Bergey's manual of systematic bacteriology

BERGEY'S MANUAL OF
Systematic
Bacteriology
Second Edition

Volume Three
The *Firmicutes*

BERGEY'S MANUAL OF
Systematic
Bacteriology
Second Edition

Volume Three
The *Firmicutes*

Paul De Vos, George M. Garrity, Dorothy Jones, Noel R. Krieg, Wolfgang Ludwig, Fred A. Rainey, Karl-Heinz Schleifer and William B. Whitman
EDITORS, VOLUME THREE

William B. Whitman
DIRECTOR OF THE EDITORIAL OFFICE

Aidan C. Parte
MANAGING EDITOR

EDITORIAL BOARD
Michael Goodfellow, Chairman, **Peter Kämpfer,** Vice Chairman,
Paul De Vos, Fred A. Rainey, Karl-Heinz Schleifer and William B. Whitman
WITH CONTRIBUTIONS FROM 165 COLLEAGUES

Springer

William B. Whitman
Bergey's Manual Trust
Department of Microbiology
527 Biological Sciences Building
University of Georgia
Athens, GA 30602-2605
USA

ISBN: 978-0-387-95041-9   e-ISBN: 978-0-387-68489-5
DOI: 10.1007/b92997
Springer Dordrecht Heidelberg London New York

Library of Congress Control Number: 2009933884

Printed on acid-free paper.

Springer is part of Springer Science+Business Media (www.springer.com)

This volume is dedicated to our colleagues
James T. Staley and George M. Garrity,
who retired from the Board of Trustees of Bergey's Manual Trust
during preparation of this volume.
We deeply appreciate their efforts as editors, authors and officers of the Trust.
They have devoted many years to helping
the Trust meet its objectives.

Preface to Volume Three of the Second Edition of *Bergey's Manual of Systematic Bacteriology*

A number of important changes occurred at Bergey's Manual Trust during the preparation of this volume. In 2006, George Garrity retired from the Trust, and the Trust moved its offices from Michigan State University to the University of Georgia. We are deeply indebted to Professor Garrity, under whose supervision much of this volume was prepared. James T. Staley's wise council guided this transition until he retired from the Trust in 2008 after 32 years of service.

The officers of the Trust have also changed during this time. Barny Whitman became Treasurer and Director of the Editorial Office in 2006. Michael Goodfellow succeeded Professor Staley as Chair in 2008 and Peter Kämpfer succeeded Professor Goodfellow as Vice-Chair in 2008. The Trust was also fortunate to acquire the services of Dr Aidan Parte as Managing Editor in 2007.

Much as things have changed, prokaryotic systematics has remained a vibrant and exciting field of study, one of challenges and opportunities, great discoveries and gradual advances. To honor the leaders of our field, the Trust presented the Bergey Award in recognition of outstanding contributions to the taxonomy of prokaryotes to Jean Paul Euzéby (2005), David P. Labeda (2006), and Jürgen Wiegel (2008). In recognition of life-long contributions to the field of prokaryotic systematics, the Bergey Medal was presented to Richard W. Castenholz (2005), Kazau Komagata (2005), Klaus P. Schaal (2006), Fergus Priest (2008), and James T. Staley (2008).

Acknowledgements

The Trust is indebted to all of the contributors and reviewers, without whom this work would not be possible. The Editors are grateful for the time and effort that each has expended on behalf of the entire scientific community. We also thank the authors for their good grace in accepting comments, criticisms, and editing of their manuscripts.

The Trust recognizes its enormous debt to Dr Aidan Parte, whose enthusiasm and professionalism have made this work possible. His expertise and good judgment have been extremely valued.

We also recognize the special efforts of Dr Jean Euzéby and Professor Aharon Oren for their assistance on the nomenclature and etymologies.

We also thank the Department of Microbiology at Michigan State University and especially Connie Williams, for her assistance in bring this volume to completion, and Walter Esselman, the Chair of the Department of Microbiology and Molecular Genetics, who facilitated our move to the University of Georgia. We thank our current copyeditors, proofreaders and other staff, including Susan Andrews, Joanne Auger, Frances Brenner, Robert Gutman, Judy Leventhal, Linda Sanders and Travis Dean, whose hard work and attention to detail have made this volume possible. Lastly, we thank the Department of Microbiology at the University of Georgia for its assistance and encouragement in thousands of ways.

William B. (Barny) Whitman

Contents

Contributors

Didier Alazard
IRD, UMR 180, Universités de Provence et de la Méditerranée, ESIL case 925, 163 avenue de Luminy, 13288 Marseille cedex 09, France
didier.alazard@univmed.fr

Luciana Albuquerque
Department of Zoology and Center for Neurosciences and Cell Biology, University of Coimbra, 3004-517 Coimbra, Portugal
luciana@cnc.uc.pt

Marie Asao
Southern Illinois University, Department of Microbiology, Mail Stop 6508, Carbondale, IL 62901-4399, USA
asao@micro.siu.edu

Sandra Baena
Departamento de Biología, Pontificia Universidad Javeriana, AA 56710 SantaFe de Bogotá, Colombia
baena@javeriana.edu.co

Georges Barbier
Université Européenne de Bretagne/Université de Brest, EA3882 Laboratoire Universitaire de Biodiversité et Ecologie Microbienne, IFR148 ScInBioS, ESMISAB, Technopôle de Brest Iroise, 29280 Plouzané, France
georges.barbier@univ-brest.fr

Julia A. Bell
Food Safety and Toxicology Center, Michigan State University, East Lansing, MI 48824, USA
bellj@msu.edu

Yoshimi Benno
Japan Collection of Microorganisms, Microbe Divison, RIKEN BioResource Center, 2-1 Hirosawa, Wako, Saitama 351-0198, Japan
benno@jcm.riken.jp

Teruhiko Beppu
Life Science Research Center, College of Bioresource Sciences, Nihon University, 1866 Kameino, Fujisawa 252-8510, Japan
beppu@wmail.brs.nihon-u.ac.jp

Hanno Biebl
National Research Centre for Biotechnology, Mascheroder Weg 1, GBF German Research Centre for Biotechnology, D-38124 Braunschweig, Germany
hannobiebl@web.de

George W. Bird
Department of Entomology, 243 Natural Science, E. Lansing, MI 48824, USA
bird@msue.msu.edu

Johanna Björkroth
Department of Food and Environmental Hygiene, Faculty of Veterinary Medicine, Helsinki University, P.O. Box 57, FIN-00014 Helsinki, Finland
johanna.bjorkroth@helsinki.fi

Michael Blaut
Department of Gastrointestinal Microbiology, Arthur-Scheunert-Allee 114-116, German Institute of Human Nutrition, D-14553 Bergholz-Rehbrücke, Germany
blaut@mail.dife.de

Monica Bonilla-Salinas
Laboratory of Microbiology, Wageningen University, Dreijenplein 10, 6703 HB Wageningen, The Netherlands

Philipp P. Bosshard
Institute of Medical Microbiology, Gloriastrasse 30/32, University of Zürich, CH-8028 Zürich, Switzerland
philboss@immv.unizh.ch

Wolfgang Buckel
Max-Planck-Institut für terrestrische Mikrobiologie, Karl-von-Frisch-Strasse, D-35043 Marburg, Germany
buckel@staff.uni-marburg.de

Hans-Jürgen Busse
Institut für Bakteriologie, Mykologie und Hygiene, Veterinarmedizinische Universitat, Veterinarplatz 1, A-1210 Wien, Austria
Hans-Juergen.Busse@vu-wien.ac.at

Ercole Canale-Parola
Department of Microbiology, University of Massachusetts, Amherst, MA 01003-0013, USA

Jean-Philippe Carlier (Deceased)
Centre National de Référence Pour les Bactéries Anaèrobies et le Botulisme, Institut Pasteur, 25-28 Rue du Dr. Roux, 75724 Paris Cedex 15, France

Jean-Luc Cayol
Laboratoire de Microbiologie IRD, UMR 180, Universités de Provence et de la Méditerranée, 163 Avenue de Luminy, Case 925, 13288 Marseille Cedex 09, France
jean-luc.cayol@univmed.fr

Dieter Claus
Chemnitzer Strasse 3, 37085 Göttingen, Germany
dclaus@gmx.net

Matthew D. Collins
Department of Food Science and Technology, The University of Reading, P.O. Box 226, Whiteknights, Reading RG6 6AP, UK

Gregory M. Cook
Department of Microbiology and Immunology,
Otago School of Medical Sciences, University of Otago,
P.O. Box 56, Dunedin, New Zealand
greg.cook@stonebow.otago.ac.nz

Nancy A. Cornick
Veterinary Microbiology and Preventative Medicine,
2180 Veterinary Medicine, College of Veterinary Medicine,
Iowa State University, Ames, IA 50011, USA
ncornick@iastate.edu

Michael A. Cotta
USDA/ARS, National Center for Agricultural Utilization
Research, 1815 N. University Street, Peoria, IL 61604, USA
Mike.Cotta@ars.usda.gov

Milton S. da Costa
Departamento de Bioquímica, Universidade de Coimbra,
3001-401 Coimbra, Portugal
milton@ci.uc.pt

Elke De Clerck
Milliken Europe N.V., Ham 18–24, B-9000 Gent, Belgium

Paul De Vos
Laboratory for Microbiology, University of Ghent,
K. L. Ledeganckstraat, 35, B-9000 Ghent, Belgium
Paul.Devos@ugent.be

Luc A. Devriese
Laboratory of Veterinary Bacteriology and Mycology,
Faculty of Veterinary Medicine, University of Ghent,
Salisburylaan 133, B-9820 Merelbeke, Belgium
devriese.okerman@skynet.be

Dhiraj P. Dhotre
National Centre for Cell Science, Pune University Campus,
Ganeshkhind, Pune 411 007, India
dheerajdhotre@gmail.com

Leon M. T. Dicks
Department of Microbiology, Private Bag X1, 7602 Matieland
(Stellenbosch), South Africa
LMTD@sun.ac.za

Donald W. Dickson
Bldg. 970, Natural Area Drive, University of Florida,
Gainesville, FL 32611-0620, USA
dwd@ufl.edu

Abhijit S. Dighe
Orthopaedic Surgery Research Center, Room B 035,
Cobb Hall, P.O. Box 800374, University of Virginia (UVa),
Charlottesville, VA 22908, USA
asd2n@virginia.edu

Anna E. Dinsdale
Department of Biological and Biomedical Sciences,
Glasgow Caledonian University, Cowcaddens Road,
Glasgow G4 0BA, UK
Anna.Dinsdale@gcal.ac.uk

Xiuzhu Dong
No. 3A, Datun Road, Chaoyang District, Beijing 100101, China
dongxz@sun.im.ac.cn

Julia Downes
Floor 28, Guy's Tower, Department of Microbiology, King's
College London, Guy's Hospital, London SE1 9RT, UK
julie.downes@kcl.ac.uk

Harold L. Drake
Department of Ecological Microbiology, University of Bayreuth,
Dr.-Hans-Frisch-Strasse 1-3, D-95440 Bayreuth, Germany
hld@uni-bayreuth.de

Sylvia H. Duncan
Microbial Ecology Group, Rowett Institute of Nutrition and
Health, University of Aberdeen, Greenburn Road, Bucksburn,
Aberdeen AB21 9SB, UK
S.Duncan@rowett.ac.uk

Dieter Ebert
Evolutionsbiologie, Zoologisches Institut, Universität Basel,
CH 4051 Basel, Switzerland
dieter.ebert@unibas.ch

Erkki Eerola
Department of Medical Microbiology, University of Turku,
Turku, Finland
erkki.eerola@utu.fi

Jean P. Euzéby
Ecole Nationale Veterinaire, 23 chemin des Capelles,
B.P. 87614, 31076 Toulouse cedex 3, France
euzeby@bacterio.org

Takayuki Ezaki
Department of Microbiology, 40 Tsukasa-machi,
Gifu University School of Medicine, Gifu 500-8705, Japan
tezaki@gifu-u.ac.jp

Enevold Falsen
Guldhedsgatan 10, University of Goteburg,
S-41346 Goteborg, Sweden
falsen@ccug.gu.se

Marie-Laure Fardeau
Laboratoire de Microbiologie IRD, UMR 180, Universités
de Provence et de la Méditerranée, 163 Avenue de Luminy,
Case 925, 13288 Marseille Cedex 09, France
marie-laure.fardeau@univmed.fr

Harry J. Flint
Microbial Ecology Group, Rowett Institute of Nutrition
and Health, University of Aberdeen, Greenburn Road,
Bucksburn, Aberdeen AB21 9SB, UK
H.Flint@rowett.ac.uk

Charles M. A. P. Franz
Max Rubner Institute, Department or Safety and Quality of
Fruits and Vegetables, Haid-und-Neu-Strasse 9,
D-76131 Karlsruhe, Germany
Charles.Franz@mri.bund.de

Michael W. Friedrich
Max Planck Institute for Terrestrial Microbiology,
Karl-von-Frisch-Strasse, D-35043 Marburg, Germany
michael.friedrich@mpi-marburg.mpg.de

Dagmar Fritze
DSM - Deutsche Sammlung von Mikroorganismen und
Zellkulturen, Inhoffenstrasse 7 B, D-38124 Braunschweig,
Germany
dfr@dsmz.de

Tateo Fujii
4-5-7 konan, Minato-ku, Tokyo University of Marine Science
and Technology, Tokyo 108-8477, Japan
ttfujii@tokyo-u-fish.ac.jp

Jean-Louis Garcia
Laboratoire de Microbiologie IRD, UMR 180, Universités
de Provence et de la Méditerranée, 163 Avenue de Luminy,
Case 925, 13288 Marseille Cedex 09, France
garcia@esil.univ-mrs.fr

Elena S. Garnova
Laboratory of Relict Microbial Communities, Institute
of Microbiology, Russian Academy of Science (RAS),
Prospect 60-let Oktyabrya 7/2, 117312 Moscow, Russia
egarnova@yahoo.com

Robin M. Giblin-Davis
University of Florida IFAS, Ft. Lauderdale Research and
Education Center, 3205 College Avenue, Ft. Lauderdale,
FL 33314-7799, USA
giblin@ufl.edu

Michael Goodfellow
Department of Microbiology, The Medical School,
University of Newcastle-upon-Tyne, Framlington Place,
Newcastle-upon-Tyne NE1 7RU, UK
M.Goodfellow@newcastle.ac.uk

Anita S. Gössner
Department of Ecological Microbiology, University of Bayreuth,
Dr.-Hans-Frisch-Strasse 1-3, D-95440 Bayreuth, Germany
a.goessner@uni-bayreuth.de

Isabelle Grech-Mora
Laboratoire de Microbiologie IRD, UMR 180, Universités
de Provence et de la Méditerranée, 163 Avenue de Luminy,
Case 925, 13288 Marseille Cedex 09, France

Auli Haikara
Biotekniikan Laboratory, P.O. Box 1500, Valtion Teknillinen
Tutkimuskeskus, Tietotie 2, Espoo FIN-02044 VTT, Finland
auli.haikara@kolumbus.fi

Walter P. Hammes
Talstr. 60/1, D-70794 Filderstadt, Germany
hammeswp@uni-hohenheim.de

Satoshi Hanada
Institute for Biological Resources and Functions, National
Institute of Advanced Industrial Science and Technology
(AIST), Tsukuba Central 6, 1-1-1 Higashi,
Tsukuba 305-8566, Japan
s-hanada@aist.go.jp

Theo A. Hansen
Department of Microbiology, Groningen Biomolecular
Sciences and Biotechnology, University of Groningen,
Kerklaan 30, NL 9751 NN Haren, The Netherlands
T.A.Hansen@rug.nl

Jeremy M. Hardie
Queen Mary University of London, Barts & The London
School of Medicine and Dentistry, Institute of Dentistry,
Turner Street, London E1 2AD, UK
jeremy.hardie@btinternet.com

Guadalupe Hernandez-Eugenio
Laboratoire de Microbiologie IRD, UMR 180, Universités
de Provence et de la Méditerranée, 163 Avenue de Luminy,
Case 925, 13288 Marseille Cedex 09, France

Christian Hertel
German Institute of Food Technology (DIL e.V.),
Professor-von-Klitzing-Strasse 7, D-49610 Quakenbrück,
Germany
c.hertel@dil-ev.de

Jeroen Heyrman
Ghent University, Department BFM (WE10V), Laboratory
of Microbiology, K.-L. Ledeganckstraat 35, B-9000 Gent,
Belgium
Jeroen.Heyrman@UGent.Be

Hans Hippe
Zur Scharfmuehle 46, 37083 Göttingen, Germany
hhi@t-online.de

Becky Jo Hollen
Department of Biological Sciences, 202 Life Sciences Building,
Louisiana State University, Baton Rouge, LA 70001, USA
bholle1@lsu.edu

Christof Holliger
EPFL LBE, Laboratory for Environmental Biotechnology,
CH C3 425 (Bâtiment Chimie), Station 6, CH-1015 Lausanne,
Switzerland
christof.holliger@epfl.ch

Kim Holmstrøm
Bioneer A/S, Kogle Allé 2, DK-2970 Hørsholm, Denmark
kho@bioneer.dk

Wilhelm H. Holzapfel
School of Life Sciences, Handong Global University,
Pohang, Gyeongbuk, 791-708, South Korea
wilhelm@woodapple.net

John V. Hookey
Department for Bioanalysis and Horizon Technologies,
Health Protection Agency, Centre for Infections,
61 Colindale Avenue, London NW9 5EQ, UK

Robert Huber
Kommunale Berufsfachschule für biologisch-technische
Assistenten, Stadtgraben 39, D-94315 Straubing, Germany
robert.huber@bta-straubing.de

Philip Hugenholtz
Microbial Ecology Program, Joint Genome Institute,
2800 Mitchell Drive, Walnut Creek, CA 94598, USA
hugenholtz@lbl.gov

Morio Ishikawa
Department of Fermentation Science, Faculty of Applied
Bio-Science, Tokyo University of Agriculture, 1-1 Sakuragaoka
1-chome, Setagaya-ku, Tokyo 156-8502, Japan
m1ishika@nodai.ac.jp

Jari Jalava
National Public Health Institute, Department of Microbiology,
Turku University, Kiinamyllynkatu 13, Turku, FIN-20520, Finland
jari.jalava@utu.fi

Peter H. Janssen
Grasslands Research Centre, AgResearch, Private Bag 11008,
Palmerston North 4442, New Zealand
peter.janssen@agresearch.co.nz

Graeme N. Jarvis
Rumen Biotechnology, Grasslands Research Centre,
AgResearch, Private Bag 11008, Palmerston North 4442,
New Zealand
Graeme.Jarvis@agresearch.co.nz

Pierre Juteau
Département d'assainissement, Cégep de Saint-Laurent,
625 avenue Sainte-Croix, Montréal QC, Canada H4L 3X7
pjuteau@cegep-st-laurent.qc.ca

Riikka Juvonen
VTT Biotechnology, P.O. Box 1500, Espoo, FI-02044 VTT, Finland
riikka.juvonen@vtt.fi

Akiko Kageyama
Kitasato Institute for Life Sciences, Kitasato University,
5-9-1 Shirokane, Minato-ku, Tokyo 108-8641, Japan
kageyama@nihs.go.jp

Yoichi Kamagata
Research Institute of Genome-Based Biofactory, National
Institute of Advanced Industrial Science and Technology
(AIST), Sapporo, Hokkaido 062-8517, Japan
y.kamagata@aist.go.jp

Peter Kämpfer
Institut für Angewandte Mikrobiologie,
Justus-Liebig-Universität Giessen, Heinrich-Buff-Ring 26-32
(IFZ), D-35392 Giessen, Germany
Peter.Kaempfer@umwelt.uni-giessen.de

Yoshiaki Kawamura
Department of Microbial-Bioinformatics, Regeneration and
Advanced Medical Science, Gifu University Graduate School
of Medicine, 40 Tsukasa-machi, Gifu 500-8705, Japan
kawamura@cc.gifu-u.ac.jp

Byung-Chun Kim
Biological Resources Center, KRIBB, Daejeon, 305-806,
Republic of Korea

Bon Kimura
Department of Food Science Technology, Kounan 4-5-7
Minato-ku, Tokyo University of Fisheries, Tokyo, 108 8477, Japan
kimubo@tokyo-u-fish.ac.jp

Oleg R. Kotsyurbenko
Helmholtz Centre for Infection Research, Environmental
Microbiology Laboratory, Inhoffenstrasse 7, D-38124
Braunschweig, Germany
kotsor@hotmail.com

Lee R. Krumholz
Department of Botany and Microbiology and Institute
for Energy and the Environment, 770 Van Vleet Oval,
The University of Oklahoma, Norman, OK 73019, USA
krumholz@ou.edu

Jan Kuever
Department of Microbiology, Bremen Institute for Materials
Testing, Foundation Institute for Materials Science,
Paul-Feller-Strasse 1, D-28199 Bremen, Germany
kuever@mpa-bremen.de

Paul A. Lawson
Department of Botany and Microbiology, George Lynn Cross
Hall, 770 Van Vleet Oval, The University of Oklahoma,
Norman, OK 73019-0245, USA
paul.lawson@ou.edu

Ute Lechner
Martin-Luther-University Halle-Wittenberg, Institute of
Biology/Microbiology, Kurt-Mothes-Str. 3, 06099 Halle, Germany
ute.lechner@mikrobiologie.uni-halle.de

Yong-Jin Lee
Department of Microbiology, Biological Sciences Building,
University of Georgia, Cedar Street, Athens, GA 30602, USA
yjlee01@gmail.com

Jørgen J. Leisner
Department of Veterinary Pathobiology, Faculty of Life
Sciences, University of Copenhagen, Grønnegårdsvej 15,
DK-1870 Frederiksberg C. (Copenhagen), Denmark
jjl@life.ku.dk

Niall A. Logan
Department of Biological and Biomedical Sciences, Glasgow
Caledonian University, Cowcaddens Road, Glasgow G4 0BA,
UK
nalo@gcal.ac.uk

Wolfgang Ludwig
Lehrstuhl für Mikrobiologie, Technische Universität
München, Am Hochanger 4, D-85350 Freising, Germany
ludwig@mikro.biologie.tu-muenchen.de

Heinrich Lünsdorf
HZI - Helmholtz Zentrum für Infektionsforschung, Abtlg.
Vakzinologie/Elektronenmikroskopie, Inhoffenstrasse 7,
D-38124 Braunschweig, Germany
heinrich.luensdorf@helmholtz-hzi.de

Bogusław Lupa
Department of Microbiology, 527 Biological Sciences Building,
University of Georgia, Cedar Street, Athens, GA 30602, USA
lupa@uga.edu

Michael T. Madigan
Southern Illinois University, Department of Microbiology,
Mail Stop 6508, Carbondale, IL 62901-4399, USA
madigan@micro.siu.edu

Michel Magot
Université de Pau et des Pays de l'Adour, Environnement
et Microbiologie, IBEAS - BP1155, 64013 Pau, France
michel.magot@univ.pau.fr

Hélène Marchandin
Laboratoire de Bactériologie, Hôpital Arnaud de Villeneuve,
371 Avenue du Doyen Gaston Giraud, 34295 Montpellier
Cedex 5, France
h-marchandin@chu-montpellier.fr

James McLauchlin
Food Water and Environmental Microbiology Network,
Health Protection Agency Regional Microbiology Network,
7th Floor Holborn Gate, 330 High Holborn, London WC1V
7PP, UK
Jim.McLauchlin@HPA.org.uk

Tahar Mechichi
Laboratoire des Bioprocédés, Centre de Biotechnologie
de Sfax BP "K", 3038 Sfax, Tunisia
mechichi.tahar@cbs.rnrt.tn

Encarnación Mellado
Departamento de Microbiología y Parasitología, Facultad
de Farmacia, Universidad de Sevilla, Spain
emellado@us.es

Noha M. Mesbah
Department of Microbiology, University of Georgia,
527 Biological Sciences Bldg., Cedar Street, Athens,
GA 30602-2605, USA
nmesbah@uga.edu

Elizabeth Miranda-Tello
Departamento de Biotecnología Ambiental, Ecología
Microbiana Aplicada y Contaminación, El Colegio de la
Frontera Sur, Unidad Chetumal, Av. del Centenario km 5.5,
Col. Calderitas, C.P. 77900 Chetumal, Quintana Roo,
Mexico
emiranda@ecosur.mx

Koji Mori
NITE Biological Resource Center (NBRC),
National Institute of Technology and Evaluation (NITE),
2-5-8 Kazusakamatari, Kisarazu, Chiba 292-0818,
Japan
mori@nbrc.nite.go.jp

Youichi Niimura
Department of Bioscience, Tokyo University of Agriculture,
1-1-1 Setagaya-ku, Tokyo 156-85027, Japan
niimura@nodai.ac.jp

Gregory R. Noel
Department of Crop Sciences, USDA ARS,
University of Illinois, Urbana, IL 61801, USA
g-noel1@illinois.edu

Spyridon Ntougias
Institute of Kalamata, National Agricultural Research
Foundation, Lakonikis 87, 24100 Kalamata, Greece
sntougias@in.gr

Kiyofumi Ohkusu
Department of Microbiology, Regeneration and Advanced
Medical Science, Gifu University Graduate School of
Medicine, Yanagido, Gifu 501-1194, Japan
ohkusu@cc.gifu-u.ac.jp

Bernard Ollivier
Laboratoire de Microbiologie IRD, UMR 180, Universités
de Provence et de la Méditerranée, 163 Avenue de Luminy,
Case 925, 13288 Marseille cedex 09, France
bernard.ollivier@univmed.fr

Rob U. Onyenwoke
Room 8105 Neuroscience Research Building, UNC School
of Medicine Campus Box 7250, 115 Mason Farm Road,
Chapel Hill, NC 27599-7250, USA
onyenwok@email.unc.edu

Ronald S. Oremland
US Geological Survey, Bldg. 15, McKelvey Building, MS 480,
345 Middlefield Road, Menlo Park, CA 94025, USA
roremlan@usgs.gov

Aharon Oren
Department of Plant and Environmental Sciences,
The Institute of Life Sciences, The Hebrew University
of Jerusalem, Jerusalem, Israel
orena@shum.cc.huji.ac.il

Ro Osawa
Department of Bioscience, Graduate School of Science,
Kobe University, Rokkodai 1-1, Nada-ku,
Kobe City 657-8501, Japan
osawa@ans.kobe-u.ac.jp

Yong-Ha Park
Korean Institute of Science and Technology, Bioinformatics &
Systematics Laboratory, Korean, Collection for Type Cultures,
Korea Inst. of Sci. & Tech., Daeduk Science Park,
Republic of Korea
peter@yumail.ac.kr

Sofiya N. Parshina
Winogradsky Institute of Microbiology, Russian Academy
of Sciences, 7/2, Prospekt 60-letiya Oktyabrya 117312,
Moscow, Russia
sonjaparshina@mail.ru

Bharat K. C. Patel
Microbial Discovery Research Unit, School of Biomolecular
Sciences, Griffith University, Nathan Campus, Kessels Road,
Brisbane, Queensland 4111, Australia
B.Patel@griffith.edu.au

Milind S. Patole
National Centre for Cell Science, Pune University Campus,
Ganeshkhind, Pune 411 007, India
patole@nccs.res.in, milindpatole@hotmail.com

Elena V. Pikuta
Astrobiology Laboratory, room 4247, National Space Science
and Technology Center, 320 Sparkman Drive, Huntsville,
AL 35805, USA
pikutae@UAH.edu

Caroline M. Plugge
Laboratory of Microbiology, Wageningen University,
Dreijenplein 10, 6703 HB Wageningen, The Netherlands
Caroline.Plugge@wur.nl

Gérard Prensier
CNRS, UMR 6023 Biologie des Protistes, Complexe
Scientifique des Cézeaux, 63177 Aubière cedex, France
Gerard.Prensier@univ-bpclermont.fr

James F. Preston III
Department of Microbiology and Cell Science,
University of Florida, Gainesville, FL 32611, USA
jpreston@ufl.edu

Fergus G. Priest
School of Life Sciences, Heriot Watt University, Edinburgh
EH14 4AS, UK
f.g.priest@hw.ac.uk

Rüdiger Pukall
DSM - Deutsche Sammlung von Mikroorganismen und
Zellkulturen, Inhoffenstrasse 7 B, D-38124 Braunschweig,
Germany
rpu@dsmz.de

Fred A. Rainey
Department of Biological Sciences, 202 Life Sciences Building,
Louisiana State University, Baton Rouge, LA 70001, USA
frainey@lsu.edu

Dilip R. Ranade
Microbial Sciences Division, Agharkar Research Institute,
G. G. Agarkar Road, Pune 411004, India
drranade@gmail.com, drranade@aripune.org

Gilles Ravot
Protéus SA,70, allée Graham Bell, Parc Georges Besse,
30000 Nimes, France
g.ravot@proteus.fr

Catherine E. D. Rees
School of Biosciences, University of Nottingham, Sutton
Bonnington Campus, Loughborough, Leicestershire
LE12 5RD, UK
cath.rees@nottingham.ac.uk

Kathryn L. Ruoff
Pathology Department, Dartmouth Hitchcock Medical Center,
One Medical Center Drive, Lebanon, NH 03756, USA
kathryn.l.ruoff@hitchcock.org

James B. Russell
Department of Microbiology, 157 Wing Hall,
Cornell University, Ithaca, NY 14853-8101, USA
jbr8@cornell.edu

Nicholas J. Russell
Imperial College London, Wye, Ashford, Kent TN25 5AT, UK
nicholas.russell@imperial.ac.uk

Masataka Satomi
2-12-4 Fukuura, Kanazawa-ku, National Research Institute
of Fisheries & Science, Yokohama 236-8648, Japan
msatomi@affrc.go.jp

Bernhard Schink
Lehrstuhl für Mikrobielle Ökologie, Fakultät für Biologie,
Universität Konstanz, Fach M 654, D-78457 Konstanz,
Germany
Bernhard.Schink@uni-konstanz.de

Karl-Heinz Schleifer
Lehrstuhl für Mikrobiologie, Technische Universität
München, Am Hochanger 4, D-85350 Freising, Germany
schleife@mikro.biologie.tu-muenchen.de

Heinz Schlesner
Institut für Allgemeine Mikrobiologie, Christian-Albrechts-
Universität, Am Botanischen Garten 1-9, D-24118 Kiel,
Germany
hschlesner@t-online.de

Yuji Sekiguchi
Bio-Measurement Research Group, Institute for Biological
Resources and Functions, National Institute of Advanced
Science and Technology (AIST), AIST Tsukuba Central 6,
Ibaraki 305-8566, Japan
y.sekiguchi@aist.go.jp

Haroun N. Shah
Molecular Identification Services Unit, Department
for Bioanalysis and Horizon Technologies, Health Protection
Agency, Centre for Infections, 61 Colindale Avenue,
London NW9 5EQ, UK
haroun.shah@hpa.org.uk

Sisinthy Shivaji
Centre for Cellular and Molecular Biology (CCMB),
Uppal Road, Hyderabad 500 007, India
shivas@ccmb.res.in

Yogesh S. Shouche
Microbial Culture Collection (DBT), National Centre for
Cell Science, Pune University Campus, Ganeshkhind,
Pune 411 007, India
yogesh@nccs.res.in, yogesh.shouche@gmail.com

Maria V. Simankova
Winogradsky Institute of Microbiology, Russian Academy
of Sciences, 7/2, Prospekt 60-letiya Oktyabrya 117312,
Moscow, Russia
msimankova@mail.ru

Alexander Slobodkin
Winogradsky Institute of Microbiology, Russian Academy
of Sciences, 7/2, Prospekt 60-letiya Oktyabrya 117312,
Moscow, Russia
aslobodkin@hotmail.com

Alanna M. Small
Department of Medicine, Tulane University School of
Medicine, 430 Tulane Ave., SL-50, New Orleans,
LA 70112, USA

Peter H. A. Sneath
Department of Infection, Immunity and Inflammation,
University of Leicester, Leicester LE1 9HN, UK
phas1@le.ac.uk

Tatyana G. Sokolova
Winogradsky Institute of Microbiology, Russian Academy
of Sciences, 7/2, Prospekt 60-letiya Oktyabrya 117312,
Moscow, Russia
tatso@mail.ru

Mark D. Spanevello
Microbial Discovery Research Unit, School of Biomolecular
Sciences, Griffith University, Nathan Campus, Kessels Road,
Brisbane, Queensland 4111, Australia

Stefan Spring
Microbiology Department, DSM - Deutsche Sammlung von
Mikroorganismen und Zellkulturen, Inhoffenstrasse 7 B,
D-38124 Braunschweig, Germany
ssp@dsmz.de

Erko Stackebrandt
DSM - Deutsche Sammlung von Mikroorganismen und
Zellkulturen, Inhoffenstrasse 7 B, D-38124 Braunschweig,
Germany
erko@dsmz.de

Alfons J. M. Stams
Laboratory of Microbiology, Wageningen University,
Dreijenplein 10, Building no. 316, 6703 HB Wageningen,
The Netherlands
fons.stams@wur.nl

Thaddeus B. Stanton
Agricultural Research Service – Midwest Area, National
Animal Disease Center, United States Department of
Agriculture, P.O. Box 70, 2300 Dayton Road, Ames,
IA 50010-0070, USA
Thad.Stanton@ars.usda.gov

John F. Stolz
Bayer School of Natural and Environmental Sciences,
Duquesne University, Pittsburgh, PA 15282, USA
stolz@mail.duq.edu

Carsten Strömpl
Finanzabteilung, Helmholtz Zentrum für Infektionsforschung
HZI, Inhoffenstrasse 7, D-38124 Braunschweig, Germany
cst@gbf.de

Ken-ichiro Suzuki
NITE Biological Resource Center (NBRC), Department
of Biotechnology, National Institute of Technology and
Evaluation, 5-8, Kazusakamatari 2-chrome, Kisarazu-shi,
Chiba 292-0818, Japan
suzuki-ken-ichiro@nite.go.jp

Pavel Švec
Masaryk University, Faculty of Science, Department
of Experimental Biology, Czech Collection of Microrganisms,
Tvrdého 14, 602 00 Brno, Czech Republic
mpavel@sci.muni.cz

Ken Takai
Subground Animalcule Retrieval (SUGAR) Program,
Japan Agency for Marine-Earth Science & Technology, 2-15
Natsushima-cho, Yokosuka 237-0061, Japan
kent@jamstec.go.jp

David Taras
Department of Gastrointestinal Microbiology,
German Institute of Human Nutrition, Bergholz-Rehbrücke,
Germany

Michael Teuber
Labor für Lebensmittelmikrobiolgie, ETH-Zentrum, Institute
fuer Lebensmittelwissenschaft, Raemistrasse 101, 8092 Zurich,
Switzerland
michael.teuber@ilw.agrl.ethz.ch

Brian J. Tindall
DSM - Deutsche Sammlung von Mikroorganismen und
Zellkulturen, Inhoffenstrasse 7 B, D-38124 Braunschweig,
Germany
bti@dsmz.de

Jean Pierre Touzel
NRA, UMR 614 Fractionnement des Agro-ressources
et Environnement, 8 rue Gabriel-Voisin, B.P. 316,
51688 Reims cedex 2, France
touzel@reims.inra.fr

Kenji Ueda
Life Science Research Center, College of Bioresource
Sciences, Nihon University, 1866 Kameino,
Fujisawa 252-8510, Japan
ueda@brs.nihon-u.ac.jp

Marc Vancanneyt
BCCM/LMG Bacteria Collection, Faculty of Sciences,
Ghent University, K.L. Ledeganckstraat 35, B-9000 Ghent,
Belgium
Marc.Vancanneyt@UGent.be

Antonio Ventosa
Departamento de Microbiologia y Parasitologia, Facultad
de Farmacia, Universidad de Sevilla, Apdo. 874, 41080 Sevilla,
Spain
ventosa@us.es

William G. Wade
Infection Research Group, King's College London Dental
Institute, Floor 28, Tower Wing, Guy's Campus, London SE1
9RT, UK
william.wade@kcl.ac.uk

Nathalie Wery
INRA, UR50, Laboratoire de Biotechnologie de
l'Environnement, Avenue des Etangs, 11100 Narbonne, France
weryn@supagro.inra.fr

Robert A. Whiley
Queen Mary University of London, Barts & The London
School of Medicine and Dentistry, Institute of Dentistry,
Turner Street, London E1 2AD, UK
r.a.whiley@qmul.ac.uk

Terence R. Whitehead
USDA/ARS, National Center for Agricultural Utilization
Research, 1815 N. University Street, Peoria, IL 61604,
USA
Terry.Whitehead@ars.usda.gov

William B. Whitman
Department of Microbiology, University of Georgia,
527 Biological Sciences Building, Cedar Street, Athens,
GA 30602-2605, USA
whitman@uga.edu

Juergen Wiegel
Department of Microbiology, 211–215 Biological Sciences
Building, University of Georgia, Cedar Street, Athens,
GA 30602-2605, USA
jwiegel@uga.edu

Anne Willems
Laboratorium voor Microbiologie, Vakgroep Biochemie,
Fysiologie en Microbiologie, Universiteit Gent,
K. L. Ledeganckstraat 35, B-9000 Gent, Belgium
Anne.Willems@rug.ac.be

Kazuhide Yamasato
Faculty of Applied Bio-Science, Department of Fermentation
Science, 1-1 Sakuragaoka 1-chome, Setagayaku,
Tokyo 156-8502, Japan
yamasato@ka5.koalanet.ne.jp

Fujitoshi Yanagida
Institute of Enology and Viticulture, University of Yamanashi,
1-13-1, Kitashin, Kofu, Yamanashi 400-0005, Japan
yanagida@mail.yamanashi.ac.jp

Jung-Hoon Yoon
Laboratory of Microbial Function, Korea Research Institute
of Bioscience and Biotechnology (KRIBB), P.O. Box 115,
Yusong, Taejon, South Korea
jhyoon@kribb.re.kr

George A. Zavarzin
Winogradsky Institute of Microbiology, Russian Academy
of Sciences, 7/2, Prospekt 60-letiya Oktyabrya 117312,
Moscow, Russia
zavarzin@inmi.host.ru

Daria G. Zavarzina
Winogradsky Institute of Microbiology, Russian Academy
of Sciences, 7/2, Prospekt 60-letiya Oktyabrya 117312,
Moscow, Russia
zavarzinatwo@mail.ru

Tatjana N. Zhilina
Winogradsky Institute of Microbiology, Russian Academy
of Sciences, 7/2, Prospekt 60-letiya Oktyabrya 117312,
Moscow, Russia
zhilinat@mail.ru

On using the *Manual*

NOEL R. KRIEG AND GEORGE M. GARRITY

Citation

The *Systematics* is a peer-reviewed collection of chapters, contributed by authors who were invited by the Trust to share their knowledge and expertise of specific taxa. Citations should refer to the author, the chapter title, and inclusive pages rather than to the Editors.

Arrangement of the Manual

As in the previous volumes of this edition, the *Manual* is arranged in phylogenetic groups based upon the analyses of the 16S rRNA presented in the introductory chapter "Revised road map to the phylum *Firmicutes*". These groups have been substantially modified since the publication of volume 1 in 2001, reflecting both the availability of more experimental data and a different method of analysis. Since volume 3 includes only the phylum *Firmicutes*, taxa are arranged by class, order, family, genus and species. Within each taxon, the nomenclatural type is presented first and indicated by a superscript T. Other taxa are presented in alphabetical order without consideration of degrees of relatedness.

Articles

Each article dealing with a bacterial genus is presented wherever possible in a definite sequence as follows:

a. Name of the genus. Accepted names are in **boldface**, followed by "defining publication(s)", i.e. the authority for the name, the year of the original description, and the page on which the taxon was named and described. The superscript AL indicates that the name was included on the Approved Lists of Bacterial Names, published in January 1980. The superscript VP indicates that the name, although not on the Approved Lists of Bacterial Names, was subsequently validly published in the *International Journal of Systematic and Evolutionary Microbiology* (or the *International Journal of Systematic Bacteriology*). Names given within quotation marks have no standing in nomenclature; as of the date of preparation of the *Manual* they had not been validly published in the *International Journal of Systematic and Evolutionary Microbiology*, although they may have been "effectively published" elsewhere. Names followed by the term "nov." are newly proposed but will not be validly published until they appear in a Validation List in the *International Journal of Systematic and Evolutionary Microbiology*. Their proposal in the *Manual* constitutes only "effective publication", not valid publication.

b. Name of author(s). The person or persons who prepared the Bergey's article are indicated. The address of each author can be found in the list of Contributors at the beginning of the *Manual*.

c. Synonyms. In some instances a list of some synonyms used in the past for the same genus is given. Other synonyms can be found in the *Index Bergeyana* or the *Supplement to the Index Bergeyana*.

d. Etymology of the name. Etymologies are provided as in previous editions, and many (but undoubtedly not all) errors have been corrected. It is often difficult, however, to determine why a particular name was chosen, or the nuance intended, if the details were not provided in the original publication. Those authors who propose new names are urged to consult a Greek and Latin authority before publishing in order to ensure grammatical correctness and also to ensure that the meaning of the name is as intended.

e. Salient features. This is a brief resume of the salient features of the taxon. The most important characteristics are given in **boldface**. The DNA G+C content is given.

f. Type species. The name of the type species of the genus is also indicated along with the defining publication(s).

g. Further descriptive information. This portion elaborates on the various features of the genus, particularly those features having significance for systematic bacteriology. The treatment serves to acquaint the reader with the overall biology of the organisms but is not meant to be a comprehensive review. The information is normally presented in the following sequence:

Colonial morphology and pigmentation
Growth conditions and nutrition
Physiology and metabolism
Genetics, plasmids, and bacteriophages
Phylogenetic treatment
Antigenic structure
Pathogenicity
Ecology

h. Enrichment and isolation. A few selected methods are presented, together with the pertinent media formulations.

i. Maintenance procedures. Methods used for maintenance of stock cultures and preservation of strains are given.

j. Procedures for testing special characters. This portion provides methodology for testing for unusual characteristics or performing tests of special importance.

k. Differentiation of the genus from other genera. Those characteristics that are especially useful for distinguishing the genus from similar or related organisms are indicated here, usually in a tabular form.

l. Taxonomic comments. This summarizes the available information related to taxonomic placement of the genus and indicates the justification for considering the genus a distinct taxon. Particular emphasis is given to the methods of molecular biology used to estimate the relatedness of the genus to other taxa, where such information is available. Taxonomic information regarding the arrangement and status of the various species within the genus follows. Where taxonomic controversy exists, the problems are delineated and the various alternative viewpoints are discussed.

m. Further reading. A list of selected references, usually of a general nature, is given to enable the reader to gain access to additional sources of information about the genus.

n. Differentiation of the species of the genus. Those characteristics that are important for distinguishing the various species within the genus are presented, usually with reference to a table summarizing the information.

o. List of species of the genus. The citation of each species is given, followed in some instances by a brief list of objective synonyms. The etymology of the specific epithet is indicated. Descriptive information for the species is usually presented in tabular form, but special information may be given in the text. Because of the emphasis on tabular data, the species descriptions are usually brief. The type strain of each species is indicated, together with the collection(s) in which it can be found. (Addresses of the various culture collections are given in the article in Volume 1 entitled Culture Collections: An Essential Resource for Microbiology.) The 16S rRNA gene sequence used in phylogenetic analysis and placement of the species into the taxonomic framework is given, along with the GenBank (or other database) accession number. Additional comments may be provided to point the reader to other well-characterized strains of the species and any other known DNA sequences that may be relevant.

p. Species *incertae sedis*. The List of Species may be followed in some instances by a listing of additional species under the heading "Species *Incertae Sedis*" or "Other organisms". The taxonomic placement or status of such species is questionable, and the reasons for the uncertainty are presented.

q. References. All references given in the article are listed alphabetically at the end of the family chapter.

Tables

In each article dealing with a genus, there are generally three kinds of table: (a) those that differentiate the genus from similar or related genera, (b) those that differentiate the species within the genus, and (c) those that provide additional information about the species (such information not being particularly useful for differentiation). The meanings of symbols are as follows:

+: 90% or more of the strains are positive

d: 11–89% of the strains are positive

−: 90% or more of the strains are negative

D: different reactions occur in different taxa (e.g., species of a genus or genera of a family)

v: strain instability (NOT equivalent to "d")

w: weak reaction.

ND, not determined or no data.

These symbols, and exceptions to their use, as well as the meaning of additional symbols, are given in footnotes to the tables.

Use of the *Manual* for determinative purposes

Many chapters have keys or tables for differentiation of the various taxa contained therein. For identification of species, it is important to read both the generic and species descriptions because characteristics listed in the generic descriptions are not usually repeated in the species descriptions.

The index is useful for locating the articles on unfamiliar taxa or in discovering the current classification of a particular taxon. Every bacterial name mentioned in the *Manual* is listed in the index. In addition, an up-to-date outline of the taxonomic framework is provided in the introductory chapter "Revised road map to the phylum *Firmicutes*".

Errors, comments, suggestions

As in previous volumes, the editors and authors earnestly solicit the assistance of all microbiologists in the correction of possible errors in *Bergey's Manual of Systematic Bacteriology*. Comments on the presentation will also be welcomed as well as suggestions for future editions. Correspondence should be addressed to:

Editorial Office
Bergey's Manual Trust
Department of Microbiology
University of Georgia
Athens, GA 30602-2605
USA
Tel: +1-706-542-4219; fax +1-706-542-2674
e-mail: bergeys@uga.edu

Revised road map to the phylum *Firmicutes*

WOLFGANG LUDWIG, KARL-HEINZ SCHLEIFER AND WILLIAM B. WHITMAN

Starting with the Second Edition of *Bergey's Manual of Systematic Bacteriology*, the arrangement of content follows a phylogenetic framework or "road map" based largely on analyses of the nucleotide sequences of the ribosomal small-subunit RNA rather than on phenotypic data (Garrity et al., 2005). Implicit in the use of the road map are the convictions that prokaryotes have a phylogeny and that phylogeny matters. However, the reader should be aware that phylogenies, like other experimentally derived hypotheses, are not static but may change whenever new data and/or improved methods of analysis become available (Ludwig and Klenk, 2005). Thus, the large increases in data since the publication of the taxonomic outlines in the preceding volumes have led to a re-evaluation of the road map. Not surprisingly, the taxonomic hierarchy has been modified or newly interpreted for a number of taxonomic units of the *Firmicutes*. These changes are described in the following paragraphs.

The taxonomic road map proposed in Volume 1 and updated and emended in Volume 2 was derived from phylogenetic and principal-component analyses of comprehensive datasets of small-subunit rRNA sequences. A similar approach is continued here. Since the introduction of comparative rRNA sequencing (Ludwig and Klenk, 2005; Ludwig and Schleifer, 2005), there has been a continuous debate concerning the justification and power of a single marker molecule for elucidating and establishing the phylogeny and taxonomy of organisms, respectively. Although generally well established in taxonomy, the polyphasic approach cannot be currently applied for sequence-based analyses due to the lack of adequate comprehensive datasets for alternative marker molecules. Even in the age of genomics, the datasets for non-rRNA markers are poor in comparison to more than 300,000 rRNA primary structures available in general and special databases (Cole et al., 2007; Pruesse et al., 2007). Nevertheless, the data provided by the full genome sequencing projects allow defining a small set of genes representing the conserved core of prokaryotic genomes (Cicarelli et al., 2006; Ludwig and Schleifer, 2005). Furthermore, comparative analyses of the core gene sequences globally support the small-subunit rRNA derived view of prokaryotic evolution. Although the tree topologies reconstructed from alternative markers differ in detail, the major groups (and taxa) are verified or at least not disproved (Ludwig and Schleifer, 2005). Consequently, the structuring of this volume is based on updated and curated (http://www.arb-silva.de; Ludwig et al., 2004) databases of processed small-subunit rRNA primary structures.

Data analysis

The current release of the integrated small-subunit rRNA database of the SILVA project (Pruesse et al., 2007) provided the basis for these phylogenetic analyses of the *Firmicutes*. The tools of the ARB software package (Ludwig et al., 2004) were used for data evaluation, optimization and phylogenetic inference. The alignment of sequences comprising at least 1000 monomers was manually evaluated and optimized for all representatives of the phylum. Phylogenetic treeing was performed with all of the approximately 14,000 sequences from *Firmicutes* which contain at least 1400 nucleotides and an additional 1000 sequences from representatives of the other phyla and domains. For recognizing and avoiding the influences of chimeric sequences, all calculations were performed twice, once including and once excluding environmental clone data. The datasets also varied with respect to the inclusion of highly variable sequence positions, which were eliminated in some analyses (Ludwig and Klenk, 2005). The consensus tree used for evaluating or modifying the taxonomic outline was based on maximum-likelihood analyses (RAXML, implemented in the ARB package; Stamatakis et al., 2005) and further evaluated by maximum-parsimony and distance matrix analyses with the respective ARB tools (Ludwig et al., 2004). In the case that type strains were only represented by partial sequences (less than 1400 nucleotides), the respective data were inserted by a special ARB-tool allowing the optimal positioning of branches to the reference tree without admitting topology changes.

Taxonomic interpretation

The phylogenetic conclusions were used for evaluating and modifying the taxonomic outline of the *Firmicutes*. In order to ensure applicability and promote acceptance, the proposed modifications were made following a conservative procedure. The overall organization follows the type 'taxon' principle as applied in the previous volumes. Taxa defined in the outline of the preceding volumes were only unified, dissected or transferred in the cases of strong phylogenetic support. This approach is justified by the well-known low significance of local tree topologies (also called "range of unsharpness" around the nodes; Ludwig and Klenk, 2005). Thus, many of the cases of paraphyletic taxa found were maintained in the current road map if the respective (sub)-clusters rooted closely together, even if they were separated by intervening clusters representing other taxa. While reorganization of these taxa may be warranted, it was not performed in the absence of confirmatory evidence. The names of validly published but phylogenetically misplaced type strains are also generally maintained. These strains are mentioned in the context of the respective phylogenetic groups. In case of paraphyly, all concerned species or higher taxa are assigned to the respective (sub)-groups. New higher taxonomic ranks are only proposed if species or genera — previously assigned to different higher taxonomic units — are significantly unified in a monophyletic branch.

The taxonomic backbone of the *Firmicutes*

In the current treatment, the phylum *Firmicutes* contains three classes, "*Bacilli*", "*Clostridia*" and "*Erysipelotrichia*". This organization is similar to that of Garrity et al. (2005). However, the *Mollicutes* were removed from the phylum given the general low support by alternative markers (Ludwig and Schleifer, 2005) and its unique phenotypic properties, in particular the lack of rigid cell walls (see Emended description of *Firmicutes*, this volume). The family *Erysipelotrichaceae*, which includes wall-forming Gram-positive organisms previously classified with the *Mollicutes*, was retained in the *Firmicutes* as a novel class, "*Erysipelotrichia*", and order, "*Erysipelotrichales*".

While the bipartition of the classes "*Clostridia*" and "*Bacilli*" is corroborated by the new analyses, some of the taxa previously assigned to the "*Clostridia*" tend to root outside the *Firmicutes* and may represent separate phyla. These include taxa previously classified within the "*Thermoanaerobacterales*" and

Syntrophomonadaceae (Garrity et al., 2005), which may contain a number of phylogenetic clades that are distinct at the phylum level. However, given the absence of corroboration by other phylogenetic markers for many of these assignments and a clear consensus on the definition of a phylum, these taxa were retained within the *Firmicutes* for the present.

Class *"Bacilli"*

Compared to Garrity et al. (2005), only minor restructuring of the "*Bacilli*" is indicated by this new analysis of the rRNA data. The separation into two orders, *Bacillales* and "*Lactobacillales*", is well supported (Figure 1). However, a number of paralogous

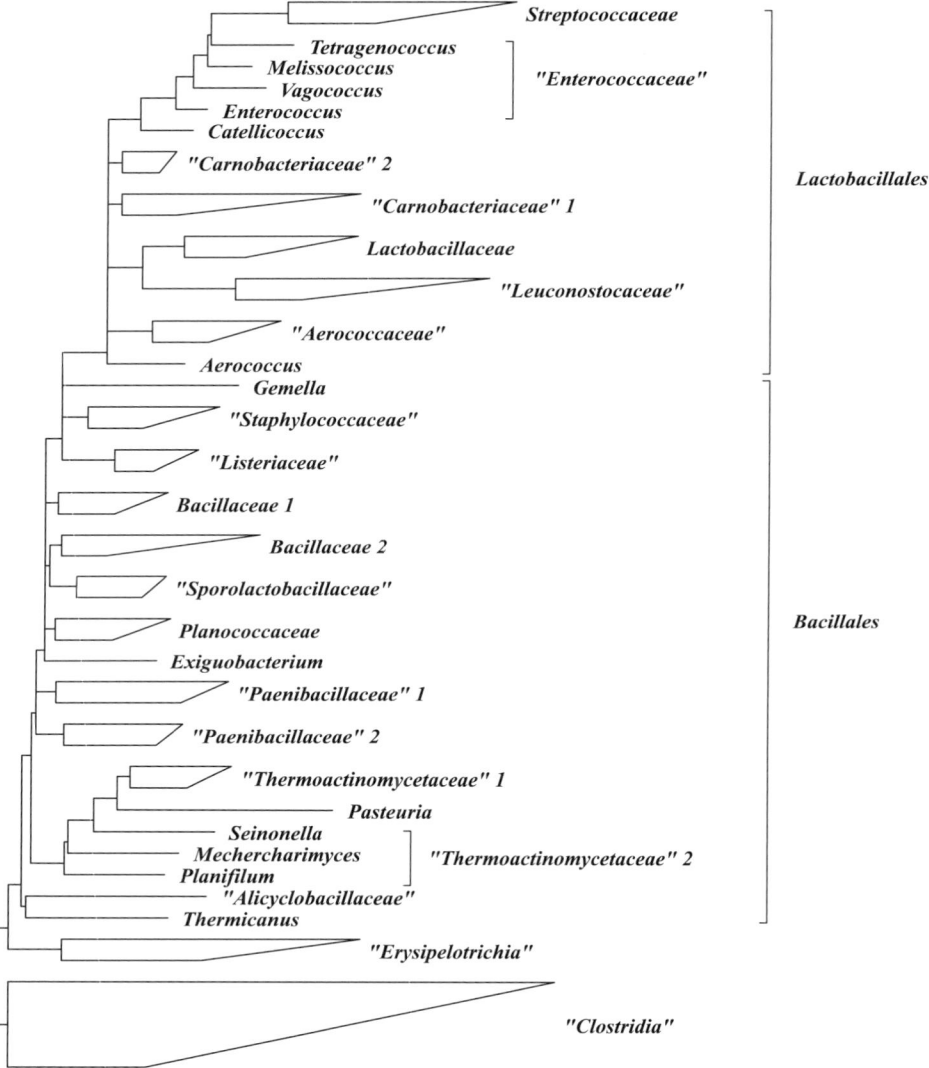

FIGURE 1. Consensus dendrogram reflecting the phylogenetic relationships of the classes "*Bacilli*" and "*Erysipelotrichia*" within the *Firmicutes*. The tree is based on maximum-likelihood analyses of a dataset comprising about 5000 almost full-length high-quality 16S rRNA sequences from representatives of the *Firmicutes* and another 1000 representing the major lines of decent of the three domains *Bacteria*, *Archaea*, and *Eucarya*. The topology was evaluated by distance matrix and maximum-parsimony analyses of the dataset. In addition, maximum-parsimony analyses of all currently available almost complete small-subunit rRNA sequences (137,400 of ARB-SILVA release 92, Prüsse et al., 2007) were performed. Only alignment positions invariant in at least 50% of the included primary structures from *Firmicutes* were included for tree reconstruction. Multifurcations indicate that a common relative branching order was not significantly supported applying alternative treeing methods. The (horizontal) branch lengths indicate the significance of the respective node separation.

groups are found within the "*Bacilli*", some of which have been reclassified.

Order *Bacillales*

The definition and taxonomic organization of the order *Bacillales* is as outlined in the previous volumes (Figure 2). Of the ten families proposed in Garrity et al. (2005), eight are retained. Upon transfer of the type genus *Caryophanon* to the *Planococcaceae*, the family *Caryophanaceae* was removed. Although the family *Caryophanaceae* Peskoff 1939[AL] has priority over *Planococcaceae* Krassilnikov 1949[AL], the former is confusing because it is a misnomer, meaning 'that which has a conspicuous nucleus', and was based upon misinterpretation of staining results (Trentini, 1986). Similarly, upon transfer of the type genus *Turicibacter*

to the family "*Erysipelotrichaceae*", the family "*Turicibacteraceae*" was removed. In addition, the genus *Pasteuria* was transferred out of the family "*Alicyclobacillaceae*" to the family *Pasteuriaceae*. As described below, a number of genera were also moved to families *incertae sedis* in recognition of the ambiguity of their phylogeny and taxonomic assignments.

Family *Bacillaceae*

The 16S rRNA-based phylogenetic analyses indicate that the family *Bacillaceae* is paraphyletic and composed of species misassigned to the genus *Bacillus* as well as genera misassigned to the family (Figure 2). Reclassification of some taxa is proposed to correct some of these problems. However, the complete reorganization of this old and well-abused taxon is outside the scope of this work.

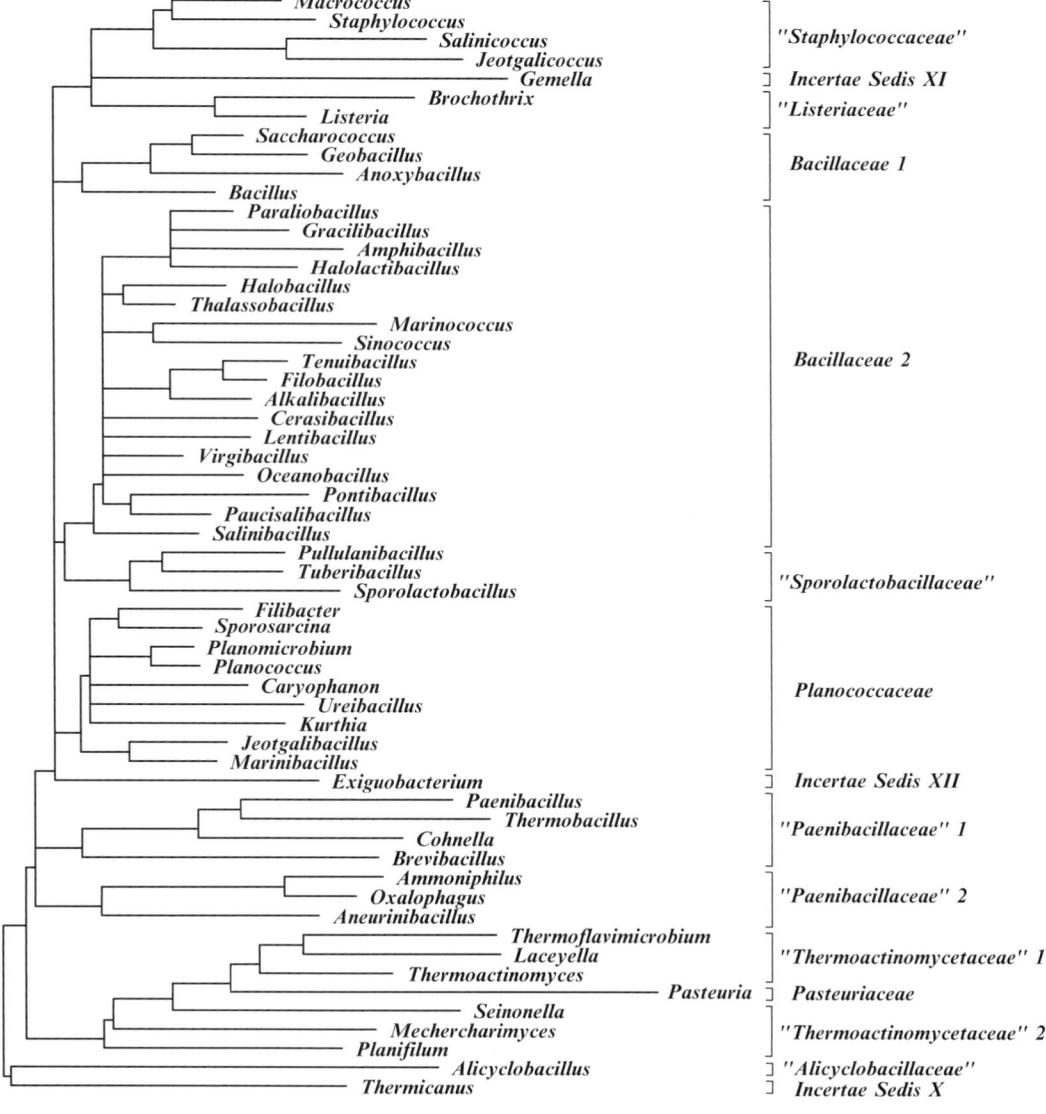

FIGURE 2. Consensus dendrogram reflecting the phylogenetic relationships of the order *Bacillales* within the class "*Bacilli*". Analyses were performed as described for Figure 1.

Genus *Bacillus*

The majority of the *Bacillus* species with validly published names are phylogenetically grouped into subclusters within this genus. However, some validly named species of *Bacillus* are not phylogenetically related to the type species, *B. subtilis*, and are more closely related to other genera. The phylogenetic subclusters within the genus *Bacillus* are:

a: *Bacillus subtilis, amyloliquefaciens, atrophaeus, mojavensis, licheniformis, sonorensis, vallismortis,* including the very likely misclassified *Paenibacillus popilliae.*

b: *Bacillus farraginis, fordii, fortis, lentus, galactosidilyticus*

c: *Bacillus asahii, bataviensis, benzoevorans, circulans, cohnii, firmus, flexus, fumarioli, infernus, jeotgali, luciferensis, megaterium, methanolicus, niacini, novalis, psychrosaccharolyticus, simplex, soli, vireti*

d: *Bacillus anthracis, cereus, mycoides, thuringiensis, weihenstephanensis*

e: *Bacillus aquimaris, marisflavi*

f: *Bacillus badius, coagulans, thermoamylovorans, acidicola, oleronius, sporothermodurans*

g: *Bacillus alcalophilus, arsenicoselenatis, clausii, gibsonii, halodurans, horikoshii, krulwichiae, okhensis, okuhidensis, pseudoalcaliphilus, pseudofirmus*

h: *Bacillus arsenicus, barnaricus, gelatini, decolorationis,*

i: *Bacillus carboniphilus, endophyticus, smithii,*

j: *Bacillus pallidus,*

k: *Bacillus funiculus, panaciterrae*

The *Bacillus* cluster contains three additional groups of related genera: *Anoxybacillus, Geobacillus,* and *Saccharococcus.*

In addition to these taxa, which compose the family *Bacillaceae sensu stricto,* other phylogenetic groups have been assigned to this family (Garrity et al., 2005). Although the largest group appears to warrant elevation to a novel family, it is retained within the *Bacillaceae* in the present outline. This cluster comprises the genera *Alkalibacillus* (new; Jeon et al., 2005), *Amphibacillus, Cerasibacillus* (new; Nakamura et el., 2004), *Filobacillus, Gracilibacillus, Halobacillus* (new; Spring et al., 1996), *Halolactibacillus* (new; Ishikawa et al., 2005), *Lentibacillus, Oceanobacillus, Paraliobacillus, Paucisalibacillus* (new; Nunes et al., 2006); not described in the current volume), *Pontibacillus, Salibacillus* (not described in the current volume), *Tenuibacillus, Thalassobacillus* (new; Garcia et al., 2005), and *Virgibacillus.* The type strains of other species are positioned phylogenetically among the members of this lineage and merit taxonomical emendation: *Bacillus halophilus* and *Bacillus thermocloacae, Sinococcus,* and *Marinococcus.* For this reason, *Marinococcus* was transferred from the *Sporolactobacillaceae* in the current outline.

In addition, the genera *Ureibacillus, Marinibacillus, Jeotgalibacillus,* and *Exiguobacterium* were previously assigned to the *Bacillaceae* (Garrity et al., 2005). *Ureibacillus* falls within the clade represented by *Planococcaceae,* and it was reassigned to that family. *Marinibacillus* and *Jeotgalibacillus* are closely related to each other as well as to *Bacillus aminovorans.* This group is distantly related to the *Planococcaceae,* and they are also assigned to that family. Lastly, *Exiguobacterium* is not closely related to any of the described families, and it is assigned to a Family XII *Incertae Sedis* in the current road map.

Bacillus schlegelii and *Bacillus solfatarae* represent their own deeply branching lineage of the "*Bacilli*" and warrant reclassification.

Family "*Alicyclobacillaceae*"

Only the type genus *Alicyclobacillus* is retained in this family, and two genera previously classified with the *Alicyclobacillaceae* have been reclassified (Garrity et al., 2005). According to the new 16S rRNA sequence analyses, *Sulfobacillus* represents a deep branch of the "*Clostridia*", and it is now placed within Family XVII *Incertae Sedis* of the *Clostridiales.* *Pasteuria,* which was also previously classified within this family, is an obligate parasite of invertebrates. While it can be cultivated within the body of its prey, it has not been cultured axenically. Because of the substantial phenotypic differences and low 16S rRNA sequence similarity with *Alicyclobacillus,* it is now classified within its own family, *Pasteuriaceae* (see below). Lastly, *Alicyclobacillus* possesses a moderate relationship to *Bacillus tusciae,* which could be reclassified to this family.

Family "*Listeriaceae*"

The monophyletic family "*Listeriaceae*" combines the genera *Listeria* and *Brochothrix* as in the previous outline.

Family "*Paenibacillaceae*"

The members of the family "*Paenibacillaceae*" are distributed between two phylogenetic clusters. *Paenibacillus, Brevibacillus, Cohnella* (new; Kämpfer et al., 2006) and *Thermobacillus* share a common origin and represent the first group. Some validly named *Bacillus* species are found among the *Paenibacillus* species: *Bacillus chitinolyticus, edaphicus, ehimensis,* and *mucilaginosus.* The second group comprises the genera *Aneurinibacillus, Ammoniphilus,* and *Oxalophagus.* Although not clearly monophyletic, these two clusters are often associated together in several types of analyses. Thus, in the absence of clear evidence for a separation, the second cluster is retained within the family. In contrast, *Thermicanus,* which was classified within this family by Garrity et al. (2005), appears to represent a novel lineage of the *Bacilli.* In recognition of its ambiguous status, it was reclassified within Family X *Incertae Sedis.*

Family *Pasteuriaceae*

This family contains *Pasteuria,* an obligate parasite of invertebrates which has not yet been cultivated outside of its host. Although this genus was previously classified within the "*Alicyclobacillaceae*", the current analyses suggest that it is more closely associated with the "*Thermoactinomycetaceae*". In spite of the similarities in morphology and rRNA sequences between *Pasteuria* and *Thermoactinomycetes,* these genera were not combined into a single family for two reasons. First, in the absence of an axenic culture of *Pasteuria,* additional phenotypic and genotypic evidence for combining these organisms into a single family are not available. Second, the obligately pathogenic nature of *Pasteuria* was judged to be distinctive enough to warrant a unique classification in the absence of evidence to the contrary.

Family *Planococcaceae*

The family *Planococcaceae* is a clearly monophyletic unit that contains the genera *Planococcus, Filibacter, Kurthia, Planomicrobium,* and *Sporosarcina* as well as three genera transferred from the *Bacillaceae* (*Jeotgalibacillus, Marinibacillus,* and *Ureibacillus*) and *Caryophanon.* *Caryophanon* is the only genus of the *Caryophanaceae* in the previous outline and is transferred to the *Planococcaceae* based upon its rRNA-based phylogeny. Thus, the family *Caryophanaceae* is not used in the current outline. Again, some validly named species of *Bacillus* are found in the *Planococcaceae* radiation: *Bacillus fusiformis, sphaericus, massiliensis, psychrodurans,* and *psychrotolerans.*

Family *"Sporolactobacillaceae"*

Given that *Marinococcus* is transferred to the *Bacillaceae*, the family *"Sporolactobacillaceae"* is now composed of only the genus *Sporolactobacillus.* A moderate relationship to this genus is found for some validly named *Bacillus* species: *Bacillus agaradhaerens, clarkii, selenitireducens,* and *vedderis* as well as two recently described genera, *Tuberibacillus* and *Pullulanibacillus,* which are not included in this volume (Hatayama et al., 2006).

Family *"Staphylococcaceae"*

The family *"Staphylococcaceae"*, as defined in the taxonomic outline of the previous volumes, is paraphyletic (Garrity et al., 2005). Whereas the four genera *Staphylococcus, Jeotgalicoccus, Macrococcus,* and *Salinicoccus* are clearly monophyletic, the genus *Gemella* represents a separate unit paraphyletic to the first cluster. Moreover, *Gemella* is distinguished from *"Staphylococcaceae"* stricto sensu because it is catalase- and oxidase-negative and possesses predominantly straight-chained, saturated and monounsaturated rather than branched-chain

membrane lipids (K. Bernard, personal communication). Thus, *Gemella* is transferred to Family XI *Incertae Sedis* within the *Bacillales.*

Family *"Thermoactinomycetaceae"*

The family now contains six newly described genera in addition to the original genus *Thermoactinomyces.* The new genera are *Laceyella, Mechercharimyces, Planifilum, Seinonella, Shimazuella,* and *Thermoflavimicrobium.*

Order *"Lactobacillales"*

As in the previous outline, this order is composed of six families (Figure 3).

Family *Lactobacillaceae*

In agreement with the previous outlines, the *Lactobacillaceae* is a monophyletic group that harbors three genera: *Lactobacillus, Paralactobacillus,* and *Pediococcus.*

Family *"Aerococcaceae"*

Two paraphyletic groups are combined in the family *"Aerococcaceae"*. The majority of the genera are unified in a phylogenetically tight group comprising *Abiotrophia, Dolosicoccus, Eremococcus, Facklamia, Globicatella,* and *Ignavigranum.* Only the type genus *Aerococcus* represents a separate lineage.

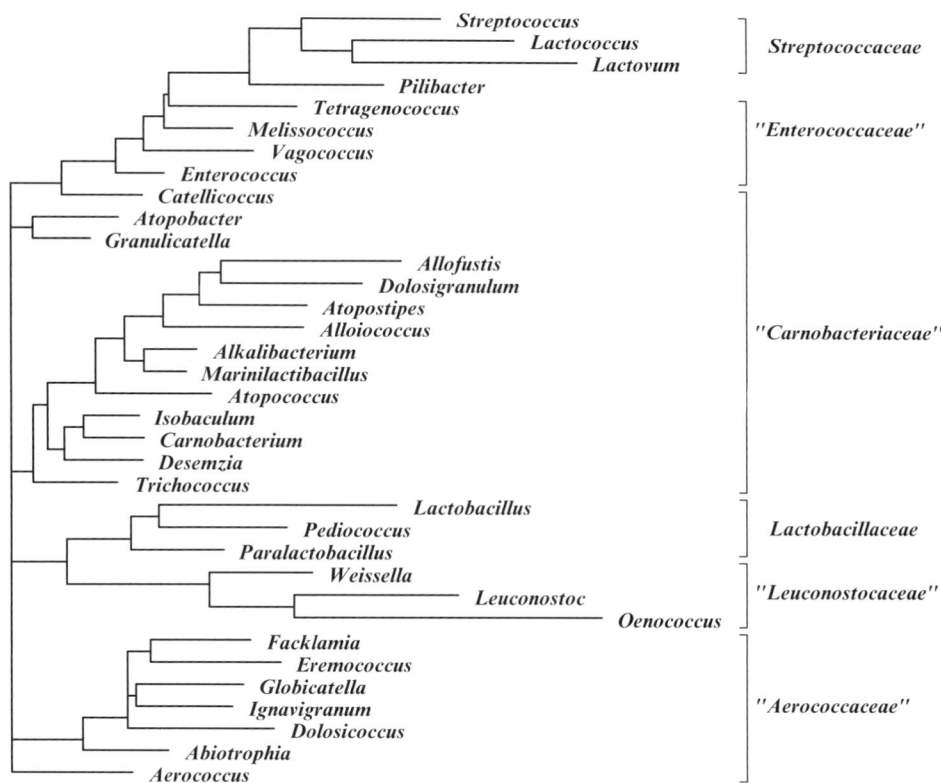

FIGURE 3. Consensus dendrogram reflecting the phylogenetic relationships of the order *"Lactobacillales"* within the class *"Bacilli"*. Analyses were performed as described for Figure 1.

Family *"Carnobacteriaceae"*

The members of the family *"Carnobacteriaceae"* are found in two paraphyletic clusters. *Carnobacterium* together with *Alkalibacterium, Allofustis, Alloiococcus, Atopococcus* (new; Collins et al., 2005), *Atopostipes, Desemzia, Dolosigranulum Isobaculum, Marinilactibacillus,* and *Trichococcus* represent the most comprehensive group. *Granulicatella* and *Atopobacter* (formerly in the *"Enterococcaceae"*) are in the second group. However, the phylogenetic position of these genera remains ambiguous, and reassignment may be warranted as more information becomes available.

Family *"Enterococcaceae"*

Four genera remain within the family *"Enterococcaceae"*: *Enterococcus, Melissococcus, Tetragenococcus,* and *Vagococcus. Atopobacter* was transferred to the *"Carnobacteriaceae"* (see above). The recently described genus *Catellicoccus,* which is not described in this volume, phylogenetically represents a sister group to the *"Enterococcaceae".*

Family *"Leuconostocaceae"*

No changes of the taxonomic organization are made for the *"Leuconostocaceae",* which unifies three phylogenetically related genera: *Leuconostoc, Oenococcus,* and *Weissella.*

Family *Streptococcaceae*

In addition to the genera *Streptococcus* and *Lactococcus* in the previous road map, the family *Streptococcaceae* comprises a third, recently discovered, genus, *Lactovum* (Matthies et al., 2004).

Families *incertae sedis*

In the current road map, *Thermicanus, Gemella,* and *Exiguobacterium* have been reclassified into different families *incertae sedis* in recognition of their ambiguous taxonomic assignments (see above). The genera *Oscillospira* and *Syntrophococcus,* which were classified in this category in the previous road map (Garrity et al., 2005), have been transferred to the *"Ruminococcaceae"* and *"Lachnospiraceae"* in the *Clostridiales,* respectively (see below).

Class *"Clostridia"*

The class *"Clostridia"* is comprised of three orders, *Clostridiales, Halanaerobiales,* and *"Thermoanaerobacterales".* This organization is similar to the previous roadmap (Garrity et al., 2005) and unites the orders *Clostridiales* Prevot 1953[AL] and *Eubacteriales* Buchanan 1917[AL]. Preference is given to *Clostridiales* because of the priority of its type genus. Moreover, because many of the species united in this group were previously classified with the genus *Clostridium,* this classification is the least likely to cause confusion. While the order *Halanaerobiales* is monophyletic, the remaining two orders are paraphyletic and each include taxa with only low similarity to the majority of the *Firmicutes* (Figure 4).

Order *Clostridiales*

In the previous road map (Garrity et al., 2005), the order *Clostridiales* was composed of eight families, many of which were paraphyletic. While it was not possible to fully address this problem, the current road map increases the number of families to ten and notes nine additional families as *incertae sedis* (Figures 5 and 6). This is only a first step, and significant further reorganization is warranted, especially as new data and concepts are applied to these taxa.

Seven of the eight original families are retained in the current outline. However, the family *"Acidaminococcaceae"* was not used in recognition of the priority of *Veillonellaceae* 1971[AL]. In addition, the family *"Ruminococcaceae"* is proposed to accommodate a large number of genera transferred from other families. A new family, *"Gracilibacteraceae",* is also proposed for a newly discovered genus, *Gracilibacter.*

Family *Clostridiaceae*

The family *Clostridiaceae* comprises 13 genera in the current outline (Figure 5). Phylogentically, these genera are distributed among three paraphyletic clusters and a fourth clade represented by a single genus, *Caminicella.* In addition, seven genera were transferred to other families. Three genera (*Acetivibrio, Faecalibacterium,* and *Sporobacter*) were transferred to the newly named family *"Ruminococcaceae"* (see below), unifying phylogenetically related former members of the *Clostridiaceae* and *"Lachnospiraceae".* The genus *Coprobacillus* was transferred to the *Erysipelotrichaceae.* The genus *Dorea* was transferred to the *"Lachnospiraceae".* The genus *Tepidibacter* was transferred to the *"Peptostreptococcaceae".* Lastly, the genus *Acidaminobacter* was transferred to Family XII *Incertae Sedis.*

The first clostridial cluster is composed of the genera *Clostridium, Anaerobacter, Caloramator, Oxobacter Sarcina,* and *Thermobrachium* Despite intense restructuring, the genus *Clostridium* is still partly paraphyletic, comprising a large collection of validly published species and species groups. Species whose common ancestry with the type species *Clostridium butyricum* is highly supported by the rRNA data remain in this genus. They are (in alphabetical not phylogenetic order): *Clostridium absonum, acetobutylicum, acetireducens, acidisoli, akagii, algidicarnis, argentinense, aurantibutyricum, baratii, beijerinckii, botulinum, bowmanii, butyricum, carnis, cellulovorans, chartatabidum, chauvoei, cochlearium, colicanis, collagenovorans, cylindrosporum, diolis, disporicum, esthertheticum, fallax, felsineum, frigoris, frigidicarnis, gasigenes, grantii, haemolyticum, histolyticum, homopropionicum, intestinale, kluyveri, lacusfryxellense, limosum, lundense, novyi, paraputrificum, pascui, pasteurianum, peptidivorans, perfringens, proteolyticum, puniceum, putrificum, putrefaciens, quinii, roseum, saccharobutylicum, saccharoperbutylacetonicum, sardiniense, sartagoforme, scatologenes, septicum, sporogenes, subterminale, tertium, tetani, tetanomorphum, thermopalmarium, thermobutyricum, thiosulfatireducens, tyrobutyricum, uliginosum,* and *vincentii.* Species which according to phylogenetic relationships should be assigned to other taxonomic units are mentioned below. Some species previously classified with *Eubacterium* also belong into the radiation of *Clostridium sensu stricto: Eubacterium budayi, combesii, moniliforme, nitritogenes,* and *tarantellae.*

Other genera of this clade, *Anaerobacter, Oxobacter,* and *Sarcina,* are partly intermixed with *Clostridium* species, indicating that further reorganization of this cluster remains to be done.

The second *Clostridiaceae* cluster comprises the genera *Alkaliphilus, Anoxynatronum* (new; Garnova et al., 2003), *Natronincola, Tindallia* as well as the *Clostridium* species *alcalibutyricum, felsineum, formicoaceticum,* and *halophilum.* The genera *Thermohalobacter–Caloranaerobacter* (Wery et al., 2001) represent the third cluster. *Caminicella,* represents a fourth paraphyletic lineage

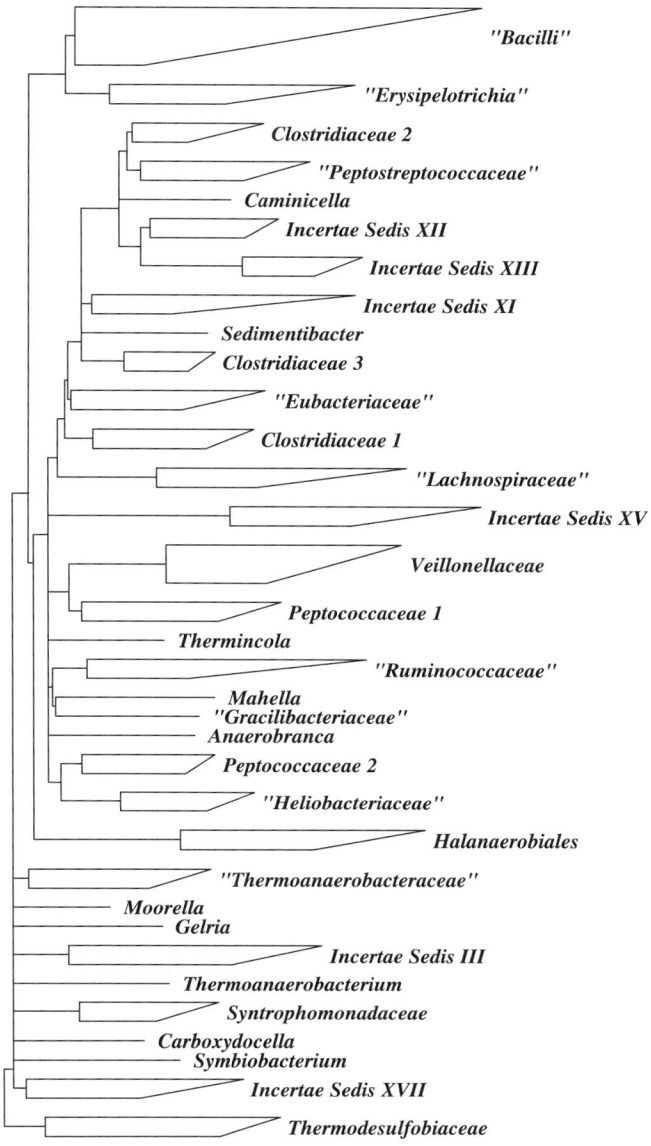

FIGURE 4. Consensus dendrogram reflecting the phylogenetic relationships of the class "*Clostridia*" within the *Firmicutes*. Analyses were performed as described for Figure 1.

previously classified within this family. Based upon the rRNA analyses, reclassification of these groups into other families may be warranted.

Family "*Eubacteriaceae*"

The six genera of the family "*Eubacteriaceae*" are monophyletic. They comprise the species of *Eubacterium stricto sensu* (*Eubacterium limosum, aggregans, barkeri, callanderi*) as well as the genera *Acetobacterium, Alkalibacter, Anaerofustis, Garciella,* and *Pseudoramibacter.* It is noteworthy that *Garciella* is the only thermophile among this group, and its assignment is the least strongly supported by the rRNA analyses reported here. Other analyses suggest a closer affiliation for this genus to the thermophiles *Thermohalobacter* and *Caloranaerobacter* (*Clostridiaceae* group 3,

D. Alazard, personal communication). Therefore, reclassification may be warranted in the future. The genera *Anaerovorax* and *Mogibacterium* have been transferred to Family XIII *Incertae Sedis.* Additional *Eubacterium* species (*Eubacterium infirmum, minutum, nodatum,* and *sulci*) are closely related to these other genera.

Family "*Gracilibacteraceae*"

The family "*Gracilibacteraceae*" is proposed to encompass the newly described genus *Gracilibacter* (Lee et al., 2006).

Family "*Heliobacteriaceae*"

The family "*Heliobacteriaceae*" is maintained as defined in the previous volumes. It comprises four genera: *Heliobacterium, Heliobacillus, Heliophilum,* and *Heliorestis.*

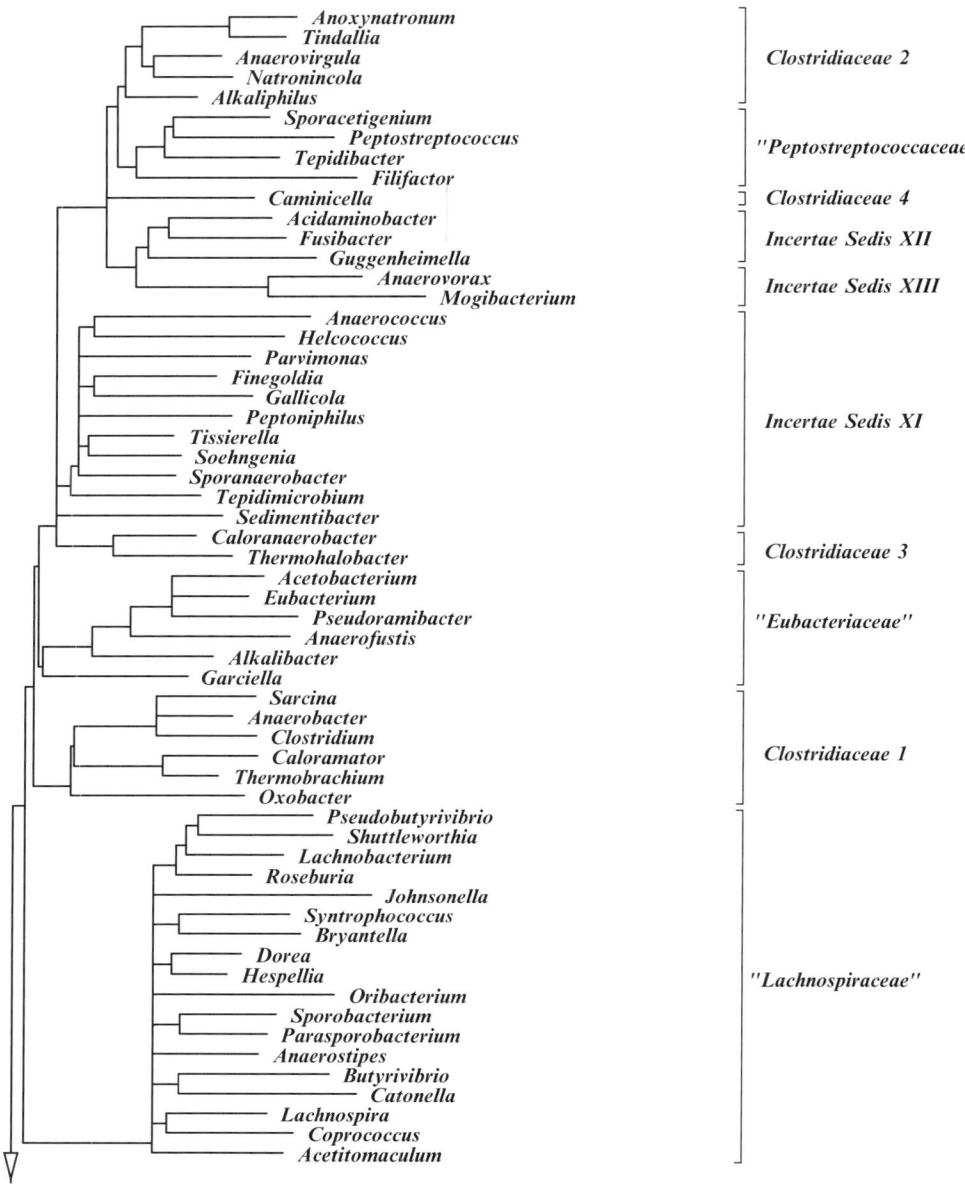

FIGURE 5. Consensus dendrogram reflecting the phylogenetic relationships of the order *Clostridiales* (part one) within the class "*Clostridia*". Analyses were performed as described for Figure 1.

Family "*Lachnospiraceae*"

The family "*Lachnospiraceae*" currently comprises 19 genera. The family is monophyletic, although a number of subclusters can be recognized. The "*Lachnospiraceae*" genera are: *Acetitomaculum*, *Anaerostipes*, *Bryantella*, *Butyrivibrio*, *Catonella*, *Coprococcus*, *Dorea* (formerly in the *Clostridiaceae*), *Hespellia*, *Johnsonella*, *Lachnobacterium*, *Lachnospira*, *Moryella* (new; Carlier et al., 2007), *Oribacterium* (Carlier et al., 2004), *Parasporobacterium*, *Pseudobutyrivibrio*, *Roseburia*, *Shuttleworthia*, *Sporobacterium*, and *Syntrophococcus* (formerly *incertae sedis* within the "*Lactobacillales*").

A number of *Clostridium* species are found within the radiation of "*Lachnospiraceae*": *Clostridium aerotolerans*, *algidixylanolyticum*, *aminophilum*, *aminovalericum*, *amygdalinum*, *bolteae*, *celerecrescens*, *coccoides*, *colinum*, *fimetarium*, *glycyrrhizinilyticum*, *hathewayi*, *herbivorans*, *hylemonae*, *indolis*, *lactatifermentans*, *lentocellum*, *methoxybenzovorans*, *neopropionicum*, *nexile*, *oroticum*, *piliforme*, *polysaccharolyticum*, *populeti*, *propionicum*, *proteoclasticum*, *scindens*, *sphenoides*, *saccharolyticum*, *symbiosum*, *xylanolyticum*, and *xylanovorans*. Furthermore, *Eubacterium cellulosolvens*, *eligens*, *hallii*, *ramulus*, *rectale*, *ruminantium*, *uniforme*, *ventriosum*, and *xylano-*

philum; Ruminococcus gnavus, hansenii, hydrogenotrophicus, obeum, productus, schinkii, and *torques,* as well as *Desulfotomaculum guttoideum* phylogenetically belong to this group.

Family *Peptococcaceae*

The current members of the family *Peptococcaceae* occupy two paraphyletic groups within the radiation of the *Clostridiales* (Figure 6).

The first cluster includes the genera *Peptococcus, Dehalobacter, Desulfitibacter, Desulfitobacterium, Desulfonispora, Desulfosporosinus,* and *Syntrophobotulus,* which form a tight monophyletic group. The second group is not closely related and includes *Cryptanaerobacter, Desulfotomaculum,* and *Pelotomaculum* species. Given its significant relationship to these genera, *Sporotomaculum* was transferred from the family "*Thermoanaerobacteraceae*".

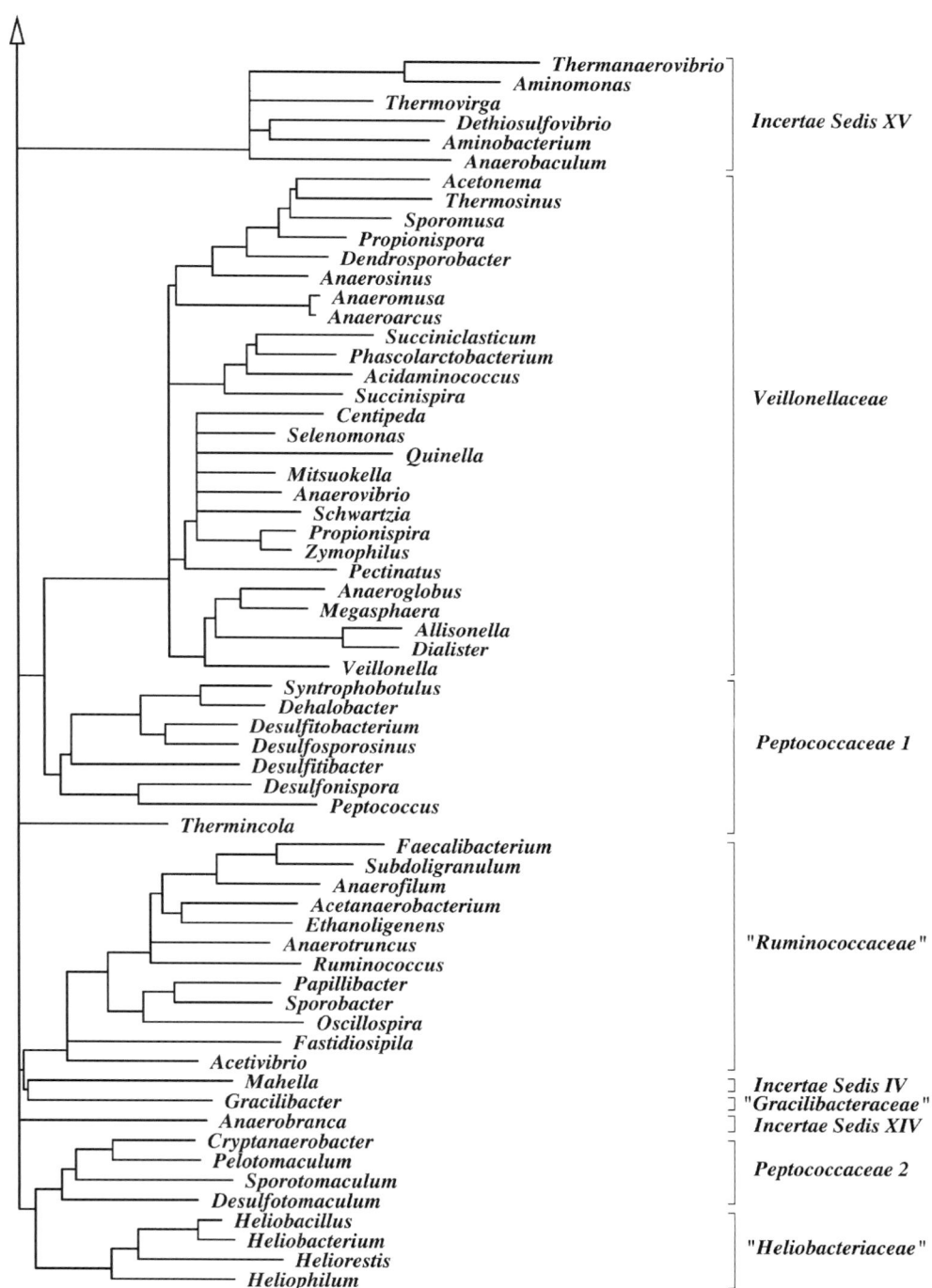

FIGURE 6. Consensus dendrogram reflecting the phylogenetic relationships of the order *Clostridiales* (part two) within the class "*Clostridia*". Analyses were performed as described for Figure 1. Family *Incertae Sedis* IV is from the "*Thermoanaerobacterales*".

Lastly, *Thermincola* is currently classified within this group even though rRNA analyses suggest only a weak relationship to the *Peptococcaceae sensu stricto*.

Family "*Peptostreptococcaceae*"

The family "*Peptostreptococcaceae*" comprised 12 genera distributed over a number of paraphyletic groups in the original road map (Garrity et al., 2005). Most of these genera are now transferred to Family XI *Incertae Sedis* (Figure 5). The current family is greatly circumscribed and monophyletic. It includes the genera *Peptostreptococcus*, *Filifactor*, *Sporacetigenium*, and *Tepidibacter* together with a number of validly published *Clostridium* and *Eubacterium* species: *Clostridium bartlettii* (new; Song et al., 2004), *bifermentans*, *difficile*, *ghoni*, *glycolicum*, *hiranonis*, *irregularis*, *litorale*, *lituseburense*, *mangenotii*, *paradoxum*, *sordellii*, *sticklandii*, and *Eubacterium tenue*, and *yurii*.

Family "*Ruminococcaceae*"

A new family is proposed for a monophyletic lineage comprising 11 genera (Figure 6). Concerning their assignment in the outline of the previous volumes, this group represents a mixture of former "*Acidaminococcaceae*", *Clostridiaceae*, and "*Lachnospiraceae*" as well as newly described genera: *Acetanaerobacterium* (new; Chen and Dong, 2004), *Acetivibrio* (formerly in the *Clostridiaceae*), *Anaerofilum* (formerly in the "*Lachnospiraceae*"), *Anaerotruncus*, *Ethanoligenens* (new; Xing et al., 2006), *Faecalibacterium* (formerly in the *Clostridiaceae*), *Fastidiosipila*, *Oscillospira*, *Papillibacter* (formerly in the "*Acidaminococacceae*"), the type species of the genus *Ruminococcus* (*R. flavefaciens* and the three species *Ruminococcus albus*, *bromii*, and *callidus*; the other *Ruminococcus* species remain within the family "*Lachnospiraceae*"),

Sporobacter (formerly in the *Clostridiaceae*), and *Subdoligranulum* (new; Holmstrom et al., 2004). A number of validly published *Clostridium* species belong to this lineage according to their phylogentic relationships: *Clostridium aldrichii*, *alkalicellulosi* (new; Zhilina et al., 2005), *cellobioparum*, *cellulolyticum*, *hungatei*, *josui*, *leptum*, *methylpentosum*, *orbiscindens*, *papyrosolvens*, *stercorarium*, *straminisolvens* (new; Kato et al., 2004), *termitidis*, *thermocellum*, *thermosuccinogenes*, and *viride*. *Eubacterium siraeum* also belongs to this lineage.

Family *Syntrophomonadaceae*

The genera previously assigned to the family *Syntrophomonadaceae* are widely dispersed and not closely related (Garrity et al., 2005). On the basis of the current rRNA-based phylogenetic analyses, many of these genera may not even be members of the phylum *Firmicutes*. Despite these tendencies, these taxa are maintained in the current volume owing to their taxonomic history and the lack of additional data demanding an official description of new phyla. The four very deep groups include the following: the *Syntrophomonadaceae sensu stricto* comprise the genera *Syntrophomonas*, *Pelospora*, *Syntrophospora*, *Syntrophothermus*, and *Thermosyntropha* (Figure 7). These five genera are retained within the family *Syntrophomonadaceae* in the current road map. Another lineage is represented by *Aminobacterium*, *Aminomonas*, *Anaerobaculum*, *Dethiosulfovibrio*, *Thermanaerovibrio*, and *Thermovirga* (new; Dahle et al., 2006). This group has been assigned to Family XV *Incertae Sedis* (Figure 6). The genera *Thermaerobacter* and *Sulfobacillus* (formerly assigned to the family "*Alicyclobacillaceae*" in the class *Bacilli*) share a common ancestor and represent a third deep lineage, now assigned to Family XVII *Incertae Sedis*. *Caldicellulosiruptor* represents the final deep lineage. Because of similarities to genera already classified

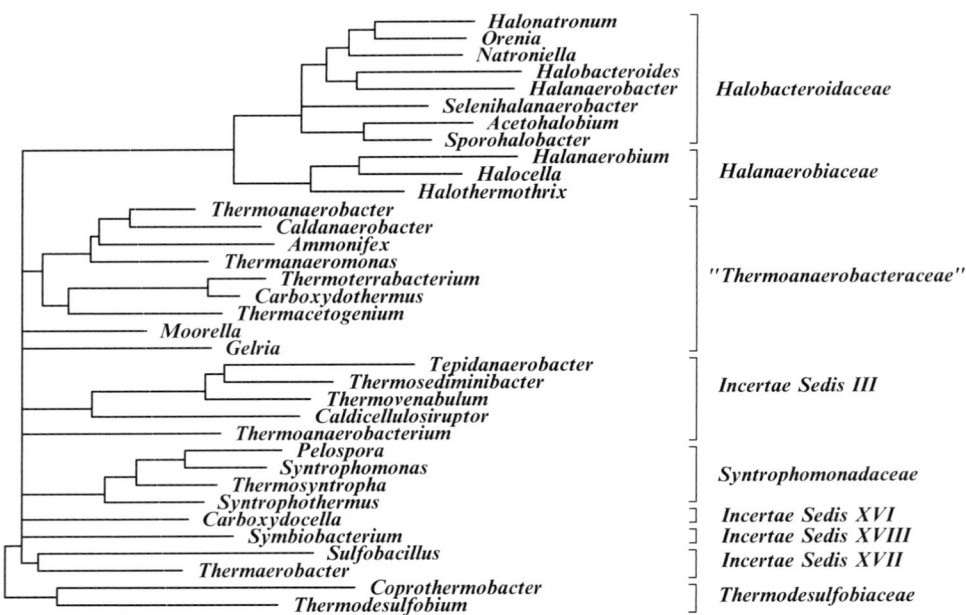

FIGURE 7. Consensus dendrogram reflecting the phylogenetic relationships of the orders *Halanaerobiales* and "*Thermoanaerobacterales*" as well as some deep branches of the *Clostridiales* within the class "*Clostridia*". Analyses were performed as described for Figure 1.

within the "*Thermoanaerobacterales*", it is reclassified to Family III *Incertae Sedis* within that order (Figure 7). Lastly, two genera (*Anaerobranca* and *Carboxydocella*) appear to represent lineages of the *Firmicutes*, although they are not closely related to each other. They are reclassified into Family XIV *Incertae Sedis* and Family XVI *Incertae Sedis*, respectively (Figures 5 and 7).

Family *Veillonellaceae*

The genera previous classified within the family "*Acidaminococcaceae*" were reclassified into the family *Veillonellaceae* due to the priority of this name. After the transfer of the genus *Papillibacter* to the new family "*Ruminococcaceae*", this family became monophyletic (Figure 6). The 26 genera currently harbored by the family *Veillonellaceae* are: *Acetonema, Acidaminococcus, Allisonella, Anaeroarcus, Anaeroglobus, Anaeromusa, Anaerosinus, Anaerovibrio, Centipeda, Dendrosporobacter, Dialister, Megasphaera, Mitsuokella, Pectinatus, Phascolarctobacterium, Propionispira, Propionispora, Quinella, Schwartzia, Selenomonas, Sporomusa, Succiniclasticum, Succinispira, Thermosinus, Veillonella,* and *Zymophilus.*

Families *Incertae Sedis*

Nine families *incertae sedis* were created to recognize some of the ambiguities remaining in the current classification.

Family XI *Incertae Sedis* contains a tight monophyletic cluster comprised of the genera *Anaerococcus, Finegoldia, Gallicola, Helcococcus, Parvimonas,* and *Peptoniphilus* which were transferred from the family "*Peptostreptococcaceae*". *Soehngenia, Sporanaerobacter, Tepidimicrobium* (new; Slobodkin et al., 2006) and *Tissierella* represent four additional genera associated with this group. The genus *Tissierella* is closely related to *Clostridium hastiforme* and *ultunense. Sedimentibacter,* which was also previously classified within the "*Peptostreptococcaceae*", appears to represent a separate but neighboring lineage in the phylogenetic tree. Thus, it remains classified with other genera transferred from this family in the current outline.

Family XII *Incertae Sedis* includes *Guggenheimella* (new; Wyss et al., 2005), a newly described genus that, while clearly a member of the *Clostridiales,* cannot be assigned to any of the defined families. It possesses moderate relationships to *Acidaminobacter* (previously classified with the *Clostridiaceae*) and *Fusibacter* (previously classified with the "*Peptostreptococcaceae*").

Family XIII *Incertae Sedis* contains the genera *Anaerovorax* and *Mogibacterium,* which were transferred from the "*Eubacteriaceae*". This group possesses a weak relationship with Family XII *Incertae Sedis.*

Family XIV *Incertae Sedis* is comprised of *Anaerobranca,* which was previously classified within the *Syntrophomonadaceae,* however, it is not closely related to any other previously described family.

Family XV *Incertae Sedis* is comprised of *Aminobacterium, Aminomonas, Anaerobaculum, Dethiosulfovibrio,* and *Thermanaerovibrio.* These genera form a monophyletic clade that was previously classified within the *Syntrophomonadaceae,* however, they possess only low relatedness to the type genus of that family and, thus, warrant reclassification.

Family XVI *Incertae Sedis* is comprised of *Carboxydocella,* which was previously classified within the *Syntrophomonadaceae.* Because it is not closely related to any other previously described family, it has been classified within its own group.

Family XVII *Incertae Sedis* is comprised of the genera *Sulfobacillus* and *Thermaerobacter.* Formerly classified with the "*Alicyclobacillaceae*" and *Syntrophomonadaceae,* respectively, these genera represent either a very deep group of the phylum *Firmicutes* or, perhaps, a novel phylum.

Family XVIII *Incertae Sedis* is comprised of *Symbiobacterium.* This genus also appears to represent either a very deep group of the Phylum *Firmicutes* or another novel phylum.

Family XIX *Incertae Sedis* includes *Acetoanaerobium,* whose rRNA has not been sequenced but whose phenotypic properties suggest an affiliation to this order.

Order *Halanaerobiales*

The taxonomic organization of this order remains as outlined in the previous volumes (Garrity et al., 2005). It contains two families, *Halanaerobiaceae* and *Halobacteroidaceae* (Figure 7).

Family *Halanaerobiaceae*

The monophyletic family *Halanaerobiaceae* currently is comprised of three genera, *Halanaerobium, Halocella,* and *Halothermothrix.*

Family *Halobacteroidaceae*

Eight genera are unified in a tight monophyletic cluster of the family *Halobacteroidaceae. Halobacteroides, Acetohalobium, Halanaerobacter, Halonatronum, Natroniella, Orenia, Selenihalanaerobacter,* and *Sporohalobacter* represent this family.

Order "*Thermoanaerobacterales*"

The order "*Thermoanaerobacterales*" is used in preference to "*Thermoanaerobacteriales*" (Garrity et al., 2005) in recognition of the priority of *Thermoanaerobacter* over *Thermoanaerobacterium* as the type genus. As in the case of the *Syntrophomonadaceae,* the diversity within this order is very large, and some members may represent novel phyla (Mori et al., 2003); Figure 7). Therefore, reclassification within this group is expected in the near future, and the current classification is of limited biological significance.

Family "*Thermoanaerobacteraceae*"

Eight genera are currently classified within this family, but they are not closely related to each other. The type genus *Thermoanaerobacter* forms part of a monophyletic cluster comprised of *Ammonifex, Carboxydibrachium* (now reclassified within *Caldanaerobacter*), *Caldanaerobacter, Carboxydothermus* (previously classified within the *Peptococcaceae*), *Thermacetogenium,* and *Thermanaeromonas.* Two additional genera, *Gelria* and *Moorella,* are neighbors of this cluster.

Family *Thermodesulfobiaceae*

A second family is comprised of the genera *Coprothermobacter* and *Thermodesulfobium.* Although not particularly closely related to each other, they are both much more distantly related to other members of the phylum *Firmicutes.*

Families *Incertae Sedis*

Family III Incertae Sedis contains a monophyletic cluster that is weakly related to both the "*Thermoanaerobacteraceae*" and *Syntrophomonadaceae* (Figure 7). It is comprised of the genera *Caldicellulosiruptor* (from the *Syntrophomonadaceae*), *Thermosediminibacter,* and *Thermovenabulum.* While *Thermoanaerobacterium* is not a member of this cluster, it appears as a neighboring group in phylogenetic trees. Therefore, it is included in this family until additional analyses warrant a more informed reclassification.

Family IV *Incertae Sedis* includes the genus *Mahella.* While it possesses some relatedness to members of the "*Thermoanaerobacteraceae*" and Family III *Incertae Sedis* (above), its distinctive position in the rRNA gene tree, near the "*Ruminococcaceae*" within the *Clostridiales,* and phenotypic properties suggest that it represents a novel family (Figure 6).

Class *"Erysipelotricha"*

With the elevation of the *Mollicutes* to the phylum *Tenericutes,* the creation of a separate class within the *Firmicutes* was warranted for the family *Erysipelotrichaceae.* This new classification recognizes the low similarity of the rRNA of this group with other members or the phylum as well as the similarity in cell wall and other phenotypic features. This class comprises a single order, "*Erysipelotrichales*", and the family *Erysipelotrichaceae.*

Family *Erysipelotrichaceae*

In comparison to the outlines of the previous volumes, the family *Erysipelotrichaceae* was extended by the inclusion of four additional genera. Besides one newly described genus (*Allobaculum*), three genera were transferred from other families. The family is now organized in eight genera: *Erysipelothrix, Allobaculum* (new; Greetham et al., 2004), *Bulleidia, Catenibacterium* (formerly in the "*Lachnospiraceae*"), *Coprobacillus* (formerly in the *Clostridia*), *Holdemania, Solobacterium,* and *Turicibacter* (formerly *incertae sedis* among *Bacillales*). Furthermore, a number of type species validly published as members of genera not assigned to the *Erysipelotrichaceae* should also be classified within this family. These include *Clostridium catenaformis, cocleatum, innocuum, ramosum,* and *spiroforme; Eubacterium biforme, cylindroides, dolichum,* and *tortuosum; Lactobacillus catenaformis* and *vitulinus;* and *Streptococcus pleomorphus.*

References

Carlier, J.-P., G. K'Ouas, I. Bonne, A. Lozniewski and F. Mory. 2004. *Oribacterium sinus* gen. nov., sp. nov., within the family '*Lachnospiraceae*' (phylum *Firmicutes*). Int. J. Syst. Evol. Microbiol. *54*: 1611–1615.

Carlier, J.-P., G. K'Ouas and X.Y. Han. 2007. *Moryella indoligenes* gen. nov., sp. nov., an anaerobic bacterium isolated from clinical specimens. Int. J. Syst. Evol. Microbiol. *57*: 725–729.

Chen, S. and X. Dong. 2004. *Acetanaerobacterium elongatum* gen. nov., sp. nov., from paper mill waste water. Int. J. Syst. Evol. Microbiol. *54*: 2257–2262.

Cicarelli, F.D., T. Doerks, C. von Mering, C.J. Creevey, B. Snel and P. Bork. 2006. Toward automatic reconstruction of a highly resolved tree of life. Science *311*: 1283–1287.

Cole, J.R., B. Chai, R.J. Farris, Q. Wang, A.S. Kulam-Syed-Mohideen, D.M. McGarrell, A.M. Bandela, E. Cardenas, G.M. Garrity and J.M. Tiedje. 2007. The ribosomal database project (RDP-II): introducing myRDP space and quality controlled public data. Nucleic Acids Res *35*: D169–172.

Collins, M.D., A. Wiernik, E. Falsen and P.A. Lawson. 2005. *Atopococcus tabaci* gen. nov., sp. nov., a novel Gram-positive, catalase-negative, coccus-shaped bacterium isolated from tobacco. Int. J. Syst. Evol. Microbiol. *55*: 1693–1696.

Dahle, H. and N.K. Birkeland. 2006. *Thermovirga lienii* gen. nov., sp. nov., a novel moderately thermophilic, anaerobic, amino-acid-degrading

bacterium isolated from a North Sea oil well. Int. J. Syst. Evol. Microbiol. *56*: 1539–1545.

Garcia, M.T., V. Gallego, A. Ventosa and E. Mellado. 2005. *Thalassobacillus devorans* gen. nov., sp. nov., a moderately halophilic, phenol-degrading, Gram-positive bacterium. Int J Syst Evol Microbiol *55*: 1789–1795.

Garnova, E.S., T.N. Zhilina, T.P. Tourova and A.M. Lysenko. 2003. *Anoxynatronum sibiricum* gen.nov., sp.nov. alkaliphilic saccharolytic anaerobe from cellulolytic community of Nizhnee Beloe (Transbaikal region). Extremophiles *7*: 213–220.

Garrity, G.M., J.A. Bell and T. Lilburn. 2005. The Revised Road Map to the Manual. *In* Brenner, Krieg, Staley and Garrity (ed.), Bergey's Manual of Systematic Bacteriology, 2nd edn, vol. 2, The *Proteobacteria,* Part A, Introductory Essays. Springer, New York, pp. 159–220.

Greetham, H.L., G.R. Gibson, C. Giffard, H. Hippe, B. Merkhoffer, U. Steiner, E. Falsen and M.D. Collins. 2004. *Allobaculum stercoricanis* gen. nov., sp. nov., isolated from canine faeces. Anaerobe *10*: 301–307.

Hatayama, K., H. Shoun, Y. Ueda and A. Nakamura. 2006. *Tuberibacillus calidus* gen. nov., sp. nov., isolated from a compost pile and reclassification of *Bacillus naganoensis* Tomimura *et al.* 1990 as *Pullulanibacillus naganoensis* gen. nov., comb. nov. and *Bacillus laevolacticus* Andersch *et al.* 1994 as *Sporolactobacillus laevolacticus* comb. nov. Int. J. Syst. Evol. Microbiol. *56*: 2545–2551.

Holmstrom, K., M.D. Collins, T. Moller, E. Falsen and P.A. Lawson. 2004. *Subdoligranulum variable* gen. nov., sp. nov. from human feces. Anaerobe *10*: 197–203.

Ishikawa, M., K. Nakajima, Y. Itamiya, S. Furukawa, Y. Yamamoto and K. Yamasato. 2005. *Halolactibacillus halophilus* gen. nov., sp. nov. and *Halolactibacillus miurensis* sp. nov., halophilic and alkaliphilic marine lactic acid bacteria constituting a phylogenetic lineage in *Bacillus* rRNA group 1. Int J Syst Evol Microbiol *55*: 2427–2439.

Jeon, C.O., J.M. Lim, J.M. Lee, L.H. Xu, C.L. Jiang and C.J. Kim. 2005. Reclassification of *Bacillus haloalkaliphilus* Fritze 1996 as *Alkalibacillus haloalkaliphilus* gen. nov., comb. nov. and the description of *Alkalibacillus salilacus* sp. nov., a novel halophilic bacterium isolated from a salt lake in China. Int J Syst Evol Microbiol *55*: 1891–1896.

Kämpfer, P., R. Rosselló-Mora, E. Falsen, H.J. Busse and B.J. Tindall. 2006. *Cohnella thermotolerans* gen. nov., sp. nov., and classification of '*Paenibacillus hongkongensis*' as *Cohnella hongkongensis* sp. nov. Int. J. Syst. Evol. Microbiol. *56*: 781–786.

Kato, S., S. Haruta, Z.J. Cui, M. Ishii, A. Yokota and Y. Igarashi. 2004. *Clostridium straminisolvens* sp. nov., a moderately thermophilic, aerotolerant and cellulolytic bacterium isolated from a cellulose-degrading bacterial community. Int. J. Syst. Evol. Microbiol. *54*: 2043–2047.

Lee, Y.J., C.S. Romanek, G.L. Mills, R.C. Davis, W.B. Whitman and J. Wiegel. 2006. *Gracilibacter thermotolerans* gen. nov., sp. nov., an anaerobic, thermotolerant bacterium from a constructed wetland receiving acid sulfate water. Int. J. Syst. Evol. Microbiol. *56*: 2089–2093.

Ludwig, W., O. Strunk, R. Westram, L. Richter, H. Meier, Yadhukumar, A. Buchner, T. Lai, S. Steppi, G. Jobb, W. Förster, I. Brettske, S. Gerber, A.W. Ginhart, O. Gross, S. Grumann, S. Hermann, R. Jost, A. König, T. Liss, R. Lüßmann, M. May, B. Nonhoff, B. Reichel, R. Strehlow, A. Stamatakis, N. Stuckmann, A. Vilbig, M. Lenke, T. Ludwig, A. Bode and K.H. Schleifer. 2004. ARB: A software environment for sequence data. Nucleic Acids Res. *32*: 1363–1371.

Ludwig, W. and H.P. Klenk. 2005. Overview: a phylogenetic backbone and taxonomic framework for procaryotic systematics. *In* Brenner, Krieg, Staley and Garrity (ed.), Bergey's Manual of Systematic Bacteriology, 2nd edn, vol. 2, The *Proteobacteria,* Part A, Introductory Essays. Springer, New York, pp. 49–65.

Ludwig, W. and K.H. Schleifer. 2005. Molecular phylogeny of bacteria based on comparative sequence analysis of conserved genes. *In* Sapp (ed.), Microbial phylogeny and evolution, concepts and controversies. Oxford University Press, New York, pp. 70–98.

Matthies, C., A. Gößner, G. Acker, A. Schramm and H.L. Drake. 2004. *Lactovum miscens* gen. nov., sp. nov., an aerotolerant, psychrotolerant, mixed-fermentative anaerobe from acidic forest soil. Res Microbiol *155*: 847–854.

Mori, K., H. Kim, T. Kakegawa and S. Hanada. 2003. A novel lineage of sulfate-reducing microorganisms: *Thermodesulfobiaceae* fam. nov., *Thermodesulfobium narugense*, gen. nov., sp. nov., a new thermophilic isolate from a hot spring. Extremophiles *7*: 283–290.

Nakamura, K., S. Haruta, S. Ueno, M. Ishii, A. Yokota and Y. Igarashi. 2004. *Cerasibacillus quisquiliarum* gen. nov., sp. nov., isolated from a semi-continuous decomposing system of kitchen refuse. Int J Syst Evol Microbiol *54*: 1063–1069.

Nunes, I., I. Tiago, A.L. Pires, M.S. da Costa and A. Verissimo. 2006. *Paucisalibacillus globulus* gen. nov., sp. nov., a Gram-positive bacterium isolated from potting soil. Int. J. Syst. Evol. Microbiol. *56*: 1841–1845.

Prüsse, E., C. Quast, K. Knittel, B. Fuchs, W. Ludwig, J. Peplies and F.O. Glöckner. 2007. SILVA: a comprehensive online resource for quality checked and aligned rRNA sequence data compatible with ARB. Nucleic Acids Res *35*: 7188–7196.

Slobodkin, A.I., T.P. Tourova, N.A. Kostrikina, A.M. Lysenko, K.E. German, E.A. Bonch-Osmolovskaya and N.-K. Birkeland. 2006. *Tepidimicrobium ferriphilum* gen. nov., sp. nov., a novel moderately thermophilic, Fe(III)-reducing bacterium of the order *Clostridiales*. Int. J. Syst. Evol. Microbiol. *56*: 369–372.

Song, Y.L., C.X. Liu, M. McTeague, P. Summanen and S.M. Finegold. 2004. *Clostridium bartlettii* sp. nov., isolated from human faeces. Anaerobe *10*: 179–184.

Spring, S., W. Ludwig, M.C. Marquez, A. Ventosa and K.-H. Schleifer. 1996. *Halobacillus* gen. nov., with descriptions of *Halobacillus litoralis* sp. nov. and *Halobacillus trueperi* sp. nov., and transfer of *Sporosarcina halophila* to *Halobacillus halophilus* comb. nov. Int. J. Syst. Bacteriol. *46*: 492–496.

Stamatakis, A.P., T. Ludwig and H. Meier. 2005. RAxML-II: A program for sequential, parallel & distributed inference of large phylogenetic trees. Concur. Comput: Prac. Exper. *17*: 1705–1723.

Trentini, W.C. 1986. Genus *Caryophanon* Peshkoff 1939, 244[AL]. *In* Sneath, Mair, Sharpe and Holt (ed.), Bergey's Manual of Systematic Bacteriology, vol. 2. The Williams & Wilkins Co., Baltimore, pp. 1259–1260.

Wery, N., J.M. Moricet, V. Cueff, J. Jean, P. Pignet, F. Lesongeur, M.A. Cambon-Bonavita and G. Barbier. 2001. *Caloranaerobacter azorensis* gen. nov., sp. nov., an anaerobic thermophilic bacterium isolated from a deep-sea hydrothermal vent. Int. J. Syst. Evol. Microbiol. *51*: 1789–1796.

Wyss, C., F.E. Dewhirst, B.J. Paster, T. Thurnheer and A. Luginbühl. 2005. *Guggenheimella bovis* gen. nov., sp. nov., isolated from lesions of bovine dermatitis digitalis. Int. J. Syst. Evol. Microbiol. *55*: 667–671.

Xing, D., N. Ren, Q. Li, M. Lin, A. Wang and L. Zhao. 2006. *Ethanoligenens harbinense* gen. nov., sp. nov., isolated from molasses wastewater. Int. J. Syst. Evol. Microbiol. *56*: 755–760.

Zhilina, T.N., V.V. Kevbrin, T.P. Tourova, A.M. Lysenko, N.A. Kostrikina and G.A. Zavarzin. 2005. *Clostridium alkalicellum* sp. nov., an obligately alkaliphilic cellulolytic bacterium from a soda lake in the Baikal region. Mikrobiologiya *74*: 557–566.

Taxonomic outline of the phylum *Firmicutes*

WOLFGANG LUDWIG, KARL-HEINZ SCHLEIFER AND WILLIAM B. WHITMAN

All taxa recognized within this volume of the rank of genus and above are listed below. The nomenclatural type is listed first within each taxon followed by the remaining taxa in alphabetical order.

Phylum XIII. *Firmicutes*
- Class I. *"Bacilli"*
  - Order I. *Bacillales*[AL (T)]
    - Family I. *Bacillaceae*[AL]
      - Genus I. *Bacillus*[AL (T)]
      - Genus II. *Alkalibacillus*[VP]
      - Genus III. *Amphibacillus*[VP]
      - Genus IV. *Anoxybacillus*[VP]
      - Genus V. *Cerasibacillus*[VP]
      - Genus VI. *Filobacillus*[VP]
      - Genus VII. *Geobacillus*[VP]
      - Genus VIII. *Gracilibacillus*[VP]
      - Genus IX. *Halobacillus*[VP]
      - Genus X. *Halolactibacillus*[VP]
      - Genus XI. *Lentibacillus*[VP]
      - Genus XII. *Marinococcus*[VP]
      - Genus XIII. *Oceanobacillus*[VP]
      - Genus XIV. *Paraliobacillus*[VP]
      - Genus XV. *Pontibacillus*[VP]
      - Genus XVI. *Saccharococcus*[VP]
      - Genus XVII. *Tenuibacillus*[VP]
      - Genus XVIII. *Thalassobacillus*[VP]
      - Genus XIX. *Virgibacillus*[VP]
    - Family II. *"Alicyclobacillaceae"*
      - Genus I. *Alicyclobacillus*[VP (T)]
    - Family III. *"Listeriaceae"*
      - Genus I. *Listeria*[AL (T)]
      - Genus II. *Brochothrix*[AL]
    - Family IV. *"Paenibacillaceae"*
      - Genus I. *Paenibacillus*[VP (T)]
      - Genus II. *Ammoniphilus*[VP]
      - Genus III. *Aneurinibacillus*[VP]
      - Genus IV. *Brevibacillus*[VP]
      - Genus V. *Cohnella*[VP]
      - Genus VI. *Oxalophagus*[VP]
      - Genus VII. *Thermobacillus*[VP]
    - Family V. *Pasteuriaceae*[AL]
      - Genus I. *Pasteuria*[AL (T)]
    - Family VI. *Planococcaceae*[AL]
      - Genus I. *Planococcus*[AL (T)]
      - Genus II. *Caryophanon*[VP]
      - Genus III. *Filibacter*[VP]
      - Genus IV. *Jeotgalibacillus*[VP]
      - Genus V. *Kurthia*[AL]
      - Genus VI. *Marinibacillus*[VP]
      - Genus VII. *Planomicrobium*[VP]
      - Genus VIII. *Sporosarcina*[AL]
      - Genus IX. *Ureibacillus*[VP]
    - Family VII. *"Sporolactobacillaceae"*
      - Genus I. *Sporolactobacillus*[AL (T)]
    - Family VIII. *"Staphylococcaceae"*
      - Genus I. *Staphylococcus*[AL (T)]
      - Genus II. *Jeotgalicoccus*[VP]
      - Genus III. *Macrococcus*[VP]
      - Genus IV. *Salinicoccus*[VP]
    - Family IX. *"Thermoactinomycetaceae"*
      - Genus I. *Thermoactinomyces*[AL (T)]
      - Genus II. *Laceyella*[VP]
      - Genus III. *Mechercharimyces*[VP]
      - Genus IV. *Planifilum*[VP]
      - Genus V. *Seinonella*[VP]
      - Genus VI. *Shimazuella*[VP]
      - Genus VII. *Thermoflavimicrobium*[VP]
    - Family X. *Incertae Sedis*
      - Genus I. *Thermicanus*[VP]
    - Family XI. *Incertae Sedis*
      - Genus I. *Gemella*[AL]
    - Family XII. *Incertae Sedis*
      - Genus I. *Exiguobacterium*[VP]
  - Order II. *"Lactobacillales"*
    - Family I. *Lactobacillaceae*[AL]
      - Genus I. *Lactobacillus*[AL (T)]
      - Genus II. *Paralactobacillus*[VP]
      - Genus III. *Pediococcus*[AL]
    - Family II. *"Aerococcaceae"*
      - Genus I. *Aerococcus*[AL (T)]
      - Genus II. *Abiotrophia*[VP]
      - Genus III. *Dolosicoccus*[VP]
      - Genus IV. *Eremococcus*[VP]
      - Genus V. *Facklamia*[VP]
      - Genus VI. *Globicatella*[VP]
      - Genus VII. *Ignavigranum*[VP]
    - Family III. *"Carnobacteriaceae"*
      - Genus I. *Carnobacterium*[VP (T)]
      - Genus II. *Alkalibacterium*[VP]
      - Genus III. *Allofustis*[VP]
      - Genus IV. *Alloiococcus*[VP]
      - Genus V. *Atopobacter*[VP]
      - Genus VI. *Atopococcus*[VP]
      - Genus VII. *Atopostipes*[VP]

          Genus VIII. *Desemzia*[VP]
          Genus IX. *Dolosigranulum*[VP]
          Genus X. *Granulicatella*[VP]
          Genus XI. *Isobaculum*[VP]
          Genus XII. *Marinilactibacillus*[VP]
          Genus XIII. *Trichococcus*[VP]
        Family IV. "*Enterococcaceae*"
          Genus I. *Enterococcus*[VP (T)]
          Genus II. *Melissococcus*[VP]
          Genus III. *Tetragenococcus*[VP]
          Genus IV. *Vagococcus*[VP]
        Family V. "*Leuconostocaceae*"
          Genus I. *Leuconostoc*[AL (T)]
          Genus II. *Oenococcus*[VP]
          Genus III. *Weissella*[VP]
        Family VI. *Streptococcaceae*[AL]
          Genus I. *Streptococcus*[AL (T)]
          Genus II. *Lactococcus*[VP]
          Genus III. *Lactovum*[VP]
    Class II. "*Clostridia*"
      Order I. *Clostridiales*[AL (T)]
        Family I. *Clostridiaceae*[AL]
          Genus I. *Clostridium*[AL (T)]
          Genus II. *Alkaliphilus*[VP]
          Genus III. *Anaerobacter*[VP]
          Genus IV. *Anoxynatronum*[VP]
          Genus V. *Caloramator*[VP]
          Genus VI. *Caloranaerobacter*[VP]
          Genus VII. *Caminicella*[VP]
          Genus VIII. *Natronincola*[VP]
          Genus IX. *Oxobacter*[VP]
          Genus X. *Sarcina*[AL]
          Genus XI. *Thermobrachium*[VP]
          Genus XII. *Thermohalobacter*[VP]
          Genus XIII. *Tindallia*[VP]
        Family II. "*Eubacteriaceae*"
          Genus I. *Eubacterium*[AL (T)]
          Genus II. *Acetobacterium*[AL]
          Genus III. *Alkalibacter*[VP]
          Genus IV. *Anaerofustis*[VP]
          Genus V. *Garciella*[VP]
          Genus VI. *Pseudoramibacter*[VP]
        Family III. "*Gracilibacteraceae*"
          Genus I. *Gracilibacter*[VP (T)]
        Family IV. "*Heliobacteriaceae*"
          Genus I. *Heliobacterium*[VP (T)]
          Genus II. *Heliobacillus*[VP]
          Genus III. *Heliophilum*[VP]
          Genus IV. *Heliorestis*[VP]
        Family V. "*Lachnospiraceae*"
          Genus I. *Lachnospira*[AL (T)]
          Genus II. *Acetitomaculum*[VP]
          Genus III. *Anaerostipes*[VP]
          Genus IV. *Bryantella*[VP]
          Genus V. *Butyrivibrio*[AL]
          Genus VI. *Catonella*[VP]
          Genus VII. *Coprococcus*[AL]
          Genus VIII. *Dorea*[VP]
          Genus IX. *Hespellia*[VP]
          Genus X. *Johnsonella*[VP]

          Genus XI. *Lachnobacterium*[VP]
          Genus XII. *Moryella*[VP]
          Genus XIII. *Oribacterium*[VP]
          Genus XIV. *Parasporobacterium*[VP]
          Genus XV. *Pseudobutyrivibrio*[VP]
          Genus XVI. *Roseburia*[VP]
          Genus XVII. *Shuttleworthia*[VP]
          Genus XVIII. *Sporobacterium*[VP]
          Genus XIX. *Syntrophococcus*[VP]
        Family VI. *Peptococcaceae*[AL]
          Genus I. *Peptococcus*[AL (T)]
          Genus II. *Cryptanaerobacter*[VP]
          Genus III. *Dehalobacter*[VP]
          Genus IV. *Desulfitobacterium*[VP]
          Genus V. *Desulfonispora*[VP]
          Genus VI. *Desulfosporosinus*[VP]
          Genus VII. *Desulfotomaculum*[AL]
          Genus VIII. *Pelotomaculum*[VP]
          Genus IX. *Sporotomaculum*[VP]
          Genus X. *Syntrophobotulus*[VP]
          Genus XI. *Thermincola*[VP]
        Family VII. "*Peptostreptococcaceae*"
          Genus I. *Peptostreptococcus*[AL (T)]
          Genus II. *Filifactor*[VP]
          Genus III. *Tepidibacter*[VP]
        Family VIII. "*Ruminococcaceae*"
          Genus I. *Ruminococcus*[AL (T)]
          Genus II. *Acetanaerobacterium*[VP]
          Genus III. *Acetivibrio*[VP]
          Genus IV. *Anaerofilum*[VP]
          Genus V. *Anaerotruncus*[VP]
          Genus VI. *Faecalibacterium*[VP]
          Genus VII. *Fastidiosipila*[VP]
          Genus VIII. *Oscillospira*[AL]
          Genus IX. *Papillibacter*[VP]
          Genus X. *Sporobacter*[VP]
          Genus XI. *Subdoligranulum*[VP]
        Family IX. *Syntrophomonadaceae*[VP]
          Genus I. *Syntrophomonas*[VP (T)]
          Genus II. *Pelospora*[VP]
          Genus III. *Syntrophospora*[VP]
          Genus IV. *Syntrophothermus*[VP]
          Genus V. *Thermosyntropha*[VP]
        Family X. *Veillonellaceae*[VP]
          Genus I. *Veillonella*[AL (T)]
          Genus II. *Acetonema*[VP]
          Genus III. *Acidaminococcus*[AL]
          Genus IV. *Allisonella*[VP]
          Genus V. *Anaeroarcus*[VP]
          Genus VI. *Anaeroglobus*[VP]
          Genus VII. *Anaeromusa*[VP]
          Genus VIII. *Anaerosinus*[VP]
          Genus IX. *Anaerovibrio*[AL]
          Genus X. *Centipeda*[VP]
          Genus XI. *Dendrosporobacter*[VP]
          Genus XII. *Dialister*[VP]
          Genus XIII. *Megasphaera*[AL]
          Genus XIV. *Mitsuokella*[VP]
          Genus XV. *Pectinatus*[AL]
          Genus XVI. *Phascolarctobacterium*[VP]

Genus XVII. *Propionispira*^{VP}
Genus XVIII. *Propionispora*^{VP}
Genus XIX. *Quinella*^{VP}
Genus XX. *Schwartzia*^{VP}
Genus XXI. *Selenomonas*^{AL}
Genus XXII. *Sporomusa*^{VP}
Genus XXIII. *Succiniclasticum*^{VP}
Genus XXIV. *Succinispira*^{VP}
Genus XXV. *Thermosinus*^{VP}
Genus XXVI. *Zymophilus*^{VP}
Family XI. *Incertae Sedis*
Genus I. *Anaerococcus*^{VP}
Genus II. *Finegoldia*^{VP}
Genus III. *Gallicola*^{VP}
Genus IV. *Helcococcus*^{VP}
Genus V. *Parvimonas*^{VP}
Genus VI. *Peptoniphilus*^{VP}
Genus VII. *Sedimentibacter*^{VP}
Genus VIII. *Soehngenia*^{VP}
Genus IX. *Sporanaerobacter*^{VP}
Genus X. *Tissierella*^{VP}
Family XII. *Incertae Sedis*
Genus I. *Acidaminobacter*^{VP}
Genus II. *Fusibacter*^{VP}
Genus III. *Guggenheimella*^{VP}
Family XIII. *Incertae Sedis*
Genus I. *Anaerovorax*^{VP}
Genus II. *Mogibacterium*^{VP}
Family XIV. *Incertae Sedis*
Genus I. *Anaerobranca*^{VP}
Family XV. *Incertae Sedis*
Genus I. *Aminobacterium*^{VP}
Genus II. *Aminomonas*^{VP}
Genus III. *Anaerobaculum*^{VP}
Genus IV. *Dethiosulfovibrio*^{VP}
Genus V. *Thermanaerovibrio*^{VP}
Family XVI. *Incertae Sedis*
Genus I. *Carboxydocella*^{VP}
Family XVII. *Incertae Sedis*
Genus I. *Sulfobacillus*^{VP}
Genus II. *Thermaerobacter*^{VP}
Family XVIII. *Incertae Sedis*
Genus I. *Symbiobacterium*^{VP}
Family XIX. *Incertae Sedis*
Genus I. *Acetoanaerobium*^{VP}

Order II. *Halanaerobiales*^{VP}
Family I. *Halanaerobiaceae*^{VP}
Genus I. *Halanaerobium*^{VP (T)}
Genus II. *Halocella*^{VP}
Genus III. *Halothermothrix*^{VP}
Family II. *Halobacteroidaceae*^{VP}
Genus I. *Halobacteroides*^{VP (T)}
Genus II. *Acetohalobium*^{VP}
Genus III. *Halanaerobacter*^{VP}
Genus IV. *Halonatronum*^{VP}
Genus V. *Natroniella*^{VP}
Genus VI. *Orenia*^{VP}
Genus VII. *Selenihalanaerobacter*^{VP}
Genus VIII. *Sporohalobacter*^{VP}
Order III. "*Thermoanaerobacterales*"
Family I. "*Thermoanaerobacteraceae*"
Genus I. *Thermoanaerobacter*^{VP (T)}
Genus II. *Ammonifex*^{VP}
Genus III. *Caldanaerobacter*^{VP}
Genus IV. *Carboxydothermus*^{VP}
Genus V. *Gelria*^{VP}
Genus VI. *Moorella*^{VP}
Genus VII. *Thermacetogenium*^{VP}
Genus VIII. *Thermanaeromonas*^{VP}
Family II. *Thermodesulfobiaceae*^{VP}
Genus I. *Thermodesulfobium*^{VP (T)}
Genus II. *Coprothermobacter*^{VP}
Family III. *Incertae Sedis*
Genus I. *Caldicellulosiruptor*^{VP}
Genus II. *Thermoanaerobacterium*^{VP}
Genus III. *Thermosediminibacter*^{VP}
Genus IV. *Thermovenabulum*^{VP}
Family IV. *Incertae Sedis*
Genus I. *Mahella*^{VP}
Class III. "*Erysipelotrichia*"
Order I. "*Erysipelotrichales*^(T)"
Family I. *Erysipelotrichaceae*^{VP}
Genus I. *Erysipelothrix*^{AL (T)}
Genus II. *Allobaculum*^{VP}
Genus III. *Bulleidia*^{VP}
Genus IV. *Catenibacterium*^{VP}
Genus V. *Coprobacillus*^{VP}
Genus VI. *Holdemania*^{VP}
Genus VII. *Solobacterium*^{VP}
Genus VIII. *Turicibacter*^{VP}

Phylum XIII. *Firmicutes* Gibbons and Murray 1978, 5 (*Firmacutes* [sic] Gibbons and Murray 1978, 5)

KARL-HEINZ SCHLEIFER

Fir.mi.cu′tes. L. adj. *firmus* strong, stout; L. fem. n. *cutis* skin; N.L. fem. pl. n. *Firmicutes* division with strong (and thick) skin, to indicate Gram-positive type of cell wall.

Gibbons and Murray (1978) described the division *Firmicutes* encompassing all of Gram-positive *Bacteria* ("bacteria with a Gram-positive type of cell wall"). Thus, both Gram-positive bacteria with a low DNA mol% G + C as well as Gram-positive with a high DNA mol% G + C were included in the division whereas the *Mollicutes* were placed in a separate division. In (2001), Garrity and Holt described the new phylum *Firmicutes* encompassing both Gram-positive bacteria with a low DNA mol% G + C (classes "*Clostridia*" and "*Bacilli*") and the class *Mollicutes*. However, the *Mollicutes* are phenotypically so different (no cell walls, no peptidoglycan, no muramic acid, flexible and highly pleomorphic cells) from the typical *Firmicutes* that we propose to remove them from the phylum *Firmicutes*. Moreover, comparative sequence analysis of various genes encoding conserved proteins (*atp*β, *hsp60*, *rpoB*, *rpoC*, *aspRS*, *trpRS*, *valRS*) studied so far have shown that *Mollicutes* are also phylogenetically quite distinct from typical *Firmicutes* (Ludwig and Schleifer, 2005). This was supported in a more recent study by Case et al. (2007). They found that the *Firmicutes* compared in their study formed a monophyletic group in the RpoB protein tree well-separated from the *Mollicutes*. The *Mollicutes* should be excluded from the phylum *Firmicutes* and classified in the phylum *Tenericutes* (Murray, 1984). There are only a few members of the *Firmicutes* (*Erysipelotrichaceae*) which, based on their 16S rRNA and 23S rRNA gene sequences, reveal a closer phylogenetic relatedness to *Mollicutes*. Unfortunately, there are currently no sequence data on conserved protein-coding genes available for members of *Erysipelotrichaceae*. There are also some organisms which are still listed in this volume within the phylum *Firmicutes* but they root in the 16S rRNA tree outside the *Firmicutes* and may represent separate phyla. These are members of the *Thermoanaerobacteraceae* and *Syntrophomonadaceae*.

The phylum *Firmicutes* consists of at least 26 families and 223 genera. All members possess a rigid cell wall. The members studied so far all contain muramic acid in their cell walls. Some contain a cell wall teichoic acid. Most members are Gram-positive, but there are also some that stain Gram-negative (e.g., *Veillonellaceae*, *Syntrophomonadaceae*). The phylum is phenotypically diverse. Cells may be spherical, or straight, curved, and helical rods or filaments, with or without flagella and with or without heat-resistant endospores. They are aerobes, facultative or strict anaerobes. Some members of the *Firmicutes* are thermophiles and/or halophiles. Most of them are chemo-organotrophs, a few are anoxygenic photoheterotrophs. Most grow at neutral pH, while some are acidophiles or alkaliphiles. The mol% G + C content of DNA is generally <50.

Type order: Bacillales Prévot 1953, 692.

References

Case, R.J., Y. Boucher, I. Dahllof, C. Holmstrom, W.F. Doolittle and S. Kjelleberg. 2007. Use of 16S rRNA and *rpoB* genes as molecular markers for microbial ecology studies. Appl Environ Microbiol 73: 278–288.

Garrity, G.M. and J.G. Holt. 2001. The Road Map to the *Manual. In* Boone, Castenholz and Garrity (Editors), Bergey's Manual of Systematic Bacteriology, vol. 1, The *Archaea* and the Deeply Branching and Phototrophic *Bacteria*. Springer, New York, pp. 119–166.

Gibbons, N.E. and R.G.E. Murray. 1978. Proposals concerning the higher taxa of bacteria. Int J Syst Bacteriol 28: 1–6.

Ludwig, W. and K.H. Schleifer. 2005. Molecular phylogeny of bacteria based on comparative sequence analysis of conserved genes. *In* Sapp (Editor), Microbial phylogeny and evolution, concepts and controversies. Oxford University Press, New York, pp. 70–98.

Murray, R.G.E. 1984. The higher taxa, or, a place for everything…? *In* Krieg and Holt (Editors), Bergey's Manual of Systematic Bacteriology, vol. 1. Williams and Wilkins, Baltimore, pp. 31–34.

Prévot, A.R. 1953. Dictionnaire des Bactéries Pathogènes, 2nd edn. Masson, Paris.

Class I. **Bacilli** class. nov.

WOLFGANG LUDWIG, KARL-HEINZ SCHLEIFER AND WILLIAM B. WHITMAN

Ba.cil′li. N.L. masc. n. *Bacillus* type genus of the type order of the class; N.L. masc. pl. n. *Bacilli* the *Bacillus* class.

The class *Bacilli* is circumscribed for this volume on the basis of the phylogenetic analyses of the 16S rRNA sequences and includes the order *Bacillales* and its close relatives. This class includes the orders *Bacillales* and *Lactobacillales* (ord. nov., this volume). Members of this class generally stain Gram-positive and form a Gram-positive type of cell wall. They may or may not form endospores. Many also grow aerobically or microaerophilically. Some are facultative anaerobes.

Type order: **Bacillales** Prévot 1953, 60[AL].

Reference

Prévot, A.R. 1953. Dictionnaire des Bactéries Pathogènes. *In* Hauduroy, Ehringer, Guillot, Magrou, Prévot, Rossetti and Urbain (Editors), 2nd edn. Masson, Paris, pp. 692.

Order I. **Bacillales** Prévot 1953, 60[AL]

PAUL DE VOS

Ba.cil.la′les. N.L. masc. n. *Bacillus* type genus of the order; suff. *-ales* ending denoting an order; N.L. fem. pl. n. *Bacillales* the *Bacillus* order.

The delineation of the order of the *Bacillales* is based on 16S sequence analysis and consists of the families of *Bacillaceae* (which contains the type genus), *Alicyclobacillaceae* (fam. nov., this volume), *Listeriaceae* (fam. nov., this volume), *Paenibacillaceae* (fam. nov., this volume), *Pasteuriaceae, Planococcaceae, Sporolactobacillaceae* (fam. nov., this volume), *Staphylococcaceae* (fam. nov., this volume), *Thermoactinomycetaceae* (Matsuo et al., 2006), and three family *incertae sedis*. Most of the family names were not validly published, but their allocation is in agreement with the phylogenetic outline given in this volume. One of the few general characteristics of the order is the formation of **endospores** that are formed in representatives of many genera of this order, although exceptions exist. Furthermore, the cell wall stains generally **Gram-positive for young cells**. The staining is supported by the chemical composition of the cell wall. **If known,** most genera allocated to the above families have **menaquinone**

7 (MK-7) although various exceptions are found, in particular amongst the *Thermoactinomycetaceae* where menaquinones ranging from MK- 6 to MK-11 have been reported.

Type genus: **Bacillus** Cohn 1872, 174[AL].

References

Cohn, F. 1872. Untersuchungen über Bakterien. Bertr. Biol. Pflanz. *1 (Heft II)*: 127–224.

Matsuo, Y., A. Katsuta, S. Matsuda, Y. Shizuri, A. Yokota and H. Kasai. 2006. *Mechercharimyces mesophilus* gen. nov., sp. nov. and *Mechercharimyces asporophorigenens* sp. nov., antitumour substance-producing marine bacteria, and description of *Thermoactinomycetaceae* fam. nov. Int. J. Syst. Evol. Microbiol. *56*: 2837–2842.

Prévot, A.R. 1953. Dictionnaire des Bactéries Pathogènes. *In* Hauduroy, Ehringer, Guillot, Magrou, Prévot, Rossetti and Urbain (Editors), 2nd edn. Masson, Paris, pp. 692.

Family I. **Bacillaceae**

NIALL A. LOGAN AND PAUL DE VOS

Ba.cil.la′ce.ae. N.L. masc. n. *Bacillus* type genus of the family; suff. *-aceae* ending to denote a family; N.L. fem. pl. n. *Bacillaceae* the *Bacillus* family.

The family *Bacillaceae* was circumscribed for this volume on the basis of phylogenetic analysis of 16S rRNA gene sequences; the family contains *Bacillus* and 18 other genera. The majority of taxa are aerobic or facultatively anaerobic chemo-organotrophic rods that possess Gram-positive type cell-wall structures and that form endospores. However, there are exceptions to all of these characteristics; the family includes strict anaerobes, autotrophs, cocci, and organisms that do not form endospores. Although the commonest Gram reaction is Gram-positive, many species of the family are Gram-variable or frankly Gram-negative. Table 1 summarizes the diversity of some routine phenotypic characters among the genera of this family.

Type genus: **Bacillus** Cohn 1872, 174[AL].

Family I. *Bacillaceae*[AL]
Genus I. *Bacillus*[AL (T)]
Genus II. *Alkalibacillus*[VP]
Genus III. *Amphibacillus*[VP]
Genus IV. *Anoxybacillus*[VP]
Genus V. *Cerasibacillus*[VP]

Genus VI. *Filobacillus*[VP]
Genus VII. *Geobacillus*[VP]
Genus VIII. *Gracilibacillus*[VP]
Genus IX. *Halobacillus*[VP]
Genus X. *Halolactibacillus*[VP]
Genus XI. *Lentibacillus*[VP]
Genus XII. *Marinococcus*[VP]
Genus XIII. *Oceanobacillus*[VP]
Genus XIV. *Paraliobacillus*[VP]
Genus XV. *Pontibacillus*[VP]
Genus XVI. *Saccharococcus*[VP]
Genus XVII. *Tenuibacillus*[VP]
Genus XVIII. *Thalassobacillus*[VP]
Genus XIX. *Virgibacillus*[VP]

The following genera were published later than the deadline for inclusion: *Caldalkalibacillus, Halalkalibacillus, Lysinibacillus, Ornithinibacillus, Paucisalibacillus, Pelagibacillus, Piscibacillus, Salimicrobium, Salirhabdus, Salsuginibacillus, Terribacillus,* and *Vulcanibacillus*.

Genus I. **Bacillus** Cohn 1872, 174[AL]

Niall A. Logan and Paul De Vos

Ba.cil′lus. N.L. masc. n. *Bacillus* a rodlet.

Cells rod-shaped, straight or slightly curved, occurring singly and in pairs, some in chains, and occasionally as long filaments. Endospores are formed, no more than one to a cell; these spores are very resistant to many adverse conditions. Gram-positive, or Gram-positive only in early stages of growth, or Gram-negative. A *meso*-DAP direct murein cross-linkage type is commonest, but L-Lys-D-Glu, Orn-D-Glu and L-Orn-D-Asp have occasionally been reported. Motile by means of peritrichous or degenerately peritrichous flagella, or nonmotile. Aerobes or facultative anaerobes, but a few species are described as strictly anaerobic. The terminal electron acceptor is oxygen, replaceable by alternatives in some species. Most species will grow on routine media such as nutrient agar and blood agar. Colony morphology and size very variable between and within species. A wide diversity of physiological abilities is exhibited, ranging from psychrophilic to thermophilic, and acidophilic to alkaliphilic; some strains are salt tolerant and some are halophilic. Catalase is produced by most species. Oxidase-positive or -negative. Chemo-organotrophic; two species are facultative chemolithotrophs: prototrophs to auxotrophs requiring several growth factors. Mostly isolated from soil, or from environments that may have been contaminated directly or indirectly by soil, but also found in water, food and clinical specimens. The resistance of the spores to heat, radiation, disinfectants, and desiccation results in species being troublesome contaminants in operating rooms, on surgical dressings, in pharmaceutical products and in foods. Most species have little or no pathogenic potential and are rarely associated with disease in humans or other animals; an exception is *Bacillus anthracis*, the agent of anthrax; several other species may cause food poisoning and opportunistic infections, and strains of *Bacillus thuringiensis* are pathogenic to invertebrates.

DNA G + C content (mol%): 32–66 (T_m).

Type species: **Bacillus subtilis** Cohn 1872, 174[AL].

Further descriptive information

Phylogeny. A phylogenetic tree, based on 16S rDNA sequences, is shown in Figure 8. The tree includes 142 named Bacillus species as listed in this chapter (but excludes *Bacillus laevolacticus* and *Bacillus tequilensis*). *Bacillus tusciae* and *Bacillus schlegelii* lie at the edge of the tree, and their respective closest neighbors, on the basis of 16S rDNA gene sequence comparisons, are an unknown *Alicyclobacillus* species and *Aneurinibacillus*.

It is well known that 16S rDNA sequences do not always allow species to be discriminated, and that DNA–DNA hybridizations may be needed for this. However, sequences of other genes (the so-called core genes) may be more appropriate for discriminating these relatively recent branchings of the evolutionary tree that correspond to bacterial species. The ad hoc committee for the re-evaluation of the species definition in bacteriology (Stackebrandt et al., 2002) advised that genetic differences of the so-called core genes should be explored in order to come to a finer "bacterial species concept" in the future. The groupings (phylogenetic trees) that are obtained from comparisons either of sequences of individual core genes, or of concatenated gene sequences of several core genes, need to be validated against the phylogenetic species concept (Wayne et al., 1987). Recent data (Wang et al., 2007a) clearly show that in the *Bacillus subtilis* group, within which species delineation is very difficult, core genes such as *gyrB* allow differentiation on a genetic basis. A debate began recently concerning the impact of these new findings of genome analysis on bacterial taxonomy (Buckley and Roberts, 2007). Analysis of whole-genome sequences showed that about 80% of an individual genome may be shared by all pathogenic isolates of *Streptococcus agalactiae* (Tettelin et al., 2005), indicating that in closely related strains belonging to the same species, at least, a vast amount of the genetic information is shared. The interested reader is referred to the literature (e.g., Kunin et al., 2005, 2007; Dagan and Martin, 2006).

Cell morphology. *Bacillus* cells may occur singly and in pairs, in chains (which may be of great length), and as filaments. Trichome-forming "*Arthromitus*" strains from sow bug or wood louse (*Porcellio scaber*) guts, with endospore-forming filaments over 100 μm long and up to 180 cells per filament in animals cultivated in darkness, have been identified as *Bacillus cereus* (Jorgensen et al., 1997) and similar filamentous organisms have been isolated from moths, roaches and termites (Margulis et al., 1998; see *Habitats*, below). The rod-shaped cells of *Bacillus* species are usually round-ended, but the cells of members of the *Bacillus cereus* group have often been described as squared. Cell diameters range from 0.4 to 1.8 μm and lengths from 0.9 to 10.0 μm, but the cells of a particular strain are usually quite regular in size, and individual species normally have dimensions within fairly narrow limits. For example, cells of *Bacillus pumilus* are typically 0.6–0.7 by 2.0–3.0 μm, while those of *Bacillus megaterium* are usually 1.2–1.5 by 2.0–5.0 μm. Pleomorphism, showing as cells and filaments with swollen regions, and entirely swollen cells, may be observed in cultures grown in suboptimal conditions; this is seen, for example, in cultures of *Bacillus fumarioli* grown on relatively rich media (Logan et al., 2000), and such stressed cultures sporulate poorly. *Bacillus* cytoplasm may stain uniformly or be vacuolate; vacuolation (the presence of inclusions is visible by phase-contrast microscopy as areas less refractive than spores, and in Gram-stained preparations by unstained globules) is enhanced in some species (*Bacillus cereus* and *Bacillus megaterium*, for example) by cultivation on an agar medium containing a fermentable carbohydrate such as glucose, so that copious storage material is produced.

Sporangial morphologies are characteristic of species, and so often valuable in identification (see *Life cycle*, below), but an individual strain may show some variation and produce, for example, both oval and spherical spores. The commonest spore shape is ellipsoidal or oval, but shapes range from frankly cylindrical through ellipsoidal to spherical, and irregular forms such as kidney- or banana-shaped spores may be seen in some species. The position of the spore is also characteristic; the most

TABLE 1. Phenotypic characteristics of genera belonging to the family *Bacillaceae*[a]

Genus key: 1. *Bacillus*; 2. *Alkalibacillus*; 3. *Amphibacillus*; 4. *Anoxybacillus*; 5. *Cerasibacillus*; 6. *Filobacillus*; 7. *Geobacillus*; 8. *Gracilibacillus*; 9. *Halobacillus*; 10. *Halolactibacillus*; 11. *Lentibacillus*; 12. *Marinococcus*; 13. *Oceanobacillus*; 14. *Paraliobacillus*; 15. *Pontibacillus*; 16. *Saccharococcus*; 17. *Tenuibacillus*; 18. *Thalassobacillus*; 19. *Virgibacillus*

Characteristic	1	2	3	4	5	6	7	8	9	10	11	12	13	14	15	16	17	18	19
Number of species in genus	141	4	3	10	1	1	17	4	4	2	4	3	3	1	2	1	1	1	9
Gram reaction	+/v/−	+	+	+	+	−	+/−	+	+	+	v	+	+	+	+	+	+	+	+
Predominant cell shape:																			
Rods	+	+	+	+	+	+	+	+	+	+	+		+	+	+		+	+	+
Cocci												+				+			
Mean cell width:																			
<0.5 mm	+					+		+		+				+			+		+
0.5–1.0 mm	+	+	+	+	+	+	+	+	+		+		+	+	+			+	+
1.0 or >1.0 mm	+/−	+	+	+			+		+			+							+
Motility	+/−	+	+	+/−			+/−	+	+/−		+/−	v	+	+					+
Spore formation	+/−	+	+	+	+	+	+	+	+		+	−	+	+	+	−	+	+	+
Spore shape:																			
Ellipsoidal	+	−	+	+			+	+	+	+	+		+	+	+			+	+
Cylindrical	+			+			+												
Spherical	+	+		+	+	+		+	+		+		+	+	+		+		+
Oxygen requirements:																			
Aerobes	+	+	+	+	+	+	+	+	+	+	+	+	+	+	+	+	+	+	+
Facultative anaerobes	+	−	+	+	+	+	+	+	−	+	−	−	+	+	−	−		−	+
Strict anaerobes	+				−				−	+					−				
Growth in media with added NaCl:																			
0%	+	−		+	+	−	+	v	−	+	−	v	+	−	−		−	−	v
5%	+	+	+	+	+	+	+	+	+	+	+	+	+	+	+		+	+	+
10%	+	+		−	−	+	+	+	+	+	+	+	+	+	+		+	+	+
20%	+	+			−	+		+	+	+	+	+	+	+	−				+
Growth at:																			
10 °C	+							+	+	+	+	+	+						+
20 °C	+	+	+	+	+		+	+	+	+	+	+	+	+	+	+	+	+	+
30 °C	+	+	+	+	+	+	+	+	+	+	+	+	+	+	+	+	+	+	+
40 °C	+	+	+	+	+	+	+	+	+	+	+	+	+				+	+	+
50 °C	+	+	+	+			+			+						+			+
60 °C	+		−	+			+			+						+			
Growth at pH:																			
5	+			+			+	+	+		+		+	+	+				+
6	+	−	+	+		+	+	+	+	+	+		+	+	+		+	+	+
7	+	+	+	+	+	+	+	+	+	+	+	+	+	+	+	+	+	+	+
8	+	+	+	+	+	+	+	+	+	+	+	+	+	+	+		+	+	+
9	+	+	+	+	+		+	+	+	+	+		+	+		+		+	+
10	+	+	+	+						+								+	+
Catalase	+/−	+	−	+/−	+	+	+/−	+	+	−	+	+	+	+	+	+	+	+	+
Oxidase	+/−	−	−	+/−	−	−	v	+	+	−	v	−	v	+	+/−	+/−	+	−	+

[a]Symbols: +, at least one species within the genus gives a positive reaction; +/−, some species within the genus are positive, some are negative; −, negative; v, varies within the genus; w, weak.

FIGURE 8. Unrooted neighbor-joining phylogenetic tree of *Bacillus* species based on 16S rRNA gene sequences. Alignment of sequences was performed using CLUSTALX, BIOEDIT and TREECON. Bootstrap values above 70% are shown (based on 1000 replications) at the branch points. Sequence accession numbers for each strain are given in parentheses.

frequently observed is a subterminal placement, and position can range from central through paracentral and subterminal to terminal. An individual strain may exhibit a range of spore positions. In small sporangia it is sometimes difficult to categorize spore positions with confidence. In just over half of the validly published *Bacillus* species the spores swell the sporangia slightly or appreciably, while in the remainder sporangial swelling has not been observed, but both swollen and unswollen sporangia may be observed within a single strain. The sporangia of *Bacillus thuringiensis* are characterized by their parasporal inclusions of crystalline protein known as δ-endotoxins, which are often toxic to insects and other invertebrates. Insecticidal strains of *Bacillus sphaericus* also produce crystalline parasporal inclusions; these are less prominent than those of *Bacillus thuringiensis*, but are generally visible with the aid of a good phase-contrast microscope (Priest, 2002).

L-form *Bacillus* cells have been reported from both humans, other animals and plants. Several authors have found L-forms in the blood of normal and arthritic persons in association with erythrocytes (Bisset and Bartlett, 1978; Pease, 1970, 1974), in

other body fluids such as synovial fluids of arthritic patients (Pease, 1969), in association with neoplasms (Livingston and Alexander-Jackson, 1970), and in chickens and turkeys with infectious synovitis (Livingston and Alexander-Jackson, 1970; Roberts, 1964). As demonstrated by Bisset and Bartlett (1978), these organisms often revert to small, acid-fast diphtheroids, and on prolonged (up to 25 months) primary culture or subculture, and especially when grown in the presence of agents known to stimulate reversion of L-forms, some of them increase in size, lose their acid-fastness, and become Gram-positive endospore-forming rods. These organisms produce licheniform colonies on agar media, like the "*Bacillus endoparasiticus*" of Benedek (1955) from arthritic patients. The fully reverted isolates of Bisset and Bartlett (1978) were phenotypically similar to *Bacillus licheniformis* in other respects, and they named them "*Bacillus licheniformis* var. *endoparasiticus*."

Symbiotic associations between L-form bacteria and plants have been observed (Paton and Innes, 1991), and this has encouraged the induction and characterization of a stable L-form of *Bacillus subtilis* (Allan, 1991; Allan et al., 1993). Artificially induced symbiosis of this stable L-form of *Bacillus subtilis* in strawberry plants has been demonstrated by ELISA (Ferguson et al., 2000), and a symbiosis of the same strain in Chinese cabbage seedlings has been shown to inhibit the germination of *Botrytis cinerea* conidia (Walker et al., 2002).

Cell-wall composition. Information on murein structure is known for only about half of the valid species of *Bacillus* (Table 2), but Bacher et al. (2001) have shown that matrix-assisted laser desorption/ionization time-of-flight mass spectrometry with nano-electrospray ionization quadrupole ion-trap mass spectrometry allows the ready determination of peptidoglycan structure in *Bacillus subtilis* vegetative cells and *Bacillus megaterium* spores. The vegetative cells of the majority of *Bacillus* species that have been studied have the most common type of cross-linkage in which a peptide bond is formed between the diamino acid in position 3 of one subunit and the D-Ala in position 4 of the neighboring peptide subunit, so that no interpeptide bridge is involved. The diamino acid in most *Bacillus* species is *meso*-diaminopimelic acid (*meso*-DAP), and this cross-linkage is now usually known as DAP-direct (A1γ in the classification of Schleifer and Kandler 1972). Where the structure is known, this cross-linkage is also typical of the examined representatives of several genera whose species were previously accommodated in *Bacillus*: *Alkalibacillus, Brevibacillus, Geobacillus, Gracilibacillus, Paenibacillus, Salibacillus,* and *Virgibacillus* (Table 2).

A different type of cross-linkage is found in the spherical-spored members of the genus (informally known as the *Bacillus sphaericus* group) and in other genera containing spherical-spored organisms. *Bacillus sphaericus* and its close relatives typically have the cross-linkage type A4α (L-Lys-D-Asp or L-Lys-D-Glu), with L-Lys in position 3 of the peptide subunit with bridging to the D-Ala in position 4 of the neighboring peptide subunit by D-Asp or D-Glu. *Bacillus sphaericus* and *Bacillus fusiformis* have accordingly been transferred to the new genus *Lysinibacillus* (Ahmed et al., 2007c), but information on the peptidoglycan structure of other potential members of this genus is awaited). Three members of the *Bacillus sphaericus* group, however, *Bacillus insolitus, Bacillus psychrodurans* and *Bacillus psychrotolerans,* have L-Orn in position 3 of the peptide subunit with bridging by D-Glu to the D-Ala in position 4 of the neighboring peptide subunit (type A4β, or L-Orn-D-Glu) (Abd El-Rahman et al., 2002; Stackebrandt et al., 1987), and this structure is also found in the halophile *Filobacillus milensis* (Schlesner et al., 2001), which, although it bears spherical spores, is not related to any of the other spherical-spored groups but lies closest to *Bacillus haloalkaliphilus* (now reclassified into *Alkalibacillus*). Some spherical-spored species formerly classified in *Bacillus* have been transferred to other genera: *Bacillus globisporus, Bacillus pasteurii* and *Bacillus psychrophilus* have been transferred to *Sporosarcina* (Yoon et al., 2001b) and they share with the type species of that genus, *Sporosarcina ureae,* A4α cross-linking based on L-Lys in position 3 of the peptide subunit with interpeptide bridges of D-Asp, D-Glu, L-Ala-D-Asp, or Gly-D-Glu. *Bacillus thermosphaericus,* which has been transferred to the new genus *Ureibacillus,* also has a L-Lys-D-Asp structure (Fortina et al., 2001b). Two other, monospecific, genera of spherical-spored species have been proposed (Yoon et al., 2001c): *Bacillus marinus* has been transferred to *Marinibacillus,* and *Jeotgalibacillus alimentarius* accommodates a single isolate from a traditional food; both species have a direct L-Lys cross-linkage. *meso*-DAP has been found in the peptidoglycan of spores of *Bacillus sphaericus* and *Bacillus pasteurii* (Ranftl and Kandler, 1973).

Other than the absence of DAP from the walls of *Bacillus horikoshii* (Nielsen et al., 1994), no information is available about cross-linkage in members of the phylogenetically distinct group of alkaliphilic or alkalitolerant species which contains this species and *Bacillus agaradhaerens, Bacillus alcalophilus, Bacillus clarkii, Bacillus clausii, Bacillus gibsonii, Bacillus*

TABLE 2. Murein cross-linkage types found in *Bacillus* species and in former *Bacillus* species that have been transferred to other genera

	Murein cross-linkage[a]	Reference
Bacillus		
B. subtilis	*meso*-DAP direct	Schleifer and Kandler (1972)
B. anthracis	(*meso*-DAP direct)	Schleifer and Kandler (1972)
B. aquimaris	*meso*-DAP[b]	Yoon et al. (2003a)
B. barbaricus	DAP[b]	Taubel et al. (2003)
B. badius	*meso*-DAP direct	Schleifer and Kandler (1972)
B. cereus	*meso*-DAP direct	Schleifer and Kandler (1972)
B. coagulans	*meso*-DAP direct	Schleifer and Kandler (1972)
B. fastidiosus	*meso*-DAP direct	Claus and Berkeley (1986)
B. firmus	(*meso*-DAP direct)	Schleifer and Kandler (1972)
B. funiculus	DAP[b]	Ajithkumar et al. (2002)
B. halophilus	*meso*-DAP direct	Ventosa et al. (1989)
B. hwajinpoensis	*meso*-DAP[b]	Yoon et al. (2004b)
B. horti	*meso*-DAP[b]	Yumoto et al. (1998)
B. indicus[c]	L-Orn-D-Asp	Suresh et al. (2004)
B. jeotgali	*meso*-DAP direct	Yoon et al. (2001a)
B. lentus	(*meso*-DAP direct)	Schleifer and Kandler (1972)
B. licheniformis	*meso*-DAP direct	Schleifer and Kandler (1972)
B. marisflavi	*meso*-DAP[b]	Yoon et al. (2003a)

(continued)

TABLE 2. (continued)

	Murein cross-linkage[a]	Reference
B. megaterium	(meso-DAP direct)	Schleifer and Kandler (1972)
B. methanolicus	meso-DAP direct	Arfman et al. (1992)
B. mycoides	meso-DAP direct	Claus and Berkeley (1986)
B. oleronius	meso-DAP direct	Kuhnigk et al. (1995)
B. pumilus	meso-DAP direct	Schleifer and Kandler (1972)
B. schlegelii	meso-DAP direct	Krüger and Meyer (1984)
B. smithii	DAP[b]	Nakamura et al. (1988)
B. thermocloacae	meso-DAP direct	Demharter and Hensel (1989b)
B. thuringiensis	meso-DAP direct	Schleifer and Kandler (1972)
B. vietnamensis	meso-DAP[b]	Noguchi et al. (2004)
Alkaliphilic and alkalitolerant *Bacillus* species		
B. cohnii	L-Orn-D-Asp	Spanka and Fritze (1993)
B. halmapalus	No DAP	Nielsen et al. (1994)
Alkaliphilic species in 6th 16S rRNA group of Nielsen et al. (1994)		
B. horikoshii	No DAP	Nielsen et al. (1994)
Spherical-spored *Bacillus* species		
B. fusiformis[d]	L-Lys-D-Asp	Ahmed et al. (2007c)
B. insolitus	Orn-D-Glu	Stackebrandt et al. (1987)
B. neidei	L-Lys-D-Glu	Nakamura et al. (2002)
B. psychrodurans	Orn-D-Glu	Abd El-Rahman et al. (2002)
B. psychrotolerans	Orn-D-Glu	Abd El-Rahman et al. (2002)
B. pycnus	L-Lys-D-Glu	Nakamura et al. (2002)
B. silvestris	L-Lys-D-Glu	Rheims et al. (1999)
B. sphaericus[d]	L-Lys-D-Asp	Schleifer and Kandler (1972)
Alkalibacillus		
A. haloalkaliphilus	meso-DAP direct	Fritze (1996b)
Brevibacillus		
Br. Brevis	meso-DAP direct	Schleifer and Kandler (1972)
Br. laterosporus	meso-DAP direct	Schleifer and Kandler (1972)
Geobacillus		
G. stearothermophilus	(meso-DAP direct)	Schleifer and Kandler (1972)
G. thermoleovorans	DAP[b]	Zarilla and Perry (1987)
G. pallidus	meso-DAP direct	Scholz et al. (1987)
Gracilibacillus		
Gr. dipsosauri	meso-DAP direct	Lawson et al. (1996)
Marinibacillus		
M. marinus	L-Lys-direct	Yoon et al. (2001b)
Paenibacillus		
P. polymyxa	(meso-DAP direct)	Schleifer and Kandler (1972)
P. alvei	meso-DAP direct	Schleifer and Kandler (1972)
P. amylolyticus[e]	(meso-DAP direct)	Schleifer and Kandler (1972)
P. lentimorbus	meso-DAP direct	Schleifer and Kandler (1972)
P. macerans	meso-DAP direct	Schleifer and Kandler (1972)
Sporolactobacillus		
S. laevolacticus	meso-DAP direct	Andersch et al. (1994)
Sporosarcina		
S. ureae	L-Lys-Gly-D-Glu	Stackebrandt et al. (1987)
S. globisporus	L-Lys-D-Glu	Stackebrandt et al. (1987)
S. psychrophilus	L-Lys-D-Glu	Stackebrandt et al. (1987)
S. pasteurii	L-Lys-D-Asp	Ranftl and Kandler (1973)
Ureibacillus		
U. thermosphaericus	L-Lys-D-Asp	Andersson et al. (1995)
Virgibacillus		
V. pantothenticus	meso-DAP direct	Schleifer and Kandler (1972)
V. halodenitrificans	meso-DAP direct	Denariaz et al. (1989)
V. marismortui	meso-DAP[b]	Arahal et al. (1999)
V. salexigens	meso-DAP[b]	Garabito et al. (1997)

[a]Data in parentheses were not obtained from the type strain of the species.
[b]Configuration not determined.
[c]This neutrophilic species is closely related to the alkaliphilic species *Bacillus cohnii* and *Bacillus halmapalus*.
[d]Ahmed et al. (2007c) proposed the transfer of these species to the new genus *Lysinibacillus*.
[e]The strain analyzed by Schleifer and Kandler (1972) as *Bacillus circulans* (ATCC 9966) has been reallocated to *Paenibacillus amylolyticus*.

pseudalcaliphilus, *Bacillus pseudofirmus*, and *Bacillus vedderi*. The two closely related species *Bacillus cohnii* (alkaliphilic) and *Bacillus halmapalus* (alkalitolerant) do not belong in this phylogenetic group, and lie nearer to *Bacillus cereus*: the cross-linkage of *Bacillus cohnii* is L-Orn-D-Asp (Spanka and Fritze, 1993), while *Bacillus halmapalus* has been shown to lack DAP (Nielsen et al., 1994).

Other cell-wall polymers have attracted less attention than murein, and the small amounts of reported data for a few strains do not allow the taxonomic values, if any, of these components to be recognized; the subject has been reviewed by Naumova and Shashkov (1997). Teichoic acids have been found in *Bacillus coagulans*, *Bacillus licheniformis* and *Bacillus subtilis*, and teichuronic acids have been found in *Bacillus licheniformis*, *Bacillus megaterium* and *Bacillus subtilis*. Aono and Ohtani (1990) and Aono et al. (1993) found the acidic polymers teichuronic acid and teichuronopeptide in the cell walls of alkaliphilic *Bacillus* strains and suggested that these components might be important in alkalophily as mutants deficient in them grew poorly at high pH. Fox et al. (1998) described the use of gas chromatography-mass spectrometry and liquid chromatography-mass spectrometry in the investigation of teichoic acids and teichuronic acids in *Bacillus* species.

Naumova and Shashkov (1997) also reviewed studies on sugar-phosphate polymers (found in *Bacillus pumilus* and *Bacillus subtilis*) and anionic polysaccharides (found in *Bacillus cereus* and *Bacillus megaterium*), but again the information is too sparse to reveal any taxonomic implications.

Capsules. Gram-positive bacteria may produce two kinds of capsule, composed of polyglutamic acid or polysaccharide, but their production by *Bacillus* species has not appeared to be of much taxonomic value. Although most *Bacillus subtilis* strains do not produce significant capsular material in the laboratory, the genome sequence of strain 168 indicates that this organism possesses the genes encoding both types of capsule (Foster and Popham, 2002). The production of poly-γ-glutamic acid by "*Bacillus subtilis* var. *natto*" during the stationary phase of growth is economically important in the manufacture of the fermented soybean product natto (Ueda, 1989).

The poly-γ-D-glutamic acid capsule of *Bacillus anthracis* is encoded by the three plasmid pXO2 genes *capA*, *capB*, and *capC*, and it is an important virulence factor for this organism as non-capsulate strains are avirulent (see *Pathogenicity*, below). The sequences of the enzymes encoded by the three genes suggest that they are membrane-associated (Mock and Fouet, 2001). The capsule is produced *in vivo* and when grown in appropriate conditions in the laboratory (see *Procedures for testing special characters*, below). *Bacillus anthracis* is a member of the *Bacillus cereus* group of closely related species, but none of the species besides *Bacillus anthracis* appears to produce this capsule. Although homologs of *Bacillus anthracis* virulence plasmid pXO1 genes were found in half of a set of 19 other members of the *Bacillus cereus* group in hybridization experiments, few pXO2 genes were found that hybridized with genomic DNA from the 19 *Bacillus cereus* group strains (Read et al., 2003). The capsule of *Bacillus anthracis* was reviewed by Mock and Fouet (2001). Other *Bacillus* species, outside the *Bacillus cereus* group, are known to produce poly-γ-glutamic acid. Synthesis by *Bacillus licheniformis* is carried out by a membrane-associated complex

that catalyzes glutamic acid racemization, polymerization, and membrane translocation (Gardner and Troy, 1979); as with "*Bacillus subtilis* var. *natto*", production of the capsular material is induced during the stationary phase (Foster and Popham, 2002). While D-glutamic acid is the predominant stereoisomer incorporated into the polymer, the ratio of D- and L-glutamic acids may vary according to the rate at which D-glutamic acid is being formed in the *Bacillus subtilis* cell (Aschiuchi et al., 1999), but in *Bacillus licheniformis* two glutamyl polypeptides are formed, one of each isomer, and the ratio is influenced by the concentrations of certain metal ions in the growth medium (Thorne, 1993). *Bacillus megaterium* is also known to produce poly-γ-glutamic acid, and can form a capsule comprising both polysaccharide and polypeptide, with the former at the cell poles and equators and the latter located laterally (Guex-Holzer and Tomcsik, 1956). Applications of bacterial poly-γ-glutamic acid are reviewed by Shih and Van (2001).

Carbohydrate polymers are formed by several *Bacillus* species, dextrans and levans being produced extracellularly by *Bacillus licheniformis* and *Bacillus subtilis* from sucrose (Claus and Berkeley, 1986), but true polysaccharide capsules have not been reported for *Bacillus subtilis*. The *Bacillus subtilis* genome contains two operons and some additional genes that show great similarity to capsule synthesis loci in *Staphylococcus aureus* and *Streptococcus pneumoniae*, but it is not known if they are truly genes for capsule synthesis (Foster and Popham, 2002). The extracellular polysaccharides of *Bacillus licheniformis* and *Bacillus subtilis* are of economic importance in the spoilage of bread and alcoholic beverages by "ropiness." Analysis of the polysaccharide of a *Bacillus licheniformis* from ropy cider found that it was a heteropolymer containing over 80% mannose (Larpin et al., 2002). Aubert (1951) assumed that a heteropolysaccharide of D-glucose, D-galactose and D-ribose extractable from *Bacillus megaterium* KM with hot water was probably capsular material, but Cassity and Kolodziej (1984) concluded that a heteropolysaccharide of D-glucose, D-xylose, D-galactose, and L-arabinose produced by another strain of this species was intracellular and that it was used as a source of carbon and energy during sporulation. Several polysaccharides from *Bacillus* strains have been found to cross-react with antisera to capsules from other genera: *Bacillus mycoides* with *Streptococcus pneumoniae* type III, and *Bacillus pumilus* with *Haemophilus influenzae* type b and with *Neisseria meningitidis* group A (Myerowitz et al., 1973).

Flagella. Many species of *Bacillus* are motile by means of peritrichous flagella, which are not usually numerous and may be very few in number. Flagellation has not been considered a particularly useful taxonomic character for the genus, but the presence or absence of motility continues to be indicated in most species descriptions, and it is of some value in identification. For example, *Bacillus anthracis* and *Bacillus mycoides* are nonmotile, while most *Bacillus cereus* strains are motile. The flagella of *Bacillus thuringiensis* may bind to insect cells and be important in virulence (Zhang et al., 1995). The value of H-antigens in the typing of *Bacillus cereus*, *Bacillus thuringiensis* and *Bacillus sphaericus*, and other aspects of *Bacillus* flagellar antigens, are discussed in Antigens and vaccines, below. The flagella of *Bacillus subtilis* are well characterized, and reviews may be found in Sonenshein et al. (1993) and in Aizawa et al. (2002).

S-layers. Surface or S-layers are two-dimensional arrays composed of protein or glycoprotein molecules. The S-layer proteins assemble themselves into very stable structures which have oblique, square or hexagonal lattice symmetries, are 5–25 nm thick, and contain pores of 2–8 nm in diameter (Sleytr et al., 2001). The phylogenetic origins of the S-layers of some *Bacillus cereus* group strains was investigated by Mignot et al. (2001), and the possession of an S-layer was found to be largely restricted to a genetically clustered subgroup of clinical and insect isolates, suggesting a role in pathogenicity and the influence of ecological pressures to maintain the layer. It has been shown that the S-layer of *Bacillus cereus* is involved in the adhesion of the organism to host cell molecules, and polymorphonuclear leukocytes, as well as enhancing the organism's radiation resistance (Kotiranta et al., 2000). However, S-layers are apparently of no value as taxonomic markers, as in some species, including *Bacillus cereus*, their presence is strain-dependent (Kotiranta et al., 1998; Sleytr et al., 2001). The S-layer of *Bacillus anthracis* is reviewed by Mock and Fouet (2001).

Colony characteristics. *Bacillus* species show a very wide range of colonial morphologies, both within and between species, and of course medium composition and other incubation conditions have a strong influence. Despite this diversity, however, *Bacillus* colonies on routine media are not generally difficult to recognize. Some species have characteristic yet seemingly infinitely variable colonial morphologies: colonies of *Bacillus cereus* and relatives are very variable, but readily recognized (Figure 9a, b, h): they are characteristically large (2–7 mm in diameter) and vary in shape from circular to irregular, with entire to undulate, crenate or fimbriate edges; they have matt or granular textures, but smooth and moist colonies are not uncommon. Although colonies of *Bacillus anthracis* and *Bacillus cereus* can be similar in appearance, those of the former are generally smaller, non-hemolytic, may show more spiking or tailing along the lines of inoculation streaks, and are very tenacious as compared with the usually more butyrous consistency of *Bacillus cereus* and *Bacillus thuringiensis* colonies, so that they may be pulled into standing peaks with a loop. The colonies of *Bacillus mycoides* differ from those of other members of the *Bacillus cereus* group; they are characteristically rhizoid or hairy-looking and adherent, and they readily cover the whole agar surface (Figure 9d).

The colonies of other species vary from moist and glossy (Figure 9c, e, f) through granular to wrinkled (Figure 9h); shapes vary from round to irregular, sometimes spreading, with entire through undulate or crenate to fimbriate edges. After 24–48 h incubation, colonial sizes of mesophilic strains typically range from 1 to 5 mm; color commonly ranges from buff or creamy-gray to off-white, but occasional strains may produce black, brown, orange, pink or yellow pigments; such pigmentation tends to be characteristic of species or subspecies. Elevations range from effuse through raised to convex. Consistency is usually butyrous, but mucoid and dry, adherent colonies are not uncommon. Hemolysis may be absent, slight or marked, partial or complete. *Bacillus subtilis* (Figure 9g) and *Bacillus licheniformis* produce similar colonies which are exceptionally variable in appearance and often appear to be mixed cultures – the colonies are irregular in shape and of moderate (2–4 mm) diameter, and range in consistency from moist and butyrous

or mucoid (with margins varying from undulate to fimbriate), through membranous with an underlying mucoid matrix (with or without mucoid beading at the surface), to a rough and crusty appearance as they dry. The "licheniform" colonies of *Bacillus licheniformis* tend to be quite adherent. Rotating and migrating microcolonies (Figure 9i), which may show spreading growth (the V morphotype, see below), were observed macroscopically in about 13% of strains received as *Bacillus circulans* (Logan et al., 1985) but this very heterogeneous species has undergone radical taxonomic revision, and organisms producing motile microcolonies are now allocated to *Paenibacillus cookii, Paenibacillus glucanolyticus, Paenibacillus lautus*, and some unidentified *Paenibacillus* species (Alexander and Priest, 1989; Logan et al., 2004a). Most of the colonial morphologies illustrated here are shown in color by Logan and Turnbull (2003).

Matsushita et al. (1998, 1999) have constructed a mathematical model to explain some of the morphological variation

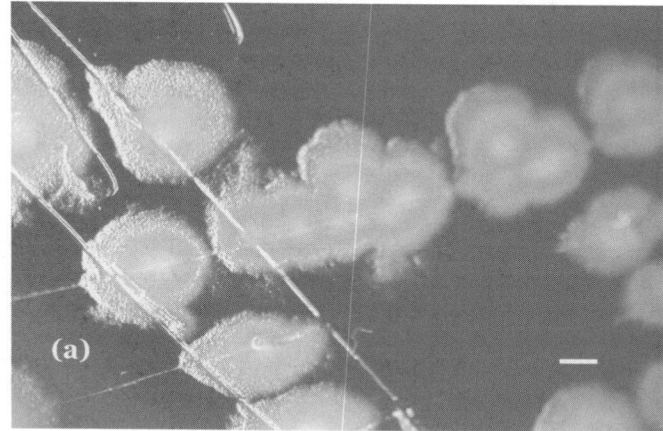

FIGURE 9. Colonies of endospore-forming bacteria on blood agar [parts (a)–(c), (e)–(f), (h)] and nutrient agar [parts (d), (g), (i)] after 24–36 h at 37 °C. These figures illustrate some of the diversity of colonial appearance within the genus, but the appearances shown should not be regarded as necessarily typical of the species illustrated. Bars for parts (a)–(f) and (h)–(i) = 2 mm; bar for (g) = 4 mm. (a) *Bacillus anthracis*: circular to irregular colonies with entire to undulate, crenate and fimbriate edges, and granular surface textures; (b) *Bacillus cereus*: irregular, with undulate, crenate and fimbriate edges, and matt or granular textures; (c) *Bacillus megaterium*: glossy, round to irregular colonies with entire to undulate margins; (d) *Bacillus mycoides*: rhizoid or hairy-looking, adherent colonies which may readily cover the whole agar surface; (e) *Bacillus pumilus*: wrinkled, irregular colonies with undulate margins; (f) *Bacillus sphaericus*: smooth, glossy, round to irregular colonies with entire to undulate margins; (g) *Bacillus subtilis*: irregular colonies that may give the appearance of a mixed culture. They range in consistency from moist through butyrous or mucoid to membranous, with an underlying mucoid matrix (with or without mucoid beading at the surface), and become rough and crusty in appearance as they dry. Margins vary from undulate to fimbriate; (h) *Bacillus thuringiensis*: circular to irregular colonies with entire or undulate edges, and matt to granular surface textures; (i) Motile, spreading microcolonies sometimes seen in strains that were previously assigned to *Bacillus circulans*, but which are now usually allocated to *Paenibacillus* species (see text). Photographs prepared by N. A. Logan.

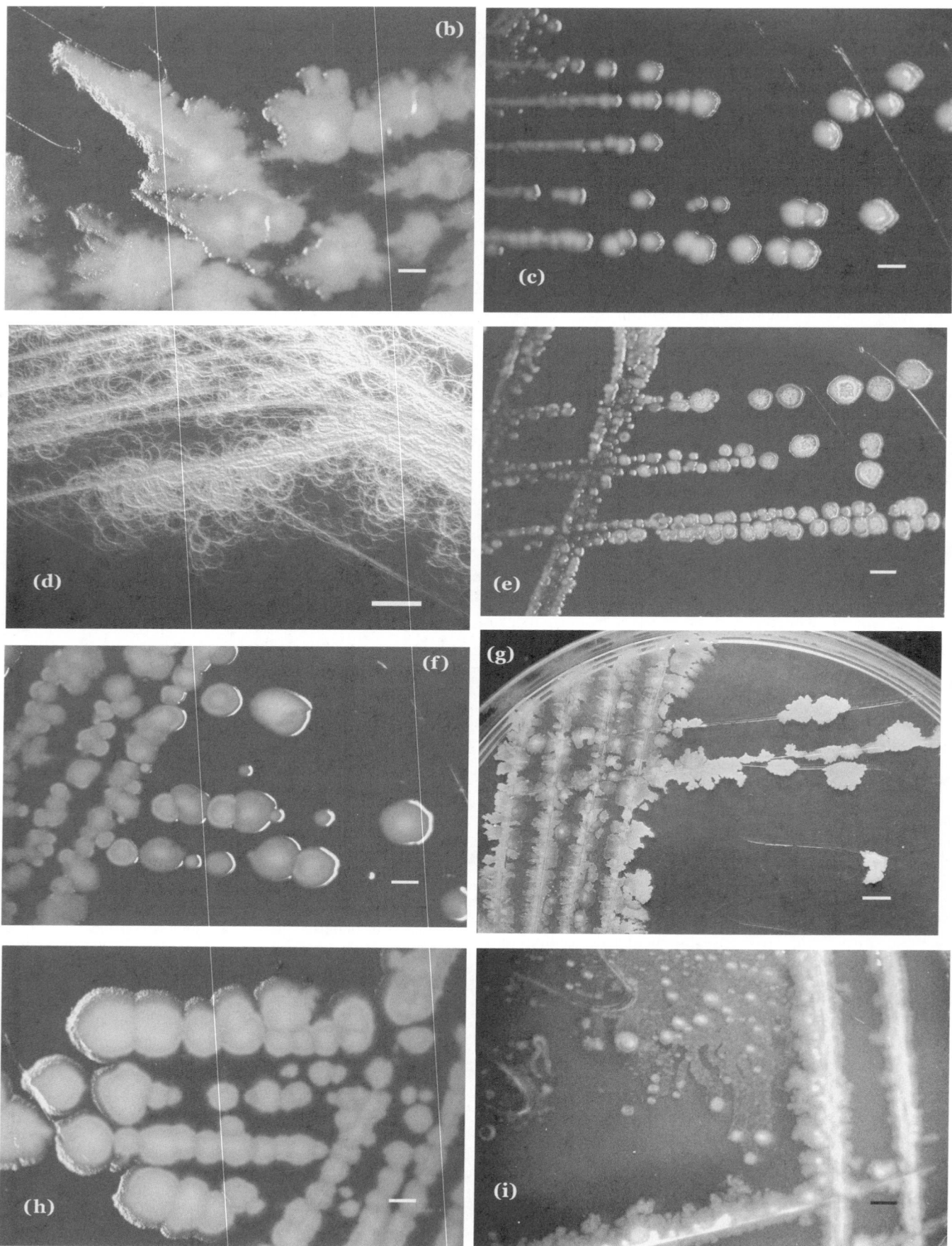

FIGURE 9. (continued)

seen in colonies of *Bacillus subtilis*: it is a diffusion-reaction-type model, where colony patterns are influenced by substrate softness and nutrient concentration, and colonies comprise active and inactive cells. The active cells grow, divide and move, while inactive cells do not. Concentric ring-like colonies reflect alternate periods of advance and rest of the growing interface, which consists of the active cells. Active cells also form the tips of the growing branches of "dense branching morphology" colonies. Ben-Jacob et al. (1998) combined a detailed study of bacterial colony development with pattern-formation concepts derived from non-living systems to construct a model which suggested that cooperative cellular behavior, involving long-range chemorepulsion and short-range chemoattraction, occurs. They defined three colonial "morphotypes": tip-splitting or branching (T), chiral (C), where the thin branches all have a same-handed twist, and vortex (V), where the tip of each branch bears a leading droplet containing many bacteria. The T morphotype is seen in *Paenibacillus dendritiformis*, the C morphotype particularly where a rapid growth transition from the T type occurs on softer agar (the reverse transition, C to T, occurring on harder agar), and the V morphotype is characteristic of some *Paenibacillus* strains formerly classfied as *Bacillus circulans*. Stecchini et al. (2001) found that the radial growth rate of *Bacillus cereus* colonies diminished as the agar content increased, and that colony density decreased during the incubation period, being lowest at the lower agar concentrations because the liquid film was thicker. Delprato et al. (2001) found that the bacteria in the central regions of *Bacillus subtilis* colonies migrated to the colony edge and formed a ring pattern following exposure of the whole colony to UV radiation, and that cells grew both inwards and outwards when the irradiation ceased; they proposed a diffusion-reaction model in which the radiation initiates a waste-limited chemotaxis.

Sporulation is strongly associated with the spatial development of the bacterial community; in *Bacillus subtilis* biofilms, sporulating aerial structures (primitive fruiting bodies) may be formed by motile cells that align themselves to form chains of attached cells (Branda et al., 2001). *Bacillus subtilis* uses an elaborate peptide quorum-sensing system to choose between the competent (i.e., for exogenous DNA uptake) state and the sporulation process, and sporulation occurs only poorly at low cell densities, even if the cells are starved (Miller and Bassler, 2001). To explain differences in the architectures of colonies grown from vegetative cells and those grown from spores, characterized by different glycocalyx wetting angles, Puzyr et al. (2002) suggested that germinating spores and vegetative cells of *Bacillus subtilis* adopt different strategies of substrate colonization.

Life cycle. Cohn (1876), Koch (1876) and Tyndall (1877) independently discovered that certain bacteria could spend part of their lives as the dormant cellular structures now known as endospores. The first two of these authors recognized the significance of these structures in the epidemiology of anthrax, and Koch's study of the life history of *Bacillus anthracis* proved the germ theory of disease and so marked the genesis of clinical bacteriology. Although Pasteur (1870) had previously figured endospores in a work on silkworm diseases, he did not clearly attribute the longevity of the pathogens to their spores.

The ability to form endospores in aerobic conditions has been a defining character of the genus *Bacillus* since the 1920s, and has been applied in all editions of the *Manual*.

Spore formation is most important in identification to genus level. Before attempting to identify to species level it is important to establish that the isolate really is an aerobic endospore-former, and that other inclusions are not being mistaken for spores.

Endospores are so named because they are formed intracellularly, and they differ from their parent vegetative cells in many ways: they are optically refractile, and are highly resistant to chemical and physical stresses that are lethal to vegetative cells. These properties are conferred by the spores' special chemical composition and ultrastructure, and much effort has been expended over many years in order to elucidate the processes of spore formation and germination, and the molecular mechanisms that make endospores the hardiest form of life known on Earth. Although endospores are to be found in other genera, *Clostridium* for example, it is the spores of *Bacillus subtilis* that have been the most intensively studied, especially those of strain 168.

Under suitable nutritional, temperature, pH, gaseous and other conditions, *Bacillus* cells will grow and divide by binary fission, with the dividing septum traversing the middle of the cell. Depending on species, strain, and cultural conditions, daughter cells may separate so that the culture appears to be composed of single cells and pairs of dividing cells when viewed by phase-contrast microscopy. In other cases, daughter cells may remain attached to each other, so that chains of cells are seen. Filaments may also be observed, and these can often be symptomatic of a stressed culture. An organism that exists predominantly as regular rods in optimal growth conditions may produce swollen, pleomorphic, unhealthy-looking cells when stressed.

Endospores are formed at the end of the exponential growth phase, and at least two kinds of environmental factors have been implicated in the induction of sporulation. One trigger for sporulation is nutritional deprivation, for example when an actively growing culture is transferred from a rich to a poor growth medium. Many other factors are known to affect endospore formation, including growth temperature, environmental pH, aeration, presence of certain minerals, and carbon, nitrogen and phosphorus sources and their concentrations. Another influence is population density: as the mass of a culture increases, there is an extracellular accumulation of a secreted peptide (competence and sporulation factor, or CSF), which acts as an autoinducer for quorum sensing (Miller and Bassler, 2001). When this peptide reaches a concentration that relates to a particular cell density, high intracellular levels of CSF lead to an increase of the phosphorylated form of a response regulator (SpoOA), which leads to derepression of various stationary-phase genes, some of which are needed for sporulation (Sonenshein, 2000). Studies of *Bacillus subtilis* biofilms have shown that the cells do not behave as strictly unicellular organisms, but that sporulation is also tightly linked with the spatial development of the microbial community. Motile cells may form aligned chains of attached cells that produce aerial structures; these can be seen as primitive fruiting bodies, as they are the preferred sites of sporulation (Branda et al., 2001).

Sporulation is closely tied to the cell cycle, and a round of DNA replication must be initiated as a prerequisite for the sporulation pathway being activated (Michael, 2001). The cell division of vegetative growth is symmetrical, and yields two similar cells. During sporulation, however, cell division is asymmetrical and two quite different kinds of cells, the small forespore and the larger mother cell, are produced, each with its own copy of the chromosome. The two different kinds of division are believed to use essentially the same protein machinery (Errington, 2001). At the commencement of sporulation, the chromosomes form an elongated structure called the axial filament, with migration of a specific region of the chromosomes towards the poles, and polar septation bisects one end of this filament so that only part of the nucleoid lies within the forespore; the remainder of the chromosome is then transferred into the forespore from the mother cell (Errington, 2001; Levin and Grossman, 1998). The process of sporulation may be divided into seven morphologically recognizable stages following vegetative growth: I, preseptation, with the DNA forming the axial filament; II, asymmetric septation, the membrane of the developing spore surrounds the spore protoplast and becomes detached from the membrane of the mother cell; III, the forespore so formed becomes surrounded by the cytoplasm of the mother cell and so is contained within two membranes of opposing polarity; IV, spore cortex formation commences, with a primordial cell wall being laid down between the membranes, next to the forespore inner membrane; the cortex (a thicker layer of electron-transparent peptidoglycan, unique to bacterial endospores) is laid down on the outside of this primordial cell wall; an exosporium, a thin and delicate proteinaceous outermost covering, may be formed at this stage; V, proteinaceous spore coats are synthesized and begin to be deposited outside the cortex; VI, the spore matures, and acquires its refractility and heat resistance; VII, the sporangium lyses and releases the mature spore (Foster, 1994). In a laboratory culture of *Bacillus subtilis*, the whole process of sporulation may take about 8 h. The genetics of sporulation are reviewed by Piggot and Losick (2002).

Endospores are metabolically extremely dormant and do not contain ATP; this dormancy is the key to their resistance to many agents, including heat, radiation and chemicals, and their survival over long periods. Spore structure, resistance and germination are reviewed by Atrih and Foster (2001). The spore cortex is essential for spore dehydration (10–30% of the water content of the vegetative cell) and so for the maintenance of dormancy, and for the spore's heat resistance; the temperature of sporulation influences the mature spore's heat resistance (Nicholson et al., 2000). The mechanisms of spore resistance to chemical agents and radiation are not well understood, but saturation of the chromosome by protective small acid-soluble proteins (SASPs) is believed to play a part, while the spore coats are believed to prevent access of peptidoglycan-lytic enzymes to the spore cortex, and are also known to protect from hydrogen peroxide and UV radiation (Riesenman and Nicholson, 2000); coat assembly and composition are reviewed by Takamatsu and Watabe (2002). SASPs appear to be more important than the low core water content in protecting DNA from heat and oxidative damage (Setlow, 1995). Spore core and coat proteins are reviewed by Driks

(2002). Pyridine-2,6-dicarboxylic acid (dipicolinic acid; DPA) is a unique and quantitatively important spore component (comprising 5–14% of the spore dry weight), and Ca^{2+} and other divalent cations are chelated by it, but precisely how it contributes to spore resistance is unclear (Slieman and Nicholson, 2001). DPA may be used as a marker for detecting the presence of spores by Curie-point pyrolysis mass spectrometry and by Fourier-transform infrared spectroscopy (Goodacre et al., 2000). Mechanisms of spore resistance have been reviewed by Nicholson et al. (2000) in the contexts of survival both in extreme terrestrial conditions and during travel through extraterrestrial environments. The function of the exosporium is not known; the exosporium of *Bacillus cereus* has been characterized by Charlton et al. (1999).

Conversion from the dormant spore to vegetative cell involves the three steps: activation, germination, and outgrowth. Dormancy may be broken by heat treatment at a time and sublethal temperature appropriate to the organism concerned, and by ageing at low temperatures, but endospores of many species do not require such activation. The heat treatment procedure used to assist the isolation of *Bacillus* species, by destruction of all kinds of vegetative cells, is often effective in activation (see *Enrichment* and *isolation procedures*, below). Following dormancy, with or without activation, the spore may encounter conditions that trigger germination; the cortex is rapidly hydrolyzed, SASPs are quickly degraded, and refractility is lost in a matter of minutes. The germinated spore protoplast then outgrows: it visibly swells owing to water uptake, biosynthesis recommences (taking advantage of the nutrients released by germination as well as those available in its new environment), and a new vegetative cell emerges from the broken spore coat; another period of vegetative reproduction ensues. It seems remarkable that a metabolically dormant spore can monitor its external environment in order to trigger germination within seconds in suitable conditions, and that this triggering mechanism can escape the constraints of dormancy while being resistant to damaging agents. Germination can be induced by exposure to nutrients such as amino acids and sugars, by mixtures of these, by non-nutrients such as dodecylamine, and by enzymes and high hydrostatic pressure; for many species, L-alanine is an important germinant, while D-alanine can bind at the same site as L-alanine and acts as a competitive inhibitor (Foster and Johnstone, 1990; Johnstone, 1994). The mechanism of germination has been most studied for *Bacillus megaterium*. Although spore structure is very similar between species, and cortex peptidoglycan structure is highly conserved (Atrih and Foster, 2001), there are many different germinant receptor specificities; nonetheless, the underlying mechanism of germination may be universal (Foster and Johnstone, 1990). Recent developments in the understanding of spore germination process and the spore components required for it are reviewed by Moir et al. (2002) and Paidhungat and Setlow (2002).

Various aspects of spores have been considered as taxonomic characters. Spore antigens are considered under *Antigens* and *vaccines*, below. The spore coat may open by splitting polarly, equatorially, transversely, or by expansion with the halves of the coat at each end of the outgrowing cell, or by the coat lysing. Small-celled organisms such as *Bacillus subtilis* tend to leave well-defined spore coat residues, while large-celled species such as *Bacillus cereus* and *Bacillus megaterium* may not. Lamana (1940a, 1954) studied modes of spore germination for nine species and found it to be of potential value for differentiation between the small-celled species and between this group and the large-celled species, but, with two exceptions (Burdon, 1956; Gould, 1962), little further attention has been paid to this character.

Bradley and Franklin (1958) showed that most of the 20 species they studied could be distinguished by electron microscopy of carbon replicas of spore surface patterns. Bulla et al. (1969) found that scanning electron microscopy gave inadequate resolution for such studies, but Murphy and Campbell (1969) achieved good resolution of *Bacillus polymyxa* spores by this method, and Gray and Hull (1971) considered this approach to be promising in the study of the *Bacillus circulans* complex. Later authors have sometimes described spore surface structure in proposals for new species, but too few such descriptions are available to judge the taxonomic value of spore surface characteristics across the genus.

Electron microscopy has revealed sword-shaped appendages radiating from one end of the exosporium of the spores of two phylloplane strains of *Bacillus cereus* (Mizuki et al., 1998). The proteinaceous spore appendages of 10 *Bacillus cereus* strains isolated from food-borne illness outbreaks and food industry sources showed some antigenic relationship, but when subjected to SDS-PAGE analysis none showed identical patterns (Stalheim and Granum, 2001). Smirnova et al. (1991) found that hemagglutination patterns of fimbriated *Bacillus thuringiensis* spores correlated with the subspecies of the strains rather than with their flagellar serovars. Song et al. (2000) reported that under strictly standardized growth conditions, spore fatty acid profiles, like those of vegetative cells, are stable and potentially of taxonomic value.

The microscopic morphologies of *Bacillus* species, especially of their sporangia, are well established as valuable characters. Smith et al. (1946, 1952) and Gordon et al. (1973) used cell size, appearance of cytoplasm and sporangial morphology as the basis of their division of the genus into three groups of species, and this arrangement still correlates quite well with the present classification of the aerobic endospore-formers. Sporangial morphology, and cell size, shape and cytoplasmic appearance, remain useful characters in polyphasic taxonomic studies, and sporangial characters are particularly valuable in identification. Spore shapes vary from cylindrical (Figure 10a) through ellipsoidal (Figure 10b–e, g) to spherical (Figure 10f); bean- or kidney-shaped, curved-cylindrical, and pear-shaped spores are also seen occasionally. Spores may be terminally (Figure 10f), subterminally (Figure 10a–g), paracentrally (Figure 10b, d, e, g) or centrally (Figure 10f) positioned within sporangia and may distend them (Figure 10c–f). Despite within-species and within-strain variation, sporangial morphologies tend to be characteristic of species, and for some species may allow tentative identification by the experienced worker. Routine recognition of *Bacillus thuringiensis* is largely dependent on observation of its cuboid or diamond-shaped parasporal crystals in sporangia (Figure 10h).

Nutrition and growth conditions. Despite the very wide diversity of the genus, most *Bacillus* species will grow well on routine media such as nutrient agar or trypticase soy agar, and most will grow on blood agar. However, some isolates, particularly those from nutritionally poor environments, may grow poorly if at all on these standard media and so require weaker formulations; for example, strains of *Bacillus thuringiensis* (Forsyth and Logan, 2000) isolated from Antarctic soils required *Bacillus fumarioli* agar or a half-strength formulation of this medium for reliable cultivation, and they would not grow consistently on trypticase soy agar.

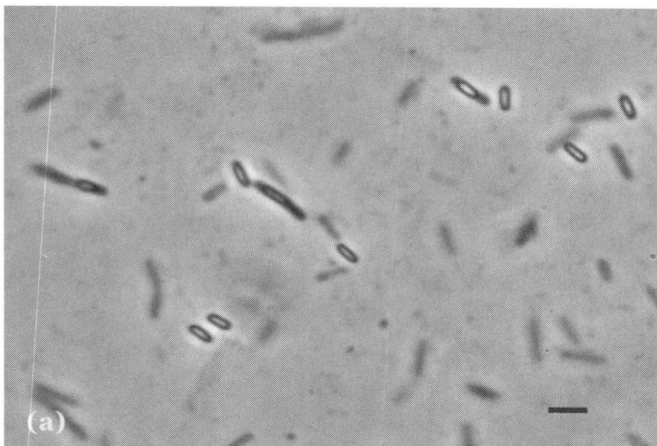 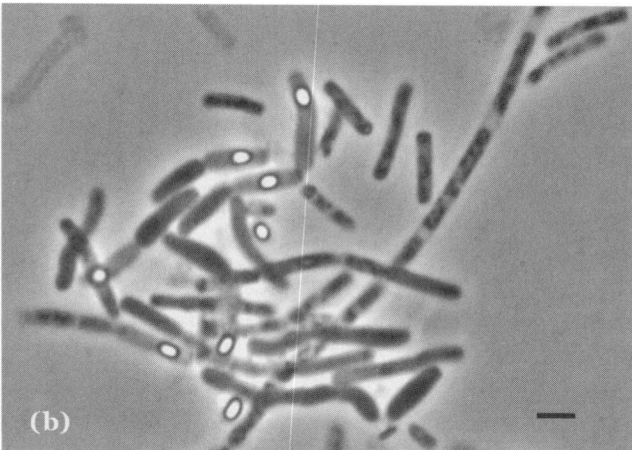

FIGURE 10. Photomicrographs of *Bacillus* species viewed by phase-contrast microscopy. Bars = 2 μm. (a) *Bacillus pumilus*: slender cells with cylindrical, subterminal spores, not swelling the sporangia; (b) *Bacillus cereus*: broad cells with ellipsoidal, paracentral and subterminal spores, not swelling the sporangia and showing some poly-β-hydroxybutyrate inclusions, which are smaller and less phase-bright than the spores; (c) *Bacillus circulans*:

(continued)

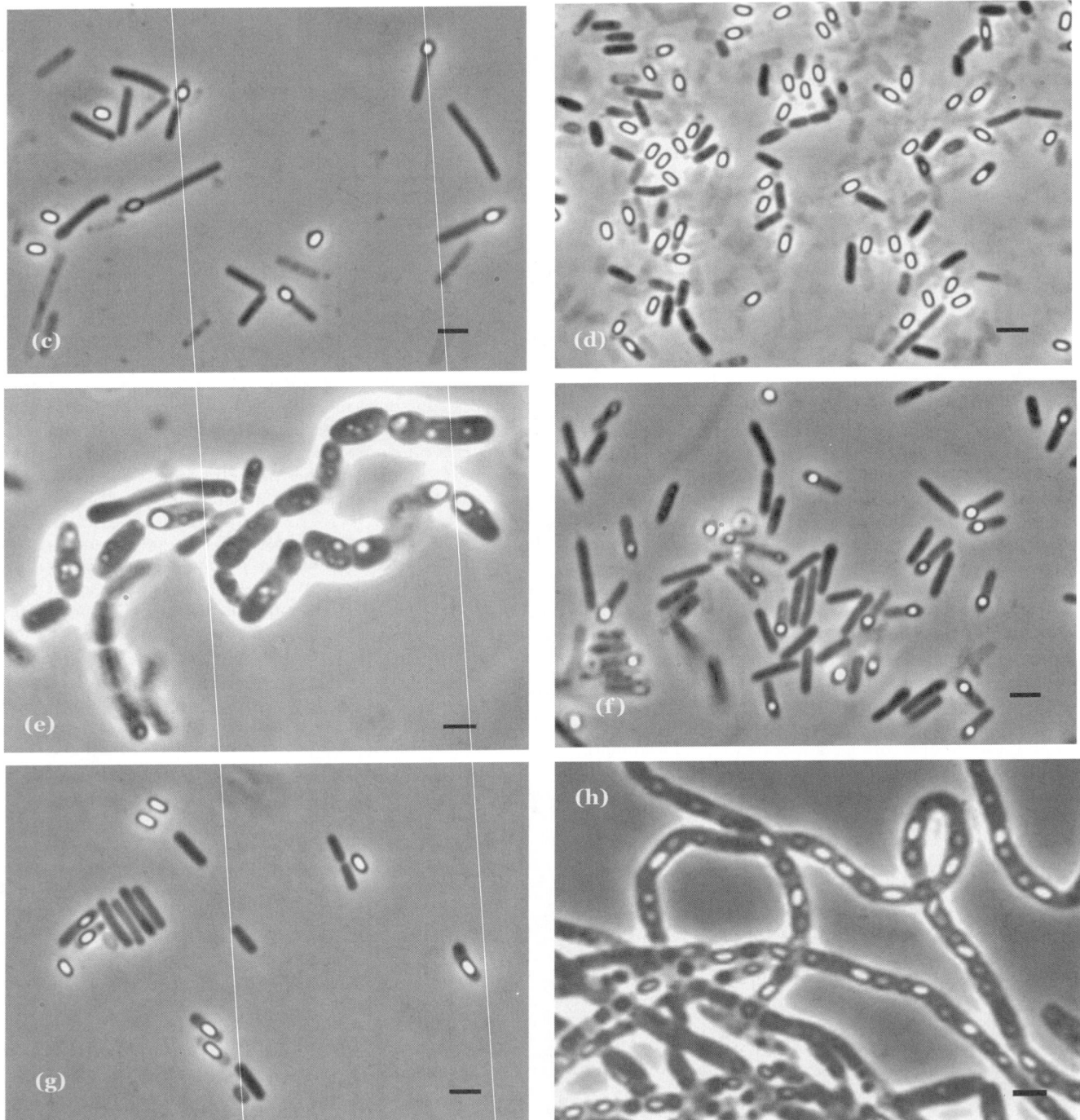

FIGURE 10. (continued) ellipsoidal, subterminal spores, swelling the sporangia; (d) *Bacillus licheniformis*: some chaining of cells evident; ellipsoidal, central and subterminal spores, not swelling the sporangia; (e) *Bacillus megaterium*: broad cells with ellipsoidal to spherical, subterminal and terminal spores, not swelling the sporangia, and showing poly-β-hydroxybutyrate inclusions, which are smaller and mostly less phase-bright than the spores; (f) *Bacillus sphaericus*: spherical, terminal spores, swelling the sporangia; (g) *Bacillus subtilis*: ellipsoidal, central, paracentral and subterminal spores, not swelling the sporangia; (h) *Bacillus thuringiensis*: broad cells with ellipsoidal, subterminal spores, not swelling the sporangia, and showing parasporal crystals of insecticidal toxin, which are less phase-bright than the spores. Photomicrographs prepared by N. A. Logan.

In the First Edition of this *Manual*, Claus and Berkeley (1986) listed five of their 34 valid species that would not grow on nutrient agar. Three of these (*Bacillus larvae*, *Bacillus lentimorbus* and *Bacillus popilliae*) have been transferred to *Paenibacillus*, and one (*Bacillus pasteurii*) has been transferred to *Sporosarcina*, leaving *Bacillus fastidiosus* as the only exception. However, of the 68 *Bacillus* species newly described or validated in the two decades following the preparation of the First Edition of this *Manual*, some 24 grow poorly or not at all on nutrient agar because of its neutral pH, and/or insufficient salinity, or because it is nutritionally too weak or too rich. *Bacillus benzoevorans* does not grow on peptone or tryptone media, but may be cultivated on yeast extract media containing sodium acetate or benzoate (Pichinoty et al., 1984). *Bacillus fastidiosus* strains usually need allantoic acid, allantoin or uric acid as sole carbon, nitrogen and energy sources, but some strains will grow on certain peptones, especially at high concentrations. *Bacillus laevolacticus* requires glucose or other carbohydrate for growth (Andersch et al., 1994). *Bacillus psychrodurans* and *Bacillus psychrotolerans* do not grow, or grow only weakly, on nutrient agar or in nutrient broth, and require a rich medium such as casein-peptone soymeal-peptone agar (Abd El-Rahman et al., 2002). *Bacillus sporothermodurans* also grows weakly on nutrient agar but grows on Brain heart Infusion Agar or in nutrient agar supplemented with vitamin B_{12} (Pettersson et al., 1996). Both *Bacillus fumarioli* and *Bacillus naganoensis* are moderately acidophilic, and will not grow at pH 7.0 (Logan et al., 2000; Tomimura et al., 1990); also, *Bacillus fumarioli* sporulates poorly on trypticase soy agar even when adjusted to its optimum pH of 5.5, and requires a weaker medium such as *Bacillus fumarioli* agar or half-strength *Bacillus fumarioli* agar. *Bacillus aeolius*, *Bacillus halodenitrificans*, *Bacillus halophilus*, *Bacillus horti* (the type strain) and *Bacillus jeotgali* do not grow in routine media without added NaCl (Denariaz et al., 1989; Gugliandolo et al., 2003a; Ventosa et al., 1989; Yoon et al., 2001a; Yumoto et al., 1998). The majority of *Bacillus* species that do not grow on routine media, however, are alkaliphiles: *Bacillus alcalophilus*, *Bacillus agaradhaerens*, *Bacillus clarkii*, *Bacillus cohnii*, *Bacillus krulwichiae*, *Bacillus pseudoalcalophilus*, *Bacillus pseudofirmus*; *Bacillus haloalkaliphilus* (which also needs NaCl; now reclassified in *Alkalibacillus*), *Bacillus thermocloacae* and *Bacillus vedderi* will not grow at pH 7.0 (Agnew et al., 1995; Demharter and Hensel, 1989b; Fritze, 1996a; Nielsen et al., 1995a; Spanka and Fritze, 1993; Yumoto et al., 2003), while the alkalitolerant organisms *Bacillus clausii*, *Bacillus gibsonii*, *Bacillus halmapalus*, *Bacillus horikoshii* and *Bacillus okuhidensis*, will all grow at pH 7.0. The two arsenate- and selenate-reducing species *Bacillus arseniciselenatis* and *Bacillus selenitireducens* are both obligately alkaliphilic and halophilic (Switzer Blum et al., 1998). Table 6 shows differential characters of species with pH optima for growth of 8 or above.

Chemically defined media have been developed for several species, often with the optimization of industrial processes in mind. Minimal growth requirements have been established for rather few species, may be influenced by environmental conditions, and further emphasize the diversity of the genus.

Most species will use glucose and/or other fermentable carbohydrates as sole sources of carbon and energy. Patterns of acid production from, or assimilation of, carbon substrates are of great value in the characterization and identification of *Bacillus*

species (Logan, 2002), but some species do not appear to utilize carbohydrates at all. *Bacillus azotoformans* uses a range of organic acids as carbon sources and does not attack carbohydrates; *Bacillus badius* and *Bacillus benzoevorans* assimilate certain amino acids and organic acids and do not produce acid from glucose and other carbohydrates. As indicated above, *Bacillus fastidiosus* usually uses allantoic acid, allantoin or uric acid as its sole carbon and energy source. The spherical-spored species *Bacillus fusiformis*, *Bacillus neidei*, *Bacillus pycnus*, *Bacillus silvestris* and *Bacillus sphaericus* do not produce acid or gas from D-glucose or other carbohydrates; *Bacillus fusiformis* utilizes acetate, citrate, formate, lactate and succinate. *Bacillus carboniphilus*, *Bacillus insolitus*, *Bacillus siralis* and *Bacillus thermocloacae* do not produce acid or gas from glucose or a range of other carbohydrates; the growth of *Bacillus carboniphilus* is promoted by activated carbon and graphite. *Bacillus schlegelii* and *Bacillus tusciae* will grow chemolithoautotrophically, using H_2 as electron donor and CO_2 as carbon source, and for the former species CO will satisfy both requirements. When growing chemoorganoheterotrophically, *Bacillus schlegelii* utilizes acetate, butyrate, fumarate, propionate, succinate, phenol, 1-propanol and a small number of amino acids, and *Bacillus tusciae* uses a few alcohols, amino acids and organic acids, as their sole carbon sources, but neither species metabolizes carbohydrates. *Bacillus methanolicus* can grow on methanol, and some strains will also grow on ethanol.

Bacillus subtilis is attracted by many sugars (Ordal et al., 1979); following the genome sequencing of this species, its carbohydrate uptake and metabolism have been reviewed by Deutscher et al. (2002). In a theoretical analysis of metabolic fluxes, the capacity of *Bacillus licheniformis* for the production of certain industrial enzymes was found to be affected by the carbon sources used (Calik and Özdamar, 2001).

Bacillus species may use inorganic and organic sources of nitrogen. Many species will utilize an ammonium salt as their sole nitrogen source, amino acids are widely utilized, and strains of some species can use urea. The two facultative autotrophs *Bacillus schlegelii* and *Bacillus tusciae* can utilize ammonium ions, asparagine and urea as sole nitrogen sources. In the presence of molybdate, *Bacillus niacini* can use nicotinate as its sole source of carbon, nitrogen and energy. A soil isolate identified as *Bacillus coagulans* was found to use pyridine as sole carbon, nitrogen and energy source (Uma and Sandhya, 1997). Strains of *Bacillus pumilus* resistant to and able to utilize cyanide have been isolated (1983; Meyers et al., 1991; Skowronski and Strobel, 1969) and a cyanide-degrading enzyme purified and characterized (Meyers et al., 1991, 1993). In studies of the chemotaxis and motility of *Bacillus subtilis*, all 20 common amino acids have been found to attract the organism (Garrity and Ordal, 1995). Leucine, threonine and valine were found to be essential for growth and emetic toxin production by *Bacillus cereus* (Agata et al., 1999). Although Achouak et al. (1999) concluded that nitrogen fixation among aerobic endospore-formers is restricted to certain species of *Paenibacillus*, nitrogen fixation has been demonstrated in several *Bacillus* isolates from soil, including strains of *Bacillus azotoformans*, *Bacillus cereus*, *Bacillus licheniformis*, *Bacillus megaterium* (Rózycki et al., 1999) and *Bacillus sphaericus*. Some *Bacillus* species may stimulate the nitrogen-fixing activities of unrelated organisms, and so perhaps benefit from the nitrogen so fixed: a *Bacillus firmus* strain growing in association with a strain of *Klebsiella terrigena* was found to increase nitrogen fixation by the latter, probably owing to the protection of nitrogenase by the phenolic compounds it excreted (Zlotnikov et al., 2001); a *Bacillus cereus* strain was found to stimulate nodulation in legumes, so enhancing nitrogen fixation by bradyrhizobia (Vessey and Buss, 2002).

Little comprehensive information is available on the vitamin requirements of individual *Bacillus* species. Many do not require such growth factors, but yeast extract will often stimulate better growth. Adams and Stokes (1968) studied the requirements of the psychrophiles *Bacillus insolitus* and *Bacillus psychrosaccharolyticus:* the former required biotin and thiamine, while the latter needed niacin and thiamine, and biotin was essential or stimulator, depending upon the strain. Among spherical-spored species, *Bacillus neidei* and *Bacillus sphaericus* require both biotin and thiamin for growth, but *Bacillus pycnus* does not. In the presence of molybdate, *Bacillus niacini* can use nicotinate (niacin) as sole source of carbon, nitrogen and energy. *Bacillus sporothermodurans* and *Bacillus subterraneus* require biotin and thiamin for growth, but neither require cystine. For some species, such as *Bacillus thermoamylovorans*, vitamins and nucleic acid derivatives will stimulate growth, but are not essential.

Growth temperature ranges vary appreciably between the strains of species, and maxima and minima may be extended beyond the usual limits of a species for strains found in unusually hot or cold environments. Isolates of *Bacillus licheniformis* and *Bacillus megaterium* from an Antarctic geothermal lake, for example, were found to have maxima of 68 °C and 63 °C, 13 °C and 18 °C, respectively, higher than the previously published limits for these species (Llarch et al., 1997). The vast majority of established species are mesophiles, with optima between 25 °C and 40 °C and typically around 30 °C, minima in the range 5–20 °C, and maxima of 35–55 °C. Several species, *Bacillus coagulans*, *Bacillus fumarioli*, *Bacillus infernus*, *Bacillus methanolicus*, *Bacillus okuhidensis*, *Bacillus smithii*, *Bacillus thermoamylovorans* and *Bacillus tusciae*, have higher growth temperature optima, ranging from 40 °C to 55 °C and above, with minima in the range 25–40 °C and maxima of 55–65 °C, and may be regarded as only moderately thermophilic. With minimum temperatures for growth of 37 °C and above, optima in the range 55–70 °C and maxima of 65–75 °C, *Bacillus schlegelii* and *Bacillus thermocloacae* may be regarded as true thermophiles. *Bacillus psychrodurans*, *Bacillus psychrosaccharolyticus*, and *Bacillus psychrotolerans* grow and sporulate around 0 °C and have maximum growth temperatures between 30 °C and 35 °C, while *Bacillus insolitus*, with a maximum growth temperature of 25 °C, an optimum of 20 °C and a minimum below 0 °C, is a true psychophile. Growth temperature ranges and optima are given for most species in the *List of species of the genus*, below, and the differential characters of species with optimum temperatures of 50 °C and above are shown in Table 8.

Although aerobic growth has long been a defining character of members of the genus, some 20 species are facultatively anaerobic, and the definition was undermined by the discoveries of *Bacillus infernus* and *Bacillus arseniciselenatis*, which are strictly anaerobic (Boone et al., 1995; Switzer Blum et al., 1998). Nitrate respiration is a common property in the genus. Although *Bacillus subtilis* has long been regarded as a strict aerobe, which will like many *Bacillus* species, however, grow anaerobically using nitrate or nitrite as an electron acceptor, it has recently been shown to grow by fermentation in the absence of electron acceptors (Clements et al., 2002; Nakano and Zuber, 2002) (see *Metabolism* and *metabolic pathways*, below).

Survival. Spores are readily formed by strains of many species, but it is a mistake to assume that a primary culture or subculture in or on a routine growth medium will automatically yield spores if stored on the bench or in the incubator. *Bacillus* strains will not sporulate under all cultural conditions, and if conditions are not suitable for sporulation the culture may die (see *Life cycle*, above). Most strains will sporulate if grown for a few days on a routine, solid growth medium supplemented with 5 mg/l manganese sulfate; failure to sporulate on such a medium may be addressed by cultivating on a nutritionally weaker, manganese-supplemented, medium. Repeated subculture of a strain sometimes leads to the production of fewer spores or the complete loss of ability to sporulate; some strains, however, appear able to survive for long periods in refrigerated cultures, even though they have not sporulated.

It is best to grow the organism on nutrient agar containing manganese for a few days, and refrigerate when microscopy shows that most cells have sporulated. For most species sporulated cultures, sealed after incubation, can survive in a refrigerator for many years.

Metabolism and metabolic pathways. The majority of information on the metabolism and biochemistry of *Bacillus* species relates to *Bacillus subtilis* alone or to comparisons of this with other species, and further valuable information has been forthcoming from studies aimed at the optimization of various industrial processes employing several other aerobic endospore-forming species.

It is now established that *Bacillus subtilis*, which was long regarded as a strict aerobe, is capable of growing anaerobically, not only with nitrate as electron acceptor but also by fermentation in the absence of electron acceptors. This species and its close relatives apparently cannot use other electron acceptors such as dimethyl sulfoxide, fumarate and trimethylamine N-oxide, and have been considered to lie in an intermediate position between the true facultative anaerobes now allocated to *Paenibacillus* and the aerobes of the *Bacillus sphaericus* group (Priest, 1993), which are strictly oxidative. *Bacillus cereus*, *Bacillus licheniformis* and *Bacillus thuringiensis* can ferment carbohydrates in the absence of exogenous electron acceptors, and many *Bacillus* species can use nitrate as an electron acceptor in the absence of oxygen, but several species such as *Bacillus megaterium* and *Bacillus pumilus* are unable to do this. *Bacillus subtilis* uses pyruvate dehydrogenase for conversion of pyruvate to acetyl-coenzyme A in anaerobic as well as in aerobic conditions and fermentation is stimulated by pyruvate. The fermentation is of the mixed acid-butanediol type, and products include acetate, acetoin, 2,3 butanediol, ethanol and lactate (Nakano et al., 1997); *Bacillus licheniformis* also carries out a mixed acid fermentation (Shariati et al., 1995). During nitrate respiration, *Bacillus subtilis* reduces nitrate to nitrite and ammonium, and, unlike the denitrifier *Bacillus licheniformis*, it does not produce the gaseous products NO, N_2O and N_2 (Nakano and Zuber, 2002). A homolog of the *Bacillus subtilis* gene encoding membrane-bound respiratory nitrate reductase is found in *Bacillus anthracis* (Nakano and Zuber, 2002). *Bacillus licheniformis* shows poor anaerobic growth on fumarate, but it can grow in the presence of arginine using the arginine deiminase pathway (Maghnouj et al., 1998); *Bacillus cereus* also possesses genes for this pathway but *Bacillus anthracis* does not (Ivanova et al., 2003; Read et al., 2003). In the First Edition of this Manual (Claus and Berkeley, 1986), ability to grow and sporulate in air was implicit in the definition of *Bacillus*, but the proposals of the species *Bacillus infernus* and *Bacillus arseniciselenatis*, which are strictly anaerobic, undermine this long-held element of the genus definition. *Bacillus arseniciselenatis* and *Bacillus selenitireducens* are two alkaliphiles isolated from a lakewater containing unusually high levels of arsenic, and they will grow by respiratory (dissimilatory) reduction of As(V) to As(III) (arsenate to arsenite) and oxidation of lactate to acetate and CO_2. The former will also grow by dissimulatory reduction of Se(VI) to Se(IV) (selenate to selenite) and the latter will reduce Se(IV) to Se(0), so that co-cultures will reduce selenate to elemental selenium (Switzer Blum et al., 1998). Such organisms or their enzymes are of interest for the bioremediation of environments contaminated with toxic oxyanions of arsenic and selenium. Lindblow-Kull et al. (1982) isolated a *Bacillus* strain from the seeds of the selenium-accumulating plant *Astragalus crotalariae*. It grew optimally in the presence of 3–100 mM selenite in nutrient broth, giving a strong red color owing to elemental selenium, and growth also occurred with selenate or tellurate. *Bacillus infernus*, a strict anaerobe, was isolated from a deep terrestrial subsurface environment and it can use Fe^{3+} and MnO_2, as well as trimethylamine N-oxide and nitrate, as electron acceptors (Boone et al., 1995), while *Bacillus subterraneus*, which is a facultative anaerobe isolated from a deep subsurface thermal aquifer, also uses Fe^{3+} and MnO_2, as well as fumarate, nitrate and nitrite as electron acceptors (Kanso et al., 2002).

The respiratory cytochromes and other heme proteins of *Bacillus subtilis* and relatives have been reviewed by von Wachenfeldt and Hederstedt (2002).

The natural habitat of *Bacillus subtilis* is soil, which contains a wide range of carbohydrates and polysaccharides from microorganisms, plants and animals, and so it can utilize a wide range of such substrates and possesses a large number of enzymes which degrade polysaccharides. Carbohydrates are taken into the cell by a range of means, including ATP-binding cassette (ABC) transporters and phosphotransferase systems (PTS); there are 77 putative ABC transporters and at least 16 PTS sugar transporters encoded in the genome (Kunst et al., 1997); 75 ABC transporter/ATP-binding proteins are encoded by the *Bacillus halodurans* genome (Takami et al., 2000). ABC transporters are important in Gram-positive organisms, given their single-membrane cell envelope, as they allow them to escape the toxic actions of many compounds. *Bacillus anthracis* has reduced numbers of PTS and other types of sugar transporters and lacks pathways for catabolism of several sugars compared with *Bacillus subtilis* (Read et al., 2003). Carbohydrate uptake and metabolism in *Bacillus subtilis* has been reviewed by Deutscher et al. (2002) and the regulation of carbon catabolism in *Bacillus* species was reviewed by Stulke and Hillen (2000).

Because many *Bacillus* species grow aerobically and produce acid from carbohydrates by oxidation rather than fermentation, they normally produce smaller amounts of acid from carbohydrates in comparison with most *Paenibacillus* species. Also, because the ammonia they produce from peptones may neutralize the small amount of acid produced, it is necessary to use a medium of low protein to carbohydrate ratio, and a sensitive indicator such as phenol red in order to detect acid production.

Most members of the *Bacillus sphaericus* group will not use carbohydrates as carbon or energy sources, and use certain organic acids and amino acids instead.

The genome of *Bacillus subtilis* encodes an Embden–Meyerhof–Parnas glycolytic pathway, coupled to a functional tricarboxylic acid (Krebs) cycle (Kunst et al., 1997), and the *Bacillus halodurans* genome is little different to that of *Bacillus subtilis* in this respect (Takami et al., 2000). *Bacillus subtilis* appears to have no glyoxylate shunt, but some *Bacillus* species, including *Bacillus halodurans* and *Bacillus anthracis*, produce glyoxylate shunt (or bypass) enzymes and/or have genes encoding components of this shunt, which allows acetate or fatty acids to be used as sole sources of carbon (Sonenshein, 2002). Inactivating mutations in the Krebs cycle genes of *Bacillus subtilis* cause defects in sporulation, and although most such defects are attributable to the conventional roles of the affected enzymes, other defects cannot be explained in this way and their mechanisms are unclear (Sonenshein, 2002). It appears that some Krebs cycle proteins may have regulatory as well as enzymic activities: the E2 subunit of the pyruvate dehydrogenase complex of *Bacillus thuringiensis* can bind to DNA, and in so doing has been implicated in regulation of the expression of a gene for toxin production (Walter and Aronson, 1999).

Bacillus subtilis can use ammonium, nitrate, amino acids, some purines, urea, uric acid, allantoin, and peptides as sole nitrogen sources. Glutamine, followed by arginine, is the best source for rapid growth. In order to utilize the nitrogen compounds that permit optimal growth rates. this organism, like other Gram-positive bacteria, regulates nitrogen metabolism genes by mechanisms very different to the pathway found in enteric bacteria. *Bacillus subtilis* controls gene expression in response to nitrogen availability with the three proteins GlnR, TnrA and CodY. Also, although σ^{54} factors were initially believed to be present only in Gram-negative bacteria, the *Bacillus subtilis* sigL regulon was found to contain a homolog of σ^{54}, and is now known to contain genes involved in carbon and nitrogen source utilization (Fisher and Débarbouillé, 2002). *Bacillus subtilis* also possesses many genes involved in the degradation of opines and related molecules derived from plants (Kunst et al., 1997). Although *Bacillus subtilis*, *Bacillus anthracis*, *Bacillus cereus* and *Bacillus halodurans* have broad similarities in their metabolisms, *Bacillus anthracis* and *Bacillus cereus* have greater capacities for the utilization of amino acids and peptides. *Bacillus anthracis* and *Bacillus cereus* have wider ranges of coding sequences for secreted proteases, 48 and 51 respectively, compared with *Bacillus subtilis*, which has only 30, and wider ranges of peptidases too (Ivanova et al., 2003). The *Bacillus anthracis* genome also encodes 17 ABC-type peptide binding proteins, has nine homologs of the BrnQ branched chain amino-acid transporter, and has six LysE/Rht amino-acid efflux systems, whereas *Bacillus subtilis* has only four, two and two respectively. *Bacillus anthracis* and *Bacillus cereus* thus appear to be adapted to protein-rich environments such as animal matter (Ivanova et al., 2003; Read et al., 2003). *Bacillus* proteases are of considerable economic value, especially as detergent additives, are intensely studied with a view to enhancing their behaviors in industrial processes, and the search for new strains producing enzymes with novel properties continues (Outtrup and Jørgensen, 2002).

Although iron is an essential nutrient for most organisms, free iron availability is severely restricted in neutral, aerobic environments, including animal tissues, owing to its insolubility in such conditions. Bacteria secrete siderophores into their environments in order to chelate iron, and the ferri-siderophore complexes can then be assimilated. *Bacillus subtilis* regulates iron uptake by members of the ferric uptake regulator (Fur) family of proteins, and possesses three such homologs, Fur, PerR (peroxide stress response) and Zur (zinc uptake regulation). The *Bacillus subtilis* homolog BsuFur shows only 33% sequence similarity to *Escherichia coli* Fur (EcoFur) and, unlike EcoFur, it does not respond to Mn(II) *in vivo* (Herbig and Helmann, 2002). *Bacillus anthracis* has a wider range of iron-acquisition genes than does *Bacillus subtilis*; it possesses 15 ABC uptake systems for iron siderophores or chelates, and two clusters of genes for the biosynthesis of siderophores. There are in the sequence of *Bacillus anthracis* two genes involved in the synthesis of an aerobactin-like siderophore that are not found in the sequenced strains of *Bacillus subtilis* and *Bacillus cereus* (Read et al., 2003). *Bacillus anthracis* carries genes for two sphere-like proteins which have internal cavities and act as ferritins, and so are involved in iron uptake and regulation; the immunogenicities of these proteins make them of interest in the development of new anthrax vaccines (Papinutto et al., 2002). *Bacillus cereus* can use hemoglobin, heme and heme-albumin complex as its iron sources, but does not appear to use other iron-binding proteins such as lactoferrin and transferrin; it does not digest these two proteins, but it will digest heme-protein complexes to elicit release of heme which may then be captured as an iron source (Sato et al., 1999).

Halophilic species and some alkaliphilic *Bacillus* strains have obligate requirements for Na$^+$. The strain of the alkaliphile *Bacillus halodurans* that was subjected to complete genome sequencing requires Na$^+$ for growth in alkaline conditions, where the environmental sodium ions are essential for solute transport through the cytoplasmic membrane. ATP metabolism through the action of ATPases is considered to be important in generating a proton-motive force across the cytoplasmic membrane by extrusion of H$^+$; *Bacillus halodurans* possesses genes for four types of ATPases which are well conserved between the genome of this species and *Bacillus subtilis* (Takami et al., 2000). The *Bacillus halodurans* genome was also found to carry protein coding sequences that are candidates for Na$^+$/H$^+$ antiporter genes, at least some of which are involved in halotolerance and alkali-tolerance and allow the organism to maintain an intracellular pH lower than the environmental pH (Takami et al., 2000). The *Bacillus subtilis* 168 genome possesses a single ABC-type putative Na$^+$ efflux system (Saier et al., 2002).

The abilities of some *Bacillus* strains to metabolize and transform complex organic compounds are of interest in both bioremediation and pharmaceutical production, and studies of isolates from special environments and searches for activities of potential value in biotechnological applications have revealed a number of unfamiliar substrates.

Reports of *Bacillus* strains or their enzymes capable of metabolizing environmental pollutants include: a *Bacillus* sp. capable of oxidizing H$_2$S in chicken feces (Nakada and Ohta, 1998); a *Bacillus sphaericus* isolate from agricultural soil which oxidizes *p*-nitrophenol (Kadiyala et al., 1998); the naphthalene-degrading "*Bacillus naphthovorans*" from oil-contaminated tropical marine sediments (Zhuang et al., 2002; a *Bacillus* sp. that can utilize dimethylphthalate as sole carbon source (Niazi et al., 2001); a *Bacillus* sp. capable of using 4-chlorobiphenyl as sole carbon source, metabolizing it to 4-chlorobenzoic acid (Sàágua et al., 1998); and an engineered *Bacillus megaterium* cytochrome P450 that degrades polycyclic aromatic hydrocarbons (Carmichael and Wong, 2001).

The uricase of *Bacillus fastidiosus* catalyzes the oxidation of uric acid into the more soluble allantoin, and conjugates of this enzyme with soluble polymers to reduce antigenicity are of value in the therapy of gout, and of hyperuricemias associated with blood malignancies and chemotherapy (Schiavon et al., 2000). Some strains of *Bacillus cereus*, *Bacillus megaterium* and *Bacillus sphaericus* are capable of biotransformations of inexpensive natural steroidal substrates into high-value therapeutic compounds (Manosroi et al., 1999; Wadhwa and Smith, 2000). A *Bacillus subtilis* isolate from soil has been reported to produce the aroma compound vanillin by degradation of the phenylpropanoid isoeugenol, which it could use as sole carbon source (Shimoni et al., 2000). Decarboxylation of the abundant ferulic acid into the useful aromatic compound 4-vinylguaiacol has been described for *Bacillus pumilus* (Lee et al., 1998), while another strain of this species has been reported to be able to use phenols and cresols as sole carbon sources (Günther et al., 1995).

Genetics. Aerobic endospore-formers have been and continue to be important in many fields of basic research, and long-term studies of the sporulation process in *Bacillus subtilis* have led to its being probably the best understood developmental system. Although endospores are to be found in other genera, it is the spores of *Bacillus subtilis* that have been the most intensively studied, especially those of strain 168. Burkholder and Giles (1947) produced auxotrophic mutants of the Marburg (i.e., type) strain by exposure to UV light and X-rays in the 1940s, and their tryptophan auxotrophic strain 168 was chosen by Spizizen (1958) as the recipient in his demonstration of transformation of this species by bacterial DNA in the 1950s. Because studies on individual genes and gene products have been performed mainly on *Bacillus subtilis*, and to a lesser extent on pathogenic members of the *Bacillus cereus* group, such properties have as yet made little contribution to our understanding of the phylogeny of the genus *Bacillus*, but this will of course change as more species have their genomes sequenced (Stackebrandt and Swiderski, 2002).

The same laboratory strain, *Bacillus subtilis* 168, was also the first *Bacillus*, indeed the first Gram-positive bacterium, to have its genome sequenced (Kunst et al., 1997), and the implications of this knowledge in our understanding of the cellular architecture, chromosomal replication, cellular division, metabolism and metabolic regulation, macromolecular synthesis, adaption and differentiation of this organism have been reviewed by Sohenshein et al. (2002). As these authors observe, it has become clear from comparisons with the genome sequences of other organisms that the proteins of macromolecular synthesis, and the enzymes of biosynthesis and biodegradation, are widely conserved among prokaryotes, but that Gram-positive and Gram-negative organisms regulate gene expression and the activities of their gene products somewhat differently. The *Bacillus subtilis* genome is similar in size to that of *Escherichia coli*, and these two organisms have orthologous counterpart genes representing about one-quarter of their genomes (Kunst et al., 1997). The same authors found that of the 450 genes encoded by *Mycoplasma genitalium*, some 300 had products similar to proteins of *Bacillus subtilis*, and this is particular of interest given the belief that mycoplasmas are derived from Gram-positive bacteria.

Subsequent to this pioneering work on the *Bacillus subtilis* genome, those of three other *Bacillus* species, *Bacillus halodurans*, *Bacillus anthracis* and *Bacillus cereus*, have been sequenced (Ivanova et al., 2003; Read et al., 2003; Takami et al., 2000), so that comparisons can be made and both common and specific features of these organisms (a soil bacterium, an alkaliphile, a pathogen of humans and other animals, and an opportunistic pathogen, respectively) can be identified.

The genome of *Bacillus subtilis* has 4,214,810 bp comprising 4,100 protein-coding sequences (CDSs); although the mean mol% G+C is 43.5 for this organism, the ratio varies greatly throughout the chromosome. There are many gene duplications, including rRNA genes, and a particularly conspicuous duplication is a 190 bp element that is repeated 10 times, with five repeats lying each side of the origin of replication; similar sequences have been found in the closely related species *Bacillus licheniformis* (Kunst et al., 1997). The genome of *Bacillus halodurans* has 4,202,353 bp containing 4,066 predicted CDSs, its mean mol% G+C is 43.7, and 16S rDNA sequence analysis shows it to be a close relative of *Bacillus subtilis*. The principal apparent difference between the two organisms is the alkaliphily of *Bacillus halodurans*, and so it was naturally of interest to identify differences between the genomes and to try and correlate these with phenotype. Both genomes showed substantial conservation of a common region comprising, amongst others, the functions of cell division, DNA replication, RNA modification, nucleotide and nucleic acid metabolism, metabolism of enzymes and prosthetic groups, glycolytic pathways and the TCA cycle, protein secretion, motility and chemotaxis (Takami et al., 2000). *Bacillus halodurans* was found to carry 112 CDSs which showed similarity with transposases or recombinases from other prokaryotes, indicating their important evolutionary roles in horizontal gene transfer. *Bacillus subtilis*, on the other hand, has only ten transposons and transposon-related proteins (Takami et al., 2000). However, the genome of *Bacillus subtilis* was found to contain at least 10 prophages or remnants of prophages, which suggest that horizontal gene transfer by bacteriophages may have played an important part in the evolution of this organism (Kunst et al., 1997); *Bacillus halodurans*, on the other hand, has no intact prophage. The σ factors required for sporulation are well conserved between the two genomes, but of 11 σ factors belonging to the extracytoplasmic function (ECF) only one is found in *Bacillus subtilis* and 10 are unknown outside *Bacillus halodurans*; these unique σ factors may play parts in alkaliphily given the roles of ECF σ factors in the control of specific molecule or ion uptake or secretion, or of various extracellular stress signals (Takami et al., 2000). Other differences between the organisms concern genes affecting competence, the control of sporulation, and cell-wall components. The last of these includes teichuronopeptide (a compound known to contribute to alkaliphily) in *Bacillus halodurans*, and the genome of this organism also possesses five candidates for Na$^+$/H$^+$ antiporter genes that may relate to its haloduric and alkaliphilic phenotype.

The chromosomes of *Bacillus anthracis* and *Bacillus subtilis* encode similar sporulation machineries, and metabolic and transport genes, and both encode numbers of predicted drug efflux pumps common in soil bacteria. Particular differences include the extended capacity of *Bacillus anthracis* for amino acid and peptide utilization, including more peptide binding proteins, secreted proteases and peptidases, and amino-acid efflux systems, and *Bacillus cereus* is likewise well equipped with proteolytic enzymes, peptide and amino acid transporters and amino-acid degradation pathways (Ivanova et al., 2003). These may correlate with their being adapted to protein-rich environments, and they have lesser capacities than *Bacillus subtilis* for sugar utilization. *Bacillus subtilis* carries 41 genes for degradation of carbohydrate polymers, whereas *Bacillus anthracis* has 15 and *Bacillus cereus* only 14 (Ivanova et al., 2003). The *Bacillus anthracis* genome encodes several detoxification functions for which homologs are not apparent in *Bacillus subtilis*; one of these is cytoplasmic Cu-Zn superoxide dismutase (SodC) which counteracts nitric oxide-mediated killing in the macrophage and has an important role in several other intracellular bacteria.

Bacillus anthracis has 5,227,293 bp and 5,508 CDSs (5,503,799 bp and 5,838 CDSs when the virulence plasmids are included), while *Bacillus cereus* carries 5,426,909 bp and 5,366 CDSs. It is well established that the genes on its virulence plasmids are essential to the virulence of *Bacillus anthracis*, as strains cured of one or both plasmids are avirulent. Study of the genome sequence shows that this organism has chromosomally encoded proteins, including hemolysins, phospholipases and iron-acquisition proteins which might contribute to pathogenicity, and surface proteins which might have potentials as drug and vaccine targets. Nearly all of these potential virulence factors and surface proteins have homologs in *Bacillus cereus*, even the sequenced *Bacillus cereus* strain which is considered to be non-pathogenic (Ivanova et al., 2003), underlining yet again the well-established close relationship between these two organisms (Turnbull et al., 2002). The chromosome of *Bacillus anthracis* carries most of its housekeeping functions, and these mostly have homologs in the sequence of *Bacillus cereus*, while the two virulence plasmids pXO1 (toxic complex) and pXO2 (capsule) carry transposons, and genes of unknown function in addition to the key virulence determinants. *Bacillus anthracis* also chromosomally encodes proteins with homology to virulence factors of *Listeria monocytogenes*, and these may be significant for intracellular survival, multiplication and escape (Read et al., 2003). The genes for a complex of three non-hemolytic enterotoxins and two channel-forming hemolysins that have roles in the pathogenicities of *Bacillus cereus* and *Bacillus thuringiensis* have homologs in *Bacillus anthracis*. *Bacillus anthracis* also carries two (and *Bacillus cereus* three) homologs of *Bacillus thuringiensis* immune inhibitor A protease which has a role in virulence to insects, and both it and *Bacillus cereus* encode a homolog of the metalloprotease enhancin which boosts viral infectivity in insect guts. It has been suggested that these genes may be evidence that the *Bacillus cereus* group had an insect-infecting ancestor, and their possession of genes for chitinolytic enzymes is consistent with this idea (Ivanova et al., 2003; Read et al., 2003).

Comparative genome hybridization of 19 *Bacillus cereus* and *Bacillus thuringiensis* strains against a *Bacillus anthracis* microarray revealed 66–92% homology of chromosomal genes, and several major differences between *Bacillus anthracis* and *Bacillus cereus* reflect altered gene expression as opposed to gene gains or losses. In the *Bacillus cereus* and *Bacillus thuringiensis* strains were very few homologs of genes found on the virulence plasmid pXO2, but about half of the 19 strains carried homologs of genes (but not of the anthrax toxin genes) found in the virulence plasmid pXO1. There are many mobility genes on pXO1 (Okinaka et al., 1999), and plasmid transfer is known to occur within the *Bacillus cereus* group, but there is little localized variation in the G+C and dinucleotide contents of the *Bacillus anthracis* chromosome and virulence plasmids which suggests that most of the genes are native to this group of species (Read et al., 2003).

The great phylogenetic heterogeneity of *Bacillus sensu lato* was long evident from its wide mol% G+C range of 43–68 (Claus and Berkeley, 1986), and this heterogeneity has clearly been demonstrated by the 16S rRNA and rDNA sequence analyses that have followed. The impact of such analyses on the taxonomy of *Bacillus sensu lato* is discussed in *Taxonomic comments* (below) and in Stackebrandt and Swiderski (2002). Although 16S rDNA sequence comparisons are valuable in the determination of approximate phylogenetic relationships at the generic level and higher, they are not appropriate for the classification of strains at the species level (Stackebrandt and Goebel, 1994). Xu and Cote (2003) compared the sequences of the 16S–23S internal transcribed spacer region (ITS) of representatives of 27 *Bacillus* species and 19 strains representing five other endospore-forming genera. Although they found general agreement with polyphasic taxonomies incorporating 16S rDNA sequence comparisons, they also found support for the division of *Bacillus* into further new genera, and revealed unexpected groupings. For example, *Bacillus coagulans* was found to lie nearer to *Geobacillus* strains than to the other *Bacillus* species, *Bacillus laevolacticus* grouped with *Virgibacillus pantothenticus*, and *Bacillus badius* with *Marinibacillus marinus*, yet *Bacillus circulans* remained ungrouped (Xu and Cote, 2003). The ITS region is hypervariable in comparison with the more conserved 16S rRNA coding region, and ITS-PCR fingerprints have been used to investigate the relationships of members of the genus. Daffonchio et al. (1998a) were able to separate several species of *Bacillus* by this approach, but distinctions of very closely related species were not possible, and single-strand conformation polymorphism analysis was used to distinguish members of the "*Bacillus subtilis* group," while *Bacillus mycoides* could be separated from *Bacillus cereus/Bacillus thuringiensis* by restriction analysis. When ITS-PCR, analysis of the regions between tRNA genes (tDNA-PCR), and RAPD were applied to *Bacillus licheniformis* the 10 strains studied fell into two clusters by all three fingerprinting methods. With *Bacillus cereus*, on the other hand, it was found that ITS-PCR and tDNA-PCR gave virtually identical profiles among the 21 strains, but that these strains showed great diversity in RAPD analysis and in plasmid profiles (Daffonchio et al., 1998b). Part of the ITS region has been used as a probe for the detection of *Bacillus sporothermodurans* (de Silva et al., 1998).

De Vos (2002) reviewed several other approaches to the analysis of nucleic acids that cover a wide range of taxonomic levels, and these are summarized here. As direct sequencing of 16S rDNA is still relatively expensive and not available to all microbiologists, indirect, fast and less expensive methods such as ARDRA, to characterize the 16S rDNA part of the ribosomal operon via restriction analysis, have offered a good alternative for sequencing aerobic endospore-forming bacteria (Heyndrickx et al., 1996c; Logan et al., 2000). This method also has the advantages that both computerized interpretation of the data and database construction are possible.

Because *Bacillus sensu lato* members contain 9–12 rRNA operons (e.g., Johansen et al., 1996; Okamoto et al., 1993), ribotyping of the aerobic spore-formers has been considered as a potentially useful approach to unravel their taxonomic structure. At present, the number of studies in which ribotyping has been used to characterize members of *Bacillus* is rather limited, and published reports deal mainly with intraspecific variation. In these studies the investigators try to find a correlation between the intraspecific distribution of ribopatterning in correlation with, for example, (i) food poisoning with *Bacillus licheniformis* (Salkinoja-Salonen et al., 1999), (ii) toxin production by members of the *Bacillus cereus* group, including strains of *Bacillus cereus* from food poisoning incidents (Pirttijärvi et al., 1999), (iii) tracing of certain *Bacillus thuringiensis* types (Akhurst et al., 1997) and (iv) differentiation between toxic and nontoxic *Bacillus sphaericus* strains (Aquino de Muro et al., 1992).

Although comparative data with other fine DNA fingerprinting methods are somewhat scarce for representatives of *Bacillus*, at least one study has revealed that randomly amplified polymorphic DNA (RAPD) analysis is only slightly more discriminative than the automated ribotyping (riboprinting) for *Bacillus cereus* isolates (Andersson et al., 1999a).

Denaturing-gradient gel electrophoresis (DGGE) and temperature-gradient gel electrophoresis (TGGE) are based on similar principles, and allow discrimination at the subspecies level and often at the strain level (De Vos, 2002). As patterns from both techniques can be obtained after amplification of target DNA taken from non-purified biological material such as soil or water samples, the methods allow visualization of the genetic biodiversity, including the non-cultivable bacterial components of biotopes. Comparison of the sequences of the dominant bands visualized by these approaches with databases such as EMBL may be indicative for the dominant bacterial component of the biotope under study. Using TGGE, for example, an unknown group of *Bacillus* members has been discovered as the main bacterial component in Drentse grassland in the Netherlands (Felske et al., 1998; Felske et al., 1999).

As identical 16S rDNA sequences do not guarantee species identity (Fox et al., 1992), DNA:DNA hybridizations are needed when 16S rDNA sequences between strains show 97% similarity or more with existing taxa. The study of DNA relatedness by different techniques has been widely applied to *Bacillus*, but only two methods are currently used for species delineation within the genus: the liquid renaturation method (De Ley et al., 1970, or a variant) and the later microplate method of Ezaki et al. (1989). Data obtained by both methods have been evaluated and compared (Goris et al., 1998). Neither of these methods allows the determination of Δthermostability (expressed as ΔT_m) of the hybrid, but differences in ΔT_m between the hybrid and the homologous duplex are important and can be decisive for taxonomic conclusions (Grimont et al., 1982).

Several typing methods are based on indirect comparative analysis of nucleic acid characteristics, and were originally developed for the discrimination of species, subspecies and even strains for epidemiological studies. Restriction Fragment Length Polymorphism (RFLP) analysis of whole bacterial genomes yielded very complex patterns of DNA fragments that are difficult to compare because of their smear-like appearance. The use of restriction enzymes that cut infrequently drastically reduces the number of the DNA fragments, the high molecular mass of which required the development of a specific technique known as pulsed field gel electrophoresis (PFGE) for the satisfactory separation of fragments on agarose gels. The method has been applied to differentiate between strains of *Bacillus sphaericus* (Zahner et al., 1998), and between very closely related species such as *Bacillus anthracis, Bacillus cereus, Bacillus mycoides* and *Bacillus thuringiensis* (Carlson et al., 1994; Harrell et al., 1995; Helgason et al., 2000; Liu et al., 1997). The last two of these studies dealt with infrequently reported clinical infections by *Bacillus cereus*.

Several other genomic typing techniques overcome the problem of interpreting complex banding patterns by visualizing only selected parts of bacterial genomes that have been amplified using the PCR. The banding patterns obtained using RAPD, in which oligonucleotides of about 10–20 bp are used as primers, are not always very reproducible, so that databases are of limited use and data exchanged between laboratories have to be interpreted with great care. Nonetheless, RAPD has been applied to the discrimination of *Bacillus thuringiensis* (Brousseau et al., 1993), *Bacillus sphaericus* (Woodburn et al., 1995) and thermophilic (now mostly assigned to *Geobacillus*) and mesophilic *Bacillus* members (Ronimus et al., 1997). A second group of PCR-based typing methods uses repetitive element primers and so is known as rep-PCR. It is based upon the observation that repetitive elements are dispersed throughout genomes of bacteria, and consensus motifs deduced from the sequence of these repetitive elements can be used as primers. The electrophoretic patterns revealed allow discrimination at the within-species level and sometimes at the strain level, and have been used to investigate the genetic diversity of novel species (Heyrman et al., 2003a; Heyrman et al., 2003b; Heyrman et al., 2004; Logan et al., 2002b), to unravel the genetic diversity of *Bacillus sphaericus* (da Silva et al., 1999; Miteva et al., 1999), and to demonstrate the presence of the thermoresistant organism *Bacillus sporothermodurans* in UHT treated milk (Klijn et al., 1997). Amplified Fragment Length Polymorphism (AFLP) is based upon a specific combination of PCR and restriction methodologies (Zabeau and Vos, 1993), and although much more complex than RAPD and Rep-PCR methods, it is also much more reproducible. It has been used in epidemiological studies of *Bacillus cereus* (Mantynen and Lindstrom, 1998; Ripabelli et al., 2000; Schraft et al., 1996) and for the genetic comparison of *Bacillus anthracis* and its closest relatives (Jackson et al., 1999; Keim et al., 1997; Turnbull et al., 2002). Further molecular characterization showed that variable number tandem repeats (VNTR), which are short, tandemly repeated sequences which undergo very rapid mutational change, were responsible for the variations seen by AFLP. Multiple-locus VNTR analysis (MVLA) thus offers greater discriminatory power than AFLP and should be useful for investigating the ecology and epidemiology of anthrax (Turnbull et al., 2002).

Hansen et al. (2001) developed a PCR assay for the detection of members of the *Bacillus cereus* group, using a 16S rRNA probe. Real-time PCR assays, which use primer and fluorescently labeled gene probe systems to allow the rapid and sensitive detection of genes specific for *Bacillus anthracis*, have been developed in several laboratories. Makino and Cheun (2003) described an assay that targeted genes for capsule and PA and allowed a single spore to be detected in 100 l of air in 1 h. Drego et al. (2002) outlined an assay targeting fragments of a chromosomal gene (*rpo*) for detecting the organism in clinical samples. Hoffmaster et al. (2002) evaluated and validated a three-target assay, with primers for capsule, PA and *rpo*, in order to test suspect isolates and to screen environmental samples during the outbreak that followed the 2001 bioterrorist attack in the USA, and a similar approach was evaluated by Ellerbrok et al. (2002).

Antigens and vaccines. Despite the promising findings of some early studies of somatic and spore antigens from a range of species (Doak and Lamanna, 1948; Lamana, 1940a, c, 1942), of flagellar, somatic and spore antigens of *Bacillus* (now *Paenibacillus*) *polymyxa* (Davies, 1951), and the potential taxonomic value of spore precipitinogens reported by Norris and Wolf (1961) following an extension of Davies' work to a wider range of species, serological studies have been taken little further for classification and identification of members of this genus. Paradoxically, however, although the *Bacillus cereus* group appeared to one of the least tractable in these early studies, where species-specific antigens were sought, the H-antigens of *Bacillus cereus* and *Bacillus thuringiensis* are now used with considerable success for typing purposes; *Bacillus anthracis* does not possess H-antigens as it is nonmotile. Smith et al. (1952) reviewed the earliest work, following their own disappointing results with antisera to vegetative cells of a range of species, and Berkeley et al. (1984) reviewed the application of serological methods to the identification of *Bacillus* species.

Somatic antigens have been little used for the identification of *Bacillus* strains. Investigations into the O-antigens of a *Bacillus cereus* and *Bacillus licheniformis* (Norris and Wolf, 1961) found them too strain-specific to be of taxonomic value. (Walker and Wolf, 1971) and Wolf and Sharp (1981) found the O-antigens of *Bacillus* (now *Geobacillus*) *stearothermophilus* to show some correlation with the three biochemical and physiological subgroups of this species that they recognized. Serotyping of *Bacillus thuringiensis* strains has been attempted on the basis of extracellular heat-stable somatic antigens (HSSAs; Ueda et al., 1989; Ohba et al., 1992), by observing the formation of immunoprecipitation haloes around colonies on antiserum-agar plates; they found a lack of correlation with H-antigen serogroups, while field isolates showed little HSSA variation within a single H-serovar. Concerns about the potential of *Bacillus anthracis* as a biological weapon have emphasized the need for a rapid method for the identification of *Bacillus anthracis* and diagnosis of anthrax. Polyclonal antibodies lack the desired specificity, because they react with other members of the *Bacillus cereus* group. Phillips and Ezzell (1999) were able to identify *Bacillus anthracis* by raising polyclonal antibodies against extracted vegetative cell antigens, absorbing with *Bacillus cereus* and *Bacillus thuringiensis*, and detecting reactions by immunofluorescence or immunoblotting. A monoclonal antibody specific to the *Bacillus anthracis* cell-wall polysaccharide antigen is effective in identification (Ezzell and Welkos, 1999), but this antigen may be masked *in vivo* by the organism's poly-γ-D-glutamic acid capsule. De et al. (2002) therefore developed a two-component direct fluorescent-antibody assay that allows rapid, sensitive and specific detection of the cell wall and capsule of *Bacillus anthracis* in clinical specimens.

Flagellar antigens have been more widely used in *Bacillus* typing than any other kind of antigen, as they provide the highest strain specificity, and valuable serotyping schemes have been developed for *Bacillus thuringiensis* and *Bacillus cereus*. High frequencies of H-antigen sharing between *Bacillus cereus* and *Bacillus thuringiensis* have been reported; in one study of *Bacillus cereus* strains from soils, phylloplanes and animal feces the seropositivity of the isolates with *Bacillus thuringiensis* H-antisera was 60–77% (Shisa et al., 2002). The common flagellar antigen of *Bacillus cereus* has been shown by SDS-PAGE and immunoblot assay to be due to a 61-kDa protein, and monoclonal antibody studies showed that the common antigenic epitope of the 61-kDa protein also exists in the flagella of *Bacillus thuringiensis* (Murakami et al., 1993).

Sixty-nine serotypes and 13 subantigenic groups of *Bacillus thuringiensis* have been recognized, giving 82 serovars (Lecadet et al., 1999). New strains of the species are screened by reference H-antisera and antisera are prepared against any strains that do not agglutinate. New antisera are then screened with all the known H-antigens, and a new serovar is recognized if cross-reactions do not occur or if new subfactors can be demonstrated by the antiserum-saturation technique (de Barjac, 1981). New serovars are registered at the International Entomopathogenic *Bacillus* Centre (IEBC) Collection at the Institute Pasteur, Paris, France; this laboratory was the international reference for *Bacillus thuringiensis* since 1965. Distinct serovars of *Bacillus thuringiensis* are given names and abbreviations, such as finitimus (FIN, H-antigen 2), sotto (SOT, H4a4b), tolworthi (TOL, H9) and pirenaica (PIR, H57). The first two of these two names were formerly used as the specific epithets of distinct species (Gordon et al., 1973). Although these serovar names have often been informally regarded as subspecific epithets, they do not represent validated subspecies of *Bacillus thuringiensis* and should instead be regarded as varieties. These varieties do not, unfortunately, show much correlation with toxicity to invertebrates; for example, serovar morrisoni (MOR, H8a8b) includes strains pathogenic to mosquitoes (Diptera), Coleoptera and Lepidoptera (Lecadet et al., 1999), while the invertebrate toxicity, if any, of many serovars (especially the recently recognized ones) is unknown.

A strain differentiation system for *Bacillus cereus* based on H-antigens is available at the Food Hygiene Laboratory, Central Public Health Laboratory, Colindale, London, UK, for investigations of food-poisoning outbreaks or other *Bacillus cereus*-associated clinical problems (Kramer and Gilbert, 1992). This system was developed by Taylor and Gilbert (1975) for the investigation of food poisoning outbreaks and recognized 18 serovars, but many strains from outbreaks were untypable (Gilbert and Parry, 1977). Terayama et al. (1978) extended the system and recognized further serovars in foods. Some of the serotypes show some correlation with pathogenicity and biotype; the distinction is not absolute, however, and it is probable that some organisms can produce both diarrheal and emetic toxins. Serovar 1, 3, 5 and 8 strains comprise a biotype distinguishable from other *Bacillus cereus* serovars and untypable strains, and these former four serovars have often been isolated in connection with cases of the emetic form of food poisoning (Logan and Berkeley, 1984; Logan et al., 1979), a form especially associated with cooked rice. Gilbert and Parry (1977) found that strains of serovar 1 are found more frequently in cooked rice than in uncooked rice, while strains of serovar 17 are commoner in uncooked rice than they are in cooked rice; they later showed that serovar 1 strains form more heat resistant spores than do serovar 17 strains (Parry and Gilbert, 1980).

Flagellar serotyping was developed for *Bacillus sphaericus* (de Barjac et al., 1985) independently of the recognition of DNA relatedness groups, so that the DNA relatedness group IIA (see *Bacillus sphaericus* in *List of Species*, below), the group in which the mosquitocidal strains of this species lie, is divided into 9 non-consecutively numbered serovars. The serotyping scheme for *Bacillus sphaericus* shows good agreement with a phage typing scheme (Yousten, 1984) for this species, but, as with *Bacillus thuringiensis*, the types do not always concur with pathogenicity (Priest, 2002).

The study of H-antigens of *Bacillus subtilis* (Simon et al., 1977) by agglutination tests was complicated by the tendency of the cells to clump spontaneously, and so double-diffusion in agar or complement fixation methods were applied. At least five distinct serovars were recognized, and although cross-reactions made the establishment of a practical serotyping scheme difficult, this approach was considered to be potentially useful in taxonomic studies.

As spores are so characteristic of the genus *Bacillus*, it is not surprising that their antigens have attracted much interest. Early workers (reviewed by Norris, 1962) recognized: (i) that any attempt to use living spores as antigens might be complicated by their germination in the host animal, giving rise to antisera against vegetative cells as well as spores; (ii) the need for a rapid method to overcome any problems of germination during the agglutination test itself; (iii) the tendency of spores to auto-agglutinate owing to their hydrophobic surfaces; and (iv) the need to remove vegetative cell debris from spore suspensions– by using media promoting complete sporulation, or by autolysis with lysozyme or thiomersalate, or by autoclaving. Once these problems are overcome, spore antigens can be useful for serological identification, and a revival of interest has been stimulated by the need to detect the spores of *Bacillus anthracis* in the contexts of biowarfare and bioterrorism (Iqbal et al., 2000). Several antigens common to endospores from different genera may be found on the exosporium, and an antibody to one of these antigens also reacted with vegetative cells of *Bacillus cereus* and *Clostridium sporogenes* (Quinlan and Foegeding, 1997).

Lamana (1940c) was able to differentiate between spores of several of the small-celled species by their antigens, but again was less successful with the large-celled species (Lamana, 1940b), and Lamana and Eisler (1960) were unable to separate *Bacillus anthracis* from *Bacillus cereus* using spore agglutinogens. Norris and Wolf (1961) reported a similar picture from their study of spore agglutinogens, but found that spore precipitinogens were of some taxonomic value, and the latter observation was confirmed for the *Bacillus circulans* complex by Wolf and Chowhury (1971a). Walker and Wolf (1971) found that spore agglutinogens supported the main biochemical subdivisions of their *Bacillus* (now *Geobacillus*) *stearothermophilus* strains, but could not detect spore precipitinogens. Smirnova et al. (1991) considered that hemagglutination patterns of fimbriated *Bacillus thuringiensis* spores might be of taxonomic value.

In order to circumvent cell-clumping problems, Kim and Goepfert (1972) developed a fluorescent antibody technique for confirming identifications of *Bacillus cereus* from food poisoning cases, but could not distinguish between spores of this species and *Bacillus thuringiensis*; it has subsequently been recognized that strains of the latter species may also cause food poisoning (Damgaard et al., 1997). Phillips and Martin (1983a, b) used radiolabeled polyclonal antibody to probe for *Bacillus anthracis* spores attached to solid supports. Spores would not attach reliably to microtiter plates, but although attachment to glass slides was better, the background signal in the immunoradiometric assay (IRMA) was higher; the best sensitivity was achieved with an indirect assay, but specificity was only moderate. Fluorescein-conjugated polyclonal antibodies to *Bacillus anthracis* spore surface antigens were found to cross-react with spores of several other *Bacillus* species, but these cross-reactions could be absorbed with strains of *Bacillus cereus*; however, spores of the Vollum strain (=type strain) did not react with antibodies to spores of most of the other strains of *Bacillus anthracis* tested (Phillips and Martin, 1988). A monoclonal antibody to viable and heat-killed spores of *Bacillus anthracis* was used in an immunofluorescence assay and achieved higher specificity but lower sensitivity than the IRMA approaches, but the epitope recognized by the antibody appeared to be unstable in spores stored for long periods (Phillips et al., 1988).

Anthrax vaccine The Sterne attenuated live spore vaccine, based on a toxigenic but non-capsulate strain, was introduced for animal vaccination in the late 1930s and spores of this strain remain in use as the basis of livestock vaccines in most parts of the world today. As this vaccine can show some slight virulence for certain animals, it is not considered suitable for human protection in the West, but live spore preparations are used in China and Russia. The former USSR vaccine was developed in the 1930s and 1940s, and licensed for administration by injection in 1959. It was based on two avirulent, non-capsulate *Bacillus anthracis* strains TI-I and 3, which were derived from virulent agents at the Sanitary-Technical Institute (STI), in Kirov (now Viatka). It was reported that in 30 years of use no adverse effects were associated with this vaccine, and so reconsideration of the suitability of live spore vaccines for human use has been suggested (Shiyakhov and Rubinstein, 1994). The UK vaccine is an alum-precipitated culture filtrate of the Sterne strain; it was first formulated in 1954, introduced for workers at risk in 1965, and licensed for human use in 1979. The current human vaccine in the USA is an aluminum hydroxide-adsorbed vaccine strain culture filtrate containing a relatively high proportion of protective antigen (PA) and relatively low amounts of lethal factor and edema factor; it was licensed in 1972 (Turnbull, 2000). Concerns about the lack of efficacy and safety data on the long-established UK and US vaccines, especially following the Sverdlovsk incident, and allegations that anthrax vaccination contributed to Gulf War syndrome in military personnel, have led to demands for new vaccines that would necessarily undergo stricter testing than was customary in the past. Favored active ingredients of these next-generation vaccines are whole-length recombinant PA or a mutant (non-toxic) portion of this molecule (Turnbull, 2000).

Antibiotic sensitivity. Most strains of *Bacillus anthracis* are susceptible to penicillin, there being few authenticated reports of resistant isolates (Lalitha and Thomas, 1997); consequently this antibiotic has been the mainstay of treatment and there have been few studies on the organism's sensitivity to other antibiotics. Mild and uncomplicated cutaneous infections may

be treated with oral penicillin V, but the treatment usually recommended is intramuscular procaine penicillin or benzyl penicillin (penicillin G). In severe cases, and gastrointestinal and inhalational infections, the recommended therapy has been penicillin G by slow intravenous injection or infusion until the fever subsides, followed by intramuscular procaine penicillin; the organism is normally susceptible to streptomycin, which may act synergistically with penicillin (Turnbull et al., 1998). The use of an adequate dose of penicillin is important, as Lightfoot et al. (1990) found that strains grown in the presence of subinhibitory concentrations of flucloxacillin *in vitro* became resistant to penicillin and amoxycillin. The study of Lightfoot et al. (1990) on 70 strains, and that of Doganay and Aydin (1991) on 22 isolates, found that most strains were sensitive to penicillins, with minimal inhibitory concentrations of 0.03 mg/l or less; however, the former authors found that two resistant isolates from a fatal case of inhalational infection had MICs in excess of 0.25 mg/l. *Bacillus anthracis* is resistant to many cephalosporins. Coker et al. (2002) found that of 25 genetically diverse, mainly animal and human isolates from around the world, five strains were resistant to the "second generation" cephalosporin cefuroxime, and 19 strains showed intermediate susceptibility to this agent; all strains were sensitive to the "first generation" cephalosporin cephalexin, and to the "second generation" cefaclor, and three were resistant to penicillin but were negative for β-lactamase production. Mohammed et al. (2002) studied 50 historical isolates from humans and animals and 15 clinical isolates from the 2001 bioterrorist attack in the USA; the majority of their strains could be regarded as nonsusceptible to the "third generation" cephalosporin ceftriaxone, and three strains were resistant to penicillin. Genomic sequence data indicate that *Bacillus anthracis* possesses two β-lactamases: a potential penicillinase (class A) and a cephalosporinase (class B) which is expressed (Bell et al., 2002). Tetracyclines, chloramphenicol, gentamicin and erythromycin are suitable for the treatment of patients allergic to penicillin; tests in primates showed doxycycline to be effective, a finding confirmed by Coker et al. (2002), and indicated the suitability of ciprofloxacin (Turnbull et al., 1998). Mohammed et al. (2002) found that most of their strains showed only intermediate susceptibility to erythromycin. Esel et al. (2003) found that ciprofloxacin and the newer quinolone gatifloxacin had a good *in vitro* activity against 40 human isolates collected in Turkey, but that for another new quinolone, levofloxacin, it was observed that minimum inhibitory concentrations were high for 10 strains. Because human cases tend to be sporadic, clinical experience of alternative treatment strategies was sparse until the bioterrorist attack occurred in the US in late 2001. The potential and actual use of *Bacillus anthracis* as a bioweapon has also emphasized the need for post-exposure prophylaxis; recommendations include ciprofloxacin or doxycycline, with amoxycillin as an option for the treatment of children and pregnant or lactating women, given the potential toxicity of quinolones and tetracyclines; however, β-lactams do not penetrate macrophages well, and these are the sites of spore germination (Bell et al., 2002). Combination therapy, begun early, with a fluoroquinolone such as ciprofloxacin and at least one other antibiotic to which the organism is sensitive, appears to improve survival (Jernigan et al., 2001). Following the 2001 outbreak in the US, the recommendation for initial treatment of inhalational anthrax is ciprofloxacin or doxycycline along with one or more agents to which the organism is normally sensitive; given supportive sensitivity testing, a penicillin may be used to complete treatment. The same approach is recommended for cutaneous infections (Bell et al., 2002). Doxycycline does not penetrate the central nervous system well, and so is not appropriate for the treatment of meningitis.

Despite the well-established importance of *Bacillus cereus* as an opportunistic pathogen, there have been rather few studies of its antibiotic sensitivity, and most information has to be gleaned from the reports of individual cases or outbreaks. *Bacillus cereus* and *Bacillus thuringiensis* produce a broad spectrum β-lactamase and are thus resistant to penicillin, ampicillin, and cephalosporins; they are also resistant to trimethoprim. An *in vitro* study of 54 isolates from blood cultures by disk diffusion assay found that all strains were susceptible to imipenem and vancomycin and that most were sensitive to chloramphenicol, ciprofloxacin, erythromycin and gentamicin (with 2, 2, 6 and 7% strains, respectively, showing moderate or intermediate sensitivities), while 22 and 37% of strains showed only moderate or intermediate susceptibilities to clindamycin and tetracycline, respectively (Weber et al., 1988); in the same study, microdilution tests showed susceptibility to imipenem, vancomycin, chloramphenicol, gentamicin and ciprofloxacin with MICs of 0.25–4, 0.25–2, 2.0–4.0, 0.25–2 and 0.25–1.0 mg/l, respectively. A plasmid carrying resistance to tetracycline in *Bacillus cereus* has been transferred to a strain of *Bacillus subtilis* and stably maintained (Bernhard et al., 1978).

Although strains are almost always susceptible to clindamycin, erythromycin, chloramphenicol, vancomycin, and the aminoglycosides and are usually sensitive to tetracycline and sulfonamides, there have been several reports of treatment failures with some of these drugs: a fulminant meningitis which did not respond to chloramphenicol (Marshman et al., 2000); a fulminant infection in a neonate which was refractory to treatment that included vancomycin, gentamicin, imipenem, clindamycin, and ciprofloxacin (Tuladhar et al., 2000); failure of vancomycin to eliminate the organism from cerebrospinal fluid in association with a fluid shunt infection (Berner et al., 1997); persistent bacteremias with strains showing resistance to vancomycin in two hemodialysis patients (A. von Gottberg and W. van Nierop, personal communication). Oral ciprofloxacin has been used successfully in the treatment of *Bacillus cereus* wound infections. Clindamycin with gentamicin, given early, appears to be the best treatment for ophthalmic infections caused by *Bacillus cereus*, and experiments with rabbits suggest that intravitreal corticosteroids and antibiotics may be effective in such cases (Liu et al., 2000).

Information is sparse on treatment of infections with other *Bacillus* species. Gentamicin was effective in treating a case of *Bacillus licheniformis* ophthalmitis and cephalosporin was effective against *Bacillus licheniformis* bacteremia/septicemia. Resistance to macrolides appears to occur naturally in *Bacillus licheniformis* (Docherty et al., 1981). *Bacillus subtilis* endocarditis in a drug abuser was successfully treated with cephalosporin, and gentamicin was successful against a *Bacillus subtilis* septicemia. Penicillin, or its derivatives, or cephalosporins probably form the best first choices for treatment of infections attributed to other *Bacillus* species. In the study by Weber et al. (1988), isolates of *Bacillus megaterium* (13 strains), *Bacillus pumilus* (4), *Bacillus subtilis* (4), *Bacillus circulans* (3), *Bacillus amyloliquefaciens* (2)

and *Bacillus licheniformis* (1), along with five strains of *Bacillus* (now *Paenibacillus*) *polymyxa* and three unidentified strains from blood cultures, over 95% of isolates were susceptible to imipenem, ciprofloxacin and vancomycin; while between 75% and 90% were susceptible to penicillins, cephalosporins and chloramphenicol. Isolates of "*Bacillus polymyxa*" and *Bacillus circulans* were more likely to be resistant to the penicillins and cephalosporins than strains of the other species – it is possible that some or all of the strains identified as *Bacillus circulans* might now be accommodated in *Paenibacillus*, along with "*Bacillus polymyxa*." An infection of a human bite wound with an organism identified as *Bacillus circulans* did not respond to treatment with amoxycillin and flucloxacillin, but was resolved with clindamycin (Goudswaard et al., 1995). A recurrent septicemia with *Bacillus subtilis* in an immunocompromised patient yielded two isolates, both of which could be recovered from the probiotic preparation that the patient had been taking; one isolate was resistant to penicillin, erythromycin, rifampin and novobiocin, while the other was sensitive to rifampin and novobiocin but resistant to chloramphenicol (Oggioni et al., 1998).

A strain of *Bacillus circulans* showing vancomycin resistance has been isolated from an Italian clinical specimen (Ligozzi et al., 1998). Vancomycin resistance was reported for a strain of *Bacillus* (now *Paenibacillus*) *popilliae* in 1965, and isolates of this species dating back to 1945 have been shown to carry a *vanA*- and *vanB*-like gene, that is to say a gene resembling those responsible for high-level vancomycin resistance in enterococci. Vancomycin-resistant enterococci (VRE) were first reported in 1986, and so it has been suggested that the resistance genes in *Bacillus popilliae* and VRE may share a common ancestor, or even that the gene in *Bacillus popilliae* itself may have been the precursor of those in VRE; *Bacillus popilliae* has been used for over 50 years as a biopesticide, and no other potential source of *vanA* and *vanB* has been identified (Rippere et al., 1998). Of two South African vancomycin-resistant clinical isolates, one was identified as *Paenibacillus thiaminolyticus* and the other was unidentified but considered to be related to *Bacillus lentus* (Forsyth and Logan, unpublished); the latter was isolated from a case of neonatal sepsis, and has been shown to have inducible resistance to vancomycin and teicoplanin; this is in contrast to the *Bacillus circulans* and *Paenibacillus thiaminolyticus* isolates mentioned above, in which expression of resistance was found to be constitutive (A. von Gottberg and W. van Nierop, personal communication).

Isolates of novel *Bacillus* species from pristine Antarctic environments showed sensitivity to: ampicillin, chloramphenicol, colistin sulfate, kanamycin, nalidixic acid (*Bacillus fumarioli* resistant), nitrofurantoin, streptomycin and tetracycline (Logan et al., 2000, 2002b, and unpublished information).

Pathogenicity. The majority of *Bacillus* species apparently have little or no pathogenic potential and are rarely associated with disease in humans or other animals. The principal exceptions to this are *Bacillus anthracis* (anthrax), *Bacillus cereus* (food poisoning and opportunistic infections), and *Bacillus thuringiensis* (pathogenic to invertebrates), but a number of other species, particularly *Bacillus licheniformis*, have been implicated in food poisoning and other human and animal infections. The resistance of the spores to heat, radiation, disinfectants, and desiccation also results in *Bacillus* species being troublesome contaminants in the operating room, on surgical dressings, in pharmaceutical products and in foods.

Bacillus anthracis. Anthrax is primarily a disease of herbivores, and before an effective veterinary vaccine became available in the late 1930s, it was one of the foremost causes worldwide of mortality in cattle, sheep, goats, and horses. In 1945 an outbreak in Iraq killed 1 m sheep. The development and application of veterinary and human vaccines together with improvements in factory hygiene and sterilization procedures for imported animal products, and the increased use of man-made alternatives to animal hides or hair, have resulted over the past half century in a marked decline in the incidence of the disease in both animals and humans. Nevertheless, the disease continues to be endemic in several countries of Africa, Asia, and central and southern Europe, particularly those that lack an efficient vaccination policy, and nonendemic regions must be constantly on the alert for the arrival of *Bacillus anthracis* in imported products of animal origin. Sites where these materials were formerly handled, such as disused tanneries, may be sources of infection when they are disturbed during redevelopment. Likewise, anthrax carcasses can remain infectious for many years, even when buried with quicklime. The cycle of infection is as follows: the spores are ingested by a grazing animal and may gain access to the lymphatics, and so to the spleen, though abrasions in the alimentary canal; following several days of the organism multiplying and producing toxin in the spleen, the animal suffers a sudden and fatal septicemia and collapses; hemorrhagic exudates escape from the mouth, nose and anus and contaminate the soil, where the vegetative cells sporulate in the air. The spores remain viable in soil for many years and their persistence does not depend on animal reservoirs, so that *Bacillus anthracis* is exceedingly difficult to eradicate from an endemic area. *Bacillus anthracis* continues to be generally regarded as an obligate pathogen. Its continued existence in the ecosystem appears to depend on a periodic multiplication phase within an animal host, with its environmental presence reflecting contamination from an animal source at some time (Lindeque and Turnbull, 1994); however, some authorities believe that self-maintenance may occur within certain soil environments (Cherkasskiy, 1999). Direct animal-to-animal transmission within a species (that is to say, excluding carnivorous scavenging of meat from anthrax carcasses) is very rare.

Bacillus anthracis has long been considered a potential agent for biological warfare or bioterrorism. It is believed its first use was against livestock during World War I (Barnaby, 2002; Christopher et al., 1997). It has been included in various development and offensive programmes in several countries since (Alibek, 1999; Barnaby, 2002; Mangold and Goldberg, 1999; Mikkola et al., 2000; Zilinskas, 1997) and it has also been used in terrorist attacks (Christopher et al., 1997; Lane and Fauci, 2001; Takahashi et al., 2004). In public consciousness, *Bacillus anthracis* is associated more with warfare and terrorism than with a disease of herbivores, and it is feared accordingly. Apart from artificial attacks, humans almost invariably contract anthrax directly or indirectly from animals. It is a point-source type of disease, and direct human-to-human transmission is exceedingly rare. Circumstantial evidence shows that humans are moderately resistant to anthrax as compared with obligate herbivores; infectious doses in the human inhalational and intestinal forms are generally very high (LD_{50} 2,500 to 55,000 spores). Naturally acquired human anthrax may result from close contact with infected animals or their carcasses after death from the disease, or be acquired by those employed in processing wool, hair, hides, bones, or other animal products. Most cases (about 99%) are cutaneous infections, but *Bacillus anthracis* meningitis and gastrointestinal anthrax are occasionally reported. In industrial settings, inhalation of spore-laden dust may also occur; anthrax weapons are normally intended to cause the inhalational form, but are likely to cause cutaneous cases as well. There have been a few reports of laboratory-acquired infections, none of them recent (Collins, 1988), but a major outbreak of anthrax occurred in April 1979 in the city of Sverdlovsk (now Yekaterinburg) in the Urals as a result of the accidental release of spores from a military production facility; 77 cases were recorded and 66 patients died (Meselson et al., 1994).

Cutaneous infection occurs through a break in the skin, and as *Bacillus anthracis* is not invasive the lesions generally occur on exposed regions of the body; this includes the eyelids. Before the availability of antibiotics and vaccines, 10–20% of untreated cases of cutaneous anthrax were fatal, and the rare fatalities seen today are due to obstruction of the airways by the edema that accompanies lesions on the face or neck, and sequelae of secondary cellulitis or meningitis. Inhalational anthrax cases are more often fatal, because they go unrecognized until too late for effective therapy, but undiagnosed, low-grade infections with recovery may occur. The number of recorded cases of inhalational anthrax is lower than might be expected from the high profile given to this condition. In the 20th century there were just 18 reported cases (two of them laboratory-acquired) in the USA, 16 (88.9%) fatal (Brachman and Kaufmann, 1998); figures in the UK showed a similar picture. In 11 confirmed cases of inhalational anthrax that followed a bioterrorist attack, in which spores were delivered in mailed letters and packages, early recognition and treatment helped a survival level of 55% to be achieved (Bell et al., 2002; Jernigan et al., 2001). Oropharyngeal and gastrointestinal anthrax are not uncommon in regions of the world where animal anthrax is endemic and socio-economic conditions are poor, and people eat the meat of animals that have died suddenly; such cases are greatly underreported (Anonymous, 1994; Dietvorst, 1996). Gastrointestinal infections are mainly characterized by gastroenteritis, and asymptomatic infections and symptomatic infections with recovery are not uncommon (CDC, 2000). The symptoms of oropharyngeal infections are fever, toxemia, inflammatory lesions in the oral cavity and oropharynx, cervical lymph node enlargement, and edema, and there is a high case-fatality rate (Sirisanthana and Brown, 2002). Meningitis can develop from any of the forms of anthrax. The emergence of clinical signs is rapidly followed by unconsciousness, and the prognosis is poor. Outbreaks of primary anthrax meningoencephalitis have been reported from India and elsewhere (George et al., 1994; Kwong et al., 1997).

Infection occurs when endospores enter the body from the environment, and the spore is the primary infectious form of the organism (Hanna and Ireland, 1999). Spores are rapidly phagocytosed by macrophages, some of which undergo lysis, and in cases of inhalational anthrax the surviving macrophages are carried towards the mediastinal lymph nodes by the lymphatics. Phagocytosed spores may not germinate for up to 60 d, and so incubation of the inhalational form of the disease may take between 2 d and 6–8 weeks; this latency does not appear to occur in the cutaneous form of the disease. By analogy with other *Bacillus* species, germination is presumed to be triggered by a specific chemical germinant, but this remains unidentified. In several *Bacillus* species L-alanine is a germinant, and its binding to a receptor causes loss of spore refractility and resistance, the cortex swells, and metabolic activity commences; triggering of the germinant receptor is believed to activate endogenous proteolytic activity which converts the proenzyme of a germination-specific cortex-lytic enzyme to its active form which allows hydrolysis of the cortex, uptake of water, and all the other events associated with germination. The elevated CO_2 level and body temperature of the host cause the organism to transcriptionally activate the capsule and toxin genes. These genes are carried on two plasmids: plasmid pXO1 encodes the toxin genes, and plasmid pXO2 encodes the capsule genes; loss of either plasmid effectively renders the organism avirulent. Spores germinating in the presence of serum and elevated levels of CO_2 release blebs of capsular material through openings in the spore surface; the capsule of poly-γ-D-glutamic acid is purported to resist phagocytosis by virtue of its negative charge (Ezzell and Welkos, 1999). The anthrax toxin complex comprises three components: edema factor (EF), protective antigen (PA), and lethal factor (LF), none of which is toxic alone; EF and LF are active in binary combinations with PA and have different activities. PA molecules bind to molecules of a particular host cell membrane protein (anthrax toxin receptor or ATR; Bradley et al., 2001) and form ring-shaped prepores of heptameric oligomers (aggregates of seven); PA is then cleaved and so activated by a furin-like protease on the surface of the cell under attack. An active PA heptamer can then bind one or more molecules of EF, LF or both. The complex passes into the cell by receptor-mediated endocytosis and into an acidified endosome; following conformational change of the heptamer in the low-pH environment, the complex escapes directly to the cytosol by insertion of the heptamer into the endosomal membrane. The PA-EF binary toxin interacts with the abundant host protein calmodulin (CaM; the major intracellular calcium receptor) and becomes an active adenylyl cyclase in most cell types; this elevates levels of the secretogogue cAMP and leads to hypovolaemic shock. The crystal structure of EF in complex with CaM has been elucidated (Drum et al., 2002). The PA-LF binary toxin is a zinc metalloprotease which cleaves members of the mitogen-activated protein kinase kinase family, so affecting certain signaling pathways, and levels of shock-inducing cytokines. This toxin primarily affects macrophages, and removal of macrophages from mice using silica renders them insensitive to the toxin (Hanna, 1999); however, the process which leads to macrophage lysis is unclear (Pannifer et al., 2001). The importance of any interaction between EF and LF awaits clarification. The molecular pathogenesis of infection with *Bacillus anthracis* was reviewed by Little and Ivins (1999).

Bacillus cereus. *Bacillus cereus* is next in importance to *Bacillus anthracis* as a pathogen of humans (and other animals), causing food-borne illness and opportunistic infections, and its ubiquity ensures that cases are not uncommon. In relation to food-borne illness, *Bacillus cereus* is the etiological agent of two distinct food poisoning syndromes (Kramer and Gilbert, 1992): (i) the diarrheal-type, characterized by abdominal pain with diarrhea 8–16 h after ingestion of the contaminated food and associated with a diversity of foods from meats and vegetable dishes to pastas, desserts, cakes, sauces, and milk, and (ii) the

emetic-type characterized by nausea and vomiting 1–5 h after eating the offending food, predominantly oriental rice dishes, although occasionally other foods such as pasteurized cream, milk pudding, pastas, and reconstituted formulas have been implicated. One outbreak followed the mere handling of contaminated rice in a children's craft activity (Briley et al., 2001), and fulminant liver failure associated with the emetic toxin has been reported (Mahler et al., 1997). Both syndromes arise as a direct result of the fact that *Bacillus cereus* spores can survive normal cooking procedures. Under improper storage conditions after cooking, the spores germinate and the vegetative cells multiply. In diarrheal illness, the toxin(s) responsible are produced by organisms in the small intestine (infective doses 10^4–10^9 cells per gram of food), while the emetic toxin is preformed and ingested in food (about 10^5–10^8 cells per gram in order to produce sufficient toxin). Variations in infective dose of the diarrheal illness will reflect the proportion of ingested cells that are sporulated, and so can survive the acid barrier of the stomach. The capacity of the strain concerned to produce toxin(s) will, of course, influence the infective or intoxicating dose in both types of illness. It is likely that cases showing both diarrheal and emetic symptoms are caused by organisms producing both diarrheal and emetic toxins. Strains of *Bacillus thuringiensis*, which are close relatives of *Bacillus cereus*, may also produce the diarrheal toxin, and *Bacillus thuringiensis* has indeed been implicated in cases of gastroenteritis (Damgaard et al., 1997); strains of this species commonly carry genes for *Bacillus cereus* enterotoxins (Rivera et al., 2000) and Fletcher and Logan (1999) found that strains of both *Bacillus mycoides* and *Bacillus thuringiensis* were positive in commercial tests for enterotoxin and in a cytotoxicity assay. Cases of illness caused by *Bacillus thuringiensis* may have been diagnosed as caused by *Bacillus cereus*, as the former may not produce its characteristic insecticidal toxin crystals when incubated at 37 °C, owing to the loss of the plasmids carrying the toxin genes (Granum, 2002). The safety of using *Bacillus thuringiensis* as a biopesticide on crop plants has been reviewed by Bishop (2002); Bishop et al. (1999) found that the main pesticide strains that they assayed produced low titers of enterotoxin.

The toxigenic basis of *Bacillus cereus* food poisoning and other *Bacillus cereus* infections has begun to be elucidated, and a complex picture is emerging (Beecher, 2001; Granum, 2002). *Bacillus cereus* is known to produce six toxins, four of which are enterotoxins, and the emetic toxin. The enterotoxins are (i) Hemolysin BL (Hbl), a 3-component proteinaceous toxin which also has dermonecrotic and vascular permeability activities, and causes fluid accumulation in ligated rabbit ileal loops; Hbl is produced by about 60% of strains tested (Granum, 2002), and it has been suggested that is a primary virulence factor in *Bacillus cereus* diarrhea, but the mechanism of its enterotoxic activity is unclear (Granum, 2002); (ii) Non-hemolytic enterotoxin (Nhe) is another 3-component proteinaceous toxin which is produced by most strains tested (Granum, 2002), and whose components show some similarities to Hbl; (iii) and (iv) Enterotoxin T (BceT) and Enterotoxin FM (EntFM) are single-component proteinaceous toxins whose roles and characteristics are not known; also, Cytotoxin K (CytK) is similar to the β-toxin of *Clostridium perfringens* and was associated with a French outbreak of necrotic enteritis in which three people died (Lund et al., 2000). The genetics of toxin production are summarized by Granum (2002).

The emetic toxin, cereulide, is a dodecadepsipeptide comprising a ring of four amino- and/or oxy-acids: [D-O-Leu-D-Ala-L-O-Val-L-Val] thrice repeated; chemically speaking, it is closely related to the potassium ionophore valinomycin (Agata et al., 1994). It is resistant to heat, pH and proteolysis, but it is not antigenic (Kramer and Gilbert, 1989). Cereulide is probably an enzymically synthesized peptide rather than a direct genetic product; it is produced in larger amounts at lower incubation temperatures, its production does not appear to be connected with sporulation (Finlay et al., 2000), and it is produced in aerobic, and microaerobic, but not in anaerobic conditions (Finlay et al., 2002). Its mechanism of action is unknown, but it has been shown to stimulate the vagus afferent through binding to the 5-HT$_3$ receptor (Agata et al., 1995). The earliest detection system for emetic toxin involved monkey-feeding tests (Logan et al., 1979), but a semi-automated metabolic staining assay has now been developed (Finlay et al., 1999).

Bacillus cereus is also a destructive ocular pathogen. Endophthalmitis may follow penetrating trauma of the eye, intraocular surgery, or hematogenous spread, and it may evolve very rapidly. Loss both of vision and the eye is likely if appropriate treatment is instituted too late (Das et al., 2001; Davey and Tauber, 1987). *Bacillus cereus* keratitis associated with contact lens wear has also been reported (Pinna et al., 2001) Other *Bacillus cereus* infections occur mainly, though not exclusively, in persons predisposed by neoplastic disease, immunosuppression, alcoholism and other drug abuse, or some other underlying condition, and fatalities occasionally result. Reported conditions include bacteremia, septicemia, fulminant sepsis with hemolysis, meningitis, brain hemorrhage, ventricular shunt infections, infections associated with central venous catheters, endocarditis, pseudomembranous tracheobronchitis, pneumonia, empyema, pleurisy, lung abscess, brain abscess, osteomyelitis, salpingitis, urinary tract infection, dermatolymphangioadenitis associated with filarial lymphedema, and primary cutaneous infections. Wound infections, mostly in otherwise healthy persons, have been reported following surgery, road traffic and other accidents, scalds, burns, plaster fixation, drug injection (including a case associated with contaminated heroin; (Dancer et al., 2002) and close-range gunshot and nail bomb injuries; some became necrotic and gangrenous. A fatal inflammation was caused by a blank firearm injury; blank cartridge propellants are commonly contaminated with the organism (Rothschild and Leisenfeld, 1996). Neonates also appear to be particularly susceptible to *Bacillus cereus*, especially with umbilical stump infections; respiratory tract infections associated with contaminated ventilation systems have also occurred (Van Der Zwet et al., 2000). Other infections reported in neonates include intestinal perforation, meningoencephalitis, and bacteremia refractory to therapy. There have been reports of wound, burn, and ocular infections with *Bacillus thuringiensis* (Damgaard et al., 1997), but there is as yet no evidence of infections associated with the use of this organism as an insecticide.

Bacillus cereus also causes infections in domestic animals. It is a well-recognized agent of mastitis and abortion in cattle, and can cause these conditions in other livestock (Blowey and Edmondson, 1995).

Other species. Reports of infections with non-*Bacillus cereus* group species are comparatively rare, but very diverse (Berkeley and Logan, 1997; Logan, 1988), and there have been several hospital pseudoepidemics associated with contaminated blood culture systems. *Bacillus licheniformis* has been reported from ventriculitis following the removal of a meningioma, cerebral abscess after penetrating orbital injury, septicemia following arteriography, bacteremia associated with indwelling central venous catheters (Blue et al., 1995), bacteremia during pregnancy with eclampsia and acute fibrinolysis, peritonitis in a CAPD patient and in a patient with volvulus and small-bowel perforation, ophthalmitis, and corneal ulcer after trauma. There have also been reports of L-form organisms, phenotypically similar to *Bacillus licheniformis*, occurring in blood and other body fluids of patients with arthritis, patients with neoplasms, clinically normal persons, and in association with infectious synovitis in birds (see *Cell morphology*, above). Although some authors have claimed a relationship between these organisms and diseases with postulated immunological elements, and higher isolations from the synovial fluids and membranes of arthritic patients have been reported, Bartlett and Bisset (1981) were unable to confirm the latter association. *Bacillus licheniformis* can cause food-borne diarrheal illness, and has been associated with an infant fatality (Mikkola et al., 2000). A toxin possibly associated with *Bacillus licheniformis* food poisoning has been identified (Mikkola et al., 2000), and toxigenic strains of *Bacillus pumilus* have been isolated in association with food-borne illness and from clinical and environmental specimens (Suominen et al., 2001). *Bacillus licheniformis* is frequently associated with bovine abortion and has been reproduced by experimental infection of cows, which demonstrated the tropism of the organism for the bovine placenta (Agerholm et al., 1997); this species has also been associated with abortion in water buffalo (Galiero and De Carlo, 1998), and is occasionally associated with bovine mastitis (Blowey and Edmondson, 1995). Many of these types of *Bacillus licheniformis* and *Bacillus cereus* infections are associated with wet and dirty conditions during winter housing, particularly when the animals lie in spilled silage (Blowey and Edmondson, 1995); in one outbreak, a water tank contaminated with *Bacillus licheniformis* was implicated (Parvanta, 2000).

The name *Bacillus subtilis* was often used to mean any aerobic, endospore-forming organism, but since 1970 there have been reports of infection in which identification of this species appears to have been made accurately. They include bacteremias associated with immunosuppression, surgical intervention, neoplastic disease, and trauma (de Boer and Diderichsen, 1991); other cases associated with neoplastic disease include: fatal pneumonia and bacteremia, a septicemia and an infection of a necrotic axillary tumour in breast cancer patients; breast prosthesis and ventriculo-atrial shunt infections; endocarditis in a drug abuser; meningitis following a head injury; cholangitis associated with kidney and liver disease; and isolations from dermatolymphangioadenitis associated with filarial lymphedema (Olszewski et al., 1999), and from surgical wound-drainage sites. *Bacillus subtilis* has also been associated with cases of bovine mastitis and of ovine abortion (Logan, 1988).

Bacillus subtilis has been implicated in food-borne illness: vomiting has been the commonest symptom, but with accompanying diarrhea frequently reported, the onset periods have been short (ranging from 10 min to 14 h; median 2.5 h), the bacterial loads of the organism were high (10^5–10^9 c.f.u./g), and the implicated foods were often prepared dishes in which meat or fish were served with cereal-based components such as bread, pastry, rice or stuffing (Kramer and Gilbert, 1989).

A probiotic preparation labeled as containing strains *Bacillus subtilis* led to a fatal septicemia in an immunocompromised patient (Oggioni et al., 1998); subsequently, the organisms concerned were identified as *Bacillus clausii* (Spinosa et al., 2000). These authors reported another *Bacillus clausii* infection, cholangitis in polycystic kidney disease in a 15-year-old French boy who had undergone renal transplant. The original authors (Wallet et al., 1996) had identified the organism as *Bacillus subtilis*; their patient had not been taking a probiotic preparation and the source of the infecting *Bacillus clausii* was unclear (Spinosa et al., 2000).

Organisms identified as *Bacillus circulans* have been isolated from cases of meningitis, a cerebrospinal fluid shunt infection, endocarditis, a wound infection in a cancer patient, a bite wound, and endophthalmitis (Tandon et al., 2001). Roy et al. (1997) reported epidemic endophthalmitis associated with isolates identified as *Bacillus circulans* that contaminated a product used during cataract surgery. It must be noted, however, that many isolates previously identified as *Bacillus circulans* might have been misallocated (see comments on *Bacillus circulans* in the *List of species*, below). *Bacillus coagulans* has been isolated from corneal infection, bacteremia and bovine abortion. *Bacillus pumilus* has been found in cases of pustule and rectal fistula infection, and in association with bovine mastitis. *Bacillus sphaericus* has been implicated in a fatal lung pseudotumour, and meningitis. Among 18 cancer patients with 24 bacteremic episodes, Banerjee et al. (1988) isolated *Bacillus cereus* (eight cases), *Bacillus circulans* (3), *Bacillus subtilis* (2), *Bacillus coagulans* (1), *Bacillus licheniformis* (1), *Bacillus pumilus* (1), *Bacillus sphaericus* (1) and six unidentified aerobic endospore-formers.

Fish. There have been several reports of *Bacillus* infections among farmed fish. *Bacillus mycoides* was isolated from necrotic muscular lesions in channel catfish (*Ictalurus punctatus*) during an epizootic in a commercial pond in Alabama, USA, and similar lesions could be reproduced by subcutaneous injection of the isolate (Goodwin et al., 1994). An unidentified aerobic endospore-former was associated with a septicemic condition affecting a variety of widely cultivated freshwater fish in Nigeria, and its etiological role was confirmed by reinfection trials (Oladosu et al., 1994). Ferguson et al. (2001) isolated an unidentified *Bacillus* from a severe multi-focal, necrotizing and granulomatous infection of the intensively reared catfish *Pangasius hypophthalmus* in the Mekong delta; the condition was reproduced experimentally, but subsequently the pathogenicity of this organism has not been demonstrable (M. Crumlish, personal communication).

Insect pathogens. *Bacillus thuringiensis* strains produce crystalline, proteinaceous, parasporal bodies within the sporangia (Figure 10h), and the insecticidal activities of many of these δ-endotoxins have made the organism (often referred to as Bt) one of the most widely produced and studied bacteria in biotechnology. The δ-endotoxins are produced as the organism begins to sporulate (in most cases; Sekar, 1988), and may represent up to 30% of sporangial dry weight by the completion of sporulation (Baum and Malvar, 1995). The δ-endotoxin genes are designated with the "*cry*" prefix to indicate that their product proteins are crystalline; they are nearly always located on large conjugative plasmids (Aronson, 1993), and transposons and insertion elements have been found associated with them.

During infection, plasmids may be transferred between strains by conjugation (Thomas et al., 2000). There are over 80 different classes and subclasses of Cry proteins, representing at least four distinct protein families, and they have their own nomenclatural system (Crickmore et al., 1998). Related proteins are produced by *Clostridium bifermentans* (Barloy et al., 1996) and *Paenibacillus popilliae* (Zhang et al., 1997). Some strains of Bt which are active against Diptera produce structurally unrelated crystalline proteins known as Cyt toxins (Ellar, 1997). The full list of Cry and Cyt toxins is maintained at: http://www.biols.susx.ac.uk/Home/Neil_Crickmore/Bt/. An individual sporangium may carry a single type of insecticidal crystal protein, or several different types, comprising one or more parasporal bodies. Despite the name, the targets of insecticidal crystal proteins are not restricted to insects such as Lepidoptera (butterflies and moths), Coleoptera (beetles) and Diptera (flies); some mites, nematodes, flatworms and protozoa are also susceptible (Feitelson et al., 1992). The protein structure and mode of action of the δ-endotoxins remain subjects of intensive study, which has been reviewed by Ellar (1997), Schnepf et al. (1998) and Aronson and Shai (2001). The δ-endotoxins have three-domain structures (Groschulski et al., 1995; Li et al., 1991), with molecular masses usually of around 70 kDa or 130–140 kDa. Their phylogenetic relationships have been investigated by Bravo (1997). Domain I forms pores in susceptible gut epithelia, while Domain II has receptor-binding properties which influence the toxin's spectrum of activity (reviewed by Dean et al., 1996). Several roles have been suggested for Domain III; it is believed to play a part in receptor binding – perhaps an initial and reversible binding which is followed by an irreversible interaction mediated by Domain II (Lee et al., 1999b). An aminopeptidase N (Knight et al., 1994) and a cadherin-like glycoprotein (Vadlamudi et al., 1995; Vladmudi et al., 1995) have been identified as δ–endotoxin-binding proteins in insect larval gut epithelia.

The ingested protoxin dissolves in the alkaline midgut of a susceptible insect larva and undergoes proteolytic activation. Binding to the specific receptor on the brush-border cell is followed by insertion into the membrane to form a pore (Knowles, 1994) through which small molecules and ions can pass, so that water is taken up by osmosis and the cells swell and lyse (Ellar, 1997). It has been suggested that the concomitant fall in midgut pH is followed by spore germination and a fatal septicemia, while the micro-organism benefits from the nutrients so liberated (Ellar, 1990). With neutral or acidic guts, proteolytic nicking may allow protoxin solubilization (Carroll et al., 1997), and pore characteristics may vary with pH (Schwarz et al., 1993). The applications and development of Bt pesticides, and their safety, have been reviewed by Bishop (2002), while Van Rie (2002) has reviewed the development of transgenic crop plants.

Bacillus sphaericus has been divided into six taxa on the basis of DNA relatedness, supported by other molecular studies, and some strains belonging to group IIA (Rippere et al., 1997) are pathogenic to mosquitoes and have been exploited for biological control purposes. The parasporal crystal proteins synthesized by pathogenic strains of *Bacillus sphaericus* share no homology with those produced by *Bacillus thuringiensis* (Baumann et al., 1991; Charles et al., 1996; Porter et al., 1993). Early isolates, recovered in California, had mosquitocidal activities too low to be of practical use (Kellen et al., 1965), but further isolates, strain numbers 1,593 from Indonesia, 2,362 from Nigeria and 2,297 from Sri Lanka, showed toxicities high enough to be of value in vector control programmes (reviewed by Baumann et al., 1991; Charles et al., 1996). These later, highly toxic, strains all produced parasporal toxin crystals which were absent from the earlier, poorly toxic, isolates. The crystals, although smaller than those seen in *Bacillus thuringiensis*, are visible by phase-contrast microscopy (Priest, 2002). Screening programmes around the world have yielded over 560 further mosquitocidal isolates, and these are held by the International Entomopathogenic *Bacillus* Centre at the Institute Pasteur, Paris. Toxicity for different species of mosquito vary: there is no activity against *Aedes aegypti*; other *Aedes* species are moderately susceptible; intermediate activity against *Anopheles* species has attracted interest for malaria control; high activity against *Culex quinquefasciatus* is used to combat Japanese encephalitis and filariasis (Priest, 2002). Insects other than mosquitoes, and mammals, are unaffected by the toxins.

In all cases, it is the mosquito larvae that are attacked. As they feed, they ingest the spore/crystal complex, and symptoms follow rapidly; the spores then germinate so that the organism can take advantage of the rich nutrient supply. The parasporal crystal produced by high-toxicity strains of *Bacillus sphaericus* lies within the exosporium and is a binary toxin (Bin) reminiscent of cholera and diphtheria toxins. Its *bin* genes are highly conserved, only four variants being known (Priest, 2002). When ingested by a mosquito larva the component polypeptides, BinA and BinB, dissolve in the alkaline conditions of the midgut; BinA is slowly reduced from 42 kDa to an active form of about 39 kDa by host proteases, while BinB is more rapidly reduced from 51 kDa to an active form of about 43 kDa (Aly et al., 1989; Broadwell and Baumann, 1987). The exact mechanism of toxicity is not clear, but it is known that BinB is responsible for specific binding while BinA effects channel formation (Priest, 2002). Following mitochondrial swelling, large vacuoles appear in cells of the gastric caecum and posterior midgut (Davidson, 1981), peristalsis stops, and the larvae may die within 6 h. The Mtx1 protein, which is formed within the vegetative cells, is responsible for the toxicity of some strains that lack parasporal crystals. This protein of about 100 kDa is broken into subunits of 27 kDa and 70 kDa which respectively resemble an ADP-ribosylating toxin and a glycoprotein-binding protein (Hazes and Read, 1995; Thanabalu et al., 1993). Although highly toxic to a wide range of mosquitoes, it is not highly expressed and is subject to proteolytic attack within the bacterium (Ahmed et al., 1995; Wati et al., 1997). Other mosquitocidal toxin genes include *mtx2* (Liu et al., 1996) and *mtx3* (Thanabalu and Porter, 1996), whose related products act by pore formation. As some isolates that lack all the toxins described above show weak pathogenicity, it is likely that further toxins await discovery (Priest, 2002).

Ecology. Most *Bacillus* species are saprophytes widely distributed in the natural environment, but some species are opportunistic or obligate pathogens of animals, including humans, other mammals, and insects. *Bacillus anthracis* is, to all intents and purposes, an obligate pathogen of animals and humans. The habitats of most species are soils of all kinds, ranging from acid through neutral to alkaline, hot to cold, and fertile to desert, and the water columns and bottom deposits of fresh and marine waters. Their endospores readily survive distribution in soils, dusts and aerosols from these natural environments to a wide variety of other habitats, and Nicholson et al. (2000) has considered the roles of *Bacillus* spores in the natural environment. Some species appear to be ubiquitous contaminants of man, other animals, their foodstuffs, water and environments, natural, domestic, industrial and hospital. Their wide distribution is in part owing to the extraordinary longevity of their endospores, which show much greater resistance to physical and chemical agents, such as heat, cold, desiccation, radiation, disinfectants, antibiotics and other toxic agents, than their counterpart vegetative cells. Endospores are typically more resistant to heat than vegetative cells by a factor of 10^5 or more, while resistance to UV and ionizing radiation may be 100-fold or more. If protected from radiation, spores may survive for very long periods. Striking claims include the isolation of a viable strain of *Bacillus sphaericus* from an extinct bee preserved in 25- to 40-million-year-old amber (Cano and Borucki, 1995), and the recovery of an aerobic endospore-former related to *Salibacillus* and *Virgibacillus* from a water droplet trapped in a Permian salt crystal for an estimated 250 million years (Vreeland et al., 2000); the latter report has been contested on account of the apparent modernity of its DNA (Graur and Pupko, 2001; Nickle et al., 2002), and both isolations need to be confirmed by independent laboratories. *Bacillus infernus* was isolated from Triassic shales, lying at depths of around 2.7 km below the land surface, that may have been hydrologically isolated for 140 million years, and although they have not been demonstrated for this organism, endospores are likely to have contributed to its survival (Boone et al., 1995). Another organism from a deep subsurface environment is *Bacillus subterraneus*, which was isolated from the Great Artesian Basin of Australia, a thermal aquifer up to 2 km deep (Kanso et al., 2002); spores have not been demonstrated for this species either.

The commonly isolated species, such as *Bacillus subtilis* and *Bacillus cereus* are very widely distributed worldwide; *Bacillus thuringiensis* has been isolated from all continents, including Antarctica (Forsyth and Logan, 2000); as well as being a common organism in the natural environment and foods, *Bacillus cereus* frequently contaminates domestic kitchen environments (Beumer and Kusumaningrum, 2003). In studies of indoor and outdoor air and dusts, *Bacillus* species commonly dominate the cultivable flora or form a large part of it (Aldagal and Fung, 1993; Andersson et al., 1999b; Dutkiewicz et al., 2001; Marafie and Ashkanani, 1991; Schaffer and Lighthart, 1997; Venkateswaran et al., 2001), and honeybees have been found to scavenge airborne spores electrostatically (Lighthart et al., 2000). The presence of *Bacillus fumarioli* strains showing similar phenotypic behavior and substantial genotypic similarity from Candlemas Island in the South Sandwich archipelago and from volcanoes some 5600 km distant on continental Antarctica, is most convincingly explained by their carriage in the air as free spores or spores attached to plant propagules, as no birds are known to visit the latter sites (Logan et al., 2000). Strains indistinguishable from these Antarctic isolates have been cultivated from gelatin production plants in Belgium, France and the USA (De Clerck et al., 2004a).

Bacillus species are often isolated following heat treatment of specimens in order to select for spores, and presence of spores in a particular environment does not necessarily indicate that the organism is metabolically active there; however, it is reasonable to assume that large numbers of endospores in a given environment reflects former or current activity of vegetative *Bacillus* cells there. Indeed, the ease with which strains of a close relative of *Bacillus pumilus* and strains of *Bacillus thuringiensis* have been isolated from pristine environments along the Victoria Land coast of Antarctica (Forsyth and Logan, 2000; and unpublished observations) cannot easily be explained as widespread and chance contamination with endospores from another source; the organisms almost certainly undergo some multiplication in those environments. *Bacillus fumarioli* was found as both spores and vegetative cells at geothermal sites in Antarctica where soil temperatures ranged from 3.4 °C to 62.5 °C; the proportions of sporulated cells tended to be higher at the temperature extremes and lower at temperatures approaching the optimum growth temperature (50 °C) of the organism (Logan et al., 2000). Isolation of organisms showing special adaptions to the environments in which they are found, such as acidophily, alkaliphily, halophily, psychrophily, and thermophily, suggests that these organisms must be metabolically active in these niches, but it tells us little about the importances of their roles in the ecosystems, and nothing about their interactions with other members of the flora. *Bacillus thermantarcticus*, which warrants transfer to *Geobacillus* (see *Species Incertae Sedis*, below), was also found in the geothermal soil of Cryptogam Ridge, Mount Melbourne, Antarctica (Nicolaus et al., 1996), a site from which *Bacillus fumarioli* was isolated (Logan et al., 2000).

Many *Bacillus* species will degrade biopolymers, with versatilities varying according to species, and it is therefore assumed that they have important roles in the biological cycling of carbon and nitrogen; it is further assumed that their activities in food spoilage and biodegradation reflect the contamination of these materials by endospores derived from dusts and other vehicles. Valid though these assumptions may be, the ever-increasing diversity of known *Bacillus* species and their apparent primary habitats implies that such generalizations may deserve reconsideration in some cases, and that certain species may have quite specialized activities.

Habitats. Although isolates of many of the established species have been derived from soil, or from environments that may have been contaminated directly or indirectly by soil, the range of isolation sources is very wide, and includes, in addition to temperate, acidic, neutral and alkaline soils, fresh and marine waters, foods and clinical specimens: air (*Bacillus carboniphilus*), arsenic-rich sediments (*Bacillus arseniciselenatis*, *Bacillus selenitireducens*) and arsenic-contaminated mud and water (*Bacillus indicus*, *Bacillus macyae*), bauxite-processing waste (*Bacillus vedderi*), brine (*Bacillus haloalkaliphilus*), compost (*Bacillus circulans*, *Bacillus coagulans*, *Bacillus licheniformis*, *Bacillus sphaericus*, *Bacillus subtilis*), emperor moth caterpillars ("phane"; *Bacillus cereus*, *Bacillus circulans*, *Bacillus licheniformis*, *Bacillus megaterium*, *Bacillus mycoides*, *Bacillus pumilus*, *Bacillus subtilis*), feathers (*Bacillus cereus*, *Bacillus licheniformis*, *Bacillus pumilus*, *Bacillus subtilis*), feces (*Bacillus alkalophilus*, *Bacillus badius*, *Bacillus cohnii*, *Bacillus flexus*, *Bacillus halodurans*, *Bacillus megaterium*, *Bacillus pseudofirmus*), geothermally heated soils (*Bacillus fumarioli*, *Bacillus luciferensis*, *Bacillus schlegelii*), honey bee and greater wax moth frass (*Bacillus cereus*, *Bacillus megaterium*, *Bacillus sphaericus*), inner tissues of plants (*Bacillus amyloliquefaciens*, *Bacillus cereus*, *Bacillus endophyticus*, *Bacillus insolitus*, *Bacillus licheniformis*, *Bacillus megaterium*, *Bacillus pumilus*, *Bacillus subtilis*), invertebrates (*Bacillus oleronius*, *Bacillus sphaericus*, *Bacillus*

thuringiensis), leather (*Bacillus cereus*, *Bacillus firmus*, *Bacillus licheniformis*, *Bacillus megaterium*, *Bacillus pumilus*, *Bacillus sphaericus*, *Bacillus subtilis*), milk (*Bacillus cereus*, *Bacillus coagulans*, *Bacillus licheniformis*, *Bacillus smithii*, *Bacillus sporothermodurans*, *Bacillus weihenstephanensis*), naturally heated waters (*Bacillus methanolicus*, *Bacillus okuhidensis*), poultry litter and manure (*Bacillus cereus*, *Bacillus fastidiosus*, *Bacillus halodurans*, *Bacillus pumilus*, *Bacillus subtilis*), paper and paperboard (*Bacillus amyloliquefaciens*, *Bacillus cereus*, *Bacillus circulans*, *Bacillus coagulans*, *Bacillus firmus*, *Bacillus flexus*, *Bacillus halodurans*, *Bacillus licheniformis*, *Bacillus megaterium*, *Bacillus mycoides*, *Bacillus pumilus*, *Bacillus sphaericus*, *Bacillus subtilis*, *Bacillus thuringiensis*), recycled paper pulp (*Bacillus pumilus*), seaweed (*Bacillus algicola*), saline and hypersaline environments (*Bacillus alcalophilus*, *Bacillus firmus*, *Bacillus halodenitrificans*, *Bacillus halophilus*, *Bacillus megaterium*), sewage and wastewater treatment processes (*Bacillus funiculus*, *Bacillus thermocloacae*), sheep fleece (*Bacillus cereus*, *Bacillus thuringiensis*), silage (*Bacillus coagulans*, *Bacillus siralis*), soda lakes (*Bacillus agaradhaerens*, *Bacillus cohnii*, *Bacillus pseudofirmus*, *Bacillus vedderi*), solfatara (*Bacillus tusciae*), gemstones (*Bacillus badius*, *Bacillus cereus*, *Bacillus circulans*, *Bacillus coagulans*, *Bacillus firmus*, *Bacillus lentus*, *Bacillus licheniformis*, *Bacillus mycoides*, *Bacillus subtilis*; Khan et al., 2001), stone surfaces of ancient monuments (*Bacillus licheniformis*, *Bacillus megaterium*, *Bacillus mycoides*, *Bacillus subtilis*; Turtura et al., 2000), subterranean soil and water (*Bacillus infernus*, *Bacillus subterraneus*), and wall paintings (*Bacillus decolorationis*, Heyrman et al., 2003a; *Bacillus barbaricus*, Taubel et al., 2003).

Most species of *Bacillus* are heterotrophic organisms that have been isolated on complex organic media. Relatively few attempts have been made to isolate aerobic endospore-formers which can utilize inorganic sources of carbon and energy, or to demonstrate that established heterotrophic species are capable of facultative autotrophy. The two thermophiles *Bacillus schlegelii* and *Bacillus tusciae* remain the only species in the genus shown to be facultatively chemolithoautotrophic. Nitrogen fixation is well established for certain species in *Paenibacillus* (*Paenibacillus azotofixans*, *Paenibacillus macerans*, *Paenibacillus polymyxa*), but less is known about *Bacillus* species utilizing atmospheric nitrogen; although *Bacillus edaphicus* and *Bacillus mucilaginosus* were isolated on nitrogen-free media, these species are actually members of *Paenibacillus* (see *Species Incertae Sedis*, below). However, several studies have demonstrated nitrogen fixation by strains of *Bacillus cereus*, *Bacillus licheniformis*, *Bacillus megaterium*, *Bacillus sphaericus*, and unidentified strains (some of which may have been *Paenibacillus* species) isolated from rhizospheres and phylloplanes, and from endophytic sites and mycorrhizae (Rózycki et al., 1999). A comparative phylogenetic study, however, concluded that nitrogen fixation among aerobic endospore-formers is restricted to certain species of *Paenibacillus* (Achouak et al., 1999). Nitrogen-fixing *Bacillus* and *Paenibacillus* growing in the rhizosphere may help to promote plant growth; other ways in which aerobic endospore-formers may promote the growth of plants include (Chanway, 2002): the production of phytohormones, increasing nutrient availability (to the plant or to other, nitrogen-fixing, bacteria; Zlotnikov et al., 2001), the suppression of ethylene production by the plant in its rhizosphere, interactions with symbiotic bacteria and fungi (Medina et al., 2003), enhancement of root nodulation, and biological control of plant pathogens by various mechanisms including the production of antibiotics. In one study, *Bacillus subtilis* and *Bacillus mycoides* were found to dominate the rhizosphere of tea bushes (Pandey and Palni, 1997), a strain of the latter species having antifungal activity. Epiphytic *Bacillus* strains can have protective roles in the phyllosphere (Collins and Jacobsen, 2003; Jock et al., 2002). Representatives of several species, including *Bacillus amyloliquefaciens*, *Bacillus cereus*, *Bacillus endophyticus*, *Bacillus insolitus*, *Bacillus licheniformis*, *Bacillus megaterium*, *Bacillus pumilus* and *Bacillus subtilis*, have been isolated from the inner tissues of healthy plants, including cotton, grape, pea, spruce and sweet corn, and some strains appear to have important roles in growth promotion and plant protection (Reva et al., 2002). These endophytes and epiphytes can have potential as agents for the biocontrol of plant diseases, and their spores offer advantages in the formulation of such preparations (Emmert and Handelsman, 1999). Hosford (1982) and Leary et al. (1986), on the other hand, reported *Bacillus* species showing pathogenicity for plants.

Bacillus species are known to have roles in the postharvest processing and flavor development of cocoa (Schwan et al., 1995), coffee (Silva et al., 2000), tobacco (English et al., 1967) and vanilla (Röling et al., 2001), in the production of natural fibers and other vegetable products, in several traditional fermented foods based on leaves and seeds (and poultry eggs) (often dominated by *Bacillus subtilis*; Wang and Fung, 1996; Beaumont, 2002; Sarkar et al., 2002), and in composting (Blanc et al., 1999; Strom, 1985).

Bacillus species may cause deterioration of hides intended for leather production (Birbir and Ilgaz, 1996). Keratinolytic strains of *Bacillus cereus*, *Bacillus licheniformis*, *Bacillus pumilus*, *Bacillus subtilis* and of unidentified *Bacillus* species have roles in the degradation of feathers in poultry waste and may be found in the plumage of many bird species (Burtt and Ichida, 1999; Kim et al., 2001). *Bacillus* species also play a part in the degradation of chitin. This activity has been demonstrated for strains of *Bacillus circulans*, *Bacillus coagulans*, *Bacillus lentus*, *Bacillus licheniformis*, *Bacillus megaterium*, *Bacillus pumilus* and *Bacillus thuringiensis* (Clements et al., 2002), and strains of *Bacillus amyloliquefaciens*, *Bacillus cereus*, *Bacillus megaterium*, *Bacillus sphaericus* and *Bacillus subtilis* which utilize the chitin in crustacean wastes have been isolated (Sabry, 1992; Wang and Hwang, 2001). The value of chitinolysis to insect-pathogenic strains of *Bacillus thuringiensis* is evident, and an exochitinase from a strain of "*B thuringiensis* subsp. *pakistani*" was found to be toxic to *Aedes aegypti* larvae (Thanthiankul et al., 2002). Chitinases from a soil isolate of *Bacillus amyloliquefaciens* have been found to have antifungal properties (Wang et al., 2002). (The two species *Bacillus chitinolyticus* and *Bacillus ehimensis*, which were isolated using a chitin medium, are members of *Paenibacillus*).

Strains of several *Bacillus* species have been found to accumulate metal ions non-enzymically by adsorption to their cell surfaces and this can be of importance in waste treatment and natural environments: *Bacillus licheniformis* cells can accumulate cerium, cobalt and copper ions from aqueous and simulated waste solutions (Hafez et al., 2002), *Bacillus subtilis* may accumulate aluminum, cadmium, iron and zinc, and aluminosilicates (Urrutia and Beveridge, 1995), and an unidentified *Bacillus* strain bound chromium, copper and lead ions (Nourbakhsh et al., 2002). *Bacillus megaterium* biomass was found to bioreduce ions of the precious metals gold, palladium, platinum, rhodium

and silver (Lin et al., 2001). *Bacillus arseniciselenatis* and *Bacillus selenitreducens* can use oxyanions of the two highly toxic elements arsenic and selenium as terminal electron acceptors in anaerobic respiration, and the environmental impact of such activity is becoming appreciated (Stolz and Oremland, 1999).

The trichome-forming bacteria "*Anisomitus*", "*Arthromitus*", "*Entomitus*", "*Coleomitus*", "*Metabacterium*", and "*Sporospirillum*", which occur in the alimentary tracts of animals, and which have been reported to form endospores, were listed as *Genera Incertae Sedis* in the First Edition of this *Manual* (Claus and Berkeley, 1986). Since that time, molecular methods have allowed considerable progress to be made in the taxonomy of some of these organisms. A cultivable "*Arthromitus*" strain from sow bug or wood louse (*Porcellio scaber*) has been identified as *Bacillus cereus*, and similar organisms have been isolated from moths, roaches and termites (Jorgensen et al., 1997; Margulis et al., 1998). *Bacillus oleronius* was first isolated from the hindgut of the termite *Reticulitermes santonensis*, and cellulolytic strains of the *Bacillus cereus* group and *Bacillus megaterium* have been found in the gut of another termite, *Zootermopsis angusticollis* (Wenzel et al., 2002). An "*Arthromitus*"-like endospore has been reported from a Miocene termite preserved in amber (Wier et al., 2002). On the other hand, several nonculturable, segmented, filamentous bacteria from chickens, mice, rats and trout have been shown to represent a distinct subline of the *Clostridium* subphylum, which has been proposed as "*Candidatus* Arthromitus" (Snel et al., 1995; Urdaci et al., 2001). Strains of "*Metabacterium polyspora*" from guinea pig cecum are closely related to the extremely large, viviparous, intestinal symbionts of the surgeonfish, *Epulopiscium* species, and also belong to the *Clostridium* subphylum (Angert et al., 1996).

Cellular fatty acids. This approach was recently reviewed by Kämpfer (2002). On the basis of the fatty acid compositions of 19 *Bacillus sensu lato* species Kaneda (1977) recognized six groups (A-F). All except group D were found to contain major amounts of branched chain acids, while within groups A–C only insignificant amounts (<3%) of unsaturated fatty acids were found; these latter three groups could be separated on the basis of their predominant fatty acids. Group A contained species now allocated to *Bacillus* (*Bacillus circulans*, *Bacillus licheniformis*, *Bacillus megaterium*, *Bacillus pumilus* and *Bacillus subtilis*), *Brevibacillus* and *Paenibacillus* and contained $C_{14:0 \text{ anteiso}}$ (26–60%) and $C_{15:0 \text{ iso}}$ (13–30%) acids, with chain lengths of 14–17. Among group B strains, now all allocated to *Paenibacillus*, $C_{15:0 \text{ anteiso}}$ acid predominated (39–62%), with chain lengths between 14 and 17. In group C species, now accommodated in *Geobacillus*, $C_{15:0 \text{ iso}}$ acid was predominant. In group D, now *Alicyclobacillus*, a unique fatty acid pattern with up to 70% cyclohexane fatty acids of chain length 17–19 was found. Group E comprised members of the *Bacillus cereus* group of species which, unlike the other groups, always had small proportions (7–12%) of unsaturated fatty acids present; the predominant fatty acids (19–21%) were of the $C_{15:0 \text{ iso}}$ type. The psychrophiles *Bacillus* (now *Sporosarcina*) *globisporus* and *Bacillus insolitus* formed group F and contained large proportions (17–28%) of unsaturated fatty acids, and the predominant branched-chain fatty acids in these two species were $C_{15:0 \text{ anteiso}}$ acids. The work of Kaneda was largely confirmed, and was supplemented, by the comprehensive study of Kämpfer (1994). For many species of *Bacillus sensu stricto* (but excepting *Bacillus badius*, the *Bacillus cereus* group, *Bacillus circulans*,

Bacillus coagulans, Bacillus simplex and *Bacillus smithii*) profiles (with ranges as percentage of total given in parentheses) were $C_{15:0\ anteiso}$ (25–66%), $C_{15:0\ iso}$ (22–47%), and $C_{17:0\ anteiso}$ (2–12%). For the *Bacillus cereus* group (*Bacillus anthracis, Bacillus cereus, Bacillus mycoides, Bacillus pseudomycoides, B thuringiensis*), levels of $C_{15:0\ anteiso}$ were lower (3–7%), and amounts of unsaturated fatty acids were generally higher (> 10%). *Bacillus smithii* and *Bacillus coagulans* showed higher amounts of $C_{17:0\ anteiso}$ (17–42%) and lower amounts of $C_{15:0\ anteiso}$ and $C_{15:0\ iso}$, and they also form a separate lineage within the *Bacillus* rRNA group. Other findings were low amounts of $C_{15:0\ anteiso}$ (<10%) and relatively high amounts of unsaturated acids in *Bacillus badius*, and very low amounts of $C_{15:0\ iso}$ in *Bacillus circulans*. These results indicated that the *Bacillus* rRNA group is still heterogeneous, and it was predicted that further taxonomic rearrangements would follow (Kämpfer, 2002).

With the division of *Bacillus sensu lato* into several phylogenetically distinct genera, the re-evaluation of the ability of fatty acid data to differentiate taxa has become possible. Numerical analyses of fatty acid data have shown that, in terms of level of taxonomic resolution, profiles vary largely within the species (Kämpfer, 1994). Although *Bacillus* species often cannot be differentiated by fatty acid analysis, especially in cases when large numbers of strains are examined, distinction of individual species, or even subspecies in certain cases, can be possible if small numbers of strains are studied (because possible intraspecific variation is not detected and hence cannot influence the interpretation of the results), or if genomically well-characterized groups of strains such as *Bacillus cereus* and its relatives are being investigated. Thus, given highly standardized growth conditions to achieve reproducible results, and a reliable database containing information on genomically homogeneous strains, fatty acid patterns can be used for *Bacillus* species identification. Also, whole-cell fatty acid analysis is valuable as a rapid and fairly inexpensive screening and identification method as part of polyphasic taxonomic studies (Kämpfer, 2002).

A GLC system developed for the identification of bacteria and yeasts by fatty acid methyl ester (FAME) analysis is marketed as the Microbial Identification System (MIDI, Newark, DE, USA). It requires highly standaridized cultivation and extraction procedures and provides a species-specific fatty acid database. Fatty acid peaks can be named by comparing retention times with those of a known mixture, but definitive identification can be made only by mass spectrometry. Species identification within *Bacillus* is not always possible by this approach, and additional phenotypic and/or genotypic characterization is often necessary (Kämpfer, 2002).

Sodium-Dodecylsulfate-Polyacrylamide Gel Electrophoresis (SDS-PAGE) of whole-cell proteins. Whole-cell protein patterning (SDS-PAGE) is a rapid and cost-effective method for the comparison of large groups of bacteria, and is a valuable initial step in polyphasic characterization. It requires highly standardized conditions of growth, combined with a rigorously standardized procedure for analysis, and normalization of the data for computer-assisted comparison of the results. Normalization between different electrophoretic runs can be achieved by the inclusion of a carefully chosen bacterial pattern in which a protein extract of a standard organism is loaded into the outer lanes and the central lane of the gel. If a molecular mass marker is loaded as well, the molecular mass of the protein

bands can be estimated easily. This method was reviewed by De Vos (2002). It has made important contibutions to polyphasic taxonomic studies within *Bacillus* (De Clerck and De Vos, 2002) and to the recognition of novel *Bacillus* species such as *Bacillus fumarioli, Bacillus luciferensis* and *Bacillus shackletonii* (Logan et al., 2000, 2002b, 2004b).

Whole-cell spectrometric analysis. A variety of instrument-based physico-chemical techniques for whole-cell analysis, including pyrolysis mass spectrometry (PyMS), matrix-assisted laser desorption-ionization time-of-flight mass spectrometry (MALDI-TOF-MS) and Fourier transform infrared spectrometry (FT-IR), have been applied to *Bacillus*. These "fingerprinting" methods require expensive instruments and complex data handling, but offer simple sample preparation, speed of analysis, high throughput, and low processing costs, and they yield data that can discriminate at genus, species and strain level. They can be valuable in polyphasic taxonomic studies, but better long-term reproducibility of spectra is required for the building of databases for routine identification applications. These approaches are reviewed by Magee and Goodacre (2002).

Bacteriophage. The potential for phage typing of *Bacillus* species has been little investigated. The γ phage has long been used as a specific test for identifying *Bacillus anthracis*, but preparations of this phage are not widely available, and the specific molecular tests for this species now exist. Ackermann et al. (1995) developed phage typing schemes for *Bacillus subtilis* and *Bacillus thuringiensis*. They examined 200 *Bacillus subtilis* strains and were able to distinguish 29 phagovars using 10 phages; with 500 strains of *Bacillus thuringiensis*, 10 phages allowed the recognition of 25 phagovars, but the phagovars showed no correlation with H-serovars. Pantasicocaldas et al. (1992) investigated the population dynamics of *Bacillus subtilis* and its phages in soil.

Enrichment and isolation procedures

Detection of *Bacillus* strains by cultivation and isolation relies mainly upon resistance of their endospores to heat (and to other conditions) which are lethal to vegetative cells of most kinds of bacteria. Only endospores and extremely themophilic, non-spore-forming bacteria are able to survive the heat treatment, so that subsequent aerobic culture at lower temperatures yields only isolates germinating from the surviving spores. Such an approach may also heat-activate the spores and so enhance their germination. However, an obvious disadvantage is that vegetative cells of *Bacillus* will be lost along with the other, non-spore-forming bacteria; as a result, it will not be known whether the isolation of a *Bacillus* strain from a particular habitat reflects its multiplication in and colonization of that niche, or mere survival of a dormant, contaminant spore. Furthermore, spores formed in natural environments may have heat sensitivities that are different to those formed in laboratory cultures, spores may be present in very small numbers or even absent, and heat treatment may be mutagenic. It is therefore valuable to reserve part of the specimen or sample for cultivation from an unheated control preparation for study in parallel with the heat-treated sample. With many environments this will allow isolates arising from vegetative cells to be obtained – problems of heavy bacterial load can, of course, be addressed by dilution of the sample. Repeated and consistent isolations of similar strains from heat-treated samples taken from pristine or sparsely populated environments would suggest that the organisms must be colonizing the habitat rather than just be reflecting heavy contamination, and it seems likely that populations of *Bacillus* strains in many habitats will exist as both vegetative and sporulated cells (see the comments on *Bacillus fumarioli* in the *Ecology* section of *Further descriptive information*, above). Notwithstanding this, it is clear that spores may be carried for long distances by water or the air, and locally large numbers of spores may be regularly deposited in sites that are apparently unsuitable for their germination and continued development; it is well known, for example, that thermophilic endospore-formers may easily be isolated from cold environments (Marchant et al., 2002).

Heat treatment may vary from 60 °C to 80 °C for 10 min or longer; 80 °C for 10 min is widely used. The given time allows for a period of heat penetration of the sample followed by a sufficient holding period at temperature; this assumes that the specimen is in an aqueous suspension in a water bath, and is adequately immersed. Solid samples may be emulsified in sterile, deionized water, 1:2 (w/v) prior to heating; the unheated control is prepared in the same way, but is unheated, or else the suspension intended for heating may be sampled for cultivation prior to heating. Direct plate cultures are made on appropriate solid media by spreading up to 250 µl volumes from undiluted, and 10-, 100- and 1000-fold dilutions of the treated sample.

Enrichment culture of samples from habitats expected to be sparsely populated with bacteria can be done by heat-treating suspensions in a suitable broth medium (perhaps 1–5 g of soil in 10 ml broth, for example), then incubating them and streak-diluting onto solid media as soon as they show signs of turbidity, with further subcultures from the reincubated broths being made at intervals thereafter. This approach, using different incubation temperatures and a range of pH, and with paired cultures from samples that had not been subjected to heat treatment, yielded a variety of isolates from pristine soils (Logan et al., 2000). Broths for particular kinds of organisms (nitrogen-free broths – in flasks so as to give a large broth surface/air interface – to isolate nitrogen fixers, for example) can be made selective or elective as desired.

Using a disinfectant to inactivate vegetative cells is for most such agents constrained by the difficulty of removing or neutralizing the agent prior to cultivation. Ethanol, however, may easily be removed by evaporation, and it does not have the potentially mutagenic effect of heat treatment. Filter-sterilized 95–100% ethanol is added to a final concentration of 1:1 (v/v) and held for 30–60 min at room temperature. The ethanol must first be filter-sterilized, because some commercial batches may be contaminated with spores.

For soil and certain other kinds of samples, desiccation by air drying will destroy many vegetative cells, but inactivation of cyst-forming bacteria such as *Azotobacter* may require an extended period of drying.

Although useful for the detection and isolation of well-represented species, the approaches outlined above are rather crude for the isolation of organisms present only in relatively small numbers, and for the slow-growing, less vigorous types. Methods that have been described for the isolation of individual species are outlined below. For isolation of species from clinical cases, most strains will survive carriage in freshly collected specimens or in a standard transport medium (for safety concerning

Bacillus anthracis, see below); heat treatment is not appropriate for fresh clinical specimens, where spores are usually sparse or absent. All the clinically significant isolates reported to date are of species that grow, and often sporulate, on routine laboratory media at 37 °C. It seems unlikely that many clinically important, but more fastidious, strains are being missed for the want of special collection, media or growth conditions.

Safety. With the exception of *Bacillus anthracis*, species of aerobic endospore-forming bacteria that may be isolated from clinical specimens can be handled safely on the open bench, and no special precautions are required for specimen collection. Isolation and presumptive identification of *Bacillus anthracis* can be performed safely in the routine clinical microbiology laboratory, provided that normal good laboratory practice is observed; vaccination is not required for minimal handling of the organism. If aerosols are likely to be generated, the work should be performed in a safety cabinet. Anthrax is not highly contagious; cutaneous anthrax is readily treated and is only life-threatening in exceptional cases, and the infectious doses in the human inhalation and intestinal forms (also treatable if recognized) are generally very high ($LD_{50} > 10,000$ spores). Precautions, therefore, need to be sensible, not extreme. When collecting specimens related to suspected anthrax, disposable gloves, disposable apron or overalls, and boots which can be disinfected after use should be worn; for dusty samples that might contain many spores, the use of head-gear and dust masks should be considered. Discard disposable items into suitable containers for autoclaving followed by incineration. Non-autoclavable items should be immersed overnight in 10% formalin (5% formaldehyde solution). Glutaraldehyde (5%) is also effective. Items that cannot be immersed should be bagged and sent for formaldehyde fumigation. Ethylene oxide and hydrogen peroxide vapor are also effective fumigants, but the latter is inappropriate if organic matter is being treated. The best disinfectant for specimen spillages is again formalin; where this is considered impractical, 10% hypochlorite solution can be used, although its limitations should be appreciated; it is rapidly neutralized by organic matter, and it corrodes metals. Other strong oxidizing agents, such as hydrogen peroxide (5%) and peracetic acid (1%) are also effective but with the same organic matter limitations. Specimen collection and handling from humans and animals is described by Logan and Turnbull (2003), Logan et al. (2007) and Turnbull et al. (1998); non-clinical materials associated with attempts at deliberate release of the organism as a weapon may be very hazardous and no attempt to sample or process them should be made without the appropriate instructions from the correct authorities.

Bacillus aeolius was isolated from a 45 °C water sample collected from a shallow (15 m depth) marine hot spring, by aerobic enrichment in marine broth at 65 °C for 3 d, followed by plating on marine agar (Gugliandolo et al., 2003a).

Bacillus agaradhaerens, Bacillus alcalophilus and other alcaliphiles. These may be isolated by direct plating of heat-treated samples onto alkaline nutrient media, or by enrichment for 1 or more days in alkaline liquid nutrient media, with shaking, at the chosen incubation temperature, followed by plating. Media may be adjusted to a pH between 8.6 and 11.0, by using sodium hydrogen carbonate, sodium carbonate, sodium sesquicarbonate, trisodium phosphate or sodium perborate. A modification (Grant and Tindall, 1980) of the widely used medium of

Horikoshi (1971) is: glucose, 10.0 g; peptone, 5.0 g; yeast extract, 5.0 g; KH_2PO_4, 1.0 g; $MgSO_4 \cdot 7H_2O$, 0.2 g; distilled water 900 ml. Agar is added at the recommended concentration if a solid medium is required. After autoclaving, 100 ml of 20% (w/v) $Na_2CO_3 \cdot 10H_2O$, which has been sterilized by autoclaving separately, is added (at 60 °C if the medium contains agar); the final pH of the medium is 10.5, and the concentration of the Na_2CO_3 solution can be reduced to give lower pH values. Other carbon sources, such as starch, may be substituted for the glucose. Most alkaliphilic *Bacillus* species require Na^+ ions for growth, germination and sporulation (Horikoshi, 1998). Optimal growth of *Bacillus agaradhaerens* is at pH 10.0 or above, no growth occurs at pH 7.0, and up to 16% NaCl is tolerated; optimal growth of *Bacillus alcalophilus* is about pH 10.0, no growth occurs at pH 7.0, and 5–8% NaCl is tolerated (Nielsen et al., 1995a).

Bacillus algicola was isolated from 5 g seaweed allowed to decay in 200 ml sterilized natural sea water for 2 months at 22 °C, supplemented with and aqueous ethanol extract of a protein inhibitor of endo(1→3)-β-D-glucanases (Yermakova et al., 2002). 0.1 ml of suspension was then plated onto plates of Marine Agar 2216 (Difco) and on plates of medium B (peptone, 2.0 g; casein hydrolysate, 2.0 g; yeast extract, 1.0 g; glucose, 1.0 g; KH_2PO_4, 0.02 g; $MgSO_4 \cdot 7H_2O$, 0.05 g; agar, 15 g; natural sea water, 500 ml; distilled water at pH 7.5–7.8, 500 ml), which was also used as the maintenance medium.

Bacillus amyloliquefaciens. This organism has been isolated from a wide range of environments, including industrial amylase fermentations, foods and soil. It is phenotypically similar to *Bacillus subtilis.* Welker and Campbell (1967) isolated strains of this organism from commercial α-amylase concentrates as follows: 1 ml or 0.5 g of α-amylase concentrate is placed in a tube containing 5 ml of sterile distilled water and heated at 80 °C for 10 min.; the heated suspension is then streaked onto plates of tryptose blood agar base containing 1% starch; after incubation at 37 °C for 18–20 h, colonies showing wide haloes of starch hydrolysis are purified by restreaking on fresh plates of the same medium. The screening method of Effio et al. (2000), which uses dye-linked starch to reveal colonies of amylolytic bacteria, may also be of value.

Bacillus anthracis. For isolation of this organism from old carcasses, animal products or environmental specimens, the organisms will mostly be present as spores. The heat treatment recommended for isolating this species is 62.5 °C for 15 min; this will both heat-shock the spores to enhance germination, and effectively destroy nonspore-forming contaminants [solid samples should first be emulsified in sterile, deionized water, 1:2 (w/v)]. Direct plate cultures are made on blood, nutrient, or, selective agars, as appropriate, by spreading up to 250 μl volumes from undiluted, and 10- and 100-fold dilutions of the treated sample. There is no effective enrichment method for *Bacillus anthracis* in old animal specimens or environmental samples; isolation from these is best done with polymyxin-lysozyme EDTA-thallous acetate (PLET) agar (Knisely, 1966). Aliquots (250 μl) of the undiluted and 1:10 and 1:100 dilutions of heat-treated suspension of the specimen are spread across PLET plates which are read after incubation for 36–40 h at 37 °C. Roughly circular, creamy-white colonies, 1–3 mm diameter, with a ground-glass texture must be distinguished from those of other members of the *Bacillus cereus* group; they are subcultured on (i), blood agar plates to test for γ phage and penicillin susceptibility, and for

hemolysis, and (ii), directly or subsequently in blood to look for capsule production using M'Fadyean's stain. PCR-based methods are being used increasingly for confirming the identity of isolates (Turnbull et al., 1998).

Bacillus aquimaris and *Bacillus marisflavi* were isolated from sea water of a tidal flat by dilution plating onto marine agar (Yoon et al., 2003a).

Bacillus arseniciselenatis. The only reported isolate of this organism was isolated from the arsenic-rich sediment of Mono Lake, California. Lake sediment (15 ml) was added to 45 ml supplemented, autoclaved lake water and sealed under N_2 in a 160 ml serum bottle. Supplements were 10 mM sodium lactate, 10 mM Na_2SeO_4, 0.5 g yeast extract, in 1 l lake water, with 2.5% cysteine-sulfide reducing agent. The slurry was incubated statically in the dark at 20 °C for 18 d, by which time red deposits of elemental selenium were evident. A 1.0 ml aliquot was transferred to a smaller serum bottle containing 20 ml of the supplemented lake water and incubated in the same way for 1 week. A stable enrichment culture was achieved by weekly transfers into 10 ml medium in anaerobic, crimp-sealed test tubes for over 1 year. Manipulation was done in an anaerobic glove box using standard anaerobic methods. The enrichment was streaked onto solidified supplemented lake water medium and the plates incubated under N_2 for 1 week, after which red colonies were seen. Colonies were inoculated into further sealed test tubes of supplemented lake water, and non-spore-forming organisms were destroyed by heat treatment at 70 °C for 20 min (Switzer Blum et al., 1998).

Bacillus asahii was cultivated from soil using PY-1 medium (peptone, 8 g; yeast extract, 3 g; agar, if required, 15 g, distilled water, 1 l).

Bacillus atrophaeus is essentially a variant of *Bacillus subtilis* which forms a soluble black pigment on media containing utilizable carbohydrates. Most strains have been isolated from soil. No method for its specific enrichment or selective isolation are known to the authors, but its pigment formation is evidently of value in its recognition on plate cultures; it forms bluish-black colonies on glycerol-glutamate agar (Arai and Mikami, 1972).

Bacillus azotoformans. For selective isolation, soil samples are heat-treated (80 °C, 10 min) in peptone broth, and incubated under an atmosphere of pure N_2O at 32 °C; gassy and foaming cultures are subcultured several times in fresh medium under the same conditions. Pure cultures are obtained by streak plating from the enrichment cultures onto the same medium solidified with agar, and incubating at 32 °C in air (Pichinoty et al., 1976). The enrichment medium contains: peptone, 10 g; $Na_2HPO_4 \cdot 12H_2O$, 3.6 g; KH_2PO_4, 1.0 g; NH_4Cl, 0.5 g; $MgSO_4 \cdot 7H_2O$, 0.03 g; trace elements solution (Pichinoty et al., 1977), 0.2 ml; distilled water, 1 l; adjust pH to 7.0 and autoclave at 121 °C for 15 min.

Bacillus badius. No methods for the specific enrichment or selective isolation of this species are known to the authors.

Bacillus barbaricus was isolated from samples of an experimental wall painting by suspending in sterile saline, shaking for 1 h, then plating dilutions onto PYES agar and incubating at room temperature. PYES agar contained: peptone from casein, 3 g; yeast extract, 3 g; disodium succinate, 2.3 g; agar to solidify; distilled water, 1 l; pH 7.2 (Taubel et al., 2003).

Bacillus bataviensis was isolated from soil by plating two dilution series, one unheated and the other heated at 80 °C for

15 min, onto 35 different agar media, most of which were mineral media with different complex components as sole carbon sources (Felske et al., 1999; Heyrman et al., 2004).

Bacillus benzoevorans was originally isolated by aerobic enrichment culture of heat-treated soil samples at 32 °C for a week in a minimal medium containing benzoate, *p*-hydroxybenzoate or cyclohexane carboxylate as carbon and energy sources (Pichinoty, 1983). The enrichment medium comprises: Na$_2$HPO$_4$·12H$_2$O, 3.575 g; KH$_2$PO$_4$, 0.98 g; MgSO$_4$·7H$_2$O, 0.03 g; NH$_4$Cl, 0.5 g; trace element solution (Pichinoty et al., 1977), 0.2 ml; benzoate, *p*-hydroxybenzoate or cyclohexane carboxylate, 2 g; yeast extract (except in case of *p*-hydroxybenzoate), 0.1 g; distilled water to 1 l; pH adjusted to 7.0 with NaOH.

Bacillus carboniphilus was isolated from the air using a otherwise nonpermissive medium which had been spotted with sterilized graphite suspension (Fujita et al., 1996). The basal medium was Penassay Broth (Bacto Antibiotic Medium 3; Difco), the potassium and sodium levels of which were stressful to the organism: beef extract, 1.5 g; yeast extract, 1.5 g; peptone, 5.0 g; glucose, 1.0 g; NaCl, 3.5 g; K$_2$HPO$_4$, 3.68 g; KH$_2$PO$_4$, 1.32 g; agar 15 g; distilled water, 1 l. 200 μl of a thick suspension of graphite in 1% Triton X-100, sterilized by heating for 1 min at about 700 °C on a gas burner, was spotted on the surface of each agar plate, and plates were incubated at 44 °C for 48–96 h. Colonies appearing around the graphite spots were purified by streaking on trypticase soy agar plates.

Bacillus cereus. With food specimens submitted for the investigation of food-poisoning incidents, heating part of the specimen at 62.5 °C for 15 min will both heat-shock the spores and effectively destroy nonspore-forming contaminants [solid samples should first be emulsified in sterile, deionized water, 1:2 (w/v)]. The other part of the specimen is cultivated without heat treatment in case spores are very heat-sensitive, or absent. Direct plate cultures are made on blood, nutrient, or, selective agars, as appropriate, by spreading up to 250 μl volumes from undiluted, and 10-, 100- and 1000-fold dilutions of the treated sample. Heat treatment is not suitable for human specimens where spores are usually sparse or absent. Enrichment procedures are generally inappropriate for isolations from clinical specimens, but when searching for *Bacillus cereus* in stools ≥3 d after a food-poisoning episode, nutrient or tryptic soy broth with polymyxin (100,000 U/l) may be added to the heat-treated specimen. Several media have been designed for isolation, identification and enumeration of *Bacillus cereus*. They exploit the organism's egg-yolk reaction positivity and acid-from-mannitol negativity; pyruvate and polymyxin may be included for selectivity. Three satisfactory formulations are MEYP (Mannitol, Egg Yolk, Polymyxin B agar), PEMBA (Polymyxin B, Egg yolk, Mannitol, Bromthymol blue Agar) and BCM (*Bacillus cereus* Medium) (van Netten and Kramer, 1992).

Bacillus circulans. Although several methods for isolating this organism have been described (Claus and Berkeley, 1986), recent developments in the taxonomy of this species (which had long been known as a complex of species) indicate that very few of the strains that have been allocated to this taxon on the basis of phenotypic properties are actually closely related to the type strain. It is not possible, therefore, to recommend a method for its specific isolation.

Bacillus clarkii. Optimal growth of this alkaliphilic organism is above pH 10.0, no growth occurs at pH 7.0, and up to 16% NaCl

is tolerated (Nielsen et al., 1995a); see *Bacillus agaradhaerens, Bacillus alcalophilus* and other alcaliphiles.

Bacillus clausii. Optimal growth of this alkalitolerant organism is above pH 8.0, good growth occurs at pH 7.0, and 8–10% NaCl is tolerated (Nielsen et al., 1995a); see *Bacillus agaradhaerens, Bacillus alcalophilus* and other alcaliphiles.

Bacillus coagulans. Methods for the specific isolation of this species are based upon its thermotolerance and acid tolerance; it may be isolated from soil and other sources by enriching and then subculturing at pH 5.5–6.0 with incubation at around 50 °C. Allen (1953) used an enrichment medium of glucose, 10 g; yeast extract, 5.0 g; distilled water to 1 l, adjusted with lactic acid to pH 5.5 and autoclaved at 121 °C for 15 min.; flasks containing 200 ml are inoculated with 1 ml of heat-treated suspension of soil or other material and incubated for 24 h at 50 °C; 0.1 ml aliquots are then transferred to fresh medium and incubated for 24 h at 50 °C; loopfuls are streaked onto nutrient agar and incubated at 45 °C for 48 h. Small. round, whitish. opaque or opalescent colonies are likely to be *Bacillus coagulans*. A similar approach was used by Emberger (1970): heat-treated soil samples were incubated anaerobically at 54 °C in nutrient broth which had been supplemented with 1% (w/v) glucose and adjusted to pH 6.0. Loopfuls were then streaked onto plates of the same medium solidified with agar and incubated aerobically at 45 °C; this yielded colonies of *Bacillus coagulans* and *Bacillus licheniformis*.

Bacillus cohnii. Samples of alkaline or possibly alkaline soil from horse meadows, and from horse feces that have been lying for some time, or other potential sources, are heat treated at 80 °C for 10 min and 1–2 drop inocula are added to nutrient broth which has been adjusted to pH 9.7 by adding sodium sesquicarbonate (Na$_2$CO$_3$) to a final concentration of 0.1 mol/l (Spanka and Fritze, 1993). Cultures are incubated aerobically at 45 °C for 1–2 d, and streaked onto plates of the same medium solidified with agar, with incubation at 45 °C or 28–30 °C for 2–7 d. Small, cream-white, seldom-occurring colonies are picked for purification.

Bacillus decolorationis was isolated from scrapings of biofilms on ancient mural paintings by homogenizing the samples in physiological water, preparing a dilution series, and then plating on five different media: (i) trypticase soy broth (BBL) solified with Bacto agar (Difco); (ii) the aforementioned trypticase soy agar supplemented with 10% NaCl; (iii) R2A agar (Difco); (iv) R2A agar supplemented with 10% NaCl; (v) starch-casein medium, which contained starch, 10 g; casein, 0.3 g; KNO$_3$, 2 g; K$_2$HPO$_4$, 2 g; NaCl, 2 g; MgSO$_4$·7H$_2$O, 0.05 g; CaCO$_3$, 0.02 g; FeSO$_4$·7H$_2$O, 0.01 g; agar, 20 g; water, 1 l. All media were supplemented with 0.03% cycloheximide to inhibit fungal growth. Plates were incubated aerobically at 28 °C for 3 weeks (Heyrman et al., 1999).

Bacillus drentensis – see *Bacillus bataviensis*.

Bacillus endophyticus. Strains of this species were isolated from the inner tissues of healthy cotton plants (Reva et al., 2002). Pieces of stem about 1 cm in diameter were flamed with ethanol and the outer layers removed with a sterile scalpel; slices of inner stem were placed on nutrient agar plates and incubated for 48 h at 30 °C. Bacterial growth associated with the stems was then purified by repeated streak cultivation on plates of the same medium.

Bacillus farraginis, Bacillus fordii and *Bacillus fortis* were isolated from samples of raw milk, feed concentrate and green fodder, and samples talen from milking apparatus, by heat treating at 100 °C for 30 min followed by plating on brain heart infusion (Oxoid) supplemented with bacteriological agar no. 1 (Oxoid) and filter-sterilized vitamin B_{12} (1 mg/l) (Sigma).

Bacillus fastidiosus uses uric acid, allantoin and allantoic acid as sole carbon, nitrogen and energy sources, and this property may be used in its enrichment and isolation. The method of Fahmy (personal communication to Claus and Berkeley, 1986) uses the following enrichment medium: K_2HPO_4, 0.8 g; KH_2PO_4, 0.2 g; $MgSO_4 \cdot 7H_2O$, 0.5 g; $CaCl_2 \cdot 2H_2O$, 0.05 g; $FeSO_4 \cdot 7H_2O$, 0.015 g; $MnSO_4 \cdot H_2O$, 0.01 g; uric acid, 10 g; distilled water to 1 l. The pH is not adjusted. 300 ml Erlemneyer flasks containing 30 ml of this medium are inoculated with 5 ml of a soil suspension in water that has been heat treated for 5 min at 80 °C, and the inoculated flasks are shaken at 30 °C for 24 h. One milliliter quantities are transferred to fresh flasks of medium for further incubation under the same conditions. Loopfuls, or 0.1 ml quantities, of serial dilutions of the enrichment culture are streaked or spread on plates of the following medium: a layer of the medium described above, from which the uric acid has been omitted but which has been solidified with agar, is overlaid with uric acid agar; the latter is prepared by sterilizing a 3% (w/v) agar solution and a 2% (w/v) uric acid suspension separately and then combining equal parts and pouring to a depth of about 3 mm. Colonies of uric acid-degrading organisms produce clear haloes in the milky layer of uric acid agar, owing to utilization of the acid and its solubilization in the rising pH caused by splitting of the urea that is formed from the uric acid. *Bacillus fastidiosus* colonies usually show rhizoid outgrowths on this medium. Suspect colonies are suspended in nutrient broth and streaked on uric acid agar and nutrient agar; isolates developing only on the former medium are likely to be this species, and they may be purified on plates of the mineral agar to which 2% (w/v) allantoin has been added before autoclaving.

Bacillus firmus. Methods for isolating members of this species complex, particularly pigmented organisms from salt marshes and marine environments, have been described (Claus and Berkeley, 1986), but none of them are specific, and recent developments in the taxonomy of the group have revealed that few authentic strains of *Bacillus firmus* originate from salty environments or are strongly pigmented; most strains have been isolated from soil, as laboratory contaminants, and as contaminants of food and pharmaceutical production environments. Until the species is more tightly defined, it is not possible to recommend a method for its specific isolation.

Bacillus flexus was revived by Priest et al. (1988) to accommodate two strains that showed low DNA homology to *Bacillus megaterium* but which showed phenotypic similarity to that species. No method has been described for its specific isolation.

Bacillus fordii and *Bacillus fortis* – see *Bacillus farraginis*.

Bacillus fumarioli. Strains of this species were isolated from soil samples that were collected from geothermal sites in the Antarctic and transported and stored both chilled and frozen. 1 g quantities of soil are added to 9 ml *Bacillus fumarioli* broth (BFB) in duplicate at pH 5.5, and one of each pair is heat treated at 80 °C for 10 min to kill vegetative cells. Broths are incubated at 50 °C in air (bottles loose-capped), and subcultured by streaking onto *Bacillus fumarioli* agar (BFA, which is BFB containing 5 mg/l $MnSO_4 \cdot 4H_2O$, to enhance sporulation, and 18 g/l agar) and incubated at 50 °C (with the plates in loosely closed polythene bags to avoid drying-out) for 1–2 d (Logan et al., 2000). BFB is an adaption of medium B of Nicolaus et al. (1998) and contains yeast extract, 4 g; $(NH_4)_2SO_4$, 2 g; KH_2PO_4, 3 g and 4 ml each of solutions A and B in distilled water, 1 l; adjust to pH 5.5. Solution A: $(NH_4)_2SO_4$, 12.5 g; $MgSO_4 \cdot 7H_2O$, 5.0 g; distilled water, 100 ml; Solution B, $CaCl_2 \cdot 2H_2O$, 6.25 g; distilled water, 100 ml. Growth and sporulation may be enhanced by enriching and subculturing on 1/2 BFB and 1/2 BFA, in both of which all components excepting the water, and the agar in the latter, are reduced by half. Using the same medium and conditions, further isolates have been obtained from gelatin production plants in France, Belgium and the USA (De Clerck et al., 2004a).

Bacillus funiculus was described on the basis of a single isolate from a domestic wastewater treatment tank (Ajithkumar et al., 2001). Water from a sludge-circulating tank was diluted to 10^4 in 0.5% NaCl solution and plated onto the following medium: nutrient broth (Oxoid CM-1), 4 g; potato starch, 5 g; glucose, 8 g; NaCl, 5 g; yeast extract, 0.5 g; agar, 15 g; distilled water, 1 l. Plates were incubated at 32 °C for 24 h, and then at 20 °C for 1 week, and colonies of this species were white, opaque, round and umbonate.

Bacillus fusiformis was revived for four strains, three of which were urease-positive, that had been assigned to *Bacillus sphaericus*. No method for its specific enrichment or isolation have been described, but urease positivity may be of value when screening colonies from *Bacillus sphaericus* isolations (see below); however, other spherical-spored organisms, including mosquito pathogenic strains of *Bacillus sphaericus* (Krych et al., 1980) and *Sporosarcina psychrophila* (formerly *Bacillus psychrophilus*), are also urease-positive.

Bacillus galactosidilyticus. Two strains of this species were isolated from raw milk after a heat treatment of 80 °C for 10 min in one case, and 100 °C for 30 min in the other, in order to pasteurize the samples and activate any endospores present; samples were then plated on brain heart infusion (BHI) (Oxoid) solidified with Bacteriological Agar no. 1 (15 g/l) (Oxoid) and supplemented with filter-sterilized vitamin B_{12} (1 mg/l), incubating at 37 °C for 48 h (Heyndrickx et al., 2004).

Bacillus gelatini. Strains were isolated from samples of gelatin batches from a gelatin production plant by enrichment of 30 g of sample in 70 ml Trypticase Soy Broth (Oxoid) at 45 °C and 55 °C for 24 h, and plating on Trypticase Soy Agar (Oxoid) and Brain heart Infusion Agar (BBL), supplemented with 1 mg vitamin B_{12}/l and Nutrient Agar supplemented with 1.2% gelatin at 45 °C and 55 °C (De Clerck et al., 2004c).

Bacillus gibsonii. Optimal growth of this alkalitolerant organism is about pH 8.0, growth occurs at pH 7.0, and up to 9% NaCl is tolerated (Nielsen et al., 1995a); see *Bacillus agaradhaerens*, *Bacillus alcalophilus* and other alcaliphiles.

Bacillus halmapalus. Optimal growth of this alkalitolerant organism is about pH 8.0, and good growth occurs at pH 7.0, but 5% NaCl is not tolerated (Nielsen et al., 1995a); see *Bacillus agaradhaerens*, *Bacillus alcalophilus* and other alcaliphiles.

Bacillus halodurans. Optimal growth of this alkaliphilic organism is pH 9.0–10.0, most strains grow at pH 7.0, and good growth is obtained at up to 12% NaCl (Nielsen et al., 1995a); see *Bacillus agaradhaerens*, *Bacillus alcalophilus* and other alcaliphiles.

Bacillus halophilus. This species is based upon one strain isolated from rotting wood on a seashore (Ventosa et al., 1989) by enrichment in Sehgal and Gibbons complex medium

(Onishi et al., 1980): NaCl, 234 g; vitamin-free Casamino acids (Difco), 7.5 g; yeast extract, 10 g; sodium citrate, 3 g; KCl, 2 g; $MgSO_4 \cdot 7H_2O$, 2 g; $FeCl_3 \cdot nH_2O$, 2.3 mg; distilled water 1 l; pH 6.6), followed by inoculation onto 15% (w/v) salt MH medium: yeast extract, 10 g; proteose peptone, 5 g; glucose, 1 g; agar, 20 g; 15% (w/v) NaCl solution 1 l.

Bacillus horikoshii. Optimal growth of this alkalitolerant organism is about pH 8.0, growth occurs at pH 7.0, and 8–9% NaCl is tolerated (Nielsen et al., 1995a); see *Bacillus agaradhaerens*, *Bacillus alcalophilus* and other alcaliphiles.

Bacillus horti. This species is based upon two strains isolated from garden soil in Japan using an alkaline peptone-yeast extract agar (Yumoto et al., 1998); details of sample preparation and inoculation were not reported. The isolation medium comprised: peptone, 8 g; yeast extract, 3 g; K_2HPO_4, 1 g; $FeSO_4 \cdot 7H_2O$, 10 mg; EDTA, 3.5 mg; $ZnSO_4 \cdot 7H_2O$, 3 mg; $Co(NO_3)_2 \cdot 6H_2O$, 2 mg; $MnSO_4 \cdot nH_2O$, 2 mg; $CuSO_4 \cdot 5H_2O$, 1 mg; H_3BO_3, 1 mg; agar, 15 g; $NaHCO_3 / Na_2CO_3$ buffer (100 mM; pH 10) in deionized water, 1 l.

Bacillus hwajinpoensis was isolated from sea water of the East Sea in Korea by dilution plating onto marine agar (Yoon et al., 2004b).

Bacillus indicus was isolated from a sand sample from an arsenic-contaminated aquifer by cultivation on nutrient agar containing 5% sodium arsenate. Nutrient agar without sodium arsenate was used for growth and maintenance.

Bacillus infernus. This anaerobic species was isolated from shale taken from a depth of about 2.7 km below the land surface, where conditions were estimated to be anoxic, thermic (60 °C), and brackish (1.2% NaCl). Core sample material was placed in an inert atmosphere, pared to remove potential surface contamination, and transferred to the laboratory on ice. Enrichment was done by adding 50 mg pieces of rock to MSA medium (Boone et al., 1995) and incubating for 40 d at 50 °C in pure nitrogen at pH 8.2. The medium comprised: yeast extract, 0.5 g; peptone, 0.5 g; NaCl, 10 g; NaOH, 4 g; NH_4Cl, 1 g; $MgCl_2 \cdot 6H_2O$, 1 g; K_2HPO_4, 0.4 g; $CaCl_2 \cdot 2H_2O$, 0.4 g; $Na_2S \cdot 9H_2O$, 250 mg; disodium $EDTA \cdot 2H_2O$, 5 mg; $CoCl_2 \cdot 6H_2O$, 1.5 mg; resazurin, 1 mg, $MnCl_2 \cdot 4H_2O$, 1 mg; $FeSO_4 \cdot 7H_2O$, 1 mg; $ZnCl_2$, 1 mg; $AlCl_3 \cdot 6H_2O$, 0.4 mg; $Na_2WO_4 \cdot 2H_2O$, 0.3 mg; $CuCl_2 \cdot 2H_2O$, 0.2 mg, $NiSO_4 \cdot 6H_2O$, 0.2 mg; H_2SeO_3, 0.1 mg; H_3BO_3, 0.1 mg; $Na_2MoO_4 \cdot 2H_2O$, 0.1 mg; 20 mM formate, 20 mM acetate, 20 mM MnO_2 (MnO_2 prepared by mixing a 9.5% $KMnO_4$ solution with an equal volume of a 17.8% $MnCl_2 \cdot 4H_2O$ solution) in deionized water, 1 l. Ingredients excepting sulfide were added together and solution equilibrated under pure N_2; the medium was dispensed into bottles sealed to exclude air. Sulfide was added from O_2-free stock solution 1 h before use. The enrichment culture was serially diluted and inoculated onto MSA medium lacking the NaCl, but solidified with 18 g agar per liter, in roll tubes. Pinpoint colonies became visible in the zones of clearing of the MnO_2 in 3–4 weeks, and they were picked, suspended in MSA medium, serially diluted and reinoculated into the roll tube medium to purify.

Bacillus insolitus. This psychrophilic organism, along with other organisms, was isolated from marshy and normal soil by enrichment in trypticase soy broth at 0 °C for 2 weeks; cultures showing turbidity were streaked on trypticase soy agar and incubated at 0 °C for 2 weeks, and isolates were purified on nutrient agar (Larkin and Stokes, 1966). Cells of this species appear coccoid when grown on nutrient agar, but rod-shaped on richer media such as trypticase soy agar (Larkin and Stokes, 1967).

Bacillus jeotgali. Strains of this species were isolated by dilution plating of jeotgal, a Korean traditional fermented seafood, on trypticase soy agar supplemented with artificial sea water (per liter: NaCl, 24 g; $MgSO_4 \cdot 7H_2O$, 7 g; $MgCl_2 \cdot 6H_2O$, 5.3 g; KCl, 0.7 g; $CaCl_2$, 0.1 g) at pH 7.5 and incubated at 30 °C. Colonies were cream-yellow or light orange-yellow, smooth, flat and irregular (Yoon et al., 2001a).

Bacillus krulwichiae was isolated from garden soil contaminated with aromatic compounds. Soil samples were added to alkaline mineral basal salt medium (AMBS) and incubated aerobically for 48 h at 30 °C; 0.5 ml amounts were transferred to fresh medium and incubated for 24 h, then plated onto ABMS agar plates, purified five times, and maintained on peptone-yeast extract-alkaline (PYA) agar at 27 °C. AMBS contained: yeast extract, 0.2 g; hydroxybenzoate, 3 g; NH_4NO_3, 2.5 g; K_2HPO_4, 1.5 g; Na_2HPO_4, 1.5 g; $MgSO_4 \cdot 7H_2O$, 0.5 g; $CaCl_2 \cdot 2H_2O$, 20 mg; $FeSO_4 \cdot 7H_2O$, 10 mg; $MnSO_4 \cdot nH_2O$, 1 mg; $ZnSO_4 \cdot 7H_2O$, 0.5 mg; Na_2CO_3, 10 g; distilled water, 1 l; pH 10. PYA agar contained: peptone, 8 g; yeast extract, 3 g; K_2HPO_4, 1 g; $FeSO_4 \cdot 7H_2O$, 10 mg; EDTA, 3.5 mg; $ZnSO_4 \cdot 7H_2O$, 3 mg; $Co(NO_3)_2 \cdot 6H_2O$, 2 mg; $MnSO_4 \cdot nH_2O$, 2 mg; $CuSO_4 \cdot 5H_2O$, 1 mg; H_3BO_3, 1 mg; $NaHCO_3 / Na_2CO_3$ buffer (100 mM in deionized water; pH 10), 1 l (Yumoto et al., 2003).

Bacillus lentus. Gibson originally isolated this species from soil by plating dilutions of soil suspensions on nutrient agar supplemented with 10% urea (peptone, 10 g; meat extract, 10 g; agar, 15 g; distilled water to 1 l; adjusted to pH 7.0–7.5 and autoclaved at 121 °C for 15 min; 100 g crystalline urea is added to molten agar immediately before use and steamed for 10 min prior to cooling and pouring) and incubating at 25 °C. However, recent work on the taxonomy of the species (Logan and De Vos, unpublished observations) has revealed that few authentic strains of *Bacillus lentus* are available for study and that many unrelated strains have been assigned to the species. Until the species is more tightly defined, it is not possible to recommend a method for its specific isolation.

Bacillus licheniformis. Strains of this species may be obtained from soil by anaerobic enrichment in peptone-meat extract-KNO_3 medium (Claus, 1965): soil suspensions are heat treated at 80 °C for 10 min, and 2 ml quantities are added to glass-stoppered bottles of a medium containing peptone, 5 g; meat extract, 3 g; KNO_3, 80 g; distilled water 1 l; pH 7.0. The bottles are filled completely, without trapping bubbles of air, and incubated at 40–45 °C for 48 h, whereupon most will show turbidity and gas production. Loopfuls are streaked onto plates of glucose mineral base agar: glucose, 10 g; $(NH_4)_2SO_4$, 1 g; K_2HPO_4, 0.8 g; KH_2PO_4, 0.2 g; $MgSO_4 \cdot 7H_2O$, 0.5 g; $CaSO_4 \cdot 2H_2O$, 0.05 g; $FeSO_4 \cdot 7H_2O$, 0.01 g; agar, 12 g; distilled water to 1 l; adjust to pH 6.8. Colonies are often reddish and have lobes and mounds of slime.

Bacillus luciferensis. Strains of this species were isolated from soil samples that were collected from a geothermal site on Candlemas Island, South Sandwich archipelago, and transported and stored both chilled and frozen. 1 g quantities of soil were added to 9 ml trypticase soy broths in duplicate at pH 6.5 and one of each pair was heat treated at 80 °C for 10 min to kill vegetative cells. Spread plates were inoculated with 0.1 ml soil suspension on trypticase soy agar at pH 6.5 (containing 5 mg/l

$MnSO_4$ to enhance sporulation) and incubated at 30 °C. The suspensions remaining were incubated at the same temperature in a waterbath, and streaked onto the solid medium as soon as they became turbid (Logan et al., 2000).

Bacillus macyae was isolated from arsenic-contaminated mud from a gold mine, as described by Santini et al. (2002). The medium for *Bacillus macyae* is NaCl, 1.2 g; KCl, 0.3 g; NH_4Cl, 0.3 g; KH_2PO_4, 0.2 g; Na_2SO_4, 0.3 g; $MgCl_2·6H_2O$, 0.4 g; $CaCl_2·2H_2O$, 0.15 g; $NaHCO_3$, 0.6 g; trace element solution with 5.2 g/l Na_2-EDTA and pH 6.5, 1.00 ml; resazurin, 0.5 mg; yeast extract, 0.8 g; Na-lactate, 1.1 g; KNO_3, 0.5 g; distilled water 1000.00 ml. Prepare the medium anaerobically under 100% N_2. Add sodium lactate (10 mM; electron donor) and sodium nitrate (or sodium arsenate; electron acceptors) to 5 mM from sterile, anaerobic stock solutions. Adjust final medium pH to 7.4–7.8. The trace element solution contains: HCl (25%; 7.7 M); $FeCl_2·4H_2O$, 1.5 g; $CoCl_2·6H_2O$, 190 mg; $MnCl_2·4H_2O$, 100 mg; $ZnCl_2$, 70 mg; H_3Bo_3, 6 mg; $Na_2MoO_4·2H_2O$, 36 mg; $CuCl_2·2H_2O$, 2 mg; $NiCl_2·6H_2O$, 24 mg; distilled water, 990 ml; dissolve the $FeCl_2$ in the HCl, then dilute in water, add and dissolve other salts, and make up to 1 l. Incubate at 28 °C.

Bacillus marisflavi – see *Bacillus aquimaris*.

Bacillus megaterium. The method of Claus (1965) may be used to isolate strains of this species from soil. Plate 0.1 ml volumes of dilutions of heat-treated soil suspensions on glucose mineral base agar: glucose, 10 g; $(NH_4)_2SO_4$ or KNO_3, 1 g; K_2HPO_4, 0.8 g; KH_2PO_4, 0.2 g; $MgSO_4·7H_2O$, 0.5 g; $CaSO_4·2H_2O$, 0.05 g; $FeSO_4·7H_2O$, 0.01 g; agar, 12 g; distilled water to 1 l; adjust to pH 7.0. Plates are incubated at 30 °C. On the nitrate medium, white, round, smooth and shiny colonies 1–3 mm in diameter may develop in 36–48 h. On the ammonium medium (necessary, because not all strains can use nitrate), a variety of colonies may develop in 24–36 h, but colonies of *Bacillus megaterium* can be detected by their appearance. Suspect colonies from either medium should be observed microscopically for the typically large cells of this species, then purified on nutrient agar or trypticase soy agar.

Bacillus methanolicus. Dijkhuizen et al. (1988) isolated their thermotolerant, methanol-utilizing strains using the following enrichment medium: filter-sterilized methanol to 50 mMol; yeast extract, 0.5 g; Casamino acids, 0.5 g; peptone, 0.5 g; $(NH_4)_2SO_4$, 1.5 g; K_2HPO_4, 4.65 g; NaH_2PO_4 H_2O, 1.5 g; $MgSO_4·7H_2O$, 0.2 g; trace element solution, 0.2 ml; vitamin solution, 1 ml; distilled water, 1 l; the medium was adjusted to pH 7.0. The trace element solution contained: EDTA, 50 g; $ZnSO_4·7H_2O$, 22 g; $CaCl_2$, 5.54 g; $MnCl_2·4H_2O$, 5.06 g; $FeSO_4·7H_2O$, 4.99 g; $(NH_4)_6Mo_7O_{24}·4H_2O$, 1.10 g; $CuSO_4·5H_2O$, 1.57 g; $CoCl_2·6H_2O$, 1.61 g; distilled water, 1 l; pH adjusted to 6.0 with KOH. The vitamin solution contained biotin, 100 mg; thiamin, HCl, 100 mg; riboflavin, 100 mg; pyridoxalphosphate, 100 mg; pantothenate, 100 mg; nicotinic acid amide, 100 mg; *p*-aminobenzoic acid, 20 mg; folic acid, 10 mg; vitamin B12, 10 mg; lipoic acid, 10 mg; distilled water, 1 l. To conical flasks containing 25 ml of the medium, 1–5 ml of liquid samples or 1–5 g of soil samples (both previously heat treated at 80 °C for 10 min, the soil samples suspended in 5 ml mineral medium) were added, and incubated at 50–55 °C. Dense growth usually appeared within 72 h, and this was subcultured onto methanol agar plates (the enrichment medium solidified with 1.5% agar), which were incubated at 50 °C. Colonies were transferred to the liquid medium after 24 h incubation. Cells apparently lysed rapidly on plates, and

so repeated transfers (about 10) to liquid culture, followed by plating, were used to select variants less susceptible to lysing on solid medium. Cultures were then serially diluted in mineral medium and plated on methanol agar.

Bacillus mojavensis. This species is phenotypically virtually indistinguishable from *Bacillus subtilis*. Original strains were isolated from desert soil using a non-specific method for *Bacillus* species (see *Bacillus subtilis*, below), and the emerging colonies screened for the *Bacillus subtilis* phenotype (Roberts et al., 1994). See also *Bacillus subtilis*.

Bacillus mycoides. Strains of this species are readily isolated from a wide variety of sources, including soil, and are easily recognized by the characteristic rhizoid morphologies of their colonies (Figure 9d). Small drops of heat-treated or untreated soil suspensions (or soil crumbs, or appropriate preparations of other source material) are placed in the centers of nutrient or trypticase soy agar plates and incubated at 28 °C. A *Bacillus mycoides* colony will develop as rhizoid growth which may spread to cover the agar surface within 2–3 d of incubation. Purification is best attempted by subculturing from the outer edge of the colony, and is aided by suspending growth (which may be quite adherent) in a nutrient broth, shaking vigorously, and then serially diluting prior to plating.

Bacillus naganoensis. The single original strain of this species was isolated on a medium containing colored starch and pullulan, designed to screen for organisms producing α-amylases and pullulanases (Tomimura et al., 1990). The initial screening medium contained: tryptone, 2 g; $(NH_4)_2SO_4$, 1 g; KH_2PO_4, 0.3 g; $MgSO_4·7H_2O$, 0.2 g; $CaCl_2·2H_2O$, 0.2 g; $FeSO_4·7H_2O$, 0.01 g; $MnCl_2·4H_2O$, 0.001 g; agar, 20 g; soluble starch, 10 g; blue-colored soluble starch (Rinderknecht et al., 1967), 3 g; red-colored pullulan, 7.5 g; distilled water to 1 l; adjust to pH 4.0 using 0.2 N sulfuric acid. The colored pullulan was prepared by dissolving 100 g pullulan in 2 l distilled water, heating to 50 °C and adding 10 g Brilliant Red and 100 ml of 10% Na_3PO_4; after 75 min incubation the product was precipitated with 1600 ml 99.5% ethanol, collected by decantation, washed twice with 60% ethanol, washed once with 99.5% ethanol, and air-dried (Tomimura et al., 1990). A 2 g portion of soil sample was suspended in 10 ml water and a 0.1 ml volume spread onto the screening medium and incubated at 30 °C. Discrete colonies surrounded by blue zones (indicating pullulan, but not starch, hydrolysis), were subcultured onto a variant of the screening medium in which amylopectin (10 g/1 l) replaced the starches and pullulan; following incubation, pullulanase producing colonies were revealed by the dark blue zones surrounding them, against a light-purple background, when exposed to iodine vapor.

Bacillus nealsonii was isolated from a spacecraft-assembly plant by exposing 2.5 × 5 cm (0.05–0.08 cm thick) stainless steel witness plates, which had been cleaned by ultrasonication and solvent treatment, then sterilized by heating at 175 °C for 2 h, on 2-m-high stands for 9 months. Each plate was placed in 30 ml sterile phosphate-buffered rinse solution (pH 7.2) and sonicated (25 kHz, 0.35 W/cm²) for 2 min. Each rinse was divided into two equal parts and one was heat treated at 80 °C for 15 min, the other not. Total aerobic counts were determined in tryptic soy agar pour plates, incubated at 32 °C for 3–7 d (Venkateswaran et al., 2003).

Bacillus neidei. This species contains soil isolates previously allocated to *Bacillus sphaericus*, and is distinguished from that species by a small number of phenotypic characters. Unlike *Bacillus sphaericus*, it has a requirement for cystine (Nakamura et al., 2002). No method for its specific enrichment or isolation has been described. See *Bacillus sphaericus*.

Bacillus niacini. Strains of this and other nicotinate-utilizing species were recognized by their production of blue or brown haloes on nicotinate medium. About 1 g of sample, which may have been heat treated (60 °C or 80 °C for 10 min), was suspended in 20 ml nicotinate medium (Nagel and Andreesen, 1989) and in 20 ml of the same medium supplemented with 0.05% yeast extract. This medium comprised: nicotinic acid or 6-hydroxynicotinic acid, 40 mM; $CaCl_2$, 0.18 mM; $MnSO_4$, 0.14 mM; $MgSO_4$, 2.0 mM; NH_4Cl, 5,6 mM; NaCl, 0.85 mM; $FeSO_4$, 0.036 mM; potassium phosphate buffer (0.6 M, pH 7.5), 5 ml; Tris/HCl (1.0 M, pH 7.5), 100 ml; trace element solution, 1 ml; filter-sterilized vitamin solution, 5 ml; distilled water to 1 l; pH adjusted to 7.5. After 3–5 d incubation at 30 °C on a rotary shaker, aliquots are plated onto nicotinate-yeast extract medium solidified with agar. The trace element solution contained: HCl (25% = 7.7 M), 12.5 ml; $FeSO_4 \cdot 7H_2O$, 2.1 g; H_3BO_3, 0.03 g; $MnCl_2 \cdot 4H_2O$, 0.1 g; $CoCl_2 \cdot 6H_2O$, 0.19 g; $NiCl_2 \cdot 6H_2O$, 0.024 g; $CuCl_2 \cdot 2H_2O$, 0.002 g; $ZnSO_4 \cdot 7H_2O$, 0.144 g; $Na_2MoO_4 \cdot 2H_2O$, 0.036 g; distilled water, 987 ml; autoclaved in sealed bottles with about ⅓ head space of air. The vitamin solution comprised: lipoic acid, 60 mg; *p*-aminobezoic acid, 50 mg; calcium-D-pantothenate, 50 mg; cyanocobalamin, 50 mg; nicotinic acid, 50 mg; riboflavin, 50 mg; thiamine hydrochloride, 50 mg; biotin, 20 mg; folic acid, 20 mg; pyridoxal hydrochloride, 10 mg; distilled water,1 l. Colonies with blue or brown haloes were streaked on the same medium to purify, then single colonies were suspended in saline and plated onto tryptone-yeast extract-succinate medium (tryptone, 10 g; yeast extract, 5 g; disodium succinate, 2 g; agar, 20 g; distilled water, 1 l; pH 7.5) or other complex medium, allowed to sporulate, and subjected to heat treatment then further cultivation on the same medium (Nagel and Andreesen, 1991).

Bacillus novalis – see *Bacillus bataviensis*.

Bacillus odysseyi. The description of this species (La Duc et al., 2004) indicates that the two isolates were isolated from two spa waters in Japan; one at 59 °C and pH 6.4 and the other at 51 °C and pH 8.1; both isolates showed optimal growth between 45 °C and 50 °C at pH 10.5 in heart infusion broth.

Bacillus okuhidensis. The description of this species (Li et al., 2002) indicates that the two isolates were isolated from two spa waters in Japan; one at 59 °C and pH 6.4 and the other at 51 °C and pH 8.1; both isolates showed optimal growth between 45 °C and 50 °C at pH 10.5 in heart infusion broth.

Bacillus oleronius. The original isolate of this species was cultivated from the hindgut contents of a termite (*Reticulitermes santonensis*) using the following enrichment medium: yeast extract, 1 g; vitamin and trace element solution (Balch et al., 1979), 10 ml; distilled water,1 l; NaCl, 24.1 mM; KCl, 21.5 mM; K_2HPO_4, 10.8 mM; KH_2PO_4, 6.9 mM; $MgSO_4 \cdot 5.3$ mM; $CaCl_2$, 0.53 mM; adjusted to pH 7.2; after sterilization 1 mM of each of the following aromatic substrates were added: benzoic acid, coumaric acid, ferulic acid, 4-hydroxybenzoixc acid and vanillic acid. After enrichment, the organism was maintained on trypiticase soy agar, with monthly transfers. Further strains of this species have been isolated from animal feed concentrate, raw milk and dairy plant using a method devised for the isolation of *Bacillus sporothermodurans* (Scheldeman et al., 2002): samples

were heat treated for 30 min at 100 °C, followed by plating on brain heart infusion broth solidified with bacteriological agar and supplemented with vitamin B_{12} (1 mg/l, filter-sterilized); plates were incubated for 48 h at 37 and 55 °C, but 37 °C is the optimal temperature for growth and isolation of this species.

Bacillus pseudalcalophilus. Optimal growth of this alkaliphilic organism is about pH 10.0, no growth occurs at pH 7.0, and up to 10% NaCl is tolerated (Nielsen et al., 1995a); see *Bacillus agaradhaerens*, *Bacillus alcalophilus* and other alcaliphiles.

Bacillus pseudofirmus. Optimal growth of this alkaliphilic organism is about pH 9.0, most strains do not grow at pH 7.0, and 16–17% NaCl is tolerated (Nielsen et al., 1995a); see *Bacillus agaradhaerens*, *Bacillus alcalophilus* and other alcaliphiles.

Bacillus pseudomycoides is indistinguishable from *Bacillus mycoides* by conventional characters, and will probably be isolated along with strains of that species. It was proposed on the basis of differences in fatty acid composition and DNA relatedness between strains of *Bacillus mycoides* (Nakamura, 1998). See *Bacillus mycoides*.

Bacillus psychrosaccharolyticus. This psychrophilic organism was isolated by the same methods as were used for *Bacillus insolitus*. Cells of this species appear granular when grown on nutrient agar and lightly stained, and are larger and vacuolate when grown on glucose agar (Larkin and Stokes, 1967).

Bacillus pumilus. There is no specific method of enrichment or isolation of this species that is known to the authors; however, Knight and Proom (1950) found that when their suspensions of soil in distilled water were incubated at 37 °C for 3 d, then plated on nutrient agar and incubated at 37 °C, the mixed collection of colonies arising included strains of this species.

Bacillus pycnus. Like *Bacillus neidei*, this species contains soil isolates previously allocated to *Bacillus sphaericus*, and is distinguished from that species by a small number of phenotypic characters. Unlike *Bacillus sphaericus*, it does not have a requirement for biotin and thiamin (Nakamura et al., 2002). No method for its specific enrichment or isolation has been described. See *Bacillus sphaericus*.

Bacillus schlegelii. Aragno (1978) isolated this thermophilic, hydrogen-oxidizing organism by adding superficial samples (0.5 g) of sediment from a eutrophic lake in Switzerland to 100 ml bottles containing 20 ml of a basal mineral medium and Bonjour et al. (1988) made further isolations from the same lake, from air and from geothermal sites in Iceland and Italy using this three-part medium: Solution 1: $Na_2HPO_4 \cdot 2H_2O$, 4.5 g; KH_2PO_4, 1.5 g; NH_4Cl, 1 g; $MgSO_4 \cdot 7H_2O$, 0.2 g; trace elements solution, 1 ml; distilled water, 1 l; Solution II: $CaCl_2 \cdot 2H_2O$, 100 mg; ferric ammonium citrate, 50 mg; distilled water, 100 ml; Solution III: $NaHCO_3$, 5 g; distilled water, 100 ml; the three solutions were autoclaved separately and mixed in the proportions I, 1 l; II, 10 ml; III, 10 ml after cooling; pH 7.0. The trace elements solution contained: H_3BO_3, 300 mg; $CoCl_2 \cdot 6H_2O$, 200 mg; $ZnSO_4 \cdot 7H_2O$, 100 mg; $MnCl_2 \cdot 4H_2O$, 30 mg; $Na_2MoO_4 \cdot 2H_2O$, 30 mg; $NiCl_2 \cdot 6H_2O$, 20 mg; $CuCl_2 \cdot 2H_2O$, 10 mg; distilled water, 1 l. Cultures were incubated in desiccators at 65 °C under an atmosphere of 0.05 atm O_2 + 0.1 atm CO_2 + 0.45 atm H_2 (partial pressure measured at room temperature). Dense growth was apparent after 4 d, and it was twice subcultured in fresh medium using loopful inocula. Pure cultures were obtained by streaking on plates of the same medium solidified with agar and incubating in the same conditions as before.

Further strains have been obtained by Krüger and Meyer (1984) from sludge samples of a sugar factory settling pond, by, and by Hudson et al. (1988) from geothermally heated Antarctic soil. Krüger and Meyer used carbon monoxide as the sole carbon and energy source for enrichment, and cultivated at 65 °C in the mineral medium of Meyer and Schlegel (1983) under an atmosphere of (v/v) 5% CO_2, 35% CO and 60% air. After 1 week of incubation, positive enrichments were serially subcultured in fresh medium using a 10% inoculum. Hudson et al. (1988) isolated their strain from a soil sample taken from Mount Erebus, Antarctica. The temperature of the sample site was 37 °C, and the sample was kept unfrozen; 1 g was added to 100 ml of a thiosulfate medium: $Na_2S_2O_3$, 10 g; $NaHCO_3$, 2 g; $NaNO_3$, 0.689 g; NH_4Cl, 0.4 g; Na_2HPO_4, 0.111 g; KNO_3, 0.103 g; $MgSO_4 \cdot 7H_2O$, 0.1 g; $CaSO_4 \cdot 2H_2O$, 0.06 g; NaCl, 0.008 g; nitrilotriacetic acid, 0.1 g; (also, for later work, yeast extract, 0.1 g); phenol red, 0.024 g; distilled water to 1 l; pH 7.0. Trace elements and ferric chloride, and vitamins were also added. Incubation was at 60 °C, and growth was evident from a fall in pH shown by the phenol red indicator. pure cultures were obtained by streaking on the same medium (without yeast extract) solidified with agar.

Bacillus selenitireducens. The single reported strain of this medium was isolated using the same method employed for *Bacillus arseniciselenatis* (see above), but it was found not to use selenate as the electron acceptor and could use arsenate in the lake water medium or selenite instead (Switzer Blum et al., 1998).

Bacillus shackletonii – see *Bacillus luciferensis.*

Bacillus silvestris. The single strain upon which this species is based was isolated from a 10 g sample of beech forest soil by storing it in humid conditions and incubating at 40 °C for 15 min followed by suspension in 100 ml of a germination solution comprising: malt extract, 1 g; glucose, 0.4 g; yeast extract, 0.4 g; $CaCO_3$, 0.2 g; L-asparagine, 1 g; distilled water, 1 l; filter-sterilized cycloheximide solution (0.1 g in 10 ml water) was added after cooling to room temperature (Rheims et al., 1999). Dilutions in the germination medium were prepared to 10^{-3} and this final dilution was stirred at room temperature for 15 min and then aliquots of 100 μl were streak diluted on various media and incubated at different temperatures. The strain was isolated on a plate of tryptic soy broth supplemented with 0.3% yeast extract and solidified with agar; incubation was at 25 °C for 1 d.

Bacillus simplex. This species is based on two soil isolates received as "*Bacillus simplex*" and "*Bacillus teres*" (Priest et al., 1988), and no method for isolating strains of the species is known to the authors.

Bacillus siralis. The original strain of this species was isolated from silage by heat treating the sample at 100 °C for 60 min and plating onto brain heart infusion (BHI) agar (de Silva et al., 1998). Pettersson et al. (2000) made further isolations by suspending 1 g of lyophilized silage in 1 ml sterile water, heating at 100 °C for 90 min, then diluting to 10^{-2} in sterile water; 50 ml aliquots were spread on plates of BHI agar and on plates of the following medium: yeast extract, 0.5 g; salts solution, 1 ml; nutrient agar, 100 ml. The salts solution comprised: $MgCl_2$, 1×10^{-3} M; $CaCl_2$, 7×10^{-4} M; $MnCl_2$, 5×10^{-5} M. Plates were incubated at 24–48 h at 37 °C, and colonies resembling those of the original isolate were purified on plates of the supplemented nutrient agar.

Bacillus smithii. This species contains isolates from evaporated milk, canned foods, cheese, and sugar beet juice extrac-

tion plant which were previously assigned to *Bacillus coagulans*; no methods for its specific enrichment or isolation are known to the authors – see *Bacillus coagulans.*

Bacillus soli – see *Bacillus bataviensis.*

Bacillus sonorensis. This species is phenotypically similar to *Bacillus licheniformis*, although yellow/cream pigmentation has been given as a distinguishing character, and colonies are bright yellow on a medium containing glucose 40 g; neopeptone, 10 g; agar 15 g, distilled water, 1 l; pH 5.6; whereas those of *Bacillus licheniformis* are cream (Palmisano et al., 2001). Original strains were isolated from desert soil using a general method for *Bacillus* species (see *Bacillus subtilis* below).

Bacillus sphaericus. Members of this species and closely related, spherical-spored organisms may be enriched using a method described by Beijerinck and Minkman (1910): 10–20 g of casein or its sodium salt is added to 1 l tap water, and the medium is dispensed in 30 ml quantities in 300 ml Erlenmeyer flasks; about 1 g of soil is added to each and the suspensions are heated to boiling, cooled and then incubated at 37 °C for 3 d. Spherical-spored organisms will usually predominate in the resulting mixture of sporulating organisms. The flasks are heated to 80 °C for 5 min and loopfuls are streaked on plates of nutrient agar at pH 7.0 and incubated at 30 °C; colonies yielding sporangia swollen with spherical spores are further purified. Massie et al. (1985) described a medium containing sodium acetate as the only major source of carbon, for isolating *Bacillus sphaericus* from soil: $Na_2HPO_4 \cdot 12H_2O$, 11.2 g; KH_2PO_4, 2.4 g; $(NH_4)_2SO_4$, 2.0 g; $MgSO_4 \cdot 7H_2O$, 50 mg; $MnCl_2 \cdot 4H_2O$, 4 mg; $FeSO_4 \cdot 7H_2O$, 2.8 mg; sodium acetate·$3H_2O$, 5.0 g; trisodium citrate·$2H_2O$, 20 mg; distilled water to 1 l; pH 7.2. The basal medium, without salts of Mg^{2+}, Mn^{2+} and Fe^{2+} and the carbon source, was autoclaved for 10 min at 121 °C. The salts were dissolved together at ×50 strength in 0.005 M-H_2SO_4 and autoclaved separately. The carbon source was autoclaved separately as a 25% w/v aqueous solution. The medium was solidified with agar. The medium was used unsupplemented or else it was supplemented with (per liter): glutamate, 1 g; thiamine, 10 mg; biotin 0.001 mg; in the combinations: biotin plus thiamine, thiamine plus glutamate, glutamate alone. The medium was also prepared with the acetate concentration doubled, and this version also used supplemented or unsupplemented. About 1 g moist soil was suspended in 5 ml sterile water and heated at 60 °C for 30 min; 50 ml medium in a 250 ml flask was inoculated with 1 g soil suspension and incubated at 30 °C with shaking; the next day, a loopful of culture was streaked onto solid medium, of the same composition as that in the original flask, and incubated at 30 °C.

Strains of the species pathogenic to mosquito larvae may be naturally resistant to chloramphenicol and streptomycin (Burke and McDonald, 1983) and a selective, differential medium containing streptomycin has been described by Hertlein et al. (1979). The BATS medium of Yousten et al. (1985) uses arginine as the sole carbon and nitrogen source and uses streptomycin as the selective agent; it was found to inhibit the growth of 68% the nonpathogenic *Bacillus sphaericus* strains tested: Na_2HPO_4, 5.57 g; KH_2PO_4, 2.4 g; $MgSO_4 \cdot 7H_2O$, 50 mg; $MnCl_2 \cdot 4H_2O$, 4 mg; $FeSO_4 \cdot 7H_2O$, 2.8 mg; $CaCl_2 \cdot 2H_2O$, 1.5 mg; L-arginine, 5 g; thiamine, 20 mg; biotin, 2 μg; streptomycin sulfate, 100 mg; agar, 20 g; distilled water, 1 l. The arginine, thiamine, biotin and streptomycin are prepared as a filter-sterilized stock solution. The Mg^{2+}, Mn^{2+}, Fe^{2+} and Ca^{2+} salts are prepared as an acidified

[0.03% (v/v) concentration of H_2SO_4], autoclaved stock solution. Stock solutions are added to the autoclaved phosphate salts-agar medium when it has cooled to 50 °C after autoclaving.

Bacillus sporothermodurans. Strains of this species are characterized by their highly heat-resistant spores (HHRS), and they were first detected in ultrahigh-temperature (UHT) treated milk which is heated to 135–142 °C for a few seconds; further strains were isolated from silage. The organism has been sought in milk, feed and silage by heating samples to 100 °C for 60 min, plating onto brain heart infusion (BHI) agar, and incubating at 37 °C for 2 d. Colonies resembling *Bacillus sporothermodurans* are selected and purified by streaking on plates of BHI agar or nutrient agar which have been supplemented with 1 mg per liter of vitamin B_{12} (Pettersson et al., 1996). The approach described is not specific for this species, and a range of other HHRS is likely to be isolated. Further strains of this species have been isolated from raw, UHT and sterilized milks, animal feed concentrate, and soy meal using the following method (Scheldeman et al., 2002): samples were heat treated for 30 min at 100 °C, followed by plating on brain heart infusion broth solidified with bacteriological agar and supplemented with vitamin B_{12} (1 mg/l, filter-sterilized); plates were incubated for 48 h at 20, 37 and 55 °C, but 37 °C is the optimal temperature for growth and isolation of this species.

Bacillus subterraneus. The single isolate of this species was cultivated from deep subterranean thermal water (71 °C, pH 7.8) collected in a sterile glass container from a bore well tapping the Great Artesian Basin of Australia; the containers were completely filled and sealed with air-proof enclosures. For enrichment, 1 ml of water was injected into 10 ml of metal reduction (MR) medium, supplemented with 0.016 g iron oxide, in Hungate tubes and incubated at 40 °C for 1 week (Kanso et al., 2002). MR medium was prepared as follows: yeast extract, 2 g; NH_4Cl, 1 g; $K_2HPO_4 \cdot 3H_2O$, 0.08 g; $MgCl_2 \cdot 6H_2O$, 4.5 g; $CaCl_2 \cdot 2H_2O$, 0.375 g; NaCl, 20 g; vitamin solution (Patel et al., 1985), 10 ml; trace element solution, 1 ml; distilled water to 1 l. The trace element solution contained: nitrilotriacetic acid neutralized to pH 6.5 with KOH, 12.5 g; $FeCl_3 \cdot 4H_2O$, 0.2 g; $MnCl_2 \cdot 4H_2O$, 0.01 g; $CoCl_2 \cdot 6H_2O$, 0.017 g; $CaCl_2 \cdot 2H_2O$, 0.1 g; $ZnCl_2$, 0.1 g; $CuCl_2$, 0.02 g; H_3BO_3, 0.01 g; $Na_2MoO_4 \cdot 2H_2O$, 0.01 g; NaCl, 1.0 g; Na_2SeO_3, 0.02 g; distilled water, 1 l. The medium was boiled and then cooled under a stream of oxygen-free N_2 gas to about 50 °C, 3.6 g of $NaHCO_3$ were added, and the pH adjusted to 7.1; 10 ml aliquots were dispensed under N_2 into Hungate tubes (Hungate, 1969), and this gas phase was subsequently replaced with N_2/CO_2 (80:20) and the medium autoclaved at 121 °C for 15 min. The enrichment culture that developed could reduce Fe(III), and nitrate (20 mM) was used as an alternative electron acceptor; serially diluted enrichments in MR medium with nitrate were purified on plates of the same medium solidified with 2% agar and incubated at 40 °C for 1 week.

Bacillus subtilis. This species is readily isolated from dried grass, and has long been known as the "hay bacillus." Zopf (1885) recommended the following method: hay is soaked in water for 4 h at 36 °C, using as small a volume of water as possible, and the fluid is decanted and diluted to a specific gravity of 1.004; the pH is adjusted to 7.0 and 500 ml of the suspension is transferred to a sterile 1-l Erlenmeyer flask, which is plugged with cotton wool and then boiled for 1 h. The flask is incubated at 36 °C for about 28 h, and the pellicle that usually forms will often be found to yield only *Bacillus subtilis.* Knight and Proom (1950) added air-dried soil to 10 ml of nutrient broth to a give total volume of 15 ml, shook the vessel well, plated loopfuls of the suspension on nutrient agar, and then incubated at 37 °C for 2 d; the resulting flora was predominantly *Bacillus subtilis.* Roberts and Cohan (1995) and their colleagues isolated the *Bacillus subtilis* strains which they subsequently allocated to *Bacillus mojavensis*, *Bacillus vallismortis* and *Bacillus subtilis* subsp. *spizizenii* (and close relatives of *Bacillus licheniformis* allocated to *Bacillus sonorensis*) by suspending 1 g quantities of desert soils in 5 ml amounts of sterile water, heat-treating suspensions at 80 °C for 10 min, vortex mixing for 1 min and allowing to settle for a further minute, then plating 50–100 µl onto tryptone blood agar base plates and incubating at 37 °C for 48 h (F.M. Cohan, personal communication).

Bacillus thermoamylovorans. The single original strain of this species was isolated from palm wine (Combet-Blanc et al., 1995) using Hungate's anaerobic methods (Hungate, 1969; Macy et al., 1972; Miller and Wolin, 1974). Palm wine, 10 ml; 10% $NaHCO_3$, 1 ml; 2% $Na_2S \cdot 9H_2O$, 0.2 ml were added to a 60 ml serum bottle containing 20 ml of basal medium supplemented with 0.3% glucose. Bottles were incubated at 50 °C for 24 h, and then a 1-ml sample was transferred to a fresh serum bottle of basal medium. This step was twice repeated. A pure culture was obtained by the roll-tube method of Hungate, using Hungate tubes containing 4.5 ml of basal medium supplemented with 0.3% glucose. Basal medium contained: yeast extract, 5 g; Biotrypcase (bioMérieux), 5 g; KH_2PO_4, 1 g; NH_4Cl, 1 g; $MgCl_2 \cdot 6H_2O$, 0.4 g; $FeSO_4 \cdot 7H_2O$, 5 mg; mineral solution, 25 ml; trace element solution, 1 ml; Tween 80, 1 ml; distilled water, 1 l; pH adjusted to 7.5 with 10 M KOH. The mineral solution contained: KH_2PO_4, 6 g; $(NH_4)_2SO_4$, 6 g; NaCl, 12 g; $MgSO_4 \cdot 7H_2O$, 2.6 g; $CaCl_2 \cdot 2H_2O$, 0.16 g; distilled water, 1 l. The trace element solution contained: nitrilotriacetic acid (neutralized to pH 6.5 with KOH), 1.5 g; $MgSO_4 \cdot 7H_2O$, 3 g; $MnSO_4 \cdot 2H_2O$, 0.5 g; NaCl, 1 g; $FeSO_4 \cdot 7H_2O$, 0.1 g; $CoCl_2$ or $CoSO_4$, 0.1 g; $CaCl_2 \cdot 2H_2O$, 0.1 g; $ZnSO_4$, 0.1 g; $CuSO_4 \cdot 5H_2O$, 0.01 g; $AlK(SO_4)_2$, 0.1 g; H_3BO_3, 0.01 g; $Na_2MoO_4 \cdot 2H_2O$, 0.01 g; distilled water, 1 l; pH adjusted to 7.0 with KOH. The medium was boiled, and cooled under a stream of O_2-free N_2 at room temperature, distributed into serum bottles and Hungate tubes, and autoclaved at 110 °C for 45 min. Energy sources were injected into the bottles and tubes from separately sterilized stock solutions.

Bacillus thermocloacae. Strains of this species were isolated from an aerobic culture at 60 °C of municipal sewage sludge in a laboratory fermenter (Demharter and Hensel, 1989b). Samples were taken from the fermenter, homogenized, diluted and plated on Ottow's (1974) medium (glucose, 1.0 g; peptone, 7.5 g; meat extract, 5.0 g; yeast extract, 2.5 g; Casamino acids, 2.5 g; NaCl, 5.0 g; agar, 13 g; tap water, 1 l; pH 7.2–7.4.) supplemented with 100 ml per liter of aqueous sludge extract, adjusted to a final pH of 8.5, and incubated for 3–4 d at 60 °C.

Bacillus thuringiensis. The isolation of *Bacillus thuringiensis* from environmental samples can be very laborious. After heat treatment and plate cultivation, it requires the microscopic examination of material from all colonies with *Bacillus cereus*-type morphologies for the parasporal crystals characteristic of *Bacillus thuringiensis.* Therefore, several methods have been devised in order to increase the proportion of such colonies which are *Bacillus thuringiensis.* Travers et al. (1987) described a method which uses acetate to inhibit germination of *Bacillus thuringiensis* spores, followed by heat to kill germinated cells of other spore-formers and cells of non-spore-forming bacteria: L broth (tryptone, 10 g; yeast extract. 5 g; NaCl, 5 g; distilled water, 1 l) is buffered with 0.25 M sodium acetate. 0.5 g of sample is added to 10 ml of this broth in a 125 ml triple-baffled flask, and the flask is shaken at 250 r.p.m. at 30 °C for 4 h. A sample of this suspension is then heat treated for 3 min at 80 °C in a flow-through heat treater, and used to inoculate plates of L agar (as L broth, but solidified with agar; sodium acetate is not included). Carozzi et al. (1991) used a similar approach, but with heat treatment at 65 °C for 10 min. Johnson and Bishop (1996) enriched *Bacillus thuringiensis* in penicillin broth, followed by plating on penicillin agar, and found their method to be superior to those of Travers et al. (1987) and Carozzi et al. (1991): 0.25 g of sample is placed in a tube containing 2 ml of nutrient broth which has been supplemented with 1 ml/l CCY salts and 20 IU/ml penicillin G. The CCY salts solution (to aid sporulation) contains: acid casein hydrolysate, 1 g/l; enzymic casein hydrolysate, 1 g; glycerol, 0.6 g; enzymic yeast extract, 0.4 g; glutamine, 20 mg; distilled water, 1 l; and the following salts: K_2HPO_4, 26 mM; KH_2PO_4, 13 mM; $MgCl_2 \cdot 6H_2O$, 0.5 mM; $CaCl_2 \cdot 6H_2O$, 0.2 mM; $FeCl_3 \cdot 6H_2O$, 0.05 mM; $ZnCl_2$, 0.05 mM; $MnCl_2 \cdot 4H_2O$, 0.01 mM. The suspension is heat-shocked at 70 °C for 10 min and then added to 50 ml of the same medium in a 250 ml flask. The flask is shaken at 200 r.p.m. at 30 °C until sporulation is complete. The suspension is centrifuged at 3,600 r.p.m. for 1 h, and the pellet resuspended in 2 ml of fresh medium and the heat treatment and shaking procedure repeated; serial dilutions of the suspension are then inoculated onto plates of the supplemented nutrient broth which has been solidified with agar, and incubated at 30 °C until sporulation is complete (as judged by microscopy), and observed for the characteristic sporangial morphology of *Bacillus thuringiensis*. Forsyth and Logan (2000) found that a penicillin-based method was unsuitable for the isolation of *Bacillus thuringiensis* from Antarctic soils, as their isolates from these environments showed some penicillin sensitivity. Heat treatment and enrichment in BFB (see *Bacillus fumarioli*, above), followed by plating from turbid enrichments onto BFA and observation for parasporal crystals gave best results; over all the samples studied, however, acetate selection methods gave yields no better than the BFB method (Logan and Grieg, unpublished observations).

Bacillus tusciae. Strains of this hydrogen-oxidizing thermoacidophile were isolated from ponds in a geothermally heated area in Italy (Bonjour and Aragno, 1984). The enrichment medium was the mineral medium used by Bonjour et al. (1988) (see *Bacillus schlegelii*, above): 0.5 g per liter of $NaHCO_3$ were added for autotrophic growth, and the medium was acidified to pH 3.5 using 5 M HCl. About 1 g amounts of pond sediment mixed with overlying water were added to 20 ml of medium in 100 ml Pyrex screw-capped bottles, and cultures were incubated in desiccators under a gas mixture of 0.05 atm O_2 + 0.1 atm CO_2 + 0.45 H_2 (partial pressures measured at room temperature; the reduced total pressure of 0.6 atm allowed incubation at high temperatures without overpressure) for 5 d at 55 °C. For isolation, cultures were subcultured by streaking onto plates of the same medium solidified with agar and incubated under the same conditions as for enrichment.

Bacillus vallismortis. This species is phenotypically virtually indistinguishable from *Bacillus subtilis*, and original strains were isolated from desert soil using a non-specific method for *Bacillus* species (see *Bacillus subtilis*, above).

Bacillus vedderi. Samples of mud from a bauxite-processing red mud tailing pond were inoculated into alkaline oxalate medium and incubated at room temperature or 45 °C for 2 weeks (Agnew et al., 1995). Bottles showing turbidity were subcultured into bottles of fresh medium which, once turbid, were subcultured by streaking on plates of the same medium solidified with 2% (w/v) agar. Isolates were routinely grown at 37 °C. The alkaline oxalate medium comprised: Na_2CO_3, 2.65 g; $(NH_4)_2SO_4$, 1 g; K_2HPO_4, 0.17 g; $MgSO_4$, 0.15 g; distilled water, 1 l; pH adjusted to 10.5 with 5 M NaOH before autoclaving. After cooling, the following additions were made from sterile stock solutions to the final concentrations shown: sodium oxalate, 0.67% (w/v); yeast extract, 0.1% (w/v); mineral solution, 0.2% (v/v). The mineral solution contained: nitrilotriacetic acid, 1.5 g; $MgSO_4 \cdot 7H_2O$, 3 g; $MnSO_4 \cdot 2H_2O$, 0.5 g; NaCl, 1 g; $FeSO_4 \cdot 7H_2O$, 0.1 g; $CaCl_2 \cdot 2H_2O$, 0.1 g; $CoCl_2$, 0.1 g; $ZnSO_4$, 0.1 g; $CuSO_4 \cdot 5H_2O$, 0.01 g; $AlK(SO_4)_2$, 0.01 g; H_3BO_3, 0.01 g; $Na_2MoO_4 \cdot 2H_2O$, 0.01 g; distilled water, 1 l.

Bacillus vietnamensis was isolated from *nuoc mam* (Vietnamese fish sauce); trypticase soy agar was used as basal and maintenance medium.

Bacillus vireti – see *Bacillus bataviensis*.

Bacillus weihenstephanensis. This species is phenotypically similar to *Bacillus cereus* and only distinguishable from it by ability to grow at 7 °C, inability to grow at 43 °C, and by certain 16S rDNA signature sequences (Lechner et al., 1998), but not all psychrotolerant organisms resembling *Bacillus cereus* are *Bacillus weihenstephanensis*. Strains were isolated from German dairies by plating pasteurized milk samples on plate count agar and incubating at 7 ± 0.5 °C for 10–16 d. Psychrotolerance was confirmed by inoculating purified colonies into liquid culture of plate count medium and incubating at 7 ± 0.5 °C with agitation until growth was visible. Plate count medium contained: peptone, 5 g; yeast extract, 2.5 g; glucose, 1 g; distilled water, 1 l; pH, 7.0.

Maintenance procedures

Bacillus strains may be preserved on slopes of a suitable growth medium that encourages sporulation, such as nutrient agar or trypticase soy agar containing 5 mg/l of $MnSO_4 \cdot 7H_2O$. Slopes should be checked microscopically for spores before sealing, to prevent drying out, and storage in a refrigerator; on such sealed slopes the spores should remain viable for many years. For longer-term preservation, lyophilization and liquid nitrogen may be used, as long as cryoprotectants are added.

Procedures for testing special characters

Introduction. The ubiquity and huge diversity of *Bacillus* species presents a huge diagnostic challenge. In the First Edition of this *Manual*, the genus *Bacillus* was essentially defined as: "aerobic, endospore-forming, Gram-positive rod-shaped bacteria," and the methods of Gordon et al. (1973) could confidently be recommended for identification of the majority of the 34 valid species (five further species were added in proof) listed by Claus and Berkeley (1986). The scheme of Smith et al. (1952) split the species into three groups according to their sporangial morphologies, and then further divided them by biochemical and physiological tests, and this system culminated in the monograph of Gordon et al. (1973). This approach was effective for some years, but *Bacillus* identification was still generally perceived as complicated, the chief difficulties being the need for special media, and between-strain variation. Much of the latter was a reflection of unsatisfactory taxonomy, but, as the studies of Logan and Berkeley (1981) revealed, test inconsistency exacerbated the problems. Logan and Berkeley (1984) addressed these problems with a large database for 38 clearly defined taxa (species) using miniaturized tests in the API 20E and 50CHB Systems (bioMérieux, Marcy l'Etoile, France), and this scheme, with updates, remains in common use. Since the development of the Gordon et al. and Logan and Berkeley schemes, the genus has been radically changed, and the task of identification made more complicated, by: (i) the proposal of many new species (frequently from exotic habitats, and often primarily on the basis of molecular analyses), (ii) the allocation of strict anaerobes (*Bacillus arseniciselenatis, Bacillus infernus*) and organisms in which spores have not been observed (*Bacillus infernus, Bacillus thermoamylovorans*) to the genus, and (iii) the transfer of many species to the new genera, *Alicyclobacillus, Aneurinibacillus, Brevibacillus, Geobacillus, Gracilibacillus, Marinibacillus, Paenibacillus, Salibacillus* (now merged with *Virgibacillus), Ureibacillus* and *Virgibacillus*, and to the long-established genus *Sporosarcina*.

Unfortunately, taxonomic progress has not revealed readily determinable features characteristic of each genus. Also, many recently described species represent genomic groups disclosed by DNA–DNA pairing experiments, and routine phenotypic characters for distinguishing some of them are very few and of unproven value. Furthermore, several recently described species were proposed on the basis of very few strains so that the within-species diversities of such taxa, and so their true boundaries, remain unknown. It is clear, therefore, that the identification of *Bacillus* species has become increasingly difficult since the publication of the schemes of Gordon et al. and Logan and Berkeley, a period during which demands to identify such organisms have greatly increased, especially in the medical and biotechnological fields.

The identification scheme of Gordon et al. (1973) embraced 18 species, only half of which now remain in *Bacillus*. The genus now contains 90 species, and encompasses acidophiles, alkaliphiles, neutrophiles, halophiles, mesophiles, psychrophiles and thermophiles, in addition to the largely neutrophilic and mesophilic species studied by Gordon et al. (1973), so that their methods can no longer be expected, even with substantial modifications and an expanded database, to allow recognition of all the species now allocated to the genus. Identification of *Bacillus* species with routine phenotypic tests must therefore call upon a variety of characterization methods, and a unified approach is no longer possible. Despite this, the traditional characterization tests used by Gordon et al. retain their place in *Bacillus* identification, because the most commonly encountered species are still distinguishable by these methods.

Fritze (2002) recommended a stepwise approach to identification of the aerobic endospore-formers: (i) differential cultivation at a range of temperature (say 5, 30 and 55 °C) and pH conditions (say pH 4.5, 7–7.2 and 9) [to which we may add salt concentrations appropriate to halotolerant and halophilic organisms], (ii) selection of spores by heat treatment or alcohol treatment (the latter being preferred, because not all spores are sufficiently heat-resistant to survive the former procedure), and (iii) characterization using the appropriate media and incubation temperatures.

Reva et al. (2001) described a phenotypic identification scheme for aerobic endospore-formers based on 115 characters and a key. Their database included just 69 strains of validly published *Bacillus* species, with numbers of representatives of each species ranging from one for *Bacillus badius, Bacillus mycoides, Bacillus lentus* and *Bacillus thuringiensis* to 27 for *Bacillus subtilis*. Four of the 13 *Bacillus* species they included were not represented by their type strains. They also included strains of two *Brevibacillus* species, four *Paenibacillus* species and one *Virgibacillus* species. Identification schemes are naturally limited by their databases, and schemes based on keys are prone to failure with atypical isolates. The strengths and weaknesses of the Reva et al. (2001) scheme can only be revealed by usage; the characters they employed were investigated by the traditional procedures described by Gordon et al. (1973) and Claus and Berkeley (1986), and the discriminative efficiencies they calculated for these characters might be valuable in the construction of "home-grown" identification systems.

Reference strains. In the following accounts of media and methods, the above-mentioned constraints must be borne in mind, and it is recommended that the original and emended descriptions of the more recently described species are consulted wherever possible, and that cultures of those organisms are obtained for comparison. It should also be remembered that 16S rDNA sequencing is not always reliable as a stand-alone tool for identification, and that a polyphasic taxonomic approach is advisable for the identification of some of the more rarely encountered species and the confident recognition of new taxa. It must be appreciated that the species descriptions given in the text and tables below are largely lifted from authors' descriptions of their proposed species, and that a number of such descriptions were based upon few strains, so that within-species diversities in such cases are unknown. Characterization methods and their interpretation vary, and typographic errors in the compilation of descriptions are bound to occur (several being encountered in the preparation of this review), so original descriptions should never be relied upon entirely. Nomenclatural types exist for a good reason and are usually easily available; there is no substitute for direct laboratory comparisons with authentic reference strains. However, collections of *Bacillus* species in laboratories around the world harbor many misnamed strains. This is not necessarily a reflection on the competence of those assembling the collections; it is more a symptom of the unsatisfactory state of the classification of the organisms at the time the cultures were acquired, and this underlies the difficulty that many bacteriologists frequently encounter with *Bacillus* identification.

Unfortunately, several of the groups whose taxonomies are the most complex are the ones whose members are frequently submitted to reference laboratories. Such organisms are frequently included in the databases of commercial kits, but it can be difficult to obtain sufficient authentic strains of some these species to allow a satisfactory database entry to be made. Ideally, an entry in the database should reflect at least ten representative strains of the species, but for some taxa, and particularly for the new species which have been based upon just one strain or very few strains, this can be impossible. It can also be a problem for some of the older-established species, such as *Bacillus circulans* and *Bacillus lentus*, as the representative strains found in culture collections around the world are sometimes the widely dispersed subcultures of a few original isolates. These problems emphasize the importance of basing proposals for new taxa on adequate numbers of strains to reflect the diversities of the taxa.

A further problem has emerged with the splitting of well-established (but not necessarily homogeneous) species or groups into large numbers of new taxa over a short period. *Bacillus circulans* was long referred to as a complex rather than a species, but the revision of the taxonomy of this group, and consequent proposals for several new species to be derived from it, have led to difficulties in identification. Although the proposals were mostly based upon polyphasic taxonomic studies, initial recognition of the new taxa depended largely upon DNA relatedness data. A DNA:DNA reassociation study of *Bacillus circulans* strains yielded *Bacillus circulans sensu stricto*, *Bacillus amylolyticus*, *Bacillus lautus*, *Bacillus pabuli* and *Bacillus validus* and evidence for the existence of five other species (Nakamura, 1984a; Nakamura and Swezey, 1983). *Bacillus amylolyticus*, *Bacillus lautus*, *Bacillus pabuli* and *Bacillus validus* are now accommodated in *Paenibacillus*, and these species and *Bacillus circulans* are difficult to distinguish using routine phenotypic tests.

Such radical taxonomic revisions have left many culture collections worldwide with few representatives of *Bacillus circulans sensu stricto*, but with numerous misnamed strains of this species, which may or may not belong to one of the newly proposed taxa. The curators will normally not be able to know which are which without considerable expenditure in scholarship and experimental work, and in many cases a collection will hold only one authentic strain, the type strain, of a species - be it an old or new species.

When attempting to construct an identification scheme for *Bacillus* the implication of such rapid taxonomic progress is huge. Accessing authentic strains of many species, even well-established ones, may require much time and effort, and for several of them the strains available may be too few to allow the diversities of the taxa to be adequately reflected in the identification scheme. Smith et al. (1952) and Gordon et al. (1973) showed commendable restraint in their concept of a bacterial species, saying in the latter monograph: "When only a few strains of a group are available, as often happens, their species descriptions must remain tentative until verified by the study of more strains". Just as taxonomists can only be as good as their culture collections, so identification systems can only be as good as their databases.

Although many new characterization methods have been developed over the last 30 years, the principle of identification remains the same; identifications cannot be achieved, strictly speaking; the best that can be done is to seek the taxon to which the unknown strain probably belongs. The outcome is expressed as a probability and, as with the classification upon which the scheme is based, the answer cannot be final (Logan, 2002). It should not be assumed that, because traditional approaches for identifying *Bacillus* are perceived as being difficult and unreliable, any newer approach is likely to be superior regardless of the size and quality of its database; whatever characterization method is used, considerable amounts of time, money and expertise need to be invested in the construction of

reliable and detailed databases, which must be founded upon wide diversities of authentic reference strains.

Standardization. Whatever methods are used to generate the characters upon which identifications are based, standardization of methodologies and inclusion of reference strains is crucial. The methods used to generate the characters included in species descriptions have not usually been standardized between laboratories, and the test results shown in differentiation tables often include information lifted from the literature, so that data are often not comparable. As the number of valid *Bacillus* species increases, the task of studying related and reference organisms in parallel becomes more demanding for classification and identification work, and authors may be tempted to lean ever more heavily on data presented in the literature. In addition, standardization of methodologies for many phenotypic tests is inherently impossible between those organisms whose temperatures and pH ranges for growth do not overlap. Miniaturized versions of traditional biochemical tests (API kits, VITEK cards, and Biolog plates) offer standardized methods for a range of biochemical characters; the first named offers some versatility in temperature and pH, while the last named can be incubated at a range of temperatures.

Media. The most widely used solid media for cultivating neutrophilic *Bacillus* species are nutrient agar and trypticase soy agar; these media may be adjusted to higher or lower pH for cultivating alcaliphiles and acidiphiles, and NaCl may be added for the cultivation of halophiles. Sporangial morphology remains an important character, and although many strains will sporulate on these media within a few days of incubation, the addition of 5 mg/l of $MnSO_4 \cdot xH_2O$ is recommended for encouraging sporulation. Gordon et al. (1973) recommended soil extract agar to encourage sporulation for the purpose of strain maintenance, but this approach is unnecessarily laborious as the addition of manganese ions appears to serve just as well. Rich media such as blood agar may not yield sporulated cells and the culture might die without sporulating. Many of the more recently described *Bacillus* species have been found in unusual environments and/or have been isolated and studied using special media; these media are described in the descriptions of these species (see below). Most of the media (or revisions described by Claus and Berkeley, 1986) used by Gordon et al. (1973) for separation of the nine species that they listed, and which remain in the genus, are given below. Other media employed by Gordon et al., such as litmus milk, are not listed because they gave very poorly reproducible results (Logan and Berkeley, 1981) and/or are now rarely used for some other reason.

Anaerobic agar. Trypticase (trypsin hydrolysate of casein), 20 g; glucose, 10 g; NaCl, 5 g; sodium thioglycolate (mercaptoacetic acid, sodium salt), 2 g; sodium formaldehydesulfoxylate (hydroxymethanesulfinic acid, monosodium salt dihydrate), 1 g; agar 15 g; distilled water, 1 l; pH 7.2. Distribute into 15 mm diameter glass tubes to 75 mm depth and autoclave at 121 °C for 20 min. Several commercial preparations are available. (Note: Gordon et al. omitted the indicator and glucose from the usual formulation of this medium).

Citrate and propionate utilization media. Trisodium citrate·2H$_2$O, 1 g (or sodium propionate, 2 g); MgSO$_4$·7H$_2$O, 1.2 g; (NH$_4$)$_2$HPO$_4$, 0.5 g; KCl, 1 g; trace element solution, 40 ml; phenol red (0.04% w/v solution), 20 ml; agar, 15 g; distilled water, 920 ml; pH 6.8;

distribute in tubes, autoclave at 121 °C for 20 min and set as slopes. Trace element solution: ethylenediaminetetraacetate, 500 mg; $FeSO_4 \cdot 7H_2O$, 200 mg; H_3BO_3, 30 mg; $CoCl_2 \cdot 6H_2O$, 20 mg; $ZnSO_4 \cdot 7H_2O$, 10 mg; $MnCl_2 \cdot 4H_2O$, 3 mg; $Na_2MoO_4 \cdot 2H_2O$, 3 mg; $NiCl_2 \cdot 6H_2O$, 2 mg; $CuCl_2 \cdot 2H_2O$, 1 mg; distilled water, 1 l.

Egg-yolk reaction medium. Tryptone (trypsin hydrolysate of casein), 10 g; Na_2HPO_4, 5 g; KH_2PO_4, 1 g; NaCl, 2 g; $MgSO_4 \cdot 7H_2O$, 0.1 g; glucose, 2 g; distilled water, 1 l; pH 7.6; autoclave at 121 °C for 20 min. Add sterile, commercially prepared egg-yolk emulsion (at the concentration recommended by the manufacturer), or 1.5 ml egg yolk (aseptically aspirated from a hen's egg) to 100 ml of basal medium and allow to stand overnight in a refrigerator. Dispense the supernatant in 2.5 ml amounts; basal medium without yolk is similarly dispensed. Modern practice replaces this medium with egg-yolk agar, without a noticeable difference in sensitivity: add 10 ml commercially prepared egg-yolk emulsion to 90 ml molten nutrient agar held at 45–50 °C; mix and pour as plates.

Glucose agar. D-Glucose, anhydrous, 10 g; nutrient agar, 1 l; pH 6.8; mix thoroughly and autoclave at 115 °C for 20 min.

Medium for acid production from carbohydrates. Basal medium: $(NH_4)_2HPO_4$, 1 g; KCl, 0.2 g; $MgSO_4 \cdot 7H_2O$, 0.2 g; yeast extract, 0.2 g; agar 15 g; distilled water, 1 l; adjust to pH 7.0; bromcresol purple (0.04% w/v solution), 15 ml; autoclave at 121 °C for 20 min. Aqueous solutions (10% w/v) of carbohydrates are filter-sterilized or may be autoclaved at 121 °C for 20 min. Gordon et al. (1973) used L-arabinose, D-glucose, D-mannitol and D-xylose, but on the basis of API 50CHB tests (see *Miniaturized biochemical test systems*, below) the following other carbohydrates may be valuable for differentiation of certain species: N-acetylglucosamine, D-mannose, D-tagatose, galactose, gluconate, glycerol, glycogen, inulin, melezitose, methyl α-D-mannoside, β-methylxyloside, salicin and starch. Aseptically add carbohydrate solution to molten base to a final concentration of 0.5% w/v, and set the medium as slopes. Media for testing acidophilic and alkaliphilic strains are described in *Methods, pH* below.

Milk agar. Skim milk powder, 5 g in 50 ml of distilled water; agar, 1 g in 50 ml of distilled water. Autoclave separately at 121 °C for 20 min, cool to 45 °C, mix, and pour into Petri dishes. Dry the surfaces of the plates before use.

Nitrate broth. Peptone (trypsin hydrolysate of meat), 5 g; beef extract, 3 g; KNO_3, 1 g; distilled water, 1 l; pH 7.0. Distribute into test tubes containing inverted Durham's tubes, and autoclave at 121 °C for 20 min.

Gelatin medium. Gelatin, 120 g; distilled water, 1 l; pH 7.0; autoclave at 121 °C for 20 min; distribute in test tubes. Alternatively, tubes of commercially prepared nutrient gelatin may be used, or plates of nutrient agar supplemented with 0.4% gelatin.

Phenylalanine agar. NaCl, 5 g; yeast extract, 3 g; DL-phenylalanine, 2 g; Na_2HPO_4, 1 g; agar, 12 g; distilled water, 1 l; pH 7.3; distribute in tubes, autoclave at 121 °C for 20 min and set as slopes. Claus and Berkeley (1986) favored the commercially prepared product available from BBL (www.voightglobal.com).

Resistance to lysozyme medium. Prepare a solution of lysozyme containing 10,000 enzyme units/ml in distilled water and sterilize it by filtration. Add 1 ml of this medium to 99 ml sterile nutrient broth and distribute the mixture in 2.5 ml amounts.

Sabouraud dextrose broth. Neopeptone (enzyme digest of casein and meat), 10 g; D-glucose, 40 g; agar, 15 g; distilled water, 1 l; pH 5.7; dispense into test tubes, sterilize by autoclaving at 121 °C for 20 min, and set as slopes. The broth is prepared by reducing the glucose content to 20 g, omitting the agar, and distributing in tubes.

Sodium chloride media. Tubes of nutrient broth are prepared with 0, 5, 7 and 10% (w/v) NaCl.

Starch agar. Suspend 1 g of potato starch in 10 ml cold distilled water and mix with 100 ml nutrient agar. Autoclave at 121 °C for 20 min, cool to 45 °C, mix thoroughly, and pour into Petri dishes.

Tyrosine agar. L-tyrosine, 0.5 g; distilled water, 10 ml; autoclave at 121 °C for 20 min. Mix aseptically with 100 ml sterile, molten nutrient agar, cool to about 50 °C and pour into Petri dishes, taking care to achieve a uniform distribution of the tyrosine crystals. Dry plates before use.

Voges – Proskauer broth. Proteose peptone (enzyme digest of meat), 7 g; glucose, 5 g; NaCl, 5 g; distilled water, 1 l; pH 6.5; dispense 5 ml amounts into 20 mm test tubes and sterilize by autoclaving at 115 °C for 20 min.

Methods. Current schemes for identifying *Bacillus* species may be roughly divided into four categories according to the kinds of characters they use: (i) traditional biochemical, morphological and physiological characters, (ii) miniaturized versions of traditional biochemical tests (API kits, VITEK cards, and Biolog plates), (iii) chemotaxonomic characters [such as fatty acid methyl ester (FAME) profiles, and pyrolysis mass spectrometry], and (iv) genomic characters (ribotyping, and nucleic acid probes). However, as early as the work of Smith et al. (1952) it was becoming clear that no one phenotypic technique would be suitable for identifying all *Bacillus* species. The problems have mounted up as further species from extreme environments have subsequently been proposed, and the potentials of chemotaxonomic analyses and studies of nucleic acids have therefore been investigated. The sections that follow outline these approaches and summarize their current contributions to identification for *Bacillus* and its relatives. However, it is impossible to devise standard conditions to accommodate the growth of strains of all species for chemotaxonomic work, and it remains unknown to the taxonomist if differences between taxa are consequences of genetic or environmental factors. The need to substantiate each characterization method by other techniques (be they phenotypic or genotypic) has become increasingly important as new techniques emerge. This need is satisfied by the polyphasic approach now usual for the better classification studies, and the same approach may sometimes be necessary in order to identify strains from some of the less familiar species.

Sporulation and microscopic appearance. Before attempting to identify to species level it is important to establish that the isolate really is an aerobic endospore-former. Isolates of large, aerobic Gram-positive rods have often been submitted to reference laboratories as *Bacillus* species, even though sporulation had not been observed, or because PHB granules or other storage inclusions had been mistaken for spores. It should also be borne in mind that *Bacillus* species do not always stain Gram-positive. See *Media* (above) for comments on suitable sporulation media; cultures grown on rich media may lyse and die rather than sporulate. Sporulation has not been observed in several recently described species (*Bacillus infernus, Bacillus thermoamylovorans*), but the potential to form endospores may be detected using a PCR method based upon certain genes for sporulation (Brill and Wiegel, 1997).

A Gram-stained smear showing cells with unstained areas suggestive of spores can be stripped of oil with acetone/alcohol, washed, and then stained for spores. Spores are stained in heat-fixed smears by flooding with 10% aqueous malachite green for up to 45 min. (without heating), followed by washing and counterstaining with 0.5% aqueous safranin for 30 s; spores are green within pink/red cells at 1000 × magnification. Phase-contrast (at 1000 × magnification) should be used if available, as it is superior to spore-staining and more convenient. Spores are larger, more phase-bright, and more regular in shape, size and position than other kinds of inclusion such as polyhydroxybutyrate (PHB) granules (Figure 10e), and sporangial appearance is valuable in identification (Figure 10). Members of the *Bacillus cereus* group and *Bacillus megaterium* will produce large amounts of storage material when grown on carbohydrate media such as glucose agar, but on routine media this vacuolate or foamy appearance is rarely sufficiently pronounced to cause confusion (Figure 10e).

General morphology should be studied in relatively young (18–24 h at 30 °C) cultures grown in nutrient broths aerated by shaking. Morphologies of cells raised on nutrient agar plates or slopes may be heterogeneous owing to varying conditions of oxygen supply within colonies. Wet preparations may be viewed by phase-contrast microscopy at 1000 × magnification, and observed for cell size (diameter), shape, shapes of ends of cells (rounded, squared, tapered), chains, filaments, and motility; for cells grown on glucose agar observe for storage inclusions (use the type strains of *Bacillus cereus* and *Bacillus subtilis* as positive and negative controls, respectively). Study cultures grown for 24 h and up to 7 d on medium supplemented with 5 mg/l $MnSO_4$ for spores: observe for spore shape (spherical, cylindrical, ellipsoidal), position (central or paracentral, subterminal, terminal), presence of parasporal bodies (use the type strain of *Bacillus thuringiensis* as positive control), and for swelling of the sporangium. Cells in wet preparations may be immobilized by coating clean slides with a thin (0.5 mm) layer of sterile 2% water agar; a drop of turbid suspension of the organism is placed on the solidified agar, overlaid with a coverslip, and viewed by phase-contrast microscopy in the normal way.

Capsule formation by Bacillus anthracis.   The capsule of virulent *Bacillus anthracis* can be demonstrated on nutrient agar containing 0.7% sodium bicarbonate incubated overnight under 5–7% CO_2 (candle jars perform well). Colonies of the capsulated *Bacillus anthracis* appear mucoid, and the capsule can be visualized by staining smears with M'Fadyean's polychrome methylene blue or India ink, or by indirect fluorescent antibody staining (Logan and Turnbull, 2003). More simply, 2.5 ml of blood (defibrinated horse blood seems best; horse or fetal calf serum are quite good) can be inoculated with a pinhead quantity of growth from the suspect colony, incubated statically for 6–18 h at 37 °C, and M'Fadyean stained. The M'Fadyean stain is preferable to other capsule staining methods, as it is more specific for *Bacillus anthracis* capsules. As *Bacillus anthracis* is suspected, safety precautions must be taken throughout capsule staining procedures; all materials coming into contact with the specimen, including spent reagents and rinsings, must either be discarded

into a disinfectant effective against endospores or autoclaved. For the M'Fadyean stain, make a thin smear from the specimen, and also from a positive control, on a clean slide and allow to dry. Fix by immersion in 95% or absolute alcohol for 30–60 s. Put a large drop (approx. 50 µl) of polychrome methylene blue (M'Fadyean stain) on the smear and ensure all the smear is covered by spreading the stain with an inoculating loop ("flooding" the slide is wasteful, unnecessary and ecologically undesirable). Leave for one minute and wash the stain off with water (into 10% hypochlorite solution). Blot the slide and allow to dry. At 100–400× magnification, the organisms can be seen as fine short threads; at 1,000× magnification (oil immersion), if virulent *Bacillus anthracis* is present, the capsule should be seen as a clearly demarcated zone around the blue-black, often square-ended rods which lie in short chains of two to a few cells in number. The positive control can be prepared by culturing a virulent strain in defibrinated horse blood as described above. For India ink negative-staining, place a large loopful of undiluted India ink on a cleaned slide, and mix in a small portion of the bacterial colony or a small loopful of the deposit from a centrifuged liquid culture. Drop a cleaned cover glass on, avoiding air bubbles, and press firmly between two sheets of blotting paper. When examined at 1000× under oil, the capsules appear as haloes around the highly refractive outlines of the bacterial cells. When capsules are absent, the ink particles directly abut the cell wall, and the cells are not easily seen. Phase-contrast is superior to bright-field microscopy, as the bacterial cells can be seen clearly in all cases.

Gamma phage sensitivity of Bacillus anthracis.   Schuch et al. (2002) found that the PlyG lysin of γ phage may be used to detect *Bacillus anthracis* by luminescence, and that the same lysin could kill vegetative cells and germinating spores. blood agar plates to test for γ phage and penicillin susceptibility. Enquiries about gamma phage and indirect fluorescent antibody capsule staining should be addressed to the Diagnostics Systems Division, USAMRIID, Fort Detrick, Frederick, MD 21702-5011, USA.

Incubation temperature and time.   Select an incubation temperature that matches the optimum growth temperature of the organism(s) as closely as is practical. Convenient temperatures are 20 °C for psychrophiles, 30 °C for mesophiles, and 45, 50 or 55 °C for strains growing up to 55–65 °C. Claus and Berkeley (1986) suggested that the incubation temperature should lie 10–15 °C below the maximum growth temperature, and this remains sound advice; however, it should be noted that some new isolates of well-established species and strains of certain more recently described species may have unusually narrow or wide growth temperature ranges, so that this rule of thumb can be difficult to apply. Although Gordon et al. (1973) stipulated incubation periods of up to 14 d for many tests, most strains will have realized their potentials on these test media within 7 d.

pH.   The media and the methods used by Gordon et al. (1973), and presented here in updated form, were mainly developed for mesophilic, neutrophilic species. They will not be applicable to acidophilic and alkaliphilic organisms. Certain media and methods may be adapted by adjusting the acidity as far as pH 6 for moderate acidophiles; for tests based upon acid production from carbohydrates, an indicator with a lower end point such as bromcresol purple will need to be selected (see Logan et al., 2000, and *Miniaturized biochemical test systems*, below).

For alkaliphilic organisms, the methods described by Fritze et al. (1990) may be recommended: the alkalinity of the casein, gelatin, nitrate and phenylalanine media, and media for determining growth temperatures and salt tolerance may be raised as far as pH 9.5–10 by adding 100 ml/l 1 M sodium sesquicarbonate after autoclaving; the phenylalanine medium is prepared with only 50 ml/l 1 M sodium sesquicarbonate, and it and the nitrate reduction medium should be acidified at the time of reading the results. Acid production from carbohydrates may be detected by using thymol blue in the following basal medium: K_2HPO_4, 7 g; NaCl, 5 g; KH_2PO_4, 2 g; $(NH_4)_2SO_4$, 1 g; $MgSO_4 \cdot 7H_2O$, 0.1 g; vitamin solution, 1 ml; distilled water, 1 l; adjust to pH 8.9–9.1 with NaOH. The vitamin solution contained: pyridoxine HCl, 100 mg; *p*-aminobenzoic acid, 50 mg; calcium D-pantothenate, 50 mg; nicotinic acid, 50 mg; riboflavin, 50 mg; thiamin HCl, 50 mg; D-biotin, 20 mg; folic acid, 20 mg; vitamin B12, 1 mg; distilled water, 1 l. The same basal medium, containing glucose as the carbon source and with pH adjusted with sodium sesquicarbonate, is used to test for ability to grow in neutral and alkaline conditions. Other tests useful for characterizing alkaliphiles are: diaminopimelic acid in cell walls, glucuronidase, pullulanase, Tween hydrolysis, and urease. The presence of diaminopimelic acid in cell walls is tested as follows: hydrolyze 1 mg dried cells with 1 ml 6 N HCl in a sealed, hard-glass tube held at 100 °C for 18 h; cool; filter the sample through paper and wash with 1 ml H_2O; remove HCl by drying 2–3 consecutive times under reduced pressure at 40 °C on a rotary evaporator; take up residue with 0.3 ml H_2O and spot 5 ml onto a cellulose-coated (microcrystalline) thin layer chromatography plate; separate the amino acids using the solvent mixture methanol-water-10 N HCl-pyridine (80:17.5:2.5:10, by volume), and detect them using acetonic ninhydrin spray (0.1% w/v) followed by heating at 100 °C for 2 min. DAP spots are olive green fading to yellow, while other amino acids give purple spots. β-glucuronidase is detected by 4-methyl-umbelliferone glucuronide (MUG) agar: tryptose, 20 g; NaCl, 5 g; cysteine·HCl, 1 g; agar, 12 g; distilled water, 1 l; pH 9.7; autoclave and hold molten at 45–50 °C; dissolve MUG in warm water and filter through a 0.22 µm membrane or autoclave the solution; add this solution to the agar base at 100 µg/ml and distribute in microtiter plate wells; when solidified, stab inoculate the wells and seal the microtiter plate with plate tape; observe after overnight incubation for fluorescence at about 366 nm. For pullulanase, add 0.3% (w/v) pullulan to a minimal medium at pH 9.7, and supplement with sterile solutions of peptone (0.5%), yeast extract (0.05%), and vitamin supplements; inoculate, incubate and reveal the reaction as for starch hydrolysis (below). For hydrolysis of Tween 20, 40 and 60, use the following medium: peptone, 10 g; NaCl, 5 g; $CaCl_2 \cdot H_2O$, 0.1 g; agar, 1 l water; pH 9.7; autoclave at 121 °C for 20 min and cool to 45–50 °C; add Tween to final concentration of 1% and pour plates; inoculate as single streaks, incubate up to 7 d, and observe for opaque halo. For urease detection, strains are cultivated on slopes of nutrient agar and growth is washed off at 3 and 7 d with 2 ml distilled water into a test tube; a drop of phenol red indicator is added and the reaction brought to pH 7 with dilute HCl; the suspension is equally divided and one tube has 0.1 g crystalline urea added (the other tube is a negative control) and allowed to stand; an alkaline reaction demonstrates the presence of urease. Nielsen et al. (1995a) determined carbohydrate utilization profiles of alkaliphiles using the API 50CH gallery (see *Miniaturized biochemical test systems*, below).

Salinity. Halotolerant and halophilic organisms may be tested in the Gordon et al. (1973) media, but supplemented with up to 10% NaCl; reactions of these salt-loving organisms in routine tests may often be very weak. Alternatively, media prepared in a marine broth or marine salts base may be used. Halophiles may often also be akaliphilic.

Traditional characterization tests. *Bacillus cereus*, *Bacillus circulans*, *Bacillus coagulans*, *Bacillus firmus*, *Bacillus licheniformis*, *Bacillus megaterium*, *Bacillus pumilus*, *Bacillus sphaericus* and *Bacillus subtilis* are the only species listed and distinguished by Gordon et al. (1973) that remain in *Bacillus* after later taxonomic revisions, and the tests outlined below are still valuable for distinguishing between these commonly encountered species.

Inocula. Use 1 drop inocula of overnight (for mesophiles at 30 °C) nutrient broth cultures delivered with Pasteur pipettes for liquid and sloped media. For plate media, use the same culture, but apply the inoculum with a moderate sized (2–3 mm outside diameter) loop.

Maximum and minimum growth temperatures. Use slopes of nutrient agar or some other suitable growth medium for the organism, immerse bottles to their necks in a waterbath at the chosen temperature and allow to equilibrate prior to inoculation. Intervals of 5 °C are recommended. Take care to maintain the water levels in the waterbaths throughout the test. Observe for growth after 3 d for temperatures of 55 °C or higher, after 5 d at 30–50 °C, after 14 d at 20 and 25 °C, and after 21 d for temperatures below 20 °C.

Acid production from carbohydrates. Inoculate, incubate, and observe for growth and acid production (shown by the indicator passing from mauve through gray to yellow) for up to 14 d. Use the type strain of *Bacillus subtilis* as positive control and the type strain of *Bacillus sphaericus* as negative control for L-arabinose, D-glucose, D-mannitol and D-xylose.

Anaerobic growth. Inoculate a tube of anaerobic agar with a small loopful of broth culture by stabbing to the bottom of the tube, or use a Pasteur pipette to seed molten medium held at about 40 °C and then allow the agar to solidify. Incubate mesophiles for 3–7 d. Use the type strains of *Bacillus cereus* and *Bacillus megaterium* as positive and negative controls, respectively.

Casein decomposition. Inoculate plates of milk agar with one-streak inocula, incubate, and observe for zones of clearing around the growth over 7 d and up to 14 d. At the termination of the test, scrape growth aside with a loop and observe for weak reactions which may have occurred beneath the colony. Use the type strains of *Bacillus megaterium* and *Paenibacillus macerans* as positive and negative controls, respectively.

Citrate and propionate utilization. Inoculate slants of the citrate and propionate utilization media and incubate for up to 7 d. Observe for a red (alkaline) reaction which indicates utilization of the substrate as sole carbon source. Use the type strains of *Bacillus subtilis* and *Bacillus badius* as positive and negative controls for citrate, and the type strains of *Bacillus licheniformis* and *Bacillus subtilis* as positive and negative controls for propionate, respectively.

Egg-yolk reaction medium. Inoculate into an egg-yolk broth and a control broth lacking egg yolk, incubate for up to 7 d, observing the egg-yolk broth at 1 or 2-d intervals for a heavy white precipitate in or on the surface of the medium. If using plates of egg-yolk agar, apply one-streak inocula, incubate as for tubes, and observe for a zone of whitish opacity in the medium around the growth. Use the type strains of *Bacillus cereus* and *Bacillus megaterium* as positive and negative controls, respectively.

Gelatin hydrolysis. Inoculate tubes of nutrient gelatin and incubate at 28 °C; at 2- to 3-d intervals, for up to 4 weeks, hold the tubes at 20 °C for 4 h and observe for liquefaction. If using plate medium, inoculate with a single streak and incubate for 3–5 d, scrape some growth aside with a loop to reveal weak reactions which may have occurred beneath the colony, then flood the plate with 10 ml of 1 N H_2SO_4 saturated with Na_2SO_4; unchanged gelatin forms an opaque precipitate within 1 h, and a clear zone indicates hydrolysis. Use the type strains of *Bacillus cereus* and *Bacillus coagulans* as positive and negative controls, respectively.

Growth at pH 5.7 in Sabouraud media. Inoculate a slope of Sabouraud dextrose agar and a tube of Sabouraud dextrose broth, and a tube of nutrient broth as a control, incubate, and observe for growth in either or both Sabouraud media for up to 14 d. Use the type strains of *Bacillus cereus* and *Bacillus badius* as positive and negative controls, respectively.

Nitrate reduction. Inoculate nitrate broths and incubate. After 3 and 7 d, observe for gas in the Durham tube, indicating reduction of nitrate through nitrite to nitrogen gas), and touch a loopful of culture onto a strip of potassium iodide/starch paper which has been moistened with a few drops of 1 N hydrochloric acid and observe for a purple color which indicates the presence of nitrite. Strains negative at 7 d are tested at 14 d by mixing 1 ml culture with 3 drops of each of: (i) sulfanilic acid, 0.8 g; 5 N acetic acid (glacial acetic acid and water 1:2.5), 100 ml; (ii) dimethyl-α-naphthylamine, 0.6 ml; acetic acid, 100 ml. A red or yellow (=high concentration) color indicates the presence of nitrite. If still negative, add 4–5 mg zinc dust to the tube; if a red color develops (owing to reduction of nitrate to nitrite by the zinc) it indicates the absence of nitrate reduction by the organism, and confirms that rapid reduction of nitrate to nitrogen gas (not trapped by the Durham tube) has not occurred within the first 3 d of incubation. Use the type strains of *Bacillus cereus* and *Bacillus megaterium* as positive and negative controls, respectively.

Phenylalanine deamination. Inoculate duplicate slopes of phenylalanine agar and incubate for 7 d. Pipette 4–5 drops of 10% (w/v) ferric chloride solution over the slope and observe for a green color beneath the growth; this indicates the formation of phenylpyruvic acid from the phenylalanine. If negative, the second tube is tested after 14 d further incubation. Use the type strains of *Bacillus megaterium* and *Bacillus cereus* as positive and negative controls, respectively.

Resistance to lysozyme. Lightly inoculate a tube of resistance-to-lysozyme medium and a control tube of 2.5 ml nutrient broth and incubate. Observe for growth or its absence in the lysozyme medium after 7–14 d. Use the type strains of *Bacillus cereus* and *Bacillus megaterium* as positive and negative controls, respectively.

Sodium chloride tolerance. Lightly inoculate NaCl broths, incubate at a slant in order to enhance aeration, and observe for growth after 7 and 14 d.

Starch hydrolysis. Inoculate duplicate plates of starch agar and incubate. At 3 and 5 d scrape some growth aside with a loop to reveal weak reactions which may have occurred beneath the colony and flood the plates with 95% ethanol in order to make the unchanged starch turn white and opaque; observe for a clear zone around and under the growth which indicates starch hydrolysis. Use the type strains of *Bacillus cereus* and *Bacillus sphaericus* as positive and negative controls, respectively.

Tyrosine decomposition. Use one-streak inocula on plates of tyrosine agar and incubate; protect from drying during incubation. Observe for clearing of the tyrosine crystals around and below the growth after 7 and 14 d. Use the type strains of *Bacillus cereus* and *Bacillus sphaericus* as positive and negative controls, respectively.

Voges–Proskauer reaction. Inoculate Voges–Proskauer broths in triplicate and test for acetyl methyl carbinol production after incubation for 3, 5 and 7 d by adding 3 ml of 40% (w/v) NaOH to the culture and adding 0.5–1 mg creatine. Vortex mix to aerate and observe for the production of a red color after 30–60 min at room temperature. Use the type strains of *Bacillus cereus* and *Bacillus megaterium* as positive and negative controls, respectively.

Chemotaxonomic characters. Chemotaxonomic fingerprinting techniques applied to aerobic endospore-formers include FAME profiling (Kämpfer, 2002), polyacrylamide gel electrophoresis (PAGE) analysis (De Vos, 2002), pyrolysis mass spectrometry, and Fourier-transform infra-red spectroscopy (Magee and Goodacre, 2002). Only one of these approaches, FAME analysis, is supported by a commercially available database for routine identification. Fatty acid analysis can play a very useful part in polyphasic taxonomic studies of *Bacillus*. However, fatty acid profiles across the aerobic endospore-forming genera do not, given frequent and considerable within-species heterogeneity, form the basis of a reliable, stand-alone identification scheme (Kämpfer, 1994, 2002). A further difficulty is the need for a standardized media and incubation temperature for preparing isolates for FAME analysis, making databases for acidophiles, alkaliphiles, neutrophiles, mesophiles, psychrophiles and thermophiles incompatible. The commercially available Microbial Identification System software (MIDI, Newark, Delaware, USA) includes a FAME database for the identification of aerobic endospore-formers. Although it cannot be expected to give an accurate or reliable identification with every isolate, it is certainly a valuable screening tool when used with caution.

Serology. See *Antigenic structure* in *Further descriptive information*, above.

Genotypic methods. As with other groups of bacteria, studies of 16S rDNA and of DNA have very valuable applications in the classification of aerobic endospore-formers (De Vos, 2002). Nucleic acid fingerprinting techniques are also of great potential for typing work, of course. A good example is the ability to differentiate *Bacillus anthracis* strains by amplified fragment length polymorphism (AFLP) analysis (Keim et al., 1997) on account of variable number tandem repeats (VNTR; Keim et al., 2000; Turnbull et al., 2002), as the distinction of isolates of this species for epidemiological or strategic purposes has long been a challenge. AFLP also shows promise for the epidemiological typing of *Bacillus cereus* (Ripabelli et al., 2000).

At present, however, nucleic acid analyses are not entirely suitable for the routine identification of aerobic endospore-formers; their value in classification does not necessarily make them suitable as routine diagnostic tools at the species level. Amplified rDNA restriction analysis (ARDRA), for example, has been and continues to be exceptionally effective in the classification of *Bacillus* (Logan et al., 2000). It is a very powerful technique for recognizing new taxa and can be used to screen large numbers of strains much faster than is reasonably possible with 16S rDNA sequencing, but it is not always capable of distinguishing closely related species (Logan et al., 2002b). Sequencing of 16S rDNA is not always capable of resolving species either; not only are the sequences within and adjacent to this gene almost identical among *Bacillus anthracis*, *Bacillus cereus* and *Bacillus thuringiensis*, but the variable sites within these sequences can differ among multiple rRNA cistrons within a single strain (Turnbull et al., 2002).

Other fingerprinting methods such as ribotyping, which is commercially available, are presently limited by the appropriateness of the restriction enzymes they use, and by the sizes of the databases available to those developing them – both in terms of the numbers of species included and of the numbers of authentic strains representing those species.

The use of gene probes in conjunction with the PCR, to allow rapid and sensitive detection, is covered in *Genetics*, above.

Miniaturized biochemical test systems. Over 50 years after Smith et al. (1946) published their first identification scheme, the most widely used commercially available methods for identifying members of the genus *Bacillus* and its relatives are still based upon miniaturized developments of traditional, routine biochemical tests: the API 20E and 50CHB Systems (bioMérieux, Marcy l'Etoile, France), the VITEK System (bioMérieux, Hazelwood, Missouri, USA) and Biolog (Biolog Inc., Hayward, California, USA) (Logan et al., 2000).

The API 20E/50CHB kits contain miniaturized and standardized versions of conventional biochemical tests, the API 50CHB comprising 48 tests for acid production from carbohydrates, esculin hydrolysis, and a negative control. They can be used for distinguishing between a number of well-established *Bacillus* species, can also recognize biotypes within the *Bacillus cereus* group (Logan et al., 1979) and may be used for the presumptive distinction of *Bacillus anthracis* from other members of the *Bacillus cereus* group within 48 h. The overall findings of an international reproducibility trial employing code-numbered *Bacillus sensu lato* strains showed that better test reproducibility could be achieved with API tests than with the conventional tests of Gordon et al. (1973), even when the latter had been carefully standardized by the Subcommittee on the Taxonomy of the Genus *Bacillus* of the International Committee on Systematic Bacteriology (Logan and Berkeley, 1981). API tests have proved valuable in the characterization of several novel species of aerobic endospore-formers (Heyndrickx et al., 1998, 1999; Heyrman et al., 2003a, b, 2004; Logan et al., 2000, 2002b, 2004b). The API 50CH gallery offers some flexibility in incubation temperature and in pH and salinity of the suspension medium. Deinhard et al. (1987a) used an ammonium salts-yeast extract medium at pH 4, with bromphenol blue as indicator, to investigate acid production from carbohydrates in

the API 50CH gallery by strains of *Bacillus* (now *Alicyclobacillus*) *acidoterrestris*, and Logan et al. (2000) used a similar medium at pH 6 with 0.033% bromcresol purple as indicator to characterize strains of *Bacillus fumarioli*: KH_2PO_4, 3 g; $MgSO_4 \cdot 7H_2O$, 0.5 g; $CaCl_2 \cdot 2H_2O$, 0.25 g; $(NH_4)_2SO_4$, 0.2; yeast extract, 0.5 g; trace element solution, 1 ml; bromcresol purple, 0.33 g; distilled water, 1 l; pH 6.0. Trace element solution contained: $FeSO_4 \cdot 7H_2O$, 0.05 g; $ZnSO_4 \cdot 7H_2O$, 0.05 g; $MnSO_4 \cdot 3H_2O$, 0.05 g; distilled water, 100 ml; 0.1 M H_2SO_4, 1 ml. Nielsen et al. (1995a) used the 50CH gallery for testing carbohydrate utilization by alkaliphilic *Bacillus* species. Heyrman et al. (2003b) added up to 10% NaCl to the API 50CHB suspension medium in order to detect acid production by strains of halotolerant and halophilic *Bacillus* and *Virgibacillus* strains.

The Vitek system arose from the Auto Microbial System (AMS) which was used for the direct identification of microbes from urine samples (Aldridge et al., 1977) and which led from the Microbial Load Monitor (MLM), which was an instrumental system developed for NASA in the 1960s for the detection of specific micro-organisms in a space-craft environment. bioMérieux offers a *Bacillus* card for the VITEK automated identification system; identifications are automatically attempted hourly between 6 h of incubation and the final reading at 15 h.

As many new species have been proposed since the API and Vitek schemes were established, updated databases are being prepared following the study of a large number of strains which have been carefully authenticated by polyphasic taxonomic study. Biolog also offers a database for the aerobic endospore-forming bacteria. The Biolog system is based on carbon source utilization patterns, indicated by the reduction of a tetrazolium dye, using a 96-well MicroPlate that is inoculated with a standardized suspension of a pure culture and incubated as appropriate. Unlike the API 50CH, the system is based on the process of metabolism itself rather than the release of metabolic by-products such as acid. For *Bacillus* identification the first release of the Biolog Gram-positive panel was dogged by problems of false positive results (Baillie et al., 1995), and this problem has been addressed by using a more viscous suspension medium to reduce flocculation and pellicle formation. The MicroPlates can be read after 4–6 h incubation, then reincubated for a further 12–18 h if an acceptable similarity threshold has not been reached.

The effectiveness of such kits can vary with the genera and species of aerobic endospore formers concerned, but they are improving with continuing development and expanding databases (Logan, 2002), and many of the characters they test are valuable in polyphasic taxonomic studies. It is stressed that their use for identification should always be preceded by the basic characterization tests, especially endospore formation, described above.

Differentiation from closely related taxa

The 142 species of *Bacillus* present such a wide diversity of routine phenotypic features that there are no characters that reliably allow distinction of *Bacillus* species from members of the other 18 genera in the family *Bacillaceae*, or from some genera of aerobic endospore-forming bacteria in other families. Table 1 summarizes some routine phenotypic characters that may be of value for differentiating between members of the family.

Taxonomic comments

Ferdinand Cohn established the genus *Bacillus* in 1872 to include the three species of rod-shaped bacteria, *Bacillus subtilis* (type species), *Bacillus anthracis* and *Bacillus ulna*, without taking motility or sporulation into account. Ehrenberg had described *Vibrio subtilis* in 1835, and "subtilis" is one of the earliest bacterial species epithets still in use. Davaine (1868) had proposed the genus *Bacteridium* to accommodate the nonmotile organism that causes anthrax. Cohn illustrated spores, and in later publications he discussed the resistance of spores and their significance in anthrax epidemiology, but Winter (1880) was the first to include "propagation through spores" in the description of the genus, and Prazmowski (1880) was the first to use sporulation as a differential (i.e., taxonomic) characteristic; he proposed the genus name *Clostridium* for organisms that differed from *Bacillus* in having spindle-shaped sporangia. At this time some workers still believed that all bacteria existed in several morphological and physiological forms, and that classification of the "fission fungi" was of no scientific value – Buchner (1882) claimed that shaking cultures of *Bacillus subtilis* at different temperatures could yield *Bacillus anthracis*!

The definition of the genus *Bacillus* as rod-shaped, aerobic or facultatively anaerobic organisms forming resistant endospores has been used for many years, and it remains of practical value despite the proposal of the strictly anaerobic species *Bacillus infernus* (Boone et al., 1995), *Bacillus arsenicoselenatis* and *Bacillus selenitireducens* (Switzer Blum et al., 1998). However, it was not until the 1880s that the name *Bacillus* and sporulation were brought together, and not until 1920 that aerobic growth became a defining character. From the 1880s to the 1900s *Bacillus* was variously used to contain rod-shaped organisms, all the rods except those in *Clostridium*, all the spore-formers, only those spore-formers producing unswollen sporangia, all motile rods, peritrichously flagellate rods (a classification that became the best-established in the American literature up to 1920), and nonmotile spore-formers with unswollen sporangia. The term *Bacillus* has thus been used in two senses: as a genus name, and, as "bacillus," as a general reference to shape; unfortunately the latter remains the most widely accepted definition of the term, especially by medical bacteriologists. As early as 1913, Vuillemin (1913) considered the name *Bacillus* so vulgarized by its various applications that it should lose nomenclatural status.

Although some took physiological characters as well as morphological ones into account in their classifications, it was not until the Committee of the Society of American Bacteriologists on Characterization and Classification of Bacterial Types reported in the early volumes of the *Journal of Bacteriology* (Winslow et al., 1917, 1920), that satisfactory and largely uncontested definitions of the bacterial groups emerged. The family *Bacillaceae* was defined as "Rods producing endospores, usually Gram-positive. Flagella when present peritrichic. Often decompose protein media actively through the agency of enzymes" and *Bacillus* was described as "Aerobic forms. Mostly saprophytes. Liquefy gelatin. Often occur in long threads and form rhizoid colonies. Form of rod usually not greatly changed at sporulation". The Committee also used the requirement of oxygen and sporangial shape for differentiation between *Bacillus* and *Clostridium*, the other genus in the family *Bacillaceae*, and this description was applied in the first and second editions of *Bergey's Manual of Determinative Bacteriology* (Bergey et al., 1923,

1925). So, as noted by Gordon (1981), the definition of the genus *Bacillus* as aerobic endospore-forming rods had become widely established by the 1920s.

Identification remained difficult, however, and the early editions of *Bergey's Manual* were not practical bench books for many taxa. Although the commonest spore-forming bacteria were accurately described in papers published early in the 20th century, Ford and his coworkers found that identification of their fresh isolates from milk remained difficult. This stimulated an extensive investigation of many strains from various environments in order to test the classification devised for the milk strains (Laubach et al., 1916; Lawrence and Ford, 1916), and 26 species were recognized – four of them new.

In 1937 it was agreed that "the genus *Bacillus* should be so defined as to exclude bacterial species which do not produce endospores" (Nomenclature Committee of the International Society for Microbiology, 1937; St. John-Brooks and Breed, 1937), over 100 years after Ehrenberg first described what is now the type species of the genus.

Some late-19th century classification schemes excluded motile organisms from *Bacillus*, and so abandoned *Bacillus subtilis* as the type species of the genus. Although other schemes retained *Bacillus subtilis*, in the late 1890s and early 1900s some confusion was emerging about the identity of the type. When describing *Bacillus cereus*, Frankland and Frankland (1887) noted its similarity both to *Bacillus anthracis* and a culture of *Bacillus subtilis* received from Koch. That could not have been *Bacillus subtilis* as recognized nowadays, and two very different type strains seemed to exist: one bore small spores and germinated equatorially, and the other formed much larger spores with germination occurring at the pole. The former was a strain from the University of Marburg and the latter was the Michigan type originating from the laboratory of Koch in 1888, and then maintained at the University of Michigan.

Following studies of strains from various culture collections, and after finding that the small-spored type tended to overgrow the large-spored type in cultures that mimicked the methods used by Cohn and other earlier workers, Conn (1930) suggested that the Marburg type should be called *Bacillus subtilis* Cohn. The Nomenclature Committee of the International Society for Microbiology sought the opinions of its members (Breed and St. John-Brooks, 1935), and in 1936, at the Second International Congress for Microbiology in London, the Marburg strain of *Bacillus subtilis* was officially adopted as the generic type (Nomenclature Committee of the International Society for Microbiology, 1937; St. John-Brooks and Breed, 1937).

With the resolution of the type strain controversy, the discovery that many pathogenic "*Bacillus subtilis*" strains were in fact *Bacillus cereus* stimulated a large taxonomic study of the genus (Clark, 1937) followed by a grouping of the mesophilic species in the form of a diagnostic key (Smith and Clark, 1937). In another study with a key, Gibson and Topping (1938) considered the *Bacillus circulans* and *Bacillus fusiformis* groups to be species complexes exhibiting several variations, and the problems so presented remain incompletely resolved to this day. Smith's team meticulously characterized their cultures and then emphasized the similarities rather than the differences between their strains, so "lumping" their taxa rather than splitting them. In a report published a decade after the study began (Smith et al., 1946), they recognized three groups of species: Group

One comprised those with oval to cylindrical spores without definite swelling of the sporangia, and included the *Bacillus cereus* and *Bacillus subtilis* groups, *Bacillus pumilus*, *Bacillus lentus*, *Bacillus megaterium* and *Bacillus firmus*. Group Two contained those with oval spores and swollen sporangia, and included *Bacillus alvei*, *Bacillus brevis*, *Bacillus circulans*, *Bacillus laterosporus*, *Bacillus macerans* and *Bacillus polymyxa*. Group Three consisted of *Bacillus pasteurii* and *Bacillus sphaericus*, both of which produced round spores with distinct swelling of the sporangia. The first truly workable diagnostic key designed for *Bacillus* identification emerged as a result of this work and appeared in the 6th edition of *Bergey's Manual* (Breed et al., 1948). The effectiveness of the scheme was demonstrated by Knight and Proom (1950); in their study of 296 strains all but 51 could be allocated to the species or groups previously described.

Smith et al. (1952) published a revision of their 1946 report, another classic example of painstaking and objective work. It was based on the study of 1134 strains, and such were the problems of synonymy in the genus that although 491 of these had 158 species names on receipt, all but 20 could be assigned to only 19 species. The classification outlined was used in the 7th edition of *Bergey's Manual* (Breed et al., 1957) and was adopted by most bacteriologists working with *Bacillus*. The designation of *Bacillus anthracis* as a variety of *Bacillus cereus* resulted in much controversy, but Smith et al. (1946, 1952) cited several reports on the loss of pathogenicity by *Bacillus anthracis* and considered such strains to be indistinguishable from *Bacillus cereus*. These studies from Smith's laboratory shaped the future of *Bacillus* taxonomy, and various research groups began applying existing techniques and new methods to the taxonomy of the genus *Bacillus*.

Numerical taxonomic methods were first applied to *Bacillus* by Sneath, 1962) using the data of Smith et al. (1952), and a phenogram was constructed which "largely agreed" with the 1952 classification. Although current classifications were criticized by Bonde (1975) on the grounds that their systems depended more on laboratory culture collection strains than on fresh isolates, his own classification generally agreed with that of Gordon et al. (1973).

During the 1970s several new approaches to characterization such as serology, enzyme and other molecular studies, and pyrolysis gas-liquid chromatography (Oxborrow et al., 1977) began to emerge as potentially useful taxonomic tools in bacterial taxonomy as a whole but despite this, the taxonomy of *Bacillus* remained relatively untouched, with few new species or subspecies being described and names validated. With *Bacillus* the emphasis during this time was on facilitating the identification of members of the genus.

One of the most influential and significant studies of the genus *Bacillus* was published in by Gordon et al. (1973), 1,134 strains were included in a classification which formed the basis for the *Bacillus* section in the Eighth Edition of *Bergey's Manual* (Gibson and Gordon, 1974) but numerical methods were not used in the analysis. The classification was very similar to that proposed by Smith et al. (1946, 1952), the main difference being that subgrouping was not attempted and the species were arranged as a spectrum of morphological characteristics. Given the period at which the work was done, it is surprising that numerical analysis was not attempted, and so the success of the arrangement that was made is all the more impressive. The characterization tests applied by Gordon et al. (1973) were used by the International Committee on Systematic Bacteriology Subcommittee on the Taxonomy of the Genus *Bacillus* as standard methods, but international reproducibility trials found that, even in the hands of *Bacillus* experts, the miniaturized, highly standardized tests in the API System gave more rapid and consistent results (Logan and Berkeley, 1981). The increasing incidence of *Bacillus* isolations from clinical environments emphasized the need for a rapid identification scheme, which could only follow an improved taxonomy of the genus. Consequently, Logan and Berkeley (1984) developed a *Bacillus* identification scheme based upon API tests.

It was also the classification of Gordon et al. (1973) with support from the work of Logan and Berkeley (1981) that formed the basis of the list of *Bacillus* included in the Approved Lists of Bacterial Names published by the International Committee of Systematic Bacteriology (Skerman et al., 1980). These lists marked a new starting date for bacterial nomenclature, the previous date being that of Linnaeus' monumental classification work, *Species Plantarum*, which was published in 1753. Since that time, many synonyms had been inadvertently been proposed, and the number of *Bacillus* species described fluctuated greatly through the eight editions of *Bergey's Manual*, ranging from a peak of 146 species in 1939 to the smallest number of 22 in 1974 (a further 26 appeared as *species incertae sedis*, and many of these were represented by very few strains).

The need for subdividing *Bacillus* had long been recognized from its DNA base composition range of 32–69 mol% G + C and the arrangements that had emerged from numerical taxonomies of phenotypic data (Logan, 1994). Early phylogenetic studies based on 16S rRNA cataloging (e.g., Fox et al., 1977, 1981; Stackebrandt et al., 1987) confirmed the evolutionary heterogeneity of *Bacillus*, and were found to be entirely consistent with the distribution of murein types among the species (Stackebrandt et al., 1987), but it was considered too early to split *Bacillus* to provide a more "natural" classification, as very few species had yet been analyzed by 16S rRNA cataloguing or full sequence comparative studies. The genus was thus still kept as one taxonomic entity in *Bergey's Manual of Systematic Bacteriology* (Claus and Berkeley, 1986).

Ash et al. (1991) recognized five phylogenetically distinct clusters among 51 type strains of *Bacillus* species on the basis of 16S rRNA sequence similarities. Rössler et al. (1991) published the results of a similar study in the same year, finding four major clusters which showed high correlation with those described by Ash et al. A weakness of both studies was their reliance upon single (type) strains of each species, so that within-species diversities were not indicated and the authenticities of the strains were not controlled. Indeed, in the Ash et al. study, *Bacillus acidoterrestris* and *Bacillus lautus* were misplaced owing, presumably, to contaminants (Heyndrickx et al., 1996b; Wisotzkey et al., 1992). A sixth rRNA group containing *Bacillus alcalophilus* was recognized by Nielsen et al. (1994) but insufficient phenotypic and genotypic data were available to allow the proposal of a new genus.

With the accumulation of further 16S rRNA (rDNA) sequence data, *Bacillus* has been divided into more manageable and better-defined groups. So far, nine new genera have been established. However, this taxonomic progress has not revealed readily determinable features characteristic of each genus, and they show wide ranges of sporangial morphologies and phenotypic test patterns.

The proposal of *Alicyclobacillus* (Wisotzkey et al., 1992) initiated the splitting of the genus *Bacillus* and extensive reclassification ensued. *Alicyclobacillus* contains 19 species of thermoacidophiles, including organisms formerly called *Bacillus acidocaldarius* (Darland and Brock, 1971), *Bacillus acidoterrestris* (Deinhard et al., 1987a) and *Bacillus cycloheptanicus* (Deinhard et al., 1987b). These organisms exhibit ω-alicyclic fatty acids as the major natural membranous lipid components, a phenotypic trait not found in other *Bacillus* species.

Ash et al. (1993) proposed the genus *Paenibacillus* to encompass their previously described Group 3 (Ash et al., 1991) as they considered these organisms to be phylogenetically "so removed" from the cluster which contained *Bacillus subtilis* as to warrant such an action. Data from their 1991 publication were used along with previously published phenotypic characters, and a gene probe based on 16S rRNA. *Paenibacillus* contains 95 species and includes organisms formerly called *Bacillus alginolyticus* (Nakamura, 1987), *Bacillus alvei* (Cheshire and Cheyne, 1885), *Bacillus amylolyticus* (Nakamura, 1984a), *Bacillus chondroitinus* (Nakamura, 1987), *Bacillus curdlanolyticus* (Kanzawa et al., 1995), *Bacillus glucanolyticus* (Alexander and Priest, 1989), *Bacillus gordonae* (Pichinoty et al., 1986), a synonym of *Bacillus validus*, now *Paenibacillus validus* (Heyndrickx et al., 1995), *Bacillus kobensis* (Kanzawa et al., 1995), *Bacillus larvae* (White, 1906) and *Bacillus pulvifaciens* (Nakamura, 1984c) (now both subspecies of *Paenibacillus larvae* (Heyndrickx et al., 1996c), *Bacillus lautus* (Nakamura, 1984a), *Bacillus lentimorbus* (Dutky, 1940), *Bacillus macerans* (Schardinger, 1905), *Bacillus macquariensis* (Marshall and Ohye, 1966), *Bacillus pabuli* (Nakamura, 1984a), *Bacillus peoriae* (Montefusco et al., 1993) *Bacillus polymyxa* (Prazmowski, 1880), *Bacillus popilliae* (Dutky, 1940), *Bacillus thiaminolyticus* (Nakamura, 1990) and *Bacillus validus* (Nakamura, 1984a). It was perhaps surprising to find that *Bacillus circulans sensu stricto* was recovered within the cluster containing *Bacillus subtilis*, as it is phenotypically similar to *Paenibacillus polymyxa* and, *Paenibacillus macerans*; however, strains originally named *Bacillus amylolyticus*, *Bacillus pabuli* and *Bacillus validus* had been members of the *Bacillus circulans* "complex" (Gibson and Topping, 1938) until these three species were revived by Nakamura (1984c). Another member of this complex, *Bacillus lautus*, was also revived by Nakamura (1984c) and later transferred to *Paenibacillus* (Heyndrickx et al., 1996c). Unfortunately, there is no phenotypic character which might allow the differentiation of the genus *Paenibacillus* from other genera of aerobic endospore-forming genera, and so the taxonomic gain in phylogenetic accuracy was not accompanied by easier routine identification.

Following 16S rRNA gene sequence analyses of type strains of species closely related to *Bacillus brevis* and *Bacillus aneurinolyticus* Shida et al. (1996) recognized two distinct clusters and proposed the *Bacillus brevis* cluster as a new genus *Brevibacillus* and the *Bacillus aneurinolyticus* cluster as *Aneurinibacillus*. *Brevibacillus* (Shida et al., 1996), contains the 10 species formerly known as *Bacillus agri* (Nakamura, 1993), *Bacillus borstelensis* (Shida et al., 1995), *Bacillus brevis* (Migula, 1900), *Bacillus centrosporus* (Nakamura, 1993), *Bacillus choshinensis* (Takagi et al., 1993), *Bacillus formosus* (Shida et al., 1995), *Bacillus laterosporus* (Laubach et al., 1916), *Bacillus parabrevis* (Takagi et al., 1993), *Bacillus reuszeri* (Shida et al., 1996) and *Bacillus thermoruber* (Manachini et al., 1985). *Bacillus galactophilus* (Takagi et al., 1993) is a synonym of *Bacillus agri* (Shida et al., 1994). *Aneurinibacillus* (Shida et al., 1996) emend. Heyndrickx et al. (1997) accommodates the three species *Bacillus aneurinilyti-*

cus (Shida et al., 1994), *Bacillus migulanus* (Takagi et al., 1993), and *Bacillus thermoaerophilus* (MeierStauffer et al., 1996).

Subsequent developments included proposals for three genera of halotolerant and halophilic species: *Virgibacillus* (Heyndrickx et al., 1998) containing *Bacillus pantothenticus* (Proom and Knight, 1950); *Gracilibacillus* (Wainø et al., 1999) containing *Bacillus dipsosauri* (Lawson et al., 1996); *Salibacillus* (Wainø et al., 1999) containing *Bacillus marismortui* (Arahal et al., 1999, 2000) and *Bacillus salexigens* (Garabito et al., 1997).

It is possible that the boundaries between some of the new genera, in terms of 16S rDNA sequence differences, may become obscured as and when further species are discovered; indeed, Heyrman et al. (2003b) transferred the species of *Salibacillus* to *Virgibacillus*.

Several of the familiar *Bacillus* thermophiles comprise a distinct evolutionary line, and so the genus *Geobacillus* (Nazina et al., 2001) accommodates species formerly called *Bacillus kaustophilus* (Priest et al., 1988), *Bacillus stearothermophilus* (Donk, 1920), *Bacillus thermocatenulatus* (Golovacheva et al., 1975), *Bacillus thermodenitrificans* (Manachini et al., 2000), *Bacillus thermoglucosidasius* (Suzuki et al., 1983) and *Bacillus thermoleovorans* (Zarilla and Perry, 1987). However, the longer-established species in this genus await polyphasic taxonomic study and circumscription, as many misnamed strains are known to exist in collections.

The thermophilic, round-spored, organism *Bacillus thermosphaericus* (Andersson et al., 1996) has been accommodated within the new genus *Ureibacillus* (Fortina et al., 2001b). Several other round-spored species, *Bacillus globisporus* (Larkin and Stokes, 1967), *Bacillus pasteurii* (Chester, 1898) and *Bacillus psychrophilus* (Nakamura, 1984b), have been transferred (Yoon et al., 2001b) to *Sporosarcina* (Kluyver and van Neil, 1936) which was established to accommodate the motile, spore-forming coccus *Sporosarcina ureae* (Beijerinck, 1901). As this proposal placed spore-forming rods and cocci in the same genus, the definition of *Sporosarcina* had to be emended. Two further new genera of round-spored organisms were subsequently proposed to accommodate single species, but they were only represented by single strains in the study concerned: *Jeotgalibacillus alimentarius* (Yoon et al., 2001c), and *Marinibacillus* to which *Bacillus marinus* was transferred (Yoon et al., 2001c).

Bacillus continues to accommodate the best-known species such as *Bacillus subtilis* (the type species), *Bacillus anthracis*, *Bacillus cereus*, *Bacillus licheniformis*, *Bacillus megaterium*, *Bacillus pumilus*, *Bacillus sphaericus* and *Bacillus thuringiensis*. It still remains a large genus, with 90 species (May 2004), since losses of species to other genera have been balanced by proposals for new *Bacillus* species. Seventy-one of the present members of *Bacillus* were proposed after the treatment of the genus for the first edition of the *Systematics* was compiled, and of the 34 species described in that edition, only 18 remain in *Bacillus*.

Members of the *Bacillus cereus* group, *Bacillus anthracis Bacillus cereus Bacillus thuringiensis*, are really pathovars of a single species (Turnbull et al., 2002), and yet the phylogenetic and phenetic distinction of this group probably support generic status. The internal division of the different 16S rRNA groups within the genus *Bacillus sensu lato* is presently far from clear. Many of its species fall into several apparently distinct rRNA sequence groups such as the "*Bacillus subtilis* group", the "*Bacillus cereus* group", and the "*Bacillus sphaericus* group", but although such divisions may also be phenotypically distinguishable, intermediate organisms may make satisfactory subdivision difficult.

Numerous *Bacillus* species mentioned in the literature over the years are not now recognized because of synonymy, incomplete characterization, or the lack of type strains, or even the complete loss of original isolates. Some invalid names persist, however, and can cause much confusion. Good examples are "*Bacillus subtili* var. *niger*" and "*Bacillus globigii*". Strains bearing these names are used for sterilization control and other purposes, but the names themselves are invalid. Most of these currently used strains may be regarded as *Bacillus subtilis*, and are listed as such in the catalogues of several culture collections; all three names may appear on the packaging of the commercially available biological indicator products that use them. *Bacillus subtilis* var. *niger* is a name that was applied to strains that produce a distinctive black pigment, hence their popularity as biological indicators. However, N. R. Smith's strain 1221A of *Bacillus subtilis* var. *niger*, which is a standard strain for sterility testing, is noted for its red pigment, and is also known as *Bacillus globigii*. Gordon et al. (1973) found their strains labeled as *Bacillus globigii* were *Bacillus circulans (sensu lato)*, *Bacillus licheniformis*, *Bacillus pumilus* or *Bacillus subtilis* var. *niger*, and they commented: "strains of *Bacillus subtilis*, *Bacillus pumilus* and *Bacillus circulans* labeled *Bacillus globigii* are extant, and some have been widely distributed. As a result, the name *Bacillus globigii* is meaningless".

New species and subspecies of aerobic endospore formers are regularly described: 182 new species were proposed between the publication of the First Edition of this *Manual* in 1986 and the time of preparing the present edition (May 2004); of these, 121 were initially assigned to *Bacillus*, 43 were then transferred to new genera, and three others await transfer (these are *Bacillus edaphicus* and *Bacillus mucilaginosus*, which belong in *Paenibacillus*, and *Bacillus thermantarcticus*, which belongs in *Geobacillus*). The remaining new species were distributed among the 11 new genera mentioned in the preceding sentence and ten further new genera; 15 of the new genera contain a single species. During that period only six proposals for merging species (Heyndrickx et al., 1996a; Heyndrickx et al., 1995; Rosado et al., 1997; Shida et al., 1994; Sunna et al., 1997b), and one proposal for merging genera (Heyrman et al., 2003b) had been made. At the time of writing there are 440 valid species of aerobic endospore-formers among 47 genera. Unfortunately, new taxa have often been proposed on the basis of very few strains (48 of the present *Bacillus* species are based upon the study of a single strain, and nine on the basis of only two strains), so that their within-species variations are unknown. It is of small comfort that the proportion of new *Bacillus* species proposed on the basis of a single isolate is, at 33%, less than the proportion of single-isolate taxa described for prokaryotes overall between 1990 and 2000 (Christensen et al., 2001).

Another regrettable circumstance is that many recently described species of *Bacillus*, and species of the genera recently derived from it, represent genomic groups disclosed by DNA–DNA pairing experiments, and routine phenotypic characters for distinguishing some of them are very few and of unproven value. An extreme example of this kind of problem is the splitting of strains of *Bacillus subtilis* into two subspecies and three new species: *Bacillus atrophaeus* (Nakamura, 1989), *Bacillus mojavensis* (Roberts et al., 1994), *Bacillus vallismortis* (Roberts et al., 1996), *Bacillus subtilis* subsp. *spizizenii* (Nakamura et al., 1999) and *Bacillus subtilis* subsp. *subtilis*. These proposals were based principally upon DNA relatedness studies (the 70% relatedness threshold for species being rigorously applied),

with distinctions between these "cryptic" individual taxa being supported by a miscellany of approaches which included small differences in fatty acid compositions, multilocus enzyme electrophoresis, restriction digest analysis of selected genes, and transformation resistance. The only distinctive phenotypic character cited among these proposals was the production of brown pigment by *Bacillus atrophaeus* on media containing tyrosine, and so the recognition of the four new taxa appears to be of little practical value.

It is known (Fox et al., 1992; Stackebrandt and Goebel, 1994) that rDNA sequence analysis alone does not allow unequivocal differentiation at the species or finer taxonomic levels, and may even lead to completely mistaken conclusions on the exact phylogenetic positions of certain strains (Clayton et al., 1995), especially when taxa are proposed on the basis of lone strains. Christensen et al. (2001) proposed an addition to Recommendation 30b of the *Bacteriological Code* (1990 Revision), which advises that proposals for new taxa be based upon at least five strains from different sources, and that descriptions be based upon comparative studies including reference strains. Adoption of this suggestion, and its application by reviewers of proposals for new taxa, would do much to improve the practical usefulness of future developments in *Bacillus* taxonomy.

An important proposal, published at the time that this account was going to press, was for the transfer of *Bacillus sphaericus* and *Bacillus fusiformis* to the new genus *Lysinibacillus* (Ahmed et al., 2007c). Other transfers of species listed below include a proposal by Jeon et al. (2005b) for *Bacillus haloalkaliphilus* to be reclassified in a new genus *Alkalibacillus*, and a proposal by Hatayama et al. (2006) for the transfer of *Bacillus laevolacticus* to *Sporolactobacillus*, and the reclassification of *Bacillus halophilus* as *Salimicrobium halophilum* (Yoon et al., 2007b) and *Bacillus arvi*, *Bacillus arenosi* and *Bacillus neidei* into a new genus, *Viridibacillus* (Albert et al., 2007).

It is clear that the taxonomic reshuffling of the genus has not yet come to an end. Furthermore, none of the other new genera of aerobic endospore formers, *Sulfobacillus*, *Amphibacillus*, *Halobacillus*, *Ammoniphilus*, *Thermobacillus*, *Filobacillus*, *Oceanobacillus*, *Lentibacillus*, *Paraliobacillus*, *Cerasibacillus*, *Halolactibacillus*, *Pontibacillus*, *Tenuibacillus*, *Salinibacillus*, *Pullulanibacillus*, *Tuberibacillus*, *Caldalkalibacillus*, *Ornithinibacillus*, *Paucisalibacillus*, *Vulcanibacillus*, *Pelagibacillus*, and *Piscibacillus* (Golovacheva and Karavaiko, 1978; Hatayama et al., 2006; Ishikawa et al., 2002; Ishikawa et al., 2005; Kim et al., 2007b; L'Haridon et al., 2006; Lim et al., 2005b; Lu et al., 2001; Mayr et al., 2006; Nakamura et al., 2004b; Niimura et al., 1990; Nunes et al., 2006; Ren and Zhou, 2005a, b; Schlesner et al., 2001; Spring et al., 1996; Tanasupawat et al., 2007; Touzel et al., 2000; Xue et al., 2006; Yoon et al., 2002; Zaitsev et al., 1998) contain species formerly assigned to *Bacillus*. "*Bacillus flavothermus*" is accommodated in *Anoxybacillus* as *Anoxybacillus flavithermus* (Pikuta et al., 2000a).

Further reading

Berkeley, R.C.W., M. Heyndrickx, N.A. Logan and P. De Vos. 2002. Applications and Systematics of *Bacillus* and Relatives. Blackwell Science, Oxford.

Differentiation of the species of the genus *Bacillus*

Differential characteristics of the species of the genus *Bacillus* are shown in Table 3, and additional data are shown in Table 4.

List of species of the genus *Bacillus**

1. **Bacillus subtilis** (Ehrenberg 1835) Cohn 1872, 174[AL.] *Nom. cons.* Nomencl. Comm. Intern. Soc. Microbiol. 1937, 28; Opin. A. Jud. Comm. 1955, 39 (*Vibrio subtilis* Ehrenberg 1835, 279.)

sub'ti.lis. L. adj. *subtilis* slender.

Aerobic, Gram-positive, motile rods, forming ellipsoidal to cylindrical spores which lie centrally, paracentrally and subterminally in unswollen sporangia (Figure 10g). Cells grown on glucose agar stain evenly. Cells 0.7–0.8 by 2.0–3.0 μm, occurring singly and in pairs, seldom in chains. Colonial morphology is exceptionally variable, within and between strains, and may give the appearance of a mixed culture. Colonies are round to irregular in shape and of moderate (2–4 mm) diameter, with margins varying from undulate to fimbriate; they become opaque, with surfaces that are dull and which may become wrinkled; color is whitish, and may become creamy or brown; textures range from moist and butyrous or mucoid, through membranous with an underlying mucoid matrix, with or without mucoid beading at the surface, to rough and crusty as they dry (Figure 9g). Pigments, varying from cream through yellow, orange, pink and red, to brown or black, may be formed on potato or agar media containing glucose; strains forming brown or black pigment were often formerly called "*Bacillus subtilis* var. *aterrimus.*" Strains forming brownish-black pigment on tyrosine (and so often evident on the crude media available to earlier workers), and often formerly called "*Bacillus subtilis* var. *niger*," have been split from *Bacillus subtilis* as *Bacillus atrophaeus.*

Optimum growth temperature 28–30 °C, with minimum of 5–20 °C and maximum of 45–55 °C. Growth occurs between pH 5.5 and 8.5, but limits have not been recorded. Some restricted anaerobic growth may occur in complex media with glucose or (less effectively) nitrate. Growth occurs on minimal medium with glucose and an ammonium salt as sole sources of carbon and nitrogen. Grows in presence of up to 7% NaCl, some strains will tolerate 10% NaCl. Catalase-positive, oxidase variable. Casein, esculin, gelatin and starch are hydrolyzed, phenylalanine and urea are not hydrolyzed. Pectin and polysaccharides of plant tissues are decomposed. Dextran and levan are formed extracellularly from sucrose. Citrate is utilized as sole carbon source by most strains; propionate is not utilized. Nitrate is reduced to nitrite. Voges–Proskauer-positive. Acid without gas is produced from glucose and from a wide range of other carbohydrates.

The practical values of the distinction of the subspecies of *Bacillus subtilis* and of the species *Bacillus mojavensis* and *Bacillus vallismortis* are questionable. The distinction of

Bacillus atrophaeus from the *Bacillus subtilis* subspecies and from *Bacillus mojavensis* and *Bacillus vallismortis* is dependent upon brownish-black pigment production on tyrosine agar by strains of *Bacillus atrophaeus.* See Table 5.

Endospores are very widespread in soil, dust and on vegetation, and in many other environments. The vegetative organisms participate in the early stages of the breakdown of organic matter. Causative agent of ropy (slimy) bread.

1a. **Bacillus subtilis subsp. subtilis** Nakamura, Roberts and Cohan 1999, 1214[VP]

Description is that given above for the species.

Phenotypically similar to *Bacillus atrophaeus* and distinguishable from that species only by the pigmentation of the latter. Not distinguishable from *Bacillus mojavensis, Bacillus subtilis* subsp. *spizizenii* and *Bacillus vallismortis* by conventional phenotypic tests.

DNA G + C content (mol%): 41.5–47.5 (T_m) for 31 strains, 41.8–46.3 (Bd) for 34 strains, and 42.9 (T_m) for the type strain.

Type strain: ATCC 6051, IAM 12118, CCM 2216, DSM 10, IFO 12210, NCIMB 3610, NCTC 3610, NRRL NRS-744.

EMBL/GenBank accession (16S rRNA gene): AB042061 (IAM 12118).

Additional remarks: Strains designated by Gibson (1944) and Smith et al. (1946) as synonyms of *Bacillus subtilis* included *Bacillus aterrimus, Bacillus mesentericus, Bacillus natto, Bacillus niger, Bacillus nigrificans* and *Bacillus panis.* "*Bacillus natto*" is a name given to *Bacillus subtilis* strains associated with natto, a Japanese food made by fermenting soybeans with these organisms. Strains formerly designated "*Bacillus amyloliquefaciens*" or "*Bacillus subtilis* var. *amyloliquefaciens*" are now accommodated within *Bacillus amyloliquefaciens.*

1b. **Bacillus subtilis subsp. spizizenii** Nakamura, Roberts and Cohan 1999, 1214[VL.]

spi.zi.ze'ni.i. L. gen. n. *spizizenii* named after the American bacteriologist J. Spizizen.

Phenotypically indistinguishable from *Bacillus subtilis* subsp. *subtilis,* and separated from that taxon only by DNA relatedness values of 58–68% with 12 strains of that subspecies, by the presence of ribitol in the cell wall, and by transformation studies.

The type strain was isolated from Tunisian soil.

Type strain: NRRL B-23049, DSM 15029, LMG 19156, KCTC 3705.

EMBL/GenBank accession number (16S rRNA gene): AF074970 (NRRL B-23049).

2. **Bacillus aeolius** Gugliandolo, Maugeri, Caccamo and Stackebrandt 2003b, 1701[VP] (Effective publication: Gugliandolo, Maugeri, Caccamo and Stackebrandt 2003a, 175.)

ae.o'li.us. L. adj. *aeolius* pertaining to the Eolian Island (Insulae Aeoliae) where the organism was isolated from a shallow marine hydrothermal vent.

Aerobic, Gram-positive, motile rods, 0.5 μm by 2.0 μm, forming terminal, oval endospores. Description is based upon a single isolate. Catalase-negative, oxidase-positive. Growth

---

*Type strains distributed by international culture collections may be subjected to quality control, and the user should establish whether or not this is the case. The type strain accession numbers shown for *Bacillus* species were chosen to give an acceptable geographic spread. Sometimes more than one 16S rRNA gene sequence is available from EMBL/GenBank for different subcultures of the same type strain; we indicate alongside the EMBL/GenBank accession number which sequence has been used to construct the tree (Figure 8). Other sequences may or may not be reliable and users of these should check carefully.

TABLE 3. Differential characteristics of the species of the genus *Bacillus*[a,b]

Characteristic	1. B. subtilis[c]	2. B. aeolius	3. B. agaradhaerens[d]	4. B. alcalophilus[d]	5. B. algicola[d]	6. B. amyloliquefaciens[c]	7. B. anthracis[e]	8. B. aquimaris	9. B. arseniciselenatis[d]	10. B. asahii	11. B. atrophaeus[c]	12. B. azotoformans	13. B. badius	14. B. barbaricus	15. B. bataviensis	16. B. benzeovorans	17. B. carboniphilus	18. B. cereus[e]
Pigmented colonies:																		
Yellow-pink-red	−h	−		−	+h	−	−	+h	−h		−		−	−	−	−	+h	−h
Dark brown/black	−h	−		−	−	−	−	−	−		+h		−	−	−	−	−	−
Motility	+	+			+	+	−	+		+	+	+	+	−	+	+	+	+
Cell diameter >1.0 μm	−	−	−	−	−	−	+	−	−	−	−	−	−	v	−	d	+	+
Spore formation:	+	+	+	+	+	+	+	+	+	+	+	+	+	+	+	+	+	+
Ellipsoidal	+	+	+	+	+	+	+	+	+		+	+	+	+	+	+	+	+
Cylindrical	v	+																d
Spherical	−	−	−	−	−	−	−	−		−	−	−	−	d	−	−	−	−
Borne terminally	−	+	−	−	+	−	−	−		+	−	+	v	−	−	−	−	−
Swell sporangia	−		+	−	−	−	−	+		−		+	−	+	+		−	−
Parasporal crystals	−	−		−	−	−	−	−	−	−	−		−	−	−		−	−
Catalase	+	−		+	w		+	+	+	+	+	−	+	+		d	+	+
Aerobic growth	+	+	+	+	+	+	+	+	−	+	+	+	+	+	+	+	+	+
Anaerobic growth	−	−		−	−	+	+	−	+		−	−	−	+	+	+	−	+
Voges-Proskauer	+	+	ng	−		+	+		ng	−	+		−			−	−	+
Acid from:																		
L-Arabinose	+	+	ng	+		d	−	−			+		−		−		−	−
D-Glucose	+	+	ng	+		+	+	+	−		+	−	−	+	+	−	−	+
Glycogen	+	−	ng			+	+	+					−	+	−	−		+
D-Mannitol	+	+	ng	+		+	−	−		−	+		−	−	+	−		
D-Mannose	+	+	ng			d	−	−		−	d	−	−		+	−		−
Methyl β-xyloside	−	−	ng			−	−	−			−		−	−	−	−		−
Salicin	+	+	ng			+	−	−			+		−	−	w	−	−	d[j]
Starch	+	+	ng			+	+	+					−	+	d/w	−	−	+[j]
D-Xylose	+	+	ng	+		d	−	−		−	+		−	−	−	−		
Hydrolysis of:																		
Casein	+	+	+	+	−	+	+	+		+	+		+		−		+	+
Gelatin	+	+	+	+	+	+	+			−	+		+		+	−	+	+
Starch	+	+	+	+	+	+	+	+		w	+		−	−			+	+
Utilization of:																		
Citrate	+	+		−	−	+	d[i]		+		+	+	+	−	−		−	+
Propionate	−			−	−	−			−		−	−				+	−	
Egg yolk reaction	−				−		+				−	−						+
Nitrate reduction	+	−	+	−	w	+	+		+	w	+	+	−		+	+	−	d
Growth at pH:																		
5		−	−	−	−			−	−					−	d	−		
6	+	−	−	−			+	+	+	+	+	+		w	+		+	+
7	+	+	−	−	+	+	+	+	−	+	+	+	+	+[q]	+	+	+	+
8	+	+	+	+	+	−		+	w	+				+[q]	+			
9		+	+	+	+	−		−	+	+				+[q]	+			
10	−	+	d	+	−		−	+	−					d				
Growth in NaCl:																		
2%	+	+	+	+	+	+	+	+	−		+			w			+	+
5%	+	+	+	+	−	+	+	+	+		+	−	+	−		−	+	+
7%	+	−	+	+		+	+	+			+	−	+	−		−	+	d
10%	d	−	+	−		d	−	+	+		−	−	−			−		
NaCl required for growth	−	+	+		−	−	+[x]	+		−			−			−		−
Growth at:																		
5 °C	d	−		−	−		−				d	−					−	
10 °C	d	−	+	+	+	−	−	+			+		−				−	d
20 °C	+	−	+	+	+	+	+	+	+	+	+		+	+			+	+
30 °C	+	+	+	+	+	+	+	+		+	+	+	+	+	+	+	+	+
40 °C	+	+	+	+	+	+	+	+		+	+	+	+	+	+	d	+	+
50 °C	d	+	−	−	−	+	−			−	+	−	+	−	+		−	
55 °C	d	+	−	−		−	−				d	−	−	−	d		−	
65 °C	−	+	−	−		−	−				−		−				−	
Growth with lysozyme present	d			−			+				−	−	−				−	+
Respiratory growth with As(V)									+									
Respiratory growth with Se(IV) or Se(VI)									+									
Autotrophic with H₂ + CO₂ or CO	−		−		−		−	−			−		−				−	−
Degradation of tyrosine	−		−		−		−	−			−	−	+				−	+
Deamination of phenylalanine	−		−		−		−	−			−	−	−				−	−
Allantoin or urate required	−	−	−	−	−	−	−	−			−	−	−	−	−	−	−	−

(continued)

TABLE 3. (continued)

Characteristic	19. B. circulans	20. B. clarkii[d]	21. B. clausii	22. B. coagulans	23. B. cohnii[d]	24. B. decolorationis	25. B. drentensis	26. B. endophyticus	27. B. farraginis	28. B. fastidiosus	29. B. firmus	30. B. flexus	31. B. fordii	32. B. fortis	33. B. fumarioli[f]	34. B. funiculus	35. B. fusiformis[g]	36. B. galactosidilyticus	37. B. gelatini
Pigmented colonies:																			
Yellow-pink-red	−	+[h]	−	−	−	−	−	+[h]	−	d[h]	+[h]	−	−	−	−	−	−	−	−
Dark brown/black	−	−	−	−	−	−	−	−	−	−	−	−	−	−	−	−	−	−	−
Motility	+			+	+	+	+	−	+	+	+		+	+	+	+	+	+	+
Cell diameter >1.0 μm	−	−	−	−	−	−	d	v	−	+	−	−	−	−	−	+	−	−	−
Spore formation:	+	+	+	+	+	+	+	+	+	+	+	+	+	+	+	+	+	+	+
Ellipsoidal	+	+	+	+	+	+	+		+	+	+	+	+	+	+	+	−	+	+
Cylindrical	−	−	−	−	−	−	−		−	−	d	−			+	−	−	−	−
Spherical	−	−	−	v	−	v	+	−		−	−	−			−	−	+	−	
Borne terminally	v	−	−	v	+	−	−	v		v	−	−			−	−	d	−	
Swell sporangia	+	d	v	+	+	+	+	−	v	−	v	−	v	+	−	−	+	+	−
Parasporal crystals	−	−	−	−	−	−	−	−		−	−	−							
Catalase	+			+	+	+		+	+	+	+		+	+	+	+		+	+
Aerobic growth	+	+	+	+	+	+	+	+	+	+	+	+	+	+	+	+	+	+	+
Anaerobic growth	+			+	−		+	−		−	+	−	−	−	−			+	−
Voges-Proskauer	−	ng		d	ng	−	−/w	−	−	ng	−	−	−	−	+	+	−	−	−
Acid from:																			
L-Arabinose	+	ng		d	ng	−	−	+	−	ng	−	−	−	−	−k		−	d/w	−
D-Glucose	+	ng		+	ng	w	w	+	−	ng	+	+	−	−	+k	+	−	+	+
Glycogen	+	ng		−	ng	−	−	−	−	ng	d		−		−k		−	−	−
D-Mannitol	+	ng		−	ng	d/w	−	+	−	ng	+	+	−	−	+k		−	d/w	+
D-Mannose	+	ng		+	ng	w	d/w	+	−	ng	−	−	−	−	+k		−	d/w	+
Methyl β-xyloside	+	ng		−	ng	−	−	−		ng	−		−		−k		−	−	−
Salicin	+	ng		d	ng	w	w	−		ng	−	d	−	−	−k		−	d/w	−
Starch	+	ng		+	ng	−	d/w	−		ng	+		−		−k		−	d/w	−
D-Xylose	+	ng		d	ng	−	d/w	d	−	ng	d	−	−	−	−k		−	d/w	+
Hydrolysis of:																			
Casein	w	+	+	−	d	+	−	−	−	−	w	+	−	−	−	−	+	w	+
Gelatin	d	+	+	d	+	+	−	−	−	−	+	+	−	−	+	−	+	−	+
Starch	w	−	+	+	+					−	+	+	−			+	−		
Utilization of:																			
Citrate	−			−		−	−	+	−	−	−	+	−	−	−	−	+	−	−
Propionate								−	−	−	−		d/w	−					
Egg yolk reaction	−			−						−	−	−					−		
Nitrate reduction	d	+	+	−	+	+	d	−	−	−	+	−	−	−	−	+	−	+	−
Growth at pH:																			
5	d	−	−	+	−	−	−		−	−	+	+	+	+	+	+	+	+	+
6	+	−	+	+	−	d	+		+	−	+	+	+	+	−	+	+	+	+
7	+	−	+	+		+	+	+	+	d	+	+	+	+	−	+	+	+	+
8	+		+	+		−	+		+	+	+	+	+	+	−	+	+	+	+
9	+	+		+	+			+	+	+	+	+	+	+	−	+	+	+	+
10	d	+	−	+	+	−	d				+				−	−	−	+	d
Growth in NaCl:																			
2%			+	+	+	+		+	+	+	+	+	+	+			+		+
5%		+	+	−	+	+		+	+	+	+	+	+	+			+		+
7%		+	+	−		+		+	+	−	+	+	+	+			+		+
10%		+	d	−	−	+		+	−		+								+
NaCl required for growth	−	+	−																
Growth at:																			
5 °C		−	−		−		+	−		−		−				−	−		
10 °C		−	−		+		+	+		+		−				−	−		
20 °C		+	+		+		+	+	w	+		+		w	w	−	+	+	
30 °C	+	+	+	+	+	+	+	+	+	+	+	+	+	+	+	+	+	+	
40 °C	+	+	+	+	+	+	+	+	+	d	+		+	+	+	+		+	+
50 °C	+	−	+	d	−	−	+	−		−	d	−			+		−	−	+
55 °C	d	−	−	d	−	−	d	−		−	−	−			+	−	−	−	+
65 °C	−	−	−	−	−	−	−	−		−	−				−	+	−	−	
Growth with lysozyme present								+									+		
Respiratory growth with As(V)																			
Respiratory growth with Se(IV) or Se(VI)																			
Autotrophic with $H_2 + CO_2$ or CO	−	−	−					−			−				−		−		
Degradation of tyrosine										−	−								
Deamination of phenylalanine		+	−							−	−							d	
Allantoin or urate required	−	−	−	−	−	−	−	−	−	+	−	−	−	−	−	−	−	−	−

(continued)

TABLE 3. (continued)

Characteristics	38. B. gibsonii	39. B. halmapalus	40. B. halodurans[d]	41. B. halophilus	42. B. horikoshii	43. B. horti[d]	44. B. hwajinpoensis	45. B. indicus	46. B. infernus[f]	47. B. insolitus[g]	48. B. jeotgali	49. B. krulwichiae[d]	50. B. lentus	51. B. licheniformis[c]	52. B. luciferensis	53. B. macyae	54. B. marisflavi	55. B. megaterium	56. B. methanolicus[f]
Pigmented colonies:																			
Yellow-pink-red	+[h]	−	−	−	−	−	+[h]	+[h]	−	−	+[h]	−	−	−	−	−	+[h]	−[h]	−
Dark brown/black	−	−	−	−	−	−	−	−	−	−	−	−	−	−	−	−	−	−[h]	−
Motility				+		+	−	−	−	+	+	+	+	+	+	+	+	+	−
Cell diameter >1.0 μm	−	−	−	−	−	−	+	+	−	v	v	−	−	−	−	−	−	+	d
Spore formation:	+	+	+	+	+	+	+	+	−	+	+	+	+	+	+	+	+	+	+
Ellipsoidal	+	+	+	+	+	+	+			+	+	+	+	+	+	+	+	+	+
Cylindrical	−	−	−	−	−	−	−			+	−	−	−	v			−	−	−
Spherical	−	−	−	−	−	−	−			+	−						−	v	−
Borne terminally	−	−	−	−	−	−	v			+	−	−	−	v			−	−	−
Swell sporangia	−	−	+	−	v	+	+	+	−	−	+	−	−	v	−	v	−	+	+
Parasporal crystals			−	−	−	−	−	−											
Catalase				+		+	+	+	−	+	+	+	+	+	w	+	+	+	+
Aerobic growth	+	+	+	+	+	+	+	+	−	+	+	+	+	+	+	−	+	+	+
Anaerobic growth				−		−	−		+	−	+	+	+	−	+	+	+	−	−
Voges-Proskauer								−						+	+			−m	
Acid from:																			
L-Arabinose							−		−	−	−			+/w	+	−		−	+
D-Glucose				+		+	+		+	−	+	+		+/w	+	+		+	+	+
Glycogen											+		−	+	−		−	+	−
D-Mannitol				−			+			−	−		+	+/w	+	−		+	+	+
D-Mannose				+			+			−		−	−	+/w	+			+	
Methyl β-xyloside											−			+/w	−		−	−	
Salicin				+							−			+/w	+	+		+	+	−
Starch											+			+/w	+	w		−	+	−
D-Xylose				+		+	−			−	−		+	+/w	+		−	+	+
Hydrolysis of:																			
Casein	+	+	+	−	+	+	+		−	ng	−	d	−	+	w		+	+	−
Gelatin	+	+	+	−	+	+	+	+	−	−	+	d	−	+	+				+
Starch	−	+	+	+	+	+	+	+			+	+	+	+			−	+	d
Utilization of:																			
Citrate				−					−	−	−		d	+	−	−		+	
Propionate				−						−			−	+					
Egg yolk reaction				−									−	−	−			−	
Nitrate reduction	d	−	d	−	−	+	+	−	+	−	−	+	+	+	+	−	+	d	−
Growth at pH:																			
5	−	−	−	−	−	−	+	−	−		−	−	d		−		+		
6	−	−	−	+	−	−	+	+	−		+	−	+	+	+		+	+	−
7	+	+	+	+	+	+	+	+	+	+	+	−	+	+	+	+	+	+	+
8	+	+	+	+	+	+				+	+	+	+	+		+	+	+	
9	−	−	+	−	+					−		+	+				+		
10	−	−	+	−	+					−		+	d				−		
Growth in NaCl:																			
2%	+		+	−	+	d	+	+	+	d	+	+	+	+		+	+	+	d
5%	+	−	+	+	+	+	+	−	+	−	+	+	+	+		−	+	+	−
7%	+	−	+	+	+	+	+		+	−	+	+	+	+		−	+	+	
10%	d	−	+	+	−	+	+		+	−	+	+				−	+	−	
NaCl required for growth	−	−	−	+	−	d	+		−	−	−	−	−					−	
Growth at:																			
5 °C	−	−	−	−	−	−				+	−			−			−	d	−
10 °C	+	+	−	−	+	−	+			+	+	−		+	−		+	d	−
20 °C	+	+	+	+	+	+	+	+		+	+			+	+		+	+	
30 °C	+	+	+	+	+	+	+	+		+	+			+	+	+	+	+	
40 °C	−	+	+	+	+	+	+		−	−	+	+	−	+	+		+	d	+
50 °C	−	−	+	+	−	−	−		+	−	−	−		+			−	d	+
55 °C	−	−	+	−	−	−	−		+	−	−	−		d	−		−	d	+
65 °C	−	−	−	−	−	−	−							−	−		−	−	−
Growth with lysozyme present										−				d				−	
Respiratory growth with As(V)																	+		
Respiratory growth with Se(IV) or Se(VI)																	−		
Autotrophic with H₂ + CO₂ or CO	−	−	−	−										−				−	−
Degradation of tyrosine				−						−	−			−			−	d	
Deamination of phenylalanine	−	−	−	−	−	−				d				−				+	
Allantoin or urate required	−	−	−	−	−	−							−					−	

(continued)

TABLE 3. (continued)

Characteristic	57. B. mojavensis [c]	58. B. mycoides [e]	59. B. naganoensis	60. B. nealsonii	61. B. neidei [g]	62. B. niacini	63. B. novalis	64. B. odysseyi	65. B. okuhidensis [d]	66. B. oleronius	67. B. pseudalcalophilus [d]	68. B. pseudofirmus [d]	69. B. pseudomycoides [e]	70. B. psychrodurans [g]	71. B. psychrosaccharolyticus	72. B. psychrotolerans [g]	73. B. pumilus [c]	74. B. pycnus [g]	75. B. schlegelii [f]
Pigmented colonies:																			
Yellow-pink-red	–	–	–	–	–	–	–	–	–	–	–	+[h]	–	–	–	–	–	–	–
Dark brown/black	–	–	–	–	–	–	–	–	–	–	–	–	–	–	–	–	–	–	–
Motility	+	–	–	+	+	d	+	+	+	–		–		–		+	+	+	+
Cell diameter >1.0 µm	–	+	–	–	–	v	v	–	–	–	–	–	v	–	v	–	–	+	–
Spore formation:																			
Ellipsoidal	+	+	+	+	–	+	+	+	+	+	+	+	+	v[o]	+	v[o]	+	–	–
Cylindrical	–	–	–		–	–	–	–	–	–	–	–	–	–	–	–	–	–	–
Spherical	–	–	–	–	+	–	–	+	–	+	–	–	–	v[o]	–	v[o]	–	+	+
Borne terminally	–	–	–	+	–	–	+	–	–	–	–	–	–	+	–	+	–	+	+
Swell sporangia	–	–	+	–	+	v	+	+	v	+	+	–	–	+	+	+	–	+	+
Parasporal crystals	–	–	–	–	–	–	–	–	–	–	–	–	–	–	–	–	–	–	–
Catalase	+	+	+	+	+	+		+	+	+				+	+	+	+	+	+
Aerobic growth	+	+	+	+	+	+	+	+	+	+	+	+	+	+	+	+	+	+	+
Anaerobic growth	–	+	–	+	–	–	+	–	–	–			+	+	+	–	–	–	
Voges-Proskauer	+	+	–		–	–	d			d	ng			+				+		
Acid from:																			
L-Arabinose	+	–	+[m]	+	–	–	–		–		ng		–	d/w	+	d/w	+	–	–
D-Glucose	+	+	+[m]	+	–	+	+		+		ng		+	d/w	+	d/w	+	–	–
Glycogen		+		–		d			–		ng						–		
D-Mannitol	+	–	+[m]	+	–	–	d/w		+		ng		–	d/w	+	d/w	+		
D-Mannose	+			+	–	+	+		d/w		ng						+	–	
Methyl β-xyloside		–		–		–	–		–		ng						–		
Salicin	+	d		+	–	d	–		d/w		ng						+	–	
Starch		+		+	–	+	–		d/w		ng						–		
D-Xylose	+	–	+[m]	+	–	+	–		–		ng		–	d/w	+	d/w	+		–
Hydrolysis of:																			
Casein	+	+	–		–	–	+		+		+	+	+	–	+	–	+		w
Gelatin	+	+	–	+		+	+		+		w	+	+	d	+	d	+		–
Starch	+	+	+	–		d	–		+		d/w	+	+	+	+	+	+		–
Utilization of:																			
Citrate	+	d	–		–	d	–			–				d	–	–	–	+	–
Propionate	–	–	–		–					–				–				–	+
Egg yolk reaction	–	+	–		–		–			–				+				–	
Nitrate reduction	+	d	–		+	+	–		+	+	–			+	+	+	–	–	+
Growth at pH:																			
5			+	–		–		+	–	–			–		–	–	d		
6	+	+	+	+	+		+	+	+	+		–	–	+		d	+	+	+
7	+	+	–	+	+	+	+	+	+	+	–	d	+	+	+	+	+	+	+
8			–	+		+	+	+	+		+	+		+			+		
9			–	+		–	+	+	+		+	+		+			+		
10			–	+		–	d	+	+		+			–			–		
Growth in NaCl:																			
2%	+		+	+	+	–		+	+					+	+	+	+		+
5%	+	–		+	+	–		+	+	+	+	+		d	d	d	+	–	–
7%	+	d	–	+		–		–	+	+	+	+	+[n]	–	–	–	+	–	–
10%	+		–	–		–		–	+		+	+					+	–	–
NaCl required for growth	–	–		–		–							–				–		
Growth at:																			
5 °C	d	–		–	d	–			–					+	+	+	d	d	–
10 °C	+	d	–		+	+			–				–	+	+	+	d	+	–
20 °C	+	+	–		+	+			–				+	+	+	+	+	+	–
30 °C	+	+	+	+	+	+	+		+	+		+	+	+	+	+	+	+	–
40 °C	+	d	+	+	+	+	+	+	+	+		+	+	–	–	d	+	+	–
50 °C	+	–	–	+	–	–	+		+	+		–	–	–	–	–	d	–	+
55 °C	d	–	–	+	–	–	d		+			–		–	–	–	–	–	+
65 °C	–	–	–	–		–	–		–								–		+
Growth with lysozyme present	–	+	–		–				–					+	–		–	d	–
Respiratory growth with As(V)																			
Respiratory growth with Se(IV) or Se(VI)																			
Autotrophic with H₂ + CO₂ or CO	–	–		–		–											–	–	+
Degradation of tyrosine	–	d	–		–									+	–		–	–	
Deamination of phenylalanine	–	–	–		–				–			+		–	–		–	–	
Allantoin or urate required	–	–	–		–									–			–	–	

(continued)

TABLE 3. (continued)

Characteristic	76. B. seleniireducens[d]	77. B. shackletonii	78. B. silvestris[g]	79. B. simplex	80. B. siralis	81. B. smithii	82. B. soli	83. B. sonorensis[c]	84. B. sphaericus[g]	85. B. sporothermodurans	86. B. subterraneus	87. B. thermoamylovorans[f]	88. B. thermocloacae[d,f]	89. B. thuringiensis[e]	90. B. tusciae[f]	91. B. vallismortis[c]	92. B. vedderi[d]	93. B. vietnamensis	94. B. vireti	95. B. weihenstephanensis[e]
Pigmented colonies:																				
Yellow-pink-red	−	−	−	−	−	−	−	w[h]	−	−	+[h]	−	−	−	−	−	−	−	−	−
Dark brown/black	−	−	−	−	−	−	−	−	−	−	−	−	−	−	−	−	−	−	−	−
Motility	−	+	+	+		+	+	+	+	d	+	+		+	+	+	+	+	+	+
Cell diameter >1.0 μm	−	−	−	−	−	−	v	−	−	−	−	−	−	+	−	−	−	−	−	+
Spore formation:	−	+	+	+	+	+	+	+	+	Rarely	−	−	d	+	+	+	+	+	+	+
Ellipsoidal	+	−	+	+	v	+	+		+		+		+	+	+	+	v	+	+	+
Cylindrical	−	−	−	−	v	−	−	−	−	−	−	−	d	−	−	−	−	−	−	d
Spherical	−	+	d	−	−	−	−	+	d		−		−	−	−	−	v	−	−	−
Borne terminally	−	+	−	v	v	−	−	+	d		−		+	−	−	−	+	−	−	−
Swell sporangia	+	+	−	+	v	v	−	+	v		−		+	−	+	−	+	−	v	−
Parasporal crystals	−	−	−	−	−	−	−	−	−p		−		−	+	−	−	−	−	−	−
Catalase	+	+	+	+	+	+		+	+	+	+	+	+	+	w	+	+	+		+
Aerobic growth	−	+	+	+	+	+	+	+	+	+	+	+	+	+	+	+	+	+	+	+
Anaerobic growth	+	−	d/w	−	+	+	+	−	−	+	+	−	+	+	−	−	+		+	+
Voges-Proskauer	d[r]	−	−	−	−	−	−	+	−	d	−		−	+		+	ng		−	+
Acid from:																				
L-Arabinose	−	−	d/w	−	d	−	+	−	−		+		−	+		+	ng	−	−	−
D-Glucose	+	−	+	−	+	+	+	−	+		+		−	+	−	+	ng	+	+	+
Glycogen	−	−	−	−		+	−	−	−		+		−	+	−		ng	+	+	+
D-Mannitol	w	−	d/w	−	d	−	+	−	d		−		−	−		+	ng	+	+	−
D-Mannose	w	−	−	−	d	+		−	d		+		d	+		+	ng	d	+	−
Methyl β-xyloside	−	−	−		−	−		−	−		−		−	−			ng			−
Salicin	+	−	d/w	−	d	−		−	d		+		d	−		+	ng	−	−	d
Starch	−	−	−	−	−	+		−	d		+		+	+	−		ng	+	+	+
D-Xylose	−	−	d/w	−	d	−	+	−	−		+		−	−		+	ng	−	−	−
Hydrolysis of:																				
Casein	w	−	d	+		+	+	d	−	−		−		+		+	−	+	+	+
Gelatin	d[r]	−	d	+	−	+		d	+	+		−		+			w	+	+	+
Starch	−	−	+	−	w	−		+	−	−	+	+		+	−	+	−	+		+
Utilization of:																				
Citrate	−	d[r]	−	d	−	d	−	+	+	d				+	−	+		−	−	+
Propionate	−		−	+		d		+								−				
Egg yolk reaction			−	−		−		−	−					+		−				+
Nitrate reduction	+	−	−	+	+	−	+	+	−	+	+	−		+	+	+		−	+	d
Growth at pH:																				
5	−	+	d			+		−	+		−			+		−		+		
6	−	+			+	+	+	d	+	−	+	−		w	+	−	−	+		+
7	−	+	+	+	+	+	+	+	+	+	+	+		+	−	+		+	+	+
8	−	+		+		−		+	+	+	+	+	+	+			−	+		
9	+	d		+		−		+	+	+	+	−	+			−	+	+		
10	+	−				−		−								−	+	+		
Growth in NaCl:																				
2%	+		+	+	+			+	+	+	+		+	+	−	+	+	+		+
5%	+		+	d	+	−		−	+	+	+		w	+	−	+	+	+		+
7%	+		−	−	+			−	+	−	+		−	+	−	+	+	+		d
10%	+		−	−	−			−	−		−			−		+	−	+		
NaCl required for growth	+		−	−	−			−	−		−			−		−		−		−
Growth at:																				
5 °C		−				−		−	−		−		−	−		d				d
10 °C		−	+			−		−	d	−	−		−	d		+		+		+
20 °C	+	+	+	+		−		+	+	+	+		−	+	−	+		+		+
30 °C		+	+	+		+	+	+	+	+	+		−	+	−	+		+	+	+
40 °C		+	+			+	+	+	d	+	+		+	+	−	+	+	+	+	d
50 °C		+	−	−	+	+	−	+	−	+	−	+	+	−	+	+	d	d	−	
55 °C		d	−	−	+	−	+	−	d	−	+	+	−	+	−	−	−	−		
65 °C		−	−	−	d	−	−	−	−	−	+	+	−	−	−	−		−		
Growth with lysozyme present		−				−		−	d					+		−		+		+
Respiratory growth with As(V)	+																			
Respiratory growth with Se(IV) or Se(VI)	+																			
Autotrophic with H₂ + CO₂ or CO	−				−			−	−					−	+					−
Degradation of tyrosine			+		−			−						+				+		+
Deamination of phenylalanine			−		−				+					−		−				−
Allantoin or urate required	−	−	−	−	−	−	−	−	−	−	−	−	−	−	−	−	−	−	−	−

(continued)

TABLE 3. (continued)

aSymbols: +, >85% positive; d, different strains give different reactions (16–84% positive); –, 0–15% positive; v, variation within strains; w, weak reaction; –/w, negative or weak reaction; d/w, different strains give different reactions and reactions are weak when positive; ng, no growth in the test medium; no entry indicates that no data are available.

bCompiled from Larkin and Stokes (1967); Nakayama and Yanoshi (1967); Gordon et al. (1973), Pichinoty et al. (1976, 1983, 1984); Aragno (1978); Schenk and Aragno (1979); Pichinoty (1983); Bonjour and Aragno (1984); Logan and Berkeley (1984), Claus and Berkeley (1986), Priest et al. (1987, 1988); Nakamura et al. (1988, 1999, 2002); Demharter and Hensel (1989b); Denariaz et al. (1989); Nakamura (1989, 1998); Ventosa et al. (1989); Tomimura et al. (1990); Nagel and Andreesen (1991); Arfman et al. (1992); Spanka and Fritze (1993); Andersch et al. (1994); Roberts et al. (1994, 1996); Agnew et al. (1995); Boone et al. (1995); Combet-Blanc et al. (1995); Nielsen et al. (1995a); Fritze (1996a); Fujita et al. (1996); Kuhnigk et al. (1995); Kuroshima et al. (1996); Pettersson et al. (2000, 1996); Shelobolina et al. (1997); Lechner et al. (1998); Switzer Blum et al. (1998); Yumoto et al. (2004c, 2003, 1998); Rheims et al. (1999); Logan et al. (2002a, b, 2000, 2004b); Palmisano et al. (2001); Yoon et al. (2001a, 2003a); Abd El-Rahman et al. (2002); Ajithkumar et al. (2002); Kanso et al. (2002); Li et al. (2002); Reva et al. (2002); Venkateswaran et al. (2003); Gugliandolo et al. (2003a); Heyrman et al. (2003a, 2005a, 2004); Taubel et al. (2003); De Clerck et al. (2004b, 2004c); Heyndrickx et al. (2004); La Duc et al. (2004); Ivanova et al. (2004a); Noguchi et al. (2004); Santini et al. (2004); Scheldeman et al. (2004); Suresh et al. (2004).

cSee Table 5 for a comparison of the subspecies of *Bacillus subtilis* and the closely related species *Bacillus subtilis, Bacillus atrophaeus, Bacillus mojavensis* and *Bacillus vallismortis*, and the species *Bacillus amyloliquefaciens, Bacillus licheniformis, Bacillus pumilus*, and *Bacillus sonorensis*.

dSee Table 6 for a comparison of alkaliphilic species: *Bacillus agaradhaerens, Bacillus alcalophilus, Bacillus algicola, Bacillus arseniciselenatis, Bacillus clarkii, Bacillus cohnii, Bacillus halodurans, Bacillus horti, Bacillus krulwichiae, Bacillus okuhidensis, Bacillus pseudoalcaliphilus, Bacillus pseudofirmus, Bacillus selenitireducens, Bacillus thermocloacae, Bacillus vedderi*.

eSee Table 7 for comparison of the closely related species *Bacillus anthracis, Bacillus cereus, Bacillus mycoides, Bacillus pseudomycoides, Bacillus thuringiensis*, and *Bacillus weihenstephanensis*.

fSee Table 8 for comparison of the thermophilic species (optimum growth at 50 °C or above): *Bacillus aeolius, Bacillus fumarioli, Bacillus infernus, Bacillus methanolicus, Bacillus schlegelii, Bacillus thermoamylovorans, Bacillus thermocloacae*, and *Bacillus tusciae*.

gSee Table 9 for comparison of the neutrophilic, non-thermophilic species that form spherical spores: *Bacillus fusiformis, Bacillus insolitus, Bacillus neidei, Bacillus psychrodurans, Bacillus psychrotolerans, Bacillus pycnus*, and *Bacillus sphaericus*.

hPigmentation: *Bacillus subtilis* may form pigments, varying from cream through yellow, orange, pink and red, to brown or black, on potato or agar media containing glucose, and strains forming brown or black pigment were often formerly called "*Bacillus subtilis* var. *aterrimus*"; *Bacillus algicola* produces semitransparent, creamy, slightly yellowish colonies; *Bacillus aquimaris* colonies are pale orange-yellow; *Bacillus arseniciselenatis* and *Bacillus selenitireducens* will produce red colonies, owing to elemental selenium precipitation, on selenium oxide media; *Bacillus atrophaeus* forms a dark brownish-black soluble pigment in 2–6 d on media containing tyrosine or other organic nitrogen source; *Bacillus carboniphilus* produces grayish yellow pigment on nutrient agar and brownish red pigment on trypto-soya agar; some strains of *Bacillus cereus* may produce a yellowish-green fluorescent pigment on various media, some strains may produce a pinkish brown diffusible pigment on nutrient agar, and on starch-containing media containing sufficient iron some strains produce the red pigment pulcherrimin; *Bacillus clarkii* colonies may be cream-white to pale yellow in color, and one of the three strains described produces dark yellow colonies with age; *Bacillus endophyticus* colonies may be white or pink-red, even on the same plate, and media containing ampicillin and lysozyme commonly yield red colonies; colonies of *Bacillus fastidiosus* on uric acid medium may become yellowish; *Bacillus firmus* colonies are creamy-yellow to pale orangey-brown after 3 d on TSA at 30 °C; *Bacillus gibsonii* colonies are yellow; *Bacillus indicus* colonies are yellowish-orange; *Bacillus hwajinpoensis* colonies are light yellow; *Bacillus jeotgali* colonies are cream-yellow to light orange-yellow; many strains of *Bacillus licheniformis* can produce red pigment (assumed to be pulcherrimin) on carbohydrate media containing sufficient iron, and colonies on glycerol/glutamate medium are reddish-brown; *Bacillus marisflavi* colonies are pale yellow; *Bacillus megaterium* colonies may become yellow and then brown or black on long incubation; *Bacillus pseudofirmus* colonies are yellow; *Bacillus sonorensis* colonies are yellowish-cream on routine media, and bright yellow on pH 5.6 agar; *Bacillus subterraneus* colonies are dark yellow to orange on tryptic soy agar.

iCitrate test results may vary according to the test method used; Gordon et al. (1973) found citrate utilization to be a variable property among 23 strains of *Bacillus anthracis*, while Logan and Berkeley (1984) and Logan et al. (1985) obtained negative results for 37 strains using the API 20E test method. For *Bacillus badius*, Gordon et al. (1973) obtained negative results with two strains, while Logan and Berkeley (1984) obtained positive results for two strains using the API 20E test method.

jStrains of *Bacillus cereus* of serovars 1, 3, 5, and 8, which are particularly associated with outbreaks of emetic-type food poisoning, do not produce acid from salicin and starch, whereas strains of *Bacillus cereus* of other serotypes are usually positive for these reactions. See Table 7.

kFor *Bacillus fumarioli*, acid production from carbohydrates is tested at pH 6 – see Logan et al. (2000) and *Testing for special characters*.

lGordon et al. (1973) found the Voges–Proskauer reaction to be negative for 60 strains of *Bacillus megaterium*, while Logan and Berkeley (1984) obtained positive results for all but one of 33 strains using the API 20E test method.

mAcid production from carbohydrates by *Bacillus naganoensis* is slow, and shows only after extended (>14 d) incubation.

nThe published description of *Bacillus pseudomycoides* (Nakamura, 1998) records that 7% salt is tolerated, but the differentiation table in that publication indicates the opposite result.

oSpores of *Bacillus psychrodurans* and *Bacillus psychrotolerans* are rarely formed; on casein-peptone soymeal-peptone agar spores are predominantly spherical, but on marine agar they are predominantly ellipsoidal.

pMosquitocidal strains of *Bacillus sphaericus* produce parasporal toxin crystals which are smaller than those produced by *Bacillus thuringiensis*, but which are nonetheless visible by phase-contrast microscopy.

qGrowth occurs within the range pH 7.2–9.5 in media adjusted with NaOH and HCl, but not in media where the pH is adjusted using buffered systems as described by Nielsen et al. (1995a).

rNegative when incubated at 30 °C, but may become positive slowly when incubated at 40 °C.

sGrowth is poor in the absence of NaCl.

TABLE 4. Additional data for differentiation of *Bacillus* species[a,b]

Characteristic	1. B. subtilis	2. B. aeolius	3. B. agaradhaerens	4. B. alcalophilus	5. B. algicola	6. B. amyloliquefaciens	7. B. anthracis	8. B. aquimaris	9. B. arseniciselenati	10. B. asahii	11. B. atrophaeus	12. B. azotoformans	13. B. badius	14. B. barbaricus	15. B. bataviensis	16. B. benzoevorans	17. B. carboniphilus
Oxidase	d	+			−			−	+	+	−	+		−		−	+
β-Galactosidase	+					d	−			−		−	−	−	+		−
Lysine decarboxylase	−	−				−	−										−
Ornithine decarboxylase	−	−				−	−								−		−
Arginine dihydrolase	−	−				−	−					−			−		−
Acid from:																	
N-Acetyl-D-glucosamine	d	+				d	+	−					−	+	+		
Adonitol	−					−	−	−					−	−	−		−
Amygdalin	+					+	−	−					−	−	d/w		
D-Arabinose	−					−	−	−		−			−	−	−		
D-Arabitol	−					−	−	−					−	−	−		
L-Arabitol	−					−	−	−					−	−	−		
Arbutin	+					+	d	−					−	−	d/w		
Cellobiose	+	+				+	−	−			d		−	−	+		−
Dulcitol	−					−	−	−					−	−	+		
Erythritol	−					−	−	−					−	−	+		
Fructose	+	+				+	d	+		−	+	−		d	+		−
D-Fucose	−					−	−	−					−	−	−		
L-Fucose	−					−	−	−						−	d/w		
Galactose	d					d	−	−		−	d	−		d	+		−
β-Gentiobiose	+					+	−	−					−	−	+		
Gluconate	−					−	−	−					−	−	−		
Glycerol	+	+		+		+	−	−		−			−	−	w		−
meso-Inositol	+	+				+	−	−		−			−	−	−		−
Inulin	d					−	−	−					−	−	d/w		
2-Ketogluconate	−	+				−	−	−					−	−	−		
5-Ketogluconate	−	+				−	−	+					−	−	−		
Lactose	d			+		d	−	−		−	−	−	−	d	+		−
Lyxose	−					−	−	−					−	−	−		
Maltose	+	+		+		+	+	+		−	d	−	−	+	+		−
Melezitose	−	+				−	−	−					−	−	+		
Melibiose	d	+				d	−	−			−		−	−	d/w		
Methyl α-D-mannoside	−					−	−	−					−	−	d/w		
Methyl β-D-xyloside	−					−	−	−					−	−	−		
Raffinose	+	+				+	−	−		−			−	−	+		−
Rhamnose	−					−	−	−		−			−	−	−		−
Ribose	+	+				+	+	+			d	−	−	−	w		−
Sorbitol	+			+		+	−	−		−	d		−	−	−		
Sorbose	−					−	−	−					−	−	−		−
Stachyose								−									
Sucrose	+	+		+		+	+	+		−	+		−	d	d/w		−
Tagatose	−					−	−	−					−	−	−		
Trehalose	+	+				d	+	+		−	+		−	+	+		−
Turanose	+	+				d	−	−					−	d	+		
Xylitol	−					−	−	−					−	−	−		
L-Xylose	−					−	−	−		−	+		−	−	−		
Utilization of:																	
Acetate					−				−			+				+	−
N-Acetyl-D-glucosamine		+	+	+	−								−	+			
Alanine					−								+	+			
Arabinose		+	+	+	−												
Arabitol					−												

(continued)

TABLE 4. (continued)

Characteristic	1. *B. subtilis*	2. *B. aeolius*	3. *B. agaradhaerens*	4. *B. alcalophilus*	5. *B. algicola*	6. *B. amyloliquefaciens*	7. *B. anthracis*	8. *B. aquimaris*	9. *B. arseniciselenati*	10. *B. asahii*	11. *B. atrophaeus*	12. *B. azotoformans*	13. *B. badius*	14. *B. barbaricus*	15. *B. bataviensis*	16. *B. benzoevorans*	17. *B. carboniphilus*
Aspartate					−				−			+	+	−			
Benzoate																+	
Ethanol									−								
Formate					−				−								
Fructose			+	+	+				+				−	+			
Fumarate													+			d	−
Galactose			+	+	−				−					−			
Gentiobiose				+	−												
Gluconate		+	−	−										−			
Glucose		+	+	+	+				−					+			
Glutamate					−				−			+	+				−
Glycerol			+	+	+								+				
Glycine									−								
Glycogen			+	+	−												
Histidine					−									+			
m-Hydroxybenzoate														−		d	
Inositol			−	d	−									−			
Lactate					−				+			+	+	−			−
Lactose			d	+	−												
Malate		+							+			+		+		+	−
Maltose		+	+	+	+								−	+			
Mannitol		+			−								−				
Mannose		+	+	d	+									+			
Melibiose			+	+	−				−								
Methanol									−								
Phenol																+	
Phenylacetate		+											+			+	
Phenylpropionate																+	
Proline					−									+		+	
Pyruvate												+		+		+	
Rhamnose			d	+	−									−			
Ribose			+	+										+			
Salicin			+	+										−			
Serine									−					+			
Sorbitol			d	−	−												
Succinate					−				−			+					−
Trehalose			+	+	+								−	+			
Turanose			+	+	−								−				
Valine																	
Xylitol			−	−	−												
Xylose			+	d													
Hydrolysis of:																	
Cellulose			+														
DNA						d					+						d
Elastin						+											
Esculin	+	−				+	+	−					−	−	+	+	+
Hippurate			−	−										+		+	+
4-Methylumbelliferone glucuronide			−	−													
Pectin	+				−												
Pullulan			d	+													
Tributyrin						+											
Tween 20			−	d		+					+				−		
Tween 40			+	+	−	+					+						
Tween 60			d	+		+					+						
Tween 80			−	−	−				+			−		−		+	+
Urea	−	−		−	+	−		−	−			−		−		+	−
Xylan			+														

(continued)

TABLE 4. (continued)

Characteristic	18. B. cereus	19. B. circulans	20. B. clarkii	21. B. clausii	22. B. coagulans	23. B. cohnii	24. B. decolorationis	25. B. drentensis	26. B. endophyticus	27. B. farraginis	28. B. fastidiosus	29. B. firmus	30. B. flexus	31. B. fordii	32. B. fortis	33. B. fumarioli	34. B. funiculus	35. B. fusiformis	36. B. galactosidilyticus	37. B. gelatini
Oxidase	−						+		+	+	+		−	+	+		−	d		−
β-Galactosidase	−	+			d	+	d[d]	+	+	−		−	+	−	−	−		−	+	−
Lysine decarboxylase	−	−		−			−	−	−			−				−			−	−
Ornithine decarboxylase	−	−		−			−	−				−				−			−	−
Arginine dihydrolase	d	−			d		−	−	−	−		−				−			−	−
Acid from:																				
N-Acetyl-D-glucosamine	+	+			+		w[d]	+		d/w			−	−	w				+	w
Adonitol	−	d			−		−	−	−			−	−	−	−			−	−	−
Amygdalin	−	+			d		−	d/w	−			−	−	−	−				−	−
D-Arabinose	−	−			−		−	−	−			−	−	−	−				−	−
D-Arabitol	−	d			d		−	−	−			−	−	−	−	−			−	−
L-Arabitol	−	+			−		−	−	−			−	−	−	−				−	−
Arbutin	+[c]	+			d		d/w[d]	d/w	−			−	−	−	−				d/w	−
Cellobiose	d	+			d		w[d]	−	−			−	−	−	−			−	d/w	d
Dulcitol	−	−			−		−	−	−			−	−	−	−				−	−
Erythritol	−	−			−		−	−	−			−	−	−	−				−	−
Fructose	+	+			+		w[d]	+	−			d	+	−	+				+	+
D-Fucose	−	−			−		−	−	−			−	−	−	−				−	−
L-Fucose	−	−			−		−	−	−			−	−	−	−				−	−
Galactose	−	+			+		d/w[d]	d/w	−	−		−	+	−	−	d		−	d/w	d
β-Gentiobiose	d	+			d		w[d]	−	−			−	−	−	−				d/w	
Gluconate	d	+			d		−	d/w	−			−	−	−	−					
Glycerol	+[c]	+			+		d/w[d]	−	−	−		d/w	+	−	−	d		−	−	+
meso-Inositol	−	+			d		−	−	+			−	d	−	−			−	−	d
Inulin	−	+			−		−	d/w	−			−	−	−	−				d/w	−
2-Ketogluconate	−	d			−		−	−	−			−	−	−	−				−	−
5-Ketogluconate	−	d			−		w[d]	−	−			−	−	−	−				−	d
Lactose	−	+			d		d/w[d]	+	−	−		−	+	−	−	d		−	d/w	−
Lyxose	−	d			−		−	−	−			−	−	−	−				−	
Maltose	+	+			+		w[d]	+	−			+	+	−	−	d		−	d/w	w
Melezitose	−	+			−		−	d/w	−			−	−	−	−	d			d/w	d
Melibiose	−	+			+		−	+	+			−	−	−	−	d			d/w	−
Methyl α-D-mannoside	−	+			d		−	−	−			−	−	−	−	−			−	−
Methyl β-D-xyloside	−	+			d		−	−				−	−	−	−	−			−	−
Raffinose	−	+			d		−	d/w	d			−	+	−	−	d		−	d/w	−
Rhamnose	−	d			d		−	−	+			−	−	−	−	−		−	d/w	−
Ribose	+	d			d		w[d]	d/w	+			−	−	−	−	d			d/w	w
Sorbitol	−	+			d		−	−	−			−	−	−	−				−	−
Sorbose	−	−			−		−	−				−	−	−	−					
Stachyose																				
Sucrose	d	+			d		w[d]	d/w	+			−	+	−	−	+		d	d/w	−
Tagatose	−	−			−		−	−	−			−				−			−	−
Trehalose	+	+			+		w[d]	d/w	−			d/w	+	−	−	w		−	d/w	+
Turanose	d	+			d		−	d/w	−			−	−	−	−	d			d/w	d
Xylitol	−	+			−		−	−	−			−	−	−	−				−	
L-Xylose	−	−			−		−	−				−				−	−		−	−
Utilization of:																				
Acetate	d	+							−					+				+		
N-Acetyl-D-glucosamine					+					−		+		−	−					
Alanine										+		+		−	−					
Arabinose					+					d		−		−			−			
Arabitol					d					−		−		−	−					

(continued)

TABLE 4. (continued)

Characteristic	18. B. cereus	19. B. circulans	20. B. clarkii	21. B. clausii	22. B. coagulans	23. B. cohnii	24. B. decolorationis	25. B. drentensis	26. B. endophyticus	27. B. farraginis	28. B. fastidiosus	29. B. firmus	30. B. flexus	31. B. fordii	32. B. fortis	33. B. fumarioli	34. B. funiculus	35. B. fusiformis	36. B. galactosidilyticus	37. B. gelatini
Aspartate						−				d			−	−	−					
Benzoate													−	−	−					
Ethanol																				
Formate	d												+					+		
Fructose				+						−				−	−			+		
Fumarate										+			−	−	−					
Galactose				d						d			−	−	−					
Gentiobiose				d						−			−	−	−					
Gluconate				d					+	−			+	−	d			d		
Glucose				+						−			+	−	d			+		
Glutamate										+			+	−	d					
Glycerol				+						−			+	−	−			+		
Glycine																				
Glycogen				+																
Histidine										+			d	d	+					
m-Hydroxybenzoate										d			−	−	d					
Inositol				d						d			−	−	−			−		
Lactate	d									+			+	−	d		d	+		
Lactose				d						−			−	−	−			−		
Malate										+			+	−	d			−		
Maltose				+						−			+	−	−			+		
Mannitol										d			+	−	−					
Mannose				+						−			−	−	−					
Melibiose				+						−			−	−	d					
Methanol										−										
Phenol																				
Phenylacetate										−			−	−	d					
Phenylpropionate										−			−	−	−					
Proline										+			+	−	d					
Pyruvate																				
Rhamnose				+						d			−	−	−					
Ribose				+						d			−	−	d					
Salicin				+														−		
Serine										−			+	−	−					
Sorbitol				+						d			−	−	−					
Succinate	d									+		+	+	−	−			+		
Trehalose				+						−			+	−	−			+		
Turanose				+						−				−	−					
Valine																				
Xylitol				+						−				−	−			−		
Xylose				+						−			−	−	−			−		
Hydrolysis of:																				
Cellulose																				
DNA													+					+		
Elastin													+					d		
Esculin	+	+		d			+	+				d	−			−	+	−	+	+
Hippurate			+	−		+														
4-Methylumbelliferone glucuronide			−	−		+														
Pectin													−					−		
Pullulan			−	−		+							+					−		
Tributyrin																				
Tween 20			−										+					+		
Tween 40			+	−																
Tween 60			+	−		+														
Tween 80			−	−									d				−	+		
Urea	d	d			−	−	−	−	−	−	+	−	+	−	−	−	+	d	d	−
Xylan																				

(continued)

TABLE 4. (continued)

Characteristic	38. B. gibsonii	39. B. halmapalus	40. B. halodurans	41. B. halophilus	42. B. horikoshii	43. B. horti	44. B. hwajinpoensis	45. B. indicus	46. B. infernus	47. B. insolitus	48. B. jeotgali	49. B. krulwichiae	50. B. lentus	51. B. licheniformis	52. B. luciferensis	53. B. macyae	54. B. marisflavi	55. B. megaterium	56. B. methanolicus
Oxidase				+		+	−			+	−	+		d		−	−	d	+
β-Galactosidase				+		+				+		−	+	+	−			+	
Lysine decarboxylase				−						−			−	−	−			−	
Ornithine decarboxylase				−						−			−	−	−			−	
Arginine dihydrolase				−				+		−			−	+	−			−	
Acid from:											c								
N-Acetyl-D-glucosamine										+	−		+	+	+		−	+	
Adonitol							−			−	−		−	−	−		−	−	
Amygdalin										−	−		+/w	−	−		−	−	−
D-Arabinose				−		−				−	−		−	−	−		−	+	
D-Arabitol										d	−		−	−	−		−	d	
L-Arabitol										−	−						−		
Arbutin										+	−		+/w	+	+		+	+	
Cellobiose						+			−	+	d	−	+/w	+	+		+	+	
Dulcitol										−	−		−	−	−			−	
Erythritol				−						−	−		−	−	−			−	
Fructose				+		+				+	+	+	+/w	+	+		+	+	d
D-Fucose										−	−		−	−	−		−	−	
L-Fucose										−	−		d/w	−	−			−	
Galactose						+			−	d	−	+	+/w	+			w	+	−
β-Gentiobiose										+	d		+/w	+	+		+	+	
Gluconate										−	d/w		−	−	−		−	d	
Glycerol				+						+		+	−	+			+	+	−
meso-Inositol				−		−	−			d	−	−	d/w	d	−		−	+	−
Inulin										d	−		−	d	−		−	+	−
2-Ketogluconate										−	−		−	−	−			−	
5-Ketogluconate										−	d		−	−	−			−	
Lactose				−	d/w		−			d	−	−	+	+	−		−	+	−
Lyxose										d	−		−	−	−			−	
Maltose				−			+			+	+	+	+/w	+	+		+	+	d
Melezitose							−			d	−		+/w	−	+		−	d	
Melibiose					−		+			d	−	−	+/w	d	−		+	+	
Methyl α-D-mannoside										d	−		+/w	d	−		+	−	
Methyl β-D-xyloside										d	−		+/w	−	−		−	−	
Raffinose					−		+			d	−	−	+/w	d	−		w	+	d
Rhamnose							−			d	−	−	+/d	d	−		−	−	
Ribose						+	+			+	−	+	+/w	+	−		+	+	d
Sorbitol							−		−	d	−		d/w	+	−		−	d	d
Sorbose										−	−		−	−	−			−	
Stachyose							+										+		
Sucrose				+		−	+		−	+	d	+	+	+	+		+	+	−
Tagatose										d	−		−	+	−		−	−	
Trehalose				+			+			d	+	+	+	+	+		+	+	d
Turanose										+	−		d/w	+	+		−	+	
Xylitol										−	−		−	−	−			d	
L-Xylose										−	−		−	−	−		−		−
Utilization of:																			
Acetate										−							+	d	d
N-Acetyl-D-glucosamine	−	+	+	−	+								−						
Alanine				−									−						
Arabinose	d	−	d	+	−				−				−						
Arabitol	−	−		d						−			−	−	−	−			

(continued)

TABLE 4. (continued)

Characteristic	38. *B. gibsonii*	39. *B. halmapalus*	40. *B. halodurans*	41. *B. halophilus*	42. *B. horikoshii*	43. *B. horti*	44. *B. hwajinpoensis*	45. *B. indicus*	46. *B. infernus*	47. *B. insolitus*	48. *B. jeotgali*	49. *B. krulwichiae*	50. *B. lentus*	51. *B. licheniformis*	52. *B. luciferensis*	53. *B. macyae*	54. *B. marisflavi*	55. *B. megaterium*	56. *B. methanolicus*
Aspartate				−									−					+	
Benzoate				−								+	−						
Ethanol				−					−										d
Formate									+										
Fructose	+	+	+	−	+								−				−		
Fumarate				−				−					−						
Galactose	d	−	+		−								−						
Gentiobiose	+	−	d		d								−						
Gluconate	−	d	+		d								−						
Glucose	+	+	+	+	+								−						
Glutamate				+									−				+		
Glycerol	+	+	+	−	+								−						
Glycine				−															
Glycogen	−	+	+		+								−						
Histidine				−									−						
m-Hydroxybenzoate												+	−						
Inositol	−	−	+	−				+					−						
Lactate				+					+				−					+	
Lactose	+	−	+	−	−			+					−						
Malate													−						
Maltose	+	+	+	−	+			+					−						
Mannitol				−									−						+
Mannose	+	+	+	−	d			+					−						
Melibiose	+	−	+	−				+					−						
Methanol	+	−																	+
Phenol																			
Phenylacetate													−						
Phenylpropionate													−						
Proline				−				−					−						
Pyruvate								−									+		d
Rhamnose	d	−	+	−	−			+					−						
Ribose	+	−	+	−	d			+					−						
Salicin	+	d	+	−	−														
Serine				−				−					−						
Sorbitol	−	−	−										−						
Succinate				−					−								+		−
Trehalose	+	+	+	−	+								−						
Turanose	+	+	+		d														
Valine				−															
Xylitol	−	−	d		−								−						
Xylose	d	d	+	−	d								−				−		
Hydrolysis of:																			
Cellulose																			
DNA				+		+						+							
Elastin																			
Esculin				+			+	+		+	+		+	+	+			+	+
Hippurate	−	+	−		+	d						+		−					−
4-Methylumbelliferone glucuronide	+	−	−	−															
Pectin																+			
Pullulan	−	+	+		+														
Tributyrin																			
Tween 20	−	−	−		−	−	+	−				+							
Tween 40	d	−	+		d	−	+					+							
Tween 60	d	−	+		d	−	+					+							
Tween 80	−	−	−		+	−	+				d	+						−	
Urea				+			−	−	−		d	+		+	d	−			−
Xylan																			

(continued)

TABLE 4. (continued)

Characteristic	57. *B. mojavensis*	58. *B. mycoides*	59. *B. naganoensis*	60. *B. nealsonii*	61. *B. neidei*	62. *B. niacini*	63. *B. novalis*	64. *B. odysseyi*	65. *B. okuhidensis*	66. *B. oleronius*	67. *B. pseudalcalophilus*	68. *B. pseudofirmus*	69. *B. pseudomycoides*	70. *B. psychrodurans*	71. *B. psychrosaccharolyticus*	72. *B. psychrotolerans*	73. *B. pumilus*	74. *B. pycnus*	75. *B. schlegelii*	76. *B. selenitireducens*
Oxidase	+	−		−		d		−		+			−	+	d	+	d		w	−
β-Galactosidase		d		+		+	−			−					+		+			
Lysine decarboxylase	−	d		−		−	−	−		−							−			
Ornithine decarboxylase	−	d		−		−	−	−		−							−			
Arginine dihydrolase	−	d		−		−	−	−		−							−			
Acid from:																				
N-Acetyl-D-glucosamine		+		+		+	+			+							+			
Adonitol		−		−		−										−	−			
Amygdalin		d				+	d/w			−							+			
D-Arabinose		−				−	−			−							−			
D-Arabitol		−		+		−				−						−				
L-Arabitol		−				−	−			−							−			
Arbutin		d				d	d/w			−							+			
Cellobiose	+	d				d	d/w			+					d		+			
Dulcitol		−				d	−			−						−	−			
Erythritol		−				−	−									−	−			
Fructose	+	d				+	+			+					d		+			
D-Fucose		−				−	−			−										
L-Fucose		−				−	−			−							−			
Galactose	+	d		+		+	w			d/w					−		+			
β-Gentiobiose		−		+		d	d/w			−							+			
Gluconate		−		+		−	d/w			−							−			
Glycerol		+		+		−	d/w			d/w					d		+			
meso-Inositol		−		+		d	−			−					−		d			
Inulin		−				d	−			−							−			
2-Ketogluconate		−		+		−	−			−							−			
5-Ketogluconate		−				−	d/w			−							−			
Lactose	−	−	w	+		d	−			−					−		d			
Lyxose		−		+			d/w			−							−			
Maltose	+	+				+	+			d/w					d		d			
Melezitose		−		+		d	−			−							−			
Melibiose		−		+		+	−			−							d			
Methyl α-D-mannoside		−				d				−							+			
Methyl β-D-xyloside						d				−							−			
Raffinose		−		+		d	−			−					−		+			
Rhamnose	+	−		+		d	−			−							−			
Ribose	+	d		−		d	d/w			d/w							+			
Sorbitol	+	−		+		d	d/w			−					d		d			
Sorbose		−					−			−										
Stachyose																				
Sucrose	+	d				+	−								d		+			
Tagatose		−		+		−	−			+							d			
Trehalose	+	+				+	+			d/w					d		+			
Turanose		d		+		d	−			−							d			
Xylitol		−				d	−			−							−			
L-Xylose		−				−	−			−							−			
Utilization of:																				
Acetate		d				+		+		−					d				+	−
N-Acetyl-D-glucosamine									+	+	−	+								
Alanine							−	+		−									−	+
Arabinose								w	+	−	d	d							−	+
Arabitol								+		−	−									

(continued)

TABLE 4. (continued)

Characteristic	57. B. mojavensis	58. B. mycoides	59. B. naganoensis	60. B. nealsonii	61. B. neidei	62. B. niacini	63. B. novalis	64. B. odysseyi	65. B. okuhidensis	66. B. oleronius	67. B. pseudalcalophilus	68. B. pseudofirmus	69. B. pseudomycoides	70. B. psychrodurans	71. B. psychrosaccharolyticus	72. B. psychrotolerans	73. B. pumilus	74. B. pycnus	75. B. schlegelii	76. B. selenitireducens
Aspartate						d													–	–
Benzoate																			–	
Ethanol																			–	–
Formate		d				d									d				–	–
Fructose										+	+	+							–	+
Fumarate		d				+				w									+	
Galactose									+	–	+	–							–	+
Gentiobiose									+		+	–								
Gluconate											d	d		–						
Glucose											+	+							–	+
Glutamate					+	+												+	+	–
Glycerol										+	+	+					–		–	
Glycine																				
Glycogen									+		+	d								
Histidine										w									–	
m-Hydroxybenzoate																				
Inositol									+		–	–								
Lactate		d				d				–					d				+	+
Lactose									+		+	–								
Malate						+				+									+	
Maltose											+	+							–	
Mannitol				+					+										–	
Mannose									–		–	d							–	
Melibiose									+	–	–	–							–	
Methanol									–										–	
Phenol																			+	
Phenylacetate																				
Phenylpropionate																				
Proline																			+	
Pyruvate						+		+										+	+	+
Rhamnose											d	–							–	
Ribose										+	+	+								
Salicin									+	–	+	d								
Serine										–									–	–
Sorbitol									+		–	–								
Succinate		–				+				–					–				+	–
Trehalose										w	+	+							–	
Turanose										–	+	d								
Valine																			–	
Xylitol									+		–	–								
Xylose									+	w	+	d							–	
Hydrolysis of:																				
Cellulose														–		–				
DNA														+	d	+				
Elastin															d					
Esculin		+					+			+				–	d	–	+			
Hippurate			–								–	–			–		–			
4-Methylumbelliferone glucuronide										–	–	–								
Pectin														–	–	–				
Pullulan			+								+	d		–	d	d				
Tributyrin																				
Tween 20										–	–	–		+	d	+				
Tween 40					–					–	d	+		+		+			–	
Tween 60										–	d	+		+		+				
Tween 80	–				–					–	–			d	d	d			–	
Urea	–	d		–		d	–	–	–					–	d	–			–	–
Xylan																	–			

TABLE 4. (continued)

Characteristic	77. *B. shackletonii*	78. *B. silvestris*	79. *B. simplex*	80. *B. siralis*	81. *B. smithii*	82. *B. soli*	83. *B. sonorensis*	84. *B. sphaericus*	85. *B. sporothermodurans*	86. *B. subterraneus*	87. *B. thermoamylovorans*	88. *B. thermocloacae*	89. *B. thuringiensis*	90. *B. tusciae*	91. *B. vallismortis*	92. *B. vedderi*	93. *B. vietnamensis*	94. *B. vireti*	95. *B. weihenstephanensis*
Oxidase		−	−	+	+			+	+	−	−	+	−	+	+	+	+		
β-Galactosidase	+	−						−	−	−	−	+	−	+	+	+	+	d	
Lysine decarboxylase		−	−					−	−				−				+	d	
Ornithine decarboxylase	−		−		−	−		−	−				+					−	
Arginine dihydrolase	−	−	−	d	−	−		−	−	−	+	−	+				−	−	
Acid from:																			
N-Acetyl-D-glucosamine	+	−	+		−	+		d	+		+		+				+	+	
Adonitol	−	−	−		−			−	−		−		−				+	+	
Amygdalin	−	−			−			−	−		−		−				−	−	
D-Arabinose	+	−			−			−	d		+		−				−	−	
D-Arabitol	−	−		−	−			−	−		−		−				−	−	
L-Arabitol	−	−			−			−	−		−		−				−	−	
Arbutin	−	−			−			−	−		−		−				−	−	
Cellobiose	w	−	−		−			−	d		+		+				−	−	
Dulcitol	+	−	d/w	−	−			−	d		+	−	d	+			−	−	
Erythritol	−	−			−			−	−		−		−				−	−	
Fructose	−	−			−			−	−		−		−				−	−	
D-Fucose	w	−	+	−	+	+		−	+		+	−	+		+		+	+	
L-Fucose	−	−			−			−	−		−		−				−	−	
Galactose	w	−	−		−	d/w		−	−		−		−				−		w
β-Gentiobiose	w	−			−			−	−	−	+	−			+		−		w
Gluconate	w	−			−			−	−		+		−				−	−	
Glycerol	−		d/w	−	d			−	d		+		−				−		d
meso-Inositol	−		d/w	−	d			−	d		−		+				+	−	
Inulin	−		d/w	−				−	−	−	−		−				−		d
2-Ketogluconate	−		+					−	−		−		−				+	−	
5-Ketogluconate	−							−	−				−				−	−	
Lactose	w							−	−		+	−	−				−	−	
Lyxose	−							−	−		+	−		−			−		
Maltose	w	−	d/w	−	+	w		−	+		+		+		+		+	+	
Melezitose	−	−			−			−	d		+		−				−	−	
Melibiose	−	−			−			−	d		+		−				−	−	
Methyl α-D-mannoside	−	−			−			−	−	−	−		−		−		−	−	
Methyl β-D-xyloside	−	−	−		−			−	−		−		−				−		d
Raffinose	−	−	d/w	−	−			−	−		−		−				−	−	
Rhamnose	−	−	−	−	d			−	−	−	+		−		+		−		d
Ribose	w	−	d/w	−	d	w		−	−				+		+		+		d
Sorbitol	w	−	d/w	−	d	w		−	−	−		−	+		+		+	w	
Sorbose	−	−	d/w	−	d			−	−	−			−		+		−	−	
Stachyose	−	−	−		−			−	−		−		−				−	−	
Sucrose	−	−	+	−	−	d/w		−	+		+	−	d		+		+	+	
Tagatose	w	−			−			−	d		−		−				−	−	
Trehalose	w	−	+	−	+	w		−	+		+		+		+		+	+	
Turanose	−	−						−	d		+		−				−	−	
Xylitol	−	−	−		−			−	d(w)		−		−				−	−	
L-Xylose	−	−	−		−			−			−		−				−	−	
Utilization of:			f					−											
Acetate					+			+											
N-Acetyl-D-glucosamine			+					+			−		d	+		−			
Alanine			+														+		
Arabinose			+										−						
Arabitol			−														−	−	

(continued)

TABLE 4. (continued)

Characteristic	77. *B. shackletonii*	78. *B. silvestris*	79. *B. simplex*	80. *B. siralis*	81. *B. smithii*	82. *B. soli*	83. *B. sonorensis*	84. *B. sphaericus*	85. *B. sporothermodurans*	86. *B. subterraneus*	87. *B. thermoamylovorans*	88. *B. thermocloacae*	89. *B. thuringiensis*	90. *B. tusciae*	91. *B. vallismortis*	92. *B. vedderi*	93. *B. vietnamensis*	94. *B. vireti*	95. *B. weihenstephanensis*
Aspartate			+										+	−					
Benzoate			+											−					
Ethanol										+				+					
Formate													d	−					
Fructose			+							+									
Fumarate			+		+			−					d						
Galactose			−																
Gentiobiose			−																
Gluconate			+											−			+		
Glucose			+							+				−		+	+		
Glutamate			+					−					+	−					
Glycerol		+	+					−		+				−			−		
Glycine								−						d					
Glycogen														−					
Histidine			+																
m-Hydroxybenzoate			+																
Inositol			+																
Lactate			+							+			d						
Lactose			−														−		
Malate			+		+									+			+		
Maltose			+											−		+			
Mannitol			+													+	+		
Mannose			−														−		
Melibiose			−														−		
Methanol														−					
Phenol																			
Phenylacetate			+																
Phenylpropionate			+																
Proline			+											−					
Pyruvate								−					d	−					
Rhamnose			+																
Ribose		+	+											−					
Salicin																			
Serine			−											−					
Sorbitol			+																
Succinate			+		+			−	−	−			−	+			−		
Trehalose			+											−		+			
Turanose			−																
Valine														+					
Xylitol			−																
Xylose			+													+			
Hydrolysis of:																			
Cellulose																			
DNA			d		+			+											
Elastin			−					d											
Esculin	+	−		+	d	+		−	+	+		−	+				+	+	
Hippurate			+		+			−											
4-Methylumbelliferone glucuronide																			
Pectin			−					−								+			
Pullulan			−		d			−											
Tributyrin												−							
Tween 20			+					+											
Tween 40																			
Tween 60																			
Tween 80		−	d					d							w				
Urea	−		−		−			d	−	−			−	+			−	−	
Xylan																+			

(continued)

TABLE 4. (continued)

[a]Symbols: +, >85% positive; d, variable (16–84% positive); –, 0–15% positive; w, weak reaction; d/w, variable and weak when positive; no entry indicates that no data are available.

[b]Compiled from Larkin and Stokes (1967); Nakayama and Yanoshi (1967); Gordon et al. (1973), Pichinoty et al. (1984, 1976, 1983); Aragno (1978); Schenk and Aragno (1979); Pichinoty (1983); Bonjour and Aragno (1984); Logan and Berkeley (1984); Claus and Berkeley (1986); Priest et al. (1987, 1988); Nakamura et al. (1988, 1999, 2002) Demharter and Hensel (1989b)<qu ref=81>; Denariaz et al. (1989); Nakamura (1989, 1998); Ventosa et al. (1989); Tomimura et al. (1990); Nagel and Andreesen (1991); Arfman et al. (1992); Spanka and Fritze (1993); Andersch et al. (1994); Roberts et al. (1994, 1996); Agnew et al. (1995); Boone et al. (1995); Combet-Blanc et al. (1995); Nielsen et al. (1995a); Fritze (1996a); Fujita et al. (1996); Kuhnigk et al. (1995); Kuroshima et al. (1996); Pettersson et al. (2000, 1996); Shelobolina et al. (1997); Lechner et al. (1998); Switzer Blum et al. (1998); Yumoto et al. (1998, 2003, 2004c); Rheims et al. (1999); Logan et al. (2002a, b, 2000, 2004b); Palmisano et al. (2001); Yoon et al. (2001a); Abd El-Rahman et al. (2002); Ajithkumar et al. (2002); Kanso et al. (2002); Li et al. (2002); Reva et al. (2002); Venkateswaran et al. (2003); Gugliandolo et al. (2003a); Heyrman et al. (2003a, 2005a, 2004); Taubel et al. (2003); Yoon et al. (2003a); De Clerck et al. (2004b, 2004c); Heyndrickx et al. (2004); La Duc et al. (2004); Ivanova et al. (2004a); Noguchi et al. (2004); Santini et al. (2004); Scheldeman et al. (2004); Suresh et al. (2004).

[c]Reactions differ between strains of the emetic biotype of *Bacillus cereus* for these substrates.

[d]Results obtained when inocula are supplemented with 7% NaCl.

[e]Results obtained when grown at pH 10.

[f]Assimilation data for *Bacillus simplex* are for the type strain only.

TABLE 5. Differentiation of *Bacillus subtilis* from closely related *Bacillus* species[a,b]

Characteristic	1. *B. subtilis* subspp. *subtilis* and *spizizenii*[c]	6. *B. amyloliquefaciens*	11. *B. atrophaeus*	51. *B. licheniformis*	57. *B. mojavensis*	73. *B. pumilus*	83. *B. sonorensis*	91. *B. vallismortis*
Pigmented colonies								
Yellow-pink-red	–[d]	–	–	–	–	–	w[e]	–
Dark brown/black	–[d]	–	+[f]	–	–	–	–	–
Anaerobic growth	–	–	–	+	–	–	+	–
Acid from:								
Glycogen	+	+	nd	+	nd	–	nd	nd
Methyl α-D-mannoside	–	–	nd	–	nd	+	nd	nd
Starch	+	+	nd	+	nd	–	nd	nd
Hydrolysis of starch	+	+	+	+	+	–	+	+
Utilization of propionate	–	–	–	+	–	–	+	–
Nitrate reduction	+	+	+	+	+	–	+	+
Growth in NaCl:								
5%	+	+	+	+	+	+	–	+
7%	+	nd	+	+	+	+	–	+
10%	nd	d	nd	nd	+	+	–	+
Growth at:								
5 °C	–	–	d	–	d	d	–	d
10 °C	d	–	+	–	+	d	–	+
50 °C	d	+	+	+	+	d	+	+
55 °C	–	–	d	d	d	–	+	–
65 °C	–	–	–	–	–	–	–	–

[a]Symbols: +, >85% positive; d, different strains give different reactions (16–84% positive); –, 0–15% positive; w, weak reaction; nd, no data are available.

[b]Compiled from Claus and Berkeley (1986), Priest et al. (1987), Nakamura (1989), Roberts et al. (1994, 1996), Nakamura et al. (1999), and Palmisano et al. (2001).

[c]The subspecies of *Bacillus subtilis* are not distinguishable by routine phenotypic tests.

[d]*Bacillus subtilis* may form pigments, varying from cream through yellow, orange, pink and red, to brown or black, on potato or agar media containing glucose; strains forming brown or black pigment were often formerly called "*Bacillus subtilis* var. *aterrimus*."

[e]Colonies are yellowish-cream on routine media, and bright yellow on pH 5.6 agar.

[f]This species accommodates strains forming brownish-black pigment on tyrosine (and so often evident on the crude media available to earlier workers), and often formerly called "*Bacillus subtilis* var. *niger*."

temperature range is 37–65 °C, with an optimum growth temperature of 55 °C. pH range for growth 7–9, with an optimum of pH 8.0. Grows in the range 0.5–5% NaCl, with an optimum of 2% NaCl. Acid is produced from glucose and a wide range of other carbohydrates. The following may be utilized as carbon sources: arabinose, N-acetylglucosamine, citrate, glucose, gluconate, malate, maltose, mannitol, mannose, phenylacetate. Produces acetoin but not H_2S or indole. Nitrate is not reduced. Casein, gelatin and starch are hydrolyzed, but esculin and urea are not. Arginine dihydrolase, and lysine and ornithine decarboxylases negative. Exopolysaccharides are produced in mineral medium supplemented with sucrose. See Table 8.

Source: a shallow marine hydrothermal vent, Vulcano Island, Eolian Islands, Italy.

DNA G + C content (mol%): 40.8 (T_m) (for methods, see Maugeri et al., 2001).

Type strain: 4-1, DSM 15804, and CIP 107628.

EMBL/GenBank accession number (16S rRNA gene): AJ504797 (4-1).

3. **Bacillus agaradhaerens** Nielsen, Fritze and Priest 1995b, 879[VP] (Effective publication: Nielsen, Fritze and Priest 1995a, 1758.)

a.gar.ad'hae.rens. Malayan n. *agar* gelling polysaccharide from brown algae; L. adj. *adhaerens* adherent; N.L. adj. *agaradhaerens* adhering to the agar.

Strictly alkaliphilic organisms forming ellipsoidal spores which lie subterminally in swollen sporangia. Cells 0.5–0.6 by 2.0–5.0 μm. Colonies are adherent, white and rhizoid with filamentous margins. Growth temperature range 10–45 °C. Optimal growth at pH 10.0 or above; no growth at pH 7.0. Grows (sometimes only weakly) in presence of up to 16% NaCl. Nitrate is reduced to nitrite. Casein, cellulose, gelatin, starch, Tween 40 and xylan are hydrolyzed. Tween 60 is hydrolyzed by most strains. Hippurate, 4-methylumbelliferone glucuronide, Tween 20 and 80 are not hydrolyzed; phenylalanine is not deaminated. Glucose and a range of other carbohydrates can be utilized as sole sources of carbon. See Table 6.

Source: soil.

DNA G + C content (mol%): 39.3–39.5 (HPLC).

Type strain: PN-105, ATCC 700163, DSM 8721, LMG 17948.

EMBL/GenBank accession number (16S rRNA gene): X76445 (DSM 8721).

4. **Bacillus alcalophilus** Vedder 1934, 141[AL] (emend. Nielsen, Fritze and Priest 1995a, 1758.)

al.cal.o.phil'us. N.L. *alcali* En. alkali from the Arabic *al* the; *qaliy* soda ash; Gr. adj. *philos* loving; N.L. adj. *alcalophilus* liking alkaline (media).

Alkaliphilic organisms forming ellipsoidal spores which lie subterminally in unswollen sporangia. Cells 0.5–0.7 by 3.0–5.0 μm. Colonies are white, circular, smooth and shiny, sometimes with darker centers. Growth temperature range 10–40 °C. Optimal growth at pH 9.0–10.0; no growth at pH 7.0. Maximum NaCl concentration tolerated ranges from less than 5% up to 8%. Nitrate is usually not reduced to nitrite. Casein, gelatin, pullulan, starch, and Tween 40 and 60 are hydrolyzed. Hippurate, 4-methylumbelliferone glucuronide, Tween 20 (usually) and 80 are not hydrolyzed; phenylalanine is not deaminated. Glucose and a range of other carbohydrates can be utilized as sole sources of carbon. See Table 6.

Source: a variety of materials after enrichment at pH 10.

DNA G + C content (mol%): 36.2–38.4 (HPLC analysis) and 37.0 (T_m), 36.7 (Bd) for the type strain.

Type strain: Vedder 1, ATCC 27647, DSM 485, JCM 5262, LMG 17938, NCIMB 10436.

EMBL/GenBank accession number (16S rRNA gene): X76436 (DSM 485).

5. **Bacillus algicola** Ivanova, Alexeeva, Zhukova, Gorshkova, Buljan, Nicolau, Mikhailov and Christen 2004b, 1425[VP] (Effective publication: Ivanova, Alexeeva, Zhukova, Gorshkova, Buljan, Nicolau, Mikhailov and Christen 2004a, 304.)

al.gi'co.la. L. fem. n. *alga -ae*, alga; L. suff. *-cola* (from L. masc. n. *incola -ae* inhabitat, dweller); N.L. masc. n. *algicola* algae-dweller.

Gram-positive cells (0.5–0.9 μm in diameter and 1.8–5.0 μm long) are aerobic, filamentous with "cross-like" branching, and produce subterminally located ellipsoidal spores (0.5–0.7 μm by 0.7–1.0 μm). Description is based upon a single isolate. Colonies are semitransparent, creamy, and slightly yellowish in color. Growth occurs between 10 °C and 45 °C with optimum at 28–30 °C. No growth is detected at 4 °C and at 50 °C. Alkalitolerant, growing at pH 7–10. Growth occurs at 0–3% NaCl. Anaerobic growth and oxidase are negative. Catalase and nitrate reduction are weak. Urea, alginate, starch, and gelatin are hydrolyzed. Do not decompose agar and casein. According to Biolog, utilizes dextrin, cellobiose, D-fructose, α-D-glucose, maltose, D-mannose, sucrose, D-trehalose, pyruvic acid methyl ester, β-hydroxybutyric acid, α-ketobutyric acid, inosine, uridine, thymidine, glycerol, and DL-α-glycerol phosphate. The predominant cellular fatty acids are $C_{14:0\ iso}$, $C_{15:0\ iso}$, $C_{15:0\ anteiso}$, $C_{16:0\ iso}$; and $C_{17:0\ anteiso}$. See Table 6.

Source: degraded thallus of brown alga *Fucus evanescens* collected from Kraternaya Bight, Pacific Ocean.

DNA G + C content (mol%): 37.4 (T_m).

Type strain: KMM 3737, CIP 107850.

GenBank/EMBL accession number (16S rRNA gene): AY228462 (KMM 3737).

6. **Bacillus amyloliquefaciens** Priest, Goodfellow, Shute and Berkeley 1987, 69[VP]

am.yl.o.li.que.fac'i.ens. L. n. *amylum* starch; L. part. adj. *liquefaciens* dissolving; N.L. part. adj. *amyloliquefaciens* starch-digesting.

Strictly aerobic, Gram-positive, motile rods, 0.7–0.9 by 1.8–3.0 μm, often occurring in chains, and forming ellipsoidal spores (0.6–0.8 by 1.0–1.4 μm) which lie centrally, paracentrally and subterminally in unswollen sporangia. No growth below 15 °C or above 50 °C; optimum growth temperature 30–40 °C. Casein, elastin, esculin, gelatin, starch and Tween 20, 40 and 60 are degraded, but adenine, cellulose, guanine, hypoxanthine, pectin, testosterone, tyrosine, urea and xanthine are not. Nitrate is reduced to nitrite. Voges–Proskauer-positive. Citrate is utilized as sole carbon source, propionate is not. Growth occurs in presence of 5% NaCl, and most strains tolerate 10% NaCl. Acid without gas is produced from glucose and a range of other carbohydrates. This species is important as a source of α-amylase and protease for industrial applications. See Table 5.

Source: soil and industrial amylase fermentations.

TABLE 6. Differentiation of alkaliphilic *Bacillus* species[a,b]

Characteristic	3. *B. agaradhaerens*	4. *B. alcalophilus*	5. *B. algicola*	9. *B. arseniciselenatis*	20. *B. clarkii*	23. *B. cohnii*	40. *B. halodurans*	43. *B. horti*	49. *B. krulwichiae*	65. *B. okuhidensis*	67. *B. pseudalcalophilus*	68. *B. pseudofirmus*	76. *B. selenitireducens*	88. *B. thermocloacae*	92. *B. vedderi*
Colonies pigmented	−	−	+[c]	−[c]	+[c]	−	−	−	−	−	−	+[c]	−[c]	−	−
Motility	nd	nd	+	−	nd	+	nd	+	+	+	nd	nd	−	−	+
Spores:															
Ellipsoidal	+	+	+	nd	+	+	−	+	+	+	+	+	nd	+	v
Spherical	−	−	−	nd	−	−	+	−	−	−	−	−	nd	−	v
Borne terminally	−	−	+	nd	−	+	+	−	−	−	−	−	nd	−	v
Swollen sporangia	+	−	−	nd	d	+	+	+	−	v	+	−	nd	+	+
Catalase	nd	+	w	+	nd	+	nd	+	+	+	nd	nd	+	+	+
Anaerobic growth	nd	−	−	+[d]	nd	nd	nd	−	+	nd	nd	nd	+	−	+
Hydrolysis of:															
Casein	+	+	−	nd	+	+	d/w	+	d	+	+	+	nd	−	−
Gelatin	+	+	+	nd	+	+	+	+	d	+	+	+	nd	−	w
Starch	+	+	+	nd	−	+	w	+	+	+	+	+	nd	−	−
Nitrate reduction	+	−	w	nd	+	+	−	+	+	+	−	−	nd	−	nd
Growth at pH:															
6	−	−	−	−	−	−	−	−	−	+	−	−	−	−	−
7	−	−	+	−	−	−	+	+	−	+	−	d	−	−	−
8	+	+	+	w	nd	nd	+	+	+	+	+	+	−	−	−
9	+	+	+	+	+	+	+	+	+	+	+	+	−	+	+
10	+	d	+	+	+	+	+	+	+	+	+	nd	+	−	+
Optimum pH	10	9–10	9	8.5–10	10	9.7	9–10	8–10	8–10	10.5	10	9	8.5–10	8–9	10
Growth in NaCl:															
2%	+	+	+	−	nd	+	+	d	+	+	nd	nd	+	+	+
5%	+	+	−	+	+	+	+	+	+	+	+	+	+	w	+
7%	+	+	nd	+	+	nd	+	+	+	+	+	+	+	−	+
10%	+	−	nd	+	+	−	+	+	+	+	+	+	+	−	−
NaCl required for growth	+	−	−	+	+			d	−	−	−	−	+	−	−
Growth at:															
10 °C	+	+	+	nd	−	+	−	−	−	−	+	+	nd	−	nd
20 °C	+	+	+	+	+	+	+	+	nd	−	+	+	+	−	nd
30 °C	+	+	+	nd	+	+	+	+	nd	+	+	+	nd	−	
40 °C	+	+	+	nd	+	+	+	+	+	+	+	+	nd	+	+
50 °C	−	−	−	nd	−	−	−	−	−	+	−	−	nd	+	+
55 °C	−	−	−	nd	−	−	−	−	−	+	−	−	nd	+	d
Deamination of phenylalanine	−	−	nd	nd	+	−	−	−	nd	−	+	−	nd	+	−
Respiratory growth with As(V)	nd	nd	nd	+	nd	nd	nd	nd	nd	nd	nd	nd	+	nd	nd
Respiratory growth with Se(IV) or Se(VI)	nd	nd	nd	+	nd	nd	nd	nd	nd	nd	nd	nd	+	nd	nd

[a]Symbols: +, >85% positive; d, different strains give different reactions (16–84% positive); −, 0–15% positive; v, variation within strains; w, weak reaction; d/w, d, different strains give different reactions, but positive reactions are weak; nd, no data are available.

[b]Compiled from Claus and Berkeley (1986), Demharter and Hensel (1989b); Spanka and Fritze (1993); Agnew et al. (1995); Nielsen et al. (1995a); Fritze (1996a); Yumoto et al. (1998, 2003); Switzer Blum et al. (2001); Li et al. (2002); Ivanova et al. (2004a).

[c]*Bacillus algicola* produces semitransparent, creamy, slightly yellowish colonies; *Bacillus arseniciselenatis* and *Bacillus selenitireducens* will produce red colonies, owing to elemental selenium precipitation, on selenium oxide media; *Bacillus clarkii* colonies may be cream-white to pale yellow in color, and one of the three strains described produces dark yellow colonies with age; *Bacillus pseudofirmus* colonies are yellow.

[d]*Bacillus arseniciselenatis* does not grow aerobically; *Bacillus selenitireducens* grows weakly in microaerobic conditions.

DNA G + C content (mol%): 44.35 ± 0.38 (T_m) for eight strains; 44.2 ± 0.7 (Bd) with a range of 44–46; the mol% G + C of the type strain is 44.6.

Type strain: Fukumoto strain F, ATCC 23350, DSM 7, LMG 9814, NCIMB 12077, NRRL B-14393.

EMBL/GenBank accession number (16S rRNA gene): X60605 (ATCC 23350).

7. **Bacillus anthracis** Cohn 1872, 177[AL]

an'thra.cis. Gr. n. *anthrax* charcoal, a carbuncle; N.L. n. *anthrax* the disease anthrax; N.L. gen. n. *anthracis* of anthrax.

Phenotypically similar to *Bacillus cereus* (see below and Table 7) except in the characters undernoted. Colonies of *Bacillus anthracis* (Figure 9a) are similar to those of *Bacillus cereus*, but those of the former are generally smaller, non-hemolytic, may show more spiking or tailing along the lines of inoculation streaks, and are very tenacious as compared with the usually more butyrous consistency of *Bacillus cereus* and *Bacillus thuringiensis* colonies, so that they may be pulled into standing peaks with a loop. Nonmotile. Usually susceptible to penicillin. Susceptible to gamma phage (see Logan and Turnbull, 2003, Logan et al., 2007). Produces a glutamyl-polypeptide capsule *in vivo* and when grown on nutrient agar containing 0.7% sodium bicarbonate incubated overnight under 5–7% CO_2. Colonies of the capsulate *Bacillus anthracis* appear mucoid, and the capsule can be visualized by staining smears with M'Fadyean's polychrome methylene blue or India Ink (Turnbull et al., 1998). An isolate showing the characteristic phenotype but unable to produce capsules may be an avirulent form lacking either or both capsule or toxin genes (Turnbull et al., 1992). Virulent and avirulent strains may be distinguished from other members of the *Bacillus cereus* group using tests in the API System (Logan et al., 1985). Virulence genes are carried by plasmids pX01 (toxins) and pX02 (capsule); these plasmids may be transmissible to other members of the *Bacillus cereus* group (Turnbull et al., 2002). Primer sequences are now available for confirming the presence of the toxin and capsule genes, and hence the virulence of an isolate. Genetically very closely related to *Bacillus cereus* and other members of the *Bacillus cereus* group (Turnbull, 1999); *Bacillus anthracis* may be distinguished from other members of the *Bacillus cereus* group by amplified fragment length polymorphism (AFLP) analysis (Keim et al., 2000; Turnbull et al., 2002). A 277 bp DNA sequence (Ba813) has been described as a specific chromosomal marker for *Bacillus anthracis* and in combination with sequencing of parts of *lef* and *cap* genes its sequence allows the identification of virulent strains (application note 209 of Pyrosequencing AB). Isolates of *Bacillus anthracis* show considerable molecular homogeneity, and the species may derive from a relatively recent common ancestor.

Causative agent of the disease anthrax in herbivorous and other animals and man. Widely studied and developed

TABLE 7. Differentiation of *Bacillus cereus* from closely related *Bacillus* species[a,b].

Characteristic	7. *B. anthracis*	18. *B. cereus*	60. *B. cereus (emetic biovar)*[c]	58. *B. mycoides*	69. *B. pseudomycoides*	89. *B. thuringiensis*	95. *B. weihenstephanensis*
Motility	−	+	+	−	−	+	+
Rhizoid colonies	−	−	−	+	+	−	−
Cell diameter >1.0 μm	+	+	+	+	v	+	+
Parasporal crystals	−	−	−	−		+	−
Acid from:							
Glycerol	−	+	d	+	nd	+	nd
Glycogen	+	+	−	+	nd	+	+
Salicin	−	d	−	d	nd	d	d
Starch	+	+	−	+	nd	+	+
Arginine dihydrolase	−	d	d	d	nd	+	nd
Utilization of citrate	d[d]	+	+	d	d	+	+
Nitrate reduction	+	d	+	d	+	+	d
Growth at:							
5 °C	−	−	nd	−	−	−	+
10 °C	−	d	nd	d	−	d	+
40 °C	+	+	nd	d	+	+	−
Degradation of tyrosine	−	+	nd	d	+	+	+

[a]Symbols: +, >85% positive; d, different strains give different reactions (16–84% positive); −, 0–15% positive; v, variation within strains; nd, no data are available.

[b]Compiled from Gordon et al. (1973); Logan and Berkeley (1984); Logan et al. (1985); Claus and Berkeley (1986); Lechner et al. (1998); Nakamura (1998).

[c]Strains of *Bacillus cereus* of serovars 1, 3, 5 and 8, which are particularly associated with outbreaks of emetic-type food poisoning.

[d]Citrate test results may vary according to the test method used; Gordon et al. (1973) found citrate utilization to be a variable property among 23 strains of *Bacillus anthracis*, while Logan and Berkeley (1984) and Logan et al. (1985) obtained negative results for 37 strains using the API 20E test method.

as a biological weapon. Generally considered to be an obligate pathogen; if it ever multiplies in the environment, it probably only does so rarely. Spores remain viable in soil for many years and their persistence does not depend on animal reservoirs.

Source: blood of animals and humans suffering from anthrax, from anthrax carcasses, and from animal products and soil contaminated with spores of the organism.

DNA G + C content (mol%): 32.2–33.9 (T_m) for five strains, and 33.2 (T_m) for the type strain.

Type strain: Vollum strain, ATCC 14578, NCIB 9377, NCTC 10340.

The 16S rRNA (or rDNA) gene sequence of the type strain is not available in the EMBL/GenBank database. However, 16S rDNA sequences of 98 strains of this species in EMBL are nearly all identical. Accession number AF176321 corresponds with the strain "Sterne."

8. **Bacillus aquimaris** Yoon, Kim, Kang, Oh and Park 2003a, 1301[VP]

a.qui.ma'ris. L. n. *aqua* water; L. gen. n. *maris* of the sea; N.L. gen. n. *aquimaris* of the water of the sea.

Aerobic, Gram-variable rods, 0.5–0.7 by 1.2–3.5 μm, motile by means of peritrichous flagella. Description is based on a single isolate. Ellipsoidal endospores are borne centrally in large, swollen sporangia. Colonies are pale orange-yellow, circular to slightly irregular, slightly raised, and 2–4 mm in diameter after 3 d at 30 °C on marine agar. Optimal growth temperature is 30–37 °C. Growth occurs at 10 and 44 °C, but not at 4 or above 45 °C. Optimal growth pH is 6.0–7.0, and no growth is observed at pH 9.0 or 4.5. Optimal growth occurs in the presence of 2–5% (w/v) NaCl. Growth is poor in the absence of NaCl, but occurs in the presence of up to 18% (w/v) NaCl. Catalase-positive, oxidase- and urease-negative. Casein, starch and Tween 80 are hydrolyzed. Esculin, hypoxanthine, tyrosine and xanthine are not hydrolyzed. Acid is produced from D-fructose, D-glucose, glycogen, 5-ketogluconate, maltose, D-ribose, starch, sucrose and D-trehalose. The cell-wall peptidoglycan contains *meso*-diaminopimelic acid. The predominant menaquinone is MK-7. The major fatty acids are $C_{15:0\ iso}$ and $C_{15:0,\ anteiso}$.

Source: sea water of a tidal flat of the Yellow Sea in Korea.

DNA G + C content (mol%) of the type strain is: 38 (HPLC).

Type strain: TF-12, JCM 11545, KCCM 41589.

GenBank accession number (16S rRNA gene): AF483625 (TF-12).

9. **Bacillus arseniciselenatis** (nom. corrig. *Bacillus arsenicoselenatis* [sic]) Switzer Blum, Burns Bindi, Buzzelli, Stolz and Oremland 2001, 793[VP] (Effective publication: Switzer Blum, Burns Bindi, Buzzelli, Stolz and Oremland 1998, 28.)

ar.se.ni.ci.se.le.na'tis. L. n. *arsenicum* arsenic; N.L. n. *selenas* -*atis* selenate; N.L. gen. n. *arseniciselenatis* of arsenic (and) selenate.

Strictly anaerobic, nonmotile, spore-forming, Gram-positive rods which show respiratory growth with Se(VI) (selenate), As(V) (arsenate), Fe(III), nitrate and fumarate as electron acceptors. Cells are 0.5–1.0 by 3–10 μm. Description is based upon a single strain. Catalase- and oxidase-positive. Colonies are formed on lactate/selenate/

yeast extract-supplemented lakewater medium incubated anaerobically at 20 °C. Grows fermentatively on fructose. Uses lactate, malate, fructose, starch and citrate as electron donors. Moderately halophilic, with optimum salinity of 60 g/l NaCl, and a requirement for NaCl for growth. Moderately alkaliphilic, with optimum growth in the range pH 8.5–10. See Table 6.

Source: arsenic-rich sediment of Mono Lake, California.

DNA G + C content (mol%) of the type strain: 40.0% (T_m).

Type strain: E1H, ATCC 700614, DSM 15340.

EMBL/GenBank accession number (16S rRNA gene): AF064705 (E1H).

10. **Bacillus asahii** Yumoto, Hirota, Yamaga, Nodasaka, Kawasaki, Matsuyama and Nakajima 2004c, 1999[VP]

as.a.hi'i. N.L. gen. n. *asahii* of Asahi; named after Asahi Kasei Co., a researcher from which isolated the bacterium.

Cells are Gram-positive peritrichously flagellated straight rods (1.4–3.0 × 0.4–0.8 μm) and produce terminally or centrally located ellipsoidal spores. Description is based upon a single isolate. Utilizes butyrate as carbon source for growth. Spores do not cause swelling of sporangium. Colonies are circular and white. Catalase and oxidase reactions are positive. Nitrate reduction to nitrite is weakly positive. Negative for indole production, Voges–Proskauer test, methyl red test, growth on MacConkey agar and H_2S production. Trypsin, esterase (C4) and esterase/lipase (C8) are positive. Alkaline phosphatase, valine arylamidase, cystine arylamidase, chymotrypsin, acid phosphatase, β-glucosidase, β-glucosidase and *N*-acetyl-β-glucosaminidase are negative. Growth occurs at pH 6–9; growth at pH 5 is variable. Growth occurs at 0–1% NaCl but not at 2% NaCl. Growth occurs at 15–45 °C, but not above 50 °C. No acid is produced carbohydrates. Hydrolysis of casein, DNA, and Tween 20, 40 and 60 is observed but hydrolysis of gelatin is not. Hydrolysis of starch is weak. $C_{15:0\ iso}$ (39.0%) and $C_{15:0\ anteiso}$ (27.8%) represent the main fatty acids produced during growth in PY-1 medium.

Source: a soil sample obtained from Tagata-gun, Shizuoka, Japan.

DNA G + C content (mol%): 39.4 (HPLC).

Type strain: FERM BP-4493, MA001, JCM 12112, NCIMB 13969, CIP 108638.

GenBank/EMBL accession number (16S rRNA gene): AB109209 (FERM BP-4493).

11. **Bacillus atrophaeus** Nakamura 1989, 299[VP]

a.tro.phae'us. L. adj. *ater* black; Gr adj. *phaeus* brown; N.L. adj. *atrophaeus* dark brown.

Aerobic, Gram-positive, motile rods, forming ellipsoidal spores which lie centrally or paracentrally in unswollen sporangia. Cells 0.5–1.0 by 2.0–4.0 μm, occurring singly and in short chains. Colonies are opaque, smooth circular and entire and up to 2 mm in diameter after 2 d at 28 °C, and form a dark brownish-black soluble pigment in 2–6 d on media containing tyrosine or other organic nitrogen source. Optimum growth temperature 28–30 °C, with minimum of 5–10 °C and maximum of 50–55 °C. Catalase-positive, oxidase-negative.

Includes strains formerly called "*Bacillus subtilis* var. *niger*"; some strains so designated are used for autoclave sterility testing. Phenotypically distinguishable from *Bacillus mojavensis*, *Bacillus subtilis* subsp. *spizenii* and *Bacillus subtilis* subsp. *subtilis* only by pigment production and a negative oxidase reaction. Phenotypically indistinguishable from *Bacillus vallismortis*. See Table 5.

Source: isolated mainly from soil.

DNA G + C content (mol%): 41.0–43.0 (Bd).

Type strain: NRRL NRS-213, ATCC 49337, DSM 7264, JCM 9070, LMG 17795, NCIMB 12899.

EMBL/GenBank accession number (16S rRNA gene): AB021181 (JCM 9070).

12. **Bacillus azotoformans** Pichinoty, de Barjac, Mandel and Asselineau 1983, 660[VP]

a.zo.to.for′mans. Fr. n. *azote* nitrogen; L. part. adj. *formans* forming; N.L. part. adj. *azotoformans* nitrogen-forming.

Gram-negative, peritrichously motile rods (0.5–0.8 μm by 3–7 μm), forming ellipsoidal, subterminal and terminal spores which swell the sporangia. Nitrate, nitrite and nitrous oxide are denitrified with the production of N_2. For anaerobic growth, nitrate, nitrite, nitrous oxide, tetrathionate and fumarate act as terminal electron acceptors. Growth requirements are complex; non-fermentative, carbohydrates are not attacked, a range of organic acids is utilized as carbon sources. Colonies circular and partially translucent, with entire margins, on yeast extract agar. Oxidase-positive, catalase-negative. Gelatin, starch and Tween 80 not hydrolyzed. Maximum growth temperature 42–46 °C.

Source: garden soil by enrichment culture in peptone broth under N_2O.

DNA G + C content (mol%): 39.0–43.9 (mean of 39.8 for 17 strains) (Bd) and 39.0 for the type strain.

Type strain: Pichinoty 1, ATCC 29788, NRRL B-14310, DSM 1046, LMG 9581, NCIMB 11859.

EMBL/GenBank accession number (16S rRNA gene): D78309 (DSM 1046). This sequence seems to be somewhat more reliable than the one reported for ATCC 29788. Both sequences differ and contain inadequately determined bases.

13. **Bacillus badius** Batchelor 1919, 23[AL]

ba.di′us. L. adj. *badius* chestnut brown.

Aerobic, Gram-positive, motile rods, cells 0.8–1.2 by 2.5–5.0 μm, occurring singly and in pairs and chains, forming ellipsoidal spores which are located subterminally, and sometimes paracentrally or terminally, and which do not swell the sporangia. Growth occurs between 15 °C and 50 °C, with the optimum around 30 °C. Catalase-positive. Casein and gelatin are hydrolyzed; starch is not hydrolyzed. Tyrosine is degraded. Grows in presence of up to 7% NaCl. Citrate may be utilized as sole carbon source. Nitrate is not reduced. Acid is not produced from glucose and other carbohydrates. Assimilates certain amino acids and organic acids.

Source: feces, dust, marine sources, foods, antacids and gelatin production plant.

DNA G + C content (mol%): 43.8 (T_m) and 43.5 (Bd) for the type strain.

Type strain: ATCC 14574, DSM 23, LMG 7122, NCIMB 9364, NRRL NRS-663, IAM 11059.

EMBL/GenBank accession number (16S rRNA gene): X77790 (ATCC 14574).

14. **Bacillus barbaricus** Taubel, Kämpfer, Buczolits, Lubitz and Busse 2003, 729[VP]

bar.ba′ri.cus. L. adj. *barbaricus* strange, foreign, referring to the strange behavior towards growth at different pH levels.

Facultatively anaerobic, Gram-positive, nonmotile rods, 0.5 μm wide and 4–5 μm long. Oval endospores are borne subterminally in swollen sporangia. Colonies are brownish, opaque, circular, flat and 3–7 mm in diameter when grown on peptone-yeast extract-succinate (PYES) agar. Catalase-positive, oxidase- and urease-negative. Nitrate is not reduced. Indole and H_2S are not produced. Alkalitolerant; growth is weak at pH 6.0, but strong at pH 7.2, 8.0 and 9.5 in PYES medium adjusted with HCl or NaOH before autoclaving, but no growth in buffered media at pH 7.0–11.0. Good growth occurs at temperatures ranging from 18 °C to 37 °C, with no growth at 4 or 47 °C. Weak growth occurs in presence of 2% NaCl, and no growth in 5% NaCl. Hippurate is decomposed. Esculin is not hydrolyzed. Citrate is not utilized. Acid is produced from D-glucose, N-acetyl-glucosamine, maltose, trehalose, starch and glycogen. Acid production is variable from D-fructose (type strain weakly positive), galactose, methyl-D-glucoside, lactose, sucrose and D-turanose (type strain negative). The cell-wall diamino acid is diaminopimelic acid and MK-7 is the predominant menaquinone. The polar lipid profile is composed of the major compounds phosphatidylethanolamine, phosphatidylglycerol and diphosphatidylglycerol. The fatty acid profile consists of the predominant compounds $C_{15:0\ anteiso}$ and $C_{15:0\ iso}$; $C_{14:0\ iso}$ and $C_{16:0\ iso}$ are present in moderate amounts.

Source: an experimental wall painting exposed in the Virgilkapelle in Vienna, Austria.

DNA G + C content (mol%): not reported.

Type strain: V2-BIII-A2, DSM 14730, CCM 4982.

EMBL/GenBank accession number (16S rRNA gene): AJ422145 (V2-BIII-A2).

15. **Bacillus bataviensis** Heyrman, Vanparys, Logan, Balcaen, Rodríguez-Díaz, Felske and De Vos 2004, 55[VP]

ba.ta.vi.en′sis. L. adj. *bataviensis* pertaining to Batavia, the name with which Julius Caesar described The Netherlands.

Facultatively anaerobic, Gram-positive or Gram-variable (at 24 h), motile, slightly tapered rods, 0.7–1.2 μm in diameter, occurring singly, and in pairs and short chains. Endospores are mainly ellipsoidal but may be spherical, and lie centrally, paracentrally and occasionally subterminally in slightly swollen sporangia. Colonies on TSA are butyrous, cream-colored, and produce a soft-brown pigment that diffuses in the agar; they are slightly raised and umbonate, have regular margins and smooth or rough, eggshell-textured surfaces. The optimum temperature for growth is 30 °C, and the maximum growth temperature lies between 50 °C and 55 °C. The optimum pH for growth is 7.0–8.0, and the pH range for growth is from 4.0–6.0 to 9.5–10.0. Casein is not hydrolyzed. In the API 20E strip, *o*-nitrophenyl-β-D-galactopyranoside hydrolysis is positive, gelatin is hydrolyzed by most strains, and nitrate reduction is positive; Voges–Proskauer reaction is negative, and reactions for arginine dihydrolase (one strain positive), lysine decarboxylase, ornithine decarboxylase, citrate utilization, hydrogen sulfide production, urease, tryptophan deaminase, indole production are negative. Hydrolysis of esculin is positive. Acid without gas is produced from the following carbohydrates in the API 50 CH gallery using the CHB suspension medium: N-acetyl-D-glucosamine, D-cellobiose, D-fructose, galactose, β-gentiobiose, D-glucose, glycerol (weak), lactose, maltose, D-mannitol, D-mannose, D-melezitose, raffinose, ribose (weak), salicin (weak), D-trehalose and D-turanose. The following reactions are variable between strains and, when positive, are usually weak: amygdalin, arbutin, L-fucose, inulin, D-melibiose, starch and sucrose; type strain is positive but weak for: arbutin, L-fucose, inulin, D-melibiose, methyl α-D-glucoside, methyl α-D-mannoside and sucrose. The major cellular fatty acids are $C_{15:0 \text{ iso}}$ and $C_{15:0 \text{ anteiso}}$, present at a level of about 37 and 21%, respectively, while $C_{16:1 \omega 11c}$ accounts for about 11% of the total fatty acids.

Source: soil in the Drentse A agricultural research area, The Netherlands.

DNA G + C content (mol%): 39.6–40.1 (type strain 40.1) (HPLC).

Type strain: LMG 21833, DSM 15601.

EMBL/GenBank accession number (16S rRNA gene): AJ542508 (LMG 21833).

16. **Bacillus benzoevorans** Pichinoty, Asselineau and Mandel 1987, 179[VP] (Effective publication: Pichinoty, Asselineau and Mandel 1984, 215.)

ben.zo.e.vor'ans. L. part. adj. *acidum benzoicum.* benzoic acid; L. *vorans* devouring; N.L. part. adj. *benzoevorans* devourer of benzoic acid.

Prototrophic, facultatively anaerobic, Gram-variable, large (1.8 μm diameter) filaments and rods which use aromatic acids and phenols, but not carbohydrates and amino acids (except glycine) as carbon and energy sources. Do not grow in media containing only peptone or tryptone; grow rapidly in media containing yeast extract and sodium acetate or benzoate. Filamentous growth on solid and in stationary liquid media, with motile rods appearing in shaken liquid culture. Form ellipsoidal spores which do not swell the sporangia. Colonies circular, flat, off-white and opaque with matt surface. Nitrate, but not nitrite, reduced. Optimum growth temperature 32 °C; maximum 39–45 °C.

Source: pasteurized soil by aerobic enrichment in minimal medium containing benzoate, p-hydroxybenzoate or cyclohexane carboxylate.

DNA G + C content (mol%) of the type strain: 41.3% (T_m).

Type strain: Pichinoty strain B1, ATCC 49005, DSM 5391, LMG 20225, NCIMB 12555, NRRL B-14535, CCM 3364.

EMBL/GenBank accession number (16S rRNA gene): X60611 (NCIMB 12555).

17. **Bacillus carboniphilus** Fujita, Shida, Takagi, Kunugita, Pankrushina and Matsuhashi 1996, 118[VP]

car.bo.ni'phi.lus. L. n. *carbo* coal, carbon; Gr adj. *philos* loving; N.L. adj. *carboniphilus* carbon-loving.

Aerobic, Gram-positive, peritrichously motile rods, forming ellipsoidal spores which lie centrally or terminally in unswollen sporangia. Cells 0.5–0.9 by 3.0–5.0 μm. Growth promoted by activated carbon and graphite. Colonies on nutrient agar are circular, flat, smooth, and grayish yellow; brown-red pigment is produced on trypto-soya agar. Growth temperature range 17–47 °C. Strictly aerobic; nitrate not reduced. Grows in presence of 7% NaCl. Catalase- and oxidase-positive; casein, gelatin, hippurate, starch and Tween 80 are hydrolyzed. Acid and gas are not produced from glucose and a range of other carbohydrates.

Source: air, using antibiotic-containing medium spotted with sterile graphite.

DNA G + C content (mol%): 37.8–38.1 (T_m), and 37.9 for the type strain.

Type strain: strain Matsuhashi Kasumi 6, JCM 9731, ATCC 700100, LMG 18001, NCIMB 13460.

EMBL/GenBank accession number (16S rRNA gene): AB021182 (JCM 9731).

18. **Bacillus cereus** Frankland and Frankland 1887, 257[AL]

ce're.us. L. adj. *cereus* waxen, wax-colored.

Facultatively anaerobic, Gram-positive, usually motile rods 1.0–1.2 by 3.0–5.0 μm, occurring singly and in pairs and long chains, and forming ellipsoidal, sometimes cylindrical, subterminal, sometimes paracentral, spores which do not swell the sporangia (Figure 10b); spores may lie obliquely in the sporangia. Cells grown on glucose agar produce large amounts of storage material, giving a vacuolate or foamy appearance. Colonies are very variable in appearance, but nevertheless distinctive and readily recognized: they are characteristically large (2–7 mm in diameter) and vary in shape from circular to irregular, with entire to undulate, crenate or fimbriate edges; they usually have matt or granular textures, but smooth and moist colonies are not uncommon (Figure 9b). Colonies are usually whitish to cream in color, but some strains may produce a pinkish brown pigment, and some strains produce a yellow diffusible pigment or a yellowish-green fluorescent pigment. Fresh plate cultures commonly have a "mousy" smell. Minimum temperature for growth is usually 10–20 °C, and the maximum 40–45 °C, with the optimum about 37 °C. Psychrotolerant strains growing at 6 °C have been isolated. Egg yolk reaction is positive. Catalase-positive, oxidase-negative. Casein, gelatin and starch are hydrolyzed. Voges–Proskauer-positive. Citrate is utilized as sole carbon source. Nitrate is reduced by most strains. Tyrosine is decomposed. Phenylalanine is not deaminated. Resistant to 0.001% lysozyme. Acid without gas is produced from glucose and a limited range of other carbohydrates. Most strains produce acid from salicin and starch, but strains of serovars 1, 3, 5 and 8 (which include strains associated with emetic food poisoning) do not produce acid from these substrates. Extracellular products include hemolysins, enterotoxins, heat-stable emetic toxin, cytotoxin, proteolytic enzymes and phospholipase; psychrotolerant strains may produce toxins (Stenfors and Granum, 2001).

Bacillus cereus has been divided into serovars on the basis of H-antigens (Kramer and Gilbert, 1992); 42 serovars are presently recognized (Ripabelli et al., 2000). Plasmid banding patterns and amplified fragment length polymorphism analysis may be of value in distinguishing between strains of the same serotype (Nishikawa et al., 1996; Ripabelli et al., 2000).

Endospores are very widespread in soil, in milk and other foods, and in many other environments. The vegetative organisms may multiply readily in a variety of foods and may cause diarrheal and emetic food poisoning syndromes. Growth in milk may result in "bitty cream defect". Occasionally causes opportunistic infections in man and other animals. Certain endospore-forming, trichome-forming bacteria that occur in the alimentary tracts of animals, some of which have been called "*Arthromitus*", have been identified as *Bacillus cereus*; see *Cell morphology* and *Habitats*, in *Further descriptive information*, above.

DNA G + C content (mol%): 31.7–40.1 (T_m) for 11 strains, 34.7–38.0 (Bd), and 35.7 (T_m), 36.2 (Bd) for the type strain.

Type strain: ATCC 14579, DSM 31, JCM 2152, LMG 6923, NCIMB 9373, NRRL B-3711, IAM 12605.

EMBL/GenBank accession number (16S rRNA gene): D16266 (IAM 12605).

Additional remarks: Phenotypically similar to other members of the *Bacillus cereus* group: *Bacillus anthracis*, *Bacillus mycoides*, *Bacillus thuringiensis* and *Bacillus weihenstephanensis*. For distinguishing characters see the individual species descriptions and Table 7. Another member of the group, *Bacillus pseudomycoides*, is separated from *Bacillus cereus* only by DNA relatedness and some differences in fatty acid composition. Genetic evidence supports the recognition of members of the *Bacillus cereus* group as one species, given that differentiation often relies on the presence of virulence characters which are carried by extrachromosomal mobile genetic elements (Turnbull et al., 2002), but practical considerations argue against such a move.

19. **Bacillus circulans** Jordan 1890, 821[AL]

cir′cu.lans. L. part. adj. *circulans* circling.

For many years this species accommodated a wide variety of phenotypically unrelated strains. It was referred to by Gibson and Topping (1938) as a complex rather than a species, and later investigators agreed with this description. Strains were frequently allocated to this species on account of their distinctive motile microcolonies (Figure 9i); however, Jordan named his isolate for the circular motion that he saw in the interior of colonies observed under low magnification, rather than because of motile microcolonies. Jordan's original strain is considered lost, but Ford's isolate 26, that he believed to be of the same species as Jordan's strain, is available. Smith and Clark (1938) observed the rotary motion within the colonies of Ford's strain and noted also the production of motile microcolonies. Despite a few discrepancies between Jordan's and Ford's descriptions of their strains, Smith et al. (1952) considered that Ford's strain 26 could be accepted as authentic and this became the type strain. The production of motile microcolonies is more characteristic of strains now allocated to *Paenibacillus* (see *Further descriptive information, Colony characteristics*, above).

Further grounds for the allocation of later isolates to this species were the production of sporangia swollen by subterminal to terminal ellipsoidal spores, and their being very active in the production of acid from a very wide range of carbohydrates. DNA relatedness studies revealed at least 10 homology groups among strains labeled *Bacillus circulans*, and it became clear that the phenotypic and genotypic heterogeneity of the complex had resulted from the allo-

cation of unrelated strains to the species (Nakamura and Swezey, 1983). This work led to the allocation of members of several of the homology groups to new or revived species which were subsequently assigned to *Paenibacillus*: *Paenibacillus amylolyticus*, *Paenibacillus lautus*, *Paenibacillus pabuli* and *Paenibacillus validus*. A further group of strains previously assigned to *Bacillus circulans* was proposed as the new species *Bacillus* (now *Paenibacillus*) *glucanolyticus* on the basis of a numerical taxonomic study (Alexander and Priest, 1989). However, many misnamed strains remain allocated to *Bacillus circulans* and await reallocation, and authentic strains of this species are in the minority in most collections.

The description which follows is based upon the type strain and several other strains which have been shown by amplified rDNA restriction analysis, polyacrylamide gel electrophoresis of whole-cell proteins, and various phenotypic characters (De Vos, Logan and colleagues, unpublished data) to be closely related to the type strain. Phylogenetic studies indicate that *Bacillus circulans*, *Bacillus firmus* and *Bacillus lentus* are related.

Facultatively anaerobic, motile, straight, round-ended, occasionally slightly tapered and curved rods 0.6–0.8 μm in diameter, appearing singly or in pairs and occasionally short chains. Endospores are ellipsoidal and lie terminally or subterminally in swollen sporangia (Figure 9c). Colonies grown for 2 d on TSA at 30 °C are 1–3 mm in diameter, opaque, cream-colored, slightly convex, with eggshell surface textures and irregular margins that may spike along the streak lines. Optimum temperature lies between 30 °C and 37 °C; maximum temperature for growth lies between 50 °C and 55 °C. The optimum pH for growth is 7.0. Minimum pH for growth lies between 4.0 and 5.0. The maximum pH lies between 9 and 10. Casein and starch are weakly hydrolyzed. In the API 20E strip, *o*-nitrophenyl-β-D-galactopyranoside hydrolysis is positive and urease production, hydrolysis of gelatin and nitrate reduction are occasionally positive. Arginine dihydrolase, lysine decarboxylase and ornithine decarboxylase production, citrate utilization, hydrogen sulfide, tryptophan deaminase and indole production and Voges–Proskauer reaction are negative. In the API 50CH gallery using the CHB suspension medium, hydrolysis of esculin is positive, and acid without gas is produced from a very wide range of carbohydrates: Production of acid without gas is variable for: adonitol, D-arabitol, 2-keto- and 5-keto-D-gluconate, rhamnose and ribose; the type strain is positive for adonitol, rhamnose and ribose. Acid production is negative for the following substrates: D-arabinose, dulcitol, erythritol, D-fucose, L-fucose, L-sorbose, D-tagatose and L-xylose. In the variable results, the type strain scores positive for: adonitol, rhamnose and ribose. Occasional strains may produce acid without gas from D-lyxose.

Source: sewage, soil, food and infant bile.

DNA G + C content (mol%): 35.7 (T_m), 36.2 (Bd) for the type strain.

Type strain: ATCC 4513, DSM 11, JCM 2504, LMG 13261, IAM 12462.

EMBL/GenBank accession number (16S rRNA gene): D78312 (IAM 12462).

20. **Bacillus clarkii** Nielsen, Fritze and Priest 1995b, 879[VP] (Ef-

fective publication: Nielsen, Fritze and Priest 1995a, 1758.)

clar'ki.i. N.L. gen. n. *clarkii* of Clark, named after the American bacteriologist Francis E. Clark.

Strictly alkaliphilic and moderately halophilic organisms forming ellipsoidal spores which lie subterminally. The sporangia of the type strain are distinctly swollen, those of the other two strains characterized by Nielsen et al. were not swollen. Cells 0.6–0.7 by 2.0–5.0 μm. Colonies are circular and smooth, creamy-white to pale yellow or (with age) dark yellow, and with entire margins. Growth temperature range 15–45 °C. Optimal growth at pH 10.0 or above; no growth at pH 7.0. Grows in presence of up to 16% NaCl; unable to grow in the absence of sodium ions. Nitrate is reduced to nitrite. Casein, hippurate, gelatin, and Tween 40 and 60 are hydrolyzed. Pullulan, starch, and Tween 20 and 80 are not hydrolyzed; phenylalanine is not deaminated. See Table 6.

Source: mud and soil.

DNA G + C content (mol%): 42.4–43.0 (HPLC analysis).

Type strain: PN-102, ATCC 700162, DSM 8720, LMG 17947.

EMBL/GenBank accession number (16S rRNA gene): X76444 (DSM 8720).

21. **Bacillus clausii** Nielsen, Fritze and Priest 1995b, 879[VP] (Effective publication: Nielsen, Fritze and Priest 1995a, 1759.)

clau'si.i. N.L. gen. n. *clausii* of Claus, named after the German bacteriologist Dieter Claus.

Alkalitolerant organisms forming ellipsoidal spores which lie paracentrally to subterminally in sporangia which may be slightly swollen. Cells 0.5–0.7 by 2.0–4.0 μm. Colonies are white and filamentous with filamentous margins. Growth temperature range 15–50 °C. Optimal growth at pH 8.0; good growth at pH 7.0. Grows in presence of up to 8–10% NaCl. Nitrate is reduced to nitrite. Casein, gelatin and starch are hydrolyzed. Hippurate, pullulan, and Tween 20, 40, 60 and 80 are not hydrolyzed; phenylalanine is not deaminated. Glucose and a wide range of other carbohydrates can be utilized as sole sources of carbon. Strains in this species were formerly assigned to *Bacillus lentus* type II by Gordon and Hyde (1982).

Source: clay and soil.

DNA G + C content (mol%): 42.8–45.5 (HPLC analysis).

Type strain: PN-23, ATCC 700160, DSM 8716, LMG 17945, NCIMB 10309.

EMBL/GenBank accession number (16S rRNA gene): X76440 (DSM 8716).

22. **Bacillus coagulans** Hammer 1915, 119[AL]

co.a'gu.lans. L. part. adj. *coagulans* curdling, coagulating.

Moderately thermophilic, aciduric, facultatively anaerobic, Gram-positive, motile rods. The cell diameter is 0.6–1.0 μm. Spores are ellipsoidal but sometimes appear spherical; they lie subterminally and occasionally paracentrally or terminally in slightly swollen sporangia; some strains do not sporulate readily. After 2 d incubation on TSA at 40 °C, colonies are <1 to 3 mm in diameter, white, convex with entire margins and smooth surfaces; they become cream-colored with age. Growth occurs at 30 °C, the optimum growth temperature lies between 40 °C and 57 °C, and the maximum temperature for growth lies between 57 °C and 61 °C. The optimum

pH for growth is 7.0; cells are able to grow at pH 4 and the maximum pH for growth lies between 10.5 and 11. Does not grow in presence of 5% NaCl. Minimal nutritional requirements are variable, and may include several amino acids and vitamins. Catalase-positive. Starch is hydrolyzed. Tyrosine is not decomposed. Casein is not hydrolyzed. In the API 20E strip, strains give variable results for arginine dihydrolase, gelatin liquefaction (type strain positive), nitrate reduction, ONPG (type strain positive) and the Voges–Proskauer test (type strain weak positive); all the strains are negative for lysine decarboxylase and ornithine decarboxylase reactions, citrate utilization, hydrogen sulfide production, urease, tryptophan deaminase, and indole production. In the API 50CH gallery using the CHB suspension medium, hydrolysis of esculin is variable (most strains positive), and acid without gas is produced from the following carbohydrates by more than 85% of strains: D-galactose, D-fructose, D-glucose, glycerol, maltose, D-mannose, D-melibiose, N-acetylglucosamine, starch and D-trehalose. Acid production from the other substrates varies between strains (see Tables 3 and 4, and De Clerck et al., 2004b).

Bacillus coagulans is economically important as a food spoilage agent, as a producer of commercially valuable products such as lactic acid, thermostable enzymes, and the antimicrobial peptide coagulin, and as a probiotic for chickens and piglets, but several taxonomic studies revealed considerable diversity within the species. De Clerck et al. (2004b) carried out a polyphasic taxonomic study of 30 strains, and found that although individual characterization methods revealed subgroups of strains, these intraspecies groupings were not sufficiently consistent among the different methods to support the proposal of subspecies, nor were there any features to suggest such a division, and DNA–DNA relatedness data and 16S rDNA sequence comparisons upheld the accommodation of all the strains in one species.

Source: soil, canned foods, tomato juice, gelatin, milk, medical preparations and silage.

DNA G + C content (mol%): 44.3–50.3 (T_m) for seven strains, 45.4–56.0 (Bd) for three strains, and 47.4 (HPLC), 47.1 (T_m), 44.5 (Bd) for the type strain.

Type strain: ATCC 7050, DSM 1, JCM 2257, LMG 6326, NCIMB 9365, NRRL NRS-609, IAM 12463.

EMBL/GenBank accession number (16S rRNA gene): D16267 (IAM 12463).

23. **Bacillus cohnii** Spanka and Fritze 1993, 155[VP]

coh'ni.i. N.L. gen. n. *cohnii* of Cohn; named after the German bacteriologist Ferdinand Cohn.

Alkaliphilic, Gram-positive, peritrichously motile rods, forming ellipsoidal spores which lie subterminally to terminally in swollen sporangia. In the cell wall, diaminopimelic acid is replaced by ornithine, and aspartic acid forms the interpeptide bridge. Colonies are creamy white, and 1–2 mm in diameter after 2 d at 45 °C. Growth temperature range 10–47 °C. Growth in presence of 5% but not 10% NaCl. Grows at pH 9.7. Nitrate is reduced. Catalase- and oxidase-positive. Gelatin, hippurate, starch and Tween 60 are hydrolyzed; hydrolysis of casein, pullulan and Tween 80 usually positive. Urea is not hydrolyzed, phenylalanine is not deaminated. See Table 6.

Source: soil and feces.

DNA G + C content (mol%): 33.9–35.0, and that of the type strain is 34.6 (T_m).

Type strain: RSH, ATCC 51227, DSM 6307, LMG 16678, IFO 15565.

EMBL/GenBank accession number (16S rRNA gene): X76437 (DSM 6307).

24. **Bacillus decolorationis** Heyrman, Balcaen, Rodríguez-Díaz, Logan, Swings and De Vos 2003a, 462[VP]

de.co.lo.ra.ti.on'is. L. gen. n. *decolorationis* of discoloration.

Aerobic, Gram-variable, motile, rods and coccoid rods, 0.5–0.8 μm wide and 1.0–4.0 μm long, that occur singly, in pairs or short chains. Spores are produced slowly and in small numbers in culture; they are ellipsoidal, sometimes nearly spherical, central to subterminal and swell the sporangia slightly. Colonies on TSA are cream-colored to beige, circular with a smooth to slightly irregular margin, low-convex with a glistening and rough surface. Oxidase- and catalase-positive. The temperature range for growth is 5–40°C with optimal growth at 25–37°C. The NaCl concentration for growth is 0–10% (w/v), with an optimum of 4–7% (w/v). Casein hydrolysis is positive within 4 d incubation. In the API 20E strip, conversion of nitrates to nitrite and dinitrogen is positive and gelatin hydrolysis occurs with or without added salt, but only with large inocula. Reactions are negative for arginine dihydrolase, lysine decarboxylase, ornithine decarboxylase, citrate utilization, hydrogen sulfide production, urease, tryptophan deaminase, indole production and Voges–Proskauer. The ONPG reaction is negative without added NaCl and variable (type strain positive) when supplemented with 7% NaCl. Acid is produced weakly and without gas from the following carbohydrates in the API 50 CH gallery using the CHB suspension medium supplemented with 7% NaCl: cellobiose, D-fructose, gentiobiose, D-glucose, 5-keto-D-gluconate, maltose, D-mannose, N-acetylglucosamine, ribose, salicin, sucrose and trehalose. Esculin hydrolysis is positive with or without added NaCl. Results are variable amongst strains for weak acid production from arbutin, galactose, glycerol, lactose and D-mannitol; type strain is positive for arbutin, glycerol and D-mannitol. The major fatty acid is $C_{15:0\ anteiso}$, present at about 68%; $C_{17:0\ anteiso}$ accounts for about 11% of the total.

Source: mural paintings, discolored by microbial growths.

DNA G + C content (mol%) of the type strain: 39.8 (HPLC).

Type strain: LMG 19507, DSM 14890.

EMBL/GenBank accession number (16S rRNA gene): AJ315075 (LMG 19507).

25. **Bacillus drentensis** Heyrman, Vanparys, Logan, Balcaen, Rodríguez-Díaz, Felske and De Vos 2004, 56[VP]

dren.ten'sis. N.L. adj. *drentensis* of Drente, a province in The Netherlands.

Facultatively anaerobic, Gram-positive or Gram-variable, motile, tapered rods, 1.5–3.5 μm in diameter, occurring singly and in pairs. Cells show pleomorphism (narrow and broad cells, the latter showing swellings) and produce intracellular storage products (possibly PHB) on TSA. Endospores

are spherical or ellipsoidal and lie in paracentral or occasionally subterminal positions in swollen sporangia. Colonies are slightly convex with regular margins when small, and sometimes wrinkled with irregular margins and prominent centers when larger. Colonies are cream-colored and produce a brownish soluble pigment; consistency is butyrous, with an eggshell-like surface texture. The optimum temperature for growth is 30°C, and the maximum growth temperature lies between 50°C and 55°C. The optimum pH for growth is 7.0–8.0, and growth occurs from pH 5.5–6.0 to 9.5–10.0. Casein is not hydrolyzed. In the API 20E strip, o-nitrophenyl-β-D-galactopyranoside hydrolysis is positive, Voges–Proskauer reaction is variable (most strains negative, positive strains weak), and nitrate reduction is variable; reactions for arginine dihydrolase, lysine decarboxylase, ornithine decarboxylase, citrate utilization, hydrogen sulfide production, urease, tryptophan deaminase, indole production and gelatin hydrolysis are negative. Hydrolysis of esculin is positive. Acid without gas is produced from the following carbohydrates in the API 50 CH gallery using the CHB suspension medium: N-acetyl-D-glucosamine, D-fructose, D-glucose (some strains, including the type strain, weak), lactose, maltose, D-melibiose and salicin (some strains, including the type strain, weak). The following reactions are variable between strains and, when positive, are usually weak: amygdalin, arbutin, galactose, gluconate, inulin, D-mannose, D-melezitose, α-methyl-D-glucoside, raffinose, ribose, starch, sucrose, D-trehalose, D-turanose and D-xylose; type strain is positive for inulin, D-mannose, D-melezitose, sucrose and weak for raffinose, ribose, starch and D-turanose. The major cellular fatty acids are $C_{15:0\ iso}$ and $C_{15:0\ anteiso}$, present at a level of about 32 and 22%, respectively, while $C_{16:1\ \omega11c}$ accounts for about 13% of the total fatty acids.

Source: soil in the Drentse an agricultural research area, The Netherlands.

DNA G + C content (mol%): 39.3–39.4 (HPLC) and 39.4 for the type strain.

Type strain: LMG 21831, DSM 15600.

EMBL/GenBank accession number (16S rRNA gene): AJ542506 (LMG 21831).

26. **Bacillus endophyticus** Reva, Smirnov, Pettersson and Priest 2002, 106[VP]

en.do.phy'ti.cus. Gr. *endo* within; Gr. n. *phyton* plant.; L. masc. suff. -*icus* adjectival suffix used with the sense of belonging to; N.L. adj. *endophyticus* within plant; originally isolated from plant tissues.

Strictly aerobic, Gram-positive, nonmotile rods, forming ellipsoidal spores which lie subterminally or terminally in unswollen sporangia. Cells 0.5–1.5 by 2.5–3.5 μm, occurring singly and in short or long chains, the latter appearing filamentous. Vacuoles are formed in the cytoplasm of cells grown on media containing 2% glucose. Colonies are circular, 1–3 mm in diameter, with entire or slightly indented margins, and may be slimy or rough; they are usually white, but pink and red pigmentation is occasionally seen. Growth temperature range 10–45°C; optimum about 28°C. Grows in presence of 10% NaCl. Nitrate is not reduced to nitrite. Catalase- and oxidase-positive. Casein, gelatin, starch and urea are not hydrolyzed. Acid without gas produced from D-glucose and a range of other carbohydrates. Citrate and gluconate are utilized; acetate, propionate and tartrate are not.

Source: inner tissues of healthy cotton plants.

DNA G + C content (mol%): not reported.

Type strain: 2DT, ATCC 29604, NRRL NRS- 1705, LMG 7124, NCIMB 11326, CIP 106778, JCM 9331.

EMBL/GenBank accession number (16S rRNA gene): AF295302 (2DT).

27. **Bacillus farraginis** Scheldeman, Rodríguez-Díaz, Goris, Pil, De Clerck, Herman, De Vos, Logan and Heyndrickx 2004, 1362[VP]

far.ra.gin'is. L. gen. fem. n. *farraginis* from mixed fodder for cattle, referring to feed concentrate for dairy cattle as the principal isolation source.

Cells are long, straight, round-ended, motile, strictly aerobic, Gram-negative rods, occurring singly, in pairs or filaments. Cell diameter is 0.5–0.8 μm and cell length 1.2–4 μm. Spores are ellipsoidal and occur paracentrally or subterminally in occasionally slightly swollen sporangia. Colonies grown for 3 d at 30 °C on nutrient agar are cream-colored or translucent, slightly raised, with irregular margins and granular, glossy surfaces. Colony diameter is no greater than 1 mm. Good growth occurs at 30 and 45 °C and weak growth occurs at 20 °C. Some strains are capable of growth at pH 9 but none grows at pH 5. Growth is not inhibited by 7% (w/v) NaCl. Hydrolysis of starch and casein is not observed within 7 d of incubation at 30 °C, and growth in casein agar is poor or negative. Catalase- and oxidase are positive. All strains are unreactive in the API 20E and API 50CHB test kits. In the Biotype100 kit using the Biotype 2 medium, nearly all strains (>83%) belonging to the species are able to use the following substrates as sole carbon sources: 4-aminobutyrate, 5-aminovalerate, D- and L-alanine, fumarate, L-glutamate, glutarate, L-histidine, 3-hydroxybutyrate, 2-oxoglutarate, D- and L-malate, DL-lactate, L-proline, putrescine, succinate, L-tryptophan and L-tyrosine. Many strains (>45%) are able to use L-aspartate, dulcitol, *m*-hydroxybenzoate, malonate, D-mannitol, D-ribose and D-sorbitol. Other substrates are used less frequently (17–44%): L-arabinose, D-galactose, gentisate, D-glucuronate, *p*-hydroxybenzoate, *myo*-inositol and α-L-rhamnose.

The type strain utilizes the following substrates as sole carbon sources: 4-aminobutyrate, 5-aminovalerate, D- and L-alanine, L-aspartate, dulcitol, fumarate, D-glucosamine, L-glutamate, glutarate, histamine, L-histidine, *m*-hydroxybenzoate, 3-hydroxybutyrate, 2-oxoglutarate, D- and L-malate, malonate, DL-lactate, L-proline, putrescine, D-ribose, D-sorbitol, succinate, *meso*-tartrate, L-tryptophan and L-tyrosine. The major cellular fatty acids (>5% of total cellular fatty acids) are $C_{15:0\ iso}$, $C_{15:0\ anteiso}$, $C_{17:0\ anteiso}$, $C_{16:0\ iso}$ and $C_{16:1}\ \omega_{7c}$ alcohol.

Source: cattle feed concentrate, milking clusters, hay, silage, grass, lucerne and green fodder.

DNA G + C content (mol%): 43.7 (HPLC).

Type strain: R-6540, MB 1885, LMG 22081, DSM 16013.

GenBank/EMBL accession number (16S rRNA gene): AY443034 (R-6540).

28. **Bacillus fastidiosus** den Dooren de Jong 1929, 344[AL]

fas.tid'i.os.us. L. adj. *fastidiosus* disdainful, fastidious.

Strictly aerobic rods about 1.3 μm in diameter, forming ellipsoidal spores which usually lie centrally, paracentrally and subterminally, occasionally terminally, in unswollen sporangia. Colonies on 1% uric acid agar become opaque and are usually unpigmented but may become yellowish; margins are often ragged and have hair-like outgrowths, or the colonies may be rhizoid. Colonies are surrounded by zones of clearing and the reaction becomes strongly alkaline. Growth occurs at 10 °C and 40 °C, but not at 5 °C or 50 °C. Does not grow at pH 6.8 or below. Grows in presence of 5% NaCl. Grows on allantoic acid, allantoin or uric acid as sole carbon, nitrogen and energy sources. Some strains will grow on certain peptones, especially at high concentrations. Growth factors not required, Nitrate is not reduced to nitrite. Catalase- and oxidase-positive. Urea is hydrolyzed; casein, gelatin and starch are not hydrolyzed. Acid and gas are not produced from D-glucose and other carbohydrates; there is no growth in the media used to test these characters. Citrate and propionate are not utilized.

Source: soil and poultry litter.

DNA G + C content (mol%): 34.3–35.1 (T_m) for 17 strains, and 35.1 (T_m), 35.1 (Bd) for the type strain.

Type strain: Delft LMD 29-14, ATCC 29604, DSM 91, LMG 7124, NCIMB 11326, NRRL NRS-1705, KCTC 3393.

EMBL/GenBank accession number (16S rRNA gene): X60615 (DSM 91).

29. **Bacillus firmus** Bredemann and Werner in Werner 1933, 446[AL]

fir'mus. L. adj. *firmus* strong, firm.

This species has for long been genetically heterogeneous, and many strains have been incorrectly assigned to it. Strains received as *Bacillus firmus* show phenotypic profiles that appear to overlap with those of strains assigned to *Bacillus lentus*, so that Gordon et al. (1977) raised the question of whether *Bacillus firmus–Bacillus lentus* represented a single species or a series of strains.

The description which follows is based upon the type strain and 17 other strains which have been shown by amplified rDNA restriction analysis, polyacrylamide gel electrophoresis of whole-cell proteins, and various phenotypic characters (De Vos, Logan and colleagues, unpublished data) to be closely related to the type strain. Phylogenetic studies indicate that *Bacillus circulans*, *Bacillus firmus* and *Bacillus lentus* are related.

Facultatively anaerobic, straight, round-ended, motile rods, 0.8–0.9 μm in diameter, that occur singly, in pairs, or occasionally as short chains. Endospores are ellipsoidal or cylindrical, lie subterminally, paracentrally or centrally, and may swell the sporangia slightly. Colonies grown for 3 d on TSA at 30 °C are 1–12 mm in diameter, creamy-yellow to pale orangey-brown in color, are of butyrous consistency, have margins that vary from entire to finely rhizoidal and surface appearances that are egg-shell to glossy, sometimes with granular or zoned areas in center. Maximum growth temperature is 40–50 °C, the optimum temperature lies between 30 °C and 40 °C, and growth occurs at 20 °C. The optimum pH for growth is 7.0–9.0; the minimum is 6.0–7.0, and the maximum lies between 11 and 11.5. Grows in presence of 7% NaCl. Catalase-positive. Casein is weakly hydrolyzed and a pale to dark honey-brown diffusible pigment is produced on it. Starch is hydrolyzed, but strength of reaction varies among strains. Citrate and propionate are not utilized. In the API 20E strip, gelatin is partially or completely hydrolyzed by most strains and nitrates are totally or partially reduced. *o*-nitrophenyl-β-D-galactopyranoside is not hydrolyzed, arginine dihydrolase, lysine decarboxylase and ornithine decarboxylase are not produced, citrate is not utilized, hydrogen sulfide, urease, tryptophan deaminase and indole are not produced and Voges–Proskauer reaction is negative. In the API 50CH gallery using the CHB suspension medium, hydrolysis of esculin is variable and acid without gas is positive or weakly positive from the following carbohydrates: D-glucose, maltose, mannitol, starch and sucrose. In the API Biotype 100 kit the following substrates are utilized as sole carbon sources: D- and L-alanine, D-gluconate, D-glucosamine, α-D-glucose, L-glutamate, glycerol, 2-oxoglutarate, DL-lactate, L-malate, maltose, maltotriose, D-mannitol, N-acetyl-D-glucosamine, L-proline, sucrose, L-serine, succinate and D-trehalose. The type strain and some other strains are positive or weak for: glycerol, N-acetylglucosamine and D-trehalose.

According to their patterns of acid production from other carbohydrates, and use of other substrates as sole carbon sources, 2 biotypes may be recognized, with Biovar 1 containing the type strain; Biovar 2 strains may represent a distinct species and are distinct from Biovar 1 strains in their slightly stronger acid production from the above-mentioned carbohydrates and their acid production from: D-fructose, glycogen and, although variable among Biovar 2 strains, D-xylose. Only Biovar 2 strains are able to utilize: *cis*-aconitate, citrate, β-D-fructose, DL-glycerate, 2-keto-D-gluconate, maltitol, 3-methyl-D-glucopyranose, methyl α-D-glucopyranoside, tricarballylate, trigonelline, L-tryptophan and D-xylose. In assimilation tests giving variable results, the type strain is positive for: L-histidine and 3-hydroxybutyrate.

Source: soil and other environments.

DNA G + C content (mol%): 41.4 (T_m), 40.7 (Bd).

Type strain: ATCC 14575, DSM 12, JCM 2512 (D78314), LMG 7125, NCIMB 9366 NRRL B-14307, IAM 12464.

EMBL/GenBank accession number (16S rRNA gene): D16268 (IAM 12464).

30. **Bacillus flexus** (*ex* Batchelor 1919) Priest, Goodfellow and Todd 1989, 93[VP] (Effective publication: Priest, Goodfellow and Todd 1988, 1878.)

fle'xus. L. adj. *flexus* flexible.

Strictly aerobic, Gram-variable rods, forming ellipsoidal spores which lie centrally or paracentrally in unswollen sporangia. Description is based upon two strains. Mean cell width 0.9 μm. Colonies are opaque and smooth. Growth occurs at 17–37 °C, but not at 5 °C or 50 °C. Grows between pH 4.5 and 9.5. Grows in presence of 10% NaCl. Nitrate is not reduced to nitrite. Oxidase-positive. Casein, elastin, gelatin, pullulan, starch and urea are hydrolyzed; esculin is not. Acid without gas is produced from D-glucose and a range of other carbohydrates; acid is not produced from pentoses. Acetate, citrate, formate and succinate are utilized; gluconate, lactate and malonate are not.

Source: feces and soil.

DNA G + C content (mol%): 35 and 36 (T_m) for two strains.

Type strain: ATCC 49095, DSM 1320, LMG 11155, NCIMB 13366, NRRL NRS-665, IFO 15715.

EMBL/GenBank accession number (16S rRNA gene): AB021185 (IFO 15715).

31. **Bacillus fordii** Scheldeman, Rodríguez-Díaz, Goris, Pil, De Clerck, Herman, De Vos, Logan and Heyndrickx 2004, 1363[VP]

for'di.i. N.L. gen. n. *fordii* named after W. W. Ford, an American microbiologist working on aerobic spore-forming bacteria at the beginning of the twentieth century.

Cells are long, straight, round-ended, motile, strictly aerobic, Gram-negative rods, occurring singly or in pairs. Cell diameter is 0.6–0.8 μm and length 1.6–3.5 μm. Spores are ellipsoidal and occur paracentrally or subterminally in, occasionally, slightly swollen sporangia. Colonies grown on nutrient agar at 30 °C for 3 d are cream-colored, raised, with entire margins and smooth glossy surfaces. Their maximum diameter is 2 mm. Good growth occurs at 30 and 45 °C and weak growth occurs at 20 °C. Growth occurs at pH 9 and some strains grow at pH 5. Growth is not inhibited by 7% (w/v) NaCl. Hydrolysis of starch and casein is not observed within 7 d of incubation at 30 °C. Growth on casein agar is colored faint pink. Catalase and oxidase are positive. All strains are unreactive in the API 20E and API 50CHB test kits. In the Biotype100 kit using the Biotype 2 medium, all strains show very good production of biomass using malonate and L-tyrosine as sole carbon sources. Variable results with good production of biomass are obtained for L-histidine, 2-oxoglutarate, glutarate, DL-lactate, 5-aminovalerate and L-tryptophan. All or most strains are capable of weak growth from the following carbon sources: esculin, gentisate, protocatechuate, *meso*-tartrate, D-glucosamine, DL-glycerate, quinate, ethanolamine, D-glucuronate, *p*-hydroxybenzoate, hydroxyquinoline β-glucuronide and 2- and 5-keto-D-gluconate. Growth occurs seldom and weakly from the following substrates: adonitol, L-alanine, L-arabinose, L-arabitol, benzoate, fumarate, D-galacturonate, D-gluconate, α-D-glucose, histamine, 3-hydroxybutyrate, D-lyxose, D-malate, methyl α-galactopyranoside, methyl β-D-glucopyranoside, N-acetyl d-glucosamine, phenylacetate, propionate, putrescine, D-raffinose, D-ribose, D-saccharate, L-sorbose, succinate, D-tagatose, L-tartrate, tricarballylate, trigonelline and D-xylose.

Where results are variable, the type strain uses the following substrates: esculin, 5-aminovalerate, L-arabinose, benzoate, ethanolamine, D-galacturonate, gentisate, D-glucuronate, glutarate, DL-glycerate, histamine, *p*-hydroxybenzoate, 3-hydroxybutyrate, hydroxyquinoline-β-glucuronide, 2- and 5-keto-D-gluconate, 2-oxoglutarate, DL-lactate, D-lyxose, N-acetyl D-glucosamine, phenylacetate, protocatechuate, putrescine, quinate, D-raffinose, *meso*-tartrate, tricarballylate, L-tryptophan and D-xylose. The major cellular fatty acids (>5% of total cellular fatty acids) are $C_{15:0\ iso}$, $C_{15:0\ anteiso}$, $C_{17:0\ anteiso}$, $C_{16:1\ \omega11c}$ and $C_{17:0\ iso}$.

Source: cattle feed concentrate, milking clusters, filter cloths, and raw milk.

DNA G + C content (mol%): 41.9 (HPLC).

Type strain: R-7190, MB 1878, LMG 22080, DSM 16014.

GenBank/EMBL accession number (16S rRNA gene): AY443034 (R-7190).

32. **Bacillus fortis** Scheldeman, Rodríguez-Díaz, Goris, Pil, De Clerck, Herman, De Vos, Logan and Heyndrickx 2004, 1362[VP]

for'tis. L. adj. *fortis* strong, referring to the fact that the strains were isolated after heat treatment for 30 min at 100 °C.

Cells are straight, round-ended, motile, strictly aerobic, Gram-negative rods, occurring singly or in pairs. Cell diameter is 0.6–0.8 μm and length 1.0–3.5 μm. Spores are oval and occur centrally or paracentrally in slightly swollen sporangia. Colonies grown on nutrient agar at 30 °C for 3 d are cream-colored, raised with entire margins and smooth, glossy surfaces. Their maximum diameter is 1 mm. Good growth occurs at 30 and 45 °C and weak growth occurs at 20 °C. Growth does not occur at pH 9 or 5. Growth is not inhibited by 7% (w/v) NaCl. Hydrolysis of starch and casein is not observed within 7 d of incubation at 30 °C, and growth on casein agar is poor or negative. Catalase and oxidase are positive. All strains are unreactive in the API 20E and API 50CHB test kits. In the Biotype100 kit using the Biotype 2 medium, strains use L-tryptophan and L-histidine as sole carbon sources. Most strains (>50%) use the following substrates as sole carbon sources: 4-aminobutyrate, 5-aminovalerate, esculin, ethanolamine, glutarate, hydroxyquinoline β-glucuronide, 2-oxoglutarate, DL-lactate, malonate, phenylacetate, L-proline, putrescine, D-ribose and L-tyrosine. Some strains (<50%) use the following substrates: citrate, erythritol, D-gluconate, D-glucosamine, α-D-glucose, L-glutamate, DL-glycerate, histamine, *m*-hydroxybenzoate, 2- and 5-keto-D-gluconate, L- and D-malate, α-D-melibiose, protocatechuate, L-sorbose and L-tartrate. Where results are variable, the type strain uses the following substrates: esculin, D-gluconate, D-glucosamine, D-glucuronate, DL-glycerate, hydroxyquinoline-β-glucuronide, 2-oxoglutarate, 2- and 5-keto-D-gluconate, DL-lactate, malonate, phenylacetate, protocatechuate, L-tartrate and L-tyrosine. The major cellular fatty acids (>5% of total cellular fatty acids) are $C_{15:0\ iso}$, $C_{15:0\ anteiso}$, $C_{16:0\ iso}$, $C_{15:0}$ and $C_{17:0\ anteiso}$.

Source: cattle feed concentrate, milking clusters, soy, and raw milk.

DNA G + C content (mol%): 44.3 (HPLC).

Type strain: R-6514, LMG 22079, DSM 16012.

GenBank/EMBL accession number (16S rRNA gene): AY443034 (R-6514).

33. **Bacillus fumarioli** Logan, Lebbe, Hoste, Goris, Forsyth, Heyndrickx, Murray, Syme, Wynn-Williams and De Vos 2000, 1751[VP]

fum.a.rio'li. nemt. L. gen. n. *fumariolum* a smoke hole; L. gen. n. *fumarioli* of a smoke hole, whence fumarole, a hole emitting gases in a volcanic area.

Moderately thermoacidophilic and strictly aerobic, feebly motile, Gram-positive organisms growing and sporulating best at pH 5.5 and 50 °C on nutrient-weak media such as BFA (Logan et al., 2000) and BFA at half nutrient-strength, but also growing and sporulating weakly on trypticase soy agar containing 5 mg/l MnSO₄. Colonies 3–10 mm in diameter, low convex, circular and slightly irregular, glossy, creamy-brown, and butyrous. Spores ellipsoidal to cylindrical, lying paracentrally and subterminally, and not swelling

the sporangia. Temperature limits for growth: 25–30 °C and 55 °C; optimum temperature is about 50 °C. Limits of pH for growth: 4–5 and 6–6.5. Catalase-positive. Nitrate is not reduced. Gelatin is hydrolyzed, but esculin and casein are not. Acid without gas produced from D-fructose, D-glucose, mannitol, D-mannose, N-acetylglucosamine (weak), sucrose, D-trehalose (weak). Acid production from galactose, glycerol, lactose, maltose, D-melibiose, D-melezitose, methyl-α-D-glucoside, D-raffinose, ribose and D-turanose varies between strains. See Table 8.

Source: geothermal soils and active and inactive fumaroles in continental and maritime Antarctica, and from gelatin production plants in Belgium, France, and the USA.

DNA G + C content (mol%): 40.7% (T_m).

Type strain: Logan B1801, LMG 19448 (replaces LMG 17489), NCIMB 13771, KCTC 3851.

EMBL/GenBank accession number (16S rRNA gene): AJ250056 (LMG 17489).

34. **Bacillus funiculus** Ajithkumar, Ajithkumar, Iriye and Sakai 2002, 1143[VP]

fu.ni'cu.lus. L. masc. n. *funiculus* string, rope; referring to the filamentous appearance of the cells.

Aerobic, Gram-variable, motile rods 0.8–2.0 by 4.0–6.0 μm, which form filamentous trichomes by cellular binding. Description is based upon a single isolate. Ellipsoidal spores lie centrally in unswollen sporangia. On prolonged incubation heat-resistant, spore-like resting cells, which outgrow by budding, are formed. Colonies are round, opaque, and off-white to colorless. Growth temperature range 20–40 °C, optimum about 30 °C. The pH range for growth is 5.0–9.0 with optimum at 7.0–8.0. Catalase-positive, oxidase-negative. Voges–Proskauer reaction is positive. Citrate utilization negative. Nitrate is reduced to nitrite. Esculin, starch and urea are hydrolyzed; casein, gelatin and Tween 80 are not hydrolyzed. Acid without gas is produced from glucose. Glucose and a range of other carbohydrates are utilized as sole carbon sources.

Source: activated sewage sludge.

DNA G + C content (mol%): 37.2 (HPLC).

Type strain: NAF001, DSM 15141, JCM 11201, CIP 107128, KCTC 3796.

EMBL/GenBank accession number (16S rRNA gene): AB049195 (NAF001).

35. **Bacillus fusiformis** (*ex* Smith, Gordon and Clark 1946) comb. nov. *Bacillus sphaericus* var. *fusiformis* Smith, Gordon and Clark 1946), Priest, Goodfellow and Todd 1989, 93[VP] (Effective publication: Priest, Goodfellow and Todd 1988, 1878.)

fus.i.form'is. L. n. *fusus* spindle; L. suff. *-formis* of the shape of; N.L. adj. *fusiformis* spindle-shaped.

Strictly aerobic, Gram-variable rods, forming spherical spores which lie centrally or terminally in swollen sporangia. Mean cell width ≤0.9 μm. Colonies are opaque and smooth. Growth occurs at 17–37 °C, but not at 5 °C or 50 °C. Grows between pH 6.0 and 9.5. Growth in presence of 2–5% NaCl varies. Nitrate is not reduced to nitrite. Casein and gelatin are hydrolyzed; urea hydrolysis varies; esculin is not hydrolyzed. Acid and gas are not produced from D-glucose or other carbohydrates. Acetate, citrate, formate, lactate and

TABLE 8. Differentiation of thermophilic *Bacillus* species[a,b]

Characteristic	2. *B. aeolius*	33. *B. fumarioli*	46. *B. infernus*	56. *B. methanolicus*	75. *B. schlegelii*	87. *B. thermoamylovorans*	88. *B. thermocloacae*	90. *B. tusciae*
Motility	+	+	−	−	+	+	−	+
Spore formation:	nd	+	−	+	+	−	d	+
Ellipsoidal	+	+	nd	+	−	nd	+	+
Cylindrical	+	+	nd	−	−	nd	−	−
Spherical	−	−	nd	−	+	nd	−	−
Borne terminally	+	−	nd	−	+	nd	+	−
Swollen sporangia	nd	−	nd	+	+	nd	+	+
Catalase	−	+	−	+	+	+	+	w
Aerobic growth	+	+	−	+	+	+	+	+
Anaerobic growth	−	−	+	−	−	+	−	−
Voges–Proskauer	+	+	nd	nd	−	nd	−	nd
Acid from:								
L-Arabinose	+	−[c]	−	nd	−	+	−	−
D-Glucose	+	+[c]	+	nd	−	+	−	−
Glycogen	−	−[c]	nd	−	nd	+	nd	−
D-Mannitol	+	+[c]	nd	+	−	−	nd	−
D-Mannose	+	+[c]	−	nd	nd	+	−	−
Salicin	+	−[c]	nd	−	nd	+	nd	−
Starch	+	−[c]	nd	−	nd	+	nd	−
D-Xylose	+	−[c]	−	nd	−	+	−	−
Hydrolysis of:								
Casein	+	−	−	−	w	nd	−	nd
Gelatin	+	+	−	nd	−	nd	−	nd
Starch	+	nd	−	d	−	+	−	−
Nitrate reduction	−	−	+	−	+	−	−	+
Growth at pH:								
5	−	+	−	nd	nd	−	−	+
6	−	+	−	−	+	+	−	w
7	+	−	+	+	+	+	−	−
8	+	−	+	nd	nd	+	+	−
9	+	−	−	nd	nd	−	+	−
10	−	−	−	nd	nd	−	−	−
Growth in NaCl:								
2%	+	nd	+	d	+	nd	+	−
5%	+	nd	+	−	−	nd	w	−
7%	−	nd	+	nd	−	nd	−	−
10%	−	nd	+	nd	−	nd	−	−
Growth at:								
30 °C	−	+	−	−	−	nd	−	−
40 °C	+	+	−	+	−	nd	+	−
50 °C	+	+	+	+	+	+	+	+
55 °C	+	+	+	+	+	+	+	+
60 °C	+	−	+	+	+	−	+	nd
65 °C	+	−	−	−	+	−	+	−
70 °C	−	−	−	−	+	nd	−	−
Optimum growth temperature (°C)	55	50	61	55	70	50	55–60	55
Autotrophic with $H_2 + CO_2$ or CO	−	−	−	−	+	−	−	+

[a]Symbols: +, >85% positive; d, different strains give different reactions (16–84% positive); −, 0–15% positive; w, weak reaction; nd, no data are available.

[b]Compiled from Schenk and Aragno (1979); Bonjour and Aragno (1984); Demharter and Hensel (1989b); Arfman et al. (1992); Boone et al. (1995); Combet-Blanc et al. (1995); Logan et al. (2000); Gugliandolo et al. (2003a).

[c]For *Bacillus fumarioli*, acid production from carbohydrates is tested at pH 6 – see Logan et al. (2000) and *Testing for special characters.*

succinate are utilized; gluconate and malonate utilization varies between strains.

Phenotypically similar to *Bacillus sphaericus*, and according to Priest et al. (1988) distinguishable from that organism by urease positivity, ability to grow in presence of 7% NaCl, and sensitivity to 1 μg/ml tetracycline; however, the data reported in that study indicated that only three of the four strains assigned to this species were urease-positive, and only one strain could grow at 2 or 5% NaCl. In a study of 12 strains belonging to the *Bacillus fusiformis* DNA homology group (Krych et al., 1980), all strains could grow in 7% NaCl and only one strain failed to degrade urea, but members of other *Bacillus sphaericus* homology groups were also positive for these characters. See Table 9.

Ahmed et al. (2007c) proposed the transfer of this species to the new genus *Lysinibacillus*.

Source: soil.

DNA G + C content (mol%): 35–36 (T_m).

Type strain: ATCC 7055, DSM 2898, LMG 9816, NRRL NRS-350, IFO 15717.

EMBL/GenBank accession number (16S rRNA gene): AJ310083 (DSM 2898).

36. **Bacillus galactosidilyticus** Heyndrickx, Logan, Lebbe, Rodríguez-Díaz, Forsyth, Goris, Scheldeman and De Vos 2004, 619^{VP}

ga.lac.to.si.di.ly′ti.cus N.L. neut. n. *galactosidum* galactoside, N.L. adj. *lyticus* lysing, dissolving; N.L. adj. *galactosidilyticus* referring to positive ONPG test revealing β-galactosidase activity.

Facultatively anaerobic, Gram-positive or Gram-variable, small, plump, round-ended rods 0.7–0.9 μm by 2–5 μm, with tumbling motility, occurring singly and in pairs, and occasionally in short chains. Ellipsoidal endospores are borne in central, paracentral and subterminal positions within slightly swollen sporangia. After 2 d on TSA, the creamy or off-white colonies have opaque centers and are approximately 1 mm in diameter, smooth, flat and butyrous; the margins are usually irregular with pointed projections that may spread and become rhizoid in older cultures. Catalase-positive. Growth occurs at 30 and 40 °C but not at 50 °C. Alkalitolerant; growth occurs between pH 6 and 10.5, but not at pH 5 or below. Casein hydrolysis is very weak. In the API 20E strip, *o*-nitrophenyl-β-D-galactopyranoside is hydrolyzed, nitrate is reduced to nitrite, urease production is variable (type strain is negative), arginine dihydrolase, lysine decarboxylase and ornithine decarboxylase are negative, citrate is not utilized, hydrogen sulfide is not produced, the Voges–Proskauer reaction is negative, indole is not produced, and gelatin is not hydrolyzed. Hydrolysis of esculin is positive. In the API 50 CHB gallery, acid without gas is produced, often weakly, from N-acetylglucosamine, D-fructose and D-glucose. Acid production from the following carbohydrates is variable, and when positive is usually very weak: amygdalin, L-arabinose, arbutin, D-cellobiose, galactose, gentiobiose, inulin, lactose, maltose, mannitol, D-mannose, D-melezitose, D-melibiose, methyl-D-glucoside, D-raffinose, rhamnose, ribose, salicin, starch, sucrose, D-trehalose, D-turanose and D-xylose; the type strain is positive for arbutin, D-cellobiose, D-melibiose, D-melezitose, D-raffinose, starch, sucrose and D-trehalose.

The major cellular fatty acids are: $C_{15:0\ anteiso}$ (33% of total), $C_{16:0}$ (27%), $C_{15:0\ iso}$ (13%) and $C_{14:0}$ (8%).

Source: raw milk, partially decomposed wheat grain and infant bile.

DNA G + C content (mol%): 35.7–38.2 (HPLC), and for the type strain is 37.7.

Type strain: LMG 17892, DSM 15595.

EMBL/GenBank accession number (16S rRNA gene): AJ535638 (LMG 17892).

37. **Bacillus gelatini** De Clerck, Rodríguez-Díaz, Vanhoutte, Heyrman, Logan and De Vos 2004c, 944^{VP}

ge.la.ti′ni. N.L. gen. neut. n. *gelatini* from gelatin.

Strictly aerobic, Gram-variable, feebly motile, round-ended, straight rods, 0.5–0.9 μm by 4–10 μm, which form long chains and occasionally appear singly. Endospores are oval, lie paracentrally and subterminally, and do not swell the sporangia. Colonies on TSA incubated at 30 °C for 4 d are smooth, cream-colored but darker in the center, have slightly irregular borders, and are waxy in appearance, with eggshell-textured surfaces. Colonies are slightly convex, but older colonies are flatter with concave, transparent centers, and diameters range from 1 to 4 mm. The maximum temperature for growth lies between 58 °C and 60 °C and the optimum temperature lies between 40 °C and 50 °C. Good growth occurs at pH 5–8; the minimum pH for growth is 4–5 and the maximum is 9–10. Good growth occurs in nutrient broth with 15% NaCl added. Catalase-positive, oxidase-negative. Casein is hydrolyzed. In the API 20E strip, hydrolysis of gelatin is positive. All the strains are negative for *o*-nitrophenyl-β-D-galactopyranoside hydrolysis, arginine dihydrolase, lysine decarboxylase and ornithine decarboxylase reactions, citrate utilization, hydrogen sulfide production, urease, tryptophan deaminase, indole production, Voges–Proskauer reaction, and nitrate reduction. In the API 50CH gallery using the CHB suspension medium, hydrolysis of esculin is positive, and acid without gas is produced, often weakly, from the following carbohydrates: D-fructose, D-glucose, glycerol, mannitol, D-mannose, D-trehalose and D-xylose. Most strains show a very weak production of acid from N-acetylglucosamine, maltose, and ribose. The more reactive strains may also produce acid from D-cellobiose, D-galactose, 5-keto-D-gluconate, D-melezitose, *meso*-inositol, methyl D-glucoside and D-turanose. The major cellular fatty acids are $C_{15:0\ iso}$, $C_{17:0\ iso}$ and $C_{17:0\ anteiso}$ (respectively representing about 60, 13 and 10% of total fatty acid). The following fatty acids are present in smaller amounts: $C_{15:0\ anteiso}$, $C_{16:0\ iso}$ and $C_{16:0}$ (respectively representing about 9, 4 and 2% of total fatty acid).

Source: gelatin production plants.

DNA G + C content (mol%): 41.5% (HPLC).

Type strain: LMG 21880, DSM 15865.

EMBL/GenBank accession number (16S rRNA gene): AJ551329 (LMG 21880).

38. **Bacillus gibsonii** Nielsen, Fritze and Priest 1995b, 879^{VP} (Effective publication: Nielsen, Fritze and Priest 1995a, 1759.)

gib.so′ni.i. N.L. gen. *gibsonii* of Gibson, named after the British bacteriologist Thomas Gibson.

TABLE 9. Differentiation of spherical-spored *Bacillus* species[a,b]

Characteristic	35. *B. fusiformis*	47. *B. insolitus*	61. *B. neidei*	64. *B. odysseyi*	70. *B. psychrodurans*	72. *B. psychrotolerans*	74. *B. pycnus*	75. *B. schlegelii*	78. *B. silvestris*	84. *B. sphaericus*
Cell diameter >1.0 mm	−	v	−	−	−	−	+	−	−	−
Spores:										
Ellipsoidal	−	+	−	−	+[c]	+[c]	−	−	−	−
Spherical	+	v	+	+	+[c]	+[c]	+	+	+	+
Borne terminally	d	+	+	+	+	+	+	+	+	+
Sporangia swollen	+	−	+	+	+	+	+	+	+	+
Parasporal crystals	−	−	−	−	−	−	−	−	−	−[d]
Hydrolysis of:										
Casein	+	ng	−	−	nd	−	−	w	−	d
Gelatin	+	−	nd	−	d	d	nd	−	−	d
Starch	−	−	−	−	+	+	−	−	−	−
Utilization of citrate	+	−	−	nd	−	−	−	−	−	d
Nitrate reduction	−	−	−	−	+	−	−	+	−	−
Growth at pH:										
5	−	nd	nd	−	−	−	nd	nd	nd	−
6	+	nd	+	+	nd	nd	+	+	nd	d
7	+	+	+	+	+	+	+	+	+	+
8	+	nd	nd	+	nd	nd	nd	nd	nd	+
9	+	nd	nd	+	nd	nd	nd	nd	nd	+
10	−	nd	nd	+	nd	nd	nd	nd	nd	−
Growth in NaCl:										
2%	+	d	+	+	+	+	nd	+	+	+
5%	+	−	+	+	d	d	−	−	+	+
7%	+	−	nd	−	−	−	−	−	−	−
NaCl required for growth	nd	−	−	nd	nd	nd	nd	nd	nd	
Growth at:										
5 °C	−	+	d	−	+	+	d	−	−	−
10 °C	nd	+	+	−	+	+	+	−	+	d
20 °C	+	+	+	−	+	+	+	−	+	+
30 °C	+	−	+	+	+	+	+	−	+	+
40 °C	nd	−	+	+	−	d	+	−	+	d
50 °C	nd	−	−	nd	nd	nd	nd	+	−	nd
Deamination of phenylalanine	d	d	nd	nd	−	−	nd	nd	−	+
Autotrophic with H₂ + CO₂ or CO	nd	nd	nd	nd	nd	nd	nd	+	nd	nd

[a]Symbols: +, >85% positive; d, different strains give different reactions (16–84% positive); −, 0–15% positive; d/w, different strains give different reactions, but positive reactions are weak; v, variation within strains; w, weak reaction; ng, no growth in test medium; nd, no data are available.

[b]Compiled from Larkin and Stokes (1967); Schenk and Aragno (1979); Logan and Berkeley (1984); Claus and Berkeley (1986); Priest et al. (1988); Fritze (1996a); Rheims et al. (1999); Abd El-Rahman et al. (2002); Nakamura et al. (2002); Priest (2002); La Duc et al. (2004).

[c]Spores of *Bacillus psychrodurans* and *Bacillus psychrotolerans* are rarely formed; on casein-peptone soymeal-peptone agar spores are predominantly spherical, but on marine agar they are predominantly ellipsoidal.

[d]Mosquitocidal strains of *Bacillus sphaericus* produce parasporal toxin crystals which are smaller than those produced by *Bacillus thuringiensis*, but which are nonetheless visible by phase-contrast microscopy.

Alkalitolerant organisms forming ellipsoidal spores which lie subterminally and, in ageing cultures paracentrally and occasionally laterally, in unswollen sporangia. Cells 0.6–1.0 by 2.0–3.0 μm. Colonies are, yellow, smooth, shiny and circular. Growth temperature range 10–30–37 °C. Optimal growth at pH 8.0; growth occurs at pH 7.0. Grows in presence of up to 9–12% NaCl. Nitrate reduction varies between strains. Casein, and gelatin are hydrolyzed. Hippurate, pullulan, starch and Tween 20 are not hydrolyzed; phenylalanine is not deaminated. Glucose and a range of other carbohydrates can be utilized as sole sources of carbon.

Source: soil.

DNA G + C content (mol%): 40.6–41.7 (HPLC analysis).

Type strain: Nielsen PN-109, ATCC 700164, DSM 8722, LMG 17949.

EMBL/GenBank accession number (16S rRNA gene): X76446 (DSM 8722).

39. **Bacillus halmapalus** Nielsen, Fritze and Priest 1995b, 879[VP] (Effective publication: Nielsen, Fritze and Priest 1995a, 1759.)

hal.ma'pa.lus. Gr. n. *halme* brine; Gr adj. *hapalos* delicate; N.L. adj. *halmapalus* sensitive to brine.

Alkalitolerant organisms forming ellipsoidal spores which lie paracentrally to subterminally in unswollen sporangia. Description is based upon two isolates. Cells 0.6–1.0 by 3.0–4.0 μm. Colonies are small, circular, shiny and creamy-white with entire margins. Growth temperature range 10–40 °C. Optimal growth at pH 8.0; growth occurs at pH 7.0. No growth in presence of 5% NaCl. Nitrate is not reduced. Casein, gelatin, hippurate, pullulan and starch are hydrolyzed. Tween 20, 40, 60 and 80 are not hydrolyzed; phenylalanine is not deaminated. Glucose and a narrow range of other carbohydrates can be utilized as sole sources of carbon. It is distinguished from *Bacillus horikoshii* by its larger cell size, lower salt tolerance, and DNA relatedness.

Source: soil.

DNA G + C content (mol%): 38.6 (HPLC analysis).

Type strain: Nielsen PN-118, ATCC 700165, DSM 8723, LMG 17950.

EMBL/GenBank accession number (16S rRNA gene): X76447 (DSM 8723).

40. **Bacillus halodurans** Nielsen, Fritze and Priest 1995b, 879[VP] (Effective publication: Nielsen, Fritze and Priest 1995a, 1759.)

ha.lo.du'rans. Gr. n. *hals* salt; L. pres. part. *durans* enduring; N.L. adj. *halodurans* salt-enduring.

Alkaliphilic and moderately halotolerant organisms forming ellipsoidal spores which lie subterminally in slightly swollen sporangia. Cells 0.5–0.6 by 3.0–4.0 μm, and occur in long chains even when sporulated. Colonies are white and circular with slightly filamentous margins. Growth temperature range 15–55 °C. Optimal growth at pH 9–10; most strains grow at pH 7.0. Growth in presence of up to 12% NaCl. Most strains do not reduce nitrate. Casein, gelatin pullulan, starch, and Tween 40 and 60 are hydrolyzed, but most strains do not hydrolyze Tween 20, and Tween 80 is not hydrolyzed. Most strains do not hydrolyze hippurate. Phenylalanine is not deaminated. Glucose and a wide range of other carbohydrates can be utilized as sole sources of carbon. The type strain was previously named "*Bacillus alcalophilus* subsp. *halodurans*" (Boyer et al., 1973). Strains in this species were formerly assigned to *Bacillus lentus* type III by Gordon and Hyde (1982). See Table 6.

Source: soil.

DNA G + C content (mol%) ranges from 42.1–43.9 (HPLC).

Type strain: PN-80, ATCC 27557, DSM 497, LMG 7121, NRRL B-3881.

EMBL/GenBank accession number (16S rRNA gene): AJ302709 (DSM 497).

41. **Bacillus halophilus** Ventosa, García, Kamekura, Onishi and Ruiz-Berraquero 1990a, 105[VP] (Effective publication: Ventosa, García, Kamekura, Onishi and Ruiz-Berraquero 1989, 164.)

hal.o.phi'lus. Gr. n. *hals* salt; Gr. adj. *philos* loving: N.L. adj. *halophilus* salt-loving.

Halophilic, strictly aerobic, Gram-positive, motile rods, 0.5–1.0 by 2.5–9.0 μm, occurring singly and in pairs or chains, and forming ellipsoidal spores which lie centrally in unswollen sporangia. Description is based upon a single isolate. Colonies on 15% NaCl medium are circular, smooth, entire, opaque and unpigmented. Grows at between 3% and 30% total salts with optimal growth at about 15% salts. Growth occurs between 15 °C and 50 °C, and is optimal at 37 °C. The pH range for growth is 6.0–8.0, with the optimum at 7.0. Catalase- and oxidase-positive. Chemo-organotroph. Acid is produced without gas from glucose and a range of other carbohydrates. Nitrate is not reduced. Esculin, DNA and urea are hydrolyzed; casein, gelatin, starch and tyrosine are not. Negative for arginine dihydrolase, lysine and ornithine decarboxylases, Voges–Proskauer, and indole. Utilizes a range of amino acids, carbohydrates and organic acids as sole carbon and energy sources, and utilizes a small range of amino acids as sole carbon, nitrogen and energy sources.

Source: rotting wood on seashore.

DNA G + C content (mol%): 51.5 (T_m).

Type strain: Kamekura N23–2, ATCC 49085, DSM 4771, LMG 17942, KCTC 3566.

EMBL/GenBank accession number (16S rRNA gene): AB021188 (DSM 4771).

42. **Bacillus horikoshii** Nielsen, Fritze and Priest 1995b, 879[VP] (Effective publication: Nielsen, Fritze and Priest 1995a, 1760.)

ho.ri.ko'shi.i. N.L. gen. n. *horikoshii* of Horikoshi; named after the Japanese microbiologist Koki Horikoshi.

Alkalitolerant organisms forming ellipsoidal spores which lie subterminally in sporangia which may be slightly swollen. Cells 0.6–0.7 by 2.0–4.0 μm. Colonies are small, circular, shiny and creamy-white with entire margins. Growth temperature range 10–40 °C. Optimal growth at pH 8.0; growth occurs at pH 7.0. 8–9% NaCl is tolerated. Nitrate is not reduced. Casein, gelatin, hippurate, pullulan, starch and Tween 80 are hydrolyzed. Tween 40 and 60 are hydrolyzed by most strains. Tween 20 is not hydrolyzed; phenylalanine is not deaminated. Glucose and a narrow range of other carbohydrates can be utilized as sole sources of carbon. Distinguished from *Bacillus halmapalus* by having a smaller cell size, higher salt tolerance, and by DNA relatedness.

Source: soil.

DNA G + C content (mol%): 41.1–42.0 (HPLC).

Type strain: Nielsen PN-121, ATCC 700161, DSM 8719, LMG 17946.

EMBL/GenBank accession number (16S rRNA gene): AB043865 (DSM 8719).

43. **Bacillus horti** Yumoto, Yamazaki, Sawabe, Nakano, Kawasaki, Ezura and Shinano 1998, 570[VP]

hor'ti. L. masc. n. *hortus* garden; L. gen. n. *horti* from the garden.

Alkaliphilic, strictly aerobic, Gram-negative, motile rods, 0.6–0.8 by 1.5–6.0 µm, forming ellipsoidal, subterminal spores in swollen sporangia. Description is based upon two isolates. Colonies on complex medium at pH 10 are white. Grow occurs at pH 7, with optimum growth at pH 8–10. Grows in presence of 3–11% NaCl but not at 12% NaCl. Growth occurs between 15 °C and 40 °C; no growth at 10 and 45 °C. Catalase- and oxidase-positive. Nitrate is reduced to nitrite, *o*-nitrophenyl-β-D-galactopyranoside is hydrolyzed and H_2S is produced at pH 7. Acid is produced without gas from glucose and a narrow range of other carbohydrates. Casein, gelatin, starch and DNA are hydrolyzed; Tween 20, 40, 60 and 80 and urea are not. See Table 6.

Source: garden soil in Japan.

DNA G + C content (mol%): 40.9% for the type strain and 40.2 for another strain (HPLC).

Type strain: K13, ATCC 700778, DSM 12751, JCM 9943, LMG 18497.

EMBL/GenBank accession number (16S rRNA gene): D87035 (K13).

44. **Bacillus hwajinpoensis** Yoon, Kim, Kang, Oh and Park 2004b, 807[VP]

hwa.jin.po.en'sis. N.L. adj. *hwajinpoensis* of Hwajinpo, a beach of the East Sea in Korea, where the type strain was isolated.

Aerobic, nonmotile rods, 1.0–1.3 µm in diameter and 2.5–4.0 µm long. Gram-positive, but Gram-variable in older cultures. Description is based on a single isolate. Ellipsoidal endospores are borne centrally or terminally in swollen sporangia. Colonies are smooth, circular to slightly irregular, slightly raised, light yellow in color and 2–4 mm in diameter after 3 d cultivation at 30 °C on marine agar. Optimum growth temperature is 30–35 °C. Growth occurs at 10 and 40 °C but not at 4 °C or above 41 °C. Optimum pH for growth is 6.0–7.0. Growth is observed at pH 5.0, but not at pH 4.5. NaCl is required for growth. Optimal growth occurs in the presence of 2–5% NaCl. Growth occurs in the presence of 19% NaCl but is inhibited by 20% NaCl. No anaerobic growth on marine agar. Esculin is hydrolyzed. Hypoxanthine, tyrosine, urea and xanthine are not hydrolyzed. Acid is produced from D-mannitol and stachyose. Cell-wall peptidoglycan contains *meso*-diaminopimelic acid. Predominant menaquinone is MK-7. Major fatty acid is $C_{15:0 \text{ anteiso}}$.

Source: sea water of the East Sea in Korea.

DNA G + C content (mol%): 40.9 (HPLC).

Type strain: SW-72, KCCM 41641, JCM 11807.

EMBL/GenBank accession number (16S rRNA gene): AF541966 (SW-72).

45. **Bacillus indicus** Suresh, Prabagaran, Sengupta and Shivaji 2004, 1374[VP]

in'di.cus. L. masc. adj. *indicus* pertaining to India, Indian.

Cells are aerobic, Gram-positive, nonmotile rods measuring approximately 0.9–1.2 µm wide and 3.3–5.3 µm long. Description is based upon a single isolate. Produces subterminal endospores in a slightly swollen sporangium. Colonies on nutrient agar are yellowish-orange pigmented, circular, raised, smooth, convex and 3.0–4.0 mm in diameter. The pigment in acetone exhibits three absorption maxima at 404, 428 and 451 nm, characteristic of carotenoids. Grows in the range of 15–37 °C (optimum 30 °C) but not at 40 °C. Grows between pH 6 and 7 and tolerates up to 2.0% (w/v) NaCl. Positive for catalase, gelatinase, amylase, arginine dihydrolase and esculin. Does not hydrolyze Tween 20 or urea. Does not reduce nitrate to nitrite and is negative for indole production, Voges–Proskauer test and citrate utilization. Utilizes D-cellobiose, *meso*-erythritol, inositol, lactose, D-melibiose, D-maltose, D-mannose, sucrose, L-rhamnose, D-ribose, raffinose, L-arginine, L-tryptophan and L-tyrosine as sole carbon sources. The major fatty acids are $C_{14:0 \text{ iso}}$ (10.9%), $C_{15:0 \text{ iso}}$ (33.5%), $C_{15:0 \text{ anteiso}}$ (19.3%), $C_{16:0 \text{ iso}}$ (11.0%), $C_{16:0}$ (5.9%) and $C_{17:0 \text{ iso}}$ (10.8%). The main proportion of the polar lipids consists of phosphatidylglycerol, diphosphatidylglycerol and phosphatidylethanolamine. The major respiratory quinone is MK-7. The cell wall is an A4β-murein with ornithine as the diamino acid and aspartic acid as the interpeptide bridge.

Source: sand of an arsenic-contaminated aquifer in West Bengal, India.

DNA G + C content (mol%): 41.2 (T_m).

Type strain: Sd/3, MTCC 4374, DSM 15820.

GenBank/EMBL accession number (16S rRNA gene): AJ583158 (Sd/3).

46. **Bacillus infernus** Boone, Liu, Zhao, Balkwill, Drake, Stevens and Aldrich 1995, 447[VP]

in.fer'nus. N.L. adj. *infernus* that which comes from below (the ground).

Strictly anaerobic, thermophilic, nonmotile rods 0.7–0.8 by 4–8 µm. Cell-wall morphology is Gram-positive but the Gram reaction is ambiguous. Endospores not observed, but their presence has been inferred from the survival of heat-treated cultures. Growth occurs at 45–60 °C but not at 40 or 65 °C; optimum temperature for growth is about 61 °C. Optimum pH for growth is about 7.3; grows well at pH 8.1 but does not grow at pH 9.2. The type strain is halotolerant; other strains have not been tested for this property. Grows fermentatively with glucose as substrate, but not with a range of other carbohydrates, alcohols and organic acids. Respiratory growth uses formate or lactate as electron donors and MnO_2, Fe^{3+}, trimethylamine oxide, and nitrate as electron acceptors. Nitrate is reduced to nitrite but not to ammonia or dinitrogen. Sulfate and thiosulfate are not reduced. Casein, gelatin and starch are not hydrolyzed. See Table 8.

Source: a shale core taken from 2.7 km below the land surface in the Taylorsville Triassic Basin in Virginia, USA.

DNA G + C content (mol%): not reported.

Type strain: Boone TH-23, DSM 10277, SMCC/W 479.

EMBL/GenBank accession number (16S rRNA gene): U20385 (Boone TH-23).

47. Bacillus insolitus Larkin and Stokes 1967, 891[AL].

in.so.li′tus. L. adj. *insolitus* unusual.

Strictly aerobic, Gram-positive, nonmotile cocci and motile rods and coccoid rods 1.0–1.5 by 1.6–2.7 μm on nutrient agar, and 0.7–0.9 by 2.4–5.3 μm on trypticase soy agar, occurring singly and in pairs. Description is based upon two isolates. Spores vary in shape from spherical to cylindrical, and from 0.7 to 1.4 μm in diameter, depending on the growth medium; terminal ellipsoidal and cylindrical spores are formed in rod-shaped cells. Sporangia are not appreciably swollen. Motile at 5 °C and 20 °C by one polar and one subpolar flagellum. Colonies on nutrient agar are small, soft, off-white and irregular. Optimum growth temperature about 20 °C, minimum below 0 °C, maximum 25 °C; sporulates and germinates at 0 °C. Tolerance of 2% NaCl varies; 4% NaCl is not tolerated. Catalase- and oxidase-positive. Gelatin and starch are not hydrolyzed; one of the two strains hydrolyzes urea (Logan and Berkeley, 1984). No growth on milk agar. Nitrate is not reduced to nitrate. Citrate not utilized as sole carbon source. No acid or gas produced from glucose or a range of other carbohydrates. See Table 9.

Source: normal and marshy soil.

DNA G + C content (mol%): 35.9 (T_m), 36.1 (Bd) for the type strain and 41.0 (T_m) for another strain.

Type strain: W 16B, ATCC 23299, DSM 5, LMG 17757, NCIMB 11433, KCTC 3737.

EMBL/GenBank accession number (16S rRNA gene): X60642 (DSM 5). This sequence displays 41 nucleotides as N-hits, indicating the weak quality of the sequence analysis. A further partial sequence of ATCC 23299 is available under accession number AF478084.

48. Bacillus jeotgali Yoon, Kang, Lee, Kho, Choi, Kang and Park 2001a, 1091[VP]

je.ot.ga′li. N.L. gen. n. *jeotgali* of jeotgal, Korean traditional fermented seafood.

Facultatively anaerobic, Gram-variable, motile rods 0.8–1.1 by 4.0–6.0 μm, forming ellipsoidal spores in swollen sporangia. Description is based upon two isolates. Colonies are cream-yellow or light orange-yellow, smooth and flat with irregular margins. Growth occurs at 10 and 45 °C but not at 55 °C; optimum growth temperature 30–35 °C. Growth occurs at pH 7.0–8.0. Tolerates up to 13% NaCl. Growth poor on nutrient agar and trypticase soy agar without added salts. Catalase-positive, oxidase-negative. Esculin, gelatin, starch and urea are hydrolyzed; casein, hypoxanthine, tyrosine and xanthine are not hydrolyzed. Acid without gas is produced from glucose and a narrow range of other carbohydrates.

Source: jeotgal, a Korean traditional fermented seafood.

DNA G + C content (mol%): 41.0 (HPLC).

Type strain: YKJ-10, AF221061, JCM 10885, CIP 107104, KCCM 41040.

EMBL/GenBank accession number (16S rRNA gene): AF221061 (YKJ-10).

49. Bacillus krulwichiae Yumoto, Yamaga, Sogabe, Nodasaka, Matsuyama, Nakajima and Suemori 2003, 1534[VP]

krul.wich.i′ae. N.L. fem. gen. n. *krulwichiae* of Krulwich; named after American microbiologist Terry A. Krulwich who made fundamental contributions to the study of alkaliphilic bacteria.

Alkaliphilic, facultatively anaerobic, Gram-positive, peritrichously flagellated straight rods, 0.5–0.7 by 1.5–2.6 μm. Ellipsoidal endospores are borne subterminally and do not cause swelling of sporangia. Description is based on two isolates. Colonies are circular and colorless. Catalase and oxidase reactions are positive. Negative for indole production, ONPG hydrolysis, and H_2S production. Growth occurs at pH 8–10, but no growth occurs at pH 7. Grows in presence of 14% NaCl, but not at higher concentrations. Nitrate is reduced to nitrite. Acid, but no gas, is produced from D-xylose, D-glucose, D-fructose, D-galactose, D-ribose, maltose, sucrose, trehalose, glycerol and mannitol when grown at pH 10. Positive for hydrolysis of starch, DNA, hippurate and Tween 20, 40, 60 and 80. Hydrolysis of casein and gelatin is variable among strains. Utilizes benzoate and m-hydroxybenzoate as sole carbon sources. The major isoprenoid quinones are menaquinone-5, -6 and -7. The main fatty acids produced during growth in an alkaline medium (pH 10) are $C_{15:0\ iso}$ (17.1–19.2%) and $C_{15:0\ anteiso}$ (45.6–49.0%). See Table 6.

Source: a soil sample obtained from Tsukuba, Ibaraki, Japan.

DNA G + C content (mol%): 40.6–41.5 (HPLC).

Type strain: AM31D, IAM 15000, NCIMB 13904, JCM 11691.

EMBL/GenBank accession number (16S rRNA gene): AB086897 (AM31D).

50. Bacillus lentus Gibson 1935, 364[AL]

len′tus. L. adj. *lentus* slow.

As with *Bacillus firmus*, strains allocated to this species are genetically heterogeneous, and many strains have been incorrectly assigned to it. Strains received as *Bacillus lentus* show phenotypic profiles that appear to overlap with those of strains assigned to *Bacillus firmus*, so that Gordon et al. (1977) raised the question of whether *Bacillus firmus-Bacillus lentus* represented a single species or a series of strains.

The description which follows is based upon the type strain and a small number of other strains which have been shown by amplified rDNA restriction analysis, polyacrylamide gel electrophoresis of whole-cell proteins, and various phenotypic characters (Logan, De Vos and colleagues, unpublished data) to be closely related to the type strain. Phylogenetic studies indicate that *Bacillus circulans*, *Bacillus firmus* and *Bacillus lentus* are related.

Strictly aerobic, Gram-positive, straight or slightly curved, round-ended, motile rods 0.7–0.8 μm in diameter that occur singly, in pairs and occasionally in short chains. Endospores are ellipsoidal, lie subterminally or paracentrally, and may swell the sporangia slightly. After 2 d on TSA at 30 °C, colonies are 1–2 mm in diameter, whitish, opaque and flat, with glossy surfaces and entire margins. Optimum growth temperature is 30 °C, minimum temperature is 10 °C and maximum lies below 40 °C. The optimum pH is 8.0; the minimum pH lies between 5.0 and 6.0 and the maximum between 9.5 and 10. Catalase-positive. Grows in presence of 5% NaCl. Starch hydrolysis is positive but casein is not hydrolyzed. In the API 20E strip, *o*-nitrophenyl-β-D-galactopyranoside hydrolysis, urease production and nitrate reduction are positive. Citrate utilization is variable. Arginine dihydrolase, lysine decarboxylase, ornithine decarboxylase, hydrogen sulfide production, tryptophan deaminase, indole production, gelatin hydrolysis, and Voges–Proskauer reaction are negative. Acid without gas is produced, often weakly, from D-glucose and from a wider range of other carbohydrates than is attacked by strains of *Bacillus firmus*. In the API 50CH gallery using the CHB suspension medium, hydrolysis of esculin is positive, and production of acid without gas is positive for lactose, *N*-acetylglucosamine, sucrose, and D-trehalose. Acid production is weak or positive for amygdalin, L-arabinose, arbutin, D-cellobiose, D-fructose, galactose, gentiobiose, D-glucose, maltose, mannitol, D-mannose, D-melezitose, D-melibiose, methyl-xyloside, D-raffinose, rhamnose, ribose, salicin, starch and D-xylose. In the API Biotype 100 kit, hydroxyquinoline-β-glucuronide is hydrolyzed and D-glucosamine, D-glucuronate and 2-keto-D-gluconate are assimilated.

Source: soil.

DNA G + C content (mol%): 36.3 (T_m), 36.4 (Bd).

Type strain: ATCC 10840, AF478107, DSM 9, JCM 2511, LMG 16798, NCIMB 8773.

EMBL/GenBank accession number (16S rRNA gene): AB021189 (NCIMB 8773).

51. **Bacillus licheniformis** (Weigmann 1898) Chester 1901, 287[AL] (*Clostridium licheniforme* Weigmann 1898, 822)

li.che.ni.for′mis. Gr. n. *lichen* lichen; L. adj. suff. *-formis* -like, in the shape of; N.L. adj. *licheniformis* lichen-shaped.

Facultatively anaerobic, Gram-positive, motile rods, forming ellipsoidal to cylindrical spores which lie centrally, paracentrally and subterminally in unswollen sporangia (Figure 10d). Cells grown on glucose agar stain evenly. Cells 0.6–0.8 by 1.5–3.0 μm, occurring singly and in pairs, and chains. Colonial morphology is variable, within and between strains, and, as with *Bacillus subtilis*, may give the appearance of a mixed culture. Colonies are round to irregular in shape and of moderate (2–4 mm) diameter, with margins varying from undulate to fimbriate; they become opaque, with surfaces that are dull and which may become wrinkled; color is whitish, and may become creamy or brown (perhaps red on carbohydrate media containing sufficient iron); textures range from moist and butyrous or mucoid, through membranous with an underlying mucoid matrix, with or without mucoid beading at the surface, to rough and crusty as they dry; these "licheniform" colonies tend to be quite adherent to the agar. Minimum growth temperature 15 °C, maximum 50–55 °C; an isolate from a geothermal environment with a maximum growth temperature of 68 °C has been reported (Llarch et al., 1997). Growth occurs at pH 5.7 and 6.8, but limits have not been reported. Grows in presence of up to 7% NaCl. Catalase-positive, oxidase variable. Casein, esculin, gelatin and starch are hydrolyzed; occasional strains will hydrolyze urea; phenylalanine is not deaminated. Usually arginine dihydrolase-positive. Pectin and polysaccharides of plant tissues are decomposed. Dextran and levan are formed extracellularly from sucrose. Citrate and propionate are utilized as sole carbon sources by most strains. Nitrate is reduced to nitrite.

Voges–Proskauer-positive. Acid without gas is produced from glucose and from a wide range of other carbohydrates.

Widely distributed in soil and many other environments, including milk and other foods, and clinical and veterinary specimens. Vegetative growth may occur readily in foods held at 30–50 °C. Occasionally reported as an opportunistic pathogen in man and other animals, and as a cause of food poisoning.

DNA G + C content (mol%): 42.9–49.9 (T_m) for 12 strains, 44.9–46.4 (Bd) for 19 strains, and 46.4 (T_m), 44.7 (Bd) for the type strain.

Type strain: ATCC 14580, CCM 2145, DSM 13, LMG 12363, IFO 12200, NCIMB 9375.

EMBL/GenBank accession number (16S rRNA gene): X68416 (DSM 13).

Additional remarks: Gibson (1944) considered *Bacillus globigii* to be a synonym of *Bacillus licheniformis*, but as Gordon et al. (1973) pointed out, strains of *Bacillus circulans*, *Bacillus pumilus* and *Bacillus subtilis* labeled *Bacillus globigii* have been widely circulated, so that the name is meaningless. Strains named *Bacillus globigii* were formerly popular for tracing studies, including those associated with the development of biological weapons.

52. **Bacillus luciferensis** Logan, Lebbe, Verhelst, Goris, Forsyth, Rodríguez-Díaz, Heyndrickx and De Vos 2002b, 1988[VP]

lu.cif.er.en′sis. N.L. adj. *luciferensis* referring to Lucifer Hill, a volcano on Candlemas Island, South Sandwich Islands, the soil of which yielded the organism.

Motile rods (0.4–0.8 by 3–6 μm) occurring singly and in pairs and showing pleomorphism. Gram-positive, but becoming Gram-negative within 24 h of culture at 30 °C. Ellipsoidal endospores lie subterminally and occasionally terminally, and may swell the sporangia slightly. Colonies are 1–5 mm in diameter, creamy-gray, raised, translucent, glossy, moist and loosely butyrous, with irregular margins and surfaces. The growth temperature range lies between 15–20 °C and 35–45 °C, with an optimum of 30 °C. The pH range for growth is from 5.5–6.0 to 8.0–8.5, with an optimum of 7.0. Organisms are facultatively anaerobic and weakly catalase-positive. Esculin and gelatin are hydrolyzed, casein is weakly hydrolyzed. Nitrate is not reduced. Acid without gas is produced from glucose and a range of other carbohydrates. The major cellular fatty acids are $C_{15:0\ anteiso}$ and $C_{15:0\ iso}$ (representing about 25% and 50% of total fatty acid, respectively).

Source: geothermal soil taken from an active fumarole on Lucifer Hill, a volcano on Candlemas Island, South Sandwich archipelago.

DNA G + C content (mol%): 33.0 (T_m) for the type strain.

Type strain: Logan SSI061, LMG 18422, CIP 107105.

EMBL/GenBank accession number (16S rRNA gene): AJ419629 (LMG 18422).

53. **Bacillus macyae** Santini, Streimann and vanden Hoven 2004, 2244[VP]

ma.cy′ae. N.L. fem. gen. n. *macyae* of Macy, named after the late Joan M. Macy, La Trobe University, Australia, in tribute to her research in the area of environmental microbiology.

Cells are Gram-positive, motile rods (2.5–3 μm long and 0.6 μm wide) and produce subterminally located ellipsoidal spores. Spores do not cause swelling of sporangia. Colonies are round and white. Catalase reaction is positive and oxidase is negative. Strict anaerobe that respires with arsenate and nitrate as terminal electron acceptors. Arsenate is reduced to arsenite and nitrate to nitrite. The electron donors used for anaerobic respiration are acetate, lactate, pyruvate, succinate, malate, glutamate and hydrogen (with acetate as carbon source). Growth occurs at 28–37 °C, pH 7–8.4 and 0.12–3% NaCl.

Source: arsenic-contaminated mud from a gold mine in Bendigo, Victoria, Australia.

DNA G + C content (mol%): 37 (HPLC).

Type strain: JMM-4, DSM 16346, JCM 12340.

GenBank/EMBL accession number (16S rRNA gene): AY032601 (JMM-4).

54. **Bacillus marisflavi** Yoon, Kim, Kang, Oh and Park 2003a, 1301[VP]

ma.ris.fla'vi. L. gen. neut. n. *maris* of the sea; L. masc. adj. *flavus* yellow; N.L. gen. masc. n. *marisflavi* of the Yellow Sea.

Aerobic rods, 0.6–0.8 μm by 1.5–3.5 μm, motile by means of a single polar flagellum. Gram-positive, but Gram-variable in older cultures. Description is based on a single isolate. Ellipsoidal endospores lie centrally or subterminally in swollen sporangia. Colonies are pale yellow, smooth, circular to slightly irregular, slightly raised, and 2–4 mm in diameter after 3 d at 30 °C on marine agar. Optimal growth temperature is 30–37 °C. Growth occurs at 10 and 47 °C, but not at 4 or above 48 °C. Optimal growth pH is 6.0–8.0. Growth is observed at pH 4.5, but not at pH 4.0. Growth occurs in the presence of 0–16% (w/v) NaCl, and optimal growth occurs at 2–5% (w/v) NaCl. Catalase-positive, oxidase- and urease-negative. Esculin and casein are hydrolyzed. Hypoxanthine, starch, Tween 80, tyrosine and xanthine are not hydrolyzed. Acid is produced from arbutin, D-cellobiose, D-fructose, gentiobiose, D-glucose, glycerol, maltose, D-mannitol, D-mannose, melibiose, methyl-D-mannoside, D-ribose, salicin, stachyose, sucrose, D-trehalose and D-xylose and produced weakly from D-galactose and D-raffinose. The cell-wall peptidoglycan contains *meso*-diaminopimelic acid. The predominant menaquinone is MK-7. The major fatty acids are $C_{15:0\ anteiso}$ and $C_{15:0\ iso}$.

Source: sea water of a tidal flat of the Yellow Sea in Korea.

DNA G + C content (mol%): 49 (HPLC).

Type strain: TF-11, KCCM 41588, JCM 11544.

EMBL/GenBank accession number (16S rRNA gene): AF483624 (TF-11).

55. **Bacillus megaterium** de Bary 1884, 499[AL]

me.ga.te'ri.um. Gr. adj. *mega* large; Gr. n. *teras, teratis* monster, beast; N.L. n. *megaterium* big beast.

Aerobic, Gram-positive, motile rods, large cells 1.2–1.5 by 2.0–5.0 μm, occurring singly and in pairs and chains, forming ellipsoidal and sometimes spherical spores which are located centrally, paracentrally or subterminally, and which do not swell the sporangia (Figure 10e). Cells grown on glucose agar produce large amounts of storage material, giving a vacuolate or foamy appearance. Colonies are glossy, round to irregular, and have entire to undulate margins (Figure 9c). Minimum temperature for growth 3–15 °C, maximum 35–45 °C, with the optimum around 30 °C. The temperature range of a water isolate from an Antarctic geothermal island was 17–63 °C, with an optimum of 60 °C (Llarch et al., 1997). Catalase-positive. Casein, gelatin and starch are hydrolyzed. Phenylalanine is deaminated by most strains; tyrosine degradation is variable. Most strains grow in presence of 7% NaCl, but none grow at 10% NaCl. Citrate is utilized as sole carbon source. Most strains do not reduce nitrate. Acid without gas is produced from glucose and a wide range of other carbohydrates.

Source: soil, cow feces, foods and clinical specimens.

DNA G + C content (mol%): 37.0–38.1 (T_m) for 12 strains, and 37.2 (T_m) for the type strain.

Type strain: ATCC 14581, CCM 2007, DSM 32, NCIMB 9376, NCTC 10342, LMG 7127, IAM 13418.

EMBL/GenBank accession number (16S rRNA gene): D16273 (IAM 13418).

Additional remarks: Gordon et al. (1973) found that their 60 cultures of *Bacillus megaterium* formed two merging aggregates of strains, and Hunger and Claus (1981) recognized three DNA relatedness groups among 21 strains labeled as *Bacillus megaterium*, with the type strain lying within relatedness group A; Priest et al. (1988) revived the names *Bacillus simplex* for strains of DNA relatedness group B and *Bacillus flexus* for two strains which showed low homology with these two relatedness groups.

56. **Bacillus methanolicus** Arfman, Dijkhuizen, Kirchhof, Ludwig, Schleifer, Bulygina, Chumakov, Govorhukina, Trotsenko, White and Sharp 1992, 444[VP]

me.tha'no.li.cus. N.L. n. *methanol* methanol; L. masc. suff. *-icus* adjectival suffix used with the sense of belonging to; N.L. masc. adj. *methanolicus* relating to methanol.

Methylotrophic, thermotolerant, strictly aerobic, nonmotile, Gram-positive rods, usually occurring singly. Filamentous cells may be seen, especially in older cultures. Ellipsoidal endospores lie centrally to subterminally and swell the sporangia. Grows on methanol, some strains also grow on ethanol. Colonies on tryptone soya agar are circular, and usually have rough surfaces and crenated, undulating margins. The growth temperature range lies between 35 °C and 60 °C, with an optimum of 55 °C. Catalase- and oxidase-positive. Casein and hippurate are not hydrolyzed; starch hydrolysis varies between strains. Nitrate is not reduced. Acid without gas is produced from glucose and a narrow range of other carbohydrates. See Table 8.

Source: soil, aerobic (and thermophilic) wastewater treatment systems and volcanic hot springs.

DNA G + C content (mol%): 48–50 (T_m).

Type strain: Dijkhuizen PB1, NCIMB 13113, LMG 16799, KCTC 3735.

EMBL/GenBank accession number (16S rRNA gene): X64465; this is for strain C1 (=NCIMB 13114) which is not the type strain.

57. **Bacillus mojavensis** Roberts, Nakamura and Cohan 1994, 263[VP]

mo.hav.en'sis. N.L. masc. adj. *mojavensis* from the Mojave Desert.

Aerobic, Gram-positive, motile rods, forming ellipsoidal spores which lie centrally or paracentrally in unswollen sporangia. Cells 0.5–1.0 by 2.0–4.0 μm, occurring singly and in short chains. Colonies are opaque, smooth, circular and entire and 1.0–2.0 mm in diameter after 2 d at 28 °C. Optimum growth temperature 28–30 °C, with minimum of 5–10 °C and maximum of 50–55 °C. Catalase-positive, oxidase-positive. Casein, gelatin and starch are hydrolyzed; Tween 80, tyrosine and urea are not. Nitrate is reduced to nitrite. Acid without gas is produced from glucose and a range of other carbohydrates.

Phenotypically indistinguishable from *Bacillus subtilis* subsp. *subtilis* and *Bacillus subtilis* subsp. *spizizenii* and distinguished from those organisms principally by DNA relatedness and resistance to transformation. Phenotypically indistinguishable from *Bacillus vallismortis*, and distinguished from that organism by DNA relatedness, restriction digestion analysis, and fatty acid analysis.

Phenotypically distinguishable from *Bacillus atrophaeus* only by failure to produce dark brown pigmented colonies on media containing tyrosine or other organic nitrogen source.

Source: desert soils.

DNA G + C content (mol%): 43.0 (T_m).

Type strain: Cohan RO-H-1, ATCC 51516, NRRL B-14698, DSM 9205, LMG 17797, NCIMB 13391, IFO 15718.

EMBL/GenBank accession number (16S rRNA gene): AB021191 (IFO 15718).

58. **Bacillus mycoides** Flügge 1886, 324[AL]

my.co.i′des. Gr. n. *myces* fungus; Gr. *eidus* form, form, shape; N.L. adj. *mycoides* fungus-like.

Facultatively anaerobic, Gram-positive, nonmotile organisms forming ellipsoidal spores which lie paracentrally to subterminally in unswollen sporangia. Cells 1.0–1.2 by 3.0–5.0 μm, occurring singly and in chains. Cells grown on glucose agar produce large amounts of storage material, giving a vacuolate or foamy appearance. Colonies are white to cream, opaque, and characteristically rhizoid; this ability to form rhizoid colonies may be lost. Minimum growth temperature 10–15 °C, maximum 35–40 °C. Grows at pH 5.7, and in 0.001% lysozyme. Ability to grow in presence of 7% NaCl varies between strains. Catalase-positive, oxidase-negative. Lecithinase and Voges–Proskauer reactions are positive. Citrate utilization variable; propionate not utilized. Nitrate reduction is variable. Casein and starch are hydrolyzed; decomposition of tyrosine variable. Acid without gas is produced from glucose and a limited range of other carbohydrates.

Phenotypically similar to other members of the *Bacillus cereus* group: *Bacillus anthracis, Bacillus cereus, Bacillus thuringiensis* and *Bacillus weihenstephanensis. Bacillus mycoides* is distinguished by its characteristic rhizoid colonies Figure 9d) and absence of motility. Smith et al. (1952) and Gordon et al. (1973) considered *Bacillus mycoides* to be a variety of *Bacillus cereus.* For distinguishing characters within the *Bacillus cereus* group, see the individual species descriptions and Table 7. Those strains which have been proposed as *Bacillus pseudomycoides* can only be separated from *Bacillus mycoides* by DNA relatedness and some differences in fatty acid composition.

Source: mainly from soil.

DNA G + C content (mol%): 32.5–38.4 (T_m) for nine strains, 35.2–39.0 (Bd) for four strains, and 34.2 (T_m), 34.1 (Bd) for the type strain.

Type strain: ATCC 6942, DSM 2048, LMG 7128, NRRL B-14811, NRS 273, NCIMB 13305, KCTC 3453.

EMBL/GenBank accession number (16S rRNA gene): AB021192 (ATCC 6462).

59. **Bacillus naganoensis** Tomimura, Zeman, Frankiewicz and Teague 1990, 124[VP]

na.ga.no.en′sis. N.L. masc. adj. *naganoensis* of the Japanese Prefecture Nagano.

Aerobic, moderately acidophilic, Gram-positive, nonmotile rods, forming ellipsoidal spores which lie subterminally in swollen sporangia. Description is based upon a single isolate. Cells are 0.5–1.0 by 2.1–10.0 μm, have rounded or square ends, and occur singly or in chains. Colonies are opaque, smooth, glistening, circular and entire, and reach 2.0–3.0 mm in diameter. Optimum growth temperature 28–33 °C, with minimum above 20 °C and maximum below 45 °C. The pH range for growth is about 4.0–6.0. Catalase-positive. A thermostable pullulanase is produced. Casein, gelatin and starch are hydrolyzed. Hippurate and tyrosine are not decomposed. Citrate and propionate are not utilized as sole sources of carbon. Nitrate is not reduced. Acid without gas is produced from glucose and a small range of other carbohydrates after extended (>14 d) incubation. Not pathogenic to mice.

Source: soil by selection using a pullulan-containing medium.

DNA G + C content (mol%): 45 ± 2 (T_m).

Type strain: ATCC 53909, DSM 10191, LMG 12887, KCTC 3742.

EMBL/GenBank accession number (16S rRNA gene): AB021193 (ATCC 53909).

Additional remark: this species has recently (Hatayama et al., 2006) been classified as *Pullulanibacillus naganoensis.*

60. **Bacillus nealsonii** Venkateswaran, Kempf, Chen, Satomi, Nicholson and Kern 2003, 171[VP]

neal′son.i.i. N.L. gen. n. *nealsonii* referring to Kenneth H. Nealson, an American microbiologist.

Facultatively anaerobic, Gram-positive, motile rods 1.0 by 4.0–5.0 μm. The ellipsoidal spores are 0.5 by 1.0 μm, and bear an additional extraneous layer similar to an exosporium. Description is based upon a single isolate. Young colonies on TSA are 3–4 mm in diameter, irregular, rough and umbonate with undulate or lobate edges and are of a beige color. Sodium ions are not essential for growth; up to 8% NaCl is tolerated. Growth occurs at pH 6–10, with an optimum of pH 7. Optimum growth temperature is 30–35 °C, with minimum of 25 °C and maximum of 60 °C. Catalase and β-galactosidase are produced, but gelatinase, arginine dihydrolase, lysine and ornithine decarboxylases, lipase, amylase and alginase are not. H_2S is not produced from thiosulfite. Denitrification does not occur. Acid is produced from glucose and a wide range of other carbohydrates.

Source: dust particles collected at a spacecraft-assembly facility.

DNA G + C content (mol%): not reported.

Type strain: FO-92, ATCC BAA-519, DSM 15077.

EMBL/GenBank accession number (16S rRNA gene): AF234863 (FO-92).

61. **Bacillus neidei** Nakamura, Shida, Takagi and Komagata 2002, 504[VP]

nei′de.i. N.L. gen. n. *neidei* of the early microbiologist Neide.

Aerobic, Gram-positive, motile rods, forming spherical spores which lie terminally in swollen sporangia. Cells are about 1.0 by 3.0–5.0 μm. Colonies are translucent, thin, smooth, circular and entire, and reach 1 mm in diameter after 24 h of incubation at 28 °C. Optimum growth temperature 28–33 °C, with minimum 5–10 °C and maximum 40–45 °C. Catalase-positive. Biotin, thiamin and cystine are required for growth. Casein, starch, Tween 40 and 80, tyrosine and urea are not hydrolyzed. L-alanine, citrate β-hydroxybutyrate, propionate and pyruvate are not oxidized. Grows in presence of 5% NaCl, but sensitive to 0.001% lysozyme. Nitrate is not reduced to nitrite. No acid or gas produced from glucose and other common carbohydrates. Cell-wall peptidoglycan type is L-Lys-D-Glu. Phenotypically similar to *Bacillus fusiformis*, *Bacillus pycnus* and *Bacillus sphaericus*, and separable from these species by growth factor requirements, several substrate oxidation and decomposition tests, and differences in fatty acid compositions. See Table 9.

Source: soil.

DNA G + C content (mol%): 35 (T_m).

Type strain: NRRL Bᴅ-87, JCM 11077, LMG 21635.

EMBL/GenBank accession number (16S rRNA gene): AF169520 (NRRL Bᴅ-87).

Additional remark: this species has recently been classified as *Viridibacillus neidei* by Albert et al. (2007).

62. **Bacillus niacini** Nagel and Andreesen 1991, 137[VP]

ni.a.ci′ni. N.L. n. *niacinum* niacin or nicotinic acid; N.L. gen. n. *niacini* of nicotinic acid.

Aerobic rods 0.9–1.4 by 3–5.6 μm, forming central, and sometimes subterminal, ellipsoidal spores which may swell the sporangia slightly. Cells may be pleomorphic and increase in width. Long chains may be formed when grown on complex media. Gram-variable when grown in nutrient broth, and Gram-positive when grown on nicotinate agar. Some strains motile. Colonies are smooth and have light beige centers surrounded by translucent areas of variable extension, and are about 3–5 mm in diameter. In the presence of molybdate, nicotinate can be used as sole source of carbon, nitrogen and energy. Grows at 10–40 °C. Catalase-positive or weakly positive; oxidase usually positive. Some strains are indole-positive. Gelatin is hydrolyzed, sometimes weakly; starch is hydrolyzed by some strains, urease is usually negative; casein, phenylalanine and tyrosine are not decomposed. Utilization of aspartate, citrate, formate and lactate as sole carbon sources varies between strains. Optimum pH for growth between 7 and 8. Nitrate is usually reduced to nitrite. Acid without gas is produced from glucose and from a number of other carbohydrates, depending upon the strain.

Source: soil.

DNA G + C content (mol%): 37–39 (T_m).

Type strain: IFO 15566, DSM 2923, LMG 16677.

EMBL/GenBank accession number (16S rRNA gene): AB021194 (IFO 15566).

63. **Bacillus novalis** Heyrman, Vanparys, Logan, Balcaen, Rodríguez-Díaz, Felske and De Vos 2004, 52[VP]

no.va′lis. L. gen. n. *novalis* of fallow land.

Facultatively anaerobic, Gram-positive, motile, slightly curved, round-ended rods, 0.6–1.2 μm in diameter, occurring singly and in pairs, and occasionally in short chains or filaments. Endospores are mainly ellipsoidal, and lie in subterminal and occasionally paracentral positions in slightly swollen sporangia. When grown on TSA, colonies are raised, butyrous, cream-colored, produce a soft brown pigment that diffuses in the agar, and have slightly irregular margins and smooth or eggshell-textured surfaces; they sometimes have iridescent centers when viewed by low-power microscopy. Optimal growth occurs at 30–40 °C, and the maximum growth temperature lies between 50 °C and 55 °C. Growth occurs from pH 4.0–5.0–9.5–10.0, and the optimum pH for growth is 7.0–9.0. Casein is hydrolyzed. In the API 20E strip, Voges–Proskauer reaction is negative, gelatin is hydrolyzed by most strains, and nitrate reduction is positive (sometimes weakly); reactions for *o*-nitrophenyl-β-D-galactopyranoside hydrolysis, arginine dihydrolase, lysine decarboxylase, ornithine decarboxylase, citrate utilization, hydrogen sulfide production, urease, tryptophan deaminase, indole production, are negative. Hydrolysis of esculin positive. Acid without gas is produced (weakly by some strains) from the following carbohydrates in the API 50 CH gallery using the CHB suspension medium: *N*-acetyl-D-glucosamine, D-fructose, galactose (always weak), D-glucose, maltose, D-mannose, D-trehalose. The following reactions are variable between strains and, when positive, are usually weak: amygdalin, arbutin, D-cellobiose, β-gentiobiose, gluconate, glycerol, 5-keto-D-gluconate, D-lyxose, D-mannitol, ribose, sorbitol, D-xylose; the type strain is positive for sorbitol and D-xylose and weakly positive for: amygdalin, D-cellobiose, β-gentiobiose, D-mannitol. The major cellular fatty acids are $C_{15:0\ iso}$ and $C_{15:0\ anteiso}$, present at a level of about 44 and 31% of the total fatty acid content, respectively.

Source: soil in the Drentse A agricultural research area, The Netherlands.

DNA G + C content (mol%): 40.0–40.5 (HPLC) and 40.5 for the type strain.

Type strain: LMG 21837, DSM 15603.

EMBL/GenBank accession number (16S rRNA gene): AJ542512 (LMG 21837).

64. **Bacillus odysseyi** La Duc, Satomi and Venkateswaran 2004, 200[VP]

o.dys.se′yi. L. n. *Odyssea* the Odyssey; N.L. gen. n. *odysseyi* pertaining to the Mars Odyssey spacecraft, from which the organism was isolated.

Strictly aerobic, Gram-positive motile rods, 4–5 μm in length and 1 μm in diameter. Forms spherical endospores which are borne terminally and swell the sporangia. Spores show an additional exosporium layer. Description is based upon a single isolate. Colonies on TSA are round, smooth, flat with entire edges and beige in color. Sodium ions are not essential for growth; growth occurs in 0–5% NaCl. Grows at pH 6–10 (optimum at pH 7) and 25–42°C (optimum 30–35°C). With the exception of arabinose, breakdown of sugars to acids does not occur following prolonged incubation. Glucose is not utilized as sole carbon source. Pyruvate, amino acids, purine or pyrimidine bases and related compounds are preferred as carbon and energy sources. Catalase-positive, but does not produce gelatinase, arginine dihydrolase, lysine or ornithine decarboxylase, lipase, amylase or alginase. Does not produce H_2S from thiosulfite and is not involved in denitrification. Closely related to species that have been transferred to the novel genus *Lysinibacillus* (Ahmed et al., 2007c), but data on peptidoglycan composition and polar lipids are not available for *Bacillus odysseyi*, and so it has not been transferred to the new genus. See Table 9.

Source: the surface of the Mars Odyssey spacecraft.

DNA G + C content (mol%): not reported.

Type strain: 34hs-1, ATCC PTA-4993, NRRL B-30641, NBRC 100172.

EMBL/GenBank accession number (16S rRNA gene): AF526913 (34hs-1).

65. **Bacillus okuhidensis** Li, Kawamura, Shida, Yamagata, Deguchi and Ezaki 2002, 1208[VP]

o.ku.hid.en′sis. N.L. masc. adj. *okuhidensis* referring to Okuhida in Gifu, Japan, where the strains were originally isolated.

Alkaliphilic, weakly Gram-positive rods, 0.5–1.0 by 5–7 μm, forming ellipsoidal, subterminal spores that may swell the sporangia slightly. Description is based upon two isolates. Motile by means of peritrichous flagella. Cells stain slightly Gram-positive in the exponential growth phase and Gram-negative in the stationary phase. Colonies are circular, convex, smooth, and yellowish. Growth temperature range 30–60°C; optimum 45–50°C. Optimal growth at pH 10.5; pH range 6.0–11.0. Grows in presence of 10% NaCl. Catalase- and oxidase-positive. Casein, starch and gelatin are hydrolyzed; hippurate, and Tween 20, 40 and 60 are not. Phenylalanine is not deaminated. Nitrate is reduced to nitrite. A range of carbohydrates can be utilized as sole sources of carbon. The major cellular fatty acids are $C_{15:0\,iso}$ (43.75% ± 0.7%) and $C_{15:0\,anteiso}$ (25.8% ± 0.6%). See Table 6.

Source: hot spa water.

DNA G + C content (mol%): 40.0–41.1 (HPLC).

Type strain: GTC 854, JCM 10945, DSM 13666.

EMBL/GenBank accession number (16S rRNA gene): AB047684 (GTC 854).

66. **Bacillus oleronius** Kuhnigk et al. 1996, 625[VP] (Effective publication: Kuhnigk et al. 1995, 704.)

o.le.ro′ni.us. N.L. adj. *oleronius* of Île de Oléron, France, where the termite host thrives.

Cells are nonmotile, Gram-negative, medium-sized rods, that occur singly and in pairs, and sometimes form short chains of 3–4 cells. They bear ellipsoidal endospores that lie in subterminal and paracentral positions within swollen sporangia. After 2 d on TSA colonies are approximately 1–2 mm diameter, circular, entire, shiny, beige or cream and butyrous with slightly translucent edges. Organisms are strictly aerobic and catalase-positive. Growth may occur between 30°C and 50°C, with an optimum of 37°C. Casein is not hydrolyzed and starch is sometimes hydrolyzed weakly. Nitrate is reduced to nitrite, the Voges–Proskauer reaction is variable (type strain positive), citrate is not utilized, hydrogen sulfide and indole are not produced, and the ONPG reaction is negative. Esculin is hydrolyzed, gelatin is weakly hydrolyzed and urea is not hydrolyzed. Acid without gas is produced from N-acetylglucosamine, D-cellobiose, D-fructose, D-glucose, mannitol and D-tagatose. Acid production from the following carbohydrates is variable, and when positive it is weak: galactose, glycerol, maltose, D-mannose, ribose salicin, starch and D-trehalose; the type strain produces acid from: glycerol, maltose, ribose, starch and D-trehalose. The major cellular fatty acids (mean percentage + standard deviation of total fatty acids) after 24 h growth on brain heart infusion supplemented with vitamin B_{12} at 37°C are: $C_{15:0\,iso}$ (39.24 ± 1.38), $C_{15:0\,anteiso}$ (22.89 ± 2.22) and $C_{17:0\,anteiso}$ (20.78 ± 0.85).

Source: the hindgut of termite *Reticulitermes santonensis* (Feytaud), and from raw milk and cattle feed concentrate.

DNA G + C content (mol%): 35.2–34.7 (HPLC) and 35.2 for the type strain.

Type strain: Kuhnigk RT 10, DSM 9356, ATCC 700005, LMG 17952, CIP 104972.

EMBL/GenBank accession number (16S rRNA gene): X782492 (DSM 9356).

67. **Bacillus pseudalcaliphilus** Nielsen, Fritze and Priest 1995b, 879 (Effective publication: Nielsen, Fritze and Priest 1995a, 1760.)

pseu.dal.ca.li′phi.lus. Gr. adj. *pseudes* false; N.L. adj. *alcalophilus* a specific epithet; N.L. adj. *pseudalcaliphilus* false *alcalophilus* because it is phenotypically closely related to *Bacillus alcalophilus* but phylogenetically distinct.

Alkaliphilic organisms forming ellipsoidal spores which lie paracentrally to subterminally in swollen sporangia. Cells 0.5–0.6 by 2.0–4.0 μm. Colonies are white and circular with undulate margins. Growth temperature range 10–40°C. Optimal growth at about pH 10.0; no growth at pH 7.0. Maximum NaCl concentration tolerated is 10%. Nitrate is not reduced. Casein, gelatin, pullulan and starch are hydrolyzed. Hippurate and Tween 20 are not hydrolyzed; phenylalanine is not deaminated. Glucose and a range of other carbohydrates can be utilized as sole sources of carbon.

Phenotypically similar to *Bacillus alcalophilus*, but phylogenetically distinct. See Table 6.

Source: soil.

DNA G + C content (mol%): 38.2–39.0 (HPLC).

Type strain: Nielsen PN-137, DSM 8725, ATCC 700166, LMG 17951, CIP 105304.

EMBL/GenBank accession number (16S rRNA gene): X76449 (DSM 8725).

68. **Bacillus pseudofirmus** Nielsen, Fritze and Priest 1995b, 879[VP] (Effective publication: Nielsen, Fritze and Priest 1995a, 1760.)

pseu.do.fir′mus. Gr. adj. *pseudes* false; L. adj. *firmus* a specific epithet; N.L. adj. *pseudofirmus* false *firmus* referring to physiological similarities to *Bacillus firmus.*

Alkaliphilic and halotolerant organisms forming ellipsoidal spores which lie paracentrally to subterminally in unswollen sporangia. Cells 0.6–0.8 by 3.0–6.0 μm. Colonies are yellow and circular with irregular margins. Growth temperature range 10–45 °C. Optimal growth at about pH 9.0; no growth at pH 7.0 for most strains. Maximum NaCl concentration tolerated is 16–17%. Nitrate is not reduced. Casein, gelatin, starch and Tween 40 and 60 are hydrolyzed; some strains can hydrolyze pullulan. Hippurate and Tween 20 are not hydrolyzed. Phenylalanine is deaminated. Glucose and a range of other carbohydrates can be utilized as sole sources of carbon.

Phenotypically similar to *Bacillus firmus*, but alkalophilic and phylogenetically distinct. See Table 6.

Source: soil and animal manure.

DNA G + C content (mol%): 39.0–40.8 (HPLC).

Type strain: PN-3, DSM 8715, NCIMB 10283, LMG 17944, ATCC 700159.

EMBL/GenBank accession number (16S rRNA gene): X76439 (DSM 8715).

69. **Bacillus pseudomycoides** Nakamura 1998, 1035[VP]

pseu.do.my.co.i′des. Gr adj. *pseudes* false; N.L. adj. *mycoides* fungus-like; N.L. adj. *pseudomycoides* false fungus-like.

Facultatively anaerobic, Gram-positive, nonmotile organisms forming ellipsoidal spores which lie paracentrally to subterminally in unswollen sporangia. Cells 1.0 by 3.0–5.0 μm, occurring singly and in short chains. Cells grown on glucose agar produce large amounts of storage material, giving a vacuolate or foamy appearance. Colonies are white to cream, opaque, and usually rhizoid. Growth temperature range 15–40 °C, optimum 28 °C. Grows at pH 5.7, in 7% NaCl, and in 0.001% lysozyme. Catalase-positive, oxidase-negative. Lecithinase and Voges–Proskauer reactions are positive. Citrate utilization variable; propionate not utilized. Nitrate is reduced to nitrite. Casein, starch and tyrosine are hydrolyzed. Acid without gas is produced from glucose and a limited range of other carbohydrates.

Phenotypically similar to *Bacillus cereus* and indistinguishable from *Bacillus mycoides* by conventional characters; distinguished from them by DNA relatedness and some differences in fatty acid composition. For distinguishing characters within the *Bacillus cereus* group, see the individual species descriptions and Table 7.

Source: mainly from soil.

DNA G + C content (mol%): 34.0–36.0 (T_m).

Type strain: NRRL B-617, DSM 12442, LMG 18993.

EMBL/GenBank accession number (16S rRNA gene): AF013121 (NRRL B-617).

70. **Bacillus psychrodurans** Abd El-Rahman, Fritze, Spröer and Claus 2002, 2132[VP]

psy.chro.dur′ans. Gr. adj. *psychros* cold; L. pres. part. *durans* enduring; N.L. part. adj. *psychrodurans* cold-enduring.

Aerobic, Gram-positive, psychrotolerant rods 0.5–0.6 by 2.0–5.0 μm. Sporulation infrequently observed; spores lie terminally in swollen sporangia and are predominantly spherical in casein-peptone soymeal-peptone agar cultures and predominantly ellipsoidal in marine agar cultures. No growth or very poor growth in/on nutrient agar/broth. Minimum growth temperature –2–0 °C and maximum 30–35 °C. Catalase-positive. Starch, Tween 20, 40, 60 and 80 (type strain negative for Tween 80) are hydrolyzed. Gelatin usually hydrolyzed. Esculin, casein and urea not hydrolyzed. Grows in presence of 3 and usually 5%, but not 7% NaCl. Sensitive to 0.001% lysozyme. Will grow anaerobically with KNO$_3$. Nitrate is reduced to nitrite. No acid or very weak acid production from glucose and other common carbohydrates; no gas produced. Cell-wall peptidoglycan type is L-Orn-D-Glu. Phenotypically similar to *Bacillus insolitus* and *Bacillus psychrotolerans*, and separable from these species by anaerobic growth with KNO$_3$, nitrate reduction, and NaCl tolerance. See Table 9.

Source: garden soil in Egypt.

DNA G + C content (mol%): 36–37 (Bd) and 36.3 for the type strain.

Type strain: 68E3, DSM 11713, NCIMB 13837, KCTC 3793.

EMBL/GenBank accession number (16S rRNA gene): AJ277984 (DSM 11713).

71. **Bacillus psychrosaccharolyticus** (*ex* Larkin and Stokes 1967) Priest, Goodfellow and Todd 1989, 93[VP] (Effective publication: Priest, Goodfellow and Todd 1988, 1879.)

psy.chro.sacch′ar.o.lyt.ic.us. Gr. adj. *psychros* cold; Gr. n. *sakchâron* sugar; Gr. adj. *lutikos* able to dissolve; N.L. adj. *psychrosaccharolyticus* cold (adapted), sugar-fermenting.

Facultatively anaerobic, Gram-positive or variable, peritrichously motile, pleomorphic rods, which vary from coccoid to elongate, forming ellipsoidal spores which lie centrally or paracentrally in swollen sporangia; the spore may fill most of the sporangium and may lie laterally. Description is based on three strains. Cells range from 0.6–1.5 μm by 1.5–3.5 μm; normal size for cells grown on nutrient agar is 0.9–1.0 μm by 2.5–3.0 μm. Colonies are opaque and smooth. Growth occurs at 0–30 °C; sporulates and germinates at 0 °C. Grows between pH 6.0–7.2 and 9.5. Growth in presence of 2–5% NaCl varies. Nitrate is reduced. Casein, esculin, gelatin, pullulan, starch and urea are hydrolyzed. Acid without gas is produced from D-glucose and some other carbohydrates. Acetate, citrate, gluconate, malonate and succinate are not utilized.

Source: soil and lowland marshes.

DNA G + C content (mol%): 43–44 (T_m).

Type strain: NRRL-B3394, DSM 13778, LMG 9580, NCIMB 11729, ATCC 23296, KCTC 3399.

EMBL/GenBank accession number (16S rRNA gene): AB021195 (ATCC 23296).

72. **Bacillus psychrotolerans** Abd El-Rahman, Fritze, Spröer and Claus 2002, 2131[VP]

psy.chro.tol′er.ans. Gr. adj. *psychros* cold; L. pres. part. *tolerans* tolerating; N.L. part. adj. *psychrotolerans* cold-tolerating.

Strictly aerobic, Gram-positive, psychrotolerant rods 0.4–1.0 by 2.0–7.0 μm. Sporulation infrequently observed; spores lie terminally in swollen sporangia and are predominantly spherical in cultures on casein-peptone soymeal-peptone agar containing Mn$^+$ and predominantly ellipsoidal in marine agar cultures. No growth or very poor growth in/on nutrient agar/broth. Minimum growth temperature –2–0 °C and maximum 30–40 °C. Catalase-positive. Starch, and Tween 20, 40 and 60 are hydrolyzed. Tween 80 usually hydrolyzed. Gelatin usually not hydrolyzed. Esculin, casein and urea not hydrolyzed. Grows in presence of 3, usually not at 5%, and not 7% NaCl. Sensitive to 0.001% lysozyme. Will not grow anaerobically with KNO$_3$. Nitrate is not reduced to nitrite. No acid or very weak acid production from glucose and other common carbohydrates; no gas produced. Cell-wall peptidoglycan type is L-Orn-D-Glu. Phenotypically similar to *Bacillus insolitus* and *Bacillus psychrotolerans*, and separable from these species by strict aerobic growth, inability to reduce nitrate, and NaCl tolerance. See Table 9.

Source: field soil in Germany.

DNA G + C content (mol%): 36–38 (Bd) and 36.5 for the type strain.

Type strain: 3H1, DSM 11706, NCIMB 13838, KCTC 3794.

EMBL/GenBank accession number (16S rRNA gene): AJ277983 (DSM 11706).

73. **Bacillus pumilus** Meyer and Gottheil in Gottheil 1901, 680[AL]

pu'mi.lus. L. adj. *pumilus* little.

Aerobic, Gram-positive or Gram-variable, motile, small rods 0.6–0.7 by 2.0–3.0 µm, occurring singly and in pairs, and forming cylindrical to ellipsoidal spores which lie centrally, paracentrally and subterminally in unswollen sporangia (Figure 10a). Cells grown on glucose agar stain evenly. Colonial morphology is variable; colonies may be wrinkled and irregular (Figure 9e), and they are unpigmented and most are smooth and opaque. Minimum growth temperature >5–15 °C, maximum 40–50 °C. Growth occurs at pH 6.0 and 9.5; some strains will grow at pH 4.5. Grows in presence of 10% NaCl. Catalase-positive. Casein, esculin and gelatin are hydrolyzed; starch is not hydrolyzed. Phenylalanine is not deaminated. Citrate is utilized as sole carbon source; propionate is not. Nitrate is not reduced. Voges–Proskauer-positive. Acid without gas is produced from glucose and from a wide range of other carbohydrates.

Source: soil and many other environments, including foods, and clinical and veterinary specimens.

DNA G + C content (mol%): 39.0–45.1 (T_m) for 12 strains, 40.0–46.9 (Bd) for 25 strains, and to be 41.9 (T_m), 40.7 (Bd) for the type strain.

Type strain: NCDO 1766, ATCC 7061, DSM 27, JCM 2508, NCIMB 9369.

EMBL/GenBank accession number (16S rRNA gene): X60637 (NCDO 1766).

Isolates of *Bacillus pumilus* from Antarctic soils and penguin rookeries show some phenotypic distinction from other strains of the species, including the production of a diffusible yellow pigment by some strains on initial culture (Logan and Forsyth, unpublished observations).

74. **Bacillus pycnus** Nakamura, Shida, Takagi and Komagata 2002, 504[VP]

pyc'nus. Gr adj. *pyknos* thick; N.L. adj. *pycnus* thick, referring to thick cells.

Aerobic, Gram-positive, motile rods, forming spherical spores which lie terminally in swollen sporangia. Cells are about 1.0–1.5 by 3.0–5.0 µm. Colonies are translucent, thin, smooth, circular and entire, and reach 1 mm in diameter after 24 h of incubation at 28 °C. Optimum growth temperature 28–33 °C, with minimum 5–10 °C and maximum 40–45 °C. Catalase-positive. Biotin, thiamin and cystine are not required for growth. Casein, starch, Tween 40 and 80, tyrosine and urea are not hydrolyzed. β-Hydroxybutyrate and pyruvate are oxidized; L-alanine, citrate and propionate are not oxidized. Does not grow in the presence of 5% NaCl or 0.001% lysozyme. Nitrate is not reduced to nitrite. No acid or gas produced from glucose and other common carbohydrates. Cell-wall peptidoglycan type is L-Lys-D-Glu. Phenotypically similar to *Bacillus fusiformis*, *Bacillus neidei* and *Bacillus sphaericus*, and separable from these species by growth factor requirements, several substrate oxidation and decomposition tests, and differences in fatty acid compositions. See Table 9.

Source: soil.

DNA G + C content (mol%): 35 (T_m).

Type strain: NRRL NRS-1691, JCM 11075, DSM 15030, LMG 21634.

EMBL/GenBank accession number (16S rRNA gene): AF169531 (NRS-1691).

75. **Bacillus schlegelii** Schenk and Aragno 1981, 215[VP] (Effective publication: Schenk and Aragno 1979, 338.)

schle.gel'i.i. N.L. gen. n. *schlegelii* of Schlegel, named after H. G. Schlegel, a German bacteriologist.

Facultatively chemolithoautotrophic, thermophilic, strictly aerobic, motile, Gram-variable rods 0.6 by 2.5–5 µm, forming terminally located, spherical spores which swell the sporangia. Colonies are cream-colored, circular or spreading. No growth factors are required. Optimum growth temperature about 70 °C; no growth at 37 or 80 °C. Optimum pH for growth 6–7. Grows in presence of 3% but not 5% NaCl. Strictly respiratory, with oxygen as terminal electron acceptor. Nitrate is reduced to nitrite, but nitrate respiration does not occur. Catalase-positive and oxidase weakly positive. Grows chemolithoautotrophically, using H$_2$ as electron donor and CO$_2$ as carbon source, or CO which satisfies both requirements, or chemoorganoheterotrophically. Can also grow autotrophically on thiosulfate (Hudson et al., 1988). Hydrogenase is constitutive and has a temperature optimum between 70 °C and 75 °C. Carbohydrates are not metabolized. Utilizes acetate, butyrate, fumarate, propionate, succinate, phenol, 1-propanol and a small number of amino acids as sole carbon sources. Ammonium ions, asparagine and urea can be utilized as sole nitrogen sources. Casein is weakly hydrolyzed; gelatin, starch and urea are not hydrolyzed. It is unclear from phylogenetic studies as to whether this species still belongs in the genus *Bacillus*. See Tables 8 and 9.

Source: lake sediment, geothermal soils, and sugar factory sludge.

DNA G + C content (mol%): 62.3–65.4 (T_m) on the basis of two studies, and 67.1–67.7 (Bd) in one study; that of the type strain is 64.4 (T_m), 67.7 (Bd).

Type strain: Aragno MA-48, DSM 2000, LMG 7133, ATCC 43741, NCIMB 13107.

EMBL/GenBank accession numbers (16S rRNA gene): Z26934 (DSM 2000) and AB042060 (ATCC 43741);

these 16S rDNA sequences differ considerably from each other, showing only 98.2% similarity.

76. **Bacillus selenitireducens** Switzer Blum, Burns Bindi, Buzzelli, Stolz and Oremland 2001, 29[VP]

se.le.ni.ti.re.du'cens. M.L. masc. gen. n. *selenitis* of selenite; L. part. adj. *reducens* reducing; M.L. part. adj. *selenitireducens* reducing selenite.

Facultatively anaerobic, nonmotile, non-spore-forming, Gram-positive rods 2–6 μm by 0.5 μm, which show weak microaerobic growth and anaerobic respiratory growth with Se(IV) (selenite), As(V) (arsenate), nitrate, nitrite, trimethylamine oxide and fumarate as electron acceptors. Description is based upon a single isolate. Quantitatively reduces Se(IV) to Se (0) (elemental selenium) during growth. Red colonies formed on lactate/selenite/yeast extract-supplemented lakewater medium incubated anaerobically at 20 °C. Grows fermentatively with fructose, glucose or starch. Uses lactate, glucose and pyruvate as electron donors. Moderately halophilic, with salinity optimum of 24–60 g/l NaCl. Moderately alkaliphilic, with optimum growth in the range pH 8.5–10. See Table 6.

Source: arsenic-rich sediment of Mono Lake, California.

DNA G + C content (mol%) of the type strain: 49.0 (T_m).

Type strain: MLS10, ATCC 700615, DSM 15326.

EMBL/GenBank accession number (16S rRNA gene): AF064704 (MLS10).

77. **Bacillus shackletonii** Logan, Lebbe, Verhelst, Goris, Forsyth, Rodríguez-Díaz, Heyndrickx and De Vos 2004b, 375[VP]

sha.ckle.ton'i.i. N.L. adj. *shackletonii* of Shackleton, referring to *R.R.S. Shackleton*, the ship used by the first British scientific expedition to visit Candlemas Island, the vessel being named in honor of the celebrated Antarctic explorer Sir Ernest Shackleton.

Aerobic, Gram-variable, motile, round-ended rods 0.7–0.9 by 2.5–4.5 μm occurring singly. Gram-positive reactions are only seen in cultures of 18 h or less at 30 °C. Endospores are ellipsoidal, lie subterminally and occasionally paracentrally, and usually swell the sporangia. After 2 d on TSA colonies are 2–5 mm in diameter, have a granular appearance and butyrous texture, with opaque, cream-colored centers and translucent irregular margins. Minimum temperature for growth lies between 15 °C and 20 °C, the optimum temperature for growth is 35–40 °C, and the maximum growth temperature is 50–55 °C. Minimum pH for growth lies between 4.5 and 5.0, the optimum pH for growth is 7.0, and the maximum pH for growth lies between 8.5 and 9.0. Catalase-positive. Do not grow readily on casein agar, but when they do grow on it they may hydrolyze the casein. Starch is not hydrolyzed. At 30 °C in the API 20E strip, *o*-nitrophenyl-β-D-galactopyranoside is hydrolyzed slowly, reactions for arginine dihydrolase, lysine decarboxylase, ornithine decarboxylase, citrate utilization, hydrogen sulfide production, urease, tryptophan deaminase, indole production, Voges–Proskauer reaction, gelatin hydrolysis, and nitrate reduction are negative. (In the API 20E strip incubated at 40 °C, citrate may be utilized slowly, gelatin may be hydrolyzed slowly, and the Voges–Proskauer reaction may be positive). In the API 50 CH gallery hydrolysis

of esculin is positive. Acid without gas is produced from the following carbohydrates: amygdalin, cellobiose, D-glucose, *N*-acetylglucosamine and salicin; weak acid reactions were detected for arbutin, D-fructose, galactose, β-gentiobiose, lactose, maltose, D-mannitol, D-mannose, ribose, D-tagatose and D-trehalose. The major cellular fatty acids are $C_{15:0\ anteiso}$, $C_{15:0\ iso}$, $C_{16:0\ iso}$ and $C_{17:0\ anteiso}$ (respectively representing about 35, 31, 6 and 18% of total fatty acids).

Source: unheated volcanic soil taken from the eastern lava flow of Candlemas Island, South Sandwich archipelago.

DNA G + C content (mol%): 35.4 (type strain) to 36.8 (HPLC).

Type strain: Logan SSI024, LMG 18435, CIP 107762.

EMBL/GenBank accession number (16S rRNA gene): AJ250318 (LMG 18435).

78. **Bacillus silvestris** Rheims, Frühling, Schumann, Rohde and Stackebrandt 1999, 800[VP]

sil.ves'tris. L. masc. adj. *Silvestris* of or belonging to a wood or forest, isolated from a forest.

Aerobic, Gram-positive, motile rods 0.5–0.7 by 0.9–2.0 μm, forming spherical spores which lie terminally in swollen sporangia. Description is based upon a single isolate. Colonies are whitish and shiny. Optimum growth temperature 20–30 °C, with minimum of 10 °C and maximum of 40 °C. Catalase-positive, oxidase-negative. Casein, esculin, gelatin, starch and Tween 80 and tyrosine are not hydrolyzed. Citrate and propionate are not utilized as sole carbon sources. Grows in the presence of up to 5% NaCl. Does not grow in the presence of lysozyme. Nitrate is not reduced to nitrite. No acid or gas produced from, and no utilization of, glucose and other common carbohydrates. Cell-wall peptidoglycan contains lysine, glutamic acid and alanine. This cell-wall composition differentiates this species from members of the novel genus *Lysinibacillus* that has been proposed to accommodate *Bacillus fusiformis*, *Bacillus sphaericus*, and the novel species *Lysinibacillus boronitolerans* (Ahmed et al., 2007c). See Table 8.

Source: forest soil.

DNA G + C content (mol%): 39.3 (HPLC).

Type strain: HR3-23, DSM 12223, LMG 18991.

EMBL/GenBank accession number (16S rRNA gene): AJ006086 (HR3-23).

79. **Bacillus simplex** (*ex* Priest, Goodfellow and Todd 1989) Heyrman, Logan, Rodríguez-Díaz, Scheldeman, Lebbe, Swings, Heyndrickx and De Vos 2005a, 129[VP]

sim'plex. L. adj. *simplex* simple.

Rods are straight, 0.7–0.9 μm in diameter, round-ended or occasionally slightly tapered and occur in chains and sometimes singly or in pairs. Motile. Endospores are ellipsoidal, occasionally spherical, lie centrally, paracentrally or subterminally, and do not obviously swell the sporangia. Gram reaction is variable. Colonies on nutrient agar at 30 °C, are 3–6 mm in diameter after 2 d, cream-colored, glossy, with irregular margins, slightly raised and umbonate. Most strains are strictly aerobic, although some strains may grow weakly on nutrient agar in anaerobic conditions. They grow at 20° and 30 °C but are not able to grow at

45 °C. Strains grow well at pH 7 and pH 9; growth at pH 5 is variable. Casein hydrolysis is variable and the medium becomes tinted brown. Starch is hydrolyzed. Tolerance of 5% NaCl (w/v) is variable and no growth occurs with 7% NaCl (w/v). Oxidase-negative, catalase-positive. ONPG, arginine dihydrolase, lysine and ornithine decarboxylase, hydrogen sulfide production, urease, indole and Voges–Proskauer are negative; citrate utilization is negative, but type strain is positive in API Biotype 100 citrate assimilation test. Gelatin hydrolysis variable. Nitrate is reduced to nitrite. Hydrolysis of esculin is variable and weak. Acid without gas is produced weakly, from D-fructose, N-acetylglucosamine, D-glucose, inulin, D-trehalose and sucrose. Acid is produced weakly and variably from salicin. Two biovars may be recognized: strains belonging to *Bacillus simplex* Biovar 1 produce acid weakly and variably from L-arabinose, mannitol, D-raffinose, ribose and sorbitol, while acid production is always negative from D-cellobiose, glycerol, maltose, *meso*-inositol, and D-xylose; strains of *Bacillus simplex* Biovar 2, produce acid weakly from L-arabinose, mannitol, D-raffinose, ribose, sorbitol, and D-xylose, and are variable for weak acid production from D-cellobiose, glycerol, maltose and *meso*-inositol. For the variable characters the type strain shows: weak or moderate acid production for L-arabinose, mannitol, ribose and sorbitol, and no acid from D-raffinose. The major cellular fatty acids are $C_{15:0\ anteiso}$ and $C_{15:0\ iso}$, present at on mean 59.03 (±5.88) and 15.55 (±2.95)% of the total fatty acids, respectively.

Heyrman et al. (2005a) considered that strains of "*Bacillus carotarum*" and its suggested synonyms "*Bacillus capri*," "*Bacillus cobayae*" and "*Bacillus musculi*," and strains of "*Bacillus maroccanus*" and "*Bacillus macroides*" NCIMB 8796 (=NCDO=LMG 18508), should be reclassified as *Bacillus simplex*.

Source: soil.

DNA G + C content (mol%): 39.5–41.8 (T_m).

Type strain: ATCC 49097, DSM 1321, LMG 11160, NRRL-NRS 960, IFO 15720.

EMBL/GenBank accession number (16S rRNA gene): AJ439078 (DSM 1321).

80. **Bacillus siralis** Pettersson, de Silva, Uhlén and Priest 2000, 2186[VP]

si.ra'lis. L. masc. n. *sirus* grain pit, silo; N.L. adj. *siralis* belonging to the silo.

Aerobic, Gram-positive rods 0.5–0.8 by 2.0–3.0 μm, forming ellipsoidal spores which lie subterminally to terminally in swollen sporangia. Colonies on brain heart infusion agar after 24 h are 3–5 mm in diameter, and are circular and entire, light brown to brown in color, with shiny, glistening and granular surfaces; on nutrient agar the colonies are smaller, pale and opaque. Maximum growth temperature 50 °C. Catalase- and oxidase-positive. Grows in presence of 7% NaCl but not 10%. Nitrate is reduced to nitrite but not to dinitrogen; nitrate respiration positive. Casein, esculin and gelatin are hydrolyzed; starch is not. Citrate is not used as sole carbon source. Acid is not produced from glucose and other carbohydrates. Contains characteristic inserts of 49 bases in the distal region of the 16S rRNA genes.

Source: silage.

DNA G + C content (mol%): not reported.

Type strain: 171544, NCIMB 13601, CIP 106295, DSM 13140.

EMBL/GenBank accession number (16S rRNA gene): AF071856 (171544).

81. **Bacillus smithii** Nakamura, Blumenstock and Claus 1988, 70[VP]

smi'thi.i. N.L. gen. n. *smithii* named after Nathan R. Smith, American bacteriologist and *Bacillus* taxonomist.

Facultatively anaerobic, facultatively thermophilic, Gram-positive, motile rods 0.8–1.0 by 5.0–6.0 μm, forming ellipsoidal to cylindrical spores which lie terminally or subterminally in unswollen or slightly swollen sporangia. Colonies are unpigmented, translucent, thin, smooth, circular, entire, and about 2 mm in diameter. Growth temperature range 25–60 or 65 °C. Catalase- and oxidase-positive. No growth in presence of 3% NaCl or 0.001% lysozyme. Nitrate is not reduced to nitrite. DNA and hippurate are hydrolyzed; starch is weakly hydrolyzed; esculin and pullulan hydrolysis is variable; casein, chitin, gelatin, tyrosine and urea are not hydrolyzed. Utilization of citrate and propionate as sole carbon sources is variable. Acid without gas is produced from glucose and a variable range of other carbohydrates.

Source: evaporated milk, canned foods, cheese, and sugar beet juice.

DNA G + C content (mol%): 38.1–40.4 (Bd), 38.7–39.7 (T_m); that of the type strain is 40.2 (Bd).

Type strain: NRRL NRS-173, JCM 9076, LMG 12526, DSM 4216.

EMBL/GenBank accession number (16S rRNA gene): Z26935 (DSM 4216).

82. **Bacillus soli** Heyrman, Vanparys, Logan, Balcaen, Rodríguez-Díaz, Felske and De Vos 2004, 55[VP]

so'li. L. gen. n. *soli* of soil.

Facultatively anaerobic, Gram-positive or Gram-variable, motile, round-ended rods 0.6–1.2 μm in diameter, sometimes curved, occurring singly and in pairs and chains. Ellipsoidal endospores are borne paracentrally, and may swell the sporangia. On TSA, colonies are butyrous, cream-colored, low, slightly umbonate, and have entire margins and glossy or eggshell textured surfaces. The optimum temperature for growth is 30 °C, and the maximum growth temperature lies between 40 °C and 45 °C. The optimum pH for growth is 7.0–8.0, and growth occurs from pH 5.0–4.0 to 9.0–9.5. Hydrolysis of casein is positive. In the API 20E strip, gelatin is hydrolyzed and nitrate reduction is positive; reactions for *o*-nitrophenyl-β-D-galactopyranoside hydrolysis, arginine dihydrolase, lysine decarboxylase, ornithine decarboxylase, citrate utilization, hydrogen sulfide production, urease, tryptophan deaminase, indole production, and Voges–Proskauer are negative. Hydrolysis of esculin positive. Acid without gas is produced from the following carbohydrates in the API 50 CH gallery using the CHB suspension medium: N-acetyl-D-glucosamine, D-fructose, D-glucose, glycogen, maltose (weak), D-mannose, ribose (weak), starch and D-trehalose (weak). Acid production from galactose and sucrose is variable, and weak when positive; type strain is weakly positive for galactose and negative for sucrose. The major cellular fatty acids are $C_{15:0\ iso}$ and $C_{15:0\ anteiso}$, present at a level of about 43 and 34%, respectively.

Source: soil of the Drentse A agricultural research area, The Netherlands.

DNA G + C content (mol%): 40.1–40.4 (type strain, 40.1) (HPLC).

Type strain: LMG 21838, DSM 15604.

EMBL/GenBank accession number (16S rRNA gene): AJ542513 (LMG 21838).

83. **Bacillus sonorensis** Palmisano, Nakamura, Duncan, Istock and Cohan 2001, 1678[VP]

so.no.ren'sis. N.L. adj. *sonorensis* of the Sonoran, named after the Sonoran Desert, where the organism was found.

Facultatively anaerobic, Gram-positive, motile rods, forming ellipsoidal spores which lie subterminally in unswollen sporangia. Cells 1.0 by 2.0–5.0 μm, occurring singly and in pairs and short chains. Colonies are yellowish cream, with mounds and lobes of amorphous slime, and 2–4 mm in diameter after 2 d at 30 °C; colonies on tyrosine agar are brown. Minimum growth temperature about 15 °C and maximum about 55 °C. Growth is inhibited by 5% NaCl and by 0.001% lysozyme. Citrate and propionate are utilized. Catalase-positive. Casein and starch are hydrolyzed. Nitrate is reduced to nitrite. Acid without gas is produced from glucose and other carbohydrates.

Phenotypically similar to *Bacillus licheniformis* and distinguishable from that species mainly by pigment production on tyrosine agar, certain gene sequences, enzyme electrophoresis, and DNA relatedness.

Source: desert soil.

DNA G + C content (mol%): 46.0 (T_m).

Type strain: L87-10, NRRL B-23154, DSM 13779.

EMBL/GenBank accession number (16S rRNA gene): AF302118 (NRRL B-23154).

84. **Bacillus sphaericus** Meyer and Neide in Neide 1904, 337[AL]

sphae'ri.cus. L. adj. *sphaericus* spherical.

Aerobic, Gram-positive, motile rods, forming spherical spores which lie terminally in swollen sporangia (Figure 10f). Cells are about 1.0 by 1.5–5.0 μm. Colonies are opaque, unpigmented, smooth and often glossy, and usually entire. Minimum growth temperature 10–15 °C and maximum 30–45 °C. Grows at pH 7.0–9.5; some strains grow at pH 6.0. Catalase- and oxidase-positive. Biotin and thiamin are required for growth; cystine is not required. Tween 20 is hydrolyzed; casein, gelatin, Tween 80 and urea hydrolysis variable; starch, and tyrosine are not hydrolyzed. Phenylalanine is deaminated. Citrate is utilized as sole carbon source. Grows in the presence of 5% NaCl, but not in 7% NaCl. Nitrate is not reduced to nitrite. No acid or gas produced from glucose and other common carbohydrates. Cell-wall peptidoglycan type is L-Lys-D-Asp. See Table 9.

Ahmed et al. (2007c) proposed the transfer of this species to the new genus *Lysinibacillus*.

Source: soil and water, and a variety of other environments including foods, clinical specimens and mosquitoes.

DNA G + C content (mol%): 37.3 (T_m), 38.3 (Bd) for the type strain.

Type strain: IAM 13420, ATCC 14577, CCM 2120, DSM 28, NCIMB 9370, LMG 7134.

EMBL/GenBank accession number (16S rRNA gene): AJ310084 (DSM 28). Two other 16S rDNA sequences in the EMBL/GenBank database, L14010 (ATCC 14577) and X60639 (NCDO 1767), are of poor quality.

Additional remarks: Nucleic acid studies have shown that *Bacillus sphaericus* is genetically heterogeneous, and have revealed six DNA relatedness groups (Krych et al., 1980; Rippere et al., 1997) and seven 16S rDNA sequence similarity groups (Nakamura et al., 2002); strains of three groups have been allocated to the species *Bacillus fusiformis* (Priest et al., 1988), *Bacillus neidei* and *Bacillus pycnus* (Nakamura et al., 2002). *Bacillus sphaericus* is phenotypically similar to *Bacillus fusiformis*, *Bacillus neidei* and *Bacillus pycnus*, and only separable from these species by growth factor requirements, several substrate oxidation and decomposition tests, and differences in fatty acid compositions. It is this lack of diagnostic characters that has hindered the recognition of the various molecularly defined groups as taxa of species rank.

Strains insecticidal for mosquitoes are found in DNA homology group IIA of Krych et al. (1980) (Rippere et al., 1997), and other taxonomic studies (see Priest, 2002) have confirmed the distinctness of the group. Serotyping (de Barjac et al., 1985) and phage typing (Yousten, 1984) schemes have been developed for group IIA. It must be emphasized that many members of the group are not mosquitocidal. Although strains in this group represent a distinct taxon, the lack of defining phenotypic characters has discouraged the proposal of a new species, and they remain allocated to *Bacillus sphaericus*; however, the name "*B. culicivorans*" has been suggested for the group (Priest, 2002). Recently, this species has been reclassified as *Lysinibacillus sphaericus* Ahmed et al. (2007c).

85. **Bacillus sporothermodurans** Pettersson, Lembke, Hammer, Stackebrandt and Priest 1996, 763[VP]

spo.ro.ther.mo.du'rans. Gr. n. *sporos* seed, spore; Gr. adj. *thermos* warm, hot; L. adj. part. *durans* resisting. N.L. adj. part. *sporothermodurans* with heat-resisting spores.

Aerobic, Gram-positive cells that usually occur as motile, thin rods in chains. Strains require vitamin B_{12} (cyanocobalamin) for satisfactory growth. After 2 d on brain heart infusion (BHI) agar supplemented with 5 mg/l $MnSO_4$ and with 1 mg/l vitamin B_{12}, colonies are 1–2 mm diameter, flat, circular, entire, beige or cream and smooth or glossy in appearance. They bear spherical to ellipsoidal endospores which lie in paracentral and subterminal, sometimes terminal, positions within slightly swollen and unswollen sporangia; the spores of the type strain, though scanty, are ellipsoidal, terminal, and do not swell the sporangia. Sporulation is infrequent but can be enhanced by using BHI-soil extract agar supplemented with vitamin B_{12} and $MnSO_4$. Strains isolated from ultrahigh-temperature (UHT) treated (135–142 °C for several seconds) milk grow poorly and sporulate poorly, but their spores show very high heat resistance and have the ability to survive ultra-heat treatment. This very high heat resistance may decrease upon subculture. Isolates from farm environments may grow more readily than UHT milk isolates but be less heat resistant. Oxidase- and catalase-positive. Casein and starch are not hydrolyzed. Nitrate is reduced to nitrite, the Voges–Proskauer reaction is variable (type strain positive), citrate utilization is variable (type strain negative), hydrogen sulfide and indole are not produced, and the ONPG reaction is negative. Gelatin and esculin are hydrolyzed, urea is not. Growth may occur between 20 °C and 55 °C, with an optimum of about 37 °C. Growth occurs between pH 5 and 9, and NaCl is tolerated up to 5%. Acid without gas is produced from N-acetylglucosamine, D-glucose, D-fructose, maltose, and from sucrose and D-trehalose by most strains, but reactions may be weak. Acid production from the following carbohydrates is variable: amygdalin, arbutin, D-cellobiose, gentiobiose, glycerol, mannitol, D-mannose, D-melezitose, methyl-D-glucoside, salicin, starch, D-tagatose, D-turanose and xylitol (weak). The type strain produces acid without gas from arbutin, D-cellobiose, glycerol, mannitol, D-melezitose, salicin, D-tagatose, D-turanose and xylitol, but not from amygdalin, gentiobiose, D-mannose, methyl-D-glucoside and starch.

Source: UHT-treated milk and dairy farm environments.

DNA G + C content (mol%): 36 (HPLC).

Type strain: M215, DSM 10559, LMG 17894, NCIMB 13600, KCTC 3777.

EMBL/GenBank accession number (16S rRNA gene): U49078 (M215).

86. **Bacillus subterraneus** Kanso, Greene and Patel 2002, 873[VP]

sub.ter.ra′ne.us. L. adj. *subterraneus* underground, subterranean, referring to the isolation source.

Facultatively anaerobic, Gram-negative, non-spore-forming, motile, curved rods, 0.5–0.8 by 2.0–25.0 μm, occurring singly and also in pairs and chains. Description is based upon a single isolate. After 2 d incubation at 40 °C, colonies on nutrient agar are 0.5–1.2 mm in diameter, translucent and convex, with undulating irregular edges, while on tryptic soy agar they are dark yellow to orange, mucoid and rhizoid. Optimum growth temperature 37–40 °C, temperature range for growth about 20–45 °C. pH range for growth 6.5–9.0. Utilizes amorphous iron (III), manganese (IV), nitrate, nitrite and fumarate as electron acceptors in the presence of yeast extract, or certain carbohydrates, ethanol or lactate. Electron acceptors are not required for growth, but growth is better in the presence of nitrate. Yeast extract can be used as sole carbon and energy source. Growth occurs in the presence of up to 9% NaCl. Catalase-positive, oxidase-negative. Esculin, gelatin and starch are hydrolyzed; casein and urea are not hydrolyzed.

Source: deep subterranean waters of the Great Artesian Basin of Australia.

DNA G + C content (mol%): 43 ± 1 (T_m).

Type strain: COOI3B, ATCC BAA-136, DSM 13966.

EMBL/GenBank accession number (16S rRNA gene): AY672638 (COOI3B).

87. **Bacillus thermoamylovorans** Combet-Blanc, Ollivier, Streicher, Patel, Dwivedi, Pot, Presnier and Garcia 1995, 15[VP]

ther.mo.a.my.lo.vo′rans. Gr. adj. *thermos* hot; Gr. n. *amylum* starch; L. v. *vorare* to devour; N.L. adj. *thermoamylovorans* utilizing starch at high temperature.

Facultatively anaerobic, moderately thermophilic, Gram-positive, slightly motile rods, 0.45–0.5 μm by 3.0–4.0 μm. Description based upon a single isolate. Endospores have

not been detected; cells killed by heating at 80 °C for 5 min. Colonies are white and lenticular, and 2–3 mm in diameter after 2 d. Optimum growth temperature about 50 °C; maximum 58 °C. Grows between pH 5.4 and 8.5, with optimum pH 6.5–7.5. Catalase-positive, oxidase-negative. Amylolytic. Nitrate is not reduced to nitrite. Vitamins and nucleic acid derivatives will stimulate growth, but are not essential. Acid without gas is produced from glucose, starch and a range of other carbohydrates; heterolactic fermentation of hexoses yields acetate, formate, lactate and ethanol. See Table 8.

Source: Senegalese palm wine.

DNA G + C content (mol%): 38.8 ± 0.2 mol% (HPLC).

Type strain: CNCM I-1378, strain DKP, LMG 18084.

EMBL/GenBank accession number (16S rRNA gene): L27478 (CNCM I-1378).

88. **Bacillus thermocloacae** Demharter and Hensel 1989a, 495 (Effective publication: Demharter and Hensel 1989b, 274.)

ther.mo.clo′a.cae. Gr. n. *therme* heat; L. n. *cloaca* sewer; N.L. gen. n. *thermocloacae* of a heated sewer.

Aerobic, moderately alkaliphilic and thermophilic, Gram-positive, nonmotile rods, 0.5–0.8 μm by 3.0–8.0 μm. Description is based upon three isolates. Spore formation only detected in one strain; ellipsoidal spores lie subterminally and terminally in swollen sporangia. Colonies are flat to convex, pale, transparent to opaque, and circular with entire or slightly lobed margins, and reach 2–5 mm in diameter after 1–2 d at 60 °C. Optimum growth temperature 55–60 °C; minimum 37 °C and maximum 70 °C. Optimum pH 8–9; no growth at pH 7. Grows in presence of up to 5% NaCl, but growth with 5% NaCl is weak. Catalase- and oxidase-positive. Casein, esculin, gelatin, starch and tributyrin not hydrolyzed. Voges–Proskauer-negative. Nitrate is not reduced to nitrite. No acid or gas are produced from glucose and other carbohydrates. See Tables 6 and 8.

Source: heat-treated sewage sludge.

DNA G + C content (mol%): 42.8–43.7 (T_m), 41.7–42.1 (HPLC).

Type strain: S 6025, DSM 5250.

EMBL/GenBank accession number (16S rRNA gene): Z26939 (DSM 5250).

89. **Bacillus thuringiensis** Berliner 1915, 29[AL]

thur.in.gi.en′sis. N.L. masc. adj. *thuringiensis* of Thuringia, the German province from where the organism was first isolated.

Facultatively anaerobic, Gram-positive, usually motile rods 1.0–1.2 by 3.0–5.0 μm, occurring singly and in pairs and chains, and forming ellipsoidal, sometimes cylindrical, subterminal, sometimes paracentral, spores which do not swell the sporangia; spores may lie obliquely in the sporangia. Sporangia carry parasporal bodies adjacent to the spores; these crystalline protein inclusions (Figure 10h) may be bipyramidal, cuboid, spherical to ovoid, flat-rectangular, or heteromorphic in shape. They are formed outside the exosporium and readily separate from the liberated spore. They are known as δ-endotoxins or insecticidal crystal proteins, and are protoxins which may be toxic for certain insects and other invertebrates including flatworms, mites, nematodes and protozoa. The ability to synthesize parasporal bodies is plasmid borne, has been transferred to strains of *Bacillus cereus* and even to *Bacillus pumilus* (Selinger et al., 1998), and may be lost on subculture. Cells grown on glucose agar produce large amounts of storage material, giving a vacuolate or foamy appearance. Like those of *Bacillus cereus*, colonies are very variable in appearance, but nevertheless distinctive and readily recognized (Figure 9h): they are usually whitish to cream in color, large (2–7 mm in diameter), and vary in shape from circular to irregular, with entire to undulate, crenate or fimbriate edges; they usually have matt or granular textures, but smooth and moist colonies are not uncommon. Minimum temperature for growth is 10–15 °C, and the maximum 40–45 °C. Egg yolk reaction is positive. Catalase-positive, oxidase-negative. Casein, gelatin and starch are hydrolyzed. Voges–Proskauer-positive. Citrate is utilized as sole carbon source. Nitrate is reduced. Tyrosine is decomposed. Phenylalanine is not deaminated. Resistant to 0.001% lysozyme. Acid without gas is produced from glucose and a limited range of other carbohydrates. Some strains can produce diarrheal enterotoxin.

Phenotypically similar to other members of the *Bacillus cereus* group: *Bacillus anthracis*, *Bacillus cereus*, *Bacillus mycoides*, *Bacillus pseudomycoides* and *Bacillus weihenstephanensis*. *Bacillus thuringiensis* is distinguished by its characteristic parasporal crystals. Smith et al. (1952) and Gordon et al. (1973) considered *Bacillus thuringiensis* to be a variety of *Bacillus cereus*. For distinguishing characters within the *Bacillus cereus* group, see the individual species descriptions and Table 7.

Endospores are very widespread in soil and many other environments, and this organism has been isolated from all continents, including Antarctica. Although numerous strains are toxic to invertebrates, this property has not been demonstrated in many other strains. Natural epizootics do not seem to occur, and it has been suggested that the natural habitat of this organism is soil.

DNA G + C content (mol%): 33.5–40.1 (T_m) for two strains; 35.7–36.7 (Bd) for four strains, and 33.8 (T_m), 34.3 (Bd) for the type strain.

Type strain: IAM 12077, ATCC 10792, NRRL NRS-996, DSM 2046, LMG 7138, NCIMB 9134.

EMBL/GenBank accession number (16S rRNA gene): D16281 (IAM 12077).

Additional remarks: Bacillus thuringiensis has been divided on the basis of flagellar (H) antigens into 69 serotypes with 13 subantigenic groups, giving a total of 82 serovars (Lecadet et al., 1999); see *also Antigenic Structure*, above), but there is little correlation between serotype and insecticidal toxicity, the latter being mainly encoded by plasmids. Ribotyping data have shown good correlation with serotypes for 10 well known serovars (Priest et al., 1994); other approaches to subspecies analysis of *Bacillus thuriengiensis* are discussed by Lecadet et al. (1999).

90. **Bacillus tusciae** Bonjour and Aragno 1985, 223[VP] (Effective publication: Bonjour and Aragno 1984, 400.)

tus'cia.e. L. gen. n. *tusciae* from Tuscia, the Roman name for the region of central Italy where the organism was found.

Facultatively chemolithoautotrophic, moderately thermophilic, strictly aerobic, motile (by one lateral flagellum), Gram-positive rods 0.8 by 4–5 μm, forming subterminal,

ellipsoidal spores which swell the sporangia. Description is based upon two isolates. The spreading colonies are creamy-white and chalky. Heavy autotrophic cultures form a yellow, water-soluble pigment. No growth factors are required. Strictly respiratory, with oxygen as terminal electron acceptor. Nitrate is reduced to nitrite, but nitrate respiration does not occur. Grows chemolithoautotrophically, using H_2 as electron donor and CO_2 as carbon source, or chemoorganoheterotrophically. Optimum growth temperature about 55 °C; no growth at 35 or 65 °C. Optimum pH for growth 4.2–4.8; weak growth at pH 3.5 and 6.0. No growth in presence of 1% NaCl. Catalase weakly positive and oxidase-positive. Carbohydrates are not metabolized. Starch is not hydrolyzed. Utilizes a few alcohols, amino acids and organic acids as sole carbon sources, with ammonium as the nitrogen source. Ammonium ions, asparagine and urea can be utilized as sole nitrogen sources. See Table 8.

Source: an acidic pond in a solfatara in Italy.

DNA G + C content (mol%): 57–58 (T_m), and for the type strain 57.5 (T_m).

Type strain: Aragno T2, DSM 2912, LMG 17940, IFO 15312.

EMBL/GenBank accession number (16S rRNA gene): AB042062 (IFO 15312).

91. **Bacillus vallismortis** Roberts, Nakamura and Cohan 1996, 474[VP]

val.lis.mor'tis. L. n. *vallis* valley; L. fem. n. *mors* death; N.L. gen. fem. n. *vallismortis* of Death Valley.

Aerobic, Gram-positive, motile rods, forming ellipsoidal spores which lie centrally or paracentrally in unswollen sporangia. Cells 0.8–1.0 by 2.0–4.0 μm, occurring singly and in short chains. Colonies are opaque, smooth, circular and entire and 1.0–2.0 mm in diameter after 2 d at 28 °C. Optimum growth temperature 28–30 °C, with minimum of 5–10 °C and maximum of about 50 °C. Catalase-positive, oxidase-positive. Citrate is utilized as a sole carbon source; propionate is not. Casein and starch are hydrolyzed. Tween 80 is decomposed weakly, phenylalanine and tyrosine are not decomposed. Nitrate is reduced to nitrite. Acid without gas is produced from glucose and a range of other carbohydrates.

Indistinguishable from *Bacillus mojavensis*, *Bacillus subtilis* subsp. *subtilis* and *Bacillus subtilis* subsp. *spizizenii* by conventional phenotypic tests, and distinguished from those organisms principally by DNA relatedness, by data from restriction digestion analyses of certain genes, and by fatty acid analysis. Phenotypically distinguishable from *Bacillus atrophaeus* only by failure to produce dark brown pigmented colonies on media containing tyrosine or other organic nitrogen source. See Table 5.

Source: desert soil.

DNA G + C content (mol%): 43.0 (T_m).

Type strain: DV1-F-3, NRRL B-14890, DSM 11031, LMG 18725, KCTC 3707.

EMBL/GenBank accession number (16S rRNA gene): AB021198 (DSM 11031).

92. **Bacillus vedderi** Agnew, Koval and Jarrell 1996, 362 (Effective publication: Agnew, Koval and Jarrell 1995, 229.)

ved'der.i. M.L. gen. n. *vedderi* of Vedder, named after A. Vedder, the Dutch microbiologist who described *Bacillus alcalophilus* in 1934.

Alkaliphilic, facultatively anaerobic, Gram-positive, motile, narrow rods forming ellipsoidal to spherical spores which lie terminally in swollen sporangia. Description is based upon a single isolate. Colonies are white, flat and circular, and 1.5 mm in diameter after 2 d growing on alkaline oxalate medium at 37 °C. Optimum growth temperature 40 °C; maximum 45–50 °C. Optimal growth at pH 10.0; pH range 8.9–10.5. Grows in presence of 7.5% NaCl, but not 10% NaCl. Growth stimulated by presence of vitamins (in yeast extract). Catalase- and oxidase-positive. Pectin and birchwood xylan are hydrolyzed; gelatin and carboxymethylcellulose are weakly hydrolyzed; casein, starch and oakwood xylan are not hydrolyzed. Glucose and a small range of other carbohydrates can be utilized as sole sources of carbon. See Table 6.

Source: red mud bauxite-processing waste, using alkaline oxalate enrichment.

DNA G + C content (mol%): 38.3 (T_m).

Type strain: JaH, DSM 9768, ATCC 7000130, LMG 17954, NCIM B 13458.

EMBL/GenBank accession number (16S rRNA gene): Z48306 (JaH).

93. **Bacillus vietnamensis** Noguchi, Uchino, Shida, Takano, Nakamura and Komagata 2004, 2119[VP]

vi.et.nam.en'sis. N.L. adj. *vietnamensis* referring to Vietnam, the country where the type strain was isolated.

Cells are rod-shaped, measuring 0.5–1.0 by 2.0–3.0 μm, Gram-positive and aerobic. They are motile with peritrichous flagella. Ellipsoidal spores develop centrally in the cells and sporangia are not swollen. Catalase and oxidase are produced. Nitrate reduction, indole production, arginine dihydrolase and urease are negative. Growth occurs in the presence of lysozyme. Casein, starch, DNA, esculin, gelatin, *p*-nitrophenyl β-D-galactopyranoside and tyrosine are hydrolyzed. Production of hydrogen sulfide is not detected on trypticase soy agar. Acid is produced from glycerol, D-ribose, D-glucose, D-fructose, mannitol, *N*-acetyl D-glucosamine, maltose, sucrose, trehalose, inulin, starch and glycogen; no acid is produced from erythritol, D-arabinose, L-arabinose, D-xylose, L-xylose, adonitol, methyl β-D-xyloside, galactose, D-mannose (NRRL B-14850 produces acid from this sugar), L-sorbose, rhamnose, dulcitol, inositol, sorbitol, methyl α-D-mannoside, methyl β-D-glucoside, amygdalin, arbutin, salicin, cellobiose, lactose, melibiose, melezitose, D-raffinose, xylitol, β-gentiobiose, D-turanose, D-lyxose, D-tagatose, D-fucose, L-fucose, D-arabitol, L-arabitol, D-gluconate, 2-ketogluconate and 5-ketogluconate. Assimilation is positive for glucose, D-mannitol, *N*-acetyl D-glucosamine, maltose, gluconate and DL-malic acid, and negative for L-arabinose, D-mannose, n-capric acid, citrate and adipic acid. Growth occurs at 0–15% (w/v) NaCl (optimum at 1%). The isolates are regarded as moderately halotolerant bacteria. Growth occurs at 10–40 °C (optimum at 30–40 °C) (16–3 and NRRL B-14850 grow at 50 °C). Growth occurs at pH 6.5–10.0 but not at pH 6.0. DNA G + C content is 43–44 mol% (HPLC). The major fatty acid is $C_{15:0}$ anteiso (48.3 ± 11.9%), with lesser $C_{15:0}$ iso (16.2 ± 4.4%). The major quinone is MK-7. *meso*-Diaminopimelic acid is found in the cell walls. Strains have been isolated from Vietnamese fish sauce and from the Gulf of Mexico. Major cellular fatty acids are $C_{15:0\ anteiso}$ (51.4%) and $C_{15:0\ iso}$ (19.8%).

Source: Vietnamese fish sauce.
DNA G + C content (mol%): 43 (HPLC).
Type strain 15-1, JCM 11124, NRIC 0531, NRRL 23890.
GenBank/EMBL accession number (16S rRNA gene): AB099708.

94. **Bacillus vireti** Heyrman, Vanparys, Logan, Balcaen, Rodríguez-Díaz, Felske and De Vos 2004, 54[VP]

vi.re'ti. L. gen. n. *vireti* of a field.

Facultatively anaerobic, Gram-negative, motile, slightly curved, round-ended rods, 0.6–0.9 μm in diameter, occurring singly and in pairs. Do not produce endospores on TSA supplemented with 5 mg/l $MnSO_4$, but sporulate on *Bacillus fumarioli* agar at pH 7 after 48 h. Endospores are ellipsoidal, lie in central, paracentral, or sometimes subterminal positions, and may swell the sporangia slightly; the ends of the sporangia may be slightly tapered. After 3 d of growth on TSA, colonies are dark cream-colored, circular, raised and up to 4 mm in diameter, with entire edges. Colonies have loose biomass and egg-shell textured surfaces. The optimum temperature for growth is 30 °C, and the maximum growth temperature lies between 40 °C and 45 °C. Growth occurs from pH 4.0–5.0 to 7.0–7.5; the optimum lies at the upper end of this range. Casein is hydrolyzed. In the API 20E strip, gelatin is hydrolyzed and nitrate reduction is positive; *o*-nitrophenyl-β-D-galactopyranoside hydrolysis is variable, reactions for arginine dihydrolase, lysine decarboxylase, ornithine decarboxylase, citrate utilization, hydrogen sulfide production, urease, tryptophan deaminase, indole production, and Voges–Proskauer are negative. Hydrolysis of esculin positive. Acid without gas is produced from the following carbohydrates in the API 50 CH gallery using the CHB suspension medium: *N*-acetyl-D-glucosamine, D-fructose, L-fucose (weak), galactose (weak), D-glucose, glycogen, maltose, D-mannitol, D-mannose, methyl α-D-glucoside (weak), ribose (weak), starch, sucrose and D-trehalose. The following reactions are variable between strains and, when positive, are usually weak: gluconate, *meso*-inositol, methyl α-D-mannoside, rhamnose; type strain is weak for gluconate and methyl α-D-mannoside and negative for *meso*-inositol and rhamnose. The major cellular fatty acids are $C_{15:0\ iso}$ and $C_{15:0\ anteiso}$, present at a level of about 47 and 34%, respectively.

Source: soil of Drentse A agricultural research area, The Netherlands.
DNA G + C content (mol%): 39.8–40.3 (type strain, 40.2) (HPLC).
Type strain: LMG 21834, DSM 15602.
EMBL/GenBank accession number (16S rRNA gene): AJ542509 (LMG 21834).

95. **Bacillus weihenstephanensis** Lechner, Mayr, Francis, Prüss, Kaplan, Wiessner-Gunkel, Stewart and Scherer 1998, 1380[VP]

we'ihen.ste'phan.en.sis. N.L. masc. adj. *weihenstephanensis* referring to Freising-Weihenstephan in Southern Germany, where the type strain was isolated.

Phenotypically similar to *Bacillus cereus* and distinguished from it by ability to grow at 7 °C, inability to grow at 43 °C, and by certain 16S rDNA signature sequences. Distinguished from *Bacillus anthracis*, *Bacillus mycoides*, B, *pseudomycoides* and *Bacillus thuringiensis* by the same characters that differentiate those species from *Bacillus cereus*. For distinguishing characters within the *Bacillus cereus* group, see the individual species descriptions and Table 7.

Source: pasteurized milk.

DNA G + C content (mol%): not reported, but can be expected to lie within the range reported for *Bacillus cereus*.

Type strain: DSM 11821, WSCB 10204, LMG 18989.

EMBL/GenBank accession number (16S rRNA gene): AB021199 (DSM 11821).

Additional remarks: Although pychrotolerance is an important distinguishing character of *Bacillus weihenstephanensis*, it must be appreciated that not all psychrotolerant organisms resembling *Bacillus cereus* are *Bacillus weihenstephanensis* (Stenfors and Granum, 2001), and the practical value of recognizing this close relative of *Bacillus mycoides* may be questioned.

Species *Incertae Sedis*

Claus and Berkeley (1986) listed 26 *species incertae sedis* in the First Edition of this *Manual*. 12 of these have been revived since then and, in some cases, transferred to other genera; either at the times of their revivals or later. Details are given in the species listings of the appropriate genera as follows: *Bacillus amyloliquefaciens* (Priest et al., 1987); *Bacillus flexus* (Priest et al., 1989); *Bacillus laevolacticus* (Andersch et al., 1994); *Bacillus psychrosaccharolyticus* (Priest et al., 1989); *Aneurinibacillus aneurinilyticus* (Heyndrickx et al., 1997; Shida et al., 1996); *Anoxybacillus flavithermus* (Pikuta et al., 2000a); *Geobacillus thermocatenulatus* (Golovacheva et al., 1975; Nazina et al., 2001); *Geobacillus thermodenitrificans* (Manachini et al., 2000; Nazina et al., 2001); *Paenibacillus agarexedens* (Uetanabaro et al., 2003); *Paenibacillus apiarius* (Nakamura, 1996); *Paenibacillus larvae* subsp. *pulvifaciens* (Heyndrickx et al., 1996c; Nakamura, 1984c); *Paenibacillus thiaminolyticus* (Nakamura, 1990; Shida et al., 1997). Many other names that have been proposed in the past for *Bacillus* species were discussed by Smith et al. (1952) and Gordon et al. (1973), and many were considered by these authors to be synonyms of established species. For comprehensive listings of such names the reader is referred to *Index Bergeyana* (Buchanan et al., 1966; Gibbons et al., 1981).

White et al. (1993) proposed the merger of the two species "*Bacillus caldotenax*" and "*Bacillus caldovelox*" (both Heinen and Heinen, 1972) within a revived *Bacillus caldotenax*, but found "*Bacillus caldolyticus*" (Heinen and Heinen, 1972) to show low homology with these two species. Sunna et al. (1997b) identified "*Bacillus caldolyticus*", "*Bacillus caldotenax*" and "*Bacillus caldovelox*," as members of *Bacillus thermoleovorans* on the basis of DNA homology studies, but this proposal has not been validated. In any case these species belong in the genus *Geobacillus* (see the chapter on *Geobacillus*). Polyphasic studies of "*Bacillus longisporus*," "*Bacillus nitritollens*" and "*Bacillus similibadius*" (all Delaporte, 1972) have not revealed homogeneous groupings among strains inherited from Delaporte (Logan and De Vos, unpublished). Heyrman et al. (2005a) found "*Bacillus carotarum*" and "*Bacillus maroccanus*" to be synonyms of *Bacillus simplex*.

Of the species remaining, the following may be considered for revival, and some may warrant transfer to other genera, after more detailed studies have been performed or after additional strains have been obtained and examined:

a. "*Bacillus agrestis*" Werner 1933, 468

The species was accepted as a synonym of *Bacillus megaterium* by Gordon et al. (1973) but differs from typical members of that species in that its cells are smaller (mean diameter <1.0 μm), poly-β -hydroxybutyrate is not formed, esculin is not hydrolyzed, and phenylalanine is not deaminated. This strain is not closely related to *Bacillus megaterium sensu stricto* (Hunger and Claus, 1981).

DNA G + C content (mol%): 37.4 (T_m).

Representative strain: NRS 602 (DSM 1316).

b. "*Bacillus aminovorans*" den Dooren de Jong 1926, 157

Strictly aerobic, motile rods, 0.8–1.5 × 1.5–5.0 μm, Gram-positive in young cultures. Spherical endospores are borne centrally and paracentrally, and do not swell the sporangia. Maximum growth temperature 37 °C. Feeble growth on nutrient agar, and better growth on trypticase peptone medium. Utilizes mono-, di- and trimethylamine and glucose. Fructose and maltose used by some strains. No growth with other carbohydrates, organic acids, or amino acids, except gluconate, glutamate, 3-hydroxybenzoate and citrate. Some strains utilize betaine. Farrow et al. (1994) did not find a specific relationship between this organism and the spherical spore-formers of Group 2 of Ash et al. (1991).

Source: Soil.

DNA G + C content (mol%): 40.4–41.8 (T_m) for about 20 strains.

Original strain: ATCC 7046 (DSM 1314).

c. "*Bacillus freudenreichii*" (Miquel) Chester 1898, 110

This species was described by Claus and Berkeley (1986) as very similar to *Bacillus* (now *Brevibacillus*) *brevis* in nutritional requirements (Bornside and Kallio, 1956), morphology and physiology, but differing in that it produces a considerable titratable alkalinity in urea broth and is less tolerant of acid. Additionally, growth occurs in 5% NaCl broth and phenylalanine is deaminated.

Source: Soil, river water and sewage.

Representative strain: ATCC 7053.

d. "*Bacillus macroides*" Bennett and Canale-Parola 1965, 204

Originally described as "*Lineola longa*" (Pringsheim, 1950). This is organism is unreactive in routine phenotypic tests, and Claus and Berkeley (1986) considered that its characters conform to those of *Bacillus sphaericus* (with the one exception that its spore is frankly oval and scarcely distends the sporangium) and to *Bacillus badius*. Minimal nutritional requirements are a carbon energy source, NH_4-N, thiamine, biotin and, in one strain, guanine. Carbon sources include various amino acids and C2–C5 *n*-fatty acids, but not sugars. Proteolytic action is not detected within 3 weeks. By comparing the sequences of the 16S–23S internal transcribed spacer region of representatives of 27 *Bacillus* species and 19 strains representing five other endospore-forming genera, Xu and Cote (2003) found that "*Bacillus macroides*" strain ATCC 12905 (=LMG 18508=NCDO 1661) showed phylogenetic relationships with *Bacillus fusiformis* and *Bacillus sphaericus* (now both reclassified in *Lysinibacillus*; Ahmed et al., 2007c). Heyrman et al. (2005a) reclassified "*Bacillus macroides*" strain NCIMB 8796 (=DSM 54=LMG 18474) as a strain of *Bacillus simplex*, and on the basis of subsequent work Heyrman et al. (unpublished) propose the revival of "*Bacillus macroides*" as *Lysinibacillus macroides*, for the single strain ATCC 12905.

Source: Cow dung, plant material decaying in water.

DNA G + C content (mol%): 37.6–38.9 (T_m).

Representative strain: ATCC 12905, LMG 18474, DSM 54.

e. "*Bacillus pacificus*" Delaporte 1967, 3071

Cells oval, exceptionally large, measuring 1.5–2.1 × 2.7–3.4 μm, with capsules and lipid inclusions. Motile, with one or two flagella inserted at or near one pole, or both poles. Spores are ellipsoidal and 1.3–1.5 × 2.7–3.4 μm in size. Best medium reported is 0.1% tryptone in sea water; no growth occurs on ordinary nutrient agar. Growth good at 28–40 °C but none at 4 °C. Glucose broth reaches pH 6 in 10 d; no acetoin is formed. Gelatin is slowly liquefied. Nitrate is reduced to nitrite. Catalase-positive. Grows in 10% NaCl.

Source: Shore sand, Pacific ocean, California.

Original strain: ATCC 25098 NCIMB 1862.

f. "*Bacillus xerothermodurans*" Bond and Favero 1977, 159

Strictly aerobic rods, 0.7–1.2 by 1.6–2.8 μm, which are pleomorphic, especially in older cultures. Spores are spherical to oval, and swell the sporangia. Scanning electron microscopy of spores revealed a surface honeycomb pattern of polygonal depressions surrounded by straight ridges. Unusual ultrastructure with an irregular, thick outer spore coat composed of globular subunits and laminated inner spore coat containing up to nine distinct layers. Catalase-positive. Growth occurs in 10% NaCl broth. Potato starch is hydrolyzed. Reactions in other properties studied are negative. Cleaned spore preparations show extreme resistance to dry heat, the strain was isolated after heating dry samples at 125 °C for 48 h.

Source: Sandy soil, Cape Kennedy, Florida.

Original strain: ATCC 27380, DSM 520.

Misclassified species

16S rRNA gene sequence comparison studies indicate that the following species are currently misclassified within *Bacillus*. They await formal proposals of transfer to other genera:

To transfer to *Geobacillus*

Bacillus thermantarcticus (nom. corrig. *Bacillus thermoantarcticus* [sic]) Nicolaus, Lama, Esposito, Manca, di Prisco and Gambacorta (1996). corrig. *Bacillus thermantarcticus* Nicolaus, Lama, Esposito, Manca, di Prisco and Gambacorta 2002, 3[VP] therm'ant.arct'ic.us. Gr. n. *therme* heat, N.L. adj. *antarcticus* from Antarctica, from Antarctic geothermal soil.

Aerobic, Gram-positive, motile rods, 0.6–2.0 μm wide and 3.0–5.0 μm long, with oval endospores which are borne terminally. Description is based on a single isolate. Colonies are opaque, flat and circular with entire margins. In stationary phase of growth an exopolysaccharide is produced. Catalase-negative, oxidase-positive. Temperature range for growth is 37–65 °C, and optimal growth occurs at 63 °C. Growth occurs in the pH range 5.5–9.0, and optimum is pH 6.0. Growth is weak in the presence of 2% NaCl but inhibited by 5% NaCl. Growth occurs on yeast extract. Glucose, trehalose and xylose can be utilized as sole carbon sources. Citrate and propionate are not utilized. Nitrate is not reduced. Gelatin and starch are hydrolyzed, but casein is not hydrolyzed. Hippurate and tyrosine are not degraded. Exo- and endocellular α-glucosidases, an intracellular alcohol dehydrogenase and an exocellular xylanase are produced. The major fatty acids at 60 °C are $C_{17:0\ anteiso}$ (36% of total), $C_{17:0\ iso}$ (27%), $C_{15:0\ iso}$ (15%) and $C_{16:0\ iso}$ (13%) (Nicolaus et al., 1995).

DNA G + C content (mol%): 53.7.

Type strain: DSM 9572, strain M1.

EMBL/GenBank accession number (16S rRNA gene): not available.

To transfer to *Paenibacillus*

Bacillus edaphicus Shelobolina et al. 1998, 631[VP] (Effective publication: Shelobolina et al. 1997, 688.)

e.daph'ic.us Gr. n. *edaphos* ground; L. masc. suff. *-icus* adjectival suffix used with the sense of belonging to; N.L. adj. *edaphicus* living in soil.

Strictly aerobic, chemo-organotrophic, nonmotile regular rods 1–1.5 by 4–10 μm in size. Produce mucous, smooth, transparent, convex colonies with even edges, 0.8–2.0 cm in diameter, on synthetic media with carbohydrates but devoid of nitrogen sources. Cells grown on such media are surrounded by mucous capsules 7–12 μm thick. Colonies are smooth, moist, light and flat, but raised in the center, and 0.5 cm in diameter, and with even edges when grown on potato agar and on synthetic media with carbohydrates and ammonium nitrogen. No growth occurs on nutrient agar or in nutrient broth. On media containing ammonium nitrogen, ellipsoidal endospores (1–1.2 × 1.7–2.0 μm) with eight to ten longitudinal ridge-like protusions are formed. Cell-wall structure is of Gram-positive type. Catalase-positive. Glucose is not fermented, and nitrate is not reduced. Starch and tyrosine are hydrolyzed but gelatin is not liquefied. Sugars and polyols are used as carbon and energy sources, and acid is produced from a range of carbohydrates. Cells are resistant to lysozyme. Palmitic, anteisopentadecanoic and stearic acids are the most frequent in the cellular lipids. Temperature range for growth is 7–45 °C.

Source: soil.

DNA G + C content (mol%): 54.6–56.5 (type strain, 56.4) (T_m).

Type strain: DSM 12974, VKPM B-7517, strain T7.

EMBL/GenBank accession number (16S rRNA gene): AF006076.

Bacillus mucilaginosus Avakyan et al. 1998, 631[VP] (Effective publication: Avakyan et al. 1986, 480; emend. Shelobolina et al. 1998, 631[VP]; effective publication: Shelobolina et al. 1997, 688.)

mu.ci.la.gi.no'sus. L. masc. adj. *mucilaginosus* slimy.

Strictly aerobic, chemo-organotrophic, nonmotile, regular, round-ended rods, borne singly and 1–1.2 by 4–7 μm in size; they are surrounded by capsules. On potato agar colonies are light gray, smooth, even-edged, wet and shiny; they do not exceed 0.5 cm in diameter. On Ashby sucrose agar and on synthetic media with carbohydrates and ammonium nitrogen, colonies are convex, semitransparent, mucous, even-edged, of viscous consistency, and 0.5–1 cm in diameter. No growth occurs on nutrient agar or nutrient gelatin, or in nutrient broth. On media containing ammonium nitrogen, and on potato agar, oval endospores (1–1.2 × 1.7–2.0 μm) with nine longitudinal ridge-like protusions are formed. Spores are borne centrally and subterminally to give a fusiform appearance to sporangia. Sporulation does not occur on media lacking a nitrogen source. Cell-wall structure is of Gram-positive type, but Gram staining may yield varying results. Catalase-positive. Glucose is not fermented, and nitrate is not reduced. Starch is hydrolyzed, but tyrosine is not, and gelatin is not liquefied. Carbohydrates, polyols, and some organic acids are used as carbon and energy sources, and acid is produced from a range of carbohydrates. Cells are resistant to lysozyme. Anteisopentadecanoic, palmitic and stearic acids are the most frequent in the cellular lipids. Temperature range for growth is 10–45 °C.

Source: soil.

DNA G + C content (mol%): 55.8 (T_m).

Type strain: VKM B-1480D, VKPM B-7519.

EMBL/GenBank accession number (16S rRNA gene): AF006077.

Note added in proof

Between the completion of the manuscript and tables (in late 2004) and the time of going to press (late 2007), the following new *Bacillus* species names were validly published, and with one exception (on account of a short sequence for *Bacillus tequilensis*) they have been included in Figure 8:

Bacillus acideceler Peak et al. 2007, 2035^{VP}

a.ci.de'ce.ler. N.L. neut. n. *acidum*, acid, L. masc. adj. *celer*, fast, N.L. masc. adj. *acidiceler*, fast-growing in acid.

Gram-positive, endospore-forming rod, isolated from a forensic specimen considered a credible threat of harboring anthrax.

DNA G + C content (mol%): 37.3.

Type strain: strain CBD 119, DSM 18954 and NRRL B-41736.

GenBank/EMBL/DDBJ accession number (16S rRNA gene): DQ374637.

Bacillus acidicola Richard, Archambault, Rosselló-Mora and Tindall 2005, 2129^{VP}

a.ci.di'co.la. N.L. n. *acidum* an acid; L. suff. *-cola* an inhabitant of a place, a resident; N.L. masc. n. *acidicola* an inhabitant of acidic environments.

Cells occur singly or in chains, and in liquid culture can form filamentous rods that are 1.0–1.3 μm wide.

Source: acidic *Sphagnum* peat bog.

DNA G + C content (mol%): 42.3.2 (HPLC).

Type strain: 10, DSM 14745, ATCC BAA-366, NRRL B-23453.

GenBank/EMBL/DDBJ accession number (16S rRNA gene): AF547209 (10).

Bacillus aerius Shivaji, Chaturvedi, Suresh, Reddy, Dutt, Wainwright, Narlikar and Bhargava 2006, 1471^{VP}

ae'ri.us. L. masc. adj. *aerius* pertaining to the air, aerial.

Shows high 16S rRNA gene sequence similarity with *Bacillus aerophilus, Bacillus licheniformis, Bacillus sonorensis* and *Bacillus stratosphericus.*

Source: air sample collected at high altitude.

DNA G + C content (mol%): 45 T_m.

Type strain 24K, MTCC 7303, JCM 13348.

GenBank/EMBL/DDBJ accession number (16S rRNA gene): AJ831843 (24K).

Bacillus aerophilus Shivaji, Chaturvedi, Suresh, Reddy, Dutt, Wainwright, Narlikar and Bhargava 2006, 1471^{VP}

ae.ro.phi'lus. Gr. n. *aêr* air; Gr. adj. *philos* loving; N.L. masc. adj. *aerophilus* air-loving.

Shows high 16S rRNA gene sequence similarity with *Bacillus aerius, Bacillus licheniformis, Bacillus sonorensis* and *Bacillus stratosphericus.*

Source: air sample collected at high altitude.

DNA G + C content (mol%): 44 T_m.

Type strain 28K, MTCC 7304, JCM 13347.

GenBank/EMBL/DDBJ accession number (16S rRNA gene): AJ831844 (28K).

Bacillus akibai Nogi, Takami and Horikoshi 2005, 2314^{VP}

a.ki.ba'i. N.L. gen. n. *akibai* of Akiba, named after the Japanese microbiologist Teruhiko Akiba, who made fundamental contributions to the study of alkaliphilic bacteria.

Related to *Bacillus krulwichiae.*

Source: preparation of carboxymethyl cellulase.

DNA G + C content (mol%): 34.4 (HPLC).

Type strain 1139, JCM 9157, ATCC 43226.

GenBank/EMBL/DDBJ accession number (16S rRNA gene): AB043858 (1139).

Bacillus altitudinis Shivaji, Chaturvedi, Suresh, Reddy, Dutt, Wainwright, Narlikar and Bhargava 2006, 1472^{VP}

al.ti'tu.di.nis. L. fem. n. *altitudo* altitude; L. fem. gen. n. *altitudinis* of altitude.

Shows high 16S rRNA gene sequence similarity with *Bacillus pumilus.*

Source: air sample collected at high altitude.

DNA G + C content (mol%): 43 (T_m).

Type strain 41KF2b, MTCC 7306, JCM 13350.

GenBank/EMBL/DDBJ accession number (16S rRNA gene): AJ831842 (41KF2b).

Bacillus alveayuensis Bae, Lee and Kim 2005, 1214^{VP}

al.ve.a.yu.en'sis. L. n. *alveus* trough; N.L. masc. adj. *ayuensis* pertaining to Ayu (as a locality); N.L. masc. adj. *alveayuensis* pertaining to the Ayu Trough in the Pacific Ocean.

Thermophile, growing at up to 65 °C and growing optimally at 3% NaCl but inhibited by 5% NaCl.

Source: deep-sea sediment.

DNA G + C content (mol%): 38.7 (T_m).

The type strain is TM1, KCTC 10634, JCM 12523.

GenBank/EMBL/DDBJ accession number (16S rRNA gene): AY605232 (TM1).

Bacillus arenosi Heyrman, Rodríguez-Díaz, Devos, Felske, Logan and De Vos 2005b, 115^{VP}

ar.en.o'si. L. gen. n. *arenosi* of a sandy place.

Closely related to *Bacillus arvi.* Spherical endospores are borne terminally and swell the sporangia slightly; largely unreactive in routine biochemical tests.

Source: soil.

DNA G + C content (mol%): 35 (HPLC).

Type strain LMG 22166, DSM 16319.

GenBank/EMBL/DDBJ accession number (16S rRNA gene): AJ627212 (LMG 22166).

Comment: this species has recently been reclassified as *Viridibacillus arenosi* (Albert et al., 2007).

Bacillus arsenicus Shivaji, Suresh, Chaturvedi, Dube and Sengupta 2005, 1126^{VP}

ar.sen.i'cus. N.L. masc. adj. *arsenicus* pertaining to arsenic.

Grows in the presence of 20 mM arsenate and 0.5 mM arsenite.

Source: arsenic ore.

DNA G + C content (mol%): 35 (T_m).

Type strain Con a/3, MTCC 4380, DSM 15822, JCM 12167.

GenBank/EMBL/DDBJ accession number (16S rRNA gene): AJ606700 (Con a/3).

Bacillus arvi Heyrman, Rodríguez-Díaz, Devos, Felske, Logan and De Vos 2005b, 115[VP]

ar'vi. L. gen. n. *arvi* of a field.

Closely related to *Bacillus arenosi*. Spherical endospores are borne terminally and swell the sporangia slightly; acid is produced from few carbohydrates.

Source: soil.

DNA G + C content (mol%): 35 (HPLC).

Type strain LMG 22165, DSM 16317.

GenBank/EMBL/DDBJ accession number (16S rRNA gene): AJ627211 (LMG 22165).

Comment: this species has recently been reclassified as *Viridibacillus arvi* (Albert et al., 2007).

Bacillus axarquiensis Ruiz-Garcia, Quesada, Martínez-Checa, Llamas, Urdaci and Béjar 2005b, 1282[VP]

a.xar.qui.en'sis. N.L. adj. masc. *axarquiensis* pertaining to Axarquia, the Arabic name for the region surrounding the city of Málaga in Southern Spain.

Halotolerant, biosurfactant producer.

Source: brackish river sediment.

According to Wang et al. (2007b), *Bacillus axarquiensis* and *Bacillus malacitensis* are later heterotypic synonyms of *Bacillus mojavensis*.

DNA G + C content (mol%): 42.5 (T_m).

Type strain CR-119, CECT 5688, LMG 22476.

GenBank/EMBL/DDBJ accession number (16S rRNA gene): AY603657 (CR-119).

Bacillus bogoriensis Vargas, Delgado, Hatti-Kaul and Mattiasson 2005, 901[VP]

bo.gor.i.en'sis. N.L. adj. *bogoriensis* pertaining to Lake Bogoria, a soda lake in Kenya.

Grows in pH range 1 and tolerates 2 M NaCl.

Source: soda lake.

DNA G + C content (mol%): 37.5 (HPLC).

Type strain: LBB3, ATCC BAA-922, LMG 22234.

GenBank/EMBL/DDBJ accession number (16S rRNA gene): AY376312 (LBB3).

Bacillus boroniphilus Ahmed, Yokota and Fujiwara 2007a, 893[VP] (Effective publication: Ahmed, Yokota and Fujiwara 2007c, 222.)

boron.i.phi'lus. N.L. n. *boron -onis* boron; Gr. adj. *philos* loving; N.L. masc. adj. *boroniphilus* boron-loving.

Boron is required for growth and more than 450 mM is tolerated. Also tolerates up to 7.0% (w/v) NaCl in the presence of 50 mM B in agar medium but grows optimally without NaCl.

From naturally boron-containing soil of Hisarcik area in the Kutahya Province, Turkey.

DNA G + C content (mol%): 41.1–42.2.

Type strain: T-15Z, ATCC BAA-1204, DAM 17376, IAM 15287, JCM 21738.

GenBank/EMBL/DDBJ accession number (16S rRNA gene): AB198719 (T-15Z).

Bacillus cellulosilyticus Nogi, Takami and Horikoshi 2005, 2314[VP]

cell.u.lo.si.ly'ti.cus. N.L. neut. n. *cellulosum* cellulose; Gr. adj. *lutikos* able to loosen, able to dissolve; N.L. masc. adj. *cellulosilyticus* cellulose-dissolving.

Grows at pH 0 with optimum of pH 0, and tolerates up to 12% NaCl.

Source: a cellulase preparation.

DNA G + C content (mol%): 39.6 (HPLC).

Type strain: N4, DSM 2522, JCM 9156, ATCC 21833, CCRC 15439, CIP 109017.

GenBank/EMBL/DDBJ accession number (16S rRNA gene): AB043852 (N4).

Bacillus cibi Yoon, Lee and Oh 2005c, 735[VP]

ci'bi. L. n. *cibus -i* food; L. gen. n. *cibi* of food.

Produces orange-yellow colonies, and ellipsoidal endospores are borne centrally or subterminally in swollen sporangia.

Source: a fermented seafood.

DNA G + C content (mol%): 45 (HPLC).

Type strain: JG-30, KCTC 3880, DSM 16189.

GenBank/EMBL/DDBJ accession number (16S rRNA gene): AY550276 (JG-30).

Bacillus chagannorensis Carrasco, Marquez, Xue, Ma, Cowan, Jones, Grant and Ventosa 2007, 2087[VP]

N.L. masc. adj. *chagannorensis* pertaining to Lake Chagannor.

A Gram-positive, moderately halophilic, spore-forming bacterium isolated from a soda lake, Lake Chagannor, in the Inner Mongolia Autonomous Region, China.

DNA G + C content (mol%): 53.8.

Type strain: CG-15, CCM 7371, CECT 7153, CGMCC 1.6292 and DSM 18086.

GenBank/EMBL/DDBJ accession number (16S rRNA gene): AM492159.

Bacillus decisifrondis Zhang, Xu and Patel 2007, 977[VP]

de.ci.si.fron'dis. L. part. adj. *decisus* thrown off, dead, died; L. n. *frons frondis* of/from foliage; N.L. gen. n. *decisifrondis* from thrown off decayed foliage.

Produces cream, round, smooth colonies, and cells are motile rods, producing subterminal spherical spores in swollen sporangia.

Source: soil underlying the decaying leaf litter of a slash pine forest located in south east Queensland, Australia.

DNA G + C content (mol%): 41±1 (T_m).

Type strain: E5HC-32, DSM 17725, JCM 13601.

GenBank/EMBL/DDBJ accession number (16S rRNA gene): DQ465405 (E5HC-32).

Bacillus foraminis Tiago, Pires, Mendes, Morais, da Costa and Veríssimo 2006, 2573[VP]

fo.ra'mi.nis. L. n. *foramen -inis* a hole; L. gen. n. *foraminis* from a hole.

Spores not observed and cells do not exhibit resistance to 80 °C for 8 min.

Source: highly alkaline, non-saline groundwater.

DNA G C content (mol%): 43.1 (HPLC).

Type strain: CV53, LMG 23174, CIP 108889.

GenBank/EMBL/DDBJ accession number (16S rRNA gene): AJ717382 (CV53).

Bacillus hemicellulosilyticus Nogi, Takami and Horikoshi 2005, 2312[VP]

125

hem.i.cell.u.lo.si.ly′ti.cus. N.L. neut. n. *hemicellulosum* hemicellulose; Gr. adj. *lutikos* able to loosen, able to dissolve; N.L. masc. adj. *hemicellulosilyticus* hemicellulose-dissolving.

Grows at pH 1 with optimum of pH 10, and tolerates up to 12% NaCl.

Source: a hemicellulase preparation.

DNA G + C content (mol%): 36.8 (HPLC).

Type strain: C-11, JCM 9152, DSM 16731.

GenBank/EMBL/DDBJ accession number (16S rRNA gene): AB043846 (C-11).

Bacillus herbersteinensis Wieser, Worliczek, Kämpfer and Busse 2005, 2122^{VP}

her.ber.stein′en.sis. N.L. masc. adj. *herbersteinensis* pertaining to Castle Herberstein in Styria, in which the chapel with the medieval wall painting is located from which the type strain was isolated.

Wide ranges of carbohydrates and organic acids are assimilated, but many amino acids are not assimilated and acid is not produced from most carbohydrates.

Source: medieval wall painting.

DNA G + C content (mol%): 36.2–36.9 (HPLC).

Type strain: D-1,5a, DSM 16534, CCM 7228.

GenBank/EMBL/DDBJ accession number (16S rRNA gene): AJ781029 (D-1,5a).

Bacillus humi Heyrman, Rodríguez-Díaz, Devos, Felske, Logan and De Vos 2005b, 116^{VP}

hu′mi. L. gen. n. *humi* of earth, soil.

Ellipsoidal and sometimes spherical endospores are borne terminally and swell the sporangia slightly; acid is produced from a few carbohydrates.

Source: soil.

DNA G + C content (mol%): 37.5 (HPLC).

Type strain: LMG 22167, DSM 16318.

GenBank/EMBL/DDBJ accession number (16S rRNA gene): AJ627210 (LMG 22167).

Bacillus idriensis Ko, Oh, Lee, Lee, Lee, Peck, Lee and Song 2006, 2543^{VP}

id.ri.en′sis. N.L. masc. adj. *idriensis* arbitrary specific epithet pertaining to IDRI, the Infectious Disease Research Institute, where this study was performed.

Related to *Bacillus cibi.*

Source: blood of a neonate with sepsis.

DNA G + C content (mol%): 41.2 (T_m).

Type strain SMC 435, KCCM 90024, JCM 13437.

GenBank/EMBL/DDBJ accession number (16S rRNA gene): AY904033 (SMC 435).

Bacillus infantis Ko, Oh, Lee, Lee, Lee, Peck, Lee and Song 2006, 2543^{VP}

in.fan′tis. L. gen. n. *infantis* of an infant, baby, the putative source of the type strain.

Related to *Bacillus firmus.*

Source: blood of a neonate with sepsis.

DNA G + C content (mol%): 40.8 (T_m).

Type strain: SMC 435, KCCM 90025, JCM 13438.

GenBank/EMBL/DDBJ accession number (16S rRNA gene): AY904032 (SMC 435).

Bacillus koreensis Lim et al. 2006b, 62^{VP}

ko.re.en′sis. N.L. masc. adj. *koreensis* pertaining to Korea.

Closest relative is *Bacillus flexus.*

Source: rhizosphere of willow.

DNA G + C content (mol%): 36 (HPLC).

Type strain BR030, KCTC 3914, DSM 16467.

GenBank/EMBL/DDBJ accession number (16S rRNA gene): AY667496 (BR030).

Bacillus kribbensis Lim et al. 2007, 2914^{VP}

krib.ben′sis. N.L. masc. adj. *kribbensis* arbitrary name formed from the acronym of the Korea Research Institute of Bioscience and Biotechnology, KRIBB, where taxonomic studies on this species were performed.

Source: a field used for potato cultivation in Jeju, Korea.

DNA G + C content (mol%): 43.3 (HPLC).

Type strain: BT080, KCTC 13934, DSM 17871).

GenBank/EMBL/DDBJ accession number (16S rRNA gene): DQ280367 (BT080).

Bacillus lehensis Ghosh, Bhardwaj, Satyanarayana, Khurana, Mayilraj and Jain 2007, 241^{VP}

le.hen′sis. N.L. masc. adj. *lehensis* pertaining to Leh, in India, where the type strain was isolated.

Colonies are circular, convex, smooth and pigmented creamish-yellow, and cells are aerobic, Gram-positive, motile rods producing subterminal oval spores in unswollen sporangia.

Source: soil collected from Leh, India.

DNA G + C content (mol%): 41, 4 (T_m).

Type strain: MLB2, MTCC 7633, JCM 13820.

GenBank/EMBL/DDBJ accession number (16S rRNA gene): AY793550 (MLB2).

Bacillus litoralis Yoon and Oh 2005, 1947^{VP}

li.to.ra′lis. L. masc. adj. *litoralis* of the shore.

Optimal growth in 2–3% NaCl; no growth without NaCl or with >11% NaCl.

Source: tidal sediment of Yellow Sea in Korea.

DNA G + C content (mol%): 35.2 (HPLC).

Type strain: SW-211, KCTC 3898, DSM 16303.

GenBank/EMBL/DDBJ accession number (16S rRNA gene): AY608605 (SW-211).

Bacillus macauensis Zhang, Fan, Hanada, Kamagata and Fang 2006, 352^{VP}

ma.cau.en′sis. N.L. masc. adj. *macauensis* pertaining to Macau, the city where the type strain was isolated. Forms long, unbranched chains of cells; related to unnamed deep-sea isolates, *Bacillus barbaricus* and *Bacillus megaterium.*

Source: a drinking water treatment plant.

DNA G + C content (mol%): 40.8 (HPLC).

Type strain: ZFHKF-1, JCM 13285, DSM 17262.

GenBank/EMBL/DDBJ accession number (16S rRNA gene): AY373018 (ZFHKF-1).

Bacillus malacitensis Ruiz-Garcia, Quesada, Martínez-Checa, Llamas, Urdaci and Béjar 2005b, 1282^{VP}

ma.la.ci.ten′sis. L. adj. masc. *malacitensis* pertaining to Flavia Malacita, the Roman name for Málaga in southern Spain.

Halotolerant; surfactant producing.

Source: brackish river sediment.

DNA G + C content (mol%): 41 (T_m).

Type strain: CR-95, CECT 5687, LMG 22477.

GenBank/EMBL/DDBJ accession number (16S rRNA gene): AY603656 (CR-95).

Comment: this species is considered as a hetrotypic synonym of *Bacillus mojavensis* (Wang et al., 2007b).

Bacillus mannanilyticus   Nogi, Takami and Horikoshi 2005, 2314[VP]

mann.an.i.ly'ti.cus. N.L. neut. n. *mannanum* mannan; Gr. adj. *lutikos* able to loosen, able to dissolve; N.L. masc. adj. *mannanilyticus* mannan-dissolving.

Produces yellow colonies; pH range for growth is 0 with optimum of pH 9.

Source: a β-mannosidase and β-mannanase preparation.

DNA G + C content (mol%): 37.4 (HPLC).

Type strain: AM-001, JCM 10596, DSM 16130.

GenBank/EMBL/DDBJ accession number (16S rRNA gene): AB043864 (AM-001).

Bacillus massiliensis   Glazunova, Raoult and Roux 2006, 1487[VP]

mas.si.li.en'sis. L. masc. adj. *massiliensis* of Massilia, the ancient Greek and Roman name for Marseille, France, where the type strain was isolated.

Member of *Bacillus sphaericus* group, forming terminal spherical spores that swell the sporangia. Closely related to species that have been transferred to the novel genus *Lysinibacillus* (Ahmed et al., 2007c), but data on peptidoglycan composition and polar lipids are not available for *Bacillus massiliensis*, and so it has not been transferred to the new genus.

Source: cerebrospinal fluid.

DNA G + C content (mol%) not reported.

Type strain: 4400831, CIP 108446, CCUG 49529.

GenBank/EMBL/DDBJ accession number (16S rRNA gene): AY677116 (4400831).

Bacillus muralis   Heyrman, Logan, Rodríguez-Díaz, Scheldeman, Lebbe, Swings, Heyndrickx and De Vos 2005a, 129[VP]

mu.ra'lis. L. masc. adj. *muralis* pertaining or belonging to walls.

Related to *Bacillus simplex*.

Source: mural painting in a church in Germany.

DNA G + C content (mol%): 41.2 (HPLC).

Type strain: LMG 20238, DSM 16288.

GenBank/EMBL/DDBJ accession number (16S rRNA gene): AJ628748 (LMG 20238).

Bacillus murimartini   Borchert, Nielsen, Graber, Kaesler, Szewyck, Pape, Antrnikian and Schäfer 2007, 2892[VP]

mu.ri.mar.ti'ni. L. n. *murus* wall; N.L. gen. n. *martini* of Martin (masc. name of a saint); N.L. gen. n. *murimartini* from the wall of the (St) Martin church in Greene-Kreiensen, Germany.

Source: a church wall mural painting in Germany.

DNA G + C content (mol%): 39.6.

Type strain: type strain LMG 21005 andNCIMB 14102.

GenBank/EMBL/DDBJ accession number (16S rRNA gene): AJ880003.

Bacillus niabensis   Kwon, Lee, Kim, Weon, Kim, Go and Lee 2007, 1910[VP]

niab.en'sis. N.L. masc. adj. *niabensis* arbitrary name formed from NIAB, the acronym for the National Institute of Agricultural Biotechnology, Korea, where taxonomic studies on this species were performed.

Colonies are yellowish-white, 2–3 mm in diameter, and circular with clear margins, and cells are motile, by means of single polar flagella. Forms ellipsoidal or oval spores that lie subterminally or terminally in swollen sporangia.

Source: cotton-waste composts in Suwon, Korea.

DNA G + C content (mol%): 37.7–40.9 (HPLC).

Type strain: 4T19, KACC 11279, DSM 17723.

GenBank/EMBL/DDBJ accession number (16S rRNA gene): AY998119 (4T19).

Bacillus okhensis   Nowlan, Dodia, Singh and Patel 2006, 1076[VP]

ok.hen'sis. N.L. masc. adj. *okhensis* pertaining to Port Okha, a port of the Dwarka region in India, where the type strain was isolated.

Halotolerant and related to *Bacillus krulwichiae*; bears a subterminal tuft of flagella, but spores have not been detected.

Source: soil of natural saltpan.

DNA G + C content (mol%): 41 (T_m).

Type strain: Kh101, JCM 13040, ATCC BAA-1137.

GenBank/EMBL/DDBJ accession number (16S rRNA gene): DQ026060 (ATCC BAA-1137).

Bacillus oshimensis   Yumoto, Hirota, Goto, Nodasaka and Nakajima 2005a, 910[VP]

o'shi.men.sis. N.L. masc. adj. *oshimensis* from Oshima, the region where the micro-organism was isolated.

Grows in 0–20% NaCl, with 7% NaCl optimal; grows from pH 7, with pH 10 optimal.

Source: soil.

DNA G + C content (mol%): 40.8 (HPLC).

Type strain: K11, JCM 12663. NCIMB 14023.

GenBank/EMBL/DDBJ accession number (16S rRNA gene): AB188090 (K11).

Bacillus panaciterrae   Ten, Baek, Im, Liu, Aslam and Lee 2006, 2864[VP]

pa.na.ci.ter'rae. N.L. n. *Panax -acis* scientific name for ginseng; L. n. *terra* soil; N.L. gen. n. *panaciterrae* of soil of a ginseng field.

Utilizes a wide range of carbohydrates, amino acids and organic acids, and hydrolyzes chitin; forms ellipsoidal endospores centrally in swollen sporangia.

Source: soil.

DNA G + C content (mol%): 47.8 (HPLC).

Type strain: Gsoil 1517, KCTC 13929, CCUG 52470, LMG 23408.

GenBank/EMBL/DDBJ accession number (16S rRNA gene): AB245380 (Gsoil 1517).

Bacillus patagoniensis   Olivera et al. 2005, 446[VP]

pa.ta.go'ni.en.sis. N.L. masc. adj. *patagoniensis* pertaining to Patagonia, in Argentina, where the type strain was isolated.

Alkalitolerant and halotolerant.

Source: desert soil rhizosphere.

DNA G + C content (mol%): 39.7 (HPLC).

Type strain: PAT 05, DSM 16117, ATCC BAA-965.

GenBank/EMBL/DDBJ accession number (16S rRNA gene): AY258614 (PAT 05).

Bacillus plakortidis Borchert, Nielsen, Graber, Kaesler, Szewyck, Pape, Antranikian and Schäfer 2007, 2892[VP]

Source: material from the sponge *Plakortis simplex* that was obtained from the Sula-Ridge, Norwegian Sea.

DNA G + C content (mol%): 41,1.

Type strain: P203,DSM 19153 and NCIMB 14288.

GenBank/EMBL/DDBJ accession number (16S rRNA gene): AJ880003.

Bacillus pocheonensis Ten, Baek, Im, Larin, Lee, Oh and Lee 2007, 2535[VP]

N.L. masc. adj. *pocheonensis* pertaining to Pocheon Province in South Korea.

A Gram-positive, nonmotile, endospore-forming rod.

Source: soil of a ginseng field in Pocheon Province, South Korea.

DNA G + C content (mol%): 44.9.

Type strain: Gsoil 420, KCTC 13943 and DSM 18135.

GenBank/EMBL/DDBJ accession number (16S rRNA gene): AB245377.

Bacillus qingdaonensis Wang, Li, Liu, Cao, Li and Guo 2007c, 1146[VP]

qing.da.o.nen′sis. N.L. masc. adj. *qingdaonensis* pertaining to Qingdao, the name of the place from which the type strain was isolated.

A moderately haloalkaliphilic, aerobic, rod-shaped, nonmotile, Gram-positive organism capable of growth at salinities of 2.5–20% (w/v) NaCl. Spores were not observed.

Source: a crude sea-salt sample collected near Qingdao in eastern China.

DNA G + C content (mol%): 48 (HPLC).

Type strain: CM1, CGMCC 1.6134, JCM 14087.

GenBank/EMBL/DDBJ accession number (16S rRNA gene): DQ115802 (CM1).

Bacillus ruris Heyndrickx, Scheldeman, Forsyth, Lebbe, Rodríguez-Díaz, Logan and De Vos 2005, 2553[VP]

ru′ris. L. neut. n. *rus* the country, the farm; L. gen. n. *ruris* from the country, the farm.

Related to *Bacillus galactosidilyticus.*

Source: raw milk and dairy cattle feed concentrate.

DNA G + C content (mol%): 39.2 (HPLC).

Type strain: LMG 22866. DSM 17057.

GenBank/EMBL/DDBJ accession number (16S rRNA gene): AJ535639 (LMG 22866).

Bacillus safensis Satomi, La Duc and Venkateswaran 2006, 1739[VP]

sa.fen′sis. N.L. masc. adj. *safensis* arbitrarily derived from SAF, the spacecraft-assembly facility at the Jet Propulsion Laboratory, Pasadena, CA, USA, from where the organism was first isolated.

Closely related to *Bacillus pumilus* on basis of 16S rRNA and *gyrB* gene sequences.

Source: a spacecraft-assembly plant.

DNA G + C content (mol%): 41.1.4 (HPLC).

Type strain: FO-36b, ATCC BAA-1126, NBRC 100820.

GenBank/EMBL/DDBJ accession number (16S rRNA gene): AF234854 (FO-36b).

Bacillus salarius Lim et al. 2006c, 376[VP]

sa.la′ri.us. L. masc. adj. *salarius* of or belonging to salt. Member of the alkaliphilic group (Group 6 of Nielsen et al., 1994) of *Bacillus*; grows at 0% NaCl, with optimum of 12% NaCl, and at pH 6.5 with optimum pH of 8.

Source: sediment of a salt lake.

DNA G + C content (mol%): 43 (HPLC).

Type strain: BH169, KCTC 3912, DSM 16461.

GenBank/EMBL/DDBJ accession number (16S rRNA gene): AY667494 (BH169).

Bacillus saliphilus Romano, Lama, Nicolaus, Gambacorta and Giordano 2005a, 162[VP]

sal.i.phi′lus. L. n. *sal* salt; Gr. adj. *philos* loving; N.L. masc. adj. *saliphilus* salt-loving.

A coccoid member of the alkaliphilic group (Group 6 of Nielsen et al., 1994) of *Bacillus*; grows at 5% NaCl, with optimum of 16% NaCl, and at pH 0 with optimum pH of 9. Spores not reported.

Source: green algal mat in a mineral pool.

DNA G + C content (mol%): 48.4 (HPLC).

Type strain: 6AG, DSM 15402, ATCC BAA-957.

GenBank/EMBL/DDBJ accession number (16S rRNA gene): AJ493660 (6AG).

Bacillus selenatarsenatis Yamamura, Yamashita, Fujimoto, Kuroda, Kashiwa, Sei, Fujita and Ike 2007, 1063[VP]

se′le.nat.ar.se.na′tis. N.L. gen. n. *selenatis* of selenate; N.L. gen. n. *arsenatis* of arsenate; N.L. gen. n. *selenatarsenatis* of selenate and arsenate.

Gram-positive, spore-forming, motile rods. Colonies are round and white. Selenate is reduced to elemental selenium via the intermediate selenite, arsenate to arsenite and nitrate to ammonia via the intermediate nitrite.

Source: an effluent drain in a glass-manufacturing plant in Japan.

DNA G + C content (mol%): 42.8 (HPLC).

Type strain: SF-1, JCM 14380, DSM 18680.

GenBank/EMBL/DDBJ accession number (16S rRNA gene): AB262082 (SF-1).

Bacillus seohaeanensis Lee, Lim, Park, Jeon, Li and Kim 2006a, 1896[VP]

seo.hae.an.en′sis. N.L. masc. adj. *seohaeanensis* of Seohaean, the Korean name for the west coast of Korea, where the type strain was isolated.

Related to *Bacillus aquimaris* and *Bacillus marisflavi.*

Source: a solar saltern.

DNA G + C content (mol%): 39 (HPLC).

Type strain: BH724, KCTC 3913, DSM 16464.

GenBank/EMBL/DDBJ accession number (16S rRNA gene): AY667495 (BH724).

Bacillus stratosphericus Shivaji, Chaturvedi, Suresh, Reddy, Dutt, Wainwright, Narlikar and Bhargava 2006, 1471[VP]

stra.to.sphe.ri′cus. N.L. fem. n. *stratosphera* stratosphere; L. suff. *-icus* adjectival suffix used with the sense of belonging to; N.L. masc. adj. *stratosphericus* belonging to the stratosphere.

Shows high 16S rRNA gene sequence similarity with *Bacillus aerius, Bacillus aerophilus, Bacillus licheniformis* and *Bacillus sonorensis.*

Source: air sample collected at high altitude.

DNA G + C content (mol%): 44 (T_m).

Type strain: 41KF2a, MTCC 7305, JCM 13349.

GenBank/EMBL/DDBJ accession number (16S rRNA gene): AJ831841 (41KF2a).

Bacillus taeanensis Lim, Jeon and Kim 2006a, 2905^{VP}

tae.an.en'sis. N.L. masc. adj. *taeanensis* belonging to Taean, where the organism was isolated.

Neutrophilic and halotolerant, with optimum growth at 2–5% NaCl.

Source: solar saltern.

DNA G + C content (mol%): 36 (HPLC).

Type strain: BH030017, KCTC 3918, DSM 16466.

GenBank/EMBL/DDBJ accession number (16S rRNA gene): AY603978 (BH030017).

Bacillus tequilensis Gatson, Benz, Chandrasekaran, Satomi, Venkateswaran and Hart 2006, 1481^{VP}

te.qui.len'sis. N.L. masc. adj. *tequilensis* referring to Tequila, Mexico.

Member of the *Bacillus subtilis* group.

Source: Mexican shaft tomb sealed in approximately 74 AD.

DNA G + C content (mol%): not reported.

Type strain: 10b, ATCC BAA-819, NCTC 13306.

GenBank/EMBL/DDBJ accession number (16S rRNA gene): AY197613: AY197613 (10b); the sequence contains only 549 bp and was therefore not included in the phylogenetic tree (Figure 8).

Bacillus thioparans Pérez-Ibarra, Flores and Garica-Varela 2007a, 1933^{VP} (Effective publication: Pérez-Ibarra, Flores and Garica-Varela 2007b, 295.)

thi.o'parus. Gr. n. *thios* sulfur; L. v. *paro* to produce; M.L. adj. *thioparus* sulfur-producing.

Source: a continuous wastewater treatment culture system operating with a bacterial consortium. Gram-variable, aerobic, moderately halotolerant, motile and endospore-forming rods.

DNA G + C content (mol%): 43.8 (T_m).

Type strain: BMP-1, BM-B-436 and CECT 7196.

GenBank/EMBL/DDBJ accession number (16S rRNA gene): DQ371431.

Bacillus velezensis Ruiz-Garcia, Béjar, Martínez-Checa, Llamas and Quesada 2005a, 195^{VP}

vel.e.zen'sis. N.L. adj. masc. *velezensis* pertaining to Vélez, named thus for being first isolated from the river Vélez in Málaga, southern Spain.

Member of the *Bacillus subtilis* group.

Source: mouth of River Vélez, Spain.

DNA G + C content (mol%): 46.6.4 (T_m).

Type strain: CR-502, CECT 5686, LMG 22478.

GenBank/EMBL/DDBJ accession number (16S rRNA gene): AY603658 (CR-502).

Bacillus wakoensis Nogi, Takami and Horikoshi 2005, 2312^{VP}

wa.ko.en'sis. N.L. masc. adj. *wakoensis* of Wako, a city in Japan.

Related to *Bacillus krulwichiae.*

Source: preparation of cellulase.

DNA G + C content (mol%): 38.1 (HPLC).

Type strain: N-1, JCM 9140, DSM 2521.

GenBank/EMBL/DDBJ accession number (16S rRNA gene): AB043851 (N-1).

Genus II. **Alkalibacillus** Jeon, Lim, Lee, Xu, Jiang and Kim 2005b, 1894^{VP}

PAUL DE VOS

Al.ka.li.ba.cil´lus. N.L. n. *alkali* alkali; L. n. *bacillus* rod; N.L. masc. n. *Alkalibacillus* bacillus living under alkaline conditions.

Rod-shaped, Gram-positive (may be Gram-variable in older cultures) bacterium. Mostly long cells, 0.8–1.6 μm in diameter and 2.0–7.0 μm in length. Endospores are spherical and located terminally in swollen sporangia. Cell-wall peptidoglycan is of the A1γ type and *meso*-diaminopimelic acid is the diamino acid. Obligately aerobic. Cells are motile by peritrochous or polar flagella. MK-7 is always present as the major isoprenoid quinone; in one species, DeMK-6 is also found. The predominant fatty acids are $C_{15:0\ iso}$, $C_{15:0\ ante}$, and $C_{17:0\ ante}$. Phylogenetically, the genus belongs to the family *Bacillaceae.*

DNA G + C content (mol%): 37–41.

Type species: **Alkalibacillus haloalkaliphilus** (Fritze 1996a) Jeon, Lim, Lee, Xu, Jiang and Kim 2005b, 1894^{VP} (*Bacillus haloalkaliphilus* Fritze 1996a, 100).

Further descriptive information

Representatives of this genus were first described by Fritze (1996a). Colonies are creamy to white on salt-containing media. Except for strains of one species, no growth occurs on media without NaCl; optimal NaCl concentration for growth is about 10%. The pH range for growth is 7.0–10.0 for most species, except for one which grows at pH 7.0–9.0; the optimal pH varies from 8.0 to 9.7 depending on the species. The temperature range for growth is from about 15 to 50 °C, with optimal growth between 30 °C and 37 °C. Enrichment media have been used in isolation campaigns. Media compositions can be found in the literature (Fritze, 1996a; Jeon et al., 2005b; Ren and Zhou, 2005b; Romano et al., 2005b; Usami et al., 2007), but always contain complex organic mixtures such as Casamino acids and yeast extract. Although nearly all strains need a relatively high concentration of salt, some were isolated from non-salty environments such as forest soil. In these particular cases, their role in the ecosystem is unclear. Members of the genus are widespread because they have been retrieved from various geographical regions.

Taxonomic comments

The genus *Alkalibacillus* now encompasses four species and was separated from the genus *Bacillus* by the reclassification of *Bacillus haloalkaliphilus.* Phylogenetically, members of the genus *Alkalibacillus* are most closely related to those of the moderately halophilic genera *Tenuibacillus* and *Filobacillus*. A fifth species, "*Alkalibacillus halophilus*", which accommodates isolates from hypersaline soil in China has been isolated, but its name has not yet been validly published (Tian et al., 2007). 16S rRNA gene sequence analysis revealed that the closest phylogenetic relatives of the type species, *Alkalibacillus haloalkaliphilus*, are members of the recently described genera *Tenuibacillus* (Ren and Zhou, 2005b) and *Filobacillus* (Schlesner et al., 2001). The physiological characteristics of these genera are very similar. Differences at the generic level are observed in cell-wall composition: in *Alkalibacillus*, the murein is of the A1γ type, whereas it is of the A4β type in *Filobacillus*. The microbiologist who wants to apply physiological tests for comparative analysis at the species/strain level should be aware that different incubation conditions may result in different characteristics. The question may be raised whether these genera, members of which differ by less than 4% in their 16S rRNA gene sequences, should not be reclassified into a single genus.

List of species of the genus *Alkalibacillus*

1. **Alkalibacillus haloalkaliphilus** (Fritze 1996a) Jeon, Lim, Lee, Xu, Jiang and Kim 2005b, 1894[VP] (*Bacillus haloalkaliphilus* Fritze 1996a, 100).

   ha.l.o.al.ka.li'phi.lus. Gr. n. *hals* salt; N.L. n. *alkali* alkali; Gr. adj. *philos* loving; M.L. adj. *haloalkaliphilus* loving briny and alkaline media.

   Strictly alkaliphilic, halophilic, and extremely halotolerant motile rods and filaments. Forms spherical spores that lie terminally and swell the sporangia. Cells are 0.3–0.5 × 3.0–8.0 μm. Colonies on alkaline nutrient agar supplemented with 5–10% NaCl are creamy-white, whereas with 20% NaCl they are yellowish. Grows at 15 and 40 °C. Optimal growth occurs at pH 9.7 or above; no growth is seen at pH 7.0. Grows in 20–25% NaCl; no growth or only very weak growth occurs without added NaCl and optimal sporulation occurs at 5% NaCl. Nitrate is not reduced to nitrite. Hippurate and 4-methyl-umbelliferone-glucuronide are hydrolyzed. Gelatin and starch are hydrolyzed weakly. Casein hydrolysis may be positive, weak or negative. Pullulan, Tween 20, and 80, and urea are not hydrolyzed. Strains have been isolated from brine, camel dung, loam, mud, and salt at Wadi Natrun, Egypt.

   DNA G + C content (mol%): 37 (Bd) to 38 (HPLC).

   Type strain: WN13, ATCC 700606, DSM 5271, LMG 17943, NCIMB 13457.

   EMBL/GenBank accession number (16S rRNA gene): X72876 (DSM 5271).

2. **Alkalibacillus filiformis** Romano, Lama, Nicolaus, Gambacorta and Giordano 2005b, 2397[VP]

   fi.li.for'mis. L. neut. n. *filum* a thread; L. suff. *-formis* -like, of the shape of; N.L. masc. adj. *filiformis* thread-shaped.

   Cells are Gram-positive, sporulating rods, 0.25–0.30 μm in width and 9.0–11.0 μm long. The terminally spherical endospores are located in swollen sporangia. Smooth, convex, and regular circular, white to transparent colonies are obtained on enrichment medium 1 (Romano et al., 2005b). Growth occurs between 15 °C and 45 °C, with optimal growth at 30 °C in media at pH 7–10 (optimal pH 9.0). Growth occurs in media without added NaCl, but is optimal in media with 10% salt added; one strain is known to tolerate up to 18% NaCl. In addition to the characteristics that allow species differentiation (Table 10), *Alkalibacillus filiformis* is able to grow on glucose as sole carbon source, does not hydrolyze phenylalanine, and does not reduce nitrate. Catalase reaction is weak. Shows α-glucosidase activity.

Menaquinones found are MK-7 (70%) and DeMK-6 (30%); phosphatidylglycerol and diphosphatidylglycerol are the predominant lipids. The major fatty acids are $C_{15:0\ iso}$, $C_{15:0\ ante}$, $C_{16:0\ iso}$, $C_{16:0}$, $C_{17:0\ iso}$, and $C_{17:0\ ante}$. The cell wall is of type A1γ (*meso*-diaminopimelic acid directly cross-linked). Accumulates glycine betaine (major component) and glutamate (minor component) for osmoprotection. The closest neighbor phylogenetically is *Alkalibacillus haloalkaliphilus* on the basis of 16S rRNA gene sequence analysis. Antibiotic sensitivity data are given in Romano et al. (2005b). The type strain was isolated from water of a small mineral pool with gas bubbles at the Malvizza site (Montecalvo Irpino, Campania Region, Italy).

   DNA G + C content (mol%): 39.5 (HPLC).

   Type strain: 4AG, DSM 15448, ATCC BAA-956.

   EMBL/GenBank accession number (16S rRNA gene): AJ493661 (4AG).

3. **Alkalibacillus salilacus** Jeon, Lim, Lee, Xu, Jiang and Kim 2005b, 1895[VP]

   sa.li.lac'us. L. n. *sal* salt; L. n. *lacus* lake; N.L. gen. masc. n. *salilacus* of a salt lake.

   Gram-positive motile rods of approximately 0.4–0.5 μm wide and 1.6–3.0 μm long. Spherical spores are formed terminally in swollen sporangia. Growth occurs at 15–40 °C, pH 7.0–9.0 and 5–20% (w/v) NaCl. Colonies on marine agar (MA) supplemented with 10% NaCl are cream, smooth, low convex, and circular/slightly irregular. Catalase-positive and oxidase-negative. Nitrate is reduced to nitrite. In addition to the characteristics given in Table 10, does not hydrolyze Tween 80, L-tyrosine, hypoxanthine, xanthine, or urea. Furthermore, in addition to the sugars and sugar alcohols mentioned in Table 10, acid is produced from L-arabinose, D-ribose, and α-D-lactose, but not from D-glucose, glycerol, D-xylose, L-rhamnose, adonitol, D-raffinose, arbutin, D-salicin, D-melibiose, or D-mannose. The predominant cellular fatty acids are $C_{15:0\ ante}$, $C_{15:0\ iso}$, $C_{17:0\ ante}$, and $C_{16:0\ iso}$. The type strain was isolated from a salt lake in the Xinjiang province of China.

   DNA G + C content (mol%): 41.0 (HPLC).

   Type strain: BH163, KCTC 3916, DSM 16460.

   EMBL/GenBank accession number (16S rRNA gene): AY671976 (BH163).

4. **Alkalibacillus silvisoli** Usami, Echigo, Fukushima, Mizuki, Yoshida and Kamekura 2007, 773[VP]

   sil.vi.so'li. L. n. *silva* forest; L. n. *solum* soil; N.L. gen. n. *silvisoli* of forest soil.

TABLE 10. Differential characteristics of *Alkalibacillus* species[a]

Characteristic	1. *A. haloalkaliphilus*	2. *A. filiformis*	3. *A. salilacus*	4. *A. silvisoli*
Motility	+	−	+	+
Gram stain	−	+	+	+
NaCl range for growth (%, w/v)	>0–25.0	0–18.0	5.0–20.0	5.0–25
pH optimum	9.7	9.0	8.0	9.0–9.5
Growth temperature range (°C)	>50	15–45	15–40	20–50
Catalase	+	w	+	+
Oxidase	+	−	−	−
Acid production from:				
D-Fructose	−	−	+	−
D-Galactose	−	−	−	+
Maltose	−	−	−	+
Trehalose	−	−	−	+
D-Mannitol	−	−	−	+
Hydrolysis of:				
Starch	w	−	−	−
Casein	−/w	−	−	+
Gelatin	+	+	−	+
Hippurate	+	−	−	−
Esculin	+	−	+	−
Nitrate reduction	−	−	−	+

[a]Data are based on Table 1 in Usami et al. (2007). w, Weak.

Cells are Gram-positive (may become Gram-variable in old cultures) motile rods of 0.3–0.5 × 4.0–7.0 μm in size with a single polar flagella. Endospores are located terminally in swollen sporangia. The A1γ type of peptidoglycan is present with *meso*-diaminopimelic acid. Colonies in media with 10% NaCl are creamy and opaque. Growth occurs at NaCl concentrations between 5.0% and 25.0% (w/v) for the type strain, although a second strain can grow in media without added NaCl. Optimum concentration is again strain dependent and is 10.0–15.0% (w/v) for the type strain. The optimal pH range for growth may also be strain dependent and is between pH 8.5 and 9.5; growth is observed at pH 7.0–10.0. The temperature range observed varies from 20°C to 50°C (optimal 30–37°C). Strictly aerobic. In addition to sugars given in Table 10, acid is produced from sucrose, but not from D-glucose or D-xylose. Catalase-positive

and oxidase-negative. Does not hydrolyze DNA, pullulan, or Tween 80. Does not reduce nitrate and gas production is not observed. Antibiotic sensitivity data are given in Usami et al. (2007). The predominant isoprenoid quinone is MK-7. The major cellular fatty acids are, in order of importance, $C_{15:0\ iso}$, $C_{17:0\ ante}$, $C_{15:0\ ante}$, $C_{16:0\ iso}$, $C_{17:0\ iso}$, and $C_{16:0\ ante}$. The type strain was isolated from non-saline surface soil from a forest in Kawagoe, Saitama Prefecture, Japan. A second strain, HN_2 (which has not yet been deposited in a culture collection), was isolated from non-saline surface soil from a forest in Yachiyo, Chiba Prefecture, Japan.

DNA G + C content (mol%): 37.0 (HPLC).

Type strain: BM2, JCM 14193, DSM 18495.

EMBL/GenBank accession number (16S rRNA gene): AB264528 (BM2).

Genus III. **Amphibacillus** Niimura, Koh, Yanagida, Suzuki, Komagata and Kozaki 1990, 299[VP] emend. An, Ishikawa, Kasai, Goto and Yokota 2007b, 2492

YOUICHI NIIMURA AND KEN-ICHIRO SUZUKI

Am.phi.ba.cil'lus. Gr. pref. *amphi* both sides or double; L. dim. n. *bacillus* a small rod; N.L. masc. n. *Amphibacillus* rod capable of both aerobic and anaerobic growth.

Cells are rods, occurring singly, in pairs or, sometimes, in short chains. Gram-positive, or Gram-positive in the very early stages of growth and loosely Gram-positive in the stationary growth phase. Cells are motile by means of flagella or nonmotile. Heat-resistant. Oval endospores are formed in terminal or center position, but sometimes sporangia are rapidly lysed and the spores are liberated. Growth is good in both well-aerated and strictly anaerobic liquid media, and also on aerobic agar plates and anaerobic plates. Cell yields and growth rates

are almost the same under all these conditions. Growth does not occur in the absence of glucose under either aerobic or anaerobic conditions. Chemo-organotrophic. Main products from glucose are ethanol, acetic acid, and formic acid under anaerobic conditions and acetic acid under aerobic conditions. Lactic acid is sometimes produced under aerobic and anaerobic conditions. Alkaliphilic, sometimes halophilic or halotolerant. Respiratory quinones, cytochromes, and catalase are absent. Located within the phylogenetic group composed of halophilic/halotolerant/alkaliphilic and/or alkalitolerant genera in *Bacillus* rRNA group 1.

DNA G + C content (mol%): 36–41.5 (T_m).

Type species: **Amphibacillus xylanus** Niimura, Koh, Yanagida, Suzuki, Komagata and Kozaki 1990, 300[VP].

Further descriptive information

The catalase test is an important and useful test for characterization and differentiation of facultatively anaerobic spore-forming bacilli. The presence of catalase is generally determined by the formation of bubbles from cells put in 3% H_2O_2. When cells of *Amphibacillus fermentum* and *Amphibacillus tropicus* are cultured in high concentrations of Na_2CO_3, small bubbles that are not caused by the catalase reaction are sometimes formed in the catalase test 2–3 min after mixing. The catalase reaction can be clearly distinguished from the reactions like that shown above by using a positive control, i.e., a catalase-positive strain such as *Bacillus subtilis*, because the true catalase reaction produces a lot of bubbles rapidly as soon as cells are put into H_2O_2 solution. Spectrophotometric analysis of catalase using cell-free extract also showed that *Amphibacillus fermentum* and *Amphibacillus tropicus* clearly lack catalase. Spectrophotometric analysis of catalase is useful for determination of the catalase reaction in cells cultured in high concentrations of Na_2CO_3. Therefore, strains of the genus *Amphibacillus* are catalase-negative, as stated above in the genus description; however, the original descriptions of *Amphibacillus fermentum* and *Amphibacillus tropicus* indicate that they are catalase-positive (Zhilina et al., 2001a).

Amphibacillus xylanus, the type species of the genus, in spite of lacking a respiratory system and catalase, grows well on plates and in liquid cultures under both strictly anaerobic and aerobic conditions. *Amphibacillus xylanus* is distinctive in this trait from other facultative anaerobes. This characteristic is due to the presence of anaerobic and aerobic pathways producing similar amounts of ATP (Figure 11) (Nishiyama et al., 2001). Accordingly, it has been suggested that *Amphibacillus fermentum* and *Amphibacillus tropicus* have two major metabolic systems, as observed in *Amphibacillus xylanus*. However, it has also been suggested that these two species

differ from *Amphibacillus xylanus* in that they have a side enzymic pathway that produces lactic acid under both aerobic and anaerobic conditions (Y. Niimura and others, unpublished data).

During anaerobic metabolism in *Amphibacillus xylanus*, NADH formed from NAD^+ in the glycolytic pathway is reoxidized by NAD-linked aldehyde dehydrogenase and NAD-linked alcohol dehydrogenase. NADH produced by both the glycolytic pathway and pyruvate metabolism of the aerobic pathway should be oxidized to NAD during the reduction of oxygen to water by the NADH oxidase-Prx system, which provides metabolic balance in the aerobic pathway in *Amphibacillus xylanus*. The NADH oxidase-Prx system, which shows extremely high turnover numbers and low K_m values for peroxides, is induced markedly in the presence of hydrogen peroxide. Thus, the *Amphibacillus xylanus* NADH oxidase-Prx system plays an important role not only as an NAD-regenerating system, but also in removing peroxides in the bacterium, which lacks both a respiratory chain and catalase, conventional peroxide-removing enzymes. Enzyme assays and immunoblot analysis revealed that NADH oxidase participates in oxygen metabolism instead of the respiratory system in *Amphibacillus fermentum* and *Amphibacillus tropicus*, as in *Amphibacillus xylanus*, and the NADH oxidase-Prx system may also function as a peroxide-reduction system in *Amphibacillus tropicus* as well as in *Amphibacillus xylanus* (Y. Niimura and others, unpublished data).

Amphibacillus species are alkaliphilic, as they grow optimally at pH values above 8.5. In contrast to *Amphibacillus xylanus*, *Amphibacillus fermentum* and *Amphibacillus tropicus* are halophiles. The optimum NaCl concentrations for growth of *Amphibacillus fermentum* and *Amphibacillus tropicus* are 10.8% and between 5.4% and 10.8%, respectively. *Amphibacillus fermentum* and *Amphibacillus tropicus* do not require chloride ions, but need sodium ions and carbonate, having an obligate requirement for sodium carbonate: growth is observed after the replacement of NaCl with an equimolar amount of Na_2CO_3 + $NaHCO_3$.

Amphibacillus species require carbohydrates represented by glucose for growth in both aerobic and anaerobic conditions. A wide variety of mono-, oligo-, and polysaccharides is utilized under both conditions.

Enrichment and isolation procedures

The strains of the genus *Amphibacillus* isolated so far are alkaliphilic and show good growth on sugars under anaerobic conditions (Niimura et al., 1987, 1989). Therefore, alkaline media containing carbohydrates are used for isolation. Cultivation for enrichment and isolation is carried out anaerobically. Strains of *Amphibacillus xylanus* were isolated from alkaliphilic composts of manure with grass and rice straw in Japan. *Amphibacillus fermentum* and *Amphibacillus tropicus* were isolated during dry and rainy periods, respectively, from bottom sediment of a coastal lagoon of Lake Magadi, Kenya (Zhilina et al., 2001a). Successive enrichment cultures were applied under strictly anaerobic conditions with nitrogen and a reducing agent, titanium (III) citrate for *Amphibacillus xylanus* and Na_2S for *Amphibacillus fermentum* and *Amphibacillus tropicus*. Titanium (III) citrate is added just before use. The enrichment medium (I) for *Amphibacillus xylanus* is alkaliphilic (pH 10) and contains Na_2CO_3 solution and the enrichment medium (II) for *Amphibacillus fermentum* and *Amphibacillus tropicus* is also alkaliphilic (pH 10), containing 2.65 M Na^+ (0.9 M NaCl, 0.6 M Na_2CO_3, and 0.55 M $NaHCO_3$). Alkaline solution (Na_2CO_3/$NaHCO_3$) is sterilized separately by filtration and mixed with the basal medium, which is sterilized just before use.

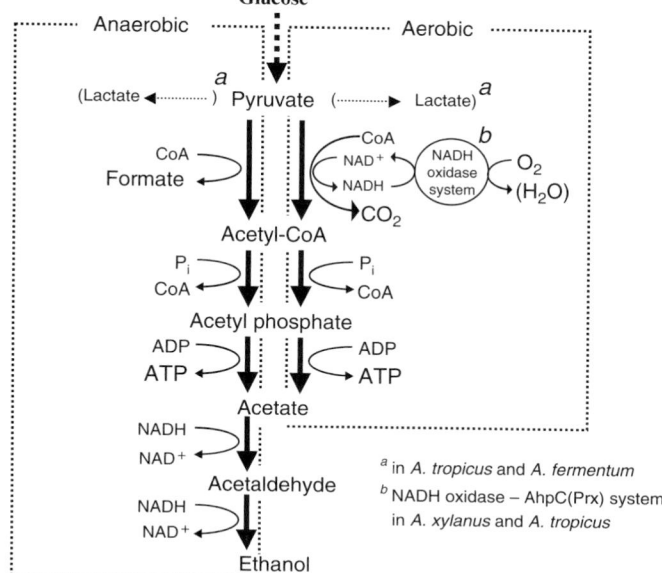

FIGURE 11. Predicted metabolic pathway of *Amphibacillus* species.

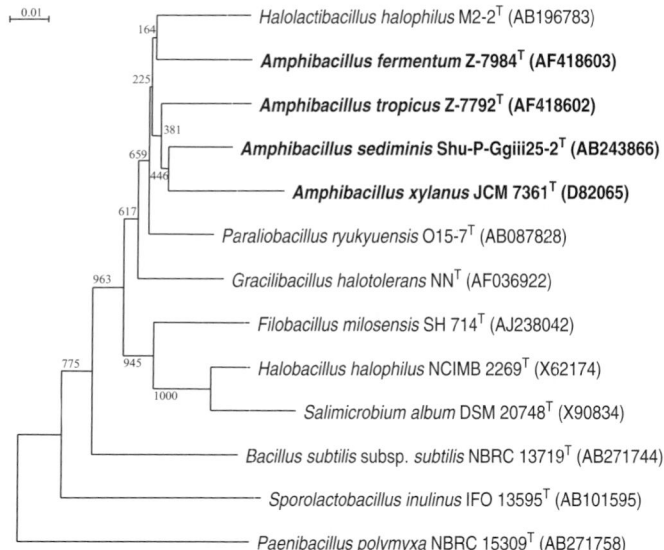

FIGURE 12. Phylogenetic tree of the genus *Amphibacillus* and related genera based on 16S rRNA gene sequences. Scale bar indicates the K_{nuc} values calculated from the nucleotide sequences. Numbers at branches indicate the confidence limits estimated by bootstrap analysis with 1,000 resampling trials (shown only for the major clusters). Tree courtesy of M. Miyashita.

Maintenance procedures

Strains of *Amphibacillus xylanus* are maintained on anaerobic culture medium containing xylan as a carbon source and stored at 5 °C at 1- to 2-month intervals. For long-term preservation, freezing and freeze-drying are used. Cells are harvested and resuspended in neutral media containing 10% glycerol or DMSO for freezing preservation and stored at –80 °C or lower. For drying, cells are suspended in 10% skim milk containing 1% monosodium L-glutamate for freeze-drying or in 0.1 M potassium phosphate buffer (pH 7.0) containing 3% monosodium L-glutamate, 1.5% adonitol, and 0.05% HCl–L-cysteine for liquid drying. Handling of cell suspensions under anaerobic conditions is not necessary.

Differentiation of the genus *Amphibacillus* from other genera

The genus *Amphibacillus* is classified in the family *Bacillaceae*, a large family of aerobic Gram-positive low-G + C-containing bacteria. The three species of the genus *Amphibacillus* constitute an independent line of descent within the group composed of halophilic/halotolerant/alkaliphilic and/or alkalitolerant members in rRNA group 1 of the genus *Bacillus* based on 16S rRNA gene sequences and occupy a phylogenetic position closely related to the genera *Halolactibacillus*, *Paraliobacillus*, and *Gracilibacillus* (Figure 12).

The similarity values of the type strain of *Amphibacillus xylanus* to the type strains of *Gracilibacillus dipsosauri*, *Gracilibacillus halotolerans*, *Halolactibacillus halophilus*, *Halolactibacillus miurensis*, and *Paraliobacillus ryukyuensis* are ~92.9–93.9%. In contrast, the similarity values of the type strain of *Amphibacillus xylanus* to those of *Amphibacillus fermentum* and *Amphibacillus tropicus* are 93.4% and 94.0%, respectively.

The three species of the genus *Amphibacillus* share physiological, biochemical, and chemotaxonomic characteristics in common and can be distinguished clearly from members of the genera *Gracilibacillus* and *Paraliobacillus* in the HA group by their lack of catalase, cytochromes, and quinones (Table 11) (Ishikawa et al., 2002). The *Amphibacillus* species can be differentiated from *Halolactibacillus* by spore formation, main metabolic products, and good growth under both anaerobic and aerobic conditions (Ishikawa et al., 2005). The main metabolic product of *Halolactibacillus* species is lactic acid, in contrast to the main products of *Amphibacillus xylanus*, which are formic acid, acetic acid, and ethanol. In addition to production of formic acid, acetic acid, and ethanol, *Amphibacillus fermentum* and *Amphibacillus tropicus* also produce lactic acid like *Halolactibacillus* species (Y. Niimura and others, unpublished data). This phenotypic trait in metabolism and 16S rRNA gene sequence similarity indicate that the positions of *Amphibacillus fermentum* and *Amphibacillus tropicus* may be rather closer to members of the genus *Halolactibacillus* than those of the genus *Amphibacillus*.

List of species of the genus *Amphibacillus*

1. **Amphibacillus xylanus** Niimura, Koh, Yanagida, Suzuki, Komagata and Kozaki 1990, 300[VP]

xy.la′nus. N.L. adj. *xylanus* pertaining to xylan.

The characteristics are as described for the genus and listed in Table 11. Cells are motile by means of flagella or nonmotile (type strain is nonmotile). In aerobic and anaerobic cultures, colonies on glucose agar are small, circular, smooth, convex, entire, and white after 1 d of incubation. Cells are rods, 0.3–0.5 µm in diameter and 0.9–1.9 µm long. Oval endospores are formed under both aerobic and anaerobic conditions.

Heat-resistant. Good growth occurs in both well-aerated and anaerobic cultures (E_h, –370 mV; pH 10) when titanium (III) citrate is used as a reducing agent. Cell yields and growth rates are the same under aerobic and anaerobic conditions. Growth occurs: between pH 8.0 and 10.0, but not at pH 7.0; in the presence of 3% NaCl, but not in 6% NaCl; and between 25 °C and 45 °C, but not at 50 °C. Negative for nitrate reduction, H_2S production, and indole production.

Growth is not observed in nutrient broth. Citrate utilization and hydrolysis of gelatin are negative. D-Xylose, L-arabinose, D-ribose, D-glucose, D-fructose, esculin, salicin, maltose, sucrose, cellobiose, trehalose, soluble starch, pectin, and xylan (oat spelt) are utilized. Ethanol, acetic acid, and formic acid are produced from glucose under anaerobic conditions and acetic acid is produced under aerobic conditions. Lactic acid is not produced under either condition. The fermentation product from xylan is acetic acid in aerobic culture; formic acid, ethanol, and acetic acid are produced in anaerobic culture.

TABLE 11. Differential characteristics of *Amphibacillus* species and members of closely related genera in the phylogenetic tree[a]

Characteristic	*Amphibacillus xylanus*[b]	*Amphibacillus fermentum*[c]	*Amphibacillus tropicus*[c]	*Halolactibacillus halophilus*[d]	*Halolactibacillus miurensis*[d]	*Paraliobacillus ryukyuensis*[e]	*Gracilibacillus halotolerans*[f]	*Gracilibacillus dipsosauri*[f]
Spore formation	+	+	+	−	−	+	+	+
Anaerobic growth	+	+	+	+	+	+	−	−
Catalase	−	−	−	−	−	+	+	+
Cytochromes	−	−	−	−	−	+	+	+
Quinones	−	−	−	−	−	+	+	+
Products formed during anaerobic growth:								
Acectate	+	+	+	+	+	+	−	−
Formate	+	+	+	+	+	+	−	−
Ethanol	+	+	+	+	+	+	−	−
Lactate	−	+	w	+	+	+	−	−
Pyruvate	w	−[g]	−[g]	−	−	−	−	−
Mol% G + C	36–38	42	39	40	39	36	38	39

[a]Symbols: +, positive; −, negative; w, weakly positive. All species grow aerobically.
[b]Niimura et al. (1989).
[c]Zhilina et al. (2001a).
[d]Ishikawa et al. (2005).
[e]Ishikawa et al. (2002).
[f]Wainø et al. (1999).
[g]Pyruvate is produced under aerobic conditions.

The cell wall contains *meso*-diaminopimelic acid. The predominant cellular fatty acids are $C_{15:0 \text{ anteiso}}$, $C_{16:0}$, $C_{16:0 \text{ iso}}$, $C_{14:0}$, and $C_{15:0 \text{ iso}}$.

The type strain was isolated from an alkaline compost of manure with grass and rice straw.

DNA G + C content (mol%): 36–38 (T_m).
Type strain: Ep01, DSM 6626, JCM 7361, NBRC 15112.
GenBank accession number (16S rRNA gene): D82065, AJ496807.

2. **Amphibacillus fermentum** Zhilina, Garnova, Tourova, Kostrikina and Zavarzin 2002, 685[VP] (Effective publication: Zhilina, Garnova, Tourova, Kostrikina and Zavarzin 2001a, 720)

fer.men.tum. L. n. *fermentum* that which causes fermentation.

Cells are motile by means of one subterminal flagellum. Cells are rod-shaped, 0.5–0.75 μm in diameter and 1.5–4 μm long and occurring singly, in pairs, or sometimes in short chains. Spores are not observed, but cells are heat-resistant.

Good growth occurs in both well-aerated and anaerobic cultures. Strictly alkaliphilic. Growth occurs at pH 7.0–10.5, with optimum growth at pH 8.0–9.5. Growth is obligately dependent on the CO_3^{2-} ion. Growth occurs at a total mineralization of 0.17–3.3 M Na+ with an optimum of 1.87 M Na+ (in the form of sodium carbonates). The Cl− ion is not required. Mesophilic. Growth is observed between 18 °C and 56 °C; optimal growth occurs at 36–38 °C. Chemo-organotrophic. Yeast extract is required for anabolic growth (the obligate requirement is methionine). Sulfur is used as an electron acceptor. Tolerant to sulfide. Sulfur reduction is not coupled to energy generation.

D-Glucose, maltose, mannose, xylose, fructose, sucrose, maltose, cellobiose, and trehalose are utilized anaerobically. In addition, ribose, arabinose, galactose, lactose, and N-acetylglucosamine are utilized aerobically. Ethanol, acetic acid, formic

acid, and lactic acid are produced anaerobically from glucose. Acetic acid and pyruvic acid, along with small amounts of lactic and formic acids, are produced aerobically from glucose.

The major cellular fatty acids are anteiso-, iso-branched, and straight-chain acids.

The type strain was isolated from the bottom sediment of a coastal lagoon of Lake Magadi, Kenya.

DNA G + C content (mol%): 41.5 (T_m).
Type strain: Z-7984, DSM 13869, UNIQEM 210.
GenBank accession number (16S rRNA gene): AF418603.

3. **Amphibacillus tropicus** Zhilina, Garnova, Tourova, Kostrikina and Zavarzin 2002, 685[VP] (Effective publication: Zhilina, Garnova, Tourova, Kostrikina and Zavarzin 2001a, 720)

tro.pi.cus. L. adj. *tropicus* tropical, an organism isolated from a tropical lake.

Cells are motile by means of peritrichous flagella. Cells are thin rods, 0.4–0.5 μm in diameter and 2–6 μm long and occurring singly or in pairs. Heat-resistant. Oval endospores are formed terminally.

Good growth occurs in both well-aerated and anaerobic cultures. Strictly alkaliphilic. Growth occurs at pH 8.0–11.5, with optimum growth between pH 9.5 and 9.7. Growth is obligately dependent on the CO_3^{2-} ion. Growth occurs at a total mineralization of 0.17–3.6 M Na+, with an optimum of 1–1.87 M Na+ (in the form of sodium carbonates). The Cl− ion is not required. Mesophilic. Growth is observed between 18 °C and 56 °C; optimal growth is at 38 °C. Chemo-organotrophic. Yeast extract is required for anabolic growth (the obligate requirement is methionine). Sulfur is used as an electron acceptor. Sulfur reduction is not coupled to energy generation. High concentrations of sulfide are inhibitory.

D-Glucose, maltose, sucrose, cellobiose, trehalose, melibiose, peptone, yeast extract, and, at a low rate, Tween

80 are utilized anaerobically. Starch, glycogen, and xylan are hydrolyzed. In addition, xylose, fructose, and lactose are utilized aerobically. Yeast extract, peptone, and Tween 80 are not utilized aerobically. Ethanol, acetic acid, formic acid, and a small amount of lactic acid are produced anaerobically from glucose. Acetic acid, pyruvic acid, and lactic acid are produced aerobically from glucose. The major cellular fatty acids are anteiso-, iso-branched, and straight-chain acids.

The type strain was isolated from the bottom sediment of a coastal lagoon of Lake Magadi, Kenya.

DNA G + C content (mol%): 39.2 (T_m).

Type strain: Z-7792, DSM 13870, UNIQEM 212.

GenBank accession number (16S rRNA gene): AF418602.

Genus IV. **Anoxybacillus** Pikuta, Lysenko, Chuvilskaya, Mendrock, Hippe, Suzina, Nikitin, Osipov and Laurinavichius 2000a, 2114[VP] emend. Pikuta, Cleland and Tang 2003a, 1561

ELENA V. PIKUTA

An.o.xy.ba.cil´lus. Gr. pref. *an* without; Gr. adj. *oxys* acid or sour and in combined words indicating oxygen; L. masc. n. *bacillus* small rod; N.L. masc. n. *Anoxybacillus* small rod living without oxygen

Cells are rod-shaped and straight or slightly curved, sometimes with angular division and Y-shaped cells, 0.4–1.5 × 2.5–9.0 μm in size, often in pairs or short chains, with rounded ends. Gram-positive. Motile or nonmotile. Endospores are round, oval or cylindrical and resistant to heating and freezing. Spores are located at the end of the cells. There is not more than one spore per cell. Aerobes, facultative aerobes or facultative anaerobes; catalase-variable. Alkaliphilic, alkalitolerant or neutrophilic, moderately thermophilic. Chemoorganotrophic, with a fermentative or oxygen respiration metabolism.

DNA G + C content (mol%): 42–57.

Type species: **Anoxybacillus pushchinoensis** corrig. Pikuta Lysenko, Chuvilskaya, Mendrock, Hippe, Suzina, Nikitin, Osipov and Laurinavichius 2000a, 2116[VP] emend. Pikuta, Cleland and Tang 2003a, 1561.

Further descriptive information

Phylogenetic analysis indicated that the closest neighbor of *Anoxybacillus pushchinoensis* (with 98.8% similarity) was the bacterium designated "*Bacillus flavothermus*" (Pikuta et al., 2000a). Previous work had shown that the invalidly named species "*Bacillus flavothermus*" was phylogenetically distinct from other members of genus *Bacillus* (with 7–16% differences) and suggested that it be recognized as a separate genus (Rainey et al., 1994). Consequently, the new genus, *Anoxybacillus* was created to contain the species *Anoxybacillus pushchinoensis* and "*Bacillus flavothermus*," reclassified as *Anoxybacillus flavithermus* comb. nov., with the name correction. The name *Anoxybacillus* was chosen because of the ability of both species to live without oxygen.

Cell morphology of most species is the same; the cells are straight rods, except for *Anoxybacillus contaminans*, which is curved or curled. Cells of *Anoxybacillus pushchinoensis* and *Anoxybacillus kamchatkensis* may also have Y-shaped rods due to angular division, which begins in the exponential growth phase. One cell can have 2, 3, or 4 branched ends at the same time, but more often two branches of equal length arise from one cell at a pole. Some species are motile, but *Anoxybacillus pushchinoensis* and *Anoxybacillus voinovskiensis* are not. All species have terminal, spherical, oval, or cylindrical endospores. All species have a Gram-positive cell-wall structure with a thick layer of peptidoglycan, except for *Anoxybacillus contaminans*, which is Gram-variable. In the case of *Anoxybacillus pushchinoensis* K1, the outer and inner S-layers are clearly visible in ultrathin section. Old cells of some cultures have cytoplasm with dark granulations and light regions. Colonies have different characteristics according to species. *Anoxybacillus pushchinoensis* has white colonies with a yellowish center, circular shape (lens-shaped in deep agar), granular surface, and uneven edges. Colonies of *Anoxybacillus flavithermus* are round in shape with a smooth surface and bright yellow in color (as a result of high concentration of flavins). Colonies of the species *Anoxybacillus gonensis* are cream-colored with irregular shape and rough edges. Colonies of *Anoxybacillus contaminans* are circular with regular margins and raised centers and edges, and are opaque, glossy, and cream-colored. *Anoxybacillus voinovskiensis* colonies are circular with faint cream color and colonies of *Anoxybacillus ayderensis* and *Anoxybacillus kestanbolensis* are regular circle-shaped with round edges and cream color. Colonies of *Anoxybacillus kamchatkensis* grown aerobically are pinpoint, yellowish translucent, round with even edges; anaerobically grown colonies of this species are white, opaque, round with even edges and flat surface. Strictly aerobic cells of *Anoxybacillus rupiensis* form whitish colonies, about 5 mm in diameter with irregular margin. Colonies of *Anoxybacillus amylolyticus* are circular, cream, and smooth.

Cells multiply by binary fusion with the formation of two daughter cells.

Many members of the genus *Anoxybacillus* are alkaliphilic, but not all of them are obligate alkaliphiles. Most of species can grow at neutral pH and are not dependent upon carbonate ions. The only slightly acidophilic species of this genus that grow optimally at pH 5.6 is *Anoxybacillus amylolyticus*. Some species require specific carbonate-containing media because of obligate dependence on carbonates, as in the case of *Anoxybacillus pushchinoensis*[*]. The highest maximum pH for growth (pH 11.0) was observed for *Anoxybacillus ayderensis*. The common characteristic of all *Anoxybacillus* species is independence from NaCl and a comparatively low resistance to salt (5–6% NaCl inhibits growth). All species are moderately thermophilic bacteria with an optimal temperature for growth of 50–62 °C.

The genus contains saccharolytic and proteolytic species and, in natural communities, they perform the function of

---

[*](per liter): NaCl, 5 g; Na_2CO_3, 2.76 g; $NaHCO_3$, 10.0 g; KCl, 0.2 g; K_2HPO_4, 0.2 g; $MgCl_2 \cdot 6H_2O$, 0.1 g; NH_4Cl, 1.0 g; $Na_2S \cdot 9H_2O$, 0.5 g; resazurin, 0.001 g; yeast extract, 0.02 g; glucose, 5.0 g; vitamin solution (Wolin et al., 1963), 2 ml; trace mineral solution 1 ml (mg per 200 ml water: $MnCl_2 \cdot 4H_2O$, 720; $Fe(NH_4)(SO_4)_2 \cdot 12H_2O$, 400; $FeSO_4 \cdot 7H_2O$, 200; $CoCl_2 \cdot 6H_2O$, 200, $ZnSO_4 \cdot 7H_2O$, 200; $Na_2MoO_4 \cdot 2H_2O$, 20; $NiCl_2$, 100; $CuSO_4 \cdot 5H_2O$, 20; $AlK(SO_4)_2 \cdot 12H_2O$, 20; H_3BO_3, 20; and 5 ml HCl concentrated), and the final pH was adjusted to 9.5 with 6M NaOH. High purity nitrogen was used for the gas phase.

primary anaerobes or aerobes in the trophic chains of organic matter decomposition. They produce low energy products such as acetate, ethanol, and hydrogen that are used by secondary anaerobes or aerobes as electron donors. *Anoxybacillus flavithermus*, *Anoxybacillus kamchatkensis*, and *Anoxybacillus gonensis* have the capacity to hydrolyze both sugars and proteolysis products (amino acids in peptone and yeast extract), but *Anoxybacillus pushchinoensis* and *Anoxybacillus voinovskiensis* are saccharolytic and cannot grow on peptone, casein, gelatin, or yeast extract (yeast extract is used only as a source of carbon and vitamins).

Anoxybacillus species can use oxygen or nitrate as electron acceptors (except for *Anoxybacillus gonensis* and *Anoxybacillus rupiensis*, which do not reduce nitrate) and, without electron acceptors, they perform fermentation by the Embden–Meyerhof pathway. *Anoxybacillus pushchinoensis* is an aerotolerant anaerobe and prefers anaerobic conditions, but other species are facultative anaerobes. The only strictly aerobic species of this genus is *Anoxybacillus rupiensis*.

The percentage similarity as indicated by 16S rDNA sequence analysis is as follows: *Anoxybacillus pushchinoensis* and *Anoxybacillus flavithermus*, 98.9%; *Anoxybacillus flavithermus* and *Anoxybacillus gonensis*, 97%; and *Anoxybacillus pushchinoensis* and *Anoxybacillus gonensis*, 96%. *Anoxybacillus contaminans* has less then 97% similarity with *Anoxybacillus pushchinoensis*, *Anoxybacillus flavithermus*, and *Anoxybacillus gonensis*. *Anoxybacillus voinovskiensis* has 95.7% similarity with *Anoxybacillus flavithermus*, 94.8% with *Anoxybacillus gonensis*, and 94.5% with *Anoxybacillus pushchinoensis*. *Anoxybacillus ayderensis* has more than 98% similarity to the sequences of *Anoxybacillus gonensis* and *Anoxybacillus flavithermus*, and 97% similarity to *Anoxybacillus pushchinoensis*. *Anoxybacillus kestanbolensis* exhibits 97% similarity to *Anoxybacillus flavithermus* and higher than 96% similarity to *Anoxybacillus gonensis* and *Anoxybacillus pushchinoensis*. *Anoxybacillus amylolyticus* shows 98.2% similarity to *Anoxybacillus voinovskiensis* and 98.1% to *Anoxybacillus contaminans*; it is also shared a similarity of 96%, and 94% with *Anoxybacillus ayderensis* and *Anoxybacillus kestanbolensis*, respectively; it has 97.5% similarity with *Geobacillus tepidamans*. The level of 16S rRNA gene sequence similarity between *Anoxybacillus kamchatkensis* and the type strains of *Anoxybacillus* species (*Anoxybacillus pushchinoensis*, *Anoxybacillus flavithermus*, *Anoxybacillus gonensis*, *Anoxybacillus ayderensis*, and *Anoxybacillus kestanbolensis*) are correspondingly following: 97.7%, 98.7%, 98.9%, 99.2%, and 97.6%. For *Anoxybacillus rupiensis* and *Geobacillus tepidamans* it shows 96.8%.

DNA–DNA hybridization between *Anoxybacillus pushchinoensis* and *Anoxybacillus flavithermus* showed 58.8% homology; *Anoxybacillus flavithermus* and *Anoxybacillus gonensis* showed 53.4% homology, and *Anoxybacillus pushchinoensis* and *Anoxybacillus gonensis* showed 45% homology. DNA–DNA hybridization between *Anoxybacillus ayderensis* and *Anoxybacillus gonensis* showed 68.6% homology, and between *Anoxybacillus kestanbolensis* and *Anoxybacillus flavithermus* showed 60.4% homology. DNA–DNA hybridization between *Anoxybacillus ayderensis* and *Anoxybacillus pushchinoensis* showed 45.1% homology, between *Anoxybacillus kestanbolensis* and *Anoxybacillus flavithermus* it was 42.9%, and between *Anoxybacillus ayderensis* and *Anoxybacillus kestanbolensis* it was 40.5%. Hybridization *Anoxybacillus amylolyticus* with *Anoxybacillus voinovskiensis* showed 32%, with *Anoxybacillus contaminans* 30.7%, and with *Geobacillus tepidamans* 30.2%. Homology of *Anoxybacillus kamchatkensis* with *Anoxybacillus pushchinoensis*, *Anoxybacillus flavithermus*, *Anoxybacillus gonensis*, and *Anoxybacillus ayderensis* is 53%, 55%, 51%, and 51%, respectively. For *Anoxybacillus rupiensis* and *Geobacillus tepidamans* it shows 32%.

Anoxybacillus pushchinoensis is sensitive to the antibiotic bacitracin (100 μg/ml), but not to penicillin, vancomycin, ampicillin, streptomycin (all at 250 μg/ml), or chloramphenicol (100 μg/ml). The growth of *Anoxybacillus gonensis* is inhibited by chloramphenicol, ampicillin, streptomycin (25 μg/ml), and tetracycline (12.5 μg/ml). For *Anoxybacillus ayderensis* and *Anoxybacillus kestanbolensis*, the inhibition of growth by ampicillin (25 μg/ml), streptomycin (25 μg/ml), kanamycin (10 μg/ml), tetracycline (12.5 μg/ml), and gentamicin (10 μg/ml) was described. *Anoxybacillus amylolyticus* is sensitive to: kanamycin (5 μg/ml), penicillin G, ampicillin, gentamicin, chloramphenicol, tylosin, fusid acid (10 μg/ml), lincomycin (15 μg/ml), streptomycin (25 μg/ml), novobiocin, and tetracycline (30 μg/ml). Growth of *Anoxybacillus rupiensis* cells is inhibited by tetracycline, gentamicin, streptomycin, erythromycin, carbenicillin, and chloramphenicol, but the cells are resistant to ampicillin, oxacillin, penicillin, and nalidixic acid.

Most species of the genus have been isolated from hot springs. *Anoxybacillus flavithermus* was isolated from a hot spring in New Zealand; *Anoxybacillus gonensis*, *Anoxybacillus ayderensis*, and *Anoxybacillus kestanbolensis* were respectively isolated from the Gonen, Ayder, and Kestanbol hot springs in Turkey; *Anoxybacillus voinovskiensis* and *Anoxybacillus kamchatkensis* both were isolated from a hot spring on the Kamchatka peninsula in Russia. *Anoxybacillus amylolyticus* was isolated from geothermal soils of Mount Rittmann on Antarctica. *Anoxybacillus rupiensis* was isolated from hot springs in the region of Rupi basin in Bulgaria. *Anoxybacillus contaminans* was isolated as a contaminant of gelatin production plant in Belgium. The situation with *Anoxybacillus puschinensis* is not completely understood. The type strain K1 was isolated from manure (equine and porcine) that was collected 20 years previously from farms in the Moscow region of Russia and stored in a cold room at "Laboratory of Anaerobic Processes" of the Institute of Biochemistry and Physiology of Microorganisms in Pushchino, Russian Academy of Sciences. The original source of this bacterium may have been soils that survived passage through the digestive tract, or the manure itself may have been the natural ecosystem. It is known that the temperature of manure during long-term storage can increase spontaneously to the range appropriate for moderately thermophilic micro-organisms. Perhaps microniches were created with optimal pH and redox potential by microbial activity that could have provided suitable conditions for this organism. Microscopy of the manure sample before culture on laboratory media showed strain K1 cells were the dominant forms in the manure. Pathogenicity, antigenic structure, mutants, plasmids, phages, and phage typing have not been studied.

Enrichment and isolation procedures

Aerobic or anaerobic techniques can be used for cultivation of *Anoxybacillus* strains. Isolation of *Anoxybacillus flavithermus* and *Anoxybacillus gonensis* was performed in nutrient broth incubated at 60–70 °C and individual colonies were obtained on agar by the streak plate method. Isolation and purification of *Anoxybacillus pushchinoensis* was performed by dilution methods in Hungate tubes under anaerobic conditions (medium described above). Colonies of *Anoxybacillus pushchinoensis* were obtained on 3% Difco agar (w/v), to which the carbonate solution was added separately after sterilization by the roll-tube method. Antibiotics can be used for isolating clean cultures from enrichments; the pure culture of *Anoxybacillus pushchinoensis* was isolated by adding 500 μg/ml penicillin and 500 μg/ml streptomycin to an enrichment culture.

Maintenance procedures

Most species of *Anoxybacillus* can be maintained on liquid or agar media (nutrient broth). For cultivation of *Anoxybacillus pushchinoensis*, the previously described medium is used, but culture of this species can be achieved by aerobic procedures. The best method for the long-term preservation of *Anoxybacillus* cultures is lyophilization.

Procedures for testing special characters

The determination of the moderately thermophilic nature, alkali-tolerance or alkaliphilic nature, dependence upon carbonate ions, relationship to oxygen, and spore formation does not require specific procedures. Isolation of DNA and amplification were performed by the usual methods, i.e., phenol/chloroform extraction and PCR (by *Thermus aquaticus* thermostable DNA polymerase).

Differentiation of the genus *Anoxybacillus* from other genera

Characters that distinguish *Anoxybacillus* from closely related taxa are listed in Table 12.

Taxonomic comments

The genus *Anoxybacillus* includes 10 species, *Anoxybacillus pushchinoensis*, *Anoxybacillus amylolyticus*, *Anoxybacillus ayderensis*, *Anoxybacillus contaminans*, *Anoxybacillus flavithermus*, *Anoxybacillus gonensis*, *Anoxybacillus kamchatkensis*, *Anoxybacillus kestanbolensis*, *Anoxybacillus rupiensis*, and *Anoxybacillus voinovskiensis*. On the basis of 16S rDNA sequencing studies, all ten species of the genus are closely related (94.5–99.2%), but the percentage homology by DNA–DNA hybridization (30.7–68.6%) and their different physiological properties indicate that they are ten distinct species. The species epithets were corrected during the reclassification of *Anoxybacillus flavithermus* (formerly "*Bacillus flavothermus*") and emendation of the description of *Anoxybacillus pushchinoensis* (formerly *Anoxybacillus pushchinensis*). A tree showing the phylogenetic relationships of *Anoxybacillus* is shown in Figure 13.

The special attributes of the genus are as follows: moderate thermophilia, alkaliphilia, or alkali tolerance; spore formation; Gram-positive staining; and capability of growth in both aerobic and anaerobic conditions with sugars. Differentiation of species of the genus *Anoxybacillus* is shown in Table 13. Detailed characteristics of the species are presented in Table 14

TABLE 12. Distinctive characteristics of the genus *Anoxybacillus* and other endospore-forming genera[a]

Feature	*Anoxybacillus*[d]	*Alkaliphilus*[e]	*Amphibacillus*[f]	*Anaerovirgula*[g]	*Bacillus*[h]	*Clostridium*[i]	*Desulfotomaculum*[j]	*Sporolactobacillus*[k]	*Tindallia*[l,m]
Motility	+/−	+	+	+	+	+	+/−	+	+/−
Gram reaction	+	+	+	+	+	+/−	+/−	+	+
Relation to O₂	An/Aer	oblig An	f Aer	Oblig An	An/f Aer	An (atl)	oblig An	An/f Aer	oblig An
Reduction of:									
SO₄²⁻ to H₂S	−	−[b]	−	−	−	−	+	−	−
NO₃²⁻ to NO₂⁻	+/−	−	−	−	+/−	+/−	−	−	−
Activity of:									
Catalase	+/−	ND	−	−	+	−	−	−	−
Oxidase	+/−	ND	−	−	+/−	−	−	−	−
NaCl (3–12%) requirement	−	−	−	−	+/−	+/−	+/−	−	+
CO₃²⁻ requirement	+/−	−[c]	−	−	−	−	+/−	−	−
Lactate as sole end product	−	ND	−	−	−	−	−	+	−

[a]Symbols: +, >85% positive; d, different strains give different reactions (16–84% positive); −, 0–15% positive; w, weak reaction; ND, not determined. Abbreviations: An, anaerobe; An (atl), aerotolerant anaerobe; f Aer, facultative aerobe; oblig An, obligative anaerobe.
[b]Reduction of sulfur, thiosulfate, and fumarate.
[c]Takai and Fredrickson, personal communication.
[d]Data from Pikuta et al. (2000a).
[e]Data from Takai et al. (2001).
[f]Data from Pikuta et al. (2000a).
[g]Data from Pikuta et al. (2006).
[h]Data from Pikuta et al. (2000a).
[i]Data from Claus and Berkeley (1986).
[j]Data from Pikuta et al. (2000a, 2000b).
[k]Data from Pikuta et al. (2000a).
[l]Data from Pikuta et al. (2003b).
[m]Data from Kevbrin et al. (1998).

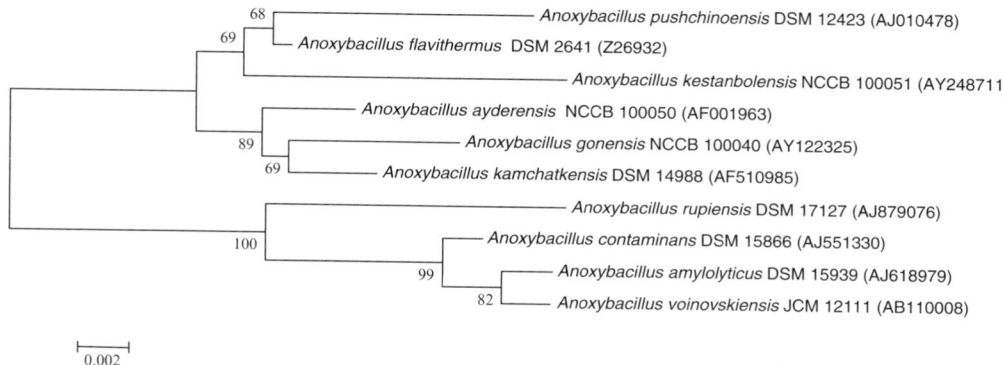

FIGURE 13. Evolutionary relationships within the genus *Anoxybacillus*. The evolutionary history was inferred using the neighbor-joining method. The bootstrap consensus tree inferred from 2,000 replicates is taken to represent the evolutionary history of the taxa. The percentage of replicate trees in which the associated taxa clustered together in the bootstrap test (2,000 replicates) are shown next to the branches. The tree is drawn to scale, with branch lengths in the same units as those of the evolutionary distances used to infer the phylogenetic tree. The evolutionary distances were computed using the Jukes–Cantor method and are in the units of the number of base substitutions per site. All positions containing gaps and missing data were eliminated from the dataset (Complete deletion option). There were a total of 1,298 positions in the final dataset. Phylogenetic analyses were conducted in MEGA software version 4.

TABLE 13. Diagnostic characteristics for species of genus *Anoxybacillus*[a]

Characteristic	1. *A. pushchinoensis*[b]	2. *A. amylolyticus*[c]	3. *A. ayderensis*[d]	4. *A. contaminans*[e]	5. *A. flavithermus*[f]	6. *A. gonensis*[g]	7. *A. kamchatkensis*[h]	8. *A. kestanbolensis*[i]	9. *A. rupiensis*[j]	10. *A. voinovskiensis*[k]
Yellow colonies	–	–	–	–	+	–	–	–	–	–
Motility	–	+	+	+	+	+	+	+	+	–
Gram reaction	+	+	+	Variable	+	+	+	+	+	+
CO_3^{2-} requirement	+	–	–	–	–	–	ND	–	–	–
Catalase	–	+	+	+	+	+	–	+	+	+
NO_3^{-} reduction	+	+	+	+	+	–	ND	+	–	+
Growth on:										
Peptone	–	ND	+	ND	+	+	+	+	+	+
Xylose	–	–	+	+	ND	+	–	–	+	+
Gelatin hydrolysis	–	–	+	+	–	+	–	–	+	–
Casein hydrolysis	–	–	–	–	+	+	–	–	+	–

[a]Symbols: +, >85% positive; d, different strains give different reactions (16–84% positive); –, 0–15% positive; w, weak reaction; ND, not determined.
[b]Data from Pikuta et al. (2000a).
[c]Data from Poli et al. (2006).
[d]Data from Dulger et al. (2004c).
[e]Data from De Clerck et al. (2004c).
[f]Data from Heinen et al. (1982).
[g]Data from Belduz et al. (2003).
[h]Data from Kevbrin et al. (2005).
[i]Data from Dulger et al. (2004).
[j]Data from Derekova et al. (2007).
[k]Data from Yumoto et al. (2004a).

TABLE 14. Descriptive table of *Anoxybacillus* species[a]

Characteristic	1. *A. pushchinoensis*[d]	5. *A. flavithermus*[e]	6. *A. gonensis*[f]	4. *A. contaminans*[g]	10. *A. voinovskiensis*[h]	3. *A. aydemensis*[i]	8. *A. kestanbolensis*[j]	2. *A. amylolyticus*[k]	9. *A. rupiensis*[l]	7. *A. kamchatkensis*[m]
Cell diameter, length (μm)	0.5–0.6, 3.0–5.0	0.85, 2.3–7.1	0.75, 5.0	0.7–1.0, 4.0–10.0	0.4–0.6, 1.5–5.0	0.55, 4.6	0.65, 4.75	0.5, 2.0–2.5	0.7–1.5, 3.3–7.0	1.0, 2.5–8.8
Gram reaction	+	+	+	Variable	+	+	+	+	+	+
Motility	–	+	+		+	+	+	+	+	+
Spore location	Terminal	Terminal	Terminal	Terminal/subterminal	–[b]	Terminal	Terminal	Terminal	Terminal	Terminal
Temperature range (optimum) (°C)	37–65 (62)	30–72 (60)	40–70 (55–60)	40–60 (50)	30–64 (54)	30–70 (50)	40–70 (50–55)	45–65 (61)	35–67 (55)	38–67 (60)
pH range (optimum)	8.0–10.5 (9.5–9.7)	6–9	6–10.0 (7.5–8.0)	4.5–10.0 (7.0)	7–8	6.0–11.0 (7.5–8.5)	6–10.5 (7.5–8.5)	5.6 (ND)	5.5–8.5 (6.0–6.5)	5.7–9.9 (6.8–8.5)
NaCl range (optimum) (%, w/v)	0–3.0 (0.5–1.0)	0–2.5	0–4.0 (2.0)	0–5 (0.5)	0–3 (ND)	0–2.5 (1.5)	0–4 (2.5)	0.6 (ND)	ND (ND)	ND (ND)
Relation to O₂	Aerotolerant anaerobe	Facultative anaerobe	Facultative anaerobe	Facultative anaerobe	Facultative anaerobe	Facultative anaerobe	Facultative anaerobe	Facultative anaerobe	Strict aerobe	Facultative aerobe
Catalase	–	+	+	+	+	+	+	+	–	–
Oxidase	ND	+	+	–	+	+	+	–	ND	+
NO₃⁻ reduction	+	+	–	+	+	+	+	–	–	ND
Substrates:										
D-Glucose	+	+	+	+	+	+	+	+	+	+
D-Fructose	+	ND	+	+	+	+	+	ND	+	+
Starch	+	+	+	+	–	+	–	+	–	–
Peptone	–	+	+	–	+[c]	+	+	ND	ND	+
Gelatin	–	–	–	+	–	–	+	–	+	–
Casein	–	+	–	–	–	–	–	–	–	–
Antibiotics:										
Sensitive to	B	ND	A, S, T, C	ND	ND	A, S, K, T, G	A, S, K, T, G	A,P,C,K,F,G,S,T,N,L	T,G,S,E,C	ND
Resistant to	P, A, V, S, C	ND	ND	ND	ND	ND	ND	ND	A,O,P	ND
G + C mol%	42.2	41.6	57.0	44.4	43.9	54.0	50.0	43.5	41.7	42.3
Source	Manure, Russia	Hot spring, New Zealand	Hot spring, Turkey	Gelatin batches, Belgium	Hot spring, Russia	Hot spring, Turkey	Hot spring, Turkey	Geothermal soil, Antarctica	Hot spring, Bulgaria	Hot spring, Russia

[a]Symbols: +, >85% positive; d, different strains give different reactions (16–84% positive); –, 0–15% positive; w, weak reaction; ND, not determined. Antibiotics: A, ampicillin; B, bacitracin; C, chloramphenicol; E, erytromycin; F, fusidic acid; G, gentamicin; P, penicillin; O, oxacillin; S, streptomycin; V, vancomycin; K, kanamycin; T, tetracycline; N, novobiocin; L, linkomycin.

[b]Spores were never observed.

[c]Personal communication.

[d]Data from Pikuta et al. (2000a).

[e]Data from Heinen et al. (1982).

[f]Data from Belduz et al. (2003).

[g]Data from De Clerck et al. (2004c).

[h]Data from Yumoto et al. (2004a).

[i]Data from Dulger et al. (2004c).

[j]Data from Dulger et al. (2004c).

[k]Data from Poli et al. (2006).

[l]Data from Derekova et al. (2007).

[m]Data from Kevbrin et al. (2005).

List of species of the genus *Anoxybacillus*

1. **Anoxybacillus pushchinoensis** corrig. Pikuta, Lysenko, Chuvilskaya, Mendrock, Hippe, Suzina, Nikitin, Osipov and Laurinavichius 2000a, 2116^{VP} emend. Pikuta, Cleland and Tang 2003a, 1561

push.chi.noen′sis. N.L. masc. adj. *pushchinoensis* pertaining to Pushchino, a research center near Moscow, Russia, where the organism was isolated.

Data are from Pikuta et al. (2000a) and Pikuta et al. (2003a). Straight rods, 0.4–0.5 × 2.5–3.0 μm in size, single, in pairs, sometimes in irregular curved chains. Gram-positive. Nonmotile. Y-shaped cells occur at angular division. Forms round endospores. Aerotolerant anaerobe, chemoheterotrophic, alkaliphilic, moderately thermophilic. Grows at 37–65 °C, with an optimum at 62 °C. Obligate alkaliphile that cannot grow at pH 7.0; grows in a pH range of 8.0–10.5 with an optimum of 9.5–9.7. CO_3^{2-} is obligately required. Optimal growth at 1% NaCl; tolerant to 3% NaCl. Growth substrates are D-glucose, sucrose, D-fructose, D-trehalose, and starch. The major fermentation products are hydrogen and acetic acid. Nitrate is reduced to nitrite. Sulfate, sulfite, thiosulfate, and sulfur are not reduced. Yeast extract stimulates growth. Vitamins are required. Catalase-negative. Gelatin and casein are not hydrolyzed.

Source: cow and pig manure with neutral pH.
DNA G + C content (mol%): 42.2 ± 0.2 (HPLC).
Type strain: K1, ATCC 700785, DSM 12423, VKM B-2193.
GenBank accession number (16S rRNA gene): AJ010478.

2. **Anoxybacillus amylolyticus** Poli, Esposito, Lama, Orlando, Nicolaus, de Appolonia, Gambacorta and Nicolaus 2006, 1459^{VP} (Effective publication: Poli, Esposito, Lama, Orlando, Nicolaus, de Appolonia, Gambacorta and Nicolaus 2006, 305.)

a.mi.lo.ly.ti.cus. Gr. n. *amulon* starch; connecting vowel -*o*-; Gr. adj. *luticos* able to dissolve; N.L. masc. adj. *amylolyticus* starch-dissolving.

Data from Poli et al. (2006). Cells are Gram-positive, motile, straight rods, 0.5 × 2.0 – 2.5 μm in size. Spores are terminal, ellipsoidal to cylindrical endospores. Colonies are circular, smooth, and cream in color. Facultative anaerobe. Catalase-positive but oxidase-negative. Reduces nitrate to nitrite. Hydrolysis hippurate and starch. Utilizes D-galactose, D-trehalose, D-maltose, raffinose, and sucrose when the medium supplemented with 0.06% of yeast extract. It is positive for tyrosine decomposition. Sensitive to lysozyme. Negative with respect to casein and gelatin hydrolysis and phenylalanine deamination. Does not produce indole. On sugar media it is able to produce exopolysaccharide, possesses a constitutive extracellular amylase activity. Does not grow on media without yeast extract, but acetate (as source of carbon) with D-glucose, D-lactose, D-fructose, D-arabinose, D-cellobiose, D-mannose, D-ribose, D-xylose, D-sorbose, and glycerol. It is slightly acidophilic, growing at pH 5.6. It is thermophile, growth occurs between 45 °C and 65 °C with optimum temperature at 61 °C. Grow at NaCl 0.6% but not at concentrations higher than 3%. Antibiotics inhibited growth: kanamycin (5 μg), penicillin G, ampicillin, gentamicin, chloramphenicol, tylosin, fusid acid (10 μg), lincomycin (15 μg), streptomycin (25 μg), novobiocin, tetracycline (30 μg).

Source: geothermal soil of Mount Rittmann on Antarctica.
DNA G + C content (mol%): 43.5 (HPLC).
Type strain: MR3C, ATCC BAA-872, DSM 15939, CIP 108338.
GenBank accession number (16S rRNA gene): AJ618979.

3. **Anoxybacillus ayderensis** Dulger, Demirbag and Belduz 2004, 1503^{VP}

ay.de.ren.sis. N.L. masc. adj. *ayderensis* pertaining to Ayder, a hot spring in the province of Rize, Turkey, from where organism was isolated.

Data from Dulger et al. (2004). Cells are rod-shaped, Gram-positive, spore-forming, 0.55 × 4.60 μm in size. Location of spherical spores is terminal. Colonies are 1–2 mm in diameter, regular circle shaped with round edges, and cream color. Facultatively anaerobic, alkalitolerant, and moderately thermophilic chemo-organotroph. Catalase- and oxidase-positive. Nitrate reduced to nitrite. Starch and gelatin, but not caseine hydrolyzed. Urease, indole, and hydrogen sulfide not produced. D-glucose, D-raffinose, D-sucrose, D-xylose, D-fructose, L-arabinose, maltose, D-mannose, and peptone utilized. Temperature range for growth 30–70 °C; optimum growth at 50 °C. Growth range at 0–2.5% NaCl and optimum growth occurs at 1.5%. Optimum pH is 7.5–8.5; pH range for growth is 6–11. Growth was inhibited in the presence of ampicillin, streptomycin, tetracycline, gentamicin, and kanamycin.

Source: Ayder hot spring, Turkey.
DNA G + C content (mol%): 54 (T_m).
Type strain: AB04, NCIMB 13972, NCCB 100050.
GenBank accession number (16S rRNA gene): AF001963.

4. **Anoxybacillus contaminans** De Clerck, Rodríguez-Díaz, Vanhoutte, Heyrman, Logan and De Vos 2004c, 944^{VP}

con.ta′mi.nans. L. part. adj. *contaminans* contaminating.

Data are from De Clerck et al. (2004c). Cells are curved or frankly curled, round ended, Gram-variable, feebly motile rods that occur singly, in pairs, or short chains. Sizes of cells are 0.7–1.0 × 4.0–10.0 μm. Endospores are oval and located subterminally or terminally within slightly swelled sporangia. Colonies are circular, 1–2 mm in diameter, with regular margins and raised centers and edges, and are opaque, glossy, and cream-colored. Facultatively anaerobic, catalase-positive, but oxidase-negative. Nitrate is reduced to nitrite. Moderately thermophilic. The temperature range for growth is 40–60 °C with the optimum at 50 °C. Alkalitolerant, grows with pH optimum at 7.0, minimum at pH 4–5, and maximum at pH 9–10. Doesn't require NaCl for growth; NaCl range for growth is 0–5%. Chemoheterotrophic. Gelatin, but not *o*-nitrophenyl-β-D-galactopyranoside or casein is hydrolyzed. All strains are negative for arginine dihydrolase, lysine decarboxylase, ornithine decarboxylase, citrate utilization, hydrogen sulfide production, urease, tryptophan deaminase, indole production, and the Voges–Proskauer reaction. Hydrolysis of esculin is weak. Small amounts of acid without gas are produced from L-arabinose, D-fructose, D-galactose, D-glucose, glycerol, glycogen, maltose, D-mannose, D-melezitose, methyl-D-glucoside, N-acetylglucosamine, D-raffinose, ribose, starch, sucrose, D-trehalose, D-turanose, and D-xylose. Production of acid is negative for adonitol, amygdalin, D-arabinose, D- and L-arabitol, arbutin, D-cellobiose, dulcitol, D- and L-fucose, gentiobiose, gluconate, inulin, 2- and 5-keto-D-gluconate, lactose, D-lyxose, mannitol, D-melibiose, *meso*-inositol, methyl-D-mannoside, methyl-xyloside, rhamnose, salicin, sorbitol, L-sorbose, D-tagatose, L-xylose, and xylitol.

The major cellular fatty acids are $C_{15:0\,iso}$, $C_{16:0}$, and $C_{17:0\,iso}$ (52, 11 and 12% of total fatty acid respectively). The following fatty acids are presented in smaller amounts: $C_{14:0}$, $C_{15:0\,anteiso}$, $C_{16:0\,iso}$, and $C_{17:0\,anteiso}$ (3, 7, 5 and 7% of total fatty acids, respectively).

Source: gelatin sample from gelatin production plant.
DNA G + C content (mol%): 44.4 (HPLC).
Type strain: DSM 15866, LMG 21881.
GenBank accession number (16S rRNA gene): AJ551330.

5. **Anoxybacillus flavithermus** (Heinen, Lauwers and Mulders 1982) Pikuta, Lysenko, Chuvilskaya, Mendrock, Hippe, Suzina, Nikitin, Osipov and Laurinavichius 2000a, 2116[VP] (*Bacillus flavothermus* Heinen, Lauwers and Mulders 1982, 270.)

fla.vi.ther′mus. L. adj. *flavus* yellow; Gr. adj. *thermos* warm; N.L. adj. *flavithermus* to indicate a yellow thermophilic organism

Data are from Heinen et al. (1982), Claus and Berkeley (1986), Sharp et al. (1992), Rainey et al. (1994), and Pikuta et al. (2000a). Rods, 0.85×2.3–$7.1\,\mu m$. Motile. Gram-positive. Terminal spores. Colonies are round, smooth, yellow. Facultatively anaerobic. Catalase-positive. Oxidase-positive. Starch, but not gelatin, hydrolyzed. Grows in peptone-yeast extract media. Glucose, mannose, maltose, sucrose, arabinose, rhamnose, and sorbitol utilized. Positive for acetoin, arginine dihydrolase, lysine decarboxylase, tryptophan deaminase, and β-galactosidase. Nitrate reduced to nitrite. Urease, ornithine decarboxylase, indole, and H_2S not produced. Growth in 2.5% NaCl broth, but not in 3% NaCl. Optimal pH for growth of 6–9. No growth at pH 5.0. Temperature range for growth 30–72°C; optimal growth at 60°C (aerobic) and 65°C (anaerobic). *Source:* hot spring, New Zealand.

DNA G + C content (mol%): 41.6 (HPLC).
Type strain: d.y., DSM 2641, NBRC 15317, LMG 18397.
GenBank accession number (16S rRNA gene): AF004589, Z26932.

6. **Anoxybacillus gonensis** Belduz, Dulger and Demirbag 2003, 1319[VP]

go.nen.sis. N.L. masc. adj. *gonensis* pertaining to Gonen, a hot spring in the province of Balikesir, Turkey, where organism was isolated.

Data are from Belduz et al. (2003). Rod-shaped, Gram-positive, motile, spore-forming, measuring $0.75 \times 5.0\,\mu m$. Terminal spherical endospores are formed. Colonies rough, cream-colored. Facultatively anaerobic. Catalase weak-positive. Oxidase-positive. Starch and gelatin hydrolyzed. Glucose, glycogen, raffinose, sucrose, xylose, mannitol, and peptone utilized. Nitrate not reduced to nitrite. Urease, indole, and H_2S not produced. Growth in 4% NaCl broth. Alkalitolerant: pH range for growth 6.0–10.0 and optimal pH for growth of 7.5–8.0. Moderate thermophile with temperature range from 40°C to 70°C and optimum at 55–60°C.

Source: Gonen hot spring, Turkey.
DNA G + C content (mol%): 57 (T_m).
Type strain: G2, NCIMB 139330, NCCB 100040.
GenBank accession number (16S rRNA gene): AY122325.

7. **Anoxybacillus kamchatkensis** Kevbrin, Zengler, Lysenko and Wiegel 2005, 397[VP]

kam.chat.ken′sis. L. adj. *kamchatkensis* pertaining to Kamchatka penninsula, Russia, where the organism was isolated.

Data are from Kevbrin et al. (2005). Cells are straight rods, 1.0×2.5–$8.8\,\mu m$ in size, single or in pairs. Gram-positive with Gram-positive-type cell wall. Motile by peritrichous flagella. Y-shaped (or branched) cells are infrequently observed. Forms terminal oval spores. Aerobically grown colonies are pinpoint, yellowish translucent, round with an even edge; anaerobically grown colonies are white, opaque, round with even edges and flat surface.

Facultative aerobe. Catalase- and oxidase-negative. Alkalitolerant moderate thermophile with an optimum growth temperature at 60°C and range between 38°C and 67°C (no growth at or below 37°C and at or above 68°C). Grows in a $pH^{25°C}$ range of 5.7–9.9 with an optimum of 6.8–8.5. Yeast extract (0.1 g/l) and B_{12} vitamin are required for growth on carbohydrates. Aerobic utilization of ribose, glucose, fructose, galactose, mannitol, maltose, trehalose, sucrose, pyruvate, yeast extract, peptone, tryptone, Casamino acids, and pectin. Anaerobic utilization of glucose, fructose, mannitol, maltose, trehalose, sucrose, and yeast extract. Starch, casein, and gelatin are not hydrolyzed. Fermentation products of glucose are lactate as main product and acetate, formate, and ethanol as minor products. End products of glucose oxidation are lactate, acetate, and traces of fumarate, succinate, and ethanol. Does not grow on: arabinose, xylose, xylitol, mannose, sorbitol, inositol, lactose, raffinose, and gluconate.

Source: volcanic thermal fields of the Geyser Valley in Kamchatka, Russia.
DNA G + C content (mol%): 42.3 (T_m).
Type strain: JW/VK-KG4, ATCC BAA-549, DSM 14988.
GenBank accession number (16S rRNA gene): AF510985.

8. **Anoxybacillus kestanbolensis** Dulger, Demirbag and Belduz 2004, 1503[VP]

kes.tan.bo.len.sis. N.L. masc. adj. *kestanbolensis* pertaining to Kestanbol, a hot spring in the province of Canakkale, Turkey, from where organism was isolated.

Data from Dulger et al. (2004). Cells are rod-shaped, Gram-positive, motile, spore-forming, with sizes $0.65 \times 4.75\,\mu m$. Location of spherical endospores is terminal. Colonies are 1–1.5 mm in diameter, regular circle shaped with round edges, and cream color. Facultatively anaerobic, alkalitolerant, and moderately thermophilic chemo-organotroph. Catalase- and oxidase-positive. Nitrate is reduced to nitrite. Starch, but not gelatin and caseine, is hydrolyzed. Urease, indole, and hydrogen sulfide are not produced. D-Glucose, D-raffinose, D-sucrose, D-fructose, maltose, D-mannitol, D-mannose, and peptone are utilized. Temperature range for growth is 40–70°C and optimum at 50–55°C. Grows in the absence of NaCl; growth range from 0 to 4% NaCl and optimum growth occurs at 2.5%. pH range for growth from 6.0

to 10.5 and optimum pH is 7.5–8.5. Growth was inhibited in presence of ampicillin, streptomycin, tetracycline, gentamicin, and kanamycin.

Source: Kestanbol hot spring, Turkey.

DNA G + C content (mol%): 50 (T_m).

Type strain: K4, NCIMB 13971, NCCB 100051.

GenBank accession number (16S rRNA gene): AY248711.

9. **Anoxybacillus rupiensis** Derekova, Sjøholm, Mandeva, and Kambourova, 2007 581[VP]

ru.pi.en'sis (N.L. masc. adj. *rupiensis* pertaining to Rupi Basin, the place of isolation of the type strain).

Data are from Derekova et al. (2007). Straight, motile rods, 0.7–1.5 × 3.3–7.0 μm in size, single, in pairs, sometimes in chains. Gram-positive. Forms terminal ellipsoidal or cylindrical endospores. Colonies are whitish, 5 mm in diameter with irregular margin. Thermophile with range of growth between 35 °C and 67 °C (optimum 55 °C). pH range from 5.5 to 8.5 (optimum 6.0–6.5). Utilizes sugars, polysaccharides and polyols in the presence of proteinaceous substrates or inorganic nitrogen. Growing on ribose, xylose, fructose, glucose, and maltose. It does not grow on galactose, L-ramnose, raffinose, sucrose, lactose, phenyl-alanine, tyrosine, and citrate. Hydrolyzing starch, casein, and xylan but not salicin, inulin, gelatin, olive oil, and pectin. Growth supported by mannitol but not by ribitol, galactitol and sorbitol. Indole is not produced, the Voges–Proskauer reaction and methyl red test are negative. Obligate aerobe, catalase-positive. Does not reduce nitrate to nitrite.

The major cellular fatty acids are $C_{15:0\,iso}$ and $C_{17:0\,iso}$.

Source: terrestrial hot spring at Rupi Basin.

DNA G + C content (mol%): 41.7 (HPLC).

Type strain: R270, DSM 17127, NBIMCC 8387.

GenBank accession number (16S rRNA gene): AJ879076.

10. **Anoxybacillus voinovskiensis** Yumoto, Hirota, Kawahara, Nodasaka, Okuyama, Matsuyama, Yokota, Nakajima and Hoshino 2004a, 1242[VP]

vo.ino.vskien'sis. N.L. adj. *voinovskiensis* from Voinovskie, named after the Voinovskie Springs from where the microorganism was isolated.

Data from Yumoto et al. (2004a). Cells are Gram-positive, nonmotile, straight rods, 0.4–0.6 × 1.5–5.0 μm in size. Spores were never observed. Colonies are circular and faint cream in color. Facultatively anaerobic (prefers aerobic conditions). Catalase- and oxidase-positive. Nitrate is reduced to nitrite. Negative for hydrogen sulfide production and hydrolysis of casein, gelatin, starch, DNA, and Tween 20 and 80. Hydrolyzes Tween 40 and 60. Growth occurs at pH 7–8, but not at 9–10. Growth occurs at 30–64 °C with optimum temperature at 54 °C. Growth in the absence of NaCl but not at concentrations higher than 3%. Acid is produced from D-glucose, D-xylose, D-arabinose, D-fructose, maltose, D-mannose, sucrose, sorbitol, and cellobiose in aerobic conditions. Grows on peptone without sugars. No acid is produced from D-galactose, raffinose, melibiose, inositol, mannitol, trehalose, L-rhamnose, and lactose in aerobic conditions.

Source: hot spring in Kamchatka, Russia.

DNA G + C content (mol%): 43.9 (HPLC).

Type strain: TH13, JCM 12111, NCIMB 13956.

GenBank accession number (16S rRNA gene): AB110008.

Species Candidatus

The 16S rDNA sequence of "*Anoxybacillus hidirlerensis*" CT1Sari[T] was deposited in GenBank with accession number EF433758 (Inan et al., unpublished) and "*Anoxybacillus bogroviensis*" with accession number AM409184 (Atanassova et al., unpublished).

Acknowledgements

The author is thankful to Dr Damien Marsic and Mr Richard Hoover for assistance.

Genus V. **Cerasibacillus** Nakamura, Haruta, Ueno, Ishii, Yokota and Igarashi 2004b, 1067[VP]

PAUL DE VOS

Ce.ras.i.ba.cil'lus. L. neut. n. *cerasum* a cherry; L. masc. n. *bacillus* small rod; N.L. masc. n. *cerasibacillus* a cherry *Bacillus*, as the appearance of its sporangium is cherry-like.

Rod-shaped, **Gram-positive** bacterium, 0.8 μm in diameter and 2.5–5.0 μm in length, occurring as single rods, in pairs, or in short chains. Terminal **spherical endospores** are formed. Strictly aerobic. Growth occurs at 30–55 °C (optimal growth at 50 °C) and in the pH range 7.5–10 (optimum 8–9). Good growth occurs at low NaCl concentrations, but no growth is observed in TSB with 10% NaCl (assessed after 6d incubation). **Catalase-positive**, does not reduce nitrate, and negative for the Voges–Proskauer test and indole production (API strip). **Casein is not hydrolyzed.** Acid production from xylose has been observed. Peptidoglycan of the *meso*-diaminopimelic acid type is present in the cell wall. The major cellular fatty acid is $C_{15\,iso}$ and **MK-7** is the main menaquinone type.

DNA G + C content (mol%): 33.9–41.8 (HPLC).

Type species: **Cerasibacillus quisquiliarum** Nakamura, Haruta, Ueno, Ishii, Yokota and Igarashi 2004b, 1067[VP].

Further information

Colonies on TSA plates at 37 °C are pigmented (light yellowish-brown), round, and opaque. Longer incubation at 37 °C and/or growth on TSA at 50 °C reveals amorphous, translucent colonies. The single strain (BLx) of the single species of the taxon was isolated from a decomposing system of kitchen refuse (Nakamura et al., 2004b) after its presence had been detected by a non-cultural based approach (denaturing-gradient gel electrophoresis, DGGE) (Haruta et al., 2002) and molecular characterization of the dominant bands. Nakamura et al. (2004a) showed that the abundance of strain BLx increased following a measured increase in gelatinase activity in the composting process, which may directly link the ecological/metabolic role of the strain in the composting process. This hypothesis was further supported by the observation that at least one of the gelatinases of strain BLx has a very similar N-terminal amino acid sequence to the gelatinase found in the compost under study.

Enrichment procedures

Specific enrichment conditions for the vegetative cells are unknown, but spore formation seems to be stimulated by adding the following trace elements to a general medium: $MgSO_4$ (1 mM), $Ca(NO_3)_2$ (1 mM), $MnCl_2$ (10 μM), and $FeSO_4$ (1 μM).

Differentiation of the genus *Cerasibacillus* from other genera

Phylogenetically, the organism belongs to the *Virgibacillus–Lentibacillus* lineage as a clearly separated branch (16S rRNA gene

sequence analysis). Differentiation from these phylogenetically adjacent genera is possible based on morphological, physiological, and biochemical characteristics (Nakamura et al., 2004b).

Taxonomic comments

Due to its rather general habitat, it is likely that representatives are far more abundant in various thermophilic environments than is presently known. The reason for our lack of awareness of their metabolic role in various decay processes of organic compounds in nature and/or man-made processes is most probably linked to the particular growth conditions of the composting process of kitchen refuse, as well as the incubation conditions.

List of species of the genus *Cerasibacillus*

1. **Cerasibacillus quisquiliarum** Nakamura, Haruta, Ueno, Ishii, Yokota and Igarashi 2004b, 1067[VP]

quis.qui.li.a'rum. L. gen. pl. n. *quisquiliarum* of kitchen refuse.

This species is the only species described so far and its general characteristics conform to those of the genus. Further characteristics were determined using the API 50CHB system (bioMérieux), which measures: acid production from carbohydrates; hydrolysis of esculin, gelatin, starch, and urea; H_2S production; and nitrate reduction. This analysis demonstrated that gelatin is hydrolyzed, whereas starch, esculin, and urea are not. Acid is produced from D-ribose, L-sorbose, D-tagatose, and 5-ketogluconate, but not from

glycerol, erythritol, D-arabinose, L-arabinose, L-xylose, adonitol, methyl β-D-xylose, galactose, glucose, fructose, mannose, rhamnose, dulcitol, inositol, mannitol, sorbitol, methyl α-D-mannose, methyl α-D-glucose, *N*-acetylglucosamine, amygdalin, arbutin, salicin, cellobiose, maltose, lactose, melibiose, sucrose, trehalose, inulin, raffinose, glycogen, xylitol, gentiobiose, turanose, D-fucose, L-fucose, D-arabitol, L-arabitol, gluconate, or 2-ketogluconate.

DNA G + C content (mol%): 33.9–41.8 (HPLC).

Type strain: BLx, DSM 15825, IAM 15044, KCTC 3815.

EMBL/GenBank accession number (16S rRNA gene): AB107894 (BLx).

Genus VI. **Filobacillus** Schlesner, Lawson, Collins, Weiss, Wehmeyer, Völker and Thomm 2001, 430[VP]

HEINZ SCHLESNER

Fi.lo.ba.cil'lus. L. neut. n. *filum* thread; L. n. bacillus rod; N.L. masc. n. *Filobacillus* a thread-like rod.

Rod-shaped cells, **0.3–0.4 × 3–7 μm**. Spores are spherical and are located terminally. Sporangium is swollen. Motile by one laterally inserted flagellum. The cells **stain Gram-negative, but the cell wall is of the Gram-positive type**. KOH test negative. Colonies are white, smooth, and round with entire margins. Mesophilic. Optimum temperature for growth is between 30 °C and 38 °C. No growth at 42 °C.

Aerobic and chemo-organotrophic. Glucose is not fermented. No dissimilatory nitrate reduction. **Requires 2% NaCl for growth and tolerates up to 23%**; Optimum concentration 8–14%. pH range for growth 6.5–8.9; pH optimum 7.3–7.8. The cell wall contains an **L-Orn-D-Glu type murein** (variation A4β).

DNA G + C content (mol%): 35 (HPLC).

Type species: **Filobacillus milosensis** corrig. Schlesner, Lawson, Collins, Weiss, Wehmeyer, Völker, Thomm 2001, 430[VP].

Further descriptive information

Cells of *Filobacillus* are morphologically very similar to cells of a number of aerobic spore-forming bacteria, however, can easily be distinguished by several characteristics (Table 15).

16S rRNA analysis indicates that this organism is a member of the rRNA group 1 of *Bacillus*-like bacteria according to Ash et al. (1991).

Enrichment and isolation procedures

The only strain was isolated from the beach of Palaeochori Bay near a shallow water hydrothermal vent area, Milos, Greece. At a hot spot (surface temperature 62 °C) in front of the waterline, a hole of about 10 cm in depth was dug, and a sample was taken with a 20 ml syringe from the accumulating interstitial water. 100 μl of sample was transferred to 50 ml medium M36M: Casein after Hammarsten (Merck), 1.0 g; yeast extract, 0.25 g; gelatin, 1.0 g; Hutner's basal salts medium (HBM), 20 ml; vitamin solution no. 6 (Staley, 1968), 10 ml; artificial sea water (ASW, Lyman and Fleming, 1940), 3.5-fold concentrated, 970 ml. ASW was modified by addition of the following salts (per liter): $MnCl_2$, 0.4 g; Na_2SiO_3, 0.57 g; $(NH_4)_2SO_4$, 0.26 g. After an incubation of three weeks at 37 °C, upcoming colonies were streaked on medium M13(3x) + 10% NaCl: peptone, 0.75 g; yeast extract, 0.75 g; glucose, 0.75 g; NaCl, 100 g; HBM, 20 ml; vitamin solution no. 6, 10 ml; 0.1 M Tris/HCl buffer, pH 7.5 for liquid media, pH 8.5 for agar solid media, 50 ml; ASW, 250 ml; distilled water to 1 l.

TABLE 15. Characteristics differentiating *Filobacillus* from other physiologically or morphologically similar taxa[a]

Characteristic	*Filobacillus milensis*[b]	*Bacillus agaradhaerens*[c]	*Bacillus haloalkaliphilus*[c]	*Bacillus halophilus*[d]	*Bacillus marismortui*[e]	*Bacillus pseudofirmus*[c]	*Gracilibacillus dipsosauri*[f]	*Gracilibacillus halotolerans*[g]	*Halobacillus halophilus*[g]	*Halobacillus litoralis*[h]	*Marinococcus albus*[g]	*Salibacillus salexigens*[g]	*Virgibacillus pantothenticus*[g]
Spore shape	S	E	S	E	E	E	S	E	S	E/S	–	E	E/S
Sporangium position	T	ST	T	C	T/ST	ST	T	T	C/T	C/ST	NA	C/ST	T
Gram reaction	–	+	–[a]	+	+	+	+	+	+	+	+	+	+
Growth in the presence of 20% NaCl	+	–	+	+	+	–	+	+	+	+	+	+	+
Growth at 50°C	–	–	–	+	+	ND	+	+	–	–	ND	–	–
Murein type	Orn-D-Glu	ND	m-Dpm	m-Dpm	m-Dpm	ND	m-Dpm	m-Dpm	Orn-D-Asp	Orn-D-Asp	m-Dpm	m-Dpm	m-Dpm
Acid produced from:													
Glucose	–	+	–	+	+	+	+	+	–	+	–	+	+
Trehalose	–	+	–	+	–	+	+	+	–	+	–	+	+
Xylose	–	–	–	+	–	–	+	+	–	+	–	–	–
Casein	–	+	–	–	+	+	–	–	+	–	–	+	+
Hydrolysis of:													
Gelatin	–	+	+	–	+	+	+	+	+	+	–	+	+
Starch	–	+	–	–	–	–	+	+	+	–	–	+	+
Nitrate reduction	+	+	+	–	+	+	+	+	+	–	–	–	+
G + C content (mol%)	35	39.3–39.5	37–38	51.5	40.7	39–40.8	39.4	38	40.1–40.9	42	44.9	39.5	36.9

[a]Symbols: +, >85% positive; d, different strains give different reactions (16–84% positive); –, 0–15% positive; w, weak reaction; ND, not determined. Abbreviations: E, ellipsoidal; S, spherical; C, central; ST, subterminal; T, terminal; NA, not applicable; ND, no data.
[b]Data from Schlesner et al. (2001).
[c]Data from Fritze (1996a).
[d]Data from Ventosa et al. (1989).
[e]Data from Arahal et al. (1999).
[f]Data from Lawson et al. (1996).
[g]Data from Waino et al. (1999).
[h]Data from Spring et al. (1996).
[i]Data from Hao et al. (1984).
[j]Data from Garabito et al. (1997).
[k]Data from Heyndrickx et al. (1998).

Maintenance procedures

When grown on slants (M13(3*x*) + 10%NaCl), the strain can be kept at 4–5 °C for at least three months. It is easily revived from lyophilized cultures and can be stored at −70 °C in a solution of 50% glycerol in M13(3*x*) + 10% NaCl.

Differentiation of the genus *Filobacillus* from closely related taxa

Table 15 lists the major features which differentiate *Filobacillus* from other genera of spore-forming bacteria.

List of species of the genus *Filobacillus*

1. **Filobacillus milosensis** corrig. Schlesner, Lawson, Collins, Weiss, Wehmeyer, Völker and Thomm 2001, 430[VP]

   mi.los.en′sis. N.L. adj. *milosensis* from the island Milos, Greece, where the organism was isolated.

   Description as for the genus. Further characteristics are given in Tables 15 and 16.

DNA G + C content (mol%): 35 (HPLC).

Type strain: SH 714, ATCC 700960, CIP 107088, DSM 13259, JCM 12288.

GenBank accesssion number (16S rRNA gene): AJ238042.

TABLE 16. Characteristics of *Filobacillus milensis*[a]

Characteristic	Result
Acid from carbohydrates:	
Glucose	−
Galactose	−
Fructose	−
Maltose	−
Mannitol	−
Sucrose	−
Trehalose	−
Xylose	−
Hydrolysis of:	
Casein	−
DNA	+
Esculin	−
Gelatin	−
Hippurate	+
Pullulan	−
Starch	−
Tributyrin	+
Sensitive to:	
Ampicillin	+
Chloramphenicol	+
Kanamycin	−
Streptomycin	+
Tetracycline	+
Vancomycin	+
Voges–Proskauer reaction	−
Production of:	
Catalase	+
Cytochromoxidase	−
L-Alanine aminopeptidase	−
Phosphatase	−

[a]Symbols: +, >85% positive; −, 0–15% positive.

Genus VII. **Geobacillus** Nazina, Tourova, Poltaraus, Novikova, Grigoryan, Ivanova, Lysenko, Petrunyaka, Osipov, Belyaev and Ivanov 2001, 442[VP]

NIALL A. LOGAN, PAUL DE VOS AND ANNA DINSDALE

Ge.o.ba.cil′lus. Gr. n. *Ge* the Earth; L. dim. n. *bacillus* small rod; N.L. masc. n. *Geobacillus* earth or soil small rod.

Obligately thermophilic. Vegetative cells are rod-shaped and **produce one endospore per cell.** Cells occur either singly or in short chains and are motile by means of peritrichous flagella or they are nonmotile. The cell-wall structure is Gram-positive, but the Gram-stain reaction may vary between positive and negative. **Ellipsoidal or cylindrical endospores** are located terminally or subterminally in slightly swollen or non-swollen sporangia. Colony morphology and size are variable; pigments may be produced on certain media. Chemo-organotrophic. Aerobic or facultatively anaerobic. Oxygen is the terminal electron acceptor, replaceable in some species by nitrate. The temperature range for growth is 35–75 °C, with an **optimum at 55–65 °C.** Neutrophilic. Growth occurs at pH 6.0–8.5, with optimal growth at pH 6.2–7.5. Growth factors, vitamins, NaCl, and KCl are not required by most species. **Most species can utilize n-alkanes as carbon and energy sources.** Most species produce acid, but not gas from fructose, glucose, maltose, mannose, and sucrose. Most species produce catalase. Oxidase reaction varies. Phenylalanine is not deaminated, tyrosine is not degraded, and indole is not produced. The major cellular fatty acids are $C_{15:0\ iso}$,

$C_{16:0\ iso}$, and $C_{17:0\ iso}$, which make up more than 60% of the total. The main menaquinone type is MK-7. The lowest level of 16S rRNA gene sequence similarity between all *Geobacillus* species is around 93%, which indicates that at least some species need to be reclassified at the genus level. Species are widely distributed in nature, in heated and unheated environments.

DNA G + C content (mol%): 48.2–58 (T_m).

Type species: **Geobacillus stearothermophilus** (Donk 1920) Nazina, Tourova, Poltaraus, Novikova, Grigoryan, Ivanova, Lysenko, Petrunyaka, Osipov, Belyaev and Ivanov 2001, 443[VP] (*Bacillus stearothermophilus* Donk 1920, 373).

Further descriptive information

Phylogeny. A phylogenetic tree, based on 16S rRNA gene sequences, is shown in Figure 14. The tree includes all *Geobacillus* species with validly published names; *species incertae sedis* are omitted, but are listed below with some additional information. *Geobacillus debilis* holds a separate position and, on the basis of 16S rRNA gene sequence analysis, is most probably not

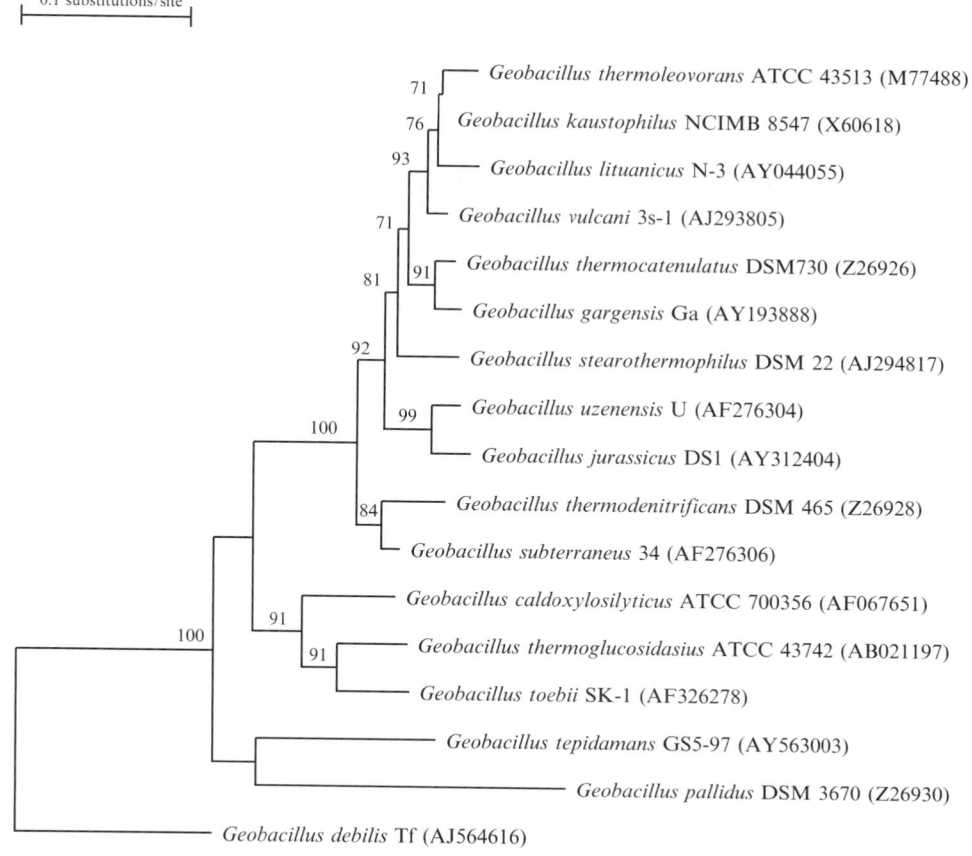

0.1 substitutions/site

71 — *Geobacillus thermoleovorans* ATCC 43513 (M77488)

76 — *Geobacillus kaustophilus* NCIMB 8547 (X60618)

93 — *Geobacillus lituanicus* N-3 (AY044055)

71 — *Geobacillus vulcani* 3s-1 (AJ293805)

81 / 91 — *Geobacillus thermocatenulatus* DSM730 (Z26926)

Geobacillus gargensis Ga (AY193888)

92 — *Geobacillus stearothermophilus* DSM 22 (AJ294817)

99 — *Geobacillus uzenensis* U (AF276304)

100 — *Geobacillus jurassicus* DS1 (AY312404)

84 — *Geobacillus thermodenitrificans* DSM 465 (Z26928)

Geobacillus subterraneus 34 (AF276306)

Geobacillus caldoxylosilyticus ATCC 700356 (AF067651)

91 — *Geobacillus thermoglucosidasius* ATCC 43742 (AB021197)

91 — *Geobacillus toebii* SK-1 (AF326278)

Geobacillus tepidamans GS5-97 (AY563003)

Geobacillus pallidus DSM 3670 (Z26930)

100

Geobacillus debilis Tf (AJ564616)

FIGURE 14. Unrooted neighbor-joining tree of *Geobacillus* type strains based on 16S rRNA gene sequences. Alignment of sequences was performed using CLUSTALX, BIOEDIT, and TREECON. Bootstrap values above 70% are shown (based on 1000 replications) at the branch points. Accession numbers for each strain are given in parentheses.

an authentic *Geobacillus* species. This is most likely also the case for *Geobacillus pallidus* and *Geobacillus tepidamans*. Phylogenetic analysis shows that on the basis of 16S rRNA gene sequences, *Geobacillus tepidamans* belongs to the genus *Anoxybacillus*, whereas *Geobacillus pallidus* does not seem to be closely related to any valid taxa.

It is well known that 16S rRNA gene sequences do not always allow species to be discriminated and that DNA–DNA hybridization data may be needed for this. However, sequences of other genes (the so-called core genes) may be more appropriate for discriminating these relatively recent branchings of the evolutionary tree that correspond to bacterial species (Stackebrandt et al., 2002).

Cell morphology. Vegative cells are rod-shaped and produce one endospore per cell. Cells occur either singly or in short chains and are motile by means of peritrichous flagella or nonmotile. The cell-wall structure is Gram-positive, but the Gram-stain reaction may vary between positive and negative.

Cell-wall composition and fine structure. Vegetative cells of the majority of *Bacillus* species that have been studied, and of the examined representatives of several genera whose species were previously accommodated in *Bacillus*, have the most common type of cross-linkage in which a peptide bond is formed between the diamino acid in position 3 of one subunit and the D-Ala in position 4 of the neighboring peptide subunit, so that

no interpeptide bridge is involved. The diamino acid in the two *Geobacillus* species for which it has been determined, *Geobacillus stearothermophilus* (Schleifer and Kandler, 1972) and *Geobacillus thermoleovorans* (Zarilla and Perry, 1987), is diaminopimelic acid and the configuration has been determined for the former as *meso*-diaminopimelic acid (*meso*-DAP); this cross-linkage is now usually known as DAP-direct (A1γ in the classification of Schleifer and Kandler, 1972).

Organisms growing at high temperatures need enzyme adaptions to give molecular stability as well as structural flexibility (Alvarez et al., 1999; Kawamura et al., 1998; Perl et al., 2000), heat-stable protein-synthesizing machinery, and adaptions of membrane phospholipid composition. They differ from their mesophilic counterparts in the fatty acid and polar headgroup compositions of their phospholipids. The effect of temperature on the membrane composition of *Geobacillus stearothermophilus* has been studied intensively. Phosphatidylglycerol (PG) and cardiolipin (CL) comprise about 90% of the phospholipids, but as the growth temperature rises the PG content increases at the expense of the CL content. The acyl-chain composition of all the membrane lipids also alters; the longer, saturated-linear and iso fatty acids with relatively high melting points increase in abundance, whereas anteiso fatty acids and unsaturated components with lower melting points decrease. As a result, the organism is able to maintain nearly constant membrane fluidity across its whole growth temperature range; this has been termed homeoviscous adaption (Martins et al., 1990; Tolner et al., 1997). An alternative theory, homeophasic adaption, considers that maintenance of the liquid-crystalline phase is more important than an absolute value of membrane fluidity in *Bacteria* (Tolner et al., 1997).

Amino acid transport in *Geobacillus stearothermophilus* is Na$^+$-dependent, which is unusual for neutrophilic organisms such as these, but common among marine bacteria and alkalophiles; however, the possession of primary and secondary Na$^+$-transport systems may be advantageous to the organism by allowing energy conversion via Na$^+$-cycling when the phospholipid adaptions needed to give optimal membrane fluidity at the organism's growth temperature also result in membrane leakiness (de Vrij et al., 1990; Tolner et al., 1997).

Colonial characteristics and life cycle. Colony morphologies and sizes are variable; pigments may be produced on certain media. Ellipsoidal or cylindrical endospores are located terminally or subterminally in slightly swollen or non-swollen sporangia. For details of sporulation, see the treatment of *Bacillus*.

Nutrition and growth conditions. All species of *Geobacillus* are obligately thermophilic chemo-organotrophs. They are aerobic or facultatively anaerobic and oxygen is the terminal electron acceptor, replaceable in some species by nitrate. Temperature ranges for growth generally lie between 37 °C and 75 °C, with optima between 55 °C and 65 °C. They are neutrophilic and grow within a relatively narrow pH range of 6.0–8.5 and their optima lie within the pH range 6.2–7.5. For the species tested, growth factors, vitamins, NaCl, and KCl are not required and most strains will grow on routine media such as nutrient agar. A wide range of substrates is utilized, including carbohydrates, organic acids, peptone, tryptone, and yeast extract; the ability to utilize hydrocarbons as carbon and energy sources is a widely distributed property in the genus (Nazina et al., 2001). A strain of *Geobacillus thermoleovorans* has been found to have extracellular lipase activity and high growth rates on lipid substrates such as olive oil, soybean oil, mineral oil, tributyrin, triolein, Tween 20, and Tween 40 (Lee et al., 1999a). A solvent-tolerant *Geobacillus pallidus* strain that degrades 2-propanol was reported by Bustard et al. (2002).

Pathogenicity. The body temperatures of humans and other animals lie at or near the minimum temperatures for growth for species of *Geobacillus* and there have been no reports of infections with these organisms.

Habitats. Although thermophilic aerobic endospore-formers and other thermophilic bacteria might be expected to be restricted to hot environments, they are also very widespread in cold environments and appear to be ubiquitously distributed in soils worldwide. Strains with growth temperature ranges of 40–80 °C can be isolated from soils whose temperatures never exceed 25 °C (Marchant et al., 2002); indeed, Weigel (1986) described how easy it is to isolate such organisms from cold soils and even from Arctic ice. That the spores of endospore-formers may survive in such cool environments without any metabolic activity is understandable, but their wide distribution and contribution of up to 10% of the cultivable flora suggests that they do not merely represent contamination from hot environments (Marchant et al., 2002). It has been suggested that the direct heating action of the sun on the upper layers of the soil and local heating from the fermentative and putrefactive activities of mesophiles might be sufficient to allow multiplication of thermophiles (Norris et al., 1981). The first described strains of the species now called *Geobacillus stearothermophilus* were isolated from spoiled, canned corn and string beans. This organism and other *Bacillus* species have long been important in the canned food and dairy industries and are responsible for "flat sour" spoilage of canned foods and products such as evaporated milk (Kalogridou-Vassilliadu, 1992). The organisms may thrive in parts of the food-processing plant and their contaminating spores may survive the canning or dairy process and then outgrow in the product if it is held for any time at an incubating temperature. This is a particular problem for foods such as military rations that may need to be stored in tropical climates (Llaudes et al., 2001). *Geobacillus stearothermophilus* may represent up to a third of thermophilic isolates from foods (Deák and Temár, 1988) and approaching two-thirds of the thermophiles in milk (Chopra and Mathur, 1984). Other sources of *Geobacillus stearothermophilus* include geothermal soil, rice soils (Garcia et al., 1982), desert sand, composts (Blanc et al., 1997), water, ocean sediments, and shallow marine hydrothermal vents (Caccamo et al., 2001). *Geobacillus vulcani* was isolated from a shallow (3 m below sea level) hydrothermal vent at Vulcano Island in the Eolian Islands, Italy (Caccamo et al., 2000). *Geobacillus caldoxylosilyticus* was found in Australian soils, and subsequently in uncultivated soils from China, Egypt, Italy, and Turkey, and in central heating system water (Obojska et al., 2002). *Geobacillus debilis* was found in cool soils in Northern Ireland. *Geobacillus thermodenitrificans* has been isolated from soils from Australia, Asia, and Europe, from shallow marine hydrothermal vents (Caccamo et al., 2001), from sugar beet juice, and, along with *Geobacillus thermoglucosidasius*, in other soils (Mora et al., 1998) and hot composts (Blanc et al., 1997). *Geobacillus tepidamans* was also isolated from sugar beet juice and geothermally heated soil in Yellowstone National Park. *Geobacillus toebii* was found in hot hay compost. *Geobacillus pallidus* has been found in compost, sewage, and wastewater treatment processes. *Geobacillus thermoleovorans* was first cultivated from soil, mud, and activated sludge collected in the USA and further isolations have been made from shallow marine hydrothermal vents (Maugeri et al., 2001), deep subterranean petroleum reservoirs (Kato et al., 2001), and Japanese, Indonesian, and Icelandic hot springs (Lee et al., 1999a; Markossian et al., 2000; Sunna et al., 1997a). *Geobacillus kaustophilus* was first isolated from pasteurized milk and other strains have been found in spoiled, canned food, and in geothermal and temperate soils from Iceland, New Zealand, Europe, and Asia (White et al., 1993). *Geobacillus jurassicus*, *Geobacillus subterraneus*, and *Geobacillus uzenensis* were all isolated from the formation waters of high-temperature oilfields in China, Kazakhstan, and Russia, whereas *Geobacillus lituanicus* was found in crude oil in Lithuania and *Geobacillus thermocatenulatus* was isolated from a slimy bloom on the inside surface of a pipe in a steam and gas thermal bore-hole in thermal zone of Yangan-Tau mountain in the South Urals. *Geobacillus gargensis* was isolated from a microbial mat that formed in the Garga hot spring in the Transbaikal region of Russia (Nazina et al., 2004). Unidentified strains belonging to the genus *Geobacillus* have been reported from deep-sea hydrothermal vents lying at 2000 to 3500 m (Marteinsson et al., 1996) and from sea mud of the Mariana Trench at 10,897 m below the surface (Takami et al., 1997).

Enrichment and isolation procedures

Thermophiles may be obtained easily by incubating environmental or other samples in routine cultivation media at 65 °C and above. As for other aerobic endosporeformers, it is useful to heat-treat the specimens to select for endospores and encourage their germination (see *Enrichment and isolation procedures*, in *Bacillus*, above). Allen (1953) described enrichment methods for strains belonging to particular physiological groups. A selective procedure for the isolation of flat sour organisms from food was described by Shapton and Hindes (1963) using yeast-glucose-tryptone agar, which contains peptone (5 g), beef extract (3 g), tryptone (2.5 g), yeast extract (1 g), and glucose (1 g) in distilled water (1000 ml). The method is as follows: dissolve medium components by heating; adjust to pH 8.4; simmer for 10 min then pass through coarse filter paper if necessary; cool and make back up to 1000 ml; adjust to pH 7.4; add sufficient agar to solidify and 2.5 ml of 1% aqueous solution bromocresol purple; sterilize by autoclaving; prepare food sample in 1/4 strength Ringer's solution and pasteurize with molten medium at 108 °C (8 p.s.i. or 55 kPa) for 10 min; reduce temperature to 100 °C and maintain for 20 min; cool to 50 °C; pour plates and allow to set; incubate at 55 °C for 48 h and observe for yellow colonies.

The following procedures are those used in the isolation of strains of *Geobacillus* species, but do not necessarily represent methods especially designed to enrich or select for those species. *Geobacillus caldoxylosilyticus* was isolated from Australian soil by adding 0.1–0.2 g sample to minimal medium and incubating at 65 °C for up to 24 h (Ahmad et al., 2000b). Minimal medium contained: xylose, 10 g; K_2HPO_4, 4 g; KH_2PO_4, 1 g; NH_4NO_3, 1 g; NaCl, 1 g; $MgSO_4$, 0.25 g; trace mineral solution, 10 ml; water to 1000 ml; pH, 6.8; 1.5% agar was added when a solid medium was desired. Trace mineral solution contained: EDTA, 5.0 g; $CaCl_2·2H_2O$, 6.0 g; $FeSO_4·7H_2O$, 6.0 g; $MnCl_2·4H_2O$, 1.15 g; $CoCl_2·6H_2O$, 0.8 g; $ZnSO_4·7H_2O$, 0.7 g; $CuCl_2·2H_2O$, 0.3 g; H_3BO_3, 0.3 g; $(NH_4)_6Mo_7O_{24}·4H_2O$, 0.25 g; water, 1000 ml. After two transfers of 1 ml culture into fresh medium, enrichments were plated on solidified minimal medium and incubated at 65 °C for 24 h. Further isolations from soils taken from China, Egypt, Italy, and Turkey were made by heating samples at 90 °C for 10 min, plating on CESP agar and incubating at 65 °C for 24 h (Fortina et al., 2001a). CESP agar contained: casitone, 15 g; yeast extract, 5 g; soytone, 3 g; peptone, 2 g; $MgSO_4$, 0.015 g; $FeCl_3$, 0.007 g; $MnCl_2·4H_2O$, 0.002 g; water, 1000 ml; pH, 7.2. *Geobacillus gargensis* was isolated from the upper layer of a microbial mat from the Garga spring, Eastern Siberia, by serial dilutions and inoculation onto the agar medium described by Adkins et al. (1992) supplemented with 15 mM sucrose: TES [*N*-tris(hydroxymethyl)methyl-2-amino-ethanesulfonic acid], 10 g; NH_4Cl, 1 g; NaCl, 0.8 g; $MgSO_4·7H_2O$, 0.2 g; $CaCO_3$ (precipitated chalk), 0.2 g; KCl, 0.1 g; K_2HPO_4, 0.1 g; $CaCl_2·2H_2O$, 0.02 g; yeast extract, 0.2 g; trace metal solution, 5 ml; vitamin solution, 10 ml; water to 1000 ml, pH, 7.0; agar was added to solidify. Trace metal solution (Tanner, 1989) contained: nitrilotriacetic acid (2 g, pH adjusted to 6 with KOH); $MnSO_4·H_2O$, 1 g; $Fe(NH_4)_2(SO_4)_2·6H_2O$, 0.8 g; $CoCl_2·6H_2O$, 0.2 g; $ZnSO_4·7H_2O$, 0.2 g; $CuCl_2·2H_2O$, 0.02 g; $NiCl_2·6H_2O$, 0.02 g; $Na_2MoO_4·2H_2O$, 0.02 g; Na_2SeO_4, 0.02 g; Na_2WO_4, 0.02; water, 1,000 ml. Vitamin solution (Tanner, 1989) contained: pyridoxine.HCl, 10 mg; thiamine.HCl, 5 mg; riboflavin, 5; calcium pantothenate, 5 mg; thioctic acid, 5 mg; *p*-aminobenzoic acid, 5 mg; nicotinic acid,

5 mg; vitamin B_{12}, 5 mg; biotin, 2 mg; folic acid, 2 mg; water, 1000 ml. Plates were incubated at 60 °C. Prickett (1928) isolated *Geobacillus kaustophilus* from uncooled pasteurized milk by plating on peptonized milk agar, followed by subculturing on the same medium or on nutrient agar supplemented with 1% yeast extract, 0.25% tryptophan broth, and 0.05% glucose. Donk (1920) reported finding *Bacillus* (now *Geobacillus) stearothermophilus* in spoiled cans of corn and string beans, but the method of isolation was not described; the reader is referred to the method described above for the isolation of flat sour organisms. Original strains of *Geobacillus pallidus* were isolated from heat-treated municipal and yeast factory wastes that had been held at 60 °C in an aerated laboratory fermenter. Dilutions of the homogenized effluents were cultivated for 3–5 d at 60 °C on the enriched nutrient agar medium of Ottow (1974): glucose, 1.0 g; peptone, 7.5 g; meat extract, 5.0 g; yeast extract, 2.5 g; Casamino acids, 2.5 g; NaCl, 5.0 g; agar, 13 g; tap water, 1000 ml; pH 7.2–7.4. *Geobacillus thermoleovorans* was isolated by adding soil, mud, and water samples to L-salts basal medium supplemented with 0.1% (v/v) n-heptadecane and incubated at 60 °C for 1–2 weeks, followed by transfer from turbid cultures to fresh medium of the same composition; after several such transfers, pure cultures were obtained by streaking on plates of L-salts basal medium supplemented with 0.2% (v/v) n-heptadecane and solidified with 2% agar (Merkel et al., 1978; Zarilla and Perry, 1987). L-salts (Leadbetter and Foster, 1958) contained: $NaNO_3$, 2.0 g; $MgSO_4·7H_2O$, 0.2 g; NaH_2PO_4, 0.09 g; KCl, 0.04 g; $CaCl_2$, 0.015 g; $FeSO_4·7H_2O$, 1.0 mg; $ZnSO_4·7H_2O$, 70.0 μg; H_3BO_3, 10.0 μg; $MnSO_4·5H_2O$, 10.0 μg; MoO_3, 10.0 μg; $CuSO_4·5H_2O$, 5.0 μg; deionized water, 1000 ml. *Geobacillus subterraneus* and *Geobacillus uzenensis* were isolated from serial dilutions of thermophilic hydrocarbon-oxidizing enrichments taken from oilfields; the enrichments were inoculated onto the medium described by Zarilla and Perry (1987) supplemented with 0.1% n-hexadecane and incubated at 55–60 °C (Nazina et al., 2001). *Geobacillus jurassicus* was isolated from oilfield formation water by diluting enrichment cultures grown in a modification of the medium of Adkins et al. (1992) (NH_4Cl, 1 g; KCl, 0.1 g; KH_2PO_4, 0.75 g; K_2HPO_4, 1.4 g; $MgSO_4·7H_2O$, 0.2 g; $CaCl_2·2H_2O$, 0.02 g; NaCl, 1.0 g; water, 1000 ml; pH 7.0) supplemented with 4% (v/v) crude oil, incubated at 60 °C, and plated on the same medium solidified with 2% agar. *Geobacillus thermocatenulatus* was isolated from a slimy bloom at about 60 °C on the inside surface of a pipe in a steam and gas thermal bore-hole in the thermal zone of Mount Yangan-Tau in the South Urals using potato-peptone and meat-peptone media (Golovacheva et al., 1965; Golovacheva et al., 1975). Mora et al. (1998) isolated novel strains of *Geobacillus thermodenitrificans* from soil by suspending 1 g soil sample in 5 ml sterile distilled water and heat treating at 90 °C for 10 min, then plating 1 ml on nutrient agar and incubating at 65 °C for 24 h. *Geobacillus thermoglucosidasius* was isolated from Japanese soil by adding 0.1 g sample to 5 ml medium I in large (1.8 × 19 cm) test tubes and incubating at 65 °C for 18 h with the tubes leaning at an angle of about 10°, followed by further enrichments in tubes of the same medium and then purification of plates of medium I solidified with 3% agar (Suzuki et al., 1976). Medium I contained: peptone, 5 g; meat extract, 3 g; yeast extract, 3 g; K_2HPO_4, 3 g; KH_2PO_4, 1 g; water, 1000 ml, pH 7.0. For *Geobacillus debilis*, 100 mg basalt till soil sample, taken at a depth of 50 mm below the surface, was suspended in 50 ml sterile Ringer's solution containing 0.1% Triton

X-100 and placed in a sonicating bath for 10 min. A sample (1 ml) was serially diluted in Ringer's solution and spread plates were prepared on nutrient broth at pH 6.8–7.2 solidified with 0.8% Gellan Gelrite gum and incubated at 70 °C for 24 h under aerobic conditions. Resulting colonies were isolated either onto specialized *Bacillus* medium (nutrient broth, 16 g; $MgSO_4·7H_2O$, 0.5 g; KCl, 2.0 g; 10^{-3} M $Ca(NO_3)_2$; 10^{-4} M $MnCl_2$; 10^{-6} M $FeSO_4$; glucose, 1 g; water, 1000 ml; Leighton and Doi, 1971), or trypticase soy broth solidified with agar or Gelrite Gellan gum, and further purified before being stored as stock cultures either at room temperature or at 4 °C. *Geobacillus lituanicus* was isolated using tenfold serial dilutions of crude oil. The dilutions were inoculated onto Czapek agar and plates were incubated aerobically at 60 °C for 48 h. *Geobacillus tepidamans* was enriched from sugar beet extraction juice samples in SVIII/glc medium (peptone, 10 g; yeast extract, 5 g; meat extract, 5 g; glucose, 3 g; $K_2HPO_4·3H_2O$, 1.3 g; $MgSO_4·7H_2O$, 0.1 g; water, 1000 ml; pH 7.2 ± 0.2) at 55 °C and subcultured in SVIII/glc broth until pure cultures were obtained (Schäffer et al., 1999). Another strain of *Geobacillus tepidamans* was isolated from a high-temperature soil that was collected aseptically and transported back to the laboratory in sterile tubes suspended in heated water contained in a Thermos bottle. Soil samples were serially diluted in 0.1 M NH_4PO_4 buffer (pH 6.0; 65 °C), and aliquots from each dilution were spread onto 0.1% yeast extract agar. Single colonies were repeatedly subcultured until a pure culture was obtained. *Geobacillus toebii* was isolated from a suspension of hay compost plated onto solid modified basal medium and incubated at 60 °C for 3 d (Sung et al., 2002). The medium contained: polypeptone, 5 g; K_2HPO_4, 6 g; KH_2PO_4, 2 g; yeast extract, 1 g; $MgSO_4·7H_2O$, 0.5 g; L-tyrosine, 0.5 g; agar to solidify; and deionized water, 1000 ml. *Geobacillus vulcani* was isolated from a marine sediment sample by inoculation into Bacto Marine Broth (Difco) and Medium D (Castenholz, 1969; Degryse et al., 1978) and incubating aerobically for 3 d at 65 °C, followed by plating positive cultures onto Bacto Marine Agar (Difco).

Identification

There are rather few routine phenotypic characters that can be used reliably to distinguish between the members of *Geobacillus*. Characters testable by the API system (bioMérieux), especially acid production from a range of carbohydrates, that are valuable for differentiating between *Bacillus* species, show relatively little variation in pattern between several *Geobacillus* species. Most species show 16S rRNA gene sequence similarities higher than 96.5%, and so they cluster together quite closely in trees based on such data. They may also show high similarities in other phenotypic analyses. The distinction of six species by Nazina et al. (2001) was mainly supported by DNA–DNA relatedness data, and their differentiation table for eight species was compiled from the literature for all of the six previously established species, so that the characterization methods used were not strictly comparable; furthermore, data were incomplete for these species. The same is true of the differentiation table that accompanied the description of *Geobacillus toebii* (Sung et al., 2002). The species *Geobacillus kaustophilus*, *Geobacillus stearothermophilus*, *Geobacillus thermocatenulatus*, *Geobacillus thermoglucosidasius*, and *Geobacillus thermoleovorans*, especially, need to be characterized alongside the recently described and other revived species in order to allow their descriptions to be emended where necessary.

16S rRNA gene sequencing is not reliable as a stand-alone tool for identification and a polyphasic taxonomic approach is advisable for the identification of *Geobacillus* species and the confident recognition of suspected new taxa. Species descriptions accompanying proposals of novel species will be based on differing test methods and reference strains of established taxa are often not included for comparison, so original descriptions should never be relied upon entirely. Nomenclatural types exist for a good reason and are usually easily available; there is no substitute for direct laboratory comparisons with authentic reference strains. Differentiation characteristics of *Geobacillus* species are given in Tables 17 and 18 provides additional information on biochemical characteristics.

Taxonomic comments

In the first edition of this *Manual*, Claus and Berkeley (1986) listed only three strict thermophiles (that is to say growing at 65 °C and above) in the genus *Bacillus*: *Bacillus acidocaldarius*, *Bacillus schlegelii*, and *Bacillus stearothermophilus*. The last-named of these species had been established for many years (Donk, 1920) and had become and remained something of a dumping-ground for any thermophilic, aerobic endospore-formers; the original strain of the species was thought to have been lost, but Gordon and Smith (1949) considered Donk's description to match most of the obligately thermophilic *Bacillus* strains that they studied. Walker and Wolf (1971), however, believed that two of their cultures [NRRL 1170 and NCA 26 (=ATCC 12980^T)] represented Donk's original strain. Mishustin (1950) renamed *Denitrobacterium thermophilum* of Ambroz (1913) as *Bacillus thermodenitrificans*, Heinen and Heinen (1972) proposed *Bacillus caldolyticus*, *Bacillus caldotenax*, and *Bacillus caldovelox*, and Golovacheva et al. (1975) described *Bacillus thermocatenulatus*, but all of these species were listed as *species incertae sedis* by Claus and Berkeley (1986). However, the heterogeneity of *Bacillus stearothermophilus* was widely appreciated. Walker and Wolf (1971) found that their strains of *Bacillus stearothermophilus* formed three groups on the basis of biochemical tests (Walker and Wolf, 1961, 1971) and serology (Walker and Wolf, 1971; Wolf and Chowhury, 1971b), and their division of the species was further supported by studies of esterase patterns (Baillie and Walker, 1968), further studies with routine phenotypic characters (Logan and Berkeley, 1981), and polar lipids (Minnikin et al., 1977). Klaushofer and Hollaus (1970) also recognized these three major subdivisions within the thermophiles. Walker and Wolf (1971) regarded the recognition of only one thermophilic species of *Bacillus* in the 7th edition of *Bergey's Manual of Determinative Bacteriology* (Breed et al., 1957) as a "dramatic restriction," and in the 8th edition of *Bergey's Manual of Determinative Bacteriology* Gibson and Gordon (1974) commented that the species was "markedly heterogeneous" and that "the emphasis on ability to grow at 65 °C has the effect of excluding organisms that have temperature maxima between 55 °C and 65 °C, although they have not so far been distinguished from *Bacillus stearothermophilus* by any other property." They concluded, "As yet there has been no agreement on how classification in this part of the genus might be improved." Early studies on *Bacillus* thermophiles were comprehensively reviewed by Wolf and Sharp (1981), who also used the scheme of Walker and Wolf (1971) to allocate several thermophilic species to the three previously established groups.

TABLE 17. Differentiation of *Geobacillus* species[a,b]

Characteristic	1. G. stearothermophilus	2. G. caldoxylosilyticus	3. G. debilis	4. G. gargensis	5. G. jurassicus	6. G. kaustophilus	7. G. lituanicus	8. G. pallidus	9. G. subterraneus	10. G. lepidamans	11. G. thermocatenulatus	12. G. thermodenitrificans	13. G. thermoglucosidasius	14. G. thermoleovorans	15. G. toebii	16. G. uzenensis	17. G. vulcani
Motility	+	+	+	+	+	+[d]	+	+	+	+	+	+[d]	+	−	+	+	+
Spore shape:																	
Ellipsoidal	+	+	+[d]	+	+	+	+	+	+	+	+[d]	+	+	+	+[d]	+	+
Cylindrical	d	d[d]	+[d]	−[d]	−[d]	+	−[d]	+	−[d]	−	+	−[d]	d[d]	−[d]	−[d]	−[d]	−
Spore position:																	
Subterminal	+	−[d]	−	−[d]	+[d]	+	+	+	+	−[d]	+[d]	+	+[d]	+[d]	+	+[d]	d[d]
Terminal	d	+	+	+	+	+	+[d]	+	+	+	+	+	+	+	+	+	+
Sporangia swollen	d	+	−[c]	+[f]	+	+[g]	+[f]	+	−	+	+[f]	+	+	+[f]	+	d	d[d]
Aerobic growth	+	+	+	+	+	+	+	+	+	+	+	−	+	+	+	+	+
Anaerobic growth	−	d/w	−[c]	−[c]	w[d]	+[d]	+	−	+[d]	−[c]	+	d	−[c]	−[c]	w[d]	+[d]	−[c]
Acid from:																	
ʟ-Arabinose	d	+	d/w	−	+	d	+[f]	−	−	+[d]	−	+	−[c]	−	−	+[f]	−
Cellobiose	−	+	−[c]	+	+	d	+	−	+	+[d]	+	+	+	+[f]	−	+	+
Galactose	−	+	−[c]	+	+	d	+	−	+	+	+	+[f]	+	+	−[c]	+	+
Glycerol	+	w[d]	−	+	w[d]	d	+[d]	+[d]	+[d]	+[d]	+	+[f]	+	d[e]	+	+	+
Glycogen	+	+[d]	−[d]	−[d]	+	−[d]	−[d]	−[d]	+	+[d]	−[d]	−[d]	+	+[d]	−[d]	+	+[c]
myo-Inositol	−	−[d]	−[d]	−	−	d	+[d]	+[d]	−	+	−	−	−	−[c]	−[d]	−	−[c]
Lactose	−	+	−	+	+	−[g]	−[d]	−[d]	+[d]	w	−	+[f]	+	−	+	+	−[c]
ᴅ-Mannitol	d	−[d]	−[d]	−	−	+	+	−[c]	+[d]	+	+	w[d]	+	+	−[c]	+	+
Mannose	+	+[d]	−	+	+	+	+	−[h]	−	+	+	+	+	+	+[d]	+	+
ʟ-Rhamnose	−	d	−[d]	−	−	−	−[d]	−	−	+	−	d	+	−	−	−	−
Sorbitol	−	−[d]	d/w	+	+	−	−	+[d]	+[c]	+	+[f]	−	+	−	−[c]	+	−
Trehalose	+	+	+	+	+[c]	+	+[d]	d	+	+	+	+	+	+	+[d]	+	+
ᴅ-Xylose	d	+	−	+	+	+	+	−	+	+	+	+	+	d[e]	−[d]	−[c]	+
Utilization of:																	
n-Alkanes	+		d/w	+	+	+			+		−	+	+/−[c]	+	−	+	+
Acetate	+	d	−	+	+	+		−	+	−	+			+	+	+	+
Citrate	−		−/w	−	−	d			+		+	d		+	−	+	−
Formate	−			−	−	−			+		−				−	+	−
Lactate				+	+	d			+		+	−			−	+	−
Hydrolysis of:																	
Casein	d/w	+	d	+[f]	−[c]	+	+	+	−	−	w	−/w	+[f]	+[d]	−	−	−
Esculin	d	+[d]	+[d]	+	+	d	+[d]	d[e]	−[c]	+[d]	+	+[d]	+	+	+[f]	+	+
Gelatin	+	+[d]	d	−[c]	+	d	+[d]	−	+	−	−[c]	w[d]	−[c]	−[c]	−	−[c]	+
Starch	+	+	−	+	−	+	+	w	w	w	−	d	+[g]	+	−[c]	+	+[c]
Urea	−	−	−	−	−[c]	−	−[d]	−	−[f]	−[f]	+	−	+[f]	−	−[c]	+[f]	−[g]
Nitrate reduction	d	+	−d	+	−[c]	+[g]	+[d]	−	+	+[f]	+[f]	+	+	+	+	+	−[c]

(continued)

TABLE 17. (continued)

Characteristic	1. G. stearothermophilus	2. G. caldoxylosilyticus	3. G. debilis	4. G. gargensis	5. G. jurassicus	6. G. kaustophilus	7. G. tibiancus	8. G. pallidus	9. G. subterraneus	10. G. lepidamans	11. G. thermocatenulatus	12. G. thermodenitrificans	13. G. thermoglucosidasius	14. G. thermoleovorans	15. G. toebii	16. G. uzenensis	17. G. vulcani
Growth at pH:																	
6	+	+	+[d]	+	-	+	+[d]	-	+	+	-	+	-	-	+	-[c]	+
8	+	+	+[d]	+	-	+[d]	+[d]	+	-[c]	+	+	+	+	-	+	-[c]	+
8.5	-[c]	d	+[d]	+	-	+[d]	+[d]	+	-[c]	+	+	d	+	-	+	-[c]	+
Growth in NaCl:																	
1%	+	-[d]	w[d]	+	+	+[d]	-	+	+	+	+	+	+	+	+	+	+
2%	+[f]	-[d]	-[d]	-	+	d	-[d]	+	+	d	+	+[f]	-	-	+	+	+[f]
3%	+[f]	-	-[d]	-	+	-[d]	-[d]	+	+	-	+	+[f]	-[d]	-	+	+	+[f]
4%	+[f]	-[d]	-[d]	-	+	-[d]	-[d]	+	d	-[d]	+	-	-[d]	-	+	+	-
5%	d[f]	-[d]	-[d]	-	+	-	-[d]	+	d	-[d]	+[d]	-	-[d]	-	-[c]	-	-
Growth at:																	
35°C	d	-	-[d]	-	-	-	-	-	-	-	+	-	-	-	-	-	-
40°C	+	-	-[d]	-[c]	+	+	-	+	-	+	+	-	-	-	-	-[c]	+
45°C	+	+	-	+	+	+	-	+	+	+	+	d	+	+	+	+	+
70°C	+[f]	+	+	+	-	d	+	d	d	-	+	d	+	+	+	-[c]	+
75°C	d[f]	+	-	-	+	-	-	+	-	-[d]	+	-	+	-	-	-	-

[a]Symbols: +, >85% positive; d, variable (16–84% positive); –, 0–15% positive; +/w, positive or weakly positive; d/w, variable, but weak when positive; –/w, occasional strains are weakly positive; no entry indicates that no data are available.

[b]Compiled from: Golovacheva et al. (1965, 1975); Suzuki et al. (1983); Logan and Berkeley (1984); Claus and Berkeley (1986); Zarilla and Perry (1987); Priest et al. (1988); Scholz et al. (1988); White et al. (1993); Sunna et al. (1997b); Ahmad et al. (2000b); Caccamo et al. (2000); Manachini et al. (2000); Fortina et al. (2001a); Nazina et al. (2004, 2005b, 2001); Sung et al. (2002); Banat et al. (2004); Kuisiene et al. (2004); Schaffer et al. (2004).

[c]Tests as Christensen's citrate-positive and Simmons' citrate-negative (Suzuki et al., 1983).

[d]Data from Dinsdale and Logan (unpublished); all species were negative for lysine decarboxylase, ornithine decarboxylase, H₂S production, and acid from D-arabinose, L-arabitol, dulcitol, erythritol, D-fucose, gluconate, 2-keto-D-gluconate, D-lyxose, and methyl-xyloside.

[e]Positive according to Dinsdale and Logan (unpublished).

[f]Negative according to Dinsdale and Logan (unpublished).

[g]Variable according to Dinsdale and Logan (unpublished).

[h]Weak according to Dinsdale and Logan (unpublished).

TABLE 18. Additional characters of the species of the genus *Geobacillus*[a,b]

Characteristic	1. G. stearothermophilus	2. G. caldoxylosilyticus	3. G. debilis	4. G. gargensis	5. G. jurassicus	6. G. kaustophilus	7. G. titanicus	8. G. pallidus	9. G. subterraneus	10. G. tepidamans	11. G. thermocatenulatus	12. G. thermodenitrificans	13. G. thermoglucosidasius	14. G. thermoleovorans	15. G. toebii	16. G. uzenensis	17. G. vulcani
Voges–Proskauer	d[e]	−[c]	−	−[c]	−[c]		w[d]	−	−[c]	−[c]	−[c]	−[g]	−[g]	−[c]	+	−[c]	+
ONPG[d]	−	−	+	−	−	d	+	−	−	+	+	−	−	−	−	−	−
Arginine dihydrolase[d]	−	−	+	−	−	+	−	−	+	−	−	−	−	−	−	−	−
Methyl red	d	−	−	−	−	+	−	−	+	−	−	−	w	−	−	−	−
Utilization of:																	
Benzoate					−				+							+	
Butyrate				+	+				+							+	
Fumarate				+	+				+							+	
Malate					+				+					+		+	
Phenylacetate									−							+	+
Propionate														d			+
Pyruvate				+	+				+					+		+	
Succinate				+	+				+					+		+	
Butanol				−					+					+		+	
Ethanol				−					+								
Phenol				−	−				+							+	
Acid from:																	
N-Acetylglucosamine	−	−[d]	+[d]	−[d]	−[d]	−[d]	w[d]	−[d]	−[d]	+[d]	+/w[d]	d[d]	+[d]	w[d]	−[d]	−[d]	+[d]
Adonitol[d]	−	−	+	−	−	−	−	+	−	+	+	−	+	−	−	−	
Amygdalin[d]	−	+	+	+	−	−	−	+	−	+	+	+	+	+	−	−	d
D-Arabitol[d]	−	+	w	+	−	−	−	+	−	+	+	+	+	+	+	w	−
Arbutin[d]	+	+	−[e]	+	+	−	+	+	+	+	+	+	+	+	+	+	+
Fructose	+	+	+[e]	−	+	−	+	+	d	+	+	+	+	+	+	+	+
Gentiobiose[d]	+	−	+[e]	+	+	−	−	+	+	+	+	+	+	+	+	+	+
D-Glucose	+	+	−	+	+	−	+	+	+	+	+	+	+	+	+	+	+
Inulin[d]	+	+	−	+	+	−	+	+	+	+	+	+	+	+	+	−	+
Maltose	+	+	−[d]	+	+	−[d]	w[d]	d	+[d]	+[d]	+[d]	+	−[d]	+	+[d]	+[d]	+[d]
D-Melezitose[d]	+	−	−	+	+	−	w[d]	−[d]	−[d]	+[d]	+[d]	d[d]	d[d]	+/w[d]	−[d]	−[d]	+[d]
D-Melibiose	+	+[d]	−	−[d]	+[d]	−	+[d]	+[d]	−[d]	+[d]	+[d]	+[d]	+[d]	+[d]	+[d]	−[d]	+[d]
Methyl D-glucoside	+	w[d]	−[d]	−	w[d]	d[d]	+[d]	+[d]	−[d]	−	w[d]	−[d]	+[d]	−/w[d]	+[d]	+[d]	d[d]
Methyl D-mannoside[d]	d	+[d]	w[d]	−[d]	+[d]	d	+[d]	w[d]	−[d]	+[d]	w[d]	−/w[d]	+[d]	−/w[d]	−[d]	+[d]	d[d]
Raffinose	+	+[d]	+[f]	−[c]	−	−	−	−[c]	−	+[d]	w[d]	+	d	+[d]	w[d]	−	+[d]
Ribose	−	+	d/w	+	+	d	+	−[c]	−[f]	w[f]	+	−/w[d]	−[c]	+	+[d]	−[d]	+
Salicin	−	+[d]	+[d]	+[d]	+	d	+	d	+[d]	+[d]	+[d]	d[d]	+	+	+	+[d]	+[d]
L-Sorbose[d]	+	−	−	−	−	d	−	d	−	−	−	+[d]	d	w[d]	w	−	+
Starch	+	+[d]	+[d]	+	w[d]	d[d]	+[d]	+	+[d]	+[d]	+	+[d]	+	−	+[d]	+	+[d]
Sucrose	+	+	−	+	+	d	+	+	+[d]	−	+	+[g]	d	+	w	+	+
D-Turanose[d]	d	+[d]	w[d]	−[d]	+[d]	−[d]	−[d]	w[d]	−[d]	+[d]	w[d]	d[d]	+[d]	w[d]	−[d]	+[d]	+[d]
Xylitol[d]	−	−	−	−	−	−	−	+	+	−	+	+	−	−	−	−	−
Phenylalanine deamination												+				−	
Gas from nitrate		d		−	+	−			+	−	+	+	−	−	+	−	

[a]See Table 17 for explanation of footnotes.

Group 1 was the most heterogeneous assemblage. It comprised strains that produced gas from nitrate, hydrolyzed starch only weakly, and produced slightly to definitely swollen sporangia with cylindrical to oval spores; it was divided into five subgroups on the basis of growth temperature maxima and minima. This group accommodated the majority of strains received as *Bacillus stearothermophilus*, as well as *Bacillus caldotenax, Bacillus caldovelox, Bacillus kaustophilus* (Prickett, 1928), and *Bacillus thermodenitrificans*.

Group 2 contained strains of *Bacillus stearothermophilus* that were described as "relatively inert" and showed lower growth temperature ranges than members of the other groups, but which had greater salt tolerance. They produced definitely swollen sporangia with oval spores.

Group 3 strains hydrolyzed starch strongly and produced definitely swollen sporangia with cylindrical to oval spores. They were divided into four subgroups on the basis of certain biochemical characters and growth temperatures. Group 3 included the type strain of *Bacillus stearothermophilus, Bacillus calidolactis* (Galesloot and Labots, 1959; Grinsted and Clegg, 1955), and *Bacillus thermoliquefaciens* (Galesloot and Labots, 1959).

Having considered a wide range of evidence, Wolf and Sharp (1981) concluded that the earlier "restrictive attitude" (i.e., regarding *Bacillus stearothermophilus* as the only obligate thermophile in the genus) was "no longer tenable," and regretted that differences in sporangial morphologies among thermophiles were disregarded by Gibson and Gordon (1974). Wolf and Sharp (1981) also showed the spread of DNA G + C content of 44–69 mol% among the *Bacillus* thermophiles, but they did not emphasize the taxonomic significance of this broad range. Claus and Berkeley (1986) were unable to take the taxonomy of the *Bacillus stearothermophilus* group any further, but noted that the heterogeneity of the species was indicated by the wide range of DNA base composition.

Following the pioneering work of the late 1960s and early 1970s, several novel thermophilic species were described, but the overall taxonomy of the group languished for some years, despite the continuing considerable interest in the biology of the thermophiles and the potential applications of their enzymes. Of the novel species described in the decade before 1980, when the Approved Lists of Bacterial Names were published (Skerman et al., 1980), only *Bacillus acidocaldarius* (Darland and Brock, 1971) was included; *Bacillus caldolyticus, Bacillus caldotenax, Bacillus caldovelox, Bacillus thermocatenulatus,* and *Bacillus thermodenitrificans* were excluded. Subsequent proposals of thermophilic taxa included: *Bacillus flavothermus* (Heinen et al., 1982), *Bacillus thermoglucosidasius* (Suzuki et al., 1983), *Bacillus tusciae* (Bonjour and Aragno, 1984), *Bacillus acidoterrestris* (Deinhard et al., 1987a; Demharter and Hensel, 1989b), *Bacillus cycloheptanicus* (Deinhard et al., 1987b), *Bacillus pallidus* (Scholz et al., 1987), *Bacillus thermoleovorans* (Zarilla and Perry, 1987), *Bacillus thermocloacae* (Demharter and Hensel, 1989b), *Bacillus thermoaerophilus* (MeierStauffer et al., 1996), *Bacillus thermoamylovorans* (Combet-Blanc et al., 1995), *Bacillus thermosphaericus* (Andersson et al., 1995), *Bacillus thermantarcticus* (corrig. Nicolaus et al. (2002); *Bacillus thermoantarcticus* [sic], Nicolaus et al., 1996), and *Bacillus vulcani* (Caccamo et al., 2000). Also, some species that had been excluded from the Approved Lists were revived: *Bacillus thermoruber (ex* Guicciardi et al., 1968; Manachini

et al., 1985), *Bacillus kaustophilus (ex* Prickett, 1928; Priest et al., 1988), and *Bacillus thermodenitrificans (ex* Klaushofer and Hollaus, 1970; Manachini et al., 2000). Several of these species were subsequently allocated to new genera as: *Alicyclobacillus acidocaldarius, Alicyclobacillus acidoterrestris, Alicyclobacillus cycloheptanicus* (Wisotzkey et al., 1992), *Brevibacillus thermoruber* (Shida et al., 1996), *Aneurinibacillus thermoaerophilus* (Heyndrickx et al., 1997), *Anoxybacillus flavithermus* (Pikuta et al., 2000a), and *Ureibacillus thermosphaericus* (Fortina et al., 2001b).

De Bartolemeo et al. (1991) subjected moderately and obligately thermophilic species of *Bacillus* to numerical taxonomic analysis and found four groups within *Bacillus stearothermophilus*; three groups corresponded with those previously recognized by Walker and Wolf (1971) and other authors, whereas the fourth group comprised biochemically inert strains of high G + C content that were incapable of growing above 65 °C. White et al. (1993) carried out a polyphasic, numerical taxonomic study on a large number of thermophilic *Bacillus* strains and recommended the revival of *Bacillus caldotenax* and *Bacillus thermodenitrificans* and proposed an emended description of *Bacillus kaustophilus*. However, the clusters they found in their numerical analysis revealed considerable heterogeneity within the species and species groups, and these clusters were often only separated from each other by small margins, indicating that separation of some of the species by routine tests would probably be difficult.

Ash et al. (1991) included strains of *Bacillus stearothermophilus, Bacillus acidoterrestris, Bacillus kaustophilus,* and *Bacillus thermoglucosidasius* in their comparison of the 16S rRNA gene sequences of the type strains of 51 *Bacillus* species. Their strain of *Bacillus acidoterrestris* was later found to have been a contaminant or misnamed culture (Wisotzkey et al., 1992), but their strains of *Bacillus stearothermophilus, B kaustophilus,* and *Bacillus thermoglucosidasius* grouped together in an evolutionary line (called group 5) that was distinct from other *Bacillus* species, implying that these thermophiles might represent a separate genus. Rainey et al. (1994) compared the 16S rRNA gene sequences of 16 strains of 14 thermophilic *Bacillus* species and found that strains of *Bacillus caldolyticus, Bacillus caldotenax, Bacillus caldovelox, Bacillus kaustophilus, Bacillus thermocatenulatus, Bacillus thermodenitrificans,* and *Bacillus thermoleovorans* grouped with *Bacillus stearothermophilus* at similarities of greater than 98%, whereas *Bacillus thermoglucosidasius* joined the group at 97% similarity. This group thus constituted group 5 *sensu* Ash et al. (1991), a coherent and phylogenetically distinct group of thermophilic *Bacillus* species that did not, however, include all the obligate thermophiles in the genus. Studholme et al. (1999) examined whether transformability is a trait associated with a particular phylogenetic group of thermophilic *Bacillus*. Two of their three transformable strains, all received as *Bacillus stearothermophilus*, were more closely related to *Bacillus thermodenitrificans* and *Bacillus thermoglucosidasius* when their 16S rRNA gene sequences were compared; it was concluded therefore that although transformability might be strain-specific, it is not limited to a single thermophilic *Bacillus* species.

Sunna et al. (1997b) identified *Bacillus kaustophilus* and *Bacillus thermocatenulatus* as well as "*Bacillus caldolyticus*," "*Bacillus caldotenax*," and "*Bacillus caldovelox*" as members of *Bacillus thermoleovorans* on the basis of DNA–DNA relatedness ranging from 73% to 88% between the type and reference strains of all these species. They proposed the merger of these species and gave an emended description of *Bacillus thermoleovorans*, but this proposal has not been validated. However, Nazina et al. (2004) found DNA–DNA reassociation values of only 47–54% between *Bacillus kaustophilus*, *Bacillus thermocatenulatus*, and *Bacillus thermoleovorans*.

Following the discovery of two novel thermophilic, aerobic endospore-formers in petroleum reservoirs, Nazina et al. (2001) proposed that the valid species of Ash et al. Group 5 should be accommodated in a new genus, *Geobacillus*, along with the novel species *Geobacillus subterraneus* and *Geobacillus uzenensis*. The new genus thus contained eight species: *Geobacillus stearothermophilus* (type species), *Geobacillus kaustophilus*, *Geobacillus subterraneus*, *Geobacillus thermocatenulatus*, *Geobacillus thermodenitrificans*, *Geobacillus thermoglucosidasius*, *Geobacillus thermoleovorans*, and *Geobacillus uzenensis*, whereas *Bacillus pallidus*, *Bacillus schlegelii*, *Bacillus thermantarcticus*, *Bacillus thermoamylovorans*, *Bacillus thermocloacae*, *Bacillus tusciae*, and *Bacillus vulcani* remained in *Bacillus*. Zeigler (2005) analyzed the full-length *recN* and 16S rRNA gene sequences of the type strains of *Geobacillus subterraneus* and *Geobacillus uzenensis*, along with those of two other isolates described as belonging to *Geobacillus subterraneus* and found that they clustered within the same similarity group. It was not clear, however, whether the close relationship shown with these methods was due to sequencing errors in one or both GenBank entries or that the strain for *Geobacillus uzenensis* used in the study by Zeigler (2005) was not in fact the same as the type strain studied by Nazina et al. (2001).

It was clear from the phylogenetic analyses accompanying their proposals (Caccamo et al., 2000; Nicolaus et al., 1996) that *Bacillus thermantarcticus* and *Bacillus vulcani* belonged to *Geobacillus* and should be transferred to that genus; Nazina et al. (2004) formally proposed the transfer of the latter species. Zeigler (2005) recommended the transfer of *Bacillus thermantarcticus* to the genus *Geobacillus* on the basis of full-length *recN* and 16S rRNA gene sequences, but its transfer awaits formal proposal and this species is currently covered in *Bacillus*. Although *Bacillus schlegelii* and *Bacillus tusciae* remain in *Bacillus*, they lie at some distance from other members of the genus (see *Bacillus*). Following the proposal of *Geobacillus*, *Saccharococcus caldoxylosilyticus* (Ahmad et al., 2000b) has been transferred to the genus as *Geobacillus caldoxylosilyticus* (Fortina et al., 2001a), *Bacillus pallidus* has been transferred as *Geobacillus pallidus* (Banat et al., 2004), and six novel species, *Geobacillus toebii* (Sung et al., 2002), *Geobacillus gargensis* (Nazina et al., 2004), *Geobacillus debilis* (Banat et al., 2004), *Geobacillus lituanicus* (Kuisiene et al., 2004), *Geobacillus tepidamans* (Schaffer et al., 2004), and *Geobacillus jurassicus* (Nazina et al., 2005a; Nazina et al., 2005b) have been described.

This progress, however, leaves the long-established species and type species of the genus, *Geobacillus stearothermophilus*, without a modern description based upon a polyphasic taxonomic study. The description given by Nazina et al. (2001) for *Geobacillus stearothermophilus* is largely based upon the one given by Claus and Berkeley (1986) at a time when the species was essentially all-embracing for thermophilic *Bacillus* strains, albeit generally recognized as being heterogeneous. Also, strains of several revived species, such as *Geobacillus kaustophilus* and *Geobacillus thermodenitrificans*, might formerly have been classified within "*Bacillus stearothermophilus*" *sensu lato*, yet no emended description of

Geobacillus stearothermophilus has been published following these proposed revivals; it is clear, therefore, that *Geobacillus stearothermophilus* is without a practically useful definition at present.

Two new taxa were proposed during preparation of this article, but neither has been validated. A single strain of "*Geobacillus caldoproteolyticus*" (Chen et al., 2004) was isolated from sewage sludge in Singapore and deposited as DSM 15730 and ATCC BAA-818. Another proposal based upon a single isolate is "*Geobacillus thermoleovorans* subsp. *stromboliensis*" (Romano et al., 2005c), which was isolated from a geothermal environment in the Eolian Islands in Italy and deposited as DSM 15393 and ATCC BAA-979.

Miscellaneous comments

The major cellular fatty acid components of *Geobacillus* species following incubation at 55 °C are (percentages of total are given in parentheses) $C_{15:0\ iso}$ (20–40%; mean 29%), $C_{16:0\ iso}$ (6–39%; mean 25%), and $C_{17:0\ iso}$ (7–37%; mean 19.5%), which account for 60–80% of the total. As minor components, $C_{15:0\ ante}$ (0.6–6.4%; mean 2.3%), $C_{16:0}$ (1.7–11.2%; mean 5.8%), and $C_{17:0\ ante}$ (3.1–18.7%; mean 7.3%) are detected (Nazina et al., 2001). The figures given by Fortina et al. (2001a) for *Geobacillus caldoxylosilyticus* and Sung et al. (2002) for *Geobacillus toebii* generally lie within these ranges, with the exception that strains of the former species showed 45–57% $C_{15:0\ iso}$.

Direct comparison of profiles between the obligately thermophilic *Geobacillus* species and mesophilic aerobic endospore-formers is not normally possible as the assays of members of the two groups have not usually been done at the same temperature. For many species of *Bacillus sensu stricto*, fatty acid profiles obtained following incubation at 30 °C are $C_{15:0\ ante}$ (25–66%), $C_{15:0\ iso}$ (22–47%), and $C_{17:0\ ante}$ (2–12%). For the *Bacillus cereus* group, levels of $C_{15:0\ ante}$ were lower (3–7%) and amounts of unsaturated fatty acids were generally higher (>10%) (Kämpfer, 1994). The thermotolerant species *Bacillus coagulans* and *Bacillus smithii* showed higher amounts of $C_{17:0\ ante}$ (means of 28 and 42%, respectively), and generally lower amounts of $C_{15:0\ ante}$ (means of 55 and 12%, respectively) and $C_{15:0\ iso}$ (means of 9 and 19%, respectively), but for these data, the former was incubated at 30 °C and the latter at 57 °C.

Llarch et al. (1997) compared the fatty acid profiles of aerobic endospore-formers isolated from Antarctic geothermal environments; their six isolates had temperature ranges with minima between 17 °C and 45 °C and maxima between 62 °C and 73 °C, with optima of 60–70 °C. Two strains (growth temperature ranges of 37–70 °C and 45–73 °C) were found to lie nearest to *Bacillus stearothermophilus* in a phenotypic analysis, whereas two other isolates could be identified as strains of *Bacillus licheniformis* (temperature range 17–68 °C) and *Bacillus megaterium* (temperature range 17–63 °C), whose maximum growth temperatures were extended beyond those seen in strains from temperate environments. The fatty acid profiles for all of these strains were compared following incubation at 45 °C; the two *Bacillus stearothermophilus*-like strains showed profiles of $C_{15:0\ iso}$ (19 and 40%), $C_{16:0\ iso}$ (47 and 5%), and $C_{17:0\ iso}$ (7.5 and 23%), which accounted for 55–73% of the total, whereas for minor components the patterns were $C_{15:0\ ante}$ (2.6 and 9.6%), $C_{16:0}$ (4 and 5.8%), and $C_{17:0\ ante}$ (4.6 and 8.7%); these profiles are consistent with those reported for *Geobacillus*. The profiles for the *Bacillus licheniformis* and *Bacillus megaterium* strains were $C_{15:0\ iso}$ [38.4 and 20.5%, respectively (values for mesophilic strains of these species, from Kämpfer (1994), were 33–38 and 15–48%, respectively)], $C_{16:0\ iso}$ [5.2 and 1.9% (mesophiles 2 and 0.9–2.4%)], and $C_{17:0\ iso}$ [24.4 and 2.3% (mesophiles 10 and 0.5–1.7%)], which accounted for 25–68% of the total, whereas for other components, the patterns were $C_{15:0\ ante}$ [10.2 and 50% (mesophiles 30 and 32–67%)], $C_{16:0}$ [12.4 and 3.3% (mesophiles 2 and 1.5–2.8%)], and $C_{17:0\ ante}$ [6 and 7.7% (mesophiles 10 and 1.7–3%)]. These profiles suggest that any potential distinctions between the rather variable fatty acid profiles of *Geobacillus* species and *Bacillus* species are largely lost when strains of each group are incubated at the same temperature.

Maintenance procedures

Geobacillus strains may be preserved on slopes of a suitable growth medium that encourages sporulation, such as nutrient agar or trypticase soy agar containing 5 mg/l $MnSO_4 \cdot 7H_2O$. Slopes should be checked microscopically for spores, before sealing to prevent drying out, and stored in a refrigerator; on such sealed slopes, the spores should remain viable for many years. For longer-term preservation, lyophilization and liquid nitrogen may be used, as long as cryoprotectants are added.

List of species of the genus *Geobacillus*

1. **Geobacillus stearothermophilus** (Donk 1920) Nazina, Tourova, Poltaraus, Novikova, Grigoryan, Ivanova, Lysenko, Petrunyaka, Osipov, Belyaev and Ivanov 2001, 443[VP] (*Bacillus stearothermophilus* Donk 1920, 373)

ste.a.ro.ther.mo′phi.lus. Gr. n. *stear* fat; Gr. n. *therme* heat; Gr. adj. *philos* loving; N.L. adj. *stearothermophilus* (presumably intended to mean) heat- and fat-loving.

Aerobic, Gram-positive, Gram-variable, and Gram-negative motile rods, varying from 2 to 3.5 μm in length and 0.6 to 1.0 μm in diameter, normally present as single cells or in short chains. Ellipsoidal, occasionally cylindrical, spores are located subterminally and sometimes terminally within sporangia that are usually swollen (Figure 15). Colonies are circular and usually convex, and may be smooth and crenate. Minimum growth temperatures lie in the range 30–45 °C; maximum growth temperatures in the range 70–75 °C. Most strains grow in the range 40–70 °C. Grows in pH 6–8. Starch and gelatin are hydrolyzed; casein hydrolysis is usually positive, but often weak. Nitrate reduction is variable. Growth occurs in the presence of 2% NaCl and sometimes in 5%, but not 7% NaCl. Inhibited by 0.001% lysozyme. Acetate is utilized, but citrate is not usually utilized. Catalase variable and usually oxidase-negative. Acid without gas is produced

from a range of carbohydrates. Hydrocarbons (C_{10}, C_{11}) may be used as carbon and energy sources. Major fatty acids are $C_{15:0\ iso}$, $C_{16:0\ iso}$, and $C_{17:0\ iso}$, which comprise more than 60% of the total fatty acids.

Source: soil, hot springs, desert sand, Arctic waters, ocean sediments, food, and compost.

Bacillus stearothermophilus was for a long period regarded by many as the only obligate thermophile in the genus (see *Taxonomic comments*, above). In the First Edition of this *Manual*, Claus and Berkeley (1986) noted that the heterogeneity of the species was indicated by the wide range of DNA base composition, but they were unable to resolve the taxonomy of the group. Notwithstanding the proposal of the genus *Geobacillus*, several taxonomic studies of thermophilic aerobic endospore-formers, and the proposals of a number of new and revived species, the type species still awaits an emended, modern description based upon polyphasic study of a number of authentic strains following delineation of *Bacillus stearothermophilus sensu stricto*. The description given above is based upon data reported by Gordon et al. (1973), Logan and Berkeley (1984), Claus and Berkeley (1986), White et al. (1993), Kämpfer (1994), and Nazina et al. (2001).

DNA G + C content (mol%): 43.5–62.2 (T_m) and 46.0–52.0 (Bd); 51.5 and 51.9 (T_m) and 51.5 (Bd) for the type strain.

Type strain: ATCC 12980, DSM 22, NCIMB 8923, NCTC 10339.

EMBL/GenBank accession number (16S rRNA gene): AJ294817 (DSM 22).

2. **Geobacillus caldoxylosilyticus** (Ahmad et al., 2000b) Fortina, Mora, Schumann, Parini, Manachini and Stackebrandt 2001a, 2069[VP] (*Saccharococcus caldoxylosilyticus* Ahmad, Scopes, Rees and Patel 2000b)

cal.do.xy.lo.si.ly′ti.cus. L. adj. *caldus* hot; N.L. neut. n. *xylosum* xylose; N.L. adj. *lyticus* dissolving, degrading; N.L. adj. *caldoxylosilyticus* hot and xylose-degrading.

Gram-positive, motile, catalase-positive rods, 4–6 μm in length and 0.5–1.0 μm in diameter, normally present as single cells or in short chains. Oval spores are located terminally within swollen sporangium. Colonies are small, flat, regular in shape, and off-white to beige in color. Type strain grows anaerobically, but other strains may grow only weakly in anaerobic conditions. The optimal growth temperature lies between 50 °C and 65 °C (range 42–70 °C). Optimum pH for growth is 6.5; no growth occurs at pH 5 or 11. Growth is inhibited by 3% NaCl. Nitrate reduction is positive, with weak

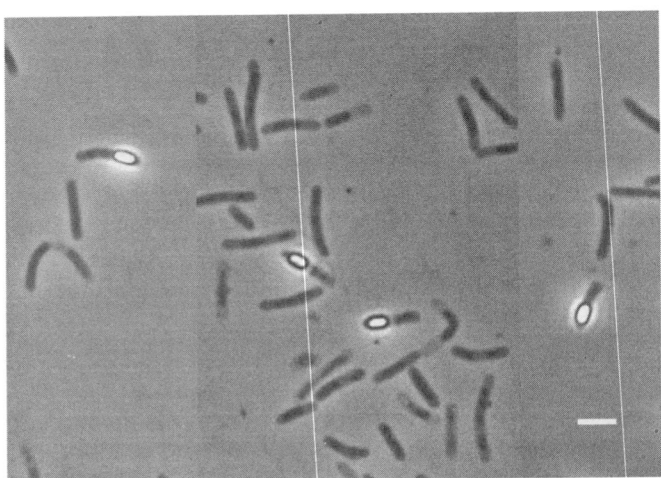

FIGURE 15. Photomicrograph of cells of the type strain of *Geobacillus stearothermophilus* viewed by phase-contrast microscopy, showing ellipsoidal, subterminal spores that slightly swell the sporangia. Bar = 2 μm. Photomicrograph prepared by N.A. Logan.

anaerobic production of gas from nitrate. Produces acid from glucose, fructose, sucrose, maltose, trehalose, lactose, galactose, cellobiose, arabinose, ribose, and xylose. Strains vary in their utilization of rhamnose and citrate. Starch and casein are hydrolyzed. All strains are negative for indole and urease production, and the Voges–Proskauer test. The main cellular fatty acids are $C_{15:0\ iso}$ and $C_{17:0\ iso}$.

Source: soil from Australia, China, Egypt, Italy, and Turkey.

DNA G + C content (mol%): 44.0–50.2; 44.4 or 45.8 in two determinations (T_m) for the type strain.

Type strain: S1812, ATCC 700356, DSM 12041.

EMBL/GenBank accession number (16S rRNA gene): AF067651 (ATCC 700356).

3. **Geobacillus debilis** Banat, Marchant and Rahman 2004, 2199[VP]

de'bil.is. L. masc. adj. *debilis* weak or feeble, referring to the restricted substrate range for this species.

Gram-negative rods, 0.5–1.0 μm wide by 1.0–14.0 μm long, and motile. Spores are produced sparsely and are terminal, sporangium not swollen. Colonies are flat and cream-colored with smooth margins. Description is based upon two isolates. Growth occurs at 50–70 °C, with an optimum above 60 °C. Obligate aerobe. Positive for catalase and oxidase. Produces acid from raffinose and trehalose; some strains also produce acid from ribose, sorbitol, and arabinose. Does not utilize starch, but casein and gelatin are used in some strains; very limited use of alkanes. The species is phylogenetically isolated and belongs most probably to a new genus.

Source: undisturbed subsurface soil in Northern Ireland.

DNA G + C content (mol%): 49.9 (HPLC).

Type strain: Tf, DSM 16016, NCIMB 13995.

EMBL/GenBank accession number (16S rRNA gene): AJ564616 (Tf).

4. **Geobacillus gargensis** Nazina, Lebedeva, Poltaraus, Tourova, Grigoryan, Sokolova, Lysenko and Osipov 2004, 2023[VP]

gar.gen'sis. N.L. masc. adj. *gargensis* of Garga, pertaining to the Garga hot spring located in Eastern Siberia (Russia), from which the type strain was isolated.

Aerobic, Gram-positive rods, motile by means of peritrichous flagella and produces terminally located ellipsoidal spores in slightly swollen sporangia. Description is based upon one isolate. Cells are 1.0–1.5 μm wide and 6–12 μm long. Chemo-organotrophic metabolism. Acid but no gas is produced from a wide range of carbohydrates. Utilizes hydrocarbons (C12–C16), acetate, butyrate, lactate, pyruvate, fumarate, succinate, peptone, tryptone, nutrient broth, potato agar, and yeast extract as carbon and energy sources. No growth occurs on methanol, ethanol, propanol, butanol, isobutanol, formate, or citrate (Simmons'). Nitrate is reduced to nitrite. Unable to grow autotrophically on H_2 + CO_2. Unable to produce NH_3 from peptone. Catalase-positive. Starch, esculin, and casein are hydrolyzed, but gelatin is not. Urea is not decomposed and H_2S and indole are not produced. The egg-yolk lecithinase, Voges–Proskauer, and methyl red tests are negative. Growth factors, vitamins, NaCl, and KCl are not required for growth. The temperature range for growth is 45–70 °C, with optimum growth at 60–65 °C. Growth occurs at pH 5.5–8.5, with optimum

growth at pH 6.5–7.0. Able to grow in both the absence of NaCl and the presence of 1% (w/v) NaCl. Major fatty acids are the iso-branched saturated fatty acids $C_{15:0\ iso}$, $C_{16:0\ iso}$, and $C_{17:0\ iso}$.

Source: the Garga hot spring, Transbaikal region, Russia.

DNA G + C content (mol%): 52.9 (T_m).

Type strain: Ga, DSM 15378, VKM B-2300.

EMBL/GenBank accession number (16S rRNA gene): AY193888 (Ga).

5. **Geobacillus jurassicus** Nazina, Sokolova, Grigoryan, Shestakova, Mikhailova, Poltaraus, Tourova, Lysenko, Osipov and Belyaev 2005b, 983[VP] (Effective publication: Nazina, Sokolova, Grigoryan, Shestakova, Mikhailova, Poltaraus, Tourova, Lysenko, Osipov and Belyaev 2005b, 50)

ju.ras.si'cus. N.L. masc. adj. *jurassicus* of Jurassic, referring to the geological period of oil-bearing formation, from where the strains were isolated.

Cells are rod-shaped, motile due to peritrichous flagella, and produce terminally located ellipsoidal spores in slightly swollen sporangia. Description is based upon two isolates. Cell wall is Gram-positive. Colonies grown on nutrient agar are round, mucous, colorless, and have a diameter of about 2 mm. Aerobic and chemo-organotrophic. Acid, but not gas, is produced from a wide range of carbohydrates. The utilized carbon and energy sources are hydrocarbons (C_6, C_{10}, C_{11}, C_{14}, and C_{16}), methane-naphthenic oil, acetate, butyrate, pyruvate, lactate, benzoate, fumarate, succinate, malate, ethanol, peptone, tryptone, and yeast extract. Can grow in nutrient broth and on potato agar. Cannot grow on methanol, propanol, butanol, isobutanol, phenol, phenylacetate, alanine, glutamate, serine, formate, propionate, or Simmons' citrate agar. Poor growth is observed on asparagine and glutamine. Catalase-positive and produces NH_3 from peptone. Urea and tyrosine are not degraded. Phenylalanine is not deaminated. H_2S, indole, and dihydroxyacetone are not produced. Esculin, gelatin, and starch are hydrolyzed. Casein is not hydrolyzed. Does not grow autotrophically on H_2 + CO_2 and does not ferment glucose with gas production. The egg-yolk lecithinase, Voges–Proskauer, and methyl red tests are negative. Fe^{2+} is not used as an electron acceptor. Nitrate is not reduced to nitrite or dinitrogen. Grows at temperatures between 45 °C and 65 °C, with optimum growth at 58–60 °C. Grows at pH 6.4–7.8, with optimum growth at pH 7.0–7.2. Can grow in the absence of NaCl and in the presence of 5–5.5% (w/v) NaCl. Major cellular fatty acids are $C_{15:0\ iso}$, $C_{16:0\ iso}$, and $C_{17:0\ iso}$.

Source: the formation water of a high-temperature oilfield.

DNA G + C content (mol%): 53.8–54.5 and 54.5 for the type strain (T_m).

Type strain: DS1, VKM B-2301, DSM 15726.

EMBL/GenBank accession number (16S rRNA gene): AY312404 (DS1).

6. **Geobacillus kaustophilus** (Priest, Goodfellow and Todd 1988) Nazina, Tourova, Poltaraus, Novikova, Grigoryan, Ivanova, Lysenko, Petrunyaka, Osipov, Belyaev and Ivanov 2001, 444[VP] ("*Bacillus kaustophilus*" Prickett 1928, 38; *Bacillus kaustophilus* nom. rev. Priest, Goodfellow and Todd 1988, 1879; emend. White, Sharp and Priest 1993, PAGE)

kau.sto'phil.us. Gr. adj. n. *kaustos* burnt, red hot; Gr. adj. *philos* loving; N.L. masc. adj. *kaustophilus* loving intense heat.

Priest et al. (1988) proposed the revival of this species on the basis of five strains, and White et al. (1993) emended the description after study of 14 strains. As there are several discrepancies between the two descriptions, the later (White et al., 1993) description is followed where such discrepancies occur; however, there are also some discrepancies between the description given by White et al. (1993) and the characters presented in the table accompanying the description.

Facultatively aerobic, motile, Gram-positive rods, 0.7–1.5 μm in diameter by 2.0–4.5 μm in length. Ellipsoidal spores are located terminally and subterminally within sporangia that are usually swollen. Colonies are 2–3 mm in diameter, circular to irregular, smooth, convex and entire. Translucent, but with faint brown color when incubated at 37–45 °C, becoming more intensely reddish-brown with prolonged incubation. Pigment is produced on tyrosine. Minimum growth temperature is 37 °C, optimum 60–65 °C, maximum 68–72 °C. Catalase- and oxidase-positive. Nitrate is reduced to nitrite by some strains. Casein and starch are hydrolyzed; most strains hydrolyze tributyrin, gelatin, and esculin. Voges–Proskauer-negative. Acid without gas is produced from a range of carbohydrates. Growth in the presence of 2% NaCl varies; usually no growth occurs in 5% NaCl. Note Sunna et al. (1997b) considered that this species should be merged with *Bacillus thermoleovorans*, but this proposal has not been validated.

Source: pasteurized milk and soil.

DNA G + C content (mol%): 52–58, and 53.9 for the type strain (T_m).

Type strain: ATCC 8005, LMG 9819, NCIMB 8547.

EMBL/GenBank accession number (16S rRNA gene): X60618 (NCIMB 8547).

7. **Geobacillus lituanicus** Kuisiene, Raugalas and Chitavichius 2004, 1993[VP]

li.tu.a'ni.cus. M.L. adj. *lituanicus* of Lithuania, referring to the Lithuanian oilfield from where the type strain was isolated.

Cells are rod-shaped, occurring in chains, motile by means of peritrichous flagella, varying in length from 4.4 to 5.8 μm and in diameter from 1.1 to 1.4 μm. Description is based upon one isolate. Oval subterminal endospores are produced within the slightly distended sporangia. Gram staining is positive. Colonies are small, round, tawny, convex, opaque, and shiny. Obligately thermophilic, the optimal growth temperature ranges between 55 °C and 60 °C, with a minimum of 55 °C and a maximum of 70 °C. Aerobic/facultatively anaerobic chemo-organotroph; nitrate is the terminal electron acceptor under anaerobic conditions. Inhibited by lysozyme. Proteolytic. *Geobacillus thermoleovorans* is the closest phylogenetic neighbor.

Source: the crude oil of a high-temperature oilfield.

DNA G + C content (mol%): 52.5 (according to DNASTAR software; Kuisiene et al., 2004).

Type strain: N-3, DSM 15325, VKM B-2294.

EMBL/GenBank accession number (16S rRNA gene): AY044055 (N-3).

8. **Geobacillus pallidus** Scholz, Demharter, Hensel and Kandler (1987) Banat, Marchant and Rahman 2004, 2200[VP] (*Bacillus pallidus* Scholz, Demharter, Hensel and Kandler 1987, PAGE)

pa'l.li.dus. L. masc. adj. *pallidus* pale, referring to the pale colony color.

Thermophilic, aerobic, Gram-positive, motile rods 0.8–0.9 × 2–5 μm, occurring singly and in pairs and chains. Forms central to terminal ellipsoidal to cylindrical spores that swell the sporangia slightly. Colonies are flat to convex, circular or lobed, smooth and opaque, and reach diameters of 2–4 mm after 4 d incubation at 55 °C. Optimum growth occurs at 60–65 °C with a minimum of 37 °C and a maximum of 65–70 °C. Grows at pH 8.0–8.5. Catalase- and oxidase-positive. Casein, gelatin, and urea are not hydrolyzed; starch is weakly hydrolyzed; tributyrin is hydrolyzed. Grows in the presence of up to 10% NaCl. Citrate is not utilized as sole carbon source. Nitrate is not reduced. Acid without gas is produced from glucose and from a small number of other carbohydrates. Phylogenetically, *Geobacillus pallidus* most probably does not belong to this genus. However, further research is needed before reclassification can be established.

Source: heat-treated sewage sludge.

DNA G + C content (mol%): 39–41 (T_m).

Type strain: H12, DSM 3670, LMG 11159, KCTC 3564.

EMBL/GenBank accession number (16S rRNA gene): Z26930 (DSM 3670).

9. **Geobacillus subterraneus** Nazina, Tourova, Poltaraus, Novikova, Grigoryan, Ivanova, Lysenko, Petrunyaka, Osipov, Belyaev and Ivanov 2001, 443[VP]

sub.ter.ra'ne.us. L. adj. *subterraneus* subterranean, below the Earth's surface.

Gram-positive, rod-shaped cells, 0.8–1.5 μm by 4.6–8.0 μm, motile by means of peritrichous flagella and producing subterminally or terminally located ellipsoidal spores in unswollen sporangia. Description is based upon three isolates. Colonies are round, mucous, and colorless. Growth temperature range is between 45–48 °C and 65–70 °C. pH range for growth is 6.0–7.8. Growth occurs without NaCl and in the presence of 3–5% NaCl. Acid, but no gas, is produced from cellobiose, galactose, glycerol, mannose, and ribose. No acid is produced from adonitol, arabinose, inositol, raffinose, rhamnose, sorbitol, or xylose. Utilizes the following carbon and energy sources: hydrocarbons (C_{10}–C_{16}), methane-naphthenic and naphthenic-aromatic oil, acetate, benzoate, butanol, butyrate, ethanol, formate, fumarate, lactate, phenol, phenylacetate, pyruvate, succinate, peptone, tryptone, nutrient broth, potato agar, and yeast extract. Nitrate is reduced to dinitrogen. Does not grow autotrophically on $H_2 + CO_2$. Casein, gelatin, and urea are not hydrolyzed. Esculin and starch are degraded. Phenylalanine is not deaminated, Fe^{3+} is not reduced, and tyrosine is not decomposed. Dihydroxyacetone, H_2S, and indole are not produced. The egg-yolk lecithinase and Voges–Proskauer reactions are negative. Major fatty acids are $C_{15:0\ iso}$, $C_{16:0\ iso}$, and $C_{17:0\ iso}$, which comprise more than 80% of the total fatty acids.

Source: formation waters of high-temperature oilfields.

DNA G + C content (mol%): 49.7–52.3 (T_m).

Type strain: 34, DSM 13552, VKM B-2226, AS 12763.

EMBL/GenBank accession number (16S rRNA gene): AF276306 (34).

10. **Geobacillus tepidamans** Schaffer, Franck, Scheberl, Kosma, McDermott and Messner 2004, 2366[VP]

te.pid.a'mans. L. adj. *tepidus* (luke) warm; L. part. adj. *amans* loving; N.L. part. adj. *tepidamans* loving warm (conditions).

Straight rods, 3.9–4.7 × 0.9–1.2 μm in size, single cells, sometimes in short chains. Description is based upon two isolates. Gram-positive, motile. Moderately thermophilic. Forms oval terminal endospores that swell the sporangia. Positive or weakly positive for catalase. Requires oxygen as an electron acceptor. Covered with an oblique *S*-layer lattice, composed of identical *S*-layer glycoprotein protomers. Grows at 39–67 °C, with an optimum of 55 °C; the pH range for optimal growth is 6–9. Type strain is inhibited by 3% NaCl. Inhibited by 0.001% lysozyme. DNA and Tween 80 are hydrolyzed. Negative for the Voges–Proskauer reaction (pH 6.5–7) and acid production from basal medium. The major cellular fatty acids are $C_{15:0\ iso}$ and $C_{17:0\ iso}$. The type strain of this species is phylogenetically very close to *Anoxybacillus* and most probably belongs to this genus.

Source: a sugar-beet factory in Austria and from geothermally heated soil, Yellowstone National Park, USA.

DNA G + C content (mol%): 42.4–43.2, and 43.2 for the type strain (HPLC).

Type strain: GS5-97, ATCC BAA-942, DSM 16325.

EMBL/GenBank accession number (16S rRNA gene): AY563003 (GS5-97).

11. **Geobacillus thermocatenulatus** (Golovacheva, Loginova, Salikhov, Kolesnikov and Zaitseva 1975) Nazina, Tourova, Poltaraus, Novikova, Grigoryan, Ivanova, Lysenko, Petrunyaka, Osipov, Belyaev and Ivanov 2001, 444[VP] (*Bacillus thermocatenulatus* Golovacheva, Loginova, Salikhov, Kolesnikov and Zaitseva 1975, 230)

ther.mo.ca.ten'ul.at.us. Gr. n. *therme* heat; N.L. adj. *catenulatus* chain-like; N.L. adj. *thermocatenulatus* thermophilic, chain-like, referring to two of the organism's features.

Facultatively anaerobic, Gram-positive, peritrichously motile rods, 0.9 × 6–8 μm, forming long chains. Description is based on a single isolate. The cylindrical spores are 1.0 × 1.7–2.0 μm in size, are borne terminally and swell the sporangia slightly. Colonies on malt-peptone and potato-peptone agars are round and raised, with entire margins, and a yellowish bloom; they show slight concentric striation and have paste-like consistency. Minimum growth temperature is about 35 °C, optimum is 65–75 °C on agar and 55–60 °C in broth, and maximum growth temperature is about 78 °C. Casein is hydrolyzed weakly, but gelatin, starch, and urea are not hydrolyzed. Grows in presence of 4% NaCl. Nitrate is reduced to gaseous nitrogen. Citrate is utilized. Acetoin, H_2S, and indole are not produced. Acid without gas is produced from cellobiose, fructose, galactose, glucose, glycerol, mannitol, sucrose, and trehalose. Maltose, mannose, and a number of hydrocarbons (C_{10}–C_{16}) are utilized as carbon and energy sources. The major cellular fatty acids are $C_{15:0\ iso}$, $C_{16:0\ iso}$, and $C_{17:0\ iso}$, making up more than 75% of the total. *Note* Sunna et al. (1997b) considered that this species should be merged with *Bacillus thermoleovorans*, but this proposal has not been validated.

Source: a slime layer inside a hot-gas bore-hole pipe.

DNA G + C content (mol%): 69 (TLC and paper chromatography; as reported by Golovacheva et al. (1975) or 55.2 (T_m, as reported by Nazina et al., 2001).

Type strain: DSM 730, VKM B-1259, strain 178.

EMBL/GenBank accession number (16S rRNA gene): Z26926 (DSM 730).

12. **Geobacillus thermodenitrificans** (Manachini, Mora, Nicastro, Parini, Stackebrandt, Pukrall and Fortina 2000) Nazina, Tourova, Poltaraus, Novikova, Grigoryan, Ivanova, Lysenko, Petrunyaka, Osipov, Belyaev and Ivanov 2001, 444[VP] ("*Denitrobacterium thermophilum*" Ambroz 1913; "*Bacillus thermodenitrificans*" Mishustin 1950; "*Bacillus thermodenitrificans*" Klaushofer and Hollaus, 1970; *Bacillus thermodenitrificans* nom. rev. Manachini, Mora, Nicastro, Parini, Stackebrandt, Pukall and Fortina 2000, 1336)

ther.mo.de.ni.tri'fi.cans. Gr. n. *therme* heat; N.L. part. adj. *denitrificans* denitrifying; *thermodenitrificans* thermophilic denitrifying, referring to two of the organism's features.

Gram-positive, straight rods, 0.5–1 × 1.5–2.5 μm; endospores are oval, subterminal or terminal, and do not distend the sporangium. Colonies are flat with lobate margins and off-white to beige in color. Growth occurs at 50–65 °C and some strains, including the type strain, are capable of growth at 45 and 70 °C. The optimum pH values for growth lie between 6 and 8. Catalase-positive. Grows in the presence of 3% NaCl. Resistant to phenol concentrations of 10–20 mM. Nitrate and nitrite are reduced to gas and gas is produced anaerobically from nitrate. Most strains are able to hydrolyze starch and a few strains degrade casein weakly. Anaerobic growth in glucose broth, growth at pH 9, and citrate utilization are variable between strains. Negative for indole and urease production, and the Voges–Proskauer reaction. Arabinose, cellobiose, fructose, galactose, glucose, lactose, maltose, mannose, ribose, trehalose, and xylose are utilized as sole carbon sources. The main fatty acids are $C_{15:0\ iso}$ and $C_{17:0\ iso}$, which account for over 66–69% of the total; minor fatty acids are $C_{16:0}$, $C_{16:0\ iso}$, and $C_{17:0\ ante}$.

Source: soil.

DNA G + C content (mol%): 48.4–52.3, and 50.3 for the type strain (T_m).

Type strain: DSM 465, LMG 17532, ATCC 29492.

EMBL/GenBank accession number (16S rRNA gene): Z26928 (DSM 465).

13. **Geobacillus thermoglucosidasius** (Suzuki, Kishigami, Inoue, Mizoguchi, Eto, Takagi and Abe 1983) Nazina, Tourova, Poltaraus, Novikova, Grigoryan, Ivanova, Lysenko, Petrunyaka, Osipov, Belyaev and Ivanov 2001, 444[VP] (*Bacillus thermoglucosidasius* Suzuki, Kishigami, Inoue, Mizoguchi, Eto, Takagi and Abe 1983, 493)

ther'mo.glu.co.si.da'si.us. Gr. n. *therme* heat; N.L. adj. *glucosidasius* of glucosidase; N.L. adj. *thermoglucosidasius* indicating the production of heat-stable glucosidase.

Strictly aerobic, Gram-positive, motile rods, 0.5–1.2 × 3.0–7.0 μm; the ellipsoidal endospores are borne terminally and swell the sporangium. Colonies are 0.8–1.2 mm in diameter, flat, smooth, translucent, glistening, circular, entire, and faintly brown in color. Smooth, viscid pellicles are formed in broth. Growth occurs between 42 °C and

67–69 °C; optimal growth temperature is 61–63 °C. Growth occurs at initial pH values of 6.5–8.5. Positive for catalase and oxidase. Grows in the presence of 0.5%, but not 2% NaCl. Nitrate is reduced to nitrite, but denitrification does not occur. H_2S is produced. Citrate utilization is positive in Christensen's medium, but negative in Simmons' medium. Casein and starch are hydrolyzed, but gelatin is not. Urease-positive. Indole production and Voges–Proskauer reaction are negative. Acid without gas is produced from cellobiose, fructose, glucose, glycerol, maltose, mannose, mannitol, rhamnose, salicin, sorbitol, starch, sucrose, trehalose, and xylose. Exo-oligo-1,6-glucosidase is synthesized in large amounts. The main cellular fatty acids are $C_{15:0\ iso}$, $C_{16:0\ iso}$, $C_{16:0}$, $C_{17:0\ iso}$, and $C_{17:0\ ante}$, making up 90% of the total.

Source: Japanese soil.

DNA G + C content (mol%): 45–46 (T_m).

Type strain: LMG 7137, ATCC 43742, DSM 2542.

EMBL/GenBank accession number (16S rRNA gene): AB021197 (ATCC 43742).

14. **Geobacillus thermoleovorans** (Zarilla and Perry 1987) Nazina, Tourova, Poltaraus, Novikova, Grigoryan, Ivanova, Lysenko, Petrunyaka, Osipov, Belyaev and Ivanov 2001, 444[VP] (*Bacillus thermoleovorans* Zarilla and Perry 1987, 263)

therm'o.le.o.vo'rans. Gr. n. *therme* heat; L. n. *oleum* oil; L. v. *vorare* to devour; N.L. pres. part. *thermoleovorans* indicating heat-requiring bacteria capable of utilizing oil (hydrocarbons).

Obligately thermophilic, strictly aerobic, generally Gram-negative, nonmotile, rod-shaped cells, 1.5–3.5 μm in length. The oval endospores are borne terminally and swell the sporangia slightly. Colonies are not pigmented. Growth factors are not required. Growth occurs at temperatures between 45 °C and 70 °C, with optimum growth occurring between 55 °C and 65 °C. Grows within the pH range 6.2–7.5. Catalase-positive. Starch is usually hydrolyzed. Voges–Proskauer reaction is negative. A variety of compounds serve as carbon and energy sources, including casein, yeast extract, nutrient broth, peptone, tryptone, acetate, butyrate, pyruvate, cellobiose, galactose, glucose, glycerol, maltose, mannitol, mannose, ribose, sucrose, trehalose, xylose, and n-alkanes (C_{13}–C_{20}). Ammonium salts can be used as nitrogen sources and most strains can utilize $NaNO_3$. The major dibasic amino acid in the peptidoglycan is diaminopimelic acid. The major cellular fatty acids are $C_{15:0\ iso}$, $C_{16:0\ iso}$, and $C_{17:0\ iso}$, making up more than 60% of the total fatty acids.

Source: soil near hot water effluent, non-thermal muds and activated sludge.

DNA G + C content (mol%): 52–58 (T_m).

Type strain: ATCC 43513, DSM 5366, strain LEH-1.

EMBL/GenBank accession number (16S rRNA gene): M77488 (ATCC 43513).

Note: "*Geobacillus thermoleovorans* subsp. *stromboliensis*" has been proposed by Romano et al. (2005c).

15. **Geobacillus toebii** Sung, Kim, Bae, Rhee, Jeon, Kim, Kim, Hong, Lee, Yoon, Park and Baek 2002, 2254VP

to.e'bi.i. N.L. neut. gen. n. *toebii* derived from toebi, a special farmland compost in Korea, from which the organism was isolated.

Aerobic, Gram-positive, motile rods, 2.0–3.5 μm long and 0.5–0.9 μm wide. Ellipsoidal spores are located subterminally to terminally in swollen sporangia. Description is based on a single isolate. Growth occurs at 45–70 °C with optimal growth at 60 °C. No growth is observed at 80 °C. Growth at 60 °C occurs between pH 6.0 and 9.0, with an optimum pH of about 7.5. No growth is observed in the presence of 0.02% azide or 5% NaCl. Catalase-positive. Acid is produced from D-glucose and inositol, but not from xylose or mannitol. Casein is hydrolyzed, but esculin, gelatin, and starch are not. n-Alkanes are utilized, but acetate, formate, and lactate are not. Denitrifier. The Voges–Proskauer test is positive. Cell-wall peptidoglycan contains *meso*-diaminopimelic acid. The major cellular fatty acids are $C_{15:0\ iso}$, $C_{16:0\ iso}$, and $C_{17:0\ iso}$, which comprise over 85% of the total. Produces factors that stimulate the growth of *Symbiobacterium toebii*.

Source: farmland hay compost in Kongju, Korea.

DNA G + C content (mol%): 43.9 (HPLC).

Type strain: SK-1, DSM 14590, KCTC 0306BP.

EMBL/GenBank accession number (16S rRNA gene): AF326278 (SK-1).

16. **Geobacillus uzenensis** Nazina, Tourova, Poltaraus, Novikova, Grigoryan, Ivanova, Lysenko, Petrunyaka, Osipov, Belyaev and Ivanov 2001, 443[VP]

u.ze.nen'sis. N.L. adj. *uzenensis* of Uzen, referring to the Uzen oilfield, Kazakhstan, from where the type strain was isolated.

Gram-positive or negative, rod-shaped cells, 0.9–1.7 × 4.7–8.5 μm, motile by means of peritrichous flagella and producing terminally located ellipsoidal spores in swollen or non-swollen sporangia. Description is based on two isolates. Colonies are round, mucous and colorless. Growth temperature range is 45–65 °C. pH range for growth is 6.2–7.8. Growth occurs without NaCl and in the presence of up to 4% NaCl. Acid, but no gas, is produced from arabinose, cellobiose, galactose, glycerol, maltose, mannitol, mannose, ribose, and trehalose. No acid is produced from adonitol, inositol, raffinose, rhamnose, sorbitol, or xylose. Utilizes the following carbon and energy sources: hydrocarbons (C_{10}–C_{16}), methane-naphthenic and naphthenic-aromatic oil, acetate, benzoate, butanol, butyrate, ethanol, fumarate, lactate, malate, phenol, phenylacetate, propionate, pyruvate, succinate, peptone, tryptone, nutrient broth, potato agar, and yeast extract. Nitrate is reduced to nitrite. Does not grow autotrophically on $H_2 + CO_2$. Esculin, gelatin, and starch are hydrolyzed, but not casein. Phenylalanine is not deaminated. Fe^{3+} is not reduced. Tyrosine and urea are not decomposed. Dihydroxyacetone, H_2S, and indole are not produced. The egg-yolk lecithinase, Voges–Proskauer, and methyl red tests are negative. Major fatty acids are $C_{15:0\ iso}$, $C_{16:0\ iso}$, $C_{16:0}$, $C_{17:0\ iso}$, and $C_{17:0\ ante}$, which comprise more than 80% of the total fatty acids.

Source: formation waters of high-temperature oilfields.

DNA G + C content (mol%): 50.4–51.5 (T_m).

Type strain: U, DSM 13551, VKM B-2229, AS 12764.

EMBL/GenBank accession number (16S rRNA gene): AF276304 (U).

17. **Geobacillus vulcani** (Caccamo et al., 2000) Nazina, Lebedeva, Poltaraus, Tourova, Grigoryan, Sokolova, Lysenko and Osipov 2004, 2023[VP] (*Bacillus vulcani* Caccamo, Gugliandolo, Stackebrandt and Maugeri 2000, 2011)

vul.ca′ni. N.L. gen. masc. n. *vulcani* of the volcano, pertaining to the Eolian Island volcano with a shallow marine hydrothermal vent, from where the organism was isolated.

Aerobic, Gram-positive, motile rods, 4–7 µm long and 0.6–0.8 µm wide, with oval endospores that are borne terminally. Description is based on a single isolate. Negative for catalase and oxidase. Temperature range for growth is 37–72 °C, and optimal growth occurs at 60 °C. Growth occurs in the pH range 5.5–9.0, and optimum is pH 6.0. Growth occurs in the presence of 0–3% NaCl, with optimal growth at 2% NaCl. The following carbon sources support growth: adipate, cellobiose, citrate, fructose, galactose, gluconate, glucose, lactose, malate, maltose, mannitol, mannose, phenylacetate, sucrose, and trehalose. Acid is produced from a wide range of carbohydrates. Positive for Voges–Proskauer test. Indole and H_2S are not produced. Nitrate is not reduced. Esculin, gelatin, starch, and Tween 20 are hydrolyzed, but casein, Tween 80, and urea are not. The major fatty acids are $C_{17:0\ iso}$ (21% of total), $C_{15:0\ iso}$ (16.6%), $C_{16:0\ iso}$ (14.6%), and $C_{17:0\ ante}$ (11.4%); unsaturated acids are absent but $nC_{18:0}$ are present (13%).

DNA G + C content (mol%): 53.0 (HPLC).

Type strain: 3s-1, CIP 106305, DSM 13174.

EMBL/GenBank accession number (16S rRNA gene): AJ293805 (3s-1).

Species Incertae Sedis

Of the thermophilic, aerobic endospore-forming bacteria that were listed as *species incertae sedis* by Claus and Berkeley (1986), "*Bacillus thermocatenulatus*" and "*Bacillus thermodentrificans*" have been validly published and transferred to the genus *Geobacillus*. The remaining invalid, thermophilic species listed by Claus and Berkeley (1986) were "*Bacillus caldolyticus*," "*Bacillus caldotenax*," "*Bacillus caldovelox*," and "*Bacillus flavothermus*."

"*Bacillus flavothermus*" (Heinen et al., 1982) has been revived and accommodated within *Anoxybacillus* (Pikuta et al., 2000a) as *Anoxybacillus flavithermus*; members of this genus will grow aerobically, but prefer anaerobic conditions (Pikuta et al., 2003a).

Sharp et al. (1980) found high DNA–DNA relatedness between "*Bacillus caldotenax*" and "*Bacillus caldovelox*," but low relatedness between these two species and "*Bacillus caldolyticus*." White et al. (1993) therefore merged "*Bacillus caldotenax*" and "*Bacillus caldovelox*" (Heinen and Heinen, 1972) into one species as "*Bacillus caldotenax*" but this proposal has not been validated; they left "*Bacillus caldolyticus*" as a *species incertae sedis*, although DNA–DNA relatedness of 93% between this organism and "*Bacillus caldotenax*" was found. Sharp et al. (1992) also investigated the relatedness of "*Bacillus caldolyticus*," "*Bacillus caldotenax*," "*Bacillus caldovelox*," and *Bacillus kaustophilus* using metabolic studies and phage typing. Rainey et al. (1994) found that "*Bacillus caldovelox*," "*Bacillus caldolyticus*," and "*Bacillus caldotenax*" shared almost identical 16S rRNA gene sequences with *Bacillus kaustophilus* and *Bacillus thermoleovorans* and concluded that DNA pairing studies would be required in order to determine whether all five species should be combined into one species. As mentioned above (see *Taxonomic comments*),

Sunna et al. (1997b) identified *Bacillus kaustophilus* and *Bacillus thermocatenulatus*, as well as "*Bacillus caldolyticus*," "*Bacillus caldotenax*," and "*Bacillus caldovelox*," as members of *Bacillus thermoleovorans*, and they proposed the merger of all these species as *Bacillus thermoleovorans*, but this proposal has not been validated. The taxonomic position of "*Bacillus caldovelox*," "*Bacillus caldolyticus*," and "*Bacillus caldotenax*," along with *Bacillus caldoxylosilyticus* and *Bacillus thermantarcticus*, was also questioned by Nazina et al. (2001) following biochemical and physiological characterization, fatty acid analysis, DNA–DNA hybridization, and 16S rRNA gene sequence comparisons. Zeigler (2005) also supported the findings of Sunna et al. (1997b), with *Bacillus kaustophilus*, *Bacillus thermocatenulatus*, and *Bacillus thermoleovorans* clustering together, alongside the type strains of *Bacillus vulcani* and *Bacillus lituanicus*. The status of each of the three species "*Bacillus caldolyticus*," "*Bacillus caldotenax*," and "*Bacillus caldovelox*" therefore still remains unclear, as the findings and conclusions of Sharp et al. (1980) and White et al. (1993) are not in complete accord with those of Rainey et al. (1994), Sunna et al. (1997b), and Zeigler (2005). Outline descriptions of these species, taken from Heinen and Heinen (1972), Logan and Berkeley (1984), Sharp et al. (1989), and White et al. (1993), therefore follow.

a. "*Bacillus caldolyticus*" Heinen and Heinen, 1972, 17.

Cell diameter is 0.7 µm. Motile. Spores are ellipsoidal to cylindrical, subterminal and terminal, swelling the sporangium. Aerobic, with weak growth in glucose broth under anaerobic conditions. Catalase-negative; oxidase-positive. Casein, gelatin, and starch are hydrolyzed. Acid is produced without gas from cellobiose, fructose, galactose, glucose, glycerol, glycogen, maltose, mannitol, mannose, melezitose, raffinose, ribose, salicin, starch, sucrose, and trehalose. Grows in 2% NaCl broth. No growth at pH 5.7 or in the presence of sodium azide. Optimum pH range 6.0–8.0. Grows at 55 °C, optimum growth temperature is 72 °C, maximum temperature for growth is 82 °C. According to Sharp et al. (1980), nitrate is reduced to nitrite, but Logan and Berkeley (1984) reported a negative reaction in this character. Sharp et al. (1980) found only 38% DNA–DNA relatedness with "*Bacillus caldotenax*," but White et al. (1993) found 93% relatedness with "*Bacillus caldotenax*."

Source: a hot natural pool in the USA.

DNA G + C content (mol%): 52.3 (T_m).

Type strain: DSM 405, IFO 15313, LMG 17975.

EMBL/GenBank accession number (16S rRNA gene): Z26924 (DSM 405).

b. "*Bacillus caldotenax*" Heinen and Heinen 1972, 17.

Cell diameter is 0.5 µm. Motile. Spores are ellipsoidal, subterminal and terminal, swelling the sporangium. Aerobic, with weak growth in glucose broth under anaerobic conditions. Catalase-negative; oxidase-positive. Casein, gelatin, and starch are hydrolyzed. Acid is produced without gas from cellobiose, fructose, galactose, glucose, glycerol, glycogen, maltose, mannitol, mannose, melezitose, raffinose, ribose, salicin, starch, and sucrose. Claus and Berkeley (1986) indicated that this organism does not produce acid from trehalose, but Logan and Berkeley (1984) and White et al. (1993) reported that the reference strain was positive for this character. Grows in 2% NaCl broth. No growth at pH 5.7 or in the presence of sodium azide. Optimum pH range 7.5–8.5. Grows at 55 °C, optimal growth temperature is 80 °C, and maximum growth temperature is 85 °C. According to Sharp et al. (1980), nitrate is reduced to nitrite, but Logan and Berkeley (1984) reported a negative reaction in this character. Sharp et al. (1980) found only 38% DNA–DNA relatedness with "*Bacillus caldolyticus*," but high relatedness with "*Bacillus caldovelox*," whereas White et al. (1993) found 93% relatedness between "*Bacillus caldotenax*" and "*Bacillus caldolyticus.*"

Source: superheated pool water in the USA.

DNA G + C content (mol%): 54.6 (T_m).

Type strain: DSM 406, IFO 15314, LMG 17974.

EMBL/GenBank accession number (16S rRNA gene): Z26922 (DSM406).

c. "*Bacillus caldovelox*" Heinen and Heinen 1972, 17.

Cell diameter is 0.6 µm. Motile. Spores are ellipsoidal, subterminal and terminal, swelling the sporangium. Aerobic, with weak growth in glucose broth under anaerobic conditions. Catalase-negative; oxidase-positive. Casein, gelatin, and starch are hydrolyzed. Acid is produced without gas

from cellobiose, fructose, galactose, glucose, glycerol, glycogen, maltose, mannitol, mannose, melezitose, raffinose, ribose, salicin, starch, sucrose, and trehalose. Grows in 2% NaCl broth. No growth at pH 5.7 or in the presence of sodium azide. Optimum pH range 6.3–8.5. Grows at 55 °C, optimum growth temperature is 60–70 °C, maximum growth temperature is 76 °C (Sharp et al., 1980). According to Sharp et al. (1980), nitrate is reduced to nitrite, but Logan and Berkeley (1984) reported a negative reaction in this character. Sharp et al. (1980) found high DNA relatedness between "*Bacillus caldovelox*" and "*Bacillus caldotenax.*"

Source: superheated pool water in the USA.

DNA G + C content (mol%): 65.1 (Tm; Sharp et al., 1980), but this value may be an error (Sharp et al., 1992).

Type strain: DSM 411, IFO 15315, LMG 14463.

EMBL/GenBank accession number (16S rRNA gene): Z26925 (DSM 411).

Note: The proposals for "*Geobacillus caldoproteolyticus*" (Chen et al., 2004) and "*Geobacillus thermoleovorans* subsp. *stromboliensis*" (Romano et al., 2005c) await validation; see Taxonomic comments, above.

Genus VIII. **Gracilibacillus** Wainø, Tindall, Shumann and Ingvorsen 1999, 829[VP]

THE EDITORIAL BOARD

Gra.ci.li.ba.cil'lus. L. adj. gracilis slender; L. masc. n. bacillus a rod; N.L. masc. n. Gracilibacillus the slender bacillus/rod.

Gram-positive, spore-forming rods (mostly thin) or filaments. Terminal ellipsoidal and/or spherical endospores. Motile. Halotolerant with growth occurring in 0–20% (w/v) NaCl. Strains produce acid from D-glucose. Catalase-positive. Starch and esculin are hydrolyzed. Tests for arginine dihydrolase, lysine, and ornithine decarboxylases and indole production are negative. The predominant cellular fatty acids are $C_{15:0\ anteiso}$, $C_{15:0\ iso}$, $C_{17:0\ anteiso}$, and $C_{16:0}$. The major polar lipids are phosphatidylglycerol and diphosphatidylglycerol. The predominant menaquinone type is MK-7. The main cell-wall peptidoglycan contains meso-diaminopimelic acid and is directly cross-linked (peptidoglycan type A1γ). Gracilibacillus can be distinguished from other closely related genera by 16S rRNA gene sequences.

DNA G + C content (mol%): 35.8–39.4.

Type species: **Gracilibacillus halotolerans** Wainø, Tindall, Shumann and Ingvorsen 1999, 829[VP].

Further descriptive information

Gracilibacillus currently contains four species. Three species, Gracilibacillus halotolerans, Gracilibacillus boraciitolerans, and Gracilibacillus orientalis were originally placed in the genus Gracilibacillus (Wainø et al., 1999; Ahmed et al., 2007b; and Carrasco et al., 2006, respectively). Gracilibacillus dipsosauri (originally described as Bacillus dipsosauri by Lawson et al., 1996) was transferred from Bacillus to Gracilibacillus by Wainø et al. (1999). Table 19 gives characteristics helpful for differentiating the species of Gracilibacillus.

Sources.   Strains of Gracilibacillus were isolated from diverse sources. The type strain, Gracilibacillus halotolerans, was isolated

from surface mud of Great Salt Lake, Utah, USA. Gracilibacillus boraciitolerans was isolated from soil naturally high in boron minerals in the Hisarcik area of the Kutahya Province of Turkey. Gracilibacillus diposauri was isolated from the salt glands of a desert iguana found near Las Vegas, Nevada, USA. Gracilibacillus orientalis was isolated from two salt lakes located in Inner Mongolia, China. There are three reported strains of Gracilibacillus orientalis, but only one of the other three species.

Cell and colony morphology and growth requirements.   Strains are spore-forming, rod-shaped organisms occurring as single cells, usually thin cells, and filaments. Cells are 0.3–0.9 × 1.8–10 µm. Colonies of the four species are entire and 0.3–3.0 mm in diameter. Gracilibacillus boraciitolerans colonies are viscous and contain a light pink to red pigment. Colonies of the other three species are white to cream and smooth. The optimal temperature of incubation ranges from 25 °C to 47 °C. The range of growth for the four species in NaCl is 0–20%; Gracilibacillus orientalis is the only species that requires NaCl for growth; however, Gracilibacillus dipsosauri grows more poorly without it. Gracilibacillus dipsosauri can grow under anaerobic conditions, whereas Gracilibacillus halotolerans and Gracilibacillus orientalis are obligate aerobes (data not available for Gracilibacillus boraciitolerans). The optimal pH for growth is 7.5–8.5 for Gracilibacillus boraciitolerans and 7.5 for the other three species.

Phenotypic analysis.   Phenotypic characteristics are given in the genus description, in Table 19, and in the individual species descriptions. Casein is not hydrolyzed by the three species tested (no data available for Gracilibacillus boraciitolerans). Gracilibacillus halotolerans is the only H_2S-, urease-, and Voges–Proskauer-positive species. Gracilibacillus halotolerans, Gracilibacillus boraciitolerans, and *Gracilibacillus orientalis* produce β-galactosidase (data not available for *Gracilibacillus dipsosauri*).

Whole-cell fatty acid profiles for the four species are similar. The ranges of the predominant cellular fatty acids found in the four species (Ahmed et al., 2007b) grown in marine broth (Difco) or, for *Gracilibacillus orientalis*, marine broth plus 1.5% (w/v) NaCl are $C_{15:0\ anteiso}$ (30–46%), $C_{15:0\ iso}$ (6.4–28%), $C_{17:0\ anteiso}$ (9.9–19%), $C_{16:0}$ (5.3–16%), $C_{16:0\ iso}$ (1.9–7.1%), and $C_{15:0}$ (1.9–6.6%). $C_{17:0\ iso}$ is present in *Gracilibacillus boraciitolerans*, *Gracilibacillus dipsosauri*, and *Gracilibacillus orientalis* (3.2–6.8%), but not in *Gracilibacillus halotolerans*.

Genotypic analysis. The G + C content of the type strains of each species ranges from 35.8 to 39.4 mol%. DNA–DNA hybridization among three strains of the species of *Gracilibacillus orientalis* is 94–98% (Carrasco et al., 2006). Hybridization of the DNA of the *Gracilibacillus boraciitolerans* type strain to the type strains of the other three *Gracilibacillus* species and to the type strain of *Paraliobacillus ryukuensis* (Ishikawa et al., 2002), a close relative, was 13–26% (Ahmed et al., 2007b).

Phylogenetic analysis. The 16S rRNA gene sequence of the type strain of *Gracilibacillus boraciitolerans* clustered with *Gracilibacillus orientalis* at 96.7%, *Gracilibacillus halotolerans* at 95.5%, and *Gracilibacillus dipsosauri* at 95.4%. In addition, there was a 95.7% similarity with *Paraliobacillus ryukuensis* (Ahmed et al., 2007b).

Enrichment and isolation procedures

Enrichment and isolation procedures for each species are described in the following publications: *Gracilibacillus halotolerans* (Wainø et al., 1999), *Gracilibacillus boraciitolerans* (Ahmed et al., 2007a), *Gracilibacillus dipsosauri* (Deutch, 1994), and *Gracilibacillus orientalis* (Ventosa et al., 1983).

Maintenance procedures

Strains grow on marine agar and broth (Difco) with or without 1.5% (w/v) NaCl and on 10% MH (moderate halophile) medium (Garabito et al., 1997). Isolates can be maintained by lyophilization and as a glycerol (35% w/v) stock at –80 °C.

Differentiation of the genus *Gracilibacillus* from other genera

The closest relatives to *Gracilibacillus* are given in the phylogenetic tree (Figure 2) generated by Ludwig et al., this volume. It differs from *Paraliobacillus ryukuensis*, *Amphibacillus xylanus* (Niimura et al., 1990), and *Halolactibactillus halophilus* (Ishikawa et al., 2005), its closest relatives, because they are fermenters and require glucose for growth under aerobic conditions. In addition, *Halolactibacillus halophilus* does not form spores and does not contain respiratory quinones. Table 20 gives major characteristics for differentiating the type strains of these genera.

List of species of the genus *Gracilibacillus*

1. **Gracilibacillus halotolerans** Wainø, Tindall, Schumann and Ingvorsen 1999, 829[VP]

ha.lo'to.le.rans. Gr.n. *hals* salt; L. adj. *tolerans* tolerating; N.L. adj. *halotolerans* salt-tolerating.

Characteristics of *Gracilibacillus halotolerans* are given in the genus description and in Table 19. Cells are thin rods, 0.4–0.6 × 2–5 μm (filamentous forms also occur) and are motile with peritrichous flagella. Cells grow at 6–50 °C with optimum growth at 47 °C. Growth occurs in 0–20% (w/v) NaCl with optimal growth at 0% NaCl. The pH range for growth is 5–10 with optimal growth at approximately pH 7.5.

Gracilibacillus halotolerans is differentiated from the other species by its production of urease and H_2S (Table 19). Tween 80 is hydrolyzed and alkaline phosphatase is produced. Phenylalanine deaminase, chitinase, and lecithinase are not produced. Acid is produced from D-glucose and D-xylose, but they are not fermented (Wainø et al., 1999).

Using Tris-medium (10% NaCl) and 0.2% (w/v) substrates (Wainø et al., 1999), the following compounds are used for growth: amylose, DL-arabinose, D-cellobiose, D-fructose, D-galactose, glycogen, inulin, lactose, maltose, D-mannose, D-melibiose, D-melezitose, raffinose, L-rhamnose, starch, D-trehalose, D-xylose, glycerol, L-ascorbic acid, D-galacturonic acid, D-gluconic acid, D-glucuronic acid, L-malate, oxoglutaric acid, *N*-acetylglucosamine, trimethylamine, and Tween 80. Growth does not occur on fucose, butanol, ethanol,

TABLE 19. Differential characteristics for *Gracilibacillus* species[a,b]

Characteristic	1. *G. halotolerans*	2. *G. boraciitolerans*	3. *G. dipsosauri*	4. *G. orientalis*
Growth pigment	Creamy white	Light pink to red	White	Cream
Anaerobic growth	–	ND	+	–
Optimal growth temperature,°C	47	25–28	45	37
Spore shape	E	S/E	S	S
NaCl growth range (% w/v)	0–20	0–11	0–15	1–20
Boron tolerance (mM)	0–50	0–450	0–150	ND
Oxidase test	+	+	+	–
Gelatin hydrolysis	+	–	+	–
Urea hydrolysis	+	–	–	+
H_2S production	+	–	–	–
Nitrate reduction to nitrite	+	–	+	–
Voges–Proskauer test	–	+	–	ND
DNA G + C content (mol%)	38	35.8	39.4	37.1

[a]Taken from Wainø et al. (1999), Ahmed et al. (2007b), Lawson et al. (1996), and Carrasco et al. (2006).
[b]Symbols: +, positive; –; negative; E, ellipsoidal; S, spherical; ND, no data.

TABLE 20. Characteristics helpful in differentiating *Gracilibacillus* from closely related genera based on type strain reactions[a,b]

Characteristic	*Gracilibacillus halotolerans*	*Paraliobacillus ryukyuensis*	*Amphibacillus xylanus*	*Halolactibacillus halophilus*
Spore formation	+	+	+	–
Anaerobic growth	–[c]	+(F)	+(F)	+(F)
Glucose required for aerobic growth	–	+	+	+
Major isoprenoid quinones	MK-7	MK-7	None	None

[a]Taken from Isikawa et al. (2005).
[b]Symbols: +, positive, –; negative; F, fermentation.
[c]*Gracilibacillus dipsosauri* exhibits anaerobic respiration (see species description).

methanol, pentanol, propanol, D-sorbitol, acetate, adipic acid, anisic acid, benzoate, butyrate, caproic acid, caprylate, citrate, formate, fumarate, glutaric acid, glycolate, glyoxylate, lactate, nicotinate, picolinic acid, propionate, pyruvate, succinate, valerate, L-alanine, L-arginine, L-aspartate, betaine, L-cysteine, L-glutamate, L-lysine, L-methionine, L-ornithine, L-phenylalanine, L-proline, L-serine, L-threonine, trytophan, acetamide, benzamide, sulfanilamide, ethanolamine, and methylamine.

Cells are susceptible to bacitracin, carbenicillin, erythromycin, novobiocin, penicillin, and rifampin, but resistant to gentamicin, kanamycin, nalidixic acid, neomycin, and tetracycline.

In addition to the major polar lipids, phosphatidylglycerol and diphosphatidylglycerol, two phospholipids of unknown structure were detected. The primary cellular fatty acid composition (discussed in the genus description) is similar to the other species except that $C_{17:0\ iso}$ is not detected (Ahmed et al., 2007b).

Genotypic and phylogenetic data are given in the genus description. In addition, using 16S rRNA gene sequencing, Wainø et al. (1999) reported that the type strain showed a 96% similarity to *Gracilibacillus dipsosauri.*

DNA G + C content (mol%): 38 (HPLC).

Type strain: DSM 11805, ATCC 700849.

GenBank accession number (16S rRNA gene): AF036922.

2. **Gracilibacillus boraciitolerans** Ahmed, Yokota and Fujiwara 2007b, 800[VP]

bo.ra′ci.i.to′le.rans. N.L. n. *boracium* boron; L. part. adj. *tolerans* tolerating; N.L. part. adj. *boraciitolerans* boron-tolerating.

Characteristics of *Gracilibacillus boraciitolerans* are given in the genus description and Table 19. Individual cells are 0.3–0.9 × 2.0–4.5 μm, occurring singly and occasionally in pairs; filaments also occur. Spherical endospores are produced in non-swollen or slightly swollen sporangia and are in a terminal or subterminal position. Cells are motile by a long, filamentous, monotrichous flagellum. Ahmed et al. (2007b) studied colonies on BUG agar medium (pH 7.5) at 30°C for 4 d. Young colonies are dirty white, but become pink and then red in several days. The pink or red pigments may diffuse into the agar after several days. They are 2–3 mm in diameter, circular with entire margins, slightly convex, translucent, and viscous. Growth occurs at 16–37°C with an optimum temperature of 25–28°C. There is no growth at 45°C. The NaCl tolerance range is 0–11% (w/v) with an optimal range of 0.5–3.0%. The pH range for growth was 6.0–10.0 with optimal pH of 7.5–8.5. The type strain tolerates 0–450 mM boron, but grows optimally without boron.

Gracilibacillus boraciitolerans can be differentiated from the other species because it is gelatinase-negative and Voges–Proskauer-positive (Table 19). Further phenotypic tests indicate that *Gracilibacillus boraciitolerans* gives a positive O-nitrophenyl β-D-galactopyranoside (ONPG) test, but is negative for tryptophan deaminase and citrate utilization (API 20E; BioMérieux). Cells can produce acid from L-arabinose, D-ribose, glucose, D-mannose, esculin, D-cellobiose, D-maltose, D-lactose, D-melibiose, and D-trehalose; weak acid is produced from D-xylose, methyl β-D-xylopyranoside, D-fructose, D-mannitol, and D-sorbitol (API 50CHB). Strong enzyme activity (API ZYM) occurs for alkaline phosphatase, β-galactosidase, and α- and β-glucosidase, whereas weak enzyme activity occurs with α-glactosidase, esterase (C8), esterase lipase (C8), and leucine arylamidase.

Using the Biolog system, cells can oxidize 3-methyl glucose, amygdalin, arbutin, D-cellobiose, dextrin, D-fructose, D-galactose, D-mannitol, D-mannose, D-melizitose, D-melibiose, D-psicose, D-raffinose, D-ribose, D-sorbitol, D-trehalose, D-xylose, gentiobiose, glycerol, lactulose, L-arabinose, maltose, maltotriose, palatinose, salicin, sucrose, turanose, α-D-glucose, α-D-lactose, methyl α-D-galactoside, methyl β-D-galactoside, methyl α-D-glucoside, methyl β-D-glucoside, DL-lactic acid, D-glucuronic acid, gluconic acid, pyruvic acid, and α-ketobutyric acid.

Cells are resistant to penicillin, amoxycillin, and metronidazole (ATB-VET Strip; BioMérieux).

In addition to the major polar lipids phosphatidylglycerol and diphosphatidylglycerol, moderate to minor amounts of an unknown aminolipid and three polar lipids were detected. The primary cellular fatty acids (discussed in the genus description) are similar to those of the other species, especially *Gracilibacillus orientalis* (GC-based Microbial Identification System; MIDI).

Genetic and phylogenetic data are given in the genus description.

DNA G + C content (mol%): 35.8 (HPLC).

Type strain: T-16X, DSM 17256, IAM 15263, ATCC BAA-1190.

GenBank accession number (16S rRNA gene): AB197126.

3. **Gracilibacillus dipsosauri** Wainø, Tindall, Schumann and Ingvorsen 1999, 829[VP] (*Bacillus dipsosauri* Lawson, Deutch and Collins 1996, 112)

dip.so.sau′ri. N.L. zool. name *Dipsosaurus* the desert iguana. N.L. gen. n. *dipsosauri* of the desert iguana because it was first isolated from the nasal salt glands of the desert iguana.

Dipsosaurus dorsalis, a desert iguana, has salt glands in its nasal cavities that allow it to excrete a concentrated brine of KCl during osmotic stress (Deutch, 1994). *Gracilibacillus dipsosauri* (formerly *Bacillus dipsosauri*) is the first reported Gram-positive, spore-forming halophile isolated from an animal with salt glands. Its characteristics suggested that it belonged in the genus *Bacillus* (which contains bacilli that form round, terminal spores). However, phylogenetic studies (Lawson et al., 1996) showed that it displayed relatively low sequence similarities with all members of the genus *Bacillus* tested (approximate range 84–93%). It was placed in the genus *Bacillus* as a matter of convenience. In 1999 Wainø et al. transferred it to the new genus, *Gracilibacillus*, because it clustered with the new species, *Gracilibacillus halotolerans* (96% similarity).

Characteristics of *Gracilibacillus dipsosauri* are given in the genus description and Table 19. Deutch (1994) observed that in trypticase soy broth (TSB) cultures containing 0.5–1.5 M (2.9–8.8% w/v) NaCl, cells were thin motile rods 0.3 × 2–3 μm. With 2 M (11.7 w/v) NaCl, cells were more filamentous and nonmotile. Spherical endospores within swollen terminal sporangia were formed after 2–3 d on TSB agar containing 1 M (7.5% w/v) KCl. Longer filaments containing several spherical endospores were sometimes observed. Sporulation occurred less frequently in liquid cultures. Growth temperature ranges from 28 °C to 50 °C with an optimal temperature of 45 °C. No growth was observed at 2.5 M (14.6%) NaCl (Deutch, 1994), but it did grow well at 2.5 M (18.75%) KCl. Carrasco et al. (2006) reported that *Gracilibacillus dipsosauri* grows in 0–15% (w/v) NaCl with an optimum of 3%. Deutch (1994) reported that although growth was observed without added salt, there was a longer lag time and growth rate was slower than normal. The pH range for growth is 6.5–10 with an optimum of 7.5. Cells grow aerobically and anaerobically in media containing nitrate or nitrite as terminal electron acceptors; however, they behave as strict aerobes in thioglycollate broth. Colonies are smooth, white, circular, and 2.0 mm in diameter when grown at 37 °C on TSB agar.

Gracilibacillus dipsosauri can be distinguished from the other species because it can grow anaerobically, it hydrolyzes gelatin and reduces nitrate (Table 19). In addition, using phenol red fermentation broth, weak acid is produced in glucose, sucrose, mannitol, and dulcitol. No acid or gas was seen during fermentation in API test strips or in Bromcresol purple fermentation broths. Triacylglycerides, ONPG, and *p*-nitrophenylgalactoside are hydrolyzed. Phospholipids and red blood cells are not hydrolyzed. Methyl red test is negative.

Cells are sensitive to chloramphenicol, kanamycin, and triple sulfa, but resistant to ampicillin, bacitracin, and streptomycin.

In addition to the major polar lipids, phosphatidylglycerol and diphosphatidylglycerol, two phospholipids of unknown structure were detected. The primary cellular fatty acids are similar to those of the other species (discussed above).

Genotypic and phylogenetic data are given in the genus description.

DNA G + C content (mol%): 39.4.

Type strain: ATCC 700347, DD1, NCFB 3027, DSM 11125.
GenBank accession number (16S rRNA gene): X82436.

4. **Gracilibacillus orientalis** Carrasco, Márquez, Yanfen, Ma, Cowan, Jones, Grant and Ventosa 2006, 56[VP]

o.ri.en.ta'lis. L. adj. *orientalis* eastern, bacterium inhabiting the East.

Characteristics of *Gracilibacillus orientalis* are given in the genus description and Table 19. Individual cells are 0.7–0.9 × 2.0–10.0 μm. Spherical endospores are formed in swollen sporangia at a terminal position. Colonies are 0.3–0.6 mm in diameter and cream-colored and opaque with entire margins when cultivated for 2 d on agar containing 10% MH medium. Growth occurs at 4–45 °C with an optimum temperature of 37 °C. *Gracilibacillus orientalis* is moderately halotolerant; the NaCl tolerance range is wide, 1–20% (w/v), with an optimal growth at 10.0%. The pH range for growth is 5.0–9.0 with an optimum of 7.5.

Gracilibacillus orientalis can be differentiated from other species because it requires NaCl for growth and is oxidase-negative (Table 19). Further phenotypic characteristics include acid production from arabinose, galactose, glycerol, D-fructose, D-lactose, D-mannitol, D-xylose, maltose, D-trehalose, and sucrose. Phosphatase test is positive. Tween 80 is not hydrolyzed and methyl red, phenylalanine deaminase, and Simmons citrate tests are negative. Compounds used as sole carbon and energy sources are acetate, citrate, formate, fumarate, D-fucose, lactose, propanol, D-sorbitol, and valerate. Compounds not utilized as sole carbon and energy sources are D-arabinose, D-cellobiose, D-galactose, maltose, D-mannose, D-melibiose, D-melezitose, L-raffinose, D-trehalose, D-xylose, butanol, ethanol, methanol, benzoate, propionate, and succinate. Compounds not used as sole carbon, nitrogen and energy sources are L-alanine, L-arginine, aspartic acid, L-cysteine, phenylalanine, glutamic acid, DL-lysine, L-methionine, L-ornithine, L-threonine, tryptophan, and L-serine.

Cells are susceptible to bacitracin, chloramphenicol, erythromycin, and rifampin. They are resistant to ampicillin, gentamicin, kanamycin, nalidixic acid, neomycin, novobiocin, and penicillin.

In addition to the major polar lipids, phosphatidylglycerol and diphosphatidylglycerol, *Gracilibacillus orientalis* contains phosphatidylethanolamine and a phospholipid and two amino phospholipids of unknown structure. The cellular fatty acid profile of the type strain is similar to those reported for the other three species (discussed in the genus description). It is especially similar to *Gracilibacillus boraciitolerans* in that it contains the same components in slightly differing amounts (Ahmed et al., 2007b).

Genotypic and phylogenetic data are given in the genus description. In addition, using 16S rRNA gene sequencing, Carrasco et al. (2006) reported that the type strain clustered with the type strain of *Gracilibacillus halotolerans* (95.4% similarity) and the type strain of *Gracilibacillus dipsosauri* (95.4%). The other closely related relative was *Paraliobacillus ryukyuensis* (94.8%).

DNA G + C content (mol%): 37.1 (T_m).
Type strain: XH-63, CCM 7326, AS 1.4250 CECT 7097.
GenBank accession number (16S rRNA gene): AM040716.

Genus IX. **Halobacillus** Spring, Ludwig, Marquez, Ventosa and Schleifer 1996, 495[VP]

STEFAN SPRING

Ha.lo.ba.cil′lus. Gr. n. *hals* salt; L. n. *bacillus* rod; N.L. masc. n. *Halobacillus* salt (-loving) rod.

Spherical to oval cells, 1.0–2.5 μm in diameter, occurring singly, in pairs, or aggregates (packets of four or more cells) or straight rod-shaped cells with pointed ends having a width of 0.5–1.4 μm and a length of 2.0–4.5 μm, occurring singly, in pairs, or short chains. The length of rod-shaped cells can be up to 20 μm under some culture conditions. Gram-positive. Endospores are formed. Motility, if present, is tumbling and conferred by one or more flagella. Colonies are circular, smooth, slightly raised, and opaque. Pigmentation by a nondiffusible pigment is variable ranging from cream-white or pale yellow to bright orange.

Chemo-organotrophic. Strictly aerobic, respiratory metabolism. Moderately halophilic. Growth is optimal at salt concentrations between 5% and 10%, temperatures of 30–38 °C, and pH values between 7.0 and 8.0. Catalase and oxidase are produced. The cell wall contains peptidoglycan of the Orn-D-Asp type (A4β type according to the murein key of Schleifer and Kandler, 1972). The cellular fatty acid pattern is characterized by major amounts of branched fatty acids, especially $C_{15:0\ anteiso}$, and a significant amount of $C_{16:1\ \omega7c}$ alcohol.

Widely distributed in a variety of hypersaline environments ranging from salt marsh soils and sediments to fermented food and mural paintings.

DNA G + C content (mol%): 40–43.

Type species: **Halobacillus halophilus** (Claus, Fahmy, Rolf and Tosunoglu 1983) Spring, Ludwig, Marquez, Ventosa and Schleifer 1996, 495[VP] (*Sporosarcina halophila* Claus, Fahmy, Rolf and Tosunoglu 1983, 503.)

Further descriptive information

The type species of the genus, *Halobacillus halophilus*, is characterized by coccoid cell morphology (Figure 16a), whereas all other known species (e.g., *Halobacillus trueperi*) are rod-shaped (Figure 16b). It is unlikely that the absence of a uniform morphotype within the genus indicates a fundamental difference between the species. It probably reflects an ongoing development of cell morphologies from spherical to rod-shaped, or vice versa, with common transition forms like oval or wedge-shaped cells.

Upon division, cells of *Halobacillus halophilus* are hemispherical but have a tendency to elongate giving rise to oval cells. Abnormally large cells occur often. Division into two or three perpendicular planes in *Halobacillus halophilus* can lead to the formation of threes, tetrads, or packets of eight or more cells. In contrast, strains of the other known *Halobacillus* species are characterized by rod-shaped cells that form short chains. The morphology of the rod-shaped cells is usually not very regular. The cells have pointed ends that often taper towards one end, resulting in shapes resembling an elongated egg or wedge.

Spores are either spherical and located centrally or laterally (*Halobacillus halophilus*, Figure 16a) or, more frequently, ellipsoidal with a subterminal or central position (all other described species, Figure 16b). They are highly refractile, survive heating at 75 °C for at least 10 min, and have a size in the range 0.5 to 1.5 μm. The walls of endospores of *Halobacillus halophilus*

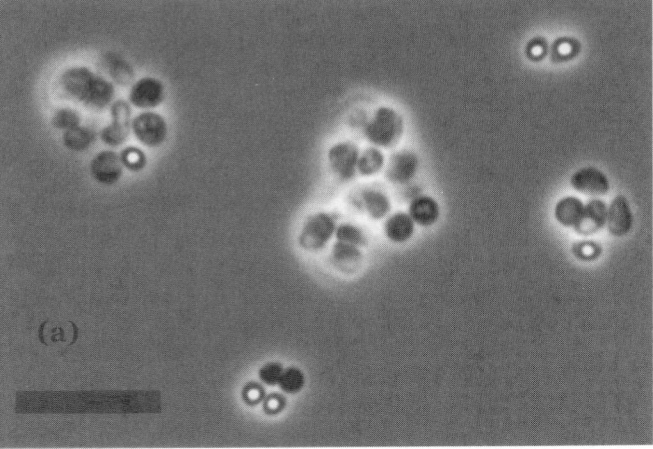

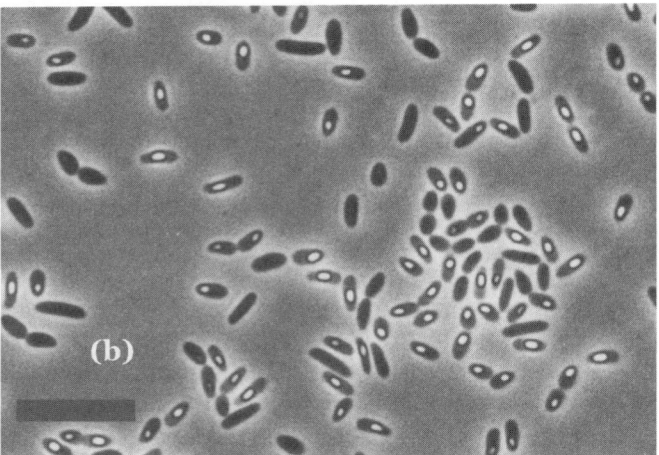

FIGURE 16. Phase-contrast photomicrographs of sporulating cultures of (a) *Halobacillus halophilus* DSM 2266[T] and (b) *Halobacillus trueperi* DSM 10404[T]. Bars = 10 μm.

contain diaminopimelic acid which is not found in walls of vegetative cells (Claus et al., 1983). The spores of this species also contain dipicolinic acid (Fahmy et al., 1985).

With the exception of *Halobacillus karajensis* all *Halobacillus* species are motile. Motility is difficult to detect in some strains and depends on culture conditions and growth phase. Flagella can be exceedingly long and are inserted predominantly as tufts at both poles and sometimes laterally.

Colonies may be bright orange, pale yellow, or cream-white. Pigmentation varies among strains and depends on salt concentration and incubation time. The pigment is water insoluble and nondiffusible.

Halobacillus halophilus is an obligate, moderate halophilic bacterium that requires sodium, magnesium, and chloride ions for growth. Poor or no growth occurs at NaCl concentrations below 3% and MgCl$_2$ concentrations below 0.5%. The chloride dependence of growth in *Halobacillus halophilus* has been studied extensively; flagella synthesis, endospore germination, and glycine betaine transport have been found to be dependent on the chloride concentration (Dohrmann and Muller, 1999; Rossler and Muller, 1998, 2001; Rossler et al., 2000). In contrast, the salt requirement of *Halobacillus trueperi* and *Halobacillus litoralis* is less pronounced; both of these species show good growth in medium supplemented with only 0.5% NaCl (Table 21).

All known members of the genus *Halobacillus* are positive for the hydrolysis of gelatin and DNA; they are negative for nitrate reduction, Voges–Proskauer reaction, and hydrolysis of urea and Tween 80.

Whole-cell fatty acid compositions of the validly published *Halobacillus* species are shown in Table 22 (R. M. Kroppenstedt, personal communication). In general, the fatty acid patterns are very similar among the species of the genus and

TABLE 21. Differential characteristics of *Halobacillus* species and phylogenetically closely related taxa[a]

Characteristic	*Halobacillus halophilus*[b]	*Halobacillus karajensis*[b]	*Halobacillus litoralis*[b]	*Halobacillus trueperi*[b]	*Bacillus halophilus*[c]	*Marinococcus albus*[d]
Morphology	Coccoid	Rod	Rod	Rod	Rod	Coccus
Flagella	Peritrichous	None	Peritrichous	Peritrichous	Peritrichous	1 or 2
Pigmentation	Orange	Colorless	Orange	Orange	Colorless	Colorless
Spores	Spherical	Ellipsoidal (spherical)	Ellipsoidal (spherical)	Ellipsoidal (spherical)	Ellipsoidal	None
NaCl range (%)	2–20	1–24	0.5–25	0.5–30	3–30	5–20
Temperature range (°C)	15–40	10–49	10–43	10–44	15–50	ND
Nitrate reduction	–	–	–	–	–	+
Acid from:						
D-Galactose	–	–	–	+	ND	–
Glucose	–	+	+	+	+	–
Maltose	–	+	+	+	–	–
D-Xylose	–	–	+	–	+	–
Hydrolysis of:						
Casein	+	+		–	–	–
Gelatin	+	+	+	+	–	–
Esculin	–	+	–	–	+	–
Starch	+	+	–	–	–	–
Urea	–	–	–	–	+	+
Cell-wall type	Orn-D-Asp	Orn-D-Asp	Orn-D-Asp	Orn-D-Asp	*m*-Dpm	*m*-Dpm
G+C content (mol%)	40.1–40.9	41.3	42	43	51.5	44.9
Source of isolation	Salt marsh soil	Hypersaline soil	Hypersaline sediment	Hypersaline sediment	Seashore drift wood	Solar saltern

[a]Symbols: +, >85% positive; d, different strains give different reactions (16–84% positive); –, 0–15% positive; w, weak reaction; ND, not determined.
[b]Data from Amoozegar et al. (2003).
[c]Data from Ventosa et al. (1989).
[d]Data from Hao et al. (1984).

TABLE 22. Fatty acid composition of type strains of *Halobacillus* species after growth on MB agar (DIFCO 2216) at 28 °C for 48 h prior to analysis[a]

Equivalent chain-length	Fatty acid[b]	*Halobacillus halophilus* (DSM 2266[T])	*Halobacillus karajensis* (DSM 14948[T])	*Halobacillus litoralis* (DSM 10405[T])	*Halobacillus trueperi* (DSM 10404[T])
13.618	$C_{14:0\ iso}$	10.6	2		23.2
14.623	$C_{15:0\ iso}$	7.4	11.3	16.3	6.6
14.715	$C_{15:0\ anteiso}$	42.1	42.4	45.6	19.2
15	$C_{15:0}$	0.8	0.3		
15.387	$C_{16:1\ \omega7c\ OH}$	8.8	9.7	2.6	12.2
15.627	$C_{16:0\ iso}$	14.2	6.9	1.2	28
15.756	$C_{16:1\ \omega11c}$	0.6	0.9	1.3	0.7
15.998	$C_{16:0}$	1	1.1	0.9	1
16.388	$C_{17:1\ \omega10c\ iso}$		1.1	2.5	
16.478	Summed feature 4[c]	0.9	3.3	6.8	1.2
16.631	$C_{17:0\ iso}$	1.7	5	7.8	2.8
16.724	$C_{17:0\ anteiso}$	11.6	16	15	5.1

[a]Values are percentages of total fatty acids.
[b]The position of the double bond in unsaturated fatty acids is located by counting from the methyl (ω) end of the carbon chain; *cis* and *trans* isomers are indicated by the suffixes c and t, respectively.
[c]Summed feature 4 contained one or more of the following fatty acids: $C_{17:1\ iso\ I}$ and/or $C_{17:1\ anteiso}$.

differences are mainly due to varying quantities of some fatty acids. Branched fatty acids of the iso- and anteiso-type with a chain length of 15:0, 16:0 and 17:0 are clearly dominant as in many other species of Gram-positive, spore-forming halophilic or halotolerant bacilli (e.g., Niimura et al., 1990; Heyndrickx et al., 1998; Wainø et al., 1999). In contrast, the occurrence of the unsaturated fatty acids $C_{16:1\ \omega7c}$ alcohol and $C_{16:1\ \omega11c}$ seems to be a typical characteristic of the genus *Halobacillus*. The polar lipid pattern of members of the genus *Halobacillus* resembles that of *Marinococcus albus* and is comprised of phosphatidyl glycerol, diphosphatidyl glycerol, and an unknown glycolipid similar to that found in *Salinicoccus roseus* (Wainø et al., 1999).

The menaquinone system has been determined only in *Halobacillus halophilus* and *Halobacillus karajensis*. In these species MK-7 is the predominant menaquinone (Amoozegar et al., 2003; Claus et al., 1983).

Based on comparative analyses of 16S rRNA gene sequences, the genus *Halobacillus* is located phylogenetically at the periphery of the *Bacillus* rRNA group 1 as defined by Ash et al. (1991). The rRNA group 1 is also known as the core cluster of the genus *Bacillus* comprising the true *Bacillus* species. Members of the genus *Halobacillus* form a distinct branch within this phylogenetic group along with other phenotypically diverse species that display various traits that are not in accord with the characteristics of *Bacillus subtilis*, the type species of the genus *Bacillus*.

Consequently, several of these newly isolated species that were only loosely associated with the rRNA group 1 were placed in novel genera (e.g., *Amphibacillus* gen. nov., Niimura et al., 1990; *Halobacillus* gen. nov., Spring et al., 1996; and *Filobacillus* gen. nov., Schlesner et al., 2001). Several species which were closely related to the newly described taxa but originally affiliated to the genus *Bacillus* or *Sporosarcina*, in the case of *Halobacillus*, were relocated into newly proposed genera as new combinations (e.g., *Gracilibacillus dipsosauri* comb. nov.; Wainø et al., 1999) and *Virgibacillus pantothenticus* comb. nov.; Heyndrickx et al., 1998). In the phylogenetic tree in Figure 17, the location of *Halobacillus* species among other related taxa of Gram-positive, moderately halophilic bacteria is shown. The species most closely related to members of the genus *Halobacillus* are *Marinococcus albus* and *Bacillus halophilus* which together form a stable cluster in most trees independent of the calculation method used (Figure 17). The similarity values of almost complete 16S rRNA gene sequences among members of this group range from 94.2% to 99.3%.

Enrichment and isolation procedures

Most members of the genus *Halobacillus* have been isolated by plating serial diluted suspensions of particulate matter from hypersaline sites on solid media. Selective enrichment methods in liquid media are not available. Suitable media for isolating *Halobacillus* species are Bacto Marine Agar 2216 (Difco)

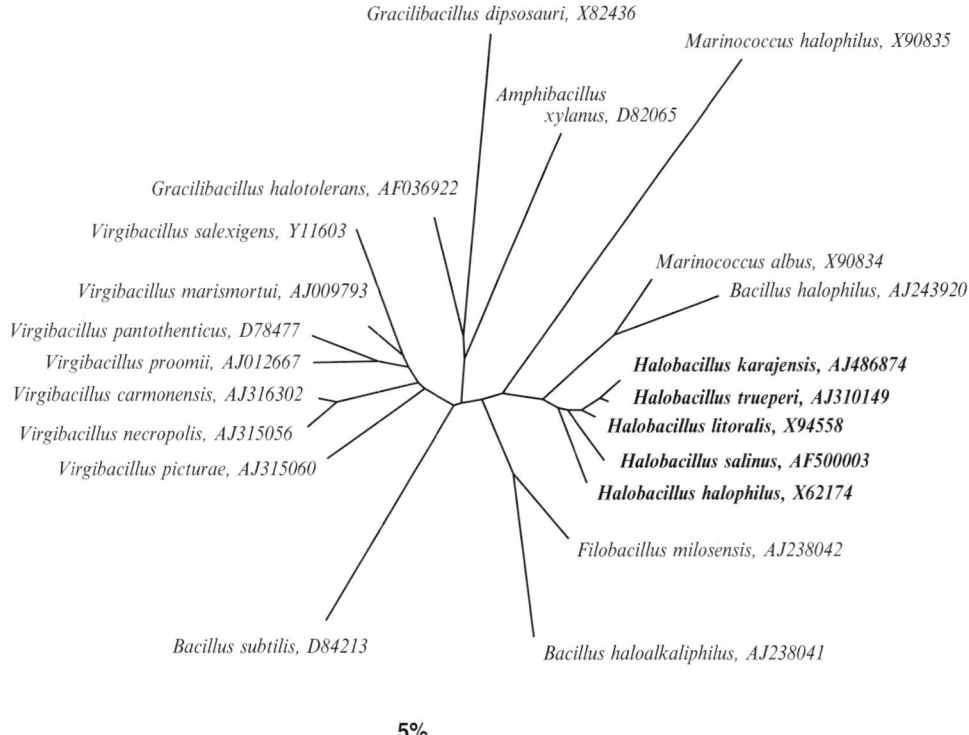

FIGURE 17. Phylogenetic tree based on almost complete 16S rRNA gene sequences showing the position of members of the genus *Halobacillus* among their closest relatives. The GenBank/EMBL accession number for each sequence is shown in parentheses. The tree was reconstructed using the ARB program package (Ludwig and Strunk, 1997). It is derived from a distance matrix on a selection of 16S rRNA sequences using the neighbor-joining method of Saitou and Nei (1987). Phylogenetic distances were calculated as described by Jukes and Cantor (1969). The sequence of *Bacillus subtilis* was used as an outgroup. The bar indicates 5% estimated sequence divergence.

or nutrient agar supplemented with double concentrated sea water (Claus et al., 1992). Several methods have been reported to reduce background growth of undesired microbial species thriving under similar conditions. Claus and Fahmy (1986) heated suspensions of soil samples at 70°C for 10 min in order to kill vegetative cells and enrich for spore-formers. They noted that it is important to plate heat treated samples on nutrient agar medium supplemented with double concentrated sea water to ensure efficient germination of spores. The number of undesired colonies may be further reduced by increasing the NaCl concentration of the isolation agar to 20% or more, which is tolerated by most strains of *Halobacillus* but inhibitory to several related genera (Spring et al., 1996). Pinar et al. (2001) supplemented enrichment media with 50 μg cycloheximide per ml to avoid fungal growth.

Colonies on agar plates normally appear after 3 d of incubation at 30°C in the dark. In most cases, colonies of *Halobacillus* strains will develop a pale-yellow to orange pigmentation, which is only seen in older colonies reaching a diameter of 1–2 mm. In addition, one species has been reported to show only a cream-white pigmentation (Amoozegar et al., 2003). Hence, a clear affiliation of novel isolates to the genus *Halobacillus* is not possible solely on the basis of morphological characteristics. For purification, cell material taken from appropriate colonies should be resuspended in a drop of nutrient broth and restreaked onto agar plates of suitable composition.

Maintenance procedures

Vegetative cultures of *Halobacillus*, grown on nutrient agar slants supplemented with 3–10% NaCl and 0.5% MgCl₂, are viable usually for about 6 months if tightly sealed to avoid drying and stored between 4°C and 10°C in the dark. Viability is increased to several years if sporulated cultures are kept at 4 to 20°C in screw-capped tubes. Fahmy et al. (1985) recommended the following medium to obtain good sporulation in *Halobacillus halophilus* at an incubation temperature below 25°C: peptone, 5.0 g; yeast extract, 1.0 g; NaCl, 24.32 g; MgCl₂·6H₂O, 10.99 g; Na₂SO₄, 4.06 g; CaCl₂·2H₂O, 1.51 g; KCl, 0.69 g; NaHCO₃, 0.20 g; KBr, 0.10 g; SrCl₂·6H₂O, 0.042 g; H₃BO₃, 0.027 g; Na₂SiO₃·9H₂O, 0.005 g; NaF, 0.003 g; NH₄NO₃, 0.002 g; FePO₄·4H₂O, 0.10 g; MnCl₂, 0.01 g; agar, 15 g; distilled water, 1,000 ml.

For the long-term preservation of *Halobacillus* strains, freeze-drying of vegetative cells or of spores is recommended. As a protective menstrum, skim-milk (20% w/v) or serum containing 5% *meso*-inositol is suitable. Both vegetative cells and spores can be successfully preserved for long periods in liquid nitrogen without severe loss in survival using glycerol (10%) or dimethylsulfoxide (5%) as cryoprotective agents.

Differentiation of the genus *Halobacillus* from other genera

The genus *Halobacillus*, despite its intrageneric variability, can be distinguished easily from most members of the other related taxa shown in the phylogenetic tree in Figure 10. Within this group, only the species *Virgibacillus salexigens*, *Virgibacillus marismortui*, and *Bacillus halophilus* are phenotypically quite similar. They are also Gram-positive, spore-forming, obligately aerobic, grow at neutral pH values, and show a requirement for salt in the medium. The differentiation from the genus *Halobacillus* is, however, possible by analyzing the murein type of the cell wall, which is *m*-Dpm in these species in contrast to Orn-D-Asp in members of the genus *Halobacillus*. *Filobacillus milosensis* is the only representative of this phylogenetic group with a similar murein structure to that of *Halobacillus* (A4β), but cells stain Gram-negative and the cell-wall murein contains Orn-D-Glu instead of Orn-D-Asp. Several species related to *Halobacillus* are facultatively anaerobic; they include *Amphibacillus* species, *Gracilibacillus* species, *Virgibacillus pantothenticus*, and *Virgibacillus proomii*. Other taxa can be discriminated by their obligate alkaliphilic growth (*Amphibacillus*, *Bacillus haloalkaliphilus*) or the absence of spores (*Marinococcus*). *Gracilibacillus halotolerans* is the only member of this phylogenetic group which is halotolerant rather than moderately halophilic, growing optimally in media without salt.

Taxonomic comments

A novel representative of the genus *Halobacillus* was isolated from fish fermentation tanks in Thailand. Strain fs-1 was selected for the secretion of proteinases that are thought to accelerate the liquefaction of fish necessary for the production of fish sauce. A formal description of this strain as *Halobacillus thailandensis* was published by Chaiyanan et al. (1999). According to the given description, this species is phylogenetically and phenotypically quite similar to *Halobacillus litoralis* and *Halobacillus trueperi*, however neither the type strain fs-1 nor its 16S rRNA gene sequence has been deposited in a public culture collection or database. Therefore, the name of this species has never been validated.

A novel *Halobacillus* species, *Halobacillus salinus*, has been isolated from a salt lake of the East Sea in Korea (Yoon et al., 2003b). At the time of chapter preparation, only the 16S rRNA gene sequence of this newly described species was available. The phylogenetic position is shown in Figure 10.

Acknowledgements

I am grateful to R. M. Kroppenstedt for providing data on whole-cell fatty acid patterns of *Halobacillus* strains.

List of species of the genus *Halobacillus*

In addition to the description given for the genus, several traits of *Halobacillus* species that are useful for their differentiation are summarized in the species descriptions. In Table 21, distinguishing characteristics of *Halobacillus* species are listed.

1. **Halobacillus halophilus** (Claus, Fahmy, Rolf and Tosunoglu 1983) Spring, Ludwig, Marquez, Ventosa and Schleifer 1996, 495^VP (*Sporosarcina halophila* Claus, Fahmy, Rolf and Tosunoglu 1983, 503.)

ha.lo′phi.lus. Gr. masc. n. *hals, halos* salt; Gr. adj. *philos* loving; N.L. masc. adj. *halophilus* salt-loving.

Spherical or oval cells, occurring singly, in pairs, triads, tetrads, or packages. Spherical cells 1.0–2.5 μm in diameter, oval cells 1.0–2.0 by 2.0–3.0 μm. Motile by one or more randomly spaced flagella on each cell. Endospores round, 0.5–1.5 μm, located centrally or laterally. Colonies round, smooth, opaque, and forming an orange, nondiffusible pigment. Casein, gelatin, pullulan, and starch are hydrolyzed, but esculin, Tween 80, and tyrosine are not hydrolyzed. Generally no acid produced from glucose or other sugars. Salinity range for growth between 2% and 15% NaCl; temperature range between 15 °C and 37 °C; pH range between 7.0 and 9.0.

Source: salt marsh soils.

DNA G + C content (mol%): 40.1–40.9 (T_m).

Type strain: 3, ATCC 35676, DSM 2266.

GenBank accession number (16S rRNA gene): X62174.

2. **Halobacillus karajensis** Amoozegar, Malekzadeh, Malik, Schumann and Spröer 2003, 1062[VP]

ka.ra.jen'sis. N.L. adj. *krajensis* from the region of Karaj, Iran, where the organism was isolated.

Cells are rod-shaped, 0.8–0.9 by 2.5–4.0 μm, occurring singly, in pairs, or in short chains. Filamentous cells can be observed under suboptimal conditions for growth. Nonmotile. Endospores are ellipsoidal or spherical and located at a central or subterminal position. Colonies round, smooth, opaque, and with a white or cream color. Gelatin, casein, esculin, and starch are hydrolyzed, but Tween 80 and tyrosine are not hydrolyzed. Acid is produced from D-fructose, D-glucose, maltose, mannitol, mannose, and raffinose, but not from D-arabinose, D-galactose, sucrose, and D-xylose. Salinity range for growth between 1.0% and 24% NaCl; temperature range between 10 °C and 49 °C; pH range 6.0–9.6.

Source: saline soil near Karaj (Iran).

DNA G + C content (mol%): 41.3 (T_m).

Type strain: MA-2, DSM 14948, LMG 21515.

GenBank accession number (16S rRNA gene): AJ486874.

3. **Halobacillus litoralis** Spring, Ludwig, Marquez, Ventosa and Schleifer 1996, 495[VP]

li. to.ra'lis. L. masc. adj. *litoralis* pertaining to the shore.

Cells are rod-shaped, 0.7–1.1 by 2.0–4.5 μm, occurring singly, in pairs, or in short chains. Sometimes filamentous cells up to 20 μm long can be observed. Motile by means of several flagella inserted at both poles or laterally. Endospores are ellipsoidal or sometimes spherical and located at a central or subterminal position. Colonies round, smooth, opaque, and forming an orange, nondiffusible pigment. Gelatin is

hydrolyzed, but casein, esculin, pullulan, starch, Tween 80, and tyrosine are not hydrolyzed. Acid is produced from D-fructose, D-glucose, maltose, D-mannitol, sucrose, D-trehalose, and D-xylose, but not from D-galactose. Salinity range for growth between 0.5% and 25% NaCl; temperature range between 10 °C and 43 °C; pH range between 6.0 and 9.5.

Source: sediment obtained from the Great Salt Lake (Utah).

DNA G + C content (mol%): 42 (T_m).

Type strain: SL-4, ATCC 700076, CIP 104798, DSM 10405, LMG 17438.

GenBank accession number (16S rRNA gene): X94558.

4. **Halobacillus trueperi** Spring, Ludwig, Marquez, Ventosa and Schleifer 1996, 495[VP]

true'per.i. N.L. gen. n. *trueperi* of Trueper, in honor of Hans G. Trüper, a German microbiologist.

Cells are rod-shaped, 0.7–1.4 by 2.0–4.5 μm, occurring singly, in pairs, or in short chains. Sometimes cells up to 20 μm long are present. Motile by means of several flagella, which are inserted at both poles or laterally. Endospores are ellipsoidal or sometimes spherical and located at a central or subterminal position. Colonies round, smooth, opaque and forming an orange, nondiffusible pigment. Gelatin and pullulan are hydrolyzed, but casein, esculin, starch, Tween 80, and tyrosine are not hydrolyzed. Acid is produced from D-fructose, D-galactose, D-glucose, maltose, D-trehalose, and sucrose, but not from D-mannitol and D-xylose. Salinity range for growth between 0.5% and 30% NaCl; temperature range between 10 °C and 44 °C; pH range between 6.0 and 9.5.

Source: sediment obtained from the Great Salt Lake (Utah).

DNA G + C content (mol%): 43 (T_m).

Type strain: SL-5, ATCC 700077, CIP 104797, DSM 10404, LMG 17437.

GenBank accession number (16S rRNA gene): AJ310149.

Note added in proof: Since this chapter was prepared, the following new species have been validly published: *Halobacillus aidingensis* (Liu et al., 2005), *Halobacillus campisalis* (Yoon et al., 2007a), *Halobacillus dabanensis* (Liu et al., 2005), *Halobacillus faecis* (An et al., 2007a), *Halobacillus kuroshimensis* (Hua et al., 2007), *Halobacillus locisalis* (Yoon et al., 2004a), *Halobacillus mangrovi* (Soto-Ramirez et al., 2008), *Halobacillus profundi* (Hua et al., 2007), and *Halobacillus yeomjeoni* (Yoon et al., 2005a).

Genus X. **Halolactibacillus** Ishikawa, Nakajima, Itamiya, Furukawa, Yamamoto and Yamasato 2005, 2435[VP]

Morio Ishikawa and Kazuhide Yamasato

Ha.lo.lac'ti.ba.cil'lus. Gr. n. *hals* salt; L. n. *lac* lactis milk; L. masc. n. *bacillus* stick, a small rod; N.L. masc. n. *Halolactibacillus* salt (-loving) lactic acid rodlet.

Cells are **Gram-positive, nonspore-forming, straight rods,** occurring singly, in pairs, or in short chains, and elongated. **Motile with peritrichous flagella. Catalase- and oxidase-negative.** Nitrate is not reduced. Starch and casein are hydrolyzed.

Growth does not occur in the absence of sugars. Slightly halophilic and highly halotolerant. Alkaliphilic. Mesophilic. **In anaerobic cultivation, L-lactic acid is the major end product from glucose. In addition to lactate, considerable amounts of**

formate, acetate, and ethanol are produced in a molar ratio of approximately 2:1:1 without gas production. Carbohydrates and related compounds are aerobically metabolized to acetate and pyruvate without production of lactate, formate, and ethanol. The cell-wall peptidoglycan is *meso*-diaminopimelic acid. Cellular fatty acids are of the straight-chain, anteiso-branched saturated, iso-branched saturated, and monounsaturated acids. Major cellular fatty acids are $C_{13:0\ ante}$ and $C_{16:0}$. Respiratory quinones and cytochromes are absent. Located within the phylogenetic group composed of halophilic/halotolerant/alkaliphilic and/or alkalitolerant genera in *Bacillus* rRNA group 1.

DNA G + C content (mol%): 38.5–40.7.

Type species: **Halolactibacillus halophilus** Ishikawa, Nakajima, Itamiya, Furukawa, Yamamoto and Yamasato 2003b, 2437[VP].

Further descriptive information

Descriptive information is based on the descriptions of *Halolactibacillus halophilus* (six strains) and *Halolactibacillus miurensis* (five strains).

The genus *Halolactobacillus* is a lactic acid bacterium belonging to family *Bacillaceae*, order *Bacillales*, class "*Bacilli*" in phylum "*Firmicutes*."

Phylogenetic position of the genus *Halolactibacillus* based on 16S rRNA gene sequence analysis is given in Figure 18. The characteristics of *Halolactibacillus halophilus* and *Halolactibacillus miurensis* are listed in Table 23, Table 24, and Table 25.

Lactate is produced in yields of 50–60% of the amount of glucose consumed at an optimum pH under anaerobic cultivation. The other end products are formate, acetate, and ethanol in a molar ratio of μ2:1:1. No gas is produced (Table 24). The L-isomer of lactate is 80–95% of the total lactate produced. The amount of lactate relative to that of the other three products is markedly affected by the initial pH of the fermentation medium. The lactate yield increases at acidic pH values and decreases at the more alkaline pH values. At all pH values, carbon recovery from glucose consumed is about 100%, and the 2:1:1 molar ratio of formate, acetate, and ethanol produced is retained (Table 24).

The similar alkaliphilic lactic acid bacteria, *Marinilactibacillus psychrotolerans* and *Alkalibacterium olivapovliticus*, likewise produce formate, acetate, and ethanol at a molar ratio of 2:1:1 (in addition to lactate) under anaerobic conditions. Their product ratios relative to lactate are similarly affected by the initial pH of the fermentation medium (Ishikawa et al., 2003b). Pyruvate is converted to lactate by lactate dehydrogenase and to formate, acetate, and ethanol by pyruvate-formate lyase. It is considered that the product balance depends on the relative activities of the two enzymes involved (Ishikawa et al., 2003b; Janssen et al., 1995; Rhee and Pack, 1980). *Halolactibacillus*, as well as *Marinilactibacillus psychrotolerans* and *Alkalibacterium olivapovliticus*, are lactic acid bacteria in which pyruvate-formate lyase would be active (especially in *Halolactibacillus*) in the pH range that results in normal growth.

Halolactibacillus metabolizes glucose oxidatively, though it lacks respiratory quinones and cytochromes. Products from glucose under aerobic cultivation conditions are acetate, pyruvate, and lactate, but formate and ethanol are not produced (Table 25). The imbalance in carbon recovery can be ascribed to CO_2 generation, if *Amphibacillus xylanus*, a facultative anaerobe lacking in catalase, respiratory quinones, and cytochromes (Niimura et al., 1989; Niimura et al., 1990) and some homofermentative lactic acid bacteria (*Lactobacillus*, *Pediococcus*, and *Streptococcus*; Sakamoto and Komagata, 1996) are similar to *Halolactibacillus* in the aerobic metabolism of glucose. The oxidative pathway of glucose in these bacteria is mediated by the NADH oxidase/peroxidase system to produce acetate and CO_2 from pyruvate using O_2 as an electron acceptor. Assuming that equimolar amounts of acetate and CO_2 are produced in *Halolactibacillus* and *Marinilactibacillus psychrotolerans*, carbon recovery under aerobic conditions can be calculated as 93–97% (nearly 100%).

Halolactibacillus requires glucose for growth even under aerobic conditions. Under aerobic conditions, growth in 2.5% NaCl GYPF broth (GYPF (GYPB) broth: (per liter) 10 g glucose, 5 g yeast extract, 5 g peptone, 5 g fish extract (beef extract), 1 g K_2HPO_4, 1 g sodium thioglycolate, and 5 ml salt solution (per liter, 40 mg $MgSO_4{\cdot}7H_2O$, 2 mg $MnSO_4{\cdot}4H_2O$, 2 mg $FeSO_4{\cdot}7H_2O$, pH 8.5) is weak when the initial glucose concentration is decreased to 0.1%, and does not occur when glucose is omitted. In GCY broth (composed of 1.0% glucose, 0.5% Vitamin assay Casamino acids (Difco), 0.05% yeast extract, and inorganic components of GYPF broth, pH 8.5), the final OD_{660} of the culture is about 0.20 and in the absence of glucose is less than 0.02.

Halolactibacillus is slightly halophilic (Kushner, 1978; Kushner and Kamekura, 1988) and highly halotolerant. The optimum NaCl concentrations for growth are 2.0% (0.34 M) to 3.0% (0.51 M) for *Halolactibacillus halophilus* and 2.5% (0.43 M) to 3.0% for *Halolactibacillus miurensis*. The maximum specific growth rates, μ_{max} (h^{-1}), of *Halolactibacillus halophilus* IAM 15242[T] are 0.18 in 0%, 0.22 in 0.5%, 0.30 in 1.0%, 0.46 in 1.5%, 0.48 in 2.0%, 0.54 in 2.5%, 0.40 in 3.0%, 0.40 in 3.75%, and 0.40 in 5.0% NaCl. Those of *Halolactibacillus miurensis* IAM 15247[T] are 0.40 in 0%, 0.44 in 0.5%, 0.56 in 1.0%, 0.56 in 1.5%, 0.56 in 2.0%, 0.70 in 2.5%, 0.60 in 3.0%, 0.48 in 3.75%, and 0.48 in 5.0% NaCl.

Halolactibacillus halophilus is able to grow between 0 and 23.5–24.0% (4.02–4.11 M) NaCl and *Halolactibacillus miurensis* in 0–25.5% (4.36 M) NaCl.

Halolactibacillus is alkaliphilic, as it grows optimally at pH values above 8.0 (Jones et al., 1994) (8.0–9.0 for *Halolactibacillus halophilus* and 9.5 for *Halolactibacillus miurensis*). For *Halolactibacillus halophilus*

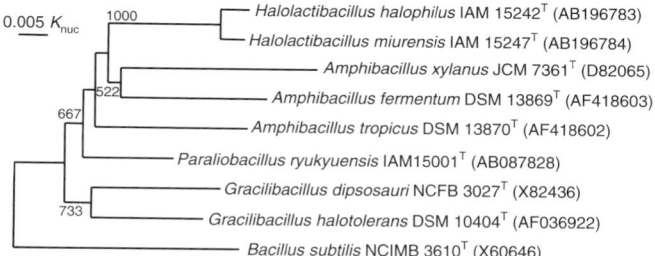

FIGURE 18. Phylogenetic relationships between *Halolactibacillus* and some other related bacteria belonging to the phylogenetic group composed of halophilic/halotolerant/alkaliphilic and/or alkalitolerant genera in *Bacillus* rRNA group 1. The tree, reconstructed using the neighbor-joining method, is based on a comparison of μ1,400 nucleotides. *Bacillus subtilis* NCIMB 3610[T] is used as an outgroup. Bootstrap values, expressed as a percentage of 1,000 replications, are given at branching points.

TABLE 23. Characteristics differentiating *Halolactibacillus* species[a,b]

Characteristic	*H. halophilus*	*H. miurensis*
NaCl optima (%)	2.0–3.0	2.5–3.0
NaCl range (%)	0–23.5 to 24.0	0–25.5
pH optima	8.0–9.0	9.5
pH range	6.5–9.5	6.0–6.5 to 10.0
Temperature optimum (°C)	30–37	37–40
Temperature range (°C)	5–10 to 40	5–45
Casein hydrolysis	–	–
Gelatin hydrolysis	–	–
Starch hydrolysis	w	w
Nitrate reduction	–	–
NH$_3$ from arginine	–	–
Dextran from sucrose	–	–
DNase	–	–
Fermentation of:		
D-Glucose, D-fructose, D-mannose, D-galactose, maltose, sucrose, D-cellobiose, lactose, melibiose, D-trehalose, D-raffinose, D-mannitol, starch, α-methyl-D-glucoside, D-salicin, gluconate	+	+
Glycerol	+	w
D-Ribose	(+)	+
D-Arabinose, D-rhamnose	–	(–)
Adonitol, myo-inositol, dulcitol, D-sorbitol	–	–
L-Arabinose, D-xylose, D-melezitose, inulin	–	+
Gas from gluconate	–	–
Yields of lactate from glucose (%)	50–60	50–60
Major fatty acid composition (% of total):[c]		
C$_{12:0}$	2.5	2.1
C$_{13:0\ iso}$	6.5	5.7
C$_{13:0\ ante}$	19.1	18.8
C$_{14:0\ iso/ante}$	–	0.5
C$_{14:0}$	4.1	4.0
C$_{15:0\ iso}$	3.5	3.7
C$_{15:0\ ante}$	6.2	7.6
C$_{15:0}$	1.4	1.6
C$_{16:0\ iso}$	0.9	0.8
C$_{16:0}$	43.1	37.2
C$_{16:1}$	–	1.3
C$_{16:1\ \omega7}$	–	0.7
C$_{17:0\ iso}$	1.5	3.5
C$_{17:0\ ante}$	–	2.5
C$_{17:0}$	–	0.9
C$_{18:0}$	4.6	5.5
C$_{18:1\ \omega9}$ (oleic acid)	2.7	2.2
C$_{18:2}$	1.1	1.2

[a]Symbols: +, all strains positive; (+), most strains positive; w, weakly positive; (–), most strains negative; –, all strains negative.

[b]Data from Ishikawa et al. (2005).

[c]Fatty acid compositions are of strain IAM 15242[T] (*Halolactibacillus halophilus*) and of strain IAM 15247[T] (*Halolactibacillus miurensis*).

IAM 15242[T], the μ_{max} values (h^{-1}) are 0.38 at pH 7.0, 0.40 at pH 7.5, 0.42 at pH 8.0, 0.52 at pH 8.5, 0.50 at pH 9.0, and 0.14 at pH 9.5. For *Halolactibacillus miurensis* IAM 15247[T], they are 0.46 at pH 7.0, 0.46 at pH 7.5, 0.46 at pH 8.0, 0.46 at pH 8.5, 0.48 at pH 9.0, 0.68 at pH 9.5, and 0.40 at pH 10.0. The final pH of cultures in 2.5% NaCl GYPF broth reaches 5.2–6.0, which is 0.5–1.3 pH units lower than the minimum pH required to initiate growth.

Halolactibacillus is mesophilic. The optimum growth temperatures of *Halolactibacillus halophilus* and *Halolactibacillus miurensis* are 30–37 °C and 37–40 °C, respectively. The μ_{max} values (h^{-1}) of *Halolactibacillus halophilus* IAM 15242[T] are 0.48 at 25 °C, 0.58 at 30 °C, 0.60 at 37 °C, 0.42 at 40 °C, and 0.06 at 42.5 °C. Those of *Halolactibacillus miurensis* IAM 15247[T] are 0.62 at 25 °C,

0.64 at 30 °C, 0.74 at 37 °C, 0.74 at 40 °C, and 0.18 at 42.5 °C. Growth occurs at 5–10 °C to 40 °C and at 5–45 °C for *Halolactibacillus halophilus* and *Halolactibacillus miurensis*, respectively. One exceptional strain, *Halolactibacillus miurensis* IAM 15249, is able to grow at –1.8 °C.

A fairly wide range of hexoses, disaccharides, trisaccharides, and related compounds are fermented. Among 6 sugar alcohols, mannitol and glycerol are fermented. D-Ribose is fermented by both species. Three other pentoses are fermented by *Halolactibacillus miurensis* but not by *Halolactibacillus halophilus*. Gluconate is fermented without production of gas by both species.

The G + C contents of the DNA of *Halolactibacillus* fall into narrow ranges: 39.6–40.7 mol% for *Halolactibacillus halophilus*

TABLE 24. Effect of initial pH of culture medium on the product balance of glucose fermentation in *Halolactibacillus* species[a]

	H. halophilus IAM 15242[T]			*H. miurensis* IAM 15247[T]		
Initial pH of culture medium	7	8	9	7	8	9
End products [mol/(mol glucose)]:						
Acetate	0.27	0.37	0.74	0.28	0.45	0.51
Ethanol	0.16	0.46	0.47	0.18	0.53	0.32
Formate	0.73	0.81	1.84	0.76	0.81	1.28
Lactate	1.50	1.13	0.45	1.30	1.13	0.73
Lactate yield from consumed glucose (%)	75	57	22	65	57	37
Carbon recovery (%)	101	98	93	93	103	86

[a]Data from Ishikawa et al. (2005).

TABLE 25. Products from glucose under aerobic and anaerobic cultivation conditions for *Halolactibacillus* species[a,b]

	H. halophilus IAM 15242[T]		*H. miurensis* IAM 15247[T]	
Cultivation	Aerobic	Anaerobic	Aerobic	Anaerobic
Glucose consumed (mM)	25.0	24.8	21.0	27.9
Products (mM):				
Acetate	13.4	9.8	12.9	9.7
Ethanol	ND	12.1	ND	12.8
Formate	ND	22.3	ND	24.9
Lactate	19.0	27.6	13.0	29.4
Pyruvate	15.9	ND	13.8	ND
Carbon recovery (%)	88	100	84	94

[a]ND, not detected.

[b]Data from Ishikawa et al. (2005).

and 38.5–40.0 mol% for *Halolactibacillus miurensis*. DNA–DNA relatedness values among the strains of *Halolactibacillus halophilus* are 82–92% and those among the strains of *Halolactibacillus miurensis* are 79–96%. DNA–DNA relatedness values between the two species is 20–40%. Levels of DNA–DNA relatedness between the type strains of *Halolactibacillus* species and the strains of the phylogenetically related genera, *Amphibacillus*, *Gracilibacillus*, and *Paraliobacillus*, are 2–24%. The sequence similarities of 16S rRNA genes (1491 bases in length and covering positions 41–1508) between the type strains of the two species of *Halolactibacillus* is 99.1%. The similarity values of *Halolactibacillus* to *Paraliobacillus*, *Amphibacillus*, *Gracilibacillus*, and *Virgibacillus marismortui* are 94.8–95.1%, 92.9–94.3%, 93.7–94.1%, and 93.8–94.2%, respectively. *Halolactibacillus* constitutes an independent line of descent within the group composed of halophilic/halotolerant/alkaliphilic and/or alkalitolerant genera (henceforth referred to as the HA group in this chapter) in rRNA group 1 of the phyletic group classically defined as the genus *Bacillus* (Ash et al., 1991), and occupies a phylogenetic position that is closely related to the genera *Paraliobacillus*, *Gracilibacillus*, and *Amphibacillus* (Figure 18).

Halolactibacillus strains were isolated from decaying algae and living sponge collected from Oura beach on the Miura Peninsula in the middle of the Japanese mainland. *Halolactibacillus* is a marine inhabitant. It was isolated from marine organisms and is slightly halophilic, halotolerant, and alkaliphilic which are physiological properties consistent with the physico-chemical conditions found in sea water [total salt concentration 3.2–3.8% (w/v), pH 8.2–8.3 (surface)]. *Marinilactibacillus psychrotolerans*, isolated from marine organisms, is also a marine-inhabiting lactic acid bacterium and slightly halophilic, halotolerant, and alkaliphilic (Ishikawa et al., 2003b). For such organisms, Ishikawa et al. (2003b) proposed the term "marine lactic acid

bacteria." *Halolactibacillus*, as well as *Marinilactibacillus psychrotolerans*, is a marine lactic acid bacterium on the basis of habitat, physiological properties, and lactic acid fermentation.

Enrichment and isolation procedures

Halolactibacillus can be isolated from marine materials by successive enrichment cultures in 7% NaCl GYPF or GYPB (glucose-yeast extract-peptone-fish or beef extract) isolation broth, pH 9.5 or 10.0 at 30 °C under anaerobic cultivation conditions. The first enrichment culture in which the pH has decreased below 7.0 is selected and subcultured. The second enrichment culture incubated at 30 °C is pour-plated with an agar medium supplemented with $CaCO_3$, overlaid with an agar medium containing 0.1% sodium thioglycolate, and incubated anaerobically. Prolonged incubation in enrichment culture should be avoided as cells in culture tend to autolyse. The compositions of the media and the procedures were described by Ishikawa et al. (2003b).

Maintenance procedures

Halolactibacillus species are maintained by serial transfer in a stab culture stored at 5–10 °C at 1–2-month intervals. The medium is 7% NaCl GYPF or GYPB agar supplemented with 12 g Na_2CO_3, 3 g $NaHCO_3$, and 5 g $CaCO_3$ per liter. Solutions of main components, buffer compounds, and $CaCO_3$ are autoclaved separately and mixed aseptically. *Halolactibacillus* species can be maintained in 2.5% NaCl GYPF or GYPB agar, pH 9.0, supplemented with 5 g $CaCO_3$ per liter. To prepare this medium, a double-strength solution of the main components is adjusted to pH 9.0, sterilized by filtration, and aseptically mixed with an equal volume of autoclaved 2.6% agar solution. Then, autoclaved $CaCO_3$ (as a slurry with a small amount of water) is added. Strains are maintained by freezing at –80 °C or below in 2.5% GYPFK or GYPBK broth (GYPF (GYPB)

TABLE 26. Characteristics differentiating *Halolactibacillus* from other related members of the halophilic/halotolerant/alkaliphilic and/or

Characteristic	*Halolactibacillus*[b]	*Alkalib-acterium*[c,d,e,f]	*Amphibacillus xylanus*[g]
Spore formation	−	−	+
Anaerobic growth	+ (F)	+ (F)	+ (F)
Catalase	−	−	−
Glucose requirement in aerobic cultivation	+	ND	+
NaCl (range, %)	0–25.5	0–17	3, +; 6, −
NaCl (optimum, %)	2–3	2–13	ND
pH (range)	6–10	8.5–12	8–10
pH (optimum)	8–9.5	9.0–10.5	ND
Major isoprenoid quinones	None	None	None
Peptidoglycan type	*m*-Dpm	Orn-D-Asp, Lys-D-Asp, Lys(Orn)-D-Asp	*m*-Dpm
G+C content (mol%)	38.5–40.7	39.7–43.2	36–38
Major cellular fatty acids:			
$C_{13:0\ ante}$	+	−	−
$C_{15:0\ iso}$	−	−	−
$C_{15:0\ ante}$	−	−	+
$C_{16:0}$	+	+	+
$C_{16:0\ iso}$	−	−	+
$C_{16:1\ \omega7}$	−	+	−
$C_{16:1\ \omega9}$	−	+	−
$C_{17:0\ ante}$	−	−	−
$C_{18:1\ \omega9}$	−	+	−
Isolation source	Decaying marine algae, living sponge	Wash-waters of edible olives, polygonum indigo fermentation liquor	Alkaline manure with rice and straw

[a]Symbols: +, positive; −, negative; ND, no data; F, fermentation; ANR, anaerobic respiration; *m*-Dpm, *meso*-diaminopimelic acid; Orn, ornithine; Asp, aspartic acid; Glu, glutamic acid.
[b]Data from Ishikawa et al. (2005).
[c]Data from Ishikawa et al. (2003b).
[d]Data from Ntougias and Russell (2001).
[e]Data from Yumoto et al. (2004b).
[f]Data from Nakajima et al. (2005).
[g]Data from Niimura et al. (1990).
[h]Data from Zhilina et al. (2001a).
[i]Data from Wainø et al. (1999).
[j]Data from Deutch (1994).
[k]Data from Lawson et al. (1996).
[l]Data from Ishikawa et al. (2002).
[m]Data from Heyndrickx et al. (1998, 1999).
[n]Spore formation was not observed but culture survived heating.
[o]Produced in aerobic cultivation.

broth in which the concentration of K_2HPO_4 is increased to 1%) supplemented with 10% (w/v) glycerol. Strains are kept by L-drying in an adjuvant solution composed of (per liter) 3 g sodium glutamate, 1.5 g adonitol, and 0.05 g cysteine hydrochloride in 0.1 M phosphate buffer (KH_2PO_4-K_2HPO_4), pH 7.0 (Sakane and Imai, 1986). Strains can be kept by freeze-drying with a standard suspending fluid containing an appropriate concentration of NaCl.

Differentiation of *Halolactibacillus* species from other related genera and species

Halolactibacillus is distinguished from facultatively anaerobic and/or phylogenetically close members of the HA group by the combination of physiological, biochemical, and chemot-

axonomic characteristics (Table 26). Among these characteristics, catalase, respiratory quinones, cytochromes, and major fatty acids are of high differentiating value. *Halolactibacillus* conforms to two genera in the group of typical lactic acid bacteria, *Marinilactibacillus* and *Alkalibacterium*, with respect to the phenotypic properties of cellular morphology, motility, halophilic and halotolerant properties, and lactic acid fermentation pattern. In addition to lactate, *Halolactibacillus halophilus*, *Halolactibacillus miurensis*, *Marinilactibacillus psychrotolerans*, and *Alkalibacterium olivapovlyticus* anaerobically produce formate, acetate, and ethanol with a molar ratio of 2:1:1 and the ratio of the three products to lactate is affected by the pH of cultivation medium. They share the ability to metabolize glucose aerobically to produce pyruvate and acetate. However, alkalitolerant group in *Bacillus* rRNA group 1, *Marinilactibacillus psychrotolerans*, and *Alkalibacterium species*

Amphibacillus fermentum[b]	*Amphibacillus tropicus*[b]	*Gracilibacillus halotolerans*[i]	*Gracilibacillus dipsosauri*[j,k]	*Marinilactibacillus psychrotolerans*[c]	*Paraliobacillus ryukyuensis*[l]	*Virgibacillus pantothenticus*[m]
+[n]	+	+	+	−	+	+
+ (F)	+ (F)	−	+ (ANR)	+ (F)	+ (F)	+ (F)
+	+	+	+	−	+[o]	+
+	+	−	−	+	+	−
0.98–19.7	0.98–20.9	0–20	0–18.6 (KCl)	0–20	0–22	0–10≤
10.8	5.4–10.8	0	3.7 (KCl)	2.0–3.75	0.75–3	4
7–10.5	8.5–11.5	5–10	6–10≤	6.0–10.0	5.5–9.5	ND
8.5–9	9.5–9.7	7.5	7.5	8.5–9.0	7–8.5	7
ND	ND	MK-7	MK-7	None	MK-7	MK-7
ND	ND	*m*-Dpm	*m*-Dpm	Orn-D-Glu	*m*-Dpm	*m*-Dpm
41.5	39.2	38	39.4	34.6–36.2	35.6	38.3
ND	ND	−	−	−	ND	−
ND	ND	−	+	−	ND	+
ND	ND	+	+	−	ND	+
ND	ND	+	+	+	ND	−
ND	ND	−	−	−	ND	−
ND	ND	−	−	+	ND	−
ND	ND	−	−	−	ND	−
ND	ND	+	+	−	ND	+
ND	ND	−	−	+	ND	−
Sediment, soda lake	Sediment, soda lake	Surface mud, Great Salt Lake	Nasal salt glands of a desert iguana	Decaying marine algae, living sponge, raw Japanese ivory shell	Decaying marine alga	Soils

Halolactibacillus can be distinguished from these lactic acid bacteria by the chemotaxonomic characteristics of peptidoglycan type and cellular fatty acid composition. *Halolactibacillus* is phenotypically similar to *Paraliobacillus ryukyuensis* which has a lactic acid fermentation pattern similar to that described above, but is differentiated from this bacterium by the lack of spore formation, catalase, respiratory quinones, and cytochromes.

Halolactibacillus halophilus and *Halolactibacillus miurensis* can be distinguished on the basis of fermentation pattern of carbon compounds: *Halolactibacillus halophilus* does not ferment L-arabinose, D-xylose, D-melezitose, or inulin but ferment glycerol, whereas *Halolactibacillus miurensis* ferments these carbohydrates and weakly ferments glycerol (Table 23).

Taxonomic comments

Halolactibacillus possesses all the essential characteristics of lactic acid bacteria that have been attributed to the most typical lactic acid bacteria including production of lactic acid through the Embden–Meyerhof pathway and lack of catalase, quinones, cytochromes, and respiratory metabolism, but is discrete in the phylogenetic group in which it belongs. Typical lactic acid bacteria can be considered to have evolved retrogressively from facultative anaerobes as close ancestors (Whittenbury, 1964). *Halolactibacillus* also may have evolved as a lactic acid bacterium by following independent but similar evolutionary processes within the HA group while retaining physiological characteristics consistent with the physico-chemical factors of salt concentration and pH that prevail in marine environments.

List of species of the genus *Halolactibacillus*

1. **Halolactibacillus halophilus** Ishikawa, Nakajima, Itamiya, Furukawa, Yamamoto and Yamasato 2005, 2437[VP]

ha.lo.phi′lus. Gr. n. *hals* salt; Gr. adj. *philos* loving; N.L. masc. adj. *halophilus* salt-loving.

The characteristics are as described for the genus and as listed in Table 23, Table 24, and Table 25. The morphology is as shown in Figure 19. Deep colonies in 2.5% NaCl GYPF agar medium are pale yellow, and lenticular, with diameters of 2–4 mm after 3 d at 30 °C. Surface colonies are round, convex, entire, pale yellow, and transparent, with diameters of 0.8–1.0 mm after 3 d at 30 °C. Cells are 0.6–0.9 × 3.6–4.5 μm, occurring singly, in pairs, or in short chains, and elongated. The density and size of colonies that develop on semisolid medium that is evenly inoculated are uniform from the surface to the bottom.

Source: decaying marine algae and a living sponge. The G + C content of the type strain is 40.2 mol%.

DNA G + C content (mol%): 39.6–40.7 (HPLC).

Type strain: M2-2, DSM 17073, IAM 15242, JCM 21694, NBRC 100868, NRIC 0628.

GenBank accession number (16S rRNA gene): AB196783.

2. **Halolactibacillus miurensis** Ishikawa, Nakajima, Itamiya, Furukawa, Yamamoto and Yamasato 2005, 2437[VP]

mi.u.ren′sis. N.L. masc. adj. *miurensis* from the Miura Peninsula, Japan, where the strains were isolated.

The characteristics are as described for the genus and as listed in Table 23, Table 24, and Table 25. The morphology is as shown in Figure 19. Deep colonies in 2.5% NaCl GYPF agar medium are pale yellow and lenticular, with diameters of 2–4 mm after 3 d at 30 °C. Surface colonies are round, convex, entire, pale yellow, and transparent, with diameters of 1.0–1.5 mm after 3 d at 30 °C. Cells are 0.6–0.9 × 3.6–4.5 μm, occurring singly, in pairs, or in short chains, and elongated. The density and size of colonies that develop on semisolid medium that is evenly inoculated are uniform from the surface to the bottom. The G + C content of the type strain is 38.5 mol%.

Source: decaying marine alga.

DNA G + C content (mol%): 38.5–40.0 (HPLC).

Type strain: M23-1, DSM 17074, IAM 15247, JCM 21699, NBRC 100873, NRIC 0633.

GenBank accession number (16S rRNA gene): AB196784.

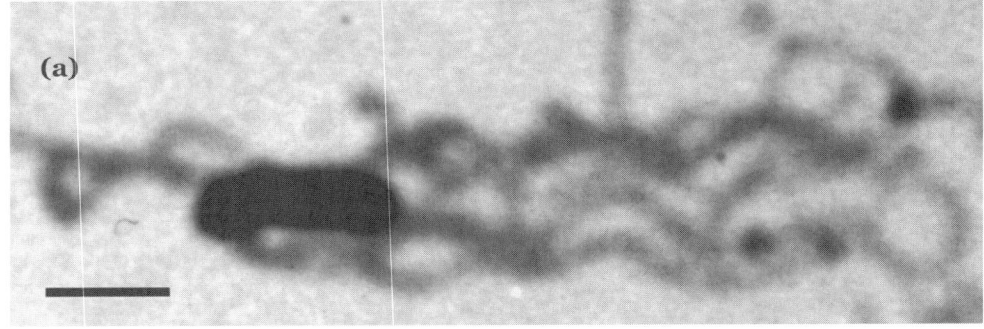

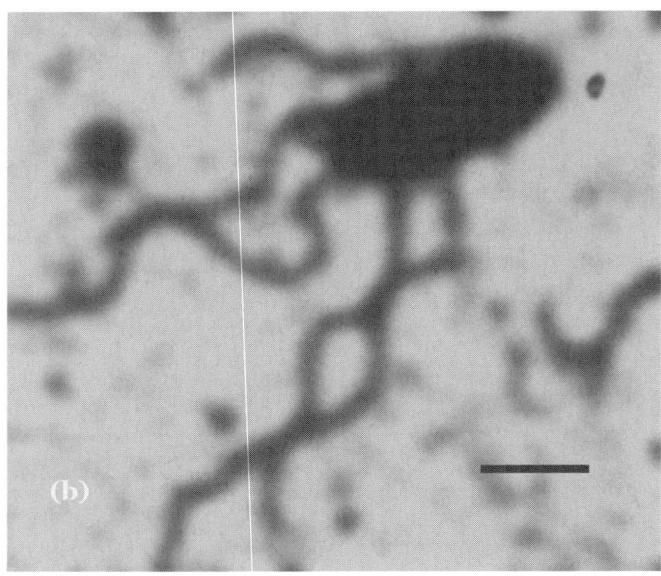

FIGURE 19. Photomicrographs of cells and peritrichous flagella of (a) *Halolactibacillus halophilus* IAM 15242[T] and (b) *Halolactibacillus miurensis* IAM 15247[T] grown anaerobically at 30 °C for 2 d on NaCl GYPFK agar. Bars = 2 μm.

Genus XI. **Lentibacillus** Yoon, Kang and Park 2002, 2047^{VP} emend. Jeon, Lim, Lee, Lee, Lee, Xu, Jiang and Kim 2005a, 1342

JEROEN HEYRMAN AND PAUL DE VOS

Len.ti.ba.cil'lus. L. adj. *lentus* slow; L. dim. n. *bacillus* small rod; N.L. masc. n. *Lentibacillus* slowly growing bacillus/rod.

Rod-shaped cells, forming terminal endospores that swell the sporangia. Gram-variable, motile or nonmotile. Colonies are white to cream-colored, smooth and circular to slightly irregular. Catalase-positive, oxidase variable, and urease-negative. Unable to hydrolyze starch, tyrosine, or xanthine. No acid production from D-melibiose, raffinose, or L-rhamnose. **Moderately to extremely halophilic. The major fatty acid is C$_{15:0\ anteiso}$ and branched saturated fatty acids account for 95% total fatty acids.** The cell-wall peptidoglycan contains *meso*-diaminopimelic acid at position 3 of the peptide subunit. The predominant menaquinone is MK-7. The major polar lipids are diphosphatidylglycerol and phosphatidylglycerol.

DNA G + C content (mol%): 42.0–44.0.

Type species: **Lentibacillus salicampi** Yoon, Kang and Park 2002, 2047^{VP}.

Further descriptive information

As also discussed for the genus *Oceanobacillus*, *Lentibacillus* is part of a quite recently described lineage of halotolerant or halophilic genera in the *Bacillus sensu lato*-group, which may undergo further taxonomic changes in the future. *Lentibacillus*, which was first proposed for a single strain described as *Lentibacillus salicampi* (Yoon et al., 2002), has been extended with the addition of four further species, namely *Lentibacillus juripiscarius* (Namwong et al., 2005), *Lentibacillus salarius* (Jeon et al., 2005a), *Lentibacillus lacisalsi* (Lim et al., 2005c), and *Lentibacillus halophilus* (Tanasupawat et al., 2006). Additional strains of the type species *Lentibacillus salicampi* have also been isolated (Namwong et al., 2005).

Cells are rods of 0.2–0.7 × 1.0–6.0 µm that form spherical or oval endospores. Of all the species in the genus, only cells of *Lentibacillus juripiscarius* are nonmotile. The genus name is based on the slow growth observed for *Lentibacillus salicampi* (Yoon et al., 2002). Strains belonging to the other *Lentibacillus* species generally show slow growth on media with low NaCl content [e.g., Marine Agar (MA)]; however, they grow well on media with higher NaCl content. For *Lentibacillus salicampi* (Yoon et al., 2002), optimal growth occurs at NaCl concentrations of 4–8% (w/v), no growth occurs without NaCl, and the upper NaCl (w/v) limit is 23%, according to Yoon et al. (2002), or 25%, according to Namwong et al. (2005). *Lentibacillus juripiscarius* (Namwong et al., 2005) requires 3–30% (w/v) NaCl, with an optimum of 10%. *Lentibacillus lacisalsi* (Lim et al., 2005c) requires 5–25% (w/v) NaCl for growth and grows optimally in the range 12–15%. *Lentibacillus salarius* (Jeon et al., 2005a) grows in 1–20% (w/v) NaCl and shows optimal growth at 12–14%. *Lentibacillus halophilus* is an extreme halophile, showing an NaCl range for growth of 12–30% (w/v), with an optimum of 20–26%. Unlike other members of the genus, *Lentibacillus lacisalsi* is not able to grow at a pH below 7. *Lentibacillus salicampi* has been described as strictly aerobic (Yoon et al., 2002). However, with the isolation of additional strains, Namwong et al. (2005) demonstrated that *Lentibacillus salicampi* and *Lentibacillus juripiscarius* are able to grow anaerobically in medium containing nitrate (1%, w/v). *Lentibacillus halophilus* was unable to grow under anaerobic conditions on media with or without added nitrate (1%, w/v). For *Lentibacillus lacisalsi* and *Lentibacillus salarius*, anaerobic growth was only tested on media without added nitrate. Additional discriminative characteristics are summed up in Table 27. For the type strain of *Lentibacillus salicampi*, conflicting results have been reported for acid production from carbohydrates. Namwong et al. (2005) isolated additional strains of *Lentibacillus salicampi* and analyzed them together with the type strain. In their analysis, the type strain of *Lentibacillus salicampi* (JCM 11462^T) produced acid from cellobiose (weak), D-fructose, D-galactose (weak), D-glucose, D-mannose (weak), D-ribose, and xylose (sometimes weak). These results were reproducible. According to Yoon et al. (2002), strain SF-20^T (=JCM 11462^T) did not produce acid from any of these sugars. Whether these conflicting results are due to examination of a different subculture of the type strain is not clear. Fatty acids that can occur in amounts above 20% are C$_{15:0\ anteiso}$, C$_{16:0\ iso}$, and C$_{17:0\ anteiso}$. Additional fatty acids that may occur in moderate amounts (±5–20%) are C$_{14:0\ iso}$ and C$_{15:0\ iso}$. However, it is not possible to compare the profiles for the different species, as both the growth media (MA, MA + 10% NaCl, JCM medium no. 377) and the incubation time (2, 3, 7d) were different. It is known that culture conditions can have a major influence on the fatty acid profile (Drucker, 1981). Furthermore, Jeon et al. (2005a) determined the fatty acid profile of *Lentibacillus salarius* from cells grown on MA for 5d and MA + 10% NaCl for 2d. The obtained fatty acid profiles differed markedly. In order to use fatty acid profiles as a distinguishing character within *Lentibacillus*, strains need to be grown and analyzed under strictly standardized conditions. Also, for comparison with closely related genera (e.g., *Oceanobacillus* and *Virgibacillus*) standard conditions are necessary.

Enrichment and isolation procedures

Lentibacillus salicampi strain SF-20^T (Yoon et al., 2002), *Lentibacillus salarius* (Jeon et al., 2005a), and *Lentibacillus lacisalsi* (Lim et al., 2005c) were all isolated from sediment from a salt field (Korea) or lake (China) by plating on MA (Difco) supplemented with salt (8.1, 15, and 20%, w/v, respectively). Additional strains of *Lentibacillus salicampi* and *Lentibacillus juripiscarius* were isolated from fish sauce by plating on JCM medium no. 377, designated *Lentibacillus* medium (composition per liter: 100 g NaCl, 5 g Casamino acids, 5 g yeast extract, 1 g glutamic acid, 2 g KCl, 3 g trisodium citrate, 20 g MgSO$_4$, 36 mg FeCl$_2$·4H$_2$O, 0.36 mg MnCl$_2$·4H$_2$O, 20 g agar; pH 7.2). *Lentibacillus halophilus* was isolated from JCM medium no. 168, which is identical to medium no. 377, except for the addition of 200 g NaCl instead of 100 g.

No specific selective isolation procedures have been described for *Lentibacillus*, but representatives of the genus might be selected for, together with other halotolerant/halophilic bacteria, by using media containing 15% (w/v) NaCl. Inoculation of such media could be preceded by a heating step (5–10 min at 80 °C) in order to select for spore-formers.

TABLE 27. Differentiation data for *Lentibacillus* species[a,b]

Characteristic	1. *L. salicampi*	2. *L. halophilus*	3. *L. juripiscarius*	4. *L. lacisalsi*	5. *L. salarius*
Motility	+	+	–	+	+
Growth at:					
pH 6.0	+	+	+	–	+
pH 9.0	NG	–	+	+	–
Temperature range (°C)	15–40	15–42	10–45	15–40	15–50
Reduction of nitrate	+	–	+	+	+
NaCl range:					
5%	+	–	+	+	+
10%	+	–	+	+	+
25%	–	+	+	+	–
Oxidase	+	+	+	+	–
Hydrolysis of:					
Casein	+	–	+	–	–
Tween 80	+	–	+	–	–
Acid production from:					
L-Arabinose	–	–	–	+	+
D-Glucose	CR	–	+	+	+
D-Fructose	CR	–	+	+	+
Glycerol	+	–	+	–	+
Lactose	–	–	–	–	+
Maltose	–	–	–	–	w
Mannitol	–	–	–	–	+
D-Mannose	CR	–	–	–	+
D-Ribose	CR	–	+	+	+
Trehalose	–	–	–	–	w
D-Xylose	CR	–	+	+	+

[a]Symbols: +, positive; –, negative; w, weak reaction; NG, not given; CR, conflicting results by different researchers (negative according to Yoon et al., 2002; positive according to Namwong et al., 2005).

[b]Data compiled from Yoon et al. (2002), Namwong et al. (2005), Jeon et al. (2005a), Lim et al. (2005c), and Tanasupawat et al. (2006).

Maintenance procedures

Lentibacillus strains can be preserved in the refrigerator in tubes containing broth medium or agar slopes, after checking the culture microscopically for sporulation. For long-term preservation, lyophilization or liquid nitrogen may be used under cryoprotection.

Procedures for testing of special characters

Lentibacillus strains were described using standard methodology, except for the addition of NaCl to the culture media.

Differentiation from closely related taxa

Phylogenetically, the genera most closely related to *Lentibacillus* are *Virgibacillus* and *Oceanobacillus*. *Lentibacillus* is not readily distinguishable from *Virgibacillus* and *Oceanobacillus* can only be differentiated from *Lentibacillus* by its slightly lower G + C content (35.8–40.1 and 42.0–44.0 mol% for *Oceanobacillus* and *Lentibacillus*, respectively) (see also section on *Oceanobacillus*).

The distinction between *Lentibacillus* and other halophilic endospore-forming genera of the *Bacillaceae* is also not straightforward, as discussed in more detail for the genus *Oceanobacillus*. *Lentibacillus* has peptidoglycan that contains *meso*-diaminopimelic acid and the predominant menaquinone is MK-7, which differentiates it from the genera *Halobacillus*, *Filobacillus*, *Jeotgalibacillus*, *Amphibacillus*, and *Halolactibacillus*. Differentiation from other halophilic endospore-forming genera is, when possible, based on only minor phenotypic differences. Furthermore, as many of the remaining halophilic endospore-forming genera are represented by a single species, it is likely that the discovery

of additional species or strains within these genera/species will result in an even less pronounced differentiation.

Taxonomic comments

The main reason for the creation of a separate genus status for *Lentibacillus salicampi* was its phylogenetic position as determined by 16S rRNA gene sequence analysis. In a neighbor-joining tree (Saitou and Nei, 1987) constructed by Yoon et al. (2002), *Lentibacillus salicampi* diverged at the bottom of a cluster including *Virgibacillus pantothenticus*, *Virgibacillus proomii*, *Virgibacillus salexigens* (formerly *Salibacillus salexigens*) and *Virgibacillus marismortui* (formerly *Salibacillus marismortui*). Since at the time of the description *Virgibacillus* and *Salibacillus* were still separate genera, Yoon et al. (2002) concluded that their isolate could not be attributed to one or the other and described it as a novel genus. Phenotypically, this description was supported by the slow growth and fatty acid profile of the strain. Since the time of the description, *Salibacillus* has been transferred to *Virgibacillus* (Heyrman et al., 2003<qu ref=58>), five novel *Virgibacillus* species have been described (Heyrman et al., 2003b; Lee et al., 2006b; Yoon et al., 2005b), and *Bacillus halodenitrificans* has been transferred to *Virgibacillus* as *Virgibacillus halodenitrificans* (Yoon et al., 2004c). Furthermore, *Lentibacillus* has been expanded with the description of *Lentibacillus juripiscarius* (Namwong et al., 2005), *Lentibacillus salarius* (Jeon et al., 2005a), *Lentibacillus lacisalsi* (Lim et al., 2005c), and *Lentibacillus halophilus* (Tanasupawat et al., 2006). Additionally, the closely related genus *Oceanobacillus* has been described to accommodate the species *Oceanobacillus iheyensis*

(Lu et al., 2002; Lu et al., 2001) and later expanded by the description of *Oceanobacillus oncorhynchi* (Yumoto et al., 2005b) and the transfer of *Virgibacillus picturae* (Heyrman et al., 2003b) to *Oceanobacillus* as *Oceanobacillus picturae* (Lee et al., 2006b). With every addition of a novel species in the genera *Lentibacillus*, *Virgibacillus*, and *Oceanobacillus*, the phylogenetic relationships between these genera changed and the phenotypic differences originally differentiating them disappeared. As the majority of species belonging to *Lentibacillus*, *Virgibacillus*, and

Oceanobacillus have been described in the last five years, future species descriptions in the neighborhood of these genera can be expected. The description of two novel species within the genus *Lentibacillus*, namely *Lentibacillus kapialis* and *Lentibacillus halodurans*, and two novel *Oceanobacillus* species, namely *Oceanobacillus chironomi* and *Oceanobacillus profundus*, are in press. These data will probably allow better assessment of whether the current situation is satisfactory or whether further rearrangements are necessary.

List of species of the genus *Lentibacillus*

1. **Lentibacillus salicampi** Yoon, Kang and Park 2002, 2047[VP]

sa.li.cam'pi. L. n. *sal* salt; L. n. *campus* field; N.L. gen. n. *salicampi* of a salt field.

Morphology and general characters are as for the generic description and further descriptive information.

Cells are Gram-variable rods, 0.4–0.7 × 2.0–4.0 µm, motile by a single flagellum. Optimal growth temperature is 30 °C. Growth occurs at 15 and 40 °C, but not at 10 or above 41 °C. Optimal pH for growth is 6.0–8.0 and no growth is observed at pH 5.0. Esculin and hypoxanthine are not hydrolyzed. Acid is produced from stachyose, but not from adonitol, lactose, D-melezitose, *myo*-inositol, D-sorbitol, or sucrose. Conflicting results are reported in literature for the acid production from, e.g., cellobiose and D-galactose.

Source: a salt field of the Yellow Sea in Korea and fish sauce (Thailand).

DNA G + C content (mol%): 44.0 (reverse-phase HPLC).

Type strain: SF-20, ATCC BAA-719, CIP 107807, JCM 11462, KCCM 41560, KCTC 3792.

GenBank/EMBL/DDBJ accession number (16S rRNA gene): AY057394.

2. **Lentibacillus halophilus** Tanasupawat, Pakdeeto, Namwong, Thawai, Kudo and Itoh 2006, 1862[VP]

ha.lo'phi.lus. Gr. n. *hals, halos* salt; Gr. adj. *philos* loving; N.L. masc. adj. *halophilus* salt-loving.

Morphology and general characters are as for the generic description and further descriptive information.

Cells are Gram-positive rods, mostly 0.4–0.6 × 1.0–3.0 µm, though longer cells (up to 6 µm) or short filaments are also observed. Growth occurs between 15 °C (weakly) and 42 °C, but not at 10, 45, or 50 °C. Optimum temperature range is 30–37 °C. Growth is observed between pH 6 and 8, but not at pH 5 or 9; optimum pH is 7.0–7.5. Does not hydrolyze esculin, arginine, gelatin, phenylalanine, or hypoxanthine. Acid is not produced from cellobiose, D-galactose, D-melezitose, *myo*-inositol, salicin, sorbitol, or sucrose.

Source: fish-sauce fermentation in Thailand.

DNA G + C content (mol%): 42.1–43.1 (reverse-phase HPLC).

Type strain: PS11-2, JCM 12149, TISTR 1549, PCU 240.

GenBank/EMBL/DDBJ accession number (16S rRNA gene): AB191345.

3. **Lentibacillus juripiscarius** Namwong, Tanasupawat, Smitinont, Visessanguan, Kudo and Itoh 2005, 319[VP]

ju.ris.pis'ca.ri.us. L.n. *jus, juris* sauce; L. adj. *piscarius -a -um* of, or belonging to, fish; N.L. masc. adj. *juripiscarius* of a fish sauce.

Morphology and general characters are as for the generic description and further descriptive information.

Cells are Gram-positive, nonmotile rods, 0.4–0.5 × 1.5–6.0 µm. Growth range is 10–45 °C, with an optimum at 37 °C. Grows at pH 5.0–9.0, with an optimum at pH 7.0. Hydrolyzes gelatin, but not arginine, hypoxanthine, phenylalanine, or tributyrin. Negative results for Voges–Prouskauer reaction, methyl red test, and indole and H₂S formation. Produces acid from sucrose, but not from amygdalin, cellobiose, D-galactose, inulin, melezitose, methyl α-D-glucoside, *myo*-inositol, salicin, or sorbitol.

Source: fish sauce (Thailand).

DNA G + C content (mol%): 43.0 (reverse-phase HPLC).

Type strain: IS40–3, CIP 108664, DSM 16577, JCM 12147, PCU 229, TISTR 1535.

GenBank/EMBL/DDBJ accession number (16S rRNA gene): AB127980.

4. **Lentibacillus lacisalsi** Lim, Jeon, Song, Lee, Ju, Xu, Jiang and Kim 2005c, 1807[VP]

la.ci.sal'si. L. masc. n. *lacus* lake; L. adj. *salsus -a -um* salted, salt; N.L. gen. n. *lacisalsi* of a salt lake.

Morphology and general characters are as for the generic description and further descriptive information. Description is based on a single strain.

Cells are 0.4–0.6 × 1.2–3.0 µm, motile with peritrichous flagella. Growth occurs at 15–40 °C and pH 7.0–9.5, with optimum growth at 30–32 °C and pH 8.0. Does not hydrolyze esculin or hypoxanthine. Does not produce acid from adonitol, arbutin, or D-salicin.

Source: a salt lake in China.

DNA G + C content (mol%): 44.0 (reverse-phase HPLC).

Type strain: BH260, KCTC 3915, DSM 16462.

GenBank/EMBL/DDBJ accession number (16S rRNA gene): AY667497.

5. **Lentibacillus salarius** Jeon, Lim, Lee, Lee, Lee, Xu, Jiang and Kim 2005a, 1342[VP]

sa.la'ri.us. L. masc. adj. *salarius* of, or belonging to, salt, because of the isolation of this micro-organism from saline sediment.

Morphology and general characters are as for the generic description and further descriptive information. Description is based on a single strain.

Cells are Gram-positive rods, $0.2–0.3 \times 1.5–3.0\,\mu m$, motile by flagella. Growth occurs at $15–50\,°C$ and pH 6.0–8.5, with optimum growth at $30–35\,°C$ and pH 7.0–7.5. Able to hydrolyze esculin, but not hypoxanthine. Does not produce acid from adonitol, arbutin, or D-salicin.

Source: saline soil in China.
DNA G + C content (mol%): 43.0 (reverse-phase HPLC).
Type strain: BH139, KCTC 3911, DSM 16459.
GenBank/EMBL/DDBJ accession number (16S rRNA gene): AY667493.

Genus XII. **Marinococcus** Hao, Kocur and Komagata 1985, 535[VP] (Effective publication: Hao, Kocur and Komagata 1984, 456.)

ANTONIO VENTOSA

Ma.ri.no.coc′cus. Gr. adj. *marino* marine; Gr. n. *kokkos* a grain or berry; N. L. masc. n. *Marinococcus* marine coccus.

Gram-positive, spherical cells, $1.0–1.2\,\mu m$ in diameter, occurring singly, in pairs, tetrads, or clumps. Motile. The motile cells usually have one or two flagella. Non-spore-forming. Colonies are circular and smooth and may be either orange, yellowish orange or creamy white. Moderately halophilic. Growth occurs in media with 5 to 20% NaCl. Optimum temperature for growth is $28–37\,°C$. Chemo-organotrophic. Metabolism respiratory. Strictly aerobic. Catalase-positive. Acid may or may not be produced from sugars. The cell wall contains peptidoglycan of *meso*-diaminopimelic acid type. The major menaquinone is MK-7. Found in sea water, solar salterns, and saline soils.

DNA G + C content (mol%): 43.9–48.5.

Type species: **Marinococcus halophilus** Hao, Kocur and Komagata 1985, 535[VP] (Effective publication: Hao, Kocur and Komagata 1984, 456.) (*Planococcus halophilus* Novitsky and Kushner 1976.).

Further descriptive information

The major cellular fatty acids of *Marinococcus halophilus* and *Marinococcus albus* are $C_{15:0\ anteiso}$ acid and $C_{17:0\ anteiso}$ acid (Hao et al., 1984). Similar results were obtained by Monteoliva-Sanchez et al. (1989) for a group of isolates belonging to *Marinococcus halophilus*. For *Marinococcus halotolerans* the major fatty acids are $C_{15:0\ anteiso}$, $C_{17:0\ anteiso}$ and $C_{16:0\ iso}$ (Li et al., 2005). The quinone system of *Marinococcus* is menaquinone, with MK-7 as the major component (Hao et al., 1984; Li et al., 2005; Marquez et al., 1992). The phospholipids of *Marinococcus halotolerant* are phosphatidylinositol and diphosphatidylglycerol (Li et al., 2005). Species of the genus *Marinococcus* have low extracellular hydrolytic activity, except *Marinococcus halophilus*, which has proteolytic activity. In a screening focused on the isolation of moderately halophilic bacteria with hydrolytic activities from several hypersaline environments, a *Marinococcus* strain able to produce a lipase has been reported (Sanchez-Porro et al., 2003).

The ability of *Marinococcus halophilus* and *Marinococcus albus* to precipitate carbonates has been studied in several strains isolated from the Salar de Atacama (Chile). The biolith precipitated were spherical and varied with the salinity; they were of magnesium calcite, with Mg content increasing with increasing salinity (Rivadeneyra et al., 1999). Several plasmids have been detected in *Marinococcus halophilus*; the complete nucleotide sequence (3874 bp) of one of these plasmids, designated pPL1, has been determined. Plasmids have not been detected in *Marinococcus albus* (Louis and Galinski, 1997a).

The species of the genus *Marinococcus* are moderately halophilic, which are defined as those micro-organisms that grow optimally in media with 3–15% NaCl. To grow over a wide range of salt concentrations, moderately halophilic bacteria accumulate organic osmotic solutes (Ventosa et al., 1998). In the species of the genus *Marinococcus* the main osmotic solutes are ectoine and hydroxyectoine (Ventosa et al., 1998). The genes responsible for the synthesis of the compatible solute ectoine have been identified and sequenced. The three genes (*ectA*, *ectB*, and *ectC*) of the biosynthetic pathway of ectoine were cloned by functional expression in *Escherichia coli*; these genes were not only expressed, but also osmoregulated in *Escherichia coli* (Louis and Galinski, 1997b). A stress-inducible promoter region from *Marinococcus halophilus* has been investigated upstream of the *ectA* gene, using the green fluorescent protein as a reporter molecule (Bestvater and Galinski, 2002). *Marinococcus halophilus* does not entirely rely on ectoine synthesis for osmoadaption, and similarly to other halophilic bacteria, it can also take up compatible solutes from the external medium. To allow for the uptake of external solutes, *Marinococcus halophilus* is equipped with osmoregulated transport systems, similarly to nonhalophilic bacteria; two transporters for compatible solutes belonging to the betaine-carnitine-choline transporter family have been reported for *Marinococcus halophilus* (Vermeulen and Kunte, 2004). Two structural genes encoding a betaine transporter named BetM, which also accepts ectoine as an additional substrate, and a transport system specific for the uptake of ectoines named EctM have been identified and characterized (Vermeulen and Kunte, 2004). The protein stabilization by several naturally occurring osmolytes has been investigated. Knapp et al. (1999) showed that the osmolyte hydroxyectoine purified from *Marinococcus* is a very efficient stabilizer and they suggest that this compatible solute could be an interesting stabilizer in biotechnological processes in which enzymes are applied in the presence of denaturants or at high temperature.

Enrichment and isolation procedures

Organisms of the genus *Marinococcus* have been isolated from different hypersaline or saline environments, such as water of ponds of salterns, saline soils, or sea water. Specific enrichment or selective isolation media have not been described. They grow well in complex culture media supplemented with a mixture of salts. The strains can be isolated either by direct inoculation on plates or by diluting the samples in sterile salt solution and then plating them on the isolation medium. The medium described by Ventosa et al. (1983) can be used. The composition of this medium is as follows (in g/l): NaCl, 178.0; $MgSO_4 \cdot 7H_2O$, 1.0; $CaCl_2 \cdot 2H_2O$, 0.36; KCl, 2.0; $NaHCO_3$, 0.06; NaBr, 0.23; $FeCl_3 \cdot 6H_2O$, trace; proteose-peptone no. 3 (Difco), 5.0; yeast extract (Difco), 10.0; glucose, 1.0; Bacto-agar (Difco), 20.0. The pH is adjusted to 7.2. Plates are incubated aerobically at 30–37 °C for 7–15 d. Colonies of *Marinococcus halophilus* and *Marinococcus halotolerant* are yellow-orange or orange, water-insoluble pigmented, while the colonies of *Marinococcus albus* are creamy white. Recently, *Marinococcus* has been isolated from marine sponges growing at a depth of about 300 m on the Sula Ridge close to the Norwegian coast; a rapid identification of *Marinococcus* and other bacterial isolates has been described, based on a rapid proteometric clustering using Intact-Cell MALDI-TOF (ICM) mass spectrometry (Dieckmann et al., 2005).

Maintenance procedures

Strains belonging to *Marinococcus* can be maintained by the standard procedures such as freeze-drying or storage at −80 °C or under liquid nitrogen. Slant cultures can be conserved several months at room temperature by using a medium with 10% salts. Nutrient agar plus a mixture of salts or MH medium can be used. The composition of MH medium is (in g/l): NaCl, 81.0; $MgCl_2$, 7.0; $MgSO_4$, 9.6; $CaCl_2$, 0.36; KCl, 2.0; $NaHCO_3$, 0.06; NaBr, 0.026; proteose-peptone no. 3 (Difco), 5.0; yeast extract (Difco), 10.0; glucose, 1.0; Bacto-agar (Difco), 20.0 (Ventosa et al., 1982; Ventosa et al., 1983).

Differentiation of the genus *Marinococcus* from other genera

The genus *Marinococcus* can be differentiated from other Gram-positive cocci by comparative analysis of the 16S rRNA sequence as well as by several phenotypic and chemotaxonomic features. The species of *Marinococcus* are moderately halophilic, growing in media with 5–20% NaCl; besides, they are motile Gram-positive cocci, in contrast to other related halophilic cocci of the genera *Salinicoccus* (Ventosa et al., 1990b), *Nesterenkonia* (Stackebrandt et al., 1995), or *Jeotgalicoccus* (Yoon et al., 2003c) that are nonmotile. *Marinococcus* has MK-7 as the characteristic predominant menaquinone system, similarly to *Jeotgalicoccus* (Yoon et al., 2003c), in contrast to *Salinicoccus*, which has MK-6 (Ventosa et al., 1990b), and *Nesterenkonia*, which has MK-8, MK-7, and MK-6 (Stackebrandt et al., 1995). Another differential feature of *Marinococcus* is that its cell wall contains murein of the *meso*-diaminopimelic acid type, while *Salinicoccus* contains murein of the L-Lys-Gly$_5$ type (Ventosa et al., 1990b), *Nesterenkonia* has murein of the L-Lys-Gly L-Glu type (Stackebrandt et al., 1995), and *Jeotgalicoccus* has murein of the L-Lys-Gly$_{3-4}$-L-Ala(Gly) type (Yoon et al., 2003c).

Taxonomic comments

Novitsky and Kushner (1976) studied a motile, halophilic coccus obtained from the culture collection of the National Research Council of Canada (NRCC), designated NRCC 14033. This organism was probably a contaminant in the original culture of *Micrococcus* sp. H$_5$, originally isolated from salted mackerel by Venkataraman and Sreenivasan (1954) and they proposed to place it as a new species of the genus *Planococcus*, as *Planococus halophilus*, on the basis of its salt

requirements (it cannot grow on nutrient medium at 30 °C without added salt) and the different cell-wall peptidoglycan, possessing *meso*-diaminopimelic acid (Novitsky and Kushner, 1976). This species was included in the Approved Lists of Bacterial Names (Skerman et al., 1980). Ventosa et al. (1983) studied 38 moderately halophilic Gram-positive cocci isolated from several saline soils and the ponds of a saltern in Spain; the 25 strains of group I were assigned to the species *Planococcus halophilus*, the 10 strains of group II were similar to the species *Sporosarcina halophila* (currently named as *Halobacillus halophilus*), and group III comprised three strains that differed from other previously described species and were tentatively designated *Planococcus* sp. On the basis of the phenotypic features as well as the chemotaxonomic data of the type strain of *Planococcus halophilus* as well as several representative strains of the study of Ventosa et al. (1983), Hao et al. (1984) proposed to place *Planococcus halophilus* in a new genus, *Marinococcus*, as *Marinococcus halophilus*, and the three strains of group III of the study of Ventosa et al. (1983) in the new species *Marinococcus albus*. An extensive study of the type strain of *Marinococcus halophilus* and another 55 additional strains isolated from hypersaline soils and salterns located in different areas of Spain was carried out by Marquez et al. (1992); besides the phenotypic data reported by Hao et al. (1984), they reported the results for many phenotypic features, including the growth on different compounds as the sole source of carbon and energy or carbon, nitrogen, and energy, as well as their antibiotic susceptibility. The G + C content of the studied strains ranged from 46.6 to 48.8 mol% (Marquez et al., 1992).

A numerical taxonomic study based on the phenotypic features of 22 strains of moderately halophilic motile cocci isolated from the Salar de Atacama (Chile) showed that they were closely related to *Marinococcus* (Valderrama et al., 1991). The chemotaxonomic data as well as the results of the DNA–DNA hybridization studies showed that strains included in phenons A and B of the previous study constitute additional strains of the species *Marinococcus albus* and *Marinococcus halophilus*, respectively (Márquez et al., 1993). The species *Marinococcus hispanicus*, proposed by Marquez et al. (1990) to accommodate five moderately halophilic Gram-positive non-motile cocci, was recently transferred to the genus *Salinicoccus*, as *Salinicoccus hispanicus* (Ventosa et al., 1992), on the basis of the chemotaxonomic results and nucleic acids hybridization studies. Recently, a third species of the genus *Marinococcus*, *Marinococcus halotolerant* has been isolated from a saline soil located in Qinghai, north-west China (Li et al., 2005). Phylogenetic analysis of *Marinococcus halophilus* and *Marinococcus halotolerans* based on the 16S rRNA showed that they belong to the low G + C Gram-positive branch but are not closely related to any other species (Farrow et al., 1992; Li et al., 2005). Currently, it is accepted that the species *Marinococcus albus* is not phylogenetically related to the other two *Marinococcus* species, constituting an incoherent phylogenetic cluster, and these data clearly support the placement of *Marinococcus albus* in a different genus.

Differentiation of the species of the genus *Marinococcus*

Some differential features of the species of the genus *Marinococcus* are given in Table 28.

TABLE 28. Differential characteristics of the species of the genus *Marinococcus*[a]

Characteristic	1. *M. halophilus*	2. *M. albus*	3. *M. halotolerans*
Colony pigmentation	Yellowish orange	Creamy white	Orange
Growth without salt	–	–	+
Oxidase	–	+	–
Acid production from:			
D-Glucose	+	–	+
Glycerol	+	–	ND
Maltose	+	–	ND
D-Mannitol	+	–	+
Sucrose	+	–	ND
D-Trehalose	+	–	ND
D-Xylose	+	–	ND
Nitrate reduction	–	+	+
Hydrolysis of:			
Casein	+	–	–
Gelatin	+	–	–
G+C content (mol%)	46.4	44.9	48.5

[a]Symbols: +, >85% positive; –, 0–15% positive; ND, not determined.

List of species of the genus *Marinococcus*

1. **Marinococcus halophilus** Hao, Kocur and Komagata 1985, 535[VP] (Effective publication: Hao, Kocur and Komagata Hao 1984, 456) (*Planococcus halophilus* Novitsky and Kushner (1976)

Hal.o.phi'lus. Gr. n. *hals* the sea, salt; Gr. adj. *philos* loving; N.L. masc. adj. *halophilus* salt-loving.

See the generic description for many features.

Colonies on nutrient agar plate with 5–20% NaCl are circular, entire, glistening, convex, smooth, and yellow orange. In salt broth or nutrient broth containing NaCl, slight turbidity is formed with sediment.

Halophilic. Optimum growth in media with 5–15% NaCl. Catalase-positive. Benzidine test for porphyrine positive. Acid but not gas is produced from D-glucose, glycerol, D-xylose, D-trehalose, maltose, sucrose, and D-mannitol in MOF medium of Leifson. Acid is not produced from lactose, D-arabinose, D-galactose, and D-fructose. Gelatin, casein, and esculin are hydrolyzed.

The following tests are negative: oxidase, urease, extracellular DNase, production of acetoin, production of H$_2$S, production of indole, hydrolysis of starch, tyrosine, and Tween 80, nitrate reduction, arginine dihydrolase, lysine and ornithine decarboxylases, phenylalanine deaminase, phosphatase, egg yolk reaction, hemolysis, growth on nutrient agar without salt, and growth on Simmons citrate agar.

Source: sea water, solar salterns, and saline soils.

DNA G + C content (mol%) 46.4 (*T*$_m$).

Type strain: strain HK 718, CCM 2706, IAM 12844, JCM 2479, ATCC 27964, DSM 20408, LMG 17439.

GenBank accession number (16S rRNA gene): X90835.

2. **Marinococcus albus** Hao, Kocur and Komagata 1985, 535[VP] (Effective publication: Hao, Kocur and Komagata 1984, 456.)

al'bus. L. adj. *albus* white.

See the generic description for many features.

Colonies on nutrient agar plate with 5–20% NaCl are round, smooth, opaque, and nonpigmented. Halophilic. Optimum growth in media with 5–15% NaCl.

The following tests are positive: Benzidine test for porphyrine, oxidase, nitrate reduction, urease, and DNase. The following tests are negative: production of acid from glucose and other sugars, hydrolysis of gelatin, casein, starch, esculin, and Tween 80, production of acetoin, production of H$_2$S, production of indole, arginine dihydrolase, lysine and ornithine decarboxylases, phenylalanine deaminase, phosphatase, egg yolk reaction, hemolysis, growth on nutrient agar without salt, and growth on Simmons citrate agar.

Source: solar salterns.

DNA G + C content (mol%) 44.9 (*T*$_m$).

Type strain: strain HK 733, CCM 3517, IAM 12845, JCM 2574, ATCC 49811, DSM 20748, LMG 17430.

GenBank accession number (16S rRNA gene): X90834.

3. **Marinococcus halotolerans** Li, Schumann, Zhang, Chen, Tian, Xu, Stackebrandt and Jiang 2005, 1803[VP]

Ha.lo.to'le.rans. Gr. n. *hals* salt; L. pres. part. *tolerans* tolerating; N.L. part. adj. *Halotolerans* referring to the ability of the organism to tolerate high salt concentrations.

See the generic description for many features.

Colonies are circular, opaque and orange pigmented. Halophilic. The optimum concentration of MgCl$_2$·6H$_2$O for growth is 10% (this salt can be substituted by NaCl or KCl). The optimum growth pH and temperature are 7.0–7.5 and 28 °C, respectively. Catalase-positive and oxidase-negative. Acid is produced from esculin, glucose and mannitol. Nitrate is reduced.

The following tests are negative: hydrolysis of gelatin, casein, Tween 80 and starch, methyl red, Voges–Proskauer, production of melanin, indole, and H$_2$S, arginine dihydrolase and ornithine decarboxylase. The following substrates are utilized: maltose, mannitol, glucose, mannose, fructose, cellobiose, salicin, acetamide, galactose, xylose and dextrin; adonitol, arabinose, arabitol, rhamnose, inositol and sorbitol are not utilized.

Source: a saline soil from Qinghai, north-west China.

DNA G + C content (mol%): 48.5 (HPLC).

Type strain: YIM 70157, DSM 16375, KCTC 19045.

GenBank accession number (16S rRNA gene): AY817493.

Genus XIII. **Oceanobacillus** Lu, Nogi and Takami 2002, 687[VP] (Effective publication: Lu, Nogi and Takami 2001, 296) emend. Yumoto, Hirota, Nodasaka and Nakajima 2005b, 1523 emend. Lee, Lim, Lee, Lee, Park, Kim 2006b, 256

JEROEN HEYRMAN AND PAUL DE VOS

O.ce.a.no.ba.cil′lus. Gr. n. *okeanos* the ocean; L. dim. n. *bacillus* a small rod; N.L. masc. n. *Oceanobacillus* the ocean bacillus (rod).

Rod-shaped cells, forming ellipsoidal subterminal or terminal endospores that swell the sporangia. Gram-positive and motile by peritrichous flagella. Colonies are circular and white to beige. Obligatory aerobic or facultative anaerobic. **Facultative or obligatory alkaliphilic** and mesophilic. **Halotolerant or halophytic, optimal growth at NaCl concentrations 3–10% (w/v) and able to grow in concentrations up to 20% (w/v).** Catalase-positive; oxidase-variable. Negative for urease and indole production. **The major cellular fatty acid is $C_{15:0\ anteiso}$.** The main menaquinone type is MK-7.

DNA G + C content (mol%): 35.8–40.1.

Type species: **Oceanobacillus iheyensis** Lu, Nogi and Takami 2002, 687[VP] (Effective publication: Lu, Nogi and Takami 2001, 296.).

Further descriptive information

Oceanobacillus is part of a phylogenetic cluster of halotolerant or halophilic genera in the *Bacillus sensu lato* group, also including the genera *Alkalibacillus, Amphibacillus, Cerasibacillus, Filobacillus, Gracilibacillus, Halobacillus, Halolactibacillus, Jeotgalibacillus, Lentibacillus, Marinibacillus, Paraliobacillus, Pontibacillus, Salinibacillus, Tenuibacillus, Thalassobacillus,* and *Virgibacillus.* Thirteen out of 17 of these genera were described since the year 2000, indicating that this is a quite recent lineage that is expected to undergo further taxonomic changes based on a better understanding of the phylogeny. *Oceanobacillus,* first described with the single species *Oceanobacillus iheyensis* by Lu et al. (2001), has been extended recently with two species, *Oceanobacillus oncorhynchi* (Yumoto et al., 2005b) and *Oceanobacillus picturae* comb. nov. (Lee et al., 2006b; basonym *Virgibacillus picturae,* Heyrman et al., 2003b).

Romano et al. (2006) subdivided *Oceanobacillus oncorhynchi* into two subspecies, each containing a single strain: *Oceanobacillus oncorhynchi* subsp. *oncorhynchi* and *Oceanobacillus oncorhynchi* subsp. *incaldanensis.* Whether or not *Oceanobacillus oncorhynchi* subsp. *incaldanensis* should be transferred to a separate species is debatable (see *Taxonomic comments*).

Cells are rods of 0.4–0.8 × 1.1–6μm that mainly occur singly. Spore formation was not observed in *Oceanobacillus oncorhynchi* subsp. *incaldanensis* strain 20AG (DSM 16557, ATCC BAA-954) (Romano et al., 2006). However, the authors did not check whether or not the strain grew after a heating step (10 min at 80°C) or whether spores were observed after growth on medium supplemented with manganese to enhance spore formation (Charney et al., 1951). *Oceanobacillus* stains are extremely halotolerant, being able to grow in medium containing up to 21% (w/v) NaCl. Most *Oceanobacillus* strains can grow in the absence of NaCl. Strain 20AG (DSM 16557, ATCC BAA-954), the only representative of *Oceanobacillus oncorhynchi* subsp. *incaldanensis,* is the only member of *Oceanobacillus* that is not able to grow without added salt (the minimal requirement is 5% NaCl). The optimal NaCl concentration for cell growth lies between 3% and 10% (w/v) for the different members of *Oceanobacillus.* For *Oceanobacillus iheyensis,* it was determined that salt tolerance was pH dependent (lower salt tolerance at higher pH) and that the lag phase and generation is extended with increasing NaCl concentration (Lu et al., 2001). All *Oceanobacillus* stains are alkalitolerant, *Oceanobacillus iheyensis* has a pH range of 6.5–10.0 (Lu et al., 2001), *Oceanobacillus picturae* of 7.0–10.0 (Lee et al., 2006b), and *Oceanobacillus oncorhynchi* subsp. *incaldanensis* of 6.5–9.5 (Romano et al., 2006). *Oceanobacillus oncorhynchi* subsp. *oncorhynchi* is the only alkaliphilic member of *Oceanobacillus,* not able to grow below pH 9.0 (Yumoto et al., 2005b). The subspecies of *Oceanobacillus oncorhynch* show differences in spore formation, pH, and NaCl range and further differ in their ability to grow under anaerobic conditions. *Oceanobacillus oncorhynchi* subsp. *oncorhynchi* is the only member of *Oceanobacillus* showing growth under anaerobic conditions. All strains are mesophilic, cell growth occurs at temperatures of 5–42°C with an optimum between 25°C and 37°C. In addition, *Oceanobacillus iheyensis* can grow at pressures of up to 30 MPa, which corresponds to the pressure at a depth of 3,000 m. This is not unexpected, since *Oceanobacillus iheyensis* was isolated at a depth of 1,050 m from the Iheya Ridge of the Nansei Islands (27°47.18′ N, 126°54.15′ E). A strain with 100% 16S rDNA sequence similarity to *Oceanobacillus picturae* has been isolated from coastal sediment (Francis et al., 2001); the dormant spores of this strain enzymically oxidized soluble Mn(II) to insoluble Mn(IV) oxides and, therefore, Francis et al. (2001) cautiously speculated that the production of Mn(II)-oxidizing spores by *Virgibacillus picturae* may have contributed to the biodeterioration of the mural paintings.

Acid production (not tested for *Oceanobacillus oncorhynchi* subsp. *incaldanensis*) is (weakly) positive for D-glucose, D-fructose, and D-mannose, and negative for amygdalin, D-arabinose, *myo*-inositol, and 5-keto-D-gluconate. In the Biolog (Biolog Inc.) test, *Oceanobacillus iheyensis* assimilates only four substrates, namely α-D-glucose, maltose, D-mannose, and turanose (Lu et al., 2001). Hydrolysis of starch (not tested for *Oceanobacillus picturae*) is negative. Differential characters are listed in Table 29.

Oceanobacillus iheyensis strain HTE831[T] (Lu et al., 2001) and *Oceanobacillus oncorhynchi* subsp. *incaldanensis* strain 20AG (Romano et al., 2006) were both susceptible to ampicillin, bacitracin, chloramphenicol, and penicillin G. *Oceanobacillus iheyensis* is further resistant to erythromycin, nalidixic acid, and spectinomycin, and is susceptible to carbenicillin, gentamicin, kanamycin, novobiocin, rifampin, and tetracycline. *Oceanobacillus oncorhynchi* subsp. *incaldanensis* is resistant to kanamycin and tetracycline, and susceptible to erythromycin, lincomycin, streptomycin, tetracycline, and vancomycin.

Major cellular fatty acids for *Oceanobacillus iheyensis* and *Oceanobacillus oncorhynchi* subsp. *oncorhynchi* (Lee et al., 2006b; Yumoto et al., 2005b) are $C_{15:0\ iso}$ (31.6–34.3% and 22.7%, respectively) and $C_{15:0\ anteiso}$ (25.3–38.7% and 49.3%, respectively). *Oceanobacillus picturae* shows dominant amounts of $C_{15:0\ anteiso}$ (53.1–59.2%) and moderate amounts of $C_{14:0\ iso}$ (8.8–10.7%), $C_{16:0\ iso}$ (7.0–10.4%), and $C_{17:0\ anteiso}$ (11.9–14.3%) (Heyrman et al., 2003b; Lee et al., 2006b). For *Oceanobacillus oncorhynchi* subsp. *incaldanensis* (Romano et al., 2006), it is only stated that the fatty acids $C_{14:0\ anteiso}$, $C_{15:0\ iso}$, and $C_{15:0\ anteiso}$ account for about

TABLE 29. Differential characteristics of *Oceanobacillus* species[a,b]

Characteristic	1. *O. iheyensis*	2. *O. oncorhynchi* subsp. *oncorhynchi*	3. *O. oncorhynchi* subsp. *incaldanensis*	4. *O. picturae*
Spore shape	E	E	–	E(S)
Spore position	ST	S	–	T
pH range	6.5–10	9–10	6.5–9.5	7–10
Anaerobic growth	–	+	–	–
Temperature range (°C)	15–42	15–40	10–40	5–40
Reduction of nitrate	–	+	+	+
Growth at 0.5% NaCl	+	+	–	w
Gelatin hydrolysis	+	–	–	–/w
Casein hydrolysis	+	–	–	+/w
Acid production from:				
D-Melibiose	–	+	NG	w
D-Trehalose	–	+	NG	d
Growth on:				
D-Fructose	+	NG	+	–
D-Xylose	–	NG	+	–

[a]Symbols: +, positive; –, negative; +/w, weak to moderately positive reaction; –/w, negative or weakly positive reaction; w, weak reaction; NG, not given. Spore shape abbreviations: E, ellipsoidal; S, spherical. Spore position abbreviations: T, terminal; ST, subterminal.

[b]Data compiled from Lu et al. (2001); Yumoto et al. (2005b); Lee et al. (2006b); Romano et al. (2006).

90% of the total fatty acid content. Lu et al. (2001) estimated the genome size of *Oceanobacillus iheyensis* around 3.6 Mb by pulsed field gel electrophoresis after digestion with *Apa*I or *Sse*8387I. The whole genome sequence of *Oceanobacillus iheyensis* was determined (Takami et al., 2002). It is a single circular chromosome consisting of 3,630,528 bp with a mean G +C content of 35.7% (36.1% and 31.8% in the coding and noncoding region, respectively). From the total genome, 3,496 protein coding sequences were found, covering 85% of the chromosome. The mean size of the predicted proteins in *Oceanobacillus iheyensis* is 32.804 kDa, ranging from 2.714 to 268.876 kDa. Comparative analysis of the genome sequence of *Oceanobacillus iheyensis* with those of *Bacillus subtilis*, *Bacillus halodurans*, *Staphylococcus aureus*, and *Clostridium acetobutylicum* (Takami et al., 2002) revealed that 838 out of 3,496 (24.0%) putative proteins identified in the *Oceanobacillus iheyensis* genome have no orthologous relationship to proteins encoded in the four other genomes. Further, 354 proteins (10.1%) were identified as common proteins only among *Bacillus*-related species and 243 putative proteins (7.0%) were *shared only between the two alkaliphiles*, *Oceanobacillus iheyensis* and *Bacillus halodurans* (Brown et al., 2003) studied the presence of genes resembling *Staphylococcus aureus* isoleucyl-transfer-RNA synthetase 2 (IleRS2) and *Streptococcus pneumoniae* methionyl-tRNA synthetase 2 (MetRS2) in genomes of Gram-positive bacteria. These genes render an organism resistant to antibiotics inhibiting IleRS1 and MetRS1. The genome of *Oceanobacillus iheyensis* contains an IleRS2 homolog similar to that of *Staphylococcus aureus*, no IleRS1, and a typical MetRS1.

Enrichment and isolation procedures

HTE831[T], the only representative of *Oceanobacillus iheyensis*, was isolated from a deep-sea mud sample on a medium consisting of 1% polypeptone, 0.5% yeast extract, 0.1% K$_2$HPO$_4$, and 0.02% MgSO$_4$·7H$_2$O supplemented with 3% NaCl (Lu et al., 2001). The strain grows well in Marine Broth (Difco). R-2[T], the only representative of *Oceanobacillus oncorhynchi* subsp. *oncorhynchi*, was obtained by enriching µl ml of a viscous liquid of rainbow trout skin in 250 ml PYA broth (pH 10) for 30 h at 27 °C, shaken at 140

r.p.m. (Yumoto et al., 2005b). PYA medium contains 8 g peptone, 3 g yeast extract, 1 g K$_2$HPO$_4$, 3.5 mg EDTA, 3 mg ZnSO$_4$·7H$_2$O, 10 mg FeSO$_4$·7H$_2$O, 2 mg MnSO$_4$·5H$_2$O, 1 mg CuSO$_4$·5H$_2$O, 2 mg Co(NO$_3$)$_2$·6H$_2$O, and 1 mg H$_3$BO$_3$ in 1 l NaHCO$_3$/Na$_2$CO$_3$ buffer (100 mM in deionized water, pH 10). Strain 20AG, the only representative of *Oceanobacillus oncorhynchi* subsp. *incaldanensis*, was enriched from an algal mat in the following medium (per liter, pH 9.0): 3.0 g Na$_2$CO$_3$, 2.0 g KCl, 1.0 g MgSO$_4$·7H$_2$O, 100.0 g NaCl, 3.0 g Na$_3$-citrate, 10.0 g yeast extract, 0.36 mg MnCl$_2$·4H$_2$O, and 50 mg FeSO$_4$ (Romano et al., 2006). Na$_2$CO$_3$ and NaCl were autoclaved separately. Strains of *Oceanobacillus picturae* were isolated from standard media (TSA and R2A) with 10% salt added and were further subcultured on Marine Agar.

Oceanobacillus strains might be selected for, together with other halotolerant and alkaliphilic bacteria, on media supplemented with 15% NaCl and adjusted to pH 9.5. Further, inoculation of such media could be preceded by a heating step (5–10 min at 80 °C) in order to kill vegetative cells and so select for endospore-formers. Whether *Oceanobacillus oncorhynchi* subsp. *incaldanensis* (Romano et al., 2006) survives such a heating step should be checked.

Maintenance procedures

Oceanobacillus strains can easily be preserved in the refrigerator in tubes containing broth medium or agar slopes after checking the culture microscopically for sporulation. For long term preservation, lyophilization or liquid nitrogen may be used with the addition of a proper cryoprotectant.

Procedures for testing special characters

All *Oceanobacillus* species were described by using standard methodology, without genus-specific alterations of the protocols.

Differentiation from closely related taxa

Phylogenetically, the genera most closely related to *Oceanobacillus* are *Virgibacillus* and *Lentibacillus*. *Oceanobacillus* cannot be distinguished from *Virgibacillus* on the genus level. *Oceanobacillus* can only be differentiated from *Lentibacillus* by its slightly lower G +C content (35.8–40.1 and 42.0–44.0 mol%, respectively).

The distinction between *Oceanobacillus* and other halophilic endospore-forming genera of the *Bacillaceae* is also not straightforward. *Halobacillus* (Amoozegar et al., 2003; Liu et al., 2005; Spring et al., 1996; Yoon et al., 2004a; Yoon et al., 2005a; Yoon et al., 2003b) is the only genus that has been shown to contain peptidoglycan of the Orn-D-Asp type (A4β), however, this feature was not tested for *Oceanobacillus*. *Gracilibacillus* (Carrasco et al., 2006; Wainø et al., 1999) tests positive for starch hydrolysis (not tested for *Oceanobacillus picturae*). Members of *Amphibacillus* (Niimura et al., 1990; Zhilina et al., 2001b) lack isoprenoid quinones and are catalase-negative. Amphibacillus shares these features with Halolactibacillus (Ishikawa et al., 2005).

Halolactibacillus strains further grow at NaCl concentrations of 25% and show $C_{13:0\ anteiso}$ and $C_{16:0}$ as the major fatty acids. Strain YKJ-13[T], the only representative of *Jeotgalibacillus alimentarius* (Yoon et al., 2001c) contains MK-7 and MK-8 as the predominant menaquinones and the peptidoglycan type is L-Lys-direct (A1α). *Oceanobacillus* differs from *Marinibacillus* (Rüger, 1983; Rüger et al., 2000; Yoon et al., 2004a; Yoon et al., 2001c) in that members of the latter genus require sea water for growth. *Filobacillus* (Schlesner et al., 2001), represented by a single strain, contains peptidoglycan of the Orn-D-Glu type (variation A4β) and does not produce acid from D-glucose (not tested for *Oceanobacillus oncorhynchi* subsp. *incaldanensis*). O15-7[T], the only representative of *Paraliobacillus* (Ishikawa et al., 2002), has a glucose requirement during aerobic cultivation. However, *Oceanobacillus oncorhynchi* subsp. *oncorhynchi*, the only representative of *Oceanobacillus* able to grow under anaerobic conditions, was not tested for this feature. *Cerasibacillus* (Nakamura et al., 2004b) and *Tenuibacillus* (Ren and Zhou, 2005a), represented by one and two strains respectively, do not produce acid from D-fructose, D-glucose, or D-mannose (not tested for *Oceanobacillus oncorhynchi* subsp. *incaldanensis*). *Thalassobacillus* (Garcia et al., 2005), yet another one-strain genus, has endospores in central position and a slightly higher G + C content (42.8 in comparison with 35.8–40.1 for *Oceanobacillus*). *Alkalibacillus* (Jeon et al., 2005b; Romano et al., 2005b), *Pontibacillus* (Lim et al., 2005a; Lim et al., 2005b), and *Salinibacillus* (Ren and Zhou, 2005a) are not readily distinguishable from *Oceanobacillus*. As many of the above-mentioned genera are represented by a single species which is often only represented by one or two strains, it is possible that the discovery of additional species or strains within these genera will result in even fewer discriminatory characteristics.

Taxonomic comments

Since the description of *Oceanobacillus*, several taxonomic rearrangements have occurred. Lu et al. (2001) created a separate genus status for *Oceanobacillus iheyensis* as the species diverged at the bottom of a cluster including *Bacillus halodenitrificans* (currently *Virgibacillus halodenitrificans*), *Virgibacillus pantothenticus*, *Virgibacillus proomii*, *Salibacillus marismortui*, and *Salibacillus salexigens* (currently *Virgibacillus marismortui* and *Virgibacillus salexigens*; Heyrman et al., 2003b).

Since at the time of description, *Virgibacillus* and *Salibacillus* were still separate genera, Lu et al. (2001) concluded that their isolate could not be attributed to one or the other and proposed a novel genus. Further, a DNA–DNA relatedness study was performed between *Oceanobacillus iheyensis* and *Virgibacillus pantothenticus*, *Virgibacillus salexigens*, *Virgibacillus marismortui*,

and *Bacillus halodenitrificans*. All homology values were below 30%. Lu et al. (2001) concluded from these low values that strain HTE831[T] (*Oceanobacillus iheyensis*) should not be categorized as a member of a new species within a known genus. Such a conclusion is false since no comments on genus affiliation can be deduced from DNA–DNA relatedness values. Since the description of *Oceanobacillus iheyensis*, the species of *Salibacillus* were transferred to *Virgibacillus* (Heyrman et al., 2003b), *Bacillus halodenitrificans* was transferred to *Virgibacillus* (Yoon et al., 2004c), and five novel *Virgibacillus* species have been described including *Virgibacillus carmonensis*, *Virgibacillus necropolis*, *Virgibacillus picturae* (basonym of *Oceanobacillus picturae*; Heyrman et al., 2003b), *Virgibacillus dokdonensis* (Yoon et al., 2005b) and *Virgibacillus koreensis* (Lee et al., 2006b). Yumoto et al. (2005b) described *Oceanobacillus oncorhynchi* as a novel species. Further, the closely related genus *Lentibacillus* was described by Yoon et al. (2002) and expanded with four novel species (Jeon et al., 2005a; Lim et al., 2005c; Namwong et al., 2005; Tanasupawat et al., 2006). With every addition of a novel species in the genera *Oceanobacillus*, *Virgibacillus*, and *Lentibacillus*, the phylogenetic relationships between these genera have changed and the phenotypic differences originally differentiating them have disappeared. This led for example to the transfer of *Virgibacillus picturae* to *Oceanobacillus* on the basis of a 16S rDNA tree (Lee et al., 2006b). The majority of species belonging to *Virgibacillus*, *Oceanobacillus*, and *Lentibacillus* were described in the last five years. Therefore, future species descriptions in the neighborhood of these genera can be expected. The following new species descriptions were "in press" at the time of writing but have since been published (albeit too late for discussion in this section): *Lentibacillus kapialis* (Pakdeeto et al., 2007), *Lentibacillus halodurans* (Yuan et al., 2007), *Oceanobacillus chironomi* (Raats and Halpern, 2007), and *Oceanobacillus profundus* (Kim et al., 2007a). These descriptions will probably allow better assessment of whether the three genera should remain as currently described or whether further rearrangements are necessary.

Romano et al. (2006) isolated a bacterial strain from an algal mat that showed 99.5% 16S rDNA sequence similarity to *Oceanobacillus oncorhynchi*. This strain differed from *Oceanobacillus oncorhynchi* in that no spore formation was observed and the cells were larger, while growth occurred at acidic pH and no anaerobic growth occurred without added NaCl. Furthermore, hydrolysis of PNPG and differences in fatty acid composition were reported. DNA–DNA relatedness of this strain to the type strain of *Oceanobacillus oncorhynchi* was 59.0%. Referring to the new guidelines on the species concept for prokaryotes (Rosselló-Mora and Amann, 2001; Stackebrandt et al., 2002), Romano et al. (2006) concluded that the nearly identical 16S rRNA gene sequence (99.5% similarity), DNA–DNA relatedness above 50%, and few phenotypic differences did not justify a classification of this single strain into a separate species. Therefore, they described this strain as a subspecies of *Oceanobacillus oncorhynchi*, namely *Oceanobacillus oncorhynchi* subsp. *incaldanensis*. Looking at the data given by Romano et al. (2006), the taxonomic status of *Oceanobacillus oncorhynchi* subsp. *incaldanensis* is debatable. Though it is true that the current bacterial species concept (Rosselló-Mora and Amann, 2001; Stackebrandt et al., 2002) states that the value for species delineation on the basis of DNA–DNA relatedness (set at 70%) should be interpreted in a more relaxed manner.

However, in this particular case, the DNA relatedness value is clearly below 70% and there are several phenotypic differences. Of course, as both subspecies of *Oceanobacillus oncorhynchi* are only represented by a single strain, it is difficult to assess whether the phenotypic variation found is intra- or interspecies variation. This clearly demonstrates that the analysis of several strains per (sub)species is necessary to allow clear distinction.

List of species of the genus *Oceanobacillus*

1. **Oceanobacillus iheyensis** Lu, Nogi and Takami 2002, 687[VP] (Effective publication: Lu, Nogi and Takami 2001, 296.)

i.he.yen'sis. N.L. masc. adj. *iheyensis* pertaining to the Iheya Ridge, Okinawa Trough, Japan.

Morphology and general characters as for generic description.

Cells are $0.6–0.8 \times 2.5–3.5\,\mu m$. Obligatory aerobe. Growth occurs at 0–21% (w/v) NaCl, with optimum growth at 3% NaCl. Growth occurs at temperatures of 15–42 °C (optimum 30 °C). The pH range for growth is 6.5–10 (optimum 7.0–9.5). Nitrate reduction to nitrite is negative. Hydrolyzes gelatin, casein, Tween 40, and Tween 60. Does not hydrolyze starch. Acid is produced from glycerol, but not from galactose, glucitol, D-melibiose, rhamnose, and D-trehalose. The following four substrates are oxidized in the Biolog test: α-D-glucose, maltose, D-mannose, and turanose. The major cellular fatty acids are $C_{15:0\ anteiso}$, $C_{15:0\ iso}$, and $C_{14:0\ iso}$.

Source: mud at a depth of 1050 m on the Iheya Ridge.

DNA G + C content (mol%): 35.8 (reverse-phase HPLC).

Type strain: HTE831, CIP 107618, DSM 14371, JCM 11309.

GenBank accession number (16S rRNA gene): AB010863.

2. **Oceanobacillus oncorhynchi** Yumoto, Hirota, Nodasaka and Nakajima 2005b, 1523[VP] emend. Romano, Lama, Nicolaus, Poli, Gambacorta and Giordano 2006, 809.

on.co.rhyn'chi. N.L. gen. n. *oncorhynchi* of *Oncorhynchus*, named after the rainbow trout *Oncorhynchus mykiss.*

Morphology and general characters as for generic description.

Growth occurs at 10–40 °C, with the optimum at 30–37 °C. Does not hydrolyze casein, gelatin, and starch. Oxidase-positive. Reduces nitrate to nitrite.

DNA G + C content (mol%): 38.5 (reverse-phase HPLC).

Type strain: R-2, JCM 12661, NCIMB 14022.

GenBank accession number (16S rRNA gene): AB188089.

2a. **Oceanobacillus oncorhynchi subsp. oncorhynchi** Yumoto, Hirota, Nodasaka and Nakajima 2005b, 152[VP] emend. Romano, Lama, Nicolaus, Poli, Gambacorta and Giordano 2006, 809.

Cells are $0.4–0.6 \times 1.1–1.4\,\mu m$. Facultative anaerobe. Growth occurs at 0–22% (w/v) NaCl with the optimum at 7% (w/v). The pH range for growth is 9.0–10.0; no growth at 7.0–8.0. Hydrolyzes Tween 40. Does not hydrolyze lipid (tributyrin) or Tweens 20, 60, and 80. Reactions for ONPG hydrolysis and deamination of phenylalanine are negative. Acid is produced from D-melibiose, sucrose, raffinose, D-galactose, and trehalose. No acid is produced from D-arabinose, *myo*-inositol, and sorbitol. The major fatty acids are $C_{15:0\ anteiso}$ (49.3%), $C_{15:0\ iso}$ (22.7%), and $C_{17:0\ anteiso}$ (18.0%).

Source: the skin of the rainbow trout (*Oncorhynchus mykiss*).

DNA G + C content (mol%): 38.5 (reverse-phase HPLC).

Type strain: R-2, JCM 12661, NCIMB 14022.

GenBank accession number (16S rRNA gene): AB188089.

2b. **Oceanobacillus oncorhynchi subsp. incaldanensis** Romano, Lama, Nicolaus, Poli, Gambacorta and Giordano 2006, 808[VP]

in. cald. en'sis. N.L. masc. adj. *incaldensis* pertaining to the Incaldana site, southern Italy, where the type strain was isolated

Cells are $0.5–0.8 \times 1.2–2.0\,\mu m$. Nonspore-forming. Strictly aerobic. Growth occurs at 5–20% (w/v) NaCl, with the optimum at 10% (w/v). The pH range for growth is 6.5–9.5 with an optimum at pH 9.0. Grows on D-arabinose, D-cellobiose, D-fructose, D-galactose, D-glucose, D-lactose, D-maltose, D-mannose, D-ribose, D-sorbose, D-sucrose, D-trehalose, D-xylose, glycerol, Na-acetate, and Na-citrate as sole carbon sources. KOH positive. Negative for aminopeptidase. The main fatty acids are $C_{14:0\ anteiso}$, $C_{15:0\ iso}$ and $C_{15:0\ anteiso}$, accounting for about 90% of the total fatty acid composition. The predominant polar lipids are phosphatidyl glycerol and diphosphatidyl choline.

Source: an algal mat collected from a sulfurous spring in the Santa Maria Incaldana site (Mondragone, Italy).

DNA G + C content (mol%): 40.1 (reverse-phase HPLC).

Type strain: 20AG, ATCC BAA-954, DSM 16557.

GenBank accession number (16S rRNA gene): AJ640134.

3. **Oceanobacillus picturae** (Heyrman et al., 2003b) Lee, Lim, Lee, Lee, Park and Kim 2006b, 256[VP] (*Virgibacillus picturae* Heyrman, Logan, Busse, Balcaen, Lebbe, Rodriguez-Diaz, Swings and De Vos 2003b, 509.)

pictur.ae. L. gen. n. *picturae* of a painting.

Morphology and general characters as for generic description.

Cells are $0.5–0.7 \times 2.0–6.0\,\mu m$. Obligatory aerobe. Weak growth without added NaCl and optimal growth at NaCl (w/v) concentrations of 5 and 10%. Temperature range for growth is 5–40 °C with optimal growth at 25–35 °C. Positive results for ONPG and nitrate reduction. Negative for arginine dihydrolase, lysine decarboxylase, ornithine decarboxylase, citrate utilization, hydrogen sulfide production, urease, tryptophan deaminase, and Voges–Proskauer. Growth on different sugars as sole carbon source is weakly positive for raffinose and negative for D-arabinose, cellobiose, D-fructose, and D-xylose. The major fatty acids are $C_{15:0\ anteiso}$ (59.2%), $C_{15:0\ anteiso}$ (11.9%), and $C_{14:0\ iso}$ (10.7%). The polar lipid pattern contains predominant amounts of diphosphatidyl glycerol and moderate amounts of phosphatidyl glycerol.

Source: mural paintings in Austria and Spain.

DNA G + C content (mol%): 39.5 (reverse-phase HPLC).

Type strain: DSM 14867, LMG 19492, KCTC 3821.

GenBank accession number (16S rRNA gene): AJ315060.

Genus XIV. **Paraliobacillus** Ishikawa, Ishizaki, Yamamoto and Yamasato 2003a, 627[VP] (Effective publication: Ishikawa, Ishizaki, Yamamoto and Yamasato 2002, 275.)

KAZUHIDE YAMASATO AND MORIO ISHIKAWA

Pa.ra.li.o.ba.cil′lus. Gr. adj. *paralios* littoral; L. n. *bacillus* rod; N.L. masc. n. *Paraliobacillus* rod inhabiting littoral (marine) environment.

Cells are Gram-positive, **endospore-forming rods** that are **motile by peritrichous flagella** (Figure 20). **Facultatively anaerobic. Catalase-positive when cultivated aerobically.** Pseudocatalase-negative. **Requires carbohydrate, sugar alcohol, or related compounds for growth in both aerobic and anaerobic conditions. Glucose is aerobically metabolized to produce acetate and pyruvate as main organic acids. In anaerobic cultivation, lactate, formate, acetate, and ethanol are the end products from glucose, with a molar ratio of approximately 2:1:1 for the latter three products, without gas production. Slightly halophilic and extremely halotolerant. Slightly alkaliphilic.** Contains *meso*-diaminopimelic acid in cell-wall peptidoglycan. **The menaquinone type is menaquinone-7. Cytochromes are present.** Occupies an independent lineage within the halophilic/halotolerant/alkaliphilic and/or alkalitolerant group[*] in rRNA group 1 of the phyletic group classically defined as the genus *Bacillus* (Ash et al., 1991).

DNA G + C content (mol%): 35.6 (HPLC).

Type species: **Paraliobacillus ryukyuensis** Ishikawa, Ishizaki, Yamamoto and Yamasato 2003a, 627[VP] (Effective publication: Ishikawa, Ishizaki, Yamamoto and Yamasato 2002, 275.).

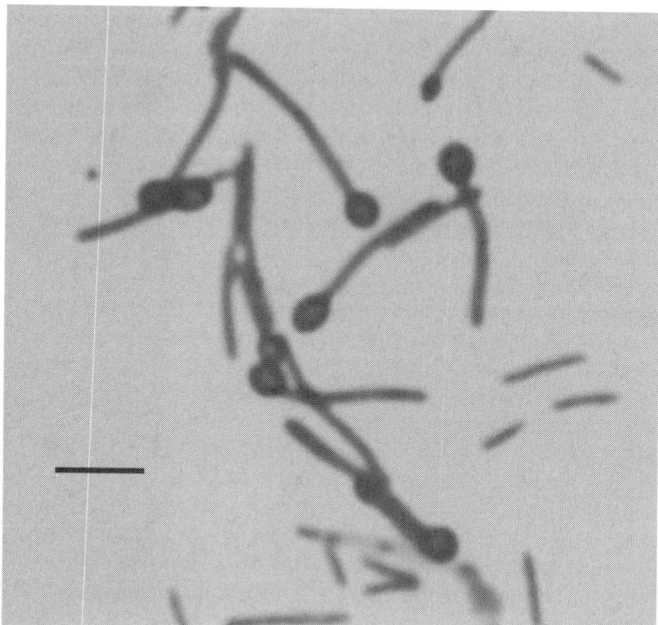

FIGURE 20. Photomicrograph of cells and peritrichous flagella of *Paraliobacillus ryukyuensis* IAM 15001[T] grown anaerobically on 2.5% NaCl GYPFK agar[††] at 20°C for 2 d. Bar = 2 μm.

Further descriptive information

Descriptive information is based on one strain of *Paraliobacillus ryukyuensis* (Ishikawa et al., 2002). The phylogenetic position of *Paraliobacillus* based on 16S rRNA gene sequence analysis is given in Figure 18.

The characteristics of *Paraliobacillus ryukyuensis* are listed in Table 30, Table 31, and Table 32. *Paraliobacillus ryukyuensis* utilizes a wide range of carbohydrates and related compounds: pentoses, hexoses, disaccharides, trisaccharides, polysaccharides, sugar alcohols, and related carbon compounds. Ethanol, formate, lactate, pyruvate, and components of the TCA cycle are not utilized.

Paraliobacillus ryukyuensis is slightly halophilic, as it grows optimally in NaCl concentrations between 0.75% (w/v) and 3.0% (Kushner, 1978; Kushner and Kamekura, 1988). The specific growth rates, μ_{max} (h^{-1}), are 0.22 in 0–0.5% (w/v), 0.36–0.38 in 0.75–3.0%, 0.31 in 4.0%, 0.28 in 5.0%, and 0.20 in 7.0% NaCl. This bacterium is highly tolerant to an elevated NaCl concentration, growing at 22% which is comparable to the extremely halotolerant members in the halophilic/halotolerant/alkaliphilic and/or alkalitolerant group[1] in rRNA group 1 (HA group).

With respect to growth responses to pH of cultivation medium, the μ_{max} (h^{-1}) values are 0.18 at pH 5.5, 0.36 at pH 6.0, 0.38 at pH 6.5, 0.40 at pH 7.0–8.5, and 0.20 at pH 9.0. *Paraliobacillus ryukyuensis* can be characterized as slightly alkaliphilic according to Jones et al. (1994) who defined alkaliphiles as organisms that grow optimally at a pH greater than 8. The final pH is 4.5 in both aerobic and anaerobic cultivations, which is 1.0 pH unit lower than the minimum pH for growth initiation (pH 5.5).

Generation of catalase is induced by oxygen; *Paraliobacillus ryukyuensis* produces catalase on an agar plate or in aerated broth medium, but no catalase activity is detected for anaerobically grown cells.

Paraliobacillus ryukyuensis requires carbohydrate (and related carbon compounds) for growth. Under anaerobic conditions, growth does not occur in 2.5% NaCl GYPF broth[†] without glucose. Under aerobic conditions, absorbances at 660 nm of stationary cultures in broth media of 2.5% NaCl GP[‡] and 2.5% NaCl GCY[§] are 0.29 and 0.38, respectively, while those in the absence of glucose are below 0.03.

---

et al., 2001), *Paraliobacillus* (Ishikawa et al., 2002,) *Lentibacillus* (Yoon et al., 2002), *Cerasibacillus* (Nakamura et al., 2004b), *Alkalibacillus* (Fritze, 1996a; Jeon et al., 2005b), *Halolactibacillus* (Ishikawa et al., 2005), *Tenuibacillus* (Ren and Zhou, 2005b), *Pontibacillus* (Lim et al., 2005a). *Salinibacillus* (Ren and Zhou, 2005a), *Thalassobacillus* (Garcia et al., 2005), *Ornithinibacillus* (Mayer et al., 2006), *Paucisalibacillus* (Nunes et al., 2006), and *Pelagibacillus* (Kin et al., 2007b).

[†] GYPF (GYPB) broth is composed of (per liter) 10 g glucose, 5 g yeast extract, 5 g peptone, 5 g fish extract (beef extract), 1 g K$_2$HPO$_4$, 1 g sodium thioglycolate, and 5 ml salt solution (per ml, 40 mg MgSO$_4$·7H$_2$O, 2 mg MnSO$_4$·4H$_2$O, 2 mg FeSO$_4$·7H$_2$O), pH 8.5. Sterilized by filtration. Sodium thioglycolate is omitted for aerobic cultivation.

[‡] 2.5% GP broth is composed of (per liter) 10 g glucose, 10 g peptone, and inorganic components of GYPF[†] broth, pH 8.5. Sterilized by filtration.

[§] 2.5% NaCl GCY broth is composed of (per liter) 10 g glucose, 5 g Vitamin assay Casamino acids (Difco), 0.5 g yeast extract, and inorganic components of GYPF[2] broth[2], pH 8.5. Sterilized by filtration.

---

[*] A monophyletic subgroup within rRNA group 1 *Bacillus*, which is composed of halophilic, halotolerant, alkaliphilic and/or alkalitolerant organisms (Ishikawa et al., 2003b). The present composing genera or species are *Bacillus halophilus* (Ventosa et al., 1989), *Amphibacillus* (Niimura et al., 1990), *Halobacillus* (Claus et al., 1983; Spring et al., 1996), *Virgibacillus* (Heyndrickx et al., 1998; Heyndrickx et al., 1999; Heyrman et al., 2003b), *Gracilibacillus* (Deutch, 1994; Lawson et al., 1996; Wainø et al., 1999), *Filobacillus* (Schlenser et al., 2001), *Oceanobacillus* (Lu

TABLE 30. Phenotypic characteristics of *Paraliobacillus ryukyuensis*[a,b]

Characteristic	
NaCl optima (%)	0.75–3.0
NaCl range (%)	0–22
pH optima	7.0–8.5
pH range	5.5–9.5
Temperature optima (°C)	37–40
Temperature range (°C)	10–47.5
Gelatin hydrolysis	–
Starch hydrolysis	+
Nitrate reduction	–
NH₃ from arginine	–
Utilization of carbon compounds:	
D-Ribose, L-arabinose, D-xylose, D-glucose, D-fructose, D-mannose, maltose, sucrose, D-cellobiose, lactose, D-trehalose, D-raffinose, D-melezitose, glycerol, D-mannitol, D-sorbitol, myo-inositol, starch, inulin, α-methyl-D-glucoside, D-salicin, gluconate	+
Adonitol, dulcitol	w
D-Arabinose, D-galactose, D-rhamnose, melibiose	–
Ethanol, formate, acetate, lactate, pyruvate, succinate, malate, fumarate, oxaloacetate, 2-oxoglutarate, citrate	–

[a]Symbols: +, positive; w, weakly positive; –, negative.
[b]Data from Ishikawa et al. (2002).

TABLE 31. Products from glucose in aerobic and anaerobic cultivations of *Paraliobacillus ryukyuensis* IAM 15001[Ta,b]

Cultivation[c]	Cultivation time (h)	Absorbance at 660 nm	Glucose consumed (mM)	Products (mM)					Carbon recovery (%)
				Pyruvate	Lactate	Formate	Acetate	Ethanol	
Aerobic	11	0.48	5.66	0.55	0.18	ND	5.05	ND	36
	14	1.09	11.32	5.83	0.59	ND	12.16	ND	64
Anaerobic	12	0.36	4.00	ND	0.97	6.90	3.29	3.27	96
	13	0.48	5.33	ND	1.85	9.45	4.59	4.30	103

[a]Symbols: ND, not detected.
[b]Data from Ishikawa et al. (2005).
[c]*Paraliobacillus ryukyuensis* IAM 15001[T] was cultivated in GYPF broth, pH 8.0, under aerobic cultivation by shaking and under anaerobic cultivation.

TABLE 32. Effect of the initial pH of the medium on the composition of products of glucose fermentation by *Paraliobacillus ryukyuensis* IAM 15001[Ta,b]

Initial pH	End products (mol/mol glucose)				Lactate yield from consumed glucose (%)	Carbon recovery (%)
	Lactate	Formate	Acetate	Ethanol		
6.5	1.03	0.94	0.41	0.37	51	93
7.0	1.00	0.84	0.43	0.48	50	94
8.0	0.41	1.60	0.67	0.65	21	91
9.0	0.16	1.85	0.88	0.93	8	100

[a]*Paraliobacillus ryukyuensis* IAM 15001[T] was cultivated in heavily buffered 2.5% NaCl GYPF broths with different initial pH values. Decrease in pH during cultivation was 0.5 units or less.
[b]Data from Ishikawa et al. (2005).

Under aerobic conditions, acetate, pyruvate, and a small amount of lactate are produced without production of formate and ethanol. The carbon recovery varies from 36% to 64% depending on growth phase; at the exponential phase it is about twofold that at the stationary phase, accompanying much increased production of pyruvate and acetate (Table 31). In anaerobic cultivation, lactate, formate, acetate, and ethanol are produced from glucose without gas production, with a well-balanced carbon recovery (96–103%). The molar ratio for formate, acetate, and ethanol is approximately 2:1:1 (Table 31). The amount of lactate produced relative to the total amount of the other three products is markedly affected by the pH during cultivation (Table 32). As the initial pH of the medium is lowered, the relative amount of lactate increases, while the relative total amount of the other three products decreases. The opposite occurs when the initial pH is increased. For each of the initial pH, the molar ratio for the three products relative to the lactate produced is substantially retained. *Exiguobacterium aurantiacum* (Collins et al., 1983; Gee et al., 1980), a facultative anaerobe in *Bacillaceae, Trichococcus (Lactosphaera*; Janssen et al., 1995), *Marinilactibacillus* (Ishikawa et al., 2003b), *Halolactibacillus* (Ishikawa et al., 2005), and *Alkalibacterium* (Ishikawa et al., 2003b) exhibit the same behavior in glucose fermentation with respect to the effect of pH of the cultivation medium. The product balance is considered to depend on the relative activities of the two enzymes involved in pyruvate metabolism, pyruvate-formate lyase and lactate dehydrogenase (Gunsalus and Niven, 1942; Janssen et al., 1995; Rhee and Pack, 1980). The ratio of the L(+) isomer to the total amount of lactate produced was 52%.

Both aerobically and anaerobically grown cells possess menaquinone-7, the concentration of which is much greater in aerobic cultivation.

The sequence similarities of 16S rRNA genes of *Paraliobacillus*, approximately 1500 bases in length covering the positions 32–1510 (*Escherichia coli* numbering system), to *Gracilibacillus*, *Amphibacillus*, and *Halolactibacillus*, the phylogenetically closest genera, are 94.7–95.8%, 92.6–94.3%, and 93.7–94.1%, respectively. The similarity values between *Paraliobacillus* and other members of the HA group are below 95.1%.

Paraliobacillus ryukyuensis was isolated from a decomposing marine alga taken at a foreshore site near the Oujima islet adjacent to the main island of Okinawa, Japan. Because this bacterium was isolated from a marine substrate and possesses slightly halophilic, highly halotolerant, and slightly alkaliphilic properties, the organism is considered to be well-adapted to marine environments as have been other members of the HA group inhabiting saline environments. Since only one strain has been isolated to date, the habitat and ecology of *Paraliobacillus* cannot be generalized.

Enrichment and isolation procedures

Since only one strain has been isolated, generalized method for isolation cannot be given. Strain IAM 15001[T] was isolated by successive enrichment culture in 7% and 18% NaCl GYPFSK isolation broths[**] (each was cultured for 21 d and 15 d, respectively) and was pour-plated with 12% NaCl GYPFSK isolation agar (2.0% agar) supplemented with 5 g (per liter) $CaCO_3$.

Maintenance procedures

Paraliobacillus ryukyuensis is maintained by serial transfer in a stab culture stored at 5–10 °C at 1-month intervals. The medium is 7% NaCl GYPF or GYPB agar, pH 7.5, supplemented with 5 g/l $CaCO_3$, autoclaved at 110 °C for 10 min. Marine agar 2216 (Difco) supplemented with glucose and $CaCO_3$ can be used. The strain has been maintained by freezing at –80 °C or below in 2.5% NaCl GYPFK or GYPBK broth[††], pH 7.5, supplemented with 10% (w/v) glycerol. Marine broth 2216 (Difco) supplemented with glucose and 10% (w/v) glycerol can also be used. The strain can be preserved by L-drying in an adjuvant solution composed of (per liter) 3 g sodium glutamate, 1.5 g adonitol, and 0.05 g cysteine hydrochloride in 0.1 M phosphate buffer (KH_2PO_4–K_2HPO_4), pH 7.0 (Sakane and Imai, 1986) or by lyophilization with a standard cryoprotectant containing an appropriate concentration of NaCl.

Procedures for testing special characters

Spores are abundantly formed on yeast extract salts agar which is composed of (per liter) 5 g yeast extract, 20 g NaCl, 5 g $MgCl_2 \cdot 7H_2O$, 0.2 g $CaCl_2$, 0.1 g KH_2PO_4, 10 ml salt solution[†], and

15 g agar, pH 7.5, autoclaved at 110 °C for 10 min and Marine agar 2216 (Difco). Spore formation is scarce on an agar medium which contains glucose and low concentrations of Mg^{2+} and Ca^{2+}, such as 2% NaCl GYPF or GYPB agar[†], pH 7.5.

For anaerobic cultivation in pH experiments, AnaeroPack-Keep (Mitsubishi Gas Chemical Company, Tokyo) which absorbs O_2 but does not generate CO_2 is adequate.

Differentiation of *Paraliobacillus* from other related genera and species

Paraliobacillus is differentiated from other members of the HA group by the combination of morphological, physiological, biochemical, and chemotaxonomic features. Comparison of characteristics of closely related/phenotypically similar bacteria is summarized in Table 33. *Paraliobacillus* is differentiated from aerobes in the HA group by acid-producing fermentation. This bacterium is distinguished from facultative anaerobes in the HA group, from *Virgibacillus* (*Virgibacillus pantothenticus* and *Virgibacillus proomii*) by glucose requirement under aerobic conditions, and from *Amphibacillus* by the formation of catalase and the possession of respiratory quinone and cytochromes. *Paraliobacillus* is phenotypically similar to *Halolactibacillus* in the HA group, and *Marinilactibacillus* and *Alkalibacterium* in typical lactic acid bacteria group in lactic acid fermentation pattern (see *Further descriptive information*) and similar to *Marinilactibacillus* and *Halolactibacillus* in glucose requirement under aerobic conditions. *Paraliobacillus* is differentiated from *Halolactibacillus*, *Marinilactibacillus*, and *Alkalibacterium* by the formation of spores and catalase and the possession of respiratory quinones, from *Halolactibacillus* and *Marinilactibacillus* further by the possession of cytochromes (no data for *Alkalibacterium*), and from *Marinilactibacillus* and *Alkalibacterium* further by peptidoglycan type and major cellular fatty acid composition.

Paraliobacillus ryukyuensis shares a requirement for glucose under aerobic conditions with *Amphibacillus xylanus* (Niimura et al., 1990), *Amphibacillus tropicus* (Zhilina et al., 2001a), and *Amphibacillus fermentum* (Zhilina et al., 2001a). From these bacteria, *Paraliobacillus ryukyuensis* is differentiated by the characteristics described above and the optimum and range of pH for growth.

Taxonomic comments

Induction of catalase by oxygen is considered to be characteristic of *Paraliobacillus ryukyuensis*, as it has not yet been observed for facultative anaerobes. Oxygen or hydrogen peroxide induction of catalase reported for facultative anaerobes (Whittenbury, 1964; Finn and Condon, 1975; Hassan and Fridovich, 1978; Loewen et al., 1985) is involved in stimulated or increased production of catalase which is produced at low level when not subjected to oxidizing agent.

Echigo et al. (2005) isolated five strains of spore-formers which are highly halotolerant from non-saline Japanese soils, which can grow at 20–25% NaCl and have 95.5–96% similarities of partial 16S rRNA gene sequences to *Paraliobacillus ryukyuensis*. These authors discussed the possibility that the isolates are vagrants that have accumulated in soils and were transported by Asian dust storms over open seas, having survived for long periods in a dormant state as spores. Though their habitats are unclear and detailed taxonomic characterization was not done, these spore-formers may be related to *Paraliobacillus*.

[**] GYPFSK isolation broth is GYPF broth[†], pH 7.5, to which 50 ml soy sauce and 10 mg (per liter) cycloheximide are added and in which concentration of K_2HPO_4 is increased to 1%. Autoclaving is at 110°C for 10 min.

[††] GYPFK (GYPBK) broth is GYPF (GYPF) broth is which concentration of K_2HPO_4 is increased to 1%.

TABLE 33. Characteristics differentiating *Paraliobacillus* from related members of the halophilic/halotolerant/alkaliphilic and/or alkalitolerant group in *Bacillus* rRNA group 1, *Alkalibacterium* species and *Marinilactibacillus psychrotolerans*[a]

Characteristic	*Paraliobacillus ryukyuensis*[b]	*Amphibacillus xylanus*[c]	*Amphibacillus fermentum*[c]	*Amphibacillus tropicus*[d]	*Gracilibacillus halotolerans*[e]	*Gracilibacillus dipsosauri*[e,f,g]	*Halolactibacillus*[h]	*Virgibacillus pantothenticus*[i]	*Alkalibacterium*[j,k,l]	*Marinilactibacillus psychrotolerans*[m]
Spore formation	+[n]	+	+[o]	+	+	+	−	+	−	−
Anaerobic growth	+ (F)	+ (F)	+ (F)	+ (F)	−	+ (ANR)	+ (F)	+ (F)	+ (F)	+ (F)
Catalase	+	−	+	+	+	+	−	+	−	+
Glucose requirement in aerobic cultivation	+	+	+	+	−	−	+	−	ND	+
NaCl (range, %)	0–22	<6	0.98–19.7	0.98–20.9	0–20	0–18.6 (KCl)	0–25.5	0 to ≥10	0–17	0–20
NaCl (optimum, %)	0.75–3	ND	10.8	5.4–10.8	0	3.7 (KCl)	2–3	4	2–13	2.0–3.75
pH (range)	5.5–9.5	8–10	7–10.5	8.5–11.5	5–10	6 to ≥10	6–10	ND	8.5–12	6.0–10.0
pH (optimum)	7–8.5	ND	8.5–9	9.5–9.7	7.5	7.5	8–9.5	7	9.0–10.5	8.5–9.0
Major isoprenoid quinones	MK-7	None	ND	ND	MK-7	MK-7	None	MK-7	None	None
Peptidoglycan type	*m*-Dpm	*m*-Dpm	ND	ND	*m*-Dpm	*m*-Dpm	*m*-Dpm	*m*-Dpm	Orn-D-Asp, Lys-D-Asp, Orn (Lys)-D-Asp	Orn-D-Glu
G+C content (mol%)	35.6	36–38	41.5	39.2	38	39.4	38.5–40.7	38.3	39.7–43.2	34.6–36.2
Major cellular fatty acids:										
$C_{13:0\ ante}$	ND	−	ND	ND	−	−	+	−	−	−
$C_{15:0\ ante}$	ND	+	ND	ND	+	+	−	+	−	−
$C_{15:0\ iso}$	ND	−	ND	ND	−	+	−	+	−	−
$C_{16:0}$	ND	+	ND	ND	+	+	+	−	+	+
$C_{16:0\ iso}$	ND	+	ND	ND	−	−	−	−	−	−
$C_{16:1\ \omega7}$	ND	−	ND	ND	−	−	−	−	+	+
$C_{16:1\ \omega9}$	ND	−	ND	ND	−	−	−	−	+	−
$C_{17:0\ ante}$	ND	−	ND	ND	+	+	−	+	−	−
$C_{18:1\ \omega9}$	ND	−	ND	ND	−	−	−	−	+	+
Phylogenetic position	HA group	HA group	HA group	HA group	HA group	HA group	HA group	HA group	Typical lactic bacteria group	Typical lactic bacteria group
Source	Decaying marine alga	Alkaline manure with rice and straw	Sediment, soda lake	Sediment, soda lake	Surface mud, Great Salt Lake	Nasal salt glands of a desert iguana	Decaying marine algae, living sponge	Soils	Washwaters of edible olives, polygonum indigo fermentation liquor	Decaying marine algae, living sponge, raw Japanese ivory shell

[a]Symbols: +, positive; −, negative; ND, no data; F, fermentation; ANR, anaerobic respiration; *m*-Dpm, *meso*-diaminopimelic acid; Orn, ornithine; Asp, aspartic acid; Glu, glutamic acid.
[b]Data from Ishikawa et al. (2002).
[c]Data from Niimura et al. (1990).
[d]Data from Zhilina et al. (2001a).
[e]Data from Wainø et al. (1999).
[f]Data from Deutch (1994).
[g]Data from Lawson et al. (1996).
[h]Data from Ishikawa et al. (2005).
[i]Data from Heyndrickx et al. (1998, 1999).
[j]Data from Ntougias and Russell (2001).
[k]Data from Yumoto et al. (2004b).
[l]Data from Nakajima et al. (2005).
[m]Data from Ishikawa et al. (2003b).
[n]Produced in aerobic cultivation.
[o]Spore formation was not observed but culture survived heating.

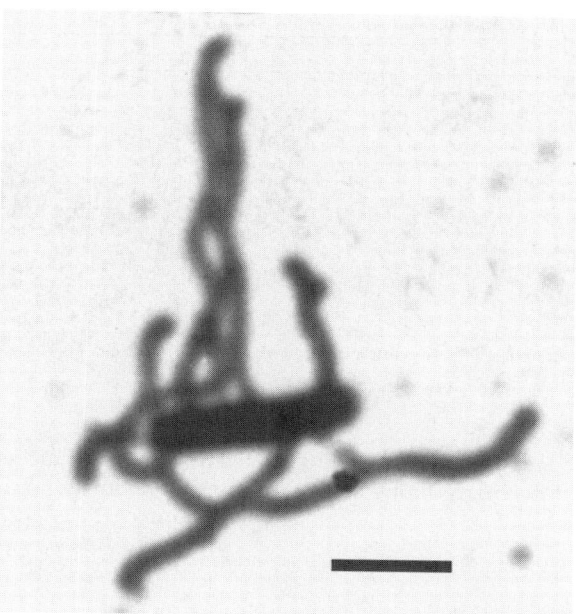

FIGURE 21. Photomicrograph of spores and sporangia of *Paraliobacillus ryukyuensis* IAM 15001[T] grown on yeast extract salts agar (composition, see *Procedures for testing special characters*) at 37 °C for 3 d. Bar = 5 μm.

List of species of the genus *Paraliobacillus*

1. **Paraliobacillus ryukyuensis** Ishikawa, Ishizaki, Yamamoto and Yamasato 2003a, 627[VP] (Effective publication: Ishikawa, Ishizaki, Yamamoto and Yamasato 2002, 276.)

ryu.kyu.en′sis. N.L. adj. *ryukyuensis* from the Ryukyu Islands, Japan, where the type strain was isolated.

The characteristics are as described for the genus and as listed in Table 30, Table 31, and Table 32. Colonies on 2.5% NaCl GYPF agar are round, convex, entire, yellow, and transparent, with diameters of 1.2–1.5 mm after 3 d growth at 30 °C. Deep colonies are creamy white, opaque, and lenticular, with diameters of 0.5–1.5 mm. Cells are 0.4–0.5 × 2.3–4.5 μm occurring singly, in pairs, or in chains. Endospores are spherical to ellipsoidal (predominant form) and terminal, measuring 0.9–1.0 × 0.9–1.4 μm (Figure 21). Sporangia are definitely swollen (Figure 21).

Source: a decomposing marine alga.

DNA G + C content (mol%): 35.6 (HPLC).

The type strain: O15-7, DSM 15140, IAM 15001, JCM 21472, NBRC 10,0001, NRIC 0520.

GenBank accession number *(16S rRNA gene):* AB087828.

Genus XV. **Pontibacillus** Lim, Jeon, Song and Kim 2005a, 168[VP] emend. Lim, Jeon, Park, Kim, Yoon and Kim 2005b, 1030

Paul De Vos

Pon.ti.ba.cil′lus. L. n. *pontus* the sea; L. masc. n. *bacillus* a small staff, a wand; N.L. masc. n. *Pontibacillus* bacillus pertaining to the sea.

Gram-positive, spore-forming rods, 0.4–0.9 μm × 2.3–4 μm. Motile by means of peritrichous flagella. Terminal, spherical endospores in swollen sporangia. Catalase-positive. Urease-negative. Oxidase is reported to be different in the different species. Strictly aerobic. No growth occurs in media without NaCl or with more than 15% NaCl. Cell-wall peptidoglycan contains A1γ type *meso*-DAP. Major isoprenoid quinone is MK-7. Predominant cellular fatty acids are $C_{15:0\,ante}$, $C_{15:0\,iso}$, and $C_{16:0\,iso}$ on Marine Agar (MA). Phylogenetically, the genus belongs to the family *Bacillaceae*.

DNA G + C content (mol%): 40.8–42.

Type species: **Pontibacillus chungwhensis** Lim, Jeon, Song and Kim 2005b, 169[VP].

Further descriptive information

After 2 d of incubation on MA colonies are creamy-yellow, slightly irregular in shape. Optimal growth occurs in media containing 2.0–9.0% NaCl. Both species grow in the pH range of 6.5–9.0 with slight differences per species. All tests were performed at 35 °C. Phylogenetically, *Pontibacillus* belongs to the *Bacillaceae* with, as close neighbors, other salt-tolerant taxa of the *Bacillaceae* such as *Gracilibacillus*, *Virgibacillus*, *Halobacillus*, *Filobacillus*, and *Lentibacillus* at the level of 93.3–95.9% 16S rDNA similarity.

Strains of *Pontibacillus* have been isolated from a solar saltern in Korea on media that must contain mineral salts and trace elements.

Differentiation of the genus *Pontibacillus* from other genera

Apart from the sequence analysis of the 16S rDNA (see above), *Pontibacillus* can be discriminated at the generic level by a number of biochemical characteristics as given in Table 1 in Lim et al. (2005b). The interested researcher must be aware that often the data given are limited to the type strains of the phy-

TABLE 34. Characteristics differentiating *Pontibacillus* species (type strains)[a,b]

Characteristic	1. *P. chungwhensis*	2. *P. marinus*
Colony pigmentation on MA	Yellow	Creamy
NaCl concentration range for growth (%)	1–15	1–9
Oxidase reaction	–	+
Nitrate reduction	–	+
Hydrolysis of esculin	–	+
Hydrolysis of casein	+	–
Hydrolysis of starch	+	–
Acid production from D-glucose	+	–
Acid production from D-fructose	–	+
Major fatty acids	$C_{15:0\ iso}$, $C_{15:0\ ante}$, $C_{16:0\ iso}$	$C_{15:0\ iso}$, $C_{15:0\ ante}$, $C_{16:1\ \omega7c\ OH}$

[a]Symbols: +, >85% positive; d, different strains give different reactions (16–84% positive); –, 0–15% positive.

[b]Data as given in Lim et al. (2005a).

logenetically closest taxa and that the methodology applied to obtain these phenotypic data may be different, resulting in less reliable comparisons.

Taxonomic comments

The genus *Pontibacillus* now encompasses two species, *Pontibacillus chunghwensis* and *Pontibacillus marinus*, that were both iso-lated from the same niche (solar saltern) and are most probably not reflecting the real ecological, biochemical, and physiological variation of the genus.

List of species of the genus *Pontibacillus*

1. **Pontibacillus chungwhensis** Lim, Jeon, Song and Kim 2005b, 169[VP]

chung.when'sis. N.L. masc. adj. *chungwhensis* belonging to Chungwha, where the organism was isolated.

Cells are µ0.6–0.9 µm × 2.3–3.0 µm. The colonies on MA are yellow, low convex, smooth, and circular (slightly irregular). Temperature for growth ranges from 15 °C to 45 °C with an optimum from 35 °C to 40 °C in the pH range of 6.5–9.5 (with optimum pH 7.5–8.5). Optimal growth occurs in media containing 2–5% w/v NaCl and no growth occurs without NaCl or in the presence of more than 15% (w/v) NaCl. Nitrate is not reduced to nitrite. Oxidase-negative. Casein, starch, and Tween 80 are hydrolyzed while esculin, L-tyrosine, hypoxanthine, xanthine, and gelatin are not. Acids are produced from D-glucose, D-ribose, maltose, glycerol, and D-trehalose, but not from D-xylose, L-arabinose, L-rhamnose, α-D-lactose, adonitol, D-raffinose, D-mannitol, D-fructose, arbutin, D-salicin, D-melibiose, or D-mannose. Cellular fatty acid profiles on MA and MA plus 3% (w/v) NaCl can be found from the species differentiation table (Table 34).

DNA G + C content (mol%): 41 (HPLC).

Type strain: BH030062, DSM 16287, KCTC 3890.

GenBank accession number (16S rRNA gene): AY553296.

2. **Pontibacillus marinus** Lim, Jeon, Park, Kim, Yoon and Kim 2005a, 1030[VP]

ma'rin.us. L. masc. adj. *marinus* of the sea.

Rods of 0.4–0.9 µm × 3.3–4.0 µm. Strictly aerobic. Catalase- and oxidase-positive. Nitrate is reduced to nitrite. Creamy colonies on MA, flat, smooth, and circular to slightly irregular. Growth occurs at 15–40 °C with an optimum at 30 °C, pH 6.0–9.0 with an optimum pH of 7.0–7.5 and 1–9% (w/v) NaCl with an optimum of 2–5%. Tween 80 and esculin are hydrolyzed. Hydrolysis of casein, starch, gelatin, L-tyrosine, hypoxanthine, xanthine, and urea is not observed. Production of acid is observed from sucrose, D-melibiose, D-trehalose, D-raffinose, D-fructose, D-ribose, and maltose, but not from D-glucose, glycerol, D-xylose, L-arabinose, L-rhamnose, α-D-lactose, adonitol, D-mannitol, inositol, or D-mannose. Major cellular fatty acids on MA are $C_{15:0\ iso}$, $C_{15:0\ ante}$, and $C_{16:1\ \omega7c}$ alcohol. Further information is given in the species differentiation table (Table 34).

DNA G + C content (mol%): 42.0 (HPLC).

Type strain: BH030004, DSM 16465, KCTC 3917.

GenBank accession number (16S rRNA gene): AY603977.

Genus XVI. **Saccharococcus** Nystrand 1984a, 503[VP] (Effective publication: Nystrand 1984b, 217.)

EDITORIAL BOARD

Sac.cha.ro.coc'cus. Gr. n. *sakchâr* sugar; L. n. *coccus* a grain, berry; N.L. masc. n. *Saccharococcus* the sugar coccus, a coccus isolated from beet sugar extraction.

Spherical cells of 1–2 µm in diameter occurring in irregular clusters. Nonmotile, nonspore-forming. Gram-positive; the cell wall contains peptidoglycan with mesodiaminopimelic acid as diamino acid. Theichoic acids are not present. The cells are lysed by 100 µg/ml of egg white lysozyme. Aerobic and heterotrophic. **Catalase and oxidase are produced. Thermophilic** with an optimal range of 68–70 °C and a maximum range of 75–78 °C. Acid, but no gas is produced from most mono- and disaccharides. L(+)Lactic acid is the main metabolite from the carbohydrate degradation.

DNA G+C content (mol%): 48.

Type species: **Saccharococcus thermophilus** Nystrand, 1984a, 503[VP] (Effective publication: Nystrand, 1984b, 217.).

Further descriptive information

Various strains of cocci have been isolated from five different sugar beet extraction plants in Sweden (Nystrand, 1984b). At that time, bacterial contamination pointed to members of the genus *Bacillus*, mainly *Bacillus stearothermophilus* (now *Geobacillus thermophilus*, see *Geobacillus*, above). The cell morphology and the lack of endospore formation favored the new generic description. Biochemically, the isolates were characterized on the basis of the API panels that were available at that time. Complete details are given in Nystrand (1984b). Strains were resistant to polymyxin B and sensitive to chloramphenicol, tetracycline, erythromycin, neomycin, penicillin-G, streptomycin, and bacitracin. The strains are also sensitive to a number of disinfectants but relatively tolerant to formaldehyde. This observation has practical implications concerning disinfection of the sugar production plants where the bacteria were at the basis of important loss of sucrose. The contamination source can be found in the contaminating soil. At least some strains contain plasmids that are quite similar in length to the plasmids of *Geobacillus stearothermophilus*.

Enrichment and isolation procedures

The members of *Saccharococcus thermophilus* could only be isolated from fresh samples immediately after their removal from the sugar beet extraction process. Preservation occurs by freeze-drying in a medium of 10% skim milk and 2% sucrose. Strains were isolated on Tryptone Glucose Extract Agar with an addition of 5 g/l sucrose after incubation at 70 °C for 16–24 h.

The strains require regular (daily) transfer to fresh medium to ensure their viability.

Differentiation of the genus *Saccharococcus* from closely related genera

Sequence analysis of the 16S rRNA gene revealed that *Saccharococcus thermophilus* falls in the immediate vicinity of members of the *Geobacillus* species with sequence similarity values of around 97% (Fortina et al., 2001a). These authors did not reclassify *Saccharococcus thermophilus* as a member of *Geobacillus* most probably because of its morphological and biochemical differences from the geobacilli. Phylogenetically, *Saccharococcus* is a member of *Geobacillus* which was already discovered by Rainey and Stackebrandt (1993) and is supported by the presence of the unusual tertiary polyamine, N4 (aminopropyl) spermidine (Hamana et al., 1993). Polyamines are regarded as evolutionary taxonomic markers and most often support phylogenetic relationships demonstrated via 16S rRNA gene sequence analysis.

Taxonomic comments

The genus *Saccharococcus* belongs in the same phylogenetic group as the genus *Geobacillus*. Reclassification at the generic level would have important nomenclatural implications because of the priority rule. Indeed, the genus *Geobacillus* should be renamed "*Saccharococcus*" which would mean that "coccus" as trivial characteristic would be linked to "bacillus"-like morphology as is the case for the present members of *Geobacillus*.

List of species of the genus *Saccharococcus*

1. **Saccharococcus thermophilus** Nystrand, 1984a, 503[VP] (Effective publication: Nystrand, 1984b, 217.)

ther.mo′phi.lus. Gr. n. *therme* heat; Gr. adj. *philos* loving; N.L. adj. *thermophilus* heat-loving.

Characteristics are the same as that for the genus. Biochemical characteristics are mentioned in detail by Nystrand (1984b). API systems reveal that a large variety of sugars (mono-, di-, and trisaccharides) can be used. Further positive reactions are reported for the sugar alcohols mannitol and sorbitol, *N*-acetylglucosamine, and glycerol. The presence of chymotrypsin, esterase, esterase lipase, α-galactosidase, α-glucosidase, various phosphatases, and leucine arylamidase are reported via the API ZYM; tetrathionate reductase and lipase are reported via API 50E and gelatinase via the API 20E methodologies. Methyl red-positive and Voges–Proskauer-negative. Phenotypic differentiation between *Saccharococcus thermophilus* and most closely related *Geobacillus* members is given in Table 35.

Source: Swedish beet sugar extraction at 70 °C where the redox potential is more than −300 mV.

DNA G + C content (mol%): 47.8 (T_m).

Type strain: R Nystrandt 657, ATCC 43125, CCM 3586, DSM 4749.

GenBank accession number (16S rRNA gene): L09227, X70430.

Genus XVII. **Tenuibacillus** Ren and Zhou 2005b, 98[VP]

PAUL DE VOS

Te.nu.i.ba.cil′lus. L. adj. *tenuis* slender, thin, slim; L. n. *bacillus* small rod; N.L. masc. n. *Tenuibacillus* a slender rod.

Gram-positive, aerobic, organotrophic rods of about 0.3–0.5 µm wide and 2.0–6.0 µm long. Motile by a single polar flagellum. Terminal spherical spores in a swollen sporangium are formed. NaCl is needed for growth. Nitrate is not reduced to nitrite. Positive for catalase and oxidase; negative for phosphoesterase and cellulase. Produces H_2S, but not NH_3. Methyl red and Voges–Proskauer tests are negative. *meso*-Diamin- opimelic acid is the peptidoglycan component. Phylogenetically, *Tenuibacillus* belongs to the *Bacillaceae* and closest 16S rRNA gene sequence similarities are found with members of the genera *Filobacillus* and *Alkalibacillus*. Other moderate halophiles/halotolerants such as *Lentibacillus*, *Gracilibacillus*, and *Virgibacillus* are more distantly related phylogenetically. The major fatty acids are $C_{15:0\ iso}$, $C_{15:0\ ante}$, $C_{17:0\ iso}$, and $C_{16:0\ iso}$.

TABLE 35. Characteristics differentiating *Saccharococcus thermophilus* from the phylogenetically closest *Geobacillus* species[a]

Characteristic	*Saccharococcus thermophilus*[b]	*Geobacillus caldoxylosilyticus*[c]	*Geobacillus stearothermophilus*[d]	*Geobacillus thermodenitrificans*[e]	*Geobacillus thermoglucosidasius*[f]
Cell morphology	Cocci	Rods	Rods	Rods	Rods
Spore formation	−	+	+	+	+
Growth in the presence of 3% NaCl	ND	−	v	+	−
Starch hydrolysis	−	+	+	W	+
Casein hydrolysis	ND	+	v	+	+
Citrate reaction	−	+	−	−	−
Anaerobic growth	−	+	−	+	−
Nitrate reduction	v	+	v	+	+
Denitrification (gas formation)	v	−	−	−	−
Urease	−	v	−	+	−
Utilization of:					
Arabinose	+	+	−	+	−
Galactose	+	+	v	+	−
Lactose	−	+	v	+	−
Rhamnose	+	v	−	−	+
Xylose	+	+	−	+	+

[a]Symbols: +, >85% positive; d, different strains give different reactions (16–84% positive); −, 0–15% positive; v, variable; w, weak reaction; ND, not determined.
[b]Data from Nystrandt (1984b).
[c]Data from Fortina et al. (2001a) and from Ahmad et al. (2000a).
[d]Data from Logan and Berkeley (1984) and from Claus and Berkeley (1986).
[e]Data from Manachini et al. (2000).
[f]Data from Suzuki et al. (1983).

DNA G + C content (mol%): 36.5–37 (HPLC).

Type species: **Tenuibacillus multivorans** Ren and Zhou 2005b, 98[VP].

Further descriptive information and differentiation from other genera

The optimal growth temperature reported is between 36 °C and 41 °C in media that contain optimally 5–8% (w/v) NaCl at pH 7. *Tenuibacillus* representatives were isolated on Halophiles Moderate (HM) agar plates (Ventosa et al., 1982) from a neutral hypersaline lake in China (Ren and Zhou, 2005b). The phenotypic and phylogenetic diversity of the genus is unknown as only two strains of the species, which show 99.8% 16S rRNA gene sequence similarity to each other in the most variable region of the cistron, have been reported and also because the most closely related genus phylogenetically (*Filobacillus*) also comprises only one species. The classic phenotypic characteristics of moderate halophilic sporeformers are generally very similar and, hence, their discrimination is based merely on genetic diversity. Although the isolation source of *Tenuibacillus* (hypersaline lake in China) may at first sight seem to be a relatively isolated ecosystem, it is to be expected that more representatives are present in similar environments. Further differences between *Filobacillus* and *Tenuibacillus* can be found in murein type (L-ornithine for cross-linking in *Filobacillus*), fatty acid composition (the ratio of $C_{15:0\,iso}$ to $C_{15:0\,ante}$ differs in both taxa), oxidase reaction, position of the flagella, and the Gram-staining reaction (for further details, see Ren and Zhou, 2005b).

List of species of the genus *Tenuibacillus*

1. **Tenuibacillus multivorans** Ren and Zhou 2005b, 98[VP]

mul.ti.vor′ans. L. part. adj. *multus* many, numerous; L. v. *voro* to devour, swallow; N.L. part. adj. *multivorans* devouring numerous kinds of substrates.

Gram-positive in fresh cultures and Gram-variable in aged cultures. The description complies with the genus description and is based on two strains. On HM medium, the circular, translucent, convex colonies (about 1–2 mm in diameter after 2 d) become brown from the center outwards upon ageing. NaCl range for growth is 1–20%; no growth is observed at either 37 °C or 20 °C in rich media without NaCl. Temperature range for growth is 21–42 °C (optimum 36–41 °C). pH range for growth is 6.5–9.0 (optimum pH 7.0–8.0). General physiological and biochemical testing according to methods described by Smibert and Krieg (1981) reveals that the saccharides, polysaccharides, and sugar alcohols tested, including arabinose, xylose, D-fructose, glucose, D-mannose, rhamnose, DL-sorbose, cellobiose, D-galactose, lactose, maltose, melibiose, sucrose, trehalose, melezitose, D-raffinose, inulin, dulcitol, erythritol, glycerol, inositol, mannitol, and salicin, are metabolized, but acid is not produced from them. Gelatin, casein, esculin, and Tween 40 and 60 are hydrolyzed, but starch, and Tween 20 and 80 are not.

Source: a hypersaline lake, Ai-Ding Lake, in the Xin-Jiang Uigur autonomous area of North-West China.

DNA G + C content (mol%): 36.5–37 (HPLC).

Type strain: 28-1, AS 1.344, NBRC 100370.

EMBL/GenBank accession number (16S rRNA gene): AY319933 (28-1).

Genus XVIII. **Thalassobacillus** Garcia, Gallego, Ventosa and Mellado 2005, 1793[VP]

ANTONIO VENTOSA, ENCARNACIÓN MELLADO AND PAUL DE VOS

Tha.las′so.ba.cil′lus. Gr. fem. n. *thalassa* sea; L. masc. n. *bacillus* rod; N.L. masc. n. *Thalassobacillus* rod from the sea.

Gram-positive rods forming ellipsoidal endospores exhibiting a central position. Motile. The catalase reaction is positive, whereas oxidase and urease are negative. Nitrate reduction to nitrite is positive. Members are moderately halophilic and do not grow in media without NaCl. Peptidoglycan type A1γ with *meso*-diaminopimelic acid is found as major component of the cell wall. Major fatty acids are $C_{15:0\,ante}$, $C_{16:0\,iso}$, and $C_{15:0\,iso}$. MK-7 is the predominant menaquinone. Based on comparative 16S rRNA gene sequence analysis, belongs phylogenetically to the family *Bacillaceae* in a separate lineage close to the genus *Halobacillus*; other genera that include moderately halophilic species, for example, *Virgibacillus, Lentibacillus, Gracilibacillus*, and *Pontibacillus*, etc., are more distantly related.

DNA G + C content (mol%): 42.4 (T_m, type strain of the type species).

Type species: **Thalassobacillus devorans** Garcia, Gallego, Ventosa and Mellado 2005, 1793[VP].

Further descriptive information

This genus currently contains only *Thalassobacillus devorans*, represented only by the type strain, indicating that the genomic

and phenotypic variability of this taxon is unknown. The strain was obtained after an isolation campaign to search for aromatic-hydrocarbon-degrading halophiles in Southern Spain. It is well known that these components, either natural or andropogenic, are recalcitrant compounds that lead to health and environmental problems. In the bioremediation processes, salt is often a limiting factor (Oren et al., 1992). Other described moderately halophilic degraders of aromatic compounds include members of the genera *Halomonas, Marinobacter,* and *Arhodomonas* (Adkins et al., 1993; Alva and Peyton, 2003; Garcia et al., 2004; Hedlund et al., 2001; Huu et al., 1999; Muñoz et al., 2001; Nicholson and Fathepure, 2004). Isolated after enrichment in a medium containing KOH (0.075 M), KH_2PO_4 (0.1 M), $(NH_4)_2SO_4$ (0.015 M), and 1% (v/v) $MgSO_4/FeSO_4$ solution ($MgSO_4$, 1.6 mM; $FeSO_4$, 39 μM), to which 10% NaCl, 0.005% yeast extract, and 0.05% (w/v) phenol were added (Garcia et al., 2005). Differentiation from other genera is merely on differences in 16S rRNA gene sequences. Furthermore, members of the genus *Halobacillus*, the closest relative phylogenetically, contain Orn-D-Asp in the cell wall, whereas the only *Thalassobacillus* species contains *meso*-diaminopimelic acid.

List of species of the genus *Thalassobacillus*

1. **Thalassobacillus** devorans Garcia, Gallego, Ventosa and Mellado 2005, 1793[VP]

   de.vo′rans. L. v. *devorare* to devour; L. part. adj. *devorans* devouring (organic compounds).

   Cells are rods, 1.0–1.2 μm wide and 2.0–4.0 μm long. Motile by means of flagella. On complex medium with 10% salts, the cream-colored colonies are uniformly round, circular, regular, and convex. Moderately halophilic: grows in media with salt concentrations between 0.5% and 20% (w/v) NaCl, with optimal growth in 7.5–10% (w/v) NaCl. No growth occurs in media without NaCl. No other salt requirements determined. Growth occurs between 15 °C and 45 °C and at pH 6.0–10.0, with optimal growth at 37 °C and pH 7.0. Strictly aerobic. Indole, methyl red, and Voges–Proskauer tests are negative. Gelatin and Tween

80 are hydrolyzed, but starch, casein, and esculin are not. Acid production is observed from D-glucose, trehalose, D-mannose, and D-fructose. The metabolic fingerprint has been determined using the Biolog GP panel. Only a limited number of substrates were respired, including sugars, sugar alcohols, and a limited number of acids. Most compounds of the Biolog GP microplates were not metabolized (for details, see Garcia et al., 2005). Other characteristics are as for the genus. The type strain was isolated from a saline soil in southern Spain.

DNA G + C content (mol%): 42.4 (T_m).

Type strain: G-19.1, DSM 16966, CECT 7046, CCM 7282.

EMBL/GenBank accession number (16S rRNA gene): AJ717299 (G-19.1).

Genus XIX. **Virgibacillus** Heyndrickx, Lebbe, Kersters, De Vos, Forsyth and Logan 1998, 104[VP] emend. Wainø, Tindall, Schumann and Ingvorsen 1999, 830 emend. Heyrman, Logan, Busse, Balcaen, Lebbe, Rodríguez-Díaz, Swings and De Vos 2003b, 510

JEROEN HEYRMAN, PAUL DE VOS AND NIALL LOGAN

Vir.gi.ba.cil′lus. L. n. *virga* a green twig, transf., a branch in a family tree; L. dim. n. *bacillus* a small rod and also a genus of aerobic endospore-forming bacteria; N.L. masc. n. *Virgibacillus* a branch of the genus *Bacillus*.

Motile, Gram-positive rods, 0.3–0.8 μm × 2–8 μm in size. Occur singly, in pairs, chains or, especially in older cultures, filaments. **Bear spherical to ellipsoidal endospores that lie in terminal (sometimes subterminal or paracentral) positions in**

swollen sporangia. Aerobic or weakly facultatively anaerobic. Catalase-positive. Colonies are of small to moderate diameter (0.5–5 mm after 2 d on marine agar or trypticase soy agar), circular and slightly irregular, smooth, glossy or sometimes matt, low convex, and slightly transparent to opaque; usually unpigmented or creamy to yellowish white, but one species produces a pink pigment on marine agar. In the API 20E strip and in conventional tests, the Voges–Proskauer reaction is negative, indole is not produced, nitrate reduction to nitrite is variable. Casein is hydrolyzed by all species, and esculin and gelatin are hydrolyzed, sometimes weakly, by most species. In the API 20E strip, urease and hydrogen sulfide are usually not produced, but a few strains give weak positive reactions for the latter; a few strains also give positive reactions for arginine dihydrolase, citrate utilization, and o-nitrophenyl-β-D-galactoside. **Growth is stimulated by 4–10% NaCl.** Growth may occur between 10 °C and 50 °C, with an optimum of about 28 °C or 37 °C. **For the species examined, raffinose can be used as sole carbon source;** no growth on D-arabinose, D-fructose, or D-xylose. The major fatty acid is $C_{15:0 \, ante}$. The major polar lipids are diphosphatidyl glycerol and phosphatidyl glycerol. Five phospholipids and one polar lipid of unknown structure are present in all species. The presence of phosphatidyl ethanolamine and other lipids varies. The main menaquinone type is MK-7, with minor to trace amounts of MK-6 and MK-8. In the species examined, the cell wall contains peptidoglycan of the *meso*-diaminopimelic acid type (Arahal et al., 1999; Claus and Berkeley, 1986). Inhabitants of soil and other, especially salty, environments; also isolated from food, water, and clinical specimens.

DNA G + C content (mol%): 36–43.

Type species: **Virgibacillus pantothenticus** (Proom and Knight 1950) Heyndrickx, Lebbe, Kersters, De Vos, Forsyth and Logan 1998, 105[VP] emend. Heyndrickx, Lebbe, Kersters, Hoste, De Wachter, De Vos, Forsyth and Logan 1999, 1089 (*Bacillus pantothenticus* Proom and Knight, 1950, 539.).

Further descriptive information

Virgibacillus cells are motile, salt-tolerant, Gram-positive rods, 0.3–0.8 μm × 2–8 μm. They occur singly, in pairs, or short chains or filaments. They bear oval to ellipsoidal endospores that lie in swollen sporangia. On marine agar, colonies are small, circular, low-convex, and slightly transparent to opaque. Although some species, notably *Virgibacillus pantothenticus*, *Virgibacillus proomii* and *Virgibacillus dokdonensis*, will grow on routine cultivation media such as nutrient agar and trypticase soy agar, all species grow on media supplemented with 4–10% NaCl, and all grow well on marine agar. Marine agar and broth may be prepared by making up a solution of an ocean salts mixture (as sold for keeping tropical fish or available from some chemical suppliers) according to the supplier's instructions, and adding nutrients such as peptone and/or yeast extract as required; marine agar and broth are also available commercially from bacteriological media suppliers. Several species will tolerate salt concentrations of 20–25%. Two species, *Virgibacillus marismortui* and *Virgibacillus salexigens*, will not grow in the absence of salt, and *Virgibacillus carmonensis* and *Virgibacillus necropolis* grow poorly, if at all, without salt. *Virgibacillus dokdonensis*, *Virgibacillus koreensis*, *Virgibacillus halodenitrificans*, *Virgibacillus pantothenticus*, and *Virgibacillus proomii* show some growth in anaerobic conditions, but the other species of the genus are strictly aerobic.

The different members of the genus show a wide range of activities in routine phenotypic tests, and perhaps this reflects undiscovered requirements for growth factors and/or special environmental conditions. *Virgibacillus carmonensis* and *Virgibacillus necropolis* give very weak reactions in routine phenotypic tests even when the test media have been supplemented with 7% salt (Heyrman et al., 2003b). Although one strain of *Virgibacillus proomii* was isolated from a clinical specimen and another from processed food, there have been no reports of *Virgibacillus* strains being associated with disease or food spoilage.

The fatty acid profile of *Virgibacillus* strains shows a dominance of branched fatty acids. The major fatty acid is $C_{15:0 \, ante}$, accounting for 58, 52, 43, 34, 68, 55, 38, and 31% of the total fatty acids in *Virgibacillus carmonensis*, *Virgibacillus halodenitrificans*, *Virgibacillus koreensis*, *Virgibacillus marismortui*, *Virgibacillus necropolis*, *Virgibacillus pantothenticus*, *Virgibacillus proomii*, and *Virgibacillus salexigens*, respectively (Lee et al., 2006b). For *Virgibacillus dokdonensis* (Yoon et al., 2005b), the amount of $C_{15:0 \, ante}$ (tested under different culture conditions) was 34%. The importance of using standardized growth conditions for chemotaxonomic characterization has been shown for *Virgibacillus* by Wainø et al. (1999). The major polar lipids are diphosphatidyl glycerol and, in smaller amounts, phosphatidyl glycerol. Five phospholipids and one polar lipid of unknown structure are present in all species. The presence of phosphatidyl ethanolamine varies; it is present in *Virgibacillus dokdonensis*, *Virgibacillus pantothenticus*, and *Virgibacillus marismortui*, and to a lesser extent in *Virgibacillus proomii*, but it is present only at trace levels or is undetectable in the other species (Heyrman et al., 2003b; Yoon et al., 2005b). The presence of other polar lipids also varies between species and, overall, the polar lipid profiles of the genus appear to be species- or strain-specific. The main menaquinone type is MK-7, occurring at levels of 94–99%, with minor amounts of MK-6 and trace to minor amounts of MK-8 (Heyrman et al., 2003b; Lee et al., 2006b; Yoon et al., 2005b; Yoon et al., 2004c); this is similar to the profile seen in most other aerobic, endospore-forming bacteria. The peptidoglycans of *Virgibacillus dokdonensis*, *Virgibacillus halodenitrificans*, *Virgibacillus koreensis*, *Virgibacillus pantothenticus*, and *Virgibacillus marismortui* contain *meso*-diaminopimelic acid (Arahal et al., 1999; Claus and Berkeley, 1986; Lee et al., 2006b; Yoon et al., 2005b; Yoon et al., 2004c).

Virgibacillus strains generally have an aerobic respiratory type of metabolism, though *Virgibacillus dokdonensis*, *Virgibacillus koreensis*, *Virgibacillus pantothenticus*, and *Virgibacillus proomii* can also grow in an anaerobic chamber (Heyndrickx et al., 1999; Lee et al., 2006b; Yoon et al., 2005b). *Virgibacillus halodenitrificans* can grow in the presence of nitrate as alternative e-acceptor, indicating the capacity of anaerobic respiration. Members of the genus are catalase-positive and are usually able to hydrolyze gelatin, esculin, and casein (Heyrman et al., 2003b; Lee et al., 2006b; Yoon et al., 2005b). *Virgibacillus pantothenticus* has a nutritional requirement for pantothenic acid, thiamine, and biotin (Proom and Knight, 1950). The type strain of *Virgibacillus halodenitrificans* requires organic (reduced) sulfur (Denariaz et al., 1989). Other than the features given in species descriptions accompanying the proposals of new taxa, *Virgibacillus* strains have rarely been tested for their physiological properties. Kuhlmann and Bremer (2002) studied various aerobic endospore-formers for the synthesis of compatible solutes (organic osmolytes that organisms may produce and accumulate in very high concentrations in order to adjust to high-osmolality environments). Endogenous synthesis of ectoine and glutamate was observed in the type strains of both *Virgibacillus pantothenticus* and

Virgibacillus salexigens; the former also produced proline and the latter hydroxyectoine. Of the thirteen members of *Bacillaceae* tested, these strains were the only ones producing three compatible solutes, and *Virgibacillus salexigens* was the only one that produced hydroxyectoine.

Virgibacillus occurs in various salty and non-salty habitats. *Virgibacillus pantothenticus* was originally isolated from soils in southern England (Proom and Knight, 1950) and has also been found in hypersaline soil and salterns (Garabito et al., 1998), antacids (Claus and Berkeley, 1986), canned chicken (Heyndrickx et al., 1999), and the alimentary tracts of worker honey bees (Gilliam and Valentine, 1976); *Virgibacillus proomii* has been isolated from soil, infant bile, and a water supply (Heyndrickx et al., 1999); *Virgibacillus salexigens* was isolated from Spanish salterns and hyper saline soils (Garabito et al., 1997); *Virgibacillus halodenitrificans* was isolated from a French solar evaporation pond (Denariaz et al., 1989); *Virgibacillus marismortui* was found in water samples that had been taken from the Dead Sea by B.E. Volcani in 1936 and held as sealed enrichment cultures for 57 years (Arahal et al., 1999); *Virgibacillus necropolis* and *Virgibacillus carmonensis* were isolated from samples of biofilms taken from damaged mural paintings in the Roman Servilia tomb in the necropolis of Carmona, Spain (Heyrman et al., 2003b). *Virgibacillus koreensis* was isolated from a salt field near Taean-Gun on the Yellow Sea in Korea (Lee et al., 2006b), while *Virgibacillus dokdonensis* has been isolated from sea water collected at Dokdo, a Korean island in the East Sea (Yoon et al., 2005b).

The international databases EMBL and GenBank contain 16S rDNA sequences of all members of *Virgibacillus* as it is currently described. In addition, partial sequences of the *recA* and *splB* (spore photoproduct lyase) genes are available for *Virgibacillus salexigens* and *Virgibacillus marismortui*. These genes were sequenced by Maughan et al. (2002), to comment on the ancestry between "*Bacillus permians*" and *Virgibacillus marismortui* (see *Taxonomic comments*). For *Virgibacillus pantothenticus*, a sequence of the betaine/carnitine/choline transporter (*ectT*) gene involved in osmoregulation is available. In addition, *ectA*, *ectB* (diaminobutyric acid aminotransferase), and *ectC* (ectoine synthase) are also available for *Virgibacillus pantothenticus* and *Virgibacillus salexigens*.

Enrichment and isolation procedures

No specific selective isolation procedures for *Virgibacillus* species have been described, but representatives of the genus might be selected for, together with other halotolerant/halophilic bacteria, by using media with 5–10% NaCl added. Inoculation of such media could be preceded by a heating step (e.g., 5–10 min at 80 °C) in order to kill vegetative cells and so select for endospore-formers.

Virgibacillus pantothenticus and *Virgibacillus proomii* strains were originally isolated on routine media, such as nutrient agar (Heyndrickx et al., 1999; Proom and Knight, 1950), but in the case of the former, isolation followed enrichment in 4% NaCl broth. All other species were originally isolated on media containing 7–10% NaCl (Arahal et al., 1999; Garabito et al., 1997; Heyrman et al., 2003b; Lee et al., 2006b) or on Marine Agar (Yoon et al., 2005b). *Virgibacillus* strains all grow well on marine agar. Optimal growth temperatures are between 25 °C and 30 °C for *Virgibacillus carmonensis*, *Virgibacillus koreensis*, and *Virgibacillus necropolis*, and 37 °C for the other species.

Virgibacillus pantothenticus was originally isolated from samples of soil collected in southern England by adding the soil to 10 ml broth (presumably nutrient broth) that had been supplemented with 4% NaCl to give a volume of 15 ml, followed by shaking to mix and incubation at 28 °C or 37 °C for 2 d. A loopful of enrichment was then plated on nutrient agar and incubated at 37 °C (Knight and Proom, 1950; Proom and Knight, 1950).

The description of *Bacillus halodenitrificans* is based upon study of one of four strains isolated from a solar evaporation pond (Denariaz et al., 1989) by injecting samples into Hungate tubes containing a yeast extract medium (Na_2HPO_4, 3.8 g, or NaH_2PO_4, 1.43 g, or KH_2PO_4, 1.3 g; $(NH_4)_2SO_4$, 1 g; yeast extract, 1 g; $Mg(NO_3)_2 \cdot 6H_2O$, 1 g) supplemented with 1.06 M $NaNO_3$ and either 1% sodium acetate, 1% tryptone, or 10 ml sodium lactate per 1000 ml, and incubating for 1–2 weeks at 37 °C. The cultures produced N_2O and were inoculated into fresh tubes of medium, incubated for 1 d, and streaked onto the acetate-supplemented medium containing 1.7 M NaCl and 1.5% agar for aerobic incubation at 37 °C.

Arahal et al. (1999) isolated *Bacillus marismortui* from water samples that had been taken from the surface level of the Northern basin of the Dead Sea near the mouth of the River Jordan by B.E. Volcani in 1936; 1% peptone had been added and the closed 500 ml bottles had been stored in the dark at 18–20 °C and left unopened for 57 years. The three isolation media used were formulated as follows: SW-10 (Nieto et al., 1989) medium contained yeast extract, 5 g; NaCl, 81 g; $MgSO_4$, 9.6 g; $MgCl_2$, 7 g; KCl, 2 g; $CaCl_2$, 0.36 g; $NaHCO_3$, 0.06; NaBr, 0.026 g; water 1000 ml; pH 7.2. HM medium (Ventosa et al., 1982) contained: yeast extract, 10 g; proteose peptone, 5 g; glucose, 1 g; NaCl, 178.0 g; $MgSO_4 \cdot 7H_2O$, 1.0 g; $CaCl_2 \cdot 2H_2O$, 0.36 g; KCl, 2.0 g; $NaHCO_3$, 0.06 g; NaBr, 0.23 g; $FeCl_3 \cdot 6H_2O$, trace; water 1000 ml; pH 7.2 adjusted with KOH. M4 medium (Arahal et al., 1999) contained NaCl, 103 g; $MgSO_4 \cdot 7H_2O$, 18 g; $CaCl_2 \cdot 2H_2O$, 0.25 g; KCl, 0.19 g; peptone, 0.5 g; water 1000 ml; pH 7.2. Cultures were incubated at 37 °C in an orbital shaker.

Virgibacillus salexigens was isolated from ponds of solar salterns and from hyper saline soils by direct inoculation, or following dilution of samples in sterile salt solution, onto plates of HM medium (described above for *Virgibacillus marismortui*) solidified with 20 g/l agar; plates were incubated at 37 °C (Garabito et al., 1997; Ventosa et al., 1983).

Virgibacillus carmonensis and *Virgibacillus necropolis* were isolated from scrapings of biofilms on ancient mural paintings by homogenizing the samples in physiological water, preparing a dilution series, and then plating on five different media: (i) trypticase soy broth (BBL) solidified with Bacto agar (Difco); (ii) the aforementioned trypticase soy agar supplemented with 10% NaCl; (iii) R2A agar (Difco); (iv) R2A agar supplemented with 10% NaCl; (v) starch-casein medium, which contained starch, 10 g; casein, 0.3 g; KNO_3, 2 g; K_2HPO_4, 2 g; NaCl, 2 g; $MgSO_4 \cdot 7H_2O$, 0.05 g; $CaCO_3$, 0.02 g; $FeSO_4 \cdot 7H_2O$, 0.01 g; agar, 20 g; water 1,000 ml. All media were supplemented with 0.03% cycloheximide to inhibit fungal growth. Plates were incubated aerobically at 28 °C for three weeks (Heyrman et al., 1999).

Virgibacillus koreensis (Lee et al., 2006b) was isolated on nutrient agar containing artificial sea water including 7% NaCl and was further subcultured on marine agar (Difco). *Virgibacillus dokdonensis* was isolated from sea water on marine agar.

Maintenance procedures

Virgibacillus strains can easily be preserved on slopes of a suitable growth medium that encourages sporulation; slopes should be checked microscopically for spores before sealing and storage in a refrigerator. For long term preservation, lyophilization and liquid nitrogen may be used as long as a cryoprotectant is added.

Procedures for testing special characters

Reliable and relatively rapid identification at the genus level is possible by comparative 16S rDNA sequencing since sequences from reference strains are available for comparison in international databases. For more rapid identification, the sequencing can be restricted to the first ±500 bp since this part contains four hypervariable regions (of 20–30 bp). The remaining ± 1,000 bp sequence only contains two hypervariable regions. Determination of the nearly complete 16S rDNA sequence may give an idea about the species rank of a novel isolate, however, DNA–DNA hybridization experiments should be performed to verify this assumption, as demanded by the current bacterial species concept (Stackebrandt et al., 2002).

Identification is also possible using polyacrylamide gel electrophoresis of cellular proteins, amplified rDNA restriction analysis (Heyndrickx et al., 1999), and fatty acid analysis (Heyndrickx et al., 1999; Heyrman et al., 2003b; Lee et al., 2006b). However, these methods depend on the availability of reference profiles and have not been tested for all species of the genus under the same culture conditions.

Identification solely on the basis of phenotypic data are not straightforward because other taxa of aerobic endospore-forming bacteria share many phenotypic properties with *Virgibacillus*. Also, as already noted, *Virgibacillus carmonensis* and *Virgibacillus necropolis* give very weak reactions in routine phenotypic tests, even when the test media have been supplemented with 7% salt. It is useful to supplement all test media in this way (excepting, of course, salt tolerance test media), and to follow the test protocols recommended for *Bacillus* species; suspension media for miniaturized tests in the API system (bioMérieux, France) may also be supplemented with NaCl. However, it must be appreciated that mere growth in these media does not signify the presence of *Virgibacillus* as many other taxa will grow in these conditions, and that data obtained with such media are not strictly comparable with data from test media not supplemented with NaCl. Phenotypic characteristics differentiating between the currently described *Virgibacillus* species are given in Table 36. Additional data are given in Table 37.

Taxonomic comments

Knight and Proom (1950) attempted to identify 296 *Bacillus* isolates from soil and investigated the nutritional requirements of 220 strains. They found that 11 of their isolates had nutritional requirements different from those of other *Bacillus* species. Their initial growth was stimulated by the addition of 4% NaCl, and (for most strains) pantothenic acid "satisfied a nutritional requirement". They accordingly proposed the new species *Bacillus pantothenticus* (Proom and Knight, 1950).

Garabito et al. (1997) isolated six moderately halophilic strains from salterns and hypersaline soils and proposed them as the novel species *Bacillus salexigens*. In a dendrogram based on 16S rDNA sequences, *Bacillus salexigens* formed a phylogenetic branch with *Halobacillus halophilus*, *Halobacillus litoralis*, *Bacil-*

lus dipsosauri (now *Gracilibacillus dipsosauri*; Wainø et al., 1999), and *Bacillus pantothenticus*. Within this branch, *Bacillus salexigens* showed highest sequence similarities to *Halobacillus litoralis* and *Bacillus pantothenticus* (95.8 and 95.5%, respectively). However, the mol% G + C of *Bacillus salexigens* (39.5 mol%) and *Halobacillus halophilus* (51.1 mol%) differed. *Bacillus salexigens* was also found to differ from all the members of *Halobacillus* in the composition of its cell-wall peptidoglycan which is the *meso*-diaminopimelic acid type in *Bacillus salexigens* and the Orn-D-Asp type in *Halobacillus* (Spring et al., 1996). Garabito et al. (1997) therefore allocated their novel species to *Bacillus* rather than to *Halobacillus*. They further observed that *Bacillus salexigens* was most closely related to *Bacillus pantothenticus*, a species that was allocated to phylogenetic group 1 (*Bacillus sensu stricto*) by Ash et al. (1991) and they suggested that both species might represent a phylogenetic group in the genus *Bacillus sensu lato* that is distinct from the six phylogenetic groups recognized at that time. Heyndrickx et al. (1998) performed a polyphasic study on 12 strains received as *Bacillus pantothenticus* using amplified 16S rDNA restriction analysis, fatty acid methyl ester analysis, SDS-PAGE of whole-cell proteins, and routine diagnostic characters based on the API system and morphological observations. Since *Bacillus pantothenticus* was known to lie at the border of rRNA group 1 of Ash et al. (1991), Heyndrickx et al. (1998) compared their data with those obtained for representative *Bacillus* species belonging to both rRNA groups 1 and 2, other related genera, and *Bacillus dipsosauri*. Unfortunately, a species only very recently published at that time, *Bacillus salexigens* (Garabito et al., 1997) was not studied by Heyndrickx et al. From the phylogenetic point of view, *Amphibacillus*, *Halobacillus*, and *Marinococcus* were found to be the most closely related genera. However, the *Bacillus pantothenticus* strains differed from these genera in various respects including their ability to grow on media without salt, their pH-ranges (*Amphibacillus* contains obligate alkalophiles that lack isoprenoid quinones; Niimura et al., 1990), cell-wall type (*Halobacillus* has a different murein type, as mentioned above), and cell morphology (*Marinococcus* has a coccoid cell morphology and does not form endospores; Hao et al., 1984). Based on these findings it was proposed to accommodate the *Bacillus pantothenticus* strains in a new genus, *Virgibacillus* (Heyndrickx et al., 1998), and nine of the strains investigated were renamed *Virgibacillus pantothenticus*; additional strains were subsequently proposed as the new species *Virgibacillus proomii* (Heyndrickx et al., 1999)

Earlier, Arahal et al. (1999) investigated a group of 91 isolates from the Dead Sea. On the basis of 16S rDNA sequence data, this group of isolates appeared to be most closely related to *Virgibacillus pantothenticus* and *Bacillus salexigens* (with 16S rDNA sequence similarities of 97.8 and 96.6%, respectively). The authors argued that there were sufficient differentiating characteristics between their isolates and *Virgibacillus pantothenticus* to warrant the proposal of a separate species, namely their isolates' inabilities to grow without NaCl or to grow under anaerobic conditions and their higher G + C content of 39.0–42.8 mol% compared with 36.9–38.3 for *Virgibacillus pantothenticus*. The Dead Sea isolates were therefore proposed as the new *Bacillus* species, *Bacillus marismortui*. In the same volume of the *International Journal of Systematic Bacteriology*, Wainø et al. (1999) had proposed the transfer of *Bacillus salexigens* to the novel genus *Salibacillus* and differentiated this genus from *Virgibacillus* by the characteristics described above (Garabito et al., 1997) and by differences in polar lipid and fatty acid patterns.

TABLE 36. Characteristics differentiating the species of the genus *Virgibacillus*[a,b]

Characteristic	1. V. pantothenticus	2. V. carmonensis	3. V. dokdonensis	4. V. halodenitrificans	5. V. koreensis	6. V. marismortui	7. V. necropolis	8. V. proomii	9. V. salexigens
Pink-pigmented colonies	−	+	−	−	−	−	−	−	−
Spore position:									
Central	−	−	−	−	−	−	+	−	+
Subterminal	(+)	+	+	+	−	+	+	(+)	+
Terminal	+	−	+	+	+	+	+	+	+
Spore shape[c]	ES	E(S)	ES	E	E	E	E	ES	E
Anaerobic growth	+	−	+	+	+	−	−	+	−
Acid from:									
N-Acetylglucosamine	+		+	−	+	(w)	+	−	w
Amygdalin	+	−	−	+	−	−	−	−	w
D-Arabinose	+	−	−	−	−	−	−	−	
D-Fructose	+[d]	−	+	+	+	+	(+)	+	−
L-Fucose	d	−	−	−	−	−	−	−	+
Galactose	−	−	+	+	−	−	−	+	−[d]
D-Glucose	+[d]	−	+	+	w	+[d]	(w)	+	+
Glycerol	+	−	w	−	+	(w)	−	+[d]	
Glycogen	−	−	−	−	−	−	d	−	
meso-Inositol	−	−	+	−	−	−	−	−	
D-Mannitol	−	−	−	d	−	−	−	+	−
D-Mannose	+[d]	−	+	+	−	+	−	−	+[d]
D-Melibiose	−	−	+	−	−	+	(w)	+	+
L-Rhamnose	+	−	−	−	−	−	−	(−)	−
D-Trehalose	+	−	−	+	w	−	(w)	+	−
D-Turanose	+	−	−	−	−	−	−	+	−
Hydrolysis of:									
Casein	+	+	+	+	+	+	+	+	
Esculin	+	(w)	+	−	+	+	+	+	
Gelatin	+	−	+	+	−	+	(w)	+	+
H₂S production	(−)	−	−	−	−	CR	−	+	+
Nitrate reduction	d	+	−	+	−	+	+	−	+
NaCl required for growth	−	+	−	d	−	+	−	−	+
Growth in 25% NaCl	−	−	−	d	−	−	−	−	w
Growth at:									
5 °C	−	−	−	−	−	−	−	−	−
10 °C	−	+	−	+	+	−	−	−	−
15 °C	+	+	+	+	+	−	+	−	−
40 °C	+	+	+	+	+	+	+	+	+
45 °C	+	−	+	+	+	+	+	+	+
50 °C	+	−	+	−	+	+	−	+	−

[a]Symbols: +, >85% positive; (+), 75–84% positive; d, variable (26–74% positive); (−), 16–25% positive; −, 0–15% positive; w/+, weak to moderately positive reaction; w, weak reaction; (w), very weak reaction.

[b]Data from Arahal et al. (1999); Garabito et al. (1997); Heyndrickx et al. (1998, 1999); Heyrman et al. (2003b); Proom and Knight (1950), Yoon et al. (2005b, 2004c), and Lee et al. (2006b).

[c]Abbreviations: E, ellipsoidal; S, spherical; CR, conflicting results between the table and the species description in Arahal et al. (1999).

[d]For the type strain the opposite reaction was reported by Heyrman et al. (2003b).

Vreeland et al. (2000) claimed to have isolated a spore-former from within a 250 million-year-old primary salt crystal, a strain that is closely related to *Bacillus marismortui* and *Virgibacillus pantothenticus* according to 16S rDNA sequence analysis, and the authors submitted their sequence data to the GenBank database under the name "*Bacillus permians*" (accession no. AF166093). However, this isolate showed 99% 16S rDNA sequence similarity to *Bacillus marismortui*, representing a divergence smaller than that seen between two strains of *Bacillus marismortui*. On this account, because of its very low rate of nucleotide substitution (perhaps four orders of magnitude lower than the typical rate for prokaryotes), and because it does not occupy an ancestral position in a tree based upon 16S rRNA sequence comparisons, its antiquity has been questioned

TABLE 37. Additional characteristics differentiating the species of *Virgibacillus*[a,b]

Characteristic	1. *V. pantothenticus*	2. *V. carmonensis*	3. *V. dokdonensis*	4. *V. halodenitrificans*	5. *V. koreensis*	6. *V. marismortui*	7. *V. necropolis*	8. *V. proomii*	9. *V. salexigens*
Arginine dihydrolase	−	−	−	−	−	−	−	d	−
β-Galactosidase	d	−		d	+		−	−	−
Lysine decarboxylase	−	−	−	−			−	−	−
Ornithine decarboxylase	−	−	−	−			−	−	−
Oxidase			+	+	+	+			+
Acid from:									
Adonitol	−	−		−			−	−	
D-Arabitol	−	−					−	−	
L-Arabitol	−	−					−	+	
Arbutin	+	−					−	+	
Cellobiose	d	−	+	−	+		−	d	
Dulcitol	−	−					−	−	
Erythritol	−	−					−	−	
D-Fucose		−					−	−	
β-Gentiobiose	d	−					−	−	
Gluconate	d	−					−	d	
Inulin	−	−					−	−	
2-Ketogluconate	−	−					−	−	
Lactose	d	−	+	d		−	−	−	d
Lyxose	−	−					−		
Maltose	+	−	+	+	+	+	−	+	d
Melezitose	−	−	−	−			−	−	
Methyl α-D-mannoside	d	−					−	−	
Methyl β-D-xyloside	−	−					−	−	
Raffinose	−	−	−	−			−	−	
Ribose	+	−	+	+			(w)	+	
Sorbitol	d	−	+	−			−	−	
Sorbose	−	−					−	−	
Sucrose	+	−	+	+		−	−	d	
D-Tagatose	+	−					(w)	+	
Xylitol	−	−					−	−	
L-Xylose	−	−					−	−	
Utilization of:									
Acetate			+	+					d
N-Acetyl-D-glucosamine								+	
Alanine				+					
Arabinose		−		−					
Arabitol				−					d
Arginine				+					
Asparagine				+					
Aspartate				−					
Cellobiose	w	w	+	−			+	w	+
Citrate	d	−	+	−		−		d	−
Cysteine				−					
Ethanol				−					
Formate			−	+					
Fructose		−		+		+			+
Fumarate				+					
Galactose			−	+					
Gluconate				+					
Glucose	w	−		+			+	w	+
Glutamate			−	+					

TABLE 37. (continued)

Characteristic	1. *V. pantothenticus*	2. *V. carmonensis*	3. *V. dokdonensis*	4. *V. halodenitrificans*	5. *V. koreensis*	6. *V. marismortui*	7. *V. necropolis*	8. *V. proomii*	9. *V. salexigens*
Glutamine				+					
Glycerol				−					
Glycine				+					+
Inulin				+					
Isoleucine				+		+			
Lactate				+					
Lactose	−	−		+			w	w	+
Leucine				−					d
Lysine				+					
Malate			−	+					
Maltose			+	+					d
Mannitol				+		+			+
Mannose			+	+		+			d
Melibiose	+	+		−					
Methanol				−			+	+	
Methionine				−					
Phenylalanine				+					
Proline				−					
Propionate				+					
Pyruvate			+			+			
Raffinose	+	+				+			
Rhamnose						+	+	+	
Ribose						+			d
Salicin			+	+					+
Serine				−					+
Sorbitol				+					
Succinate			−	+		+			−
Sucrose	−	+		+			+	+	d
Trehalose	−	+		+			+	+	d
Turanose				+					
Valine				+					
D-Xylose			−						d
Hydrolysis of:									
DNA				−		+			+
Tween 80	+		+			−			
Urea	−	−	−	−		+		−	−

[a]Symbols: +, >85% positive; (+), 75–84% positive; d, variable (26–74% positive); (−), 16–25% positive; −, 0–15% positive; w/+, weak to moderately positive reaction; w, weak reaction; (w), very weak reaction; blanks indicate no data available.

[b]Data from Arahal et al. (1999); Garabito et al. (1997); Heyndrickx et al. (1998, 1999); Heyrman et al. (2003b); Proom and Knight (1950), Yoon et al. (2005b, 2004c), Lee et al. (2006b).

(Graur and Pupko, 2001; Nickle et al., 2002). Maughan et al. (2002) sequenced a part of the *recA* and *splB* genes of "*Bacillus permians*" and its closest relatives to further investigate their ancestry. Although these authors were cautious in making conclusions, the results further question the authenticity of the finding of Vreeland et al. (2000).

After the creation of the genus *Salibacillus*, Arahal et al. (2000) transferred *Bacillus marismortui* to *Salibacillus* and differentiated this genus from *Virgibacillus* (*Virgibacillus pantothenticus* and *Virgibacillus proomii*) on the basis of spore shape, the inability of *Salibacillus* species to grow under anaerobic conditions or to grow without NaCl, their production of H_2S, their failure to produce acid from D-galactose and D-trehalose, and their failure to hydrolyze starch and Tween 80 (although no data for these last two characters were available for *Virgibacillus proomii*). They further claimed distinction on the basis of the optimal growth of *Salibacillus* strains in 10% NaCl (although this was never tested for *Virgibacillus pantothenticus* and *Virgibacillus proomii*), and the differing G + C contents of the type strains (although there was overlap in these when the G + C data for other strains were also taken into account).

Yoon et al. (2002) described the novel genus *Lentibacillus* on the basis of a single isolate which was proposed as the novel species *Lentibacillus salicampi*. This organism clustered close to *Salibacillus* and *Virgibacillus* in a phylogenetic tree based on 16S rDNA sequence comparisons, but it showed slower growth and a higher proportion of the fatty acid $C_{16:0 \text{ iso}}$ in comparison with *Virgibacillus* and *Salibacillus*, and therefore the authors considered that it warranted separate genus status. Lu et al. (2002, 2001) described another halophilic endospore-forming genus, *Oceanobacillus*, based on a single strain isolated from the deep-sea. Since at the time of description *Virgibacillus* and *Salibacillus* were still separate genera, Lu et al. (2001) concluded that their isolate could not be attributed to one or the other and proposed a novel genus, despite the limited distinguishing characteristics between their isolate and the members of both existing genera.

The properties distinguishing *Virgibacillus* and *Salibacillus* were reinvestigated by Heyrman et al. (2003b) when they characterized three potential new species isolated from walls and mural paintings. In a phylogenetic tree based on 16S rDNA sequences, the novel isolates formed a monophyletic branch with the members of *Virgibacillus* and *Salibacillus* and were positioned approximately equidistantly from these two genera. The mural isolates and the type strains of *Virgibacillus* and *Salibacillus* species were compared in their growths at different salt concentrations, and it was found that all strains grew optimally at salt concentrations of 5–10%. They also investigated the polar lipid patterns of all type strains, since Wainø et al. (1999) distinguished *Virgibacillus pantothenticus* and *Salibacillus salexigens* on the basis that only the pattern of the former species contains phosphatidylethanolamine, an aminophospholipid, and a glycolipid. The polar lipid pattern

of *Salibacillus marismortui* also showed a moderate amount phosphatidylethanolamine, and the other polar lipid patterns did not allow a satisfactory distinction between members of *Virgibacillus* and *Salibacillus*. Heyrman et al. (2003b) concluded that the remaining characteristics for differentiating the genera were not convincing enough to maintain their separation, and they proposed the transfer of the two species of *Salibacillus* to *Virgibacillus*. They also discovered three novel *Virgibacillus* species among their isolates from walls and mural paintings: *Virgibacillus carmonensis*, *Virgibacillus necropolis*, and *Virgibacillus picturae*. Unfortunately, the authors did not include the single strain genera *Lentibacillus* and *Oceanobacillus*, which had been described only recently. *Virgibacillus* thus contained seven species but, on the basis of 16S rDNA sequence comparisons, as already noted, *Bacillus halodenitrificans* also fell within the genus, and Yoon et al. (2004c) have proposed its transfer to this genus. Lee et al. (2006b) isolated a novel *Virgibacillus* strain from a Korean salt field and described it as *Virgibacillus koreensis*. The genus *Oceanobacillus* had since then expanded by one species, *Oceanobacillus oncorhynchi* (Yumoto et al., 2005b), and *Lentibacillus* by three species *Lentibacillus juripiscarius* (Namwong et al., 2005), *Lentibacillus salarius* (Jeon et al., 2005a), and *Lentibacillus lacisalsi* (Lim et al., 2005c). In a phylogenetic tree including members of *Virgibacillus*, *Oceanobacillus*, and *Lentibacillus* (lacking *Lentibacillus lacisalsi*), Lee et al. (2006b) observed that *Virgibacillus picturae* clustered closer to *Oceanobacillus* and transferred the species as *Oceanobacillus picturae*. In the same year, Yoon et al. (2005b) described *Virgibacillus dokdonensis*.

As discussed above, the majority of species belonging to *Virgibacillus*, *Oceanobacillus*, and *Lentibacillus* were described in the last 5 years. In addition, major changes such as the transfer of *Salibacillus* to *Virgibacillus* and the descriptions of *Oceanbacillus* and *Lentibacillus* were all made in a short time interval; as a result, the different studies were not able to compare results. With every addition of a novel species in the genera *Oceanobacillus*, *Virgibacillus*, and *Lentibacillus*, the phylogenetic relationships between these genera changed and the phenotypic differences originally differentiating them disappeared. Therefore, future species descriptions in the neighborhood of these genera can be expected. This will probably allow better assessment of whether the three genera should remain as currently described or whether further rearrangements are necessary. Two novel additions to the genus *Virgibacillus* are already listed as in press, *Virgibacillus halophilus* and *Virgibacillus olivae*.

Further reading

Arahal, D.R. and A. Ventosa, 2002. Moderately halophilic and halotolerant species of *Bacillus* and related genera. *In* Berkeley, Heyndrickx, Logan and De Vos (Editors), Applications and Systematics of *Bacillus* and Relatives, Blackwell Science, Oxford, pp. 83–99.

List of species of the genus *Virgibacillus*

1. **Virgibacillus pantothenticus** (Proom and Knight 1950) Heyndrickx, Lebbe, Kersters, De Vos, Forsyth and Logan 1998, 105[VP] emend. Heyndrickx, Lebbe, Kersters, Hoste, De Wachter, De Vos, Forsyth and Logan 1999, 1089 (*Bacillus pan-*

tothenticus Proom and Knight, 1950, 539.)

pan.to.then'tic.us. N.L. *acidum pantothenticum* pantothenic acid; N.L. adj. *pantothenticus* relating to pantothenic acid.

Cells are motile, Gram-positive, usually long, rods (0.5–0.7 × 2–8 μm) that sometimes (especially in older cultures) form chains and/or filaments. They bear spherical to ellipsoidal endospores which lie in terminal, sometimes subterminal, positions in swollen sporangia. After 2 d on trypticase soy agar, colonies are 1–4 mm in diameter, low convex, circular and slightly irregular, butyrous (sometimes slightly tenacious when cells form filaments), creamy-gray, and almost opaque with an eggshell or glossy appearance; the appearance under 10× magnification is reminiscent of white soap flakes in a grayish matrix. After 4 d, colonies smell of ammonia and are 5–10 mm in diameter with lobed and/or fimbriate margins. Organisms are facultatively anaerobic. They have a nutritional requirement for pantothenic acid, thiamine, biotin, and amino acids. Hydrogen sulfide is usually not produced, but a few strains give weak positive reactions in the API 20E strip; a few strains also give positive reactions for arginine dihydrolase, citrate utilization, and *o*-nitrophenyl-β-D-galactoside in the API 20E strip. Nitrate reduction to nitrite is variable. Esculin and casein are hydrolyzed; gelatin is usually hydrolyzed. Growth is stimulated by 4% NaCl and not inhibited by 10% NaCl. Growth may occur between 15 °C and 50 °C with an optimum of about 37 °C. Acid without gas is produced from the following carbohydrates: *N*-acetylglucosamine, D-arabinose, arbutin, D-fructose, galactose, D-glucose, glycerol, maltose, D-mannose, α-methyl-D-glucoside, rhamnose, ribose, salicin, starch, sucrose, D-tagatose, trehalose, and D-turanose; acid production from the following carbohydrates is variable: amygdalin, cellobiose, L-fucose, β-gentiobiose, gluconate, lactose, α-methyl-D-mannoside, and sorbitol. The major cellular fatty acids (>10%) are $C_{15:0\ ante}$ and $C_{17:0\ ante}$.

Source: soils, hyper saline soil and salterns, antacids, canned chicken, and the alimentary tracts of honey bees.

DNA G + C content (mol%): 36.9 (T_m).

Type strain: B0018, ATCC 14576, CCUG 7424, CFBP 4270, CIP 51.24, DSM 26, HAMBI 476, JCM 20334, LMG 7129, NCIMB 8775, NCDO 1765, NCTC 8162, NRRL NRS-1321, VKM B-507.

GenBank accession number (16S rRNA gene): X60627.

2. **Virgibacillus carmonensis** Heyrman, Logan, Busse, Balcaen, Lebbe, Rodríguez-Díaz, Swings and De Vos 2003b, 507[VP]

car.mo.nen′sis. N.L. adj. *carmonensis* of Carmona referring to the mural paintings of the necropolis at Carmona, Spain, from where the strains were isolated.

Cells are motile, Gram-positive rods (0.5–0.7 × 2–7 μm) that mostly occur singly, sometimes in pairs and short chains. They bear ellipsoidal, sometimes nearly spherical, endospores that lie in subterminal positions in swollen sporangia. After 24 h on marine agar, colonies are 0.5–1.0 mm in diameter, low convex, circular with slightly irregular margins, smooth, and transparent with the larger colonies having a pink tint. After 2 d the colonies turn bright pink and opaque. Organisms do not grow in an anaerobic chamber at 37 °C. The temperature range for growth is 10–40 °C with optimal growth 25–30 °C. No growth without added salt and optimal growth at NaCl concentrations of 5 and 10%. In the API 20E kit, strains give positive results for nitrate reduction, and negative results for *o*-nitrophenyl-β-D-galactosidase, arginine dihydrolase, lysine decarboxylase, ornithine decarboxylase, citrate utilization, hydrogen sulfide production, urease, tryptophan deaminase, and gelatinase. Casein hydrolysis positive. No hemolysis on 5% horse blood agar and no growth on this medium when supplemented with 7% NaCl. With the exception of very weak reactions in esculin hydrolysis and acid production from 5-keto-D-gluconate, strains are unreactive in the API 50CHB gallery even when the CHB suspension medium is supplemented with 7% NaCl. In marine broth, in which the peptone component is replaced by the tested sugar 1% (w/v), growth is observed for the following sugars: cellobiose (weak growth), D-melibiose, raffinose, sucrose, and D-trehalose. No growth on D-arabinose, D-fructose, D-glucose, DL-lactose, and D-xylose. The major cellular fatty acids (>10%) are $C_{15:0\ ante}$ and $C_{17:0\ ante}$.

Source: samples of biofilms taken from damaged ancient mural paintings.

DNA G + C content (mol%): 38.9 (HPLC).

Type strain: DSM 14868, LMG 20964.

GenBank accession number (16S rRNA gene): AJ316302.

3. **Virgibacillus dokdonensis** Yoon, Kang, Lee, Lee and Oh 2005b, 1836[VP]

dok.do.nen′sis. N.L. masc. adj. *dokdonensis* of Dokdo, a Korean island.

Description is based on a single strain. Cells are Gram-variable rods, 0.6–0.8 × 2.5–5.0 μm, motile by peritrichous flagella. They bear ellipsoidal or spherical endospores that lie in terminal or subterminal positions in swollen sporangia. After 2 d on marine agar (MA) supplemented with 3% (w/v) NaCl, colonies are 3–5 mm in diameter, flat, irregular, translucent and milky white in color. Growth occurs under anaerobic conditions on MA and MA supplemented with nitrate. Growth may occur between 15 °C and 50 °C with an optimum around 37 °C. Optimal growth in the presence of 4–5% (w/v) NaCl. Growth occurs without NaCl and in the presence of 23% (w/v) NaCl, but not in the presence of >24%. In the API 20E kit, the strain gives negative results for nitrate reduction, hydrogen sulfide production, urease, arginine dihydrolase, lysine decarboxylase, and ornithine decarboxylase. Oxidase-positive. Hydrolysis of esculin, casein, gelatin, starch, Tween 20, 40, 60, and 80. Hypoxanthine, xanthines and tyrosine are not hydrolyzed. When assayed with the API ZYM system, alkaline phosphatase, esterase (C4), esterase lipase (C8), α-chymotrypsin, naphthol-AS-BI-phosphohydrolase, and α-glucosidase are present. Acid is produced from D-cellobiose, D-galactose, D-fructose, *myo*-inositol, lactose, maltose, D-mannose, D-ribose, sucrose, and D-sorbitol. The following substrates are utilized: acetate, D-cellobiose, citrate, D-glucose, D-fructose, D-mannose, maltose, pyruvate, salicin, and sucrose. The major fatty acids (>10%) are $C_{15:0\ ante}$, $C_{15:0\ iso}$, $C_{17:0\ ante,\ and}$ $C_{16:0\ iso}$.

Source: sea water collected at the island Dokdo (Korea).

DNA G + C content (mol%): 36.9 (HPLC).

Type strain: DSW-10, DSM 16826, CIP 109001, KCTC 3933.

GenBank accession number (16S rRNA gene): AY822043.

4. **Virgibacillus halodenitrificans** (Denariaz, Payne and Le Gall 1989) Yoon, Oh and Park 2004c, 2166^VP (*Bacillus halodenitrificans* Denariaz, Payne and Le Gall 1989, 150.)

ha.lo.de.ni.tri′fi.cans. Gr n. *hals* salt, the sea; N.L. v. *denitrifico* to denitrify; N.L. part. adj. *halodenitrificans* salt-requiring denitrifier.

Description is based upon two strains. Moderately halophilic, denitrifying, Gram-variable, rods 0.6–0.8 × 2.5–4.0 μm, occurring singly or in short chains and forming flexuous elongated (15 μm) cells in anaerobic culture. Motile by means of single or peritrichous flagella; motility in young cultures shows frequent spinning and twisting; only a few flagella are observed, in lateral and polar positions. Forms terminal or subterminal ellipsoidal spores that swell the sporangia. Anaerobic growth on marine agar only occurs when nitrate is present; nitrate is reduced to nitrite (Yoon et al., 2004c); Denariaz et al. (1989) reported a reduction to N_2O, but not to N_2. Colonies on marine agar are circular to slightly irregular, raised, translucent, and cream-colored, with diameters of μ 2–3 mm, after 3 d incubation at 37 °C. Grows at temperatures ranging from 10 °C to 45 °C, with an optimum of 35–40 °C. The pH range for growth is 5.8–9.6 with an optimum of around 7.4. Grows between 2% and 23% NaCl (2–25% for the type strain), optimum 3–7% NaCl; the strain described by Yoon et al. (2004c) grew in the presence of 0.5% NaCl. Ammonia and glutamine are utilized as nitrogen sources; arginine, nitrate, and urea are not. Uses a range of amino acids, carbohydrates, and organic acids as carbon and energy sources. Oxidase-positive. Hydrolyzes casein and gelatin, but not esculin, starch, Tween 80, or urea. Arginine dihydrolase, lysine and ornithine decarboxylase-negative. The type strain contains large amounts of type *b* and *c* cytochromes which confer a pink-orange color to concentrated cell suspensions. The major fatty acids (>10%) are $C_{15:0\ ante}$, $C17:0_{ante}$, and $C_{16:0\ iso}$. For further information on the type strain, the reader should consult also Denariaz et al. (1989).

Source: a saltern in France.

DNA G + C content (mol%): 38 (chemical determination and HPLC).

Type strain: ATCC 49067, DSM 10037, JCM 12304, KCTC 3790, LMG 9818.

GenBank accession number (16S rRNA gene): AB021186, AY543169.

5. **Virgibacillus koreensis** Lee, Lim, Lee, Lee, Park and Kim 2006b, 254^VP

ko.re.en′sis. N.L. masc. adj. *koreensis* for Korea, where the type strain was isolated.

Description based on a single strain. Facultatively anaerobic, Gram-positive rods, 0.5–0.7 × 2.0–7.0 μm, motile by peritrichous flagella. Ellipsoidal spores are produced in terminal positions. On marine agar, colonies are circular, low-convex, smooth, semi-translucent and cream-colored. Growth occurs optimally at a NaCl concentration of 5–10% (w/v) and no growth occurs at concentrations of more than 20%. Growth occurs in the temperature range of 10–45 °C, with an optimum at 25 °C. The strain tests positive for β-galactosidase and negative for nitrate reduction, indole production, arginine hydrolysis, hydrogen sulfide production, and urease. Hydrolyzes esculin but not gelatin. Oxidase-positive. Acid is produced from esculin, amygdalin, L-arabinose, cellobiose, D-fructose, D-glucose (weak), maltose, D-trehalose (weak) and D-xylose. The major fatty acids (>10%) are $C_{15:0\ ante}$ and $C_{16:0\ iso}$.

Source: salt field near Taean-Gun on the Yellow Sea in Korea.

DNA G + C content (mol%): 41 (HPLC).

Type strain: BH30097, CIP 109159, JCM 12387, KCTC 3823.

GenBank accession number (16S rRNA gene): AY616012.

6. **Virgibacillus marismortui** (Arahal, Márquez, Volcani, Schleifer and Ventosa 1999) Heyrman, Logan, Busse, Balcaen, Lebbe, Rodríguez-Díaz, Swings and De Vos 2003b, 510^VP (*Bacillus marismortui* Arahal, Márquez, Volcani, Schleifer and Ventosa 1999, 529; *Salibacillus marismortui* Arahal, Márquez, Volcani, Schleifer and Ventosa 2000, 1503.)

ma.ris.mor′tu.i. L. gen. n. *maris* of the sea; L. adj. *mortuus* dead; N.L. gen. n. *marismortui* of the Dead Sea.

Strictly aerobic, motile, Gram-positive rods, 0.5–0.7 × 2.0–3.6 μm, occurring singly, in pairs or in short chains. Oval endospores are produced in terminal or subterminal positions in swollen sporangia. Colonies are cream-colored, circular, entire, and opaque. Growth occurs in 5–25% (w/v) total salts with optimal growth at 10% (w/v) salts. No growth occurs in the absence of salts. Growth occurs in the temperature range 15–50 °C, with an optimum temperature of 37 °C, and in the pH range 6.0–9.0 with an optimum of pH 7.5. Oxidase-positive. Acid is produced from D-fructose, D-glucose, glycerol, and maltose, but not from D-arabinose, D-galactose, lactose, D-mannitol, sucrose, D-trehalose, or D-xylose. Casein, DNA, and gelatin are hydrolyzed. Starch and Tween 80 are not hydrolyzed. Urease-positive. Methyl red test is positive. Nitrate is reduced to nitrite, but nitrite is not reduced. Arginine dihydrolase, Simmons citrate, H_2S production and phenylalanine deaminase tests are negative. D-Fructose, inulin, maltose, D-mannitol, D-mannose, pyruvate, D-raffinose, D-rhamnose, and succinate are utilized as sole carbon and energy sources. Susceptible to chloramphenicol, erythromycin, penicillin G, streptomycin, and tetracycline; resistant to nalidixic acid, neomycin, novobiocin, and rifampin. The major fatty acids (>10%) are $C_{15:0\ ante}$ and $C_{15:0\ iso}$.

Source: stored enrichment cultures of water taken from the Dead Sea.

DNA G + C content (mol%): 40.7 (T_m).

Type strain: 123, ATCC 700626, CECT 5066, CIP 105609, DSM 12325, LMG 18992.

GenBank accession number (16S rRNA gene): AJ009793.

7. **Virgibacillus necropolis** Heyrman, Logan, Busse, Balcaen, Lebbe, Rodríguez-Díaz, Swings and De Vos 2003b, 507^VP

ne.cro.po′lis. N.L. adj. *necropolis* of the necropolis, referring to the mural paintings of the necropolis of Carmona, Spain, where the strain was isolated.

Description based on a single strain. Strictly aerobic, motile, Gram-positive rods 0.5–0.7 × 2–5 µm, and coccoid rods, which occur singly, and in pairs or short chains; the cells in the chains tend to lie at angles to each other. Ellipsoidal endospores lie in terminal or subterminal positions in the rods, or centrally in coccoid cells, and they swell the sporangia. After 24 h on marine agar, the colonies are 0.2–0.5 mm in diameter, low convex, circular with entire margins, and smooth, cream-colored, and slightly transparent (but becoming opaque after 2 d growth). The temperature range for growth is 10–40 °C with optimal growth at 25–35 °C. Growth without added salt is weak, and optimal growth occurs at NaCl-concentrations of 5 and 10%. In the API 20E kit, it gives a positive result for nitrate reduction, a very weak reaction for gelatinase, and negative results for *o*-nitrophenyl-β-D-galactosidase, arginine dihydrolase, lysine decarboxylase, ornithine decarboxylase, citrate utilization, hydrogen sulfide production, urease, and tryptophan deaminase. Casein hydrolysis is positive. Partial hemolysis on 5% horse blood agar, but no growth on this medium when supplemented with 7% NaCl. Generally unreactive in the API 50CHB gallery, using CHB suspension medium supplemented with 7% NaCl, but very weak reactions which did not qualify as positive results are seen in the following tests: glycerol, ribose, D-glucose, D-fructose, D-mannose, *N*-acetylglucosamine, D-trehalose, D-tagatose, and 5-keto-D-gluconate. In marine broth, in which the peptone component is replaced by the tested sugar at 1% (w/v), growth is observed for the following sugars: cellobiose, D-glucose, DL-lactose (weak growth), D-melibiose, raffinose, sucrose, and D-trehalose. No growth on D-arabinose, D-fructose, and D-xylose. The major fatty acids (>10%) are $C_{15:0\ ante}$ and $C_{17:0\ ante}$.

Source: samples of biofilms taken from damaged ancient mural paintings.

DNA G + C content (mol%): 37.3 (HPLC).

Type strain: DSM 14866, LMG 19488.

GenBank accession number (16S rRNA gene): AJ315056.

8. **Virgibacillus proomii** Heyndrickx, Lebbe, Kersters, Hoste, De Wachter, De Vos, Forsyth and Logan 1999, 1087[VP]

proom.i.i. N.L. gen. n. *proomii* of Proom, referring to Harold Proom, the person who, with B.C.J.G. Knight, first isolated a member of this species and who described *Bacillus pantothenticus.*

Motile, Gram-positive, usually long, rods 0.5–0.7 × 2–8 µm which sometimes, especially in older cultures, form chains and/or filaments. They bear spherical to ellipsoidal endospores which lie in terminal, sometimes subterminal, positions in swollen sporangia. After 2 d on trypticase soy agar, colonies are 1–4 mm in diameter, low convex, circular, and slightly irregular, butyrous (sometimes slightly tenacious when cells form filaments), creamy-gray and almost opaque with an eggshell or matt appearance; the appearance under 10× magnification

is reminiscent of white soap flakes in a grayish matrix. After 4 d, colonies smell of ammonia and are 5–10 mm in diameter with lobed and/or fimbriate margins. Organisms are weakly facultatively anaerobic. A few strains may give positive reactions for arginine dihydrolase and citrate utilization in the API 20E strip. Hydrogen sulfide is not produced. Nitrate is not reduced to nitrite. Hydrolysis of esculin and casein are positive; hydrolysis of gelatin usually positive. Growth may occur between 15 °C and 50 °C with an optimum of about 37 °C. Acid without gas is produced from the following carbohydrates: *N*-acetylglucosamine, arbutin, D-fructose, galactose, D-glucose, inositol, maltose, D-mannose, ribose, salicin, D-tagatose, and trehalose; acid production from the following carbohydrates is variable: amygdalin, cellobiose, and gluconate (these three usually negative), glycogen, α-methyl-D-glucoside, rhamnose, starch, and sucrose. The major fatty acids (>10%) are $C_{15:0\ ante}$, $C_{17:0\ ante}$, and $C_{15:0\ iso}$.

Source: soil, water, and a clinical specimen of infant bile.

DNA G + C content (mol%): 37 (HPLC).

Type strain: BO413, F 2737/77, CIP 106304, DSM 13055, LMG 12370.

GenBank accession number (16S rRNA gene): AJ012667.

9. **Virgibacillus salexigens** (Garabito, Arahal, Mellado, Márquez and Ventosa 1997) Heyrman, Logan, Busse, Balcaen, Lebbe, Rodríguez-Díaz, Swings and De Vos 2003b, 509[VP] (*Bacillus salexigens* Garabito, Arahal, Mellado, Márquez and Ventosa 1997, 739; *Salibacillus salexigens* Wainø, Tindall, Schumann and Ingvorsen 1999, 830).

sal.ex′i.gens. L. n. *sal* salt; L. v. *exigo* to demand; N.L. part. adj. *salexigens* salt-demanding.

Strictly aerobic, motile, Gram-positive rods, 0.3–0.6 × 1.5–3.5 µm, occurring singly, in pairs, or in short chains. Oval endospores are produced in central or subterminal positions in swollen sporangia. Colonies are unpigmented, smooth, circular, convex, and entire. Growth occurs in the presence of 7–20% (w/v) total salts with optimal growth at 10% (w/v) salts. According to the original description, no growth occurred with 25% NaCl, but Heyrman et al. (2003b) reported weak growth with this salt concentration. No growth occurs in the absence of salts. Growth occurs in the temperature range 15–45 °C, with an optimum temperature of 37 °C, and in the pH range 6.0–11.0 with an optimum of pH 7.5. Oxidase-positive. Acid is produced from D-fructose, D-glucose, glycerol, maltose, D-mannitol, and mannose, but not from dulcitol, D-galactose, lactose, D-melibiose, L-rhamnose, D-trehalose, or D-xylose. Esculin, casein, DNA, and gelatin are hydrolyzed. Starch, Tween 80, and tyrosine are not hydrolyzed. H_2S is produced. Arginine dihydrolase, lysine and ornithine decarboxylase, phenylalanine deaminase, and Simmons citrate test negative. *N*-acetyl-D-glucosamine, *N*-acetyl-glutamate, adenosine, L-alanyl-glycine, cellobiose, 2′-deoxyadenosine, dextrin, D-fructose, D-glucose, glycerol, glycyl-L-glutamate, inosine, L-lactate, maltose, maltotriose, D-mannose, D-melibiose, pyruvate, D-psicose, D-raffinose, D-ribose, salicin, sucrose, thymidine, uridine, and D-xylose are utilized as sole carbon and energy sources. *p*-Hydroxyphenylacetic acid and succinic acid are not utilized as sole carbon and energy sources. Susceptible to ampicillin, carbenicillin, cephalotin, chloramphenicol, erythromycin, novobiocin, and penicillin G; resistant to nalidixic acid and polymyxin. The major fatty acids (>10%) are $C_{15:0\ iso}$ and $C_{15:0\ ante}$. Habitats are solar salterns and saline soils.

DNA G + C content (mol%): 39.5 (T_m).

Type strain: C-20Mo, ATCC 700290, CCM 4646, CCUG 52350, CIP 105608, DSM 11483, LMG 21520.

GenBank accession number (16S rRNA gene): Y11603.

References

Abd El-Rahman, H.A., D. Fritze, C. Spröer and D. Claus. 2002. Two novel psychrotolerant species, *Bacillus psychrotolerans* sp. nov. and *Bacillus psychrodurans* sp. nov., which contain ornithine in their cell walls. Int. J. Syst. Evol. Microbiol. *52*: 2127–2133.

Achouak, W., P. Normand and T. Heulin. 1999. Comparative phylogeny of *rrs* and *nifH* genes in the *Bacillaceae*. Int. J. Syst. Bacteriol. *49*: 961–967.

Ackermann, H.W., R.R. Azizbekyan, R.L. Bernier, H. de Barjac, S. Saindoux, J.R. Valero and M.X. Yu. 1995. Phage typing of *Bacillus subtilis* and *B. thuringiensis*. Res. Microbiol. *146*: 643–657.

Adams, J.C. and J.L. Stokes. 1968. Vitamin requirements of psychrophilic species of *Bacillus*. J. Bacteriol. *95*: 239–240.

Adkins, J.P., L.A. Cornell and R.A. Tanner. 1992. Microbial composition of carbonate petroleum reservoir fluids. Geomicrobiol. J. *10*: 87–97.

Adkins, J.P., M.T. Madigan, L. Mandelco, C.R. Woese and R.S. Tanner. 1993. *Arhodomonas aquaeolei* gen. nov., sp. nov., an aerobic, halophilic bacterium isolated from a subterranean brine. Int. J. Syst. Bacteriol. *43*: 514–520.

Agata, N., M. Mori, M. Ohta, S. Suwan, I. Ohtani and M. Isobe. 1994. A novel dodecadepsipeptide, cereulide, isolated from *Bacillus cereus* causes vacuole formation in HEp-2 cells. FEMS Microbiol. Lett. *121*: 31–34.

Agata, N., M. Ohta, M. Mori and M. Isobe. 1995. A novel dodecadepsipeptide, cereulide, is an emetic toxin of *Bacillus cereus*. FEMS Microbiol. Lett. *129*: 17–20.

Agata, N., M. Ohta, M. Mori and K. Shibayama. 1999. Growth conditions of and emetic toxin production by *Bacillus cereus* in a defined medium with amino acids. Microbiol. Immunol. *43*: 15–18.

Agerholm, J.S., N.E. Jensen, V. Dantzer, H.E. Jensen and F.M. Aarestrup. 1997. Experimental infection of pregnant cows with *Bacillus licheniformis* bacteria. Vet. Pathol. *36*: 191–201.

Agnew, M.D., S.F. Koval and K.F. Jarrell. 1995. Isolation and characterization of novel alkaliphiles from bauxite-processing waste and description of *Bacillus vedderi* sp. nov., a new obligate alkaliphile. Syst. Appl. Microbiol. *18*: 221–230.

Agnew, M.D., S.F. Koval and K.F. Jarrell. 1996. *In* Validation of the publication of new names and new combinations previously effectively published outside the IJSB. List no. 56. Int. J. Syst. Bacteriol<qu ref=73>.

Ahmad, S., R.K. Scopes, G.N. Rees and B.K. Patel. 2000a. *Saccharococcus caldoxylosilyticus* sp. nov., an obligately thermophilic, xylose-utilizing, endospore-forming bacterium. Int. J. Syst. Evol. Microbiol. *50*: 517–523.

Ahmad, S., R.K. Scopes, G.N. Rees and B.K.C. Patel. 2000b. *Saccharococcus caldoxylosilyticus* sp. nov., an obligately thermophilic, xylose-utilizing, endospore-forming bacterium. Int. J. Syst. Evol. Microbiol. *50*: 517–523.

Ahmed, H.K., W.J. Mitchell and F.G. Priest. 1995. Regulation of mosquitocidal toxin synthesis in *Bacillus sphaericus*. Appl. Microbiol. Biotechnol. *43*: 310–314.

Ahmed, I., A. Yokota and T. Fujiwara. 2007a. A novel highly boron tolerant bacterium, *Bacillus boroniphilus* sp. nov., isolated from soil, that requires boron for its growth. Extremophiles. *11*: 217–224.

Ahmed, I., A. Yokota and T. Fujiwara. 2007b. *Gracilibacillus boraciitolerans* sp. nov., a highly boron-tolerant and moderately halotolerant bacterium isolated from soil. Int. J. Syst. Evol. Microbiol. *57*: 796–802.

Ahmed, I., A. Yokota, A. Yamazoe and T. Fujiwara. 2007c. Proposal of *Lysinibacillus boronitolerans* gen. nov. sp. nov., and transfer of *Bacillus fusiformis* to *Lysinibacillus fusiformis* comb. nov. and *Bacillus sphaericus* to *Lysinibacillus sphaericus* comb. nov. Int. J. Syst. Evol. Microbiol. *57*: 1117–1125.

Aizawa, S.-I., I.B. Zhulin, L. Márquez-Magaña and G.W. Ordal. 2002. *Bacillus subtilis* and its closest relatives. *In* Sonenshein, Hoch and Losick (Editors), Chemotaxis and motility. ASM Press, Washington D.C., pp. 437–452.

Ajithkumar, V.P., B. Ajithkumar, K. Mori, K. Takamizawa, R. Iriye and S. Tabata. 2001. A novel filamentous *Bacillus* sp., strain NAF001, forming endospores and budding cells. Microbiology. *147*: 1415–1423.

Ajithkumar, V.P., B. Ajithkumar, R. Iriye and T. Sakai. 2002. *Bacillus funiculus* sp. nov., novel filamentous isolates from activated sludge. Int. J. Syst. Evol. Microbiol. *52*: 1141–1144.

Akhurst, R.J., E.W. Lyness, Q.Y. Zhang, D.J. Cooper and D.E. Pinnock. 1997. A 16S rRNA gene oligonucleotide probe for identification of *Bacillus thuringiensis* isolates from sheep fleece. J. Invertebr. Pathol. *69*: 24–30.

Albert, R.A., J. Archambault, M. Lempa, B. Hurst, C. Richardson, S. Gruenloh, M. Duran, H.L. Worliczek, B.E. Huber, R. Rossello-Mora, P. Schumann and H.J. Busse. 2007. Proposal of *Viridibacillus* gen.nov. and reclassification of *Bacillus arvi*, *Bacillus arenosi* and *Bacillus neidei* as *Viridibacillus arvi* gen. nov., comb. nov., *Viridibacillus arenosi* comb. nov. and *Viridibacillus neidei* comb. nov. Int. J. Syst. Evol. Microbiol. *57*: 2729–2739.

Aldagal, M. and Y.C. Fung. 1993. Aeromicrobiology- an assessment of a new meat research complex. J. Environ. Health *56*: 7–14.

Aldridge, C., P.W. Jones, S. Gibson, J. Lanham, M. Meyer, R. Vannest and R. Charles. 1977. Automated microbiological detection/identification system. J. Clin. Microbiol. *6*: 406–413.

Alexander, B. and F.G. Priest. 1989. *Bacillus glucanolyticus*, a new species that degrades a variety of beta-glucans. Int. J. Syst. Bacteriol. *39*: 112–115.

Alibek, K. 1999. Biohazard. Hutchinson, London.

Allan, E.J. 1991. Induction and cultivation of a stable L-form of *Bacillus subtilis*. J. Appl. Bacteriol. *70*: 339–343.

Allan, E.J., F. Amijee, R.H. Tyson, J.A. Strang, C.M. Innes and A.M. Paton. 1993. Growth and physiological characteristics of *Bacillus subtilis* L-forms. J. Appl. Bacteriol. *74*: 588–594.

Allen, M.B. 1953. The thermophilic aerobic sporeforming bacteria. Bacteriol. Rev. *17*: 125–173.

Alva, V.A. and B.M. Peyton. 2003. Phenol and catechol biodegradation by the haloalkaliphile *Halomonas campisalis*: influence of pH and salinity. Environ. Sci. Technol. *37*: 4397–4402.

Alvarez, M., J. Wouters, D. Maes, V. Mainfroid, F. Rentier-Delrue, L. Wyns, E. Depiereux and J.A. Martial. 1999. Lys13 plays a crucial role in the functional adaptation of the thermophilic triose-phosphate isomerase from *Bacillus stearothermophilus* to high temperatures. J. Biol. Chem. *274*: 19181–19187.

Aly, C., M.S. Mulla and B.A. Federici. 1989. Ingestion, dissolution, and proteolysis of the *Bacillus sphaericus* toxin by mosquito larvae. J. Invertebr. Pathol. *53*: 12–20.

Ambroz, A. 1913. *Denitrobacterium thermophilum* spec nova, ein Beitrag zur Biologie der thermophilen Bakterien. Zentbl. Bakteriol. Parasitenkd. Infektionskr. Hyg. Abt. II *37*: 3–16.

Amoozegar, M.A., F. Malekzadeh, K.A. Malik, P. Schumann and C. Sproer. 2003. *Halobacillus karajensis* sp. nov., a novel moderate halophile. Int. J. Syst. Evol. Microbiol. *53*: 1059–1063.

An, S.-Y., K. Kanoh, H. Kasai, K. Goto and A. Yokota. 2007a. *Halobacillus faecis* sp. nov., a spore-forming bacterium isolated from a mangrove area on Ishigaki Island, Japan. Int. J. Syst. Evol. Microbiol. *57*: 2476–2479.

An, S.Y., S. Ishikawa, H. Kasai, K. Goto and A. Yokota. 2007b. *Amphibacillus sediminis* sp. nov., an endospore-forming bacterium isolated from lake sediment in Japan. Int. J. Syst. Evol. Microbiol. *57*: 2489–2492.

Andersch, I., S. Pianka, D. Fritze and D. Claus. 1994. Description of *Bacillus laevolacticus* (ex Nakayama and Yanoshi 1967) sp. nov., nom. rev. Int. J. Syst. Bacteriol. *44*: 659–664.

Andersson, M., M. Laukkanen, E.L. Nurmiaholassila, F.A. Rainey, S.I. Niemela and M. Salkinojasalonen. 1995. *Bacillus thermosphaericus* sp. nov. a new thermophilic ureolytic bacillus isolated from air. Syst. Appl. Microbiol. *18*: 203–220.

Andersson, M., M. Laukkanen, E.-L. Nurmiaho-Lassila, F.A. Rainey, S.I. Niemelä and M. Salkinoja-Salonen. 1996. *B. thermosphaericus* sp. nov. a new thermophilic ureolytic *Bacillus* isolated from air. Syst. Appl. Microbiol. *18*: 203–220.

Andersson, A., B. Svensson, A. Christiansson and U. Ronner. 1999a. Comparison between automatic ribotyping and random amplified polymorphic DNA analysis of *Bacillus cereus* isolates from the dairy industry. Int. J. Food Microbiol. *47*: 147–151.

Andersson, A.M., N. Weiss, F. Rainey and M.S. Salkinoja-Salonen. 1999b. Dust-borne bacteria in animal sheds, schools and children's day care centres. J. Appl. Microbiol. *86*: 622–634.

Angert, E.R., A.E. Brooks and N.R. Pace. 1996. Phylogenetic analysis of *Metabacterium polyspora*: clues to the evolutionary origin of daughter cell production in *Epulopiscium* species, the largest bacteria. J. Bacteriol. *178*: 1451–1456.

Anonymous. 1994. Anthrax control and research, with special reference to national programme development in Africa: memorandum from a WHO meeting. Bull. World Health Organ. *72*: 13–22.

Aono, R. and M. Ohtani. 1990. Loss of alkalophily in cell-wall-component-defective mutants derived from alkalophilic *Bacillus* C-125. Isolation and partial characterization of the mutants. Biochem. J. *266*: 933–936.

Aono, R., M. Ito and K. Horikoshi. 1993. Occurrence of teichuronopeptide in cell walls of group 2 alkaliphilic *Bacillus* spp. J. Gen. Microbiol. *139*: 2739–2744.

Aquino de Muro, M., W.J. Mitchell and F.G. Priest. 1992. Differentiation of mosquito-pathogenic strains of *Bacillus sphaericus* from non-toxic varieties by ribosomal RNA gene restriction patterns. J. Gen. Microbiol. *138*: 1159–1166.

Aragno, M. 1978. Enrichment, isolation and preliminary characterization of a thermophilic, endospore-forming hydrogen bacterium. FEMS Microbiol. Lett. *3*: 13–15.

Arahal, D.R., M.C. Marquez, B.E. Volcani, K.H. Schleifer and A. Ventosa. 1999. *Bacillus marismortui* sp. nov., a new moderately halophilic species from the Dead Sea. Int. J. Syst. Bacteriol. *49*: 521–530.

Arahal, D.R., M.C. Marquez, B.E. Volcani, K.H. Schleifer and A. Ventosa. 2000. Reclassification of *Bacillus marismortui* as *Salibacillus marismortui* comb. nov. Int. J. Syst. Evol. Microbiol. *50*: 1501–1503.

Arai, T. and Y. Mikami. 1972. Chromogenicity of *Streptomyces*. Appl. Microbiol. *23*: 402–406.

Arfman, N., L. Dijkhuizen, G. Kirchhof, W. Ludwig, K.H. Schleifer, E.S. Bulygina, K.M. Chumakov, N.I. Govorukhina, Y.A. Trotsenko, D. White and R.J. Sharp. 1992. *Bacillus methanolicus* sp. nov., a new species of thermotolerant, methanol-utilizing, endospore-forming bacteria. Int. J. Syst. Bacteriol. *42*: 439–445.

Aronson, A.I. 1993. The two faces of *Bacillus thuringiensis*: Insecticidal proteins and post-exponential survival. Mol. Microbiol. *7*: 489–496.

Aronson, A.I. and Y. Shai. 2001. Why *Bacillus thuringiensis* insecticidal toxins are so effective: unique features of their mode of action. FEMS Microbiol. Lett. *195*: 1–8.

Aschiuchi, M., K. Soda and H. Misono. 1999. A poly-gamma-glutamate synthetic system of *Bacillus subtilis* IFO 3336: Gene cloning and biochemical analysis of poly-gamma-glutamate produced by *Escherichia coli* clone cells. Biochem. Biophys. Res. Comm. *263*: 6–12.

Ash, C., J.A.E. Farrow, S. Wallbanks and M.D. Collins. 1991. Phylogenetic heterogeneity of the genus *Bacillus* revealed by comparative analysis of small subunit ribosomal RNA sequences. Lett. Appl. Microbiol. *13*: 202–206.

Ash, C., F.G. Priest and M.D. Collins. 1993. Molecular identification of rRNA group 3 bacilli (Ash, Farrow, Wallbanks and Collins) using a PCR probe test. Proposal for the creation of a new genus *Paenibacillus*. Antonie Leeuwenhoek *64*: 253–260.

Atrih, A. and S.J. Foster. 2001. Analysis of the role of bacterial endospore cortex structure in resistance properties and demonstration of its conservation amongst species. J. Appl. Microbiol. *91*: 364–372.

Aubert, J.P. 1951. Biochemical study of the material output of growth of an aerobic bacterium: *Bacillus megatherium*. Ann. Inst. Pasteur (Paris) *80*: 644–658.

Avakyan, Z.A., T.A. Pivovarova and G.I. Karavaiko. 1986. Properties of a new species, *Bacillus mucilaginosus* (Russian). Mikrobiologiya. *55*: 477–482.

Avakyan, Z.A., T.A. Pivovarova and G.I. Karavaiko. 1998. Validation List no. 66. Int. J. Syst. Bacteriol. *48*: 631–632.

Bacher, G., R. Korner, A. Atrih, S.J. Foster, P. Roepstorff and G. Allmaier. 2001. Negative and positive ion matrix-assisted laser desorption/ionization time-of-flight mass spectrometry and positive ion nano-electrospray ionization quadrupole ion trap mass spectrometry of peptidoglycan fragments isolated from various *Bacillus* species. J. Mass Spectrom. *36*: 124–139.

Bae, S.S., J.H. Lee and S.J. Kim. 2005. *Bacillus alveayuensis* sp. nov., a thermophilic bacterium isolated from deep-sea sediments of the Ayu Trough. Int. J. Syst. Evol. Microbiol. *55*: 1211–1215.

Baillie, A. and P.D. Walker. 1968. Enzymes of thermophilic aerobic sporeforming bacteria. J. Appl. Bacteriol. *31*: 114–119.

Baillie, L.W., M.N. Jones, P.C. Turnbull and R.J. Manchee. 1995. Evaluation of the Biolog system for the identification of *Bacillus anthracis*. Lett. Appl. Microbiol. *20*: 209–211.

Balch, W.E., G.E. Fox, L.J. Magrum, C.R. Woese and R.S. Wolfe. 1979. Methanogens: Reevaluation of a unique biological group. Microbiol. Rev. *43*: 260–296.

Banat, I.M., R. Marchant and T.J. Rahman. 2004. *Geobacillus debilis* sp. nov., a novel obligately thermophilic bacterium isolated from a cool soil environment, and reassignment of *Bacillus pallidus* to *Geobacillus pallidus* comb. nov. Int. J. Syst. Evol. Microbiol. *54*: 2197–2201.

Banerjee, C., C.I. Bustamante, R. Wharton, E. Talley and J.C. Wade. 1988. *Bacillus* infections in patients with cancer. Arch. Intern. Med. *148*: 1769–1774.

Barloy, F., A. Delecluse, L. Nicolas and M.M. Lecadet. 1996. Cloning and expression of the first anaerobic toxin gene from *Clostridium bifermentans* subsp. *malaysia*, encoding a new mosquitocidal protein with homologies to *Bacillus thuringiensis* delta-endotoxins. J. Bacteriol. *178*: 3099–3105.

Barnaby, W. 2002. The Plague Makers: The Secret World of Biological Warfare. Vision, London.

Bartlett, R. and K.A. Bisset. 1981. Isolation of *Bacillus licheniformis* var. *endoparasiticus* from the blood of rheumatoid arthritis patients and normal subjects. J. Med Microbiol. *14*: 97–105.

Batchelor, M.D. 1919. Aerobic spore-bearing bacteria in the intestinal tract of children. J. Bacteriol. 4: 23–24.

Baum, J.A. and T. Malvar. 1995. Regulation of insecticidal crystal protein production in *Bacillus thuringiensis*. Mol. Microbiol. 18: 1–12.

Baumann, P., M.A. Clark, L. Baumann and A.H. Broadwell. 1991. *Bacillus sphaericus* as a mosquito pathogen: properties of the organism and its toxins. Microbiol. Rev. 55: 425–436.

Beaumont, M. 2002. Flavouring composition prepared by fermentation with *Bacillus* spp. Int. J. Food Microbiol. 75: 189–196.

Beecher, D.J. 2001. The *Bacillus* cereus group. *In* Sussman (Editor), Molecular Medical Microbiology. Academic Press, New York, pp. 1161–1190.

Beijerinck, M.W. 1901. Anhäufungsversuche mit Ureumbakterien, Uremspaltung durch Urease und durch Katabolismus. Zentbl. Bakteriol. Parasitenkd. Infektionskr. Hyg. Abt. II 7: 33–61.

Beijerinck, M.W. and D.C.J. Minkman. 1910. Bildung und Verbrauch con Stickstoffoxydul durch Bakterien. Zentbl. Bakteriol. Parasitenkd. Infektionskr. Hyg. Abt. II 25: 30–63.

Belduz, A.O., S. Dulger and Z. Demirbag. 2003. *Anoxybacillus gonensis* sp. nov., a moderately thermophilic, xylose-utilizing, endospore-forming bacterium. Int. J. Syst. Evol. Microbiol. 53: 1315–1320.

Bell, D.M., P.E. Kozarsky and D.S. Stephens. 2002. Clinical issues in the prophylaxis, diagnosis, and treatment of anthrax. Emerging Infect. Dis. 8: 222–225.

Ben-Jacob, E., I. Cohen and D.L. Gutnick. 1998. Cooperative organization of bacterial colonies: from genotype to morphotype. Annu. Rev. Microbiol. 52: 779–806.

Benedek, T. (Editor) 1955. Rheumotoid Arthritis and Psoriasis Vulgaris. Edward Bros. Inc., Ann Arbor, Michigan, p. 8.

Bennett, J.F. and E. Canale-Parola. 1965. The taxonomic status of *Lineola longa*. Arch. Mikrobiol. 52: 197–205.

Bergey, D.H., F.C. Harrison, R.S. Breed, B.W. Hammer and F.M. Huntoon. 1923. Bergey's Manual of Determinative Bacteriology, 1st ed. The Williams & Wilkins Co., Baltimore.

Bergey, D.H., F.C. Harrison, R.S. Breed, B.W. Hammer and F.M. Huntoon. 1925. Bergey's Manual of Determinative Bacteriology, 2nd ed. The Williams & Wilkins Co., Baltimore.

Berkeley, R.C.W., N.A. Logan, L.A. Shute and A.G. Capey. 1984. Identification of *Bacillus* species. *In* Bergan (Editor), Methods in Microbiology, vol. 16. Academic Press, London, pp. 291–328.

Berkeley, R.C.W. and N.A. Logan. 1997. *Bacillus, Alicyclobacillus* and *Paenibacillus*. *In* Emmerson, Hawkey and Gillespie (Editors), Principles and Practice of Clinical Bacteriology. John Wiley, Chichester, pp. 185–204.

Berkeley, R.C.W., M. Heyndrickx, N.A. Logan and P. De Vos. 2002. Applications and Systematics of *Bacillus* and Relatives. Blackwell Science, p. 317.

Berliner, E. 1915. Über die Schlaffsucht der Mehlmottenraupe (Ephestia kühnniella Zell) und ihren Erreger *Bacillus thuringiensis* n. sp. Z. Angew. Entomol. 2: 29–56.

Berner, R., F. Heinen, K. Pelz, V. van Velthoven, M. Sauer and R. Korinthenberg. 1997. Ventricular shunt infection and meningitis due to *Bacillus* cereus. Neuropediatrics 28: 333–334.

Bernhard, K., H. Schrempf and W. Goebel. 1978. Bacteriocin and antibiotic resistance plasmids in *Bacillus cereus* and *Bacillus subtilis*. J. Bacteriol. 133: 897–903.

Bestvater, T. and E.A. Galinski. 2002. Investigation into a stress-inducible promoter region from *Marinococcus halophilus* using green fluorescent protein. Extremophiles 6: 15–20.

Beumer, R.R. and H. Kusumaningrum. 2003. Kitchen hygiene in daily life. Int. Biodeterior. Biodegrad. 51: 299–302.

Birbir, M. and A. Ilgaz. 1996. Isolation and identification of bacteria adversely affecting hide and leather quality. J. Soc. Leather Technol. Chemists 80: 147–153.

Bishop, A.H., J. Johnson and M. Perani. 1999. The safety of *Bacillus thuringiensis* to mammals investigated by oral and subcutaneous dosage. World J. Biotechnol. 15: 375–380.

Bishop, A.H. 2002. *Bacillus thuringiensis* insecticides. *In* Berkeley, Heyndrickx, Logan and De Vos (Editors), Applications and Systematics of *Bacillus* and Relatives. Blackwell Science, Oxford, pp. 160–175.

Bisset, K.A. and R. Bartlett. 1978. The isolation and characters of L-forms and reversions of *Bacillus licheniformis* var. Endoparasiticus (Benedek) associated with the erythrocytes of clinically normal persons. J. Med. Microbiol. 11: 335–349.

Blanc, M., L. Marilley, T. Beffa and M. Arango. 1997. Thermophilic bacterial communities in hot composts as revealed by most probable number counts and molecular (16S rDNA) methods. FEMS Immunol. Med. Microbiol. 28: 141–149.

Blanc, M., L. Marilley, T. Beffa and M. Aragno. 1999. Thermophilic communities in hot composts as revealed by most probable number counts and molecular (16S rDNA) methods. FEMS Microbiol. Ecol. 28: 141–149.

Blowey, R. and P. Edmondson. 1995. Mastitis Control in Dairy Herds. An Illustrated and Practical Guide. Farming Press Books, Ipswich.

Blue, S.R., V.R. Singh and M.A. Saubolle. 1995. *Bacillus licheniformis* bacteremia: Five cases associated with indwelling central venous catheters. Clin. Infect. Dis. 20: 629–633.

Bond, W.W. and M.S. Favero. 1977. *Bacillus xerothermodurans* sp. nov., a species forming endospores extremely resistant to dry heat. Int. J. Syst. Bacteriol. 27: 157–160.

Bonde, G.J. 1975. The genus *Bacillus*. An experiment with cluster analysis. Dan. Med. Bull. 22: 41–61.

Bonjour, F. and M. Aragno. 1984. *Bacillus tusciae*, a new species of thermoacidophilic, facultatively chemolithoautotrophic, hydrogen oxidizing sporeformer from a geothermal area. Arch. Microbiol. 139: 397–401.

Bonjour, F. and M. Aragno. 1985. *In* Validation of the publication of new names and new combinations previously effectively published outside the IJSB. List no. 14. Int. J. Syst. Bacteriol. 35: 223–225.

Bonjour, F., A. Graber and M. Aragno. 1988. Isolation of *Bacillus schlegelii*, a thermophilic, hydrogen oxidizing aerobic autotroph from geothermal and nongeothermal environments. Microb. Ecol. 16: 331–337.

Boone, D.R., Y.T. Liu, Z.J. Zhao, D.L. Balkwill, G.R. Drake, T.O. Stevens and H.C. Aldrich. 1995. *Bacillus infernus* sp. nov., an Fe(III)-reducing and Mn(IV)-reducing anaerobe from the deep terrestrial subsurface. Int. J. Syst. Bacteriol. 45: 441–448.

Borchert, M.S., P. Nielsen, I. Graeber, I. Kaesler, U. Szewzyk, T. Pape, G. Antranikian and T. Schäfer. 2007. *Bacillus plakortidis* sp. nov. and *Bacillus murimartini* sp. nov., novel alkalitolerant members of rRNA group 6. Int. J. Syst. Evol. Microbiol. 57: 2888–2893.

Bornside, G.H. and R.E. Kallio. 1956. Urea-hydrolyzing bacilli. II. Nutritional profiles. J. Bacteriol. 71: 655–660.

Boyer, E.W., M.B. Ingle and G.D. Mercer. 1973. *Bacillus alcalophilus* subsp. *halodurans* subsp. nov., alkaline amylase-producing, alkalophilic organism. Int. J. Syst. Bacteriol. 23: 238–242.

Brachman, P. and A. Kaufmann. 1998. Anthrax. *In* Evans and Brachman (Editors), Bacterial Infections in Humans. Plenum Medical Book Company, New York, pp. 95–107.

Bradley, D.E. and J.G. Franklin. 1958. Electron microscope survey of the surface configuration of spores of the genus *Bacillus*. J. Bacteriol. 76: 618–630.

Bradley, K.A., J. Mogridge, M. Mourez, R.J. Collier and J.A. Young. 2001. Identification of the cellular receptor for anthrax toxin. Nature. 414: 225–229.

Branda, S.S., J.E. Gonzalez-Pastor, S. Ben-Yehuda, R. Losick and R. Kolter. 2001. Fruiting body formation by *Bacillus subtilis*. Proc. Natl. Acad. Sci. USA 98: 11621–11626.

Bravo, A. 1997. Phylogenetic relationships of *Bacillus thuringiensis* delta-endotoxin family proteins and their functional domains. J. Bacteriol. 179: 2793–2801.

Breed, R.S. and R. St. John-Brooks. 1935. Report on proposals submitted by R.E. Buchanan and H.J. Conn relative to the conservation of *Bacillus* as a bacterial generic name, fixing of the type species and of type or standard culture. Zentbl. Bakteriol. Parasitenkd. Infektkrankh. Hyg. Abt. II *92*: 481–490.

Breed, R.S., E.G.D. Murray and A.P. Hitchens. 1948. Bergey's Manual of Determinative Bacteriology, 6th ed. The Williams & Wilkins Co., Baltimore.

Breed, R.S., E.G.D. Murray and N.R. Smith (Editors). 1957. Bergey's Manual of Determinative Bacteriology. The Williams & Wilkins Co., Baltimore.

Briley, R.T., J.H. Teel and J.P. Fowler. 2001. Nontypical *Bacillus* cereus outbreak in a child care center. J. Environ. Health *63*: 9–11.

Brill, J.A. and J. Wiegel. 1997. Differentiation between sporeforming and asporogenous bacteria by a PCR and Southern hybridization based method. J. Microbiol. Methods. *31*: 29–36.

Broadwell, A.H. and P. Baumann. 1987. Proteolysis in the gut of mosquito larvae results in further activation of the *Bacillus* sphaericus toxin. Appl. Environ. Microbiol. *53*: 1333–1337.

Brousseau, R., A. Saint-Onge, G. Prefontaine, L. Masson and J. Cabana. 1993. Arbitrary primer polymerase chain reaction, a powerful method to identify *Bacillus thuringiensis* serovars and strains. Appl. Environ. Microbiol. *59*: 114–119.

Brown, J.R., D. Gentry, J.A. Becker, K. Ingraham, D.J. Holmes and M.J. Stanhope. 2003. Horizontal transfer of drug-resistant aminoacyl-transfer-RNA synthetases of anthrax and Gram-positive pathogens. EMBO Rep. *4*: 692–698.

Buchanan, R.E., J.G. Holt and E.F. Lessel, Jr. 1966. Index Bergeyana. The Williams & Wilkins Co., Baltimore.

Buchner, H. 1882. Über die experimentelle Erzengung des Milzbrandcontagiums aus den Heupilzen. *In* Nägeli (Editor), Untersuchungen über niedere Pilze. R. Olderbourg, München and Leipzeg, pp. 140–177.

Buckley, M. and R.J. Roberts. 2007. Reconciling Microbial systematics & genetics. Colloquium report American Society for Microbiology – March 2007 – http://www.asm.org/Academy/index.asp?bid=49252.

Bulla, L.A., G. St Julian, R.A. Rhodes and C.W. Hesseltine. 1969. Scanning electron and phase-contrast microscopy of bacterial spores. Appl. Microbiol. *18*: 490–495.

Burdon, K.L. 1956. Useful criteria for the identification of *Bacillus anthracis* and related species. J. Bacteriol. *71*: 25–42.

Burke, W.F. and K.O. McDonald. 1983. Naturally occurring antibiotic resistance in *Bacillus sphaericus* and *Bacillus licheniformis*. Curr. Microbiol. *9*: 69–72.

Burkholder, P.R. and N.H. Giles. 1947. Induced biochemical mutations in *Bacillus subtilis*. Am. J. Bot. *34*: 345–348.

Burtt, E.H. and J.M. Ichida. 1999. Occurrence of feather-degrading bacilli in the plumage of birds. Auk. *116*: 364–372.

Bustard, M.T., S. Whiting, D.A. Cowan and P.C. Wright. 2002. Biodegradation of high-concentration isopropanol by a solvent-tolerant thermophile, *Bacillus pallidus*. Extremophiles. *6*: 319–323.

Caccamo, D., C. Gugliandolo, E. Stackebrandt and T.L. Maugeri. 2000. *Bacillus vulcani* sp. nov., a novel thermophilic species isolated from a shallow marine hydrothermal vent. Int. J. Syst. Evol. Microbiol. *50*: 2009–2012.

Caccamo, D., T.L. Maugeri and C. Gugliandolo. 2001. Identification of thermophilic and marine bacilli from shallow thermal vents by restriction analysis of their amplified 16S rDNA. J. Appl. Microbiol. *91*: 520–524.

Calik, P. and T.H. Özdamar. 2001. Carbon sources affect metabolic capacities of *Bacillus* species for the production of industrial enzymes: Theoretical analyses for serine and neutral proteases and alpha-amylase. Biochem. Eng. J. *8*: 61–81.

Cano, R.J. and M.K. Borucki. 1995. Revival and identification of bacterial spores in 25- to 40-million-year-old Dominican amber. Science. *268*: 1060–1064.

Carlson, C.R., D.A. Caugant and A.B. Kolsto. 1994. Genotypic diversity among *Bacillus cereus* and *Bacillus thuringiensis* strains. Appl. Environ. Microbiol. *60*: 1719–1725.

Carmichael, A.B. and L.L. Wong. 2001. Protein engineering of *Bacillus megaterium* CYP102. The oxidation of polycyclic aromatic hydrocarbons. Eur. J. Biochem. *268*: 3117–3125.

Carozzi, N.B., V.C. Kramer, G.W. Warren, S. Evola and M.G. Koziel. 1991. Prediction of insecticidal activity of *Bacillus thuringiensis* strains by polymerase chain reaction product profiles. Appl. Environ. Microbiol. *57*: 3057–3061.

Carrasco, I.J., M.C. Márquez, X. Yanfen, Y. Ma, D.A. Cowan, B.E. Jones, W.D. Grant and A. Ventosa. 2006. *Gracilibacillus orientalis* sp. nov., a novel moderately halophilic bacterium isolated from a salt lake in Inner Mongolia, China. Int. J. Syst. Evol. Microbiol. *56*: 599–604.

Carrasco, I.J., M.C. Márquez, Y. Xue, Y. Ma, D.A. Cowan, B.E. Jones, W.D. Grant and A. Ventosa. 2007. *Bacillus chagannorensis* sp. nov., a moderate halophile from a soda lake in Inner Mongolia, China. Int. J. Syst. Evol. Microbiol. *57*: 2084–2088.

Carroll, J., D. Convents, J. Van Damme, A. Boets, J. Van Rie and D.J. Ellar. 1997. Intramolecular proteolytic cleavage of *Bacillus thuringiensis* Cry3A delta-endotoxin may facilitate its coleopteran toxicity. J. Invertebr. Pathol. *70*: 41–49.

Cassity, T.R. and B.J. Kolodziej. 1984. Isolation, partial characterization and utilization of a polysaccharide from *Bacillus megaterium* ATCC 19213. J. Gen. Microbiol. *130*: 535–539.

Castenholz, R.W. 1969. Thermophilic blue-green algae and the thermal environment. Bacteriol. Rev. *33*: 476–504.

CDC. 2000. Human ingestion of *Bacillus anthracis*-contaminated meat–Minnesota, August 2000. JAMA. *284*: 1644–1646.

Chaiyanan, S., T. Maugel, A. Huq, F.R. Robb and R.R. Colwell. 1999. Polyphasic taxonomy of a novel *Halobacillus, Halobacillus thailandensis* sp. nov. isolated from fish sauce. Syst. Appl. Microbiol. *22*: 360–365.

Chanway, C.P. 2002. Plant growth promotion by *Bacillus* and relatives. *In* Berkeley, Heyndrickx, Logan and De Vos (Editors), Applications and Systematics of *Bacillus* and Relatives. Blackwell Science, Oxford, pp. 219–235.

Charles, J.F., C. Nielson-LeRoux and A. Delecluse. 1996. *Bacillus sphaericus* toxins: Molecular biology and mode of action. Annu. Rev. Entomol. *41*: 451–472.

Charlton, S., A.J. Moir, L. Baillie and A. Moir. 1999. Characterization of the exosporium of *Bacillus cereus*. J. Appl. Microbiol. *87*: 241–245.

Charney, J., W.P. Fisher and C.P. Hegarty. 1951. Manganese as an essential element for sporulation in the genus *Bacillus*. J. Bacteriol. *62*: 145–148.

Chen, X.G., O. Stabnikova, J.H. Tay, J.Y. Wang and S.T.L. Tay. 2004. Thermoactive extracellular proteases of *Geobacillus caldoproteolyticus*, sp. nov., from sewage sludge. Extremophiles. *8*: 489–498.

Cherkasskiy, B.L. 1999. A national register of historic and contemporary anthrax foci. J. Appl. Microbiol. *87*: 192–195.

Cheshire, F.R. and W.W. Cheyne. 1885. The pathogenic history and history under cultivation of a new bacillus (*B. alvei*), the cause of a disease of the hive bee hitherto known as foul brood. R. Micros. Soc. J. Ser. II *5*: 581–601.

Chester, F.D. 1898. Report of the mycologist: Bacteriological work. Del. Agric. Exp. Stn. Bull. *10*: 47–137.

Chester, F.D. 1901. A Manual of Determinative Bacteriology. The Macmillan Co., New York.

Chopra, A.K. and D.K. Mathur. 1984. Isolation, screening and characterization of thermophilic *Bacillus* species isolated from dairy products. J. Appl. Bacteriol. *57*: 263–271.

Christensen, H., M. Bisgaard, W. Frederiksen, R. Mutters, P. Kuhnert and J.E. Olsen. 2001. Is characterization of a single isolate sufficient for valid publication of a new genus or species? Proposal to modify Recommendation 30b of the Bacteriological Code (1990 Revision). Int. J. Syst. Evol. Microbiol. *51*: 2221–2225.

Christopher, G.W., T.J. Cieslak, J.A. Pavlin and E.M. Eitzen, Jr. 1997. Biological warfare. A historical perspective. J. Am. Med. Assoc. *278*: 412–417.

Clark, F.E. 1937. The relationship of *Bacillus siamensis* and similar pathogenic spore-forming bacteria to *Bacillus subtilis.* J. Bacteriol. *33*: 435–443.

Claus, D. 1965. Anreicherung und Direktisolierung aerober sporenbildender Bakterien. *In* Schlegel (Editor), Anreicherungskultur und Mutantenauslese. Gustav Fischer Verlag, Stuttgart, pp. 337–362.

Claus, D., F. Fahmy, H.J. Rolf and N. Tosunoglu. 1983. *Sporosarcina halophila* sp. nov., an obligate, slightly halophilic bacterium from salt marsh soils. Syst. Appl. Microbiol. *4*: 496–506.

Claus, D. and R.C.W. Berkeley. 1986. Genus *Bacillus* Cohn 1872. *In* Murray, Sneath, Sharpe and Holt (Editors), Bergey's Manual of Systematic Bacteriology, 9th ed., vol. 2. The Williams & Wilkins Co., Baltimore, pp. 1105–1139.

Claus, D. and F. Fahmy. 1986. Genus *Sporosarcina* Kluyver and van Niel 1936. *In* Murray, Sneath, Sharpe and Holt (Editors), Bergey's Manual of Systematic Bacteriology, vol. 2. The Williams and Wilkins Co., Baltimore, pp. 1202–1206.

Claus, D., D. Fritze and M. Kocur. 1992. Genera related to the genus *Bacillus*. *In* Balows, Trüper, Dworkin, Harder and Schleifer (Editors), The prokaryotes, 2nd ed., vol. 2. Springer-Verlag, New York, pp. 1769–1791.

Clayton, R.A., G. Sutton, P.S. Hinkle, Jr., C. Bult and C. Fields. 1995. Intraspecific variation in small-subunit rRNA sequences in GenBank: why single sequences may not adequately represent prokaryotic taxa. Int. J. Syst. Bacteriol. *45*: 595–599.

Clements, L.D., B.S. Miller and U.N. Streips. 2002. Comparative growth analysis of the facultative anaerobes *Bacillus subtilis, Bacillus* licheniformis, and *Escherichia coli.* Syst. Appl. Microbiol. *25*: 284–286.

Cohn, F. 1872. Untersuchungen über Bakterien. Bertr. Biol. Pflanz. *1*(Heft II): 127–224.

Cohn, F. 1876. Untersuchungen über Bakterien. IV. Berträge zur Biologie der Bacillen. Bertr. Biol. Pflanz. *2*: 249–276.

Coker, P.R., K.L. Smith and M.E. Hugh-Jones. 2002. Antimicrobial susceptibilities of diverse *Bacillus anthracis* isolates. Antimicrob. Agents Chemother. *46*: 3843–3845.

Collins, C.H. 1988. Laboratory Acquired Infections, 2nd ed. Butterworths, London, pp. 16.

Collins, D.P. and B.J. Jacobsen. 2003. Optimizing a *Bacillus subtilis* isolate for biological control of sugar beet cercospora leaf spot. Biol. Control. *26*: 153–161.

Collins, M.D., B.M. Lund, J.A.E. Farrow and K.H. Schleifer. 1983. Chemotaxonomic study of an alkalophilic bacterium, *Exiguobacterium aurantiacum* gen. nov., sp. nov. J. Gen. Microbiol. *129*: 2037–2042.

Combet-Blanc, Y., B. Ollivier, C. Streicher, B.K.C. Patel, P.P. Dwivedi, B. Pot, G. Prensier and J.L. Garcia. 1995. *Bacillus thermoamylovorans* sp. nov., a moderately thermophilic and amylolytic bacterium. Int. J. Syst. Bacteriol. *45*: 9–16.

Conn, H.J. 1930. The identity of *Bacillus subtilis.* J. Infect. Dis. *46*: 341–350.

Crickmore, N., D.R. Zeigler, J. Feitelson, E. Schnepf, J. Van Rie, D. Lereclus, J. Baum and D.H. Dean. 1998. Revision of the nomenclature for the *Bacillus thuringiensis* pesticidal crystal proteins. Microbiol. Mol. Biol. Rev. *62*: 807–813.

da Silva, K.R., L. Rabinovitch and L. Seldin. 1999. Phenotypic and genetic diversity among *Bacillus sphaericus* strains isolated in Brazil, potentially useful as biological control agents against mosquito larvae. Res. Microbiol. *150*: 153–160.

Daffonchio, D., S. Borin, A. Consolandi, D. Mora, P.L. Manachini and C. Sorlini. 1998a. 16S-23S rRNA internal transcribed spacers as molecular markers for the species of the 16S rRNA group I of the genus *Bacillus*. FEMS Microbiol. Lett. *163*: 229–236.

Daffonchio, D., S. Borin, G. Frova, P.L. Manachini and C. Sorlini. 1998b. PCR fingerprinting of whole genomes: The spacers between the 16S and 23S rRNA genes and of intergenic tRNA gene regions reveal a different intraspecific genomic variability of *Bacillus cereus* and *Bacillus* licheniformis [corrected]. Int. J. Syst. Bacteriol. *48*: 107–116.

Dagan, T. and W. Martin. 2006. The tree of one percent. Genome Biol. *7*: 118.

Damgaard, P.H., P.E. Granum, J. Bresciani, M.V. Torregrossa, J. Eilenberg and L. Valentino. 1997. Characterization of *Bacillus thuringiensis* isolated from infections in burn wounds. FEMS Immunol. Med. Microbiol. *18*: 47–53.

Dancer, S.J., D. McNair, P. Finn and A.B. Kolsto. 2002. *Bacillus cereus* cellulitis from contaminated heroin. J. Med Microbiol. *51*: 278–281.

Darland, G. and T. Brock. 1971. *Bacillus acidocaldarius* sp. nov., an acidophilic thermophilic spore-forming bacterium. J. Gen. Microbiol. *67*: 9–15.

Das, T., K. Choudhury, S. Sharma, S. Jalali and R. Nuthethi. 2001. Clinical profile and outcome in *Bacillus endophthalmitis.* Ophthalmology. *108*: 1819–1825.

Davaine, C. 1868. Dictionnaire Encyclopedie des Sciences Medicales, Ser I, vol. 8, pp. 21.

Davey, R.T. Jr., and W.B. Tauber. 1987. Posttraumatic endophthalmitis: The emerging role of *Bacillus cereus* infection. Rev. Infect. Dis. *9*: 110–123.

Davidson, E.W. 1981. A review of the pathology of bacilli infecting mosquitoes, including an ultrastructural study of larvae fed *Bacillus sphaericus* 1593 spores. Dev. Ind. Microbiol. *22*: 69–81.

Davies, S.N. 1951. The serology of *Bacillus polymyxa.* J. Gen Microbiol. *5*: 807–816.

de Barjac, H. 1981. Identification of H-serotypes of *Bacillus thuringiensis. In* Burges (Editor), Microbial Control of Pests and Plant Diseases 1970–80. Academic Press, London, pp. 35–43.

de Barjac, H., I. Larget-Thiery, V.C. Cosmao Dumanoir and H. Ripouteau. 1985. Serological classification of *Bacillus sphaericus* in relation to toxicity to mosquito larvae. Appl. Microbiol. Biotech. *21*: 85–90.

De Bartolemeo, A., F. Trotta, F. La Rosa, G. Saltalamacchia and V. Mastrandrea. 1991. Numerical analysis and DNA base compositions of some thermophilic *Bacillus* species. Int. J. Syst. Bacteriol. *41*: 502–509.

de Bary, A. 1884. Vergleichende Morphologie und Biologie der Pilze, Mycetozoen und Bakterien. Wilhelm Engelmann, Leipzig.

De, B.K., S.L. Bragg, G.N. Sanden, K.E. Wilson, L.A. Diem, C.K. Marston, A.R. Hoffmaster, G.A. Barnett, R.S. Weyant, T.G. Abshire, J.W. Ezzell and T. Popovic. 2002. A two-component direct fluorescent-antibody assay for rapid identification of *Bacillus anthracis.* Emerg. Infect. Dis. *8*: 1060–1065.

de Boer, A.S. and B. Diderichsen. 1991. On the safety of *Bacillus subtilis* and B. amyloliquefaciens: A review. Appl. Microbiol. Biotechnol. *36*: 1–4.

De Clerck, E. and P. De Vos. 2002. Study of the bacterial load in a gelatine production process focussed on *Bacillus* and related endospore-forming genera. Syst. Appl. Microbiol. *25*: 611–617.

De Clerck, E., D. Gevers, K. Sergeant, M. Rodriguez-Diaz, L. Herman, N.A. Logan, J. Van Beeumen and P. De Vos. 2004a. Genomic and phenotypic comparison of *Bacillus fumarioli* isolates from geothermal Antarctic soil and gelatine. Res. Microbiol. *155*: 483–490.

De Clerck, E., M. Rodriguez-Diaz, G. Forsyth, L. Lebbe, N.A. Logan and P. DeVos. 2004b. Polyphasic characterization of *Bacillus coagulans* strains, illustrating heterogeneity within this species, and emended description of the species. Syst. Appl. Microbiol. *27*: 50–60.

De Clerck, E., M. Rodriguez-Diaz, T. Vanhoutte, J. Heyrman, N.A. Logan and P. De Vos. 2004c. *Anoxybacillus contaminans* sp. nov. and *Bacillus gelatini* sp. nov., isolated from contaminated gelatin batches. Int. J. Syst. Evol. Microbiol. *54*: 941–946.

De Ley, J., H. Cattoir and A. Reynaerts. 1970. The quantitative measurement of DNA hybridization from renaturation rates. Eur. J. Biochem. *12*: 133–142.

de Silva, S., B. Petterson, M. Aquino de Muro and F.G. Priest. 1998. A DNA probe for the detection and identification of *Bacillus sporothermodurans* using the 16S-23S rDNA spacer region and phylogenetic analysis of some field isolates of *Bacillus* which form highly heat resistant spores. Syst. Appl. Microbiol. *21*: 398–407.

De Vos, P. 2002. Nucleic acid analysis and SDS-PAGE of whole-cell proteins in *Bacillus* taxonomy. *In* Berkeley, Heyndrickx, Logan and De Vos (Editors), Applications and Systematics of *Bacillus* and Relatives. Blackwell Science, Oxford, pp. 141–159.

de Vrij, W., G. Speelmans, R.I.R. Heyne and W.N. Konings. 1990. Energy transduction and amino acid transport in thermophilic aerobic and fermentative bacteria. FEMS Microbiol. Rev. *75*: 183–200.

Deák, T. and É. Temár. 1988. Simplified identification of aerobic spore-formers in the investigation of foods. Int. J. Food Microbiol. *6*: 115–125.

Dean, D.H., F. Rajamohan, M.K. Lee, S.J. Wu, X.J. Chen, E. Alcantara and S.R. Hussain. 1996. Probing the mechanism of action of *Bacillus thuringiensis* insecticidal proteins by site-directed mutagenesis-a mini-review. Gene *179*: 111–117.

Degryse, E., N. Glansdorff and A. Pierard. 1978. A comparative analysis of extreme thermophilic bacteria belonging to the genus *Thermus*. Arch. Microbiol. *117*: 189–196.

Deinhard, G., P. Blanz, K. Poralla and E. Altan. 1987a. *Bacillus acidoterrestris* sp. nov., a new thermotolerant acidophile isolated from different soils. Syst. Appl. Microbiol. *10*: 47–53.

Deinhard, G., J. Saar, W. Krischke and K. Poralla. 1987b. *Bacillus cycloheptanicus* sp. nov., a new thermoacidophile containing omega-cycloheptane fatty acids. Syst. Appl. Microbiol. *10*: 68–73.

Delaporte, B. 1967. A new bacteria of the Pacific Ocean: *Bacillus pacificus* n. sp. C. R. Acad. Sci. Hebd. Seances Acad. Sci. D *264*: 3068–3071.

Delaporte, B. 1972. Three new species of *Bacillus: Bacillus similibadius* n. sp., *Bacillus longisporus* n. sp. and *Bacillus nitritollens* n. sp. Ann. Inst. Pasteur (Paris) *123*: 821–834.

Delprato, A.M., A. Samadani, A. Kudrolli and L.S. Tsimring. 2001. Swarming ring patterns in bacterial colonies exposed to ultraviolet radiation. Phys. Rev. Lett. *87*: 158102.

Demharter, W. and R. Hensel. 1989a. *In* Validation of the publication of new names and new combinations previously effectively published outside the IJSB. List no. 31. Int. J. Syst. Bacteriol. *39*: 495–497.

Demharter, W. and R. Hensel. 1989b. *Bacillus thermocloaceae* sp. nov., a new thermophilic species from sewage sludge. Syst. Appl. Microbiol. *11*: 272–276.

den Dooren de Jong, L.E. 1926. Bijdrage tot de kennis van het Mineralisatieproces. Nijgh En Van Ditmar Uitgeversmaatschappij, Rotterdam.

den Dooren de Jong, L.E. 1929. Über *Bacillus fastidiosus*. Zentbl. Bakteriol. Parasitenkd. Infektionskr. Hyg. Abt. II *79*: 344–353.

Denariaz, G., W.J. Payne and J. Legall. 1989. A halophilic denitrifier, *Bacillus halodenitrificans* sp. nov. Int. J. Syst. Bacteriol. *39*: 145–151.

Derekova, A., C. Sjoholm, R. Mandeva and M. Kambourova. 2007. *Anoxybacillus rupiensis* sp. nov., a novel thermophilic bacterium isolated from Rupi basin (Bulgaria). Extremophiles. *11*: 577–583.

Deutch, C.E. 1994. Characterization of a novel salt-tolerant *Bacillus* sp. from the nasal cavities of desert iguanas. FEMS Microbiol. Lett. *121*: 55–60.

Deutscher, J., A. Galinier and I. Martin-Verstraete. 2002. Carbohydrate uptake and metabolism. *In* Sonenshein, Hoch and Losick (Editors), *Bacillus subtilis* and its Closest Relatives. ASM Press, Washington, D.C., pp. 129–150.

Dieckmann, R., I. Graeber, I. Kaesler, U. Szewzyk and H. von Dohren. 2005. Rapid screening and dereplication of bacterial isolates from marine sponges of the sula ridge by intact-cell-MALDI-TOF mass spectrometry (ICM-MS). Appl. Microbiol. Biotechnol. *67*: 539–548.

Dietvorst, D.C.E. 1996. Farmers' attitudes towards the control and prevention of anthrax in Western Province, Zambia. Salisbury Med. Bull. No 87, Special supplement: 102–103.

Dijkhuizen, L., N. Arfman, M.M. Attwood, A.G. Brooke, W. Harder and E.M. Watling. 1988. Isolation and initial characterization of thermotolerant methylotrophic *Bacillus* strains. FEMS Microbiol. Lett. *52*: 209–214.

Doak, B.W. and C. Lamanna. 1948. On the antigenic structure of the bacterial spore. J. Bacteriol. *55*: 373–380.

Docherty, A., G. Grandi, R. Grandi, T.J. Gryczan, A.G. Shivakumar and D. Dubnau. 1981. Naturally occurring macrolide-lincosamide-streptogramin B resistance in *Bacillus licheniformis*. J. Bacteriol. *145*: 129–137.

Doganay, M. and N. Aydin. 1991. Antimicrobial susceptibility of *Bacillus anthracis*. Scand. J. Infect. Dis. *23*: 333–335.

Dohrmann, A.B. and V.V. Muller. 1999. Chloride dependence of endospore germination in *Halobacillus halophilus*. Arch. Microbiol. *172*: 264–267.

Donk, P.J. 1920. A highly resistant thermophilic organism. J. Bacteriol. *5*: 373–374.

Drego, L., A. Lombardi, E.D. Vecchi and M.R. Gismondo. 2002. Real-time PCR assay for rapid detection of *Bacillus anthracis* spores in clinical samples. J. Clin. Microbiol. *40*: 4399.

Driks, A. 2002. Proteins of the spore core and coat. *In* Sonenshein, Hoch and Losick (Editors), *Bacillus subtilis* and its Closest Relatives. ASM Press, Washington, D.C., pp. 527–535.

Drucker, D.B. 1981. Microbiological Applications of Gas Chromatography. Cambridge University Press, Cambridge.

Drum, C.L., S.Z. Yan, J. Bard, Y.Q. Shen, D. Lu, S. Soelaiman, Z. Grabarek, A. Bohm and W.J. Tang. 2002. Structural basis for the activation of anthrax adenylyl cyclase exotoxin by calmodulin. Nature. *415*: 396–402.

Dulger, S., Z. Demirbag and A.O. Belduz. 2004. *Anoxybacillus ayderensis* sp. nov. and *Anoxybacillus kestanbolensis* sp. nov. Int. J. Syst. Evol. Microbiol. *54*: 1499–1503.

Dutkiewicz, J., E. Krysinska-Traczyk, C. Skorska, J. Sitkowska, Z. Prazmo and M. Golec. 2001. Exposure to airborne microorganisms and endotoxin in herb processing plants. Ann. Agric. Environ. Med. *8*: 201–211.

Dutky, S.R. 1940. Two new spore-forming bacteria causing milky disease of Japanese beetle larvae. J. Agric. Res. *61*: 57–68.

Echigo, A., M. Hino, T. Fukushima, T. Mizuki, M. Kamekura and R. Usami. 2005. Endospores of halophilic bacteria of the family *Bacillaceae* isolated from non-saline Japanese soil may be transported by Kosa event (Asian dust storm). Saline Syst. *1*: 8.

Effio, P.C., E.F. Silva and M.T. Pueyo. 2000. A simple and rapid method for screening amylolytic bacteria. Biochem. Ed. *28*: 47–49.

Ehrenberg, C.G. 1835. Dritter Beitrag zur Erkemtiss grosser Organisation in der Richtung des kleinsten Raumes. Abh. Preuss. Akad. Wiss. Phys. K1 Berlin aus den Jahre. *1833–1835*: 143–336.

Ellar, D.J. 1990. Pathogenic determinants of entomopathogenic bacteria, pp. 298–302, Fifth International Colloquium on Invertebrate Pathology and Microbial Control: 1990, San Diego and UK.

Ellar, D.J. 1997. The structure and function of *Bacillus thuringiensis* delta-endotoxins and prospects for biopesticide improvement. *In* Microbial Insecticides: Novelty or Necessity? Symposium Proceedings, no. 68. British Crop Protection Council, Farnham, U.K., pp. 83–100.

Ellerbrok, H., H. Nattermann, M. Ozel, L. Beutin, B. Appel and G. Pauli. 2002. Rapid and sensitive identification of pathogenic and apathogenic *Bacillus anthracis* by real-time PCR. FEMS Microbiol. Lett. *214*: 51–59.

Emberger, O. 1970. Cultivation methods for the detection of aerobic sporeforming bacteria. Zentbl. Bakteriol. Parasitenkd. Infektionskr. Hyg. Abt. II *125*: 555–565.

Emmert, E.A. and J. Handelsman. 1999. Biocontrol of plant disease: A (gram-) positive perspective. FEMS Microbiol. Lett. *171*: 1–9.

English, C.F., E.J. Bell and A.J. Berger. 1967. Isolation of thermophiles from broadleaf tobacco and effect of pure culture inoculation on cigar aroma and mildness. Appl. Microbiol. *15*: 117–119.

Errington, J. 2001. Septation and chromosome segregation during sporulation in *Bacillus subtilis*. Curr. Opin. Microbiol. *4*: 660–666.

Esel, D., M. Doganay and B. Sumerkan. 2003. Antimicrobial susceptibilities of 40 isolates of *Bacillus anthracis* isolated in Turkey. Int. J. Antimicrob. Agents *22*: 70–72.

Ezaki, T., Y. Hashimoto and E. Yabuuchi. 1989. Fluorometric DNA-DNA hybridization in microdilution wells as an alternative to membrane filter hybridization in which radioisotopes are used to determine genetic relatedness among bacterial strains. Int. J. Syst. Bacteriol. *39*: 224–229.

Ezzell, J.W. and S.L. Welkos. 1999. The capsule of *Bacillus anthracis*, a review. J. Appl. Microbiol. *87*: 250.

Fahmy, F., F. Mayer and D. Claus. 1985. Endospores of *Sporosarcina halophila*: Characteristics and ultrastructure. Arch. Microbiol. *140*: 338–342.

Farrow, J.A., C. Ash, S. Wallbanks and M.D. Collins. 1992. Phylogenetic analysis of the genera *Planococcus*, *Marinococcus* and *Sporosarcina* and their relationships to members of the genus *Bacillus*. FEMS Microbiol. Lett. *72*: 167–172.

Farrow, J.A., S. Wallbanks and M.D. Collins. 1994. Phylogenetic interrelationships of round-spore-forming bacilli containing cell walls based on lysine and the non-spore-forming genera *Caryophanon*, *Exiguobacterium*, *Kurthia*, and *Planococcus*. Int. J. Syst. Bacteriol. *44*: 74–82.

Feitelson, J.S., J. Payne and L. Kim. 1992. *Bacillus thuringiensis*: Insects and beyond. Bio./Technol. *10*: 271–275.

Felske, A., A.D.L. Akkermans and W.M. De Vos. 1998. In situ detection of an uncultured predominant bacillus in Dutch grassland soils. Appl. Environ. Microbiol. *64*: 4588–4590.

Felske, A., A. Wolterink, R. van Lis, W.M. de Vos and A.D.L. Akkermans. 1999. Searching for predominant soil bacteria: 16S rDNA cloning versus strain cultivation. FEMS Microbiol. Ecol. *30*: 137–145.

Ferguson, C.M., N.A. Booth and E.J. Allan. 2000. An ELISA for the detection of *Bacillus subtilis* L-form bacteria confirms their symbiosis in strawberry. Lett. Appl. Microbiol. *31*: 390–394.

Ferguson, H.W., J.F. Turnbull, A. Shinn, K. Thompson, T.T. Dung and M. Crumlish. 2001. Bacillary necrosis in farmed *Pangasius hypophthalmus* (Sauvage) from the Mekong Delta, Vietnam. J. Fish Dis. *24*: 509–513.

Finlay, W.J., N.A. Logan and A.D. Sutherland. 1999. Semiautomated metabolic staining assay for *Bacillus cereus* emetic toxin. Appl. Environ. Microbiol. *65*: 1811–1812.

Finlay, W.J., N.A. Logan and A.D. Sutherland. 2000. *Bacillus cereus* produces most emetic toxin at lower temperatures. Lett. Appl. Microbiol. *31*: 385–389.

Finlay, W.J., N.A. Logan and A.D. Sutherland. 2002. *Bacillus cereus* emetic toxin production in relation to dissolved oxygen tension and sporulation. Food Microbiol. *19*: 423–430.

Finn, G.J. and S. Condon. 1975. Regulation of catalase synthesis in *Salmonella typhimurium*. J. Bacteriol. *123*: 570–579.

Fischer, A. 1895. Untersuchungen über bakterien. Jahrbuch für Wissenschaftliche Botanik *27*: 1–163.

Fisher, S.H. and M. Débarbouillé. 2002. Nitrogen source utilization and its regualtion. *In* Sonenshein, Hoch and Losick (Editors), *Bacillus subtilis* and its Closest Relatives. ASM Press, Washington, D.C., pp. 181–191.

Fletcher, P. and N.A. Logan. 1999. Improved cytotoxicity assay for *Bacillus cereus* diarrhoeal enterotoxin. Lett. Appl. Microbiol. *28*: 394–400.

Flügge, C. 1886. Die Mikroorganismen. F.C.W. Vogel, Leipzig.

Forsyth, G. and N.A. Logan. 2000. Isolation of *Bacillus thuringiensis* from Northern Victoria Land, Antarctica. Lett. Appl. Microbiol. *30*: 263–266.

Fortina, M.G., D. Mora, P. Schumann, C. Parini, P.L. Manachini and E. Stackebrandt. 2001a. Reclassification of *Saccharococcus caldoxylosilyticus* as *Geobacillus caldoxylosilyticus* (Ahmad *et al.* 2000) comb. nov. Int. J. Syst. Evol. Microbiol. *51*: 2063–2071.

Fortina, M.G., R. Pukall, P. Schumann, D. Mora, C. Parini, P.L. Manachini and E. Stackebrandt. 2001b. *Ureibacillus* gen. nov., a new genus to accommodate *Bacillus thermosphaericus* (Andersson *et al.* 1995), emendation of *Ureibacillus thermosphaericus* and description of *Ureibacillus terrenus* sp. nov. Int. J. Syst. Evol. Microbiol. *51*: 447–455.

Foster, S.J. 1994. The role and regulation of cell wall structural dynamics during differentiation of endospore-forming bacteria. J. Appl. Bacteriol. *76*: 25S–39S.

Foster, S.J. and K. Johnstone. 1990. Pulling the trigger: The mechanism of bacterial spore germination. Mol. Microbiol. *4*: 137–141.

Foster, S.J. and D.L. Popham. 2002. Structure and synthesis of cell wall, spore cortex, teichoic acids, S-layers, and capsules. *In* Sonenshein, Hoch and Losick (Editors), *Bacillus Subtilis* and its Closest Relatives. ASM Press, Washington, D.C., pp. 21–41.

Fox, G.E., K.R. Pechman and C.R. Woese. 1977. Comparative cataloguing of 16S ribosomal ribonucleic acid: molecular approach to prokaryotic systematics. Int. J. Syst. Bacteriol. *27*: 44–57.

Fox, G.E., E. Stackebrandt, R.B. Hespell, J. Gibson, J. Maniloff, T.A. Dyer, R.S. Wolfe, W.E. Balch, R.S. Tanner, L.J. Magrum, L.B. Zablen, R. Blakemore, R. Gupta, L. Bonen, B.J. Lewis, D.A. Stahl, K.R. Luehrsen, K.N. Chen and C.R. Woese. 1981. The phylogeny of prokaryotes. Science *209*: 457–463.

Fox, G.E., J.D. Wisotzkey and P. Jurtshuk. 1992. How close is close: 16S ribosomal RNA sequence identity may not be sufficient to guarantee species identity. Int. J. Syst. Bacteriol. *42*: 166–170.

Fox, G.E., D.S. Wunschel, A. Fox and G.C. Stewart. 1998. Complementarity of GC-MS and LC-MS analyses for determination of carbohydrate profiles of vegetative cells and spores of bacilli. J. Microbiol. Methods *33*: 1–11.

Francis, C.A., E.M. Co and B.M. Tebo. 2001. Enzymatic manganese(II) oxidation by a marine alpha-proteobacterium. Appl. Environ. Microbiol. *67*: 4024–4029.

Frankland, G.C. and P.F. Frankland. 1887. Studies on some new microorganisms obtained from air. Phil. Trans. Roy. Soc. London Ser. B Biol. Sci. *178*: 257–287.

Fritze, D. 1996a. Bacillus haloalkaliphilus sp. nov. Int. J. Syst. Bacteriol. *46*: 98–101.

Fritze, D. 1996b. Reclassification of *Bacillus haloalkaliphilus* as *Alkalibacillus haloalkaliphilus* gen. nov., comb. nov. and the description of *Alkalibacillus salilacus* sp. nov., a novel halophilic bacterium isolated from a salt lake in China. Int. J. Syst. Evol. Microbiol. *55*: 1891–1896.

Fritze, D. 2002. *Bacillus* identification- traditional approaches. *In* Berkeley, Heyndrickx, Logan and De Vos (Editors), Applications and Systematics of *Bacillus* and Relatives. Blackwell Science, Oxford, pp. 100–122.

Fritze, D., J. Flossdorf and D. Claus. 1990. Taxonomy of alkaliphilic *Bacillus* strains. Int. J. Syst. Bacteriol. *40*: 92–97.

Fujita, T., O. Shida, H. Takagi, K. Kunugita, A.N. Pankrushina and M. Matsuhashi. 1996. Description of *Bacillus carboniphilus* sp. nov. Int. J. Syst. Bacteriol. *46*: 116–118.

Galesloot, T.E. and H. Labots. 1959. Thermofiele sporevormers in melk, vooral met beyrekking tot de bereiding van gesteriliseerde melk en chocolademelk. Nederland Melk-en Zuiveltijdschrift *13*: 448–465.

Galiero, G. and E. De Carlo. 1998. Abortion in water buffalo (*Bubalus bubalis*) associated with *Bacillus licheniformis*. Vet. Rec. *143*: 640.

Garabito, M.J., D.R. Arahal, E. Mellado, M.C. Marquez and A. Ventosa. 1997. *Bacillus salexigens* sp. nov, a new moderately halophilic *Bacillus* species. Int. J. Syst. Bacteriol. *47*: 735–741.

Garabito, M.J., M.C. Marquez and A. Ventosa. 1998. Halotolerant *Bacillus* diversity in hypersaline environments. Can. J. Microbiol. *44*: 95–102.

Garcia, J.L., S. Roussos, M. Bensoussan, A. Bianchi and M. Mandel. 1982. Numerical taxonomy of a thermophilic "*Bacillus*" species isolated from West African rice soils (author's transl). Ann. Microbiol. (Paris) *133*: 471–488.

Garcia, M.T., E. Mellado, J.C. Ostos and A. Ventosa. 2004. *Halomonas organivorans* sp. nov., a moderate halophile able to degrade aromatic compounds. Int. J. Syst. Evol. Microbiol. *54*: 1723–1728.

Garcia, M.T., V. Gallego, A. Ventosa and E. Mellado. 2005. *Thalassobacillus devorans* gen. nov., sp. nov., a moderately halophilic, phenol-degrading, Gram-positive bacterium. Int. J. Syst. Evol. Microbiol. *55*: 1789–1795.

Gardner, J.M. and F.A. Troy. 1979. Chemistry and biosynthesis of the poly(gamma-D-glutamyl) capsule in *Bacillus licheniformis*. Activation, racemization, and polymerization of glutamic acid by a membranous polyglutamyl synthetase complex. J. Biol. Chem. *254*: 6262–6269.

Garrity, L.F. and G.W. Ordal. 1995. Chemotaxis in *Bacillus subtilis*: How bacteria monitor environmental signals. Pharmacol. Ther. *68*: 87–104.

Gatson, J.W., B.F. Benz, C. Chandrasekaran, M. Satomi, K. Venkateswaran and M.E. Hart. 2006. *Bacillus tequilensis* sp. nov., isolated from a 2000-year-old Mexican shaft-tomb, is closely related to *Bacillus subtilis*. Int. J. Syst. Evol. Microbiol. *56*: 1475–1484.

Gee, J.M., B.M. Lund, G. Metcalf and J. Peel. 1980. Properties of a new group of alkalophilic bacteria. J. Gen. Microbiol. *117*: 9–18.

George, S., D. Mathai, V. Balraj, M.K. Lalitha and T.J. John. 1994. An outbreak of anthrax meningoencephalitis. Trans. R. Soc. Trop. Med. Hyg. *88*: 206–207.

Ghosh, A., M. Bhardwaj, T. Satyanarayana, M. Khurana, S. Mayilraj and R.K. Jain. 2007. *Bacillus lehensis* sp. nov., an alkalitolerant bacterium isolated from soil. Int. J. Syst. Evol. Microbiol. *57*: 238–242.

Gibbons, N.E., K.B. Pattee and J.G. Holt. 1981. Supplement to Index Bergeyana. The Williams & Wilkins Co., Baltimore.

Gibson, T. 1935. The urea-decomposing microflora of soils. I. Description and classification of the organisms. Zentbl. Bakteriol. Parasitenkd. Infektionskr. Hyg. Abt. II *92*: 364–380.

Gibson, T. 1944. A study of *Bacillus subtilis* and related organisms. J. Dairy Res. *13*: 248–260.

Gibson, T. and R.E. Gordon. 1974. *Bacillus* Cohn 1872. *In* Buchanan and Gibbons (Editors), Bergey's Manual of Determinative Bacteriology. The Williams & Wilkins Co., Baltimore, pp. 529–550.

Gibson, T. and L.E. Topping. 1938. Further studies of the spore-forming bacilli. Proc. Soc. Agric. Bacteriol. 43–44.

Gilbert, R.J. and J.M. Parry. 1977. Serotypes of *Bacillus cereus* from outbreaks of food poisoning and from routine foods. J. Hyg. Camb. *78*: 69–74.

Gilliam, M. and D.K. Valentine. 1976. Bacteria isolated from the intestinal contents of foraging worker honey bees, *Apis mellifera*: the genus *Bacillus*. J. Invert. Pathol. *28*: 275–276.

Glazunova, O.O., D. Raoult and V. Roux. 2006. *Bacillus massiliensis* sp. nov., isolated from cerebrospinal fluid. Int. J. Syst. Evol. Microbiol. *56*: 1485–1488.

Golovacheva, R.S. and G.I. Karavaiko. 1978. New genus of thermophilic spore-forming bacteria, *Sulfobacillus*. Microbiology (En. transl. from Mikrobiologiya) *47*: 658–665.

Golovacheva, R.S., L.A. Egorova and L.G. Loginova. 1965. Ecology and systematics of aerobic obligately thermophilic bacteria isolated from thermal locations on Mount Yangan-Tau and Kunaschir Isle of the Kuril chain. Microbiology (En. transl. from Mikrobiologiya) *34*: 693–698.

Golovacheva, R.S., L.G. Loginova, T.A. Salikhov, A.A. Kolesnikov and G.N. Zaitseva. 1975. New thermophilic species, *Bacillus thermocatenulatus* nov. sp. Microbiology (En. transl. from Mikrobiologiya) *44*: 230–233.

Goodacre, R., B. Shann, R.J. Gilbert, E.M. Timmins, A.C. McGovern, B.K. Alsberg, D.B. Kell and N.A. Logan. 2000. Detection of the dipicolinic acid biomarker in *Bacillus* spores using Curie-point pyrolysis mass spectrometry and Fourier transform infrared spectroscopy. Anal. Chem. *72*: 119–127.

Goodwin, A.E., J.S. Roy, J.M. Grizzle and M.T. Goldsby. 1994. *Bacillus mycoides* – A bacterial pathogen of channel catfish. Dis. Aquatic Org. *18*: 173–179.

Gordon, R.E. 1981. One hundred years with the genus *Bacillus*. *In* Berkeley and Goodfellow (Editors), Classification and identification of the aerobic endospore-forming bacteria. Academic Press, London.

Gordon, R.E. and N.R. Smith. 1949. Aerobic sporeforming bacteria capable of growth at high temperatures. J. Bacteriol. *58*: 327–341.

Gordon, R.E. and J.L. Hyde. 1982. The *Bacillus firmus–Bacillus lentus* complex and the pH variants of some alkaliphilic strains. J. Gen. Microbiol. *128*: 1109–1116.

Gordon, R.E., W.C. Haynes and C.H.-N. Pang. 1973. The genus *Bacillus*. Agriculture Handbook no. 427. United States Department of Agriculture, Washington, D.C.

Gordon, R.E., J.L. Hyde and J.A. Moore. 1977. *Bacillus firmus–Bacillus lentus*: A series or one species? Int. J. Syst. Bacteriol. *27*: 265–262.

Goris, J., K.I. Suzuki, P. De Vos, T. Nakase and K. Kersters. 1998. Evaluation of a microplate DNA-DNA hybridization method compared with the initial renaturation method. Can. J. Microbiol. *44*: 1148–1153.

Gottheil, O. 1901. Botanische Beschreibung einiger bodenbakterien. Zentbl. Bakteriol. Parasitenkd. Infektionskr. Hyg. Abt. II *7*: 680–691.

Goudswaard, W.B., M.H. Dammer and C. Hol. 1995. *Bacillus circulans* infection of a proximal interphalangeal joint after a clenched-fist injury caused by human teeth. Eur. J. Clin. Microbiol. Infect. Dis. *14*: 1015–1016.

Gould, G.W. 1962. Microscopical observations on the mergence of cells of *Bacillus* spp. from spores under different cultural conditions. J. Appl. Bacteriol. *25*: 35–41.

Grant, W.D. and B.J. Tindall. 1980. The isolation of alkaliphilic bacteria. *In* Gould and Corry (Editors), Microbial Growth in Extremes of Environment. Academic Press, London, pp. 27–38.

Granum, P.E. 2002. *Bacillus cereus* and food poisoning. *In* Berkeley, Heyndrickx, Logan and De Vos (Editors), Applications and Systematics of *Bacillus* and Relatives. Blackwell Science, Oxford, pp. 37–46.

Graur, D. and T. Pupko. 2001. The Permian bacterium that isn't. Mol. Biol. Evol. *18*: 1143–1146.

Gray, T.R.G. and D.A. Hull. 1971. Taxonomy of *Bacillus circulans* with special reference to spore morphology. *In* Barker, Gould and Wolf (Editors), Spore Research 1971. Academic Press, London, pp. 219–226.

Grimont, P.A.D., K. Irino and F. Grimont. 1982. The *Serratia liquefaciens-S. proteomaculans-S. grimesii* complex: DNA relatedness. Curr. Microbiol. *7*: 63–68.

Grinsted, E. and L.F.L. Clegg. 1955. Spore-forming organisms in commercial sterilized milk. J. Dairy Res. *22*: 178–190.

Groschulski, P., L. Masson, S. Borisova, M. Pusztaicarey, J.L. Schwartz, R. Brousseau and M. Cygler. 1995. *Bacillus thuringiensis* CrylA(a) insecticidal toxin: Crystal structure and channel formation. J. Mol. Biol. *254*: 447–464.

Guex-Holzer, S. and J. Tomcsik. 1956. The isolation and chemical nature of capsular and cell-wall haptens in a *Bacillus* species. J. Gen. Microbiol. *14*: 14–25.

Gugliandolo, C., T.L. Maugeri, D. Caccamo and E. Stackebrandt. 2003a. *Bacillus aeolius* sp. nov. a novel thermophilic, halophilic marine *Bacil-*

lus species from Eolian islands (Italy). Syst. Appl. Microbiol. *26*: 172–176.

Gugliandolo, C., T.L. Maugeri, D. Caccamo and E. Stackebrandt. 2003b. *In* Validation of the publication of new names and new combinations previously effectively published outside the IJSEM. List no. 94. Int. J. Syst. Evol. Microbiol. *53*: 1701–1702.

Guicciardi, A., M.R. Biffi, P.L. Manachini, A. Craveri, C. Scolastico, B. Ridone and R. Craveri. 1968. Ricerche preliminari su un nuovo schizomicete termofilo del genere *Bacillus* e caratterizzazione del pigmento rosso prodotto. Ann. Microbiol. (Milan) *18*: 191–205.

Gunsalus, I.C. and C.F.J. Niven. 1942. The effect of pH on the lactic acid fermentaion. J. Biol. Chem. *145*: 131–136.

Günther, K., D. Schlosser and W. Fritsche. 1995. Phenol and cresol metabolism in *Bacillus pumilus* isolated from contaminated groundwater. J. Basic Microbiol. *35*: 83–92.

Hafez, M.B., A. Fouad and W. El-Dezouky. 2002. Accumulation of metal ions on *Bacillus licheniformis*. J. Radioanal. Nuc. Chem. *251*: 249–252.

Hamana, K., H. Hamana, M. Niitsu, K. Samejima, T. Sakane and A. Yokota. 1993. Tertiary and quaternary branched polyamines distributed in thermophilic *Saccharococcus* and *Bacillus*. Microbios. *75*: 23–32.

Hammer, B.W. 1915. Bacteriological studies on the coagulation of evaporated milk. Iowa Agr. Exp. Sta. Res. Bull. *19*: 119–131.

Hanna, P. 1999. Lethal toxin actions and their consequences. J. Appl. Microbiol. *87*: 285–287.

Hanna, P. and J.A.W. Ireland. 1999. Understanding *Bacillus anthracis* pathogenesis. Trends Microbiol. *7*: 180–182.

Hansen, B.M., T.D. Leser and N.B. Hendriksen. 2001. Polymerase chain reaction assay for the detection of *Bacillus cereus* group cells. FEMS Microbiol. Lett. *202*: 209–213.

Hao, M.V., M. Kocur and K. Komagata. 1984. *Marinococcus* gen. nov., a new genus for motile cocci with *meso*-diaminopimelic acid in the cell wall and *Marinococcus albus* sp. nov. and *Marinococcus halophilus* (Novitsky and Kushner) comb. nov. J. Gen. Appl. Microbiol. *30*: 449–459.

Hao, M.V., K. Kocur and K. Komagata. 1985. *In* Validation of the publication of new names and new combinations previously effectively published outside the IJSB. List no. 19. Int. J. Syst. Bacteriol. *35*: 535.

Harrell, L.J., G.L. Andersen and K.H. Wilson. 1995. Genetic variability of *Bacillus anthracis* and related species. J. Clin. Microbiol. *33*: 1847–1850.

Haruta, S., M. Kondo, K. Nakamura, H. Aiba, S. Ueno, M. Ishii and Y. Igarashi. 2002. Microbial community changes during organic solid waste treatment analyzed by double gradient-denaturing gradient gel electrophoresis and fluorescence in situ hybridization. Appl. Microbiol. Biotechnol. *60*: 224–231.

Hassan, H.M. and I. Fridovich. 1978. Regulation of the synthesis of catalase and peroxidase in *Escherichia coli*. J. Biol. Chem. *253*: 6445–6450.

Hatayama, K., H. Shoun, Y. Ueda and A. Nakamura. 2006. Tuberi*bacillus* calidus gen. nov., sp. nov., isolated from a compost pile and reclassification of *Bacillus naganoensis* Tomimura *et al.* 1990 as *Pullulanibacillus naganoensis* gen. nov., comb. nov. and *Bacillus laevolacticus* Andersch *et al.* 1994 as *Sporolactobacillus laevolacticus* comb. nov. Int. J. Syst. Evol. Microbiol. *56*: 2545–2551.

Hazes, B. and R.J. Read. 1995. A mosquitocidal toxin with a ricin-like cell-binding domain. Nat. Struct. Biol. *2*: 358–359.

Hedlund, B.P., A.D. Geiselbrecht and J.T. Staley. 2001. *Marinobacter* strain NCE312 has a *Pseudomonas*-like naphthalene dioxygenase. FEMS Microbiol. Lett. *201*: 47–51.

Heinen, U.J. and W. Heinen. 1972. Characteristics and properties of a caldo-active bacterium producing extracellular enzymes and two related strains. Arch. Mikrobiol. *82*: 1–23.

Heinen, W., A.M. Lauwers and J.W.M. Mulders. 1982. *Bacillus flavothermus*, a newly isolated facultative thermophile. Antonie Leeuwenhoek J. Microbiol. *48*: 265–272.

Helgason, E., D.A. Caugant, I. Olsen and A.B. Kolsto. 2000. Genetic structure of population of *Bacillus cereus* and *B. thuringiensis* isolates associated with periodontitis and other human infections. J. Clin. Microbiol. *38*: 1615–1622.

Herbig, A.F. and J.D. Helmann. 2002. Metal ion uptake and oxidative stress. *In* Sonenshein, Hoch and Losick (Editors), *Bacillus Subtilis and its Closest Relatives*. ASM Press, Washington, D.C., pp. 405–414.

Hertlein, B.H., R. Levy and T.W. Miller. 1979. Recycling potential and selective retrieval of *Bacillus sphaericus* from soil in a mosquito habitat. J. Invertebr. Pathol. *33*: 217–221.

Heyndrickx, M., K. Vandemeulebroecke, P. Scheldeman, B. Hoste, K. Kersters, P. Devos, N.A. Logan, A.M. Aziz, N. Ali and R.C.W. Berkeley. 1995. *Paenibacillus* (formerly *Bacillus*) *gordonae* (Pichinoty *et al.* 1986) Ash *et al.* 1994 is a later subjective synonym of *Paenibacillus* (formerly *Bacillus*) *validus* (Nakamura 1984) Ash *et al.* 1994, emended description of *P. validus*. Int. J. Syst. Bacteriol. *45*: 661–669.

Heyndrickx, M., K. Vandemeulebroecke, B. Hoste, P. Janssen, K. Kersters, P. DeVos, N.A. Logan, N. Ali and R.C.W. Berkeley. 1996a. Reclassification of *Paenibacillus* (formerly *Bacillus*) *pulvifaciens* (Nakamura 1984) Ash *et al.* 1994, a later subjective synonym of *Paenibacillus* (formerly *Bacillus*) *larvae* (White 1906) Ash *et al.* 1994, as a subspecies of *P. larvae*, with emended descriptions of *P. larvae* as *P. larvae* subsp. *larvae* and *P. larvae* subsp. *pulvifaciens*. Int. J. Syst. Bacteriol. *46*: 270–279.

Heyndrickx, M., K. Vandemeulebroecke, P. Scheldeman, K. Kersters, P. DeVos, N.A. Logan, A.M. Aziz, N. Ali and R.C.W. Berkeley. 1996b. A polyphasic reassessment of the genus *Paenibacillus*, reclassification of *Bacillus lautus* (Nakamura 1984) as *Paenibacillus lautus* comb. nov. and of *Bacillus peoriae* (Montefusco *et al.* 1993) as *Paenibacillus peoriae* comb. nov., and emended descriptions of *P. lautus* and of *P. peoriae*. Int. J. Syst. Bacteriol. *46*: 988–1003.

Heyndrickx, M., L. Vauterin, P. Vandamme, K. Kersters and P. DeVos. 1996c. Applicability of combined amplified ribosomal DNA restriction analysis (ARDRA) patterns in bacterial phylogeny and taxonomy. J. Microbiol. Methods *26*: 247–259.

Heyndrickx, M., L. Lebbe, M. Vancanneyt, K. Kersters, P. DeVos, N.A. Logan, G. Forsyth, S. Nazli, N. Ali and R.C.W. Berkeley. 1997. A polyphasic reassessment of the genus *Aneurinibacillus*, reclassification of *Bacillus thermoaerophilus* (Meier-Stauffer *et al.* 1996) as *Aneurinibacillus thermoaerophilus* comb. nov, and emended descriptions of *A. aneurinilyticus* corrig., *A. migulanus*, and *A. thermoaerophilus*. Int. J. Syst. Bacteriol. *47*: 808–817.

Heyndrickx, M., L. Lebbe, K. Kersters, P. De Vos, C. Forsyth and N.A. Logan. 1998. *Virgibacillus*: A new genus to accommodate *Bacillus pantothenticus* (Proom and Knight 1950). Emended description of *Virgibacillus pantothenticus*. Int. J. Syst. Bacteriol. *48*: 99–106.

Heyndrickx, M., L. Lebbe, K. Kersters, B. Hoste, R. De Wachter, P. De Vos, G. Forsyth and N.A. Logan. 1999. Proposal of *Virgibacillus proomii* sp. nov. and emended description of *Virgibacillus pantothenticus* (Proom and Knight 1950) Heyndrickx *et al.* 1998. Int. J. Syst. Bacteriol. *49*: 1083–1090.

Heyndrickx, M., N.A. Logan, L. Lebbe, M. Rodriguez-Diaz, G. Forsyth, J. Goris, P. Scheldeman and P. De Vos. 2004. *Bacillus galactosidilyticus* sp. nov., an alkali-tolerant beta-galactosidase producer. Int. J. Syst. Evol. Microbiol. *54*: 617–621.

Heyndrickx, M., P. Scheldeman, G. Forsyth, L. Lebbe, M. Rodriguez-Diaz, N.A. Logan and P. De Vos. 2005. *Bacillus ruris* sp. nov., from dairy farms. Int. J. Syst. Evol. Microbiol. *55*: 2551–2554.

Heyrman, J., J. Mergaert, R. Denys and J. Swings. 1999. The use of fatty acid methyl ester analysis (FAME) for the identification of heterotrophic bacteria present on three mural paintings showing severe damage by microorganisms. FEMS Microbiol. Lett. *181*: 55–62.

Heyrman, J., A. Balcaen, M. Rodriguez-Diaz, N.A. Logan, J. Swings and P. De Vos. 2003a. *Bacillus decolorationis* sp. nov., isolated from biodeteriorated parts of the mural paintings at the Servilia tomb (Roman necropolis of Carmona, Spain) and the Saint-Catherine chapel (Castle Herberstein, Austria). Int. J. Syst. Evol. Microbiol. *53*: 459–463.

Heyrman, J., N.A. Logan, H.J. Busse, A. Balcaen, L. Lebbe, M. Rodriguez-Diaz, J. Swings and P. De Vos. 2003b. *Virgibacillus carmonensis* sp nov., *Virgibacillus necropolis* sp nov and *Virgibacillus picturae* sp nov., three novel species isolated from deteriorated mural paintings, transfer of the species of the genus *Salibacillus* to *Virgibacillus*, as *Virgibacillus marismortui* comb. nov and *Virgibacillus salexigens* comb. nov., and emended description of the genus *Virgibacillus*. Int. J. Syst. Evol. Microbiol. *53*: 501–511.

Heyrman, J., B. Vanparys, N.A. Logan, A. Balcaen, M. Rodriguez-Diaz, A. Felske and P. De Vos. 2004. *Bacillus novalis* sp. nov., *Bacillus vireti* sp. nov., *Bacillus soli* sp. nov., *Bacillus bataviensis* sp. nov. and *Bacillus drentensis* sp. nov., from the Drentse A grasslands. Int. J. Syst. Evol. Microbiol. *54*: 47–57.

Heyrman, J., N.A. Logan, M. Rodriguez-Diaz, P. Scheldeman, L. Lebbe, J. Swings, M. Heyndrickx and P. De Vos. 2005a. Study of mural painting isolates, leading to the transfer of 'Bacillus maroccanus' and 'Bacillus carotarum' to *Bacillus simplex*, emended description of *Bacillus simplex*, re-examination of the strains previously attributed to 'Bacillus macroides' and description of *Bacillus muralis* sp. nov. Int. J. Syst. Evol. Microbiol. *55*: 119–131.

Heyrman, J., M. Rodriguez-Diaz, J. Devos, A. Felske, N.A. Logan and P. De Vos. 2005b. *Bacillus arenosi* sp. nov., *Bacillus arvi* sp. nov. and *Bacillus humi* sp. nov., isolated from soil. Int. J. Syst. Evol. Microbiol. *55*: 111–117.

Hoffmaster, A.R., R.F. Meyer, M.D. Bowen, C.K. Marston, R.S. Weyant, K. Thurman, S.L. Messenger, E.E. Minor, J.M. Winchell, M.V. Rassmussen, B.R. Newton, J.T. Parker, W.E. Morrill, N. McKinney, G.A. Barnett, J.J. Sejvar, J.A. Jernigan, B.A. Perkins and T. Popovic. 2002. Evaluation and validation of a real-time polymerase chain reaction assay for rapid identification of *Bacillus anthracis*. Emerg. Infect. Dis. *8*: 1178–1182.

Horikoshi, K. 1971. Production of alkaline enzymes by alkalophilic microorganisms. Part I. Alkaline protease produced by *Bacillus* No. 221. Agric. Biol. Chem. (Tokyo) *35*: 1404–1407.

Horikoshi, K. 1998. Alkaliphiles. *In* Horikoshi and Grant (Editors), Extremophiles. Wiley-Liss, New York, pp. 155–179.

Hosford, R.M. 1982. White blotch incited in wheat by *Bacillus megaterium* pv. cerealis. Phytopathol. *72*: 1453–1459.

Hua, N.P., A. Kanekiyo, K. Fujikura, H. Yasuda and T. Naganuma. 2007. *Halobacillus profundi* sp. nov. and *Halobacillus kuroshimensis* sp. nov., moderately halophilic bacteria isolated from a deep-sea methane cold seep. Int. J. Syst. Evol. Microbiol. *57*: 1243–1249.

Hudson, J.A., R.M. Daniel and H.W. Morgan. 1988. Isolation of a strain of *Bacillus schlegelii* from geothermally heated antarctic soil. FEMS Microbiol. Lett. *51*: 57–60.

Hungate, R.E. 1969. A roll tube method for cultivation of strict anaerobes. *In* Norris and Ribbons (Editors), Methods in Microbiology, vol. 3B. Academic Press, London and New York, pp. 117–132.

Hunger, W. and D. Claus. 1981. Taxonomic studies on *Bacillus megaterium* and on agarolytic strains. *In* Berkeley and Goodfellow (Editors), Classification and identification of the aerobic endospore-forming bacteria. Academic Press, London, pp. 217–239.

Huu, N.B., E.B.M. Denner, D.T.C. Ha, G. Wanner and H. Stan-Lotter. 1999. *Marinobacter aquaeolei* sp. nov., a halophilic bacterium isolated from a Vietnamese oil-producing well. Int. J. Syst. Bacteriol. *49*: 367–375.

Iqbal, S.S., M.W. Mayo, J.G. Bruno, B.V. Bronk, C.A. Batt and J.P. Chambers. 2000. A review of molecular recognition technologies for detection of biological threat agents. Biosens. Bioelectron. *15*: 549–578.

Ishikawa, M., S. Ishizaki, Y. Yamamoto and K. Yamasato. 2002. *Paraliobacillus ryukyuensis* gen. nov., sp. nov., a new Gram-positive, slightly halophilic, extremely halotolerant, facultative anaerobe isolated from a decomposing marine alga. J. Gen. Appl. Microbiol. *48*: 269–279.

Ishikawa, M., S. Ishizaki, Y. Yamamoto and K. Yamasato. 2003a. *In* Validation of the publication of new names and new combinations previously effectively published outside the IJSEM. List no. 91. Int. J. Syst. Evol. Microbiol. *53*: 627–628.

Ishikawa, M., K. Nakajima, M. Yanagi, Y. Yamamoto and K. Yamasato. 2003b. *Marinilactibacillus psychrotolerans* gen. nov., sp. nov., a halophilic and alkaliphilic marine lactic acid bacterium isolated from marine organisms in temperate and subtropical areas of Japan. Int. J. Syst. Evol. Microbiol. *53*: 711–720.

Ishikawa, M., K. Nakajima, Y. Itamiya, S. Furukawa, Y. Yamamoto and K. Yamasato. 2005. *Halolactibacillus halophilus* gen. nov., sp. nov. and *Halolactibacillus miurensis* sp. nov., halophilic and alkaliphilic marine lactic acid bacteria constituting a phylogenetic lineage in *Bacillus* rRNA group 1. Int. J. Syst. Evol. Microbiol. *55*: 2427–2439.

Ivanova, N., A. Sorokin, I. Anderson, N. Galleron, B. Candelon, V. Kapatral, A. Bhattacharyya, G. Reznik, N. Mikhailova, A. Lapidus, L. Chu, M. Mazur, E. Goltsman, N. Larsen, M. D'Souza, T. Walunas, Y. Grechkin, G. Pusch, R. Haselkorn, M. Fonstein, S.D. Ehrlich, R. Overbeek and N. Kyrpides. 2003. Genome sequence of *Bacillus cereus* and comparative analysis with *Bacillus anthracis*. Nature *423*: 87–91.

Ivanova, E.P., Y.A. Alexeeva, N.V. Zhukova, N.M. Gorshkova, V. Buljan, D.V. Nicolau, V.V. Mikhailov and R. Christen. 2004a. *Bacillus algicola* sp. nov., a novel filamentous organism isolated from brown alga *Fucus evanescens*. Syst. Appl. Microbiol. *27*: 301–307.

Ivanova, E.P., Y.A. Alexeeva, N.V. Zhukova, N.M. Gorshkova, V. Buljan, D.V. Nicolau, V.V. Mikhailov and R. Christen. 2004b. Validation of publication of new names and new combinations previously effectively published outside the IJSEM. Int. J. Syst. Evol. Microbiol. *54*: 1425–1426.

Jackson, P.J., K.K. Hill, M.T. Laker, L.O. Ticknor and P. Keim. 1999. Genetic comparison of *Bacillus anthracis* and its close relatives using amplified fragment length polymorphism and polymerase chain reaction analysis. J. Appl. Microbiol. *87*: 263–269.

Janssen, P.H., S. Evers, F.A. Rainey, N. Weiss, W. Ludwig, C.G. Harfoot and B. Schink. 1995. *Lactosphaera* gen. nov., a new genus of lactic-acid bacteria, and transfer of *Ruminococcus pasteurii* Schink 1984 to *Lactosphaera pasteurii* comb. nov. Int. J. Syst. Bacteriol. *45*: 565–571.

Jeon, C.O., J.M. Lim, J.C. Lee, G.S. Lee, J.M. Lee, L.H. Xu, C.L. Jiang and C.J. Kim. 2005a. *Lentibacillus salarius* sp. nov., isolated from saline sediment in China, and emended description of the genus *Lentibacillus*. Int. J. Syst. Evol. Microbiol. *55*: 1339–1343.

Jeon, C.O., J.M. Lim, J.M. Lee, L.H. Xu, C.L. Jiang and C.J. Kim. 2005b. Reclassification of *Bacillus haloalkaliphilus* Fritze 1996 as *Alkalibacillus haloalkaliphilus* gen. nov., comb. nov. and the description of *Alkalibacillus salilacus* sp. nov., a novel halophilic bacterium isolated from a salt lake in China. Int. J. Syst. Evol. Microbiol. *55*: 1891–1896.

Jernigan, J.A., D.S. Stephens, D.A. Ashford, C. Omenaca, M.S. Topiel, M. Galbraith, M. Tapper, T.L. Fisk, S. Zaki, T. Popovic, R.F. Meyer, C.P. Quinn, S.A. Harper, S.K. Fridkin, J.J. Sejvar, C.W. Shepard, M. McConnell, J. Guarner, W.J. Shieh, J.M. Malecki, J.L. Gerberding, J.M. Hughes and B.A. Perkins. 2001. Bioterrorism-related inhalational anthrax: The first 10 cases reported in the United States. Emerg. Infect. Dis. *7*: 933–944.

Jock, S., B. Volksch, L. Mansvelt and K. Geider. 2002. Characterization of *Bacillus* strains from apple and pear trees in South Africa antagonistic to *Erwinia amylovora*. FEMS Microbiol. Lett. *211*: 247–252.

Johansen, T., C.R. Carlson and A.B. Kolsto. 1996. Variable numbers of rRNA gene operons in *Bacillus cereus* strains. FEMS Microbiol. Lett. *136*: 325–328.

Johnson, C. and A.H. Bishop. 1996. A technique for the effective enrichment and isolation of *Bacillus thuringiensis*. FEMS Microbiol. Lett. *142*: 173–177.

Johnstone, K. 1994. The trigger mechanism of spore germination – current concepts. J. Appl. Bacteriol. *76*: S17–S24.

Jones, B.E., W.D. Grant, N.C. Collins and W.E. Mwatha. 1994. Alkaliphiles: Diversity and identification. *In* Priest, Ramos-Cormenzana and Tindall (Editors), Bacterial Diversity and Systematics. Plenum, New York, pp. 195–230.

Jordan, E.O. 1890. A report on certain species of bacteria observed in sewage, Experimental investigations upon the purification of sewage by filtration and by chemical precipitation and upon the intermittent filtration of water. Rep. Mass. Bd. Publ. Hlth. 821–844.

Jorgensen, J., S. Dolan, A. Haselton and R. Kolchinsky. 1997. Isolation and cultivation of spore-forming filamentous bacteria from *Porcellio scaber*. Can. J. Microbiol. *43*: 129–135.

Jukes, T.H. and C. Cantor. 1969. Evolution of protein molecules. *In* Murano (Editor), Mammalian Protein Metabolism. Academic Press, New York, pp. 21–132.

Kadiyala, V., B.F. Smets, K. Chandran and J.C. Spain. 1998. High affinity p-nitrophenol oxidation by *Bacillus sphaericus* JS905. FEMS Microbiol. Lett. *166*: 115–120.

Kalogridou-Vassilliadu, D. 1992. Biochemical activities of *Bacillus* species isolated from flat sour evaporated milk. J. Dairy Sci. *75*: 2681–2686.

Kämpfer, P. 1994. Limits and possibilities of total fatty acid analysis for classification and identification of *Bacillus* species. Syst. Appl. Microbiol. *17*: 86–98.

Kämpfer, P. 2002. Whole-cell fatty acid analysis in the systematics of *Bacillus* and related genera. *In* Berkeley, Heyndrickx, Logan and De Vos (Editors), Applications and Systematics of *Bacillus* and Relatives. Blackwell Science Ltd., Oxford, Berlin, pp. 271–299.

Kaneda, T. 1977. Fatty acids of the genus *Bacillus*: An example of branched-chain preference. Bacteriol. Rev. *41*: 391–418.

Kanso, S., A.C. Greene and B.K. Patel. 2002. *Bacillus subterraneus* sp. nov., an iron- and manganese-reducing bacterium from a deep subsurface Australian thermal aquifer. Int. J. Syst. Evol. Microbiol. *52*: 869–874.

Kanzawa, Y., A. Harada, M. Takeuchi, A. Yokota and T. Harada. 1995. *Bacillus curdlanolyticus* sp. nov and *Bacillus kobensis* sp. nov, which hydrolyze resistant curdlan. Int. J. Syst. Bacteriol. *45*: 515–521.

Kato, T., M. Haruki, T. Imanaka, M. Morikawa and S. Kanaya. 2001. Isolation and characterization of long-chain-alkane degrading *Bacillus thermoleovorans* from deep subterranean petroleum reservoirs. J. Biosci. Bioeng. *91*: 64–70.

Kawamura, S., Y. Abe, T. Ueda, K. Masumoto, T. Imoto, N. Yamasaki and M. Kimura. 1998. Investigation of the structural basis for thermostability of DNA-binding protein HU from *Bacillus stearothermophilus*. J. Biol. Chem. *273*: 19982–19987.

Keim, P., A. Kalif, J. Schupp, K. Hill, S.E. Travis, K. Richmond, D.M. Adair, M. Hugh-Jones, C.R. Kuske and P. Jackson. 1997. Molecular evolution and diversity in *Bacillus anthracis* as detected by amplified fragment length polymorphism markers. J. Bacteriol. *179*: 818–824.

Keim, P., L.B. Price, A.M. Klevytska, K.L. Smith, J.M. Schupp, R. Okinaka, P.J. Jackson and M.E. Hugh-Jones. 2000. Multiple-locus variable-number tandem repeat analysis reveals genetic relationships within *Bacillus anthracis*. J. Bacteriol. *182*: 2928–2936.

Kellen, W.R., T.B. Clark, J.E. Lindegren, B.C. Ho, M.H. Rogoff and S. Singer. 1965. *Bacillus sphaericus* Neide as a pathogen of mosquitoes. J. Invertebr. Pathol. *7*: 442–448.

Kevbrin, V.V., T.N. Zhilina, F.A. Rainey and G.A. Zavarzin. 1998. *Tindallia magadii* gen. nov., sp. nov.: an alkaliphilic anaerobic ammonifier from soda lake deposits. Curr. Microbiol. *37*: 94–100.

Kevbrin, V.V., K. Zengler, A.M. Lysenko and J. Wiegel. 2005. *Anoxybacillus kamchatkensis* sp. nov., a novel thermophilic facultative aerobic bacterium with a broad pH optimum from the Geyser valley, Kamchatka. Extremophiles *9*: 391–398.

Khan, M.R., M.L. Saha and H. Afroz. 2001. Microorganisms associated with gemstones. Banglad. J. Bot. *30*: 93–96.

Kim, H.U. and J.M. Goepfert. 1972. Efficacy of a fluorescent-antibody procedure for identifying *Bacillus cereus* in foods. Appl. Microbiol. *24*: 708–713.

Kim, J.M., W.J. Lim and H.J. Suh. 2001. Feather-degrading *Bacillus* species from poultry waste. Proc. Biochem. *37*: 287–291.

Kim, Y.G., D.H. Choi, S. Hyun and B.C. Cho. 2007a. *Oceanobacillus profundus* sp. nov., isolated from a deep-sea sediment core. Int. J. Syst. Evol. Microbiol. *57*: 409–413.

Kim, Y.G., C.Y. Hwang, K.W. Yoo, H.T. Moon, J.H. Yoon and B.C. Cho. 2007b. *Pelagibacillus goriensis* gen. nov., sp. nov., a moderately halo-

tolerant bacterium isolated from coastal water off the east coast of Korea. Int. J. Syst. Evol. Microbiol. *57*: 1554–1560.

Klaushofer, H. and F. Hollaus. 1970. Zur Taxonomie der hoch-thermophilen, in Zukerfabriksäften vorkommenden aeroben sporebildner. Z. Zuckerrind. *20*: 465–470.

Klijn, N., L. Herman, L. Langeveld, M. Vaerewijck, A.A. Wagendorp, I. Huemer and A.H. Weerkamp. 1997. Genotypical and phenotypical characterization of *Bacillus sporothermodurans* surviving UHT sterilization. Int. Dairy J. *7*: 421–428.

Kluyver, A.J. and C.B. van Neil. 1936. Prospects for a natural classification of bacteria. Zentbl. Bacteriol. Parasitenkd. Infektionskr. Hyg. Abt. II *94*: 369–403.

Knapp, S., R. Ladenstein and E.A. Galinski. 1999. Extrinsic protein stabilization by the naturally occurring osmolytes beta-hydroxyectoine and betaine. Extremophiles *3*: 191–198.

Knight, B.C. and H. Proom. 1950. A comparative survey of the nutrition and physiology of mesophilic species in the genus *Bacillus*. J. Gen. Microbiol. *4*: 508–538.

Knight, P.J., N. Crickmore and D.J. Ellar. 1994. The receptor for *Bacillus thuringiensis* CrylA(c) delta-endotoxin in the brush border membrane of the lepidopteran *Manduca sexta* is aminopeptidase N. Mol. Microbiol. *11*: 429–436.

Knisely, R.F. 1966. Selective medium for *Bacillus anthracis*. J. Bacteriol. *92*: 784–786.

Knowles, B.H. 1994. Mechanism of action of *Bacillus thuringiensis* insecticidal delta-endotoxin. Adv. Insect Physiol. *24*: 275–308.

Ko, K.S., W.S. Oh, M.Y. Lee, J.H. Lee, H. Lee, K.R. Peck, N.Y. Lee and J.H. Song. 2006. *Bacillus infantis* sp. nov. and *Bacillus idriensis* sp. nov., isolated from a patient with neonatal sepsis. Int. J. Syst. Evol. Microbiol. *56*: 2541–2544.

Koch, R. 1876. Untersuchungen über Bakterien. V. Die Aetiologi der Milzbrand Krankheit, begrandet auf Entwicklurgs geschichte des *Bacillus anthracis*. Bertr. Biol. Pflanz. *2*: 277–308.

Kotiranta, A., M. Haapasalo, K. Kari, E. Kerosuo, I. Olsen, T. Sorsa, J.H. Meurman and K. Lounatmaa. 1998. Surface structure, hydrophobicity, phagocytosis, and adherence to matrix proteins of *Bacillus cereus* cells with and without the crystalline surface protein layer. Infect. Immun. *66*: 4895–4902.

Kotiranta, A., K. Lounatmaa and M. Haapasalo. 2000. Epidemiology and pathogenesis of *Bacillus cereus* infections. Microbes Infect. *2*: 189–198.

Kramer, J.M. and R.J. Gilbert. 1989. *Bacillus cereus* and other *Bacillus* species. *In* Doyle (Editor), Foodborne Bacterial Pathogens. Marcel Dekker, New York and Basel, pp. 21–70.

Kramer, J.M. and R.J. Gilbert. 1992. *Bacillus cereus* gastroenteritis. *In* Tu (Editor), Food Poisoning. Handbook of Natural Toxins, vol. 7. Marcel Dekker, New York, pp. 119–153.

Krüger, B. and O. Meyer. 1984. Thermophilic bacilli growing with carbon monoxide. Arch. Microbiol. *139*: 402–408.

Krych, V.K., J.L. Johnson and A.A. Yousten. 1980. Deoxyribonucleic acid homologies among strains of *Bacillus sphaericus*. Int. J. Syst. Bacteriol. *30*: 476–484.

Kuhlmann, A.U. and E. Bremer. 2002. Osmotically regulated synthesis of the compatible solute ectoine in *Bacillus pasteurii* and related *Bacillus* spp. Appl. Environ. Microbiol. *68*: 772–783.

Kuhnigk, T., E.M. Borst, A. Breunig, H. Konig, M.D. Collins, R.A. Hutson and P. Kämpfer. 1995. *Bacillus oleronius* sp. nov., a member of the hindgut flora of the termite *Reticulitermes santonensis* (Feytaud). Can. J. Microbiol. *41*: 699–706.

Kuhnigk, T., E.M. Borst, A. Breunig, H. Konig, M.D. Collins, R.A. Hutson and P. Kämpfer. 1996. *In* Validation of the publication of new names and new combinations previously effectively published outside the IJSB. List no. 57. Int. J. Syst. Bacteriol. *46*: 625–626.

Kuisiene, N., J. Raugalas and D. Chitavichius. 2004. *Geobacillus lituanicus* sp. nov. Int. J. Syst. Evol. Microbiol. *54*: 1991–1995.

Kunin, V., D. Ahren, L. Goldovsky, P. Janssen and C.A. Ouzounis. 2005. Measuring genome conservation across taxa: divided strains and united kingdoms. Nucleic Acids Res. *33*: 616–621.

Kunin, V., L. Goldovsky, N. Darzentas and C.A. Ouzounis. 2007. The net of life: reconstructing the microbial phylogenetic network. Genome Res. *15*: 954–959.

Kunst, F., N. Ogasawara, I. Moszer, A.M. Albertini, G. Alloni, V. Azevedo, M.G. Bertero, P. Bessieres, A. Bolotin, S. Borchert, R. Borriss, L. Boursier, A. Brans, M. Braun, S.C. Brignell, S. Bron, S. Brouillet, C.V. Bruschi, B. Caldwell, V. Capuano, N.M. Carter, S.K. Choi, J.J. Codani, I.F. Connerton, A. Danchin. and et al. 1997. The complete genome sequence of the gram-positive bacterium *Bacillus subtilis*. Nature *390*: 249–256.

Kuroshima, K., T. Sakane, R. Takata and A. Yokota. 1996. *Bacillus ehimensis* sp. nov. and *Bacillus chitinolyticus* sp. nov., new chitinolytic members of the genus *Bacillus*. Int. J. Syst. Bacteriol. *46*: 76–80.

Kushner, D.J. 1978. Life in high salt and solute concentrations: halophilic bacteria. *In* Kushner (Editor), Microbial Life in Extreme Environments, Academic Press, London, pp. 318–346.

Kushner, D.J. and K. Kamekura. 1988. Physiology of halophilic eubacteria. *In* Rodriguez-Valera (Editor), Halophilic Bacteria, vol. 1. CRC Press, Boca Raton, FL, pp. 109–140.

Kwon, S.W., S.Y. Lee, B.Y. Kim, H.Y. Weon, J.B. Kim, S.J. Go and G.B. Lee. 2007. *Bacillus niabensis* sp. nov., isolated from cotton-waste composts for mushroom cultivation. Int. J. Syst. Evol. Microbiol. *57*: 1909–1913.

Kwong, K.L., T.L. Que, S.N. Wong and K.T. So. 1997. Fatal meningoencephalitis due to *Bacillus anthracis*. J. Paediatr. Child Health *33*: 539–541.

L'Haridon, S., M.L. Miroshnichenko, N.A. Kostrikina, B.J. Tindall, S. Spring, P. Schumann, E. Stackebrandt, E.A. Bonch-Osmolovskaya and C. Jeanthon. 2006. *Vulcanibacillus modesticaldus* gen. nov., sp. nov., a strictly anaerobic, nitrate-reducing bacterium from deep-sea hydrothermal vents. Int. J. Syst. Evol. Microbiol. *56*: 1047–1053.

La Duc, M.T., M. Satomi and K. Venkateswaran. 2004. *Bacillus odysseyi* sp. nov., a round-spore-forming bacillus isolated from the Mars Odyssey spacecraft. Int. J. Syst. Evol. Microbiol. *54*: 195–201.

Lalitha, M.K. and M.K. Thomas. 1997. Penicillin resistance in *Bacillus anthracis*. Lancet *349*: 1522.

Lamana, C. 1940a. The taxonomy of the genus *Bacillus*. I. Modes of spore germination. J. Bacteriol. *40*: 347–359.

Lamana, C. 1940b. The taxonomy of the genus *Bacillus*. III. Differentiation of the large celled species by means of spore antigens. J. Infect. Dis. *67*: 205–212.

Lamana, C. 1940c. The taxonomy of the genus *Bacillus*. II. Differentiation of small celled species by means of spore antigens. J. Infect. Dis. *67*: 193–204.

Lamana, C. 1942. The status of *Bacillus subtilis*, including a note on the separation of precipitinogens from bacterial spores. J. Bact. *44*: 611–617.

Lamana, C. 1954. The problem of the nomenclature of *Bacillus subtilis* and *Bacillus* vulgatus. Int. Bull. Bact. Nomen. Taxon. *4*: 133–139.

Lamana, C. and D. Eisler. 1960. Comparative study of the agglutinogens of the endospores of *Bacillus anthracis* and *Bacillus cereus*. J. Bacteriol. *79*: 435–441.

Lane, H.C. and A.S. Fauci. 2001. Bioterrorism on the home front: a new challenge for American medicine. J. Am. Med. Assoc. *286*: 2595–2597.

Larkin, J.M. and J.L. Stokes. 1966. Isolation of psychrophilic species of *Bacillus*. J. Bacteriol. *91*: 1667–1671.

Larkin, J.M. and J.L. Stokes. 1967. Taxonomy of psychrophilic strains of *Bacillus*. J. Bacteriol. *94*: 889–895.

Larpin, S., N. Sauvageot, V. Pichereau, J.M. Laplace and Y. Auffray. 2002. Biosynthesis of exopolysaccharide by a *Bacillus licheniformis* strain isolated from ropy cider. Int. J. Food Microbiol. *77*: 1–9.

Laubach, C.A., J.L. Rice and W.W. Ford. 1916. Studies on aerobic spore-bearing non-pathogenic bacteria, Part II. J. Bacteriol. *1*: 493–533.

Lawrence, J.S. and W.W. Ford. 1916. Studies on aerobic spore-bearing non-pathogenic bacteria. Part I. J. Bacteriol. *1*: 273–320.

Lawson, P.A., C.E. Deutch and M.D. Collins. 1996. Phylogenetic characterization of a novel salt-tolerant *Bacillus* species: description of *Bacillus dipsosauri* sp. nov. J. Appl. Bacteriol. *81*: 109–112.

Leadbetter, E.R. and J.W. Foster. 1958. Studies on some methane-utilizing bacteria. Arch. Mikrobiol. *30*: 91–118.

Leary, J.V., N. Nelson, B. Tisserat and E.A. Allingham. 1986. Isolation of pathogenic *Bacillus circulans* from callus cultures and healthy off-shoots of date palm (*Phoenix dactylifera* L.). Appl. Environ. Microbiol. *52*: 1173–1176.

Lecadet, M.M., E. Frachon, V.C. Dumanoir, H. Ripouteau, S. Hamon, P. Laurent and I. Thiery. 1999. Updating the H-antigen classification of *Bacillus thuringiensis*. J. Appl. Microbiol. *86*: 660–672.

Lechner, S., R. Mayr, K.P. Francis, B.M. Pruss, T. Kaplan, E. Wiessner-Gunkel, G.S. Stewartz and S. Scherer. 1998. *Bacillus weihenstephanensis* sp. nov. is a new psychrotolerant species of the *Bacillus cereus* group. Int. J. Syst. Bacteriol. *48*: 1373–1382.

Lee, I.Y., T.G. Volm and J.P.N. Rosazza. 1998. Decarboxylation of ferulic acid to 4-vinylguaiacol by *Bacillus pumilis* in aqueous-organic solvent two-phase systems. Enzyme Microb. Technol. *23*: 261–266.

Lee, D., Y. Koh, K. Kim, B. Kim, H. Choi, D. Kim, M.T. Suhartono and Y. Pyun. 1999a. Isolation and characterization of a thermophilic lipase from *Bacillus thermoleovorans* ID-1. FEMS Microbiol. Lett. *179*: 393–400.

Lee, M.K., T.H. You, F.L. Gould and D.H. Dean. 1999b. Identification of residues in domain III of *Bacillus thuringiensis* Cry1Ac toxin that affect binding and toxicity. Appl. Environ. Microbiol. *65*: 4513–4520.

Lee, J.C., J.M. Lim, D.J. Park, C.O. Jeon, W.J. Li and C.J. Kim. 2006a. *Bacillus seohaeanensis* sp. nov., a halotolerant bacterium that contains L-lysine in its cell wall. Int. J. Syst. Evol. Microbiol. *56*: 1893–1898.

Lee, J.S., J.M. Lim, K.C. Lee, J.C. Lee, Y.H. Park and C.J. Kim. 2006b. *Virgibacillus koreensis* sp. nov., a novel bacterium from a salt field, and transfer of *Virgibacillus picturae* to the genus *Oceanobacillus* as *Oceanobacillus picturae* comb. nov. with emended descriptions. Int. J. Syst. Evol. Microbiol. *56*: 251–257.

Leighton, T.J. and R.H. Doi. 1971. The stability of messenger ribonucleic acid during sporulation in *Bacillus subtilis*. J. B Chem. *246*: 3189–3195.

Levin, P.A. and A.D. Grossman. 1998. Cell cycle and sporulation in *Bacillus subtilis*. Curr. Opin. Microbiol. *1*: 630–635.

Li, J.D., J. Carroll and D.J. Ellar. 1991. Crystal structure of insecticidal delta-endotoxin from *Bacillus thuringiensis* at 2.5 Å resolution. Nature *353*: 815–821.

Li, W.J., P. Schumann, Y.Q. Zhang, G.Z. Chen, X.P. Tian, L.H. Xu, E. Stackebrandt and C.L. Jiang. 2005. *Marinococcus halotolerans* sp. nov., isolated from Qinghai, north-west China. Int. J. Syst. Evol. Microbiol. *55*: 1801–1804.

Li, Z.Y., Y. Kawamura, O. Shida, S. Yamagata, T. Deguchi and T. Ezaki. 2002. *Bacillus okuhidensis* sp. nov., isolated from the Okuhida spa area of Japan. Int. J. Syst. Evol. Microbiol. *52*: 1205–1209.

Lightfoot, N.F., R.J.D. Scott and P.C.B. Turnbull. 1990. Antimicrobial susceptibility of *B. anthracis*. Salisbury Med. Bull. (Special Suppl.) *60*: 95–98.

Lighthart, B., K. Prier, G.M. Loper and J. Bromenshenk. 2000. Bees scavenge airborne bacteria. Microb. Ecol. *39*: 314–321.

Ligozzi, M., G. Lo Cascio and R. Fontana. 1998. *vanA* gene cluster in a vancomycin-resistant clinical isolate of *Bacillus circulans*. Antimicrob. Agents Chemother. *42*: 2055–2059.

Lim, J.M., C.O. Jeon, D.J. Park, H.R. Kim, B.J. Yoon and C.J. Kim. 2005a. *Pontibacillus marinus* sp. nov., a moderately halophilic bacterium from a solar saltern, and emended description of the genus *Pontibacillus*. Int. J. Syst. Evol. Microbiol. *55*: 1027–1031.

Lim, J.M., C.O. Jeon, S.M. Song and C.J. Kim. 2005b. *Pontibacillus chungwhensis* gen. nov., sp. nov., a moderately halophilic Gram-positive bacterium from a solar saltern in Korea. Int. J. Syst. Evol. Microbiol. *55*: 165–170.

Lim, J.M., C.O. Jeon, S.M. Song, J.C. Lee, Y.J. Ju, L.H. Xu, C.L. Jiang and C.J. Kim. 2005c. *Lentibacillus lacisalsi* sp. nov., a moderately halophilic bacterium isolated from a saline lake in China. Int. J. Syst. Evol. Microbiol. *55*: 1805–1809.

Lim, J.M., C.O. Jeon and C.J. Kim. 2006a. *Bacillus taeanensis* sp. nov., a halophilic Gram-positive bacterium from a solar saltern in Korea. Int. J. Syst. Evol. Microbiol. *56*: 2903–2908.

Lim, J.M., C.O. Jeon, J.C. Lee, Y.J. Ju, D.J. Park and C.J. Kim. 2006b. *Bacillus koreensis* sp. nov., a spore-forming bacterium, isolated from the rhizosphere of willow roots in Korea. Int. J. Syst. Evol. Microbiol. *56*: 59–63.

Lim, J.M., C.O. Jeon, S.M. Lee, J.C. Lee, L.H. Xu, C.L. Jiang and C.J. Kim. 2006c. *Bacillus salarius* sp. nov., a halophilic, spore-forming bacterium isolated from a salt lake in China. Int. J. Syst. Evol. Microbiol. *56*: 373–377.

Lim, J.M., C.O. Jeon, J.R. Lee, D.J. Park and C.J. Kim. 2007. *Bacillus kribbensis* sp. nov., isolated from a soil sample in Jeju, Korea. Int. J. Syst. Evol. Microbiol. *57*: 2912–2916.

Lin, Z.Y., K.J. Fu, J.M. Wu, Y.Y. Liu and H. Cheng. 2001. Preliminary study on the mechanism of non-enzymatic bioreduction of precious metal ions. Acta Phys.-Chim. Sinica *17*: 477–480.

Lindblow-Kull, C., A. Shrift and R.L. Gherna. 1982. Aerobic, selenium-utilizing *Bacillus* isolated from seeds of *Astragalus crotalariae*. Appl. Environ. Microbiol. *44*: 737–743.

Lindeque, P.M. and P.C. Turnbull. 1994. Ecology and epidemiology of anthrax in the Etosha National Park, Namibia. Onderstepoort J. Vet. Res. *61*: 71–83.

Little, S.F. and B.E. Ivins. 1999. Molecular pathogenesis of *Bacillus anthracis* infection. Microbes Infect. *1*: 131–139.

Liu, J.W., A.G. Porter, B.Y. Wee and T. Thanabalu. 1996. New gene from nine *Bacillus sphaericus* strains encoding highly conserved 35.8-kilodalton mosquitocidal toxins. Appl. Environ. Microbiol. *62*: 2174–2176.

Liu, P.Y., S.C. Ke and S.L. Chen. 1997. Use of pulsed-field gel electrophoresis to investigate a pseudo-outbreak of *Bacillus cereus* in a pediatric unit. J. Clin. Microbiol. *35*: 1533–1535.

Liu, S.M., T. Way, M. Rodrigues and S.M. Steidl. 2000. Effects of intravitreal corticosteroids in the treatment of *Bacillus cereus* endophthalmitis. Arch. Ophthalmol. *118*: 803–806.

Liu, W.Y., J. Zeng, L. Wang, Y.T. Dou and S.S. Yang. 2005. *Halobacillus dabanensis* sp. nov. and *Halobacillus aidingensis* sp. nov., isolated from salt lakes in Xinjiang, China. Int. J. Syst. Evol. Microbiol. *55*: 1991–1996.

Livingston, V.W. and E. Alexander-Jackson. 1970. A specific type of organism cultivated from malignancy: bacteriology and proposed classification. Ann. N. Y. Acad. Sci. *174*: 636–654.

Llarch, À., N.A. Logan, J. Castellví, M.J. Prieto and J. Guinea. 1997. Isolation and characterization of thermophilic *Bacillus* spp. from geothermal environments on Deception Island, South Shetland archipelago. Microb. Ecol. *34*: 58–65.

Llaudes, M.K., L. Zhao, S. Duffy and D.W. Schaffner. 2001. Simulation and modelling of the effect of small inoculum size on time to spoilage by *Bacillus stearothermophilus*. Food Microbiol. *18*: 395–405.

Loewen, P.C., J. Switala and B.L. Triggs-Raine. 1985. Catalases HPI and HPII in *Escherichia coli* are induced independently. Arch. Biochem. Biophys. *243*: 144–149.

Logan, N.A. 1988. *Bacillus* species of medical and veterinary importance. J. Med. Microbiol. *25*: 157–165.

Logan, N.A. 1994. Bacterial Systematics. Blackwell Scientific Publications, Oxford.

Logan, N.A. 2002. Modern methods for identification. *In* Berkley, Heyndrickx, Logan and De Vos (Editors), Applications and Systematics of *Bacillus* and Relatives. Blackwell Science, Oxford, pp. 123–140.

Logan, N.A. and R.C.W. Berkeley. 1981. Classification and identification of members of the genus *Bacillus* using API tests. *In* Berkeley and Goodfellow (Editors), The Aerobic Endospore-Forming Bacteria, vol. 106–140. Academic Press, London.

Logan, N.A. and R.C. Berkeley. 1984. Identification of *Bacillus* strains using the API system. J. Gen. Microbiol. *130*: 1871–1882.

Logan, N.A. and P.C.B. Turnbull. 2003. *Bacillus* and other aerobic endospore-forming bacteria. *In* Murray, Baron, Jorgensen, Pfaller and Yolken (Editors), Manual of Clinical Microbiology, 8th ed., vol. 1. ASM, Washington, D.C., pp. 445–460.

Logan, N.A., B.J. Capel, J. Melling and R.C.W. Berkeley. 1979. Distinction between emetic and other strains of *Bacillus cereus* using the API system and numerical-methods. FEMS Microbiol. Lett. *5*: 373–375.

Logan, N.A., J.A. Carman, J. Melling and R.C. Berkeley. 1985. Identification of *Bacillus anthracis* by API tests. J. Med. Microbiol. *20*: 75–85.

Logan, N.A., L. Lebbe, B. Hoste, J. Goris, G. Forsyth, M. Heyndrickx, B.L. Murray, N. Syme, D.D. Wynn-Williams and P. De Vos. 2000. Aerobic endospore-forming bacteria from geothermal environments in northern Victoria Land, Antarctica, and Candlemas Island, South Sandwich archipelago, with the proposal of *Bacillus fumarioli* sp. nov. Int. J. Syst. Evol. Microbiol. *50*: 1741–1753.

Logan, N.A., G. Forsyth, L. Lebbe, J. Goris, M. Heyndrickx, A. Balcaen, A. Verhelst, E. Falsen, A. Ljungh, H.B. Hansson and P. De Vos. 2002a. Polyphasic identification of *Bacillus* and *Brevibacillus* strains from clinical, dairy and industrial specimens and proposal of *Brevibacillus invocatus* sp. nov. Int. J. Syst. Evol. Microbiol. *52*: 953–966.

Logan, N.A., L. Lebbe, A. Verhelst, J. Goris, G. Forsyth, M. Rodriguez-Diaz, M. Heyndrickx and P. De Vos. 2002b. *Bacillus luciferensis* sp. nov., from volcanic soil on Candlemas Island, South Sandwich archipelago. Int. J. Syst. Evol. Microbiol. *52*: 1985–1989.

Logan, N.A., E. De Clerck, L. Lebbe, A. Verhelst, J. Goris, G. Forsyth, M. Rodriguez-Diaz, M. Heyndrickx and P. De Vos. 2004a. *Paenibacillus cineris* sp. nov. and *Paenibacillus cookii* sp. nov., from Antarctic volcanic soils and a gelatin-processing plant. Int. J. Syst. Evol. Microbiol. *54*: 1071–1076.

Logan, N.A., L. Lebbe, A. Verhelst, J. Goris, G. Forsyth, M. Rodriguez-Diaz, M. Heyndrickx and P. De Vos. 2004b. *Bacillus shackletonii* sp. nov., from volcanic soil on Candlemas Island, South Sandwich archipelago. Int. J. Syst. Evol. Microbiol. *54*: 373–376.

Logan, N.A., T.J. Popovitch and A. Hoffmaster. 2007. *Bacillus* and related genera. *In* Murray, Baron, Jorgensen, Pfaller and Yolken (Editors), Manual of Clinical Microbiology, 9th ed., vol. 1. ASM, Washington, pp. 455–473.

Louis, P. and E.A. Galinski. 1997a. Identification of plasmids in the genus *Marinococcus* and complete nucleotide sequence of plasmid pPL1 from *Marinococcus halophilus*. Plasmid *38*: 107–114.

Louis, P. and E.A. Galinski. 1997b. Characterization of genes for the biosynthesis of the compatible solute ectoine from *Marinococcus halophilus* and osmoregulated expression in *Escherichia coli*. Microbiology *143*: 1141–1149.

Lu, J., Y. Nogi and H. Takami. 2001. *Oceanobacillus iheyensis* gen. nov., sp. nov., a deep-sea extremely halotolerant and alkaliphilic species isolated from a depth of 1050 m on the Iheya Ridge. FEMS Microbiol. Lett. *205*: 291–297.

Lu, J., Y. Nogi and H. Takami. 2002. *In* Validation of the publication of new names and new combinations previously effectively published outside the IJSEM. List no. 85. Int. J. Syst. Evol. Microbiol. *52*: 685–690.

Ludwig, W. and O. Strunk. 1997. Posting date. The ARB project. http://www.arb-home.de. [Online.].

Lund, T., M.L. De Buyser and P.E. Granum. 2000. A new cytotoxin from *Bacillus cereus* that may cause necrotic enteritis. Mol. Microbiol. *38*: 254–261.

Lyman, J. and R.H. Fleming. 1940. Composition of sea water. J. Mar. Res. *3*: 134–146.

Macy, J.M., J.E. Snellen and R.E. Hungate. 1972. Use of syringe methods for anaerobiosis. Am. J. Clin. Nutr. *25*: 1318–1323.

Magee, J.T. and R. Goodacre. 2002. Fingerprint spectrometry methods in *Bacillus* systematics. *In* Berkeley, Heyndrickx, Logan and De Vos

(Editors), Applications and Systematics of *Bacillus* and Relatives. Blackwell Science, Oxford, pp. 254–270.

Maghnouj, A., T.F. de Sousa Cabral, V. Stalon and C. Vander Wauven. 1998. The *arcABDC* gene cluster, encoding the arginine deiminase pathway of *Bacillus licheniformis*, and its activation by the arginine repressor argR. J. Bacteriol. *180*: 6468–6475.

Mahler, H., A. Pasi, J.M. Kramer, P. Schulte, A.C. Scoging, W. Bar and S. Krahenbuhl. 1997. Fulminant liver failure in association with the emetic toxin of *Bacillus cereus*. N. Engl. J. Med. *336*: 1142–1148.

Makino, S. and H.I. Cheun. 2003. Application of the real-time PCR for the detection of airborne microbial pathogens in reference to the anthrax spores. J. Microbiol. Methods *53*: 141–147.

Manachini, P.L., M.G. Fortina, C. Parini and R. Craveri. 1985. *Bacillus thermoruber* sp. nov., nom. rev., a red-pigmented thermophilic bacterium. Int. J. Syst. Bacteriol. *35*: 493–496.

Manachini, P.L., D. Mora, G. Nicastro, C. Parini, E. Stackebrandt, R. Pukall and M.G. Fortina. 2000. *Bacillus thermodenitrificans* sp. nov., nom. rev. Int. J. Syst. Evol. Microbiol. *50*: 1331–1337.

Mangold, T. and J. Goldberg. 1999. Plague Wars: A True Story of Biological Warfare. Macmillan, London.

Manosroi, J., M. Abe and A. Manosroi. 1999. Biotransformation of steroidal drugs using microorganisms screened from various sites in Chiang Mai, Thailand. Bioresour. Technol. *69*: 67–73.

Mantynen, V. and K. Lindstrom. 1998. A rapid PCR-based DNA test for enterotoxic *Bacillus cereus*. Appl. Environ. Microbiol. *64*: 1634–1639.

Marafie, S.M.R.H. and L. Ashkanani. 1991. Airborne bacteria in Kuwait (1986–1988). Grana *30*: 472–476.

Marchant, R., I.M. Banat, T.J. Rahman and M. Berzano. 2002. What are high-temperature bacteria doing in cold environments? Trends Microbiol. *10*: 120–121.

Margulis, L., J.Z. Jorgensen, S. Dolan, R. Kolchinsky, F.A. Rainey and S.C. Lo. 1998. The Arthromitus stage of *Bacillus cereus*: intestinal symbionts of animals. Proc. Natl. Acad. Sci. USA *95*: 1236–1241.

Markossian, S., P. Becker, H. Markl and G. Antranikian. 2000. Isolation and characterization of lipid-degrading *Bacillus thermoleovorans* IHI-91 from an icelandic hot spring. Extremophiles *4*: 365–371.

Marquez, M.C., A. Ventosa and F. Ruiz-Berraquero. 1990. *Marinococcus hispanicus*, a new species of moderately halophilic Gram-positive cocci. Int. J. Syst. Bacteriol. *40*: 165–169.

Marquez, M.C., A. Ventosa and F. Ruizberraquero. 1992. Phenotypic and chemotaxonomic characterization of *Marinococcus halophilus*. Syst. Appl. Microbiol. *15*: 63–69.

Márquez, M.C., E. WQuesada, V. Bejar and A. Ventosa. 1993. A chemotaxonomic study of some moderately halophilic Gram-positive isolates. J. Appl. Bacteriol. *75*: 604–607.

Marshall, B.J. and D.F. Ohye. 1966. Bacillus macquariensis n. sp., a psychrotrophic bacterium from sub-Antarctic soil. J. Gen. Microbiol. *44*: 41–46.

Marshman, L.A., C. Hardwidge and P.M. Donaldson. 2000. *Bacillus cereus* meningitis complicating cerebrospinal fluid fistula repair and spinal drainage. Br. J. Neurosurg. *14*: 580–582.

Marteinsson, V.G., J.L. Birrien, C. Jeanthon and D. Prieur. 1996. Numerical taxonomic study of thermophilic *Bacillus* isolated from three geographically separated deep-sea hydrothermal vents. FEMS Microbiol. Ecol. *21*: 255–266.

Martins, L.O., A.S. Jurado and V.M. Madeira. 1990. Composition of polar lipid acyl chains of *Bacillus stearothermophilus* as affected by temperature and calcium. Biochim. Biophys. Acta *1045*: 17–20.

Massie, J., G. Roberts and P.J. White. 1985. Selective isolation of *Bacillus sphaericus* from soil by use of acetate as the only major source of carbon. Appl. Environ. Microbiol. *49*: 1478–1481.

Matsushita, M., J. Wakita, H. Itoh, I. Ràfols, T. Matsuyama, H. Sakaguchi and M. Mimura. 1998. Interface growth and pattern formation in bacterial colonies. Physica A: Stat. Theoret. Phys. *249*: 517–524.

Matsushita, M., J. Wakita, H. Itoh, K. Watanabe, T. Arai, T. Matsuyama, H. Sakaguchi and M. Mimura. 1999. Formation of colony patterns by a bacterial cell population. Physica A: Stat. Mech. Appl. *274*: 190–199.

Maugeri, T.L., C. Gugliandolo, D. Caccamo and E. Stackebrandt. 2001. A polyphasic taxonomic study of thermophilic bacilli from shallow, marine vents. Syst. Appl. Microbiol. *24*: 572–587.

Maughan, H., C.W. Birky, Jr., W.L. Nicholson, W.D. Rosenzweig and R.H. Vreeland. 2002. The paradox of the "ancient" bacterium which contains "modern" protein-coding genes. Mol. Biol. Evol. *19*: 1637–1639.

Mayr, R., H.J. Busse, H.L. Worliczek, M. Ehling-Schulz and S. Scherer. 2006. *Ornithinibacillus* gen. nov., with the species *Ornithinibacillus bavariensis* sp. nov. and *Ornithinibacillus californiensis* sp. nov. Int. J. Syst. Evol. Microbiol. *56*: 1383–1389.

Medina, A., A. Probanza, F.J.G. Manero and R. Azcon. 2003. Interactions of arbuscular-mycorrhizal fungi and *Bacillus* strains and their effects on plant growth, microbial rhizosphere activity (thymidine and leucine incorporation) and fungal biomass (ergosterol and chitin). Appl. Soil Ecol. *22*: 15–28.

MeierStauffer, K., H.J. Busse, F.A. Rainey, J. Burghardt, A. Scheberl, F. Hollaus, B. Kuen, A. Makristathis, U.B. Sleytr and P. Messner. 1996. Description of *Bacillus thermoaerophilus* sp. nov., to include sugar beet isolates and *Bacillus brevis* ATCC 12990. Int. J. Syst. Bacteriol. *46*: 532–541.

Merkel, G.J., W.H. Underwood and J.J. Perry. 1978. Isolation of thermophilic bacteria capable of growth solely in long-chain hydrocarbons. FEMS Microbiol. Lett. *3*: 81–83.

Meselson, M., J. Guillemin, M. Hughjones, A. Langmuir, I. Popova, A. Shelokov and O. Yampolskaya. 1994. The Sverdlovsk anthrax outbreak of 1979. Science *266*: 1202–1208.

Meyer, O. and H.G. Schlegel. 1983. Biology of aerobic carbon monoxide oxidizing bacteria. Ann. Rev. Microbiol. *37*: 277–310.

Meyers, P.R., P. Gokool, D.E. Rawlings and D.R. Woods. 1991. An efficient cyanide-degrading *Bacillus pumilus* strain. J. Gen. Microbiol. *137*: 1397–1400.

Meyers, P.R., D.E. Rawlings, D.R. Woods and G.G. Lindsey. 1993. Isolation and characterization of a cyanide dihydratase from *Bacillus pumilus* C1. J. Bacteriol. *175*: 6105–6112.

Michael, W.M. 2001. Cell cycle: Connecting DNA replication to sporulation in *Bacillus*. Curr. Biol. *11*: R443–R445.

Mignot, T., B. Denis, E. Couture-Tosi, A.B. Kolsto, M. Mock and A. Fouet. 2001. Distribution of S-layers on the surface of *Bacillus cereus* strains: phylogenetic origin and ecological pressure. Environ. Microbiol. *3*: 493–501.

Migula, W. 1900. System der Bakterien. Handbuch der Morphologie, Entwicklungsgeschichte und Systematik der bacterien, vol. 2; p. 583. G. Fischer Verlag Jena.

Mikkola, R., M. Kolari, M.A. Andersson, J. Helin and M.S. Salkinoja-Salonen. 2000. Toxic lactonic lipopeptide from food poisoning isolates of *Bacillus* licheniformis. Eur. J. Biochem. *267*: 4068–4074.

Miller, T.L. and M.J. Wolin. 1974. A serum bottle modification of the Hungate technique for cultivating obligate anaerobes. Appl. Microbiol. *27*: 985–987.

Miller, M.B. and B.L. Bassler. 2001. Quorum sensing in bacteria. Annu. Rev. Microbiol. *55*: 165–199.

Minnikin, D.E., H. Abdolrahimzadeh and J. Wolf. 1977. Taxonomic significance of polar lipids in some thermophilic members of *Bacillus*. *In* Barker, Wolf, Ellar, Dring and Gould (Editors), Spore Research, 1976. Academic Press, London, pp. 879–893.

Mishustin, E.N. 1950. Termofilnie Mikroorganiszmi w Prirode I Praktike. Akademi Nauk SSSR, Moskwa.

Miteva, V., S. Selenska-Pobell and V. Mitev. 1999. Random and repetitive primer amplified polymorphic DNA analysis of *Bacillus sphaericus*. J. Appl. Microbiol. *86*: 928–936.

Mizuki, E., M. Obha, T. Ichimatsu, S.H. Hwang, K. Higuchi, H. Saitoh and T. Akao. 1998. Unique appendages associated with spores of *Bacillus cereus* isolates. J. Basic Microbiol. *38*: 33–39.

Mock, M. and A. Fouet. 2001. Anthrax. Annu. Rev. Microbiol. *55*: 647–671.

Mohammed, M.J., C.K. Marston, T. Popovic, R.S. Weyant and F.C. Tenover. 2002. Antimicrobial susceptibility testing of *Bacillus anthracis*: Comparison of results obtained by using the National Committee for Clinical Laboratory Standards broth microdilution reference and Etest agar gradient diffusion methods. J. Clin. Microbiol. *40*: 1902–1907.

Moir, A., B.M. Corfe and J. Behravan. 2002. Spore germination. Cell Mol. Life Sci. *59*: 403–409.

Montefusco, A., L.K. Nakamura and D.P. Labeda. 1993. *Bacillus peoriae* sp. nov. Int. J. Syst. Bacteriol. *43*: 388–390.

Monteoliva-Sanchez, M., A. Ventosa and A. Ramos-Cormenzana. 1989. Cellular fatty acid composition of moderately halophilic cocci. Syst. Appl. Microbiol. *12*: 141–144.

Mora, D., M.G. Fortina, G. Nicastro, C. Parini and P.L. Manachini. 1998. Genotypic characterization of thermophilic bacilli: A study on new soil isolates and several reference strains. Res. Microbiol. *149*: 711–722.

Muñoz, J.A., B. Perez-Esteban, M. Esteban, S. de la Escalera, M.A. Gomez, M.V. Martinez-Toledo and J. Gonzalez-Lopez. 2001. Growth of moderately halophilic bacteria isolated from sea water using phenol as the sole carbon source. Folia Microbiol. (Praha) *46*: 297–302.

Murakami, T., K. Hiraoka, T. Mikami, T. Matsumoto, S. Katagiri, K. Shinagawa and M. Suzuki. 1993. Analysis of common antigen of flagella in *Bacillus cereus* and *Bacillus thuringiensis*. FEMS Microbiol. Lett. *107*: 179–183.

Murphy, J.A. and L.L. Campbell. 1969. Surface features of *Bacillus polymyxa* spores as revealed by scanning electron microscopy. J. Bacteriol. *98*: 737–743.

Myerowitz, R.L., R.E. Gordon and J.B. Robbins. 1973. Polysaccharides of the genus *Bacillus* cross-reactive with the capsular polysaccharides of *Diplococcus pneumoniae* type 3, *Haemophilus influenzae* type b, and *Neisseria meningitidis* group A. Infect. Immun. *8*: 896–900.

Nagel, M. and J.R. Andreesen. 1989. Molybdenum-dependent degradation of nicotinic acid by *Bacillus* sp. DSM 2923. FEMS Microbiol. Lett. *59*: 147–152.

Nagel, M. and J.R. Andreesen. 1991. *Bacillus niacini* sp. nov., a nicotinate-metabolizing mesophile isolated from soil. Int. J. Syst. Bacteriol. *41*: 134–139.

Nakada, Y. and Y. Ohta. 1998. Hydrogen sulfide removal by a deodorant bacterium *Bacillus* BN53-1. J. Ferment. Bioeng. *84*: 614.

Nakajima, K., K. Hirota, Y. Nodasaka and I. Yumoto. 2005. *Alkalibacterium iburiense* sp. nov., an obligate alkaliphile that reduces an indigo dye. Int. J. Syst. Evol. Microbiol. *55*: 1525–1530.

Nakamura, L.K. 1984a. *Bacillus amylolyticus* sp.nov., nom. rev., *Bacillus lautus* sp. nov., nom. rev., *Bacillus pabuli* sp. nov., nom. rev, and *Bacillus validus* sp. nov, nom. rev. Int. J. Syst. Bacteriol. *34*: 224–226.

Nakamura, L.K. 1984b. *Bacillus psychrophilus* sp. nov., nom. rev. Int. J. Syst. Bacteriol. *34*: 121–123.

Nakamura, L.K. 1984c. *Bacillus pulvifaciens* sp. nov, nom. rev. Int. J. Syst. Bacteriol. *34*: 410–413.

Nakamura, L.K. 1987. *Bacillus alginolyticus* sp. nov. and *Bacillus chondroitinus* sp. nov., two alginate-degrading species. Int. J. Syst. Bacteriol. *37*: 284–286.

Nakamura, L.K. 1989. Taxonomic relationship of black-pigmented *Bacillus subtilis* strains and a proposal for *Bacillus atrophaeus* sp. nov. Int. J. Syst. Bacteriol. *39*: 295–300.

Nakamura, L.K. 1990. *Bacillus thiaminolyticus* sp. nov., nom. rev. Int. J. Syst. Bacteriol. *40*: 242–246.

Nakamura, L.K. 1993. DNA relatedness of *Bacillus brevis* Migula 1900 strains and proposal of *Bacillus agri* sp. nov., nom. rev., and *Bacillus centrosporus* sp. nov., nom. rev. Int. J. Syst. Bacteriol. *43*: 20–25.

Nakamura, L.K. 1996. *Paenibacillus apiarius* sp. nov. Int. J. Syst. Bacteriol. *46*: 688–693.

Nakamura, L.K. 1998. *Bacillus pseudomycoides* sp. nov. Int. J. Syst. Bacteriol. *48*: 1031–1035.

Nakamura, L.K. and J. Swezey. 1983. Taxonomy of *Bacillus circulans* Jordan 1890: base compostition and reassociation of deoxyribonucleic acid. Int. J. Syst. Bacteriol. *33*: 46–52.

Nakamura, L.K., I. Blumenstock and D. Claus. 1988. Taxonomic study of *Bacillus coagulans* Hammer 1915 with a proposal for *Bacillus smithii* sp. nov. Int. J. Syst. Bacteriol. *38*: 63–73.

Nakamura, L.K., M.S. Roberts and F.M. Cohan. 1999. Relationship of *Bacillus subtilis* clades associated with strains 168 and W23: A proposal for *Bacillus subtilis* subsp. *subtilis* subsp. nov. and *Bacillus subtilis* subsp. *spizizenii* subsp. nov. Int. J. Syst. Bacteriol. *49*: 1211–1215.

Nakamura, L.K., O. Shida, H. Takagi and K. Komagata. 2002. *Bacillus pycnus* sp. nov. and *Bacillus neidei* sp. nov., round-spored bacteria from soil. Int. J. Syst. Evol. Microbiol. *52*: 501–505.

Nakamura, K., S. Haruta, H.L. Nguyen, M. Ishii and Y. Igarashi. 2004a. Enzyme production-based approach for determining the functions of microorganisms within a community. Appl. Environ. Microbiol. *70*: 3329–3337.

Nakamura, K., S. Haruta, S. Ueno, M. Ishii, A. Yokota and Y. Igarashi. 2004b. *Cerasibacillus quisquiliarum* gen. nov., sp. nov., isolated from a semi-continuous decomposing system of kitchen refuse. Int. J. Syst. Evol. Microbiol. *54*: 1063–1069.

Nakano, M.M., Y.P. Dailly, P. Zuber and D.P. Clark. 1997. Characterization of anaerobic fermentative growth of *Bacillus subtilis*: identification of fermentation end products and genes required for growth. J. Bacteriol. *179*: 6749–6755.

Nakano, M.M. and P. Zuber. 2002. Anaerobiosis. *In* Sonenshein, Hoch and Losick (Editors), *Bacillus* and its Closest Relatives. ASM Press, Washington, D.C., pp. 393–404.

Nakayama, O. and M. Yanoshi. 1967. Spore-bearing lactic acid bacteria isolated from rhizosphere. I. Taxonomic studies on *Bacillus laevolacticus* nov. sp. and *Bacillus racemilacticus* nov. sp. J. Gen. Appl. Microbiol. *13*: 139–153.

Namwong, S., S. Tanasupawat, T. Smitinont, W. Visessanguan, T. Kudo and T. Itoh. 2005. Isolation of *Lentibacillus salicampi* strains and *Lentibacillus juripiscarius* sp. nov. from fish sauce in Thailand. Int. J. Syst. Evol. Microbiol. *55*: 315–320.

Naumova, I.B. and A.S. Shashkov. 1997. Anionic polymers in cell walls of Gram-positive bacteria. Biochemistry (Mosc.) *62*: 809–840.

Nazina, T.N., T.P. Tourova, A.B. Poltaraus, E.V. Novikova, A.A. Grigoryan, A.E. Ivanova, A.M. Lysenko, V.V. Petrunyaka, G.A. Osipov, S.S. Belyaev and M.V. Ivanov. 2001. Taxonomic study of aerobic thermophilic bacilli: descriptions of *Geobacillus subterraneus* gen. nov., sp. nov. and *Geobacillus uzenensis* sp. nov. from petroleum reservoirs and transfer of *Bacillus stearothermophilus*, *Bacillus thermocatenulatus*, *Bacillus thermoleovorans*, *Bacillus kaustophilus*, *Bacillus thermoglucosidasius* and *Bacillus thermodenitrificans* to *Geobacillus* as the new combinations *G. stearothermophilus*, *G. thermocatenulatus*, *G. thermoleovorans*, *G. kaustophilus*, *G. thermoglucosidasius* and *G. thermodenitrificans*. Int. J. Syst. Evol. Microbiol. *51*: 433–446.

Nazina, T.N., E.V. Lebedeva, A.B. Poltaraus, T.P. Tourova, A.A. Grigoryan, D. Sokolova, A.M. Lysenko and G.A. Osipov. 2004. *Geobacillus gargensis* sp. nov., a novel thermophile from a hot spring, and the reclassification of *Bacillus vulcani* as *Geobacillus vulcani* comb. nov. Int. J. Syst. Evol. Microbiol. *54*: 2019–2024.

Nazina, T.N., D.Sh. Sokolova, A.A. Grigoryan, N.M. Shestakova, E.M. Mikhailova, A.B. Poltaraus, T.P. Tourova, A.M. Lysenko, G.A. Osipov and S.S. Belyaev. 2005a. *In* Validation of the publication of new names and new combinations previously effectively published outside the IJSEM. List no. 103. Int. J. Syst. Evol. Microbiol. *55*: 983–985.

Nazina, T.N., D.S. Sokolova, A.A. Grigoryan, N.M. Shestakova, E.M. Mikhailova, A.B. Poltaraus, T.P. Tourova, A.M. Lysenko, G.A. Osipov and S.S. Belyaev. 2005b. *Geobacillus jurassicus* sp. nov., a new thermophilic bacterium isolated from a high-temperature petroleum *Geobacillus* species reservoir, and the validation of the *Geobacillus* species. Syst. Appl. Microbiol. *28*: 43–53.

Neide, E. 1904. Botanische Beschreibung einiger sporenbildenden Bakterien. Zentbl. Bakteriol. Parasitenkd. Infektionskr. Hyg. Abt. II *12*: 337–352.

Niazi, J.H., D.T. Prasad and T.B. Karegoudar. 2001. Initial degradation of dimethylphthalate by esterases from *Bacillus* species. FEMS Microbiol. Lett. *196*: 201–205.

Nicholson, C.A. and B.Z. Fathepure. 2004. Biodegradation of benzene by halophilic and halotolerant bacteria under aerobic conditions. Appl. Environ. Microbiol. *70*: 1222–1225.

Nicholson, W.L., N. Munakata, G. Horneck, H.J. Melosh and P. Setlow. 2000. Resistance of *Bacillus* endospores to extreme terrestrial and extraterrestrial environments. Microbiol. Mol. Biol. Rev. *64*: 548–572.

Nickle, D.C., G.H. Learn, M.W. Rain, J.I. Mullins and J.E. Mittler. 2002. Curiously modern DNA for a "250 million-year-old" bacterium. J. Mol. Evol. *54*: 134–137.

Nicolaus, B., M.C. Manca, L. Lama, E. Esposito and A. Gambacorta. 1995. Effects of growth temperature on the polar lipid pattern and fatty acid composition of seven thermophilic isolates from the antarctic continent. Syst. Appl. Microbiol. *18*: 32–36.

Nicolaus, B., L. Lama, E. Esposito, M.C. Manca, G. diPrisco and A. Gambacorta. 1996. "*Bacillus thermoantarcticus*" sp. nov., from Mount Melbourne, Antarctica: a novel thermophilic species. Polar Biol. *16*: 101–104.

Nicolaus, B., R. Improta, M.C. Manca, L. Lama, E. Esposito and A. Gambacorta. 1998. Alicyclobacilli from an unexplored geothermal soil in Antarctica: Mount Rittmann. Polar Biol. *19*: 133–141.

Nicolaus, B., L. Lama, E. Esposito, M.C. Manca, G. di Prisco and A. Gambacorta. 2002. *In* Validation of the publication of new names and new combinations previously effectively published outside the IJSEM. List no. 84. Int. J. Syst. Evol. Microbiol. *52*: 3–4.

Nielsen, P., F.A. Rainey, H. Outtrup, F.G. Priest and D. Fritze. 1994. Comparative 16S rDNA sequence-analysis of some alkaliphilic bacilli and the establishment of a 6th rRNA group within the genus *Bacillus*. FEMS Microbiol. Lett. *117*: 61–65.

Nielsen, P., D. Fritze and F.G. Priest. 1995a. Phenetic diversity of alkaliphilic *Bacillus* strains: proposal for nine new species. Microbiology *141*: 1745–1761.

Nielsen, P., D. Fritze and F.G. Priest. 1995b. *In* Validation of the publication of new names and new combinations previously effectively published outside the IJSB. List no. 55. Int. J. Syst. Bacteriol. *45*: 879–880.

Nieto, J.J., R. Fernandez-Castillo, M.C. Marquez, A. Ventosa, E. Quesada and F. Ruiz-Berraquero. 1989. Survey of metal tolerance in moderately halophilic eubacteria. Appl. Environ. Microbiol. *55*: 2385–2390.

Niimura, Y., F. Yanagida, T. Uchimura, N. Ohara, K. Suzuki and M. Kozaki. 1987. A new facultative anaerobic xylan-using alkalophile lacking cytochrome, quinone, and catalase. Agric. Biol. Chem. (Tokyo) *51*: 2271–2275.

Niimura, Y., E. Koh, T. Uchimura, N. Ohara and M. Kozaki. 1989. Aerobic and anaerobic metabolism in a facultative anaerobe Ep01 lacking cytochrome, quinone and catalase. FEMS Microbiol. Lett. *61*: 79–83.

Niimura, Y., E. Koh, F. Yanagida, K.I. Suzuki, K. Komagata and M. Kozaki. 1990. *Amphibacillus xylanus* gen. nov., sp. nov., a facultatively anaerobic spore-forming xylan-digesting bacterium which lacks cytochrome, quinone, and catalase. Int. J. Syst. Bacteriol. *40*: 297–301.

Nishikawa, Y., J.M. Kramer, M. Hanaoka and A. Yasukawa. 1996. Evaluation of serotyping, biotyping, plasmid banding pattern analysis, and HEp-2 vacuolation factor assay in the epidemiological investigation of *Bacillus cereus* emetic-syndrome food poisoning. Int. J. Food Microbiol. *31*: 149–159.

Nishiyama, Y., V. Massey, K. Takeda, S. Kawasaki, J. Sato, T. Watanabe and Y. Niimura. 2001. Hydrogen peroxide-forming NADH oxidase belonging to the peroxiredoxin oxidoreductase family: existence and physiological role in bacteria. J. Bacteriol. *183*: 2431–2438.

Nogi, Y., H. Takami and K. Horikoshi. 2005. Characterization of alkaliphilic *Bacillus* strains used in industry: proposal of five novel species. Int. J. Syst. Evol. Microbiol. *55*: 2309–2315.

Noguchi, H., M. Uchino, O. Shida, K. Takano, L.K. Nakamura and K. Komagata. 2004. *Bacillus vietnamensis* sp. nov., a moderately halotolerant, aerobic, endospore-forming bacterium isolated from Vietnamese fish sauce. Int. J. Syst. Evol. Microbiol. *54*: 2117–2120.

Nomenclature Committee of the International Society for Microbiology. 1937. Resolutions of the Nomenclature Committee. *In* Second International Congress for Microbiology, London 1936. Report of Proceedings. *In* St.-John-Brooks (Editor), International Society for Microbiology, London, pp. 28–29.

Norris, J.R. 1962. Bacterial spore antigens: a review. J. Gen Microbiol. *28*: 393–408.

Norris, J.R. and J. Wolf. 1961. A study of antigens of the aerobic spore-forming bacteria. J. Appl. Bacteriol. *24*: 42–56.

Norris, J.R., R.C.W. Berkeley, N.A. Logan and A.G. O'Donnell. 1981. The genera *Bacillus* and *Sporolactobacillus*. *In* Starr, Trüper, Balows and Schlegel (Editors), The Prokaryotes: a Handbook on Habitats, Isolation and Identification of Bacteria, vol. 2, Springer-Verlag, Berlin and Heidelberg, pp. 1711–1742.

Nourbakhsh, M.N., S. Kiliçarslan, S. Ilhan and H. Ozdag. 2002. Biosorption of Cr^{6+}, Pb^{2+}, and Cu^{2+} ions in industrial wastewater on *Bacillus* sp. Chem. Eng. J. *85*: 351–355.

Novitsky, T.J. and D.J. Kushner. 1976. *Planococcus halophilus* sp. nov., a facultatively halophilic coccus. Int. J. Syst. Bacteriol. *26*: 53–57.

Nowlan, B., M.S. Dodia, S.P. Singh and B.K. Patel. 2006. *Bacillus okhensis* sp. nov., a halotolerant and alkalitolerant bacterium from an Indian saltpan. Int. J. Syst. Evol. Microbiol. *56*: 1073–1077.

Ntougias, S. and N.J. Russell. 2001. *Alkalibacterium olivoapovliticus* gen. nov., sp. nov., a new obligately alkaliphilic bacterium isolated from edible-olive wash-waters. Int. J. Syst. Evol. Microbiol. *51*: 1161–1170.

Nunes, I., I. Tiago, A.L. Pires, M.S. da Costa and A. Verissimo. 2006. *Paucisalibacillus globulus* gen. nov., sp. nov., a Gram-positive bacterium isolated from potting soil. Int. J. Syst. Evol. Microbiol. *56*: 1841–1845.

Nystrand, R. 1984a. *In* Validation of the publication of new names and new combinations previously effectively published outside the IJSB. List no.16. Int. J. Syst. Evol. Microbiol. *34*: 503–504.

Nystrand, R. 1984b. *Saccharococcus thermophilus* gen. nov., sp. nov. isolated from beet sugar extraction. Syst. Appl. Microbiol. *5*: 204–219.

Obojska, A., N.G. Ternan, B. Lejczak, P. Kafarski and G. McMullan. 2002. Organophosphonate utilization by the thermophile *Geobacillus caldoxylosilyticus* T20. Appl. Environ. Microbiol. *68*: 2081–2084.

Oggioni, M.R., G. Pozzi, P.E. Valensin, P. Galieni and C. Bigazzi. 1998. Recurrent septicemia in an immunocompromised patient due to probiotic strains of *Bacillus subtilis*. J. Clin. Microbiol. *36*: 325–326.

Ohba, M., K. Ueda and K. Aizawa. 1992. Serotyping of *Bacillus thuringiensis* environmental isolates by extracellular heat-stable somatic antigens. Can. J. Microbiol. *38*: 694–695.

Okamoto, K., P. Serror, V. Azevedo and B. Vold. 1993. Physical mapping of stable RNA genes in *Bacillus subtilis* using polymerase chain reaction amplification from a yeast artificial chromosome library. J. Bacteriol. *175*: 4290–4297.

Okinaka, R.T., K. Cloud, O. Hampton, A.R. Hoffmaster, K.K. Hill, P. Keim, T.M. Koehler, G. Lamke, S. Kumano, J. Mahillon, D. Manter, Y. Martinez, D. Ricke, R. Svensson and P.J. Jackson. 1999. Sequence and organization of pXO1, the large *Bacillus anthracis* plasmid harboring the anthrax toxin genes. J. Bacteriol. *181*: 6509–6515.

Oladosu, G.A., O.A. Ayinla and M.O. Ajiboye. 1994. Isolation and pathogenicity of a *Bacillus* sp. associated with a septicemic condition in some tropical freshwater fish species. J. Appl. Ichthyol. *10*: 69–72.

Olivera, N., F. Sineriz and J.D. Breccia. 2005. *Bacillus patagoniensis* sp. nov., a novel alkalitolerant bacterium from the rhizosphere of Atriplex lampa in Patagonia, Argentina. Int. J. Syst. Evol. Microbiol. *55*: 443–447.

Olszewski, W.L., S. Jamal, G. Manokaran, S. Pani, V. Kumaraswami, U. Kubicka, B. Lukomska, F.M. Tripathi, E. Swoboda, F. Meisel-Mikolajczyk, E. Stelmach and M. Zaleska. 1999. Bacteriological studies of blood, tissue fluid, lymph and lymph nodes in patients with acute dermatolymphangioadenitis (DLA) in course of 'filarial' lymphedema. Acta Trop. *73*: 217–224.

Onishi, H., H. Fuchi, K. Konomi, O. Hidaka and M. Kamekura. 1980. Isolation and distribution of a variety of halophilic bacteria and their classification by salt-response. Agric. Biolog. Chem. *44*: 1253–1258.

Ordal, G.W., D.P. Villani and M.S. Rosendahl. 1979. Chemotaxis towards sugars by *Bacillus subtilis*. J. Gen. Microbiol. *115*: 167–172.

Oren, A., P. Gurevich, M. Azachi and Y. Henis. 1992. Microbial degradation of pollutants at high salt concentrations. Biodegradation *3*: 387–398.

Ottow, J.C. 1974. Detection of hippurate hydrolase among *Bacillus* species by thin layer chromatography and other methods. J. Appl. Bacteriol. *37*: 15–30.

Outtrup, H. and S.T. Jørgensen. 2002. The importance of *Bacillus* species in the production of industrial enzymes. *In* Berkeley, Heyndrickx, Logan and De Vos (Editors), Applications and Systematics of *Bacillus* and Relatives. Blackwell Science, Oxford, pp. 206–218.

Oxborrow, G.S., N.D. Fields and J.R. Puleo. 1977. Pyrolysis gas-liquid chromatography of the genus *Bacillus*: effect of growth media on pyrochromatogram reproducibility. Appl. Environ. Microbiol. *33*: 865–870.

Paidhungat, M. and P. Setlow. 2002. Spore germination and outgrowth. *In* Sonenshein, Hoch and Losick (Editors), *Bacillus subtilis* and its Closest Relatives. ASM Press, Washington, D.C., pp. 537–548.

Pakdeeto, A., S. Tanasupawat, C. Thawai, S. Moonmangmee, T. Kudo and T. Itoh. 2007. *Lentibacillus kapialis* sp. nov., from fermented shrimp paste in Thailand. Int. J. Syst. Evol. Microbiol. *57*: 364–369.

Palmisano, M.M., L.K. Nakamura, K.E. Duncan, C.A. Istock and F.M. Cohan. 2001. *Bacillus sonorensis* sp. nov., a close relative of *Bacillus licheniformis*, isolated from soil in the Sonoran Desert, Arizona. Int. J. Syst. Evol. Microbiol. *51*: 1671–1679.

Pandey, A. and L.M. Palni. 1997. *Bacillus* species: the dominant bacteria of the rhizosphere of established tea bushes. Microbiol. Res. *152*: 359–365.

Pannifer, A.D., T.Y. Wong, R. Schwarzenbacher, M. Renatus, C. Petosa, J. Bienkowska, D.B. Lacy, R.J. Collier, S. Park, S.H. Leppla, P. Hanna and R.C. Liddington. 2001. Crystal structure of the anthrax lethal factor. Nature *414*: 229–233.

Pantasicocaldas, M., K.E. Duncan, C.A. Istock and J.A. Bell. 1992. Population dynamics of bacteriophage and *Bacillus subtilis* in soil. Ecology *75*: 1888–1902.

Papinutto, E., W.G. Dundon, N. Pitulis, R. Battistutta, C. Montecucco and G. Zanotti. 2002. Structure of two iron-binding proteins from *Bacillus anthracis*. J. Biol. Chem. *277*: 15093–15098.

Parry, J.M. and R.J. Gilbert. 1980. Studies on the heat resistance of *Bacillus cereus* spores and growth of the organism in boiled rice. J. Hyg. (Lond.) *84*: 77–82.

Parvanta, M.F. 2000. Abortion in a dairy herd associated with *Bacillus licheniformis*. Tierarztliche Umschau *55*: 126.

Pasteur, L. 1870. Études sur la maladie des vers à soie, moyen pratique assuré de la comnattre et d'en prévenir le retour. Gauthier Villars, Paris.

Patel, B.K.C., H.W. Morgan and R.M. Daniel. 1985. A simple and efficient method for preparing anaerobic media. Biotechnol. Lett. *7*: 227–228.

Paton, A.M. and C.M.J. Innes. 1991. Methods for the establishment of intracellular associations of L-forms with higher plants. J. Appl. Bacteriol. *71*: 59–64.

Peak, K.K., K.E. Duncan, W. Veguilla, V.A. Luna, D.S. King, L. Heller, L. Heberlein-Larson, F. Reeves, A.C. Cannons, P. Amuso and J. Cattani. 2007. *Bacillus acidiceler* sp. nov., isolated from a forensic specimen, containing *Bacillus anthracis* pX02 genes. Int. J. Syst. Evol. Microbiol. *57*: 2031–2036.

Pease, P. 1969. Bacterial L-forms in the blood and joint fluids of arthritic subjects. Ann. Rheum. Dis. *28*: 270–274.

Pease, P. 1970. Morphological appearances of a bacterial L-form growing in association with the erythrocytes of arthritic subjects. Ann. Rheum. Dis. *29*: 439–444.

Pease, P. 1974. Identification of bacteria from blood and joint fluids of human subjects as *Bacillus licheniformis*. Ann. Rheum. Dis. *33*: 67–69.

Pérez-Ibarra, B.M., M.E. Flores and M. García-Varela. 2007a. List of new names and new combinations previously effectively, but not validly, published. List no. 117. Int. J. Syst. Evol. Microbiol. *57*: 1933–1934.

Pérez-Ibarra, B.M., M.E. Flores and M. García-Varela. 2007b. Isolation and characterization of *Bacillus thioparus* sp. nov., chemolithoautotrophic, thiosulfate-oxidizing bacterium. FEMS Microbiol. Lett. *271*: 289–296.

Perl, D., U. Mueller, U. Heinemann and F.X. Schmid. 2000. Two exposed amino acid residues confer thermostability on a cold shock protein. Nat. Struct. Biol. *7*: 380–383.

Pettersson, B., F. Lembke, P. Hammer, E. Stackebrandt and F.G. Priest. 1996. *Bacillus sporothermodurans*, a new species producing highly heat-resistant endospores. Int. J. Syst. Bacteriol. *46*: 759–764.

Pettersson, B., S.K. de Silva, M. Uhlen and F.G. Priest. 2000. *Bacillus siralis* sp. nov., a novel species from silage with a higher order structural attribute in the 16S rRNA genes. Int. J. Syst. Evol. Microbiol. *50*: 2181–2187.

Phillips, A.P. and K.L. Martin. 1983a. Comparison of direct and indirect immunoradiometric assays (IRMA) for *Bacillus anthracis* spores immobilised on multispot microscope slides. J. Appl. Bacteriol. *55*: 315–324.

Phillips, A.P. and K.L. Martin. 1983b. Comparison of immunoradiometric assays of *Bacillus anthracis* spores immobilised on multispot slides and on microtitre plates. J. Immunol. Methods *62*: 273–282.

Phillips, A.P. and K.L. Martin. 1988. Investigation of spore surface antigens in the genus *Bacillus* by the use of polyclonal antibodies in immunofluorescence tests. J. Appl. Bacteriol. *64*: 47–55.

Phillips, A.P., A.M. Campbell and R. Quinn. 1988. Monoclonal antibodies against spore antigens of *Bacillus anthracis*. FEMS Microbiol. Immunol. *1*: 169–178.

Pichinoty, F. 1983. Isolation of a large aerobic sporulated sheathed bacteria from soil by elective culture. Ann. Microbiol. (Paris) *134B*: 443–446.

Pichinoty, F., H. de Berjac, M. Mandel, B. Greenway and J.L. Garcia. 1976. A new, sporulating, denitrifying, mesophilic bacterium: *Bacillus azotoformans* n. sp. (author's transl.)]. Ann. Microbiol. (Paris) *127B*: 351–361.

Pichinoty, F., M. Mandel, B. Greenway and J.L. Garcia. 1977. Isolation and properties of a denitrifying bacterium related to *Pseudomonas lemoignei*. Int. J. Syst. Bacteriol. *27*: 346–348.

Pichinoty, F., H. Debarjac, M. Mandel and J. Asselineau. 1983. Description of *Bacillus azotoformans* sp. nov. Int. J. Syst. Bacteriol. *33*: 660–662.

Pichinoty, F., J. Asselineau and M. Mandel. 1984. Biochemical characterization of *Bacillus benzoevorans* sp. nov., a new filamentous, sheathed mesophilic species, degrading various aromatic acids and phenols. Ann. Microbiol. (Paris) *135B*: 209–217.

Pichinoty, F., J.B. Waterbury, M. Mandel and J. Asselineau. 1986. *Bacillus gordonae* sp. nov., a new species belonging to the second morphological group, degrading various aromatic compounds. Ann. Inst. Pasteur Microbiol. *137A*: 65–78.

Pichinoty, F., J.B. Waterbury, M. Mandel and J. Asselineau. 1987. *In* Validation of the publication of new names and new publications previously effectively published outside the IJSB. List no. 23. Int. J. Syst. Bact. *37*: 179–180.

Piggot, P.J. and R. Losick. 2002. Sporulation genes and intercompartmental regulation. *In* Sonenshein, Hoch and Losick (Editors), *Bacillus subtilis* and its Closest Relatives. ASM Press, Washington, D.C., pp. 483–517.

Pikuta, E., A. Lysenko, N. Chuvilskaya, U. Mendrock, H. Hippe, N. Suzina, D. Nikitin, G. Osipov and K. Laurinavichius. 2000a. *Anoxybacillus pushchinensis* gen. nov., sp. nov., a novel anaerobic, alkaliphilic, moderately thermophilic bacterium from manure, and description of *Anoxybacillus falvithermus* comb. nov. Int. J. Syst. Evol. Microbiol. *50*: 2109–2117.

Pikuta, E., A. Lysenko, N. Suzina, G. Osipov, B. Kuznetsov, T. Tourova, V. Akimenko and K. Laurinavichius. 2000b. *Desulfotomaculum alkaliphilum* sp. nov., a new alkaliphilic, moderately thermophilic, sulfate-reducing bacterium. Int. J. Syst. Evol. Microbiol. *50*: 25–33.

Pikuta, E., D. Cleland and J. Tang. 2003a. Aerobic growth of *Anoxybacillus pushchinoensis* K1ᵀ: emended descriptions of *A. pushchinoensis* and the genus *Anoxybacillus*. Int. J. Syst. Evol. Microbiol. *53*: 1561–1562.

Pikuta, E.V., R.B. Hoover, A.K. Bej, D. Marsic, E.N. Detkova, W.B. Whitman and P. Krader. 2003b. *Tindallia californiensis* sp. nov., a new anaerobic, haloalkaliphilic, spore-forming acetogen isolated from Mono Lake in California. Extremophiles *7*: 327–334.

Pikuta, E.V., T. Itoh, P. Krader, J. Tang, W.B. Whitman and R.B. Hoover. 2006. *Anaerovirgula multivorans* gen. nov., sp. nov., a novel spore-forming, alkaliphilic anaerobe isolated from Owens Lake, California, USA. Int. J. Syst. Evol. Microbiol. *56*: 2623–2629.

Pinar, G., C. Ramos, S. Rolleke, C. Schabereiter-Gurtner, D. Vybiral, W. Lubitz and E.B.M. Denner. 2001. Detection of indigenous *Halobacillus* populations in damaged ancient wall paintings and building materials: Molecular monitoring and cultivation. Appl. Environ. Microbiol. *67*: 4891–4895.

Pinna, A., L.A. Sechi, S. Zanetti, D. Usai, G. Delogu, P. Cappuccinelli and F. Carta. 2001. *Bacillus cereus* keratitis associated with contact lens wear. Ophthalmology *108*: 1830–1834.

Pirttijärvi, T.S.M., M.A. Amdersson, A.C. Scoging and M.S. Salkinoja-Salonen. 1999. Evaluation of methods for recognizing strains of the *Bacillus cereus* group with food poisoning potential among industrial and environmental contaminants. Syst. Appl. Microbiol. *22*: 133–144.

Poli, A., E. Esposito, L. Lama, P. Orlando, G. Nicolaus, F. de Appolonia, A. Gambacorta and B. Nicolaus. 2006. *Anoxybacillus amylolyticus* sp. nov., a thermophilic amylase producing bacterium isolated from Mount Rittmann (Antarctica). Syst. Appl. Microbiol. *29*: 300–307.

Porter, A.G., E.W. Davidson and J.W. Liu. 1993. Mosquitocidal toxins of bacilli and their genetic manipulation for effective biological control of mosquitoes. Microbiol. Rev. *57*: 838–861.

Prazmowski, A. 1880. Untersuchung über die Entwickelungsgeschichte und Fermentwirking einiger Bacterien-Arten. Hugo Voigt, Leipzig.

Prickett, P.S. 1928. Thermophilic and thermoduric microorganisms with special reference to species isolated from milk. V. Description of spore-forming types. New York Agric. Exp. Stat. Tech. Bull. *147*.

Priest, F.G. 1993. Systematics and ecology of *Bacillus*. *In* Sonenshein, Hoch and Losick (Editors), *Bacillus subtilis* and other Gram-positive Bacteria. ASM Press, Washington, D.C., pp. 3–16.

Priest, F.G. 2002. *Bacillus sphaericus* and its insecticidal toxins. *In* Berkeley, Heyndrickx, Logan and De Vos (Editors), Applications and Systematics of *Bacillus* and its Relatives. Blackwell Science, Oxford, pp. 190–205.

Priest, F.G., M. Goodfellow, L.A. Shute and R.C.W. Berkeley. 1987. *Bacillus amyloliquefaciens* sp. nov, nom. rev. Int. J. Syst. Bacteriol. *37*: 69–71.

Priest, F.G., M. Goodfellow and C. Todd. 1988. A numerical classification of the genus *Bacillus*. J. Gen. Microbiol. *134*: 1847–1882.

Priest, F.G., M. Goodfellow and C. Todd. 1989. *In* Validation of the publication of new names and new combinations previously effectively published outside the IJSB. List no. 28. Int. J. Syst. Bacteriol. *39*: 93–94.

Priest, F.G., D.A. Kaji, Y.B. Rosato and V.P. Canhos. 1994. Characterization of *Bacillus thuringiensis* and related bacteria by ribosomal RNA gene restriction fragment length polymorphisms. Microbiology *140*: 1015–1022.

Pringsheim, E.G. 1950. The bacterial genus *Lineola*. J. Gen. Microbiol. *4*: 198–209.

Proom, H. and B.C. Knight. 1950. *Bacillus pantothenticus* (n.sp.). J. Gen. Microbiol. *4*: 539–541.

Puzyr, A.P., O.A. Mogil'naia, T.Y. Krylova and L.Y. Popova. 2002. The colony architectonics in *Bacillus subtilis* 2335. Mikrobiologiia *71*: 66–74.

Quinlan, J.J. and P.M. Foegeding. 1997. Monoclonal antibodies for use in detection of *Bacillus* and *Clostridium* spores. Appl. Environ. Microbiol. *63*: 482–487.

Raats, D. and M. Halpern. 2007. *Oceanobacillus chironomi* sp. nov., a halotolerant and facultatively alkaliphilic species isolated from a chironomid egg mass. Int. J. Syst. Evol. Microbiol. *57*: 255–259.

Rainey, F.A. and E. Stackebrandt. 1993. Phylogenetic evidence for the relationship of *Saccharococcus thermophilus* to *Bacillus stearothermophilus*. Syst. Appl. Microbiol. *16*: 224–226.

Rainey, F.A., D. Fritze and E. Stackebrandt. 1994. The phylogenetic diversity of thermophilic members of the genus *Bacillus* as revealed by 16S rDNA analysis. FEMS Microbiol. Lett. *115*: 205–211.

Ranftl, H. and O. Kandler. 1973. ᴅ-Aspartyl-ʟ-alanine an interpeptide bridge of murein of *Bacillus pasteurii* Migula. Z. Naturforsch. Sect. C J. Biosci. *C28*: 4–8.

Read, T.D., S.N. Peterson, N. Tourasse, L.W. Baillie, I.T. Paulsen, K.E. Nelson, H. Tettelin, D.E. Fouts, J.A. Eisen, S.R. Gill, E.K. Holtzapple, O.A. Okstad, E. Helgason, J. Rilstone, M. Wu, J.F. Kolonay, M.J. Beanan, R.J. Dodson, L.M. Brinkac, M. Gwinn, R.T. DeBoy, R. Madpu, S.C. Daugherty, A.S. Durkin, D.H. Haft, W.C. Nelson, J.D. Peterson, M. Pop, H.M. Khouri, D. Radune, J.L. Benton, Y. Mahamoud, L. Jiang, I.R. Hance, J.F. Weidman, K.J. Berry, R.D. Plaut, A.M. Wolf, K.L. Watkins, W.C. Nierman, A. Hazen, R. Cline, C. Redmond, J.E. Thwaite, O. White, S.L. Salzberg, B. Thomason, A.M. Friedlander, T.M. Koehler, P.C. Hanna, A.B. Kolsto and C.M. Fraser. 2003. The genome sequence of *Bacillus anthracis* Ames and comparison to closely related bacteria. Nature *423*: 81–86.

Ren, P.G. and P.J. Zhou. 2005a. *Salinibacillus aidingensis* gen. nov., sp. nov. and *Salinibacillus kushneri* sp. nov., moderately halophilic bacteria isolated from a neutral saline lake in Xin-Jiang, China. Int. J. Syst. Evol. Microbiol. *55*: 949–953.

Ren, P.G. and P.J. Zhou. 2005b. *Tenuibacillus multivorans* gen. nov., sp. nov., a moderately halophilic bacterium isolated from saline soil in Xin-Jiang, China. Int. J. Syst. Evol. Microbiol. *55*: 95–99.

Reva, O.N., I.B. Sorokulova and V.V. Smirnov. 2001. Simplified technique for identification of the aerobic spore-forming bacteria by phenotype. Int. J. Syst. Evol. Microbiol. *51*: 1361–1371.

Reva, O.N., V.V. Smirnov, B. Pettersson and F.G. Priest. 2002. *Bacillus endophyticus* sp. nov., isolated from the inner tissues of cotton plants (*Gossypium* sp.). Int. J. Syst. Evol. Microbiol. *52*: 101–107.

Rhee, S.K. and M.Y. Pack. 1980. Effect of environmental pH on chain length of *Lactobacillus bulgaricus*. J. Bacteriol. *144*: 865–868.

Rheims, H., A. Fruhling, P. Schumann, M. Rohde and E. Stackebrandt. 1999. *Bacillus silvestris* sp. nov., a new member of the genus *Bacillus* that contains lysine in its cell wall. Int. J. Syst. Bacteriol. *49*: 795–802.

Richard, A.A., J. Archambault, R. Rosselló-Mora, B.J. Tindall and M. Matheny. 2005. *Bacillus acidicola* sp. nov., a novel mesophilic, acidophilic species isolated from acidic *Sphagnum* peat bogs in Wisconsin. Int. J. Syst. Evol. Microbiol. *55*: 2125–2130.

Riesenman, P.J. and W.L. Nicholson. 2000. Role of the spore coat layers in *Bacillus subtilis* spore resistance to hydrogen peroxide, artificial UV-C, UV-B, and solar UV radiation. Appl. Environ. Microbiol. *66*: 620–626.

Rinderknecht, H., P. Wilding and B.J. Haverback. 1967. A new method for the determination of alpha-amylase. Experientia *23*: 805.

Ripabelli, G., J. McLauchlin, V. Mithani and E.J. Threlfall. 2000. Epidemiological typing of *Bacillus cereus* by amplified fragment length polymorphism. Lett. Appl. Microbiol. *30*: 358–363.

Rippere, K., J.L. Johnson and A.A. Yousten. 1997. DNA similarities among mosquito-pathogenic and nonpathogenic strains of *Bacillus sphaericus*. Int. J. Syst. Bacteriol. *47*: 214–216.

Rippere, K., R. Patel, J.R. Uhl, K.E. Piper, J.M. Steckelberg, B.C. Kline, F.R. Cockerill, 3rd and A.A. Yousten. 1998. DNA sequence resembling *vanA* and *vanB* in the vancomycin-resistant biopesticide *Bacillus popilliae*. J. Infect. Dis. *178*: 584–588.

Rivadeneyra, M.A., G. Delgado, M. Soriano, A. Ramos-Cormenzana and R. Delgado. 1999. Biomineralization of carbonates by *Marinococcus albus* and *Marinococcus halophilus* isolated from the Salar de Atacama (Chile). Curr. Microbiol. *39*: 53–57.

Rivera, A.M.G., P.E. Granum and F.G. Priest. 2000. Common occurrence of enterotoxin genes and enterotoxicity in *Bacillus thuringiensis*. FEMS Microbiol. Lett. 190: 151–155.

Roberts, D.D. 1964. L-phase bacterial forms associated with infectious synovitis in chickens and turkeys. Res. Vet. Sci. 5: 441.

Roberts, M.S. and F.M. Cohan. 1995. Recombination and migration rates in natural populations of *Bacillus subtilis* and *Bacillus mojavensis*. Evolution *49*: 1081–1094.

Roberts, M.S., L.K. Nakamura and F.M. Cohan. 1994. *Bacillus mojavensis* sp. nov., distinguishable from *Bacillus subtilis* by sexual isolation, divergence in DNA-sequence, and differences in fatty acid composition. Int. J. Syst. Bacteriol. *44*: 256–264.

Roberts, M.S., L.K. Nakamura and F.M. Cohan. 1996. *Bacillus vallismortis* sp. nov., a close relative of *Bacillus subtilis*, isolated from soil in Death Valley, California. Int. J. Syst. Bacteriol. *46*: 470–475.

Röling, W.F., J. Kerler, M. Braster, A. Apriyantono, H. Stam and H.W. van Verseveld. 2001. Microorganisms with a taste for vanilla: microbial ecology of traditional Indonesian vanilla curing. Appl. Environ. Microbiol. *67*: 1995–2003.

Romano, I., L. Lama, B. Nicolaus, A. Gambacorta and A. Giordano. 2005a. *Bacillus saliphilus* sp. nov., isolated from a mineral pool in Campania, Italy. Int. J. Syst. Evol. Microbiol. *55*: 159–163.

Romano, I., L. Lama, B. Nicolaus, A. Gambacorta and A. Giordano. 2005b. *Alkalibacillus filiformis* sp. nov., isolated from a mineral pool in Campania, Italy. Int. J. Syst. Evol. Microbiol. *55*: 2395–2399.

Romano, I., A. Poli, L. Lama, A. Gambacorta and B. Nicolaus. 2005c. *Geobacillus thermoleovorans* subsp. *stromboliensis* subsp. nov., isolated from the geothermal volcanic environment. J. Gen. Appl. Microbiol. *51*: 183–189.

Romano, I., L. Lama, B. Nicolaus, A. Poli, A. Gambacorta and A. Giordano. 2006. *Oceanobacillus oncorhynchi* subsp. *incaldanensis* subsp. nov., an alkalitolerant halophile isolated from an algal mat collected from a sulfurous spring in Campania (Italy), and emended description of *Oceanobacillus oncorhynchi*. Int. J. Syst. Evol. Microbiol. *56*: 805–810.

Ronimus, R.S., L.E. Parker and H.W. Morgan. 1997. The utilization of RAPD-PCR for identifying thermophilic and mesophilic *Bacillus* species. FEMS Microbiol. Lett. *147*: 75–79.

Rosado, A.S., J.D. vanElsas and L. Seldin. 1997. Reclassification of *Paenibacillus durum* (formerly *Clostridium durum* Smith and Cato 1974) Collins *et al.* 1994 as a member of the species *P. azotofixans* (formerly *Bacillus azotofixans* Seldin et al. 1984) Ash et al. 1994. Int. J. Syst. Bacteriol. *47*: 569–572.

Rosselló-Mora, R. and R. Amann. 2001. The species concept for prokaryotes. FEMS Microbiol. Rev. *25*: 39–67.

Rössler, D., W. Ludwig, K.H. Schleifer, C. Lin, T.J. McGill, J.D. Wisotzkey, P. Jurtshuk, Jr. and G.E. Fox. 1991. Phylogenetic diversity in the genus *Bacillus* as seen by 16S rRNA sequencing studies. Syst. Appl. Microbiol. *14*: 266–269.

Rossler, M. and V. Muller. 1998. Quantative an physiological analyses of chloride dependence of growth of *Halobacillus halophilus*. App. Environ. Microbiol. *64*: 3813–3817.

Rossler, M. and V. Muller. 2001. Chloride dependence of glycine betaine transport in *Halobacillus halophilus*. FEBS Lett. *489*: 124–128.

Rossler, M., G. Wanner and V. Muller. 2000. Motility and flagellum synthesis in *Halobacillus halophilus* are chloride dependent. J. Bacteriol. *182*: 532–535.

Rothschild, M.A. and O. Leisenfeld. 1996. Is the exploding powder from blank cartridges sterile? Forensic Sci. Int. *83*: 1–13.

Roy, M., J.C. Chen, M. Miller, D. Boyaner, O. Kasner and E. Edelstein. 1997. Epidemic bacillus endophthalmitis after cataract surgery. 1. Acute presentation and outcome. Ophthalmology *104*: 1768–1772.

Rózycki, H., H. Dahm, E. Strzelczyk and C.Y. Li. 1999. Diazotrophic bacteria in root-free soil and in the root zone of pine (*Pinus sylvestris* L.) and oak (*Quercus robur* L.). Appl. Soil Ecol. *12*: 239–250.

Rüger, H.J. 1983. Differentiation of *Bacillus globisporus*, *Bacillus marinus* comb. nov., *Bacillus aminovorans*, and *Bacillus insolitus*. Int. J. Syst. Bacteriol. *33*: 157–161.

Rüger, H.J., D. Fritze and C. Sproer. 2000. New psychrophilic and psychrotolerant *Bacillus marinus* strains from tropical and polar deep-sea sediments and emended description of the species. Int. J. Syst. Evol. Microbiol. *50*: 1305–1313.

Ruiz-Garcia, C., V. Bejar, F. Martinez-Checa, I. Llamas and E. Quesada. 2005a. *Bacillus velezensis* sp. nov., a surfactant-producing bacterium isolated from the river Velez in Malaga, southern Spain. Int. J. Syst. Evol. Microbiol. *55*: 191–195.

Ruiz-Garcia, C., E. Quesada, F. Martinez-Checa, I. Llamas, M.C. Urdaci and V. Bejar. 2005b. *Bacillus axarquiensis* sp. nov. and *Bacillus malacitensis* sp. nov., isolated from river-mouth sediments in southern Spain. Int. J. Syst. Evol. Microbiol. *55*: 1279–1285.

Sàágua, M.C., G. Vieira, H. Paveia and A. Anselmo. 1998. Isolation and preliminary characterization of *Bacillus* sp. MCS, a Gram-positive 4-chlorobiphenyl degrading bacterium. Int. Biodeterior. Biodegrad. *42*: 39–43.

Sabry, S.A. 1992. Microbial degradation of shrimp-shell waste. J. Basic Microbiol. *32*: 107–111.

Saier, M.H., S.R. Goldman, R.R. Maile, M.S. Moreno, W. Weyler, N. Yang and I.T. Paulsen. 2002. Overall transport capabilities of *Bacillus subtilis*. *In* Sonenshein, Hoch and Losick (Editors), *Bacillus subtilis* and its Closest Relatives. ASM Press, Washington, D.C., pp. 113–128.

Saitou, N. and M. Nei. 1987. The neighbor-joining method: a new method for reconstructing phylogenetic trees. Mol. Biol. Evol. *4*: 406–425.

Sakamoto, M. and K. Komagata. 1996. Aerobic growth and activities of NADH oxidase and NADH peroxidase in lactic acid bacteria. J. Ferment. Bioeng. *82*: 210–216.

Sakane, K. and K. Imai. 1986. Preservation of various bacteria by L-drying, Part II. Jpn. J. Freez. Dry. *32*: 47–53.

Salkinoja-Salonen, M.S., R. Vuorio, M.A. Andersson, P. Kämpfer, M.C. Andersson, T. Honkanen-Buzalski and A.C. Scoging. 1999. Toxigenic strains of *Bacillus* licheniformis related to food poisoning. Appl. Environ. Microbiol. *65*: 4637–4645.

Sanchez-Porro, C., S. Martin, E. Mellado and A. Ventosa. 2003. Diversity of moderately halophilic bacteria producing extracellular hydrolytic enzymes. J. Appl. Microbiol. *94*: 295–300.

Santini, J.M., J.F. Stolz and J.M. Macy. 2002. Isolation of a new arsenate-respiring bacterium–physiological and phylogenetic studies. Geomicrobiol. J. *19*: 41–52.

Santini, J.M., I.C. Streimann and R.N. vanden Hoven. 2004. *Bacillus macyae* sp. nov., an arsenate-respiring bacterium isolated from an Australian gold mine. Int. J. Syst. Evol. Microbiol. *54*: 2241–2244.

Sarkar, P.K., B. Hasenack and M.J. Nout. 2002. Diversity and functionality of *Bacillus* and related genera isolated from spontaneously fermented soybeans (Indian Kinema) and locust beans (African Soumbala). Int. J. Food Microbiol. *77*: 175–186.

Sato, N., S. Ikeda, T. Mikami and T. Matsumoto. 1999. *Bacillus cereus* dissociates hemoglobin and uses released heme as an iron source. Biol. Pharm. Bull. *22*: 1118–1121.

Satomi, M., M.T. La Duc and K. Venkateswaran. 2006. *Bacillus safensis* sp. nov., isolated from spacecraft and assembly-facility surfaces. Int. J. Syst. Evol. Microbiol. *56*: 1735–1740.

Schaffer, B.T. and B. Lighthart. 1997. Survey of culturable airborne bacteria at four diverse locations in Oregon: Urban, rural, forest, and coastal. Microb. Ecol. *34*: 167–177.

Schaffer, C., W.L. Franck, A. Scheberl, P. Kosma, T.R. McDermott and P. Messner. 2004. Classification of isolates from locations in Austria and Yellowstone National Park as *Geobacillus tepidamans* sp. nov. Int. J. Syst. Evol. Microbiol. *54*: 2361–2368.

Schäffer, C., N. Muller, R. Christian, M. Graninger, T. Wugeditsch, A. Scheberl and P. Messner. 1999. Complete glycan structure of the S-layer glycoprotein of *Aneurinibacillus thermoaerophilus* GS4–97. Glycobiology *9*: 407–414.

Schardinger, F. 1905. *Bacillus macerans*, ein Aceton bildender Rottebacillus. Zentralbl. Bakteriol. Parasitenkd. Infektionskr. Hyg. Abt. II *14*: 772–781.

Scheldeman, P., L. Herman, J. Goris, P. De Vos and M. Heyndrickx. 2002. Polymerase chain reaction identification of *Bacillus sporothermodurans* from dairy sources. J. Appl. Microbiol. *92*: 983–991.

Scheldeman, P., M. Rodriguez-Diaz, J. Goris, A. Pil, E. De Clerck, L. Herman, P. De Vos, N.A. Logan and M. Heyndrickx. 2004. *Bacillus farraginis* sp. nov., *Bacillus fortis* sp. nov. and *Bacillus fordii* sp. nov., isolated at dairy farms. Int. J. Syst. Evol. Microbiol. *54*: 1355–1364.

Schenk, A. and M. Aragno. 1979. *Bacillus schlegelii*, a new species of thermophilic, facultatively chemolithoautotrophic bacterium oxidizing molecular hydrogen. J. Gen. Microbiol. *115*: 333–341.

Schenk, A. and M. Aragno. 1981. *In* Validation of the publication of new names and new combinations previously effectively published outside the IJSB. List no. 6. Int. J. Syst. Bacteriol. *31*: 215–218.

Schiavon, O., P. Caliceti, P. Ferruti and F.M. Veronese. 2000. Therapeutic proteins: A comparison of chemical and biological properties of uricase conjugated to linear or branched poly(ethylene glycol) and poly(*N*-acryloylmorpholine). Farmaco *55*: 264–269.

Schleifer, K.H. and O. Kandler. 1972. Peptidoglycan types of bacterial cell walls and their taxonomic implications. Bacteriol. Rev. *36*: 407–477.

Schlesner, H., P.A. Lawson, M.D. Collins, N. Weiss, U. Wehmeyer, H. Volker and M. Thomm. 2001. *Filobacillus milensis* gen. nov., sp. nov., a new halophilic spore-forming bacterium with Orn-D-Glu-type peptidoglycan. Int. J. Syst. Evol. Microbiol. *51*: 425–431.

Schnepf, E., N. Crickmore, J. Van Rie, D. Lereclus, J. Baum, J. Feitelson, D.R. Zeigler and D.H. Dean. 1998. *Bacillus thuringiensis* and its pesticidal crystal proteins. Microbiol. Mol. Biol. Rev. *62*: 775–806.

Scholz, T., W. Demharter, R. Hensel and O. Kandler. 1987. *Bacillus pallidus* sp. nov., a new thermophilic species from sewage. Syst. Appl. Microbiol. *9*: 91–96.

Scholz, T., W. Demharter, R. Hensel and O. Kandler. 1988. *In* Validation of the publication of new names and new combinations previously effectively published outside the IJSB. List no. 24. Int. J. Syst. Bacteriol. *38*: 136–137.

Schraft, H., M. Steele, B. McNab, J. Odumeru and M.W. Griffiths. 1996. Epidemiological typing of *Bacillus* spp. isolated from food. Appl. Environ. Microbiol. *62*: 4229–4232.

Schuch, R., D. Nelson and V.A. Fischetti. 2002. A bacteriolytic agent that detects and kills *Bacillus anthracis*. Nature *418*: 884–889.

Schwan, R.F., A.H. Rose and R.G. Board. 1995. Microbial fermentation of cocoa beans, with emphasis on enzymatic degradation of the pulp. J. Appl. Bacteriol. *79*: S96–S107.

Schwarz, J., L. Garneau, D. Savaria, L. Masson, R. Brouseau and E. Pousseau. 1993. Lepidopteran-specific crystal toxins from *Bacillus thuringiensis* form cation- and anion-selective channels in planar lipid bilayers. J. Membrane Biol. *132*: 53–62.

Sekar, V. 1988. The insecticidal crystal protein gene is expressed in vegetative cells of *Bacillus thuringiensis* var. tenebrionis. Curr. Microbiol. *17*: 347–349.

Selinger, L.B., G.G. Khachatourians, J.R. Byers and M.F. Hynes. 1998. Expression of a *Bacillus thuringiensis* delta-endotoxin gene by *Bacillus pumilis*. Can. J. Microbiol. *44*: 259–269.

Setlow, P. 1995. Mechanisms for the prevention of damage to DNA in spores of *Bacillus* species. Annu. Rev. Microbiol. *49*: 29–54.

Shapton, D.A. and W.R. Hindes. 1963. The standardization of a spore count technique. Chem. Ind. *41*: 230–234.

Shariati, P., W.J. Mitchell, A. Boyd and F.G. Priest. 1995. Anaerobic metabolism in *Bacillus licheniformis* NCIB 6346. Microbiology *141*: 1117–1124.

Sharp, R.J., K.J. Bown and A. Atkinson. 1980. Phenotypic and genotypic characterization of some thermophilic species of *Bacillus*. J. Gen. Microbiol. *117*: 201–210.

Sharp, R.J., M. Munster, A. Vivian, S. Ahmed and T. Atkinson. 1989. Taxonomic and genetic studies of *Bacillus* thermophiles. *In* Da Costa, Duarte and Williams (Editors), Microbiology of Extreme Environments and its Biotechnological Potential. Elsevier Applied Science, London, pp. 62–81.

Sharp, R.J., P.W. Riley and D. White. 1992. Heterotrophic thermophilic bacilli. *In* Kristjansson (Editor), Thermophilic Bacteria. CRC Press, Boca Raton, FL, pp. 19–50.

Shelobolina, E.S., Z.A. Avakyan, E.S. Bulygina, T.P. Turova, A.M. Lysenko, G.A. Osipov and G.I. Karavaiko. 1997. Description of a new species of mucilaginous bacteria, *Bacillus edaphicus* sp. nov., and confirmation of the taxonomic status of *Bacillus mucilaginosus* Avakyan *et al.* 1986 based on data from phenotypic and genotypic analysis. Microbiology (En. transl. from Mikrobiologiya) *66*: 681–689.

Shelobolina, E.S., Z.A. Avakyan, E.S. Bulygina, T.P. Turova, A.M. Lysenko, G.A. Osipov and G.I. Karavaiko. 1998. *In* Validation of the publication of new names and new combinations previously effectively published outside the IJSB. List no. 66. Int. J. Syst. Evol. Microbiol. *51*: 631–632.

Shida, O., H. Takagi, K. Kadowaki, H. Yano, M. Abe, S. Udaka and K. Komagata. 1994. *Bacillus aneurinolyticus* sp. nov. nom. rev. Int. J. Syst. Bacteriol. *44*: 143–150.

Shida, O., H. Takagi, K. Kadowaki, S. Udaka, L.K. Nakamura and K. Komagata. 1995. Proposal of *Bacillus reuszeri* sp. nov., *Bacillus formosus* sp. nov., nom. rev., and *Bacillus borstelensis* sp. nov., nom. rev. Int. J. Syst. Bacteriol. *45*: 93–100.

Shida, O., H. Takagi, K. Kadowaki and K. Komagata. 1996. Proposal for two new genera, *Brevibacillus* gen. nov. and *Aneurinibacillus* gen. nov. Int. J. Syst. Bacteriol. *46*: 939–946.

Shida, O., H. Takagi, K. Kadowaki, L.K. Nakamura and K. Komagata. 1997. Transfer of *Bacillus alginolyticus*, *Bacillus chondroitinus*, *Bacillus curdlanolyticus*, *Bacillus glucanolyticus*, *Bacillus kobensis*, and *Bacillus thiaminolyticus* to the genus *Paenibacillus* and emended description of the genus *Paenibacillus*. Int. J. Syst. Bacteriol. *47*: 289–298.

Shih, I.L. and Y.T. Van. 2001. The production of poly-(gamma-glutamic acid) from microorganisms and its various applications. Bioresour. Technol. *79*: 207–225.

Shimoni, E., U. Ravid and Y. Shoham. 2000. Isolation of a *Bacillus* sp. capable of transforming isoeugenol to vanillin. J. Biotechnol. *78*: 1–9.

Shisa, N., N. Wasano, A. Ohgushi, D.H. Lee and M. Ohba. 2002. Extremely high frequency of common flagellar antigens between *Bacillus thuringiensis* and *Bacillus cereus*. FEMS Microbiol. Lett. *213*: 93–96.

Shivaji, S., K. Suresh, P. Chaturvedi, S. Dube and S. Sengupta. 2005. *Bacillus arsenicus* sp. nov., an arsenic-resistant bacterium isolated from a siderite concretion in West Bengal, India. Int. J. Syst. Evol. Microbiol. *55*: 1123–1127.

Shivaji, S., P. Chaturvedi, K. Suresh, G.S. Reddy, C.B. Dutt, M. Wainwright, J.V. Narlikar and P.M. Bhargava. 2006. *Bacillus aerius* sp. nov., *Bacillus aerophilus* sp. nov., *Bacillus stratosphericus* sp. nov. and *Bacillus altitudinis* sp. nov., isolated from cryogenic tubes used for collecting air samples from high altitudes. Int. J. Syst. Evol. Microbiol. *56*: 1465–1473.

Shiyakhov, E.N. and E. Rubinstein. 1994. Human live anthrax vaccine in the former USSR. CVaccine. *12*: 727–730.

Silva, C.F., R.F. Schwan, E.S. Dias and A.E. Wheals. 2000. Microbial diversity during maturation and natural processing of coffee cherries of *Coffea arabica* in Brazil Int. J. Food Microbiol. *60*: 251–260.

Simon, M.I., S.U. Emerson, J.H. Shaper, P.D. Bernard and A.N. Glazer. 1977. Classification of *Bacillus subtilis* flagellins. J. Bacteriol. *130*: 200–204.

Sirisanthana, T. and A.E. Brown. 2002. Anthrax of the gastrointestinal tract. Emerg. Infect. Dis. *8*: 649–651.

Skerman, V.B.D., V. McGowan and P.H.A. Sneath. 1980. Approved lists of bacterial names. Int. J. Syst. Bacteriol. *30*: 225–420.

Skowronski, B. and G.A. Strobel. 1969. Cyanide resistance and cyanide utilization by a strain of *Bacillus pumilus*. Can. J. Microbiol. *15*: 93–98.

Sleytr, U.B., M. Sára, D. Pum and B. Schuster. 2001. Characterization and use of crystalline bacterial cell surface layers. Prog. Surface Sci. *68*: 231–278.

Slieman, T.A. and W.L. Nicholson. 2001. Role of dipicolinic acid in survival of *Bacillus subtilis* spores exposed to artificial and solar UV radiation. Appl. Environ. Microbiol. *67*: 1274–1279.

Smibert, R.M. and N.R. Krieg. 1981. General characterization. *In* Gerhardt, Murray, Costilow, Nester, Wood, Krieg, and Phillips (Editors), Manual of Methods for General Bacteriology. Am. Soc. Microbiol., Washington, D.C., pp. 409–443.

Smirnova, T.A., L.I. Kulinich, M. Galperin and R.R. Azizbekyan. 1991. Subspecies-specific haemagglutination patterns of fimbriated *Bacillus thuringiensis* spores. FEMS Microbiol. Lett. *69*: 1–4.

Smith, N.R. and F.E. Clark. 1937. A proposed grouping of the mesophilic, aerobic, spore-forming bacilli. Soil Sci. Soc. Am. Proc. *2*: 255.

Smith, N.R. and F.E. Clark. 1938. Motile colonies of *Bacillus alvei* and other bacteria. J. Bacteriol. *35*: 59–60.

Smith, N.R., R.E. Gordon and F.E. Clark. 1946. Aerobic Mesophilic Spore-Forming Bacteria. Miscellaneous Publication 559. United States Department of Agriculture, Washington, D.C.

Smith, N.R., R.E. Gordon and F.E. Clark. 1952. Aerobic Spore-Forming Bacteria. Monograph No. 16. United States Department of Agriculture, Washington, D.C.

Sneath, P.H.A. 1962. The construction of taxonomic groups. *In* Ainsworth and Sneath (Editors), *In* Microbial Classification. Cambridge University Press, pp. 289–322.

Snel, J., P.P. Heinen, H.J. Blok, R.J. Carman, A.J. Duncan, P.C. Allen and M.D. Collins. 1995. Comparison of 16S rRNA sequences of segmented filamentous bacteria isolated from mice, rats, and chickens and proposal of "*Candidatus* Arthromitus". Int. J. Syst. Bacteriol. *45*: 780–782.

Sonenshein, A.L. 2000. Control of sporulation initiation in *Bacillus subtilis*. Curr. Opin. Microbiol. *3*: 561–566.

Sonenshein, A.L. 2002. The Krebs citric acid cycle. *In* Sonenshein, Hoch and Losick (Editors), *Bacillus subtilis* and its Closest Relatives. ASM Press, Washington, D.C., pp. 151–162.

Sonenshein, A.L., J.A. Hoch and R. Losick (Editors). 1993. *Bacillus subtilis* and other Gram-positive Bacteria. ASM Press, Washington, D.C.

Sonenshein, A.L., J.A. Hoch and R. Losick (Editors). 2002. *Bacillus subtilis* and its Closest Relatives: From Genes to Cells. ASM Press, Washington, D.C.

Song, Y., R. Yang, Z. Guo, M. Zhang, X. Wang and F. Zhou. 2000. Distinctness of spore and vegetative cellular fatty acid profiles of some aerobic endospore-forming bacilli. J. Microbiol. Methods *39*: 225–241.

Soto-Ramirez, N., C. Sanchez-Porro, S. Rosas-Padilla, K. Almodovar, G. Jimenez, M. Machado-Rodriguez, M. Zapata, A. Ventosa and R. Montalvo-Rodriguez. 2008. *Halobacillus mangrovi* sp. nov., a moderately halophilic bacterium isolated from the black mangrove Avicennia germinans. Int. J. Syst. Evol. Microbiol. *58*: 125–130.

Spanka, R. and D. Fritze. 1993. *Bacillus cohnii* sp. nov., a new, obligately alkaliphilic, oval-spore-forming bacillus species with ornithine and aspartic acid instead of diaminopimelic acid in the cell wall. Int. J. Syst. Bacteriol. *43*: 150–156.

Spinosa, M.R., F. Wallet, R.J. Courcol and M.R. Oggioni. 2000. The trouble in tracing opportunistic pathogens: cholangitis due to *Bacillus* in a French hospital caused by a strain related to an Italian product? Microb. Ecol. Health Dis. *12*: 99–101.

Spizizen, J. 1958. Transformation of biochemically deficient strains of *Bacillus subtilis* by deoxyribonucleate. Proc. Natl. Acad. Sci. USA. *75*: 1072–1078.

Spring, S., W. Ludwig, M.C. Marquez, A. Ventosa and K.H. Schleifer. 1996. *Halobacillus* gen. nov., with descriptions of *Halobacillus litoralis* sp nov and *Halobacillus trueperi* sp. nov., and transfer of *Sporosarcina halophila* to *Halobacillus halophilus* comb. nov. Int. J. Syst. Bacteriol. *46*: 492–496.

St. John-Brooks, R. and R.S. Breed. 1937. Actions taken by the second international microbiology congress in London, 1936 regarding bacteriological nomenclature. J. Bacteriol. *33*: 445–447.

Stackebrandt, E. and B.M. Goebel. 1994. Taxonomic note: a place for DNA-DNA reassociation and 16S rRNA sequence analysis in the present species definition in bacteriology. Int. J. Syst. Bacteriol. *44*: 846–849.

Stackebrandt, E. and J. Swiderski. 2002. From phylogeny to systematics: The dissection of the genus *Bacillus*. *In* Berkeley, Heyndrickx, Logan and De Vos (Editors), Applications and Systematics of *Bacillus* and Relatives. Blackwell Science, Oxford, pp. 8–22.

Stackebrandt, E., W. Ludwig, M. Weizenegger, S. Dorn, T.J. Mcgill, G.E. Fox, C.R. Woese, W. Schubert and K.H. Schleifer. 1987. Comparative 16S ribosomal RNA oligonucleotide analyses and murein types of round spore-forming bacilli and non-spore-forming relatives. J. Gen. Microbiol. *133*: 2523–2529.

Stackebrandt, E., C. Koch, O. Gvozdiak and P. Schumann. 1995. Taxonomic dissection of the genus *Micrococcus kocuria* gen. nov., *Nesterenkonia* gen. nov., *Kytococcus* gen. nov., *Dermacoccus* gen. nov., and *Micrococcus* Cohn 1872 gen. emend. Int. J. Syst. Bacteriol. *45*: 682–692.

Stackebrandt, E., W. Frederiksen, G.M. Garrity, P.A. Grimont, P. Kämpfer, M.C. Maiden, X. Nesme, R. Rosselló-Mora, J. Swings, H.G. Trüper, L. Vauterin, A.C. Ward and W.B. Whitman. 2002. Report of the ad hoc committee for the re-evaluation of the species definition in bacteriology. Int. J. Syst. Evol. Microbiol. *52*: 1043–1047.

Stalheim, T. and P.E. Granum. 2001. Characterization of spore appendages from *Bacillus cereus* strains. J. Appl. Microbiol. *91*: 839–845.

Stecchini, M.L., M. Del Torre, S. Donda, E. Maltini and S. Pacor. 2001. Influence of agar content on the growth parameters of *Bacillus cereus*. Int. J. Food Microbiol. *64*: 81–88.

Stenfors, L.P. and P.E. Granum. 2001. Psychrotolerant species from the *Bacillus cereus* group are not necessarily *Bacillus weihenstephanensis*. FEMS Microbiol. Lett. *197*: 223–228.

Stolz, J.F. and R.S. Oremland. 1999. Bacterial respiration of arsenic and selenium. FEMS Microbiol. Rev. *23*: 615–627.

Strom, P.F. 1985. Identification of thermophilic bacteria in solid-waste composting. Appl. Environ. Microbiol. *50*: 906–913.

Studholme, D.J., R.A. Jackson and D.J. Leak. 1999. Phylogenetic analysis of transformable strains of thermophilic *Bacillus* species. FEMS Microbiol. Lett. *172*: 85–90.

Stulke, J. and W. Hillen. 2000. Regulation of carbon catabolism in *Bacillus* species. Annu. Rev. Microbiol. *54*: 849–880.

Sung, M.H., H. Kim, J.W. Bae, S.K. Rhee, C.O. Jeon, K. Kim, J.J. Kim, S.P. Hong, S.G. Lee, J.H. Yoon, Y.H. Park and D.H. Baek. 2002. *Geobacillus toebii* sp. nov., a novel thermophilic bacterium isolated from hay compost. Int. J. Syst. Evol. Microbiol. *52*: 2251–2255.

Sunna, A., S.G. Prowe, T. Stoffregen and G. Antranikian. 1997a. Characterization of the xylanases from the new isolated thermophilic xylan-degrading *Bacillus thermoleovorans* strain K-3d and *Bacillus flavothermus* strain LB3A. FEMS Microbiol. Lett. *148*: 209–216.

Sunna, A., S. Tokajian, J. Burghardt, F. Rainey, G. Antranikian and F. Hashwa. 1997b. Identification of *Bacillus kaustophilus*, *Bacillus thermocatenulatus* and *Bacillus* strain HSR as members of *Bacillus thermoleovorans*. Syst. Appl. Microbiol. *20*: 232–237.

Suominen, I., M.A. Andersson, M.C. Andersson, A.M. Hallaksela, P. Kämpfer, F.A. Rainey and M. Salkinoja-Salonen. 2001. Toxic *Bacillus pumilus* from indoor air, recycled paper pulp, Norway spruce, food poisoning outbreaks and clinical samples. Syst. Appl. Microbiol. *24*: 267–276.

Suresh, K., S.R. Prabagaran, S. Sengupta and S. Shivaji. 2004. *Bacillus indicus* sp. nov., an arsenic-resistant bacterium isolated from an aquifer in West Bengal, India. Int. J. Syst. Evol. Microbiol. *54*: 1369–1375.

Suzuki, Y., T. Kishigami and S. Abe. 1976. Production of extracellular alpha-glucosidase by a thermophilic *Bacillus* species. Appl. Environ. Microbiol. *31*: 807–812.

Suzuki, Y., T. Kishigami, K. Inoue, Y. Mizoguchi, N. Eto, M. Takagi and S. Abe. 1983. *Bacillus thermoglucosidasius* sp. nov. a new species of obligately thermophilic bacilli. Syst. Appl. Microbiol. *4*: 487–495.

Switzer Blum, J., A. Burns Bindi, J. Buzzelli, J.F. Stolz and R.S. Oremland. 1998. *Bacillus arsenicoselenatis*, sp. nov., and *Bacillus selenitireducens*, sp. nov. two haloalkaliphiles from Mono Lake, California that respire oxyanions of selenium and arsenic. Arch. Microbiol. *171*: 19–30.

Switzer Blum, J., A. Burns Bindi, J. Buzzelli, J.F. Stolz and R.S. Oremland. 2001. *In* Validation of the publication of new names and new combinations previously effectively published outside the IJSEM. List no. 80. Int. J. Syst. Evol. Micorbiol. *51*: 793–794.

Takagi, H., O. Shida, K. Kadowaki, K. Komagata and S. Udaka. 1993. Characterization of *Bacillus brevis* with descriptions of *Bacillus migulanus* sp. nov., *Bacillus choshinensis* sp. nov., *Bacillus parabrevis* sp. nov., and *Bacillus galactophilus* sp. nov. Int. J. Syst. Bacteriol. *43*: 221–231.

Takahashi, H., P. Keim, A.F. Kaufmann, C. Keys, K.L. Smith, K. Taniguchi, S. Inouye and T. Kurata. 2004. *Bacillus anthracis* incident, Kameido, Tokyo, 1993. Emerg. Infect. Dis. *10*: 117–120.

Takai, K., D.P. Moser, T.C. Onstott, N. Spoelstra, S.M. Pfiffner, A. Dohnalkova and J.K. Fredrickson. 2001. *Alkaliphilus transvaalensis* gen. nov., sp. nov., an extremely alkaliphilic bacterium isolated from a deep South African gold mine. Int. J. Syst. Evol. Microbiol. *51*: 1245–1256.

Takamatsu, H. and K. Watabe. 2002. Assembly and genetics of spore protective structures. Cell Mol. Life Sci. *59*: 434–444.

Takami, H., A. Inoue, F. Fuji and K. Horikoshi. 1997. Microbial flora in the deepest sea mud of the Mariana Trench. FEMS Microbiol. Lett. *152*: 279–285.

Takami, H., K. Nakasone, Y. Takaki, G. Maeno, R. Sasaki, N. Masui, F. Fuji, C. Hirama, Y. Nakamura, N. Ogasawara, S. Kuhara and K. Horikoshi. 2000. Complete genome sequence of the alkaliphilic bacterium *Bacillus halodurans* and genomic sequence comparison with *Bacillus subtilis*. Nucleic Acids Res. *28*: 4317–4331.

Takami, H., Y. Takaki and I. Uchiyama. 2002. Genome sequence of *Oceanobacillus iheyensis* isolated from the Iheya Ridge and its unexpected adaptive capabilities to extreme environments. Nucleic Acids Res. *30*: 3927–3935.

Tanasupawat, S., A. Pakdeeto, S. Namwong, C. Thawai, T. Kudo and T. Itoh. 2006. *Lentibacillus halophilus* sp. nov., from fish sauce in Thailand. Int. J. Syst. Evol. Microbiol. *56*: 1859–1863.

Tanasupawat, S., S. Namwong, T. Kudo and T. Itoh. 2007. *Piscibacillus salipiscarius* gen. nov., sp. nov., a moderately halophilic bacterium from fermented fish (pla-ra) in Thailand. Int. J. Syst. Evol. Microbiol. *57*: 1413–1417.

Tandon, A., M.L. Tay-Kearney, C. Metcalf and L. McAllister. 2001. *Bacillus circulans* endophthalmitis. Clin. Exp.Ophthalmol. *29*: 92–93.

Tanner, R.S. 1989. Monitoring sulfate-reducing bacteria: comparison of enumeration media. J. Microbiol. Methods. *10*: 83–90.

Taubel, M., P. Kämpfer, S. Buczolits, W. Lubitz and H.A. Busse. 2003. *Bacillus barbaricus* sp. nov., isolated from an experimental wall painting. Int. J. Syst. Evol. Microbiol. *53*: 725–730.

Taylor, A.J. and R.J. Gilbert. 1975. *Bacillus cereus* food poisoning: a provisional serotyping scheme. J. Med. Microbiol. *8*: 543–550.

Ten, L.N., S.H. Baek, W.T. Im, M. Lee, H.W. Oh and S.T. Lee. 2006. *Paenibacillus panacisoli* sp. nov., a xylanolytic bacterium isolated from soil in a ginseng field in South Korea. Int. J. Syst. Evol. Microbiol. *56*: 2677–2681.

Ten, L.N., S.H. Baek, W.T. Im, L.L. Larina, J.S. Lee, H.M. Oh and S.T. Lee. 2007. *Bacillus pocheonensis* sp. nov., a moderately halotolerant, aerobic bacterium isolated from soil of a ginseng field. Int. J. Syst. Evol. Microbiol. *57*: 2532–2537.

Terayama, T., M. Shingaki, S. Yamada, H. Ushioda, H. Igarashi, S. Sakai and H.Z. Yoji. 1978. Incidence of *Bacillus cereus* in commercial foods and serological typing of isolates. J. Food Hyg. Soc. Japan. *19*: 98–104.

Tettelin, H., V. Masignani, M.J. Cieslewicz, C. Donati, D. Medini, N.L. Ward, S.V. Angiuoli, J. Crabtree, A.L. Jones, A.S. Durkin, R.T. Deboy, T.M. Davidsen, M. Mora, M. Scarselli, I. Margarit y Ros, J.D. Peterson, C.R. Hauser, J.P. Sundaram, W.C. Nelson, R. Madupu, L.M. Brinkac, R.J. Dodson, M.J. Rosovitz, S.A. Sullivan, S.C. Daugherty, D.H. Haft, J. Selengut, M.L. Gwinn, L. Zhou, N. Zafar, H. Khouri, D. Radune, G. Dimitrov, K. Watkins, K.J. O'Connor, S. Smith, T.R. Utterback, O. White, C.E. Rubens, G. Grandi, L.C. Madoff, D.L. Kasper, J.L. Telford, M.R. Wessels, R. Rappuoli and C.M. Fraser. 2005. Genome analysis of multiple pathogenic isolates of *Streptococcus agalactiae*: Implications for the microbial "pan-genome". Proc. Natl. Acad. Sci. USA. *102*: 13950–13955.

Thanabalu, T. and A.G. Porter. 1996. A *Bacillus sphaericus* gene encoding a novel type of mosquitocidal toxin of 31.8 kDa. Gene *170*: 85–89.

Thanabalu, T., C. Berry and J. Hindley. 1993. Cytotoxicity and ADP-ribosylating activity of the mosquitocidal toxin from *Bacillus sphaericus* SSII-1: possible roles of the 27- and 70-kilodalton peptides. J. Bacteriol. *175*: 2314–2320.

Thanthiankul, S., S. Suan-Ngay, S. Tantimavanich and W. Panbangred. 2002. Chitinase from *Bacillus thuringiensis* subsp. *pakistani*. Appl. Microbiol. Biotechnol. *56*: 395–401.

Thomas, D.J., J.A. Morgan, J.M. Whipps and J.R. Saunders. 2000. Plasmid transfer between the *Bacillus thuringiensis* subspecies *kurstaki* and *tenebrionis* in laboratory culture and soil and in lepidopteran and coleopteran larvae. Appl. Environ. Microbiol. *66*: 118–124.

Thorne, C.B. 1993. *Bacillus Anthracis*. *In* Sonenshein, Hoch and Losick (Editors), *Bacillus subtilis* and other Gram-positive Bacteria. ASM Press, Washington, D.C., pp. 113–124.

Tiago, I., C. Pires, V. Mendes, P.V. Morais, M.S. da Costa and A. Verissimo. 2006. *Bacillus foraminis* sp. nov., isolated from a non-saline alkaline groundwater. Int. J. Syst. Evol. Microbiol. *56*: 2571–2574.

Tian, X.P., S.G. Dastager, J.C. Lee, S.K. Tang, Y.Q. Zhang, D.J. Park, C.J. Kim and W.J. Li. 2007. *Alkalibacillus halophilus* sp. nov., a new halophilic species isolated from hypersaline soil in Xin-Jiang province, China. Syst. Appl. Microbiol. *30*: 268–272.

Tolner, B., B. Poolman and W.N. Konings. 1997. Adaptation of microorganisms and their transport systems to high temperatures. Comp. Biochem. Physiol. *118*: 423–428.

Tomimura, E., N.W. Zeman, J.R. Frankiewicz and W.M. Teague. 1990. Description of *Bacillus naganoensis* sp. nov. Int. J. Syst. Bacteriol. *40*: 123–125.

Touzel, J.P., M. O'Donohue, P. Debeire, E. Samain and C. Breton. 2000. *Thermobacillus xylanilyticus* gen. nov., sp. nov., a new aerobic thermophilic xylan-degrading bacterium isolated from farm soil. Int. J. Syst. Evol. Microbiol. *50*: 315–320.

Travers, R.S., P.A. Martin and C.F. Reichelderfer. 1987. Selective process for efficient isolation of soil *Bacillus* spp. Appl. Environ. Microbiol. *53*: 1263–1266.

Tuladhar, R., S.K. Patole, T.H. Koh, R. Norton and J.S. Whitehall. 2000. Refractory *Bacillus cereus* infection in a neonate. Int. J. Clin. Pract. *54*: 345–347.

Turnbull, P.C., R.A. Hutson, M.J. Ward, M.N. Jones, C.P. Quinn, N.J. Finnie, C.J. Duggleby, J.M. Kramer and J. Melling. 1992. *Bacillus anthracis* but not always anthrax. J. Appl. Bacteriol. *72*: 21–28.

Turnbull, P.C. 1999. Definitive identification of *Bacillus anthracis*: a review. J. Appl. Microbiol. *87*: 237–240.

Turnbull, P.C. 2000. Current status of immunization against anthrax: old vaccines may be here to stay for a while. Curr. Opin. Infect. Dis. *13*: 113–120.

Turnbull, P.C.B., R. Böhm, O. Cosivi, M. Doganay, M.E. Hugh-Jones, D.D. Joshi, M.K. Lalitha and V. de Vos. 1998. Guidelines for the Surveillance and Control of Anthrax in Humans and Animals, 3rd ed. World Health Organization, Geneva.

Turnbull, P.C.B., P.J. Jackson, K.K. Hill, P. Keim, A.B. Kolstø and D.J. Beecher. 2002. Longstanding taxonomic enigmas within the 'Bacillus cereus group' are on the verge of being resolved by far-reaching molecular developments: Forecasts on the possible outcome by an ad hoc team. *In* Berkeley, Heyndrickx, Logan and De Vos (Editors), Applications and Systematics of *Bacillus* and Relatives. Blackwell Science, Oxford, pp. 23–36.

Turtura, G.C., A. Perfetto and P. Lorenzelli. 2000. Microbiological investigation on black crusts from open-air stone monuments of Bologna (Italy). New Microbiol. *23*: 207–228.

Tyndall, J. 1877. Further reaches on the department and vital persistence of putrefactive and infective organisms from a physical point of view. Phil. Trans. R. Soc. *167*: 149–206.

Ueda, S. 1989. Industrial application of B. subtilis. *In* Maruo and Yoshikawa (Editors), *Bacillus subtilis*, Molecular Biology, and Industrial Application. Kodansha, Tokyo, pp. 143–161.

Uetanabaro, A.P., C. Wahrenburg, W. Hunger, R. Pukall, C. Sproer, E. Stackebrandt, V.P. de Canhos, D. Claus and D. Fritze. 2003. *Paenibacillus agarexedens* sp. nov., nom. rev., and *Paenibacillus agaridevorans* sp. nov. Int. J. Syst. Evol. Microbiol. *53*: 1051–1057.

Uma, B. and S. Sandhya. 1997. Pyridine degradation and heterocyclic nitrification by *Bacillus coagulans*. Can. J. Microbiol. *43*: 595–598.

Urdaci, M.C., B. Regnault and P.A.D. Grimont. 2001. Identification by in situ hybridization of selected filamentous bacteria in the intestine of diarrheic rainbow trout (*Oncorhyncus mykiss*). Res. Microbiol. *152*: 67–73.

Urrutia, M.M. and T.J. Beveridge. 1995. Formation of short-range ordered aluminosilicates in the presence of bacterial surface (*Bacillus subtilis*) and organic ligands. Geoderma *65*: 149–165.

Usami, R., A. Echigo, T. Fukushima, T. Mizuki, Y. Yoshida and M. Kamekura. 2007. *Alkalibacillus silvisoli* sp. nov., an alkaliphilic moderate halophile isolated from non-saline forest soil in Japan. Int. J. Syst. Evol. Microbiol. *57*: 770–774.

Vadlamudi, R.K., E. Weber, I. Ji, T.H. Ji and L.A. Bulla, Jr. 1995. Cloning and expression of a receptor for an insecticidal toxin of *Bacillus thuringiensis*. J. Biol. Chem. *270*: 5490–5494.

Valderrama, M.J., B. Prado, A. Del Moral, R. Ríos, A. Ramos-Cormenzana and V. Campos. 1991. Numerical taxonomy of moderately halophilic Gram-positive cocci isolated from the Salar de Atacama (Chile). Microbiologia SEM *7*: 35–41.

Van Der Zwet, W.C., G.A. Parlevliet, P.H. Savelkoul, J. Stoof, A.M. Kaiser, A.M. Van Furth and C.M. Vandenbroucke-Grauls. 2000. Outbreak of *Bacillus cereus* infections in a neonatal intensive care unit traced to balloons used in manual ventilation. J. Clin. Microbiol. *38*: 4131–4136.

van Netten, P. and J.M. Kramer. 1992. Media for the detection and enumeration of *Bacillus cereus* in foods: a review. Int. J. Food Microbiol. *17*: 85–99.

Van Rie, J. 2002. Bt crops – a novel insect control tool. *In* Berkeley, Heyndrickx, Logan and De Vos (Editors), Applications and Systematics of *Bacillus* and Relatives. Blackwell Science, Oxford, pp. 176–189.

Vargas, V.A., O.D. Delgado, R. Hatti-Kaul and B. Mattiasson. 2005. *Bacillus bogoriensis* sp. nov., a novel alkaliphilic, halotolerant bacterium isolated from a Kenyan soda lake. Int. J. Syst. Evol. Microbiol. *55*: 899–902.

Vedder, A. 1934. *Bacillus alcalophilus* n. sp.; benevens enkele ervaringen met sterk alcalische voedingsbodems. Antonie Leeuwenhoek J. Microbiol. Serol. *1*: 141–147.

Venkataraman, R. and S. Sreenivasan. 1954. Studies on the red halophilic bacteria from salted fish and salt. Proc. Indian Acad. Sci. *39b*: 17–23.

Venkateswaran, K., M. Satomi, S. Chung, R. Kern, R. Koukol, C. Basic and D. White. 2001. Molecular microbial diversity of a spacecraft assembly facility. Syst. Appl. Microbiol. *24*: 311–320.

Venkateswaran, K., M. Kempf, F. Chen, M. Satomi, W. Nicholson and R. Kern. 2003. *Bacillus nealsonii* sp. nov., isolated from a spacecraft-assembly facility, whose spores are gamma-radiation resistant. Int. J. Syst. Evol. Microbiol. *53*: 165–172.

Ventosa, A., E. Quesada, F. Rodríguez-Valera, F. Ruiz-Berraquero and A. Ramos-Cormenzana. 1982. Numerical taxonomy of moderately halophilic Gram-negative rods. J. Gen. Microbiol. *128*: 1959–1968.

Ventosa, A., A. Ramos-Cormenzana and M. Kocur. 1983. Moderately halophilic Gram-positive cocci from hypersaline environments. Syst. Appl. Microbiol. *4*: 564–570.

Ventosa, A., M.T. Garcia, M. Kamekura, H. Onishi and F. Ruiz-Berraquero. 1989. *Bacillus halophilus* sp. nov., a moderately halophilic *Bacillus* species. Syst. Appl. Microbiol. *12*: 162–166.

Ventosa, A., M.T. Garcia, M. Kamekura, H. Onishi and F. Ruiz-Berraquero. 1990a. *In* Validation of the publication of new names and new combinations previously effectively published outside the IJSB. List no. 32. Int. J. Syst. Bacteriol. *40*: 105–106.

Ventosa, A., M.C. Marquez, F. Ruiz-Berraquero and M. Kocur. 1990b. *Salinicoccus roseus* gen. nov., sp. nov., a new moderately halophilic Gram-positive coccus. Syst. Appl. Microbiol. *13*: 29–33.

Ventosa, A., M.C. Marquez, N. Weiss and B.J. Tindall. 1992. Transfer of *Marinococcus hispanicus* to the genus *Salinicoccus* as *Salinicoccus hispanicus* comb. nov. Syst. Appl. Microbiol. *15*: 530–534.

Ventosa, A., J.J. Nieto and A. Oren. 1998. Biology of moderately halophilic aerobic bacteria. Microbiol. Mol. Biol. Rev. *62*: 504–544.

Vermeulen, V. and H.J. Kunte. 2004. *Marinococcus halophilus* DSM 20408^T encodes two transporters for compatible solutes belonging to the betaine-carnitine-choline transporter family: identification and characterization of ectoine transporter EctM and glycine betaine transporter BetM. Extremophiles *8*: 175–184.

Vessey, J.K. and T.J. Buss. 2002. *Bacillus cereus* UW85 inoculation effects on growth, nodulation, and N accumulation in grain legumes: controlled environmental studies. Can. J. Plant Sci. *82*: 283–290.

Vladmudi, R.K., E. Weber, T. Ji and J.L.A. Bulla. 1995. Cloning and expression of a receptor for an insecticidal toxin of *Bacillus thuringiensis*. J. Biol. Chem. *270*: 5490–5494.

von Wachenfeldt, C. and L. Hederstedt. 2002. Respiratory cytochromes, other heme proteins, and heme biosynthesis. *In* Sonenshein, Hoch and Losick (Editors), *Bacillus subtilis* and its Closest Relatives. ASM Press, Washington D.C., pp. 151–162.

Vreeland, R.H., W.D. Rosenzweig and D.W. Powers. 2000. Isolation of a 250 million-year-old halotolerant bacterium from a primary salt crystal. Nature *407*: 897–900.

Vuillemin, P. 1913. Genera Schizomycetum. Ann. Mycol. *11*: 512–527.

Wadhwa, L. and K.E. Smith. 2000. Progesterone side-chain cleavage by *Bacillus sphaericus*. FEMS Microbiol. Lett. *192*: 179–183.

Wainø, M., B.J. Tindall, P. Schumann and K. Ingvorsen. 1999. *Gracilibacillus* gen. nov., with description of *Gracilibacillus halotolerans* gen. nov., sp. nov.; transfer of *Bacillus dipsosauri* to *Gracilibacillus dipsosauri* comb. nov., and *Bacillus salexigens* to the genus *Salibacillus* gen. nov., as *Salibacillus salexigens* comb. nov. Int. J. Syst. Bacteriol. *49*: 821–831.

Walker, P.D. and J. Wolf. 1961. Some properties of aerobic thermophiles growing at 65. J. Appl. Bacteriol. *24*: iv–v.

Walker, P.D. and J. Wolf. 1971. The taxonomy of *Bacillus stearothermophilus*. *In* Barker, Gould and Wolf (Editors), Spore Research 1971. Academic Press, London, pp. 247–262.

Wallet, F., V. Crunelle, M. Roussel-Delvallez, A. Fruchart, P. Saunier and R.J. Courcol. 1996. *Bacillus subtilis* as a cause of cholangitis in polycystic kidney and liver disease. Am. J. Gastroenterol. *91*: 1477–1478.

Walter, T. and A. Aronson. 1999. Specific binding of the E2 subunit of pyruvate dehydrogenase to the upstream region of *Bacillus thuringiensis* protoxin genes. J. Biol. Chem. *274*: 7901–7906.

Wang, J. and D.Y. Fung. 1996. Alkaline-fermented foods: A review with emphasis on pidan fermentation. Crit. Rev. Microbiol. *22*: 101–138.

Wang, S.L. and J.R. Hwang. 2001. Microbial remediation of shellfish wastes for the production of chitinases. Enzyme Microb. Technol. *28*: 376–382.

Wang, S.L., I.L. Shih, T.W. Liang and C.H. Wang. 2002. Purification and characterization of two antifungal chitinases extracellularly produced by *Bacillus amyloliquefaciens* V656 in a shrimp and crab shell powder medium. J. Agric. Food Chem. *50*: 2241–2248.

Wang, L.T., F.L. Lee, C.J. Tai and H. Kasai. 2007a. Comparison of *gyrB* gene sequences, 16S rRNA gene sequences and DNA–DNA hybridization in the *Bacillus subtilis* group. Int. J. Syst. Evol. Microbiol. *57*: 1846–1850.

Wang, L.T., F.L. Lee, C.J. Tai, A. Yokota and H.P. Kuo. 2007b. Reclassification of *Bacillus axarquiensis* Ruiz-Garcia *et al.* 2005 and *Bacillus malacitensis* Ruiz-Garcia *et al.* 2005 as later heterotypic synonyms of *Bacillus mojavensis* Roberts *et al.* 1994. Int. J. Syst. Evol. Microbiol. *57*: 1663–1667.

Wang, Q.F., W. Li, Y.L. Liu, H.H. Cao, Z. Li and G.Q. Guo. 2007c. *Bacillus qingdaonensis* sp. nov., a moderately haloalkaliphilic bacterium isolated from a crude sea-salt sample collected near Qingdao in eastern China. Int. J. Syst. Evol. Microbiol. *57*: 1143–1147.

Wati, M.R., T. Thanabalu and A.G. Porter. 1997. Gene from tropical *Bacillus sphaericus* encoding a protease closely related to subtilisins from Antarctic bacilli. Biochim. Biophys. Acta *1352*: 56–62.

Wayne, L.G., D.J. Brenner, R.R. Colwell, P.A.D. Grimont, O. Kandler, M.I. Krichevsky, L.H. Moore, W.E.C. Moore, R.G.E. Murray, E. Stackebrandt, M.P. Starr and H.G. Trüper. 1987. Report of the ad hoc committee on the reconciliation of approaches to bacterial systematics. Int. J. Syst. Evol. Microbiol. *37*: 463–464.

Weber, D.J., S.M. Saviteer, W.A. Rutala and C.A. Thomann. 1988. In vitro susceptibility of *Bacillus* spp. to selected antimicrobial agents. Antimicrob. Agents Chemother. *32*: 642–645.

Weigel, J. 1986. Methods for isolation and study of thermophiles. *In* Brock (Editor), Thermophiles: General, Molecular and Applied Microbiology. Wiley, New York, pp. 17–37.

Weigmann, H. 1898. Ueber zwei an der Käsefereifung beteiligte Bakterien. Zentbl. Bakteriol. Parasitenkd. Infektionskr. Hyg. Abt. II *4*: 820–834.

Welker, N.E. and L.L. Campbell. 1967. Unrelatedness of *Bacillus amyloliquefaciens* and *Bacillus subtilis*. J. Bacteriol. *94*: 1124–1130.

Wenzel, M., I. Schonig, M. Berchtold, P. Kämpfer and H. Konig. 2002. Aerobic and facultatively anaerobic cellulolytic bacteria from the gut of the termite *Zootermopsis angusticollis*. J. Appl. Microbiol. *92*: 32–40.

Werner, W. 1933. Botanische Beschreibung haüfinger am Buttersäureabbau beteiligter sporenbildender Bakterienspezies. Zentbl. Bakteriol. Parasitenkd. Infektionskr. Hyg. Abt. II *87*: 446–475.

White, G.F. 1906. The bacteria of the apiary, with special reference to bee diseases. Bureau of Entomology Technical Series No. 14. U.S. Department of Agriculture, Washington, D.C.

White, D., R.J. Sharp and F.G. Priest. 1993. A polyphasic taxonomic study of thermophilic bacilli from a wide geographical area. Antonie Leeuwenhoek. *64*: 357–386.

Whittenbury, R. 1964. Hydrogen peroxide formation and catalase activity in the lactic acid bacteria. J. Gen. Microbiol. *35*: 13–26.

Wier, A., M. Dolan, D. Grimaldi, R. Guerrero, J. Wagensberg and L. Margulis. 2002. Spirochete and protist symbionts of a termite (*Mastotermes electrodominicus*) in Miocene amber. Proc. Natl. Acad. Sci. USA. *99*: 1410–1413.

Wieser, M., H. Worliczek, P. Kämpfer and H.J. Busse. 2005. *Bacillus herbersteinensis* sp. nov. Int. J. Syst. Evol. Microbiol. *55*: 2119–2213.

Winslow, C.E., J. Broadhurst, R.E. Buchanan, C. Krumwiede, L.A. Rogers and G.H. Smith. 1917. The families and genera of the bacteria: preliminary report of the committee of the Society of American Bacteriologists on characterization and classification of bacterial types. J. Bacteriol. *2*: 505–566.

Winslow, C.E., J. Broadhurst, R.E. Buchanan, C. Krumwiede, L.A. Rogers and G.H. Smith. 1920. The families and genera of the bacteria: final report of the committee of the Society of American Bacteriologists on characterization and classification of bacterial types. J. Bacteriol. *5*: 191–229.

Winter, G. 1880. Die Pilze. *In* Rabenhorst (Editor), Kryptogamen-Flora, 2nd ed., vol. 1. Eduard Kummer, Leipzig, pp. 38.

Wisotzkey, J.D., P. Jurtshuk, G.E. Fox, G. Deinhard and K. Poralla. 1992. Comparative sequence analyses on the 16S ribosomal RNA (rDNA) of *Bacillus acidocaldarius*, *Bacillus acidoterrestris*, and *Bacillus cycloheptanicus* and proposal for creation of a new genus, *Alicyclobacillus* gen. nov. Int. J. Syst. Bacteriol. *42*: 263–269.

Wolf, J. and M.S.U. Chowhury. 1971a. The *Bacillus circulans* complex: Biochemical and immunological studies. *In* Barker Gould and, Wolf (Editors), Spore Research 1971. Academic Press, London, pp. 227–245.

Wolf, J. and M.S.U. Chowhury. 1971b. Taxonomy of *B. circulans* and *B. stearothermophilus*. *In* Barker, Gould and Wolf (Editors), Spore Research 1971. Academic Press, London, pp. 349–350.

Wolf, J. and R.J. Sharp. 1981. Taxonomic and related aspects of thermophiles within the genus *Bacillus*. *In* Berkeley and Goodfellow (Editors), The Aerobic Endospore-Forming Bacteria. Academic Press, London, pp. 251–296.

Wolin, E.A., M.G. Wolin and R.S. Wolfe. 1963. Formation of methane by bacterial extracts. J. Biol. Chem. *238*: 2882–2886.

Woodburn, M.A., A.A. Yousten and K.H. Hilu. 1995. Random amplified polymorphic DNA fingerprinting of mosquito-pathogenic and nonpathogenic strains of *Bacillus sphaericus*. Int. J. Syst. Bacteriol. *45*: 212–217.

Xu, D. and J.C. Cote. 2003. Phylogenetic relationships between *Bacillus* species and related genera inferred from comparison of 3′ end 16S rDNA and 5′ end 16S–23S ITS nucleotide sequences. Int. J. Syst. Evol. Microbiol. *53*: 695–704.

Xue, Y., X. Zhang, C. Zhou, Y. Zhao, D.A. Cowan, S. Heaphy, W.D. Grant, B.E. Jones, A. Ventosa and Y. Ma. 2006. *Caldalkalibacillus thermarum* gen. nov., sp. nov., a novel alkalithermophilic bacterium from a hot spring in China. Int. J. Syst. Evol. Microbiol. *56*: 1217–1221.

Yamamura, S., M. Yamashita, N. Fujimoto, M. Kuroda, M. Kashiwa, K. Sei, M. Fujita and M. Ike. 2007. *Bacillus selenatarsenatis* sp. nov., a selenate- and arsenate-reducing bacterium isolated from the effluent drain of a glass-manufacturing plant. Int. J. Syst. Evol. Microbiol. *57*: 1060–1064.

Yermakova, S.P., V.V. Sova and T.N. Zvyagintseva. 2002. Brown seaweed protein as an inhibitor of marine mollusk endo-(1→3)-beta-D-glucanases. Carbohydr. Res. *337*: 229–237.

Yoon, J.-H., S.-J. Kang, Y.-T. Jung and T.-K. Oh. 2007a. *Halobacillus campisalis* sp. nov., containing *meso*-diaminopimelic acid in the cell-wall peptidoglycan, and emended description of the genus *Halobacillus*. Int. J. Syst. Evol. Microbiol. *57*: 2021–2025.

Yoon, J.-H., S.-J. Kang and T.-K. Oh. 2007b. Reclassification of *Marinococcus albus* Hao *et al.* 1985 as Salimicrobium album gen. nov., comb. nov. and *Bacillus halophilus* Ventosa *et al.* 1990 as *Salimicrobium halophilum* comb. nov., and description of *Salimicrobium luteum* sp. nov. Int. J. Syst. Evol. Microbiol. *57*: 2406–2411.

Yoon, J.H., S.S. Kang, K.C. Lee, Y.H. Kho, S.H. Choi, K.H. Kang and Y.H. Park. 2001a. *Bacillus jeotgali* sp. nov., isolated from jeotgal, Korean traditional fermented seafood. Int. J. Syst. Evol. Microbiol. *51*: 1087–1092.

Yoon, J.H., K.C. Lee, N. Weiss, Y.H. Kho, K.H. Kang and Y.H. Park. 2001b. *Sporosarcina aquimarina* sp. nov., a bacterium isolated from seawater in Korea, and transfer of *Bacillus globisporus* (Larkin and Stokes 1967), *Bacillus psychrophilus* (Nakamura 1984) and *Bacillus pasteurii* (Chester

1898) to the genus *Sporosarcina* as *Sporosarcina globispora* comb. nov., *Sporosarcina psychrophila* comb. nov. and *Sporosarcina pasteurii* comb. nov., and emended description of the genus *Sporosarcina*. Int. J. Syst. Evol. Microbiol. *51*: 1079–1086.

Yoon, J.H., N. Weiss, K.C. Lee, I.S. Lee, K.H. Kang and Y.H. Park. 2001c. *Jeotgalibacillus alimentarius* gen. nov., sp. nov., a novel bacterium isolated from jeotgal with L-lysine in the cell wall, and reclassification of *Bacillus marinus* Ruger 1983 as *Marinibacillus marinus* gen. nov., comb. nov. Int. J. Syst. Evol. Microbiol. *51*: 2087–2093.

Yoon, J.H., K.H. Kang and Y.H. Park. 2002. *Lentibacillus salicampi* gen. nov., sp nov., a moderately halophilic bacterium isolated from a salt field in Korea. Int. J. Syst. Evol. Microbiol. *52*: 2043–2048.

Yoon, J.H., I.G. Kim, K.H. Kang, T.K. Oh and Y.H. Park. 2003a. *Bacillus marisflavi* sp. nov. and *Bacillus aquimaris* sp nov., isolated from sea water of a tidal flat of the Yellow Sea in Korea. Int. J. Syst. Evol. Microbiol. *53*: 1297–1303.

Yoon, J.H., H.K.G. Kook and Y.H. Park. 2003b. *Halobacillus salinus* sp. nov., isolated from a salt lake on the coast of the East Sea in Korea. Int. J. Syst. Evol. Microbiol. *53*: 687–693.

Yoon, J.H., K.C. Lee, N. Weiss, K.H. Kang and Y.H. Park. 2003c. *Jeotgalicoccus halotolerans* gen. nov., sp. nov. and *Jeotgalicoccus psychrophilus* sp. nov., isolated from the traditional Korean fermented seafood jeotgal. Int. J. Syst. Evol. Microbiol. *53*: 595–602.

Yoon, J.H., K.H. Kang, T.K. Oh and Y.H. Park. 2004a. *Halobacillus locisalis* sp. nov., a halophilic bacterium isolated from a marine solar saltern of the Yellow Sea in Korea. Extremophiles. *8*: 23–28.

Yoon, J.H., I.G. Kim, K.H. Kang, T.K. Oh and Y.H. Park. 2004b. *Bacillus hwajinpoensis* sp. nov. and an unnamed *Bacillus* genomospecies, novel members of *Bacillus* rRNA group 6 isolated from sea water of the East Sea and the Yellow Sea in Korea. Int. J. Syst. Evol. Microbiol. *54*: 803–808.

Yoon, J.H., T.K. Oh and Y.H. Park. 2004c. Transfer of *Bacillus halodenitrificans* Denariaz *et al.* 1989 to the genus *Virgibacillus* as *Virgibacillus halodenitrificans* comb. nov. Int. J. Syst. Evol. Microbiol. *54*: 2163–2167.

Yoon, J.H., S.J. Kang, C.H. Lee, H.W. Oh and T.K. Oh. 2005a. *Halobacillus yeomjeoni* sp. nov., isolated from a marine solar saltern in Korea. Int. J. Syst. Evol. Microbiol. *55*: 2413–2417.

Yoon, J.H., S.J. Kang, S.Y. Lee, M.H. Lee and T.K. Oh. 2005b. *Virgibacillus dokdonensis* sp. nov., isolated from a Korean island, Dokdo, located at the edge of the East Sea in Korea. Int. J. Syst. Evol. Microbiol. *55*: 1833–1837.

Yoon, J.H., C.H. Lee and T.K. Oh. 2005c. *Bacillus cibi* sp. nov., isolated from jeotgal, a traditional Korean fermented seafood. Int. J. Syst. Evol. Microbiol. *55*: 733–736.

Yoon, J.H. and T.K. Oh. 2005. *Bacillus litoralis* sp. nov., isolated from a tidal flat of the Yellow Sea in Korea. Int. J. Syst. Evol. Microbiol. *55*: 1945–1948.

Yousten, A.A. 1984. Bacteriophage typing of mosquito-pathogenic strains of *Bacillus sphaericus*. J. Invert. Pathol. *43*: 124–125.

Yousten, A.A., S.B. Fretz and S.A. Jelley. 1985. Selective medium for mosquito-pathogenic strains of *Bacillus sphaericus*. Appl. Environ. Microbiol. *49*: 1532–1533.

Yuan, S., P. Ren, J. Liu, Y. Xue, Y. Ma and P. Zhou. 2007. *Lentibacillus halodurans* sp. nov., a moderately halophilic bacterium isolated from a salt lake in Xin-Jiang, China. Int. J. Syst. Evol. Microbiol. *57*: 485–488.

Yumoto, I., K. Yamazaki, T. Sawabe, K. Nakano, K. Kawasaki, Y. Ezura and H. Shinano. 1998. *Bacillus horti* sp. nov., a new Gram-negative alkaliphilic *Bacillus*. Int. J. Syst. Bacteriol. *48*: 565–571.

Yumoto, I., S. Yamaga, Y. Sogabe, Y. Nodasaka, H. Matsuyama, K. Nakajima and A. Suemori. 2003. *Bacillus krulwichiae* sp. nov., a halotolerant obligate alkaliphile that utilizes benzoate and m-hydroxybenzoate. Int. J. Syst. Evol. Microbiol. *53*: 1531–1536.

Yumoto, I., K. Hirota, T. Kawahara, Y. Nodasaka, H. Okuyama, H. Matsuyama, Y. Yokota, K. Nakajima and T. Hoshino. 2004a. *Anoxybacillus voinovskiensis* sp. nov., a moderately thermophilic bacterium from a hot spring in Kamchatka. Int. J. Syst. Evol. Microbiol. *54*: 1239–1242.

Yumoto, I., K. Hirota, Y. Nodasaka, Y. Yokota, T. Hoshino and K. Nakajima. 2004b. *Alkalibacterium psychrotolerans* sp. nov., a psychrotolerant obligate alkaliphile that reduces an indigo dye. Int. J. Syst. Evol. Microbiol. *54*: 2379–2383.

Yumoto, I., K. Hirota, S. Yamaga, Y. Nodasaka, T. Kawasaki, H. Matsuyama and K. Nakajima. 2004c. *Bacillus asahii* sp. nov., a novel bacterium isolated from soil with the ability to deodorize the bad smell generated from short-chain fatty acids. Int. J. Syst. Evol. Microbiol. *54*: 1997–2001.

Yumoto, I., K. Hirota, T. Goto, Y. Nodasaka and K. Nakajima. 2005a. *Bacillus oshimensis* sp. nov., a moderately halophilic, non-motile alkaliphile. Int. J. Syst. Evol. Microbiol. *55*: 907–911.

Yumoto, I., K. Hirota, Y. Nodasaka and K. Nakajima. 2005b. *Oceanobacillus oncorhynchi* sp. nov., a halotolerant obligate alkaliphile isolated from the skin of a rainbow trout (*Oncorhynchus mykiss*), and emended description of the genus *Oceanobacillus*. Int. J. Syst. Evol. Microbiol. *55*: 1521–1524.

Zabeau, M. and P. Vos. 1993. Selective restriction fragment amplification: A general method for DNA fingerprinting. European Patent Office Publ. 0534 858 A1, bulletin 93/13.

Zahner, V., H. Momen and F.G. Priest. 1998. Serotype H5a5b is a major clone within mosquito-pathogenic strains of *Bacillus sphaericus*. Syst. Appl. Microbiol. *21*: 162–170.

Zaitsev, G.M., I.V. Tsitko, F.A. Rainey, Y.A. Trotsenko, J.S. Uotila, E. Stackebrandt and M.S. Salkinoja-Salonen. 1998. New aerobic ammonium-dependent obligately oxalotrophic bacteria: description of *Ammoniphilus oxalaticus* gen. nov., sp. nov. and *Ammoniphilus oxalivorans* gen. nov., sp. nov. Int. J. Syst. Bacteriol. *48*: 151–163.

Zarilla, K.A. and J.J. Perry. 1987. *Bacillus thermoleovorans*, sp. nov., a species of obligately thermophilic hydrocarbon utilizing endospore-forming bacteria. Syst. Appl. Microbiol. *9*: 258–264.

Zeigler, D.R. 2005. Application of a *recN* sequence similarity analysis to the identification of species within the bacterial genus *Geobacillus*. Int. J. Syst. Evol. Microbiol. *55*: 1171–1179.

Zhang, M.Y., A. Lovgren and R. Landen. 1995. Adhesion and cytotoxicity of *Bacillus thuringiensis* to cultured Spodoptera and *Drosophila* cells. J. Invertebr. Pathol. *66*: 46–51.

Zhang, J., T.C. Hodgman, L. Krieger, W. Schnetter and H.U. Schairer. 1997. Cloning and analysis of the first cry gene from *Bacillus popilliae*. J. Bacteriol. *179*: 4336–4341.

Zhang, T., X. Fan, S. Hanada, Y. Kamagata and H.H. Fang. 2006. *Bacillus macauensis* sp. nov., a long-chain bacterium isolated from a drinking water supply. Int. J. Syst. Evol. Microbiol. *56*: 349–353.

Zhang, L., Z. Xu and B.K. Patel. 2007. *Bacillus decisifrondis* sp. nov., isolated from soil underlying decaying leaf foliage. Int. J. Syst. Evol. Microbiol. *57*: 974–978.

Zhilina, T.N., E.S. Garnova, T.P. Tourova, N.A. Kostrikina and G.A. Zavarzin. 2001a. *Amphibacillus fermentum* sp. nov. and *Amphibacillus tropicus* sp. nov., new alkaliphilic, facultatively anaerobic, saccharolytic bacilli from Lake Magadi. Microbiology (En. transl. from Mikrobiologiya). *70*: 711–722.

Zhilina, T.N., E.S. Garnova, T.P. Turova, N.A. Kostrikina and G.A. Zavarzin. 2001b. [*Amphibacillus fermentum* sp. nov., *Amphibacillus tropicus* sp. nov. – new alkaliphilic, facultatively anaerobic, saccharolytic Bacilli from Lake Magadi]. Mikrobiologiia. *70*: 825–837.

Zhilina, T.N., E.S. Garnova, T.P. Tourova, N.A. Kostrikina and G.A. Zavarzin. 2002. *In* Validation of the publication of new names and new combinations previously effectively published outside the IJSEM. List no. 85. Int. J. Syst. Evol. Microbiol. *52*: 685–690.

Zilinskas, R.A. 1997. Iraq's biological weapons. The past as future? J. Am. Med. Assoc. *278*: 418–424.

Zlotnikov, A.K., Y.N. Shapovalova and A.A. Makarov. 2001. Association of *Bacillus* firmus E3 and *Klebsiella terrigena* E6 with increased ability for nitrogen fixation. Soil Biol. Biochem. *33*: 1525–1530.

Zopf, W. 1885. Die Spaltpilze, 3rd ed. Edward Trewendt, Breslau.

Family II. **Alicyclobacillaceae** fam. nov.

MILTON S. DA COSTA AND FRED A. RAINEY

A.li.cy.clo.ba.cil.la'ce.ae. N.L. masc. n. *Alicyclobacillus* type genus of the family; -aceae ending to denote a family; N.L. fem. pl. n. *Alicyclobacillaceae* the *Alicyclobacillus* family.

Cells are straight rods of variable length, generally nonmotile. Terminal or subterminal ovoid endospores are formed. The majority of the species stain Gram-positive. Strains are non-pigmented. **Aerobic** with a strictly respiratory type of metabolism, but a few strains reduce nitrate to nitrite and some reduce Fe^{3+}. Mesophilic, slightly thermophilic, and thermophilic, and acidophilic. Menaquinone-7 is the predominant respiratory quinone. Many species possess ω-cyclohexane or ω-cycloheptane, but some do not. Branched chain iso- and anteiso- fatty acids and straight chain fatty acids are present in all species. Most species are chemoorganotrophic; some species are mixotrophic. Organic compounds are used as sole carbon and energy sources; acid is produced from several carbohydrates. Mixotrophic species utilize Fe^{2+}, S^0, and sulfide minerals in the presence of yeast extract or sole organic compounds. The species of the genus *Alicyclobacillus*, the sole genus of the family *Alicyclobacillaceae*, form a distinct phylogenetic lineage based on 16S rRNA gene sequence comparisons (Figure 22). Found in soils and water of geothermal areas, soils, fruit juices, and ores.

DNA G+C content (mol%): 49–62.

Type genus: **Alicyclobacillus** Wisotzkey, Jurtshuk, Fox, Deinhard and Poralla 1992, 267[VP] emend. Goto, Mochida, Asahara, Suzuki, Kasai and Yokota 2003, 1542 emend. Karavaiko, Bogdanova, Tourova, Kondrat'eva, Tsaplina, Egorova, Krasil'nikova and Zakharchuk 2005, 946.

Genus I. **Alicyclobacillus** Wisotzkey, Jurtshuk, Fox, Deinhard and Poralla 1992, 267[VP] emend. Goto, Mochida, Asahara, Suzuki, Kasai and Yokota 2003, 1542 emend. Karavaiko, Bogdanova, Tourova, Kondrat'eva, Tsaplina, Egorova, Krasil'nikova and Zakharchuk 2005, 946

MILTON S. DA COSTA, FRED A. RAINEY AND LUCIANA ALBUQUERQUE

A.li.cy.clo.ba.cil'lus. Gr. adj. *aliphos* fat; Gr. n. *kyklos* circle; L. n. *bacillus* small rod; N.L. masc. n. *Alicyclobacillus* small rod with ω-alicyclic fatty acids.

Straight rods, 0.3–1.1 μm ×1.5–6.3 μm. Terminal or subterminal ovoid endospores are formed. In some strains the sporangium is swollen. The majority of the species stain Gram-positive; one species stains Gram-negative. Many of the strains are **nonmotile, but others have motility. Colonies are not pigmented. Aerobic** with a strictly respiratory type of metabolism, but a few strains reduce nitrate to nitrite, and some reduce Fe^{3+}. The strains of these organisms have variable oxidase and catalase reactions. **Mesophilic, slightly thermophilic,** and **thermophilic,** with an optimum growth temperature 35–65°C; the temperature range for growth 4–70°C. **Acidophilic;** the pH range for growth is from about 0.5–6.5; the optimum is from pH 1.5–5.5. **Menaquinone-7** is the predominant respiratory quinone. Two phospholipids, one aminoglycolipid, one glycolipid, and a sulfonolipid are generally present. Fatty acids are predominantly ω-**cyclohexane or** ω-**cycloheptane**; three species do not possess these fatty acids. Branched chain iso- and anteiso-fatty acids and straight-chain fatty acids are also present. Some strains are known to possess hopanoids. Most species are **chemoorganotrophic;** two species are **mixotrophic.** Monosaccharides, disaccharides, amino acids, and organic acids are used as sole carbon and energy sources; acid is produced from several carbohydrates. Mixotrophic species utilize Fe^{2+}, S^0, and sulfide minerals in the presence of yeast extract or sole organic compounds. Some strains require yeast extract or cofactors for growth. Found in **soil and water of geothermal areas, nongeothermal soils, fruit juices, and ores.**

DNA G+C content (mol%): 48.7–62.5.

Type species: **Alicyclobacillus acidocaldarius** (Darland and Brock, 1971) Wisotzkey, Jurtshuk, Fox, Deinhard and Poralla 1992, 267 (*Bacillus acidocaldarius* Darland and Brock 1971, 9.)

Further descriptive information

The first strains of *Alicyclobacillus acidocaldarius* were classified in the genus *Bacillus* by Darland and Brock (1971) because these aerobic acidothermophiles were Gram-positive and produced endospores. Moreover, another slightly thermophilic and acidophilic endospore former, *Bacillus coagulans*, had been described, and it was logical to include the new species in a ubiquitous genus that also included thermophilic species (Becker and Pederson, 1950). More than a decade later, two other acidophilic and slightly thermophilic species named *Bacillus acidoterrestris* and *Bacillus cycloheptanicus* were described (Deinhard et al., 1987a, 1987b). By this time it had been recognized that *Bacillus acidocaldarius*, *Bacillus acidoterrestris*, and *Bacillus cycloheptanicus* possessed unique ω-alicyclic fatty acids and that the three species were closely related (De Rosa et al., 1971b; Poralla and Konig, 1983). With the application of 16S rRNA gene sequence analysis, it became evident that the genus *Bacillus* was heterogeneous (Ash et al., 1991a). However, an early phylogenetic study of Gram-positive bacilli showed that *Bacillus acidoterrestris* did not cluster with the other two species that possessed ω-alicyclic fatty acids (Ash et al., 1991b). The reason for this anomaly was resolved when it was discovered that the strain supplied by the Deutsche Sammlung von Mikroorganismen und Zellkulturen (DSMZ), Braunschweig, Germany, was not the type strain of *Bacillus acidoterrestris* (Wisotzkey et al., 1992). These authors, using the correct strain, compared the 16S rRNA gene sequences of the three acidothermophilic organisms leading to the proposal that these organisms did not belong to the genus *Bacillus*, as defined at the time, and proposed the genus *Alicyclobacillus* to accommodate the three

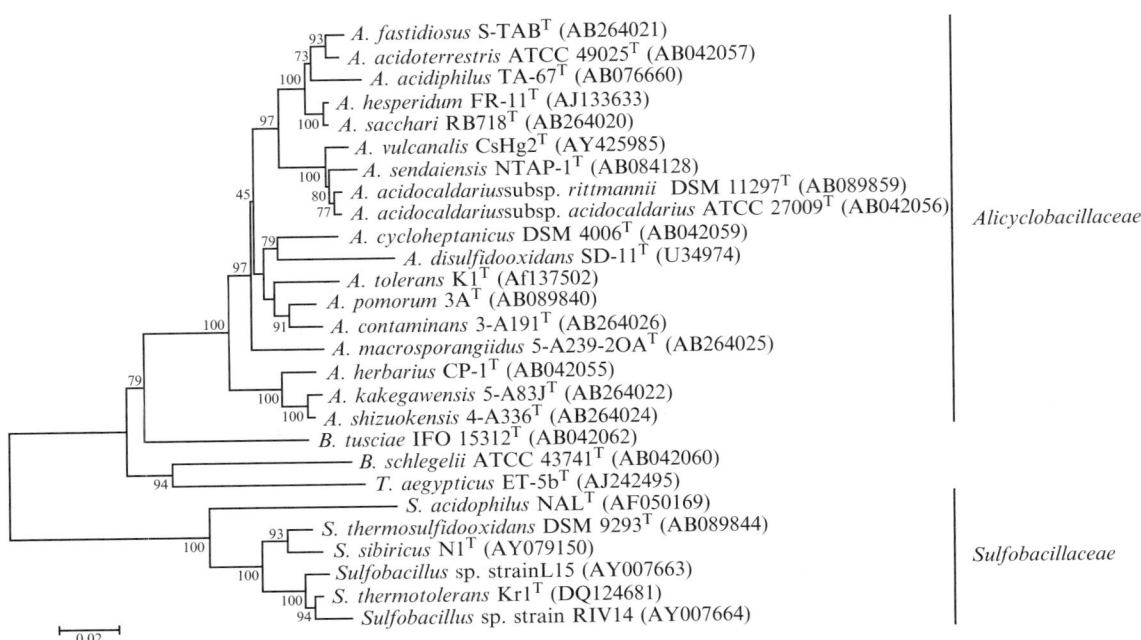

FIGURE 22. 16S rRNA gene sequence-based phylogeny of members of the genera *Alicyclobacillus* and *Sulfobacillus* and related taxa. Bar = 2 inferred nucleotide substitutions per 100 nucleotides.

species already described (Wisotzkey et al., 1992). After more than a decade, 14 new species have been described or reclassified as members of this genus.

All strains examined produce endospores which tend to be terminal or subterminal and have an ovoid morphology; some species have swollen sporangia, but others do not. The strains of the genus *Alicyclobacillus* generally stain Gram-positive, but *Alicyclobacillus sendaiensis* stains Gram-negative (Tsuruoka et al., 2003), reflecting a thin cell wall found in many strains of the genus. Many of the species of this genus are not motile, but motility has been reported in others. The type strains of *Alicyclobacillus herbarius* (Goto et al., 2002a), *Alicyclobacillus acidiphilus* (Matsubara et al., 2002), *Alicyclobacillus pomorum* (Goto et al., 2003), *Alicyclobacillus contaminans*, *Alicyclobacillus kakegawensis*, *Alicyclobacillus macrosporangiidus*, *Alicyclobacillus sacchari*, and *Alicyclobacillus shizuokensis* (Goto et al., 2007) possess motility, however, the position of the flagella on the cells has not been reported (Table 38).

The species of the genus *Alicyclobacillus* can be grouped into three categories in terms of their growth temperature ranges. The strains of the species *Alicyclobacillus acidocaldarius* and of two unnamed genomic species have a growth temperature range of about 45–70°C, with an optimum growth temperature of about 65°C (Albuquerque et al., 2000; Goto et al., 2002b; Karavaiko et al., 2005; Nicolaus et al., 1998). Another group of species, *Alicyclobacillus acidiphilus*, *Alicyclobacillus acidoterrestris*, *Alicyclobacillus cycloheptanicus*, *Alicyclobacillus herbarius*, *Alicyclobacillus hesperidum*, *Alicyclobacillus sendaienensis*, *Alicyclobacillus vulcanalis*, *Alicyclobacillus tolerans*, *Alicyclobacillus pomorum*, *Alicyclobacillus contaminans*, *Alicyclobacillus fastidiosus*, *Alicyclobacillus kakegawensis*, *Alicyclobacillus macrosporangiidus*, and *Alicyclobacillus shizuokensis* have a lower temperature range, 20–65°C, with an optimum for growth of 40–55°C. The species *Alicyclobacillus disulfidooxidans* and *Alicyclobacillus tolerans*, formerly classified in the genus *Sulfobacillus*, have lower growth temperature ranges of about 4–55°C and optimum growth temperatures of about 35–42°C (Table 38). All strains of the species of the genus *Alicyclobacillus* are acidophilic with pH optima for growth in the range 1.5–5.5 and pH optima of 0.5–6.5, the most acidophilic strains being those of the mesophilic species *Alicyclobacillus tolerans* and *Alicyclobacillus disulfidooxidans* which have pH optima for growth of about 1.5–2.5 (Karavaiko et al., 2005).

Most strains of the genus *Alicyclobacillus* appear to be strictly chemoorganotrophic, although the vast majority of these have not been examined for mixotrophic metabolism found in the species *Alicyclobacillus tolerans* or *Alicyclobacillus disulfidooxidans*. Some species of the genus *Alicyclobacillus* assimilate a large variety of carbohydrates, polyols, organic acids, and amino acids for growth, although most descriptions of the species of this genus report only the formation of acid from sugars, and little is known about the assimilation of noncarbohydrate carbon sources (Albuquerque et al., 2000). The strains of *Alicyclobacillus tolerans* and *Alicyclobacillus disulfidooxidans* are facultatively chemoorganotrophic and mixotrophic. For example, *Alicyclobacillus tolerans* strain K1[T] prefers mixotrophic growth conditions rather than growth on organic substrates alone (Karavaiko et al., 2005). Higher growth rates and higher biomass yields are obtained when this organism is grown on Fe^{2+}, glucose, and yeast extract than when they are grown on glucose and yeast extract alone. Autotrophic growth on Fe^{2+} is poor and ceases after two transfers of the culture; strain $K1^T$ fixes CO_2 very weakly. Mixotrophic growth also takes place with elemental sulfur, sulfide minerals, and yeast extract (Karavaiko et al., 2005). The species *Alicyclobacillus disulfidooxidans* has been reported to grow autotrophically (Dufresne et al., 1996).

The pathways involved in carbon and energy metabolism have not been reported in most strains of the genus *Alicyclobacillus*. In fact, only one strain, formerly classified as a species of the genus *Sulfobacillus*, has been examined in this respect. The type strain of *Alicyclobacillus tolerans* has enzymes of the Embden–Meyerhof pathway, but not those of the Entner–Doudoroff or pentose-phosphate pathways under mixotrophic growth (Karavaiko et al., 2001). This same strain is known to have a complete tricarboxylic acid cycle (Karavaiko et al., 2002).

The most notable characteristic of the alicyclobacilli is, without doubt, the high levels of ω-cyclohexane and ω-cycloheptane fatty acids (De Rosa et al., 1971b) which gave the organisms their generic epithet. These terminal cyclic fatty acids are found in 14 of the 17 validly named species. ω-Cyclohexane fatty acids, predominantly ω-cyclohexylundecanoic acid ($C_{17:0\ \omega cyclohexane}$) and ω-cyclohexyltridecanoic acid ($C_{19:0\ \omega cyclohexane}$), constitute the major acyl chains of most species, and $C_{17:0\ \omega cyclohexane}$ can reach levels as high as 80% of the total fatty acids in *Alicyclobacillus acidophilus* (Table 39). The major fatty acids of *Alicyclobacillus cycloheptanicus*, *Alicyclobacillus kakegawensis*, *Alicyclobacillus shizuokensis*, and *Alicyclobacillus herbarius* have been identified as the very rare ω-cycloheptane fatty acids, namely ω-cycloheptylundecanoic acid ($C_{18:0\ \omega cycloheptane}$) and ω-cycloheptyltridecanoic acid ($C_{20:0\ \omega cycloheptane}$), where the former may reach about 86% of the total fatty acids in *Alicyclobacillus cycloheptanicus* (Deinhard et al., 1987b). The type strain of *Alicyclobacillus tolerans* also possesses $C_{17:0\ \omega cyclohexane\ 2OH}$ that reaches about 11.3% of the total fatty acids (Karavaiko et al., 2005). The ω-cycloheptane fatty acid-containing alicyclobacilli also possess lower amounts (2.3–3.5% of the total fatty acids) of $C_{18:0\ \omega cycloheptane\ 2OH}$ (Deinhard et al., 1987b; Goto et al., 2002a). All strains also possess low amounts of straight-chain and branched-chain iso- and anteiso-fatty acids. However, the type strains of *Alicyclobacillus pomorum*, *Alicyclobacillus contaminans*, and *Alicyclobacillus macrosporangiidus*, unlike other strains of the genus *Alicyclobacillus*, do not possess ω-cyclohexane or ω-cycloheptane fatty acids. These organisms possess only straight-chain and branched-chain iso- and anteiso-fatty acids. The lack of ω-alicyclic fatty acids in these organisms indicate that they are not necessary for growth of the alicyclobacilli at high temperature and low pH, as was sometimes hypothesized. These fatty acids remain, however, a hallmark characteristic of the bacteria of the genus *Alicyclobacillus*, although ω-alicyclic fatty acids are also found in species of the genus *Sulfobacillus* (M. S. da Costa, F. A. Rainey, and L. Albuquerque, unpublished results) and in the unrelated bacteria *Curtobacterium pusillum* (Suzuki et al., 1981) and *Propionibacterium cyclohexanicum* (Kusano et al., 1997), where they are also the major fatty acids.

Hopanoids are pentacyclic triterpenoids that are found in the membranes of many bacteria (Ourisson et al., 1987). The hopanoids are structurally similar to sterols and cause condensation of phospholipids in model membranes (Poralla et al., 1980). These lipids were identified in *Alicyclobacillus acidocaldarius* as 1,2,3,4-tetrahydroxypentane-29-hopane and 1-(O-β-N-acylglucosaminyl)-2,3,4-tetrahydroxypentane-29-hopane (Langworthy et al., 1976; Langworthy and Mayberry, 1976; Poralla et al., 1980). Hopanoids have also been found in *Alicyclobacillus acidoterrestris*, but not in *Alicyclobacillus cycloheptanicus*. Other strains have not been examined (Deinhard et al., 1987a, 1987b).

The first definite isolation of strains of *Alicyclobacillus* was by Brock and Darland (1970) and Darland and Brock (1971) from acidic and thermal environments in Yellowstone National Park. Brock and Darland described the type species of the genus, although strains of these organisms were probably isolated earlier by Uchino and Doi (1967) who reported acidophilic sporeformers from Japanese hot springs. Many strains of alicyclobacilli have been isolated from such environments all over the world, namely from the solfatara at Pisciarelli in Italy (De Rosa et al., 1971a), acidic soils associated with geothermal activity in Japan (Goto et al., 2002b; Hiraishi et al., 1997; Tsuruoka et al., 2003), the Furnas area on the Island of São Miguel in the Azores (Albuquerque et al., 2000), Coso Hot Springs in California (Simbahan et al., 2004), and the Antarctic (Nicolaus et al., 1991, 1998). Alicyclobacilli appear to be ubiquitous colonists of acidic geothermal environments. However, many strains have now been isolated from other environments, some of them quite unexpected. Strains of *Alicyclobacillus acidoterrestris* and *Alicyclobacillus cycloheptanicus* were isolated from soils that were not closely associated with geothermal activity such as garden soil, soil from woods, and apple juice (Cerny et al., 1984; Deinhard et al., 1987a, 1987b; Hippchen et al., 1981). Recently, strains of six new species have been isolated from farm soils and beverages in Japan (Goto et al., 2007). The type strains of the mesophilic species *Alicyclobacillus tolerans* and *Alicyclobacillus disulfidooxidans* were isolated from lead-zinc ores in Uzbekistan and wastewater sludge in Canada, respectively (Dufresne et al., 1996; Kovalenko and Malakhova, 1983).

Strains of seven species of alicyclobacilli have been isolated from unusual environments, namely beverages. As noted previously, one strain classified as *Alicyclobacillus acidoterrestris* was isolated from apple juice (Cerny et al., 1984; Hippchen et al., 1981). Later isolates of this same species were recovered from spoiled acidic beverages where they cause off-flavor, particularly in fruit juices (Jensen, 1999; Pettipher et al., 1997; Splittstoesser et al., 1994; Yamazaki et al., 1996). These organisms are believed to colonize soil and then contaminate the juices via the fruits used to make them (Eguchi et al., 1999) where they can grow and produce guaiacol and halophenols (Matsubara et al., 2002). The spores survive the pasteurization processes, germinate, and grow in fruit juices, giving rise to off-flavors. These organisms have been recognized as important spoilage organisms of fruit juices (Chang and Kang, 2004). Recently, several strains classified as new species of this genus have been isolated from herbal tea and fruit juices in Japan (Goto et al., 2002a, 2003, 2007; Matsubara et al., 2002).

Enrichment, isolation and growth conditions

The original strains of Darland and Brock (1971) were isolated with a liquid medium composed of 1.0 or 10.0 g glycerol, or glucose, or ribose, 1.0 g yeast extract, 0.2 g $(NH_4)_2SO_4$, 0.5 g $MgSO_4 \cdot 7H_2O$, 0.25 g $CaCl_2 \cdot 2H_2O$, 3.0 g KH_2PO_4 in 1 l of water

TABLE 38. Characteristics of the type strains of the species of the genus *Alicyclobacillus*[a]

Characteristic	1. *A. acidocaldarius* 104-1A[b,c,d]	2. *A. acidiphilus* TA-67[e]	3. *A. acidoterrestris* GD3B[c,d,f]	4. *A. contaminans* 3-A191[g]	5. *A. cycloheptanicus* SCH[c,h,i,j]	6. *A. disulfidooxidans* SD-11[j,k]	7. *A. fastidiosus* S-TAB[g]	8. *A. herbarius* CP-1[i]
Cell size (µm)	0.7–0.8 × 2.0–3.0	0.9–1.1 × 4.8–6.3	0.6–0.8 × 2.9–4.3	0.8–0.9 × 4.0–5.0	0.4–0.6 × 2.5–4.5	0.3–0.5 × 0.9–3.6	0.9–1.0 × 4.0–5.0	ND
Motility	ND	Motile	ND	Motile	Nonmotile	Nonmotile	Nonmotile	Motile
Spores	Ellipsoidal, terminal to subterminal	Oval, terminal to subterminal	Oval, terminal to subterminal	Ellipsoidal, subterminal	Oval, subterminal	Oval, subterminal	Ellipsoidal, subterminal	Oval, sub-terminal
Sporangia	Not swollen	Swollen	Slightly swollen or not swollen	Swollen	Slightly swollen	Swollen	Swollen	Swollen
Gram reaction	+	+	+	+ to v	+	v	+ to v	+
Anaerobic growth	–	–	–	–	+	+	–	
Optimum temperature (°C)	60–65	50	42–53	50–55	48	35	40–45	55–60
Growth temperature range (°C)	45–70	20–55	35–55	35–60	40–53	4–40	20–55	35–65
Optimum pH	3.0–4.0	3	4	4.0–4.5	3.5–4.5	1.5–2.5	4.0–4.5	4.5–5.0
Growth pH range	2.0–6.0	2.5–5.5	2.2–5.8	3.5–5.5	3.0–5.5	0.5–6.0	2.5–5.0	3.5–6.0
Growth in NaCl:								
1%	+	+	+	+	+	ND	+	+
2%	+	+	+	+	+	ND	+	+
3%	–	+	+	ND	+	ND	ND	+
4%	–	+	+	ND	+	ND	ND	+
5%	–	–	–		+	ND	–	+
Growth factors	Not required	Not required	Not required	Not required	Methionine or vitamin B$_{12}$, pantothenate, isoleucine	Yeast extract	Not required	Not required
Nitrate reduction to nitrite	–	–	–	–	–	ND	–	+
Presence of:								
Oxidase	–	–	–	–	+	ND	–	–
Catalase	w	+	w	–	+	ND	+	+
Mineral substrates	ND	ND	ND	ND	ND	S^0, Fe^{2+} and Fe$_2$S	ND	ND
Acid production from:								
N-Acetyl-glucosamine	–	–	–	ND	ND	–	ND	–
Adonitol	–	–	–	ND	ND	–	ND	–
Amygdalin	–	–	–	–	–	–	–	+
D-Arabinose	–	+	–	–	+	–	+	+
L-Arabinose	+	+	+	+	ND	–	+	+
D-Arabitol	–	–	–	–	+	–	–	–
L-Arabitol	–	–	–	ND	ND	–	ND	–
Arbutin	–	+	–	+	–	–	–	+
Cellobiose	+	+	+	+	–	–	ND	+
Dulcitol	–	–	–	ND	+	–	+	+
D-Fructose	+	+	+	+	+	–	+	+
D-Fucose	–	–	–	–	–	–	+	+
L-Fucose	–	–	–	–	+	–	+	–
D-Galactose	+	+	+	+	–	–	+	+
β-Gentiobiose	–	+	–	v	–	–	–	+
Gluconate	–	–	–	ND	ND	+	ND	+
D-Glucose	+	+	+	+	+	–	–	+
Glycerol	+	–	+	+	–	–	–	+
Glycogen	–	–	–	–	–	–	–	–
Erythritol	–	–	+	–	–	–	–	–
Esculin	–	+	–	+	+	–	+	+
Inositol	–	–	+	–	+	–	+	–
Inulin	–	–	–	–	ND	–	–	–
2-Keto-gluconate	–	–	–	ND	ND	–	ND	–
5-Keto-gluconate	–	–	–	–	+	–	–	+
Lactose	+	+	+	v	–	–	–	+

9. *A. hesperidum* FR-11[d]	10. *A. kakegawaensis* 5-A83J[g]	11. *A. macrosporangiidus* 5-A239-2O-A[g]	12. *A. pomorum* 3A[m]	13. *A. sacchari* RB718[g]	14. *A. sendaiensis* NTAP-1[l]	15. *A. shizuokensis* 4-A336[g]	16. *A. tolerans* K1[k]	17. *A. vulcanalis* CsHg2[m]
0.5–0.7 × 2.1–3.9	0.6–0.7 × 0–5.0	0.7–0.8 × 5.0–6.0	0.8–1.0 × 2.0–4.0	0.6–0.7 × 4.0–5.0	0.8 × 2.0–3.0	0.7–0.8 × 4.0–5.0	0.9–1.0 × 3.0–6.0	0.4–0.7 × 1.5–2.5
Nonmotile Terminal	Motile Oval, subterminal	Motile Oval, terminal	Motile Oval, subterminal	Motile Ellipsoidal, subterminal	Nonmotile Round, terminal	Motile Oval, subterminal	Nonmotile Oval, terminal to subterminal	ND Terminal
Not swollen	Swollen	Swollen	Swollen	Swollen	Swollen	Swollen	Swollen	ND
+	+ to v	+ to v	+ to v	+ to v	−	+ to v	+	+
−	−	−	−	−	−	−	−	−
50–53	50–55	50–55	45–50	45–50	55	45–50	37–42	55
40–55	40–60	35–60	30–60	30–55	40–65	35–60	20–55	35–65
3.5–4.0	4.0–4.5	4.0–4.5	4.5–5.0	4.0–4.5	5.5	4.0–4.5	2.5–2.7	4
2.5–5.5	3.5–6.0	3.5–6.0	3.0–6.0	2.5–5.5	2.5–6.5	3.5–6.0	1.5–5.0	2.0–6.0
+	+	+	+	+	+	+	ND	+
+	+	+	+	+	+	+	ND	+
+	ND	+	−	ND	+	+	ND	−
−	ND	+	−	ND	+	+	ND	−
Not required	Not required	Not required	Not required	Not required	Not required	Not required	Not required	Not required
−	−	−	−	−	+	−	−	ND
−	−	−	+	−	−	−	w	−
w	w	w	+	−	−	−	w	−
ND	ND	ND	ND	ND	ND	ND	S^0, Fe^{2+}, sulfide minerals	ND
−	ND	ND	ND	ND	−	ND	−	−
−	ND	ND	ND	ND	−	ND	−	−
−	+	−	+	−	−	−	+	−
−	+	+	−	−	−	−	w	−
+	+	+	−	+	+	+	w	+
−	+	+	ND	−	−	−	−	−
−	ND	ND	ND	ND	−	ND	−	−
+	+	−	−	+	+	+	−	w
−	ND	ND	ND	ND	−	ND	−	+
+	+	+	+	+	+	+	+	−
−	−	−	−	−	−	−	−	+
−	−	−	ND	−	−	−	−	−
+	+	+	−	+	+	+	−	+
w	+	+	−	+	−	+	+	+
−	ND	ND	ND	ND	−	ND	−	−
+	+	+	+	+	+	+	+	+
+	−	+	+	+	+	−	w	+
+	−	−	−	+	+	−	+	+
−	v	−	−	−	−	−	+	−
−	+	+	+	−	−	+	+	−
−	v	−	−	−	−	−	−	+
−	−	−	ND	+	−	−	−	−
−	ND	ND	ND	ND	−	ND	−	−
−	−	−	+	−	−	−	+	w
+	+	+	−	+	+	−	w	−

(continued)

TABLE 38. (continued)

Characteristic	1. A. acidocaldarius 104-1A[b,c,d]	2. A. acidiphilus TA-67[e]	3. A. acidoterrestris GD3B[e,d,f]	4. A. contaminans 3-A191[g]	5. A. cycloheptanicus SCH[e,h,j]	6. A. disulfidooxidans SD-11[j,k]	7. A. fastidiosus S-TAB[g]	8. A. herbarius CP-1[i]
D-Lyxose	−	−	−	−	+	−	+	+
Maltose	+	+	+	+	+	−	+	+
D-Mannose	+	+	+	+	+	−	+	+
Mannitol	+	−	+	+	+	−	+	+
Melibiose	+	−	w	−	−	−	−	+
Melezitose	−	+	−	−	−	−	−	+
Methyl α-D-glucoside	−	+	−	−	−	−	−	+
Methyl α-D-mannoside	−				ND	−	+	−
Methyl β-xyloside	−	−	−	−		−	+	+
D-Raffinose	+	+	−	−	−	−	+	+
Rhamnose	−	−	+	v	+	−	+	+
Ribose	+	+	+	+	+	−	−	+
Salicin	−	+	−	v	−	−	−	−
Sorbitol	−	+	+	−	+	−	−	−
L-Sorbose	−	+	−	v	+	−	−	−
Starch	−	−	−	−	−	−	−	+
Sucrose	+	+	+	+	−	−	+	−
D-Tagatose	−	−	−	v	+	−	+	+
Trehalose	+	+	+	+	−	−	−	+
D-Turanose	−	+	−	−	−	−	−	+
Xylitol	−	+	+	−	−	−	−	−
D-Xylose	−	+	−	+	+	−	+	+
L-Xylose	−	−	−	−	+	−	−	−
Presence of hopanoids	+	ND	+	ND	−	ND	ND	ND
Presence of a sulfonolipid	+	ND	+	ND	+	ND	ND	ND
Major menaquinone	MK-7	MK-7	MK-7	MK-7	MK-7	MK-7	MK-7	MK-7
Mol% G + C	60.3	54.1	51.6	60.1–60.6	55.6	53	53.9	56.2

[a]Symbols: +, positive; −, negative; w, weakly positive; v, variable; ND, not determined.

[b]Results from Darland and Brock (1971).

[c]Results from Wisotzkey et al. (1992).

[d]Results from Albuquerque et al. (2000).

[e]Results from Matsubara et al. (2002).

[f]Results from Deinhard et al. (1987a).

[g]Results from Goto et al. (2007).

[h]Results from Deinhard et al. (1987b).

[i]Results from Goto et al. (2002a).

[j]Results from Dufresne et al. (1996).

[k]Results from Karavaiko et al. (2005).

[l]Results from Tsuruoka et al. (2003).

[m]Results from Goto et al. (2003).

[n]Results from Simbahan et al. (2004).

with the pH adjusted to 2–5 with H_2SO_4. Turbid cultures were streaked on the same medium solidified with 2.0% agar and adjusted to pH 3.5–4.0. Later formulations of this medium containing 5.0 g/l of glucose also included, in a few cases, the addition of 1 ml of the trace element solution of Farrand et al. (1983). The medium with or without the trace element solution is commonly called *Bacillus acidocaldarius* Medium (BAM) and is used extensively for the isolation and growth of chemoorganotrophic strains of alicyclobacilli. The incuba-

tion temperature of the enrichments and the cultures depends on the organisms sought. Other modifications of the original medium include the addition of 0.02 g/l of $FeCl_3 \cdot 6H_2O$ and 0.35 g/l of tryptone (Simbahan et al., 2004). Several investigators have used completely different media; one medium used by Tsuruoka et al. (2003) contains 1.5% gelatin, 0.01% yeast extract, 0.85% NaCl, 0.3% KH_2PO_4, and 0.001% $MgSO_4 \cdot 7H_2O$ at pH 4.8. Another medium used by Goto et al. (2003) for the isolation of alicyclobacilli from juices contained 2.0 g yeast

9. *A. hesperidum* FR-11[d]	10. *A. kakegawaensis* 5-A83J[g]	11. *A. macrosporangiidus* 5-A239-2O-A[g]	12. *A. pomorum* 3A[m]	13. *A. sacchari* RB718[g]	14. *A. sendaiensis* NTAP-1[l]	15. *A. shizuokensis* 4-A336[g]	16. *A. tolerans* K1[k]	17. *A. vulcanalis* CsHg2[9n]
+	+	+	−	−	−	−	−	−
+	+	+	+	+	+	+	w	+
+	+	−	+	+	+	+	+	+
−	−	−	−	+	+	+	−	+
−	v	−	−	+	+	−	+	+
−	+	+	+	+	+	−	+	w
								−
−	−	−	ND	+	−	−	−	−
−	−	−	−	+	+	−	w	+
+	+	+	+	+	−	−	−	+
−	+	+	+	+	+	+	+	+
−	+	+	−	+	+	+	−	−
−	+	+	+	−	−	−	−	−
w	−	−	−	+	ND	−	+	−
+	+	+	+	+	+	−	+	+
−	v	−	+	−	−	+	+	−
+	+	+	+	+	+	−	w	+
+	+	−	+	+	+	+	+	+
−	+	+	−	−	−	−	w	+
−	+	+	−	+	+	+	+	+
ND	ND	ND	ND	ND	ND	+	+	−
+	ND	ND	ND	ND	ND	ND	ND	ND
MK-7	MK-7	MK-7	MK-7	MK-7	MK-7	MK-7	MK-7	ND
53.3	61.3–61.7	62.5	53.1	56.6	62.3	60.5	48.7	62

extract, 1.0 g glucose, 2.0 g soluble starch, and 15 g agar in 1 l of water adjusted to pH 3.7. Yeast extract is generally added to the media, although many strains do not require it for growth. The species *Alicyclobacillus cycloheptanicus* is known to require isoleucine, methionine or vitamin B_{12}, and pantothenate (Deinhard et al., 1987b).

The strains of mixotrophic species, such as *Alicyclobacillus tolerans*, are sensitive to high concentrations of organic compounds and have, therefore, been isolated and grown on different media with low amounts of organic carbon. The strains of this species have been grown in Manning medium (Manning, 1975;

DSMZ medium 1023) containing 0.2 g of yeast extract, 33.4 g $FeSO_4 \cdot 7H_2O$, 6.0 g $(NH_4)_2SO_4$, 0.2 g KCl, 1.0 g $MgSO_4 \cdot 7H_2O$, 0.02 g $Ca(NO_3)_2$, and 0.2 g K_2HPO_4 in 1 l of water adjusted to pH 2.5–2.7 with 0.05 M H_2SO_4 and incubated at 40°C. The species *Alicyclobacillus disulfidooxidans* has been routinely grown on a medium composed of Solution A, which contains 0.1 g of yeast extract, 3.0 g $(NH_4)_2SO_4$, 0.1 g KCl, 0.5 g $MgSO_4 \cdot 7H_2O$, 0.5 g KH_2PO4, and 0.1 g $Ca(NO_3)_2 \cdot 4H_2O$, in 1 l of water adjusted to pH 2.5 with H_2SO_4. Solution A is autoclaved, and filter-sterilized solution B containing 2.5 g glutathione in 10.0 ml of water is added to 1.0 liter of solution A (ATCC medium 2091).

TABLE 39. Fatty acid composition of the type strains of the species of the genus *Alicyclobacillus* grown at optimum temperature[a,b]

Fatty acid	1. *A. acidocaldarius* 104-1[c]	2. *A. acidiphilus* TA-67[d]	3. *A. acidoterrestris* GD3B[e]	4. *A. contaminans* 3-A191[e]	5. *A. cycloheptanicus* SCH[f]	6. *A. disulfidooxidans* SD-11[g,h]	7. *A. fastidiosus* S-TAB[e]	8. *A. herbarius* CP-1[f]	9. *A. hesperidum* FR-11[c]	10. *A. kakegawensis* 5-A83[f]	11. *A. macrosporangiidus* 5-A239-2O-A[f]	12. *A. pomorum* 3A[f]	13. *A. sacchari* RB718[g]	14. *A. sendaiensis* NTAP-1[j]	15. *A. shizuokensis* 4-A336[f]	16. *A. tolerans* K1[h]	17. *A. vulcanalis* CsHg2[k]
$C_{14:0}$	–	–	–	–	0.8	ND	–	–	–	–	–	0.5	–	–	–	ND	–
$C_{14:0}$ ante	–	–	1	–	–	ND	–	–	–	–	–	–	–	–	–	ND	–
$C_{14:0}$ iso	2.2	–	–	1.8	0.5	ND	2.5	–	5.4	–	2.1	19.9	0.8	1.1	0.7	ND	1.5
$C_{15:0}$ iso	–	0.3	–	0.5	–	ND	1.9	–	6.6	–	1.2	9.9	0.5	1.5	–	ND	0.7
$C_{15:0}$ ante	–	–	–	–	0.6	ND	–	–	–	1.2	–	0.5	–	–	0.5	ND	0.5
$C_{15:0}$	1.1	–	–	16.9	1	ND	8.3	3.8	0.9	0.9	44.2	18.3	–	–	1.9	ND	4.2
$C_{16:0}$ iso	0.7	1.5	1.1	0.9	2.4	ND	1.3	5.1	2.1	8.8	6.1	1.4	0.5	1.6	6.1	ND	7.8
$C_{16:0}$	10.8	0.6	–	29.3	2.7	ND	5.3	5.6	4.9	0.5	16.7	13.3	2.4	4.9	4	ND	8.3
$C_{17:0}$ iso	3	2.6	0.6	43.8	1.3	ND	5.4	3	10.3	1.2	25.2	34.2	4.6	4.5	1.6	ND	8.5
$C_{17:0}$ ante	–	–	–	0.6	–	ND	–	0.5	–	–	–	–	–	10.5	–	ND	–
$C_{17:0}$	–	–	–	4.7	–	ND	–	3.2	–	–	4.5	–	–	0.6	–	ND	–
$C_{18:0}$ iso	–	–	–	0.6	–	ND	–	–	–	–	–	–	–	0.9	–	ND	–
$C_{18:0}$	48.9	82.6	71.6	–	1.3	Mfa	61.1	2.4	56.8	–	–	1.3	64.4	44.1	–	60	46.2
$C_{17:0}$ ω-cyclohexane	–	–	–	–	–	ND	–	–	–	–	–	–	–	–	–	11.3	–
$C_{17:0}$ ω-cyclohexane 2OH	–	–	–	–	–	ND	–	–	–	–	–	–	–	–	–	1.2	–
$C_{18:0}$ ω-cyclohexane	–	–	–	–	–	Mfa	13.9	–	13.3	–	–	–	26.9	30.2	–	7.9	22.8
$C_{19:0}$ ω-cyclohexane	–	–	–	–	86.8	ND	–	67.1	–	65.6	–	–	–	–	65.4	ND	–
$C_{18:0}$ ω-cycloheptane	–	–	–	–	2.3	ND	–	3.5	–	15	–	–	–	–	12.4	ND	–
$C_{18:0}$ ω-cycloheptane 2OH	–	–	–	–	–	ND	–	–	–	–	–	–	–	–	–	ND	–
$C_{20:0}$ ω-cycloheptane	–	–	–	–	–	ND	–	4.5	–	6.4	–	–	–	–	7.4	ND	–

[a]Symbols: –, not detected; ND, not determined; Mfa, major fatty acid.
[b]Values for fatty acids presents at levels of less than 0.5% in all strains are not shown.
[c]Results from Albuquerque et al. (2000).
[d]Results from Matsubara et al. (2002).
[e]Results from Goto et al. (2007)
[f]Results from Goto et al. (2002a); Deinhard et al. (1987b) detected 20:0 ω-cycloheptane.
[g]Results from Dufresne et al. (1996).
[h]Results from Karavaiko et al. (2005).
[i]Results from Goto et al. (2003)
[j]Results from Tsuruoka et al. (2003)
[k]Results from Simbahan et al. (2004).

Most isolates of the genus *Alicyclobacillus* have been obtained by enrichment in BAM or Manning medium. Water or biofilm samples are inoculated into liquid medium and incubated at 40–65°C for 2–3 d. Turbid cultures are spread on the same medium solidified with agar (2–3%) and incubated at the same temperature until the organisms can be isolated. Alternatively, samples are directly spread on solid media. Membrane filtration methods have also been used for the isolation of strains of *Alicyclobacillus* and offer the advantage of recovering a larger number of different colonial types and minor populations than liquid enrichments which tend to select clones that grow better in the media used or that constitute the major populations of the samples. The membrane filtration method can be used with water or soil. Adequate volumes of the samples or dilutions are filtered through 47-mm-diameter cellulose nitrate or acetate membrane filters with pore sizes of 0.22 or 0.45 μm. The filters are placed on the surface of plates of BAM medium or a similar low-nutrient medium solidified with 2–3% agar; the plates are inverted and incubated for 2–7 d wrapped in plastic film at the appropriate temperature. Soil is resuspended in water, large particles are allowed to sediment, and the suspensions or dilutions are filtered through membrane filters (Albuquerque et al., 2000). Nonpigmented colonies can easily be observed and picked for further purification.

The selective isolation of alicyclobacilli has also been achieved by heating spores to inactivate vegetative cells of other acidophilic organisms. Heating at 80°C for 10 minutes resulted in a very high recovery of viable spores of *Alicyclobacillus acidoterrestris* and the selection of high proportions of alicyclobacilli from fruit juices (Walls and Chuyate, 2000).

Maintenance procedures

All strains grow well on one of the media described above. During incubations, Petri plates should be wrapped in plastic film to prevent evaporation. Cultures on solid medium can be maintained for a few weeks in the dark at room temperature. For long term storage, cultures can be frozen at −70°C in cryotubes containing broth supplemented with a final concentration of 15% (v/v) glycerol. Freeze-dried and liquid nitrogen storage cultures have been maintained for several years without loss of viability.

Taxonomic comments

The genus *Alicyclobacillus* presently comprises 17 species and 2 genomic species. The majority of the species are easily distinguished from each other by phenotypic and chemotaxonomic characteristics (Table 38 and Table 39). In the past, the presence of high levels of ω-alicyclic fatty acids, spore formation, and the thermoacidophilic nature of the organisms was considered sufficient for the presumptive identification of the chemoorganotrophic bacteria of this genus. The descriptions of *Alicyclobacillus pomorum*, *Alicyclobacillus contaminans*, and *Alicyclobacillus macrosporangiidus*, which do not possess ω-alicyclic, made it more difficult to recognize strains of alicyclobacilli without resorting to 16S rRNA gene sequence analysis. The absence of ω-alicyclic fatty acid also inevitably led to the emendation of the genus *Alicyclobacillus* (Goto et al., 2003). The mesophilic nature of *Alicyclobacillus tolerans* and *Alicyclobacillus disulfidooxidans*, which grow at lower temperatures than the other members of the genus, also make it

more difficult to recognize and distinguish some members of this genus from those of the unrelated genus *Sulfobacillus*. The description of two genomic species that could not be distinguished from the type strain of *Alicyclobacillus acidocaldarius*, the description of a subspecies of *Alicyclobacillus acidocaldarius*, and the reclassification of two new species formerly classified in the genus *Sulfobacillus* indicate that further discussion of the taxonomy of these organisms is merited.

Two unnamed genomic species, whose phenotypes were difficult to distinguish from that of the type strain of *Alicyclobacillus acidocaldarius*, were designated genomic species 1 and genomic species 2 by Albuquerque et al. (2000) and Goto et al. (2002b). Genomic species 1 was initially represented by strains FR-3 (DSM 11983) and FR-6 (DSM 11984^T) from the Furnas area of the Island of São Miguel in the Azores. Later Goto et al. (2002b), who described the second genomic species from soils in Japan, also isolated several strains designated KHA 31, MIH 2, UZ 1, and KHC 3 with 16S rRNA gene sequences that were identical to the Azorean genomic species 1 strains. The phenotypic characteristics of the Azorean and Japanese strains were extremely variable, while the fatty acid composition was too similar to that of the type strain of *Alicyclobacillus acidocaldarius* to allow differentiation. On the other hand, DNA–DNA reassociation studies showed that strains FR-6 and the Japanese strains shared over 89% homology with each other, but only 56–66% homology with *Alicyclobacillus acidocaldarius*. Another strain, designated MIH 332 (DSM 14672) and examined by Goto et al. (2002b), also had phenotypic and chemotaxonomic characteristics that were similar to those of the type strain of *Alicyclobacillus acidocaldarius* but a DNA–DNA reassociation value of only 39% with the type strain of the species and 44% with the DNA of strains of genomic species 1. In view of the similar phenotypes of the strains of genomic species 1, genomic species 2, and the type strain of *Alicyclobacillus acidocaldarius*, new species could not be formally described (Albuquerque et al., 2000; Goto et al., 2002b). The studies of Albuquerque et al. (2000) and Goto et al. (2002b) clearly showed large variations in the production of acid from individual carbohydrates that made it very difficult to define a common phenotype within the *Alicyclobacillus acidocaldarius* group of strains. Moreover, the isolates and the type strain had very similar fatty acid compositions that did not distinguish the organisms from each other.

The description of the subspecies *Alicyclobacillus acidocaldarius* subsp. *rittmannii* by Nicolaus et al. (1998), led to the automatic classification of the type strain of Darland and Brock (1971) as *Alicyclobacillus acidocaldarius* subsp *acidocaldarius*. The description of this subspecies was based on the close phylogenetic relationship of strain MR-1^T (DSM 11297^T) isolated from the Antarctic, with the type strain of *Alicyclobacillus acidocaldarius* with which it shared 99.1% 16S rRNA gene sequence similarity, a DNA–DNA reassociation value of 69.7%, and a distinctive fatty acid composition. However, the fatty acid composition of strain MR-1 was not compared with the type strain of *Alicyclobacillus acidocaldarius* but with a strain designated Pisciarelli (De Rosa et al., 1971a), which may or may not belong to this species. To our knowledge, 16S rRNA gene sequence analysis has not been performed with strain Pisciarelli nor has this organism been characterized in detail.

Strain MR-1 has a fatty acid composition that is different from strain Pisciarelli and which is also different from that of the type strain of *Alicyclobacillus acidocaldarius* (ATCC 27009[T], DSM 446[T]), although the latter fatty acid composition was obtained in a different study (Albuquerque et al., 2000). The simple fact that the subspecies is based on one strain alone and that other investigators have found a bewildering diversity of acid production patterns from carbohydrates argues against the classification of strain MR-1 into a new subspecies. Moreover, the fatty acid composition was compared with a strain that is not the type strain of the species. It is therefore difficult to accept the validity of *Alicyclobacillus acidocaldarius* subsp. *rittmanni*. We are, therefore, of the opinion that *Alicyclobacillus acidocaldarius* subsp. *rittmannii* is not a valid subspecies of the species *Alicyclobacillus acidocaldarius*.

Two strains, now assigned as species of *Alicyclobacillus*, were formally classified as *Sulfobacillus*. Strain SD-11[T] was classified as *Sulfobacillus disulfidooxidans* on the basis of its ability to utilize elemental sulfur and pyrite as sole source of energy (Dufresne et al., 1996). Phylogenetic analysis showed that strain SD-11[T] fell within the radiation of the species of the genus *Alicyclobacillus* (Goto et al., 2002a, 2003; Matsubara et al., 2002) and was later classified as *Alicyclobacillus disulfidooxidans* (Karavaiko et al., 2005). Another strain, designated K1[T] was isolated from lead-zinc ores and named "*Sulfobacillus thermosulfidooxidans* subsp. *thermotolerans*" because it also had the ability to oxidize iron, elemental sulfur, and sulfides, but the name of this taxon was never validly published (Kovalenko and Malakhova, 1983). Strain K1[T] was recently reclassified as *Alicyclobacillus tolerans* because of its close phylogenetic relationship to other species of *Alicyclobacillus* (Karavaiko et al., 2005). Strain K1[T] is closely related, based on 16S rRNA gene sequence analysis, to other unclassified organisms, some of which share several characteristics with this strain, namely strains SC, AGC-2,

GSM, and CLG, and may or may not belong to the species *Alicyclobacillus tolerans*. It was necessary, therefore, to emend the description of the genus *Alicyclobacillus* to include species with mixotrophic metabolism that had not been previously recognized in this genus (Karavaiko et al., 2005).

The phylogenetic relationships of the species of the genus *Alicyclobacillus* based on 16S rRNA gene sequence comparisons are shown in Figure 23. The 16S rRNA gene sequence similarity within the genus ranges from 90.7–99.6 based on the comparison of 1352 nucleotide positions. Although the species of the genus fall into four subclusters, only three of these subclusters are supported by significant bootstrap values (Figure 23). One cluster comprises the species *Alicyclobacillus acidocaldarius*, *Alicyclobacillus sendaiensis*, and *Alicyclobacillus vulcanalis*, of which the 16S rRNA gene sequences share 98.4% similarity. The second cluster comprises the species *Alicyclobacillus hesperidum*, *Alicyclobacillus sacchari*, *Alicyclobacillus fastidiosus*, *Alicyclobacillus acidoterrestris*, and *Alicyclobacillus acidiphilus*, sharing 16S rRNA gene sequence similarities in the range 96.7–99.7%. The most closely related *Alicyclobacillus* species based on 16S rRNA gene sequence comparisons are *Alicyclobacillus hesperidum* and *Alicyclobacillus sacchari* at 99.7% similarity. The branching of these two subclusters is also supported by a bootstrap value of 97%. The third phylogenetic cluster (*Alicyclobacillus cycloheptanicus*, *Alicyclobacillus disulfidooxidans*, *Alicyclobacillus tolerans*, *Alicyclobacillus pomorum*, and *Alicyclobacillus contaminans*) as shown in Figure 23 is not supported on the basis of bootstrap analyses. In all analyses, cluster IV (*Alicyclobacillus herbarius*, *Alicyclobacillus kakegawensis*, and *Alicyclobacillus shizuokensis*) falls outside the main *Alicyclobacillus* species group and shares 91.6–93.4% 16S rRNA gene sequence similarity to the other species of the genus. The species *Alicyclobacillus macrosporangiidus* does not cluster with any of the other species of the genus but is the outgroup of clusters I, II, and III.

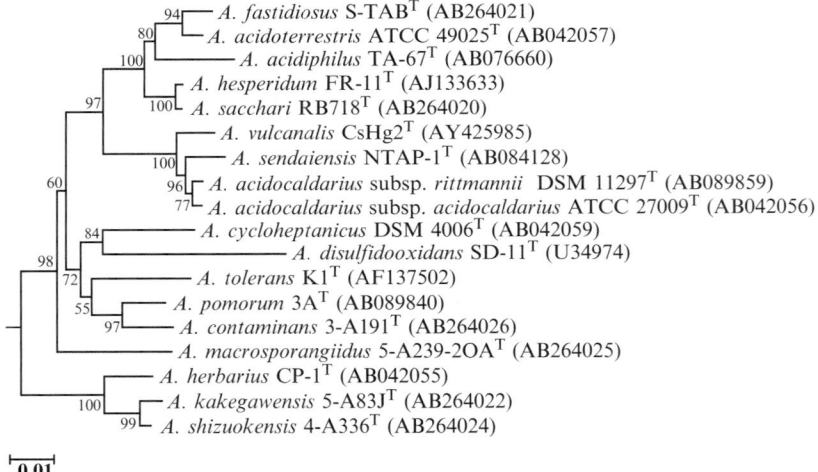

FIGURE 23. Phylogenetic tree indicating the relationships of the *Alicyclobacillus* species based on 16S rRNA gene sequence comparisons. The scale bar represents one inferred nucleotide change per 100 nucleotides. The numbers on the branching points are bootstrap values from 1000 replicate analyses. *Bacillus subtilis* was used as an outgroup organism.

List of species of the genus *Alicyclobacillus*

1. **Alicyclobacillus acidocaldarius** (Darland and Brock 1971) Wisotzkey, Jurtshuk, Fox, Deinhard and Poralla 1992, 267[VP] (*Bacillus acidocaldarius* Darland and Brock 1971, 9.)

a.ci.do.cal.da´ri.us. L. adj. *acidus* acid; L. adj. *caldarius* hot; N.L. adj. *acidocaldarius* pertaining to acid thermal environments.

The strains of this species are Gram-positive and form rod-shaped cells 2–3 μm × 0.7–0.8 μm. Short chains are present. The sporangia are not swollen; ellipsoidal endospores located subterminally or terminally. Colonies are nonpigmented. The predominant membrane acyl chains are C$_{17:0\ \omega\text{cyclohexane}}$ and C$_{19:0\ \omega\text{cyclohexane}}$; straight-chain and branched-chain fatty acids are also present. Hopanoids and a sulfonolipid are present. The major respiratory qui-none is menaquinone-7 (MK-7). Aerobic and chemo-organotrophic. Anaerobic growth does not occur in medium containing nitrate. Nitrate is not reduced to nitrite. The pH range for growth is 2.0–6.0; optimum pH around 3.0–4.0. The temperature range for growth is 45–70°C; optimum temperature is around 60–65°C. Oxidase-negative and weakly catalase-positive. Strains utilize hexoses, disaccharides, organic acids, and amino acids for growth. Acid is produced from sugars. Growth factors are not required. Strains of this species have been isolated from acidic thermal soils and water. The type strain was isolated from Yellowstone National Park.

DNA G+C content (mol%): 60.3 (*T$_m$*) or 62.3 (Bd).

Type strain: 104-1A, ATCC 27009, BCRC 14685, CCUG 28521, CIP 106131, DSM 446, HAMBI 2073, HAMBI 2071, NBRC 15652, JCM 5260, KCTC 1825, LMG 7119, NCCB 89167, NCIMB 11725, NRRL B-14509.

GenBank accession number (16S rRNA gene): AB042056, AJ496806, X60742.

2. **Alicyclobacillus acidiphilus** Matsubara, Goto, Matsumura, Mochida, Iwaki, Niwa and Yamasoto 2002, 1684[VP]

a.ci.di´phi.lus. L. n. *acidum* acid; Gr. adj. *philos* loving; N.L. adj. *acidophilus* acid-loving.

Rod-shaped cells that are 4.8–6.3 μm × 0.9–1.1 μm. Gram stain is positive. The cells are motile. Subterminal or terminal oval spores are formed; sporangia are swollen. Colonies are nonpigmented. Growth occurs at 20–55°C; the optimum growth temperature is about 50°C. The optimum pH is about 3.0; growth occurs at pH 2.5–5.5. Strains are cytochrome-oxidase-negative and catalase is positive. Aerobic and chemoorganotrophic. Yeast extract or growth factors are not required for growth. Nitrate is not reduced to nitrite. Gelatin, starch, phenylalanine, and tyrosine are not hydrolyzed. Voges–Proskauer test is weakly positive and indole production is negative. The predominant membrane fatty acids are C$_{17:0\ \omega\text{cyclohexane}}$ and C$_{19:0\ \omega\text{cyclohexane}}$; straight-chain and branched-chain fatty acids are also present. Major respiratory quinone is menaquinone-7 (MK-7). Acid is produced from several sugars. The type strain of this species was isolated from spoiled fruit juice.

DNA G+C content (mol%): 54.1 (HPLC).

Type strain: TA-67, DSM 14558, IAM 14935, JCM 21417, NBRC 100859, NRIC 6496.

GenBank accession number (16S rRNA gene): AB059677, AB076660.

3. **Alicyclobacillus acidoterrestris** (Deinhard, Blanz, Poralla and Altan 1987a) Wisotzkey, Jurtshuk, Fox, Deinhard and Poralla 1992, 268[VP] (*Bacillus acidoterrestris* Deinhard, Blanz, Poralla and Altan 1987a, 52.)

a.ci.do ter.res´tris. L. n. *acidum* acid; L. adj. *terrestris* soil; N.L. adj. *acidoterrestris* acid-loving and isolated from soil.

The strains of this species stain Gram-positive and form rod-shaped cells 2.9–4.3 μm × 0.6–0.8 μm. The sporangia are not generally swollen; oval endospores located subterminally or terminally. Colonies are nonpigmented. The predominant membrane fatty acids are C$_{17:0\ \omega\text{cyclohexane}}$ and C$_{19:0\ \omega\text{cyclohexane}}$; straight-chain and branched-chain fatty acids are also present. Hopanoids and sulfonolipids are present. The major respiratory quinone is menaquinone-7 (MK-7). Aerobic and chemoorganotrophic. Nitrate is not reduced to nitrite. The pH range for growth is 2.2–5.8; optimum pH around 4.0. The temperature range for growth is about 35–55°C; optimum temperature is 42–53°C. Oxidase-negative and weakly catalase-positive. Strains utilize hexoses, disaccharides, organic acids, and amino acids for growth. Acid is produced from sugars. Growth factors are not required. Strains of this species have been isolated from soils and apple juice. The type strain was isolated from garden soil in Germany.

DNA G + C content (mol%): 51.6 (*T$_m$*).

Type strain: GD3B, ATCC 49025, CIP 106132, DSM 3922, LMG 16906.

GenBank accession number (16S rRNA gene): AB042057, AJ133631, AY573797.

4. **Alicyclobacillus contaminans** Goto, Mochida, Kato, Asahara, Fujita, An, Kasai and Yokoto 2007, 1281[VP]

con.ta´mi.nans. L. part. adj. *contaminans* contaminating, referring to contamination of fruit juice.

Strains of this species stain Gram-positive or Gram-variable for old cultures. Cells are straight rods 4.0–5.0 μm × 0.8–0.9 μm with rounded ends. Motile. Endospores are ellipsoidal and subterminal with swollen sporangia. Colonies on BAM agar after 48 h are nonpigmented circular, opaque, entire, umbonate, and 3–5 mm in diameter. The predominant fatty acids are C$_{16:0\ iso}$, C$_{17:0\ iso}$, and C$_{17:0\ ante}$. The major respiratory quinone is menaquinone-7 (MK-7). Strictly aerobic and chemoorganotrophic. The pH optimum for growth is 4.0–4.5 with no growth at pH 3.0 or 6.0; the temperature range for growth is 35–60°C; optimum growth temperature is 50–55°C. Growth occurs in the presence of 0–2% (w/v) NaCl but not 5% (w/v) NaCl. Oxidase- and catalase-negative. Nitrate is not reduced to nitrite. Voges–Proskauer test and indole production are negative. Gelatin and esculin are hydrolyzed, but phenylalanine, starch, and tyrosine are not; arbutin hydrolysis is variable. Acid is produced from a number of sugars and sugar alcohols. The type strain of this species was isolated from soil of a crop field in Fuji city. An additional strain, E-8 (=IAM 15228), was isolated from orange juice.

DNA G + C content (mol%): 60.6 (HPLC).

Type strain: 3-A191, DSM 17975, IAM 15224, JCM 21678, NBRC 103102.

GenBank accession number (16S rRNA gene): AB264026.

5. **Alicyclobacillus cycloheptanicus** (Deinhard, Saar, Krischke and Poralla 1987b) Wisotzkey, Jurtshuk, Fox, Deinhard and Poralla 1992, 268[VP] (*Bacillus cycloheptanicus* Deinhard, Saar, Krischke and Poralla 1987b, 72.)

cy.clo.hep.ta′ni.cus. Gr. n. *kyclos* circle; Gr. n. *hepta* seven; N.L. adj. *cycloheptanicus* referring to the ω-cycloheptyl fatty acids.

The strains of this species stain Gram-positive and form nonmotile rod-shaped cells 2.5–4.5 μm × 0.4–0.6 μm. Short chains during exponential phase. The sporangia are slightly swollen; endospores are oval and subterminal. Colonies are creamy white and opaque. About 90% of the fatty acids are $C_{18:0\ \omega cycloheptane}$, $C_{20:0\ \omega cycloheptane}$, and $C_{18:0\ \omega cycloheptane\ 2OH}$. The major respiratory quinone is menaquinone-7 (MK-7). A sulfonolipid is present. Aerobic and chemoorganotrophic. Nitrate is not reduced to nitrite. The pH range for growth is 3.0–5.5; optimum pH 3.5–4.5; the temperature range for growth is 40–53°C; optimum temperature is about 48°C. Acid is produced from sugars. Methionine or vitamin B_{12}, panthotenate, and isoleucine are required for growth. Strains of this species have been isolated from soil. The type strain was isolated from soil in Germany.

DNA G + C content (mol%): 55.6 (T_m).

Type strain: SCH, ATCC 49028, ATCC BAA-2, CIP 106133, DSM 4006, HAMBI 2074, LMG 17941, NBRC 15310.

GenBank accession number (16S rRNA gene): AB042059, X52489.

6. **Alicyclobacillus disulfidooxidans** (Dufresne, Bousquent, Bassinot and Guay 1996) Karavaiko, Bogdanova, Tourova, Kondrat′eva, Tsaplina, Egorova, Krasil′nikova Zakharchuk 2005, 946[VP] (*Sulfobacillus disulfidooxidans* Dufresne, Bousquent, Bassinot and Guay 1996, 1063.)

di.sul.fi.do.ox′i.dans. N.L. n. *disulfidum* disulfide; N.L. v. *oxido* oxidize; N.L. adj. *disulfidooxidans* disulfide-oxidizing.

Aerobic Gram-variable rods, 0.3–0.5 μm × 0.9–3.6 μm. Endospores are oval and produced subterminally; sporangium swollen. Optimum growth temperature is about 35°C; the growth temperature range is 4–40°C. Optimum pH for growth is 1.5–2.5; pH range for growth is 0.5–6.0. The predominant membrane fatty acids are $C_{17:0\ \omega cyclohexane}$ and $C_{19:0\ \omega cyclohexane}$. Menaquinone-7 (MK-7) is the major respiratory quinone. Mixotrophic. Glucose, glycerol, glutamate, glutathione, cysteine, cystine, cystamine, dithio(bis)benzothiazole, S^0, Fe^{2+}, and Fe_2S are used for growth. Acid is not produced from sugars. Yeast extract is required. The type strain of this species was isolated from wastewater sludge in Canada.

DNA G + C content (mol%): 53.0 (T_m).

Type strain: SD-11, ATCC 51911, DSM 12064.

GenBank accession number (16S rRNA gene): AB089843, U34974.

7. **Alicyclobacillus fastidiosus** Goto, Mochida, Kato, Asahara, Fujita, An, Kasai and Yokoto 2007, 1281[VP]

fas.ti.di.o′sus. L. masc. adj. *fastidiosus* fastidious, referring to its fastidious character.

Strains of this species stain Gram-positive or Gram-variable for old cultures. Cells are straight rods, 4.0–5.0 μm × 0.9–1.0 μm with rounded ends. Nonmotile. Endospores are ellipsoidal and subterminal with swollen sporangia. Colonies on BAM agar after 48 h are nonpigmented circular, opaque, entire, flat, and 3–4 mm in diameter. The predominant fatty acids are $C_{17:0\ \omega cyclohexane}$ and $C_{19:0\ \omega cyclohexane}$. The major respiratory quinone is menaquinone-7 (MK-7). Strictly aerobic and chemoorganotrophic. The pH optimum for growth is 4.0–4.5 with no growth at pH 2.0 or 5.5. The temperature

range for growth is 20–55°C; optimum growth temperature is 40–45°C. Growth occurs in the presence of 0–2% (w/v) NaCl but not 5% (w/v) NaCl. Oxidase-negative and catalase-positive. Nitrate is not reduced to nitrite. Voges–Proskauer test and indole production are negative. Gelatin is hydrolyzed, but esculin, arbutin, phenylalanine, starch, and tyrosine are not. Acid is produced from a number of sugars and sugar alcohols. The type strain of this species was isolated from apple juice.

DNA G + C content (mol%): 53.9 (HPLC).

Type strain: S-TAB, DSM 17978, IAM 15229, JCM 21683, NBRC 103109.

GenBank accession number (16S rRNA gene): AB264021.

8. **Alicyclobacillus herbarius** Goto, Matsubara, Mochida, Matsumura, Hara, Niwa and Yamasoto 2002a, 112[VP]

her.ba′ri.us. N.L. adj. *herbarius* pertaining to herb, from which the organism was isolated.

Rod-shaped cells that stain Gram-positive. The cells are motile. Oval endospores located subterminally are formed; sporangia are swollen. Colonies are nonpigmented. Growth occurs at 35–65°C; the optimum growth temperature is 55–60°C. The optimum pH is 4.5–5.0; growth does not occur at pH 3.0 or pH 6.5. Aerobic and chemoorganotrophic. Oxidase-negative and catalase-positive. Yeast extract or growth factors are not required for growth. The major fatty acid is $C_{18:0\ \omega cycloheptane}$; $C_{18:0\ \omega cycloheptane\ 2OH}$, $C_{20:0\ \omega cycloheptane}$, and branched-chain fatty acids are also present. Major respiratory quinone is menaquinone-7 (MK-7). Reduce nitrate to nitrite. Voges–Proskauer test and indole production are negative. Gelatin and starch are not hydrolyzed. Acid is produced from several sugars. The type strain of this species was isolated from herbal tea made from dried hibiscus flowers.

DNA G + C content (mol%): 56.2 (HPLC).

Type strain: CP-1, DSM 13609, IAM 14883, JCM 21376, NBRC 100860, NRIC 0477.

GenBank accession number (16S rRNA gene): AB042055.

9. **Alicyclobacillus hesperidum** Albuquerque, Rainey, Chung, Sunna, Nobre, Grote, Antranikian and da Costa 2000, 454[VP]

hes.pe′ri.dum. L. fem. pl. n. *hesperidum* of the Hesperides, mythological figures whom the Greeks believed to have lived at the western edge of the Earth, interpreted by the authors as being located in the Azores.

Rod-shaped cells that are 2.1–3.9 μm × 0.5–0.7 μm. Gram stain is positive. The cells are nonmotile. Terminal spores are formed; sporangia are not swollen. Colonies are nonpigmented. Growth occurs at 40–55°C; the optimum growth temperature is 50–53°C. The optimum pH is 3.5–4.0; growth does not occur at pH 2.0 or pH 6.0. Cytochrome oxidase-negative and catalase weakly positive or negative. Yeast extract or growth factors are not required for growth. The major fatty acids are $C_{17:0\ \omega cyclohexane}$ and $C_{19:0\ \omega cyclohexane}$; branched-chain fatty acids are also present in large proportions. Major respiratory quinone is menaquinone-7 (MK-7). Aerobic and chemoorganotrophic. Nitrate is not reduced to nitrite. Gelatin, hide powder, and starch are hydrolyzed, but elastin and fibrin are not. Utilize many hexoses and disaccharides, but pentoses and polyols, with the exception of mannitol and glycerol, are not utilized as single carbon sources. Acid is produced from virtually the same sugars that are utilized as single carbon sources. Strains of this species have been isolated from sol-fataric soils at Furnas, Island of São Miguel, the Azores.

DNA G + C content (mol%): 53.3 (HPLC).

Type strain: FR-11, DSM 12489.

GenBank accession number (16S rRNA gene): AB059678, AJ133633.

10. **Alicyclobacillus kakegawensis** Goto, Mochida, Kato, Asahara, Fujita, An, Kasai and Yokoto 2007, 1283[VP]

ka.ke.ga.wa.en.sis. N.L. masc. adj. *kakegawensis* pertaining to Kakegawa, a city in Shizuoka Prefecture, Japan, where the type strain was isolated.

Strains of this species stain Gram-positive or Gram-variable for old cultures. Cells are straight rods 4.0–5.0 μm × 0.6–0.7 μm with rounded ends. Motile. Endospores are oval and subterminal with swollen sporangia. Colonies on BAM agar after 48 h are nonpigmented circular, opaque, entire, flat, and 2–3 mm in diameter. The predominant fatty acids are $C_{18:0\ \omega cyclohexane}$, $C_{18:0\ \omega cyclohexane\ 2OH}$, and $C_{20:0\ \omega cyclohexane}$. The major respiratory quinone is menaquinone-7 (MK-7). Strictly aerobic and chemoorganotrophic. The pH optimum for growth is 4.0–4.5 with no growth at pH 3.0 or 6.5. The temperature range for growth is 40–60°C; optimum growth temperature is 50–55°C. Growth occurs in the presence of 0–2% (w/v) NaCl but not 5% (w/v) NaCl. Oxidase-negative. Catalase weakly positive. Nitrate is not reduced to nitrite. Voges–Proskauer test and indole production are negative. Esculin and arbutin are hydrolyzed, but gelatin, phenylalanine, starch, and tyrosine are not. Acid is produced from a number of sugars and sugar alcohols. The type strain of this species was isolated from soil of a crop field in Kakegawa city.

DNA G + C content (mol%): 61.3 (HPLC).

Type strain: 5-A83J, DSM 17979, IAM 15227, JCM 21681, NBRC 103104.

GenBank accession number (16S rRNA gene): AB264022.

11. **Alicyclobacillus macrosporangiidus** Goto, Mochida, Kato, Asahara, Fujita, An, Kasai and Yokoto 2007, 1283[VP]

ma.cro.spo.ran′gi.i.dus. Gr. adj. *macros* big; N.L. n. *sporangium* sporangia; L. suff. *-idus* suffix expressing a quality or tendency; N.L. masc. adj. *macrosporangiidus* having large sporangia.

Strains of this species stain Gram-positive or Gram-variable for old cultures. Cells are straight rods 5.0–6.0 μm × 0.7–0.8 μm with rounded ends. Motile. Endospores are oval and terminal with swollen sporangia. Colonies on BAM agar after 48 h are nonpigmented circular, opaque, entire, convex, and 2–4 mm in diameter. The predominant fatty acids are $C_{16:0\ iso}$, $C_{17:0\ iso}$, and $C_{17:0\ ante}$. The major respiratory quinone is menaquinone-7 (MK-7). Strictly aerobic and chemoorganotrophic. The pH optimum for growth is 4.0–4.5 with no growth at pH 3.0 or 6.5; the temperature range for growth is 35–60°C; optimum growth temperature is 50–55°C. Growth occurs in the presence of 0–5% (w/v) NaCl but not 7% (w/v) NaCl. Oxidase-negative. Catalase weakly positive. Nitrate is not reduced to nitrite. Voges–Proskauer test and indole production are negative. Esculin is hydrolyzed, but arbutin, gelatin, phenylalanine, starch, and tyrosine are not. Acid is produced from a number of sugars and sugar alcohols. The type strain of this species was isolated from soil of a crop field in Fujieda city.

DNA G + C content (mol%): 62.5 (HPLC).

Type strain: 5-A239–2O-A, DSM 17980, IAM 15370, JCM 21814.

GenBank accession number (16S rRNA gene): AB264025.

12. **Alicyclobacillus pomorum** Goto, Mochida, Asahara, Suzuki, Kasai and Yokota 2003, 1542[VP]

po.mo′rum. L. neut. n. *pomum* fruit; L. gen. pl. neut. n. *pomorum* of fruits.

The strains of this species strain Gram-positive, but are Gram-variable in older cultures, and form motile rod-shaped cells 2.0–4.0 μm × 0.8–1.0 μm. The sporangia are swollen; oval endospores located subterminally. Colonies are nonpigmented. Growth does not occur at pH 2.5 or 6.5, the optimum pH for growth is 4.5–5.0. The temperature range for growth is 30–60°C; the optimum temperature for growth is about 45–50°C. Chemoorganotrophic and strictly aerobic. Does not reduce nitrate to nitrite. The predominant membrane acyl chains are iso- and anteiso-branched; ω-cyclohexane fatty acids are not detected. The major respiratory quinone is menaquinone-7 (MK-7). Oxidase-positive and catalase-positive. Esculin, gelatin, and starch are hydrolyzed, but arbutin, phenylalanine, and tyrosine are not. Acid is produced from several sugars. Growth factors are not required. The type strain of this species was isolated from spoiled mixed fruit juice.

DNA G + C content (mol%): 53.1 (HPLC).

Type strain: 3A, DSM 14955, IAM 14988, JCM 21459, NBRC 100861.

GenBank accession number (16S rRNA gene): AB089840.

13. **Alicyclobacillus sacchari** Goto, Mochida, Kato, Asahara, Fujita, An, Kasai and Yokoto 2007, 1283[VP]

sac′cha.ri. L. gen. n. *sacchari* of sugar, referring to the source of isolation.

Strains of this species stain Gram-positive or Gram-variable for old cultures. Cells are straight rods 4.0–5.0 μm × 0.6–0.7 μm with rounded ends. Motile. Endospores are ellipsoidal and subterminal with swollen sporangia. Colonies on BAM agar after 48 h are nonpigmented circular, opaque, entire, umbonate, and 3–5 mm in diameter. The predominant fatty acids are $C_{17:0\ \omega cyclohexane}$ and $C_{19:0\ \omega cyclohexane}$. The major respiratory quinone is menaquinone-7 (MK-7). Strictly aerobic and chemoorganotrophic. The pH optimum for growth is 4.0–4.5 with no growth at pH 2.0 or 6.0. The temperature range for growth is 30–55°C; optimum growth temperature is 45–50°C. Growth occurs in the presence of 0–2% (w/v) NaCl but not 5% (w/v) NaCl. Oxidase- and catalase-negative. Nitrate is not reduced to nitrite. Voges–Proskauer test and indole production are negative. Arbutin, gelatin, and starch are hydrolyzed, but esculin, phenylalanine, and tyrosine are not. Acid is produced from a number of sugars and sugar alcohols. The type strain of this species was isolated from liquid sugar.

DNA G + C content (mol%): 56.6 (HPLC).

Type strain: RB718, DSM 17974, IAM 15230, JCM 21684, NBRC 103105.

GenBank accession number (16S rRNA gene): AB264020.

14. **Alicyclobacillus sendaiensis** Tsuruoka, Isono, Shida, Hemmi, Nakayama and Nishino 2003, 1084[VP]

sen.dai.en′sis. N.L. adj. *sendaiensis* of Sendai, a city in Myagi Perfecture, Japan, where the type strain was isolated.

Nonmotile rod-shaped cells that are 2.0–$3.0\,\mu m \times 0.8\,\mu m$. Gram stain is negative. Terminal round spores. Colonies are nonpigmented. Growth occurs at 40–65°C; the optimum growth temperature is about 55°C. The optimum pH is about 5.5; growth occurs at pH 2.5–6.5. Oxidase- and catalase-negative. Yeast extract or growth factors are not required for growth. The predominant membrane fatty acids are $C_{17:0 \,\omega cyclohexane}$ and $C_{19:0 \,\omega cyclohexane}$; straight-chain and branched-chain fatty acids are also present. Major respiratory quinone is menaquinone-7 (MK-7). Aerobic and chemoorganotrophic. Nitrate is reduced to nitrite. Voges–Proskauer test positive. Gelatin is hydrolyzed. Acid is produced from several sugars. The type strain of this species was isolated from soil in Japan.

DNA G + C content (mol%): 62.3 (HPLC).

Type strain: NTAP-1, ATCC-BAA 609, JCM 11817, NBRC 100866.

GenBank accession number (16S rRNA gene): AB084128, AB222247.

15. **Alicyclobacillus shizuokensis** Goto, Mochida, Kato, Asahara, Fujita, An, Kasai and Yokoto 2007, 1283[VP]

shi.zu.o.ken'sis. N.L. masc. adj. *shizuokensis* pertaining to Shizuoka, a city in Shizuoka Prefecture, Japan, where the type strain was isolated.

Strains of this species stain Gram-positive or Gram-variable for old cultures. Cells are straight rods 4.0–$5.0\,\mu m \times 0.7$–$0.8\,\mu m$ with rounded ends. Motile. Endospores are oval and subterminal with swollen sporangia. Colonies on BAM agar after 48 h are nonpigmented circular, opaque, entire, convex, and 1–2 mm in diameter. The predominant fatty acids are $C_{18:0 \,\omega cyclohexane}$, $C_{18:0 \,\omega cyclohexane \, 2OH}$, and $C_{20:0 \,\omega cyclohexane}$. The major respiratory quinone is menaquinone-7 (MK-7). Strictly aerobic and chemoorganotrophic. The pH optimum for growth is 4.0–4.5 with no growth at pH 3.0 or 6.5. The temperature range for growth is 35–60°C; optimum growth temperature is 45–50°C. Growth occurs in the presence of 0–5% (w/v) NaCl but not 7% (w/v) NaCl. Oxidase-negative and catalase-positive. Nitrate is not reduced to nitrite. Voges–Proskauer test and indole production are negative. Esculin and arbutin are hydrolyzed, but gelatin, phenylalanine, starch, and tyrosine are not. Acid is produced from a number of sugars and sugar alcohols. The type strain of this species was isolated from soil of a crop field in Shizuoka city.

DNA G + C content (mol%): 60.5 (HPLC).

Type strain: 4-A336, DSM 17981, IAM 15226, JCM 21680, NBRC 103103.

GenBank accession number (16S rRNA gene): AB264024.

16. **Alicyclobacillus tolerans** (Kovalenko and Malakhova, 1983) Karavaiko, Bogdanova, Tourova, Kondrat'eva, Tsaplina, Egorova, Krasil'nikova and Zakharchuk 2005, 946[VP] ("*Sulfobacillus thermosulfidooxidans* subsp. *thermotolerans*" Kovalenko and Malakhova 1983, 763.)

to.le'rans. L. adj. *tolerans* tolerant to changes in growth temperature and pH.

Cells are nonmotile rod-shaped cells 3–$6\,\mu m \times 0.9$–$1.0\,\mu m$. Gram stain is positive. Terminal or subterminal oval spores are formed; sporangia are swollen. Growth occurs at about 20–55°C; the optimum growth temperature is about 37–42°C. The optimum pH is around 2.5–2.7; growth occurs at pH 1.5–5.0. Mixotrophic; Fe^{2+}, S^0, and sulfide minerals are oxidized in the presence of organic substrates. Fe^{3+} is also reduced. Facultative chemoorganotrophic. Oxidase weakly positive. Catalase weakly positive. Nitrate is not reduced to nitrite. Yeast extract enhances growth but is not required. The predominant membrane acids are $C_{17:0 \,\omega cyclohexane}$, $C_{19:0 \,\omega cyclohexane}$, and $C_{17:0 \,\omega cyclohexane \, 2OH}$; branched-chain fatty acids were also detected. Acid is produced from several sugars. The type strain of this species was isolated from oxidizable lead-zinc ores in Uzbekistan.

DNA G + C content (mol%): 48.7 (HPLC).

Type strain: K1, DSM 16297, VKM B-2304.

GenBank accession number (16S rRNA gene): AB222265, AF137502.

17. **Alicyclobacillus vulcanalis** Simbahan, Drijber and Blum 2004, 1706[VP]

vul.ca.na'lis. N.L. masc. adj. *vulcanalis* of Vulcan, belonging to Vulcan, Roman god of fire and metal making.

Rod-shaped cells that are 1.5–$2.5\,\mu m \times 0.4$–$0.7\,\mu m$. Gram stain is positive. Terminal spores are formed. Colonies are nonpigmented. Growth occurs at 35–65°C; the optimum growth temperature is about 55°C. The optimum pH is about 4.0; growth occurs at pH 2.0–6.0. Cytochrome oxidase-negative and catalase-negative. Aerobic and chemoorganotrophic. Yeast extract or growth factors are not required for growth. Starch is hydrolyzed. The predominant membrane fatty acids are $C_{17:0 \,\omega cyclohexane}$ and $C_{19:0 \,\omega cyclohexane}$; straight-chain and branched-chain fatty acids are also present. The major respiratory quinone is menaquinone-7 (MK-7). Acid is produced from several sugars. The type strain of this species was isolated from a geothermal pool in California.

DNA G + C content (mol%): 62.0 (HPLC).

Type strain: CsHg2, ATCC-BAA 915, DSM 16176.

GenBank accession number (16S rRNA gene): AB222267, AY425985.

References

Albuquerque, L., F.A. Rainey, A.P. Chung, A. Sunna, M.F. Nobre, R. Grote, G. Antranikian and M.S. da Costa. 2000. *Alicyclobacillus hesperidum* sp. nov. and a related genomic species from solfataric soils of Sao Miguel in the Azores. Int. J. Syst. Evol. Microbiol. *50*: 451–457.

Ash, C., J.A.E. Farrow, M. Dorsch, E. Stackebrandt and M.D. Collins. 1991a. Comparative analysis of *Bacillus anthracis*, *Bacillus cereus*, and related species on the basis of reverse-transcriptase sequencing of 16S ribosomal RNA. Int. J. Syst. Bacteriol. *41*: 343–346.

Ash, C., J.A.E. Farrow, S. Wallbanks and M.D. Collins. 1991b. Phylogenetic heterogeneity of the genus *Bacillus* revealed by comparative analysis of small subunit ribosomal RNA sequences. Lett. Appl. Microbiol. *13*: 202–206.

Becker, M.E. and C.S. Pederson. 1950. The physiological characters of *Bacillus coagulans* (*Bacillus thermoacidurans*). J. Bacteriol *59*: 717–725.

Brock, T.D. and G.K. Darland. 1970. Limits of microbial existence: temperature and pH. Science *169*: 1316–1318.

Cerny, G., W. Hennlich and K. Poralla. 1984. [Spoilage of fruit juice by bacilli: isolation and characterization of the spoiling microorganisms]. Z Lebensm Unters Forsch *179*: 224–227.

Chang, S.S. and D.H. Kang. 2004. *Alicyclobacillus* spp. in the fruit juice industry: history, characteristics, and current isolation/detection procedures. Crit Rev Microbiol *30*: 55–74.

Darland, G. and T. Brock. 1971. *Bacillus acidocaldarius* sp. nov., an acidophilic thermophilic spore-forming bacterium. J. Gen. Microbiol. *67*: 9–15.

De Rosa, M., A. Gambacorta and J.D. Bu'Lock. 1971a. An isolate of *Bacillus acidocaldarius*, and acidophilic thermophilic with unusual lipids. Giorn. Microbiol. *19*: 145–154.

De Rosa, M., A. Gambacorta, L. Minale and J.D. Bu'Lock. 1971b. Cyclohexane fatty acids from a thermophilic bacterium. J. Chem. Soc. London, Chem. Comm. 1334.

Deinhard, G., P. Blanz, K. Poralla and E. Altan. 1987a. *Bacillus acidoterrestris* sp. nov., a new thermotolerant acidophile isolated from different soils. Syst. Appl. Microbiol. *10*: 47–53.

Deinhard, G., J. Saar, W. Krischke and K. Poralla. 1987b. *Bacillus cycloheptanicus* sp. nov., a new thermoacidophile containing omega-cycloheptane fatty acids. Syst. Appl. Microbiol. *10*: 68–73.

Dufresne, S., J. Bousquet, M. Boissinot and R. Guay. 1996. *Sulfobacillus disulfidooxidans* sp. nov., a new acidophilic, disulfide-oxidizing, gram-positive, spore-forming bacterium. Int. J. Syst. Bacteriol. *46*: 1056–1064.

Eguchi, S.Y., G.P. Manfio, M.E. Pinhatti, E. Azuma and S.F. Variane. 1999. Acidothermophilic spore-forming bacteria (ATSB) in orange juices: detection methods, ecology, and involvement in the deterioration of fruit juices. Abecitrus, Ribeirão Preto, Brazil.

Farrand, S.G., J.D. Linton, R.J. Stephenson and W.V. Maccarthy. 1983. The use of response surface analysis to study growth of *Bacillus acidocaldarius* throughout the growth range of temperature and pH. Arch. Microbiol. *135*: 272–275.

Goto, K., H. Matsubara, K. Mochida, T. Matsumura, Y. Hara, M. Niwa and K. Yamasato. 2002a. *Alicyclobacillus herbarius* sp. nov., a novel bacterium containing omega-cycloheptane fatty acids, isolated from herbal tea. Int. J. Syst. Evol. Microbiol. *52*: 109–113.

Goto, K., Y. Tanimoto, T. Tamura, K. Mochida, D. Arai, M. Asahara, M. Suzuki, H. Tanaka and K. Inagaki. 2002b. Identification of thermoacidophilic bacteria and a new *Alicyclobacillus* genomic species isolated from acidic environments in Japan. Extremophiles *6*: 333–340.

Goto, K., K. Mochida, M. Asahara, M. Suzuki, H. Kasai and A. Yokota. 2003. *Alicyclobacillus pomorum* sp. nov., a novel thermo-acidophilic, endospore-forming bacterium that does not possess omega-alicyclic fatty acids, and emended description of the genus *Alicyclobacillus*. Int. J. Syst. Evol. Microbiol. *53*: 1537–1544.

Goto, K., K. Mochida, Y. Kato, M. Asahara, R. Fujita, S.Y. An, H. Kasai and A. Yokota. 2007. Proposal of six species of moderately thermophilic, acidophilic, endospore-forming bacteria: *Alicyclobacillus contaminans* sp. nov., *Alicyclobacillus fastidiosus* sp. nov., *Alicyclobacillus kakegawensis* sp. nov., *Alicyclobacillus macrosporangiidus* sp. nov., *Alicyclobacillus sacchari* sp. nov. and *Alicyclobacillus shizuokensis* sp. nov. Int. J. Syst. Evol. Microbiol. *57*: 1276–1285.

Hippchen, B., A. Röll and K. Poralla. 1981. Ocurrence in soil of thermo-acidophilic bacilli possessing ω-cyclohexane fatty acids and hopanoids. Arch. Microbiol. *129*: 53–55.

Hiraishi, A., K. Inagaki, Y. Tanimoto, M. Iwasaki, N. Kishimoto and H. Tanaka. 1997. Phylogenetic characterization of a new thermoacidophilic bacterium isolated from hot springs in Japan. J. Gen Appl Microbiol. *43*: 295–304.

Jensen, N.S. 1999. *Alicyclobacillus*–a new challenge for the food industry. Food Aust *51*: 33–36.

Karavaiko, G., E.N. Krasil'nikova, I.A. Tsaplina, T.I. Bogdanova and L.M. Zakharchuk. 2001. Growth and carbohydrate metabolism of sulfobacilli. Microbiology (En. transl. from Mikrobiologiya) *70*: 245–250.

Karavaiko, G.I., L.M. Zakharchuk, T.I. Bogdanova, M.A. Egorova, I.A. Tsaplina and E.N. Krasil'nikova. 2002. The enzymes of carbon metabolism in the thermotolerant bacillar strain K1. Microbiology (En. transl. from Mikrobiologiya) *71*: 651–656.

Karavaiko, G.I., T.I. Bogdanova, T.P. Tourova, T.F. Kondrat'eva, I.A. Tsaplina, M.A. Egorova, E.N. Krasil'nikova and L.M. Zakharchuk. 2005. Reclassification of 'Sulfobacillus thermosulfidooxidans subsp. *thermotolerans*' strain K1 as *Alicyclobacillus tolerans* sp. nov. and *Sulfobacillus disulfidooxidans* Dufresne et al. 1996 as *Alicyclobacillus disulfidooxidans* comb. nov., and emended description of the genus *Alicyclobacillus*. Int. J. Syst. Evol. Microbiol. *55*: 941–947.

Kovalenko, E.V. and P.T. Malakhova. 1983. The spore-forming iron-oxidizing bacterium *Sulfobacillus thermosulfidooxidans*. Mikrobiologiya *52*: 962–966 (in Russian).

Kusano, K., H. Yamada, M. Niwa and K. Yamasato. 1997. *Propionibacterium cyclohexanicum* sp. nov, a new acid-tolerant omega-cyclohexyl fatty acid-containing propionibacterium isolated from spoiled orange juice. Int. J. Syst. Bacteriol. *47*: 825–831.

Langworthy, T.A. and W.R. Mayberry. 1976. A 1,2,3,4-tetrahydroxy pentane-substituted pentacyclic triterpene from *Bacillus acidocaldarius*. Biochim Biophys Acta *431*: 570–577.

Langworthy, T.A., W.R. Mayberry and P.F. Smith. 1976. A sulfonolipid and novel glucosamidyl glycolipids from the extreme thermoacidophile *Bacillus acidocaldarius*. Biochim Biophys Acta *431*: 550–569.

Manning, H.L. 1975. New medium for isolating iron-oxidizing and heterotrophic acidophilic bacteria from acid mine drainage. Appl Microbiol *30*: 1010–1016.

Matsubara, H., K. Goto, T. Matsumura, K. Mochida, M. Iwaki, M. Niwa and K. Yamasato. 2002. *Alicyclobacillus acidiphilus* sp. nov., a novel thermo-acidophilic, omega-alicyclic fatty acid-containing bacterium isolated from acidic beverages. Int. J. Syst. Evol. Microbiol. *52*: 1681–1685.

Nicolaus, B., F. Marsiglia, E. Esposito, A. Trincone, L. Lama, R. Sharp, G. Diprisco and A. Gambacorta. 1991. Isolation of five strains of thermophilic eubacteria in Antarctica. Polar Biol. *11*: 425–429.

Nicolaus, B., R. Improta, M.C. Manca, L. Lama, E. Esposito and A. Gambacorta. 1998. Alicyclobacilli from an unexplored geothermal soil in Antarctica: Mount Rittmann. Polar Biol. *19*: 133–141.

Ourisson, G., M. Rohmer and K. Poralla. 1987. Prokaryotic hopanoids and other polyterpenoid sterol surrogates. Ann. Rev. Microbiol. *41*: 301–333.

Pettipher, G.L., M.E. Osmundson and J.M. Murphy. 1997. Methods for the detection and enumeration of *Alicyclobacillus acidoterrestris* and investigation of growth and production of taint in fruit juice and fruit juice-containing drinks. Lett. Appl. Microbiol. *24*: 185–189.

Poralla, K., E. Kannenberg and A. Blume. 1980. A glycolipid containing hopane isolated from the acidophilic, thermophilic *Bacillus acidocaldarius*, has a cholesterol-like function in membranes. FEBS Lett *113*: 107–110.

Poralla, K. and W.A. Konig. 1983. The occurrence of omega-cycloheptane fatty-acids in a thermo-acidophilic bacillus. FEMS Microbiol. Lett. *16*: 303–306.

Simbahan, J., R. Drijber and P. Blum. 2004. *Alicyclobacillus vulcanalis* sp. nov., a thermophilic, acidophilic bacterium isolated from Coso Hot Springs, California, USA. Int. J. Syst. Evol. Microbiol. *54*: 1703–1707.

Splittstoesser, D.F., J.J. Churey and C.Y. Lee. 1994. Growth characteristics of aciduric spore-forming bacilli isolated from fruit juices. J. Food Prot. *57*: 1080–1083.

Suzuki, K.I., K. Saito, A. Kawaguchi, S. Okuda and K. Komagata. 1981. Occurrence of ω-cyclohexyl fatty acids in *Curtobacterium pusillum*. J. Gen. Appl. Microbiol. *8*: 185–189.

Tsuruoka, N., Y. Isono, O. Shida, H. Hemmi, T. Nakayama and T. Nishino. 2003. *Alicyclobacillus sendaiensis* sp. nov., a novel acidophilic, slightly thermophilic species isolated from soil in Sendai, Japan. Int. J. Syst. Evol. Microbiol. *53*: 1081–1084.

Uchino, F. and S. Doi. 1967. Acido-thermophilic bacteria from thermal waters. Agric. Biol. Chem. (Tokyo) *31*: 817–822.

Walls, I. and R. Chuyate. 2000. Isolation of *Alicyclobacillus acidoterrestris* from fruit juices. J. AOAC Int. *83*: 1115–1120.

Wisotzkey, J.D., P. Jurtshuk, G.E. Fox, G. Deinhard and K. Poralla. 1992. Comparative sequence analyses on the 16S ribosomal RNA (rDNA) of *Bacillus acidocaldarius*, *Bacillus acidoterrestris*, and *Bacillus cycloheptanicus* and proposal for creation of a new genus, *Alicyclobacillus* gen. nov. Int. J. Syst. Bacteriol. *42*: 263–269.

Yamazaki, K., H. Teduka and H. Shinano. 1996. Isolation and identification of *Alicyclobacillus acidoterrestris* from acidic beverages. Biosci Biotechnol Biochem *60*: 543–545.

Family III. **Listeriaceae** fam. nov.

WOLFGANG LUDWIG, KARL-HEINZ SCHLEIFER AND WILLIAM B. WHITMAN

Lis.te.ri.a'ce.ae. N.L. fem. n. *Listeria* type genus of the family; suff. *-aceae* ending denoting family; N.L. fem. pl. n. *Listeriaceae* the *Listeria* family.

The family *Listeriaceae* is circumscribed for this volume on the basis of phylogenetic analyses of the 16S rRNA sequences and includes the genus *Listeria* and its close relative *Brochothrix*. Cells are short rods that may form filaments. Cells stain Gram-positive, and the cell walls contain *meso*-diaminopimelate. The major lipid components include saturated straight-chain and methyl-branched fatty acids. Endospores are not formed. Menaquinones are the sole respiratory quinone. Growth is aerobic and facultative anaerobic, and glucose is fermented to lactate and other products.

Type genus: **Listeria** Pirie 1940a, 383.

Genus I. **Listeria** Pirie 1940a, 383[AL]

JAMES MCLAUCHLIN AND CATHERINE E. D. REES

Lis.te'ri.a. N.L. fem. n. *Listeria* named after Lord Lister, English surgeon and pioneer of antisepsis.

Regular, short rods, 0.4–0.5 × 1–2 μm with parallel sides and blunt ends. Usually occur singly or in short chains. In older or rough cultures, filaments of ≥6 μm in length may develop. **Gram-positive** with even staining, but some cells, especially in older cultures, lose their ability to retain the Gram strain. **Not acid-fast. Capsules not formed.** Do not form spores. All species **motile with peritrichous flagella when cultured <30°C. Aerobic and facultative anaerobic**. Colonies (24–48 h) are 0.5–1.5 mm in diameter, round, translucent, low convex with a smooth surface and entire margin, non-pigmented with a crystalline central appearance. May be sticky when removed from agar surfaces, usually emulsify easily, and may leave a slight impression on the agar surface after removal. Older cultures (3–7 d) are larger, 3–5 mm in diameter, have a more opaque appearance, sometimes with a sunken center: rough colonial forms may develop. In 0.25% (w/v) agar, 8% (w/v) gelatin and 1.0% (w/v) glucose semi-solid medium, growth along the stab after 24 h at 37°C is followed by irregular, cloudy extensions into the medium. Growth spreads slowly through the entire medium. An umbrella-like zone of maximal growth occurs 3–5 mm below the surface. **Temperature limits of growth <0 to 45°C; optimal growth at 30–37°C. Do not survive heating at 60°C for 30 min.** Growth occurs between pH 6 and pH 9. Growth occurs in nutrient broth supplemented with up to 10% (w/v) NaCl. Catalase-positive, oxidase-negative. **Cytochromes produced. Homofermentative anaerobic catabolism of glucose results in production of L(+)-lactic acid, acetic acid plus other end products. Acid but no gas produced from other sugars. Methyl-red-positive, Voges–Proskauer-positive.** Exogenous citrate not utilized. **Organic growth factors are required.** Indole is not produced. Esculin and sodium hippurate are hydrolyzed. Urea is not hydrolyzed. Gelatin, casein, and milk are not hydrolyzed.

The cell wall contains a directly cross-linked peptidoglycan based on *meso*-diaminopimelic acid (*meso*-DAP) (variation A1γ of Schleifer and Kandler, 1972): the cell wall does not contain arabinose. Mycolic acids are not present. The long-chain fatty acids consist of predominantly straight-chain saturated, anteiso-methyl-branched-chain types. When grown at 37°C, the major fatty acids are 14-methylhexadecanoic (anteiso-$C_{17:0}$) and 12-methyltetradecanoic (anteiso-$C_{15:0}$). Menaquinones are the sole respiratory quinones; the major quinone contains seven isoprene units (MK-7).

All members of the genus *Listeria* are widely distributed in nature and have been isolated from soil, vegetation, sewage, water, animal feed, fresh and frozen poultry, slaughterhouse wastes, and the feces of healthy animals including humans. *Listeria monocytogenes* is pathogenic to man. *Listeria monocytogenes* and, to a lesser extent, *Listeria ivanovii* are pathogenic to a wide range of animals, especially sheep and goats. The disease in humans and livestock is predominantly transmitted by the consumption of food or feed contaminated by *Listeria monocytogenes*. The colonization by *Listeria monocytogenes* of specific sites within food manufacturing environments for long periods, together with its ability to grow in a wide range of foods (including those containing sodium chloride or sodium nitrate as preservative) at low temperatures makes this bacterium of particular concern as a contaminant of ready-to-eat refrigerated foods (Farber and Peterkin, 1991).

DNA G+C content (mol%): 36–42.5 (T_m).

Type species: **Listeria monocytogenes** Pirie 1940a.

Further descriptive information

Numerical taxonomic studies (Wilkinson and Jones, 1977) together with biochemical (Collins et al., 1979b; Fiedler, 1988), DNA base composition (Hartford and Sneath, 1993; Rocourt et al., 1982a) and 16S rRNA gene sequence studies (Collins et al., 1991) show that all the species of the genus *Listeria* form a homogeneous group. On the basis of the sequencing of the16S rRNA gene (as well as other genes), members of the genus *Brochothrix* (*Brochothrix thermosphacta* and *Brochothrix campestris*) are the closest relatives to *Listeria*, which is consistent with chemical and numerical taxonomic approaches (Collins et al., 1979b; Wilkinson and Jones, 1977); these two genera justify family status as *Listeriaceae*. The family *Listeriaceae* shows relatedness to other low-G+C Gram-positive bacteria, especially to species of *Bacillus* (Collins et al., 1991; Glaser et al., 2001). Indeed, the completed genome sequences of *Listeria monocytogenes* and *Listeria innocua* showed a close relationship to that from *Bacillus subtilis* and suggest a common origin (Glaser et al., 2001). *Listeria monocytogenes sensu lato* was reclassified into *Listeria monocytogenes sensu stricto*,

Listeria innocua, Listeria ivanovii, Listeria seeligeri, and *Listeria welshimeri* (Rocourt et al., 1982a); it is not always possible to be certain of the species designations cited in the older literature.

Six species (*Listeria monocytogenes, Listeria innocua, Listeria ivanovii, Listeria seeligeri, Listeria welshimeri,* and *Listeria grayi*) were recognized on the basis of DNA–DNA hybridization studies (Rocourt et al., 1982a). This classification was supported by the results of a second DNA–DNA homology study (Hartford and Sneath, 1993) and an analysis of multilocus enzyme electrophoresis (Boerlin et al., 1991). The study by Hartford and Sneath (1993) suggested a very close relationship between *Listeria monocytogenes* and *Listeria innocua* and there may be some overlap between these two species. 16S rRNA gene sequencing also supports this classification (Collins et al., 1991) together with sequencing of 23S rRNA genes (Sallen et al., 1996; Thompson et al., 1992), and 16S–23S intergeneric spacer regions (Drebot et al., 1996; Graham et al., 1997). Signature sequences in the V2 region of the 16S rRNA genes for the different *Listeria* species are: RAGUGUGGCGCGCAUGCCACGCU (*Listeria monocytogenes* and *Listeria innocua*); AGUAGUGACGCAUGUCAUCAC (*Listeria ivanovii*); AGGAGUGACGCAUGUCACUAC (*Listeria seeligeri*); AGUGGUGGCGCAUGCCACGGC (*Listeria welshimeri*); and AAUCACUCCGCAUGGAGCAGG (*Listeria grayi*) (Collins et al., 1991).

Complete genome sequences are available for *Listeria monocytogenes* (serotype 1/2a), *Listeria innocua* and *Listeria welshimeri* (Glaser et al., 2001; Hain et al., 2006; Nelson et al., 2004). However genome sequences other *Listeria monocytogenes* serotypes and for *Listeria seeligeri* and *Listeria ivanovii* have also been completed, and the genomes of further *Listeria moncytogenes* strains and *Listeria grayi* are currently being sequenced (http://www.genomesonline.org). The *Listeria* genome encode approximately 2800 potential proteins, of which two thirds have an assigned gene function, leaving some 900 genes with unknown function. Although differences bewteen the species are seen due to gene acquisition or gene deletion, very little rearrangement or inversion of genome segments is detected resulting in a highly conserved synteny in gene organization. This may be due to the generally low occurrence of transposons and insertion sequences in *Listeria* genomes.

Comparative genomic analysis has confirmed that *Listeria monocytogenes* and *Listeria innocua* are closely related and form a distinct group from *Listeria welshimeri, Listeria ivanovii* and *Listeria seeligeri,* while *Listeria grayi* forms the deepest branch within the genus (Schmid et al., 2005). Looking at the evolution and pathogenicity of members of the genus, it appears that the non-pathogenic species have evolved by genome reduction from a progenitor strain that carried the virulence genes (Buchrieser et al., 2003; Doumith et al., 2004; Hain et al., 2006).

All *Listeria* species produce regular, short rods, 0.4–0.5 × 1–2μm with parallel sides and blunt ends. Coccoid forms 0.4–0.5μm in diameter and 0.4–0.6μm in length are sometimes seen in smears from infected tissue (*Listeria monocytogenes* only) or from liquid cultures but rarely from colonies from solid media. These coccoid forms can be mistaken for streptococci when growing on microbiological media, but can be differentiated using a catalase test. Filaments up to 96μm are developed by some *Listeria monocytogenes* strains (Rowan et al., 2000) and are also seen when strains are grown in media containing high levels of NaCl; formation of these filaments is reversible

(Jorgensen et al., 1995). A murein hydrolase (p60) as been shown to have an essential role in cell division (Pilgrim et al., 2003) and is down-regulated in strains which form exceptionally long filaments (Kuhn and Goebel, 1989).

Colonies of all *Listeria* species show limited variation and usually appear as non-pigmented, round, translucent, low-convex with a smooth surface, entire margin and crystalline central appearance on fresh isolation. After several days, colonies become less translucent and creamier in color, margins become less entire, and a central depression may form. Rough forms in which individual bacteria do not septate and are unusually long occur where colonies have an undulating rough surface and an uneven edge: bacteria from these colonies usually autoagglutinate. Some of these long forms in *Listeria monocytogenes* are due to a defective p60 protein which has murein hydrolase activity and is required for normal septum formation and is essential for cell viability (Wuenscher et al., 1993). The growth of all *Listeria* species generates a characteristic sweet caramel or buttery smell due to the generation of butyric acid.

Two to six peritrichous flagella are produced by all *Listeria* species (Peel et al., 1988a, b) grown below 30°C. Tumbling motility is observed for cultures grown between 20–30°C, and expression of the structural gene for the flagellin protein (*flaA*) has been shown to be temperature regulated and repressed at 37°C (Dons et al., 1992). Repression of flagellae at 37°C is necessary for full virulence (Grundling et al., 2004; Kathariou et al., 1995). High-level expression is seen at 25°C, corresponding to the temperature at which tumbling motility is observed. Even in the absence of active motility, strong induction of flagella biosynthesis occurs below 20°C and flagellae have a role in attachment to solid surfaces (Vatanyoopaisarn et al., 2000). Synthesis of flagellae is reported to be sensitive to catabolite repression (Galsworthy et al., 1990) and to the osmolarity of the medium (Sanchez-Campillo et al., 1995). Motility tests should be performed in media that do not contain high concentrations of free carbohydrate since the pH drops rapidly to <5.5 and the organism can appear to be nonmotile. In hanging-drop preparations, characteristic tumbling and rotatory movements alternate with periods of rest. Brain heart infusion broths are recommended since motility can be detected using hanging drops after 4h: nutrient broths can also be used, but motility may not be detected until after overnight incubation. Motility can also be demonstrated in semi-solid agar in "U" tubes. Maximal turbidity in liquid and semi-solid agar is present 2–3mm below the surface in an umbrella-like pattern. After some d the organisms settle to the bottom in floccules. Some cultures within *Listeria* species other than *Listeria monocytogenes* may also elaborate flagella when grown at 37°C.

All species multiply rapidly in milk (Pine et al., 1989).

All *Listeria* species grow well on most non-selective bacteriological media including blood agar base, nutrient, tryptose, and tryptose soy or brain heart infusion agars. Growth is enhanced by the addition of a suitable fermentable carbohydrate (0.2–1% (w/v) glucose is suitable for all species) blood or serum.

All species grow best at neutral to slightly alkaline pH. Some cultures grow at pH 9.6 but usually die at pH lower than 5.5, consequently viable transfers may not be successful from cultures used in fermentation reactions where acid is produced. Two different mechanisms for adapting to acid stress have been described. Two glutamate decarboxylases (GadA and GadB), and a cognate H^+ antiporter homolog (GadC), contribute to the regulation of cytoplasmic pH by the consumption of H$^+$ ions (Cotter et al., 2001) are essential for survival in both gastric juices and acidic foods (Hill et al., 2002). GadB is induced under acid stress condition via the SigB stress regulon (Cotter et al., 2001). In addition an H$^+$ ATPase has been shown to have a role in acid survival and in the induction of an acid-tolerance response (Cotter et al., 2000). The genes encoding the ATPase (atpC and atpD) appear to be regulated by PrfA (Glaser et al., 2001). To date no mechanism for surviving alkaline stress conditions has been elucidated.

Listeria ivanovii, Listeria monocytogenes, and Listeria seeligeri are hemolytic on agars containing sheep, cow, horse, rabbit or human blood: the remaining three species are non-hemolytic. Variation may occur between the species' blood because of the presence of antibodies: it is therefore preferable to use saline-washed erythrocytes. Media containing plant-extracts, such as tryptose soy agar, should not be used due to an observed repression of the hemolysin gene by the plant sugar cellobiose (Park and Kroll, 1993). Treatment of media with activated charcoal can increase the levels of both hemolysis (Ripio et al., 1996) and lecithinase activity in plate assays (Ermolaeva et al., 2003). Prolonged heat treatment of the media will also diminish the hemolytic activity (Sheikh-Zeinoddin et al., 2000).

The zones of hemolysis produced by Listeria monocytogenes and Listeria seeligeri are narrow, have an indistinct margin, and, especially for Listeria seeligeri, may be detected only by removal of the colony from the agar surface. Listeria ivanovii produces wider zones of hemolysis (especially after 36 h incubation) with sharper edges and may also have double or multiple zones. Because of the relatively weak hemolytic reactions produced by Listeria, the use of layered blood agar plates is recommended. The Listeria monocytogenes type strain is nonhemolytic, CAMP-test-negative, and nonmotile and may be misidentified as Listeria innocua. This strain however, unlike Listeria innocua cultures, does exhibit amino acid peptidase activity. Wild-type Listeria monocytogenes occur with similar properties, these can be identified on the basis of the detection of hemolysin gene fragments by PCR and 16S rRNA gene sequencing (J. McLauchlin, unpublished data).

Agglutination of human, rabbit, guinea pig or sheep erythrocytes does not occur (Seeliger, 1961).

Listeria monocytogenes, Listeria seeligeri, and Listeria ivanovii all show positive CAMP test or enhancement of hemolysis reactions (Christie et al., 1944): Listeria monocytogenes and Listeria seeligeri are CAMP-test-positive with Staphylococcus aureus and negative with Rhodococcus equi; Listeria ivanovii is CAMP-test-positive with Rhodococcus equi and negative with Staphylococcus aureus. Listeria grayi, Listeria seeligeri, and Listeria welshimeri are all CAMP-test-negative (Rocourt et al., 1982a). Because the enhancement of hemolyis can be weak (especially between Listeria seeligeri and Staphylococcus aureus), a thin layer (2 mm) of 5% (v/v) washed sheep erythrocytes in nutrient agar poured over a nutrient agar base is recommended. In addition, not all Staphylococcus aureus strains are suitable since some produce too much lysis alone: Staphylococcus aureus NCTC 1803 and Rhodococcus equi NCTC 1621 are recommended for this test.

All Listeria species grow in complex media supplemented with 10% (w/v) NaCl. Some Listeria monocytogenes strains can tolerate 20% (w/v) NaCl: retention of viability for >1 year in 16% (w/v) NaCl at pH 6.0 has been reported (Seeliger, 1961). Listeria species can utilize a range of compounds as osmopro-

tectants: the rank order of a variety of different compounds tested was glycine > betaine > proline betaine > acetyl carnitine/carinitne/γ-butrobetaine > 3-dimethylsulfonioproprionate (Bayles and Wilkinson, 2000): the temperature of cultivation can influence the effectiveness of individual compounds. Ko et al. (1994) identified a chill-activated glycine betaine transporter which was 15 times more active at 7°C than at 30°C, and thus glycine betaine may also serve as a cryoprotectant for Listeria. Accumulation of glycine betaine and carnitine occurs via at least two glycine betaine transporters, BetL (Ko and Smith, 1999) and Gbu (Sleator et al., 1999), and one carnitine transporter encoded by the opuC operon (Fraser et al., 2000). Full resistance to osmotic stress also requires induction of a set of stress-response genes controlled by the alternate sigma factor SigB (Becker et al., 1998), including the ABC transporter OpuC (Sleator et al., 2001) and the general stress protein Ctc (Gardan et al., 2003).

All strains grow on 10% (w/v) and 40% (w/v) bile agar. Wetzler et al. (1968) reported that generally growth on 40% (w/v) bile (Bacto ox gall) was better than that obtained on 10% (w/v) bile. Slight growth occurs on MacConkey agar. A gene encoding a bile salt hydrolase (bsh) has been identified in Listeria monocytogenes and deletion of this gene resulted in decreased resistance to bile in vitro. The gene is up-regulated by virulence regulator PrfA, and contributes to the survival of the bacterium in the intestinal and hepatic phases of listeric infection (Dussurget et al., 2002). This gene is not present in the non-pathogenic species Listeria innocua.

All strains grow in the presence of 0.025% (w/v) thallous acetate; 3.75% (w/v) potassium thiocyanate; 0.04% (w/v) potassium tellurite; 0.01% (w/v) 2,3,5-triphenyltetrazolium chloride, tellurite and tetrazolium are reduced. Strains do not grow in the presence of 0.02% (w/v) sodium azide. Wetzler et al. (1968) reported no growth in the presence of potassium cyanide. No growth occurs on the medium of Gardner (1966). All species are catalase-positive when grown on the usual laboratory media but may give a negative reaction if cultures on media containing low concentrations of meat and yeast extract. Catalase-negative Listeria monocytogenes strains have been observed (Bubert et al., 1997). Friedman and Alm (1962) reported that catalase activity is depressed in media containing higher (10% (w/v)) concentrations of glucose. Catalase and superoxide dismutase production was increased in three Listeria monocytogenes strains by addition of iron to tryptose soy broth (Fisher and Martin, 1999): addition of iron to the medium will repress the production of listeriolysin (Bockmann et al., 1996). The addition of selenium resulted in increasing production of catalase and listeriolysin (Fisher and Martin, 1999).

Carbohydrate is essential for growth of Listeria strains and glucose is the usual choice. Fully chemically defined media that successfully support the growth of Listeria monocytogenes in both batch (Friedman and Roessler, 1961; Jones et al., 1995; Phan-Thanh and Gormon, 1997; Premaratne et al., 1991; Romick et al., 1996; Siddiqi and Khan, 1982, 1989; Trivett and Meyer, 1971) and continuous culture (Jones et al., 1995) have been described. The medium of Trivett and Meyer (1971) comprised: sodium and potassium phosphate, ammonium chloride, magnesium sulfate, ferric chloride, sodium hydroxide, nitriloacetic acid, L-cysteine, L-leucine, DL-isoleucine, DL-valine, DL-methionine, L-arginine, L-histidine, riboflavin, thiamine, D-biotin, α-lipoic acid, and glucose. The medium of Phan-Thanh and Gormon (1997) was similar to that of Trivett and Meyer (1971), but also included L-glutamine, L-tryptophan, L-phenylalanine, adenine, pyridoxal, para-aminobenzoic acid, and nicotinamide. In a defined medium, pyruvate, acetate, citrate, isocitrate, 2-oxoglutarate, succinate, fumarate, and malate do not support growth of *Listeria monocytogenes* in the absence of glucose, nor do they increase growth in the presence of glucose (Trivett and Meyer, 1971). The medium described by Phan-Thanh and Gormon (1997) supported all *Listeria* species in batch culture. Pyruvate, malate, succinate and 2-oxoglutarate have been reported to be oxidized at low rates by *Listeria monocytogenes* (Friedman and Alm, 1962; Kolb and Seidel, 1960). In a complex medium, pyruvate is utilized as a carbon source by some strains. *Listeria monocytogenes* appears to utilize a split noncyclic citrate pathway which has an oxidative and a reductive portion. The pathway is probably important in biosynthesis but not for a net gain in energy (Trivett and Meyer, 1971). It has been shown that *Listeria monocytogenes* can induce the enzymes required for the utilization of glucose 1-phosphate when grown under conditions which induce the virulence regulator PrfA (Ripio et al., 1997) and therefore growth conditions (especially temperature) should be considered when determining sugar-utilization profiles. Catabolism of glucose proceeds by the Embden–Meyerhof pathway both aerobically and anaerobically. Under anaerobic conditions the catabolism of glucose by all *Listeria* species is homofermentative, i.e., lactate is produced exclusively (Pine et al., 1989). Under aerobic conditions cell yields are considerably increased, and all species produce lactic, acetic, isobutyric, and isovaleric acids: there are differences between strains in the relative amounts of lactic and acetic acids produced (Pine et al., 1989). Friedman and Alm (1962) and Daneshvar et al. (1989) also reported the production of acetoin and pyruvate by *Listeria monocytogenes* under aerobic conditions. There is no evidence for the Entner–Doudoroff pathway, but glucose-6-phosphate dehydrogenase and 6-phosphogluconate dehydrogenase have been reported to be present (Miller and Silverman, 1959). Under anaerobic conditions, only hexoses and pentoses support growth, but under aerobic conditions maltose and lactose support growth of some species, but sucrose does not (Pine et al., 1989). No growth occurred with lactose under anaerobic conditions, but all species grew under aerobic conditions and *Listeria grayi* utilized both the galactose and glucose moieties but *Listeria monocytogenes* and *Listeria innocua* only the glucose (Pine et al., 1989). Analysis of cultures grown at 5°C in sterile milk suggested that glucose was the major and limiting substrate (Pine et al., 1989).

All species are Methyl red-positive and Voges–Proskauer-positive when tested at 48 h by the method of O'Meara or that of Barritt (see Barrow and Feltham, (1993), for methods).

When testing for the production of acid from carbohydrates, the composition of the basal medium and the pH indicator used are important. *Listeria* species are actively saccharolytic and a basal media which contains traces of fermentable carbohydrate and a pH indicator which changes color rather near neutrality can result in false-positive reactions. A variety of basal media and pH indicators have been used. Purple Broth Base (Difco) peptone water medium with phenol red as indicator (Rocourt et al., 1983) supplemented with 0.5 or 1% (w/v) of the carbohydrate are suitable for testing for acid production.

All carbohydrates should be sterilized by filtration and not by autoclaving.

Members of the genus *Listeria* are remarkably similar in their phenotypic characteristics (Feresu and Jones, 1988; Kämpfer, 1992; Kämpfer et al., 1991; Rocourt et al., 1985b; Rocourt et al., 1983; Wilkinson and Jones, 1977). Lists of results of phenotypic tests are shown in Table 40 and Table 41. All species produce acid from esculin, glucose, trehalose and salicin.

Exogenous citrate is not utilized. Iron is stimulatory for the growth of *Listeria monocytogenes* in stationary or aerated cultures (Sword, 1966; Trivett and Meyer, 1971), but aeration improves growth only in the presence of adequate iron (Trivett and Meyer, 1971). H$_2$S is not produced. Ornithine, lysine, glutamic acid, and arginine decarboxylases are not produced, nor is an arginine dihydrolase present. Phosphatase is produced and methylene blue is decolorized. Tributyrinase activity is absent, and the hydrolysis of Tweens 20, 40, 60, and 80 takes place slowly.

Rocourt et al. (1983) showed that the species of *Listeria* could be differentiated using a small number of tests, which was later confirmed by others (Kämpfer et al., 1991; McLauchlin, 1997). Kämpfer et al. (1992, 1991) showed that *Listeria monocytogenes* could be distinguished from other members of the genus by the absence of an arylesterase on alanine-substituted substrates. A patent has been granted for the use of arylesterase activity on glycine substituted substrates for the identification of *Listeria* (Monget, 1992), and this reaction is used in a commercially available identification kit (Bille et al., 1992; McLauchlin, 1997). A method based on detection of amino acid peptidase activity using alanine-derived substrates for the rapid differentiation of *Listeria monocytogenes* from the rest of the genus has been described (Clark and McLauchlin, 1997; McLauchlin, 1997).

The cell wall of *Listeria monocytogenes* has the appearance of a thick multilayered structure typical of Gram-positive bacteria (Ghosh and Murray, 1967). The cell-wall peptidoglycan contains *meso*-DAP as the diamino acid (variation A1γ of Schleifer and Kandler, 1972). Alanine and glutamic acid are also present (Fiedler et al., 1984; Fiedler and Ruhland, 1987; Fiedler and Seger, 1983; Kamisango et al., 1982; Robinson, 1968; Schleifer and Kandler, 1972; Srivastava and Siddique, 1973). In addition to *N*-acetylmuramic acid and *N*-acetylglucosamine, glucosamine also occurs as a component of the cell-wall polysaccharide (Fiedler and Seger, 1983; Hether et al., 1983; Ullmann and Cameron, 1969). Ribitol and lipo-teichoic acids are present in *Listeria monocytogenes* (Fiedler, 1988; Fiedler et al., 1984; Fujii et al., 1985; Hether and Jackson, 1983; Kamisango et al., 1983; Ruhland and Fiedler, 1987; Uchikawa et al., 1986b, a): these, together with flagella antigens, are responsible for the serological types (Fiedler et al., 1984; Wendlinger et al., 1996). Cell surface and secreted proteins occur, some of which are involved with the virulence of *Listeria monocytogenes* and *Listeria ivanovi*. The production of many of the surface-associated virulence genes is temperature-dependent and these are expressed above 30°C. Many, but not all, of the surface-associated virulence genes characterized to date are under the control of the central virulence regulator PrfA (Vazquez-Boland et al., 2001). L forms have been reported in *Listeria monocytogenes* (Gray and Killinger, 1966; Markova et al., 1997) and may have a role in infection.

Mycolic acids are not present (Jones et al., 1979) and MK-7 is the major menaquinone with MK-6 and MK-5 present as minor components for all species examined (Collins and Jones, 1981).

TABLE 40. Characteristics for differentiating species of the genus *Listeria*[a,b]

Characteristic	L. monocytogenes	L. innocua	L. ivanovii	L. seeligeri	L. welshimeri	L. grayi
β Hemolysis	+	−	+	+	−	−
CAMP test with:						
Rhodococcus equi	−	−	+	−	−	−
Staphylococcus aureus	+	−	−	+	−	−
Lecithinase	+		+	d		−
Acid production from:						
Gluconate	−			−	−	d
Glucose 1-phosphate	−	−	+	−	−	
D-Mannitol	−	−	−	−	−	+
Melezitose	+	d	+	+	+	−
Methyl α-D-glucososide	+	+	+	+	+	+[c]
Methyl α-D-mannoside	+	+	−	−	+	+
L-Rhamnose	+	d	−	−	d	+
Ribose	−	−	+[d]	−	−	+
Sucrose	+	+	+	+	+	−
Soluble starch	−	−	−	d	+	−
D-Xylose	−	−	+	+	+	−[e]
Nitrates reduced to nitrites	−	−	−	−	−	−[e]
Acid phosphatase	+	+	+	+	+	
Amino acid peptidase:						
D-Alanine	−	+	+	+	+	+
Lysine	−	+	+	+	+	+
Cystine arylamidase	−	−	−	−	−	+
Phosphoamidase	+	+	+	+	+	−
Tween 80 esterase	+	+	d	d	+	+[f]
Growth in the presence of 10 μg/ml trypaflavine	+	+	+	+	+	−
Growth in peptone water plus 10% (w/v) NaCl	d	+	d	d	+	+

[a]Symbols: +, >85% positive; d, different strains give different reactions (16–84% positive); −, 0–15% positive; w, weak reaction; ND, not determined.
[b]Data from Seeliger and Jones (1986), Kämpfer et al. (1991) and Rocourt and Catimel (1985).
[c]No acid produced by *Listeria grayi* subsp. *murrayi*.
[d]No acid produced by *Listeria ivanovii* subsp. *londoniensis*.
[e]Nitrates reduced to nitrites by *Listeria grayi* subsp. *murrayi*.
[f]Tween 80 esterase not produced by *Listeria grayi* subsp. *murrayi*.

The polar lipid composition is similar in *Listeria monocytogenes*, *Listeria innocua*, *Listeria welshimeri*, and *Listeria seeligeri*, and comprises phosphatidylglycerol, diphosphatidylglycerol, galactosylglucosyldiacylglycerol and L-lysylcardiolipin (Fischer and Leopold, 1999; Kosaric and Carroll, 1971; Shaw, 1974). Other more polar phospholipids were suggested to be polyprenol phosphate and glycerol-phospholipid plus a D-ananyl derivative (Fischer and Leopold, 1999). The fatty acid compositions of *Listeria monocytogenes*, *Listeria innocua*, and *Listeria ivanovii* are very similar. All contain predominantly straight-chain saturated anteiso- and iso-methyl-branched-chain types. The major fatty acids are 14-methylhexadecanoic (anteiso-$C_{17:0}$) and 12-methyltetradecanoic (anteiso-$C_{15:0}$) (Carroll et al., 1968; Feresu and Jones, 1988; Julak et al., 1989; Kosaric and Carroll, 1971; Nichols et al., 2002; Ninet et al., 1992; Raines et al., 1968; Tadayon and Carroll, 1971). The composition of the fatty acids changes under different growth conditions (Nichols et al., 2002; Puttmann et al., 1993), with a reduction of the fatty acid chain-length and alteration of branching from iso to anteiso forms with decreasing temperature in order to maintain membrane fluidity and function and to permit continued growth at lower temperatures (Annous et al., 1997).

Jones et al. (1979) reported cytochromes abb_1 in *Listeria monocytogenes* NCTC 7973, but in a later study cytochromes a_1bdo were demonstrated to be present in *Listeria monocytogenes*, *Listeria innocua* and *Listeria ivanovii* (S. B. Feresu, D. Jones and M. D. Collins, personal communication).

Plasmid DNA has been detected in all species of *Listeria*, most of which are larger than 20 MDa (Dykes et al., 1994; Fistrovici and Collins-Thompson, 1990; Flamm et al., 1984; Kolstad et al., 1990; Perez-Diaz et al., 1982; Peterkin et al., 1992; Slade and Collins-Thompson, 1990). Most of the larger plasmids in *Listeria monocytogenes* encode resistance to cadmium (Lebrun et al., 1994a; Lebrun et al., 1994b) : the proportion of cultures with these large plasmids varies markedly depending upon the serovar of *Listeria monocytogenes* (McLauchlin et al., 1997) Plasmid DNA was detected in the complete sequence of the *Listeria innocua* genome, and was shown to be 54 MDa in size with 79 genes (Glaser et al., 2001). Smaller plasmids encoding resistance to tetracycline alone (3 MDa; Poyart-Salmeron et al., 1992) and multiresistance to chloramphenicol, erythromycin, streptomycin and tetracycline (25 MDa; Hadorn et al., 1993; Poyart-Salmeron et al., 1990; Quentin et al., 1990; Tsakris et al., 1997) have been detected, although these are rare. Trimethoprim resistance due to the *dfrD* gene has been found to be encoded on a 2.5 MDa plasmid of *Listeria innocua* (Charpentier and Courvalin, 1997). Two bacteriocins have been found to be encoded on a 1.9 kDa plasmid of *Listeria innocua* (Kalmokoff et al., 2001). Transfer of native listerial plasmids has been demonstrated *in vitro* between strains of *Listeria monocytogenes*, to

TABLE 41. Additional descriptive characteristics for differentiating species of the genus *Listeria*[a,b]

Characteristic	*L. monocytogenes*	*L. grayi*	*L. innocua*	*L. ivanovii*	*L. seeligeri*	*L. welshimeri*
Gram stain	+	+	+	+	+	+
Acid production from:						
L-Arabinose	–	–	–	d	–	–
Dextrin	d	+	–	–	ND	ND
Galactose	d	+	–	d	–	–
Gluconate	–	d	–	–	d	–
D-Glucose	+	+	+	+	+	–
Glycerol	+	+	+	+	+	+
Glycogen	–	–	–	–	–	–
5-Ketogluconate	d	+	d	+	+	+
Lactose	+	+	+	+	+	+
D-Lyxose	d	d	d	+	+	+
Melezitose	d	d	d	–	–	d
α-D-Melibiose	–	d	–	d	d	d
Sucrose	d	–	d	–	–	–
Sorbitol	d	–	–	d	d	+
Sucrose	–	–	d	d	ND	ND
D-Tagatose	d	–	d	–	+	D
Trehalose	+	+	+	+	–	–
D-Turanose	+	+	+	+	+	+
D-Xylitol	+	d	+	+	+	+
Hydrolysis of:						
Esculin	+	+	+	+	+	+
Cellulose	–	–	–	–	ND	ND
Hippurate	+	–	+	+	ND	ND
Starch	d	+	d	δ	ND	ND
Voges–Proskauer	+	+	+	+	d	d
Methyl red test	+	+	+	+	+	+
Leucine esterase	d	+	+	d	+	+
Chymotrypsin	+	+	+	d	+	+
α-Glucosidase	d	+	+	d	+	+
β-Glucosidase	+	+	+	+	d	–
N-Acetyl-β-glucosamidase	d	+	d	d	–	+
Growth in peptone water plus 10% (w/v) NaCl	d	+	+	d	+	–
Pathogenicity for mice	+	–	–	+	–	–
Cell-wall type	A1γ	A1γ	A1γ	A1γ	A1γ	A1γ
Major peptidoglycan diamino acid	*meso*-DAP	*meso*-DAP	*meso*-DAP	*meso*-DAP	*meso*-DAP	*meso*-DAP
Major menaquinone	MK-7	MK-7	MK-7	MK-7	ND	ND
Mol% G+C (T_m)	37–39	41–42.5	36–38	37–38	36	36

[a]Symbols: +, >85% positive; d, different strains give different reactions (16–84% positive); –, 0–15% positive; w, weak reaction; ND, not determined. *meso*-DAP, *meso*-diaminopimelic acid.

[b]Data from Seeliger and Jones (1986), Kämpfer et al. (1991) and Rocourt and Catimel (1985).

other *Listeria* species, and to other species of bacteria including *Bacillus subtilis*, *Enterococcus faecalis*, *Streptococcus agalactiae*, and *Staphylococcus aureus* (Charpentier and Courvalin, 1999; Flamm et al., 1984; Perez-Diaz et al., 1982; Vicente et al., 1988). Cloning vectors based on the plasmid replicons pC194 (from *Staphylococcus aureus*; Sullivan et al., 1984) and pE194 (from *Enterococcus*; Chakraborty et al., 1992) have been successfully used for gene cloning experiments in *Listeria*.

Transposons *Tn1545*, *Tn916*, and *Tn917* (and their derivatives) introduced and expressed in *Listeria monocytogenes* have proved to be extremely useful tools in understanding the basis of virulence in the bacterium (Vazquez-Boland et al., 2001). Transfer of *Tn1545* has been demonstrated between *Enterococcus faecalis* and *Listeria monocytogenes* in the digestive tract of gnotobiotic mice (Doucet-Populaire et al., 1991). A transposon, similar to *Tn917* (designated *Tn5422*) has been detected in plasmid DNA of *Listeria monocytogenes* (Lebrun et al., 1994a; Lebrun et al., 1994b).

Lysogenic phage are commonly carried by *Listeria* (Audurier et al., 1977; Bannerman et al., 1996; Loessner, 1991; Loessner et al., 1994; Rocourt et al., 1985b; Rocourt et al., 1986; Rocourt et al., 1982b). They are generally morphologically similar with isometric heads, long non-contractile tails and correspond to the Myoviridae or Syphoviridae families (Loessner et al., 1994; Rocourt, 1986; Rocourt et al., 1986). The complete genome sequence of one lysogenic phage, A118, has been reported (Loessner et al., 2000)and phage integration was shown to occur in a homolog of the *Bacillus comK* gene, although no known function has yet been determined for this gene in *Listeria*. To date, no phage conversion has been reported due to toxin genes associated with prophage sequences. A lytic phage (A511) of the Myoviridae family with a double stranded DNA genome of approximately 116 kbp has also been described (Loessner et al., 1994). The phage genomes characterized to date have been linear double-stranded DNA of 35–116 kbp (Loessner et al., 1994; Rocourt et al., 1986) with G+C base compositions of 37–39 mol% (Loessner at el., 1994) (Loessner et al., 1994). Lytic properties of sets of phages have been used for subtyping *Listeria monocytogenes*. Bacteriocin (monocin) production is common within the genus (Bannerman et al., 1996; Curtis and Mitchell, 1992; Kalmokoff et al., 2001; Rocourt et al., 1986).

The cell surface of *Listeria* species contains a number of different antigens. Somatic (O factor) antigens based, at least in part, on ribitol teichoic acid substitution, together with flagella (H factor) antigens, have been characterized using highly absorbed rabbit antisera for serotyping members of the genus (Seeliger and Hohne, 1979; Table 42). Antigenic cross-reactions with other genera have been reported (Seeliger, 1961), in particular with some *Bacillus* species that share a common cell-wall structure.

Listeria species are usually sensitive to amikacin, amoxycillin, ampicillin, azlocillin, ciprofloxacin, chloramphenicol, clin-

damycin, coumermycin, doxycycline, enoxacin, erythromycin, gentamicin, imipen, netilmicin, penicillin, rifampin, trimethoprim, and vancomycin. This genus is less sensitive to norfloxacin and ofloxacin, and is resistant to the cephalosporins, phosphomycin, and polymyxin (Charpentier et al., 1995; MacGowan, 1990; Riviera et al., 1993). Tetracycline is less active against *Listeria monocytogenes*, and 2–5% of strains are highly resistant (Poyart-Salmeron et al., 1992): this rate of resistance has not changed over the past 30 years in the UK (Threlfall et al., 1998). Most of these highly resistant cultures contain a chromosomally encoded *tetM* gene (together with a transposon, int-Tn), which also confers resistance to minocycline (Poyart-Salmeron et al., 1992). The *tetM* gene has also been detected in *Listeria innocua* and *Listeria welshimeri* (Charpentier and Courvalin, 1999; Facinelli et al., 1993). Plasmid-encoded tetracycline resistance in *Listeria monocytogenes* is rarer than that chromosomally encoded. The latter is encoded by a *tetL* (minocycline-sensitive; Poyart Salmeron et al., 1992) or a *tetS* determinant (minocycline-resistant; Charpentier et al., 1993; Hadorn et al., 1993). The *tetS* gene was first described in *Listeria monocytogenes*, but has since been detected in *Listeria innocua*, *Listeria welshimeri*, and *Enterococcus faecalis* (Charpentier and Courvalin, 1999; Charpentier et al., 1994). Plasmid-encoded resistance to chloramphenicol (*cat221*), erythromycin (*ermAM*), streptomycin (*aad6*) and tetracycline (*tetS*) has been detected in *Listeria monocytogenes* (Charpentier and Courvalin, 1999; Hadorn et al., 1993; MacGowan, 1990; Poyart-Salmeron et al., 1990; Quentin et al., 1990; Threlfall et al., 1998). Other genes encoding resistance to tetracycline (*tetK*) and streptomycin (*aad6*) have been detected in *Listeria innocua* (Charpentier and Courvalin, 1999). A gene that encodes resistance to erythromycin (*ermC*) has been detected in *Listeria monocytogenes* and *Listeria innocua* (Roberts et al., 1996).

Listeria monocytogenes has long been used as a low-grade intracellular pathogen to study cellular immunity (Kaufmann, 1993), and more recent advances in molecular biology have further assisted the understanding of listeriosis at the cellular level (Vazquez-Boland et al., 2001). Of the six species of *Listeria*, only cultures of *Listeria monocytogenes* and *Listeria ivanovii* are virulent in all the models as measured by LD$_{50}$, the kinetics of bacterial growth in host tissue, survival in the liver and spleen or the death of the experimental animal. All the non-hemolytic species of *Listeria* (*Listeria innocua*, *Listeria welshimeri*, and *Listeria grayi*) and the weakly hemolytic *Listeria seeligeri* are avirulent in mouse-pathogenicity tests (Mainou-Fowler et al., 1988). *Listeria monocytogenes*, *Listeria ivanovii* and, to a lesser extent *Listeria seeligeri*, show properties of invasion and spreading in a range of mammalian cells growing *in vitro*, but other *Listeria* species do not (Farber et al., 1991; Pine et al., 1991; Van Langendonck et al., 1998).

Listeria monocytogenes enters both phagocytic and non-phagocytic cells. A listerial surface protein, internalin A (which is reminiscent of the M protein of the group A streptococci), has been shown to be involved with the initial stages of invasion of all cell types. A second cell surface protein (internalin B) is required for entry into hepatocyte-like but not into enterocyte-like cells. The receptor for internalin A is E-cadherin which is a mammalian surface protein involved in cell-to-cell adhesion. The cell-wall teichoic acid is also required for adhesion of *Listeria* to mammalian cells (Abachin et al., 2002). The listerial cell-surface protein, p60 (with murein hydrolase activity) may

TABLE 42. Somatic and flagella antigens used for serotyping *Listeria*[a]

Serovar	Somatic (O factor) antigens[b, c]	Flagella (H factor) antigens[b]
L. monocytogenes:		
1/2a	I, II, III	A, B
1/2b	I, II, III	A, B, C
1/2c	I, II, III	B, D
3a	II, III, IV, (XII), (XIII)	A, B
3b	II, III, IV, (XII), (XIII)	A, B, C
3c	II, III, IV, (XII), (XIII)	B, D
4a	III, (V), VII, IX	A, B, C
4ab	III, V, VI, VII, IX, X	A, B, C
4b	III, V, VI	A, B, C
4c	III, V, VII	A, B, C
4d	III, (V), VI, VIII	A, B, C
4e	III V, VI, (VIII), X	A, B, C
7	III, XII, XIII	A, B, C
L. grayi:		
Not designated	III, XII, XIV	A, B, C
L. innocua:[d]		
4ab	III, (V), VI, VII, IX	A, B, C
6a	III, V, (VI), (VII), (IX), XV	A, B, C
6b	III, (V), (VI), (VII), IX, XXI	A, B, C
L. ivanovii:		
5	III, V, VI, (VII) X	A, B, C
L. seeligeri:[d]		
1/2b	I, II, III	A, B, C
4c	III, V, VII	A, B, C
4d	III, (V), VI, VIII	A, B, C
6b	III, (V), (VI), (VII), IX, XXI	A, B, C
L. welshimeri:		
6a	III, V, (VI), (VII), (IX), XV	A, B, C
6b	III, (V), (VI), (VII), IX, XXI	A, B, C

[a]() Denotes factors not always detected.
[b]Data from Seeliger and Hohne (1979).
[c]Factor II is heat labile.
[d]Undesignated combinations of O factors also occur.

also be involved in the invasion of fibroblasts and a second surface-exposed autolysin (Ami) has been shown to be involved in adhesion to eukaryotic cells (Milohanic et al., 2000).

Within mammalian phagocytic cells, the majority of cells in phagocytic vacuoles are rapidly killed. However, some bacteria survive in the vacuole, and those in the membrane-bound compartment of the non-professional phagocyte (which probably confers on the bacterium some protection from host defences in the extracellular environment), mediate the dissolution of the vacuole membrane by means of a thiol-activated hemolysin (listeriolysin O), in combination with the action of a phospholipase C (phosphatidyl inositol-specific). *Listeria monocytogenes* then enters the host-cell cytoplasm where growth occurs. In the cytoplasm, the organism becomes surrounded by polymerized host-cell actin, which becomes preferentially polymerized at the older pole of the bacterium following cell division. The ability to polymerize actin confers intracellular mobility on the bacterium is dependent on the Act A protein and may be mediated by binding with host-cell profilactin and vasodilator-stimulated phosphoprotein (VASP), although the exact mode of action is not understood. The resulting comet tail-like structure can push the bacterial cell into an adjacent mammalian cell, where it again becomes encapsulated in a vacuole. At this stage the vacuole is double-membrane-bound from the two host cells. The primary role of the second phospholipase enzyme (a broad-spectrum lecithinase enzyme which hydrolyzes phosphatidylcholine or lecithin) is believed to be in the fusion and subsequent dissolution of these membranes (although again the hemolysin and phospholipase C may also contribute in this process; a distinct role may not exist for each of the Hly, PlcBC and PlcC proteins; rather, these enzymes work synergistically to achieve optimal levels of activity at different stages of the infection process). After release into the cytosol, intracellular growth and movement within the newly invaded cell is then repeated and the focus of infection continues to enlarge by this process of cell-to-cell spread (Vazquez-Boland et al., 2001).

The genes involved in the invasion and intracellular movement in mammalian cells are organized into a main virulence cluster containing adjacent operons which are all directly regulated by the central virulence regulator PrfA. The internalin genes are located in a separate gene cluster and are only partially regulated by PrfA, there being a basal level of expression even in the absence PrfA induction. Similar sets of virulence genes are also present in *Listeria ivanovii* and *Listeria seeligeri*. Virulence for *Listeria monocytogenes* is therefore a multifactorial property and at least nine genes and their products are required for infection, invasion, survival, mobility and cell-to-cell spread (Vazquez-Boland et al., 2001), although additional genes (especially those for cell-surface components) are also likely to be involved with the infectious process (Autret et al., 2001; Engelbrecht et al., 1998; Raffelsbauer et al., 1998). It is possible that the genes for intracellular parasitism have evolved for *Listeria monocytogenes*, as well as for *Listeria ivanovii* and *Listeria seeligeri*, to invade lower multicellular eukaryotic organisms resident in the environment. Internalin genes also occur in the non-pathogenic species *Listeria innocua* (Vazquez-Boland et al., 2001).

A wide range of animals are susceptible to experimental infection, but rabbits, mice and, to a lesser extent, rats are most frequently used since these animals die following inoculation of live bacteria by the intravenous or intraperitoneal route. The guinea pig is a better model of human infection than the mouse due to a single amino-acid difference between the murine and human E-cadherin molecules: this change reduces the affinity of the internalin A protein for the receptor and compromises invasion by *Listeria* in the mouse model (Lecuit et al., 1999). In guinea pigs and transgenic mice carrying the humanized E-cadherin gene, internalin was found to mediate invasion of enterocytes and crossing of the intestinal barrier (Lecuit et al., 2001). Virulence assays are usually performed using either intraperitoneal or tail-vein injection. Delivery directly into the animal overcomes the limitations of E-cadherin specificity when strains are administered by the oral route.

Enrichment and isolation procedures

Culture from blood or cerebrospinal fluid does not require special media. Tissues should be homogenized, suspended in broth and subcultured onto blood agar. The incubation of medulla homogenates at refrigeration temperatures for some weeks has been reported as necessary to obtain cultures which grow on artificial media (Gray and Killinger, 1966). For specimens such as feces, vaginal secretions, food and environmental samples, special selective media are necessary.

Before the mid-1980s, "cold enrichment", utilizing the ability of *Listeria* to outgrow competing organisms at refrigeration temperatures in non-selective broths, was the main method used for selective isolation (Gray and Killinger, 1966). When growing on transparent media illuminated by oblique transmitted light and viewed at low magnification ("Henry" illumination technique), all *Listeria* colonies have a characteristic blue color with a central "ground glass" appearance (Gray, 1957). However, because of the degree of skill required in recognizing characteristic colonies, the lack of specificity, and the slowness of these methods (some workers subcultured broths for up to 6 months), procedures have been much improved.

Media have been developed that rely on a number of selective agents; these include acriflavin, lithium chloride, colistin, ceftazidime, cefotetan, fosfomycin, moxolactam, nalidixic acid, cycloheximide, and polymyxin. Such media have resulted in the widespread ability of microbiology laboratories (especially those involved with the examination of foods) to selectively isolate *Listeria*. Numerous enrichment and selective isolation media have now been developed. Those mentioned here (or modifications of these) are used most frequently for the examination of foods. For selective broths, the US Food and Drugs Administration (FDA) method (Lovett et al., 1987), the US Department of Agriculture (USDA) method (McClain and Lee, 1988), or the Netherlands Government Food Inspection Service (NGFIS) method described by van Netten et al. (1989) are most often used. Selective agars most frequently used are those of Curtis et al. ((1989a); "Oxford" formulation) or the PALCAM agar of van Netten et al. (1989). These media are listed in internationally agreed standard methods (International Organization for Standardization, 1996, 1998).

The FDA method uses a single enrichment broth containing acriflavin, nalidixic acid, and cycloheximide. The USDA method consists of a double enrichment using first a University of Vermont primary enrichment broth 1 (UVM1; containing esculin nalidixic acid and acriflavin) which is subcultured into a Fraser Broth (containing esculin, nalidixic acid, lithium chloride and acriflavin). The NGFIS Method uses a single L-PALCAM (or Liquid-PALCAM) broth, the name of which is an acronym of the ingredients, polymyxin B, acriflavin, lithium chloride, ceftazidime, [a]esculin and mannitol.

After incubation the selective broths are subcultured onto selective agars, most usually PALCAM agar (containing the same agents as L-PALCAM broth) or Oxford Agar. The latter agar contains esculin, lithium chloride, cycloheximide, colistin, acrifalvin, cefotetan, and fosfomycin. On Oxford agar all *Listeria* species show a typical colonial appearance and hydrolyze esculin, to produce black zones around the colonies. On PALCAM agar the colonies have a cherry red background. It is beyond the scope of this contribution to describe the preparation of these media; for further details see Baird et al. (1987). For comparisons of the efficiency of the different selective techniques for the isolation of *Listeria* from foods, see Warburton et al. (1991)and Hayes et al. (1992).

All *Listeria* species are isolated by these methods and are morphologically indistinguishable from each other. However, Curtis et al. (1989b) reported differences in the minimal inhibitory concentrations of selective agents used in microbiological media and found that *Listeria ivanovii* and *Listeria seeligeri* were more sensitive to fosfomycin than *Listeria innocua*, *Listeria monocytogenes*, and *Listeria welshimeri*. Consequently, colonies of *Listeria ivanovii* and *Listeria seeligeri* grow more slowly and are smaller on Oxford agar, which uses fosfomycin as a selective agent.

To differentiate *Listeria monocytogenes* from other *Listeria* species on selective agars, substrates have been added to selective media to detect phospholipase (Notermans et al., 1991) or β-D-glucosidase and enhanced hemolysis (Beumer et al., 1997). Selective media, based on lipase and β-D-glucosidase activity which successfully differentiates *Listeria monocytogenes* from populations of other *Listeria* species are commercially available (Vlaemynck et al., 2000).

Maintenance procedures

Listeria species can be preserved for decades at room temperature in the dark on stab or sloped non-selective agars such as nutrient or tryptose soy agars provided the preservation vessels are well sealed and do not allow the agar to dehydrate. *Listeria* can also be preserved in glycerol on glass beads at less than –20°C (Feltham et al., 1978) or by lyophilization.

Differentiation of the genus *Listeria* from other genera

Table 43 lists the features most useful in differentiating the genus *Listeria* from other Gram-positive, non-spore-forming, rod-shaped bacteria. Identification of new *Listeria* isolates may be achieved by the following: examination of cellular and colonial morphology; growth at 37°C; oxygen requirements: catalase production; hydrolysis of esculin and sodium hippurate; alkaline phosphatase production; and production of acid from carbohydrates in a suitable medium. Determination of the cell-wall and lipid composition and 16S rRNA gene sequence analysis are not necessary for routine identification.

Bacteria with which members of the genus *Listeria* are most likely to be confused are those of the genera *Brochothrix*, *Erysipelothrix*, *Lactobacillus*, and *Kurthia*. As mentioned earlier, coccobacillary forms of the genus *Listeria* may be confused with streptococci, which are catalase-negative.

Listeria may be distinguished easily from the genus *Kurthia* by their different oxygen requirements. *Kurthia* species are strictly aerobic and produce little or no acid in sugar-fermentation tests. In contrast to *Listeria*, the cell walls of *Kurthia* species contain lysine (Keddie, 1981a).

Listeria species may be distinguished from *Erysipelothrix* species by the negative catalase test, good growth on the usual nutrient media, and colonial morphology; motility; possession of more vigorous saccharolytic activity; hydrolysis of esculin and sodium hippurate, and non-production of H₂S. *Listeria* species and *Erysipelothrix rhusiopathiae* are antigenically distinct. *Listeria* species contain cytochromes and menaquinones, which are absent in *Erysipelothrix rhusiopathiae*. The two genera also differ in the chemical composition of the cell-wall peptidoglycan: that of *Listeria* is based on *meso*-DAP while the cell wall of *Erysipelothrix rhusiopathiae* contains lysine.

Listeria species may be distinguished from most lactobacilli by the catalase test. There are, however, some catalase-positive lactobacilli and rare strains of *Listeria* that are catalase-negative. Members of the genus *Listeria* grow very poorly or not at all on MRS medium (De Man et al., 1960), on which lactobacilli grow very well. When investigated at the correct incubation temperature (20–25°C), *Listeria* are motile while most lactobacilli are not. Lactobacilli have a different colonial morphology to *Listeria* species, there are differences in the fermentation patterns of the two genera, and all strains of *Listeria* examined possess *meso*-DAP in the cell wall. Some lactobacilli also possess this amino acid in the cell wall but in other lactobacilli it is replaced by lysine or ornithine (Schleifer and Kandler, 1972). Serologically, the genera *Listeria* and *Lactobacillus* are distinct.

Members of the genus *Listeria* may be confused with *Brochothrix* species. Members of both genera are frequently isolated from prepackaged meats and poultry held at refrigeration temperatures. Both contain *meso*-DAP in the cell wall, MK-7 as the major isoprenoid quinone, and have an identical fatty acid composition (Feresu and Jones, 1988; Schleifer and Kandler, 1972). The inability of *Brochothrix thermosphacta* to grow at 37°C and the inability of *Listeria* species to grow on the medium of Gardner (1966) serve to distinguish the two genera. *Brochothrix thermosphacta* does not hydrolyze sodium hippurate but produces acid from a greater number of sugars. Serologically, the two taxa are distinct (Wilkinson and Jones, 1977).

Taxonomic comments

As described above, analysis of sequence data from the 16S rRNA confirms the close relationship within the genus *Listeria*, and indicates a close phylogenetic relationship (93% sequence similarity) with the genus *Brochothrix* (Collins et al., 1991). This latter genus comprises two species (*Brochothrix thermosphacta* and *Brochothrix campestris*) which have many phenotypic properties in common with *Listeria* (Collins et al., 1991; Seeliger and Jones, 1986; Sneath and Jones, 1986). These two genera justify status as the family *Listeriaceae* (Collins et al., 1991). 16S rRNA gene sequence analyses show relationships with other Gram-positive genera of low G+C content, including members of the genus *Bacillus* (Collins et al., 1991). Indeed, the completed genome sequences of *Listeria monocytogenes* and *Listeria innocua* show a close relationship to that from *Bacillus subtilis* and suggest a common origin (Glaser et al., 2001).

Although originally described as a monospecific genus containing only *Listeria monocytogenes* (Pirie, 1940a), six species of *Listeria* are now recognized (Moore et al., 1985; Skerman et al., 1980). The group of bacteria originally named *Listeria*

TABLE 43. Characteristics differentiating the genera *Listeria*, *Brochothrix*, *Erysipelothrix*, *Kurthia*, and *Lactobacillus*[a,b]

Genus	Listeria	Brochothrix	Erysipelothrix	Kurthia	Lactobacillus
Motile	+[c]	−	−	+[d]	−[e]
Oxygen requirement for growth at 35°C	Facultative	Facultative	Facultative	Aerobic	Facultative
Growth at 35°C	+	−	+	+	+
Catalase	+	+[f]	−	+	−[g]
H$_2$S production	−	−	+	+[h]	−
Acid from glucose	+	+	+	−	+
Peptidoglycan type[i]	A	A	B	A	A
Major peptidoglycan diamino acid	*meso*-DAP	*meso*-DAP	L-Lysine	L-Lysine	Lysine or *meso*-DAP, or ornithine
Major menaquinone	MK-7	MK-7	−	MK-7	−[j]
Fatty acid type[k,l]	S, A, I	S, A, I	S, A, I, U	S, A, I	S, A, (U)
Mol% G+C	36–38	35.6–36.1	36–40	36.7–37.9	34–53

[a]Symbols: +, >85% positive; d, different strains give different reactions (16–84% positive); −, 0–15% positive; w, weak reaction; ND, not determined.

[b]Data from McLean and Sulzbacher (1953), Davidson et al. (1968), Collins-Thompson et al. (1972), Tadayon and Caroll (1971), Schleifer and Kandler (1972), Stuart and Welshimer (1973, 1974), Seeliger and Welshimer (1974), Shaw (1974), Jones (1975), Collins and Jones (1981), Keddie (1981b), Sharpe (1981), Rocourt and Grimont (1983), Seeliger et al. (1984), Flossmann and Erler (1972), White and Mirikitani (1976).

[c]All species motile at 20–25°C, poorly or nonmotile at 37°C (Seeliger and Welshimer, 1974).

[d]Majority of strains motile, but nonmotile strains do occur (Keddie, 1981a).

[e]Most strains nonmotile, but a few motile strains occur (Sharpe, 1981).

[f]Catalase production dependent on medium and temperature of incubation (Davidson et al., 1968).

[g]Some strains give positive catalase reaction (Sharpe, 1981).

[h]Weak production of H$_2$S by some strains (Jones, 1975; Keddie, 1981a).

[i]Group A, cross-linkage between positions 3 and 4 of two peptide subunits; group B, cross-linkage between positions 2 and 4 of two peptide subunits (Schleifer and Kandler, 1972).

[j]*Lactobacillus mali* contains MK-8 and MK-9 as major menaquinones; a menaquinone has also been detected in *Listeria casei* subsp. *rhamnosus* (Collins and Jones, 1981).

[k]Straight-chain saturated; U, monounsaturated; A, anteiso-methyl-branched; I, iso-methyl-branched; cyclopropane-ring fatty acids.

[l]Those in parentheses may be present.

monocytogenes (*Listeria monocytogenes sensu lato*) was redefined on the basis of DNA–DNA hybridization studies (Rocourt et al., 1982a) to comprise the species *Listeria monocytogenes* (*sensu stricto*), *Listeria innocua* (Seeliger, 1981), *Listeria welshimeri* (Rocourt and Grimont, 1983), *Listeria seeligeri* (Rocourt and Grimont, 1983), and *Listeria ivanovii* (Seeliger et al., 1984). The genus includes one other species, *Listeria grayi* (Errebo Larsen and Seeliger, 1966). This classification is supported by the results of a second DNA–DNA homology study (Hartford and Sneath, 1993) and an analysis of multilocus enzyme electrophoresis (Boerlin et al., 1991). The study by Hartford and Sneath (1993) suggested a very close relationship between *Listeria monocytogenes* and *Listeria innocua* and there may be some overlap between these two species. DNA–DNA hybridization values obtained between different *Listeria* species are given in Table 44. Numbers of nucleotide differences and homology values of a continuous stretch of 1458 nucleotides of the 16S rRNA of *Listeria* species are given in Table 45. Signature sequences for different *Listeria* species within the V2 region of the 16S rRNA genes have been described (Collins et al., 1991).

On the basis of both phenotypic and genotypic characters, *Listeria grayi* shows a more distant relationship to the rest of the genus (Boerlin et al., 1991; Collins et al., 1991; Feresu and Jones, 1988; Hartford and Sneath, 1993; Kämpfer et al., 1991; Rocourt et al., 1982a; Wilkinson and Jones, 1977), but there is clear justification for the retention of *Listeria grayi* within the genus (Collins et al., 1991; Rocourt et al., 1987). There is insufficient justification for the renaming of this species as a new genus "*Murraya*" as suggested by Stuart and Welshimer (1974).

It is not anticipated that there will be major changes in the classification of this family in the future.

The species previously known as "*Listeria denitrificans*" (Prévot, 1961) is not a member of the genus *Listeria* and has been reclassified in a new genus *Jonesia denitrificans* (Rocourt et al., 1987).

Historical note. The original isolation of *Listeria monocytogenes* by Murray and colleagues (Murray et al., 1926) named this bacterium "*Bacterium monocytogenes*". The name was changed to "*Listerella monocytogenes*" following the recognition of the same bacterium by Pirie, which had been originally named as "*Listerella hepatolytica*" (Pirie, 1927). The genus was renamed *Listeria* (Pirie, 1940b) since "*Listerella*" had already been used for another genus of organisms (Pirie, 1940a). The conservation of the name *Listeria* was approved by the Judicial Commission on Bacteriological Nomenclature and Taxonomy (Judicial-Commission, 1954).

Comments on strains of *Listeria monocytogenes*

Although numerous *Listeria monocytogenes* cultures are available for study, most characterization has been performed on a small selection of strains (Table 46).

The type strain (originally designated 53 XXIII and was isolated from an animal in Cambridge, UK, in 1924; Murray et al., 1926) now shows atypical characteristics: most variants of this strain are non-hemolytic, CAMP-test-negative and some nonmotile (Kathariou and Pine, 1991), although hemolytic reactions on rabbit (Kathariou and Pine, 1991) and sheep (J. McLauchlin, unpublished data) blood have been detected. The type strain is non-virulent in laboratory models. A second culture from the 1924 outbreak (Murray et al., 1926) designated 58 XXIII was isolated from a different animal 2 d after the type strain. This second isolate is virulent in laboratory models, but generates non-hemolytic and avirulent mutants amongst the

TABLE 44. DNA–DNA hybridization (per cent relative binding at 60°C) of *Listeria* species[a]

Species	L. monocytogenes	L. ivanovii	L. innocua	L. welshimeri	L. seeligeri
L. monocytogenes	66–100				
L. ivanovii	27–34	85–100			
L. innocua	47–56	22–45	87–100		
L. welshimeri	41–46	22–38	42–49	83–100	
L. seeligeri	29–38	34–50	22–34	26–35	71–100
L. grayi	4–21	2–7	5–26	2–19	1–7

[a]Data adapted from Rocourt et al. (1982a).

TABLE 45. Numbers of nucleotide differences and homology values of a continuous stretch of 1458 nucleotides of the 16S rRNA of *Listeria* species

	No. nucleotide differences or % homology between:[a,b]				
Species	L. innocua	L. ivanovii	L. monocytogenes	L. seeligeri	L. welshimeri
L. innocua		98.8	99.2	98.7	99.1
L. ivanovii	18		98.4	99.1	99.1
L. monocytogenes	11	23		98.5	99.8
L. seeligeri	19	13	22		99.0
L. welshimeri	13	13	17	14	

[a]Data adapted from Collins et al. (1991).
[b]The number below the diagonal are numbers of nucleotide differences, and the number above the diagonal are homology values.

TABLE 46. *L. monocytogenes* strains commonly used for laboratory characterization

Strain designation	Original source	Serovar	Available from:[a]	Comments
53 XXIII	Rabbit peritoneal exudates, Cambridge, UK, 1924[b]	2-Jan	ATCC 15313, NCTC 10357, CIP 82.110, SLCC 53 SLCC 5850	Type strain, used for taxonomic and other studies[c]
58 XXIII	Guinea pig mesenteric lymph node, Cambridge UK, 1924[b]	1/2a	NCTC 7973, ATCC 35152, CIP 54.149, SLCC 2371, ATCC 43248–51[d]	Used for taxonomic, immunological, physiological, molecular biological, and virulence studies
Mackaness/EGD	Origin uncertain, obtained by G.B. Mackaness from E.G.D. Murray, Canada, in the 1960s[e]	1/2a	NCTC 12427, SLCC 5764	Extensively used for immunological, physiological, molecular biological, and virulence studies. Complete genome sequence available[f]
DPL10403S	Streptomycin-resistant variant of 10403 obtained from the laboratory of M.L. Gray, USA[g]	1/2a	Extensively used for molecular biological and virulence studies	
LO28	Human clinical isolate, Spain[h]	1/2c	Extensively used for molecular biological and virulence studies	
Scott A	Human clinical isolate, USA[i]	4b	ATCC 49594[j], ATCC 700302[j], ATCC 700301[j]	Used as control strain, especially by food microbiologists in the USA
646/86	Soft cheese associated with case of human listerial meningitis, UK[k]	4b	NCTC 11994	Used as control strain, especially by food microbiologists in the UK

[a]NCTC, National Collection of Type Cultures, London, UK; ATCC, American Type Culture Collection, Manassas, VA, USA; CIP, Institut Pasteur Collection, Paris, France; SLCC, Special Listeria Culture Collection, established by H.P.R. Seeliger in Wurzburg, now held by H. Hof, University of Heidelberg, Germany.
[b]Data from Murray et al. (1926).
[c]Culture shows atypical reactions, usually nonhemolytic, nonmotile, and CAMP-test-negative; deletions in the *prfA* gene.
[d]Spontaneously produced nonhemolytic variants with deletions in the *prfA* gene.
[e]H. Hof, University of Heidelberg, personal communication.
[f]Data from Glaser et al. (2001).
[g]Data from Bishop and Hinrichs (1987).
[h]Data from Vicente et al. (1985).
[i]Data from Fleming et al. (1985).
[j]Atypical forms, either forming petite colonies (Siragusa et al., 1990) or showing increased resistance to nisin (Mazzotta and Montville, 1997).
[k]Data from Bannister (1987).

hemolytic phenotype (Pine et al., 1987). Deletions of part of the *prfA* gene have been detected in the type strain and the non-hemolytic 58 XXIII (Gormley et al., 1989; Leimeister-Wachter and Chakraborty, 1989; Leimeister-Wachter et al., 1990). A request to substitute 58 XXIII for the type strain (53 XXIII; Jones and Seeliger, 1983) was rejected (Wayne, 1986). The type strain (53 XXIII) is serogroup 1/2 (flagella can not be detected in cultures from the NCTC or ATCC) and strain 58 XXIII serovar 1/2a. Representatives of these two cultures (53 XXIII and 58 XXIII) from various sources are indistinguishable by phage or pulsed-field gel electrophoresis and are therefore likely to be the same strain (J. McLauchlin, unpublished data)

The cultures designated EGD or Mackaness are used synonymously by some workers (Sokolovic and Goebel, 1989) and were original obtained from E.G.D. Murray by G.B. Mackaness at the Trudeau Institute (USA) in the 1960s, although the initial origin(s) of these cultures are obscure. Mackaness certainly used 58XXIII (see above) as represented by NCTC 7973 (Miki and Mackaness, 1964), but also describes the use of "a recent human isolate supplied by Dr EGD Murray" (Blanden et al., 1966). Later publications from this laboratory refer to "EGD" without reference to the exact origin of the culture (Mackaness, 1969). Although some workers use the designations EGD and Mackaness synonymously (Sokolovic and Goebel, 1989), physiological and genetic differences between EGD and Mackaness have been reported (Bubert et al., 1992; Sokolovic and Goebel, 1989). DNA from an EGD strain was shown to have the same restriction pattern around the region of the hemolysin gene as 58 XXIII (Gormley et al., 1989). EGD and Mackaness are both serovar 1/2a, and on the basis of phage typing and pulsed-field gel electrophoresis, two different types occur, either of which are designated EGD or Mackaness and these probably represent two distinct strains (J. McLauchlin, unpublished data). One of these strains is indistinguishable from those isolated from the 1924 outbreak in Cambridge which is the origin of the type strain (J. McLauchlin, unpublished data). EGD has been used extensively for immunological, and pathogenesis studies, and was the subject of much molecular biological analysis, including the production of a whole genome sequence (Glaser et al., 2001).

Two other *Listeria monocytogenes* strains have been studied using molecular biological techniques for the analysis of pathogenesis: i.e., DLP 10403S (serovar 1/2a) and LO28 (serovar 1/2c).

Two distinct strains have been used extensively by food microbiologists (Scott A and NCTC 11994), both of which are serovar 4b, which is the serotype most often involved with human infection. Scott A is principally used in the USA and NCTC 11994 in the UK. Both these strains were associated with human disease (Bannister, 1987; Fleming et al., 1985).

Further reading

Ryser, E.T., and E.H. Marth. 2007. *Listeria*, Listeriosis, and Food Safety, Third edition Marcel Dekker, New York.

Differentiation of species of the genus *Listeria*

The differential characters of the species of *Listeria* are listed in Table 40. The antigenic structure used for serotyping *Listeria monocytogenes* together with a descriptive table of characters of *Listeria* species are shown in Table 42 and Table 41, respectively.

List of species of the genus *Listeria*

1. **Listeria monocytogenes** (Murray, Webb and Swann 1926) Pirie 1940a, 383[AL] (*Bacterium monocytogenes* Murray, Webb and Swann 1926, 408)

   mo.no.cy.to′ge.nes. N.L. n *monocytum* a blood cell, monocyte; Gr. v. *gennaio* to produce; N.L. adj. *monocytogenes* monocyte-producing.

   Characteristics and differentiation from other *Listeria* species are given in Table 40. Antigenic composition for serotyping is shown in Table 42.

   Habitat: Widely distributed in nature, found in soil, mud, sewage, vegetation and in the feces of animals and man. Pathogenic for animals and man; most cases are food- or feed-borne. In humans causes systemic illness (most often septicemia and/or meningitis) in immunocompromised adults or juveniles, as well as infecting the unborn infant.

   DNA G+C content (mol%): 37–39 (T_m); 38 (Bd).

   Type strain: ATCC 15313, CIP 82.110, DSM 20600, NCTC 10357, SLCC 53, 53XXII.

   GenBank accession number (16S rRNA gene): U84148.

   Complete genome sequences: *Listeria monocytogenes* EGD (2,944,528 bp; 39 mol% G+C, 2853 protein-coding genes) CLIP 11262 (Glaser et al., 2001). Unfortunately, the *Listeria monocytogenes* type strain was not included in the above studies.

   Additional remarks: On the basis of both multilocus enzyme electrophoresis (Bibb et al., 1990; Piffaretti et al., 1989) and restriction fragment and sequence analysis of specific genes (Comi et al., 1997; Gutekunst et al., 1992; Rasmussen et al., 1991; Vines et al., 1992; Vines and Swaminathan, 1998; Wiedmann et al., 1997), three lineages of *Listeria monocytogenes* have been identified that represent distinct evolutionary branches. These lineages correspond to serovars 1/2a, 1/2c and 3c (Lineage I), serovars 4b, 1/2b and 3b (Lineage II) and serovars 4a and 4c (Lineage III). Outbreaks of disease are most commonly associated with organisms from Lineages I and II and Lineage III have almost solely been isolated from animal hosts. However particular phenotypic characters that account for these differences have not been identified.

2. **Listeria grayi** Errebo Larsen and Seeliger 1966, 19[AL] emend. Rocourt, Boerlin, Grimont, Jacquet and Piffaretti 1992, 173.

   gray′i. N.L. gen. n. *grayi* of Gray, named in honor of M. L. Gray, an American bacteriologist.

   Characteristics and differentiation from other *Listeria* species are given in Table 40. Antigenic composition for serotyping is shown in Table 42.

   Habitat: Widely distributed in nature, found in soil, mud, sewage, vegetation and in the feces of animals and man. Not pathogenic for animals or man.

   DNA G+C content (mol%): 41–42 (T_m).

   Type strain: ATCC 19120, CCUG 4983, CCUG 5118, CIP 6818, CIP 105447, DSM 20601, LMG 16490, NCAIM B.01871, NCTC 10815, V-1.

   GenBank accession number (16S rRNA gene): X56150, X98526.

Additional remarks: There are two subspecies of *Listeria grayi* (Rocourt et al., 1992): *Listeria grayi* subsp. *grayi* and *Listeria grayi* subsp. *murrayi* (previously named "*Listeria murrayi*"; Welshimer and Meredith, 1971). Differences between these two biotypes are shown in Table 47.

3. **Listeria innocua**   (*ex* Seeliger and Schoofs 1979) Seeliger 1983, 439[VP] (Effective publication: Seeliger 1981, 492).

in.noc′u.a. L. adj. *innocuus* harmless.

Characteristics and differentiation from other *Listeria* species are given in Table 40. Antigenic composition for serotyping is shown in Table 42.

Habitat: Widely distributed in nature, found in soil, mud, sewage, vegetation and in the feces of animals and man. Not pathogenic to animals or man.

DNA G+C content (mol%): 36–38 (T_m), 38 (Bd).

Type strain: ATCC 33090, CCUG 15531, CIP 80.11, DSM 20649, LMG 11387, NCTC 11288, SLCC 3379, 58/1971.

GenBank accession number (16S rRNA gene): X56152, X98527.

Complete genome sequences is available: *Listeria innocua* CLIP 11262 (3,011,209 bp; 37 mol% G+C, 2973 protein-coding genes) (Glaser et al., 2001).

4. **Listeria ivanovii** Seeliger, Rocourt, Schrettenbrunner, Grimont and Jones 1984, 336[VP]

i.van.ov′i.i. N.L. gen. n. *ivanovii* of Ivanov, named in honor of Ivan Ivanov, a Bulgarian bacteriologist.

Characteristics and differentiation from other *Listeria* species are given in Table 40. Antigenic composition for serotyping is shown in Table 42.

Habitat: Widely distributed in nature, found in soil, mud, sewage, vegetation and in the feces of animals and man. Pathogenic for animals.

DNA G+C content (mol%): 37–38 (T_m).

Type strain: ATCC 19119, CCUG 15528, CIP 78.42, CLIP 12510, DSM 20750, LMG 11388, NCTC 11846, SLCC 2379, SV5.

GenBank accession number (16S rRNA gene): X56151, X98528.

Additional remarks: There are two subspecies of *Listeria ivanovii*, *Listeria ivanovii* subsp. *ivanovii* and *Listeria ivanovii* subsp. *Iondoniensis* (Boerlin et al., 1992). Differences between these taxa are shown in Table 47.

5. **Listeria seeligeri** Rocourt and Grimont 1983, 869[VP]

see′li.ger.i. N.L. gen. n. *seeligeri* of Seeliger, named in honor of Heinz P.R. Seeliger, a German bacteriologist.

Characteristics and differentiation from other *Listeria* species are given in Table 40. Antigenic composition for serotyping is shown in Table 42.

Habitat: Widely distributed in nature, found in soil, mud, sewage, vegetation and in the feces of animals and man. Not pathogenic for animals or man.

DNA G+C content (mol%): 36 (T_m).

Type strain: Weis 1120, ATCC 35967, CCUG 15530, CIP 100100, DSM 20751, LMG 11386, NCAIM B.01873, NCTC 11856, SLCC 3594.

GenBank accession number (16S rRNA gene): X56148.

6. **Listeria welshimeri** Rocourt and Grimont 1983, 867[VP]

wel.shi′mer.i. N.L. gen. n. *welshimeri* of Welshimer, named in honor of Herbert J. Welshimer, an American bacteriologist.

Characteristics and differentiation from other *Listeria* species are given in Table 40. Antigenic composition for serotyping is shown in Table 42.

Habitat: Widely distributed in nature, found in soil, mud, sewage, vegetation and in the feces of animals and man. Not pathogenic for animals or man.

TABLE 47. Phenotypic characteristics distinguishing subspecies of *Listeria ivanovii* and *Listeria grayi*[a,b]

Characteristic	Species/subspecies	
	L. ivanovii subsp. *ivanovii*	*L. ivanovii* subsp. *londoniensis*
Production of acid from:		
N-Acetyl β-d-mannosamine[c]	−	+
Ribose	+	−
DNA–DNA hybridization	Group I	Group II
Multilocus enzyme electrophoresis cluster	Group I	Group II
Ribotype patterns using chromosomal DNA digested with *Eco*RI	EIV1 and EIV2	EIV3 and EIV4
Strains available[d]	NCTC 11846, ATCC, 19119, CIP 78.42	NCTC 12701, ATCC 49954, CIP 103466
	L. grayi subsp. *grayi*	*L. grayi* subsp. *murrayi*
Reduction of NO_2^- to NO_3^-	+	−
Acid production from:		
D-Lyxose	−	+
Methyl α-D-glucoside	+	−
L-Rhamnose	d	−
D-Turanose	−	+
Tween 80 esterase	+	−
Nonmurein composition of cell wall	Glucose, glucosamine	Rhamnose, glucose, glucosamine
Strains available	NCTC 10815, ATCC, 19120, CIP 105447	NCTC 10812, ATCC 25401, CIP 76.124

[a]Symbols: +, >85% positive; d, different strains give different reactions (16–84% positive); −, 0–15% positive; w, weak reaction; ND, not determined.

[b]Data from Boerlin et al. (1992), Fiedler et al. (1984), Kämpfer et al. (1991), Rocourt and Catimel (1985), and Rocourt et al. (1992).

[c]Fermentation after 18–24 h.

[d]NCTC = National Collection of Type Cultures, London, UK; ATCC = American Type Culture Collection, Manassas, VA, USA; CIP = Institut Pasteur Collection, Paris, France.

DNA G+C content (mol%): 36 (T_m).

Type strain: Welshimer V8, ATCC 35897, CCUG 15529, CIP 81.49, DSM 20650, LMG 11389, NCAIM B.01872, NCTC 11857, SLCC 5334.

GenBank accession number (16S rRNA gene): X56149, X98532.

Infrasubspecies divisions

The considerable recent interest in *Listeria monocytogenes* and in the epidemiology of listeriosis has resulted in the development of phenotypic and genotypic typing systems. Most of these methods were compared on a common set of cultures in a large multicentered study (Bille and Rocourt, 1996).

On the basis of agglutination reactions with absorbed rabbit antisera, *Listeria monocytogenes* can be subdivided into 13 serovars (Seeliger and Hone, 1979; Table 42). However, this system offers limited practical discrimination for epidemiological and ecological studies since the majority of strains causing disease (usually >90%) in both humans and other animals belong to serovars 1/2a, 1/2b, and 4b (Low and Donachie, 1997; McLauchlin, 1990; Seeliger and Hohne, 1979). Other phenotypic tests used for typing *Listeria monocytogenes* include: reactions with panels of lysogenic phage (lysotyping; Rocourt et al., 1985a; Loessner, 1991; McLauchlin et al., 1996); susceptibility to bacteriocins (monocins; Curtis and Mitchell, 1992); multilocus enzyme electrophoresis (Bibb et al., 1990; Piffaretti et al., 1989); and resistance to cadmium and arsenite (McLauchlin et al., 1997). Phage typing has had widespread use for epidemiological

typing, although interlaboratory comparability of results generated in different laboratories is problematic (McLauchlin et al., 1996).

The advent and widespread availability of molecular biological techniques has led to numerous molecular approaches to molecular typing, including: plasmid profiling (McLauchlin et al., 1997); restriction endonuclease digest of total chromosomal DNA (Nocera et al., 1993; Wesley and Ashton, 1991); probing Southern blots of chromosomal restriction endonuclease digests with probes for rRNA genes (ribotyping; Graves et al., 1991; Jacquet et al., 1992; Jaradt et al., 2002) or randomly cloned chromosomal DNA fragments (Ridley, 1995); separation of large DNA fragments by pulsed-field gel electrophoresis obtained by using low-frequency-cleavage restriction endonucleases (pulsed-field gel electrophoresis typing; Brosch et al., 1991); random amplification of polymorphic DNA (RAPD) analysis (Mazurier and Wernars, 1992); and amplified fragment length polymorphism (AFLP) analysis (Ripabelli et al., 2000). Some of these molecular methods allow a greater degree of interlaboratory comparability, and successful networks of laboratories have been established using standardized methods for pulsed-field gel electrophoresis with automatic pattern comparison of strain databases over the internet (Swaminathan et al., 2001). Results from methods based on random amplification of chromosomal sequences are still subject to laboratory variability. DNA sequenced based approaches are likely to become widely used in the future (Cai et al., 2002).

Genus II. **Brochothrix** Sneath and Jones 1976, 102[AL]

PETER H. A. SNEATH

Bro.cho.thr′ix. Gr. n. *brochos* a loop; Gr. n. *thrix* a *thread*; N.L. fem. n. *Brochothrix* loop(ed) thread.

Regular unbranched rods, usually 0.6–0.75 μm in diameter and 1–2 μm in length. **Occur singly, in short chains, or in long filament-like chains that fold into knotted masses.** In older cultures the rods may give rise to coccoid forms, which develop into rod forms when subcultured onto a suitable medium. **Capsules are not formed. Gram-positive,** but some cells (both rod and coccoid forms) lose the ability to retain the Gram stain. **No endospores are produced. Nonmotile. Aerobic and facultatively anaerobic.** After 24–48 h, colonies on nutrient agar are opaque, 0.75–1.00 mm in diameter, and convex with entire margin. In older cultures (>2 d) the edge of the colony often breaks up and the center may become raised to give a "fried-egg" appearance. **Nonpigmented.** Nonhemolytic. **Optimum temperature 20–25°C; growth occurs within the range 0–30°C; over 30°C growth rarely occurs. Catalase is produced. Fermentative metabolism of glucose results in the production of L(+)-lactic acid and some other products. Methyl-red-positive.** Nitrate is not reduced. Indole-negative. H_2S-negative. No growth at pH 3.9 or on acetate medium. Arginine not hydrolyzed. Acid but no gas is produced from a number of carbohydrates. Acetoin and acetate are the major end products of aerobic metabolism of glucose. **Usually Voges–Proskauer-positive. Exogenous citrate and urea not utilized. Enzymes of the tricarboxylic acid cycle are almost totally absent.** Organic growth factors are required. The cell wall contains a directly cross-linked peptidoglycan based upon *meso*-diaminopimelic acid (*meso*-DAP). Mycolic acids are not present. The long chain fatty acid compo-

sition is predominately of the straight chain saturated, iso- and anteiso- methyl-branched chain types. Menaquinones are the sole respiratory quinones.

DNA G+C content (mol%): 36–38 (T_m).

Type species: **Brochothrix thermosphacta** (McLean and Sulzbacher 1953) Sneath and Jones 1976, 103[AL] (*Microbacterium thermosphactum* McLean and Sulzbacher 1953, 432.).

Further descriptive information

Strains of *Brochothrix* grow at 4°C, but not at 37°C. Strains are positive for the following arylamidases (Talon et al., 1988): phenylalanine, histidine, glycyl-phenylalanine, seryl-tyrosine, glutamate, tryptophan, and histidyl-L-phenylalanine.

The following carbohydrates are fermented: glucose, ribose, fructose, mannose, *N*-acetylglucosamine, salicin, maltose, trehalose, and usually (>85%) glycerol, amygdalin, and cellobiose. Esculin is hydrolyzed.

Strains are sensitive to novobiocin, amikacin, tobramycin, gentamicin, ampicillin, and usually to tetracycline. Strains are resistant to oxacillin, nalidixic acid, and usually to colistin.

Enrichment and isolation procedures

Brochothrix thermosphacta is an important spoilage organism of meat and meat products stored aerobically or vacuum-packed at chill temperatures (Dainty and Hibbard, 1980; Gardner, 1981; Jones, 1992). Consequently, almost all isolation methods described for *Brochothrix thermosphacta* are concerned with its recovery from such sources.

After its first reputed isolation from pork trimmings and finished pork sausage (Sulzbacher and McLean, 1951), *Brochothrix thermosphacta* has been isolated on numerous occasions from the same sources and from a variety of animal and poultry meats or products based on them (see Gardner, 1981; Jones, 1992). Until fairly recently, there were few reports of the isolation of *Brochothrix thermosphacta* from other sources. McLean and Sulzbacher (1953) occasionally isolated it from equipment and tables used to prepare sausages, but assumed that its presence there was the result of contamination by pork trimmings since it could be isolated repeatedly from unopened barrels of such trimmings. Using a selective medium, Gardner (1966) isolated bacteria which he thought resembled *Brochothrix thermosphacta* from soil and feces. Collins-Thompson et al. (1971) suggested that some lipolytic Gram-positive bacteria isolated from dairy sources by Jayne-Williams and Skerman (1966) could be *Brochothrix thermosphacta*, but the published description of these organisms does not support this suggestion.

Brochothrix thermosphacta has been isolated from a wide variety of food products (especially meats) including fish (Gardner, 1981; Jones, 1992; Lannelongue et al., 1982; Nickelson et al., 1980). Both *Brochothrix thermosphacta* and *Brochothrix campestris* have been isolated from soil and grass, and the former has also been isolated from sheep wool and sheep feces (Talon et al., 1988). They are probably wide spread in the environment.

Enrichment is not usually performed for the isolation of *Brochothrix thermosphacta*. Wolin et al. (1957) placed irradiated sliced beef in sterile Petri dishes containing 3–5 ml sterile distilled water and incubated at 2°C in a slanting position until spoilage was evident. Samples plated out on a suitable nutrient agar yielded good growth of *Brochothrix thermosphacta*. There do not appear to be any other reports of enrichment.

Brochothrix thermosphacta is often the dominant organism in prepackaged meats and meat products stored at refrigerator temperature. The conditions prevailing during such storage selectively favor its growth. *Brochothrix thermosphacta* grows at 1–4°C and under conditions of low O_2 concentration (Gardner et al., 1967) The organism is mainly limited to the meat surface and grows at the meat/cling film interface (Ingram and Dainty, 1971). More recently, *Brochothrix* species have been shown to be dominant in salmon stored under Modified Atmosphere Packed (MAP) conditions (Rudi et al., 2004).

Swabs of various meat surfaces are directly plated onto suitable media, e.g., glycerol nutrient agar* (Gardner, 1966) or glucose nutrient agar† (Sulzbacher and McLean, 1951; Wolin et al., 1957). Alternatively, samples of macerated meat or other materials suspended in saline (0.85%, w/v) or peptone water (0.1%, w/v) are shaken vigorously and appropriate dilutions spread on to the same media (see above) to give well separated colonies. In addition to *Brochothrix thermosphacta*, such procedures can result in the recovery of a wide variety of other bacteria, e.g., micrococci, staphylococci, lactobacilli, *Kurthia* species, and pseudomonads.

Selective isolation of *Brochothrix thermosphacta* is achieved by the use of the selective medium (STAA)‡ of Gardner (1966). After incubation of appropriate material on the medium at 20–22°C for 2d, almost all the colonies are those of *Brochothrix thermosphacta*. A few pseudomonads may also grow; these may be detected by their positive oxidase reaction. Members of the genera *Lactobacillus*, *Listeria*, *Erysipelothrix*, *Streptococcus*, and *Bacillus* and those coryneform bacteria tested do not grow on this medium. *Brochothrix campestris* can also be isolated on STAA (Talon et al., 1988).

Maintenance procedures

Cultures may be preserved for short periods (months rather than years) in nutrient agar (plus 0.1% (w/v) glucose) stabs in screw-capped bottles. After overnight incubation at 25–30°C, the caps should be screwed tightly and the bottles stored at room or refrigerator temperature in the dark.

Longer-term preservation (over 7 years) may be achieved by freezing on glass beads at –70°C (Feltham et al., 1978). The organisms can also be preserved by lyophilization.

Differentiation of the genus *Brochothrix* from other genera

Identification of new isolates may be achieved by examination of cellular morphology, maximum growth temperature, oxygen requirements, catalase production, and tests such as production of acid from various carbohydrates. The chemical composition of the cell wall and the lipid composition also aid the identification of *Brochothrix* but it is not usually necessary to perform these analyses for routine identification.

Morphologically, *Brochothrix* may be confused with *Kurthia*, but the two genera can be distinguished by their different O_2 requirements. *Brochothrix* is facultatively anaerobic while *Kurthia* is strictly aerobic. In addition, *Brochothrix* strains produce acid from a wide variety of carbohydrates, whereas *Kurthia* is asaccharolytic. The presence of *meso*-diaminopimelic acid (*meso*-DAP) as the cell-wall diamino acid also distinguishes *Brochothrix* from *Kurthia* (Schleifer and Kandler, 1972). The latter contains lysine in the cell wall.

The cell-wall peptidoglycan composition of *Brochothrix* also serves to distinguish it from *Erysipelothrix* which, like *Kurthia*, contains lysine as the major cell-wall diamino acid. Other features which distinguish *Brochothrix* from *Erysipelothrix* are the inability of *Brochothrix* to grow at 35°C and the catalase reaction. If the test is carried out under the correct conditions, *Brochothrix* is always catalase-positive while *Erysipelothrix* is catalase-negative.

Strains of *Brochothrix* are most likely to be confused with some members of the genus *Lactobacillus* and with the genus *Listeria*. The distinctive morphology and inability to grow at 35°C, presence of catalase, cytochromes and menaquinones (Collins and Jones, 1981; Collins et al., 1979a; Davidson and Hartree, 1968), and inability to grow on acetate medium, together with low mol% G+C differentiate *Brochothrix* from those lactobacilli which possess a cell-wall peptidoglycan containing *meso*-DAP. The fatty acid composition of *Brochothrix* is also different from that of lactobacilli.(Shaw, 1974; Shaw and Stead, 1970)

Brochothrix shares many characters in common with *Listeria*. Both genera are facultative anaerobes, produce acid from carbo-

---

*Glycerol nutrient agar: peptone (Oxoid), 20 g; yeast extract (Oxoid), 2 g; glycerol, 15 g; K_2HPO_4, 1 g; $MgSO_4·7H_2O$, 1 g; agar (Oxoid No.3), 13 g; distilled water, 1000 ml; pH7.0. Autoclave 121°C for 15 min.

†Glucose nutrient agar: tryptone (Difco), 10 g; yeast extract (Difco), 5 g; K_2HPO_4, 5 g; NaCl, 5 g; agar (Difco), 15 g; distilled water, 1000 ml; pH7.0. Autoclave 121°C for 15 min.

---

‡STAA agar: to the molten glycerol nutrient agar (see above) add the following solutions in sterile distilled water: streptomycin sulfate (Glaxo) to a final concentration of 500 µg/ml, actidione (Upjohn) to 50 µg/ml, thallous acetate to 50 µg/ml. Mix well and dispense.

hydrates, contain catalase, possess *meso*-DAP as the cell-wall peptidoglycan, and have similar mol% G+C. They both fail to grow on acetate medium.

Colonies of *Brochothrix thermosphacta* do not show the blue-green coloration exhibited by *Listeria* when viewed by obliquely transmitted white light. *Brochothrix* is unable to grow at 35°C and is nonmotile unlike *Listeria*, and has a different morphology. No serological cross-reactions between *Brochothrix thermosphacta* and *Listeria* have been reported (Wilkinson and Jones, 1975).

Taxonomic comments

The genus *Brochothrix* is comprised of the bacteria isolated by Sulzbacher and McLean (1951) and named *Microbacterium thermosphactum* by McLean and Sulzbacher (1953) and *Brochothrix campestris* which was isolated by Talon et al. (1988). The genus

was tentatively placed in the family *Lactobacillaceae* by Sneath and Jones (1976). Strains of *Brochothrix thermosphacta* show over 85% DNA relatedness to one another, and strains of *Brochothrix campestris* show over 87% relatedness to one another; the two species show 15–40% relatedness (Talon et al., 1988).

Brochothrix and *Listeria* are closely related and form part of the super-cluster containing *Lactobacillus*, *Streptococcus*, and *Bacillus* (Collins et al., 1991; Morse et al., 1996; Stackebrandt and Woese, 1981; Talon et al., 1990). The closeness of *Brochothrix* to *Listeria* is supported by study of RNA polymerase (Morse et al., 1996).

Differentiation of species of the genus *Brochothrix*
There are two species of the genus *Brochothrix*. For the general characterization of the species, see the generic description. Additional characteristics are given under the species descriptions and the main differential characteristics of the two species are listed in Table 48.

List of species of the genus *Brochothrix*

1. **Brochothrix thermosphacta** (McLean and Sulzbacher 1953) Sneath and Jones 1976, 103[AL] (*Microbacterium thermosphactum* McLean and Sulzbacher 1953, 432.).

ther′mos.phac.ta. Gr. n. *therme* heat; Gr. adj. *sphactos* slain; N.L. fem. adj. *thermosphacta* killed by heat.

Gram-positive rods with long chains, loops, and knots; no capsules; nonhemolytic. Does not survive heating at 63°C for 5 min. Cytochromes are produced. Acetate is a major end product of aerobic metabolism of glucose. Enzymes of the tricarboxylic acid cycle are almost totally absent.

In addition to the features given in the generic description, acid is produced fermentatively from cellobiose, dextrin, dulcitol, galactose, gentiobiose, lactose, mannitol, raffinose, sucrose, tagatose, and xylose. Weak or delayed acid production from adonitol, inositol, and melibiose. Acid seldom produced from erythritol, methyl D-glucoside, rhamnose, sorbose, or starch. Variable for arabinose, arbutin, inulin, melezitose, and sorbitol. Milk is made slightly acid, otherwise unchanged. Oxidase-negative. Sodium hippurate and cellulose are not hydrolyzed. Gelatin is not liquefied; casein may be hydrolyzed. Gluconate is not oxidized. Phosphatase is produced but not sulfatase or lecithinase. Deoxyribonuclease and ribonuclease activities are absent or very weak. Tweens 20, 40, and 80 are not hydrolyzed; Tween 60 hydrolysis is variable. Cellulose, tyrosine, and xanthine are not hydrolyzed. Grows on and reduces 0.05% potassium tellurite. Slime often formed from sucrose. Often reduces 0.01% tetrazolium. Grows in the presence of 6.5% NaCl and usually in the presence of 8% NaCl. Grows in the presence of 0.1% sodium nitrite and 0.03% thallous acetate. Often resistant to furadoine. Sensitive to gentamicin, polymyxin B, erythromycin, novobiocin, and ampicillin; resistant to nalidixic acid and streptomycin.

The cell walls contain *meso*-DAP, glutamic acid, and alanine, but not arabinose or galactose. Mycolic acids are not present. Long chain fatty acids are predominantly of the straight chain saturated iso- and anteiso-methyl-branched chain types; the major ones are 12-methyltetradecanoic and 14-methylhexadecanoic acids. The major phospholipids are phosphatidylglycerol, diphosphatidylglycerol, and phosphatidylethanolamine. The glycolipid fraction contains acetylated glucose and small amounts of glycosyl diglyceride. Menaquinones are the sole respiratory quinones. The major menaquinone contains seven isoprene units (MK-7).

Two colony types are frequently present even in young (24 h) plate cultures (see Barlow and Kitchell, 1966). The colony types can appear so different that the culture may be thought to be contaminated. One type is convex, with entire edge and smooth surface; the other is flatter, with irregular edge and rough surface. On restreaking, individual colonies of both types give rise to colonies which exhibit both kinds of morphology.

Colony size is increased markedly if glucose is included in the medium; on such a medium *Brochothrix thermosphacta* has a sour odor due to the production of lactic acid, acetoin, acetic acid, and other fatty acids. Although it is facultatively anaerobic, better growth occurs in air.

There is general agreement that the temperature limits of growth are between 0°C and 30°C. However, limited growth has been noted at 35°C and one strain has been reported to grow at 37°C and at 45°C (see Gardner, 1981).

Brochothrix thermosphacta was originally classified in the genus *Microbacterium*, members of which are relatively heat-resistant; consequently, much attention has been paid to the heat resistance of this species (Gardner, 1981) because of its importance in food spoilage. All workers agree that the organism does not survive heating at 63°C for 5 min. However, Gardner (1981) has suggested that it is more useful to characterize the heat resistance by its D value (decimal reduction time), defined as the time (min) for a 10-fold reduction in the population under specified conditions.

The optimum pH for growth of *Brochothrix thermosphacta* is 7.0 but growth occurs from pH 5.0 to 9.0 (Brownlie, 1966). The ability of *Brochothrix thermosphacta* to grow in the presence of NaCl has been examined by many workers (Gardner, 1981).

TABLE 48. Differential characteristics of species of the genus *Brochothrix*

	B. campestris	*B. thermosphacta*
Growth with 8% NaCl after 2 d	−	+
Growth with and reduction of 0.05% potassium tellurite	−	+
Hippurate hydrolysis	+	−
Acid from rhamnose	+	−

All strains grow in the presence of 6.5% NaCl, but reports differ on growth at higher concentrations. Many strains tolerate 10% NaCl. Whether the differences are due to different strains or to different methodologies is not clear (Gardner, 1981).

Growth is inhibited by nitrite, but this is related to pH and temperature (Brownlie, 1966). Low pH and low temperatures increase the inhibition.

Collins-Thompson and Rodriguez-Lopez (1980) state that, unlike many lactic acid bacteria, *Brochothrix thermosphacta* does not appear to have a nitrite reductase system. It tolerates 500 ppm of SO_2 under both aerobic and anaerobic conditions (Dowdell and Board, 1971). Palmitic acid is inhibitory in liquid culture (Macaskie, 1982). Addition of nisin to meat products significantly inhibits its growth (Cutter and Siragusa, 1998). *Brochothrix thermosphacta* does not grow on the acetate medium of Rogosa et al. (1951) and only poorly on the MRS medium of De Man et al. (1960). These media were devised, respectively, for the selective isolation and for the cultivation of lactobacilli.

Sutherland et al. (1975) reported that *Brochothrix thermosphacta* attacked tributyrin but not beef fat. However, Patterson and Gibbs (1978) found that not one of their 95 isolates attacked tributyrin, an observation in agreement with the results of Davis et al. (1969). Collins-Thompson et al. (1971) demonstrated the presence of a glycerol ester hydrolase in *Brochothrix thermosphacta*. This lipase was active on tripropionin, tributyrin, tricaproin, tricaprylin and trilaurin but not tripalmitin. The temperature optimum of the lipase was 35–37°C with little or no activity below 20°C.

Brochothrix thermosphacta possesses a high level of enzymes associated with the catabolism of glucose (Collins-Thompson et al., 1972; Grau, 1983). Fermentative metabolism of glucose always results in the production of L(+)-lactic acid. Other end products appear to depend on the conditions of growth. McLean and Sulzbacher (1953) found only L(+)-lactic acid present in detectable quantities. Davidson et al. (1968) also reported L(+)-lactic acid to be the main end product of glucose fermentation, but small amounts of acetic and propionic acids were also detected. In glucose-limited continuous culture under anaerobic conditions, the end products of glucose metabolism have been identified as L(+)-lactic acid and ethanol (Hitchener et al., 1979). Grau (1983) reported that the organism ferments glucose anaerobically to L(+)-lactate, acetate, formate, and ethanol, and that the ratio of these products varies with the conditions of growth. Although McLean and Sulzbacher (1953) reported CO_2 production from fermentation of carbohydrates, this has not been confirmed in subsequent studies.

Aerobically the major end products of glucose metabolism are acetoin, and acetic, isobutyric, and isovaleric acids (Dainty and Hibbard, 1980). The relative proportions are also affected by growth conditions. Low glucose concentrations and neutral pH favor fatty acid production; high glucose and low pH favor acetoin production. Similar results were obtained with ribose and glycerol. Acetoin and probably acetic acid are derived from the carbohydrates, and isobutyric and isovaleric acids are derived from valine and leucine respectively (Dainty and Hibbard, 1980).

When *Brochothrix thermosphacta* is grown in a complex medium, enzymes of the tricarboxylic acid (TCA) cycle are almost totally absent (Collins-Thompson et al., 1972) However,

it has been suggested that in a defined, less complex medium, the TCA cycle enzymes may be sufficiently active to provide substrates for synthesis but not to provide energy (Grau, 1979).

Brochothrix thermosphacta requires cysteine, lipoate, nicotinate, pantothenate, *p*-aminobenzoate, biotin, and thiamine for aerobic growth in a glucose-mineral salts medium (Grau, 1979). The organism can also grow anaerobically in this medium. Macaskie et al. (1981) showed that most, but not all, of the yeast extract requirement of *Brochothrix thermosphacta* can be fulfilled by thiamine.

Brochothrix thermosphacta contains cytochromes and is unequivocally catalase-positive when cultured on a suitable medium and incubated at 20°C. Care must be taken when examining cultures for the presence of catalase. Production of the enzyme is dependent on both the growth medium and the temperature of incubation. Davidson et al. (1968) noted that *Brochothrix thermosphacta* strains grown on APT Medium (Difco) incubated at 20°C were always catalase-positive but that weak or negative reactions were obtained on HIA Medium (Difco) incubated at the same temperature. The same authors reported that negative results were obtained frequently if the bacteria were grown on either medium incubated at 30°C. This has been our experience, but we have found that BAB No. 2 (Difco) is a satisfactory alternative medium for APT (Difco). Davidson and Hartree (1968) showed that *Brochothrix thermosphacta* contained cytochromes aa_3b and noted the same effects of growth medium and temperature on the quantitative cytochrome content of the organism. No satisfactory explanation can be offered for the differences in cytochrome and catalase content. It does not appear to be due to a difference in the concentration of heme compounds in the different media. In APT Medium (Difco) – a medium which favors the formation of catalase and cytochromes – the concentration of heme compounds has been reported to be too low to be detected by the sensitive hemochromogen technique (Davidson and Hartree, 1968) but they noted that the APT Medium (Difco) does contain added iron (8.0 μg/ml). Gill et al. (1992) found that *Brochothrix thermosphacta* produced cytochromes of the *a*-, *b*- and *d*-types at 10–15°C. In high oxygen concentrations they were mostly of the *a*-type but at low oxygen tension these disappeared and were replaced by *d*-type cytochromes.

Little serological work has been carried out with *Brochothrix thermosphacta*. Wilkinson and Jones (1975) could demonstrate no serological relationships between *Brochothrix thermosphacta* and species of the genera *Listeria*, *Erysipelothrix*, and *Kurthia*.

Bacteriophages active on *Brochothrix thermosphacta* have been isolated from aqueous extracts of spoiled beef (Greer, 1983). Phage plaque size and plating efficiency were reported to be increased significantly when the incubation temperature was reduced from 25 to 1°C. Fourteen distinct phage lysotypes were detected. Greer (1983) suggested that phage typing may provide a rapid method of differentiating *Brochothrix thermosphacta* strains. None of the high titer lysates of any of the *Brochothrix thermosphacta* phages was capable of lysing *Corynebacterium flavescens*, *Microbacterium lacticum*, *Lactobacillus mali*, *Lactobacillus plantarum*, *Jonesia denitrificans*, or *Listeria grayi*.

All the *Brochothrix thermosphacta* strains typed by Greer (1983) appeared to form a homogeneous group on the basis of their other phenotypic characters. However, as indicated by phage typing, the species may not be as homogeneous as currently thought. An investigation of the esterases of a number of *Brochothrix thermosphacta* strains by gel electrophoresis indicated the presence of seven groups among the strains examined. There was no association between groups based on esterase patterns and source of isolation (Gardner, personal communication). It is possible that strains of *Brochothrix campestris* have been misidentified as *Brochothrix thermosphacta* (McCormick et al., 1998).

There is no evidence that *Brochothrix thermosphacta* is pathogenic. It is an economically important meat-spoilage organism because it grows in a wide variety of meats and meat products at low temperatures, and produces malodorous metabolic end products which make affected meat unpalatable. The main products that affect flavor are acetoin, diacetyl, and 3-methylbutanol (Dainty and Mackey, 1992). *Brochothrix thermosphacta* constitutes the major proportion of the microflora in packaged meats under vacuum-pack or aerobic high CO_2 concentration packing if the packaging film is of low permeability to oxygen. If the film is of high permeability, pseudomonads predominate, and *Listeria monocytogenes* may increase (Tsigarida et al., 2000). *Brochothrix thermosphacta* grows better at 5°C than 1°C; it is tolerant of carbon dioxide and can compete with lactic acid bacteria at low pH (Dainty and Mackey, 1992).

DNA G+C content (mol%): 34.6–36.2 (T_m).

Type strain: ATCC 11509, CCUG 35132, CIP 103251, DSM 20171, HAMBI 1439, NBRC 12167, LMG 17208, NCTC 10822.

GenBank accession number (16S rRNA gene): AY543023, M58798.

2. **Brochothrix campestris**   Talon, Grimont, Grimont, Gasser and Boeufgras 1988, 101.

cam.pes'tris. L. fem. adj. *campestris* from the fields.

Gram-positive rods, usually a mixture of long and short rods, found singly or in pairs. Colonies are circular, not pigmented, 0.7–1.0 mm in diameter after 48 h at 25°C.

In addition to the features given in the genus description, *Brochothrix campestris* shows the following characteristics: no growth with 8% or 10% NaCl or with 0.05% potassium tellurite; slime is not produced from sucrose; 0.01% tetrazolium is reduced; sodium hippurate is hydrolyzed; gelatin is not liquefied.

Acid but no gas is formed from arbutin and rhamnose and usually from gentobiose. Results are variable for inositol, mannitol, starch, sucrose, and tagatose. Sensitive to furadoine.

Brochothrix campestris produces a bacteriocin, brochocin-C (McCormick et al., 1998; Siragusa and Cutter, 1993). This is active on a wide range of Gram-positive bacteria, including strains of *Brochothrix thermosphacta* and species of *Carnobacterium, Kurthia, Enterococcus, Lactobacillus, Pediococcus,* and *Listeria,* but not on intact Gram-negative bacteria (although it is active on EDTA-treated cells of *Escherichia coli*). It is a class II bacteriocin which is cleaved to two peptides, BrcA and BrcB. The gene for a brochocin-C immunity protein, BrcI, is adjacent to the gene for brochocin-C. Brochocin-C may have promise as a food preservative.

DNA G+C content (mol%): 38 (T_m).

Type strain: S3, ATCC 43754, CIP 102920, DSM 4712, NBRC15547.

GenBank accession number (16S rRNA gene): AY543038, X56156.

References

Abachin, E., C. Poyart, E. Pellegrini, E. Milohanic, F. Fiedler, P. Berche and P. Trieu-Cuot. 2002. Formation of D-alanyl-lipoteichoic acid is required for adhesion and virulence of *Listeria monocytogenes.* Mol. Microbiol. *43:* 1–14.

Annous, B.I., L.A. Becker, D.O. Bayles, D.P. Labeda and B.J. Wilkinson. 1997. Critical role of anteiso-$C_{15:0}$ fatty acid in the growth of *Listeria monocytogenes* at low temperatures. Appl. Environ. Microbiol. *63:* 3887–3894.

Audurier, A., J. Rocourt and A.L. Courtieu. 1977. [Isolation and characterization of "*Listeria monocytogenes*" bacteriophages (author's transl.)]. Ann Microbiol (Paris) *128:* 185–198.

Autret, N., I. Dubail, P. Trieu-Cuot, P. Berche and A. Charbit. 2001. Identification of new genes involved in the virulence of *Listeria monocytogenes* by signature-tagged transposon mutagenesis. Infect. Immun. *69:* 2054–2065.

Baird, R.M., J.E.L. Corry and G.D.W. Curtis. 1987. Pharmocopoeia of culture media for food microbiology. Int. J. Food Microbiol. *9:* 89–128.

Bannerman, E., P. Boerlin and J. Bille. 1996. Typing of *Listeria monocytogenes* by monocin and phage receptors. Int. J. Food Microbiol. *31:* 245–262.

Bannister, B.A. 1987. *Listeria monocytogenes* meningitis associated with eating soft cheese. J. Infect. *15:* 165–168.

Barlow, J. and A.G. Kitchell. 1966. A note on the spoilage of prepacked lamb chops by *Microbacterium thermosphactum.* J. Appl. Bacteriol. *29:* 185–188.

Barrow, G.I. and R.K.A. Feltham. 1993. Cowan and Steel's Manual for the identification of medical bacteria, 3rd edn. Cambridge University Press.

Bayles, D.O. and B.J. Wilkinson. 2000. Osmoprotectants and cryoprotectants for *Listeria monocytogenes.* Lett. Appl. Microbiol. *30:* 23–27.

Becker, L.A., M.S. Cetin, R.W. Hutkins and A.K. Benson. 1998. Identification of the gene encoding the alternative sigma factor σB from *Listeria monocytogenes* and its role in osmotolerance. J. Bacteriol. *180:* 4547–4554.

Beumer, R.R., M.C. te Giffel and L.J. Cox. 1997. Optimization of haemolysis in enhanced haemolysis agar (EHA)—a selective medium for the isolation of *Listeria monocytogenes.* Lett. Appl. Microbiol. *24:* 421–425.

Bibb, W.F., B.G. Gellin, R. Weaver, B. Schwartz, B.D. Plikaytis, M.W. Reeves, R.W. Pinner and C.V. Broome. 1990. Analysis of clinical and food-borne isolates of *Listeria monocytogenes* in the United States by multilocus enzyme electrophoresis and application of the method to epidemiologic investigations. Appl. Environ. Microbiol. *56:* 2133–2141.

Bille, J., B. Catimel, E. Bannerman, C. Jacquet, M.N. Yersin, I. Caniaux, D. Monget and J. Rocourt. 1992. API Listeria, a new and promising one-day system to identify *Listeria* isolates. Appl. Environ. Microbiol. *58:* 1857–1860.

Bille, J. and J. Rocourt. 1996. WHO International Multicenter *Listeria monocytogenes* Subtyping Study–rationale and set-up of the study. Int. J. Food Microbiol. *32:* 251–262.

Bishop, D.K. and D.J. Hinrichs. 1987. Adoptive transfer of immunity to *Listeria monocytogenes:* the influence of in vitro stimulation on lymphocyte subset requirements. J. Immunol. *139:* 2005–2009.

Blanden, R.V., G.B. Mackaness and F.M. Collins. 1966. Mechanisms of acquired resistance in mouse typhoid. J. Exp. Med. *124:* 585–600.

Bockmann, R., C. Dickneite, B. Middendorf, W. Goebel and Z. Sokolovic. 1996. Specific binding of the *Listeria monocytogenes* transcriptional regulator PrfA to target sequences requires additional factor(s) and is influenced by iron. Mol. Microbiol. *22:* 643–653.

Boerlin, P., J. Rocourt and J.C. Piffaretti. 1991. Taxonomy of the genus *Listeria* by using multilocus enzyme electrophoresis. Int. J. Syst. Bacteriol. *41*: 59–64.

Boerlin, P., J. Rocourt, F. Grimont, P.A.D. Grimont, C. Jacquet and J.C. Piffaretti. 1992. *Listeria ivanovii* subsp. *londoniensis* subsp. nov. Int. J. Syst. Bacteriol. *42*: 69–73.

Brosch, R., C. Buchrieser and J. Rocourt. 1991. Subtyping of *Listeria monocytogenes* serovar 4b by use of low-frequency-cleavage restriction endonucleases and pulsed-field gel electrophoresis. Res. Microbiol. *142*: 667–675.

Brownlie, L.E. 1966. Effect of some environmental factors on psychrophilic microbacteria. J. Appl. Bacteriol. *29*: 447–454.

Bubert, A., S. Kohler and W. Goebel. 1992. The homologous and heterologous regions within the iap gene allow genus- and species-specific identification of *Listeria* spp. by polymerase chain reaction. Appl. Environ. Microbiol. *58*: 2625–2632.

Bubert, A., J. Riebe, N. Schnitzler, A. Schonberg, W. Goebel and P. Schubert. 1997. Isolation of catalase-negative *Listeria monocytogenes* strains from listeriosis patients and their rapid identification by anti-p60 antibodies and/or PCR. J. Clin. Microbiol. *35*: 179–183.

Buchrieser, C., C. Rusniok, F. Kunst, P. Cossart, P. Glaser and L. Consortium. 2003. Comparison of the genome sequences of *Listeria monocytogenes* and *Listeria innocua*: clues for evolution and pathogenicity. FEMS Immunol. Med. Microbiol. *35*: 207–213.

Cai, S., D.Y. Kabuki, A.Y. Kuaye, T.G. Cargioli, M.S. Chung, R. Nielsen and M. Wiedmann. 2002. Rational design of DNA sequence-based strategies for subtyping *Listeria monocytogenes*. J. Clin. Microbiol. *40*: 3319–3325.

Carroll, K.K., J.H. Cutts and E.G. Murray. 1968. The lipids of *Listeria monocytogenes*. Can. J. Biochem. *46*: 899–904.

Chakraborty, T., M. Leimeister-Wachter, E. Domann, M. Hartl, W. Goebel, T. Nichterlein and S. Notermans. 1992. Coordinate regulation of virulence genes in *Listeria monocytogenes* requires the product of the *prfA* gene. J. Bacteriol. *174*: 568–574.

Charpentier, E., G. Gerbaud and P. Courvalin. 1993. Characterization of a new class of tetracycline-resistance gene *tet(S)* in *Listeria monocytogenes* BM4210. Gene *131*: 27–34.

Charpentier, E., G. Gerbaud and P. Courvalin. 1994. Presence of the *Listeria* tetracycline resistance gene *tet(S)* in *Enterococcus faecalis*. Antimicrob. Agents Chemother. *38*: 2330–2335.

Charpentier, E., G. Gerbaud, C. Jacquet, J. Rocourt and P. Courvalin. 1995. Incidence of antibiotic resistance in *Listeria* species. J. Infect. Dis. *172*: 277–281.

Charpentier, E. and P. Courvalin. 1997. Emergence of the trimethoprim resistance gene dfrD in *Listeria monocytogenes* BM4293. Antimicrob. Agents Chemother. *41*: 1134–1136.

Charpentier, E. and P. Courvalin. 1999. Antibiotic resistance in *Listeria* spp. Antimicrob. Agents Chemother. *43*: 2103–2108.

Christie, R., N.E. Atkins and E. Munch-Peterson. 1944. A note on a lytic phenomenon shown by group B streptococci. Aust. J. Exp. Biol. Med. Sci *22*: 197–200.

Clark, A.G. and J. McLauchlin. 1997. Simple color tests based on an alanyl peptidase reaction which differentiate *Listeria monocytogenes* from other *Listeria* species. J. Clin. Microbiol. *35*: 2155–2156.

Collins-Thompson, D.L., T. Sørhaug, L.D. Witter and Z.J. Ordal. 1971. Glycerol ester hydrolase activity of *Microbacterium thermosphactum*. Appl. Microbiol. *21*: 9–12.

Collins-Thompson, D.L., T. Sørhaug, L.D. Witter and Z.J. Ordal. 1972. Taxonomic consideration of *Microbacterium lacticum*, *Microbacterium flavum* and *Microbacterium thermosphactum*. Int. J. Syst. Bacteriol. *22*: 65–72.

Collins-Thompson, D.L. and G. Rodriguez-Lopez. 1980. Influence of sodium nitrite, temperature, and lactic acid bacteria on the growth of *Brochothrix thermosphacta* under anaerobic conditions. Can. J. Microbiol. *26*: 1416–1421.

Collins, J.K. and D. Jones. 1981. The distribution of isoprenoid quinone structural types in bacteria and their taxonomic implications. Microbiol. Rev. *45*: 316–354.

Collins, M.D., M. Goodfellow and D.E. Minnikin. 1979a. Isoprenoid quinones in the classification of coryneform and related bacteria. J. Gen. Microbiol. *110*: 127–136.

Collins, M.D., D. Jones, M. Goodfellow and D.E. Minnikin. 1979b. Isoprenoid quinone composition as a guide to the classification of *Listeria*, *Brochothrix*, *Erysipelothrix* and *Caryophanon*. J. Gen. Microbiol. *111*: 453–457.

Collins, M.D., S. Wallbanks, D.J. Lane, J. Shah, R. Nietupski, J. Smida, M. Dorsch and E. Stackebrandt. 1991. Phylogenetic analysis of the genus *Listeria* based on reverse transcriptase sequencing of 16S rRNA. Int. J. Syst. Bacteriol *41*: 240–246.

Comi, G., L. Cocolin, C. Cantoni and M. Manzano. 1997. A RE-PCR method to distinguish *Listeria monocytogenes* serovars. FEMS Immunol. Med. Microbiol. *18*: 99–104.

Cotter, P.D., C.G. Gahan and C. Hill. 2000. Analysis of the role of the *Listeria monocytogenes* F_0F_1-ATPase operon in the acid tolerance response. Int. J. Food Microbiol. *60*: 137–146.

Cotter, P.D., C.G. Gahan and C. Hill. 2001. A glutamate decarboxylase system protects *Listeria monocytogenes* in gastric fluid. Mol. Microbiol. *40*: 465–475.

Curtis, G.D.W., R.G. Mitchell, A.F. King and E.J. Griffin. 1989a. A selective differential medium for the isolation of *Listeria monocytogenes*. Lett. Appl. Microbiol. *8*: 95–98.

Curtis, G.D.W., W.W. Nichols and T.J. Falla. 1989b. Selective agents for *Listeria* can inhibit their growth. Lett. Appl. Microbiol. *8*: 169–172.

Curtis, G.D.W. and R.G. Mitchell. 1992. Bacteriocin (monocin) interactions among *Listeria monocytogenes* strains. Int. J. Food Microbiol. *16*: 283–292.

Cutter, C.N. and G.R. Siragusa. 1998. Incorporation of nisin into a meat binding system to inhibit bacteria on beef surfaces. Lett. Appl. Microbiol. *27*: 19–23.

Dainty, R.H. and C.M. Hibbard. 1980. Aerobic metabolism of *Brochothrix thermosphacta* growing on meat surfaces and in laboratory media. J. Appl. Bacteriol. *48*: 387–396.

Dainty, R.H. and B.M. Mackey. 1992. The relationship between the phenotypic properties of bacteria from chill-stored meat and spoilage processes. J. Appl. Bacteriol. *73*: S103–S114.

Daneshvar, M.I., J.B. Brooks, G.B. Malcolm and L. Pine. 1989. Analyses of fermentation products of *Listeria* species by frequency-pulsed electron-capture gas-liquid chromatography. Can. J. Microbiol. *35*: 786–793.

Davidson, C.M. and E.F. Hartree. 1968. Cytochrome as a guide to classifying bacteria: taxonomy of *Microbacterium thermosphactum*. Nature *220*: 502–504.

Davidson, C.M., P. Mobbs and J.M. Stubbs. 1968. Some morphological and physiological properties of *Microbacterium thermosphactum*. J. Appl. Bacteriol. *31*: 551–559.

Davis, G.H. and K.G. Newton. 1969. Numerical taxonomy of some named coryneform bacteria. J. Gen. Microbiol. *56*: 195–214.

De Man, J.C., M. Rogosa and M.E. Sharpe. 1960. A medium for the cultivation of lactobacilli. J. Appl. Bacteriol. *23*: 130–135.

Dons, L., O.F. Rasmussen and J.E. Olsen. 1992. Cloning and characterization of a gene encoding flagellin of *Listeria monocytogenes*. Mol. Microbiol. *6*: 2919–2929.

Doucet-Populaire, F., P. Trieu-Cuot, I. Dosbaa, A. Andremont and P. Courvalin. 1991. Inducible transfer of conjugative transposon Tn1545 from *Enterococcus faecalis* to *Listeria monocytogenes* in the digestive tracts of gnotobiotic mice. Antimicrob. Agents Chemother. *35*: 185–187.

Doumith, M., C. Cazalet, N. Simoes, L. Frangeul, C. Jacquet, F. Kunst, P. Martin, P. Cossart, P. Glaser and C. Buchrieser. 2004. New aspects regarding evolution and virulence of *Listeria monocytogenes* revealed by comparative genomics and DNA arrays. Infect. Immun. *72*: 1072–1083.

Dowdell, M.J. and R.G. Board. 1971. The microbial associations in British fresh sausages. J. Appl. Bacteriol. *34*: 317–337.

Drebot, M., S. Neal, W. Schlech and K. Rozee. 1996. Differentiation of *Listeria* isolates by PCR amplicon profiling and sequence analysis of 16S-23S rRNA internal transcribed spacer loci. J. Appl. Bacteriol. *80*: 174–178.

Dussurget, O., D. Cabanes, P. Dehoux, M. Lecuit, C. Buchrieser, P. Glaser, P. Cossart and E.L.G. Consortium. 2002. *Listeria monocytogenes* bile salt hydrolase is a PrfA-regulated virulence factor involved in the intestinal and hepatic phases of listeriosis. Mol. Microbiol. *45*: 1095–1106.

Dykes, G.A., I. Geonara, M.A. Papathanasopolous and A. von Holy. 1994. Plasmid profiles of *Listeria* species associated with poultry processing. Food Microbiol. *11*: 519–523.

Engelbrecht, F., G. Dominguez-Bernal, J. Hess, C. Dickneite, L. Greiffenberg, R. Lampidis, D. Raffelsbauer, J.J. Daniels, J. Kreft, S.H. Kaufmann, J.A. Vazquez-Boland and W. Goebel. 1998. A novel PrfA-regulated chromosomal locus, which is specific for *Listeria ivanovii*, encodes two small, secreted internalins and contributes to virulence in mice. Mol. Microbiol. *30*: 405–417.

Ermolaeva, S., T. Karpova, S. Novella, M. Wagner, M. Scortti, I. Tartakovskii and J.A. Vazquez-Boland. 2003. A simple method for the differentiation of *Listeria monocytogenes* based on induction of lecithinase activity by charcoal. Int. J. Food Microbiol. *82*: 87–94.

Errebo Larsen, H. and H.P.R. Seeliger. 1966. Presented at the Proceedings of the 3rd International Symposium on Listeriosis, Biltoven The Netherlands.

Facinelli, B., M.C. Roberts, E. Giovanetti, C. Casolari, U. Fabio and P.E. Varaldo. 1993. Genetic basis of tetracycline resistance in food-borne isolates of *Listeria innocua*. Appl. Environ. Microbiol. *59*: 614–616.

Farber, J.M. and P.I. Peterkin. 1991. *Listeria monocytogenes*, a food-borne pathogen. Microbiol. Rev. *55*: 476–511.

Farber, J.M., J.I. Speirs, R. Pontefract and D.E. Conner. 1991. Characteristics of nonpathogenic strains of *Listeria monocytogenes*. Can. J. Microbiol. *37*: 647–650.

Feltham, R.K., A.K. Power, P.A. Pell and P.A. Sneath. 1978. A simple method for storage of bacteria at–76 degrees C. J. Appl. Bacteriol. *44*: 313–316.

Feresu, S.B. and D. Jones. 1988. Taxonomic studies on *Brochothrix, Erysipelothrix, Listeria* and atypical lactobacilli. J. Gen. Microbiol. *134*: 1165–1183.

Fiedler, F. and J. Seger. 1983. The murein types of *Listeria grayi, Listeria murrayi* and *Listeria denitrificans*. Syst. Appl. Microbiol. *4*: 444–450.

Fiedler, F., J. Seger, A. Shrettenbrunner and H.P.R. Seeliger. 1984. The biochemistry of murein and cell wall teichoic acids in the genus *Listeria*. Syst. Appl. Microbiol. *5*: 360–376.

Fiedler, F. and G.J. Ruhland. 1987. Structure of *Listeria monocytogenes* cell walls. Bull. Inst. Pasteur *85*: 287–300.

Fiedler, F. 1988. Biochemistry of the cell surface of *Listeria* strains: a locating general view. Infection *16 Suppl 2*: S92–97.

Fischer, W. and K. Leopold. 1999. Polar lipids of four *Listeria* species containing L-lysylcardiolipin, a novel lipid structure, and other unique phospholipids. Int. J. Syst. Bacteriol. *49*: 653–662.

Fisher, C.W. and S.E. Martin. 1999. Effects of iron and selenium on the production of catalase, superoxide dismutase, and listeriolysin O in *Listeria monocytogenes*. J. Food Prot. *62*: 1206–1209.

Fistrovici, E. and D.L. Collins-Thompson. 1990. Use of plasmid profiles and restriction endonuclease digest in environmental studies of *Listeria* spp. from raw milk. Int. J. Food Microbiol. *10*: 43–50.

Flamm, R.K., D.J. Hinrichs and M.F. Thomashow. 1984. Introduction of pAM beta 1 into *Listeria monocytogenes* by conjugation and homology between native *L. monocytogenes* plasmids. Infect. Immun. *44*: 157–161.

Fleming, D.W., S.L. Cochi, K.L. MacDonald, J. Brondum, P.S. Hayes, B.D. Plikaytis, M.B. Holmes, A. Audurier, C.V. Broome and A.L. Rein-

gold. 1985. Pasteurized milk as a vehicle of infection in an outbreak of listeriosis. N. Engl. J. Med. *312*: 404–407.

Flossmann, K.D. and W. Erler. 1972. Serologische, chemische und immunchemische untersuchungen an rothufbakterien. XI. Isolierung und charakterisierung von desoxyribonukleinsäuren aus rotlufbakterien. Arch. Exp. Vetinärmed. *26*: 817–824.

Fraser, K.R., D. Harvie, P.J. Coote and C.P. O'Byrne. 2000. Identification and characterization of an ATP binding cassette L-carnitine transporter in *Listeria monocytogenes*. Appl. Environ. Microbiol. *66*: 4696–4704.

Friedman, M.E. and W.G. Roessler. 1961. Growth of *Listeria monocytogenes* in defined media. J. Bacteriol. *82*: 528–533.

Friedman, M.E. and W.L. Alm. 1962. Effect of glucose concentration in the growth medium on some metabolic activities of *Listeria monocytogenes*. J. Bacteriol. *84*: 375–376.

Fujii, H., K. Kamisango, M. Nagaoka, K. Uchikawa, I. Sekikawa, K. Yamamoto and I. Azuma. 1985. Structural study on teichoic acids of *Listeria monocytogenes* types 4a and 4d. J. Biochem. (Tokyo) *97*: 883–891.

Galsworthy, S.B., S. Girdler and S.F. Koval. 1990. Chemotaxis in *Listeria monocytogenes*. ACTA Microbiol. Hung *37*.

Gardan, R., O. Duche, S. Leroy-Setrin and J. Labadie. 2003. Role of ctc from *Listeria monocytogenes* in osmotolerance. Appl. Environ. Microbiol. *69*: 154–161.

Gardner, G.A. 1966. A selective medium for the enumeration of *Microbacterium thermosphactum* in meat and meat products. J. Appl. Bacteriol. *29*: 455–460.

Gardner, G.A., A.W. Carson and J. Patton. 1967. Bacteriology of prepacked pork with reference to the gas composition within the pack. J. Appl. Bacteriol. *30*: 321–333.

Gardner, G.A. 1981. *Brochothrix thermosphacta (Microbacterium thermosphactum)* in the spoilage of meats: A review. *In* Roberts, Hobbs, Christian and Skovgaard (Editors), Psychotrophic Microorganisms in Spoilage and Pathogenicity. Academic Press, London, pp. pp. 139–173.

Ghosh, B.K. and R.G. Murray. 1967. Fine structure of *Listeria monocytogenes* in relation to protoplast formation. J. Bacteriol. *93*: 411–426.

Gil, A., R.G. Kroll and R.K. Poole. 1992. The cytochrome composition of the meat spoilage bacterium *Brochothrix thermosphacta*: identification of cytochrome *a3*-and *d*-type terminal oxidases under various conditions. Arch. Microbiol. *158*: 226–233.

Glaser, P., L. Frangeul, C. Buchrieser, C. Rusniok, A. Amend, F. Baquero, P. Berche, H. Bloecker, P. Brandt, T. Chakraborty, A. Charbit, F. Chetouani, E. Couve, A. de Daruvar, P. Dehoux, E. Domann, G. Dominguez-Bernal, E. Duchaud, L. Durant, O. Dussurget, K.D. Entian, H. Fsihi, F. Garcia-del Portillo, P. Garrido, L. Gautier, W. Goebel, N. Gomez-Lopez, T. Hain, J. Hauf, D. Jackson, L.M. Jones, U. Kaerst, J. Kreft, M. Kuhn, F. Kunst, G. Kurapkat, E. Madueno, A. Maitournam, J.M. Vicente, E. Ng, H. Nedjari, G. Nordsiek, S. Novella, B. de Pablos, J.C. Perez-Diaz, R. Purcell, B. Remmel, M. Rose, T. Schlueter, N. Simoes, A. Tierrez, J.A. Vazquez-Boland, H. Voss, J. Wehland and P. Cossart. 2001. Comparative genomics of *Listeria* species. Science *294*: 849–852.

Gormley, E., J. Mengaud and P. Cossart. 1989. Sequences homologous to the listeriolysin O gene region of *Listeria monocytogenes* are present in virulent and avirulent haemolytic species of the genus *Listeria*. Res. Microbiol. *140*: 631–643.

Graham, T.A., E.J. Golsteyn-Thomas, J.E. Thomas and V.P.J. Gannon. 1997. Inter- and intraspecies comparison of the 16S-23S rRNA operon intergenic spacer regions of six *Listeria* spp. Int. J. Syst. Bacteriol. *47*: 863–869.

Grau, F. 1979. Nutritional requirements of *Microbacterium thermosphactum*. Appl. Environ. Microbiol. *38*: 818–820.

Grau, F.H. 1983. End products of glucose fermentation by *Brochothrix thermosphacta*. Appl. Environ. Microbiol. *45*: 84–90.

Graves, L.M., B. Swaminathan, M.W. Reeves and J. Wenger. 1991. Ribosomal DNA fingerprinting of *Listeria monocytogenes* using a digoxigenin-labeled DNA probe. Eur. J. Epidemiol. *7*: 77–82.

Gray, M.L. 1957. A rapid method for the detection of colonies of *Listeria monocytogenes*. Zentralbl. Bakteriol. [Orig] *169*: 373–377.

Gray, M.L. and A.H. Killinger. 1966. *Listeria monocytogenes* and listeric infections. Bacteriol. Rev. *30*: 309–382.

Greer, G.G. 1983. Psychrotrophic *Brochothrix thermosphacta* bacteriophages isolated from beef. Appl. Environ. Microbiol. *46*: 245–251.

Grundling, A., L.S. Burrack, H.G.A. Bouwer and D.E. Higgins. 2004. *Listeria monocytogenes* regulates flagellar motility gene expression through MogR, a transcriptional repressor required for virulence. Proc. Natl. Acad. Sci. *101*: 12318–12323.

Gutekunst, K.A., B.P. Holloway and G.M. Carlone. 1992. DNA sequence heterogeneity in the gene encoding a 60-kilodalton extracellular protein of *Listeria monocytogenes* and other *Listeria* species. Can. J. Microbiol. *38*: 865–870.

Hadorn, K., H. Hächler, A. Schaffner and F.H. Kayser. 1993. Genetic characterization of plasmid-encoded multiple antibiotic resistance in a strain of *Listeria monocytogenes* causing endocarditis. Eur. J. Clin. Microbiol. Infect. Dis. *12*: 928–937.

Hain, T., C. Steinweg and T. Chakraborty. 2006. Comparative and functional genomics of *Listeria* spp. J. Biotechnol. *126*: 37–51.

Hartford, T. and P.H. Sneath. 1993. Optical DNA-DNA homology in the genus *Listeria*. Int. J. Syst. Bacteriol. *43*: 26–31.

Hayes, P.S., L.M. Graves, B. Swaminathan, G.W. Ajello, G.B. Malcolm, R.E. Weaver, R. Ransom, K. Deaver, B.P. Plikaytis, A. Schuchat, J.D. Wenger, R.W. Pinner, C.V. Broome and T.L.S. Group. 1992. Comparison of three selective enrichment methods for the isolation of *Listeria monocytogenes* from naturally contaminated foods. J. Food Prot. *55*: 952–959.

Hether, N.W., P.A. Campbell, L.A. Baker and L.L. Jackson. 1983. Chemical composition and biological functions of *Listeria monocytogenes* cell wall preparations. Infect. Immun. *39*: 1114–1121.

Hether, N.W. and L.L. Jackson. 1983. Lipoteichoic acid from *Listeria monocytogenes*. J. Bacteriol. *156*: 809–817.

Hill, C., P.D. Cotter, R.D. Sleator and C.G.M. Gahan. 2002. Bacterial stress response in *Listeria monocytogenes*: jumping the hurdles imposed by minimal processing. Int. Dairy J. *12*: 273–283.

Hitchener, B.J., A.F. Egan and P.J. Rogers. 1979. Energetics of *Microbacterium thermosphactum* in glucose-limited continuous culture. Appl. Environ. Microbiol. *37*: 1047–1052.

Ingram, M. and R.H. Dainty. 1971. Symposium on microbial changes in foods. Changes caused by microbes in spoilage of meats. J. Appl. Bacteriol. *34*: 21–39.

International Organization for Standardization, 1996. Microbiology of food and animal feeding stuffs: Horizontal method for the detection and enumeration of *Listeria monocytogenes*. Part 1 Detection Method, ISO. British Standards Institute, London, pp. 11209–11210.

International Organization for Standardization, 1998. Microbiology of food and animal feeding stuffs: Horizontal method for the detection and enumeration of *Listeria monocytogenes*. Part 2 Enumeration method, ISO. British Standards Institute, London, pp. 11290–11292.

Jacquet, C., J. Bille and J. Rocourt. 1992. Typing of *Listeria monocytogenes* by restriction polymorphism of the ribosomal ribonucleic acid gene region. Zentralbl. Bakteriol. *276*: 356–365.

Jaradat, Z.W., G.E. Schutze and A.K. Bhunia. 2002. Genetic homogeneity among *Listeria monocytogenes* strains from infected patients and meat products from two geographic locations determined by phenotyping, ribotyping and PCR analysis of virulence genes. Int. J. Food Microbiol. *76*: 1–10.

Jayne-Williams, D.J. and T.M. Skerman. 1966. Comparative studies on coryneform bacteria from milk and dairy sources. J. Appl. Bacteriol. *29*: 72–92.

Jones, C.E., G. Shama, P.W. Andrew, I.S. Roberts and D. Jones. 1995. Comparative study of the growth of *Listeria monocytogenes* in defined media and demonstration of growth in continuous culture. J. Appl Bacteriol. *78*: 66–70.

Jones, D. 1975. A numerical taxonomic study of coryneform and related bacteria. J. Gen. Microbiol. *87*: 52–96.

Jones, D., M.D. Collins, M. Goodfellow and D.E. Minnikin. 1979. Chemical studies in the classification of the genus *Listeria* and possibly related bacteria. *In* Ivanov (Editor), Problems of Listeriosis. National Agroindustriual Union, Centre for Scientific Studies, Sofia, Bulgaria.

Jones, D. and H.P.R. Seeliger. 1983. Designation of a new type strain for *Listeria monocytogenes*: request for an opinion. Int. J. Syst. Bacteriol. *33*: 429–429.

Jones, D. 1992. The genus *Brochothrix*. *In* Balows, Trüper, Dworkin, Harder and Schleifer (Editors), The Prokaryotes. A Handbook on the Biology of Bacteria: Ecophysiology, Isolation, Identification, Applications. Springer-Verlag, New York, pp. 1617–1628.

Jorgensen, F., P.J. Stephens and S. Knochel. 1995. The effect of osmotic shock and subsequent adaptation on the thermotolerance and cell morphology of *Listeria monocytogenes*. J. Appl. Bacteriol. *79*: 274–281.

Judicial-Commission. 1954. Opinion 12, conservation of *Listeria* Pirie 1940 as a generic name in bacteriology. Int. Bull. Bacteriol. Nomencl. Taxon. *4*: 150–151.

Julak, J., M. Ryska, I. Koruna and E. Mencikova. 1989. Cellular fatty acids and fatty aldehydes of *Listeria* and *Erysipelothrix*. Zentralbl. Bakteriol. *272*: 171–180.

Kalmokoff, M.L., S.K. Banerjee, T. Cyr, M.A. Hefford and T. Gleeson. 2001. Identification of a new plasmid-encoded sec-dependent bacteriocin produced by *Listeria innocua* 743. Appl. Environ. Microbiol. *67*: 4041–4047.

Kamisango, K., I. Saiki, Y. Tanio, H. Okumura, Y. Araki, I. Sekikawa, I. Azuma and Y. Yamamura. 1982. Structures and biological activities of peptidoglycans of *Listeria monocytogenes* and *Propionibacterium acnes*. J. Biochem (Tokyo) *92*: 23–33.

Kamisango, K., H. Fujii, H. Okumura, I. Saiki, Y. Araki, Y. Yamamura and I. Azuma. 1983. Structural and immunochemical studies of teichoic acid of *Listeria monocytogenes*. J. Biochem. (Tokyo) *93*: 1401–1409.

Kämpfer, P., S. Böttcher, W. Dott and H. Rüden. 1991. Physiological characterization and identification of *Listeria* species. Zentbl. Bakteriol. *275*: 423–435.

Kämpfer, P. 1992. Differentiation of *Corynebacterium* spp., *Listeria* spp., and related organisms by using fluorogenic substrates. J. Clin. Microbiol. *30*: 1067–1071.

Kathariou, S. and L. Pine. 1991. The type strain(s) of *Listeria monocytogenes*: a source of continuing difficulties. Int. J. Syst. Bacteriol. *41*: 328–330.

Kathariou, S., R. Kanenaka, R.D. Allen, A.K. Fok and C. Mizumoto. 1995. Repression of motility and flagellin production at 37 degrees C is stronger in *Listeria monocytogenes* than in the nonpathogenic species Listeria innocua. Can. J. Microbiol. *41*: 572–577.

Kaufmann, S.H.E. 1993. Immunity to intracellular bacteria. Annu. Rev. Immunol. *11*: 129–163.

Keddie, R.M. 1981a. The genus *Kurthia*. *In* Starr, Stolp, Trüper, Balows and Schlegel (Editors), The Prokaryotes. A Handbook on Habitats, Isolation, and Identification of Bacteria. Springer-Verlag, Berlin, pp. 1866–1869.

Keddie, R.M., and D. Jones. 1981b. The genus *Brochothrix* (formerly *Microbacterium thermosphactum*, McLean and Sulzbacher). *In* S. Starr, Trüper, Balows, and Schegel (Editors), A Handbook on Habitats, Isolation and Identification of Bacteria. Springer-Verlag, Berlin, pp. 1866–1869.

Ko, R., L.T. Smith and G.M. Smith. 1994. Glycine betaine confers enhanced osmotolerance and cryotolerance on *Listeria monocytogenes*. J. Bacteriol. *176*: 426–431.

Ko, R. and L.T. Smith. 1999. Identification of an ATP-driven, osmoregulated glycine betaine transport system in *Listeria monocytogenes*. Appl. Environ. Microbiol. *65*: 4040–4048.

Kolb, E. and H. Seidel. 1960. Ein beitrag zur kenntnis des stoffweschels von *Listeria monocytogenes* (typ 1) unter besonderer berucksichtigung der oxidation von kohlenhydraten und metaboliten des tricarbonsaurecyclus und deren beeinfussing durch hemmstoffe. Zentbl. Vetmed. *7*: 509–518.

Kolstad, J., L.M. Rørvik and P.E. Granum. 1990. Characterization of plasmids from *Listeria* sp. Int. J. Food. Microbiol. *12*: 123–132.

Kosaric, N. and K.K. Carroll. 1971. Phospholipids of *Listeria monocytogenes*. Biochim. Biophys. Acta *239*: 428–442.

Kuhn, M. and W. Goebel. 1989. Identification of an extracellular protein of *Listeria monocytogenes* possibly involved in intracellular uptake by mammalian cells. Infect. Immun. *57*: 55–61.

Lannelongue, M., M.O. Hanna, G. Finne, R.I. Nickelson and C. Vanderzant. 1982. Storage characteristics of finfish fillets (*Archosargus probatocephalus*) packaged in modified gas atmospheres containing carbon dioxide. J. Food Prot. *45*: 440–444.

Lebrun, M., A. Audurier and P. Cossart. 1994a. Plasmid borne cadmium resistance genes in *Listeria monocytogenes* are present on Tn*5422*, a novel transposon closely related to Tn*917*. J. Bacteriol. *176*: 3049–3061.

Lebrun, M., A. Audurier and P. Cossart. 1994b. Plasmid-borne cadmium resistance genes in *Listeria monocytogenes* are similar to cadA and cadC of *Staphylococcus aureus* and are induced by cadmium. J. Bacteriol. *176*: 3040–3048.

Lecuit, M., S. Dramsi, C. Gottardi, M. Fedor-Chaiken, B. Gumbiner and P. Cossart. 1999. A single amino acid in E-cadherin responsible for host specificity towards the human pathogen *Listeria monocytogenes*. EMBO J. *18*: 3956–3963.

Lecuit, M., S. Vandormael-Pournin, J. Lefort, M. Huerre, P. Gounon, C. Dupuy, C. Babinet and P. Cossart. 2001. A transgenic model for listeriosis: role of internalin in crossing the intestinal barrier. Science *292*: 1722–1725.

Leimeister-Wachter, M. and T. Chakraborty. 1989. Detection of listeriolysin, the thiol-dependent hemolysin in *Listeria monocytogenes*, *Listeria ivanovii*, and *Listeria seeligeri*. Infect. Immun. *57*: 2350–2357.

Leimeister-Wachter, M., C. Haffner, E. Domann, W. Goebel and T. Chakraborty. 1990. Identification of a gene that positively regulates expression of listeriolysin, the major virulence factor of *Listeria monocytogenes*. Proc. Natl. Acad. Sci. U. S. A. *87*: 8336–8340.

Loessner, M.J. 1991. Improved procedure for bacteriophage typing of *Listeria* strains and evaluation of new phages. Appl. Environ. Microbiol. *57*: 882–884.

Loessner, M.J., I.B. Krause, T. Henle and S. Scherer. 1994. Structural proteins and DNA characteristics of 14 *Listeria* typing bacteriophages. J. Gen. Virol. *75*: 701–710.

Loessner, M.J., R.B. Inman, P. Lauer and R. Calendar. 2000. Complete nucleotide sequence, molecular analysis and genome structure of bacteriophage A118 of *Listeria monocytogenes*: implications for phage evolution. Mol. Microbiol. *35*: 324–340.

Lovett, J., D.W. Francis and J.M. Hunt. 1987. *Listeria monocytogenes* in raw milk: Detection, incidence, and pathogenicity. J. Food Prot. *50*: 188–192.

Low, J.C. and W. Donachie. 1997. A review of *Listeria monocytogenes* and listeriosis. Vet. J. *153*: 9–29.

Macaskie, L.E., R.H. Dainty and P.J.F. Henderson. 1981. The role of thiamine as a factor for the growth of *Brochothrix thermosphacta*. J. Appl. Bacteriol. *50*: 267–274.

Macaskie, L.E. 1982. Inhibition of growth of *Brochothrix thermosphacta* by palmitic acid. J. Appl. Bacteriol. *52*: 339–343.

MacGowan, A.P. 1990. *Listeriosis*–the therapeutic options. J. Antimicrob. Chemother. *26*: 721–722.

Mackaness, G.B. 1969. The influence of immunologically committed lymphoid cells on macrophage activity in vivo. J. Exp. Med. *129*: 973–992.

Mainou-Fowler, T., A.P. MacGowan and R. Postlethwaite. 1988. Virulence of *Listeria* spp.: course of infection in resistant and susceptible mice. J. Med. Microbiol. *27*: 131–140.

Markova, N., L. Michailova, A. Vesselinova, V. Kussovski, T. Radoucheva, S. Nikolova and I. Paskaleva. 1997. Cell wall-deficient forms (L-forms) of *Listeria monocytogenes* in experimentally infected rats. Zentralbl. Bakteriol. *286*: 46–55.

Mazurier, S.I. and K. Wernars. 1992. Typing of Listeria strains by random amplification of polymorphic DNA. Res. Microbiol. *143*: 499–505.

Mazzotta, A.S. and T.J. Montville. 1997. Nisin induces changes in membrane fatty acid composition of *Listeria monocytogenes* nisin-resistant strains at 10 degrees C and 30 degrees C. J. Appl. Microbiol. *82*: 32–38.

McClain, D. and W.H. Lee. 1988. Development of 'USDA-FSIS' method for isolation of *Listeria monocytogenes* from raw meat and poultry. J. Assoc. Off. Anal. Chem. *71*: 660–664.

McCormick, J.K., A. Poon, M. Sailer, Y. Gao, K.L. Roy, L.M. McMullen, J.C. Vederas, M.E. Stiles and M.J. Van Belkum. 1998. Genetic characterization and heterologous expression of brochocin-C, an antibotulinal, two-peptide bacteriocin produced by *Brochothrix campestris* ATCC 43754. Appl. Environ. Microbiol. *64*: 4757–4766.

McLauchlin, J. 1990. Distribution of serovars of *Listeria monocytogenes* isolated from different categories of patients with listeriosis. Eur. J. Clin. Microbiol. Infect. Dis. *9*: 210–213.

McLauchlin, J. 1996. The relationship between *Listeria* and listeriosis. Food Con. *7*: 187–193.

McLauchlin, J., A. Audurier, A. Frommelt, P. Gerner-Smidt, C. Jacquet, M.J. Loessner, N. van der Mee-Marquet, J. Rocourt, S. Shah and D. Wilhelms. 1996. WHO study on subtyping *Listeria monocytogenes*: results of phage-typing. Int. J. Food Microbiol. *32*: 289–299.

McLauchlin, J. 1997. The identification of *Listeria* species. Int. J. Food Microbiol. *38*: 77–81.

McLauchlin, J., M.D. Hampton, S. Shah, E.J. Threlfall, A.A. Wieneke and G.D. Curtis. 1997. Subtyping of *Listeria monocytogenes* on the basis of plasmid profiles and arsenic and cadmium susceptibility. J. Appl. Microbiol. *83*: 381–388.

McLean, R.A. and W.L. Sulzbacher. 1953. Microbacterium thermosphactum, spec: nov.; a nonheat resistant bacterium from fresh pork sausage. J. Bacteriol. *65*: 428–433.

Miki, K. and G.B. Mackaness. 1964. The passive transfer of acquired resistance to *Listeria Monocytogenes*. J. Exp. Med. *120*: 93–103.

Miller, I.L. and S.J. Silverman. 1959. Glucose metabolism of *Listeria monocytogenes*. Bacteriol. Proc. *103*.

Milohanic, E., B. Pron, P. Berche and J.L. Gaillard. 2000. Identification of new loci involved in adhesion of *Listeria monocytogenes* to eukaryotic cells. Microbiology *146*: 731–739.

Monget, D. 1992. Procédé et milieu d'identification de bactéries du genre *Listeria* Europäisches Patentment.

Moore, W.E.C., E.P. Cato and L.V.H. Moore. 1985. Index of the bacterial and yeast nomenclatural changes published in the International Journal of Systematic Bacteriology since the 1980 Approved Lists of Bacterial Names (1 January 1980 to 1 January 1985). Int. J. Syst. Bacteriol. *35*: 382–407.

Morse, R., M.D. Collins, J.T. Balsdon, S. Wallbanks and P.T. Richardson. 1996. Nucleotide sequence of part of the *rpoC* gene encoding the beta' subunit of DNA-dependent RNA polymerase from some grampositive bacteria and comparative amino acid sequence analysis. Syst. Appl. Microbiol. *19*: 150–157.

Murray, E.G.D., R.A. Webb and M.B.R. Swann. 1926. A disease of rabbits characterised by a large mononuclear leucocytosis, caused by a hitherto undescribed bacillus *Bacterium monocytogenes* (n. sp.). J. Pathol. Bacteriol. *29*: 407–439.

Nelson, K.E., D.E. Fouts, E.F. Mongodin, J. Ravel, R.T. DeBoy, J.F. Kolonay, D.A. Rasko, S.V. Angiuoli, S.R. Gill, I.T. Paulsen, J. Peterson, O. White, W.C. Nelson, W. Nierman, M.J. Beanan, L.M. Brinkac, S.C. Daugherty, R.J. Dodson, A.S. Durkin, R. Madupu, D.H. Haft, J. Selengut, S. Van Aken, H. Khouri, N. Fedorova, H. Forberger, B. Tran, S. Kathariou, L.D. Wonderling, G.A. Uhlich, D.O. Bayles, J.B.

Luchansky and C.M. Fraser. 2004. Whole genome comparisons of serotype 4b and 1/2a strains of the food-borne pathogen *Listeria monocytogenes* reveal new insights into the core genome components of this species. Nucleic Acids Res. *32*: 2386–2395.

Nichols, D.S., K.A. Presser, J. Olley, T. Ross and T.A. McMeekin. 2002. Variation of branched-chain fatty acids marks the normal physiological range for growth in *Listeria monocytogenes*. Appl. Environ. Microbiol. *68*: 2809–2813.

Nickelson, R.I., G. Finne, M.O. Hanna and C. Vanderzant. 1980. Minced fish flesh from nontraditional Gulf of Mexico finfish species: bacteriology. J. Food Sci. *45*: 1321–1326.

Ninet, B., H. Traitler, J.M. Aeschlimann, I. Hormna, D. Hartman and J. Bille. 1992. Quantitative analysis of cellular fatty acids (CFAs) composition of the seven species of *Listeria*. Syst. Appl. Microbiol. *15*: 76–81.

Nocera, D., M. Altwegg, G. Martinetti Lucchini, E. Bannerman, F. Ischer, J. Rocourt and J. Bille. 1993. Characterization of *Listeria* strains from a foodborne listeriosis outbreak by rDNA gene restriction patterns compared to four other typing methods. Eur. J. Clin. Microbiol. Infect. Dis. *12*: 162–169.

Notermans, S.H., J. Dufrenne, M. Leimeister-Wachter, E. Domann and T. Chakraborty. 1991. Phosphatidylinositol-specific phospholipase C activity as a marker to distinguish between pathogenic and nonpathogenic *Listeria* species. Appl. Environ. Microbiol. *57*: 2666–2670.

Park, S.F. and R.G. Kroll. 1993. Expression of listeriolysin and phosphatidylinositol-specific phospholipase-C is repressed by the plant-derived molecule cellobiose in *Listeria monocytogenes*. Mol. Microbiol. *8*: 653–661.

Patterson, J.T. and P.l.A. Gibbs. 1978. Some microbiological considerations applying to the conditioning, aging and vacuum packing of lamb. J. Food Prot. *13*: 1–13.

Peel, M., W. Donachie and A. Shaw. 1988a. Physical and antigenic heterogeneity in the flagellins of *Listeria monocytogenes* and *L. ivanovii*. J. Gen. Microbiol. *134*: 2593–2598.

Peel, M., W. Donachie and A. Shaw. 1988b. Temperature-dependent expression of flagella of *Listeria monocytogenes* studied by electron microscopy, SDS-PAGE and western blotting. J. Gen. Microbiol. *134*: 2171–2178.

Perez-Diaz, J.C., M.F. Vicente and F. Baquero. 1982. Plasmids in *Listeria*. Plasmid *8*: 112–118.

Peterkin, P.I., M.A. Gardiner, N. Malik and E.S. Idziak. 1992. Plasmids in *Listeria monocytogenes* and other *Listeria* species. Can. J. Microbiol. *38*: 161–164.

Phan-Thanh, L. and T. Gormon. 1997. A chemically defined minimal medium for the optimal culture of *Listeria*. Int. J. Food Microbiol. *35*: 91–95.

Piffaretti, J.C., H. Kressebuch, M. Aeschbacher, J. Bille, E. Bannerman, J.M. Musser, R.K. Selander and J. Rocourt. 1989. Genetic characterization of clones of the bacterium *Listeria monocytogenes* causing epidemic disease. Proc. Natl. Acad. Sci. U. S. A. *86*: 3818–3822.

Pilgrim, S., A. Kolb-Maurer, I. Gentschev, W. Goebel and M. Kuhn. 2003. Deletion of the gene encoding p60 in *Listeria monocytogenes* leads to abnormal cell division and loss of actin-based motility. Infect. Immun. *71*: 3473–3484.

Pine, L., R.E. Weaver, G.M. Carlone, P.A. Pienta, J. Rocourt, W. Goebel, S. Kathariou, W.F. Bibb and G.B. Malcolm. 1987. *Listeria monocytogenes* ATCC 35152 and NCTC 7973 contain a nonhemolytic, nonvirulent variant. J. Clin. Microbiol. *25*: 2247–2251.

Pine, L., G.B. Malcolm, J.B. Brooks and M.I. Daneshvar. 1989. Physiological studies on the growth and utilization of sugars by *Listeria* species. Can. J. Microbiol. *35*: 245–254.

Pine, L., S. Kathariou, F. Quinn, V. George, J.D. Wenger and R.E. Weaver. 1991. Cytopathogenic effects in enterocytelike Caco-2 cells

differentiate virulent from avirulent *Listeria* strains. J. Clin. Microbiol. *29*: 990–996.

Pirie, J.H. 1940a. The genus *Listerella* Pirie. Science *91*: 383.

Pirie, J.H.H. 1927. A new disease of veld rodents, 'Tiger River Disease". Pub. S. Afr. Inst. Med. Res. *3*: 163–186.

Pirie, J.H.H. 1940b. *Listeria*: Change of name for a genus of bacteria. Nature *145*: 264.

Poyart-Salmeron, C., C. Carlier, P. Trieu-Cuot, A.L. Courtieu and P. Courvalin. 1990. Transferable plasmid-mediated antibiotic resistance in *Listeria monocytogenes*. Lancet *335*: 1422–1426.

Poyart-Salmeron, C., P. Trieu-Cuot, C. Carlier, A. MacGowan, J. McLauchlin and P. Courvalin. 1992. Genetic basis of tetracycline resistance in clinical isolates of *Listeria monocytogenes*. Antimicrob. Agents Chemother. *36*: 463–466.

Premaratne, R.J., W.J. Lin and E.A. Johnson. 1991. Development of an improved chemically defined minimal medium for *Listeria monocytogenes*. Appl. Environ. Microbiol. *57*: 3046–3048.

Prévot, A.R. 1961. *Listeria* (ed.), Traité de Systématique Bactérienne, vol. 2. Dunod, Paris, pp. 511–512.

Puttmann, M., N. Ade and H. Hof. 1993. Dependence of fatty acid composition of *Listeria* spp. on growth temperature. Res. Microbiol. *144*: 279–283.

Quentin, C., M.C. Thibaut, J. Horovitz and C. Bebear. 1990. Multiresistant strain of *Listeria monocytogenes* in septic abortion. Lancet *336*: 375.

Raffelsbauer, D., A. Bubert, F. Engelbrecht, J. Scheinpflug, A. Simm, J. Hess, S.H. Kaufmann and W. Goebel. 1998. The gene cluster inlC2DE of *Listeria monocytogenes* contains additional new internalin genes and is important for virulence in mice. Mol. Gen. Genet. *260*: 144–158.

Raines, L.J., C.W. Moss, D. Farshtchi and B. Pittman. 1968. Fatty acids of *Listeria monocytogenes*. J. Bacteriol. *96*: 2175–2177.

Rasmussen, O.F., T. Beck, J.E. Olsen, L. Dons and L. Rossen. 1991. *Listeria monocytogenes* isolates can be classified into two major types according to the sequence of the listeriolysin gene. Infect. Immun. *59*: 3945–3951.

Ridley, A.M. 1995. Evaluation of a restriction fragment length polymorphism typing method for *Listeria monocytogenes*. Res. Microbiol. *146*: 21–34.

Ripabelli, G., J. McLauchin and E.J. Threlfall. 2000. Amplified fragment length polymorphism (AFLP) analysis of *Listeria monocytogenes*. Syst. Appl. Microbiol. *23*: 132–136.

Ripio, M.T., G. Dominguez-Bernal, M. Suarez, K. Brehm, P. Berche and J.A. Vazquez-Boland. 1996. Transcriptional activation of virulence genes in wild-type strains of *Listeria monocytogenes* in response to a change in the extracellular medium composition. Res. Microbiol. *147*: 371–384.

Ripio, M.T., K. Brehm, M. Lara, M. Suarez and J.A. Vazquez-Boland. 1997. Glucose-1-phosphate utilization by *Listeria monocytogenes* is PrfA dependent and coordinately expressed with virulence factors. J. Bacteriol. *179*: 7174–7180.

Riviera, L., F. Dubini and M.G. Bellotti. 1993. *Listeria monocytogenes* infections: the organism, its pathogenicity and antimicrobial drugs susceptibility. New Microbiol. *16*: 189–203.

Roberts, M.C., B. Facinelli, E. Giovanetti and P.E. Varaldo. 1996. Transferable erythromycin resistance in *Listeria* spp. isolated from food. Appl. Environ. Microbiol. *62*: 269–270.

Robinson, K. 1968. The use of cell wall analysis and gel electrophoresis for the identification of coryneform bacteria *In* Gibbs and Shapton (Editors), Identification Methods for Microbiologists. Academic Press, Part B London, pp. 85–92.

Rocourt, J., F. Grimont, P.A.D. Grimont and H.P.R. Seeliger. 1982a. DNA relatedness among serovars of *Listeria monocytogenes* sensu lato. Curr. Microbiol. *7*: 383–388.

Rocourt, J., A. Schrettenbrunner and H.P. Seeliger. 1982b. Isolation of bacteriophages from *Listeria monocytogenes* Serovar 5 and

Listeria innocua. Zentralbl. Bakteriol. Mikrobiol. Hyg. [A] *251*: 505–511.

Rocourt, J. and P.A.D. Grimont. 1983. *Listeria welshimeri* sp. nov. and *Listeria seeligeri* sp. nov. Int. J. Syst. Bacteriol. *33*: 866–869.

Rocourt, J., A. Schrettenbrunner and H.P. Seeliger. 1983. Différenciation biochemique des groupes génomiques de *Listeria monocytogenes* (sensu lato). Ann. Microbiol. (Paris) *134A*: 65–71.

Rocourt, J., A. Audurier, A.L. Courtieu, J. Durst, S. Ortel, A. Schrettenbrunner and A.G. Taylor. 1985a. A multi-centre study on the phage typing of *Listeria monocytogenes.* Zentralbl. Bakteriol. Mikrobiol. Hyg. [A] *259*: 489–497.

Rocourt, J. and B. Catimel. 1985. [Biochemical characterization of species in the genus *Listeria*]. Zentralbl. Bakteriol. Mikrobiol. Hyg. [A] *260*: 221–231.

Rocourt, J., B. Catimel and A. Schrettenbrunner. 1985b. [Isolation of *Listeria seeligeri* and *L. welshimeri* bacteriophages. Lysotyping of *L. monocytogenes*, *L. ivanovii*, *L. innocua*, *L. seeligeri* and *L. welshimeri*]. Zentralbl. Bakteriol. Mikrobiol. Hyg. [A] *259*: 341–350.

Rocourt, J. 1986. Bactériophages et bactériocines du genre *Listeria.* Zentralbl. Bakteriol. Mikrobiol. Hyg. [A] *261*: 12–28.

Rocourt, J., M. Gilmore, W. Goebel and H.P.R. Seeliger. 1986. DNA relatedness among *Listeria monocytogenes* and *Listeria innocua* bacteriophages. Syst. Appl. Microbiol. *8*: 42–47.

Rocourt, J., U. Wehmeyer and E. Stackebrandt. 1987. Transfer of *Listeria dentrificans* to a new genus, *Jonesia* gen. nov., as *Jonesia denitrificans* comb. nov. Int. J. Syst. Bacteriol. *37*: 266–270.

Rocourt, J., P. Boerlin, F. Grimont, C. Jacquet and J.C. Piffaretti. 1992. Assignment of *Listeria grayi* and *Listeria murrayi* to a single species, *Listeria grayi*, with a revised description of *Listeria grayi*. Int. J. Syst. Bacteriol *42*: 171–174.

Rogosa, J., J.A. Mitchell and R.F. Wiseman. 1951. A selective medium for isolation and enumeration of oral and fecal lactobacilli. J. Bacteriol. *62*: 132–133.

Romick, T.L., H.P. Fleming and R.F. McFeeters. 1996. Aerobic and anaerobic metabolism of *Listeria monocytogenes* in defined glucose medium. Appl. Environ. Microbiol. *62*: 304–307.

Rowan, N.J., A.A.G. Candlish, A. Bubert, J.G. Anderson, K. Kramer and J. McLauchlin. 2000. Virulent rough filaments of *Listeria monocytogenes* from clinical and food samples secreting wild-type levels of cell-free p60 protein. J. Clin. Microbiol. *38*: 2643–2648.

Rudi, K., T. Maugesten, S.E. Hannevik and H. Nissen. 2004. Explorative multivariate analyses of 16S rRNA gene data from microbial communities in modified-atmosphere-packed salmon and coalfish. Appl. Environ. Microbiol. *70*: 5010–5018.

Ruhland, G.J. and F. Fiedler. 1987. Occurrence and biochemistry of lipoteichoic acids in the genus *Listeria*. Syst. Appl. Microbiol. *9*: 40–46.

Sallen, B., A. Rajoharison, S. Desvarenne, F. Quinn and C. Mabilat. 1996. Comparative analysis of 16S and 23S rRNA sequences of *Listeria* species. Int. J. Syst. Bacteriol. *46*: 669–674.

Sanchez-Campillo, M., S. Dramsi, J.M. Gomez-Gomez, E. Michel, P. Dehoux, P. Cossart, F. Baquero and J.C. Perez-Diaz. 1995. Modulation of DNA topology by *flaR*, a new gene from *Listeria monocytogenes*. Mol. Microbiol. *18*: 801–811.

Schleifer, K.H. and O. Kandler. 1972. Peptidoglycan types of bacterial cell walls and their taxonomic implications. Bacteriol. Rev. *36*: 407–477.

Schmid, M.W., E.Y. Ng, R. Lampidis, M. Emmerth, M. Walcher, J. Kreft, W. Goebel, M. Wagner and K.H. Schleifer. 2005. Evolutionary history of the genus *Listeria* and its virulence genes. Syst. Appl. Microbiol. *28*: 1–18.

Seeliger, H.P. 1981. [Nonpathogenic listeriae: *L. innocua* sp. n. (Seeliger et Schoofs, 1977) (author's transl.)]. Zentralbl. Bakteriol. Mikrobiol. Hyg. [A] *249*: 487–493.

Seeliger, H.P.R. 1961. Listeriosis. Karger, Basel.

Seeliger, H.P.R. and H.J. Welshimer. 1974. Genera of uncertain affiliation: Genus *Listeria*. *In* Buchanan and Gibbons (Editors), Bergey's Manual of Determinative Bacteriology, 8th edn. The Williams and Wilkins Co., Baltimore, pp. 593–596.

Seeliger, H.P.R. and K. Hohne. 1979. Serotyping of *Listeria monocytogenes* and related species. *In* Bergan and Norris (Editors), Methods in Microbiology, vol. 13. Academic Press, London, pp. 31–49.

Seeliger, H.P.R. and M. Schoofs. 1979. Serological analysis of nonhemolyzing *Listeria* strains belonging to a species different from *Listeria monocytogenes*. *In* Ivanov (Editor), Problems of Listeriosis. Proceedings of the VIIth International Symposium Varna 1977 National Agroindustrial Union Center for Scientific Information, Sofia, Bulgaria.

Seeliger, H.P.R. 1983. In Validation of the publication of new names and new combinations previously effectively published outside the IJSB. List no. 10. Int. J. Syst. Bacteriol. *33*: 438–440.

Seeliger, H.P.R., J. Rocourt, A. Schrettenbrunner, P.A.D. Grimont and D. Jones. 1984. *Listeria ivanovii* sp. nov. Int. J. Syst. Bacteriol. *34*: 336–337.

Seeliger, H.P.R. and D. Jones. 1986. Genus *Listeria*. *In* Sneath, Mair, Sharpe and Holt (Editors), Bergey's Manual of Systematic Bacteriology, vol. 2. The Williams and Wilkins Co., pp. 1235–1245.

Sharpe, M.E. 1981. The genus *Lactobacillus*. *In* Starr, Stolp, Trüper, Balows and Schlegel (Editors), The Prokaryotes. A Handbook on Habits, Isolation and Identification of Bacteria. Springer-Verlag, Berlin, pp. 1653–1679.

Shaw, N. and D. Stead. 1970. A study of the lipid composition of *Microbacterium thermosphactum* as a guide to its taxonomy. J. Appl. Bacteriol. *33*: 470–473.

Shaw, N. 1974. Lipid composition as a guide to the classification of bacteria. Adv. Appl. Microbiol. *17*: 63–108.

Sheikh-Zeinoddin, M., T.M. Perehinec, S.E. Hill and C.E. Rees. 2000. Maillard reaction causes suppression of virulence gene expression in *Listeria monocytogenes*. Int. J. Food Microbiol. *61*: 41–49.

Siddiqi, R. and M.A. Khan. 1982. Vitamin and nitrogen base requirements for *Listeria monocytogenes* and haemolysin production. Zentralbl. Bakteriol. Mikrobiol. Hyg [A] *253*: 225–235.

Siddiqi, R. and M.A. Khan. 1989. Amino acid requirement of six strains of *Listeria monocytogenes*. Zentralbl. Bakteriol. *271*: 146–152.

Siragusa, G.R., L.A. Elphingstone, P.L. Wiese, S.M. Haefner and M.G. Johnson. 1990. Petite colony formation by *Listeria monocytogenes* and *Listeria* species grown on esculin-containing agar. Can. J. Microbiol. *36*: 697–703.

Siragusa, G.R. and C.N. Cutter. 1993. Brochocin-C, a new bacteriocin produced by *Brochothrix campestris*. Appl. Environ. Microbiol. *59*: 2326–2328.

Skerman, V.B.D., V. McGowan and P.H.A. Sneath. 1980. Approved lists of bacterial names. Int. J. Syst. Bacteriol. *30*: 225–420.

Slade, P.J. and D.L. Collins-Thompson. 1990. *Listeria*, plasmids, antibiotic resistance, and food. Lancet *336*: 1004.

Sleator, R.D., C.G. Gahan, T. Abee and C. Hill. 1999. Identification and disruption of BetL, a secondary glycine betaine transport system linked to the salt tolerance of *Listeria monocytogenes* LO28. Appl. Environ. Microbiol. *65*: 2078–2083.

Sleator, R.D., J. Wouters, C.G. Gahan, T. Abee and C. Hill. 2001. Analysis of the role of OpuC, an osmolyte transport system, in salt tolerance and virulence potential of *Listeria monocytogenes*. Appl. Environ. Microbiol. *67*: 2692–2698.

Sneath, P.H.A. and D. Jones. 1976. *Brochothrix*, a new genus tentatively placed in family *Lactobacillaceae*. Int. J. Syst. Bacteriol. *26*: 102–104.

Sneath, P.H.A. and D. Jones. 1986. Genus Brochothrix. *In* Sneath, Mair, Sharp and Holt (Editors), Bergey's Manual of Systematic Bacteriology, vol. 2. The Williams and Wilkins Co., Baltimore, pp. 1249–1253.

Sokolovic, Z. and W. Goebel. 1989. Synthesis of listeriolysin in *Listeria monocytogenes* under heat shock conditions. Infect. Immun. *57*: 295–298.

Srivastava, K.K. and I.H. Siddique. 1973. Quantitative chemical composition of peptidoglycan of *Listeria monocytogenes*. Infect. Immun. *7*: 700–703.

Stackebrandt, E. and C.R. Woese. 1981. The evolution of prokaryotes. *In* Carlile, Collins and Moseley (Editors), Molecular and Cellular Aspects of Microbial Evolution. Cambridge University Press, Cambridge, pp. 1–31.

Stuart, S.E. and H.J. Welshime. 1973. Intrageneric Relatedness of *Listeria* Pirie. Int. J. Syst. Bacteriol. *23*: 8–14.

Stuart, S.E. and H.J. Welshimer. 1974. Taxonomic re-examination of *Listeria pirie* and transfer of *Listeria grayi* and *Listeria murrayi* to a new genus, Murraya. Int. J. Syst. Bacteriol. *24*: 177–185.

Sullivan, M.A., R.E. Yasbin and F.E. Young. 1984. New shuttle vectors for *Bacillus subtilis* and *Escherichia coli* which allow rapid detection of inserted fragments. Gene *29*: 21–26.

Sulzbacher, W.L. and R.A. McLean. 1951. The bacterial flora of fresh pork sausage. Food Technol. *5*: 7–8.

Sutherland, J.P., J.T. Patterson, P.A. Gibbs and J.G. Murrary. 1975. Some metabolic and biochemical characteristics of representative microbial isolates from vacuum-packaged beef. J. Appl. Bacteriol. *39*: 239–249.

Swaminathan, B., T.J. Barrett, S.B. Hunter and R.V. Tauxe. 2001. PulseNet: the molecular subtyping network for foodborne bacterial disease surveillance, United States. Emerg. Infect. Dis. *7*: 382–389.

Sword, C.P. 1966. Mechanisms of pathogenesis in *Listeria monocytogenes* infection. I. Influence of iron. J. Bacteriol. *92*: 536–542.

Tadayon, R.A. and K.K. Carroll. 1971. Effect of growth conditions on the fatty acid composition of *Listeria monocytogenes* and comparison with the fatty acids of *Erysipelothrix* and *Corynebacterium*. Lipids *6*: 820–825.

Talon, R., P.A.D. Grimont, F. Grimont, F. Gasser and J.M. Boeufgras. 1988. *Brochothrix campestris* sp. nov. Int. J. Syst. Bacteriol. *38*: 99–102.

Talon, R., M.-C. Champomier and M.-C. Montel. 1990. DNA-rRNA hybridization studies among *Brochothrix* spp. and some other Gram-positive bacteria. Int. J. Syst. Bacteriol. *40*: 464–466.

Thompson, D.E., J.T. Balsdon, J. Cai and M.D. Collins. 1992. Studies on the ribosomal RNA operons of *Listeria monocytogenes*. FEMS Microbiol. Lett. *75*: 219–224.

Threlfall, E.J., J.A. Skinner and J. McLauchlin. 1998. Antimicrobial resistance in *Listeria monocytogenes* from humans and food in the UK, 1967–96. Clin. Microbiol. Infect. *4*: 410–412.

Trivett, T.L. and E.A. Meyer. 1971. Citrate cycle and related metabolism of *Listeria monocytogenes*. J. Bacteriol *107*: 770–779.

Tsakris, A., A. Papa, J. Douboyas and A. Antoniadis. 1997. Neonatal meningitis due to multi-resistant *Listeria monocytogenes*. J. Antimicrob. Chemother. *39*: 553–554.

Tsigarida, E., P. Skandamis and G.J. Nychas. 2000. Behaviour of *Listeria monocytogenes* and autochthonous flora on meat stored under aerobic, vacuum and modified atmosphere packaging conditions with or without the presence of oregano essential oil at 5 degrees C. J. Appl. Microbiol. *89*: 901–909.

Uchikawa, K., I. Sekikawa and I. Azuma. 1986a. Structural studies on teichoic acids in cell walls of several serotypes of *Listeria monocytogenes*. J. Biochem. (Tokyo) *99*: 315–327.

Uchikawa, K., I. Sekikawa and I. Azuma. 1986b. Structural studies on lipoteichoic acids from four *Listeria* strains. J. Bacteriol. *168*: 115–122.

Ullmann, W.W. and J.A. Cameron. 1969. Immunochemistry of the cell walls of *Listeria monocytogenes*. J. Bacteriol. *98*: 486–493.

Van Langendonck, N., E. Bottreau, S. Bailly, M. Tabouret, J. Marly, P. Pardon and P. Velge. 1998. Tissue culture assays using Caco-2 cell line differentiate virulent from non-virulent *Listeria monocytogenes* strains. J. Appl. Microbiol. *85*: 337–346.

van Netten, P., I. Perales, A. van de Moosdijk, G.D. Curtis and D.A. Mossel. 1989. Liquid and solid selective differential media for the detection and enumeration of L. monocytogenes and other *Listeria* spp. Int. J. Food Microbiol. *8*: 299–316.

Vatanyoopaisarn, S., A. Nazli, C.E. Dodd, C.E. Rees and W.M. Waites. 2000. Effect of flagella on initial attachment of *Listeria monocytogenes* to stainless steel. Appl. Environ. Microbiol. *66*: 860–863.

Vazquez-Boland, J.A., M. Kuhn, P. Berche, T. Chakraborty, G. Dominguez-Bernal, W. Goebel, B. Gonzalez-Zorn, J. Wehland and J. Kreft. 2001. *Listeria* pathogenesis and molecular virulence determinants. Clin. Microbiol. Rev. *14*: 584–640.

Vicente, M.F., F. Baquero and J.C. Perezdiaz. 1985. Cloning and Expression of the *Listeria monocytogenes* hemolysin in *Escherichia coli*. FEMS Microbiol. Lett. *30*: 77–79.

Vicente, M.F., F. Baquero and J.C. Pérez-Díaz. 1988. Conjugative acquisition and expression of antibiotic resistance determinants in *Listeria* spp. J. Antimicrob. Chemother. *21*: 309–318.

Vines, A., M.W. Reeves, S. Hunter and B. Swaminathan. 1992. Restriction fragment length polymorphism in four virulence-associated genes of *Listeria monocytogenes*. Res. Microbiol. *143*: 281–294.

Vines, A. and B. Swaminathan. 1998. Identification and characterization of nucleotide sequence differences in three virulence-associated genes of *Listeria monocytogenes* strains representing clinically important serotypes. Curr. Microbiol. *36*: 309–318.

Vlaemynck, G., V. Lafarge and S. Scotter. 2000. Improvement of the detection of *Listeria monocytogenes* by the application of ALOA, a diagnostic, chromogenic isolation medium. J. Appl. Microbiol. *88*: 430–441.

Warburton, D.W., J.M. Farber, A. Armstrong, R. Caldeira, T. Hunt, S. Messier, R. Plante, N.P. Tiwari and J. Vinet. 1991. A comparative study of the 'FDA' and 'USDA' methods for the detection of *Listeria monocytogenes* in foods. Int. J. Food Microbiol. *13*: 105–117.

Wayne, L.G. 1986. Actions of the Judicial Commission of the International Committee on Systematic Bacteriology on requests for opinions published in 1983 and 1984. Int. J. Syst. Bacteriol. *36*: 357–358.

Welshimer, H.H. and A.L. Meredith. 1971. *Listeria murrayi* sp. nov.: A nitrate-reducing mannitol-fermenting *Listeria*. Int. J. Syst. Bacteriol. *21*: 3–7.

Wendlinger, G., M.J. Loessner and S. Scherer. 1996. Bacteriophage receptors on *Listeria monocytogenes* cells are the *N*-acetylglucosamine and rhamnose substituents of teichoic acids or the peptidoglycan itself. Microbiology *142*: 985–992.

Wesley, I.V. and F. Ashton. 1991. Restriction enzyme analysis of *Listeria monocytogenes* strains associated with food-borne epidemics. Appl. Environ. Microbiol. *57*: 969–975.

Wetzler, T.F., N.R. Freeman, M.L. French, L.A. Renkowski, W.C. Eveland and O.J. Carver. 1968. Biological characterization of *Listeria monocytogenes*. Health Lab. Sci. *5*: 46–62.

White, T.G. and F.K. Mirikitani. 1976. Some biological and physical chemical properties of *Erysipelothrix rhusiopathiae*. Cornell Vet. *66*: 152–163.

Wiedmann, M., J.L. Bruce, C. Keating, A.E. Johnson, P.L. McDonough and C.A. Batt. 1997. Ribotypes and virulence gene polymorphisms suggest three distinct *Listeria monocytogenes* lineages with differences in pathogenic potential. Infect. Immun. *65*: 2707–2716.

Wilkinson, B.J. and D. Jones. 1975. Some serological studies on *Listeria* and possibly related bacteria. *In* Woodbine (Editor), Problems of Listeriosis. Leicester University Press, Leicester, pp. 399–421.

Wilkinson, B.J. and D. Jones. 1977. A numerical taxonomic survey of *Listeria* and related bacteria. J. Gen. Microbiol. *98*: 399–421.

Wolin, E.F., J.B. Evans and C.F. Niven, Jr. 1957. The microbiology of fresh and irradiated beef. Food Res. *22*: 682–686.

Wuenscher, M.D., S. Kohler, A. Bubert, U. Gerike and W. Goebel. 1993. The iap gene of *Listeria monocytogenes* is essential for cell viability, and its gene product, p60, has bacteriolytic activity. J. Bacteriol. *175*: 3491–3501.

Family IV. **Paenibacillaceae** fam. nov.

PAUL DE VOS, WOLFGANG LUDWIG, KARL-HEINZ SCHLEIFER AND WILLIAM B. WHITMAN

Pae.ni.ba.cil.la'ce.ae. N.L. masc. n. *Paenibacillus* type genus of the family; suff. *-aceae* ending denoting family; N.L. fem. pl. n. *Paenibacillaceae* the *Paenibacillus* family.

The family *Paenibacillaceae* is circumscribed for this volume on the basis of phylogenetic analyses of the 16S rRNA sequences and includes the genus *Paenibacillus* and its close relatives. This family is distributed between two phylogenetic clusters. *Paenibacillus*, *Brevibacillus*, *Cohnella*, and *Thermobacillus* are monophyletic and represent the first group. The second clearly monophyletic group comprises the genera *Aneurinibacillus*, *Ammoniphilus*, and *Oxalophagus*. Although these two clusters are often associated together in several types of analyses, the evidence to unite them is not strong. However, in the absence of clear evidence for a separation, the second cluster is retained within the family. In contrast, *Thermicanus*, which was also classified within this family by Garrity et al. (2005), appears to represent a novel lineage of the *Bacilli*. In recognition of its ambiguous status, it was reclassified within Family X *Incertae Sedis*.

Cells are straight to curved rods, generally $0.5–1.0 \times 2–6\,\mu m$. While the cell-wall type is Gram-positive and contains *meso*-diaminopimilic acid, cells may stain Gram-negative, variable or positive. Oval or ellipsoidal endospores are formed, frequently swelling the sporangium. Motility with peritrichous flagellation is common, although some species are nonmotile. May be strictly aerobic, microaerophilic, facultative aerobic, or obligate anaerobic. May be catalase-positive or -negative. Organoheterotrophs, utilizing complex media, carbohydrates and amino acids. Some species utilize only oxalic acid as sole carbon and energy source. Both mesophilic and thermophilic, neutrophilic and alkaliphilic. Isolated from soil, roots, feces, blood, and other sources. Abundant fatty acids include $C_{15:0\ anteiso}$, $C_{15:0\ iso}$, $C_{16:0\ iso}$, and $C_{16:0}$. The major isoprenoid quinones are MK-7 or MK-6 (*Thermobacillus*).

DNA G+C content (mol%): 36–59.

Type genus: **Paenibacillus** Ash, Priest and Collins 1994, 852^{VP} (Effective publication: Ash, Priest and Collins 1993, 259) emend. Shida, Takagi, Kadowaki, Nakamura and Komagata 1997a, 297.

Genus I. **Paenibacillus** Ash, Priest and Collins 1994, 852^{VP} (Effective publication: Ash, Priest and Collins 1993, 259) emend. Shida, Takagi, Kadowaki, Nakamura and Komagata 1997a, 297

FERGUS G. PRIEST

Pae.ni.ba.cil'lus. L. adj. *paene* almost; bacterial name *Bacillus*; N.L. masc. n. *Paenibacillus* almost a bacillus.

Rod-shaped cells of Gram-positive structure, but usually stain variable or negative in the laboratory. Oval endospores are formed that distend the sporangium. Motile by means of peritrichous flagella. Facultatively anaerobic or strictly aerobic. Most species are catalase-positive. Colonies are generally smooth and translucent, light brown, white, or sometimes light pink in color. Optimum growth generally occurs at 28–40°C and pH 7.0, although strains of some species are alkaliphilic. Growth is inhibited by 10% NaCl. **Major fatty acid is $C_{15:0\ anteiso}$. Additional important fatty acid fractions often contain $C_{16:0}$, $C_{15:0\ iso}$, and $C_{17:0\ anteiso}$. Two forward PCR primers, PAEN515F (5′-GCTCG-GAGAGTGACGGGTACCTGAGA) and 843F (5′-TCGATAC-CCTTGGTGCCGAAGT) have been described, either of which can be used with an appropriate reverse primer, for example 1484R (TACCTTGTTACGACTTCACCCCA), for diagnostic amplification from the 16S rRNA gene.**

DNA G+C content (mol%): 40–59.

Type species: **Paenibacillus polymyxa** (Prazmowski 1880) Ash, Priest and Collins 1994, 852^{VP} (Effective publication: Ash, Priest and Collins 1993, 259) (*Clostridium polymyxa* Prazmowski 1880, 37; *Bacillus polymyxa* Macé 1889, 588).

Further descriptive information

Phylogenetic treatment. *Paenibacillus* emerged from the early phylogenetic dissection of *Bacillus sensu lato* based on 16S rRNA gene sequences (Ash et al., 1993). Originally referred to as RNA group 3, it soon became evident that this monophyletic group was distinct from other endospore-forming bacteria. Two conserved areas of the 16S rRNA gene have been exploited for identification of paenibacilli. A sequence from 843–862 (*Escherichia coli* numbering) was used to provide the original diagnostic oligonucleotide probe (5′-TCGATACCCTTGGT-GCCGAAGT) (Ash et al., 1993) and, although the genus has expanded to more than 80 species, this sequence remains highly specific. *Paenibacillus alginolyticus* and *Paenibacillus chondroitinus* show one base variation from the consensus (the A at position 5 is missing) and *Paenibacillus apiarius* contains a C at position 19; all other *Paenibacillus* species included in this chapter contain this sequence intact. A second oligonucleotide primer has been found useful for diagnostic purposes (Shida et al., 1997a). The variation in 16S rRNA gene sequences is greater in the target region of primer PAEN515F (5′-GCTCG-GAGAGTGACGGGTACCTGAGA) and it has not been tested with representatives of all *Paenibacillus* species. In particular, bases 6–10 (i.e., GAGAGTGA) are very unstable in the various species sequences. Nevertheless, in my experience, it has provided accurate and reproducible assignment of strains to the genus. Either of these primers can be used as a forward primer with a universal reverse primer for diagnostic PCR (Pettersson et al., 1999). An overall phylogenetic tree is given in Figure 24 based on 16S rRNA gene sequences. According to Goto et al. (2002), the HV region of the 16S rRNA gene is the best region for species discrimination, although only a limited number of species have been tested.

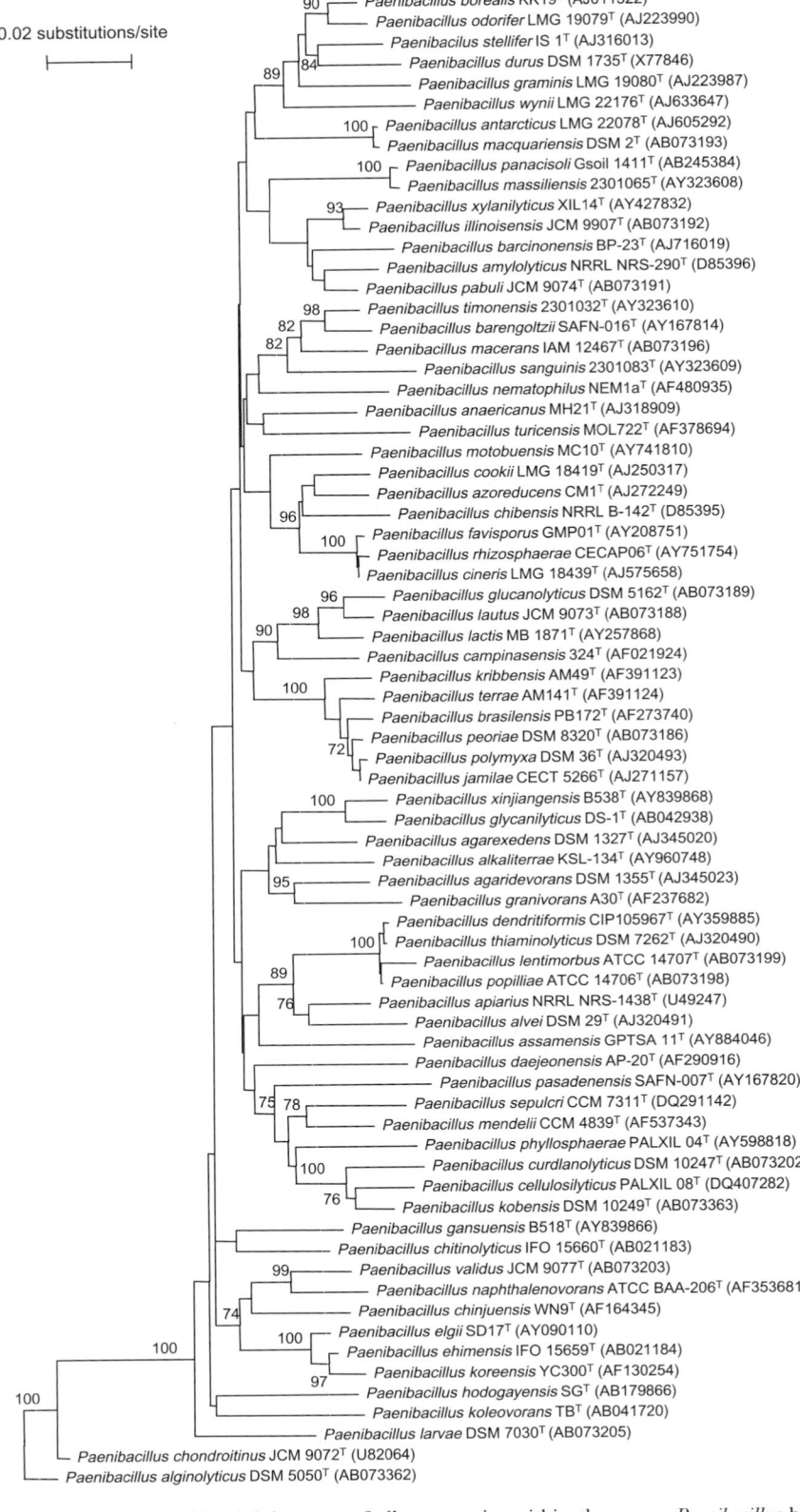

FIGURE 24. Phylogenetic neighbor-joining tree of all type strains within the genus *Paenibacillus* based on 16S rRNA gene sequences. Tree was constructed using CLUSTAL_X, BioEdit (1367 bp) and TREECON software. Bootstrap values are based on 1000 replications; values above 70% are shown at the branch nodes.

Cell structure and morphology. Paenibacilli are rod-shaped organisms generally measuring 2–5 μm in length and 0.5–0.8 μm in width. Although they have a Gram-positive wall structure, they almost invariably appear Gram-negative under the microscope, especially in older cultures. All species produce endospores that are usually of a greater diameter than the mother cell or sporangium and thus produce swelling of the sporangium (Figure 25). The spores of *Paenibacillus macerans* and *Paenibacillus polymyxa* have a heavily ridged surface and those of *Paenibacillus borealis* have a similar striped morphology (Elo et al., 2001). In transverse section, this can be confused with a spiked morphology. Most species are motile by peritrichous flagella, although in a few cases motility may be restricted to a minority of cells (e.g., *Paenibacillus popilliae*) or be absent (*Paenibacillus lentimorbus*). The cell wall peptidoglycan of those species that have been studied is invariably of the *meso*-diaminopimelic acid (DAP) type. Capsules are produced by some species under suitable growth conditions. *Paenibacillus polymyxa*, for example, synthesizes a levan capsule when grown with sucrose as carbon source. Some species produce extracellular polysaccharide (Aguilera et al., 2001; Yoon et al., 2002). S-layers are probably present in most species, although they have been reported in relatively few (e.g., *Paenibacillus alvei* and *Paenibacillus polymyxa*). The complete structure of the S-layer glycoprotein and its linkage to the peptidoglycan layer of the cell wall have been reported for *Paenibacillus alvei* (Schaffer et al., 2000).

The fatty acid composition of the paenibacilli is characteristic, containing predominantly $C_{15:0\ anteiso}$, which generally comprises around 55% total fatty acids, but ranges from 34% (*Paenibacillus naphthalenovorans*) to 80% (*Paenibacillus macquariensis*). $C_{17:0\ anteiso}$, $C_{16:0\ iso}$, and $C_{16:0}$ generally comprise the remainder of the fatty acids.

Colony characteristics. Paenibacilli usually produce small, translucent or light brown/white, sometimes pink or yellowish, colonies on agar plates. Colonies of pure cultures often show variations in opacity. This may be due to the degree of sporulation, the colonies becoming less translucent as they sporulate. Most paenibacilli are non-pigmented, the only exceptions are *Paenibacillus larvae* subsp. *pulvifaciens*, which may produce yellow-orange colonies, and the light pink colonies of *Paenibacillus chinjuensis*.

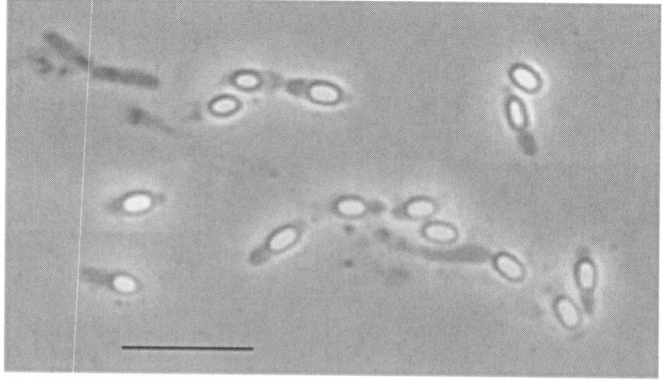

FIGURE 25. Phase-contrast micrograph of sporulating cells of *Paenibacillus polymyxa* DSM 36ᵀ. Bar = 5 μm.

Paenibacillus alvei is notable for its motile microcolonies that rapidly migrate over agar media, even on well-dried plates. Other species with motile colonies include some taxa previously included in *Bacillus circulans*, such as *Paenibacillus glucanolyticus*, and some recently described species including *Paenibacillus campinasensis* and *Paenibacillus curdlanolyticus*. Colony motility can readily be observed at a magnification of about ×50 under transmitted light.

Paenibacillus strains have also been noted for the formation of complex colonial patterns when grown under suboptimal conditions such as on low nutrient media (Cohen et al., 2000) or in the presence of antibiotics (Ben-Jacob et al., 2000). The term morphotype is used to describe the pattern-forming capacity of micro-organisms (Fujikawa and Matsushita, 1989). Different morphotypes of *Paenibacillus dendritiformis* produce both "tip-splitting" and "chiral" patterns (Figure 26) when grown under starvation conditions such as 0.2% peptone agar (Tcherpakov et al., 1999).

Nutrition and metabolism. Most paenibacilli grow on nutrient agar at neutral pH although inclusion of a fermentable carbon source (e.g., 0.5% glucose) will generally enhance growth. Tryptone soy agar (TSA) is a good alternative to nutrient agar. Exceptions include *Paenibacillus campinasensis* and *Paenibacillus daejeonensis*, which are alkaliphiles and do not grow below pH 7.5, and the fastidious insect pathogens *Paenibacillus lentimorbus* and *Paenibacillus popilliae*, which are best grown in J-broth: tryptone, 5.0 g; yeast extract, 15.0 g; K_2HPO_4, 3.0 g; glucose (sterilized separately), 2.0 g; and distilled water to 1000 ml. The pH should be adjusted to 7.3–7.5 before autoclaving. MYGGP broth (Dingman and Stahly, 1983) is an alternative and comprises: Mueller–Hinton broth, 10 g; yeast extract, 10.0 g; K_2HPO_4, 3.0 g; glucose (sterilized separately), 0.5 g; sodium pyruvate, 1.0 g; and distilled water to 1000 ml. The pH is adjusted to 7.1 before autoclaving. Some recently described species (e.g., *Paenibacillus cookii* and *Paenibacillus cineris*) need a lower pH for optimal growth. Media can be solidified by the addition of agar. *Paenibacillus larvae* will grow on nutrient agar, but is best grown on one of these more complex media. Most species grow optimally at around 30°C. However, *Paenibacillus macquariensis* is an exception to this being a psychrophile with a maximum growth temperature of 25°C and *Paenibacillus cineris* has a very wide temperature range for growth (0–50°C), probably due to dramatic changes in its natural environment (fumaroles).

Paenibacillus species do not form spores under all cultural conditions. Supplementation of nutrient media with a mixture of salts, for example $MnCl_2$ (50 μM), $CaCl_2$ (700 μM) and $MgCl_2$ (1 mM), is usually effective in enhancing sporulation.

Paenibacillus species are noted for their ability to hydrolyze a variety of carbohydrates. Carboxymethyl cellulose, chitin, chondroitin, curdlan (β-1,3-glucan), pustulan (β-1,6-glucan), β-1,4-glucan, pullulan (maltotriose units linked by α-1,6-glycosidic bonds), starch and xylan are variously hydrolyzed by different species. Indeed, *Paenibacillus glycanilyticus* was selectively isolated on the basis of its ability to hydrolyze the β-linked extracellular heteropolysaccharide from *Nostoc commune* (Dasman et al., 2002). *Paenibacillus naphthalenovorans* and *Paenibacillus validus* are atypical in their ability to degrade hydrocarbons (Daane et al., 2002), although few species have been tested in this respect so it may be more common than appreciated.

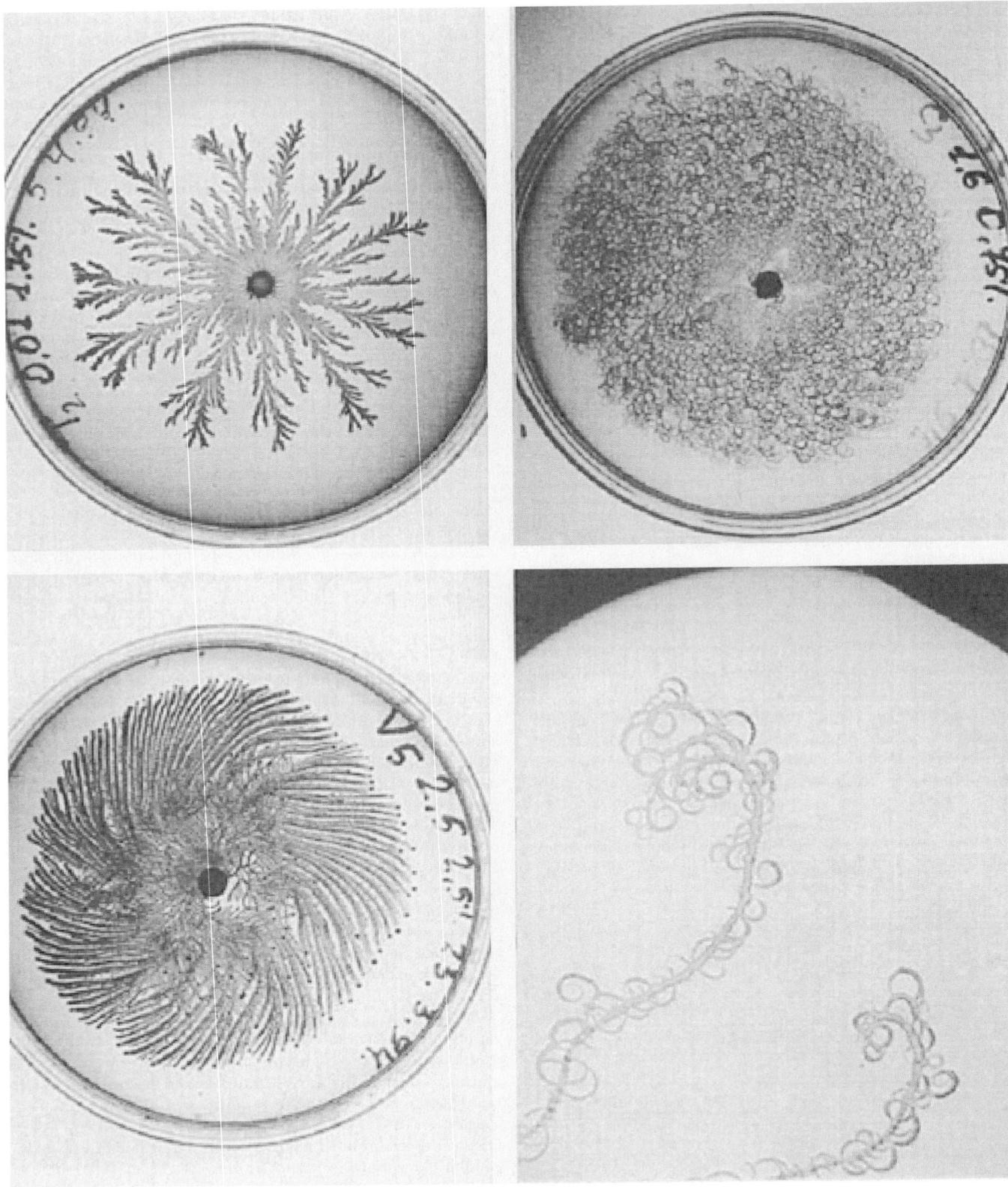

FIGURE 26. Pattern-forming colonial growth in *Paenibacillus* showing T (tip-splitting), C (chiral), and V (vortex) morphotypes. The fourth panel shows a close-up of the chiral tips. T and C are two variants of *Paenibacillus dendritiformis* that form in response to changes in the agar concentration. Vortex is an isolate of *Paenibacillus* that is closely related to *Paenibacillus lautus* (photographs courtesy of D. L. Gutnick, University of Tel Aviv).

These bacteria are predominantly facultative anaerobes but, as the genus has expanded, more strictly aerobic species have been introduced (Table 49). During growth of *Paenibacillus macerans* with glucose as carbon source, the Embden–Meyerhof pathway is used to generate pyruvate, which is converted initially to ethanol, acetic acid, and small amounts of formate. As the culture ages, the formate and acetate are catabolized to H_2, CO_2, and acetone (Weimer, 1984); hence the production of acid and gas from carbohydrates noted in the diagnostic tests (Table 49). *Paenibacillus polymyxa* is noted for its production of 2,3-butanediol during sugar catabolism, particularly at low pH (Marwoto et al., 2002; Raspoet et al., 1991). Like *Paenibacillus macerans*, it also produces hydrogen during sugar fermentation. The ability to use nitrate as an exogenous electron acceptor during anaerobic growth is a variable feature of these bacteria (Table 49).

The genus *Paenibacillus* currently includes 10 nitrogen-fixing species, although it needs to be mentioned that this character has not been tested systematically (Table 49). The *nifH* (dinitrogen reductase) gene has been detected in several of these species and is more closely related to the *nifH* of other aerobic prokaryotes than to the genes present in the anaerobic nitrogen-fixing genera such as *Clostridium* (Achouak et al., 1999). Nitrogen fixation does not seem to occur among other aerobic, endospore-forming genera.

Insect pathogenicity Members of the genus *Paenibacillus* have not been associated with human or mammalian pathogenicity, but some strains are important pathogens of insects. *Paenibacillus larvae* subsp. *larvae* causes foulbrood of honeybee (*Apis mellifera*) larvae (Chantawannakul and Dancer, 2001). The bacterium produces a fatal septicemia following ingestion of endospores. The spores germinate and grow in the larval midgut. Vegetative cells then traverse the epithelium and enter the hemocoel where they grow and sporulate in massive numbers. Infected individuals turn brown, then black, and the resultant mass becomes a hard scale of material deposited on the side of the honeycomb cell. The disease is so important that infected hives must be destroyed, generally by burning (Matheson and Reid, 1992). Methods for the recognition of *Paenibacillus larvae* subsp. *larvae* by PCR have been published (Alippi et al., 2002).

Paenibacillus larvae subsp. *pulvifaciens* has been associated with powdery scale of beehives. This material comprises the remnants of dead larvae, but it is still not clear whether the bacterium causes larval death. Indeed, it seems likely that the bacterium is non-pathogenic since reintroduction of the bacterium into healthy larvae does not induce the disease (Gilliam and Dunham, 1978). Recently both subspecies have been united again into the same species, *Paenibacillus larvae*, without subspecies discrimination based on a polyphasic study of additional strains (Genersch et al., 2006).

Paenibacillus lentimorbus and *Paenibacillus popilliae* cause milky disease in certain scarabaeid beetle larvae, notably the Japanese beetle, a common pest of lawn grass. The disease is named after the characteristic appearance of the normally translucent larvae, which become turbid due to the massive growth and sporulation of the bacteria in the hemolymph (Klein and Kaya, 1995). In brief, the larva eats the spore that germinates in the hindgut. Vegetative cells invade the hemolymph where they grow and sporulate, often reaching 10^{10} cells per ml. Death follows about 2 weeks after initial ingestion. Crystal proteins have been observed in sporulating cells of *Paenibacillus popilliae*, although their contribution to virulence has not been established. They have also been noted in a subgroup of *Paenibacillus lentimorbus* strains from South America that could be distinguished by DNA hybridization and molecular typing using random amplified polymorphic DNA (Harrison et al., 2000; Rippere et al., 1998). It is also noteworthy that crystal-forming strains have been isolated from insects other than Japanese beetle (*Popillia japonica*). Interestingly, the crystal protein of *Paenibacillus popilliae* shares some 40% sequence identity with the Cry2 polypeptides of *Bacillus thuringiensis* (Zhang et al., 1997). The two species are responsible for slightly different forms of milky disease. *Paenibacillus popilliae* causes type A milky disease and *Paenibacillus lentimorbus* has been associated with type B. The latter is characterized by the appearance of brown clots that block the circulation of hemolymph and lead to gangrenous conditions in the affected parts. The two species can be distinguished using molecular taxonomic techniques (Rippere et al., 1998).

Paenibacillus lentimorbus and *Paenibacillus popilliae* are often referred to as obligate pathogens of insects because it has proved impossible to determine *in vitro* conditions for sporulation and consequently growth in the host is necessary for the life cycle of the bacterium to be completed. This inability to culture spores *in vitro* has hampered the introduction of these bacteria for the biocontrol of Japanese beetle.

Ecology The normal habitat of the paenibacilli is the soil, particularly soils rich in humus and plant materials in which they presumably aid composting through the secretion of extracellular carbohydrases and other enzymes. Strains of several species, in particular the nitrogen-fixing species such as *Paenibacillus polymyxa*, have been associated with the rhizosphere of plants and important crop species. These bacteria enhance the growth of various plants by the production of phytohormones (Lebuhn et al., 1997; Timmusk et al., 1999) or by providing nutrients including nitrogen. They also produce antifungal compounds (Beatty and Jensen, 2002; Walker et al., 1998) and enzymes (Mavingui and Heulin, 1994) that help suppress fungal disease. *Paenibacillus polymyxa* is ubiquitous as a rhizosphere bacterium, particularly associated with grasses including crop plants such as wheat (Guemouri-Athmani et al., 2000). *Paenibacillus durus* (previously *Paenibacillus azotofixans*), on the other hand, has only been found in Brazilian and Hawaiian soils (Rosado et al., 1998). *Paenibacillus borealis* was isolated from humus in Scandinavian spruce forests, suggesting geographic localization of various species (Elo et al., 2001).

Enrichment and isolation procedures

There are no general methods for the isolation of members of the genus *Paenibacillus*. Standard procedures for selective isolation of endospores are helpful. Samples are generally heated above 70°C for 10 mins or longer to destroy vegetative cells and plated onto appropriate media for germination and outgrowth of endospores. Variations on this theme may enable the isolation of a more diverse range of bacteria. For example, exposure to 50% ethanol for 60 min will allow the survival of endospores while killing vegetative cells and may enable the recovery of spores that may be sensitive to heat. A method based on selective germination can be useful for *Paenibacillus* strains. A soil sample is heat-treated as described above, inoculated in broth for

TABLE 49. Phenotypic characteristics of *Paenibacillus* species[a]

Characteristic	1. P. polymyxa	2. P. agarexedens	3. P. agaridevorans	4. P. alginolyticus	5. P. alkaliterrae	6. P. alvei	7. P. amylolyticus	8. P. anaericanus	9. P. antarcticus	10. P. apiarius	11. P. assamensis	12. P. azoreducens	13. P. barcinonensis	14. P. barengoltzii	15. P. borealis	16. P. brasilensis	17. P. campinasensis	18. P. cellulosilyticus	19. P. chibensis
Spore shape	Oval	Oval	Oval	Oval	Oval	Oval	Oval	Oval	Oval	Oval	Oval	Oval	Oval	Oval	Oval	Oval	Oval	Oval	Oval
Swollen sporangium	+	-	+/-	+	+	+	+	+	+	+	+	+	+	+	+	+	+	+	+
Parasporal crystal	-	-	-	-	-	-	-	-	-	+	nd	-	+	-	-	-	+	-	-
Anaerobic growth	+	+	-	-	-	+	+	+	-	+	-	+	-	+	+	+	+	+	+
Catalase	+	+	+	+	+	+	+	+	+	+	+	+	+	-	+	+	+	+	+
Oxidase	-	-	+	+	+	+	-	+	+	-	+	-	-	+	-	nd	-	-	+
Nitrate reduction	+	-	-	-	+	-	+	-	-	+	-		-	+	-	+	nd	+	+
Voges–Proskauer	+	-	-	-	-	+	-	-	-	-	nd	-	-	+	+	+	nd	-	-
Hydrolysis of:																			
Casein	+	-	-	-	-	+	+(w)	nd	-	+	+	-	-	nd	+	+	+	-	-
Esculin	+	+	+	nd	-	+	nd	nd	+	nd	+	nd	nd	+	+	+	+	+	nd
Gelatin	+	-	-	nd	-	+	+	-	+	nd	+	+	+	-	-	D	+	+	-
Starch	+	+	-	+	+	D	+	+	-	+	+	+	-	nd	-	+	-	+	+
Urea	-	-	-	+	-	-	-	-	nd	+	-	-	-	nd	nd	nd	-	-	+
Degradation of:																			
Tyrosine	-	D	-	-	-	D	-	nd	-	+	-	-	-	-	nd	nd	nd	nd	-
Formation of:																				
Indole	-	-	-	-	-	+	-	-	nd	-	-	-	nd	-	nd	nd	nd	-	-
Dihydroxyacetone	+	-	-	nd	nd	+	-	-	-	-	nd	-	-	nd	-	-	-	nd	-
Utilization of:																				
Acetate	nd	nd	nd	+	-	nd	-	-	nd	nd	-	nd	nd	nd	-	nd	+	nd	-
Citrate	-	-	-	+	-	-	-	-	nd	+	nd	nd	-	nd	-	+	nd	-	-
Optimal growth temperature, °C	30	30	30	28–30	30	28	28–37	30–35	10–15	28	nd	37	30	37	28	30–32	40	28	37
Growth at:																				
50°C	-	-	-	-	-	-	-	-	-	-	-	+	-	+	-	-	-	-	+
Growth in NaCl:																				
2%	D	nd	nd	D	nd	+	+	+	+	+	+	nd	+	+	+	+	+	+	+
5%	-	-	-	-	nd	D	+	+	+	+	-	-	+	+	-	+	+	-	+
7%	-	-	-	-	-	-	+	-	-	-	-	nd	nd	-	-	-	+	-	-
Acid from:																				
L-Arabinose	++	-	-	+	nd	-	+	nd	+	-	nd	-	+	-	++	-	nd	+	+
D-Glucose	++	+	+	+	nd	+	+	+	+	+	nd	+	+	-	++	++	nd	+	+
Glycerol	++	-	-	nd	nd	+	+	nd	-	nd	nd	+	+	-	++	-	nd	nd	-
D-Mannitol	++	-	-	+	nd	-	+	nd	+	-	nd	+	+	-	++	++	nd	-	+
D-Xylose	++	-	-	+	nd	-	+	nd	-	+	nd	-	+	nd	++	-	nd	+	+
Nitrogen fixation	+	nd	nd	nd	nd	nd	nd	nd	nd	nd	nd	nd	nd	nd	+	+	nd	nd	nd
G+C (mol%)	43–46	47–49	50–52	47–49	49.4	45–47	46.3–46.6	42.6	40.7	52–54	41.2	47	45.0	nd	54	51	51	51	52.5–53.2

[a]Symbols: +, positive reaction for all or more than 90% of the strains tested. −, negative reaction for all or less than 10% of the strains tested; +/−, most strains positive for this character; ++, acid and gas production; D, various strains react differently for this character; D(+), although various strains react differently, the majority of them is positive for this test; D(w), strains react differently, but if positive, the reaction is weak; w/−, two results are reported in the literature, either weakly positive or negative; −(w), usually negative, but sometimes a weak positive reaction is reported; +(w), positive but weak reaction; nd, either not measured or not reported unambiguously in original descriptions.

TABLE 49. (continued)

Characteristic	20. P. chinjuensis	21. P. chitinolyticus	22. P. chondroitinus	23. P. cineris	24. P. cookii	25. P. curdlanolyticus	26. P. daejeonensis	27. P. dendritiformis	28. P. durus	29. P. ehimensis	30. P. elgii	31. P. favisporus	32. P. gansuensis	33. P. glucanolyticus	34. P. glycanilyticus	35. P. graminis	36. P. granivorans	37. P. hodogayensis	38. P. illinoisensis
Spore shape	Oval	Oval	Oval	Oval	Oval	Oval	Oval	Oval	Oval	Oval	Oval	Oval	Oval	Oval	Oval	Oval	Oval	nd	Oval
Swollen sporangium	+	+	+	+	+	+	+	+	+	+	+	+/–	+	+	+	+	+	–(w)	+
Parasporal crystal	+	–	–	–	–	–	–	–	–	–	–	–	–	–	–	–	–	–	–
Anaerobic growth	+	+	+	+	+	+	nd	+	+	+	+	+	–	+	+	+	+	+	+
Catalase	+	+	–	+	+	+	+	+	+	+	+	+	+	+	+	+	+	+	+
Oxidase	+	+	–	+	+	–	+	+	+	+	+	+	–	+	+	+	+	+	+
Nitrate reduction	–	+	–	+	+	+	+	–	+	+	+	–	–	+	+	+	+	–	–
Voges–Proskauer	nd	+	–	D(w)	+	–	nd	+	+	–	+	–	nd	+	–	nd	–	–	–
Hydrolysis of:																			
Casein	+	D	–	+(w)	+	–	nd	+	–	D	+	–	+	D(+)	–	nd	–	nd	+
Esculin	+	nd	nd	+	+	nd	+	nd	nd	nd	+	nd	+	+	+	+	nd	nd	nd
Gelatin	+	D	nd	–	D	nd	–	nd	–	+	nd	+	D	D(+)	–	nd	+	+	+
Starch	+	–	+	nd	nd	+	+	+	–	+	+	+	–	–	+	+	+	+	+
Urea	–	nd	+	–	–	+	–	+	nd	nd	nd	+	nd	nd	+	+	nd	–	–
Degradation of:																			
Tyrosine	–	nd	–	nd	nd	nd	nd	–	–	nd	nd	–	–	nd	–	nd	–	nd	–
Formation of:																			
Indole	nd	nd	–	–	–	–	nd	+	–	nd	+	–	nd	–	–	–	nd	nd	–
Dihydroxyacetone	nd	nd	nd	nd	nd	nd	nd	–	–	nd	nd	–	nd	nd	nd	nd	nd	nd	–
Utilization of:																			
Acetate	nd	nd	D	nd	nd	–	nd	+	nd	nd	nd	nd	nd	D	–	nd	–	nd	+
Citrate	nd	–	–	nd	nd	–	nd	–	–	–	–	+	nd	D(+)	–	nd	–	nd	–
Optimal growth temperature, °C	30–37	25–37	28–30			30	30	37	30–37	28–40	nd	37	35–40	30	28–37	28	37	30	37
Growth at:																			
50°C	–	–	–	+	+	–	nd	–	–	+	–	–	nd	–	–	–	–	–	+
Growth in NaCl:																			
2%	+	nd	+	+	+	+	nd	+	nd	nd	+	+	nd	+	+	nd	+	+	+
5%	–	nd	–	–	+	–	nd	+	–	nd	–	+(w)	nd	+	–	nd	–	–	–
7%	–	nd	–	–	–	–	nd	nd	–	nd	–	nd	nd	nd	nd	nd	–	–	–
Acid from:																			
L-Arabinose	–	–	+	+	+	+	–	+	+	+	–	nd	+	D	nd	++	+	–	+
D-Glucose	+	+	+	+	+	+	–	+	++	+	+	+	+	nd	++	++	+	+	+
Glycerol	–	nd	nd	+(w)	+	+	–	nd	–	nd	D	nd	–	D(+)	nd	++	+	+	+
D-Mannitol	–	–	+	+	+	–	+	–	++	+	+	+	+	D(+)	++	++	+	–	+
D-Xylose	–	–	+	+	nd	nd	–	nd	+	nd	D	+	+	+	nd	++	–	nd	+
Nitrogen fixation	nd	nd	nd	nd	nd	nd	nd	nd	+	nd	nd	nd	nd	nd	–	+	nd	nd	–
G+C (mol%)	53	51.3–52.8	47–48	51.5	51.6	50–52	53	55	48–53	52.9–54.9	51.7	53	50	48.1	50.5	52.1	47.8	55	47.6–48.3

(continued)

TABLE 49. (continued)

Characteristic	39. P. jamilae	40. P. kobensis	41. P. koleovorans	42. P. koreensis	43. P. kribbensis	44. P. lactis	45. P. larvae	46. P. lautus	47. P. lentimorbus	48. P. macerans	49. P. macquariensis	50. P. massiliensis	51. P. mendelii	52. P. motobuensis	53. P. naphthalenovorans	54. P. nematophilus	55. P. odorifer	56. P. pabuli
Spore shape	Oval	Oval	Oval	Oval	Oval	Oval	Oval	Oval	Oval	Oval	Oval	Oval	Oval	Oval	Oval	Oval	Oval	Oval
Swollen sporangium	+	+	+	+	+	+	+	+	+	+	+	+	+	+	+	+	+	+
Parasporal crystal	−	−	−	−	−	−	−	−	D	−	−	−	−	−	−	+(w)	−	−
Anaerobic growth	+	+	+	+	+	−	−	+	+	+	+	+	+	+	−	+	+	+
Catalase	+	+	−	+	+	+	+	+	+	+	+	+	+	+	nd	+	+	+
Oxidase	−	−	+	+	−	+	nd	−	−	nd	nd	−	−	+	D	−	+	−
Nitrate reduction	+	+	−	+	+	D	D	−	−	+	−	+	+	+	+	+	+	−
Voges–Proskauer	+	−	−	−	nd	−	−	−	−	−	−	−	−	nd	nd	−	nd	−
Hydrolysis of:																		
Casein	+	−	nd	+	+	−	+	−	nd	nd	−	nd	+	nd	−	+	nd	+
Esculin	+	nd	+	+	+	+	+	nd	nd	nd	nd	+	+	nd	nd	+	+	nd
Gelatin	+	nd	−	nd	+	−	+	nd	−	+	+	+	−	nd	−	+	nd	nd
Starch	+	+	−	+	+	+	−	+	−	+	+	+	−	nd	D	+	+	+
Urea	nd	+	−	−	−	−	nd	+	−	nd	−	−	−	nd	+	−	nd	−
Degradation of:																		
Tyrosine	nd	nd	nd	nd	nd	nd	−	−	nd	−	nd	nd	−	nd	nd	nd	nd	−
Formation of:																		
Indole	nd	−	−	−	nd	nd	−	−	−	−	−	−	−	nd	−	−	nd	−
Dihydroxyacetone	nd	nd	nd	nd	nd	nd	−	nd	−	−	−	nd	nd	nd	−	nd	nd	nd
Utilization of:																		
Acetate	nd	−	nd	+	+(w)	nd	nd	+	nd	nd	nd	nd	nd	nd	nd	nd	nd	+
Citrate	−	−	−	+		+	nd		nd	D	−	+	−	+	D	−	nd	
Optimal growth temperature, °C	30	30	30	38–40	30–37	30–40	28–30/35–37	28–30	28–30	30	20	30–37	25–30	37	30–37	30	28	28–30
Growth at:																		
50°C	−	−	−	+	−	+	−	−	−	D	−	+	+	+	nd	−	−	−
Growth in NaCl:																		
2%	+	+	nd	nd	+	nd	+	+	−	nd	+	+	nd	+	+	+	nd	+
5%	−	−	−	nd	−	nd	−	−	−	−	−	+	−	+(w)	−	−	nd	D
7%	−	−	−	−	−	nd	−	nd	−	−	−	nd	−	nd	−	−	nd	nd
Acid from:																		
L-Arabinose	+	+	−	+	+	+	+	+	−	+	+	+	+	nd	−	−	+	+
D-Glucose	+	+	−	+	+	+	−	+	−	+	+	+	+	nd	+	+	+	+
Glycerol	+	−	−	nd	+(w)	−	+	nd	nd	nd	nd	+	+	+	−	−	D	nd
D-Mannitol	+	−	−	+	+	+	−	+	−	++	+	+	nd	nd	+	−	+	+
D-Xylose	−	+	−	−	+	+	−	+	−	++	+	−	nd	nd	−	nd	+	+
Nitrogen fixation	nd	nd	nd	nd	nd	nd	nd	nd	nd	d	nd	nd	nd	nd	nd	nd	+	nd
G+C (mol%)	40.6–40.8	50–52	54–55.8	54	48	51.7	42–44	50–52	38	52–53	39	nd	50.8	47	49	44	44	48–50

(continued)

TABLE 49. (continued)

Characteristic	57. P. panacisoli	58. P. pasadenensis	59. P. peoriae	60. P. phyllosphaerae	61. P. popilliae	62. P. rhizosphaerae	63. P. sanguinis	64. P. sepulcri	65. P. stellifer	66. P. terrae	67. P. thiaminolyticus	68. P. timonensis	69. P. turicensis	70. P. validus	71. P. wynnii	72. P. xinjiangensis	73. P. xylanilyticus
Spore shape	Oval	Oval	Oval	Oval	Oval	Oval	Oval	Oval	Oval	Oval	Oval	Oval	Oval	Oval	Oval	Spherical	Oval
Swollen sporangium	+	+	+	+	+	+/−	+	+	+	+	+	+	+	+	+/−	+	+/−
Parasporal crystal	−	nd	−	−	+	−	−	−	−	−	−	−	−	−	−	−	−
Anaerobic growth	+	+	+	+	+	+	+	+	+	+	+	+	+	+	+	−	+
Catalase	+	+	+	+	−	+	+	+	+	+	+	+	+	+	+	+	+
Oxidase	+	+	−	+	−	+	+	+	+	+	+	+/−	−	+	+	−	+
Nitrate reduction	+	−	D(+)	+	−	+	−	−	−	+	+	+(w)	w/−	+	−	−	+
Voges-Proskauer	−	+	+	−	+	+	−	−	nd	nd	−	−	+	−	−	nd	+
Hydrolysis of:																	
Casein	+(w)	nd	+	−	−	−	nd	−	+	+	+	nd	nd	nd	−	−	−
Esculin	−	+	+	+	−	+	nd	+	+	+	nd	+	+	nd	+	+	nd
Gelatin	+	+	+	nd	nd	−	−	−	−	+	nd	+	−	nd	−	+	−
Starch	−	nd	+	+	−	−	nd	−	+	+	+	−	−	+	+	−	+
Urea	−	nd	−	−	nd	−	−	−	nd	−	−	−	+	−	−	−	+
Degradation of:																	
Tyrosine	nd	nd	−	nd	−	−	nd	−	−	−	+	nd	nd	+	+	nd	−
Formation of:																	
Indole	−	−	−	−	−	−	−	−	−	nd	+	−	−	−	nd	nd	−
Dihydroxyacetone	−	nd	−	nd	−	−	nd	nd	−	nd	−	nd	nd	nd	−	nd	−
Utilization of:																	
Acetate	−	nd	nd	nd	nd	nd	nd	nd	−	−	+	nd	nd	+	nd	nd	nd
Citrate	−	nd	−	−	−	−	−	−	−	−	+	−	−	−	−	−	−
Optimal growth temperature, °C	37	nd	28–30	28	28–30	28	30–37	25	28	30	28	30–37	37–42	28–30	20	30–35	37
Growth at:																	
50°C	−	nd	−	−	−	−	−	−	−	−	−	+	−	+	−	−	−
Growth in NaCl:																	
2%	+	+	+	+	+	+	+	+	+	+	+	+	+	+	+	+	+
5%	+	−	+	nd	−	+	+	−	+	+	+	−	+	−	−	−	+
7%	−	−	−	−	−	−	−	−	−	−	−	−	nd	−	−	−	−
Acid from:																	
L−Arabinose	+(w)	−	++	+	−	+	+	+	nd	+	D	+	+	+	D	−	+
D−Glucose	−	nd	++	+	+	+	+	+	nd	+	+	+	+	+	+	+	+
Glycerol	nd	nd	−	−	nd	nd	−	−	nd	+	nd	−	−	nd	D	+	+
D−Mannitol	+	nd	++	+	−	+	+	+	nd	+	D	−	−	+	+	−	−
D−Xylose	nd	nd	++	+	−	+	+	nd	nd	+	−	+	+	+	−	+	+
Nitrogen fixation	nd	nd	nd	nd	nd	nd	nd	nd	nd	nd	nd	nd	nd	nd	+	nd	nd
G+C (mol%)	53.9	nd	45–47	50.7	41	50.9	nd	50	55.6	47	52–54	nd	nd	50–52	44.6	47.0	50.5

a short period (e.g., 4 h) to allow spores to germinate and then heat-treated a second time before plating onto suitable media. Many common *Bacillus* species such as *Bacillus cereus*, *Bacillus licheniformis*, and *Bacillus subtilis* that germinate and grow vigorously can be eliminated in this way, whereas the *Paenibacillus* spores that germinate slowly survive the second pasteurization and form colonies on the plates.

In general, broths based on carbohydrate catabolism and incubated anaerobically for spore germination and outgrowth, followed by aerobic growth on plates of the same composition to eliminate anaerobic endospore-forming bacteria prove reasonably selective for *Paenibacillus* strains. For example, *Paenibacillus polymyxa* has been selectively isolated by incubating pasteurized soil samples in tubes of broth comprising peptone (10.0 g), lactose or starch (10.0 g), and distilled water to 1000 ml (pH 6.8–7) and containing a Durham tube. Culture from tubes showing gas formation is subcultured on to neutral red-peptone-starch-agar plates (prepared by adding agar and 0.005% neutral red to the broth described above). Colonies of *Paenibacillus polymyxa* accumulate the neutral red and are red-pigmented. Other carbohydrates for enrichment of specific species include alginate for *Paenibacillus alginolyticus* and *Paenibacillus chondroitinus* (Nakamura, 1987), colloidal chitin for *Paenibacillus koreensis* (Chung et al., 2000), curdlan for *Paenibacillus curdlanolyticus* and *Paenibacillus kobensis* (Kanzawa et al., 1995), and a polysaccharide derived from *Nostoc commune* for *Paenibacillus glycanilyticus* (Dasman et al., 2002). Ingenuity in the selection of carbohydrates will lead to the isolation of a great diversity of species. Enrichments with hydrocarbons have been used to isolate *Paenibacillus naphthalenovorans* (Daane et al., 2002). Nitrogen-free media are used for isolation of the nitrogen-fixing taxa (Elo et al., 2001).

Maintenance procedures

Paenibacillus species are best stored as spore preparations. Most strains produce spores readily when grown on nutrient medium supplemented with trace elements. A typical supply of trace elements includes the salts described above. If this fails, soil extract can be added to nutrient media. This involves crushing and sieving 400 g air-dried soil, which is then autoclaved in 1000 ml water for 60 min at 121°C. After the particles have settled out, the liquid is decanted and sterilized again for 30 min. This soil extract is added at 10% (by vol.) to nutrient agar or other media. Assuming that good crops of endospores can be obtained, the mixture of spores and cells can be scraped from the plate, resuspended as a turbid suspension in 20% (by vol.) sterile glycerol and frozen at –20°C. This suspension can be thawed and refrozen several times for use as a working culture. Stock cultures can be simply prepared by drying spores on a small piece of sterile filter paper and keeping it wrapped in sterile foil at room temperature. Pieces can be cut from the filter paper aseptically and immersed in broth to produce working cultures. It is sometimes preferable to store more refined spore preparations. These can be prepared by treating crude spore and cell suspensions with sterile lysozyme (100 μg/ml) for 30–60 min followed by repeated washes by centrifugation in sterile distilled water. Such spore preparations can be stored at 4°C in sterile water for many years with no appreciable loss of viability. For bacteria that do not readily differentiate into endospores, freeze-drying or storage at –80°C in 20% glycerol are alternatives.

Differentiation of the genus *Paenibacillus* from other genera

There are no diagnostic phenotypic features for the genus *Paenibacillus* although the presence of an endospore that distinctly swells the sporangium (Figure 25) is indicative. The only definitive identification to genus level is by means of PCR using the primers PAEN515F or 843F (see above) that, with a suitable reverse primer, amplify fragments from *Paenibacillus* strains exclusively, although this has not been tested thoroughly for all new species. Procedures for PCR amplification have been reported (Pettersson et al., 1999).

Taxonomic comments

The genus *Paenibacillus* was created in 1993 to accommodate a monophyletic lineage of endospore-forming bacteria previously classified in the genus *Bacillus* (Ash et al., 1993). The choice of *Paenibacillus polymyxa* as type species was challenged on the basis of priority of publication (Tindall, 2000), but *Paenibacillus polymyxa* has been retained as type species. Since 2005, the number of *Paenibacillus* species has nearly doubled from about 50 to more than 80 at this moment. The G+C content span of the overall genomic DNA at the generic level is about 20%, which is a very good indicator for the important genotypic, and hence also the phenotypic, heterogeneity of a taxon. It is generally accepted that generic variation at the genus level should not be more than 10–15%, which means that the taxonomy of *Paenibacillus* at the generic level certainly deserves to be reconsidered. As indicated above, there are no clear-cut phenotypic traits that discriminate *Paenibacillus* from other genera. Moreover, at the species level, phenotypic data are often very difficult to interpret taxonomically for obvious reasons. Firstly, the methodology that has been applied throughout the various species descriptions is often different. In our hands, the best data are obtained by a combination of API 20E, API 20NE, and API 50CH test panels combined with Biolog data because these methods are standardized, thus allowing the most comparative analysis. The Vitec approach is most probably another, although more limited, method that seems to be reliable for phenotypic analysis but has not yet been applied in a general context for *Paenibacillus* species. Secondly, the interpretation of the data, given by various authors is not always straightforward. A typical example concerns data on "utilization of carbon sources", which is often confused with "oxidation" of substrates as measured in the Biolog panel. Thirdly, and perhaps most importantly, novel species with validly described names are increasingly "one-strain species", with complete ignorance of phenotypic variability as a consequence. However, all identification procedures based on the above-mentioned commercialized phenotypic test panels are based on statistic interpretation of data! Most recent developments in bacterial taxonomy are comparative sequence analysis of house-keeping genes such as *rpoB*, *gyrA* or *gyrB*, *recA*, etc., to discriminate between species and even within species. Studies in *Paenibacillus* are very recent and mainly concentrate on typing aspects of these genes (Durak et al., 2006; Vollu et al., 2006). Although very promising taxonomic insights can be obtained with this kind of approach, more sequence data are needed before taxonomic rearrangements of *Paenibacillus* will be possible.

A final taxonomic issue concerns *Paenibacillus azotofixans*, which is regarded as a later synonym of *Paenibacillus durus* (Logan et al., 1998). The Judicial Commission recently concluded that *Paenibacillus durus* has priority over *Paenibacillus azotofixans*.

List of species of the genus *Paenibacillus*

1. **Paenibacillus polymyxa** (Prazmowski 1880) Ash, Priest and Collins 1994, 852[VP] (Effective publication: Ash, Priest and Collins 1993, 259) (*Clostridium polymyxa* Prazmowski 1880, 37; *Bacillus polymyxa* Macé 1889, 588)

po.ly.my'xa. Gr. pref. *poly* much, many; Gr. n. *myxa* slime or mucous; N.L. n. *polymyxa* much slime.

Colonies on nutrient agar are thin, often with amoeboid spreading. On glucose agar, colonies are usually heaped and mucoid with a matt surface. Levan is synthesized from sucrose, forming large capsules. Facultative anaerobe. Ferments glucose to 2,3-butanediol, ethanol, CO_2, and H_2. Reduces nitrate to nitrite. Many sugars and carbohydrates are fermented, including pectin, pullulan, starch, and xylan, but action on cellulose is weak. Most strains fix atmospheric nitrogen under anaerobic conditions. Biotin (or a trace of yeast extract) is required for growth in minimal medium. Major fatty acids are $C_{15:0\ anteiso}$ and $C_{17:0\ anteiso}$.

Isolated from decomposing plants and soil. Associated with the rhizosphere where many strains provide protection to the plant and enhance plant growth.

DNA G+C content (mol%): 43–46 (Bd).

Type strain: NRRL NRS-1105, ATCC 842, DSM 36, NCIMB 8158, KCTC 3858, LMG 13294.

EMBL/GenBank accession number (16S rDNA) of the type strain: AJ320493 (DSM 36).

2. **Paenibacilllus agarexedens** (Wieringa 1941) Uetanabaro, Wahrenburg, Hunger, Pukall, Spröer, Stackebrandt, de Canhos, Claus and Fritze 2003, 1055[VP] (*Bacillus agar-exedens* Wieringa 1941, 125)

a.gar.ex.e'dens. N.L. n. *agarum* agar; L. v. *exedere* to eat up, utilize; N.L. part. adj. *agarexedens* agar-utilizing.

Cells are Gram-negative (in young cells the Gram stain is uneven), motile, strictly aerobic rods (0.5–1.4 by 2–8 µm). Spores are ellipsoidal and most of the sporangia are not swollen. Colonies on peptone-urea agar are whitish and round with entire margins. Colonies sink into the agar within a few days, but do not liquefy it. Growth is inhibited by peptones, but this is reversed by urea. Mesophilic; maximum temperature for growth is 40°C (type strain: 35°C). No growth at pH 5.7 or in 5% NaCl. Oxidase-positive, nitrate is not reduced to nitrite, and no denitrification (i.e., gas production from nitrate) occurs. Positive for hydrolysis of agar, starch, hippurate, and esculin, but negative for dextran and pectin. Negative for: Voges–Proskauer test; urease, dextranase, DNase, and lysine decarboxylase; hydrolysis of poly-β-hydroxybutyric acid, casein, pectin, Tween 80, and chitin; and production of indole, dihydroxyacetone, and dextrin crystals.

Isolated from soil.

DNA G+C content (mol%): 47–49 (T_m).

Type strain: CIP 107437, DSM 1327.

EMBL/GenBank accession number (16S rDNA) of the type strain: AJ345020.

3. **Paenibacillus agaridevorans** Uetanabaro, Wahrenburg, Hunger, Pukall, Spröer, Stackebrandt, de Canhos, Claus and Fritze 2003, 1056[VP]

a.ga.ri.de.vo'rans. N.L. n. *agarum* agar; L. part. adj. *devorans* consuming, devouring; N.L. part. adj. *agaridevorans* agar-devouring.

Cells are Gram-negative (Gram staining is uneven in young cells), strictly aerobic, motile rods (0.6–0.8 by 2–5 µm). Spores are ellipsoidal and most of the sporangia are not swollen. Colonies on agar media sink into the agar within a few days; no liquefaction of agar occurs. Colonies on peptone-urea agar are whitish and round with entire margins. Growth is inhibited by peptones, but this may be neutralized by urea. Mesophilic, maximum growth temperature of 35°C. No growth at pH 5.7 and in media with 5% NaCl. Oxidase-positive. Nitrate is not reduced and no gas is formed from nitrate anaerobically. Hydrolyzes agar, dextran, esculin, and hippurate, but not pectin. Produces acid, but no gas, from agar and glucose. Negative for: Voges–Proskauer test; urease, DNase, and lysine decarboxylase; hydrolysis of starch, poly-β-hydroxybutyric acid, casein, pectin, Tween 80, and chitin; tyrosine degradation; deamination of phenylalanine; and production of indole, dihydroxyacetone, and dextrin crystals.

Isolated from volcanic soil, Paricutin volcano, Mexico.

DNA G+C content (mol%): 50–52 (T_m).

Type strain: CIP 107436, DSM 1355.

EMBL/GenBank accession number (16S r DNA) of the type strain: AJ345023 (DSM 1355).

4. **Paenibacillus alginolyticus** (Nakamura 1987) Shida, Takagi, Kadowaki, Nakamura and Komagata 1997a, 295[VP] (*Bacillus alginolyticus* Nakamura 1987, 285)

al.gi.no.ly'ti.cus. N.L. adj. *alginicus* pertaining to alginic acid; Gr. adj. *lyticus* dissolving; N.L. adj. *alginolyticus* alginic acid-dissolving.

Cells are Gram-positive, aerobic rods (0.5–1.0 by 4–6 µm). Motile, occurring singly and in short chains. Colonies are non-pigmented, translucent, smooth, and circular. Ellipsoidal spores are formed in swollen sporangia. Optimum growth temperature is 28–30°C with minima of 5–10°C and maxima of 35–40°C. Growth in the presence of 0.001% lysozyme occurs at pH 5.6–5.7. Growth is inhibited by 3% NaCl in the media. Oxidase is not produced and the Voges–Proskauer test is negative. Nitrate is not reduced to nitrite. Produces acid, but no gas, from a variety of sugars and hydrolyzes numerous polysaccharides including alginate, carboxymethyl cellulose and β-1,2-glucan. Can be selectively isolated by growth in minimal medium with sodium alginate as sole carbon source and a trace of yeast extract (0.1 g/l). Major fatty acids are $C_{15:0\ anteiso}$ and $C_{16:0\ iso}$.

Isolated from soil.

DNA G+C content (mol%): 47–49 (Bd).

Type strain: DSM 5050, NRRL NRS-1347.

EMBL/GenBank accession number (16S rDNA) of the type strain: AB073362 (DSM 5050).

Additional remarks: This species represents DNA homology group 3 of *Bacillus circulans sensu lato* (Nakamura and J. Swezey, 1983).

5. **Paenibacillus alkaliterrae** Yoon, Kang, Yeo and Oh 2005, 2342[VP]

al.ka.li.ter′rae. Arabic n. *alkali* (*al-qaliy*) the ashes of saltwort; L. gen. n. *terrae* of the soil or earth; N.L. gen. n. *alkaliterrae* of high-pH soil.

Gram-positive, aerobic, motile, sporulating rods (1.5–3.0 by 0.4–0.5 μm). Ellipsoidal spores are positioned centrally or subterminally in swollen sporangia. Colonies on 2-fold diluted nutrient agar (pH 7.5) are circular to slightly irregular, smooth, sticky, glistening, raised, ivory-colored, and 2.0–4.0 mm in diameter after 5 d at 30°C. Growth occurs between 10 and 37°C with optimal growth at 30°C. Grows at pH 7.0–9.5 with optimal growth at 7.5. Denitrification is not growth supportive on 2-fold diluted nutrient agar supplemented with nitrate. Esculin is hydrolyzed, but Tweens 20, 40 and 60, hypoxanthine, and xanthine are not. D-Glucose, D-fructose, D-galactose, D-cellobiose, D-mannose, trehalose, D-xylose, L-arabinose, sucrose, maltose, and salicin are utilized, but benzoate, pyruvate, formate, and L-glutamate are not. Arginine dihydrolase, lysine decarboxylase, ornithine decarboxylase, and tryptophan deaminase are absent. An API ZYM profile has been recorded. Antibiotic profiles are reported. The cell wall peptidoglycan contains *meso*-DAP and the predominant menaquinone is MK-7. The major fatty acid is $C_{15:0\ anteiso}$.

The type strain was isolated from an alkaline soil in Kwangchun, Korea.

DNA G+C content (mol%): 49.4 (HPLC).

Type strain: KSL-134, KCTC 3956, DSM 17040.

EMBL/GenBank accession number (16S rDNA) of the type strain: AY960748 (KSL-134).

6. **Paenibacillus alvei** (Cheshire and Cheyne 1885) Ash, Priest and Collins 1995, 197[VP] (Effective publication: Ash, Priest and Collins 1993, 259) (*Bacillus alvei* Cheshire and Cheyne 1885, 581)

al′ve.i. L. n. *alveus* a beehive; L. gen. n. *alvei* of a beehive.

Grows as actively motile colonies that spread across agar media. Free spores may lay side-by-side in long rows on the agar. Facultative anaerobe, but does not use nitrate as electron acceptor under anaerobic conditions. Production of indole from tryptophan and dihydroxyacetone on glycerol agar is distinctive. Minimal nutritional requirements are several amino acids plus thiamine or thiamine and biotin. Major fatty acids are $C_{15:0\ anteiso}$, $C_{15:0\ iso}$, and $C_{16:0}$.

Originally isolated from honeybee larvae suffering from European foulbrood, but not the etiological agent and not an insect pathogen. Can be isolated from beehives and soil surrounding beehives.

DNA G+C content (mol%): 45–47 (T_m).

Type strain: ATCC 6344, DSM 29, NCIMB 9371.

EMBL/GenBank accession number (16S rDNA) of the type strain: AJ320491 (DSM 29).

7. **Paenibacillus amylolyticus** (*ex* Kellerman and McBeth 1912) Ash, Priest and Collins 1995, 197[VP] (Effective publication Ash, Priest and Collins 1993, 259) (*Bacillus amylolyticus ex* Kellerman and McBeth (1912); Nakamura 1984b, 224)

emend. Shida, Takagi, Kadowaki, Nakamura and Komagata 1997b, 303

am.y.lo.ly′ti.cus. L. n. *amylum* starch; Gr. adj. *lyticus* dissolving; N.L. adj. *amylolyticus* dissolving starch.

Cells are facultative anaerobes. Rod-shaped (0.5–1.0 by 3.0–5.0 μm). Colonies on agar are grayish-white, translucent, smooth, and circular. Optimum growth temperature is 28–30°C with respective minima and maxima being 10–15 and 40–45°C. Nitrate is reduced to nitrite. Produces acid from a range of sugars and hydrolyzes starch. Major fatty acids are $C_{15:0\ anteiso}$ and $C_{16:0}$.

Habitat: soil.

DNA G+C content (mol%): 46.3–46.6 (Bd).

Type strain: ATCC 9995, DSM 3034, NRRL NRS-290.

EMBL/GenBank accession number (16S rDNA) of the type strain: D85396 (NRRL NRS-290).

Additional remarks: This species represents DNA hybridization group 11 of *Bacillus circulans sensu lato* (Nakamura and J. Swezey, 1983).

8. **Paenibacillus anaericanus** Horn, Ihssen, Matthies, Schramm, Acker and Drake 2005, 1263[VP]

an.ae.ri.ca′nus. Gr. pref. *an* no, not; Gr. n. *aer* air; Gr. adj. *ikanos* capable; N.L. masc. adj. *anaericanus* capable of anaerobic growth.

Cells are Gram-negative, facultatively anaerobic, motile rods (2.0–5.0 by 0.5–1.0 μm) that grow in chains; cells are linked by connecting filaments. Cells have a three-layered cell wall without outer membrane. Contains *b*-type cytochromes. Terminal to subterminal ellipsoidal spores are formed. Colonies are flat, smooth, circular, and entire. Grows at 5–40°C and pH 5.8–8.6, with optimum growth at 30–35°C and pH 7.7. Doubling time under optimal conditions is 5 h. Catalase- and oxidase-positive. Grows in media containing 2% NaCl, but not in 5% NaCl. Formate, acetate, and ethanol are formed when glucose is fermented. Nitrate and sulfate are not dissimilated. Low amounts (1 mM) of nitrite are reduced to N_2O.

The type strain was isolated from the gut of the earthworm *Aporrectodea caliginosa* (collected from garden soil in Bayreuth, Germany).

DNA G+C content (mol%): 42.6 (HPLC).

Type strain: MH21, DSM 15890, ATCC BAA-844.

EMBL/GenBank accession number (16S rDNA) of the type strain: AJ318909 (MH21).

9. **Paenibacillus antarcticus** Montes, Mercadé, Bozal and Guinea 2004, 1523[VP]

ant.arc′ti.cus. L. masc. adj. *antarcticus* of the Antarctic environment, where the organism was isolated.

Cells are Gram-variable, facultatively anaerobic rods (0.7 by 2.5 μm), motile by means of peritrichous flagella. Subterminal or terminal ellipsoidal spores are formed in swollen sporangia. Colonies grown on TSA are non-pigmented, circular, slightly convex, bright, and cream-colored. Growth is not inhibited by the presence of 4% NaCl or 0.001% lysozyme. Growth occurs at 4 and 31°C, but not at 0 or 32°C; optimal growth occurs at 10–15°C. Oxidase, catalase, urease, and methyl red reactions are positive. Nitrate reduction and H_2S production are negative. The predominant fatty acid is $C_{15:0\ anteiso}$.

DNA G+C content (mol%): 40.7 (HPLC).

Type strain: 20CM, LMG 22078, CECT 5836.

EMBL/GenBank accession number (16S rDNA) of the type strain: AJ605292 (LMG 22078).

10. **Paenibacillus apiarius** (Katznelson 1955) Nakamura 1996, 692[VP] (*Bacillus apiarius* Katznelson 1955, 635)

a.pi.a′ri.us. N.L. adj. *apiarius* relating to bees.

Cells are facultatively anaerobic, Gram-variable, motile rods (0.7–0.8 by 3.0–5.0 μm). Spores have thick walls; they appear to be rectangular in swollen sporangia. Agar colonies are non-pigmented, translucent, thin, smooth, and entire with a mean diameter of about 1.0 mm after 24 h incubation on TGY agar (Haynes et al., 1955) at 28°C. Optimum growth temperature is 28°C with 40°C and 15°C as minima and maxima. Growth occurs at pH 5.7, in the presence of 0.001% lysozyme, and in the presence of 5% NaCl, but is inhibited by 7% NaCl. Oxidase-negative. Nitrate is reduced to nitrite. Acid, but not gas, is produced from sugars. Hydrolyzes casein and starch. Major fatty acids are $C_{15:0 \text{ anteiso}}$, $C_{15:0 \text{ iso}}$, and $C_{17:0 \text{ anteiso}}$.

Isolated from honeybees, their larvae and their hives, but not an insect pathogen.

DNA G+C content (mol%): 52–54 (Bd).

Type strain: ATCC 29575, DSM 5581, NRRL NRS-1438.

EMBL/GenBank accession numbers (16S rDNA) of the type strain: U49427 (NRRL NRS-1438) and AJ320492 (DSM 5581).

Comment: The sequence with accession number AJ320492 is a usable sequence linked to DSM 5581, but sequence AB073201, also linked to DSM 5581, differs considerably and is likely the incorrect sequence for phylogenetic studies.

11. **Paenibacillus assamensis** Saha, Mondal, Makyilraj, Krishnamurthi, Bhattacharya and Chakrabarti 2005, 2579[VP]

as.sam.en′sis. N.L. masc. adj. *assamensis* pertaining to Assam, a north-eastern state in India, where the type strain was isolated.

Gram-variable, motile, strictly aerobic, spore-forming cells; occurs singly or in pairs. Endospores are ellipsoidal and subterminal, occurring in bulging sporangia. Cells are 1.0–2.5 μm long and 0.5–0.6 μm wide. On TSA, colonies are round, convex with undulated margins, and light yellowish-white in color, spreading as single colonies over the entire plate. Growth occurs between 20 and 37°C and at pH 6.8–12.0. Up to 2.5% NaCl is tolerated. Oxidase, catalase, gelatinase, and arginine dihydrolase reactions are positive, but urease, DNase, phenylalanine deaminase, lysine, and ornithine decarboxylase are negative. Indole and H_2S are not formed. Nitrate is not reduced and the Voges–Proskauer test is negative. Gas is not produced from glucose. Positive for the methyl red test. Cannot utilize acetate, citrate, or propionate. Hydrolyzes starch, esculin, and casein, but not tyrosine, ONPG, or Tweens 20, 40 and 80. Acid is produced from various carbohydrates. Can grow in the presence of 0.01% lysozyme, but not in 0.001% lysozyme. The major fatty acid is $C_{15:0 \text{ anteiso}}$. *meso*-DAP is reported as a cell-wall amino acid and MK-7 as the diagnostic menaquinone. PAEN 515F and PAEN 862F signature sequences in the 16S rRNA gene support the phylogenetic position of the species in the genus *Paenibacillus*.

The type strain was isolated from a warm-spring water sample from Assam, India.

DNA G+C content (mol%): 41.2 (HPLC).

Type strain: GPTSA 11, MTCC 6934, JCM 13186.

EMBL/GenBank accession number (16S rDNA) of the type strain: AY884046 (GPTSA 11).

12. **Paenibacillus azoreducens** Meehan, Bjourson and McMullan 2001, 1684[VP]

azo.re.duc′ens. Gr. n. *azo* a combining form meaning nitrogen; N.L. pres. part. *reducens* reducing; N.L. adj. *azoreducens* nitrogen-reducing, referring to the ability to decolorize azo dyes.

Cells are Gram-variable, facultatively anaerobic rods (0.5–0.8 by 3.0–6.0 μm), motile by peritrichous flagella. Ellipsoidal spores are formed in swollen sporangia. Colonies on nutrient agar containing glucose are creamish-yellow, flat, smooth, and circular. Growth occurs between 10 and 50°C with optimal growth at 37°C and pH 7. Growth is inhibited by 5% NaCl. Does not use nitrate as an electron acceptor under anaerobic conditions. Acid is produced from a range of sugars. Oxidase-negative. Positive for hydrogen sulfide production. Completely decolorizes the azo dye Remazol Black B (25–400 mg/l) within 24 h. Major fatty acids are $C_{15:0 \text{ anteiso}}$, $C_{16:0}$, and $C_{17:0 \text{ anteiso}}$.

Habitat unknown, but isolated from textile industrial waste water.

DNA G+C content (mol%): 47 (HPLC).

Type strain: CM1, DSM 13822, NCIMB 13761.

EMBL/GenBank accession number (16S rDNA) of the type strain: AJ272249 (CM1).

13. **Paenibacillus barcinonensis** Sánchez, Fritze, Blanco, Spröer, Tindall, Schumann, Kroppenstedt, Diaz and Pastor 2005, 937[VP]

bar.ci.no.nen′sis. L. masc. adj. *barcinonensis* from *Barcino*, the Roman name for Barcelona, the city in Spain where the strain was isolated.

Gram-positive, facultatively anaerobic rods (0.5–1 by 1.5–4.5 μm). Ellipsoidal endospores form in swollen sporangia at a subterminal position. Colonies are circular to slightly irregular, pale yellow in color, and 0.5 mm in diameter after 2 d growth at 30°C in nutrient broth. Growth occurs at 10–40°C and pH 5.0–10.4. Growth occurs in the presence of 5% (w/v) NaCl and 0.001% (w/v) lysozyme. Catalase-positive. Oxidase- and urease-negative. Nitrate is not reduced to nitrite or nitrogen. Casein and starch are not hydrolyzed. Citrate and propionate are not utilized. Acid is produced from various carbohydrates. The major fatty acids are $C_{15:0 \text{ anteiso}}$, $C_{16:1\omega11c}$, $C_{15:0 \text{ iso}}$, and $C_{16:0}$. The predominant menaquinone is MK-7. The major polar lipids present are diphosphatidylglycerol, phosphatidylethanolamine, and two unidentified amino-phospholipids.

The type strain was isolated from soil from a rice field in the Ebro River delta, Spain.

DNA G+C content (mol%): 45.0 (type strain, HPLC).

Type strain: BP-23, CECT 7022, DSM 15478.

EMBL/GenBank accession number (16S rDNA) of the type strain: AJ716019 (BP-23).

14. **Paenibacillus barengoltzii** Osman, Satomi and Venkateswaran 2006, 1514[VP]

ba.ren.gol′tzi.i. N.L. gen. n. *barengoltzii* of Barengoltz, referring to Jack Barengoltz, a well-known American physicist and NASA planetary protection scientist.

Cells are Gram-positive, strictly aerobic, spore-forming rods (0.5–0.8 by 3.0–5.0 µm), motile by peritrichous flagella. Ellipsoidal spores are formed in swollen sporangia. Colonies are flat, smooth, circular, entire, and brownish-yellow on nutrient agar without production of soluble pigments. Growth occurs between 10 and 50°C and pH 4.5–9.0 with optimum growth at 37°C and pH 7.0. Growth occurs in the presence of 2% NaCl and 0.001% lysozyme, but is inhibited by 5% NaCl. Positive for catalase, oxidase, and the Voges–Proskauer reaction. Hydrogen sulfide and indole are not produced. Denitrification is not reported, but nitrate is reduced to nitrite. Gelatin is not liquefied, esculin is hydrolyzed, and β-galactosidase is produced. Of the carbon substrates tested, only gluconate is utilized. Acid is not produced from D-glucose. Furthermore, the vast majority of API 20E, API 20NE, and Biolog carbohydrate tests were negative.

The type strain was isolated from clean room floors of the Jet Propulsion Laboratory Spacecraft Assembly Facility, Pasadena, CA, USA; strain SAFN-125 (=ATCC BAA-1210) is a reference strain.

DNA G+C content (mol%): not reported.

Type strain: SAFN-016, ATCC BAA-1209, NBRC 101215.

EMBL/GenBank accession number (16S rDNA) of the type strain: AY167814 (SAFN-016).

15. **Paenibacillus borealis** Elo, Suominen, Kämpfer, Juhanoja, Salkinoja-Salonen and Haahtela 2001, 542[VP]

bo.re.a′lis. N.L. adj. *borealis* pertaining to the north (wind) boreal.

Cells are Gram-negative, facultatively anaerobic, motile rods (0.75–1.0 by 3–5 µm). Ellipsoidal spores formed in swollen sporangia are subterminal or terminal. Mature spores have an unusual striped morphology when visualized by electron microscopy. Flat, smooth, opaque colonies are formed on nutrient agar. The temperature range for growth is 5–37°C with an optimum at 28°C. Grows at pH 5.6–8.0 with an optimum of pH 7. Does not tolerate 5%NaCl, 0.001%lysozyme, or 0.02% sodium azide. Nitrate is not reduced under anaerobic conditions. Acid and gas are produced from various sugars. Atmospheric nitrogen is fixed. Selective isolation can be achieved in N-free media, with or without yeast extract, and with glucose as carbon source. Major cellular fatty acids are $C_{15:0\ anteiso}$, $C_{15:0\ iso}$, and $C_{16:0}$.

Isolated from acid humus in Norway spruce forest in Finland and from the rhizoplane of birch trees.

DNA G+C content (mol%): 54 (HPLC).

Type strain: KK19, CCUG 43137, DSM 13188.

EMBL/GenBank accession number (16S rDNA) of the type strain: AJ011322 (KK19).

16. **Paenibacillus brasilensis** von der Weid, Frois Duarte, van Elsas and Seldin 2002, 2152[VP]

bra.sil.en′sis. N.L. adj. *brasilensis* referring to Brazil, the country where the strains were isolated.

Cells are Gram-positive to Gram-variable facultatively anaerobic rods. Endospores are ellipsoidal, predominantly central to subterminal, and distend the sporangium. Colonies are about 10 mm in diameter, bright yellow on TBN agar (glucose, yeast extract, biotin, thiamine medium; Seldin

et al., 1983), circular with entire margins, and adhere to the agar. Grows well on GB medium (glucose, 10 g; peptone, 10 g; NaCl, 5 g; yeast extract, 1 g; meat extract, 2 g; distilled water to 1 l and adjusted to pH 7.2 with NaOH; Seldin et al., 1983). The maximum temperature for growth is 42°C with an optimum of 30–32°C. Grows at pH 5–7 and in the presence of 2% NaCl, but not in 5% NaCl or in the presence of lysozyme. Acid and gas are produced from glucose. Acid is produced from a restricted number of carbohydrates. Reduces nitrate to nitrite. Voges–Proskauer reaction is positive. All strains fix atmospheric nitrogen.

Isolated from the rhizosphere of maize sown in the Minas Gerais area of Brazil.

DNA G+C content (mol%): not reported.

Type strain: PB172, ATCC BAA-413, DSM 14914.

EMBL/GenBank accession number (16S rDNA) of the type strain: AF273740 (PB172).

17. **Paenibacillus campinasensis** Yoon, Yim, Lee, Shin, Sato, Lee, Park and Park 1998, 836[VP]

cam.pi.na.sen′sis. N.L. adj. *campinasensis* referring to Campinas, the city in Brazil where the College of Food Engineering, State University of Brazil is located.

Cells are Gram-variable, facultatively anaerobic rods (0.6–0.9 by 3.0–6.0 µm) that are motile by means of peritrichous flagella. Ellipsoidal spores are formed in swollen sporangia. Colonies are flat, smooth, opaque, and motile, particularly on wet agar plates. Alkaliphilic, optimal growth at pH 10.0 and no growth at neutral pH. Growth in 7% NaCl is distinctive. Biochemical reactions reported are based on Biolog GN test panel results and do not provide information on acid production from carbohydrates. Gelatin, casein, esculin, and starch are hydrolyzed. Oxidase- and urease-negative. Produces crystalline cyclodextrins from starch. The major fatty acid is $C_{15:0\ anteiso}$.

DNA G+C content (mol%): 51 (HPLC).

Type strain: 324, KCTC 0364BP, BCRC 17341, JCM 11200.

EMBL/GenBank accession number (16S rDNA) of the type strain: AF021924 (324).

18. **Paenibacillus cellulosilyticus** Rivas, García-Fraile, Mateos, Martínez-Molina and Velázquez 2006, 2779[VP]

cel.lu.lo.si.ly′ti.cus. N.L. n. *cellulosum* cellulose; Gr. adj. *lutikos* able to loose, able to dissolve; N.L. adj. *lyticus -a -um* dissolving; N.L. masc. adj. *cellulosilyticus* cellulose-dissolving.

Gram-variable, aerobic or facultatively anaerobic, spore-forming rods (0.8–0.9 by 4.0–4.2 µm), motile by peritrichous flagella. Subterminal, ellipsoidal spores are formed in swollen sporangia. Chemo-organotrophic and xylanolytic. Colonies on yeast extract-glucose medium are circular, flat, white/creamy, opaque, and usually 1–3 mm in diameter after 48 h at 28°C. Growth occurs at 10–37°C with optimal growth at 28°C. Optimum pH is 7. Grows in the presence of 2% NaCl, but not in 5% NaCl. Nitrate is not reduced to nitrite. Growth in anaerobic agar is reported as negative. Cellulases, xylanases, amylases, and β-galactosidase are actively produced, but gelatinase, caseinase, arginine dihydrolase, indole, lysine decarboxylase, ornithine decarboxylase, urease, tryptophan deaminase, phenylalanine deaminase, and hydrogen sulfide are not. Esculin is hydrolyzed. Gas is not produced from D-glucose. A number of sugars and gluconate are assimilated. Acid is produced from a limited number of carbohydrates. The major quinone is MK-7. Main fatty acids are $C_{15:0\ anteiso}$ and $C_{16:0\ iso}$.

The type strain was isolated from the bract phyllosphere of *Phoenix dactylifera* in Palma de Mallorca (Spain) on XED medium (xylan, 0.7%; yeast extract, 0.3%; agar, 2.5%, w/v).

DNA G+C content (mol%): 51 (T_m).

Type strain: PALXIL08, LMG 22232, CECT 5696.

EMBL/GenBank accession number (16S rDNA) of the type strain: DQ407282 (PALXIL08).

19. **Paenibacillus chibensis** Shida, Takagi, Kadowaki, Nakamura and Komagata 1997b, 306[VP]

chi.ben'sis. N.L. adj. *chibensis* referring to Chiba, a Japanese prefecture where the research laboratory of Higeta Shoyu Co., Ltd. is located.

Cells are Gram-positive, strictly aerobic rods (0.5–0.8 by 3.0–5.0 μm), motile by means of peritrichous flagella. Ellipsoidal spores are formed in swollen sporangia Colonies are brownish-yellow, flat, smooth, circular, and entire. Growth occurs at 10–50°C and pH 4.5–9.0 with optimal growth at 37°C and pH 7.0.Growth occurs in the presence of 2% NaCl and 0.001% lysozyme, but is inhibited by 5% NaCl and 0.02% sodium azide. Catalase-positive and oxidase-negative; Voges–Proskauer reaction is negative. Hydrogen sulfide, indole, dihydroxyacetone, and lecithinase are not produced. Nitrate is reduced to nitrite; ammonium and nitrate are utilized. Acid, but no gas, is produced from glucose and a number of other sugars. Starch is hydrolyzed, but casein, gelatin, DNA, Tweens 20, 60 and 80, urea, and hippurate are not. Major fatty acids are $C_{15:0\ anteiso}$, $C_{16:0\ iso}$, and $C_{17:0\ anteiso}$.

Habitat: unknown.

DNA G+C content (mol%): 52.5–53.2 (HPLC).

Type strain: NRRL B-142, DSM 11731, JCM 9905.

EMBL/GenBank accession number (16S rDNA) of the type strain: D85395 (NRRL B-142).

20. **Paenibacillus chinjuensis** Yoon, Seo, Shin, Kho, Kang and Park 2002, 419[VP]

chin.ju.en'sis. N.L. adj. *chinjuensis* of Chinju, the city in Korea where the type strain was isolated.

Cells are Gram-positive to Gram-variable, facultatively anaerobic rods (0.8–1.1 by 3.0–5.0 μm), motile by peritrichous flagella. Ellipsoidal spores are formed in swollen sporangia. Colonies are light pink, smooth, glossy, circular, and convex with entire margins after 3–4d on TSA. No growth occurs in media containing 2% (w/v) or more NaCl. Growth occurs at 20–45°C, but not at 15 or 50°C with optimum growth at 30–37°C. The optimal pH for growth is 6.5–7.3. Growth is very slow or inhibited below pH 4.5 and above pH 9. Does not use nitrate as a terminal electron acceptor. Ferments sugars producing acid but no gas. Hydrolyzes arbutin, casein, gelatin, and starch. Major fatty acids are $C_{15:0\ anteiso}$ and $C_{15:0\ iso}$. Cell-wall peptidoglycan contains *meso*-DAP. The predominant menaquinone is MK-7.

Isolated from soil in Chinju, Korea.

DNA G+C content (mol%): 53 (HPLC).

Type strain: WN9, JCM 10939, KCTC 8951P.

EMBL/GenBank accession number (16S rDNA) of the type strain: AF164345 (WN9).

21. **Paenibacillus chitinolyticus** (Kuroshima, Sakane, Takata and Yokota 1996) Lee, Pyun and Bae 2004, 932[VP] (*Bacillus chitinolyticus* Kuroshima, Sakane, Takata and Yokota 1996, 79)

chi.ti.no.ly'ti.cus. N.L. n. *chitinum* chitin; Gr. adj. *lutikos* able to loose, able to dissolve; N.L. adj. *lyticus -a -um* dissolving; N.L. masc. adj. *chitinolyticus* decomposing chitin.

Cells are Gram-variable, aerobic rods (0.4–0.6 by 1.7–3 μm), motile by peritrichous flagella. Ellipsoidal spores are formed in swollen sporangia. Colonies are pale brown, irregular, and raised with undulate margins. The optimal growth temperature is 25–37°C, with maxima of 42–45°C and minima of 18–20°C. Oxidase is produced, Voges–Proskauer test is positive. Nitrate is reduced to nitrite. Acid, but no gas, is produced from a few carbohydrates. Variable for casein and gelatin hydrolysis. Starch is not hydrolyzed. Major fatty acid is $C_{15:0\ anteiso}$ and MK-7 is the major quinone. Decomposes chitin.

Isolated from forest soil samples obtained in Kaya, Kagoshima Prefecture, Japan.

DNA G+C content (mol%): 51.3–52.8 (HPLC).

Type strain is: EAG-3, KCTC 3791, NBRC 15660.

EMBL/GenBank accession number (16S rDNA) of the type strain: AB021183 (IFO 15660).

22. **Paenibacillus chondroitinus** (Nakamura 1987) Shida, Takagi, Kadowaki, Nakamura and Komagata 1997a, 297[VP] (*Bacillus chondroitinus* Nakamura 1987, 285)

chon.droi'ti.nus. N.L. adj. *chondroitinus* of chondroitin, pertaining to chondroitin.

Cells are Gram-positive, aerobic, motile rods (0.5–1.0 by 4.0–6 μm), occurring singly and in short chains. Ellipsoidal spores are formed in swollen sporangia. Agar colonies are non-pigmented, translucent, smooth, circular, entire, and 1–2 mm in diameter after 3 d at 28°C. Optimum growth temperature is 28–30°C with minima of 5–10°C and maxima of 35–40°C. Growth occurs at pH 5.6–5.7 and is usually inhibited by 0.001% lysozyme or 3% NaCl in the medium. Oxidase is not detected; nitrate is not reduced to nitrite. Hydrolyzes alginate and chondroitin sulfate, but not β-glucans. Acid, but no gas, is produced from sugars. Isolated following enrichment growth with alginate as major carbon source. Major fatty acids are $C_{15:0\ anteiso}$ and $C_{16:0\ iso}$.

Isolated from soil.

DNA G+C content (mol%): 47–48 (Bd).

Type strain: DSM 5051, NRRL NRS-1351, JCM 9072.

EMBL/GenBank accession number (16S rDNA) of the type strain: AB073206 (JCM 9072).

Additional remarks: This species represents DNA homology group 4 of *Bacillus circulans sensu lato* (Nakamura and Swezey, 1983).

23. **Paenibacillus cineris** Logan, De Clerck, Lebbe, Verhelst, Goris, Forsyth, Rodríguez-Díaz, Heyndrickx and De Vos 2004, 1075[VP]

ci'ne.ris. L. gen. masc. n. *cineris* of/from ash, referring to the volcanic, ash-based soil from which the type strain was isolated.

Cells are motile, Gram-negative or Gram-variable, facultatively anaerobic, round-ended rods (0.7–0.9 by

2.5–4.0 μm), occurring singly and in pairs. Endospores are formed within 2–3 d incubation on *Bacillus fumarioli* agar (BFA) at 30°C; they are ellipsoidal, paracentral and subterminal and swell the sporangia. After 2 d at 30°C, colonies are 1–5 mm in diameter, low convex, circular with slightly irregular edges, opaque, glossy, and light beige to grayish in color with paler margins. Grows on nutrient agar. Catalase- and oxidase-positive. Minimum growth temperature is 8–15°C; maximum is 50°C. Optimal pH for growth is 7.0 with minima varying from pH 5.0 to 6.5 and maxima from pH 7.5 to 11.0. Growth occurs in the presence of 3% NaCl, but is inhibited by 5% NaCl. Hydrolysis of casein is weakly positive. Nitrate is reduced. Gelatin is not hydrolyzed. Hydrolysis of esculin is positive. Acid is produced without gas from various carbohydrates. An antibiotic reaction profile has been determined. The major cellular fatty acids are $C_{15:0\ anteiso}$, $C_{16:0}$, $C_{17:0\ anteiso}$, and $C_{16:0\ iso}$ (representing about 46, 18, 10, and 9% of total fatty acids, respectively).

Isolated from the ashy soil of a cold, dead fumarole at the foot of Lucifer Hill, a volcano on Candlemas Island, South Sandwich Archipelago, Antarctica. Strain LMG 21976 is reference strain.

DNA G+C content (mol%): 51.5 (HPLC).

Type strain: Logan B1768, LMG 18439, CIP 108109.

EMBL/GenBank accession number (16S rDNA) of the type strain: AJ575658 (LMG 18439).

24. **Paenibacillus cookii** Logan, De Clerck, Lebbe, Verhelst, Goris, Forsyth, Rodríguez-Díaz, Heyndrickx and De Vos 2004, 1075[VP]

cook'i.i. N.L. gen. n. *cookii* of Cook, referring to Captain James Cook, of HMS *Resolution*, who discovered Candlemas Island on Candlemas Day (2 February), 1775.

Cells are Gram-negative or Gram-variable, round-ended, facultatively anaerobic, motile rods (0.6–0.8 by 3.0–3.5 μm). Endospores, formed within 2–3 d incubation on BFA incubated at 30°C are ellipsoidal, lie subterminally, and swell the sporangia slightly. After 2 d at 30°C, colonies are 1–4 mm in diameter, convex, yellowish, and transparent with opaque centers; motile microcolonies are formed and spread across the surface of the agar, rotating clockwise and anticlockwise. Grows on nutrient agar. Catalase-positive or weakly positive and oxidase-positive. Minimum growth temperature varies between 15 and 20°C and the maximum growth temperature is 50°C. The optimal pH for growth is 7, with respective minima and maxima of pH 5.0–5.5 and pH 7.5–10. Growth occurs in the presence of 5% NaCl, but is inhibited by 7% NaCl. Casein hydrolysis is positive, but weak. Nitrate is reduced. Hydrolysis of gelatin is variable; that of esculin is positive. Urease-negative. Acid is produced without gas from various carbohydrates. An antibiotic profile has been recorded with the disk method. The major cellular fatty acids are $C_{15:0\ anteiso}$, $C_{17:0\ anteiso}$, $C_{16:0\ iso}$, $C_{16:0}$, and $C_{15:0\ iso}$ (representing about 36, 20, 13, 11, and 6.5% of total fatty acids, respectively).

Isolated from geothermal soil taken from an active fumarole on Lucifer Hill, a volcano on Candlemas Island, South Sandwich Archipelago, Antarctica, and from a gelatin extract of bovine bones.

DNA G+C content (mol%): 51.6 (HPLC).

Type strain: LMG 18419, CIP 108110, Logan B1718.

EMBL/GenBank accession number (16S rDNA) of the type strain: AJ250317 (LMG 18419).

25. **Paenibacillus curdlanolyticus** (Kanzawa, Harada, Takeuchi, Yokota and Harada 1995) Shida, Takagi, Kadowaki, Nakamura and Komagata 1997a, 297[VP] (*Bacillus curdlanolyticus* Kanzawa, Harada, Takeuchi, Yokota and Harada 1995, 517)

curd.lan.o.ly′ti.cus. N.L. n. *curdlanum* curdlan, a polysaccharide produced by bacteria; Gr. adj. *lutikos* dissolving; N.L. adj. *curdlanolyticus* hydrolyzing curdlan.

Agar colonies are flat, smooth, opaque, and motile on minimal medium with curdlan as carbon source. Strictly aerobic, reduces nitrate to nitrite under anaerobic conditions. Acid, but no gas, is produced from sugars. Carboxymethyl cellulose, curdlan (a β-1,3-glucan), pustulan (a β-1,6-glucan), pullulan, and starch are hydrolyzed. Major fatty acids are $C_{15:0\ anteiso}$ and $C_{16:0\ iso}$.

Isolated from soil.

DNA G+C content (mol%): 50–52 (HPLC).

Type strain: IFO 15724, DSM 10247, LMG 18050, ATCC 51898.

EMBL/GenBank accession number (16S rDNA) of the type strain: AB073202 (DSM 10247).

26. **Paenibacillus daejeonensis** Lee, Lee, Chang, Hong, Oh, Pyun and Bae 2002, 2110[VP]

dae.je.on.en′sis. N.L. masc. adj. *daejeonensis* referring to Daejeon, Korea, the geographical origin of the novel species.

Cells are Gram-variable, spore-forming rods that are motile by means of peritrichous flagella. Ellipsoidal spores form in swollen sporangia. Colonies are circular, flat, smooth, opaque, and white on TSA adjusted to pH 10 with Na_2CO_3. Cells do not grow at pH 6.0, but grow at pH 7.0–13.0 with an optimum pH of 8.0. Alkaliphilic. Positive for esculin hydrolysis, β-galactosidase, oxidase, and catalase. Gelatin is not liquefied. Negative for nitrate reduction, urease, and H_2S production. Acid is produced from a restricted number of carbohydrates. The major isoprenoid quinone is menaquinone MK-7. Major fatty acids are $C_{15:0\ anteiso}$ and $C_{16:0\ iso}$, with significant amounts of $C_{15:0}$ and $C_{16:0}$. Cell-wall peptidoglycan contains *meso*-DAP.

Isolated from soil of Daejeon City, Korea. KCTC 3750 and JCM 11237 are reference strains.

DNA G+C content (mol%): 53 (HPLC).

Type strain: AP-20, JCM 11236, KCTC 3745, CIP 107805.

EMBL/GenBank accession number (16S rDNA) of the type strain: AF290916 (AP-20).

27. **Paenibacillus dendritiformis** Tcherpakov, Ben-Jacob and Gutnick 1999, 244[VP]

den.dri.ti.for′mis. Gr. n. *dendron* tree; N.L. adj. *formis* shaped; N.L. adj. *dendritiformis* tree-shaped, referring to the tree like-shapes of the colonies on agar.

Gram-negative. Cells are either small (2–3 μm long by 0.5–1.0 μm wide; T morphotype) or large (4–6 μm long by 1.0–1.5 μm wide; C morphotype), and motile. The former produces branched colony morphology on thin, peptone agar plates and the latter produces a chiral morphology on this medium. Colonies on nutrient agar are non-pigmented, translucent, thin, smooth, entire, and 1.0–2.0 mm in diameter. The transformation of C–T morphotypes and vice versa depends on the agar concentration of the medium. Spores are round, cylindrical, or ellipsoidal in swollen sporangia. Oxidase is present. Voges–Proskauer is negative. Nitrate is not reduced to nitrite. Facultatively anaerobic, producing acid from glucose. Major fatty acids are $C_{15:0\ anteiso}$ and $C_{17:0\ anteiso}$.

Habitat: soil.

DNA G+C content (mol%): 55 (method not reported).

Type strain: Gutnick strain T168, 30A1, CIP 105967, DSM 18844, LMG 21716, T168.

EMBL/GenBank accession number (16S rDNA) of the type strain: AY359885 (CIP 105967).

28. **Paenibacillus durus** (Smith and Cato 1974) Collins, Lawson, Willems, Cordoba, Fernandez-Garayzabel, Garcia, Cai, Hippe and Farrow 1994, 824^{VP} (*Clostridium durum* Smith and Cato 1974, 1396)

du'rus. L. masc. adj. *durus* hard, tough.

Colonies on GB agar are 1–2 mm in diameter, whitish, circular to slightly irregular, and convex with entire margins. Little or no growth occurs in nutrient broth under aerobic conditions, but abundant growth is observed in GB broth. Cells are capsulated and weakly motile. Facultatively anaerobic, but does not reduce nitrate to nitrite. Acid and gas are produced from sugars. Fixes nitrogen under anaerobic conditions. Selectively isolated by anaerobic incubation on nitrogen-free minimal medium containing thiamine and biotin. Major cellular fatty acids are $C_{15:0\ anteiso}$ and $C_{16:0}$.

Isolated from the rhizoplane and rhizosphere soil of various grasses and marine sediments.

DNA G+C content (mol%): 50 (T_m; a mean value of measurements ranging from 48–53).

Type strain: ATCC 27763, CIP 105291, DSM 1735, LMG 15707, VPI 6563-1.

EMBL/GenBank accession number (16S rDNA) of the type strain: X77846 (DSM 1735). More reliable 16S rRNA gene sequences are AB033195, which is linked to *Paenibacillus azotofixans* DSM 5976^T, or AJ251192, which is linked to *Paenibacillus azotofixans* ATCC 35681^T.

Additional remarks: Paenibacillus azotofixans and *Paenibacillus durus* (formerly *Clostridium durum*) were combined into one taxon by Rosado et al. (1997) as *Paenibacillus azotofixans* but, following a request to the Judicial Commission by Logan et al. (1998), *Paenibacillus durus* was ruled to have priority over *Paenibacillus azotofixans*.

29. **Paenibacillus ehimensis** (Kuroshima, Sakane, Takata and Yokota 1996) Lee, Pyun and Bae 2004, 931^{VP} (*Bacillus ehimensis* Kuroshima, Sakane, Takata and Yokota 1996, 79)

e.hi.men'sis. N.L. masc. adj. *ehimensis* of or belonging to Ehime, referring to Ehime Prefecture, Japan, the source of the soil samples from which the organisms were isolated.

Cells are aerobic rods (0.4–0.6 by 1.7–5 μm), motile by peritrichous flagella. Gram reaction is usually positive, although some strains stain Gram-negative. Ellipsoidal spores are formed in swollen sporangia. Colonies are creamy to pale brown, circular, and convex with entire margins. The optimum growth temperature ranges from 28 to 40°C with maxima of 50–53°C and minima of 18–20°C.

Catalase and oxidase are produced. Voges–Proskauer test is negative. Nitrate is reduced to nitrite. Acid, but no gas, is produced from few carbohydrates. Gelatin and starch are hydrolyzed; variable for hydrolysis of casein. Decomposes chitin.

Isolated from garden soil in Matsuyama, Ehime Prefecture, Japan.

DNA G+C content (mol%): 52.9–54.9 (HPLC).

Type strain: EAG-5, IFO 15659, KCTC 3748.

EMBL/GenBank accession number (16S rDNA) of the type strain: AB021184 (IFO 15659).

30. **Paenibacillus elgii** Kim, Bae, Jeon, Chun, Oh, Hong, Baek, Moon and Bae 2004, 2034^{VP}

el'gi.i. N.L. gen. n. *elgii* arbitrary name formed from the company name LG, where taxonomic studies on this species were performed.

Cells are facultatively anaerobic, Gram-variable rods (0.8–1.0 by 3.0–5.0 μm) that are motile with peritrichous flagella. Ellipsoidal spores are formed in swollen sporangia and mature spores have stripes on the surface. Colonies on nutrient agar are circular, flat, smooth, opaque, and white. Temperature range for growth is 20–45°C; growth occurs at pH 6.0–8.5 (optimum 7.0). Can grow in the presence of 2% NaCl. Catalase-positive, oxidase-negative. Nitrate is reduced to nitrite. Indole is produced, but H_2S is not. Casein, esculin, and starch are hydrolyzed. Acid is produced from a restricted number of sugars. Glucose, ribose, *N*-acetylglutamate, and Tween 40 are assimilated. The major isoprenoid quinone is menaquinone MK-7. The major cellular fatty acid is $C_{15:0\ anteiso}$. Cell-wall peptidoglycan contains *meso*-DAP.

Isolated from roots of *Perilla frutescens* in Seocheon County, Korea. Strain SD18 (=KCTC 3756) is a reference strain.

DNA G+C content (mol%): 51.7 (T_m).

Type strain: SD17, KCTC 10016BP, NBRC 100335.

EMBL/GenBank accession number (16S rDNA) of the type strain: AY090110 (SD17).

31. **Paenibacillus favisporus** Velázquez, de Miguel, Poza, Rivas, Rosselló-Mora and Villa 2004, 61^{VP}

fa.vi.spo'rus. L. masc. n. *favus* a honeycomb; Gr. n. *spora* a seed, spore; N.L. masc. adj. *favisporus* referring to the honeycomb form of spores.

Cells are Gram-variable, motile, aerobic or facultatively anaerobic, spore-forming rods (0.5–0.7 by 2–3 μm). Spores slightly swell the sporangia and are subterminal. Spores have an ornamentation similar to that of honeycomb. Colonies on yeast extract-glucose medium are circular, convex, white with a central brown spot, translucent, and usually 1–3 mm in diameter within 48 h at 37°C. Optimal growth temperature is 37°C; optimal growth pH is 7. Chemo-organotrophic and xylanolytic. Oxidase- and catalase-positive. Phylogenetically, most closely related to *Paenibacillus azoreducens*. The main fatty acid is $C_{15:0\ anteiso}$. Gas is not produced from D-glucose. Acid is produced from various carbohydrates. Cellulose and starch are utilized as carbon sources. In contrast, no growth occurs using L-arabinose, citrate, inositol, sorbitol, glucitol, or xylitol as carbon sources. Gelatinase, urease, and amylase are produced actively, but casein is not hydrolyzed and hydrogen sulfide is not formed. Nitrate is reduced to nitrite. An antibiotic profile has been obtained.

Isolated from old cow dung in Salamanca, Spain.

DNA G+C content (mol%): 53% (HPLC).

Type strain: GMP01, LMG 20987, CECT 5760.

EMBL/GenBank accession number (16S rDNA) of the type strain: AY208751 (GMP01).

32. **Paenibacillus gansuensis** Lim, Jeon, Lee, Xu, Jiang and Kim 2006a, 2133[VP]

gan.su.en′sis. N.L. masc. adj. *gansuensis* belonging to Gansu, from where the type strain was isolated.

Cells are Gram-positive motile rods (0.7–0.9 by 1.7–2.4 μm). Colonies are glistening, translucent, semi-sticky, irregular, slightly raised, and pale yellow on R2A agar. Growth occurs at 10–45°C (optimum 35–40°C). Nitrate is not reduced to nitrite. Catalase- and oxidase-negative. Esculin, casein, and Tween 80 are hydrolyzed. Hypoxanthine, tyrosine, starch, and xanthine are not hydrolyzed. Acids are produced from various carbohydrates. Phylogenetically, most closely related to *Paenibacillus chitinolyticus* and *Paenibacillus daejeonensis*. The 16S rRNA signatures PAEN 862F and PAEN 515F are reported to be present in the rRNA gene of this species. The major isoprenoid quinone is MK-7 and the major fatty acid is $C_{15:0\ anteiso}$; smaller, but still significant, amounts (around 10%) of $C_{16:0}$, $C_{17:0\ anteiso}$, $C_{16:0\ iso}$, and $C_{15:0\ iso}$ are also present.

Isolated from a desert soil sample from Gansu Province in China.

DNA G+C content (mol%): 50 (HPLC).

Type strain: B518, KCTC 3950, DSM 16968.

EMBL/GenBank accession number (16S rDNA) of the type strain: AY839866 (B518).

33. **Paenibacillus glucanolyticus** (Alexander and Priest 1989) Shida, Takagi, Kadowaki, Nakamura and Komagata 1997a, 297[VP] (*Bacillus glucanolyticus* Alexander and Priest 1989, 113).

glu.can.o.ly′ti.cus. N.L. n. *glucanum* glucan (a polysaccharide of D-glucose monomers) Gr. adj. *lutikos* dissolvable; N.L. adj. *glucanolyticus* hydrolyzing glucose polymers.

Cells are facultatively anaerobic, long (>3.0 μm), thin (0.9 μm) rods that produce ellipsoidal terminal spores that swell the sporangia. Colonies are flat, smooth, opaque, and motile on nutrient agar plates. Grows at pH 5.7 and 17–37°C, but not at 5 or 50°C with variable growth at 45°C. Reduces nitrate to nitrite. Voges–Proskauer reaction is negative. Produces acid, but no gas, from sugars. Hydrolyzes a range of carbohydrates including carboxymethyl cellulose, curdlan (a β-1,3-glucan), pustulan (a β-1,6-glucan), β-1,2-glucan, pullulan, and starch. Major fatty acids are $C_{15:0\ anteiso}$, $C_{16:0\ iso}$, and $C_{16:0}$.

Habitat: soil.

DNA G+C content (mol%): 48.1 (T_m).

Type strain: DSM 5162, NCIMB 12809.

EMBL/GenBank accession number (16S rDNA) of the type strain: AB073189 (DSM 5162).

34. **Paenibacillus glycanilyticus** Dasman, Kajiyama, Kawasaki, Yagi, Seki, Fukusaki and Kobayashi 2002, 1671[VP]

gly.can.i.ly′ti.cus. N.L. *glycanum* glycan, a heteropolysaccharide; Gr. adj. *lutikos* dissolving; N.L. adj. *glycanilyticus* degrading heteropolysaccharide.

Cells are Gram-positive, facultatively aerobic rods (0.5–0.8 by 3.0–5.0 μm), motile by means of peritrichous flagella. Endospores are ellipsoidal in swollen sporangia. Colonies are flat, smooth, circular, entire, and pinkish-yellow when grown on peptone-yeast extract agar. Optimal growth temperature is 28–37°C; no growth is observed at 50°C. Oxidase is not produced. Distinctive property is the ability to hydrolyze the β-1,4-linked xylogalactoglucan backbone of the heteropolysaccharide extracted from *Nostoc commune*. Major fatty acids are $C_{15:0\ anteiso}$ and $C_{16:0\ iso}$.

Habitat: soil.

DNA G+C content (mol%): 50.5 (HPLC).

Type strain: DS-1, JCM 11221, NRRL B-23455.

EMBL/GenBank accession number (16S rDNA) of the type strain: AB042938 (DS-1).

35. **Paenibacillus graminis** Berge, Guinebretière, Achouak, Normand and Heulin 2002, 613[VP]

gra′mi.nis. L. gen. neut. n. *graminis* of grass.

Cells are motile, Gram-positive, facultatively anaerobic rods (0.5–1.0 by 3.0–4.0 μm), occurring singly or in short chains. Ellipsoidal spores are formed in swollen sporangia. Nutrient agar colonies are cream-colored, smooth with regular, entire margins, and measure 1.0 to 2.0 mm in diameter after 3 d growth at 30°C on TSA. Growth temperature varies from 5–10°C (minimum) to 35–40°C (maximum). Oxidase is not produced. Acid and gas are produced from various carbohydrates. Nitrate is reduced to nitrite. Fixes nitrogen under anaerobic conditions and, at least for the type strain, the *nifH* gene has been demonstrated.

Habitat: soil, maize rhizosphere and wheat roots.

DNA G+C content (mol%): 52.1 (T_m).

Type strain: RSA19, ATCC BAA-95, LMG 19080.

EMBL/GenBank accession number (16S rDNA) of the type strain: AJ223987 (LMG 19080).

36. **Paenibacillus granivorans** van der Maarel, Veen and Wijbenga 2001, 264[VP] (Effective publication: van der Maarel, Veen and Wijbenga 2000, 347)

gra.ni.vo′rans. L. pl. n. *grani* granules; L. part. adj. *vorans* eating, devouring; N.L. part. adj. *granivorans* granules-eating, referring to its ability to hydrolyze granular starch.

Cells are Gram-positive, strictly aerobic rods (0.5–0.8 by 1.5–4.0 μm), motile by peritrichous flagella. Ellipsoidal spores are produced that do not swell the sporangia. Growth occurs optimally at 37°C (maximum of 45°C) and between pH 6–8.5 (with optimum at pH 7.0). Inhibition of growth is observed in media containing 5% NaCl. Reduces nitrate to nitrite, oxidase-negative, and produces acid from glucose and other sugars. Hydrolyzes starch, but not casein or gelatin. Has the distinctive ability to degrade native potato starch granules. Major fatty acid is $C_{15:0\ anteiso}$ (>70% of total).

Isolated from wastewater from potato starch production plant.

DNA G+C content (mol%): 47.8 (HPLC).

Type strain: A30, CBS 229.89.

EMBL/GenBank accession number (16S rDNA) of the type strain: AF237682.

37. **Paenibacillus hodogayensis** Takeda, Suzuki and Koizumi 2005, 740^{VP}

ho.do.ga.yen'sis. N.L. masc. adj. *hodogayensis* pertaining to Hodogaya, the name of a district in Yokohama, Japan, the geographical origin of isolation of the type strain.

Cells are Gram-variable or Gram-negative, aerobic, motile, spore-forming rods (1.3–1.7 by 2.3–2.8 μm). Ellipsoidal spores are formed in non-swollen or slightly swollen sporangia. Colonies on TYN agar (containing 5 g tryptone, 2.5 g yeast extract, 2.5 g NaCl, and 15 g agar in 1 l distilled water) are white, convex, and opaque. Optimum growth is at 30°C and pH 8. No growth occurs at 50°C or in the presence of 5% (w/v) NaCl. Catalase- and oxidase-positive. Nitrate is not reduced. Indole is not produced. Voges–Proskauer reaction is negative. Citrate is not utilized. Acids are produced from various carbohydrates. The predominant cellular fatty acid is $C_{15:0\ anteiso}$. The major quinone is MK-7(H_2). The PAEN515F binding site has been demonstrated in the 16S rRNA gene.

DNA G+C content (mol%): 55 (HPLC).

Type strain: SG, JCM 12520, KCTC 3919.

EMBL/GenBank accession number (16S rDNA) of the type strain: AB179866 (SG).

38. **Paenibacillus illinoisensis** Shida, Takagi, Kadowaki, Nakamura and Komagata 1997b, 304^{VP}

il.li.nois.en'sis. N.L. adj. *illinoisensis* referring to Illinois, the state where Microbial Properties Research, National Center for Agricultural Utilization Research, U.S. Department of Agriculture is located.

Cells are Gram-positive, facultatively anaerobic rods (0.5–0.8 by 3.0–5.0 μm), motile by means of peritrichous flagella. Endospores are ellipsoidal in swollen sporangia. Colonies are flat, smooth, circular, entire, and yellowish-gray. Optimal growth occurs at 37°C with 10 and 50°C as minimum and maximum, respectively. Grows at pH 4.5–9.0 with optimum at pH 7.0. Growth occurs in media with 2% NaCl, but is inhibited by 5% NaCl or 0.02% azide. Does not reduce nitrate to nitrite. Acid is produced from various sugars. Major fatty acids are $C_{15:0\ anteiso}$, $C_{16:0}$, and $C_{17:0\ anteiso}$.

Isolated from soil.

DNA G+C content (mol%): 47.6–48.3 (Bd).

Type strain: DSM 11733, IFO 15959, JCM 9907.

EMBL/GenBank accession number (16S rDNA) of the type strain: AB073192 (JCM 9907).

Additional remarks: This species represents DNA homology group 6 of *Bacillus circulans sensu lato* (Nakamura and J. Swezey, 1983).

39. **Paenibacillus jamilae** Aguilera, Monteoliva-Sánchez, Suárez, Guerra, Lizama, Bennasar and Ramos-Cormenzana 2001, 1691^{VP}

ja.mi'lae. N.L. fem. n. *jamilae* residual water of olive oil production, from *jamila* a specific term of Arabic origin commonly used in Andalusia, Spain.

Cells are Gram-variable, facultatively anaerobic rods (0.5–1.2 by 4.5–6.5 μm), motile by means of peritrichous flagella. Endospores are ellipsoidal in swollen sporangia. Colonies are convex, mucoid, opaque, and motile (microcolonies) on wet agar plates. Optimal growth occurs at 30–40°C and pH 5–12. Grows in media with 2% NaCl, but is inhibited by 5% NaCl. Reduces nitrate to nitrite. Oxidase-negative. Acid, but no gas, is produced from various sugars. Major fatty acid is $C_{15:0\ anteiso}$.

Isolated from corn-compost treated with olive mill waste water.

DNA G+C content (mol%): 40.6–40.8 (T_m).

Type strain: CECT 5266, DSM 13815.

EMBL/GenBank accession number (16S rDNA) of the type strain: AJ271157 (CECT 5266).

40. **Paenibacillus kobensis** (Kanzawa, Harada, Takeuchi, Yokota and Harada 1995) Shida, Takagi, Kadowaki, Nakamura and Komagata 1997a, 297^{VP} (*Bacillus kobensis* Kanzawa, Harada, Takeuchi, Yokota and Harada 1995, 517)

ko.ben'sis. N.L. adj. *kobensis* referring to Kobe City, Hyogo Prefecture, Japan, the source of the soil from which the organisms were isolated.

Cells are Gram-positive, facultatively aerobic rods (0.5–1.0 by 2.0–6.0 μm), motile by peritrichous flagella. Ellipsoidal spores are formed in swollen sporangia. Colonies are flat, smooth, and opaque on nutrient agar plates and are motile during growth on minimal medium with glucose or curdlan (a β-1,3-glucan) as carbon source. Grows in the presence of 0.001% lysozyme, but not in media containing 5% NaCl. No growth at 50°C. Negative for oxidase and the Voges–Proskauer reaction. Nitrate is reduced to nitrite. Curdlan, pullulan, pustulan (a β-1,6-glucan), and starch are hydrolyzed, but carboxymethyl cellulose and β-1,2-glucans are not. Acid, but no gas, is produced from sugars. Major fatty acids are $C_{15:0\ anteiso}$ and $C_{16:0\ iso}$.

Isolated from soil.

DNA G+C content (mol%): 50–52 (HPLC).

Type strain: DSM 10249, IFO 15729.

EMBL/GenBank accession number (16S rDNA) of the type strain: AB073363 (DSM 10249).

41. **Paenibacillus koleovorans** Takeda, Kamagata, Shinmaru, Nishiyama and Koizumi 2002, 1600^{VP}

ko.le.o.vo'rans. Gr. n. *koleon* sheath; L. v. *vorare* to devour; N.L. part. adj. *koleovorans* sheath-devouring.

Cells are Gram-negative, facultatively anaerobic rods (0.4–0.7 by 1–3.2 μm). Ellipsoidal spores are formed in swollen sporangia. Spores have an unusual spiked surface morphology. The creamy-gray colonies that are formed on 1% tryptone agar are of two types: flat and translucent or convex and opaque (this is not connected with degree of sporulation). Does not grow in nutrient broth. Optimal growth temperature is 30°C and optimal pH is 7. In media containing 5% (w/v) NaCl, growth is inhibited. Does not reduce nitrate to nitrite. Voges–Proskauer reaction is negative. Characteristically hydrolyzes the sheath material of *Sphaerotilus natans* and produces acid from a very restricted range of sugars. Major fatty acids are $C_{15:0\ anteiso}$, $C_{15:0\ iso}$, and $C_{16:0}$.

Isolated from soil and river water.

DNA G+C content (mol%): 54.0–55.8 (HPLC).

Type strain: TB, JCM 11186, KCTC 13912.

EMBL/GenBank accession number (16S rDNA) of the type strain: AB041720 (TB).

42. **Paenibacillus koreensis** Chung, Kim, Hwang and Chun 2000, 1499^{VP}

ko.re.en'sis. N.L. masc. adj. *koreensis* indicating Korea, the geographical origin of isolation.

Cells are Gram-variable, facultatively anaerobic rods (0.5–0.9 by 2.3–4.5 μm), motile by means of peritrichous flagella. Endospores are ellipsoidal in swollen sporangia. Colony morphology is variable on 0.1× TSA, but is typically circular, flat smooth, and opaque. Grows at 10–50°C, with an optimum at 38–48°C. Growth is inhibited by 7% NaCl in the medium. Reduces nitrate to nitrite. Positive for oxidase and negative for Voges–Proskauer reaction. Hydrolyzes casein, chitin, chitosan, esculin, and starch. Selective isolation has been achieved using colloidal chitin as carbon source. Ferments various sugars with the formation of acid, but no gas. Produces an iturin-like antifungal antibiotic. Major fatty acid is $C_{15:0\ anteiso}$.

Isolated from soil.

DNA G+C content (mol%): 54 (T_m).

Type strain: YC300, KCTC 2393, KCCM 40903.

EMBL/GenBank accession number (16S rDNA) of the type strain: AF130254 (YC300).

43. **Paenibacillus kribbensis** Yoon, Oh, Yoon, Kang and Park 2003, 299[VP]

krib.ben'sis. N.L. masc. adj. *kribbensis* arbitrary name formed from the acronym of the Korea Research Institute of Bioscience and Biotechnology, KRIBB, where taxonomic studies on this species were performed.

Cells are Gram-variable, facultatively anaerobic rods (1.3–1.8 by 4.0–4.7 μm), motile by means of peritrichous flagella. Endospores are ellipsoidal in swollen sporangia. Colonies are cream-colored, circular to slightly irregular, flat to low convex, and translucent on TSA. Growth occurs at 10–44°C, but not at 4 or 45°C, with optimum growth at 30–37°C. Growth is optimal with 0–2% NaCl in the medium, but is inhibited by 5% NaCl. Reduces nitrate to nitrite. Oxidase- and urease-negative. Hydrolyzes casein, esculin, gelatin, and starch. Produces acid, but no gas, from a variety of sugars. Major fatty acids are $C_{15:0\ anteiso}$, $C_{15:0\ iso}$, $C_{16:0}$, and $C_{17:0\ anteiso}$.

Isolated from soil.

DNA G+C content (mol%): 48 (HPLC).

Type strain: AM49, JCM 11465, KCTC 0766BP.

EMBL/GenBank accession number (16S rDNA) of the type strain: AF391123 (AM49).

44. **Paenibacillus lactis** Scheldeman, Goossens, Rodriguèz-Diàz, Pil, Goris, Herman, De Vos, Logan and Heyndrickx 2004, 889[VP]

lac'tis. L. masc. n. *lac* milk; L. gen. n. *lactis* from milk, referring to milk (and its environment on the dairy farm) as the principal isolation source.

Cells are Gram-negative or Gram-variable, straight and round-ended, aerobic, motile rods (0.6–0.9 by 3–6 μm) that may occasionally be slightly tapered and curved. Endospores are ellipsoidal or cylindrical, are located subterminally and occasionally paracentrally, and usually swell the sporangia. Colonies grown for 4 d on TSA at 30°C are opaque, cream-colored, slightly convex, and round, with rough or spreading transparent edges and with an eggshell surface texture. Motile microcolonies may spread across the surface of the agar and rotate in a clockwise direction. Colony diameter is 1–2 mm. Optimal growth temperature falls between 30 and 40°C, with maxima between 50 and 55°C. Optimum pH for

growth is 7.0 with pH 5.0–6.0 and pH 10.5–11.0 as minima and maxima, respectively. Casein is not hydrolyzed. Nitrate reduction is variable. Hydrogen sulfide production, urease reaction, and hydrolysis of gelatin are negative. Hydrolysis of esculin is positive. Acid is produced without gas from various carbohydrates. The major cellular fatty acids based on measurements of 10 strains are $C_{15:0\ anteiso}$, $C_{16:0}$, $C_{15:0\ iso}$, $C_{17:0\ anteiso}$, $C_{16:0\ iso}$, and $C_{17:0\ iso}$.

Isolated from raw and heat-treated milk in Belgium.

DNA G+C content (mol%): 51.6 (HPLC; mean value).

Type strain: MB 1871, LMG 21940, DSM 15596.

EMBL/GenBank accession number (16S rDNA) of the type strain: AY257868 (MB 1871).

45. **Paenibacillus larvae** (White 1906) Heyndrickx, Vandemeulebroecke, Hoste, Janssen, Kersters, De Vos, Logan, Ali and Berkeley 1996a, 278[VP] (*Bacillus larvae* White 1906, 42; *Paenibacillus larvae* Ash, Priest and Collins 1995, 197)

lar'vae. L. n. *larva* a larva; N.L. gen. n. *larvae* of a larva.

Cells are facultatively anaerobic rods. Spores are formed sparsely *in vitro*. Fastidious, does not survive serial transfer in nutrient broth. Grows well on Columbia blood agar containing 5% horse blood. Reduction of nitrate to nitrite is variable. Catalase is not produced or produced very weakly. Acid, but no gas, is produced from various sugars.

The species was divided into the subspecies *Paenibacillus larvae* subsp. *larvae* and *Paenibacillus larvae* subsp. *pulvifaciens* by Heyndrickx et al. (1996a), but recently the existence of two subspecies has been reconsidered by Genersch et al. (2006), who proposed reclassification into a single species based on a polyphasic study encompassing more strains than the former studies and by which the former subspecies discrepancies based on PAGE profiling of cell proteins became uncertain. Forms small (<1 mm diameter), regular, glossy, grayish colonies on Columbia blood agar after incubation for 2 to 4 d at 37°C, sometimes with a bright orange pigment. The only known host for the bacterium is the larva of the honeybee (*Apis mellifera*) in which it causes the disease American foulbrood. This disease occurs throughout the world and is a serious threat to apiary. The name is derived from the odor of the decaying larvae. Major fatty acids are $C_{15:0\ anteiso}$, $C_{15:0\ iso}$, and $C_{17:0\ anteiso}$.

Isolated from diseased honeybee larvae and the honey, wax, and pollen of affected hives and from powdery scales of honeybee larvae.

DNA G+C content (mol%): 42–43 (T_m).

Type strain: ATCC 9545, DSM 7030, LMG 15969; additional reference strains are the former type strains of *Paenibacillus* "*pulvifaciens*", i.e., ATCC 13537, DSM 3615, LMG 15974.

EMBL/GenBank accession number (16S rDNA) of the type strain: AB073205 (DSM 7030).

46. **Paenibacillus lautus** (Batchelor 1919) Heyndrickx, Vandemeulebroecke, Scheldeman, Kersters, De Vos, Logan, Aziz, Ali and Berkeley 1996b, 995[VP] ("*Bacillus lautus*" Batchelor 1919, 30; *Bacillus lautus* Nakamura 1984b, 225)

lau'tus. L. part. adj. *lautus* washed, splendid.

Grows on routine media such as nutrient agar producing circular to irregularly shaped, low, convex, grayish colonies that are 1 to 2 mm in diameter. There is a tendency for motile colonies to spread across the agar surface. Grows in 5% NaCl. Facultatively anaerobic, reduces nitrate to nitrite. Hydrolyzes a range of carbohydrates including carboxymethyl cellulose, curdlan (a β-1,3-glucan), pustulan (a β-1,6-glucan), β-1,2-glucan, and starch. Acid, but no gas, is produced from sugars. Major fatty acid is $C_{15:0\ anteiso}$.

Isolated from soil and human intestinal tracts.

DNA G+C content (mol%): 50–52 (Bd).

Type strain: DSM 3035, LMG 11157, JCM 9073.

EMBL/GenBank accession number (16S rDNA) of the type strain: AB073188 (JCM 9073).

Additional remarks: This species represents DNA hybridization group 9 of *Bacillus circulans sensu lato* (Nakamura and Swezey, 1983).

47. **Paenibacillus lentimorbus** (Dutky 1940) Pettersson, Rippere, Yousten and Priest 1999, 538[VP] (*Bacillus lentimorbus* Dutky 1940, 68)

len.ti.mor′bus. L. adj. *lentus* slow; L. n. *morbus* disease; N.L. n. *lentimorbus* the slow disease.

Cells are Gram-variable or Gram-negative, facultatively anaerobic rods (0.5–0.7 by 1.8–7 μm). Endospores are ellipsoidal in swollen sporangia. Growth occurs at 20–35°C. Nutritionally fastidious, requires J-broth (Gordon et al., 1973) or MYGPG (Costilow and Coulter, 1971) medium (*see Nutrition and metabolism*). On these media, forms cream-colored colonies less than 1 mm in diameter. Sporulation in laboratory media has not been reported. Spores are produced *in vivo* during growth in susceptible insect larvae. Spores may be accompanied by a parasporal crystal (type strain does not produce a crystal). Catalase-negative and does not reduce nitrate. Fails to grow in 2% NaCl or 1 μg/ml vancomycin, which distinguishes it from *Paenibacillus popilliae*. Causative agent of type B milky disease in Japanese beetle (*Popillia japonica* Newman) and related scarab larvae, which is characterized by brown clots that block the circulation of hemolymph. Major fatty acids are $C_{15:0\ anteiso}$ and $C_{16:0}$.

Isolated from diseased larvae of a Japanese beetle.

DNA G+C content (mol%): 38 (T_m).

Type strain: ATCC 14707.

EMBL/GenBank accession number (16S rDNA) of the type strain: AB073199 (ATCC 14707).

Additional remarks: Some strains, named *Paenibacillus lentimorbus*, have been erroneously identified at the species level.

48. **Paenibacillus macerans** (Schardinger 1905) Ash, Priest and Collins 1995, 197[VP] (Effective publication: Ash, Priest and Collins 1993, 259) (*Bacillus macerans* Schardinger 1905, 772)

ma′ce.rans. L. part. adj. *macerans* softening by steeping, retting.

Colonies on nutrient agar are thin, round to spreading. Facultatively anaerobic, reduces nitrate to nitrite. Nitrogen is generally fixed under anaerobic conditions. Hydrolyzes chitin; some strains hydrolyze pectin. Crystalline (Schardinger) dextrins are produced typically from starch. Ethanol and acetic acid are produced in the early stages of glucose fermentation followed by the disappearance of

formate and acetate with gas (H_2 and CO_2) and acetone production (Weimer, 1984). Major fatty acids are $C_{15:0\ anteiso}$, $C_{16:0}$, and $C_{16:0\ iso}$.

Major habitat is composting plant materials, less common in nutrient-poor soils.

DNA G+C content (mol%): 52–53 (T_m).

Type strain: IAM 12467, ATCC 8244, DSM 24, NCIMB 9368.

EMBL/GenBank accession number (16S rDNA) of the type strain: AB073196 (IAM 12467).

Additional remarks: The four different sequences available for the type strain differ considerably; the IAM sequence is recommended for phylogenetic studies.

49. **Paenibacillus macquariensis** (Marshall and Ohye 1966) Ash, Priest and Collins 1995, 197[VP] (Effective publication: Ash, Priest and Collins 1993, 259.) (*Bacillus macquariensis* Marshall and Ohye 1966, 41)

mac.qua.ri.en′sis. N.L. adj. *macquariensis* pertaining to Macquarie Island.

Colonies on nutrient agar are opaque and smooth with translucent fimbriate edges. Colonies are 0.5 to 1 mm in diameter after incubation for 4 d at 20°C. Facultative anaerobe. Does not reduce nitrate to nitrite. Psychrophilic, maximum growth temperature is 25°C and optimum is 15–20°C. Will grow at 0°C within 3 weeks. Major fatty acid is $C_{15:0\ anteiso}$ (80% of total).

Isolated from soil from Macquarie Island (sub Antarctic).

DNA G+C content (mol%): 39 (T_m).

Type strain: ATCC 23464, DSM 2, NCIMB 9934.

EMBL/GenBank accession number (16S rDNA) of the type strain: AB073193 (DSM 2).

50. **Paenibacillus massiliensis** Roux and Raoult 2004, 1051[VP]

mas.si.li.en′sis. L. masc. adj. *massiliensis* of Massilia, the old Greek and Roman name for Marseille, where the type strain was isolated.

Cells are Gram-positive, facultatively anaerobic rods (0.5 by 2.0–4.0 μm), motile by peritrichous flagella. Ellipsoidal endospores are formed in swollen sporangia. Grows on routine media and forms translucent, beige-colored, flat colonies after incubation for 24 h at 30°C. Catalase-positive, oxidase-negative. Optimal growth occurs at 30–37°C, but can grow at 50°C. Growth occurs in the presence of 5% (w/v) NaCl. Nitrate is reduced, but gelatin is not liquefied. Acid is produced from various carbohydrates.

Type strain was isolated from blood culture.

DNA G+C content (mol%): not reported.

Type strain: 2301065, CIP 107939, CCUG 48215.

The EMBL/GenBank accession number (16S rDNA) of the type strain: AY323608 (2301065).

51. **Paenibacillus mendelii** Šmerda, Sedláček, Páova, Durnová, Smíšková and Havel 2005, 2353[VP]

men.de′li.i. N.L. gen. n. *mendelii* of Mendel, to honor J. G. Mendel, the founder of genetics.

Cells are Gram-variable, aerobic or facultatively anaerobic, spore-forming rods. Spores are oval with a subterminal position in a swollen sporangium. Optimal growth occurs between 25 and 30°C; no growth is observed at 50°C or in media with 5% NaCl. Colonies on nutrient agar are circular, smooth, flat, bright, translucent with entire edges, and about 1–2 mm in diameter. Positive for catalase, oxidase, lecithinase, β-galactosidase, and hydrolysis of esculin. Acid is produced from various carbohydrates. Acetoin and indole are not produced. Negative for hydrolysis of casein, starch, hippurate urea, gelatin, Tween 80, and tyrosine. Citrate is not utilized. Arginine dihydrolase and DNase are not produced. Nitrate is not reduced. Hydrolysis of agar is not observed. An antibiotic profile has been reported. The major fatty acid is $C_{15:0 \, anteiso}$. The closest phylogenetic neighbors are *Paenibacillus phyllosphaerae*, *Paenibacillus curdlanolyticus*, and *Paenibacillus kobensis*.

Habitat is not known, but the type strain was isolated from surface-sterilized pea seeds.

DNA G+C content (mol%): 50.8 (HPLC).

Type strain: C/2, CCM 4839, LMG 23002.

EMBL/GenBank accession number (16S rDNA) of the type strain: AF537343 (C/2).

52. **Paenibacillus motobuensis** Iida, Ueda, Kawamura, Ezaki, Takade, Yoshida and Amako 2005, 1814[VP]

mo.to.bu.en′sis. N.L. masc. adj. *motobuensis* pertaining to Motobu in Okinawa, Japan, where the type strain was isolated.

Cells are Gram-negative, facultatively anaerobic rods (0.6–1.0 by 1.0–3 μm), motile by peritrichous flagella. Ellipsoidal spores are located terminally. Colonies are circular, flat, smooth, and opaque white. Growth occurs at 20–55°C, with an optimum of 37°C. The pH range for growth is 6.0–8.0 (optimum is pH 8.0). Growth occurs in the presence of 5% NaCl, but is inhibited by 10%. Positive catalase and oxidase reactions. Negative Voges–Proskauer reaction. Nitrate is reduced to nitrite. The major fatty acid is $C_{15:0 \, anteiso}$ (39.8%). Various compounds of the Biolog GP pallet are oxidized.

The type strain was isolated from a composting machine containing soil from Motobu, Okinawa, Japan.

DNA G+C content (mol%): 47 (HPLC).

Type strain: MC10, GTC 1835, JCM 12774, CCUG 50090.

EMBL/GenBank accession number (16S rDNA) of the type strain: AY741810 (MC10).

53. **Paenibacillus naphthalenovorans** Daane, Harjono, Barns, Launen, Palleroni and Häggblom 2002, 137[VP]

naph.tha.le.no.vo′rans. L. neut. n. *naphthalene* from Persian *neft* naphtha; L. v. *vorare* to devour; N.L. part. adj. *naphthalenovorans* naphthalene-devouring.

Cells are Gram-positive, strictly aerobic rods (0.8 by 2.8–4.0 μm), motile by means of peritrichous flagella. Ellipsoidal spores are formed in swollen sporangia. Forms white, translucent, mucoid colonies on solid media. Optimal growth occurs at 30–37°C, but does not grow at 10 or 55°C. Growth is inhibited in media with 5% NaCl and variable if 3% NaCl is present. Some strains reduce nitrate to nitrite. Acid is produced from a limited range of sugars. Able to use naphthalene as sole carbon source. Major fatty acids are $C_{15:0 \, anteiso}$, $C_{16:0}$, and $C_{16:1} \omega 11c$.

Isolated from soil from estuarine sediments and rhizosphere of salt marsh plants.

DNA G+C content (mol%): 49 (HPLC).

Type strain: ATCC BAA-206, DSM 14203.

EMBL/GenBank accession number (16S rDNA) of the type strain: AF353681 (ATCC BAA-206).

54. **Paenibacillus nematophilus** Enright, McInerney and Griffin 2003, 439[VP]

ne.ma.to′phi.lus. N.L. n. *nematoda* nematode; Gr. adj. *philos* loving or having affinity for; N.L. adj. *nematophilus* nematode-loving.

Cells are Gram-negative to Gram-variable (older cultures), motile rods (0.5–1.0 by 3.5–7.0 μm). Oval-shaped endospores are produced in swollen, spindle-shaped sporangia and lie in a central/paracentral position. Endospore is retained within the sporangium. Forms small (0.5–4.0 mm in diameter) colonies on nutrient agar that are thin, non-pigmented, regular, smooth, and slightly umbonate with an undulate edge. Grows in 2% but not 3% (w/v) NaCl. Grows optimally at 30°C. Growth occurs at 10–37°C, but not at 5 or 40°C. Grows at pH 6–11, but not at pH 5.6. Oxidase-negative. Positive for Voges–Proskauer test. Grows weakly anaerobically, does not reduce nitrate. Hydrolyzes esculin and starch, but not casein or gelatin. Produces acid, but no gas, from various sugars. Major fatty acids are $C_{15:0 \, anteiso}$ and $C_{16:0}$.

Isolated from *Heterorhabditis* species, which are insect-pathogenic nematodes. The sporangia of the bacterium adhere to the free-living, infective stage of the nematode, which carries it to new hosts in which the bacterium reproduces.

DNA G+C content (mol%): 44 (HPLC).

Type strain: NEM1a, DSM 13559.

EMBL/GenBank accession number (16S rDNA) of the type strain: AF480935 (NEM1a).

55. **Paenibacillus odorifer** Berge, Guinebretière, Achouak, Normand and Heulin 2002, 614[VP]

o.do′ri.fer. L. n. *odor* smell; L. suff. n. *-fer* carrier; L. n. *odorifer* carrier of smell.

Cells are Gram-positive, motile, facultatively anaerobic rods (0.5–1.0 by 2.0–4.0 μm), occurring singly or in short chains. Oval terminal spores are formed in swollen sporangia. Nutrient agar colonies are cream-colored, smooth with regular margins, and measure 1.5–3 mm in diameter after incubation for 3 d at 30°C on TSA. Oxidase-negative. Nitrate is reduced to nitrite. Acid is produced from various carbohydrates. Fixes nitrogen under anaerobic conditions and, at least for the type strain, the presence of the *nifH* gene has been shown. Produces a characteristic, fruity, volatile aroma on nutrient media.

Isolated from wheat roots and vegetable purées.

DNA G+C content (mol%): 44 (T_m).

Type strain: ATCC BAA-93, LMG 19079.

EMBL/GenBank accession number (16S rDNA) of the type strain: AJ223990 (LMG 19079).

56. **Paenibacillus pabuli** (*ex* Schieblich 1923) Ash, Priest and Collins 1995, 197[VP] (Effective publication: Ash, Priest and Collins 1993, 259.) (*Bacillus pabuli ex* Schieblich 1923; Nakamura 1984b, 225)

pa′bu.li. L. gen. n. *pabuli* of fodder.

Cells are 2.0–4.0 μm long by 0.5–1.0 μm wide. Facultatively anaerobic. Does not reduce nitrate to nitrite. Hydrolyzes carboxymethyl cellulose, curdlan, pectin, and starch, but not chitin, or β-1,2- and β-1,6-glucans. Acid, but no gas, is produced from various sugars. Major fatty acid is $C_{15:0\ anteiso}$ (about 70% of total).

Isolated from soil and fodder.

DNA G+C content (mol%): 48–50 (Bd).

Type strain: DSM 3036, NRRL NRS-924, JCM 9074.

EMBL/GenBank accession number (16S rDNA) of the type strain: AB073191 (JCM 9074).

Additional remarks: This species represents DNA hybridization group 5 of *Bacillus circulans sensu lato* (Nakamura and Swezey, 1983).

57. **Paenibacillus panacisoli** Ten, Baek, Im, Lee, Oh and Lee 2006, 2680^{VP}

pa.na.ci.so′li. N.L. n. *Panax -acis* scientific name of ginseng; L. n. *solum* soil; N.L. gen. n. *panacisoli* of soil of a ginseng field.

Cells are Gram-positive, facultatively anaerobic, motile, spore-forming rods (0.4–0.6 by 2.0–5.0 μm). Spores are oval and occur subterminally in swollen sporangia. On R2A, colonies are 0.5–1.0 mm in diameter, convex, irregular, undulate, non-glossy, and slightly yellowish after 1 d. Optimal growth occurs at 37°C, with 42–45°C and 15–20°C as respective maxima and minima. Optimal pH for growth is 6.5, with pH 4.5–5 and pH 8.5–9 as respective minima and maxima. Oxidase and catalase reactions are positive. Nitrate is reduced to nitrite. Tolerates 5% (w/v) NaCl, but not 7%. No growth occurs on TSA, MacConkey agar, or nutrient agar. Hydrolyzes xylan and casein (weakly), but not chitin, starch, cellulose, DNA, or esculin. Utilization of compounds for growth is limited to a few carbohydrates. Positive in API 20E for gelatin hydrolysis and tryptophan deaminase; negative for arginine dihydrolase, lysine decarboxylase, ornithine decarboxylase, β-galactosidase, urease, indole and hydrogen sulfide production, citrate utilization, and the Voges–Proskauer reaction. Acid production from carbohydrates is usually weak, if present at all. MK-7 is the predominant menaquinone. The major fatty acids are $C_{15:0\ anteiso}$, $C_{16:0\ iso}$, and $C_{16:0}$. Phylogenetically, the most closely related species are *Paenibacillus xylanilyticus*, *Paenibacillus illinoisensis*, *Paenibacillus pabuli*, and *Paenibacillus amylolyticus*.

The type strain was isolated from soil of a ginseng field of Pocheon Province, South Korea.

DNA G+C content (mol%): 53.9 (HPLC).

Type strain: Gsoil 1411, KCTC 13020, LMG 23405.

EMBL/GenBank accession number (16S rDNA) of the type strain: AB245384 (Gsoil 1411).

58. **Paenibacillus pasadenensis** Osman, Satomi and Venkateswaran 2006, 1512^{VP}

pa.sa.den.en′sis. N.L. masc. adj. *pasadenensis* referring to Pasadena, the city in which the JPL-SAF is located.

Cells are Gram-positive, spore-forming rods (0.5–0.8 by 3.0–5.0 μm), motile by peritrichous flagella. Ellipsoidal spores are formed in swollen sporangia. On nutrient agar, colonies are flat, smooth, circular, entire, and brownish-yellow, with no soluble pigment formation. Catalase and oxidase tests and the Voges–Proskauer reaction are positive. Hydrogen sulfide and indole are not produced. Nitrate is

not reduced to nitrite. Gelatin is liquefied, esculin is hydrolyzed, and β-galactosidase is produced. Growth occurs in the presence of 2% NaCl and 0.001% lysozyme. Growth is inhibited by 3% NaCl. Utilizes α-cyclodextrin, D-cellobiose, D-fructose, maltose, D-melibiose, methyl β-D-glucoside, D-ribose, pyruvic acid, L-alanyl glycine, and L-serine. Acid is not produced from D-glucose.

The type strain was isolated from the entrance floor of the JPL-SAF, Pasadena, CA, USA.

DNA G+C content (mol%): not reported.

Type strain: SAFN-007, ATCC BAA-1211, NBRC 101214.

EMBL/GenBank accession number (16S rDNA) of the type strain: AY167820 (SAFN-007).

59. **Paenibacillus peoriae** (Montefusco, Nakamura and Labeda 1993) Heyndrickx, Vandemeulebroecke, Scheldeman, Kersters, De Vos, Logan, Aziz, Ali and Berkeley 1996b, 999^{VP} (*Bacillus peoriae* Montefusco, Nakamura and Labeda 1993, 389)

pe.o′ri.ae. N.L. gen. n. *peoriae* of Peoria, named after Peoria, Illinois, where the organism was studied.

Grows on routine media such as nutrient agar producing butyrous, thin, smooth, circular, translucent colonies about 2 mm in diameter after incubation for 2 to 3 d at 30°C. Facultatively anaerobic, generally reduces nitrate to nitrite. Hydrolyzes pectin, pullulan, starch, and xylan. Acid and gas are produced from sugars. Acetylene reduction, an indication of nitrogen fixation, has been detected in some strains of this species (von der Weid et al., 2002). Major fatty acids are $C_{15:0\ anteiso}$, $C_{16:0}$, and $C_{17:0\ anteiso}$.

Isolated from soil and fodder.

DNA G+C content (mol%): 46 (Bd).

Type strain: DSM 8320, LMG 14832.

EMBL/GenBank accession number (16S rDNA) of the type strain: AB073186 (DSM 8320).

60. **Paenibacillus phyllosphaerae** Rivas, Mateos, Martínez-Molina and Velázquez 2005c, 745^{VP}

phyl.lo.sphae′rae. Gr. neut. n. *phyllon* leaf; L. fem. n. *sphaera* ball, sphere; N.L. gen. fem. n. *phyllosphaerae* of the phyllosphere.

Cells are Gram-variable, aerobic or facultatively anaerobic, spore-forming rods, motile with peritrichous flagella. Forms ellipsoidal spores in swollen sporangia in subterminal position. Chemo-organotrophic and xylanolytic. Colonies on yeast extract-glucose medium are circular, flat, whitish-cream, opaque, and 1–3 mm in diameter after 48 h growth at 28°C. Growth occurs between 10 and 37°C (optimal growth temperature is 28°C). The optimal pH for growth is 7. Oxidase- and catalase-positive. No growth in the presence of 5% NaCl. The major quinone is MK-7. The main fatty acid is $C_{15:0\ anteiso}$. Gas is not produced from D-glucose, although a number of sugars are used as carbon sources. Xylanase, cellulase, amylase, and β-galactosidase are actively produced. Caseinase, arginine dihydrolase, indole, lysine decarboxylase, ornithine decarboxylase, urease, phenylalanine deaminase, tryptophan deaminase, hydrogen sulfide, and acetoin (Voges–Proskauer medium) reactions are negative. Nitrate is reduced to nitrite.

Isolated from the phyllosphere of *Phoenix dactylifera* in Palma de Mallorca (Spain).

DNA G+C content (mol%): 50.7 (T_m).

Type strain: PALXIL04, LMG 22192, CECT 5862.

EMBL/GenBank accession number (16S rDNA) of the type strain: AY598818 (PALXIL04).

61. **Paenibacillus popilliae** (Dutky 1940) Pettersson, Rippere, Yousten and Priest 1999, 539^{VP} (*Bacillus popilliae* Dutky 1940, 68)

po.pil′li.ae. N.L. n. *Popillia* generic name of the Japanese beetle; N.L. gen. n. *popilliae* of *Popillia*.

Cells are rods (0.5–0.8 by 1.3–5.2 μm), Gram-variable to Gram-negative in exponential phase and Gram-positive when sporulating. Nutritionally fastidious, requires J-broth (Gordon et al., 1973) or MYGPG (Costilow and Coulter, 1971) medium. On these media, forms cream-colored colonies less than 1 mm in diameter. Sporulation in laboratory media is poor. Spores are produced *in vivo* during growth in susceptible insect larvae. Spores are usually accompanied by a parasporal crystal. Facultatively anaerobic, does not reduce nitrate. Catalase-negative. Most strains grow in 2% NaCl or 150 μg/ml vancomycin, which distinguishes the species from *Paenibacillus lentimorbus*; however, strains isolated in South America are vancomycin-sensitive (Harrison et al., 2000). Does not grow in media containing 5% NaCl. Causative agent of type A milky disease in Japanese beetle (*Popillia japonica* Newman) and related scarab larvae which is characterized by increasing milky whiteness of the almost translucent grub as the bacterium grows and sporulates in the hemolymph. Major fatty acids are C$_{15:0}$ anteiso and C$_{16:0}$.

Isolated from diseased larvae of Japanese beetle.

DNA G+C content (mol%): 41 (*T$_m$*).

Type strain: ATCC 14706, DSM 2047, KCTC 3766, LMG 17744.

EMBL/GenBank accession number (16S rDNA) of the type strain: AB073198 (ATCC 14706).

Additional remarks: Strains named as *Paenibacillus popilliae* are phylogenetically heterogeneous.

62. **Paenibacillus rhizosphaerae** Rivas, Gutiérrez, Abril, Mateos, Martínez-Molina, Ventosa and Velázquez 2005a, 1307^{VP}

rhi.zo.sphae′rae. Gr. fem. n. *rhiza* root; L. fem. n. *sphaera -ae* ball, any globe, sphere; N.L. gen. fem. n. *rhizosphaerae* of the rhizosphere.

Cells are Gram-positive aerobic rods (0.9–1.0 by 3.0–3.1 μm), motile by means of peritrichous flagella. Spores are subterminal and cause slight swelling of the sporangia. Colonies grown on nutrient agar (for 48 h at 28°C) are circular, convex, cream-colored, opaque, and usually 1–3 mm in size. Growth occurs at 10–37°C and pH 5–9. The optimum growth temperature is 28°C and the optimum pH is 7. Grows without NaCl and with up to 5.0% (w/v) NaCl. Oxidase- and catalase-positive. Gas is not produced from glucose. Acid is produced from various carbohydrates. Gelatinase, urease, and hydrogen sulfide are not produced. The predominant fatty acids are C$_{15:0\ anteiso}$, C$_{16:0}$, and C$_{16:0\ iso}$.

The type strain was isolated from the rhizosphere of the legume *Cicer arietinum* in Argentina.

DNA G+C content (mol%): 50.9 (*T$_m$*; type strain).

Type strain: CECAP06, LMG 21955, CECT 5831.

EMBL/GenBank accession number (16S rDNA) of the type strain: AY751754 (CECAP06).

63. **Paenibacillus sanguinis** Roux and Raoult 2004, 1052^{VP}

san′gui.nis. L. masc. gen. n. *sanguinis* of blood, referring to the fact that the type strain was isolated from a blood sample.

Gram-positive, facultatively anaerobic rods (0.5 by 2.0–3.0 μm), motile by peritrichous flagella. Ellipsoidal endospores are formed in swollen sporangia. Colonies are grayish, translucent, shiny, regular circles, and 1 mm in diameter after 24 h incubation at 30°C (the medium was not reported). Catalase- and oxidase-negative. Optimal growth occurs at 30–37°C, but no growth occurs at 50°C. Growth does not occur in the presence of 5% (w/v) NaCl. Nitrate is not reduced and gelatin is not liquefied. Acid is produced from various carbohydrates.

The type strain was isolated from blood culture.

DNA G+C content (mol%): not reported.

Type strain: 2301083, CIP 107938, CCUG 48214.

EMBL/GenBank accession number (16S rDNA) of the type strain: AY323609 (2301083).

64. **Paenibacillus sepulcri** Šmerda, Sedlá ek, Pá ová, Krej í and Havel 2006, 2343^{VP}

se.pul′cri. L. gen. n. *sepulcri* from a tomb, pertaining to the place of isolation of the type strain.

Cells are Gram-variable, facultatively anaerobic, spore-forming rods. Spores are oval, subterminally in swollen sporangia. Moderately psychrotolerant. The temperature range for growth is 10–30°C and the optimum is 25°C; no growth occurs at 37°C. Grows at pH 6.0–8.0; the optimum is between pH 7.2 and 7.4. Tolerates up to 2% NaCl, but not 3% NaCl. Colonies on nutrient agar are circular, smooth, slightly convex with complete edges, and colorless. Oxidase-positive. Acid is produced from various carbohydrates. Phylogenetically, the most closely related species is *Paenibacillus mendelii*. The predominant menaquinone is MK-7. The cell wall contains *meso*-DAP of the A1γ type with C$_{15:0}$ anteiso as major fatty acid component.

Isolated from the biodeteriorated mural paintings in the Servilia tomb (Spain).

DNA G+C content (mol%): 50 (HPLC).

Type strain: CCM 7311, LMG 19508.

EMBL/GenBank accession number (16S rDNA) of the type strain: DQ291142 (CCM 7311).

65. **Paenibacillus stellifer** Suominen, Spröer, Kämpfer, Rainey, Lounatmaa and Salkinoja-Salonen 2003, 1373^{VP}

stel′li.fer. L. masc. adj. *stellifer* star-bearing/starry, referring to the presence of star-shaped spores.

Gram-positive, facultatively anaerobic rods (0.6–0.8 by 2.5–5.0 μm), motile by means of peritrichous flagella. In swollen sporangia, terminally ellipsoidal spores are formed with spikes in electron micrographs of thin sections that are actually ribs connecting the poles of the spore. Spores as well as vegetative cells have pilus-like appendages. The temperature range for growth is 15–40°C. Does not reduce nitrate to nitrite. Hydrolyzes starch with the formation of cyclodextrins. Produces acid from various sugars. The temperature range for growth is 15–40°C. Non-hemolytic. Catalase-positive and oxidase-negative. Nitrate is not reduced to nitrite or nitrogen. Some biochemical characteristics are reported by Suominen et al. (2003) as being similar to those of *Paenibacillus durus/Paenibacillus azotofixans* without the exact data being provided, which makes a correct interpretation difficult. The major fatty acid is $C_{16:0}$ (34–45% total fatty acids) and the ratio of $C_{15:0}$ iso:anteiso (28°C) is 2.3–2.5. Produces cyclodextrins from potato starch.

Isolated from food packaging paper and board.

DNA G+C content (mol%): 55.6 (HPLC).

Type strain: IS 1, CCUG 45566, DSM 14472.

EMBL/GenBank accession number (16S rDNA) of the type strain: AJ316013 (IS 1).

66. **Paenibacillus terrae** Yoon, Oh, Yoon, Kang and Park 2003, 300[VP]

ter'rae. L. gen. n. *terrae* of the earth.

Cells are Gram-variable, facultatively anaerobic rods (1.3–1.8 by 4.0–7.0 μm), motile by means of peritrichous flagella on TSA. Ellipsoidal spores are formed in swollen sporangia. Colonies are cream-colored, irregular in shape, thin, and translucent on TSA. Growth occurs at 10–40°C with an optimum at 30°C. Optimal pH for growth is 6.5–8.0; no growth at pH 4.0. Grows optimally in the presence of 0–2% (w/v) NaCl and growth still occurs in presence of 3% (w/v), but not in the presence of 4% (w/v) NaCl. Oxidase-negative. Reduces nitrate to nitrite; one strain has been reported to denitrify (Horn et al., 2005). Hydrolyzes esculin, casein, gelatin, and starch, but not urea. Produces acid, but not gas, from a range of sugars. Major fatty acid is $C_{15:0\ anteiso}$. Predominant menaquinone is MK-7.

Isolated from a soil sample from Taejon City, Korea.

DNA G+C content (mol%): 47 (HPLC).

Type strain: AM141, KCCM 41557, JCM 11466.

EMBL/GenBank accession number (16S rDNA) of the type strain: AF391124 (AM141).

67. **Paenibacillus thiaminolyticus** (*ex* Kuno 1951) Shida, Takagi, Kadowaki, Nakamura and Komagata 1997a, 297[VP] ("*Bacillus thiaminolyticus*" ex Kuno 1951, 364; *Bacillus thiaminolyticus* Nakamura 1990, 245)

thi.am.in.o.ly'tic.us. N.L. n. *thiaminum* thiamine; N.L. adj. *lyticus* dissolving; N.L. adj. *thiaminolyticus* decomposing thiamine.

Cells are Gram-positive, facultatively anaerobic, motile rods (0.5–1.0 by 2.0–3.0 μm). Ellipsoidal spores are formed in swelling sporangia. Agar colonies are translucent, thin, smooth, and entire, measuring 1.0–2.0 mm in diameter. Optimal temperature for growth is 28°C with 20°C and 45°C as respective minimum and maximum. No growth is observed at pH 5.6 or 5.7. Most strains grow in media containing 5% NaCl, but are inhibited by 7% NaCl. Nitrate is usually reduced to nitrite. Hydrolyzes chitin, casein, pullulan, and starch. Decomposes tyrosine and thiamine. Acid, but no gas, is produced from sugars. Major fatty acids are $C_{15:0\ anteiso}$, $C_{15:0\ iso}$, and $C_{17:0\ anteiso}$.

Isolated from human fecal material and honeybee larvae.

DNA G+C content (mol%): 52–54 (Bd).

Type strain: DSM 7262, NRRL B-4156, IFO 15656.

EMBL/GenBank accession number (16S rDNA) of the type strain: AJ320490 (DSM 7262).

68. **Paenibacillus timonensis** Roux and Raoult 2004, 1053[VP]

ti.mo.nen'sis. N.L. masc. adj. *timonensis* pertaining to the Hôpital de la Timone, where the type strain was isolated.

Gram-positive, facultatively anaerobic rods (0.5 by 2.0 μm). Ellipsoidal endospores are formed in swollen sporangia. Colonies are grayish, translucent, shiny, 1.5 mm in diameter, regularly shaped, and low convex after incubation for 24 h at 30°C. Motile by peritrichous flagella. Catalase-positive, but oxidase-negative. Optimal growth occurs at 30–37°C; also grows at 50°C. Growth does not occur in the presence of 5% (w/v) NaCl. Nitrate is reduced weakly, but gelatin is not liquefied. Acid is produced from various carbohydrates.

The type strain was isolated from blood culture.

DNA G+C content (mol%): not reported.

Type strain: 2301032, CIP 108005, CCUG 48216.

EMBL/GenBank accession number (16S rDNA) of the type strain: AY323610 (2301032).

69. **Paenibacillus turicensis** Bosshard, Zbinden and Altwegg 2002, 2247[VP]

tu.ri.cen'sis. L. adj. *turicensis* referring to *Turicum*, the Latin name of Zurich, where the organism was first isolated.

Cells are Gram-positive, facultatively anaerobic, motile rods (0.5–1.1 by 5–3 μm). Oval spores are terminal to subterminal in the sporangia. Grows on sheep blood agar producing non-hemolytic, grayish-white colonies, convex with regular margins. Grows poorly on Luria–Bertani medium. Growth occurs at 15–48°C with an optimum at 37 to 42°C. Growth occurs between pH 5.5 and 9.5 with an optimum at pH 7. Grows in the presence of up to 5% NaCl and 0.1% lysozyme. Catalase- and oxidase-negative. Unable to reduce nitrate. Positive for β-galactosidase and Voges–Proskauer reaction. Ferments a variety of sugars with the production of acid, but no gas. A distinctive feature is the variability of the 16S rRNA genes. Major fatty acids are $C_{15:0\ anteiso}$, $C_{16:0}$, and $C_{14:0}$.

Isolated from the valve of a cerebrospinal shunt.

DNA G+C content (mol%): not reported.

Type strain: MOL722, DSM 14349.

EMBL/GenBank accession number (16S rDNA) of the type strain: AF378694 (MOL722).

70. **Paenibacillus validus** (Bredemann and Heigner 1935) Ash, Priest and Collins 1995, 197[VP] (Effective publication: Ash, Priest and Collins 1993, 259) emend. Heyndrickx, Vandemeulebroecke, Scheldeman, Hoste, Kersters, De Vos, Logan, Aziz, Ali, Berkeley 1995, 667 ("*Bacillus validus*" ex Bredemann and Heigener 1935; *Bacillus validus* Nakamura 1984b, 225)

val'i.dus. L. adj. *validus* strong, vigorous.

Cells are rod-shaped (0.5–1.0 by 5.0–7.0 μm). Colonies on nutrient agar are grayish-white, translucent, smooth, circular, and entire, 1–2 mm in diameter. Growth occurs optimally at 28–30°C with 5–10 and 45–50°C as respective minimum and maximum ranges. Growth is inhibited by 3% NaCl. Strictly aerobic. Does not reduce nitrate to nitrite. Acid, but no gas, is produced from sugar fermentation. Selective growth on phenanthrene has resulted in isolation of *Paenibacillus validus* strains (Daane et al., 2002). Major fatty acids are $C_{15:0\ anteiso}$, $C_{16:0\ iso}$, and $C_{16:0}$.

Isolated from soil and estuarine sediments.

DNA G+C content (mol%): 50–52 (Bd).

Type strain: JCM 9077, ATCC 43897, DSM 3037, LMG 11161.

EMBL/GenBank accession number (16S rDNA) of the type strain: AB073203 (JCM 9077).

Additional remarks: This species represents DNA hybridization group 10 of *Bacillus circulans sensu lato* (Nakamura and J. Swezey, 1983). It is synonymous with *Paenibacillus gordonae* Heyndrickx, Vandemeulebroecke, Scheldman, Hoste, Kersters, De Vos, Logan, Aziz, Ali and Berkeley (1995).

71. **Paenibacillus wynnii** Rodríguez-Díaz, Lebbe, Rodelas, Heyrman, De Vos and Logan 2005, 2097^VP

wynn'i.i. N.L. gen. n. *wynnii* of Wynn, in honor of Dr David Wynn-Williams, the Antarctic microbiologist who developed Mars Oasis as a research site.

Gram-negative, motile, facultatively anaerobic, spore-forming curved rods (0.5–0.7 by 3–5 μm) with slightly tapered ends, occurring as single cells or in pairs. Endospores are ellipsoidal or oval and lay paracentrally and subterminally in sporangia that may be swollen. On nutrient agar at 20°C, the maximum colony diameter observed after 3 d is 2 mm. Colonies are circular, convex, and glossy with entire margins. Smaller colonies are transparent and whitish, whereas larger colonies are pale yellow–orange with whitish margins and darker centers. They bear a watery biomass, but may be mucoid. Older colonies are firmly attached to the agar. The optimum temperature for growth is 20°C; some strains grow at 37°C, but no growth is observed at 40°C. Growth is observed at 4°C in broth within 7 d. Optimal pH for growth lies between 7.0 and 8.0; pH 6.0–6.5 and pH 7.0–8.0 are respective minima and maxima. Catalase-positive and oxidase-negative. Does not tolerate 3% NaCl in the medium. Growth on skimmed milk agar plates is scarce and casein is not hydrolyzed. Starch is hydrolyzed. Fixes nitrogen as demonstrated by the presence of the *nifH* gene in all strains and acetylene reduction in most of them (partial sequence of the *nifH* gene is available in EMBL under accession number AJ867247). Voges–Proskauer reaction and gelatin hydrolysis reactions are negative. Reduces nitrate. Acid without gas is produced from various carbohydrates. The main cellular fatty acids are $C_{15:0\ anteiso}$ and $C_{16:0}$, each present in about 33% of the total fatty acid content.

The type strain and at least seven other strains were from a soil sample collected from four different locations at the Mars Oasis, Antarctica.

DNA G+C content (mol%): 44.6 (HPLC).

Type strain: LMG 22176, DSM 18334, CIP 108306.

EMBL/GenBank accession number (16S rDNA) of the type strain: AJ633647 (LMG 22176).

72. **Paenibacillus xinjiangensis** Lim, Jeon, Park, Xu, Jiang and Kim 2006b, 2581^VP

xin.ji.ang.en'sis. N.L. masc. adj. *xinjiangensis* pertaining to Xinjiang in China, where the type strain was isolated.

Cells are Gram-positive, strictly aerobic, spore-forming, motile rods (0.8–1.2 by 2.0–3.2 μm). Colonies are smooth, circular to slightly irregular, slightly convex, and cream-colored on TSA. Growth occurs at 10–40°C with an optimum at 30–35°C in media containing 0–3% (w/v) NaCl (opti-

mum 0–1%) and pH 6.5–9.8 (optimum pH 8.0–8.5). Catalase-positive and oxidase-negative. Nitrate is not reduced to nitrite. Hydrolysis of esculin is positive, but casein, starch, Tweens 20 and 80, L-tyrosine, hypoxanthine, xanthine, and urea are not hydrolyzed. Acids are produced from various carbohydrates. Cell wall contains *meso*-DAP (A1γ type). The predominant menaquinone is MK-7. The major cellular fatty acids on TSA are $C_{15:0\ anteiso}$ (48.61%), $C_{16:0}$ (14.43%), $C_{16:0\ iso}$ (9.65%), and $C_{15:0\ iso}$ (9.32%).

The type strain was isolated from Xinjiang province, China.

DNA G+C content (mol%): 47.0 (HPLC).

Type strain: B538, KCTC 3952, DSM 16970.

EMBL/GenBank accession number (16S rDNA) of the type strain: AY839868 (B538).

73. **Paenibacillus xylanilyticus** Rivas, Mateos, Martínez-Molina and Velázquez 2005b, 406^VP

xy.la.ni.ly'ti.cus. N.L. neut. n. *xylanum* xylan; Gr. masc. adj. *lutikos* able to loose, dissolving; N.L. masc. adj. *xylanilyticus* xylan-dissolving.

Cells are Gram-positive, facultatively anaerobic, spore-forming rods (1.5–1.55 by 3.9–4 μm), motile by peritrichous flagella. Subterminal oval endospores are formed in slightly swollen sporangia. Xylanolytic, colonies grown for 48 h at 37°C on YNBX agar (6.7 g/l yeast extract, 7 g/l xylan, 20 g/l agar) are circular, convex, cream-colored, opaque, and usually 1–3 mm in diameter. Optimum growth temperature is 37°C and optimum pH is 7. Oxidase-negative. Nitrate is reduced to nitrite. Does not produce gas from glucose. Acid is produced from various carbohydrates. Produces xylanases, gelatinase, amylase, and β-galactosidase, but not urease, caseinase, phenylalanine deaminase, lysine decarboxylase, arginine dihydrolase, ornithine decarboxylase, tryptophan deaminase, tyrosinase, indole, dihydroxyacetone, hydrogen sulfide, or acetoin. The predominant fatty acids are $C_{15:0\ anteiso}$ and $C_{16:0}$.

The type strain was isolated from air in a research laboratory, Salamanca University, Spain.

DNA G+C content (mol%): 50.5 (T_m).

Type strain: XIL14, LMG 21957, CECT 5839.

EMBL/GenBank accession number (16S rDNA) of the type strain: AY427832 (XIL14).

The following species were described after the submission deadline.

Paenibacillus fonticola Chou, Chou, Lin, Sheu, Sheu, Arun, Young and Chen 2007, 1348^VP

fon.ti.co'la. L. masc. n. *fons fontis* a spring, fountain; L. suff. *-cola* (from L. masc. or fem. n. *incola*) an inhabitant of a place, a resident; N.L. n. *fonticola* an inhabitant of a fountain.

The type strain was isolated from a water sample collected from Jhonglun warm spring, Hiayi County, Taiwan.

DNA G+C content (mol%): 49.2 (HPLC).

Type strain: ZL, BCRC 17579, LMG 23577.

EMBL/GenBank accession number (16S rDNA) of the type strain: DQ453131 (ZL).

Paenibacillus forsythiae Ma and Chen 2008, 321^VP

for.sy'thi.ae. N.L. gen. n. *forsythiae* of Forsythia, referring to the plant *Forsythia mira*, the source of the rhizosphere soil from which the type strain was isolated.

The type strain was isolated from rhizosphere soil of the plant *Forsythia mira* in the Beijing region, China.

DNA G+C content (mol%): 50.4 (T_m).

Type strain: T98, CCBAU 10203, DSM 17842.

EMBL/GenBank accession number (16S rDNA) of the type strain: DQ338443 (T98).

Paenibacillus ginsengarvi Yoon, Ten and Im 2007, 1812[VP]

gin.seng.ar'vi. N.L. n. *ginsengum* ginseng; L. n. *arvum* a field; N.L. gen. n. *ginsengarvi* of a ginseng field, the source of the type strain).

The type strain was isolated from soil from a ginseng field in Pocheon Province, South Korea.

DNA G+C content (mol%): 48.1 (HPLC).

Type strain: Gsoil 139, KCTC 13059, DSM 18677.

EMBL/GenBank accession number (16S rDNA) of the type strain: AB271057 (Gsoil 139).

Paenibacillus ginsengisoli Lee, Ten, Baek, Im, Aslam and Lee 2007c, 1372[VP] (Effective publication: Lee, Ten, Baek, Im, Aslam and Lee 2007b, 133.)

gin.sen.gi.so'li. N.L. n. *ginsengum* ginseng; L. n. *solum* soil; N.L. gen. n. *ginsengisoli* of soil of a ginseng field, the source of the organism.

The type strain was isolated from soil of a ginseng field in Pocheon province, South Korea.

DNA G+C content (mol%): 50.7.

Type strain: Gsoil 1638, KCTC 13931, LMG 23406.

EMBL/GenBank accession number (16S rDNA) of the type strain: AB245382 (Gsoil 1638).

Paenibacillus humicus Vaz-Moreira, Faria, Nobre, Schumann, Nunes and Manaia 2007, 2270[VP]

hu'mi.cus. L. n. *humus* earth, soil and, in earth sciences or agriculture, humus; L. suff. *-icus -a -um* suffix used with the sense of belonging to; N.L. masc. adj. *humicus* pertaining to humus.

The type strain was isolated from final compost produced from poultry litter.

DNA G+C content (mol%): 58. (HPLC)

Type strain: PC-147, DSM 18784, NBRC 102415.

EMBL/GenBank accession number (16S rDNA) of the type strain: AM411528 (PC-147).

Paenibacillus provencensis Roux, Fenner and Raoult 2008, 685[VP]

pro.ven.cen'cis. N.L. masc. adj. *provencensis* pertaining to Provence, the region of France where the type strain was isolated.

The type strain was isolated from human cerebrospinal fluid.

DNA G+C content (mol%): not reported.

Type strain: 4401170, CIP 109358, CCUG 53519.

EMBL/GenBank accession number (16S rDNA) of the type strain: EF212893.

Paenibacillus sabinae Ma, Xia, Liu and Chen 2007a, 9[VP]

sa'bi.nae. N.L. gen. n. *sabinae* of Sabina, referring to the plant *Sabina squamata*, the source of the rhizosphere soil from which the type strain was isolated.

The type strain was isolated from the rhizosphere soil of *Sabina squamata* planted in Beijing, China.

DNA G+C content (mol%): 51.9 (T_m).

Type strain: T27, CCBAU 10202, DSM 17841.

EMBL/GenBank accession number (16S rDNA) of the type strain: DQ338444 (T27).

Paenibacillus soli Park, Kim, An, Yang, Oh, Chung and Yang 2007, 149[VP]

so'li. L. neut. gen. n. *soli* of soil, the source of the organism.

The type strain was isolated from a soil sample from a ginseng field in the Republic of Korea.

DNA G+C content (mol%): 56.6–57.0.

Type strain: DCY03, KCTC 13010, LMG 23604.

EMBL/GenBank accession number (16S rDNA) of the type strain: DQ309072.

Paenibacillus taiwanensis Lee, Kuo, Tai, Yokota and Lo 2007a, 1353[VP]

tai.wan.en'sis. N.L. masc. adj. *taiwanensis* of Taiwan, where the type strain was isolated.

The type strain was isolated in 2000 from farmland soil in Wu-Feng, Taiwan.

DNA G+C content (mol%): 44.6 (HPLC).

Type strain: G-soil-2-3, BCRC 17411, IAM 15414, LMG 23799, DSM 18679.

EMBL/GenBank accession number (16S rDNA) of the type strain: DQ890521.

Paenibacillus terrigena Xie and Yokota 2007, 71[VP]

ter.ri.ge'na. L. n. *terrigena* (nominative in apposition) born of, or from, the earth, earth-born.

The type strain was isolated from coastal soil from Chiba, Japan.

DNA G+C content (mol%): 48.1 (HPLC).

Type strain: A35, IAM 15291, CCTCC AB206026.

EMBL/GenBank accession number (16S rDNA) of the type strain: AB248087.

Paenibacillus urinalis Roux, Fenner and Raoult 2008, 685[VP]

u.ri.na'lis. L. masc. adj. *urinalis* pertaining to urine, urinary.

The type strain was isolated from a human urine sample.

DNA G+C content (mol%): not reported.

Type strain: 5402403, CIP 109357, CCUG 53521.

EMBL/GenBank accession number (16S rDNA) of the type strain: EF212892.

Paenibacillus woosongensis Lee and Yoon 2008, 615[VP]

woo.song.en'sis. N.L. masc. adj. *woosongensis* of Woosong, the Korean name for the university in Korea at which the organism was isolated.

The type strain was isolated from forest soil at Daejeon in Korea.

DNA G+C content (mol%): 51.7 (HPLC).

Type strain: YB-45, KCTC 3953, DSM 16971.

EMBL/GenBank accession number (16S rDNA) of the type strain: AY847463 (YB-45).

Paenibacillus zanthoxyli Ma, Zhang and Chen 2007b, 876[VP]

zan'th.ox.y.li. N.L. gen. n. *zanthoxyli* of *Zanthoxylum*, referring to the plant *Zanthoxylum simulans*, the source of the rhizosphere soil from which the type strain was isolated.

The type strain was isolated from the rhizosphere soil of *Zanthoxylum simulans* planted in Beijing, China.

DNA G+C content (mol%): 53.2 (T_m).

Type strain: JH29, CCBAU 10243, DSM 18202.

EMBL/GenBank accession number (16S rDNA) of the type strain: DQ471303 (JH29).

Genus II. **Ammoniphilus** Zaitsev, Tsitko, Rainey, Trotsenko, Uotila, Stackebrandt and Salkinoja-Salonen 1998, 161[VP]

FRED A. RAINEY

Am.mo.ni.phi′lus. N.L. neut. n. *ammonium* ammonia (NH₃); Gr. adj. *philos* loving, friendly to; N.L. masc. n. *Ammoniphilus* ammonia lover.

Straight or slightly curved rods. Motile by peritrichous flagella. Endospores are formed. Gram reaction is variable. Cell wall consists of two electron-dense layers and electron-dense granules. **Obligate aerobe.** Catalase- and oxidase-positive. **Chemo-organotroph.** Produce H_2S from cysteine. Nitrate is not reduced. Indole-negative. High concentrations of ammonium ions required for growth. Growth is optimal at $\geq 0.07\,M$ NH_4^+. Oxalate is used as the sole organic source for carbon and energy. **Obligate oxalotroph.** Temperature optimum for growth is 28–30°C; pH optimum for growth is 8.0–8.5. Cell wall contains *meso*-diaminopimelic acid. The major menaquinone is MK-7. Phylogenetically this genus belongs to the Bacilli lineage of Gram-positive bacteria. Isolated from *Rumex acetosa* roots and decaying pinewood.

DNA G+C content (mol%): 42–46 (HPLC).

Type species: **Ammoniphilus oxalaticus** Zaitsev, Tsitko, Rainey, Trotsenko, Uotila, Stackebrandt and Salkinoja-Salonen 1998, 161[VP].

Further descriptive information

The genus *Ammoniphilus* was described on the basis of properties of nine strains from a collection of fifty-eight strains. The strains used for the description were RAOx-FS[T], RAOx-RS, RAOx-1[T], RAOx-PF, RAOx-PM, RAOx-FF, RAOx-RF, RAOx-RM (isolated from the rhizosphere of sorrel [*Rumex acetosa*]), and DWOx-RM (isolated from decaying pinewood). On OM-2 medium, colonies are visible after 2 d. The colonies of the different strains show variation in color and morphology after 4–8 days incubation. Strains RAOx-1[T], RAOx-PM, RAOx-FF, and RAOx-RF have light brown colonies that are circular, convex, have an entire margin and smooth surface. Strains RAOx-RM and DWOx-RM have colonies that are light brown in the center and bright-beige around the edges while those of RAOx-PF are mucoid and bright beige. Strains RAOx-FS[T] and RAOx-RS have white colonies (0.5–2 mm in diameter). The cells of the strains of this genus are motile, straight or slightly curved rods. Cells vary in length (1.0–8.0 μm), can be single, and sometimes in pairs or chains. Strains stain Gram-negative in young cultures and Gram-positive for cultures in exponential or stationary phases of growth. The cell wall of these strains has two electron-dense layers, which is atypical for Gram-positive bacteria. The endospores, which are stable at 80°C for 10 min, are located centrally or subterminally in slightly or non-swollen sporangia. Strains RAOx-1[T] and RAOx-FS[T] have been shown to have *meso*-diaminopimelic acid in their cell walls. All nine strains examined have MK-7 as the major menaquinone. The major fatty acids found in all strains are $C_{16:0}$ (22–29%), $C_{16:1\ \omega 7c}$ (13–36%), and $-C_{15:0\ anteiso}$ (8–37%). The amounts of these fatty acids vary between strains. Smaller amounts of $C_{14:0\ iso}$, $C_{14:0}$, $C_{15:0\ iso}$, $C_{15:0}$, $C_{16:0\ iso}$, and $C_{17:0\ anteiso}$ are found in all strains.

Growth is strictly aerobic with none of the nine strains tested growing under anaerobic conditions. Strains RAOx-FS[T] and RAOx-RS grow slowly when compared to the other seven rapidly growing strains and do not grow at $\leq 14°C$. Strains RAOx-

FS[T] and RAOx-RS differ from the other strains in that they grow with 5% NaCl, require ammonium oxalate (at least 5 g/l) for growth and do not grow on mixtures of formate and glycolate or methanol and glycolate. The strains of this genus are obligately oxalotrophic, and growth is not supported by acetate, adipate, adonitol, alanine, L-arabinose, L-asparagine, L-aspartate, D-cellobiose, citrate, dulcitol, ethanol, formate, fructose, fumarate, D-galactose, gluconate, glutamate, D-glucose, glycerate, glycerol, β-glycerophosphate, *meso*-inositol, 2-oxoglutarate, L-lactate, lactose, L-leucine, D-lyxose, malate, maleate, malonate, maltose, D- mannitol, D-mannose, DL-methionine, DL-norleucine, β-phenylalanine, L-proline, L-pyruvate, raffinose, L-rhamnose, D-ribose, sarcosine, L-serine, D-sorbitol, succinate, sucrose, DL-threonine, L-tryptophan, L-tyrosine, L-valine, and D-xylose. The strains do not grow on complex media such as nutrient agar, malt extract peptone agar, potato dextrose agar, or trypticase soy agar. Replacement of ammonium oxalate by ammonium sulfate and yeast extract, malt extract, peptone or tryptone did not support growth of any of the nine strains. Some strains show weak growth on glycolate or mixtures of formate and glycolate or methanol and glycolate. These strains all grow optimally at high concentrations of oxalate (20 g/l) and good growth occurs in saturated ammonium oxalate solutions (60 g/l). All of the strains require NH_4^+ for growth and can utilize thiosulfate, sulfate, sulfite, sulfide, DMSO, methionine, cysteine, glutathione, and thiouridine as sulfur sources.

Strains RAOx-FS[T] and RAOx-RS are at the low end of the mol% G+C range with values of 42. The other seven strains have mol% G+C values in the range 45–46.

Enrichment, isolation and growth conditions

The oxalate utilizing strains of the genus *Ammoniphilus* were isolated from decaying pinewood and the rhizosphere of sorrel (*Rumex acetosa*) using OM-2 medium. Medium OM-2 contains per liter: 10–20 g of $(NH_4)_2C_2O_4$ (ammonium oxalate), 10 g of $NaHCO_3$, 10 g of $NaH_2PO_4\cdot 2H_2O$, 0.7 g of NaCl, 0.57 g of KCl, 0.1 g of $MgCl_2\cdot 6H_2O$, 0.01 g of $CaCl_2\cdot 2H_2O$, and 1 g of $Na_2S_2O_3\cdot 5H_2O$. Adjusted to pH 6.8–7.0. $(NH_4)_2C_2O_4$, $NaHCO_3$, and $NaH_2PO_4\cdot 2H_2O$ should be dissolved completely and in the order listed. The medium will be pH 8.5–9.0 after autoclaving. Bacto Agar at 1.5–2.0% (w/v) is used to prepare solid medium. For enrichment cultures, 5 g of plant roots or 10 g of decaying wood was added to 100 ml of OM-2 medium in 500-ml flasks and incubated without shaking at 28°C for 20 d. Ten-ml aliquots of enrichment cultures showing ammonia production were transferred to 90 ml of OM-2 medium in a 500-ml flask incubated and shaken at 180 r.p.m. When turbidity was observed, the culture was diluted and plated on OM-2 agar and incubated at 28°C until colonies appeared. Strains were routinely cultured on OM-2 medium. For substrate utilization tests, OM-2 medium contained ammonium sulfate (6 g/l) instead of ammonium oxalate. To examine the utilization of nitrogen sources, ammonium oxalate was replaced by sodium oxalate (10 g/l).

Taxonomic comments

The species of the genus *Ammoniphilus* belong to a distinct phylogenetic lineage within the Bacilli that includes the closely related genus *Oxalophagus*. The 16S rRNA gene sequences of *Ammoniphilus oxalaticus* and *Ammoniphilus oxalivorans* share 98.6% similarity. The closest relative is *Oxalophagus oxalicus* with around 97% 16S rRNA gene sequence similarity. Species of the genera *Bacillus* and *Aneurinibacillus* share less than 91% sequence similarity. The strains of the *Ammoniphilus* species have high 16S rRNA gene sequence similarities. 16S rRNA gene sequences (with >98.7% sequence similarity) are available for strains RAOx-FS^T (Y14580), RAOx-1^T (Y14578), and RAOx-FF (Y14579). Strains RAOx-1^T and RAOx-FF have the most similar sequences (99.3% identical). The closest phylogenetic relative of the strains of this genus is *Oxalophagus oxalicus* (96.0–96.7% 16S rRNA gene sequence similarity). DNA–DNA hybridization studies demonstrated the distinct species status of strains RAOx-FS^T and RAOx-1^T (reassociation value of 39.7%). The nine strains have been assigned to two species of the genus *Ammoniphilus*. Strains RAOx-1^T, RAOx-PF, RAOx-PM, RAOx-FF, RAOx-RF, RAOx-RM, and DWOx-RM belong to the species *Ammoniphilus oxalaticus* while the species *Ammoniphilus oxalivorans* comprises strains RAOx-FS^T and RAOx-RS.

Differentiation of the genus *Ammoniphilus* from other genera

The species of the genus *Ammoniphilus* can be differentiated from other genera on the basis of their phylogenetic position as well as phenotypic and physiological characteristics. *Ammoniphilus* is most closely related to the single species genus *Oxalophagus* to which it has 96.5% 16S rRNA gene sequence similarity. The next closest relatives are species of the genera *Bacillus* and *Aneurinibacillus* at 16S rRNA gene sequence similarity levels of below 91%. Although the 16S rRNA gene sequence similarity between the strains of the genera *Ammoniphilus* and *Oxalophagus* is high and in most cases would indicate membership in the same genus, there are very distinct physiological differences between these organisms that warrant separate genus status. In contrast to the genus *Oxalophagus*, which is catalase-negative and a strict anaerobe, strains of the genus *Ammoniphilus* are strictly aerobic and catalase-positive. The mol% G+C content of the DNA also differs between these taxa, with the value for *Oxalophagus* (35.4–37.2) being lower than that for *Ammoniphilus* strains (42–46). The use of oxalate as a sole carbon and energy source and the inability to use other organic substrates differentiates the genus *Ammoniphilus* from the distantly related species of the genus *Bacillus*. The strict requirement for NH$_4^+$ ions differentiates this genus from all other spore-forming taxa in the Bacilli lineage.

List of species of the genus *Ammoniphilus*

1. **Ammoniphilus oxalaticus** Zaitsev, Tsitko, Rainey, Trotsenko, Uotila, Stackebrandt and Salkinoja-Salonen 1998, 161^{VP}

o.xa.la'ti.cus. N.L. adj. *oxalaticus* pertaining to oxalate.

Straight or slightly curved rods. Cells are 1.0–8.0 μm in length and 0.6–1.1 μm in diameter. Cells occur singly, in pairs, or in short or long chains. Endospores are oval, centrally or subterminally located in either swollen or non-swollen sporangia, and moderately heat resistant to 80°C. Colonies on OM-2 agar after 4–8 d incubation are up to 5 mm in diameter, light brown or bright beige, convex, circular with entire margins, and smooth or mucoid. Obligate oxalotroph. Some strains grow poorly in 0.03 M NH$_4^+$ and not at all in ≤0.02 M NH$_4^+$. Growth occurs in the presence of ammonium oxalate (up to 100 g/l), with good growth occurring at 5–40 g/l, poor growth at 3–4 g/l, and no growth at ≤2 g/l. Glycolate is used by some strains as a carbon and energy source. Weak growth is supported by mixtures of formate and glyoxylate or methanol and glyoxylate. No growth occurs with other organic acids, sugars, or alcohols. Mesophilic, optimum temperature for growth is 28–30°C. Most strains grow at 10–40°C, pH 6.8–9.5, and are tolerant to 4% NaCl. Some strains slowly hydrolyze starch, gelatin, and urea. All strains form H$_2$S from cysteine but not from thiosulfate. Nitrates, nitrites, or urea are not utilized as sole nitrogen sources. No reduction of nitrate to nitrite or N$_2$; vitamins are not required; indole is not produced. The major cellular fatty acids are C$_{16:0}$ (22–29%), C$_{16:1 ω7c}$ (28–36%), and *anteiso*-C$_{15:0}$ (8–15%). Isolated from the rhizosphere of sorrel (*Rumex acetosa*) and decaying pinewood.

DNA G+C content (mol%) of the type strain: 46.0 (HPLC).

Type strain: RAOx-1, ATCC 700649, CIP 105538, DSM 11538, HAMBI 2283.

GenBank accession number (16S rRNA gene): Y14578.

2. **Ammoniphilus oxalivorans** Zaitsev, Tsitko, Rainey, Trotsenko, Uotila, Stackebrandt and Salkinoja-Salonen 1998, 161^{VP}

o.xa.li.vo'rans. N.L. neut. n. *oxalatum* oxalate; L. part. pres. *vorans* eating; N.L. part. adj. *oxalivorans* oxalate-eating.

Straight or slightly curved rods. Cells are 1.0–1.4 μm in length and 0.7–1.1 μm in diameter. Cells occur singly, in pairs or in short or long chains. Endospores are oval, and centrally or subterminally located in either slightly swollen or non-swollen sporangia. Colonies on OM-2 agar after 4–8 d incubation are up to 2 mm in diameter, white, convex, and circular with entire margins and smooth. Obligate oxalotroph. Grows poorly in 0.03–0.6 M NH$_4^+$ and not at all in ≤0.02 M NH$_4^+$. Growth occurs in ammonium oxalate (up to 100 g/l) with good growth occurring in 5–60 g/l. No growth occurs on ≤4 g/l ammonium oxalate, on glycolate, organic acids, sugars, or alcohols, or on mixtures of formate and glyoxylate or methanol and glyoxylate. Mesophilic, grows in range 20–38°C with optimum growth temperature at 28–30°C. The pH range for growth is 6.8–9.5, and strains are tolerant to 5% NaCl. Starch and gelatin are hydrolyzed. H$_2$S is formed from cysteine but not from thiosulfate. Nitrates, nitrites, or urea are not utilized as the sole nitrogen sources. No reduction of nitrate to nitrite or N$_2$; vitamins not required; urease- and indole-negative. The major cellular fatty acids are C$_{16:0}$ (26–28%), C$_{16:1ω7c}$ (13%) and *anteiso*-C$_{15:0}$ (36–37%). Isolated from the rhizosphere of sorrel (*Rumex acetosa*).

DNA G+C content (mol%) of the type strain: 42.0 (HPLC).

Type strain: RAOx-FS, ATCC 700648, CIP 105539, DSM 11537, HAMBI 2284.

GenBank accession number (16S rRNA gene): Y14580.

Genus III. **Aneurinibacillus** Shida, Takagi, Kadowaki and Komagata 1996a, 945[VP] emend. Heyndrickx, Lebbe, Vancanneyt, Kersters, De Vos, Logan, Forsyth, Nazli, Ali and Berkeley 1997, 814

NIALL A. LOGAN AND PAUL DE VOS

A.neu.ri.ni.ba.cil′lus. N.L. n. *aneurinum* thiamine; L. dim. n. *bacillus* small rod; N.L. masc. n. *Aneurinibacillus* thiamine-decomposing small rod.

Gram-positive, rod-shaped cells, 0.5–1.0 μm by 2.0–6.0 μm, and motile by peritrichous flagella. **Ellipsoidal spores**, one per cell, are **borne centrally, paracentrally, and subterminally and may swell the sporangia**. **Strictly aerobic**, but one species is microaerophilic. Grow on routine media such as nutrient agar and trypticase soy agar. Decompose thiamine. Catalase-positive, weakly positive or negative. Variable for nitrate reduction and hydrolysis of casein, gelatin, starch, and Tween 80. Urea is not hydrolyzed and indole is not produced. **Growth temperature range is from 20 to 65°C.** Growth occurs at pH 5.5 to 9.0. Growth occurs in the presence of 2–5% NaCl; some strains grow weakly at 7% NaCl. **Few carbohydrates are assimilated and acid is produced weakly, if at all, from them; amino acids and some organic acids are used as carbon sources.** The major cellular fatty acid components (ranges as percentages of total are given in parentheses) are $C_{15:0\ iso}$ (41.9–66.8), $C_{17:0\ iso}$ (1–23.8), $C_{16:0}$ (1.8–8.5), and $C_{16:0\ iso}$ (0.5–6.6). The major quinone is menaquinone 7. A specific S-layer protein is present.

DNA G+C content (mol%): 42–47.

Type species: **Aneurinibacillus aneurinilyticus** (*ex* Aoyama 1952) Heyndrickx, Lebbe, Vancanneyt, Kersters, De Vos, Logan, Forsyth, Nazli, Ali and Berkeley 1997, 815[VP] (*Bacillus aneurinolyticus* Shida, Takagi, Kadowaki, Yano, Abe, Udaka and Komagata 1994b, 146; *Aneurinibacillus aneurinolyticus* Shida, Takagi, Kadowaki and Komagata 1996a, 945).

Further descriptive information

Phylogeny. *Aneurinibacillus* belongs to the family *Bacillaceae* and is very closely related to the genus *Brevibacillus*. Both genera originated from a taxonomic rearrangement of *Bacillus* (Shida et al., 1996a). A phylogenetic tree, based on 16S rRNA gene sequences, comprising the type strains of *Aneurinibacillus* species is given in Figure 29.

The rod-shaped cells of *Aneurinibacillus* species are usually round-ended, and occur singly, in pairs, and in chains. Cell diameters range from 0.7 to 1.0 μm and lengths range from 3.0 to 6.0 μm, but the cells of particular strains are usually quite regular in size, and individual species normally have dimensions within fairly narrow limits. The spores are ellipsoidal, lie centrally, paracentrally, or subterminally, and swelling of the sporangia may be slight, moderate, or substantial, and vary within a strain (Figure 27).

Information on cell-wall composition is not available for *Aneurinibacillus* species, but *Aneurinibacillus thermoaerophilus* S-layer structure and biosynthesis have been studied intensively for a sugar beet extraction-juice isolate and a culture collection strain previously classified as *Bacillus brevis* (Kneidinger et al., 2001a, 2001b; Schäffer et al., 1999; Wugeditsch et al., 1999).

Colonies of the two mesophilic species of *Aneurinibacillus* species are flat, 0.5–3 mm in diameter after 48 h on nutrient agar at

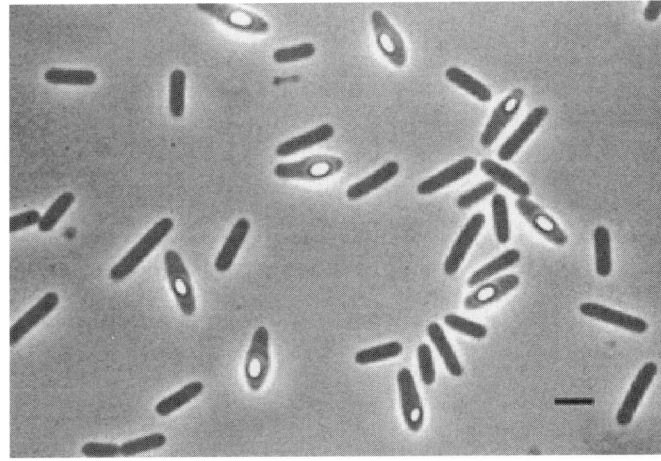

FIGURE 27. Photomicrograph of the type strain of *Aneurinibacillus aneurinilyticus* viewed by phase-contrast microscopy, showing ellipsoidal, paracentral, and subterminal spores that usually swell the sporangia. Bar = 2 μm. Photomicrograph prepared by N.A. Logan.

37°C, round or irregular in shape, with slightly crenate edges, and glossy, translucent, and creamy or yellowish gray in appearance; they become whitish and opaque as their component cells sporulate. Colonies of the thermophilic species may be larger, up to 10 mm in diameter, show a tendency to swarm across the surface of the agar, and have matt surfaces.

Aneurinibacillus species are heterotrophic and neutrophilic and will grow well on routine media such as nutrient agar or trypticase soy agar; growth may be enhanced by the addition of a small amount of yeast extract. Most strains will grow on blood agar. They will use some amino acids, carbohydrates, and organic acids as sole sources of carbon and energy. Utilization of carbohydrates may be accompanied by the production of small amounts of acid, and utilization of amino acids and organic acids may be accompanied by the production of small amounts of alkali but, generally speaking, these are not easily detected by routine test methods and some characters may prove to be inconsistent when retested. Growth temperatures range from about 20 to 50°C for the mesophiles (optima around 30–37°C) and from 40 to 60°C for the thermophilic species *Aneurinibacillus thermoaerophilus*.

Habitats. The first isolates of *Bacillus aneurinolyticus* were obtained from human feces (Aoyama, 1952); later isolations were made from human, bovine, chicken, dog, and rat feces, Japanese soil, and turban shells from the Japan Sea. The type strain of *Bacillus migulanus* was isolated from garden soil as a gramicidin producer. The original strains of the thermophilic species *Aneurinibacillus thermoaerophilus* were isolated from the high-temperature stages of beet sugar extraction and refining (Meier-Stauffer et al., 1996) and *Aneurinibacillus thermoaerophilus*, or a close relative of this species, was found to be a prominent member of the flora of hot synthetic compost (Dees and Ghiorse, 2001). The single isolate that represents *Aneurinibacillus danicus* was isolated at 45°C from a natural gas fermenter in Denmark. The geothermal soils of the Antarctic volcanoes Mount Melbourne and Mount Rittmann yielded strains of a novel, moderately thermophilic, and moderately acidophilic species, *Aneurinibacillus terranovensis*, which were isolated in small numbers along with *Bacillus fumarioli* (Allan et al., 2005; Logan et al., 2000). There are no reports of *Aneurinibacillus* species being isolated in association with infections of humans, other animals, or plants.

Enrichment and isolation

Enrichment and selective isolation methods developed especially for *Aneurinibacillus* species have not been reported. The original strains of *Aneurinibacillus aneurinilyticus* were isolated from human feces by Aoyama (1952) by adding 1 g feces to "broth", heating at 80°C for 30 min, then incubating at 37°C for 3 d. Broth (1 ml) was then added to a buffered solution (pH 7.0) of thiamin, incubated at 37°C for 30 min, and examined for evidence of thiamine degradation using the permutit-thiochrome method. Strains that could decompose thiamine were then isolated on plates of chocolate blood agar. Abe and Kimoto (1984) isolated further strains from human, bovine, chicken, dog, and rat feces, soil, and seashells, but the isolation procedure was not reported. Abe et al. (1986) described a procedure for detecting thiaminase-producing colonies: cells of the organisms under test were diluted with physiological saline and spread on a plate of nutrient agar containing 0.1% yeast extract so as to obtain no more than 100 colonies per plate, then incubated at 37°C for 4 d. The plates were then overlaid with 8 ml molten soft agar at 50–55°C containing, for detection of "*Bacillus aneurinolyticus*": thiamin, 0.8 g; agar, 5 g; 25 mM Tris/HCl buffer, 1000 ml. After the soft agar had set, the plate was covered and incubated at 60°C for 2 h, then 16 ml freshly prepared diazo-reagent was poured onto the agar surface, left for 5–10 min at room temperature, then poured off. A yellow halo surrounding the colony against a reddish-pink background revealed zones of thiamin decomposition. The diazo-reagent was prepared from three solutions: (i) 0.6 g *p*-amino-acetophenone dissolved in 9 ml concentrated HCl (11.6 M), diluted with water to 100 ml, and stored in a brown bottle; (ii) 23 g sodium nitrite dissolved in 100 ml water and stored in a brown bottle; (iii) 20 g sodium hydroxide and 28 g sodium bicarbonate dissolved in 350 ml water. The solutions could be stored at room temperature for several weeks. The reagent was prepared by adding 0.2 ml sodium nitrite solution to 0.8 ml *p*-amino-acetophenone solution in a glass tube followed by 10 ml distilled water, then 6 ml sodium hydroxide/ sodium bicarbonate buffer, which was poured gradually into the tube; the mixture was poured onto the plate immediately.

Edwards and Seddon (2000) described an isolation method for detecting colonies of gramicidin-producing *Bacillus brevis* from field trial experiments; the method utilized the ability of this strain to decompose tyrosine, so producing light-brown colonies surrounded by haloes on tyrosine agar. Isolates producing gramicidin could then be identified by paper chromatography. The type strain of *Aneurinibacillus migulanus* was isolated as a gramicidin-producing strain (Takagi et al., 1993). Tyrosine utilization is found in both mesophilic *Aneurinibacillus* species and *Brevibacillus* species (as well as many organisms outside these genera), so although this method is by no means specific or selective, it is potentially of assistance in detecting colonies of strains belonging to these two genera. Tyrosine agar contains: nutrient broth (Oxoid), 6.6 g; tyrosine, 5 g; agar, 15 g; water, 1000 ml. After autoclaving, it is stirred continuously with a magnetic stirrer (to reduce the size of the tyrosine crystals) until it has reached 50°C, whereupon it is poured immediately, giving an opaque, off-white, solid medium. *Aneurinibacillus thermoaerophilus* was originally isolated by smearing sugar beet extraction juice onto plates of TYG agar and incubating in air at about 60°C; colonies were purified by plating on the same medium. TYG agar contains: Bacto Tryptone (Difco), 5 g; yeast extract, 2.5 g; glucose, 1 g; agar, 22 g; water, 1000 ml. Dees and Ghiorse (2001) isolated a close relative of *Aneurinibacillus thermoaerophilus* from dilutions of hot synthetic compost by plating on a variety of routine media such as nutrient broth, plate count agar, and trypticase soy agar that were diluted to 1/10 of the manufacturer's recommended concentration and then amended with bacteriological agar to solidify.

Wakisaka and Koizumi (1982) noted that some aerobic endospore-formers that appeared to form minor components of soil floras showed slow and/or uneven germination compared with the more frequently encountered *Bacillus* species such as *Bacillus cereus*, *Bacillus megaterium*, *Bacillus sphaericus*, and *Bacillus subtilis*. Members of these minor populations were isolated with difficulty by the standard dilution-plate technique, but could be enriched by first removing the rapidly germinating, fast-growing members of the predominant flora. This approach has not been tried for the isolation of *Aneurinibacillus*, but it may well be of assistance (for details, see the treatment of *Brevibacillus*).

Aneurinibacillus terranovensis was isolated from volcanic soils of Mount Melbourne and Mount Rittmann, northern Victoria Land, Antarctica. Soil (1 g quantities) was added to 9 ml *Bacillus fumarioli* broth (BFB) in duplicate at pH 5.5, and one of each pair was heat-treated at 80°C for 10 min to kill vegetative cells. All broths were incubated at 50°C in water baths and inspected daily. Cultures that became turbid were subcultured by streaking onto plates of *Bacillus fumarioli* agar (BFA). Colonies appearing on streak plates were screened for vegetative and sporangial morphologies by phase-contrast microscopy; spore-forming rods were streak-purified and then transferred to slopes of the same medium for storage at 4°C after incubation and confirmation of sporulation by microscopy. BFB contains 4 g yeast extract, 2 g $(NH_4)_2SO_4$, 3 g KH_2PO_4, and 4 ml/l of each of solutions A and B [A, 125 g $(NH_4)_2SO_4$ and 50 g $MgSO_4 \cdot 7H_2O$ per liter; B, 62.5 g $CaCl_2 \cdot 2H_2O$ per liter], adjusted to pH 5.5. BFA was prepared by adding 5 mg $MnSO_4 \cdot 4H_2O$ and 18 g/l agar to BFB prior to autoclaving.

Maintenance procedures

Aneurinibacillus strains may be preserved on slopes of a suitable growth medium that encourages sporulation, such as nutrient agar or trypticase soy agar containing 5 mg/l $MnSO_4 \cdot 7H_2O$. Slopes should be checked microscopically for spores, before sealing to prevent drying out, and stored in a refrigerator; spores should remain viable on such sealed slopes for many years. For longer-term preservation, lyophilization and liquid nitrogen may be used, as long as cryoprotectants are added.

Procedures for testing special characters

Members of the genus *Aneurinibacillus* tend to be unreactive in routine biochemical tests and no special characters have been described for their differentiation; hence, their identification is difficult. They are largely unreactive in the carbohydrate utilization tests of the API 50CHB gallery (bioMérieux) and insufficiently variable in the API 20E and supplementary tests, so that the two mesophilic species in this genus are largely inseparable by these means. The API Biotype 100 gallery, which was developed as a research product for differentiating enterobacteria, proved to be of great value in differentiating species of *Brevibacillus* and *Aneurinibacillus* (Heyndrickx et al., 1997; Logan et al., 2002); it contained 99 tests for the assimilation of carbohydrates, organic acids, and amino acids, and one control tube. It was inoculated with a suspension in one of two semisolid media that differed in the number of growth factors they contained and, after incubation, the tubes were examined for turbidity. This system was adapted by using a suspension medium containing phenol red and examining the tubes for evidence of acid or alkali production (A. H. A. Albaser and N. A. Logan, unpublished results). For further guidance, the reader is referred to *Procedures for testing special characters* in the treatment of *Brevibacillus*.

Differentiation of members of the genus *Aneurinibacillus* from other genera

The genus *Aneurinibacillus* contains aerobic, endospore-forming rods that may be confused with members of other genera of aerobic endospore-formers, including *Bacillus*. The most characteristic feature of *Aneurinibacillus* species is their lack of reactivity in routine biochemical tests and their tendency to form swollen sporangia; however, members of the genus *Brevibacillus* have similar characteristics, and distinction of these two genera cannot be achieved easily on the basis of routine phenotypic tests.

Taxonomic comments

Bacillus aneurinolyticus was described by Aoyama (1952) as a thiamine-decomposing organism from human feces, but was omitted from the Approved Lists of Bacterial Names (Skerman et al., 1980) owing to a paucity of representative strains. Gordon et al. (1973) studied four strains, but they felt their experience with them was insufficient to allow delineation of the species, and Logan and Berkeley (1981) were unable to separate their strains of *Bacillus aneurinolyticus* from *Bacillus brevis*. The species was revived by Shida et al. (1994b) on the basis of a polyphasic analysis of 21 strains. Phenotypically, this organism was known to resemble *Bacillus brevis* and related taxa (Claus and Berkeley, 1986) and studies on the 16S rRNA gene sequences of the type strains of these two species suggested that *Bacillus aneurinolyticus* represented a distinct evolutionary line close to that of *Bacillus brevis* (Ash et al., 1991) or that it diverged early from the *Bacillus brevis* line (Farrow et al., 1992, 1994). Following DNA relatedness studies and chemotaxonomic analyses, the taxonomy of *Bacillus brevis* was modified by assigning some *Bacillus brevis*-group strains to the novel species *Bacillus agri* and *Bacillus centrosporus* (Nakamura, 1993), *Bacillus migulanus, Bacillus choshinensis,* and *Bacillus parabrevis* (Takagi et al., 1993), and *Bacillus reuszeri, Bacillus formosus,* and *Bacillus borstelensis* (Shida et al., 1995). The thermophilic species *Bacillus*

thermoaerophilus was described to include sugar beet isolates and *Bacillus brevis* strain ATCC 12990 (MeierStauffer et al., 1996). In the same year, Shida et al. (1996b) reported the findings of a SDS-PAGE study that supported the divergence of the *Bacillus brevis* and *Bacillus aneurinolyticus* groups, but although the *Bacillus aneurinolyticus* group species, *Bacillus aneurinolyticus* and *Bacillus migulanus,* were distinguishable by this technique, when the analysis was restricted to the type strains of individual species of the *Bacillus brevis* and *Bacillus aneurinolyticus* groups, they were not always well separated by this method. Similar problems were encountered by Logan et al. (2002). Shida et al. (1996a) created two new genera on the basis of 16S rRNA gene sequence analysis (type strains only) to contain the above-mentioned and allied species: *Aneurinibacillus* accommodated *Aneurinibacillus aneurinolyticus* and *Aneurinibacillus migulanus,* whereas *Brevibacillus* accommodated *Brevibacillus brevis* and the species derived from it (see above), *Brevibacillus laterosporus,* and *Brevibacillus thermoruber.* Meier-Stauffer et al. (1996) proposed *Bacillus thermoaerophilus* in the same year and suggested that it might, along with *Bacillus aneurinolyticus* and *Bacillus migulanus,* represent the core of a new genus, but it was not included in either of the Shida et al. (1996a, 1996b) studies.

Heyndrickx et al. (1997) carried out a polyphasic taxonomic study on 37 strains belonging to *Aneurinibacillus* and *Brevibacillus,* using amplified rDNA restriction analysis (ARDRA), fatty acid methyl ester analysis, SDS-PAGE of whole-cell proteins, pyrolysis mass spectrometry, assimilation tests, and other routine phenotypic tests. Two of the species, *Aneurinibacillus aneurinolyticus* (the type species) and *Aneurinibacillus migulanus,* were found to be quite similar phenotypically and genotypically, but distinguishable from each other by a small number of phenotypic characters. ARDRA revealed that *Aneurinibacillus aneurinolyticus, Aneurinibacillus migulanus,* and *Bacillus thermoaerophilus* formed a cluster that was quite separate from the *Brevibacillus* one, supporting the distinction of both genera, and they transferred *Bacillus thermoaerophilus* to *Aneurinibacillus.* The species epithet *aneurinolyticus* was corrected to *aneurinilyticus* at this time.

However, because *Aneurinibacillus* strains are unreactive in many of the conventionally formulated biochemical tests upon which routine identification schemes are based, identifications rely too heavily on negative test results. Distinction between most *Aneurinibacillus* and *Brevibacillus* species is thus not possible using the currently available *Bacillus* identification schemes and their separation remains difficult even with much wider selections of phenotypic tests; also, another unreactive species, *Bacillus badius,* may be easily misidentified as a member of these genera (Heyndrickx et al., 1997; Logan et al., 2002). The original recognition of the two mesophilic *Aneurinibacillus* species was based mainly upon DNA relatedness studies, molecular probing, and chemotaxonomic analyses of the relatively few available isolates (Shida et al., 1994b), using databases that remain restricted largely to reference laboratories. Such an approach to identification is unsuitable for such unreactive organisms encountered only occasionally in routine laboratories and isolates suspected of being *Aneurinibacillus* or *Brevibacillus* species may require referral to a reference laboratory.

Given these problems of identification, Goto et al. (2004) examined the hypervariable (HV) region corresponding to the 5′ end of the 16S rRNA gene (nucleotide positions 70–344 in

Bacillus subtilis numbering) in 29 strains received at the culture collections as *Brevibacillus brevis*, along with strains of *Aneurinibacillus* and *Brevibacillus* species and other, unnamed, *Brevibacillus* strains. The HV region marker had already proved to be useful taxonomically for the genera *Alicyclobacillus*, *Bacillus*, and *Paenibacillus*. They found that 14 *Brevibacillus brevis* and three *Brevibacillus* spp. strains clustered in *Aneurinibacillus*: two with *Aneurinibacillus migulanus* and 14 with *Aneurinibacillus thermoaerophilus*. One strain did not cluster with any of the existing *Aneurinibacillus* strains and was proposed as the novel species *Aneurinibacillus danicus*.

More reliable species differentiation and identification may be based in the future on genomic sequences of well selected so-called housekeeping genes.

Acknowledgements

We thank Abdul Hadi Ali Albaser for his assistance with phenotypic profiling of *Aneurinibacillus* species, and Raymond Allan for his data on *Aneurinibacillus terranovensis*.

List of species of the genus *Aneurinibacillus*

1. **Aneurinibacillus aneurinilyticus** (*ex* Aoyama 1952) Heyndrickx, Lebbe, Vancanneyt, Kersters, De Vos, Logan, Forsyth, Nazli, Ali and Berkeley 1997, 815^{VP} (*Bacillus aneurinolyticus* Shida, Takagi, Kadowaki, Yano, Abe, Udaka and Komagata 1994b, 146; *Aneurinibacillus aneurinolyticus* Shida, Takagi, Kadowaki and Komagata 1996a, 945)

a.neu.ri.no.ly'tic.us. N.L. n. *aneurinum* thiamine; N.L. adj. *lyticus* dissolving; N.L. adj. *aneurinolyticus* decomposing thiamine.

Colonies on nutrient agar after 48 h at 37°C are flat, 0.5–2 mm in diameter, round or irregular in shape, with slightly crenate edges, glossy, translucent, and creamy grayish. Vegetative cells are 0.7–0.9 μm by 3.0–5.0 μm. Colonies become whitish and opaque as their component cells sporulate. Spores are ellipsoidal, paracentral, and subterminal and may swell the sporangia. Catalase weakly positive. Growth temperatures range from 20 to 50°C. Grows at pH 5.0–9.0. Optimum temperature for growth is 37°C; optimum pH for growth is pH 7.0. Nitrate is reduced to nitrite. Casein, gelatin, and Tweens 20, 40, 60, and 80 are not hydrolyzed. Utilization of some organic compounds and other characteristics are shown in Table 50 and Table 51.

The major cellular fatty acid components (ranges as percentages of total) are $C_{14:0 \text{ iso}}$ (1.4–8.8%), $C_{14:0}$ (1.6–4.0%), $C_{15:0 \text{ iso}}$ (41.9–66.3%), $C_{16:0 \text{ iso}}$ (0.5–4.2%), $C_{16:1 \omega 11c}$ (8.0–13.5%), $C_{16:0}$ (2.2–8.5%), $C_{17:1 \omega 10c}$ (1.5–3.4%), $C_{17:0 \text{ iso}}$ (1.0–3.4%), H $C_{15:1 \text{ iso}}$ and/or I $C_{15:1 \text{ iso}}$ and/or 3-OH $C_{13:0}$ (0.9–2.8%), $C_{16:1 \omega 7c}$ and/or 2-OH $C_{15:0 \text{ iso}}$ (2.7–10.3%), and I $C_{17:1 \text{ iso}}$ and/or B $C_{17:1 \text{ anteiso}}$ (1.0–3.5%). Isolated from human, bovine, chicken, dog, and rat feces, soil, and sea shells.

DNA G+C content (mol%): 41.1–43.4 (HPLC); that of the type strain is 42.9.

Type strain: DSM 5562, LMG 15531, ATCC 12856, CIP 104007, NRRL NRS-1589.

EMBL/GenBank accession number (16S rRNA gene): X94194 (DSM 5562).

2. **Aneurinibacillus danicus** Goto, Fujita, Kato, Asahara and Yokota 2004, 425^{VP}

da'ni.cus. N.L. adj. *danicus* Danish, pertaining to Denmark.

Strictly aerobic, Gram-variable, motile rods, 0.8–1.0 μm by 4.0–6.0 μm. Ellipsoidal spores are borne subterminally in swollen sporangia. Description is based upon a single isolate. Colonies on nutrient agar are circular, entire, smooth, flat, translucent, and white, and they are 5–10 mm in diameter after 48 h. The temperature range for growth is 35–55°C and the temperature for optimum growth is 45–50°C. Optimum pH for growth is 6.5–7.0; growth occurs at pH 6.0–7.5, but not at pH 5.5 or 9.5. Growth is weak in 2% NaCl and is inhibited by 5% NaCl. Catalase- and oxidase-positive. Urease, citrate utilization, and nitrate reduction are negative. Casein, esculin, gelatin, and DNA are hydrolyzed, tyrosine is weakly hydrolyzed, and arbutin and starch are not hydrolyzed. Acid is produced from a range of carbohydrates (see Table 50). The major fatty acids are $C_{15:0 \text{ iso}}$, $C_{16:0}$, and $C_{17:0 \text{ iso}}$. The main quinone is menaquinone 7. Isolated from a natural gas fermenter.

DNA G+C content (mol%): 46.7 (HPLC).

Type strain: NCIMB 13288, IAM 15048.

EMBL/GenBank accession number (16S rRNA gene): AB112725 (NCIMB 13288).

3. **Aneurinibacillus migulanus** (Takagi, Shida, Kadowaki, Komagata and Udaka 1993) Shida, Takagi, Kadowaki and Komagata 1996a, 945^{VP} (*Bacillus migulanus* Takagi, Shida, Kadowaki, Komagata and Udaka 1993, 229)

mi.gu.la'nus. N.L. adj. *migulanus* referring to the German bacteriologist W. Migula, who contributed to bacterial taxonomy.

Colonies on nutrient agar after 48 h at 37°C are flat, 2–3 mm in diameter, with crenate edges, translucent, yellowish gray, and glossy. Vegetative cells are 0.5–1.0 μm by 2.0–6.0 μm. Colonies become creamy white and opaque as their component cells sporulate. Spores are ellipsoidal, paracentral, and subterminal and may swell the sporangia. Catalase-positive. Growth temperatures range from 20 to 50°C. Grows at pH 5.5–9.0. Nitrate is reduced to nitrite. Gelatin and starch are hydrolyzed. Casein and Tween 20, 40, 60, and 80 are not hydrolyzed. A range of carbohydrates, amino acids, and organic acids are used as carbon sources (see Table 50 and Table 51).

The major cellular fatty acid components (ranges as percentages of total) are $C_{14:0 \text{ iso}}$ (8.3–9.6%), $C_{14:0}$ (1.8–2.0%), $C_{15:0 \text{ iso}}$ (48.3–48.9%), $C_{15:0 \text{ anteiso}}$ (1.3%); $C_{15:1 \omega 6c}$ (0.8–1.0%), $C_{15:0}$ (2.3–3.3%), $C_{16:1 \omega 7c}$ alcohol (1.4%), H $C_{16:1 \text{ iso}}$ (1.3–1.7%), $C_{16:0 \text{ iso}}$ (6.5–6.6%), $C_{16:1 \omega 11c}$ (6.4–6.6%), $C_{16:1 \omega 5c}$ (1.1–1.2%), $C_{16:0}$ (3.4–3.6%), $C_{17:1 \omega 10c \text{ iso}}$ (1.7–2.2%), $C_{17:0 \text{ iso}}$ (2.1%), $C_{17:1 \omega 10c}$ (1.4–1.7%), H $C_{15:1 \text{ iso}}$ and/or I $C_{15:1 \text{ iso}}$ and/or 3-OH $C_{13:0}$ (1.3–1.6%), $C_{16:1 \omega 7c}$ and/or 2-OH $C_{15:0 \text{ iso}}$ (4.8–5.0%), and I $C_{17:1 \text{ iso}}$ and/or B $C_{17:1 \text{ anteiso}}$ (2.4–2.5%). Isolated from garden soil.

DNA G+C content (mol%): 42.5–43.2 (HPLC); that of the type strain is 42.5.

Type strain: DSM 2895, LMG 15427, ATCC 9999, CIP 103841, NCTC 7096.

EMBL/GenBank accession number (16S rRNA gene): X94195 (DSM 2895).

4. **Aneurinibacillus terranovensis** Allan, Lebbe, Heyrman, De Vos, Buchanan and Logan 2005, 1048^{VP}

terr.a.no.ven'sis. N.L. adj. *terranovensis* referring to Terra Nova Bay Station (Italy), northern Victoria Land, Antarctica, where the strains were first isolated.

TABLE 50. Differential characteristics of species of the genus *Aneurinibacillus*[a]

Characteristic	1. *A. aneurinilyticus*[b,c]	2. *A. danicus*[c]	3. *A. migulanus*[b,c]	4. *A. terranovensis*[d,e]	5. *A. thermoaerophilus*[b,f]
Hydrolysis of:					
Casein	−	+	−	ng	+
Gelatin	−	+	+	+	+
Starch	+	−	+	w	−
Growth at:					
20°C	d	−	+	d	−
30°C	+	−	+	+	−
50°C	d	+	d	+	+
55°C	−	+	−	d	+
Nitrate reduction	+	−	+	(+)	−
Acid production from:					
N-Acetylglucosamine	−	−	−	−	w
Adonitol	−	+	−	−	−
D-Arabinose	−	+	−		−
Dulcitol	−	+	−	−	−
D-Fructose	−	+	+	−[h]	w
D-Glucose	−	−	−	−[h]	+
myo-Inositol	−	−	+	−	−
D-Lyxose	−	+	−	−	+
D-Mannose	−	+	−	−[h]	−
Sorbitol	−	+	−	−	−
L-Sorbose	−	+	−	−	+
D-Tagatose	−	+	−	−	+
Xylitol	−	+	−	−	−
D-Xylose	−	+	−	−	+
Growth in NaCl:					
2%	+	w	+	−	+
5%	+	−	+	−	−
7%	w	−	w	−	−
Alkali from:[g]					
cis-Aconitate	−		−	+	
trans-Aconitate	−		+	d[h]	
Aspartate	−		+	+	
Caprylate	+		+	d[i]	
Citrate	−		+	d	
D-Gluconate	−		+	+	
Fumarate	+		+	+	
D-Galacturonate	−		+	+	
L-Glutamate	+		+	+	
D-Glucuronate	−		+	−	
DL-Lactate	+		+	d[i]	
D-Malate	+		+	d[h]	
L-Malate	+		+	d[h]	
Malonate	+		+	+	
Mucate	+		−	+	
Propionate	+		+	d[i]	
Quinate	−		−	+	
Succinate	−		+	+	
L-Tartrate	−		−	+	

[a]Symbols: +, >85% positive; (+), (75–84% positive); d, variable (16–84% positive); −, 0–15% positive; w, weak reaction; ng, no growth on test medium; no entry indicates that no data are available.

[b]From Heyndrickx et al. (1997).

[c]From Goto et al. (2004).

[d]From Allan et al. (2005).

[e]Data for this species (other than growth temperature tests) were obtained by incubating at 40°C, and (excepting acid and alkali production from carbon sources – see footnote g below) were obtained at pH 5.5.

[f]Data for this species were obtained by incubating at 55°C.

[g]Data are from A. H. A. Albaser and N. A. Logan (unpublished results) and the method of testing is described in the section *Procedures for testing special characters* in the treatment of *Brevibacillus*.

[h]Type strain gives a positive reaction.

[i]Type strain gives a negative reaction.

TABLE 51. Additional characteristics of species of the genus *Aneurinibacillus*[a]

Characteristic	1. *A. aneurinilyticus*[b]	2. *A. danicus*[c]	3. *A. migulanus*[b]	4. *A. terranovensis*[d]	5. *A. thermoaerophilus*[c]
Oxidase	+	+	+		w
Utilization of:					
N-Acetyl-D-glucosamine	−	−	−	d[e]	+
cis-Aconitate	−		d	d[e]	−
trans-Aconitate	−		d	d[f]	d[e]
D-Alanine	−		+	−	+
L-Alanine	−		+	d[e]	+
4-Aminobutyrate	−		+	−	d[e]
5-Aminovalerate	+		+	−	
L-Arabinose	−	w	−	−	−
L-Arabitol	−		−	d[e]	−
D-Arabitol	−	+	−	d[e]	d[f]
Aspartate	d		+	d[e]	+
Betaine	−		+	d[e]	+
Caprate	−		+	−	−
Caprylate	+		−	−	−
Citrate	−	−	+	d[f]	d[e]
m-Coumarate	−		−	−	−
Erythritol	−	w	−	−	−
Ethanolamine	+		+	−	+
Fructose	−	+	+	−	d[e]
Fumarate	+		+	+	d[f]
Galacturonate	−		+	d[f]	+
Gluconate	d		+	d[e]	d[f]
Glucosamine	−		+	+	−
Glucose	−	−	−	d[f]	−
L-Glutamate	−		+	+	+
Glutarate	+		−	+	+
DL-Glycerate	d		−	−	−
Glycerol	+		+	d[e]	d[f]
Histamine	+		+	+	+
L-Histidine	−		d	−	d[f]
3-Hydroxybenzoate	−		d	−	−
4-Hydroxybenzoate	−		+	−	−
β-Hydroxybutyrate	−		−	−	−
Hydroxyquinoline-β-glucuronide	+		+	d[e]	+
myo-Inositol	−		+	−	−
Itaconate	−		−	+	−
2-Ketogluconate	+		+	d[e]	−
5-Ketogluconate	−		+	d[e]	−
α-Ketoglutarate	−		+	+	
DL-Lactate	+		+	+	+
D-Malate	+		+	d[e]	+
L-Malate	+		+	d[f]	+
Malonate	+		+	−	+
Maltose	−		−	−	−
Maltotriose			−	−	−
Mannitol	−		−	−	−
Mannose	−	+	−	d	d[f]
1-0-Methyl-α-D-glucopyranoside	−		−	+	−
1-0-Methyl-β-D-glucopyranoside	−		−	−	−
Mucate	−		−	−	−
Phenylacetate	−		+	−	+
Proline	−		+	d[e]	+
Propionate	+		+	−	d[e]
Protocatechuate	−		+	−	
Putrescine	+		+	d[e]	+
Quinate	−		−	d[f]	+
Ribose	+		+	d[e]	+
Saccharate	−		+	d[e]	+
L-Serine	−		+	−	d[e]

(continued)

TABLE 51. (continued)

Characteristic	1. *A. aneurinilyticus*[b]	2. *A. danicus*[c]	3. *A. migulanus*[b]	4. *A. terranovensis*[d]	5. *A. thermoaerophilus*[c]
Succinate	+		+	d[e]	+
Sucrose	–	–	–	d[e]	–
Tagatose	d	+	–	–	–
Trehalose	–		–	d[e]	–
Tricarballylate	d		–	–	+
Trigonelline	–		+	–	d[e]
Tryptamine	–		–	–	–
Tryptophan	+		+	–	–
Turanose	–		–	–	d[f]
L-Tyrosine	+	–	+	–	+
Tween 20	–	–			
Tween 40	–	–			
Tween 60	–	–			
Tween 80	–	–			+
Growth at pH:					
5	+	–	–	+	–
6	+	+	+	+	+
7	+	+	+	d	+
8	+	–	+	–	+
9	+	–	+	–	+
Fatty acids (mean percentages of total):[g]					
$C_{15:0\ iso}$	57.2	57.7	48.6	45.4	58.5
$C_{15:0\ anteiso}$	1.3	<1.0	1.3	40.9	5.1
$C_{16:0}$	5	6.3	3.5	1.9	2.2
$C_{16:0\ iso}$	2.3	3.5	6.5	2.0	3.8
$C_{17:0\ iso}$	2	8.0	2.1	<1.0	23.6

[a]Symbols: +, >85% positive; (+), (75–84% positive); d, variable (16–84% positive); –, 0–15% positive; w, weak reaction; ng, no growth on test medium; no entry indicates that no data are available.

[b]Data are from A. H. A. Albaser and N. A. Logan (unpublished results) and the method of testing is described in the section *Procedures for testing special characters* in the treatment of *Brevibacillus*.

[c]From Goto et al. (2004).

[d]From Allan et al. (2005).

[e]Type strain gives a negative reaction.

[f]Type strain gives a positive reaction.

[g]Fatty acid data for *Aneurinibacillus terranovensis* are for cells grown on 1/2 BFA.

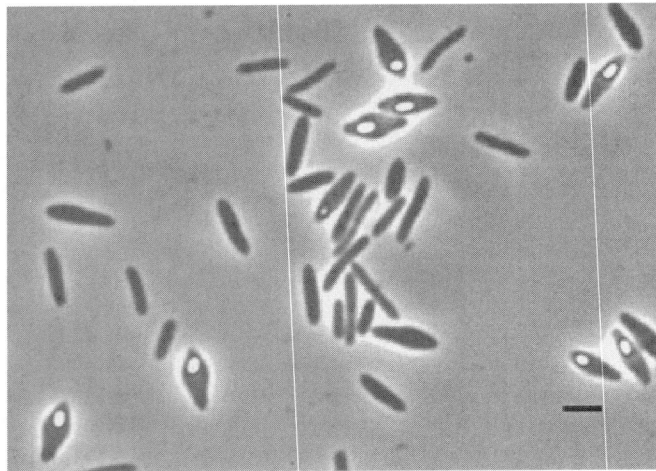

FIGURE 28. Photomicrograph of the type strain of *Aneurinibacillus terranovensis* viewed by phase-contrast microscopy, showing ellipsoidal, paracentral, and subterminal spores in very swollen sporangia. Bar = 2 μm. Photomicrograph prepared by N.A. Logan.

Cells are Gram-positive, becoming Gram-negative after 48 h, motile, round-ended rods (0.8–1 μm by 2–8 μm), occurring singly, in pairs, and in chains. Endospores are ellipsoidal, occurring paracentrally and subterminally in very swollen sporangia (Figure 28). After 48 h incubation on 1/2 BFA (pH 5.5) at 40°C, colonies are circular, flat, up to 1.5 mm in diameter, and cream-colored with a slightly glossy appearance and butyrous consistency. Minimum growth temperature lies between 20 and 25°C, with the optimum temperature for growth being between 37 and 45°C, and the maximum growth temperature lying between 50 and 55°C. Growth may occur between pH 3.5 and 7.0 and the optimum pH for growth lies between pH 5.0 and 5.5. Growth is inhibited by 2–3% NaCl. The organisms are microaerophilic and weakly positive for catalase. Gelatin is hydrolyzed slowly. Starch is hydrolyzed weakly. No growth occurs on casein agar. In the API 20E strip reactions, arginine dihydrolase, the Voges–Proskauer reaction, and nitrate reduction are positive. Citrate utilization is variable. A small range of carbohydrates, amino acids, and organic acids is utilized as sole carbon sources by all strains in the API Biotype 100 gallery (see Table 50 and Table 51). Utilization of a wider range of such substrates shows variation between strains, but the type strain utilizes *trans*-aconitate, citrate, fumarate, D-galactose, D-glucosamine, α-lactose, L-malate, and quinate; although no acid production is seen, alkaline reactions are observed from various organic compounds

Table 50). The major cellular fatty acids (from cells grown on 1/2 BFA) are $C_{15:0\ anteiso}$ and $C_{15:0\ iso}$, accounting for approximately 41 and 46% of the total fatty acid content, respectively. The following fatty acids are present in smaller amounts (at least 1.0%): $C_{14:0}$, $C_{14:0\ iso}$, $C_{16:0}$, $C_{16:0\ iso}$, and $C_{16:1\ \omega7c}$ alcohol. Isolated from volcanic soils in northern Victoria Land, Antarctica.

DNA G+C content (mol%): 43.2–44.6; that of the type strain is 43.2 (HPLC).

Type strain: LMG 22483, CIP 108308.

EMBL/GenBank accession number (16S rRNA gene): AJ715378 (LMG 22483).

5. **Aneurinibacillus thermoaerophilus** (Meier-Stauffer, Busse, Rainey, Burghardt, Scheberl, Hollaus, Kuen, Makristathis, Sleytr and Messner 1996) Heyndrickx, Lebbe, Vancanneyt, Kersters, De Vos, Logan, Forsyth, Nazli, Ali and Berkeley 1997, 816[VP] (*Bacillus thermoaerophilus* Meier-Stauffer, Busse, Rainey, Burghardt, Scheberl, Hollaus, Kuen, Makristathis, Sleytr and Messner 1996, 540)

ther.mo.aer.o'phi.lus. Gr. adj. *thermos* hot; Gr. masc. n. *aer* air; Gr. adj. *philos* loving; N.L. adj. *thermoaerophilus* loving heat air, i.e., thermophilic and aerobic.

Colonies on nutrient agar after 24 h at 55 °C are flat, 1–10 mm in diameter, irregular in shape, with a tendency to swarm across the surface of the agar; they are matt, translucent, and creamy grayish in appearance, becoming whitish and opaque in their centres. Vegetative cells are Gram-positive, motile by peritrichous flagella, and 1.0–1.2 μm by 3.5–5.5 μm. Central and paracentral spores are formed in swollen sporangia. Strictly aerobic. Catalase production variable. Growth temperatures range from 40 to 60 °C and growth occurs from pH 7.0 to 8.0. Growth occurs in presence of 3% NaCl, but not 5% NaCl. Hydrolyzes casein, gelatin, and Tween 80. Produces acid from a small range of carbohydrates. Nitrate is not reduced. A range of amino acids, carbohydrates, and organic acids is assimilated as carbon sources in the API Biotype 100 System. The major cellular fatty acid components (ranges as percentages of total) are $C_{15:0\ iso}$ (50.3–66.8%), $C_{15:0\ anteiso}$ (0–10.3%), $C_{15:0}$ (0.9–3.4%), $C_{16:0\ iso}$ (3.5–4.1%), $C_{16:0}$ (1.8–2.5%), $C_{17:0\ iso}$ (23.4–23.8%), and $C_{17:0\ anteiso}$ (0–8.3%). Isolated from sugar beet extraction juice.

DNA G+C content (mol%): 46.3–46.7 (HPLC); that of the type strain is 46.7.

Type strain: DSM 10154, LMG 17165.

EMBL/GenBank accession number (16S rRNA gene): X94196 (DSM 10154).

Genus IV. **Brevibacillus** Shida, Takagi, Kadowaki and Komagata 1996a, 942[VP]

NIALL A. LOGAN AND PAUL DE VOS

Bre.vi.ba.cil'lus. L. adj. *brevis* short; L. dim. n. *bacillus* small rod; N.L. masc. n. *Brevibacillus* short, small rod.

Gram-positive, Gram-variable, or Gram-negative, rod-shaped cells, 0.7–1.0 μm × 3.0–6.0 μm. Motile by means of peritrichous flagella. **Ellipsoidal spores are formed and swell the sporangia.** Most species grow on routine media such as nutrient agar and trypticase soy agar producing flat, smooth, yellowish-gray colonies. One species produces red pigment. **Most species are strictly aerobic, but one species is microaerophilic and one species is facultatively anaerobic.** Most species are catalase-positive. Oxidase reaction varies between species. Voges–Proskauer reaction is negative. Nitrate reduction and casein, gelatin, and starch hydrolysis varies between species. Growth is inhibited by 5% NaCl. Optimum growth occurs at pH 7.0. **Carbohydrates may be assimilated, but acid is produced weakly if at all from them by most species.** Some amino acids and organic acids may be used as carbon and energy sources. The major cellular fatty acids are $C_{15:0\ ante}$ and $C_{15:0\ iso}$.

DNA G+C content (mol%): 40.2–57.4.

Type species: **Brevibacillus brevis** (Migula 1900) Shida, Takagi, Kadowaki and Komagata 1996a, 943[VP] (*Bacillus brevis* Migula 1900, 583.).

Further descriptive information

Phylogeny. *Brevibacillus* belongs to the family *Paenibacillaceae* and a phylogenetic tree covering the present species that are represented by their type strain is given in Figure 29 as a 16S rDNA sequence based neighbor-joining tree.

Differential characteristics for members of the genus *Brevibacillus* are given in Table 52. The rod-shaped cells of *Brevibacillus* species are usually round-ended, and occur singly, in pairs, and in chains. Cell diameters range from 0.7–1.0 μm and lengths

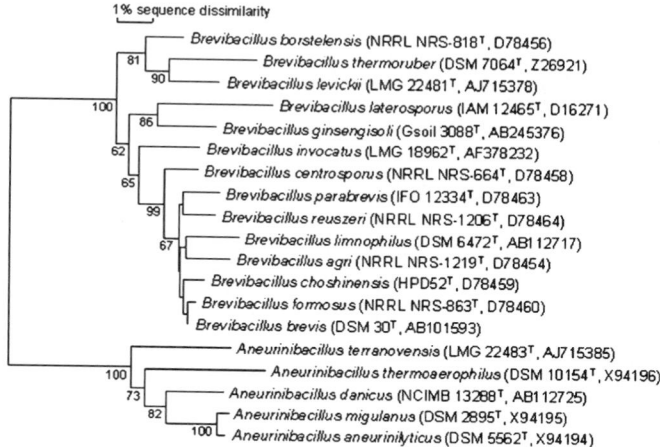

FIGURE 29. Phylogenetic neighbor-joining tree of type strains of *Brevibacillus* and *Aneurinibacillus* species based on 16S rDNA sequences. Bootstrap values are given at the branching points based on 1000 recalculations. Strain numbers with their respective EMBL accession numbers are given in parentheses.

from 3.0–6.0 μm, but the cells of a particular strain are usually quite regular in size, and individual species normally have dimensions within fairly narrow limits. Most of the species of *Brevibacillus* do not have distinctive sporangial morphologies; the spores are ellipsoidal, lie subterminally or perhaps terminally, and swell the sporangia slightly or moderately (Figure 30 and Figure 31). The notable exception is the unique sporan-

TABLE 52. Characteristics differentiating the species of the genus *Brevibacillus*[a,b]

Characteristic	*B. brevis*	*B. agri*	*B. borstelensis*	*B. centrosporus*	*B. choshinensis*	*B. formosus*	*B. ginsengisoli*	*B. invocatus*	*B. laterosporus*	*B. levickii*[c]	*B. limnophilus*	*B. parabrevis*	*B. reuszeri*	*B. thermoruber*[d]
Gram reaction	+/v	+	+	+	+	+	+	−	+/v/−	+	v	+/v	+	+
Anaerobic growth	−	−	−	−	−	−	+	−	+	−[e]	−	−	−	−
Growth at:														
20°C	d	+	+	+	+	+	+	+	+	+	+	+	+	−
50°C	−	−	+	−	−	−	−	−	d	+	−	d	−	+
55°C	−	−	−	−	−	−	−	−	−	d	−	−	−	+
NaCl tolerance:														
2%	d	+	−		−	−	+				w	d	+	
3%	−	−	−		−	−	−							−
4%								−						
5%										d	−	−		−
Hydrolysis of:														
Casein	+	+	+		−	+	+	−	+	d/w	−	+	−	+
Gelatin	+	+	+			+	+	−	+	+	−	+	−	+
ONPG	+	−	−		−	−	−	−	−	−	−	+		
Starch	−	−	−		−	−	−		−	w	−	−		+
Urea	−	−	−		−	−	+	−	−	−	−	−	−	+
Nitrate reduction	d	−	+	d	−	+	+	−	+	d	−	+		−
Acid from:[f]														
N-Acetylglucosamine	+	+	+	+	−	+			+	−			+	
D-Fructose	+	+	+	−	−	d			+	−	+	−	+	
D-Glucose	+	+	−	d	−	d			+	−	+	+	+	
Glycerol	+	+	D	d	−	+			+	−	+	+	+	
Maltose	+	+	−	−	−	−			+	−		+	+	
D-Mannitol	+	+	−	+	−	−			+	−		+	+	
D-Mannose	−	d	−	−	−	−			+	−		−		
Ribose	+	d	+	+	+	+			+	−	+	+		
D-Tagatose	−	−	+	−	−	d			−	−		−		
D-Trehalose	+	+	−	−	−	−			+	−		+	+	
D-Turanose	+	−	−	−	−	−			−	−		+		
Alkali from:[f]														
cis-Aconitate	+	d	−	+	−	+			−	d		+		
trans-Aconitate	−	d	−	−	−	−			−	d		−		
Aspartate	+	+	−	+	−	+			+	+		d	+	
Caprylate	−	−	−	d	−	−			−	d		−	+	
Citrate	+	d	−	+	d	−			−	d	−	+	+	
Fumarate	+	+	+	+	−	d		d	+	d		+		
D-Galacturonate	−	d	−	−	−	−			−	+		−	+	
D-Gluconate	+	+	−	d	d	−			−	+		−		
D-Glucuronate	−	−	−	−	−	−			−	d		−	+	
L-Glutamate	+	+	−	+	−	+		−	+	+		+	+	
DL-Lactate	+	+	−	+	−	−		d	−	d		+	+	
D-Malate	+	+	−	−	d	−			−	d		−		
L-Malate	+	+	d	+	+	d			−	+		+	+	
Malonate	+	+	+	+	d	−			−	+		d	+	
Mucate	−	−	−	−	−	−		d	−	+		d		
2-Oxoglutarate	+	+	+	+	+	+	+	+	+	+		+	+	
Propionate	+	d	+	+	+	d			−	d		d	+	
Quinate	−	−	−	−	−	−			−	+		+		
Succinate	+	+	+	+	−	+			−	+		+	+	
L-Tartrate	−	d	−	−	−	−			−	+		d		

[a]Symbols: +, >85% positive; d, results differ between strains (16–84% positive); −, 0–15% positive; +/v, positive or variable reaction within a strain; +/v/−, positive, variable or negative reaction within a strain; v, reaction varies within a strain; w, weak reaction; +/w, positive or weak positive reaction; d/w, results differ between strains, but positive reactions are weak; no entry indicates that no data are available.

[b]Data from Manachini et al. (1985), Goto et al. (2004), Heyndrickx et al. (1997), Allan et al. (2005), Albaser and Logan (unpublished results).

[c]Data for this species (other than growth temperature tests) were obtained by incubating at 40 °C, and (excepting acid and alkali production from carbon sources – see[f] below) were obtained at pH 5.5.

[d]Data for this species were obtained by incubating at 45 °C.

[e]*Brevibacillus levickii* is microaerophilic.

[f]Data for *Brevibacillus limnophilus* are from Goto et al. (2004); data for *Brevibacillus thermoruber* are from Manachini et al. (1985); data for *Brevibacillus levickii* are from Allan et al. (2005) and inoculum was at pH 7, although this is supra-optimal for this species. Data for all other species are from Albaser and Logan (unpublished results), and the method of testing is described in the section *Procedures for testing special characters.*

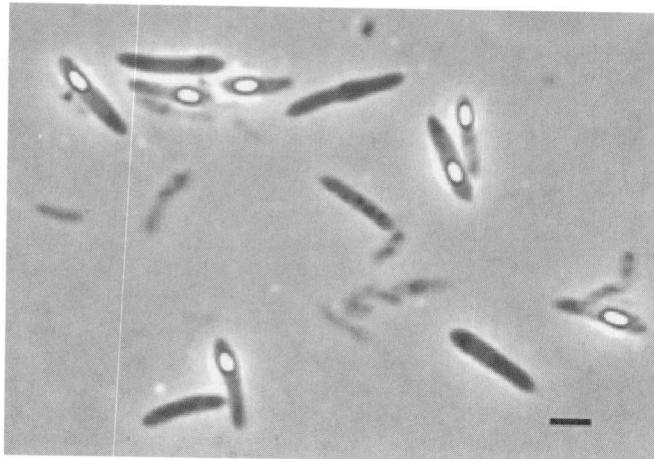

FIGURE 30. Photomicrograph of type strain of *Brevibacillus brevis* viewed by phase-contrast microscopy, showing ellipsoidal, subterminal spores that usually swell the sporangia. Bar = 2 μm. Photomicrograph prepared by N.A. Logan.

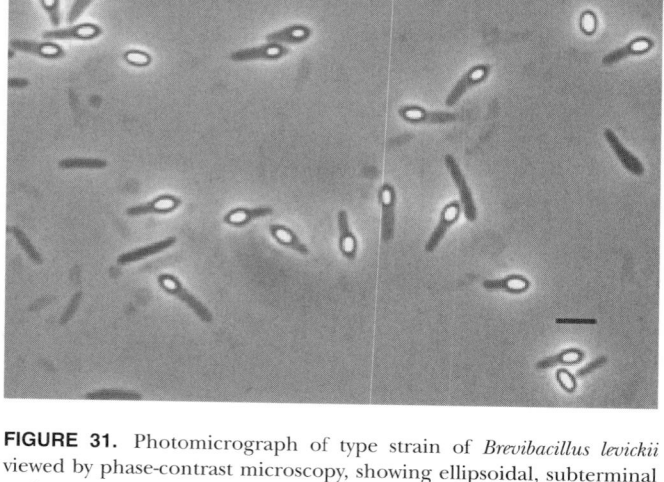

FIGURE 31. Photomicrograph of type strain of *Brevibacillus levickii* viewed by phase-contrast microscopy, showing ellipsoidal, subterminal and terminal spores in swollen sporangia. Bar = 2 μm. Photomicrograph prepared by N.A. Logan.

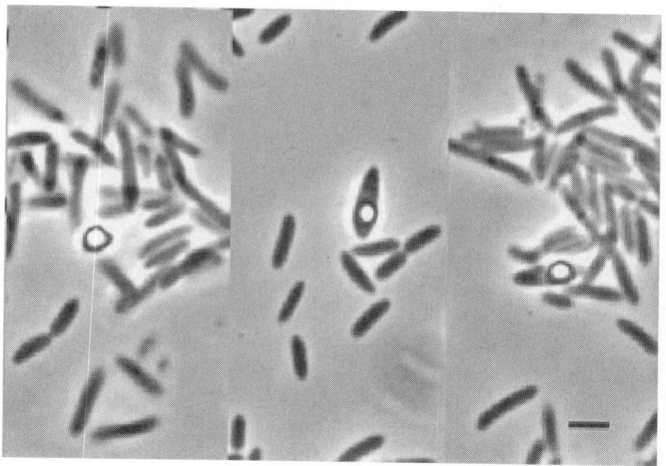

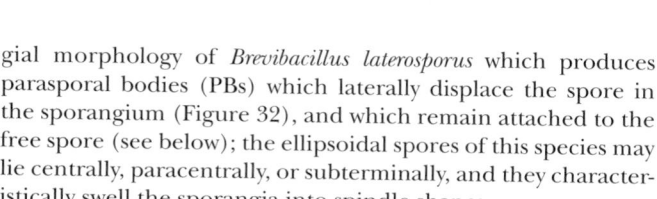

FIGURE 32. Composite photomicrograph of type strain of *Brevibacillus laterosporus* viewed by phase-contrast microscopy, The ellipsoidal spores are cradled in parasporal bodies, are borne paracentrally and subterminally, and are displaced laterally so that the sporangia are swollen into spindle shapes. Bar = 2 μm. Photomicrograph prepared by N.A. Logan.

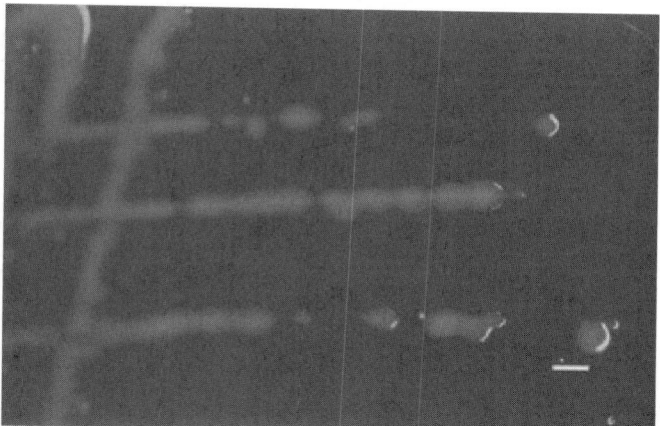

FIGURE 33. Glossy colonies of the type strain of *Brevibacillus brevis* grown on trypticase soy agar for 24–36 h. Bar = 2 mm. Photograph prepared by N.A. Logan.

gial morphology of *Brevibacillus laterosporus* which produces parasporal bodies (PBs) which laterally displace the spore in the sporangium (Figure 32), and which remain attached to the free spore (see below); the ellipsoidal spores of this species may lie centrally, paracentrally, or subterminally, and they characteristically swell the sporangia into spindle shapes.

Information on cell-wall composition is available for only two *Brevibacillus* species, *Brevibacillus brevis* and *Brevibacillus laterosporus*. They have the type of cross-linkage that is seen in the majority of *Bacillus* species for which it is known (see Table 2 in *Bacillus* section). A peptide bond is formed between the diamino acid in position 3 of one subunit and the D-Ala in position 4 of the neighboring peptide subunit so that no interpeptide bridge is involved. The diamino acid in most *Bacillus* species is *meso*-

diaminopimelic acid (*meso*-DAP), and this cross-linkage is usually known as DAP-direct (Schleifer and Kandler, 1972).

Colonies of *Brevibacillus* species are usually smooth, moist, and glossy. Their elevations are flat to slightly raised, consistencies are butyrous, and shapes vary from round to irregular (Figure 33 and Figure 34). Diameters commonly range from 1–3 mm, but sizes up to 8 mm may occur. Colony color commonly ranges from buff or creamy-gray to off-white. *Brevibacillus thermoruber* produces spreading colonies and a red, nondiffusible pigment.

Brevibacillus species are heterotrophic and neutrophilic and will grow well on routine media such as nutrient agar or trypticase soy agar. Growth may be enhanced by the addition of a small amount of yeast extract. Most strains will grow on blood agar. They will use some amino acids, carbohydrates, and organic acids as sole sources of carbon and energy (see Table 53). Utilization of carbohydrates may be accompanied by the production of small amounts of acid, and utilization of amino acids and organic

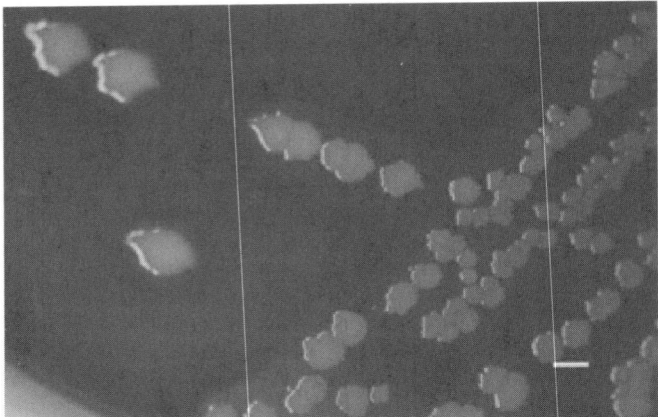

FIGURE 34. Type strain of *Brevibacillus laterosporus* grown on trypticase soy agar for 24–36 h, showing creamy-white and smooth colonies with irregular margins. Bar = 2 mm. Photograph prepared by N.A. Logan.

acids may be accompanied by the production of small amounts of alkali, but, these generally are not easily detected by the routine test methods, and some characters may prove to be inconsistent when retested. Growth temperatures vary considerably, and the descriptions of the individual species should be consulted.

Habitats. Most *Brevibacillus* strains have been isolated from the natural environment, particularly soils, where they appear to be saprophytes, but there have been some isolations from human clinical specimens and from human illness, and *Brevibacillus laterosporus* has long been associated with insect pathogenicity. As with *Bacillus*, the spores of these organisms may readily survive distribution from these natural environments to a wide variety of other habitats, and some strains have been found as contaminants in foods and pharmaceutical products. For more information on endospores in the environment, see the chapter on *Bacillus*. *Brevibacillus* species may be isolated following heat treatment of specimens in order to select for endospores. Although the presence of their spores in a given environment does not necessarily indicate that the organisms are metabolically active there, repeated and independent isolations from such a habitat make it reasonable to assume that vegetative *Brevibacillus* cells are, or have been, active there. Isolates identified as *Bacillus brevis* prior to the allocation of many strains bearing this name into the new and revived species *Brevibacillus agri*, *Brevibacillus borstelensis*, *Brevibacillus centrosporus*, *Brevibacillus choshinensis*, *Brevibacillus formosus*, *Brevibacillus* (now *Aneurinibacillus*) *migulanus*, *Brevibacillus parabrevis*, *Brevibacillus reuszeri*, *Aneurinibacillus danicus*, and strains assigned to *Bacillus brevis* by authors unaware that these nomenclatural changes had been proposed (between 1995 and 2004), may or may not be authentic strains of this species or even members of the genus. This should be borne in mind when reading the accounts of habitats in which strains have been found. Isolations of strains identified as *Bacillus brevis* have been reported from tannery processing (Birbir and Ilgaz, 1996), black crusts on open-air stone monuments in Italy (Turtura et al., 2000), and soil contaminated with hexachlorocyclohexane. A *Bacillus brevis* strain secreting an extracellular cellulase was found in another soil (Singh and Kumar, 1998), and Wenzel et al. (2002) found a cellulolytic *Brevibacillus brevis* in the gut of the termite *Zootermopsis angusti-*

collis. Isolates identified as *Brevibacillus brevis* have been reported from the airborne dust of schools and children's daycare centers (Andersson et al., 1999), food packaging products of paper and board (Pirttijarvi et al., 2000), the submerged rhizosphere of the seagrass *Vallisneria americana* (wild celery) in an estuarine environment (Kurtz et al., 2003), and in the humus of Norway spruce (*Picea abies*) (Elo et al., 2000).

The sources of the type strains of the species *Brevibacillus agri*, *Brevibacillus borstelensis*, *Brevibacillus choshinensis*, *Brevibacillus formosus*, *Brevibacillus ginsengisoli*, and *Brevibacillus reuszeri* were soils. Foodstuffs are readily contaminated by soil organisms; *Brevibacillus centrosporus* has been isolated from spinach and *Brevibacillus parabrevis* from cheese. *Brevibacillus centrosporus* has also been found in estuarine seagrass rhizosphere (Kurtz et al., 2003). Strains of *Brevibacillus invocatus* and *Brevibacillus agri* were repeatedly isolated from a pharmaceutical fermenter plant and its antibiotic raw product over a period of several months (Logan et al., 2002). *Brevibacillus agri* has also been isolated from sterilized milk, a gelatin processing plant, clinical specimens, and a public water supply where it was implicated in an outbreak of waterborne illness (Logan et al., 2002). *Brevibacillus centrosporus* was isolated from a bronchio-alveolar lavage, *Brevibacillus parabrevis* was found in a breast abscess, and both species have been isolated from human blood (Logan et al., 2002). The original strain of the thermophilic species *Brevibacillus thermoruber* was isolated from mushroom compost (Craveri et al., 1966; Guicciardi et al., 1968), *Brevibacillus borstelensis* or a close relative of this species was found to be a prominent member of the flora of hot synthetic compost (Dees and Ghiorse, 2001), and a hydrogen sulfide decomposing strain of *Brevibacillus formosus* has been isolated from pig feces compost (Nakada and Ohta, 2001). The geothermal soil of the northwest slope of Mount Melbourne, a volcano in Antarctica, yielded strains of a moderately thermophilic and moderately acidophilic species, *Brevibacillus levickii*, that were isolated in small numbers along with *Bacillus fumarioli* (Allan et al., 2005; Logan et al., 2000).

The specific epithet of *Brevibacillus laterosporus* is derived from the organism's unique sporangial morphology. It produces parasporal bodies (PBs) that displace the spore laterally in the sporangium (Figure 32); these bodies have been described as resembling canoes or the keels of ships. Montaldi and Roth (1990) examined sporangia by thin-section transmission electron microscopy and found three kinds of PB: i) a large one, associated with the spore, and of similar volume to it, with a lamellar structure of sequentially smaller layers, ii) a smaller globular or angular one of 100–200 nm in diameter that appeared at the same time as the lamellar PB but which was not attached to the spore in any way, and iii) a striated, rod-shaped PB with diameter of at least 200 nm. *Brevibacillus laterosporus* was originally isolated from water (Laubach, 1916), but McCray (1917) isolated other strains with sporangial morphologies similar to the Laubach et al. strain from the diseased larvae of bees. White (1920), who had named his bee larvae isolates as *Bacillus orpheus* in 1912 but had not described them, recognized the similarity between the two species. *Bacillus orpheus* is thus a synonym of *Brevibacillus laterosporus*. Endosporeformers named *Bacillus pulvifaciens* by Katznelson (1950) were isolated from diseased honeybee larvae (including cases of powdery scale). Gordon et al. (1973) thought that they might form a connection between *Bacillus larvae* and *Bacillus laterosporus* and

TABLE 53. Utilization of carbon compounds by *Brevibacillus* species[a]

Utilization of:[b]	B. brevis	B. agri	B. borstelensis	B. centrosporus	B. choshinensis	B. formosus	B. ginsengisoli	B. invocatus	B. laterosporus	B. levickii	B. limnophilus	B. parabrevis	B. reuszeri	B. thermoruber[d]
Acetate							+							
Adonitol	−	−	−	−	−	−	−							w
D-Alanine	+	+	+	−	−	−	+	−	−	+		−	−	
L-Alanine	+	+	+	−	−	+	+	+	−	+		+	+	
4-Aminobutyrate	d	−	+	−	−	−	−	−	−	d		+	+	
L-Arabinose	−	−	−	−	−	−	−	−	−	d		−	+	
L-Arabitol	−	−	−	−	−	−	−	−	−	+		−	−	+
Cellobiose	d	−	−	−	−	+	+	−	+	−		+	d	
Citrate	d	−	−	−	d	−	−	−	−	+	−	+	+	−
Ethanolamine	d	+	+	−	−	+		−	d	−		+	d	
D-Fructose	+	+	+	−	d	+	−	−	+	+		−	+	+
Galactose	−	−	−	−	−	−	−	−	−	d		−	−	+
Gentiobiose	−	−	−	−	−	−	−	−	−	d		−	−	+
Gluconate	d	+	+	−	d	+	−	−	+	d		+	d	
Glucosamine	−	−	−	−	−	−	−	−	−	+		d	−	
D-Glucose	+	+	−	+	d	+	+	−	+	+		+	+	
DL-Glycerate	−	−	+	−	−	−	−	−	+	+		+	+	+
Glycerol	+	+	d	−	−	+	+	−	+	+		−	+	
3-Hydroxybutyrate	+	+	+	d	−	+	+	−	+	−		+	+	+
Inositol	−	d	−	−	−	−	−	−	−	−		+	+	+
2-Ketogluconate	−	−	−	−	−	−	−	−	−	+		−	−	
5-Ketogluconate	−	−	−	−	−	−	−	−	−	+		−	−	
Lactose	−	−	−	−	−	−	−	−	−	+		−	−	
L-Malate	−	+	+	−	−	−	−	d	d	+		−	−	−
Maltitol	−	−	−	−	−	−	−	−	−	−		+	−	−
Maltose	+	+	−	−	−	+	−	−	+	+		+	−	
Maltotriose	+	+	−	−	−	+	−	−	+	+		+	−	+
Mannitol	+	+	−	+	d	+	−	+	+	+		+	−	+
Melezitose	+	−	−	−	−	+	−	+	+	+		+	+	+
1-O-Methyl-β-D-galactopyranoside	−	−	−	−	−	−	−	−	−	−		d		
1-O-Methyl-α-D-glucopyranoside	−	d	−	−	−	−	−	−	−	+		+	−	
1-O-Methyl-β-D-glucopyranoside	−	−	−	−	−	−	−	−	+	+		−	−	
Mucate	−	−	−	−	−	−	−	−	d		−	−		
2-Oxoglutarate	d	+	d	+	−	+		+	d	d		+	+	
Palatinose	+	+	−	−	−	+	−	−	−			+		
Phenylacetate	d	+	−	+	d	d	−	+	−	−		+	+	
Proline	+	+	+	−	−	+	+	−	+	+		+	+	
Putrescine	−	+	−	d	−	−		−	−	+		+	+	
Pyruvate														−
Quinate	−	−	−	−	−	−		+	−	+		−	−	
Rhamnose	−	−	−	+	−	−	−	−	−	−		−	−	
D-Ribose	+	−	+	d	+	+	−	−	+	d		−	+	+
Saccharate	−	−	−	−	−	−	−	−	−	d		−	−	+
L-Serine	+	+	+	−	−	+	+	−	+	+		+	+	
Sorbitol	−	−	−	−	−	−	−	−	−	+		−	−	
Succinate	−	+	+	+	−	d		−	+	+		−	+	w
Sucrose	+	+	−	−	−	+	−	−	−	+		+	−	w
meso-Tartrate	−	−	−	−	−	−	−	−	−	+		−	−	
Trehalose	+	+	−	−	−	+	−	−	+	d		+	−	+
Tryptophan	−	−	−	−	−	−	−	−	−	d		−	−	+
L-Tyrosine	−	d	d	d	−	+	−	−	d	−		d	−	
D-Xylose	−	−	−	−	−	−	−	−	−	−		−	−	+

[a]Symbols: +, >85% positive; d, results differ between strains (16–84% positive); -, 0–15% positive; +/v, positive or variable reaction within a strain; +/v/-, positive, variable or negative reaction within a strain; v, reaction varies within a strain; w, weak reaction; +/w, positive or weak positive reaction; d/w, results differ between strains, but positive reactions are weak; no entry indicates that no data are available.

[b]*Brevibacillus brevis*, *Brevibacillus agri*, *Brevibacillus borstelensis*, *Brevibacillus centrosporus*, *Brevibacillus choshinensis*, *Brevibacillus formosus*, *Brevibacillus invocatus*, *Brevibacillus laterosporus*, *Brevibacillus levickii*, *Brevibacillus parabrevis*, and *Brevibacillus reuszeri* were negative for utilization of: L-arabinose, D- arabitol, dulcitol, L-fucose, lactulose, lyxose, melibiose, 1–O-methyl-α-galactopyranoside, 3–O-methyl-D-glucopyranose, raffinose, sorbose, xylitol, D-xylose, histamine, trigonelline, tryptamine, 5-aminovalerate, betain, caprate, *m*-coumarate, gentisate, glutarate, 3-hydroxybenzoate, 4-hydroxybenzoate, itaconate, 3-phenylpropionate, protocatechuate, D-tartrate, L-tartrate, tricarballylate.

[c]Data for this species (other than growth temperature tests) were obtained by incubating at 40°C, and (excepting acid and alkali production from carbon sources – see [f] below) were obtained at pH 5.5.

[d]Data for this species were obtained by incubating at 45°C.

included them as unassigned strains in their listing of the latter species, however, Nakamura (1984a) revived *Bacillus pulvifaciens* as a distinct species. *Bacillus pulvifaciens* was later transferred to *Paenibacillus*. Heyndrickx et al. (1996a) showed that *Paenibacillus pulvifaciens* was a later subjective synonym of *Paenibacillus larvae* and proposed that the two species should become subspecies of *Paenibacillus larvae* because they represent distinct pathovars. In a recent taxonomic proposal, this subspecies distinction was again abandoned (Genersch et al., 2006). Although the strains of *Brevibacillus laterosporus* originally named *Bacillus orpheus* were isolated from diseased bees, it was not clear that they were clinically significant and this species is now considered to be a secondary invader. Falcon (1971) noted that *Bacillus laterosporus* had been isolated along with other bacteria in association with a natural epizootic of high mortality affecting Cabbage Moth (*Mamestra brassicae*) caterpillars on cabbage. Some *Bacillus laterosporus* strains have been convincingly shown to be pathogenic for mosquito and blackfly (*Simulium*) larvae (Favret and Yousten, 1985). Their pathogenicities are very low, and apparently are toxin-mediated but not associated with their spores. Subsequently, other strains (one of them isolated from dead insects) were found to produce crystalline inclusions during sporulation (Smirnova et al., 1996) which were toxic for larvae of the mosquito species *Aedes aegypti* and *Anopheles stephensi*, and toxic to a lesser extent for *Culex pipiens* (Orlova et al., 1998). *Bacillus laterosporus* and *Brevibacillus laterosporus* have also been isolated from a case of endophthalmitis following penetrating injury (Yabbara et al., 1977), sweet curdling milk spoilage (Heyndrickx and Scheldeman, 2002), bread dough (Bailey and von Holy, 1993), spontaneously fermenting soybeans (Sarkar et al., 2002), food packaging paper and board products (Pirttijarvi et al., 2000), tannery processing (Birbir and Ilgaz, 1996), sea water (Barsby et al., 2002), and from an estuarine seagrass rhizosphere (Kurtz et al., 2003). Garabito et al. (1998) identified some of their isolates from salterns and saline soils as *Brevibacillus laterosporus* but found that these strains showed different substrate utilization patterns from that of the type strain of this species.

Brevibacillus strains, especially of *Brevibacillus brevis*, *Brevibacillus choshinensis*, and *Brevibacillus laterosporus*, have attracted considerable interest owing to their production or transformation of valuable compounds, or their potentials as biocontrol agents. The characteristically very high productivity of heterologous polypeptides and proteins by "*Bacillus brevis*" and *Brevibacillus brevis* strains have been harnessed for the production of human growth hormone (Kajino et al., 1997), human interleukin-2 (Takimura et al., 1997), cholera toxin B subunit for use as a mucosal adjuvant (Goto et al., 2000), a thermostable alkaline protease for use as a laundry detergent additive (Banerjee et al., 1999), acetolactate decarboxylase to prevent diacetyl formation during the accelerated maturation of beer (Outtrup and Jørgensen, 2002), and artificially designed gelatins (Kajino et al., 2000). *Brevibacillus choshinensis* has been used for the production of recombinant chicken interferon-γ as a growth-promoting agent for poultry (Yashiro et al., 2001) and for the production of recombinant human epidermal growth factor multimers and their transformation into the monomeric, native form (Miyauchi et al., 1999).

An unidentified *Brevibacillus* strain isolated from petroleum-contaminated soil was found to be capable of degrading petroleum hydrocarbons (Grishchenkov et al., 2000). A strain identified as *Bacillus brevis* was isolated from soil contaminated with hexachlorocyclohexane where it degraded this polluting pesticide (Gupta et al., 2000). Another strain of this species was found to degrade the insecticide teflubenzuron (Finkelstein et al., 2001). A strain of *Brevibacillus laterosporus* was able to break down polyvinyl alcohol to acetate (Lim and Park, 2001). An isolate closely related to *Brevibacillus thermoruber* has been found to depolymerize xanthan (Nankai et al., 1999).

Brevibacillus brevis produces the broad-spectrum, topically useful peptide antibiotic gramicidin S, which attacks the lipid bilayer of the inner membranes of susceptible organisms (Prenner et al., 1999). *Brevibacillus laterosporus* was the original source of the immunosuppressive drug spergualin (Takeuchi et al., 1981), synthetic analogs of which may be used as antitumor drugs and to prevent or treat tissue rejection (Allison, 2000). Other antibiotics produced by *Brevibacillus laterosporus* include laterosporamine (Shoji et al., 1976), the anti-*Candida* basiliskamides, and tupuseleiamides (Barsby et al., 2002). *Brevibacillus* strains with antifungal properties are potentially valuable biocontrol agents. These include a *Bacillus brevis* strain active against fusarial wilt of pigeon pea (Bapat and Shah, 2000), *Brevibacillus brevis* antagonistic to *Botrytis cinerea* (Edwards and Seddon, 2001), and a *Brevibacillus laterosporus* effective against four foliar necrotrophic pathogens of wheat (Alippi et al., 2000).

Enrichment and isolation

For most species, enrichment and selective isolation methods have not been reported. An isolation method for detecting colonies of the fungicidal, gramicidin-producing *Bacillus brevis* Nagano strain was developed by Edwards and Seddon (2000) for the recovery of this organism from plants and soil in field trial experiments. It utilized the ability of this strain to decompose tyrosine, producing light-brown colonies surrounded by haloes on tyrosine agar. The Nagano strain and any other isolates producing gramicidin could then be identified by paper chromatography of ethanolic extracts with detection by ninhydrin. Tyrosine agar contains nutrient broth (Oxoid), 6.6 g; tyrosine, 5 g; agar, 15 g; water, 1000 ml . After autoclaving, it is continuously stirred with a magnetic stirrer (to reduce the size of the tyrosine crystals) until it has reached 50°C, whereupon it is poured immediately, resulting in an opaque, off-white, solid medium. Tyrosine utilization is found among other *Brevibacillus* species and in many organisms outside this genus, so this method is by no means specific for *Brevibacillus brevis*. *Brevibacillus laterosporus* may be isolated from larval remains in outbreaks of European foulbrood in honeybees where they are considered to be secondary invaders (Alippi, 1991). Wakisaka and Koizumi (1982) noted that some aerobic endosporeformers that appeared to form minor components of soil floras, including *Bacillus brevis* and *Bacillus laterosporus*, showed slow and/or uneven germination compared with the more frequently encountered *Bacillus* species such as *Bacillus cereus*, *Bacillus megaterium*, *Bacillus sphaericus*, and *Bacillus subtilis*. Members of these minor populations were isolated with difficulty by the standard dilution-plate technique, but could be enriched by removing the rapidly germinating, fast-growing members of the predominant flora even though the latter might outnumber the minor flora by 100- to 1000-fold. The dried soil sample (0.5 g) was suspended in 5 ml of sterile saline and stirred with three to five 4 mm diameter glass beads for 1 min, then placed in a vacuum desiccator for 30 min to eliminate air. The mixture was then heated at 65°C for 10 min to destroy vegetative bacteria and prompt spore germination; 1 ml of this suspension was combined with 1 ml of germination medium and incubated for 2–3 h at 30°C with gentle shaking (90 r.p.m.) followed by heating it at 65°C for 10 min to kill the newly emerged and vulnerable vegetative cells of the predominant members of the flora. After cooling, the suspension was serially diluted, and 0.5 ml quantities (of 1-, 10- and 100-fold dilutions, usually) were plated on Gly IM medium (see below), followed by incubation at 30°C for 2–12 days to recover the members of the minor population. Germination medium contained: glucose, 10 g; Casamino acids, 5 g; beef extract, 3 g; yeast extract, 1 g; DL-alanine, 1 g; water, 1000 ml; pH 6.8. Autoclave at 110°C for 15 min. Gly IM agar contained: NaCl, 3 g; beef extract, 2.5 g; polypeptone, 2.5 g; yeast extract, 2.5 g; soluble starch, 2 g; glycerol, 2 g; agar, 12.5 g; water, 1000 ml; pH 6.8.

Brevibacillus ginsengisoli was originally isolated from ginseng field soil. The sample was suspended in 50 mM phosphate buffer (pH 7.0), and the suspension was spread on one-fifth-strength modified R2A agar plates (tryptone, 0.25 g; peptone, 0.25 g; yeast extract, 0.25 g; malt extract, 0.125 g; beef extract, 0.125 g; Casamino acids, 0.25 g; soytone, 0.25 g; glucose, 0.5 g; soluble starch, 0.3 g; xylan, 0.2 g; $C_3H_3NaO_3$, 0.3 g; K_2HPO_4, 0.3 g; $MgSO_4$, 0.05 g; $CaCl_2$, 0.05 g; agar, 15 g; water) after being serially diluted with 50 mM phosphate buffer (pH 7.0). The plates were incubated for 1 month at room temperature in an anaerobic chamber. The headspace was substituted with a gas mixture comprising $N_2/CO_2/H_2$ (80:15:5, by vol). Single colonies on the plates were purified by transferring them onto new plates that were incubated using the modified R2A agar or half-strength modified R2A agar under anaerobic conditions. The organism was routinely cultured on R2A agar at 30°C.

Brevibacillus levickii was isolated from geothermal soil samples collected from the northwest slope of Mount Melbourne, northern Victoria Land, Antarctica. 1 g quantities of soil were added to 9 ml *Bacillus fumarioli* broth (BFB) in duplicate at pH 5.5, and one of each pair was heat treated at 80°C for 10 min to kill vegetative cells. All broths were incubated at 50°C in waterbaths and inspected daily. Cultures which became turbid were subcultured by streaking onto plates of *Bacillus fumarioli* agar (BFA). Colonies appearing on streak plates were screened for vegetative and sporangial morphologies by phase-contrast microscopy, and sporeforming rods were streak purified and then transferred to slopes of the same medium for storage at 4°C after incubation and confirmation of sporulation by microscopy. The recipe for BFB is given under *Enrichment and isolation* in *Aneurinibacillus*, above.

Procedures for testing special characters

From the point of view of routine diagnostic laboratories, the aerobic endosporeformers comprise two groups, the reactive ones that will give positive results in various routine biochemical tests (and which are therefore more amenable to identification by traditional methods and modern developments of such approaches) and the nonreactive ones which give few if any positive results in such tests. Members of the genus *Brevibacillus* fall into the latter category. No special characters have been described for their differentiation, so their identification is difficult. They are largely unreactive in the carbohydrate utilization tests of the API 50CHB gallery (bioMérieux, Marcy l'Etoile, France) and insufficiently variable in the API 20E and supplementary tests so that species in this genus are largely inseparable by these means. *Bacillus laterosporus* is an exception and does give useful results in this system. The API Biotype 100 gallery (bioMérieux), which was developed as a research product for differentiating enterobacteria, proved to be of great value in differentiating species of *Brevibacillus* and *Aneurinibacillus* (Allan et al., 2005; Heyndrickx et al., 1997; Logan et al., 2002); it contained 99 tests for the assimilation of carbohydrates, organic acids, and amino acids, and one control tube. It was inoculated with a suspension in one of two semi-solid media which differed in the number of growth factors they contained. After incubation, the tubes were examined for turbidity. Rigorous standardization of the suspension densities was essential. This system was adapted by using a suspension medium containing phenol red and examining the tubes for evidence of acid or alkali production (Albaser and Logan, unpublished results). The medium contained: KH_2PO_4, 1.5 g; $MgSO_4 \cdot 7H_2O$, 0.1 g; $CaCl_2 \cdot 2H_2O$, 0.125 g; $MnSO_4 \cdot 4H_2O$, 2.5 mg; phenol red 1 g; distilled water, 1000 ml; vitamin solution, 1 ml; pH 7.5. The vitamin solution contained: biotin 5 mg; thiamine, 5 mg; riboflavin, 5 mg; pyridoxal phosphate, 5 mg; pantothenate, 5 mg; nicotinic acid, 5 mg; *p*-aminobenzoic acid, 1 mg; folic acid, 0.5 mg; vitamin B_{12}, 0.5 mg; thioctic acid, 0.5 mg; deionized water, 50 ml. Several other commercially available biotyping kits have been investigated but do not give useful data for differentiating *Brevibacillus* species. Some of the special characters described in the section *Procedures for testing special characters* of *Bacillus* (see above) are applicable to *Brevibacillus*. It is strongly recommended that the original and emended descriptions of the more recently described species are consulted wherever possible and that cultures of those organisms are obtained for comparison when trying to distinguish between *Brevibacillus* species. It should be appreciated that 16S rDNA sequencing is not always reliable as a stand-alone tool for identification (see Logan et al., 2002, for example) and that a polyphasic taxonomic approach is advisable for the identification of some of the more rarely encountered species and the confident recognition of new taxa. Nomenclatural types exist for a good reason and are usually easily available. There is no substitute for direct laboratory comparisons with authentic reference strains. It must also be remembered that cultures labeled *Brevibacillus brevis* in various laboratory collections around the world may not be authentic strains of this species if they were deposited prior to the extensive splitting of this species in the mid-1990s; the use of authentic type strains is therefore essential.

Taxonomic comments

Bacillus brevis was described by Migula in 1900, and the species attracted much interest in the early 1940s due to the production of the antibiotic gramicidin. Molluscicidal activity (Singer et al., 1988) and protein overproduction (Udaka et al., 1989) were subsequently reported for some strains. Because of such interest in potential applications, numerous isolations were made, but many isolates were found to differ from the reference strains, some in particular being thermophilic, and so the delineation of the species became uncertain. Following DNA–DNA reassociation studies and chemotaxonomic analyses, the taxonomy of *Bacillus brevis* was modified by assigning some *Bacillus brevis* strains to the new or revived species *Bacillus agri* and *Bacillus centrosporus* (Nakamura, 1993), *Bacillus migulanus*, *Bacillus choshinensis*, *Bacillus parabrevis*, and *Bacillus galactophilus* (Takagi et al., 1993) and *Bacillus reuszeri*, *Bacillus formosus*, and *Bacillus borstelensis* (Shida et al., 1995), but *Bacillus galactophilus* was later recognized to be a synonym of *Bacillus agri* (Shida et al., 1994a).

Studies on the 16S rDNA sequences of the type strain of *Bacillus brevis* suggested that *Bacillus aneurinolyticus* represents a distinct evolutionary line close to that of *Bacillus brevis* (Ash et al., 1991) or that it diverged early from the *Bacillus brevis* line (Farrow et al., 1992, 1994). On the basis of a 16S rDNA gene sequence analysis of the type strains only, Shida et al. (1996a) proposed the new genera *Brevibacillus* and *Aneurinibacillus*. The former accommodated *Brevibacillus brevis* and the seven species mentioned above that were derived from it, along with *Brevibacillus laterosporus* and the thermophile *Brevibacillus thermoruber*. *Aneurinibacillus* contained "*Bacillus aneurinolyticus*" and two other species. An earlier SDS-PAGE study supported this divergence of the *Bacillus brevis* and *Bacillus aneurinolyticus* groups (Shida et al., 1996b), but the individual species of the *Bacillus brevis* group were not always well separated by this method, a problem also noticed by Logan et al. (2002). The latter authors also found that 16S rDNA sequence analysis showed low discrimination potential for the species *Brevibacillus brevis*, *Brevibacillus choshinensis*, *Brevibacillus formosus*, *Brevibacillus parabrevis*, and *Brevibacillus reuszeri*, and that most species of the genus were difficult to separate by routine phenotypic tests. A further species, *Brevibacillus invocatus*, was proposed by Logan et al. (2002), who emphasized the difficulties of distinguishing between species of this genus. In the light of these problems of identification, Goto et al. (2004) examined the hypervariable (HV) region corresponding to the 5′ end of 16S rDNA (nucleotide positions 70–344 in *Bacillus subtilis* numbering) in 52 strains of aerobic endosporeformers, 31 of which were received as *Brevibacillus*, and 3 as unidentified *Brevibacillus* species. The HV region marker had already proved to be taxonomically useful in *Alicyclobacillus*, *Bacillus*, and *Paenibacillus*, and tentative identifications by this approach were then confirmed by DNA–DNA hybridizations. They found that 14 *Brevibacillus brevis* and three *Brevibacillus* species strains clustered in *Aneurinibacillus* and proposed one of these strains as the new species *Aneurinibacillus danicus*. Of the remaining 17

strains received as *Brevibacillus brevis*, five identified as *Bacillus methanolicus*, two strains clustered close to this species but showed less than 70% DNA homology with the type strain, and one identified as *Bacillus oleronius*. Only two strains clustered with *Brevibacillus brevis*, and they showed less than 60% DNA homology with the type strain of this species. Of the remainder, three strains identified as *Brevibacillus agri*, two as *Brevibacillus parabrevis*, and 1 strain which did not cluster with an existing species of the genus was proposed as the new species *Brevibacillus limnophilus*.

Thus strains which might previously have been assigned to *Bacillus brevis* now represent some 12 mesophilic species in *Brevibacillus*. Their distinctions are based mainly upon DNA relatedness studies, molecular probing, and chemotaxonomic analyses of the relatively few available isolates using databases which are largely restricted to reference laboratories and unsuitable for organisms encountered only occasionally in routine laboratories. Distinction of most *Brevibacillus* species is not possible using the currently available *Bacillus* identification schemes, and separation remains difficult even with much wider selections of phenotypic tests (Heyndrickx et al., 1997; Logan, 2002). Also, another unreactive species, *Bacillus badius*, and species of *Aneurinibacillus* may easily be misidentified as a member of this genus. It is unfortunate that the extensive splitting proposed by the various recent taxonomic studies has not revealed characteristic phenotypic profiles that would be of value in the routine laboratory. It has been questioned whether the current taxonomy of *Brevibacillus* best serves the needs of the diagnostic bacteriologist and whether certain species might better be merged to give a more practically useful classification of this genus (Logan et al., 2002). However, new genotypic approaches based on sequences of so-called housekeeping genes may provide more straightforward (genomic) differentiation and identification.

Maintenance procedures

Brevibacillus strains may be preserved on slopes of a suitable growth medium that encourages sporulation, such as nutrient agar or trypticase soy agar containing 5 mg/l of $MnSO_4 \cdot 7H_2O$. Slopes should be checked microscopically for spores before sealing (to prevent drying out) and storage in a refrigerator; on such sealed slopes the spores should remain viable for many years. For longer term preservation, lyophilization and liquid nitrogen may be used, as long as cryoprotectants are added.

List of the species of the genus *Brevibacillus*

1. **Brevibacillus brevis** (Migula 1900) Shida, Takagi, Kadowaki and Komagata 1996a, 943[VP] (*Bacillus brevis* Migula 1900, 583.)

    bre′vis. L. adj. *brevis* short.

    Strictly aerobic, Gram positive or Gram variable, motile, rod-shaped cells, 0.7–0.9 μm × 3.0–5.0 μm, occurring singly and in pairs. The ellipsoidal spores are borne subterminally and swell the sporangia (Figure 30). Grows on routine media such as nutrient agar and trypticase soy agar, pro-

    ducing glossy, butyrous, cream-colored colonies, 1–3 mm in diameter after 24–36 h (Figure 33). Growth at 30°C may initially be slow, with more rapid growth following 24 h incubation. Catalase- and oxidase-positive. Nitrate reduction positive for most strains. Casein, DNA, gelatin, and Tween 60 are hydrolyzed; starch and urea are not hydrolyzed. Hydrolysis of Tween 80 is variable. Hydrogen sulfide and indole are not produced. Most strains do not grow at 20° or below, or above 50°C. No growth occurs at pH 5.5 and most strains do not grow at pH 9.0. D-Fructose, D-glucose, glycerol, maltose, mannitol, ribose, trehalose, and a few other carbohydrates are assimilated, and acid is produced weakly, if at all, from them. Amino acids and some organic acids are used as carbon and energy sources.

Isolated mainly from soil; also found in airborne dust, milk, rhizospheres, and paper products.

DNA G+C content (mol%): 48.7 (HPLC).

Type strain: ATCC 8246, BCRC 14682, CCM 2050, CCUG 7413, CIP 52.86, DSM 30, HAMBI 1883, NBRC 15304, JCM 2503, LMG 7123, NCCB 48009, NCIMB 9372, NCTC 2611, NRRL B-14602, NRRL NRS-604, VKM B-503, W.W. Ford 27B.

GenBank accession number (16S rRNA gene): AB101593, AB271756, D78457, X60612.

2. **Brevibacillus agri** (Laubach 1916) Shida, Takagi, Kadowaki and Komagata 1996a, 943[VP] (*Bacillus agri* (ex Laubach, Rice and Ford 1916) Nakamura 1993, 23; *Bacillus galactophilus* Takagi, Shida, Kadowaki, Komagata and Udaka 1993, 229.)

ag'ri. L. gen. n. *agri* of a field.

Strictly aerobic, Gram positive, motile, rod-shaped cells, 0.5–1.0 μm × 2.0–5.0 μm. The ellipsoidal spores swell the sporangia. Grows on routine media such as nutrient agar and trypticase soy agar, producing nonpigmented, translucent, thin, smooth, circular, entire colonies of about 2 mm in diameter. Catalase-positive, oxidase-negative. Nitrate reduction negative. Casein and gelatin are hydrolyzed; starch and urea are not hydrolyzed. Minimum growth temperature varies between 5 and 20°C, optimum temperature for growth is 28°C, and maximum growth temperature is 40°C. Growth occurs at pH 5.6 and no growth occurs at pH 9.0. Grows in presence of 2% but not 3% NaCl. D-Fructose, D-glucose, glycerol, maltose, D-mannitol, D-trehalose, and a few other carbohydrates are assimilated, and acid is produced weakly, if at all, from them. Amino acids and some organic acids are used as carbon and energy sources.

Isolated from soil, water, clinical specimens, sterilized milk, and pharmaceutical manufacturing plants.

DNA G+C content (mol%): 53.5 (HPLC).

Type strain: ATCC 51663, CCUG 31345, CIP 104002, DSM 6348, NBRC 15538, JCM 9067, LMG 15103, NRRL NRS-1219.

GenBank accession number (16S rRNA gene): AB112716, D78454.

3. **Brevibacillus borstelensis** (Stührk 1935) Shida, Takagi, Kadowaki and Komagata 1996a, 945[VP] (*Bacillus borstelensis* Shida, Takagi, Kadowaki, Udaka, Nakamura and Komagata 1995, 98.)

bor.stel.en'sis. N.L. adj. *borstelensis* referring to Borstel, Germany, where it was isolated.

Strictly aerobic, Gram positive, motile, rod-shaped cells, 0.5–0.9 μm ×2.0–5.0 μm. The ellipsoidal spores swell the sporangia. Grows on routine media such as nutrient agar and trypticase soy agar, producing flat, smooth, circular, entire colonies. One of the 16 strains contributing to the original description produced brown-red pigmentation on nutrient agar. Catalase-positive, oxidase-negative. Nitrate is reduced. Casein and gelatin are hydrolyzed; starch and urea are not hydrolyzed. Growth occurs at 20°C, the maximum is 50°C,

and the optimum temperature for growth is 30°C. Growth occurs at pH 5.5 and 5.6. Growth does not occur in the presence of 2% NaCl. D-Fructose, ribose, and a few other carbohydrates are assimilated, and acid is produced weakly, if at all, from them. D- Mannitol and D-trehalose are not assimilated. Amino acids and some organic acids, but not citrate, are used as carbon and energy sources.

Isolated from soil.

DNA G+C content (mol%): 51.3 (HPLC).

Type strain: ATCC 51668, CIP 104545, DSM 6347, NBRC 15714, JCM 9022, LMG 16009, NRRL NRS-818.

GenBank accession number (16S rRNA gene): AB112721, D78456.

4. **Brevibacillus centrosporus** (Laubach 1916) Shida, Takagi, Kadowaki and Komagata 1996a, 943[VP] (*Bacillus centrosporus* *ex* Laubach, Rice and Ford 1916) Nakamura 1993, 24.)

cen.tro.spor'us. L. n. *centrum* the center; N.L. n. *spora* spore; N.L. adj. *centrosporus* with a central spore.

Strictly aerobic, Gram positive, motile, rod-shaped cells, 0.5–1.0 μm × 2.0–6.0 μm. The ellipsoidal spores swell the sporangia. Despite the species name, the spores do not tend to lie centrally in the sporangia. Grows on routine media such as nutrient agar and trypticase soy agar, producing nonpigmented, translucent, thin, smooth, circular, entire colonies 2–3 mm in diameter. Catalase-positive, oxidase-negative. Nitrate is reduced to nitrite by some strains. Casein, gelatin, starch, and urea are not hydrolyzed. Minimum temperature for growth is 10°C, maximum is 40°C, and the optimum temperature for growth is 28°C. Growth does not occur at pH 5.6. Growth does not occur in the presence of 3% NaCl. D-Glucose, D-mannitol, ribose, and a few other carbohydrates are assimilated, and acid is produced weakly, if at all, from them. D-Fructose and trehalose are not assimilated. Some amino acids and organic acids are used as carbon and energy sources.

Isolated from child's feces, clinical specimens, spinach, and estuarine seagrass rhizosphere.

DNA G+C content (mol%): 49.8 (HPLC).

Type strain: ATCC 51661, CCUG 31347, CIP 104003, DSM 8445, NBRC 15540, JCM 9071, LMG 15106, NRRL NRS-664.

GenBank accession number (16S rRNA gene): AB112719, D78458.

5. **Brevibacillus choshinensis** (Takagi, Shida, Kadowaki, Komagata and Udaka 1993) Shida, Takagi, Kadowaki and Komagata 1996a, 943[VP] (*Bacillus choshinensis* Takagi, Shida, Kadowaki, Komagata and Udaka 1993, 229.)

cho.shi.nen'sis. N.L. adj. *choshinensis* referring to Choshi, Japan, where it was isolated.

Strictly aerobic, Gram positive, motile, rod-shaped cells, with cell diameters greater than 0.5 μm and cell lengths greater than 3.0 μm. The ellipsoidal spores swell the sporangia and lie subterminally to terminally. Grows on routine media such as nutrient agar and trypticase soy agar, producing pale yellow colonies. Catalase- and oxidase-positive. Nitrate reduction negative. Casein, gelatin, starch, and urea are not hydrolyzed. Growth occurs at 15°C, but not at 50°C. Growth does not occur at pH 5.5 and pH 9.0. Does not grow in presence of 2% NaCl. Strains are very unreactive and very few carbohydrates are assimilated by some strains while acid is produced weakly, if at all, from them. Glycerol is not assimilated. A very few amino acids, citrate, and some other organic acids may be used as carbon and energy sources.

Isolated from soil.

DNA G+C content (mol%): 48.2 (Takagi et al., 1993)–49.8 (Logan et al., 2002) (both HPLC).

Type strain: HPD52, ATCC 51359, CIP 103838, DSM 8552, NBRC 15518, JCM 8505, LMG 15968, NCIMB 13345, NRRL B-23247.

GenBank accession number (16S rRNA gene): AB112713, D78459.

6. **Brevibacillus formosus** (Heigener 1935) Shida, Takagi, Kadowaki and Komagata 1996a, 943^{VP} (*Bacillus formosus* Shida, Takagi, Kadowaki, Udaka, Nakamura and Komagata 1995, 98.)

for.mo′sus. L. adj. *formosus* beautiful.

Strictly aerobic, Gram positive, motile, rod-shaped cells, 0.5–0.9 μm × 2.0–5.0 μm. The ellipsoidal spores swell the sporangia. Description is based on the study of three strains. Grows on routine media such as nutrient agar and trypticase soy agar, producing colonies that are unpigmented, flat, smooth, circular and entire. Catalase-positive, oxidase-negative. Nitrate is reduced to nitrite. Casein and gelatin are hydrolyzed; starch and urea are not hydrolyzed. Growth occurs at 10°C and at 45°C, but not at 50°C; the optimum temperature for growth is 30°C. Growth occurs at pH 5.5 and 5.6. Growth does not occur in the presence of 2% NaCl. D-Glucose, D-fructose, glycerol, and other carbohydrates are assimilated, and acid is produced weakly, if at all, from them. A range of amino acids and organic acids may be used as carbon and energy sources.

Isolated from soil.

DNA G+C content (mol%): 47.2 (HPLC).

Type strain: F12, ATCC 51669, CIP 104544, DSM 9885, NBRC 15716, JCM 9169, LMG 16010, NRRL NRS-863.

GenBank accession number (16S rRNA gene): AB112712, D78460.

7. **Brevibacillus ginsengisoli** Baek, Im, Oh, Lee, Oh and Lee 2006, 2667^{VP}

gin.sen.gi.so′li. N.L. n. *ginsengum* ginseng; L. n. *solum* soil; N.L. gen. n. *ginsengisoli* of the soil of a ginseng field, the source of the organism.

Cells are Gram positive, aerobic or facultatively anaerobic, motile, slightly curved rods, 0.3–0.5 μm in diameter and 3.5–5.0 μm in length after 2 days culture on R2A agar. Colonies grown on R2A agar for 2 days are smooth, circular, glossy, white, and convex. Central and subterminal oval spores are formed in swollen sporangia. Grows well at 20–42°C and pH 5.0–8.5, but does not grow at 4 or 45°C. Growth occurs in the absence of NaCl and in the presence of 2.0% (w/v) NaCl but not 4% (w/v) NaCl. Grows anaerobically in denitrifying conditions. Casein and gelatin are hydrolyzed. Xylan, chitin, starch, cellulose, and DNA are not degraded. Urease, β-glucosidase, protease, and malic acid assimilation are positive in tests using API 20E and API 20NE strips. Reactions for ONPG hydrolysis, arginine dihydrolase, lysine decarboxylase, ornithine decarboxylase, citrate utilization, hydrogen sulfide production, tryptophan deaminase, indole production, acetoin production, and adipic acid assimilation are negative. Utilizes a small number of carbohydrates, amino acids, and organic acids as carbon sources. In addition to those shown in the tables, the following carbon sources are utilized in the API 50 CH and ID 32GN tests: L-histidine, salicin, and valeric acid. Utilization tests are negative for the following substrates: amygdalin, arbutin, D-arabinose, D-arabitol, D-fucose, D-lyxose, dulcitol, erythritol, L-fucose, glycogen, inulin, D-mannose, D-melibiose, methyl-α-D-mannopyranoside, methyl-β-D-xylopyranoside, D-raffinose, L-sorbose, D-tagatose, D-turanose, xylitol, L-xylose, capric acid, 3-hydroxybenzoic acid, 4-hydroxybenzoic acid, itaconic acid, propionic acid, sodium malonate, and suberic acid. MK-7 is the predominant respiratory quinone. The major cellular fatty acids are $C_{15:0\ iso}$, $C_{14:0\ iso}$, and $C_{15:0\ ante}$.

Isolated from soil from a ginseng field in Pocheon Province, South Korea.

DNA G+C content (mol%): 52.1 (HPLC).

Type strain: Gsoil 3088, KCTC 13938, LMG 23403.

GenBank accession number (16SrRNA): AB245376.

8. **Brevibacillus invocatus** Logan, Forsyth, Lebbe, Goris, Heyndrickx, Balcaen, Verhelst, Falsen, Ljungh, Hansson and De Vos. 2002, 964^{VP}

in.vo.ca′tus. L. adj. *invocatus* uninvited, referring to the isolation of strains of this organism as contaminants of an industrial fermentation.

Gram negative, motile, rod-shaped cells, 0.5–1.0 μm × 2.0–6.0 μm. Strictly aerobic. The ellipsoidal spores are borne subterminally and occasionally terminally, and swell the sporangia. Grows on routine media such as nutrient agar and trypticase soy agar. Growth at 30°C is initially slow, with more rapid growth following 24 h incubation; after 3–4 days the slightly umbonate colonies are 1–8 mm in diameter, with slightly irregular margins. Colonies are brownish-yellow, some with a single whitish concentric zone at the margin, and they are butyrous and have silky surfaces; the centers are opaque and the edges translucent. Catalase-positive. Nitrate reduction negative. Casein, gelatin, starch, and urea are not hydrolyzed; indole is not produced. Growth temperatures range from 15–35°C. Growth occurs between pH 6.0 and 8.5. Few carbohydrates are assimilated, only weakly, and acid is not produced from them; some amino acids and organic acids are used as carbon sources.

Isolated from a pharmaceutical fermentation plant and its antibiotic raw product.

DNA G+C content (mol%): 49.7 (HPLC).

Type strain: B2156, CIP 106911, JCM 12215, LMG 18962, NCIMB 13772.

GenBank accession number (16SrRNA): AB112718.

9. **Brevibacillus laterosporus** (Laubach 1916) Shida, Takagi, Kadowaki and Komagata 1996a, 945^{VP} (*Bacillus laterosporus* Laubach 1916, 505.)

la.te.ro.spor′us. L. n. *latus, lateris* the side; N.L. n. *spora* spore; N.L. adj. *laterosporus* with lateral spores.

Gram positive, Gram negative, and Gram variable, motile, rod-shaped cells, 0.5–0.9 μm × 2.0–5.0 μm. Facultatively anaerobic. Strains of this species commonly exhibit distinctive sporangial morphologies. The ellipsoidal spores are cradled in parasporal bodies (PBs) that have been described as C-shaped, or resembling canoes or the keels of ships. The spores, with their attached PBs, are borne centrally, paracentrally, and subterminally and are displaced laterally in the sporangia by the PBs, so that the sporangia are swollen into spindle shapes (Figure 32). Other species of aerobic endosporeforming bacteria may produce spores that lie laterally, but the PBs, which tend to remain firmly adherent to the spore after sporangial lysis, appear to be unique to this species. The proportion of sporangia containing PBs may vary with the strain and growth conditions. Grows on routine media such as nutrient agar and trypticase soy agar. Colonies are 1–3 mm in diameter, creamy-white, and smooth, and may have slightly irregular margins (Figure 34). Catalase-positive. Nitrate is reduced. Casein and gelatin are hydrolyzed, and starch and urea are not hydrolyzed; indole is not produced. Growth temperatures range from minima between 15 and 20°C to maxima between 35 and 50°C, with optima around 30°C. Growth does not occur at pH 5.7 but does occur at pH 6.8. D-Fructose, D-glucose, glycerol, maltose, D-mannitol, D-mannose, D-ribose, trehalose, and several other carbohydrates are assimilated, and acid is produced from them in larger quantities, so it is more readily detected than is the case with most other species of *Brevibacillus*. Some amino acids and organic acids are also used as carbon and energy sources. Some strains are pathogenic for mosquito and blackfly larvae. Parasporal toxin crystals, visible by electron microscopy, may be produced by some strains.

Isolated from soil, salterns, tap water, diseased honey bee larvae, other insects, foods, paper products, marine environments, and an eye infection.

DNA G+C content (mol%): 40.2 (T_m), 40.5 (Bd).

Type strain: ATCC 4517, ATCC 64, ATCC 8248, CCM 2116, BCRC 10607, CCUG 7421, CFBP 4222, CIP 52.83, DSM 25, HAMBI 1882, IAM 12465, NBRC 15654, JCM 2496, LMG 6931, LMG 16000, NCCB 75013, NCCB 48016, NCIMB 8213, NCIMB 9367, NCTC 6357, NRRL NRS-314, NRRL NRS-340, VKM B-499.

GenBank accession number (16S rRNA gene): AB112720, D16271, X60620.

10. **Brevibacillus levickii** Allan, Lebbe, Heyrman, De Vos, Buchanan and Logan 2005, 1048[VP]

le.vic.ki'i. N.L. gen. n. *levickii* of Levick, named after G. Murray Levick, surgeon and biological scientist of Captain R.F. Scott's Northern Party, the first scientific expedition to visit the vicinity of Mt. Melbourne in 1912.

Microaerophilic and weakly catalase-positive. Cells are Gram positive, becoming Gram negative after 48 h, motile, round-ended rods (0.7–0.8 μm × 2–5 μm) occurring singly, in pairs, and in chains. Endospores are ellipsoidal, occurring subterminally or terminally in swollen sporangia (Figure 31). After 48 h incubation at 40°C on 1/2 BFA (pH 5.5), colonies are circular, flat, up to 3.0 mm in diameter, and cream-colored with a matt appearance. Colony consistency becomes tough and difficult to break with a loop. Minimum growth temperature lies between 15 and 20°C, with the optimum temperature for growth being between 40 and 45°C, and the maximum growth temperature lying between 50 and 55°C. Growth occurs between pH 4.5 and 6.5 and the optimum pH for growth

lies between pH 5.0 and 5.5. Horse blood agar is partially hemolyzed. Gelatin is hydrolyzed, starch hydrolysis is weak, and casein hydrolysis is weak and variable. In the API 20E strip reactions for arginine dihydrolase, citrate utilization and Voges–Proskauer reaction are positive. Nitrate reduction is variable. A range of carbohydrates, amino acids, and organic acids is assimilated in the API Biotype 100 gallery as sole carbon sources. The major cellular fatty acid is $C_{15:0\ ante}$, accounting for approximately 74 % of the total fatty acid content. The following fatty acids are present in smaller amounts (at least 1 %): $C_{14:0\ iso}$, $C_{15:0\ iso}$, $C_{16:0}$, $C_{16:0\ iso}$, summed feature 4 ($C_{17:1\ iso}$ and/or $C_{17:1\ ante}$), and $C_{17:0\ ante}$.

Isolated from geothermal soil collected from the northwest slope of Mt. Melbourne, northern Victoria Land, Antarctica.

DNA G+C content (mol%): 50.3 (HPLC).

Type strain: B-1657, CIP 108307, LMG 22481.

GenBank accession number (16S rRNA gene): AJ715378.

11. **Brevibacillus limnophilus** Goto, Fujita, Kato, Asahara and Yokota 2004, 426[VP]

lim.no'phi.lus. Gr. n. *limnos* lake; Gr. adj. *philos* loving or friendly to; N.L. masc. adj. *limnophilus* lake-loving.

Strictly aerobic, Gram variable, motile rods, 0.5–0.6 μm ×2.2–4.0 μm. Ellipsoidal spores are borne subterminally in swollen sporangia. Description is based upon a single isolate. Colonies on nutrient agar are circular, entire, smooth, convex, translucent, and whitish-beige, and they are 3–4 mm in diameter after 48 h. The temperature range for growth is 20–45°C, and the temperature for optimum growth is 30–35°C. Optimum pH for growth is 7.0–7.5, growth occurs at pH 6.5–8.0 and does not occur at pH 6.0 or 8.5. Growth is weak in the presence of 2% NaCl and is inhibited by 5% NaCl. Catalase-positive and oxidase-negative. Urease, citrate utilization, and nitrate reduction are negative. Esculin is hydrolyzed, DNA is weakly hydrolyzed, and arbutin, casein, gelatin, starch, and tyrosine are not hydrolyzed. Acid is produced from L-arabinose, D-fructose, glycerol, rhamnose (weakly), and ribose. The major fatty acids are $C_{15:0\ iso}$, $C_{16:0}$, and $C_{17:0\ iso}$. The main quinone is menaquinone 7.

This organism was deposited in the ARS Culture collection as "*Bacillus limnophilus*" by Porter in 1940 and identified by Smith et al. (1952) as "*Bacillus brevis*".

Source of the type strain not reported; *Bacillus limnophilus* was described by Stührk (1935). Goto et al. (2004) did not refer to the original description, but they did indicate that the name they proposed was a revival for the strain that was deposited with this name in the ARS Culture Collection by Porter in 1940. Smith et al. (1952) considered it to be a synonym of *Bacillus brevis*.

DNA G+C content (mol%): 51.9 (HPLC).

Type strain: DSM 6472, NRRL NRS-887.

GenBank accession number (16S rRNA gene): AB112717.

12. **Brevibacillus parabrevis** (Takagi, Shida, Kadowaki, Komagata and Udaka 1993) Shida, Takagi, Kadowaki and Komagata 1996a, 943[VP] (*Bacillus parabrevis* Takagi, Shida, Kadowaki, Komagata and Udaka 1993, 229.)

pa.ra.bre'vis. Gr. prep. *para* alongside of, like; N.L. adj. *brevis* short; N.L. *parabrevis* *brevis*-like, referring to *Bacillus* (now *Brevibacillus*) *brevis*.

Strictly aerobic, Gram positive or Gram variable, motile, rod-shaped cells, 0.5–0.9 μm × 2.0–4.0 μm. The ellipsoidal spores are borne subterminally to terminally and swell the sporangia. Grows on routine media such as nutrient agar and trypticase soy agar, producing flat, smooth, yellowish-gray colonies. Catalase- and oxidase-positive. Nitrate reduction positive. Casein, DNA, gelatin, Tween 60, and Tween 80 are hydrolyzed; starch, and urea are not hydrolyzed. Hydrogen sulfide and indole are not produced. Growth occurs at 20°C, some strains grow at 15°C and most at 50°C, but no growth occurs at 55°C. Most strains will not grow at pH 5.5 or pH 9.0. Most strains will grow in presence of 2% NaCl, but not with 5% NaCl. D-Glucose, glycerol, maltose, D-mannitol, trehalose, and other carbohydrates are assimilated, but acid is produced weakly, if at all, from them. D-Fructose is not assimilated. Some amino acids and organic acids are used as carbon and energy sources.

Source of type strain not reported; other strains found in clinical specimens and cheese.

DNA G+C content (mol%): 51.8 (Takagi et al., 1993)– 52.2 (Logan et al., 2002) (both HPLC).

Type strain: ATCC 10027, CIP 103840, DSM 8376, NBRC 12334, JCM 8506, LMG 15971, NCIMB 13346, NRRL NRS-605, NRRL NRS-815.

GenBank accession number (16S rRNA gene): AB112714, D78463.

13. **Brevibacillus reuszeri** (Shida, Takagi, Kadowaki, Udaka, Nakamura and Komagata 1995) Shida, Takagi, Kadowaki and Komagata 1996a, 943[VP] (*Bacillus reuszeri* Shida, Takagi, Kadowaki, Udaka, Nakamura and Komagata 1995, 98.)

reus.ze'ri. N.L. gen. n. *reuszeri* of Reuszer, referring to H.W. Reuszer, who isolated the organism.

Strictly aerobic, Gram positive, motile, rod-shaped cells, 0.5–0.9 μm × 2.0–5.0 μm. The ellipsoidal spores swell the sporangia. Grows on routine media such as nutrient agar and trypticase soy agar, producing unpigmented, flat, smooth, circular, entire colonies. Catalase-positive, oxidase-negative. Nitrate is not reduced to nitrite. Casein, gelatin, starch, and urea are not hydrolyzed. Growth occurs at 10°C and at 45°C, and the optimum temperature for growth is 30°C. Growth occurs at pH 5.5 and 5.6. Growth occurs in the presence of 2% NaCl but not with 3% NaCl. D-Fructose, D-glucose, D-mannitol, and a few other carbohydrates are assimilated, and acid is produced weakly, if at

all, from them. Shida et al. (1995) found that this organism produced acid from glycerol and maltose, but Logan et al. (2002) found that neither was assimilated. Some amino acids and organic acids are used as carbon and energy sources.

Isolated from soil.

DNA G+C content (mol%): 46.5 (HPLC).

Type strain: H.W. Reuszer Army strain 39, ATCC 51665, CIP 104543, DSM 9887, NBRC 15719, JCM 9170, LMG 16012, NRRL NRS-1206.

GenBank accession number (16S rRNA gene): AB112715, D78464.

14. **Brevibacillus thermoruber** (Guicciardi, Biffi, Manachini, Craveri, Scolastico, Rindone and Craver 1968) Shida, Takagi, Kadowaki and Komagata 1996a, 945[VP] (*Bacillus thermoruber* (*ex* Guicciardi, Biffi, Manachini, Craveri, Scolastico, Rindone and Craver 1968) Manachini, Fortina, Parini and Craveri 1985, 495.)

ther'mo.ru.ber. Gr. n. *therme* heat; L. adj. *ruber* red; N.L. masc. adj. *thermoruber* heat-loving and red-pigment producing.

Moderately thermophilic, strictly aerobic, Gram positive, motile, rod-shaped cells, 0.8–1.0 μm × 2.5–4.8 μm. The ellipsoidal spores are borne terminally and subterminally and swell the sporangia. Description is based upon study of a single isolate. Requires biotin or thiamin for growth. Grows on glucose yeast extract agar, producing colonies that are spreading, smooth, shiny and red, with glossy, mucilaginous surfaces. The red pigment is endocellular and nondiffusible. Grows on routine media supplemented with yeast extract; biotin or thiamine required for growth. Growth in glucose-yeast extract broth is homogeneous. Catalase-negative or weakly positive. Nitrate is not reduced to nitrite. Casein, gelatin, and starch are hydrolyzed. Growth occurs between 34°C and 58°C, and the optimum temperature for growth is 45–48°C. No growth occurs in the presence of 5% NaCl. D-Fructose, D-glucose, D-mannitol, D-trehalose, and other carbohydrates are utilized as sole carbon sources according to Manachini et al. (1985), but Logan et al. (2002) were unable to reproduce these results.

Isolated from mushroom compost.

DNA G+C content (mol%): 57.0 ± 0.8 (HPLC).

Type strain: BT2, MIM 30.8.38, CIP 105255, CIP 105298, DSM 7064, HAMBI 2105, LMG 16910.

GenBank accession number (16S rRNA gene): AB112722, Z26921.

Genus V. **Cohnella** Kämpfer, Rosselló-Mora, Falsen, Busse and Tindall 2006, 784[VP]

PETER KÄMPFER, HANS-JÜRGEN BUSSE AND BRIAN J. TINDALL

Cohn.el'la. N.L. fem. dim. n. *Cohnella* named after Ferdinand Cohn, a German microbiologist who first described the bacterial genus *Bacillus* in 1872.

Spore-forming rods. Nonmotile. Gram-positive. **Aerobic.** Oxidase-positive. Good growth after 24 h on complex media such as trypticase soy agar and nutrient agar at 25–30°C. **Thermotolerant**; good growth occurs at 55°C. The major menaquinone is MK-7. The predominant polar lipids are diphosphatidylglycerol, phosphatidylglycerol, phosphatidyle-

thanolamine, and lysyl-phosphatidylglycerol. In addition, two unknown phospholipids, and four unknown amino-phospholipids are present. The main fatty acids are $C_{16:0\ iso}$, $C_{15:0\ anteiso}$, and $C_{16:0}$. Fatty acids in minor amounts are $C_{14:0}$, $C_{15:0}$, $C_{17:0\ iso}$, $C_{17:1\ iso}$, and $C_{17:0\ anteiso}$.

DNA G+C content (mol%): 57–59.

Type species: **Cohnella thermotolerans** Kämpfer, Rosselló-Mora, Falsen, Busse and Tindall 2006, 784[VP].

Further descriptive information

The 16S rRNA gene (1486 bp) of *Cohnella thermotolerans* showed the greatest degree of similarity to "*Paenibacillus hongkongensis*" (GenBank accession no. AF433165; 96.58%), described by Teng et al. (2003), but this name was not validly published. Significantly lower sequence similarities (<94.5%) were found with all other species of the genus *Paenibacillus* with validly published names (Kämpfer et al., 2006). A phylogenetic tree is shown in Figure 35.

As currently circumscribed, the genus contains only two species, *Cohnella thermotolerans* and *Cohnella hongkongensis*. Both species descriptions are based on single strains. The type strain of *Cohnella thermotolerans* was isolated on blood agar at 37°C during a routine hygienic control in a starch producing company. On trypticase soy (TS) agar, the strain was able to grow at 20–55°C, but not at 10 or 60°C. Growth at 30°C was also observed on nutrient agar and the R2A medium of Reasoner and Geldreich (1985), but not on *Salmonella–Shigella* (SS) agar. The type strain of *Cohnella hongkongensis* — originally described as "*Paenibacillus hongkongensis*" by Teng et al., (2003)—was also isolated from a blood agar culture of a 9-year-old Chinese boy with neutropenic fever and pseudobacteremia. This strain was also able to grow on complex media. On horse blood agar, the strain produced gray colonies after 24 h incubation at ambient air temperature. At 50°C it produced pinpoint colonies after 72 h incubation.

Both type strains of the *Cohnella* species display a complex polar lipid profile, which consists of diphosphatidylglycerol, phosphatidylglycerol, phosphatidylethanolamine, lysyl-phosphatidylglycerol, two unknown phospholipids, and four unknown amino-phospholipids.

The fatty acid profiles of the *Cohnella* strains (given in the species description) are similar to those of other *Paenibacillus* species, however, the amounts of $C_{16:0 \text{ iso}}$ (45.5% in *Cohnella thermotolerans*), or $C_{16:0}$ (25.3% in *Cohnella hongkongensis*) are very high (Kämpfer et al., 2006). Such high amounts of $C_{16:0 \text{ iso}}$ have not been reported to date for *Paenibacillus* species. Members of the genus *Thermobacillus*, described by Touzel et al. (2000) show a similar fatty acid content to *Cohnella thermotolerans* with respect to $C_{16:0}$ and $C_{17:0}$, but not for the other fatty acids. Kämpfer (2002) compiled fatty acid data of paenibacilli and found that the major fatty acids of the genus *Paenibacillus* (with ranges as percentages of total given in parentheses), as currently defined, are $C_{15:0 \text{ anteiso}}$ (36–80%), $C_{16:0 \text{ iso}}$ (0.5–6%), and $C_{15:0 \text{ iso}}$ (1—12%), and $C_{17:0 \text{ anteiso}}$ (6.7%). The differences in the relative amounts of $C_{16:0 \text{ iso}}$ and $C_{16:0}$, as well as the differences of more than 10% in the mol% G+C content of the DNA, may be indicative of properties that allow one not only to distinguish the two species, but (potentially) differentiate higher taxa. However, finding additional species that show similar 16S rDNA sequences and chemotaxonomic properties would be needed to test this hypothesis. Additional potential members of the genus *Cohnella* as suggested by high similarities in 16S rRNA gene sequences, which are deposited with gene banks, are apparently present in several institutional culture collections such as *Paenibacillus* sp. GP25–12 (94.6–95.4% 16S rRNA gene sequence similarity; accession no AM162342), *Cohnella* sp. T (95.2–97.3% 16S rRNA gene sequence similarity; accession no DQ333896), "*Cohnella ginsengisoli*" GR21–5 (94.4–94.9% 16S rRNA gene sequence

similarity; accession no EF368010), "*Paenibacillus panacarvi*" Gsoil 349; (94.4–94.6% 16S rRNA gene sequence similarity; accession no AB271056), and *Bacillus* sp. GL1 (94.3–94.9%16S rRNA gene sequence similarity; accession no AB024598; Nankai et al., 1999). Other strains might be also affiliated with the genus *Cohnella* but 16S rRNA gene sequences alone do not provide sufficient evidence because these strains share sequence similarities higher than 94% with only one of the established *Cohnella* species.

Enrichment and isolation procedures

No specific isolation medium has been described so far. Good growth occurs on complex media such as nutrient agar and trypticase soy agar (BBL).

Maintenance procedures

Cohnella cultures may be lyophilized by common procedures used for many bacteria. In addition, cultures can be maintained by serial transfers on solid complex media (although this is the least satisfactory of long-term storage methods). Growth on agar slants in screw-capped tubes can be kept at 4°C for about 2–4 weeks. Long-term preservation of liquid cultures supplemented with 5% dimethyl sulfoxide is recommended in liquid nitrogen in the liquid phase at –196°C.

Differentiation of the genus *Cohnella* from other genera

Members of the genus *Cohnella* may be distinguished from other genera by a combination of the cell-wall peptidoglycan, polar lipid, menaquinone, and fatty acid composition, and also the ability to grow at relatively high temperatures.

Taxonomic comments

In the field of reclassification of the bacilli, there is still a discussion as to what constitutes the genus *Bacillus* and whether other genera created over the years should be either further split or combined. In creating the genus *Cohnella* it was important to show that it could be clearly distinguished from the members of the genus *Bacillus* and also from members of the genus *Paenibacillus* (Kämpfer et al., 2006). One solution is to define the properties of the type species of these genera (i.e., *Bacillus* and *Paenibacillus*) and to create new genera where clear differences are to be found based on a combination of 16S rDNA sequence data, chemical composition, physiology, and biochemical properties.

The chemical composition of members of the bacilli is heterogeneous. This has been well documented (Minnikin and Goodfellow, 1981; O'Leary and Wilkinson, 1988). However, in recent years much emphasis has been put on the use of fatty acid composition. Although some differentiation is possible, fatty acid composition appears to be of limited value in further differentiating within the bacilli. In the case of respiratory lipoquinones, menaquinones dominate, with isoprenoid chain lengths from 6—9. The most commonly encountered menaquinone is MK-7. Closer examination of the polar lipids indicates a greater degree of variation, although this method has not been widely applied in recent years. As regards the members of *Bacillus* group 1 (Ash et al., 1991), there is chemical heterogeneity within the group (Minnikin and Goodfellow, 1981; O'Leary and Wilkinson, 1988). However, *Bacillus subtilis*—the type species of the genus *Bacillus*—produces diphosphatidylglycerol, phosphatidylglycerol, phosphatidylethanolamine, an amino-

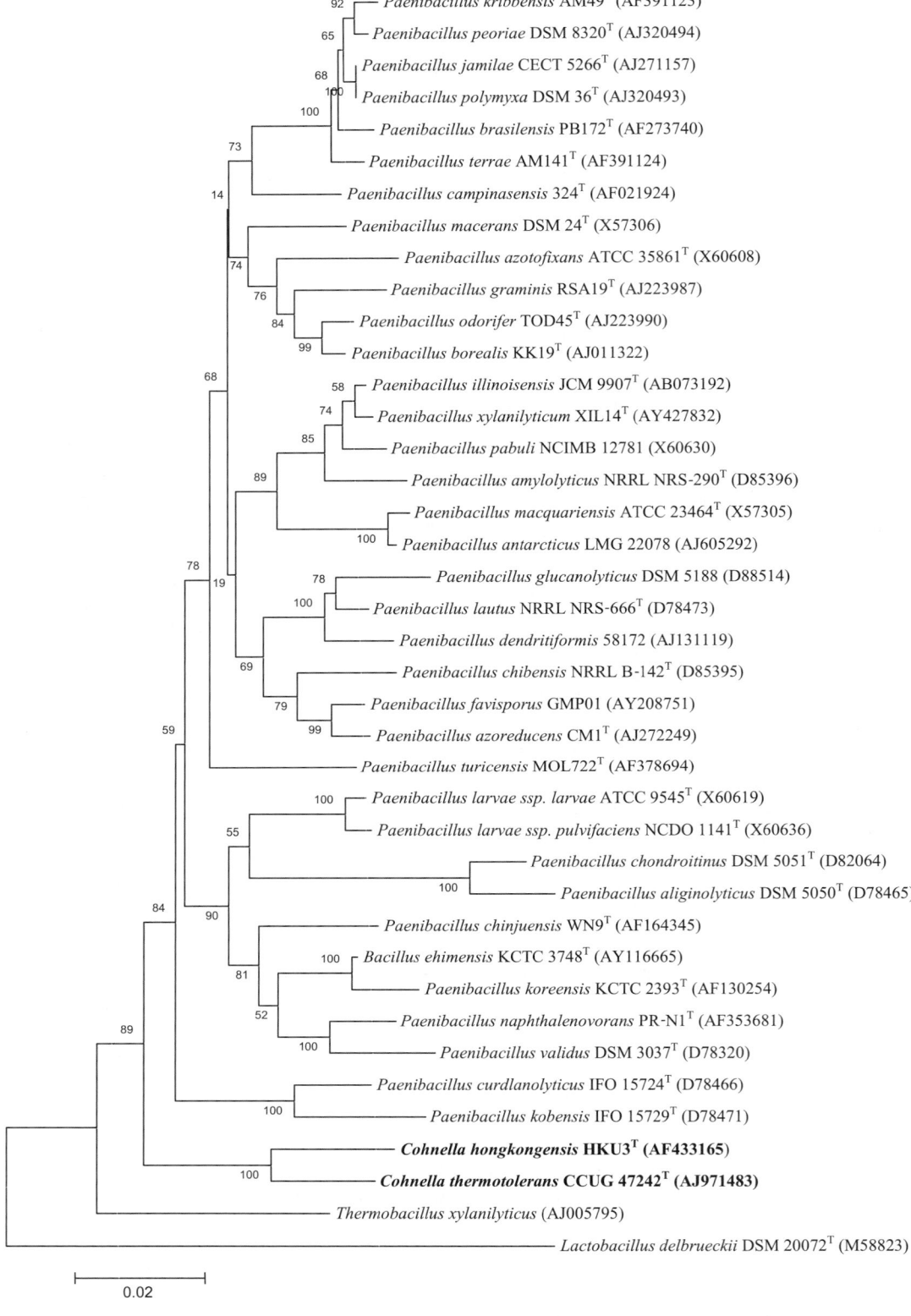

FIGURE 35. Phylogenetic analysis based on 16S rRNA gene sequences available from the European Molecular Biology Laboratory database (accession no.s are given in parentheses) constructed after multiple alignment of data by CLUSTAL_X (Thompson et al., 1997). Distances (distance options according to the Kimura-2 model) and clustering with the neighbor-joining method were performed by using the software packages MEGA (Molecular Evolutionary Genetics Analysis) version 2.1 (Kumar et al., 2001). Bootstrap values based on 1000 replications are listed as percentages at the branching points. (Data from Kämpfer et al., 2006).

acylphosphatidylglycerol, and a glycolipid that has been identified as a β-gentiobiosyldiacylglycerol (Minnikin and Goodfellow, 1981). Species sharing this polar lipid pattern, together with MK-7 and the fatty acid pattern listed in Table 54 should, in the future, constitute the "core" of the genus *Bacillus*.

On the other hand, it is still an open question which degree of variability (if any) within the polar lipid pattern will be acceptable to allow affiliation with the "core" of the genus *Bacillus*. *Bacillus herbersteinenesis* (Wieser et al., 2005), for instance, which shows a high degree of 16S rDNA sequence similarity, shares with *Bacillus subtilis* the same quinone system, diaminopimelic acid in the cell wall, a rather similar fatty acid profile, and the majority of characteristics within the polar lipid patterns, however, it can be distinguished by its lack of phosphatidylethanolamine and an unknown aminophospholipid. It is hoped that the affiliation of *Bacillus herbersteinenesis* lies with the *Bacillus* "core" and assignment to the genus *Bacillus*, or to a novel genus can be unambiguously answered by analyses of other representatives of taxa that show similar high degrees of 16S rDNA sequence similarity.

In contrast to the polar lipids of type species of the genus *Bacillus*, the polar lipid composition of the type species of the genus *Paenibacillus*, *Paenibacillus polymyxa*, is clearly different. The single glycolipid present does not have the same R_f value as β-gentiobiosyldiacylglycerol. Although the major phospholipids present are diphosphatidylglycerol, phosphatidylglycerol, and phosphatidylethanolamine, they only serve (together with the presence of menaquinones and iso/anteiso fatty acids) to confirm that this species belongs to the bacilli. The presence of a number of unidentified phospholipids serves to differentiate this species from other taxa. Preliminary work on the polar lipid composition of the genus *Paenibacillus* indicates that this group is also heterogeneous (Minnikin and Goodfellow, 1981; O'Leary and Wilkinson, 1988), and only those species sharing a similar chemical composition to the type species should be retained in the genus *Paenibacillus* in the future.

In the case of members of the genus *Cohnella*, the chemical composition also centres on the presence of menaquinones (MK-7), the dominance of iso and anteiso fatty acids, and the presence of diphosphatidylglycerol, phosphatidylglycerol,

TABLE 54. Percentage fatty acid composition of *Cohnella thermotolerans*, *Cohnella hongkongensis*, *Bacillus subtilis*, and *Paenibacillus polymyxa*[a]

Fatty acid	*C. thermotolerans*	*C. hongkongensis*	*B. subtilis*	*P. polymyxa*
$C_{13:0\ anteiso}$	0.8			
$C_{14:0\ iso}$	2.1	2.3	1	
$C_{14:0}$	1	5		
$C_{15:0\ iso}$	3.2	8.1	27	1
$C_{15:0\ anteiso}$	28.4	31.2	39	63
$C_{15:0}$	1.4	8		
$C_{16:0\ iso}$	45.5	11.9	1	6
$C_{16:1\ iso}$	1			
$C_{16:0}$	6.6	25.3	1	9
$C_{16:1\ \omega11c}$	0.9			
$C_{17:0\ iso}$	8	2		
$C_{17:0\ anteiso}$	6.7	2.6	1	17
$C_{17:1\ iso}$	1.1	1.9	3	
$C_{17:0}$	1.2			
$C_{17:1\ anteiso}$	10			
$C_{17:1\ \omega6c}$	1			
Other	8			
C18:1$_{\omega7c}$	4			

[a]Data for *Cohnella thermotolerans* and *Cohnella hongkongensis* are based on the type strains (Kämpfer et al., 2006). Data for *Bacillus subtilis* and *Paenibacillus polymyxa* were taken from Kämpfer (2002) and are compiled from several species descriptions.

and phosphatidylethanolamine as the predominant phospholipids. Among the polar lipids, however, both strains share the same properties of having a number of phospholipids and amino-phospholipids that are not present in either of the type species of the genus *Bacillus* or *Paenibacillus*. The ability to synthesize such lipids (including other amino acid derivates) must also be seen in a taxonomic and evolutionary context, being particularly prevalent in this branch of the Gram-positive Bacteria. This and previous work clearly indicate the value of chemotaxonomy within the bacilli, and further support the need for including such studies in all future taxonomic work on this group. This conclusion is also consistent with the remarks of the *ad hoc* committee (Murray et al., 1990; Wayne et al., 1987), which emphasized the need to carry out more chemotaxonomic work, particularly when defining new genera and families.

List of species of the genus *Cohnella*

1. **Cohnella thermotolerans** Kämpfer, Rosselló-Mora, Falsen, Busse and Tindall 2006, 784[VP]

ther.mo.to'ler.ans. Gr. n. *therme* heat; L. pres. part. *tolerans* tolerating; N.L. part. adj. *thermotolerans* able to tolerate high temperatures.

The description is as given for the genus, with the following additional characteristics. Good growth occurs after 24 h on TS agar and nutrient agar at 25–30°C; good growth occurs also at 55°C. The fatty acid profile of the type strain is comprised of $C_{16:0\ iso}$ (45.5%), $C_{15:0\ anteiso}$ (28.4%), $C_{17:0\ anteiso}$ (6.7%), $C_{16:0}$ (6.6%), $C_{18:1\ \omega7c}$ (4.0%), $C_{15:0\ iso}$ (3.2%), $C_{14:0\ iso}$ (2.1%), $C_{15:0}$ (1.4%), $C_{17:0\ iso}$ (1.1%), $C_{14:0}$ (1.0%), and $C_{17:1\ \omega6c}$ (1.0%). Esculin and *p*-nitrophenyl (pNP)-β-D-glucopyranoside are hydrolyzed. The following are utilized as sole carbon sources: arbutin, L-arabinose (weakly),

D-cellobiose, D-fructose, D-galactose, gluconate, D-glucose, D-maltose, D-mannose, α-D-melibiose, L-rhamnose (weakly), and D-ribose.

No acid production occurs from glucose, lactose, sucrose, D-mannitol, sulcitol, salicin, adonitol, inositol, sorbitol, L-arabinose, raffinose, rhamnose, maltose, D-xylose, trehalose, cellobiose, methyl-D-glucoside, erythritol, melibiose, D-arabitol, and D-mannose. No hydrolysis occurs of pNP-β-D-galactopyranoside, pNP-β-D-glucuronide, pNP-α-D-glucopyranoside, pNP-β-D-xylopyranoside, bis-pNP-phosphate, pNP-phenyl-phosphonate, pNP-phosphoryl-choline, 2-deoxythymidine-5'-pNP-phosphate, L-alanine-*p*-nitroanilide (pNA), L-glutamate-gamma-3-carboxy-pNA, and L-proline-pNA. The following carbon source are not utilized: *N*-acetyl-D-galactosamine, *N*-acetyl-D-glucosamine, sucrose, salicin, D-trehalose, D-xylose, adonitol, *i*-inositol, maltitol, D-mannitol, D-sorbitol, putrescine, acetate, propionate, *cis*-aconitate, *trans*-aconitate, adipate, 4-aminobutyrate, azelate, citrate, fumarate, glutarate, DL-3-hydroxybutyrate, itaconate, DL-lactate, L-malate, mesaconate, oxoglutarate, pyruvate, suberate, L-alanine, β-alanine, L-aspartate, L-histidine, L-leucine, L-ornithine, L-phenylalanine, L-proline, L-serine, L-tryptophan, 3-hydroxybenzoate, 4-hydroxybenzoate, and phenylacetate.

DNA G+C content (mol%): 59 (HPLC).

Type strain: CCUG 47242, CIP 108492, DSM 17683).

GenBank accession number (16S rRNA gene): AJ971483.

2. **Cohnella hongkongensis** Kämpfer, Rosselló-Mora, Falsen, Busse and Tindall 2006, 784^{VP} (*Paenibacillus hongkongensis* Teng, Woo, Leung, Lau, Wong and Yuen 2003, 33.)

hong.kong.en'sis. N.L. fem. adj. *hongkongensis* pertaining to Hong Kong.

The description is as given for the genus, with the following additional characteristics. Good growth occurs after 24 h on horse-blood agar at 37°C; good growth occurs also at 50°C, but not at 65°C. The fatty acid profile of the type strain is composed of $C_{16:0\ iso}$ (11.9%), $C_{15:0\ anteiso}$ (31.2%), $C_{17:0\ anteiso}$ (2.6%), $C_{16:0}$ (25.3%), $C_{15:0\ iso}$ (8.1%), $C_{13:0\ anteiso}$ (0.8%), $C_{14:0\ iso}$ (2.3%), $C_{15:0}$ (8.0%), $C_{16:1\ \omega11c}$ (0.9%), $C_{17:0\ iso}$ (1.9%), $C_{14:0}$ (5.0%), and $C_{17:0}$ (1.2%).

The type strain utilizes the following carbon sources after 7 d: *N*-acetyl-D-galactosamine, *N*-acetyl-D-glucosamine, arbutin, L-arabinose (weakly), D-cellobiose, D-fructose, D-galactose, gluconate, D-glucose, D-maltose, D-mannose, α-D-melibiose, L-rhamnose, and D-ribose. The following carbon source are not utilized: sucrose, salicin, D-trehalose, D-xylose, adonitol, *i*-inositol, maltitol, D-mannitol, D-sorbitol, putrescine, acetate, propionate, *cis*-aconitate, *trans*-aconitate, adipate, 4-aminobutyrate, azelate, citrate, fumarate, glutarate, DL-3-hydroxybutyrate, itaconate, DL-lactate, L-malate, mesaconate, oxoglutarate, pyruvate, suberate, L-alanine, β-alanine, L-aspartate, L-histidine, L-leucine, L-ornithine, L-phenylalanine, L-proline, L-serine, L-tryptophan, 3-hydroxybenzoate, 4-hydroxybenzoate, and phenylacetate.

DNA G+C content (mol%): 47.6 (T_m).

Type strain: HKU3, CCUG 49571, CIP 107898, DSM 17642.

GenBank accession number (16S rRNA gene): AF433165.

Genus VI. **Oxalophagus** Collins, Lawson, Willems, Cordoba, Fernandez-Garayzabal, Garcia, Cai, Hippe and Farrow 1994a, 822^{VP}

FRED A. RAINEY

Ox.sa.lo'pha.gus. Gr. n. *oxalis* wood sorrel (from which the name of oxalic acid is derived); Gr. masc. n. *phagos* glutton; N.L. masc. n. *Oxalophagus* oxalate eater.

Cells are Gram-positive, straight rods. Endospores are formed. Oval spores, located subterminally to centrally. Strictly anaerobic. Catalase-negative. Cytochromes are not produced. Oxalate and oxamate are decarboxylated to formate. Acetate is assimilated for cell carbon synthesis. No growth occurs with other organic acids, sugars, or alcohols. Member of the family *Paenibacillaceae* based on 16S rRNA gene sequence comparisons.

DNA G+C content (mol%): 36.3.

Type species: **Oxalophagus oxalicus** (Dehning and Schink, 1989) Collins, Lawson, Willems, Cordoba, Fernandez-Garayzabal, Garcia, Cai, Hippe and Farrow 1994, 822^{VP} (*Clostridium oxalicum* Dehning and Schink 1989, 83.).

List of species of the genus *Oxalophagus*

1. **Oxalophagus oxalicus** (Dehning and Schink 1989) Collins, Lawson, Willems, Cordoba, Fernandez-Garayzabal, Garcia, Cai, Hippe and Farrow 1994, 822^{VP} (*Clostridium oxalicum* Dehning and Schink 1989, 83.)

ox.a'li.cus. N.L. n. *acidum oxalicum* oxalic acid; N.L. masc. adj. *oxalicus* referring to the metabolism of oxalic acid.

Cells are straight rods, with rounded ends, occurring singly or in pairs, 2.5—4.8 μm long and 0.7—0.9 μm wide. Endospores are oval, 1.2—1.4 × 0.9 μm, subterminal to central and heat resistant. Motile by peritrichous flagella. Gram-positive. No cytochromes. Strictly anaerobic. Chemo-organotroph. Oxalate and oxamate decarboxylated to formate. Acetate assimilated for cell carbon synthesis. Nitrate, sulfate, sulfite, thiosulfate, and sulfur are not reduced. No growth with malonate, succinate (both 20 mM), glutamate, aspartate, glycine, fumarate + formate, glycerol, ethylene glycol, pyruvate, malate, citrate, acetoin, betaine, 1,2-propanediol, trimethoxybenzoate (all 10 mM), lactate, glycolate, methanol (all 5 mM), glyoxylate, hexamethylenetetramine, glucose, fructose, xylose, arabinose (all 2 mM), yeast extract, Casamino acids (both 0.1%), and H_2/CO_2 as substrates. Indole and catalase are not formed. Urea, gelatin, and esculin are not hydrolyzed. Growth occurs between 16–34°C, optimum 28–30°C. pH range for growth is 5.3–8.5, optimum 7.0. Enriched from pasteurized samples The type strain was isolated from anoxic freshwater sediment.

DNA G+C content (mol%): 36.3 (T_m).

Type strain: Alt Ox1, ATCC 49686, DSM 5503.

GenBank accession number (16S rRNA gene): X77840, Y14581.

Genus VII. **Thermobacillus** Touzel, O'Donohue, Debeire, Samain and Breton 2000, 318^{VP}

JEAN PIERRE TOUZEL AND GÉRARD PRENSIER

Ther.mo.ba.cil'lus. Gr. adj. *thermos* hot; N.L. dim. n. *bacillus* small rod; N.L. masc. n. *Thermobacillus* small thermophilic rod.

Cells straight rods with tapered ends 0.4–0.5 by 2.0–2.8 μm (Figure 36). Nonmotile. Ellipsoidal **endospores** are formed in swollen sporangia. **Aerobic.** Cells stain Gram-negative although the phylogenetic position is Gram-positive. Cell-wall structure is clearly of the Gram-positive type. Colonies are irregular, flat, with undulate margins. Catalase-positive and oxidase- and urease-negative. Growth occurs in the presence of 3% NaCl. Starch and esculin are hydrolyzed whereas gelatin and casein are not. Utilizes cellobiose, fructose, galactose, lactose, mannose, melezitose, melibiose, raffinose, trehalose, tributyrin, and xylose as sole carbon source for growth. Substrates which are not utilized are adonitol, casein, citrate, dextrin, dulcitol, erythritol, gelatin, inulin, salicin, sorbitol, and succinate. Nitrate is not reduced. **Thermophilic.** Grows optimally at 55°C; maximum temperature for growth is 63°C. Grows at pH 6.5–8.5; optimal pH is 7.8. **Carbon dioxide is required.** Produces large quantities of a xylan-inducible xylanase in the culture medium. The menaquinone is MK 7. The major fatty acid is $C_{16:0\ iso}$. On the basis of 16S rRNA analysis, the genus *Thermobacillus* is closely related to members of the genus *Paenibacillus* (Ash et al., 1994).

Isolated from farm soil in the north of France.

DNA G+C content (mol%): 57.5.

Type species: **Thermobacillus xylanilyticus** Touzel, O'Donohue, Debeire, Samain and Breton 2000, 319^{VP}.

Further descriptive information

The type strain is XE, which was isolated from farm soil in northern France and has been deposited with the French Collection Nationale de Cultures Microbiennes as a patent strain (CNCM-I1017).

Enrichment and isolation procedures

Thermobacillus xylanilyticus was isolated from a farm soil underlying a former manure heap in France. It is considered to be common in hot environments where plant organic matter is decaying. It is easily enriched and isolated after successive transfers in a xylan-based medium under aerobic, thermophilic (55°C) conditions in tightly closed vials that favor CO_2 accumulation (Touzel et al., 2000).

Maintenance procedures

Thermobacillus grows well in tightly closed 125 ml penicillin vials containing 10 ml liquid medium (Touzel et al., 2000) with a gas phase of air/CO_2 (90:10). The cultures are incubated overnight at 55°C on a rotary shaker. Such cultures remain viable for about 1 year when kept at 4°C. Long-term preservation is maintained in a mechanical freezer (–80°C) with 25% glycerol.

Differentiation of the genus *Thermobacillus* from other genera

Thermobacillus can be phylogenetically differentiated from *Paenibacillus* by 16S rDNA sequence comparison. Other features that distinguish these two genera include: motility, major fatty acid, and mol% G+C content of the DNA.

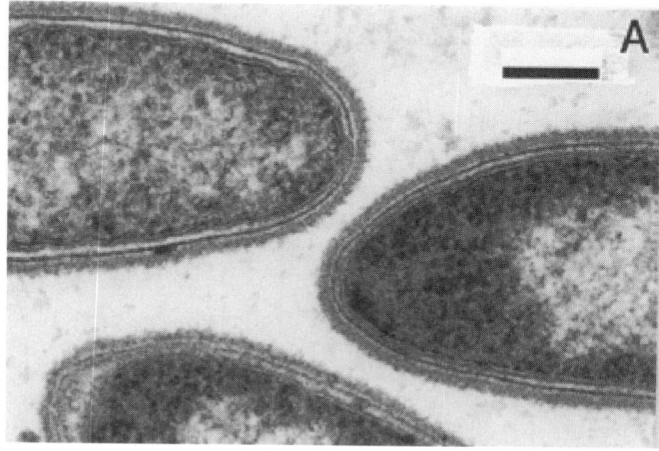
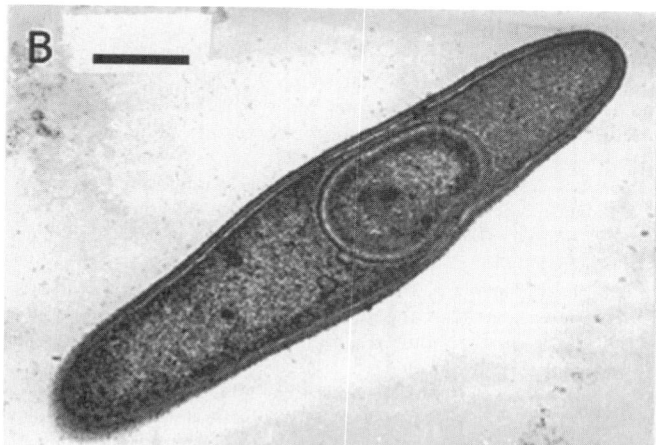

FIGURE 36. Electron micrograph of thin sections of *Thermobacillus xylanilyticus* grown in liquid medium. (A) Detail of the cell-wall structure typical of Gram-positive bacteria; bar = 0.2 μm. (B) General view of a sporulating cell; bar = 0.4 μm.

List of species of the genus *Thermobacillus*

1. **Thermobacillus xylanilyticus** Touzel, O'Donohue, Debeire, Samain and Breton 2000, 319[VP]

xy.la.ni.ly′ti.cus. Gr. n. *xylon* wood; N.L. n. *xylanum* xylan, a plant polysaccharide; Gr. adj. *lutikos* dissolving; N.L. adj. *xylanilyticus* hydrolyzing xylan.

Description is the same as for the genus.
DNA G+C content (mol%): 57.5 (T_m).
Type strain: XE, CNCM I-1017.
GenBank accession number (16S rRNA gene): AJ005795.

References

Abe, M. and M. Kimoto. 1984. Distribution of two types of regular-array particles in the cell wall of *Bacillus aneurinolyticus* (Kimura et Aoyama). Microbiol. Immunol. *28*: 841–846.

Abe, M., T. Nishimune, S. Ito, M. Kimoto and R. Hayashi. 1986. A simple method for the detection of 2 types of thiaminase-producing colonies. FEMS Microbiol. Lett. *34*: 129–133.

Achouak, W., P. Normand and T. Heulin. 1999. Comparative phylogeny of *rrs* and *nifH* genes in the *Bacillaceae*. Int. J. Syst. Bacteriol. *49*: 961–967.

Aguilera, M., M. Monteoliva-Sanchez, A. Suarez, V. Guerra, C. Lizama, A. Bennasar and A. Ramos-Cormenzana. 2001. *Paenibacillus jamilae* sp. nov., an exopolysaccharide-producing bacterium able to grow in olive-mill wastewater. Int. J. Syst. Evol. Microbiol. *51*: 1687–1692.

Alexander, B. and F.G. Priest. 1989. *Bacillus glucanolyticus*, a new species that degrades a variety of beta-glucans. Int. J. Syst. Bacteriol. *39*: 112–115.

Alippi, A.M. 1991. A comparison of laboratory techniques for the detection of significant bacteria of the honey bees, *Apis mellifera*. J. Apic. Res. *30*: 75–80.

Alippi, A.M., A.E. Perello, M.N. Sisterna, N.M. Greco and C.A. Cordo. 2000. Potential of spore-forming bacteria as biocontrol agents of wheat foliar diseases under laboratory and greenhouse conditions. J. Plant Dis. Prot. *107*: 155–169.

Alippi, A.M., A.C. Lopez and O.M. Aguilar. 2002. Differentiation of *Paenibacillus larvae* subsp. *larvae*, the cause of American foulbrood of honeybees, by using PCR and restriction fragment analysis of genes encoding 16S rRNA. Appl. Environ. Microbiol. *68*: 3655–3660.

Allan, R.N., L. Lebbe, J. Heyrman, P. De Vos, C.J. Buchanan and N.A. Logan. 2005. *Brevibacillus levickii* sp. nov. and *Aneurinibacillus terranovensis* sp. nov., two novel thermoacidophiles isolated from geothermal soils of northern Victoria Land, Antarctica. Int. J. Syst. Evol. Microbiol. *55*: 1039–1050.

Allison, A.C. 2000. Immunosuppressive drugs: the first 50 years and a glance forward. Immunopharmacology *47*: 63–83.

Andersson, A.M., N. Weiss, F. Rainey and M.S. Salkinoja-Salonen. 1999. Dust-borne bacteria in animal sheds, schools and children's day care centres. J. Appl. Microbiol. *86*: 622–634.

Aoyama, S. 1952. Studies on the thiamin-decomposing bacterium. I. Bacteriological researches of a new thiamin decomposing bacillus *Bacillus aneurinolyticus* Kimura et Aoyama. Acta. Sch. Med. Univ. Kioto *30*: 127–132.

Ash, C., J.A.E. Farrow, S. Wallbanks and M.D. Collins. 1991. Phylogenetic heterogeneity of the genus *Bacillus* revealed by comparative analysis of small subunit ribosomal RNA sequences. Lett. Appl. Microbiol. *13*: 202–206.

Ash, C., F.G. Priest and M.D. Collins. 1993. Molecular identification of rRNA group 3 bacilli (Ash, Farrow, Wallbanks and Collins) using a PCR probe test. Proposal for the creation of a new genus Paenibacillus. Antonie Leeuwenhoek *64*: 253–260.

Ash, C., F.G. Priest and M.D. Collins. 1994. *In* Validation of the publication of new names and new combinations previously effectively published outside the IJSB. List no. 51. Int. J. Syst. Bacteriol. *44*: 852.

Ash, C., F.G. Priest and M.D. Collins. 1995. *In* Validation of the publication of new names and new combinations previously effectively published outside the IJSB. List no. 52. Int. J. Syst. Bacteriol. *45*: 197–198.

Baek, S.H., W.T. Im, H.W. Oh, J.S. Lee, H.M. Oh and S.T. Lee. 2006. *Brevibacillus ginsengisoli* sp. nov., a denitrifying bacterium isolated from soil of a ginseng field. Int. J. Syst. Evol. Microbiol. *56*: 2665–2669.

Bailey, C.P. and A. von Holy. 1993. *Bacillus* spore contamination associated with commercial bread manufacture. Food Microbiol. *10*: 287–294.

Banerjee, U.C., R.K. Sani, W. Azmi and R. Soni. 1999. Thermostable alkaline protease from *Bacillus brevis* and its characterization as a laundry detergent additive. Process Biochem. *35*: 213–219.

Bapat, S. and A.K. Shah. 2000. Biological control of fusarial wilt of pigeon pea by *Bacillus brevis*. Can. J. Microbiol. *46*: 125–132.

Barsby, T., M.T. Kelly and R.J. Andersen. 2002. Tupuseleiamides and basiliskamides, new acyldipeptides and antifungal polyketides produced in culture by a *Bacillus laterosporus* isolate obtained from a tropical marine habitat. J. Nat. Prod. *65*: 1447–1451.

Batchelor, M.D. 1919. Aerobic spore-bearing bacteria in the intestinal tract of children. J. Bacteriol. *4*: 23–24.

Beatty, P.H. and S.E. Jensen. 2002. *Paenibacillus polymyxa* produces fusaricidin-type antifungal antibiotics active against *Leptosphaeria maculans*, the causative agent of blackleg disease of canola. Can. J. Microbiol. *48*: 159–169.

Ben-Jacob, E., I. Cohen, I. Golding, I.D.L. Gutnick, M. Tcherpakov, D. Helbing and I.G. Ron. 2000. Bacterial cooperative organization under antibiotic stress. Physica A *282*: 247–282.

Berge, O., M.H. Guinebretière, W. Achouak, P. Normand and T. Heulin. 2002. *Paenibacillus graminis* sp. nov. and *Paenibacillus odorifer* sp. nov., isolated from plant roots, soil and food. Int. J. Syst. Evol. Microbiol. *52*: 607–616.

Birbir, M. and A. Ilgaz. 1996. Isolation and identification of bacteria adversely affecting hide and leather quality. J. Soc. Leather Technol. Chemists *80*: 147–153.

Bosshard, P.P., R. Zbinden and M. Altwegg. 2002. *Paenibacillus turicensis* sp. nov., a novel bacterium harbouring heterogeneities between 16S rRNA genes. Int. J. Syst. Evol. Microbiol. *52*: 2241–2249.

Bredemann, G. and H. Heigner. 1935. *Bacillus validus* Bredemann and Heigner. Zentralbl. Bakteriol. Parasitenk., Infektrankh. Hyg., Abt. A *93*: 97–98.

Chantawannakul, P. and B.N. Dancer. 2001. American foulbrood in honey bees. Bee World *82*: 168–180.

Cheshire, F.R. and W.W. Cheyne. 1885. The pathogenic history and history under cultivation of a new bacillus (*B. alvei*), the cause of a disease of the hive bee hitherto known as foul brood. R. Micros. Soc. J. Ser. II *5*: 581–601.

Chou, J.H., Y.J. Chou, K.Y. Lin, S.Y. Sheu, D.S. Sheu, A.B. Arun, C.C. Young and W.M. Chen. 2007. *Paenibacillus fonticola* sp. nov., isolated from a warm spring. Int. J. Syst. Evol. Microbiol. *57*: 1346–1350.

Chung, Y.R., C.H. Kim, I. Hwang and J. Chun. 2000. *Paenibacillus koreensis* sp. nov., a new species that produces an iturin-like antifungal compound. Int. J. Syst. Evol. Microbiol. *50*: 1495–1500.

Claus, D. and R.C.W. Berkeley. 1986. Genus *Bacillus* Cohn 1872. *In* Mair Sneath, Sharpe and Holt (Editors), Bergey's Manual of Systematic Bacteriology, 9th edn, vol. 2. The Williams & Wilkins Co., Baltimore, pp. 1105–1139.

Cohen, I., I.G. Ron and E. Ben-Jacob. 2000. From branching to nebula patterning during colonial development of the *Paenibacillus alvei* bacteria. Physica A *286*: 321–336.

Collins, M.D., P.A. Lawson, A. Willems, J.J. Cordoba, J. Fernández-Garayzábal, P. Garcia, J. Cai, H. Hippe and J.A.E. Farrow. 1994. The phylogeny of the genus *Clostridium*: proposal of five new genera and eleven new species combinations. Int. J. Syst. Bacteriol. *44*: 812–826.

Costilow, R.N. and W.H. Coulter. 1971. Physiological studies of an oligosporogenous strain of *Bacillus popilliae*. Appl. Microbiol. *22*: 1076–1084.

Craveri, R., A. Guicciardi and N. Pacini. 1966. Distribution of thermophilic *Actinimycetes* in compost for mushroom production. Ann. Microbiol. (Milan) *16*: 111–113.

Daane, L.L., I. Harjono, S.M. Barns, L.A. Launen, N.J. Palleron and M.M. Haggblom. 2002. PAH-degradation by *Paenibacillus* spp. and description of *Paenibacillus naphthalenovorans* sp. nov., a naphthalene-degrading bacterium from the rhizosphere of salt marsh plants. Int. J. Syst. Evol. Microbiol. *52*: 131–139.

Dasman, S. Kajiyama, H. Kawasaki, M. Yagi, T. Seki, E. Fukusaki and A. Kobayashi. 2002. *Paenibacillus glycanilyticus* sp. nov., a novel species that degrades heteropolysaccharide produced by the cyanobacterium *Nostoc commune*. Int. J. Syst. Evol. Microbiol. *52*: 1669–1674.

Dees, P.M. and W.C. Ghiorse. 2001. Microbial diversity in hot synthetic compost as revealed by PCR-amplified rRNA sequences from cultivated isolates and extracted DNA. FEMS Microbiol Ecol *35*: 207–216.

Dehning, I. and B. Schink. 1989. Two new species of anaerobic oxalate-fermenting bacteria, *Oxalobacter vibrioformis* sp. nov. and *Clostridium oxalicum* sp. nov., from sediment samples. Arch. Microbiol. *153*: 79–84.

Dingman, D.W. and D.P. Stahly. 1983. Medium promoting sporulation of *Bacillus larvae* and metabolism of medium components. Appl. Environ. Microbiol. *46*: 860–869.

Durak, M.Z., H.I. Fromm, J.R. Huck, R.N. Zadoks and K.J. Boor. 2006. Development of molecular typing methods for *Bacillus* spp. and *Paenibacillus* spp. isolated from fluid milk products. J. Food Sci. *71*: M50–M56.

Dutky, S.R. 1940. Two new spore-forming bacteria causing milky disease of Japanese beetle larvae. J. Agric. Res. *61*: 57–68.

Edwards, S.G. and B. Seddon. 2000. Selective medium based on tyrosine metabolism for the isolation and enumeration of *Brevibacillus brevis* (*Bacillus brevis*). Lett. Appl. Microbiol. *31*: 395–399.

Edwards, S.G. and B. Seddon. 2001. Mode of antagonism of *Brevibacillus brevis* against *Botrytis cinerea in vitro*. J. Appl. Microbiol. *91*: 652–659.

Elo, S., L. Maunuksela, M. Salkinoja-Salonen, A. Smolander and K. Haahtela. 2000. Humus bacteria of Norway spruce stands: plant growth promoting properties and birch, red fescue and alder colonizing capacity. FEMS Microbiol. Ecol. *31*: 143–152.

Elo, S., I. Suominen, P. Kämpfer, J. Juhanoja, M. Salkinoja-Salonen and K. Haahtela. 2001. *Paenibacillus borealis* sp. nov., a nitrogen-fixing species isolated from spruce forest humus in Finland. Int. J. Syst. Evol. Microbiol. *51*: 535–545.

Enright, M.R., J.O. McInerney and C.T. Griffin. 2003. Characterization of endospore-forming bacteria associated with entomopathogenic nematodes, *Heterorhabditis* spp., and description of *Paenibacillus nematophilus* sp. nov. Int. J. Syst. Evol. Microbiol. *53*: 435–441.

Falcon, L.A. 1971. Use of bacteria for biological control. *In* Burges and Hussey (Editors), Microbial Control of INsects and Mites. Academic Press, London, pp. 67–95.

Farrow, J.A., C. Ash, S. Wallbanks and M.D. Collins. 1992. Phylogenetic analysis of the genera *Planococcus*, *Marinococcus* and *Sporosarcina* and their relationships to members of the genus *Bacillus*. FEMS Microbiol. Lett. *72*: 167–172.

Farrow, J.A., S. Wallbanks and M.D. Collins. 1994. Phylogenetic interrelationships of round-spore-forming bacilli containing cell walls based on lysine and the non-spore-forming genera *Caryophanon*, *Exiguobacterium*, *Kurthia*, and *Planococcus*. Int. J. Syst. Bacteriol. *44*: 74–82.

Favret, M.E. and A.A. Yousten. 1985. Insecticidal activity of *Bacillus laterosporus*. J. Invertebr. Pathol. *45*: 195–203.

Finkelstein, Z.I., B.P. Baskunov, I.M. Rietjens, M.G. Boersma, J. Vervoort and L.A. Golovleva. 2001. Transformation of the insecticide teflubenzuron by microorganisms. J. Environ. Sci. Health B *36*: 559–567.

Fujikawa, H. and M. Matsushita. 1989. Fractal growth of *Bacillus subtilis* on plates. J. Phys. Soc. Japan *58*: 3875–3878.

Garabito, M.J., M.C. Marquez and A. Ventosa. 1998. Halotolerant *Bacillus* diversity in hypersaline environments. Can. J. Microbiol. *44*: 95–102.

Garrity, G., D.J. Brenner, N.R. Krieg and J. T. Staley. 2005. Taxonomic outline of the *Archaea* and *Bacteria*. Appendix 2. The *Proteobacteria* Part A Introductory Assays. *In* Brenner, Garrity, Krieg and Staley (Editors), Bergey's Manual of Systematic Bacteriology, 2nd edn, vol. 2A. Springer, New York, pp. 207–220.

Genersch, E., E. Forsgren, J. Pentikainen, A. Ashiralieva, S. Rauch, J. Kilwinski and I. Fries. 2006. Reclassification of *Paenibacillus larvae* subsp. *pulvifaciens* and *Paenibacillus larvae* subsp. *larvae* as *Paenibacillus larvae* without subspecies differentiation. Int. J. Syst. Evol. Microbiol. *56*: 501–511.

Gilliam, M. and D.R. Dunham. 1978. Recent isolations of *Bacillus pulvifaciens* from powdery scales of honeybee, *Apis mellifera*, larvae. J. Invert. Pathol. *32*: 222–223.

Gordon, R.E., W.C. Haynes and C.H.-N. Pang. 1973. The genus *Bacillus*. Agriculture Handbook no. 427. United States Department of Agriculture, Washington, D.C.

Goto, K., Y. Kato, M. Asahara and A. Yokota. 2002. Evaluation of the hypervariable region in the 16S rDNA sequence as an index for rapid species identification in the genus *Paenibacillus*. J. Gen. Appl. Microbiol. *48*: 281–285.

Goto, K., R. Fujita, Y. Kato, M. Asahara and A. Yokota. 2004. Reclassification of *Brevibacillus brevis* strains NCIMB 13288 and DSM 6472 (=NRRL NRS-887) as *Aneurinibacillus danicus* sp. nov. and *Brevibacillus limnophilus* sp. nov. Int. J. Syst. Evol. Microbiol. *54*: 419–427.

Goto, N., J. Maeyama, Y. Yasuda, M. Isaka, K. Matano, S. Kozuka, T. Taniguchi, Y. Miura, K. Ohkuma and K. Tochikupo. 2000. Safety evaluation of recombinant cholera toxin B subunit produced by *Bacillus brevis* as a mucosal adjuvant. Vaccine *18*: 2164–2171.

Grishchenkov, V.G., R.T. Townshend, T.J. McDonald, R.L. Autenrieth, J.S. Bonner and A.M. Boronin. 2000. Degradation of petroleum hydrocarbons by facultative anaerobic bacteria under aerobic and anaerobic conditions. Process Biochem. *35*: 889–896.

Guemouri-Athmani, S., O. Berge, M. Bourrain, P. Mavingui, J.M. Thiery, T. Bhatnagar and T. Heulin. 2000. Diversity of *Paenibacillus polymyxa* populations in the rhizosphere of wheat (*Triticum durum*) in Algerian soils. Eur. J. Soil Biol. *36*: 149–159.

Guicciardi, A., M.R. Biffi, P.L. Manachini, A. Craveri, C. Scolastico, B. Ridone and R. Craveri. 1968. Ricerche preliminari su un nuovo schizomicete termofilo del genere *Bacillus* e caratterizzazione del pigmento rosso prodotto. Ann. Microbiol. (Milan) *18*: 191–205.

Gupta, A., C.P. Kaushik and A. Kaushik. 2000. Degradation of hexachlorocyclohexane (HCH; α, β, γ, and δ) by *Bacillus circulans* and *Bacillus brevis* isolated from soil contaminated with HCH. Soil Biol. Biochem. *32*: 1803–1805.

Harrison, H., R. Patel and A.A. Yousten. 2000. *Paenibacillus* associated with milky disease in Central and South American scarabs. J. Invertebr. Pathol. *76*: 169–175.

Haynes, W.C., L.J. Wickerham and C.W. Hesseltine. 1955. Maintenance of cultures of industrially important microorganisms. Appl Microbiol *3*: 361–368.

Heigener, H. 1935. Verwertung von Aminosäuren gemeinsame C- und N-Quelle durch bekannte Bodenbakterien nebst botanischer Beschreibung neu isolierter Betain- und Valin-Abbauer. Zentralbl. Bakteriol. Parasitenkd. Infektionskr. Hyg. Abt. 1 Orig Abt. II *93*: 81–113.

Heyndrickx, M., K. Vandemeulebroecke, P. Scheldeman, B. Hoste, K. Kersters, P. De Vos, N.A. Logan, A.M. Aziz, N. Ali and R.C.W. Berkeley.

1995. *Paenibacillus* (formerly *Bacillus*) *gordonae* (Pichinoty et al. 1986) Ash et al. 1994 is a later subjective synonym of *Paenibacillus* (formerly *Bacillus*) *validus* (Nakamura 1984) Ash et al. 1994, emended description of *P. validus*. Int. J. Syst. Bacteriol. *45*: 661–669.

Heyndrickx, M., K. Vandemeulebroecke, B. Hoste, P. Janssen, K. Kersters, P. De Vos, N.A. Logan, N. Ali and R.C.W. Berkeley. 1996a. Reclassification of *Paenibacillus* (formerly *Bacillus*) *pulvifaciens* (Nakamura 1984) Ash et al. 1994, a later subjective synonym of *Paenibacillus* (formerly *Bacillus*) *larvae* (White 1906) Ash et al. 1994, as a subspecies of *P. larvae*, with emended descriptions of *P. larvae* as *P. larvae* subsp. *larvae* and *P. larvae* subsp. *pulvifaciens*. Int. J. Syst. Bacteriol. *46*: 270–279.

Heyndrickx, M., K. Vandemeulebroecke, P. Scheldeman, K. Kersters, P. De Vos, N.A. Logan, A.M. Aziz, N. Ali and R.C.W. Berkeley. 1996b. A polyphasic reassessment of the genus *Paenibacillus*, reclassification of *Bacillus lautus* (Nakamura 1984) as *Paenibacillus lautus* comb. nov. and of *Bacillus peoriae* (Montefusco et al. 1993) as *Paenibacillus peoriae* comb. nov., and emended descriptions of *P. lautus* and of *P. peoriae*. Int. J. Syst. Bacteriol. *46*: 988–1003.

Heyndrickx, M., L. Lebbe, M. Vancanneyt, K. Kersters, P. De Vos, N.A. Logan, G. Forsyth, S. Nazli, N. Ali and R.C.W. Berkeley. 1997. A polyphasic reassessment of the genus *Aneurinibacillus*, reclassification of *Bacillus thermoaerophilus* (Meier-Stauffer et al. 1996) as *Aneurinibacillus thermoaerophilus* comb. nov, and emended descriptions of *A. aneurinilyticus* corrig., *A. migulanus*, and *A. thermoaerophilus*. Int. J. Syst. Bacteriol. *47*: 808–817.

Heyndrickx, M. and P. Scheldeman. 2002. Bacteria associated with spoilage in dairy products and other food. *In* Berkeley, Heyndrickx, Logan and De Vos (Editors), Applications and Systematics of *Bacillus* and Relatives. Blackwell Science, Oxford, pp. 64–82.

Horn, M.A., J. Ihssen, C. Matthies, A. Schramm, G. Acker and H.L. Drake. 2005. *Dechloromonas denitrificans* sp. nov., *Flavobacterium denitrificans* sp. nov., *Paenibacillus anaericanus* sp. nov. and *Paenibacillus terrae* strain MH72, N₂O-producing bacteria isolated from the gut of the earthworm *Aporrectodea caliginosa*. Int. J. Syst. Evol. Microbiol. *55*: 1255–1265.

Iida, K., Y. Ueda, Y. Kawamura, T. Ezaki, A. Takade, S. Yoshida and K. Amako. 2005. *Paenibacillus motobuensis* sp. nov., isolated from a composting machine utilizing soil from Motobu-town, Okinawa, Japan. Int. J. Syst. Evol. Microbiol. *55*: 1811–1816.

Kajino, T., Y. Saito, O. Asami, Y. Yamada, M. Hirai and S. Udata. 1997. Extracellular production of an intact and biologically active human growth hormone by the *Bacillus brevis* system. J. Ind. Microbiol. Biotechnol. *19*: 227–231.

Kajino, T., H. Takahashi, M. Hirai and Y. Yamada. 2000. Efficient production of artificially designed gelatins with a *Bacillus brevis* system. Appl. Environ. Microbiol. *66*: 304–309.

Kämpfer, P. 2002. Whole-cell fatty acid analysis in the systematics of *Bacillus* and related genera. *In* Berkeley, Heyndrickx, Logan and De Vos (Editors), Applications and Systematics of *Bacillus* and Relatives. Blackwell Science Ltd., Oxford, Berlin, pp. 271–299.

Kämpfer, P., R. Rosselló-Mora, E. Falsen, H.J. Busse and B.J. Tindall. 2006. *Cohnella thermotolerans* gen. nov., sp. nov., and classification of 'Paenibacillus hongkongensis' as *Cohnella hongkongensis* sp. nov. Int. J. Syst. Evol. Microbiol. *56*: 781–786.

Kanzawa, Y., A. Harada, M. Takeuchi, A. Yokota and T. Harada. 1995. *Bacillus curdlanolyticus* sp. nov and *Bacillus kobensis* sp. nov, which hydrolyze resistant curdlan. Int. J. Syst. Bacteriol. *45*: 515–521.

Katznelson, H. 1950. *Bacillus pulvifaciens* (n. sp.) an organism associated with powdery scale of honeybee larvae. J. Bacteriol. *59*: 153–155.

Katznelson, H. 1955. *Bacillus apiarius* n.sp., an aerobic spore-forming organism isolated from honeybee larvae. J. Bacteriol. *70*: 635–636.

Kellerman, K.K. and I.G. McBeth. 1912. The fermentation of cellulose. Zentralbl. Bakteriol. Parasitenkd. Infektionskr. Hyg. Abt. II *34*: 485–494.

Kim, D.S., C.Y. Bae, J.J. Jeon, S.J. Chun, H.W. Oh, S.G. Hong, K.S. Baek, E.Y. Moon and K.S. Bae. 2004. *Paenibacillus elgii* sp. nov., with broad antimicrobial activity. Int. J. Syst. Evol. Microbiol. *54*: 2031–2035.

Klein, M.G. and H.K. Kaya. 1995. *Bacillus* and *Serratia* species for *Scarab* control. Mem. Inst. Oswaldo Cruz *90*: 87–95.

Kneidinger, B., M. Graninger, G. Adam, M. Puchberger, P. Kosma, S. Zayni and P. Messner. 2001a. Identification of two GDP-6-deoxy-D-lyxo-4-hexulose reductases synthesizing GDP-D-rhamnose in *Aneurinibacillus thermoaerophilus* L420-91ᵀ. J. Biol. Chem. *276*: 5577–5583.

Kneidinger, B., M. Graninger, M. Puchberger, P. Kosma and P. Messner. 2001b. Biosynthesis of nucleotide-activated D-glycero-D-manno-heptose. J. Biol. Chem. *276*: 20935–20944.

Kumar, S., K. Tamura, I.B. Jakobsen and M. Nei. 2001. MEGA2: molecular evolutionary genetics analysis software. Bioinformatics *17*: 1244–1245.

Kuno, Y. 1951. *Bacillus thiaminolyticus*, a new thiamin-decomposing bacterium. Imp. Acad. Japan. Proc. *27*: 362–365.

Kuroshima, K., T. Sakane, R. Takata and A. Yokota. 1996. *Bacillus ehimensis* sp. nov. and *Bacillus chitinolyticus* sp. nov., new chitinolytic members of the genus *Bacillus*. Int. J. Syst. Bacteriol. *46*: 76–80.

Kurtz, J.C., D.F. Yates, J.M. Macauley, R.L. Qurles, F.J. Genthner, C.A. Chancy and R. Devereux. 2003. Effects of light reduction on growth of the submerged macrophyte *Vallisneria americana* and the community of root-associated heterotrophic bacteria. J. Exp. Marine Biol. Ecol. *291*: 199–218.

Laubach, C.A. 1916. Studies on aerobic spore-bearing non-pathogenic bacteria. Spore-bearing organisms in water. J. Bacteriol. *1*: 505–512.

Lebuhn, M., T. Heulin and A. Hartmann. 1997. Production of auxin and other indolic and phenolic compounds by *Paenibacillus polymyxa* strains isolated from different proximity to plant roots. FEMS Microbiol. Ecol. *22*: 325–334.

Lee, F.L., H.P. Kuo, C.J. Tai, A. Yokota and C.C. Lo. 2007a. *Paenibacillus taiwanensis* sp. nov., isolated from soil in Taiwan. Int. J. Syst. Evol. Microbiol. *57*: 1351–1354.

Lee, J.C. and K.H. Yoon. 2008. *Paenibacillus woosongensis* sp. nov., a xylanolytic bacterium isolated from forest soil. Int. J. Syst. Evol. Microbiol. *58*: 612–616.

Lee, J.S., K.C. Lee, Y.H. Chang, S.G. Hong, H.W. Oh, Y.R. Pyun and K.S. Bae. 2002. *Paenibacillus daejeonensis* sp. nov., a novel alkaliphilic bacterium from soil. Int. J. Syst. Evol. Microbiol. *52*: 2107–2111.

Lee, J.S., Y.R. Pyun and K.S. Bae. 2004. Transfer of *Bacillus ehimensis* and *Bacillus chitinolyticus* to the genus *Paenibacillus* with emended descriptions of *Paenibacillus ehimensis* comb. nov. and *Paenibacillus chitinolyticus* comb. nov. Int. J. Syst. Evol. Microbiol. *54*: 929–933.

Lee, M., L.N. Ten, S.H. Baek, W.T. Im, Z. Aslam and S.T. Lee. 2007b. *Paenibacillus ginsengisoli* sp. nov., a novel bacterium isolated from soil of a ginseng field in Pocheon province, South Korea. Antonie van Leeuwenhoek *91*: 127–135.

Lee, M., L.N. Ten, S.H. Baek, W.T. Im, Z. Aslam and S.T. Lee. 2007c. *In* List of new names and new combinations previously effectively, but not validly, published. List no. 116. Int. J. Syst. Evol. Microbiol. *57*: 1371–1373.

Lim, J.G. and D.H. Park. 2001. Degradation of polyvinyl alcohol by *Brevibacillus laterosporus*: metabolic pathway of polyvinyl alcohol to acetate. J. Microbiol. Biotech. *11*: 928–933.

Lim, J.M., C.O. Jeon, J.C. Lee, L.H. Xu, C.L. Jiang and C.J. Kim. 2006a. *Paenibacillus gansuensis* sp. nov., isolated from desert soil of Gansu Province in China. Int. J. Syst. Evol. Microbiol. *56*: 2131–2134.

Lim, J.M., C.O. Jeon, D.J. Park, L.H. Xu, C.L. Jiang and C.J. Kim. 2006b. *Paenibacillus xinjiangensis* sp. nov., isolated from Xinjiang province in China. Int. J. Syst. Evol. Microbiol. *56*: 2579–2582.

Logan, N.A. and R.C.W. Berkeley. 1981. Classification and identification of members of the genus *Bacillus* using API tests. *In* Berkeley and Goodfellow (Editors), The Aerobic Endospore-Forming Bacteria, pp. 106–140. Academic Press, London.

Logan, N.A., M. Heyndrickx, R.C.W. Berkeley and P. De Vos. 1998. *Paenibacillus azotofixans* (Seldin et al. 1984) Ash et al. 1995 does not have priority over *Paenibacillus durum* (Smith and Cato 1974) Collins et al. 1994: Request for an Opinion. Int. J. Syst. Bacteriol. *48*: 325–326.

Logan, N.A., L. Lebbe, B. Hoste, J. Goris, G. Forsyth, M. Heyndrickx, B.L. Murray, N. Syme, D.D. Wynn-Williams and P. De Vos. 2000. Aerobic endospore-forming bacteria from geothermal environments in northern Victoria Land, Antarctica, and Candlemas Island, South Sandwich archipelago, with the proposal of *Bacillus fumarioli* sp. nov. Int. J. Syst. Evol. Microbiol. *50*: 1741–1753.

Logan, N.A. 2002. Modern methods for identification *In* Berkley, Heyndrickx, Logan and De Vos (Editors), Applications and Systematics of *Bacillus* and Relatives. Blackwell Science, Oxford, pp. 123–140.

Logan, N.A., G. Forsyth, L. Lebbe, J. Goris, M. Heyndrickx, A. Balcaen, A. Verhelst, E. Falsen, A. Ljungh, H.B. Hansson and P. De Vos. 2002. Polyphasic identification of *Bacillus* and *Brevibacillus* strains from clinical, dairy and industrial specimens and proposal of *Brevibacillus invocatus* sp. nov. Int. J. Syst. Evol. Microbiol. *52*: 953–966.

Logan, N.A., E. De Clerck, L. Lebbe, A. Verhelst, J. Goris, G. Forsyth, M. Rodriguez-Diaz, M. Heyndrickx and P. De Vos. 2004. *Paenibacillus cineris* sp. nov. and *Paenibacillus cookii* sp. nov., from Antarctic volcanic soils and a gelatin-processing plant. Int. J. Syst. Evol. Microbiol. *54*: 1071–1076.

Ma, Y., Z. Xia, X. Liu and S. Chen. 2007a. *Paenibacillus sabinae* sp. nov., a nitrogen-fixing species isolated from the rhizosphere soils of shrubs. Int. J. Syst. Evol. Microbiol. *57*: 6–11.

Ma, Y., J. Zhang and S. Chen. 2007b. *Paenibacillus zanthoxyli* sp. nov., a novel nitrogen-fixing species isolated from the rhizosphere of *Zanthoxylum simulans*. Int. J. Syst. Evol. Microbiol. *57*: 873–877.

Ma, Y.C. and S.F. Chen. 2008. *Paenibacillus forsythiae* sp. nov., a nitrogen-fixing species isolated from rhizosphere soil of *Forsythia mira*. Int. J. Syst. Evol. Microbiol. *58*: 319–323.

Macé, E. 1889. Traité Pratique de Bactériologie, 1st edn. Ballière, Paris.

Manachini, P.L., M.G. Fortina, C. Parini and R. Craveri. 1985. *Bacillus thermoruber* sp. nov., nom. rev., a red-pigmented thermophilic bacterium. Int. J. Syst. Bacteriol. *35*: 493–496.

Marshall, B.J. and D.F. Ohye. 1966. *Bacillus macquariensis* n. sp., a psychrotrophic bacterium from sub-Antarctic soil. J. Gen. Microbiol. *44*: 41–46.

Marwoto, B., Y. Nakashimada, T. Kakizono and N. Nishio. 2002. Enhancement of (R,R)-2,3-butanediol production from xylose by *Paenibacillus polymyxa* at elevated temperatures. Biotechnol. Lett. *24*: 109–114.

Matheson, A. and M. Reid. 1992. Strategies for the prevention and control of American foulbrood. Am. Bee J. *132*: 471–475.

Mavingui, P. and T. Heulin. 1994. In vitro chitinase antifungal activity of soil, rhizosphere and rhizoplane populations of *Bacillus polymyxa*. Soil Biol. Biochem. *26*: 801–803.

McCray, A.H. 1917. Spore-forming bacteria of the apiary. J. Agric. Res. *8*: 399–420.

Meehan, C., A.J. Bjourson and G. McMullan. 2001. *Paenibacillus azoreducens* sp. nov., a synthetic azo dye decolorizing bacterium from industrial wastewater. Int. J. Syst. Evol. Microbiol. *51*: 1681–1685.

Meier-Stauffer, K., H.J. Busse, F.A. Rainey, J. Burghardt, A. Scheberl, F. Hollaus, B. Kuen, A. Makristathis, U.B. Sleytr and P. Messner. 1996. Description of *Bacillus thermoaerophilus* sp. nov., to include sugar beet isolates and *Bacillus brevis* ATCC 12990. Int. J. Syst. Bacteriol. *46*: 532–541.

Migula, W. 1900. System der Bakterien. Handbuch der Morphologie, Entwicklungsgeschichte und Systematik der bacterien, vol. 2; p. 583. G. Fischer Verlag, Jena.

Minnikin, D.E. and M. Goodfellow. 1981. Lipids in the classification of *Bacillus* and related taxa. *In* Berkeley and Goodfellow (Editors), The Aerobic Endospore-Forming Bacteria. Ninety Special Publications of the Society for General Microbiology. Academic Press, London, pp. 59–90.

Miyauchi, A., M. Ozawa, M. Mizukami, K. Yashiro, S. Ebisu, T. Tojo, T. Fujii and H. Takagi. 1999. Structural conversion from non-native to native form of recombinant human epidermal growth factor by *Brevibacillus choshinensis*. Biosci. Biotechnol. Biochem. *63*: 1965–1969.

Montaldi, F.A. and I.L. Roth. 1990. Parasporal bodies of *Bacillus laterosporus* sporangia. J. Bacteriol. *172*: 2168–2171.

Montefusco, A., L.K. Nakamura and D.P. Labeda. 1993. *Bacillus peoriae* sp. nov. Int. J. Syst. Bacteriol. *43*: 388–390.

Montes, M.J., E. Mercade, N. Bozal and J. Guinea. 2004. *Paenibacillus antarcticus* sp. nov., a novel psychrotolerant organism from the Antarctic environment. Int. J. Syst. Evol. Microbiol. *54*: 1521–1526.

Murray, R.G.E., D.J. Brenner, R.R. Colwell, P. de Vos, P. Goodfellow, P.A.D. Grimont, N. Pfennig, E. Stackebrandt and G.A. Zavarin. 1990. Report of the ad hoc committee on approaches to taxonomy within the proteobacteria. Int. J. Syst. Evol. Microbiol. *40*: 213–215.

Nakada, Y. and Y. Ohta. 2001. Rapid detection of *Brevibacillus formosus* BN53-1 in chicken feces. Can. J. Microbiol. *47*: 457–459.

Nakamura, L.K. and J. Swezey. 1983. Deoxyribonucleic acid relatedness of *Bacillus circulans* Jordan 1890 strains. Int. J. Syst. Bacteriol. *133*: 703–708.

Nakamura, L.K. 1984a. *Bacillus pulvifaciens* sp. nov, nom. rev. Int. J. Syst. Bacteriol. *34*: 410–413.

Nakamura, L.K. 1984b. *Bacillus amylolyticus* sp.nov., nom. rev., *Bacillus lautus* sp. nov., nom. rev., *Bacillus pabuli* sp. nov., nom. rev, and *Bacillus validus* sp. nov, nom. rev. Int. J. Syst. Bacteriol. *34*: 224–226.

Nakamura, L.K. 1987. *Bacillus alginolyticus* sp. nov. and *Bacillus chondroitinus* sp. nov., two alginate-degrading species. Int. J. Syst. Bacteriol. *37*: 284–286.

Nakamura, L.K. 1990. *Bacillus thiaminolyticus* sp. nov., nom. rev. Int. J. Syst. Bacteriol. *40*: 242–246.

Nakamura, L.K. 1993. DNA relatedness of *Bacillus brevis* Migula 1900 strains and proposal of *Bacillus agri* sp. nov., nom. rev., and *Bacillus centrosporus* sp. nov., nom. rev. Int. J. Syst. Bacteriol. *43*: 20–25.

Nakamura, L.K. 1996. *Paenibacillus apiarius* sp. nov. Int. J. Syst. Bacteriol. *46*: 688–693.

Nankai, H., W. Hashimoto, H. Miki, S. Kawai and K. Murata. 1999. Microbial system for polysaccharide depolymerization: enzymatic route for xanthan depolymerization by *Bacillus* sp. strain GL1. Appl. Environ. Microbiol. *65*: 2520–2526.

O'Leary, W.M. and S.G. Wilkinson. 1988. Gram-positive bacteria. *In* Ratledge and Wilkinson (Editors), Microbial Lipids. Academic Press, London, pp. 117–201.

Orlova, M.V., T.A. Smirnova, L.A. Ganushkina, V.Y. Yacubovich and R.R. Azizbekyan. 1998. Insecticidal activity of *Bacillus laterosporus*. Appl. Environ. Microbiol. *64*: 2723–2725.

Osman, S., M. Satomi and K. Venkateswaran. 2006. *Paenibacillus pasadenensis* sp. nov. and *Paenibacillus barengoltzii* sp. nov., isolated from a spacecraft assembly facility. Int. J. Syst. Evol. Microbiol. *56*: 1509–1514.

Outtrup, H. and S.T. Jørgensen. 2002. The importance of *Bacillus* species in the production of industrial enzymes. *In* Berkeley, Heyndrickx, Logan and De Vos (Editors), Applications and Systematics of *Bacillus* and Relatives. Blackwell Science, Oxford, pp. 206–218.

Park, M.J., H.B. Kim, D.S. An, H.C. Yang, S.T. Oh, H.J. Chung and D.C. Yang. 2007. *Paenibacillus soli* sp. nov., a xylanolytic bacterium isolated from soil. Int. J. Syst. Evol. Microbiol. *57*: 146–150.

Pettersson, B., K.E. Rippere, A.A. Yousten and F.G. Priest. 1999. Transfer of *Bacillus lentimorbus* and *Bacillus popilliae* to the genus *Paenibacillus* with emended descriptions of *Paenibacillus lentimorbus* comb. nov., and *Paenibacillus popilliae* comb. nov. Int. J. Syst. Bacteriol. *49*: 531–540.

Pirttijarvi, T.S., M.A. Andersson and M.S. Salkinoja-Salonen. 2000. Properties of *Bacillus cereus* and other bacilli contaminating biomaterial-based industrial processes. Int. J. Food Microbiol. *60*: 231–239.

Prazmowski, A. 1880. Untersuchung über die Entwickelungsgeschichte und Fermentwirking einiger Bacterien-Arten. Hugo Voigt, Leipzig.

Prenner, E.J., R.N. Lewis and R.N. McElhaney. 1999. The interaction of the antimicrobial peptide gramicidin S with lipid bilayer model and biological membranes. Biochim. Biophys. Acta *1462*: 201–221.

Raspoet, D., D. Pot, B. Pot, D. De Deyn, P. De Vos, K. Kersters and J. De Ley. 1991. Differentiation between 2,3-butanediol producing *Bacillus licheniformis* and *Bacillus polymyxa* strains by fermentation product profiles and whole-cell protein electrophoretic patterns. Syst. Appl. Microbiol. *14*: 1–7.

Reasoner, D.J. and E.E. Geldreich. 1985. A new medium for the enumeration and subculture of bacteria from potable water. Appl. Environ. Microbiol. *49*: 17.

Rippere, K.E., M.T. Tran, A.A. Yousten, K.H. Hilu and M.G. Klein. 1998. *Bacillus popilliae* and *Bacillus lentimorbus*, bacteria causing milky disease in Japanese beetles and related scarab larvae. Int. J. Syst. Bacteriol. *48*: 395–402.

Rivas, R., C. Gutierrez, A. Abril, P.F. Mateos, E. Martinez-Molina, A. Ventosa and E. Velazquez. 2005a. *Paenibacillus rhizosphaerae* sp. nov., isolated from the rhizosphere of Cicer arietinum. Int. J. Syst. Evol. Microbiol. *55*: 1305–1309.

Rivas, R., P.F. Mateos, E. Martinez-Molina and E. Velazquez. 2005b. *Paenibacillus xylanilyticus* sp. nov., an airborne xylanolytic bacterium. Int. J. Syst. Evol. Microbiol. *55*: 405–408.

Rivas, R., P.F. Mateos, E. Martinez-Molina and E. Velazquez. 2005c. *Paenibacillus phyllosphaerae* sp. nov., a xylanolytic bacterium isolated from the phyllosphere of *Phoenix dactylifera*. Int. J. Syst. Evol. Microbiol. *55*: 743–746.

Rivas, R., P. Garcia-Fraile, P.F. Mateos, E. Martinez-Molina and E. Velazquez. 2006. *Paenibacillus cellulosilyticus* sp. nov., a cellulolytic and xylanolytic bacterium isolated from the bract phyllosphere of *Phoenix dactylifera*. Int. J. Syst. Evol. Microbiol. *56*: 2777–2781.

Rodriguez-Diaz, M., L. Lebbe, B. Rodelas, J. Heyrman, P. De Vos and N.A. Logan. 2005. *Paenibacillus wynnii* sp. nov., a novel species harbouring the *nifH* gene, isolated from Alexander Island, Antarctica. Int. J. Syst. Evol. Microbiol. *55*: 2093–2099.

Rosado, A.S., J.D. van Elsas and L. Seldin. 1997. Reclassification of *Paenibacillus durum* (formerly *Clostridium durum* Smith and Cato 1974) Collins et al. 1994 as a member of the species *P. azotofixans* (formerly *Bacillus azotofixans* Seldin et al. 1984) Ash et al. 1994. Int. J. Syst. Bacteriol. *47*: 569–572.

Rosado, A.S., G.F. Duarte, L. Seldin and J.D. van Elsas. 1998. Genetic diversity of *nifH* gene sequences in *Paenibacillus azotofixans* strains and soil samples analyzed by denaturing gradient gel electrophoresis of PCR-amplified gene fragments. Appl. Environ. Microbiol. *64*: 2770–2779.

Roux, V. and D. Raoult. 2004. *Paenibacillus massiliensis* sp. nov., *Paenibacillus sanguinis* sp. nov. and *Paenibacillus timonensis* sp. nov., isolated from blood cultures. Int. J. Syst. Evol. Microbiol. *54*: 1049–1054.

Roux, V., L. Fenner and D. Raoult. 2008. *Paenibacillus provencensis* sp. nov., isolated from human cerebrospinal fluid, and *Paenibacillus urinalis* sp. nov., isolated from human urine. Int. J. Syst. Evol. Microbiol. *58*: 682–687.

Saha, P., A.K. Mondal, S. Mayilraj, S. Krishnamurthi, A. Bhattacharya and T. Chakrabarti. 2005. *Paenibacillus assamensis* sp. nov., a novel bacterium isolated from a warm spring in Assam, India. Int. J. Syst. Evol. Microbiol. *55*: 2577–2581.

Sanchez, M.M., D. Fritze, A. Blanco, C. Sproer, B.J. Tindall, P. Schumann, R.M. Kroppenstedt, P. Diaz and F.I. Pastor. 2005. *Paenibacillus barcinonensis* sp. nov., a xylanase-producing bacterium isolated from a rice field in the Ebro River delta. Int. J. Syst. Evol. Microbiol. *55*: 935–939.

Sarkar, P.K., B. Hasenack and M.J. Nout. 2002. Diversity and functionality of *Bacillus* and related genera isolated from spontaneously fermented soybeans (Indian Kinema) and locust beans (African Soumbala). Int. J. Food Microbiol. *77*: 175–186.

Schaffer, C., N. Muller, P.K. Mandal, R. Christian, S. Zayni and P. Messner. 2000. A pyrophosphate bridge links the pyruvate-containing secondary cell wall polymer of *Paenibacillus alvei* CCM 2051 to muramic acid. Glycoconj. J. *17*: 681–690.

Schäffer, C., N. Muller, R. Christian, M. Graninger, T. Wugeditsch, A. Scheberl and P. Messner. 1999. Complete glycan structure of the S-layer glycoprotein of *Aneurinibacillus thermoaerophilus* GS4-97. Glycobiology *9*: 407–414.

Schardinger, F. 1905. *Bacillus macerans*, ein Aceton bildender Rottebacillus. Zentralbl. Bakteriol. Parasitenkd. Infektionskr. Hyg. Abt.II *14*: 772–781.

Scheldeman, P., K. Goossens, M. Rodriguèz-Diàz, A. Pil, J. Goris, L. Herman, P. De Vos, N.A. Logan and M. Heyndrickx. 2004. *Paenibacillus lactis* sp. nov., isolated from raw and heat-treated milk. Int. J. Syst. Evol. Microbiol. *54*: 885–891.

Schieblich, M. 1923. Zwei aus Futterproben isolierte, bisher noch nicht beschrieben Bazillen. Zentralbl. Bakteriol. Parasitenkd. Infektionskr. Hyg. Abt.II *58*: 204–207.

Schleifer, K.H. and O. Kandler. 1972. Peptidoglycan types of bacterial cell walls and their taxonomic implications. Bacteriol Rev *36*: 407–477.

Seldin, L., J.D. van Elsas and E.G.C. Penido. 1983. *Bacillus* nitrogen fixers from Brazilian soils. Plant Soil *70*: 243–256.

Shida, O., H. Takagi, K. Kadowaki, S. Udaka and K. Komagata. 1994a. *Bacillus galactophilus* is a later subjective synonym of *Bacillus agri*. Int. J. Syst. Bacteriol. *44*: 172–173.

Shida, O., H. Takagi, K. Kadowaki, H. Yano, M. Abe, S. Udaka and K. Komagata. 1994b. *Bacillus aneurinolyticus* sp. nov., nom. rev. Int. J. Syst. Bacteriol. *44*: 143–150.

Shida, O., H. Takagi, K. Kadowaki, S. Udaka, L.K. Nakamura and K. Komagata. 1995. Proposal of *Bacillus reuszeri* sp. nov., *Bacillus formosus* sp. nov., nom. rev., and *Bacillus borstelensis* sp. nov., nom. rev. Int. J. Syst. Bacteriol. *45*: 93–100.

Shida, O., H. Takagi, K. Kadowaki and K. Komagata. 1996a. Proposal for two new genera, *Brevibacillus* gen. nov. and *Aneurinibacillus* gen. nov. Int. J. Syst. Bacteriol. *46*: 939–946.

Shida, O., H. Takagi, K. Kadowaki, H. Yano and K. Komagata. 1996b. Differentiation of species in the *Bacillus brevis* group and the *Bacillus aneurinolyticus* group based on the electrophoretic whole-cell protein pattern. Antonie van Leeuwenhoek *70*: 31–39.

Shida, O., H. Takagi, K. Kadowaki, L.K. Nakamura and K. Komagata. 1997a. Transfer of *Bacillus alginolyticus*, *Bacillus chondroitinus*, *Bacillus curdlanolyticus*, *Bacillus glucanolyticus*, *Bacillus kobensis*, and *Bacillus thiaminolyticus* to the genus *Paenibacillus* and emended description of the genus *Paenibacillus*. Int. J. Syst. Bacteriol. *47*: 289–298.

Shida, O., H. Takagi, K. Kadowaki, L.K. Nakamura and K. Komagata. 1997b. Emended description of *Paenibacillus amylolyticus* and description of *Paenibacillus illinoisensis* sp. nov. and *Paenibacillus chibensis* sp. nov. Int. J. Syst. Bacteriol. *47*: 299–306.

Shoji, J., R. Skazaki, Y. Wakisaka, K. Koizumi and M. Mayama. 1976. Isolation of a new antibiotic, laterosporamine. Studies on antibiotics from the genus *Bacillus*. J. Antibiot. (Tokyo) *29*: 390–393.

Singer, S., K.A. Doherty and A.D. Stambbaugh. 1988. Abstr. Annu. Meet. Am. Soc. Microbiol., vol. D34, pp. 76.

Singh, V.K. and A. Kumar. 1998. Production and purification of an extracellular cellulase from *Bacillus brevis* VS-1. Biochem. Mol. Biol. Int. *45*: 443–452.

Skerman, V.B.D., V. McGowan and P.H.A. Sneath. 1980. Approved lists of bacterial names. Int. J. Syst. Bacteriol *30*: 225–420.

Smerda, J., I. Sedlacek, Z. Pacova, E. Durnova, A. Smiskova and L. Havel. 2005. *Paenibacillus mendelii* sp. nov., from surface-sterilized seeds of *Pisum sativum* L. Int. J. Syst. Evol. Microbiol. *55*: 2351–2354.

Smerda, J., I. Sedlacek, Z. Pacova, E. Krejci and L. Havel. 2006. *Paenibacillus sepulcri* sp. nov., isolated from biodeteriorated mural paintings in the Servilia tomb. Int. J. Syst. Evol. Microbiol. *56*: 2341–2344.

Smirnova, T.A., I.B. Minenkova, M.V. Orlova, M.M. Lecadet and R.R. Azizbekyan. 1996. The crystal-forming strains of *Bacillus laterosporus*. Res. Microbiol. *147*: 343–350.

Smith, L.D.S. and E.P. Cato. 1974. *Clostridium durum*, sp.nov., the predominant organism in a sediment core from the Black Sea. Can. J. Microbiol. *20*: 1393–1397.

Smith, N.R., R.E. Gordon and F.E. Clark. 1952. Aerobic spore-forming bacteria, Monograph No. 16. U.S. Department of Agriculture, Washington, D.C.

Stührk, A. 1935. Untersuchungen über die Sporentötungszeit bei Bodenbakterien im strömenden Dampf nebst botanischer beschreibung einiger bei diesen Versuchen isolierter neuer Bakterienspezies. Zentbl. Bacteriol. Parasitenkd. Infektionskr. Hyg. Abt. II *93*: 161–198.

Suominen, I., C. Spröer, P. Kämpfer, F.A. Rainey, K. Lounatmaa and M. Salkinoja-Salonen. 2003. *Paenibacillus stellifer* sp. nov., a cyclodextrin-producing species isolated from paperboard. Int. J. Syst. Evol. Microbiol. *53*: 1369–1374.

Takagi, H., O. Shida, K. Kadowaki, K. Komagata and S. Udaka. 1993. Characterization of *Bacillus brevis* with descriptions of *Bacillus migu-*

lanus sp. nov., *Bacillus choshinensis* sp. nov., *Bacillus parabrevis* sp. nov., and *Bacillus galactophilus* sp. nov. Int. J. Syst. Bacteriol. *43*: 221–231.

Takeda, M., Y. Kamagata, S. Shinmaru, T. Nishiyama and J. Koizumi. 2002. *Paenibacillus koleovorans* sp. nov., able to grow on the sheath of *Sphaerotilus natans*. Int. J. Syst. Evol. Microbiol. *52*: 1597–1601.

Takeda, M., I. Suzuki and J. Koizumi. 2005. *Paenibacillus hodogayensis* sp. nov., capable of degrading the polysaccharide produced by *Sphaerotilus natans*. Int. J. Syst. Evol. Microbiol. *55*: 737–741.

Takeuchi, T., H. Iunuma and S. Kunimoto. 1981. A new anti-tumor antibiotic, spergualin: isolation and anti-tumore activity. J. Antibiot. (Tokyo) *34*: 1619–1621.

Takimura, Y., M. Kato, T. Ohta, H. Yamagata and S. Udaka. 1997. Secretion of human interleukin-2 in biologically active form by *Bacillus brevis* directly into culture medium. Biosci. Biotechnol. Biochem. *61*: 1858–1861.

Tcherpakov, M., E. Ben-Jacob and D.L. Gutnick. 1999. *Paenibacillus dendritiformis* sp. nov., proposal for a new pattern-forming species and its localization within a phylogenetic cluster. Int. J. Syst. Bacteriol. *49*: 239–246.

Ten, L.N., S.H. Baek, W.T. Im, M. Lee, H.W. Oh and S.T. Lee. 2006. *Paenibacillus panacisoli* sp. nov., a xylanolytic bacterium isolated from soil in a ginseng field in South Korea. Int. J. Syst. Evol. Microbiol. *56*: 2677–2681.

Teng, J.L., P.C. Woo, K.W. Leung, S.K. Lau, M.K. Wong and K.Y. Yuen. 2003. Pseudobacteraemia in a patient with neutropenic fever caused by a novel paenibacillus species: *Paenibacillus hongkongensis* sp. nov. Mol. Pathol. *56*: 29–35.

Thompson, J.D., T.J. Gibson, F. Plewniak, F. Jeanmougin and D.G. Higgins. 1997. The CLUSTAL_X windows interface: flexible strategies for multiple sequence alignment aided by quality analysis tools. Nucleic Acids Res. *25*: 4876–4882.

Timmusk, S., B. Nicander, U. Granhall and E. Tillberg. 1999. Cytokinin production by *Paenibacillus polymyxa*. Soil Biol. Biochem. *31*: 1847–1852.

Tindall, B.J. 2000. What is the type species of the genus *Paenibacillus*? Request for an Opinion. Int. J. Syst. Evol. Microbiol. *50*: 939–940.

Touzel, J.P., M. O'Donohue, P. Debeire, E. Samain and C. Breton. 2000. *Thermobacillus xylanilyticus* gen. nov., sp. nov., a new aerobic thermophilic xylan-degrading bacterium isolated from farm soil. Int. J. Syst. Evol. Microbiol. *50*: 315–320.

Turtura, G.C., A. Perfetto and P. Lorenzelli. 2000. Microbiological investigation on black crusts from open-air stone monuments of Bologna (Italy). New Microbiol. *23*: 207–228.

Udaka, S., N. Tsukagoshi and H. Yamagata. 1989. *Bacillus brevis*, a host bacterium for efficient extracellular production of useful proteins. Biotechnol. Genet. Eng. Rev. *7*: 113–146.

Uetanabaro, A.P., C. Wahrenburg, W. Hunger, R. Pukall, C. Sproer, E. Stackebrandt, V.P. de Canhos, D. Claus and D. Fritze. 2003. *Paenibacillus agarexedens* sp. nov., nom. rev., and *Paenibacillus agaridevorans* sp. nov. Int. J. Syst. Evol. Microbiol. *53*: 1051–1057.

Van der Maarel, M.J.E.C., A. Veen and D.J. Wijbenga. 2000. *Paenibacillus granivorans* sp. nov., a new *Paenibacillus* species which degrades native potato starch granules. Syst. Appl. Microbiol. *23*: 344–348.

Van der Maarel, M.J.E.C., A. Veen and D.J. Wijbenga. 2001. In Validation of the publication of new names and new combinations previously effectively published outside the IJSEM. List no. 79. Int. J. Syst. Evol. Microbiol. *51*: 263–265.

Vaz-Moreira, C.F. I., M.F. Nobre, P. Schumann, O.C. Nunes and C.M. Manaia. 2007. *Paenibacillus humicus* sp. nov., isolated from poultry litter compost. Int. J. Syst. Evol. Microbiol. *57*: 2267–2271.

Velazquez, E., T. de Miguel, M. Poza, R. Rivas, R. Rosselló-Mora and T.G. Villa. 2004. *Paenibacillus favisporus* sp. nov., a xylanolytic bacterium isolated from cow faeces. Int. J. Syst. Evol. Microbiol. *54*: 59–64.

Vollu, R.E., R. Fogel, S.C. dos Santos, F.F. da Mota and L. Seldin. 2006. Evaluation of the diversity of cyclodextrin-producing *Paenibacillus graminis* strains isolated from roots and rhizospheres of different plants by molecular methods. J. Microbiol. *44*: 591–599.

von der Weid, I., G.F. Duarte, J.D. van Elsas and L. Seldin. 2002. *Paenibacillus brasilensis* sp. nov., a novel nitrogen-fixing species isolated from the maize rhizosphere in Brazil. Int. J. Syst. Evol. Microbiol. *52*: 2147–2153.

Wakisaka, Y. and K. Koizumi. 1982. An enrichment isolation procedure for minor *Bacillus* populations. J. Antibiot. (Tokyo) *35*: 450–457.

Walker, R., A.A. Powell and B. Seddon. 1998. *Bacillus* isolates from the spermosphere of peas and dwarf French beans with antifungal activity against *Botrytis cinerea* and *Pythium* species. J. Appl. Microbiol. *84*: 791–801.

Wayne, L.G., D.J. Brenner, R.R. Colwell, P.A.D. Grimont, O. Kandler, M.I. Krichevsky, L.H. Moore, W.E.C. Moore, R.G.E. Murray, E. Stackebrandt, M.P. Starr and H.G. Trüper. 1987. Report of the ad hoc committee on the reconciliation of approaches to bacterial systematics. Int. J. Syst. Evol. Microbiol. *37*: 463–464.

Weimer, P.J. 1984. Control of product formation during glucose fermentation by *Bacillus macerans*. J. Gen. Microbiol. *130*: 103–112.

Wenzel, M., I. Schonig, M. Berchtold, P. Kämpfer and H. Konig. 2002. Aerobic and facultatively anaerobic cellulolytic bacteria from the gut of the termite *Zootermopsis angusticollis*. J. Appl. Microbiol. *92*: 32–40.

White, G.F. 1906. The bacteria of the apiary, with special reference to bee diseases. Bureau of Entomology Technical Series No. 14. U.S. Department of Agriculture, Washington, D.C.

White, G.F. 1920. European foulbrood; Bureau of Entomology Bulletin 810. U.S. Department of Agriculture, Washington, D.C.

Wieringa, K.T. 1941. *Bacillus agar-exedens*, a new species decomposing agar. Antonie van Leeuwenhoek J. Microbiol. Serol. *7*: 121–127.

Wieser, M., H. Worliczek, P. Kämpfer and H.J. Busse. 2005. *Bacillus herbersteinensis* sp. nov. Int. J. Syst. Evol. Microbiol. *55*: 2119–2213.

Wugeditsch, T., N.E. Zachara, M. Puchberger, P. Kosma, A.A. Gooley and P. Messner. 1999. Structural heterogeneity in the core oligosaccharide of the S-layer glycoprotein from *Aneurinibacillus thermoaerophilus* DSM 10155. Glycobiology *9*: 787–795.

Xie, C.H. and A. Yokota. 2007. *Paenibacillus terrigena* sp. nov., isolated from soil. Int. J. Syst. Evol. Microbiol. *57*: 70–72.

Yabbara, K.F., F. Juffali and R.M. Matossian. 1977. *Bacillus laterosporus* endophthalmitis. Arch. Ophthalmol. *95*: 2187–2189.

Yashiro, K., J.W. Lowenthal, T.E. O'Neil, S. Ebisu, H. Takagi and R.J. Moore. 2001. High-level production of recombinant chicken interferon-gamma by *Brevibacillus choshinensis*. Protein Expr. Purif. *23*: 113–120.

Yoon, J.H., D.K. Yim, J.S. Lee, K.S. Shin, H.H. Sato, S.T. Lee, Y.K. Park and Y.H. Park. 1998. *Paenibacillus campinasensis* sp. nov., a cyclodextrin-producing bacterium isolated in Brazil. Int. J. Syst. Bacteriol. *48*: 833–837.

Yoon, J.H., W.T. Seo, Y.K. Shin, Y.H. Kho, K.H. Kang and Y.H. Park. 2002. *Paenibacillus chinjuensis* sp. nov., a novel exopolysaccharide-producing bacterium. Int. J. Syst. Evol. Microbiol. *52*: 415–421.

Yoon, J.H., H.M. Oh, B.D. Yoon, K.H. Kang and Y.H. Park. 2003. *Paenibacillus kribbensis* sp. nov. and *Paenibacillus terrae* sp. nov., bioflocculants for efficient harvesting of algal cells. Int. J. Syst. Evol. Microbiol. *53*: 295–301.

Yoon, J.H., S.J. Kang, S.H. Yeo and T.K. Oh. 2005. *Paenibacillus alkaliterrae* sp. nov., isolated from an alkaline soil in Korea. Int. J. Syst. Evol. Microbiol. *55*: 2339–2344.

Yoon, M.H., L.N. Ten and W.T. Im. 2007. *Paenibacillus ginsengarvi* sp. nov., isolated from soil from ginseng cultivation. Int. J. Syst. Evol. Microbiol. *57*: 1810–1814.

Zaitsev, G.M., I.V. Tsitko, F.A. Rainey, Y.A. Trotsenko, J.S. Uotila, E. Stackebrandt and M.S. Salkinoja-Salonen. 1998. New aerobic ammonium-dependent obligately oxalotrophic bacteria: description of *Ammoniphilus oxalaticus* gen. nov., sp. nov. and *Ammoniphilus oxalivorans* gen. nov., sp. nov. Int. J. Syst. Bacteriol. *48*: 151–163.

Zhang, J., T.C. Hodgman, L. Krieger, W. Schnetter and H.U. Schairer. 1997. Cloning and analysis of the first *cry* gene from *Bacillus popilliae*. J. Bacteriol. *179*: 4336–4341.

Family V. **Pasteuriaceae** Laurent 1890, 756[AL]

Pas.teu.ri.a′ce.ae. N.L. fem. n. *Pasteuria* type genus of the family; suff. -*aceae* ending denoting family; N.L. fem. pl. n. *Pasteuriaceae* the family of *Pasteuria*.

Gram-positive, dichotomously branching, septate mycelium, the terminal hyphae of which enlarge to **form sporangia and eventually endospores. Nonmotile.** Sporangia and microcolonies are **endoparasitic in the bodies of freshwater, plant, and soil invertebrates**. Has not been cultivated axenically, but can be grown in the laboratory with its invertebrate host.

Type genus: **Pasteuria** Metchnikoff 1888, 166[AL] emend. Sayre and Starr 1985, 149; Starr and Sayre 1988a, 27.

Genus I. **Pasteuria** Metchnikoff 1888, 166[AL] emend. Sayre and Starr 1985, 149; emend. Starr and Sayre 1988a, 27 [Nom. Cons. Opin. 61 Jud. Comm. 1986, 119. Not **Pasteuria** in the sense of Henrici and Johnson (1935), Hirsch (1972), or Staley (1973); see Starr et al. (1983) and Judicial Commission (1986)]

RICHARD M. SAYRE AND MORTIMER P. STARR

revised by

DONALD W. DICKSON, JAMES F. PRESTON, III, ROBIN M. GIBLIN-DAVIS, GREGORY R. NOEL, DIETER EBERT AND GEORGE W. BIRD

Pas.teu′ri.a. N.L. gen. n. *Pasteuria* of Pasteur, named after Louis Pasteur, French savant and scientist.

Gram-positive, endospore-forming bacteria. Propagation following germination within a nematode or cladoceran host proceeds through the formation of rounded to elliptical (termed "cauliflower-like" by Metchnikoff, 1888) vegetative microcolonies from which "daughter" microcolonies may be formed. The sporogenous cells at the periphery of the colonies are usually attached by narrow "sacrificial" intercalary hyphae that lyse, resulting in developing sporangia arranged in clumps of eight or more, but more often in quartets, triplets, or doublets and, finally, as single teardrop-shaped or cup-shaped or rhomboidal mature sporangia. The rounded end of the sporangium encloses a single refractile endospore (1.0–3.0 μm in major dimension), an oblate spheroid, ellipsoidal or almost spherical in shape, usually resistant to desiccation and elevated temperatures (one species has somewhat limited heat tolerance). Nonmotile. Sporangia and microcolonies are endoparasitic in the bodies of freshwater, plant, and soil invertebrates. Axenic cultivation has not been documented, but it can be grown in the laboratory with its invertebrate host. The pathogen is horizontally transmitted via soil or waterborne spores. Infected hosts fail to reproduce.

DNA G+C content (mol%): not known.

Type species: **Pasteuria ramosa** Metchnikoff 1888, 166[AL].

Further descriptive information

A complex of errors initially confused our understanding of the genus *Pasteuria* Metchnikoff (1888). Stated briefly, Metchnikoff (1888) described an endospore-forming bacterial parasite of cladocerans, which he named *Pasteuria ramosa*. Metchnikoff presented drawings and photomicrographs of the life stages of this parasite as they occurred in the hemolymph of the water fleas *Daphnia pulex* Leydig and *Daphnia magna* Straus. He was, however, unable to cultivate the organism *in vitro*. Subsequent workers (Henrici and Johnson, 1935; Hirsch, 1972; Staley, 1973), who were looking in cladocerans for Metchnikoff's unique bacterium, reported on a different bacterium with only superficial resemblance to certain life stages of *Pasteuria ramosa*. Their investigations led to the axenic cultivation of a budding bacterium that is occasionally found on the exterior surfaces of *Daphnia* species. Unlike the organism described

by Metchnikoff, this bacterium divides by budding. It forms a major non-prosthecate appendage (a fascicle), it does not form endospores, it is not mycelial or branching, its staining reaction is Gram-negative, and it is not an endoparasite of cladocerans.

After searching for, but not finding, the bacterial endoparasite in water fleas as described by Metchnikoff, the erroneous conclusion was reached that this budding bacterium that occurs on the surfaces of *Daphnia* species was the organism Metchnikoff had described in 1888. As a result, a budding bacterium (strain ATCC 27377) was mistakenly designated (Staley, 1973) as the type culture for *Pasteuria ramosa* Metchnikoff (1888), the type (and, then, sole) species of the genus *Pasteuria*. This confusion between Metchnikoff's described cladoceran parasite and the quite different budding bacterium was resolved by Starr et al. (1983). The budding bacterium (with strain ATCC 27377 as its type culture) was named *Planctomyces staleyi* Starr, Sayre and Schmidt (1983), now named *Pirellula staleyi* (Butler et al., 2002). Conservation of the original descriptions of the genus *Pasteuria* and *Pasteuria ramosa*, as updated, was recommended (Starr et al., 1983) and approved (Judicial-Commission, 1986). Unfortunately, vestiges of this nomenclatural disorder remained for some time. For example, certain evolutionary and taxonomic inferences (Woese, 1987) regarding the genus *Pasteuria* were based upon data concerning bacteria belonging to the *Blastocaulis–Planctomyces* group of budding and non-prosthecately appendaged bacteria rather than the mycelial and endospore-forming invertebrate parasites that actually comprise the genus *Pasteuria*. A *Pasteuria ramosa*-like strain was discovered infecting the cladoceran *Moina rectirostris*, a member of the Daphniidae (Sayre et al., 1979), and this strain was used in the emendation of the species (Starr et al., 1983). Ebert et al. (1996), however, proposed that the *Daphnia*-parasitic *Pasteuria ramosa* that they had characterized from the same host as Metchnikoff (1888) be designated the neotype for *Pasteuria ramosa* Metchnikoff (1888) and that the *Moina* isolate be compared directly to the neotype in future studies.

Confusion of a different kind occurred in the nomenclature of the bacterial endoparasite of nematodes when Cobb (1906) erroneously reported the numerous highly refractive bodies infecting specimens of the nematode *Dorylaimus bulbiferous* as "perhaps monads" of a parasitic sporozoan. The incorrect placement in the protozoa of an organism now known to be a bacterial parasite of nematodes has persisted for nearly 70 years. Another incorrect placement was suggested by Micoletzky (1925), who found a nematode parasite similar in size and shape to those reported by Cobb. Micoletzky suggested that this organism belonged to the genus *Duboscqia* Perez (1908), another sporozoan group (Perez, 1908). Later, Thorne (1940) presented a detailed description of a new parasite from the nematode *Pratylenchus pratensis* (syn. *Pratylenchus brachyurus*, see Sayre et al., 1988). Thorne assumed that this organism was similar to the nematode parasite described by Micoletzky, thereby assigning it to the microsporidian genus *Duboscqia*, as *Duboscqia penetrans*.

Thorne's description and nomenclature persisted until 1975 even though other investigators (Canning, 1973; Williams, 1960), who had examined this nematode parasite in some detail, questioned this placement. It was not until the nematode parasite was re-examined using electron microscopy that its true affinities to bacteria rather than to protozoa were recognized and the name *Bacillus penetrans* (Thorne, 1940) Mankau (1975) was applied to it (Imbriani and Mankau, 1977; Mankau, 1975).

However, *Bacillus penetrans* was not included in the Approved Lists of Bacterial Names (Skerman et al., 1980), therefore it had no nomenclatural standing. Although this micro-organism forms endospores of the sort typical of the genus *Bacillus* Cohn (1872), its other traits (e.g., mycelial habit, endoparasitic associations with plant-parasitic nematodes) suggested that it did not belong in the genus *Bacillus*. Rather, it is closely related to *Pasteuria ramosa* (see Table 55 and Figure 37) and it has more properly been assigned (Sayre and Starr, 1985) to the genus *Pasteuria* Metchnikoff (1888) as *Pasteuria penetrans*.

The developmental stages that occur before sporulation are similar for all *Pasteuria* spp. Except for *Thermoactinomyces* spp., the mycelial-like proliferation during development is not found in any other endospore-forming bacteria. The morphological changes that occur during sporulation appear to be generally similar for all of the endospore-forming bacteria (Atibalentja et al., 2004a, 2004b; Chen et al., 1997b; Ebert et al., 1996, 2004; Giblin-Davis et al., 2003a, 2003b; Metchnikoff, 1888; Sayre and Starr, 1985).

Recent molecular work with 16S rRNA gene sequences support *Pasteuria* as a monophyletic clade comprising well supported lineages (Anderson et al., 1999; Atibalentja and Noel,

TABLE 55. Common characteristics of *Pasteuria ramosa* Metchnikoff (1888), *Pasteuria penetrans sensu stricto* emend. Starr and Sayre (1988a), *Pasteuria thornei* Starr and Sayre (1988a), *Pasteuria nishizawae* Sayre, Wergin, Schmidt and Starr (1991) emend. Noel, Atibalentja and Domier (2005), and "*Candidatus* Pasteuria usgae" Giblin-Davis, Williams, Bekal, Dickson, Brito, Becker and Preston (2003b)

Characterisitic	Description
Morphological similarities as observed by light microscopy:	
Vegetative cells	Microcolonies consist of dichotomously branched mycelium. Diameter of mycelial filaments similar. Mycelial filaments are seen in host tissues only during early stages of infection. Daughter microcolonies seem to be formed by lysis of "sacrificial" intercalary cells. Nearly all vegetative mycelium eventually lyses, leaving only sporangia and endospores.
Endospores	Terminal hyphae or peripheral cells of the colony elongate and swell, giving rise to sporangia. A single endospore is produced within each sporangium. Endospores are in the same general size range. Refractivity of endospores, as observed on the light microscope, increases with maturity.
Staining reaction	Gram-positive
Ultrastructural similarities:	
Vegetative cells	Mycelial cell walls are typical of Gram-positive bacteria. Mycelial filaments divided by septa. Double-layered cell walls. Where they occur, mesosomes are similar in appearance and seem to be associated with division and septum formation.
Endospores	Typical endogenous spore formation. Identical sequences in endospore formation: (a) septa form within sporangia; (b) sporangium cytoplast condenses to form forespore; (c) endospore walls form; (d) final endospore matures; and (e) "light" areas adjacent to endospore give rise to extrasporal fibers.
Similar sequences of life stages	Microcolonies. Fragmentation of microcolonies. Quartets of sporangia. Doublets of sporangia. Single sporangia. Free endospores.
Host–bacterium relationships	All parasitize invertebrates. Colonies first observed in the host are sedentary and located in the host's musculature. Growth in muscle tissue eventually leads to fragmentation and entry of microcolonies into the coelom or pseudocoelom of the respective host. Microcolonies carried passively by body fluids. Colonization of hemolymph or pseudocoelomic fluid by the parasite is extensive. Host ranges are very narrow: *Pasteuria ramosa* occurs only in cladoceran water fleas, *Pasteuria penetrans sensu stricto* in the root-knot nematode *Meloidogyne incognita*, *Pasteuria thornei* in the lesion nematode *Pratylenchus brachyurus*, *Pasteuria nishizawae* in *Heterodera* and *Globodera*, and "*Candidatus* Pasteuria usgae" in *Belonolaimus*. Host is completely utilized by the bacteria; in the end, the host becomes little more than a bag of bacterial endospores.
Survival mechanisms	Survive in field soil and at bottom of ponds. Resist desiccation. Loss of infectivity after 5 min at 70°C may indicate moderate resistant to heat.

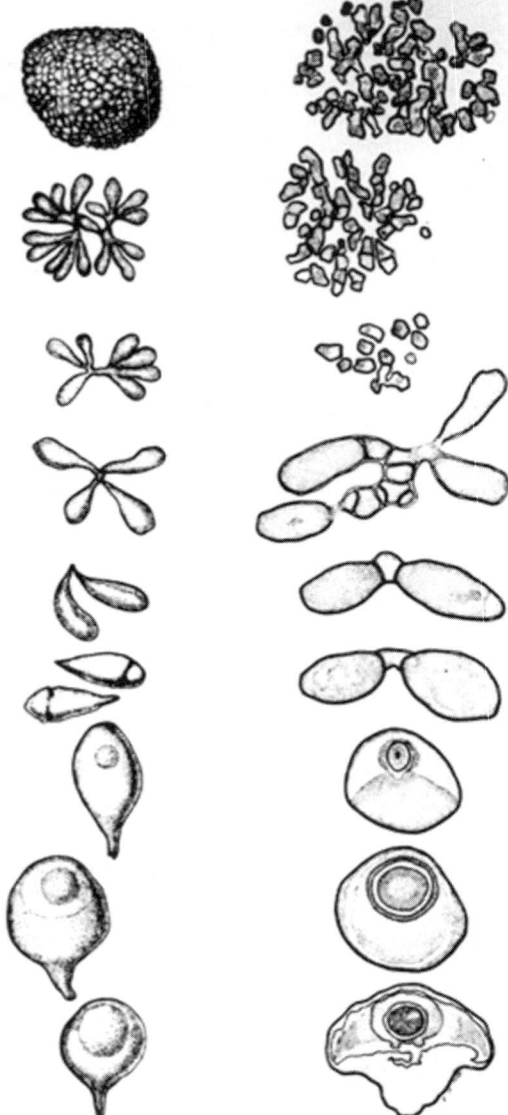

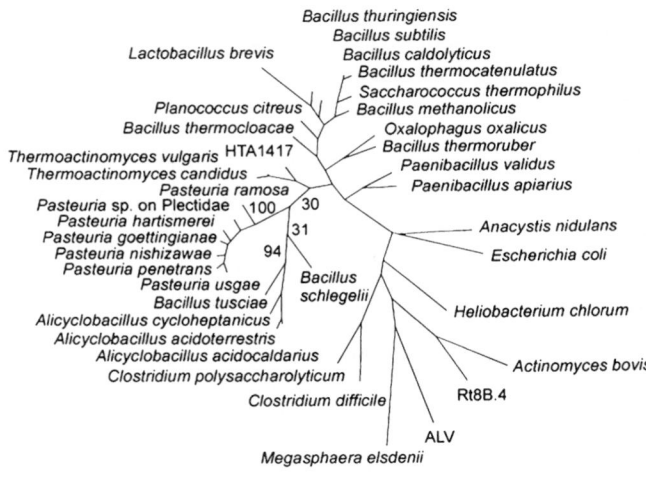

FIGURE 38. Phylogenetic position of *Pasteuria* spp. inferred from maximum-parsimony analysis of 16S rRNA gene sequences. The accession numbers of the *Pasteuria* sequences used were as follows: "*Pasteuria goettingianae*", AF515699; "*Pasteuria hartismerei*", AJ878853; *Pasteuria nishizawae*, AF134868; *Pasteuria penetrans*, AF077672; *Pasteuria* sp. parasitic on nematodes of the family Plectidae, AY652776; *Pasteuria ramosa*, U34688; and "*Candidatus* Pasteuria usgae*", AF254387. The accession numbers for the other sequences were as in Atibalentja et al. (2000). *Anacystis nidulans* (*Cyanobacteria*) and *Escherichia coli* (*Proteobacteria*) were used as outgroup taxa. Bootstrap proportions (10,000 replications) are shown, wherever possible, for nodes that are relevant for *Pasteuria* spp. The figures not shown are 98, 96, 53, 87, and 93, respectively, for branches leading to *Pasteuria* sp. on Plectidae, "*Pasteuria hartismerei*", "*Pasteuria goettingianae*", *Pasteuria nishizawae*, and *Pasteuria penetrans*. Bootstrap support was 100% for each of the branches leading to *Alicyclobacillus cycloheptanicus* and *Alicyclobacillus acidoterristris*. The alignment of the 16S rRNA gene sequences was performed with CLUSTAL X (Thompson et al., 1997) and phylogenetic analyses were conducted with PAUP* 4.0b10 (Swofford, 2003). "*Pasteuria goettingianae*", "*Pasteuria hartismerei*" and "*Candidatus* Pasteuria usgae*" are most likely common in environmental samples but require isolation and deposition of type cultures before validation of their names. From Atibalentja and Noel (2008).

FIGURE 37. Drawings of *Pasteuria penetrans sensu stricto* (left) based on electron micrographs are compared with those of *Pasteuria ramosa* (right) as drawn by Metchnikoff (1888). Starting at the top of the left column is a vegetative colony of *Pasteuria penetrans* followed by daughter colonies, quartets of sporangia, doublets of sporangia, single sporangia and, finally, at the bottom the mature endospore within the old sporangial wall. The drawings of *Pasteuria ramosa* on the right are placed in order of their occurrence in the life cycle of the parasite as reported by Metchnikoff (1888). (Reproduced by permission from R.M. Sayre, W.P. Wergin and R.E. Davis. (1977). Can. J. Microbiol. *23*: 1573–1579).

2008; Atibalentja et al., 2000; Bekal et al., 2001; Bishop et al., 2007; Ebert et al., 1996; Giblin-Davis et al., 2003b; Preston et al., 2003; Sturhan et al., 2005). Further genomic comparisons have been made through the sequences of sporulation and other genes in different isolates of *Pasteuria penetrans* and *Pasteuria ramosa* (Bird et al., 2003; Charles et al., 2005; Preston et al., 2003; Schmidt et al., 2004; Trotter and Bishop, 2003). A phylogenetic tree (Atibalentja and Noel, 2008) comparing 16S rRNA

sequences has been updated to depict the close relationship of *Pasteuria* spp. to other genera of the *Firmicutes* (Figure 38).

Comparison of *spoIIAB* gene sequences amplified by PCR from DNA from different isolates (biotypes) of *Pasteuria penetrans* identified single nucleotide polymorphisms or SNPs, indicating genetic heterogeneity within populations obtained from both individuals, as well as populations of *Meloidogyne* spp. (Nong et al., 2007). DNA sequences from three different loci derived from a single-spore isolate of *Pasteuria penetrans* were identical, supporting the need for clonal populations for definitive studies on host preference (Trotter et al., 2004).

To clarify the characteristics of the genus *Pasteuria*, the meanings of the terms "endospore" and "sporangium" must be modified slightly from their usual definition in order to be applicable to this genus. Metchnikoff observed the several stages of endosporogenesis that occurred in *Pasteuria ramosa*. In his discussion, he noted within each sporangium a single refractile body that stained with difficulty; he called this structure, as we do today, an endospore. The *Pasteuria* endospore is not entirely typical of those found in *Bacillus* or *Clostridium*.

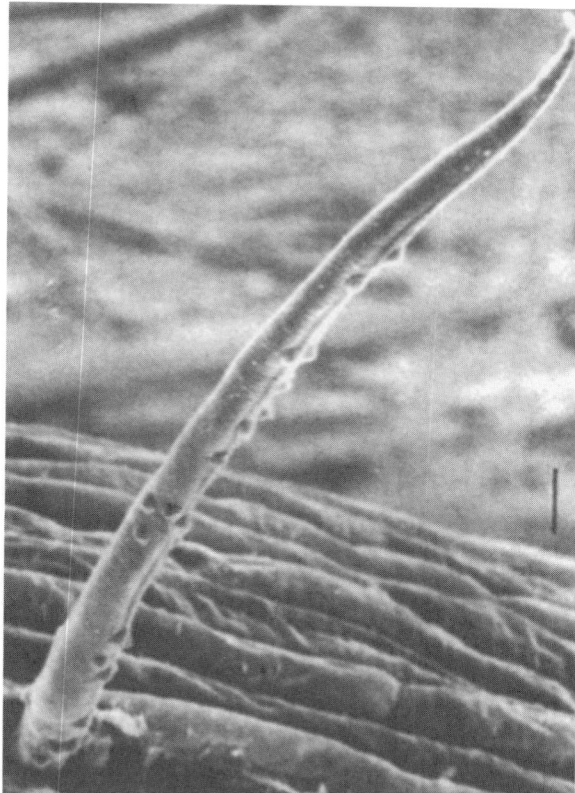

FIGURE 39. Scanning electron micrograph of endospores of *Pasteuria penetrans sensu stricto* attached to a juvenile of a root-knot nematode that has partially penetrated a tomato root. The endospores carried on the juvenile will germinate inside the plant and penetrate the developing nematode, completing their life cycle in their host. On decay of the plant root in soil, the endospores developed within the parasitized nematode are released. Bar = 10 μm.

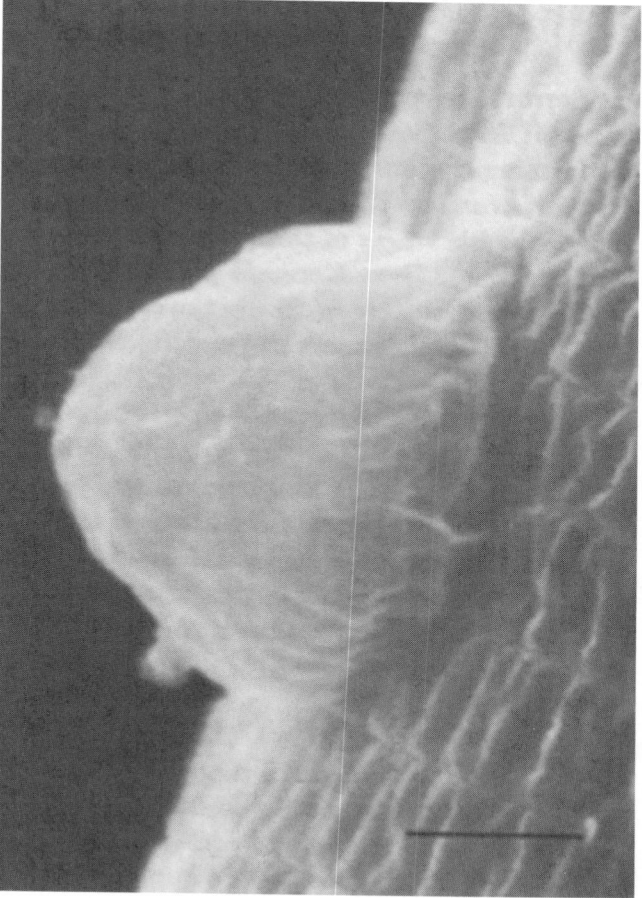

FIGURE 40. Scanning electron micrograph of an endospore of *Pasteuria penetrans sensu stricto* on cuticle of a second-stage juvenile of *Meloidogyne incognita*. This endospore has retained its exosporium, resulting in the appearance of a crinkled surface. Bar = 10 μm.

For one thing, the *Pasteuria* endospore has a mass of fibrous outgrowths emanating from the central body or core. These microfibrillar strands (usually called parasporal fibers, peripheral fibers, or perisporium), which surround the central body of the endospore, are structures comprised of glycoproteins that are presumed to function as adhesins involved in attachment of the endospore to its invertebrate host (Figure 40) (Brito et al., 2004; Davies et al., 1994; Persidis et al., 1991; Preston et al., 2003; Schmidt et al., 2004). A monoclonal antibody that recognizes a glycan-containing epitope associated with adhesins on the surfaces of endospores of *Pasteuria penetrans* (Brito et al., 2003; Schmidt et al., 2003) detects this epitope in extracts of endospores from other species of infected nematodes (Preston et al., 2003), as well as *Pasteuria ramosa* (Schmidt et al., 2008). The epitope was not detected in extracts of endospores obtained from a number of *Bacillus* spp. that were tested, but is common to endospores of all *Pasteuria* spp. that have been evaluated.

Albeit an integral part of the endospore, the parasporal fibers are arrayed differently in the different *Pasteuria* spp. discussed herein. It is difficult to include these somewhat amorphous fibrous masses in any precise measurements of endospores. For this reason, the measurements reported herein include only the major and minor axes of the central body of the endospore,

the endospore proper, and explicitly exclude the parasporal fibers. Measurements and descriptions of parasporal fibers are presented separately.

The attachment of endospores to their invertebrate hosts is mediated by the adhesins in their parasporal fibers. The attached endospore is often overlaid by seemingly nonfunctional remnants of the old sporangium (Figure 40). The presence or absence of these sporangial remnants may be due in part to the length of time the sporangium has been subjected to degradative processes in the soil or the amount of abrasion received as the nematode host moves through its environs (Figure 41). When such sporangial material was significant, it was included in the measurements of the endospores reported herein. Even though sporangial material is sometimes seen, we have adopted the convention of calling the infectious unit on the nematode exterior cuticle surface an endospore. *Pasteuria* endospores, at least in the case of the nematode parasites, also differ from those of *Bacillus* in that the former produce a germ tube that penetrates the nematode cuticle and initiates bacterial colonization within the nematode body.

The type-descriptive material (Sayre and Starr, 1985) of *Pasteuria penetrans* (*ex* Thorne (1940) Sayre and Starr (1985) refers

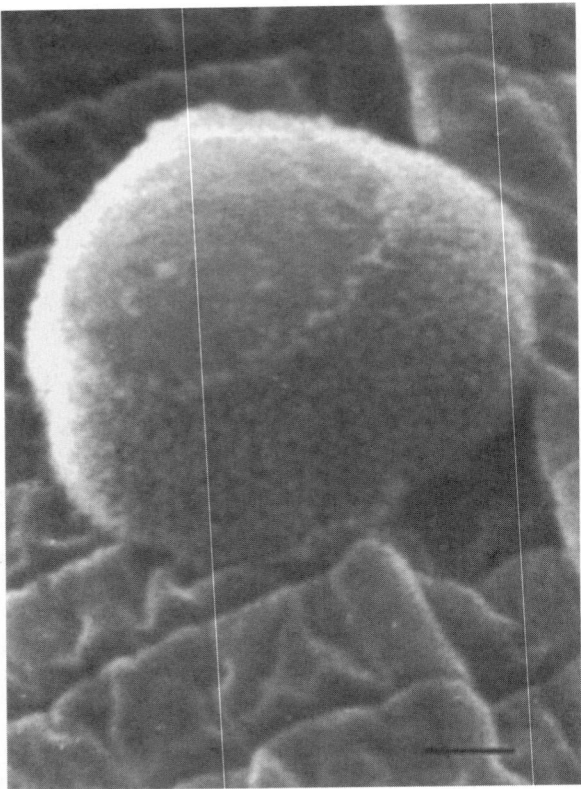

FIGURE 41. Endospore of *Pasteuria penetrans sensu stricto* attached to the cuticle along the lateral field of the juvenile of a root-knot nematode. The exosporium has been sloughed, exposing the central dome of the endospore; the peripheral fibers can be distinguished. Bar = 0.5 μm.

to the bacteria occurring on the root-knot nematode *Meloidogyne incognita*. Hence, the name *Pasteuria penetrans* must refer to that organism (Starr and Sayre, 1988a). This morphotype, however, has been isolated from other *Meloidogyne* spp. and comprises several genotypes that may defy or challenge easy species definition (Sturhan et al., 2005). Some other observed members of *Pasteuria* are demonstrably different from *Pasteuria penetrans* using a Linnaean species concept (typological) with molecular corroboration and should be assigned to other taxa. The first such assignment was made for the bacterium from the lesion nematode *Pratylenchus brachyurus*, to which the name *Pasteuria thornei* Starr and Sayre (1988a) was affixed (Starr and Sayre, 1988a). Because the obligate endoparasitic nature of *Pasteuria* currently prevents cultivation of axenic type strains, "*Candidatus*" status has been proposed for each new provisional species designation in this genus (Giblin-Davis et al., 2003b; Murray and Stackebrandt, 1995; Murray and Schleifer, 1994; Stackebrandt et al., 2002). All of the previously named species in the genus *Pasteuria* Metchnikoff (1888) (Skerman et al., 1980) have nomenclatural standing and remain as species with validly published names.

The species that remain valid include *Pasteuria ramosa* Metchnikoff (1888) [Skerman et al. (1980) as emended by Starr et al. (1983) serving as type; see Judicial Commission (1986), Wayne (1986)], *Pasteuria penetrans* (*ex* Thorne, 1940) [Sayre and Starr (1985) description and illustrations serving as type;

Validation List no. 20; (Sayre and Starr, 1986)], *Pasteuria thornei* [Sayre et al. (1988) and Starr and Sayre (1988a) description and illustrations serving as type; Validation List no. 26; (Starr and Sayre, 1988b)], and *Pasteuria nishizawae* [Sayre et al. (1991) as emended by Noel et al. (2005) description and illustrations serving as type; Validation List no. 41; (Sayre et al., 1992)]. We concur with the proposal of Ebert et al. (1996) to accept the *Daphnia* parasite that they isolated and studied, and whose 16S rRNA gene has been sequenced, as the neotype for *Pasteuria ramosa* and the genus *Pasteuria*. Unfortunately, 16S rRNA gene sequence data are not available for *Pasteuria thornei* and this species of *Pasteuria* must be rediscovered before a more complete characterization can be made.

Enrichment and isolation procedures

Attempts to devise methods for *in vitro* cultivation of *Pasteuria penetrans* using defined media have proven difficult, but some success has been reported (Bishop and Ellar, 1991; Hewlett et al., 2006). The cultivation of *Pasteuria penetrans* outside of its nematode host has been reported by an independent commercial enterprise, but complexities and undisclosed composition of the media used have precluded their application and confirmation in other laboratories. The distinctive morphology and unique relationship to invertebrate hosts shared by *Pasteuria* spp. (see Table 55) suggest that this commonality may be the harbinger of similarities in their physiological requirements for endospore germination, vegetative growth, sporulation, and their eventual axenic cultivation. Based on microscopic observations, the physiological and physical requirements for their growth *in vivo* would appear to be similar. Summarized briefly, vegetative growth of *Pasteuria* spp. seems linked through the environment provided in the coelom/pseudocoelom of their different invertebrate hosts. The hemolymph of the cladoceran or the pseudocoelomic fluid of the nematode allows for the exchange of nutrients and waste products. Also, the coelom/pseudocoelom provides space for mycelial colony development. These colonies fragment after they reach a critical size. Finally, it is reasoned that various factors, possibly the accumulation of bacterial biomass and metabolites, as well as the onset of senility in the invertebrate hosts, trigger sporulation of the bacteria. The similar physical conditions found in the separate host species suggests that the bacteria might have common nutritional requirements; hence, when the requirements for axenic cultivation of one *Pasteuria* sp. become known, they may, with slight modification, apply to other species. The unique host specificities of different *Candidatus* species (Giblin-Davis et al., 2004; Giblin-Davis et al., 2003a; Sturhan et al., 2005), as well as biotypes of *Pasteuria penetrans* defined by a marked preference for different *Meloidogyne* spp. (Oostendorp et al., 1990; Stirling, 1985), supports a role for the host in conferring virulence and host specificity. The host may play an active role in the maturation of endospores and endow it with adhesins that serve as virulence factors through their recognition of receptors on the cuticle of the host and the attachment that is required for infection (Preston et al., 2003).

Since isolation and cultivation of *Pasteuria ramosa*, apart from in its cladoceran host, has not yet been achieved, this bacterium is usually studied in field samples of the infected invertebrate host. The following procedures may increase the chances of finding *Pasteuria ramosa* in its natural habitat.

1. Cladocerans should be collected during the warmest part of the growing season. The parasite is most often found in *Daphnia magna*, but also in other *Daphnia* spp. (Ebert, 2005).

2. Because the frequency of occurrence of the parasite in a cladoceran population may be low (less than 10%), a large number of living specimens needs to be examined to increase the odds for detection. Prevalence, however, may reach up to 100% of all the adult hosts (Duncan and Little, 2007).

3. The internal parasites are most easily identified by using an inverted microscope at magnifications of ×100–250. Infected *Daphnia* are typically large and much less transparent than uninfected animals. The parasite fills the entire body cavity.

Although axenic cultivation of *Pasteuria penetrans* has been reported but not confirmed, investigations have been limited to studies on naturally or artificially infected nematode hosts (Bekal et al., 1999; Chen et al., 1997a; Chen and Dickson, 1998; Imbriani and Mankau, 1977; Mankau, 1975; Mankau and Imbriani, 1975; Sayre et al., 1983, 1988; Sayre and Wergin, 1977; Starr and Sayre, 1988a) and exploration of the bacterium's potential as a biological control agent against plant-parasitic nematode populations (Chen and Dickson, 1998). Consequently, the studies have depended on finding, maintaining, and manipulating nematode populations infected with these bacteria. Because of this direct dependence on host–nematode populations, the procedures and methods used for maintaining members of the nematode-associated *Pasteuria* are by-and-large those used in maintaining the nematodes (Southey, 1986; Zuckerman et al., 1984).

A few generalizations about *Pasteuria* spp. follow:

1. The group of nematodes that are parasitized by *Pasteuria* spp. is widespread and diverse. The bacterial parasite has been reported from about 116 nematode genera and 323 nematode species. *Pasteuria* spp. have been reported in a dozen states of the United States, as well as roughly 51 countries on five continents and on various islands in the Atlantic, Pacific, and Indian Oceans (Chen and Dickson, 1998; Ciancio et al., 1994; Sayre and Starr, 1988; Sturhan, 1988).

2. Members of nematode-associated *Pasteuria* will most likely be found in soils where nematode populations have been consistently high and are causing crop damage (in the case of plant-parasitic nematodes only). However, numerous nematode suppressive soils have been identified where *Pasteuria* causes a precipitous drop in plant nematode numbers (Chen and Dickson, 1998). Planting of susceptible crops is necessary for maintenance of the nematode populations and multiplication of *Pasteuria* spp. *Pasteuria* spp. may also be associated with plant nematodes in greenhouse situations.

3. To find the bacterial endospores, nematodes are extracted from the suspected soil, e.g., by a centrifugal-flotation method (Jenkins, 1964). Increasing the sucrose concentration used for extraction of nematodes results in a higher percentage recovery of endospore-filled specimens (Oostendorp et al., 1991). Other separation methods, relying on the nematode's mobility (e.g., Baermann, 1917), may not yield those nematodes that are heavily encumbered with endospores or endospore-filled bodies since such nematodes are partially immobilized. Addition of healthy nematodes to soils, together with subsequent extraction and examination, is the most commonly used bioassay for determining the presence of members of the genus *Pasteuria* in soil. Attachment assays have been developed that allow estimations of the number of endospores per gram of soil and the extent to which soils are suppressive for plant nematodes (Chen and Dickson, 1998; Oostendorp et al., 1990). Also, methods have been developed for the quantification of *Pasteuria* endospore concentrations in tomato root material (Chen et al., 1996), the extraction and purification of endospores (Chen et al., 2000), and for determining suppressive soils caused by *Pasteuria* (Chen et al., 1997a; Dickson et al., 1994; Stirling, 1984; Walia et al., 2004). PCR-based methods have been developed for amplifying and sequencing 16S rRNA genes from soil samples and single nematodes (Atibalentja et al., 2004b; Duan et al., 2003). Immunoassays using polyclonal (Costa et al., 2006; Fould et al., 2001) and monoclonal (Schmidt et al., 2003) antibodies directed against endospore surface proteins have been developed for the detection and quantification of bacterial endospores in soil and tissue samples. PCR-based assays for sporulation and other genes have been developed for the identification and quantification of the vegetative cells in plant tissues (Schmidt et al., 2004).

4. Occurrence of endospores on the surfaces of nematodes is most easily observed by means of an inverted microscope at ×250–400 magnification. Several stains, e.g., crystal violet, cotton blue, Brilliant Blue G, etc., are useful for visualizing external spores. The endospores also may be readily identified with a fluorescent immunoassay and monoclonal antibodies that recognize adhesin-related epitopes (Davies et al., 1994; Schmidt et al., 2003). The application of immunoassays avoids potential misidentification of endospores of *Paenibacillus* spp. that may be associated with entomopathogenic rhabditid nematodes (El-Borai et al., 2005; Enright et al., 2003).

5. Direct confirmation of the presence of members of the *Pasteuria* group inside the nematode can be made by microscopic examination of the pseudocoelom of nematodes, also at ×250–400 magnification. Both juveniles and adults should be examined for the characteristic mycelial colonies and endospores. Treating nematodes with methyl blue and lactophenol stain works well for visualizing the various developmental stages within the nematode pseudocoelom (Serracin et al., 1997). In uninfected nematodes, the anterior region of the esophagus is clear, with internal structures visible; when filled with endospores, the region will not be clear and internal structures will be masked by the bacterium.

6. Detection of the bacterium in the sedentary endoparasitic nematodes depends on manual (Thorne, 1940) and/or enzymic (Dickson et al., 1970) removal of root-tissues from around the female nematodes. The freed females are placed on glass slides, crushed, and their body contents are examined microscopically for vegetative stages and endospores of the bacterium.

Maintenance procedures

Since the recently reported method of axenic cultivation is proprietary and has not been optimized or confirmed, *Pasteuria* spp. are maintained by co-cultivation with their respective invertebrate host.

Pasteuria ramosa can be grown in the laboratory in clonal cultures of *Daphnia magna*. (Ebert et al., 2004). An endospore-filled body of *Daphnia magna* is shown in Figure 42. *Daphnia* are maintained on a diet of chemostat-grown unicellular green

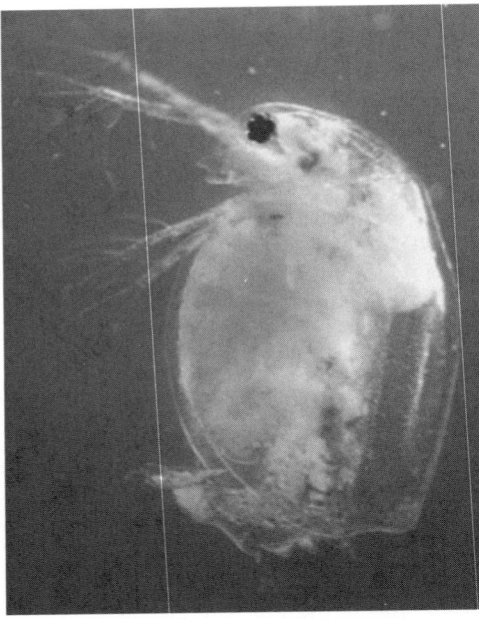

FIGURE 42. Photomicrograph of *Daphnia magna* showing body cavity filled with endospores of *Pasteuria ramosa.*

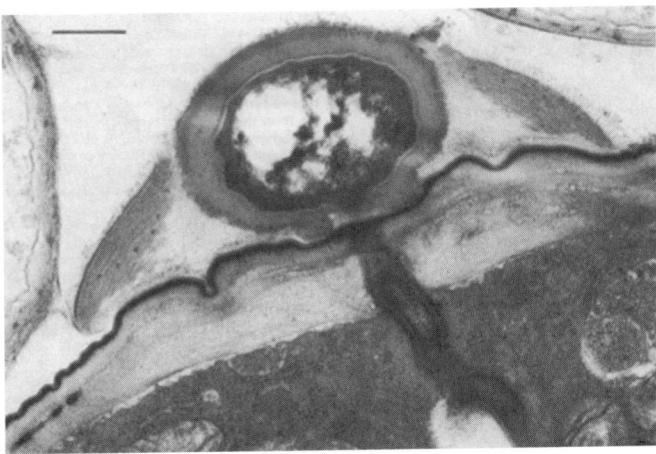

FIGURE 43. Section through a germinated endospore of *Pasteuria penetrans sensu stricto*; the penetrating germ tube follows a sinuous path as it travels the cuticle and hypodermis of the nematode. Bar = 0.5 μm.

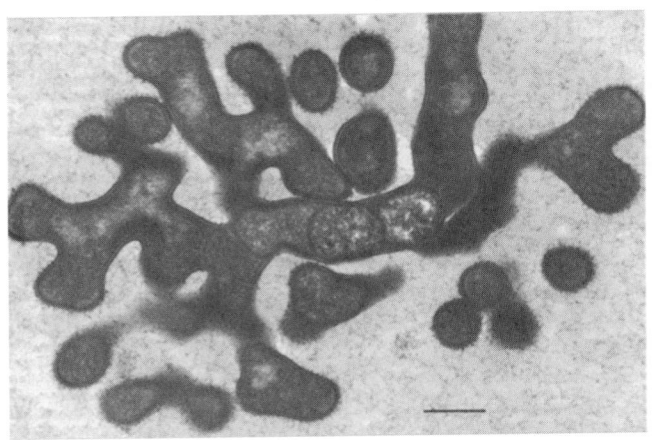

FIGURE 44. Section of a mycelial colony of *Pasteuria penetrans sensu stricto* in the pseudocoelom of the nematode. The septate hyphae appear to bifurcate at the margins of the colony. Bar = 0.5 μm.

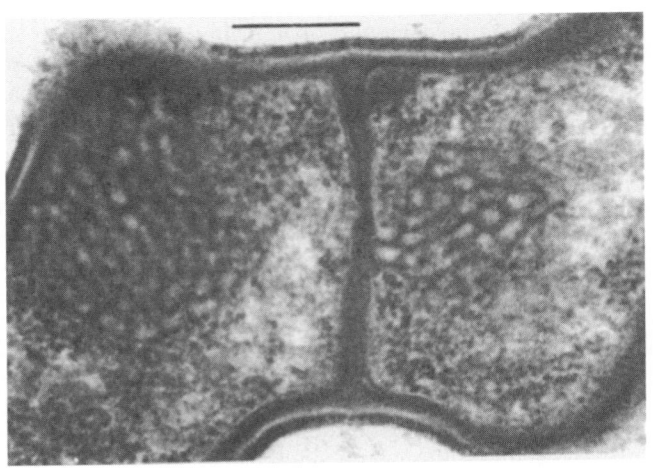

FIGURE 45. Hyphal cells of *Pasteuria penetrans sensu stricto* are bounded by a compound wall consisting of a double membrane. A mesosome is associated with the septum. Bar = 0.5 μm.

algae *Scenedesmus* sp. At 20°C, an infected *Daphnia magna* produces several million *Pasteuria* spores within about 40 d.

Endospores used to inoculate healthy *Daphnia magna* are obtained from two sources: (a) the crushed bodies of living or dead parasitized cladocerans in late-stage infections; and (b) sediments from the bottoms of aquaria (or ponds) in which dead and parasitized cladocerans have accumulated. Frozen cadavers can also be used. The infection rate is dose-dependent. High infection rates can be reached by adding 10,000 endospores to jars with 20 ml water and a single *Daphnia magna* in each jar (Regoes et al., 2003). Resistance of *Daphnia magna* clones to *Pasteuria ramosa* is widespread (Carius et al., 2001; Decaestecker et al., 2003; Little and Ebert, 1999) and may be the most common reason for a failure to cultivate the bacterium. Furthermore, there are strong host clone–parasite isolate interactions (Carius et al., 2001).

Nematode-associated *Pasteuria* can be maintained in a system consisting of the immediate nematode host and its host plant. A good example is the system consisting of *Pasteuria penetrans–Meloidogyne incognita* and tomato plants (Sayre and Wergin, 1977). To initiate and increase a bacterial population, dried bacterial endospore preparations (e.g., Stirling and Wachtel, 1980) are mixed into soils that are heavily infested with juveniles of *Meloidogyne incognita*. The juveniles become encumbered with the bacterial endospores as they move through the soil in a random fashion (Figure 39). The endospores adhering to the nematode cuticle are carried by the juvenile into tomato roots, where germination occurs after the nematode initiates feeding. An endospore penetrates the cuticle by means of a single germ tube (Figure 43), which extends into the pseudocoelomic cavity. The bacterium then enters into its vegetative endoparasitic developmental stages (Figure 44 and Figure 45). Finally, bacteria form in the mature and moribund host nematodes and in sporangia that contain endospores (Figure 46, Figure 47, and Figure 48). In summary, the developmental stages of the bacterium include recognition and attachment to a susceptible host nematode, infection of the

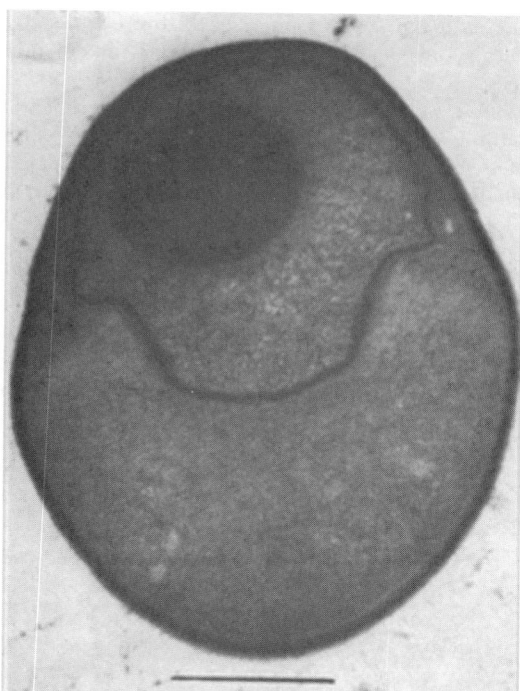

FIGURE 46. An early stage of endospore development in *Pasteuria penetrans sensu stricto* is shown in this median section through a sporangium. An electron-opaque body has formed with the forespore; the body is surrounded by membranes that will condense and contribute to the multilayered wall of the mature endospore. Bar = 0.5 μm.

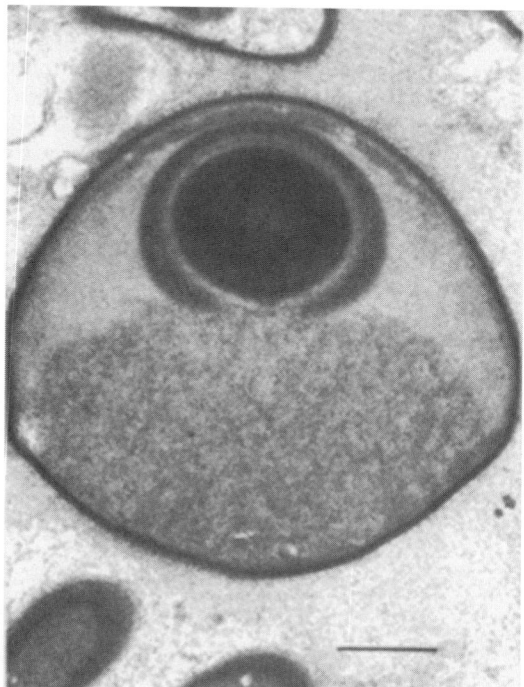

FIGURE 47. Section through a sporangium of *Pasteuria penetrans sensu stricto* with an almost mature endospore. The lateral regions (light areas) will differentiate into parasporal fibers. Bar = 0.5 μm.

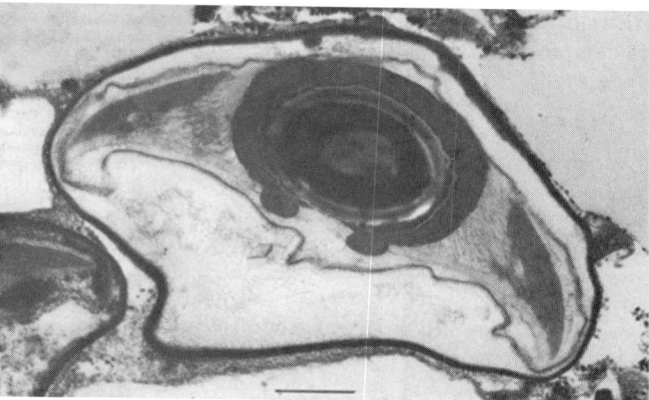

FIGURE 48. Median section through a sporangium of *Pasteuria penetrans sensu stricto* containing a fully mature endospore. Final stages of endospore differentiation include formation of an encircling membrane or exosporium and emergence of parasporal fibers within the granular material that lies laterally around the spore. Bar = 0.5 μm.

host nematode via a germ tube, followed by vegetative growth, sporulation, and maturation within the host pseudocoelom.

Endospores of *Pasteuria penetrans* growing inside *Meloidogyne incognita* can be harvested by two methods. The simplest procedure is to allow the nematode-infested plant roots to decay *in situ* in soil; during such decay, about 2×10^6 endospores are released from each female nematode. The soil containing the endospores is air-dried, mixed, and stored. Such preparations have yielded bacterial endospores that can attach to the juvenile of their respective host nematode even after several years in storage (Mankau, 1973). A second method for obtaining a more concentrated preparation of the bacterial endospores has been demonstrated (Stirling and Wachtel, 1980). Freshly hatched juveniles of *Meloidogyne incognita* may be encumbered with endospores by placing them in aqueous suspensions containing endospores. A centrifuge method can be used to help obtain consistent attachment of endospores to host nematodes (Hewlett and Dickson, 1993). These encumbered juveniles are then allowed to penetrate roots of tomato seedlings. After the life cycle of the nematode is completed in soil, the galled roots are harvested, washed, air-dried, ground into a fine powder, and stored. Such preparations have provided adequate sources of endospores for use in bioassays and other procedures. A more efficient and rapid method of obtaining endospore-filled female nematodes is by using an enzyme (cellulase and/or pectinase) preparation (Brito et al., 2003; Chen et al., 2000; Schmidt et al., 2004). Purification can be achieved by selective filtration steps and centrifugation in sucrose or renografin (sodium diatrizoate) gradients (Chen et al., 2000).

Pasteuria thornei and "*Candidatus* Pasteuria usgae" can be maintained in similar systems, except that the bacteria must be maintained on the migratory endoparasitic host nematode *Pratylenchus brachyurus* or the ectoparasitic nematode *Belonolaimus longicaudatus*, respectively. *Pratylenchus brachyurus* may be collected from roots of infected hosts, whereas *Belanolaimus longicaudatus* may be collected from around roots of infected plants. Numerous bacterial endospores are liberated upon decay of the plant roots and the cadavers of infected nematodes. Healthy juveniles or adults of either nematode migrating through such soils can become encumbered with endospores of their respective bacterial parasite and repeat the developmental cycle, with not only maintenance of the bacterium, but also a net increase in the endospore content of the soil.

At present, the only method of producing endospores of *Pasteuria nishizawae* is to collect infected females and cysts. This is difficult due to the fragile nature of infected cysts and the difficulty in identifying them. At a certain stage, the infected cysts are usually a grayish-green color, but this coloration is often difficult to observe. Infected females cannot be readily identified without crushing them.

A number of investigators have cultivated members of *Pasteuria* in three-membered systems consisting of plant-tissue cultures, gnotobiotically reared nematodes, and the desired bacterium free of contaminating microbes (Bekal et al., 1999; Chen and Dickson, 1998). Once perfected, these cultural methods are useful for maintenance of these bacteria, at least until methods for their axenic cultivation become generally available.

Differentiation of the genus *Pasteuria* from other genera

Table 55 summarizes the characteristic features of species of the genus *Pasteuria*.

Taxonomic comments

De Toni and Trevisan (1889) provided the first generic diagnosis of *Pasteuria*; it followed quickly and closely the original description of *Pasteuria ramosa* by Metchnikoff (1888). However, some other early investigators, particularly those interested in taxonomic coherence, not having observed this enigmatic organism and relying solely on descriptions, rejected both the generic and specific concepts (Lehmann and Neumann, 1896; Migula, 1904). Laurent (1890) suggested that a bacteroid-forming species from nodules on leguminous plants, together with *Pasteuria ramosa*, comprised the new family he erected, *Pasteuriaceae*. Similarly, Vuillemin (1913) believed that a generic relationship existed between *Nocardia* and *Pasteuria*. De Toni and Revisan (1889) speculated that *Pasteuria ramosa*, because of its ability to form endospores, should be placed in the subtribe *Pasteurieae* of the tribe *Bacilleae*. The unusual morphology of *Pasteuria ramosa* became the basis for numerous suppositions about its affinities to other bacterial groups (Buchanan, 1925). This speculative process continued up until recently (Sayre et al., 1983; Sayre and Starr, 1985, 1988; Starr and Sayre, 1988a). The advent of 16S rRNA gene analyses indicated that *Pasteuria* is a deep lineage within the *Bacillales* (Atibalentja et al., 2000; Ebert et al., 1996). Although it has been classified within the "*Alicyclobacillaceae*" (Garrity et al., 2005), the low sequence similarity of 85% and its distinctive phenotype suggests that it would be more properly classified within its own family, the *Pasteuriaceae*.

A significant change made since the 9th edition of the *Manual* is the reduction in the confusion between Metchnikoff's Gram-positive, endospore-forming, mycelial cladoceran parasite with certain Gram-negative, budding, non-prosthecately appendaged aquatic bacteria. Conservation of *Pasteuria ramosa sensu* Metchnikoff (1888) on the basis of type-descriptive material, as well as rejection of ATCC 27377 as the type of *Pasteuria ramosa* Metchnikoff (1888) because it actually is a quite different organism (*Planctomyces staleyi* Starr, Sayre and Schmidt (1983), have been recommended (Starr et al., 1983) and approved (Judicial-Commission, 1986). The detailed drawings, photomicrographs, and lengthy description offered by Metchnikoff (1888) provided a sound basis for comparing his species with the current cladoceran parasites. As stated

above, a *Pasteuria ramosa*-like strain was discovered infecting *Moina rectirostris* (Sayre et al., 1979) that was used in the emendation of the species (Starr et al., 1983). We support the proposal of Ebert et al. (1996) that the *Daphnia*-parasitic *Pasteuria ramosa* characterized from the same host as Metchnikoff (1888) from Europe be designated the neotype for *Pasteuria ramosa* Metchnikoff (1888) and that the *Moina* isolate of Sayre et al. (1979) be compared directly to the neotype in future studies.

One reason investigators (Henrici and Johnson, 1935; Hirsch, 1972) have been interested in re-examining *Pasteuria ramosa* stems from the questions raised by Metchnikoff's assertion that the bacterium divides longitudinally. Metchnikoff suggested that cells of *Pasteuria ramosa* undergo a longitudinal fission, giving rise to a branched structure in which the two daughter cells remain attached at their tips. Based on a hypothesis about the evolution of division patterns in bacteria, he concluded that *Pasteuria ramosa*, a fairly primitive bacterium in his view, divided longitudinally.

Observations on the *Pasteuria ramosa*-like parasite of *Moina* (Sayre et al., 1979) indicated that cleavage indeed occurs in three planes, as evidenced by the spherical mycelial colonies. However, no common plane of division was found, as illustrated in drawing 2 of the Metchnikoff (1888) paper, starting at the surface of the cauliflower-like growth and ending at its interior. The prominent bifurcations in the mycelium (Figure 49), which Metchnikoff also observed and which may have prompted his longitudinal fission theory, are not products of atypical fission but rather are probably the branched distal or terminal cells of the mycelium undergoing rapid enlargement during formation of endogenous spores. These terminal cells develop into sporangia.

Of some historical interest are a few scattered reports in the literature about organisms similar in appearance to *Pasteuria ramosa*. In these reports, each author came to a different taxonomic decision about the organism, as follows: *Torula* or other yeast species (Ruehberg, 1933); two different genera of the microsporidia (Jirovec, 1939; Weiser, 1943), and a haplosporidium (*Dermocystidium daphniae* Jirovec (1939) and a possible intermediate stage in the life cycle of a *Dermocystidium* sp. (Sterba and Naumann, 1970). These results prob-

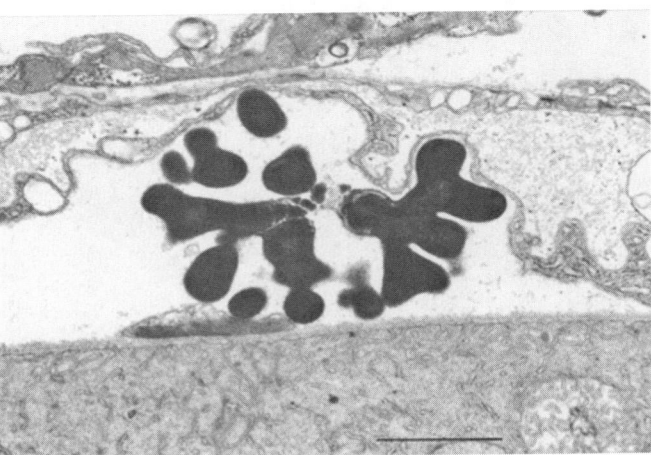

FIGURE 49. Cross-section of a fragmenting mycelium of *Pasteuria ramosa* in the body cavity of a cladoceran. Bar = 2.0 μm.

ably stem from the inability of these investigators to cultivate the particular organism they observed, taken together with their dependence solely on morphology and staining reactions as the basis for their descriptions and classifications. Surprisingly, Metchnikoff (1888), the first person to report on *Pasteuria*, recognized it correctly as a bacterium, whereas the later workers did not and, moreover, were apparently unaware of Metchnikoff's work.

Until rather recently, taxonomic studies of *Pasteuria* were carried out mainly by nematologists. Cobb (1906) erroneously designated such microbes as protozoan parasites of nematodes. At first glance, it would appear from the literature that subsequent workers (Steiner, 1938; Thorne, 1940) had independently come to the same conclusion. However, their conclusions were probably by consensus. Members of *Pasteuria* were then, and largely still are, essentially known only to the community of plant nematologists. When Thorne described *Duboscqia penetrans* as a protozoan, he could not have realized its bacterial nature because electron microscopes were not then available. Later, Williams (1960) studied a similar organism in a population of root-knot nematodes from sugarcane, presented an interpretation of its life stages, and indicated some reservations about Thorne's identification. Canning (1973) also doubted the identification as a protozoan and stressed the organism's fungal characteristics. Finally, Mankau (1975) and Imbriani and Mankau (1977) established the bacterial nature of the organism and brought its attendant taxonomic problems to the attention of bacteriologists. Some results of the ensuing interdisciplinary enterprise are summarized elsewhere (Sayre and Starr, 1988; Sayre et al., 1988; Starr and Sayre, 1988a).

Acknowledgements

This chapter is dedicated to the memory of Dr Richard Sayre and his wife, Diane, who perished on 10 June 1998 in a boating accident in the Galapagos Islands.

We thank R. E. Davis for calling our attention to the genus *Pasteuria* and for his help in initiating its study. Thanks also go to W. P. Wergin, J. R. Adams, C. Pooley, R. Reise, and S. Ochs for sustained technical advice and assistance, particularly in the preparation of the figures involving transmission and scanning electron microscopy. Thanks are due R. L. Gherna for supplying strain ATCC 27377. We are grateful to J. M. Schmidt for her corroboration during the period when the taxonomic confusion between *Pasteuria* and *Planctomyces* was being corrected. Skillful bibliographic and redactional assistance was provided by P. B. Starr.

Further reading

Chen, Z.X. and D.W. Dickson. 2004. Biological control of nematodes with bacterial antagonists. *In* Chen, Chen, and Dickson (Editors), Nematology Advances and Perspectives, Vol. 2, Nematode Management and Utilization, CABI Publishing, Cambridge, MA, pp. 1041–1082.

Gowen, S., K.G. Davies and B. Pembroke. 2008. Potential use of *Pasteuria* spp. in the management of plant parasitic nematodes. *In* Ciancio and Mukerji (Editors), Integrated Management and Biocontrol of Vegetables and Grain Crops Nematodes, Springer, Dordrecht, The Netherlands, pp. 205–219.

Differentiation and characteristics of the species and *Candidatus* species of the genus *Pasteuria*

The differential characteristics of nominal *Pasteuria* and *Candidatus* species are given in Table 56.

List of species of the genus *Pasteuria*

1. **Pasteuria ramosa** Metchnikoff 1888, 166[AL.] [Nom. Cons. Opin. 61 Jud. Comm. 1986, 119. Not *Pasteuria ramosa* in the sense of Henrici and Johnson 1935, Hirsch 1972, and Staley 1973; see Starr et al. (1983) and Judicial Commission (1986)]

ra.mo′sa. L. fem. adj. *ramosa* much-branched.

Gram-positive. Sporangia and microcolonies are parasitic in the hemocoel of cladocerans, water fleas of the genera *Daphnia* and *Moina*. Usually occur attached to one another at pointed ends of the teardrop-shaped sporangia, forming quartet, triplet, and doublet configurations. The rounded end of the sporangium encloses a single refractile endospore, having axes of 1.37–1.61 × 1.20–1.46 μm, narrowly elliptic in cross-section. Endospores are 4.2–5.4 × 4.9–6.0 μm. Nonmotile. Endospores are resistant to desiccation, but with only limited heat tolerance. They have been recovered from 30-year-old pond sediments (Decaestecker et al., 2004). Vegetative stages are cauliflower-like, septate, mycelial growths that branch dichotomously and fragment to form microcolonies. Has not been cultivated axenically, but can be grown in the laboratory with the invertebrate host. The type-descriptive material consists of descriptions and illustrations in Metchnikoff's original publication (Metchnikoff, 1888) and elsewhere (Ebert et al., 1996; Sayre et al., 1979, 1983; Starr and Sayre, 1988a; Starr et al., 1983).

DNA G+C content (mol%): not reported.

Type strain: descriptions and illustrations serving as type.

GenBank accession number (16S rRNA gene): AY762091 (Ebert et al., 1996).

2. **Pasteuria nishizawae** Sayre, Wergin, Schmidt and Starr 1992, 327[VP] (Effective publication: Sayre, Wergin, Schmidt and Starr 1991, 562.) (emend. Noel, Atibalentja and Domier 2005, 1683.)

ni.shi.za′wae. N.L. gen. n. *nishizawae* of Nishizawa, named after Tsutomu Nishizawa, a Japanese nematologist who discovered and first investigated bacterial parasites of cyst-forming nematodes.

Gram-positive vegetative cells, forms endospores. Obligate endoparasitic bacterium of the pseudocoelom of *Heterodera glycines* (soybean cyst nematode). Microcolony shape is cauliflower-like initially and later fragments into clusters of elongated grape-like immature sporangia occurring in configurations of octets, quartets, and doublets. Sporangia are cup-shaped with diameters and heights (under the light microscope) of 5.3 and 4.3 μm, respectively, and (under the transmission electron microscope) of 4.4 and 3.1 μm, respectively. Sporangial wall and mother cell matrix disintegrate at maturity leaving the exosporium as the outermost layer of the endospore. The surface of the exosporium is velutinous to hairy. The stem cell is observed occasionally. The central body is oblate spheroid, ellipsoid to narrowly elliptical. Orientation of major axis to sporangium base is horizontal with diameter and height of 2.1 and 1.7 μm, respectively (when viewed under the light microscope) or 1.6 and 1.3 μm, respectively (when viewed by transmission electron microscopy). The epicortical layer entirely surrounds the cortex. A laminar inner spore coat with alternating layers of dense and light materials occurs

TABLE 56. Comparison of *Pasteuria ramosa*, *Pasteuria nishizawae*, *Pasteuria penetrans sensu stricto* emend., *Pasteuria thornei*, and "*Candidatus* Pasteuria usgae"

Characteristic	1. *P. ramosa*	2. *P. nishizawae*	3. *P. penetrans*	4. *P. thornei*	5. "*Candidatus* P. usgae"
Colony shape	Cauliflower-like floret	Cauliflower-like floret initially, later fragments into cluster of elongated grape-like sporangia	Spherical to clusters of elongated grape-like sporangia	Small, elongate clusters	Cauliflower-like floret
Sporangia					
Shape	Teardrop	Cup	Cup	Rhomboidal	Cup to rhomboidal
Diameter, µm[a]	2.12–2.77	4.1–4.7	3.0–3.9	2.22–2.7	4.7–7.1
Height, µm[a]	3.40–4.35	2.8–3.4	2.26–2.60	1.96–2.34	2.73–4.97
Fate of sporangial wall at maturity of endospore	Remains rigidly in place; external markings divide sporangium in three parts	Sporangial wall and mother cell matrix of endospore disintegrate, leaving exosporium as the outermost layer of the endospore; no clear external markings	Basal portion collapses inward in the developed endospore; no clear external markings	Remains essentially rigid, sometimes collapsing at bases; no clear external markings	Basal portion collapses inward in the developed endospore; no clear external markings
Exosporium	Present	Present, velutinous to hairy surface	Present	Present	Present
Stem cell	Remains attached to most sporangia	Seen occasionally	Seen occasionally	Neither stem cell nor second sporangium seen	Rarely seen
Central body	Oblate spheroid, an ellipsoid, narrowly elliptic in section	Oblate spheroid, ellipsoid to narrowly elliptical in section	Oblate spheroid, ellipsoid to broadly elliptical in section	Oblate spheroid; ellipsoid, sometimes almost spherical, narrowly elliptical in section	Oblate spheroid, ellipsoid to broadly elliptical in section
Orientation of major axis to sporangium base	Vertical	Horizontal	Horizontal	Horizontal	Horizontal
Cell dimensions, µm[a]	1.37–1.61 × 1.20–1.46	1.4–1.8 × 1.2–1.4	0.99–1.21 × 1.30–1.54	0.96–1.20 × 1.15–1.43	2.40–3.91 × 1.44–2.34
Wall thickness, µm[a]	0.28–0.34		0.22–0.26	0.17–0.23	0.36–0.59
Protoplast	Contains pronounced stranded inclusions	Stranded inclusions sometimes seen	Stranded inclusions observed	Stranded inclusions observed	
Partial middle spore wall	Not observed	Surrounds endospore laterally, not in basal or polar areas	Surrounds endospore somewhat sublaterally	Surrounds endospores somewhat sublaterally	
Pore					
Occurrence	Absent	Present	Present	Present	Present
Characteristics	–	Thickness of basal wall constant and is the depth of pore	Basal annular opening formed from thickened outer wall	Basal cortical wall thins to expose inner endospore	Basal cortical wall thins to expose inner endospore
Diameter, µm[a]	–	0.3±0.1	0.28±0.11	0.13±0.01	0.29±0.1
Parasporal structures, origin and orientation	Long primary fibers arise laterally from cortical wall, bending sharply downward to yield numerous secondary fibers arrayed internally toward the granular matrix	Same as *P. penetrans* but additional layer is formed on obverse surface of endospore	Fibers arise directly from cortical wall, gradually arching downward to form an attachment layer of numerous shorter fibers	Long fibers arise directly from cortical wall, bending sharply downward to form an attachment layer of numerous shorter fibers	Same as *P. penetrans*

TABLE 56. (continued)

Characteristic	1. *P. ramosa*	2. *P. nishizawae*	3. *P. penetrans*	4. *P. thornei*	5. *"Candidatus* P. usgae"
Matrix, at maturity	Persists as fine granular material	Persists, numerous strands are formed and partial collapse may occur	Becomes coarsely granular; lysis occurs; sporangial wall collapses; base is vacuolate	Persists, but more granular; some strands are formed and partial collapse may occur	Same as *P. penetrans*
Host	Cladocerans (*Daphnia, Moina*)	Cyst nematodes (*Heterodera glycines*)	Nematodes *Meloidogyne* spp.)	Nematodes (*Pratylenchus brachyurus*)	Nematodes (*Belonolaimus longicaudatus*)
Completes life cycle in nematode juveniles	–	No, only in female and cyst	Mostly only in female, occasionally seen in 2nd stage juvenile[b]	Yes, in 2nd, 3rd, and 4th stage juveniles and adult	3rd, 4th stage juveniles and adults
Location in host	Hemocoel and musculature; sometimes found attached to coelom walls	Pseudocoelom and musculature; no attachment to pseudocoelom walls seen	Pseudocoelom and musculature; no attachment to pseudocoelom walls seen	Pseudocoelom and musculature; no attachment to pseudocoelom walls seen	Pseudocoelom and musculature; no attachment to pseudocoelom walls seen
Attachment of spores on host	Spores not observed to attach or accumulate on surface of cladoceran	Spores accumulate on juveniles, rare on male	Spores accumulate in large numbers on cuticular surface	Spores accumulate in large numbers on cuticular surface	Spores accumulate in large numbers on cuticular surface
Mode of penetration of host	Not known; suspected to occur through gut wall	Direct penetration of nematode cuticle by germ tube	Direct penetration of nematode cuticle by germ tube	Direct penetration suspected but not seen	Direct penetration of nematode cuticle by germ tube
Source of host	Pond mud, freshwater	Soil, plants	Soil, plants	Soil, plants	Soil, plants

[a]Measurements are based on preparations examined by transmission electron microscopy. Somewhat different apparent sizes are obtained by phase-contrast light microscopy and scanning electron microscopy.
[b]Rarely seen in males but, when observed, thought to be sex-reversed males (Hatz and Dickson, 1992).

between the outer spore membrane and the epicortical layer. The outer spore coat consists of several layers of electron-dense materials and is surrounded laterally and ventrally by microprojections. Including the microprojections, the outer spore coat is thickest at the top of the central body and then tapers gradually to 0.2 μm at the spore equator and to 0.1 μm or less around a 0.3 μm wide basal pore. Parasporal structures consist of long primary fibers arising laterally from the outer spore coat and bending downwards to yield numerous secondary fibers arrayed ventrally. Depending upon the extent of invagination of the basal adhesion layer, additional partial hirsute layers may be present on the obverse face of the central body. Spores attach to second-stage juveniles in the soil, but rarely to males in the soil. The cuticle and body wall are penetrated by the germ tube that develops after the infective second-stage juvenile penetrates the host plant root. The life cycle is completed only in the pseudocoelom of females, but may also be completed in the cyst (female cadaver). The only confirmed host is *Heterodera glycines*. Attachment of endospores to second-stage juveniles of *Globodera rostochiensis* (potato cyst nematode) with endospores obtained from *Heterodera elachista* (upland rice nematode), *Heterodera lespedezae* (lespedeza cyst nematode), *Heterodera schachtii* (sugarbeet cyst nematode), and *Heterodera trifolii* (clover cyst nematode) indicates that these nematode species may be hosts. Completion of the life cycle of *Pasteuria nishizawae* in these nematode species has not been confirmed.

DNA G+C content (mol%): not reported.

Type strain: descriptions and illustrations serving as type.

GenBank accession number (16S rRNA gene): AF134868 and AF516396.

3. **Pasteuria penetrans** (*ex* Thorne 1940) Sayre and Starr 1986, 355^VP (Effective publication: Sayre and Starr 1985, 163.) (emend. Sayre, Starr, Golden, Wergin and Endo 1988, 28; *Duboscqia penetrans* Thorne 1940, 51.)

pen′e.trans. L. part. adj. *penetrans* penetrating, entering.

Gram-positive vegetative cells. Mycelium is septate; hyphal strands, 0.2–0.5 μm in diameter, branch dichotomously. The sporangia, formed by expansion of hyphal tips, are cup shaped, approximately 2.26–2.60 μm in height with a diameter of 3.0–4.0 μm. Each sporangium is divided into two unequal sections. The smaller proximal body is not as refractile as the larger, rounded, cup-shaped portion, which encloses an ellipsoidal endospore broadly elliptic in section having axes of 0.99–1.21 × 1.30–1.54 μm. Endospores seem to be of the kind typical of the genus *Bacillus*; they are resistant to both heat and desiccation. Nonmotile. Sporangia and vegetative cells are found as parasites in the pseudocoelomic cavities of *Meloidogyne* spp. The epithet is now restricted to members of *Pasteuria penetrans* with cup-shaped sporangia and ellipsoidal endospores broadly elliptic in section occurring primarily as parasites of *Meloidogyne* spp. Has not been cultivated axenically; the type-descriptive material consists of the text and photographs in Sayre and Starr (1985) and Starr and Sayre (1988a). *Pasteuria penetrans* differs from other described members of *Pasteuria* in host specificity, in size and shape of sporangia and endospores, and in other morphological and developmental characteristics.

The influence of temperature on the development of *Pasteuria penetrans* in *Meloidogyne* spp. has been observed in growth chambers (Hatz and Dickson, 1992; Serracin et al., 1997; Stirling, 1981). The parasite's greatest endospore

attachment rate to second-stage juveniles was at 30°C and the bacterium developed more quickly within its nematode host at 30 and 35°C than at 25°C. Development time quickly decreases as temperature decreases, e.g., at 35, 28, and 21°C, mature endospores were detected at 28, 35, and >90 d, respectively (Hatz and Dickson, 1992; Serracin et al., 1997).

DNA G+C content (mol%): not reported.

Type strain: descriptions and illustrations serving as type.

GenBank accession number (16S rRNA gene): AF077672 and AF375881 (Anderson et al., 1999).

4. **Pasteuria thornei** Starr and Sayre 1988, 328^VP (Effective publication: Starr and Sayre 1988a, 28.)

thor′ne.i. M.L. gen. n. *thornei* of Thorne, named after Gerald Thorne, a nematologist from the United States, who described and named this parasite of *Pratylenchus* as a protozoan parasite.

Gram-positive vegetative cells. Mycelium is septate; hyphal strands, 0.2–0.5 μm in diameter, branch dichotomously. Sporangia, formed by expansion of hyphal tips, are rhomboidal in shape, approximately 2.22–2.70 μm in diameter and 1.96–2.34 μm in height. Each sporangium is divided into two almost equal units. The smaller unit, proximal to the mycelium, is not refractile and contains a granular matrix interspersed with many fibrillar strands. The refractile apical unit is cone shaped; it encloses an ellipsoidal endospore, sometimes almost spherical, having axes of 0.96–1.20 × 1.15–1.43 μm, with cortical walls about 0.13 μm in thickness. A sublateral epicortical wall gives the endospore a somewhat triangular appearance in cross-section. The tapering outer cortical wall at the base of the endospore forms an opening approximately 0.13 μm in diameter. Sporangia and endospores are found as parasites of lesion nematodes (*Pratylenchus* spp.). Has not been cultivated axenically; the type-descriptive material consists of the text and photographs in Starr and Sayre (1988a) and Sayre et al. (1988). *Pasteuria thornei* differs from *Pasteuria penetrans* and other members of *Pasteuria* in host specificity, size and shape of sporangia and endospores, and other morphological and developmental traits.

DNA G+C content (mol%): not reported.

Type strain: descriptions and illustrations serving as type.

GenBank accession number (16S rRNA gene): not reported.

5. "**Candidatus Pasteuria usgae**" Giblin-Davis, Williams, Bekal, Dickson, Brito, Becker and Preston 2003b, 197

us′gae. N.L. gen. n. *usgae* of U.S.G.A., the acronym for the United States Golf Association, in gratitude for their financial support to study this potential biological control agent against *Belonolaimus longicaudatus* in turfgrass ecosystems.

Organism is nonmotile with Gram-positive vegetative cells. Mycelium is septate; hyphal strands branch dichotomously with expansion of hyphal tip forming sporangium. With scanning electron microscopy, peripheral fibers of the mature endospore protrude around the exposed spherical outer coat of the spore creating a crenate border as opposed to other species of *Pasteuria* described from nematodes that have no scalloped border. The sporangium and central body diameters were on average at least 0.5 and 0.7 μm wider than these respective measurements for the other described species of *Pasteuria*. In lateral view with transmission electron microscopy, the shape of the central body is a rounded-rectangle to a rounded-trapezoid in transverse section that contrasts with the circular shape for

Pasteuria ramosa, the horizontally oriented elliptical shapes for *Pasteuria penetrans* and *Pasteuria nishizawae*, and the rounded-square shape for *Pasteuria thornei*. The outer spore coat is thickest laterally, thinner on top and thinnest across the bottom of the spore, being 7–8 times thicker laterally than along the bottom. These measurements contrast with all other described species having outer spore coats with relatively uniform thickness. No basal ring exists around the pore opening as in *Pasteuria penetrans*. The outer coat wall thickness at its thickest point is >15% (both walls >30%) of the diameter of the central body compared with 3 to <13% (both walls 6 to <25%) for the other described species of *Pasteuria*. The epicortical wall remnant of the mature endospore occurs between the cortex and the inner spore coat in a sublateral band, similar to *Pasteuria thornei* but different from the other three described species. The epicortical wall in the other described species is as follows: completely concentric in *Pasteuria ramosa* and *Pasteuria nishizawae*, and lateral in *Pasteuria penetrans*.

Obligate endoparasitic bacterium of the pseudocoelom of *Belonolaimus longicaudatus* that cannot be cultivated on cell-free media. Cultivated only by attachment of endospores to *Bacillus longicaudatus* and co-cultivation on excised axenic root or greenhouse plant cultures. Transmission occurs horizontally. Host infection is via cuticular penetration by attached endospores that occurs on all stages of *Belonolaimus longicaudatus* except eggs. Sporogenesis, which leads to the death of the host, occurs in the pseudocoelom of J3 through adult stage nematodes. Sporogenesis is typical of other nematode-specific *Pasteuria*. Host range appears to be limited to *Belonolaimus longicaudatus*, although attachment of endospores has been observed on *Bacillus euthychilus*, but not other soil-inhabiting nematodes.

DNA G+C content (mol%): not reported.

Type strain: descriptions and illustrations serving as type.

GenBank accession number (16S rRNA gene): AF254387.

Further information

Pasteuria ramosa. When using light microscopy, the earliest visible growth stages of *Pasteuria ramosa* in the water fleas *Daphnia magna* and *Moina rectirostris* Leydig 1860 are the cauliflower-like microcolonies usually found on the inner wall of the invertebrate's carapace Ebert et al., (1996), Sayre et al., 1979; Figure 50). The next detectable stage consists of quartets of sporangia carried in the hemolymph throughout the body of the cladoceran. Later, as the parasite develops, the hemolymph is noticeably clouded by the myriad immature sporangia of *Pasteuria ramosa*, mainly singles or doublets.

Electron micrographs of the "cauliflower" stage reveal circular patterns of septate hyphal strands of an actinomycete-like organism (Figure 49). The hyphae measure approximately 0.67 μm in width and their wall, which is fairly homogeneous in density, is 14.5–15 nm thick. The periplasmic region between the wall and the cell membrane is about 8.7–9.0 nm in width. The cell membrane measures 5.8 nm in thickness. Mesosomes (circular membrane complexes) are often found associated with the septa (Sayre et al., 1979; Figure 51).

The distal or terminal mycelial cells of the microcolonies enlarge to form teardrop-shaped sporangia that, when viewed by light microscopy, measure approximately 4.8–5.7 × 3.3–4.1 μm in diameter. Similar materials prepared for transmission electron microscopy and sectioned gave measurements of 3.40–4.35 μm in height and 2.12–2.77 μm in diameter. Early in this process, two septa form; these septa divide the sporangium into anterior, mid-

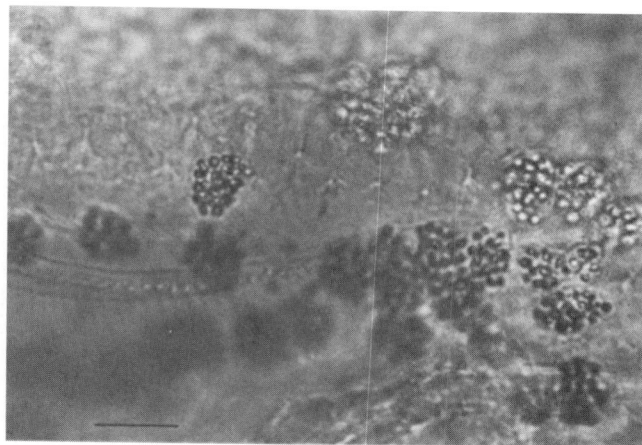

FIGURE 50. Cauliflower-like, branching, mycelial colony of *Pasteuria ramosa* attached to the inner walls of the carapace of the cladoceran *Moina rectirostris*. Bar = 10 μm.

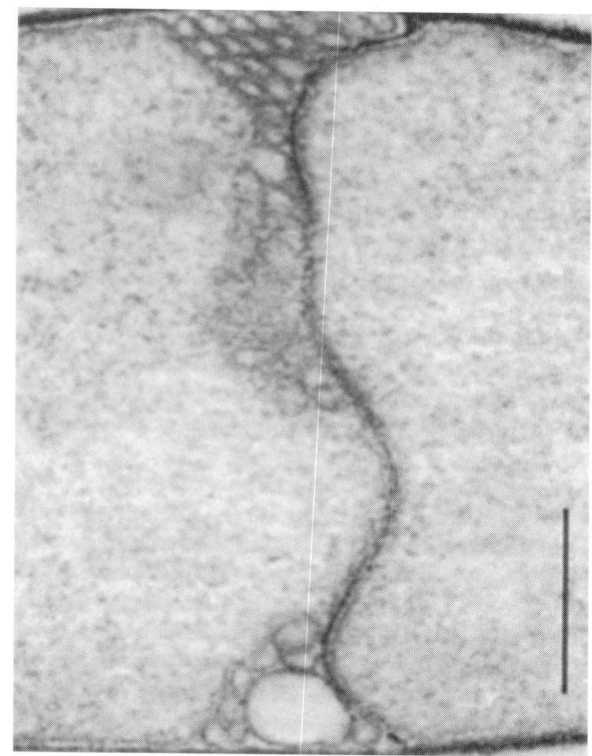

FIGURE 51. Mesosome associated with the septum in a dividing cell of *Pasteuria ramosa*. Bar = 0.5 μm.

dle, and stem sections (Metchnikoff, 1888; Figure 52). The partitioning is reflected in the outer wall of the sporangia (Figure 53).

The anterior section, the upper two-thirds of the sporangium, gives rise to the forespore. Within the anterior section, granular material condenses to form an electron-dense endospore, slightly ellipsoidal to almost spherical in shape, having axes measuring 1.37–1.61 × 1.20–1.46 μm. It comprises a multilayered central cytoplasm containing numerous doubled fibrillar strands (Figure 54). Structural changes also occur within the median section, where electron-transparent areas appear to expand and attach laterally to the multilayered endospore wall to form fibrous appendages. The mode of penetration of spores into host cladocerans is not known.

Parasitized water fleas (*Moina rectirostris*) taken from a pond in College Park, MD, USA, were found to yield about 2×10^5 *Pasteuria ramosa* sporangia per host individual. Generally, the parasite was found in mature females with no young in their egg pouches. *Daphnia* species, reported previously (Metchnikoff, 1888) as hosts of this organism, were not parasitized by this bacterial strain (Sayre et al., 1979). Neither sporangia from crushed water fleas nor sediments from the rearing aquaria of parasitized *Moina rectirostris* (Sayre and Wergin, 1977) resulted in infection when added to healthy populations of *Daphnia magna* or *Daphnia pulex*. These two *Daphnia* species were listed by Metchnikoff (1888) as hosts for *Pasteuria ramosa*. However, because *Moina rectirostris* is in the same family (Daphniidae) as these two *Daphnia* spp., this result may suggest only a very marked host specificity in the particular strain of *Pasteuria ramosa* available to us, perhaps at the level of a *forma specialis*, the situation in which one form of a parasite reproduces only in one host species and not in others that are closely related taxonomically. Recent work on *Pasteuria ramosa* from populations of *Daphnia magna* demonstrated strong specificity (Carius et al., 2001). The type-descriptive material of this taxon (Metchnikoff, 1888) is attached to the form on *Daphnia* species. If later work should show differences warranting separation at the specific

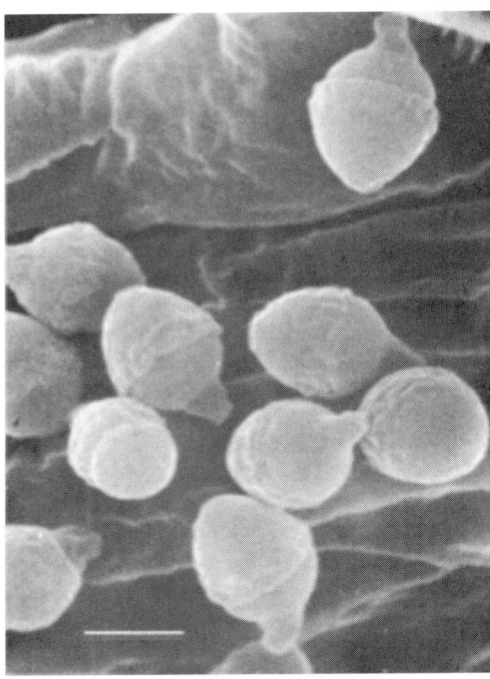

FIGURE 53. Scanning electron micrograph of sporangia of *Pasteuria ramosa*. External ridges mark boundaries of the endospore, middle section, and stem of a mature sporangium. Bar = 2.5 μm.

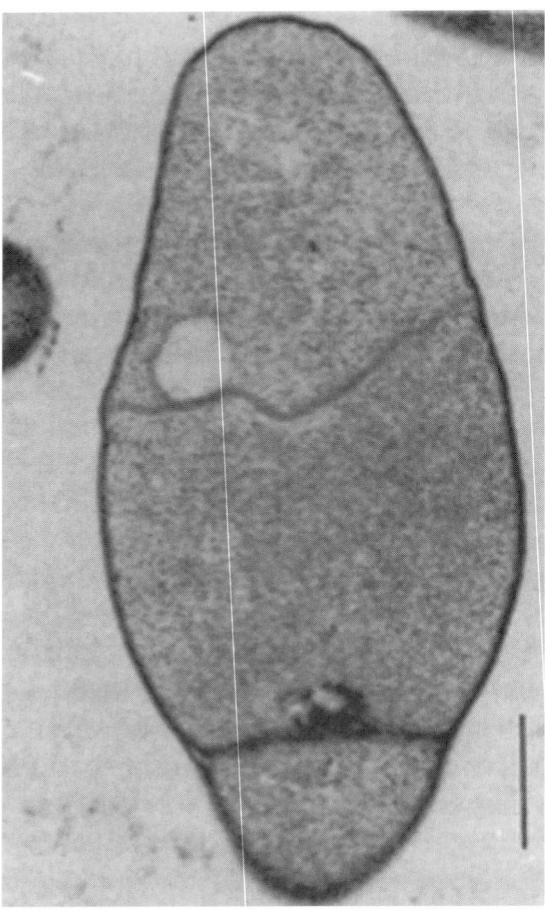

FIGURE 52. Septa divide the immature sporangium of *Pasteuria ramosa* into three parts. Bar = 0.5 μm.

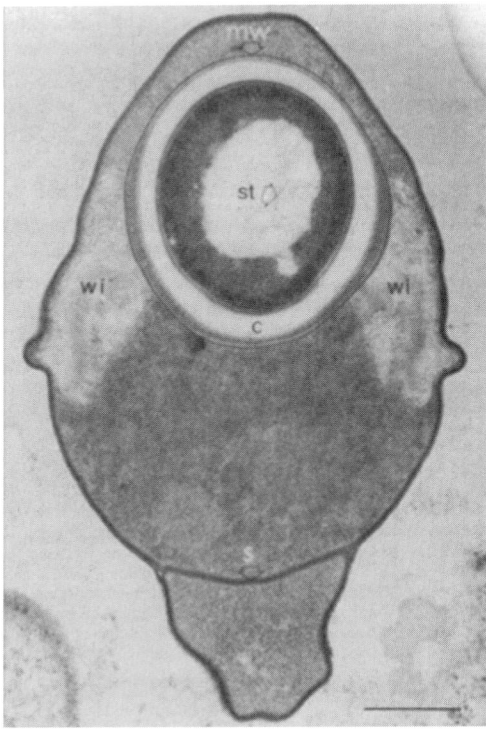

FIGURE 54. Mature sporangium of *Pasteuria ramosa* containing an endospore, made up of multilayered spore walls (mw), cortex (c), and cytoplast with stranded inclusions (st). Septum (s) separates the stem from the middle section. The function of the fibrous appendages (wi; wing-shaped light areas) is not known. Bar = 0.5 μm.

or subspecific level, the taxon on *Moina* would of course have to be given a different name from *Pasteuria ramosa.*

The influence of water temperature on the occurrence of *Pasteuria ramosa* in *Moina rectirostris* in nature was observed over a 3-year period. The parasite was not found until the surface water temperature in the pond reached 26°C or higher, usually about mid-July in the College Park (MD, USA) area. The apparent temperature requirement was confirmed in laboratory tests in which water in the aquaria was held at constant temperatures; the parasite was found in 6 and 3 d at 26° and 31°C, respectively, but not at all at 21°C. *Pasteuria ramosa* from *Daphnia magna* can be cultured from 15 to 25°C (Ebert et al., 1996).

Although endospores of *Pasteuria ramosa* appear to withstand desiccation, they have only limited resistance to heat. Air-dried sporangia, which were stored for 6 months, were capable of infecting healthy populations of cladocerans. However, when air-dried aquarium sediments were heated to 40, 60, or 80°C for 10 min and then added to cultures of healthy cladocerans, they failed to develop after treatment at 80°C (Sayre et al., 1979). Endospores from *Pasteuria ramosa* from *Daphnia* populations can survive in pond sediments for decades (Decaestecker et al., 2004).

Pasteuria penetrans. Members of nematode-associated *Pasteuria* share several morphological, ultrastructural, and ecological features Sayre and Starr, 1985; Starr et al., 1983; Starr and Sayre, 1988a; Table 55; Figure 37): all are Gram-positive; all form endospores; all form mycelia, septate and dichotomously branched, vegetative cells; and all parasitize invertebrates (Table 55). The members of the *Pasteuria penetrans* group differ in many respects from *Pasteuria ramosa*: colony shape, shape and size of sporangia (Figure 55) and endospores, and host relations (Table 56). Upon recognition (Sayre and Starr, 1985) of its relationship to the genus *Pasteuria*, the first of these nematode parasites to receive such taxonomic attention was renamed *Pasteuria penetrans* (*ex* Thorne) Sayre and Starr (1985). Subsequently, the name *Pasteuria penetrans sensu stricto* (i.e., in the strict sense) was limited in scope to the bacterium parasitic on *Meloidogyne* spp. and particularly *Meloidogyne incognita* (Starr and Sayre, 1988a). A second species, *Pasteuria thornei*, was erected for parasites of the lesion nematodes of the genus *Pratylenchus* and particularly *Pasteuria brachyurus* (Starr and Sayre, 1988a).

Much of the following information stems from studies of *Pasteuria penetrans*. *Pasteuria thornei*, which has received much less study than its relative, is similar in most respects examined (transmission electron micrographs of sporangia of *Pasteuria thornei* at various stages are shown in Figure 56, Figure 57, and Figure 58); where substantial differences have been observed, they are noted below. Cross-sections viewed by transmission electron microscopy (Imbriani and Mankau, 1977; Sayre and Starr, 1985; Sayre and Wergin, 1977; Starr and Sayre, 1988a) reveal that the endospore of *Pasteuria penetrans sensu stricto* consists of a central, highly electron-opaque core surrounded by an inner and an outer wall composed of several distinct layers (Figure 48). When observed with the transmission electron microscope, the peripheral matrix of the spore is fibrillar. Fine microfibrillar strands, about 1.5 nm thick, extend outward and downward from the sides of the endospore to the cuticle of the nematode, where they become more electron-dense.

A mature endospore of *Pasteuria penetrans sensu stricto* attaches to the surface of a nematode so that a basal ring of wall material lies flatly against the cuticle. A median section through the endospore

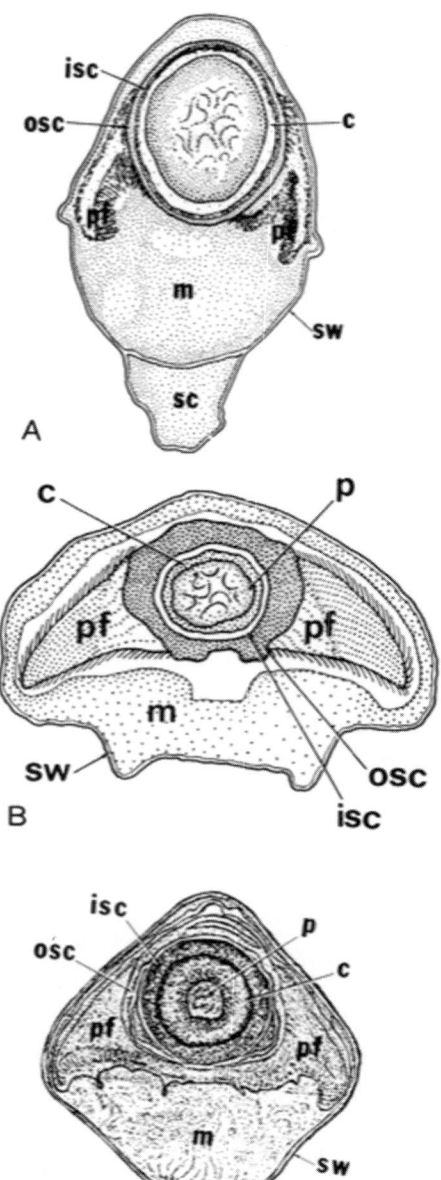

FIGURE 55. Drawings of sections through typical mature sporangia of three *Pasteuria* species emphasizing ultrastructural features useful in identification. A, *Pasteuria ramosa*, parasite of cladoceran water fleas. B, *Pasteuria penetrans sensu stricto*, parasite of *Meloidogyne incognita*. C, *Pasteuria thornei*, parasite of *Pratylenchus brachyurus*. Labels: protoplast (p) containing stranded inclusions; multilayered structures including a cortex (c) and inner (isc) and outer (osc) spore coats; parasporal fibers (pf); granular matrix (m); sporangial wall (sw); in *Pasteuria ramosa*, a stem cell (sc) remains attached to the sporangium.

and perpendicular to the surface of the nematode bisects this basal ring. As a result, the ring appears as two protruding pegs, which are continuous with the outer layer of the spore wall and rest on the cuticular surface of the nematode (Figure 43).

The peripheral fibers of the endospore also are closely associated with the nematode's cuticle. These fibers, which encircle the endospore, lie along the surface of the nematode and follow the irregularities of the cuticular annuli. They seem not

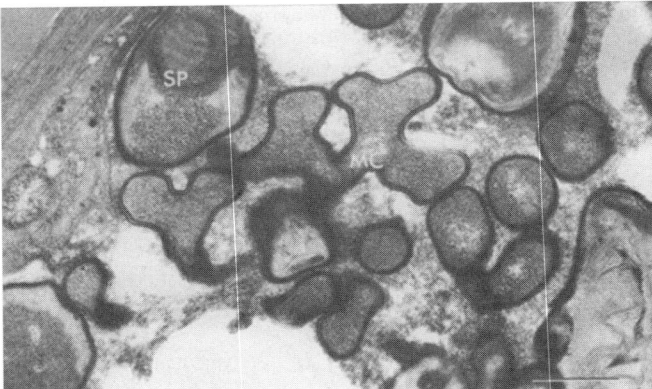

FIGURE 56. Transmission electron micrograph of *Pasteuria thornei* parasitizing a juvenile of the lesion nematode *Pratylenchus brachyurus*, showing simultaneous occurrence of vegetative microcolonies (MC) and sporangia (SP) of the bacterium in the nematode's pseudocoelomic cavity. Bar = 1.0 μm.

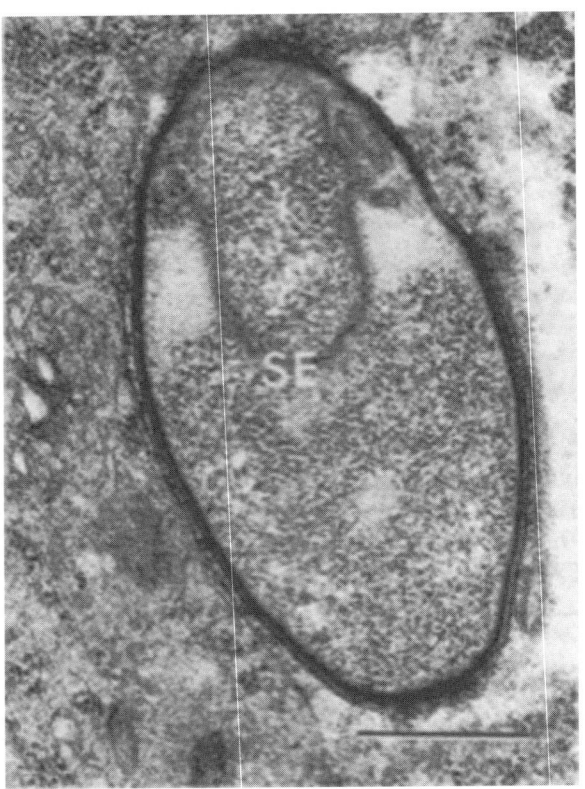

FIGURE 57. Section of immature sporangium of *Pasteuria thornei* showing the developing septum (SE) separating the polar area that condenses into an endospore from the parasporal matrix; the light areas sublateral to the polar area will develop into parasporal fibers. Bar = 0.5 μm.

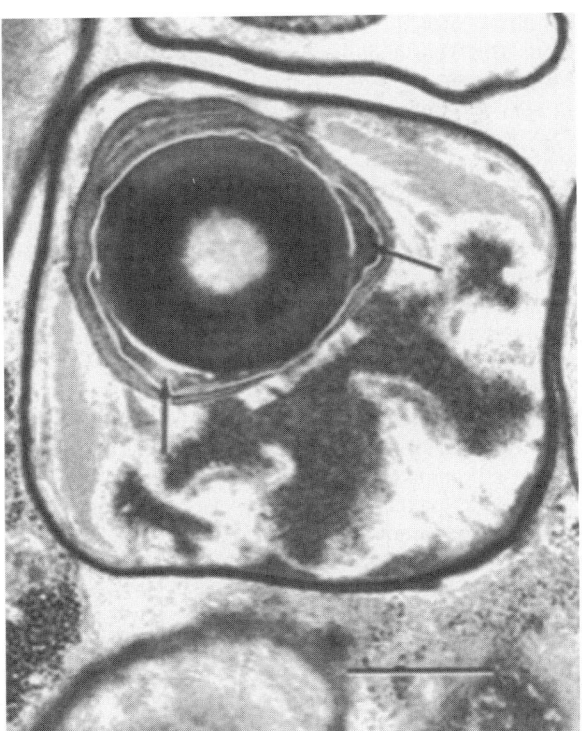

FIGURE 58. Lateral view of a mature sporangium of *Pasteuria thornei* showing inner sublateral cortical wall (arrows) that gives the endospore an angular appearance. The basal cortical wall for the endospore thins to provide a germinal pore. The basal portion of the sporangium contains an irregular granular matrix intermingled with fibrillar strands. Bar = 0.5 μm.

to penetrate the cuticle. The germ tube of the endospore of *Pasteuria penetrans* emerges through the central opening of the basal ring, penetrates the cuticle of the nematode, and enters the hypodermal tissue (Figure 43). Hyphae were initially encountered beneath the cuticle of the nematode near the site of germ-tube penetration. From this site, they apparently penetrate the hypodermal and muscle tissues and enter the pseudocoelom.

Mycelial colonies of *Pasteuria penetrans* up to 10 μm in diameter are formed in the pseudocoelom, where they are observed after the diseased juvenile penetrates plant roots (Figure 44). The hyphae comprising the colony are septate. A hyphal cell, which is 0.40–0.50 μm in cross-section, is bounded by a compound wall, 0.12 μm thick, composed of an outer and an inner membrane. The inner membrane of the wall forms the septation and delineates individual cells. In addition, this membrane is continuous with a membrane complex or mesosome that is frequently associated with the septum (Figure 45). Because of the sinuous and branching growth habit, cell length cannot be determined from thin section electron micrographs.

Sporulation of *Pasteuria penetrans* occurs in a developmental pattern that is similar to that observed in axenic cultures of *Bacillus* spp., e.g., *Bacillus subtilis*. During the controlled and semisynchronous growth of *Meloidogyne arenaria* race 1 infected with a line (P20) of *Pasteuria penetrans*, endospore maturation was coincident with the formation of spore-associated adhesin, as determined by a specific monoclonal antibody (Brito et al., 2003).

Sporulation of *Pasteuria penetrans* is initiated in the female nematode. As the process begins, the terminal hyphal cells of the mycelium bifurcate and enlarge from typical hyphal cells to ovate cells measuring 2.0 × 4.0 μm. Structure and content of the cytoplast change from a granular matrix, which contains numerous ribosomes as found in the hyphal cells, to one lacking particulate organelles. During these changes, the developing sporangia separate from their parental hyphae, which cease to grow and eventually degenerate.

After these early structural alterations, a membrane forms within the sporangium and separates the upper third of the cell or forespore from its lower or parasporal portion (Figure 46). The granular matrix confined within the membrane then condenses into an electron-opaque body, 0.6 μm in diameter, which eventually becomes encircled by a multilayered wall. The resulting discrete structure is an endospore.

Coincident with the formation of an endospore in *Pasteuria penetrans* is the emergence of parasporal fibers. These fine fibers, which form around the base of the spore, differentiate from an electron-translucent, granular substance. They appear to connect with and radiate from the external layer of the wall of the endospore (Figure 47). During development of the parasporal fibers, the formation of another membrane, the exosporium, isolates the newly formed endospore within the sporangium. At this later stage of spore development, the granular content of the paraspore becomes less dense, degenerates, and disappears. As a result, the mature sporangium contains a fully developed endospore enclosed within the exosporium (Figure 48).

The cell wall of the sporangium of *Pasteuria penetrans* remains intact until the remnants of the infected nematode are disrupted, after which event the endospores are released. The exosporium apparently remains associated with the endospore until contact is made with a new nematode and the infection cycle restarts. The vermiform juvenile stages of the nematodes are encumbered by the parasite as they migrate through soils infested by the endospores of *Pasteuria penetrans sensu stricto* and the infectious cycle is repeated.

Much morphological, ultrastructural, developmental, and host diversity is evident in *Pasteuria* (Giblin-Davis et al., 2001; Sayre and Starr, 1985; Sayre et al., 1988). We believe this diversity speaks for the existence of many taxa within the group. *Pasteuria thornei*, *Pasteuria nishizawae*, and "*Candidatus* Pasteuria usgae" represent three cases where novel species have been proposed on substantial morphological, morphometric, host range, and molecular grounds (Table 56) (Giblin-Davis et al., 2001; Giblin-Davis et al., 2003b; Sayre et al., 1988; Starr and Sayre, 1988a).

Other organisms – "Pasteuria hartismerei". *Pasteuria hartismerei* Bishop, Gowen, Pembroke and Trotter (2007) was described from *Meloidogyne ardenensis* on the basis that the endospores lacked a basal ring on their ventral side and they had a unique clumping nature inside their host (Bishop et al., 2007). Also, a unique 16S rRNA sequence was obtained which differs from that of other described *Pasteuria*. This taxon will become a *Candidatus* species when validated according to the *International Code of Nomenclature of Bacteria (1992)* (Euzéby, 1998).

References

Anderson, J.M., J.F. Preston, D.W. Dickson, T.E. Hewlett, N.H. Williams and J.E. Maruniak. 1999. Phylogenetic analysis of *Pasteuria penetrans* by 16S rRNA gene cloning and sequencing. J. Nematol. *31*: 319–325.

Atibalentja, N., G.R. Noel and L.L. Domier. 2000. Phylogenetic position of the North American isolate of *Pasteuria* that parasitizes the soybean cyst nematode, *Heterodera glycines*, as inferred from 16S rDNA sequence analysis. Int. J. Syst. Evol. Microbiol. *50*: 605–613.

Atibalentja, N., B.P. Jakstys and G.R. Noel. 2004a. Life cycle, ultrastructure, and host specificity of the North American isolate of *Pasteuria* that parasitizes the soybean cyst nematode, *Heterodera glycines*. J. Nematol. *36*: 171–180.

Atibalentja, N., G.R. Noel and A. Ciancio. 2004b. A simple method for PCR-amplification, cloning, and sequencing of *Pasteuria* 16S rDNA from small numbers of endospores. J. Nematol. *36*: 100–105.

Atibalentja, N. and G.R. Noel. 2008. Bacterial endosymbionts of plant-parasitic nematodes. Symbiosis *46*: 87–93.

Baermann, G. 1917. Eine einfache Methode zur Affindung von Ankylostomun (Nematoden) Larven in Erdproben. Geneesk. Tijdschr. Nederl. Indiue *57*: 131–137.

Bekal, S., R.M. Giblin-Davis and J.O. Becker. 1999. Gnotobiotic culture of *Pasteuria* sp. on *Belonolaimus longicaudatus*. J. Nematol. *31*: 522.

Bekal, S., J. Borneman, M.S. Springer, R.M. Giblin-Davis and J.O. Becker. 2001. Phenotypic and molecular analysis of a *Pasteuria* strain parasitic to the sting nematode. J. Nematol. *33*: 110–115.

Bird, D.M., C.H. Opperman and K.G. Davies. 2003. Interactions between bacteria and plant-parasitic nematodes: now and then. Int J Parasitol *33*: 1269–1276.

Bishop, A.H. and D.J. Ellar. 1991. Attempts to culture *Pasteuria penetrans* in vitro. Biocontrol Sci. Technol. *1*: 101–114.

Bishop, A.H., S.R. Gowen, B. Pembroke and J.R. Trotter. 2007. Morphological and molecular characteristics of a new species of *Pasteuria* parasitic on *Meloidogyne ardenensis*. J. Invertebr. Pathol. *96*: 28–33.

Brito, J.A., J.F. Preston, D.W. Dickson, R.M. Giblin-Davis, D.S. Williams, H.C. Aldrich and J.D. Rice. 2003. Temporal formation and immuno-localization of an endospore surface epitope during *Pasteuria penetrans* sporogenesis. J. Nematol. *35*: 278–288.

Brito, J.A., T.O. Powers, P.G. Mullin, R.N. Inserra and D.W. Dickson. 2004. Morphological and molecular characterization of *Meloidogyne mayaguensis* isolates from Florida. J. Nematol. *36*: 232–240.

Buchanan, R.E. 1925. General Systematic Bacteriology. The Williams & Wilkins Co., Baltimore.

Butler, M.K., J. Wang, R.I. Webb and J.A. Fuerst. 2002. Molecular and ultrastructural confirmation of classification of ATCC 35122 as a strain of *Pirellula staleyi*. Int. J. Syst. Evol. Microbiol. *52*: 1663–1667.

Canning, E.U. 1973. Protozoal parasites as agents for biological control of plant-parasitic nematodes. Nematologica *19*: 342–348.

Carius, H.J., T.J. Little and D. Ebert. 2001. Genetic variation in a host-parasite association: potential for coevolution and frequency-dependent selection. Evolution Int. J. Org. Evol. *55*: 1136–1145.

Charles, L., I. Carbone, K.G. Davies, D. Bird, M. Burke, B.R. Kerry and C.H. Opperman. 2005. Phylogenetic analysis of *Pasteuria penetrans* by use of multiple genetic loci. J. Bacteriol. *187*: 5700–5708.

Chen, S.Y., J. Charnecki, J.F. Preston and D.W. Dickson. 2000. Extraction and purification of *Pasteuria* spp. endospores. J. Nematol. *32*: 78–84.

Chen, Z.X., D.W. Dickson and T.E. Hewlett. 1996. Quantification of endospore concentrations of *Pasteuria penetrans* in tomato root material. J. Nematol. *28*: 50–55.

Chen, Z.X., D.W. Dickson, D.J. Mitchell, R. McSorley and T.E. Hewlett. 1997a. Suppression mechanisms of *Meloidogyne arenaria* race 1 by *Pasteuria penetrans*. J. Nematol. *29*: 1–8.

Chen, Z.X., D.W. Dickson, L.G. Freitas and J.F. Preston. 1997b. Ultrastructure, morphology, and sporogenesis of *Pasteuria penetrans*. Phytopathology *87*: 273–283.

Chen, Z.X. and D.W. Dickson. 1998. Review of *Pasteuria penetrans*: biology, ecology, and biological control potential. J. Nematol. *30*: 313–340.

Ciancio, A., R. Bonsignore, N. Vovlas and F. Lamberti. 1994. Host records and spore morphometrics of *Pasteuria penetrans* group parasites of nematodes. J. Invertebr. Pathol. *63*: 260–267.

Cobb, N.A. 1906. Fungus maladies of the sugar cane, with notes on associated insects and nematodes, 2nd edn. Hawaiian Sugar Planters' Assoc., Expt. Sta. Div. Path. Physiol. *6*: 163–195.

Cohn, F. 1872. Untersuchungen über Bakterien. Bertr. Biol. Pflanz. *1 (Heft II)*: 127–224.

Costa, S.R., B.R. Kerry, R.D. Bardgett and K.G. Davies. 2006. Exploitation of immunofluorescence for the quantification and characterization of small numbers of *Pasteuria* endospores. FEMS Microbiol. Ecol. *58*: 593–600.

Davies, K.G., M. Redden and T. K. Pearson. 1994. Endospore heterogeneity in *Pasteuria penetrans* related to attachment to plant-parasitic nematodes. Lett. Appl. Microbiol. *19*: 370–373.

De Toni, J.B. and V. Trevisan. 1889. Schizomycetaceae Naeg. *In* Saccardo (Editor), Sylloge Fungorum, vol. 8, pp. 923–1087.

Decaestecker, E., A. Vergote, D. Ebert and L. De Meester. 2003. Evidence for strong host clone-parasite species interactions in the *Daphnia* microparasite system. Evolution Int. J. Org. Evol. *57*: 784–792.

Decaestecker, E., C. Lefever, L. De Meester and D. Ebert. 2004. Haunted by the past: evidence for resting stage banks of microparasites and epibionts of *Daphnia*. Limnol. Oceanogr. *49*: 1355–1364.

Dickson, D.W., J.N. Sasser and D. Huisingh. 1970. Comparative disc-electrophoretic protein analyses of selected *Meloidogyne, Ditylenchus, Heterodera* and *Aphelenchus* spp. J. Nematol. *2*: 286–293.

Dickson, D.W., M. Oostendorp, R.M. Giblin-Davis and D.J. Mitchell. 1994. Control of plant-parasitic nematodes by biological antagonists. *In* Rosen, Bennett and Capinera (Editors), Pest Management in the Subtropics. Biological control – a Florida perspective, Intercept, Hampshire, UK, pp. 575–601.

Duan, Y.P., H.F. Castro, T.E. Hewlett, J.H. White and A.V. Ogram. 2003. Detection and characterization of *Pasteuria* 16S rRNA gene sequences from nematodes and soils. Int. J. Syst. Evol. Microbiol. *53*: 105–112.

Duncan, A.B. and T.J. Little. 2007. Parasite-driven genetic change in a natural population of *Daphnia*. Evolution Int. J. Org. Evol. *61*: 796–803.

Ebert, D., P. Rainey, T.M. Embley and D. Scholz. 1996. Development, life cycle, ultrastructure and phylogenetic position of *Pasteuria ramosa* Metchnikoff 1888: rediscovery of an obligate endoparasite of *Daphnia magna* Straus. Philos. Trans. R. Soc. Lond. B *351*: 1689–1701.

Ebert, D., H.J. Carius, T. Little and E. Decaestecker. 2004. The evolution of virulence when parasites cause host castration and gigantism. Am. Nat. *164*: S19–S32.

Ebert, D. 2005. Ecology, Epidemiology, and Evolution of Parasitism in *Daphnia* National Library of Medicine (US), National Center for Biotechnology Information, Bethesda, MD.

El-Borai, F.E., L.W. Duncan and J.F. Preston. 2005. Bionomics of a phoretic association between *Paenibacillus* sp. and the entomopathogenic nematode *Steinernema diaprepesi*. J. Nematol. *37*: 18–25.

Enright, M.R., J.O. McInerney and C.T. Griffin. 2003. Characterization of endospore-forming bacteria associated with entomopathogenic nematodes, *Heterorhabditis* spp., and description of *Paenibacillus nematophilus* sp. nov. Int. J. Syst. Evol. Microbiol. *53*: 435–441.

Euzéby, J.P. 1998. List of prokaryotic names with standing in nomenclature, Centre Interuniversitaire de Calcul de Toulouse, Toulouse, France.

Fould, S., A.L. Dieng, K.G. Davies, P. Normand and T. Mateille. 2001. Immunological quantification of the nematode parasitic bacterium *Pasteuria penetrans* in soil. FEMS Microbiol. Ecol. *37*: 187–195.

Garrity, G.M., J.A. Bell and T. Lilburn. 2005. The Revised Road Map to the Manual. *In* Brenner, Krieg, Staley and Garrity (Editors), Bergey's Manual of Systematic Bacteriology, 2nd edn, vol. 2, The *Proteobacteria*, Part A, Introductory Essays. Springer, New York, pp. 159–220.

Giblin-Davis, D.S.W. R.M., W.P. Wergin, D.W. Dickson, T.E. Hewlett, S. Bekal and J.O. Becker. 2001. Ultrastructure and development of *Pasteuria* sp. (S-1 strain), an obligate endoparasite of *Belonolaimus longicaudatus* (Nemata: Tylenchida). J. Nematol. *33*: 227–238.

Giblin-Davis, R.M., D.S. Williams, J.A. Brito, D.W. Dickson and J.F. Preston. 2003a. Ultrastructure and development of two *Pasteuria* species on *Hoplolaimus galeatus*. J. Nematol. *35*: 340.

Giblin-Davis, R.M., D.S. Williams, S. Bekal, D.W. Dickson, J.A. Brito, J.O. Becker and J.F. Preston. 2003b. 'Candidatus Pasteuria usgae' sp. nov., an obligate endoparasite of the phytoparasitic nematode *Belonolaimus longicaudatus*. Int. J. Syst. Evol. Microbiol. *53*: 197–200.

Giblin-Davis, R.M., B.J. Center, D.S. Williams, L.M. Schmidt, J.A. Brito, D.W. Dickson and J.F. Preston. 2004. Isolation of a *Pasteuria* sp. that is easily cultured on a bacteriovorous nematode, *Bursilla* sp. in soil. J. Nematol. *36*: 319–320.

Hatz, B. and D.W. Dickson. 1992. Effect of temperature on attachment, development, and interactions of *Pasteuria penetrans* on *Meloidogyne arenaria*. J. Nematol. *24*: 512–521.

Henrici, A.T. and D.E. Johnson. 1935. Studies of freshwater bacteria: II. Stalked bacteria, a new order of *Schizomycetes*. J. Bacteriol. *30*: 61–93.

Hewlett, T.E. and D.W. Dickson. 1993. A centrifugation method for attaching endospores of *Pasteuria* spp. to nematodes Suppl. J. Nematol. *25*: 785–788.

Hewlett, T.E., S.T. Griswold and K.S. Smith. 2006. Biological control of *Meloidogyne incognita* using in vitro produced *Pasteuria penetrans* in a microplot study. J. Nematol. *38*: 274 (Abstr.).

Hirsch, P. 1972. Re-evaluation of *Pasteuria ramosa* Metchnikoff 1888, a bacterium pathogenic for *Daphnia* species. Int. J. Syst. Bacteriol. *22*: 112–116.

Imbriani, J.L. and R. Mankau. 1977. Ultrastructure of the nematode pathogen, *Bacillus penetrans*. J. Invertebr. Pathol. *30*: 337–347.

Jenkins, W.R. 1964. A rapid centrifugal-flotation technique for separating nematodes from soil. Plant Dis. Rep *48*: 692.

Jirovec, O. 1939. *Dermocystidium vejdovskyi* n. sp., ein neuer Parasit des Hechtes, nebst einer Bemerkung ueber *Dermocystidium daphniae* (Ruehberg). Arch. Protiskenk. *92*: 137–146.

Judicial-Commission. 1986. Opinion 61. Rejection of the type strain of *Pasteuria ramosa* (ATCC 27377) and conservation of the species *Pasteuria ramosa* Metchnikoff 1888 on the basis of the type descriptive material. Int. J. Syst. Bacteriol. *36*: 119.

Lapage, S.P., P.H.A. Sneath, E.F. Lessel, V.B.D. Skerman, H.P.R. Seeliger and W.A. Clark. 1992. International Code of Nomenclature of Bacteria (1990 Revision). Bacteriological Code. Published for IUMS by the American Society for Microbiology, Washington, D.C.

Laurent, E. 1890. Sur le microbe des nodosites des legumineuses. CR Acad Sci Paris *3*: 754–756.

Lehmann, K.B. and R. Neumann. 1896. Atlas und Grundriss der Bakteriologie und Lehrbuch der speciellen bakteriologischen Diagnostik, vol. II. J.F. Lehmann, Munchen.

Little, T. and D. Ebert. 1999. Associations between parasitism and host genotype in natural populations of *Daphnia* (Crustacea: Cladocera). J. Animal Ecol. *68*: 134–149.

Mankau, R. 1973. Utilization of parasites and predators in nematode pest management ecology. Proceedings of the Tall Timbers Conference on Ecological Animal Control by Habitat Management *4*: 129–143.

Mankau, R. 1975. *Bacillus penetrans* n. comb. causing a virulent disease of plant-parasitic nematodes. J. Invertbr. Pathol. *26*: 333–339.

Mankau, R. and J.L. Imbriani. 1975. The life cycle of an endoparasite in some Tylenchid nematodes. Nematologica *9*: 40–45.

Metchnikoff, E. 1888. *Pasteuria ramosa*, un représentant des bactéries à division longitudinale Ann. Inst. Pasteur (Paris) *2*: 165–170.

Micoletzky, H. 1925. Die freilebenden Suesswasser une Moornematoden Daenmarks. Andr. Fred. Host & Son, Kobenhavn.

Migula, W. 1904. Allgemeine Morphologie, Entwicklungsgeschichte, Anatomie und Systematik der Schizomyceten. *In* Lafar (Editor), Handbuch der technischen Mykologie, 2nd edn. Gutav Fischer, Jena, pp. 29–149.

Murray, R.G. and E. Stackebrandt. 1995. Taxonomic note: implementation of the provisional status *Candidatus* for incompletely described procaryotes. Int J Syst Bacteriol *45*: 186–187.

Murray, R.G.E. and K.H. Schleifer. 1994. Taxonomic notes: a proposal for recording the properties of putative taxa of procaryotes. Int. J. Syst. Bacteriol. *44*: 174–176.

Noel, G.R., N. Atibalentja and L.L. Domier. 2005. Emended description of *Pasteuria nishizawae*. Int. J. Syst. Evol. Microbiol. *55*: 1681–1685.

Nong, G., V. Chow, L.M. Schmidt, D.W. Dickson and J.F. Preston. 2007. Multiple-strand displacement and identification of single nucleotide polymorphisms as markers of genotypic variation of *Pasteuria penetrans* biotypes infecting root-knot nematodes. FEMS Microbiol Ecol *61*: 327–336.

Oostendorp, M., D.W. Dickson and D.J. Mitchell. 1990. Host range and ecology of isolates of *Pasteuria* spp. from the southeastern United States. J. Nematol. *22*: 524–531.

Oostendorp, M., T.E. Hewlett, D.W. Dickson and D.J. Mitchell. 1991. Specific gravity of spores of *Pasteuria penetrans* and extraction of spore-filled nematodes from soil. Suppl. J. Nematol. *23*: 729–732.

Perez, C. 1908. Sur *Duboscqia legeri*, microsporidie nouvelle parasite die *Termes lucifugus* et sur la classification des microsporidies. Soc. Biol. (Paris) *65*: 631.

Persidis, A., J.G. Lay, T. Manousis, A.H. Bishop and D.J. Ellar. 1991. Characterisation of potential adhesins of the bacterium *Pasteuria penetrans*, and of putative receptors on the cuticle of Meloidogyne incognita, a nematode host. J Cell Sci *100*: 613–622.

Preston, J.F., D.W. Dickson, J.E. Maruniak, G. Nong, J.A. Brito, L.M. Schmidt and R.M. Giblin-Davis. 2003. *Pasteuria* spp.: systematics and phylogeny of these bacterial parasites of phytopathogenic nematodes. J. Nematol. *35*: 198–207.

Regoes, R.R., J.W. Hottinger, L. Sygnarski and D. Ebert. 2003. The infection rate of *Daphnia magna* by *Pasteuria ramosa* conforms with the mass-action principle. Epidemiol Infect *131*: 957–966.

Ruehberg, W. 1933. Ueber eine Hefeinfektion bei *Daphnia magna*. Arch. Protiskenk. *80*: 72–100.

Sayre, R.M. and W.P. Wergin. 1977. Bacterial parasite of a plant nematode: morphology and ultrastructure. J. Bacteriol. *129*: 1091–1101.

Sayre, R.M., W.P. Wergin and R.E. Davis. 1977. Occurrence in *Monia rectirostris* (Cladocera: Daphnidae) of a parasite morphologically similar to *Pasteuria ramosa* (Metchnikoff, 1888). Can J Microbiol *23*: 1573–1579.

Sayre, R.M., J.R. Adams and W.P. Wergin. 1979. Bacterial parasite of a cladoceran: morphology, development in vivo, and taxonomic relationships with *Pasteuria ramosa* Metchnikoff 1888. Int. J. Syst. Bacteriol. *29*: 252–262.

Sayre, R.M., R.L. Gherna and W.P. Wergin. 1983. Morphological and taxonomic reevaluation of *Pasteuria ramosa* Metchnikoff 1888 and "*Bacillus penetrans*" Mankau 1975. Int. J. Syst. Bacteriol. *33*: 636–649.

Sayre, R.M. and M.P. Starr. 1985. *Pasteuria penetrans* (ex Thorne 1940) nom. rev., comb. n., sp. n., a mycelial and endospore-forming bacterium parasitic in plant-parasitic nematodes. Proc. Helminthol. Soc. Wash. *52*: 149–165.

Sayre, R.M. and M.P. Starr. 1986. *In* Validation of the publication of new names and new combinations previously effectively published outside the IJSB, List no. 20. Int. J. Syst. Bacteriol. *36*: 354–356.

Sayre, R.M. and M.P. Starr. 1988. Bacterial diseases and antagonisms of nematodes. *In* Poinar and Jansson (Editors), Nematode Pathology, vol. 1. CRC Press, Boca Raton, Florida, pp. 65–101.

Sayre, R.M., M.P. Starr, A.M. Golden, W.P. Wergin and B.Y. Endo. 1988. Comparison of *Pasteuria penetrans* from Meloidogyne incognita with a related mycelial and endospore-forming bacterial parasite from Pratylenchus brachyurus. Proc. Helminthol. Soc. Wash. *55*: 28–49.

Sayre, R.M., W.P. Wergin, J.M. Schmidt and M.P. Starr. 1991. *Pasteuria nishizawae* sp. nov., a mycelial and endospore-forming bacterium parasitic on cyst nematodes of genera *Heterodera* and *Globodera*. Res. Microbiol. *142*: 551–564.

Sayre, R.M., W.P. Wergin, J.M. Schmidt and M.P. Starr. 1992. *In* Validation of the publication of new names and new combinations previously effectively published outside the IJSB. List no. 41. Int. J. Syst. Bacteriol. *42*: 327–329.

Schmidt, L.M., J.F. Preston, D.W. Dickson, J.D. Rice and T.E. Hewlett. 2003. Environmental quantification of *Pasteuria penetrans* endospores using in situ antigen extraction and immunodetection with a monoclonal antibody. FEMS Microbiol. Ecol. *44*: 17–26.

Schmidt, L.M., J.F. Preston, G. Nong, D.W. Dickson and H.C. Aldrich. 2004. Detection of *Pasteuria penetrans* infection in *Meloidogyne arenaria* race 1 in planta by polymerase chain reaction. FEMS Microbiol. Ecol. *48*: 457–464.

Schmidt, L.M., L. Mouton, G. Nong, D. Ebert and J.F. Preston. 2008. Genetic and immunological comparison of the cladoceran parasite *Pasteuria ramosa* with the nematode parasite *Pasteuria penetrans*. Appl. Environ. Microbiol. *74*: 259–264.

Serracin, M., A.C. Schuerger, D.W. Dickson and D.P. Weingartner. 1997. Temperature-dependent development of *Pasteuria penetrans* in *Meloidogyne arenaria*. J. Nematol. *29*: 228–238.

Skerman, V.B.D., V. McGowan and P.H.A. Sneath. 1980. Approved lists of bacterial names. Int. J. Syst. Bacteriol *30*: 225–420.

Southey, J.F. (Editor). 1986. Laboratory Methods for Work with Plant and Soil Nematodes. [Reference book 402, Ministry of Agriculture, Fisheries and Food, United Kingdom], 6th edn. Her Majesty's Stationery Office, London.

Stackebrandt, E., W. Frederiksen, G.M. Garrity, P.A. Grimont, P. Kämpfer, M.C. Maiden, X. Nesme, R. Rosselló-Mora, J. Swings, H.G. Truper, L. Vauterin, A.C. Ward and W.B. Whitman. 2002. Report of the ad hoc committee for the re-evaluation of the species definition in bacteriology. Int. J. Syst. Evol. Microbiol. *52*: 1043–1047.

Staley, J.T. 1973. Budding bacteria of the *Pasteuria-Blastobacter* group. Can. J. Microbiol. *19*: 609–614.

Starr, M.P., R.M. Sayre and J.M. Schmidt. 1983. Assignment of ATCC 27377 to *Planctomyces staleyi* sp. nov. and conservation of *Pasteuria ramosa* Metchnikoff 1888 on the basis of type descriptive material: request for an opinion. Int. J. Syst. Bacteriol. *33*: 666–671.

Starr, M.P. and R.M. Sayre. 1988a. *Pasteuria thornei* sp. nov. and *Pasteuria penetrans sensu stricto* emend., mycelial and endospore-forming bacteria parasitic, respectively, on plant-parasitic nematodes of the genera *Pratylenchus* and *Meloidogyne*. Ann. Inst. Pasteur Microbiol. *139*: 11–31.

Starr, M.P. and R.M. Sayre. 1988b. In Validation of the publication of new names and new combinations previously effectively published outside the IJSB. List no. 26. Int. J. Syst. Bacteriol. *38*: 328–329.

Steiner, G. 1938. Opuscula miscellanea nematologica VII. Proc. Helminthol. Soc. Wash. *5*: 35–40.

Sterba, G. and W. Naumann. 1970. Unteruschungen ueber *Dermocystidium granulosum* n. sp. bei *Tetraodon palembangensis* (Bleeker 1852). Arch. Protiskenk. *112*: 106–118.

Stirling, G.R. and M.F. Wachtel. 1980. Mass production of *Bacillus penetrans* for the biological control of root-knot nematodes. Nematologica *26*: 308–312.

Stirling, G.R. 1981. Effect of temperature on infection of *Meloidogyne javanica* by *Bacillus penetrans*. Nematologica *27*: 458–462.

Stirling, G.R. 1984. Biological control of *Meloidogyne javanica* with *Bacillus penetrans*. Phytopathology *74*: 55–60.

Stirling, G.R. 1985. Host specificity of *Pasteuria penetrans* within the genus *Meloidogyne*. Nematologica *31*: 203–209.

Sturhan, D. 1988. New host and geographical records of nematode-parasitic bacteria of the *Pasteuria penetrans* group. Nematologica *34*: 350–356.

Sturhan, D., T.S. Shutova, V.N. Akimov and S.A. Subbotin. 2005. Occurrence, hosts, morphology, and molecular characterisation of *Pasteuria* bacteria parasitic in nematodes of the family Plectidae. J Invertebr Pathol *88*: 17–26.

Swofford, D.L. 2003. PAUP* – Phylogenetic analysis using parsimony* and other methods, version 4, Sinauer Associates, Sunderland, MA.

Thompson, J.D., T.J. Gibson, F. Plewniak, F. Jeanmougin and D.G. Higgins. 1997. The CLUSTAL_X Windows interface: flexible strategies for multiple sequence alignment aided by quality analysis tools. Nucleic Acids Res. *25*: 4876–4882.

Thorne, G. 1940. *Duboscqia penetrans* n. sp. (Sporozoa, Microsporidia, Nosematidae), a parasite of the nematode *Pratylenchus pratensis* (de Man) Filipjev. Proc. Helminthol. Soc. Wash. *7*: 51–53.

Trotter, J.R. and A.H. Bishop. 2003. Phylogenetic analysis and confirmation of the endospore-forming nature of *Pasteuria penetrans* based on the *spo0A* gene. FEMS Microbiol. Lett. *225*: 249–256.

Trotter, J.R., D.A. Darban, S.R. Gowen, A.H. Bishop and B. Pembroke. 2004. The isolation of a single spore isolate of *Pasteuria penetrans* and its pathogenicity on Meloidogyne javanica. Nematology *6*: 463–471.

Vuillemin, P. 1913. Genera Schizomycetum. Ann. Mycol. Berlin *11*: 512–527.

Walia, R.K., T.E. Hewlett and D.W. Dickson. 2004. Microwave treatment of *Pasteuria penetrans* parasite preparation for selective elimination of undesired micro-organisms. Nematol. Mediterr. *32*: 15–17.

Wayne, L.G. 1986. Actions of the Judicial Commission of the International Committee on Systematic Bacteriology on requests for opinions published in 1983 and 1984. Int. J. Syst. Bacteriol. *36*: 357–358.

Weiser, J. 1943. Beitraege zur Entwicklungsgeschichte von *Dermocystidium daphniae* Jirovec. Zool. Anz. *142*: 200–205.

Williams, J.R. 1960. Studies on the nematode soil fauna of sugarcane fields of Mauritius. 5. Notes upon a parasite of root-knot nematodes. Nematologica *5*: 37–42.

Woese, C.R. 1987. Bacterial evolution. Microbiol. Rev. *51*: 221–271.

Zuckerman, B.M., W.F. Mai and M.B. Harrison (Editors). 1984. Laboratory Manual for Plant Nematology. Massachusetts Agricultural Experiment Station, Amherst.

Family VI. Planococcaceae Krasil'nikov 1949, 328[AL]

WOLFGANG LUDWIG, KARL-HEINZ SCHLEIFER AND WILLIAM B. WHITMAN

Plan.o.coc.ca'ce.ae. N.L. masc. n. *Planococcus* type genus of the family; L. suff. *-aceae* ending denoting family; N.L. fem. pl. n. *Planococcaceae* the *Planococcus* family.

The family *Planococcaceae* is circumscribed for this volume on the basis of phylogenetic analyses of the 16S rRNA sequences and includes the genus *Planococcus* and its close relatives, *Caryophanon, Filibacter, Jeotgalibacillus, Kurthia, Marinibacillus, Planomicrobium, Sporosarcina,* and *Ureibacillus*. Cells are cocci or rods, sometimes forming filaments or trichomes. Usually strictly aerobic heterotrophs, although some species are facultatively aerobic. Motile by flagella or gliding. Catalase-positive, oxidase-positive or negative. May or may not form endospores. The dominant fatty acids are usually $C_{15:0\ iso}$ or $C_{15:0\ anteiso}$.

DNA G+C content (mol%): 34–48.

Type genus: **Planococcus** Migula 1894, 236.

Genus I. Planococcus Migula 1894, 236[AL]

SISINTHY SHIVAJI

Plan.o.coc'cus. Gr. n. *planos* wanderer; Gr. n. *coccus* a grain, berry; N.L. masc. n. *Planococcus* motile coccus.

Cells coccoid, 1.0–1.2 μm in diameter, occurring singly, in pairs, in groups of three, or as tetrads or clumps of cells. Gram-positive to Gram-variable. **Motile**. The cells have one or two flagella. Chemo-organotrophic, respiratory metabolism. **Aerobic. Catalase-positive. Lack endospores**. Colonies are circular, slightly convex, smooth, glistening, and **yellow-orange in color**. Halotolerant; can tolerate 1–17% NaCl and are occasionally halophilic. Most strains are oxidase-negative and hydrolyze gelatin but not starch. Do not reduce nitrate. Growth factors are usually not required. Either psychrophilic or mesophilic. Distributed in sea water, marine clams, fish, shrimp, prawns, cyanobacterial mats, freshwater ponds, sulfur springs, and glacial soil.

DNA G+C content (mol%): 39–48 (T_m, Bd).

Type species: **Planococcus citreus** Migula 1894, 236[AL].

Further descriptive information

Planococcus species have been isolated using various kinds of media such as sea water agar (1.0% beef extract, 1.0% peptone, 2.0% agar, tap water 250 ml and sea water 750 ml; pH 7.2), nutrient agar containing 1–5% NaCl, Zobell marine agar 2216 (Difco Laboratories, Detroit, USA) or antarctic bacterial medium (0.5% peptone, 0.2% yeast extract, and 1.5% agar, pH 6.4) with 1.5% NaCl or in the absence of salt. Cells are coccoid, motile, and possess one or two flagella; a few cells possess three to four flagella. The flagella are often irregular, but some show a regular sine curve. Motile cells occur in both liquid and solid media.

The fine structure of the cells of planococci is similar to that of other Gram-positive, catalase-positive cocci (Novitsky and Kushner, 1976). The cell wall of *Planococcus citreus* is double layered, and its thickness varies with the age of the culture from 25–35 nm. Cell wall peptidoglycan is of the L-Lys-D-Glu type (Alam et al., 2003; Mayilraj et al., 2005; Reddy et al., 2002; Schleifer and Kandler, 1970). $C_{15:0\ ante}$ is the predominant fatty acid (Alam et al., 2003; Engelhardt et al., 2001; Junge et al., 1998; Mayilraj et al., 2005; Nakagawa et al., 1996; Reddy et al., 2002; Romano et al., 2003; Thirkell and Summerfield, 1980) followed by $C_{17:0\ ante}$ and $C_{16:0\ iso}$. Strains of psychrophilic *Planococcus antarcticus* possess significant amounts of $C_{15:1}$ and $C_{18:1}$ (Reddy et al., 2002).

All species produce a yellow-orange, water-insoluble but methanol-soluble carotenoid pigment. In methanol, the pigments exhibited three absorption maxima at about 440, 465, and 488 nm, a characteristic feature of carotenoids. The concentration of sea salt in the medium and the age of the culture appear to influence the quantity of the pigment synthesized and the type of carotenoid (Thirkell and Summerfield, 1980). Hydrostatic pressure of 20–40 MPa has no influence on the growth and pigment production of *Planococcus citreus* (Courington and Goodwin, 1955; Oppenheimer and Zobell, 1952).

MK-7 and MK-8 are the menaquinones that are normally present (Alam et al., 2003; Jeffries, 1969; Mayilraj et al., 2005; Reddy et al., 2002; Yamada et al., 1976). In *Planococcus kocurii* and *Planococcus maritimus*, in addition to MK-7 and MK-8, MK-6 is also present (Hao and Komagata, 1985; Yoon et al., 2003), and in *Planococcus rifietoensis* only MK-8 is present (Romano et al., 2003). The phospholipid pattern of planococci is similar to that of *Sporosarcina* and contains a large amount of cardiolipins (Thirkell and Summerfield, 1977; Yamada et al., 1976) including phosphatidylglycerol (PG), diphosphtaidylglycerol (DPG), and phosphatidylethanolamine (PE) as in *Planococcus antarcticus, Planococcus maritimus,* and *Planococcus stackebrandtii* (Mayilraj et al., 2005; Reddy et al., 2002; Yoon et al., 2003). In *Planococcus rifietoensis*, PE is replaced by phosphocholine (Romano et al., 2003). All the reported species are halotolerant (7–17% NaCl) and *Planococcus maitriensis* is halophilic (Alam et al., 2003; Novitsky and Kushner, 1976).

A serological examination of *Planococcus citreus* has shown no antigenic relationship to staphylococci and micrococci (Oeding, 1971). Sensitivity to antibiotics has not been studied in all the species of *Planococcus*. *Planococcus citreus, Planococcus maritimus, Planococcus maitriensis,* and *Planococcus antarcticus* are sensitive to tetracycline and chloramphenicol. The last three species were also sensitive to streptomycin, lincomycin, and bacitracin but varied in their sensitivity to erythromycin, penicillin G, gentamicin, rifampin, and ampicillin (Alam et al., 2003; Jeffries, 1969; Kocur et al., 1970; Novitsky and Kushner, 1976; Reddy et al., 2002; Yoon et al., 2003)

Planococci have been isolated from various terrestrial, aquatic, and marine habitats such as sea water (Yoon et al., 2003; Zobell and Upham, 1944), fish-brining tanks (Georgala, 1957), marine clams and fish (Hao and Komagata, 1985; Novitsky and Kushner, 1976), shrimp (Alvarez, 1982), antarctic soil (Shivaji et al., 1988), antarctic cyanobacterial mats (Alam et al., 2003; Reddy et al., 2002), sulfur spring (Romano et al., 2003), and glacial soil (Mayilraj et al., 2005).

Enrichment and isolation procedures

Various media listed above may be used for the isolation of planococci. Enrichment may be facilitated by using sea water agar or nutrient agar containing 5–7% NaCl.

Maintenance procedures

Planococci cultures can be stored in screw-capped tubes or on plates containing semisolid medium. Tubes and plates are inoculated and, after overnight growth at optimum temperature, stored at 4°C. Cultures are lyophilized for long-term preservation.

Differentiation of the genus *Planococcus* from other genera

Table 57 indicates the characteristics of *Planococcus* that distinguish it from other genera of morphologically or physiologically similar taxa.

Taxonomic comments

The genus *Planococcus* was first proposed more than hundred years ago to include Gram-positive motile cocci (Kocur et al., 1970; Migula, 1894). Despite this clear-cut criterion based on cell shape and motility (motile cocci), the genus *Planococcus* still included species which were either coccoid (*Planococcus citreus* and *Planococcus halophilus*) or rod-shaped (*Planococcus okeanokoites* and *Planococcus mcmeekinii*) (Junge et al., 1998; Migula, 1894; Nakagawa et al., 1996; Novitsky and Kushner, 1976; Zobell and Upham, 1944). In addition, motile cocci were also included in the genus *Micrococcus* (Hao and Komagata, 1985), but such motile cocci differed from species of the genera

Micrococcus and *Staphylococcus* in their motility and DNA base composition (Hao and Komagata, 1985). Taking these phenotypic traits into consideration, Boháček et al. (1967) separated the motile from the nonmotile cocci and subsequently, on the basis of DNA base composition, Kocur et al. (1970) divided the motile cocci into two groups. Group I included strains with a low mol% G+C content of DNA (39.6–42.2 mol%) and were identified as *Planococcus kocurii* (Hao and Komagata, 1985). Group II included strains with high mol% G+C content of DNA (48–52.1 mol%) and were identified as *Planococcus citreus* (Migula, 1894). Subsequent phylogenetic analysis based on 16S rRNA gene sequences confirmed that these *Planococcus* species are distinct from other cocci (Farrow et al., 1992; Farrow et al., 1994a) and exhibit no specific relationship to the genera *Micrococcus* and *Staphylococcus* (Stackebrandt and Woese, 1979). However, phylogenetically they appear to be related to the genera *Bacillus* and *Sporosarcina* (Boháček et al., 1968b; Pechman et al., 1976; Stackebrandt and Woese, 1979) and chemotaxonomic characteristics such as mol% G+C content in the DNA and cell wall composition support these observations (Boháček et al., 1968b; Kocur et al., 1970; Schleifer and Kandler, 1970) but planococci are nonsporeforming and thus are distinctly different. Phylogenetic analysis of *Planococcus mcmeekinii*, *Planococcus okeanokoites*, *Planococcus kocurii*, and *Planococcus citreus* based on 16S rRNA gene sequences indicated that the four species formed two distinct phylogenetic clusters (Yoon et al., 2001a). One of the clusters included *Planococcus okeanokoites* and *Planococcus mcmeekinii*. Both these species have MK-8 as the predominant menaquinone and MK-7 as the minor component (Nakagawa et al., 1996). *Planococcus citreus* and *Planococcus kocurii* constituted the second cluster. These two species differ in their menaquinone composition, with *Planococcus citreus* having MK-6, MK-7, and MK-8 and *Planococcus kocurii* having MK-7 and MK-8 (Hao and Komagata, 1985). Subsequent studies confirmed the preceding phylogenetic affiliations (Alam et al., 2003; Mayilraj et al., 2005; Reddy et al., 2002) and clearly

TABLE 57. Characteristics differentiating the genus *Planococcus* from other morphologically or biochemically similar taxa[a]

Characteristics	*Planococcus*	*Sporosarcina*	*Micrococcus*	*Planomicrobium*[b]
Spores	−	+	−	−
Motility	+	+	−	+
Mol% G+C of the DNA[c]	39–52	40–44	65–75	35–47
Peptidoglycan type[d]	L-Lys-D-Glu	L-Lys-Gly-D-Glu	Mostly L-Lys-peptide subunit, or L-Lys-L-Ala$_{3-4}$	L-Lys-D-Glu or L-Lys-D-Asp
Menaquinone pattern[e]	MK-6, MK-7, MK-8	MK-7	Hydrogenated MK-7, MK-8, MK-9	MK-6, MK-7, MK-8
Phosphatidyl ethanolamine[f]	+	+	−	+
Aliphatic hydrocarbons[g]	Absent	nd	Present	nd
Yellow-orange pigment	+	−	−	+
Growth on nutrient agar containing 12% NaCl	+	−	−	−
Gelatin hydrolysis	+	−	d	+
Urease	−	+	d	−
NO$_3^-$ reduced to NO$_2^-$	−	+	d	d

[a]+, Positive; −, negative; nd, not determined; d, some strains are positive.
[b]Data from Yoon et al. (2001a).
[c]Data from Boháček et al. (1968b, 1968a); Kocur et al. (1970).
[d]Data from Schleifer and Kandler (1970); Novitsky and Kushner (1976).
[e]Data from Jeffries (1969); Yamada et al. (1976).
[f]Data from Komura et al. (1975).
[g]Data from Morrison et al. (1971).

indicated a robust phylogenetic relationship between *Planococcus citreus*, *Planococcus kocurii*, *Planococcus antarcticus*, *Planococcus maritimus*, *Planococcus rifietoensis*, *Planococcus maitriensis* and *Planococcus columbae* but *Planococcus stackebrandtii* (Alam et al., 2003; Hao and Komagata, 1985; Mayilraj et al., 2005; Migula, 1894; Reddy et al., 2002; Romano et al., 2003; Suresh et al., 2007; Yoon et al., 2003) proved to be an exception and it grouped with *Planomicrobium psychrophilus* with a low bootstrap value. Thus, based on the phylogenetic inference, morphological features, and chemotaxonomic properties, it was suggested

that *Planococcus okeanokoites* and *Planococcus mcmeekinii* be classified into the new genus *Planomicrobium*, separating it from *Planococcus* as represented by *Planococcus citreus* and *Planococcus kocurii* (Yoon et al., 2001a). All of the eight species of *Planococcus* listed above are Gram-positive, motile, coccoid, orange-yellow in color, halotolerant and catalase-positive.

Differentiation and characteristics of the species of the genus Planococcus.   The differential characteristics of the species of *Planococcus* are indicated in TABLE 58. Other characteristics of the species are presented in TABLE 59 and Table 60.

TABLE 58. Characteristics differentiating the species of the genus *Planococcus*[a]

Characteristics	*P. citreus*[b,c,d,e,f]	*P. antarcticus*[g]	*P. columbae*[h]	*P. kocurii*[e,f]	*P. maitriensis*[i]	*P. maritimus*[j]	*P. rifietoensi*[k]	*P. stackebrandtii*[l]
Colony color	Orange/ yellow	Orange	Orange	Orange/ yellow	Orange	Yellow/ orange	Orange	Orange
Growth range (°C)	4–37	0–30	8–42	4–37	0–30	4–41	5–42	15–30
Growth at:								
5°C	+	+	-	+	+	+	+	nd
37°C	+	−	+	+	−	+	+	nd
NaCl requirement	No	No	No	No	Yes	No	No	No
NaCl tolerance (%)	15	12	14	10	12.5	17	15	7
Oxidase	−	−	+	−	−	−	+	−
Phosphatase	−	−	nd	−	−	nd	nd	nd
Nitrate reduction	−	−	+	−	+	−	−	−
Lipase	−	d	+	−	−	−	nd	+
Esculin hydrolysis	−	d	-	−	+	−	nd	nd
Starch hydrolysis	−	−	-	−	−	−	nd	−
Casein hydrolysis	nd	nd	-	nd	nd	+	−	−
Utilization of carbon compounds:								
Glucose	−	+	+	−	+	nd	−	nd
Glutamate	−	+	nd	−	−	nd	nd	nd
Succinate	+	+	nd	+	+	nd	nd	nd
Acid from D-glucose	+	+	-	+	−	+	nd	−
Sensitivity to:								
Amoxycillin	nd	R	nd	nd	S	nd	nd	nd
Ampicillin	nd	R	nd	nd	S	nd	S	nd
Erythromycin	S	S	nd	nd	R	nd	S	nd
Carbenicillin	nd	R	nd	nd	S	nd	nd	nd
Gentamicin	nd	S	S	nd	S	nd	R	nd
Kanamycin	nd	R	nd	nd	S	nd	S	nd
Neomycin	nd	S	S	nd	S	nd	R	nd
Nystatin	nd	R	nd	nd	nd	nd	S	nd
Penicillin G	S	S	R	nd	R	nd	R	nd
Tobramycin	nd	R	R	nd	nd	nd	nd	nd
Menaquinones	MK-7, MK-8	MK-7, MK-8	MK-7, MK-8, MK-7(H$_2$)	MK-7, MK-8	MK-7, MK-8	MK-6, MK-7, MK-8	MK-8	MK-7, MK-8
Phospholipids	PG, DPG, PE	PG, DPG, PE	PG, DPG, PC, PI	PG, DPG, PE	nd	PG, DPG, PE	PG, DPG, PC	PG, DPG, PE
Mol% G + C of DNA	48.5	41.5	50.5	39.6–42.9	39±2.5	48	47.9	40

[a]Symbols:+, positive; −, negative; nd, not determined; PG, phosphatidylglycerol; DPG, diphosphatidylglycerol; PE, phosphatidylethanolamine; PC, phosphocholine; S, sensitive; R, resistant; d, some strains are positive.
[b]Kocur (1986).
[c]Jeffries (1969).
[d]Yamada et al. (1976).
[e]Hao and Komagata (1985).
[f]Komura et al. (1975).
[g]Reddy et al. (2002).
[h]Suresh et al. (2007).
[i]Alam et al. (2003).
[j]Yoon et al. (2003).
[k]Romano et al. (2003).
[l]Mayilraj et al. (2005).

TABLE 59. Other characteristics of the species of the genus *Planococcus*[a]

Characteristics	*P. citreus*[b,c,d,e,f]	*P. antarcticus*[g]	*P. columbae*[h]	*P. kocurii*[i,e,f]	*P. maitriensis*[j]	*P. maritimus*[i]	*P. rifietoensis*[k]	*P. stackebrandtii*
Cell shape	Cocci	Cocci	Cocci	Cocci	Cocci	Cocci	Cocci	Cocci
Cell arrangement	Single, pairs, in threes, tetrads	Single, pairs, in threes, tetrads		Single, pairs, in threes, tetrads	Single, pairs, in threes, tetrads	nd	Single, pairs, tetrads, clumps	Pairs, clumps
Motile	+	+	+	+	+	+	+	+
Endospores	−	−	−	−	−	−		−
Catalase	+	+	+	+	+	+	+	+
Gelatinase	+	+	+	d	+	+	+	+
Arginine dihydrolase	−	+	−	−	+			−
Urease	−	−	+	−	−			
Indole production	−	−	−	−				
Methyl red test	−	−	−	−				
Voges–Proskauer reaction	−	−	−	−			−	−
Utilization of:								
Acetate		+	+		+			
L-Alanine		−			−			
Cellulose		−	+		−			
Dextrin		−	−		−			
Dulcitol		−	+		−			
D-Fructose		+	+		+			
D-Galactose		−	+		−			
Glucosamine		−	+		−			
L-Glutamine		−			−			
Glycerol		+			+			
L-Glycine		−			−			
Glycogen		−			−			
Inositol		+			−			
Lactic acid	−	−						
Lactose		+	+		−		−	
L-Lysine		−	+		+		−	
D-Maltose		−	+					
D-Mannitol		−	+					
D-Mannose		−	+					
L-Melibiose		+			+			
L-Methionine		−			−			
Pyruvate		+		d	+			
D-Raffinose		−			−			
D-Ribose		−	+		+			
D-Sorbitol		−			−			
Sucrose		−	+		+		−	
D-Trehalose		−			+		−	
L-Tyrosine		−			−			
D-Xylose		+	+		−		−	
Acid from:								
D-Cellobiose	−	−	−					
D-Fructose			−	d		+		+
D-Galactose		+	−			−		+

(continued)

TABLE 59. (Continued)

Characteristics	P. citreus[b,c,d,e,f]	P. antarcticus[g]	P. columbae[h]	P. kocurii[e,f]	P. maitriensis[i]	P. maritimus[j]	P. rifietoensis[k]	P. stackebrandtii[l]
D-Glucose	+	+	–	d	–	+		–
Lactose	–	–	–	–		–		+
Sucrose	–	+	+	–		–		+
D-Xylose	–		–	–		–		–
Sensitivity to:								
Bacitracin		S			S		S	
Chloramphenicol	S	S	S		S		S	
Chlortetracycline		S			S		S	
Lincomycin		S			S			
Nalidixic acid		S	S		S			
Rifampin		S	S		S			
Streptomycin		S	S		S		S	
Tetracycline	S	S			S		S	
Vancomycin					S		S	
Peptidoglycan	L-Lys-D-Glu	L-Lys-D-Glu	L-Lys-D-Glu	L-Lys-D-Glu	L-Lys-D-Glu	L-Lys-D-Glu		

[a]Symbols: +, positive; –, negative; PG, phosphatidylglycerol; DPG, diphosphatidylglycerol; PE, phosphatidylethanolamine; PC, phosphocholine; S, sensitive; R, resistant; d, some strains are positive; nd, not determined.
[b]Jeffries (1969).
[c]Yamada et al. (1976).
[d]Kocur (1986).
[e]Hao and Komagata (1985).
[f]Komura et al. (1975).
[g]Reddy et al. (2002).
[h]Suresh et al. (2007).
[i]Alam et al. (2003).
[j]Yoon et al. (2003).
[k]Romano et al. (2003).
[l]Mayilraj et al. (2005).

TABLE 60. Comparison of the fatty acid composition (%) of the species of the genus *Planococcus*[a]

Fatty acid	P. citreus[b]	P. antarcticus[c]	P. columbae[d]	P. kocurii[b]	P. maitriensis[e]	P. maritimus[f]	P. rifietoensis[g]	P. stackebrandtii[h]
$C_{12:0}$					0.4			
$C_{14:0\ iso}$	3.3	1	10.5	16.4	4.1	13.1		4.7
$C_{14:0\ ante}$							7.7	
$C_{14:0}$					0.9		6.75	
$C_{15:0\ iso}$		1.3	25.2	4.9	2.8	9.5	traces	2.9
$C_{15:0\ ante}$	61.7	43.2	35.1	41.6	27.3	30.6	50.7	49.8
$C_{15:0}$	3.7	14.2		11.3	2.5	3.1	7.2	5.5
$C_{15:1}$		9.8						
$C_{16:0\ iso}$	6	4	11.5	11.2	9.2	18.5	5	5.7
$C_{16:0}$	4.1	4.2	1.5		7.2	0.8	17.4	
$C_{16:0\ 2OH}$						0.8		
$C_{16:1}$	2.5	3		4.3				
$C_{16:1\ iso}$		1.2						
$C_{16:1\ \omega7c\ OH}$			4.7			8.9		8.5
$C_{16:1\ \omega11c}$						0.7		1.7
$C_{16:1\ \omega9c}$					1.6			
$C_{16:1\ \omega7c}$					3.8			
$C_{17:0\ iso}$		0.3	5.0			3.1	Traces	2.9
$C_{17:0\ ante}$	13.9	9.5	4.3	3.6	6.6	4.4	5.4	4.6
$C_{17:0}$		1	1.8		5.3	2.5		2.1
$C_{17:1\ iso} + C_{17:1\ ante}$		4.2				1.3		8.6
$C_{17:1\ \omega10c\ iso}$						0.8		1.7
$C_{18:0\ iso}$						2.1		
$C_{18:0}$		0.3		4				
$C_{18:1}$		1						
$C_{18:1\ \omega9c}$					4.2			

[a]Traces indicates < 0.3%.
[b]Englehardt et al. (2001).
[c]Reddy et al. (2002).
[d]Suresh et al. (2007).
[e]Alam et al. (2003).
[f]Yoon et al. (2003).
[g]Romano et al. (2003).
[h]Mayilraj et al. (2005).

List of species of the genus *Planococcus*

1. **Planococcus citreus** Migula 1894, 236[AL]

ci′tre.us. L. masc. adj. *citreus* lemon yellow.

The cell and colony morphology are as given for the genus. Physiological and biochemical characteristics are listed in Table 58, Table 59, and Table 60.

DNA G+C content (mol%): 48–52 (T_m, Bd).

Type strain: ATCC 14404, CCM 316, CIP 81.74, DSM 20549, JCM 2532, LMG 17319, NBRC 15849, NCIMB 1493, VKM B-1307.

GenBank accession number (16S rRNA gene): X62172.

2. **Planococcus antarcticus** Reddy, Prakash, Vairamani, Prabhakar, Matsumoto and Shivaji 2002, 1437[VP] (Effective publication: Reddy, Prakash, Vairamani, Prabhakar, Matsumoto and Shivaji 2002, 260.)

an.tarc′ti.cus. L. masc. adj. *antarcticus* pertaining to Antarctica.

The cell and colony morphology are as given for the genus. Physiological and biochemical characteristics are listed in Table 58, Table 59, and Table 60. This was the first psychrophilic species of *Planococcus* isolated, and it has significant quantities of $C_{15:1}$, $C_{16:1\ iso}$ and $C_{18:1}$ (Table 60).

DNA G+C content (mol%): 41.5 (T_m).

Type strain: CMS 26or, MTCC 3854, DSM 14505.

GenBank accession number (16S rRNA gene): AJ314745.

3. **Planococcus columbae** Suresh, Mayilraj, Bhattacharya and Chakrabarti 2007, 1269[VP].

co.lum′ba.e. L. gen. n. *columbae* of a pigeon, *Columba livia*.

The cell and colony morphology are as given for the genus. Physiological and biochemical characteristics are listed in Table 58, Table 59, and Table 60.

DNA G+C content (mol%): 50.5 (T_m).

Type strain: PgEx 11, ATCC 7251, DSM 17517.

4. **Planococcus kocurii** Hao and Komagata 1986, 573[VP] (Effective publication: Hao and Komagata 1985, 452.)

ko.cu′ri.i. N.L. masc. gen. n. *kocurii* named in honor of Miloslav Kocur, a Czechoslovakian bacteriologist.

The cell and colony morphology are as given for the genus. Physiological and biochemical characteristics are listed in Table 58, Table 59, and Table 60.

DNA G+C content (mol%): 39.6–42.9 (T_m).

Type strain: HK 701, AJ 3345, ATCC 43650, CCM 1849, DSM 20747, IAM 12847, JCM 2569, LMG 17320, NBRC 15850, NCIMB 629.

GenBank accession number (16S rRNA gene): X62173.

5. **Planococcus maitriensis** Alam, Singh, Dube, Reddy and Shivaji 2003, 307[VP] (Effective publication: Alam, Singh, Dube, Reddy and Shivaji 2003, 509.)

mai.tri.en'sis. N.L. masc. adj. *maitriensis* pertaining to the Indian station Maitri in Antarctica.

The cell and colony morphology are as given for the genus. Physiological and biochemical characteristics are listed in Table 58, Table 59, and Table 60.

DNA G+C content (mol%): 39 (T_m).
Type strain: S1, MTCC 4827, DSM 15305.
GenBank accession number (16S rRNA gene): AJ544622.

6. **Planococcus maritimus** Yoon, Weiss, Kang, Oh and Park 2003, 2016[VP]

ma.ri'ti.mus. L. masc. adj. *maritimus* living near the sea.

The cell and colony morphology are as given for the genus. Physiological and biochemical characteristics are listed in Table 58, Table 59, and Table 60. This species is tolerant to 17% NaCl.

DNA G+C content (mol%): 48 (HPLC).
Type strain: TF-9, JCM 11543, KCCM 41587.
GenBank accession number (16S rRNA gene): AF50,0007.

7. **Planococcus rifietoensis** Romano, Giordano, Lama, Nicolaus and Gambacorta 2003, 1701[VP] (Effective publication: corrig. Romano, Giordano, Lama, Nicolaus and Gambacorta 2003, 364.)

ri.fie.to.en'sis. N.L. masc. adj. *rifietoensis* pertaining to Rifieto Spring in Italy.

The cell and colony morphology are as given for the genus. Physiological and biochemical characteristics are listed in Table 58, Table 59, and Table 60.

DNA G+C content (mol%): 47.9 (HPLC).
Type strain: M8, ATCC BAA-790, DSM 15069.
GenBank accession number (16S rRNA gene): AJ493659.

8. **Planococcus stackebrandtii** Mayilraj, Prasad, Suresh, Saini, Shivaji and Chakrabarti 2005, 93[VP]

sta.cke.brand.ti'i. N.L. gen. n. *stackebrandtii* of Stackebrandt, to honor Erko Stackebrandt, a German microbiologist.

The cell and colony morphology are as given for the genus. Physiological and biochemical characteristics are listed in Table 58, Table 59, and Table 60.

DNA G+C content (mol%): 40 (T_m).
Type strain: K22-03, DSM 16419, JCM 12481, MTCC 6226.
GenBank accession number (16S rRNA gene): AY437845.

Acknowledgements

S. Shivaji thanks Dr G. S. N. Reddy for his inputs at the proof level.

Genus II. **Caryophanon** Peshkoff 1939, 244[AL]

DAGMAR FRITZE AND DIETER CLAUS

Ca.ry.o'pha.non. Gr. n. *karyon* nut, kernel, nucleus; Gr. adj. *phaneros* bright, conspicuous; N.L. neut. n. *Caryophanon* that which has a conspicuous nucleus.

Slightly curved to straight **multicellular rods** (trichomes) with rounded or slightly tapered ends. Cell diameter is 1.0–3.5 μm and length is 4–20 μm or more. Several trichomes may form short chains. Asporogenous, nonbranching. Gram-positive. Motile by means of peritrichous flagella. After isolation in pure culture, the typical cell morphology may be preserved only in liquid media containing cow dung. Strictly aerobic. Chemoorganotrophic with presumed respiratory metabolism. Carbohydrates are not used as substrates. Acetate and other organic acids are the only major carbon sources used. Biotin is required and thiamine is stimulatory. Catalase-positive; cytochrome oxidase-negative; indole not produced. Found **associated with cattle dung. Not** known to be **pathogenic**.

DNA G+C content (mol%): 41–46 (T_m).

Type species: **Caryophanon latum** Peshkoff 1939, 244[AL].

Further descriptive information

The original false interpretation of cross-walls, nuclear material and cytoplasm in stained trichomes led to the name *Caryophanon* (Peshkoff, 1939). The true nature of the trichomes was shown by Pringsheim and Robinow (1947) by improved cytological techniques. Studies on both species have been reviewed by Trentini (1978). Since then only a few studies have been published in which *Caryophanon* strains have been used or at least cited (only 11 retrieved in search). A table of diagnostic characters is given in Table 61 and physiological properties are given in Table 62.

The close phylogenetic relationship between *Caryophanon* and members of the genus *Bacillus* and related genera, in particular

TABLE 61. Diagnostic table for *Caryophanon* species

Characteristic	C. latum		C. tenue
Trichome width (μm)	2.3–3.5		1.0–2.0
Trichome length (μm)	6.0–20.0		4.0–10.0
Number of cells per trichome	4–15		2–3
rRNA gene sequence similarity		96.2–92.5%	
DNA–DNA hybridization		13–30%	
Mol% G + C	44.0–45.6		41.2–41.6
Genome size (×10⁶ Da)	1100–1200		900–1000
Cell wall	Lys-D-Glu		nt

nt, Not tested.

TABLE 62. Physiological properties of *Caryophanon*[a]

Characteristic	*C. latum*[b]	*C. tenue*[b]	*C. latum* DSM 14151[Tc]	*C. tenue* DSM 14152[Tc]
Catalase	+	+	nt	nt
Oxidase	–	–	nt	nt
Motility	+	+	–[d]	–[d]
Growth temperature, °C:				
5	na	na	–	–
10	+	na	+	+
15	na	na	+	+
35	na	na	+	+
37	+	na	+	+
40	na	na	+	+
45	–	na	–	–
Anaerobic growth	–	–		
Anaerobic growth (BBL)	na	na	nt	nt
Anaerobic growth (CASO)	na	na	–	–
Growth in the presence of:			–	–
5% NaCl	na	na	–	–
7% NaCl	na	na	–	–
10% NaCl	na	na	–	–
VP test	na	na	–	–
pH in VP	na	na	7.9–8.1	7.7–7.8
Growth at pH 5.7 (Sabouraud)	na	na	–	–
Resistance to lysozyme	na	na	–	–
Acid from:				
Glucose	–	–	–	–
Arabinose	'–'	'–'	–	–
Xylose	'–'	'–'	–	–
Mannitol	'–'	'–'	–	–
Hydrolysis of:				
Starch	–	–		–
Casein	–	–	–	–
Gelatin	–	–	ng	ng
Cellulose	–	–	+	+
Tributyrin	w	na	nt	nt
Tween 80	na	na	nt	nt
Tween 60	na	na	–	+
Tween 40	na	na	–	+
Tween 20	na	na	–	+
Tyrosine	na	na	–	+
Hippurate	na	na	–	–
Urea	–	na	–	–
Uric acid	–	na	–	–
Lecithin	na	na	nt	nt
Utilization of:			ng	ng
Citrate	na	na	+	+
Propionate	na	na	+	+
Acetate	+	+	nt	nt
Butyrate	+	+	nt	nt
Valerate	+	+	nt	nt
Capronate	+	+	nt	nt
Stearate	+	+	nt	nt
Methylpropionate	+	+	nt	nt
2-Methylbutyrate	+	+	nt	nt
Poly-β-hydroxybutyrate	+	na	nt	nt
Deamination of phenylalanine	na	na	–	–
Reduction of NO_3 to NO_2	–	na	–	–
Production of indole	–	–	–	–

[a]+, Positive; –, negative; na, data not available in literature; nt, not tested; ng, no growth; w, very weak; '–': literature states 'and no other sugars.'
[b]Literature data available for species only.
[c]Own data; physiological tests were performed according to Gordon et al. (1973); Gordon's J-medium with glucose replaced by propionate was used as the basic medium; urease was tested using urease test broth (BBL 11797); motility was tested on soft agar (0.1% yeast extract, 0.01% K_2HPO_4, 0.2% agar); Tween-80 was tested according to Cowan and Steel (1974). Anaerobic growth was tested in anaerobic agar (BBL 210926) and casein-peptone soymeal-peptone glucose agar (Merck CASO Broth 105459 plus 1.5% agar).
[d]Motility lost on culture media.

the round-spore-forming *Bacillus* species, was described by Stackebrandt et al. (1987). Placement of both *Caryophanon* species within the Gram-positive aerobic endospore-forming bacteria, in particular within rRNA group 2 (Ash et al., 1991), was confirmed by Farrow et al. (1994a) and by our own re-analysis. When comparing sequences of the type strain of *Caryophanon tenue* as determined by Farrow et al. (1994a) and by our own analysis, a similarity of 99.6% was determined. However, a comparison of sequences determined for the type strain of *Caryophanon latum* in both studies revealed that sequence similarities differed by more than 3% Figure 59).

Caryophanon grows in the form of trichomes. To correctly interpret the term trichome, the applicable definition of Starr and Skerman (1965) is given here: "the designation trichome, frequently misused, is intended to mean the assemblage of cells in uniseriately multicellular bacteria, irrespective of their mode of movement, in which adjacent cells have a relatively large area of close contact". Endospores, sheaths or capsules are not formed. However, Trentini and Gilleland (1974) reported that one or two superficial wall layers containing protein are present in cells of *Caryophanon latum*. Mesosomes, nucleidosomes, and analogs of mitochondria have been described by Shadrina et al. (1982).

Under optimal growth conditions, individual cells of *Caryophanon latum* exhibit a disk-like shape within a trichome. Cells are larger in width than in length, showing cross wall formation at various stages of closure. In *Caryophanon tenue*, cells within trichomes are slightly larger in length than in width. They lack the multiple septation typical of *Caryophanon latum* and show mostly only one cross septum in a trichome (Peshkov and Marek, 1972).

The size of trichomes of *Caryophanon latum* in enrichment cultures varies between 2.3 × 6.0 and 3.5 × 20 μm. The cell number within trichomes ranges from 4 in the shorter to 15 in the longer ones. The dimensions of *Caryophanon tenue* trichomes range between 1.0 × 4.0 and 2.0 × 10.0 μm and numbers of cells within trichomes usually range from 2 to 3.

On artificial media, the morphology of trichomes may differ greatly from that in enrichment cultures. Cells may be deformed, thinner, longer, or crooked. In older cultures, strains of *Caryophanon latum* often form small, spheroidal forms. It has been postulated that these are part of a normal life cycle. According to Kele (1970), these are degenerate structures. Changes in the cell morphology of both *Caryophanon* species during longer cultivation on agar media have been described by Peshkov et al. (1978).

Various solid media adjusted to pH 7.5–8.5 have been described for growing strains of *Caryophanon*. Clarified manure extract agar[1] (Peshkov, 1967), cow dung agar[2] (Smith and Trentini, 1972), peptone-yeast extract-acetate agar (Pringsheim and Robinow, 1947), cow dung agar with lactalbumin hydrolysate (Moran and Witter, 1976), or a semisynthetic medium[3] (Smith and Trentini, 1973; Trentini, 1978) support good growth of most strains of *Caryophanon latum* and *Caryophanon tenue*. However, with most of these media, the true enrichment morphology of most strains of *Caryophanon latum* cannot be maintained. For both species, biotin has been found to be essential for growth, whereas thiamine seems to be stimulatory.

Growth in liquid media like dung extract generally is scant. A defined liquid medium for *Caryophanon latum* has been developed by Kele and McCoy (1971). They noted that after some subcultures, nonmotile laboratory strains became actively motile and developed normal morphology. However, this medium did not support high density populations and some of the strains tested failed to grow.

Because growth in liquid media is scant, detailed studies of the effects of temperature and pH on growth of *Caryophanon latum* have been performed only with agar cultures (Moran and Witter, 1976). When measuring colony growth rate, an optimum for these parameters was found at pH 8.0 and 35°C. Detailed data for minimum and maximum growth temperatures for both *Caryophanon* species cannot be found in literature. With strains of *Caryophanon latum*, growth occurs at 10 and 37°C, but not at 45°C and at pH values between 6 and 8, with optimum growth at pH 7.5–8.0 (Weeks and Kelley, 1958). Data on the influence of temperature or pH on growth of *Caryophanon tenue* are not available.

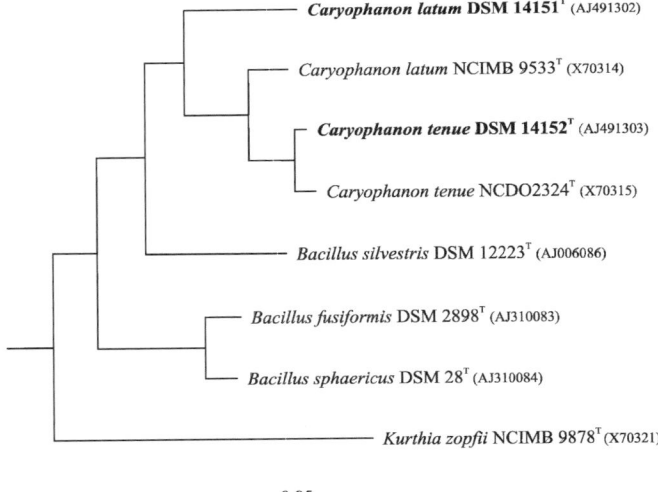

Caryophanon latum DSM 14151[T] (AJ491302)

Caryophanon latum NCIMB 9533[T] (X70314)

Caryophanon tenue DSM 14152[T] (AJ491303)

Caryophanon tenue NCDO2324[T] (X70315)

Bacillus silvestris DSM 12223[T] (AJ006086)

Bacillus fusiformis DSM 2898[T] (AJ310083)

Bacillus sphaericus DSM 28[T] (AJ310084)

Kurthia zopfii NCIMB 9878[T] (X70321)

0.05

FIGURE 59. Neighbor-joining tree showing the 16S rRNA gene sequence similarity of the type strains of *Caryophanon latum* and *Caryophanon tenue* (X70314 and X70315, deposited by Farrow et al., 1994a; AJ491302 and AJ491303, own deposits) in comparison with closest relatives belonging to the *Bacillus* rRNA group 2. The sequence of *Kurthia zopfii* was used to root the dendrogram. Bar = 0.05 sequence divergence.

[1] Clarified manure extract agar (Peshkov, 1967; modified). Clarified cow manure extract, 250.0 ml; peptone from casein, 5.0 g; sodium acetate trihydrate, 2.3 g; distilled water, 750.0 ml; agar, 15.0 g. Adjust to pH 7.8–8.0 and sterilize the medium at 121°C for 15 min. To prepare the manure extract, 250 g fresh cow dung is mixed with 750 ml distilled water, sterilized for 15 min at 120°C, and filtered through several layers of gauze. The filtrate is centrifuged at 3300 g for 15 min and the supernatant is frozen at 20°C. Due to the freezing step, colloidal particles form larger aggregates which, after thawing, can be removed by centrifugation at 3300 g. The clarified extract is stored frozen.

[2] Cow dung agar (Smith and Trentini, 1972). Fresh cow dung, 250 g; distilled water, 750 ml; agar, 15 g. Before agar is added, dung and water are mixed for a few minutes in a blender to break up large dung particles and to ensure thorough dispersion. Sterilize at 121°C for 15 min.

[3] Semisynthetic medium for *Caryophanon latum* (Trentini, 1978). Lactalbumin hydrolysate, 10.0 g; sodium acetate, 5.0 g; $MgCl_2$·$6H_2O$, 20.3 mg; $CaCl_2$, 11.1 mg; $Cu_2(SO_4)_3$, 0.4 mg; $FeSO_4$, 0.152 mg; biotin, 0.02 mg; thiamine–HCl, 0.05 mg; nitrilotriacetic acid, 19.1 mg; double-distilled water, 1000.0 ml. Adjust the medium with anhydrous Na_2CO_3 to pH 7.5 before autoclaving. The pH of the medium should be around 8.0 after sterilization.

In cell walls of *Caryophanon latum*, the peptidoglycan type is A4α (Lys-D-Glu) (Stackebrandt et al., 1987). For *Caryophanon tenue*, no data on cell wall composition are available. Cell walls are sensitive to lysozyme: septal peptidoglycan is split very quickly, whereas wall peptidoglycan is more resistant (Trentini and Murray, 1975).

Menaquinones with six isoprene units (MK-6) are the major isoprenologues in *Caryophanon latum* and *Caryophanon tenue* (Collins and Jones, 1981).

The fatty acid composition data for strains of *Caryophanon latum* and *Caryophanon tenue* are shown in Table 63.

Studies on the fine structure of *Caryophanon latum* and *Caryophanon tenue* have been published by Peshkov and Marek (1972).

Caryophanon latum and *Caryophanon tenue* are strictly aerobic species. According to Provost and Doetsch (1962), *Caryophanon latum* forms a yellowish, nondiffusible pigment on several media that shows a strong absorption peak at 430 nm and which is presumed to be a carotenoid. Pigment formation is inhibited by diphenylamine. In older cultures and in cultures with unbalanced nutrition, *Caryophanon latum* synthesizes large quantities of poly-β-hydroxybutyric acid from butyrate and acetate (Provost and Doetsch, 1962; Shekhovtsov and Zharikova, 1978).

According to Provost and Doetsch (1962), acetate has been found to be the major carbon source for *Caryophanon latum*, followed by butyrate. With the exception of β-hydroxybutyrate, numerous other fatty acids, tricarboxylic acids, sugars, sugar alcohols, amino acids, and aliphatic alcohols cannot be used as substrates. However, Rowenhagen (1987) has shown that a number of additional fatty acids (e.g., propionate, valerate, capronate, and stearate) are oxidized by several strains of both *Caryophanon* species. Strains of *Caryophanon tenue* preferred valerate and capronate to acetate. It seems that each strain is specialized to one or two fatty acids as a result of a competitive pressure of other *Caryophanon* strains in the same habitat.

The DNA base composition of 36 strains of *Caryophanon latum* collected from Canada, England, Germany, Scotland, USA, and the former USSR formed a tight group with G + C contents of 44.0–45.6 mol% (T_m), whereas three isolates of *Caryophanon tenue* clustered with 41.2–41.6 mol% G + C (T_m). DNA–DNA hybridization studies revealed intragroup relatedness values of 78–94% (*Caryophanon latum*) and 82–94% (*Caryophanon tenue*). *Caryophanon latum* and *Caryophanon tenue* showed intergroup relatedness values of only 13–30% (Adcock et al., 1976). The genome size of *Caryophanon tenue* is 900–1000×10⁶ Da.

Phages have been isolated for both *Caryophanon* species. They were specific for *Caryophanon latum* or for *Caryophanon tenue* or were active against both species (Trentini, 1978).

According to Provost and Doetsch (1962) and W.C. Trentini (unpublished results), *Caryophanon latum* and *Caryophanon tenue* show a similar antibiotic sensitivity pattern to 45 antibiotics. Both species are resistant to streptomycin, polymyxin B, nalidixic acid, and several sulfa drugs. To isolate *Caryophanon latum*, cow dung agar containing 80 µg/ml streptomycin sulfate can be used.

Caryophanon latum and *Caryophanon tenue* are not known to be pathogenic.

Caryophanon latum and *Caryophanon tenue* were both isolated from fresh cattle dung. In most successive studies, *Caryophanon latum* was found only in cattle manure or cattle manure-contaminated materials like bedding straw, barn dust, or barnyard soil. It was absent from freshly voided droppings and old, dried samples. Reports that *Caryophanon latum* has been isolated from other sources could not be verified. *Caryophanon latum* has not been found as part of the natural flora of the intestinal tract of cattle (Trentini, 1978) and seems to be a natural, specific, and temporary coprophilic resident of cattle dung. It seems to be dispersed to new droppings by contaminated air and by flying insects (Trentini, 1978; Trentini and Machen, 1973). Whether the habitat of *Caryophanon tenue* is also restricted like that of *Caryophanon latum* is not known.

Drozd et al. (1987) described a case of concrete corrosion caused by a *Caryophanon* species.

TABLE 63. Fatty acid composition (%) of *Caryophanon* species[a]

Fatty acid	C. latum DSM 14151ᵀ=NCIMB 9533ᵀ	C. latum DSM 14843ᵀ=NCIMB 9533ᵀ	C. latum DSM 14837=NCIMB 702034=NCDO 2034	C. latum DSM 14844=NCIMB 9534	C. latum DSM 484	C. tenue DSM 14152ᵀ
iso-14:0	4.28	4.93	15.56	10.30	8.25	5.35
14:0	2.46	3.30	2.02	2.26	1.13	1.19
iso-15:0	39.44	35.68	20.52	27.70	21.86	28.81
anteiso-15:0	2.98	2.94	1.25	2.26	1.65	4.67
15:0	2.14	2.25	2.81	2.32	3.55	–
16:1ω7c alcohol	8.68	8.75	21.88	19.23	16.98	12.51
iso-16:0	1.74	2.11	13.09	8.04	12.72	10.59
16:1ω11c	23.28	25.72	15.05	16.88	13.74	18.50
16:0	4.01	4.25	4.05	5.26	5.61	5.52
iso-17:1ω10c	3.67	3.36	0.91	1.16	1.80	3.48
iso-17:0	1.53	1.21	1.61	1.35	3.38	4.44
anteiso-17:0	1.37	0.85	–	–	1.47	2.46
18:1ω9c	–	–	1.26	2.10	2.32	–

[a]*Note:* Analysis through the MIDI system; as *Caryophanon* does not grow on the standard medium used for culturing organisms for fatty acid analysis, medium no. 34 from the DSMZ catalogue designed for *Caryophanon* was used; data are therefore not directly comparable with other data; values lower than 0.36% are not recorded by the system; values for anteiso-11:0, iso-13:0, iso-15:1 at 5, iso-18:0, 18:0 and 20:4ω6,9,12,15c are lower than 1% or not detectable; values for 17:0 are 1.68% for DSM 484, but are lower than 1% or not detectable for all other strains; data produced at the DSMZ.

Enrichment and isolation procedures

Caryophanon strains can be isolated only after an enrichment step. According to Pringsheim and Robinow (1947) and Weeks and Kelley (1958), samples of fresh cow dung are collected, placed in closed bottles, and kept at room temperature. After 1 to 2 d, when *Caryophanon* has multiplied in the sample, a loopful of dung is suspended in a few drops of tap water. Samples that are microscopically rich in *Caryophanon latum* are streaked on cow dung agar or peptone-yeast extract-acetate agar. Small sized granular colonies of the large bacterium, among those of other organisms, are visible after about 24 h. With the aid of a dissecting microscope colonies can be picked with a capillary pipette or needle and checked microscopically for the presence of *Caryophanon*. Pure cultures are obtained by repeated re-streaking on agar plates. Provost and Doetsch (1962) isolated *Caryophanon latum* by the following method. Fresh cow dung samples are incubated at room temperature in glass bottles and covered with aluminum foil. After 1 d, wet mount preparations are examined microscopically for typical trichomes. Part of a positive sample is sterilized at 121°C for 15 min and is then inoculated with several loops of material taken from the remaining portion. After incubation for 24 h at room temperature, the inoculated dung sample usually shows high numbers of trichomes. About 2 g enriched sample is suspended in 10 ml water. The suspension is forced through several layers of gauze to remove large particles. The filtrate is centrifuged at 50 g for 8 min and the sediment is resuspended in water. This process is repeated 10 times. The resulting sediment is streaked on plates of peptone-yeast extract-acetate agar. After about 24 h at room temperature, the plates are examined for minute, granular colonies using a dissecting microscope. Material from colonies is checked microscopically for the presence of *Caryophanon*. Pure cultures are obtained by repeated streaking.

Smith and Trentini (1972) used cow dung-streptomycin agar for the effective isolation of *Caryophanon latum*. Their isolation method is based on differential low-speed centrifugation followed by filtration and plating on cow dung agar containing 80 µg/ml filter-sterilized streptomycin sulfate. Positive colonies are purified by re-streaking on cow dung agar without streptomycin.

Whereas *Caryophanon latum* can be detected easily because of its typical morphology (Figure 60), the presence of *Caryophanon tenue* in cow dung or enrichment samples is more difficult to detect. Often, its trichomes are only slightly bigger in width and length than other rod-shaped bacteria developing in the enrichment cultures. To detect *Caryophanon tenue*, it is best to prepare samples on slides covered with 1.0 ml 5 % or 10 % (w/v) gelatin and dried at about 50°C to enhance the contrast of cross septa and to look for trichomes with only one or two cross septa per cell (Figure 61). Methods for the specific isolation of *Caryophanon tenue* have not been described up to now. Cultures have been isolated apparently by random.

Maintenance procedures

Caryophanon cultures can be maintained on clarified manure extract agar[1] or cow dung agar[2]. After incubation at about 25°C for 48 h, *Caryophanon* strains survive storage at 4°C for

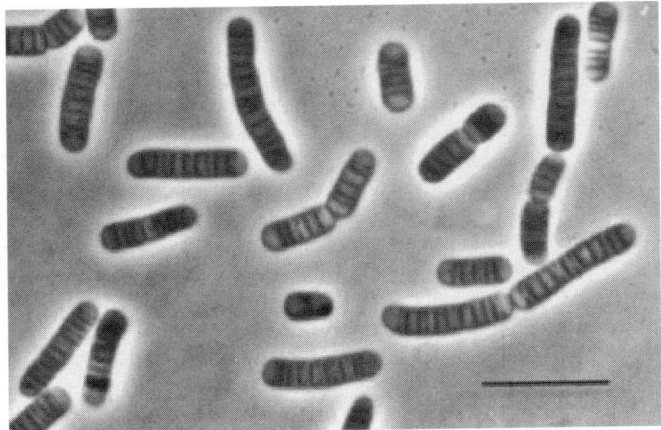

FIGURE 60. *Caryophanon latum* strain Lenglern; clarified manure extract agar; 30°C; microscopic slides covered with 0.5 ml of 5% (w/v) gelatin and dried at 50°C; bar = 10 µm.

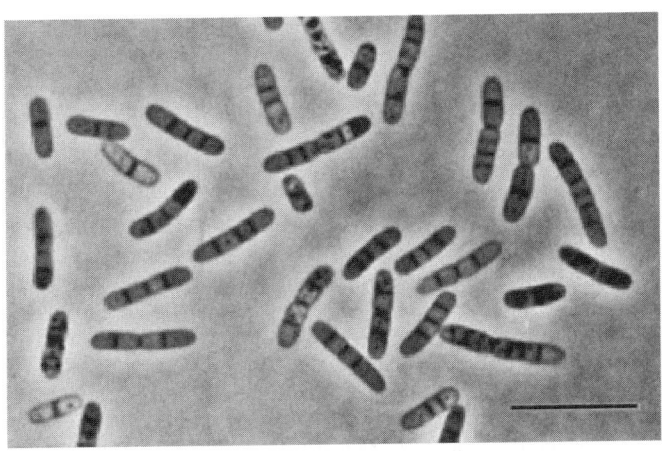

FIGURE 61. *Caryophanon tenue* strain Uelzen; clarified manure extract agar; 30°C; microscopic slides covered with 0.5 ml of 5% (w/v) gelatin and dried at 50°C; bar = 10 µm.

at least 4 weeks. For long-term storage, isolates can be readily freeze-dried or kept in or above liquid nitrogen by standard methods.

In enrichment cultures, the numbers of *Caryophanon latum* rapidly fall to zero after about 2 d at room temperature. However, actively moving trichomes can usually be maintained for some weeks by keeping the enrichment cultures (cow dung mixed with some water) at about 4°C after incubation for 1 to 2 d at room temperature.

Differentiation from closely related taxa

The two species of the genus are still today mainly defined morphologically; thus, differentiation from other bacterial genera present in cow manure is possible through their trichomous appearance. Differentiation from other trichome-forming bacteria is by the presence of flagella (Figure 62) and the absence of gliding movement.

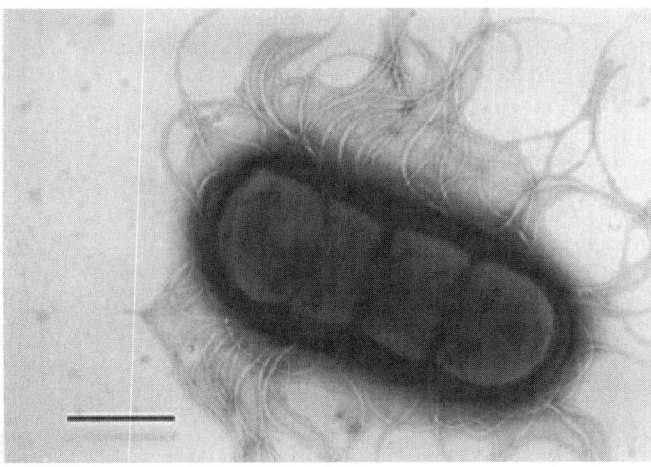

FIGURE 62. *Caryophanon tenue* strain Uelzen; clarified manure extract agar; 30°C; negative staining, uranyl acetate; bar = 10 μm.

Taxonomic comments

Whereas the data presented for *Caryophanon latum* are mostly based on several strains, those for *Caryophanon tenue* are generally based on a single isolate.

Acknowledgements

Dr R.M. Kroppenstedt (DSMZ) is thanked for doing FAME analysis, Dr M. Madkour (University of Göttingen, Institute for Microbiology and Genetics) is thanked for taking electron micrographs, J. Swiderski (DSMZ) is thanked for constructing the 16S rRNA gene sequence dendrogram, and C. Wahrenburg (DSMZ) is thanked for doing the physiological tests.

Further reading

Claus, D., D. Fritze and M. Kocur. 1992. Genera related to the genus *Bacillus* – *Sporolactobacillus*, *Sporosarcina*, *Planococcus*, *Filibacter*, and *Caryophanon*. *In* Balows, Trüper, Dworkin, Harder and Schleifer (Editors), The Prokaryotes, 2nd Ed., Vol. II, Springer-Verlag, New York, pp. 1769–1791.

List of species of the genus *Caryophanon*

1. **Caryophanon latum**   Peshkoff 1939, 244[AL]

   la′tum. L. neut. adj. *latum* broad.

   DNA G+C content (mol%): 44.0–45.6 (T_m).
   Type strain: NCIMB 9533, ATCC 33407, DSM 14151, LMG 17312, KCTC 3403, VKM B-105.
   GenBank accession number (16S rRNA gene): X70314 (Farrow et al., 1994a), AJ491302.

2. **Caryophanon tenue**   (*ex* Peshkoff 1939) Trentini 1988, 220[VP] (Effective publication: Trentini 1986, 1259.)

   te′nu.e. L. neut. adj. *tenue* slender.

   DNA G+C content (mol%): 41.2–41.6 (T_m).
   Type strain: NCIMB 9535, ATCC 33098, DSM 14152, LMG 17313, KCTC 3404, VKM B-106.
   GenBank accession number (16S rRNA gene): X70315 (Farrow et al., 1994a), AJ491303.

Genus III. **Filibacter** Maiden and Jones 1985, 375[VP] (Effective publication: Maiden and Jones 1984, 2957.)

Erko Stackebrandt

Fi.li.bac′ter. L. n. *filum* a thread; N.L. masc. n. *bacter* masculine form of Gr. neut. n. *bactron* a rod; N.L. masc. n. *Filibacter* thread rod.

Filaments composed of cylindrical cells, straight or curved. Filaments are neither sheathed nor branched. Gram stain negative. Cell junctions clearly marked by constrictions. **Filaments are flexible,** but do not show active flexing. **Motile by gliding.** Resting stages not observed. Nonspore-forming. Strictly aerobic, catalase and oxidase-positive. The mol% of the DNA is 44. **Peptidoglycan contains lysine. Murein type L-lysine-D-glutamic acid, variation A4. Major respiratory lipoquinone is menaquinone MK-7. Membrane fatty acids are dominated by** $C_{15:0\ anteiso}$ **and** $C_{17:0\ anteiso}$. **Cytochromes are predominantly of the** *c* type while *b*-type cytochromes occur in lesser amounts. Phylogenetically, a member of *Bacillus* RNA group 2, subphylum *Clostridium–Bacillus*, phylum of Gram-positive bacteria. *Bacillus globisporus*, *Bacillus pasteuri*, and *Bacillus psychrophilus* are close phylogenetic neighbors.

Type species: **Filibacter limicola** Maiden and Jones 1985, 375[VP] (Effective publication: Maiden and Jones 1984, 2957).

Further descriptive information

As the genus currently contains a single species only, the description of the genus is that of the species. Only the type strain 1SS101[T] (NCIMB 11923[T], DSM 13886[T]) has been analyzed in detail. Morphological, cultural, and most of the physiological data were take from Maiden and Jones (1984).

Filaments of up to 150 μm in length are multicellular, the cells being 3–30 μm in length. Within longer cells, septa without constrictions can be seen (Maiden and Jones, 1984). The study by Clausen et al. (1985) pointed out that Gram-negative staining behavior does not necessarily indicate the presence of the typical Gram-negative wall morphology. Electron micrographs of thin sections of glutaraldehyde fixed cells of *Filibacter limicola* showed a five-layered cell wall, including an electron-dense peptidoglycan layer of approximately 12.5 nm thickness and a surface (S) layer. An outer membrane, which is typical of Gram-negative bacteria, was absent. The cell wall of *Filibacter limicola* is about

40 nm thick, and thus within the range of 20–80 nm typical for a Gram-positive cell wall (Glauert and Thornley, 1969). Neither sheaths nor capsules were observed. Electron micrographs of unfixed cells and filaments indicated the presence of fibrillar material which often appeared as a capsule-like structure around the cells. Safranin stain of *Filibacter limicola* cells was even. Upon staining with Sudan Black one particle per cell was stained. Poly-hydroxybutyrate and volutin tests were negative (Maiden and Jones, 1984) and no intracellular sulfur globules were deposited when cells were grown in the presence of sulfide.

The growth habits of *Filibacter limicola* on solid media are similar to those of *Vitreoscilla stercoraria*, in that cells spread widely over the surface of agar producing whorls of growth and spiral colonies. This distinctive growth was interpreted as gliding (Maiden and Jones, 1984). Some filaments were reported to move slowly (5 μm/min). In wet mounts, the strain was usually immotile, but often showed vibrating movements, unlike Brownian motion. As reported by Maiden and Jones (1984), occasionally cells 10–20 μm long were seen to rotate slowly through 360° with one end appearing to be attached to the glass. Under these conditions, only rarely, was gliding motility observed.

Filibacter limicola is strictly aerobic, growing well at the surface (3–4 mm) in shake cultures. In tubes with an agar plug cells grew faintly at the surfaced of semi-solid agar. The organism is catalase and oxidase-positive. The cytochrome spectrum of *Filibacter limicola* showed a Soret peak at 409 nm in the oxidized state, which increased and was shifted to 417 nm when reduced. The reduced spectrum also showed the characteristic cytochrome *c* beta (522 nm) and alpha peaks (552 nm). Small shoulders in the alpha and beta peaks occurring in reduced minus oxidized difference spectra of the alpha and beta peaks indicated the presence of minor amounts of *b*-type cytochromes.

Filibacter limicola grows well on Tryptone soy broth (TSB) (Oxoid CM129) with a generation time of 6.6 h. It also grows with Casamino acids or the complete amino acid mixture in the presence of vitamins (Maiden and Jones, 1984). Good growth also occurred on Anaerobe Agar (BBL, Becton–Dickinson) and in CASO (casein-peptone soymeal-peptone; Merck 5458), supplemented with inosine. No growth occurred in the absence of vitamins. One of the most salient metabolic properties of this organism is its inability to utilize organic compounds in the absence of amino acids. These results were confirmed in the DSMZ laboratories. Many of the organic compounds tested in the presence of amino acids were inhibitory. The only compounds which enhanced yields were acetate, butyrate, and glycerol at 2 mg carbon atom/l with amino acids at 36.3 carbon atom/l. Stimulation of growth decreased at higher substrate concentrations and lower amino acid concentrations.

Of the amino acids that supported growth, combinations of those of the following families were found stimulating: the glutamate family (glutamate, glutamine arginine, proline) plus aspartate (aspartate, asparagines lysine, threonine isoleucine, methionine) or serine (serine, glycine, cysteine) families, and the aspartate family plus the pyruvate (alanine, valine, leucine), serine (serine, glycine, cysteine) or aromatic (tryptophan, phenylalanine, tyrosine) families. The requirements for and interdependence of particular combinations of amino acids from different families may be due to coupled oxidation-reduction reactions such as in the Stickland reaction in anaerobes. No single amino acid or mixture of amino acids from a single biosynthetic family supported growth. Histidine and members of the aromatic family (in most cases) were reported inhibitory. More detailed information on the mutual influence of amino acid families on support and inhibition of growth has been given by Jones and Maiden (1984).

Physiological reactions of strain DSM 13886^T towards the substrate panel provided by the API 20 NE (API BioMérieux, Marcy l'Etoile, France) confirmed the positive urease reaction and gelatin hydrolysis described by Maiden and Jones (1984); in addition, the strain was found to reduce nitrate and to cleave *p*-nitro-phenyl-β-D-galactopyranoside. None of the other substrates of the API 20 NE panel, none of carbohydrates (sugars and acids) provided by the API 50 CHE panel, as well as those carbohydrates used for the differentiation of *Bacillus* species (Gordon et al., 1973), supported growth within 8 days.

Filibacter limicola has only been isolated once, from sediments of Blelham Tarn, a eutrophic freshwater lake in the English Lake District (Maiden, 1983). This lake was selected for enrichment studies because here the greatest numbers and variety of filamentous bacteria had been observed (Godinho-Orlandi and Jones, 1981). The nutrition with respect to amino acids as substrates for carbon sources resembles that of the gliding organism *Vitreoscilla stercoraria*, a strain of which has also been isolated from the same source. The ecological role has been discussed by Maiden and Jones (1984). As compared to the amino acid concentration in surface waters, that of sediments is 10^3 times higher [35 μg/l ((Sepers, 1981)] and 35 mg/l glycine equivalents ((Simon and Jones, 1982), respectively], and the selective use of amino acids may reflect their relative abundance in the sediment. *Filibacter limicola* shows a positive urease reaction; urea may originate as a product of decomposing phytoplankton (Satoh, 1980).

Enrichment and isolation procedures

For isolation, sediment cores are sampled and treated according to Maiden and Jones (1984). Sediment was sampled using a Jenkins surface mud-sampler (Ohnstadt and Jones, 1982). The top 1–2 cm layer of a deep-water sample (13–14 m in depth) was diluted with filtered lake water, mixed vigorously with a vortex mixer, and used for the inoculation of isolation media. The highest number of colonies containing filamentous bacteria was observed on MYP plates at 10°C. MYP medium has the following composition (mg/l): yeast extract (Difco), 10; peptone (Oxoid L37), 100; K_2HPO_4, 28; $MgCl_2 \cdot 6H_2O$, 127; KNO_3, 4; $(NH_4)_2SO_4$, 60; $MnSO_4 \cdot 4H_2O$, 8; Ferric citrate, 6. The pH of the medium was adjusted with 1 M $KHCO_3$ before autoclaving and the final pH was 7.0–7.2. Filaments developed after 14 d incubation. Good growth, but less abundant growth, also occurred on CASO agar and Tryptone soy agar at 20°C. During incubation, plates were examined by eye, and through a dissecting microscope at a magnification of 32x. Small agar blocks containing filaments occurring away from the inoculation sample were excised, transferred to fresh agar, and subcultured until pure cultures were obtained.

Maintenance procedures

Cultures may be grown at 20°C on the isolation media, indicated under Enrichment (MYP, Oxoid CM129, Oxoid CM 131).

Cultures, grown at 20°C, can be maintained at 10°C for some weeks. Long-term preservation can be done by lyophilization or storage in liquid nitrogen.

Special characters

Because of the morphological similarity between *Filibacter limicola* and *Vitreoscilla stercoraria*, affiliation of new *Filibacter* isolates should include 16S rDNA sequence analysis and/or determination of chemotaxonomic properties, such as the presence of the L-lysine-D-glutamic acid peptidoglycan type (Schleifer and Kandler, 1972) and the isoprenoid type MK-7 (Collins, 1994). Verification of isolates which are genomically highly related to the type strain ISS101 (DSM 13886[T], NCIMB 11923[T]) should be done by RiboPrint™ analysis. The pattern of strain DSM 13886 (Figure 63) is distinct and clearly different from those of *Bacillus* species, shown to be phylogenetic neighbors by 16S rDNA analysis, e.g., *Bacillus pasteurii*, *Bacillus globisporus*, and *Bacillus psychrophilus*.

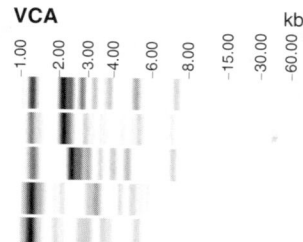

VCA   kb

Sporosarcina pasteurii DSM 33[T]

Sporosarcina ureae DSM 2281[T]

Filibacter limicola DSM 13886

Sporosarcina psychrophila DSM 3[T]

Sporosarcina globispora DSM 4[T]

FIGURE 63. Riboprint™ patterns of the DNA of *Filibacter limicola* DSM and the type strains of *Sporosarcina urea* and related *Bacillus* species of RNA group 2. The pattern was generated by the Riboprint™ robot (Qualicon, Wilmington, DE), in which DNA was cleaved by *Eco*RI, the resulting fragments separated on a membrane, and fragments of the *rrn* operons were visualized by hybridization with a fluorescent *rrn* operon probe. The scale bar on top of the fragments indicates the length (in kb) of the fluorescence-positive fragments. Fresh material (agar colonies) was analyzed except for *Sporosarcina pasteurii* DSM33[T], for which frozen material was used.

Differentiation of *Filibacter* from other taxa

Filibacter limicola resembles *Vitreoscilla stercoraria* in superficial morphological properties, such as the formation of Gram-negative, unpigmented, multicellular, and gliding filaments (see *Special characters* for differentiation of *Filibacter* from other gliding bacteria). The genomic and chemotaxonomic characters of *Filibacter* give evidence for a close relationship with members of the *Bacillus* rRNA group 2 (Ash et al., 1991), which, in addition to members of the genus *Bacillus* and *Sporosarcina*, contains non-sporeforming organisms, presently classified in *Caryophanon*, *Planococcus* and *Kurthia* (Claus et al., 1991). However, none of these taxa is as closely related to any *Sporosarcina* species as *Filibacter limicola* is to *Sporosarcina pasteurii*, *Sporosarcina globispora* and *Sporosarcina psychrophila* (Yoon et al., 2001b). An intense search for endospores in *Filibacter limicola*, using a broad range of techniques which have triggered spore formation in *Bacillus* species (Gordon et al., 1973), has failed (Maiden and Jones, 1984; studies of the DSMZ identification service).

The compilation of the main differential morphological, physiological, and biochemical properties (Table 64) indicates that *Filibacter limicola* can easily been diagnosed as a unique taxonomic entity, even if formation of endospores may be described in future studies on this organism.

Taxonomic comments

The first evidence that *Filibacter limicola* is not a member of the *Flexibacteraceae*, as proposed in the species description (Maiden and Jones, 1984), originated from 16S rRNA oligonucleotide cataloging (Clausen et al., 1985). Comparison of the catalog of 16S rRNA fragments of *Filibacter limicola* with those of gliding organisms such as *Vitreoscilla stercoraria*, *Cytophaga johnsonae*, and *Flexibacter elegans* and representatives of Gram-positive organisms, indicated a phylogenetic relationship with members of the latter taxa, especially to the round-spore-forming *Bacillus pasteurii*, *Bacillus insolitus*, *Bacillus psychrophilus* (Stackebrandt and Woese, 1981), *Sporosarcina urea* (Pechman et al., 1976), and to the asporogenous species *Planococcus citreus* (Stackebrandt and Woese, 1979). The phylogenetic relationship was supported by several common phenotypic properties, such as the presence of lysine as the diagnostic amino acid in the peptidoglycan and the occurrence of MK-7 as the major isoprenoid quinone. In a subsequent study (Stackebrandt et al., 1987), the diversity of the phylogenetically coherent but morphologically diverse group, to which *Filibacter limicola* belongs, was extended by the inclusion of members of *Caryophanon* and *Sporosarcina globispora*. With the introduction of reverse transcriptase sequencing of 16S DNA the phylogenetic heterogeneity was fully explored (Ash et al., 1991) and the group containing *Filibacter limicola* and its nearest phylogenetic neighbors was defined as RNA group 2 of the genus *Bacillus* and related taxa. In addition to the organisms indicated above, members of the genus *Kurthia* were affiliated to this group (Farrow et al., 1994b). As the 16S rDNA sequence of *Filibacter limicola* NCIMB 11923[T] (accession no. X70317) contained several ambiguous nucleotides and stretches of undetermined nucleotides, the almost complete sequence (1517 nucleotides) was redetermined for a new culture of strain NCIMB 11923[T], deposited in the DSMZ as DSM 12288[T] (accession no. AJ292316). Though the two sequences showed only 95% similarity to each other, the phylogenetic position of the strains as members of RNA group 2 remained unchanged. The new sequence placed *Filibacter limicola* next to the species *Sporosarcina psychrophila*, *Sporosarcina pasteurii*, and *Sporosarcina globispora*. Depending upon the algorithm used (DeSoete, 1983; Felsenstein, 1993), *Bacillus pasteurii* grouped either with the other two *Sporosarcina* species (consensus tree using DNAPARS) or it branched below the lineage defined by *Filibacter limicola*, *Sporosarcina psychrophila* and *Sporosarcina globispora* [consensus tree using distance matrix analyses or maximum-likelihood (ML)]. Figure 64 displays the position of *Filibacter limicola* within RNA group 2 (ML algorithm) together with organisms analyzed recently, such as members of *Planococcus* (Ash et al., 1992), and various *Bacillus* species (Rheims et al., 1999).

Acknowledgements

I wish to thank the following DSMZ staff members for their support: Peter Schumann, riboprinting analysis; Susanne Verbarg and Anja Frühling, determination of physiological properties; and Jolantha Swiderski, 16S rDNA analyses.

TABLE 64. Diagnostic characteristics distinguishing *Filibacter limicola* DSM from phylogenetically closely related taxa[a]

Characteristic	*Filibacter limicola*[b]	*Sporosarcina globispora*[c]	*Sporosarcina pasteurii*[c]	*Sporosarcina psychrophila*[c]	*Sporosarcina ureae*[d]
Morphology:					
Cell shape	Filaments, long rods, straight or curved, rounded ends	Straight rods	Straight rods	Straight rods	Spherical or oval, may develop into tetrads or packages of eight or more
Cell dimensions, width × length	1.1 × 3–30 µm	0.6–1.0 × 1.5–5 µm	0.5–1.2 × 1.3–4 µm	>1 × 3–7 µm	1.0–2.5 µm
Colonial morphology	Spreading whorls and spiral colonies	Not distinctive, usually circular and glossy, pigmentation varies on different media	Translucent, slightly raised, circular, entire, smooth, glossy	Round, smooth, opaque; cream colored, pale yellow to bright orange	
Endospores, position	–	Spherical, terminal	Spherical, terminal	Spherical	Spherical, central or lateral
Motility	Gliding	+	+	+	+
Relation to oxygen	Obligate aerobic	Aerobic, no anaerobic growth	Aerobic, no anaerobic growth	Aerobic, anaerobic growth with glucose	Obligate aerobic
Temperature response:					
Growth at 4°C	+	+	nd	+	nd
Growth at 30°C	–	d	+	+	nd
Optimum temperature	20°C	nd	nd	25°C	26°C
Hydrolysis of:					
Casein	–	d	d	–	–
Gelatin	+	+	+	+	–
Starch	–	d	–	–	–
Decomposition of urea	+	+	+	+	+
Growth on amino acids plus vitamins only	+	–	–	–	–
Utilization of other organic compounds	–	Glucose, acid from glycerol, lactose and sucrose	nd	Acetate, fumarate, malate, succinate; acid from various carbohydrates	Acetate, glutamate
Enzymes:					
Catalase	+	+	+	+	+
Oxidase	+	+	nd	+	+
DNA mol% G+C	44	40	38	44	40–41

[a]Symbols: –, 90% of strains or more are negative; +, 90% or more are positive; d, 11–89% of strains are positive; nd, not determined.
[b]According to Maiden and Jones (1984).
[c]Claus and Berkeley (1986).
[d]Claus and Fahmy (1986).

List of species of the genus *Filibacter*

1. **Filibacter limicola** Maiden and Jones 1985, 375[VP] (Effective publication: Maiden and Jones 1984, 2957.)
   li.mi′co.la L.n. *limus* mud; L. suff. *cola* dweller; N.L. masc. n. *limicola* mud-dweller.

   Filaments 1.1 µm × 8–150 µm, cells 1.1 µm × 3–30 µm, rarely less than three diameters in length. Gliding motility about 5 µm/min. Good growth at pH 7.0, no growth at pH 5.7. Optimal temperature 20°C, slow growth at 4°C and no growth above 26°C. Sugars and organic acids not utilized, amino acids only are used as carbon and energy sources. Good growth on peptone-containing media. Gelatin hydrolysed, but casein and starch not hydrolysed. Nitrate reduction in CASO broth positive. *p*-Nitrophenyl-β-D-galactopyranoside hydrolysis and urease-positive.

   DNA G+C content (mol%): 44 mol% (T_m).

   Type strain: 1SS101, ATCC 43646, NCIMB 11923, DSM 13886.

   GenBank accession number (16S rDNA): AJ292316.

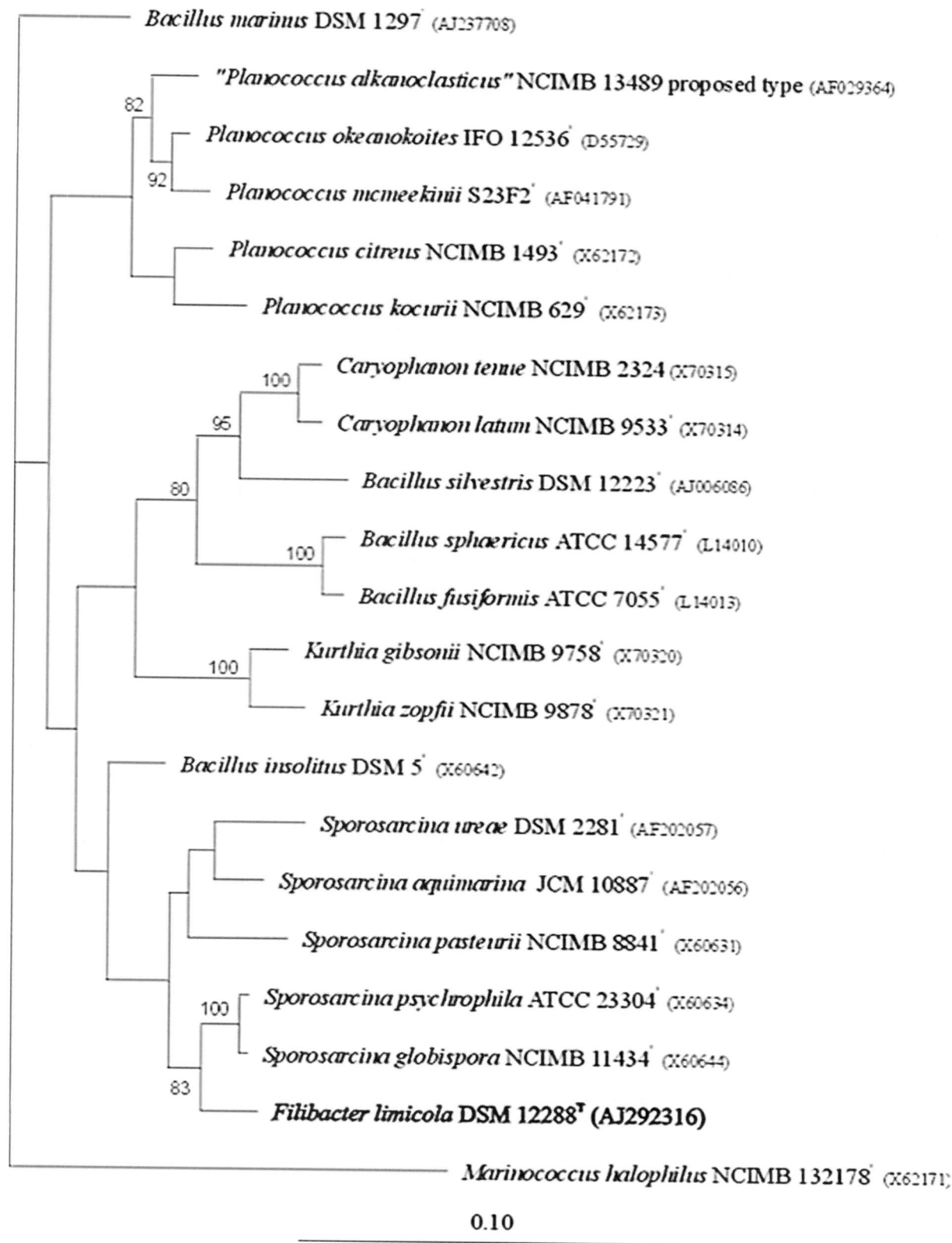

FIGURE 64. 16S rDNA sequence-based dendrogram displaying the phylogenetic position of *Filibacter limicola* and related taxa of *Bacillus* RNA group 2 by the maximum-likelihood algorithm (Felsenstein, 1993). Bootstrap values (expressed as percentages of 100 replications) of 75% and higher are indicated at the branch points. Scale bar = 10 inferred nucleotide substitutions per 100 nucleotides.

Genus IV. **Jeotgalibacillus** Yoon, Weiss, Lee, Lee, Kang and Park 2001c, 2092[VP]

ELKE DE CLERCK AND PAUL DE VOS

Je.ot.ga.li.ba.cil'lus. N.L. n. *jeotgalum* (Korean n. *jeotga*) jeotgal, traditional Korean food; L. n. *bacillus* rod; N.L. masc. n. *Jeotgalibacillus* rod from jeotgal.

Rods $1.0–1.2 \times 2.0–4.0\,\mu m$, but longer rods are often observed. Motile by peritrichous flagella. **Round endospores** are formed in swollen sporangia. Cells stain Gram-variable. **Cell wall peptidoglycan contains L-lysine at position 3 of the peptide subunit.** Facultatively anaerobic. Optimal growth temperature is 30–35°C. Optimal pH for growth is 7.0–8.0. Catalase- and oxidase-positive. Urease-negative. Nitrate is reduced to nitrite. **The predominant menaquinones are MK-7 and MK-8.** The major fatty acid is $C_{15:0\,iso}$.

DNA G+C content (mol%): 44.

Type species: **Jeotgalibacillus alimentarius** Yoon, Weiss, Lee, Lee, Kang and Park 2001c, 2092[VP].

Further descriptive information

The genus *Jeotgalibacillus* contains a single species, *Jeotgalibacillus alimentarius*, which is phylogenetically related to members of *Bacillus* rRNA group 2 and forms a coherent cluster with *Marinibacillus marinus*.

Jeotgalibacillus alimentarius contains a single strain, YKJ-13. Most cells of strain YKJ-13 are $1.0–1.2 \times 2.0–4.0\,\mu m$ in a 3-d-old culture on Marine agar at 30°C, but longer rods may also be observed. MK-7 is the predominant menaquinone (62%), but a significant amount of MK-8 (35%) also occurs. The fatty acid profile is characterized by the predominance of $C_{15:0\,iso}$ (47.6%), followed by $C_{15:0\,anteiso}$ (15.2%).

After incubation on Marine agar, colonies are smooth, glistening, irregular, flat-raised, and orange-yellow in color.

Optimal growth is observed at 30–35°C. Growth occurs at 10 and 45°C but not at 4 or 50°C. Strain YKJ-13 grows optimally at pH 7.0–8.0, and no growth occurs at pH values below 6.0. Optimal growth is observed in the presence of 3–12% NaCl (w/v); growth occurs with 19% NaCl and weakly with 20% NaCl, but not with more than 21% NaCl. Strain YKJ-13 grows under anaerobic conditions on Marine agar and shows catalase and oxidase activities, but no urease activity.

Esculin, casein, gelatin, and Tween 80 are hydrolyzed. No hydrolysis of hypoxanthine, starch, tyrosine, or xanthine is observed. Nitrate is reduced to nitrite.

Enrichment and isolation procedures

Jeotgalibacillus cells can be enriched in Marine broth at 30°C and harvested from Marine agar after incubation for 3 d.

Differentiation of the genus *Jeotgalibacillus* from other genera

Jeotgalibacillus is considered a member of *Bacillus* rRNA group 2 because of the formation of round endospores, the inclusion of L-lysine at position 3 of the peptide subunit of the peptidoglycan, and the abundance of $C_{15:0\,iso}$, the major fatty acid. Those properties have been stated as characteristic for members of this group (Rheims et al., 1999; Shida et al., 1997; Stackebrandt et al., 1987). The menaquinone profile of *Jeotgalibacillus* has not been found among other members of *Bacillus* rRNA group 2. *Jeotgalibacillus* differs from the closest related genus, *Marinibacillus*, in its ability to grow anaerobically, and in its NaCl and growth temperature tolerances. Whereas *Marinibacillus* shows slight growth with 7% NaCl and no growth with 10% NaCl, *Jeotgalibacillus* grows with 20% NaCl. Moreover, *Jeotgallibacillus alimentarius* strain YKJ-13T is a mesophile, whereas *Marinibacillus marinus*, the only species within the genus *Marinibacillus*, is a psychrophile.

Taxonomic comments

In the tree-based neighbor-joining algorithm, strain YKJ-13 is phylogenetically related to members of *Bacillus* rRNA group 2 and forms a coherent cluster with *Marinibacillus marinus* DSM 1297[T]. The relationship between strain YKJ-13, *Marinibacillus marinus* DSM 1297[T], and members of *Bacillus* rRNA group 2 is also found in trees generated by the maximum-likelihood and maximum-parsimony algorithms.

List of species of the genus *Jeotgalibacillus*

1. **Jeotgalibacillus alimentarius** Yoon, Weiss, Lee, Lee, Kang and Park 2001c, 2092[VP]

   a.li.men.ta'ri.us. L. adj. *alimentarius* relating to food.

   Cell morphology, nutrition, and physiology are as described for the genus and as listed in Table 65.

   DNA G+C content (mol%): 44 (HPLC).
   Type strain: YKJ-13, JCM 10872, KCCM 80002.
   GenBank accession number (16S rRNA gene): AF281158.

Genus V. **Kurthia** Trevisan 1885, 92[AL] Nom. cons. Opin. 13 Jud. Comm. 1954, 152

RÜDIGER PUKALL AND ERKO STACKEBRANDT

Kurth'i.a. N.L. fem. gen. n. *Kurthia* named for H. Kurth, the German bacteriologist who described the type species.

In young cultures (12–24 h): **regular, unbranched rods** with rounded ends occurring in chains that are often parallel. The rods are ~0.6–1.2 μm in diameter and vary in length according to the stage of growth ~2–5 μm long; filaments (5–10 μm) may occur. Older cultures (3–7 d) of **some species are composed of coccoid cells** formed by fragmentation of the rods, but short rods may be the dominant forms in such cultures; one species does not form coccoid cells. **Does not form endospores.**

TABLE 65. Characteristics of *Jeotgalibacillus alimentarius*[a]

Characteristic	Reaction or result
Rod-shaped cells	+
Swollen sporangia	+
Round spore shape	+
Gram staining	v
Motility	+
Pigment	+
Anaerobic growth	+
Oxidase, catalase	+
Nitrate reduction	+
Decomposition of:	
Casein, esculin, gelatin, Tween 80	+
Hypoxanthine, starch, tyrosine, urea, xanthine	–
Acid production from:	
Adonitol, arabinose, cellobiose, lactose, mannose, rhamnose, sorbitol, xylose	–
Glucose, fructose, galactose, mannitol, maltose, melibiose, raffinose, sucrose, trehalose	+
Growth in NaCl (%):	
0	
7, 10	+
20	w
21	–
Growth at (°C):	
5	–
10, 30, 37	+
Optimum growth temperature (°C)	30–35
Maximum growth temperature (°C)	45–50
Peptidoglycan type	L-Lys-direct
Predominant menaquinones	MK-7, MK-8
Major fatty acid	$C_{15:0\ iso}$
Mol% G+C of DNA	44

[a]Symbols: +, positive; –, negative; w, weakly positive; v, variable.

Gram-positive. Not acid-fast. The rods are usually motile by peritrichous flagella but nonmotile strains occur. **Strictly aerobic.** Optimum temperature 25–30°C or 20–25°C, one species grows up to 45°C. Grows well on peptone-yeast extract media at neutral pH, producing rhizoid colonies. Two species grow in 7% generally agree with previous reports on *Kurthia zopfii* and *Kurthia gibsonii* (Goodfellow et al., 1980), in that $C_{15:0\ ante}$ and $C_{15:0\ iso}$ acids are the major component. Detailed percentages are given in the species descriptions.

Phylogenetic position. The phylogenetic position of *Kurthia* was elucidated in comparison with non-spore-forming aerobic spherical and rod-shaped Gram-positive organisms with a DNA G+C content of less than 55 mol%. Members of *Filibacter* (Clausen et al., 1985), *Sporosarcina*, *Planococcus* and *Marinococcus* (Farrow et al., 1992; Ludwig et al., 1981), as well as *Kurthia*, *Caryophanon*, *Planococcus* and *Exiguobacterium* (Farrow et al., 1994a) were found to branch among members of the genus *Bacillus* (Ash et al., 1991; Stackebrandt et al., 1987; Stackebrandt and Swiderski, 2002). The genus *Bacillus* itself was found to contain several remotely related phylogenetic branches, many of which are today described as individual genera (see chapters on the family *Bacillaceae*). The core genus of *Bacillus* still forms a phylogenetically broad cluster, consisting of different groups (Ash et al., 1991), one of which, *Bacillus* group 2, embraces the round spore-forming bacilli (e.g., *Bacillus psychrodurans*, *Bacillus nedei*, *Bacillus psychrus*, *Bacillus sphaericus*, *Bacillus insolitus*, *Bacillus*

silvestris) and their non-spore-forming relatives (Stackebrandt et al., 1987).

With 16S rRNA gene sequence similarities ranging between 95.8 and 97.7%, the type strains of the three *Kurthia* species form a tight phylogenetic cluster (Figure 66). The branching is supported by high bootstrap values. *Kurthia* species are remotely related to the type strains of two *Bacillus* species, i.e., *Bacillus pycnus* JCM 11075[T] and *Bacillus neidei* JCM 11077[T] (94.2–96.2%). However, low bootstrap values indicate that the stability of their branching is not significant and may change with other sequences included in the database. These five species form a sister clade of a cluster consisting of species of *Bacillus*, *Sporosarcina*, *Filibacter*, and *Planococcus*.

Genes other than those encoding 16S rRNA have rarely been sequenced. An example is the suite of 11 biotin biosynthesis genes involved in the stepwise synthesis of biotin in *Kurthia* sp. DSM 10609 (Kiyasu et al., 2001; accession numbers AB045873–AB045875). Besides a chitinase gene of *Kurthia zopfii* (accession no. D63702), the *cpn60* (chaperonin 60) gene has partially been sequenced for all three type strains of *Kurthia* (accession nos AY123713, AY123714, AY123716).

Habitat (Keddie and Shaw, 1986). Although a number of strains of bacteria identified as "*Kurthia* species," have been isolated from various clinical materials, most commonly from the feces of patients suffering from diarrhea, there is no evidence of pathogenicity in authentic members of the genus (see

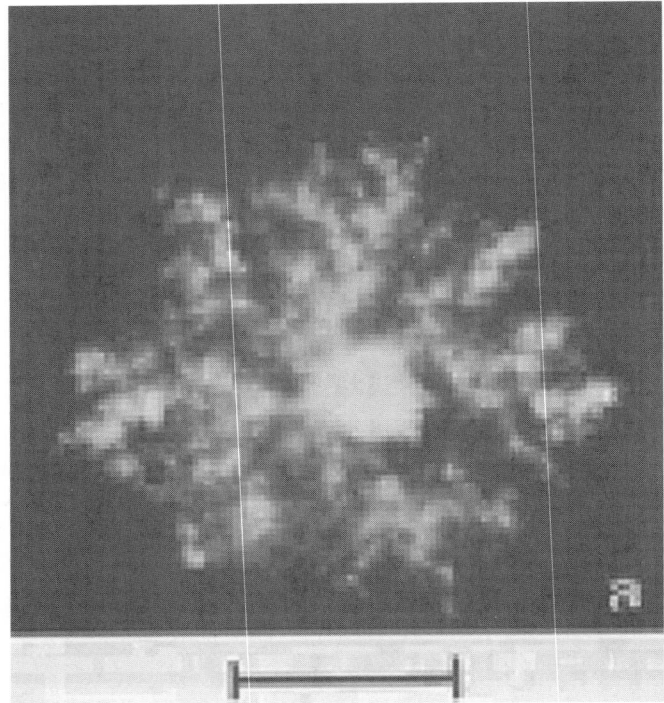

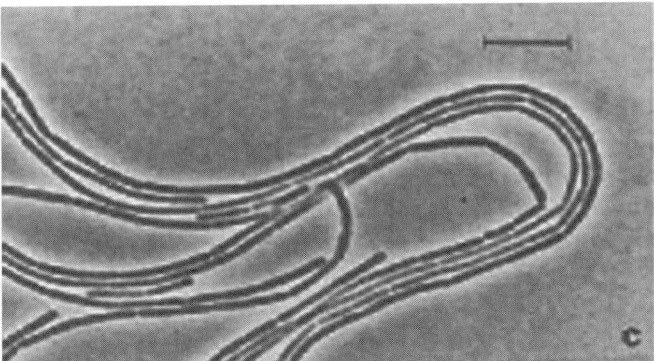

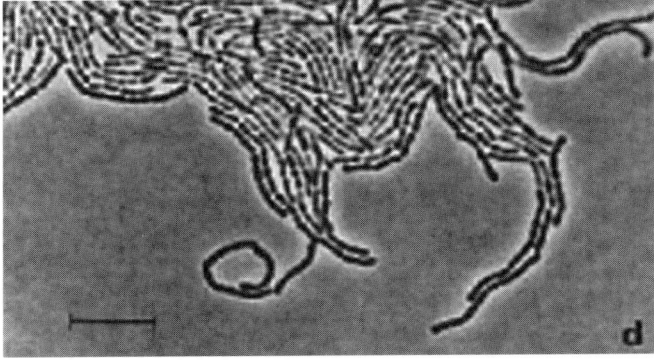

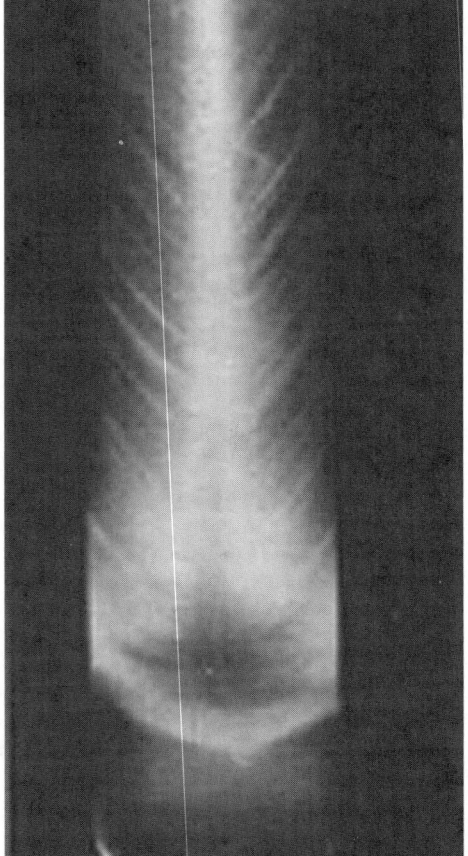

FIGURE 65. Growth behavior of *Kurthia zopfii* NCIB 9878[T]. (a) Rhizoid colony on yeast nutrient agar after 4 d incubation at 25°C; bar = 10 mm. (b) Yeast nutrient gelatin slant showing "bird's feather" type of growth; incubated 5 d at 20°C. *Kurthia zopfii* isolate: edge of colony on yeast nutrient agar incubated at 25°C. (c) After 24 h, showing long filaments composed of rods. (d) After 3 d, showing development of coccoid forms. Bar = 10 μm.

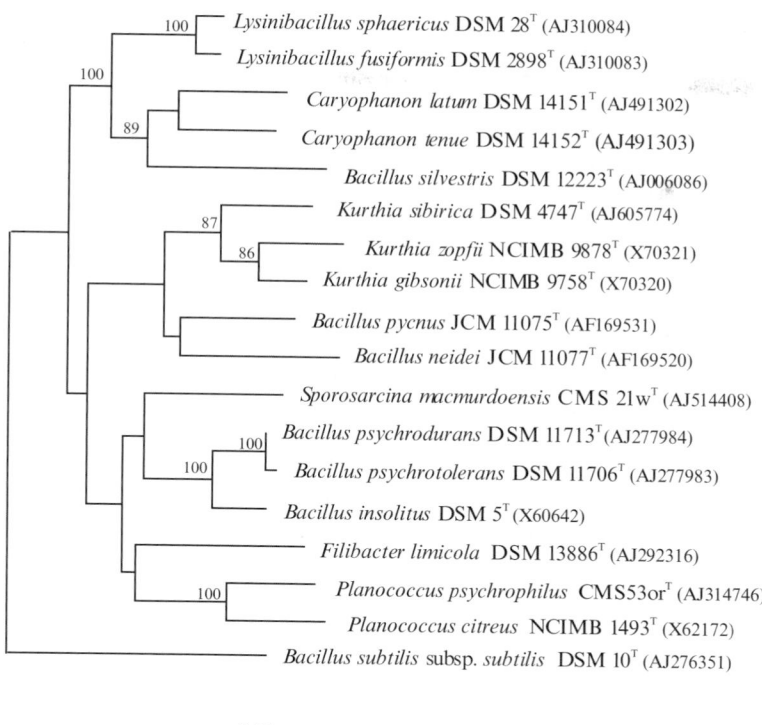

Lysinibacillus sphaericus DSM 28^T (AJ310084)

Lysinibacillus fusiformis DSM 2898^T (AJ310083)

Caryophanon latum DSM 14151^T (AJ491302)

Caryophanon tenue DSM 14152^T (AJ491303)

Bacillus silvestris DSM 12223^T (AJ006086)

Kurthia sibirica DSM 4747^T (AJ605774)

Kurthia zopfii NCIMB 9878^T (X70321)

Kurthia gibsonii NCIMB 9758^T (X70320)

Bacillus pycnus JCM 11075^T (AF169531)

Bacillus neidei JCM 11077^T (AF169520)

Sporosarcina macmurdoensis CMS 21w^T (AJ514408)

Bacillus psychrodurans DSM 11713^T (AJ277984)

Bacillus psychrotolerans DSM 11706^T (AJ277983)

Bacillus insolitus DSM 5^T (X60642)

Filibacter limicola DSM 13886^T (AJ292316)

Planococcus psychrophilus CMS53or^T (AJ314746)

Planococcus citreus NCIMB 1493^T (X62172)

Bacillus subtilis subsp. *subtilis* DSM 10^T (AJ276351)

0.01

FIGURE 66. Neighbor-joining tree (Felsenstein, 1993) generated on the basis of 16S rRNA gene sequences showing the nearest neighbors of members of the genus *Kurthia*. Bootstrap value was calculated for 500 replicate trees. The bar corresponds to a 1% difference in nucleotide sequences.

Keddie, 1981, for references). *Kurthia zopfii* and *Kurthia gibsonii* are commonly isolated from meat (particularly after storage for a few days at 16°C (Gardner, 1969) and meat products; it is likely that the meat becomes contaminated with *Kurthia* in the abattoir (Keddie, 1981). The other common source of *Kurthia zopfii* and *Kurthia gibsonii* is feces of certain farm animals, especially of chickens and pigs, several hours to a few days after they have been voided (Keddie, 1981). There are also reports of presumptive *Kurthia* species being isolated from such diverse sources as "sloughing spoilage" of ripe olives, the gut of a crab, wet stored wood, dental plaque of beagle dogs and from air at an altitude greater than 10,000 feet (see Keddie, (1981) for references), but there is considerable doubt about the accuracy of identification of many of these isolates. However, authentic *Kurthia* species have been isolated occasionally from milk, soil and surface waters, presumably as a result of contamination with animal dung (Keddie, 1981). Recent reports describe *Kurthia* species in bottled drinking water (Jeena et al., 2006) and the nasal cavity of sea lions (Hernandez-Castro et al., 2005). *Kurthia* species were isolated from the stomach and intestinal contents of the Susuman mammoth (Belikova et al., 1980). Of 13 strains referred to by these authors as *Kurthia zopfii*, four had the characters of that species, three appeared to be *Kurthia gibsonii* and the remainder, mainly psychrophiles, were described as *Kurthia sibirica* (Belikova et al., 1986).

Isolation and enrichment procedures (Keddie and Shaw, 1986)

Species of the genus may be isolated by making use of their unusual cultural properties. Nutrient gelatin plates are inoculated heavily with a single, central streak of the material to be examined, or of a suspension or macerate of solid material in sterile water. The plates are incubated at 20°C with the lids upright and examined daily. Liquefaction of the gelatin around the streak soon occurs but in successful cultures filamentous outgrowths of *Kurthia* appear beyond this zone in 2–3 d. To obtain pure cultures, a small piece of nutrient gelatin containing outgrowths is streaked out on a suitable nutrient agar medium. Isolates are then examined for the characteristic properties of *Kurthia*: useful screening tests are colony form, morphology and Gram reaction, production of "bird's feather" growth on nutrient gelatin slants and aerobic growth in glucose nutrient agar shake cultures.

The composition of the nutrient gelatin medium used is important and, in particular, the concentration and brand of gelatin used; not all brands support the typical outgrowths. A nutrient gelatin (YNG; Keddie, 1981) of the following composition may be used (per liter of distilled water): meat extract (Lab Lemco powder, Oxoid), 4 g; peptone (Difco), 5 g; yeast extract (Difco), 2.5 g; NaCl, 5 g; gelatin (BDH Chemical Co., Poole,

UK), 100 g; pH 7.0. The medium is sterilized for 30 min (for quantities up to 100 ml). The medium should be inoculated with a reference strain of *Kurthia zopfii* (NCIB 9878) to test its ability to allow good outgrowth production. The gelatin manufactured by the BDH Chemical Co., Poole, England is satisfactory but, with some batches, a lower concentration than that stated may give more satisfactory outgrowths. Although all *Kurthia* strains tested grow well in yeast nutrient broth, YNB (YNG without gelatin), YNG prepared with some batches of gelatin has given poor growth. Dissolving the constituents of YNG in mineral base E (Owens and Keddie, 1969) to give MYNG (Shaw and Keddie, 1983b) overcomes this problem. A single central streak on a suitable nutrient agar often allows the detection and isolation of *Kurthia* and should be used in addition to the gelatin streak method in case rapid liquefaction of gelatin prevents the detection of *Kurthia*. YNA is inoculated with a single, central streak as described above. Yeast extract nutrient agar (YNA) has a composition similar to YNG but the gelatin is replaced by 12 g/l of Bacto-Agar (Difco). Plates are incubated at 25°C and each day the edge of the streak is scanned by low power (100×) microscope for the characteristic skein-like outgrowths of *Kurthia*. Pure cultures are obtained by picking carefully from outgrowths and streaking on YNA.

Maintenance procedures

Kurthia species grow well on media such as YNA (described above) at 25°C. YNA cultures should remain viable for at least 6 months at room temperature (20°C) provided they are not allowed to dry out. They may be preserved for long periods by lyophilization.

Differentiation of the genus *Kurthia* from other genera

The 16S rRNA gene sequence similarities, in combination with absence of endospores, strictly aerobic growth and bird's feather-like growth on gelatin nutrient agar unambiguously assign new Gram-positive isolates to the genus *Kurthia* (Table 66). Peptidoglycan of the Lys-Asp type is another characteristic trait.

Differentiation of the species of the genus *Kurthia*

DNA–DNA similarity studies of a few strains (some atypical) support the conclusion that *Kurthia zopfii*, *Kurthia gibsonii* and *Kurthia sibirica* are distinct species with DNA similarity values ranging between 20 and 40% (Cherevach et al., 1983). Chemotaxonomically the type strains of the species can be differentiated on the basis of qualitative and quantitative differences in their fatty acid profiles. The circumscription of the genus as given in the First Edition of Bergey's Manual of Systematic Bacteriology (Keddie and Shaw, 1986) was based on the characters of *Kurthia zopfii* and *Kurthia gibsonii*. Both species were shown to resemble each other closely in the numerical phenetic study of Shaw and Keddie (1983b). In order to compare the type strains of the three species on

TABLE 66. Differentiation of *Kurthia* from morphologically similar taxa and phylogenetic relatives[a]

Characteristics	*Kurthia*[b]	*Brochothrix*[b,c]	*Filibacter*[d]	*Listeria*[b,e]	Nearest *Bacillus* species[f]	*Caryophanon*[g]	*Sporosarcina*[h]
Morphology	Rods	Rods	Rods, straight or curved	Rods	Rods	Multicellular rods	Rods, orspheres,
Endospores	–	–	–	–	Round spores	–	+
Motility	+	–	Gliding	+	+ or nd	+	+
Glucose fermentation	–	+	–	+	– or weak	–	D
Facultatively anaerobic	–	+	–	+	–	–	D
Bird's feather' growth on nutrient gelatin	+	–	nd	NA	nd	NA	nd
Peptidoglycan type	Lys-Asp	*meso*-A$_2$pm	Lys-Glu	*meso*-A$_2$pm	Lys-D-Glu or Orn-Glu	Lys-D-Glu	Lys-Glu, Lys-Als-Asp, Lys-Asp, Lys-Gly-Glu
Major fatty acids	C$_{15:0\,iso}$, C$_{15:0\,ante}$	C$_{14:0\,ante}$, C$_{15:0\,ante}$, C$_{16:0}$[h]	C$_{15:0\,ante}$, C$_{17:0\,ante}$	C$_{15:0\,ante}$, C$_{17:0\,ante}$	C$_{15:0\,iso}$, or C$_{15:0}$ + C$_{15:0\,anteiso}$ + C$_{16:1\,11cis}$	C$_{15:0}$, C$_{16:1\,\omega11cis}$	C$_{15:0\,anteiso}$
DNA mol% G+C	36–38	36	44	36–38	36–37	41–46	39–44
Major menaquinone	MK-7	MK-7	MK-7	MK-7	nd	MK-6	MK-7

[a]Symbols: –, 90% or more of strains are negative; +, 90% or more of strains are positive; D, different reactions in different species of a genus; NA, not applicable; nd, not determined.

[b]Keddie and Shaw (1986).

[c]Stackebrandt and Jones (2005).

[d]Stackebrandt (2005) is the reference to the filibacter chapter in Bergey complete reference at the bottom.

[e]Seeliger and Jones (1986).

[f]*Bacillus neidei*, *Bacillus pycnus* (Nakamura et al., 2002); *Bacillus psychrodurans* (Abd El-Rahman et al., 2002).

[g]Reddy *et al.* (2003) and this chapter.

[h]Personal data.

the same information basis, BIOLOG substrate panels were tested (Table 67).

In general, physiological reactions are weak, leading to many weak or doubtful positive reactions (Table 67; see also Keddie and Jones, 1992) Physiological similarities between *Kurthia zopfii* and *Kurthia gibsonii* support the phylogenetic adjacent position of the two type strains in the phylogenetic tree (Figure 66). Nevertheless these two species can be differentiated from each other by a few reactions, notably utilization of *N*-acety-β-D-glucosamine and γ-hydroxybutyric acid by *Kurthia zopfii*. *Kurthia sibirica* shows a more distinct metabolic pattern as it utilizes D-fructose, D-mannose and D-psicose, but fails to utilize uridine-5 -monophsphate, succinic acid monomethylester and pyruvic acid methyl ester. The riboprint patterns of the three type strains are distinct (Figure 67).

Clinical sources. Reports on the authenticity of *Kurthia* species isolated from various clinical sources (Faoagali, 1974; Severi, 1946; Yang et al., 1985) and most frequently from the feces of patients suffering from diarrhea (Elston, 1961; Jarumilinta et al., 1976) should be taken with caution as none of these strains may have been properly classified. Also, in all of these cases, the connection between the occurrence of the pre-

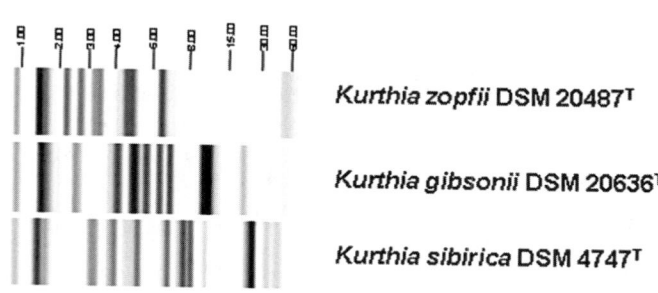

Kurthia zopfii DSM 20487[T]

Kurthia gibsonii DSM 20636[T]

Kurthia sibirica DSM 4747[T]

FIGURE 67. Diversity of normalized *Eco*RI ribotype patterns found within the type strains of the genus *Kurthia*. Automated ribotyping was carried out by using the RiboPrinter Microbial Characterization system (Qualicon). The analysis is based on the detection of polymorphism that exists within and around the *rrn* operons of the strains.

sumptive *Kurthia* and the clinical condition was, at most, tenuous. Some organisms were considered to be "*Kurthia bessonii*" or "*Kurthia variabilis*", neither of which was included in the Approved Lists of Bacterial Names (Skerman et al., 1980) and are therefore not legitimate species.

TABLE 67. Results of substrate utilization as determined by BIOLOG GP2 microtiter plates scored after incubation of 24 h at 28°C or 72 h at 25°C for *Kurthia* species[a]

Carbon source	1. *K. zopfii* DSM 20580[T]	2. *K. gibsonii* DSM 20636[T]	3. *K. sibirica* DSM 4747[T]
Glycogen	w	−	−
N-Acetyl-β-D-glucosamine	++	−	−
N-Acetyl-β-D-mannosamine	w	−	−
L-Arabinose	−	−	w
D-Fructose	−	−	++
D-Mannose	−	−	++
D-Psicose	−	−	+
D-Ribose	++	+	++
Sedoheptulosan	−	−	w
Acetic acid	+	+	+
γ-Hydroxybutyric acid	+	−	−
α-Ketovaleric acid	+	+	+
Pyruvatic acid methyl esther	+	++	−
L-Malic acid	−	w	−
Succinic acid mono-methyl ester	++	++	−
Propionic acid	−	−	−
Pyruvic acid	++	+	++
L-Alanine	+	w	−
L-Alanyl-glycine	−	−	−
L-Glutamic acid	w	−	+
L-Serine	−	w	−
Glycerol	w	+	−
2,3-Butanediol	−	w	−
Adenosine	++	+	w
2′-Deoxy adenosine	+	+	+
Inosine	+	+	−
Thymidine	++	++	+
Uridine	+	+	−
Adenosine-5′-monophosphate	w	+	−
Thymidine-5′-monophosphate	−	+	−
Uridine-5′-monophosphate	+	++	−

[a]Inocula were supplemented with 1 ml of sterile sodium salicylate (100 mM) as recommended by the manufacturer's instructions. Cavities showing a photometric value above 20% and 30% of the highest value obtained for the individual strain were scored as weak (w) or positive, respectively. ++, very strong reaction >70% of highest reading; +, strong reaction >30–<70 % of highest reading.

Further reading

Keddie, R.M. 1981. The genus *Kurthia*. *In* Starr, Stolp, Trüper, Balows and Schlegel (Editors), The Prokaryotes. A Handbook on Habitats, Isolation, and Identification of Bacteria, Springer-Verlag, Berlin, pp. 1888–1893.

Ludwig, W., E. Seewaldt, K.H. Schleifer and E. Stackebrandt. 1981. The phylogenetic status of *Kurthia zopfii*. FEMS Microbiol. Lett. *10*: 193–197.

Shaw, S. and R.M. Keddie. 1983a. A numerical taxonomic study of the genus *Kurthia* with a revised description of *Kurthia zopfii* and a description of *Kurthia gibsonii* sp. nov. Syst. Appl. Microbiol. *4*: 253–276.

List of species of the genus *Kurthia*

1. **Kurthia zopfii** (Kurth) Trevisan 1885, 92[AL] nom. cons. Opin. 13, Jud. Comm. 1954, 152 (*Bacterium zopfii* Kurth 1883, 98)

zop'fi.i. M.L. gen. n. *zopfii* of Zopf, named for W. Zopf, a German botanist.

The description is based on the emended description of Shaw and Keddie (1983b) and personal data. Morphology and general characteristics are given in the generic description. Other characteristics are given in Table 66 and Table 67.

Polar lipids are available only for one strain (NCTC 404, Collins et al., (1979; Goodfellow et al., 1980). Major fatty acids are $C_{15:0\ iso}$ (42.9%) and $C_{15:0\ ante}$ (39.3%); $C_{14:0}$ (1.7%), $C_{14:0\ iso}$ (6.8%), $C_{16:1\ \omega7cEtOH}$ (1.5%), $C_{16:1\ \omega11c}$ (2.0%), and $C_{16:0}$ (4.6%) occur in smaller amounts.

*DNA G+C content (mol%):*36–38 (T_m) (Keddie, 1981; Shaw S. and Keddie, 1984).

Type strain: NCIB 9878 (DSM 20580).

GenBank accession number (16S rRNA gene): X70321.

2. **Kurthia gibsonii** Shaw and Keddie 1983b, 672 (Effective publication: Shaw and Keddie 1983b, 268.)

gib'son.i.i. M.L. gen. n. *gibsonii* of Gibson, named for T. Gibson.

The description is based on that of Shaw and Keddie (1983b) supplemented with personal data. Morphology and general characteristics are given in the generic description. Growth occurs at 40°C; acid is formed from butane-1-ol; n-octanoate, crotonate, DL-lactate, oxalacetate, and L-alanine are utilized. There is no production of H$_2$S. Other characteristics are given in Table 66 and Table 67.

Data on the isoprenoid quinones and polar lipids are available for four strains (Collins et al., 1979; Goodfellow et al., 1980). Major fatty acids are $C_{14:0\ iso}$ (8.9%), $C_{15:0\ iso}$ (26.8%), $C_{15:0\ ante}$ (31.1%) and $C_{16:0\ iso}$ (8.8%); $C_{14:0}$ (1.2%), $C_{15:0}$ (4.3%), $C_{16:1\ \omega11c}$ (4.1%), $C_{16:0}$ (6.0%), $C_{17:0iso}$ (1.7%) and $C_{17:0\ ante}$ (4.6%) occur in smaller amounts.When supplied with a suitable source of amino acids cells require biotin and thiamine (Shaw and Keddie, 1983a).

DNA G+C content (mol%): 36–38 (T_m) (Keddie, 1981; Shaw S. and Keddie, 1984).

Type strain: NCIB 9758 (DSM 20636).

GenBank accession number (16S rRNA gene): X70320.

3. **Kurthia sibirica** Belikova, Cherevach and Kalakutskii 1988, 220[VL] (Effective publication: Belikova, Cherevach and Kalakutskii 1986, 834.)

si.bi'ri.ca. M.L. fem. adj. *sibirica* (Siberian) from Siberia where it was found in a mammoth from which the bacteria were isolated.

The description is based on the original description of Belikova et al. (1986) supplemented with personal data. General characsistics are consistant with those given in the generic description. Other characteristics are given in Table 66 and Table 67.

Cells are regular rods with rounded ends, 0.6–1.0μm in diameter. They vary in length with age: in young cultures from 5–10μm or long filamentsand in old cultures from 2–5μm, but not coccoid. Colonies and smears are gray-cream or yellow, depending on culturing conditions. A yellow pigment, probably a carotinoid (absorption maxima ranging between 400 and 480nm), is produced. Nutritional requirements are complex. It grows well in media containing peptone and yeast extract. Requires biotin, thiamine, pantothenate, nicotinic acid and pyridoxal-5-phosphate. Blood is hemolysed. Growth occurs between 5–37°C, fresh isolates may grow at 0–1°C, the optimum temperature for growth is in the 20–25°C range. Growth occurs in a pH ranging from 5.5–9.5. No growth occurs at 7% NaCl. Cells are phosphatase positive, urease and lecithinase negative, and do not hydrolyze starch. Acid is formed from fructose and glycerol. No acid is produced from glucose, ribose and ethanol. Major fatty acids are $C_{15:0\ iso}$ (65.4%), $C_{15:0\ ante}$ (12.2%) and $C_{17:1\ \omega10c}$ (6.5%); $C_{14:0\ iso}$ (2.0%) $C_{16:1\ \omega7cEtOH}$ (1.4%) $C_{16:0\ iso}$ (1.1%) $C_{16:1\ \omega11c}$ (3.6%) $C_{16:0}$ (2.4%) $C_{17:0\ iso}$ (2.5%) and $C_{17:0\ ante}$ (1.4%) occur in smaller amounts.

Kurthia sibirica is sensitive to penicillin (2 units), streptomycin (2 units), erythromycin (2μg), novobiocin (5μg) neomycin (5μg), oleandomycin (5μg), polymyxin B (300μg) and to 0.1% chlorous triphenyltetrazole added to the medium.

Isolated from the intestinal tract and the stomach of a mammoth in the Magadan provinceof Siberia.

DNA G+C content (mol%): 37.

Type strain: 13-2 (VKM B-1549, ATCC 49154, DSM 4747).

GenBank accession number (16S rRNA gene): AJ605774.

Genus VI. **Marinibacillus** Yoon, Weiss, Lee, Lee, Kang and Park 2001c, 2092[VP] emend. Yoon, Kim, Schumann, Oh and Park 2004, 1320

JEROEN HEYRMAN AND PAUL DE VOS

Ma.ri.ni.ba.cil'lus. L. adj. *marinus* of the sea; L. n. *bacillus* rod; N.L. masc. n. *Marinibacillus* rod of the sea.

Rod-shaped, **Gram-positive** (may be Gram-variable in older cultures) bacterium, 0.8–1.6μm in diameter and 2.0–7.0μm in length. Endospores are round to ellipsoidal and are located centrally, terminally, or subterminally. The sporangia are not swollen or are only slightly swollen. Strains without the ability to form endospores have been isolated. **Cell wall peptidoglycan** contains L-lysine at position 3 of the peptide subunit. Filaments present. Motile by single polar flagellum or peritrichous (degenerated) flagella. Obligately aerobic. **Strains require sea water or marine** agar for growth. The optimum NaCl concentration for growth varies between 1.2 and 3%, while growth in 10% NaCl is variable. Strains are **psychrophilic or psychrotolerant with growth at 1–4°C**; some strains may grow up to 39°C whereas others do not grow above 4–8°C. Catalase-positive. No reduction of nitrate to gas is observed. **The predominant menaquinone is MK-7** and the predominant fatty acids are $C_{15:0 \, iso}$ and $C_{15:0 \, ante}$.

DNA G+C content (mol%): 33.9–41.8.

Type species: **Marinibacillus marinus** (Rüger and Richter 1979) Yoon, Weiss, Lee, Lee, Kang and Park 2001b, 2092[VP] (*Bacillus globisporus* subsp. *marinus* Rüger and Richter 1979, 196; *Bacillus marinus* Rüger 1983, 157 emend. Rüger, Fritze and Spröer 2000, 1310.).

Further descriptive information

Colonies on sea water agar are round with entire margin, flat to raised, colorless, mostly translucent; light orange pigment may be formed. Colony diameter of 2–3 mm may be obtained after three days while a diameter of 1–2 mm is reached after 2 weeks of incubation on sea water medium.

Strains of *Marinibacillus marinus* have been isolated from sediments of the Iberian deep-sea, the tropical Atlantic, and the Arctic and Antarctic Oceans. Taking into consideration the known deep-water circulation in the polar and tropical sectors of the Atlantic Ocean (Hollister and McCave, 1984; Mantayla and Reid, 1983), it is assumed that the origin of *Marinibacillus marinus* was in the Antarctic and that the species spread from there to the north. The ability to form endospores as a survival strategy may not be necessary in environments with long periods of uniform environmental conditions, like the deep-sea. This may explain why endospores could be detected in only a few of the *Marinibacillus marinus* isolates. Additional information on the biochemical characteristics of *Marinibacillus* species can be found in Rüger, 1983; Rüger et al., (2000), and Yoon et al., (2001b), (2004).

Enrichment and isolation procedures

Strains grow on sea water medium and/or Marine Agar. Good growth occurs between 2 d and 2 weeks of incubation, but the extremely psychrophilic strains require longer incubation times of up to 4 weeks at 4°C.

Differentiation of the genus *Marinibacillus* from other genera

Sequence analysis of the 16S rDNA revealed the closest phylogenetic relation of the type strain of *Marinibacillus marinus* to be the type strain of *Jeotgalibacillus alimentarius*, the only species within the genus *Jeotgalibacillus* (95.7% similarity). However, there are some physiological differences. Whereas *Marinibacillus marinus* shows slight growth at 7% NaCl and no growth at 10% NaCl, *Jeotgalibacillus alimentarius* grows at 20% NaCl. *Marinibacillus marinus* is a psychrophile, while *Jeotgalibacillus alimentarius* is a mesophile. Also, some chemotaxonomic properties are important to distinguish *Marinibacillus* from *Jeotgalibacillus*. Whereas *Marinibacillus* has $C_{15:0 \, ante}$ as the predominant fatty acid, *Jeotgalibacillus* is characterized by the predominance of $C_{15:0 \, iso}$. Furthermore, the menaquinone profile of *Jeotgalibacillus* is characterized by the predominance of MK-7 (62%) followed by MK-8 (35%), whereas *Marinibacillus* has 87% MK-7 and no MK-8. Other round spore-forming bacilli, originally belonging to rRNA group 2, can be discriminated from *Marinibacillus* rather easily by comparative 16S rDNA analysis.

Taxonomic comments

The genus *Marinibacillus* now encompasses two species, *Marinibacillus marinus* and *Marinibacillus campisalis*, which both can be regarded as illegitimate because they belong to a genus of which the type species was not validly published. The first species was originally described as *Bacillus globisporus* subsp. *marinus* (Rüger and Richter, 1979) and later elevated to the species level as *Bacillus marinus* (Rüger, 1983). An emended description of the species has been formulated (Ruger et al., 2000) and a reclassification to *Marinibacillus marinus* was proposed (Yoon et al., 2001c).

Marinibacillus is phylogenetically related to members of *Bacillus* rRNA group 2 and forms a coherent cluster with *Jeotgalibacillus*. The phylogenetic relationship between *Marinibacillus*, *Jeotgalibacillus*, and members of *Bacillus* rRNA group 2 is found with the tree-based neighbor-joining algorithm and in trees generated by the maximum-likelihood and maximum-parsimony algorithms. Species of the *Bacillus* rRNA group 2 may deserve to be separated at the generic level from the genus *Bacillus*. The genera *Sporosarcina* (Yoon et al., 2001b), *Jeotgalibacillus* (Yoon et al., 2001c), and *Marinibacillus* (Yoon et al., 2001c) can be considered as a next step in the further splitting of the former genus *Bacillus sensu* Claus and Berkeley (1986).

List of species of the genus *Marinibacillus*

1. **Marinibacillus marinus** (Rüger and Richter 1979) Yoon, Weiss, Lee, Lee, Kang and Park 2001c, 2092[VP] (*Bacillus globisporus* subsp. *marinus* Rüger and Richter 1979, 196; *Bacillus marinus* Rüger 1983, 157 emend. Rüger, Fritze and Spröer 2000, 1310.)

ma.ri′nus. L. adj. *marinus* of the sea.

Gram-positive rods, 0.8–1.1 × 2.0–7.0 μm. Filaments present. Single or in pairs. Motile strains are peritrichously flagellated or show degenerated flagella. The spores are terminally or subterminally located. Cells without the ability to form spores have been isolated. Colonies on sea water agar are round with entire margin, flat to raised, colorless, mostly translucent with

a diameter of 1–2 mm after 2 weeks of incubation. Growth occurs at 1–4°C. The highest maximum growth temperature was 30°C and numerous strains did not grow at temperatures exceeding 4 or 8°C. Catalase-positive, aerobic acid production from glucose-positive in MOF medium. Further detailed biochemical characterization is given in Table 68.

DNA G+C content (mol%): 38 (HPLC).

Type strain: 581, ATCC 29841, CCUG 28884, CIP 103308, DSM 1297, LMG 6930, NCMB 2140, NRRL B-14321.

EMBL/GenBank accession number (16S rDNA): AB021190, AJ237708. This suggested change is unclear to me: it is well

TABLE 68. Characteristics differentiating the species of the genus *Marinibacillus*[a]

Characteristic	*M. marinus*	*M. campisalis*
Cell morphology	Rod	Rod
Spore position	Terminal or subterminal	Terminal or subterminal
Swollen sporangia	– or Slightly	Slightly
Spore shape	Round or slightly ellipsoidal	Round or ellipsoidal
Gram stain	+	+ or d
Pigment	–	Light orange on marine agar
Motility	+ (Peritrichous)	+ (Polar)
H_2S production from cysteine	d (+)	–
Anaerobic growth	–	–
Cytochrome oxidase	d(–)	–
Catalase	+	+
Nitrate reduction to nitrite	d (–)	+
Decomposition of:		
Casein	d (+)	+
Esculin	d (+)	+
Gelatin	d (+)	+
Hypoxanthine	(–)	–
Starch	–	–
Tween 80	(–)	–
Tyrosine	(–)	–
Xanthine	(–)	–
Urease activity	–(d)[b]	
Acid production from:		
Arabinose	–	–
D-Cellobiose	–	+
D-Glucose	d (+)	+
Lactose	–	–
Maltose	d (+)	+
D-Mannitol	–	+
D-Mannose	d (+)	–
Melibiose	(–)	+
D-Raffinose	–	–
L-Rhamnose	–	–
D-Sorbitol	–	–
Sucrose	–(d)[b]	+
D-Trehalose	d (+)	+
D-Xylose	d (+)	–
Growth in NaCl (%):		
0	–	–
7	d (w+)	+
10	–	+
20	–	
21		
Optimal concentration for growth	1.2–0.4%	2–3%
Growth at (°C):		
5	+	+
10	+	+
30	d (+)	+
37	–	+
Optimum growth temperature (°C)	12–23	30
Maximum growth temperature (°C)	25–30	39
Peptidoglycan type	L-Lys-direct	L-Lys-direct
Predominant menaquinone	MK-7	MK-7
Major fatty acids	$C_{15:0\ ante}$	$C_{15:0\ ante}$
G + C content (mol%)	37–41.6	41.8

[a]Symbols: +, property of the species; –, not a property of the species; w, weakly positive reaction; D, variable reaction. Data from Rüger et al., (2000) and Yoon et al., (2001c), (2004).
[b]Data in the literature show some contradictions.

known that there are often various 16S sequences available of the same strain - in this case the type strain -, but they do not always have the same quality. Although there might even be more than one sequence of good quality we have preferred here to give only one accession number with the indication between brackets of the subculture that has been reported to link with that particular accession number. This is the reason why only one accession number has been mentioned.

2. **Marinibacillus campisalis** Yoon, Kim, Schumann, Oh and Park 2004, 1320[VP]

cam.pi.sa'lis. L. n. *campus* field; L. gen. n. *salis* of salt; N.L. gen. n. *campisalis* of the field of salt.

Contains only one strain. Cells are rods, 1.3–1.6 × 2.5–4.0 μm. Gram-positive, but Gram-variable in older cultures. Round to ellipsoidal endospores lie centrally or subterminally in slightly swollen sporangia. Colonies are smooth, glistening, circular to slightly irregular, flat to raised, light orange–yellow in color and 2–3 mm in diameter after 3 d incubation on marine agar. Growth occurs at 4°C; maximum growth temperature is 39°C. Optimal pH for growth is 7.0–8.0; no growth occurs at pH 4.5.

Optimal growth occurs in the presence of 2–3% (w/v) NaCl. Growth occurs in the presence of 15% NaCl, but not 16% NaCl. Xylan is not hydrolyzed. Acid is produced from D-fructose and melibiose and weakly produced from D-melezitose. Acid is not produced from adonitol, D-galactose, *myo*-inositol, D-ribose, or stachyose. Other characteristics are given in Table 68. The peptidoglycan type is A1 (L-Lys direct). The predominant menaquinone is MK-7. The major fatty acid is $C_{15:0\ ante}$. Isolated from marine solar saltern in Korea.

DNA G+C content (mol%): 41.8 (HPLC).

Type strain: SF-57, JCM 11810, KCCM 41644.

EMBL/GenBank accession number (16S rDNA): AY190535.

Genus VII. **Planomicrobium** Yoon, Kang, Lee, Lee, Kho, Kang and Park 2001a, 1518[VP]

JUNG-HOON YOON, BYUNG-CHUN KIM AND YONG-HA PARK

Pla.no.mi.cro'bi.um. Gr. n. *planos* wanderer; Gr. adj. *micros* small; Gr. n. *bios* life; N.L. neut. n. *microbium* microbe; N.L. neut. n. *Planomicrobium* motile microbe.

Rods or rods to cocci/short rods in liquid culture and on plates usually 1.0–2.0 μm. Spores are not formed. **Gram-positive to Gram-variable. Motile by means of a single polar flagellum or peritrichous flagella.** Aerobic. Colonies on plates are circular, smooth, and **yellow to orange or pale orange** in color. Nitrate is or is not reduced to nitrite. Catalase-positive. Oxidase-negative or positive. Urease-negative. Hydrolysis of esculin, casein, starch, and Tween 80 are negative or positive. Hydrolysis of gelatin is positive. The menaquinone profile is characterized by the predominance of MK-8 followed by MK-7, or by the predominance of MK-8 followed by MK-7 and MK-6. The cellular fatty acids are mainly unsaturated and branched fatty acids (Table 69). The cellular phospholipids are phosphatidylethanolamine, phosphatidylglycerol, and bisphosphatidylglycerol. In the 16S rRNA gene sequences, two sequence signatures, cytosine and guanine, are present at positions 183 and 190 (*Escherichia coli* numbering), respectively (Dai et al., 2005).

DNA G+C content (mol%): 34.8–47 (HPLC, T_m).

Type species: **Planomicrobium koreense** Yoon, Kang, Lee, Lee, Kho, Kang and Park 2001a, 1518[VP].

Further descriptive information

Phylogenetic analysis shows that *Planomicrobium* forms a distinct cluster containing four species previously described as *Planococcus* species, and the relationship between this cluster

TABLE 69. Cellular fatty acid profiles of the type strains of genus *Planomicrobium*[a,b]

Fatty acid	P. koreense	P. alkanoclasticum	P. chinense	P. mcmeekinii	P. okeanokoites	P. psychrophilum
Saturated:						
$C_{15:0}$	2.4	–	4.9	1.6	–	–
$C_{16:0}$	–	–	1.5	1.1	1.1	4.4
$C_{17:0}$	–	–	–	–	–	6.1
$C_{18:0}$	–	–	–	–	–	1.6
Unsaturated:						
$C_{16:1\ \omega7c}$ alcohols	23.6	–	12.1	22.9	26.1	–
$C_{16:1\ \omega11c}$	1.7	6.4	4.5	2.3	2.1	3.2
$C_{18:1\ \omega9c}$	–	2.7	–	–	1.0	1.0
Branched:						
$C_{14:0\ iso}$	21.0	6.0	10.4	18.1	40.4	3.2
$C_{15:0\ iso}$	6.4	7.5	3.8	5.3	–	5.6
$C_{15:0\ anteiso}$	26.1	45.5	49.7	32.0	4.0	41.3
$C_{16:0\ iso}$	11.1	17.1	4.1	8.7	24.5	8.1
$C_{16:1\ iso}$	–	–	–	–	–	7.2
$C_{17:0\ iso}$	1.6	7.4	–	1.3	–	–
$C_{17:0\ anteiso}$	1.6	10.2	2.8	1.2	–	7.3
$C_{17:1\ iso\ \omega5c}$	2.8	–	–	–	–	–
$C_{17:1\ iso\ \omega10c}$	1.9	–	1.7	2.4	–	–
$C_{18:0\ iso}$	–	1.2	–	–	0.9	–
$C_{17:1\ iso}$ I and $C_{17:1\ anteiso}$ B	–	9.3	4.4	3.0	–	11.3

[a]Symbols: +, >85% positive; d, different strains give different reactions (16–84% positive); –, 0–15% positive; w, weak reaction; nd, not determined.

[b]Data from Engelhardt et al. (2001) for *Planomicrobium alkanoclasticum* cultured on sea water agar (NCIMB medium 209); Yoon et al. (2001a) for *Planomicrobium koreense*, *Planomicrobium mcmeekinii*, and *Planomicrobium okeanokoites* cultured on marine agar; Reddy et al. (2002) for *Planomicrobium psychrophilum*; and Dai et al. (2005) for *Planomicrobium okeanokoites* cultured in marine broth.

and other species described as *Planococcus* is supported by bootstrap analysis at a confidence level of 100%. The 16S–23S internally transcribed spacer (ITS) sequence similarity and DNA–DNA relatedness values between *Planomicrobium koreense* and the type strains of other *Planococcus* species are in the range 74.6–83.2% and 10.4–20.5%, respectively. Cell wall peptidoglycan type is L-Lys-D-Glu or L-Lys-D-Asp. *Planomicrobium alkanoclasticum* MAE2 (basonym: *Planococcus alkanoclasticus*) degrades linear and branched alkanes from C11–C33 (Engelhardt et al., 2001). *Planomicrobium okeanokoites* IFO 12536 (basonym: *Flavobacterium okeanokoites*) produces *Fok*I restriction endonuclease recognizing the nonpalindromic pentadeoxyribonucleotide 5′-GGATG-3′:5′-CATCC-3′ in duplex DNA and cleaving 9 and 13 nucleotides away from the recognition site, and adenine-N6-specific DNA-methyltransferase M.*Fok*I, an adenine-N6-specific DNA-methyltransferase which specifically methylates both adenine residues within 5′-GGATG-3′:5′-CATCC-3′ sequences (Landry et al., 1989; Sugisaki and Kanazawa, 1981). Members of the genus *Planomicrobium* are currently represented by six species including four species reclassified from *Planococcus* (*Planomicrobium okeanokoites, Planomicrobium mcmeekinii, Planomicrobium psychrophilum*, and *Planomicrobium alkanoclasticum*), and two isolates (*Planomicrobium koreense* and *Planomicrobium chinense*). Characteristics that are useful in differentiating members of the genus *Planomicrobium* are given in Table 70.

Enrichment and isolation procedures

Members of the genus *Planomicrobium* have been isolated from specimens related to marine environment such as seafood, coastal sediments, and Antarctic sea ice. For the isolation of *Planomicrobium*, media containing NaCl, marine agar (MA,Difco), or artificial sea water basal medium with 1% peptone and 0.5% yeast extract (Eguchi et al., 1996) may be used.

Maintenance procedures

Members of the genus *Planomicrobium* may be maintained on marine agar for short term storage at 4°C. Freeze-drying is recommended for long term storage of members of this genus.

Taxonomic comments

Based on 16S rRNA gene sequence data, the closest relative of *Planomicrobium* is *Planococcus*. *Planomicrobium* is distinguishable from *Bacillus* species and the other genera belonging to the *Bacillus* rRNA group 2 by menaquinone type and lack of spore formation (Rheims et al., 1999).

List of species of the genus *Planomicrobium*

1. **Planomicrobium koreense** Yoon, Kang, Lee, Lee, Kho, Kang and Park 2001a, 1518[VP]

   ko.re.en′se. N.L. neut adj. *koreense* from Korea.

   Cells are cocci or short rods in the early growth phase but soon change to rods (Figure 68). Gram-positive, and Gram-

variable in old cultures. Motile by means of a single polar flagellum (Figure 68). Colonies on plates are circular, smooth, low convex, and yellow to orange in color. Arbutin and elastin are not hydrolyzed. Growth occurs at 4–38°C, but weakly at 39°C and no growh at 40°C. The optimal growth temperature is 20–30°C. The optimal pH for growth is 7.0–8.5, and growth

TABLE 70. Characteristics differentiating type strains of the genus *Planomicrobium*[a]

Property	P. koreense	P. alkanoclasticum	P. chinense	P. mcmeekinii	P. okeanokoites	P. psychrophilum
Cell shape	Coccoid/rods	Rods	Coccoid/short rods	Coccoid/rods	Rods	Rods
Growth temperature (°C)	4–38	15–41	12–43	0–37	20–37	2–30
Growth in presence of NaCl (%)	0–7	0.8–6	0–10	0–14	0–7	0–12
Catalase	+	+	+	+	+	+
Oxidase	−	−	−	−	w	+
Urease	−	nr	nr	−	−	−
Nitrate reduction	−	−	+	+	−	−
Hydrolysis of:						
Casein	+	nr	−	+	+	nr
Esculin	+	nr	−	nr	−	+
Starch	−	+	−	−	−	−
Gelatin	+	+	+	+	+	+
Tween 80	−	−	−	−	−	+
Acid production from glucose	w	−	−	+	−	−
DNA G+C content (mol%)	47.0 (HPLC)	45.3 (HPLC)	34.8 (T_m)	35.0 (HPLC)	46.3 (HPLC)	44.5 (T_m)
Peptidoglycan type	L-Lys-D-Glu	nr	L-Lys-D-Glu	L-Lys-D-Asp	L-Lys-D-Asp	L-Lys-D-Glu
Major menaquinones	MK-8, MK-7, MK-6	MK-8, MK-7	MK-8, MK-7	MK-8, MK-7	MK-8, MK-7	MK-8, MK-7
Phospholipids[b]	PE, PG, BPG	nr	nr	nr	PE, PG, BPG	PE, PG, BPG

[a]Symbols: +, >85% positive; d, different strains give different reactions (16–84% positive); −, 0–15% positive; w, weakly positive; nr, not reported. (Data were derived from Nakagawa et al., 1996; Junge et al., 1998; Engelhardt et al., 2001; Yoon et al., 2001a; Reddy et al., 2002; and Dai et al., 2005.)

[b]PE, phosphatidylethanolamine; PG, phosphatidylglycerol; BPG, bisphosphatidylglycerol.

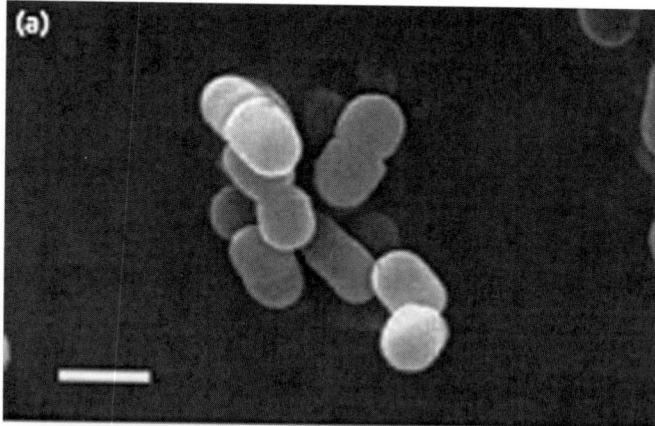

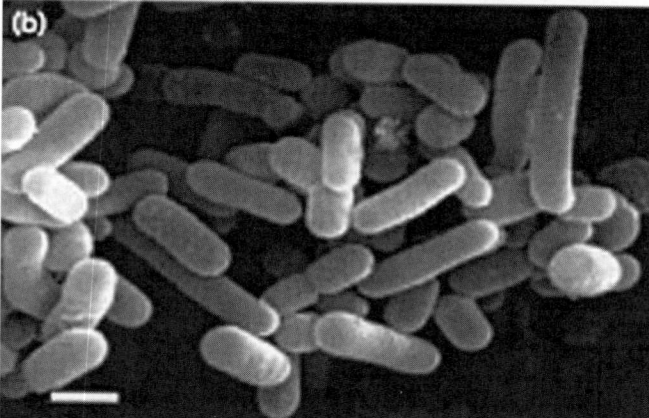

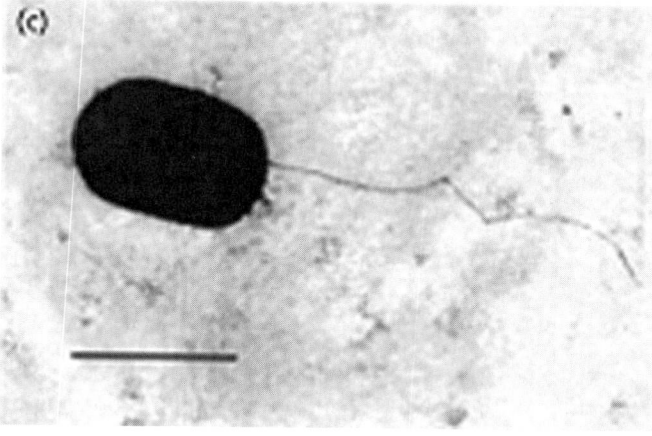

FIGURE 68. Electron microscopy of *Planomicrobium koreense* grown on IFO medium no. 326. Scanning electron micrograph of cells from (a) early growth phase and (b) stationary phase. (c) Transmission electron micrograph of cells from exponentially growing culture. Bars = 1 μm.

is inhibited at pH 5.5 and 10. Growth occurs in the presence of 0–6% NaCl but weakly in the presence of 7% NaCl and does not occur in the presence of more than 8% NaCl. The optimal concentration of NaCl for growth is 1–4%.

Acid is produced from esculin, cellobiose, maltose, lactose, melibiose, and 5-ketogluconate. No acid is produced from glycerol, erythritol, D-arabinose, L-arabinose, ribose, D-xylose, L-xylose, adonitol, β-methyl-D-xyloside, galactose, fructose, mannose, sorbose, rhamnose, dulcitol, inositol, mannitol, sorbitol, α-methyl-D-mannoside, α-methyl-D-glucoside, N-acetylglucosamine, amygdalin, arbutin, salicin, sucrose, trehalose, inulin, melezitose, raffinose, starch, glycogen, xylitol, gentiobiose, D-turanose, D-lyxose, D-tagatose, D-fucose, L-fucose, D-arabitol, L-arabitol, gluconate, or 2-ketogluconate. Isolated from the Korean traditional fermented seafood jeotgal.

DNA G+C content (mol%): 47 (HPLC).

Type strain: JG07, CIP 107134, JCM 10704, KCTC 3684.

GenBank accession number (16S rRNA gene): AF144750.

2. **Planomicrobium alkanoclasticum** (Engelhardt, Daly, Swannell and Head 2001) Dai, Wang, Wang, Liu and Zhou 2005, 702[VP] (*Planococcus alkanoclasticus* Engelhardt, Daly, Swannell and Head 2001, 245.)

al.kan.o.cla'sti.cum. N.L. n. *alkanum* alkane; Gr. adj. *clastos* broken; N.L. neut. adj. *alkanoclasticum* breaking alkanes.

The following description is taken from the original paper. Cells are rods that are 0.4–0.8 μm wide and 1.7–2.6 μm long. Gram-positive, though Gram stain reaction may be variable. Chemo-organotroph with respiratory metabolism. Colonies on marine agar or nutrient agar with 0.8% NaCl are orange pigmented. Temperature range for growth is 15–41°C. Obligate requirement for NaCl but will not grow at NaCl concentrations of 6% or greater.

Produce DNase and phosphatase. Negative tests were obtained for the utilization of α-cyclodextrin, β-cyclodextrin, dextrin, glycogen, inulin, mannan, N-acetyl-D-glucosamine, N-acetyl-D-mannosamine, amygdalin, L-arabinose, D-arabitol, arbutin, cellobiose, D-fructose, L-fucose, D-galactose, D-galacturonic acid, gentiobiose, D-gluconic acid, α-D-glucose, m-inositol, α-D-lactose, lactulose, maltose, maltotriose, D-mannitol, D-mannose, D-melezitose, D-melibiose, α-methyl-D-galactoside, β-methyl-D-galactoside, 3-methyl glucose, α-methyl-D-glucoside, β-methyl-D-glucoside, α-methyl-D-mannoside, palatinose, D-psicose, D-raffinose, L-rhamnose, D-ribose, salicin, seduheptulosan, D-sorbitol, stachyose, sucrose, D-tagatose, D-trehalose, turanose, xylitol, D-xylose, acetic acid, α-hydroxybutyric acid, β-hydroxybutyric acid, γ-hydroxybutyric acid, p-hydroxyphenylacetic acid, α-ketoglutaric acid, α-ketovaleric acid, lactamide, D-lactic acid methyl ester, L-lactic acid, D-malic acid, L-malic acid, methyl pyruvate, monomethylsuccinate, propionic acid, pyruvic acid, succinamic acid, succinic acid, N-acetyl L-glutamic acid, alaninamide, D-alanine, L-alanine, L-alanyl-glycine, L-asparagine, L-glutamic acid, glycyl-L-glutamic acid, L-pyroglutamic acid, L-serine, putrescine, 2,3-butanediol, glycerol, adenosine, 2'-deoxyadenosine, inosine, thymidine, uridine, adenosine-5'-monophosphate, thymidine-5'-monophosphate, uridine-5'-monophosphate, fructose-6-phosphate, glucose-1-phosphate, glucose-6-phosphate, and DL-α-glycerol. Isolated from intertidal beach sediment.

DNA G+C content (mol%): 45.3 ± 0.4% (HPLC, n. 4).

Type strain: MAE2, CIP 107718, NCIMB 13489.

GenBank accession number (16S rRNA gene): AF029364.

3. **Planomicrobium chinense** Dai, Wang, Wang, Liu and Zhou 2005, 701[VP]

chin.en'se. N.L. neut. adj. *chinense* from China.

Cells are coccoid or short rods, $0.8 \times 1.0\,\mu m$. Gram-positive. Motile by polar flagella. Strictly aerobic. Colonies are smooth, circular, low-convex, and yellow to orange in color when cultivated on MA. Temperature range for growth is 10–45°C, but growth does not occur at 46°C. The optimal growth temperature is 30–35°C. Optimal pH for growth is 6.0–7.0; no growth occurs below pH 5.0 or above pH 10.0.

Acid is not produced from sucrose, raffinose, lactose, arabinose, cellobiose, xylose, rhamnose, melibiose, mannose, or mannitol. Citrate does not support growth.

Isolated from coastal sediment from the Eastern China Sea in Fujian Province, China.

DNA G+C content (mol%): 34.8 (T_m).

Type strain: DX3-12, AS 1.3454, JCM 12466.

GenBank accession number (16S rRNA gene): AJ697862.

4. **Planomicrobium mcmeekinii** (Junge, Gosink, Hoppe and Staley 1998) Yoon, Kang, Lee, Lee, Kho, Kang and Park 2001a, 1519[VP] (*Planococcus mcmeekinii* Junge, Gosink, Hoppe and Staley 1998, 312.)

mc.mee.kin'i.i. N.L. gen. n. *mcmeekinii* of McMeekin named after Thomas A. McMeekin, Australian microbiologist.

Cells are rod-shaped, $0.8–1.2 \times 0.8–10\,\mu m$ long, and occur singly. Gram-positive. Aerobic. Colonies are pale orange, circular, convex, undulate on SWCm agar (Irgens et al., 1989). Growth from 0–37°C. Tolerate up to 14% NaCl. Oxidative and fermentative glucose metabolism in MOF medium of Leifson. Halotolerant. No sodium requirement.

Utilize proprionate, pyruvate, acetate, and malate as sole carbon source. Isolated from Antarctic sea ice.

DNA G+C content (mol%): 35 (HPLC).

Type strain: S23F2, ATCC 700539, CIP 105673.

GenBank accession number (16S rRNA gene): AF041791.

5. **Planomicrobium okeanokoites** (Zobell and Upham 1944) Yoon, Kang, Lee, Lee, Kho, Kang and Park 2001a, 1518[VP] (*Flavobacterium okeanokoites* ZoBell and Upham 1944; *Planococcus okeanokoites* Nakagawa, Sakane and Yokota 1996, 869.)

o.ke.a.no.ko.i'tes. Gr. masc. n. *okeanos* the ocean; Gr. fem. n. *choite* bed; N.L. fem. gen. n. *okeanokoites* of the ocean bed.

Cells are rods $0.4–0.8 \times 1.0–20\,\mu m$ long. Most cells are less than $2.8\,\mu m$ long. Gram-positive to Gram-variable in medium containing NaCl. Gram-negative in medium without NaCl. Motile by means of peritrichous flagella. Strict aerobes. The color of the cell mass is usually bright yellow to bright orange. Chemo-organotrophs. Metabolism is respiratory. The optimum temperature range for growth is 20–37°C. The optimum level of salinity for growth is 3–5% NaCl. No growth occurs in the presence of more than 7% NaCl.

The following tests are positive: arginine dihydrolase, lysine decarboxylase, and ornithine decarboxylase. The following tests are negative: methyl red, Voges–Proskauer reaction, indole production, phenylalanine deaminase, hydrolysis of cellulose, and decomposition of tyrosine. Acid is not produced from D-galactose, L-arabinose, D-xylose, sucrose, maltose, lactose, or mannitol in Hugh–Leifson O-F medium. The molar ratio of the amino acids glutamic acid, lysine, alanine, and aspartic acid in the cell wall is 1:1:2:1. Isolated from marine mud.

DNA G+C content (mol%): 46.3 (HPLC).

Type strain: ATCC 33414, CCM 320, CIP 105082, NBRC 12536, LMG 4030, NCIMB 561, VKM B-1175.

GenBank accession number (16S rRNA gene): D55729.

6. **Planomicrobium psychrophilum** (Reddy, Prakash, Vairamani, Prabhakar, Matsumoto and Shivaji 2002) Dai, Wang, Wang, Liu and Zhou 2005, 702[VP] (*Planococcus psychrophilus* Reddy, Prakash, Vairamani, Prabhakar, Matsumoto and Shivaji 2002, 260.)

psy.chro.phi'lum. Gr. n. *psychros* cold; Gr. adj. *philos* loving; N.L. neut. adj. *psychrophilum* cold-loving.

Cells are rod-shaped, single, and Gram-positive. Colonies on peptone–yeast extract medium are orange, smooth, convex, circular, and 1–2 mm in diameter. Growth occurs at 2–30°C. The optimal growth temperature is 22°C. Cultures can grow at pH 6–12, optimum growth at pH 7, and they tolerate up to 12% NaCl. Pigment is insoluble in water but soluble in methanol and exhibits absorption maxima at 440, 465, and 487.5 nm. Pigment synthesis is not dependent on the growth phase or the growth conditions.

Can hydrolyze esculin. Lipase, β-galactosidase, and arginine dihydrolase-positive, but negative with respect to phosphatase, indole production, the methyl red and Voges–Proskauer tests, and levan formation. Can utilize rhamnose, melibiose, trehalose, xylose, glycerol, lysine, sodium acetate, sodium succinate, inositol, glutamic acid, and pyruvate, but not glucose, lactose, sorbose, arabinose, cellobiose, sucrose, fructose, mannose, mannitol, raffinose, ribose, lactose, lactic acid, adonitol, maltose, glucosamine, sorbitol, melizitol, β-hydroxybutyric acid, dulcitol, dextran, PEG, glycine, sodium citrate, cellulose, inulin, alanine, phenylalanine, methionine, glutamine, arginine, serine, potassium hydrogen phthalate, myristic acid, creatinine, tyrosine, or glycogen as the sole carbon source. Further, it does not produce acid or gas from sucrose, cellobiose, lactose, sodium glutamate, or sodium thioglycolate.

It is sensitive to penicillin, chlortetracycline, chloramphenicol, neomycin, streptomycin, novobiocin, tetracycline, bacitracin, furazolidone, colistin, kanamycin, lincomycin, cotrimoxazole, ampicillin, amoxycillin, trimethoprim, erythromycin, nalidixic acid, nystatin, gentamicin, and polymyxin B, but resistant to carbenicillin, tobramycin, oxytetracycline, nitrofurazone, and nitrofurantoin. Isolated from a cyanobacterial mat sample from McMurdo Dry Valleys, Antarctica.

DNA G+C content (mol%): 44.5 (T_m).

Type strain: CMS 53or, DSM 14507, MTCC 3812.

GenBank accession number (16S rRNA gene): AJ314746.

Genus VIII. **Sporosarcina** Kluyver and van Niel 1936, 401[AL] emend. Yoon, Lee, Weiss, Kho, Kang and Park 2001b, 1085

THE EDITORIAL BOARD

Spo.ro.sar.ci′na. Gr. n. *spora* a spore; L. fem. n. *sarcina* a package, bundle; N.L. fem. n. *Sporosarcina* a sporeforming package.

Endospore forming cocci or rods. Gram positive or variable. Most species are motile. **Strict or facultative aerobes.** Catalase positive. Most species are also oxidase and urease positive. Nitrate reduction to nitrite is variable. Optimum growth temperature and pH are 20–30°C and 6.5–8, respectively. Many species grow at low temperatures <10°C. Many species are also halotolerant and grow at <3–15% NaCl. Cell wall peptidoglycan contains L-lysine at position 3 of the peptide subunit and is the A4α type. **The principal menaquinone is MK-7. Major fatty acid is C$_{15:0\ ante}$.**

DNA G + C content (mol%): 38–46.5.

Type species: **Sporosarcina ureae** (Beijerinck 1901) Kluyver and van Niel 1936, 401[VP] (*Planosarcina ureae* Beijerinck 1901, 52.).

Further descriptive information

Coherent clustering of the nine species of *Sporosarcina* is suggested by sequence analyses of the 16S rRNA gene. One study found that *Sporosarcina saromensis* forms a cluster with the type strains for *Sporosarcina aquimarina* (97.3% similarity), *Sporosarcina globispora* (96.9%), *Sporosarcina psychrophila* (96.8%), *Sporosarcina ureae* (96.8%), *Sporosarcina pasteurii* (96.0%), and *Sporosarcina macmurdoensis* (95.9%) (An et al., 2007). Another study found that *Sporosarcina koreensis* forms a cluster with the type strains for *Sporosarcina soli* (98.9%), *Sporosarcina globispora* (97.3%), *Sporosarcina aquimarina* (97.2%), and *Sporosarcina psychrophila* (96.9%) (Kwon et al., 2007).

Though similar in phenotypic and physiological characteristics, *Sporosarcina* species may also be distinguished on the basis of phenotypic differences (see Table 71). The cells of some species occur singly or in pairs (*Sporosarcina globisporus*, *Sporosarcina koreensis*, and *Sporosarcina macmurdoensis*), whereas others also form clusters or short chains (*Sporosarcina psychrophila* and *Sporosarcina soli*). Endospores can be terminal (*Sporosarcina aquimarina*, *Sporosarcina globispora*, *Sporosarcina koreensis*, *Sporosarcina pasteurii*, *Sporosarcina psychrophila*, and *Sporosarcina saromensis*), subterminal (*Sporosarcina macmurdoensis*), or central (*Sporosarcina soli*). Only two species are nonmotile (*Sporosarcina macmurdoensis* and *Sporosarcina soli*). All species tested are catalase positive. Only one species is negative for urease and oxidase (*Sporosarcina macmurdoensis*). While *Sporosarcina pasteurii* is positive for urease, it has not been tested for catalase and oxidase.

TABLE 71. Differential phenotypic and physiological characteristics of species of the genus *Sporosarcina*[a]

Characteristic	S. ureae	S. aquimarina	S. globispora	S. koreensis	S. macmurdoensis	S. pasteurii	S. psychrophila	S. saromensis	S. soli
Colony color[b]	W	O	W	O	W	W	W	B	O
Cell shape[c]	S	R	R	R	R	R	R	R	R
Spore position[d]	NA	T	T	T	ST	T	T	T	C
Motility	+	+	+	+	–	+	+	+	C
NaCl (%) tolerance	3	13	5	7	3	10	5	9	5
Optimum pH	7	6.5–7.0	7	7	7	9	7	6.5	8
Growth at (°C):									
5 and 10	NA	+	+	–	+	NA	+	+	–
30	+	+	+	+	–	+	w	+	+
40	–	–	–	+	–	w	–	+	–
Optimum growth temperature (°C)	25	25	20–25	30	20	30	25	27	30
Anaerobic growth	–	+	+	–	+	+	+	–	–
Presence of:									
Oxidase	+	+	+	+	–	NA	+	+	+
Urease	+	+	+	+	–	+	+	+	+
Hydrolysis of:									
Casein	–	–	–	–	NA	w	–	–	–
Gelatin	–	+	+	+	+	+	+	+	–
Starch	–	–	–	–	+	–	+	+	–
Tween 80	w	–	–	–	NA	NA	–	NA	–
Tyrosine	+	w	NA	–	NA	NA	–	NA	–
Nitrate reduction	+	+	+	–	–	+	+	–	+
DNA G + C content (mol%)	40–41.5	40	40	46.5	44	38.5	44.1	46	44.5

[a]Adapted from Table 1 in Kwon et al., (2007). Symbols: +, positive, –, negative,; w, weak, NA, data not available.
[b]B, beige, O, light orange; W, white.
[c]R, rod; S, spherical.
[d]C, central; ST, subterminal; T, terminal.

Isolation and enrichment procedures

Isolation is best achieved by plating soil or water dilutions on an appropriate agar. The growth of *Sporosarcina ureae* can be enhanced by addition of 3% urea. Two methods for isolating *Sporosarcina ureae* are described in detail in Claus et al. (1991) and include Claus's 1981 modification of Gibson's method and the method used by Pregerson in 1973. In the former, about 5 g of dried soil was suspended in 20 ml of sterile tap water, and the suspension was diluted 1:10 and 1:100. About 0.1 ml of each dilution was streaked on nutrient agar plates (peptone, 5 g; meat extract, 3 g; agar, 15 g; distilled water, 100 ml) supplemented with 30–100 g urea/liter. After incubation at 25°C for 3 or more days, colonies were visible under a dissecting microscope or by transmitted light. Colonies were checked microscopically for the sarcina morphology and restreaked on nutrient agar containing 1% urea for purification. Testing motility and endospore formation was used to confirm the provisional identification.

The second method used tryptic soy yeast extract agar (Difco Tryptic Soy Broth, 27.5 g; Difco Yeast Extract, 5.0 g; glucose, 5.0 g; Bacto Agar, 15.0 g; distilled water, 1000 ml), which was adjusted to pH 8.5 with NaOH before sterilization. After sterilization, a filter-sterilized urea solution was added to give a final concentration of 1% (Claus and Fahmy, 1986). If cells from picked colonies exhibited motile tetrads and/or packets under the microscope, the colonies were restreaked on the same medium for further purification.

Other species were isolated by a variety of methods. *Sporosarcina aquimarina* was isolated on trypticase soy agar (TSA) supplemented with artificial sea water (pH 7.5) (Yoon et al., 2001b). *Sporosarcina koreensis* and *Sporosarcina soli* were isolated upon dilution and plating of soil on TSA (30°C for 2 days) (Kwon et al., 2007). *Sporosarcina macmurdoensis* was isolated from a suspension of cyanobacterial mat after plating on ABM agar (0.5% peptone, 0.2% yeast extract, 1.5% agar, pH 7.2) (Reddy et al., 2003). *Sporosarcina saromensis* was isolated from surface water of Lake Saroma or marine sediment by plating on JCM57 medium (10 g of glucose, 1.0 g of asparagine, 0.5 g of K_2HPO_4, 2.0 g of yeast extract, 15 g of agar per liter of distilled water, pH adjusted to 7.3) or 1/10 diluted marine agar 2216 (An et al., 2007).

Maintenance procedures

Sporosarcina ureae vegetative cultures grown on nutrient agar are viable for up to a year when stored at 4–10°C in the dark. Endospores survive several years in screw-capped tubes under the same conditions (Claus and Fahmy, 1986). In general, strains of *Sporosarcina ureae* form spores in nutrient agar supplemented with urea (final concentration 0.2%) if incubated at 25°C. Addition of 50 mg of $MnSO_4$ H_2O per liter may enhance sporulation. Good sporulation can also be expected in the medium of MacDonald and MacDonald (1962) (see Claus and Fahmy, 1986).

Cryopreservation is recommended for long term storage. Cryoprotectants in use include skim milk (20%, w/v), serum containing 5% *meso*-inositol, glycerol (10%), and DMSO (5%).

Differentiation of the genus *Sporosarcina* from other species

The genus belongs to the family *Planococcaceae* and can be differentiated from other family members on the basis of its morphology, chemotaxonomic markers and 16S rRNA sequence. Species of *Sporosarcina* form coccoid or rod-shaped cells and round or oval endospores. Most species are motile. All species described so far possess MK-7 as the major menaquinone and an A4α peptidoglycan type in the cell wall (Reddy et al., 2003).

Taxonomic comments

Nine species are currently assigned to the genus *Sporosarcina*. In the last edition of *Bergey's Manual of Systematic Bacteriology*, this genus comprised the type species, *Sporosarcina ureae*, and *Sporosarcina halophila* (Claus and Fahmy, 1986). However, subsequent 16S rRNA sequence analyses led to the reassignment of the latter species to a new genus, *Halobacillus* (Spring et al., 1996). Further analyses of *Bacillus* species discovered close relationships of *Bacillus globisporus*, *Bacillus psychrophilus*, and *Bacillus pasteurii* to *Sporosarcina ureae*, and these organisms were reclassified as *Sporosarcina globispora*, *Sporosarcina psychrophila*, and *Sporosarcina pasteurii*, respectively (Yoon et al., 2001b). Five additional species have also been isolated in soil and water samples.

List of species of the genus *Sporosarcina*

1. **Sporosarcina ureae** (Beijerinck 1901) Kluyver and van Niel 1936, 401[AL] (*Planosarcina ureae* Beijerinck 1901, 52.)

ure.ae. N.L. n. *urea* urea; N.L. gen. n. *ureae* of urea.

The description is the same as for the genus except as follows. Cells are spherical or oval cocci (1.0–2.5 μm in diameter). They occur singly, in pairs, tetrads, and in packets of eight or more cells as a result of division in two or three perpendicular planes. Endospores are 0.5–1.5 μm in diameter. Cells are motile, and flagella are randomly spaced. Colonies are gray or cream, becoming yellowish, orange, or brown depending upon the strain and medium. Negative for acid production from D-glucose and D-xylose. In nature, it may play an active role in urea decomposition. Other characteristics are described in Table 71.

DNA G + C content (mol%): 40–42 (T_m).

Type strain: ATCC 6473, DSM 2281, JCM 2577, LMG 17366, NBRC 12699, VKM B-595.

GenBank accession number (16S rRNA gene): AF202057.

2. **Sporosarcina aquimarina** Yoon, Lee, Weiss, Kho, Kang and Park 2001b, 1084[VP]

a.qui.ma.ri′na. L. n. *aqua* water; L. adj. *marinus* of the sea; M.L. adj. *aquimarina* pertaining to sea water.

Cells are rods (0.9–1.2 μm × 2.0–3.5 μm) that form round, terminal endospores in swollen sporangia in 3-day-old cultures on trypticase soy agar. Gram variable. Motile by means of a single polar flagellum. Colonies are light orange, smooth, circular to irregular, and raised on TSA. Grows in the presence of 13% NaCl but not in the presence of more than 14% NaCl. Growth occurs at 4 and 37°C, but not at 40°C. The optimal growth temperature is 25°C. The optimal pH for growth is 6.5–7.0. Growth is inhibited below pH 5.0. Does not hydrolyze esculin, arbutin, casein, hypoxanthine, starch, Tween 80, and xanthine. Negative for deamination of arginine and production of indole. Acid is produced from N-acetylglucosamine, esculin, fructose, glycerol, 5-ketogluconate, ribose, and D-tagatose but not from erythritol, D-arabinose, L-arabinose, D-xylose, L-xylose, adonitol, β-methyl-D-xyloside, galactose, glucose, mannose, sorbose, rhamnose, dulcitol, inositol, mannitol, sorbitol, α-methyl-D-mannoside, α-methyl-D-glucoside, amygdalin, arbutin, salicin, cellobiose, maltose, lactose, melibiose, sucrose, trehalose, inulin, melezitose, raffinose, starch, glycogen, xylitol, gentiobiose, D-turanose, D-lyxose, D-fucose, L-fucose, D-arabitol, L-arabitol, gluconate, and 2-ketogluconate. The major fatty acid is $C_{15:0\,ante}$. Isolated from sea water in Korea. Other characteristics are included in Table 71 and the genus description.

DNA G + C content (mol%): 40 (HPLC).

Type strain: SW28, JCM 10887, KCCM 41039.

GenBank accession number (16S rRNA gene): AF202056.

3. **Sporosarcina globispora** (Larkin and Stokes 1967) Yoon, Lee, Weiss, Kho, Kang, and Park 2001b, 1085[VP] (*Bacillus globisporus* Larkin and Stokes 1967, 892.)

glo.bis'por.a. L. n. *globus* a sphere; N.L. n. *spora* a spore; N.L. fem. adj. *globispora* with spherical spores.

Cells are rods (usually 0.9–1.0 µm × 2.5–4.0 µm) with rounded ends. They occur singly or in pairs, stain uniformly, and are motile by means of peritrichous flagella. Gram positive or Gram variable. They form terminal to subterminal (sometimes slightly lateral), thick-walled, easily stained spores (1.0–1.1 µm in diameter). Colonies on nutrient agar are off-white, raised, and irregular with lobate to undulate margins. In glucose broth, growth occurs under anaerobic conditions. Growth is the same on nutrient agar, glucose agar, and tyrosine agar, but better on soybean agar and trypticase soy agar. No growth or scant growth on glucose nitrate agar and proteose-peptone acid agar. Moderate turbidity forms with sediment in nutrient broth. A small curd forms in litmus milk, which becomes slightly alkaline, after 3 weeks of growth. No indole production. Using ammonium salts as the source of nitrogen, cells form acid but not gas from glucose, lactose, sucrose, and glycerol but not from mannitol, arabinose, and xylose. However, no acid is formed from lactose in the presence of an organic nitrogen source. Negative for starch hydrolysis, acetylmethylcarbinol production, and citrate utilization. Maximum temperature for growth is 20–25°C. Grows and sporulates at 0°C. Isolated from soil and river water. Other characteristics are included in Table 71 and the genus description.

DNA G + C content (mol%): 39.7 (Bd), 39.8 (T_m).

Type strain: 785, Larkin and Stockes W 25, ATCC 23301, CCUG 7419, CIP 103266, DSM 4, HAMBI 471, NBRC 16082, JCM 10046, LMG 6928, NRRL NRS-1533, NRRL B-3396, VKM B-1435. *GenBank accession number (16S rRNA gene):* X54967, X68415.

4. **Sporosarcina koreensis** Kwon, Kim, Song, Weon, Schumann, Tindall, Stackebrandt and Fritze 2007, 1697[VP]

ko.re.en'sis. N.L. fem. adj. *koreensis* referring to Korea, where the isolates were collected.

The species characteristics are the same as those of the genus except as follows. Cells are sporeforming rods (0.5–0.7 µm × 2.5–3.0 µm) occurring singly or in short chains. Gram positive. They form mainly oval, terminal endospores in swollen sporangia. Colonies are light orange (after 2 days on TSA at 30°C). Growth is strictly aerobic and does not occur in >7% NaCl. The growth temperature range is 15–40°C (optimum, 30°C), and growth pH range is 6.0–9.0 (optimum, 7.0; no growth at pH 5.7 or 10.0). Negative for anaerobic growth, formation of indole and dihydroxyacetone, in the Voges–Proskauer test, for phenylalanine deamination, and acid production from D-glucose, L-arabinose, D-xylose, and D-mannitol. No utilization of citrate or propionate. The major fatty acids are $C_{15:0\,iso}$ and $C_{15:0\,ante}$. The major polar lipids are diphosphatidylglycerol, phosphatidylglycerol, phosphatidylethanolamine, an unidentified phospholipid, and two unidentified aminophospholipids. The type strain was isolated from upland soil in Suwon, Korea. Other characteristics are included in Table 71.

DNA G + C content (mol%): 46.5 (HPLC).

Type strain: F73, DSM 16921, KACC 11299.

GenBank accession number (16S rRNA gene): DQ073393.

5. **Sporosarcina macmurdoensis** Reddy, Matsumoto and Shivaji 2003, 1364[VP]

mac.mur.do.en'sis. N.L. fem. adj. *macmurdoensis* pertaining to the McMurdo Region, Antarctica, where the isolates were collected.

Cells are single, nonmotile, rods and form subterminal spores. Gram positive. Colonies (2–3 mm in diameter) are white, circular, flat, and opaque. They tolerate a maximum of 3% (w/v) NaCl and grow at pH 6–9 (pH 7 is optimum for growth) and at psychrophilic temperatures (4–25°C; 18–20°C is optimal for growth). They do not hydrolyze esculin or reduce nitrate to nitrite. They are negative for lipase, β-galactosidase, arginine dihydrolase, arginine decarboxylase, lysine decarboxylase, indole production, methyl red test, and Voges–Proskauer test. Neither acid nor gas is produced from L-arabinose, D-fructose, D-galactose, lactose, D-mannose, D-mannitol, L-rhamnose, sucrose, or D-xylose. Utilizes dulcitol, D-fructose, D-galactose, D-glucose, *meso*-inositol, lactose, D-maltose, D-mannose, pyruvate, D-raffinose, D-xylose, and L-glutamic acid as sole carbon sources but not acetate, adonitol, L-arabinose, D-cellobiose, cellulose, citrate, dextran, glucose, *meso*-erythritol, fumaric acid, glycerol, inulin, lactic acid, D-mannitol, D-melibiose, melezitose, L-rhamnose, D-ribose, sorbitol, D-sorbose, sucrose, succinic acid, trehalose, thioglycollate, L-alanine, L-arginine, L-aspartic acid, L-aspargine, L-glutamine, L-lysine, L-histidine, L-isoleucine, L-leucine, L-lysine, L-methionine, L-phenylalanine, L-proline, L-serine, L-threonine, L-tyrosine, L-tryptophan, or L-valine. Sensitive to amikacin, ampicillin, amoxycillin, bacitracin, carbenicillin, cefazoline, cefaperazone, cephotaxime, chloramphenicol, chlorotetracycline, co-trimoxazole, ciprofloxacin, erythromycin, furazolidone, furoxone, gentamicin, kanamycin, lomefloxacin, nalidixic acid, neomycin, nitro-furazone, nitrofurantoin, norfloxacin, novobiocin, nystatin, oxytetracycline, penicillin, polymyxin-B, rifampin, roxithromycin, streptomycin, tetracycline, tobramycin, trimethoprim and vancomycin, but resistant to cefuroxime, colistin, and lincomycin. Peptidoglycan type is L-Lys-D-Glu of the A4α variant. The major fatty acids include $C_{16:1 \text{ iso}}$ in addition to $C_{15:0 \text{ ante}}$. Other characteristics are included in Table 71 and the genus description.

DNA G + C content (mol%): 44 (T_m).

Type strain: CMS 21w, CIP 107784, DSM 15428, MTCC 4670.

GenBank accession number (16S rRNA gene): AJ514408.

6. **Sporosarcina pasteurii** (Miquel 1889) Yoon, Lee, Weiss, Kho, Kang and Park 2001b, 1085[VP] (*Urobacillus pasteurii* Miquel 1889, 519; *Bacillus pasteurii* Chester 1898, 47.)

pas.teur′i.i. N.L. gen. n. *pasteurii* of Pasteur; named for Louis Pasteur, French chemist and bacteriologist.

Cells are rods, 0.5–1.2 × 1.3–4 μm. Colonies are usually circular and glossy and have variable opacity and size, depending on the medium. In liquid media, growth is turbid with slimy deposits but rarely a fragile pellicle. Cultures are very adept at converting urea to ammonium carbonate, but this activity frequently decreases upon transfer in artificial media. Requires alkaline medium (optimum pH ca. 9) containing NH_3 (1% NH_4Cl). Growth in culture medium is supported by casein hydrolyzate (pH 8.5–9.5), ammonia, thiamine, and for some strains, biotin, and nicotinic acid. Isolated from soil, sewage, and incrustrations on urinals. Other characteristics are included in Table 71 and the genus description.

DNA G + C content (mol%): 38.4 (Bd), 38.5 (T_m).

Type strain: ATCC 11859, CCUG 7425, CIP 66.21, DSM 33, LMG 7130, NCCB 48021, NCIMB 8841, NCTC 4822, NRRL NRS-673, VKM B-513.

GenBank accession number (16S rRNA gene): X60631.

7. **Sporosarcina psychrophila** (Nakamura 1984) Yoon, Lee, Weiss, Kho, Kang, and Park 2001b, 1085[VP] (*Bacillus psychrophilus* Nakamura 1984, 122.)

psy.chro′phil.a. Gr. adj. *psychros* cold; Gr. adj. *philos* liking, preferring; N.L. adj. *psychrophila* preferring cold.

Rod-shaped (0.5–1.0 μm × 3.0–7.0 μm). Cells occur singly and in chains. Gram positive. Motile. Round endospores produced in swollen sporangia. Colonies (1–2 mm in diameter) are nonpigmented, translucent, slightly raised, circular, entire, smooth, and slightly glossy. Grows anaerobically in the presence of glucose. Acetylmethylcarbinol, indole, and hydrogen sulfide are not produced. Arginine-, lysine-, and ornithine-decomposing enzymes are not produced. Acetate, fumarate, malate, and succinate, but not citrate, are utilized. No growth occurs at pH 5.6 or 5.7. Litmus milk is unchanged at 7 d. Temperature for growth ranges from 0–3°C to 30°C and is optimum at 25°C. Acid but no gas is produced from D-fructose, D-galactose, D-glucose, D-mannitol, maltose, D-ribose, sucrose, trehalose, and D-xylose. Isolated from soil and river water. Other characteristics are included in Table 71 and the genus description.

DNA G + C content (mol%): 44.1 (Bd).

Type strain: ATCC 23304, CCM 2117, BCRC 11738, CCUG 28886, CIP 103267, DSM 3, IAM 12468, NBRC 15381, JCM 9075, LMG 6929, NRRL B-3397, NRRL NRS-1530, W16A.

GenBank accession number (16S rRNA gene): D16277, X60634.

8. **Sporosarcina saromensis** An, Haga, Kasai, Goto and Yokata 2007, 1870[VP]

sa.ro.men′sis. N.L. fem. adj. *saromensis* pertaining to Lake Saroma, where the type strain was collected.

Cells are sporeforming rods (0.8–1.0 μm × 2.0–3.2 μm). Gram positive. They form spherical, terminal endospores. Colonies are circular, convex, and beige on TSA medium containing 50% Herbst's artificial sea water. The growth temperature range is 5–40°C (optimum, 27°C; no growth at 45°C) and growth pH range is 5.5–9.0 (optimum, 6.5). Negative for formation of indole, H_2S, and acetoin, and tests for arginine hydrolase, lysine decarboxylase, ornithine decarboxylase, tryptophan deaminase, and citrate utilization. Acid is not produced from carbohydrates in the API 50CHI gallery. L-Lys-D-Glu is the cell wall peptidoglycan type. The major fatty acids are $C_{15:0 \text{ iso}}$ and $C_{15:0 \text{ ante}}$. The major polar lipids are diphosphatidylglycerol, phosphatidylglycerol, and phosphatidylethanolamine. The type strain was isolated from surface water in Lake Saroma. A reference strain, HG711, was isolated from sediment in Nagasuka fishery harbor (Miyagi, Japan). Other characteristics are included in Table 71 and the genus description.

DNA G + C content (mol%): 46.0 (HPLC).

Type strain: HG645, IAM 15429, JCM 23205, KCTC 13119, MBIC08270, NBRC 103571.

GenBank accession number (16S rRNA gene): AB243859.

9. **Sporosarcina soli** Kwon, Kim, Song, Weon, Schumann, Tindall, Stackebrandt and Fritze 2007, 1697[VP]

so′li. L. neut. gen. n. *soli* of soil, the source of the organism.

Cells are sporeforming rods (0.7–1.0 μm × 2.0–3.0 μm) occurring singly, in pairs, or occasionally in short chains. Gram positive. They form mainly round, central endospores in nonswollen sporangia. Colonies are light orange after 2 d on TSA at 30°C. The growth temperature range is 15–37°C (optimum, 30°C), and growth pH range is 7.0–9.0 (optimum, 8.0; no growth at 5.7 or 10.0). Negative for formation of indole and dihydroxyacetone, in the Voges–Proskauer test, for phenylalanine deamination, nitrate reduction, acid production from D-glucose, L-arabinose, D-xylose, and D-mannitol, and for hydrolysis of starch, casein, and Tween 80. No utilization of citrate or propionate. L-Lys-D-Glu is the peptidoglycan type. The major fatty acids are $C_{15:0 \text{ iso}}$ and $C_{15:0 \text{ ante}}$. The major polar lipids are diphosphatidylglycerol, phosphatidylglycerol, phosphatidylethanolamine, and an unidentified phospholipid. The type strain was isolated from upland soil in Suwon, Korea. Other characteristics are included in Table 71 and the genus description.

DNA G + C content (mol%): 44.5 (HPLC).

Type strain: strain I80, DSM 16920, KACC 11300.

GenBank accession number (16S rRNA gene): DQ073394.

Genus IX. **Ureibacillus** Fortina, Pukall, Schumann, Mora, Parini, Manachini and Stackebrandt 2001, 453[VP]

DAGMAR FRITZE AND PAUL DE VOS

Ur.e.i.ba.cil'lus. N.L. n. *urea* urea; L. dim. n. *bacillus* a rod and also the name of a genus of aerobic endospore-forming bacteria; N.L. masc. n. *Ureibacillus* a ureolytic aerobic bacillus.

Gram-stain-negative, rod-shaped, motile bacterium, 0.5–0.7 μm in diameter and 1.0–6.0 μm in length. **Endospores are spherical,** subterminal or terminal in swollen sporangia. Aerobic. Cross-linkage of the **peptidoglycan is of the L-Lys-D-Asp type.** Phosphatidylglycerol, diphosphatidylglycerol, phospolipids, and glycolipids of unknown composition are found as polar lipids. $C_{16:0\ iso}$ is the major fatty acid component. The predominant menaquinones are either **MK-7 or a mixture of MK-8 and MK-9**.

DNA G+C content (mol%): 35.7–41.5.

Type species: **Ureibacillus thermosphaericus** (Andersson et al., 1996) Fortina, Pukall, Schumann, Mora, Parini, Manachini and Stackebrandt 2001, 453[VP] (*Bacillus thermosphaericus* Andersson, Laukkanen, Nurmiaho-Lassila, Rainey, Niemelä and Salkinoja-Salonen 1996, 362.).

Further descriptive information

Colonies on CESP agar (see Fortina et al., (2001) for composition) are circular, with entire margins, transparent, and show swarming; for *Ureibacillus suwonensis,* no real colony formation has been observed. Growth occurs on complex media. For some strains, growth is limited to temperatures of 40 to about 65°C, with optimal growth at 50–60°C; others may grow at 35–37°C (*Ureibacillus thermosphaericus*). Strains may grow with 3% (w/v) NaCl in the medium and even up to 5% (w/v) NaCl at pH 9.0. Casein is never hydrolyzed and acetylmethylcarbinol formation is negative. Fatty acids that are present in substantial amounts are $C_{15:0\ iso}$, $C_{16:0\ iso}$, and $C_{16:0}$. Data from fatty acid analyses obtained after growth on non-standard media allow discrimination at the intrageneric level (Fortina et al., 2001). Remarkable intrageneric differences are found in the menaquinone composition, which varies between almost exclusively MK-7 to a mixture of major amounts of MK-8 and MK-9 with minor fractions of MK-11, MK-10, and MK-7, and possibly a minor fraction of MK-6.

Taxonomic comments

The genus *Ureibacillus* was created to accommodate a number of thermophilic spore-formers that held a separate phylogenetic position in the genus *Bacillus* and of which a number of airborne strains were initially classified and described as *Bacillus thermosphaericus* (Andersson et al., 1995; Andersson et al., 1996); three further strains, isolated at a later date (Mora et al., 1998), were found to be the closest relatives of these airborne isolates. The distinctiveness of this new phylogenetic lineage is also supported by chemotaxonomic and phenotypic characteristics. On the basis of DNA–DNA hybridization data, the four strains (one of which was the original type strain of *Bacillus thermosphaericus*) studied by Fortina et al. (2001) clearly belonged to two different species, one of which complies with the type species *Ureibacillus thermosphaericus.* Species differentiation is only partially supported by ribotyping results. A third species was recently found in compost of cotton waste in Korea. Discrimination from members of other genera is on the basis of 16S rRNA gene sequence analysis, but also on the basis of the amino acid composition of peptide bridges in the peptidoglycan, and menaquinone and fatty acid compositions.

Hydrolysis of urea (reflected in the name *Ureibacillus*) could not be confirmed for any of the three species when using stringent test systems (Kim et al., 2006).

List of species of the genus *Ureibacillus*

1. **Ureibacillus thermosphaericus** (Andersson, Laukkanen, Nurmiaho-Lassila, Rainey, Niemelä and Salkinoja-Salonen 1996) Fortina, Pukall, Schumann, Mora, Parini, Manachini and Stackebrandt 2001, 453[VP] (*Bacillus thermosphaericus* Andersson, Laukkanen, Nurmiaho-Lassila, Rainey, Niemelä and Salkinoja-Salonen 1996, 362.)

   ther.mo.sphae'ri.cus. Gr. adj. *thermos* hot; L. adj. *sphaericus* spherical; N.L. n. *thermosphaericus* the hot sphere.

   Pleomorphic cells that vary during the incubation phase between cocci, short rods and extremely long rods. Motility is enhanced upon higher incubation temperature. Peritrichously inserted flagella. Strictly aerobic. Grows between 33 and 64°C. The spherical spores lay terminally in swollen sporangia. Colonies are circular, entire, flat, transparent, and swarming. Growth occurs in the presence of 5% NaCl and at pH 9.0. Negative for the following characters: anaerobic growth; Voges–Proskauer; indole production; nitrate reduction; starch, casein, and gelatin hydrolysis; and acid production from arabinose, ribose, xylose, glucose, maltose, mannose, rhamnose, trehalose, sucrose, and citrate. Unable to use sugars as carbon and energy source. Uses esculin, urea (see comment above), and malate. Peptidoglycan type is A4α-lysine-aspartic acid. The major menaquinone is MK-7, but MK-8 is also present as a minor component. The main fatty acid is $C_{16:0\ iso}$ (58–61%). The type strain was isolated from landfill air.

   DNA G+C content (mol%): 35.7–39.2 (T_m).

   Type strain: P-11, DSM 10633, KACC 10504.

   EMBL/GenBank accession number (16S rRNA gene): AB101594 (P-11).

2. **Ureibacillus terrenus** Fortina, Pukall, Schumann, Mora, Parini, Manachini and Stackebrandt 2001, 454[VP]

   ter.re'nus. L. adj. *terrenus* from earth, referring to the habitat of the organism.

   Description is not well-documented, but characteristics are similar to those given for the genus and for *Ureibacillus thermosphaericus.* In general, cells are Gram-stain-negative rods, forming endospores in swollen sporangia that lay terminal/subterminal. Not all strains grow at pH 9 or in media with 5% NaCl. The temperature range for growth is 42–65°C; does not grow at 37°C. Major menaquinones are MK-9 and MK-8, with MK-7 as a minor component. The type strain was isolated from soil.

DNA G+C content (mol%): 39.6–41.5 (T_m).

Type strain: TU1A, DSM 12654, LMG 19470.

EMBL/GenBank accession number (16S rRNA gene): AJ276403 (DSM 12654).

3. **Ureibacillus suwonensis** Kim, Lee, Weon, Kwon, Go, Park, Schumann and Fritze 2006, 665[VP]

su.won.en'sis. N.L. masc. adj. *suwonensis* referring to Suwon Region in Korea, where the bacterium was first found.

Gram-stain-negative, spore-forming, motile (by means of peritrichously inserted flagella) rods (0.5–0.7 μm wide and 1.5–2.0 μm long), appearing singly and in chains.

Endospores are spherical to oval in subterminally or terminally swollen sporangia. Colonies appear to be smeared over the surface of solid media. Temperature for growth ranges from 35 to 60°C. Grows in the presence of 5% NaCl. Strictly aerobic. Catalase, oxidase, and arginine dihydrolase reactions are positive. Weak reaction for phenylalanine deamination. Strains test negative for the production of indole and dihydroxyacetone, and also in the Voges–Proskauer test. Nitrate reduction is not observed. Acid is not produced from glucose, arabinose, xylose, or mannitol. Esculin, starch, gelatin, casein, and urea are not hydrolyzed. The cross-linkage of peptidoglycan is as for the genus. The major cellular fatty acid is $C_{16:0\ iso}$ (65–66%). The major isoprenoid quinones are MK-9, MK-8 and MK-7. Isolated from cotton compost in Suwon, Korea.

DNA G+C content (mol%): 41.5 (HPLC).

Type strain: 6T19, KACC 11287, DSM 16752.

EMBL/GenBank accession number (16S rRNA gene): AY850379 (6T19).

References

Abd El-Rahman, H.A., D. Fritze, C. Spröer and D. Claus. 2002. Two novel psychrotolerant species, *Bacillus psychrotolerans* sp. nov. and *Bacillus psychrodurans* sp. nov., which contain ornithine in their cell walls. Int. J. Syst. Evol. Microbiol 52: 2127–2133.

Adcock, K.A., R.J. Seidler and W.C. Trentini. 1976. Deoxyribonucleic acid studies in the genus *Caryophanon*. Can. J. Microbiol. 22: 1320–1327.

Alam, S.I., L. Singh, S. Dube, G.S. Reddy and S. Shivaji. 2003. Psychrophilic *Planococcus maitriensis* sp.nov. from Antarctica. Syst. Appl. Microbiol. 26: 505–510.

Alvarez, R.J. 1982. Role of *Planococcus citreus* in the spoilage of *Penaeus* shrimp. Zentbl. Bakteriol. Mikrobiol. Hyg. Abt. I Orig. C 3: 503–512.

An, S.Y., T. Haga, H. Kasai, K. Goto and A. Yokota. 2007. *Sporosarcina saromensis* sp. nov., an aerobic endospore-forming bacterium. Int. J. Syst. Evol. Microbiol. 57: 1868–1871.

Andersson, M., M. Laukkanen, E.L. Nurmiaholassila, F.A. Rainey, S.I. Niemelä and M. Salkinoja-Salonen. 1995. *Bacillus thermosphaericus* sp. nov. a new thermophilic ureolytic bacillus isolated from air. Syst. Appl. Microbiol. 18: 203–220.

Andersson, M., M. Laukkanen, E.-L. Nurmiaho-Lassila, F.A. Rainey, S.I. Niemelä and M. Salkinoja-Salonen. 1996. *In* Validation of the publication of new names and new combinations previously effectively published outside the IJSB. List no. 56. Int. J. Syst. Bacteriol. 46: 362–363.

Ash, C., J.A.E. Farrow, S. Wallbanks and M.D. Collins. 1991. Phylogenetic heterogeneity of the genus *Bacillus* revealed by comparative analysis of small subunit ribosomal RNA sequences. Lett. Appl. Microbiol. 13: 202–206.

Ash, C., S. Wallbanks and M.D. Collins. 1992. Phylogenetic analysis of the genera *Marinococcus*, *Planococcus*, and *Sporosarcina* and their relationship to members of the genus *Bacillus*. FEMS Microbiol. Lett. 93: 167–172.

Beijerinck, M.W. 1901. Anhäufungsversuche mit Ureumbakterien, Uremspaltung durch Urease und durch Katabolismus. Zentbl. Bakteriol. Parasitenkd. Infektionskr. Hyg. Abt. II 7: 33–61.

Belikova, V.A., N.V. Cherevach and L.V. Kalakutskii. 1986. A new species of bacteria of the genus *Kurthia*, *Kurthia sibirica* sp. nov. Microbiology (En. transl. from Mikrobiologiya) 55: 668–672.

Belikova, V.L., N. V. Cherevach, L.M. Baryshnikova and L.V. Kalakutskii. 1980. Morphologic, physiologic and biochemical characteristics of *Kurthia zopfii*. Microbiologiya 49: 51–55.

Belikova, V.L., N. V. Cherevach, L. M. Baryshnikova and L.V. Kalakutskii. 1988. *In* Validation of the publication of new names and new combinations previously effectively published outside the IJSB. List no. 5. Int. J. Syst. Bacteriol. 38: 220–222.

Boháček, J., M. Kocur and T. Martinec. 1967. DNA base composition and taxonomy of some micrococci. J. Gen. Microbiol. 46: 369–376.

Boháček, J., M. Kocur and T. Martinec. 1968a. Deoxyribonucleic acid base composition of some marine and halophilic micrococci. J. Appl. Bacteriol. 31: 215–219.

Boháček, J., M. Kocur and T. Martinec. 1968b. Deoxyribonucleic acid base composition of *Sporosarcina ureae*. Arch. Mikrobiol. 64: 23–28.

Cherevach, N.V., T.P. Tourova and V.L. Belikova. 1983. DNA-DNA Homology studies among strains of *Kurthia zopfii*. FEMS Microbiol. Lett. 19: 243–245.

Chester, F.D. 1898. Report of the mycologist: bacteriological work. Del. Agric. Exp. Stn. Bull. 10: 47–137.

Claus, D. and R.C.W. Berkeley. 1986. Genus *Bacillus* Cohn 1872. *In* Mair, Sneath, Sharpe and Holt (Editors), Bergey's Manual of Systematic Bacteriology, vol. 2. The Williams & Wilkins Co., Baltimore, pp. 1105–1139.

Claus, D. and F. Fahmy. 1986. Genus *Sporosarcina* Kluyver and van Niel 1936. *In* Mair, Sneath, Sharpe and Holt (Editors), Bergey's Manual of Systematic Bacteriology, vol. 2. The Williams & Wilkins Co., Baltimore, pp. 1202–1206.

Claus, D., D. Fritze and M. Kocur. 1991. Genera related to the genus *Bacillus*–*Sporolactobacillus*, *Sporosarcina*, *Planococcus*, *Filibacter*, and *Caryophanon*. *In* Balows, Trüper, Dworkin, Harder and Schleifer (Editors), The Prokaryotes, 2nd edn, vol. 2. Springer, New York, pp. 1769–1791.

Clausen, V., J.G. Jones and E. Stackebrandt. 1985. 16S ribosomal RNA analysis of *Filibacter limicola* indicates a close relationship to the genus *Bacillus*. J. Gen. Microbiol. 131: 2659–2663.

Collins, M.D., M. Goodfellow and D.E. Minnikin. 1979. Isoprenoid quinones in the classification of coryneform and related bacteria. J. Gen. Microbiol. 110: 127–136.

Collins, M.D. and D. Jones. 1981. Distribution of isoprenoid quinone structural types in bacteria and their taxonomic implication. Microbiol Rev 45: 316–354.

Collins, M.D. 1994. Isoprenoid quinones. *In* Goodfellow and O'Donnell (Editors), Chemical Methods in Prokaryotic Systematics. John Wiley & Sons, New York, pp. 265–309.

Courington, D.P. and T.W. Goodwin. 1955. A survey of the pigments of a number of chromogenic marine bacteria, with special reference to the carotenoids. J. Bacteriol. 70: 568–571.

Cowan, S.T. and K.J. Steel. 1974. Cowan and Steel's Manual for Identification of Medical Bacteria, 2nd edn. Cambridge University Press, London.

Dai, X., Y.N. Wang, B.J. Wang, S.J. Liu and Y.G. Zhou. 2005. *Planomicrobium chinense* sp. nov., isolated from coastal sediment, and transfer of *Planococcus psychrophilus* and *Planococcus alkanoclasticus* to *Planomicrobium* as *Planomicrobium psychrophilum* comb. nov. and *Planomicrobium alkanoclasticum* comb. nov. Int. J. Syst. Evol. Microbiol. *55*: 699–702.

DeSoete, G. 1983. A least square algorithm for fitting additive trees to proximity data. Psychometrika *48*: 621–626.

Drozd, G.Y., M.A. Sobol and Y.S. Varenko. 1987. Concrete corrosion caused by a *Caryophanon* species. Mikrobiol. Zh. (Kiev). *49*: 61–64.

Eguchi, M., T. Nishikawa, K. Macdonald, R. Cavicchioli, J.C. Gottschal and S. Kjelleberg. 1996. Responses to stress and nutrient availability by the marine ultramicrobacterium *Sphingomonas* sp. strain RB2256. Appl. Environ. Microbiol. *62*: 1287–1294.

Engelhardt, M.A., K. Daly, R.P. Swannell and I.M. Head. 2001. Isolation and characterization of a novel hydrocarbon-degrading, Gram-positive bacterium, isolated from intertidal beach sediment, and description of *Planococcus alkanoclasticus* sp. nov. J. Appl. Microbiol. *90*: 237–247.

Faoagali, J.L. 1974. *Kurthia*, an unusual isolate. Amer. J. Clin. Pathol. *62*: 604–606.

Farrow, J.A., C. Ash, S. Wallbanks and M.D. Collins. 1992. Phylogenetic analysis of the genera *Planococcus*, *Marinococcus* and *Sporosarcina* and their relationships to members of the genus *Bacillus*. FEMS Microbiol. Lett. *72*: 167–172.

Farrow, J.A., S. Wallbanks and M.D. Collins. 1994a. Phylogenetic interrelationships of round-spore-forming bacilli containing cell walls based on lysine and the non-spore-forming genera *Caryophanon*, *Exiguobacterium*, *Kurthia*, and *Planococcus*. Int. J. Syst. Bacteriol. *44*: 74–82.

Farrow, J.A.E., S. Wallbanks and M.D. Collins. 1994b. Phylogenetic interrelationships of round-spore-forming bacilli containing cell walls based on lysine and the non-spore-forming genera *Caryophanon*, *Exiguobacterium*, *Kurthia*, and *Planococcus*. Int. J. Syst. Bacteriol. *44*: 74–82.

Felsenstein, D. 1993. PHYLIP (Phylogeny Inference Package) 3.57. Department of Genetics, University of Washington, Seattle.

Fortina, M.G., R. Pukall, P. Schumann, D. Mora, C. Parini, P.L. Manachini and E. Stackebrandt. 2001. *Ureibacillus* gen. nov., a new genus to accommodate *Bacillus thermosphaericus* (Andersson *et al.* 1995), emendation of *Ureibacillus thermosphaericus* and description of *Ureibacillus terrenus* sp. nov. Int. J. Syst. Evol. Microbiol. *51*: 447–455.

Gardner, G.A. 1969. Physiological and morphological characteristics of *Kurthia zopfii* isolated from meat products. J. Appl Bacteriol *32*: 371–380.

Georgala, D.L. 1957. Quantitative and qualitative aspects of the skin flora of North Sea cod and the effect thereon of handling on ship and on shore. PhD thesis, University of Aberdeen.

Glauert, A.M. and M.J. Thornley. 1969. The topography of the bacterial cell wall. Annu. Rev. Microbiol. *23*: 159–198.

Godinho-Orlandi, M.J.L. and J.G. Jones. 1981. The distribution of some genera of filamentous bacteria in littoral and profundal lake sediments. J. Gen. Microbiol. *123*: 91–101.

Goodfellow, M., M.D. Collins and D.E. Minnikin. 1980. Fatty acid and polar lipid composition in the classification of *Kurthia*. J. Appl Bacteriol *48*: 269–276.

Gordon, R.E., W.C. Haynes and C.H.-N. Pang. 1973. The genus *Bacillus*. Agriculture Handbook no. 427. United States Department of Agriculture, Washington, D.C.

Hao, M.V. and K. Komagata. 1985. A new species of *Planococcus*, *Planococcus kocurii* isolated from fish, frozen foods, and fish curing brine. J. Gen. Appl. Microbiol. *31*: 441–455.

Hao, M.V. and K. Komagata. 1986. A new species of *Planococcus*, *P. kocurii* isolated from fish, frozen foods and fish curing bring. J. Gen. Appl. Microbiol. *31*: 306–314.

Hernandez-Castro, R., L. Martinez-Chavarria, A. Diaz-Avelar, A. Romero-Osorio, C. Godinez-Reyes, A. Zavala-Gonzalez and A. Verdugo-Rodriguez. 2005. Aerobic bacterial flora of the nasal cavity in Gulf of California sea lion (*Zalophus californianus*) pups. Vet. J. *170*: 359–363.

Hollister, D.D. and I.N. McCave. 1984. Sedimentation under deep-sea storms. Nature *309*: 220–225.

Irgens, R.L., I. Suzuki and J.T. Staley. 1989. Gas vacuolated bacteria obtained from marine waters of Antarctica. Curr. Microbiol. *18*: 261—265.

Jeena, M.I., P. Deepa, K.M. Mujeeb Rahiman, R.T. Shanthi and A.A. Hatha. 2006. Risk assessment of heterotrophic bacteria from bottled drinking water sold in Indian markets. Int. J. Hyg. Environ. Health *209*: 191–196.

Jeffries, L. 1969. Menaquinones in the classification of *Micrococcaceae* with observations on the application of lysozyme and novobiocin sensitivity test. Int. J. Syst. Bacteriol. *19*: 183–187.

Junge, K., J.J. Gosink, H.G. Hoppe and J.T. Staley. 1998. *Arthrobacter*, *Brachybacterium* and *Planococcus* isolates identified from Antarctic Sea ice brine. Description of *Planococcus mcmeekinii*, sp. nov. Syst. Appl. Microbiol. *21*: 306–314.

Keddie, R.M. 1949. A study of Bacterium zopfii Kurth, Dissertation. Edinburgh School of Agriculture, Edinburgh, UK.

Keddie, R.M. 1981. The Genus *Kurthia*. *In* Starr, Stolp, Trüper, Balows and Schlegel (Editors), The Prokaryotes: a Handbook on Habitats, Isolation and Identification of Bacteria. Springer-Verlag, Berlin, pp. 1888–1893.

Keddie, R.M. and S. Shaw. 1986. Genus *Kurthia*. *In* Mair, Sneath, Sharpe and Holt (Editors), Bergey's Manual of Systematic Bacteriology, vol. 2. The Williams & Wilkins Co., Baltimore, pp. 1255–1258.

Keddie, R.M. and D. Jones. 1992. The genus *Brochothrix* (formerly *Microbacterium thermosphactum*, McLean and Sulzbacher). *In* Stolp, Starr, Trüper, Balows and Schlegel (Editors), The Prokaryotes: a Handbook on Habitats, Isolation and Identification of Bacteria. Springer-Verlag, Berlin, pp. 1654–1662.

Kele, R.A. 1970. Investigations on the nutrition, morphogenesis and habitat of *Caryophanon latum*. PhD thesis, University of Wisconsin, Madison.

Kele, R.A. and E. McCoy. 1971. Defined liquid minimal medium for *Caryophanon latum*. Appl. Microbiol. *22*: 728–729.

Kim, B.Y., S.Y. Lee, H.Y. Weon, S.W. Kwon, S.J. Go, Y.K. Park, P. Schumann and D. Fritze. 2006. *Ureibacillus suwonensis* sp. nov., isolated from cotton waste composts. Int. J. Syst. Evol. Microbiol. *56*: 663–666.

Kiyasu, T., Y. Nagahashi and T. Hoshino. 2001. Cloning and characterization of biotin biosynthetic genes of *Kurthia* sp. Gene *265*: 103–113.

Kluyver, A.J. and C.B. van Neil. 1936. Prospects for a natural classification of bacteria. Zentbl. Bacteriol. Parasitenkd. Infektionskr. Hyg. Abt. II *94*: 369–403.

Kocur, M., Z. Pacova, W. Hodgkiss and T. Martinec. 1970. The taxonomic status of the genus *Planococcus* Migula 1894. Int. J. Syst. Bacteriol. *20*: 241–248.

Kocur, M. 1986. Family I. *Micrococcaceae*, Genus III. *Planococcus* Migula 1894, 236^AL. *In* Sneath, Mair, Sharpe and Holt (Editors), Bergey's Manual of Systematic Bacteriology, vol. 2. The Williams & Wilkins Co., Baltimore, pp. 1011–1013.

Komura, I., K. Yamada and K. Komagata. 1975. Taxonomic significance of phospholipid composition in aerobic Gram positive cocci. J. Gen. Appl. Microbiol. *21*: 97–107.

Krasil'nikov, N.A. 1949. Opredelitelv Bakterii i Actinomicetov. Akad. Nauk. SSSR, Moscow.

Kurth, H. 1883. Bacterium zopfii. Ein Beitrag zur Kenntniss der Morphologie und Physiologie der Spaltpilze. Bot. Zeitung 41: 369–386, 393–405, 409–420, 425–435.

Kwon, S.W., B.Y. Kim, J. Song, H.Y. Weon, P. Schumann, B.J. Tindall, E. Stackebrandt and D. Fritze. 2007. Sporosarcina koreensis sp. nov. and Sporosarcina soli sp. nov., isolated from soil in Korea. Int. J. Syst. Evol. Microbiol. 57: 1694–1698.

Landry, D., M.C. Looney, G.R. Feehery, B.E. Slatko, W.E. Jack, I. Schildkraut and G.G. Wilson. 1989. M.FokI methylates adenine in both strands of its asymmetric recognition sequence. Gene 77: 1–10.

Larkin, J.M. and J.L. Stokes. 1967. Taxonomy of psychrophilic strains of Bacillus. J. Bacteriol. 94: 889–895.

Ludwig, W., E. Seewaldt, K.H. Schleifer and E. Stackebrandt. 1981. The phylogenetic status of Kurthia zopfii. FEMS Microbiol. Lett. 10: 193–197.

MacDonald, R.E. and S.W. MacDonald. 1962. The physiology and natural relationships of the motile, sporeforming sarcinae. Can. J. Microbiol. 8: 795–808.

Maiden, M.F. and J.G. Jones. 1984. A new filamentous, gliding bacterium, Filibacter limicola gen. nov. sp. nov., from lake sediment. J. Gen. Microbiol. 130: 2943–2959.

Maiden, M.F.J. 1983. The biology of filamentous bacteria in freshwater sediments. PhD thesis, University of Wales Institute of Science and Technology.

Maiden, M.F.J. and J.G. Jones. 1985. In Validation of the publication of new names and new combinations previously effectively published outside the IJSB. List no. 18. Int. J. Syst. Bacteriol. 35: 375–376.

Mantayla, A.W. and J.L. Reid. 1983. Abyssal characteristics of the world ocean waters. Deep-Sea Res. 30: 805–833.

Mayilraj, S., G.S. Prasad, K. Suresh, H.S. Saini, S. Shivaji and T. Chakrabarti. 2005. Planococcus stackebrandtii sp. nov., isolated from a cold desert of the Himalayas, India. Int. J. Syst. Evol. Microbiol. 55: 91–94.

Migula, W. 1894. Über ein neues System der Bakterien. Arb. Bakteriol. Inst. Karlsruhe. 1: 235–238.

Miquel, P. 1889. Étude sur la fermentation ammoniacale et sur les ferments de l'urée. Ann. Micrographie 1: 506–519.

Mora, D., M.G. Fortina, G. Nicastro, C. Parini and P.L. Manachini. 1998. Genotypic characterization of thermophilic bacilli: a study on new soil isolates and several reference strains. Res. Microbiol. 149: 711–722.

Moran, J.W. and L.D. Witter. 1976. Effect of temperature and pH on the growth of Caryophanon latum colonies. Can. J. Microbiol. 22: 1401–1403.

Morrison, S.J., T.G. Tornabene and W.E. Kloos. 1971. Neutral lipids in the study of relationships of members of the family Micrococcaceae. J. Bacteriol. 108: 353–358.

Nakagawa, Y., T. Sakane and A. Yokota. 1996. Emendation of the genus Planococcus and transfer of Flavobacterium okeanokoites Zobell and Upham 1944 to the genus Planococcus as Planococcus okeanokoites comb. nov. Int. J. Syst. Bacteriol. 46: 866–870.

Nakamura, L.K. 1984. Bacillus psychrophilus sp. nov., nom. rev. Int. J. Syst. Bacteriol. 34: 121–123.

Nakamura, L.K., O. Shida, H. Takagi and K. Komagata. 2002. Bacillus pycnus sp. nov. and Bacillus neidei sp. nov., round-spored bacteria from soil. Int. J. Syst. Evol. Microbiol. 52: 501–505.

Novitsky, T.J. and D.J. Kushner. 1976. Planococcus halophilus sp. nov., a facultatively halophilic coccus. Int. J. Syst. Bacteriol. 26: 53–57.

Oeding, P. 1971. Serological investigation of Planococcus strains. Int. J. Syst. Bacteriol. 21: 323–325.

Ohnstadt, F.R. and J.G. Jones. 1982. The Jenkin Surface-mud Sampler User Manual. Freshwater Biological Association Occasional Publication No. 15.

Oppenheimer, C.H. and C.E. Zobell. 1952. The growth and viability of sixty-three species of marine bacteria as influenced by hydrostatic pressure. J. Mar. Res. 11: 10–18.

Owens, J.D. and R.M. Keddie. 1969. The nitrogen nutrition of soil and herbage coryneform bacteria. J. Appl Bacteriol 32: 338–347.

Pechman, K.J., B.J. Lewis and C.R. Woese. 1976. Phylogenetic status of Sporosarcina ureae. Int. J. Syst. Bacteriol. 26: 305–310.

Peshkoff, M.A. 1939. Cytology, karyology and cycle of development of new microbes - Caryophanon latum and Caryophanon tenue. Compt. Rend. (Doklady) Acad. Sci. l'URRS 25: 244–247.

Peshkov, M.A. 1967. New high-productivity nutritive media for Caryophanon and a method for quantitative evaluation of bacterial yield. Microbiology (En. trans. from Mikrobiologiya) 37: 548–551.

Peshkov, M.A. and B.I. Marek. 1972. Fine structure of Caryophanon latum and Caryophanon tenue Peshkov. Microbiology (En. transl. from Mikrobiologiya) 41: 941–945.

Peshkov, M.A., V.P. Shekhovtsov and G.G. Zharikova. 1978. Morphology of the multicellular, trichome bacterium, Caryophanon, in the process of growth on an agarized medium. Mikrobiologiia 47: 539–544.

Pringsheim, E.G. and C.F. Robinow. 1947. Observation on two very large bacteria, Caryophanon latum Peshkov and Lineola longa (nomen provisorum). J. Gen. Microbiol. 1: 267–278.

Provost, P.J. and R.N. Doetsch. 1962. An appraisal of Caryophanon latum. J. Gen. Microbiol. 28: 547–557.

Reddy, G.S., G.I. Matsumoto and S. Shivaji. 2003. Sporosarcina macmurdoensis sp. nov., from a Cyanobacterial mat sample from a pond in the McMurdo Dry Valleys, Antarctica. Int. J. Syst. Evol. Microbiol. 53: 1363–1367.

Reddy, G.S.N., J.S.S. Prakash, M. Vairamani, S. Prabhakar, G.I. Matsumoto and S. Shivaji. 2002. Planococcus antarcticus and Planococcus psychrophilus spp. nov. isolated from Cyanobacterial mat samples collected from ponds in Antarctica. Extremophiles 6: 253–261.

Rheims, H., A. Fruhling, P. Schumann, M. Rohde and E. Stackebrandt. 1999. Bacillus silvestris sp. nov., a new member of the genus Bacillus that contains lysine in its cell wall. Int. J. Syst. Bacteriol. 49: 795–802.

Romano, I., A. Giordano, L. Lama, B. Nicolaus and A. Gambacorta. 2003. Planococcus rifietoensis sp. nov. isolated from algal mat collected from a sulfurous spring in Campania (Italy). Syst. Appl. Microbiol. 26: 357–366.

Rowenhagen, B. 1987. Fettsäureverwertung durch Caryophanon latum und Caryophanon tenue sowie morphologische und cytologische Unterschiede dieser Bakterien species. PhD thesis, University of Braunschweig.

Ruger, H.J., D. Fritze and C. Sproer. 2000. New psychrophilic and psychrotolerant Bacillus marinus strains from tropical and polar deep-sea sediments and emended description of the species. Int. J. Syst. Evol. Microbiol. 50: 1305–1313.

Rüger, H.J. and G. Richter. 1979. Bacillus globisporus subsp. marinus subsp. nov. Int. J. Syst. Bacteriol. 29: 196–203.

Rüger, H.J. 1983. Differentiation of Bacillus globisporus, Bacillus marinus comb. nov., Bacillus aminovorans, and Bacillus insolitus. Int. J. Syst. Bacteriol. 33: 157–161.

Satoh, Y. 1980. Production of urea by bacterial decomposition of organic matter including phytoplankton. Int. Rev. Ges. Hydrobiol. 65: 295–301.

Schleifer, K.H. and O. Kandler. 1970. Amino acid sequence of the murein of Planococcus and other Micrococcaceae. J. Bacteriol. 103: 387–392.

Schleifer, K.H. and O. Kandler. 1972. Peptidoglycan types of bacterial cell walls and their taxonomic implications. Bacteriol. Rev. 36: 407–477.

Seeliger, H.P.R. and D. Jones. 1986. Genus Listeria. In Sneath, Mair, Sharpe and Holt (Editors), Bergey's Manual of Systematic Bacteriology, vol. 2. The Williams & Wilkins Co., pp. 1235–1245.

Sepers, A.B.J. 1981. Diversity of ammonifying bacteria. Hydrobiologia 83: 343–350.

Severi, R. 1946. L'azione patogena delle Kurthia e la loro sistematica. Una nuova specie: "Kurthia variabilis". Giorn. Batteriol. Immunol. 24: 107–114.

Shadrina, I.A., A.V. Mashkovtseva, N.A. Kostrikina and V.I. Viryuzova. 1982. The heterogeneity of intracellular membranes in Caryophanon latum. Mikrobiologiya 51: 809–814.

Shaw, S. and R.M. Keddie. 1983a. The vitamin requirements of *Kurthia zopfii* and *Kurthia gibsonii*. Syst. Appl. Microbiol *4*: 439–443.

Shaw, S. and R.M. Keddie. 1983b. A numerical taxonomic study of the genus *Kurthia* with a revised description of *Kurthia zopfii* and a description of *Kurthia gibsonii* sp. nov. Syst. Appl. Microbiol. *4*: 253–276.

Shaw S. and R.M. Keddie. 1984. The genus *Kurthia*: cell wall composition and DNA base content. Syst. Appl. Microbiol *5*: 220–224.

Shekhovtsov, V.P. and G.G. Zharikova. 1978. Cytomorphology of lipid inclusion of *Caryophanon* during its growth on a solid medium. Microbiology (En. transl. from Mikrobiologiya) *47*: 590–594.

Shida, O., H. Takagi, K. Kadowaki, L.K. Nakamura and K. Komagata. 1997. Transfer of *Bacillus alginolyticus*, *Bacillus chondroitinus*, *Bacillus curdlanolyticus*, *Bacillus glucanolyticus*, *Bacillus kobensis*, and *Bacillus thiaminolyticus* to the genus *Paenibacillus* and emended description of the genus *Paenibacillus*. Int. J. Syst. Bacteriol. *47*: 289–298.

Shivaji, S., N. Shymala Rao, L. Saisree, V. Sheth, G.S.N. Reddy and P.M. Bhargava. 1988. Isolation and identification of *Micrococcus roseus* and *Planococcus* sp. from Schirmacher Oasis, Antarctica. J. Biosci. *13*: 409–414.

Simon, B.M. and J.G. Jones. 1982. Gas chromatographic determination of total amino acids and protein in sediments using the ninhydrin-CO_2 reaction. Hydrobiologia *97*: 81–86.

Skerman, V.B.D., V. McGowan and P.H.A. Sneath. 1980. Approved lists of bacterial names. Int. J. Syst. Bacteriol *30*: 225–420.

Smith, D.L. and W.C. Trentini. 1972. Enrichment and selective isolation of *Caryophanon latum*. Can. J. Microbiol. *18*: 1197–1200.

Smith, D.L. and W.C. Trentini. 1973. On the gram reaction of *Caryophanon latum*. Can. J. Microbiol. *19*: 757–760.

Spring, S., W. Ludwig, M.C. Marquez, A. Ventosa and K.H. Schleifer. 1996. *Halobacillus* gen. nov., with descriptions of *Halobacillus litoralis* sp nov and *Halobacillus trueperi* sp. nov., and transfer of *Sporosarcina halophila* to *Halobacillus halophilus* comb. nov. Int. J. Syst. Bacteriol. *46*: 492–496.

Stackebrandt, E. and C. Woese. 1979. Phylogenetic dissection of the family *Micrococcaceae*. Curr. Microbiol. *2*: 317–322.

Stackebrandt, E. and C.R. Woese. 1981. The evolution of prokaryotes. *In* Carlile, Collins and Moseley (Editors), Molecular and Cellular Aspects of Microbial Evolution. Cambridge University Press, Cambridge, pp. 1–31.

Stackebrandt, E., W. Ludwig, M. Weizenegger, S. Dorn, T.J. Mcgill, G.E. Fox, C.R. Woese, W. Schubert and K.H. Schleifer. 1987. Comparative 16S ribosomal RNA oligonucleotide analyses and murein types of round spore-forming bacilli and non-spore-forming relatives. J. Gen. Microbiol. *133*: 2523–2529.

Stackebrandt, E. and J. Swiderski. 2002. From phylogeny to systematics: the dissection of the genus *Bacillus*. *In* Berkeley, Heyndrickx, Logan and De Vos (Editors), Applications and Systematics of *Bacillus* and Relatives. Blackwell Science, Oxford, pp. 8–22.

Stackebrandt, E. and D. Jones. 2005. The genus *Brochothrix*. *In* Dworkin (Editor), The Prokaryotes: an Evolving Electronic Resource for the Microbiological Community, 3rd edn, release 3.19. Springer-Verlag, New York.

Starr, M.P. and V.B. Skerman. 1965. Bacterial diversity: The natural history of selected morphologically unusual bacteria. Annu. Rev. Microbiol. *19*: 407–454.

Sugisaki, H. and S. Kanazawa. 1981. New restriction endonucleases from *Flavobacterium okeanokoites* (*Fok*I) and *Micrococcus luteus* (*Mlu*I). Gene *16*: 73–78.

Suresh, K., S. Mayilraj, A. Bhattacharya and T. Chakrabarti. 2007. *Planococcus columbae* sp. nov., isolated from pigeon faeces. Int. J. Syst. Evol. Microbiol. *57*: 1266–1271.

Thirkell, D. and M. Summerfield. 1977. Membrane lipids of *Planococcus citreus* Migula from cells grown in presence of 3 different concentra-

tions of sea salt added to a basic medium. Antonie van Leeuwenhoek J. Microbiol. *43*: 43–54.

Thirkell, D. and M. Summerfield. 1980. Variation in pigment production by *Planococcus citreus* Migula with cultural age and with sea salt concentration in the medium. Antonie van Leeuwenhoek *46*: 51–57.

Trentini, W.C. and C. Machen. 1973. Natural habitat of *Caryophanon latum*. Can. J. Microbiol. *19*: 689–694.

Trentini, W.C. and H.E. Gilleland, Jr. 1974. Ultrastructure of the cell envelope and septation process in *Caryophanon latum* as revealed by thin section and freeze-etching techniques. Can. J. Microbiol. *20*: 1435–1442.

Trentini, W.C. and R.G.E. Murray. 1975. Ultrastructural effects of lysozymes on the cell wall of *Caryophanon latum*. Can. J. Microbiol. *21*: 164–172.

Trentini, W.C. 1978. Biology of the genus *Caryophanon*. Annu. Rev. Microbiol. *32*: 123–141.

Trentini, W.C. 1986. Genus *Caryophanon* Peshkoff 1939, 244AL. *In* Sneath, Mair, Sharpe and Holt (Editors), Bergey's Manual of Systematic Bacteriology, vol. 2. The Williams & Wilkins Co., Baltimore, pp. pp. 1259–1260.

Trentini, W.C. 1988. *In* Validation of the publication of new names and new combinations previously effectively published outside the IJSB. List no. 25. Int. J. Syst. Bacteriol. *38*: 220–222.

Trevisan, V. 1885. Caratteri di alcuni nuovi generi Batteriacee. Atti Accad. Fis-Med-Stat., vol. 3 (Ser 4), Milano.

Weeks, O.B. and L.M. Kelley. 1958. Observations on the growth of the bacterium *Caryophanon latum*. J. Bacteriol. *75*: 326–330.

Yamada, Y., G. Inouye, Y. Tahara and K. Kondo. 1976. The menaquinone system in the classification of aerobic Gram-positive cocci in the genera *Micrococcus*, *Staphylococcus*, *Planococcus* and *Sporosarcina*. J. Gen. Appl. Microbiol. *22*: 227–236.

Yang, M., S. Y, P. Ge, Q. Dong and Z.M. 1985. A case of infant septicemia caused by *Kurthia zopfii*. Chinese J. Microbiol. Immunol. *5*: 485.

Yoon, J.H., S.S. Kang, K.C. Lee, E.S. Lee, Y.H. Kho, K.H. Kang and Y.H. Park. 2001a. *Planomicrobium koreense* gen. nov., sp. nov., a bacterium isolated from the Korean traditional fermented seafood jeotgal, and transfer of *Planococcus okeanokoites* (Nakagawa *et al.* 1996) and *Planococcus mcmeekinii* (Junge *et al.* 1998) to the genus *Planomicrobium*. Int. J. Syst. Evol. Microbiol. *51*: 1511–1520.

Yoon, J.H., K.C. Lee, N. Weiss, Y.H. Kho, K.H. Kang and Y.H. Park. 2001b. *Sporosarcina aquimarina* sp. nov., a bacterium isolated from seawater in Korea, and transfer of *Bacillus globisporus* (Larkin and Stokes 1967), *Bacillus psychrophilus* (Nakamura 1984) and *Bacillus pasteurii* (Chester 1898) to the genus *Sporosarcina* as *Sporosarcina globispora* comb. nov., *Sporosarcina psychrophila* comb. nov. and *Sporosarcina pasteurii* comb. nov., and emended description of the genus *Sporosarcina*. Int. J. Syst. Evol. Microbiol. *51*: 1079–1086.

Yoon, J.H., N. Weiss, K.C. Lee, I.S. Lee, K.H. Kang and Y.H. Park. 2001c. *Jeotgalibacillus alimentarius* gen. nov., sp. nov., a novel bacterium isolated from jeotgal with L-lysine in the cell wall, and reclassification of *Bacillus marinus* Ruger 1983 as *Marinibacillus marinus* gen. nov., comb. nov. Int. J. Syst. Evol. Microbiol. *51*: 2087–2093.

Yoon, J.H., N. Weiss, K.H. Kang, T.K. Oh and Y.H. Park. 2003. *Planococcus maritimus* sp. nov., isolated from sea water of a tidal flat in Korea. Int. J. Syst. Evol. Microbiol. *53*: 2013–2017.

Yoon, J.H., I.G. Kim, P. Schumann, T.K. Oh and Y.H. Park. 2004. *Marinibacillus campisalis* sp. nov., a moderate halophile isolated from a marine solar saltern in Korea, with emended description of the genus *Marinibacillus*. Int. J. Syst. Evol. Microbiol. *54*: 1317–1321.

Zobell, C.E. and H.C. Upham. 1944. A list of marine bacteria including descriptions of sixty new species. Bull. Scripps Inst. Oceanogr. Univ. Calif. *5*: 239–292.

Family VII. **Sporolactobacillaceae** fam. nov.

Wolfgang Ludwig, Karl-Heinz Schleifer and William B. Whitman

Spo.ro.lac.to.ba.cil.la'ce.ae. N.L. masc. n. *Sporolactobacillus* type genus of the family; suff. *-aceae* ending denoting family; N.L. fem. pl. n. *Sporolactobacillaceae* the *Sporolactobacillus* family.

The family *Sporolactobacillaceae* is circumscribed for this volume on the basis of phylogenetic analyses of the 16S rRNA sequences and includes the genus *Sporolactobacillus* and its close relatives. It is composed of Gram-positive rods. The cell walls contain *meso*-diaminopimelate. Endospores are formed. Facultatively anaerobic or aerobic growth. Cells contain menaquinone as the major respiratory quinine. Mesophilic and thermophilic species.

Type genus: **Sporolactobacillus** Kitahara and Suzuki 1963, 69[AL].

Genus I. **Sporolactobacillus** Kitahara and Suzuki 1963, 69[AL]

Fujitoshi Yanagida and Ken-ichiro Suzuki

Spo.ro.lac.to.ba.cil'lus. Gr. n. *spora* seed; L. n. *lac, lactis* milk; L. dim. n. *bacillus* a small rod; N.L. masc. n. *Sporolactobacillus* sporing milk rodlet.

Cells are **straight rods**, $0.4–1.0 \times 2.0–4.0\,\mu m$, occurring singly, in pairs, and, rarely, in short chains. **Endospores are formed** that are resistant to heating at 80°C for 10 min. **Gram-stain-positive**. Mostly motile by peritrichous flagella. Facultatively anaerobic or microaerophilic; good growth occurs on media containing glucose, but poor or no growth occurs in nutrient broth; D- or DL-lactic acid is produced homofermentatively; catalase negative; does not contain cytochromes. Mesophilic. Acid is produced from glucose, fructose, galactose, mannose, maltose, sucrose, and trehalose. The cell wall contains **meso-diaminopimelic acid (meso-DAP)**. The predominant isoprenoid quinone is a **menaquinone with seven isoprene units (MK-7)**. Ubiquinone is not detected. The cellular fatty acids comprise predominantly **12-methyl tetradecanoic acid ($C_{15:0\ ante}$) and 14-methyl hexadecanoic acid ($C_{17:0\ ante}$)**, with 13-methyl tetradecanoic acid ($C_{15:0\ iso}$) and 14-methyl pentadecanoic acid ($C_{16:0\ iso}$) as minor components.

DNA G+C content (mol%): 43–50 (T_m).

Type species: **Sporolactobacillus inulinus** (Kitahara and Suzuki 1963) Kitahara and Lai 1967, 197[AL] (*Lactobacillus (Sporolactobacillus) inulinus* Kitahara and Suzuki 1963, 69).

Further descriptive information

Although strains of *Sporolactobacillus* show resistance to heat at 80°C for 10 min, spores are rarely observed microscopically. Media for sporulation contain: 0.1% (w/v) yeast extract; 0.5% (w/v) meat extract; 1.0% (w/v) $(NH_4)_2SO_4$; 0.5% (w/v) α-methyl D-glucoside; 20% (v/v) tomato serum; and an excess of $CaCO_3$ (Kitahara and Lai, 1967; Kitahara and Toyota, 1972). However, Doores and Westhoff (1983) reported improved sporulation in the absence of tomato serum. Motility occurs by means of a small number of long peritrichous flagella.

The cell wall shows a Gram-positive type of structure with peptidoglycan of the *meso*-DAP direct linkage type. In spite of the presence of polysaccharides, teichoic acid is not found (Okada et al., 1976; Weiss et al., 1967). Isoprenoid quinones are exclusively menaquinones, predominantly MK-7 and small proportions of MK-6 and MK-5 (Collins and Jones, 1979; Yanagida et al., 1997). Cellular fatty acids consist mostly of anteiso- and iso-branched acids (Uchida and Mogi, 1973; Yanagida et al., 1997). These chemotaxonomic characteristics are similar to those of members of the genus *Bacillus*.

Catalase is absent. Growth is observed under anaerobic cultivation. Very recently *Bacillus laevolacticus* was transferred to the genus *Sporolactobacillus* based on phylogenetic analysis (Hatayama et al., 2006). Hatayama and colleagues showed that strains of *Sporolactobacillus laevolacticus* did not have catalase activity, in contrast to data from previous studies (Nakayama and Yanoshi, 1967a; Yanagida et al., 1987a).

Carbohydrates are essential substrates for growth. Although acid is produced from a limited number of carbohydrates, species and subspecies of the genus vary in their substrate utilization patterns, as shown in Table 72. Lactic acid is produced from glucose by homofermentation.

Phylogenetically, the genus *Sporolactobacillus* is part of the family "*Sporolactobacillaceae*" of the order *Bacillales* based on 16S rRNA gene sequences. Previously, several DNA–DNA hybridi-zation studies have shown low relatedness values of the type strain of *Sporolactobacillus inulinus* with *Bacillus coagulans* and with *Lactobacillus* species, including *Lactobacillus plantarum* (Dellaglio et al., 1975; Miller et al., 1971). Fox et al. (1977, 1980) and Stackebrandt et al. (1987) indicated a distant relationship between *Sporolactobacillus inulinus* and *Bacillus* species based on 16S rRNA oligonucleotide cataloging. Ash et al. (1991) classified the genus *Bacillus* into five groups by comparative analysis of nearly (95%) complete 16S rRNA primary sequences; *Sporolactobacillus inulinus* was separated phylogenetically from all of the *Bacillus* species. Suzuki and Yamasato (1994) determined 16S rRNA gene sequences of 16 strains of the spore-forming lactic acid bacteria and revealed that the isolates of Nakayama and Yanoshi (1967a, 1967b) formed one cluster with *Sporolactobacillus inulinus* and that the cluster could be further divided into five subclusters, namely four catalase negative subclusters and one positive subcluster. Andersch et al. (1994) proposed *Bacillus laevolacticus* for a group of catalase positive strains. Later, Hatayama et al. (2006) transferred *B. laevolacticus* to the genus *Sporolactobacillus* based on phylogenetic analysis of the 16S rRNA gene.

TABLE 72. Differential characteristics of *Sporolactobacillus* species and subspecies[a]

Characteristic	1. *S. inulinus*	2. *S. kofuensis*	3. *S. lactosus*	4. *S. laevolacticus*	5a. *S. nakayamae* subsp. *nakayamae*	5b. *S. nakayamae* subsp. *racemicus*	6. *S. terrae*
DNA homology group[b]	1	5	6	–	2	4	3
Number of strains studied[b]	2	2	10	12	17	4	8
Growth at 15°C	–	–	d	d	d	+	d
Litmus milk:							
Acidification	–	–	d	d	d	+	d
Reduction	–	–	+	d	d	d	–
Acid production from:							
Arabinose	–	–	d	–	–	–	d
Xylose	–	–	d	–	–	–	–
Galactose	–	+	+	+	+	+	d
Cellobiose	–	–	d	d	–	d	d
Lactose	–	–	+	d	–	d	–
Melibiose	–	–	+	d	d	+	d
Starch	+	d	d	d	d	d	d
Inulin	d	+	+	d	d	+	+
Isomer of lactic acid produced	D	D	D	D	D	DL	D

[a]+, 90% or more strains are positive; –, 90% or more strains are negative; d, 11–89% strains are positive.
[b]Yanagida et al. (1997); *Sporolactobacillus laevolacticus* was not included in this study.

Enrichment and isolation procedures

Sample suspensions are incubated anaerobically at 30°C after a heat treatment of 10 min at 80°C to suppress non-spore-forming aerobic bacteria. After 4–7 d incubation, one drop of the turbid broth is spread on GYP agar containing 1% $CaCO_3$ and incubated in an anaerobic jar under a 100% CO_2 atmosphere. Acid-producing bacteria are recognized by the appearance of clear zones around colonies. They are purified by repeated isolation. The production of lactic acid is confirmed by HPLC (Yanagida et al., 1997).

For further cultivation, *Sporolactobacillus* strains grow well on GYP medium, which contains 2% glucose, 1% yeast extract, 1% peptone, 1% sodium acetate, and 0.5% (v/v) salt solution (pH 6.8). The salt solution contains 4% $MgSO_4 \cdot 7H_2O$, 0.2% $MnSO_4 \cdot 4H_2O$, 0.2% $FeSO_4 \cdot 7H_2O$, and 0.2% NaCl. MRS medium at pH 5.5 is also useful for selective isolation and enrichment. The *Sporolactobacillus* strains isolated so far are mesophilic, with optimal growth temperatures of 30–35°C.

Maintenance procedures

Sporolactobacillus strains can be lyophilized or liquid-dried by using the same procedures as used for lactobacilli. They can also be preserved by freezing at –80°C or lower in the presence of 10% (v/v) skimmed milk broth or glycerol.

Differentiation of the genus *Sporolactobacillus* from the other genera

Phenotypic characteristics of the genus *Sporolactobacillus* and related spore-forming genera are summarized in Table 73. Some characteristics, e.g., endospore formation, the presence of iso- and anteiso-branched cellular fatty acids, MK-7 as the predominant menaquinone, and *meso*-DAP-containing peptidoglycan, are shared with the genera *Bacillus*, *Brevibacillus*, and *Paenibacillus*. Generally, members of these genera are strict aerobes and catalase positive, whereas *Sporolactobacillus* species are microaerophilic and catalase negative. *Bacillus coagulans* produces lactic acid anaerobically as an end product from glucose. Members of the genus *Amphibacillus* lack menaquinones, catalase and cytochromes, and major fermentation products from glucose are ethanol, acetic acid, and formic acid (Niimura et al., 1990).

Some non-sporulating Gram-positive bacteria show similar characteristics to those of members of the genus *Sporolactobacillus*. *Lactobacillus plantarum* strains produce lactic acid homofermentatively and possess *meso*-DAP-containing peptidoglycan, as observed in *Sporolactobacillus* species, but their cellular fatty acids are straight chain acids. Like members of the genus *Sporo-lactobacillus*, species of the genus *Marinococcus* are characterized by *meso*-DAP, MK-7, and iso- and anteiso-branched chain fatty acids (Li et al., 2005). However, strains of *Marinococcus* are strictly aerobic and do not form spores.

TABLE 73. Discriminative characteristics of some genera of aerobic, endospore-forming bacteria[a]

Characteristic	Sporolactobacillus[b]	Amphibacillus[c]	Bacillus[d]	Brevibacillus[e]	Paenibacillus[f]	Sporosarcina[g]
Rod-shaped cells	+	+	+	+	+	−
Spore shape	Oval	Oval	Oval or spherical	Oval	Oval	Round
Anaerobic growth	+	+	V	V	+	−
Growth in nutrient broth	−	−	+	+	+	+ or −
Catalase activity	−	−	+	+	+	+
Production of lactic acid	+	+	V	nt	nt	nt
Major isoprenoid quinone	MK-7[h]	None	MK-7	MK-7	MK-7	MK-7
Major cellular fatty acids	$C_{15:0\ ante}$, $C_{17:0\ ante}$	$C_{15:0\ ante}$, $C_{16:0}$, $C_{16:0\ iso}$, $C_{15:0\ iso}$	V	$C_{15:0\ ante}$ and $C_{15:0\ iso}$ or $C_{15:0}$	$C_{15:0\ ante}$	$C_{15:0\ ante}$
DNA G + C content (mol%)	43–50	36–42	32–66	40–57	40–54	40–42

[a]+, Positive; −, negative; V, variable; nt, not tested.
[b]Kandler and Weiss (1986), Yanagida et al. (1997).
[c]Niimura et al. (1990).
[d]Claus and Berkeley (1986).
[e]Shida et al. (1996).
[f]Heyndrickx et al. (1996a, 1996b), Shida et al. (1997).
[g]Yoon et al. (2001).
[h]Collins and Jones (1979).

Phylogenetically, the genus *Marinococcus* is the closest neighbor of the genus *Sporolactobacillus*, whereas the genus *Lactobacillus* is distantly related.

Taxonomic comments

The genus *Sporolactobacillus* was first established by Kitahara and Suzuki (1963) to accommodate the spore-forming lactic acid bacteria as a subgenus of the genus *Lactobacillus*. Later, the subgenus was elevated to the genus level in the Approved Lists of Bacterial Names (Skerman et al., 1980). Current phylogenetic analysis based on 16S rRNA gene sequences places the genus *Sporolactobacillus* in the family *Sporolactobacillaceae* of the order *Bacillales*, whereas *Lactobacillus* falls in the order *Lactobacillales*. A tree showing members of the family *Sporolactobacillaceae* and related genera is shown in Figure 69.

Sporolactobacillus inulinus was first isolated from chicken feed (Kitahara and Lai, 1967; Kitahara and Suzuki, 1963). Later, Nakayama and Yanoshi (1967a, 1967b) isolated a number of spore-forming lactic acid bacteria from soil of plant rhizospheres and proposed four species, namely "*Bacillus laevolacticus*", "*Bacillus racemilacticus*", "*Sporolactobacillus laevus*", and "*Sporolactobacillus racemicus*" – none of these names was validly published. Yanagida et al. (1987b, 1987a) studied these spore-forming lactic acid bacteria by DNA–DNA hybridization as well as biochemical and physiological tests.

Andersch *et al.* (1994) proposed *Bacillus laevolacticus* for some catalase-positive strains. Based on the results of their previous studies, Yanagida et al. (1997) proposed four novel species and one novel subspecies in the genus *Sporolactobacillus*. Recently, *Bacillus laevolacticus* was transferred to the genus *Sporolactobacillus* as *Sporolactobacillus laevolacticus* (Hatayama et al., 2006), based on its phylogenetic position, as well as weak or non-existent catalase activity. The data for catalase activity differ in these studies. Catalase acti-vity in *Sporolactobacillus laevolacticus* exists, but is extremely weak and quite different from the strong catalase activity observed in species of *Bacillus*. Strains of other species of the genus *Sporolactobacillus* do not possess catalase activity.

Differentiation of species of the genus *Sporolactobacillus*

The species of *Sporolactobacillus* have been established on the basis of DNA–DNA hybridization with consideration of phenotypic characteristics (Yanagida et al., 1987a, 1987b, 1997). The acid production profiles from various carbohydrates and the isomers of lactic acid produced are useful phenotypic criteria for species differentiation. Molecular techniques including 16S rRNA gene sequence analysis may provide useful information to classify and identify species of the genus.

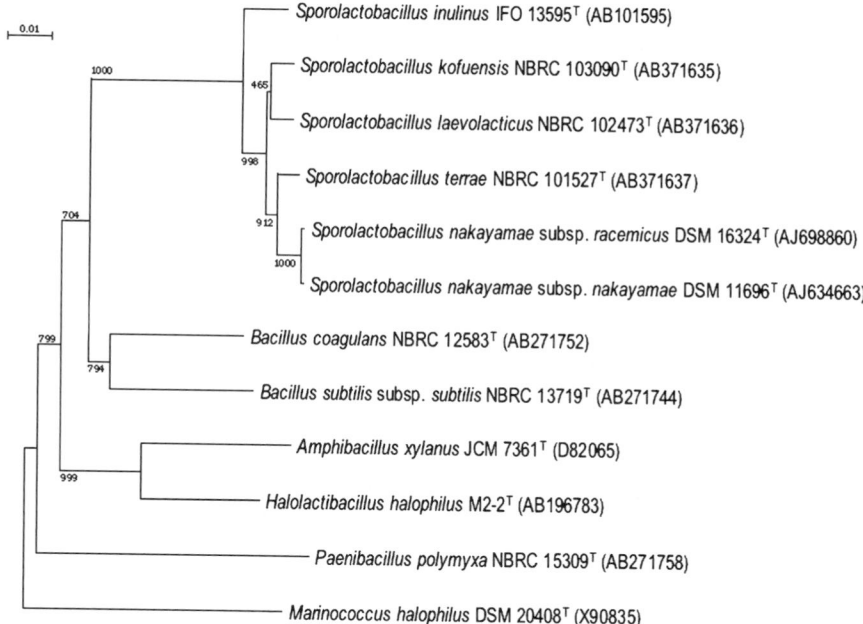

FIGURE 69. Phylogenetic tree of members of the genus *Sporolactobacillus* and related genera based on 16S rRNA gene sequences. Scale bar indicates the K_{nuc} values calculated from the nucleotide sequences. Numbers at branches indicate the confidence limits estimated by bootstrap analysis with 1000 resampling trials, which are shown only for the major clusters. (Tree and some sequences courtesy of M. Miyashita).

List of species of the genus *Sporolactobacillus*

1. **Sporolactobacillus inulinus** (Kitahara and Suzuki 1963) Kitahara and Lai 1967, 197^AL (*Lactobacillus (Sporolactobacillus) inulinus* Kitahara and Suzuki 1963, 69*)

i.nu.li′nus. N.L. n. *inulum* inulin; N.L. adj. *inulinus* pertaining to inulin.

Cell morphology and chemotaxonomic characteristics are as described for the genus. Additional species characteristics are as follows. L-Lactic acid is produced. Acid is produced from raffinose, mannitol, starch, and methyl α-D-glucoside, but not from arabinose, ribose, xylose, galactose, rhamnose, cellobiose, lactose, melibiose, or salicin. The type strain produces acid from inulin.

Source: chicken feed and soil.

DNA G+C content (mol%): 47–50 (T_m); 47 for type strain.

Type strain: ATCC 15538, DSM 20348, JCM 6014, LMG 11481, NBRC 13595, NCIMB 9743.

GenBank accession number (16S rRNA gene): D16283, AB101595.

2. **Sporolactobacillus kofuensis** Yanagida, Suzuki, Kozaki and Komagata 1997, 503^VP

ko.fu.en′sis. N.L. adj. *kofuensis* belonging to Kofu-city, Yamanashi, Japan, the place of origin of the soil from which the organism was isolated.

Cell morphology and chemotaxonomic characteristics are as described for the genus. Additional species characteristics are as follows. L-Lactic acid is produced. Acid is produced from galactose, raffinose, mannitol, and inulin, but not from arabinose, ribose, xylose, rhamnose, cellobiose, lactose, melibiose, salicin, or sorbitol.

Source: soil.

DNA G+C content (mol%): 43 (T_m; type strain).

Type strain: M-19, JCM 3419, LMG 18786 and NBRC 103090.

GenBank accession number (16S rRNA gene): AB371635.

3. **Sporolactobacillus lactosus** Yanagida, Suzuki, Kozaki and Komagata 1997, 503^VP

lac.to′sus. N.L. adj. *lactosus* of lactose, pertaining to lactose.

Cell morphology and chemotaxonomic characteristics are as described for the genus. Additional species characteristics are as follows. Grows at 45°C and most strains grow at 15°C. L-Lactic acid is produced. Litmus milk is acidified and afterwards reduced by some strains. Acid is produced from galactose, lactose, melibiose, and raffinose.

---

* Editorial note: The original designation inadvertently reduced *Lactobacillus* to subgeneric rank and was presumably was a *lapsus calami* for *Lactobacillus (Sporolactobacillus) inulinus*. In subsequent papers, Kitahara and co-workers referred to the organism as *Sporolactobacillus inulinus*, elevating *Sporolactobacillus* to generic rank. The Approved Lists of Bacterial Names cites Kitahara and Suzuki (1963) as the authors of the generic name and Kitahara and Lai (1967) as authors of the type species name, and we will follow that designation here.

Some strains produce acid from arabinose, ribose, xylose, rhamnose, cellobiose, salicin, sorbitol, starch, methyl α-D-glucoside, and inulin.

Source: soil in Japan and fermentation starters.

DNA G+C content (mol%): 43–46 (T_m); 46 for type strain.

Type strain: strain X1-1, JCM 9690 (the type strain does not seem to be available from any international culture collection and, therefore, this taxon may lose its standing in nomenclature).

4. **Sporolactobacillus laevolacticus** (Andersch, Pianka, Fritze and Claus 1994) Hatayama, Shoun, Ueda and Nakamura 2006, 2550[VP] (*Bacillus laevolacticus* Andersch, Pianka, Fritze and Claus 1994, 663; "*Bacillus laevolacticus*" Nakayama and Yanoshi 1967a, 149)

lae.vo.lac′ticus. N.L. adj. *laevolacticus* referring to D-(−)-lactic acid, the sole (levorotatory) isomer of lactic acid that is produced by the bacterium.

Cell morphology and chemotaxonomic characteristics are as described for the genus. Additional species characteristics are as follows. Litmus milk is acidified, reduced, and coagulated. DL-Lactic acid is produced. Acid is produced from galactose, lactose, melibiose, raffinose, mannitol, and inulin, but not from arabinose, ribose, xylose, rhamnose, cellobiose, salicin, or methyl α-D-glucoside. Some strains produce acid from sorbitol and starch.

Additional description by Andersch et al. (1994). No growth occurs in the presence of lysozyme or 5% NaCl. Starch and pullulan are hydrolyzed. Citrate and propionate are not utilized. Gelatin, DNA, tyrosine, and casein are not hydrolyzed. Indole is not produced. Nitrate is not reduced to nitrite. Negative for egg yolk lecithinase. No phenylalanine deaminase.

Additional description by Hatayama et al. (2006). Cell-wall peptidoglycan contains *meso*-DAP, alanine, glutamic acid, galactose, mannose, and rhamnose.

Source: soil in Japan and fermentation starters.

DNA G+C content (mol%): 43–45 (T_m); 45 for type strain.

Type strain: M-8, ATCC 23492, DSM 442, IAM 12321, JCM 2513, LMG 6329, NBRC 102473, NCIMB 10269.

GenBank accession number (16S rRNA gene): AB362642, AB371636.

5. **Sporolactobacillus nakayamae** Yanagida, Suzuki, Kozaki and Komagata 1997, 502[VP]

na.ka.ya′mae. N.L. gen. n. *nakayamae* of Nakayama, named after Ooki Nakayama, a Japanese microbiologist who isolated a number of *Sporolactobacillus* strains.

Cell morphology and chemotaxonomic characteristics are as described for the genus. In addition, most strains grow at 15°C, but not at 45°C.

Source: soil.

DNA G+C content (mol%): 43–47 (T_m).

Type strain: M-114, ATCC 700379, DSM 11696, IAM 12388, JCM 3514, LMG 18787, NBRC 101526.

GenBank accession number (16S rRNA gene): AJ634663.

a. **Sporolactobacillus nakayamae subsp. nakayamae** Yanagida, Suzuki, Kozaki and Komagata 1997, 502[VP]

na.ka.ya′mae. N.L. gen. n. *nakayamae* of Nakayama, named after Ooki Nakayama, a Japanese microbiologist who isolated a number of *Sporolactobacillus* strains.

Cell morphology and chemotaxonomic characteristics are as described for the genus. In addition to the characteristics for the genus and the species, those for the subspecies are as follows. Does not acidify Litmus milk. L-Lactic acid is produced. Acid is produced from galactose, but not from arabinose, ribose, xylose, rhamnose, or lactose. Most strains produce acid from galactose, but not from cellobiose, melibiose, salicin, starch, or inulin.

Source: soil.

DNA G+C content (mol%): 43–47 (T_m); 45 for type strain.

Type strain: M-114, ATCC 700379, DSM 11696, IAM 12388, JCM 3514, LMG 18787, NBRC 101526.

GenBank accession number (16S rRNA gene): AJ634663.

b. **Sporolactobacillus nakayamae subsp. racemicus** Yanagida, Suzuki, Kozaki and Komagata 1997, 503[VP]

ra.ce.mi′cus. N.L. adj. *racelicus* racemic; DL-lactic acid is produced.

Cell morphology and chemotaxonomic characteristics are as described for the genus. In addition to the characteristics for the genus and the species, those for the subspecies are as follows. DL-Lactic acid is produced. Litmus milk is acidified. Acid is produced from galactose, melibiose, raffinose, and inulin, but not from arabinose, ribose, or xylose.

Source: soil.

DNA G+C content (mol%): 43–46 (T_m); 45 for type strain.

Type strain: M-17, ATCC 700381, DSM 16324, IAM 12396, JCM 3417, LMG 18785, NBRC 101524.

GenBank accession number (16S rRNA gene): AJ698860.

6. **Sporolactobacillus terrae** Yanagida, Suzuki, Kozaki and Komagata 1997, 503[VP]

ter′rae. L. n. *terra* earth. L. gen. n. *terrae* of the earth.

Cell morphology and chemotaxonomic characteristics are as described for the genus. Additional species characteristics are as follows. Most strains grow at 15°C. Acid is produced from inulin, but not from ribose, xylose, rhamnose, lactose, or sorbitol. Most strains produce acid from galactose. Most strains do not produce acid from arabinose, melibiose, or starch.

Source: soil.

DNA G+C content (mol%): 43–46 (T_m); 44 for type strain.

Type strain: M-116, ATCC 700380, DSM 11697, IAM 14264, JCM 3516, LMG 18887, NBRC 101527.

GenBank accession number (16S rRNA gene): AJ634662, AB371637.

References

Andersch, I., S. Pianka, D. Fritze and D. Claus. 1994. Description of *Bacillus laevolacticus* (*ex* Nakayama and Yanoshi 1967) sp. nov., nom. rev. Int. J. Syst. Bacteriol. *44*: 659–664.

Ash, C., J.A.E. Farrow, S. Wallbanks and M.D. Collins. 1991. Phylogenetic heterogeneity of the genus *Bacillus* revealed by comparative analysis of small subunit ribosomal RNA sequences. Lett. Appl. Microbiol. *13*: 202–206.

Claus, D. and R.C.W. Berkeley. 1986. Genus *Bacillus* Cohn 1872. *In* Sneath, Mair, Sharpe and Holt (Editors), Bergey's Manual of Systematic Bacteriology, vol. 2. The Williams & Wilkins Co., Baltimore, pp. 1105–1139.

Collins, M.D. and D. Jones. 1979. Isoprenoid quinone composition as a guide to the classification of *Sporolactobacillus* and possibly related bacteria. J. Appl. Bacteriol *47*: 293–297.

Dellaglio, F., V. Bottazzi and M. Vescovo. 1975. Deoxyribonucleic acid homology among *Lactobacillus* species of subgenus *Streptobacterium* Orla-Jensen. Int. J. Syst. Bacteriol. *25*: 160–172.

Doores, S. and D.C. Westhoff. 1983. Selective method for the isolation of *Sporolactobacillus* from food and environment sources. J. Appl Bacteriol *54*: 273–280.

Fox, G.E., K.R. Pechman and C.R. Woese. 1977. Comparative cataloguing of 16S ribosomal ribonucleic acid: molecular approach to prokaryotic systematics. Int. J. Syst. Bacteriol. *27*: 44–57.

Fox, G.E., E. Stackebrandt, R.B. Hespell, J. Gibson, J. Maniloff, T.A. Dyer, R.S. Wolfe, W.E. Balch, R.S. Tanner, L.J. Magrum, L.B. Zablen, R. Blakemore, R. Gupta, L. Bonen, B.J. Lewis, D.A. Stahl, K.R. Luehrsen, K.N. Chen and C.R. Woese. 1980. The phylogeny of prokaryotes. Science *209*: 457–463.

Hatayama, K., H. Shoun, Y. Ueda and A. Nakamura. 2006. *Tuberibacillus calidus* gen. nov., sp. nov., isolated from a compost pile and reclassification of *Bacillus naganoensis* Tomimura et al. 1990 as *Pullulanibacillus naganoensis* gen. nov., comb. nov. and *Bacillus laevolacticus* Andersch *et al.* 1994 as *Sporolactobacillus laevolacticus* comb. nov. Int. J. Syst. Evol. Microbiol. *56*: 2545–2551.

Heyndrickx, M., K. Vandemeulebroecke, B. Hoste, P. Janssen, K. Kersters, P. De Vos, N.A. Logan, N. Ali and R.C.W. Berkeley. 1996a. Reclassification of *Paenibacillus* (formerly *Bacillus*) *pulvifaciens* (Nakamura 1984) Ash et al. 1994, a later subjective synonym of *Paenibacillus* (formerly *Bacillus*) *larvae* (White 1906) Ash et al. 1994, as a subspecies of *P. larvae*, with emended descriptions of *P. larvae* as *P. larvae* subsp. *larvae* and *P. larvae* subsp. *pulvifaciens*. Int. J. Syst. Bacteriol. *46*: 270–279.

Heyndrickx, M., K. Vandemeulebroecke, P. Scheldeman, K. Kersters, P. De Vos, N.A. Logan, A.M. Aziz, N. Ali and R.C.W. Berkeley. 1996b. A polyphasic reassessment of the genus *Paenibacillus*, reclassification of *Bacillus lautus* (Nakamura 1984) as *Paenibacillus lautus* comb. nov. and of *Bacillus peoriae* (Montefusco et al. 1993) as *Paenibacillus peoriae* comb. nov., and emended descriptions of *P. lautus* and of *P. peoriae*. Int. J. Syst. Bacteriol. *46*: 988–1003.

Kandler, O. and N. Weiss. 1986. Genus *Sporolactobacillus*. *In* Sneath, Mair, Sharpe and Holt (Editors), Bergey's Manual of Systematic Bacteriology, vol. 2. The Williams & Wilkins Co., Baltimore, pp. 1139–1141.

Kitahara, K. and J. Suzuki. 1963. *Sporolactobacillus* nov. subgen. J. Gen. Appl. Microbiol. *9*: 59–71.

Kitahara, K. and C.L. Lai. 1967. On the spore formation of *Sporolactobacillus inulinus*. J. Gen. Appl. Microbiol. *13*: 197–203.

Kitahara, K. and T. Toyota. 1972. Auto-spheroplastization and cell-permeation in *Sporolactobacillus inulinus*. J. Gen. Appl. Microbiol. *18*: 99–107.

Li, W.J., P. Schumann, Y.Q. Zhang, G.Z. Chen, X.P. Tian, L.H. Xu, E. Stackebrandt and C.L. Jiang. 2005. *Marinococcus halotolerans* sp. nov., isolated from Qinghai, north-west China. Int. J. Syst. Evol. Microbiol. *55*: 1801–1804.

Miller, A., 3rd, W.E. Sandine and P.R. Elliker. 1971. Deoxyribonucleic acid homology in the genus *Lactobacillus*. Can. J. Microbiol. *17*: 625–634.

Nakayama, O. and M. Yanoshi. 1967a. Spore-bearing lactic acid bacteria isolated from rhizosphere. I. Taxonomic studies on *Bacillus laevolacticus* nov. sp. and *Bacillus racemilacticus* nov. sp. J. Gen. Appl. Microbiol. *13*: 139–153.

Nakayama, O. and M. Yanoshi. 1967b. Spore-bearing lactic acid bacteria isolated from rhizosphere. II. Taxonomic studies on the catalase-negative strains. J. Gen. Appl. Microbiol. *13*: 155–165.

Niimura, Y., E. Koh, F. Yanagida, K.I. Suzuki, K. Komagata and M. Kozaki. 1990. *Amphibacillus xylanus* gen. nov., sp. nov., a facultatively anaerobic spore-forming xylan-digesting bacterium which lacks cytochrome, quinone, and catalase. Int. J. Syst. Bacteriol. *40*: 297–301.

Okada, S., T. Toyoda, M. Kozaki and K. Kitahara. 1976. Studies on the cell wall of *Sporolactobacillus inulinus*. J. Agric. Chem. Soc. Jap. *50*: 259–263.

Shida, O., H. Takagi, K. Kadowaki and K. Komagata. 1996. Proposal for two new genera, *Brevibacillus* gen. nov. and *Aneurinibacillus* gen. nov. Int. J. Syst. Bacteriol. *46*: 939–946.

Shida, O., H. Takagi, K. Kadowaki, L.K. Nakamura and K. Komagata. 1997. Transfer of *Bacillus alginolyticus*, *Bacillus chondroitinus*, *Bacillus curdlanolyticus*, *Bacillus glucanolyticus*, *Bacillus kobensis*, and *Bacillus thiaminolyticus* to the genus *Paenibacillus* and emended description of the genus *Paenibacillus*. Int. J. Syst. Bacteriol. *47*: 289–298.

Skerman, V.B.D., V. McGowan and P.H.A. Sneath. 1980. Approved lists of bacterial names. Int. J. Syst. Bacteriol *30*: 225–420.

Stackebrandt, E., W. Ludwig, M. Weizenegger, S. Dorn, T.J. Mcgill, G.E. Fox, C.R. Woese, W. Schubert and K.H. Schleifer. 1987. Comparative 16S ribosomal RNA oligonucleotide analyses and murein types of round spore-forming bacilli and non-spore-forming relatives. J. Gen. Microbiol. *133*: 2523–2529.

Suzuki, T. and K. Yamasato. 1994. Phylogeny of spore-forming lactic acid bacteria based on 16S rRNA gene sequences. FEMS Microbiol. Lett. *115*: 13–17.

Uchida, K. and K. Mogi. 1973. Cellular fatty acid spectra of *Sporolactobacillus* and some other *Bacillus-Lactobacillus* intermediates as a guide to their taxonomy. J. Gen. Appl. Microbiol. *19*: 129–140.

Weiss, N., R. Plapp and O. Kandler. 1967. Die Aminosäuresequenz des DAP-des DAP haltigen Mureins von *Lactobacillus plantarum* und *Lactobacillus inulinus*. Arch. Mikrobiol. *58*: 313–323.

Yanagida, F., K. Suzuki, T. Kaneko, M. Kozaki and K. Komagata. 1987a. Morphological, biochemical, and physiological characteristics of spore-forming lactic acid bacteria. J. Gen. Appl. Microbiol. *33*: 33–45.

Yanagida, F., K. Suzuki, T. Kaneko, M. Kozaki and K. Komagata. 1987b. Deoxyribonucleic acid relatedness among some spore-forming lactic acid bacteria. J. Gen. Appl. Microbiol. *33*: 47–55.

Yanagida, F., K.I. Suzuki, M. Kozaki and K. Komagata. 1997. Proposal of *Sporolactobacillus nakayamae* subsp. *nakayamae* sp. nov., subsp. nov., *Sporolactobacillus nakayamae* subsp. *racemicus* subsp. nov., *Sporolactobacillus terrae* sp. nov., *Sporolactobacillus kofuensis* sp. nov., and *Sporolactobacillus lactosus* sp. nov. Int. J. Syst. Bacteriol. *47*: 499–504.

Yoon, J.H., K.C. Lee, N. Weiss, Y.H. Kho, K.H. Kang and Y.H. Park. 2001. *Sporosarcina aquimarina* sp. nov., a bacterium isolated from seawater in Korea, and transfer of *Bacillus globisporus* (Larkin and Stokes 1967), *Bacillus psychrophilus* (Nakamura 1984) and *Bacillus pasteurii* (Chester 1898) to the genus *Sporosarcina* as *Sporosarcina globispora* comb. nov., *Sporosarcina psychrophila* comb. nov. and *Sporosarcina pasteurii* comb. nov., and emended description of the genus *Sporosarcina*. Int. J. Syst. Evol. Microbiol. *51*: 1079–1086.

Family VIII. **Staphylococcaceae** fam. nov.

KARL-HEINZ SCHLEIFER AND JULIA A. BELL

Staph.y.lo.coc.ca′ce.ae. N.L. masc. n. *Staphylococcus* type genus of the family; *-aceae* ending to denote family; N.L. fem. pl. n. *Staphylococcaceae* the *Staphylococcus* family.

Circumscribed on the basis of 16S rDNA sequence analysis; contains the genera *Staphylococcus*, *Jeotgalicoccus*, *Macrococcus*, and *Salinicoccus*.

DNA G+C content (mol%): 30–51 (T_m, Bd).

Type genus: **Staphylococcus** Rosenbach 1884, 18.

Genus I. **Staphylococcus** Rosenbach 1884, 18[AL] (Nom. Cons. Opin. 17 Jud. Comm. 1958, 153.)

KARL-HEINZ SCHLEIFER AND JULIA A. BELL

Staph.y.lo.coc′cus. Gr. n. *staphyle* bunch of grapes; Gr. n. *kokkos* a grain or berry; N.L. masc. n. *Staphylococcus* the grape-like coccus.

Cells spherical, 0.5–1.5 μm in diameter, occurring singly, in pairs, in tetrads, in short chains (3–4 cells), and characteristically dividing in more than one plane to form irregular grape-like clusters. **Gram-stain-positive.** Nonmotile. Resting stages not produced. **Cell wall contains peptidoglycan and teichoic acid. The diamino acid present in the peptidoglycan is L-lysine (peptidoglycan group A3).** Usually unencapsulated or limited capsule formation.

Facultative anaerobes. With the exception of *Staphylococcus aureus* subsp. *anaerobius* and *Staphylococcus saccharolyticus*, growth is more rapid and abundant under aerobic conditions. **Usually catalase-positive and oxidase-negative.** Most strains grow in the presence of 10% NaCl and between 18 and 40°C.

Chemo-organotrophs; **metabolism respiratory and fermentative.** Some species are mainly respiratory or mainly fermentative. Unsaturated menaquinones and cytochromes *a* and *b* (and *c* in *Staphylococcus fleuretti*, *Staphylococcus lentus*, *Staphylococcus sciuri*, and *Staphylococcus vitulinus*) form the electron transport system. L- and/or D-lactic acid may be produced from glucose under anaerobic conditions. Lactose or D-galactose metabolized via the D-tagatose-6-phosphate or Leloir pathways, depending upon the particular species. Carbohydrates and/or amino acids utilized as carbon and energy sources. A variety of carbohydrates may be utilized aerobically with the production of acid. For most species, the main product of glucose fermentation is lactic acid; in the presence of air, the main products are acetic acid and CO_2.

Nutritional requirements are variable. Most species require an organic source of nitrogen, i.e., certain amino acids, and B group vitamins. Others can grow with $(NH_4)_2SO_4$ as a sole source of substrate nitrogen. Uracil and/or a fermentable carbon source may be required by certain species for anaerobic growth.

Susceptible to lysis by lysostaphin, but resistant to lysis by lysozyme. Species or strains which have significant amounts of L-serine or L-alanine replacing glycine in the interpeptide bridge of the peptidoglycan are generally less susceptible to lysostaphin than those with an interpeptide bridge consisting solely of glycine residues. Some species resistant to novobiocin. Generally susceptible to furazolidone and nitrofuran, and resistant to erythromycin and bacitracin at low levels. Usually susceptible to antibacterials such as phenols and their derivatives, salicylanilides, carbanilides, and halogens (chlorine and iodine) and their derivatives.

Host for a variety of bacteriophages which may have a narrow or wide host range. Transfer of characters by transduction, transformation, and cell-to-cell contact has been demonstrated in some species.

Natural populations are mainly associated with skin, skin glands, and mucous membranes of warm-blooded animals. Host or niche range may be narrow or wide, depending upon the particular species or subspecies. Some organisms may be isolated from a variety of animal products (e.g., meat, milk, cheese) and environmental sources (e.g., fomites, soil, sand, dust, air, or natural waters). **Some species are opportunistic pathogens of humans and/or animals.** A list of staphylococcal species/subspecies together with some differential characteristics is summarized in Table 74.

DNA G+C content (mol%): 27–41 (T_m, Bd).

Type species: **Staphylococcus aureus** Rosenbach 1884, 18[AL].

Further descriptive information

Phylogeny. Based on comparative 16S rRNA sequence studies, the genus *Staphylococcus* belongs to the phylum *Firmicutes* [Gram-positive bacteria with a low (<50 mol%) DNA G+C content]. It is closely related to bacilli and other members of the phylum *Firmicutes* such as macrococci, enterococci, streptococci, lactobacilli, and listeria (Figure 70). It has been proposed to combine the genera *Staphylococcus*, *Gemella*, *Jeotgalicoccus*, *Macrococcus*, and *Salinicoccus* within the family *Staphylococcaceae* (Garrity and Holt, 2001); in the present volume, *Gemella* has been classified in Family X. *Incertae Sedis*.

The genus *Staphylococcus* is monophyletic (Figure 70) and well separated from other related genera with intergenera 16S rRNA sequence similarities of 93.4–95.3%. The intragenus similarities are significantly higher with at least 96.5% (Figure 71). Based on DNA–DNA hybridization studies, staphylococcal species can be placed in species groups (Kloos et al., 1992). The most important ones are the two coagulase-negative and novobiocin-susceptible species groups *Staphylococcus epidermidis* (e.g., *Staphylococcus capitis*, *Staphylococcus caprae*, *Staphylococcus epidermidis*, *Staphylococcus haemolyticus*, *Staphylococcus hominis*, *Staphylococcus saccharolyticus*, *Staphylococcus warneri*) and *Staphylococcus simulans* (e.g., *Staphylococcus carnosos*, *Staphylococcus simulans*); the two coagulase-negative and novobiocin-resistant species groups *Staphylococcus saprophyticus* (e.g., *Staphylococcus cohnii, Staphylococcus saprophyticus, Staphylococcus xylosus*) and *Staphylococcus sciuri* (e.g., *Staphylococcus lentus, Staphylococcus sciuri, Staphylococcus vitulinus*), and the two coagulase-positive and novobiocin-susceptible species groups *Staphylococcus intermedius* (e.g., *Staphylococcus delphini, Staphylococcus intermedius*) and *Staphylococcus aureus* (e.g., *Staphylococcus aureus, Staphylococcus aureus* subsp. *anaerobius*).

The different species are genotypically well separated and demonstrate 60% or less DNA–DNA similarity under optimal hybridization conditions and less than 30% under restrictive conditions. The only exception is *Staphylococcus pulvereri/Staphylococcus vitulinus*. The two species appear identical on the basis of DNA similarity, pulse-field gel electrophoresis, ribotyping, and other taxonomic criteria (Kloos et al., 2001; Webster et al., 1994) *Staphylococcus pulvereri* is a later synonym of *Staphylococcus vitulinus*.

Species identity is also made on the basis of a variety of phenotypic characters such as cell-wall composition, colony morphology, activities and molecular properties of various enzymes, production of acid from various carbohydrates, resistance to certain antibiotics, nutritional and oxygen requirements, cytochrome composition, oxygen requirements, and fatty acid composition.

Cell-wall composition. The ultrastructure and chemical composition of the cell wall of staphylococci is similar to that of other Gram-positive bacteria. It consists of a thick (usually 60–80 nm), rather homogeneous, and not very electron-dense layer. It is made up of peptidoglycan, teichoic acid, and protein (Schleifer, 1984). A characteristic feature of the peptidoglycan of all staphylococci, with the exception *Staphylococcus succinus* subsp. *succinus*, is the occurrence of a peptidoglycan of group A 3 with L-lysine as the diamino acid in position 3 of the peptide subunit and a glycine-rich interpeptide bridge (Schleifer and Kandler, 1972). Penta- and hexaglycine interpeptide bridges are found in about half of the staphylococcal species (peptidoglycan type: Lys–Gly$_{5-6}$). In most of the other half, a minor part of the glycine residues can be replaced with L-serine (peptidoglycan type: Lys–Gly$_4$, Ser). *Staphylococcus sciuri, Staphylococcus lentus, Staphylococcus fleuretti*, and *Staphylococcus vitulinus* have an L-alanine instead of a glycine residue bound to lysine of the peptide subunit (peptidoglycan type: Lys–Ala–Gly$_4$). *Staphylococcus succinus* subsp. *succinus*, however, has been described with a completely different peptidoglycan type with *meso*-diaminopimelic acid in position 3 of the peptide subunit and a direct cross-linkage without an interpeptide bridge (Lambert et al., 1998). The occurrence of such a peptidoglycan type would be unique for staphylococci and is rather questionable because the closely related *Staphylococcus succinus* subsp. *casei* contains a typical staphylococcal peptidoglycan type (L-Lys–Gly$_4$, L-Ser; Place et al., 2002).

Additions to the peptidoglycan of *Staphylococcus aureus* also occur in the form of covalently linked wall teichoic acid. They are attached by phosphodiester linkages on C$_6$ of some of the muramic acid residues (Hay et al., 1965). Staphylococcal cell-wall teichoic acids are water-soluble polymers containing repeating phosphodiester groups that are covalently linked to peptidoglycan. They consist of polyols (glycerol or ribitol), sugars and/or N-acetylamino-sugars. Most staphylococci contain glycerol or ribitol teichoic acids. The teichoic acids consist of polymerized polyol phosphates that are substituted with various combinations of sugars and/or N-acetylamino- sugar residues, and also ester-linked D-alanine residues. In some cases, N-acetylamino-sugar residues can also form an integral part of the polymer chain (Endl et al., 1983; Endl et al., 1984). The occurrence of the same major components does not always mean that the structure of teichoic acid is identical; for example, the teichoic acids of *Staphylococcus capitis* and *Staphylococcus hyicus* exhibit a similar composition, but their structures are quite different.

Pathogenicity. The coagulase-positive species *Staphylococcus aureus, Staphylococcus intermedius, Staphylococcus delphini*, and *Staphylococcus schleiferi* subsp. *coagulans* and the coagulase variable species *Staphylococcus hyicus* are regarded as potentially serious pathogens. *Staphylococcus aureus* is responsible for a variety of infections. In the late 1950s and early 1960s, *Staphylococcus aureus* caused considerable morbidity and mortality as a nosocomial pathogen of hospitalized patients. Among the major human infections caused by this species are furuncles, carbuncles, impetigo, toxic epidermal necrolysis (scalded skin syndrome), pneumonia, osteomyelitis, acute endocarditis, myocarditis, pericarditis, enterocolitis, mastitis, cystitis, prostatitis, cervicitis, cerebritis, meningitis, bacteremia, toxic shock syndrome, and abscesses of the muscle, skin, urogenital tract, central nervous system, and various intra-abdominal organs. In addition, staphylococcal enterotoxin is involved in food poisoning. Methicillin-resistant *Staphylococcus aureus* (MRSA) strains have emerged in the 1980s as a major clinical and epidemiological problem in hospitals. These strains are beginning to spread out of the hospitals and into communities.

Staphylococcus aureus is also capable of producing infections in a variety of other mammals and birds. The more common natural infections include mastitis, synovitis, arthritis, endometritis, furuncles, suppurative dermatitis, pyemia, and septicemia. Staphylococcal mastitis in either a clinical or subclinical form may have considerable economic consequences in the dairy industry. *Staphylococcus aureus* subsp. *anaerobius* is the etiologic agent of an abscess disease in sheep, symptomatically similar to caseous lymphadenitis (de la Fuente et al., 1985).

Staphylococcus intermedius is a serious opportunistic pathogen of dogs and may cause otitis externa, pyoderma, abscesses, reproductive tract infections, mastitis, and purulent wound infections. *Staphylococcus hyicus* has been implicated as the etiologic agent of infectious exudative epidermitis (greasy pig disease) and septic polyarthritis of pigs, skin lesions in cattle and horses, osteomyelitis in poultry and cattle, and occasionally has been associated with mastitis in cattle. *Staphylococcus delphini* has been implicated in purulent skin lesions of dolphins (Varaldo et al., 1988). *Staphylococcus schleiferi* subsp. *coagulans* is associated with the external auditory meatus of dogs suffering from external ear otitis (Igimi et al., 1990).

The coagulase-negative staphylococcal species constitute a major component of the normal microflora of the human; their role in causing nosocomial infections has been recognized and well documented over the last two decades. The increase in infections by these organisms has been correlated with the wide medical use of prosthetic and indwelling devices and the growing number of immunocompromised patients in hospitals. Within the coagulase-negative staphylococci, *Staphy-*

TABLE 74. Characteristics differentiating the species and subspecies of the genus *Staphylococcus*[a]

Characteristic	1a. S. aureus subsp. aureus[a]	1b. S. aureus subsp. anaerobius[a]	2. S. arlettae[a]	3. S. auricularis[a]	4a. S. capitis subsp. capitis[a]	4b. S. capitis subsp. urealyticus[a]	5. S. caprae[a]	6a. S. carnosus subsp. carnosus[a]	6b. S. carnosus subsp. utilis[a]	7. S. chromogenes[a]	8a. S. cohnii subsp. cohnii[a]	8b. S. cohnii subsp. urealyticus[a]	9. S. condimenti[a]	10. S. delphini[b]	11. S. epidermidis[a]	12a. S. equorum subsp. equorum[a]	12b. S. equorum subsp. linens[a]	13. S. felis[a]	14. S. fleurettii[a]	15. S. gallinarum[a]	16. S. haemolyticus[a]	17a. S. hominis subsp. hominis[a]	17b. S. hominis subsp. novobiosepticus[a]	18. S. hyicus[a]	19. S. intermedius[a]
Colony diameter (mm)	>5	1–3	6–8	<5	<5	4.3–7.1	d	>5	0.5–1.5	>5	d	>5	1–2	5–7	<5	4–6	1–2	5–8	<3/8–12	>5	>5	>5	4–6	>5	>5
Time (d)	5	2	2	5	5	5	5	5	2	5	5	5	2	5	5	2 d	2	5	Not given	5	5	5	5 d	5	5
Temperature (°C)	34–35 for 3 d; then 25 for 2 d	Not given	37	34–35 for 3 d; then 25 for 2 d	34–35 for 3 d; then 25 for 2 d	34–35 for 3 d; then 25 for 2 d	34–35 for 3 d; then 25 for 2 d	34–35 for 3 d; then 25 for 2 d	37	34–35 for 3 d; then 25 for 2 d	34–35 for 3 d; then 25 for 2 d	34–35 for 3 d; then 25 for 2 d	37	34–35 for 3 d; then 25 for 2 d	34–35 for 3 d; then 25 for 2 d	37	32	34–35 for 3 d; then 25 for 2 d	37	34–35 for 3 d; then 25 for 2 d	34–35 for 3 d; then 25 for 2 d	34–35 for 3 d; then 25 for 2 d	35 for 3 d; then 25 for 2 d	34–35 for 3 d; then 25 for 2 d	34–35 for 3 d; then 25 for 2 d
Medium	P	Blood agar	BHIA	P	P	P	P	P	P	P	P	P	P	P	P	BHIA	PC skim milk	P	P/TSA	P	P	P	P, TSA, TSA +	P	P
Pigment	+w	–	+	–	–	d	–	–	ND	+	–	d	+	–	–	–	–	d	d	d	d	–	–	–	+
Aerobic growth	+	–	+	+	+	+	+	+	+	+	+	+	+	–	+	+	+	+	+	+	+	+	+	+	+
Anaerobic growth (thioglycolate)	+	+	–	-w	(+)	(+)	+	+	+	+	+	d	(+)	+	+	–	+	+	+	+	+	(+)	-w	-w	+
Growth on NaCl agar:																									
10% (w/v)	+	+	ND	+	+	ND	ND	+	+	+	+	+	+	+	w	ND	+	+	+	ND	+	w	ND	+	+
15% (w/v)	w	d	ND	w	-w	ND	ND	+	+	-w	d	d	+	+	–	ND	w	+	ND	ND	d	–	ND	-w	d
Growth at:																									
15°C	+	ND	ND	–		ND	ND	+	+	+	+	+	+	ND	-w	ND	ND	+w	ND	ND	-w	-w	ND	+	+
45°C	+	–	ND	+	+	ND	+	–		-w	d	d	d	+	+	–	ND	+	ND	+w	+	+	ND	-w	+
Lactic acid production:																									
L(+) Isomer	+	+	w	w	+	ND	+	ND		+	w	w	ND	ND	+	w	ND	ND	ND	+	–	d	ND	+	+
D(–) Isomer	+	–	w	–	-w	ND	+	ND		–	–	w	ND	ND	+	w	ND	ND	ND	–	+	+	d	–	–
Acetoin production	+	+	ND	d	d	d	+	+		ND	–	d	d	ND	+	–	+	+	d	+	–	–	–	+	+
Alkaline phosphatase	+	+	+	–	–	–	+	+		–	+	+	–	+w	+	+	+	–	–	+	d	+	–	+	d
Arginine dihydrolase	+w	ND	–	d	d	ND	d	+		+	+	–	-w	+	+	+w	–	–	+	–	–	+	d	–	+
Clumping factor	+	–	–	–	–	–	–	ND		–	–	–	ND	–	+	–	–	ND	–	–	–	–	–	–	+
Coagulase	+	+	–	–	–	–	–	–		–	–	–	+	–	+	–	–	ND	–	–	–	–	–	d	+
Deoxyribonuclease (DNase agar)	+	+	–	-w	w	ND	+	w		ND	w	–	-w	-w	ND	w	-w	–	ND	ND	ND	ND	d	-w	ND
Fibrinolysin	+	ND	–	ND	ND		–	ND		ND	–	ND	ND	ND	–	d	ND	ND	ND	–	ND	ND	ND	d	d
β-Glucosidase	+	–	ND	–	–	–	–		–	ND	d	–	–	ND	ND	(d)	ND	ND	–	ND	+	d	–	d	d
β-Glucuronidase	–	–	ND	–	–	–	–		–	–	–	–	+	–	ND	–	+	–	–	–	–	d	–	d	–
β-Galactosidase	–	–	ND	(d)	–	–	–		+	–	-w	–	+	–	ND	–	+	–	–	-w	–	–	–	ND	d
Heat-stable nuclease	+	+	+	–	–	–	–	d		ND	-w	–	–	ND	+	+	-w	d	–	-w	ND	w	(+)	-w	–
Hemolysis[a]	+	+	–		–	-w	(d)	(+)	–	ND	–	(d)	(d)	ND	+	-w	d	–	-w	ND	w	(+)	-w	–	ND
Hyaluronidase	+	+	–	ND	ND	ND	–	ND		ND	–	ND	ND	ND	ND	–	d	–	ND	–	ND	ND	ND	+	+
Nitrate reduction	+	–	–	(d)	d	d	+	+		+	–	–	–	+	+w	+	+	+	+	+	d	d	+	+	+
Oxidase	–	–	–	–	–	–	–	–	ND	–	–	–	–	ND	ND	–	–	–	+	–	–	–	–	d	+
Urease	+w	ND	–	–	–	+	+	–		–	d	–	–	+	+	+	+	+	–	+	–	+	–	+	+
Aerobic production of acid from:																									
Arabinose	–	–	+	–	–	–	–	–		–	–	–	–	–	–	+	–	–	d	+	–	–	–	–	–
Cellobiose	–	–	–	–	–	–	–	–		–	–	–	–	ND	–	-w	–	–	+	–	–	–	–	–	–
Fructose	+	+	+	+	+	ND	–	+	+	+	+	+	+	+	+	+	+	+	+	d	+	+	+	+	+
Fucose	–	ND	w	–	–	ND	–	–	ND	–	–	–	ND	ND	–	–	ND	ND	–	w	–	–	ND	–	–
Galactose	+	–	d	–	–	ND	+	d	–	+	–	d	+	ND	d	d	–	–	d	–	+	d	d	d	+	d
Lactose	+	–	+	–	–	d	d	d	–	+	–	+	+	d	d	d	–	+	–	d	d	d	+	–	d
Maltose	+	+	+	(+)	–	+	d	–	–	d	(d)	(+)	–	ND	+	–	+	–	–	+	+	+	+	–	(w)
Mannitol	+	–	+	–	+	+	d	–	–	d	d	d	+	+	–	d	ND	d	ND	+	d	–	–	–	(d)
Mannose	+	–	+w	–	+	+	+	+	–	+	(d)	+	+	+	(+)	+	ND	+	+	+	–	–	d	d	+
Melezitose	–	–	+	–	–	ND	–	–	–	–	–	–	ND	–	(d)	–	ND	–	–	–	–	d	d	–	–
Raffinose	–	–	+	–	–	–	–	–	–	–	–	–	ND	–	–	–	ND	–	–	–	–	–	–	–	–
Ribose	+	–	+	–	–	ND	–	ND	+	–	–	–	ND	ND	d	+	–	–	-w	ND	+	d	–	+	+
Salicin	–	–	ND	–	–	ND	–	–	ND	–	–	ND	ND	–	+	+	–	+	–	d	–	+	–	+	–
Sucrose	+	+	+	d	(+)	+	–	–	–	+	–	–	–	w	+	+	+	–	d	+	+	+	(+)	+	+
Trehalose	+	–	+	(+)	–	–	+	d	d	+	+	+	–	–	–	–	+	+	+	d	d	–	+	+
Turanose	+w	ND	+	(d)	–	d	–	–	–	d	–	–	ND	d	d	w	ND	+	+	d	d	ND	–	d
Xylitol	–	–	–	–	ND	–	ND	–	(d)	(d)	ND	–	–	+	ND	–	–	d	–	–	ND	–	–	–
Xylose	–	–	+	–	–	–	–	ND	–	–	–	ND	–	–	+	–	–	d	–	–	–	+	–	–	–
Novobiocin resistance (MIC ≥ 1.6 µg/ml)	–	+	+	–	–	–	–	–	–	–	+	+	ND	–	–	+	–	–	+	+	–	–	+	–	–

20. S. kloosii[a]	21. S. lentus[b]	22. S. lugdunensis[c]	23. S. lutrae[a]	24. S. muscae[a]	25. S. nepalensis[c]	26. S. pasteuri[c]	27. S. piscifermentans[d]	28. S. pulvereri[e]	29. S. saccharolyticus[a]	30a. S. saprophyticus subsp. saprophyticus[b]	30b. S. saprophyticus subsp. bovis[a]	31a. S. schleiferi subsp. schleiferi[a]	31b. S. schleiferi subsp. coagulans[a]	32a. S. sciuri subsp. sciuri[a]	32b. S. sciuri subsp. carnaticus[a]	32c. S. sciuri subsp. rodentium[a]	33. S. simulans[a]	34a. S. succinus subsp. succinus[c]	34b. S. succinus subsp. casei[a]	35. S. vitulinus[c]	36. S. warneri[c]	37. S. xylosus[a]
3–6	<5	<5	3.5–4.5	5–6	2–6	4–7	ND	4–8	<5	>5	4	<5	>5	7±2/10±2	5±2/9±1	6±2/10±1	>5	4–6	4–6	<3/8–12	d	>5
2	5	5	3	5	2	5 d	na	5	5	5	3	5	5	5 d	5 d	5 d	5	2 d	2 d	Not given	5	5
37	34–35 for 3 d; then 25 for 2 d	37	37	37	37	34–35 for 3 d; then 25 for 2 d	na	34–35 for 3 d; then 25 for 2 d	34–35 for 3 d; then 25 for 2 d	34–35 for 3 d; then 25 for 2 d	35	34–35 for 3 d; then 25 for 2 d	34–35 for 3 d; then 25 for 2 d	34–35 for 3 d; then 25 for 2 d	34–35 for 3 d; then 25 for 2 d	34–35 for 3 d; then 25 for 2 d	34–35 for 3 d; then 25 for 2 d	25	32	35	34–35 for 3 d; then 25 for 2 d	34–35 for 3 d; then 25 for 2 d
BHIA	P	P	P	P	P	P	na	P	P	P	P	P	P	P/TSA	P/TSA	P/TSA	P	TSA	PC skim milk	P/TSA	P	P
d	d	d	–	–	–	d	–	–	–	d	–	–	d	d	d	–	–	–	–	+	d	d
+	+	+	+	+	+	+	+	+	–w	+	+	+	+	+	+	+	–	+	+	+	d	d
–	–w	+	+	+	+	+	+	(+)	+	(+)	+	+	+	(+)	–w	–w	+	–	–	–	+	d
ND	+	+	+	+	d	+	+	+	ND	+	+	+	ND	+	ND	ND	+	+	+	ND	+	+
(+)	–w	(+)	–	–	–	+	d	+	ND	d	–	+	ND	d	ND	ND	w	+	+	ND	w	d
ND	–w	ND	ND	–	–	+	ND	+	–w	+	+	ND	ND	+	ND	ND	+	–	–	ND	d	+
ND	–	+	ND	–	–	+	ND	–	+	d	–	+	ND	–w	–w	–w	+	–	–	–	+	–w
w	+	w	ND	ND	ND	+	+	ND	w	w	ND	+	ND	+	ND	ND	+	ND	ND	ND	+	w
w	–	+	ND	ND	ND	+	+	ND	–	w	ND	w	ND	–	ND	ND	d	ND	ND	ND	+	–w
ND	–	+	–	–	–	d	–	d	ND	+	d	+	+	–w	–w	–w	–w	–	–	–	+	d
+	w	–	+	+	+	–	d	ND	–	d	+	+	+	d	d	d	w	+	w	–	–	d
–	–	–	–	–	–	ND	+	d	+	–w	–	+	+	–	–	–	+	–	–	–	d	–
–	–	+	–	v	–	ND	–	–	–	–	–	+	+	d[r]	+	–	ND	ND	d	–	–	–
–	–	–	+	–	–	–	–	–	–	–	–	+	+	–	ND	ND	–	–	–	–	–	–
–	+w	ND	w	ND	ND	–	ND	ND	ND	–	ND	ND	ND	+w	ND	ND	w	ND	ND	ND	d	–w
–	ND	–	ND	–	ND	–	ND	ND	ND	ND	–	–	ND	ND	ND	ND	ND	ND	ND	ND	ND	ND
ND	+	ND	ND	ND	ND	+	+	ND	ND	d	ND	ND	ND	+	+	+	–	ND	ND	d	+	+
ND	–	–	ND	ND	+	+	–	d	ND	d	–	ND	–	–	–	–	d	+	+	–	d	d
ND	–	–	+	–	+	–	d	ND	d	d	d	ND	–	+	+	+	–	–	–	–	–	+
–	ND	–	ND	–	–	–	ND	ND	–	–	+	+	–	–	–	–	–w	ND	ND	–	–	–
d	–	w	+	+	ND	+	–	ND	–	–	–	–w	+	w	w	d	–w	–	–	+	(d)	–w
–	ND	ND	–	ND	–	ND	ND	ND	ND	ND	–	ND	–	ND	ND	ND	ND	ND	ND	ND	ND	ND
–	+	+	+	+	+	d	+	ND	+	+	+	+	+	+	+	+	–	+	+	–w	d	d
–	+	–	–	–	+	–	–	ND	–	–	–	–	–	+	+	+	+	–	+	–	–	d
d	–	d	+	–	+	+	+	d	ND	+	+	–	+	–	–	–	+	+	+	–	+	+
d	d	–	ND	–	+	–	– (d)	– (d)	–	–	d	–	–	d	d	d	–	–	–	–	–	+
–	+	–	ND	–	–	–	–	–	–	–	–	+	d	d	–	w	–	(d)	–	–	–	–
+	(+)	+	ND	+	+	+	+	+	(+)	+	+	w	+	+	+	+	–	+	+	+	–	–
–	d	ND	ND	–	–	ND	–	ND	ND	–	ND	ND	ND	+	+	+	–	+	+	ND	–	–
+w	d	ND	+	–	+	ND	d	ND	ND	–	d	ND	+	(+)	(+)	(+)	–w	w	+	–	d	d
d	d	+	+	–	+	ND	d	d	–	d	d	–	d	d	–	+	+	+	–	d	d	d
+	d	+	+	–	+	d	d	+	–	+	+	–	–	(d)	(+)	(+)	–w	+	+	–	d	+
+	+	–	ND	–	+	d	d	d	–	d	+	–	d	+	+	+	+	+	+	+	d	d
–	(+)	+	+	–	+	–	–	d	(+)	–	d	+	+	(d)	d	+	d	–	+	+	d	d
+w	–	ND	ND	–	–	ND	d	ND	ND	–	ND	ND	d	–	d	–	–	+	+	–	–	+
–w	+	–	ND	–	–	–	–	–	ND	–	–	–	–	–	–	–	–	+	+	–	d	–
+	+	–	ND	–	–	d	–	ND	–	+	–	–	+	+	+	+	d	w	+	d	d	d
ND	d	ND	ND	–	+	ND	d	ND	ND	–	ND	ND	ND	+	+	+	d	w	+	d	d	d
–	+	+	ND	+	+	+	d	+	–	+	d	–	d	+	+	+	+	+	+	+	+	+
+	+	+	+	+	+	+	+	d	–	+	+	d	–	+	+	+	+	+	+	(d)	+	+
w	–	d	ND	+	d	ND	–	–	ND	+	d	–	ND	–	–	–	–	–	+	–	d	d
–w	–	–	ND	ND	d	–	–	ND	–	d	ND	–	–	–	–	–	–	–	+	–	–	–w
–	–w	–	+	+	+	–	–	ND	–	–	–	–	–	d	+	d	–	–	w	(d)	–	+
+	+	–	–	–	+	–	–	+	–	+	+	–	–	+	+	+	–	+	+	–	–	+

TABLE 74. (continued)

[a]Symbols: +, 90% or more strains positive; –, 90% or more strains negative; d, 11–89% strains positive; ( ) delayed reaction; w, weak reaction; –w, negative to weak reaction; +w, positive to weak reaction; ND, not determined. Positive hemolytic reactions include greening of the agar as well as clearing. Media: P-agar (Kloos and George, 1991; Kloos et al., 1974); BHIA, brain heart infusion agar; PC skim agar: commercially available from Merck; TSA, tryptic soy agar; TSA+: TSA supplemented with either sheep or bovine blood: P/TSA, P and TSA, respectively.

[b]Data from Holt et al. (1994), except for bacitracin resistance data, which are from Igimi et al. (1989).

[c]Data from Probst et al. (1998).

[d]Data from Place et al. (2003a).

[e]Data from Vernozy-Rozand et al. (2000).

[f]Data from Kloos et al. (1998b).

[g]Data from Foster et al. (1997).

[h]Data from Hájek et al. (1992).

[i]Data from Spergser et al. (2003).

[j]Data from Chesneau et al. (1993).

[k]Data from Tanasupawat et al. (1992).

[l]Data from Zakrzewska-Czerwinska et al. (1995). *Staphylococcus pulvereri* is considered a junior heterotypic synonym of *Staphylococcus vitulinus* (Kloos et al., 2001; Petráš, 1998).

[m]Data from Hájek et al. (1996).

[n]Data from Kloos et al. (1997).

[o]Data from Lambert et al. (1998).

[p]Data from Place et al. (2003a).

[q]Data from Webster et al. (1994). *Staphylococcus vitulinus* is considered a senior heterotypic synonym of *Staphylococcus pulvereri* (Kloos et al., 2001; Petráš, 1998).

[r]Kloos et al. (1997) found that *Staphylococcus sciuri* subsp. *sciuri* and *Staphylococcus sciuri* subsp. *carnaticus* strains were negative when tested for Staph-A-Lex agglutination; all *Staphylococcus sciuri* subsp. *sciuri* strains were negative and some *Staphylococcus sciuri* subsp. *carnaticus* strains were positive when tested for Staph Latex agglutination. *Staphylococcus sciuri* subsp. *rodentium* was positive when tested with both systems.

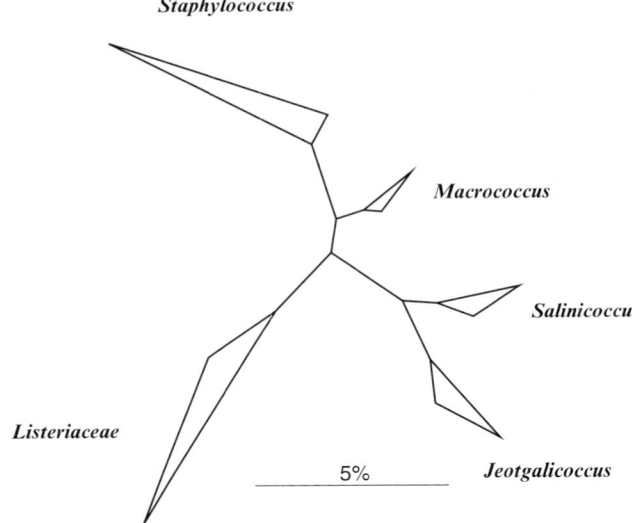

FIGURE 70. 16S rRNA based tree reflecting the phylogenetic relationships of the genus *Staphylococcus* and selected representatives of the phylum *Firmicutes*. The tree topology was reconstructed by applying the ARB-parsimony tool (Ludwig et al., 2004) upon an ARB 16SrRNA database (http://www.arb-home.de/) comprising 28,980 almost complete sequences. The tree topology was corrected according to the data obtained by applying distance and maximum-likelihood methods. Only sequence positions which shared identical nucleotides in at least 50% of all compared sequences from representatives of the *Staphylococcaceae* were used to construct the tree. Length bar indicates 5% estimated sequence divergence.

lococcus epidermidis is the species that is associated most commonly with disease. It appears to have the greatest pathogenic potential and adaptive diversity. This species has been implicated in bacteremia, native and prosthetic valve endocarditis,

osteomyelitis, pyoarthritis, peritonitis during continuous ambulatory dialysis, mediastinitis, infections of permanent pacemakers, vascular grafts, cerebrospinal fluid shunts, prosthetic joints, and a variety of orthopedic devices, and urinary tract infections including urethritis and pyelonephritis. Recent reviews have been published on the nature of human infections caused by *Staphylococcus epidermidis* and other coagulase-negative species (Crossley and Archer, 1997; Kloos and Bannerman, 1994; Rupp and Archer, 1994). Nosocomial methicillin-resistant *Staphylococcus epidermidis* (MRSE) strains became a serious clinical problem in the 1980s, especially in patients with prosthetic heart valves or who have undergone other forms of cardiac surgery (Archer and Tenenbaum, 1980; Karchmer et al., 1983). *Staphylococcus epidermidis* has also occasionally been associated with mastitis in cattle (Baba et al., 1980; Devriese and De Keyser, 1980; Holmberg, 1986).

Certain other coagulase-negative species have been associated with infections in humans and animals. *Staphylococcus haemolyticus* is the second most frequently encountered species of this group found in human clinical infections. It has been implicated in native valve endocarditis, septicemia, peritonitis, and urinary tract infections, and is occasionally associated with wound, bone, and joint infections. *Staphylococcus haemolyticus* has occasionally been associated with mastitis in cattle (Baba et al., 1980). *Staphylococcus caprae*, which has previously been misidentified as *Staphylococcus haemolyticus*, *Staphylococcus hominis*, and *Staphylococcus warneri*, is widely distributed in human clinical specimens (Kawamura et al., 1998). *Staphylococcus caprae* has been implicated in cases of infective endocarditis, bacteremia, and urinary tract infections. *Staphylococcus lugdunensis* has been implicated in native and prosthetic valve endocarditis, septicemia, brain abscess, and chronic osteoarthritis and infections of soft tissues, bone, peritoneal fluid, and catheters, especially in patients with underlying diseases. *Staphylococcus schleiferi* has been implicated in human brain empyema, osteoarithritis, bacteremia, wound infections, and infections

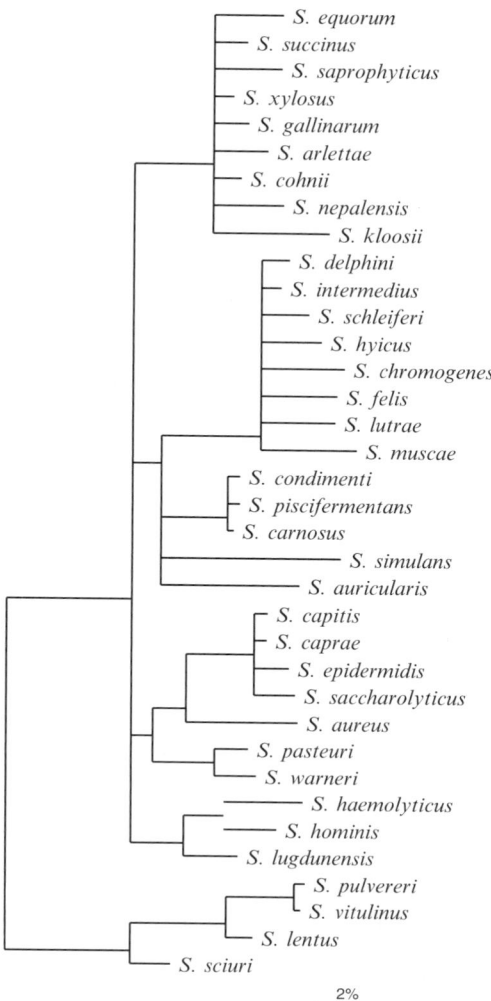

FIGURE 71. 16S rRNA based tree reflecting the relationships of the type strains of *Staphylococcus* species. The tree topology was reconstructed as described in the legend of Figure 70 in the *Staphylococcaceae* chapter. A relative branching order is shown if supported by all three treeing methods. Multifurcations indicate that a significant relative branching order could not be determined or not supported by all three treeing methods. The length bar indicates 2% estimated sequence divergence.

associated with a cranial drain and jugular catheter. However, this species occurs less frequently than *Staphylococcus lugdunensis* in the hospital environment and human infections. *Staphylococcus saprophyticus* is an important opportunistic pathogen in human urinary tract infections, especially in young, sexually active females. It is considered to be the second most common cause of urinary tract infections, such as acute cystitis or pyelonephritis, in these patients. This species has also occasionally been isolated from wound infections and septicemia (Fleurette et al., 1987; Marsik and Brake, 1982).

Staphylococcus warneri has been associated with mastitis in cattle (Devriese and De Keyser, 1980; Devriese and Derycke, 1979). *Staphylococcus simulans* has been isolated from patients with chronic osteomyelitis and from mastitis in cows. *Staphylococcus felis*, a relative of *Staphylococcus simulans*, has been associated

with skin infections of cats (Igimi et al., 1989). *Staphylococcus capitis* has been implicated in endocarditis, septicemia, and catheter infections. *Staphylococcus hominis* has been associated with human endocarditis, peritonitis, septicemia, and arthritis. *Staphylococcus cohnii* has been isolated from urinary tract infections and arthritis. Staphylococcus *chromogenes*, a close relative of *Staphylococcus hyicus*, is commonly present in the milk of cows suffering from mastitis, although its role as an etiologic agent is questionable (Devriese and De Keyser, 1980; Langlois et al., 1983; Watts et al., 1984). *Staphylococcus sciuri* subspecies have been isolated from wound, skin, and soft tissue infections (Marsou et al., 1999).

Genomes. The size range of staphylococcal genomes is about 2–3 Mbp (George and Kloos, 1994; Kloos et al., 1998b). Eight staphylococcal strains (seven *Staphylococcus aureus* and one *Staphylococcus epidermidis*) are completely sequenced. The genome size of the *Staphylococcus aureus* strains is 2.82–2.9 Mbp and that of *Staphylococcus epidermidis* 2.5 Mbp. Sequences of the genomes of *Staphylococcus epidermidis* RP62A, *Staphylococcus carnosus*, and *Staphylococcus haemolyticus* are due for completion but so far only the *Staphylococcus aureus* strains N315 and Mu50 (Kuroda et al., 2001) are annotated. Strain N315 is a methicillin-resistant *Staphylococcus aureus* and strain Mu50, in addition to being methicillin-resistant, is also vancomycin-resistant. Both strains contain one plasmid and three pathogenicity islands. The genome of *Staphylococcus aureus* N315, which causes acute infections, was compared with the genome of *Staphylococcus epidermidis* RP62A, which can cause chronic infections (Götz et al., 2003). Of the 125 selected virulence genes in *Staphylococcus aureus* N315, only 22 genes (18%) had homologous equivalents in *Staphylococcus epidermidis* RP62A. This clearly reflects the difference in the pathogenic potential of both species representatives and explains why *Staphylococcus aureus* is a rather aggressive pathogen and *Staphylococcus epidermidis* is not. Of the 40 toxin genes in *Staphylococcus aureus*, only three were identified in the *Staphylococcus epidermidis* genome. This finding is in agreement with the earlier observations of decreased toxin production of *Staphylococcus epidermidis* and corroborates well with the decreased severity of an *Staphylococcus epidermidis* infection. Compared to *Staphylococcus aureus*, *Staphylococcus epidermidis* also revealed a decreased number of other potential virulence factors such as exoenzymes and adhesins.

Enrichment and isolation procedures

Isolation of staphylococci from clinical specimens. Procedures used in the isolation and enumeration of staphylococci from clinical specimens including blood, pus, purulent fluids, sputum, urine, and feces can be found in the following texts: American Society for Microbiology (ASM) *Manual of Clinical Microbiology*, 7th edition (Murray et al., 1999), *A Guide to Specimen Management in Clinical Microbiology*, 2nd edition (Miller, 1998), and *Clinical Microbiology Procedures Handbook* (Isenberg, 1994).

Staphylococci from clinical specimens are usually isolated in primary culture on blood agar and in a fluid medium such as thioglycolate broth (Kloos and Bannerman, 1999; Murray et al., 1999). On blood agar, abundant growth of most staphylococcal species occurs within 18–24 h at 34–37°C. Colonies should be allowed to grow for at least an additional 2–3 d before the primary isolation plate is confirmed for species or strain composition (Kloos and Schleifer, 1975a).

Several basic methods are available for isolating staphylococci and other aerobic bacteria from skin and the adjacent mucous membranes (reviewed by Noble and Sommerville, 1974). The medium most widely used for the isolation and culture of natural populations of staphylococci from the skin of humans and animals is P agar (Kloos et al., 1974).

Isolation of *Staphylococcus aureus* from food samples. A nonselective enrichment for the detection of *Staphylococcus aureus* in food samples has been described by Downes and Ito (2001). Black to dark gray colonies are selected for coagulase testing. When supplemented with egg yolk tellurite enrichment, Baird-Parker agar base can also be used for the detection and enumeration of coagulase-positive staphylococci in foods.

Selective enrichments. Selective enrichment is recommended for raw food ingredients and unprocessed foods expected to contain <100 *Staphylococcus aureus* cells/g and a large population of competing species. The recommended procedures for the selective enrichment of *Staphylococcus aureus* are the most probable number technique or the direct surface plating on Baird-Parker agar (Horowitz, 2000). Instead of Baird-Parker agar, the following media for the selective isolation and enumeration of staphylococci can be used: tellurite polymyxin egg yolk agar (Crisley et al., 1964), Kalium-Rhodanid (= potassium thiocyanate)-Actidione-Natriumazid (=sodium azide)-Egg yolk-Pyruvate (KRA-NEP) agar (Sinell and Baumgart, 1966), and Schleifer-Krämer (SK) agar (Harvey and Gilmour, 1988; Schleifer and Krämer, 1980).

Ecology

Human and animal sources. Staphylococci are a major group of bacteria inhabiting the skin, skin glands, and mucous membranes of humans, other mammals, and birds. They may be found on the skin as residents or transients (Kloos and Musselwhite, 1975; Noble and Somerville, 1974). Resident bacteria are indigenous to the host, maintain relatively stable populations, and increase in numbers mainly by multiplication of those already present. Transient bacteria are derived from exogenous sources, found primarily on exposed skin, and may easily be washed away.

Staphylococcus aureus is a major species of primates, though specific ecovars or biotypes can occasionally be found living on different domestic animals or birds (Kloos, 1980; Meyer, 1967). This species is found infrequently on nonprimate wild animals. In humans, *Staphylococcus aureus* has a niche preference for the anterior nares, especially in the adult. Nasal carrier rates range from less than 10% to more than 40% in normal adult human populations residing outside of the hospital (Noble and Somerville, 1974). *Staphylococcus epidermidis* (Schleifer and Kloos, 1975) is the most prevalent and persistent *Staphylococcus* species on human skin. It is found over much of the body surface and produces the largest populations where moisture content and nutrition are high, such as in the anterior nares, axillae, inguinal and perineal area, and toe webs. This species may be found occasionally on other hosts, such as domestic animals, but it is presumably transferred there from human sources. *Staphylococcus hominis* (Kloos and Schleifer, 1975a) is also prevalent

on human skin. Its population size is usually second or equal to *Staphylococcus epidermidis* on skin sites where apocrine glands are numerous (e.g., in the axillae and inguinal and perineal areas). It can also colonize the drier regions of skin (e.g., on the extremities) more successfully than other species. *Staphylococcus haemolyticus* (Schleifer and Kloos, 1975) shares many of the habitats of *Staphylococcus hominis*, but it is usually found in smaller populations. Some individuals may carry unusually large populations of *Staphylococcus haemolyticus*. *Staphylococcus warneri* (Kloos and Schleifer, 1975a) is found usually in small numbers on human skin, though a few individuals may carry unusually large populations. Occasionally, small transient populations of *Staphylococcus haemolyticus* or *Staphylococcus warneri* may be isolated from domestic animals. *Staphylococcus capitis* (Kloos and Schleifer, 1975a) produces large populations on the human scalp following puberty. It is also found on other regions of the adult head, e.g., forehead, face, eyebrows, and external auditory meatus in moderate-sized to large populations. *Staphylococcus auricularis* is one of the major species found living in the adult human external auditory meatus and demonstrates a strong preference for this niche (Kloos and Schleifer, 1983).

Staphylococcus caprae, originally isolated from the skin of domestic goats or in their milk (Devriese et al., 1983; Poutrel, 1984), has been isolated from human clinical specimens (Kawamura et al., 1998; Vandenesch et al., 1995). *Staphylococcus hominis* subsp. *novobiosepticus*, *Staphylococcus lugdunensis*, *Staphylococcus pasteuri*, and *Staphylococcus schleiferi* are other clinically significant species isolated from human specimens (Chesneau et al., 1993; Freney et al., 1988; Kloos et al., 1998b). Their original niche preference and prevalence is undetermined. The host range of *Staphylococcus saprophyticus* and similar species varies from humans to lower mammals and birds (Kloos, 1980) (Devriese, 1986). As a group, these staphylococci are most prevalent on lower primates and mammals. *Staphylococcus saprophyticus* is found usually in small, transient populations on the skin of humans or other primates. It may also be isolated from lower mammals and environmental sources. *Staphylococcus cohnii* (Schleifer and Kloos, 1975) is found as a temporary resident or transient on human skin. *Staphylococcus cohnii* subsp. *urealyticus* (Kloos and Wolfshohl, 1991) is sometimes found on human skin, but it is often one of the major species and subspecies of nonhuman primates, especially the lower primates. *Staphylococcus xylosus* (Schleifer and Kloos, 1975) is often found as a transient on the skin of lower primates and other mammals, and occasionally on birds (Akatov et al., 1985; Devriese et al., 1985; Kloos et al., 1976b). The related species *Staphylococcus kloosii* has been found living on a variety of lower mammals including wild marsupials, rodents, and carnivores, and less frequently on domestic animals (Kloos, 1980) (Schleifer et al., 1984). *Staphylococcus arlettae* has been isolated from poultry and goats, *Staphylococcus equorum* from horses, and *Staphylococcus gallinarum* from poultry (Devriese et al., 1983; Schleifer, 1984).

Staphylococcus intermedius is a major species of domestic dogs (Hajek and Marsalek, 1976). This species can be found in relatively large populations on canine skin and can on occasion be transferred to the skin of human handlers (Kloos et al., 1976b). *Staphylococcus intermedius* also appears to be indigenous to a variety of other carnivores including mink, foxes (Hajek, 1976), and raccoons (Kloos et al., 1976b). It has also been isolated from horses and pigeons (Hajek, 1976). *Staphylococcus felis* is one of the major species of the domestic cat (Igimi et al., 1990). *Staphylococcus schleiferi* subsp. *coagulans* is a coagulase-positive species that has been isolated from the external auditory meatus of dogs with ear infections (Igimi et al., 1990). Other coagulase-positive species isolated from animals include *Staphylococcus delphini* (Varaldo et al., 1988) and *Staphylococcus lutrae* (Foster et al., 1997). *Staphylococcus sciuri*, *Staphylococcus sciuri* subsp. *carnaticus*, and *Staphylococcus sciuri* subsp. *rodentium* have been isolated from a variety of lower mammals and domestic animals (Kloos et al., 1997; Kloos et al., 1976a). In addition, the *Staphylococcus sciuri* subspecies may be isolated from human clinical specimens (Marsou et al., 1999). *Staphylococcus sciuri* appears to be a natural reservoir of methicillin resistance and staphylolytic enzyme genes (Kloos et al., 1997). Their role, if any, in the appearance of methicillin resistance in the more pathogenic species *Staphylococcus aureus* has not yet been determined. *Staphylococcus hyicus* and *Staphylococcus chromogenes* are found predominantly on domestic ungulates such as pigs, cattle, and horses (Devriese, 1986; Hajek et al., 1986). *Staphylococcus lentus* (Schleifer et al., 1983b) has been isolated in large populations from domestic sheep and goats (Kloos et al., 1976a; Kloos et al., 1976b) and occasionally from other domestic animals (Devriese et al., 1985); it is a major bacterium in saliva of rabbits (Kanda et al., 2001).

Flies of the genera *Musca*, *Fannia*, and *Stomoxys*, commonly found in human and animal habitations, can carry populations of staphylococci (e.g., *Staphylococcus muscae*; Hajek et al., 1992) and appear to be significant vectors of these organisms in an epizootiological chain (Hájek and Balusek, 1985).

Environmental sources and food. Staphylococci have been isolated sporadically from a wide variety of environmental sources such as soil, beach sand, sea water, fresh water, plant surfaces and products, feeds, meat and poultry, dairy products, and the surfaces of cooking ware, utensils, furniture, clothing, blankets, carpets, linens, paper currency, and dust and air in various inhabited areas.

Staphylococcus aureus has been confirmed to be a major causative agent of food poisoning. Other staphylococci may be involved in the production of certain fermented foods. The predominant bacterium in fermented meat is *Staphylococcus carnosus* which appears in the early literature as *Micrococcus* (Schleifer and Fischer, 1982). For more than 50 years, *Staphylococcus carnosus* has been used alone or in combination with lactobacilli or pediococci as a starter culture for the production of raw fermented sausages. During the ripening process of dry sausage, *Staphylococcus carnosus* exerts several desired functions (Götz, 1990; Liepe and Porobic, 1983). *Staphylococcus condimenti* was isolated from soy sauce mash (Probst et al., 1998), *Staphylococcus equorum* subsp. *linens* and *Staphylococcus succinus* subsp. *casei* from surface ripened cheeses (Place et al., 2002; Place et al., 2003b), *Staphylococcus fleurettii* from goats' milk cheese (Vernozy-Rozand et al., 2000), and *Staphylococcus piscifermentans* (Tanasupawat et al., 1992) from fermented shrimp and fish.

Staphylococcus sciuri and *Staphylococcus xylosus* can grow in habitats containing only an inorganic nitrogen source and thus might be more free-living than other staphylococci (Emmett and Kloos, 1979). These species have been isolated in small numbers from beach sand, natural waters, marsh grass (Kloos and Schleifer, 1981), and plant products (Pioch et al., 1988).

Identification

Conventional methods. Conventional methods for the determination of phenotypic characters were developed first and later staphylococci were examined through genotypic studies (reviewed by Kloos and Schleifer, 1986; Schleifer, 1986). A simplified system for the identification of human staphylococci has been described by Kloos and Schleifer (1975b). Key characteristics now used for species and subspecies identification include the following: colony morphology, oxygen requirements, coagulase, clumping factor, heat-stable nuclease (thermonuclease), hemolysins, catalase, oxidase, alkaline phosphatase, urease, ornithine decarboxylase, pyrrolidonyl arylamidase, β-galactosidase, acetoin production, nitrate reduction, esculin hydrolysis, aerobic acid production from a variety of carbohydrates including D-trehalose, D-mannitol, D-mannose, D-turanose, D-xylose, D-cellobiose, L-arabinose, maltose, lactose, sucrose, and raffinose, and intrinsic resistance to novobiocin and polymyxin B (reviewed by Kloos and Bannermann, 1999). Some conventional methods may require up to 3–5 d before a final result can be obtained, while others only require several hours for interpretation. Characteristic properties useful for the differentiation of the species and subspecies of the genus *Staphylococcus* are compiled in Table 74.

Rapid identification systems. Identification in the routine or clinical laboratory is carried out with rapid species identification kits or automated systems that require only a few hours to 1 d for the completion of tests. Identification of a number of the *Staphylococcus* species can be made with an accuracy of 70 to >90% using the commercial systems (Kloos et al., 1998a). *Staphylococcus aureus*, *Staphylococcus epidermidis*, *Staphylococcus capitis*, *Staphylococcus haemolyticus*, *Staphylococcus saprophyticus*, *Staphylococcus simulans*, and *Staphylococcus intermedius* can be identified reliably by most of the commercial systems now available. Some identification systems now available include the following: RAPIDEC Staph (identification of *Staphylococcus aureus*, *Staphylococcus epidermidis*, and *Staphylococcus saprophyticus*) and API STAPH (bioMérieux Vitek, Inc., Hazelwood, Mo.); VITEK, a fully automated microbiology system that uses a Gram-positive Identification (GPI) Card (bioMérieux Vitek); MicroScan Pos ID panel (read manually or on MicroScan instrumentation), MicroScan Rapid Pos ID panel (read by the WalkAway systems) (in addition, the ID panels are available with antimicrobial agents for susceptibility testing) (Dade MicroScan, Inc., West Sacramento, Calif.); Crystal Gram-positive Identification System, Crystal Rapid Gram-positive Identification System, Pasco MIC/ID Gram-positive Panel, and the Phoenix, an automated identification system (Becton Dickinson Microbiology Systems); GP MicroPlate test panel (read manually, using Biolog MicroLog system or read automatically with the Biolog MicroStation system) (Biolog, Haywood, Calif.); MIDI Sherlock Identification System Microbial Identification System (MIS) that automates microbial identification by combining cellular fatty acid analysis with computerized high-resolution gas chromatography (MIDI, Newark, Del.); and RiboPrinter Microbial Characterization System (Qualicon, Inc. Wilmington, Del.), based on ribotype pattern analysis.

Oligonucleotide probes directed against rRNA are also used for the rapid identification of the species *Staphylococcus aureus*, e.g., AccuProbe culture identification test for *Staphylococcus aureus* (Gen-Probe, Inc., San Diego, Calif.). The oligonucleotide probe tests are usually very accurate (100% specificity). Additionally, PCR analysis of the 16S–23S rRNA intergenic spacer region has preliminarily shown successful results in discriminating among 31 *Staphylococcus* species (Mendoza et al., 1998). PCR analysis allows for the identification of pure culture.

Differentiation at the infrasubspecific level. For clinical, particularly epidemiological, studies it is important to distinguish bacteria at the infrasubspecific level. Therefore, many different methods were used for the typing of *Staphylococcus aureus* and *Staphylococcus epidermidis* strains. Besides conventional techniques including plasmid profiles and phage typing, molecular techniques are applied. The most powerful methods are pulsed-field gel electrophoresis (PFGE) and multilocus sequence typing (MLST). PFGE is based on direct comparison of DNA fragment patterns of the entire genome when restricted with a rare-cutting enzyme (Cookson et al., 1996; Tenover et al., 1995). This method has proved very useful for the investigation of nosocomical outbreaks and of the spread of multi-drug-resistant *Staphylococcus aureus* strains (Cox et al., 1995; de Sousa et al., 1998). A major disadvantage of PFGE is that it is rather time-consuming and subjective in interpretation (van Belkum et al., 1998). An alternative highly discriminative method is MLST. It is based on the comparison of sequence fragments of several house-keeping genes (Maiden et al., 1998). For each gene fragment, the different sequences are assigned as distinct alleles, and each isolate is defined by the alleles at each of the house-keeping loci. Sequence data are readily compared between laboratories and easily catalogd. MLST has been used for the characterization of natural populations of *Staphylococcus aureus* (Enright et al., 2000; Grundmann et al., 2002; van Leeuwen et al., 2003) and of *Staphylococcus epidermidis* (Wang et al., 2003).

Taxonomic comments

The genus *Staphylococcus* has been considerably expanded since publication of Volume 2 of the first edition of *Bergey's Manual of Systematic Bacteriology* in which 19 species were listed. Currently, 37 species and several subspecies are recognized in the genus *Staphylococcus*. They form a coherent and well-defined group of related species that is widely divergent from those of the genus *Micrococcus*. A simple test system for the separation of staphylococci from micrococci has been described by Kloos and Schleifer (1975b). Staphylococci can be separated from macrococci on the basis of generally smaller cells, presence of cell-wall teichoic acids, a generally lower DNA G+C content (<40% mol%), 16S rRNA sequence, and unique ribotype patterns. The former *Staphylococcus caseolyticus* has been transferred to the genus *Macrococcus* as *Macrococcus caseolyticus* (Kloos et al., 1998a). Most members of the genus *Staphylococcus* are oxidase-negative whereas macroococci and salinicocci as well as jeotgalicocci are oxidase-positive. The DNA G+C content of salinicocci is significantly higher (46–51 mol%) than that of staphylococci.

List of species of the genus *Staphylococcus*

1. **Staphylococcus aureus** Rosenbach 1884, 18[AL]

au′re.us. L. masc. adj. *aureus* golden.

Nonmotile, nonspore-forming, Gram-stain-positive cocci; occur singly and form pairs and clusters. Facultatively anaerobic. Grows well in medium containing 10% NaCl, poorly in 15% NaCl. Positive reactions: alkaline phosphatase, catalase, coagulase, heat-stable nuclease, hemolysis, and hyaluronidase. Negative reactions: oxidase, β-galactosidase, β-glucuronidase. Produces acid aerobically from fructose, maltose, and sucrose. No acid produced from arabinose, cellobiose, melezitose, raffinose, salicin, xylitol, or xylose. Novobiocin-susceptible. Peptidoglycan type L-Lys–Gly$_{5-6}$. Teichoic acid contains ribitol and *N*-acetylglucosamine.

Additional characteristics of the subspecies are given below.

1a. **Staphylococcus aureus subsp. aureus** Rosenbach 1884, 18[AL]

Nonmotile, nonspore-forming, Gram-stain-positive cocci, 0.5–1.0 μm in diameter; occur singly and form pairs and clusters. Colonies raised, smooth, glistening, translucent, with entire margins. Pigmentation varies from gray to yellow to orange. The pigments triterpenoid carotenoids or derivatives of them and are located in the cell membrane. Colony diameter >5 mm. Colonies are smooth, raised, glistening, circular, entire, and translucent. May produce capsules. Encapsulated strains usually produce smaller and more convex colonies. Capsular polysaccharides have been shown to contain N-acetyl-D-amino-galacturonic acid, *N*-acetyl-D-fucosamine, and taurine. Cell membranes contain the glycolipids, mono- and diglucosyldiglyceride and the phospholipids, lysyl-phosphatidylglycerol, and cardiolipin.

Facultatively anaerobic. Growth is best under aerobic conditions. Temperature range for growth 10–45°C; optimum 30–37°C. Growth good in medium containing 10% NaCl; poor at 15% NaCl. Produces D- and L-lactate anaerobically from glucose. Addition of glucose to a growth medium suppresses the tricarboxylic acid cycle, resulting in the accumulation of acetate and carbon dioxide. D-Galactose and lactose are metabolized via the D-tagatose-6-phosphate pathway. Contain class I fructose-1,6-bisphosphate aldolase.

Protein A produced. Positive reactions: acetoin production, alkaline phosphatase, arginine dihydrolase, caseinase, catalase, clumping factor, coagulase, fibrinolysin, gelatinase, β-glucosidase, heat-stable nuclease (thermonuclease), hemolysin (at least four different ones), lipase, esterases, nitrate reduction, and urease. Negative reactions: amylase, ornithine decarboxylase, lysine decarboxylase and oxidase. Esculin and starch are not usually hydrolyzed.

Acid produced aerobically from fructose, galactose, glucose, glycerol, lactose, maltose, mannitol, mannose, ribose, sucrose, trehalose, and turanose. A few strains do not produce detectable acid from lactose, galactose, turanose, or mannitol. No acid produced from adonitol, arabitol, arabinose, cellobiose, dextrin, dulcitol, erythritol, erythrose, fucose, gentiobiose, inositol, lyxose, melezitose, melibiose, raffinose, rhamnose, salicin, sorbitol, sorbose, xylitol, and xylose. Novobiocin-susceptible. Resistance to penicillin G and methicillin is quite common. Peptidoglycan type

L-Lys–Gly$_{5–6}$. Teichoic acid contains ribitol and *N*-acetylglucosamine. The major cellular fatty acids are C$_{Br-15}$, C$_{20}$, C$_{18}$, and C$_{Br-17}$. Unsaturated menaquinones (mainly MK-8; MK-7 and MK-9 as minor components) are present.

Some strains produce an epidermolytic toxin. Many strains produce enterotoxins. At least five different types of enterotoxins have been identified (A, B, C, D, and E). In addition, three different forms of enterotoxin C are known. Enterotoxin A is the most common one. An enterotoxin-like protein (enterotoxin F, pyrogenic exotoxin C) is produced by many of the strains associated with toxic shock syndrome.

Phage typing can be used for the differentiation of *Staphylococcus aureus* strains. Chromosomal and plasmid DNA may be transferred to appropriate recipient cells via transduction or transformation.

Isolated from nasal membranes (anterior nares, nasopharynx) and skin of warm-blooded animals. Can cause infection, food poisoning, and toxic shock syndrome.

DNA–DNA hybridization studies yielded relative DNA binding values >75% among *Staphylococcus aureus* subsp. *aureus* strains (Kloos and Wolfshohl, 1982; Meyer and Schleifer, 1978).

DNA G+C content (mol%): 32–36 (T_m, Bd).

Type strain: ATCC 12600, ATCC 12600-U, CCM 885, CCUG 1800, CIP 65.8, DSM 20231, HAMBI 66, NCAIM B.01065, NCCB 72047, NCTC 8532.

GenBank accession number (16S rRNA gene): D83357, L37597, X68417.

1b. Staphylococcus aureus subsp. anaerobius De La Fuente, Suarez and Schleifer 1985, 100^{VP}

an.ae.ro′bi.us. Gr. pref. an not; Gr. n. *aer* air; Gr. n. *bios* life; N.L. adj. *anaerobius* not living in air.

Nonmotile, nonspore-forming, Gram-stain-positive cocci, 0.8–1.0 μm in diameter; occur singly and form pairs and clusters. Grows microaerobically and anaerobically; weak aerobic growth can be obtained on subculturing. Produces L-lactate, acetate, and succinate anaerobically from glucose. Colonies on blood agar white, opaque, glistening, entire, smooth, and convex. Colony diameter 1–3 mm after incubation at 2 d on blood agar. Temperature range for optimal growth 30–40°C. No growth at 20 or 45°C. All strains tolerate 10% NaCl; most do not tolerate 15% NaCl. Positive reactions: acid and alkaline phosphatases, coagulase, hemolysis, caseinase, gelatinase, hyaluronidase, both heat-labile and heat-stable nucleases, and tellurite reduction. Negative reactions: acetoin production, benzidine test, catalase, oxidase, nitrate reduction, staphylokinase, and Tween 80 hydrolysis. Acid produced aerobically from fructose, glucose, maltose, and sucrose. No acid produced aerobically from amygdalin, arabinose, cellobiose, galactose, galactitol, gentiobiose, glycerol, inositol, inulin, lactose, mannitol, mannose, melibiose, melezitose, raffinose, rhamnose, ribose, salicin, sorbitol, trehalose, xylitol, or xylose. Lysozyme-resistant; lysostaphin-sensitive. Peptidoglycan type L-Lys–Gly$_{5–6}$. Ribitol-type teichoic acid containing *N*-acetylglucosamine. Class I fructose-1,6-bisphosphate aldolase. No production of enterotoxins A, B, C, D, or E or protein A.

Distinguished from *Staphylococcus aureus* subsp. *aureus* by lack of pigment and clumping factor and by the inability to carry out anaerobic fermentation of mannitol, to grow at 45°C, to produce acetoin from glucose, to reduce nitrate, to produce β-glucosidase, and to produce acid from galactose, lactose, mannose, mannitol, ribose, and trehalose.

Found in abcesses of sheep; also pathogenic for goats. Not pathogenic for mice, rabbits, or guinea pigs. Primary isolation requires a medium supplemented with serum, egg yolk, or blood.

DNA–DNA hybridization studies yielded relative DNA binding values ≥80% between three strains of *Staphylococcus aureus* subsp. *anaerobius* and the type strain of *Staphylococcus aureus* subsp. *aureus* (de la Fuente et al., 1985).

DNA G+C content (mol%): 31.5–32.7 (T_m).

Type strain: MVF-7, ATCC 35844, CCUG 37246, CIP 103780, DSM 20714.

GenBank accession number (16S rRNA gene): D83355.

2. Staphylococcus arlettae Schleifer, Kilpper-Bälz and Devriese 1985, 224^{VP} (Effective publication: Schleifer, Kilpper-Bälz and Devriese 1984, 504..)

ar.let′tae. N.L. gen. n. *arlettae* of Arletta; named for Arlette van de Kerckhove, who has studied this and related species for many years.

Nonmotile, nonspore-forming, Gram-stain-positive cocci, 0.5–1.5 μm in diameter; occur singly and form pairs, clusters, and chains. Colonies beige or yellow, opaque, with entire margins. Colony diameter 6–8 mm after incubation for 2 d at 37°C on brain heart infusion agar. Aerobic. Positive reactions: catalase, esculin hydrolysis, and alkaline phosphatase. Produces very small amounts of lactate from glucose anaerobically. Weakly positive reactions: gelatinase and tributyrin hydrolysis. Negative reactions: coagulase, clumping factor, heat-stable nuclease, oxidase, hyaluronidase, staphylokinase, Tween 80 hydrolysis, fibrinolysin, hemolysis of sheep red blood cells, reduction of nitrate, urease, caseinase, and arginine hydrolysis. Acid produced aerobically from L-arabinose, arbutin, D-fructose, D-fucose (weak), β-gentiobiose, lactose, D-glucose, glycerol, maltose, D-mannitol, D-melezitose, raffinose, D-ribose, sucrose, trehalose, D-turanose, and D-xylose. No acid produced from amygdalin, cellobiose, L-rhamnose, D-sorbitol, or xylitol. Some strains produce acid from D-galactose, D-mannose, and D-melibiose. Peptidoglycan type Lys–Gly$_{5–6}$. Teichoic acids contain glucosamine, glycerol, and ribitol. Novobiocin-resistant.

Isolated from poultry (skin and nares) and goats (nares).

DNA–DNA hybridization studies yielded relative DNA binding values of 82–100% among *Staphylococcus arlettae* strains and 7–36% between *Staphylococcus arlettae* strains and nine other *Staphylococcus* species (Schleifer et al., 1984).

DNA G+C content (mol%): 31–32.6 (T_m).

Type strain: BP47, ATCC 43957, CCUG 32416, CIP 103501, DSM 20672, LMG 19114, NCTC 12413, NRRL B-14764

GenBank accession number (16S rRNA gene): AB009933.

3. Staphylococcus auricularis Kloos and Schleifer 1983, 9^{VP}

au.ri.cu.la′ris. L. adj. *auricularis* pertaining to the ear; named for the region of the body (external auditory meatus, external ear) where this species is commonly found.

Nonmotile, nonspore-forming, Gram-stain-positive cocci, 0.8–1.2 μm in diameter; form tetrads and pairs. Colonies white, convex, smooth, slightly glistening, opaque, and butyrous, with undulate margins that become crimped with age. Colony diameter 1.4–2.8 mm on P agar after incubation at 34–35°C for 3 d, then 25°C for 2 d. Facultatively anaerobic; anaerobic growth in thioglycolate medium slow and sparse. Grows best under aerobic conditions. Grows well in medium containing 10% NaCl, poorly in 15% NaCl. Temperature range for growth 20–45°C; optimum 30–40°C. Produces small amounts of L-lactate anaerobically from glucose. Positive reactions: catalase and lipase. Negative reactions: alkaline phosphatase, clumping factor, coagulase, heat-stable nuclease, hyaluronidase, oxidase, and urease. Acid produced aerobically from fructose, glucose, glycerol, maltose (delayed), and trehalose (delayed). No acid produced from arabinose, cellobiose, fucose, gentiobiose, inositol, mannitol, melezitose, raffinose, rhamnose, salicin, sorbitol, sorbose, xylitol, or xylose. Novobiocin-susceptible. Peptidoglycan type L-Lys–Gly$_{4.3–4.8}$, Ser$_{0–0.6}$. Teichoic acid contains N-acetylglucosamine 1-phosphate. Variable characteristics include acetoin production, arginine dihydrolase, β-galactosidase, hemolysis, and nitrate reduction.

Isolated from the external auditory meatus of primates, including humans.

DNA–DNA hybridization studies yielded relative DNA binding values of 90–99% between the type and other strains of *Staphylococcus auricularis* and 28–42% between the type strain of *Staphylococcus auricularis* and strains of 14 other *Staphylococcus* species (Kloos and Schleifer, 1983).

DNA G+C content (mol%): 38–39 (T_m).

Type strain: WK 811M, ATCC 33753, CCUG 15601, CCUG 36974, CIP 103587, DSM 20609, NCTC 12101, NRRL B-14762.

GenBank accession number (16S rRNA gene): D83358, L37598.

4. **Staphylococcus capitis** Kloos and Schleifer 1975a, 64[AL]

ca'pi.tis. L. n. *caput* head; L. gen. n. *capitis* of the head; pertaining to that part of the human body where cutaneous populations of this species are usually the largest and most frequent.

Nonmotile, nonspore-forming, Gram-stain-positive cocci. Facultative anaerobes. Grows best under aerobic conditions. Positive reactions: catalase and aerobic production of acid from mannose, mannitol, and sucrose. Negative reactions: alkaline phosphatase, coagulase, clumping factor, β-galactosidase, β-glucosidase, β-glucuronidase, heat-stable nuclease, oxidase, and production of acid from arabinose, cellobiose, raffinose, trehalose, and xylose. Novobiocin-susceptible.

Additional characteristics of the subspecies are given below.

4a. **Staphylococcus capitis subsp. capitis** Kloos and Schleifer 1975a, 64[AL]

Nonmotile, nonspore-forming, Gram-stain-positive cocci, 0.8–1.2 μm in diameter; occur in pairs and tetrads. Colonies smooth, slightly convex, glistening, opaque, and white or grayish white; after storage in the cold (4–10°C), the colonies of some strains become yellowish or yellow-orange. Colony diameter 1–3 mm on P agar. Facultatively anaerobic.

Growth is best under aerobic conditions. Produces mainly L-lactate and lesser amounts of D-lactate anaerobically from glucose. Temperature range for growth 18–45°C; optimum 30–40°C. Growth good in medium containing 10% NaCl, poor or absent in 15% NaCl. Positive reactions: benzidine test, catalase, DNase (weak, not thermostable), lipases, and hemolysis of human blood. Negative reactions: alkaline phosphatase, coagulase, clumping factor, β-galactosidase, β-glucosidase, β-glucuronidase, heat-stable nuclease, proteases, urease, and oxidase. Acid produced aerobically from fructose, glucose, glycerol, mannose, mannitol, and sucrose (delayed). No acid produced from adonitol, arabinose, arabitol, cellobiose, erythritol, erythrose, fucose, galactose, gentiobiose, inositol, lyxose, maltose, melezitose, melibiose, raffinose, rhamnose, ribose, salicin, sorbitol, sorbose, tagatose, turanose, xylitol, or xylose. Lysozyme-resistant; moderate resistance to lysostaphin. Novobiocin-susceptible. Peptidoglycan type L-Lys–Gly$_{3.5–4.3}$, L-Ser$_{0.8–1.2}$. Teichoic acid contains glycerol and N-acetylglucosamine. Some strains may contain polymeric glutamic acid in their cell walls. Major fatty acids C$_{Br-15}$, C$_{18}$, and C$_{20}$. Major menaquinone MK-7. Variable characteristics include growth on medium containing 15% NaCl, acetoin production, arginine dihydrolase, phosphatase, and nitrate reduction.

Isolated from the skin of the human head.

DNA–DNA hybridization studies yielded relative DNA binding values >90% among strains of *Staphylococcus capitis* (Kloos, 1980; Kloos and Wolfshohl, 1982).

DNA G+C content (mol%): 31–38.8 (T_m).

Type strain: LK499, ATCC 27840, CCM 2734, CCUG 7326, CIP 81.53, DSM 20326, JCM 2420, LMG 13353, NCTC 11045, NRRL B-14752.

GenBank accession number (16S rRNA gene): L37599.

4b. **Staphylococcus capitis subsp. urealyticus** Bannerman and Kloos 1991, 145[VP]

ur.e.o.ly'ti.cus. N.L. n. *urea* urea; Gr. adj. *lutikos* dissolving; N.L. masc. adj. *ureolyticus* urea-dissolving.

Nonmotile, nonspore-forming, Gram-stain-positive cocci, 0.8–1.0 μm in diameter; occur singly and form pairs and clusters. Colonies raised, opaque, glistening; some strains develop yellow pigmentation late in incubation; colonies may be either smooth or rough and have either slightly irregular or entire edges. Colony diameter 4.3–7.1 mm. Facultatively anaerobic. Positive reactions: arginine utilization, catalase, nitrate reduction, and urease. Negative reactions: alkaline phosphatase, clumping factor, coagulase, esculin hydrolysis, β-galactosidase, β-glucosidase, β-glucuronidase, heat-stable nuclease, ornithine decarboxylase, and oxidase. Acid produced aerobically from D-mannose, D-mannitol, maltose, and sucrose. Acid not produced from N-acetylglucosamine, L-arabinose, D-cellobiose, raffinose, D-trehalose, or D-xylose. Novobiocin-susceptible. Peptidoglycan type L-Lys–Gly$_{3.6}$–Ser$_{1.4}$. Major fatty acids C$_{15:0 iso}$, C$_{15:0 ante}$, C$_{18:0}$, and C$_{20:0}$. Variable characteristics include colony pigmentation, hemolysis, acetoin production, nitrate reduction, and acid production from α-lactose and D-turanose.

Distinguished from *Staphylococcus capitis* subsp. *capitis* by production of urease, ability to produce acid from maltose, and fatty acid composition.

Isolated from humans, human clinical specimens, and animals.

DNA–DNA hybridization studies yielded relative DNA binding values of 81–100% among strains of *Staphylococcus capitis* subsp. *urealyticus*, of 80–100% between strains of *Staphylococcus capitis* subsp. *urealyticus* and strains of *Staphylococcus capitis* subsp. *capitis*, and 4–23% between the type strain of *Staphylococcus capitis* subsp. *urealyticus* and strains of 24 other *Staphylococcus* species (Bannerman and Kloos, 1991).

DNA G+C content (mol%): 38.8 (T_m).

Type strain: MAW 8436, ATCC 49326, CCUG 35142, CIP 104192, DSM 6717.

GenBank accession number (16S rRNA gene): AB009937.

5. **Staphylococcus caprae** Devriese, Poutrel, Kilpper-Bälz and Schleifer 1983, 483[VP]

ca′prae. L. fem. n. *capra* goat; L. fem. gen. n. *caprae* of a goat.

Nonmotile, nonspore-forming, Gram-stain-positive cocci, 0.8–1.2 μm in diameter; occur singly and form pairs, chains, and clusters. Colonies nonpigmented, slightly convex, opaque, and glistening. Colony diameters differ; usually 5–8 mm. Facultatively anaerobic. Produces predominantly L-lactate anaerobically from glucose. FBP-aldolase is of class I type. Temperature range for growth 25–45°C, but growth is much slower at 25°C. Positive reactions: acetoin production, alkaline phosphatase, arginine dihydrolase, catalase, heat-labile DNase, hemolysis (delayed reaction), nitrate reduction, and urease. Negative reactions: caseinase, clumping factor, coagulase, esculin hydrolysis, heat-stable DNase, gelatinase, hyaluronidase, oxidase, staphylokinase, and Tween 80 hydrolysis. Acid produced aerobically from galactose, glucose, glycerol, lactose, mannose, and trehalose. No acid produced from amygdalin, arabinose, arbutin, cellobiose, fructose, fucose, β-gentiobiose, glucitol, mannitol, melezitose, melibiose, raffinose, rhamnose, ribose, salicin, sucrose, turanose, xylitol, or xylose. Novobiocin-susceptible. Lysozyme-resistant. Peptidoglycan type L-Lys–Gly$_{4–5}$, Ser$_{0.8–1.2}$. Teichoic acid contains glycerol, *N*-acetylglucosamine, and glucose. Variable characteristics include production of acid aerobically from maltose and mannitol.

Isolated from goats' milk.

DNA–DNA hybridization studies yielded relative DNA binding values of about 85% between the type strain and a second strain of *Staphylococcus caprae* and of 5–36% between the two *Staphylococcus caprae* strains and strains of 14 other *Staphylococcus* species (Devriese et al., 1983).

DNA G+C content (mol%): 35–37 (T_m).

Type strain: 143.22, ATCC 35538, CCM 3573, CCUG 15604, CIP 104000, DSM 20608, NCTC 12196, NRRL B-14757.

GenBank accession number (16S rRNA gene): AB009935, Y12593.

6. **Staphylococcus carnosus** Schleifer and Fischer 1982, 153[VP]

car.no′sus. L. adj. *carnosus* pertaining to flesh.

Nonmotile, nonspore-forming, Gram-stain-positive cocci. Facultatively anaerobic. Grows best under aerobic conditions. Grows on medium containing 15% NaCl. Grows at 15°C. Positive reactions: arginine dihydrolase and nitrate reduction. Negative reactions: coagulase, β-glucuronidase,

and urease. Acid produced aerobically from fructose. No acid produced from arabinose, cellobiose, maltose, melezitose, sucrose, turanose, or xylose. Acid production from trehalose is variable. Novobiocin-susceptible.

Additional characteristics of the subspecies are given below.

6a. **Staphylococcus carnosus subsp. carnosus** Schleifer and Fischer 1982, 153[VP]

Nonmotile, nonspore-forming, Gram-stain-positive cocci, 0.5–1.5 μm in diameter; occur singly and form pairs. Colonies gray-white, slightly raised, smooth, and slightly glistening. May produce pigment during storage. Colony diameter 1–3 mm. Facultatively anaerobic. Produces D- and L-lactate anaerobically from glucose. Temperature range for growth 15–45°C; optimum 30–40°C. Grows in medium containing 15% NaCl. Positive reactions: acetoin production, alkaline phosphatase, arginine dihydrolase, catalase, nitrate reduction, β-galactosidase, and phosphatase. Negative reactions: coagulase, hemolysis, oxidase, and urease. Acid produced aerobically from fructose, glucose, glycerol, mannitol, and mannose. No acid produced from arabinose, fucose, maltose, melibiose, melezitose, raffinose, rhamnose, salicin, sucrose, turanose, xylitol, and xylose. Lysozyme-resistant. Novobiocin-susceptible. Peptidoglycan type L-Lys–Gly$_{5–6}$. Variable characteristics include production of heat-stable DNase, and acid production from galactose, lactose, and trehalose.

Isolated from dry sausage and is used as starter culture for the production of dry sausage.

DNA–DNA hybridization studies yielded relative DNA binding values of 100% between the type strain and three other strains of *Staphylococcus carnosus* and of 10–32% between the type strain of *Staphylococcus carnosus* and the type strains of eleven other *Staphylococcus* species (Schleifer and Fischer, 1982).

DNA G+C content (mol%): 34.8–36 (T_m).

Type strain: 361, ATCC 51365, CCUG 15605, CIP 103274, DSM 20501, NRRL B-14760.

GenBank accession number (16S rRNA gene): AB009934.

6b. **Staphylococcus carnosus subsp. utilis** Probst, Hertel, Richter, Wassill, Ludwig and Hammes 1998, 657[VP]

ut.ti′lis. L. adj. *utilis* useful.

Nonmotile, nonspore-forming, Gram-stain-positive cocci, 1 μm in diameter; occur singly and form pairs and clusters. Colonies cream-colored, opaque, and entire. Colony diameter 0.5–1.5 mm after incubation at 2 d at 37°C on P agar. Temperature range for growth 15–42°C; no growth at 8°C. Tolerates 15% NaCl. Positive reactions: nitrate and nitrite reductases, catalase, and arginine dihydrolase. Negative reactions: phosphatase, β-galactosidase, β-glucuronidase, lecithinase, coagulase, and esculin hydrolysis. Acid produced aerobically from *N*-acetylglucosamine, fructose, and glucose, glycerol. No acid produced aerobically from cellobiose, galactose, lactose, maltose, mannitol, mannose, melezitose, raffinose, ribose, sorbitol, sucrose, or turanose. Lysozyme-resistant; lysostaphin-sensitive. Novobiocin-susceptible. Major menaquinone MK-7; minor menaquinones MK-6 and MK-8. Some strains produce acid from trehalose.

Distinguished from *Staphylococcus carnosus* subsp. *carnosus* by inability to grow at 45°C; lack of lipolytic activity, alkaline phosphatase, and β-galactosidase; and lack of acid production from mannose and mannitol.

Found in fermented seafood sauces.

DNA G+C content (mol%): 34.8 (no method given).

Type strain: SK 11, LTH 3728, CIP 105758, DSM 11676, JCM 6067.

GenBank accession number (16S rRNA gene): AB233329.

7. **Staphylococcus chromogenes** (Devriese, Hájek, Oeding, Meyer and Schleifer 1978) Hájek, Devriese, Mordarski, Goodfellow, Pulverer and Varaldo 1987, 179[VP] (Effective publication: Hájek, Devriese, Mordarski, Goodfellow, Pulverer and Varaldo 1986, 170.) (*Staphylococcus hyicus* subsp. *chromogenes* Devriese, Hájek, Oeding, Meyer and Schleifer 1978, 488.)

chro.mo'ge.nes. Gr. n. *chroma* color; Gr. v. *gennaio* to produce; N.L. adj. *chromogenes* producing color.

Nonmotile, nonspore-forming Gram-stain-positive cocci, 0.8–1.0 μm in diameter; form pairs, tetrads, and clusters as well as occurring singly. Colonies cream or orange, glistening, butyrous, with entire margins. Colony diameter 4–7 mm. Facultatively anaerobic. Produces L-lactate from glucose anaerobically. Optimum temperature range for growth 30–38°C. No growth at 45°C. Growth in medium containing 10% NaCl, but no growth or only weak growth at 15% NaCl. Positive reactions: arginine dihydrolase, benzidine test, caseinase, catalase, gelatinase, hippurate hydrolysis, nitrate reduction, phosphatase, and proteinase. Negative reactions: oxidase, acetoin production, clumping factor, coagulase, esculin hydrolysis, fibrinolysin, heat-stable nuclease, hemolysin, hyaluronidase, lecithinase, tellurite reduction, H_2S production, indole production, and Tween 80 hydrolysis. Produces acid aerobically from trehalose, sucrose, ribose, mannose, lactose, glycerol, glucose, galactose, and fructose. No acid produced from xylose, tagatose, salicin, raffinose, melezitose, melibiose, lyxose, β-gentiobiose, fucose, cellobiose, arabinose, xylitol, sorbitol, dulcitol, rhamnose, arabitol, amygdalin, or adonitol. Lysostaphin-sensitive; lysozyme-resistant. Novobiocin-susceptible. Peptidoglycan type L-Lys–Gly$_{4-5}$, Ser$_{0-0.3}$. Teichoic acid contains glycerol and α-N-acetylglucosamine. Major menaquinone MK-7. Variable characteristics include β-glucosidase, urease, and acid production from maltose, mannitol, and turanose.

Distinguished from *Staphylococcus hyicus* by pigmentation and lack of the following characteristics: heat-stable nuclease, hyaluronidase, Tween 80 hydrolysis, lecithinase, and pathogenicity for piglets.

Isolated from poultry, swine, and cattle.

Initial DNA–DNA hybridization studies yielded relative DNA binding values of 92–95% between two *strains of Staphylococcus chromogenes*, 30–55% between two strains of *Staphylococcus chromogenes* and three strains of *Staphylococcus hyicus*, and 9–19% between two strains of *Staphylococcus chromogenes* and three other *Staphylococcus* species (Devriese et al., 1978). Further DNA–DNA hybridization studies yielded relative DNA binding values of 84–100% between the type strain and nine other strains of *Staphylococcus chromogenes* and 43–65% between the type strain of *Staphylococcus chromogenes* and ten strains of *Staphylococcus hyicus* (Hajek et al., 1986).

DNA G+C content (mol%): 32–37 (T_m).

Type strain: A.C. Baird Parker 1462, ATCC 43764, CCM 3387, CIP 81.59, DSM 20454, NCTC 10530, NRRL B-14759.

GenBank accession number (16S rRNA gene): D83360.

Additional remarks: First described as *Staphylococcus hyicus* subsp. *chromogenes* (Devriese et al., 1978).

8. **Staphylococcus cohnii** Schleifer and Kloos 1975, 54[AL]

coh'ni.i. N.L. gen. n. *cohnii* of Cohn; named for Ferdinand Cohn, a German botanist and bacteriologist.

Nonmotile, nonspore-forming Gram-stain-positive cocci, 0.5–1.2 μm in diameter; form pairs as well as occurring singly. Facultatively anaerobic. Grows best under aerobic conditions. Produces small amounts of lactate (L and/or D) anaerobically from glucose. Growth good in medium containing up to 10% NaCl, poor at 15% NaCl. Grows at 15°C. Catalase-positive. Negative reactions: oxidase, nitrate reduction, coagulase, clumping factor, heat-stable nuclease, and β-glucosidase. Acid produced aerobically from fructose, glucose, glycerol, and trehalose. No acid produced from adonitol, arabinose, arabitol, cellobiose, dulcitol, erythritol, erythrose, fucose, gentiobiose, inositol, lyxose, melezitose, melibiose, raffinose, rhamnose, ribose, salicin, sorbitol, sorbose, sucrose, tagatose, turanose, or xylose. Novobiocin-resistant. Peptidoglycan type L-Lys–Gly$_{5-6}$. Teichoic acid contains glycerol, glucose, and N-acetylglucosamine, occasionally N-acetylgalactosamine or ribitol. Major menaquinone MK-7. Major fatty acids C_{Br-15}, C_{17}, C_{18}, and C_{20}. Variable characteristics include growth at 45°C, acetoin production, caseinase, gelatinase, and production of acid from mannitol and xylitol.

Isolated from the skin of humans and other primates.

Additional characteristics of the subspecies are given below.

8a. **Staphylococcus cohnii subsp. cohnii** Schleifer and Kloos 1975, 54[AL]

Nonmotile, nonspore-forming Gram-stain-positive cocci, 0.5–1.2 μm in diameter; form pairs as well as occurring singly. Colonies nonpigmented or yellowish, smooth, glistening, opaque, with entire margins. Colony diameters differ. Facultatively anaerobic. FBP-aldolase of class I type. Produces small amounts of L-lactate anaerobically from glucose. Catalase-positive. Negative reactions: alkaline phosphatase, arginine dihydrolase, clumping factor, coagulase, esculin hydrolysis, β-galactosidase, β-glucosidase, β-glucuronidase, heat-stable nuclease, nitrate reduction, ornithine decarboxylase, oxidase, and urease. Acid produced aerobically from β-D-fructose, D-glucose, glycerol, and D-trehalose. No acid produced from N-acetylglucosamine, L-arabinose, D-cellobiose, α-lactose, β-melezitose, L-rhamnose, ribose, sucrose, D-turanose, or D-xylose. Novobiocin-resistant. Peptidoglycan type L-Lys–Gly$_{5-6}$. Polyglycerolphosphate teichoic acids with N-acetylglucosamine. Major fatty acids $C_{15:0\ iso}$ and $C_{15:0\ ante}$. Variable characteristics include growth at 45°C, tolerates 15% NaCl, hemolysis, acetoin production, and production of acid from maltose, D-mannitol, D-mannose, and xylitol.

Isolated from the skin of humans.

DNA–DNA hybridization studies yielded mean relative DNA binding values of 94 ± 3% among eleven strains of *Staphylococcus cohnii* subsp. *cohnii*, 65 ± 5% between strains of *Staphylococcus cohnii* subsp. *cohnii* and *Staphylococcus cohnii* subsp. *urealyticus*, and 8–19% between *Staphylococcus cohnii* subsp. *cohnii* and ten other *Staphylococcus* species (Kloos and Wolfshohl, 1991).

DNA G+C content (mol%): 36–38 (T_m).

Type strain: GH137, ATCC 29974, CCUG 6463, CCUG 7322, CCUG 35144, CIP 81.54, DSM 20260, JCM 2417, NCTC 11041, NRRL B-14756.

GenBank accession number (16S rRNA gene): D83361.

8b. Staphylococcus cohnii subsp. urealyticus corrig. Kloos and Wolfshohl 1991, 287[VP]

u.re.a.ly′ti.cus. N.L. fem. n. *urea* urea; Gr. adj. *lutikos* dissolving; N.L. masc. adj. *urealyticus* urea-dissolving.

Nonmotile, nonspore-forming Gram-stain-positive cocci, 0.5–1.2 μm in diameter; form pairs as well as occurring singly. Colonies raised to convex, smooth, glistening, translucent, with entire margins and gray or grayish white concentric rings. Colony diameter 5.5–8.0 mm on P agar. Human isolates nonpigmented; isolates from other primates pigmented and may have bands of alternating colors. Facultatively anaerobic. Produces equal small amounts of L- and D-lactate anaerobically from glucose. Positive reactions: alkaline phosphatase (expression may be delayed), catalase, β-galactosidase, β-glucuronidase, and urease. Negative reactions: clumping factor, coagulase, heat-stable nuclease, ornithine decarboxylase, and oxidase. Acid produced aerobically from β-D-fructose, D-glucose, glycerol, lactose, maltose (delayed), D-mannose, and D-trehalose. No acid produced from L-arabinose, D-cellobiose, β-gentiobiose, D-melezitose, β-melibiose, raffinose, L-rhamnose, D-ribose, salicin, sucrose, D-turanose, or D-xylose. Novobiocin-resistant. Peptidoglycan type L-Lys–Gly$_{5-6}$. Polyglycerolphosphate teichoic acids with N-acetylglucosamine and glucose. Major fatty acids $C_{15:0\ iso}$ and $C_{15:0\ ante}$. Variable characteristics include growth at 45°C, tolerates 15% NaCl, hemolysis, nitrate reduction, acetoin production, β-glucosidase, arginine dihydrolase, esculin hydrolysis, and production of acid from galactose, mannitol, xylitol, and N-acetylglucosamine.

Distinguished from *Staphylococcus cohnii* subsp. *cohnii* by production of alkaline phosphatase, β-glucuronidase, β-galactosidase, and urease activity.

Isolated from the skin of humans and other primates.

DNA–DNA hybridization studies yielded mean relative DNA binding values of 92 ± 4% among 28 strains of *Staphylococcus cohnii* subsp. *urealyticus*, 68 ± 3% between strains of *Staphylococcus cohnii* subsp. *cohnii* and *Staphylococcus cohnii* subsp. *urealyticus*, and 7–15% between *Staphylococcus cohnii* subsp. *cohnii* and ten other *Staphylococcus* species (Kloos and Wolfshohl, 1991).

DNA G+C content (mol%): 37 (no method given).

Type strain: CK27, ATCC 49330, CCUG 35143, CIP 104024, DSM 6718.

GenBank accession number (16S rRNA gene): AB009936.

9. Staphylococcus condimenti Probst, Hertel, Richter, Wassill, Ludwig and Hammes 1998, 656[VP]

con.di.men′ti. L. n. *condimentum* spice; L. gen. n. *condimenti* of the spice.

Nonmotile, nonspore-forming, Gram-stain-positive cocci, 1 μm in diameter; occur singly and form pairs, chains, and clusters. Colonies cream-colored, opaque, raised, and entire.

Colony diameter 1–2 mm after incubation at 2 d at 37°C on P agar. Temperature range for growth 15–42°C. Tolerates 15% NaCl. Positive reactions: alkaline phosphatase, arginine dihydrolase, catalase, β-galactosidase, lipase, nitrate and nitrite reductases, and urease. Negative reactions: coagulase, esculin hydrolysis, β-glucuronidase, and lecithinase. Acid produced aerobically from fructose, N-acetylglucosamine, galactose, glucose, glycerol, lactose, mannitol, mannose, sorbitol, sucrose (weak), and trehalose. No acid produced from arabinose, cellobiose, raffinose, ribose, or turanose. Lysozyme-resistant; lysostaphin-sensitive. Major menaquinone MK-7; minor menaquinones MK-6 and MK-8. Growth at 45°C is variable.

Found in soy sauce mash.

DNA–DNA hybridization studies yielded relative DNA binding values of 98% between the type strain and a second strain of *Staphylococcus condimenti*, 58% between the type strain of *Staphylococcus condimenti* and the type strain of *Staphylococcus carnosus* subsp. *carnosus*, and 51% between the type strain of *Staphylococcus condimenti* and the type strain of *Staphylococcus piscifermentans* (Probst et al., 1998).

DNA G+C content (mol%): 35.2–36 (no method given).

Type strain: F-2, LTH 3734, CCUG 39902, CIP 105760, DSM 11674, JCM 6074.

GenBank accession number (16S rRNA gene): Y15750.

10. Staphylococcus delphini Varaldo, Kilpper-Bälz, Biavasco, Satta and Schleifer 1988, 437[VP]

del.phi′ni. L. gen. n. *delphini* of a dolphin.

Nonmotile, nonspore-forming, Gram-stain-positive cocci, 0.8–1.0 μm in diameter; occur singly and form clusters. Colonies nonpigmented, convex, smooth, glistening, and opaque. Colony diameter 5–7 mm. Facultatively anaerobic. Grows at 45°C; tolerate 15% NaCl. Positive reactions: arginine dihydrolase, caseinase, catalase, coagulase, hemolysis, nitrate reduction, phosphatase, and urease. Weak heat-labile DNase activity. Negative reactions: acetoin production, clumping factor, fibrinolysin, and gelatinase. Acid produced aerobically from fructose, glucose, lactose, mannitol, mannose, and sucrose. Acid not produced from arabinose, trehalose, xylitol, or xylose. Lysozyme-resistant; lysostaphin-sensitive. Novobiocin-susceptible. Peptidoglycan type Lys–Gly$_{5-6}$. Teichoic acid contains glycerol and N-acetylglucosamine.

Isolated from skin lesions of captive dolphins in an aquarium.

DNA–DNA similarity studies yielded relative DNA binding values of 99% between two strains of *Staphylococcus delphini* and 4–35% between the type strain of *Staphylococcus delphini* and the type strains of eleven other *Staphylococcus* species; the most closely related species was *Staphylococcus intermedius*.

DNA G+C content (mol%): 39 (T_m).

Type strain: Heidy, ATCC 49171, CCUG 30107, CIP 103732, DSM 20771, NCTC 12225, NRRL B-14767.

GenBank accession number (16S rRNA gene): AB009938.

11. Staphylococcus epidermidis (Winslow and Winslow 1908) Evans 1916, 449[AL] emend. Schleifer and Kloos 1975, 52 ("*Albococcus epidermidis*" Winslow and Winslow 1908, 201.)

e.pi.der′mi.dis. Gr. n. *epiderma* the outer skin; N.L. gen. n. *epidermidis* of the epidermis.

Nonmotile, nonspore-forming, Gram-stain-positive cocci, 0.8–1.0 μm in diameter; occur in pairs and tetrads, occasionally single cells. Colonies usually nonpigmented (gray or grayish white), smooth, raised, glistening, with entire margins. May be mucoid or slimy; vary in translucency; may become sticky with age. Colony diameter 2.5–4.0 mm. Facultatively anaerobic, growth is best under aerobic conditions. Acetate and carbon dioxide are the major end products of aerobic glucose metabolism. Produces predominantly L-lactate anaerobically from glucose. D-Galactose and lactose are metabolized via the D-tagatose-6-phosphate pathway. FBP-aldolase belongs to the class I type. Growth is good at NaCl concentrations up to 7.5% and relatively poor at 10% NaCl. Temperature range for growth 15–45°C; optimum 30–37°C. Positive reactions: acetoin production, alkaline phosphatase, arginine dihydrolase, nitrate reduction, and urease. Negative reactions: clumping factor, coagulase, esculin hydrolysis, β-galactosidase, β-glucuronidase, ornithine decarboxylase, oxidase, starch hydrolysis, and growth in medium containing 15% NaCl. Acid produced aerobically from fructose, glucose, glycerol, maltose, mannose (delayed), and sucrose. No acid produced from arabinose, adonitol, arabitol, cellobiose, dulcitol, erythritol, erythrose, fucose, gentiobiose, inositol, lyxose, mannitol, melibiose, raffinose, rhamnose, salicin, sorbose, tagatose, trehalose, xylitol, or xylose. Novobiocin-susceptible. Lysozyme-resistant. Peptidoglycan type L-Lys–Gly$_{4–5}$, L-Ser$_{0.7–1.5}$. Teichoic acid contains glycerol, *N*-acetylglucosamine, and glucose. Major fatty acids C$_{Br-15}$, C$_{18}$, and C$_{20}$. Major menaquinone MK-7. Variable characteristics include acid phosphatase, β-glucosidase, hyaluronidase, fibrinolysin; hemolysis (weak), DNase (weak), proteases, lipases, and production of acid from galactose, lactose, melezitose, ribose, and turanose.

Phage typing of *Staphylococcus epidermidis* has not been very successful. Chromosomal and plasmid DNA may be transferred to appropriate recipient cells via transduction.

Isolated from human skin where it is a resident; occasionally found on the skin of other mammals, particularly those living in close association with humans. Opportunistic pathogen, may colonize various indwelling medical devices, postoperative infections, urinary tract, and wound infections.

DNA–DNA hybridization studies yielded relative DNA binding values >85% among *Staphylococcus epidermidis* strains (Kloos and Wolfshohl, 1982).

DNA G+C content (mol%): 30–37 (T_m).

Type strain: Fussel, 2466, ATCC 14990, CCUG 18000 A, CCUG 39508, CIP 81.55, DSM 20044, LMG 10474, NCTC 11047.

GenBank accession number (16S rRNA gene): D83363, L37605.

12. **Staphylococcus equorum** Schleifer, Kilpper-Bälz and Devriese 1985, 224[VP] (Effective publication: Schleifer, Kilpper-Bälz and Devriese 1984, 506.)

e.quor′rum. L. gen. n. *equus* of horses.

Nonmotile, nonspore-forming, Gram-stain-positive cocci, 0.5–1.8 μm in diameter; occur singly and form pairs, clusters, and chains. Colonies white, opaque, with entire margins. Colony diameter 4–6 mm after incubation at 2 d at 37°C on brain heart infusion agar. Aerobic. Optimum growth temperature 30°C, no growth at 42°C. Positive reactions: catalase, β-glucuronidase, β-galactosidase, nitrate reduction, and urease. Weakly positive reactions: esculin and tributyrin hydrolysis. Negative reactions: oxidase, arginine hydrolysis, anaerobic growth. Acid produced aerobically from D-fructose, D-glucose, glycerol, D-mannose. Peptidoglycan type Lys–Gly$_{5–6}$. Teichoic acids contain glucosamine, glycerol, and small amounts of ribitol. Novobiocin-resistant. Variable characters include hemolysis, caseinase, and production of acid from lactose, and D-turanose.

Additional characteristics of the subspecies are given below.

12a. **Staphylococcus equorum subsp. equorum** Schleifer, Kilpper-Bälz and Devriese 1985, 224[VP] (Effective publication: Schleifer, Kilpper-Bälz and Devriese 1984, 506.)

Nonmotile, nonspore-forming, Gram-stain-positive cocci, 0.5–1.8 μm in diameter; occur singly and form pairs, clusters, and chains. Colonies white, opaque, with entire margins. Colony diameter 4–6 mm after incubation at 2 d at 37°C on brain heart infusion agar. Aerobic. Optimum growth temperature 30°C; no growth at 42°C. Positive reactions: catalase, β-glucuronidase, β-galactosidase, nitrate reduction, urease, and phosphatase. Weakly positive reactions: esculin and tributyrin hydrolysis. Negative reactions: coagulase, clumping factor, gelatinase, heat-stable nuclease, oxidase, hyaluronidase, staphylokinase, Tween 80 hydrolysis, fibrinolysin, acetoin production, and arginine hydrolysis. Acid produced aerobically from L-arabinose, D-fructose, D-galactose, D-glucose, glycerol, maltose, D-mannose, D-mannitol, D-melezitose, D-ribose, salicin, sorbitol, sucrose, trehalose, and D-xylose. No acid produced from amygdalin, D-fucose, β-gentiobiose, D-melibiose, raffinose, L-rhamnose, or xylitol. Peptidoglycan type Lys–Gly$_{5–6}$. Teichoic acids contain glucosamine, glycerol, and small amounts of ribitol. Novobiocin-resistant. Variable characters include hemolysis, caseinase, and production of acid from arbutin, cellobiose, lactose, and D-turanose.

Isolated from the skin of horses.

DNA–DNA hybridization studies yielded relative DNA binding values of 75–100% among *Staphylococcus equorum* strains and 15–33% between *Staphylococcus equorum* strains and *Staphylococcus kloosii*, *Staphylococcus arlettae*, *Staphylococcus saprophyticus*, *Staphylococcus cohnii*, *Staphylococcus gallinarum*, *Staphylococcus caprae*, *Staphylococcus epidermidis*, *Staphylococcus lentus*, and *Staphylococcus xylosus* (Schleifer, 1984).

DNA G+C content (mol%): 33.4–34.7 (T_m).

Type strain: PA231, ATCC 43958, CCUG 30109, CIP 103502, DSM 20674, NCTC 12414, NRRL B-14765.

GenBank accession number (16S rRNA gene): AB009939.

12b. **Staphylococcus equorum subsp. linens** Place, Hiestand, Gallmann and Teuber 2003b, 1219[VP] (Effective publication: Place, Hiestand, Gallmann and Teuber 2003c, 36.)

lin′ens. L. pres. part. of *linere* smearing; named because the organism was isolated from the surface of a smeared red cheese.

Nonmotile, Gram-stain-positive cocci, 0.9–1.4 μm in diameter; occur singly and form pairs and chains. Colonies white, glossy, and opaque. Colony diameter 1–2 mm after 2 d on PC skim milk agar (Merck) at 32°C. Aerobic. Tolerates 15% NaCl in the medium, although growth is poor. Temperature range for growth 6–40°C; optimum 32°C. Positive reactions: nitrate reduction, and urease. Negative reactions: acetoin production, arginine dihydrolase, ornithine decarboxylase, β-glucuronidase, β-galactosidase, alkaline phosphatase, esculin hydrolysis, hemolysis, and oxidase. Acid produced aerobically from fructose, D-mannose, and turanose (weak). No acid produced from L-arabinose, cellobiose, D-galactose, α-D-lactose, maltose, mannitol, D-raffinose D-ribose, salicin, sucrose, D-trehalose, or D-xylose. Ferments D-glucose strongly and *N*-acetyl-D-glucosamine, 3-methyl-glucose, maltotriose, methyl-pyruvate, and pyruvic acid weakly. Does not ferment D-alanine, L-alanyl-glycine, arbutin, 2,3-butanediol, β-hydroxybutyric acid, or Tween 40. Novobiocin-susceptible. Peptidoglycan type L-Lys–Gly$_5$.

Distinguished from *Staphylococcus equorum* subsp. *equorum* by novobiocin susceptibility; lack of β-galactosidase; inability to produce acid from D-galactose, α-D-lactose, maltose, D-ribose, salicin, sucrose, and D-trehalose; and inability to ferment D-alanine, L-alanyl-glycine, D-mannitol, and D-raffinose.

DNA–DNA hybridization studies yielded relative DNA binding values of 68% between the type strains of *Staphylococcus equorum* subsp. *equorum* and *Staphylococcus equorum* subsp. *linens* and 68% between the type strains of *Staphylococcus equorum* subsp. *linens* and *Staphylococcus xylosus* (Place et al., 2003c).

Isolated from a surface-ripened Swiss cheese.

DNA G+C content (mol%): 35.1 (HPLC).

Type strain: RP29, CIP 107656, DSM 15097.

GenBank accession number (16S rRNA gene): AF527483.

13. **Staphylococcus felis** Igimi, Kawamura, Takahashi and Mitsuoka 1989, 374[VP]

fe'lis. L. gen. n. *felis* of a cat.

Nonmotile, nonspore-forming, Gram-stain-positive cocci, 0.8–1.2 μm in diameter; occur singly and form pairs and clusters. Colonies low convex, nonpigmented, opaque, smooth, and glistening. Colony diameter 5–8 mm. Facultatively anaerobic. Grows in 10% NaCl in 24 h and in 15% NaCl in 72 h. Weak growth at 15 and 45°C. Weakly hemolytic; no α-, β-, or δ-hemolysin detected. Positive reactions: phosphatase, β-galactosidase, nitrate reduction, arginine dihydrolase, and urease. Negative reactions: coagulase, clumping factor, hyaluronidase, heat-stable nuclease (or very weak), acetoin production, and oxidase. Acid produced aerobically from fructose, glucose, glycerol, lactose, mannose, and trehalose. No acid produced aerobically from xylose, arabinose, cellobiose, raffinose, maltose, or xylitol. Acid production from sucrose, galactose, mannitol, and ribose varies between strains. Lysostaphin-sensitive. Novobiocin- and bacitracin-susceptible.

Isolated from various infections of cats.

DNA–DNA hybridization studies yielded relative DNA binding values of 84–93% between the type and four other

Staphylococcus felis strains and 1–9% between the type strain of *Staphylococcus felis* and the type strains of 28 other *Staphylococcus* species and subspecies (Igimi et al., 1989).

DNA G+C content (mol%): 35 (T_m).

Type strain: GD521, ATCC 49168, CCUG 32418, CCUG 38440, CIP 103366, DSM 7377, JCM 7469, NRRL B-14779.

GenBank accession number (16S rRNA gene): D83364.

14. **Staphylococcus fleurettii** Vernozy-Rozand, Mazuy, Meugnier, Bes, Lasne, Fiedler, Étienne and Freney 2000, 1523[VP]

fleu.rett'i.i. N.L. gen. n. *fleurettii* of Fleurette in honor of the French microbiologist Jean Fleurette for his contribution to the taxonomy of staphylococci.

Nonmotile, nonspore-forming, Gram-stain-positive cocci, 0.8–1.4 μm in diameter; occur singly and form chains and clumps. Colonies opaque and usually nonpigmented (may be cream colored), with irregular centres and margins. Colony diameter <3 mm on P agar or 8–12 mm on tryptic soy agar at 37°C (incubation time not given). Facultatively anaerobic. Tolerates 10% NaCl. Positive reactions: catalase, nitrate reduction, and oxidase. Negative reactions: arginine dihydrolase, coagulase, clumping factor, β-galactosidase, β-glucuronidase, heat-stable nuclease, ornithine decarboxylase, staphylocoagulase, and urease. Acid produced aerobically from D-fructose, maltose, D-mannose, sucrose, D-trehalose, and turanose. No acid produced aerobically from cellobiose, L-fucose, D-galactose, α-lactose, D-melezitose, D-raffinose, salicin, or D-xylitol. Lysozyme-resistant; lysostaphin-sensitive. Novobiocin- and bacitracin-resistant. Peptidoglycan contains lysine, alanine, and glycine. Poly-(glycerolphosphate) teichoic acids contain glucose. Lipoteichoic acids present. No production of enterotoxins SEA, SEB, SEC$_1$, SEC$_2$, SEC$_3$, SED, and SEE. Variable characteristics include colony pigmentation, acetoin production, alkaline phosphatase, esculin hydrolysis, and acid production from L-arabinose and D-xylose.

Found in goats' milk cheese.

DNA–DNA hybridization studies yielded relative DNA binding values of 86–100% between the type strain and five other *Staphylococcus fleurettii* strains and 8–44% between the *Staphylococcus fleurettii* type strain and the type strains of 36 *Staphylococcus* species and subspecies (Vernozy-Rozand et al., 2000).

DNA G+C content (mol%): 31–32 (HPLC).

Type strain: 241, ATCC BAA-274, CCUG 43834, CIP 106114, DSM 13212.

GenBank accession number (16S rRNA gene): AB233330.

15. **Staphylococcus gallinarum** Devriese, Poutrel, Kilpper-Bälz and Schleifer 1983, 481[VP]

gal.li.na'rum. L. fem. n. *gallina* hen; L. fem. gen. pl. n. *gallinarum* of hens.

Nonmotile, nonspore-forming, Gram-stain-positive cocci, 0.5–1.8 μm in diameter; occur singly and form pairs, short chains, and small groups. Colonies yellow, yellowish, or nonpigmented, flat, opaque, and dry, with lobate or crenate edges. Colony diameter 10–15 mm after 2 d on brain heart infusion agar and 5–10 mm after 2 d on tryptone soy agar. Small colony type smooth and glistening with entire edges, 4–7 mm in diameter after 2 d on brain heart infusion agar. Facultatively anaerobic, growth is better under aerobic conditions. Produces L-lactate anaerobically from glucose. FBP-aldolase is of class I type. Temperature range for growth 25–42°C. Positive reactions: alkaline phosphatase, caseinase, catalase, esculin hydrolysis, nitrate reduction, and urease. Negative reactions: acetoin production, clumping factor, coagulase, heat-stable DNase, hyaluronidase, lipase, and oxidase. Acid produced aerobically from amygdalin, L-arabinose, arbutin, cellobiose, D-fructose, D-fucose (weak), D-galactose, β-gentiobiose, D-glucitol, glycerol, maltose, D-mannitol, D-mannose, melezitose, melibiose, D-ribose, salicin, sucrose, trehalose, turanose, and D-xylose. No acid produced from L-rhamnose. Novobiocin-resistant. Peptidoglycan type L-Lys–Gly$_{4–5}$, L-Ser$_{0.7–1.0}$. Teichoic acid contains glucose, glycerol, and N-acetylglucosamine (minor amounts). Variable characteristics include hemolysis (weak) and acid production from lactose and xylitol.

Isolated from the nares and skin of poultry.

DNA–DNA hybridization studies yielded relative DNA binding values of 72–100% among *Staphylococcus gallinarum* strains and 4–25% between the type strain of *Staphylococcus gallinarum* and 13 strains of other *Staphylococcus* species (Devriese et al., 1983).

DNA G+C content (mol%): 34–35 (T_m).

Type strain: VIII1, ATCC 35539, CCM 3572, CCUG 15600, CIP 103504, DSM 20610, LMG 19121, NCTC 12195, NRRL B-14763.

GenBank accession number (16S rRNA gene): D83366.

16. **Staphylococcus haemolyticus** Schleifer and Kloos 1975, 56[AL]

hae.mo.ly'ti.cus. N.L. adj. *haemolyticus* blood-dissolving.

Nonmotile, nonspore-forming, Gram-stain-positive cocci, 0.8–1.3 μm in diameter; form pairs and tetrads. Colonies nonpigmented or yellowish, smooth, glistening, opaque, with entire margins. Colony diameter 5–9 mm. Facultatively anaerobic. Growth is best under aerobic conditions. Produces only or predominantly D-lactate anaerobically from glucose. FBP-aldolase of class I type. Grows well in medium containing 10% NaCl; growth poor or absent at 15% NaCl. Temperature range for growth 18–45°C (optimum 30–40°C); growth poor or absent at 15°C. Positive reactions: acetoin production, arginine dihydrolase, benzidine test, catalase, hemolysis (sometimes weak), and lipase. Negative reactions: coagulase, DNase, ornithine decarboxylase, phosphatase, urease, and oxidase. Produces acid aerobically from glucose, glycerol, maltose, sucrose, and trehalose. No acid produced from adonitol, arabinose, arabitol, cellobiose, dulcitol, erythritol, erythrose, fucose, gentiobiose, inositol, lyxose, mannose, melezitose melibiose, raffinose, rhamnose, salicin, sorbitol, sorbose, tagatose, xylitol, or xylose. Lysostaphin-sensitive; lysozyme-resistant. Novobiocin-susceptible. Peptidoglycan type L-Lys–Gly$_{3.3–4.0}$, L-Ser$_{0.9–1.5}$. Teichoic acid contains glycerol and N-acetylglucosamine. Major fatty acids C_{Br-15}, C_{Br-17}, C_{18}, and C_{20}. The fatty acid profile is rather similar to that of *Staphylococcus hominis* and *Staphylococcus aureus*. Major menaquinone MK-7. Variable characteristics include acetoin production, β-glucosidase, β-glucuronidase, nitrate reduction, and acid production from fructose, galactose, lactose, mannitol, ribose, and turanose.

Isolated from humans and other primates; sometimes associated with infections.

DNA–DNA hybridization studies yielded relative DNA binding values ≥75% among *Staphylococcus hemolyticus* strains isolated from humans and among *Staphylococcus hemolyticus* strains isolated from nonhuman primates, however, relative DNA binding values between *Staphylococcus hemolyticus* strains from humans and from nonhuman primates were somewhat less: 65–77% under optimal hybridization conditions (Kloos and Wolfshol, 1979).

DNA G+C content (mol%): 34–36 (T_m).

Type strain: ATCC 29970, CCUG 7323, CIP 81.56, DSM 20263, JCM 2416, LMG 13349, NCTC 11042, NRRL B-14755.

GenBank accession number (16S rRNA gene): D83367, L37600.

17. **Staphylococcus hominis** Kloos and Schleifer 1975a, 68[AL] emend. Kloos, George, Olgiate, Van Pelt, McKinnon, Zimmer, Muller, Weinstein and Mirrett 1998b, 807.

ho'mi.nis. L. gen. n. *hominis* of humans; named for the host on whose skin this species is commonly found.

Nonmotile, nonspore-forming, Gram-stain-positive cocci, 1.0–1.5 μm in diameter; occur singly and form tetrads and smaller numbers of pairs. Aerobic or facultatively anaerobic; anaerobic growth weak and delayed. Positive reactions: catalase and urease. Negative reactions: coagulase, clumping factor, alkaline phosphatase, β-galactosidase, β-glucosidase, β-glucuronidase, latex agglutination, ornithine decarboxylase, and oxidase. Acid produced aerobically from β-D-fructose, D-glucose, glycerol, maltose, and sucrose. No acid produced from L-arabinose, D-cellobiose, mannitol, D-mannose, D-raffinose, salicin, D-sorbitol, and D-xylose.

In conjunction with the description of *Staphylococcus hominis* subsp. *novobiosepticus* and the creation of *Staphylococcus hominis* subsp. *hominis*, Kloos et al. (1998b) emended the description of *Staphylococcus hominis* to include novobiocin-resistant strains as well as novobiocin-susceptible strains.

Additional characteristics of the subspecies are given below.

17a. **Staphylococcus hominis subsp. hominis** Kloos and Schleifer 1975a, 68[AL] emend. Kloos, George, Olgiate, Van Pelt, McKinnon, Zimmer, Muller, Weinstein and Mirrett 1998b, 807.

Nonmotile, nonspore-forming, Gram-stain-positive cocci, 1.0–1.5 μm in diameter; occur singly and form tetrads and smaller numbers of pairs. Colonies may be nonpigmented or cream to yellow-orange. Colonies smooth, opaque, raised to umbonate, and butyrous, with entire margins. Colony diameter 3–5 mm. Aerobic or facultatively anaerobic; anaerobic growth weak and delayed. Growth is best under aerobic conditions. Nonhemolytic on sheep or bovine blood agar. Temperature range for growth 20–45°C; optimum 30–40°C. May produce only D-lactate or both L- and D-lactate from glucose anaerobically. Class I fructose-1,6-bisphosphate aldolase. Positive reactions: catalase, lipase, and urease. Negative reactions: alkaline phosphatase, caseinase, coagulase, DNase, β-galactosidase, β-glucosidase, β-glucuronidase, gelatinase, latex agglutination, ornithine decarboxylase, staphylocoagulase. capsular polysaccharide/adhesion, and oxidase. Acid produced aerobically from β-D-fructose, D-glucose, glycerol, maltose, and sucrose. No acid produced from L-arabinose, D-cellobiose, D-mannose, D-raffinose, salicin, D-sorbitol, and D-xylose. Acid production from N-acetylglucosamine, galactose, α-lactose, D-mannitol, D-melezitose, D-trehalose, and turanose varies between strains. Novobiocin-susceptible. Peptidoglycan type L-Lys–Gly$_{3.5-4.5}$, L-Ser$_{0.6-1.3}$. Teichoic acid contains glycerol and N-acetylglucosamine. Major fatty acids are C$_{Br-15}$, C$_{18}$, and C$_{20}$. Major quinone is MK-7. Variable characteristics: acetoin production, arginine utilization, and nitrate reduction.

Isolated from the skin of humans (one of the major staphylococcal species) and clinical specimens from human infections. Appears to be quite human-specific.

DNA–DNA hybridization studies yielded relative DNA binding values >85% between *Staphylococcus hominis* strains (Kloos, 1980; Kloos and Wolfshohl, 1982). A later study yielded relative DNA binding values of 84–94% between the type strain and four other strains of *Staphylococcus hominis* subsp. *hominis*, 62–76% between the type strain of *Staphylococcus hominis* subsp. *hominis* and eight strains of *Staphylococcus hominis* subsp. *novobiosepticus*, and 63–77% between the type strain of *Staphylococcus hominis* subsp. *novobiosepticus* and five strains of *Staphylococcus hominis* subsp. *hominis* (Kloos et al., 1998b).

DNA G+C content (mol%): 30–36 (T_m).

Type strain: ATCC 27844, CCUG 35516, CIP 102258, CIP 81.57, DM 122, DSM 20328, JCM 2419, LMG 13348, NCTC 11320, NRRL B-14737.

GenBank accession number (16S rRNA gene): L37601, X66101.

17b. **Staphylococcus hominis subsp. novobiosepticus** Kloos, George, Olgiate, Van Pelt, McKinnon, Zimmer, Muller, Weinstein and Mirrett 1998b, 809[VP]

no.vo.bio.sep′ti.cus. N.L. n. *novobiocinum* novobiocini; L. adj. *septicus* septic; N.L. adj. *novobiosepticus* intended to mean resistant to novobiocin and growing in blood.

Nonmotile, nonspore-forming, Gram-stain-positive cocci, 1.0–1.5 μm in diameter; occur singly and form tetrads and smaller numbers of pairs. Colonies grayish-white, convex to umbonate, butyrous, and opaque, with entire margins. Colony diameter 4–6 mm after incubation at 34–35°C for 3 d, then 25°C for 2 d on P, tryptic soy agar, or tryptic soy agar plus sheep blood agar. Aerobes or facultative anaerobes; anaerobic growth weak and delayed. Temperature optimum 35°C. Nonhemolytic. Positive reactions: catalase and urease. Negative reactions: oxidase, coagulase, alkaline phosphatase, β-glucosidase, β-galactosidase, β-glucuronidase, and capsular polysaccharide/adhesin. Acid produced aerobically from β-D-fructose, D-glucose, glycerol, maltose, and sucrose. No acid produced aerobically from N-acetyl-D-glucosamine, L-arabinose, D-cellobiose, D-mannitol, D-mannose, D-raffinose, salicin, D-sorbitol, D-trehalose, or D-xylose. Novobiocin-resistant. Strains differ in acid production from α-lactose and D-melezitose

and in production of acetoin. No production of ammonia from arginine.

Distinguished from *Staphylococcus hominis* subsp. *hominis* by novobiocin resistance, inability to utilize arginine, and inability to produce acid aerobically from D-trehalose and N-acetylglucosamine.

Isolated from human blood and clinical specimens from other body sites.

DNA–DNA hybridization studies yielded relative DNA binding values of 82–100% between the type strain and seven other strains of *Staphylococcus hominis* subsp. *novobiosepticus*, 63–77% between the type strain of *Staphylococcus hominis* subsp. *novobiosepticus* and five strains of *Staphylococcus hominis* subsp. *hominis*, and 62–76% between the type strain of *Staphylococcus hominis* subsp. *hominis* and five strains of *Staphylococcus hominis* subsp. *novobiosepticus* (Kloos et al., 1998b).

DNA G+C content (mol%): 35 (T_m).

Type strain: R22, ATCC 700236, CCUG 42399, CIP 105719.

GenBank accession number (16S rRNA gene): AB233326.

18. **Staphylococcus hyicus** (Sompolinsky 1953) Devriese, Hájek, Oeding, Meyer and Schleifer 1978, 488[AL] ("*Micrococcus hyicus*" Sompolinsky 1953, 307; *Staphylococcus hyicus* subsp. *hyicus* Devriese et al., 1978, 488.)

hy′i.cus. Gr. n. *hyos* hog, pig; N.L. masc. adj. *hyicus* pertaining to a pig.

Nonmotile, nonspore-forming, Gram-stain-positive cocci, 0.6–1.3 μm in diameter; form pairs, tetrads, and small clusters; occasionally occur as single cells. Colonies nonpigmented, slightly convex, glistening, and opaque, with entire margins. Colony diameter 4–7 mm. Growth poor or absent at 45°C and delayed at 22°C; optimum temperature range for growth 30–35°C. Facultatively anaerobic, growth is best under aerobic conditions. Produces L-lactate anaerobically from glucose. Galactose is metabolized via the D-tagatose-6-phosphate pathway. Strains tested possess both class I and II FBP-aldolases. Positive reactions: arginine dihydrolase, benzidine test, catalase, caseinase, gelatinase, heat-stable nuclease, hippurate hydrolysis, hyaluronidase, nitrate reduction, phosphatase, tributyrin hydrolysis, and Tween hydrolysis. Negative reactions: acetoin production, clumping factor, esculin hydrolysis, growth on medium containing 15% NaCl, hemolysin, H$_2$S production, indole production, and oxidase. Acid produced aerobically from fructose, glucose, galactose, glycerol, lactose, mannose, ribose, sucrose, and trehalose. No acid produced from adonitol, amygdalin, L(+)-arabinose, arabitol, cellobiose, dulcitol, L(−)-fucose, β-gentiobiose, lyxose, mannitol, melibiose, melezitose, raffinose, L(+)-rhamnose, salicin, sorbose, tagatose, xylitol, and xylose. Lysozyme-resistant, lysostaphin-sensitive. Novobiocin-susceptible. Peptidoglycan type L-Lys–Gly$_{4-5}$, Ser$_{0-0.3}$. Teichoic acid contains glycerol and N-acetylglucosamine.

Coagulase is produced by some strains (25–56%) although it is often weak in activity and may require 18–24 h for detection in the tube test with rabbit plasma. Other variable characters include fibrinolysin, β-glucosidase, β-glucuronidase, and urease. Variation in these characteristics is associated with differences in host of origin as well as geographic region of origin.

Isolated from skin of poultry, cattle, and swine. Pig strains pathogenic. Rarely isolated from cattle with mastitis.

Initial DNA–DNA hybridization studies yielded relative DNA binding values of 48–95% among three *Staphylococcus hyicus* strains, 30–55% between three strains of *Staphylococcus hyicus* and two strains of *Staphylococcus chromogenes*, and 10–19% between *Staphylococcus hyicus* and three other *Staphylococcus* species (Devriese et al., 1978). Further DNA–DNA hybridization studies yielded relative DNA binding values of 67–95% between the type strain and nine other strains of *Staphylococcus hyicus* and 38–52% between the type strain of *Staphylococcus hyicus* and ten strains of *Staphylococcus chromogenes* (Hajek et al., 1986).

DNA G+C content (mol%): 33–34 (T_m).

Type strain: D. Sompolinsky no. 1, ATCC 11249, CCM 2368, CIP 81.58, DSM 20459, JCM 2423, LMG 13352, NCTC 10350.

GenBank accession number (16S rRNA gene): D83368.

Additional remarks: Originally described as *Staphylococcus hyicus* subsp. *hyicus* (*Staphylococcus hyicus* Group A strains); became *Staphylococcus hyicus* when *Staphylococcus hyicus* subsp. *chromogenes* was elevated to species rank as *Staphylococcus chromogenes* (Hajek et al., 1986).

19. **Staphylococcus intermedius** Hájek 1976, 405[AL]

in.ter.me′di.us. L. adj. *intermedius* in between, intermediate; intended to indicate that this species possesses some properties of *Staphylococcus aureus* and *Staphylococcus epidermidis*.

Nonmotile, nonspore-forming, Gram-stain-positive cocci, 0.5–1.5 μm in diameter; form pairs and clusters; also occur as single cells. Colonies nonpigmented, slightly convex, smooth, glistening, and butyrous, with entire margins. Colony diameter 5.0–6.5 mm. Facultatively anaerobic, growth is best under aerobic conditions. Produces L-lactate anaerobically from glucose. Contains FBP-dependent L-lactate dehydrogenase. Galactose and lactose are metabolized via the Leloir pathway and not via the D-tagatose-6-phosphate pathway as in the case of *Staphylococcus aureus*. Strains possess both class I and class II FBP-aldolases. Grows on agar medium containing up to 12.5% NaCl. Temperature range for growth 15–45°C; optimum 30–40°C. Positive reactions: acid and alkaline phosphatase, benzidine test, catalase, coagulase (rabbit and bovine plasma), gelatinase, heat-labile and heat-stable nucleases, methyl red test, nitrate reduction, and urease. Esterases are produced; their patterns are complex (polymorphic) and their electrophoretic migrations are slower than the major esterase of *Staphylococcus aureus*. Negative reactions: acetoin production, esculin hydrolysis, fibrinolysin, indole, H$_2$S, protein A, oxidase, and tellurite reduction. Acid produced aerobically from fructose, galactose, glucose, glycerol, maltose (weak), mannose, ribose, sucrose, and trehalose. No acid produced from arabinose, cellobiose, fucose, melezitose, raffinose, rhamnose, salicin, sorbitol, xylitol, or xylose. Lysozyme-resistant; lysostaphin-sensitive. Novobiocin-susceptible. Peptidoglycan type L-Lys–Gly$_{4-5}$, L-Ser$_{0.5-1.0}$. Teichoic acid contains glycerol, *N*-acetylglucosamine, and/or glucose. Major menaquinone MK-7. Variable characteristics include arginine dihydrolase (pigeon strains usually negative), clumping factor, caseinase, coagulation of human plasma, hemolysin, β-D-galactosidase,

β-glucosidase, Tween hydrolysis, and acid production from lactose, mannitol, and turanose.

Isolated from dogs, horses, mink, and pigeons. Often predominant staphylococcal species inhabiting the nasal membranes and skin of carnivora (e.g., dogs, minks, raccoons, foxes). Associated with a variety of infections in dogs. Rarely isolated from humans or other primates.

DNA–DNA hybridization studies yielded relative DNA binding values of 75–95% among six strains of *Staphylococcus intermedius*, 33–55% between six strains of *Staphylococcus intermedius* and five strains of *Staphylococcus aureus*, and 40–45% between six strains of *Staphylococcus intermedius* and the type strain of *Staphylococcus epidermidis* (Meyer and Schleifer, 1978).

DNA G+C content (mol%): 31–36 (T_m).

Type strain: H 11/68, ATCC 29663, CCM 5739, CCUG 6520, CCUG 27191, CIP 81.60, CNCTC M16/75, DSM 20373, JCM 2422, LMG 13351, NCTC 11048, NRRL B-14754.

GenBank accession number (16S rRNA gene): D83369.

20. **Staphylococcus kloosii** Schleifer, Kilpper-Bälz and Devriese 1985, 224[VP] (Effective publication: Schleifer, Kilpper-Bälz and Devriese 1984, 506.)

kloo′si.i. N.L. gen. n. *kloosii* of Kloos, named for Wesley E. Kloos, who made important contributions to the systematics of staphylococci.

Nonmotile, nonspore-forming, Gram-stain-positive cocci, 0.6–1.8 μm in diameter; occur singly and form pairs, clusters, and chains. Colonies opaque, glistening, with entire margins; yellowish or nonpigmented. Colony diameter 3–6 mm after incubation for 2 d at 37°C on brain heart infusion agar. Temperature range for growth 25–42°C. Aerobic. Positive reactions: catalase, phosphatase, and tributyrin hydrolysis. Negative reactions: coagulase, clumping factor, gelatinase, heat-stable nuclease, oxidase, hyaluronidase, staphylokinase, Tween 80 hydrolysis, fibrinolysin, and arginine hydrolysis. Acid produced aerobically from D-fructose, D-glucose, D-galactose, glycerol, maltose, D-mannitol, melizitose (weak), D-ribose, trehalose, and D-turanose (weak). No acid produced from amygdalin, arbutin, cellobiose, D-fucose, D-mannose, D-melibiose, L-rhamnose, sorbitol, sucrose or D-xylose. Peptidoglycan type Lys–Gly$_{5-6}$ or Lys–Gly$_5$, Ser$_{0.3-0.8}$. Teichoic acids contain glycerol, ribitol, and either glucosamine or galactosamine; glucose sometimes present. Novobiocin-resistant. Variable characters include hemolysis, urease, esculin hydrolysis, production of acetoin, and production of acid from L-arabinose, β-gentiobiose, lactose, D-melezitose (weak), raffinose (weak), and xylitol (weak).

Isolated from the skin of various animals, including squirrels, raccoons, opossums, swine, sheep, and horses.

DNA–DNA hybridization studies yielded relative DNA binding values of 51–100% among *Staphylococcus kloosii* strains and 20–38% between *Staphylococcus kloosii* strains and nine other *Staphylococcus* species (Schleifer et al., 1984).

DNA G+C content (mol%): 31–32.3 (T_m).

Type strain: SC210, ATCC 43959, CCUG 30110, CIP 103503, DSM 20676, NCTC 12415, NRRL B-14766.

GenBank accession number (16S rRNA gene): AB009940.

21. **Staphylococcus lentus** (Kloos, Schleifer and Smith 1976a) Schleifer, Geyer, Kilpper-Bälz and Devriese 1983a, 897[VP] (Effective publication: Schleifer, Geyer, Kilpper-Bälz and Devriese 1983b, 385) (*Staphylococcus sciuri* subsp. *lentus* Kloos, Schleifer and Smith 1976a, 30.)

len'tus. L. adj. *lentus* slow; pertaining to slow growth.

Nonmotile, nonspore-forming, Gram-stain-positive cocci, 0.7–1.2 μm in diameter; occur singly and form pairs and tetrads. Colonies convex, small, glistening, smooth, with slightly undulate margins; pigmentation grayish white, white, or cream. Some strains mucoid. Colony diameter 0.5–5 mm on P agar or 2–7 mm on tryptic soy agar after incubation at 34–35°C for 3 d, then 25°C for 2 d. Facultative anaerobes; anaerobic growth weak. Does not grow anaerobically in thioglycolate medium. Growth slow in medium containing 10% NaCl; growth poor in 15% NaCl. Produces L-lactate from glucose anaerobically. Temperature optimum 25–35°C; no growth at 15 or 45°C. Positive reactions: benzidine test, caseinase, catalase, esculin hydrolysis, gelatinase, β-glucosidase, nitrate reduction, nuclease, and oxidase (may be weak). Alkaline phosphatase activity weak or absent. Negative reactions: acetoin production, arginine dihydrolase, coagulase, elastase, β-galactosidase, β-glucuronidase, lipase, hemolysin, and urease. Acid produced aerobically from D(+)-cellobiose, β-D(−)-fructose, β-gentiobiose, D(+)-glucose, glycerol, D-mannitol, D(+)-mannose, raffinose, D(−)-ribose, sucrose, and D(+)-trehalose. No acid produced from adonitol, D-arabitol, dulcitol, *meso*-erythritol, D-erythrose, L(−)-fucose, *meso*-inositol, D-lyxose, D(+)-melezitose, L-sorbose, D(+)-tagatose, D(+)-turanose, or D(+)-xylose. Lysozyme-resistant. Resistant to 2–4 μg/ml novobiocin. Lysostaphin-sensitive. Peptidoglycan type L-Lys–L-Ala–Gly$_4$; serine sometimes present. Teichoic acids contain glycerol and glucosamine. Major menaquinone MK-6. Variable characteristics include production of staphylokinase and acid production from L(+)-arabinose, fucose, galactose, lactose, maltose, melibiose, rhamnose, salicin, sorbitol, and xylitol (weak).

Distinguished from *Staphylococcus sciuri* by production of acid from raffinose.

Isolated from udders of goats and sheep.

DNA–DNA hybridization studies yielded relative DNA binding values of 80–100% between the type strain and 14 other strains of *Staphylococcus lentus* and 35–41% between the between the type strain *Staphylococcus lentus* and four strains of *Staphylococcus sciuri* subsp. *sciuri*. The type strain of *Staphylococcus sciuri* subsp. *sciuri* was 36–43% similar to twelve strains of *Staphylococcus lentus* (Schleifer et al., 1983b).

DNA G+C content (mol%): 29.8–34.2 (T_m).

Type strain: Roguinsky K21, ATCC 29070, CCUG 15599, CIP 81.63, NCTC 12102, NRRL B-14758.

GenBank accession number (16S rRNA gene): D83370.

Additional remarks: Originally described as *Staphylococcus sciuri* subsp. *lentus*.

22. **Staphylococcus lugdunensis** Freney, Brun, Bes, Meugnier, Grimont, Grimont, Nervi and Fleurette 1988, 170[VP]

lug.dun.en'sis. L. adj. *lugdunensis* pertaining to Lugdunum, the Latin name of Lyon, a French city where the organism was first isolated.

Nonmotile, nonspore-forming, Gram-stain-positive cocci, 0.8–1.0 μm in diameter; occur singly and form pairs, clusters, and chains. Facultatively anaerobic. Colonies cream or pale yellow to golden, glistening, smooth, and entire. Colonial morphology somewhat variable. Colony diameter 1–4 mm after incubation for 72 h at 35°C on P agar. Temperature range for good growth 30–45°C; weak growth at 20°C. Tolerates 10% NaCl; delayed growth on 15% NaCl. Produces predominantly D-lactate anaerobically from glucose. Positive reactions: catalase, clumping factor, hemolysis (weak), nitrate reduction, ornithine decarboxylase, pyrrolidonyl aminopeptidase, *N*-acetylglucosaminidase, and acetoin production. Negative reactions: oxidase, coagulase, gelatinase, heat-stable nuclease, staphylokinase, alkaline phosphatase, β-glucuronidase, and β-galactosidase. Acid produced aerobically from D-fructose, D-glucose, α-methyl-D-glucoside, glycerol, lactose, maltose, D-mannose, sucrose, and trehalose. No acid produced aerobically from L-arabinose, cellobiose, D-mannitol, D-melibiose, D-raffinose, ribose, xylitol, or D-xylose. Lysostaphin-sensitive; lysozyme-resistant. Novobiocin-susceptible; bacitracin-resistant. Peptidoglycan type L-Lys–Gly$_{5–6}$. Teichoic acids contain glucosamine, glycerol, and glucose. Protein A not present. No production of TSST-1 toxin, exfoliative toxin, or enterotoxins A, B, or C. Variable characteristics include colony pigmentation, urease, and production of acid from turanose.

Isolated from clinical specimens from humans.

DNA–DNA hybridization studies yielded relative DNA binding values of 81–100% between the type and ten other strains of *Staphylococcus lugdunensis* and 1–8% between the type strain of *Staphylococcus lugdunensis* and type strains of 25 other species of *Staphylococcus* (Freney et al., 1988).

DNA G+C content (mol%): 32 (T_m).

Type strain: N860297, ATCC 43809, CCUG 25348, CIP 103642, DSM 4804, LMG 13346, NCTC 1221, NRRL B-14774.

GenBank accession number (16S rRNA gene): AB009941.

23. **Staphylococcus lutrae** Foster, Ross, Hutson and Collins 1997, 726[VP]

lu'trae. L. fem. gen. n. *lutrae* of an otter.

Nonmotile, nonspore-forming, Gram-stain-positive cocci; occur singly and form pairs and clusters. Colonies convex, nonpigmented, opaque, glistening, and smooth, with entire margins. Colony diameter 3.5–4.5 mm after incubation for 3 d at 37°C on P agar. Facultatively anaerobic. Temperature range for growth 25–42°C. Positive reactions: growth in 10% NaCl, alkaline phosphatase, β-galactosidase, hemolysis (sheep blood), catalase, coagulase, DNase (weak), nitrate reduction, and urease. Negative reactions: growth in 15% NaCl, oxidase, clumping factor, hyaluronidase, arginine dihydrolase, and acetoin production. Acid produced aerobically from galactose, glucose, lactose, maltose, mannose, trehalose, xylose, adonitol, dulcitol, inositol, and sorbitol. Novobiocin-susceptible. Lysostaphin-sensitive. Acid production from mannitol varies between strains.

Isolated from dead otters.

16S rDNA sequence analysis showed that *Staphylococcus lutrae* was most closely related to *Staphylococcus felis, Staphylococcus schleiferi, Staphylococcus intermedius,*

Staphylococcus delphini, and *Staphylococcus muscae* (Foster et al., 1997).

DNA G+C content (mol%): 34 (no method given).

Type strain: M340/94/1, ATCC 700373, CCUG 38494, CIP 105399, DSM 10244.

GenBank accession number (16S rRNA gene): AB233333.

24. **Staphylococcus muscae** Hájek, Ludwig, Schleifer, Springer, Zitzelberger, Kroppenstedt and Kocur 1992, 99[VP]

mus'cae. L. gen. n. *muscae* of a fly.

Nonmotile, nonspore-forming, Gram-stain-positive cocci (sometimes ovoid), 0.4–1.1µm in diameter; occur singly and form clumps. Colonies gray-white, opaque, convex, low, with raised centres, smooth, somewhat glistening, and butyrous; margins entire. Colony diameter 5–6 mm after incubation for 5 d at 37°C on P agar. Facultatively anaerobic; better growth under aerobic conditions. No growth at 10 or 45°C. Positive reactions: growth on 10% NaCl, catalase, alkaline phosphatase, heat-labile nuclease, hemolysis on ovine blood agar, lecithinase, nitrate reduction, and degradation of Tweens 20, 40, and 80. Negative reactions: oxidase, coagulase, gelatinase, arginine dihydrolase, urease, β-galactosidase, α- or β-hemolysins, fibrinolysin, protease, protein A, clumping factor, acetoin production, degradation of starch and esculin, and growth on 15% NaCl. Acid produced aerobically from fructose, glucose, glycerol, sucrose, trehalose, turanose, and xylose. No acid produced aerobically from arabinose, cellobiose, fucose, galactose, lactose, maltose, mannose, melezitose, melibiose, raffinose, rhamnose, ribose, sorbose, tagatose, dulcitol, inositol, mannitol, arbutin, inulin, or salicin. Lysozyme-resistant. Bacitracin-resistant; novobiocin-susceptible. Peptidoglycan type Lys–Gly$_{5-6}$ or Lys–Gly$_4$, Ser. Major fatty acids C$_{15:0 \text{ iso}}$ and C$_{16:0}$.

Isolated from flies associated with cattle but not thought to be a permanent part of the insects' microflora.

DNA–DNA hybridization studies yielded relative DNA binding values of 85–100% among four *Staphylococcus muscae* strains. 16S rDNA sequence analysis showed that the closest relative to *Staphylococcus muscae* was *Staphylococcus schleiferi* (Hajek et al., 1992).

DNA G+C content (mol%): 40–41 (T_m).

Type strain: MB4, ATCC 49910, CCUG 36972, CCM 4175, CIP 103641, DSM 7068.

GenBank accession number (16S rRNA gene): S83566.

25. **Staphylococcus nepalensis** Spergser, Wieser, Täubel, Rosselló-Mora, Rosengarten and Busse 2003, 2009[VP]

ne.pa.len'sis. N.L. masc. adj. *nepalensis* pertaining to the kingdom of Nepal, where clinical samples for bacteriological examination were collected from local goats.

Nonmotile cocci; occur singly and form pairs and clusters. Colonies white, opaque, convex, low, glistening, and smooth. Colony diameter 2–6 mm after incubation for 2 d at 37°C on P agar. Facultatively anaerobic. No growth in 15% NaCl; growth variable at lower concentrations. Temperature range for growth 20–40°C; optimum 30°C. Positive reactions: catalase, urease, alkaline phosphatase, esculin and Tween 80 hydrolysis, pyrrolidonyl arylamidase, β-galactosidase, β-glucuronidase, and nitrate reduction. Negative reactions: acetoin production, arginine dihydrolase, oxidase,

clumping factor, coagulase, heat-labile and heat-stable nucleases, indole, H$_2$S, and hyaluronidase. Acid produced aerobically from *N*-acetylglucosamine, L-arabinose, arbutin, erythritol, D-fructose, galactose, D-glucose, glycerol, lactose, maltose, mannose, mannitol, salicin, sucrose, trehalose, and D-xylose. No acid production from adonitol, amygdalin, D-arabinose, L-arabitol, D-cellobiose, dulcitol, D-fucose, L-fucose, gentiobiose, gluconate, glycogen, inositol, inulin, D-lyxose, 2-ketogluconate, 5-ketogluconate, melibiose, melezitose, methyl-α-D-glucoside, methyl-α-D-mannoside, methyl-β-D-xyloside, D-raffinose, rhamnose, ribose, L-sorbose, starch, D-tagatose, or L-xylose. Variable characteristics include aerobic production of acid from D-arabitol, sorbitol, turanose, and xylitol. Major quinone MK-7, minor quinones MK-6 and MK-8. Major fatty acids C$_{15:0 \text{ ante}}$, C$_{15:0 \text{ iso}}$, and C$_{17:0 \text{ ante}}$. Major polar lipids diphosphatidylglycerol, phosphatidylglycerol, and an unidentified lipid. Novobiocin- and bacitracin-resistant. Lysozyme-resistant; lysostaphin-sensitive.

Isolated from goats with respiratory disease.

DNA–DNA hybridization studies yielded relative DNA binding values of 42–68% between the type strain of *Staphylococcus nepalensis* and eleven other *Staphylococcus* species and subspecies. 16S rDNA sequence analysis yielded similarity values of 98.1–99% between *Staphylococcus nepalensis* and seven other *Staphylococcus* species and subspecies (Spergser et al., 2003).

DNA G+C content (mol%): 33 (HPLC).

Type strain: CW1, CCM 7045, DSM 15150.

GenBank accession number (16S rRNA gene): AJ517414.

26. **Staphylococcus pasteuri** Chesneau, Morvan, Grimont, Labischinski and El Solh 1993, 241[VP]

pas'teur.i. L. gen. n. *pasteuri* of Pasteur, honoring the French microbiologist Louis Pasteur for his contribution in 1878 to the recognition of staphylococci as pathogenic agents and also referring to the research institute, Institute Pasteur, Paris, France, where the new species was characterized.

Nonmotile, nonspore-forming, Gram-stain-positive cocci, 0.5–1.5 µm in diameter; occur singly and form pairs, tetrads, and clusters. Colonies usually yellow, raised, glistening, and smooth. Colony diameter 4–7 mm. Facultatively anaerobic. Temperature range for growth 15–45°C. Grows in 5–15% NaCl. Produces L- and D-lactate from glucose anaerobically. Positive reactions: catalase, β-glucosidase, β-glucuronidase, urease, and hemolysis. Negative reactions: oxidase, coagulase, clumping factor, β-galactosidase, fibrinolysin, DNase, ornithine decarboxylase, esculin hydrolysis, and alkaline phosphatase. Acid produced aerobically from β-D-fructose, D-glucose, sucrose, and trehalose. No acid produced aerobically from L-arabinose, D-cellobiose, D-mannose, α-D-melibiose, D-raffinose, D-ribose, D-xylose, or xylitol. Variable characteristics include acetoin production, nitrate reduction, and aerobic production of acid from maltose and mannitol. Lysostaphin-resistant. Novobiocin-susceptible; bacitracin-resistant. Peptidoglycan type L-Lys–Gly$_4$, Ser. Teichoic acid contains glycerol but not ribitol. Produces β-toxin.

Isolated from humans, animals, and food.

DNA–DNA hybridization studies yielded relative DNA binding values of 93–98% between the type strain and three other *Staphylococcus pasteuri* strains and 2–13% between the type strain of *Staphylococcus pasteuri* and the type strains of 27 other *Staphylococcus* species (Chesneau et al., 1993).

DNA G+C content (mol%): 35 (T_m).

Type strain: BM9357, ATCC 51129, CCUG 32420, CIP 103540, DSM 10656.

GenBank accession number (16S rRNA gene): AB009944, AF041361.

27. **Staphylococcus piscifermentans** Tanasupawat, Hashimoto, Ezaki, Kozaki and Komagata 1992, 578[VP]

pis.ci.fer.men'tans. L. n. *piscis* fish; L. part. adj. *fermentans* fermenting; N.L. part. adj. *piscifermentans* fish fermenting.

Nonmotile, nonspore-forming, Gram-stain-positive cocci, 1.0 μm in diameter; occur singly and form pairs, tetrads, and clusters. Colonies low, convex, white and opaque with a yellow-orange tint and entire margins. Colony diameters not reported. Facultatively anaerobic. Produces D- and L-lactate anaerobically from glucose. Temperature range for growth 18–42°C. Positive reactions: arginine hydrolysis, benzidine test, catalase, β-glucosidase, growth in 10% NaCl, nitrate reduction, and urease. Negative reactions: oxidase, coagulase, hemolysis, acetoin production, alkaline phosphatase, gelatinase, citrate utilization, β-glucuronidase, and β-galactosidase. Acid produced aerobically from D-fructose, D-glucose, and D-trehalose. No acid produced aerobically from D-arabinose, D-cellobiose, L-fucose, gluconate, D-mannose, raffinose, D-turanose, D-xylose, gluconate, or xylitol. Resistant to lysozyme; sensitive to lysostaphin;-susceptible to novobiocin. Major menaquinone MK-7; minor menaquinones MK-6 and MK-8. Peptidoglycan type L-Lys–Gly$_{4-5}$. Major fatty acids C$_{15:0\ ante}$, C$_{18:0}$, and C$_{20:0}$. Variable characteristics: growth in 15% NaCl, esculin hydrolysis, Tween 80 hydrolysis, and production of acid from D-galactose, glycerol, lactose, maltose, D-mannitol, D-melezitose, D-ribose, salicin, sucrose, and D-sorbitol.

Isolated from fermented shrimp and fish.

DNA–DNA hybridization studies yielded relative DNA binding values of 65–100% among strains of *Staphylococcus piscifermentans* and 2–32% relative DNA binding between *Staphylococcus piscifermentans* strain Ph79L and 19 other species and subspecies of *Staphylococcus* (Tanasupawat et al., 1992).

DNA G+C content (mol%): 36–37 (T_m).

Type strain: SK03, ATCC 51136, CCUG 35133, CIP 103958, DSM 7373, JCM 6057, NRIC 1817, TISTR 824.

GenBank accession number (16S rRNA gene): AB009943, Y15754.

28. **Staphylococcus pulvereri** Zakrzewska-Czerwinska, Gaszewska-Mastalarz, Lis, Gamian and Mordarski 1995, 171[VP]

pul'ver.er.i. N.L. gen. n. *pulvereri* of Pulver, in honor of the German microbiologist Gerhard Pulverer for contributions to the study of staphylococcal infections in clinical microbiology.

Nonmotile, nonspore-forming, Gram-stain-positive cocci. Colonies nonpigmented or yellowish, smooth, with entire margins. Colony diameter 4–8 mm. Facultatively anaerobic; anaerobic growth weak; no anaerobic growth in thioglycolate medium. Grows on medium containing 15% NaCl and at 10°C. No growth at 45°C. Positive reactions: catalase. Negative

reactions: clumping factor and staphylocoagulase. Acid produced aerobically from D-fructose, D-glucose, maltose, and sucrose. No acid produced aerobically from D-arabinose, D-cellobiose, raffinose, D-ribose, or D-turanose. Resistant to lysozyme and novobiocin; susceptible to lysostaphin. Major fatty acids include both iso and anteiso forms of C$_{15}$, C$_{17}$ and C$_{19}$ fatty acids. Peptidoglycan type L-Lys–Ala–Gly$_{4-5}$. Teichoic acids contain polyglycerol phosphate, ribose, and *N*-acetylglucosamine. Variable characteristics: acetoin production, alkaline phosphatase, arginine dihydrolase, β-galactosidase, β-glucuronidase, urease, and acid production from lactose, D-mannitol, D-mannose, *N*-acetylglucosamine, and D-trehalose.

Isolated from humans and animals.

DNA–DNA hybridization studies yielded relative DNA binding values of 85–100% between the type strain and four other *Staphylococcus pulvereri* strains and 3–36% between the *Staphylococcus pulvereri* type strain and 30 other species of *Staphylococcus* (Vernozy-Rozand et al., 2000). The most closely related species was *Staphylococcus vitulinus*. On the basis of DNA–DNA hybridization studies, pulse field electrophoresis, ribotyping and other taxonomic criteria, *Staphylococcus pulvereri* is identical with *Staphylococcus vitulinus* (Kloos et al., 2001).

DNA G+C content (mol%): 27–30 (T_m).

Type strain: B92/87, NT215, ATCC 51698, CCM 4481, CCUG 33938, CIP 104364, DSM 9930, PCM 2443.

GenBank accession number (16S rRNA gene): AB009942.

Additional remarks: Staphylococcus pulvereri is considered a junior heterotypic synonym of *Staphylococcus vitulinus* (Kloos et al., 2001; Petráš, 1998).

29. **Staphylococcus saccharolyticus** (Foubert and Douglas 1948) Kilpper-Bälz and Schleifer 1984, 91[VP] (Effective publication: Kilpper-Bälz and Schleifer 1981, 329.) ("*Micrococcus saccharolyticus*" Foubert and Douglas 1948, 31; *Peptococcus saccharolyticus* (Foubert and Douglas 1948) Douglas 1957, 478.)

sac.cha.ro.ly'ti.cus. Gr. n. *sacchar* sugar; Gr. adj. *lyticus* able to loose; M.L. adj. *saccharolyticus* sugar-digesting.

Nonmotile, nonspore-forming, Gram-stain-positive cocci, 0.6–1.0 μm in diameter; occur singly and form pairs, tetrads, and clusters. Colonies nonpigmented (faintly yellow on agar containing hemin), slightly convex, glistening, and opaque, with entire margins. Colony diameter 0.5–2.0 mm. Anaerobic, no growth or poor growth under aerobic conditions. Optimum temperature range for growth 30–37°C. Produces CO$_2$, ethanol, acetate, small amounts of formate, and lactate anaerobically from glucose. FBP-aldolase is of the class I type. Positive reactions: arginine dihydrolase, benzidine test, catalase (weak unless medium contains hemin), and nitrate reduction. Negative reactions: amylase, coagulase, esculin hydrolysis, gelatinase, hemolysis, H$_2$S, indole, and oxidase. Acid produced from fructose, glucose, glycerol, and mannose. No acid produced from adonitol, amygdalin, arabinose, cellobiose, galactose, inulin, lactose, maltose, mannitol, melezitose, raffinose, ribose, rhamnose, salicin, sorbitol, sucrose, trehalose, xylitol, or xylose. Lysozyme-resistant; lysostaphin-sensitive. Novobiocin-susceptible. Peptidoglycan type L-Lys–Gly$_4$, Ser$_{0.7-1.2}$. Teichoic acid contains glycerol and *N*-acetylglucosamine. Acetoin production is variable.

Isolated from human skin; occasionally found in clinical specimens.

DNA–DNA hybridization studies yielded relative DNA binding values of 93–100% between the type strain and nine other strains of *Staphylococcus saccharolyticus* (*Peptococcus saccharolyticus*) (Crosa et al., 1979), 22–24% between two *Staphylococcus saccharolyticus* strains and the type strain of *Staphylococcus aureus*, and 10–39% between two strains of *Staphylococcus saccharolyticus* and *Staphylococcus* species (Kilpper et al., 1980).

DNA G+C content (mol%): 33–34 (T_m).

Type strain: ATCC 14953, CCUG 9989, CCUG 24040, CIP 103275, DSM 20359, NCTC 11807, NRRL B-14778.

GenBank accession number (16S rRNA gene): L37602.

30. **Staphylococcus saprophyticus** (Fairbrother 1940) Shaw, Stitt and Cowan 1951, 1021[AL] emend. Schleifer and Kloos 1975, 53 (Fairbrother 1940, 87.)

sa.pro.phy'tic.us. Gr. adj. *sapros* putrid; Gr. n. *phyton* plant; N.L. adj. *saprophyticus* saprophytic, growing on dead tissues.

Nonmotile, nonspore-forming, Gram-stain-positive cocci, 0.6–1.2 µm in diameter; occur singly and form pairs. Colonies nonpigmented or slightly yellow, raised, smooth, glistening, and opaque, with entire margins. Facultatively anaerobic. Grows best under aerobic conditions. Many strains only ferment glucose weakly and would be misclassified as micrococci on the basis of the O/F-test. Produces mainly L-lactate (small amounts) anaerobically from glucose. FBP-aldolase is of the class I type. Grows in medium containing 15% NaCl. Temperature range for growth 10–40°C; optimum 28–35°C. Positive reactions: benzidine test, catalase, and urease. Negative reactions: coagulase, clumping factor, and oxidase. Acid produced aerobically from fructose, glucose, glycerol, maltose, and trehalose. No acid produced from adonitol, arabitol, cellobiose, dulcitol, erythritol, erythrose, fucose, gentiobiose, inositol, lyxose, melibiose, melezitose, raffinose, rhamnose, sorbitol, sorbose, tagatose, or xylose. Lysozyme-resistant; lysostaphin-sensitive. Novobiocin-resistant. Peptidoglycan type L-Lys–Gly$_{4-5}$, Ser$_{0.6-0.8}$. Teichoic acid contains ribitol, small amounts of glycerol, and *N*-acetylglucosamine. Major fatty acids C_{Br-15}, C_{16}, C_{18}, and C_{20}. Major menaquinone MK-7. Variable characteristics: arginine dihydrolase (weak), β-galactosidase, growth at 45°C, caseinase, gelatinase, and production of acid from fructose, lactose, trehalose, turanose, and xylitol.

Isolated from skin of humans and animals. Associated with human urinary tract infections.

Additional characteristics of the subspecies are given below.

30a. **Staphylococcus saprophyticus subsp. saprophyticus** (Fairbrother 1940) Shaw, Stitt and Cowan 1951, 1021[AL] emend. Schleifer and Kloos 1975, 53 (Fairbrother 1940, 87.)

Nonmotile, nonspore-forming, Gram-stain-positive cocci, 0.6–1.2 µm in diameter; occur singly and form pairs. Colonies nonpigmented or slightly yellow, raised, smooth, glistening, and opaque, with entire margins. Colony diameter 4.0–9.0 mm. Facultatively anaerobic. Produces mainly L-lactate (small amounts) anaerobically from glucose.

Grows in medium containing 10% NaCl. Temperature range for growth 10–40°C; optimum 28–35°C. Lactose is metabolized via the Leloir pathway (Schleifer et al., 1978). Positive reactions: acetoin production, benzidine test, catalase, and urease. Negative reactions: coagulase, DNase, esculin hydrolysis, hemolysin, nitrate reduction, ornithine decarboxylase, and phosphatase. Acid produced aerobically from fructose, glucose, glycerol, maltose, sucrose, trehalose, and turanose. No acid produced from adonitol, arabinose, arabitol, cellobiose, dulcitol, erythritol, erythrose, fucose, galactose, gentiobiose, inositol, lyxose, mannose, melibiose, melezitose, raffinose, rhamnose, ribose, salicin, sorbitol, sorbose, tagatose, or xylose. Lysozyme-resistant; lysostaphin-sensitive. Novobiocin-resistant. Peptidoglycan type L-Lys–Gly$_{4-5}$, Ser$_{0.6-0.8}$. Teichoic acid contains ribitol, small amounts of glycerol, and *N*-acetylglucosamine. Major fatty acids C_{Br-15}, C_{16}, C_{18}, and C_{20}. Major menaquinone MK-7. Variable characteristics: arginine dihydrolase (weak), β-glucosidase, β-galactosidase, growth at 45°C, growth in media containing 15% NaCl, caseinase, gelatinase, and production of acid from lactose, mannitol, and xylitol.

Isolated from skin of humans and animals. Associated with human urinary tract infections. *Staphylococcus saprophyticus* and *Staphylococcus epidermidis* are the predominant coagulase-negative staphylococcal species isolated from human urinary infections.

DNA–DNA hybridization studies yielded relative DNA binding values of >85% among strains of *Staphylococcus saprophyticus* (Kloos and Wolfshohl, 1982).

DNA G+C content (mol%): 31–36 (T_m).

Type strain: S-41, ATCC 15305, CCUG 3706, BCRC 10786, CIP 76.125, DSM 20229, IAM 12452, JCM 2427, LMG 13350, NCAIM B.01067, NCCB 73011, NCTC 7292, NRRL B-14751.

GenBank accession number (16S rRNA gene): D83371, L37596.

30b. **Staphylococcus saprophyticus subsp. bovis** Hájek, Meugnier, Bes, Brun, Fiedler, Chmela, Lasne, Fleurette and Freney 1996, 793[VP]

bo'vis. L. n. *bos* a cow; L. gen. n. *bovis* of a cow.

Nonmotile, nonspore-forming, Gram-stain-positive cocci, 0.7–1.4 µm in diameter; occur singly and form pairs, chains, and clumps. Colonies low, convex, glistening, butyrous, with entire margins; presence and color of pigmentation varies between strains. Colony diameter 4 mm after incubation for 3 d at 35°C on P agar. Facultatively anaerobic. Grows weakly at 10°C; does not grow at 45°C. Grows in medium containing 10% but not 15% NaCl. Positive reactions: benzidine test, catalase, hippurate hydrolysis, nitrate reduction, and urease. Negative reactions: acetoin production, arginine dihydrolase, clumping factor, coagulase, elastase, fibrinolysin, heat-labile and heat-stable nucleases, hyaluronidase, hemolysin, ornithine decarboxylase, esculin hydrolysis, starch hydrolysis, utilization of citrate, and production of H_2S. Acid produced aerobically from *N*-acetylglucosamine, fructose, glucose, maltose, ribose, trehalose, glycerol, and mannitol. No acid produced aerobically from cellobiose, raffinose, rhamnose, xylose, adonitol, or dulcitol. Sensitive to lysostaphin; resistant to lysozyme. Resistant to novobiocin and bacitracin. Peptidoglycan type Lys–Gly$_{3.4-4.0}$, Ser$_{0.2-0.7}$. Major fatty acids C$_{15:0\ ante}$ and C$_{17:0\ ante}$. Teichoic acids of most strains contain ribitol and N-acetylglucosamine. Variable characteristics: colony pigmentation; gelatinase; caseinase; β-galactosidase; β-glucuronidase; lecithinase; hydrolysis of Tween 20, 40, and 80; alkaline phosphatase; ability to produce acetoin; and production of acid from arabinose, galactose, lactose, mannose, sucrose, and turanose.

Distinguished from *Staphylococcus saprophyticus* subsp. *saprophyticus* by smaller colony size, ability to reduce nitrate, and fermentation of ribose.

Isolated from the nostrils of healthy cattle.

DNA–DNA hybridization studies yielded relative DNA binding values of 94–99% between the type and four other *Staphylococcus saprophyticus* subsp. *bovis* strains, 71% between the type strain of *Staphylococcus saprophyticus* subsp. *bovis* and the type strain of *Staphylococcus saprophyticus* subsp. *saprophyticus*, and 2–32% between the type strain of *Staphylococcus saprophyticus* subsp. *bovis* and 19 other *Staphylococcus* species (Hajek et al., 1996).

DNA G+C content (mol%): 31 (HPLC).

Type strain: KV 12, CCUG 36975, CCM 4410, CCUG 38042, CIP 105260.

GenBank accession number (16S rRNA gene): AB233327.

31. **Staphylococcus schleiferi** Freney, Brun, Bes, Meugnier, Grimont, Grimont, Nervi and Fleurette 1988, 171[VP]

schlei′fer.i. N.L. gen. n. *schleiferi* of Schleifer, in honor of Karl-Heinz Schleifer, German microbiologist, for his many contributions to the taxonomy of Gram-positive organisms.

Nonmotile, nonspore-forming, Gram-stain-positive cocci, 0.8–1.0 μm in diameter; occur singly and form pairs, clusters, and chains. Colonies nonpigmented, low convex, smooth, glistening, with entire margins. Facultatively anaerobic. Tolerates 10% NaCl. Temperature range for growth 20–45°C. Hemolytic. Produces mostly L-lactic acid in thioglycolate medium. Positive reactions: arginine dihydrolase, alkaline phosphatase, heat-stable nuclease, catalase, nitrate reduction, and acetoin production. Negative reactions: oxidase and staphylokinase. Acid produced aerobically from D-glucose, glycerol, and D-mannose. No acid produced from L-arabinose, D-cellobiose, α-methyl-D-glucoside, lactose, maltose, D-mannitol, D-melibiose, D-raffinose, xylitol, or D-xylose. Novobiocin-susceptible; bacitracin-resistant. Lysostaphin-sensitive; lysozyme-resistant. Peptidoglycan type L-Lys–Gly$_{5-6}$–Ser$_{0.2-0.3}$. Teichoic acids contain glucosamine, glucose, and glycerol. No production of protein A, TSST-1 toxin, exfoliative toxin, or enterotoxins A, B, or C. Variable characteristics include gelatinase and acid production from trehalose and D-fructose.

Isolated from human clinical specimens.

DNA–DNA hybridization studies yielded relative DNA binding values of 73–100% between the type and eleven other strains of *Staphylococcus schleiferi* and 2–9% between the type strain of *Staphylococcus schleiferi* and the type strains of 25 other species of *Staphylococcus* (Freney et al., 1988).

Additional characteristics of the subspecies are given below.

31a. **Staphylococcus schleiferi subsp. schleiferi** Freney, Brun, Bes, Meugnier, Grimont, Grimont, Nervi and Fleurette 1988, 171[VP]

Nonmotile, nonspore-forming, Gram-stain-positive cocci, 0.8–1.0 μm in diameter; occur singly and form pairs, clusters, and chains. Colonies nonpigmented, low convex, smooth, glistening, with entire margins. Colony diameter 3–5 mm after 4 d on P agar. Facultatively anaerobic. Tolerates 10% NaCl; growth on 15% NaCl delayed. Temperature range for growth 20–45°C. Weakly hemolytic. Produces mostly L-lactic acid in thioglycolate medium. Positive reactions: arginine dihydrolase, alkaline phosphatase, heat-stable nuclease, catalase, clumping factor, nitrate reduction, and acetoin production. Negative reactions: oxidase, coagulase, staphylokinase, ornithine decarboxylase, and urease. Acid produced aerobically from D-glucose, glycerol, and D-mannose. No acid produced from L-arabinose, D-cellobiose, α-methyl-D-glucoside, lactose, maltose, D-mannitol, D-melibiose, D-raffinose, ribose, sucrose, turanose, xylitol, or D-xylose. Novobiocin-susceptible; bacitracin-resistant. Lysostaphin-sensitive; lysozyme-resistant. Peptidoglycan type L-Lys–Gly$_{5-6}$–Ser$_{0.2-0.3}$. Teichoic acids contain glucosamine, glucose, and glycerol. No Protein A. No production of TSST-1 toxin, exfoliative toxin, or enterotoxins A, B, or C. Variable characteristics include gelatinase and acid production from trehalose and D-fructose.

Isolated from human clinical specimens.

DNA–DNA hybridization studies yielded relative DNA binding values of 73–100% between the type strain and eleven other strains of *Staphylococcus schleiferi* and 2–9% between the type strain of *Staphylococcus schleiferi* and the type strains of 25 other species of *Staphylococcus* (Freney et al., 1988).

DNA G+C content (mol%): 37 (T$_m$).

Type strain: N850274, ATCC 43808, CCUG 25351, CIP 103643, DSM 4807, LMG 13347, NCTC 12218, NRRL B-14775.

GenBank accession number (16S rRNA gene): S83568.

31b. **Staphylococcus schleiferi subsp. coagulans** Igimi, Takahashi and Mitsuoka 1990, 410[VP]

co.a′gu.lans. L. adj. *coagulans* curdling, coagulating.

Nonmotile, nonspore-forming, Gram-stain-positive cocci, 0.8–1.2 μm in diameter; occur singly and form pairs and clusters. Colonies nonpigmented, low convex, opaque, and glistening. Colony diameter 1.5–2.0 mm after 24 h on blood agar at 37°C. Facultatively anaerobic. Positive reactions: acetoin production, arginine hydrolysis, coagulase, heat-stable nuclease, hemolysis, nitrate reduction, phosphatase, and urease. Produces β-hemolysin. Negative reactions: clumping factor, hyaluronidase, and oxidase. Acid produced aerobically from fructose, galactose, glucose, glycerol, mannose, and ribose. No acid produced from adonitol, arabinose, cellobiose, dulcitol, erythritol, glycogen, inositol, maltose, melibiose, melezitose, raffinose, rhamnose, sorbitol, trehalose, xylitol, or xylose. Acid production from lactose, mannitol, and sucrose varies between strains.

Distinguished from *Staphylococcus schleiferi* subsp. *schleiferi* by production of coagulase, urease, lack of clumping factor, and the production of acid from D-ribose.

Isolated from external ear infections of dogs.

DNA–DNA hybridization studies yielded relative DNA binding values of 67–99% among *Staphylococcus schleiferi* subsp. *coagulans* strains, 55–73% between *Staphylococcus schleiferi* subsp. *coagulans* strains and the type strain of *Staphylococcus schleiferi* subsp. *schleiferi*, and 5–27% between *Staphylococcus schleiferi* subsp. *coagulans* strains and *Staphylococcus aureus* subsp. *aureus*, *Staphylococcus aureus* subsp. *anaerobius*, *Staphylococcus intermedius*, and *Staphylococcus hyicus* (Igimi et al., 1990).

DNA G+C content (mol%): 35–37 (T_m).

Type strain: GA211, ATCC 49545, CCUG 37248, CIP 104370, DSM 6628, JCM 7470.

GenBank accession number (16S rRNA gene): AB009945.

32. **Staphylococcus sciuri** Kloos, Schleifer and Smith 1976a, 23[AL] emend. Kloos, Ballard, Webster, Hubner, Tomasz, Couto, Sloan, Dehart, Fiedler, Schubert, de Lencastre, Santos Sanchez, Heath, Leblanc and Ljungh 1997, 320.

sci′ur.i. L. masc. n. *Sciurus* a squirrel and also the genus name of a squirrel on whose skin this species is commonly found in large populations; L. gen. n. *sciuri* of the squirrel.

Nonmotile, nonspore-forming, Gram-stain-positive cocci, 0.7–1.2 μm in diameter; occur singly and form pairs and tetrads. Most isolates facultatively anaerobic; some do not grow in anaerobic zone of thioglycolate medium. Grows much better under aerobic conditions. Many strains only ferment glucose weakly and would be misclassified as micrococci on the basis of the O/F-test. Possesses cytochromes aa_3 and c and therefore oxidase-positive. No growth or poor growth at 45°C. Positive reactions: benzidine test, catalase, β-glucosidase, nitrate reduction, and oxidase. Negative reactions: arginine dihydrolase, coagulase, elastase, β-galactosidase, β-glucuronidase, lipase, urease, and heat-stable nuclease. Acid produced aerobically from β-D(−)-fructose, D(+)-glucose, glycerol, D-mannitol, D(−)-ribose, and sucrose. No acid produced from *N*-acetylglucosamine, D-arabitol, dulcitol, *meso*-erythritol, D-erythrose, L(−)-fucose, D-lyxose, L-sorbose, tagatose, or xylitol. Lysozyme-resistant. Novobiocin-resistant. Peptidoglycan type L-Lys–L-Ala–Gly$_4$. Teichoic acids usually contain glycerol and glucosamine, sometimes ribitol or glucose. Major menaquinone MK-6.

Isolated from a variety of animals and animal products. Also found in soil, sand, and water.

Additional characteristics of the subspecies are given below.

32a. **Staphylococcus sciuri subsp. sciuri** Kloos, Schleifer and Smith 1976a, 23[AL] emend. Kloos, Ballard, Webster, Hubner, Tomasz, Couto, Sloan, Dehart, Fiedler, Schubert, de Lencastre, Santos Sanchez, Heath, Leblanc and Ljungh 1997, 320.

Nonmotile, nonspore-forming, Gram-stain-positive cocci, 0.7–1.2 μm in diameter; occur singly and form pairs and tetrads. Most isolates facultatively anaerobic; anaerobic growth poor or absent. Colonies large, raised, smooth, glistening, and opaque. Colony edges often undulating; center elevated in aged colonies; pigmentation gray-white to cream or yellowish. Colony diameter 7 ± 2 mm on P agar and 10 ± 2 mm on tryptic soy agar after incubation at

34–35°C for 3 d, then 25°C for 2 d. Produces small amount of L-lactate from glucose anaerobically. FBP-aldolase is of class I type. Tolerates 10% NaCl; growth poor in 15% NaCl. Optimal temperature for growth 35–40°C on tryptic soy agar and 35°C on P agar. Positive reactions: catalase, benzidine test, caseinase, gelatinase, hemolysis (weak), nitrate reduction, oxidase, and DNase. Negative reactions: acetoin production (or very weak), clumping factor, coagulase, elastase, and lipase. Protein A not produced. Acid produced aerobically from D(+)-cellobiose (sometimes weak), β-D(−)-fructose, D(+)-fucose, D(+)-galactose, β-gentiobiose, D(+)-glucose, glycerol, D-mannitol, D(−)-ribose, D-sorbitol, sucrose, and D(+)-trehalose (sometimes weak). No acid produced from adonitol, D-arabitol, dulcitol, *meso*-erythritol, D-erythrose, L(−)-fucose, *meso*-inositol, D-lyxose, D-melibiose, raffinose, L-sorbose, tagatose, D-turanose, or xylitol. Lysozyme-resistant. Novobiocin-resistant. Peptidoglycan type L-Lys–L-Ala–Gly$_4$. Major menaquinone MK-6. Variable characteristics include anaerobic growth, alkaline phosphatase, and production of acid from L-arabinose, α-lactose, maltose, D-mannose, D-melezitose, α-L-rhamnose, salicin, D-sorbitol, and D-xylose.

All strains of *Staphylococcus sciuri* subsp. *sciuri*, 58% of *Staphylococcus sciuri* subsp. *carnaticus* strains, and 67% of *Staphylococcus sciuri* subsp. *rodentium* strains produce acid aerobically from cellobiose. All strains of *Staphylococcus sciuri* subsp. *sciuri*, 29% of *Staphylococcus sciuri* subsp. *carnaticus* strains, and 67% of *Staphylococcus sciuri* subsp. *rodentium* strains possess alkaline phosphatase. No strains of *Staphylococcus sciuri* subsp. *sciuri* or *Staphylococcus sciuri* subsp. *carnaticus* strains, but approximately 95% of *Staphylococcus sciuri* subsp. *rodentium* strains were positive when tested for Staph-A-Lex agglutination; no strains of *Staphylococcus sciuri* subsp. *sciuri*, 36% of *Staphylococcus sciuri* subsp. *carnaticus* strains, and 94% of *Staphylococcus sciuri* subsp. *rodentium* strains were positive when tested for Staph Latex agglutination (Kloos et al., 1997).

Isolated from squirrels, mice, a prairie vole, opossums, a red kangaroo, horses, bottle-nosed dolphins, a pilot whale, a dog, a squirrel monkey, cattle and meat from cattle, and marsh grass.

DNA–DNA hybridization studies yielded relative DNA binding values of 81–100% among *Staphylococcus sciuri* subsp. *sciuri* strains, 61–73% between *Staphylococcus sciuri* subsp. *sciuri* strains and *Staphylococcus sciuri* subsp. *carnaticus* strains, and 49–68% between *Staphylococcus sciuri* subsp. *sciuri* strains and *Staphylococcus sciuri* subsp. *rodentium* strains (Kloos et al., 1997).

DNA G+C content (mol%): 35–36 (T_m).

Type strain: SC 116, DD 4277, ATCC 29062, CCUG 15598, CIP 81.62, DSM 20345, JCM 2425, NCTC 12103, NRRL B-14777.

GenBank accession number (16S rRNA gene): AJ421446, S83569.

32b. **Staphylococcus sciuri subsp. carnaticus** Kloos, Ballard, Webster, Hubner, Tomasz, Couto, Sloan, Dehart, Fiedler, Schubert, de Lencastre, Santos Sanchez, Heath, Leblanc and Ljungh 1997, 320[VP]

car.na′ti.cus. N.L. adj. *carnaticus* pertaining to meat.

Nonmotile, nonspore-forming, Gram-stain-positive cocci, 0.7–1.2 μm in diameter; occur singly and form pairs and tetrads. Colonies of type strain smooth and glistening with undulate, raised margins; translucent to opaque; may have nonpigmented and cream-colored pigmented sectors. Colony diameter 5 ± 2 mm on P agar and 9 ± 1 mm on tryptic soy agar after incubation at 34–35°C for 3 d, then 25°C for 2 d. Facultatively anaerobic; anaerobic growth delayed and weak. Temperature range for growth 15–40°C; optimum 35–40°C on tryptic soy agar and 35°C on P agar; no growth or poor growth at 45°C. Tolerate 10% NaCl; growth poor in 15% NaCl. Positive reactions: benzidine test, caseinase, catalase, esculin hydrolysis, gelatinase, β-glucosidase, hemolysis (weak), nitrate reduction, DNase, and oxidase. Negative reactions: acetoin production, arginine dihydrolase, coagulase, heat-stable nuclease, elastase, β-galactosidase, β-glucuronidase, lipase, urease. Acid produced aerobically from β-D(−)-fructose D-fucose, galactose, D(+)-glucose, maltose (delayed), D-mannitol, D(−)-ribose, sucrose, D-trehalose, and D-xylose,. No acid production from D-arabitol, dulcitol, meso-erythritol, D-erythrose, L(−)-fucose, D-lyxose, D-melezitose, α-L-rhamnose, L-sorbose, tagatose or xylitol. Lysozyme-resistant. Novobiocin-resistant. Peptidoglycan type L-Lys–L-Ala–Gly$_4$. Teichoic acids usually contain glycerol and glucosamine, sometimes ribitol or glucose. Major menaquinone MK-6. Variable characteristics: alkaline phosphatase, clumping factor, and production of acid from L-arabinose, cellobiose, α-lactose, mannose, salicin, and sorbitol.

All Staphylococcus sciuri subsp. carnaticus strains, 12% of Staphylococcus sciuri subsp. sciuri strains, and 17% of Staphylococcus sciuri subsp. rodentium strains produce acid aerobically from D-xylose. All Staphylococcus sciuri subsp. carnaticus strains, 50% of Staphylococcus sciuri subsp. sciuri strains, and 89% of Staphylococcus sciuri subsp. rodentium strains produce acid aerobically from maltose at least weakly. All Staphylococcus sciuri subsp. carnaticus strains, 38% of Staphylococcus sciuri subsp. sciuri strains, and 56% of Staphylococcus sciuri subsp. rodentium strains are unable to produce acid aerobically from melezitose. Seven percentage of Staphylococcus sciuri subsp. carnaticus strains, 65% of Staphylococcus sciuri subsp. sciuri strains, and 33% of Staphylococcus sciuri subsp. rodentium strains produce colonies with diameters ≥7 mm on P agar (Kloos et al., 1997). No strains of Staphylococcus sciuri subsp. sciuri or Staphylococcus sciuri subsp. carnaticus strains, but approximately 95% of Staphylococcus sciuri subsp. rodentium strains were positive when tested for Staph-A-Lex agglutination; no strains of Staphylococcus sciuri subsp. sciuri, 36% of Staphylococcus sciuri subsp. carnaticus strains, and 94% of Staphylococcus sciuri subsp. rodentium strains were positive when tested for Staph Latex agglutination (Kloos et al., 1997).

Isolated from a cavy, a bottle-nosed dolphin, cattle, and meat from cattle.

DNA–DNA hybridization studies yielded relative DNA binding values of 81–98% among Staphylococcus sciuri subsp. carnaticus strains, 61–73% between Staphylococcus sciuri subsp. carnaticus strains and Staphylococcus sciuri subsp. sciuri strains, and 53–65% between Staphylococcus

sciuri subsp. carnaticus strains and Staphylococcus sciuri subsp. rodentium strains (Kloos et al., 1997).

DNA G+C content (mol%): 36 (T_m).

Type strain: DD 791, ATCC 700058, CCUG 39509, CIP 105826.

GenBank accession number (16S rRNA gene): AB233331.

32c. **Staphylococcus sciuri subsp. rodentium** Kloos, Ballard, Webster, Hubner, Tomasz, Couto, Sloan, Dehart, Fiedler, Schubert, de Lencastre, Santos Sanchez, Heath, Leblanc and Ljungh 1997, 321[VP]

ro.den′ti.um. L. pl. gen. n. *rodentium* of rodents.

Nonmotile, nonspore-forming, Gram-stain-positive cocci, 0.7–1.2 μm in diameter; occur singly and form pairs and tetrads. Colonies of type strain smooth and glistening with undulate, raised margins; translucent, yellow-gray. Colony diameter 6 ± 2 mm on P agar and 10 ± 1 mm on tryptic soy agar after incubation at 34–35°C for 3 d, then 25°C for 2 d. Most isolates facultatively anaerobic; some do not grow in anaerobic zone of thioglycolate medium; growth of others weak and delayed. Growth temperature optimum 35–40°C on tryptic soy agar and 35°C on P agar; no growth or poor growth at 45°C. Positive reactions: benzidine test, caseinase, catalase, clumping factor, esculin hydrolysis, gelatinase, β-glucosidase, nitrate reduction, nuclease, and oxidase. Negative reactions: acetoin production, arginine dihydrolase, coagulase, elastase, β-galactosidase, β-glucuronidase, lipase, ornithine decarboxylase, urease, and heat-stable nuclease. Acid produced aerobically from β-D(−)-fructose, fucose, galactose, D(+)-glucose, glycerol, maltose, D-mannose, D-mannitol, D(−)-ribose, sucrose, and D-trehalose. No acid produced from N-acetylglucosamine, D-arabitol, dulcitol, meso-erythritol, D-erythrose, α-lactose, D-lyxose, L-sorbose, tagatose, or xylitol. Lysozyme-resistant. Novobiocin-resistant. Peptidoglycan type L-Lys–L-Ala–Gly$_4$. Teichoic acids usually contain glycerol and glucosamine, sometimes ribitol or glucose. Major menaquinone MK-6. Variable characteristics include alkaline phosphatase, hemolysis, staphylolysin, and production of acid from L-arabinose, D-cellobiose, D-melezitose, α-L-rhamnose, salicin, D-sorbitol, and D-xylose.

No strains of Staphylococcus sciuri subsp. sciuri or Staphylococcus sciuri subsp. carnaticus strains, but approximately 95% of Staphylococcus sciuri subsp. rodentium strains were positive when tested for Staph-A-Lex agglutination; no strains of Staphylococcus sciuri subsp. sciuri, 36% of Staphylococcus sciuri subsp. carnaticus strains, and 94% of Staphylococcus sciuri subsp. rodentium strains were positive when tested for Staph Latex agglutination (Kloos et al., 1997). Staphylococcus sciuri subsp. rodentium strains were also more likely to exhibit at least intermediate levels of resistance to cefazolin, methicillin, and oxacillin (Kloos et al., 1997).

Isolated from squirrels, rats, a pilot whale, and humans.

DNA–DNA hybridization studies yielded relative DNA binding values of 80–88% between the type strain and three other strains of Staphylococcus sciuri subsp. rodentium strains, 49–68% between the type strain of Staphylococcus sciuri subsp. rodentium strain and eight Staphylococcus sciuri subsp. sciuri strains, and 53–65% between the type strain of *Staphylococcus sciuri* subsp. *rodentium* strains and seven *Staphylococcus sciuri* subsp. *carnaticus* strains (Kloos et al., 1997).

DNA G+C content (mol%): 35 (T_m).

Type strain: DD 4761, R1–33, ATCC 700061, CCUG 37923, CIP 105829.

GenBank accession number (16S rRNA gene): AB233332.

33. **Staphylococcus simulans** Kloos and Schleifer 1975a, 69[AL]

sim'u.lans. L. part. adj. *simulans* imitating; named for having similarities to certain coagulase-positive staphylococci, including *Staphylococcus aureus*.

Nonmotile, nonspore-forming, Gram-stain-positive cocci, 0.8–1.5 μm in diameter; form pairs and tetrads; occasionally occur as single cells. Colonies nonpigmented (gray-white), slightly raised, smooth, glistening, translucent, with entire margins. Colony diameter 5–7.5 mm. Facultatively anaerobic. Grows better under aerobic conditions. Strains may produce either L-lactate alone or both D- and L-lactate anaerobically from glucose. Grows in medium containing 10% NaCl; growth poor or absent at 15% NaCl. FBP-aldolase of class I type. Temperature range for growth 15–45°C; optimum 25–40°C. Positive reactions: arginine dihydrolase, benzidine test, catalase (may be weak), DNase, β-galactosidase, heat-stable nuclease (some strains), lipase, nitrate reduction, and urease. Negative reactions: caseinase, coagulase, clumping factor, gelatinase, β-glucosidase, and oxidase. Acid produced aerobically from fructose, glucose, glycerol, lactose, mannitol, and sucrose. No acid production from adonitol, arabinose, arabitol, cellobiose, dulcitol, erythritol, erythrose, fucose, gentiobiose, inositol, lyxose, melezitose, melibiose, raffinose, rhamnose, salicin, sorbitol, sorbose, tagatose, turanose, xylitol, or xylose. Lysozyme-resistant; lysostaphin-sensitive. Novobiocin-susceptible. Peptidoglycan type L-Lys–L-Gly$_{5-6}$. The lysostaphin-producing strain (NRRL B-22628) designated *Staphylococcus simulans* biovar *staphylolyticus* (Sloan et al., 1982) has a peptidoglycan of the type L-Lys–Gly$_{2.3}$, L-Ser$_{1.3}$. The high serine content in the interpeptide bridge of the peptidoglycan contributes to the lysostaphin resistance of this strain. Teichoic acid contains glycerol, *N*-acetylgalactosamine, and *N*-acetylglucosamine. Major fatty acids C$_{Br-15}$, C$_{18}$, and C$_{20}$. Major menaquinone MK-7. Variable characteristics include acetoin production (weak), phosphatase (weak), β-glucuronidase, growth in medium containing 15% NaCl, hemolysis (may be weak), and acid production from galactose, maltose, mannose, ribose, and trehalose.

Staphylococcus simulans is found occasionally on the skin of humans and other primates; may be associated with infection.

DNA–DNA hybridization studies yielded relative DNA-binding values >80% among *Staphylococcus simulans* strains (Kloos and Wolfshohl, 1982). The *Staphylococcus simulans* biovar *staphylolyticus* demonstrated 100% DNA–DNA similarity with the type strain of *Staphylococcus simulans* (Sloan et al., 1982). *Staphylococcus simulans* is more closely related to *Staphylococcus carnosus* than to other species.

DNA G+C content (mol%): 34–38 (T_m).

Type strain: MK 148, ATCC 27848, CCUG 7327, CIP 81.64, DSM 20322, HAMBI 2058, JCM 2424, NCTC 11046, NRRL B-14753.

GenBank accession number (16S rRNA gene): D83373.

34. **Staphylococcus succinus** Lambert, Cox, Mitchell, Rosselló-Mora, Del Cueto, Dodge, Orkland and Cano 1998, 516[VP]

suc.cin'us. L. masc. adj. *succinum* of amber.

Nonmotile, nonspore-forming, Gram-stain-positive cocci, 0.6–1.9 μm in diameter. Occur in rosettes of 3–6 cells. Colonies may be either opaque, white, crenated, rough, and umbonate, or opaque, grayish white, glistening, smooth, entire, and raised. Temperature range for growth 25–40°C; optimum 28°C. No growth at 45°C. Aerobic. No growth in anaerobic zone of thioglycolate medium. Can grow in 15% NaCl. Positive reactions: β-galactosidase, β-glucuronidase, and urease. Acid produced aerobically from fructose, lactose, maltose, mannitol, salicin, sucrose, and trehalose. Negative reactions: acetoin production, oxidase, reduction of nitrate to nitrite, and hydrolysis of arginine. Novobiocin- and bacitracin-resistant. Additional characteristics of the subspecies are given below.

34a. **Staphylococcus succinus** subsp. **succinus** Lambert, Cox, Mitchell, Rosselló-Mora, Del Cueto, Dodge, Orkland and Cano 1998, 516[VP]

Gram-stain-positive cocci, 0.6–1.9 μm in diameter. Occur in rosettes of 3–6 cells. Colonies may be either opaque, white, crenated, rough, and umbonate or opaque, grayish white, glistening, smooth, entire, and raised. Colony diameter 4–6 mm after incubation for 2 d at 25°C on tryptic soy agar. Temperature range for growth 25–40°C; optimum 28°C. No growth in anaerobic zone of thioglycolate medium. Grows in presence of 40% bile and at pH 4.9. Acid produced aerobically from fructose, galactose (weak), lactose, maltose (weak), mannitol, salicin, sucrose, and trehalose. Does not ferment arabinose, fucose, mannose, melizitose, raffinose, turanose, xylitol, or xylose. Produces catalase, β-D-glucuronidase, β-D-galactopyranosidase, phosphatase, and urease. Hydrolyzes indoxyl phosphate. Does not produce oxidase, reduce nitrate to nitrite, or hydrolyze arginine. No acetoin production. Novobiocin- and bacitracin-resistant. DAP-type peptidoglycan (has to be confirmed). Major fatty acids are iso and anteiso forms of C$_{13:0}$, C$_{15:0}$, and C$_{17:0}$; also contains C$_{20:0}$ and tuberculostearic acid.

Found in Dominican amber.

In DNA–DNA hybridization studies, the DNA of the type strain of *Staphylococcus succinus* gave a relative binding value of 95% to the DNA of a second strain of *Staphylococcus succinus*, 6% to *Staphylococcus saprophyticus*, 23% to *Staphylococcus xylosus*, and 38% to *Staphylococcus equorum*.

DNA G+C content (mol%): 34.4–35 (HPLC).

Type strain: AMG-D1, ATCC 700337, CCUG 43571, DSM 14617.

GenBank accession number (16S rRNA gene): AF004220.

34b. **Staphylococcus succinus** subsp. **casei** Place, Hiestand, Burri and Teuber 2003b, 1[VP] (Effective publication: Place, Hiestand, Burri and Teuber 2002, 359.)

ca'se.i. L. gen. n. *casei* of cheese, named because the organism was isolated from cheese.

Nonmotile cocci, 1.1–2.0 μm in diameter that form pairs and clumps as well as occurring singly. Does not form rosettes. Nonmotile. Colonies opaque, white, crenated, rough, and umbonate. Colony diameter 4–6 mm after incubation for 2 d at 32°C on PC skim milk agar. Temperature range for growth 6–41°C; optimum 32°C. Does not grow anaerobically. Tolerant to 15% NaCl. No growth at pH 4.9. Positive reactions: nitrate reduction to nitrite, urease, β-galactosidase, β-glucuronidase, and esculin hydrolysis. Negative reactions: acetoin production, arginine dihydrolase, ornithine decarboxylase, oxidase, and hemolysis. Weak production of alkaline phosphatase. Acid produced aerobically from fructose, D-galactose, α-D-lactose, maltose, mannitol, D-mannose, D-melezitose, D-ribose, salicin, sucrose, and D-trehalose. Weak acid production from turanose and xylose. No acid produced from L-arabinose, cellobiose, L-fucose, raffinose, or xylitol. Ferments adenosine, 2′ deoxyadenosine, 2,3-butanediol, gentiobiose, inosine, α-ketovaleric acid, D-melibiose, D-sorbitol, and thymidine-5′monophosphate. Does not ferment D-alanine. Novobiocin-resistant. Peptidoglycan type L-Lys–Gly$_4$, Ser. Major fatty acids are iso and anteiso forms of C$_{15:0}$ and C$_{17:0}$. Also contains tuberculostearic acid.

Distinguished from *Staphylococcus succinus* subsp. *succinus* by ability to reduce nitrate, production of acid aerobically from D-mannose and D-melizitose, and inability to grow at pH 4.9; fermentation patterns of the two subspecies also differ, although some reactions are weak (Place et al., 2002).

Found in surface ripened cheese.

DNA–DNA hybridization studies yielded a relative DNA binding value of 72.3% between the type strain of *Staphylococcus succinus* subsp. *casei* and the type strain of *Staphylococcus succinus* subsp. *succinus* and 42–51.2% between the type strain of *Staphylococcus succinus* subsp. *casei* and the type strains of *Staphylococcus equorum*, *Staphylococcus xylosus*, and *Staphylococcus gallinarum*. 16S rDNA sequence analysis showed that the *Staphylococcus* species most closely related to *Staphylococcus succinus* subsp. *succinus* and *Staphylococcus succcinus* subsp. *casei* were *Staphylococcus saprophyticus*, *Staphylococcus xylosus*, *Staphylococcus equorum*, *Staphylococcus arlettae*, and *Staphylococcus gallinarum* (Place et al., 2002).

DNA G+C content (mol%): 34.4 (HPLC).

Type strain: SB72, DSM 15096, CIP 107658.

GenBank accession number (16S rRNA gene): AJ320272.

35. **Staphylococcus vitulinus** corrig. Webster, Bannerman, Hubner, Ballard, Cole, Bruce, Fiedler, Schubert and Kloos 1994, 458^{VP}

vit′u.lin.us. L. masc. adj. *vitulinus* pertaining to veal.

Nonmotile, nonspore-forming, Gram-stain-positive cocci; occur singly and form pairs, tetrads, and clusters. Colonies usually pigmented, opaque, raised, with irregular edges; may be sectored. Colony diameter <3 mm on P agar and 8–12 mm on tryptic soy agar after incubation at 35°C. No growth at 45°C or anaerobically in thioglycolate medium. Weak growth on sheep blood agar or tryptic soy agar in anaerobic jars. Positive reactions: catalase, oxidase, and nitrate reduction. Negative reactions: acetoin production, alkaline phosphatase, arginine utilization, β-galactosidase, β-glucuronidase, ornithine decarboxylase, staphylocoagulase,

thermonuclease, and urease. Acid produced aerobically from fructose, glycerol, mannitol, and sucrose. No acid produced from N-acetyl-D-glucosamine, L-arabinose, galactose, lactose, maltose, D-mannose, melezitose, raffinose, rhamnose, D-turanose, and xylitol. Peptidoglycan type L-Lys–L-Ala–Gly$_{3–3.4}$. Glycerol teichoic acids contain N-acetylglucosamine, N-acetylglucosamine 1-phosphate, or N-acetylgalactosamine. Novobiocin-resistant. Variable characteristics include β-glucosidase, esculin hydrolysis, clumping factor, and aerobic production of acid from D-cellobiose, D-ribose, salicin, D-trehalose, and D-xylose.

Isolated from various animals (horse, pine vole, pilot whale) and meats (veal, beef, chicken, lamb).

DNA–DNA hybridization studies yielded relative DNA binding values of 74–97% at 55°C and 75–93% at 70°C among strains of *Staphylococcus vitulinus*. Relative DNA binding values of two strains of *Staphylococcus vitulinus* to the type strains of nine other *Staphylococcus* species were 13–53% at 55°C and 5–15% at 70°C; the most closely related species were *Staphylococcus sciuri* and *Staphylococcus lentus* (Webster et al., 1994).

DNA G+C content (mol%): 34 (T_m).

Type strain: DD 756, ATCC 51145, CCM 4511, CCUG 33636, CIP 104850.

GenBank accession number (16S rRNA gene): AB009946.

Additional remarks: Staphylococcus vitulinus is considered a senior heterotypic synonym of *Staphylococcus pulvereri* (Kloos et al., 2001; Petráš, 1998).

36. **Staphylococcus warneri** Kloos and Schleifer 1975a, 63^{AL}

war.ner′i. N.L. gen. n. *warneri* of Warner; named for Arthur Warner, from whom this organism was originally isolated.

Nonmotile, nonspore-forming, Gram-stain-positive cocci, 0.5–1.2 μm in diameter; occur singly and form pairs, occasionally tetrads. Colonies sticky, raised, smooth, glistening, and opaque, with entire margins. Pigmentation varies from nonpigmented to yellow-orange; colonies may have a yellow-orange rim or a yellowish center. Colony diameters 3–6 mm. Facultatively anaerobic. Growth is best under aerobic conditions. Produces both D- and L-lactate anaerobically from glucose. FBP-aldolase is of the class I type. Grows in medium containing 10% NaCl. Temperature range for growth 15–45°C (optimum 30–40°C). Positive reactions: acetoin production, benzidine test, catalase, β-glucosidase, lipase, and urease. Negative reactions: caseinase, coagulase, clumping factor, gelatinase, ornithine decarboxylase, oxidase, and phosphatase. Acid produced aerobically from fructose, glucose, glycerol, maltose (delayed), sucrose, and trehalose. No acid produced from adonitol, arabinose, arabitol, cellobiose, dulcitol, erythritol, erythrose, fucose, gentiobiose, inositol, lyxose, mannose, melibiose, raffinose, rhamnose, salicin, sorbitol, sorbose, tagatose, xylitol, or xylose. Novobiocin-susceptible. Lysozyme-resistant; slightly lysostaphin-resistant. Peptidoglycan type L-Lys–Gly$_{3.3–4.5}$, Ser$_{0.6–1.4}$. Teichoic acid contains glycerol, glucose, and N-acetylglucosamine. Major fatty acids C$_{Br-15}$, C$_{18}$, and C$_{20}$; C$_{22}$ also produced. Major menaquinone MK-7. Variable characteristics include arginine dihydrolase, β-glucuronidase, growth at 15°C, growth in medium containing 15% NaCl, heat-labile DNase (may be weak), hemolysis (may be weak), nitrate reduction (weak), and acid production from galactose, lactose, mannitol, melezitose, ribose, and turanose.

Isolated from skin of humans and nonhuman primates. One of the major *Staphylococcus* species found living on the skin and nasal membranes of various monkeys and prosimians. Sometimes associated with infections.

DNA–DNA hybridization studies yielded relative DNA binding values of >80% both among *Staphylococcus warneri* strains from humans and among *Staphylococcus warneri* strains from nonhuman primates, but only 50–67% between human and primate strains under optimal conditions (Kloos and Wolfshol, 1979).

DNA G+C content (mol%): 34–35 (T_m).

Type strain: AW25, ATCC 27836, CCUG 7325, CIP 81.65, DSM 20316, JCM 2415, LMG 13354, NCTC 11044, NRRL B-14736.

GenBank accession number (16S rRNA gene): L37603.

37. **Staphylococcus xylosus** Schleifer and Kloos 1975, 57[AL]

xy.lo′sus. N.L. adj. *xylosus* belonging to xylose.

Nonmotile, nonspore-forming, Gram-stain-positive cocci, 0.8–1.2 μm in diameter; occur singly and form pairs, occasionally tetrads. Colony characteristics variable: may be raised to slightly convex, rough or smooth, dull to glistening, usually opaque, with entire, undulate, or crenate margins. Pigmentation ranges from white to yellowish to yellow-orange. Colony diameter 5–10 mm. Facultatively anaerobic. Grows best under aerobic conditions. Some strains ferment glucose weakly and would be misclassified as microcci on the basis of the O/F-test. Produces

small amounts of ʟ-lactate anaerobically from glucose. FBP-aldolase is of the class type I. Galactose is metabolized via the Leloir pathway (Schleifer et al., 1978). Temperature range for growth 10–40°C; optimum 25–35°C. Grows in medium containing 10% NaCl. Positive reactions: benzidine test, catalase, β-galactosidase, β-glucosidase, and urease. Negative reactions: arginine dihydrolase, coagulase, clumping factor, and hemolysis. Acid produced aerobically from arabinose, fructose, glucose, glycerol, maltose, mannose, sucrose, trehalose, and xylose. No acid produced from adonitol, arabitol, cellobiose, dulcitol, erythritol, erythrose, fucose, lyxose, melezitose, melibiose, raffinose, sorbose, or tagatose. Lysostaphin-sensitive; lysozyme-resistant. Novobiocin-resistant. Peptidoglycan type ʟ-Lys–Gly$_{5-6}$. Teichoic acid contains glycerol, ribitol, and glucosamine. Major fatty acids C$_{Br-15}$, C$_{Br-17}$, C$_{18}$, and C$_{20}$. Major menaquinone MK-7. Variable characteristics include acetoin production, β-glucuronidase, DNase (weak), growth at 45°C, growth in medium containing 15% NaCl, lipase, alkaline phosphatase, nitrate reduction, and acid production from arabinose, galactose, gentiobiose, inositol, lactose, mannitol, rhamnose, ribose, salicin, sorbitol, turanose, and xylitol (weak).

Isolated from mammals, including humans and other primates, and the environment. Rarely associated with human or animal infections.

DNA–DNA hybridization studies yielded relative DNA binding values >70% among strains of *Staphylococcus xylosus*, except for one group of strains from New World monkeys. Relative DNA binding values between these strains and the other staphylococci examined were 38–39% (Kloos and Wolfshohl, 1982; Schleifer et al., 1979).

DNA G+C content (mol%): 30–36 (T_m).

Type strain: KL162, ATCC 29971, CCUG 7324, CIP 81.66, DSM 20266, HAMBI 2057, JCM 2418, LMG 20217, NCTC 11043, NRRL B-14776.

GenBank accession number (16S rRNA gene): D83374.

Addendum

Staphylococcus pseudintermedius Devriese, Vancanneyt; Baele, Vaneechoutte, de Grafe, Snauwaert, Cleenwerck, Dawyndt, Swings, Decostere and Haesebrouck 2005, 1571

pseu.do.in.ter.me′di.us. Gr. adj. *pseudes* or *pseudos* false; L. masc. adj. *intermedius* intermediate, and also a specific epithet; N.L. masc. adj. *pseudintermedius* a false (*Staphylococcus*) *intermedius*, because of the high phenotypic similarity to *Staphylococcus intermedius*.

Nonmotile, non-spore-forming Gram-stain-positive cocci predominantly in groups. Colonies on Columbia sheep blood agar unpigmented and surrounded by double zone hemolysis. Outer band incompletely hemolytic, turns into complete hemolysis after being put at 4°C (hot–cold hemolysis), typical of staphylococcal β-hemolysin (a sphingomyelinase). Catalase-positive, coagulate rabbit plasma, clumping-factor-negative. Produce strong DNase. Positive reactions for acetoin, β-glucosidase, arginine dihydrolase, urease, nitrate reduction, pyrrolidonyl arylamidase and ONPG (β-galactosidase). Negative reaction: β-glucuronidase. Susceptible to 8 μg acriflavine ml^{-1} and to novobiocin. Resistant to deferoxamine. Acid produced from glycerol (weakly and delayed), ribose, galactose, ᴅ-glucose, ᴅ-fructose, ᴅ-mannose, mannitol (weakly and delayed), *N*-acetyl-glucosamine, maltose, lactose, sucrose, trehalose and ᴅ-turanose (weakly and delayed). No acid produced from erythritol, ᴅ-arabinose, ʟ-arabinose, ᴅ-xylose, ʟ-xylose, adonitol, methyl β-ᴅ-xyloside, ʟ-sorbose, rhamnose, dulcitol, inositol, sorbitol, methyl α-ᴅ-glucoside, methyl α-ᴅ-xyloside, amygdalin, arbutin, esculin, salicin, cellobiose, melibiose, inulin, melezitose, ᴅ-raffinose, starch, glycogen, xylitol, ᴅ-lyxose, ᴅ-tagatose, ᴅ-fucose, ʟ-fucose, ʟ-arabitol, gluconate, 2-ketogluconate or 5-ketogluconate.

Isolated from lesions in different animal host species, but the habitat and pathogenic activity are unknown.

DNA G+C content (mol%): 38 (HPLC).

Type strain: LMG 22219[T] (=ON 86[T]=CCUG 49543[T]).

GenBank accession number (16S rRNA gene): AJ780976.

Staphylococcus simiae sp. nov. Pantucek,. Sedlacek, Petras, Koukalova, Svec, Stetina, Vancanneyt, Chrasttinova, Vokurkova, Ruzickova, Doskar, Swings,and Hajek 2005, 1957

si′mi.ae. L. gen. fem. n. *simiae* of/from a monkey.

Nonmotile, non-spore-forming Gram-stain-positive cocci; 0.7–0.8 μm in diameter; singly in pairs and in irregular clusters. Colonies 1–1.5 mm in diameter on glucose/yeast extract/peptone agar after 24 h at 36.5°C; circular, smooth, flat with low-convex centres, glistening, white with continuous margins. Strains grow anaerobically in thioglycolate medium. Grow at 15 and 45°C and in 12 % NaCl. All strains susceptible to lysostaphin, novobiocin and nitrofurantoin and resistant to bacitracin and polymyxin B. Catalase-positive, oxidase-negative, coagulase-negative and clumping-factor-negative. Produce alkaline phosphatase, acid phosphatase, urease, leucine arylamidase, esterase and esterase-lipase and reduce nitrate to nitrite. Weak δ-hemolysis activity. No production of acetoin, heat-stable or heat-labile nuclease, arginine arylamidase, ornithine decarboxylase, valine arylamidase, cystine arylamidase, pyrrolidonyl arylamidase, lipase, *N*-acetyl-β-glucosaminidase, α-fucosidase, α-galactosidase, α-glucosidase, α-mannosidase or β-glucuronidase. β-galactosidase weakly positive when ONPG is used as a substrate and variable when 2-naphthyl β-D-galactopyranoside is used; the type strain is β-galactosidase-negative. Similarly, cells are β-glucosidase-positive when 2-nitrophenyl β-D-glucopyranoside is used as a substrate and negative when 6-bromo-2-naphthyl β-D-glucopyranoside is used. Acid produced aerobically from D-fructose, D-glucose, galactose, lactose, D-maltose, D-mannitol, sucrose, D-trehalose and D-turanose. Delayed positive test result (2 d) obtained for acidification of melezitose and *N*-acetylglucosamine. No aerobic production of acid from D-cellobiose, D-raffinose, D-mannose, D-melibiose, D-ribose, xylitol, methyl α-D-glu-coside and for anaerobic fermentation of D-mannitol. Variable reactions obtained for acid production from L-arabinose (3/8 positive, type strain negative), sorbitol (3/8 positive, type strain negative) and D-xylose (3/8 positive, type strain negative), for production of arginine dihydrolase (5/8 positive, type strain positive) and lecithinase (4/8 positive, type strain weakly positive) and for hydrolysis of casein (6/8 positive, type strain weakly positive), gelatin (6/8 positive, type strain positive) and Tween 80 (5/8 positive, type strain positive). No hydrolysis of elastin, esculin, starch and tyrosine.

Isolated from the feces of a squirrel monkey.

DNA G+C content (mol%): 33.8 (HPLC).

Type strain: CCM 7213[T] (=LMG 22723[T]).

GenBank accession number (16S rRNA gene): AY727530.

Genus II. **Jeotigalicoccus** Yoon, Lee, Weiss, Kang and Park 2003, 600[VP]

THE EDITORIAL BOARD

Je.ot.ga.li.coc′cus. N.L. n. *jeotgalum* (Korean n. *jeotgal*) jeotgal, traditional Korean seafood; Gr. masc. n. *kokkos* a grain or berry; N.L. masc. n. *Jeotgalicoccus* coccus from jeotgal.

Nonsporeforming cocci (0.6–1.1 μm). **Gram-positive. Nonmotile. Facultatively anaerobic.** Optimal pH for growth is 7.0–8.0 and no growth occurs at 5.5. **Catalase- and oxidase-positive.** Urease negative. Nitrate is not reduced to nitrite. Cell-wall peptidoglycan contains L-lysine at position 3 of the peptide subunit. **Peptidoglycan type is A3α** based on L-Lys–Gly$_{3-4}$–L-Ala(Gly). **The principal menaquinone is MK-7. Major fatty acids are** C$_{15:0\ ante}$ **and** C$_{15:0\ iso}$.

DNA G+C content (mol%): 38–42.

Type species: **Jeotgalicoccus halotolerans** Yoon, Lee, Weiss, Kang and Park 2003, 600[VP].

Further descriptive information

Strains grow optimally in the presence of 2–5% (w/v) NaCl. On marine agar, colonies are smooth, glistening, low convex, circular to slightly irregular, and light yellow in color. They have the peptidoglycan type A3α, based on the structure L-Lys–Gly$_{3-4}$–L-Ala(Gly). Analysis of dihydrophenylated cell walls indicated the N-terminal glycine residue of the interpeptide bridge is mostly replaced by an alanine residue. MK-7 represents 80–86% of the menaquinone. The main cellular fatty acids are C$_{15:0\ ante}$ and C$_{15:0\ iso}$. The polar fatty acids include phosphatidylglycerol, diphosphatidylglycerol, and unidentified phospholipids. No glycolipids are detected.

Differential characteristics for the species of the genus are given in Table 75.

Differentiation from closely related taxa

Jeotgalicoccus is distinguished from *Salinicoccus* by the type of menaquinone (MK-7 vs. MK-6, respectively), type of cell-wall peptidoglycan (L-Lys–Gly$_{3-4}$–L-Ala(Gly) vs. L-Lys–Gly$_5$) and colony color (light yellow vs. reddish orange). It can also be distinguished from *Nesterenkonia* on the basis predominant menaquinones and cell-wall peptidoglycan (MK-8 and MK-9; L-Lys–Gly–Glu) as well as mol% G + C of the DNA (42 vs. 72), from *Staphylococcus* on the basis of predominant menaquinones (MK-6, 7, 8), from *Macrococcus* on the basis of peptidoglycan type (L-Lys–Gly$_{3-4}$, L-Ser), from *Marinococcus* on the basis of peptidoglycan type (*meso*-diaminopimelic acid at position 3), and from *Tetragenococcus* on the basis of peptidoglycan type (L-Lys–D-Asp [A4α]), fatty acid profile (abundant C$_{18:1}$ and C$_{16:0}$), and some physiological properties (no catalase or oxidase activity, and lactic acid production in some species).

Taxonomic comments

On the basis of its 16S rRNA gene, *Jeotgalicoccus* is assigned to the family *Staphylococcaceae*. Its closest relatives within that family are *Salinicoccus roseus* DSM 5351[T] and *Salinicoccus hispanicus* DSM 5352[T] (with 92.6–93.5% sequence similarity).

List of species of the genus *Jeotgalicoccus*

1. **Jeotgalicoccus halotolerans** Yoon, Lee, Weiss, Kang and Park 2003, 600[VP]

ha.lo.to′le.rans. Gr. n. *hals* salt; L. pres. part. *tolerans* tolerating, enduring; N.L. part. adj. *halotolerans* salt-tolerating.

Characteristics are the same as described for the genus except as follows. Growth occurs at 4 and 42°C, but not at 43°C, and at pH 5.5, and is optimal at 30–35°C and at pH 7.0–8.0. Growth occurs in the presence of 0–20%, but not 21%, NaCl and is optimal in 2–5% NaCl. Grows anaerobically on marine agar. Acid is produced from L-arabinose, D-mannose, D-ribose, and D-mannitol but not from D-cellobiose, D-fructose, D-galactose, D-glucose, lactose, maltose, D-melezitose, melibiose, D-raffinose, L-rhamnose, stachyose, sucrose, D-trehalose, D-xylose, adonitol, myoinositol, or D-sorbitol.

Source: traditional Korean fermented seafood.

DNA G + C content (mol%): 42 (HPLC).

Type strain: YKJ-101, JCM 11198, KCCM 41448.

GenBank accession number (16S rRNA gene): AY028925.

TABLE 75. Characteristics differentiating the species of the genus *Jeotgalicoccus*[a]

Characteristic	1. *J. halotolerans*	2. *J. pinnipedialis*	3. *J. psychrophilus*
Growth at:			
4°C	+	−	+
37°C	+	+	−
42°C	+	+	−
Growth in NaCl:			
0%	+	−	−
14%	+	−	+
20%	+	−	−
Acid production from:			
Arabinose	+	−	−
D-Mannitol	+	−	−

[a]Symbols: +, >85% positive; −, 0–15% positive.

2. **Jeotgalicoccus pinnipedialis** Hoyles, Collins, Foster, Falsen and Schumann 2004, 747[VP]

pin.ni.ped.i.a′lis. N.L. masc. adj. *pinnipedialis* pertaining to pinnipeds.

The description of this species is identical to that of the genus except as follows. Cells appear as grape-like bunches, in pairs, or in tetrads. They are non-acid-fast. Nonhemolytic, buff or fawn, round, convex colonies (2 mm in diameter) form on blood agar after aerobic incubation for 24 h. Colonies have a similar appearance on nutrient agar. Increasing the CO$_2$ concentration does not enhance growth. Grows at 25 and 42°C, but not at 4°C. Grows in 2% and 6% NaCl, but not in 0% and 14%. Using API systems, acid is not produced from arabinose, cellobiose, glucose, glycogen, fructose, lactose, mannose, mannitol, maltose, melibiose, α-methyl D-glucoside (methyl α-D-glucopyranoside), raffinose, ribose, sucrose, trehalose, turanose, xylitol, or D-xylose. Gelatin, but not esculin or hippurate, is hydrolyzed. Acid phosphatase, phosphoamidase, pyrazinaminidase, and pyrrolidonyl arylamidase activities are detectable, but ester lipase C8 activity is weakly detectable or absent. No activity for the following is present: alkaline phosphatase, arginine dihydrolase, arginine arylamidase, chymotrypsin, cystine arylamidase, esterase C4, α-fucosidase, α-galactosidase, β-galactosidase, α-glucosidase, β-glucosidase, β-glucuronidase, leucine arylamidase, lipase C14, α-mannosidase, ornithine decarboxylase,

N-acetylglucosaminidase, trypsin, valine arylamidase, or urease. Acetoin is not produced. An additional polar lipid (phosphatidylinositol) was detected in addition to those included in the genus description. Its habitat is unknown.

Source: mouth swab of a Southern elephant seal.

DNA G + C content (mol%): 38.6 (HPLC).

Type strain: A/G14/99/10, CCUG 42722, CIP 107946.

GenBank accession number (16S rRNA gene): AJ251530.

3. **Jeotgalicoccus psychrophilus** Yoon, Lee, Weiss, Kang and Park 2003, 601[VP]

psy.chro′phil.us. Gr. adj. *psychros* cold; Gr. adj. *philos* liking, loving; N.L. adj. *psychrophilus* cold-loving.

The description of this species is identical to that of the genus except as follows. Growth occurs at 4 and 34°C, but not above 35°C, and at pH 5.5, and is optimal at 20–25°C and at pH 7.0–8.0. Growth occurs in the presence of 14%, but not at 0% or 15%, NaCl and is optimal in 2–5% NaCl. Acid is not produced from L-arabinose, D-cellobiose, D-fructose, D-galactose, D-glucose, lactose, maltose, D-mannose, D-melezitose, melibiose, D-raffinose, L-rhamnose, D-ribose, stachyose, D-trehalose, D-xylose, adonitol, D-mannitol, myo-inositol, or D-sorbitol.

Source: a traditional Korean fermented seafood.

DNA G + C content (mol%): 42 (HPLC).

Type strain: YKJ-115, JCM 11199, KCCM 41449.

GenBank accession number (16S rRNA gene): AY028926.

Genus III. **Macrococcus** Kloos, Ballard, George, Webster, Hubner, Ludwig, Schleifer, Fiedler and Schubert 1998a, 871[VP]

KARL-HEINZ SCHLEIFER

Ma.cro.coc′cus. Gr. adj. *macrus* large; Gr. masc. n. *kokkos* a grain or berry; N.L. masc. n. *Macrococcus* a large coccus.

Cells are **spherical** or **coccoid**, 0.74–2.5 μm in diameter, occurring mostly in pairs and tetrads or in irregular clusters. Nonmotile. Do not form endospores. Gram-stain-positive. Metabolism is mainly respiratory; growth is chemo-organotrophic and only marginally facultatively anaerobic. **Catalase-** and **oxidase-positive**. Resistant to bacitracin and lysozyme and susceptible to furazolidone. Do not produce acid from D-cellobiose.

Habitat: skin of animals, milk of cattle.

DNA G+C content (mol%): 38–45.

Type species: **Macrococcus equipercicus** Kloos, Ballard, George, Webster, Hubner, Ludwig, Schleifer, Fiedler and Schubert 1998a, 873[VP].

Further descriptive information

Macrococci are similar to oxidase-positive, novobiocin-resistant staphylococci (*Staphylococcus sciuri*, *Staphylococcus lentus*, and *Staphylococcus vitulinus*) but differ in their somewhat higher G+C content of DNA and generally larger cells. Strains of *Macrococcus lamae*, however, are novobiocin-sensitive. Most macrococci do not grow in the anaerobic portion of a thioglycolate semisolid medium. They are negative for coagulase, ornithine decarboxylase, β-glucuronidase, and β-galactosidase. Do not utilize arginine. No acid production from D-cellobiose, D-melezitose, D-xylose, L-arabinose, D-turanose, D-sorbitol, xylitol, salicin, and D-raffinose. Macrococci can contain *a-*, *b-*, and/or *c*-type cytochromes. The genome size is in the range of 1500–1800 kb (Kloos et al., 1998a).

Enrichment and isolation procedures

Isolation of macrococci does not require a special medium. Good growth occurs on nutrient or brain heart infusion agar at 30–35°C.

Taxonomic comments

The genus *Macrococcus* has been described on the basis of comparative 16S rRNA analysis, DNA–DNA hybridization studies, ribotype patterns, cell-wall composition, and phenotypic characteristics (Kloos et al., 1998a; Mannerova et al., 2003). Macrococci can be separated from staphylococci on the basis of a generally higher DNA G+C content (38–45 mol%), unique ribotype patterns, and usually larger cells. Members of the genus *Macrococcus* are oxidase-positive, whereas most staphylococci are oxidase-negative. The DNA G+C content of *Macrococcus caseolyticus* is 38–39 mol% (Schleifer et al., 1982) which is similar to that of several staphylococcal species.

The peptidoglycan type of the cell wall of four macrococcal species has been determined (Kloos et al., 1998a). *Macrococcus caseolyticus*, *Macrococcus equipercicus*, and *Macrococcus carouselicus* contain the Lys–Gly$_{3-4}$, L-Ser-type and *Macrococcus bovicus* contains the Lys–Gly$_3$, L-Ser-type. Small amounts of an atypical cell-wall teichoic acid (poly(*N*-acetylglucosaminylphosphate) or poly(*N*-acetylgalactosaminylphosphate)) were found in *Macrococcus caseolyticus* (Kloos et al., 1998a; Schleifer et al., 1982). A similar cell-wall teichoic acid occurs in *Staphylococcus auricularis* (Endl et al., 1983). In contrast to staphylococci, none of the other macrococcal species tested contain a cell-wall teichoic acid (Kloos et al., 1998a; Schleifer, 1986).

Based on comparative 16S rRNA gene sequence studies, the genus *Macrococcus* belongs to the phylum *Firmicutes*. They are closely related to staphylococci and members of the *Firmicutes* such as bacilli, enterococci, streptococci, lactobacilli, and listeria (Figure 72). It has been proposed to combine the genera *Staphylococcus*, *Gemella*, *Macrococcus*, and *Salinicoccus* within the family *Staphylococcaceae* (Garrity and Holt, 2001); in the present volume, *Gemella* has been classified in Family X. *Incertae Sedis*.

The genera *Staphylococcus* and *Macrococcus* are monophyletic and well separated from each other with intergeneric 16S rRNA sequence similarities of 93.4–95.3%. The intrageneric similarities are significantly higher with at least 96.5% for staphylococci and 97.7% for macrococci. A phylogenetic tree based on comparative sequence analysis of the 16S rRNA genes of the type strains of the seven macrococcal species is depicted in Figure 73.

The phylogenetic relationships of staphylococci and four of the seven macrococcal species were also inferred from comparative sequence analysis of the highly conserved gene encoding the heat-shock protein (Hsp60) (Kwok and Chow, 2003). The four macrococcal species (*Macrococcus bovicus*, *Macrococcus carouselicus*, *Macrococcus caseolyticus*, and *Macrococcus equipercicus*) are clearly separated from the staphylococcal species, but are closely related (especially *Macrococcus caseolyticus*) to the *Staphylococcus sciuri* species group, the only staphylococci

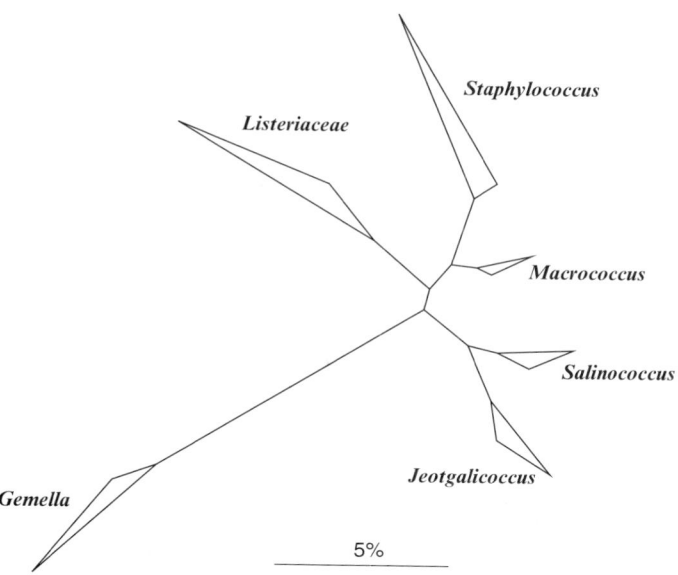

FIGURE 72. 16S rRNA tree reflecting the phylogenetic relationships of the genus *Macrococcus* and selected representatives of the phylum *Firmicutes*. The tree topology was reconstructed by applying the ARB-parsimony tool (Ludwig et al., 2004) upon an ARB 16SrRNA database (http://www.arb-home.de) comprising 28,980 almost complete sequences. The tree topology was corrected according to the data obtained by applying distance and maximum-likelihood methods. Only sequence positions which shared identical nucleotides in at least 50% of all compared sequences from representatives of the *Staphylococcaceae* were used to construct the tree. Length bar indicates 5% estimated sequence divergence.

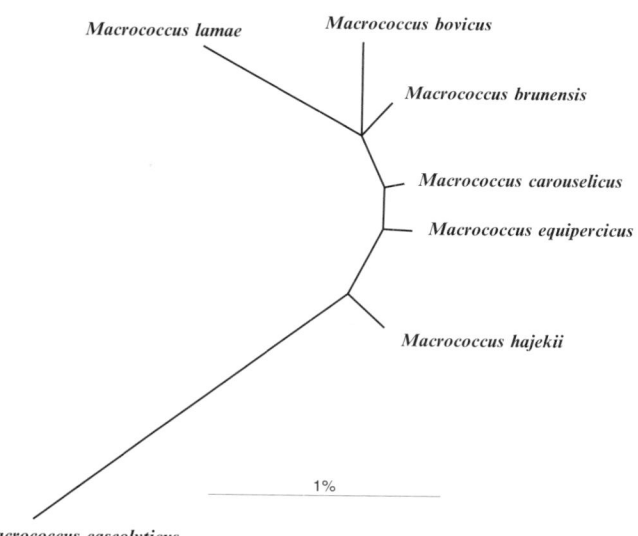

FIGURE 73. 16S rRNA based tree reflecting the relationships of the type strains of the *Macrococcus* species. The tree topology was reconstructed as described in the legend of Figure 72. The length bar indicates 1% estimated sequence divergence.

that are oxidase-positive. Pairwise sequence identity scores based on partial *hsp60* gene sequences among the four macrococcal species ranged from 82–87%, while those among the 40 staphylococcal species and subspecies ranged from 74–98% (Kwok and Chow, 2003). Thus, *hsp60* sequences appear to be more discriminatory than 16S rRNA gene sequences for differentiating macrococcal and staphylococcal species.

Based on DNA–DNA hybridization studies, the species *Macrococcus equipericus*, *Macrococcus bovicus*, and *Macrococcus carouselicus* are more closely related to one another than to *Macrococcus caseolyticus* (Kloos et al., 1998a). Both DNA–DNA hybridization studies and 16S rRNA sequence analysis indicate a closer relationship of the genus *Macrococcus* to the *Staphylococcus sciuri* species group than to other staphylococcal species.

In particular, *Macrococcus caseolyticus* is more closely related to staphylococci than the other macrococci. This agrees well with the significantly lower DNA G+C content (38–39 mol%) of *Macrococcus caseolyticus* that is shared by several of the staphylococcal species.

Acknowledgements

I thank Wolfgang Ludwig for reconstructing the 16S rDNA tree.

Differentiation of the species of the genus *Macrococcus*

Differential characteristics of the species of the genus *Macrococcus* are compiled in Table 76. Identification of macrococci at the species level on the basis of phenotypic tests alone is unreliable; molecular methods have to be applied.

List of species of the genus *Macrococcus*

1. **Macrococcus equipericus** Kloos, Ballard, George, Webster, Hubner, Ludwig, Schleifer, Fiedler and Schubert 1998a, 873[VP]

e.qui.per′ci.cus. L. gen. n. *equi* of horse; N.L. adj. *equipericus* pertaining to a horse named Percy, from which this species was first isolated.

Surface colonies about 6 mm in diameter. Convex, entire, butyrous, dull to slightly glistening, and opaque. Light to medium orange. No or only weak anaerobic growth. Optimum growth temperature is 35°C. No hemolysis. No production of acetylmethylcarbinol (acetoin). Acid produced from D-fructose and, with the exception of one strain out of 22 strains, from D-mannitol. No acid production from lactose and, with the exception of one strain, from D-ribose.

DNA G+C content (mol%): 45 (T_m).

Type strain: DD 9350, Kloos H8h3, ATCC 51831, CCUG 38363, CIP 105716.

EMBL accession number (16S rRNA gene): Y15712.

Additional remarks: Type strain was isolated from the skin of an Irish thoroughbred horse. This species has a preference for horses and ponies and is commonly found as large populations on the skin of these mammals.

2. **Macrococcus bovicus** Kloos, Ballard, George, Webster, Hubner, Ludwig, Schleifer, Fiedler and Schubert 1998a, 874[VP]

TABLE 76. Diagnostic and descriptive features of *Macrococcus* species

Characteristics	1. *M. equipericus*	2. *M. bovicus*	3. *M. brunensis*	4. *M. carouselicus*	5. *M. caseolyticus*	6. *M. hajekii*	7. *M. lamae*
Orange pigment	+	d	–	d	–	–	+
Resistance to novobiocin (5 μg)	+	+	+	+	+	+	–
Growth in 7.5% NaCl	+	+	–	+	+	–	–
Acid from:							
Glycerol	+	+	–	d	+	–	–
D-Mannitol	+	+	+	d	–	+	+
Maltose	d	d	+	–	+	+	+
Sucrose	–	d	–	d	d	+	+
D-Cellobiose	–	–	–	–	–	–	–
Esculin hydrolysis	d	d	–	+	d	–	–
Acetoin production	–	–	–	–	+	–	–
DNase	–	d	–	+	d	–	–
Alkaline phosphatase	–	–	+	–	–	+	+
Urease	d	d	–	–	–	–	–
Pyrrolidonyl arylamidase	–	–	–	–	d	–	–
Nitrate reduction	–	d	+	–	+	+	–
Mol% G+C of DNA	45	42–44	42	41	38–39	40	41

bov.ic′us. Gr. n. *bou* cow; L. gen. n. *bovis* of a cow; N.L. adj. *bovicus* pertaining to a bovine or cow, from which organism was first isolated.

Surface colonies about 4 mm in diameter. Slightly convex, entire, butyrous, glistening, and opaque. Pale yellow to medium orange. No anaerobic growth. Optimum growth temperature is 35°C. No cell-wall teichoic acid. Partial hemolysis (greening) of horse and bovine blood. No production of acetoin. Acid produced from glycerol, D-mannitol, and D-fructose. No acid production from lactose.

DNA G+C content (mol%): 42–44 (T_m).

Type strain: Kloos C2F4, DD 4516, ATCC 51825, CCUG 38365, CIP 105714.

EMBL accession number (16S rRNA gene): Y15714.

Additional remarks: Type strain was isolated from the skin of a Holstein cow. This species appears to have a preference for cattle, horses, and ponies.

3. **Macrococcus brunensis** Mannerová, Pantůek, Doška, Švec, Snauwaert, Vancanneyt, Swings and Sedláek 2003, 1652[VP]

bru.nen′sis. L. adj. *brunensis* from *Bruna*, the Roman name of the city of Brno, Czech Republic, where the type strain was isolated.

Surface colonies 2–4 mm in diameter. Circular, smooth, and glossy. No pigment produced. Facultatively anaerobic growth. Growth in 4% NaCl. No growth at 4 or 42°C. Hydrolysis of casein and gelatin, but not Tween 80, starch, lecithin, esculin, or tyrosine. No production of acetoin. No hemolysis. No activities of urease, hemolysis, arginine dihydrolase, arginine arylamidase, esterase, lipase, naphthol-AS-BI-phosphohydrolase, valine arylamidase, cystine arylamidase, α-galactosidase, *N*-acetyl-β-glucosaminidase, α-mannosidase, and α-fucosidase. Acid produced from maltose, D-mannitol, D-fructose, D-trehalose, and D-glucose. No acid production from sucrose, D-galactose, lactose, D-ribose, D-melibiose, or *N*-acetylglucosamine.

DNA G+C content (mol%): 41–42 (HPLC).

Type strain: CCM 4811, LMG 21712.

GenBank accession number (16S rRNA gene): AY119686.

Additional remarks: Type strain was isolated from llama skin.

4. **Macrococcus carouselicus** Kloos, Ballard, George, Webster, Hubner, Ludwig, Schleifer, Fiedler and Schubert 1998a, 874[VP]

car.ou.sel′i.cus. N.L. adj. *carouselicus* pertaining to carousel or merry-go-round, which has carousel horses.

Surface colonies 5–7 mm in diameter. Slightly convex, entire, butyrous, glistening, and opaque. Cream to light orange pigmentation. No anaerobic growth. Optimum growth at 35°C. No hemolysis and no cell-wall teichoic acid. No acetoin production. All strains are susceptible to lysostaphin and oxacillin. Weak acid production from D-fructose. No acid production from maltose, D-ribose, or lactose.

DNA G+C content (mol%): 41 (T_m).

Type strain: DD 9348, Kloos H8b16, ATCC 51828, CCUG 38360, CIP 105711.

EMBL accession number (16S rRNA gene): Y15713.

Additional remarks: Type strain was isolated from the skin of an Irish thoroughbred horse. This species has a prefer-

ence for horses and is commonly found as large populations on the skin of these mammals.

5. **Macrococcus caseolyticus** Kloos, Ballard, George, Webster, Hubner, Ludwig, Schleifer, Fiedler and Schubert 1998a, 871[VP]

ca.se.o.ly′ti.cus. L. n. *caseus* cheese; Gr. adj. *lutikos* able to loose; N.L. adj. *caseolyticus* casein-dissolving.

Surface colonies 3–7 mm. Slightly convex, entire, butyrous, glistening, and opaque. Unpigmented or pale yellow. No or weak anaerobic growth. Growth in 10% NaCl. Optimum growth at 35°C. Low amounts of atypical cell-wall teichoic acid: poly(*N*-acetylglucosaminylphosphate) or poly(*N*-acetylgalactosaminylphosphate).Most strains, except the type strain, produced partial hemolysis (greening) of horse blood. Produce acetoin. No urease and β-glucosidase activities. Acid production from maltose, D-fructose, D-trehalose, and D-glucose. No acid production from D-mannitol.

DNA G+C content (mol%): 38–39 (T_m).

Type strain: DD 4508, ATCC 13548, CCUG 15606, CIP 100755, DSM 20597.

EMBL accession number (16S rRNA gene): Y15711.

Additional remarks: Type strain was isolated from cow's milk, which was previously designated the type strain of *Staphylococcus caseolyticus* (Schleifer et al., 1982). This relatively uncommon species appears to have a preference for cattle, sheep, goats, and whales and may be found in their milk and meat products.

6. **Macrococcus hajekii** Mannerová, Pantůek, Doška, Švec, Snauwaert, Vancanneyt, Swings and Sedláek 2003, 1653[VP]

ha.je′ki.i. N.L. gen. n. *hajekii* of Hajek, named after Vaclav Hajek, a Czech microbial taxonomist.

Surface colonies 2–3 mm in diameter. Circular, smooth, and glossy, without pigment. No growth at 4 or 42°C or in 4% NaCl. Hydrolysis of casein and gelatin. No hydrolysis of Tween 80, starch, lecithin, esculin, and tyrosine. No acetoin production. No activities of urease, hemolysis, arginine dihydrolase, arginine arylamidase, esterase, lipase, naphthol-AS-BI-phosphohydrolase, valine arylamidase, cystine arylamidase, α-galactosidase, *N*-acetyl-β-glucosaminidase, α-mannosidase, α-fucosidase, or acid phophatase. Acid production from sucrose, maltose, D-mannitol, D-fructose, D-trehalose, and D-glucose. No acid production from xylitol, D-galactose, lactose, D-ribose, D-melibiose, *N*-acetylglucosamine, or methyl α-D-glucoside.

DNA G+C content (mol%): 40 (HPLC).

Type strain: CCM 4809, LMG 21711.

GenBank accession number (16S rRNA gene): AY119685.

Additional remarks: Type strain was isolated from llama skin.

7. **Macrococcus lamae** Mannerová, Pantůek, Doška, Švec, Snauwaert, Vancanneyt, Swings and Sedláek 2003, 1653[VP]

la′mae. N.L. fem. gen. n. *lamae* of *Lama*, the zoological genus name of llama.

Surface colonies 2–5 mm in diameter. Circular, smooth, and glossy. Orange pigment. No growth at 4 and 42°C and in 4% NaCl. Hydrolysis of casein and gelatin, but not Tween 80, starch, lecithin, esculin, or tyrosine.

No acetoin production. No activities of urease, hemolysis, arginine dihydrolase, arginine arylamidase, esterase, lipase, naphthol-AS-BI-phosphohydrolase, valine arylamidase, cystine arylamidase, α-galactosidase, N-acetyl-β-glucosaminidase, α-mannosidase, or α-fucosidase. Acid production from sucrose, maltose, D-mannitol, D-fructose, D-trehalose, and D-glucose. No acid production from xylitol, D-galactose, lactose, ribose, D-melibiose, N-acetylglucosamine, or methyl α-D-glucoside. Susceptible to novobiocin.

DNA G+C content (mol%): 41–42 (HPLC).

Type strain: CCM 4815, LMG 21713.

GenBank accession number (16S rRNA gene): AY119687.

Additional remarks: Type strain was isolated from llama skin.

Genus IV. **Salinicoccus** Ventosa, Márquez, Ruiz-Berraquero and Kocur 1990b, 320[VP] (Effective publication: Ventosa, Márquez, Ruiz-Berraquero and Kocur 1990a, 32.)

ANTONIO VENTOSA

Sa.li.ni.coc'cus. L. adj. *salinus* saline; Gr. n. *kokkos* a grain or berry; N.L. masc. n. *Salinicoccus* saline coccus.

Gram-stain-positive spherical cells, 0.5–2.5 μm in diameter, occurring singly, in pairs, tetrads, or clumps. **Nonmotile, nonspore-forming.** Colonies are round, smooth, and form **pink-red or orange, nondiffusible pigments. Strictly aerobic.** Catalase- and oxidase-positive. **Moderately halophilic.** Optimal NaCl concentration for growth is 4–10%; growth occurs in media with 0–25% NaCl. Temperature range for growth is 4–49°C; optimal temperature is 30–37°C. pH range for growth is 6–11.5; optimal pH is 7–9.5. **The predominant lipoquinone is MK-6. The cell wall contains murein of the L-Lys–Gly$_5$ type,** which corresponds to the A3α type according to Schleifer and Kandler (1972).

DNA G+C content (mol%): 45.6–51.2.

Type species: **Salinicoccus roseus** Ventosa, Márquez, Ruiz-Berraquero and Kocur 1990b, 320[VP] (Effective publication: Ventosa, Márquez, Ruiz-Berraquero and Kocur 1990a, 32.).

Further descriptive information

The polar lipids of *Salinicoccus* are phosphatidylglycerol, diphosphatidylglycerol, and a glycolipid of unknown structure (Ventosa et al., 1990a; Ventosa et al., 1993). Sprott et al. (2006) reported that a major lipid in *Salinicoccus hispanicus* is the sulfonolipid sulfoquinovosyl diacylglycerol. They observed that the ratio of sulfonolipid to phospholipid is dependent on the salinity of the growth media, when grown in media with low phosphate content. The fatty acid composition of species of *Salinicoccus* includes the presence of predominantly straight and branched-chain fatty acids (Aslam et al., 2007; Ventosa et al., 1990a; Ventosa et al., 1993; Zhang et al., 2002; Zhang et al., 2007).

In order to grow over a wide range of salt concentrations, moderately halophilic micro-organisms accumulate organic osmotic solutes (Ventosa et al., 1998). In *Salinicoccus roseus* and *Salinicoccus hispanicus* the main organic osmotic solute is proline (Ventosa et al., 1998). When grown in complex media, they accumulate glycine betaine from the medium (Ventosa et al., 1998).

The susceptibility pattern of *Salinicoccus roseus* to 16 antimicrobial compounds has been reported and shows a remarkable resistance to ampicillin, cephalotin, erythromycin, gentamicin, novobiocin, and penicillin G (Nieto et al., 1993). It must be considered that the salt concentration may influence the susceptibility of moderately halophilic bacteria to antimicrobial compounds (Coronado et al., 1995).

In general, the species of the genus *Salinicoccus* do not produce extracellular hydrolytic enzymes. In a screening focused on the isolation from several hypersaline environments of moderately halophilic strains with hydrolytic activities, two *Salinicoccus* strains able to produce a protease and a lipase, respectively, have been reported (Sanchez-Porro et al., 2003).

Enrichment and isolation procedures

Organisms of the genus *Salinicoccus* have been isolated from water of salterns and saline soils, from soda lakes, and from salted materials. Specific enrichment or isolation media have not been described. They grow well in complex culture media supplemented with a mixture of salts. For the species that grow at neutral pH values the medium described by Ventosa et al. (1983) can be used. The alkaliphilic species requires a pH of 9.0–9.5 and can be grown by using the medium described by Zhang et al. (2002). They can be grown aerobically at 32–37°C. Typically, colonies of strains of *Salinicoccus* can be differentiated in these media by their pink to red or orange pigmentation, since most moderately halophilic species are cream, white, yellow, or nonpigmented.

Maintenance procedures

Strains belonging to *Salinicoccus* can be maintained by the standard procedures such as freeze-drying, storage at –80°C or under liquid nitrogen. Slant cultures can be conserved for several months at room temperature using a medium with 10% salts, at pH 7.5 or 9.0 (for the alkaliphilic species). Nutrient agar plus a mixture of salts can be used. For species that grow at neutral pH the MH medium has been described. The composition of this medium is (in g/l): MaCl, 81.0; MgCl$_2$, 7.0; MgSO$_4$, 9.6; CaCl$_2$, 0.36; KCl, 2.0; NaHCO$_3$, 0.06; NaBr, 0.026; proteose-peptone no. 3 (Difco), 5.0; yeast extract(Difco), 10.0; glucose, 1.0; Bacto-agar, 20.0 (Ventosa et al., 1982; Ventosa et al., 1983). For haloalkaliphilic species the following medium can be used (in g/l): NaCl, 100; Na$_2$CO$_3$, 10; K$_2$HPO$_4$, 1; MgSO$_4$ · 7H$_2$O, 0.2; glucose, 10; polypeptone, 5; yeast extract, 5; agar, 20 (Zhang et al., 2002).

Differentiation of the genus *Salinicoccus* from other genera

The genus *Salinicoccus* can be differentiated from other Gram-stain-positive cocci by comparative analysis of the 16S rRNA sequence as well as by several phenotypic and chemotaxonomic features. In contrast to the members of the genus *Marinococcus* that are also moderately halophilic, the species of *Salinicoccus* are nonmotile. *Salinicoccus* has MK-6 as the characteristic predominant menaquinone system in contrast to other Gram-stain-positive cocci such as *Marinococcus* (Hao et al., 1984) and *Jeotgalicoccus* (Yoon et al., 2003) that have MK-7, and *Nesterenkonia* that has MK-6, MK-7, and MK-8 (Stackebrandt et al., 1995). Another differential and characteristic feature of the species of *Salinicoccus* is that their cell walls contain murein of the L-Lys–Gly₅ type in contrast to that found in *Marinococcus*, showing peptidoglycan of the *meso*-diaminopimelic acid type (Hao et al., 1984), in *Nesterenkonia*, that has murein of L-Lys–Gly–L-Glu type (Mota et al., 1997; Stackebrandt et al., 1995), and the peptidoglycan of *Jeotgalicoccus*, based on L-Lys–Gly₃₋₄–L-Ala(Gly) (Yoon et al., 2003). Finally, another differential feature is the DNA base composition. The G + C range for *Salinicoccus* is 45.6–51.2 mol%, while for *Marinococcus* it is 44.9–46.4 mol% (Hao et al., 1984), for *Nesterenkonia* it is 70.0–72.0 mol% (Mota et al., 1997; Stackebrandt et al., 1995), and for *Jeotgalicoccus* it is 42 mol% (Yoon et al., 2003).

Taxonomic comments

The genus *Salinicoccus* was proposed on the basis of a single Gram-stain-positive, moderately halophilic, non-motile coccus, that was proposed as the new species *Salinicoccus roseus* (Ventosa et al., 1990a). Lately, two culture collection strains, CCM 168 and CCM 1405, previously assigned to the genus *Micrococcus* were also shown to be members of this species (Ventosa et al., 1993). In a study based on chemotaxonomic features, Ventosa et al. (1992) showed that the moderately halophilic bacterium *Marinococcus hispanicus* was also a species of the genus *Salinicoccus*, *Salinicoccus hispanicus*. Laterly, Zhang et al. (2002) described a new species of the genus *Salinicoccus*, *Salinicoccus alkaliphilus*, an alkaliphilic and moderately halophilic coccus isolated from a soda lake. More recently, three new species have been proposed and several new isolates will probably represent new species of this genus. Phylogenetically, members of the genus *Salinicoccus* constitute a monophyletic branch within the *Firmicutes*.

Differentiation of the species of the genus *Salinicoccus*

Some differential features of the species of the genus *Salinicoccus* are given in Table 77.

List of species of the genus *Salinicoccus*

1. **Salinicoccus roseus** Ventosa, Márquez, Ruiz-Berraquero and Kocur 1990b, 320^VP (Effective publication: Ventosa, Márquez, Ruiz-Berraquero and Kocur 1990a, 32.)

ro′se.us. L. adj. *roseus* rose-colored.

See the generic description for many features.

Spherical cells, 1.0–2.5 μm in diameter, occurring singly, in pairs, tetrads, or clumps. Nonmotile, nonspore-forming. Colonies are round, smooth, and form a pink-red, nondiffusible pigment. In liquid medium the cultures are slightly turbid and form a viscous sediment.

Moderately halophilic. Growth occurs in media containing 0.9–25 % NaCl. Optimal growth at 10% NaCl. No growth in the absence of NaCl. Growth at 15–40°C and pH 6–9; optimal growth at 37°C and pH 8.0.

Acid is not produced from arabinose, fructose, D-galactose, D-glucose, glycerol, lactose, maltose, D-mannitol, sucrose, D-trehalose, and D-xylose. Nitrate is reduced to nitrite. Esculin is not hydrolyzed. Benzidine test is positive. The following tests are negative: urease, Simmons' citrate, arginine, lysine, ornithine decarboxylases, methyl red, Voges–Proskauer, indole, egg-yolk, β-galactosidase, H₂S production, and phenylalanine deaminase.

The following compounds are utilized as sole carbon and energy sources: D-gluconolactone, D-glucosamine, D-melibiose, sucrose, D-trahalose, erythritol, ethanol, *m*-inositol, propanol, D-sorbitol, *N*-acetylglucosamine, aconitate, α-aminobutyrate, benzoate, butyrate, citrate, fumarate, DL-glycerate, D-gluconate, hippurate, DL-malate, propionate, quinate, and D-tartrate. The following compounds are not utilized as sole carbon and energy sources: starch, D-amygdalin, L-arabinose, D-cellobiose, esculin, D-fructose, D-fucose, D-galactose, D-glucose, inulin, lactose, maltose, D-mannose, L-raffinose, L-rhamnose, ribose, salicin, D-xylose, adonitol, dulcitol, DL-glycerol, D-mannitol, acetate, caprylate, glucuronate, *p*-hydroxybenzoate, malonate, oxalate, pyruvate, saccharate, salicylate, suberate, succinate, L-aspartic acid, betaine, creatine, and ethionine.

The following compounds are utilized as sole carbon, nitrogen, and energy sources: L-alanine, allantoin, DL-arginine, L-asparagine, phenylalanine, glycine, L-glutamine, L-histidine, L-isoleucine, L-leucine, and L-proline. The following compounds are not utilized as sole carbon, nitrogen, and energy sources: L-glutamic acid, L-lysine, methionine, L-ornithine, L-serine, threonine, L-tryptophan, and L-valine.

Isolated from a marine saltern in Spain, salted meat, and salted horse hide.

DNA G+C content (mol%): 51.2 (T_m).

Type strain: 9, ATCC 49258, CCM 3516, CIP 104761, DSM 5351.

GenBank accession number (16S rRNA gene): X94559.

2. **Salinicoccus alkaliphilus** Zhang, Xue, Ma, Zhou, Ventosa and Grant 2002, 792^VP

al.ka.li′phi.lus. N.L. n. *alkali* alkali; Gr. adj. *philos* loving; N.L. masc. adj. *alkaliphilus* liking alkaline media.

See the generic description for many features.

Cells are cocci, 0.5–0.8 μm in diameter, occurring singly, in pairs, tetrads, or clumps. Nonmotile, nonspore-forming. Colonies are round, smooth, and slightly convex with pinkish color and nondiffusible pigment.

Alkaliphilic and moderately halophilic. Growth occurs at pH 6.5–11.5; optimal growth at pH 9.5. Growth occurs in media containing 0–25 % NaCl. Optimal growth at 10 % NaCl. Growth at 10–49°C; optimal growth at 32°C.

Acid is produced from D-glucose but not from maltose, galactose, fructose, sucrose, D-arabinose, glycerol, or D-xylose. Nitrate is reduced. Urease-positive. Esculin is hydrolyzed, but casein, starch, gelatin, and Tween 80, 60 and 20 are not. The following tests are negative: methyl red, Voges–Proskauer, indole, and H₂S production.

The following compounds are utilized as sole carbon and energy sources: D-arabinose, D-melibiose, creatinine, esculin, lactose, D-xylose, D-fructose, amylose, sorbose, D-α-melizitose,

TABLE 77. Characteristics differentiating the species of the genus *Salinicoccus*[a]

Characteristic	1. S. roseus	2. S. alkaliphilus	3. S. hispanicus	4. S. luteus	5. S. jeotgali	6. S. salsiraiae
Diameter of cells (μm)	1.0–2.5	0.5–0.8	1.0–2.0	0.9	1.0–2.0	1.0–2.5
Colony pigmentation	Pink-red	Pinkish	Reddish orange	Orange	Orange	Pink-red
Salt range for growth (%)	0.9–25	0–25	0.5–25	1–25	0.5–15	0–22
pH for growth:						
Range	6–9	6.5–11.5	5–9	7–11	6.5–11	6.5–9.5
Optimum	8	9.5	7.5	8–9	7	8
Temperature for growth (°C):						
Range	15–40	10–49	15–37	4–45	20–30	20–45
Optimum	37	32	37	30	30	37
Acid production from:						
Fructose	–	–	+	–	+	+
D-Galactose	–	–	+	–	–	–
D-Glucose	–	+	+	+	+	+
Glycerol	–	–	+	ND	+	+
Nitrate reduction	–	+	+	+	+	+
Methyl red	–	–	+	–	+	–
Hydrolysis of:						
Esculin	–	+	+	+	+	–
Casein	+	–	–	–	–	+
Gelatin	+	–	+	–	–	+
Tween 80	+	–	–	–	+	ND
Utilization of compounds as carbon and energy sources:						
Adonitol	–	+	–	+	ND	ND
Cellobiose	–	+	–	+	ND	ND
D-Fructose	–	+	–	+	ND	ND
D-Galactose	–	+	–	+	ND	ND
Mannose	–	+	+	+	ND	ND
Rhamnose	–	+	+	ND	+	ND
Ribose	–	+	+	ND	–	ND
Sorbitol	+	+	–	+	–	ND
G+C content (mol%)	51.2	49.6	45.6–49.3	49.7	47.0	46.2

[a]Symbols: +, >85% positive; –, 0–15% positive; ND, not determined.

erythritol, inulin, *meso*-erythritol, trehalose, D-mannose, D-sorbitol, stachyose, ribose, mannitol, D-galactose, adonitol, D-cellobiose, glycogen, dulcitol, and α-L-rhamnose.

The following compounds are utilized as sole nitrogen and energy sources: polypeptone, yeast extract, DL-tryptophan, bacto-peptone, casitone, urea, L-proline, L-methionine, L-serine, L-arginine, L-alanine, L-threonine, L-histidine, sodium L-glutamate, L-ornithine, L-leucine, L-glycine, L-glutamic acid, and DL-lysine. The following compounds are not utilized as sole nitrogen and energy sources: L-asparagine, L-tyrosine, L-isoleucine, and L-cysteine.

Isolated from Baer Soda Lake in Inner Mongolia Autonomous Region, China.

DNA G+C content (mol%): 49.6 (T_m).

Type strain: T8, AS 1.2691, JCM 11311.

GenBank accession number (16S rRNA gene): AF275710.

3. **Salinicoccus hispanicus** (Márquez, Ventosa and Ruiz-Berraquero 1990) Ventosa, Márquez, Weiss and Tindall 1993, 398[VP] (Effective publication: Ventosa, Márquez, Weiss and Tindall 1992, 533) (*Marinococcus hispanicus* Márquez, Ventosa and Ruiz-Berraquero 1990, 167.)

his.pa'ni.cus. L. adj. *hispanicus* Spanish.

See the generic description for many features.

Spherical cells, 1.0–2.0 μm in diameter, occurring singly, in pairs, tetrads, or clumps. Nonmotile, nonspore-forming. Colonies are round and smooth and form a reddish orange, nondiffusible pigment. Broth cultures are uniformly turbid.

Moderately halophilic. Growth occurs in media containing 0.5–25 % NaCl. Optimal growth at 10% NaCl. No growth in the absence of NaCl. Growth at 15–37°C and pH 5–9; optimal growth at 37°C and pH 7.5.

Acid is produced from fructose, galactose, glycerol, D-glucose, and D-mannitol, but not from arabinose, lactose, D-trehalose, and xylose. Nitrate is reduced to nitrite. Gelatin, esculin, and DNA are hydrolyzed. Casein, starch, Tween 80, and tyrosine are not hydrolyzed. Methyl red test positive. The following tests are negative: Simmons' citrate, arginine, lysine and ornithine decarboxylase, Voges–Proskauer, indole, H_2S production, and phenylalanine deaminase.

The following compounds are utilized as sole carbon and energy sources: D-glucose, D-mannose, L-raffinose, L-rhamnose, ribose, salicin, and sucrose. The following compounds are not utilized as sole carbon and energy sources: amygdalin, D-cellobiose, D-fructose, D-fucose, D-galactose, D-galactosamine, lactose, maltose, D-trehalose, D-xylose, adonitol, dulcitol, erythritol, ethanol, propanol, D-sorbitol, N-acetylglucosamine, DL-α-aminobutyrate, δ-aminovalerate, benzoate, butyrate, caprylate, lactate, DL-malate, oxalate, propionate, quinate, D-saccharate, succinate, and D-tartrate.

L-Glutamine is utilized as sole source of carbon, nitrogen, and energy. The following compounds are not utilized as sole carbon, nitrogen, and energy sources: L-allantoin, DL-arginine, L-asparagine, aspartic acid, betaine, creatine, ethionine, phenylalanine, glycine, glutamic acid, L-histidine, L-leucine, DL-lysine, L-ornithine, putrescine, sarcosine, L-serine, L-threonine, and L-valine.

Isolated from salterns and saline soils.

DNA G+C content (mol%): 45.7 (T_m).

Type strain: J-82, ATCC 49259, CCUG 43288, CIP 104760, CCM 4148, DSM 5352.

GenBank accession number (16S rRNA gene): AY028927.

Additional Remarks: Mol% G + C ranges from 45.6 to 49.3 (T_m) for various strains.

4. **Salinicoccus luteus** Zhang, Yu, Liu, Zhang, Xu and Li 2007, 1903[VP]

lu.te′us. L. masc. adj. *luteus* orange-colored.

See the generic description for many features.

Cells are nonmotile, nonspore-forming cocci, 0.9 μm in diameter. Colonies are circular, opaque and approximately 1.0 mm in diameter after 48 h at 30°C, with orange pigmentation.

Moderately halophilic. Growth occurs at pH 7.0–11.0; optimal growth at pH 8.0–9.0. Growth occurs in media containing 1–25 % NaCl. Optimal growth at 10 % NaCl. Growth at 4–45°C; optimal growth at 30°C.

Acid is produced from D-glucose, maltose, sucrose, malonate and N-acetylglucosamine. Nitrate is reduced. Ornithine and arginine decarboxylases, arginine dihydrolase and β-glucosidase-positive. Tween 20 is hydrolyzed, but casein, starch, gelatin, and Tween 80 are not. The following tests are negative: methyl red, Voges–Proskauer, H_2S and melanin production, N-acetylglucosamindase, β-glucuronidase, β-galactosidase and α-galactosidase.

The following compounds are utilized as sole carbon and energy sources: maltose, mannitol, glucose, adonitol, arabinose, arabitol, mannose, inositol, sorbitol, fructose, cellobiose, salicin, acetamide, galactose, xylose and dextrin. Rhamnose and starch are not utilized.

Isolated from a desert soil sample from Egypt.

DNA G+C content (mol%): 49.7 (HPLC).

Type strain: YIM 70202, CGMCC 1.6511, KCTC 3941.

GenBank accession number (16S rRNA gene): DQ352839.

5. **Salinicoccus jeotgali** Aslam, Lim, Im, Yasir, Chung and Lee 2007, 637[VP]

je.ot.ga′li. N.L. gen. n. *jeotgali* of jeotgal, a traditional Korean fermented seafood, from which the type strain was isolated.

See the generic description for many features.

Cells are nonmotile, nonspoeforming cocci, 1.0–2.0 μm in diameter. Colonies are smooth, circular, orange-colored with transparent edges and 1–2 mm in diameter.

Moderately halophilic. Growth occurs at pH 6.5–11.0; optimal growth at pH 7.0. Growth occurs in media containing 0.5–15 % NaCl. Optimal growth at 5 % NaCl. Growth at 20–30°C; optimal growth at 30°C.

Acid is produced from D-glucose, D-fructose, D-maltose, D-trehalose, N-acetyl-D-glucosamine and glycerol but not from D-mannitol, sucrose, D-arabinose, L-arabinose, D-ribose, D-xylose, L-xylose, D-adonitol, D-galactose, D-mannose, L-sorbose, L-rhamnose, D-sorbitol, D-cellobiose, D-lactose, D-melibiose, D-melezitose, D-raffinose, D-turanose, D-lyxose, D-tagatose, D-fucose, L-fucose, D-arabitol, L-arabitol, erythritol, methyl β-D-xylopyranoside, dulcitol, inositol, methyl α-D-mannopyranoside, methyl α-D-glucopyranoside, amygdalin, arbutin, salicin, inulin, starch, glycogen, xylitol, gentiobiose, potassium gluconate, potassium 2-ketogluconate or potassium 5-ketogluconate. Nitrate is reduced to nitrite, but not to nitrogen gas. Indole and H_2S are produced. Esculin is hydrolyzed, but casein, gelatin, and DNA are not. The following tests are negative: urease, Voges–Proskauer, arginine dihydrolase, lysine and ornithine decarboxylase, and β-glucosidase.

The following compounds are utilized as sole carbon and energy sources: salicin, L-arabinose, D-glucose, citrate, histidine, L-rhamnose, N-acetyl-D-glucosamine, D-maltose, D-lactate, L-alanine and 5-ketogluconate. The following compounds are not utilized as sole carbon and energy sources: mannitol, D-melibiose, L-fucose, D-sorbitol, propionate, caprate, valerate, 2-ketogluconate, 3-hydroxybutyrate, L-proline, D-ribose, inositol, D-sucrose, itaconate, suberate, malonate, acetate, glycogen, 3-hydroxybenzoate, 4-hydroxybenzoate and L-serine.

Isolated from jeotgal, a traditional Korean fermented seafood.

DNA G+C content (mol%): 47.0 (HPLC).

Type strain: S2R53-5, KCTC 13030, LMG 23640.

GenBank accession number (16S rRNA gene): DQ471329.

6. **Salinicoccus salsiraiae** França, Rainey, Nobre and da Costa 2007, 433[VP] (Effective publication: França, Rainey, Nobre and da Costa 2006, 535.)

sal.si.ra′i.a.e. L. adj. *salsus* salted; L. n. *raia* a ray; N.L. gen. n. *salsiraiae* of a salted ray.

See the generic description for many features.

Spherical cells, 1.0–2.5 μm in diameter, occurring singly, in pairs, tetrads, or clumps. Nonmotile, nonspore-forming. Colonies form a pink-red pigment.

Moderately halophilic. Growth occurs in media containing 0–22 % NaCl. Optimal growth at 4% NaCl. Growth at 20–45°C and pH 6.5–9.5; optimal growth at 37°C and pH 8.0.

Acid is produced from D-ribose, glycerol, D-glucose, D-fructose, N-acetylglucosamine, D-maltose, D-trehalose, 2-ketogluconate, and 5-ketogluconate. Nitrate is reduced to nitrite. Gelatin, casein, arbutin, hippurate, hide powder azure and DNA are hydrolyzed. Starch125, elastin, fibrin, and esculin are not hydrolyzed. Methyl red, Voges–Proskauer and indole production are negative.

Isolated from a salted ray (skate).

DNA G+C content (mol%): 46.2 (HPLC).

Type strain: RH1, LMG 22840, CIP 108576.

GenBank accession number (16S rRNA gene): DQ333949.

References

Akatov, A.K., V. Hájek, T. M. Samsonova and J. Balusek. 1985. Classification and drug resistance of coagulase-negative staphylococci isolated from wild birds *In* Jeljaszewicz (Editor), The Staphylococci: Proceedings of the 5th International Symposium on Staphylococci and Staphylococcal Infections, Gustav-Fischer Verlag Stuttgart, Germany, pp. 125–127

Archer, G.L. and M.J. Tenenbaum. 1980. Antibiotic-resistant *Staphylococcus epidermidis* in patients undergoing cardiac surgery. Antimicrob. Agents Chemother. *17*: 269–272.

Aslam, Z., J.H. Lim, W.T. Im, M. Yasir, Y.R. Chung and S.T. Lee. 2007. *Salinicoccus jeotgali* sp. nov., isolated from jeotgal, a traditional Korean fermented seafood. Int. J. Syst. Evol. Microbiol. *57*: 633–638.

Baba, E., T. Fukata and H. Matsumoto. 1980. Ecological studies on coagulase-negative staphylococci in and around udder. Bull. Univ. Osaka Prefect. Ser. B *32*: 70–75.

Bannerman, T.L. and W.E. Kloos. 1991. *Staphylococcus capitis* subsp. *ureolyticus* subsp. nov. from human skin. Int. J. Syst. Bacteriol. *41*: 144–147.

Chesneau, O., A. Morvan, F. Grimont, H. Labischinski and N. Elsolh. 1993. *Staphylococcus pasteuri* sp. nov., isolated from human, animal, and food specimens. Int. J. Syst. Bacteriol. *43*: 237–244.

Cookson, B.D., P. Aparicio, A. Deplano, M. Struelens, R. Goering and R. Marples. 1996. Inter-centre comparison of pulsed-field gel electrophoresis for the typing of methicillin-resistant *Staphylococcus aureus*. J. Med. Microbiol. *44*: 179–184.

Coronado, M.J., C. Vargas, H.J. Kunte, E.A. Galinski, A. Ventosa and J.J. Nieto. 1995. Influence of salt concentration on the susceptibility of moderately halophilic bacteria to antimicrobials and its potential use for genetic transfer studies. Curr. Microbiol. *31*: 365–371.

Cox, R.A., C. Conquest, C. Mallaghan and R.R. Marples. 1995. A major outbreak of methicillin-resistant *Staphylococcus aureus* caused by a new phage-type (EMRSA-16). J. Hosp. Infect. *29*: 87–106.

Crisley, F.D., R. Angelotti and M.J. Foter. 1964. Multiplication of *Staphylococcus aureus* in synthetic cream fillings and pies. Public Health Rep. *79*: 369–376.

Crosa, J.H., B.L. Williams, J.J. Jorgensen and C.A. Evans. 1979. Comparative study of deoxyribonucleic acid homology and physiological characteristics of strains of *Peptococcus saccharolyticus*. Int. J. Syst. Bacteriol. *29*: 328–332.

Crossley, K.B. and G.L. Archer. 1997. The Staphylococci in Human Disease. Churchill Livingston, New York.

de la Fuente, R., G. Suarez and K.H. Schleifer. 1985. *Staphylococcus aureus* subsp. *anaerobius* subsp. nov., the causal agent of abscess disease of sheep. Int. J. Syst. Bacteriol. *35*: 99–102.

de Sousa, M.A., I.S. Sanches, M.L. Ferro, M.J. Vaz, Z. Saraiva, T. Tendeiro, J. Serra and H. de Lencastre. 1998. Intercontinental spread of a multidrug-resistant methicillin-resistant *Staphylococcus aureus* clone. J. Clin. Microbiol. *36*: 2590–2596.

Devriese, L.A., V. Hajek, P. Oeding, S.A. Meyer and K.H. Schleifer. 1978. *Staphylococcus hyicus* (Sompolinsky 1953) comb. nov. and *Staphylococcus hyicus* subsp. *chromogenes* subsp. nov. Int. J. Syst. Bacteriol. *28*: 482–490.

Devriese, L.A. and J. Derycke. 1979. *Staphylococcus hyicus* in cattle. Res. Vet. Sci. *26*: 356–358.

Devriese, L.A. and H. De Keyser. 1980. Prevalence of different species of coagulase-negative staphylococci on teats and in milk samples from dairy cows. J. Dairy Res. *47*: 155–158.

Devriese, L.A., B. Poutrel, R. Kilpper-Bälz and K.H. Schleifer. 1983. *Staphylococcus gallinarum* and *Staphylococcus caprae*, two new species from animals. Int. J. Syst. Bacteriol. *33*: 480–486.

Devriese, L.A., K.H. Schleifer and G.O. Adegoke. 1985. Identification of coagulase negative Staphylococci from farm animals. J. Appl. Bacteriol. *58*: 45–55.

Devriese, L.A. 1986. Coagulase-negative staphylococci in animals. *In* Mardh and Schleifer (Editors), Coagulase-negative Staphylococci Almquist and Wiksell International Stockholm, Sweden, pp. 51–57.

Devriese, L.A., M. Vancanneyt, M. Baele, M. Vaneechoutte, E. De Graef, C. Snauwaert, I. Cleenwerck, P. Dawyndt, J. Swings, A. Decostere and F. Haesebrouck. 2005. *Staphylococcus pseudintermedius* sp. nov., a

coagulase-positive species from animals. Int. J. Syst. Evol. Microbiol. *55*: 1569–1573.

Douglas, H.C. 1957. Genus VI. *Peptococcus* Kluyver and van Niel 1936. *In* Breed, Murray and Smith (Editors), Bergey's Manual of Determinative Bacteriology. The Williams & Wilkins Co., Baltimore, pp. 474–480.

Downes, F.P. and K. Ito. 2001. Compendium of Methods for the Microbiological Examination of Foods. American Public Health Association.

Emmett, M. and W.E. Kloos. 1979. Nature of arginine auxotrophy in cutaneous populations of staphylococci. J. Gen. Microbiol. *110*: 305–314.

Endl, J., H.P. Seidl, F. Fiedler and K.H. Schleifer. 1983. Chemical composition and structure of cell wall teichoic acids of staphylococci. Arch. Microbiol. *135*: 215–223.

Enright, M.C., N.P.J. Day, C.E. Davies, S.J. Peacock and B.G. Spratt. 2000. Multilocus sequence typing for characterization of methicillin-resistant and methicillin-susceptible clones of *Staphylococcus aureus*. J. Clin. Microbiol. *38*: 1008–1015.

Evans, A.C. 1916. The bacteria of milk freshly drawn from normal udders. J. Infect. Dis. *18*: 437–476.

Fairbrother, R.W. 1940. Coagulase production as a criterion for the classification of the staphylococci. J. Pathol. Bacteriol. *50*: 83–88.

Fleurette, J., Y. Brun, M. Bes, M. Coulet and F. Forey. 1987. Infections caused by coagulase-negative staphylococci other than *S. epidermidis* and *S. saprophyticus*. *In* Peters (Editor), Pathogenicity and Clinical Significance of Coagulase-negative Staphylococci. Gustav Fischer-Verlag, Stuttgart, Germany, pp. 195–208.

Foster, G., H.M. Ross, R.A. Hutson and M.D. Collins. 1997. *Staphylococcus lutrae* sp. nov., a new coagulase-positive species isolated from otters. Int. J. Syst. Bacteriol. *47*: 724–726.

Foubert, E.L., Jr and H.C. Douglas. 1948. Studies on the anaerobic micrococci. I. Taxonomic considerations. J. Bacteriol. *56*: 25–34.

Franca, L., F.A. Rainey, M.F. Nobre and M.S. da Costa. 2006. *Salinicoccus salsiraiae* sp. nov.: a new moderately halophilic gram-positive bacterium isolated from salted skate. Extremophiles *10*: 531–536.

França, L., F.A. Rainey, M.F. Nobre and M.S.d. Costa. 2007. *In* List of new names and new combinations previously effectively, but not yet, published. Validation List no. 114. Int. J. Syst. Evol. Microbiol. *57*: 433–434.

Freney, J., Y. Brun, M. Bes, H. Meugnier, F. Grimont, P.A.D. Grimont, C. Nervi and J. Fleurette. 1988. *Staphylococcus lugdunensis* sp. nov. and *Staphylococcus schleiferi* sp. nov., two species from human clinical specimens. Int. J. Syst. Bacteriol. *38*: 168–172.

Garrity, G.M. and J.G. Holt. 2001. The Road Map to the *Manual*. *In* Boone, Castenholz and Garrity (Editors), Bergey's Manual of Systematic Bacteriology, vol. 1, The *Archaea* and the Deeply Branching and Phototrophic *Bacteria*. Springer, New York, pp. 119–166.

George, C.G. and W.E. Kloos. 1994. Comparison of the *Sma*I-digested chromosomes of *Staphylococcus epidermidis* and the closely-related species *Staphylococcus capitis* and *Staphylococcus caprae*. Int. J. Syst. Bacteriol. *44*: 404–409.

Götz, F. 1990. *Staphylococcus carnosus*: A new host organism for gene cloning and protein production. J. Appl. Bacteriol. Symp. Suppl. *69*: S49-S53.

Götz, F., T. Bannermann and K.H. Schleifer. 2003. The genera *Staphylococcus* and *Macrococcus*. *In* Dworkin, Falkow, Rosenberg and Stackebrandt (Editors), The Prokaryotes. Springer, New York.

Grundmann, H., S. Hori, M.C. Enright, C. Webster, A. Tami, E.J. Feil and T. Pitt. 2002. Determining the genetic structure of the natural population of *Staphylococcus aureus*: A comparison of multilocus sequence typing with pulsed-field gel electrophoresis, randomly amplified polymorphic DNA analysis, and phage typing. J. Clin. Microbiol. *40*: 4544–4546.

Hájek, V. 1976. *Staphylococcus intermedius*, a new species isolated from animals. Int. J. Syst. Bacteriol. *26*: 401–408.

Hájek, V. and E. Marsalek. 1976. Staphylococci outside the hospital. *Staphylococcus aureus* in sheep. Zentralbl Bakteriol [Orig B] *161*: 455–461.

Hájek, V., L.A. Devriese, M. Mordarski, M. Goodfellow, G. Pulverer and P.E. Varaldo. 1986. Elevation of *Staphylococcus hyicus* subsp. *chromo-*

genes (Devriese *et al.*, 1978) to species status: *Staphylococcus chromogenes* (Devriese *et al.*, 1978) comb. nov. Syst. Appl. Microbiol. *8*: 169–173.

Hájek, V., W. Ludwig, K.H. Schleifer, N. Springer, W. Zitzelsberger, R.M. Kroppenstedt and M. Kocur. 1992. *Staphylococcus muscae*, a new species isolated from flies. Int. J. Syst. Bacteriol. *42*: 97–101.

Hájek, V., H. Meugnier, M. Bes, Y. Brun, F. Fiedler, Z. Chmela, Y. Lasne, J. Fleurette and J. Freney. 1996. *Staphylococcus saprophyticus* subsp., *bovis* subsp. nov., isolated from bovine nostrils. Int. J. Syst. Bacteriol. *46*: 792–796.

Hájek, V. and J. Balusek. 1985. Presented at the The Staphylococci: Proceedings of the 5th International Symposium on Staphylococci and Staphylococcal Infections, Gustav Fischer-Verlag Stuttgart, Germany.

Hájek, V., L.A. Devriese, M. Mordarski, M. Goodfellow, G. Pulverer and P.E. Varaldo. 1987. *In* Validation of the publication of new names and new combinations previously effectively published outside the IJSB. List no. 23. Int. J. Syst. Bacteiol. *37*: 179–180.

Hao, M.V., M. Kocur and K. Komagata. 1984. *Marinococcus* gen. nov., a new genus for motile cocci with *meso*-diaminopimelic acid in the cell wall and *Marinococcus albus* sp. nov. and *Marinococcus halophilus* (Novitsky and Kushner) comb. nov. J. Gen. Appl. Microbiol. *30*: 449–459.

Harvey, J. and A. Gilmour. 1988. Isolation and characterization of staphylococci from goats milk produced in Northern Ireland. Lett. Appl. Microbiol. *7*: 79–82.

Hay, J.B., A.R. Archibald and J. Baddiley. 1965. The molecular structure of bacterial walls: The size of ribitol teichoic acids and the nature of their linkage to glycosaminopeptides. Biochem. J. *97*: 723–730.

Holmberg, O. 1986. Coagulase-negative staphylococci in bovine mastitis. *In* Mardh and Schleifer (Editors), Coagulase-negative Staphylococci. Almquist and Wiksell International, Stockholm, Sweden, pp. 203–211.

Holt, J.G., N.R. Krieg, P.H.A. Sneath, J.T. Staley and S.T. Williams (Editors). 1994. Bergey's Manual of Determinative Bacteriology, 9th edn. The Williams & Wilkins Co., Baltimore.

Horowitz, W. 2000. Official Methods of Analysis of AOAC International, Gaithersburg, Maryland.

Hoyles, L., M.D. Collins, G. Foster, E. Falsen and P. Schumann. 2004. *Jeotgalicoccus pinnipedialis* sp. nov., from a southern elephant seal (*Mirounga leonina*). Int. J. Syst. Evol. Microbiol. *54*: 745–748.

Igimi, S., S. Kawamura, E. Takahashi and T. Mitsuoka. 1989. *Staphylococcus felis*, a new species from clinical specimens from cats. Int. J. Syst. Bacteriol. *39*: 373–377.

Igimi, S., E. Takahashi and T. Mitsuoka. 1990. *Staphylococcus schleiferi* subsp. *coagulans* subsp. nov., isolated from the external auditory meatus of dogs with external ear otitis. Int. J. Syst. Bacteriol. *40*: 409–411.

Isenberg, H.D. 1994. Clinical Microbiology Procedures Handbook. American Society for Microbiology, Washington, D.C.

Judicial Commission. 1958. Opinion. 17. Conservation of the generic name *Staphylococcus* Rosenbach, designation of *Staphylococcus aureus* Rosenbach as the the nomenclatural type of the genus *Staphylococcus* Rosenbach, and designation of a neotype culture of *Staphylococcus aureus* Rosenbach. Int. J. Syst. Bacteriol. *8*: 153–154.

Kanda, M., H. Inoue, T. Fukuizumi, T. Tsujisawa, K. Tominaga and J. Fukuda. 2001. Detection and rapid increase of salivary antibodies to *Staphylococcus lentus*, an indigenous bacterium in rabbit saliva, through a single tonsillar application of bacterial cells. Oral Microbiol. Immunol. *16*: 257–264.

Karchmer, A.W., G.L. Archer and W.E. Dismukes. 1983. *Staphylococcus epidermidis* causing prosthetic valve endocarditis: Microbiologic and clinical observations as guides to therapy. Ann. Intern. Med. *98*: 447–455.

Kawamura, Y., X.G. Hou, F. Sultana, K. Hirose, M. Miyake, S.E. Shu and T. Ezaki. 1998. Distribution of *Staphylococcus* species among human clinical specimens and emended description of *Staphylococcus caprae*. J. Clin. Microbiol. *36*: 2038–2042.

Kilpper-Bälz, R. and K.H. Schleifer. 1981. Transfer of *Peptococcus saccharolyticus* Foubert and Douglas to the genus *Staphylococcus* - *Staphylococcus saccharolyticus* (Foubert and Douglas) comb. nov. Zentbl. Bakteriol. Mikrobiol. Hyg. Abt. Orig. C *2*: 324–331.

Kilpper-Bälz, R. and K.H. Schleifer. 1984. *In* Validation of the publication of new names and new combinations previously effectively published outside the IJSB. List no. 13. Int. J. Syst. Bacteriol. *34*: 91–92.

Kilpper, R., U. Buhl and K.H. Schleifer. 1980. Nucleic acid homology studies between *Peptococcus saccharolyticus* and various anaerobic and facultative anaerobic gram-positive cocci. FEMS Microbiol. Lett. *8*: 205–210.

Kloos, W.E., T.G. Tornabene and K.H. Schleifer. 1974. Isolation and characterization of micrococci from human skin, including two new species, *Micrococcus lylae* and *Micrococcus kristinae*. Int. J. Syst. Bacteriol. *24*: 79–101.

Kloos, W.E. and M.S. Musselwhite. 1975. Distribution and persistence of *Staphylococcus* and *Micrococcus* species and other aerobic bacteria on human skin. Appl. Microbiol. *30*: 381–395.

Kloos, W.E. and K.H. Schleifer. 1975a. Isolation and characterization of staphylococci from human skin. 2. Descriptions of four new species: *Staphylococcus warneri*, *Staphylococcus capitis*, *Staphylococcus hominis*, and *Staphylococcus simulans*. Int. J. Syst. Bacteriol. *25*: 62–79.

Kloos, W.E. and K.H. Schleifer. 1975b. Simplified scheme for routine identification of human *Staphylococcus* species. J. Clin. Microbiol. *1*: 82–88.

Kloos, W.E., K.H. Schleifer and R.F. Smith. 1976a. Characterization of *Staphylococcus sciuri* sp. nov. and its subspecies. Int. J. Syst. Bacteriol. *26*: 22–37.

Kloos, W.E., R.J. Zimmerman and R.F. Smith. 1976b. Preliminary studies on characterization and distribution of *Staphylococcus* and *Micrococcus* species on animal skin. Appl. Environ. Microbiol. *31*: 53–59.

Kloos, W.E. and J.F. Wolfshol. 1979. Evidence for deoxyribonucleotide sequence divergence between staphylococci living on human and other primate skin. Curr. Microbiol. *3*: 167–172.

Kloos, W.E. 1980. Natural populations of the genus *Staphylococcus*. *In* Starr, Ingraham and Balows (Editors), Anual Review of Microbiology, vol. 34. Annual Reviews, Inc., Palo Alto, CA, pp. 559–592.

Kloos, W.E. and K.H. Schleifer. 1981. The genus *Staphylococcus*. *In* Starr, Stolp, H.G. Trüper, Balows and Schegel (Editors), The Prokaryotes. Springer-Verlag, New York, pp. 1548–1569.

Kloos, W.E. and J.F. Wolfshohl. 1982. Identification of *Staphylococcus* species with the API STAPH-IDENT system. J. Clin. Microbiol. *16*: 509–516.

Kloos, W.E. and K.H. Schleifer. 1983. *Staphylococcus auricularis* sp. nov., an inhabitant of the human external ear. Int. J. Syst. Bacteriol. *33*: 9–14.

Kloos, W.E. and K.H. Schleifer. 1986. Genus 4. *Staphylococcus* Rosenbach 1984. *In* Holt, Sneath, Mair and Sharpe (Editors), Bergey's Manual of Systematic Bacteriology. The Williams & Wilkins Co., Maltimore, pp. 1013–1035.

Kloos, W.E. and C.G. George. 1991. Identification of *Staphylococcus* species and subspecies with the Microscan Pos ID and rapid Pos ID panel systems. J. Clin. Microbiol. *29*: 738–744.

Kloos, W.E. and J.F. Wolfshohl. 1991. *Staphylococcus cohnii* subspecies: *Staphylococcus cohnii* subsp. *cohnii* subsp. nov. and *Staphylococcus cohnii* subsp. urea*Lyticum* subsp. nov. Int. J. Syst. Bacteriol. *41*: 284–289.

Kloos, W.E., K.H. Schleifer and F. Götz. 1992. The Prokyarotes. *In* Balows, Trüper, Dworkin, Harder and Schleifer (Editors), The Prokaryotes. Springer-Verlag, New York, pp. 1369–1420.

Kloos, W.E. and T.L. Bannerman. 1994. Update on clinical significance of coagulase-negative staphylococci. Clin. Microbiol. Rev. *7*: 117–140.

Kloos, W.E., D.N. Ballard, J.A. Webster, R.J. Hubner, A. Tomasz, I. Couto, G.L. Sloan, H.P. Dehart, F. Fiedler, K. Schubert, H. deLencastre, I.S. Sanches, H.E. Heath, P.A. Leblanc and A. Ljungh. 1997. Ribotype delineation and description of *Staphylococcus sciuri* subspecies and their potential as reservoirs of methicillin resistance and staphylolytic enzyme genes. Int. J. Syst. Bacteriol. *47*: 313–323.

Kloos, W.E., D.N. Ballard, C.G. George, J.A. Webster, R.J. Hubner, W. Ludwig, K.H. Schleifer, F. Fiedler and K. Schubert. 1998a. Delimiting the genus *Staphylococcus* through description of *Macrococcus caseolyticus* gen. nov., comb. nov. and *Macrococcus equipercicus* sp. nov., *Macrococcus bovicus* sp. nov. and *Macrococcus carouselicus* sp. nov. Int. J. Syst. Bacteriol. *48*: 859–877.

Kloos, W.E., C.G. George, J.S. Olgiate, L. Van Pelt, M.L. McKinnon, B.L. Zimmer, E. Muller, M.P. Weinstein and S. Mirrett. 1998b. *Staphylococcus hominis* subsp. *novobiosepticus* subsp. nov., a novel trehalose- and N-acetyl-D-glucosamine-negative, novobiocin- and multiple-antibiotic-resistant subspecies isolated from human blood cultures. Int. J. Syst. Bacteriol. *48*: 799–812.

Kloos, W.E. and T.L. Bannerman. 1999. *Staphylococcus* and *Micrococcus*. *In* Murray, Baron, Pfaller, Tenover and Yolkin (Editors), Manual of Clinical Microbiology. ASM Press, Washington, D.C., pp. 264–282.

Kloos, W.E., J.M. Hardie and R.A. Whiley. 2001. International Committee on Systematic Bacteriology. Subcommittee on the taxonomy of staphylococci and streptococci. Minutes of the meetings, 17 September, 1996. Int. J. Syst. Evol. Microbiol. *51*: 717–718.

Kuroda, M., T. Ohta, I. Uchiyama, T. Baba, H. Yuzawa, I. Kobayashi, L.Z. Cui, A. Oguchi, K. Aoki, Y. Nagai, J.Q. Lian, T. Ito, M. Kanamori, H. Matsumaru, A. Maruyama, H. Murakami, A. Hosoyama, Y. Mizutani-Ui, N.K. Takahashi, T. Sawano, R. Inoue, C. Kaito, K. Sekimizu, H. Hirakawa, S. Kuhara, S. Goto, J. Yabuzaki, M. Kanehisa, A. Yamashita, K. Oshima, K. Furuya, C. Yoshino, T. Shiba, M. Hattori, N. Ogasawara, H. Hayashi and K. Hiramatsu. 2001. Whole genome sequencing of meticillin-resistant *Staphylococcus aureus*. Lancet *357*: 1225–1240.

Kwok, A.Y. and A.W. Chow. 2003. Phylogenetic study of *Staphylococcus* and *Macrococcus* species based on partial *hsp60* gene sequences. Int. J. Syst. Evol. Microbiol. *53*: 87–92.

Lambert, L.H., T. Cox, K. Mitchell, R.A. Rosselló-Mora, C. Del Cueto, D.E. Dodge, P. Orkand and R.J. Cano. 1998. *Staphylococcus succinus* sp. nov., isolated from Dominican amber. Int. J. Syst. Bacteriol. *48*: 511–518.

Langlois, B.E., R.J. Harmon and K. Akers. 1983. Identification of *Staphylococcus* species of bovine origin with the API Staph-Ident system. J. Clin. Microbiol. *18*: 1212–1219.

Liepe, H.U. and R. Porobic. 1983. Influence of storage conditions on survival rates and fermentative activity of lyophilized staphylococci. Fleischwirtschaft *63*: 1756–1757.

Ludwig, W., O. Strunk, R. Westram, L. Richter, H. Meier, Yadhukumar, A. Buchner, T. Lai, S. Steppi, G. Jobb, W. Förster, I. Brettske, S. Gerber, A.W. Ginhart, O. Gross, S. Grumann, S. Hermann, R. Jost, A. König, T. Liss, R. Lüßmann, M. May, B. Nonhoff, B. Reichel, R. Strehlow, A. Stamatakis, N. Stuckmann, A. Vilbig, M. Lenke, T. Ludwig, A. Bode and K.H. Schleifer. 2004. ARB: A software environment for sequence data. Nucleic Acids Res. *32*: 1363–1371.

Maiden, M.C., J.A. Bygraves, E. Feil, G. Morelli, J.E. Russell, R. Urwin, Q. Zhang, J. Zhou, K. Zurth, D.A. Caugant, I.M. Feavers, M. Achtman and B.G. Spratt. 1998. Multilocus sequence typing: a portable approach to the identification of clones within populations of pathogenic microorganisms. Proc. Natl. Acad. Sci. U.S.A. *95*: 3140–3145.

Mannerova, S., R. Pantucek, J. Doskar, P. Svec, C. Snauwaert, M. Vancanneyt, J. Swings and I. Sedlacek. 2003. *Macrococcus brunensis* sp. nov., *Macrococcus hajekii* sp. nov. and *Macrococcus lamae* sp. nov., from the skin of llamas. Int. J. Syst. Evol. Microbiol. *53*: 1647–1654.

Marquez, M.C., A. Ventosa and F. Ruiz-Berraquero. 1990. *Marinococcus hispanicus*, a new species of moderately halophilic Gram-positive cocci. Int. J. Syst. Bacteriol. *40*: 165–169.

Marsik, F.J. and S. Brake. 1982. Species identification and susceptibility to 17 antibiotics of coagulase-negative staphylococci isolated from clinical specimens. J. Clin. Microbiol. *15*: 640–645.

Marsou, R., M. Bes, M. Boudouma, Y. Brun, H. Meugnier, J. Freney, F. Vandenesch and J. Etienne. 1999. Distribution of *Staphylococcus sciuri* subspecies among human clinical specimens, and profile of antibiotic resistance. Res. Microbiol. *150*: 531–541.

Mendoza, M., H. Meugnier, M. Bes, J. Etienne and J. Freney. 1998. Identification of *Staphylococcus* species by 16S–23S rDNA intergenic spacer PCR analysis. Int. J. Syst. Bacteriol. *48*: 1049–1055.

Meyer, S.A. and K.H. Schleifer. 1978. Deoxyribonucleic acid reassociation in classification of coagulase-positive staphylococci. Arch. Microbiol. *117*: 183–188.

Meyer, W. 1967. A proposal for subdividing the species *Staphylococcus aureus*. Int. J. Syst. Bacteriol. *17*: 387–389.

Miller, J.M. 1998. A Guide to Specimen Management in Clinical Microbiology. American Society for Microbiology, Washington, D.C.

Mota, R.R., M.C. Marquez, D.R. Arahal, E. Mellado and A. Ventosa. 1997. Polyphasic taxonomy of *Nesterenkonia halobia*. Int. J. Syst. Bacteriol. *47*: 1231–1235.

Murray, P.R., E.J. Baron, M.A. Pfaller, F.C. Tenover and R.H. Yolken. 1999. Manual of Clinical Microbiology, 7th ed. American Society for Microbiology, Washington, D.C.

Nieto, J.J., R. Fernandez-Castillo, M.T. Garcia, E. Mellado and A. Ventosa. 1993. Survey of antimicrobial susceptibility of moderately halophilic eubacteria and extremely halophilic aerobic archaeobacteria: utilization of antimicrobial resistance as a genetic marker. Syst. Appl. Microbiol. *16*: 352–360.

Noble, W.C. and D.A. Somerville. 1974. Microbiology of Human Skin. W.B.Saunders, London.

Pantucek, R., I. Sedlacek, P. Petras, D. Koukalova, P. Svec, V. Stetina, M. Vancanneyt, L. Chrastinova, J. Vokurkova, V. Ruzickova, J. Doskar, J. Swings and V. Hajek. 2005. *Staphylococcus simiae* sp. nov., isolated from South American squirrel monkeys. Int. J. Syst. Evol. Microbiol. *55*: 1953–1958.

Petráš, P. 1998. *Staphylococcus pulvereri* equals *Staphylococcus* vitulus? Int. J. Syst. Bacteriol. *48*: 617–618.

Pioch, G., H. Heyne and W. Witte. 1988. Coagulase-negative staphylococci in combined fodder and in grain. Zentbl. Mikrobiol. *143*: 157–171.

Place, R.B., D. Hiestand, S. Burri and M. Teuber. 2002. *Staphylococcus succinus* subsp. *casei* subsp. nov., a dominant isolate from a surface ripened cheese. Syst. Appl. Microbiol. *25*: 353–359.

Place, R.B., D. Hiestand, S. Burri and M. Teuber. 2003a. *In* Validation of the publication of new names and new combinations previously effectively published outside the IJSEM. List no. 89. Int. J. Syst. Evol. Microbiol. *53*: 1–2.

Place, R.B., D. Hiestand, H.R. Gallmann and M. Teuber. 2003b. *Staphylococcus equorum* subsp. *linens*, subsp. nov., a starter culture component for surface ripened semi-hard cheeses. Syst. Appl. Microbiol. *26*: 30–37.

Place, R.B., D. Hiestand, H.R. Gallmann and M. Teuber. 2003c. *In* Validation of the publication of new names and new combinations previously effectively published outside the IJSEM. List no. 93. Int. J. Syst. Evol. Microbiol. *53*: 1219–1220.

Poutrel, B. 1984. Udder infection of goats by coagulase-negative staphylococci. Vet. Microbiol. *9*: 131–137.

Probst, A.J., C. Hertel, L. Richter, L. Wassill, W. Ludwig and W.P. Hammes. 1998. *Staphylococcus condimenti* sp. nov., from say sauce mash, and *Staphylococcus carnosus* (Schleifer and Fischer 1982) subsp. utilis subsp. nov. Int. J. Syst. Bacteriol. *48*: 651–658.

Rosenbach, F.J. 1884. Micro-organismen bei den Wund-Infections-Krankheiten des Menschen. J.F.Bergmann, Weisbaden.

Rupp, M.E. and G.L. Archer. 1994. Coagulase-negative staphylococci: pathogens associated with medical progress. Clin. Infect. Dis. *19*: 231–243.

Sanchez-Porro, C., S. Martin, E. Mellado and A. Ventosa. 2003. Diversity of moderately halophilic bacteria producing extracellular hydrolytic enzymes. J. Appl Microbiol. *94*: 295–300.

Schleifer, K.H. and O. Kandler. 1972. Peptidoglycan types of bacterial cell walls and their taxonomic implications. Bacteriol Rev *36*: 407–477.

Schleifer, K.H. and W.E. Kloos. 1975. Isolation and characterization of Staphylococci from human skin. 1. Amended descriptions of *Staphylococcus epidermidis* and *Staphylococcus* saprophytic: *Staphylococcus cohnii*, *Staphylococcus haemolyticus*, and *Staphylococcus xylosus*. Int. J. Syst. Bacteriol. *25*: 50–61.

Schleifer, K.H., A. Hartinger and F. Gotz. 1978. Occurrence of D-tagatose-6-phosphate pathway of D-galactose metabolism among staphylococci. FEMS Microbiol. Lett. *3*: 9–11.

Schleifer, K.H., S.A. Meyer and M. Rupprecht. 1979. Relatedness among coagulase-negative staphylococci: Deoxyribonucleic acid reassociation and comparative immunological studies. Arch. Microbiol. *122*: 93–101.

Schleifer, K.H. and E. Krämer. 1980. Selective medium for isolating staphylococci. Zentbl. Bakteriol. Hyg. Abt.1 Orig. C *1*: 270–280.

Schleifer, K.H. and U. Fischer. 1982. Description of a new species of the genus *Staphylococcus: Staphylococcus carnosus*. Int. J. Syst. Bacteriol. *32*: 153–156.

Schleifer, K.H., R. Kilpper-Bälz, U. Fischer, A. Faller and J. Endl. 1982. Identification of *Micrococcus* candidus ATCC 4852 as a strain of *Staphylococcus epidermidis* and of *Micrococcus* caseolyticus ATCC 13548 and *Micrococcus varians* ATCC 29750 as members of a new species, *Staphylococcus caseolyticus*. Int. J. Syst. Bacteriol. *32*: 15–20.

Schleifer, K.H., U. Geyer, R. Kilpper-Bälz and L.A. Devriese. 1983a. *In* Validation of the publication of new names and new combinations previously effectively published outside the IJSB. List no. 12. Int. J. Syst. Bacteriol. *33*: 896–897.

Schleifer, K.H., U. Geyer, R. Kilpper-Bälz and L.A. Devriese. 1983b. Elevation of *Staphylococcus sciuri* subsp. *lentus* (Kloos et al.) to species status: *Staphylococcus lentus* (Kloos et al.) comb. nov. Syst. Appl. Microbiol. *4*: 382–387.

Schleifer, K.H. 1984. The cell envelope. *In* Easmon and Adlam (Editors), Staphylococci and Sraphylococcal Infections, vol. 2. Academic Press, London, pp. 358–428.

Schleifer, K.H., R. Kilpper-Bälz and L.A. Devriese. 1984. *Staphylococcus arlettae* sp. nov., *Staphylococcus equorum* sp. nov. and *Staphylococcus kloosii* sp. nov.: three new coagulase-negative, novobiocin-resistant species from animals. Syst. Appl. Microbiol. *5*: 501–509.

Schleifer, K.H., R. Kilpper-Bälz and L.A. Devriese. 1985. *In* Validation of the publication of new names and new combinations previously effectively published outside the IJSB. List no. 17. Int. J. Syst. Bacteriol. *35*: 223–225.

Schleifer, K.H. 1986. Taxonomy of coagulase-negative staphylococci. *In* Mardh and Schleifer (Editors), Coagulase-negative Staphylococci. Almquist and Wiksell International, Stockholm, Sweden, pp. 11–26.

Shaw, J., M. Stitt and S.T. Cowan. 1951. Staphylococci and their classification. J. Gen. Microbiol. *5*: 1010–1023.

Sinell, H.J. and J. Baumgart. 1966. Selektivnährböden zur Isolierung von Staphylokokken aus Lebensmitteln. Zentbl. Bakteriol. Abt. 1 Orig. *197*: 447–461.

Sloan, G.L., J.M. Robinson and W.E. Kloos. 1982. Identification of *Staphylococcus staphylolyticus* NRRL B-2628 as a biovar of *Staphylococcus simulans*. Int. J. Syst. Bacteriol. *32*: 170–174.

Sompolinsky, D. 1953. De l'impetigo contagiosa suis et du *Micrococcus hyicus* n. sp. Schweiz. Arch. Tierheikd. *95*: 302–309.

Spergser, J., M. Wieser, M. Taubel, R.A. Rosselló-Mora, R. Rosengarten and H.J. Busse. 2003. *Staphylococcus nepalensis* sp. nov., isolated from goats of the Himalayan region. Int. J. Syst. Evol. Microbiol. *53*: 2007–2011.

Sprott, G.D., L. Bakouche and K. Rajagopal. 2006. Identification of sulfoquinovosyl diacylglycerol as a major polar lipid in *Marinococcus halophilus* and *Salinicoccus hispanicus* and substitution with phosphatidylglycerol. Can. J. Microbiol. *52*: 209–219.

Stackebrandt, E., C. Koch, O. Gvozdiak and P. Schumann. 1995. Taxonomic dissection of the genus *Micrococcus: Kocuria* gen. nov., *Nesterenkonia* gen. nov., *Kytococcus* gen. nov., *Dermacoccus* gen. nov., and *Micrococcus* Cohn 1872 gen. emend. Int. J. Syst. Bacteriol. *45*: 682–692.

Tanasupawat, S., Y. Hashimoto, T. Ezaki, M. Kozaki and K. Komagata. 1992. *Staphylococcus piscifermentans* sp. nov., from fermented fish in Thailand. Int. J. Syst. Bacteriol. *42*: 577–581.

Tenover, F.C., R.D. Arbeit, R.V. Goering, P.A. Mickelsen, B.E. Murray, D.H. Persing and B. Swaminathan. 1995. Interpreting chromosomal DNA restriction patterns produced by pulsed-field gel electrophoresis: criteria for bacterial strain typing. J. Clin. Microbiol. *33*: 2233–2239.

van Belkum, A., W. van Leeuwen, M.E. Kaufmann, B. Cookson, F. Forey, J. Etienne, R. Goering, F. Tenover, C. Steward, F. O'Brien, W. Grubb, P. Tassios, N. Legakis, A. Morvan, N. El Solh, R. de Ryck, M. Struelens, S. Salmenlinna, J. Vuopio-Varkila, M. Kooistra, A. Talens, W. Witte and H. Verbrugh. 1998. Assessment of resolution and intercenter reproducibility of results of genotyping *Staphylococcus aureus* by pulsed-field gel electrophoresis of *Sma*I macrorestriction fragments: a multicenter study. J. Clin. Microbiol. *36*: 1653–1659.

van Leeuwen, W.B., C. Jay, S. Snijders, N. Durin, B. Lacroix, H.A. Verbrugh, M.C. Enright, A. Troesch and A. van Belkum. 2003. Multilocus sequence typing of *Staphylococcus aureus* with DNA array technology. J. Clin. Microbiol. *41*: 3323–3326.

Vandenesch, F., S.J. Eykyn, M. Bes, H. Meugnier, J. Fleurette and J. Etienne. 1995. Identification and ribotypes of *Staphylococcus caprae* isolates isolated as human pathogens and from goat milk. J. Clin. Microbiol. *33*: 888–892.

Varaldo, P.E., R. Kilpper-Bälz, F. Biavasco, G. Satta and K.H. Schleifer. 1988. *Staphylococcus delphini* sp. nov., a coagulase-positive species isolated from dolphins. Int. J. Syst. Bacteriol. *38*: 436–439.

Ventosa, A., E. Quesada, F. Rodríguez-Valera, F. Ruiz-Berraquero and A. Ramos-Cormenzana. 1982. Numerical taxonomy of moderately halophilic Gram-negative rods. J. Gen. Microbiol. *128*: 1959–1968.

Ventosa, A., A. Ramos-Cormenzana and M. Kocur. 1983. Moderately halophilic Gram-positive cocci from hypersaline environments. Syst. Appl. Microbiol. *4*: 564–570.

Ventosa, A., M.C. Marquez, F. Ruiz-Berraquero and M. Kocur. 1990a. *Salinicoccus roseus* gen. nov., sp. nov., a new moderately halophilic Gram-positive coccus. Syst. Appl. Microbiol. *13*: 29–33.

Ventosa, A., M.C. Marquez, F. Ruiz-Berraquero and M. Kocur. 1990b. *In* Validation of the publication of new names and new combinations previously effectively published outside the IJSB. List no. 34. Int. J. Syst. Bacteriol. *40*: 320–321.

Ventosa, A., M.C. Marquez, N. Weiss and B.J. Tindall. 1992. Transfer of *Marinococcus hispanicus* to the genus *Salinicoccus* as *Salinicoccus hispanicus* comb. nov. Syst. Appl. Microbiol. *15*: 530–534.

Ventosa, A., M.C. Marquez, N. Weiss and B.J. Tindall. 1993. *In* Validation of the publication of new names and new combinations previously effectively published outside the IJSB. List no. 45. Int. J. Syst. Bacteriol. *43*: 398–399.

Ventosa, A., J.J. Nieto and A. Oren. 1998. Biology of moderately halophilic aerobic bacteria. Microbiol Mol Biol Rev *62*: 504–544.

Vernozy-Rozand, C., C. Mazuy, H. Meugnier, M. Bes, Y. Lasne, F. Fiedler, J. Etienne and J. Freney. 2000. *Staphylococcus fleurettii* sp. nov., isolated from goat's milk cheeses. Int. J. Syst. Evol. Microbiol. *50*: 1521–1527.

Wang, X.M., L. Noble, B.N. Kreiswirth, W. Eisner, W. McClements, K.U. Jansen and A.S. Anderson. 2003. Evaluation of a multilocus sequence typing system for *Staphylococcus epidermidis*. J. Med. Microbiol. *52*: 989–998.

Watts, J.L., J.W. Pankey and S.C. Nickerson. 1984. Evaluation of the Staph-Ident and STAPHase systems for identification of staphylococci from bovine intramammary infections. J. Clin. Microbiol. *20*: 448–452.

Webster, J.A., T.L. Bannerman, R.J. Hubner, D.N. Ballard, E.M. Cole, J.L. Bruce, F. Fiedler, K. Schubert and W.E. Kloos. 1994. Identification of the *Staphylococcus sciuri* species group with EcoRI fragments containing ribosomal RNA sequences and description of *Staphylococcus vitulus* sp. nov. Int. J. Syst. Bacteriol. *44*: 454–460.

Winslow, C.E.A. and A. Winslow. 1908. The Systematic Relationships of the Coccaceae. John Wiley and Sons, New York.

Yoon, J.H., K.C. Lee, N. Weiss, K.H. Kang and Y.H. Park. 2003. *Jeotgalicoccus halotolerans* gen. nov., sp. nov. and *Jeotgalicoccus psychrophilus* sp. nov., isolated from the traditional Korean fermented seafood jeotgal. Int. J. Syst. Evol. Microbiol. *53*: 595–602.

Zakrzewska-Czerwinska, J., A. Gaszewska-Mastalarz, B. Lis, A. Gamian and M. Mordarski. 1995. *Staphylococcus pulvereri* sp. nov., isolated from human and animal specimens. Int. J. Syst. Bacteriol. *45*: 169–172.

Zhang, W., Y. Xue, Y. Ma, P. Zhou, A. Ventosa and W.D. Grant. 2002. *Salinicoccus alkaliphilus* sp. nov., a novel alkaliphile and moderate halophile from Baer Soda Lake in Inner Mongolia Autonomous Region, China. Int. J. Syst. Evol. Microbiol. *52*: 789–793.

Zhang, Y.Q., L.Y. Yu, H.Y. Liu, Y.Q. Zhang, L.H. Xu and W.J. Li. 2007. *Salinicoccus luteus* sp. nov., isolated from a desert soil. Int. J. Syst. Evol. Microbiol. *57*: 1901–1905.

Family IX. **Thermoactinomycetaceae** Matsuo, Katsuta, Matsuda, Shizuri, Yokota and Kasai 2006, 2840[VP]

MICHAEL GOODFELLOW AND AMANDA L. JONES

Ther'mo.ac.ti.no.my.ce.ta'ce.ae. N.L. masc. n. *Thermoactinomyces* type genus of the family; *-aceae* ending to denote a family; N.L. fem. pl. n. *Thermoactinomycetaceae* the *Thermoactinomyces* family.

Aerobic, Gram-stain-positive, chemo-organotroph. **Substrate mycelium well-developed, branched, and septate.** Usually forms an abundant white or yellow aerial mycelium. **Spores formed singly on aerial and substrate hyphae, sessile or on simple or branched sporophores with the structure and properties of bacterial endospores. Mesophilic or thermophilic. Wall peptidoglycan contains *meso*-diaminopimelic acid but no characteristic sugars.** Major menaquinones unsaturated with seven or nine isoprene units. Rich in straight-chain and iso- and anteiso-fatty acids. Members of the family have been isolated from air, humidifier and air conditioning units, soils, muds, marine sediments, moldy and decaying plant materials, and composts, often with spontaneous heating.

DNA G+C content (mol%): 40.0–60.3.

Type genus: **Thermoactinomyces** Tsiklinsky 1899, 501 emend. Yoon, Kim, Shin and Park 2005, 398.

Further descriptive information

Phylogeny. The classification of thermoactinomycetes has undergone significant changes since the First Edition of *Bergey's Manual of Systematic Bacteriology* where they were assigned to the genus *Thermoactinomyces* (Lacey and Cross, 1989), a taxon which mainly encompassed thermophilic organisms. The genus *Thermoactinomyces* and its type species, *Thermoactinomyces vulgaris*, were described by Tsiklinsky in 1899. Subsequently, additional *Thermoactinomyces* species were proposed, including *Thermoactinomyces peptonophilus* Nonomura and Ohara 1971, *Thermoactinomyces sacchari* Lacey (1971a), *Thermoactinomyces candidus* Kurup et al. 1975, *Thermoactinomyces intermedius* Kurup et al. 1980, *Thermoactinomyces thalpophilus* Unsworth and Cross (1980), and *Thermoactinomyces putidus* Lacey and Cross 1989. In addition, *Thermoactinomyces dichotomicus* was proposed by Cross and Goodfellow (1973) for organisms previously classified as "*Actinobifida dichotomica*" by Krasil'nikov and Agre (1964). It was recognized that members of the species were closely related though the standing of some of the species was questioned (Lacey and Cross, 1989).

More recently, *Thermoactinomyces candidus* and *Thermoactinomyces thalpophilus* were reclassified as synonyms of *Thermoactinomyces vulgaris* and *Thermoactinomyces sacchari*, respectively, based on DNA–DNA relatedness and 16S rRNA gene sequence data (Yoon et al., 2000). Additional comparative studies on representatives of validly published *Thermoactinomyces* species showed that they belong to four phyletic lines, the taxonomic standings of which were supported by chemotaxonomic and phenotypic data (Yoon et al., 2005). Three new genera were proposed on the basis of these data. *Thermoactinomyces putidus* and *Thermoactinomyces sacchari* were assigned to the genus *Laceyella* as *Laceyella putida* and *Laceyella sacchari*, *Thermoactinomyces peptonophilus* to the genus *Seinonella* as *Seinonella peptonophila*, and *Thermoactinomyces dichotomicus* to the genus *Thermoflavimicrobium* as *Thermoflavimicrobium dichotomicum*. The revised genus *Thermoactinomyces* was left as a home for *Thermoactinomyces intermedius* and *Thermoactinomyces vulgaris*.

Similar studies on additional thermoactinomycetes led to proposals for the recognition of three additional taxa, the genera *Mechercharimyces* Matsuo et al. 2006, *Planifilum* Hatayama et al. 2005b, and *Shimazuella* Park et al. 2007. The genus *Mechercharimyces* contains two species, *Mechercharimyces mesophilus*, the type species, and *Mechercharimyces asporophorigenens*. The genus *Planifilum* also contains two species, *Planifilum fimeticola*, the type species, and *Planifilum fulgidum*. The genus *Shimazuella* contains a single species, *Shimazuella kribbensis*. Members of the genera *Mechercharimyces* and *Shimazuella* are mesophiles whereas those assigned to the genus *Planifilum* are thermophiles.

The family *Thermoactinomycetaceae* was. proposed by Matsuo et al. (2006) to accommodate the genera *Laceyella*, *Mechercharimyces*, *Planifilum*, *Seinonella*, *Thermoactinomyces*, and *Thermoflavimicrobium*. This taxon is one of the families that constitute the order *Bacillales*. The relationships between the constituent genera of the family *Thermoactinomyces* and between them and the type strains of the type species of some of the genera classified in the order *Bacillales* are shown in Figure 74.

General comments. The following sections provide further descriptive information on the properties of species assigned to the genus *Thermoactinomyces sensu lato*, along the lines described in the previous edition of the *Systematics* (Lacey and Cross, 1989). In contrast, nothing is known about the properties of the species classified in the genera *Mechericharimyces* and *Planifilum* beyond the details published in the original species descriptions. In addition to this unavoidable limitation, the details of the better-known species of the genus *Thermoactinomyces sensu lato* are given under names of the taxa used in the earlier literature though the names in the Figures have been updated. It has been shown that *Thermoactinomyces putidus*, *Thermoactinomyces sacchari*, and *Thermoactinomyces vulgaris* strains share little DNA hybridization with each other (Hirst et al., 1991).

Cell morphology. The substrate mycelium consists of stable, branched, septate hyphae from which aerial hyphae arise. They first arise vertically but, in most species, then form a loose network of almost straight hyphae over the substrate (Locci, 1972). Intercalary growth of the primary substrate mycelium has been reported for *Thermoactinomyces thalpophilus* (Kretschmer, 1984c). Aerial hyphae of *Thermoactinomyces vulgaris* are hydrophobic and may be found within 6.5 h of spore germination (Kretschmer, 1984a). The aerial hyphae may autolyze (within 2–4 d in *Thermoactinomyces vulgaris* or more rapidly in *Thermoactinomyces sacchari*) depositing spores on the agar surface (Küster and Locci, 1963; Lacey, 1971a). Spores are formed singly on both substrate (Figure 75) and aerial mycelium and may be either sessile or on hyphae called sporophores (Figure 76, Figure 77, Figure 78, Figure 79, and Figure 80). In *Thermoactinomyces dichotomicus*, both sporophores and mycelium may be dichotomously branched (Figure 80). The spores are spheroidal, 0.5–1.5 µm in diameter, with a ridged surface that gives an angular appearance by

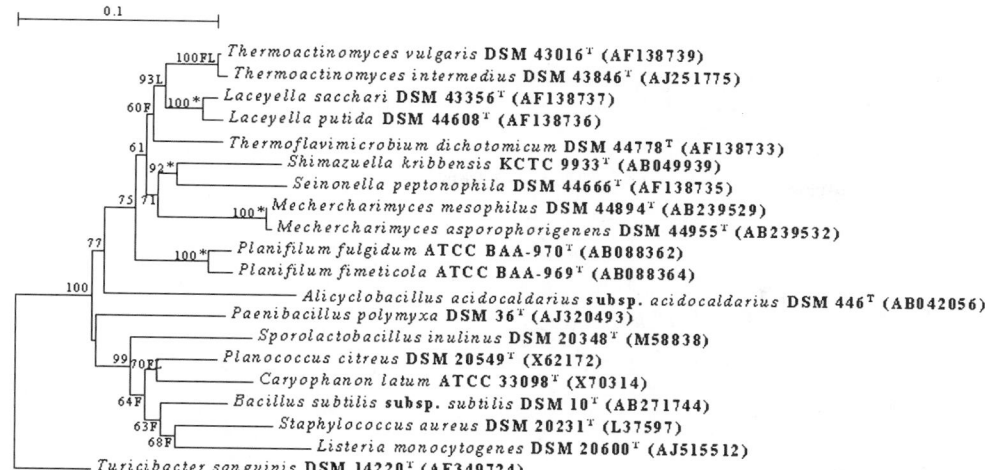

FIGURE 74. Neighbor-joining tree showing the positions of genera and species of the family *Thermoactinomycetaceae* and their relationship to phylogenetically close organisms classified in the order *Bacillales*. The asterisks indicate phyletic lines that were also recovered using the least-squares, maximum-likelihood, and maximum-parsimony tree-making algorithms. When lineages were not recovered in all three algorithms, F, L, and P indicate branches that were recovered using the least-squares, maximum-likelihood and maximum-parsimony methods, respectively. The numbers at the nodes are percentage bootstrap values based on a neighbor-joining analysis of 1000 resampled datasets; only values above 50% are given. The scale bar indicates 0.01 substitutions per nucleotide position.

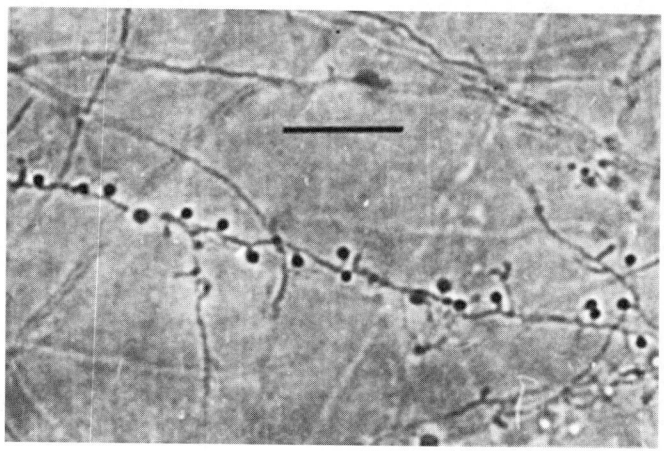

FIGURE 75. Substrate mycelium spores of *Laceyella* (*Thermoactinomyces*) *sacchari*. Half-strength nutrient agar, incubation at 55°C. Bar = 10 μm.

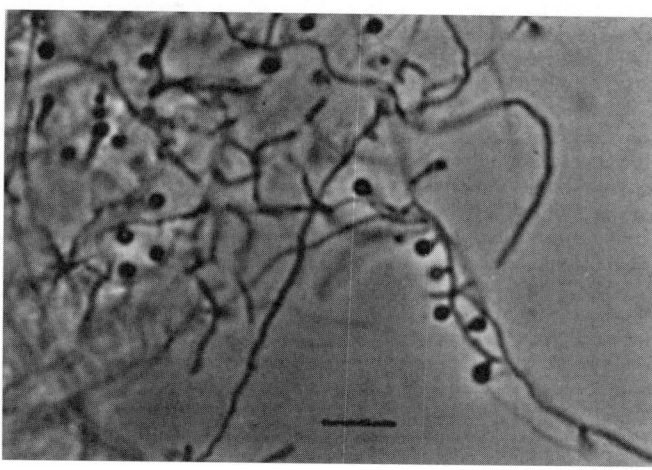

FIGURE 76. Inclined coverslip preparation of growth of *Thermoactinomyces vulgaris*. CYC agar, incubation at 50°C. Bar = 5 μm.

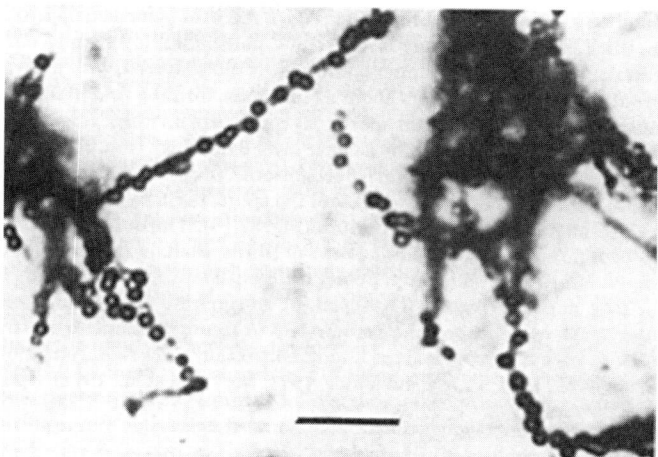

FIGURE 77. Aerial mycelium spores of *Laceyella sacchari* (previously *Thermoactinomyces thalpophilus*). Half-strength nutrient agar, incubation at 55°C. Bar = 10 μm.

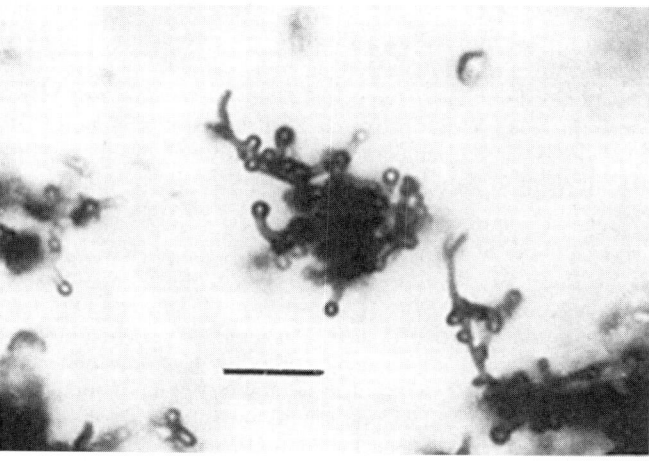

FIGURE 78. Aerial mycelium spores of *Laceyella* (*Thermoactinomyces*) *sacchari*. Yeast-malt agar, incubation at 55°C. Bar = 5 μm.

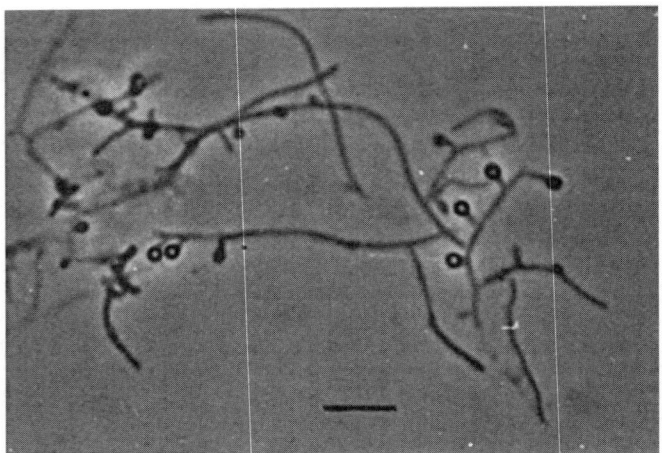

FIGURE 79. Inclined coverslip preparation of *Laceyella* (*Thermoactinomyces*) *putida*. CYC agar, incubation at 50°C. Bar = 10 μm.

FIGURE 80. Aerial mycelium spores of *Thermoflavimicrobium* (*Thermoactinomyces*) *dichotomicus*. Half-strength nutrient agar, incubation at 55°C. Bar = 20 μm.

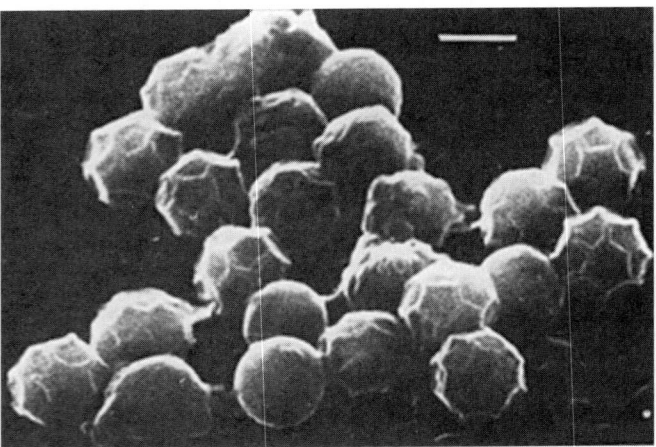

FIGURE 81. Scanning electron micrograph of spores of *Laceyella* (*Thermoactinomyces*) *sacchari*. Bar = 1 μm.

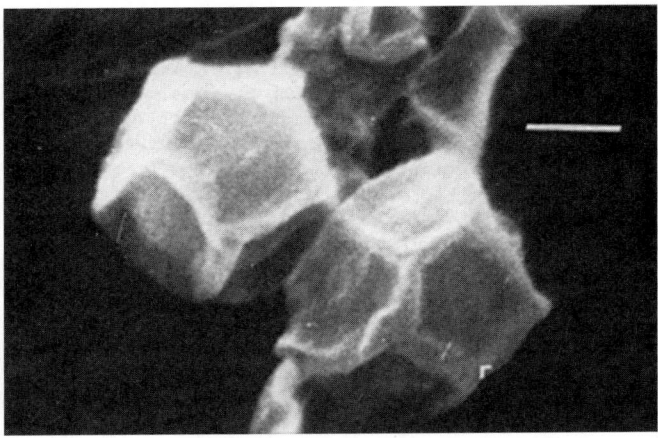

FIGURE 82. Scanning electron micrograph of spores of *Thermoflavimicrobium* (*Thermoactinomyces*) *dichotomicum*. Bar = 0.5 μm.

transmission electron microscopy. By scanning electron microscopy, the ridges can be seen to form pentagonal and hexagonal areas on the spore surface. Most species resemble that studied by McVittie et al. (1972) in having approximately 12 pentagonal and 12 hexagonal faces (Figure 81), although *Thermoactinomyces dichotomicus* has only approximately 12 faces in total (Figure 82). The spores are refractile and phase-bright by light microscopy, staining only with endospore stains.

Cell-wall composition. The wall peptidoglycan contains *meso*-A$_2$pm but no diagnostic sugars (wall chemotype III; Becker et al., 1965). Menaquinones are unsaturated, with seven (*Thermoactinomyces dichotomicus*, *Thermoactinomyces sacchari*, and *Thermoactinomyces vulgaris*) or nine (*Thermoactinomyces putidus* and *Thermoactinomyces thalpophilus*) isoprene units in major amounts (Collins et al., 1982). Organisms contain straight-chain and iso- and anteiso-branched fatty acids, but phospholipids of diagnostic value have not been studied (Goodfellow and Cross, 1984). As with bacterial endospores, spores of *Thermoactinomyces* species contain dipicolinic acid (6.5–7% (w/w) in *Thermoactinomyces vul-*

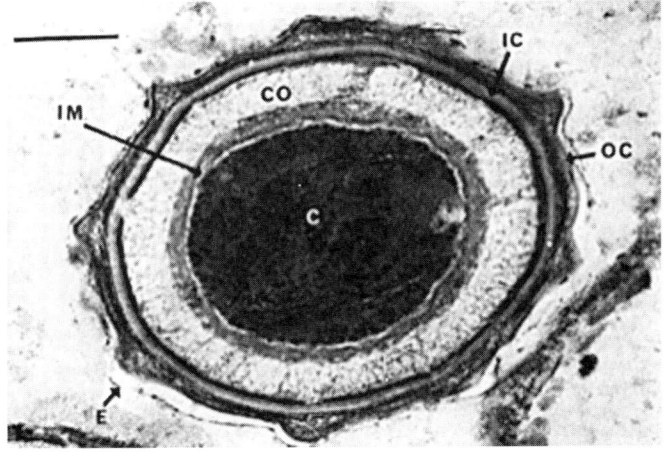

FIGURE 83. Thin-section of *Laceyella* (*Thermoactinomyces*) *sacchari* endospore. Bar = 0.1 μm. C, core; CO, cortex; IC, inner spore coat; IM, inner forespore membrane; OC, outer spore coat; E, possible exosporium.

garis and *Thermoactinomyces sacchari*, 3.6% in *Thermoactinomyces dichotomicus*, but only 0.6% in *Thermoactinomyces peptonophilus*) (Attwell, 1978; Cross, 1968; Lacey, 1971a). They also contain large amounts of calcium and magnesium ions (Kalakoutskii and Agre, 1973, 1976; Kalakoutskii et al., 1969). The DNA base composition of *Thermoactinomyces* species is similar to that of other endospore-forming bacteria. Values of mol% G+C (T_m) of 52.0 (Fritzsche, 1967), 53.4–54.8 (Craveri and Manachini, 1966), 54.1–54.8 (Craveri et al., 1966b), and 48.7–52.2 (Hirst et al., 1991) have been reported.

Fine structure. In thin section, hyphae of *Thermoactinomyces* species are 0.3–0.6 μm in diameter and bounded by a wall about 22 nm thick, similar to that of other Gram-stain-positive bacteria. The cytoplasm is bounded by a plasma membrane; a typical unit membrane about 10 nm thick that bears tubular vesicular mesosomes 0.1–0.3 μm in diameter. Septa are formed by a single ingrowth of plasma membrane and hyphal wall, often bearing mesosomes (type 1; Williams et al., 1973). Nuclear material occurs as an axial filament, and ribosomes, about 12 nm in diameter, are usually present. Such structures have been observed in *Thermoactinomyces peptonophilus*, *Thermoactinomyces sacchari*, and *Thermoactinomyces thalpophilus* (Attwell, 1978; Cross et al., 1968b; Dorokhova et al., 1968, 1970b, a; Lacey and Vince, 1971), but in *Thermoactinomyces sacchari*, although the cytoplasm is uniformly dense in young cells, it soon becomes coarsely granular and less dense and is then released by lysis of the cells (Lacey and Vince, 1971).

Spore formation follows the same stages of development as with *Bacillus* endospores. The first stage is the formation of a double membrane across the cell forming a spore septum about 18 h after inoculation. The septum lengthens as the area of attachment to the plasma membrane progressively decreases, engulfing cytoplasm and nuclear material from the mother cell. Mesosomes are closely associated with this process. Eventually this membrane closes and breaks away from the plasma membrane to form a forespore, one to a cell, which moves to the hyphal wall and out into a lateral sporangium that is sessile in *Thermoactinomyces thalpophilus* and terminal on a short sporophore in *Thermoactinomyces sacchari* (Figure 83). A cortex is formed up to 0.17 μm thick, with a dense inner layer, a thick, pale middle layer, and a dark, granular outer layer. The inner spore coat is multilayered, with 6–10 alternating light and dark layers about 27 nm thick forming first in discrete zones. The two-layer outer spore coat consists of an inner, ridged layer 25–70 nm thick surrounded by a dense layer 8 nm thick (Figure 84). After maturity, often no sporangium is recognizable (probably due to lysis), but a thin, dark membrane may surround the spore, perhaps the remains of the sporangium or of an exosporium (Figure 84). The outer coat of the mature spore consists of parallel rows of fibrils, each measuring about 5 nm. Nuclear material is often poorly differentiated from the cytoplasm but may appear irregular, U-shaped, or as separate areas at opposite poles of the spore. Ribosome granules up to 15 nm in diameter are frequent in the cytoplasm, and the middle cortex may show radial striations (Lacey and Vince, 1971; McVittie et al., 1972). Spore development in *Thermoactinomyces dichotomicus* and *Thermoactinomyces peptonophilus* is essentially the same, differing chiefly in the degree of sporophore development and in the possible presence of an exosporium in *Thermoactinomyces dichotomicus* (Attwell, 1978; Cross et al., 1968b; Dorokhova et al., 1970b).

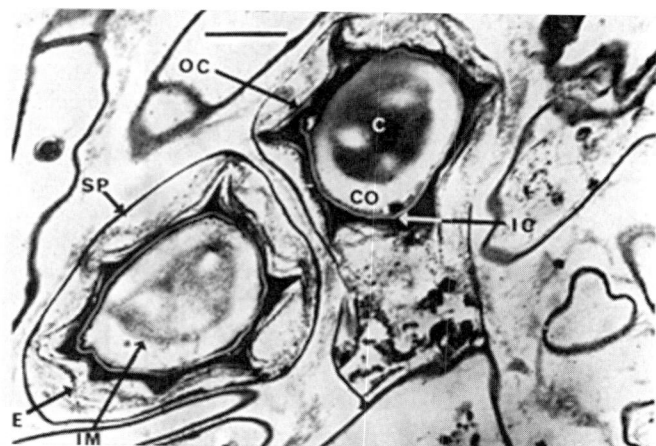

FIGURE 84. Thin-section of *Laceyella* (*Thermoactinomyces*) *dichotomicum* endospore. Bar = 0.4 μm. C, core; CO, cortex; IC, inner spore coat; IM, inner forespore membrane; OC, outer spore coat; E, possible exosporium, SP, sporangial wall.

Spore germination may occur within 3 h of incubation but frequently takes longer. Enhanced CO_2 levels (Kretschmer and Jacob, 1983) and a pH greater than 7.0 are reported to be essential to germination. Optimum pH is 8.0–10.0 (Foerster, 1975). Activation and germination may be stimulated by calcium dipicolinate, L-alanine, L-leucine, L-α-aminobutyric acid, inosine, and adenosine; by heating briefly to 100°C; or by cooling to 20°C. However, heating may also deactivate spores (Agre et al., 1972a, b; Attwell et al., 1972; Foerster, 1975; Kalakoutskii and Agre, 1976; Kirillova et al., 1974). Agre et al. (1972a, 1972b) observed variation in the ability of spores to germinate, identifying in aqueous spore suspensions three types consisting of those germinating without previous activation, those requiring mild activation treatment, and those needing severe activation treatment. Activated spores either had germinated, were activated, or could be deactivated by heating at 80°C. Ability to germinate was decreased by drying. Kokina and Agre (1977b, 1977a) found that cultures degenerated with repeated subculture, resulting in less sporulation and poor spore germination.

On germination, the cortex disappears, the core swells to fill the space, and U-shaped fibrillar areas of nuclear material arise. The core swells to about 2 μm diameter with destruction of the spore coats, although the outermost layer may remain more or less intact. Finally, a germ tube is formed with an axial nuclear area of coarsely granular material and the spore contents then senesce (Dorokhova et al., 1968; Lacey and Vince, 1971).

Cultural characteristics. Colonies are usually fast-growing at 55°C on nutrient or Czapek-yeast-casein (CYC)* agars, flat or lightly ridged, usually with white aerial mycelium. However, in *Thermoactinomyces sacchari* aerial mycelia are often transient, and colonies may appear "bacterial". In contrast, in *Thermoactinomyces dichotomicus* the aerial mycelium is yellow. In *Thermoactinomyces putidus* and *Thermoactinomyces sacchari*, lysis may deposit spores on the

---

*CYC agar: Czapek-Dox agar powder (Oxoid CM 97), 33.4 g; yeast extract, 2.0 g; vitamin-free Casamino acids, 6.0 g; pH 7.2. Novobiocin (Albamycin, Upjohn; 25 μg/ml) and cycloheximide (Acti-dione, Koch-Light; 50 μg/ml) added after autoclaving.

agar surface (Lacey, 1971a; Unsworth, 1978). *Thermoactinomyces peptonophilus* is exceptional in being mesophilic and requiring peptone and B vitamins for growth. Aerial mycelium production is favored by 0.02% (w/v) glycerol or glucose as supplements in glycerol-asparagine, oatmeal, yeast-starch, and peptone-yeast extract (PY)* agars (Nonomura and Ohara, 1971). Soluble pigments are produced by some species, and *Thermoactinomyces intermedius*, *Thermoactinomyces putidus*, and *Thermoactinomyces thalpophilus* produce brown, water-soluble melanin pigments with 0.5% L-tyrosine in CYC agar (Kurup et al., 1980; Lacey, 1971a; Unsworth, 1978).

Life cycles. Alternative life cycles for a *Thermoactinomyces* species, *Thermoactinomyces thalpophilus*, have been suggested (Kretschmer, 1984a). The primary mycelium formed on germination, with septa about every 10 µm, continues to grow without producing endospores, until growth conditions become limiting. At that time, two types of secondary mycelium, substrate or aerial, may form depending on the composition of the substrate, while older hyphae may lyse. Hydrophobic aerial mycelium may develop as soon as nutrients become limiting, while secondary substrate mycelium develops after a lag phase. Coincidentally, the intervals between septa decreases to 0.75–0.8 µm and endospores form. Whether secondary substrate or aerial mycelia are formed depends on the nature of the limitation and differs between different agar media. The low phosphate in corn steep liquor (CSL) agar† favors secondary aerial mycelium production, whereas the low nitrogen in Czapek-Dox (CD) agar‡ favors the production of secondary substrate mycelium. Changes in gene expression occur during the lag period before the secondary substrate mycelium is formed, resulting in the production of extracellular proteases. Since spore formation starts as soon as the aerial mycelium is produced, mature spores are produced by this route approximately 5 h earlier than by the secondary substrate mycelium route. Intercalary extension of cells may sometimes be observed giving a zig-zag appearance, but separation of cells (as in *Nocardia*) never occurs (Kretschmer, 1984b).

Nutrition and growth conditions. Nutritional requirements of *Thermoactinomyces* species are incompletely known. All the thermophilic species tested degrade casein and, except for some strains of *Thermoactinomyces sacchari*, gelatin, but none degrades hypoxanthine or xanthine. *Thermoactinomyces peptonophilus* degrades neither gelatin nor casein. *Thermoactinomyces vulgaris* isolates differ greatly in their ability to utilize individual carbon sources. Few isolates of any species reduce nitrate to nitrite. Water-soluble melanin pigments are produced by *Thermoactinomyces intermedius*, *Thermoactinomyces putidus*, and *Thermoactinomyces thalpophilus* on media containing tyrosine provided that these do not contain inhibitory peptones (Unsworth, 1978). Biotin and methionine are reported to be essential for growth of at least some *Thermoactinomyces* isolates, and no aerial growth is obtained in the absence of methionine (Webley, 1958).

Most *Thermoactinomyces putidus* isolates produce acid from sucrose, and up to 30% of *Thermoactinomyces thalpophilus* strains produce acid from fructose, glycerol, mannitol, mannose, ribose, sucrose, and trehalose. A few isolates of *Thermoactinomyces vulgaris* produce acid from these sugars, whereas only 20% of *Thermoactinomyces sacchari* produce acid from glycerol but no other sugar. Acid production by other species has not been tested (Unsworth, 1978). *Thermoactinomyces dichotomicus*, *Thermoactinomyces intermedius*, *Thermoactinomyces sacchari*, *Thermoactinomyces thalpophilus*, and *Thermoactinomyces vulgaris* are all resistant to lysozyme, but the resistance of other species is not known (Goodfellow and Pirouz, 1982; Kurup et al., 1980).

Thermophilic *Thermoactinomyces* species all grow between 35° and 58°C, but *Thermoactinomyces putidus* and *Thermoactinomyces thalpophilus* grow down to 29°C, *Thermoactinomyces sacchari* grows up to 60°C, and *Thermoactinomyces dichotomicus* and *Thermoactinomyces vulgaris* up to 62°C. *Thermoactinomyces peptonophilus* grows poorly at 25°C and optimally at 35°C, and fails to grow at 45°C (Lacey, 1971a; Nonomura and Ohara, 1971; Unsworth, 1978).

Production of aerial mycelium may be inhibited when air is replaced by pure oxygen, perhaps through the inactivation of essential -SH groups in thiol enzymes (Webley, 1954). Respiration is greatest in 1–2-d-old cultures bearing aerial mycelium. Vegetative mycelium more than 1 d old and spores respire little (Erikson and Webley, 1953), perhaps a consequence of low cytochrome α content (Taptykova et al., 1969).

Enzyme production. Because of their activity at high temperatures, enzymes of *Thermoactinomyces* species, especially their amylases, lipases, and proteases, have attracted much interest. Thermitase is an extracellular endopeptidase that is reportedly well-suited for use in the food industry. It is the principal component of proteases produced in culture medium; temperature of optimal activity against peptide esters is 60°C, against peptide p-nitroanilides is 65–75°C, and against casein is 90°C, although it is rapidly inactivated at this temperature (Behnke et al., 1982). Optimum production of thermitase is at 50–52°C with a pH of 6.6–7.3 (Leuchtenberger and Ruttloff, 1983). The crystal structure of thermitase from *Thermoactinomyces vulgaris* has been determined (Teplyakov et al., 1989).

Protease production begins at the transition from exponential to linear growth phases after about 5 h of incubation. During linear growth, up to 45% of the hyphae are lysed. After 22 h of fermentation, the enzyme comprises three components forming 33%, 64%, and 3% of the total, differing in their retention in Sephadex gel and also in the protein bands on polyacrylamide gel electrophoresis. Enzyme production ceases after 10 h of growth when easily utilizable carbohydrates are exhausted, leading to decreased respiration (Kretschmer et al., 1982; Taufel et al., 1979). Enzyme synthesis is promoted by adding rape oil to the medium as an antifoam agent because lysis of the mycelium is decreased (Leuchtenberger and Ruttloff, 1983). Autolysis is closely associated with heat inactivation of the enzyme thermitase.

Serine proteases reported by Roberts et al. (1977) from both *Thermoactinomyces thalpophilus* and *Thermoactinomyces vulgaris* were similar, but *Thermoactinomyces sacchari* shows no proteolytic activity. However, zymograms of the first two species differ, with *Thermoactinomyces vulgaris* showing two intense proteolytic bands at rA 0.3 and 1.4 and *Thermoactinomyces thalpophilus* 3–4 bands between rA 0.1 and 0.4. Proteolytic enzymes with molecular weights of 27,500 and 23,800 are produced optimally by some

---

*PY agar: peptone, 20 g; yeast extract, 20 g; glycerol, 2 g; MgSO$_4$ 7H$_2$O, 0.3 g; agar, 20 g; water 1 l; pH 7.6.

†CSL agar: corn steep liquor (50% dry matter), 5 g: vitamin-free Casamino acids (Difco), 3 g; sucrose, 15 g; corn starch, 5 g; NaNO$_3$, 1 g; NaCl, 2.5 g; KCl, 0.025 g; CaCl$_2$, 0.25 g; FeSO$_4$ 7H$_2$O, 0.005 g; magnesium glycerophsphate, 0.25 g; agar, 20 g; distilled water, 1 l; pH 7.2.

‡CD agar: sucrose, 10 g; vitamin-free Casamino acids (Difco), 1.5 g; casein peptone (Serval), 1.5 g; NaCl, 1.0 g; Kh$_2$PO$_4$, 1.2 g; Na$_2$HPO$_4$, 1.2 g; Na$_2$HPO$_4$ 2H$_2$O, 2.8 g; CaCO$_3$, 0.005 g; FeSO$_4$ 7H$_2$O, 0.01 g; MgSO$_4$, 0.5 g; agar, 20 g; water, 1 l; pH 6.3.

Thermoactinomyces isolates at 60–70°C and pH 9.0 and cause lysis of heat-inactivated cells of fungi, mycobacteria, and *Firmicutes* and especially, *Proteobacteria* (Desai and Dhala, 1966; Desai and Dhala, 1967; Desai and Dhala, 1969; Golovina et al., 1973). Strong lipolytic activity may be found in *Thermoactinomyces thalpophilus* (Elwan et al., 1978a; Elwan et al., 1978b). Lipase is produced optimally at 55°C and pH 6.8 in shake culture in a Czapek medium containing corn oil and 0.2% (w/v) yeast extract, incubated for 24–36h. Activity of the isolated enzyme is greatest at 55°C and pH 8.0. Inactivation occurs in 45min at 80°C and in 5min at 90°C and 100°C.

Kuo and Hartman (1966) first reported α-amylase activity by a *Thermoactinomyces* species, probably *Thermoactinomyces thalpophilus*, and subsequently this was confirmed by others (Allam et al., 1975; Obi and Odibo, 1984; Shimizu et al., 1978). Although the organism used by Allam et al. (1975) was described as *Thermomonospora vulgaris*, illustrations of it in Hussein et al. (1975) are consistent with *Thermoactinomyces thalpophilus*. However, more than one enzyme may be involved. Kuo and Hartman (1966, 1967) found a neutral α-amylase with optimum activity at 60°C and pH 5.9–7.0, whereas other α-amylases with different temperature and pH optima of 70°C and pH 5.0 and 80°C and pH 7.0, respectively, were found in other studies (Obi and Odibo, 1984; Shimizu et al., 1978). The amylase of Shimizu et al. (1978) also had pullulan-hydrolyzing ability, giving panose as the main product, unlike other pullulanases. Cellulolytic acitivity has been reported in a *Thermoactinomyces* isolate by Hägerdal et al. (1978) but not by Fergus (1969), nor has cellulose utilization ever been found in taxonomic tests. It is possible that the isolate used by Hägerdal et al. (1978) was in fact a species of *Thermomonospora*, a genus in which cellulolytic activity is well known (McCarthy and Cross, 1984a).

Enzymic profiles of double-dialysis antigens (Edwards, 1972) on API ZYM strips (APL Systems) show similarities between *Thermoactinomyces sacchari* and *Thermoactinomyces vulgaris* in that both possess alkaline phosphatase, C_4 esterase, and C_8 esterase-lipase, all of which were absent from *Thermoactinomyces thalpophilus* antigens. P. Boiron (personal communication) has additionally found C_{14} lipase, leucine arylamidase, acid phosphatase, and naphthol AS-BI phosphohydrolase in *Thermoactinomyces sacchari* preparations. However, *Thermoactinomyces thalpophilus* is the only species to possess α-glucosidase. Weak acid phosphatase activity is present in *Thermoactinomyces vulgaris* and phosphoamidase activity in *Thermoactinomyces thalpophilus*. C_4 esterase, leucine aminopeptidase, and chymotrypsin activity are present in *Thermoactinomyces thalpophilus* when whole cells are used (Hollick, 1982).

Antimicrobial activity. Antimicrobial activity has been reported in cultures of *Thermoactinomyces dichotomicus* and *Thermoactinomyces thalpophilus* (Craveri et al., 1964; Cross and Unsworth, 1976; Krasil'nikov and Agre, 1964). Both these species can inhibit growth of *Thermoactinomyces vulgaris*, which accounts for the so-called autoinhibition phenomenon (Locci, 1963) observed between isolates when *Thermoactinomyces vulgaris* was considered to be a single variable species. *Thermomonospora chromogena* (Krasil'nikov and Agre 1965) McCarthy and Cross 1984b is also inhibited by *Thermoactinomyces dichotomicus*, whereas thermorubin (Moppett et al., 1972) isolated from cultures of *Thermoactinomyces thalpophilus* (then named *Thermoactinomyces antibioticus*) is more inhibitory to Gram-stain-positive

than to Gram-stain-negative bacteria, but it is also highly toxic to mammals (Kosmachev, 1962; Terao et al., 1965).

Genetics. Recombination, with substitution of small fragments of homologous segments of genetic material, has been observed with mutants obtained from a *Thermoactinomyces thalpophilus* isolate (Hopwood and Ferguson, 1970). It was also shown that *Thermoactinomyces thalpophilus*, like other prokaryotes, has partially diploid zygotes. Most recombinants differed from one or another parent strain by only a single marker, irrespective of the coupling of the markers. The process was characterized as transformation that occurred when agar containing a constant amount of DNA was tested with >10^5 spores leading to confluent mycelial growth (Hopwood and Wright, 1972). The transformation frequency has been estimated at 1×10^{-3} to 1×10^{-4} (Kretschmer and Sarfert, 1980). This high transformation frequency in *Thermoactinomyces thalpophilus* implies that there is little extracellular DNase activity. Transformation was inhibited if DNase was added during the first 6–8h of incubation but not if it was added after 7–9h of growth. Competence develops at the time when aerial mycelium first appears (Hopwood and Wright, 1972).

Phages. Phages to *Thermoactinomyces* species have been reported frequently (Agre, 1961; Kretschmer, 1982; Kretschmer and Sarfert, 1980; Kurup and Heinzen, 1978; Patel, 1969; Prauser and Momirova, 1970; Sarfert et al., 1979; Treuhaft, 1977), mostly from *Thermoactinomyces thalpophilus* but also from *Thermoactinomyces sacchari* and *Thermoactinomyces vulgaris*. The size and structure of phages from various sources differ, having heads 60–72 × 62–84nm and tails 5–20 × 90–120nm (Agre, 1961; Kurup and Heinzen, 1978). Some of the tails had a helical structure with about 29 turns (Patel, 1969); in two others, the DNA content was 28.8×10^6 and 37×10^6 daltons per phage (Kretschmer, 1982). Infectivity was lost after 10min at 70°C and pH 3.6 or when treated with H_2O_2 (Kurup and Heinzen, 1978).

Phages differed in their species specificity. Those from *Thermoactinomyces sacchari* were species specific, but some from *Thermoactinomyces thalpophilus* and *Thermoactinomyces vulgaris* could infect the other species (Kurup and Heinzen, 1978; Treuhaft, 1977). None infected other genera. Plaque morphology differed among hosts. When *Thermoactinomyces thalpophilus* phage infected *Thermoactinomyces thalpophilus*, confluent lysis was characteristic, but, when infecting *Thermoactinomyces vulgaris*, only small plaques were formed (Treuhaft, 1977). Seven host range/ plaque type groups were distinguished using two isolates of *Thermoactinomyces thalpophilus* and *Thermoactinomyces vulgaris* (Treuhaft, 1977) and three types of interaction between phage and host were found in *Thermoactinomyces thalpophilus* (Kretschmer, 1982). Multiplication of phages occurred only in the primary mycelium, and they decreased in number in secondary mycelium and during sporulation. The phage genome was incorporated into spores early in their formation in a heat-stable state and only multiplied on germination. Growing secondary substrate mycelium was competent to take up exogenous DNA, but transfection did not occur (Kretschmer, 1980).

Antigenicity. Species of *Thermoactinomyces* differ antigenically, although there are common components that give some cross-reactivity. *Thermoactinomyces dichotomicus*, *Thermoactinomyces putidus*, *Thermoactinomyces sacchari*, and *Thermoactinomyces vulgaris* are serologically homogeneous, whereas *Thermoactinomyces thalpophilus* isolates are heterogeneous (Arden-Jones and

Cross, 1980). *Thermoactinomyces intermedius* and *Thermoactinomyces sacchari* both cross-react with *Thermoactinomyces thalpophilus* and *Thermoactinomyces vulgaris* but not with one another (Kurup et al., 1980; Kurup et al., 1976b; Lacey, 1971a). Distinctive protein bands are found on polyacrylamide gel electrophoresis of double-dialysis antigens (Edwards, 1972). *Thermoactinomyces vulgaris* gives 10–16 bands, five of which are glycoprotein. There are major bands at rA 0.42, 0.66, and 1.32 that differ from a more diffuse band at rA 0.54 and a prominent solitary band ar rA 0.97 in *Thermoactinomyces thalpophilus*. *Thermoactinomyces sacchari* combined the features of both preceding species but two bands are distinctive (Hollick et al., 1979; Roberts et al., 1977). Pyridine extracts revealed only 8–11 protein bands, three of which were glycoprotein. Crossed immunoelectrophoresis of *Thermoactinomyces vulgaris* antigen revealed 15 immunogens when tested against homologous antiserum and 19 bands with isoelectric points between 3.5 and 5.7 during flat bed isoelectric focusing. Most were heat labile and unaffected by Pronase (Hollick and Larsh, 1979). Most components of *Thermoactinomyces sacchari* antigen are heat labile and partially sensitive to Pronase (Lehrer and Salvaggio, 1978), perhaps because this species lacks serine protease (Roberts et al., 1977).

Antibiotic sensitivity. There is little information on antibiotic sensitivity of *Thermoactinomyces* species. However, all of the thermophilic species, but not the mesophilic *Thermoactinomyces peptonophilus*, are tolerant to nalidixic acid and up to 200 µg novobiocin/ml (Cross, 1968; Cross and Attwell, 1975). Most isolates are sensitive to ampicillin, benzylpenicillin, cephaloridine, chloramphenicol, colistin sulfate, demethylchlortetracycline, erythromycin, gentamicin, kanamycin, neomycin, nitrofurantoin, oleandomycin, penicillin, streptomycin, and tetracycline; *Thermoactinomyces dichotomicus* isolates are also sensitive to lincomycin and vancomycin and *Thermoactinomyces vulgaris* is sensitive to chloramphenicol. Isolates of some species differ in their sensitivity (Goodfellow and Pirouz, 1982).

Pathogenicity. *Thermoactinomyces* species have often been implicated as causes of extrinsic allergic alveolitis (hypersensitivity pneumonitis). *Thermoactinomyces dichotomicus*, *Thermoactinomyces thalpophilus*, and *Thermoactinomyces vulgaris* have all been implicated in farmers' lung disease, although *Faenia rectivirgula* (Cross et al., 1968a) is the major source of the antigen. *Thermoactinomyces putidus* has been identified from a lung biopsy of a patient (Cross and Unsworth, 1976; Molina, 1974; Pepys et al., 1963; Pether and Greatorex, 1976; Terho and Lacey, 1979; Unsworth, 1978; Wenzel et al., 1967; Wenzel et al., 1974). *Thermoactinomyces sacchari* is a principal source of bagassosis antigen (Lacey, 1971a), and *Thermoactinomyces vulgaris* is reported to cause humidifier fever (Banaszak et al., 1970; Sweet et al., 1971). Often the role of individual species has not been clear because *Thermoactinomyces thalpophilus* and *Thermoactinomyces vulgaris* have not been differentiated. More farmers have been found with precipitins to *Thermoactinomyces vulgaris* than to *Thermoactinomyces thalpophilus* (Greatorex and Pether, 1975; Terho and Lacey, 1979), but most screening has been done with antigens prepared from *Thermoactinomyces thalpophilus*. Although *Thermoactinomyces vulgaris* is much more abundant than *Thermoactinomyces thalpophilus* in hay (Terho and Lacey, 1979), isolates of *Thermoactinomyces thalpophilus* were chosen for antigen production because they were regarded, at the time,

as more vigorously growing variants of *Thermoactinomyces vulgaris*, following the species concept of Küster and Locci (1964). Commercial antigens labeled "*Thermoactinomyces vulgaris*" have represented both species and it is necessary that these be identified and the role of *Thermoactinomyces* species in farmer's lung be re-evaluated (Lacey, 1981). *Thermoactinomyces thalpophilus* and *Thermoactinomyces vulgaris*, at least, should be present in panels of antigens used for screening (Terho and Lacey, 1979). A 16S rRNA primer set is now available for the identification of *Thermoactinomyces senso lato* strains associated with allergic alveolitis and pneumonitis (Xu et al., 2002).

Ecology. *Thermoactinomyces* species are most abundant in moldy fodders and other vegeTable matter including straw cereal grains, cotton, composts, hay, and manures (Craveri et al., 1966a; Desai and Dhala, 1966; Fergus, 1964; Forsyth and Webley, 1948; Gregory and Lacey, 1963; Gregory et al., 1963; Henssen, 1957; Lacey, 1973; Lacey, 1978; Lacey and Lacey, 1987). They are favored by spontaneous heating to temperatures up to 70°C, often resulting in production of more than 10^7 spores/g dry weight. The spores easily become airborne when the substrate is disturbed. However, growth may be limited where aeration is restricted. Growth in agar cultures was halved by decreasing the oxygen concentrations in air to 1% (v/v). Although some growth occurred with 0.1% oxygen, little or no sporulation occurred with less than 1% (Deploey and Fergus, 1975). *Thermoactinomyces vulgaris* is usually the most abundant species, but *Thermoactinomyces thalpophilus* is also common (Terho and Lacey, 1979), and *Thermoactinomyces dichotomicus* has been isolated from mushroom composts. *Thermoactinomyces sacchari* is most abundant in heated sugar cane bagasse where it occupies a niche similar to that of *Thermoactinomyces thalpophilus* and *Thermoactinomyces vulgaris* in moldy hay. All three species have also been isolated from soil and peat, although usually in small numbers seldom exceeding 10^4/g dry weight of soil (Goodfellow and Cross, 1974; Küster and Locci, 1963). Many originate from manure, sewage, or dung added to the soil (Cross, 1968; Diab, 1978), but deposition of airborne spores from moldy hay on farms is also possible. Some growth may also occur on vegetation heated by the sun. Even in temperate regions, solar heating may raise the temperature of soil and litter to more than 30°C (Eggins et al., 1972).

Erosion of soil may result in the accumulation of spores in lake muds and marine sediments, giving counts of 10^4–10^6 spores/g dry weight (Cross and Johnston, 1971). The presence of thermoactinomycetes in marine environments is a reliable and established indicator of terrestrial wash-in as spore germination and growth do not occur at low *in situ* temperatures (Attwell and Colwell, 1984; Goodfellow and Haynes, 1984; Pathom-aree et al., 2006). The occurrence of *Thermoactinomyces* species in deep mud cores and in archaeological excavations suggests that spores may remain viable for thousands of years and hence may be useful as palaeoindicators in studies on the agricultural history of soils (Cross and Attwell, 1974; Jackson et al., 1997; Nilsson and Renberg, 1990; Seaward et al., 1976; Unsworth et al., 1977). *Thermoactinomyces peptonophilus* and *Thermoactinomyces putidus* have also been isolated from soil (Nonomura and Ohara, 1971), but *Thermoactinomyces intermedius* has been found only in air conditioners, humidifiers, house dust, and grass compost (Kurup et al., 1980) where it occurs with other *Thermoactinomyces* species (Kurup et al., 1976a).

Miscellaneous. Spores of *Thermoactinomyces* species are characteristically heat resistant, surviving up to 4 h at 100°C in sucrose solution or 15 h of dry heat at this temperature (Fergus, 1967). Survival curves and $D_{100°C}$ values have been calculated for *Thermoactinomyces dichotomicus* (77 min), *Thermoactinomyces sacchari* (59 min), and *Thermoactinomyces thalpophilus* (11 min) (Cross et al., 1968b; Lacey, 1971a). A report that heat resistance may be lost in 24 h at 4°C (Kirillova et al., 1973) was not confirmed by Foerster (1978). Heat shock at 100°C or low-temperature storage may also sometimes decrease germination or kill spores (Attwell and Cross, 1973; Ensign, 1978; Foerster, 1978; Kirillova et al., 1973).

Enrichment and isolation procedures

Isolation of most thermophilic *Thermoactinomyces* species may be achieved on agar media containing 25 μg novobiocin/ml and 50 μg cycloheximide/ml incubated at 50–55°C. SuiTable media include CYC agar and half-strength nutrient or tryptone soya-casein agars (Lacey and Dutkiewicz, 1976b). Samples may be suspended in an aqueous diluent containing gelatin (0.5 g/l) or agar (0.2 g/l) and suiTable dilutions spread on agar in pre-poured plates. Alternatively, spores may be suspended in the air of a small wind tunnel or sedimentation chamber and plated using an Andersen sampler (Gregory and Lacey, 1963; Lacey and Dutkiewicz, 1976a).

Maintenance procedures

Thermophilic species may be maintained on the same media as used for isolation, incubating for 2–3 d at 50–55°C. Incubation may be continued for up to 1 week if plates are enclosed in poly-ethylene bags or in sealed containers with some water. Transfer of *Thermoactinomyces sacchari* is aided by transfer of agar bearing the culture and by sealing the Petri dish with a broad rubber band (Lacey, 1971a). Cultures can be maintained on agar slopes in screw-capped bottles at room temperature or 40°C, but for long-term preservation lyophilization is preferred with spores suspended in double-strength skim milk or other media.

Differentiation of *Thermoactinomyces sensu latu* from other genera

Some *Thermomonospora* species often appear similar to *Thermoactinomyces* species on isolation plates, growing well at 55°C, producing white aerial mycelium, and having a chemotype III wall (Becker et al., 1965). However, the sporophores of the latter show differing degrees of dichotomous branching, causing the spores to appear clustered, although they are formed singly, which gives the colony a granular appearance. Also, *Thermomonospora* spores are usually ovoid with a smooth or spiny surface, they are not endospores, and they are killed at 70°C (Cross and Lacey, 1970; McCarthy and Cross, 1984a). *Thermomonospora* species will not grow in the presence of 25 μg novobiocin/ml. *Saccharomonospora viridis* (Schuurmans et al., 1956) Nonomura and Ohara, 1971 also produces single oval spores that are not endospores and are heat sensitive. Colonies of this taxon are characteristically blue-green but may remain white when grown at suboptimal temperatures. This genus has a chemotype IV wall (Becker et al., 1965).

Differentiation of the genera of the family *Thermoactinomycetaceae*

Phenotypic characteristics that differentiate the genera are shown in Table 78.

TABLE 78. Phenotypic characteristics differentiating member genera of the family *Thermoactinomycetaceae*[a,b]

Characteristic	*Thermoactinomyces*	*Laceyella*	*Mechercharimyces*	*Planifilum*	*Seinonella*	*Shimazuella*	*Thermoflavimicrobium*
Aerial mycelium	White	White	White	None[c]	White	White	Yellow
Dichotomously branched sporophores	–	–	–	–	–	–	+
Degradation of:							
Casein	+	+	+	+	–	+	+
Gelatin	+	+	+	nd	–	–	–
Hypoxanthine	–	–	–	–	–	–	+
Starch	–	+	–	+	–	+	+
Xanthine	–	–	–	–	–	–	+
Optimal temperature for growth (°C)	50–55	48–55	30	53–63	35	32	55
Growth on 25 μg/ml novobiocin	+	+	+	nd	–	+	+
Predominant menaquinone	MK-7	MK-9	MK-9	MK-7	MK-7	MK-9	MK-7
Other menaquinones making up >10% peak area ratio	MK-8 or MK-9	MK-7 – MK8 or MK-10	MK-8	None[d]	MK8, MK-9, MK-10	MK-10	None
Major fatty acids	$C_{15:0\ iso}$, $C_{17:0\ iso}$, $C_{15:0\ ante}$	$C_{15:0\ iso}$, $C_{15:0\ ante}$	$C_{15:0\ iso}$, $C_{17:0}$ $_{iso-\omega11c}$, $C_{15:0\ ante}$	$C_{17:0\ iso}$, $C_{17:0\ ante}$ ($C_{15:0\ iso}$ or $C_{16:0}$)	$C_{14:0\ iso}$, $C_{15:0\ ante}$, $C_{16:0\ iso}$	$C_{16:0\ iso}$, $C_{15:0\ ante}$, $C_{16:0\ iso}$	$C_{15:0\ iso}$, $C_{15:0\ ante}$, $C_{16:0\ iso}$
DNA G+C content (mol%)	48	48–49	44.9–45.2	58.7–60.3	40	39.4	43

[a]Symbols: +, positive; –, negative; nd, not determined.
[b]Data from Hatayama et al. (2005b), Yoon et al. (2005), Matsuo et al. (2006) and Park et al. (2007).
[c]Aerial mycelia not observed on Bacto nutrient, Czapek-Dox-yeast extract, Luria–Bertani or starch-yeast agar plates (Hatayama et al., 2005b).
[d]MK-8 detected at a trace level (Hatayama et al., 2005b).

Taxonomic comments

Until recently, the genus *Thermoactinomyces* was considered to be an actinomycete, mainly because of its ability to form aerial mycelium when cultured on solid media. However, a wealth of taxonomic data, including the ability to produce dipicolinic acid-containing endospores (Cross et al., 1968b; Lacey and Vince, 1971), low G+C content of DNA (Lacey and Cross, 1989), menaquinone composition (Collins et al., 1982; Tseng et al., 1990), 5S rRNA (Park et al., 1993) and 16S rRNA gene sequences (Stackebrandt and Woese, 1981; Yoon and Park, 2000), and comparative ribosomal AT-L30 protein analyzes (Ochi, 1994) showed that thermoactinomycetes were closely related to other endospore-forming bacteria. Consequently, it is now accepted that the genus is a member of the order *Bacillales*. Nevertheless, because of their morphological properties, thermoactinomycetes were considered with the actinomycetes in the last edition of the *Systematics* (Lacey and Cross, 1989).

The Eighth Edition of the *Determinative* (Küster, 1974) listed only two species of *Thermoactinomyces* (*Thermoactinomyces sacchari* and *Thermoactinomyces vulgaris*), the Approved Lists of Bacterial Names (Skerman et al., 1980) included five species, and the last edition of the *Systematics* seven (Lacey and Cross, 1989). Changes since the Eighth Edition include the transfer of *Thermoactinomyces dichotomicus* from the genus *Actinobifida* on the basis of endospore formation, the description of *Thermoactinomyces candidus*, *Thermoactinomyces intermedius*, *Thermoactinomyces peptonophilus*, and *Thermoactinomyces putidus* as new species, and the revival of *Thermoactinomyces thalpophilus*. Numerical studies have shown the genus *Thermoactinomyces* to be defined at the 70% similarity (S_{SM} coefficient) level, although *Thermoactinomyces intermedius* and *Thermoactinomyces peptonophilus* were not included (Unsworth, 1978). Individual species were defined at the 79% similarity (S_{SM}) level or greater, while in another study, *Thermoactinomyces sacchari*, *Thermoactinomyces thalpophilus*, and *Thermoactinomyces vulgaris* formed an aggregate cluster at the 85% S_{SM} level of similarity (Goodfellow and Pirouz, 1982).

The status of *Thermoactinomyces candidus*, *Thermoactinomyces thalpophilus*, and *Thermoactinomyces vulgaris* has been a source of confusion; indeed the epithet *vulgaris* has been used in the literature in three senses: synonymous with *Thermoactinomyces candidus*, synonymous with *Thermoactinomyces thalpophilus*, and for an aggregate species comprising all the thermophilic taxa except *Thermoactinomyces dichotomicus* and *Thermoactinomyces sacchari*. The confusion arose because of changing concepts of *Thermoactinomyces* species. Prior to 1964, six species had been described. Of these, three including "*Thermoactinomyces glaucus*" Henssen 1957, "*Thermoactinomyces thermophilus*" (Berestnev) Waksman 1961, and "*Thermoactinomyces monosporus*" (Lehmann and Schutze) Waksman 1953 (in Waksman and Corke, 1953), are *nomina dubia*; one, "*Thermoactinomyces viridis*" Schuurmans et al., 1956, is now placed in the genus *Saccharomonospora* Nonomura and Ohara 1971; and the remaining two, *Thermoactinomyces thalpophilus* and *Thermoactinomyces vulgaris*, were placed in synonymy by Küster and Locci 1964. *Thermoactinomyces vulgaris* was considered to be a variable species, a conclusion supported by Flockton and Cross (1975), but it was concluded that the variation was insufficient to justify creation of additional taxa. As a consequence, *Thermoactinomyces vulgaris* acquired characters that were not present in the original concept of Tsiklinsky 1899. A prime example is the ability to utilize starch. Tsiklinsky (1899) stated clearly "il ne donne pas d'amylase", but later Kuo and Hartman (1966) described isolates that produced amylase, and this character is found in the description of *Thermoactinomyces vulgaris* in the Eighth Edition of the *Determinative*. Kurup et al. (1975) placed isolates producing amylase or not into two species, supported also by differences in their ability to utilize arbutin, esculin, hypoxanthine, and tyrosine. They named isolates lacking amylase as *Thermoactinomyces candidus* and those producing amylase as *Thermoactinomyces vulgaris* and also noted that isolates of *Thermoactinomyces candidus* produced spores on short sporophores while those of *Thermoactinomyces vulgaris* were mostly sessile. Sporophores are also described by Tsiklinsky 1899 in her description of *Thermoactinomyces vulgaris*. It is therefore appropriate that such isolates should remain the type species of *Thermoactinomyces* rather than those considered by Kurup et al. 1975 to be *Thermoactinomyces vulgaris*.

Type cultures of *Thermoactinomyces vulgaris* are not extant and the neotype proposed for the genus *Thermoactinomyces* and for *Thermoactinomyces vulgaris* is the oldest strain. This was isolated by Erikson 1953 as "*Micromonospora vulgaris*" strain D, and is listed in the Approved Lists of Bacterial Names as KCC A-0162. This corresponds to Tsiklinsky's original concept of *Thermoactinomyces vulgaris*, as does also *Thermoactinomyces candidus*. The two species should therefore be regarded as synonymous and, in accordance with the Code of Bacteriological Nomenclature, the oldest legitimate epithet retained. Thus *Thermoactinomyces vulgaris* remains the legitimate name for this taxon. Isolates that Kurup et al. (1975) named as *Thermoactinomyces vulgaris* are thus left without a name, but correspond to the aggregate cluster for which Unsworth and Cross (1980) proposed reviving the name *Thermoactinomyces thalpophilus* Waksman and Corke 1953. The original strains of Waksman and Corke are not extant, but a strain isolated by Henssen 1957, which she considered identical with strains from Waksman, has been designated the neotype. However, this occurs at the margin of the *Thermoactinomyces thalpophilus* phenon of Unsworth and Cross (1980), and additional reference strains have been specified. Isolations from hay, cotton, and other substrates suggest that isolates unable to produce amylase and with the characters of *Thermoactinomyces vulgaris sensu* Unsworth and Cross (1980) are more abundant than those producing amylase and therefore most likely to be the type isolated by Tsikinsky (1899) (Cross and Unsworth, 1981; Lacey and Lacey, 1987; Terho and Lacey, 1979). "*Thermoactinomyces antibioticus*" Craveri, Coronelli, Pagani and Sensi 1964 is a synonym of *Thermoactinomyces thalpophilus* and "*Thermoactinomyces albus*" Orlowska (1969) of *Thermoactinomyces vulgaris*. In contrast, there is evidence that additional species of *Thermoactinomyces sensu lato* remain to be described (Goodfellow and Pirouz, 1982; Song et al., 2001).

Genus I. **Thermoactinomyces** Tsiklinsky 1899, 501[AL] emend. Yoon, Kim, Shin and Park 2005, 398[VP]

MICHAEL GOODFELLOW AND AMANDA L. JONES

Ther.mo.ac.ti.no.my′ces. Gr. adj. *thermos* hot; Gr. n. *actis, actinos* a ray; Gr. n. *myces* fungus; N.L. masc. n. *Thermoactinomyces* heat (loving) ray fungus.

Aerobic, Gram-stain-positive, non-acid-fast chemo-organotroph. **Aerial mycelium is abundant and white.** Well-developed, branched and septate substrate mycelium is formed. **Endospores are sessile and produced singly on aerial and substrate hyphae or on unbranched short sporophores. Thermophilic.** Growth occurs at 55°C, but not at 30°C. Wall peptidoglycan contains *meso*-diaminopimelic acid but no characteristic sugars. **The predominant menaquinone is MK-7. The major fatty acid is $C_{15:0\ iso}$; significant amounts of $C_{17:0\ iso}$ are present.** The phylogenetically nearest neighbor is the genus *Laceyella*.

DNA G+C content (mol%): 48.

Type species: **Thermoactinomyces vulgaris** Tsiklinsky 1899, 501[AL]. (*Thermoactinomyces albus* Orlowska 1969, 25; *Thermoactinomyces candidus* Kurup, Barboriak, Fink and Lechevalier 1975, 152.).

Further descriptive information

Thermoactinomyces strains degrade arbutin and esculin, but not chitin (Lacey and Cross, 1989). Additional shared phenotypic features are given in Table 78. Ohshima et al. (1994) purified, characterized, cloned, and sequenced a gene expressing a thermostable leucine dehydrogenase from *Thermoactinomyces intermedius*. *Thermoactinomyces vulgaris* strains degrade elastin, DNA, RNA, and Tweens 20, 40, 60, and 80, but not adenine, cellulose, guanine, keratin, testosterone, or xylan, use D-glucose as a sole carbon source but not starch, grow in the presence of lysozyme and at 1% (w/v) sodium chloride but not in the presence of demethylchlortetracycline (500), gentamicin (50), kanamycin (100), neomycin (50), or vancomycin (50); the Figures in parentheses indicate the concentrations of antimicrobial compounds (µg/ml) used to soak filter paper discs (Goodfellow and Pirouz, 1982; Lacey and Cross, 1989). *Thermoactinomyces vulgaris* strain R-47 produces α-amylases that hydrolyze cyclodextrins, pullulan, and starch (Abe et al., 2005; Ohtaki et al., 2006). Additional phenotypic properties of *Thermoactinomyces vulgaris* strains can be found in the corresponding

section under the family *Thermoactinomycetaceae*. An isolate of *Thermoactinomyces vulgaris* recovered from compost was shown to be pathogenic for mice (Unaogu and Gugnani, 1999).

Enrichment and isolation procedures

Isolation of members of the genus can be achieved using CYC, half-strength nutrient, and tryptic soy agars supplemented with 25 µg novobiocin/ml and 50 µg cycloheximide/ml, and incubating for 2–3 d at 50–55°C (Lacey and Dutkiewicz, 1976c).

Maintenance procedures

Cultures can be maintained on the same media used for selective isolation, incubating at 50–55°C for up to 3d. Long-term storage can be accomplished by freezing at –80°C or in liquid nitrogen or by lyophilization with spores suspended in double strength skim milk.

Differentiation of the genus *Thermoactinomyces* from other genera

The genus *Thermoactinomyces* is phylogenetically distinct from neighboring genera based on 16S rRNA gene sequences (Figure 74). The type strains of the two constituent species share a 16S rRNA gene similarity of 99.4%, a value that corresponds to 9 nucleotide differences at 1437 locations. Members of the two species can be distinguished from those in other genera classified in the family *Thermoactinomycetaceae* using a range of phenotypic properties (Table 78).

Differentiation of the species of the genus *Thermoactinomyces*

Few phenotypic characteristics have been highlighted for the differentiation of *Thermoactinomyces intermedius* and *Thermoactinomyces vulgaris*. However, unlike *Thermoactinomyces intermedius*, *Thermoactinomyces vulgaris* strains are unable to degrade tyrosine or produce melanin pigments on CYC agar supplemented with 0.5% (w/v) tyrosine (Lacey and Cross, 1989).

List of species of the genus *Thermoactinomyces*

1. **Thermoactinomyces vulgaris** Tsiklinsky 1899, 501[AL] (*Thermoactinomyces albus* Orlowska 1969, 25; *Thermoactinomyces candidus* Kurup, Barboriak, Fink and Lechevalier 1975, 152.)

   vul.ga′ris. L. adj. *vulgaris* common.

   Colonies fast-growing, flat on nutrient and CYC agars at 55°C, with a moderate covering of white mycelium and often, a feathery margin on CYC agar. Endospores are produced on short, unbranched sporophores (Figure 75 and Figure 76). The colony reverse is white or cream, never pink or brown. Soluble pigments are not formed. Does not produce amylase or degrade L-tyrosine. Grows on CYC agar + 5% (w/v) NaCl.

   Frequently isolated from soils and muds, vegeTable composts, hay, straw, cereal grains, sugar cane bagasse, cotton, mushroom compost, humidifiers and air conditioning units, and from air.

   A probable cause of extrinsic allergic alveolitis (hypersensitivity pneumonitis), but its importance has probably been underestimated because isolates used in testing patients'

   antisera have often been *Laceyella sacchari* (previously *Thermoactinomyces thalpophilus*).

   DNA G+C content (mol%): 48.0 (HPLC).

   Type strain: ATCC 43649, CBS 505.77, DSM 43016, JCM 3162, KCC A-0162, KCTC 9076, NBRC 13606, VKM Ac-1195.

   GenBank accession number (16S rRNA gene): AF138739.

2. **Thermoactinomyces intermedius** Kurup, Hollick and Pagan, 1981, 216[VP] (Effective publication: Kurup, Hollick and Pagan 1980, 107.)

   in.ter.me′di.us. L. masc. adj. *intermedius* intercalated, intermediate.

   Colonies have white aerial mycelium and yellowish to yellowish-brown substrate mycelium. Brown, water-soluble melanin pigments produced on CYC agar + 0.5% (w/v) L-tyrosine. Degrades L-tyrosine. Endospores are sessile or produced on short sporophores. Growth is good at 50–55°C but poor at 37°C. The maximum temperature for growth has not been determined. Few nutritional and physiological characters have been determined. Isolated from air conditioner filters.

DNA G+C content (mol%): unknown.

Type strain: T-323, ATCC 33205, DSM 43846, DSM 44011, JCM 3312, KCTC 9646, NBRC 14230, NRRL B-16979, VKM Ac-1427.

GenBank accession number (16S rRNA gene): AF138734, AJ251775.

Genus II. **Laceyella** Yoon, Kim, Shin and Park 2005, 398[VP]

MICHAEL GOODFELLOW AND AMANDA L. JONES

La.cey.el′la. N.L. dim. fem. n. *Laceyella* named to honor John Lacey, an English microbiologist, for his contributions to the taxonomy of the genus *Thermoactinomyces* and actinomycetes.

Aerobic, Gram-stain-positive, non-acid-fast chemo-organotroph. Aerial and substrate mycelia are formed. **Aerial mycelium is white.** Yellow-brown or grayish-yellow soluble pigment may be produced. **Endospores produced on short or long sporophores. Thermophilic.** Wall peptidoglycan contains *meso*-diaminopimelic acid but no characteristic sugars. **The predominant menaquinone is MK-9. The major fatty acids are $C_{15:0\ iso}$ and $C_{15:0\ ante}$.** The phylogenetic nearest neighbor is the genus *Thermoactinomyces*.

DNA G+C content (mol%): 48–49.

Type species: **Laceyella sacchari** (Waksman and Cork 1953) Yoon, Kim, Shin and Park 2005, 398[VP] (*Thermoactinomyces thalpophilus* Waksman and Corke 1953, 378; *Thermoactinomyces sacchari* Lacey 1971a, 327; *Thermoactinomyces thalpophilus* Lacey and Cross 1989, 2582.)

Further descriptive information

Laceyella strains use D-glucose but not D-mannose as sole carbon sources, produce alkaline phosphatase, C_4 esterase, and C_8 lipase but not α- or β- glucosidase or β-glucurosidase (API ZYM tests) and are sensitive to ampicillin (25), chloramphenicol (50), colistin sulfate (10), erythromycin (10), nitrofurantoin (200), oleandomycin (5), streptomycin (100), and sulfafurazole (500) but not to nalidixic acid (30) or sulfafurazole (100); the Figures in parentheses indicate the concentration of antimicrobial agent (μg/ml) used to soak filter paper discs (Goodfellow and Pirouz, 1982; Lacey and Cross, 1989). They are also sensitive to penicillin (10 IU). Additional shared phenotypic features are given in Table 78 and in the corresponding section under the family *Thermoactinomycetaceae*. *Laceyella sacchari* strains degrade elastin, DNA, RNA, and Tweens 20, 40, 60, and 80 but not adenine, cellulose, chitin, or guanine, use L-arabinose but not *meso*-inositol, D-mannose, L-rhamnose, or D-xylose as sole carbon sources, and are sensitive to filter paper discs soaked in cephaloridine (100), demethylchlortetracycline (500), gentamicin (50), kanamycin (100), neomycin (50), and vancomycin (50); as

outlined above (Goodfellow and Pirouz, 1982; Lacey and Cross, 1989).

Enrichment and isolation procedures

Isolation of *Laceyella* species may be achieved on CYC agar supplemented with 25 μg novobiocin/ml and 50 μg cycloheximide/ml incubated at 50–55°C for up to 3 d (Lacey and Cross, 1989). Isolation of *Laceyella sacchari* is best achieved on yeast-malt agar (Shirling and Gottlieb, 1966) supplemented 25 μg novobiocin and 50 μg cycloheximide/ml and incubated at 55°C (Lacey, 1971b).

Maintenance procedures

Cultures can be maintained on the same media used for selective isolation with plates incubated at 50–55°C for up to 3 d. Long-term storage can be accomplished by freezing at −80°C, or in liquid nitrogen, or by lyophilization with spores suspended in double strength skim milk.

Differentiation of the genus *Laceyella* from other genera

The genus *Laceyella* is phylogenetically distinct from neighboring genera based on 16S rRNA gene sequences (Figure 74). The type strains of the two constituent species share a 16S rRNA gene similarity of 98.3%, a value that corresponds to 25 nucleotide differences at 1458 locations. Members of the two species can be distinguished from those classified in the family *Thermoactinomycetaceae* using a range of phenotypic properties (Table 78).

Differentiation of the species of the genus *Laceyella*

Members of the *Laceyella* species can be distinguished using a range of phenotypic characters (Lacey and Cross, 1989). Only the *Laceyella putida* strains degrade L-tyrosine, produce acid phosphatase, chymotrypsin, and leucine arylamidase, use D-sucrose as a sole carbon source, and form melanin pigments on CYC agar supplemented with 0.5%, w/v tyrosine. In contrast, only *Laceyella sacchari* strains degrade DNA and use D-fructose and D-mannose as sole carbon sources.

List of species of the genus *Laceyella*

1. **Laceyella sacchari** (Waksman and Cork 1953) Yoon, Kim, Shin and Park 2005, 398[VP] (*Thermoactinomyces thalpophilus* Waksman and Corke 1953, 378; *Thermoactinomyces sacchari* Lacey 1971a, 327; *Thermoactinomyces thalpophilus* Lacey and Cross 1989, 2582.)

   sac′cha.ri. N.L. n. *saccharum* generic name of sugar cane; N.L. gen. n. *sacchari* of sugar cane.

   Produces olive-buff, lightly ridged colonies which are "bacterial-like" in appearance. A sparse, transient, tufted aerial mycelium rapidly autolyzes depositing endospores in a thick layer on the surface of yeast malt or nutrient agar supplemented with 1% (w/v) glucose. Growth on nutrient agar is poor, restricted and thin with no aerial mycelium and few spores. Endospores are produced on sporophores up to

3 μm long (Figure 78). Yellow-brown soluble pigments may be formed. Grows at 55°C but growth is variable at 30°C. Water-soluble melanin may be produced on CYC agar supplemented with 0.5% (w/v) L-tyrosine. Elastin, DNA, RNA, and Tween 20, 40, 60, and 80 are degraded but not adenine, cellulose, guanine, or keratin. Degradation of esculin, arbutin, chitin and tyrosine is variable. D-fructose, D-glucose, and D-mannitol are used as carbon sources but not cellulose, *meso*-inositol, D-raffinose, L-rhamnose, or D-xylose. Does not grow in the presence of 5% (w/v) NaCl; growth variable in the presence of 1% (w/v) NaCl. Isolated from sugar cane, self-heated sugar cane bagasse, filter press muds, sugar mills, and soil.

DNA G+C content (mol%): 48 (HPLC).

Type strain: ATCC 27375, CCUG 7967, DSM 43356, JCM 3137, JCM 3214, KCTC 9790, NBRC 13920, NCIMB 10486, NCTC 10721, NRRL B-16981, VKM Ac-1360.

GenBank accession number (16S rRNA gene): AF138737, AJ251779.

2. **Laceyella putida** (Lacey and Cross 1989) Yoon, Kim, Shin and Park 2005, 399[VP] (*Thermoactinomyces putidus* Lacey and Cross 1989, 2582.)

pu'ti.da. L. fem. adj. *putida* stinking, fetid.

Colonies are usually highly wrinkled and puckered with endospores formed on short and unbranched sporophores (Figure 79). Aerial mycelium is white, but may appear cream, pale yellow or yellowish-brown due to a yellowish-brown substrate mycelium. During sporulation, hyphae lyse quickly, leaving spores on the surface of agar. A grayish-yellow soluble pigment may be produced; a brown, water-soluble melanin pigment is formed on CYC agar supplemented with 0.5% (w/v) L-tyrosine. Grows between 36°C and 58°C, optimally at 48°C. Sensitive to 1% (w/v) NaCl. Cultures characteristically produce a distinctive, unpleasant smell. Tyrosine is degraded, but not DNA. D-glucose and D-sucrose are used as sole carbon sources, but not D-fructose, glycerol, D-mannitol, D-mannose, D-ribose, or D-trehalose. Isolated from deep mud cores, soil, and a lung biopsy of a patient with farmer's lung.

DNA G+C content (mol%): 49 (HPLC).

Type strain: ATCC 49853, DSM 44608, KCTC 3666, JCM 8091, NCIMB 12324.

GenBank accession number (16S rRNA gene): AF138736, AJ251776.

Genus III. **Mechercharimyces** Matsuo, Katsuta, Matsuda, Shizuri, Yokota and Kasai 2006, 2840[VP]

MICHAEL GOODFELLOW AND AMANDA L. JONES

Me.cher.cha.ri'my.ces. N.L. n. *Mecherchar* a marine lake located on Mechechar Island in the Republic of Palau, where the organisms were isolated; Gr. masc. n. *mukes* fungus; N.L. masc. n. *Mechercharimyces* a fungus of Mecherchar.

Aerobic, Gram-stain-positive, chemo-organotroph. Aerial and substrate mycelia are formed. **Aerial mycelium is abundant and white.** Well-developed, branched and septate substrate mycelium is formed on marine agar 2216. Soluble pigments are not produced. **Endospores formed singly on short, unbranched sporophores. Mesophilic.** Cell-wall peptidoglycan contains *meso*-diaminopimelic acid, alanine, and glutamic acid but no characteristic sugars. **The predominant menaquinone is MK-9. The major fatty acid is C$_{15:0 iso}$.** The phylogenetically nearest neighbor is the genus *Seinonella*.

DNA G+C content (mol%): 45.

Type species: **Mechercharimyces mesophilus** Matsuo, Katsuta, Matsuda, Shizuri, Yokota and Kasai 2006, 2840[VP].

Further descriptive information

The chemotaxonomic characteristics of the type strains of *Mechercharimyces mesophilus* (YM3-251[T]) and *Mechercharimyces asporophorigenens* (YM11-542[T]) were determined by using biomass cultured in marine broth 2216 at the exponential phase of growth. The cell-wall peptidoglycan contained *meso*-diaminopimelic acid, glutamic acid, and alanine in the ratio 1.1:1:2.7 for strain YM3-251[T] and 1.2:1:2.9 for strain YM11-542[T], but no characteristic sugars. MK-8 and MK-9 were observed in strains YM3-251[T] (21.7% and 73.0%, respectively) and YM11-542[T] (23.9% and 76.1%, respectively). This profile is similar to those of members of the genus *Laceyella* Yoon et al. 2005.

In API ZYM tests, alkaline phosphatase activity was observed in the two type strains and in the additional strains of *Mechercharimyces mesophilus* (YM-653 and YM-671). Degradation of substrates was determined using marine agar 2216 supplemented with 0.1% esculin, 0.15% chitin, 1% gelatin, 0.5% hypoxanthine, 0.5% tyrosine, and 0.5% xanthine. All four strains degraded casein

and gelatin, but not the remaining substrates. They also formed a dark brown pigment on the L-tyrosine-containing marine agar. Reproducible results were not obtained using GP2 Microplates (Biolog) because of the weak growth of the strains. The type strain of *Mechercharimyces mesophilus* produces two cytotoxic substances, namely mechercharmycins A and B (Kanoh et al., 2005).

Enrichment and isolation procedures

Mechercharimyces mesophilus strains YM3-251[T], YM3-653 and YM3-671 were isolated from sediment samples collected from a marine lake on Mecherchar Island, Republic of Palau, by using, respectively 1/10 MYGS-AF medium*, skim milk medium†, 1/10 PYGS-AF medium‡ (Matsuo et al., 2006). The *Mechercharimyces*

---

*1/10 MYGS-AF medium: Malt extract 1 g, yeast extract 500 mg, glucose 500 mg, seawater 1 l, agar 20 g; pH 7.8–8.0; cycloheximide 100 mg, nystatin 50 mg, griseofulvin 20 mg.

†Skim milk medium: skim milk (Difco) 5 g, distilled water 200 ml, yeast extract 500 mg, seawater 800 ml, agar 20 g, pH 7.8–8.0.

‡1/10 PYGS-AF medium:, Metals mix X[1] 250 ml, distilled water 750 ml, agar 20 g, bacto peptone 1 g, yeast extract 0.5 g, C Soln[2] 5 ml, cycloheximide 50 mg, griseofluvin 25 mg, nalidixic acid 20 mg, aztreonam 40 mg.

[1]Metal mix X: NaCl 500 g, MgSO$_4$ · 7H$_2$O 180 g, CaCl$_2$ · 2H$_2$O 2.8 g, KCl 14 g, Na$_2$HPO$_4$ · 12H$_2$O 5 g, FeSO$_4$ · 7H$_2$O 200 mg, PII metals[3] 600 ml, S2 metals[4] 100 ml, distilled water, 4300 ml, pH 7.6.

[2]C soln: sodium pyruvate 25 g, mannitol 50 g, glucose 50 g, distilled water 500 ml pH7.5, sterilized by filtration.

[3]PII metals: Na$_2$-EDTA 1 g, H$_3$BO$_3$ 1.13 g, Fe soln 1 ml (FeCl$_3$ · 6H$_2$O [2.42 g/50 ml]), Mn soln 1 ml (MnCl$_2$ · 4H$_2$O [7.2 g/50 ml]), Zn soln 1 ml (ZnCL$_2$ [0.52 g/50 ml (+HCl)]), Co soln 1 ml (CoCl$_2$ · 6H$_2$O [0.2 g/50 ml]), distilled water 996 ml, pH 7.5.

[4]S2 metals: NaBr 1.28 g, Mo soln 10 ml (Na$_2$MoO$_4$ · 2H$_2$O [0.63 g/50 ml]), Sr soln 10 ml (SrCl$_2$ · 6H$_2$O [3.04 g/50 ml]), Rb soln 10 ml (RbCl [141.5 mg/50 ml]), Li soln 10 ml (LiCl l [0.61 g/50 ml]), I soln 10 ml (KI [6.55 mg/50 ml]), V soln 10 ml (V$_2$O$_5$ [1.785 mg/50 ml (+ NaOH)]), distilled water 940 ml, pH 7.5.

phorigenens strain was isolated from a marine lake in the northern part of Urukthapel Island, Republic of Palau, by using 1/10 PYGS-AF medium. For the isolation of strain YM3-251[T], the sediment sample was heated at 52°C for 2 h prior to inoculating the medium. Strain YM3-653 was isolated from a sediment sample that had been dried on filter paper for 24 h and then on silica gel for a month. The two remaining strains were isolated from sediment samples without preheating.

Maintenance procedures

Specific information on suitable procedures for the maintenance of *Mechercharimyces* species is not available. It can be assumed that the procedures satisfactory for other members of the family *Thermoactinomycetaceae* will be suitable for the maintenance of *Mechercharimyces* isolates.

Differentiation of the genus *Mechercharimyces* from other genera

The genus *Mechercharimyces* is phylogenetically distinct from neighboring genera based on 16S rRNA gene sequences (Figure 74). The four constituent members of the clade have highly conserved sequences as they share 16S rRNA similarities within the range 99.4–99.9%, values that correspond to 2–9 nucleotide differences at 1428 locations. The organisms can also be readily distinguished from representatives of the genera *Laceyella*, *Seinonella*, *Thermoactinomyces*, and *Thermoflavimicrobium* in phylogenetic and molecular evolutionary analyzes based on *gyrB* sequences. They share an optimal growth temperature that distinguishes them from members of these taxa (Table 78). Together with *Laceyella* strains they are characterized by having MK-9 as the predominant isoprenologue.

Differentiation of the species of the genus *Mechercharimyces*

Members of the two *Mechercharimyces* species can be distinguished on the basis of 16S rRNA and *gyrB* gene sequence similarities, the level of DNA–DNA relatedness, and the presence of endospores.

List of species of the genus *Mechercharimyces*

1. **Mechercharimyces mesophilus** Matsuo, Katsuta, Matsuda, Shizuri, Yokota and Kasai 2006, 2840[VP]

me.so.phi'lus. Gr. adj. *mesos* middle; Gr. adj. *philos* loving; N.L. masc. adj. *mesophilus* middle (temperature) -loving, mesophilic.

Colonies are fast-growing, lightly ridged, with a moderate covering of white aerial hyphae and a feathery margin on marine agar 2216 at 27°C. Endospores are formed singly on short, unbranched sporophores. Produces a dark-brown pigment on marine agar supplemented with L-tyrosine. Growth occurs at 15–37°C, optimally at 30°C. Shows trypsin activity in API ZYM tests. Casein and gelatin are degraded, but not hypoxanthine, starch, L-tyrosine, or xanthine. Growth occurs in the presence of 25 μg novobiocin/ml. Major cellular fatty acids are $C_{15:0\ iso}$, $C_{16:0\ iso}$, $C_{17:1\ iso\ \omega11c}$, and $C_{17:0\ iso}$. Isolated from sediment samples collected from a marine lake in Mecherchar Island, Republic of Palau.

DNA G+C content (mol%): 45.1 (HPLC).

Type strain: YM3-251, DSM 44894, MBIC06230.

GenBank accession number (16S rRNA gene): AB239529.

2. **Mechercharimyces asporophorigenens** Matsuo, Katsuta, Matsuda, Shizuri, Yokota and Kasai 2006, 2840[VP]

a.spo'ro.pho.ri.gen.ens. Gr. prep. *a* not; Gr. n. *sporophora* sporophore; L. part. adj. *genens* producing; N.L. part. adj. *asporophorigenens* sporophore nonproducing.

Colonies are fast-growing, lightly ridged, with a moderate covering of white aerial hyphae and a feathery margin on marine agar 2216 at 27°C. Oval shaped endospores are borne on aerial and substrate hyphae. Produces a dark-brown pigment on marine agar supplemented with L-tyrosine. Growth occurs at 20–37°C, optimally at 30°C. Does not show trypsin activity in API ZYM tests. Casein and gelatin are degraded but not hypoxanthine, starch, L-tyrosine or xanthine. Growth occurs in the presence of 25 μg/ml novobiocin. Major cellular fatty acids are $C_{15:0\ iso}$, $C_{16:0\ iso}$, $C_{17:1\ iso\ \omega11c}$, and $C_{17:0\ iso}$. Isolated from a sediment sample collected from a marine lake in the northern part of Urukthapel Island, Republic of Palau.

DNA G+C content (mol%): 45.2 (HPLC).

Type strain: YM11-542, DSM 44955, MBIC06487.

GenBank accession number (16S rRNA gene): AB239532.

Genus IV. **Planifilum** Hatayama, Shoun, Ueda and Nakamura 2005b, 2104[VP]

MICHAEL GOODFELLOW AND AMANDA L. JONES

Pla.ni.fi'lum. L. adj. *planus* flat; L.neut. n. *filum* a thread; N.L. neut. n. *Planifilum* a flat thread.

Aerobic, Gram-stain-positive organism **which forms a substrate mycelium, but not an aerial mycelium on CYC, LB, and SY agars. Single endospores are borne on substrate hyphae. Thermophilic.** Casein and starch are degraded but not hypoxanthine or xanthine. The cell-wall peptidoglycan contains *meso*-diaminopimelic acid, alanine, and glutamic acid but no diagnostic sugars. **The predominant menaquinone is MK-7. The major fatty acids are $C_{17:0\ iso}$, $C_{17:0\ ante}$, and either $C_{15:0\ iso}$ or $C_{16:0\ iso}$.** The phylogenetic nearest neighbor is the genus *Thermoactinomyces*.

DNA G+C content (mol%): 58.7–60.3.

Type species: **Planifilum fimeticola** Hatayama, Shoun, Ueda and Nakamura 2005b, 2104[VP].

Further descriptive information

Does not degrade 1% avicel, 1% carboxymethylcellulose, or 1% xylan. Utilization of sugars by the methods of Shirling and Gottlieb (1966) and Fotina et al. (2001) did not yield reproducible results due to weak growth of the strains.

Enrichment and isolation procedures

The two species were isolated from samples taken from a hyperthermal composting process (Hatayama et al., 2005a) by cultivation on Luria–Bertani agar plates at 65°C.

Maintenance procedures

Specific information on suitable procedures for the maintenance of *Planifilum* species is not available. It can be assumed that the procedures satisfactory for other members of the family *Thermoactinomycetaceae* will also be suitable for the maintenance of *Planifilum* isolates.

Differentiation of the genus *Planifilum* from other genera

Planifilum isolates can be distinguished from members of the other genera classified in the family *Thermoactinomycetaceae* based on the absence of aerial mycelia, growth temperatures, DNA G+C content, and cellular fatty acid and menaquinone composition (Table 78). Phylogenetic analyzes show that *Planifilum* isolates form a distinct branch in the *Thermoactinomycetaceae* 16S rRNA gene tree (Figure 74).

Differentiation of the species of the genus *Planifilum*

Members of the two *Planifilum* species can be distinguished on the basis of 16S rRNA gene sequence similarities, levels of DNA–DNA relatedness, differences in cellular fatty acid profiles, and on the ability to degrade L-tyrosine (Hatayama et al., 2005b).

List of species of the genus *Planifilum*

1. **Planifilum fimeticola** Hatayama, Shoun, Ueda and Nakamura 2005b, 2104[VP]

   fi.me.ti.co′la. L. n. *fimetum* a dung-hill and, by extension, compost; L. masc. suffix *-cola* inhabitant; N.L. masc. n. *fimeticola* inhabitant of compost, referring to the habitat of the type strain.

   Colonies are lustrous, cream-yellow with radial wrinkles. Growth occurs at 50–65°C, optimally at 55–63°C. L-Tyrosine is degraded. The major cellular fatty acids are $C_{16:0\ iso}$, $C_{17:0\ iso}$, and $C_{17:0\ ante}$. Isolated from a hyperthermal composting process plant in Okinawa Prefecture, Japan.

   DNA G+C content (mol%): 60.3% (HPLC).

   Type strain: H0165, ATCC BAA-969, JCM 12507.

   GenBank accession number (16S rRNA gene): AB088364.

2. **Planifilum fulgidum** Hatayama, Shoun, Ueda and Nakamura 2005b, 2104[VP]

   ful′gi.dum. L. neut. adj. *fulgidum* lustrous, referring to the colony character.

   Colonies are lustrous, cream-yellow with radial wrinkles. Growth occurs at 50–67°C, optimally at 60–65°C. L-tyrosine is not degraded. The major cellular fatty acids are $C_{17:0\ iso}$, $C_{17:0\ ante}$, and $C_{15:0\ iso}$. Isolated from a hyperthermal composting process plant in Okinawa Prefecture, Japan.

   DNA G+C content (mol%): 58.7–60 (HPLC).

   Type strain: 500275, ATCC BAA-970, JCM 12508.

   GenBank accession number (16S rRNA gene): AB088362.

Genus V. **Seinonella** Yoon, Kim, Shin and Park 2005, 399[VP]

Michael Goodfellow and Amanda L. Jones

Sei.no.nel′la. N.L. dim. fem. n. *Seinonella* named to honor Akiro Seino, a Japanese microbiologist, for his contributions to the genus *Thermoactinomyces* and actinomycetes.

Aerobic, Gram-stain-positive, non-acid-fast chemo-organotroph. The substrate mycelium is white to yellowish-brown and the aerial mycelium white. **Sessile endospores are produced on the substrate mycelium and on filamentous branches of the aerial mycelium. Growth at 25–45°C, and optimally at 35°C.** The wall peptidoglycan contains *meso*-diaminopimelic acid. **The predominant menaquinone is MK-7**, significant amounts of MK-8, MK-9, and MK 10 are also present. **The major fatty acids are $C_{14:0\ iso}$ and $C_{15:0\ ante}$.** The phylogenetic nearest neighbor is the genus *Mercheracharimyces*.

DNA G+C content (mol%): 40.

Type species: **Seinonella peptonophila** (Nonomura and Ohara 1971) Yoon, Kim, Shin and Park 2005, 400[VP] (*Thermoactinomyces peptonophilus* Nonomura and Ohara 1971, 902.)

Further descriptive information

Endospores are less heat-resistant ($D_{90°C}$ = 45 min) than those of other thermoactinomyces (Attwell, 1978). Does not grow below pH 5.0.

Enrichment and isolation procedures

Seinonella peptonophila has been isolated from dry heat treated (100°C) soil samples using MGA-SE agar*. However, growth on this medium occurred only in the presence of actinomycete colonies because of the restricted nutritional requirments of the organism (Nonomura and Ohara, 1971).

---

*MGA-SE agar: glucose 2.0 g, L-asparagine 1.0 g, K_2HPO_4 0.5 g, $MgSO_4$·$7H_2O$ 0.5 g, soil extract 200 ml; penicillin 1 mg, polymixin β 5 mg, cycloheximide 50 mg, nystatin 50 mg, agar 20 g, water 800 ml, pH 8.0. Soil extract: 1000 g soil autoclaved with 1 l water for 30 mins, decanted, and filtered.

Maintenance procedures

Seinonella peptonophila must be grown on glycerol-asparagine agar (Shirling and Gottlieb, 1966) supplemented with 10 g yeast extract or on PY agar with incubation at 35°C (Nonomura and Ohara, 1971). Long-term storage can be achieved by freezing at −80°C or in liquid nitrogen or by lyophilization with spores suspended in double strength skim milk.

Differentiation of the genus *Seinonella* from other genera

The genus *Seinonella* is phylogenetically distinct from neighboring genera based on 16S rRNA gene sequences (Figure 74). The organism can be distinguished from those classified in the family *Thermoactinomycetaceae* using a combination of phenotypic properties (Table 78), notably by its low optimal temperature for growth.

List of species of the genus *Seinonella*

1. **Seinonella peptonophila** (Nonomura and Ohara 1971) Yoon, Kim, Shin and Park 2005, 400[VP] (*Thermoactinomyces peptonophilus* Nonomura and Ohara 1971, 902.)

   pep.to.no′phi.la. Gr. adj. *peptos* cooked; Gr. adv. *philos* loving; N.L. adj. *peptonophila* peptone loving (Note: Rule 61 of the *Bacteriological Code* prevents the correction of the epithet to *peptoniphila*).

   B vitamins and high concentrations of peptone (3%, w/v) are essential for growth. Aerial mycelium production is favored by glycerol or glucose (0.2%, w/v) and is best on supplemented glycerol-asparagine, oatmeal, yeast-starch, and PY agars at 35°C. Grows poorly at 25°C and not at all at 45°C. Optimum pH for growth is 7.0–8.0; no growth at pH 5.0. Nitrate reduction is negative. Tyrosine is not degraded. Isolated from soil.

   DNA G+C content (mol%): 40 (HPLC).

   Type strain: ATCC 27302, DSM 44666, JCM 10113, KCTC 9740.

   GenBank accession number (16S rRNA gene): AF 138735.

Genus VI. **Shimazuella** Park, Dastagar, Lee, Yeo, Yoon and Kim 2007, 2663[VP]

MICHAEL GOODFELLOW AND AMANDA L. JONES

Shi.maz ue′lla. N.L. fem. dim. n. *Shimazuella* of Shimazu, named after Akira Shimazu of Tokyo University for his contributions to prokaryotic taxonomy.

Aerobic, Gram-stain-positive, mesophilic, chemorganotroph. Aerial and substrate mycelia are formed. Aerial mycelium is abundant and white. Well developed, branched and septate mycelium is formed on Bennett's and yeast extract-malt extract agars. Soluble pigments are not produced. **Single endospores formed on both aerial and substrate mycelium with sizes ranging from 1.0–1.4 to 0.7–0.9 μm. Extensively branched sporophores are formed on aerial hyphae with sizes ranging from 0.3–0.6 μm in length. Spore surface spiny.** Grows at 20–50°C and optimally at 32°C. Cell-wall peptidoglycan contains *meso*-diaminopimelic acid, alanine, and glutamic acid but no characteristic sugars. Predominant menaquinone is MK-9. Major fatty acid is $C_{15:0\ iso}$. The phylogenetically nearest neighbor is the genus *Laceyella*.

DNA G+C content (mol%): 39.4.

Type species: **Shimazuella kribbensis** Park, Dastagar, Lee, Yeo, Yoon and Kim 2007, 2663[VP].

Further descriptive information

Has chemotaxonomic and morphological features consistent with its classification in the family *Thermoactinomycetaceae*. Grows on glycerol-asparagine, inorganic salts-starch, oatmeal, tyrosine, and yeast extract-malt extracts agar (Shirling and Gottlieb, 1966). Phosphatidylethanolamine is the diagnostic phospholipid. Contains MK-9 and MK-10 in the ratio of 7:3.

Isolation procedures

Strain A 9500[T] was isolated from a soil sample collected from Mount Sobaek, Republic of Korea, by plating soil suspensions onto Bennett's agar (Atlas, 1993) and incubating at 30°C for 2 weeks.

Maintenance procedures

The organism can be maintained on Bennett's agar plates but additional information on suitable procedures for the maintenance of *Shimazuella* strains is not available. It can be assumed that the procedures satisfactory for other membes of the family *Thermoactinomycetaceae* will be suitable for the maintenance of *Shimazuella* isolates.

Differentiation of the genus *Shimazuella* from other genera

The genus *Shimazuella* is phylogenetically distinct from neighboring genera based on 16S rRNA gene sequences (Figure 74). Members of the taxon can be distinguished from genera classified in the family *Thermoactinomycetaceae* using a range of phenotypic properties (Table 78), notably by its menaquinone profile.

List of species of the genus *Shimazuella*

1. **Shimazuella kribbensis** Park, Dastagar, Lee, Yeo, Yoon and Kim 2007, 2663[VP]

   krib.ben′sis. N.L. fem. adj. *kribbensis* pertaining to KRIBB, an arbitrary adjective formed from the acronym of the Korea Research Institute of Bioscience and Biotechnology (KRIBB), where the taxonomic studies on the organism were undertaken.

   Fast-growing, lightly ridged colonies with a pale yellow substrate mycelium, a feathery margin, and an abundant white aerial mycelium are formed on Bennett's agar. Casein and starch are degraded, but not gelatin, hypoxanthine, L-tyrosine, or xanthine. Growth occurs in the presence of 25 μg novobiocin/ml. The fatty acid profile consists of $C_{15:0 \text{ ante}}$ (43.4%), $C_{16: \text{ iso}}$ (14.2), $C_{16:0}$ (7.9%), $C_{15:0 \text{ iso}}$ (7.4%), $C_{17:0 \text{ ante}}$ (7.2%), $C_{14:0 \text{ iso}}$ (6.1%), $C_{14:0}$ (6.1%), $C_{15:0}$ (4.0%), and $C_{16:0 \text{ ante}}$ (3.8%). Isolated from a soil sample collected from Mount Sobaek, Republic of Korea.

DNA G+C content (mol%): 39.4 (HPLC).

Type strain: A9500, KCTC 9933, KCCM 41585.

GenBank accession number (16S rRNA gene): AB049939.

Genus VII. **Thermoflavimicrobium** Yoon, Kim, Shin and Park 2005, 399[VP]

MICHAEL GOODFELLOW AND AMANDA L. JONES

Ther'mo.fla.vi.mi.cro'bi.um. Gr. adj. *thermos* hot; L. adj. *flavus* yellow; Gr. adj. *mikros* small; Gr. n. *bios* life; N.L. neut. n. *Thermoflavimicrobium* a thermophilic yellow-colored microbe.

Aerobic, Gram-stain-positive, non-acid-fast chemo-organotroph. Aerial and substrate mycelia are formed. **Aerial mycelium is abundant and yellow. Sessile endospores are produced on dichotomously branched sporophores. Thermophilic.** Growth occurs at 55°C but not at 30°C. The wall peptidoglycan contains *meso*-diaminopimelic acid but no characteristic sugars. **Predominant menaquinone is MK-7. Major fatty acids are $C_{15:0 \text{ iso}}$, $C_{15:0 \text{ ante}}$, and $C_{16:0 \text{ iso}}$.** The phylogenetically nearest neighbor is the genus *Planofilum*.

DNA G+C content (mol%): 43.

Type species: **Thermoflavimicrobium dichotomicum** (Krasil'nikov and Agre, 1964) Yoon, Kim, Shin and Park 2005, 399[VP] (*Actinobifida dichotomica* Krasil'nikov and Agre 1964, 939; *Thermoactinomyces dichotomicus* corrig. Cross and Goodfellow 1973, 77.).

Further descriptive information

Thermoflavimicrobium strains grow up to 62°C, produce alkaline phosphatase, C_4 esterase, and C_8 lipase but not chymotrypsin, α- or β- glucosidase, β-glucuronidase, or leucine arylamidase (API ZYM tests) and are sensitive to cephaloridine (100), demethylchlortetracycline (500), gentamicin (100), kanamycin (100), lincomycin (100), neomycin (100), streptomycin (100), tobramycin (100), and vancomycin (100) but not to novobiocin (50); the Figures in parentheses indicate the concentrations of antimicrobial compounds (μg/ml) used to soak filter paper discs (Goodfellow and Pirouz, 1982; Lacey and Cross, 1989).

The organism is also sensitive to penicillin (10 IU) and NaCl (0.5%, w/v).

Enrichment and isolation procedures

Isolation may be achieved on half-strength nutrient agar supplemented with 25 μg novobiocin/ml and 50 μg cycloheximide/ml following incubation for up to 3 d at 50–55°C (Lacey and Cross, 1989). Colonies on isolation plates are recognized by their bright yellow color.

Maintenance procedures

Agre (1964) recommended a maize-starch medium[*]. Cultures can be maintained on agar slopes in screw-capped bottles at room temperature or at 4°C. Long-term storage can be accomplished by freezing at −80°C or in liquid nitrogen or by lyophilization with spores suspended in double strength skim milk.

Differentiation of the genus *Thermoflavimicrobium* from other genera

The genus *Thermoflavimicrobium* is phylogenetically distinct from neighboring genera based on 16S rRNA gene sequences (Figure 74). Members of the genus can be distinguished from those classified in the family *Thermoactinomycetaceae* using a range of phenotypic properties (Table 78), notably by their ability to produce single spores on dichotomously branched sporophores (Figure 80) and yellow pigmented colonies.

List of species of the genus *Thermoflavimicrobium*

1. **Thermoflavimicrobium dichotomicum** (Krasil'nikov and Agre 1964) Yoon, Kim, Shin and Park 2005, 399[VP] (*Actinobifida dichotomica* Krasil'nikov and Agre 1964, 939; *Thermoactinomyces dichotomicus* corrig. Cross and Goodfellow 1973, 77.)

di.chot'o.mi.cus. Gr. adj. *dichotomos* cut in two parts, forked; L. suff. *-icus* suffix used with several meanings; N.L. neut. adj. *dichotomicus* dichotomous.

Distinctive, fast-growing yellow to orange colonies with dichotomously branched mycelium and sporophores are produced on nutrient and CYC agars at 55°C; margins are entire on CYC agar. The presence of an exosporium surrounding the spore has been suggested. Elastin, DNA, guanine, RNA, and Tween 20, 40, 60, and 80 are degraded, but not esculin, adenine, arbutin, cellulose, hippurate, keratin, or tyrosine. Growth occurs in the presence of 0.5% (w/v) NaCl,

but not in the presence of 1.0% (w/v) NaCl. L-arabinose, D-galactose, D-glucose, glycerol, D-lactose, D-maltose, D-mannitol, *meso*-inositol, D-raffinose, L-rhamnose, D-sorbitol, starch, sucrose, and D-xylose are used as sole carbon sources. Isolated from soil and mushroom compost.

DNA G+C content (mol%): 43 (HPLC).

Type strain: ATCC 49854, DSM 44778, JCM 9688, KCTC 3667, NCIMB 10211, VKM Ac-1435.

GenBank accession number (16S rRNA gene): AF138733, L16902.

---

[*]Maize-starch medium: split maize 50 g, boiled in 1 l water for 30 min, then filtered before adding starch 10 g, NaCl 5 g, $CaCl_2$ 0.5 g, peptone (Oxoid) 5 g, agar 20 g, pH 7.2.

References

Abe, A., H. Yoshida, T. Tonozuka, Y. Sakano and S. Kamitori. 2005. Complexes of *Thermoactinomyces vulgaris* R-47 alpha-amylase 1 and pullulan model oligossacharides provide new insight into the mechanism for recognizing substrates with alpha-(1,6) glycosidic linkages. FEBS J. *272*: 6145–6153.

Agre, N.S. 1961. The phage of the thermophilic *Micromonospora* vulgaris. Mikrobiologiya *30*: 414–417.

Agre, N.S. 1964. A contribution to the technique of isolation and cultivation of thermophilic actinomycetes (in Russian). Mikrobiologiya *33*: 913–917.

Agre, N.S., I.P. Kirillova and L.V. Kalakoutskii. 1972a. Spore germination in thermophilic actinomycetes. I. Preliminary observations with *Thermoactinomyces vulgaris* and *Actinobifida dichotomica*. Zentralbl. Bakteriol. Parasitenkd. Infektionskr. Hyg. Abt. II *127*: 525–538.

Agre, N.S., I.P. Kirillova and L.V. Kalakoutskii. 1972b. Spore germination in thermophilic actinomycetes. II. Germinal changes in water suspensions of *Thermoactinomyces vulgaris* spore. Zentralbl. Bakteriol. Parasitenkd. Infektionskr. Hyg. Abt. II *127*: 539–544.

Allam, A.M., A.M. Hussein and A.M. Ragab. 1975. Amylase of a thermophilic actinomycete *Thermoactinomyces vulgaris*. Z. Allg. Mikrobiol. *15*: 393–398.

Arden-Jones and M.T. Cross. 1980. Antigenic variation within and between species of *Thermoactinomyces*. Abstr. 89. Annu. Meet. Am. Soc. Microbiol. 1980.

Atlas, R.M. 1993. Handbook of Microbiological Media. CRC Press Inc, Florida, USA.

Attwell, R.W., T. Cross and G.W. Gould. 1972. Germination of *Thermoactinomyces vulgaris* endospores: microscopic and optical density studies showing the influences of germinants, heat treatment, strain differences and antibiotics. J. Gen. Microbiol. *73*: 471–481.

Attwell, R.W. and T. Cross. 1973. Germination of actinomycete spores. *In* Sykes and Skinner (Editors), *Actinomycetales*: Characteristics and Practical Importance. Soc. Appl. Bacteriol. Symp. Ser. No. 2. Society for Applied Bacteriology, London, pp. 197–207.

Attwell, R.W. 1978. The spores of *Thermoactinomyces peptonophilus*. Actinomycetes *13*: 30.

Attwell, R.W. and R.R. Colwell. 1984. Thermoactinomycetes as terrestrial indicators for marine and estuarine waters. *In* Ortiz-Ortiz, Bojalil and Yakoleff (Editors), Biological Biochemical and Biomedical Aspects of Actinomycetes. Academic Press, Orlando, pp. 441–452.

Banaszak, E.F., W.H. Thiede and J.N. Fink. 1970. Hypersensitivity pneumonitis due to contamination of an air conditioner. N. Engl. J. Med. *283*: 271–276.

Becker, B., M.P. Lechevalier and H.A. Lechevalier. 1965. Chemical composition of cell-wall preparations from strains of various form-genera of aerobic actinomycetes. Appl. Microbiol. *13*: 236–243.

Behnke, U., H. Ruttloff and R. Kleine. 1982. Preparation and characterization of proteases from *Thermoactinomyces vulgaris*. V. Investigations on autolysis and thermostability of the purified protease. Z. Allg. Mikrobiol. *22*: 511–519.

Collins, M.D., G.C. Mackillop and T. Cross. 1982. Menaquinone Composition of Members of the Genus *Thermoactinomyces*. FEMS Microbiol. Lett. *13*: 151–153.

Craveri, R., C. Coronelli, H. Pagani and P. Sensi. 1964. Thermorubin, a new antibiotic from a thermoactinomycete. Clin. Med. *71*: 511–521.

Craveri, R., A. Guicciardi and N. Pacini. 1966a. Distribution of thermophilic actinomycetes in compost for mushroom production. Ann. Microbiol. (Milan) *16*: 111–113.

Craveri, R. and P.L. Manachini. 1966. Base composition of DNA in *Streptomyces argenteolus* and *Thermoactinomyces vulgaris* cultivated at different temperatures (in Italian). Ann. Microbiol. Enzimol. *16*: 1–3.

Craveri, R., P.L. Manachini and N. Pacini. 1966b. Deoxyribonucleic acid base composition of actinomycetes with different temperature requirements for growth. Ann. Micobiol. Enzimol. *16*: 115–117.

Cross, T. 1968. Thermophilic actinomycetes. J. Appl. Bacteriol. *31*: 36–53.

Cross, T., A. Maciver and J. Lacey. 1968a. The thermophilic actinomycetes in mouldy hay: *Micropolyspora faeni* sp. nov. J. Gen. Microbiol. *50*: 351–359.

Cross, T., P.D. Walker and G.W. Gould. 1968b. Thermophilic actinomycetes producing resistant endospores. Nature *220*: 352–354.

Cross, T. and J. Lacey. 1970. Studies on the genus Thermonospora. *In* Prauser (Editor), The *Actinomycetales*, VEB Gustav Fischer Verlag, Jena, pp. 211–219.

Cross, T. and D.W. Johnston. 1971. *Thermoactinomyces vulgaris*. II. Distribution in natural habitats. *In* Barker, Gould and Wolf (Editors), Spore Research 1971. Academic Press, London, pp. 315–330.

Cross, T. and M. Goodfellow. 1973. Taxonomy and classification of the actinomycetes. *In* Sykes and Skinner (Editors), *Actinomycetales*: Characteristics and Practical Importance. Academic Press, London, New York, pp. 11–112.

Cross, T. and R.W. Attwell. 1974. Recovery of viable thermoactinomycete endospores from deep mud cores. *In* Barker, Gould and Wolf (Editors), Spore Research 1973. Academic Press, London, pp. 11–20.

Cross, T. and R.W. Attwell. 1975. Actinomycete spores. *In* Gerhardt, Costilow and Sadoff (Editors), Spores. VI. American Society for Microbiology, Washington, D.C, pp. 3–13.

Cross, T. and B.A. Unsworth. 1976. Farmer's lung: A neglected antigen. Lancet Infect. Dis. *1*: 958–959.

Cross, T. and B.A. Unsworth. 1981. Taxonomy of the endospore-forming actinomycetes. *In* Berkeley and Goodfellow (Editors), The Aerobic Endospore Forming Bacteria: Classification and Identification. Academic Press, London, pp. 17–32.

Deploey, J.J. and C.L. Fergus. 1975. Growth and sporulation of thermophilic fungi and actinomycetes in O_2-N_2 atmosphere. Mycologia *67*: 780–797.

Desai, A.J. and S.A. Dhala. 1966. Isolation and study of thermophilic actinomycetes from soil, manure and compost from Bombay. Indian J. Microbiol. *6*: 54–58.

Desai, A.J. and S.A. Dhala. 1967. Bacteriolysis by thermophilic actinomycetes. Antonie van Leeuwenhoek *33*: 56–62.

Desai, A.J. and S.A. Dhala. 1969. Purification and properties of proteolytic enzymes from thermophilic actinomycetes. J. Bacteriol. *100*: 149–155.

Diab, A. 1978. Studies on thermophilic microorganisms in certain soils in Kuwait. Zentralbl. Bakteriol. Naturwiss. *133*: 579–587.

Dorokhova, L.A., N.S. Agre, L.V. Kalakoutskii and N.A. Krasil'nikov. 1968. Fine structure of spores in a thermophilic actinomycete, *Micromonospora vulgaris*. J. Gen. Appl. Microbiol. *14*: 295–303.

Dorokhova, L.A., N.S. Agre, L.V. Kalakoutskii and N.A. Krasil'nikov. 1970a. Electron microscopic study of spore formation in *Micromonospora vulgaris*. Mikrobiologiya *39*: 680–684.

Dorokhova, L.A., N.S. Agre, L.V. Kalakoutskii and N.A. Krasil'nikov. 1970b. Comparative study on spores of some actinomycetes with special reference to their thermoresistance. *In* Prauser (Editor), The *Actinomycetales*, Gustav Fischer Verlag, Jena, pp. 227–232.

Edwards, J.H. 1972. The double dialysis method of producing farmer's lung antigens. J. Lab. Clin. Med. *79*: 683–688.

Eggins, H.O.W., A.v. Szilvinyi and D. Allsopp. 1972. The isolation of actively growing thermophilic fungi from insulated soils. Int. Biodeterior. Bull. *8*: 53–58.

Elwan, S.H., S.A. Mostafa, A.A. Khodair and O. Ali. 1978a. Lipase productivity of lipolytic strain of *Thermoactinomyces vulgaris*. Zentralbl. Bakteriol. Parasitenkd. Infektionskr. Hyg. Abt. II *133*: 706–712.

Elwan, S.H., S.A. Mostafa, A.A. Khodair and O. Ali. 1978b. Identity and lipase activity of an isolate of *Thermoactinomyces vulgaris*. Zentralbl. Bakteriol. Parasitenkd. Infektionskr. Hyg. Abt. II *133*: 713–722.

Ensign, J.C. 1978. Formation, properties, and germination of actinomycete spores. Annu. Rev. Microbiol. *32*: 185–219.

Erikson, D. and D.M. Webley. 1953. The respiration of a thermophilic actinomycete, *Micromonospora vulgaris*. J. Gen. Microbiol. *8*: 455–463.

Fergus, C.L. 1964. Thermophilic and thermotolerant molds and actinomycetes of mushroom compost during peak heating. Mycologia *56*: 267–284.

Fergus, C.L. 1967. Resistance of spores of some thermophilic actinomycetes to high temperature. Mycopath. Mycol. Appl. *32*: 205–208.

Fergus, C.L. 1969. The cellulolytic activity of thermophilic fungi and actinomycetes. Mycologia *61*: 120–129.

Flockton, H.O. and T. Cross. 1975. Variability in *Thermoactinomyces vulgaris*. J. Appl. Bacteriol. *38*: 309–313.

Foerster, H.F. 1975. Germination characteristics of some of the thermophilic actinomycete spores. *In* Gerhardt, Costilow and Sadoff (Editors), Spores, Vol. VI. American Society for Microbiology, Washington, D.C., pp. 36–43.

Foerster, H.F. 1978. Effects of temperature on the spores of thermophilic actinomycetes. Arch. Microbiol. *118*: 257–264.

Forsyth, W.G.C. and D.M. Webley. 1948. The microbiology of composting. II. A study of the aerobic thermophilic bacterial flora developing in grass composts. Proc. Soc. Appl. Bacteriol. *3*: 34–39.

Fortina, M.G., D. Mora, P. Schumann, C. Parini, P.L. Manachini and E. Stackebrandt. 2001. Reclassification of *Saccharococcus caldoxylosilyticus* as *Geobacillus caldoxylosilyticus* (Ahmad et al. 2000) comb. nov. Int. J. Syst. Evol. Microbiol. *51*: 2063–2071.

Fritzsche, H. 1967. Infrared studies of deoxyribonucleic acids, their consituents and analogues. II. Deoxyribonucleic acids with different base composition. Biopolymers *5*: 863–870.

Golovina, L.G., E.P. Guzhova, T.I. Bogdanov and L.G. Loginova. 1973. Lytic enzymes produced by thermophilic actinomycetes *Micromonospora vulgaris* PA 11-4. Mikrobiologiya *62*: 620–626.

Goodfellow, M. and T. Cross. 1974. actinomycetes. *In* Dickinson and Pugh (Editors), Biology of Plant Litter Decomposition. Academic Press, London, pp. 269–302.

Goodfellow, M. and T. Pirouz. 1982. Numerical classification of sporoactinomycetes containing *meso*-diaminopimelic acid in the cell wall. J. Gen. Microbiol. *128*: 503–527.

Goodfellow, M. and T. Cross. 1984. Classification. *In* Goodfellow, Mordarski and Williams (Editors), The Biology of the Actinomycetes. Academic Press, London, pp. 7–164.

Goodfellow, M. and J.A. Haynes. 1984. Actinomycetes in marine sediments. *In* Ortiz-Ortiz, Bojalil and Yakoleff (Editors), Biological, Biochemical and Biomedical Aspects of Actinomycetes. Academic Press, Orlando, pp. 453–472.

Greatorex, F.B. and J.V.S. Pether. 1975. Use of a serologically distinct strain of *Thermoactinomyces* vulgaris in the diagnosis of farmers' lung disease. J. Clin. Pathol. *28*: 1000–1002.

Gregory, P.H. and M.E. Lacey. 1963. Mycological examination of dust from mouldy hay associated with farmer's lung disease. J. Gen. Microbiol. *30*: 75–88.

Gregory, P.H., M.E. Lacey, G.N. Festerstein and F.A. Skinner. 1963. Microbial and biochemical changes during the moulding of key. J. Gen. Microbiol. *33*: 147–174.

Hägerdal, B.G.R., J.D. Ferchak and E.K. Pye. 1978. Cellulolytic enzyme system of *Thermoactinomyces* sp. grown on microcrystalline cellulose. Appl. Environ. Microbiol. *36*: 606–612.

Hatayama, K., S. Kawai, H. Shoun, Y. Ueda and A. Nakamura. 2005a. *Pseudomonas azotifigens* sp. nov., a novel nitrogen-fixing bacterium isolated from a compost pile. Int. J. Syst. Evol. Microbiol. *55*: 1539–1544.

Hatayama, K., H. Shoun, Y. Ueda and A. Nakamura. 2005b. *Planifilum fimeticola* gen. nov., sp. nov. and *Planifilum fulgidum* sp. nov., novel members of the family 'Thermoactinomycetaceae' isolated from compost. Int. J. Syst. Evol. Microbiol. *55*: 2101–2104.

Henssen, A. 1957. Beiträge zur Morphologie und Systematik der thermophilen Actinomyceten. Arch. Mikrobiol. *26*: 373–414.

Hirst, J.M., C.R. Bailey and F.G. Priest. 1991. Deoxyribonucleic acid sequence homology among some strains of *Thermoactinomyces*. Lett. Appl. Microbiol. *13*: 35–38.

Hollick, G.E., N.K. Hall and H.W. Larsh. 1979. Chemical and seriological comparison of two antigen extracts of *Thermoactinomyces candidus*. Mykosen *22*: 49–59.

Hollick, G.E. and H.W. Larsh. 1979. Crossed immunoelectrophoretic analysis of two antigen extracts from *Thermoactinomyces candidus*. Infect. Immun. *26*: 1057–1064.

Hollick, G.E. 1982. Enzymatic profiles of selected thermophilic actinomycetes. Microbios *35*: 187–196.

Hopwood, D.A. and H.M. Ferguson. 1970. Genetic recombination in a thermophilic actinomycete, *Thermoactinomyces vulgaris*. J. Gen. Microbiol. *63*: 133–136.

Hopwood, D.A. and H.M. Wright. 1972. Transformation in *Thermoactinomyces vulgaris*. J. Gen. Microbiol. *71*: 383–398.

Hussein, A.M., A.M. Allam and A.M. Ragab. 1975. Taxonomical studies on thermophilic actinomycetes of some soils in Egypt. Ann. Microbiol. (Milan) *25*: 19–28.

Jackson, A.M., P.R. Poulton and A.S. Ball. 1997. Importance of farming practice on the isolation frequency of *Thermoactinomyces* species. Soil Biol. Biochem. *29*: 207–210.

Kalakoutskii, L.V., N.I. Nikitina and O.I. Artamonova. 1969. Spore germination in actinomycetes (in Russian). Mikrobiologiya *38*: 834–841.

Kalakoutskii, L.V. and N.S. Agre. 1973. Endospores of *Actinomyces*: dormancy and germination. *In* Sykes and Skinner (Editors), *Actinomycetales*: Characteristics and Practical Importance, Soc. Appl. Bacteriol. Symp. Ser. No. 2. Society for Applied Bacteriology, London, pp. 179–195.

Kalakoutskii, L.V. and N.S. Agre. 1976. Comparative aspects of development and differentiation in actinomycetes. Bacteriol. Rev. *40*: 469–524.

Kanoh, K., Y. Matsuo, K. Adachi, H. Imagawa, M. Nichizawa and Y. Shizuri. 2005. Mechercharmycins A and B cytotoxic substances from marine – derived *Thermoactinomyces* sp. YM3-251. J. Antibiot. *58*: 289–292.

Kirillova, I.P., N.S. Agre and L.V. Kalakoutskii. 1973. Conditions for initiation of *Thermoactinomyces vulgaris* spores (in Russian). Mikrobiologiya *42*: 867–872.

Kirillova, I.P., N.S. Agre and L.V. Kalakoutskii. 1974. Spore initiation and minimum temperature for growth of *Thermoactinomyces vulgaris*. Z. Allg. Mikrobiol. *14*: 69–72.

Kokina, V.Y. and N.S. Agre. 1977a. Investigation of spores of *Thermoactinomyces vulgaris* degenerated cultures (in Russian) Mikrobiologiya *46*: 378–380.

Kokina, V.Y. and N.S. Agre. 1977b. Factors causing degeneration of *Thermoactinomyces vulgaris* cultures (in Russian). Mikrobiologiya *46*: 304–310.

Kosmachev, A.E. 1962. A thermophilic *Micromonospora* and its production of antibiotic T-12 under conditions of surface and submerge fermentation at 50°C–60°C (in Russian). Mikrobiologiya *31*: 66–71.

Krasil'nikov, N.A. and N.S. Agre. 1964. A new actinomycete genus – Actinobifida n. gen. yellow group – Actinobifida dichotomica n. sp. (in Russian). Mikrobiologiya *33*: 935–943.

Krasil'nikov, N.A. and N.S. Agre. 1965. The brown group of Actinobifida chromogena n. sp. (in Russian). Mikrobiologiya *34*: 284–291.

Kretschmer, S. 1980. Transinfection in *Thermoactinomyces vulgaris*. Z. Allg. Mikrobiol. *20*: 73–75.

Kretschmer, S. and E. Sarfert. 1980. Transfection in *Thermoactinomyces vulgaris*. Z. Allg. Mikrobiol. *20*: 73–75.

Kretschmer, S. 1982. Alteration of interaction with virulent bacteriophage Ta 1 during differentiation of *Thermoactinomyces vulgaris*. Z. Allg. Mikrobiol. *22*: 629–637.

Kretschmer, S., D. Körner, G. Stohbach, P. Klingenberg, H.-E. Jacob, J. Gumpet and H. Tuttloff. 1982. Physiological and cell biological characteristics of protease-forming *Thermoactinomyces vulgaris* during prolonged culture in a fermenter (in German). Z. Allg. Mikrobiol. *22*: 693–703.

Kretschmer, S. and H.-E. Jacob. 1983. Autolysis of *Thermoactinomyces vulgaris* spores lacking carbon dioxide during germination Z. Allg. Mikrobiol. *23*: 27–32.

Kretschmer, S. 1984a. Alternative life cycles in *Thermoactinomyces vulgaris*. Z. Allg. Mikrobiol. *24*: 93–100.

Kretschmer, S. 1984b. Intercalary growth of *Thermoactinomyces vulgaris*. Z. Allg. Mikrobiol. *24*: 211–215.

Kretschmer, S. 1984c. Characterization of aerial mycelium of *Thermoactinomyces vulgaris*. Z. Allg. Mikrobiol. *24*: 101–111.

Kuo, M.J. and P.A. Hartman. 1966. Isolation of amylolytic strains of *Thermoactinomyces vulgaris* and production of thermophilic actinomycete amylases. J. Bacteriol. *92*: 723–726.

Kuo, M.J. and P.A. Hartman. 1967. Purification and partial characterization of *Thermoactinomyces vulgaris* amylases. Can. J. Microbiol. *13*: 1157–1163.

Kurup, V.P., J.J. Barboriak, J.N. Fink and M.P. Lechevalier. 1975. *Thermoactinomyces candidus*, a new species of thermophilic actinomycetes. Int. J. Syst. Bacteriol. *25*: 150–154.

Kurup, V.P., J.N. Fink and D.M. Bauman. 1976a. Thermophilic actinomycetes from the environment. Mycologia *68*: 662–666.

Kurup, V.P., J.J. Barboriak, J.N. Fink and G. Scribner. 1976b. Immunologic cross reactions among thermophilic actinomycetes associated with hypersensitivity pneumonitis. J. Allergy Clin. Immunol *57*: 417–421.

Kurup, V.P. and R.J. Heinzen. 1978. Isolation and characterization of actinophages of *Thermoactinomyces* and *Micropolyspora*. Can. J. Microbiol. *24*: 794–797.

Kurup, V.P., G.E. Hollick and E.F. Pagan. 1980. *Thermoactinomyces intermedius*, a new species of amylase negative thermophilic actinomycetes. Science-Ciencia Bol. Cien. Sur *7*: 104–108.

Kurup, V.P., G.E. Hollick and E.F. Pagan. 1981. *In* Validation of the publication of new names and new combinations previously effectively published outside the IJSB. List no. 6. Int. J. Syst. Bacteriol. *31*: 215–218.

Küster, E. and R. Locci. 1963. Studies on peat and peat microorganisms. I. Taxonomic studies on thermophilic actinomycetes isolated from peat. Arch. Mikrobiol. *45*: 188–197.

Küster, E. and R. Locci. 1964. Taxonomic studies on the genus *Thermoactinomyces*. Int. Bull. Bacteriol. Nomencl. Taxon. *14*: 109–114.

Küster, E. 1974. Genus II. *Thermoactinomyces* Tsiklinsky 1899, 501. *In* Buchanan and Gibbons (Editors), Bergey's Manual of Determinative Bacteriology, 8th edn. The Williams & Wilkins Co., Baltimore, pp. 855–856.

Lacey, J. 1971a. *Thermoactinomyces sacchari* sp. nov., a thermophilic actinomycete causing bagassosis. J. Gen. Microbiol. *66*: 327–338.

Lacey, J. 1971b. The microbiology of moist barky storage in unsealed silos. Ann. Appl. Biol. *69*: 187–212.

Lacey, J. and D.A. Vince. 1971. Endospore formation and germination in a new *Thermoactinomyces* species. *In* Barker, Gould and Wolf (Editors), Spore Research 1971. Academic Press, London, pp. 181–187.

Lacey, J. 1973. Actinomycetes in soils, composts and fodders. *In* Sykes and Skinner (Editors), *Actinomycetales*: Characteristics and Practical Importance, Society for Applied Bacteriology, Ser. No. 2. Soc. Appl. Bacteriol. Symp., London, pp. 231–251.

Lacey, J. and J. Dutkiewicz. 1976a. Isolation of actinomycetes and fungi from mouldy hay using a sedimentation chamber. J. Appl Bacteriol. *41*: 315–319.

Lacey, J. and J. Dutkiewicz. 1976b. Methods for examining the microflora of mouldy hay. J. Appl. Bacteriol. *41*: 13–27.

Lacey, J. and J. Dutkiewicz. 1976c. Methods for examining the microflora of mouldy hay. J. Appl. Bacteriol *41*: 13–27.

Lacey, J. 1978. Ecology of actinomycetes in fodders and related substances. Zentralbl. Bakteriol. Parasitenkd. Infektionskr. Hyg. Abt. I *Suppl. 6*: 161–170.

Lacey, J. 1981. Airborne actinomycete spores as respiratory allergens. Zentralbl. Bakteriol. Mikrobiol. Hyg. I Abt *Suppl. 11*: 243–250.

Lacey, J. and M.E. Lacey. 1987. Micro-organisms in the air of cotton mills. Ann. Occup. Hyg. *31*: 1–19.

Lacey, J. and T. Cross. 1989. Genus *Thermoactinomyces* Tsiklinsky 1899, 501AL. *In* Williams, Sharpe and Holt (Editors), Bergey's Manual of Systematic Bacteriology, vol. 4. The Williams & Wilkins Co., Baltimore, pp. 2574–2585.

Lehrer, S.B. and J.E. Salvaggio. 1978. Characterization of *Thermoactinomyces sacchari* antigens. Infect. Immun. *20*: 519–525.

Leuchtenberger, A. and H. Ruttloff. 1983. Effect of oil and fatty acids on growth and enzyme formation by *Thermoactinomyces vulgaris*. III. Influence of culture vessel, strain and medium composition (in German). Z Allg. Mikrobiol. *23*: 635–644.

Locci, R. 1963. The phenomenon of autoinhibition in *Thermoactinomyces vulgaris* (in Italian). G. Microbiol. *11*: 183–189.

Locci, R. 1972. On the spore formation process in actinomycetes. IV. Examination by scanning electron microscopy of the genera *Thermoactinomyces*, *Actinobifida* and *Thermomonospora*. Riv. Patol. Veg *4* 63–80.

Matsuo, Y., A. Katsuta, S. Matsuda, Y. Shizuri, A. Yokota and H. Kasai. 2006. *Mechercharimyces mesophilus* gen. nov., sp. nov. and *Mechercharimyces asporophorigenens* sp. nov., antitumour substance-producing marine bacteria, and description of *Thermoactinomycetaceae* fam. nov. Int. J. Syst. Evol. Microbiol. *56*: 2837–2842.

McCarthy, A.J. and T. Cross. 1984a. A taxonomic study of *Thermomonospora* and other monosporic actinomycetes. J. Gen. Microbiol. *130*: 5–25.

McCarthy, A.J. and T. Cross. 1984b. *In* Validation of the publication of new names and new combinations previously effectively published outside the IJSB. List no. 15. Int. J. Syst. Bacteriol. *34*: 355–357.

McVittie, A., H. Wildermuth and D.A. Hopwood. 1972. Fine structure and surface topography of endospores of *Thermoactinomyces vulgaris*. J. Gen. Microbiol. *71*: 367–381.

Molina, C. 1974. Farmer's lung in France. *In* de Haller and Suter (Editors), Aspergillosis and Farmer's Lung in Man and Animal. Hans Huber Publishers, Bern, pp. 205–206.

Moppett, C.E., D.T. Dix, F. Johnson and C. Coronelli. 1972. Structure of thermorubin A, the major orange-red antibiotic of *Thermoactinomyces antibioticus*. J. Am Chem Soc *94*: 3269–3272.

Nilsson, M. and I. Renberg. 1990. Viable endospores of *Thermoactinomyces vulgaris* in lake sediments as indicators of agricultural history. Appl Environ Microbiol *56*: 2025–2028.

Nonomura, H. and Y. Ohara. 1971. Distribution of actinomycetes in soil. X. New genus and species of monosporic actinomycetes in soil. J. Ferment. Technol *49*: 895–903.

Obi, S.K.C. and F.J.C. Odibo. 1984. Some properties of a highly thermostable α-amylase from a *Thermoactinomyces* sp. Can. J. Microbiol. *30*: 780–785.

Ochi, K. 1994. Phylogenetic diversity in the genus *Bacillus* and comparative ribosomal protein AT-L30 analyses of the genus *Thermoactinomyces* and relatives. Microbiology *140*: 2165–2171.

Ohshima, T., N. Nishida, S. Bakthavatsalam, K. Kataoka, H. Takada, T. Yoshimura, N. Esaki and K. Soda. 1994. The purification, characterization, cloning and sequencing of the gene for a halostable and thermostable leucine dehydrogenase from *Thermoactinomyces intermedius*. Eur. J. Biochem *222*: 305–312.

Ohtaki, A., M. Mizuno, H. Yoshida, T. Tonozuka, Y. Sakano and S. Kamitori. 2006. Structure of a complex of *Thermoactinomyces vulgaris* R-47 alpha-amylase 2 with maltohexaose demonstrates the important role of aromatic residues at the reducing end of the substrate binding cleft. Carbohydr Res *341*: 1041–1046.

Orlowska, B. 1969. *Thermoactinomyces albus*, a new species of thermophilic actinomycetes. Annual Report of the Ludwig Hirszfeld Institute for Immunol. Exp. Ther.: 25–26.

Park, D.J., S.G. Dastager, J.C. Lee, S.H. Yeo, J.H. Yoon and C.J. Kim. 2007. *Shimazuella kribbensis* gen. nov., sp. nov., a mesophilic representative of the family *Thermoactinomycetaceae*. Int. J. Syst. Evol. Microbiol. *57*: 2660–2664.

Park, Y.-H., D.-G. E. Kim, Y.-H.K. Yim, T.-I. Mheen and M. Goodfellow. 1993. Suprageneric classification of *Thermoactinomyces vulgaris* by nucleotide sequencing of 5S ribosomal RNA. Zentbl. Bakteriol. *278*: 469–478.

Patel, J.J. 1969. Phages of lysogenic *Thermoactinomyces vulgaris*. Arch. Mikrobiol. *69*: 294–300.

Pathom-aree, W., J.E.M. Stach, A.C. Ward, K. Horikoshi, A.T. Bull and M. Goodfellow. 2006. Diversity of actinomycetes isolated from Challenger Deep sediment (10,898) from the Mariana Trench. Extremophiles *10*: 181–189.

Pepys, J., P.A. Jenkins, G.N. Festenstein, P.H. Gregory, M.E. Lacey and F.A. Skinner. 1963. Farmer's lung: Thermophilic actinomycetes as a source of "farmer's lung hay" antigen. Lancet Infect. Dis.: 607–611.

Pether, J.V.S. and F.B. Greatorex. 1976. Farmer's lung disease in Somerset. Br. J. Ind. Med. *33*: 265–268.

Prauser, H. and S. Momirova. 1970. [Phage sensitivity, cell wall composition and taxonomy of various thermophilic actinomycetes]. Z. Allg. Mikrobiol. *10*: 219–222.

Roberts, R.C., D.P. Zais, J.J. Marx and M.W. Treuhaft. 1977. Comparative electrophoresis of the proteins and proteases in thermophilic actinomycetes. J. Lab. Clin. Med. *90*: 1076–1085.

Sarfert, E., S. Kretschmer, H. Triebel and G. Luck. 1979. Properties of *Thermoactinomyces vulgaris* phage Ta, and its extracted DNA. Z. Allg. Mikrobiol. *19*: 203–210.

Schuurmans, D.M., B.H. Olson and C.L.S. Clemente. 1956. Production and isolation of thermoviridin an antibiotic produced by *Thermoactinomyces viridis* n. sp. Appl. Microbiol. *4*: 61–66.

Seaward, M.R.D., T. Cross and B.A. Unsworth. 1976. Viable bacterial spores recovered from an archaeological excavation. Nature (Lond.) *261*: 407–408.

Shimizu, M., M. Kanno, M. Tamura and M. Suckane. 1978. Purification and some properties of a novel α-amylase produced by a strain of *Thermoactinomyces vulgaris*. Agric. Biol. Chem. *42*: 1681–1688.

Shirling, E.B. and D. Gottlieb. 1966. Methods for characterization of *Streptomyces* species. Int. J. Syst. Bacteriol. *16*: 313–340.

Skerman, V.B.D., V. McGowan and P.H.A. Sneath. 1980. Approved lists of bacterial names. Int. J. Syst. Bacteriol *30*: 225–420.

Song, J., H.Y. Weon, S.H. Yoon, D.S. Park, S.J. Go and J.W. Suh. 2001. Phylogenetic diversity of thermophilic actinomycetes and *Thermoactinomyces* spp. isolated from mushroom composts in Korea based on 16S rRNA gene sequence analysis. FEMS Microbiol. Lett. *202*: 97–102.

Stackebrandt, E. and C.R. Woese. 1981. Towards a phylogeny of the actinomycetes and related organisms. Curr. Microbiol. *5*: 197–202.

Sweet, L.C., J.A. Anderson, Q.C. Callies and E.O. Coates. 1971. Hypersensitivity pneumonitis related to home furnace humidifier. J. Allergy Clin. Immunol. *48*: 171–178.

Taptykova, S.D., L.V. Kalakoutskii and N.S. Agre. 1969. Cytochromes in spores of actinomycetes. J. Gen. Appl. Microbiol. *15*: 383–386.

Taufel, A., U. Behnke and H. Ruttloff. 1979. Isolation and characterisation of proteases from *Thermoactinomyces vulgaris*. IV. Extracellular protease spectrum during the course of culture (in German). Z. Allg. Mikrobiol. *19*: 129–138.

Teplyakov, A.V., I.P. Kuranova and E.H. Harutyunyan. 1989. Crystal structure of thermitase from *Thermoactinomyces vulgaris* at 2.2A resolution. FEBS Lett. *244*: 208–218.

Terao, M., K. Furuya and R. Emokita. 1965. Studies on antibiotics from thermophilic actinomycetes. I. An antibiotic produced by strain no. BT3-4. Annu. Rep. Sankyo. Res. Lab. *17*: 110–117.

Terho, E.O. and J. Lacey. 1979. Microbiological and serological studies on farmer's in Finland. Clin. Allergy *9*: 43–52.

Treuhaft, M.W. 1977. Isolation of bacteriophage from *Thermoactinomyces*. J. Clin. Microbiol. *6*: 420–424.

Tseng, M., T. Kudo and A. Seino. 1990. Identification of thermophilic actinomycetes isolated from mushroom compost in Taiwan. Bull JFCC *6*: 6–13.

Tsiklinsky, P. 1899. On the thermophilic moulds (in French). Ann. Inst. Pasteur *13*: 500–505.

Unaogu, I.C. and H.C. Gugnani. 1999. Pathogenicity of *Thermoactinomyces vulgaris* for laboratory mice. J. Mycol. Med. *9*: 237–238.

Unsworth, B.A., T. Cross, M.R. Seaward and R.E. Sims. 1977. The longevity of thermoactinomycete endospores in natural substrates. J. Appl. Bacteriol. *42*: 45–52.

Unsworth, B.A. 1978. The genus *Thermoactinomyces*. Tsiklinsky, PhD thesis. University of Bradford, U.K.

Unsworth, B.A. and T. Cross. 1980. Thermophilic actinomycetes implicated in Farmer's Lung: numerical taxonomy of *Thermoactinomyces* species. *In* Goodfellow and Board (Editors), Microbiological Classification and Identification. Academic Press, London, pp. 389–390.

Waksman, S.A. and C.T. Cork. 1953. *Thermoactinomyces* Tsiklinsky, a genus of thermophilic actinomycetes. J. Bacteriol. *66*: 377–378.

Waksman, S.A. 1961. The actinomycetes. Classification, Identification and Descriptions of Genera and Species, Volume 2, The Williams & Wilkins Co., Baltimore, pp. 1–363.

Webley, D.M. 1954. The effect of oxygen on the growth and metabolism of the aerobic thermophilic actinomycete *Micromonospora vulgaris*. J. Gen. Microbiol *19*: 402–406.

Webley, D.M. 1958. A defined medium for the growth of the thermophilic actinomycete *Micromonospora vulgaris*. J. Gen. Microbiol *19*: 402–406.

Wenzel, F.J., D.A. Emanuel and B.R. Lawton. 1967. Pneumonitis due to *Micromonospora vulgaris* (farmer's lung). Am. Rev. Respir. Dis. *95*: 652–655.

Wenzel, F.J., R.L. Gray, R.C. Roberts and D.A. Emanuel. 1974. Serologic studies in farmer's lung. Precipitins to the thermophilic actinomycetes. Am. Rev. Respir. Dis. *109*: 464–468.

Williams, S.T., G.P. Sharples and R.M. Bradshaw. 1973. The fine structure of the *Actinomycetales*. *In* Sykes and Skinner (Editors), *Actinomycetales*: Characteristics and Practical Importance. Academic Press, London, pp. 113–130.

Xu, J., J.R. Rao, B.C. Millar, J.S. Eborn, J. Evans, J.G. Barr and J.E. Moore. 2002. Improved molecular identification of *Thermoactinomyces* spp. associated with mushroom worker's lung by 16S rDNA sequence typing. J. Med. Microbiol. *51*: 1117–1127.

Yoon, J.H. and Y.H. Park. 2000. Phylogenetic analysis of the genus *Thermoactinomyces* based on 16S rDNA sequences. Int. J. Syst. Evol. Microbiol. *50*: 1081–1086.

Yoon, J.H., Y.K. Shin and Y.H. Park. 2000. DNA-DNA relatedness among *Thermoactinomyces* species: *Thermoactinomyces candidus* as a synonym of *Thermoactinomyces vulgaris* and *Thermoactinomyces thalpophilus* as a synonym of *Thermoactinomyces sacchari*. Int. J. Syst. Evol. Microbiol. *50*: 1905–1908.

Yoon, J.H., I.G. Kim, Y.K. Shin and Y.H. Park. 2005. Proposal of the genus *Thermoactinomyces* sensu stricto and three new genera, *Laceyella*, *Thermoflavimicrobium* and *Seinonella*, on the basis of phenotypic, phylogenetic and chemotaxonomic analyses. Int. J. Syst. Evol. Microbiol. *55*: 395–400.

Family X. **Incertae Sedis**

Previously assigned to the "*Paenibacillaceae*" by Garrity et al. (2005), subsequent analyses suggest that the genus *Thermicanus* represents a very deep group that is only distantly related to any of the previously described families within the *Bacillales*. In view of its ambiguous status, it has been assigned to its own family *incertae sedis*.

Genus I. **Thermicanus** Gössner, Devereux, Ohnemüller, Acker, Stackebrandt and Drake 2000, 423^{VP} (Effective publication: Gössner, Devereux, Ohnemüller, Acker, Stackebrandt and Drake 1999, 5131.)

HAROLD L. DRAKE

Therm.i.ca′nus. Gr. adj. *thermos* hot; Gr. adj. *hikanos*; N.L. masc. adj. *icanus* capable; N.L. masc. n. *Thermicanus* the capable thermophile.

Thermophilic, chemotrophic, and **facultative microaerophile.** Cells are **weakly Gram-stain-positive** and **rod-shaped** (Figure 85).
DNA G+C content (mol%): 50.3.

Type species: **Thermicanus aegyptius** Gössner, Devereux, Ohnemüller, Acker, Stackebrandt and Drake 2000, 423^{VP} (Effective publication: Gössner, Devereux, Ohnemüller, Acker, Stackebrandt and Drake 1999, 5131.).

Further descriptive information

Members of the genus *Thermicanus* are currently represented by a single species, i.e., the type species *Thermicanus aegyptius*. *Thermicanus aegyptius* is phylogenetically distantly related to the genera *Paenibacillus*, *Oxalophagus*, *Bacillus*, and *Thermoactinomyces*. Sequence similarity of the 16S rRNA gene of *Thermicanus aegyptius* to that of its closest relative approximates 88% (Gössner et al., 1999).

Enrichment and isolation procedures

The type species, *Thermicanus aegyptius*, was isolated from soil obtained from Egypt (Gössner et al., 1999). A near pH neutral medium designed for the enrichment of acetogens was inoculated with soil and incubated at 55°C. An isolated colony obtained from an anoxic streak plate prepared from an acetogenic enrichment contained two organisms designated ET-5a (a new strain of the acetogen *Moorella thermoacetica*, DSM 12797) and ET-5b (*Thermicanus aegyptius*). *Thermicanus aegyptius* was subsequently enriched and isolated from the co-culture by selective cultivation on cellobiose under oxic conditions. The type species was derived from an isolated colony.

Maintenance procedures

The type species is easily maintained in the carbonate-buffered, yeast extract medium described by Gössner et al. (1999). Stability under long-term storage conditions has not been determined.

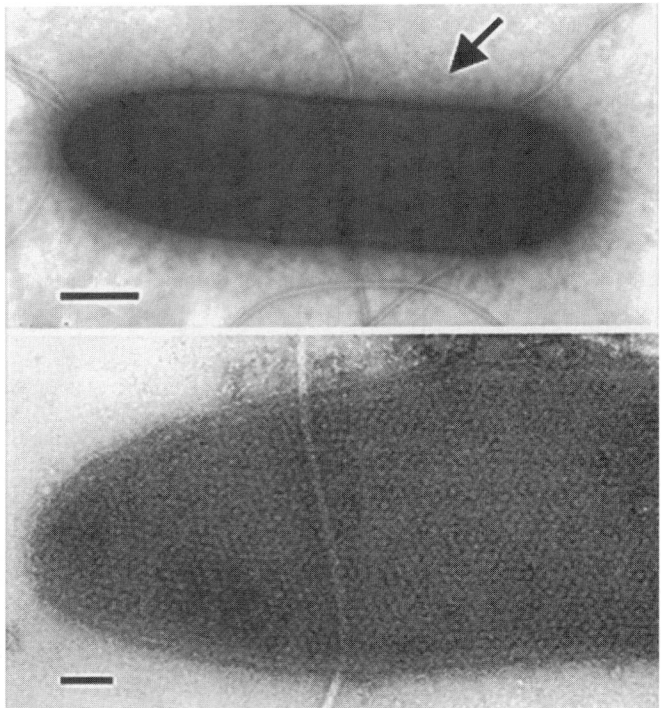

FIGURE 85. Transmission electron micrographs of *Thermicanus aegyptius* (DSM 12793). (top) The arrow indicates fibrillar structures in the capular domain. The large filaments are flagella. Bar = 0.3 μm. (bottom) Hexagonal subunits of the S-layer. Bar = 0.1 μm. (Reproduced with permission from Gössner et al.; Applied and Environmental Microbiology *65*: 5124–5133, 1999, ©American Society for Microbiology.)

List of species of the genus *Thermicanus*

1. **Thermicanus aegyptius** Gössner, Devereux, Ohnemüller, Acker, Stackebrandt and Drake 2000, 423^{VP} (Effective publication: Gössner, Devereux, Ohnemüller, Acker, Stackebrandt and Drake 1999, 5131.)

ae.gyp′ti.us. L. adj. *aegyptius* Egyptian or from Egypt (to indicate the origin of the type species).

Cells are approximately 2.5 × 0.5 μm with an S-layer and outer and cytoplasmic membranes; nonspore-forming. Cells are flagellated and motile. Colonies on solidified media are beige. Growth is optimal at 55–60°C and pH 6.5–7; doubling times approximate 1.5–2 h. Prefers anoxic or microaerophilic conditions. Substrates include stachyose, raffinose, maltose, sucrose, cellobiose, lactose, galactose, glucose, fructose, mannose, and xylose. Acetate, formate, and succinate are only utilized under oxic conditions. Fermentation products are acetate, succinate, ethanol, formate, lactate, and H_2; fermentation products are also formed under oxic conditions. Nitrate, sulfate, and thiosulfate are not dissimilated; Fe^{3+} is reduced to Fe^{2+} as a side reaction. Particulate and soluble fractions

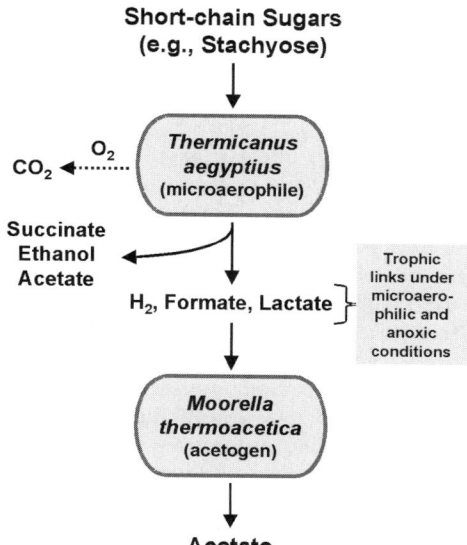

Short-chain Sugars (e.g., Stachyose)

CO_2 ← O_2 ← **Thermicanus aegyptius** (microaerophile)

Succinate Ethanol Acetate

H_2, Formate, Lactate

Trophic links under microaerophilic and anoxic conditions

Moorella thermoacetica (acetogen)

Acetate

FIGURE 86. Scheme illustrating the trophic interaction of *Thermicanus aegyptius* and *Moorella thermoacetica*. (Reproduced with permission from Gössner et al.; Applied and Environmental Microbiology 65: 5124–5133, 1999, ©American Society for Microbiology.)

have *b*-type cytochromes. The thermophilic acetogen *Moorella thermoacetica* ET-5a (DSM 12797) grows by symbiotic interaction with *Thermicanus aegyptius* on oligosaccharides via the commensal transfer of H_2, formate, and lactate (Figure 86).

DNA G+C content (mol%): 50.3 (HPLC).

Type strain: ET-5b, ATCC 700890, DSM 12793.

GenBank accession number (16S rRNA gene): AJ242495.

References

Garrity, G.M., J.A. Bell and T. Lilburn. 2005. The Revised Road Map to the *Manual. In* Brenner, Krieg, Staley and Garrity (Editors), Bergey's Manual of Systematic Bacteriology, Second Edition, Vol. 2, The *Proteobacteria,* Part B, The *Gammaproteobacteria.* Springer, New York, pp. 159–206.

Gössner, A.S., R. Devereux, N. Ohnemuller, G. Acker, E. Stackebrandt and H.L. Drake. 1999. *Thermicanus aegyptius* gen. nov., sp. nov., isolated from oxic soil, a fermentative microaerophile that grows commensally with the thermophilic acetogen *Moorella thermoacetica.* Appl. Environ. Microbiol. *65:* 5124–5133.

Gössner, A., R. Devereux, N. Ohnemüller, G. Acker, E. Stackebrandt and H.L. Drake. 2000. *In* Validation of the publication of new names and new combinations previously effectively published outside the IJSEM. List no. 73. Int. J. Syst. Evol. Microbiol. *50:* 423–424.

Family XI. **Incertae Sedis**

Previously assigned to the "*Staphylococcaceae*" by Garrity et al. (2005), subsequent analyses suggest that the genus *Gemella* is outside the radiation of this family. Moreover, *Gemella* is distinguished from the *Staphylococcaceae stricto sensu* because it is catalase- and oxidase-negative and possesses predominantly straight-chained saturated and monounsaturated rather than branched-chain fatty acids. In view of its ambiguous status, it has been assigned to its own family *incertae sedis.*

Genus I. **Gemella** Berger, 1960, 253[AL]

MATTHEW D. COLLINS AND ENEVOLD FALSEN

Ge.mel'la. L. n. *gemellus* a twin; N.L. fem. n. *Gemella* a little twin.

Ovoid in shape and arranged in pairs, tetrads, small clusters, and sometimes in short chains. Cells stain Gram-positive, but some strains may decolorize readily and appear Gram-negative. Nonmotile and nonsporeforming. **No growth at 10 or 45°C or in broth containing 6.5% NaCl.** Negative bile-esculin reaction. **Facultatively anaerobic. Catalase- and oxidase-negative. Gas is not produced in MRS broth. Glucose and some other carbohydrates are fermented.** Esculin, gelatin, and hippurate are not hydrolyzed. **Pyrrolidonyl arylamidase is produced by most strains.** Leucine aminopeptidase may or may not be produced. Nitrate is not reduced. **Vancomycin-sensitive.**

DNA G + C content (mol%): 30–34.

Type species: **Gemella haemolysans** (Thjøtta and Bøe, 1938) Berger, 1960, 253[AL] (*Neisseria haemolysans* Thjøtta and Bøe 1938, 531.).

Further descriptive information

The cell walls of gemellae are of a Gram-positive type. They generally stain Gram-positive, but some decolorize easily and give Gram-variable or Gram-negative reactions. The walls of gemellae are relatively thin (10–20 nm; Mills et al., 1984; Reyn et al., 1970) which probably accounts for their Gram-variable character. Gemellae divide in two planes, generally at right angles to each other. Cells are cocci arranged in pairs often with adjacent sides flattened, or arranged in tetrads, clusters, or short chains. Pleomorphism may be observed; elongate and rod-shaped forms occur. Morphology varies with strain and cultural conditions (Berger, 1992). Size of the cells may vary considerably; diameter varies from about 0.5 mm to more than 1 mm and "giant cells" have been observed (Berger, 1992). In *Gemella morbillorum* pleomorphism may be very pronounced; coccal forms are frequently elongate, and cells may be of unequal size. Elongate cells are generally 0.5 by 1.2 μm (Holdeman and Moore, 1974), but longer cells (up to 2–3 μm) have been reported (Berger, 1992; Prévot, 1933).

Growth of *Gemella* species is slow. Colonies are small, circular, entire, low convex, translucent to opaque, nonpigmented, smooth, and occasionally mucoidal. On blood agar media, some strains of gemellae are hemolytic (Berger, 1992). On trypticase soy sheep blood agar, most strains are α- or non-hemolytic. The expression of hemolysis can depend on the type of blood and agar base used. Wide zone (β) hemolysis may be observed with some strains, especially *Gemella haemolysans*, on blood agar bases containing rabbit blood. Examination for β-hemolysis is best performed on Mueller–Hinton agar supplemented with rabbit blood (Berger, 1992).

Gemellae are facultatively anaerobic. Some strains may require anaerobic conditions on primary isolation but become more aerotolerant after transfer to suitable media. While incubation in elevated CO_2 concentrations may stimulate growth, *Gemella haemolysans* grows better in the presence of free O_2. Strains of *Gemella morbillorum* prefer an anaerobic atmosphere. *Gemella* species are cytochrome oxidase, catalase and peroxidase-negative. The optimum growth temperature of gemellae is 35 to 37°C; all species grow over a wide range of temperatures, but none grow at 10 or 45°C. Gemellae are fermentative; acid is produced from some carbohydrates. Acid formation in carbohydrate broth, using conventional tests, can be somewhat varied with some substrates. There is sometimes poor correlation between commercially available miniaturized tests (e.g., API systems) and conventional tube methods. Glucose is degraded fermentatively by *Gemella* species. Major end products of anaerobic glucose metabolism by *Gemella haemolysans* and *Gemella morbillorum* are L-lactic and acetic acids (Brooks et al., 1971; Holdeman and Moore, 1974). Acid is produced from mannitol and sorbitol by some *Gemella* strains, but these can be variable characteristics depending on the test method. Using the API rapid ID 32Strep system, *Gemella sanguinis*, *Gemella cuniculi*, and most strains of *Gemella morbillorum* ferment these substrates whereas *Gemella haemolysans* and *Gemella palaticanis* do not. Some strains of *Gemella bergeri* produce acid from mannitol but fail to ferment sorbitol. *Gemella haemolysans*, *Gemella morbillorum*, *Gemella palaticanis*, and *Gemella sanguinis* produce acid from maltose and sucrose whereas *Gemella bergeri* and *Gemella cuniculci* do not. *Gemella palaticanis* produces acid from lactose whereas other gemellae do not. Using the same test system, none of the *Gemella* species produces acid from D-arabitol, L-arabinose, cyclodextrin, glycerol, melibiose, melezitose, methyl-β-D-glucopyranoside, pullulan, raffinose, ribose, or tagatose. Gemellae do not hydrolyze esculin, gelatin, or hippurate. They are urease-negative and do not reduce nitrate. Some species produce acid and alkaline phosphatases. Activities for both of these enzymes are detected in *Gemella cuniculi*, *Gemella hemolyticus*, and *Gemella sanguinis* using the API rapid ID 32Strep and API ZYM systems but not in the other species. All gemellae are arginine dihydrolase, α-galactosidase, β-galactosidase, β-glucuronidase, α-mannosidase, α-fucosidase, lipase C14, trypsin, *N*-acetyl-β-glucosaminidase, and valine arylamidase-negative.

The reported DNA G+C contents of *Gemella haemolysans*, *Gemella bergeri*, *Gemella morbillorum*, *Gemella palaticanis*, and *Gemella sanguinis* are 33.5 ± 1.6 (Bd), 32.5 (T_m), 30 (T_m), 32 (T_m), and 31 (T_m) mol%, respectively (Collins et al., 1998a, 1998b, 1999b; Kilpper-Bälz and Schleifer, 1988). The cell-wall murein of *Gemella morbillorum* is of the L-Lysine–Ala$_{1-3}$ type (Kilpper-Bälz and Schleifer, 1988).

Information on the antimicrobial susceptibilities of gemellae is somewhat fragmentary (Berger, 1992; Buu-Hoi et al., 1982). *Gemella haemolysans* and *Gemella morbillorum* are susceptible to penicillins, cephalosporins, tetracyclines, chloramphenicol, and lincomycins. *Gemella haemolysans* is also strongly inhibited by macrolide antibiotics (erythromycin, spiramycin, oleandomycin), vancomycin, ristocetin, novobiocin, and tyrothricin. Some strains are, however, resistant to erythromycin and tetracycline. *Gemella haemolysans* is resistant to sulfonamides and trimethoprim and also displays low-level resistance to aminoglycosides (streptomycin, kanamycin, gentamicin, tobramycin, amikacin, and neomycin). Synergy between penicillin G and either gentamicin or streptomycin, and between vancomycin and the aforementioned aminoglycosides is also observed (Buu-Hoi et al., 1982). The antimicrobial susceptibilities of gemellae resemble those of viridans streptococci.

Gemellae are residents of the mucous membranes of humans and some other animals (Berger, 1992). In healthy people, *Gemella haemolysans* has been found in the oral cavity and upper respiratory tract whereas *Gemella morbillorum* is, in addition, found as a component of the normal human intestinal flora (Berger, 1985, 1992; Facklam and Elliott, 1995; Holdeman and Moore, 1974). *Gemella haemolysans* and *Gemella morbillorum*, like many other commensal bacteria of the human microbiota, are opportunistic pathogens causing severe localized and generalized infection, particularly in immunocompromised patients (Durak et al., 1983; Eggelmeijer et al., 1992; Etienne et al., 1984; Mitchell and Teddy, 1985; Petit et al., 1993; Pradeep et al., 1997). *Gemella haemolysans* has been isolated from blood cultures of patients with endocarditis (Brack et al., 1991; Fresard et al., 1993; Kaufhold et al., 1989; Morea et al., 1991) and from cerebrospinal fluid (CSF) cultures of patients with meningitis (Aspevall et al., 1991; May et al., 1993; Mitchell and Teddy, 1985; Petit et al., 1993). *Gemella morbillorum* has been recovered from blood cultures of patients with endocarditis (Maxwell, 1989; Omran and Wood, 1993), from cultures of synovial fluid from septic arthritis (von Essen et al., 1993) and from CSF cultures of patients with meningitis (Debast et al., 1993). Little is known about the distribution of the more recently described *Gemella* species. Both *Gemella bergeri* and *Gemella sanguinis* have been isolated from blood cultures of persons with subacute bacterial endocarditis (Collins et al., 1998a, 1998b; Shukla et al., 2002). *Gemella palaticanis* was originally isolated from the oral cavity of a dog (Collins et al., 1999b) whereas the only known strain of *Gemella cuniculi* was recovered, in mixed culture, from a submandibular abscess of a rabbit (Hoyles et al., 2000).

Isolation procedures

Gemella species can be grown on blood agar at 37°C in air (with or without additional CO_2, 5–10%), with the exception of *Gemella morbillorum*, for which oxygen has to be excluded (Berger and Pervanidis, 1986). While incubation in elevated CO_2 concentrations stimulates the growth of gemellae, the aerotolerant *Gemella morbillorum* prefers an anaerobic atmosphere.

For recovering *Gemella haemolysans* from oropharyngeal swabs, organisms should be streaked onto blood agar, where they form smooth, nonhemolytic or α-hemolytic colonies that resemble viridans streptococci. The expression of hemolysis, an important characteristic for the presumptive identification of *Gemella haemolysans*, depends on the nature of the growth medium (choice of blood and agar base). β-Hemolysis is only consistent on Mueller–Hinton agar supplemented with 5% rabbit blood. This is the best-suited medium for the isolation of colonies suspected of being *Gemella haemolysans*. For isolating *Gemella haemolysans* from septicemic infections, freshly drawn blood can be transferred to commercially available blood culture media and incubated aerobically or anaerobically. The liquid medium generally becomes slightly turbid after 3 d of incubation after which culture fluid can be streaked onto blood (rabbit, sheep, or horse) agar plates incubated aerobically or under increased CO_2 levels. The isolation of *Gemella haemolysans* from cerebrospinal fluid (CSF) can be achieved by streaking out a small amount of CSF deposit onto blood agar (Mitchell and Teddy, 1985). *Gemella haemolysans* has also been recovered from dental plaque (see De Jong and van der Hoeven (1987) for a description of the method).

Gemella morbillorum has been isolated from a wider range of clinical sources than *Gemella haemolysans*. From swabs and pus, *Gemella morbillorum* can be isolated by plating onto blood (usually 5% defibrinated sheep blood) agar plates. Minute colonies of *Gemella morbillorum* are generally either nonhemolytic or are surrounded by a zone of α-hemolysis (greening) (Facklam, 1977; Holdeman and Moore, 1974). In *Gemella morbillorum*, hemolysis is not a constant character. Hemolysis may or may not be influenced by blood source (Berger and Pervanidis, 1986; Facklam and Wilkinson, 1981). Hemolysis may also be influenced by the strain, presence/absence of oxygen, and CO_2 levels. *Gemella morbillorum* can be recovered from dental plaque. Samples taken with a sterile curette are immediately transferred to thioglycolate broth (Difco) containing 20% beef infusion, and dilutions streaked onto Columbia blood agar incubated in an atmosphere containing H_2, CO_2, and N_2 (1:1:8; Kolenbrander and Williams, 1983). *Gemella morbillorum* has been isolated from the human intestinal tract using rumen-fluid-glucose-cellobiose agar (RGCA) used for culturing anaerobes (Holdeman et al., 1976; Holdeman and Moore, 1974). The composition of RGCA is given in the VPI Anaerobic Laboratory Manual (Holdeman and Moore, 1975).

Maintenance procedures

Gemellae grow poorly on media without serum or blood. They can be routinely maintained on a variety of agar media, such as heart infusion agar, Mueller–Hinton agar, trypticase soy agar, and Columbia agar, supplemented with blood (5–7% v/v), serum (5–10% v/v) or ascitic fluid (10% v/v). Plates can be incubated aerobically (with or without additional CO_2). Fresh isolates of *Gemella morbillorum*, which are not adapted to these conditions, should be cultured anaerobically. Heart infusion broth, brain heart infusion broth, and trypticase soy broth enriched with serum (10%

v/v) serve as suitable liquid media. *Gemella morbillorum* has been preserved in chopped meat-glucose broth at room temperature (Kannangara et al., 1981). For long-term preservation, strains can be stored on cryogenic beads at −70°C, or lyophilized.

Taxonomic comments

Thjøtta and Bøe (1938) originally isolated the bacterium, which is now referred to as *Gemella haemolysans*, from the sputum of a patient with chronic bronchitis and assigned it to the genus *Neisseria*, as *Neisseria haemolysans*. The species differed markedly from other members of the *Neisseria* genus, and Berger (1960) therefore created the monospecific genus *Gemella* to accommodate the species. Berger (1960, 1961) originally proposed *Gemella* as an aerobic, oxidase-negative, catalase-negative genus within the family *Neisseriaceae*, but later studies by Reyn et al. (1970, 1966) showed *Gemella haemolysans* possessed a Gram-positive type cell wall, and it is now known that phylogenetically the species belongs to the order *Bacillales* of the *Firmicutes* (Collins et al., 1998a, 1998b, 1999b; Ludwig et al., 1988; Reyn et al., 1970; Whitney and O'Connor, 1993). A second species, *Gemella morbillorum*, was added to the genus *Gemella* by Kilpper-Bälz and Schleifer (1988). *Gemella morbillorum* has had a checkered taxonomic history. The species was first isolated by Tunnicliff in 1917 and named "*Diplococcus rubeolae*" (Tunnicliff, 1933) but this name was withdrawn (Tunnicliff, 1936) and Prévot's epithet for the same organism, *Diplococcus morbillorum*, was adopted until the species was transferred to the genus *Peptostreptococcus*, as *Peptostreptococcus morbillorum* by Smith (1957). Holdeman and Moore (1974) considered the species to be an anaerobic to aerotolerant streptococcus, and reclassified the species as *Streptococcus morbillorum*. The species was finally transferred to the genus *Gemella* on the basis of physiological and phylogenetic evidence (Kilpper-Bälz and Schleifer, 1988), a placement confirmed by comparative 16S rRNA sequencing (Collins et al., 1998a, 1998b, 1999b; Whitney and O'Connor, 1993). In the past few years, four other new species have been assigned to the *Gemella* genus including *Gemella bergeri* and *Gemella sanguinis* from human clinical sources (Collins et al., 1998a, 1998b), *Gemella palaticanis* from the oral cavity of a dog (Collins et al., 1999b) and *Gemella cuniculi* from a rabbit (Hoyles et al., 2000). All six *Gemella* species form a robust rRNA cluster within the *Bacillales* and do not display a particularly close phylogenetic affinity with any described Gram-positive, catalase-negative genus (Collins et al., 1999b; Hoyles et al., 2000; Whitney and O'Connor, 1993).

Differentiation of the genus *Gemella* from other genera

The identification of *Gemella* species in the routine laboratory can be problematic. Gemellae grow poorly on blood agar and often take 48 h to grow. On commonly used blood agar media, gemellae form small, nonhemolytic or α-hemolytic colonies that resemble viridans streptococci. β-Hemolysis (some strains) is only evident on Mueller–Hinton agar supplemented with rabbit blood. Cellular morphology may be helpful in differentiating gemellae from streptococci. Cells of *Gemella* species may appear

TABLE 79. Characteristics distinguishing species of the genus *Gemella*[a,b]

Characteristic	1. *G. haemolysans*	2. *G. bergeri*	3. *G. cuniculi*	4. *G. morbillorum*	5. *G. palaticanis*	6. *G. sanguinis*
Acid from:						
Lactose	−	−	−	−	+	−
Maltose	+	−	−	+	+	+
Mannitol	−	d	+	d	−	+
Sorbitol	−	−	+	d	−	+
Sucrose	+	−	−	+	+	+
Trehalose	−	−	−	−	+	−
Production of:						
APPA	−	d	−	d	+	+
GTA	d	−	−	−	+	−
PAC	+	−	+	−	−	+
PAL	+	−	+	−	d	+
Voges–Proskauer test	−	−	−	−	−	+

[a]Symbols: +, >85% positive; d, different strains give different reactions (16–84% positive); −, 0–15% positive.

[b]Tests performed using API Rapid ID32 Strep system except for production of PAC which is performed using API ZYM kit. Abbreviations: PAC, acid phosphatase; PAL, alkaline phosphatase; APPA, alanyl phenylalanine proline arylamidase; GTA, glycyl tryptophan arylamidase.

Gram-variable. They divide in two planes, generally at right angles to each other. Cocci are often arranged in pairs with adjacent sides flattened, or are arranged in tetrads, small clusters or short chains. Gemellae isolates fail to grow in broth containing 6.5% NaCl, fail to give a positive bile-esculin reaction, and do not grow at 10 or 45°C. Most *Gemella* strains are pyrolydonyl-arylamidase-positive, but the only known isolate of *Gemella cuniculi* is pyrolydonyl-arylamidase-negative.

Differentiation of the species of the genus *Gemella*

Numerous biochemical tests have been proposed to differentiate the various *Gemella* species, but some of these are controversial and dependent on the test method. There is sometimes poor correlation between results obtained by conventional methods and those obtained by miniaturized test systems (such as API kits). *Gemella haemolysans* can, however, usually be distinguished from *Gemella morbillorum* by producing alkaline phosphatase and by failing to produce acid from mannitol and sorbitol. Most strains of *Gemella morbillorum* are alkaline-phosphatase-negative and most pro-

duce acid from mannitol and sorbitol (Berger, 1961; Berger and Pervanidis, 1986). The reduction of nitrite is also a typical characteristic of *Gemella haemolysans* but is negative for *Gemella morbillorum* (Berger and Pervanidis, 1986). *Gemella sanguinis* and *Gemella bergeri*, also associated with human clinical sources, can be readily distinguished biochemically from each other and the aforementioned species by using the API rapid ID 32Strep system. *Gemella sanguinis* is similar to *Gemella morbillorum* in forming acid from mannitol and sorbitol but differs from the latter species in producing alkaline phosphatase. Similarly, *Gemella bergeri* differs from *Gemella haemolysans*, *Gemella morbillorum*, and *Gemella sanguinis* by failing to produce acid from maltose and sucrose (Collins et al., 1998b). It further differs from *Gemella haemolysans* and *Gemella sanguinis* by being alkaline-phosphatase-negative (Collins et al., 1998a). Biochemical tests using the commercially available API rapid ID 32Strep system, which are useful in distinguishing between the various *Gemella* species, are shown in Table 79.

All six *Gemella* species possess characteristic 16S rRNA gene sequences. 16S rRNA gene sequencing is probably the quickest and most reliable tool for identifying *Gemella* isolates to the species level.

List of species of the genus *Gemella*

1. **Gemella haemolysans** (Thjøtta and Bøe 1938) Berger 1960, 253 (*Neisseria haemolysans* Thøtta and Bøe 1938, 531.)
hae.mo.ly′sans. Gr. n. *haema* blood; Gr. v. *lyo* dissolve, break up; N.L. part. adj. *haemolysans* dissolving blood.
Type strain: ATCC 10379, CCUG 37985, CIP 101126, LMG 18984, NCTC 12968.
GenBank accession number (16S rRNA gene): L14326, M58799.

2. **Gemella bergeri** Collins, Hutson, Falsen, Sjödén and Facklam 1998c, 631[VP] (Effective publication: 1998b, 1292.)
ber.ger.i. N.L. gen. n. *bergeri* of Berger, named after Ulrich Berger in recognition of his contributions to the microbiology of gemellae.

Type strain: 617-93, ATCC 700627, CCUG 37817, CIP 105584, LMG 18983.
GenBank accession number (16S rRNA gene): Y13365.

3. **Gemella cuniculi** Hoyles, Foster, Falsen and Collins 2000, 2039[VP]
cu.ni′cu.li. L. gen. masc. n. *cuniculi* of the rabbit.
Type strain: M60449/99/1, ATCC BAA-287, CCUG 42726, CIP 106481.
GenBank accession number (16S rRNA gene): AJ251987.

4. **Gemella morbillorum** (Prévot 1933) Kilpper-Bälz and Schleifer 1988, 442[VP] (*Diplococcus morbillorum* Prévot 1933, 148; *Streptococcus morbillorum* Holdeman and Moore 1974, 269.)

mor.bil′lor.um. N.L. gen. n. *morbillorum* of measles; once considered to be associated with measles.

Type strain: Prévot 2917B, ATCC 27824, CCUG 15561, CCUG 18164, CIP 81.10, DSM 20572, LMG 18985, NCTC 11323, VPI 5424.

GenBank accession number (16S rRNA gene): L14327.

5. **Gemella palaticanis** Collins, Rodriguez Jovita, Foster, Sjödén and Falsen 1999b, 1525[VP]

pa.la.ti.ca′nis. L. neut. n. *palatum* gum; L. masc. n. *canis* dog; N.L. gen. masc. n. *palaticanis* of the gum of a dog.

Type strain: M663-98-1, ATCC BAA-58, CCUG 39489, CIP 106318.

GenBank accession number (16S rRNA gene): Y17280.

6. **Gemella sanguinis** Collins, Hutson, Falsen, Sjödén and Facklam 1999a, 1[VP] (Effective publication: 1998a, 3092.)

san′gui.nis. L. gen. n. *sanguinis* of the blood.

Type strain: 2045-94, ATCC 700632, CCUG 37820, CIP 105929, LMG 18986.

GenBank accession number (16S rRNA gene): Y13364.

References

Aspevall, O., E. Hillebrant, B. Linderoth and M. Rylander. 1991. Meningitis due to *Gemella haemolysans* after neurosurgical treatment of trigeminal neuralgia. Scand. J. Infect. Dis. *23*: 503–505.

Berger, U. 1960. *Neisseria* haemolysans (Thjøtta and Bøe, 1938): studies on its place in the system. Z. Hyg. Infektionskr. *146*: 253–259.

Berger, U. 1961. A proposed new genus of Gram-negative cocci: *Gemella*. Int. Bull. Nomencl. Taxon. *11*: 17–19.

Berger, U. 1985. Prevalence of *Gemella haemolysans* on the pharyngeal mucosa of man. Med. Microbiol. Immunol. *174*: 267–274.

Berger, U. and A. Pervanidis. 1986. Differentiation of *Gemella haemolysans* (Thjøtta and Bøe 1938) Berger 1960, from *Streptococcus morbillorum* (Prevot 1933) Holdeman and Moore 1974. Zentralbl Bakteriol Mikrobiol Hyg [A] *261*: 311–321.

Berger, U. 1992. The genus *Gemella*. *In* Balows, Trüper, Dworkin, Harder and Schleifer (Editors), The Prokaryotes. A Handbook on the Biology of Bacteria: Ecophysiology, Isolation, Identification, Applications, 2nd edn. Springer-Verlag, New York, pp. 1643–1653.

Brack, M.J., P.G. Avery, P.J. Hubner and R.A. Swann. 1991. *Gemella haemolysans*: a rare and unusual cause of infective endocarditis. Postgrad. Med. J. *67*: 210.

Brooks, J.B., D.S. Kellogg, L. Thacker and E. Turner. 1971. Analysis by gas chromatography of fatty acids found in whole cultural extracts of *Neisseria* species. Can. J. Microbiol. *17*: 531–543.

Buu-Hoi, A., A. Sapoetra, C. Branger and J.F. Acar. 1982. Antimicrobial susceptibility of *Gemella haemolysans* isolated from patients with subacute endocarditis. Eur. J. Clin. Microbiol. *1*: 102–106.

Collins, M.D., R.A. Hutson, E. Falsen, B. Sjödén and R.R. Facklam. 1998a. Description of *Gemella sanguinis* sp. nov., isolated from human clinical specimens. J. Clin. Microbiol. *36*: 3090–3093.

Collins, M.D., R.A. Hutson, E. Falsen, B. Sjödén and R.R. Facklam. 1998b. *Gemella bergeriae* sp. nov., isolated from human clinical specimens. J. Dermatol. Tokyo *25*: 1290–1293.

Collins, M.D., R.A. Hutson, E. Falsen, B. Sjödén and R.R. Facklam. 1998c. *In* Validation of the publication of new names and new combinations previously effectively published outside the IJSB. List no. 48. Int. J. Syst. Bacteriol. *48*: 631–632.

Collins, M.D., R.A. Hutson, E. Falsen, B. Sjödén and R.R. Facklam. 1999a. *In* Validation of the publication of new names and new combinations previously effectively published outside the IJSB. List no. 68. Int. J. Syst. Bacteriol. *49*: 1–3.

Collins, M.D., M.R. Jovita, G. Foster, B. Sjoden and E. Falsen. 1999b. Characterization of a *Gemella*-like organism from the oral cavity of a dog: description of *Gemella palaticanis* sp. nov. Int. J. Syst. Bacteriol. *49*: 1523–1526.

De Jong, M.H. and J.S. Van der Hoeven. 1987. The growth of oral bacteria on saliva. J. Dent. Res. *66*: 498–505.

Debast, S.B., R. Koot and J.F. Meis. 1993. Infections caused by *Gemella morbillorum*. Lancet *342*: 560.

Durak, D.T., E.L. Kaplan and A.L. Bisno. 1983. Apparent failures of endocarditis prophylaxis. Analysis of 52 cases submitted to a national registry. JAMA *250*: 2318–2322.

Eggelmeijer, F., P. Petit and B.A. Dijkmans. 1992. Total knee arthroplasty infection due to *Gemella haemolysans*. Br. J. Rheumatol. *31*: 67–69.

Etienne, J., M.E. Reverdy, L.D. Gruer, V. Delorme and J. Fleurette. 1984. Evaluation of the API 20 STREP system for species identification of streptococci associated with infective endocarditis. Eur. Heart. J. *5*: 25–27.

Facklam, R. and J.A. Elliott. 1995. Identification, classification, and clinical relevance of catalase-negative, gram-positive cocci, excluding the streptococci and enterococci. Clin. Microbiol. Rev. *8*: 479–495.

Facklam, R.R. 1977. Physiological differentiation of viridans streptococci. J. Clin. Microbiol. *5*: 184–201.

Facklam, R.R. and H.W. Wilkinson. 1981. The family *Streptococcaceae* (medical aspects). *In* Starr, Stolp, Trüper, Balows and Schlegel (Editors), The Prokaryotes: A Handbook on Habitats, Isolation, and Identification of Bacteria, Vol. 2. Springer Verlag, Berlin, pp. pp. 1572–1597.

Fresard, A., V.P. Michel, X. Rueda, G. Aubert, G. Dorche and F. Lucht. 1993. *Gemella haemolysans* endocarditis. Clin. Infect. Dis. *16*: 586–587.

Garrity, G.M., J.A. Bell and T. Lilburn. 2005. The Revised Road Map to the *Manual*. *In* Brenner, Krieg, Staley and Garrity (Editors), Bergey's Manual of Systematic Bacteriology, 2nd edn, Vol. 2, The *Proteobacteria*, Part B, The *Gammaproteobacteria*. Springer, New York, pp. 159–206.

Holdeman, L.V. and W.E.C. Moore. 1974. New genus, *Coprococcus*, twelve new species, and emended descriptions of four previously described species of bacteria from human feces. Int. J. Syst. Bacteriol. *24*: 260–277.

Holdeman, L.V. and W.E.C. Moore. 1975. Anaerobe Laboratory Manual, 3rd edition, Blacksburg, Virginia.

Holdeman, L.V., I.J. Good and W.E. Moore. 1976. Human fecal flora: variation in bacterial composition within individuals and a possible effect of emotional stress. Appl. Environ. Microbiol. *31*: 359–375.

Hoyles, L., G. Foster, E. Falsen and M.D. Collins. 2000. Characterization of a *Gemella*-like organism isolated from an abscess of a rabbit: description of *Gemella cuniculi* sp. nov. Int. J. Syst. Evol. Microbiol. *50*: 2037–2041.

Kannangara, D.W., H. Thadepalli, V.T. Bach and D. Webb. 1981. Animal model for anaerobic lung abscess. Infect. Immun. *31*: 592–597.

Kaufhold, A., D. Franzen and R. Lutticken. 1989. Endocarditis caused by *Gemella haemolysans*. Infection *17*: 385–387.

Kilpper-Bälz, R. and K.H. Schleifer. 1988. Transfer of *Streptococcus morbillorum* to the genus *Gemella* as *Gemella morbillorum* comb. nov. Int. J. Syst. Bacteriol. *38*: 442–443.

Kolenbrander, P.E. and B.L. Williams. 1983. Prevalence of viridans streptococci exhibiting lactose-inhibitable coaggregation with oral actinomycetes. Infect. Immun. *41*: 449–452.

Ludwig, W., M. Weizenegger, R. Kilpper-Bälz and K.H. Schleifer. 1988. Phylogenetic relationships of anaerobic streptococci. Int. J. Syst. Bacteriol. *38*: 15–18.

Maxwell, S. 1989. Endocarditis due to *Streptococcus morbillorum*. J. Infect. *18*: 67–72.

May, T., C. Amiel, C. Lion, M. Weber, A. Gerard and P. Canton. 1993. Meningitis due to *Gemella haemolysans*. Eur. J. Clin. Microbiol. Infect. Dis. *12*: 644–645.

Mills, J., L. Pulliam, L. Dall, J. Marzouk, W. Wilson and J.W. Costerton. 1984. Exopolysaccharide production by viridans streptococci in experimental endocarditis. Infect. Immun. *43*: 359–367.

Mitchell, R.G. and P.J. Teddy. 1985. Meningitis due to *Gemella haemolysans* after radiofrequency trigeminal rhizotomy. J. Clin. Pathol. *38*: 558–560.

Morea, P., M. Toni, M. Bressan and P. Stritoni. 1991. Prosthetic valve endocarditis by *Gemella haemolysans*. Infection *19*: 446.

Omran, Y. and C.A. Wood. 1993. Endovascular infection and septic arthritis caused by *Gemella morbillorum*. Diagn. Microbiol. Infect. Dis. *16*: 131–134.

Petit, J.Y., J.C. Layre, I. Lamaury, C. Perez, O. Jonquet and F. Janbon. 1993. *Gemella haemolysans* purulent meningitis. Presse Med. *22*: 444.

Pradeep, R., M. Ali and C.F. Encarnacion. 1997. Retropharyngeal abscess due to *Gemella morbillorum*. Clin. Infect. Dis. *24*: 284–285.

Prévot, A.R. 1933. Etudes de systematique bactérienne. I. Lois générales. II. Cocci anaérobies. Ann. Sci. Nat. Bot. *15*: 23–261.

Reyn, A., A. Birch-Andersen and S.P. Lapage. 1966. An electron microscope study of thin sections of *Haemophilus vaginalis* (Gardner and Dukes) and some possibly related species. Can. J. Microbiol. *12*: 1125–1136.

Reyn, A., A. Birch-Andersen and U. Berger. 1970. Fine structure and taxonomic position of *Neisseria* haemolysans (Thjøtta and Bøe 1938) or *Gemella haemolysans* (Berger 1960). Acta Pathol. Microbiol. Scand. [B] Microbiol. Immunol. *78*: 375–389.

Shukla, S.K., T. Tak, R.C. Haselby, C.S. McCauley, Jr and K.D. Reed. 2002. Second case of infective endocarditis caused by *Gemella sanguinis*. Wisconsin Med. J. *101*: 37–39.

Smith, L.D.S. 1957. *Peptostreptococcus* Kluyver and Van Neil, 1936. *In* Breed, Murray and Smith (Editors), Bergey's Manual of Determinative Bacteriology, 7th edn. The Williams & Wilkins Co., Baltimore, pp. 533–541.

Thjøtta, T. and J. Bøe. 1938. *Neisseria* haemolysans: a hemolytic species of *Neisseria* Trevisan. Acta Pathol. Microbiol. Scand. Suppl. *37*: 527–531.

Tunnicliff, R. 1933. Colony formation of *Diplococcus rubeolae*. J. Infect. Dis. *52*: 39–53.

Tunnicliff, R. 1936. Opsonins for *Diplococcus morbillorum* and for *Streptococcus scarlatinae* in convalescent measles serum, convalescent scarlet fever and placental extract. J. Infect. Dis. *58*: 1–4.

von Essen, R., M. Ikavalko and B. Forsblom. 1993. Isolation of *Gemella morbillorum* from joint fluid. Lancet *342*: 177–178.

Whitney, A.M. and S.P. O'Connor. 1993. Phylogenetic relationship of *Gemella morbillorum* to *Gemella haemolysans*. Int. J. Syst. Bacteriol. *43*: 832–838.

Family XII. **Incertae Sedis**

Previously assigned to the *Bacillaceae* by Garrity et al. (2005), subsequent analyses suggest that the genus *Exiguobacterium* is only distantly related to this and any of the previously described families within the *Bacillales*. Its inability to form endospores, which is a common property within the *Bacillaceae*, further supports its assignment to its own family *incertae sedis*.

Genus I. **Exiguobacterium** Collins, Lund, Farrow and Schleifer 1984, 91[VP] (Effective publication: Collins, Lund, Farrow and Schleifer 1983, 2039.)

MATTHEW D. COLLINS

Ex.ig.uo.bac.te'ri.um. L. adj. *exiguus* short, small; L. n. *bacillus* rod; N.L. neut. n. *Exiguobacterium* small rod

Cells may be ovoid or short rods occurring singly, in pairs, or chains. Gram-stain-positive and not acid-fast. Nonspore-forming. Motile. Facultatively anaerobic and catalase-positive. No growth at 10 or 45°C. Gas is not produced. Acid is produced from D-glucose and some other sugars. **The cell-wall murein is based on L-lysine. Menaquinones are the sole respiratory quinones, with menaquinone-7 predominating.** Mycolic acids are not present. **The long-chain cellular fatty acids consist of a mixture of straight-chain saturated, anteiso- and iso-methyl branched, and monounsaturated types.** The polar lipids are diphosphatidylglycerol, phosphatidylglycerol, phosphatidylserine and phosphatidylethanolamine; phosphatidylinositol and unidentified phospholipids may be present. **The G+C content of the DNA is 46.6–55.8 mol% (T_m).**

Type species: **Exiguobacterium aurantiacum** Collins, Lund, Farrow and Schleifer 1984, 91[VP] (Effective publication: Collins, Lund, Farrow and Schleifer 1983, 2040).

Further descriptive information

Cells vary in shape from short rods to coccoid forms; cells may occur singly, in pairs or in chains. Cells vary in length from about 1–4 μm; long distorted rods may be observed during exponential growth of *Exiguobacterium aurantiacum* at pH >10. Exiguobacteria are motile by means of peritrichous flagella. In cellular morphology they may somewhat resemble some coryneforms. Colonies of *Exiguobacterium aurantiacum* on PPYG agar (Gee et al., 1980) are 2–3 mm in diameter after 3 d at 25°C, and are low convex, orange, opaque, butyrous and easily emulsified. Colonies on heart infusion agar are normally flatter and fainter orange. The orange pigment does not diffuse into the medium; pigment production does not occur anaerobically. Colonies of *Exiguobacterium acetylicum* on PYE agar are flat with irregular edge and are yellow-orange in color; the pigment does not diffuse into the medium. Surface colonies of *Exiguobacte-*

rium antarcticum and *Exiguobacterium undae* on tryptone soy agar are 2–3 mm in diameter, orange in color, convex, entire, and shiny after 2 d at 25°C. The optimum growth temperature of *Exiguobacterium acetylicum* is between 25 and 30°C and for *Exiguobacterium aurantiacum* it is approximately 37°C; growth occurs at 10 and 40°C. The temperature range for growth of *Exiguobacterium aurantiacum* is from 7 to 43°C. *Exiguobacterium acetylicum* grows slowly at 5°C but not at 45°C. Neither *Exiguobacterium undae* nor *Exiguobacterium antarcticum* grow at 45°C. *Exiguobacterium aurantiacum* is alkalophilic whereas *Exiguobacterium acetylicum* is not. The pH range for growth of *Exiguobacterium aurantiacum* at 25°C is approximately 6.5–11.5, with two maxima, at pH 8.5 and 9.5 (Gee et al., 1980).

Exiguobacteria grow under aerobic and anaerobic conditions. All species are catalase-negative. *Exiguobacterium aurantiacum* is oxidase-negative whereas *Exiguobacterium acetylicum*, *Exiguobacterium antarcticum* and *Exiguobacterium undae* are oxidase-positive. All species produce acid from glucose, fructose, glycogen, β-gentiobiose, maltose, salicin, sucrose and trehalose but not from adonitol, D-arabitol, L-arabitol, D-arabinose, L-arabinose, dulcitol, erythritol, D-fucose, inositol, inulin, 2-keto-gluconate, 5-keto-gluconate, lactose, D-lyxose, melezitose, rhamnose, sorbitol, L-sorbose, D-xylose, L-xylose, D-turinose or xylitol. All species hydrolyze casein, DNA, esculin and starch; *Exiguobacterium acetylicum*, *Exiguobacterium undae* and *Exiguobacterium antarcticum* also hydrolyzes gelatin but *Exiguobacterium aurantiacum* may or may not. None of the species hydrolyzes cellulose or Tweens 40 and 80. *Exiguobacterium aurantiacum* does not attack dextran, tributyrin, or pectin. Reports vary on the ability of *Exiguobacterium aurantiacum* to reduce nitrate (Collins et al., 1983; Fritze et al., 1990) but *Exiguobacterium acetylicum*, *Exiguobacterium undae*, and *Exiguobacterium antarcticum* do not reduce nitrate. Using API systems, both *Exiguobacterium acetylicum* and *Exiguobacterium aurantiacum* produce α-glucosidase, esterase C-4, ester lipase C8, and pyrazinamidase. Alkaline phosphatase is produced by *Exiguobacterium acetylicum* but may or may not be produced by *Exiguobacterium aurantiacum*. β-Galactosidase is detected in *Exiguobacterium acetylicum* but not in *Exiguobacterium aurantiacum*. Neither species produces acid phosphatase, chymotrypsin, trypsin, α-fucosidase, α-galactosidase, β-glucosidase, β-glucuronidase, lipase C14, leucine arylamidase, α-mannosidase, N-acetyl-β-glucosaminidase, pyrolydonyl arylamidase, valine arylamidase, or urease.

The cell-wall murein of *Exiguobacterium aurantiacum*, *Exiguobacterium undae*, and *Exiguobacterium antarcticum* is type L-lysine-glycine (Collins et al., 1983; Fruhling et al., 2002) whereas that of *Exiguobacterium acetylicum* is reported to be type L-lysine-D-aspartic acid (Schleifer and Kandler, 1972). The major polar lipids of exiguobacteria are diphosphatidylglycerol, phosphatidylglycerol, phosphatidylserine, and phosphatidylethanolamine; phosphatidylinositol and other unidentified phospholipids may be produced but glycolipids are not present (Fruhling et al., 2002; Yamada and Komagata, 1970). It is not known how much the phospholipid composition of exiguobacteria is influenced with growth conditions such as media and pH. The fatty acids of exiguobacteria are characterized by mixtures of straight-chain saturated, anteiso- and iso-methyl-branched chain, and monounsaturated types. Species of exiguobacteria display differences in their fatty acid profiles but there are also some quantitative differences between different reports in the literature (Collins and Kroppenstedt, 1983; Fruhling et al., 2002). The G+C content of DNA of *Exiguobacterium aurantiacum* is within the range 53.2–55.8 mol% (T_m) whereas that of the type strain of *Exiguobacterium acetylicum*, ATCC 953, is reported to be 46.6 mol% (T_m) (Yamada and Komagata, 1970). There are no data on the G+C contents of *Exiguobacterium antarcticum* or *Exiguobacterium undae*.

There is little information on susceptibilities of exiguobacteria to antimicrobial agents. The growth of *Exiguobacterium aurantiacum* is inhibited by chloramphenicol (10 μg per disc), erythromycin (10 μg per disc), novobiocin (5 μg per disc), oleandomycin (5 μg per disc), penicillin (1 μg per disc) and tetracycline (10 μg per disc) but not by sulfafurazole (100 μg per disc) (Gee et al., 1980).

Exiguobacterium aurantiacum was isolated from potato-processing effluent (Gee et al., 1980) whereas the type strain of *Exiguobacterium acetylicum*, ATCC 953, was recovered from creamery waste (Levine and Soppeland, 1926). *Exiguobacterium undae* has been isolated from garden pond water in Germany whereas *Exiguobacterium antarcticum* has been recovered from a microbial mat from Lake Fryxell, Antarctica (Fruhling et al., 2002).

Isolation procedures

Exiguobacterium aurantiacum has been recovered from potato-processing effluent by enrichment in PPYG medium (composition g/l: peptone (Difco), 5; yeast extract (Difco), 1.5; glucose, 5; $Na_2HPO_4 \cdot 12H_2O$, 1.5; NaCl, 1.5; $MgCl_2 \cdot 6H_2O$ 0.1; Na_2CO_3, 5.03. Solutions of glucose and Na_2CO_3 are sterilized separately by autoclaving; final pH of medium 10.5–11.0) at 20°C followed by plating on PPYG agar (Gee et al., 1980) (Gee et al., 1980). *Exiguobacterium undae* has been isolated from water by streaking onto GS medium (composition g/l: glucose, 0.15; yeast extract, 1.0; $(NH_4)_2SO_4$, 0.5; $CaCO_3$, 0.1; $Ca(NO_3)_2$, 0.1; KCl, 0.05; K_2PO_4, 0.05; $MgSO_4 \cdot 7H_2O$, 0.05; $Na_2S \cdot 9H_2O$, 0.2; 10 ml of vitamin cocktail; GS medium no. 851 DSMZ, 1998) (DSMZ, 1998). Isolated colonies developing after 2 days at room temperature were purified on tryptone soy agar (Difco) (Fruhling et al., 2002). *Exiguobacterium antarcticum* has been retrieved from anaerobic enrichments from a microbial mat from an Antarctic Lake (see Brambilla et al., (2001) for details of the methods used).

Maintenance procedures

Exiguobacteria grow well on ordinary laboratory growth media such as nutrient agar, tryptone soy agar, or heart infusion agar at 30–37°C. The pH of the medium should be adjusted to about 8.5 for *Exiguobacterium aurantiacum* as this species prefers alkaline conditions. For long-term preservation, strains can be stored on cryogenic beads at −70°C or lyophilized.

Taxonomic comments

The genus *Exiguobacterium* was created in 1983 (Collins et al., 1983) to accommodate an alkalophilic bacterium originally isolated from potato-processing effluent by (Gee et al., 1980). The genus originally contained a single species, *Exiguobacterium aurantiacum*. This species showed some morphological similarities with some coryneform bacteria but chemotaxonomic investigations, most notably G+C, murein and menaquinone determinations, indicated a closer affinity to some low-G+C-content taxa, such as *Bacillus pasteurii*, *Bacillus sphaericus*, and *Kurthia* (Collins et al., 1983). *Exiguobacterium acetylicum* was assigned to the genus *Exiguobacterium* by Farrow *et al.* (1994). This species was originally isolated by Levine and Soppeland (1926) from skimmed milk and designated "*Flavobacterium acetylicum*". The species was subsequently classified in the genus *Brevibacterium* by Breed (1957). Chemotaxonomic studies by Collins and Kroppenstedt (1983) however revealed that the type strain of the species was incompatible with the genus *Brevibacterium*, and a very close chemical similarity with *Exiguobacterium* became apparent. The affinity between *Brevibacterium acetylicum* and *Exiguobacterium aurantiacum* was unequivocally demonstrated by comparative 16S rRNA sequencing (Farrow et al., 1994), with the two species forming a distinct phylogenetic group within the overall radiation of genera associated with *Bacillus* (including genera such as *Kurthia* and *Planococcus*). Although *Exiguobacterium aurantiacum* differed markedly from *Exiguobacterium acetylicum* in being alkalophilic, the two species display only about 5.5% 16S rRNA sequence divergence, and treeing analysis shows their association is statistically significant (Farrow et al., 1994). Recently two further species, *Exiguobacterium undae* and *Exiguobacterium antarcticum*, were assigned to the genus by (Fruhling et al., 2002). Phylogenetically *Exiguobacterium*

undae and *Exiguobacterium antarcticum* are more closely related to each other than to *Exiguobacterium acetylicum* or *Exiguobacterium aurantiacum*. The former two species show about 2.1–2.8% 16S rRNA gene sequence divergence to *Exiguobacterium acetylicum* and about 6.2–6.7% divergence to *Exiguobacterium aurantiacum*.

Differentiation of the genus *Exiguobacterium* from other genera

Traits which are useful in differentiating exiguobacteria from closely related genera include their rod-shaped morphology, motility, peritrichous flaggelation, absence of endospores, facultatively anaerobic nature, and catalase- and oxidase-positive reactions. Useful chemical characteristics include presence of a cell-wall murein based on L-Lysine, MK-7 as the major respiratory quinone, long-chain fatty acids of the straight-chain saturated, iso- and anteiso- methyl-branched and monounsaturared types, phospholipids comprising diphosphatidylglycerol, phosphatidylglycerol, phosphatidylserine, and phosphatidylethanolamine (also phosphatidylinositol and other unidentified phospholipids may be produced) and a G+C range of approximately 46–56 mol%.

Differentiation of the species of the genus *Exiguobacterium*

Available comparative biochemical data on the different *Exiguobacterium* species is currently based on the analysis of relatively few strains (in some cases single strains). Although there are numerous tests which "potentially" serve to distinguish between species, until a larger number of strains of each species are examined, it is difficult to assess how reliable, if at all, these reported differential tests are. Tests

TABLE 80. Tests differentiating *Exiguobacterium* species using the API 50CHE system[a,b]

Test	*E. acetylicum* DSM 20416	*E. aurantiacum* DSM 6208	*E. undae* strains L1–L4[c]	*E. antarcticum* DSM 14480
Acid from:				
Glycerol	–	+	+	+
Ribose	–	–	+	+
Galactose	–	–	+	+
D-Mannose	+	–	+	+
Mannitol	+	+	+	–
Amygdalin	–	+	d	+
Arbutin	–	+	d	+
Cellobiose	+	–	d	+
Melibiose	–	–	+	–
D-Raffinose	–	–	d	–
Methyl α-D-glucoside	–	+	d	–

[a]Symbols: +, >85% positive; d, different strains give different reactions (16–84% positive); –, 0–15% positive.
[b]Data from Frühling et al. (2002).
[c]Strains L1–L4 include the type strain L2 = DSM 14481.

TABLE 81. Tests differentiating *Exiguobacterium* species using the Biolog GP Microplate system[a,b]

Test	*E. acetylicum* DSM 20416	*E. aurantiacum* DSM 6208	*E. undae* strains L1–L4[c]	*E. antarcticum* DSM 14480
Utilization of:				
Glycogen	+	–	+	+
N-Acetyl glucosamine	–	+	+	+
N-Acetyl mannosamine	–	–	w+	w+
Cellobiose	+	–	+	+
D-Galactose	–	–	+	+
D-Mannitol	+	–	+	–
D-Raffinose	–	–	+	+
D-Ribose	–	–	+	+
D-Sorbose	+	–	–	–
Acetic acid	w+		+	+
γ-Hydroxy butyrate	–	–	w+	w+
Methyl pyruvate	–	+	–	–
Methyl succinate	–	–	w+	w+
L-Alanine	+	–	w+	w+
2,3-Butanediol	–	–	d	+
Adenosine 5'-monophosphate	–	–	–	+
Thymidine 5'-monophosphate	–	–	w+	+
Uridine 5'-monophosphate	–	–	–	+
Fructose 6-phosphate	+	–	–	–
Glucose 1-phosphate	+	–	–	–
Glucose 6-phosphate	+	–	–	–
DL-α-Glycerol phosphate	+	–	–	–

[a]Symbols: +, >85% positive; –, 0–15% positive; w, weak reaction.
[b]Data from Frühling et al. (2002).
[c]Strains L1–L4 includes the type strain L2 = DSM 14481.

which appear to be useful in distinguishing *Exiguobacterium* species in the literature, using the API 50CHE system, are shown in Table 80. *Exiguobacterium* species are also purported to be distinguished from each other by using the Biolog GP Microplate system (see Table 81 and Frühling et al., 2002).

List of species of the genus *Exiguobacterium*

1. **Exiguobacterium aurantiacum** Collins, Lund, Farrow and Schleifer 1984, 91[VP] (Effective publication: Collins, Lund, Farrow and Schleifer 1984, 2040.)
au.ran.ti′ac.um. L. neut. n. *aurum* gold; N.L. neut. n. *aurantium* generic name of the orange; N.L. neut. adj. *aurantiacum* orange-colored.

   DNA G+C content (mol%): 53.2–55.8 (T_m).
   Type strain: ATCC 35652, BL 77/1, CIP 103353, NCIMB 11798, DSM 6208, CCUG 44910, NBRC 14763, NCDO 2321, 1005, IMET 11072.
   GenBank accession number (16S rRNA gene): X70316.

2. **Exiguobacterium acetylicum** (Levine and Soppeland 1926) Farrow, Wallbanks and Collins 1994, 81[VP] (Levine and Soppeland 1926, 46.)
a.ce.ty′li.cum. L. neut. n. *acetum* vinegar; N.L. neut. n. *acetylum* the organic radical acetyl; N.L. adj. *acetylicus* pertaining to acetyl.

   DNA G+C content (mol%): 46.6 (T_m).

   Type strain: ATCC 953, CCUG 32630, DSM 20416, NCIMB 9889, CIP 82.109, HAMBI 2009, NBRC 12146, JCM 1968.
   GenBank accession number (16S rRNA gene) for strain C/ C-aer/b: X70313.

3. **Exiguobacterium antarcticum** Frühling, Schumann, Hippe, Sträubler and Stackebrandt 2002, 1175[VP]
ant.arc′ti.cum. N.L. gen. n. *antarcticum* of Antarctica.

   DNA G+C content (mol%): not reported.
   Type strain: H2, CIP 107163, DSM 14480.
   GenBank accession number (16S rRNA gene): AJ297437.

4. **Exiguobacterium undae** Frühling, Schumann, Hippe, Sträubler and Stackebrandt 2002, 1173[VP]
un′dae. L. fem. n. *unda* water; L. gen. n. *undae* of the water.

   DNA G+C content (mol%): not reported.
   Type strain: L2, CIP 107162, DSM 14481.
   GenBank accession number (16S rRNA gene): AJ344151.

   Editorial note: Since this chapter was accepted for publication, eight more species have been described. They are:

Exiguobacterium aestuarii (Kim et al., 2005), *Exiguobacterium artemiae* (Lopez-Cortes et al., 2006), *Exiguobacterium indicum* (Chaturvedi and Shivaji, 2006), *Exiguobacterium marinum* (Kim et al., 2005), *Exiguobacterium mexicanum* (Lopez-Cortes et al., 2006), *Exiguobacterium oxidotolerans* (Yumoto et al., 2004), *Exiguobacterium profundum* (Crapart et al., 2007), and *Exiguobacterium sibiricum* (Rodrigues et al., 2006).

References

Brambilla, E., H. Hippe, A. Hagelstein, B.J. Tindall and E. Stackebrandt. 2001. 16S rDNA diversity of cultured and uncultured prokaryotes of a mat sample from Lake Fryxell, McMurdo Dry Valleys, Antarctica. Extremophiles 5: 23–33.

Breed, R.S. 1957. Family IX. *Brevibacteriaceae* Breed, 1953. *In* Breed, Murray and Smith (ed.), Bergey's Manual of Determinative Bacteriology, 7th Ed. edn. The Williams & Wilkins Co., Baltimore, pp. pp. 490–503.

Chaturvedi, P. and S. Shivaji. 2006. *Exiguobacterium indicum* sp. nov., a psychrophilic bacterium from the Hamta glacier of the Himalayan mountain ranges of India. Int. J. Syst. Evol. Microbiol. 56: 2765–2770.

Collins, M.D. and R.M. Kroppenstedt. 1983. Lipid composition as a guide to the classification of some coryneform bacteria containing an A-4-alpha type peptidoglycan. Syst. Appl. Microbiol. 4: 95–104.

Collins, M.D., B.M. Lund, J.A.E. Farrow and K.H. Schleifer. 1983. Chemotaxonomic study of an alkalophilic bacterium, *Exiguobacterium aurantiacum* gen. nov., sp. nov. J. Gen. Microbiol. 129: 2037–2042.

Collins, M.D., B.M. Lund, J.A.E. Farrow and K.H. Schleifer. 1984. *In* Validation of the publication of new names and new combinations previously effectively published outside the IJSB. List no. 13. Int. J. Syst. Bacteriol. 34: 91–92.

Crapart, S., M.L. Fardeau, J.L. Cayol, P. Thomas, C. Sery, B. Ollivier and Y. Combet-Blanc. 2007. *Exiguobacterium profundum* sp. nov., a moderately thermophilic, lactic acid-producing bacterium isolated from a deep-sea hydrothermal vent. Int. J. Syst. Evol. Microbiol. 57: 287–292.

DSMZ. 1998. Catalog of Strains, 5th edition, Braunschweig: DSMZ.

Farrow, J.A.E., S. Wallbanks and M.D. Collins. 1994. Phylogenetic interrelationships of round-spore-forming bacilli containing cell walls based on lysine and the non-spore-forming genera *Caryophanon*, *Exiguobacterium*, *Kurthia*, and *Planococcus*. Int. J. Syst. Bacteriol. 44: 74–82.

Fritze, D., J. Flossdorf and D. Claus. 1990. Taxonomy of alkaliphilic *Bacillus* strains. Int. J. Syst. Bacteriol. 40: 92–97.

Fruhling, A., P. Schumann, H. Hippe, B. Straubler and E. Stackebrandt. 2002. *Exiguobacterium undae* sp. nov. and *Exiguobacterium antarcticum* sp. nov. Int. J. Syst. Evol. Microbiol. 52: 1171–1176.

Garrity, G.M., J.A. Bell and T. Lilburn. 2005. The Revised Road Map to the Manual. *In* Brenner, Krieg, Staley and Garrity (Editors), Bergey's Manual of Systematic Bacteriology, 2nd edn, vol. 2, The Proteobacteria, Part B, The Gammaproteobacteria Springer, New York, pp. 159–206.

Gee, J.M., B.M. Lund, G. Metcalf and J. Peel. 1980. Properties of a new group of alkalophilic bacteria. J. Gen. Microbiol. 117: 9–18.

Kim, I.G., M.H. Lee, S.Y. Jung, J.J. Song, T.K. Oh and J.H. Yoon. 2005. *Exiguobacterium aestuarii* sp. nov. and *Exiguobacterium marinum* sp. nov., isolated from a tidal flat of the Yellow Sea in Korea. Int. J. Syst. Evol. Microbiol. 55: 885–889.

Levine, M. and L. Soppeland. 1926. Bacteria in creamery waste. Bull. Iowa State Agr. Coll. 77: 1–72.

Lopez-Cortes, A., P. Schumann, R. Pukall and E. Stackebrandt. 2006. *Exiguobacterium mexicanum* sp. nov. and *Exiguobacterium artemiae* sp. nov., isolated from the brine shrimp *Artemia franciscana*. Syst Appl Microbiol 29: 183–190.

Rodrigues, D.F., J. Goris, T. Vishnivetskaya, D. Gilichinsky, M.F. Thomashow and J.M. Tiedje. 2006. Characterization of *Exiguobacterium* isolates from the Siberian permafrost. Description of *Exiguobacterium sibiricum* sp. nov. Extremophiles 10: 285–294.

Schleifer, K.H. and O. Kandler. 1972. Peptidoglycan types of bacterial cell walls and their taxonomic implications. Bacteriol Rev 36: 407–477.

Yamada, K. and K. Komagata. 1970. Taxonomic studies on coryneform bacteria. III. DNA base composition of coryneform bacteria. J. Gen. Appl. Microbiol. 16: 215–224.

Yumoto, I., M. Hishinuma-Narisawa, K. Hirota, T. Shingyo, F. Takebe, Y. Nodasaka, H. Matsuyama and I. Hara. 2004. *Exiguobacterium oxidotolerans* sp. nov., a novel alkaliphile exhibiting high catalase activity. Int. J. Syst. Evol. Microbiol. 54: 2013–2017.

Order II. **Lactobacillales** ord. nov.

WOLFGANG LUDWIG, KARL-HEINZ SCHLEIFER AND WILLIAM B. WHITMAN

Lac.to.ba.cil.la'les. N.L. masc. n. *Lactobacillus* type genus of the order; suff. *-ales* ending to denote an order; N.L. fem. pl. n. *Lactobacillales* the *Lactobacillus* order.

The order *Lactobacillales* is circumscribed for this volume on the basis of the phylogenetic analyses of the 16S rRNA sequences and includes the family *Lactobacillaceae* and its close relatives "*Aerococcaceae*", "*Carnobacteriaceae*", "*Enterococcaceae*", "*Leuconostocaceae*", and *Streptococcaceae*. It is composed of Gram-stain-positive rods and cocci. Endospores are not formed. Usually facultatively anaerobic and catalase-negative.

Type genus: **Lactobacillus** Beijerinck 1901, 212[AL].

References

Beijerinck, M.W. 1901. Sur les ferments lactiques de l'industrie. Arch. Néer. Sci. (sect. 2) 6: 212–243.

Family I. **Lactobacillaceae** Winslow, Broadhurst, Buchanan, Krumwiede, Rogers and Smith 1917, familia

Lac.to.ba.cil.la′ce.ae. N.L. masc. n. *Lactobacillus* type genus of the family; suff. *-aceae* ending denoting family; N.L. fem. pl. n. *Lactobacillaceae* the *Lactobacillus* family.

The family *Lactobacillaceae* is circumscribed on the basis of phylogenetic analyses of the 16S rRNA gene sequences and includes the genera *Lactobacillus*, *Paralactobacillus* and *Pediococcus*.

Cells vary from long and slender, sometimes bent rods to short coryneform coccobacilli or spherical cells. Chain formation common; with the exception of pediococci which form pairs or tetrads. Nonsporeforming, Gram-positive, fermentative metabolism, obligately saccharo-clastic. At least half of the end product carbon is lactate. Additional products may be acetate, ethanol, CO_2, formate, or succinate. Oxidase- and cytochrome-negative. Catalase usually negative, although some strains may produce pseudocatalase. Complex nutritional requirements for amino acids, peptides, nucleic acid derivatives, vitamins, salts, fatty acids or fatty acid esters, and fermentable carbohydrates.

Type genus: **Lactobacillus** Beijerinck 1901, 212[AL].

Genus I. **Lactobacillus** Beijerinck 1901, 212[AL]

WALTER P. HAMMES AND CHRISTIAN HERTEL

Lac.to.ba.cil′lus. L. n. *lac*, *lactis* milk; L. n. *bacillus* a rod; N.L. masc. n. *Lactobacillus* milk rodlet.

Cells vary from **long and slender, sometimes bent rods** to **short, often coryneform coccobacilli;** chain formation common. **Motility** uncommon; when present, by peritrichous flagella. **Nonspore-forming. Gram-stain-positive.** Some strains exhibit bipolar bodies, internal granulations, or a barred appearance with the Gram-reaction or methylene blue stain.

Fermentative metabolism; obligately saccharoclastic. At least half of **end product** carbon is **lactate.** Lactate is usually not fermented. Additional products may be acetate, ethanol, CO_2, formate, or succinate.

Facultatively anaerobic; surface growth on solid media generally enhanced by anaerobiosis or reduced oxygen pressure and 5–10% CO_2. Strictly aerobic conditions are commonly growth inhibitory; some are anaerobes on isolation.

Nitrate reduction unusual; if present, only when terminal pH is poised above 6.0 and/or heme is added to the growth medium. **Gelatin not liquefied.** Casein not digested, but small amounts of soluble nitrogen produced by most strains. Indole and H_2S not produced.

Catalase and cytochrome negative (porphyrins absent), however, a few strains in several species decompose peroxide by a pseudocatalase or by true catalase when heme is present. Benzidine reaction negative.

Pigment production rare; if present, yellow or orange to rust or brick red.

Complex nutritional requirements for amino acids, peptides, nucleic acid derivatives, vitamins, salts, fatty acids or fatty acid esters, and fermentable carbohydrates. Nutritional requirements are generally characteristic for each species, often particular strains only.

Growth temperature range 2–53°C, optimum generally 30–40°C.

Aciduric, optimal pH usually 5.5–6.2; growth generally occurs at pH 5.0 or less; the growth rate is often reduced in neutral or initially alkaline conditions.

Found in dairy products, grain products, meat and fish products, beer, wine, fruits and fruit juices, pickled vegetables, mash, sauerkraut, silage, sourdough, water, soil, and sewage; they are a part of the normal flora in the mouth, intestinal tract, and vagina of humans and many animals. Pathogenicity is absent or, in rare cases, restricted to persons with an underlying disease.

DNA G+C content (mol%): 32–55 (Bd, T_m).

Type species: **Lactobacillus delbrueckii** (Leichmann 1896b) Beijerinck 1901, 229[AL] (*Bacillus Delbrücki* (*sic*) Leichmann 1896b, 284.

Further descriptive information

Introductory remarks. In the previous edition of *Bergey's Manual* (Sneath et al., 1986), only 44 *Lactobacillus* species were listed. Subsequently, the genus *Weissella* was created from a group of heterofermentative species (Collins et al., 1993) and three species were no longer included in the genus *Lactobacillus* (*Lactobacillus vitulinus*, *Lactobacillus maltaromicus*, *Lactobacillus bavaricus*). At a rate of six species *per annum*, the number of new *Lactobacillus* species grew up during the last 10 years to the present number of 96 species and 16 subspecies. The character of the species description has changed in so far as morphological, physiological, and biochemical properties became reduced, and more genotypic data, especially sequences of the 16S rRNA genes, have been used to support the species rank of new isolates. Therefore, genotypic characteristics will increasingly be needed to allot an isolate to one of the newly described (and often not known for long) species. In addition, in several cases only a single isolate was obtained and simple properties have only a limited value for correct identification. To facilitate access to the species of the genus, the *Lactobacillus* species are listed in Table 82. Each species is classified as belonging to one of three fermentation-type groups (A, B, or C, see below), the habitat or origin of the isolates is indicated, and the phylogenetic group is given.

TABLE 82. List of the species of the genus *Lactobacillus*

Species	Type of glucose fermentation[a]	Main habitat or source of isolation[b]	Phylogenetic group[c]
Lactobacillus delbrueckii subsp. *delbrueckii*	A	F	de
L. delbrueckii subsp. *bulgaricus*	A	F	de
L. delbrueckii subsp. *lactis*	A	F	de
L. delbrueckii subsp. *indicus*	A	F	de
L. acetotolerans	B	D	de
L. acidifarinae	C	F	u
L. acidipiscis	B	F	sl
L. acidophilus	A	I,F	de
L. agilis	B	S,I	sl
L. algidus	B	D	sl
L. alimentarius	B	F, D	u
L. amylolyticus	A	F	de
L. amylophilus	A	F	de
L. amylovorus	A	F	de
L. animalis	B	I	sl
L. antri	C	I	u
L. aviarius subsp. *aviarius*	A	I	sl
L. aviarius subsp. *araffinosus*	A	I	sl
L. bifermentans	B	D	u
L. brevis	C	F, D	u
L. buchneri	C	F, D	u
L. casei	B	F	u
L. coleohominis	C	I	re
L. collinoides	C	D	u
L. coryniformis subsp. *coryniformis*	B	F	u
L. coryniformis subsp. *torquens*	B	F	u
L. crispatus	A	I,F	de
L. curvatus	B	I, F, D	u
L. cypricasei	B	F	sl
L. diolivorans	C	F	u
L. durianis	C	F	u
L. equi	A	I	sl
L. farciminis	A	F	u
L. ferintoshensis	C	F	u
L. fermentum	C	F, D	re
L. fornicalis	B	I	de
L. fructivorans	C	D	u
L. frumenti	C	F	re
L. fuchuensis	B	D	u
L. gallinarum	A	I	de
L. gasseri	A	I	de
L. gastricus	C	I	re
L. graminis	B	F	u
L. hammesii	C	F	u
L. hamsteri	B	I	de
L. helveticus	A	F	de
L. hilgardii	C	D	u
L. homohiochii	B	D	u
L. iners	A	I	de
L. ingluviei	C	I	re
L. intestinalis	B	I	de
L. jensenii	B	I	de
L. johnsonii	A	I, F	de
L. kalixensis	A	I	de
L. kefiranofaciens subsp. *kefiranofaciens*	A	F	de
L. kefiranofaciens subsp. *kefirgranum*	A	F	de
L. kefir	C	F	u
L. kimchii	B	F	u
L. kitasatonis	A	I	de
L. kunkeei	C	D	u

(continued)

TABLE 82. (continued)

Species	Type of glucose fermentation[a]	Main habitat or source of isolation[b]	Phylogenetic group[c]
L. lindneri	C	D	u
L. malefermentans	C	D	u
L. mali	A	D	sl
L. manihotivorans	A	F	u
L. mindensis	A	F	u
L. mucosae	C	I, F	re
L. murinus	B	I,F	sl
L. nagelii	A	D	sl
L. oris	C	I	re
L. panis	C	F	re
L. pantheris	A	I	u
L. parabuchneri	C	I	u
L. paracasei subsp. *paracasei*	B	I, D, F	u
L. paracasei subsp. *tolerans*	B	D	u
L. paracollinoides	C	F	u
L. parakefiri	C	F	u
L. paralimentarius	B	F	u
L. paraplantarum	B	D, I	u
L. pentosus	B	F, S	u
L. perolens	B	D	u
L. plantarum subsp. *plantarum*	B	F, D	u
L. plantarum subsp. *argentoratensis*	B	F, D	u
L. pontis	C	F	re
L. psittaci	A	(P)	de
L. reuteri	C	I, F	re
L. rhamnosus	B	F	u
L. rossii	C	F	re
L. ruminis	A	I	sl
L. saerimneri	A	I	sl
L. sakei subsp. *sakei*	B	I, F, D	u
L. sakei subsp. *carnosus*	B	F, D	u
L. salivarius subsp. *salivarius*	A	I	sl
L. salivarius subsp. *salicinius*	A	I	sl
L. sanfranciscensis	C	F	u
L. satsumensis	A	F	sl
L. sharpeae	A	S	u
L. spicheri	C	F	u
L. suebicus	C	F	u
L. suntoryeus	A	F	de
L. thermotolerans	C	I	re
L. ultunensis	A	I	de
L. vaccinostercus	C	I	u
L. vaginalis	C	I	re
L. versmoldensis	B	F	u
L. zeae	B	F	u
L. zymae	C	F	u

[a]Abbreviations: A, obligately homofermentative; B, facultatively heterofermentative; C, obligately heterofermentative.

[b]D, food associated, usually involved in spoilage; F, involved in fermentation of food and feed; I, associated with humans and/or animals, e.g., oral cavity, intestines, vagina; S, sewage; (P), recovered from diseased parrot, safe status not known.

[c]Determined by C. Hertel, unpublished results. de, *Lactobacillus delbrueckii* group; re, *Lactobacillus reuteri* group; sl, *Lactobacillus salivarius* group; u, unique. To relate species to phylogenetic groups, sequences of species have been considered for which at least 90% of the complete 16S rDNA sequences have been published. These species were considered for construction of the phylogenetic trees (Figure 90, Figure 91, and Figure 92).

Cell morphology. The variability of lactobacilli from long, straight, or slightly crescent rods to coryneform coccobacilli is depicted in Figure 87. The length of the rods and the degree of curvature depends on the age of the culture, the composition of the medium (e.g., availability of oleic acid esters) (Jacques et al., 1980), and the oxygen tension. However, the main morphological differences between the species usually remain clearly recognizable. Some species of the gas-producing lactobacilli (e.g., *Lactobacillus fermentum*, *Lactobacillus brevis*) always exhibit a mixture of long and short rods (Figure 87E).

Cell division occurs only in one plane. The tendency toward chain formation varies between species and even strains. It depends on the growth phase and the pH of the medium (Rhee and Pack, 1980). The asymmetrical development of cells during cell division in coryneform lactobacilli (Figure 88) leads to wrinkled chains or even ring formation. Irregular involution forms

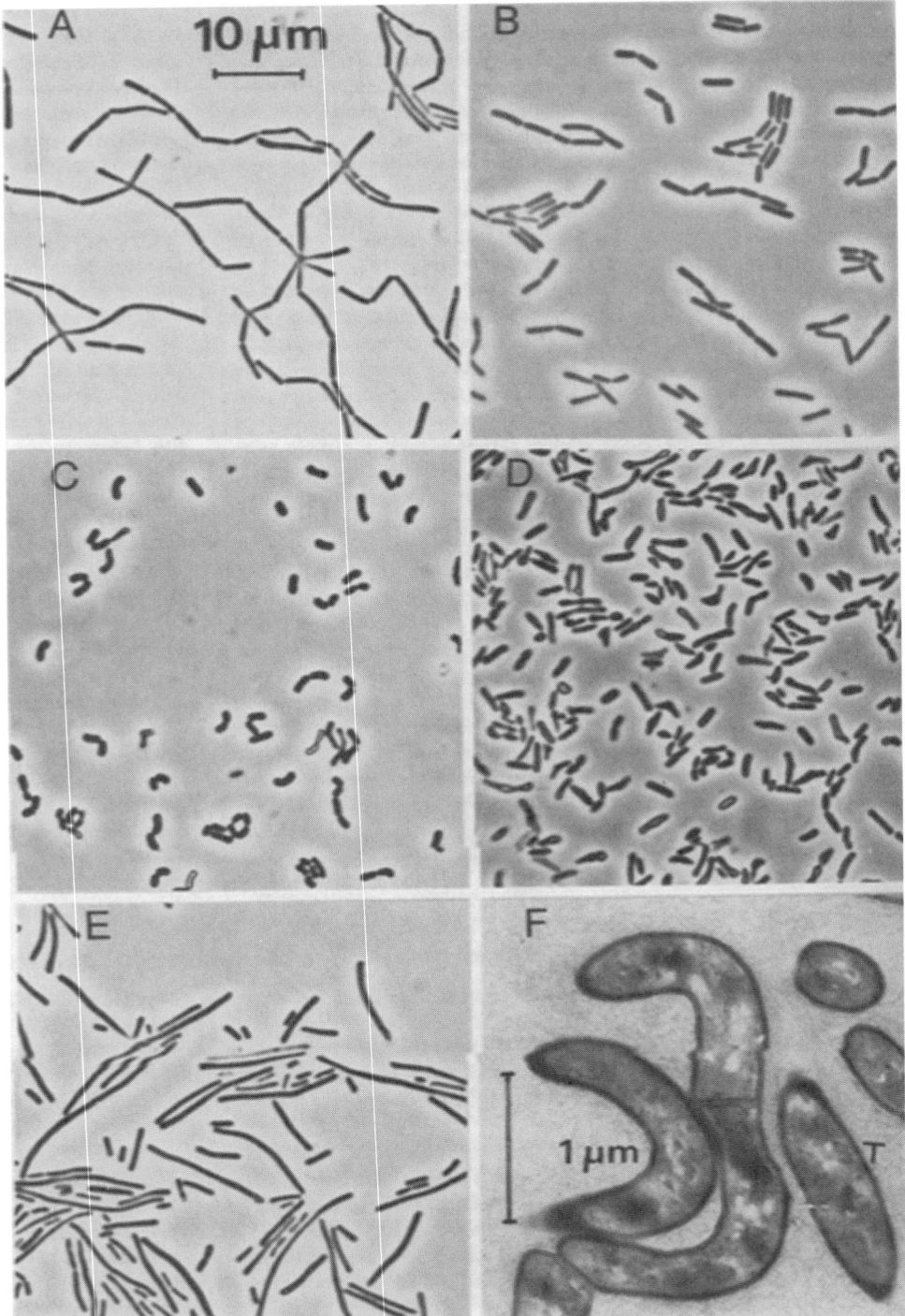

FIGURE 87. Phase contrast (A–E) and electron (F) micrographs showing different cell morphology of lactobacilli: A, *Lactobacillus gasseri*; B, *Lactobacillus agilis*; C, *Lactobacillus curvartus*; D, *Lactobacillus minor*; E, *Lactobacillus fermentum*; and F, involution form of lactobacilli in a thin section of a kefir grain.

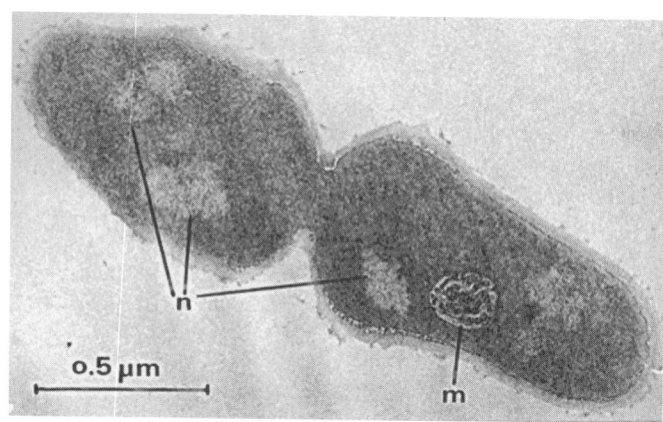

FIGURE 88. Electron micrograph of a dividing cell of *Lactobacillus coryniformis* showing asymmetric growth (m, mesosome; n, nucleoid).

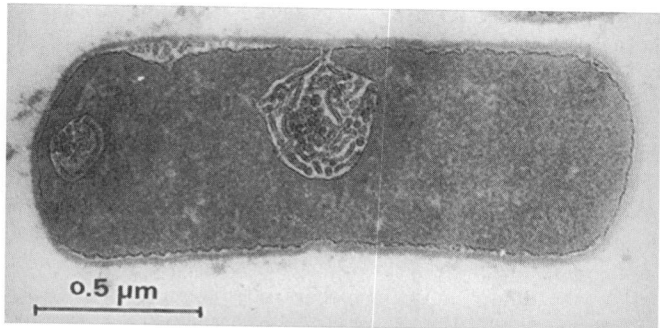

FIGURE 89. Electron micrograph of *Lactobacillus acidophilus* showing a mesosome connected with the cytoplasmic membrane.

may be observed under symbiotic growth, e.g., in kefir grains (Figure 87F) or under the influence of high concentrations of glycine, D-amino acids, or cell wall-active antibiotics (Hammes et al., 1973; Schleifer et al., 1976). Motility by peritrichous flagellation is observed in only a few species. It is highly dependent on the medium and the age of the culture and is sometimes observed only during isolation, but lost after several transfers on artificial media.

All lactobacilli stain clearly Gram-stain-positive. Dying cells may give variable results. Internal granulation is often revealed by Gram or methylene blue stain especially in the homofermentative long rods. The large bipolar bodies probably contain polyphosphate and appear very electron-dense in electron microscopy.

Cell wall and fine structure. Electron micrographs of thin sections reveal a typical Gram-stain-positive cell wall profile (Figure 88 and Figure 89). The cell wall contains peptidoglycan (murein) of various chemotypes of the cross-linkage group A. The Lys–D-Asp type is the most widespread peptidoglycan type (Schleifer and Kandler, 1972) (some species have the Orn–D-Asp type), followed by the *meso*-Dpm-direct type. The species *Lactobacillus sanfranciscensis* and *Lactobacillus rossiae* are unique as they possess the Lys–Ala and Lys–Ser–Ala$_2$ types, respectively. The cell wall also contains polysaccharides attached to peptidoglycan by phosphodiester bonds (Knox and Hall, 1964). Membrane-bound teichoic acid is present in all species (Archibald and Baddiley, 1966); cell wall-bound teichoic acid is present only in some of the species (Coyette and Ghuysen, 1970; Knox and Wicken, 1973).

S-Layers that have been detected in the group of lactic acid bacteria were found exclusively in several species of the genus *Lactobacillus*. In these species, the proteinaceous S-layers are characterized by small size (40–60 kDa) and high predicted pI value (pH >9) (Avall-Jaaskelainen and Palva, 2005).

Extracellular polysaccharides (EPS) are formed by strains of numerous *Lactobacillus* species and can be divided into two classes of compounds: (i) extracellularly synthesized homopolysaccharides (mainly dextran, glucan, and levan) and (ii) heteropolysaccharides with and without regularly repeating units that are synthesized from nucleotide-activated precursors. Homosaccharides are mainly produced by heterofermentative

species, especially those occurring in cereal environments, e.g., *Lactobacillus pontis*, *Lactobacillus frumenti*, *Lactobacillus sanfranciscensis*, and *Lactobacillus reuteri* (Tieking et al., 2003). The production is greatly enhanced by addition of sucrose to the medium. Heteropolysaccharides (mw 4×10^4–6×10^6) are produced in small amounts, usually 0.1–1.5 g/l by homofermentative and facultatively heterofermentative species (e.g., *Lactobacillus kefiranofaciens*, *Lactobacillus delbrueckii* subsp. *bulgaricus*, *Lactobacillus paracasei*, *Lactobacillus rhamnosus*, *Lactobacillus helveticus*, and *Lactobacillus sakei*) that, for example, are isolated from dairy or meat environments (De Vuyst and Degeest, 1999).

EPS-producing strains of *Lactobacillus delbrueckii* subsp. *bulgaricus* are utilized for production of dairy products such as yogurt as they positively affect the rheological properties with regard to texture, shear stability of the gel, decreased syneresis, and viscosity. The polysaccharide from *Lactobacillus kefiranofaciens* appears to constitute a matrix, so named kefir grains, acting as an ecological niche (Cheirsilp et al., 2003) for the microbial associations containing yeasts and lactobacilli. These grains are harvested and serve as inocula for kefir production. In meat products, the slime-forming potential is seen in spoiled refrigerated material.

In addition to nucleoids and ribosomes typical of all prokaryotes, electron micrographs of thin sections frequently show large mesosomes (Figure 89). They are formed by invaginations of the cytoplasmic membrane and are filled with tubuli, probably derived from secondary membrane invaginations (Schötz et al., 1965).

Colony and cultural characteristics. Colonies on agar media are usually small (2–5 mm) with entire margins, convex, smooth, glistening, and commonly opaque without pigment. In rare cases they are yellowish or reddish. A triterpeneoid carotenoid 4,4′-diaponneuroponeurosporene was identified as pigment in *Lactobacillus plantarum* (Breithaupt et al., 2001). Some species form rough colonies. Clearing zones caused by exoenzymes are usually not observed when grown on agar media containing dispersed protein or fat. However, most strains exhibit slight proteolytic activity due to cell-wall-bound or cell-wall-released proteases and peptidases (Kunji et al., 1996) and a weak lipolytic activity due to predominantly intracellular lipases. Distinct starch degradation leading to clearing zones on starch plates is only observed in a few species (e.g., *Lactobacillus amylophilus*, *Lactobacillus amylovorans*, and *Lactobacillus fermentum*). Growth in liquid media generally occurs throughout the liquid, but the cells settle soon after growth ceases. The sediment is smooth and homogeneous, rarely granular or slimy. Aggregation of cells in culture can be mediated by factors of proteinaceous nature (Schachtsiek et al., 2004). Clumping of cells of the same strain is called autoaggregation, whereas in coaggregation, aggregates form with genetically distinct cells. Both types occur in lactobacilli, e.g., in *Lactobacillus crispatus, Lactobacillus gasseri, Lactobacillus reuteri*, and *Lactobacillus coryniformis*. Aggregation plays a role in colonization of the oral cavity and the urogenital tract as well as in genetic exchange via conjugation.

Lactobacilli do not develop characteristic odors when grown in common media. When growing in food, they produce numerous volatile compounds that contribute either to food spoilage or to the desired, pleasant aroma of fermented food. Diacetyl, acetic acid, and acetaldehyde are examples of compounds derived from carbohydrate metabolism, and H_2S, amines, carbonyl compounds, cresol, skatol, benzaldehyde, and methanethiol are derived from amino acid catabolism (Margalith, 1981; Tanous et al., 2002). Their concentration is strongly affected by the potential of the strain, the microbial coassociation, and the environmental factors such as pH, water activity, temperature, and presence of electron acceptors (e.g., oxygen, fructose).

Nutrition and growth conditions. Lactobacilli are extremely fastidious organisms, adapted to complex organic substrates. They not only require carbohydrates as energy and carbon sources, but also require nucleotides (Elli et al., 2000), amino acids, and vitamins. Although pantothenic acid and nicotinic acid are, with the exception of a few strains, required by all species, thiamine is only necessary for the growth of the heterofermentative lactobacilli. The requirements for folic acid, riboflavin, pyridoxal phosphate, and *p*-aminobenzoic acid are scattered among the various species with riboflavin being the most frequently required compound. Biotin and vitamin B_{12} are required by only a few strains. Although the pattern of vitamin auxotrophy is considered to be characteristic of particular species (Rogosa et al., 1961), deviating strains are common (Abo-Elnaga and Kandler, 1965a; Ledesma et al., 1977). Vitamin-dependent strains are commonly used for bioassays of vitamins and are listed in the catalogues of most culture collections. The pattern of amino acid requirements also differs among species and even strains. By sequential mutagenesis, quintuple mutants of *Lactobacillus casei* were obtained which had lost their requirement for 5 amino acids (Morishita et al., 1974). However, the mutants grew significantly slower and reverted frequently to their amino-acid-dependent state when transferred back to the complete medium. Corresponding results were also obtained with four other species (Morishita et al., 1981).

These studies show that many of the nutritional requirements of lactobacilli can be the result of numerous minor defects within the genome and that much of the information encoding the various biosynthetic pathways is still present in the chromosome. On the other hand, the comparison of the complete genomes of *Lactobacillus plantarum* and *Lactobacillus johnsonii* revealed remarkable differences in the genetic endowments of these two species which are adapted to a great variety of environments and to the intestinal tract, respectively (Boekhorst et al., 2004). Although the *Lactobacillus plantarum* genome encodes 90 proteins predicted to be involved in transport and metabolism of vitamins and cofactors, the *Lactobacillus johnsonii* genome encodes only 30. In addition, 268 proteins are pre-

dicted to be involved in metabolism and transport of amino acids in *Lactobacillus plantarum*, whereas only 125 are encoded in the *Lactobacillus johnsonii* genome. *Lactobacillus plantarum* encodes the enzymes required for biosynthesis of all amino acids with the exception of leucine, isoleucine, and valine, whereas *Lactobacillus johnsonii* is predicted to be incapable of synthesizing most, if not all, of the 20 standard amino acids. A similar reduction of the genetic endowment was also found for proteins predicted to be involved in fatty acid biosynthesis and carbohydrate transport and metabolism.

The various requirements for essential nutrients are normally met when the media contain fermentable carbohydrate, peptone, and meat and yeast extracts. Supplementations with tomato juice, manganese, acetate and oleic acid esters, and especially Tween 80, are stimulatory or even essential for most species. Therefore, these compounds are included in the widely used MRS medium (De Man et al., 1960). Lactobacilli that are adapted to very particular substrates may require special growth factors. For instance, D-mevalonic acid is necessary for rice wine (sake) spoilage organisms (Tamura, 1956), and a small peptide isolated from freshly prepared yeast extract was found to be required for luxurious growth of *Lactobacillus sanfranciscensis* (Berg et al., 1981), a sourdough organism. To meet the requirement of a still unknown growth factor, some of the original substrate must be added.

Lactobacilli grow best in slightly acidic media with an initial pH of 6.4–5.4. Growth ceases commonly when pH 3.6–4.0 is reached, depending on the species and strain. Exceptions are found in isolates from fruit mashes, e.g., *Lactobacillus suebicus* grows even at pH 2.8 (Kleynmans et al., 1989). A similar pH tolerance is also found in strains of *Lactobacillus casei* and *Lactobacillus plantarum*. Lactobacilli can adapt to growth at an initial pH of 10.0 by gradually increasing the starting pH during subsequent enrichment test runs (Suhaimi et al., 1987). Although most strains are fairly aerotolerant, optimal growth is achieved under microaerophilic or anaerobic conditions. Increased CO_2 concentration (~5%) may stimulate growth.

Most lactobacilli grow best at mesophilic temperatures with an upper limit of around 40°C. Some also grow below 15°C and some strains even below 5°C. The thermophilic lactobacilli may have an upper limit of 55°C and do not grow below 15°C. Extremely thermophilic lactobacilli growing above 55°C are as yet unknown.

Metabolism. Metabolically, lactobacilli are at the threshold of anaerobic-to-aerobic life. They possess efficient carbohydrate fermentation pathways coupled with substrate level phosphorylation. A second substrate level phosphorylation site is the conversion of carbamyl phosphate to CO_2 and NH_3 the final step of arginine fermentation. However, only some of the species forming NH_3 from arginine are able to grow on arginine as the only energy source. In addition to substrate level phosphorylation, energy is generated by secondary transport systems including uniporters, proton-solute symporters, and antiporters (Konings, 2002), all contributing to the generation of a proton motive force (PMF). These systems are of special importance for cell survival under stress conditions, for example, as they occur after consumption the carbohydrate substrate, accumulation of lactic acid, and corresponding decrease of pH. Lactobacilli, except *Lactobacillus mali* and *Lactobacillus rhamnosus*, contain no isoprenoid quinones (Collins and Jones, 1981) and no cytochrome systems to perform oxidative phosphorylation. However, they possess flavine-containing oxidases and peroxidases to carry out the oxidation of $NADH_2$ with O_2 as the final electron acceptor. When nitrite is used as an electron acceptor for lactate oxidation (Wolf et al., 1990), ammonia is formed in a reaction catalyzed by a heme-dependent nitrite reductase present in *Lactobacillus plantarum* and *Lactobacillus pentosus*. In strains of *Lactobacillus delbrueckii* subsp. *lactis*, *Lactobacillus sakei*, *Lactobacillus farciminis*, *Lactobacillus brevis*, *Lactobacillus buchneri*, and *Lactobacillus suebicus*, a heme-independent enzyme is active and forms NO and N_2O as end products. Lactobacilli are also able to perform a manganese-catalyzed scavenging of superoxide (Archibald and Fridovich, 1981; Götz et al., 1980), although they are generally devoid of superoxide dismutase and usually catalase. However, superoxide dismutase activity has been described for an oxygen tolerant strain of *Lactobacillus sakei* (Amanatidou et al., 2001).

The main fermentation pathways for hexoses are the Embden-Meyerhof pathway converting 1 mol of hexose to 2 mol of lactic acid (homolactic fermentation) and the 6-phosphogluconate pathway, resulting in 1 mol CO_2, 1 mol ethanol (or acetic acid), and 1 mol lactic acid (heterolactic fermentation). Under aerobic conditions, most strains are able to reoxidize $NADH_2$ with oxygen serving as the final electron acceptor, thus acetyl-CoA is not, or at least not completely, reduced to ethanol. Consequently, additional ATP is formed by substrate level phosphorylation and varying ratios of acetic acid and ethanol are found, depending on the oxygen supply.

Pyruvate, intermediately formed in both pathways, may partly undergo several alternative conversions yielding either the well-known aroma compound diacetyl and its derivatives or acetic acid (ethanol); with hexose limitation, the latter pathway may become dominant and the homolactic fermentation may be changed to a heterofermentation with acetic acid, ethanol, and formic acids as the main products (de Vries et al., 1970; Thomas et al., 1979). Even lactate may be partially oxidized and broken down to acetic acid and formate or CO_2 by various little known mechanisms (Kandler et al., 1983). The conversion of glycerol to 1,3-propanediol with glucose serving as electron donor is observed in *Lactobacillus brevis* (Schütz and Radler, 1984), *Lactobacillus buchneri*, and *Lactobacillus reuteri*. In this pathway, glycerol serves as electron acceptor and is converted to 3-hydroxypropionaldehyde. The reaction is catalyzed by a vitamin B_{12}-dependent glycerol dehydratase. Remarkably, strains of *Lactobacillus reuteri* can synthesize cobalamine and are thus vitamin B_{12}-independent (Taranto et al., 2003). *Lactobacillus reuteri* can accumulate and excrete 3-hydroxypropionaldehyde (Talarico and Dobrogosz, 1989), which has an antimicrobial potential and has become known as reuterin. Strains of *Lactobacillus buchneri* and *Lactobacillus parabuchneri* can produce 1,2-propanediol from lactate in a sequence of reactions presumably catalyzed by lactaldehyde dehydrogenase and an NAD-linked 1,2-propanediol-dependent oxidoreductase (Krooneman et al., 2002). Remarkably, *Lactobacillus diolivorans* can utilize 1,2-propanediol converting it (still hypothetically) to propione aldehyde that becomes disproportioned to 1-propanol and propionic acid (Oude Elferink et al., 2001). The antimycotic potential of propionic acid is well established and can contribute to the stability of silage.

At the enzyme level, homo- and heterofermentative lactobacilli differ with respect to the presence or absence of FDP aldolase or phosphoketolase. Whereas the heterofermentative lactobacilli possess phosphoketolase but no aldolase, the obligately homofermentative ones possess FDP aldolase but no phosphoketolase. They are thus unable to ferment any of the pentoses which are broken down by the heterofermenters via phosphoketolase yielding equimolar amounts of lactic acid and acetic acid. This basic rule no longer appears to be absolutely valid, as homofermentation with fructose and a fructose-inducible FDP aldolase were detected in a strain of *Lactobacillus brevis* (Saier et al., 1996). Correspondingly, fructose phosphoketolase was found in a ribose-fermenting strain of *Lactobacillus acidophilus* (Biddle and Warner, 1993). These findings have strong implications for the traditional grouping of the lactobacilli (see below). Homofermentative lactobacilli, traditionally named "Streptobacteria" (Orla-Jensen, 1919), possess an inducible phosphoketolase with pentoses acting as inducers. They are thus able to ferment pentoses upon adaption to lactic acid and acetic acid, whereas hexoses are homofermentatively metabolized. Therefore, these lactobacilli must be called facultative heterofermenters (group B; see below). The ability to ferment pentoses does not necessarily mean that the organism can grow on these carbohydrates. Because hexoses are required for biosynthesis of peptidoglycan and other building blocks, these sugars have to be synthesized from the C_5 compounds. It was shown (Westby et al., 1993) that an arabinose, xylose, and ribose fermenting strain of *Lactobacillus plantarum* does not grow on the pentoses and is devoid of fructose-1,6-diphosphatase which plays a key role in gluconeogenesis. A homolactic fermentation of pentoses was detected in an unidentified thermophilic *Lactobacillus* (Barre, 1969) and is also found in *Lactobacillus murinus* (D. Hemme, personal communication). In these organisms an as yet unknown pathway appears to exist which does not involve phosphoketolase. Such fermentations may involve the transformation of pentoses to hexoses via transaldolase and transketolase reactions followed by glycolysis (Kandler, 1983) with lactic acid being the only fermentation product.

Carbohydrates may also contribute to other reactions. Sucrose is not only a substrate for fermentation, but also for the formation of dextrans and levans (slime) catalyzed by dextran and levan sucrases, respectively (see *Cell wall and fin structure*, above). Fructose serves not only as a substrate for fermentation, but also as an electron acceptor and becomes reduced to mannitol by most heterofermentative lactobacilli. Correspondingly, glycerol is formed from triosephosphate and excreted into the medium by some heterofermentative strains and even erythritol is formed in an unspecific reaction by *Lactobacillus pontis* (Hammes et al., 1996).

Solutes such as amino acids and mono- and oligosaccharides are taken up by lactic acid bacteria by the different transport systems common to bacteria. As described by Poolman (1993, 2002), the energy- providing principles used in the translocation processes are: (i) secondary transport (e.g., PMF-driven), (ii) primary transport (e.g., ATP-driven), and (iii) group translocation (chemical modification with concomitant transport), by means of the phosphoenolpyruvate:sugar phosphotransferase system (sugar PTS). Most information on these transport systems in lactic acid bacteria have been obtained from studies with *Lactococcus lactis*, and rather little is known about these systems in lactobacilli. Oligosaccharides are split inside the cell by the respective glycosidases prior to the phosphorylation of the resulting monosaccharides. However, at least lactose and galactose are taken up by some lactobacilli via the phosphoenolpyruvate-dependent phosphotransferase system (Chassy and Thompson, 1983). The lactose phosphate formed is split to glucose and D-galactose-6-phosphate. The latter is then metabolized via the D-tagatose-6-phosphate pathway (Kandler, 1983). The hierarchical utilization of available carbohydrates is governed by a transcriptional control mechanism that has evolved in low-G+C Gram-stain-positive bacteria including lactobacilli (Poolman, 2002; Titgemeyer and Hillen, 2002). The gene expression is regulated by the pleiotropic control protein CcpA that binds the seryl-phosphorylated form of the phosphotransferase HPr (HPr-ser-P). Operons and genes under control of this mechanism have been characterized in *Lactobacillus delbrueckii* subsp. *lactis*, *Lactobacilluscasei*, *Lactobacillus plantarum*, and *Lactobacillus pentosus*. HPr-ser-P can also inhibit carbohydrate permeases and thus trigger inducer exclusion and mediate inducer expulsion. In this mechanism of carbon catabolite repression, the histidyl-phosphorylated HPr that is generally involved in transport of many carbon sources via the phosphotransferase system controls further regulatory reactions and catabolic enzymes.

Organic acids such as citric, tartaric, succinic, and malic acids, are degraded by lactobacilli. These compounds are constituents of raw plant materials used for the fermentative production of, for example, wine and cider. Citric acid is also contained in milk. The conversion of these compounds has profound effects on sensory properties. The degradation proceeds commonly via oxaloacetic acid and pyruvate to CO_2 and lactic or acetic acid (Radler, 1986; Radler and Brohl, 1984). The reactions result in a change of charges that can be used to generate metabolic energy (Konings, 2002). Malate is converted by lactobacilli to CO_2 and lactate. Detailed studies on the catalytic and regulatory properties of the DNA-dependent malic enzymes of lactic acid bacteria were performed by London et al. (1971). Alternatively, malic acid is split to CO_2 and L(+)-lactic acid in many lactobacilli by a multifunctional malolactic enzyme with all intermediates remaining tightly bound to the enzyme complex (Radler, 1975). *Lactobacillus fermentum* possesses neither the malic nor the malolactic enzyme and converts malate via fumarate to succinate (Whiting, 1975).

Several amino acids, e.g., lysine, ornithine, histidine, phenylalanine, and tyrosine, are decarboxylated and excreted by lactobacilli (Straub et al., 1995). The resulting biogenic amines are of hygienic concern, and strains used in starter cultures should lack the potential of their formation. The reaction can again be used by the lactobacilli to generate metabolic energy.

Phenolic compounds are converted in several reactions. Chlorogenic acid and *p*-coumarylquinic acid are metabolized after malic acid fermentation is completed. They are hydrolyzed and the resulting (−)-quinic acid is reduced to (−)-dehydroshikimic acid by heterofermentative lactobacilli. It is further reduced to dihydroxycyclohexane-c-1-carboxylic acid by homofermenters. Shikimic acid may be reduced to catechol by *Lactobacillus plantarum* which also converts *p*-coumaric acid to *p*-ethylphenol. The electron source of these reactions is lactate which becomes oxidized to CO_2 and acetic acid (Whiting, 1975). Lactobacilli isolated from whisky fermentations decarboxylate *p*-coumaric acid and/or ferulic acid with the production of 4-vinylphenol and/or 4-vinylguaiacol, respectively (van Beek and Priest, 2002). Isolates from food and from human and animal intestines possess tannase (tannin acylhydrolase). Hydrolyzable tannins such as gallotannin and ellagitannin are widely distributed in plants and are considered effective antinutritive compounds. The activity was described for *Lactobacillus plantarum*, *Lactobacillus paraplantarum*, *Lactobacillus pentosus*, *Lactobacillus murinus*, and *Lactobacillus animalis* (Nishitani et al., 2004; Sasaki et al., 2005).

The lactic acid formed by the various fermentation pathways possesses either the L- or the D-configuration depending on the stereospecificity of the lactated dehydrogenase present in the cells. Racemate may be formed when both L- and D-lactate dehydrogenase are present in the same cell or, in rare cases, by the action of an inducible lactate racemase in combination with a constitutive L-lactate dehydrogenase (Stetter and Kandler, 1973). Lactate dehydrogenases of the various species often differ from each other considerably, e.g., with respect to their electrophoretic mobility and their kinetic properties. Most enzymes are nonallosteric, but some species contain allosteric L-lactate dehydrogenases with FDP and Mn^{2+} acting as effectors (Garvie, 1980; Hensel et al., 1977).

Phages. *Lactobacillus* phages can interfere with the performance of their hosts in fermentation processes and have been studied under this practical aspect (Hammes and Hertel, 2003). Phages inhibiting processes in the dairy field are of greatest importance and, therefore, most knowledge is available for dairy *Lactobacillus* phages, including those added to the process to achieve probiotic effects (Brussow, 2001). All *Lactobacillus* phages belong to the families *Myoviridae* and *Siphoviridae*, have double stranded DNA, and are *cos*-site or *pac*-site phages. The genome size ranges between 40 kbp (*Lactobacillus casei* phage pL1) and 133 kbp (*Lactobacillus plantarum* phage fri). Complete sequences are available for *Lactobacillus delbrueckii* subsp. *lactis*, *Lactobacillus casei*, *Lactobacillus johnsonii*, and *Lactobacillus plantarum* (Desiere et al., 2002). No sequence similarities were detected between phages infecting distinct *Lactobacillus* species. The incidence of lysogeny is very high and DNA/DNA hybridization revealed a close relatedness between the virulent and temperate phages of a species (Lahbib-Mansais et al., 1988; Mata and Ritzenthaler, 1988).

Antagonistic compounds. Lactobacilli share with other lactic acid bacteria the potential to inhibit the growth of competing micro-organisms and, thus, to prevent food spoilage and the growth of undesired micro-organisms in association with humans and animals. The primary effect of pH reduction is of major importance, but this effect is supported by formation of numerous compounds that may be produced depending on environmental factors and on the genetic endowment of a species or strain (Hammes and Tichaczek, 1994; Ouwehand, 1998). At low pH, lactic acid and acetic acid act as weakly dissociated compounds which are inhibitory per se as is the gaseous fermentation product CO_2. H_2O_2 is formed in reactions catalyzed by flavoproteins. In substrates such as milk, lactoperoxidase catalyzes a reaction that uses H_2O_2 and thiocyanate and forms hypothianite ($OSCN^-$) which inhibits membrane functions (Kamau et al., 1990) and several reactions of glycolysis (Condon, 1987). The effect of H_2O_2 on Gram-stain-positive organisms is bacteriostatic, whereas it is bacteriocidal against Gram-stain-negative bacteria. Many low molecular mass compounds with antimicrobial properties are produced by lactobacilli. Reuterin, already introduced as 3-hydroxypropanal (see above), is excreted by

Lactobacillus reuteri, *Lactobacillus buchneri*, *Lactobacillus brevis*, *Lactobacillus collinoides*, and *Lactobacillus coryniformis* (Schnürer and Magnusson, 2005). It inhibits thiol enzymes and has a broad antimicrobial spectrum including fungi, protozoa, and bacteria. Pyroglutamic acid (2-pyrolidone-5 carboxylic acid) with activity against *Bacillus subtilis*, *Panothoea agglomerans*, and *Pseudomonas* species is produced by strains of *Lactobacillus casei* (Chen and Russell, 1989; Huttunen et al., 1995), *Lactobacillus helveticus*, *Lactobacillus delbrueckii* subsp. *bulgaricus*, and *Lactobacillus delbrueckii* subsp. *lactis* (Mucchetti et al., 2002). Strains of *Lactobacillus plantarum* produce benzoic acid, methylhydantoin, mevalonolactone, and cyclo (glycyl–L-leucyl). The producer strain inhibits *Panothoea agglomerans* and *Fusarium avenaceum*, and the single compounds each exhibit a fraction of the total effect. *Lactobacillus plantarum* MiLAB 393 produces cyclo (L-Phe–L-Pro), cyclo (L-Phe–*trans*-4-OH–L-Pro) and phenyl lactic acid exhibiting activity especially against molds and yeasts. The latter compound as well as 4-hydroxyphenyllactic acid appears to be a rather common product in food-fermenting lactobacilli (Valerio et al., 2004) and is especially active against molds causing spoilage of baked goods (Lavermicocca et al., 2003). *Lactobacillus reuteri* produces a tetramic acid derivative named reutericyclin (Ganzle et al., 2000). The compound is also formed in sourdough (Ganzle and Vogel, 2003) and inhibits Gram-stain-positive bacteria. Bacteriocins are proteinaceous compounds that have been divided into four classes (Klaenhammer, 1993). Class I includes lantibiotics; Class II includes small (<10 kDa), moderate (100°C) to high (121°C) heat-stable non-lanthionine-containing membrane-active peptides. Subgroups have been defined within this class: IIa are *Listeria*-active peptides with a -Y-G-IV-G-V-X-C-sequence near the amino terminus. IIb are two-peptide bacteriocins, and IIc are thiol-activated peptides. Class III are large (>30 kDa) heat-labile proteins, and Class IV contains complex bacteriocins, e.g., proteins with lipid and/or carbohydrate. Groups I and II are by far most studied for their biosynthesis, genetics regulation, and mode of action (Nes et al., 1996; Twomey et al., 2002). Lactobacilli are most often cited for production of bacteriocins and produce all four classes of bacteriocins (Klaenhammer, 1993). In general, their inhibitory spectrum is narrow and includes closely related species. Depending on environmental conditions (Ganzle et al., 1999) and type of compound, the spectrum may also include rather unrelated organisms including *Proteobacteria* and even *Archaea* (Hammes et al., 1979). Bacteriocinogenic strains have been found among homo- and heterofermentative species. Class I bacteriocins are ribosomally synthesized polypeptides that are characterized by containing ring structures introduced post-translationally by lanthionine or methyllanthionine bridges. Further modifications may include presence of D-alanine, dehydroalanine, dehydroaminobutyrate, and N-terminal bound pyruvate (lactocin S) or lactate. The compounds partially interfere with membrane function by pore formation and partially inhibit cell wall synthesis by complexing with lipid II (Reisinger et al., 1980). Well studied examples of lantibiotics (Twomey et al., 2002) are plantaricin C (Turner et al., 1999), plantaricin W (Holo et al., 2001), and lactocin S (Mortvedt et al., 1991). Each compound belongs to one specific lantibiotic subgroup. Plantaricin W is a two-component bacteriocin characterized by an inherent activity of each component. Plwα and Plwβ group with the mersacidin and ltnAZ compounds, respectively. Lactocin S

is produced by *Lactobacillus sakei* L 45 and constitutes a separate group. Plantaricin C is included in the lacticin 481 group.

Resistance to chemotherapeutics. Initially studies on the sensitivity or resistance pattern of lactobacilli toward antibiotics originated mainly from problems created by the presence of antibiotics in milk resulting from mastitis therapy (Sozzi and Smiley, 1980). The general increase of antibiotic resistance in pathogenic bacteria and the recognition that some *Lactobacillus* species had been involved in cases of bacteremia led to new and special interest in the sensitivity of lactobacilli to chemotherapeutics. Sensitivity testing of lactobacilli is still problematic as several species require complex media with compounds and pH conditions that interfere with the test substance. To determine antibiotic concentrations needed in plasma levels, the differentiation between minimal inhibitory concentration (MIC) and minimal bactericidal concentration (MBC) is essential. In a study (Bayer et al., 1978) of the effects of penicillin, ampicillin, clindamycin, cephalotin, cefoxitin, and metronidazol on 40 lactobacilli belonging to seven species, it was observed that the MBC:MIC ratios are high, ranging from 30:1 for cephalotin to 266:1 for ampicillin. For cefoxitin and metronidazol an achievable MIC and/or MBC were observed with 87.5–100% of the strains. In a study of 15 antibiotics with major use in veterinary practice and with 43 lactobacilli from nine species mainly of food and partially of clinical origin it has been observed (Klein, 1992) that intrinsic antibiotic resistance exists against colistin sulfate and sulfonamide. In this study, MRS medium had been used in agar diffusion and suspension tests. Thirty-one strains of *Lactobacillus delbrueckii* subsp. *bulgaricus* used in yogurt cultures had an intrinsic resistance toward mycostatin, nalidixic acid, neomycin, polymyxin B, trimethoprim, colimycin, sulfomethoxazol, and sulfonamides (Sozzi and Smiley, 1980). Intrinsic resistance exists, furthermore, for most lactobacilli against vancomycin because of the presence of D-alanine:D-alanine ligase related enzymes (Elisha and Courvalin, 1995). The acquisition of antibiotic resistance by horizontal gene transfer has become of great concern (Teuber et al., 1999). Resistance genes were located on conjugative or mobilizable plasmids and transposons in isolates from humans or animals and in food-associated lactobacilli. The evolution of antibiotic resistance is enhanced by these genetic elements, by possession of integrons and insertion elements, and by lytic and temperate phages. All elements occur in lactobacilli. Lactobacilli associated with food may become part of the intestinal association. These bacteria together with the resident lactobacilli may acquire and exchange resistance genes and contribute to their spread in human-associated bacteria. For that reason, it is necessary not to use lactobacilli that carry acquired antibiotic resistance genes in starter cultures.

Pathogenicity. Based on their numbers in human food, lactobacilli are the most important group of living micro-organisms ingested. From long-term experience, it can be concluded that lactobacilli are safe organisms. However, an increasing number of reports of cases of lactobacillemia led to some doubts on the validity of that general statement. Infections caused by lactobacilli are very rare and have been estimated to represent 0.05–0.48% of all cases of infective endocarditis and bacteremia (Gasser, 1994; Husni et al., 1997; Saxelin et al., 1996). In the vast majority of these cases an underlying disease indicated a predisposition of the patients (Hammes and Hertel, 2003). Lactobacilli were involved in endocarditis, bloodstream infection, and local infection (Aguirre and Collins, 1993; Gasser, 1994). The most frequently isolated species were *Lactobacillus rhamnosus*, *Lactobacillus paracasei*, and *Lactobacillus plantarum*, and with lower incidence *Lactobacillus brevis*, *Lactobacillus delbrueckii*, *Lactobacillus gasseri*, *Lactobacillus jensenii*, *Lactobacillus johnsonii*, *Lactobacillus salivarius*, *Lactobacillus acidophilus*, *Lactobacillus casei*, and *Lactobacillus fermentum* were found. The correct identification of the latter three species may be questioned (see below) because of the insufficient identification methods available at the time of the early reports and, in the case of *Lactobacillus casei*, changes in nomenclature at the time of the description of *Lactobacillus paracasei* and *Lactobacillus rhamnosus* (Collins et al., 1989). All species are part of the human intestinal association. With the exception of *Lactobacillus gasseri*, *Lactobacillus jensenii*, and *Lactobacillus salivarius* they also occur in food association and are partially used in starter cultures in food fermentation or as probiotics. Thus, the clinical isolates may have originated from the indigenous body association (mouth, vagina, intestines) or from ingested food. Pathogenicity factors are unknown for lactobacilli. Their application in food should exclude any avoidable negative effect on human health. Clearly, transferable antibiotic resistance should be absent. Additional undesired properties (Hammes and Hertel, 2003) are formation of biogenic amines in food and, for short-bowl patients, the formation of D-lactic acid is a disadvantageous property, as the lactobacilli (e.g., in probiotics) may overgrow commensal bacteria and cause D-lactate acidosis (Bongaerts et al., 1997; Coronado et al., 1995). Binding to extracellular matrices (e.g., collagen) or serum proteins (e.g., fibrinogen and fibronectin) as well as aggregation of human platelets were also considered as critical properties (Anonymous, 2000). Isolates from infective endocarditis (IE) patients (five strains each of the species *Lactobacillus paracasei* and *Lactobacillus rhamnosus*) were investigated for these properties by (Harty et al., 1994). Ten control strains of the same species and various oral isolates were used as comparators. Platelets aggregated with *Lactobacillus rhamnosus* IE-isolates and half of the control strains. A positive reaction with *Lactobacillus paracasei* was observed for two out of five IE-isolates and two out of nine control strains. Platelet aggregation was also observed with oral strains of *Lactobacillus acidophilus* (1/1), *Lactobacillus salivarius* (2/3), *Lactobacillus plantarum* (3/5), and *Lactobacillus fermentum* (2/3). These results show that platelet aggregation is dispersed among lactobacilli and not a characteristic of clinical isolates. Virtually the same result was obtained in the study of the potential of the strains to bind to fibrinogen, fibronectin, and collagen. A similarly dispersed distribution was observed in preceding studies (Harty et al., 1993) in which just a significant tendency of the IE-isolates to show higher hydrophobicity, hydroxyapatite adhesion, and salivary aggregation values was observed. These results are consistent with the conclusion that a specific pathogenicity related property cannot be found in isolates from diseased persons.

Ecology, habitats, and biotechnology. Lactobacilli grow under anaerobic conditions or at least under reduced oxygen tension in all habitats providing ample carbohydrates, breakdown products of protein and nucleic acids, and vitamins. A mesophilic to slightly thermophilic temperature range is favorable. However, strains of some species (e.g., *Lactobacillus sakei*, *Lactobacillus curvatus*, *Lactobacillus fuchuensis*, *Lactobacillus algidus*, *Lactobacillus plantarum*) grow, albeit slowly, even at temperatures close to freezing point, e.g., in refrigerated meat (Kato et al., 2000; Kitchell and Shaw, 1975; Sakala et al., 2002) and fish (Ringo and Gatesoupe, 1998; Schröder et al., 1980). Lactobacilli are, in general, aciduric or acidophilic, and decrease the pH of their substrates to below 4.0 by lactic acid formation, thus preventing, or at least severely delaying, growth of virtually all competitors except other lactic acid bacteria and yeasts. These properties make lactobacilli valuable inhabitants of the intestinal tract of humans and animals and important contributors to food technology. However, they are also potent spoilage organisms as they may affect the sensory properties through flavor, texture, color, slime, cloudiness, and formation of biogenic amines.

Several individual species have adapted to specific ecological niches and generally are not found outside their specialized habitats. The relative ease with which such species can be reisolated from their respective sources since their first discovery, sometimes almost 100 years ago, indicates that these niches are, in fact, their natural habitats, although sometimes man-made.

Plant sources. Lactobacilli occur in nature in low numbers on all plant surfaces (Keddie, 1959; Mundt and Hammer, 1968), and together with other lactic acid bacteria grow luxuriously in all decaying plant material, especially decaying fruits. Hence, lactobacilli are important for the production as well as the spoilage of fermented vegetable feed and food (e.g., silage, sauerkraut, kimchi, olives, mixed pickles) and beverages (e.g., beer, wine, juices). The chiefly isolated species include *Lactobacillus plantarum*, *Lactobacillus brevis*, *Lactobacillus coryniformis*, *Lactobacillus paracasei*, *Lactobacillus curvatus*, *Lactobacillus sakei*, and *Lactobacillus fermentum* (Carr et al., 1975; Hammes and Hertel, 2003; Kandler, 1984; Sharpe, 1981).

Fermented products from plants can be grouped according to the nature of the fermentation substrates. Starchy substrates are cereals (Hammes et al., 2005), potatoes, and cassava; products of their fermentation are, e.g., sourdough, beer, spirits, cassava sour starch, and fufu. Roughly 30 *Lactobacillus* species have been isolated from sourdough, including highly adapted species described recently, *Lactobacillus sanfranciscensis*, *Lactobacillus pontis*, *Lactobacillus panis*, *Lactobacillus mindensis*, *Lactobacillus hammesii*, *Lactobacillus acidifarinae*, *Lactobacillus spicheri*, *Lactobacillus paralimentarius*, and *Lactobacillus frumenti* (De Vuyst and Neysens, 2005; Ehrmann and Vogel, 2005). In addition, species also common in other substrates occur, e.g., *Lactobacillus plantarum*, *Lactobacillus reuteri*, *Lactobacillus fermentum*, *Lactobacillus delbrueckii*, *Lactobacillus fructivorans*, *Lactobacillus alimentarius*, *Lactobacillus parabuchneri*, *Lactobacillus farciminis*, *Lactobacillus brevis*, *Lactobacillus reuteri*, *Lactobacillus homohiochii*, *Lactobacillus hilgardii*, and *Lactobacillus amylovorus*, and, remarkably, species that are known as inhabitants of the human and/or animal tract, such as *Lactobacillus gasseri*, *Lactobacillus amylovorus*, *Lactobacillus pontis*, *Lactobacillus sakei*, *Lactobacillus delbrueckii* subsp. *lactis*, *Lactobacillus murinus*, *Lactobacillus acidophilus*, *Lactobacillus reuteri*, *Lactobacillus johnsonii*, and *Lactobacillus mucosae*. These lactobacilli often occur together with *Weissella* species, pediococci, leuconostocs, and yeasts (Hammes et al., 2005). The composition of the association is governed by the technology of dough preparation (Ganzle et al., 1998). Traditionally propagated doughs contain mainly 2–3 obligately heterofermentative species (*Lactobacillus sanfranciscensis*, *Lactobacillus pontis*, and *Lactobacillus mindensis*).

In beer, lactobacilli are the most important agents of spoilage. Adapted species tolerate ethanol (commonly 4–5%), pH values of 3.8–4.7, high CO_2 concentration, and bitter hops compounds (Sakamoto and Konings, 2003). The following species have been identified in spoiled beer: *Lactobacillus brevis*, *Lactobacillus buchneri*, *Lactobacillus paracasei*, *Lactobacillus coryniformis*, *Lactobacillus curvatus*, *Lactobacillus lindneri*, *Lactobacillus malefermentans*, *Lactobacillus parabuchneri*, *Lactobacillus plantarum*, *Lactobacillus paraplantarum*, *Lactobacillus paracollinoides*, and *Lactobacillus collinoides*. The common hop sensitivity of Gram-stain-positive bacteria is overcome by lactobacilli, e.g., *Lactobacillus brevis*, by active extrusion of the drug mediated by a (multi) drug resistance pump located in the cytoplasmic membrane (Sakamoto et al., 2001). This mechanism is supported by the ability of the resistant strains to maintain a larger ΔpH (Simpson and Fernandez, 1994) and to create an ATP pool that is larger than in hop-sensitive strains. In lactic-acid containing beer, such as the Belgian Rodenbach or Gueuze beer, the Berliner Weisse beer, the Russian kwass (Verachtert and Iserentant, 1995), and maheu (bantu beer), less hops are commonly used and lactobacilli contribute to the acid taste of the alcoholic drinks. In the case of maheu, the thermophilic *Lactobacillus delbrueckii* subsp. *delbrueckii*, which was first isolated from potato and grain mashes fermented at 40–55°C, is used (Henneberg, 1903). The organism is also used for production of lactic acid (Buchta, 1983).

In whisky production, the wort is not boiled before yeasts start the alcoholic fermentation and therefore, lactobacilli can grow and are considered to contribute favorably to the flavor (Pedersen et al., 2004; van Beek and Priest, 2002). The following species have been found in the fermenting whisky wort: *Lactobacillus plantarum*, *Lactobacillus fermentum*, *Lactobacillus brevis*, *Lactobacillus paracasei*, *Lactobacillus mucosae*, *Lactobacillus ferintoshensis*, *Lactobacillus sanfranciscensis*-like bacteria, and *Lactobacillus suntoreyus* (Cachat and Priest, 2005).

In the seed mash of sake fermentation (moto), *Lactobacillus sakei* and *Lactobacillus plantarum* have been described as acidifying lactobacilli (Kitahara et al., 1957), and *Lactobacillus homohiochii*, *Lactobacillus fermentum*, *Lactobacillus plantarum*, and *Lactobacillus fructivorans* were found in spoiled sake (hiochi) (Katagiri et al., 1934). *Lactobacillus satsumensis* was found in the mash of shochu, a traditional distilled Japanese spirit (Endo and Okada, 2005).

The second group of fermented plant products is raw material containing soluble carbohydrates such as vegetables and fruits for food fermentation, and grass, clover, maize, etc. for silage production. The fermentation processes are commonly not controlled by starter cultures and are characterized by successions in the microbial associations involved. An example is the fermentation of cabbage to produce sauerkraut (Hammes and Hertel, 2003; Pedersen, 1979; Splittstoesser et al., 1975; Stamer, 1975). In an initial phase, for 5 d *Leuconostoc mesenteroides* represents approximately 90% of the lactic acid bacteria involved and *Lactobacillus sakei*, *Lactobacillus curvatus*, *Lactobacillus plantarum* and *Lactobacillus brevis* are found within the remaining 10%. Thereafter "betabacteria" evolve (mainly *Lactobacillus brevis* and *Lactobacillus buchneri*), and finally homofermenters dominate (mainly *Lactobacillus plantarum*, *Lactobacillus curvatus*, and *Lactobacillus sakei*). Similar successions occur in the Korean kimchi, made from Chinese cabbage, radish, cucumber, onion, garlic,

pepper, etc. *Lactobacillus kimchii* was isolated from fermenting kimchi (Yoon et al., 2000).

Successions occur also in fermenting silage (Hammes and Hertel, 2003). In silage made from grass and red clover, pediococci, *Lactobacillus plantarum*, and *Lactobacillus graminis* became dominant after an initial phase characterized by the presence of leuconostocs and "*Lactobacillus coprophilus*" (Beck et al., 1987).

In wine, lactobacilli have to tolerate high levels of alcohol (e.g., up to 15–22% in desert wines), low pH (3.0–4.0), and the presence of SO_2. In addition to *Oenococcus oeni*, pediococci, and leuconostocs, the following lactobacilli have been isolated from wine at the various production stages: *Lactobacillus brevis*, *Lactobacillus hilgardii*, *Lactobacillus plantarum*, *Lactobacillus fermentum*, *Lactobacillus buchneri*, *Lactobacillus fructivorans*, *Lactobacillus mali*, *Lactobacillus jensenii*, *Lactobacillus kunkeei*, and *Lactobacillus nageli* (Edwards et al., 2000). The latter two species have been shown to slow down alcoholic fermentation of grape musts and to be responsible for sluggish or stuck alcohol fermentation which is a well-known problem in wine making (Edwards et al., 2000; Edwards et al., 1998). The presence of the lactobacilli is undesirable because they may affect the flavor and cause cloudiness. *Lactobacillus fructivorans* and *Lactobacillus hilgardii* can even spoil fortified wines (Lonvaud-Funel, 1999). They may also form biogenic amines as shown for *Lactobacillus higardii* (Farias et al., 1993) and *Lactobacillus buchneri* (Liu et al., 1994). Heterofermentative lactobacilli, e.g., *Lactobacillus buchneri*, may also produce the ethylcarbamate precursor (carbamylphosphate) from arginine. The lactobacilli association in wine is characterized by population successions and preferential growth at the end of malolactic fermentation which is desirably performed by *Oenococcus oeni*.

Lactobacillus hilgardii and *Lactobacillus fructivorans* (Fornachon et al., 1949) are typical organisms of acidic and alcoholic beverages; *Lactobacillus collinoides* (Carr and Davies, 1972) and *Lactobacillus mali* (Carr and Davies, 1970; Carr et al., 1977) are found in cider and other fruit juices. Fruit mashes for production of fruit brandy may be acidified with sulfuric acid to achieve pH values <3.0. Highly acid-resistant lactobacilli were isolated from mashes that had sometimes been stored for a year (Heinzl and Hammes, 1986). Strains of *Lactobacillus plantarum* and *Lactobacillus suebicus* grow at pH 2.5 in the presence of 12 and 14% ethanol, respectively. *Lactobacillus brevis* and *Lactobacillus hilgardii* are additional species that are found in numbers of >10^6 c.f.u./ml.

Milk and dairy products. Milk contains no lactobacilli when it leaves the udder, but is very easily contaminated with lactobacilli by dust, dairy utensils, etc. Because streptococci grow faster, the number of lactobacilli usually remains fairly low even in spontaneously soured milk. Only after prolonged incubation do lactobacilli take over because of their higher acid tolerance. In sour whey, the most acid-tolerant and thus typical species is *Lactobacillus helveticus* which produces as much as 3% lactic acid. It is traditionally used in starters for the production of Swiss cheese and other types of hard cheeses, e.g., Grana, Gorgonzola, and Parmesan (Bottazzi et al., 1973). Nowadays, *Lactobacillus delbrueckii* subsp. *bulgaricus* and subsp. *lactis* are also used (Auclair and Accolas, 1983; Biede et al., 1976; Teuber, 2000). Artisanal starter cultures are still in use, in Italy and Switzerland, for example. In addition to enterococci, these contain, *Streptococcus thermophilus*, lactococci, *Lactobacillus fermentum*,

Lactobacillus salivarius, *Lactobacillus helveticus*, and *Lactobacillus delbrueckii* subsp. *lactis*. In all types of cheese with ripening periods longer than about 14 d, several mesophilic lactobacilli (*Lactobacillus plantarum*, *Lactobacillus brevis*, *Lactobacillus casei*, etc.) originating from the milk or the dairy environment reach levels as high as 10^6–10^8 c.f.u./g cheese (Abo-Elnaga and Kandler, 1965b; Sharpe, 1962; Van Kerken and Kandler, 1966). *Lactobacillus cypricasei* has been isolated from Halloumi cheese, a semi-hard cheese of Cyprus (Lawson et al., 2001a).

Lactobacilli that are very specifically adapted for the production of sour milks include *Lactobacillus delbrueckii* subsp. *bulgaricus* and subsp. *indicus* which are components present in the yogurt and Indian dahi flora, respectively (Davis, 1975; Dellaglio et al., 2005) and also *Lactobacillus kefiri* and *Lactobacillus parakefiri* (Kandler and Kunath, 1983; Takizawa et al., 1994) as well as *Lactobacillus kefiranofaciens* subsp. *kefiranofaciens* and subsp. *kefirgranum* (Fujisawa et al., 1988; Vancanneyt et al., 2004) which are the heterofermentative and homofermentative components, respectively, of the Caucasian sour milk kefir. These two sour milks are the only known habitats of these lactobacilli.

Several species of lactobacilli may contribute to spoilage of dairy products by slime or gas production. The formation of biogenic amines by *Lactobacillus buchneri*, for example, is of hygienic concern, and *Lactobacillus bifermentans* has been found to cause the blowing of Edam cheese (Pette and van Beynum, 1943).

Meat and meat products. Lactobacilli play an important role in fermented sausages. The most common naturally occurring species found in ripening raw sausages are *Lactobacillus sakei*, *Lactobacillus curvatus*, and with minor incidence, *Lactobacillus versmoldensis Lactobacillus plantarum*, *Lactobacillus brevis*, *Lactobacillus farciminis*, *Lactobacillus alimentarius*, *Weissella* species, pediococci, and leuconostocs (Hammes et al., 1990; Krockel et al., 2003). Because meat does not contain appreciable amounts of fermentable carbohydrates, glucose, and/or sucrose is added to the meat mixture together with spices and curing salt. The pH drops during fermentation to values ranging between 4.8 and 5.4, depending on the sausage type. Starters added to the meat mix usually contain, in addition to micrococci/staphylococci, *Lactobacillus sakei*, *Lactobacillus curvatus*, *Lactobacillus plantarum*, or pediococci (Hammes and Hertel, 1998; Lücke, 2000). Various species of lactobacilli multiply during cold storage of meat and meat products such as sausage and cooked ham, especially when packaged air tight. They may delay spoilage by proteolytic bacteria, but may also lead to spoilage by producing off-flavor, acid taste, gas, slime, or greening (Egan, 1983). In addition to leuconostocs, *Weissella* species, and carnobacteria, *Lactobacillus sakei*, *Lactobacillus curvatus*, *Lactobacillus algidus*, *Lactobacillus fucuensis*, *Lactobacillus plantarum*, and *Lactobacillus brevis* have been isolated frequently. In meat packed in oxygen-impermeable bags and stored refrigerated for 16 weeks, a strain of *Lactobacillus delbrueckii* grew to dominant numbers in the lactic acid bacterial association (Jones, 2004).

Fish and marinated fish. Lactobacilli have been found to occur in the intestines of fish (Ringo and Gatesoupe, 1998), however, they do not belong to the dominant lactic acid bacteria association. For example, they represent only 0.44% in wild brown trout (Gonzalez et al., 2000). *Lactobacillus plantarum*-like strains have been identified. In Asian regions numerous types of food are produced by fermentation of fish and shellfish.

Lactobacilli and species of all other lactic acid bacteria except carnobacteria are involved in that process (Paludan-Muller et al., 1999; Tanasupawat et al., 1998). *Lactobacillus acidipiscis* had been isolated from pla-ra and pla-chom in Thailand (Tanasupawat et al., 2000). Homo- and heterofermentative lactobacilli play an important role in the spoilage of marinated fish (Blood, 1975). In these and similar types of food, acetic acid, sometimes in combination with lactic acid, is used as an acidulans. Acetic acid is a weak acid (pK 4.75) and a traditional food preservative. It is suggested that the acetic acid added to herring provides the necessary acid environment for the action of proteinases present in the fish muscle (Meyer, 1962). The free amino acids thus liberated then provide the energy source for acetic-acid-tolerant and salt-tolerant lactobacilli which are able to decarboxylate amino acids and form biogenic amines. The CO_2 formed is the first indication of spoilage. In carbohydrate-containing marinades, the carbohydrates may be the source of CO_2 liberated by heterofermentative lactobacilli. Meyer (1956) distinguished between a "carbohydrate swell" and a "protein swell". Lactobacilli isolated from marinated herring were mainly allotted to *Lactobacillus plantarum*, *Lactobacillus brevis*, and *Lactobacillus buchneri*. However, reinvestigation of such isolates by employing modern biochemical and genomic characteristics is necessary to elucidate their true taxonomic positions. *Lactobacillus plantarum* and *Lactobacillus casei* have been identified as causatives of ropy appearance in cooked marinades (Priebe, 1970).

Mayonnaise, dressings, and salads are further examples of food preserved mainly by acetic acid. They are sensitive to spoilage by yeasts and lactic acid bacteria. *Lactobacillus plantarum*, *Lactobacillus buchneri*, *Lactobacillus brevis*, *Lactobacillus delbrueckii*, *Lactobacillus casei*, and *Lactobacillus fructivorans* were isolated from spoiled food of these types (Baumgart et al., 1983). The highest resistance to acetic acid was found for *Lactobacillus acetotolerans* (Entani et al., 1986). The organism grows even in fermenting rice vinegar broth and tolerates 4–5 % acetic acid at pH 3.5.

Humans and animals. Humans and warm-blooded animals harbor lactobacilli in the oral cavity, the intestines, and the vagina (Hammes and Hertel, 2003).

Oral cavity. Lactobacilli constitute <0.1% of cheek and tongue bacteria, <0.005% of the intragingival plaque, and <1% of the saliva and gingival crevice bacteria (Marsh and Martin, 1984). In saliva specimens of 130 school children, *Lactobacillus casei* and *Lactobacillus fermentum* were identified in 59 and 45% of samples, respectively (Rogosa et al., 1953). With lower incidence, *Lactobacillus acidophilus* (22%), *Lactobacillus brevis* (17%) and *Lactobacillus buchneri*, *Lactobacillus salivarius*, and *Lactobacillus plantarum* were found. *Lactobacillus oris* was isolated from human saliva and described as new species by (Farrow and Collins, 1988). *Lactobacillus rhamnosus* was dominant in softened and hard dentin of molars (Kneist et al., 1988).

The comparison of the saliva-*Lactobacillus* association between 12 individuals with open caries (group A) and 12 no-caries individuals (group B) revealed that of 153 isolates obtained, 82% (group A) and 90% (group B) were identified as homofermentative species (Botha et al., 1998). With decreasing incidence, *Lactobacillus paracasei* (A, 39%; B, 30%) and *Lactobacillus rhamnosus* (A, 31%; B, 41%) were identified. The heterofermentative species identified were *Lactobacillus fermentum* (A, 68%; B, 100%) and *Weissella* species.

Similar studies were performed with the direct dental carious lesions and use of culture-dependent as well as independent identification methods. In the predominant association, *Lactobacillus rhamnosus* constituted 8%, whereas in the teeth with caries, *Lactobacillus gasseri/Lactobacillus johnsonii* and *Lactobacillus rhamnosus* represented 14 and 8%, respectively, of the total association. Lactobacilli are commonly recovered from dental lesions but do not appear to be involved in the progression of dental caries (Hammes and Hertel, 2003). According to the "dental plaque hypothesis" (Marsh, 1994), carbohydrates cause the selective growth of lactic acid bacteria and especially lactobacilli at the expense of less acid-tolerant species. The low pH in carious cavities is initially the result of streptococcal metabolism which is followed in a succession by the more acid-resistant lactobacilli growing to high numbers in the cavity.

Intestinal tract of humans and animals. The anatomical differences that exist between the intestines of humans and animals strongly affect the numbers and composition of the microorganisms, especially lactobacilli. The stomach is lined with a glandular mucosa in humans, whereas in pigs, mice, and rats it is partly lined with a nonglandular, stratified squamous epithelium (Tannock, 1992). This type of epithelium is also present in the crop of birds and in the stomach of horses. Lactobacilli adhere directly to these epithelia named *pars oesophagea* in pigs and forestomach in mice and rats, and form a cell layer. Correspondingly, they colonize these epithelia in birds and horses. From these places, the lactobacilli continuously inoculate the digesta resulting in large numbers in the proximal part of the digestive tract. The *Lactobacillus* association in ruminants, especially of the rumen of adult animals, has not been investigated in depth. Lactic acid bacteria constitute a minor fraction in adult sheep and cattle (Stewart, 1992). They dominate, however, in young animals and animals fed with a high grain starch diet (Hespell et al., 1997). In these animals, lactobacilli and *Streptococcus bovis* grow up to >10^9 c.f.u./ml and decrease the pH to 4.5. These conditions can cause acidosis in the animals.

In the human stomach, the numbers of lactobacilli are low (<10^3 c.f.u./ml) and only species adapted to the acidic environment (pH 2.2–4.2) can use this ecological niche. Two obligately heterofermentative species (*Lactobacillus gastricus* and *Lactobacillus antri*) and two obligately homofermentative species (*Lactobacillus kalixensis* and *Lactobacillus ultunensis*) were isolated from biopsies of healthy volunteers (Roos et al., 2005). The numbers of bacteria increase from >10^4 c.f.u./ml in the digesta of the duodenum to 10^8–10^9 in the ileum. Lactobacilli together with enterococci dominate in the duodenum and in the jejunum. In the ileum, the composition of the microbial association approaches that of the large intestines. Most knowledge of the latter has been obtained from analysis of feces. Lactobacilli can be found in feces in numbers ranging from 0 to <10^9 c.f.u./g. They constitute only a minor fraction within the >10^{11} bacteria/g feces which make up approximately 55% of the total mass. In animals, e.g., pigs, dogs, chicken, mice, rats, and hamsters, lactobacilli represent a greater fraction of the population (Mitsuoka, 1992). Species detected in the intestinal tract of humans and animals are compiled in Table 83 which also includes also species that have been detected by culture-independent methods (Leser et al., 2002; Vaughan et al., 2002). It is assumed that numerous species are not yet culturable (Vaughan et al., 2002) and that specifically adjusted conditions are needed for

the detection of others by culturing. For example, *Lactobacillus sakei* does not grow on the commonly used acetate-rich media but requires an acetate-free medium, incubation at 30°C, and an atmosphere with 2% (Dal Bello and Hertel, 2006). Changes in taxonomy and the use of genotypic identification methods indicate that the conclusions with regard to dominant species drawn from studies between 1960 and 1980 need certain modifications. *Lactobacillus acidophilus* isolates have to be allotted mainly to *Lactobacillus gasseri* and *Lactobacillus crispatus*. *Lactobacillus fermentum* strains belong to *Lactobacillus reuteri* and the *Catenabacterium catenaforme* isolates were identified as *Lactobacillus ruminis* (Mitsuoka, 1992; Reuter, 2001). The definition of autochthonous human *Lactobacillus* species is still a matter of controversial discussion (Hammes and Hertel, 2003). As *Lactobacillus salivarius* and *Lactobacillus ruminis* strains were recovered from some individuals for at least 18 months, these species may be autochthonous (Tannock et al., 2000). However, they are not present in all individuals and, in fact, this is true for all *Lactobacillus* species. Furthermore, lactobacilli cannot be recovered at all from some individuals. Moreover, ecological studies indicate that most *Lactobacillus* species found in the human gastrointestinal tract are likely to be transient, originating from either the oral cavity (Dal Bello and Hertel, 2006) or food (Dal Bello et al., 2003; Heilig et al., 2002; Walter et al., 2001).

In accordance with the probiotic concept (Guarner and Schaafsma, 1998), lactic acid bacteria (mainly lactobacilli and bifidobacteria) are used in food, preferentially in fermented milk products and as feed additives, to achieve an improvement of health (Klaenhammer, 2001). Species used in food and having research documentation for defined strains include the following lactobacilli: *Lactobacillus acidophilus*, *Lactobacillus casei*, *Lactobacillus delbrueckii* subsp. *bulgaricus*, *Lactobacillus fermentum*, *Lactobacillus johnsonii*, *Lactobacillus paracasei*, *Lactobacillus plantarum*, *Lactobacillus reuteri*, *Lactobacillus rhamnosus*, and *Lactobacillus salivarius* (Sanders and Huis in 't Veld, 1999). Cultures for farm animals contain *Lactobacillus acidophilus*, *Lactobacillus brevis*, *Lactobacillus casei*, *Lactobacillus delbrueckii* subsp. *bulgaricus*, *Lactobacillus fermentum*, *Lactobacillus helveticus*, *Lactobacillus plantarum*, *Lactobacillus reuteri*, and *Lactobacillus rhamnosus* (Fuller, 1999). These lactobacilli usually persist during the intestinal transit but do not colonize for an extended time.

Vertebrates acquire their intestinal association at or after birth. Lactobacilli can be cultured from human feces 1–3 d after birth. The numbers detected in the following periods increase (Conway, 1997) (from weeks 1–19 to approximately 10^7 c.f.u./g feces) depending on individual factors and on the diet (Mitsuoka, 1992). The food may contain lactobacilli that persist on the intestinal transit and might be recovered from feces. This may at least partially explain the fluctuation of the *Lactobacillus* association observed in individuals of all ages.

Vaginal lactobacilli. Similar to intestinal lactobacilli, the presence of lactobacilli in the vagina is subject to great variation with regard to the numbers as well as species and strains. Healthy subjects may contain none or virtually 100% lactobacilli in their vaginal association (Hyman et al., 2005). Lactobacilli are believed to maintain low pH at values ranging throughout the menstrual cycle between 3.5 and 4.5, and to exert a protective effect against colonization of undesired micro-organisms (Hill et al., 1985). This concept may be modified because certain subjects do not harbor lactobacilli and because the vaginal

TABLE 83. *Lactobacillus* species in intestines of human and animal[a,b]

Species	Human	Pig	Birds	Cattle	Dog	Mouse	Rat	Hamster	Horse	Jaguar[c]
L. delbrueckii	+	+				+	+			
L. acidophilus	+	+	+	+		+	+			
L. agilis		+	+						+	
L. amylovorus		M		+					+	
L. antri	+									
L. aviarius			+							
L. brevis	+	+	+	+						
L. casei	+									
L. crispatus	M	+	M						+	
L. curvatus	+									
L. equi									M	
L. fermentum	+	+	+	+		+	+			
L. fructivorans	+									
L. gallinarum		M								
L. gasseri	M			+		+				
L. gastricus	+									
L. hamsteri								M		
L. ingluviei			+							
L. intestinalis						M	M			
L. johnsonii	+	+	M	M					+	
L. kalixensis	+									
L. kitasatonis			+							
L. mucosae		+								
L. murinus/animalis			+		M	M	+			
L. oris	+									
L. panis		+								
L. pantheris										+
L. paracasei	+			+						
L. paraplantarum	+									
L. plantarum	+	+		+					+	
L. pontis		+								
L. reuteri	M	M	M	M	M	M	M	M	M	
L. rhamnosus	+									
L. ruminis	M	+		M						
L. saerimneri		+								
L. sakei	+									
L. salivarius	M	M	M						M	
L. sharpae		+								
L. thermotolerans			+							
L. ultunensis	+									
L. vaginalis	+	+								

[a]Symbols: M, major component of *Lactobacillus* species; +, occasionally recovered.
[b]Modified from Hammes and Hertel, (2003). References include Mitsuoka, 1992; Tannock, 1992; Walter et al., (2001), Tannock et al., 2000; Roos et al., 2000; Gusils et al., 1999; Rubio et al., 1998; Marounek et al., 1988; Stewart, 1992; Serra et al., (1992), Walter et al., 2001; Yuki et al., 2000; Morotomi et al., 2002; Liu and Dong, 2002; Vaughan et al., 2002; Leser et al., 2002; Dal Bello et al., 2003; Mukai et al., 2003; Niamsup et al., 2003; Roos et al., (2005).
[c]The species composition has not been studied so far.

secretion is already acidic at birth, when the vagina is still sterile (Hammes and Hertel, 2003; Redondo-Lopez et al., 1990). Early descriptions of vaginal lactobacilli suffer from the inaccuracy of the identification methods of their time (Döderlein, 1894; Lenzner, 1966; Rogosa and Sharpe, 1960; Wylie and Henderson, 1969). At later times, species were isolated from the female urogenital tract that can be considered to be adapted to this environment. *Lactobacillus jensenii* was isolated from vaginal discharge (Gasser et al., 1970) and belongs together with *Lactobacillus gasseri* and *Lactobacillus crispatus* to the most often detected lactobacilli. *Lactobacillus gasseri* is a predominantly detected organism which is also part of the intestinal association in humans and animals, and occurs also in fermented food (Lauer and Kandler, 1980b). Similarily, *Lactobacillus crispatus* (Cato et al.,

1983) is commonly associated with humans, animals (Table 83), and fermented food.

Adaption comparable to *Lactobacillus jensenii* can be attributed to *Lactobacillus vaginalis* which was originally isolated from patients suffering from trichomoniasis (Embley et al., 1989), *Lactobacillus fornicalis* isolated from the posterior fornix of a healthy woman attending pre- and postnatal clinics (Dicks et al., 2000), *Lactobacillus iners* isolated from the urine and vaginal discharge of adult women (Falsen et al., 1999), and *Lactobacillus coleohominis*, isolated from the vaginas of young women (Nikolaitchouk et al., 2001). Extensive studies of vaginal lactobacilli have been performed using phenotypic as well as genotypic methods (Antonio et al., 1999; Hyman et al., 2005; Vasquez et al., 2002; Zhou et al., 2004), showing or confirming that

Lactobacillus crispatus, Lactobacillus gasseri, Lactobacillus jensenii, and *Lactobacillus iners* are the most commonly occurring species. Numerous additional species were recovered from the human body association (*Lactobacillus reuteri, Lactobacillus ruminis, Lactobacillus oris, Lactobacillus vaginalis*) (Antonio et al., 1999; Burton et al., 2003) and mainly from food association (*Lactobacillus rhamnosus, Lactobacillus paracasei, Lactobacillus fermentum* and *Lactobacillus plantarum*). Solely using gene-targeted methods, rather unexpected, tightly food-associated bacteria have been identified, such as *Lactobacillus kefiranofaciens, Lactobacillus delbrueckii* subsp. *delbrueckii,* subsp. *lactis,* and subsp. *bulgaricus.* Remarkably, in one subject 20 contigs allotted to *Lactobacillus crispatus* were identified showing that they are far from clonal. Lactobacilli with defined properties have been used therapeutically in the treatment of bacterial vaginosis (McLean and Rosenstein, 2000; Reid and Heinemann, 1999) with the aim of restoring the normal *Lactobacillus* association.

Sewage and manure.   Sewage and manure are secondary habitats of all lactobacilli found in the intestine or other primary habitats in more or less direct contact. *Lactobacillus coryniformis* and *Lactobacillus curvatus* are frequently found in manure (Abo-Elnaga and Kandler, 1965b). *Lactobacillus vaccinostercus* has only been found in cow dung (Okada et al., 1979).

In municipal sewage, levels of 10^4–10^5 lactobacilli/ml have been found (Weiss et al., 1981). The heterofermentative strains (approx. 25% of the isolates) have been classified as *Lactobacillus fermentum, Lactobacillus reuteri,* and *Lactobacillus brevis* and, to a lesser extent, as *Lactobacillus confusus.* The homofermentative strains (approx. 75% of the isolates) belonged to a great number of species. However, about 10% of the strains could not be allotted to any of the known species. They have been described as representatives of the two new species *Lactobacillus sharpeae* and *Lactobacillus agilis.*

Enrichment and isolation procedures

Procedures for the isolation of lactobacilli must take into account their aciduric or acidophilic nature, their complex nutritional requirements, and their preference for microaerophilic conditions. When lactobacilli are the predominant flora in the source material, the rather nonselective MRS agar (De Man et al., 1960) or the somewhat similar APT agar (Evans and Niven, 1951) may be used for isolation. APT agar and sorbic acid agar (Reuter, 1968) are commonly used for isolating lactobacilli from meat products. When lactobacilli occur as part of a complex population, selective media are required. Most lactobacilli from many different sources have been successfully isolated on the widely used acetate medium (SL) of Rogosa et al., (1951). SL medium is not completely selective for lactobacilli as other lactic acid bacteria, e.g., leuconostocs, pediococci, enterococci, *Weissella* species, bifidobacteria, and yeasts may also grow. Thus, colonies may have to be further examined. Yeasts can be eliminated by the addition of cycloheximide at a concentration of 100 mg/l. Intestinal lactobacilli are commonly isolated on MRS, SL, or LAMVAP agar (Hartemink et al., 1997). For the isolation of anaerobic lactobacilli from intestinal sources, 0.05% (w/v) cysteine should be added, and it may be necessary to pre-reduce poured, dried plates by overnight incubation in an anaerobic jar.

Lactobacillus iners (Falsen et al., 1999), which was isolated from the human vagina, does not grow on MRS or SL agar but needs to be cultured on blood agar. Depending on the source of isolation, other lactobacilli, mainly from quite specialized environments, require minor modifications of SL medium such as supplementing with more or less specific growth factors, e.g., meat extract, tomato juice, fresh yeast extract, malt extract, ethanol, mevalonic acid, or even some of the natural substrate (beer, fruit juices, cheese whey). Replacement of glucose, either completely or partially, by other carbohydrates such as maltose, fructose, sucrose, or arabinose is recommended in some cases, especially where heterofermentative lactobacilli play an important role. An example is Homohiochii Medium (Kleynmans et al., 1989). For the detection of beer-spoiling bacteria including nutritionally fastidious lactobacilli, a special medium (NBB medium) has been described by Back (1980). Several other media are also in use, which have been reviewed by Sakamoto and Konings (2003). An extensive literature survey of the use of culture media for lactic acid bacteria, including lactobacilli, was given by Carr et al. (2002).

Because most lactobacilli generally grow better either anaerobically or in the presence of increased CO_2 tension, agar plates should be incubated in jars that have been evacuated and filled with 90% N_2 or H_2 + 10% CO_2 or in anaerobic jars (BBL, Oxoid) using H_2 + CO_2 generating kits.

Maintenance procedures

For short-term preservation, cultures are preferably inoculated into MRS or optimal medium agar stabs, incubated until growth becomes visible, stored at 4–7°C, and transferred monthly. Some species or strains, however, die out quite rapidly within a series of transfers. Alternatively, cultures grown to the early stationary growth phase may be deep frozen in the growth medium and stored at –20°C for several months. The cultures can be kept for years by adding 20% glycerol and storing at –80°C in screw caps.

The method of choice for long-term preservation is lyophilization. Cells grown to the late exponential growth phase are collected by centrifugation, resuspended in sterile skim milk or horse serum containing 7.5% (w/v) glucose, and lyophilized. Ampules are sealed under vacuum and stored at 5–8°C. Most strains preserved by this method are still viable after 10–20 years, although some require more frequent relyophilization. Strains may also be kept for long periods (over 30 years) in liquid nitrogen.

Procedures for testing special characters

The development and increasingly common use of genotypic identification methods (see below) has led progressively to a reduction of the physiological and biochemical characterization of newly described species. An example is *Lactobacillus suntoreyus* whose description included the basic properties, i.e., morphology, culture conditions, motility, spore formation potential, and was furthermore limited by the carbohydrate fermentation pattern, sequencing of the 16S rDNA, and DNA/DNA hybridization. Formerly included properties (Kandler and Weiss, 1986; Vandamme et al., 1996) such as configuration of lactate, G+C content, cell wall composition (peptidoglycan type, teichoic acids), electrophoretic pattern of lactate dehydrogenases, and cellular fatty acids are thus no longer known. The peptidoglycan type is a especially reliable criterion for the identification of certain species, e.g., *Lactobacillus sanfranciscensis* (Lys–Ala) and *Lactobacillus rossiae* (Lys–Ser–Ala$_2$). It is also of great value when obligately heterofermentative lactobacilli that can be confused with weissellas have to be identified. The interpeptide bridge of the latter genus is (Lys– mono-amino mono-carboxylic acids).

Carbohydrate fermentation. MRS broth without meat extract and glucose and with 0.05% (w/v) chlorphenol red is generally used as basal medium. Filter-sterilized solutions of the test carbohydrates are added to a final concentration of 1%. Tests are incubated at the optimum growth temperature and results recorded up to 7d. In a few cases, e.g., for some strains of *Lactobacillus delbrueckii*, the addition of 0.2% meat extract broadens the pattern of fermented carbohydrates somewhat, and the fermentation of glucose is distinctly improved. For strains that will not grow reasonably in MRS broth, the optimal growth medium should be used as basal medium. The use of this procedure delivers clear results with regard to the growth and fermentation potential, i.e., whether or not the fermented carbohydrate supports the growth of the strain. However, as it is a time and labor consuming procedure, the use of the API system is presently the standard. As discussed previously, the results obtained are not always consistent with growth of the organisms on the fermentable carbohydrate.

Lactic acid configuration. The amount of the isomers of lactic acid produced is best determined enzymically using D-lactate (Gawehn and Bergmeyer, 1974) and L-lactate dehydrogenase (Gutmann and Wahlefeld, 1974). Corrections must be made for the lactic acid content of the medium before inoculation. Care must be taken to analyze cultures after they have reached the stationary growth phase, since some DL-lactate-formers produce predominantly L(+)-lactic acid or, in a few cases, D(−)-lactic acid during the early growth phase.

Differentiation from closely related taxa

Lactobacilli are metabolically very similar to the other genera of lactic acid bacteria. Only their rod shape readily distinguishes them from the coccal genera. Strains of *Lactoococcus* that form atypically elongated cells may also be confused with coccoid rods of lactobacilli. Here, differentiation may require genotypic identification as in the case of *Lactobacillus xylosus* and "*Lactobacillus hordniae*", both of which have been shown to belong to the genus *Lactococcus* (Garvie et al., 1981; Kilpper-Bälz et al., 1982).

The rod-shaped bifidobacteria, which until the eighth edition of *Bergey's Manual* had long been included in the genus *Lactobacillus* as "*Lactobacillus bifidus*", may be differentiated from lactobacilli on the basis of their characteristic hexose fermentation pathway which yields lactic acid and acetic acid at a molar ratio of 2:3, but no CO_2, instead of lactic acid, acetic acid (or ethanol), and CO_2 at a molar ratio of 1:1:1 which is the pattern of fermentation products typical of obligately heterofermentative lactobacilli. Carnobacteria can be differentiated from lactobacilli by their reduced acid tolerance and their potential to grow at pH values up to 9.5. The differentiation of lactobacilli from weissellas is problematic, but possible by determination of the peptidoglycan type (see above). The differentiation from *Paralactobacillus* is best made by using genotypic identification methods.

Genotypic identification methods. For identification of lactobacilli, genotypic methods are increasingly applied and the polyphasic approach is considered to provide the most reliable identification results. The classical genotypic methods are based on the determination of DNA base composition (mol% G+C)

and DNA–DNA similarity. The whole genome DNA–DNA similarity study is still an important criterion in the current species concept (Rosselló-Mora and Amann, 2001). Its usefulness for delineation of *Lactobacillus* species is reviewed by Vandamme et al. (1996). However, this approach was mainly employed to identify new species or to clarify relationships among existing species. It plays only a minor role in rapid identification of unknown isolates. For this purpose, comparative analysis of 16S and/or 23S rRNA gene sequences can be used, but it should be taken into account that several *Lactobacillus* species show very high 16S rRNA sequence similarity, e.g., *Lactobacillus plantarum*, *Lactobacillus pentosus*, and *Lactobacillus paraplantarum* (99.7–99.9%), *Lactobacillus kimchii* and *Lactobacillus paralimentarius* (99.9%), *Lactobacillus mindensis* and *Lactobacillus farciminis* (99.9%), and *Lactobacillus animalis* and *Lactobacillus murinus* (99.7%). In such cases multilocus sequence analysis (MLSA) of conserved protein-coding loci may help to overcome this limitation. Recently, the applicability of this technique to identifying species of lactic acid bacteria was demonstrated for the genus *Enterococcus* (Naser et al., 2005). Nevertheless, the analysis of 16S rRNA sequences (especially of the first 900 bases) is still a fast tool that can be used to gain insight into the taxonomic position of an unknown *Lactobacillus* isolate.

Comparative analysis of rRNA sequences reveals evolutionarily less conserved regions that are diagnostic for species, genus, or groups of phylogenetically related organisms (Amann et al., 1995; Schleifer et al., 1993). These regions serve as target sites for specific probes and primers (Amann and Ludwig, 2000; Coeuret et al., 2003; Satokari et al., 2003) which can be used in combination with a variety of hybridization and PCR techniques, e.g., *in situ* colony hybridization (Brockmann et al., 1996; Hertel et al., 1992), reverse dot blot hybridization (Ehrmann et al., 1994), fluorescent *in situ* hybridization (Blasco et al., 2003; Lick et al., 2000; Matte-Tailliez et al., 2001), species-specific PCR (Bunte et al., 2000; Dickson et al., 2005), and multiplex PCR (Muller et al., 2000; Settanni et al., 2005; Song et al., 2000; Yost and Nattress, 2000). In addition, the DNA spacer sequences between the rRNA genes of lactobacilli are highly variable but sufficiently conserved to construct species-specific PCR primers (Berthier and Ehrlich, 1998; Chen et al., 2000; Tannock et al., 1999; Tilsala-Timisjarvi and Alatossava, 1997). It was shown that 16S–23S rRNA gene intergenic spacer region sequence comparison can especially be useful in identifying closely related species, e.g., *Lactobacillus curvatus*, *Lactobacillus sakei*, and *Lactobacillus graminis* as well as *Lactobacillus paraplantarum*, *Lactobacillus plantarum*, and *Lactobacillus pentosus* (Berthier and Ehrlich, 1998). The amplified rDNA restriction analysis (ARDRA), which is mostly based on the restriction length polymorphism of 16S rRNA gene fragments amplified by PCR, is also a useful tool for species identification (Bouton et al., 2002; Giraffa et al., 1998; Moschetti et al., 1997; Roy et al., 2000; Ventura et al., 2000).

Methods have been developed for culture-independent analysis of the diversity of complex microbial communities. Denaturing gradient gel electrophoresis (DGGE) of 16S rDNA amplicons has been demonstrated to be a suitable tool to identify *Lactobacillus* species in and to monitor the diversity and dynamics of a microbiota in which lactobacilli belong to the dominating organisms (ben Omar and Ampe, 2000; Cocolin et al., 2001; Ercolini et al., 2001; Meroth et al., 2004; Meroth et al.,

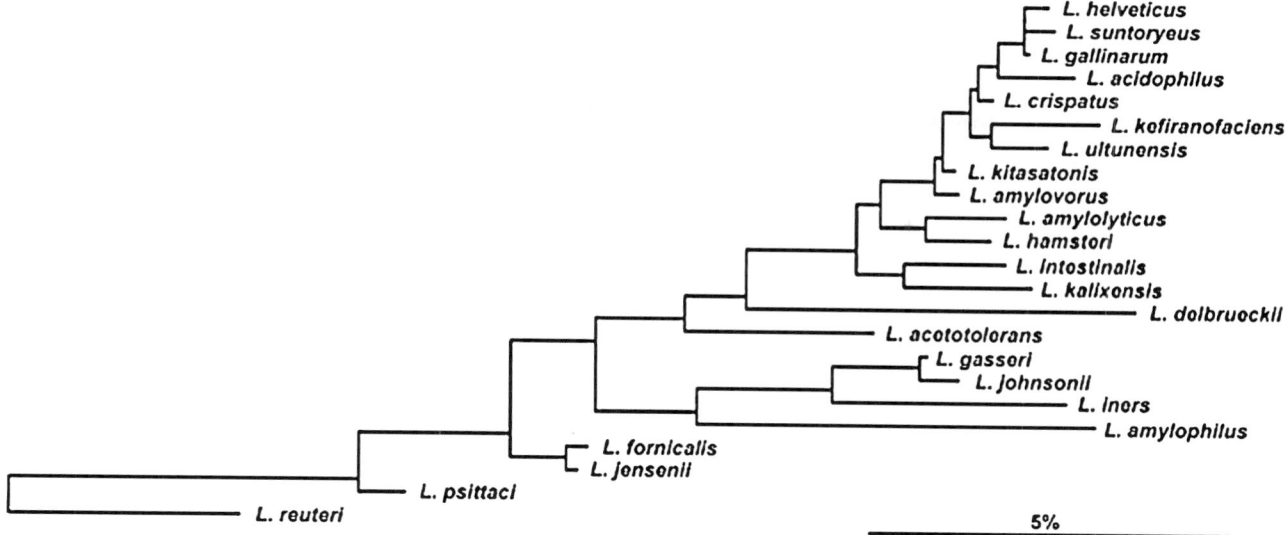

FIGURE 90. Maximum likelihood tree reflecting the relationship among members of the *Lactobacillus delbrueckii* group. The tree is based on analyses of all available at least 90% complete 16S rRNA sequences of *Lactobacillaceae*. Alignment positions that share identical residues in at least 50% of all sequences of the genera *Lactobacillus* and *Pediococcus* were considered. The bars indicate 5% estimated sequence divergence.

2003; Randazzo et al., 2002). For detection of *Lactobacillus* species in a microbiota in which they represent only a minor part of the population (e.g., the intestinal microbiota), the DGGE has to be combined with a *Lactobacillus* group-specific PCR (Heilig et al., 2002; Walter et al., 2001). The usefulness of PCR-DGGE was demonstrated by the identification of *Lactobacillus iners* as the most common *Lactobacillus* species in the vaginal bacterial microbiota (Burton et al., 2003).The important role was difficult to detect because this species does not grow on the common selective media used for isolation of lactobacilli (Falsen et al., 1999).

Taxonomic comments

The species of the genera *Lactobacillus* belong together with the genera *Paralactobacillus* and *Pediococcus* to the family *Lactobacillaceae* which is a member of the order "*Lactobacillales*", one of the major phylogenetic groups of the *Firmicutes*. Although the pediococci are highly related to the lactobacilli, they are treated in separate chapters. Using 16S rRNA gene sequence analysis, the phylogeny of lactic acid bacteria has been studied extensively in the past (reviewed by Hammes and Hertel, 2003). It has become evident that only little correlation exists between the traditional classification and the phylogenetic relatedness of lactic acid bacteria. As a result of these studies, it has been proposed (Hammes and Hertel, 2003) to subdivide the genus *Lactobacillus* into seven phylogenetic groups (*Lactobacillus buchneri* group, *Lactobacillus casei* group, *Lactobacillus delbrueckii* group, *Lactobacillus plantarum* group, *Lactobacillus reuteri* group, *Lactobacillus sakei* group, and *Lactobacillus salivarius* group). Following this concept of grouping, the phylogenetic relatedness of lactobacilli was reinvestigated by using all available 16S rRNA gene sequences deposited in public databases (C. Hertel, unpublished results). Various datasets differing with respect to the selection of sequences and sequence positions were used for phylogenetic analyzes applying three methods, namely, distance matrix, maximum-parsimony, and maximum-likelihood

(Ludwig and Klenk, 2001). The analyzes revealed that just three distinct phylogenetic groups can be defined. These are depicted in Figure 90, Figure 91, and Figure 92 as *Lactobacillus delbrueckii* group, *Lactobacillus reuteri* group, and *Lactobacillus salivarius* group, respectively. The definition of further groups of lactobacilli was no longer possible as the discriminatory power of the 16S rRNA gene sequences is limited, and the large increase of new 16S rRNA gene sequences in recent years abolished formerly recognized phylogenetic lines of relationship. The newly described *Lactobacillus* species branch deeply within the previously defined seven phylogenetic groups. For example, in the case of *Lactobacillus algidus*, the positioning varies with the sequence positions selected and the methods used for tree construction. In just a few cases of tree construction, the species could be allotted to the *Lactobacillus salivarius* group as shown in Figure 92. Thus, a more reliable insight into the evolutionary relationships of the numerous *Lactobacillus* species must still be obtained by applying novel approaches that are based on comparative analysis of conserved macromolecules or even whole-genome sequences (Coenye et al., 2005).

The *Lactobacillus delbrueckii* group (Figure 90) is consistent with the formerly described *Lactobacillus delbrueckii* group (Collins et al., 1991; Hammes and Hertel, 2003; Schleifer and Ludwig, 1995a, b). As *Lactobacillus delbrueckii* is the type species of the genus *Lactobacillus*, we use the designation *Lactobacillus delbrueckii* group instead of *Lactobacillus acidophilus* group used by Schleifer and Ludwig (1995b). It contains mainly obligate homofermenters but also some facultative heterofermenters. The G+C content within this group ranges rather widely from 34 up to 51 mol% which may be explained by changes in the codon usages originating from the degeneracy of the genetic code (Schleifer and Ludwig, 1995b). The peptidoglycan type of all species in this group is of the Lys–D-Asp type (Schleifer and Kandler, 1972). *Lactobacillus delbrueckii* contains four subspecies and *Lactobacillus kefiranofaciens* two subspecies that cannot be reliably differentiated by rRNA sequence analysis.

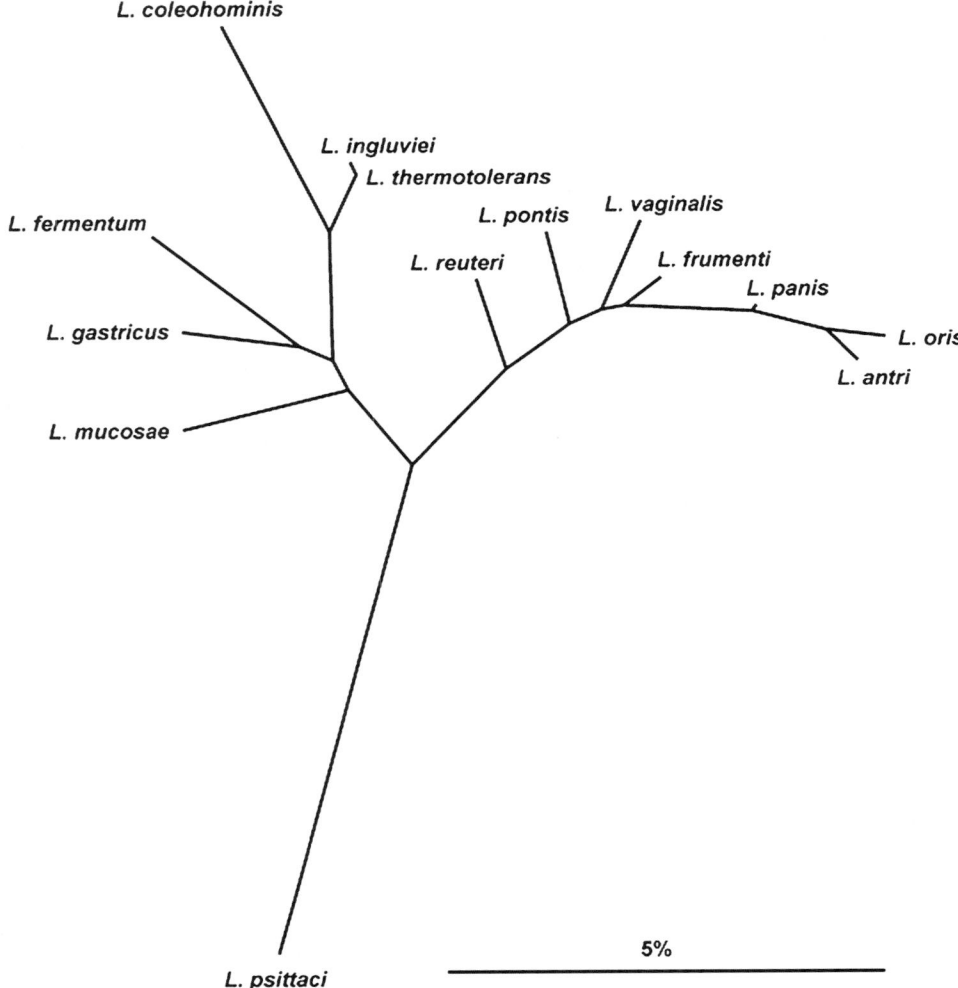

FIGURE 91. Maximum likelihood tree reflecting the relationship among members of the *Lactobacillus reuteri* group. The tree is based on analyses of all available at least 90% complete 16S rRNA sequences of *Lactobacillaceae*. Alignment positions that share identical residues in at least 50% of all sequences of the genera *Lactobacillus* and *Pediococcus* were considered. The bars indicate 5% estimated sequence divergence.

The *Lactobacillus reuteri* group (Figure 91) exclusively contains obligate heterofermenters, and its members show again a wide range in the G+C content of their DNA (38–54 mol%). In contrast to the previously defined *Lactobacillus reuteri* group (Hammes and Hertel, 2003), the species *Lactobacillus durianis*, *Lactobacillus rossiae*, *Lactobacillus suebicus*, and *Lactobacillus vaccinostercus* could no longer be allotted to the *Lactobacillus reuteri* group. The split into two evolutionary lines containing as prominent species *Lactobacillus reuteri* and *Lactobacillus fermentum*, respectively, was evident in most of the phylogenetic analyzes. The peptidoglycan types in this group are Lys–D-Asp, Orn–D-Asp.

The *Lactobacillus salivarius* group (Figure 92) contains obligate homofermenters and facultative heterofermenters, and again the G+C content ranges widely within the species (32–44%). The peptidoglycan types occurring in this group are Lys–D-Asp and *meso*-Dpm-direct. Certain species show high 16S rRNA sequence similarities, e.g., *Lactobacillus animalis* and *Lactobacillus murinus* (99.7%), *Lactobacillus cypricasei* and *Lactobacillus acidipiscis* (99.7%). A similar situation can also be found among other species of the genus *Lactobacillus*, e.g., in *Lactobacillus*

plantarum, *Lactobacillus pentosus*, and *Lactobacillus paraplantarum* (99.7–99.9%). The species *Lactobacillus aviarius* and *Lactobacillus salivarius* each contain two subspecies which cannot be differentiated by rRNA sequence analysis.

The history of the genus *Lactobacillus* starts with the proposal of a new genus by Beijerinck (1901). In the following years, new species were described which increased to the present number of 96 recognized species. Their taxonomic treatment was and still is a challenge for taxonomists, and their task is by far not finished. The taxonomic tools and the availability of genomes of lactic acid bacteria (reviewed by Klaenhammer et al., 2005) will provide a basis for a more conclusive classification. Several scientists contributed to the present status of *Lactobacillus* taxonomy (Kandler and Weiss, 1986; Orla-Jensen, 1919, 1942, 1943; Rogosa, 1970, 1974; Sharpe, 1979). Their bases for grouping the large number of species were and still are the characteristics of morphology, fermentation of carbohydrates, nutritional requirements, growth dependence on temperature, agglutination properties, and the potential for performing homo- or heterofermentation. Kandler and Weiss

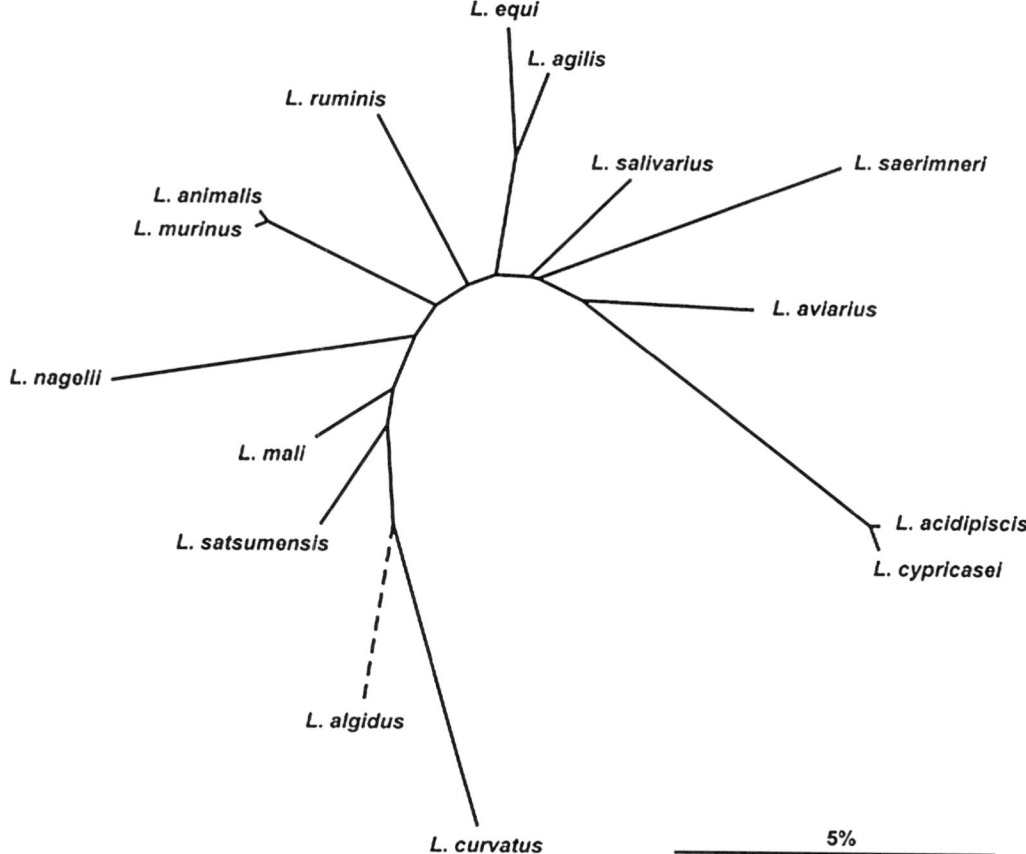

FIGURE 92. Maximum likelihood tree reflecting the relationship among members of the *Lactobacillus salivarius* group. The tree is based on analyses of all available at least 90% complete 16S rRNA sequences of *Lactobacillaceae*. Alignment positions that share identical residues in at least 50% of all sequences of the genera *Lactobacillus* and *Pediococcus* were considered. The positioning of *Lactobacillus acidipiscis* is based on partial sequence data and may be subject to changes. *Lactobacillus algidus* is included into this tree, although its positioning depends on the method applied for tree reconstruction. The bars indicate 5% estimated sequence divergence.

(1986) have already recognized that their scheme of allotting the species to groups (I-III) is not supported by phylogenetic data. We follow their way of grouping, which no longer includes the formerly used criteria of morphology and the range of growth temperature, but rests solely on the fermentation types. It was recognized that a substantial number of species allotted by Kandler and Weiss (1986) to the obligately homofermentative group (I) bacteria had to be transferred to the new group of facultatively heterofermentative species (B). This change is remarkable as it especially concerns species of the *Lactobacillus delbrueckii* and the *Lactobacillus salivarius* groups, which are now found in two groups (A and B). The allotment of the species to groups named A, B, and C is indicated in Table 82, and the differentiation characteristics for *Lactobacillus* species shown Table 84, Table 85, and Table 86 mainly rest on that grouping. The groups are defined as follows:

Group A, obligately homofermentative lactobacilli: Hexoses are fermented almost exclusively to lactic acid by the Embden–Meyerhof pathway; pentoses or gluconate are not fermented.

It can be assumed that for certain strains of group A species, the formation of acid from pentoses will be discovered, as occurred recently. In the hands of Kandler and Weiss (1986), a number of species, which are now placed in group B, did not grow on pentoses, despite a reported potential to ferment pentoses. As has been discussed previously in the metabolism section (see above), the formation of acid from a carbohydrate does not necessarily mean that the organism can grow on it. Thus, there is clearly a need for further studies of the metabolic potential of these organisms.

Group B, facultatively heterofermentative lactobacilli: Hexoses are fermented almost exclusively to lactic acid by the Embden-Meyerhof pathway or, at least by some species, to lactic acid, acetic acid, ethanol, and formic acid under glucose limitation; pentoses are fermented to lactic acid and acetic acid via an inducible phosphoketolase.

Group C, obligately heterofermentative lactobacilli: Hexoses are fermented to lactic acid, acetic acid (ethanol), and CO_2 via the phosphogluconate pathway. Pentoses are fermented to lactic acid and acetic acid by the related pentose phosphate pathway.

In general, these pathways involve phosphoketolase. *Lactobacillus bifermentans* probably possesses other pathways for carbohydrate breakdown leading to production of gas from hexoses and has therefore tentatively also been included in group C. *Lactobacillus bifermentans* ferments glucose homofermentatively to DL-lactic acid, but, depending on pH, the lactic acid formed

TABLE 84. Key characteristics of Group A lactobacilli (obligately homofermentative)[a]

Species	L. delbrueckii subsp. delbrueckii	L. delbrueckii subsp. bulgaricus	L. delbrueckii subsp. lactis	L. delbrueckii subsp. indicus	L. acidophilus	L. amylolyticus	L. amylophilus	L. amylovorus	L. aviarius subsp. aviarius	L. aviarius subsp. araffinosus	L. crispatus	L. farciminis	L. gallinarum	L. gasseri	L. helveticus
Phylogenetic group	de	de	de	de	de	de	de	de	sl	sl	de	u	de	de	de
Peptidoglycan type	Lys–D-Asp	Lys–D-Asp	Lys–D-Asp	Lys–D-Asp	Lys–D-Asp	Lys–D-Asp	Lys–D-Asp	Lys–D-Asp	Lys–D-Asp	Lys–D-Asp	Lys–D-Asp	Lys–D-Asp	Lys–D-Asp	Lys–D-Asp	Lys–D-Asp
G+C content (mol%)	49–51	49–51	49–51	49–51	34–37	39	44–46	40.3	39–43	41.3	35–38	34–36	36–37	33–35	37
Lactic acid isomer(s)	d	d	d	d	dl	dl	l	dl	dl	l(d)	dl	l(d)	dl	dl	dl
Growth (°C) 15/45	−/+	−/+	−/+	−/+	−/+	−/+	+/−	−/+	−/ND	−/ND	−/+	+/−	+/+	−/+	−/+
Carbohydrates fermented:															
Amygdalin	−	−	+	−	+	d	−	+	d	d	+	+	+	+	−
Cellobiose	−	d	d	−	+	−	−	+	+	d	+	+	+	+	−
Galactose	−	−	d	−	+	+	+	+	d	−	+	+	+	+	+
Lactose	−	+	+	+	+	−	−	−	d	−	+	+	d	d	+
Maltose	d	−	+	−	+	+	+	+	+	+	+	+	+	d	d
Mannitol	−	−	−	−	−	−	−	−	−	−	−	−	−	−	−
Mannose	+	−	+	+	+	+	+	+	+	+	+	+	+	+	d
Melibiose	−	−	−	−	d	d	−	−	d	−	−	−	+	d	−
Raffinose	−	−	−	−	d	d	−	−	+	−	−	−	+	d	−
Salicin	−	−	+	−	+	d	−	+	+	d	+	+	+	+	−
Sucrose	+	−	+	d	+	+	−	+	+	+	+	−	+	+	+
Trehalose	d	−	+	−	d	−	−	+	+	+	−	+	−	d	d

[a]Symbols and abbreviations: +, 90% or more of strains are positive; −, 90% or more are negative; d, 11–89% of strains are positive; w, weak positive reaction; ND, no data available; (), isomers in parentheses indicate <15% of total lactic acid; mDpm, *meso*-diaminopimelic acid; de, *Lactobacillus delbrueckii*-group; sl, *Lactobacillus salivarius*-group; re, *Lactobacillus reuteri*-group; u, unique.

[b]W. P. Hammes, unpublished results.

	L. iners	*L. johnsonii*	*L. kalixensis*	*L. kefiranofaciens* subsp. *kefiranofaciens*	*L. kefiranofaciens* subsp. *kefirgranum*	*L. kitasatonis*	*L. mali*	*L. manihotivorans*	*L. mindensis*	*L. nagelii*	*L. pantheris*	*L. psittaci*	*L. ruminis*	*L. saerimneri*	*L. salivarius* subsp. *salivarius*	*L. salivarius* subsp. *salicinius*	*L. satsumensis*	*L. sharpeae*	*L. suntoryeus*	*L. ultunensis*
	de	de	de	de	de	de	sl	u	u	sl	u	de	sl	sl	sl	sl	sl	u	de	de
	Lys–D-Asp	Lys–D-Asp	Lys–D-Asp	Lys–D-Asp[b]	Lys–D-Asp	ND	mDpm	Lys–D-Asp	Lys–D-Asp	mDpm[b]	mDpm[b]	Lys–D-Asp[b]	mDpm	mDpm	Lys–D-Asp	Lys–D-Asp	mDpm	mDpm	ND	Lys–D-Asp
	34.4	33–35	35.5	34–39	34–39	37–40	32.5	48.4	37.5	ND	52.7	ND	44–47	42.9	34–36	34–36	39–41	53	ND	35.7
	l	dl	dl	d(l)	d(l)	d/l	l	l	l	dl	d	nd	l	dl	l	l	l	l	nd	dl
	–/ND	+/+	–/+	–/–	w/–	–/+	+/d	+	+/–	+/+	+/–	ND	–/d	w/+	–/+	–/+	+/+	+/–	–/+	–/–
	ND	+	+	–	–	–	+	+	d	+	–	ND	+	–	–	–	ND	+	+	–
	ND	+	+	–	d	d	d	+	+	+	+	ND	+	–	–	–	–	+	+	+
	ND	+	+	+	+	+	d	+	–	+	+	ND	+	–	+	+	d	+	d	+
	–	d	+	+	+	ND	d	+	–	–	+	–	d	–	+	+	–	+	ND	+
	d	+	+	+	+	+	d	+	+	+	+	ND	+	–	+	+	ND	+	+	+
	–	–	d	–	–	ND	–	–	–	+	–	–	–	–	+	+	+	–	–	–
	ND	+	+	ND	+	+	+	+	+	+	+	ND	+	+	–	–	+	–	+	+
	–	d	+	+	+	–	d	+	–	–	–	–	+	–	+	+	–	–	ND	d/w
	–	d	+	+	+	–	d	+	–	–	–	+	+	–	+	+	–	–	ND	d
	ND	+	+	–	d	d	+	+	+	+	+	ND	+	–	–	+	+	+	+	+
	–	+	+	+	d	+	+	+	–	+	–	+	+	+	+	+	+	–	+	+
	–	d	+	–	d	–	+	+	–	–	+	ND	–	+	+	+	+	–	ND	+

TABLE 85. Key characteristics of Group B lactobacilli (facultatively heterofermentative)[a]

Species	L. acetotolerans	L. acidipiscis	L. agilis	L. algidus	L. alimentarius	L. animalis	L. bifermentans	L. casei	L. coryniformis subsp. coryniformis	L. coryniformis subsp. torquens	L. curvatus	L. cypricasei	L. equi	L. formicalis	L. fuchuensis	L. graminis
Phylogenetic group	de	sl	sl	sl	u	sl	u	u	u	u	u	sl	sl	de	u	u
Peptidoglycan type	Lys–D–Asp	Lys–D–Asp	mDpm	mDpm	Lys–D–Asp	Lys–D–Asp	Lys–D–Asp	Lys–D–Asp	Lys–D–Asp	Lys–D–Asp	Lys–D–Asp	Lys–D–Asp[b]	Lys–D–Asp[b]	ND	Lys–D–Asp[b]	Lys–D–Asp
G+C content (mol%)	35–36.5	39–42	43–44	36.8 ± 3	36–37	41–44	45	45–47	45	45	42–44	ND	38–40	37	41–41.7	41–43
Lactic acid isomer(s)	DL	L	L	L	L(D)	L	DL	L	D(L)	D	DL	ND	DL	DL	L(D)	DL
Growth (°C) 15/45	–/–	–/–	–/+	+/–	+/–	–/+	+/–	+/–	+/–	+/–	+/–	–/–	–/+	–/–	+/–	+/–
Carbohydrates fermented:																
Amygdalin	–	d	+	d	ND	d	–	+	–	–	d	ND	–	+	+	+
Arabinose	–	d	–	+	d	d	–	–	–	–	–	d	d	–	–	–
Cellobiose	d	–	+	d	+	+	–	+	–	–	+	+	–	+	+	+
Esculin	+	–	+	+	+	+	–	+	d	–	+	+	d	+	+	+
Gluconate	–	–	–	–	+	–	–	+	+	+	–	ND	–	ND	+	–
Mannitol	d	d	+	–	–	–	+	+	+	+	–	–	+	+	–	–
Melezitose	–	–	+	–	–	–	–	+	–	–	–	–	+	–	–	–
Melibiose	–	–	+	d	–	+	–	–	d	–	–	–	+	–	–	–
Raffinose	–	–	+	d	–	d	–	–	d	–	–	–	+	–	–	–
Ribose	d	+	+	+	+	d(w)	+	+	–	–	+	d	d	+	+	–
Sorbitol	–	–	d	–	–	–	+	+	d	–	–	–	d	+	–	–
Sucrose	–	d	+	d	+	ND	–	+	+	+	d	d	+	d	–	+
Xylose	–	–	–	–	–	–	–	–	–	–	–	–	d	–	–	+

[a]Symbols and abbreviations: +, 90% or more of strains are positive; –, 90% or more are negative; d, 11–89% of strains are positive; w, weak positive reaction; ND, no data available; (), isomers in parentheses indicate <15% of total lactic acid; mDpm, *meso*-diaminopimelic acid; de, *Lactobacillus delbrueckii*-group; sl, *Lactobacillus salivarius*-group; re, *Lactobacillus reuteri*-group; u, unique.

[b]W. P. Hammes, unpublished results.

[c]Strains formerly designated *L. casei* subsp. *pseudoplantarum* produce DL-lactic acid.

[d]According to Carlsson and Gothefors (1975), 60 out of 64 strains ferment ribose.

L. hamsteri	*L. homohiochii*	*L. intestinalis*	*L. jensenii*	*L. kimchii*	*L. murinus*	*L. paracasei* subsp. *paracasei*	*L. paracasei* subsp. *tolerans*	*L. paralimentarius*	*L. paraplantarum*	*L. pentosus*	*L. perolens*	*L. plantarum* subsp. *plantarum*	*L. plantarum* subsp. *argentoratensis*	*L. rhamnosus*	*L. sakei* subsp. *sakei*	*L. sakei* subsp. *carnosus*	*L. versmoldensis*	*L. zeae*
de	u	de	de	u	sl	u	u	u	u	u	u	u	u	u	u	u	u	u
Lys–D-Asp 33–35	Lys–D-Asp 35–38	Lys–D-Asp 33–35	Lys–D-Asp 35–37	Lys–D-Asp[b] 35	Lys–D-Asp 43–44	Lys–D-Asp 45–47	Lys–D-Asp 45–47	Lys–D-Asp[b] 37–38	mDpm 44–45	mDpm 46–47	Lys–D-Asp 49–53	mDpm 44–46	mDpm 44–46	Lys–D-Asp 45–47	Lys–D-Asp 42–44	Lys–D-Asp 42–44	ND 40.5	Lys–D-Asp 48–49
DL	DL	DL	D	DL	L	L[c]	L	ND	DL	DL	L	DL	DL	L	DL/L(D)	DL/L(D)	L	L(D)
−/ND	+/−	−/+	−/+	+/−	−/+	+/d	+/−	+/+	+/−	+/−	+/−	+/−	+/−	+/+	+/−	+/−	+/−	+/+
ND	−	−	+	+	d	+	−	+	+	+	+	+	+	+	−	−	−	+
+	−	−	−	+	+	−	−	−	d	+	d	d	d	d	d	d	−	−
+	d	d	+	+	+	+	−	+	+	+	+	+	+	+	d	d	−	+
+	ND	−	+	+	+	+	−	+	+	ND	+	+	+	+	+	+	−	+
+	−	ND	−	+	−	+	w	−	+	+	+	+	+	+	+	+	ND	+
+	d	+	d	−	d	+	−	−	+	+	−	+	+	+	−	−	−	+
−	−	−	−	+	−	+	−	−	+	d	+	+	+	−	+	−	−	+
+	−	d	−	−	+	−	−	−	+	+	+	+	+	−	+	+	+	−
+	−	d	−	−	+	−	−	−	d	+	+	+	+	−	−	−	−	−
+	d	d	+[d]	+	+	+	−	+	+	+	−	+	+	+	+	+	+	+
+	−	−	−	−	−	d	−	−	d	+	d	+	+	+	−	−	−	−
+	−	+	+	+	+	+	−	+	+	+	+	+	+	+	+	+	−	+
d	−	−	−	+	−	−	−	−	−	+	d	d	d	−	−	−	−	−

TABLE 86. Characteristics differentiating Group C lactobacilli (obligately heterofermentative)[a]

Species	L. acidifarinae	L. antri	L. brevis	L. buchneri	L. coleohominis[b]	L. collinoides	L. diolivorans	L. durianis	L. ferintoshensis	L. fermentum	L. fructivorans	L. frumenti	L. gastricus	L. hammesii	L. hilgardii	L. ingluviei	L. kefiri
Phylogenetic group	u	re	u	u	re	u	u	u	u	re	u	re	re	u	u	re	u
Peptidoglycan type	ND	Lys–D-Asp	Lys–D-Asp	Lys–D-Asp	mDpm	Lys–D-Asp	Lys–D-Asp[c]	ND	mDpm[c]	Orn–D-Asp	Lys–D-Asp	Lys–D-Asp	Orn–D-Asp	Lys–D-Asp	Lys–D-Asp	ND	Lys–D-Asp
G+C content (mol%)	51	44.9	44–47	44–46	ND	46	40	43.3	ND	52–54	38–40	43–45	41.3	52.6	39–41	49	41–42
Growth (°C) 15/45	+/–	–/+	+/–	+/–	–/+	+/–	+/–	+/–	+/–	–/+	+/–	–/+	–/–	+/–	+/–	–/ND	+/–
NH₃ from arginine	+	ND	+	+	–	+	ND	–	+	+	+	+?	ND	–	+	ND	+
Carbohydrates fermented:																	
Arabinose	+	+	+	+		+	+	+	+	d	–	d	d	+	–	+	d
Cellobiose	–	–	–	–		–	–	–	d	d	–	+	+	+	–	–	–
Esculin	–	W	d	d	–	+	–	d	+	–	–	+	–	–	–	d	–
Galactose	+	+	d	d	–	+	+	d	+	+	–	+	+	+	d	–	
Maltose	+	+	+	+	d	+	+	–	+	+	d	+	+	+	+	d	+
Mannose	–	–	–	–	–	–	–	–	+	w	–	+	+	+	–	d	–
Melezitose	–	–	–	+	–	+	–	–	d	–	–	d	–	–	d	–	–
Melibiose	–	+	+	+	–	+	+	–	d	+	–	+	+	–	–	d	+
Raffinose	–	+	d	d	–	–	–	–	d	+	–	+	+	–	–	d	–
Ribose	+	+	+	+	+	+	+	+	+	+	w	+	d	–	+	+	+
Sucrose	–	+	d	d	–	–	–	–	+	+	d	+	+	–	d	+	–
Trehalose	–	–	–	–	–	–	–	–	+	d	–	+	+	+	–	–	–
Xylose	–	d	d	d	–	+	+	+	+	d	–	–	–	+	+	+	–

[a]Symbols and abbreviations: +, 90% or more of strains are positive; –, 90% or more are negative; d, 11–89% of strains are positive; w, weak positive reaction; ND, no data available; (), isomers in parentheses indicate <15% of total lactic acid; mDpm, *meso*-diaminopimelic acid; de, *Lactobacillus delbrueckii*-group; sl, *Lactobacillus salivarius*-group; re, *Lactobacillus reuteri*-group; u, unique.

[b]*Lactobacillus coleohominis* has been found to produce gas from glucose in our laboratory (W. P. Hammes and C. Hertel, unpublished results).

[c]W. P. Hammes, unpublished results.

L. kunkeei	*L. lindneri*	*L. malefermentans*	*L. mucosae*	*L. oris*	*L. panis*	*L. parabuchneri*	*L. paracollinoides*	*L. parakefiri*	*L. pontis*	*L. reuteri*	*L. rossiae*	*L. sanfranciscensis*	*L. spicheri*	*L. suebicus*	*L. thermotolerans*	*L. vaccinostercus*	*L. vaginalis*	*L. zymae*
u	u	u	re	re	re	u	u	u	re	re	u	u	u	u	re	u	re	u
Lys–D-Asp	Lys–D-Asp	Lys–D-Asp	Lys–D-Asp	Lys–D-Asp	Lys–D-Asp	Lys–D-Asp	ND	Lys–D-Asp	Orn-D-Asp	Lys–D-Asp	Lys–Ser–Ala2	Lys–Ala	Lys–D-Asp	mDpm	ND	mDpm	Orn-D-Asp	ND
ND	35	41–42	46.5	49–51	48	44	44.8	41–42	53–55	40–42	44.6	36–38	55	40.4	50.5	36	38–41	53–54
+/–	+/–	+/–	–/+	–/d	–/+	+/ND	+/–	+/+	+/+	–/+	+/–	+/–	+/–	+/d	–/+	–/–	–/+	+/–
–	–	+	+	–	–	+	ND	+	+	+	+	–	+	ND	+	–	ND	+
ND	–	–	d	+	+	+	d	+	–	+	+	–	–	+	+	+	–	+
–	–	–	–	d	–	–	–	–	–	–	–	–	d	ND	w	–	–	
–	–	–	+	d	+	–	ND	–	–	ND	–	ND	–	–	+	–	d	–
–	–	–	d	+	+	+	ND	+	d	+	W	d	–	+	–	w	+	d
–	+	+	+	+	+	+	+	+	+	+	+	+	+	+	ND	+	+	+
–	–	–	–	d	+	ND	–	–	–	–	w	–	–	ND	ND	–	+	–
–	–	–	–	–	–	+	–	d	–	–	–	–	–	–	–	–	–	–
–	–	–	d	+	+	+	+	d	d	+	d	–	–	d	d	–	+	–
w	–	–	d	+	+	+	–	–	d	+	–	–	–	–	d	–	+	–/w
–	–	+	+	+	+	+	+	+	+	+	+	d	+	+	+	+	d	+
+	–	–	+	+	+	+	–	–	+	+	–	d	–	d	ND	–	+	–
–	–	–	–	d	–	–	–	ND	–	–	–	–	–	–	ND	–	–	–
–	–	–	d	+	+	–	+	–	–	–	d	–	+	+	+	+	–	d

is more or less completely split into acetic acid, CO_2, and H_2 (Kandler et al., 1983; Pette and van Beynum, 1943).

Further reading

Pot, B., W. Ludwig, K. Kersters, and K. Schleifer. (1994). Taxonomy of lactic acid bacteria. *In* De Vuyst and Vandamme (Editors). Bacteriocins of lactic acid bacteria. Microbiology, genetics and application, Blackie Academic and Professional, London, pp. 13–90.

Salminen, S., and A. von Wright. (1998). Lactic Acid Bacteria, Microbiology and Functional Aspects. 2nd edn. Marcel Dekker, Inc., New York, Basel.

Wood, B. J., and P. J. Warner. (2003). Genetics of Lactic Acid Bacteria. Springer-Verlag, New York.

Differentiation and characteristics of the species of the genus *Lactobacillus*

The key characteristics for differentiation of the species of the genus *Lactobacillus* are compiled in Table 84, Table 85, and Table 86.

List of species of the genus *Lactobacillus**

1. **Lactobacillus delbrueckii** (Leichmann 1896b) Beijerinck 1901, 229^{AL} (*Bacillus Delbrücki* (*sic*) Leichmann 1896b, 284.) del.bruec′ki.i. N.L. gen. n. *delbrueckii* of Delbrück; named for M. Delbrück, a German bacteriologist.

Cells are nonmotile rods with rounded ends (0.5–0.8 × ~2–9 μm) occurring singly and in short chains. Often containing internal granulation demonstrable with methylene blue stain. Surface growth is greatly enhanced by reduced O_2-tension or anaerobiosis. Obligately homofermentative. Good growth at 45°C or even at 48–52°C. Indole is not formed. Growth factor requirements: pantothenic acid and niacin generally essential; riboflavin, folic acid, vitamin B_{12}, and thymidine are essential for particular strains; thiamine, pyridoxin, biotin, and *p*-aminobenzoic acid are not required. The species contains four subspecies.

Phylogenetic position: Lactobacillus delbrueckii group.

DNA G+C content (mol%): 49–51 (Bd, T_m).

Additional remarks: Because of the high phenotypic and genomic similarities between *Lactobacillus delbrueckii, Lactobacillus leichmannii, Lactobacillus lactis,* and *Lactobacillus bulgaricus,* only *Lactobacillus delbrueckii* is retained here as a separate species. Both *Lactobacillus lactis* and *Lactobacillus leichmannii* are treated as *Lactobacillus delbrueckii* subsp. *lactis* and *Lactobacillus bulgaricus* as *Lactobacillus delbrueckii* subsp. *bulgaricus* (see Weiss et al., 1983b, 1984).

1a. **Lactobacillus delbrueckii subsp. delbrueckii** (Leichmann 1896a) Weiss, Schillinger and Kandler 1984, 270^{VP} (Effective publication: Weiss, Schillinger and Kandler 1983b, 556.) (*Bacillus Delbrücki* (*sic*) Leichmann 1896b, 284.)

Arginine hydrolysis is strain dependent. Distinguishing physiological and biochemical properties are compiled in Table 84.

Isolated mainly from plant material fermented at high temperatures (40–53°C).

Type strain: ATCC 9649, CCUG 34222, CIP 57.8, DSM 20074, NBRC 3202, JCM 1012, LMG 6412, NCIMB 8130, NRRL B-763, VKM B-1596.

DNA G+C content (mol%): 49–51 (Bd, T_m)

GenBank accession number (16S rRNA gene): AY050172, M58814, X52654.

1b. **Lactobacillus delbrueckii subsp. bulgaricus** (Orla-Jensen 1919) Weiss, Schillinger and Kandler 1984, 270^{VP} (Effective publication: Weiss, Schillinger and Kandler 1983b, 556.) (*Thermobacterium bulgaricum* Orla-Jensen 1919, 164; *Lactobacillus bulgaricus* Rogosa and Hansen 1971, 181.)

bul.ga′ri.cus. N.L. adj. *bulgaricus* Bulgarian.

Arginine hydrolysis negative.

Ferments only a few carbohydrates. The distinguishing physiological and biochemical properties are compiled in Table 84.

Isolated from yogurt and cheese.

Type strain: ATCC 11842, CCUG 41390, CIP 101027, DSM 20081, NBRC 13953, JCM 1002, LMG 6901, LMG 13551, NCTC 12712, VKM B-1923.

DNA G+C content (mol%): 49–51 (Bd, T_m).

GenBank accession number (16S rRNA gene): AY050171.

1c. **Lactobacillus delbrueckii subsp. lactis** (Orla-Jensen 1919) Weiss, Schillinger and Kandler 1984, 270^{VP} (Effective publication: Weiss, Schillinger and Kandler 1983b, 556) (*Thermobacterium lactis* Orla-Jensen 1919, 164; *Lactobacillus leichmannii* (Henneberg) Bergey, Harrison, Breed, Hammer and Huntoon 1923, 249.)

lac′tis. L. n. *lac* milk; L. gen. n. *lactis* of milk.

Arginine hydrolysis is strain dependent. The distinguishing physiological and biochemical properties are compiled in Table 84.

Isolated from milk, cheese, compressed yeast, and grain mash.

Type strain: ATCC 12315, CCUG 31454, CIP 101028, DSM 20072, JCM 1248, LMG 7942, NRRL B-4525.

DNA G+C content (mol%): 49–51 (Bd, T_m).

GenBank accession number (16S rRNA gene): AY050173, M58823.

---

* Editorial note: Since this chapter was accepted for publication, 24 new species have been validly published. They are: *Lactobacillus apodemi* Osawa et al. (2006), *Lactobacillus camelliae* Tanasupawat et al. (2007), *Lactobacillus ceti* Vela et al. (2008), *Lactobacillus composti* Endo and Okada (2007a), *Lactobacillus concavus* Tong and Dong (2005b), *Lactobacillus crustorum* Scheirlinck et al. (2007b), *Lactobacillus cypricasei* Lawson et al. (2001a), *Lactobacillus durianis* Leisner et al. (2002), *Lactobacillus equigenerosi* Endo et al. (2008), *Lactobacillus farraginis* Endo and Okada (2007b), *Lactobacillus ghanensis* Nielsen et al. (2007), *Lactobacillus harbinensis* Miyamoto et al. (2006), *Lactobacillus hayakitensis* Morita et al. (2007), *Lactobacillus namurensis* Scheirlinck et al. (2007a), *Lactobacillus nantensis* Valcheva et al. (2006), *Lactobacillus parabrevis* Vancanneyt et al. (2006), *Lactobacillus parafarraginis* Endo and Okada (2007b), *Lactobacillus rennini* Chenoll et al. (2006), *Lactobacillus secaliphilus* Ehrmann et al. (2007), *Lactobacillus senmaizukei* Hiraga et al. (2008), *Lactobacillus siliginis* Aslam et al. (2006), *Lactobacillus sobrius* Konstantinov et al. (2006), *Lactobacillus thailandensis* Tanasupawat et al. (2007), and *Lactobacillus vini* Rodas et al. (2006).

1d. Lactobacillus delbrueckii subsp. indicus Dellaglio, Felis, Castioni, Torriani and Germond 2005, 403[VP]

in′di.cus. L. masc. adj. *indicus* from India, referring to the geographical origin of the strains.

Growth is observed in MRS plus 2.5% (w/v) NaCl, except for strain NCC780; none of the strains grows in MRS plus 5% (w/v) NaCl. All the strains grow in MRS at pH 3, 4, and 5, but not at pH 7.8. Ferments only a few carbohydrates. Distinguishing physiological and biochemical properties are compiled in Table 84.

Genotypic analysis showed that the subspecies represents a coherent cluster differentiating it from all other subspecies.

Isolated from a traditional Indian dairy fermented (type Dahi) product.

Type strain: NCC725, DSM 15996, LMG 22083.

DNA G+C content (mol%): 49–51 (Bd, T_m).

GenBank accession number (16S rRNA gene): AY421720.

2. **Lactobacillus acetotolerans** Entani, Masai and Suzuki 1986, 547[VP]

a.ce.to.tole.rans. L. n. *acetum* vinegar; L. pres. part. *tolerans* tolerating, enduring; M.L. part. adj. *acetotolerans* vinegar tolerating.

Cells are nonmotile rods (0.4–0.5 × 1.1–3.4 μm) occurring singly, in pairs, or occasionally in short chains. The colonies are 0.3–1.5 mm in diameter, circular to irregular, convex, opaque, yellowish white, rough, and undulate when grown at 30°C for 14 d on Briggs agar (pH 5.0). Growth is generally observed at pH 3.3–6.6 and at 23–40°C. Resistant to 4–5% and 9–11% acetic acid at pH 3.5 and 5.0, respectively. Facultatively heterofermentative (some strains utilize ribose). No gas is produced from D-glucose or D-gluconate. Arginine and starch not hydrolyzed. Additional physiological and biochemical characteristics are presented in Table 85. Milk is not curdled. Riboflavin, pantothenic acid, folic acid, uracil, and Tween 80 are required for growth, but thiamine, *p*-aminobenzoic acid, biotin, adenine, guanine, xanthine, mevalonic acid, acetic acid, and ethanol are not required. Some strains require pyridoxal, nicotinic acid, vitamin B_{12}, and peptides.

Isolated from fermented vinegar broth.

Phylogenetic position: Lactobacillus delbrueckii group.

Type strain: ATCC 43578, CCUG 32229, CIP 103180, DSM 20749, JCM 3825, LMG 10751, NBI 3014.

DNA G+C content (mol%): 35.3–36.5 (T_m).

GenBank accession number (16S rRNA gene): e.g., M58801.

Additional remarks: Two biovars were described. Strains of biovar I produce acid from D-mannitol but not from cellobiose, whereas strains of biovar II produce acid from cellobiose but not from D-mannitol. Strains of biovar II are more fastidious than those of biovar I. Only the strains of biovar II require nicotinic acid, vitamin B_{12}, and peptides in addition to common growth factors.

3. **Lactobacillus acidifarinae** Vancanneyt Neysens, De Wachter, Engelbeen, Snauwaert, Cleenwerck, Van der Meulen, Hoste, Tsakalidou, De Vuyst and Swings 2005, 619[VP]

a.ci.di.fa′ri.nae. L. adj. *acidus* sour; N.L. n. *farina* flour; N.L. gen. n. *acidifarinae* of a sour flour.

Cells are nonmotile rods (1.0 × 2–20 μm) occurring singly, in pairs, or in chains. After 24 h, colonies are beige, circular with a rough and wrinkled surface, and approximately 1 mm in diam-

eter. Growth occurs at 5% NaCl. Obligately heterofermentative. Gas is produced from glucose and gluconate. Additional physiological and biochemical characteristics are presented in Table 86.

Isolated from a Belgian artisan wheat sourdough.

Phylogenetic position: unique.

Type strain: R-19065, CCM 7240, LMG 22200.

DNA G+C content (mol%): 51 (HPLC).

GenBank accession number (16S rRNA gene): AJ632158.

4. **Lactobacillus acidipiscis** Tanasupawat, Shida, Okada and Komagata 2000, 1481[VP]

a.ci.di′pis.cis. L. adj. *acidus* sour; L. n. *piscis* fish; N.L. gen. n. *acidipiscis* of a sour fish, an isolation source of strains of this species.

Cells are nonmotile rods (0.4–0.6 × 1.2–5.0 μm) occurring singly, in pairs, and in chains. Colonies on MRSH agar plate are circular, slightly convex with an entire margin, and nonpigmented. Does not produce gas from glucose. Arginine is not hydrolyzed; no formation of slime from sucrose. Most strains show no reaction in litmus milk. Grows at 25–37°C but not at 15 or 42°C. Does not grow at pH 4.0 or pH 8.5. Grows in 10% NaCl. Some strains grow in the presence of 12% NaCl. Facultatively heterofermentative. Additional physiological and biochemical characteristics are presented in Table 85. Requires niacin and calcium pantothenate for growth. Major cellular fatty acids are straight-chained $C_{16:0}$ and $C_{18:0}$.

Isolated from fermented fish (pla-ra and pla-chom) in Thailand.

Phylogenetic position: Lactobacillus salivarius group.

Type strain: FS60–1, CIP 106750, HSCC 1411, JCM 10692, NRIC 0300, PCU 207, TISTR 1386.

DNA G+C content (mol%): 38.6–41.5 (HPLC).

GenBank accession number (16S rRNA gene): AB023836.

5. **Lactobacillus acidophilus** (Moro 1900) Hansen and Mocquot 1970, 326[AL] (*Bacillus acidophilus* Moro 1900, 115.)

a.ci.do′phi.lus. N.L. n. *acidum* acid; Gr. adj. *philos* loving; N.L. adj. *acidophilus* acid-loving.

Cells are nonmotile rods with rounded ends (0.6–0.9 × 1.5–6 μm) occurring singly, in pairs, and in short chains. The colonies on trypticase-glucose-agar are rough, showing fuzzy outlines. Starch is fermented by most strains. Obligately homofermentative. Additional physiological and biochemical characteristics are presented in Table 84. Nutritional requirements: calcium pantothenate, folic acid, niacin, and riboflavin are essential; pyridoxal, thiamine, thymidine, and vitamin B_{12} are not required.

Isolated from the intestinal tract of humans and animals, human mouth, vagina, sourdough, and wine.

Phylogenetic position: Lactobacillus delbrueckii group.

Type strain: ATCC 4356, CCUG 5917, CIP 76.13, DSM 20079, NBRC 13951, JCM 1132, LMG 9433, LMG 13550, NCTC 12980, NRRL B-4495, VKM B-1660.

DNA G+C content (mol%): 34–37 (Bd, T_m).

GenBank accession number (16S rRNA gene): M58802, X61138.

Additional remarks: Lactobacillus acidophilus cannot be differentiated reliably from *Lactobacillus gasseri*, *Lactobacillus crispatus*, and *Lactobacillus amylovorus* by any simple phenotypic test; electrophoretic analysis of soluble cellular proteins or lactate dehydrogenases, detailed cell wall studies or, preferentially, genotypical identification methods are necessary. Synonymous with *Lactobacillus acidophilus* group A-1 of Johnson et al. (1980).

6. **Lactobacillus agilis** Weiss, Schillinger, Laternser and Kandler 1982, 266^{VP} (Effective publication: Weiss, Schillinger, Laternser and Kandler 1981, 252.)

a'gi.lis. L. adj. *agilis* agile, motile.

Cells are rods with rounded ends (0.7–1.0 × 3–6 μm) occurring singly, in pairs, and in short chains. Motile with peritrichous flagella; motility normally easy to demonstrate in MRS broth. Arginine not hydrolyzed. Facultatively heterofermentative. Additional physiological and biochemical characteristics are presented in Table 85.

Habitat is the crop of pigeons (Baele et al., 2001b); isolated from municipal sewage.

Phylogenetic position: Lactobacillus salivarius group.

Type strain: CIP 101264, CCUG 31450, DSM 20509, JCM 1187, LMG 9186, NRRL B-14856.

DNA G+C content (mol%): 43–44 (T_m).

GenBank accession number (16S rRNA gene): M58803.

Additional remarks: "*Lactobacillus plantarum* var. *mobilis*" isolated from feces of turkey (Harrison and Hansen, 1950) was only tentatively named and therefore omitted from the Approved Lists of Bacterial Names (Skerman et al., 1980). According to the original description and later investigations (Sharpe et al., 1973) this organism may belong to *Lactobacillus agilis*.

7. **Lactobacillus algidus** Kato, Sakala, Hayashidani, Kiuchi, Kaneuchi and Ogawa 2000, 1148^{VP}

al'gi.dus. L. adj. *algidus* cold, referring to the ability to grow at low temperature.

The cellular morphology depends on growth medium: coccobacilli or short rods in MRS broth and rods (0.3–0.7 × 1.1–1.9 μm) on MRS agar. Nonmotile. Colonies are circular, convex, smooth, cream-white, and 1.0–1.5 mm in diameter after 48 h incubation at 20°C on MRS agar. Facultatively heterofermentative. Growth occurs at 0–25°C. No growth at 30°C. No dextran is produced from sucrose. Additional physiological and biochemical characteristics are presented in Table 85.

The major cellular fatty acids produced by strain M6A9^T are the straight-chain mono-unsaturated oleic acid ($C_{18:1\ \omega9}$) and straight-chain saturated palmitic acid ($C_{16:0}$).

Isolated from vacuum-packaged refrigerated beef, stored at low temperature.

Phylogenetic position: Lactobacillus salivarius group.

Type strain: M6A9, CIP 106688, JCM 10491.

DNA G+C content (mol%): 36.8 ± 3 (HPLC).

GenBank accession number (16S rRNA gene): AB033209.

8. **Lactobacillus alimentarius** Reuter 1983a, 672^{VP} (Effective publication: Reuter 1983b, 278.)

a.li.men.ta'ri.us. L. adj. *alimentarius* pertaining to food.

Cells are nonmotile, short, slender rods (0.6–0.8 × 1.5–2.5 μm). Growth in the presence of 10% NaCl. Acetoin is produced from glucose. Facultatively heterofermentative. Additional physiological and biochemical characteristics are presented in Table 85.

Isolated from marinated fish products, meat products (fermented sausages, sliced prepacked sausages), and sourdough.

Phylogenetic position: unique.

Type strain: ATCC 29643, CCUG 30672, CIP 102986, DSM 20249, JCM 1095, LMG 9187.

DNA G+C content (mol%): 36–37 (T_m).

GenBank accession number (16S rRNA gene): M58804.

9. **Lactobacillus amylolyticus** Bohak, Back, Richter, Ehrmann, Ludwig and Schleifer 1999, 1^{VP} (Effective publication: Bohak, Back, Richter, Ehrmann, Ludwig and Schleifer 1998, 363.)

a.my.lo.ly'ti.cus. Gr. n. *amylum* starch, Gr. adj. *lyticus* able to loosen; N.L. adj. *amylolyticus* starch-digesting.

Cells are nonmotile rods with rounded ends (0.7–0.9 μm × 5–20 μm) occurring singly, in pairs, or in short chains. Colonies on MRS flat, rough, dull surface, beige to dirty-white color; diameter 2–3 mm. Growth in agar stabs occurs throughout the stab but not up to the surface. Can grow up to 52°C; optimum at 45°C-48°C; no growth at 20°C. Optimum at pH 5–5.5; no growth below pH 3.5 or above pH 6. Obligately homofermentative. Esculin is hydrolyzed in some cases; arginine is not hydrolyzed. Urease is not produced. Additional physiological and biochemical characteristics are presented in Table 84. Cell walls contain teichoic acid.

Isolated from brewery, unhoped wort, malt-mash, and malt.

Phylogenetic position: Lactobacillus delbrueckii group.

Type strain: LA 5, CCUG 39901, DSM 11664, JCM 12529, LMG 18796.

DNA G+C content (mol%): 39 (T_m).

GenBank accession number (16S rRNA gene): Y17361.

10. **Lactobacillus amylophilus** Nakamura and Crowell 1981, 216^{VP} (Effective publication: Nakamura and Crowell 1979, 539.)

a.my.lo'phi.lus. Gr. n. *amylum* starch; Gr. adj. *philos* loving; N.L. adj. *amylophilus* starch-loving.

Cells are nonmotile thin rods (0.5–0.7 × 2–3 μm) occurring singly and in short chains. Actively ferments starch and displays extracellular amylolytic enzyme activity. Obligately homofermentative. Additional physiological and biochemical characteristics are presented in Table 84. Growth factor requirements: riboflavin, pyridoxal, pantothenic acid, niacin, and folic acid are essential; thiamine is not required.

Isolated from swine waste-corn fermentation.

Phylogenetic position: Lactobacillus delbrueckii group.

Type strain: ATCC 49845, CCUG 30137, CIP 102988, DSM 20533, NBRC 15881, JCM 1125, LMG 6900, NCAIM B.01457, NRRL B-4437, NRRL B-4476.

DNA G+C content (mol%): 44–46 (Bd).

GenBank accession number (16S rRNA gene): M58806.

11. **Lactobacillus amylovorus** Nakamura 1981, 61^{VP}

a.my.lo.vo'rus. Gr. n. *amylum* starch; L. v. *vorare* to devour; N.L. adj. *amylovorus* starch-devouring.

Cells are nonmotile rods (1 × 3–5 μm) occurring singly and in short chains. Good growth at 45°C. Actively ferments starch and displays extracellular amylolytic enzyme activity. Obligately homofermentative. Additional physiological and biochemical characteristics are presented in Table 84.

Growth factor requirements: niacin, pantothenic acid, folic acid, and riboflavin are essential; thiamine is not required.

Isolated from cattle waste-corn fermentation.

Phylogenetic position: Lactobacillus delbrueckii group.

Type strain: ATCC 33620, CCUG 27201, CIP 102989, DSM 20531, JCM 1126, LMG 9496, NCAIM B.01458, NRRL B-4540.

DNA G+C content (mol%): 40.3 ± 0.1 (Bd).

GenBank accession number (16S rRNA gene): AY944408, M58805.

Additional remarks: Synonymous with *Lactobacillus acidophilus* group A 3 of Johnson et al. (1980). See comments on *Lactobacillus acidophilus.*

12. **Lactobacillus animalis** Dent and Williams 1983, 439[VP] (Effective publication: Dent and Williams 1982, 384.)

a.ni.ma′lis. L. n. *animal* animal; L. gen. n. *animalis* of an animal.

Cells are nonmotile rods (1.0–1.2 × 3–6 μm) with rounded ends; exponential growth phase cells in MRS broth occur singly or in pairs. Good growth at 45°C. Arginine not hydrolyzed. Facultatively heterofermentative. Additional physiological and biochemical characteristics are presented in Table 85.

Isolated from dental plaques and intestines of animals.

Phylogenetic position: Lactobacillus salivarius group.

Type strain: 1535, ATCC 35046, CCUG 33906, CIP 103152, DSM 20602, NBRC 15882, JCM 5670, LMG 9843, NCIMB 13278, NRRL B-14176.

DNA G+C content (mol%): 41–44 (T_m).

GenBank accession number (16S rRNA gene): M58807, X61133.

Additional remarks: Some of the strains on which the description of *Lactobacillus animalis* was based ferment arabinose and ribose weakly thus resembling *Lactobacillus murinus.* The comparison of the sequences of >1400 bp of the 16S rRNA reveals 99.7% identity. DNA/DNA similarity studies should be directed towards establishing the genomic relationship of the different strains of *Lactobacillus animalis* among each other and with *Lactobacillus murinus.*

13. **Lactobacillus antri** Roos, Engstrand and Jonsson 2005, 80[VP] an′tri. L. gen. n. *antri* of a cave (the antrum region of the stomach).

Cells are nonmotile rods (1 × 1.2–2 μm) occurring singly or in pairs. After anaerobic growth at 37°C for 48 h, colonies on MRS agar are 2–3.5 mm in diameter; they are white, smooth, and slightly convex. Growth on MRS agar under aerobic conditions is very weak. In MRS broth, growth occurs at 25 and 45°C, but not at 20°C. Obligately heterofermentative. Gas is produced from glucose. Additional physiological and biochemical characteristics are presented in Table 86.

Isolated from a biopsy of the healthy human gastric mucosa.

Phylogenetic position: Lactobacillus reuteri group.

Type strain: Kx146A4, CCUG 48456, DSM 16041, LMG 22111.

DNA G+C content (mol%): 44 (HPLC).

GenBank accession number (16S rRNA gene): AY253659.

14. **Lactobacillus aviarius** Fujisawa, Shirasaka, Watabe and Mitsuoka 1985, 223[VP] (Effective publication: Fujisawa, Shirasaka, Watabe and Mitsuoka 1984, 419.)

a.vi.a′ri.us. L. adj. *aviarius* pertaining to birds.

Cells nonmotile, short to coccoid rods (0.5–1.0 × 0.5–1.6 μm) with rounded ends, occurring singly or in short chains. Surface colonies on BL agar after 2 d of anaerobic incubation are 0.3–1.2 μm in diameter, round, globular, yellowish-white to reddish-brown, with a smooth surface and entire edge. Strictly anaerobic. No gas from glucose. The final pH of glucose broth is 3.9–4.0. Obligately homofermentative. Two subspecies are currently recognized.

Phylogenetic position: Lactobacillus salivarius group.

DNA G+C content (mol%): 39–43 (T_m).

Additional remarks: On Validation List no. 17, the type strain, strain 75 (DSM 20655) was incorrectly cited.

14a. **Lactobacillus aviarius subsp. aviarius** Fujisawa, Shirasaka, Watabe and Mitsuoka 1985, 223[VP] (Effective publication: Fujisawa, Shirasaka, Watabe and Mitsuoka 1984, 419.)

Distinguishing physiological and biochemical properties are compiled in Table 84.

Isolated from chicken and duck alimentary tract and feces.

Type strain: 75, ATCC 43234, DSM 20655, CCUG 32230, CIP 103144, JCM 5666, LMG 10753.

DNA G+C content (mol%): 39–43.

GenBank accession number (16S rRNA gene): M58808.

14b. **Lactobacillus aviarius subsp. araffinosus** Fujisawa, Shirasaka, Watabe and Mitsuoka 1985, 223[VP] (Effective publication: Fujisawa, Shirasaka, Watabe and Mitsuoka 1984, 419.)

a.raffi.no′sus. Gr. pref. *a* not; N.L.adj. *raffinosus* of raffinose; N.L. adj. *araffinosus* not fermenting raffinose.

Distinguishing physiological and biochemical properties are compiled in Table 84.

Isolated from chicken duodenum.

Type strain: DSM 20653.

DNA G+C content (mol%): 41.3.

GenBank accession number (16S rRNA gene): unavailable.

Additional remarks: The effective date of validation for this subspecies is that of Validation List no. 17 from which it was accidentally omitted (see Fujisawa et al., 1986, and Validation List no. 20, footnote d.)

15. **Lactobacillus bifermentans** Kandler, Schillinger and Weiss 1983, 896[VP] (Effective publication: Kandler, Schillinger and Weiss 1983, 409.)

bi.fer.men′tans. L. pref. *bis* twice; L. part. *fermentans* leavening; N.L. part. adj. *bifermentans* doubly fermenting.

Cells are nonmotile irregular rods with rounded or often tapered ends (0.5–1.0 × 1.5–2.0 μm) occurring singly, in pairs, or irregular short chains, often forming clumps. No growth at 42°C. Homofermentative production of DL-lactic acid in media containing more than 1% fermentable hexoses. Lactic acid is fermented to acetic acid, ethanol, traces of propionic acid, CO_2, and H_2 at pH >4.0. Arginine not hydrolyzed. Additional physiological and biochemical characteristics are presented in Table 85. In contrast to all other lactobacilli, *Lactobacillus bifermentans* ferments lactate and produces free H_2 and was therefore put on the list of species *incertae sedis* in the eighth edition of the *Manual*.

Isolated from spoiled Edam and Gouda cheeses where it forms undesired small cracks ("Boekelscheuren", Pette and van Beynum (1943).

Phylogenetic position: unique.

Type strain: ATCC 35409, CCUG 32234, CIP 102811, DSM 20003, JCM 1094, LMG 9845.

DNA G+C content (mol%): 45 (T_m).

GenBank accession number (16S rRNA gene): M58809.

16. **Lactobacillus brevis** (Orla-Jensen 1919) Bergey, Breed, Hammer, Huntoon, Murray and Harrison 1934, 312[AL] (*Betabacterium breve* Orla-Jensen 1919, 175.)

bre'vis. L. adj. *brevis* short.

Cells are rods with rounded ends, generally 0.7–1.0 × 2–4 µm, occurring singly and in short chains. Growth factor requirements: calcium pantothenate, niacin, thiamine, and folic acid are essential; riboflavin, pyridoxal, and vitamin B_{12} not required. Obligately heterofermentative. Additional physiological and biochemical characteristics are presented in Table 86.

Isolated from milk, cheese, sauerkraut, sourdough, silage, cow manure, feces, mouth, and intestinal tract of humans and rats.

Phylogenetic position: unique.

Type strain: ATCC 14869, CCUG 30670, CIP 102806, DSM 20054, JCM 1059, LMG 6906, LMG 7944, NRRL B-4527.

DNA G+C content (mol%): 44–47 (Bd, T_m).

GenBank accession number (16S rRNA gene): M58810, X61134.

Additional remarks: Lactobacillus brevis is often difficult to distinguish clearly from *Lactobacillus buchneri*, *Lactobacillus hilgardii*, *Lactobacillus collinoides*, *Lactobacillus kefiri*, *Lactobacillus spicheri*, or *Lactobacillus acidifarnae* by simple physiological tests, especially carbohydrate fermentation reactions. Identification by using genotypical methods is most suitable (see *Lactobacillus acidophilus*).

17. **Lactobacillus buchneri** (Henneberg 1903) Bergey, Harrison, Breed, Hammer and Huntoon 1923, 251[AL] (*Bacillus Buchneri* (*sic*) Henneberg 1903, 163.)

buch'ne.ri. N.L. gen. n. *buchneri* of Buchner; named for E. Buchner, a German bacteriologist.

Cells are nonmotile rods with rounded ends (0.7–1.0 × 2–4 µm) occurring singly and in short chains. *Lactobacillus buchneri* is identical in almost all characteristics with *Lactobacillus brevis*. Obligately heterofermentative. Additional physiological and biochemical characteristics are presented in Table 86.

Isolated from pressed yeast, milk, cheese, fermenting plant material, and human mouth.

Phylogenetic position: unique.

Type strain: ATCC 4005, CCUG 21532, CIP 103023, DSM 20057, JCM 1115, LMG 6892, NCAIM B.01145, NRRL B-1837, VKM B-1599.

DNA G+C content (mol%): 44–46 (Bd, T_m).

GenBank accession number (16S rRNA gene): AB205055, M58811, X61139.

18. **Lactobacillus casei** (Orla-Jensen 1916) Hansen and Lessel 1971, 71[AL] (*Streptobacterium casei* Orla-Jensen 1919, 166.)

ca'se.i. L. n. *caseus* cheese; L. gen. n. *casei* of cheese.

Cells are nonmotile rods (0.7–1.1 × 2.0–4.0 µm) often with square ends and tending to form chains. Facultatively heterofermentative. L-LDH is activated by FDP and Mn^{2+}. Growth factor requirements: riboflavin, folic acid, calcium pantothenate, and niacin are essential; pyridoxal or pyridoxamine is essential or stimulatory; thiamine, vitamin B_{12}, and thymidine are not required. Additional physiological and biochemical characteristics are presented in Table 85.

Isolated from milk, cheese, and intestinal tract.

Phylogenetic position: unique.

Type strain: ATCC 393, BCRC 10697, CCUG 21451, CECT 475, CIP 103137, DSM 20011, IAM 12473, NBRC 15883, JCM 1134, KCTC 3109, LMG 6904, NCIMB 11970, NCIMB 11970, NRRL B-1922.

DNA G+C content (mol%): 45–47 (Bd).

GenBank accession number (16S rRNA gene): AF469172, AY773945, D16551, M23928, X61135.

Additional remarks: The taxonomic status of *Lactobacillus casei* is characterized by certain inconsistencies (Dellaglio et al., 2002; Dicks et al., 1996). The type strain ATCC 393 is closely related to *Lactobacillus zeae* and does not appear to represent the majority of strains isolated from multitude of habitats as they are described in the previous edition of *Bergey's Manual*. Instead only few strains are known to which the type strain ATCC 393 is genetically closely related. When *Lactobacillus paracasei* had been described by Collins et al. (1989), the 16S rRNA sequences of crucial strains were not included in their investigation. Thus, *Lactobacillus paracasei* reflects the ecological distribution that was formerly attributed to *Lactobacillus casei*. Based on numerous arguments, Dellaglio et al. (2002) have presented a Request for an Opinion to the Judical Commission asking to abolish the species *Lactobacillus paracasei* and to transfer the present type strain of *Lactobacillus casei* to *Lactobacillus zeae*.

19. **Lactobacillus coleohominis** Nikolaitchouk, Wacher, Falsen, Andersch, Collins and Lawson 2001, 2084[VP]

co.le.o.ho'mi.nis. Gr. n. *koleos* vagina; L. gen. n. *hominis* of humans; N.L. gen. n. *coleohominis* of the vagina of humans.

Cells are rods. On Columbia horse-blood agar, its colonies are small, entire, non-pigmented, and do not produce an odor. In the original description it was stated that glucose was fermented without gas formation. In our laboratory (Hammes and Hertel, unpublished results), the type strain produced gas from glucose. Ammonia is not produced from arginine. Obligately heterofermentative. Additional physiological and biochemical characteristics are presented in Table 86. Activity is detected for acid phosphatase (weak reaction), β-galactosidase, β-glucuronidase, and phosphoamidase (weak reaction). Activity is not detected for alanine phenylalanine proline arylamidase, alkaline phosphatase, cystine arylamidase, N-acetyl-β-glucosaminidase, chymotrypsin, ester lipase C8, lipase C14, -fucosidase, glycyl tryptophan arylamidase, α-mannosidase, β-mannosidase, pyroglutamic acid arylamidase, valine arylamidase, trypsin, or urease. The reactions are variable for arginine dihydrolase, esterase C-4, α-glucosidase, β-glucosidase, and leucine arylamidase. Hippurate are hydrolyzed. Voges–Proskauer-negative.

Isolated from human vagina and urinary tract.

Phylogenetic position: Lactobacillus reuteri group.

Type strain: CCUG 44007, CIP 106820, DSM 14060, JCM 11550.

DNA G+C content (mol%): not determined.

GenBank accession number (16S rRNA gene): AJ293530, AM113776.

20. **Lactobacillus collinoides** Carr and Davies 1972, 470[AL]

col.li.no.i′des. L. adj. *collinus* hilly; Gr. suff. *eides* resembling, similar; N.L. adj. *collinoides* hill-shaped, pertaining to colony form.

Cells are rods with rounded ends (0.6–0.8 × 3–5 μm); tendency to form long filaments, occurring singly, in palisades, and irregular clumps. Growth in MRS broth is distinctly improved by the addition of 20% tomato juice and by replacement of glucose by maltose. Mannitol formed from fructose. Obligately heterofermentative. Additional physiological and biochemical characteristics are presented in Table 86.

Isolated from cider.

Phylogenetic position: unique.

Type strain: ATCC 27612, CCUG 32259, CIP 103008, DSM 20515, JCM 1123, LMG 9194.

DNA G+C content (mol%): 46 (T_m).

GenBank accession number (16S rRNA gene): AB005893.

21. **Lactobacillus coryniformis** Abo-Elnaga and Kandler 1965c, 18[AL]

co.ry′ni.for′mis. Gr. n. *coryne* a club; L. adj. *formis* shaped; N.L. adj. *coryniformis* club-shaped.

Cells are short, often coccoid rods; frequently somewhat pear-shaped (0.8–1.1 × 1–3 μm) occurring singly, in pairs, or in short chains. Facultatively heterofermentative. Additional physiological and biochemical characteristics are presented in Table 85. Growth factor requirements: pantothenic acid, niacin, riboflavin, biotin, and *p*-aminobenzoic acid are essential for all or the majority of the strains tested; folic acid, pyridoxin, thiamine, and vitamin B_{12} are not required.

Isolated from silage, cow dung, dairy barn air, and sewage. Two subspecies are recognized within *Lactobacillus coryniformis.*

Phylogenetic position: unique.

DNA G+C content (mol%): 45 (T_m).

21a. **Lactobacillus coryniformis subsp. coryniformis** Abo-Elnaga and Kandler 1965c, 18[AL]

The lactic acid produced from glucose contains substantial amounts of the L-isomer (15–20% of total lactic acid).

Type strain: ATCC 25602, CIP 103133, DSM 20001, CCUG 30666, JCM 1164, LMG 9196, NRRL B-4391.

DNA G+C content (mol%): 45 (T_m).

GenBank accession number (16S rRNA gene): M58813.

21b. **Lactobacillus coryniformis subsp. torquens** Abo-Elnaga and Kandler 1965c, 19[AL]

tor′quens. L. pres. part. *torquens* twisting.

Exclusively D(−)-lactic acid is produced.

Type strain: ATCC 25600, CCUG 30667, CIP 103134, DSM 20004, JCM 1166, LMG 9197, NRRL B-4390.

DNA G+C content (mol%): 45 (T_m).

GenBank accession number (16S rRNA gene): not available.

22. **Lactobacillus crispatus** (Brygoo and Aladame 1953) Moore and Holdeman (1970), 15[AL] emend. Cato, Moore and Johnson 1983, 15[AL] (*Eubacterium crispatum* Brygoo and Aladame 1953, 641.)

cris.pa′tus. L. part. adj. *crispatus* curled, crisped, referring to morphology observed originally in broth media.

Cells are straight to slightly curved rods with rounded ends (0.8–1.6 × 2.3–11 μm) occurring singly and in short chains. Esculin and starch hydrolyzed. Arginine not hydrolyzed. Obligately homofermentative. Additional physiological and biochemical characteristics are presented in Table 84.

Isolated from human feces, vagina, and buccal cavities, crops and ceca of chicken; also found in patients with purulent pleurisy, leucorrhea, and urinary tract infection.

Phylogenetic position: Lactobacillus delbrueckii group.

Type strain: ATCC 33820, CCUG 30722, CIP 102990, CIPP II, DSM 20584, JCM 1185, LMG 9479, VPI 3199.

DNA G+C content (mol%): 35–38 (T_m).

GenBank accession number (16S rRNA gene): AF257097, Y17362.

Additional remarks: See comments on *Lactobacillus acidophilus.* Synonymous with *Lactobacillus acidophilus* group A2 of Johnson et al. (1980).

23. **Lactobacillus curvatus** (Troili-Petersson 1903) Abo-Elnaga and Kandler 1965c, 19[AL] (*Bacterium curvatum* Troili-Petersson 1903, 137.)

cur.va′tus. L. v. *curvare* to curve; L. past. part. *curvatus* curved.

Cells are curved, bean-shaped rods with rounded ends (0.7–0.9 × 1–2 μm) occurring in pairs and short chains; closed rings of usually four cells or horseshoe forms frequently observed. Some strains at first motile; motility lost on subculture. Some strains tested grow even at 2–4°C. L-LDH is activated by FDP and Mn^{2+}. Possesses lactic acid racemase whose biosynthesis is induced by L(+)-lactic acid. Racemase induction generally not repressed by acetate. Arginine not hydrolyzed. Formation of catalase in the presence of heme is strain dependent. Facultatively heterofermentative. Additional physiological and biochemical characteristics are presented in Table 85.

Isolated from cow dung, milk, silage, sauerkraut, prepacked finished dough, sourdough, pressed yeast, and meat products.

Phylogenetic position: unique.

Type strain: ATCC 25601, CCUG 30669, CIP 102992, DSM 20019, NBRC 15884, JCM 1096, LMG 9198, LMG 13553, NRRL B-4562.

DNA G+C content (mol%): 42–44 (T_m).

GenBank accession number (16S rRNA gene): AJ621550, AM113777.

Additional remarks: Two subspecies were validly published by Torriani et al. (1996). Based on molecular studies, it was shown by Koort et al. (2004) that the type strain of subspecies *melibiosus* was synonymous with *Lactobacillus sakei* subsp. *carnosus.*

24. **Lactobacillus cypricasei** Lawson, Papademas, Wacher, Falsen, Robinson and Collins 2001a, 48[VP]

cy.pri.ca'se.i. N.L. masc. gen. n. *cypricasei* of cheese from Cyprus, referring to the original isolation source.

Cells are rods with rounded ends occurring as single cells, in pairs, or in short chains (0.6–0.8 × 3–5 μm). When grown on MRS agar, colonies are small, entire, and cream colored. No growth is observed at 15 or 45°C in MRS broth after 48 h incubation. The optimum pH range is 5.5–7.5. Facultatively heterofermentative. No gas is produced from glucose metabolism. Additional physiological and biochemical characteristics are presented in Table 85. Positive reactions are obtained for arginine dihydrolase, β-galactosidase, β-galacturonidase, α-glucosidase, β-glucosidase, leucine arylamidase, pyroglutamic acid arylamidase (weak reaction), and valine arylamidase. Negative reactions are obtained for alanine phenylalanine proline arylamidase, alkaline phosphatase, chymotrypsin, esterase C-4, ester lipase C8, α-fucosidase, β-glucuronidase, glycine-tryptophan arylamidase, lipase C14, β-mannosidase, trypsin, and urease. Variable reactions are obtained for acid phosphatase, N-acetyl-β-glucosaminidase, cystine arylamidase, α-galactosidase, α-mannosidase, and phosphoamidase. Hippurate is not hydrolyzed. Acetoin is produced.

Isolated from Halloumi cheese, a cheese prevalent in Cyprus.

Phylogenetic position: Lactobacillus salivarius group.

Type strain: LMK3, ATCC BAA-288, CCUG 42961, CIP 106393, DSM 15353, JCM 11551.

DNA G+C content (mol%): not determined.

GenBank accession number (16S rRNA gene): AJ251560.

25. **Lactobacillus diolivorans** Krooneman, Alderkamp, Oude Elferink, Driehuis, Cleenwerck, Swings, Gottschal and Vancanneyt 2002, 645[VP]

di.o.li.vo'rans. N.L. *diol* from 1.2-propanediol; L. v. *vorare* to devour; N.L. adj. *diolivorans* devouring diols.

Cells are nonmotile rods that occur as single cells, in pairs, or occasionally in short chains. On MRS-MOD medium with 1.2-propanediol as substrate, cells are 1 μm × 2 μm whereas, on glucose, the cells are longer, up to 10 μm. Colonies are off-white on MRS medium. Obligately heterofermentative. Fermentative growth on 1,2-propanediol under anoxic conditions, producing propanol and propionic acid. Optimal temperature and pH for growth on 1,2-propanediol are 30–32°C and pH 5.7. No growth is obtained at 12 or 42°C. Growth at NaCl concentrations of 2% (w/v) at 30°C; no growth at 4%. Additional physiological and biochemical characteristics are presented in Table 86.

Isolated from maize silage in the Netherlands.

Phylogenetic position: unique.

Type strain: JKD6, DSM 14421, JCM 12183, LMG 19667.

DNA G+C content (mol%): 40 (HPLC).

GenBank accession number (16S rRNA gene): AF264701.

26. **Lactobacillus durianis** Leisner, Vancanneyt, Lefebvre, Vandemeulebroecke, Hoste, Euras Vilalta, Rusul and Swings 2002, 929[VP]

du.ri.an'is. N.L. gen. n. *durianis* of the durian fruit.

Cells are nonmotile rods occurring singly, as pairs, or in chains (fewer than five cells) in APT broth supplemented with glucose, D-xylose, or both components. Surface colonies on APT agar with D-xylose or glucose after 3 d of microaerophilic incubation at 30°C are 1 mm or less in diameter, round, and with smooth surfaces. The colonies are smaller with glucose than with D-xylose (growth is enhanced by replacing glucose with D-xylose). Obligately heterofermentative. Gas is produced from gluconic acid but not from glucose. None of the three strains reduces tetrazolium on SBM agar, nor do they grow in Clark & Lubbs broth used in the Voges–Proskauer test. Able to grow on acetate agar but unable to acidify La-broth below pH 4.15. No growth in the presence of 6.5 or 8% NaCl. Resistant towards 30 μg vancomycin/g. Additional physiological and biochemical characteristics are presented in Table 86.

Isolated from a Malaysian acid-fermented condiment, tempoyak, made from the durian fruit.

Phylogenetic position: unique.

Type strain: CCUG 45405, JCM 12184, LMG 19193.

DNA G+C content (mol%): 43.2–43.3 (HPLC).

GenBank accession number (16S rRNA gene): AJ315640.

27. **Lactobacillus equi** Morotomi, Yuki, Kado, Kushiro, Shimazaki, Watanabe and Yuyama 2002, 214[VP]

e'qui. L. n. *equus* horse; L. gen. n. *equi* of the horse.

Cells are nonmotile rods (0.6–0.8 × 1.3–3.5 μm) occurring singly and in pairs. Colonies on MRS agar are white, smooth, convex, and approximately 2 mm in diameter. Some strains contain filamentous cells. Catalase-negative. Facultatively heterofermentative, as several pentoses are fermented. Additional physiological and biochemical characteristics are presented in Table 85. Acid is not produced from glycerol. Reactions of esculin hydrolysis and acid production from rhamnose, sorbitol, N-acetylglucosamine, ribose, mannose, and D-turanose are variable.

Isolated from feces of horses.

Phylogenetic position: Lactobacillus salivarius group.

Type strain: YIT 0455, ATCC BAA-261, JCM 10991.

DNA G+C content (mol%): 38.9 ± 0.8 (HPLC).

GenBank accession number (16S rRNA gene): AB048833.

28. **Lactobacillus farciminis** Reuter 1983a, 672[VP] (Effective publication: Reuter 1983b, 278.)

far.ci'mi.nis. L. n. *farcimen* sausage; L. gen. n. *farciminis* of sausage.

Cells are nonmotile, slender rods (0.6–0.8 × 2–6 μm) occurring singly and in short chains. Grows in the presence of 10% NaCl and occasionally 12% NaCl. Arginine is hydrolyzed. Obligately homofermentative. Additional physiological and biochemical characteristics are presented in Table 84.

Isolated from meat products (fermented sausages) and sourdough.

Phylogenetic position: unique.

Type strain: ATCC 29644, DSM 20184, CCUG 30671, CIP 103136, JCM 1097, LMG 9200, NRRL B-4566.

DNA G+C content (mol%): 34–36 (T_m).

GenBank accession number (16S rRNA gene): M58817.

29. **Lactobacillus ferintoshensis** Simpson, Pettersson and Priest 2002a, 1075[VP] (Effective publication: Simpson, Pettersson and Priest 2001, 1015.)

fe.rin.to.shen'sis. N.L. adj. *ferintoshensis* from Ferintosh, a Scottish estate famous for its whisky.

Cells are nonmotile rods (3–4 μm × 1 μm) occurring singly, in pairs, or in short chains. After 48 h incubation on modified MRS agar (MRS agar supplemented with vitamins), colonies are 2–5 mm in diameter, circular, shiny, and creamy white in color. Obligately heterofermentative. Additional physiological and biochemical characteristics are presented in Table 86.

Isolated from malt whisky fermentations.

Phylogenetic position: unique

Type strain: R7-84, CIP 106749, JCM 12511.

DNA G+C content (mol%): not determined.

GenBank accession number (16S rRNA gene): AF275311.

30. **Lactobacillus fermentum** Beijerinck 1901, 233[AL]

fer.men′tum. L. n. *fermentum* ferment, yeast.

Cells are nonmotile rods (0.5–0.9 μm thick and highly variable in length), occurring singly or in pairs. Obligately heterofermentative. Additional physiological and biochemical characteristics are presented in Table 86. Growth factor requirements: calcium pantothenate, niacin, and thiamine are essential; riboflavin, pyridoxal, and folic acid not required.

Isolated from yeast, milk products, sourdough, fermenting plant material, manure, sewage, and mouth and feces of humans.

Phylogenetic position: Lactobacillus reuteri group.

Type strain: ATCC 14931, CCUG 30138, CIP 102980, DSM 20052, NBRC 15885, JCM 1173, LMG 6902, NCCB 46038, NCIMB 11840, NRRL B-4524.

DNA G+C content (mol%): 52–54 (Bd, T_m).

GenBank accession number (16S rRNA gene): M58819, X61142.

Additional remarks: Lactobacillus fermentum cannot be definitely distinguished from *Lactobacillus reuteri* by simple physiological tests. The genotypic methods used provide clear results (Dellaglio et al., 2004). *Lactobacillus cellobiosus* Rogosa et al. 1953, 693[AL] has been reclassified as a later synonym of *Lactobacillus fermentum* Beijerinck (1901) by Dellaglio et al. (2004).

31. **Lactobacillus fornicalis** Dicks, Silvester, Lawson and Collins 2000, 1258[VP]

for.nic.a′lis. N.L. adj. *fornicalis* pertaining to the posterior fornix.

Cells are nonmotile rods. Colonies on MRS agar are round, smooth, white, and approximately 1 mm in diameter. Grows well on the surface of MRS agar when not incubated under microaerophilic conditions. Facultatively heterofermentative, with no gas production from glucose or gluconate, but ribose is fermented. Voges–Proskauer-negative. Indole is not formed. Polysaccharides are not produced from sucrose. Additional physiological and biochemical characteristics are presented in Table 85.

Isolated from the posterior fornix fluid of the human vagina.

Phylogenetic position: Lactobacillus delbrueckii group.

Type strain: TV 1018, ATCC 700934, CCUG 43621, CIP 106679, DSM 13171, JCM 12512.

DNA G+C content (mol%): 37 (T_m).

GenBank accession number (16S rRNA gene): Y18654.

32. **Lactobacillus fructivorans** Charlton, Nelson And Werkman 1934, 1[AL]

fruc.ti.vo′rans. L. n. *fructus* fruit; L. v. *vorare* to eat; N.L. pres. part. *fructivorans* fruit-eating, intended to mean fructose-devouring.

Cells are nonmotile rods (0.5–0.8 × 1.5–4 μm) with rounded ends, occurring singly, in pairs, and in chains; very long, more or less curved or coiled filaments often observed. Acidophilic; favorable pH is 5.0–5.5; no growth at an initial pH higher than 6.0. Nutritionally very exacting, at least on primary isolation. Depending on the source of isolation, mevalonic acid, tomato juice, and/or ethanol are required for growth. Some strains, especially those isolated from non-alcohol-containing sources, often become less fastidious during laboratory transfers and grow well in MRS broth. Obligately heterofermentative. Additional physiological and biochemical characteristics are presented in Table 86.

Isolated from spoiled mayonnaise, salad dressings, vinegar preserves, spoiled sake, dessert wines, and aperitifs.

Phylogenetic position: unique.

Type strain: ATCC 8288, CCUG 32260, CIP 103042, DSM 20203, NBRC 13954, JCM 1117, LMG 9201, NRRL B-1841.

DNA G+C content (mol%): 38–40 (T_m).

GenBank accession number (16S rRNA gene): M58818, X76330.

Additional remarks: Lactobacillus trichodes Fornachon et al. 1949, 129[AL] and *Lactobacillus heterohiochii* Kitahara et al. 1957, 117[AL] have been reclassified as later synonyms of *Lactobacillus fructivorans* by Weiss et al. (1983a).

33. **Lactobacillus frumenti** Müller, Ehrmann and Vogel 2000b, 2132[VP]

fru.men′ti. L. gen. n. *frumenti* from cereal.

Cells are nonmotile rods occurring singly or in pairs (seldom in chains). Obligately heterofermentative. Additional physiological and biochemical characteristics are presented in Table 86.

Isolated from sourdough.

Phylogenetic position: Lactobacillus reuteri group.

Type strain: TMW 1.666, CIP 106922, DSM 13145, JCM 11122, LMG 19473.

DNA G+C content (mol%): 43–45 (HPLC).

GenBank accession number (16S rRNA gene): AJ250074.

34. **Lactobacillus fuchuensis** Sakala, Kato, Hayashidani, Murakami, Kaneuchi and Ogawa 2002, 1153[VP]

fu.chu.en′sis. N.L. adj. *fuchuensis* of Fuchu, the city in which the bacterium was originally isolated.

Cells are nonmotile, straight, and curved rods (0.5–0.75 μm × 2–6 μm) occurring singly, in pairs, or short chains. Colonies on MRS agar plates after 48 h are small, about 1.0–2.0 mm in diameter, smooth, entire, convex, and cream colored. No growth is observed at 37°C in MRS broth after 48 h incubation or on acetate agar (pH 5.4). No gas is produced from glucose or gluconate. Facultatively heterofermentative. Additional physiological and biochemical characteristics are presented in Table 85. Alanine-phenylalanine-proline arylamidase, arginine dehydrolase, α-glucosidase, N-acetyl-β-glucosaminidase, glycyltryptophan arylamidase, and β-mannosidase activities are detected. No activity for β-galactosidase, β-glucuronidase, α-galactosidase, alkaline phosphatase, or pyrrolidonyl arylamidase is detected. Hippurate is not hydrolyzed. Acetoin is produced weakly. Urease is not produced. *m*-Dmp is not detected in the cell wall.

Isolated from vacuum-packaged refrigerated beef. Habitat is not known.

Phylogenetic position: unique.

Type strain: B5M10, DSM 14340, JCM 11249.

DNA G+C content (mol%): 41–41.7 (HPLC).

GenBank accession number (16S rRNA gene): AB063479.

35. **Lactobacillus gallinarum** Fujisawa, Benno, Yaeshima and Mitsuoka 1992, 489[VP]

gallin.ar'um. L.n. *gallina* the hen. L. gen. pl. *gallinarum* of hens.

Cells from BL agar plate cultures are short to long rods (0.5–1.5 × 1.5–10 μm) occurring singly, in pairs, and sometimes in short chains. Colonies on BL agar are 0.5–2.0 mm in diameter, circular to slightly irregular, entire, grayish brown to reddish brown, and rough. Obligately homofermentative. Additional physiological and biochemical characteristics are presented in Table 84. Strains are tolerant to 4.0% NaCl. (This property differentiates the species from *Lactobacillus acidophilus*, *Lactobacillus amylovorus*, and *Lactobacillus crispatus*).

Isolated from chicken intestine.

Phylogenetic position: Lactobacillus delbrueckii group.

Type strain: ATCC 33199, CCUG 30724, CIP 103611, DSM 10532, JCM 2011, LMG 9435, VPI 1294.

DNA G+C content (mol%): 35.9–37.2 (T_m).

GenBank accession number (16S rRNA gene): AJ242968, AJ417737.

Additional remarks: Synonymous with *Lactobacillus acidophilus* group A4 of Johnson et al. (1980).

36. **Lactobacillus gasseri** Lauer and Kandler 1980a, 601[VP] (Effective publication: Lauer and Kandler 1980b, 77.)

gas'se.ri. N.L. gen. n. *gasseri* of Gasser, named for F. Gasser, a French bacteriologist.

Cells are rods with rounded ends (0.6–0.8 × 3.0–5.0 μm) occurring singly and in chains. Formation of mini-cells and snakes is frequently observed. Starch is fermented by most strains. Unlike all other lactobacilli, the d-alanyl-d-alanine termini of peptide subunits of peptidoglycan not involved in cross-linkage are preserved because of the lack of D,D-carboxypeptidase action. Arginine is not hydrolyzed. Obligately homofermentative. Additional physiological and biochemical characteristics are presented in Table 84.

Isolated from the human mouth and vagina and from the intestinal tract of man and animals; also found in wounds, urine, blood, and pus of patients suffering from septic infections.

Phylogenetic position: Lactobacillus delbrueckii group.

Type strain: 63 AM of Gasser, ATCC 33323, CCUG 31451, CIP 102991, DSM 20243, JCM 1131, LMG 9203, NRRL B-14168, NRRL B-4240.

DNA G+C content (mol%): 33–35 (T_m).

GenBank accession number (16S rRNA gene): AF519171, M58820, X61137.

Additional remarks: Synonymous with *Lactobacillus acidophilus* group B-1 of Johnson et al. (1980). See comments on *Lactobacillus acidophilus*.

37. **Lactobacillus gastricus** Roos, Engstrand and Jonsson 2005, 80[VP]

gas'tri.cus. N.L. masc. adj. *gastricus* from Gr. adj. *gastrikos* of the stomach.

Cells are nonmotile rods (0.9 × 1.2 μm) occurring as single cells or in pairs. After anaerobic growth at 37°C for 48 h, colonies on MRS agar are 2 mm in diameter; they are white, smooth, and convex. Growth on MRS agar under aerobic conditions is very weak. In MRS broth, growth occurs at 25 and 42°C, but not at 20 or 45°C. Gas is produced from glucose. Obligately heterofermentative. Additional physiological and biochemical characteristics are presented in Table 86.

Isolated from a biopsy of the healthy human gastric mucosa.

Phylogenetic position: Lactobacillus reuteri group.

Type strain: Kx156A7, DSM 16045, CCUG 48454, LMG 22113.

DNA G+C content (mol%): 41.3 (HPLC)

GenBank accession number (16S rRNA gene): AY253658.

38. **Lactobacillus graminis** Beck, Weiss and Winter 1989, 93[VP] (Effective publication: Beck, Weiss and Winter 1988, 282.)

gra'mi.nis. L. n. *gramen* grass, N.L. gen. *graminis* of grass.

Cells are nonmotile, slightly curved rods with rounded ends (0.7–1 × 1.5–2 μm) occurring singly, in pairs, or in short chains, slightly curved. Colonies smooth, round, nonpigmented. Flocculant sediment after 3 d of growth in MRS broth. Facultatively heterofermentative. Additional physiological and biochemical characteristics are presented in Table 85. Cell wall teichoic acid not detectable.

Habitat: grass silage.

Phylogenetic position: unique.

Type strain: G90 (1), ATCC 51150, CCUG 32238, CIP 105164, DSM 20719, JCM 9503, LMG 9825, NRRL B-14857.

DNA G+C content (mol%): 41–43 (T_m).

GenBank accession number (16S rRNA gene): AJ621551, AM113778.

39. **Lactobacillus hammesii** Valcheva, Korakli, Onno, Prévost, Ivanova, Ehrmann, Dousset, Gänzle and Vogel 2005, 766[VP]

ham.me.sii. N.L. gen. n. *hammesii* of Hammes, named for Walter P. Hammes, a German microbiologist who contributed to the microbiological and technological development of wheat and rye sourdough research.

Cells are nonmotile straight rods (0.5 × 2–4 μm) occurring as single in pairs or occasionally in short chains. Colonies on MRS agar appeared white circular with a smooth surface and edges (1–1.5 mm after 2 d of growth). Good growth in liquid or solid MRS in aerobiosis. The optimal initial pH ranges from 4.7 to 7.2. In MRS4 at pH 6.2 and at 30°C the specific growth rate of strain LP38 is 0.42 ± 0.01/h. Strain LP38 grows well up to a NaCl content of 2%, and at a salt content of 6.6% the specific growth rate was 26% (100% without NaCl). Obligately heterofermentative.

Additional physiological and biochemical characteristics are presented in Table 86. Fructose is either used as an energy source or as an electron acceptor and is reduced to mannitol. Strain LP38 grows better in media containing electron acceptors such as fructose in addition to a carbon source (maltose or glucose). Strain LP38 produces lactic acid and ethanol from glucose or maltose and in the presence of fructose produces lactic and acetic acid.

Isolated from sourdough.

Phylogenetic position: unique.

Type strain: LP38, CIP 108387, DSM 16381, TMW 1.1236.

DNA G+C content (mol%): 52.6 (HPLC).

GenBank accession number (16S rRNA gene): AJ632219.

40. **Lactobacillus hamsteri.** Mitsuoka and Fujisawa, 1988, 220[VP] (Effective publication: Mitsuoka and Fujisawa, 1987, 272.)

hams'te.ri. N.L. gen. n. *hamsteri* of the hamster from which the isolate was derived.

Cells are nonmotile, long stout rods (1.0–1.3 μm × 5.0–10.0 μm) occurring singly, in pairs, and in short chains. Surface colonies on BL agar (2 d anaerobic incubation) are 0.7–3.0 mm in diameter, round, umbonate, brown in color, with a rough surface and erose edge. Strictly anaerobic. Final pH of glucose broth is 3.7. Facultatively heterofermentative. Additional physiological and biochemical characteristics are presented in Table 85.

Habitat: hamster intestine and feces. Isolated from hamster feces.

Phylogenetic position: Lactobacillus delbrueckii group.

Type strain: Ha5F1, ATCC 43851, CCUG 30726, CIP 103365, DSM 5661, JCM 6256, LMG 10754.

DNA G+C content (mol%): 33.1–35.1 (T_m).

GenBank accession number (16S rRNA gene): AJ306298.

41. **Lactobacillus helveticus** (Orla-Jensen 1919) Bergey, Harrison, Breed, Hammer and Huntoon 1925, 184[AL] (*Thermobacterium helveticum* Orla-Jensen 1919, 164.)

hel.ve'ti.cus. L. adj. *helveticus* Swiss.

Cells are nonmotile rods (0.7–0.9 to 6.0 μm) occurring singly and in chains. Lactose agar colonies are small, grayish, and viscid. Maximum growth temperature 50–52°C. Arginine is not hydrolyzed. Obligately homofermentative. Additional physiological and biochemical characteristics are presented in Table 84. Growth factor requirements: calcium pantothenate, niacin, riboflavin, pyridoxal, or pyridoxamine is essential; thiamine, folic acid, vitamin B_{12}, and thymidine are not required.

Isolated from sour milk, cheese starter cultures, and cheese, particularly Emmental and Gruyère cheese.

Phylogenetic position: Lactobacillus delbrueckii group.

Type strain: ATCC 15009, CCUG 30139, CIP 103146, DSM 20075, NBRC 15019, JCM 1120, LMG 6413, LMG 13555, NRRL B-4526.

DNA G+C content (mol%): 37–40 (Bd, T_m).

GenBank accession number (16S rRNA gene): AM113779, X61141.

42. **Lactobacillus hilgardii** Douglas and Cruess 1936, 115[AL]

hil.gar'di.i. N.L. gen. n. *hilgardii* of Hilgard, named for Eugene W. Hilgard, a German-American scientist.

Cells are nonmotile rods with rounded ends (0.5–0.8 × 2–4 μm) occurring singly, in short chains, and frequently in long filaments. Growth on solid media is poor, colonies are punctiform, white, glistening, translucent, with edges entire. Optimal initial pH for growth and carbohydrate fermentation reactions is in the range of 4.5–5.5. Grows in the presence of 15–18% ethanol. End products from fructose are lactic acid, acetic acid, and CO_2. Mannitol is not formed. Obligately heterofermentative. Additional physiological and biochemical characteristics are presented in Table 86.

Isolated from California table wines; widely distributed in wines of different origin.

Phylogenetic position: unique.

Type strain: ATCC 8290, CCUG 30140, CIP 103007, DSM 20176, NBRC 15886, JCM 1155, LMG 6895, NRRL B-1843.

DNA G+C content (mol%): 39–41 (T_m).

GenBank accession number (16S rRNA gene): M58821.

43. **Lactobacillus homohiochii** Kitahara, Kaneko and Goto 1957, 118[AL]

ho'mo.hi.o'chi.i. Gr. adj. *homos* like, equal; Japanese n. *hiochi* spoiled sake; N.L. gen. n. *homohiochii* probably intended to mean homofermentative lactobacillus of hiochi.

Cells are rods with rounded ends (0.7–0.8 × 2–4 μm or, occasionally, 6 μm) Does not grow in MRS broth. In Rogosa SL broth supplemented with DL-mevalonic acid (30 mg/liter) and ethanol (40 mg/liter), copious growth is obtained at 30°C after a marked lag phase of 4–7 d. No growth at 45°C and at an initial pH higher than 5.5. Resistant to 13–16% ethanol. Facultatively heterofermentative. Additional physiological and biochemical characteristics are presented in Table 85. Growth factor requirements: D-mevalonic acid is essential or highly stimulatory; ethanol is promotive.

Isolated from spoiled sake.

Phylogenetic position: unique.

Type strain: ATCC 15434, CCUG 32247, CIP 103141, DSM 20571, NBRC 15887, JCM 1199, JCM 7793, LMG 9478, NRRL B-4559.

DNA G+C content (mol%): 35–38 (T_m).

GenBank accession number (16S rRNA gene): AJ621552, AM113780.

44. **Lactobacillus iners** Falsen, Pascual, Sjödén, Ohlén and Collins 1999, 220[VP]

L. adj. *iners* inert, lazy.

Cells are nonmotile rods and occur singly, in pairs, or in short chains. After 24 h incubation on blood agar under anaerobic conditions, colonies are small (<1 mm diameter), smooth, circular, translucent, and nonpigmented. Old cultures display a "fried egg" appearance. Grows on blood agar but not MRS agar or Rogosa agar. Grows in the presence of 1.5% NaCl, but not 3% NaCl. Obligately homofermentative. Additional physiological and biochemical characteristics are presented in Table 84. Positive for alanine-phenylalanine-proline arylamidase, esterase C4, α-glucosidase, leucine arylamidase, and phosphoamidase. Negative for alkaline phosphatase, arginine dihydrolase, chymotrypsin, α-fucosidase, α-galactosidase, β-galactosidase, β-galacturonidase, β-glucosidase, β-glucuronidase, glycyl-tryptophan arylamidase, α-mannosidase, β-mannosidase, lipase C14, trypsin, and urease. Acid phosphatase, ester lipase C8, and pyroglutamic acid arylamidase activity is detected in some strains. Hippurate is hydrolyzed. Voges–Proskauer-negative.

Isolated from human clinical specimens (urine, vagina) and medical care products.

Phylogenetic position: Lactobacillus delbrueckii group.

Type strain: CCUG 28746, CIP 105923, DSM 13335, JCM 12513, LMG 18914.

DNA G+C content (mol%): 34.4 (T_m).

GenBank accession number (16S rRNA gene): Y16329.

45. **Lactobacillus ingluviei** Baele, Vancanneyt, Devriese, Lefebvre, Swings and Haesebrouck 2003, 135[VP]

in.glu′vi.ei. L. n. *ingluvies* crop sac; L. gen. n. *ingluviei* of a crop sac.

Cells are nonmotile, very short, plump rods, rapidly decolorizing in the Gram-stain procedure. Cells mostly occur singly or in pairs, and some appear to be slightly longer than others. Colonies are white and smooth or crumbly and dry. Growth is enhanced under anaerobic conditions and also slightly in the presence of 5% CO_2, compared with aerobic growth. Better growth is obtained at 42°C than at 37°C. No growth occurs at 25°C and growth is poor at 30°C. The strains grow as nonhemolytic streptococcus-like colonies on Columbia blood agar with diameters of up to 0.5 mm. Obligately heterofermentative. Additional physiological and biochemical characteristics are presented in Table 86. The characteristic tDNA-PCR fingerprint is composed of fragments with lengths of 162.5, 176.5, 185.5, and 255.2 bp, as determined by fluorescent capillary electrophoresis.

Habitat is pigeon crop and intestines.

Phylogenetic position: Lactobacillus reuteri group.

Type strain: KR3, CCUG 45722, JCM 12531, LMG 20380.

DNA G+C content (mol%): 49 (HPLC).

GenBank accession number (16S rRNA gene): AF333975.

46. **Lactobacillus intestinalis** (*ex* Hemme 1974) Fujisawa, Itoh, Benno and Mitsuoka 1990, 303[VP]

in.tes.tin.al′is. N.L. adj. *intestinalis* pertaining to the intestine.

Cells are nonmotile rods (0.6–0.8 × 2.0–6.0 μm) occurring singly, in pairs, or occasionally in short chains. Surface colonies on glucose-blood-liver agar after 2 d of anaerobic incubation at 37°C are 0.7–2.5 mm in diameter, round, flat, and light brown with rough surfaces and erose edges. No gas from glucose and ribose is fermented. Facultatively heterofermentative. The final pH of glucose broth is 3.7. Additional physiological and biochemical characteristics are presented in Table 85.

The habitat of the species is the intestines of mice and rats.

Phylogenetic position: Lactobacillus delbrueckii group.

Type strain: Th4, ATCC 49335, CCUG 30727, CIP 104793, DSM 6629, JCM 7548, LMG 14196.

DNA G+C content (mol%): 32.5–35.4 (T_m).

GenBank accession number (16S rRNA gene): AJ306299.

Additional remarks: First described as *Thermobacterium* Group Th4 by Raibaud et al. (1973) and as *Lactobacillus intestinalis* (Hemme, 1974) but not validly published.

47. **Lactobacillus jensenii** Gasser, Mandel and Rogosa 1970, 221[AL]

jen.se′ni.i. N.L. gen. n. *jensenii* of Jensen; named for S. Orla-Jensen, a Danish microbiologist.

Cells are rods with rounded ends (0.6–0.8 × 2.0–4.0 μm) occurring singly and in short chains. Arginine is hydrolyzed. Facultatively heterofermentative. Additional physiological and biochemical characteristics are presented in Table 85. Folic acid, vitamin B_{12}, nicotinic acid, and calcium pantothenate required for growth.

Isolated from human vaginal discharge and blood clot.

Phylogenetic position: Lactobacillus delbrueckii group.

Type strain: ATCC 25258, CCUG 21961, CCUG 35572, CIP 69.17, DSM 20557, JCM 1146, LMG 6414, NRRL B-4550.

DNA G+C content (mol%): 35–37 (Bd).

GenBank accession number (16S rRNA gene): AF243176.

Additional remarks: Lactobacillus jensenii is indistinguishable from *Lactobacillus delbrueckii* by simple physiological tests.

48. **Lactobacillus johnsonii** Fujisawa, Benno, Yaeshima and Mitsuoka 1992, 489[VP]

john.so′ni.i. M.L. gen. n. *johnsonii* of Johnson; named for J.L. Johnson, an American microbiologist.

Cells from BL agar plate cultures are short to long rods (0.5–1.5 × 1.5–10 μm) occurring singly, in pairs, and sometimes in short chains. Colonies on BL agar are 0.5–2.0 mm in diameter, circular to slightly irregular, entire, grayish brown to reddish brown, and rough. Obligately homofermentative. Tolerant to 4% NaCl. Additional physiological and biochemical characteristics are presented in Table 84.

Isolated from human blood and feces of chicken, mice, calf, and pig.

Phylogenetic position: Lactobacillus delbrueckii group.

Type strain: ATCC 33200, CCUG 30725, CIP 103620, DSM 10533, JCM 2012, VPI 7960.

DNA G+C content (mol%): 32.7–34.8 (T_m).

GenBank accession number (16S rRNA gene): AJ002515.

Additional remarks: Synonymous with *Lactobacillus acidophilus* group B2 of Johnson et al. (1980). See comments on *Lactobacillus acidophilus*.

49. **Lactobacillus kalixensis** Roos, Engstrand and Jonsson 2005, 81[VP]

ka.lix.en′sis. N.L. masc. adj. *kalixensis* pertaining to Kalix, a town in northern Sweden where the gastric biopsies were sampled.

Cells are nonmotile rods (1 × 1.5–10 μm) occurring as singly, in pairs, or in chains. After anaerobic growth at 37°C for 48 h, colonies on MRS agar are 2 mm in diameter; they are white, smooth, and convex. Growth on MRS agar under aerobic conditions is weak. In MRS broth, growth occurs at 25 and 45°C, but not at 20°C. Gas is not produced from glucose. Obligately homofermentative. Additional physiological and biochemical characteristics are presented in Table 84.

Isolated from a biopsy of the healthy human gastric mucosa.

Phylogenetic position: Lactobacillus delbrueckii group.

Type strain: Kx127A2, CCUG 48459, DSM 16043, LMG 22115.

DNA G+C content (mol%): 35.5 (HPLC).

GenBank accession number (16S rRNA gene): AY253657.

50. **Lactobacillus kefiranofaciens** Fujisawa, Adachi, Toba, Arihara and Mitsuoka 1988, 13[VP] emend. Vancanneyt, Mengaud, Cleenwerck, Vanhonacker, Hoste, Dawyndt, Degivry, Ringuet, Janssens and Swings 2004, 555.

ke.fi.ra.no.fa'ci.ens. N.L. n. *kefiran* a polysaccharide of kefir grain, kefiran; L. v. *facio* produce; N.L. part. adj. *kefiranofaciens* kefiran-producing.

Cells are nonmotile rods (0.5–1.2 × 3.0–20 μm) occurring singly, in pairs, or occasionally in short chains. Colony morphology is subspecies-dependent (see below). Obligately homofermentative. Habitat of the species is kefir grains. The species contains two subspecies.

Phylogenetic position: Lactobacillus delbrueckii group.

Type strain: WT-2B, ATCC 43761, CCUG 32248, CIP 103307, DSM 5016, JCM 6985, LMG 19149.

DNA G+C content (mol%): 37.3–38.2 (HPLC) or 34.3–38.6 (T_m).

50a. **Lactobacillus kefiranofaciens subsp. kefiranofaciens** (Fujisawa, Adachi, Toba, Arihara and Mitsuoka 1988) Vancanneyt, Mengaud, Cleenwerck, Vanhonacker, Hoste, Dawyndt, Degivry, Ringuet, Janssens and Swings 2004, 555[VP] (Fujisawa, Adachi, Toba, Arihara and Mitsuoka 1988)

The description is as that for the species, with the following additional characteristics. Cells are capsulated. On modified KPL agar (pH 5.5) at 30°C after 10 d, colonies are circular or irregular, 0.5–3.0 mm in diameter, convex, transparent to translucent, white, smooth to rough, and ropy. On MLR medium under anaerobic conditions after 7–14 d incubation at 25 or 30°C, colonies are transparent, glossy, convex, and extremely slimy. Large amounts of polysaccharides are produced. Hydrolysis of esculin is negative. Additional physiological and biochemical characteristics are presented in Table 84.

Habitat of the species is kefir grains.

Type strain: WT-2B, ATCC 43761, CCUG 32248, CIP 103307, DSM 5016, JCM 6985, LMG 19149.

DNA G+C content (mol%): 37.3–38.2 (HPLC) or 34.3–38.6 (T_m).

GenBank accession number (16S rRNA gene): AJ575279, AM113781.

50b. **Lactobacillus kefiranofaciens subsp. kefirgranum** (Takizawa, Kojima, Tamura, Fujinaga, Benno and Nakase 1994) Vancanneyt, Mengaud, Cleenwerck, Vanhonacker, Hoste, Dawyndt, Degivry, Ringuet, Janssens and Swings 2004, 555[VP] (*Lactobacillus kefirgranum* Takizawa, Kojima, Tamura, Fujinaga, Benno and Nakase 1994, 438.)

ke.fir.gra'num. Turkish n. *kefir* Caucasian sour milk; L. n. *granum* grain; N.L. adj. *kefirgranum* kefir grain.

The description is as for the species, with the following additional characteristics. On R-CW agar (Kojima et al.,

1993) at 30°C for 5 d, colonies are 0.5–3.0 mm in diameter, circular to irregular, convex, opaque, white, and smooth to rough. On MLR medium under anaerobic conditions after 7–14 d incubation at 25 or 30°C, colonies are white, dry, compact, dull, and bulging. Flocculus or powdery sediment is formed in broth. Weak growth occurs at 15°C. Hydrolysis of esculin is positive. Additional physiological and biochemical characteristics are presented in Table 84.

Habitat of the species is kefir grains.

Type strain: GCL 1701, ATCC 51647, CCUG 39467, CIP 104241, DSM 10550, JCM 8572, LMG 15132.

DNA G+C content (mol%): 37.3–38.2 (HPLC) or 34.3–38.6 (T_m).

GenBank accession number (16S rRNA gene): AJ575742, AJ575261, AM113782.

51. **Lactobacillus kefiri** corrig. Kandler and Kunath 1983, 672[VP] (Effective publication: Kandler and Kunath 1983, 292.)

ke'fir.i. Latinized Turkish n. *kefirum* kefir, a Caucasian sour milk (Trüper and de Clari, 1997). N.L. gen. n. *kefiri* of kefir.

Cells are nonmotile rods with rounded ends (0.6–0.8 × 3.0–15 μm) with tendency to form chains of short rods or long filaments, often containing polyphosphate granules usually terminal. Colonies on MRS agar are grayish, smooth and flat, 2–4 mm in diameter. Obligately heterofermentative. Additional physiological and biochemical characteristics are presented in Table 86.

Isolated from kefir grains and drink kefir.

Phylogenetic position: unique.

Type strain: A/K, ATCC 35411, CCUG 30673, CIP 103006, DSM 20587, NBRC 15888, JCM 5818, LMG 9480.

DNA G+C content (mol%): 41–42 (T_m).

GenBank accession number (16S rRNA gene): AB024300, AJ621553, AY579584.

52. **Lactobacillus kimchii** Yoon, Kang, Mheen, Ahn, Lee, Kim, Park, Kho, Kang and Park 2000, 1794[VP]

kim'chi.i. N.L. gen. n. *kimchii* of kimchi, a Korean fermented-vegetable food.

Cells are short, slender, nonmotile rods (0.6–0.8 × 1.5–3.0 μm) occurring singly, in pairs, or occasionally in short chains. After 3 d incubation on MRS agar, colonies are white, circular to slightly irregular, convex, smooth, opaque, and approximately 0.8–1.5 mm in diameter. Catalase-negative. Casein is hydrolyzed but starch and urea are not. Arginine is not hydrolyzed. Indole is not produced. Facultatively heterofermentative. Optimal temperature and pH for growth are approximately 30°C and 6.0–7.0. Grows in the presence of 8% NaCl but not in the presence of 10% NaCl. Additional physiological and biochemical characteristics are presented in Table 85. The major fatty acid is the unsaturated fatty acid $C_{18:1 \omega 9c}$.

Isolated from kimchi, a Korean fermented vegetable food.

Phylogenetic position: unique.

Type strain: MT-1077, ATCC BAA-131, CIP 107019, DSM 13961, JCM 10707, KCTC 8903P.

DNA G+C content (mol%): 35 (HPLC).

GenBank accession number (16S rRNA gene): AF183558.

53. **Lactobacillus kitasatonis** Mukai, Arihara, Ikeda, Nomura, Suuki and Ohori 2003, 2057[VP]

ki.ta.sa.to'nis. L. gen. n. *kitasatonis* referring to Shibasaburo Kitasato, the founder of Kitasato Institute, the father of Japanese bacteriology.

Cells are nonmotile rods (0.6–1.0 × 2.0–20.0 μm) occurring singly, in pairs, or occasionally in short chains. When the organism is grown on MRS agar at 37°C for 2 d, colonies are 1.2–2.1 mm in diameter, circular to slightly irregular, convex, opaque, rough, and white. Obligately homofermentative. Milk is not curdled. Additional physiological and biochemical characteristics are presented in Table 84.

Habitat is the intestine of chickens.

Phylogenetic position: Lactobacillus delbrueckii group.

Type strain: JCM 1039, KCTC 3155.

DNA G+C content (mol%): 37–40 (HPLC).

GenBank accession number (16S rRNA gene): AB107638.

54. **Lactobacillus kunkeei** Edwards, Haag, Collins, Hutson and Huang 1998, 1083[VP] (Effective publication: Edwards, Haag, Collins, Hutson and Huang 1998, 700.)

kun'ke.ei L. gen. n. *kunkeei* of Kunkee. Named after Ralph E. Kunkee, University of California, Davis, CA, USA, for his contributions to the microbiology of wines.

Cells are rods (0.5 × 1–1.5 μm). Colonies on MR agar appear opaque, concave, and approximately 1–2 mm in diameter after 3 d of growth at 25°C. Catalase-positive (weakly). Obligately heterofermentative. Acid and gas are produced from glucose. Additional physiological and biochemical characteristics are presented in Table 86. Citrate or malate is utilized in the presence of glucose. Growth in MR media at pH 3.7, 4.5, or 8.0 (25°C). Weak growth in 5% NaCl.

Isolated from partially fermented grape juice.

Phylogenetic position: unique.

Type strain: YH-15, ATCC 700308, DSM 12361.

DNA G+C content (mol%): not determined.

GenBank accession number (16S rRNA gene): Y11374.

55. **Lactobacillus lindneri** (*ex* Henneberg 1901) Back, Bohak, Ehrmann, Ludwig and Schleifer 1997, 601[VP] (Effective publication: Back, Bohak, Ehrmann, Ludwig and Schleifer 1996, 324.)

lind'ne.ri. L. gen. n. *lindneri* of Lindner, referring to P. Lindner, a German bacteriologist.

Cells are nonmotile rods (0.9–5 μm), occurring singly, in pairs, or in chains. Depending on the medium, some strains show a strong tendency to alter their cell length and chain length. Colonies are usually small (2 mm), smooth, low convex, flat, and white on NBB-agar. Obligately heterofermentative. Growth occurs at 15–30°C but not above, optimum at 22–25°C; growth occurs at inital pH value of 5.8, some strains at 5.4; optimum at pH 4.6–5.2; no growth at or above pH 6.5. Acid and gas are produced from glucose. Additional physiological and biochemical characteristics are presented in Table 86. Urease is not produced. Cell wall does not contain teichoic acid.

Isolated from spoiled beer.

Phylogenetic position: unique.

Type strain: KPA, CIP 102983, DSM 20690, JCM 11027, LMG 14528.

DNA G+C content (mol%): 35 (T_m).

GenBank accession number (16S rRNA gene): X95421.

56. **Lactobacillus malefermentans** (*ex* Russell and Walker 1953) Farrow, Phillips and Collins 1989, 371[VP] (Effective pubication: Farrow, Phillips and Collins 1988, 165.)

ma.le.fer.men'tans. L. adj. *malus* bad. L. part. adj. *fermentans* fermenting; N.L. adj. *malefermentans* badly fermenting, referring to spoiled beer.

Cells are nonmotile rods (0.8–1.0 × 1.5–2.5 μm) ocurring singly, in pairs, or in chains. Gas is produced from glucose. Obligately heterofermentative. Additional physiological and biochemical characteristics are presented in Table 86. Urease is not produced. Voges–Proskauer-negative.

Isolated from beer.

Phylogenetic position: unique.

Type strain: D2 MF1, ATCC 49373, CCUG 32206, CIP 103367, DSM 5705, NBRC 15905, JCM 12497, LMG 11455, NCIMB 701410.

DNA G+C content (mol%): 41–42 (T_m).

GenBank accession number (16S rRNA gene): AJ575743, AM113783.

57. **Lactobacillus mali** Carr and Davies, 1970, 769[AL] emend. Kaneuchi, Seki and Komagata 1988, 272.

ma'li. L. n. *malus* apple; L. gen. n. *mali* of the apple.

Cells are rods (0.5–0.8 × 1.2–1.3 μm) occurring singly, in pairs, and in short chains. Nonmotile or weakly motile with a few peritrichous flagella. Colonies are white, smooth, and glistening after a few days. Liquid cultures are turbid after few days, with subsequent clearing and sediment. Obligately homofermentative (see below). Additional physiological and biochemical characteristics are presented in Table 84. No change in litmus milk. Dextran is produced from sucrose. Mannitol is not produced from fructose. Arginine is not hydrolyzed. Esculin is hydrolyzed. Acetoin is produced. H_2S, indol, lecithinase, lipase, and urease not formed. Malic acid is decomposed to lactic acid and CO_2. Most strains exhibit pseudocatalase activity when grown on MRS agar containing 0.1% (w/v) glucose. The following characteristics are negative: gas production from citrate; growth in ethanol (15 and 30%), 4% taurocholate, and sodium lauryl sulfate (0.005%); The following characteristics are variable: growth in 6% NaCl and at 45°C, catalase production.

The type strains of *Lactobacillus mali*, and *Lactobacillus yamanashiensis* are negative for the latter two reactions, indicating a heterofermentative potential in the species.

Isolated from wine must, fermenting cider, and fermented molasses.

Phylogenetic position: Lactobacillus salivarius group.

Type strain: ATCC 27053, CCUG 30141, CCUG 32228, CIP 103142, DSM 20444, JCM 1116, LMG 6899, NCIMB 10560, NRRL B-4563, VKM B-1600.

DNA G+C content (mol%): 32.5 (T_m).

GenBank accession number (16S rRNA gene): M58824.

58. **Lactobacillus manihotivorans** Morlon-Guyot, Guyot, Pot, Jacobe de Haut and Raimbault 1998, 1107[VP]

ma.ni.ho.ti.vo'rans. L. n. *manihot* cassava; L. v. *vorare* to devour; N.L. adj. *manihotivorans* cassava-devouring.

Cells are nonmotile rods that occur in short chains or occasionally as single cells. Optimal temperature, 30°C; optimal pH for growth, 6.0. Grows at NaCl concentrations of 20–65 g/l and does not grow at 100 g/l NaCl. Obligately homofermentative. Starch is utilized. Strain OND 32[T] has an extracellular amylase activity and strain YAM 1 shows cell-linked amylase activity. Additional physiological and biochemical characteristics are presented in Table 84.

Isolated from sour cassava starch fermentation.

Phylogenetic position: unique.

Type strain: OND 32, CCUG 42894, CIP 105851, DSM 13343, JCM 12514, LMG 18010.

DNA G+C content (mol%): 48.4 ± 0.2 (T_m).

GenBank accession number (16S rRNA gene): AF000162.

59. **Lactobacillus mindensis** Ehrmann, Müller and Vogel 2003, 11[VP]

min.den'sis. N.L. adj. *mindensis* pertaining to the city of Minden, Germany, from where the first strain of this species was isolated.

Cells are nonmotile rods (0.9–5 μm), occurring singly, in pairs, or in chains. Colonies are usually small (2 mm), smooth, low convex, flat, and white on MRS agar. Obligately homofermentative. Growth occurs at 15–30°C but not above. Growth optimum is at pH 4.6–5.2; no growth at or above pH 6.5. Additional physiological and biochemical characteristics are presented in Table 84. Arginine is not hydrolyzed. Urease is not produced. The cell wall does not contain teichoic acid.

Isolated from commercial sourdough starter preparations and from bakeries' sourdough after continuous propagations for long periods.

Phylogenetic position: unique.

Type strain: TMW 1.80, DSM 14500, JCM 12532, LMG 21508.

DNA G+C content (mol%): 37.5 (T_m).

GenBank accession number (16S rRNA gene): AJ313530.

60. **Lactobacillus mucosae** Roos, Karner, Axelsson and Jonsson 2000, 256[VP]

mu'co.sae. N.L. gen. n. *mucosae* of mucosa.

Cells are nonmotile rods (1 × 2–4 μm) occurring singly, in pairs, or as short chains. After anaerobic growth at 37°C for 2 d, colonies on MRS agar are 1.2 mm in diameter. They are white, smooth, and convex. Obligately heterofermentative. Additional physiological and biochemical characteristics are presented in Table 86.

Isolated from the small intestines of pigs and from sourdough.

All strains of this species so far isolated have a homolog to the mucus-binding protein, Mub, and bind mucus *in vitro*.

Phylogenetic position: Lactobacillus reuteri group.

Type strain: S32, CCUG 43179, CIP 106485, DSM 13345, JCM 12515.

DNA G+C content (mol%): 46.5 ± 0.2 (HPLC).

GenBank accession number (16S rRNA gene): AF126738.

61. **Lactobacillus murinus** Hemme, Raibaud, Ducluzeau, Galpin, Sicard and Van Heijenoort 1982, 384[VP] (Effective publication: Hemme, Raibaud, Ducluzeau, Galpin, Sicard and Van Heijenoort 1980, 306.)

mu.ri'nus. L. adj. *murinus* of mice.

Cells are nonmotile rods with rounded ends (0.8–1.0 × 2.0–4.0 μm), frequently in chains. Ribose and arabinose slowly fermented. L-LDH is activated by FDP and Mn^{2+}. Facultatively heterofermentative. Grows in the presence of 6 % (w/v) NaCl. Urea and hippurate not hydrolyzed; malate is decarboxylated. Additional physiological and biochemical characteristics are presented in Table 85. Growth factor requirement: riboflavin is essential; thiamine, and vitamin B_{12} not required.

Isolated from the intestinal tract of mice and rats and from sourdough.

Phylogenetic position: Lactobacillus salvarius group.

Type strain: 313, ATCC 35020, CCUG 33904, CIP 104818, CNRZ 220, DSM 20452, NBRC 14221, JCM 1717, LMG 14189.

DNA G+C content (mol%): 43.4–44.3 (T_m).

GenBank accession number (16S rRNA gene): AJ621554, M58826.

Additional remarks: See *Lactobacillus animalis*.

62. **Lactobacillus nagelii** Edwards, Collins, Lawson and Rodriguez 2000, 700[VP]

na'gel.i.i. N.L. gen. n. *nagelii* of Nagel, after Charles W. Nagel, Washington State University, WA, USA, for his contributions to the science of wines.

Cells are rods (0.5 × 1–1.5 μm). Colonies on MR agar appear opaque with smooth edges and are approximately 2 mm diameter after 4–5 d growth at 25°C. Glucose fermented without gas formation. Mannitol is not formed from fructose. Both citrate and malate are utilized in the presence of glucose. Arginine is not hydrolyzed. Dextran is formed from sucrose. Obligately homofermentative. Additional physiological and biochemical characteristics are presented in Table 84.

Growth in MR broth containing 5% (w/v) NaCl (pH 4.5) and at pH 3.7, 4.5, and 8.0 (25°C). Weak growth at 5°C (pH 4.5).

Isolated from partially fermented grape juice.

Phylogenetic position: Lactobacillus salvarius group.

Type strain: LuE$_{10}$, ATCC 700692, CCUG 43575, DSM 13675, JCM 12492.

DNA G+C content (mol%): not determined.

GenBank accession number (16S rRNA gene): AB162131, Y17500.

63. **Lactobacillus oris** Farrow and Collins 1988, 116[VP]

or'is. L. gen. n. *oris* of the mouth.

Cells are nonmotile rods (0.8–1.0 × 2–4 μm) occurring singly, in pairs, or short chains. Colonies are small, raised, semirough, or rough. Grows at 30 and 40°C but not at 22 or 48°C; some strains grow at 45°C. Obligately heterofermentative. Gas is produced from glucose. Additional physiological and biochemical characteristics are presented in Table 86. Some strains hydrolyze hippurate. Some strains produce urease. Grows in 4% NaCl.

Isolated from human saliva.

Phylogenetic position: Lactobacillus reuteri group.

Type strain: 5A1 of Hayward, ATCC 49062, CCUG 37396, CIP 103255, CIP 105162, DSM 4864, JCM 7507, JCM 11028, LMG 9848, NCIMB 8831.

DNA G+C content (mol%): 49.3–50.7 *(Tm)*.

GenBank accession number (16S rRNA gene): X61131, X94229.

64. **Lactobacillus panis** Wiese, Strohmar, Rainey and Diekmann 1996, 452[VP]

pa'nis. L. masc. gen. n. *panis* of bread.

Cells are nonmotile rods shaped (0.7–0.9 × 2.5–6 mm) occurring in chains or clusters. Colonies are white to gray, convex, smooth, and up to 2 mm in diameter. Anaerobic. Arginine is not hydrolyzed. Oligately heterofermentative. Additional physiological and biochemical characteristics are presented in Table 86.

Isolated from long-fermented rye sourdough.

Phylogenetic position: Lactobacillus reuteri group.

Type strain: CCUG 37482, DSM 6035, JCM 11053.

DNA G+C content (mol%): 48.0–48.3 (HPLC).

GenBank accession number (16S rRNA gene): X94230.

65. **Lactobacillus pantheris** Liu and Dong 2002, 1747[VP]

pan'ther.is. L. gen. n. *pantheris* of the panther, referring to the isolation of the strains from jaguar feces.

Cells are nonmotile rods (0.5 × 1.5–2.5 μm) after 24 h incubation in TPYG liquid medium.

Colonies are convex, opaque, and rough with irregular edges and less than 1 mm in diameter after 24 h cultivation on TPYG plates. No gas is produced from glucose fermentation. Obligately homofermentative. The optimum pH for growth is 6.0–6.4 and growth can occur at pH 4.5, but not at pH values above 7.5. Arginine is hydrolyzed and esculin is hydrolyzed. Additional physiological and biochemical characteristics are presented in Table 84.

Isolated from the feces of a jaguar in Beijing Zoo.

Phylogenetic position: unique.

Type strain: A24-2-1, AS 1.2826, JCM 12539, LMG 21017.

DNA G+C content (mol%): 52.7 *(T*m*)*.

GenBank accession number (16S rRNA gene): AF413523.

66. **Lactobacillus parabuchneri** Farrow, Phillips and Collins 1989, 371[VP] (Effective publication: Farrow, Phillips and Collins 1988, 165.)

pa.ra.buch.ne'ri. Gr. prep. *para* resembling; N.L. gen. n. *buchneri* specific epithet; N.L. gen. n. *parabuchneri* resembling *Lactobacillus buchneri.*

Cells are nonmotile rods (0.8–1.0 × 2–4 μm) occurring singly, in pairs, or short chains. Obligately heterofermentative. Additional physiological and biochemical characteristics are presented in Table 86. Urease is not produced. Voges–Proskauer-negative.

Isolated from human saliva, cheese, and contaminated brewery yeasts.

Phylogenetic position: unique.

Type strain: ATCC 49374, CCUG 32261, CIP 103368, DSM 5707, JCM 12493, LMG 11457, NCIMB 8838.

DNA G+C content (mol%): 44 *(T*m*)*.

GenBank accession number (16S rRNA gene): AB205056, AJ970317, AY026751.

67. **Lactobacillus paracasei** Collins, Phillips and Zanoni 1989, 107[VP]

pa.ra.ca'se.i. Gr. prep. *para* resembling; N.L. gen. n. *casei* a specific epithet; N.L. gen. n. *paracasei* resembling *Lactobacillus casei.*

Cells are rods (0.8–1.0 × 2.0–4.0 μm), often with square ends, occurring singly or in chains. Growth at 10 and 40°C. Some strains grow at 5 and 45°C. Facultatively heterofermentative. Two subspecies are validly published.

Phylogenetic position: unique.

DNA G+C content (mol%): 45–47 (Bd).

Further comments: see *Lactobacillus casei.*

67a. **Lactobacillus paracasei subsp. paracasei** Collins, Phillips and Zanoni 1989, 107[VP]

Cells are nonmotile rods (0.8–1.0 × 2.0–4.0 μm), often with square ends, occurring singly or in chains. L(+)-Lactic acid is produced; a few strains formerly described as "*Lactobacillus pseudoplantarum*" produce inactive lactic acid due to the activity of L-lactic acid racemase. Additional physiological and biochemical characteristics are presented in Table 85. Ammonia is not produced from arginine. Urease-negative.

Isolated from dairy products, sewage, silage, humans, and clinical sources.

Type strain: ATCC 25302, CCUG 32212, CIP 103918, DSM 5622, NBRC 15889, JCM 8130, LMG 13087, NCIMB 700151.

DNA G+C content (mol%): 45–47 (Bd).

GenBank accession number (16S rRNA gene): D79212.

67b. **Lactobacillus paracasei subsp. tolerans** (Abo-Elnaga and Kandler 1965c) Collins, Phillips and Zanoni 1989, 108[VP] (*Lactobacillus casei* subsp. *tolerans* Abo-Elnaga and Kandler 1965c, 26.)

to.le'rans. L. pres. part. *tolerans* tolerating, enduring; means survival during the pasteurization of milk.

Cells are nonmotile rods (0.8–1.0 × 2.0–4.0 μm), often with square ends, occurring singly or in chains. Grows at 10 and 37°C; no growth at 40°C. Survives heating at 72°C for 40 s. Arginine is not hydrolyzed. Urease-negative. Additional physiological and biochemical characteristics are presented in Table 85.

Isolated from dairy products.

Type strain: ATCC 25599, CCUG 34829, CIP 102994, CIP 103024, DSM 20258, NBRC 15906, JCM 1171, LMG 9191, NCIMB 9709.

DNA G+C content (mol%): 45–47 (Bd).

GenBank accession number (16S rRNA gene): AB181950, D16550.

Additional remarks: Lactobacillus paracasei subsp. *tolerans* does not ferment ribose and gluconate and therefore does not fit the definition of group B, however, the high DNA/DNA similarity with *Lactobacillus paracasei* subsp. *paracasei* indicates that the lack of these characteristics is caused by minor genomic differences.

68. **Lactobacillus paracollinoides** Suzuki, Funahashi, Koyanagi and Yamashita 2004, 116[VP]

pa.ra.col.li.noi'des. Gr. pref. *para* beside; N.L. masc. adj. *collinoides* hill-shaped, referring to the colony form of *Lac-*

tobacillus collinoides; N.L. masc. adj. *paracollinoides* beside *collinoides*, referring to the close relationship to *Lactobacillus collinoides*.

Cells are nonmotile rods, occurring singly or in short chains. Obligately heterofermentative. Additional physiological and biochemical characteristics are presented in Table 86.

Isolated from brewery environments.

Phylogenetic position: unique.

Type strain: LA2, DSM 15502, JCM 11969.

DNA G+C content (mol%): 44.8 (T_m).

GenBank accession number (16S rRNA gene): AJ786665.

Additional remarks: "*Lactobacillus pastorianus*" (Van Laer, 1892) was not validly published and was identified as *Lactobacillus paracollinoides* (Ehrmann and Vogel, 2005).

69. **Lactobacillus parakefiri** corrig. Takizawa, Kojima, Tamura, Fujinaga, Benno and Nakase 1994, 439[VP]

pa.ra.ke'fir.i. Gr. prep. *para* resembling; N.L. n. *kefiri* the specific epithet of *Lactobacillus kefiri*; N.L. gen. n. *parakefiri*, resembling *Lactobacillus kefiri*.

Cells are nonmotile rods (0.5–1.2 × 1.0–3.5 µm) occurring singly, in pairs, or occasionally in short chains. When the organism is grown on R-CW agar at 30°C for 5 d, colonies are 0.5–2.0 mm in diameter, circular to irregular, flat, opaque, white, and rough. Obligately heterofermentative. Gas is produced from glucose, but not from gluconate. Milk is curdled. Additional physiological and biochemical characteristics are presented in Table 86.

Isolated from kefir grains.

Phylogenetic position: unique.

Type strain: GCL 1731, ATCC 51648, CCUG 39468, CIP 104242, DSM 10551, NBRC 15890, JCM 8573, LMG 15133.

DNA G+C content (mol%): 41.4–42.0 (T_m).

GenBank accession number (16S rRNA gene): AY026750.

70. **Lactobacillus paralimentarius** Cai, Okada, Mori, Benno and Nakase 1999b, 1455[VP]

pa.ra.li.men.ta'ri. us. Gr. pref. *para* beside; N.L. adj. *alimentarius* a specific epithet; N.L. adj. *paralimentarius* beside [*Lactobacillus*] *alimentarius*.

Cells are nonmotile rods (0.7–1.0 × 4.0–7.0 µm). Growth occurs at pH 4.5, 40°C, and 6.5% NaCl, but not above 45°C. Grows well anaerobically on MRS agar at 37°C. Facultatively heterofermentative. Additional physiological and biochemical characteristics are presented in Table 85.

Isolated from sourdough.

Phylogenetic position: unique.

Type strain: TB 1, CCUG 43349, CIP 106794, DSM 13238, JCM 10415, LMG 19152.

DNA G+C content (mol%): 37.2–38.0 (HPLC).

GenBank accession number (16S rRNA gene): AB018528, AJ417500.

72. **Lactobacillus paraplantarum** Curk, Hubert and Bringel 1996, 598[VP]

pa.ra.plan.tar'um. Gr. prep. *para* resembling; N.L. gen. n. *plantarum* specific epithet of *Lactobacillus plantarum*; N.L. gen. pl. n. *paraplantarum* resembling *Lactobacillus plantarum*.

Cells are nonmotile rods occurring singly, in pairs, and sometimes in short chains. Colonies grown on MRS agar are round, smooth, dome shaped, and creamy colored. Grows in MRS medium at pH 5 and 7, but not at pH 4 and tolerates NaCl up to a concentration of 8%. Facultatively heterofermentative. Arginine is not hydrolyzed. Additional physiological and biochemical characteristics are presented in Table 85.

Isolated from beer and human feces in France.

Phylogenetic position: unique

Type strain: ATCC 700211, CCUG 35983, CIP 104668, CNRZ 1885, CST 10961, DSM 10667, JCM 12533, LMG 16673, NRRL B-23115.

DNA G+C content (mol%): 44–45 (E_{280}, pH 3).

GenBank accession number (16S rRNA gene): AJ306297.

72. **Lactobacillus pentosus** (ex Fred, Peterson and Anderson 1921) Zanoni, Farrow, Phillips and Collins 1987, 339[VP]

pen.to'sus. N.L. adj. *pentosus* of pentose, pertaining to pentoses.

Cells are nonmotile rods (1–1.2 µm × 2–5 µm) with rounded ends, straight, and may occur singly, in pairs, or in short chains. Facultatively heterofermentative. Arginine is not hydrolyzed. Voges–Proskauer-positive. Acid and clot produced in litmus milk. Acid is produced from glycerol. Additional physiological and biochemical characteristics are presented in Table 85.

Isolated from corn silage, fermenting olives, and sewage.

Phylogenetic position: unique.

Type strain: ATCC 8041, CCUG 33455, CIP 103156, DSM 20314, JCM 1558, LMG 10755, NCAIM B.01727, NCCB 32014, NCIMB 8026, NRRL B-227, NRRL B-473.

DNA G+C content (mol%): 46.1–47.2 (T_m).

GenBank accession number (16S rRNA gene): D79211.

73. **Lactobacillus perolens** Back, Bohak, Ehrmann, Ludwig, Pot, Kersters and Schleifer 2000, 3[VP] (Effective publication: Back, Bohak, Ehrmann, Ludwig, Pot, Kersters and Schleifer 1999, 358.)

per.o'lens. L. pref. *per* through, penetrating; L. part. adj. *olens* having an odor; N.L. part. adj. *perolens* offensive-smelling.

Cells are nonmotile rods with rounded ends (0.7–2.5 µm) occurring singly, in pairs, or in short chains. Colonies on MRS flat or little convex, wavy, or dentated with rough, dull surface, beige to dirty-white; mean diameter 2.5 mm. Can grow up to 42°C with an optimum at 28–32°C, no growth below 15°C. Optimum growth at pH 5.5–6.5, no growth below pH 3.7. Facultatively heterofermentative. Arginine is not hydrolyzed. Urease not produced.

Additional physiological and biochemical characteristics are presented in Table 85.

Isolated from spoiled soft drinks and from brewery environment.

Phylogenetic position: unique.

Type strain: L 532, DSM 12744, JCM 12534, LMG 18936.

DNA G+C content (mol%): 49–53 (T_m).

GenBank accession number (16S rRNA gene): Y19167.

74. **Lactobacillus plantarum** (Orla-Jensen 1919) Bergey, Harrison, Breed, Hammer and Huntoon 1923, 250[AL] (*Streptobacterium plantarum* Orla-Jensen 1919, 174.)

plan.ta′rum. L. fem. n. *planta* a plant; N.L. gen. pl. n. *plantarum* of plants.

Cells are nonmotile rods with rounded ends, straight (0.9–1.2 μm × 3–8 μm) occurring singly, in pairs, or in short chains. Some strains reduce nitrate provided the concentration of glucose in the medium is limited and the pH poised at 6.0 or higher. Some strains exhibit pseudocatalase activity, especially if grown under glucose limitation, or true catalase when heme is present. Cell walls contain either ribitol or glycerol teichoic acid. Facultatively heterofermentative. Growth factor requirements: calcium pantothenate and niacin required; thiamine, pyridoxal or pyridoxamine, folic acid, vitamin B_{12}, thymidine, or deoxyribosides not required; riboflavin generally not required.

Phylogenetic position: unique. The species contains two subspecies.

DNA G+C content (mol%): 44–46 (Bd, T_m).

74a. **Lactobacillus plantarum subsp. plantarum** (Bergey, Harrison, Breed, Hammer and Huntoon 1923) Bringel, Castioni, Olukoya, Felis, Torriani and Dellaglio 2005, 1633[VP] (*Lactobacillus plantarum* Bergey, Harrison, Breed, Hammer and Huntoon 1923, 250.)

The description is that of the species. Additional physiological and biochemical characteristics are presented in Table 85.

Isolated from dairy products and environments, silage, sauerkraut, pickled vegetables, sourdough, cow dung, and the human mouth, intestinal tract and stools, and from sewage.

Type strain: ATCC 14917, CCUG 30503, CIP 103151, DSM 20174, NBRC 15891, JCM 1149, LMG 6907, NCIMB 11974, NRRL B-4496.

DNA G+C content (mol%): 44–46 (Bd, T_m).

GenBank accession number (16S rRNA gene): D79210.

Additional remarks: The differentiation from the related species *Lactobacillus paraplantarum*, *Lactobacillus pentosus*, and *Lactobacillus plantarum* subsp. *argentoratensis* is problematic when using physiological and biochemical characteristics. The use of genotypic methods provides clear identification results (Bringel et al., 2005).

74b. **Lactobacillus plantarum subsp. argentoratensis** Bringel, Castioni, Olukoya, Felis, Torriani and Dellaglio 2005, 1633[VP]

ar.gen.to.ra.ten′sis. L. masc. adj. *argentoratensis* of or pertaining to Argentoratus, the Roman name of the City of Strasbourg in Alsace, France.

The description is that of the species. Additional physiological and biochemical characteristics are presented in Table 85. The only difference from the other subspecies is the absence of melizitose fermentation in 14 strains of subsp. *argentoratensis*.

Isolated from starchy food and fermenting food of plant origin.

Type strain: DKO 22, CIP 108320, DSM 16365.

DNA G+C content (mol%): 44–46 (Bd, T_m).

GenBank accession number (16S rRNA gene): AJ640078.

Additional remarks: Fort further information, see subsp. *plantarum*.

75. **Lactobacillus pontis** Vogel, Böcker, Stolz, Ehrmann, Fanta, Ludwig, Pot, Kersters, Schleifer and Hammmes 1994, 228[VP]

pon′tis L. gen. n. *pontis* of the bridge, referring to BRIDGE, which is the abbreviation for the Commission of the European Communities Research Programme entitled Biotechnology Research, for Innovation Development and Growth in Europe. During this program, the organism was isolated and characterized by the joined efforts of three laboratories.

Cells are nonmotile slender rods (0.3–0.6 × 4–6 μm) occurring singly, in pairs, and in chains. Some strains have a strong tendency to form chains of long bent rods or even resembling a vine tendril. Colonies on sanfrancisco agar are 1–2 mm in diameter, rough circular plateaus with irregular borders and smooth convex centers, translucent, and grayish after 2–5 d of anaerobic incubation at 30°C. The terminal pH in sanfrancisco-medium ranges from 3.9 to 4.2. The main fermentation products from maltose or fructose are lactate, acetate, ethanol, glycerol, and CO_2. Obligately heterofermentative. Additional physiological and biochemical characteristics are presented in Table 86.

Isolated from rye sourdough.

Phylogenetic position: Lactobacillus reuteri group.

Type strain: LTH 2587, ATCC 51518, CCUG 33456, CIP 104232, DSM 8475, JCM 11051, LMG 14187.

DNA G+C content (mol%): 53–55 (HPLC).

GenBank accession number (16S rRNA gene): AJ422032, X76329.

76. **Lactobacillus psittaci** Lawson, Wacher, Hansson, Falsen and Collins 2001b, 969[VP]

psit.ta′ci. L. gen. masc. n. *psittaci* of the parrot, from which the organism was first isolated.

Cells are coccibacilli to rod-shaped, occurring singly or in pairs. After 24 h incubation on Columbia horse blood agar, colonies are 1–2 mm in diameter, colorless, and do not produce an odor. α-Hemolytic. Gas is produced from glucose (Gibson's semisolid medium). Growth is observed on acetate agar (pH 5.4). Optimum pH range is 5.5–7.5. Grows in 2% NaCl, but not in 6% NaCl. Obligately homofermentative (see below). Additional physiological and biochemical characteristics are presented in Table 84. Positive reactions are obtained for acid phosphatase, alanine-phenylalanine-proline arylamidase, alkaline phosphatase, cystine arylamidase, *a*-galactosidase, α-glucosidase, β-glucosidase, leucine arylamidase, phosphoamidase, and valine arylamidase (weak reaction). Negative reactions are observed for *N*-acetyl-β-glucosaminidase, arginine dihydrolase, chymotrypsin, esterase C-4, ester lipase C8, lipase C14, -fucosidase, β-galactosidase, β-glucuronidase, glycyl-tryptophan arylamidase, α-mannosidase, β-mannosidase, methyl β-D-glucopyranosidase, pyrazinamidase, pyroglutamic acid arylamidase, pyrrolidonyl arylamidase, trypsin, and urease. Esculin is hydrolyzed, but hippurate is not. Voges–Proskauer test is negative.

Isolated from the lungs of a dead hyacinth macaw (*Anodorhynchus hyacinthinus*). Habitat is not known.

Phylogenetic position: Lactobacillus delbrueckii group.

Type strain: CCUG 42378, CIP 106492, DSM 15354, JCM 11552.

DNA G+C content (mol%): not determined.

GenBank accession number (16S rRNA gene): AJ272391.

Additional remarks: The formation of gas from glucose on Gibson's semisolid medium would place *Lactobacillus psittaci* to the group of obligately heterofermentative species. The phylogenetic data are not consistent with that biochemical characteristic.

77. **Lactobacillus reuteri** Kandler, Stetter and Köhl 1982, 266[VP] (Effective publication: Kandler, Stetter and Köhl 1980, 267.) (*Lactobacillus fermentum* Type II Lerche and Reuter 1962, 462.)

reu'te.ri. N.L. gen. n. *reuteri* of Reuter; named for G. Reuter, a German bacteriologist.

Cells are nonmotile, slightly irregular, bent rods with rounded ends (0.7–1.0 × 2.0–5.0 μm), occurring singly, in pairs, and in small clusters. In the original description, it was mistakenly stated that ammonia is not produced from arginine. Obligately heterofermentative. Additional physiological and biochemical characteristics are presented in Table 86.

Isolated from human and animal feces, sourdough, and meat products.

Phylogenetic position: Lactobacillus reuteri group.

Type strain: ATCC 23272, CCUG 33624, CIP 101887, DSM 20016, NBRC 15892, JCM 1112, LMG 9213, LMG 13557, NRRL B-14171.

DNA G+C content (mol%): 40–42.3 (Bd, T_m).

GenBank accession number (16S rRNA gene): L23507, X76328.

Additional remarks: Synonymous with *Lactobacillus fermentum* Type II Lerche and Reuter 1962, 462. (Lerche and Reuter, 1962). *Lactobacillus reuteri* cannot be definitely distinguished from *Lactobacillus fermentum* by simple physiological tests. Determination of diamino acid of peptidoglycan or, preferentially, genotypical methods clearly separate the two species.

78. **Lactobacillus rhamnosus** (Hansen 1968) Collins, Phillips and Zanoni 1989, 108[VP] (*Lactobacillus casei* subsp. *rhamnosus* Hansen 1968, 76.)

rham.no'sus. N.L. adj. *rhamnosus* pertaining to rhamnose.

Cells are nonmotile rods (0.8–1.0 × 2.0–4.0 μm) often with square ends, occurring singly or in chains. Facultatively heterofermentative. Additional physiological and biochemical characteristics are presented in Table 85. Arginine is not hydrolyzed. Urease-negative.

Isolated from dairy products, sewage, humans, and clinical source.

Phylogenetic position in the genus: unique.

Type strain: ATCC 7469, CCUG 21452, CIP A157, DSM 20021, NBRC 3425, JCM 1136, LMG 6400, NCAIM B.01147, NCCB 46033, NCIMB 6375, NCTC 12953, NRRL B-442, VKM B-574.

DNA G+C content (mol%): 45–47 (Bd).

GenBank accession number (16S rRNA gene): D16552.

79. **Lactobacillus rossiae** corrig. Corsetti, Settanni, Sinderen, Felis, Dellaglio and Gobbetti 2005, 39[VP]

ros.si'ae. N.L. gen. n. *rossiae* of Rossi, to honor Jone Rossi, University of Perugia, Perugia, Italy, for her major contribution to dairy and sourdough microbiology.

Cells are nonmotile rods (0.5 × 1–1.5 μm.), occurring singly, in pairs, and in short chains. After growth for 3 d at 30°C on MRS agar plates, colonies are 1–1.5 mm in diameter, white, smooth, and circular or convex. Obligately heterofermentative. Additional physiological and biochemical characteristics are presented in Table 86. Peptidoglycan structure is A3α (L-Lys–L-Ser–L-Ala$_2$) type.

Isolated from wheat sourdough.

Phylogenetic position: unique.

Type strain: CS1, ATCC BAA-822, DSM 15814.

DNA G+C content (mol%): 44.6 (HPLC).

GenBank accession number (16S rRNA gene): AJ564009.

80. **Lactobacillus ruminis** Sharpe, Latham, Garvie, Zirngibl and Kandler 1973, 47[AL]

ru'mi.nis. L n. *rumen* rumen; N.L. gen. n. *ruminis* of rumen.

Cells are rods (0.6–0.8 × 3–5 μm), occurring singly, in pairs, and in short chains. Motile by peritrichous flagella; motility not always easy to demonstrate and often sluggish, best demonstrated as stab cultures in semisolid media containing low concentrations of glucose. Anaerobic. Surface growth is obtained only under reduced oxygen pressure; growth in liquid media is improved by the addition of cysteine-HCl. Unlike the strains isolated from the rumen, many strains from sewage were nonmotile and failed to grow at 45°C. Arginine is not hydrolyzed. Obligately homofermentative. Additional physiological and biochemical characteristics are presented in Table 84.

Isolated from rumen of cow and from sewage.

Phylogenetic position: Lactobacillus salivarius group.

Type strain: ATCC 27780, CCUG 39465, CIP 103153, DSM 20403, JCM 1152, LMG 10756, NRRL B-14853.

DNA G+C content (mol%): 44–47 (T_m).

GenBank accession number (16S rRNA gene): M58828.

81. **Lactobacillus saerimneri** Pedersen and Roos 2004, 1367[VP]

sae.rim'ne.ri. N.L. gen. masc. n. *saerimneri* of Saerimner, a pig occurring in Nordic mythology, because the organism was isolated from pigs.

Cells are nonmotile rods (1 × 1.5–4 μm) occurring singly or in pairs. After anaerobic growth at 37°C for 48 h, colonies on MRS agar are 2–3 mm in diameter; they are white with an opaque border, smooth, and convex. Growth on MRS agar also occurs under aerobic conditions, but at a considerably lower rate. Gas is not produced from glucose. Obligately homofermentative. Additional physiological and biochemical characteristics are presented in Table 84. Esculin is not hydrolyzed.

Isolated from pig feces. Habitat is intestines of pigs.

Phylogenetic position: Lactobacillus salivarius group.

Type strain: GDA154, CCUG 48462, DSM 16049, LMG 22087.

DNA G+C content (mol%): 42.9 (HPLC).

GenBank accession number (16S rRNA gene): AY255802.

82. **Lactobacillus sakei** corrig. Katagiri, Kitahara and Fukami 1934, 157[AL]

sa′ke.i. Latinized Japanese n. *sakeum* sake, a Japanese rice wine, N.L. gen. n. *sakei* of sake.

Cells are nonmotile rods (0.6–0.8 × 2–3 μm) with rounded ends, occurring singly and in short chains; frequently slightly curved and irregular, especially during stationary growth phase. Many of the strains tested grow even at 2–4°C. L-LDH is activated by FDP and Mn^{2+}. Possesses lactic acid racemase; induction of racemase in most strains is repressed by acetate. Therefore, the majority of strains produce L(+)-lactic acid in MRS broth, whereas DL-lactic acid is produced in cabbage press juice. A few strains, whose identity with *Lactobacillus sakei* is confirmed by DNA/DNA similarity, however, produce inactive lactic acid also in MRS broth. Facultatively heterofermentative.

Originally isolated from sake starter; regularly found in sauerkraut and other fermented plant material, meat products, prepacked finished dough, and human feces.

Phylogenetic position: unique.

Type strain: ATCC 15521, CCUG 30501, CIP 103139, DSM 20017, NBRC 15893, JCM 1157, LMG 9468, LMG 13558.

DNA G+C content (mol%): 42–44 (T_m).

Additional remarks: Because of the difficulty in differentiating between strains of the species *Lactobacillus curvatus* and *Lactobacillus sakei*, an extraordinarily great number of strains has been subjected to detailed taxonomic studies (Berthier and Ehrlich, 1999; Torriani et al., 1996). Based on numerical analysis of total soluble cell protein patterns, biochemical test data, and RAPD-PCR, profiles of two subspecies were defined for each species. This grouping had to be withdrawn for the *Lactobacillus curvatus* subspecies because Koort et al. (2004) showed that the type strain of *Lactobacillus curvatus* subspecies *melibiosus* was actually a strain of *Lactobacillus sakei*. More clear-cut grouping of data for both species was provided by Berthier and Ehrlich (1999). Strains of *Lactobacillus sakei* were allotted to four phenotypic groups (4, 5, 6, 7) within two RAPD groups. All *Lactobacillus sakei* strains hydrolyze arginine (all *Lactobacillus curvatus* strains do not). Further key properties in phylogenetic grouping of the *Lactobacillus sakei* strains are melibiose fermentation (restricted to *Lactobacillus sakei*), formation of catalase in the presence of heme (found in 3 of 4 phenotypes), and production of D-lactate. Both *Lactobacillus sakei* and *Lactobacillus curvatus* contain strains that are negative in that property which had been formerly allotted to the invalid species "*Lactobacillus bavaricus*" (Kagermeier-Callaway and Lauer, 1995). In addition, PCR-primers were derived that permitted the identification of strains of each subgroup.

82a. **Lactobacillus sakei subsp. sakei** (corrig. Katagiri, Kitahara and Fukami 1934) Torriani, Van Reenen, Klein, Reuter, Dellaglio and Dicks 1996, 1162[VP] (corrig. Katagiri, Kitahara and Fukami 1934, 157.)

The key characteristics of facultatively heterofermentative species, shown in Table 85, do not permit differentiation of the two subspecies. Ammonia is always produced from arginine. Most strains grow in the presence of 10% NaCl and produce acetoin from glucose.

Isolated from sake starter. Regularly found in fermented meat products, vacuum-packaged meat, sauerkraut and other fermented plant material, and human feces.

Type strain: ATCC 15521, CCUG 30501, CIP 103139, DSM 20017, NBRC 15893, JCM 1157, LMG 9468, LMG 13558.

DNA G+C content (mol%): 42–44 (T_m).

GenBank accession number (16S rRNA gene): AM113784, AY204893, M58829.

82b. **Lactobacillus sakei subsp. carnosus** Torriani, Van Reenen, Klein, Reuter, Dellaglio and Dicks 1996, 1162[VP]

car.no′sus. L. adj. *carnosus* pertaining to meat.

The physiological and biochemical characteristics are presented in Table 85. They do not permit the differentiation of the two subspecies.

Isolated from fermented meat products. Regularly found in vacuum-packaged meat, sauerkraut, and other fermented plant material.

Type strain: R 14b/a, CCUG 31331, CIP 105422, JCM 11031, LMG 17302.

DNA G+C content (mol%): 42–44 (T_m).

GenBank accession number (16S rRNA gene): AY204892.

83. **Lactobacillus salivarius** Rogosa, Wiseman, Mitchell and Disraely 1953, 691[AL]

sa.li.va′ri.us. L. adj. *salivarius* salivary.

Cells are rods with rounded ends (0.6–0.9 × 1.5–5 μm) occurring singly and in chains of varying length. Arginine is not hydrolyzed. Obligately homofermentative.

Isolated from the mouth and intestinal tract of humans, hamsters, and chickens. Two subspecies are recognized.

Phylogenetic position: Lactobacillus salivarius group.

DNA G+C content (mol%): 34–36 (Bd).

83a. **Lactobacillus salivarius subsp. salivarius** Rogosa, Wiseman, Mitchell and Disraely 1953, 691[AL]

Description as for the species. Ferments rhamnose but not salicin and esculin. Additional physiological and biochemical characteristics are presented in Table 84.

Type strain: ATCC 11741, CCUG 31453, CIP 103140, DSM 20555, JCM 1231, LMG 9477, NRRL B-1949.

DNA G+C content (mol%): 34–36 (Bd).

GenBank accession number (16S rRNA gene): AF089108.

83b. **Lactobacillus salivarius subsp. salicinius** Rogosa, Wiseman, Mitchell and Disraely 1953, 691[AL]

sa.li.ci′ni.us. N.L. adj. *salicinius* pertaining to salicin, a glycoside.

Description as for the species. Ferments salicin and esculin but not rhamnose. Additional physiological and biochemical characteristics are presented in Table 84.

Type strain: ATCC 11742, CCUG 39464, CIP 103155, DSM 20554, JCM 1150, LMG 9476, NRRL B-1950.

DNA G+C content (mol%): 34–36 (Bd).

GenBank accession number (16S rRNA gene): M59054.

84. **Lactobacillus sanfranciscensis** corrig. Weiss and Schillinger 1984a, 503[VP] (Effective publication: Weiss and Schillinger 1984b, 231.)

san.fran.cis.cen′sis. N.L. adj. *sanfranciscensis* from San Francisco, named after the city where the sourdough from which the organism was first isolated had been propagated for more than 100 years.

Cells are rods with rounded ends (0.6–0.8 × 2–4 μm) occurring singly and in pairs. On primary isolation some strains do not grow reasonably in MRS broth unless freshly prepared yeast extract is added and the initial pH is lowered to 5.6. A small peptide isolated from yeast extract was found responsible for the growth-promoting effect (Berg et al., 1981). In our laboratory, all strains grew well on MRS5 (Meroth et al., 2003). Obligately heterofermentative. Additional physiological and biochemical characteristics are presented in Table 86. The peptidoglycan is of the Lys–Ala-type.

Isolated from sourdough.

Phylogenetic position: unique.

Type strain: L-12, ATCC 27651, CCUG 30143, CIP 103252, DSM 20451, JCM 5668, LMG 16002, NRRL B-3934.

DNA G+C content (mol%): 36–38 (T_m).

GenBank accession number (16S rRNA gene): AY459586, M58830, X61132, X76327.

Additional remarks: Kline and Sugihara (1971) proposed the name *Lactobacillus sanfrancisco* with reservation as to results of pending DNA/DNA similarity studies. Later on they briefly confirmed the proposal and designated a type strain (Sugihara and Kline, 1975). The name, however, was omitted from the Approved Lists of Bacterial Names and consequently has no standing in bacteriological nomenclature and has been revived (Weiss and Schillinger, 1984b).

85. **Lactobacillus satsumensis** Endo and Okada 2005, 85[VP]

sat.su.men′sis. N.L. masc. adj. *satsumensis*, pertaining to Satsuma (old name for the southern part of Kyushu in Japan), from where the type strain was isolated.

Cells are rods (0.6–0.8 × 1.0–1.5 μm) occurring singly or in pairs. Motile with peritrichous flagella. Colonies on MRS agar are white, smooth, and approximately 2 mm in diameter after incubation for 2 d. Obligately homofermentative. Additional physiological and biochemical characteristics are presented in Table 84. Dextran is formed from sucrose. Growth is observed in MRS broth containing 5% (w/v) NaCl and at pH 3.5 in MRS broth. No growth is observed in MRS broth containing 10% (v/v) ethanol.

Isolated from shochu mashes collected at a shochu distillery in the South Kyushu district of Japan.

Phylogenetic position: Lactobacillus salvarius group.

Type strain: DSM 16230, JCM 12392, NRIC 0604.

DNA G+C content (mol%): 39–41 (HPLC).

GenBank accession number (16S rRNA gene): AB154519.

86. **Lactobacillus sharpeae** Weiss, Schillinger, Laternser and Kandler 1982, 266[VP] (Effective publication: Weiss, Schillinger, Laternser and Kandler 1981, 251.)

shar′pe.ae. M.L. gen. n. *sharpeae* of Sharpe; named for M. Elisabeth Sharpe, an English bacteriologist.

Cells are nonmotile rods with rounded ends (0.6–0.8 × 3–8 μm) with a pronounced tendency to form "snakes" and, after prolonged incubation, long characteristically wrinkled chains. In broth cultures, a flocculent sediment is observed. Colonies are grayish and flat. Obligately homofermentative. Additional physiological and biochemical characteristics are presented in Table 84. Indole, lipase, and urease are not produced. Arginine is not hydrolyzed.

Habitat unknown; isolated from municipal sewage.

Phylogenetic position: unique.

Type strain: ATCC 49974, CIP 101266, DSM 20505, JCM 1186, LMG 9214, NRRL B-14855.

DNA G+C content (mol%): 53 (T_m).

GenBank accession number (16S rRNA gene): M58831.

87. **Lactobacillus spicheri** Meroth, Hammes and Hertel 2004, 631[VP] (Effective publication: Meroth, Hammes and Hertel 2004, 157.)

spi′cher.i. N.L. gen. n. *spicheri* of Spicher, honoring Gottfried Spicher, a German researcher who contributed to the chemical, microbial, and technological characterization of wheat and rye sourdoughs.

Cells are nonmotile rods (1.5 × 3–5 μm) occurring singly or in pairs and seldom in short chains. Catalase-negative. After 3 d of incubation on MRS5 agar, colonies are white, circular with a smooth surface and border, and approximately 0.5–1 mm in diameter. Grows well in MRS5 medium up to pH 5.0 or 5.5% NaCl but not with 7.0% NaCl. Obligately heterofermentative. Additional physiological and biochemical characteristics are presented in Table 86. Fermentation of glucose is weak, but it is strong in combination with fructose. Fructose is reduced to mannitol.

Isolated from an industrial processed rice sourdough.

Phylogenetic position: unique.

Type strain: LTH 5753, DSM 15429, LMG 21871.

DNA G+C content (mol%): 55 (T_m).

GenBank accession number (16S rRNA gene): AJ534844.

88. **Lactobacillus suebicus** Heinzl and Hammes 1989, 495[VP] (Effective publication: Kleynmans, Heinzl and Hammes 1989, 270.)

suè.bi.cus. L. n. *suebia* Swabia a region in southern Germany; N.L. adj. *suebicus* pertaining to Swabia.

Cells are nonmotile rods (0.6–0.8 × 2–3 μm) with rounded ends, occurring singly or pairs, seldom in short chains. Colonies are white, round, convex, smooth, slimy, and 2 mm in diameter on Homohiochi agar after incubation for 4 d at 30°C in an atmosphere 95% N_2 and 5% CO_2. Obligately heterofermentative. Additional physiological and biochemical characteristics are presented in Table 82. No reaction in litmus milk; no slime production from sucrose. The lowest limits for growth are 10°C and pH 2.8. The upper limit for growth is 14% ethanol at pH 3.3.

Habitats are stored apple and pear mashes.

Phylogenetic position: unique.

Type strain: I, ATCC 49375, CCUG 32233, CIP 103411, DSM 5007, JCM 9504, LMG 11408.

DNA G+C content (mol%): 40.4 ± 0.4 (T_m).

GenBank accession number (16S rRNA gene): AJ306403, AJ575744, AM113785.

89. **Lactobacillus suntoryeus** Cachat and Priest 2005, 33[VP]

sun.to.ry.e′us. N.L. masc. adj. *suntoryeus* occurring in Suntory malt whisky fermentations.

Cells are nonmotile rods varying in length from 2 μm to long filaments. After 48 h incubation on MRS agar (supplemented with maltose and brain heart infusion broth), colonies are circular, shiny, creamy white in color, and 2–5 mm in diameter. Growth is poor in unsupplemented

MRS medium. Obligately homofermentative. Additional physiological and biochemical characteristics are presented in Table 84.

Isolated from whisky fermentations.

Phylogenetic position: Lactobacillus delbrueckii group.

Type strain: SA, LMG 22464, NCIMB 14005.

DNA G+C content (mol%): not determined.

GenBank accession number (16S rRNA gene): AY445815.

90. **Lactobacillus thermotolerans** Niamsup, Sujaya, Tanaka, Sone, Hanada, Kamagata, Lumyong, Assavanig, Asano, Tomita and Yokota 2003, 267[VP]

therm.o.tol.er′ans. Gr. n. *thermos* heat; L. pres. part. *tolerans* tolerating; N.L. part. adj. *thermotolerans* heat-tolerating.

Cells are nonmotile rods (1 × 2–3 μm) occurring singly, in pairs, or in short chains. After anaerobic growth at 42°C for 2 d, colonies on MRS agar are 1–1.5 mm in diameter, white, circular, convex, smooth, and opaque. Anaerobic. Obligately heterofermentative. Grows up to 50°C, but not at 15°C; optimum temperature for growth is 42°C. No dextran is produced from sucrose. Additional physiological and biochemical characteristics are presented in Table 86.

Major cellular fatty acid is a straight-chain, unsaturated acid, $C_{18:1}$.

Isolated from the feces of chickens in Thailand.

Phylogenetic position: Lactobacillus reuteri group.

Type strain: G 35, DSM 14792, JCM 11425.

DNA G+C content (mol%): 50.5 (HPLC).

GenBank accession number (16S rRNA gene): AF317702.

91. **Lactobacillus ultunensis** Roos, Engstrand and Jonsson 2005, 81[VP]

ul.tun.en′sis. N.L. masc. adj. *ultunensis* pertaining to Ultuna, the site of Swedish University of Agricultural Sciences in Uppsala, Sweden.

Cells are nonmotile rods (1 × 2–30 μm) occurring as single cells, pairs, or filaments. After anaerobic growth at 37°C for 48 h, colonies on MRS agar are 2–3 mm in diameter; they are white, irregular, and have a dry appearance. Growth on MRS agar under aerobic conditions is very weak. Anaerobic. In MRS broth, growth occurs at 25 and 42°C, but not at 20 or 45°C. Gas is not produced from glucose. Obligately homofermentative. Esculin is hydrolyzed. Additional physiological and biochemical characteristics are presented in Table 84.

Isolated from a biopsy of the healthy human gastric mucosa.

Phylogenetic position: Lactobacillus delbrueckii group.

Type strain: Kx146C1, CCUG 48460, DSM 16047, LMG 22117.

DNA G+C content (mol%): 35.7 (HPLC).

GenBank accession number (16S rRNA gene): AY253660.

92. **Lactobacillus vaccinostercus** Okada, Suzuki and Kozaki 1983, 439[VP] (Effective publication: Okada, Suzuki and Kozaki 1979, 217.)

vac.ci.no.ster′cus. L. adj. *vaccinus* from cows; L. n. *stercus* dung; N.L. adj. *vaccinostercus* from cow dung.

Cells grown in xylose-YP broth are nonmotile rods (0.9–1.0 × 1.5–2.5 μm) with rounded ends, occurring mostly in pairs. Colonies on xylose-YP agar for 48h at 30°C are 2 mm in diameter. Obligately heterofermentative. Litmus milk is not changed. Grows in the range from 20–40°C. Additional physiological and biochemical characteristics are presented in Table 86. Growth factor requirements: thiamine, pantothenic acid, niacin, and biotin are essential; pyridoxal, *p*-aminobenzoic acid, and folic acid not required.

Isolated from cow dung.

Phylogenetic position: unique.

Type strain: X-94, ATCC 33310, CCUG 30723, CIP 102807, DSM 20634, JCM 1716, LMG 9215, TUA 055B.

DNA G+C content (mol%): 36 (T_m).

GenBank accession number (16S rRNA gene): AB212087, AJ621556, AM113786.

93. **Lactobacillus vaginalis** Embley, Faquir, Bossart and Collins 1989, 368[VP]

va.gi.na′lis. N.L. adj. *vaginalis* of the vagina.

Cells are nonmotile rods (0.5–0.8 ×1.5–25 μm) occurring singly, in pairs, or in short chains. Colonies are white to gray, 1–5 mm in diameter, and semirough, often with raised areas. Grows at 30 and 45°C but not at 22 or 48°C. Obligately heterofermentative. Additional physiological and biochemical characteristics are presented in Table 86. Produces straight-chain and monounsaturated long-chain fatty acids of the vaccenic series and *cis*-11,12-methylene-octadecenoic (lactobacillic) acid.

Isolated from the vagina of patients suffering from trichomoniasis.

Phylogenetic position: Lactobacillus reuteri group.

Type strain: ATCC 49540, CCUG 31452, CIP 105932, DSM 5837, JCM 9505, LMG 12891, NCTC 12197.

DNA G+C content (mol%): 38–41 (T_m).

GenBank accession number (16S rRNA gene): AF243177, X61136.

94. **Lactobacillus versmoldensis** Kröckel, Schillinger, Franz, Bantleon and Ludwig 2003, 516[VP]

vers.mold.en′sis. N.L. masc. adj. *versmoldensis* pertaining to Versmold, the town in Germany where the strains were isolated.

Cells are nonmotile rods with rounded ends, (0.9 × 1.6–6.0 μm) occurring singly, in pairs, and in chains of generally four cells. Cells aggregate during growth in MRS broth. Colonies on MRS agar after 3 d incubation at 30°C are small (up to 1 mm in diameter), circular, convex with entire edges, and grayish-white. Facultatively heterofermentative. Gas is not produced from glucose. Arginine is not hydrolyzed. Maximum NaCl tolerance for growth in MRS broth is in the range 8–14%. Additional physiological and biochemical characteristics are presented in Table 85.

Isolated from poultry salami.

Phylogenetic position: unique.

Type strain: KU-3, ATCC BAA-478, DSM 14857, NCCB 100034.

DNA G+C content (mol%): 40.5 (T_m).

GenBank accession number (16S rRNA gene): AJ496791.

95. **Lactobacillus zeae** (Kuznetsov 1959) Dicks, Du Plessis, Dellaglio and Lauer 1996, 340[VP] (*Lactobacterium zeae* Kuznetsov 1959, 368.)

ze.ae. L. gen. n. *zeae* pertaining to *Zea mays*, corn.

Cells are nonmotile rods (0.5–0.6 ×1.2–2.4 μm) occurring singly or in chains, depending on growth conditions. Surface colonies on MRS agar are smooth, glistening, white, and 1–2 mm in diameter. Grows at 10 and 45°C. Arginine is not hydrolyzed. Facultatively heterofermentative. Additional physiological and biochemical characteristics are presented in Table 85. Strain ATCC 15820[T] ferments rhamnose, whereas strain ATCC 393[T] does not. Gas is produced from gluconate. Benzidine, urease, and indole tests are negative.

Isolated from corn steep liquor.

Phylogenetic position: unique.

Type strain: ATCC 15820, CCUG 35515, DSM 20178, JCM 11302, LMG 17315, NCIMB 9537.

DNA G+C content (mol%): 48–49.

GenBank accession number (16S rRNA gene): D86516.

Additional remarks: See *Lactobacillus casei*.

96. **Lactobacillus zymae** Vancanneyt, Neysens, De Wachter, Engelbeen, Snauwaert, Cleenwerck, Van der Meulen, Hoste, Tsakalidou, De Vuyst and Swings 2005, 619[VP]

zy.mae. N.L. n. *zyma* (from Gr. n. *zume*), leaven, sourdough; N.L. gen. n. *zymae* of sourdough.

Cells are nonmotile rods (1.0 × 2–20 μm) occurring singly, in pairs, and in chains After 24 h, colonies are beige, circular with a rough and wrinkled surface, and approximately 1 mm in diameter. Growth occurs at 7% NaCl. Obligately heterofermentative. Gas is produced from glucose and gluconate. Additional physiological and biochemical characteristics are presented in Table 86.

Isolated from a Belgian artisan wheat sourdough.

Phylogenetic position: unique.

Type strain: R-18615, CCM 7241, LMG 22198.

DNA G+C content (mol%): 53–54 (HPLC).

GenBank accession number (16S rRNA gene): AJ632157.

Genus II. **Paralactobacillus** Leisner, Vancanneyt, Goris, Christensen and Rusul 2000, 22[VP]

JØRGEN J. LEISNER AND MARC VANCANNEYT

Pa.ra.lac.to.ba.cill'us. Gr. prep. *para* resembling; N.L. n. *Lactobacillus* a bacterial genus; N.L. masc. n. *Paralactobacillus* resembling the genus *Lactobacillus*.

Cells straight slender **rods. Nonmotile. Do not form spores. Gram-positive.** Aerobic and facultatively anaerobic. Grows slowly at 15°C but not at 45°C. Able to grow on acetate agar. **Catalase-negative,** oxidase-negative. **Homofermentative. D- and L-lactic acid produced from glucose.** Nitrate not reduced. **Ammonia not produced from arginine.** Esculin is hydrolyzed.

DNA G+C content (mol%): 46 (HPLC).

Type species: **Paralactobacillus selangorensis** Leisner, Vancanneyt, Goris, Christensen and Rusul 2000, 23.

Further descriptive information

Based upon 16S rDNA analysis, *Paralactobacillus* forms a separate branch of lactic acid bacteria and is most related to the genera *Lactobacillus* and *Pediococcus*. Highest sequence similarities are found with the *Lactobacillus casei/Pediococcus* cluster ranging from 90.1 to 91.7% (Leisner et al., 2000) and up to 93.1% with *Lactobacillus pantheris* (Liu and Dong, 2002). GenBank accession numbers are AF049745 and AF049742 for strains LMG 17710[T] and LMG 17714, respectively.

Cells are nonspore-forming, straight, slender rods 2.5–6.5 μm long and <1.0 μm wide, usually occurring singly or as pairs but sometimes in older cultures as short chains. Colonies < 1–1.5 mm in diameter, round, and with a gray color after growth on All Purpose Tween (APT) agar incubated at 30°C for 5 d. Visible growth at 15°C was observed only after 5–10 d incubation. Variable growth at 37°C. No growth above 37°C or below 15°C. Growth at 30°C is slow in sterilized chili bo and in Man Rogosa Sharpe (MRS) broth at initial pH 4.0–5.0 compared with strains of *Lactobacillus plantarum* or *Pediococcus acidilactici* (unpublished results). No growth in the presence of 6.5 % NaCl. No growth on cattle blood agar after 5 d incubation at 30°C (unpublished results). No growth on SBM agar[*], supplemented with 0.01% tetrazolium (unpublished results). Able to lower pH to < 4.15 in La-broth[†].

Homofermentative with no gas production from glucose. No growth with gluconic acid as substrate. Catabolism of glucose results in production of D- and L-lactic acid in the approximate ratio 1.6–2:1 (D:L) for static cultures in MRS broth. Only trace amounts of acetic acid and ethanol are produced. Acid is produced from D-glucose, D-fructose, D-mannose, *N*-acetylglucosamine, salicin, and maltose after 4 d of incubation at 30°C. No acid is produced from ribose, D-xylose, mannitol, lactose, inulin, D-raffinose, or starch. The cellular fatty acid composition is similar to that for the genus *Lactobacillus*, consisting mostly of straight-chain saturated and monounsaturated types. DNA–DNA hybridization values within the single described species of the genus range between 99% and 104% among the five isolates tested.

None of 22 strains examined show antagonistic activity towards *Lactobacillus plantarum* ATCC 14917 or *Lactobacillus pentosus* LMG 17682. Twenty-one of 22 strains examined by the spot-on-lawn method were resistant towards nisin produced by *Lactococcus lactis* ATCC 11454. LMG 17710[T] exhibits sensitivity towards nisin produced by *Lactococcus lactis* ATCC 11454 when examined by the deferred inhibition test (Moreno et al., 2002). Resistant towards 30 μg vancomycin/g MRS agar.

Genus was isolated from a Malaysian food ingredient, chili bo. The primary source of these bacteria is not known.

Enrichment and isolation procedures

Paralactobacillus may be isolated on media traditionally employed for isolation of species of *Lactobacillus*. These may include APT agar, MRS agar or acetate agar. There is no selective medium available that allows exclusive growth of *Paralactobacillus* or any indicative medium that allows ready identification of this genus in mixed cultures. Therefore, additional tests are necessary in order to obtain a proper identification. *Paralactobacillus* usually exhibits slow growth during

---

[*]SBM agar, g/l: peptone, 15; yeast extract, 1; Tween 80 1; K$_2$HPO$_4$, 2; MgSO$_4$·7H$_2$O, 0.2; MnSO$_4$·4H$_2$O, 0.05; agar 10; pH 7.0.

[†]La-broth: MRS broth without phosphate buffer and with 0.3% (w/v) sodium citrate replacing ammonium citrate and adjusted to an initial pH of 6.8.

microaerophilic incubation at 30°C. It is therefore necessary to incubate the agar plates for at least 3 d in order to secure growth of visible colonies.

Maintenance procedures

Storage in frozen medium at −80°C or as a lyophilized culture is effective for long-term maintenance.

Differentiation of the genus *Paralactobacillus* from other genera

Differentiation of the genus from other related rod-shaped taxa is indicated in Table 87. Within the family of *Lactobacillaceae*, *Paralactobacillus* is distinguished from obligatory heterofermentative *Lactobacillus* species other than *Lactobacillus rogosae* by its inability to produce gas from glucose. *Lactobacillus rogosae* is differentiated by production of acetic acid from fructose and little or no fermentation of other carbohydrates.

Paralactobacillus is distinguished from facultative heterofermentative *Lactobacillus* species, except for some strains of *Lactobacillus acetotolerans*, *Lactobacillus coleohominis*, *Lactobacillus cypricasei*, *Lactobacillus equi*, and *Lactobacillus homohiochii*, by lack of production of gas from gluconic acid and lack of production of acid from ribose or xylose. *Lactobacillus acetotolerans*, *Lactobacillus coleohominis*, and *Lactobacillus cypricasei* are distinguished from *Paralactobacillus* by no growth at 15°C; *Lactobacillus equi*

is distinguished by no growth at 15°C and production of acid from raffinose and lactose, and *Lactobacillus homohiochii* is distinguished by lack of growth in MRS broth.

The differentiation of *Paralactobacillus* from homofermentative *Lactobacillus* species is more problematic and requires the use of a combination of characters. *Paralactobacillus* is distinguished from homofermentative *Lactobacillus* species by a minimum requirement of the following combination of characteristics: nonmotile, growth at 15°C but not 45°C, able to grow under aerobic conditions, no acid from L-arabinose, raffinose, D-ribose or lactose, and production of both D- (more than 25% of total amount) and L-isomers of lactic acid.

Paralactobacillus is readily distinguished from the third genus of *Lactobacillaceae*, *Pediococcus*, by possessing a rod-shaped cell morphology.

Paralactobacillus is distinguished from other genera of lactic acid bacteria as follows: from obligatory heterofermentative *Weissella*, *Oenococcus*, and *Leuconostoc* by no production of gas from glucose; from rod-shaped *Allofustis*, *Atopostipes*, *Carnobacterium*, *Desemzia*, *Isobaculum*, *Marinilactibacillus* and *Pilibacter* by the criteria listed in Table 87; from rod-shaped *Alkalibacterium* by the criteria listed in Table 87 and by no growth with 6.5% NaCl; from the pleiomorphic *Melissococcus* by no growth at 45°C; and from *Aerococcus*, *Abiotrophia*, *Agiticoccus*, *Alloiococcus*, *Atopobacter*, *Atopococcus*, *Catellicoccus*, *Dolosicoccus*, *Dolosigranulum*,

TABLE 87. Differentiation of the genus *Paralactobacillus* from other rod-shaped genera within the order of "*Lactobacillales*"[a]

Characteristic	*Paralactobacillus*	*Allofustis*	*Atopostipes*	*Alkalibacterium*[b]	*Carnobacterium*	*Desemzia*	*Isobaculum*	Heterofermentative *Lactobacillus*[c]	Homofermentative *Lactobacillus*[d]	*Leuconostoc*	*Marinilactibacillus*	*Pilibacter*	*Weissella*[e]
G + C content (mol%)	46	39	44	39–41	33–37	40	39	33–55	32–53	37–44	34–36	38	37–47
Motility	−		−	+	d	+[f]	−	d	d	−	+		−
Growth on acetate agar	+			−				+	+				
Growth at 15°C	+			+	+			d	d	+	+		+
Growth at 45°C	−		−		−	−	−	d	d			−	d
Gas from glucose	−			d				+	−	+	−		+
Acetate from glucose	−		+	+			+	+	−	+	w		+
NH₄ from arginine	−	+	−[f]	d		−[f]	+	d	d	−	w		d
Lactic acid isomer	DL			L	L	L	L[f]	L, D, DL	L, D, DL	D	L		D, DL
Acid from lactose	−	−[f]	+[f]	−	d	+[f]	−[f]	d	d	d	d	+[f]	
Acid from raffinose	−	−[f]	+[f]	−		+[f]	−[f]	d	d	d	d	−[f]	d

[a]Symbols: +, >85% positive; d, different strains give different reactions (16–84% positive); −, 0–15% positive; w, weak reaction.

[b]Additional information is listed in the text.

[c]Include facultatively and obligatory heterofermentative species. Additional information is listed in the text.

[d]Differentiation of *Paralactobacillus* from homofermentative *Lactobacillus* requires the use of a combination of characteristics as listed in the text.

[e]Cell morphology ranges from irregular rods to short coccoid rods or cocci.

[f]Information is given for the type species which constitutes the only species of the genus.

Enterococcus, *Eremococcus*, *Facklamia*, *Globicatella*, *Granulicatella*, *Ignavigranum*, *Lactococcus*, *Lactovum*, *Streptococcus*, *Tetragenococcus*, *Trichococcus*, and *Vagococcus* by possessing a distinctive rod-shaped cell morphology.

Taxonomic comments

The genus *Paralactobacillus* was described to accommodate a Gram-positive rod originating from the Malaysian food ingredient chili bo. The genus is phylogenetically related to the genera *Lactobacillus* and *Pediococcus*.

The genera *Lactobacillus* and *Pediococcus* are phylogenetically complex and there is extensive intermixing of species of the two genera. The genus *Paralactobacillus* forms a distinct clade related to the *Lactobacillus*/*Pediococcus* group. The branching point between *Paralactobacillus* and this group has moderate support as demonstrated by a bootstrap value of 48%.

List of species of the genus *Paralactobacillus*

1. **Paralactobacillus selangorensis** Leisner, Vancanneyt, Goris, Christensen and Rusul 2000, 23[VP]

sel.an.gor'en.sis. N.L. adj. *selangorensis* belonging to the province of Selangor, Malaysia.

The characteristics are as described for genus with the following additional information. Some strains produce acid from galactose, rhamnose, α-methyl-D-glucoside, amygdalin, arbutin, cellobiose, sucrose, trehalose, β-gentiobiose, D-turanose, and D-tagatose after 4 d or more (up to 2 weeks) incubation at 30°C.

No acid is produced from glycerol, erythritol, D-arabinose, L-arabinose, L-xylose, adonitol, β-methyl-xyloside, L-sorbose, dulcitol, inositol, sorbitol, α-methyl-D-mannoside, melibiose, melizitose, glycogen, xylitol, D-lyxose, D-fucose, L-fucose, D-arabitol, L-arabitol, gluconate, 2-ketogluconate, or 5-ketogluconate.

DNA G+C content (mol%): 46 (HPLC).

Type strain: ATCC BAA-66, CCUG 43347, CIP 106482, DSM 13344, LMG 17710.

GenBank accession number (16S rRNA gene): AF049745.

Genus III. **Pediococcus** Claussen 1903, 68[AL]

WILHELM. H. HOLZAPFEL, CHARLES M.A.P. FRANZ, WOLFGANG LUDWIG AND LEON M.T. DICKS

Pe.di.o.coc'cus. Gr. n. *pedium* a plane surface; Gr. n. *kokkos* a grain or berry; N.L. masc. n. *Pediococcus* coccus growing in one plane.

Nine species are recognized within this genus, including *Pediococcus acidilactici* (Lindner, 1887; Skerman et al., 1980), *Pediococcus claussenii* (Dobson et al., 2002), *Pediococcus cellicola* (Zhang et al., 2005), *Pediococcus damnosus* (Balcke, 1884; Claussen, 1903; Skerman et al., 1980), *Pediococcus dextrinicus* (Back, 1978a; Coster and White, 1964; Skerman et al., 1980), *Pediococcus inopinatus* (Back, 1978a, 1988), *Pediococcus parvulus* (Günther and White, 1961; Skerman et al., 1980), *Pediococcus pentosaceus* (Mees, 1934; Skerman et al., 1980) with the subspecies *Pediococcus pentosaceus* subsp. *pentosaceus* and *Pediococcus pentosaceus* subsp. *intermedius*, and *Pediococcus stilesii* (Franz et al., 2006). The *Pediococcus* type species is *Pediococcus damnosus*

(Claussen, 1903). Judicial Opinion 52 states that this generic name is conserved over *Pediococcus* (Balcke, 1884) and all earlier objective synonyms. Erroneously, the genus name *Pediococcus* (Balcke, 1884) has been cited in the Approved Lists of Bacterial Names (Skerman et al., 1980) and in the Amended Edition of the Approved Lists of Bacterial Names (Euzéby, 1998; Skerman et al., 1989). These species can clearly be distinguished on the basis of DNA–DNA similarity (Table 88).

The species *Pediococcus dextrinicus* is considered atypical for the genus *Pediococcus*; it produces L(+)-lactic acid from glucose via a fructose-1,6-diphosphate FDP inducible L-lactate dehydrogenase (L-LDH; Back, 1978b) and may represent a new genus

TABLE 88. Percentage DNA–DNA similarity among pediococci and with reference also to *Tetragenococcus* and *Aerococcus* (*Pediococcus urinaeequi* – Syn. *Aerococcus*)[a]

Species	P. acidilactici	P. damnosus	P. dextrinicus	P. inopinatus	T. halophilus	P. parvulus	P. pentosaceus	P. pentosaceus subsp. intermedius	P. urinaeequi	P. cellicola	P. stilesii
P. acidilactici	100										
P. damnosus	0–7	100									
P. dextrinicus	0–5	4–5	100								
P. inopinatus	0–7	41–54	7	100							
Tetragenococcus halophilus	0–2	0–2	6	3–5	100						
P. parvulus	0–7	34–36	8	30–40	4	100					
P. pentosaceus	5–35	0–18	6	7–8	4	7	100				
P. pentosaceus subsp.*intermedius*	17–19	0–7	5	6–7	3	6	88–97	100			
P. urinaeequi	0.0E+01	0.0E+01	0.0E+01	0.0E+01	0.0E+01	0.0E+01	0.0E+01	0.0E+01	100		
P. cellicola	–	15–23	–	30–40	–	26–36	–	–	–	100	
P. stilesii	14.5	–	–	–	–	–	21	–	–	–	100

[a]Data compiled from Back and Stackebrandt (1978), Dellaglio et al. (1981), Dellaglio and Torriani (1986), Franz et al. (2006), and Zhang et al. (2005).

TABLE 89. Differentiation of the genus *Pediococcus* from the atyptical species *Pediococcus dextrinicus* and from *Aerococcus* and *Tetragenococcus* (Holzapfel et al., 2005)[a,b]

Characteristic	*Pediococcus*	*Pediococcus dextrinicus*	*Aerococcus*	*Tetragenococcus*
Growth at pH 5	+	+	−	−
Growth at pH 9	−	−	+	+
Growth tolerance to 18% NaCl	−	−	+	−
Facultative aerobic	+	+	+	−
Gas from gluconate	−	+	−	−
Configuration of lactate from glucose	DL/L(+)	L(+)	L(+)	L(+)
Catalase	−	−	−	+/−
Hippurate hydrolysis	−	−	−	+
Starch fermentation	−	+	−	−
Peptidoglycan type	Lys–D-Asp	Lys–D-Asp	Lys–D-Asp	Lys-direct

[a]Symbols: +, present; −, absent; +/−, variable.
[b]From Weiss (1992), Simpson and Taguchi (1995), and Dobson et al. (2002).

(Ludwig and Back, personal communication) (Table 89). A new *Lactobacillus* species (*Lactobacillus concavus* sp. nov.) is closely related to *Pediococcus dextrinicus* JCM 5887[T], with 97.9% similarity (Tong and Dong, 2005a). The former *Pediococcus halophilus* was renamed to *Tetragenococcus halophilus* by Collins et al. (1990) and is synonymous to "*Pediococcus soyae*" (Weiss, 1992). The genus *Tetragenococcus* exhibits a low 16S rRNA sequence similarity (≤90%) with the true pediococci (see chapter by Dicks et al. on *Tetragenoccus*, below). The taxonomic status of *Pediococcus urinaeequi* was described by Felis et al. (2005), who proposed the transfer of this species to the genus *Aerococcus* with the name *Aerococcus urinaeequi* comb. nov. on the basis of comparative analysis of the 16S rRNA gene sequence and DNA–DNA hybridization data.

The cells of typical representatives are nonmotile, spherical, occasionally ovoid, and divide to form pairs or divide alternately in two perpendicular directions to form tetrads (Axelsson, 1998) but never chains. Cells may occur singly and in pairs, especially during early or mid-exponential growth. Individual cells measure 0.5–1.0 μm. The cells are Gram-positive, asporogenic, and oxidase-negative. Cytochromes are absent, and catalase is not produced, although some strains of *Pediococcus pentosaceus* have been reported to produce catalase or pseudocatalase (Simpson and Taguchi, 1995). More than 50% of the *Pediococcus pentosaceus* strains isolated from goat milk, Feta cheese, and Kaseri cheese were reported to have a weak catalase activity (Tzanetakis and Litopoulou-Tzanetaki, 1989). Arginine hydrolysis is rare but has been recorded for *Pediococcus acidilactici* and *Pediococcus pentosaceus* (Simpson and Taguchi, 1995). Homofermentative, producing lactic acid but no CO_2 from glucose. Chemo-organotrophic. Facultatively anaerobic. Growth at pH 5 but not at pH 9, except for *Pediococcus stilesii*. Optimum growth temperature is 25–35°C, but is species dependent. All species except *Pediococcus claussenii* produce DL-lactic acid from glucose. Nitrate not reduced. *Peptidoglycan type:* Lys–D-Asp (Holzapfel et al., 2005).

DNA G+C content (mol%): 35–44 (T_m).

Type species: **Pediococcus damnosus** Claussen 1903, 68[AL].

Further descriptive information

The cell morphology of the *Pediococcus* species is similar to that of *Tetragenococcus* species, and members of these two genera are the only lactic acid bacteria (LAB) dividing in two perpendicular directions resulting in the formation of pairs and tetrads, but never of typical chains. The coccoid cells are perfectly round and rarely ovoid, in contrast to other coccus-shaped lactic acid bacteria such as *Leuconostoc*, *Lactococcus*, and *Enterococcus* species. The cell size may vary from 0.6–1.0 μm diameter, depending on the strain and species, but remains uniform within a culture.

The genus *Pediococcus* forms part of the *Clostridium* branch of descent with a low DNA base composition of less than 50 mol% G + C, whereas on the basis of 16S rDNA similarity, Collins et al. (1991) grouped this genus in the *Lactobacillus* branch of LAB. This could be confirmed by comparative analysis of the currently available sequence dataset comprising 950 almost complete 16S rRNA primary structures from representatives of the *Lactobacillales*. Overall, 16S rRNA sequence similarities for pediococci and closely to moderately related *Lactobacillus* species range from 85–94%. The corresponding values for other genera of the *Lactobacillales*, i.e., *Leuconostoc/Oenococcus*, *Enterococcus*, *Lactococcus*, *Vagococcus*, *Tetragenococcus*, and *Carnobacterium* are 80–90%, 84–91%, 84–87%, 88–90%, 88–91%, and 85–87%, respectively. Small-subunit rRNA-based phylogenetic analyses support a monophyletic group comprising eight *Pediococcus* species (*Pediococcus acidilactici*, *Pediococcus cellicola*, *Pediococcus damnosus*, *Pediococcus dextrinicus*, *Pediococcus inopinatus*, *Pediococcus parvulus*, *Pediococcus pentosaceus*, and *Pediococcus stilesii*) to the exclusion of only distantly related *Pediococcus dextrinicus* (Figure 93). The close relationship within the *Pediococcus* cluster is reflected by overall 16S rRNA sequence similarities greater than 94%, whereas the remote position of *Pediococcus dextrinicus* is documented by lower values (90.2–92.4% 16S rRNA sequence similarity for *Pediococcus dextrinicus* and other members of the cluster).

The *Pediococcus* species exhibit a low 16S rRNA sequence similarity with the genus *Tetragenococcus*. Similarity values to *Tetragenococcus halophilus* were determined for *Pediococcus acidilactici* (89.7%), *Pediococcus pentosaceus* (88.3%), *Pediococcus damnosus* (88.7%), *Pediococcus parvulus* (87.4%), and *Pediococcus dextrinicus* (88.6%) (Collins et al., 1990). The phylogenetic relatedness of *Aerococcus*, *Enterococcus*, *Vagococcus*, *Carnobacterium*, *Pediococcus*, *Tetragenococcus*, and *Lactococcus* is indicated in Figure 114 and Figure 115 in the chapter on *Tetragenococcus* by Dicks et al. in this edition.

Identification at species level within the genus *Pediococcus* may be achieved by 16S rRNA gene sequencing (Barney et al., 2001; Collins et al., 1990; Kurzak et al., 1998; Omar et al., 2000),

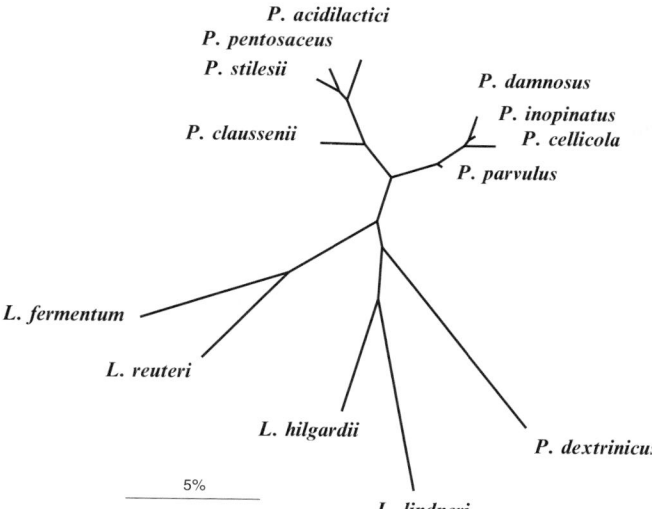

FIGURE 93. 16S rRNA-based phylogeny of the validly published *Pediococcus* and selected *Lactobacillus* reference species. The tree is based upon a maximum-parsimony analysis of 41,230 almost complete small-subunit rRNA primary structures using the ARB databases and software (Ludwig et al., 2004). The tree topology was evaluated by applying alternative treeing approaches (maximum-likelihood and distance matrix) as well as various positional filters based on conservation profiles (Ludwig and Klenk, 2001).

by the use of specific DNA target probes (Lonvaud-Funel et al., 1993, 2000a; Mora et al., 1997, 1998; Rodrigues et al., 1997), ribotyping (Barney et al., 2001; Jager and Harlander, 1992; Santos et al., 2005; Satokari et al., 2000), randomly amplified polymorphic DNA (RAPD) PCR (Fujii et al., 2005; Kurzak et al., 1998; Mora et al., 2000b; Nigatu et al., 1998; Simpson et al., 2002b; Tamang et al., 2005), and pulsed-field gel electrophoresis (PFGE) (Barros et al., 2001; Dellaglio et al., 1981; Luchansky et al., 1992). A rapid PFGE protocol was developed by Simpson et al. (2006) for species and strain-level verification of pediococci within less than 3 d. A solution phase hybridization PCR-ELISA was used for the detection and quantification of *Enterococcus faecalis* and *Pediococcus pentosaceus* in Nurmi-type cultures (Waters et al., 2005). Total genomic DNA–DNA hybridization remains the most reliable method of differentiating among species (Back and Stackebrandt, 1978; Dellaglio et al., 1981; Franz et al., 2006) and is considered the gold standard for genotypic delineation (Mehlen et al., 2004). Real-time PCR is increasingly being used for quantitative detection of *Pediococcus* species among other lactic acid bacteria (Delaherche et al., 2004; Lan et al., 2004; Stevenson et al., 2005).

Glucose is metabolized homofermentatively to lactic acid, probably by the Embden–Meyerhof pathway. For typical *Pediococcus* species, it appears that CO_2 is not produced from gluconate. *Pediococcus dextrinicus* is an exception but is also considered an atypical species (see taxonomic comments). Strains of *Pediococcus* species possess varying degrees of β- and α-D-glucopyranosidase activities which are influenced differently by ethanol, sugars, temperature, and pH. Strains with these characteristics may influence the glycoside composition of wine (Grimaldi et al., 2005).

Graham and McKay (1985) were the first to describe plasmids for the genus *Pediococcus*. Many genes regulating the fermentation of sugars are plasmid encoded, e.g., the fermentation of raffinose, melibiose, and sucrose by *Pediococcus pentosaceus* (Gonzalez and Kunka, 1986; Hoover et al., 1988), and the fermentation of sucrose for some strains of *Pediococcus acidilactici* (Hoover et al., 1988). Besides sugar fermentation, genes encoding antibiotic resistance may also be plasmid encoded. For example, an erythromycin resistance is encoded on a 40-MDa plasmid in *Pediococcus acidilactici* (Torriani et al., 1987). According to Gonzalez and Kunka (1983) plasmids are frequently transferred between *Pediococcus* and *Enterococcus*, *Streptococcus*, and *Lactococcus* species.

Limited information is available on bacteriophages of pediococci. The bacteriophages of *Pediococcus acidilactici* were studied by Caldwell et al. (1999), and were temperate, but could be induced with mitomycin C and were classified into two genetic groups (Caldwell et al., 1999).

Most of the genetic studies conducted on *Pediococcus* species revolve around the genes encoding different bacteriocins. Most bacteriocins (pediocins) produced by pediococci are grouped into Class IIa, i.e., small (less than 10 kDa), non-lanthionine-containing and *Listeria*-active peptides with a YGNGV (tyrosine-glycine-asparagine-glycine-valine) -consensus sequence in the N terminus (Nes et al., 1996). Pediocins are membrane active and resistant to temperatures of up to 100°C; a few pediocins withstand autoclaving (15 min at 121°C). Among all pediocins and other bacteriocins within Class IIa, pediocin PA-1 is the best characterized (Gonzalez and Kunka, 1987; Rodrigues et al., 2002). Pediocins share many sequence similarities with bacteriocins produced by *Lactobacillus* species, e.g., curvacin A, sakacin P, bavaricin A, and bavaricin MN; *Leuconostoc* species, e.g., leukocin A and mesentericin Y105; *Streptococcus* species, e.g., mundticin; *Enterococcus* species, e.g., enterocin A; and *Carnobacterium* species, e.g., carnobacteriocin B2 and piscicolin 126 (Holzapfel et al., 2005).

Some strains of pediococci have been associated with infections in humans and may be considered opportunistic pathogens (Barros et al., 2001; Barton et al., 2001; Colman and Efstratiou, 1987; Golledge et al., 1990; Green et al., 1991; Mastro et al., 1990; Riebel and Washington, 1990; Sarma and Mohanty, 1998). They may cause infections in individuals debilitated as a result of trauma or underlying disease (Barton et al., 2001; Facklam and Elliott, 1995; Mastro et al., 1990; Sarma and Mohanty, 1998). Some strains are resistant to vancomycin and teicoplanin (Swenson et al., 1990; Tankovic et al., 1993) which may provide them with a competitive advantage in medical care environments (Colman and Efstratiou, 1987; Golledge et al., 1990; Green et al., 1991; Mastro et al., 1990; Ruoff et al., 1988; Swenson et al., 1990). Some pediococci are also resistant to quinolone antibiotics and tetracycline (Sarma and Mohanty, 1998; Tankovic et al., 1993), but they are generally sensitive to antibiotics such as penicillin, ampicillin, and aminoglycosides (especially gentamicin and netilmicin) and moderately sensitive towards chloramphenicol (Barton et al., 2001; Mastro et al., 1990; Swenson et al., 1990; Tankovic et al., 1993).

Enrichment and isolation procedures

Members of the genus *Pediococcus* share several physiological properties with the genera *Lactobacillus*, *Leuconostoc*, and

Weissella. Together they may be called the LLPW physiological group, and they generally respond in the same way to conditions or compounds inhibitory to nonlactics (Schillinger and Holzapfel, 2003). This is because they often share the same habitat and show relatively similar growth requirements (Holzapfel, 1992). Therefore, most culture media developed for the detection of *Pediococcus* species are not completely selective for this genus. Examples of semiselective media include Selective SL Medium for Isolating Pediococci (Rogosa et al., 1951) and Acetate Agar for Isolation of *Leuconostoc* and *Pediococcus* (Whittenbury, 1965). Simpson et al. (2006) evaluated a *Pediococcus* Selective Medium (PSM) for the enumeration of *Pediococcus acidilactici* and *Pediococcus pentosaceus* from probiotic animal feed and silage inoculants. This medium is based on MRS supplemented with cysteine hydrochloride, novobiocin, vancomycin, and nystatin. Inoculation in the latter medium and incubation at 37°C under anaerobic conditions for 24 h stimulates the growth of pediococci in samples that also contain bacilli, bifidobacteria, enterococci, lactobacilli, lactococci, propionibacteria, streptococci, and yeasts. Growth of *Lactobacillus plantarum* and *Lactobacillus casei* is inhibited by the addition of ampicillin (PSM+A).

Several growth media have been developed to detect spoilage strains of *Pediococcus damnosus* and *Pediococcus claussenii* in beer. One such medium is MRS agar adjusted to pH 5.5. Back (1978a) used a combination of MRS broth and beer (1:1 v/v) and incubated at 22°C in the presence of 90% N_2 + 10% CO_2. Taguchi et al. (1990) used a complex Kirin-Ohkochi-Taguchi (KOT) medium, consisting of beer, malt extract, liver concentrate, maltose, L-malic acid, cytidine, thymidine, actidione, and sodium azide, in addition to other typical LAB growth medium components. Beer is frequently used as a major component of media (Back, 1978a; Nakagawa, 1964), thereby selectively promoting the growth of pediococci that are adapted to beer and hops. Apart from hops, selective agents commonly added to media may include cycloheximide, actidione, or sorbic acid. Gram-negative bacteria are usually inhibited by polymyxin B, acetic acid (together with reduced pH), and thallous acetate (Schillinger and Holzapfel, 2003). The European Brewery Convention has recommended three media for the simultaneous detection of lactobacilli and pediococci in beer, namely, MRS agar, supplemented with cycloheximide to prevent growth of aerobes such as yeasts and molds; Raka-Ray medium, supplemented with cycloheximide; and VLB (Versuchs- und Lehranstalt für Brauerei, Berlin, Germany) (Holzapfel et al., 2005). NBB may be one of the most successful and most appropriate media for detection and cultivation of beer pediococci. The most rapid and intensive growth rate was recorded on this medium (Dachs, 1981), whereas the high beer-specific selectivity results in the inhibition of yeasts and Gram-negative bacteria (Back, 1994, 2000).

Maintenance procedures

For short-term storage, cultures may be kept as stabs in appropriate growth media supplemented with 0.3% $CaCl_2$ as pH neutralizer, at 4°C, with monthly transfers to freshly prepared medium. For long-term storage, exponential-phase cultures may be mixed with an equal volume of 80% (v/v) sterile glycerol and stored in cryotubes at −20 or −80°C. Lyophilization with sterile skim milk or storage in liquid nitrogen is recommended for longer-term storage. Vancomycin may be added (Björkroth and Holzapfel, 2003). Strains are maintained as described in the chapter for *Leuconostoc*. For more information, see Schillinger and Holzapfel (2003).

Differentiation from closely related taxa

The true pediococci may be distinguished from closely related or morphologically similar genera and groups by a number of phenotypic characteristics, as is shown in Table 89. In general, they are characterized by their nonhalophilic but aciduric nature and facultative aerobic growth (Holzapfel et al., 2005; Weiss, 1992), which distinguishes them from *Tetragenococcus*, *Aerococcus*, and *Pediococcus dextrinicus*. The latter, in addition, is able to produce CO_2 from gluconate and to hydrolyze starch, in contrast to the true pediococci and tetragenococci.

Habitat

Pediococcus species share common habitats with other LAB, and especially with *Lactobacillus*, *Leuconostoc*, and *Weissella* species. *Pediococcus acidilactici*, *Pediococcus parvulus*, *Pediococcus inopinatus*, *Pediococcus stilesii*, and *Pediococcus pentosaceus* may be naturally associated with plants and fruits (Back, 1978a; Dellaglio et al., 1981; Mundt et al., 1969; Wilderdyke et al., 2004) and also with the fermentation of plant materials, such as silage, maize, cucumbers, olives, or cereal gruels. Some strains also contribute to meat fermentations. *Pediococcus damnosus*, *Pediococcus dextrinicus*, *Pediococcus inopinatus*, and *Pediococcus cellicola* are often associated with the brewery environment or environments in which alcoholic beverages are produced (Back, 1978a; Simpson and Taguchi, 1995; Zhang et al., 2005). Pediococci may also be important constituents of the nonstarter lactic acid bacteria involved in ripening of cheese (Olson, 1990). Some reports suggest the association of pediococcal strains with human and animal hosts, e.g., with human saliva (Sims, 1986) and with the digestive tract of humans and animals and in human feces (Heilig et al., 2002; Ruoff et al., 1988; Walter et al., 2001). *Pediococcus acidilactici* and *Pediococcus pentosaceus* were also detected in the gastrointestinal (GI) tracts of poultry (Juven et al., 1991) and ducks (Kurzak et al., 1998) and in freshwater prawns (*Macrobrachium rosenbergii*) in Thailand (Cai et al., 1999a). *Pediococcus pentosaceus* was reported to form part of the Gram-positive tonsillar and nasal microbial population of piglets (Baele et al., 2001a; Martel et al., 2003).

The association of pediococci with the clinical environment has been discussed (see above). The genomic diversity within the genus *Pediococcus* has been investigated by Simpson et al. (2002b) using RAPD-PCR and PFGE. These techniques proved useful for the rapid classification of *Pediococcus* strains isolated from sources such as food, feed, silage, beer, and human clinical samples. Comparative information may be found in the chapter on *Tetragenococcus* by Dicks et al.

TABLE 90. Phenotypic characteristics differentiating *Pediococcus* species[a,b]

Characteristic	P. damnosus	P. acidilactici	P. cellicola	P. claussenii	P. dextrinicus	P. inopinatus	P. parvulus	P. pentosaceus	P. stilesii
Growth at:									
pH 4.5	+	+	+	+	−	+	+	+	+[c]
pH 7.0	−	+	+	+[c]	+	d	d	+	+[c]
pH 8.0	−	+	+	+[c]	+	−	−	+	+[c]
pH 9.0	−	−	−	−	−	−	−	−	+[c]
35°C	−	+	+	+	+	+	+	+	+[c]
40°C	−	+	+	+[c]	+	d	−	d	+[c]
45°C	−	+	+	−	−	−	−	d	+[c]
48°C	−	+	−	−	−	−	−	d	+[c]
Lactic acid configuration	DL	DL	DL	L(+)[c]	L(+)	DL	DL	DL	DL
Acid from:									
Arabinose	−	d	d	−	−	−	−	+	−[c]
Galactose	+	+	+	−	d	+	d	+	+[c]
Lactose	−	d	+	−	d	+	−	+	−[c]
Maltose	d	−	+	d	+	+	d	+	+[c]
Melezitose	d	−	−	−	−	−	−	−	−[c]
Ribose	−	+	+	+	−	−	−	+	+[c]
Xylose	−	+	+	−	−	−	−	d	−[c]
Max. NaCl conc. for growth	5	10	?	5	6	8	8	10	8[c]
DNA G + C content (mol%) (T_m method)	38.5	42	38	40.5	40–41	39.5	41	38	38

[a]Symbols: +, 90% or more strains positive; −, 90% or more strains negative; d, 11–89% of strains positive.
[b]Data partially adapted from Simpson & Taguchi (1995), Franz et al. (2006), Holzapfel et al. (2005), and Zhang et al. (2005).
[c]Data have only been obtained with the type strain of *Pediococcus claussenii* and *Pediococcus stilesii*.

List of species of the genus *Pediococcus*

Table 90 gives some conditions under which the different species will grow and phenotypic features for a preliminary discrimination at species level.

1. **Pediococcus damnosus** Claussen 1903, 68[AL]

dam.no′sus. L. adj. *damnosus* destructive.

Morphology as general description. Growth is often slow and 2–3 d incubation at 22°C may be required for good growth. Broth cultures will develop without anaerobic conditions, but cysteine added to the medium improves growth. On an agar surface, colony development is poor unless cultures are incubated anaerobically. The final pH in broth is about 4.0. Optimum pH for growth is 5.5; cells do not grow at pH 8.5. Growth occurs in the range of 8–30°C. No growth occurs at 35°C; some strains grow in the presence of 4% NaCl. Although some strains grow in the presence of 5% (w/v) NaCl (Table 90), none grow in the presence of 5.5% NaCl (Simpson and Taguchi, 1995). Acetoin or diacetyl are readily produced and cause sarcina (buttery) odor in spoiled beer. Slime-forming strains are reported (Carr, 1970; Shimwell, 1948). The exopolysaccharide produced by a ropy strain of *Pediococcus damnosus* was shown to be a 2-substituted (1–3)-β-D-glucan (Duenas-Chasco et al., 1997). Hops humulone may cause the formation of giant cells of 5–15 µm (Nakagawa and Kitahara, 1962). *Pediococus damnosus* strain NCFB1832 produces the bacteriocin pedio-

cin PD-1 with activity against various food spoilage and food pathogenic bacteria (Bauer et al., 2005; Green et al., 1997). Fatty acid analysis using pyrolysis mass spectrometry showed that the major FAME profiles for *Pediococcus damnosus* were palmitic acid methyl ester ($C_{16:0 \, FAME}$), oleic acid methyl ester ($C_{18:1 \, FAME}$), and cyclopropyl nonadecanoic acid methyl ester ($C_{19:0 \, FAME \, cyclo}$).

Phenotypically, the inability to hydrolyze arginine separates *Pediococcus damnosus* from *Pediococcus acidilactici* and *Pediococcus pentosaceus*. Furthermore, *Pediococcus damnosus* has low genomic relatedness to *Pediococcus acidilactici* and *Pediococcus pentosaceus* (Dellaglio et al., 1981; Simpson and Taguchi, 1995). In contrast, *Pediococcus damnosus* shows a high degree of genomic relatedness to *Pediococcus inopinatus* and *Pediococcus parvulus* (34–36%) in DNA–DNA hybridization experiments (Table 88). Such a close relationship is also reflected in the phylogenetic relatedness based on 16S rDNA sequences (Dobson et al., 2002). Cells are readily destroyed by heat, i.e., 60°C for 10 min.

The type strain was isolated from lager beer yeast.

DNA G+C content (mol%): 37–42 (T_m) (Kocur et al., 1971).

Type strain: Be.1, ATCC 29358, CCUG 32251, CIP 102264, DSM 20331, JCM 5886, LMG 11484, NCIMB 12010, VKM B-1602.

GenBank accession number (16S rRNA gene): AJ318414, D87678.

2. **Pediococcus acidilactici** Lindner 1887, 440[AL]

a.ci.di.lac.ti'ci. N.L. gen. n. *acidilactici* of lactic acid.

Morphological, cultural, and physiological properties do not readily separate *Pediococcus acidilactici* from *Pediococcus pentosaceus*. Growth occurs at pH 4.2 and 8.0 and sometimes at pH 8.5. Final pH in MRS broth is between 3.5 and 3.8. Optimum temperature of growth is 40°C, while the maximum growth temperature is 50–53°C. All strains grow at 50°C. Cells grow in the presence of 9–10% (w/v) NaCl (Simpson and Taguchi, 1995) (Table 90). Cells are heat tolerant, but are killed after 10 min at 70°C. Some strains may be even more heat tolerant, particularly when newly isolated. *Pediococcus acidilactici* and *Pediococcus pentosaceus* are difficult to separate using physiological tests, suggesting a high degree of relatedness. These two species also show a higher relatedness to each other than to other pediococci when considering their phylogenetic relationship based on 16S rDNA analysis (Holzapfel et al., 2005). However, on the genome level they actually share low DNA–DNA relatedness (5–35% similarity, Table 88) (Dellaglio and Torriani, 1986). At the phenotypic level, the inability of *Pediococcus acidilactici* to ferment maltose and its ability to grow at 50°C differentiates it from *Pediococcus pentosaceus*. Some strains of *Pediococcus acidilactici* produce the bacteriocin pediocin PA-1/AcH (Marugg et al., 1992; Mora et al., 2000a).

The type strain was isolated from barley.

DNA G+C content (mol%): 38–44 (T_m) (Simpson and Taguchi, 1995).

Type strain: Back S213C, B213c, CCUG 32235, CIP 103408, DSM 20284, LMG 11384, NCIMB 12174.

GenBank accession number (16S rRNA gene): AJ305320, M58833.

3. **Pediococcus cellicola** Zhang, Tong and Dong 2005, 2169[VP]

cel.li.co'la. L. n. *cella* a storeroom for wine and food; L. suff. *cola* from L. n. *incola* an inhabitant, dweller; N.L. n. *cellicola* an inhabitant of a storeroom, indicating that the strain was isolated from a distilled-spirit-fermenting cellar.

Morphology as general description. Strains are facultatively anaerobic and produce DL-lactate. Growth at 13–44°C with an optimum at 30–37°C. Optimum pH for growth is pH 6.0–6.8. No growth at pH 3.8 and pH 8.2. No growth in 10% NaCl (Table 90). Utilizes rhamnose, ribose, D-xylose, and L-xylose as sole carbon energy source which can be used to differentiate *Pediococcus cellicola* from the closely related species *Pediococcus damnosus*, *Pediococcus inopinatus*, and *Pediococcus parvulus* that cannot ferment these sugars. Tolerant to 10% (v/v) ethanol. A phylogenetic study based on 16S rDNA sequence analysis indicated a close relationship with *Pediococcus damnosus*, *Pediococcus inopinatus*, and *Pediococcus parvulus* (Zhang et al., 2005).

The type strain was isolated from a cellar for spirit fermentation.

DNA G+C content (mol%): 37–39 (T_m) (Zhang et al., 2005).

Type strain: Z-8, AS 1.3787, LMG 22956.

GenBank accession number (16S rRNA gene): AY956788.

4. **Pediococcus claussenii** Dobson, Deneer, Lee, Hemmingsen, Glaze and Ziola 2002, 2009[VP]

claus'sen.i.i. N.L. gen. n. *Claussenii* of claussen, in honor of N.H. Claussen.

Morphology as general description. Some strains form slime. Unlike other pediococci (with the exception of *Pediococcus dextrinicus*) *Pediococcus claussenii* produces L(+)-lactate (Table 90). However, so far only the type strain of *Pediococcus claussenii* was tested (Holzapfel et al., 2005) for this trait. Should this remain a reliable phenotypic characterization trait, it may serve as a distinguishing feature for this species, as well as for the other L(+)-lactate-producing *Pediococcus* species, *Pediococcus dextrinicus*. However, *Pediococcus dextrinicus* is quite different from other pediococci at the 16S rDNA level and, as mentioned above, will probably be reclassified as a new genus in the future. Dobson et al. (2002) used three different comparative phylogenetic sequence analyses, comprising the 16S rDNA, 16S–23S internally transcribed spacer region, and the heat-shock protein HSP60 gene for investigating the phylogenetic relationships among pediococci and showed that *Pediococcus claussenii* could be well separated from other *Pediococcus* species. (As mentioned above, these investigations also confirmed other observations that *Pediococcus pentosaceus* and *Pediococcus acidilactici* are closely related.) The type strain was isolated from beer.

DNA G+C content (mol%): 40.0–41.7 (T_m) (Dobson et al., 2002).

Type strain: P06, ATCC BAA-344, DSM 14800, KCTC 3811, LMG 21948.

GenBank accession number (16S rRNA gene): AJ621555.

5. **Pediococcus dextrinicus** (Coster and White 1964) Back 1978b, 523[AL] (*Pediococcus cerevisiae* subsp. *dextrinicus* Coster and White 1964, 29.)

dex.trin'i.cus. N.L. n. *dextrinum* dextrin; N.L. masc. adj. *dextrinicus* relating to dextrin.

Morphology as general description. Less anaerobic than previously described species. Colonies develop on agar aerobically, but growth is improved in an atmosphere of H_2 + 10%CO_2. In MRS broth, the final pH is about 4.4. Growth does not occur at pH 4.2, but does at pH 8.0. Optimum pH for growth is 6.5. Optimum temperature 30–35°C. Cells grow in the presence of 6% (w/v) NaCl. In contrast to all other pediococci, *Pediococcus dextrinicus* utilizes starch and produces acid and gas from gluconate. *Pediococcus dextrinicus* produces L(+)-lactate from glucose (Table 89 and Table 90). 16S RNA sequencing has shown that the species is not closely related to other pediococci (Collins et al., 1991; Dobson et al., 2002). Thus, although this species is still classified as belonging to the genus *Pediococcus* (Euzéby, 1998; updated March 01, 2005: http://www.bacterio.cict.fr/almr.html), genetic and phenotypic data (see above) support its allocation to a new genus. Growth requirements have not been studied. Growth occurs in weakly hopped beer; its habitat is fermenting vegetables and beer.

The type strain was isolated from silage.

DNA G+C content (mol%): 40–41 (T_m) (Simpson and Taguchi, 1995).

Type strain: ATCC 33087, CCUG 18834, CIP 103407, DSM 20335, JCM 5887, LMG 11485, NCIMB 701561, VKM B-1603.

GenBank accession number (16S rRNA gene): D87679.

6. **Pediococcus inopinatus** Back 1988, 221[VP] (Effective publication: Back 1978a, 245.)

in.o.pin.a'tus. L. adj. *inopinatus* unexpected.

Morphology as general description. On agar, growth is slow and colonies may take 3–5 d to develop. The final pH in MRS broth is about 4.0. There is a close similarity between *Pediococcus inopinatus* and *Pediococcus parvulus* which both occur in the same habitat. The maximum pH value for growth is 7.5, while the maximum temperature for growth is 37–40°C. The optimum growth temperature lies between 30 and 32°C. Cells grow in the presence of 6–8% (w/v) NaCl (Table 90). *Pediococcus inopinatus* can be differentiated from *Pediococcus pentosaceus* and *Pediococcus acidilactici* by its inability to utilize pentoses and hydrolyze arginine (Table 90). It differs from *Pediococcus dextrinicus* in being unable to grow on starch or produce acid and gas from gluconate. Some strains form slime (Simpson and Taguchi, 1995). Habitat: fermenting vegetables and beverages such as beer and wine. The mol% G+C of the DNA is close to that of the other *Pediococcus* species (Table 90). Separation of *Pediococcus parvulus* and *Pediococcus inopinatus* was initially obtained by the electrophoresis of the L-(+)- and D-(−)-LDHs (Back, 1978b) and later confirmed by DNA–DNA hybridization and comparison of 16S rDNA sequences (Simpson and Taguchi, 1995).

The type strain was isolated from brewery yeast.

DNA G+C content (mol%): 39–40 (T_m) (Simpson and Taguchi, 1995).

Type strain: ATCC 49902, CCUG 38496, CIP 102406, DSM 20285, JCM 12518, LMG 11409, NCIMB 12564.

GenBank accession number (16S rRNA gene): AJ271383.

7. **Pediococcus parvulus** Günther and White 1961, 195[AL]

par'vu.lus. L. dim. adj. *parvulus* very small.

Morphology as general description. Colonies on tomato juice agar were originally reported to be very small, but improved growth can be obtained particularly when anaerobic incubation is used. Larger colonies usually develop after 48 h at 30°C. Broth cultures are improved by the addition of cysteine and Tween 80, but 48 h incubation may be needed (Garvie and Gregory, 1961). Some strains require asparagine (Garvie et al., 1961) but not folic acid. The final pH in broth is about 4.0 for well-grown strains. The optimal growth pH is 6.5 with the upper limit for growth between pH 7.0 and 7.5. The optimum growth temperature is 30°C. Maximum temperature for growth is 37–39°C. Growth occurs in the presence of 5.5–8% (w/v) NaCl (Simpson and Taguchi, 1995). Maximum temperature is 37–39°C. The inability of *Pediococcus parvulus* to utilize pentoses (Table 90) separates it from *Pediococcus pentosaceus* and *Pediococcus acidilactici* (Simpson and Taguchi, 1995). Some strains form slime and some are reported to produce high levels of histamine during wine fermentation (Landete et al., 2005). *Pediococcus damnosus* strain ATO77 produces the bacteriocin pediocin PA-1 (Bennik et al., 1997). Habitat: fermenting vegetables and wine fermentations.

The type strain was isolated from silage. Tolerance to hop antiseptic is unknown.

DNA G+C content (mol%): 40.5–41.6 (T_m) (Simpson and Taguchi, 1995).

Type strain: ATCC 19371, CCUG 28439, CIP 102262, DSM 20332, JCM 5889, LMG 11486, NBRC 100673, NCIMB 9447, VKM B-1604.

GenBank accession number (16S rRNA gene): D88528.

8. **Pediococcus pentosaceus** Mees 1934, 96[AL]

pen.to.sa'ce.us. N.L. neut. n. *pentosum* a pentose sugar; N.L. adj. *pentosaceus* relating to a pentose.

Morphology as general description. Anaerobic incubation is not necessary, and colonies should be visible on agar after aerobic incubation for 24 h at 30°C. Litmus milk reactions are variable and may be related to growth requirements. The requirement for folic acid varies between strains. A limited study of the aldolases found in pediococci has shown that *Pediococcus pentosaceus* and *Pediococcus acidilactici* are closely related species which are separate from *Pediococcus parvulus*. There is some evidence that not all strains of *Pediococcus pentosaceus* have the same aldolase (London and Chace, 1976).

In broth, growth can be very rapid and the final pH in MRS broth is usually below 4.0. Optimal growth pH is between 6.0 and 6.5, but growth may occur at pH 8.0 (Simpson and Taguchi, 1995) (Table 90). Optimal growth temperature is between 28 and 32°C and not above 39–45°C. Cells grow in the presence of 9–10% (w/v) NaCl (Simpson and Taguchi, 1995). *Pediococcus pentosaceus* is less heat resistant than *Pediococcus acidilactici* and is killed after 8 min. at 65°C. The sugar fermentation reactions of *Pediococcus pentosaceus* resemble those of *Pediococcus acidilactici*. However, the ability to ferment maltose and a lower growth temperature differentiates *Pediococcus pentosaceus* from *Pediococcus acidilactici* (although not invariably). *Pediococcus pentosaceus* could also be confused with micrococci as it can form small colonies on sugar-free agar and can grow at pH 9.0. *Pediococcus pentosaceus* can also be weakly catalase-positive when grown in medium with low glucose content (Whittenbury, 1965). The rapid growth, low final pH, and absence of cytochromes distinguish *Pediococcus pentosaceus* from micrococci.

Various strains of *Pediococcus pentosaceus* have been reported to produce bacteriocins such as pediocin PA-1 (Miller et al., 2005), pediocin ST18 (Todorov and Dicks, 2005), pediocin – A (Piva and Headon, 1994), and pediocin ACCEL (Wu et al., 2004). The type strain was isolated from dried American beer yeast.

DNA G+C content (mol%): 35–39 (T_m) (Simpson and Taguchi, 1995).

Type strain: ATCC 33316, CCUG 32205, CIP 102260, DSM 20336, JCM 5890, LMG 11488, NCIMB 12012, NCTC 12956.

GenBank accession number (16S rRNA gene): AJ305321, M58834.

9. **Pediococcus stilesii** Franz, Vancanneyt, Vandemeulebroecke, De Wachter, Cleenwerck, Hoste, Schillinger, Holzapfel and Swings 2006, 332[VP]

stiles.i.i. N.L. gen. n. *stilesii* of Stiles, in honor of M.E. Stiles, a microbiologist.

Morphology as general description. Currently only one strain known. Produces DL-lactate. Acid produced from ribose, galactose, fructose, maltose, and mannose. Unlike most other pediococci, *Pediococcus stilesii* can grow at high pH, even pH 9.6. Grows at 35–45°C, no growth at 48°C. Cells grow in the presence of 8% NaCl (Table 90). Using 16S rDNA sequencing for phylogenetic comparison, *Pediococcus pentosaceus*, *Pediococcus acidilactici*, *Pediococcus stilesii*, and *Pediococcus claussenii* were each shown to occupy a distinct branch, yet were also more closely related with each other than with other *Pediococcus* species (Franz et al., 2006). However, using DNA–DNA hybridization, only low similarity was observed with DNA from *Pediococcus acidilactici* (14.5%) and *Pediococcus pentosaceus* (21%). Pheno-

typically, *Pediococcus stilesii* is distinguished from *Pediococcus pentosaceus* by its inability to ferment arabinose and lactose, from *Pediococcus acidilactici* by its inability to ferment xylose, and from *Pediococcus claussenii* by its ability to produce acid from galactose and DL-lactate from glucose. The type strain was isolated from steeped maize grains.

DNA G+C content (mol%): 38 (T_m) (Franz et al., 2006).

Type strain: LMG 23082, BFE 1652, CCUG 51290, FAIR-E 180.

GenBank accession number (16S rRNA gene): AJ973157.

References

Abo-Elnaga, I.G. and O. Kandler. 1965a. [On the taxonomy of the species *Lactobacillus* Beijerinck. 3. The need for vitamins]. Zentralbl Bakteriol Parasitenkd Infektionskr Hyg *119*: 661–672.

Abo-Elnaga, I.G. and O. Kandler. 1965b. On the taxonomy of the genus *Lactobacillus* Beijerinck. II. Subgenus Betabacterium Orla Jensen. Zentbl. Bakteriol. Parasitenkd. Infektionskr. Hyg. *119*: 117–129.

Abo-Elnaga, I.G. and O. Kandler. 1965c. On the taxonomy of genus *Lactobacillus* Beijerinck. I. Subgenus Streptobacterium Orla Jensen. Zentbl. Bakteriol. Parasitenkd. Infektionskr. Hyg. *119*: 1–36.

Aguirre, M. and M.D. Collins. 1993. Lactic acid bacteria and human clinical infection. J. Appl. Bacteriol. *75*: 95–107.

Amanatidou, A., M.H. Bennik, L.G. Gorris and E.J. Smid. 2001. Superoxide dismutase plays an important role in the survival of *Lactobacillus sake* upon exposure to elevated oxygen. Arch. Microbiol. *176*: 79–88.

Amann, R. and W. Ludwig. 2000. Ribosomal RNA-targeted nucleic acid probes for studies in microbial ecology. FEMS Microbiol. Rev. *24*: 555–565.

Amann, R.I., W. Ludwig and K.H. Schleifer. 1995. Phylogenetic identification and in situ detection of individual microbial cells without cultivation. Microbiol. Rev. *59*: 143–169.

Anonymous. 2000. Probiotische Mikroorganismenkulturen in Lebensmitteln. Ernährungsumschau *47*: 191–195.

Antonio, M.A., S.E. Hawes and S.L. Hillier. 1999. The identification of vaginal *Lactobacillus* species and the demographic and microbiologic characteristics of women colonized by these species. J. Infect. Dis. *180*: 1950–1956.

Archibald, A.R. and J. Baddiley. 1966. The teichoic acids. Adv. Carbohydr. Chem. Biochem. *21*: 323–375.

Archibald, F.S. and I. Fridovich. 1981. Manganese and defenses against oxygen toxicity in *Lactobacillus plantarum*. J. Bacteriol. *145*: 442–451.

Aslam, Z., W.T. Im, L.N. Ten, M.J. Lee, K.H. Kim and S.T. Lee. 2006. *Lactobacillus siliginis* sp. nov., isolated from wheat sourdough in South Korea. Int. J. Syst. Evol. Microbiol. *56*: 2209–2213.

Auclair, J. and J.P. Accolas. 1983. Use of thermophilic lactic starters in the dairy industry. Antonie van Leeuwenhoek *49*: 313–326.

Avall-Jaaskelainen, S. and A. Palva. 2005. *Lactobacillus* surface layers and their applications. FEMS Microbiol. Rev. *29*: 511–529.

Axelsson, L. 1998. Lactic acid bacteria: Classification and physiology. *In* Salminen and von Wright (Editors), Lactic Acid Bacteria. Microbiology and Functional Aspects. Marcel Dekker, Inc., New York, pp. 1–72.

Back, W. 1978a. Zur Taxonomie der Gattung *Pediococcus*. Phäanotypische and genotypische Abgrenzung der bisher bekannten Arten sowie Beschreibung einer neuen bierschädlichen Art: *Pediococcus inopinatus*. Brauwissenschaft *31*: 237–250, 312–320, 336–343.

Back, W. 1978b. Elevation of *Pediococcus cerevisiae* subsp. *dextrinicus* coster and white to species status *Pediococcus dextrinicus* (Coster and White) comb. nov. Int. J. Syst. Bacteriol. *28*: 523–527.

Back, W. and E. Stackebrandt. 1978. DNA-DNA homology studies on genus *Pediococcus*. Arch. Microbiol. *118*: 79–85.

Back, W. 1980. Bierschädliche Bakterien. Nachweis und Kultivierung bierschädlicher Bakterien im Betriebslabor. Brauwelt *120*: 1562–1569.

Back, W. 1988. *In* Validation of the publication of new names and new combinations previously effectively published outside the IJSB. List no. 25. Int. J. Syst. Bacteriol. *38*: 220–222.

Back, W. 1994. Farbatlas und Handbuch der Getränkebiologie Teil 1. Verlag Hans Carl, Nürnberg.

Back, W., I. Bohak, M. Ehrmann, W. Ludwig and K.H. Schleifer. 1996. Revival of the species *Lactobacillus lindneri* and the design of a species specific oligonucleotide probe. Syst. Appl. Microbiol. *19*: 322–325.

Back, W., I. Bohak, M. Ehrmann, W. Ludwig and K.H. Schleifer. 1997. *In* Validation of the publication of new names and new combinations previously effectively published outside the IJSB. List no. 61. Int. J. Syst. Bacteriol. *47*: 601–602.

Back, W., I. Bohak, M. Ehrmann, T. Ludwig, B. Pot and K.H. Schleifer. 1999. *Lactobacillus perolens* sp. nov., a soft drink spoilage bacterium. Syst. Appl. Microbiol. *22*: 354–359.

Back, W. 2000. Farbatlas und Handbuch der Getränkebiologie Teil 2. Verlag Hans Carl, Nürnberg.

Back, W., I. Bohak, M. Ehrmann, T. Ludwig, B. Pot and K.H. Schleifer. 2000. *In* Validation of the publication of new names and new combinations previously effectively published outside the IJSB. List no. 72. Int. J. Syst. Bacteriol. *50*: 3–4.

Baele, M., K. Chiers, L.A. Devriese, H.E. Smith, H.J. Wisselink, M. Vaneechoutte and F. Haesebrouck. 2001a. The Gram-positive tonsillar and nasal flora of piglets before and after weaning. J. Appl. Microbiol. *91*: 997–1003.

Baele, M., L.A. Devriese and F. Haesebrouck. 2001b. *Lactobacillus agilis* is an important component of the pigeon crop flora. J. Appl. Microbiol. *91*: 488–491.

Baele, M., M. Vancanneyt, L.A. Devriese, K. Lefebvre, J. Swings and F. Haesebrouck. 2003. *Lactobacillus ingluviei* sp. nov., isolated from the intestinal tract of pigeons. Int. J. Syst. Evol. Microbiol. *53*: 133–136.

Balcke, J. 1884. Über häufig vorkommende Fehler in der Bierbereitung. Wochenschrift für Brauerei *1*: 181–184.

Barney, M., A. Volgyi, A. Navarro and D. Ryder. 2001. Riboprinting and 16S rRNA gene sequencing for identification of brewery *Pediococcus* isolates. Appl. Environ. Microbiol. *67*: 553–560.

Barre, P. 1969. Numerical taxonomy of lactobacilli isolated from wines. Arch. Mikrobiol. *68*: 74–86.

Barros, R.R., M.G. Carvalho, J.M. Peralta, R.R. Facklam and L.M. Teixeira. 2001. Phenotypic and genotypic characterization of *Pediococcus* strains isolated from human clinical sources. J. Clin. Microbiol. *39*: 1241–1246.

Barton, L.L., E.D. Rider and R.W. Coen. 2001. Bacteremic infection with *Pediococcus*: vancomycin-resistant opportunist. Pediatrics *107*: 775–776.

Bauer, R., M.L. Chikindas and L.M.T. Dicks. 2005. Purification, partial amino acid sequence and mode of action of pediocin PD-1, a bacteriocin produced by *Pediococcus damnosus* NCFB 1832. Int. J. Food Microbiol. *101*: 17–27.

Baumgart, J., B. Weber and B. Hanekamp. 1983. Mikrobiologische Stabilität von Feinkosterzeugnissen. Fleischwirtschaft *63*: 93–94.

Bayer, A.S., A.W. Chow, N. Concepcion and L.B. Guze. 1978. Susceptibility of 40 lactobacilli to six antimicrobial agents with broad gram-positive anaerobic spectra. Antimicrob. Agents Chemother. *14*: 720–722.

Beck, R., F. Gross and T. Beck. 1987. Untersuchungen zur Kenntnis der Gärfuttermikroflora. Das Wirtschaftseigene Futter *33*: 13–33.

Beck, R., N. Weiss and J. Winter. 1988. *Lactobacillus graminis* sp. nov., a new species of facultatively heterofermentative lactobacilli surviving at low pH in grass silage. Syst. Appl. Microbiol. *10*: 279–283.

Beck, R., N. Weiss and J. Winter. 1989. *In* Validation of the publication of new names and new combinations previously effectively published outside the IJSB. List no. 28. Int. J. Syst. Bacteriol. *39*: 93–94.

Beijerinck, M.W. 1901. Sur les ferments lactiques de l'industrie. Arch. Néer. Sci. *6*: 212–243.

ben Omar, N. and F. Ampe. 2000. Microbial community dynamics during production of the Mexican fermented maize dough pozol. Appl. Environ. Microbiol. *66*: 3664–3673.

Bennik, M., E.J. Smid and L. Gorris. 1997. Vegetable-associated *Pediococcus parvulus* produces pediocin PA-1. Appl. Environ. Microbiol. *63*: 2074–2076.

Berg, R.W., W.E. Sandine and A.W. Anderson. 1981. Identification of a growth stimulant for *Lactobacillus sanfrancisco*. Appl. Environ. Microbiol. *42*: 786–788.

Bergey, D.H., F.C. Harrison, R.S. Breed, B.W. Hammer and F.M. Huntoon. 1923. Bergey's Manual of Determinative Bacteriology, 1st edn. The Williams & Wilkins Co., Baltimore.

Bergey, D.H., F.C. Harrison, R.S. Breed, B.W. Hammer and F.M. Huntoon. 1925. Bergey's Manual of Determinative Bacteriology, 2nd edn. The Williams & Wilkins Co., Baltimore.

Bergey, D.H., R.S. Breed, B.W. Hammer, F.M. Huntoon, E.G. Murray and F.C. Harrison. 1934. Bergey's Manual of Determinative Bacteriology, 4th edn. The Williams & Wilkins Co., Baltimore.

Berthier, F. and S.D. Ehrlich. 1998. Rapid species identification within two groups of closely related lactobacilli using PCR primers that target the 16S/23S rRNA spacer region. FEMS Microbiol. Lett. *161*: 97–106.

Berthier, F. and S.D. Ehrlich. 1999. Genetic diversity within *Lactobacillus sakei* and *Lactobacillus curvatus* and design of PCR primers for its detection using randomly amplified polymorphic DNA. Int. J. Syst. Bacteriol. *49*: 997–1007.

Biddle, A.D. and P.J. Warner. 1993. Induction of the 6-phosphogluconate pathway is correlated with change of colony morphology in *Lactobacillus acidophilus*. FEMS Microbiol. Rev. *12*: P47.

Biede, S.L., G.W. Reinbold and E.G. Hammond. 1976. Influence of *Lactobacillus bulgaricus* on microbiology and chemistry of Swiss cheese. J. Dairy Sci. *59*: 854–858.

Björkroth, J. and W. Holzapfel. 2003. Genera *Leuconostoc*, *Oenococcus*, and *Weissella*. *In* Dworkin, Falkow, Rosenberg, Schleifer and Stackbrandt (Editors), The Prokaryotes: An Evolving Electronic Resource for the Microbiological Community, 3rd edn. Springer, New York.

Blasco, L., S. Ferrer and I. Pardo. 2003. Development of specific fluorescent oligonucleotide probes for in situ identification of wine lactic acid bacteria. FEMS Microbiol. Lett. *225*: 115–123.

Blood, R.M. 1975. Lactic acid bacteria in marinated herring. *In* Carr, Cutting and Whiting (Editors), Lactic Acid Bacteria in Beverages and Food. Academic Press, London, New York, San Francisco, pp. 195–208.

Boekhorst, J., R.J. Siezen, M.C. Zwahlen, D. Vilanova, R.D. Pridmore, A. Mercenier, M. Kleerebezem, W.M. de Vos, H. Brussow and F. Desiere. 2004. The complete genomes of *Lactobacillus plantarum* and *Lactobacillus johnsonii* reveal extensive differences in chromosome organization and gene content. Microbiology *150*: 3601–3611.

Bohak, I., W. Back, L. Richter, M. Ehrmann, W. Ludwig and K.H. Schleifer. 1998. *Lactobacillus amylolyticus* sp. nov., isolated from beer malt and beer wort. Syst. Appl. Microbiol. *21*: 360–364.

Bongaerts, G.P., J.J. Tolboom, A.H. Naber, W.J. Sperl, R.S. Severijnen, J.A. Bakkeren and J.L. Willems. 1997. Role of bacteria in the pathogenesis of short bowel syndrome-associated D-lactic acidemia. Microb. Pathog. *22*: 285–293.

Botha, S.J., S.C. Boy, F.S. Botha and R. Senekal. 1998. *Lactobacillus* species associated with active caries lesions. J. Dent. Assoc. S. Afr. *53*: 3–6.

Bottazzi, V., M. Vescovo and F. Dellaglio. 1973. Microbiology of Grana cheese. IX. Characteristics and distribution of *Lactobacillus helveticus* biotypes in natural whey cheese starter. Sci. Tech. Latt.-casear *24*: 23–29.

Bouton, Y., P. Guyot, E. Beuvier, P. Tailliez and R. Grappin. 2002. Use of PCR-based methods and PFGE for typing and monitoring homofermentative lactobacilli during Comte cheese ripening. Int. J. Food Microbiol. *76*: 27–38.

Breithaupt, D.E., W. Schwack, G. Wolf and W.P. Hammes. 2001. Characterization of the triterpenoid 4,4'-diaponeurosporene and its isomers in food-associated bacteria. Eur. Food Res. Technol. *213*: 231–233.

Bringel, F., A. Castioni, D.K. Olukoya, G.E. Felis, S. Torriani and F. Dellaglio. 2005. *Lactobacillus plantarum* subsp. *argentoratensis* subsp. nov., isolated from vegetable matrices. Int. J. Syst. Evol. Microbiol. *55*: 1629–1634.

Brockmann, E., B.L. Jacobsen, C. Hertel, W. Ludwig and K.H. Schleifer. 1996. Monitoring of genetically modified *Lactococcus lactis* in gnotobiotic and conventional rats by using antibiotic resistance markers and specific probe or primer based methods. Syst. Appl. Microbiol. *19*: 203–212.

Brussow, H. 2001. Phages of dairy bacteria. Ann. Rev. Microbiol. *55*: 283–303.

Brygoo, E.R. and N. Aladame. 1953. Study of a new strictly anaerobic species of the genus *Eubacterium*: *Eubacterium crispatum* n. sp. Ann. Inst. Pasteur (Paris) *84*: 640–641.

Buchta, K. 1983. Lactic acid. *In* Reed (Editor), Biotechnology, 3. Verlag Chemie, Weinheim, pp. 409–417.

Bunte, C., C. Hertel and W.P. Hammes. 2000. Monitoring and survival of *Lactobacillus paracasei* LTH 2579 in food and the human intestinal tract. Syst. Appl. Microbiol. *23*: 260–266.

Burton, J.P., P.A. Cadieux and G. Reid. 2003. Improved understanding of the bacterial vaginal microbiota of women before and after probiotic instillation. Appl. Environ. Microbiol. *69*: 97–101.

Cachat, E. and F.G. Priest. 2005. *Lactobacillus suntoryeus* sp. nov., isolated from malt whisky distilleries. Int. J. Syst. Evol. Microbiol. *55*: 31–34.

Cai, Y.M., S. Kumai, M. Ogawa, Y. Benno and T. Nakase. 1999a. Characterization and identification of *Pediococcus* species isolated from forage crops and their application for silage preparation. Appl. Environ. Microbiol. *65*: 2901–2906.

Cai, Y.M., H. Okada, H. Mori, Y. Benno and T. Nakase. 1999b. *Lactobacillus paralimentarius* sp. nov., isolated from sourdough. Int. J. Syst. Bacteriol. *49*: 1451–1455.

Caldwell, S.L., D.J. McMahon, C.J. Oberg and J.R. Broadbent. 1999. Induction and characterization of *Pediococcus acidilactici* temperate bacteriophage. Syst. Appl. Microbiol. *22*: 514–519.

Carlsson, J. and L. Gothefors. 1975. Transmission of *Lactobacillus jensenii* and *Lactobacillus acidophilus* from mother to child at time of delivery. J Clin Microbiol *1*: 124–128.

Carr, F.J., D. Chill and N. Maida. 2002. The lactic acid bacteria: a literature survey. Crit. Rev. Microbiol. *28*: 281–370.

Carr, J.G. 1970. Tetrad-forming cocci in ciders. J. Appl. Bacteriol. *33*: 371–379.

Carr, J.G. and P.A. Davies. 1970. Homofermentative lactobacilli of ciders including *Lactobacillus mali* nov. spec. J. Appl. Bacteriol. *33*: 768–774.

Carr, J.G. and P.A. Davies. 1972. The ecology and classification of strains of *Lactobacillus collinoides* nov. spec.: a bacterium commonly found in fermenting apple juice. J. Appl. Bacteriol. *35*: 463–471.

Carr, J.G., C.V. Cutting and G.C. Whiting. 1975. University of Bristol, Sept. 9–21, 1973.

Carr, J.G., P.A. Davies, F. Dellaglio, M. Vescovo and R.A.D. Williams. 1977. The relationship between *Lactobacillus mali* from cider and *Lactobacillus yamanashiensis* from wine. J. Appl. Bacteriol. *42*: 219–228.

Cato, E.P., W.E.C. Moore and J.L. Johnson. 1983. Synonym of strains of "*Lactobacillus acidophilus*" groups A2 (Johnson et al. 1980) with the type strain of *Lactobacillus crispatus* (Brygoo and Aladame 1953) Moore and Holdeman Moore 1970. Int. J. Syst. Bacteriol. *33*: 426–428.

Charlton, D.B., M.E. Nelson and C.H. Werkman. 1934. Physiology of *Lactobacillus fructivorans* sp. nov. isolated from spoiled salad dressing. Iowa State J. Sci. *9*: 1–11.

Chassy, B.M. and J. Thompson. 1983. Regulation and characterization of the galactose-phosphoenolpyruvate-dependent phosphotransferase system in *Lactobacillus casei*. J. Bacteriol. *154*: 1204–1214.

Cheirsilp, B., H. Shimizu and S. Shioya. 2003. Enhanced kefiran production by mixed culture of *Lactobacillus kefiranofaciens* and *Saccharomyces cerevisiae*. J. Biotechnol. *100*: 43–53.

Chen, G.J. and J.B. Russell. 1989. Transport of glutamine by *Streptococcus bovis* and conversion of glutamine to pyroglutamic acid and ammonia. J. Bacteriol. *171*: 2981–2985.

Chen, H., C.K. Lim, Y.K. Lee and Y.N. Chan. 2000. Comparative analysis of the genes encoding 23S-5S rRNA intergenic spacer regions of *Lactobacillus casei*-related strains. Int. J. Syst. Evol. Microbiol. *50*: 471–478.

Chenoll, E., M.C. Macian and R. Aznar. 2006. *Lactobacillus rennini* sp. nov., isolated from rennin and associated with cheese spoilage. Int. J. Syst. Evol. Microbiol. *56*: 449–452.

Claussen, N.H. 1903. Études sur les bactéries dites sarcines et sur les maladies quelles provoquent dans la bière. C.R. Trav. Lab. Carlsberg *6*: 64–83.

Cocolin, L., M. Manzano, C. Cantoni and G. Comi. 2001. Denaturing gradient gel electrophoresis analysis of the 16S rRNA gene V1 region to monitor dynamic changes in the bacterial population during fermentation of Italian sausages. Appl. Environ. Microbiol. *67*: 5113–5121.

Coenye, T., D. Gevers, Y. Van de Peer, P. Vandamme and J. Swings. 2005. Towards a prokaryotic genomic taxonomy. FEMS Microbiol. Rev. *29*: 147–167.

Coeuret, V., S. Dubernet, M. Bernardeau, M. Gueguen and J.P. Vernoux. 2003. Isolation, characterisation and identification of lactobacilli focusing mainly on cheeses and other dairy products. LAIT *83*: 269–306.

Collins, J.K. and D. Jones. 1981. The distribution of isoprenoid quinone structural types in bacteria and their taxonomic implications. Microbiol. Rev. *45*: 316–354.

Collins, M.D., B.A. Phillips and P. Zanoni. 1989. Deoxyribonucleic acid homology studies of *Lactobacillus casei*, *Lactobacillus paracasei* sp. nov., subsp. *paracasei* and subsp. *tolerans*, and *Lactobacillus rhamnosus* sp. nov., comb. nov. Int. J. Syst. Bacteriol. *39*: 105–108.

Collins, M.D., A.M. Williams and S. Wallbanks. 1990. The phylogeny of *Aerococcus* and *Pediococcus* as determined by 16S rRNA sequence analysis: description of *Tetragenococcus* gen. nov. FEMS Microbiol. Lett. *58*: 255–262.

Collins, M.D., U. Rodrigues, C. Ash, M. Aguirre, J. A. Farrow, A. Martina-Murcia, B. A. Phillips, A. M. Williams and S. Wallbanks. 1991. Phylogenetic analysis of the genus *Lactobacillus* and related lactic acid bacteria as determined by reverse transcriptase sequencing of 16S rRNA. FEMS Microbiol. Lett. *77*: 5–12.

Collins, M.D., J. Samelis, J. Metaxopoulos and S. Wallbanks. 1993. Taxonomic studies on some *Leuconostoc*-like organisms from fermented sausages: description of a new genus *Weissella* for the *Leuconostoc paramesenteroides* group of species. J. Appl. Bacteriol. *75*: 595–603.

Colman, G. and A. Efstratiou. 1987. Vancomycin-resistant leuconostocs, lactobacilli and now pediococci. J. Hosp. Infect. *10*: 1–3.

Condon, S. 1987. Responses of lactic acid bacteria to oxygen. FEMS Microbiol. Rev. *46*: 269–280.

Conway, P.L. 1997. Development of intestinal microbiota. *In* Mackie, White and Isaacson (Editors), Gastrointestinal Microbiology, 2. Chapman and Hall, London, pp. 3–38.

Coronado, B.E., S.M. Opal and D.C. Yoburn. 1995. Antibiotic-induced D-lactic acidosis. Ann. Intern. Med. *122*: 839–842.

Corsetti, A., L. Settanni, D. van Sinderen, G.E. Felis, F. Dellaglio and M. Gobbetti. 2005. *Lactobacillus rossii* sp. nov., isolated from wheat sourdough. Int. J. Syst. Evol. Microbiol. *55*: 35–40.

Coster, E. and H.R. White. 1964. Further studies of the genus *Pediococcus*. J. Gen. Microbiol. *37*: 15–31.

Coyette, J. and J.M. Ghuysen. 1970. Structure of the walls of *Lactobacillus acidophilus* strain 63 AM Gasser. Biochemistry *9*: 2935–2943.

Curk, M.C., J.C. Hubert and F. Bringel. 1996. *Lactobacillus paraplantarum* sp. nov., a new species related to *Lactobacillus plantarum*. Int. J. Syst. Bacteriol. *46*: 595–598.

Dachs, E. 1981. NBB-Nachweismedium für vierschädliche Bakterien. Brauwelt. *121*: 1778–1782, 1784.

Dal Bello, F., J. Walter, W.P. Hammes and C. Hertel. 2003. Increased complexity of the species composition of lactic acid bacteria in human feces revealed by alternative incubation conditions. Microb. Ecol. *45*: 455–463.

Dal Bello, F. and C. Hertel. 2006. Oral cavity as natural reservoir for intestinal lactobacilli. Syst. Appl. Microbiol *29*: 69–76.

Davis, J.G. 1975. The microbiology of yoghurt. *In* Carr, Cutting and Whiting (Editors), Lactic Acid Bacteria in Beverages and Food. Academic Press, London, New York, San Francisco, pp. 245–263.

De Man, J.C., M. Rogosa and M.E. Sharpe. 1960. A medium for the cultivation of lactobacilli. J. Appl. Bacteriol. *23*: 130–135.

de Vries, W., W.M. Kapteijn, E.G. van der Beek and A.H. Stouthamer. 1970. Molar growth yields and fermentation balances of *Lactobacillus casei* L3 in batch cultures and in continuous cultures. J. Gen. Microbiol. *63*: 333–345.

De Vuyst, L. and B. Degeest. 1999. Heteropolysaccharides from lactic acid bacteria. FEMS Microbiol. Rev. *23*: 153–177.

De Vuyst, L. and P. Neysens. 2005. The sourdough microflora: biodiversity and metabolic interactions. Trends Food Sci. Technol. *16*: 43–56.

Delaherche, A., O. Claisse and A. Lonvaud-Funel. 2004. Detection and quantification of *Brettanomyces bruxellensis* and 'ropy' *Pediococcus damnosus* strains in wine by real-time polymerase chain reaction. J. Appl. Microbiol. *97*: 910–915.

Dellaglio, F., L.D. Trovatelli and P.G. Sarra. 1981. DNA-DNA homology among representative strains of the genus *Pediococcus*. Zentbl. Bakteriol. Mikrobiol. Hyg. I Abteilung Orig. C *2*: 140–150.

Dellaglio, F. and S. Torriani. 1986. DNA–DNA homology, physiological characteristics and distribution of lactic acid bacteria isolated from maize silage. J. Appl. Bacteriol. *60*: 83–92.

Dellaglio, F., G.E. Felis and S. Torriani. 2002. The status of the species *Lactobacillus casei* (Orla-Jensen 1916) Hansen and Lessel 1971 and *Lactobacillus paracasei* Collins *et al.* 1989. Request for an opinion. Int. J. Syst. Evol. Microbiol. *52*: 285–287.

Dellaglio, F., S. Torriani and G.E. Felis. 2004. Reclassification of *Lactobacillus cellobiosus* Rogosa *et al.* 1953 as a later synonym of *Lactobacillus fermentum* Beijerinck 1901. Int. J. Syst. Evol. Microbiol. *54*: 809–812.

Dellaglio, F., G.E. Felis, A. Castioni, S. Torriani and J.E. Germond. 2005. *Lactobacillus delbrueckii* subsp. *indicus* subsp. nov., isolated from Indian dairy products. Int. J. Syst. Evol. Microbiol. *55*: 401–404.

Dent, V.E. and R.A.D. Williams. 1982. *Lactobacillus animalis* sp. nov., a new species of *Lactobacillus* from the alimentary canal of animals. Zentbl. Bakteriol. Mikrobiol. Hyg. I Abt. Orig. C *3*: 377–386.

Dent, V.E. and R.A.D. Williams. 1983. *In* Validation of the publication of new names and new combinations previously effectively published outside the IJSB. List no. 10. Int. J. Syst. Bacteriol. *33*: 438–440.

Desiere, F., S. Lucchini, C. Canchaya, M. Ventura and H. Brussow. 2002. Comparative genomics of phages and prophages in lactic acid bacteria. Antonie Leeuwenhoek *82*: 73–91.

Dicks, L.M.T., E.M. DuPlessis, F. Dellaglio and E. Lauer. 1996. Reclassification of *Lactobacillus casei* subsp. *casei* ATCC 393 and *Lactobacillus rhamnosus* ATCC 15820 as *Lactobacillus zeae* nom. rev., designation of ATCC 334 as the neotype of *L. casei* subsp. *casei*, and rejection of the *Lactobacillus paracasei*. Int. J. Syst. Bacteriol. *46*: 337–340.

Dicks, L.M.T., M. Silvester, P.A. Lawson and M.D. Collins. 2000. *Lactobacillus fornicalis* sp. nov., isolated from the posterior fornix of the human vagina. Int. J. Syst. Evol. Microbiol. *50*: 1253–1258.

Dickson, E.M., M.P. Riggio and L. Macpherson. 2005. A novel species-specific PCR assay for identifying *Lactobacillus fermentum*. J. Med. Microbiol. *54*: 299–303.

Dobson, C.M., H. Deneer, S. Lee, S. Hemmingsen, S. Glaze and B. Ziola. 2002. Phylogenetic analysis of the genus *Pediococcus*, including *Pediococcus claussenii* sp. nov., a novel lactic acid bacterium isolated from beer. Int. J. Syst. Evol. Microbiol. *52*: 2003–2010.

Döderlein, A. 1894. Die Scheidensekretuntersuchungen. Zentbl. Gynäkol. *18*: 10–14.

Douglas, H.C. and W.V. Cruess. 1936. A *Lactobacillus* from California wine: *Lactobacillus hilgardii*. Food Res. *1*: 113–119.

Duenas-Chasco, H.T., M.A. Rodriguez-Carvajal, P. Tejero Mateo, G. Franco-Rodriguez, J.L. Espartero, A. Irastorza-Iribas and A.M. Gil-Serrano. 1997. Structural analysis of the exopolysaccharide produced by *Pediococcus damnosus*. Carbohydr. Res. *303*: 453–458.

Edwards, C.G., K.M. Haag, M.D. Collins, R.A. Hutson and Y.C. Huang. 1998. *Lactobacillus kunkeei* sp. nov.: a spoilage organism associated with grape juice fermentations. J. Appl. Microbiol. *84*: 698–702.

Edwards, C.G., M.D. Collins, P.A. Lawson and A.V. Rodriguez. 2000. *Lactobacillus nagelii* sp. nov., an organism isolated from a partially fermented wine. Int. J. Syst. Evol. Microbiol. *50*: 699–702.

Egan, F.A. 1983. Lactic acid bacteria of meat and meat products. Antonie Leeuwenhoek J. Microbiol. Serol. *49*: 327–336.

Ehrmann, M., W. Ludwig and K.H. Schleifer. 1994. Reverse dot blot hybridization: a useful method for the direct identification of lactic acid bacteria in fermented food. FEMS Microbiol. Lett. *117*: 143–149.

Ehrmann, M.A., M.R.A. Muller and R.F. Vogel. 2003. Molecular analysis of sourdough reveals *Lactobacillus mindensis* sp. nov. Int. J. Syst. Evol. Microbiol. *53*: 7–13.

Ehrmann, M.A. and R.F. Vogel. 2005. Taxonomic note "*Lactobacillus pastorianus*" (Van Laer, 1892) a former synonym for *Lactobacillus paracollinoides*. Syst. Appl. Microbiol. *28*: 54–56.

Ehrmann, M.A., M. Brandt, P. Stolz, R.F. Vogel and M. Korakli. 2007. *Lactobacillus secaliphilus* sp. nov., isolated from type II sourdough fermentation. Int. J. Syst. Evol. Microbiol. *57*: 745–750.

Elisha, B.G. and P. Courvalin. 1995. Analysis of genes encoding D-alanine:D-alanine ligase-related enzymes in *Leuconostoc mesenteroides* and *Lactobacillus* spp. Gene *152*: 79–83.

Elli, M., R. Zink, A. Rytz, R. Reniero and L. Morelli. 2000. Iron requirement of *Lactobacillus* spp. in completely chemically defined growth media. J. Appl Microbiol. *88*: 695–703.

Embley, T.M., N. Faquir, W. Bossart and M.D. Collins. 1989. *Lactobacillus vaginalis* sp. nov. from the human vagina. Int. J. Syst. Bacteriol. *39*: 368–370.

Endo, A. and S. Okada. 2005. *Lactobacillus satsumensis* sp. nov., isolated from mashes of shochu, a traditional Japanese distilled spirit made from fermented rice and other starchy materials. Int. J. Syst. Evol. Microbiol. *55*: 83–85.

Endo, A. and S. Okada. 2007a. *Lactobacillus composti* sp. nov., a lactic acid bacterium isolated from a compost of distilled shochu residue. Int. J. Syst. Evol. Microbiol. *57*: 870–872.

Endo, A. and S. Okada. 2007b. *Lactobacillus farraginis* sp. nov. and *Lactobacillus parafarraginis* sp. nov., heterofermentative lactobacilli isolated from a compost of distilled shochu residue. Int. J. Syst. Evol. Microbiol. *57*: 708–712.

Endo, A., S. Roos, E. Satoh, H. Morita and S. Okada. 2008. *Lactobacillus equigenerosi* sp. nov., a coccoid species isolated from faeces of thoroughbred racehorses. Int J Syst Evol Microbiol *58*: 914–918.

Entani, E., H. Masai and K.I. Suzuki. 1986. *Lactobacillus acetotolerans*, a new species from fermented vinegar broth. Int. J. Syst. Bacteriol. *36*: 544–549.

Ercolini, D., G. Moschetti, G. Blaiotta and S. Coppola. 2001. Behavior of variable V3 region from 16S rDNA of lactic acid bacteria in denaturing gradient gel electrophoresis. Curr. Microbiol. *42*: 199–202.

Euzéby, J.P. 1998. Necessary corrections according to judicial opinions 16, 48 and 52. Int. J. Syst. Bacteriol. *48*: 613.

Evans, J.B. and C.F. Niven, Jr. 1951. Nutrition of the heterofermentative lactobacilli that cause greening of cured meat proucts. J. Bacteriol. *62*: 599–603.

Facklam, R. and J.A. Elliott. 1995. Identification, classification, and clinical relevance of catalase-negative, gram-positive cocci, excluding the streptococci and enterococci. Clin. Microbiol. Rev. *8*: 479–495.

Falsen, E., C. Pascual, B. Sjoden, M. Ohlen and M.D. Collins. 1999. Phenotypic and phylogenetic characterization of a novel *Lactobacillus* species from human sources: description of *Lactobacillus iners* sp. nov. Int. J. Syst. Bacteriol. *49*: 217–221.

Farias, J.A.E., M.C. Manca de Nadra, G.C. Rollan and A.M. Strasser de Saad. 1993. Histidine decarboxylase activity in lactic acid bacteria from wine. J. Int. Sci. Vigne Vin *27*: 191–199.

Farrow, J.A.E. and M.D. Collins. 1988. *Lactobacillus oris* sp. nov. from the human oral cavity. Int. J. Syst. Bacteriol. *38*: 116–118.

Farrow, J.A.E., B.A. Phillips and M.D. Collins. 1988. Nucleic acid studies on some heterofermentative lactobacilli: description of *Lactobacillus malefermentans* sp. nov. and *Lactobacillus parabuchneri* sp. nov. FEMS Microbiol. Lett. *55*: 163–168.

Farrow, J.A.E., B.A. Phillips and M.D. Collins. 1989. *In* Validation of the publication of new names and new combinations previously effectively published outside the IJSB. List no. 30. Int. J. Syst. Bacteriol. *39*: 371.

Felis, G.E., S. Torriani and F. Dellaglio. 2005. Reclassification of *Pediococcus urinaeequi* (*ex* Mees 1934) Garvie 1988 as *Aerococcus urinaeequi* comb. nov. Int. J. Syst. Evol. Microbiol. *55*: 1325–1327.

Fornachon, J.C.M., H.C. Douglas and R.H. Vaughn. 1949. *Lactobacillus trichodes* nov. sp., a bacterium causing soilage in appetizer and dessert wines. Higardia *19*: 119–132.

Franz, C.M., M. Vancanneyt, K. Vandemeulebroecke, M. De Wachter, I. Cleenwerck, B. Hoste, U. Schillinger, W.H. Holzapfel and J. Swings. 2006. *Pediococcus stilesii* sp. nov., isolated from maize grains. Int. J. Syst. Evol. Microbiol. *56*: 329–333.

Fred, E.B., W.H. Peterson and J.A. Anderson. 1921. The characteristics of certain pentose destroying bacteria, especially as concerns their action on arabinose. J. Biol. Chem. *48*: 385–412.

Fujii, T., K. Nakashima and N. Hayashi. 2005. Random amplified polymorphic DNA-PCR based cloning of markers to identify the beer-spoilage strains of *Lactobacillus brevis*, *Pediococcus damnosus*, *Lactobacillus collinoides* and *Lactobacillus coryniformis*. J. Appl. Microbiol. *98*: 1209–1220.

Fujisawa, T., S. Shirasaka, J. Watabe and T. Mitsuoka. 1984. *Lactobacillus aviarius* sp. nov.: a new species isolated from the intestine of chickens. Syst. Appl. Microbiol. *5*: 414–420.

Fujisawa, T., S. Shirasaka, J. Watabe and T. Mitsuoka. 1985. *In* Validation of the publication of new names and new combinations previously effectively published outside the IJSB. List no.17. Int. J. Syst. Bacteriol. *35*: 223–225.

Fujisawa, T., S. Shirasaka, J. Watabe and T. Mitsuoka. 1986. *In* Validation of the publication of new names and new combinations previously effectively published outside the IJSB. List no. 20. Int. J. Syst. Bacteriol. *36*: 354–356.

Fujisawa, T., S. Adachi, T. Toba, K. Arihara and T. Mitsuoka. 1988. *Lactobacillus kefiranofaciens* sp. nov. isolated from kefir grains. Int. J. Syst. Bacteriol. *38*: 12–14.

Fujisawa, T., K. Itoh, Y. Benno and T. Mitsuoka. 1990. *Lactobacillus intestinalis* (*ex* Hemme 1974) sp. nov., nom. rev, isolated from the intestines of mice and rats. Int. J. Syst. Bacteriol. *40*: 302–304.

Fujisawa, T., Y. Benno, T. Yaeshima and T. Mitsuoka. 1992. Taxonomic study of the *Lactobacillus acidophilus* group, with recognition of *Lactobacillus gallinarum* sp. nov. and *Lactobacillus johnsonii* sp. nov. and synonymy of *Lactobacillus acidophilus* group-A3 (Johnson et al. 1980) with the type strain of *Lactobacillus amylovorus* (Nakamura 1981). Int. J. Syst. Bacteriol. *42*: 487–491.

Fuller, R. 1999. Probiotics for farm animals. *In* Tannock (Editor), Probiotics: a Critical Review. Horizon Scientific Press, Norfolk, UK, pp. 15–22.

Ganzle, M.G., M. Ehmann and W.P. Hammes. 1998. Modeling of growth of *Lactobacillus sanfranciscensis* and Candida milleri in response to process parameters of sourdough fermentation. Appl. Environ. Microbiol. *64*: 2616–2623.

Ganzle, M.G., S. Weber and W.P. Hammes. 1999. Effect of ecological factors on the inhibitory spectrum and activity of bacteriocins. Int. J. Food Microbiol. *46*: 207–217.

Ganzle, M.G., A. Holtzel, J. Walter, G. Jung and W.P. Hammes. 2000. Characterization of reutericyclin produced by *Lactobacillus reuteri* LTH2584. Appl. Environ. Microbiol. *66*: 4325–4333.

Ganzle, M.G. and R.F. Vogel. 2003. Contribution of reutericyclin production to the stable persistence of *Lactobacillus reuteri* in an industrial sourdough fermentation. Int. J. Food Microbiol. *80*: 31–45.

Garvie, E.I. and M.E. Gregory. 1961. Folinic acid requirment of strains of the genus *Pediococcus*. Nature *190*: 563–564.

Garvie, E.I., M.E. Gregory and L.A. Mabbitt. 1961. The effect of asparagine on the growth of a Gram-positive coccus. J. Gen. Microbiol. *24*: 25–30.

Garvie, E.I. 1980. Bacterial lactate dehydrogenases. Microbiol. Rev. *44*: 106–139.

Garvie, E.I., J.A. Farrow and B.A. Phillips. 1981. A taxonomic study of some strains of streptococci which grow at 10° but not at 45° including *Streptococcus lactis* and *Streptococcus cremoris*. Zentbl. Bakteriol. Hyg. I Abt. Orig. C *2*: 151–165.

Gasser, F., M. Mandel and M. Rogosa. 1970. *Lactobacillus jensenii* sp. nov., a new representative of subgenus *Thermobacterium*. J. Gen. Microbiol. *62*: 219–222.

Gasser, F. 1994. Safety of lactic acid bacteria and their occurence in human clinical infections. Bull. Inst. Pasteur *92*: 45–67.

Gawehn, K. and H.U. Bergmeyer. 1974. D(–)Lactat. *In* Bergmeyer (Editor), Methoden der Enzymatischen Analyse. Verlag Chemie, Weinheim, pp. 1538–1541.

Giraffa, G., P. De Vecchi and L. Rossetti. 1998. Note: Identification of *Lactobacillus delbrueckii* subspecies *bulgaricus* and subspecies *lactis* dairy isolates by amplified rDNA restriction analysis. J. Appl. Microbiol. *85*: 918–924.

Golledge, C.L., N. Stingemore, M. Aravena and D. Joske. 1990. Septicemia caused by vancomycin-resistant *Pediococcus acidilactici*. J. Clin. Microbiol. *28*: 1678–1679.

Gonzalez, C. and B.S. Kunka. 1987. Plasmid associated bacteriocin production and sucrose fermentation in *Pediococcus acidilactici*. Appl. Environ. Microbiol. *53*: 2534–2538.

Gonzalez, C.F. and B.S. Kunka. 1983. Plasmid transfer in *Pediococcus* spp.: intergeneric and intrageneric transfer of pIP501. Appl. Environ. Microbiol. *46*: 81–89.

Gonzalez, C.F. and B.S. Kunka. 1986. Evidence for plasmid linkage of raffinose utilization and associated alpha-galactosidase and sucrose

hydrolase activity in *Pediococcus pentosaceus*. Appl. Environ. Microbiol. *51*: 105–109.

Gonzalez, C.L., J.P. Enicinas, M.L. García López and A. Otero. 2000. Characterization and identification of lactic acid bacteria from freshwater fishes. Food Microbiol. *17*: 383–391.

Götz, F., E.F. Elstner, B. Sedewitz and E. Lengfelder. 1980. Oxygen utilization by *Lactobacillus plantarum*. II. Superoxide and superoxide dismutation. Arch. Microbiol. *125*: 215–220.

Graham, D. and L.L. McKay. 1985. Plasmid DNA in strains of *Pediococcus pentosaceus*. Appl. Environ. Microbiol. *50*: 532–534.

Green, G., L.M.T. Dicks, G. Bruggeman, E.J. Vandamme and M.L. Chikindas. 1997. Pediocin PD-1, a bactericidal antimicrobial peptide from *Pediococcus damnosus* NCFB 1832. J. Appl. Microbiol. *82*: 127–132.

Green, M., K. Barbadora and M. Michaels. 1991. Recovery of vancomycin-resistant gram-positive cocci from pediatric liver transplant recipients. J. Clin. Microbiol. *29*: 2503–2506.

Grimaldi, A., E. Bartowsky and V. Jiranek. 2005. Screening of *Lactobacillus* spp. and *Pediococcus* spp. for glycosidase activities that are important in oenology. J. Appl. Microbiol. *99*: 1061–1069.

Guarner, F. and G.J. Schaafsma. 1998. Probiotics. Int. J. Food Microbiol. *39*: 237–238.

Günther, H.L. and H.R. White. 1961. The cultural and physiological characters of the pediococci. J. Gen. Microbiol. *26*: 185–197.

Gusils, C., S.N. Gonzalez and G. Oliver. 1999. Some probiotic properties of chicken lactobacilli. Can. J. Microbiol. *45*: 981–987.

Gutmann, I. and A.W. Wahlefeld. 1974. L(+)-Lactat. Bestimmung mit Lactat-Dehydrogenase und NAD. *In* Bergmeyer (Editor), Methoden der Enzymatischen Analyse. Verlag Chemie, Weinheim, pp. 1510–1514.

Hammes, W.P., K.H. Schleifer and O. Kandler. 1973. Mode of action of glycine on the biosynthesis of peptidoglycan. J. Bacteriol. *116*: 1029–1053.

Hammes, W.P., J. Winter and O. Kandler. 1979. The sensitivity of the pseudomurein-containing genus *Methanobacterium* to inhibitors of murein synthesis. Arch. Microbiol. *123*: 275–279.

Hammes, W.P., A. Bantleon and S. Min. 1990. Lactic acid bacteria in meat fermentation. FEMS Microbiol. Rev. *87*: 165–173.

Hammes, W.P. and P.S. Tichaczek. 1994. The potential of lactic acid bacteria for the production of safe and wholesome food. Z. Lebensm. Unters Forsch. *198*: 193–201.

Hammes, W.P., P. Stolz and M.G. Ganzle. 1996. Metabolism of lactobacilli in traditional sourdoughs. Adv. Food Sci. (CMTL) *5/6*: 176–184.

Hammes, W.P. and C. Hertel. 1998. New developments in meat starter cultures. Meat Sci. *49*: S125–S138.

Hammes, W.P. and C. Hertel. 2003. The genera *Lactobacillus* and *Carnobacterium*. *In* Dworkin (Editor), The Prokaroyes, 3rd edn. Springer-Verlag, New York, pp. release 3.15.

Hammes, W.P., M.J. Brandt, K.L. Francis, J. Rosenheim, M.F.H. Seitter and S.A. Vogelmann. 2005. Microbial ecology of cereal fermentations. Trends in Food Sci. and Technol. *16*: 4–11.

Hansen, P.A. 1968. Type strains of *Lactobacillus* species. A report by the taxonomic subcommittee on lactobacilli and closely related organisms. International Committee on Nomenclature of Bacteria of the International Association of Microbiological Societies. American Type Culture Collection, Rockville, Maryland.

Hansen, P.A. and G. Mocquot. 1970. *Lactobacillus acidophilus* (Moro) comb. nov. Int. J. Syst. Bacteriol. *20*: 325–327.

Hansen, P.A. and E.F. Lessel. 1971. *Lactobacillus casei* (Orla Jensen) comb. nov. Int. J. Syst. Bacteriol. *21*: 69–71.

Harrison, A.P., Jr. and P.A. Hansen. 1950. A motile *Lactobacillus* from the cecal feces of turkeys. J. Bacteriol. *59*: 444–446.

Hartemink, R., V.R. Domenech and F.M. Rombouts. 1997. LAMVAB - A new selective medium for the isolation of lactobacilli from faeces. J. Microbiol. Methods *29*: 77–84.

Harty, D.W., M. Patrikakis and K.W. Knox. 1993. Identification of *Lactobacillus* strains isolated from patients with infective endocarditis and comparison of their surface-associated properties with those of other strains of the same species. Microb. Ecol. Health Dis. *6*: 191–201.

Harty, D.W., H.J. Oakey, M. Patrikakis, E.B. Hume and K.W. Knox. 1994. Pathogenic potential of lactobacilli. Int. J. Food Microbiol. *24*: 179–189.

Heilig, H.G., E.G. Zoetendal, E.E. Vaughan, P. Marteau, A.D. Akkermans and W.M. de Vos. 2002. Molecular diversity of *Lactobacillus* spp. and other lactic acid bacteria in the human intestine as determined by specific amplification of 16S ribosomal DNA. Appl. Environ. Microbiol. *68*: 114–123.

Heinzl, H. and W.P. Hammes. 1986. Die mikroaerophile Bakterienflora von Obstmaischen. Chem. Microbiol. Technol. Lebensm. *10*: 106–109.

Hemme, D. 1974. Taxonomie des lactobacilles homofermentaires du tube digestif du rat. Description de deux nouvelles espèces *Lactobacillus intestinalis*, *Lactobacillus murini*. University of Paris VII, Paris.

Hemme, D., P. Raibaud, R. Ducluzeau, J.V. Galpin, P. Sicard and J. van Heijenoort. 1980. *Lactobacillus murinus* n. sp. a new species of the autochtoneous dominant flora of the digestive tract of rat and mouse. Annales De Microbiologie *A131*: 297–308.

Hemme, D., P. Raibaud, R. Ducluzeau, J.V. Galpin, P. Sicard and J. van Heijenoort. 1982. *In* Validation of the publication of new names and new combinations previously effectively published outside the IJSB. List No. 9. Int. J. Syst. Bacteriol. *32*: 384–385.

Henneberg, W. 1901. Zur Kenntnis der Milchsäurebakterien der Brennereimaische der Milch, und des Bierres. Wochenshrift Brau *18*: 381–384.

Henneberg, W. 1903. Kenntnis der Milchsäurebakterien der Brennereimaische, der Milch, des Bieres, der Presshefe, der Melasse, des Sauerkohls, der sauren Gurken und des Sauerteiges, sowie einige Bemerkungen über die Milchsäurebakterien des menschlichen Magens. Zeitschrift der Spiritusindustrie *26*: 329–332.

Hensel, R., U. Mayr, K.O. Stetter and O. Kandler. 1977. Comparative studies of lactic acid dehydrogenases in lactic acid bacteria. I. Purification and kinetics of the allosteric L-lactic acid dehydrogenase from *Lactobacillus casei* ssp. casei and *Lactobacillus curvatus*. Arch. Microbiol. *112*: 81–93.

Hertel, C., W. Ludwig and K.H. Schleifer. 1992. Introduction of silent mutations in a proteinase gene of *Lactococcus lactis* as a useful marker for monitoring studies. Syst. Appl. Microbiol. *15*: 447–452.

Hespell, R.B., D.E. Aiken and B.A. Dehority. 1997. Bacteria, fungi, and protozoa of the rumen. *In* Mackie, White and Isaacson (Editors), Gastrointestinal Microbiology, 2. Chapman and Hall, London, pp. 59–141.

Hill, G.B., D.A. Eschenbach and K.K. Holmes. 1985. Bacteriology of the vagina. Scand. J. Urol. Nephrol. *86*: S23–S29.

Hiraga, K., Y. Ueno, S. Sukontasing, S. Tanasupawat and K. Oda. 2008. *Lactobacillus senmaizukei* sp. nov., isolated from Japanese pickle. Int. J. Syst. Evol. Microbiol. *58*: 1625–1629.

Holo, H., Z. Jeknic, M. Daeschel, S. Stevanovic and I.F. Nes. 2001. Plantaricin W from *Lactobacillus plantarum* belongs to a new family of two-peptide lantibiotics. Microbiology *147*: 643–651.

Holzapfel, W.H. 1992. Culture media for non-sporulating gram-positive food spoilage bacteria. Int. J. Food Microbiol. *17*: 113–133.

Holzapfel, W.H., C.M.A.P. Franz, W. Ludwig, W. Back and L.M.T. Dicks. 2005. Genera *Pediococcus* and *Tetragenococcus*. *In* Dworkin, Falkow, Rosenberg, Schleifer and Stackbrandt (Editors), The Prokaryotes: An Evolving Electronic Resource for the Microbiological Community, 3rd edn.

Hoover, D.G., P.M. Walsh, K.M. Kolactis and M.M. Daly. 1988. A bacteriocin produced by *Pediococcus* species associated with a 5.5 MDa plasmid. J. Food Protect. *51*: 29–31.

Husni, R.N., S.M. Gordon, J.A. Washington and D.L. Longworth. 1997. *Lactobacillus* bacteremia and endocarditis: review of 45 cases. Clin. Infect. Dis. *25*: 1048–1055.

Huttunen, E., K. Noro and Z.N. Yang. 1995. Purification and identification of antimicrobial substances produced by two *Lactobacillus casei* strains. Int. Dairy J. *5*: 503–513.

Hyman, R.W., M. Fukushima, L. Diamond, J. Kumm, L.C. Giudice and R.W. Davis. 2005. Microbes on the human vaginal epithelium. Proc. Natl. Acad. Sci. U. S. A. *102*: 7952–7957.

Jacques, N.A., L. Hardy, K.W. Knox and A.J. Wicken. 1980. Effect of Tween 80 on the morphology and physiology of *Lactobacillus salivarius* strain IV CL-37 grown in a chemostat under glucose limitation. J. Gen. Microbiol. *119*: 195–201.

Jager, K. and S. Harlander. 1992. Characterization of a bacteriocin from *Pediococcus acidilactici* PC and comparison of bacteriocin-producing strains using molecular typing procedures. Appl. Microbiol. Biotechnol. *37*: 631–637.

Johnson, J.L., C.F. Phelps, C.S. Cummings and L. London. 1980. Taxonomy of the *Lactobacillus acidophilus* group. Int. J. Syst. Bacteriol. *30*: 53–68.

Jones, R.J. 2004. Observations on the succession dynamics of lactic acid bacteria populations in chill-stored vacuum-packaged beef. Int. J. Food Microbiol. *90*: 273–282.

Juven, B.J., R.J. Meinersmann and N.J. Stern. 1991. Antagonistic effects of lactobacilli and pediococci to control intestinal colonization by human enteropathogens in live poultry. J. Appl. Bacteriol. *70*: 95–103.

Kagermeier-Callaway, A.S. and E. Lauer. 1995. *Lactobacillus sake* Katagiri, Kitahara, and Fukami 1934 is the senior synonym for *Lactobacillus bavaricus* Stetter and Stetter 1980. Int. J. Syst. Bacteriol. *45*: 398–399.

Kamau, D.N., S. Doores and K.M. Pruitt. 1990. Enhanced thermal destruction of *Listeria monocytogenes* and *Staphylococcus aureus* by the lactoperoxidase system. Appl. Environ. Microbiol. *56*: 2711–2716.

Kandler, O., K.O. Stetter and R. Kohl. 1980. *Lactobacillus reuteri* sp. nov., a new species of heterofermentative lactobacilli. Zentbl. Bakteriol. Mikrobiol. Hyg. I Abt. Orig. C *1*: 264–269.

Kandler, O., K.O. Stetter and R. Kohl. 1982. *In* Validation of the publication of new names and new combinations previously effectively published outside the IJSB. List no.8. Int. J. Syst. Bacteriol. *32*: 266–268.

Kandler, O. 1983. Carbohydrate metabolism in lactic acid bacteria. Antonie Leeuwenhoek *49*: 209–224.

Kandler, O. and P. Kunath. 1983. *Lactobacillus kefir* sp. nov., a component of the microflora of kefir. Syst. Appl. Microbiol. *4*: 286–294.

Kandler, O., U. Schillinger and N. Weiss. 1983. *Lactobacillus bifermentans* sp. nov., nom. rev., an organism forming CO_2 and H_2 from lactic-acid. Syst. Appl. Microbiol. *4*: 408–412.

Kandler, O. 1984. Current taxonomy of lactobacilli. Ind. Microbiol. *25*: 109–123.

Kandler, O. and N. Weiss. 1986. Regular non-sporing Gram-positive rods. *In* Sneath, Mair, M.E. Sharpe and Holt (Editors), Bergey's Manual of Systematic Bacteriology, Vol. 2. The Williams & Wilkins Co., Baltimore, pp. 1208–1234.

Kaneuchi, C., M. Seki and K. Komagata. 1988. Taxonomic study of *Lactobacillus mali* Carr and Davis 1970 and related strains: validation of *Lactobacillus mali* Carr and Davis 1970 over *Lactobacillus yamanashiensis* Nonomura 1983. Int. J. Syst. Bacteriol. *38*: 269–272.

Katagiri, H., K.T. Kitahara and K. Fukami. 1934. The characteristics of the lactic acid bacteria isolated from moto, yeast mashes for sake manufacture. IV. The classification of the lactic acid bacteria. Bull. Agr. Chem. Soc. Japan *10*: 156–157.

Kato, Y., R.M. Sakala, H. Hayashidani, A. Kiuchi, C. Kaneuchi and M. Ogawa. 2000. *Lactobacillus algidus* sp. nov., a psychrophilic lactic acid bacterium isolated from vacuum-packaged refrigerated beef. Int. J. Syst. Evol. Microbiol. *50*: 1143–1149.

Keddie, R.M. 1959. The properties and classification of lactobacilli isolated from grass and silage. J. Appl. Bacteriol. *22*: 403–416.

Kilpper-Bälz, R., G. Fischer and K.H. Schleifer. 1982. Nucleic acid hybridization of group N and group D streptococci. Curr. Microbiol. *7*: 245–250.

Kitahara, K.T., T. Kaneko and O. Goto. 1957. Taxonomic studies on the hiochi-bacteria, specific saprophytes of sake. II. Identification and classification of hiochi-bacteria. J. Gen. Appl. Microbiol. *3*: 111–120.

Kitchell, A.G. and B.G. Shaw. 1975. Lactic acid bacteria in fresh and cured meat. *In* Carr, Cutting and Whiting (Editors), Lactic Acid Bacteria in Beverages and Food. Academic Press, london, New York, San Francisco, pp. 209–220.

Klaenhammer, T.R. 1993. Genetics of bacteriocins produced by lactic acid bacteria. FEMS Microbiol. Rev. *12*: 39–85.

Klaenhammer, T.R. 2001. Probiotics and prebiotics. *In* Doyle, Beauchat and Montville (Editors), Food Microbiology. ASM Press, Washington, D.C., pp. 797–812.

Klaenhammer, T.R., R. Barrangou, B.L. Buck, M.A. Azcarate-Peril and E. Altermann. 2005. Genomic features of lactic acid bacteria effecting bioprocessing and health. FEMS Microbiol. Rev. *29*: 393–409.

Klein, G. 1992. Charakterisierung von biotechnologisch eingesetzten Laktobazillenstämmen und einigen klinischen Isolaten durch Bestimmung der Resistenz gegen Antibiotoka und Chemotherapeutika und der Plasmidmuster. Inauguraldissertation zur Erlangung des Grades eines Doktors der Veterinämedizin. Freie Universität Berlin, Berlin.

Kleynmans, U., H. Heinzl and W.P. Hammes. 1989. *Lactobacillus suebicus* sp. nov., an obligately heterofermentative *Lactobacillus* species isolated from fruit mashes. Syst. Appl. Microbiol. *11*: 267–271.

Kline, L. and T.F. Sugihara. 1971. Microorganisms of the San Francisco sour dough bread process. II. Isolation and characterization of undescribed bacterial species responsible for the souring activity. Appl. Microbiol. *21*: 459–465.

Kneist, S., R. Heinrich and W. Kunzel. 1988. The presence of lactobacilli in carious dentin. Zahn. Mund. Kieferheilkd. Zentralbl. *76*: 123–127.

Knox, K.W. and E.A. Hall. 1964. The relationship between the capsular and cell wall polysaccharides of strains of *Lactobacillus*. J. Gen. Microbiol. *37*: 433–438.

Knox, K.W. and A.J. Wicken. 1973. Non-specific inhibition of the precipitin reaction between teichoic acids and antisera. Immunochemistry *10*: 93–98.

Kocur, M., T. Bergan and N. Mortensen. 1971. DNA base composition of Gram-positive cocci. J. Gen. Microbiol. *69*: 167–183.

Kojima, S., S. Takizawa, S. Tamura, S. Fujinaga, Y. Benno and T. Nakase. 1993. An improved medium for the isolation of lactobacilli from kefir grains. Biosci. Biotech. Biochem. *57*: 119–120.

Konings, W.N. 2002. The cell membrane and the struggle for life of lactic acid bacteria. Antonie Leeuwenhoek *82*: 3–27.

Konstantinov, S.R., E. Poznanski, S. Fuentes, A.D. Akkermans, H. Smidt and W.M. de Vos. 2006. *Lactobacillus sobrius* sp. nov., abundant in the intestine of weaning piglets. Int. J. Syst. Evol. Microbiol. *56*: 29–32.

Koort, J., P. Vandamme, U. Schillinger, W. Holzapfel and J. Bjorkroth. 2004. *Lactobacillus curvatus* subsp. *melibiosus* is a later synonym of *Lactobacillus sakei* subsp. *carnosus*. Int. J. Syst. Evol. Microbiol. *54*: 1621–1626.

Krockel, L., U. Schillinger, C.M. Franz, A. Bantleon and W. Ludwig. 2003. *Lactobacillus versmoldensis* sp. nov., isolated from raw fermented sausage. Int. J. Syst. Evol. Microbiol. *53*: 513–517.

Krooneman, J., F. Faber, A.C. Alderkamp, S.J. Elferink, F. Driehuis, I. Cleenwerck, J. Swings, J.C. Gottschal and M. Vancanneyt. 2002. *Lactobacillus diolivorans* sp. nov., a 1,2-propanediol-degrading bacterium isolated from aerobically stable maize silage. Int. J. Syst. Evol. Microbiol. *52*: 639–646.

Kunji, E.R., I. Mierau, A. Hagting, B. Poolman and W.N. Konings. 1996. The proteolytic systems of lactic acid bacteria. Antonie Leeuwenhoek *70*: 187–221.

Kurzak, P., M.A. Ehrmann and R.F. Vogel. 1998. Diversity of lactic acid bacteria associated with ducks. Syst. Appl. Microbiol. *21*: 588–592.

Kuznetsov, V.D. 1959. A new species of lactobacillus. Mikrobiologiia *28*: 368–373.

Lahbib-Mansais, Y., M. Mata and P. Ritzenthaler. 1988. Molecular taxonomy of *Lactobacillus* phages. Biochimie *70*: 429–435.

Lan, Y., S. Sun, S. Tamminga, B.A. Williams, M.W. Verstegen and G. Erdi. 2004. Real-time PCR detection of lactic acid bacteria in cecal contents of *Eimeria tenella*-infected broilers fed soybean oligosaccharides and soluble soybean polysaccharides. Poult. Sci. *83*: 1696–1702.

Landete, J.M., S. Ferrer and I. Pardo. 2005. Which lactic acid bacteria are responsible for histamine production in wine? J. Appl. Microbiol. *99*: 580–586.

Lauer, E. and O. Kandler. 1980a. *In* Validation of the publication of new names and new combinations previously effectively published outside IJSB. List no. 4. Int. J. Syst. Bacteriol. *30*: 601.

Lauer, E. and O. Kandler. 1980b. *Lactobacillus gasseri* sp. nov, a new species of the subgenus Thermobacterium. Zentbl. Bakteriol. Mikrobiol. Hyg. I Abt. Orig. C *1*: 75–78.

Lavermicocca, P., F. Valerio and A. Visconti. 2003. Antifungal activity of phenyllactic acid against molds isolated from bakery products. Appl. Environ. Microbiol. *69*: 634–640.

Lawson, P.A., P. Papademas, C. Wacher, E. Falsen, R. Robinson and M.D. Collins. 2001a. *Lactobacillus cypricasei* sp. nov., isolated from Halloumi cheese. Int. J. Syst. Evol. Microbiol. *51*: 45–49.

Lawson, P.A., C. Wacher, I. Hansson, E. Falsen and M.D. Collins. 2001b. *Lactobacillus psittaci* sp. nov., isolated from a hyacinth macaw (*Anodorhynchus hyacinthinus*). Int. J. Syst. Evol. Microbiol. *51*: 967–970.

Ledesma, O.V., A.P. De Ruiz Holgado, G. Oliver, G.S. De Giori, P. Raibaud and J.V. Galpin. 1977. A synthetic medium for comparative nutritional studies of lactobacilli. J. Appl. Bacteriol. *42*: 123–133.

Leichmann, G. 1896a. Über die freiwillige Säurung der Milch. Zentralblatt Bakteriologie, Parasitenkunde, Infektionskrankheiten und Hygiene, Abteilung 2 *2*: 777–780.

Leichmann, G. 1896b. Über die im Brennereiprozess bei der Bereitung der Kunsthefe auftretende spontane Milchsäuregärung. Zentbl. Bakteriol. II Abt. *2*: 281–285.

Leisner, J.J., M. Vancanneyt, J. Goris, H. Christensen and G. Rusul. 2000. Description of *Paralactobacillus selangorensis* gen. nov., sp. nov., a new lactic acid bacterium isolated from chili bo, a Malaysian food ingredient. Int. J. Syst. Evol. Microbiol. *50*: 19–24.

Leisner, J.J., M. Vancanneyt, K. Lefebvre, K. Vandemeulebroecke, B. Hoste, N.E. Vilalta, G. Rusul and J. Swings. 2002. *Lactobacillus durianis* sp. nov., isolated from an acid-fermented condiment (tempoyak) in Malaysia. Int. J. Syst. Evol. Microbiol. *52*: 927–931.

Lenzner, A.A. 1966. Some results of the investigation of lactobacilli of human microflora. Surv. Res. Med. Tartu State Univ., 1940–1965 *191*: 69–75.

Lerche, M. and G. Reuter. 1962. Das Vorkommen aerob wachsender Gram-positiver Stäbchen Des Genus *Lactobacillus Beijerinck* im Darminhalt erwachsener Menschen. Zentbl. Bakteriol. Parasitenkd. Infektionskr. Hyg. Abt. 1 *185*: 446–481.

Leser, T.D., J.Z. Amenuvor, T.K. Jensen, R.H. Lindecrona, M. Boye and K. Moller. 2002. Culture-independent analysis of gut bacteria: the pig gastrointestinal tract microbiota revisited. Appl. Environ. Microbiol. *68*: 673–690.

Lick, S., E. Brockmann and K.J. Heller. 2000. Identification of *Lactobacillus delbrueckii* and subspecies by hybridization probes and PCR. Syst. Appl. Microbiol. *23*: 251–259.

Lindner, P. 1887. Über ein neues in Matzmaischen vorkommendes, milchsäurebildendes. Ferment. Wschr. Brau. *4*: 437–440.

Liu, B. and X. Dong. 2002. *Lactobacillus pantheris* sp. nov., isolated from faeces of a jaguar. Int. J. Syst. Evol. Microbiol. *52*: 1745–1748.

Liu, S.O., G.G. Pritchard, J. Hardman and G. Pilone. 1994. Citruline production and ethylcarbomate (urethane) precursor formation from argnine degradation by wine lactic acid bacteria *Leuconostoc oenos* and *Lactobacillus buchneri*. Am. J. Enol. Vitic. *45*: 235–242.

London, J., E.Y. Meyer and S. Kulczyk. 1971. Comparative biochemical and immunological study of malic enzyme from two species of lactic acid bacteria: evolutionary implications. J. Bacteriol. *106*: 126–137.

London, J. and N.M. Chace. 1976. Aldolases of the lactic acid bacteria. Demonstration of immunological relationships among eight genera of Gram positive bacteria using an anti-pediococcal aldolase serum. Arch. Microbiol. *110*: 121–128.

Lonvaud-Funel, A., Y. Guilloux and A. Joyeux. 1993. Isolation of a DNA probe for identification of glucan-producing *Pediococcus damnosus* in wines. J. Appl. Bacteriol. *74*: 41–47.

Lonvaud-Funel, A. 1999. Lactic acid bacteria in the quality improvement and depreciation of wine. Antonie Leeuwenhoek *76*: 317–331.

Luchansky, J.B., K.A. Glass, K.D. Harsono, A.J. Degnan, N.G. Faith, B. Cauvin, G. Baccus-Taylor, K. Arihara, B. Bater, A.J. Maurer and R.B. Cassens. 1992. Genomic analysis of *Pediococcus* starter cultures used to control *Listeria monocytogenes* in turkey summer sausage. Appl. Environ. Microbiol. *58*: 3053–3059.

Lücke, F.K. 2000. Fermented meats. *In* Lund, Baird-Parker and Gould (Editors), The Microbiological Safety and Quality of Food, 1. Aspen Publishers, Inc., Gaithersburg, pp. 420–444.

Ludwig, W. and H.P. Klenk. 2001. Overview: a phylogenetic backbone and taxonomic framework for prokaryotic systematics. *In* Brenner, Krieg, Staley and Garrity (Editors), Bergey's Manual of Systematic Bacteriology, 2nd edn, Vol. 1. Springer, New York, pp. 49–66.

Ludwig, W., O. Strunk, R. Westram, L. Richter, H. Meier, Yadhukumar, A. Buchner, T. Lai, S. Steppi, G. Jobb, W. Förster, I. Brettske, S. Gerber, A.W. Ginhart, O. Gross, S. Grumann, S. Hermann, R. Jost, A. König, T. Liss, R. Lüßmann, M. May, B. Nonhoff, B. Reichel, R. Strehlow, A. Stamatakis, N. Stuckmann, A. Vilbig, M. Lenke, T. Ludwig, A. Bode and K.H. Schleifer. 2004. ARB: A software environment for sequence data. Nucleic Acids Res. *32*: 1363–1371.

Margalith, P.Z. 1981. Flavor Microbiology. Charles C. Thomas, Springfield, Illinois.

Marounek, M., K. Jehlickova and V. Kmet. 1988. Metabolism and some characteristics of lactobacilli isolated from the rumen of young calves. J Appl Bacteriolcteriol *65*: 43–47.

Marsh, P. and M. Martin. 1984. Oral Microbiology, 2nd edn, Aspects of Microbiology, 1. Van Nostrand Reinhold, London.

Marsh, P.D. 1994. Microbial ecology of dental plaque and its significance in health and disease. Adv. Dent. Res. *8*: 263–271.

Martel, A., V. Meulenaere, L.A. Devriese, A. Decostere and F. Haesebrouck. 2003. Macrolide and lincosamide resistance in the grampositive nasal and tonsillar flora of pigs. Microb. Drug Resist. *9*: 293–297.

Marugg, J.D., C.F. Gonzalez, B.S. Kunka, A.M. Ledeboer, M.J. Pucci, M.Y. Toonen, S.A. Walker, L.C. Zoetmulder and P.A. Vandenbergh. 1992. Cloning, expression, and nucleotide sequence of genes involved in production of pediocin PA-1, and bacteriocin from *Pediococcus acidilactici* PAC1.0. Appl. Environ. Microbiol. *58*: 2360–2367.

Mastro, T.D., J.S. Spika, P. Lozano, J. Appel and R.R. Facklam. 1990. Vancomycin-resistant *Pediococcus acidilactici*: nine cases of bacteremia. J. Infect. Dis. *161*: 956–960.

Mata, M. and P. Ritzenthaler. 1988. Present state of lactic acid bacteria phage taxonomy. Biochimie *70*: 395–400.

Matte-Tailliez, O., P. Quenee, R. Cibik, J. van Opstal, F. Dessevre, O. Firmesse and P. Tailliez. 2001. Detection and identification of lactic acid bacteria in milk and industrial starter culture with fluorescently labeled rRNA-targeted peptide nucleic acid probes. Lait *81*: 237–248.

McLean, N.W. and I.J. Rosenstein. 2000. Characterisation and selection of a *Lactobacillus* species to re-colonise the vagina of women with recurrent bacterial vaginosis. J. Med. Microbiol. *49*: 543–552.

Mees, R.H. 1934. Onderzoekingen over de Biersarcina. Technical University, Delft, Holland.

Mehlen, A., M. Goeldner, S. Ried, S. Stindl, W. Ludwig and K.H. Schleifer. 2004. Development of a fast DNA-DNA hybridization method based on melting profiles in microplates. Syst. Appl. Microbiol. *27*: 689–695.

Meroth, C.B., J. Walter, C. Hertel, M.J. Brandt and W.P. Hammes. 2003. Monitoring the bacterial population dynamics in sourdough fermentation processes by using PCR-denaturing gradient gel electrophoresis. Appl. Environ. Microbiol. *69*: 475–482.

Meroth, C.B., W.P. Hammes and C. Hertel. 2004. Characterisation of the microbiota of rice sourdoughs and description of *Lactobacilius spicheri* sp. nov. Syst. Appl. Microbiol. *27*: 151–159.

Meyer, V. 1956. Die Bestimmung der Bomage-Arten bei Fischkonserven. Fischwirtschaft *8*: 212.

Meyer, V. 1962. Problem des Verderbens von Fischkonserven in Dosen. VII. Untersuchungen über die Entstehung der Aminosäuren beim Marinieren von Heringen. Veroeff. Inst. Meeresforsch Bremerhaven *8*: 21.

Miller, K.W., P. Ray, T. Steinmetz, T. Hanekamp and B. Ray. 2005. Gene organization and sequences of pediocin AcH/PA-1 production operons in *Pediococcus* and *Lactobacillus* plasmids. Lett. Appl. Microbiol. *40*: 56–62.

Mitsuoka, T. and T. Fujisawa. 1987. *Lactobacillus hamsteri*, a new species from the intestine of hamsters. Proc. Jpn. Acad. Ser. B Phys. Biol. Sci. *63*: 269–272.

Mitsuoka, T. and T. Fujisawa. 1988. *In* Validation of the publication of new names and new combinations previously effectively published outside the IJSB. List no. 25. Int. J. Syst. Bacteriol. *38*: 220–222.

Mitsuoka, T. 1992. The human gastrointestinal tract. *In* Wood (Editor), The Lactic Acid Bacteria, 1. The Lactic Acid Bacteria in Health and Disease. Elsevier Applied Science Publishers, London, pp. 69–114.

Miyamoto, M., Y. Seto, D.H. Hao, T. Teshima, Y.B. Sun, T. Kabuki, L.B. Yao and H. Nakajima. 2006. Validation List no. 107. Int. J. Syst. Evol. Microbiol. *56*: 1–6.

Moore, W.E.C. and L.V. Holdeman. 1970. *Propionibacterium, Arachnia, Actinomyces, Lactobacillus* and *Bifidobacterium*. *In* Cato, Cummings, Holdeman, Johnson, Moore, Smibert and Smith (Editors), Outline of Clinical Methods in Anaerobic Bacteriology, 2nd edn. Virginia Polytechnic Institute Anaerobe Laboratory, Blacksburg, Virginia, pp. 15–22.

Mora, D., M.G. Fortina, C. Parini and P.L. Manachini. 1997. Identification of *Pediococcus acidilactici* and *Pediococcus pentosaceus* based on 16S rRNA and *ldhD* gene-targeted multiplex PCR analysis. FEMS Microbiol. Lett. *151*: 231–236.

Mora, D., C. Parini, M.G. Fortina and P.L. Manachini. 1998. Discrimination among Pediocin AcH/PA-1 producer strains by comparison of *pedB* and *pedD* amplified genes and by multiplex PCR assay. Syst. Appl. Microbiol. *21*: 454–460.

Mora, D., M.G. Fortina, C. Parini, D. Daffonchio and P.L. Manachini. 2000a. Genomic subpopulations within the species *Pediococcus acidilactici* detected by multilocus typing analysis: relationships between pediocin AcH/PA-1 producing and non-producing strains. Microbiology *146*: 2027–2038.

Mora, D., C. Parini, M.G. Fortina and P.L. Manachini. 2000b. Development of molecular RAPD marker for the identification of *Pediococcus acidilactici* strains. Syst. Appl. Microbiol. *23*: 400–408.

Moreno, M.R., J.J. Leisner, L.K. Tee, C. Ley, S. Radu, G. Rusul, M. Vancanneyt and L. De Vuyst. 2002. Microbial analysis of Malaysian tempeh, and characterization of two bacteriocins produced by isolates of *Enterococcus faecium*. J. Appl. Microbiol. *92*: 147–157.

Morishita, T., T. Fukada, M. Shirota and T. Yura. 1974. Genetic basis of nutritional requirements in *Lactobacillus casei*. J. Bacteriol. *120*: 1078–1084.

Morishita, T., Y. Deguchi, M. Yajima, T. Sakurai and T. Yura. 1981. Multiple nutritional requirements of lactobacilli: genetic lesions affecting amino acid biosynthetic pathways. J. Bacteriol. *148*: 64–71.

Morita, H., C. Shiratori, M. Murakami, H. Takami, Y. Kato, A. Endo, F. Nakajima, M. Takagi, H. Akita, S. Okada and T. Masaoka. 2007. *Lactobacillus hayakitensis* sp. nov., isolated from intestines of healthy thoroughbreds. Int. J. Syst. Evol. Microbiol. *57*: 2836–2839.

Morlon-Guyot, J., J.P. Guyot, B. Pot, I.J. de Haut and M. Raimbault. 1998. *Lactobacillus manihotivorans* sp. nov., a new starch-hydrolysing lactic acid bacterium isolated during cassava sour starch fermentation. Int. J. Syst. Bacteriol. *48*: 1101–1109.

Moro, E. 1900. Über die nach Gram färbbaren Bacillen des Säuglingsstuhles. Wien. Klin. Wochenschr. *13*: 114–115.

Morotomi, M., N. Yuki, Y. Kado, A. Kushiro, T. Shimazaki, K. Watanabe and T. Yuyama. 2002. *Lactobacillus equi* sp. nov., a predominant intestinal *Lactobacillus* species of the horse isolated from faeces of healthy horses. Int. J. Syst. Evol. Microbiol. *52*: 211–214.

Mortvedt, C.I., J. Nissen-Meyer, K. Sletten and I.F. Nes. 1991. Purification and amino acid sequence of lactocin S, a bacteriocin produced by *Lactobacillus sake* L45. Appl. Environ. Microbiol. *57*: 1829–1834.

Moschetti, G., G. Blaiotta, M. Aponte, G. Mauriello, F. Villani and S. Coppola. 1997. Genotyping of *Lactobacillus delbrueckii* subsp. *bulgaricus* and determination of the number and forms of rrn operons in *L. delbrueckii* and its subspecies. Res. Microbiol. *148*: 501–510.

Mucchetti, G., F. Locci, P. Massara, R. Vitale and E. Neviani. 2002. Production of pyroglutamic acid by thermophilic lactic acid bacteria in hard-cooked mini-cheeses. J. Dairy Sci. *85*: 2489–2496.

Mukai, T., K. Arihara, A. Ikeda, K. Nomura, F. Suzuki and H. Ohori. 2003. *Lactobacillus kitasatonis* sp. nov., from chicken intestine. Int. J. Syst. Evol. Microbiol. *53*: 2055–2059.

Muller, M.R.A., M.A. Ehrmann and R.F. Vogel. 2000. Multiplex PCR for the detection of *Lactobacillus pontis* and two related species in a sourdough fermentation. Appl. Environ. Microbiol. *66*: 2113–2116.

Mundt, J.O. and J.L. Hammer. 1968. Lactobacilli on plants. Appl. Microbiol. *16*: 1326–1330.

Mundt, J.O., W.G. Beattie and F.R. Wieland. 1969. Pediococci residing on plants. J. Bacteriol. *98*: 938–942.

Nakagawa, A. and K. Kitahara. 1962. Pleomorphism in bacteria cells. II. Giant cell formation in *Pediococcus cerevisiae* induced by hop resins. J. Gen. Appl. Microbiol. *8*: 142–148.

Nakagawa, A. 1964. A simple method for the detection of beer-sarcinae. Bullet. Brew. Sci. *10*: 7–10.

Nakamura, L.K. and C.D. Crowell. 1979. *Lactobacillus amylophilus*, a new starch-hydrolyzing species from swine waste-corn fermentation. Dev. Ind. Microbiol. *20*: 531–540.

Nakamura, L.K. 1981. *Lactobacillus amylovorus*, a new starch-hydrolyzing species from cattle waste-corn fermentations. Int. J. Syst. Bacteriol. *31*: 56–63.

Naser, S., F.L. Thompson, B. Hoste, D. Gevers, K. Vandemeulebroecke, I. Cleenwerck, C.C. Thompson, M. Vancanneyt and J. Swings. 2005. Phylogeny and identification of enterococci by *atpA* gene sequence analysis. J. Clin. Microbiol. *43*: 2224–2230.

Nes, I.F., D.B. Diep, L.S. Håvarstein, M.B. Brurberg, V. Eijsink and H. Holo. 1996. Biosynthesis of bacteriocin in lactic acid bacteria. Antonie Leeuwenhoek *70*: 113–128.

Niamsup, P., I.N. Sujaya, M. Tanaka, T. Sone, S. Hanada, Y. Kamagata, S. Lumyong, A. Assavanig, K. Asano, F. Tomita and A. Yokota. 2003. *Lactobacillus thermotolerans* sp. nov., a novel thermotolerant species isolated from chicken faeces. Int. J. Syst. Evol. Microbiol. *53*: 263–268.

Nielsen, D.S., U. Schillinger, C.M. Franz, J. Bresciani, W. Amoa-Awua, W.H. Holzapfel and M. Jakobsen. 2007. *Lactobacillus ghanensis* sp. nov., a motile lactic acid bacterium isolated from Ghanaian cocoa fermentations. Int. J. Syst. Evol. Microbiol. *57*: 1468–1472.

Nigatu, A., S. Ahrne, B.A. Gashe and G. Molin. 1998. Randomly amplified polymorphic DNA (RAPD) for discrimination of *Pediococcus pentosaceus* and *Pediococcus acidilactici* and rapid grouping of *Pediococcus* isolates. Lett. Appl. Microbiol. *26*: 412–416.

Nikolaitchouk, N., C. Wacher, E. Falsen, B. Andersch, M.D. Collins and P.A. Lawson. 2001. *Lactobacillus coleohominis* sp. nov., isolated from human sources. Int. J. Syst. Evol. Microbiol. *51*: 2081–2085.

Nishitani, Y., E. Sasaki, T. Fujisawaz and R. Osawa. 2004. Genotypic analyses of lactobacilli with a range of tannase activities isolated from human feces and fermented foods. Syst. Appl. Microbiol. *27*: 109–117.

Okada, S., Y. Suzuki and M. Kozaki. 1979. New heterofermentative *Lactobacillus* species with meso-diaminopimelic acid in peptidoglycan, *Lactobacillus vaccinostercus* Kozaki and Okada sp. nov. J. Gen. Appl. Microbiol. *25*: 215–221.

Okada, S., Y. Suzuki and M. Kozaki. 1983. *In* Validation of the publication of new names and new combinations previously effectively published outside the IJSB. List no. 10. Int. J. Syst. Bacteriol. *33*: 438–440.

Olson, N.F. 1990. The impact of lactic acid bacteria on cheese flavor. FEMS Microbiol. Rev. *87*: 131–147.

Omar, N.B., F. Ampe, M. Raimbault, J.P. Guyot and P. Tailliez. 2000. Molecular diversity of lactic acid bacteria from cassava sour starch (Colombia). Syst. Appl. Microbiol. *23*: 285–291.

Orla-Jensen, S. 1916. Maelkeri-Bakteriologi, Schonberske Forlag, Copenhagen. Orla-Jensen, S. 1919. The Lactic Acid Bacteria. Host & Son, Copenhagen.

Orla-Jensen, S. 1942. The lactic acid bacteria. *In* Munksgaard (Editor), The Lactic Acid Bacteria, 2nd edn, Copenhagen, pp. 106–107.

Orla-Jensen, S. 1943. The lactic acid bacteria. Ergänzungband. Mém. Acad. Roy. Sci Danemark, Stect. Sci. Biol. 2 (3), Copenhagen.

Osawa, R., T. Fujisawa and R. Pukall. 2006. *Lactobacillus apodemi* sp. nov., a tannase-producing species isolated from wild mouse faeces. Int. J. Syst. Evol. Microbiol. *56*: 1693–1696.

Oude Elferink, S.J., J. Krooneman, J.C. Gottschal, S.F. Spoestra, F. Faber and F. Driehuis. 2001. Anaerobic conversion of lactic acid to acetic acid and 1,2-propanediol by *Lactobacillus buchneri*. Appl. Environ. Microbiol. *67*: 125–132.

Ouwehand, A.C. 1998. Antimicrobial components from lactic acid bacteria. *In* Salminen and von Wright (Editors), Lactic Acid Bacteria. Marcel Dekker, Inc., New York, pp. 139–159.

Paludan-Muller, C., H.H. Huss and L. Gram. 1999. Characterization of lactic acid bacteria isolated from a Thai low-salt fermented fish product and the role of garlic as substrate for fermentation. Int. J. Food Microbiol. *46*: 219–229.

Pedersen, C., H. Jonsson, J.E. Lindberg and S. Roos. 2004. Microbiological characterization of wet wheat distillers' grain, with focus on isolation of lactobacilli with potential as probiotics. Appl. Environ. Microbiol. *70*: 1522–1527.

Pedersen, C. and S. Roos. 2004. *Lactobacillus saerimneri* sp. nov., isolated from pig faeces. Int. J. Syst. Evol. Microbiol. *54*: 1365–1368.

Pedersen, C.S. 1979. Microbiology of Food Fermentations, 2nd edn. AVI Publishing, Westport.

Pette, J.W. and J. van Beynum. 1943. Boekelscheurbacterien Rijkslandbauwproefstation te hoorn. Versl. Landbouwkd. Onderz. *49C*: 315–346.

Piva, A. and D.R. Headon. 1994. Pediocin A, a bacteriocin produced by *Pediococcus pentosaceus* FBB61. Microbiology *140* 697–702.

Poolman, B. 1993. Energy transduction in lactic acid bacteria. FEMS Microbiol. Rev. *12*: 125–148.

Poolman, B. 2002. Transporters and their roles in LAB cell physiology. Antonie Leeuwenhoek *82*: 147–164.

Pot, B., W. Ludwig, K. Kersters and K. Schleifer. 1994. Taxonomy of lactic acid bacteria. *In* De Vuyst and Vandamme (Editors), Bacteriocins of Lactic Acid Bacteria. Microbiology, Genetics, and Application. Blackie Academic and Professional, London, pp. 13–90.

Priebe, K. 1970. Untersuchungen zur Ursache und zur Vermeidung des Aufretens von Fadenziehen bei Bratheringsmarinaden. Arch. Lebensmittelhyg. *2*: 1–23.

Radler, F. 1975. The metabolism of organic acids. *In* Carr, Cutting and Whiting (Editors), Lactic Acid Bacteria in Beverages and Food. Academic Press, London, New York, San Francisco, pp. 17–27.

Radler, F. and K. Brohl. 1984. The metabolism of several carboxylic acids by lactic acid bacteria. Z. Lebensm. Unters. Forsch. *179*: 228–231.

Radler, F. 1986. Microbial Biochemistry. Experientia *42*: 884–893.

Raibaud, P., J.V. Galpin, R. Ducluzeau, G. Mocquot and G. Oliver. 1973. [The "*Lactobacillus*" genus in the digestive tract of rats. I. Characteristics of homofermentative strains isolated from holo- and gnotoxenic rats]. Ann. Microbiol. (Paris) *124*: 83–109.

Randazzo, C.L., S. Torriani, A.D. Akkermans, W.M. de Vos and E.E. Vaughan. 2002. Diversity, dynamics, and activity of bacterial communities during production of an artisanal Sicilian cheese as evaluated by 16S rRNA analysis. Appl. Environ. Microbiol. *68*: 1882–1892.

Redondo-Lopez, V., R.L. Cook and J.D. Sobel. 1990. Emerging role of lactobacilli in the control and maintenance of the vaginal bacterial microflora. Rev. Infect. Dis. *12*: 856–872.

Reid, G. and C. Heinemann. 1999. The role of the microflora in bacterial vaginosis. *In* Tannock (Editor), Medical Importance of the Normal Microflora. Kluwer Academic Publishers, Dordrecht, The Netherlands, pp. 477–486.

Reisinger, P., H. Seidel, H. Tschesche and W.P. Hammes. 1980. The effect of nisin on murein synthesis. Arch. Microbiol. *127*: 187–193.

Reuter, G. 1968. Erfahrungen mit Nährböden für die selektive mikrobiologische Analyse von Fleischerzeugnissen. Arch. Lebensmittelhyg. *19*.

Reuter, G. 1983a. *In* Validation of the publication of new names and new combinations previously effectively published outside the IJSB. List no. 11. Int. J. Syst. Bacteriol. *33*: 672–674.

Reuter, G. 1983b. *Lactobacillus alimentarius* sp. nov., nom. rev. and *Lactobacillus farciminis* sp. nov., nom. rev. Syst. Appl. Microbiol. *4*: 277–279.

Reuter, G. 2001. The *Lactobacillus* and *Bifidobacterium* microflora of the human intestine: composition and succession. Curr. Issues Intest. Microbiol. *2*: 43–53.

Rhee, S.K. and M.Y. Pack. 1980. Effect of environmental pH on chain length of *Lactobacillus bulgaricus*. J. Bacteriol. *144*: 865–868.

Riebel, W.J. and J.A. Washington. 1990. Clinical and microbiologic characteristics of pediococci. J. Clin. Microbiol. *28*: 1348–1355.

Ringo, E. and F.J. Gatesoupe. 1998. Lactic acid bacteria in fish: a review. Aquaculture *160*: 177–203.

Rodas, A.M., E. Chenoll, M.C. Macian, S. Ferrer, I. Pardo and R. Aznar. 2006. *Lactobacillus vini* sp. nov., a wine lactic acid bacterium homofermentative for pentoses. Int. J. Syst. Evol. Microbiol. *56*: 513–517.

Rodrigues, J.M., L.M. Cintas, P. Casaus, M.I. Suarez and P.E. Hernandez. 1997. Detection of pediocin PA-1-producing pediococci by rapid molecular biology techniques. Food Microbiol. *14*: 363–371.

Rodrigues, J.M., M.I. Martinez and J. Kok. 2002. Pediocin PA-1, a widespectrum bacteriocin from lactic acid bacteria. Crit. Rev. Food Sci. Nutr. *42*: 91–121.

Rogosa, J., J.A. Mitchell and R.F. Wiseman. 1951. A selective medium for isolation and enumeration of oral and fecal lactobacilli. J. Bacteriol. *62*: 132–133.

Rogosa, M., R.F. Wiseman, J.A. Mitchell, M.N. Disraely and A.J. Beaman. 1953. Species differentiation of oral lactobacilli from man including description of *Lactobacillus salivarius* nov. spec. and *Lactobacillus cellobiosus* nov. spec. J. Bacteriol. *65*: 681–699.

Rogosa, M. and M.E. Sharpe. 1960. Species differentiation of human vaginal lactobacilli. J. Gen. Microbiol. *23*: 197–201.

Rogosa, M., J.G. Franklin and K.D. Perry. 1961. Correlation of the vitamin requirements with cultural and biochemical characters of *Lactobacillus* spp. J. Gen. Microbiol. *25*: 473–482.

Rogosa, M. 1970. Characters used in the classification of lactobacilli. Int. J. Syst. Bacteriol. *20*: 519–533.

Rogosa, M. and P.A. Hansen. 1971. Nomenclatural considerations of certain species of *Lactobacillus Beijerinck*. Int. J. Syst. Bacteriol. *21*: 177–186.

Rogosa, M. 1974. Genus III. *Bifidobacterium* Orla-Jensen. *In* Buchanan and Gibbons (Editors), Bergey's Manual of Determinative Bacteriology, 8th edn. The Williams & Wilkins Co., Baltimore, pp. 669–676.

Roos, S., F. Karner, L. Axelsson and H. Jonsson. 2000. *Lactobacillus mucosae* sp. nov., a new species with in vitro mucus-binding activity isolated from pig intestine. Int. J. Syst. Evol. Microbiol. *50*: 251–258.

Roos, S., L. Engstrand and H. Jonsson. 2005. *Lactobacillus gastricus* sp. nov., *Lactobacillus antri* sp. nov., *Lactobacillus kalixensis* sp. nov. and *Lactobacillus ultunensis* sp. nov., isolated from human stomach mucosa. Int. J. Syst. Evol. Microbiol. *55*: 77–82.

Rosselló-Mora, R. and R. Amann. 2001. The species concept for prokaryotes. FEMS Microbiol. Rev. *25*: 39–67.

Roy, D., P. Ward, D. Vincent and F. Mondou. 2000. Molecular identification of potentially probiotic lactobacilli. Curr. Microbiol. *40*: 40–46.

Rubio, L.A., A. Brenes, I. Setien, G. de la Asuncion, N. Duran and M.T. Cutuli. 1998. Lactobacilli counts in crop, ileum and caecum of growing broiler chickens fed on practical diets containing whole or dehulled sweet lupin (*Lupinus angustifolius*) seed meal. Br. Poult. Sci. *39*: 354–359.

Ruoff, K.L., D.R. Kuritzkes, J.S. Wolfson and M.J. Ferraro. 1988. Vancomycin-resistant Gram-positive bacteria isolated from human sources. J. Clin. Microbiol. *26*: 2064–2068.

Russell, C. and T.K. Walker. 1953. *Lactobacillus malefermentans* n.sp., isolated from beer. J. Gen. Microbiol. *8*: 160–162.

Saier, M.H., Jr., J.J. Ye, S. Klinke and E. Nino. 1996. Identification of an anaerobically induced phosphoenolpyruvate-dependent fructose-specific phosphotransferase system and evidence for the Embden–Meyerhof glycolytic pathway in the heterofermentative bacterium *Lactobacillus brevis*. J. Bacteriol. *178*: 314–316.

Sakala, R.M., Y. Kato, H. Hayashidani, M. Murakami, C. Kaneuchi and M. Ogawa. 2002. *Lactobacillus fuchuensis* sp. nov., isolated from vacuum-packaged refrigerated beef. Int. J. Syst. Evol. Microbiol. *52*: 1151–1154.

Sakamoto, K., A. Margolles, H.W. van Veen and W.N. Konings. 2001. Hop resistance in the beer spoilage bacterium *Lactobacillus brevis* is mediated by the ATP-binding cassette multidrug transporter HorA. J. Bacteriol. *183*: 5371–5375.

Sakamoto, K. and W.N. Konings. 2003. Beer spoilage bacteria and hop resistance. Int. J. Food Microbiol. *89*: 105–124.

Salminen, S. and A. von Wright. 1998. Lactic Acid Bacteria, Microbiology and Functional Aspects. Marcel Dekker, Inc., New York.

Sanders, M.E. and J. Huis in't Veld. 1999. Bringing a probiotic-containing functional food to the market: microbiological, product, regulatory and labeling issues. Antonie van Leeuwenhoek *76*: 293–315.

Santos, E.M., I. Jaime, J. Rovira, U. Lyhs, H. Korkeala and J. Björkroth. 2005. Characterization and identification of lactic acid bacteria in "morcilla de Burgos". Int. J. Food Microbiol. *97*: 285–296.

Sarma, P.S. and S. Mohanty. 1998. *Pediococcus acidilactici* pneumonitis and bacteremia in a pregnant woman. J. Clin. Microbiol. *36*: 2392–2393.

Sasaki, E., T. Shimada, R. Osawa, Y. Nishitani, S. Spring and E. Lang. 2005. Isolation of tannin-degrading bacteria isolated from feces of the Japanese large wood mouse, Apodemus speciosus, feeding on tannin-rich acorns. Syst. Appl. Microbiol. *28*: 358–365.

Satokari, R., T. Mattila-Sandholm and M.L. Suihko. 2000. Identification of pediococci by ribotyping. J. Appl. Microbiol. *88*: 260–265.

Satokari, R.M., E.E. Vaughan, H. Smidt, M. Saarela, J. Matto and W.M. de Vos. 2003. Molecular approaches for the detection and identification of bifidobacteria and lactobacilli in the human gastrointestinal tract. Syst. Appl. Microbiol. *26*: 572–584.

Saxelin, M., N.H. Chuang, B. Chassy, H. Rautelin, P.H. Makela, S. Salminen and S.L. Gorbach. 1996. Lactobacilli and bacteremia in southern Finland, 1989–1992. Clin. Infect. Dis. *22*: 564–566.

Schachtsiek, M., W.P. Hammes and C. Hertel. 2004. Characterization of *Lactobacillus coryniformis* DSM 20001ᵀ surface protein Cpf mediating coaggregation with and aggregation among pathogens. Appl. Environ. Microbiol. *70*: 7078–7085.

Scheirlinck, I., R. Van der Meulen, A. Van Schoor, I. Cleenwerck, G. Huys, P. Vandamme, L. De Vuyst and M. Vancanneyt. 2007a. *Lactobacillus namurensis* sp. nov., isolated from a traditional Belgian sourdough. Int. J. Syst. Evol. Microbiol. *57*: 223–227.

Scheirlinck, I., R. Van der Meulen, A. Van Schoor, G. Huys, P. Vandamme, L. De Vuyst and M. Vancanneyt. 2007b. *Lactobacillus crustorum* sp. nov., isolated from two traditional Belgian wheat sourdoughs. Int. J. Syst. Evol. Microbiol. *57*: 1461–1467.

Schillinger, U. and W.H. Holzapfel. 2003. Culture media for lactic acid bacteria. *In* Corry, Curtis and Baird (Editors), Handbook of Culture Media for Food Microbiology, Vol. 37. Elsevier, Amsterdam, pp. 127–140.

Schleifer, K.H. and O. Kandler. 1972. Peptidoglycan types of bacterial cell walls and their taxonomic implications. Bacteriol. Rev. *36*: 407–477.

Schleifer, K.H., W.P. Hammes and O. Kandler. 1976. Effect of endogenous and exogenous factors on the primary structures of bacterial peptidoglycan. Adv. Microb. Physiol. *13*: 245–292.

Schleifer, K.H., W. Ludwig and R. Amann. 1993. Nucleic acid probles. *In* Goodfellow and McDonnell (Editors), Handbook of New Bacterial Systematics. Academic Press, London, New York, pp. 463–510.

Schleifer, K.H. and W. Ludwig. 1995a. Phylogenetic relationships of lactic acid bacteria. *In* Wood and Holzapfel (Editors), The Genera of Lactic Acid Bacteria. Blackie Academic & Professional, London, Glasgow, Weinheim, New York, Tokyo, Melbourne, Madras, pp. 7–18.

Schleifer, K.H. and W. Ludwig. 1995b. Phylogeny of the genus *Lactobacillus* and related genera. Syst. Appl. Microbiol. *18*: 461–467.

Schnürer, J. and J. Magnusson. 2005. Antifungal lactic acid bacteria as biopreservatives. Trends Food Sci. Technol. *16*: 70–78.

Schötz, F., I.G. Abo-Elnaga and O. Kandler. 1965. Zur Struktur Der Mesosomen Bei *Lactobacillus* Corynoides. Z. Naturforschung Part B *20*: 790–794.

Schröder, K., E. Clausen, A.M. Sandbreg and J. Raa. 1980. Psychrotrophic *Lactobacillus plantarum* from fish and its ability to produce antibiotic substances. *In* Connell (Editor), Control of Fish Quality. Fishing News Books Ltd., Farnham, Surrey.

Schütz, H. and F. Radler. 1984. Anaerobic reduction of glycerol to propanediol-1.3 by *Lactobacillus brevis* and *Lactobacillus buchneri*. Syst. Appl. Microbiol. *5*: 169–178.

Serra, P.G., L. Morelli and V. Bottazzi. 1992. The lactic microflora of fowl. *In* Wood (Editor), The lactic acid bacteria, vol. 1. The lactic acid bacteria in health and disease. Elsevier Applied Science, London, U.K.

Settanni, L., D. van Sinderen, J. Rossi and A. Corsetti. 2005. Rapid differentiation and in situ detection of 16 sourdough *Lactobacillus* species by multiplex PCR. Appl. Environ. Microbiol. *71*: 3049–3059.

Sharpe, M.E. 1962. Enumeration and studies of lactobacilli in food products. Dairy Sci. Abstr. *24*: 165–171.

Sharpe, M.E., M.J. Latham, E.I. Garvie, J. Zirngibl and O. Kandler. 1973. Two new species of *Lactobacillus* isolated from bovine rumen, *Lactobacillus ruminis* sp. nov. and *Lactobacillus vitulinus* sp. nov. J. Gen. Microbiol. *77*: 37–49.

Sharpe, M.E. 1979. Identification of the lactic acid bacteria. *In* Lovelock (Editor), Identification Methods for Bacteriologists. Society for Applied Bacteriology Technical Series 14. Academic Press, London, pp. 233–259.

Sharpe, M.E. 1981. The genus *Lactobacillus*. *In* Starr, Stolp, Trüper, Balows and Schegel (Editors), The Prokaryotes. A Handbook on

Habits, Isolation and Identification of Bacteria. Springer-Verlag, Berlin, pp. 1653–1679.

Shimwell, J.L. 1948. A study of ropiness in beer. Part II. Ropiness due to tetrad forming cocci. J. Inst. Brew. *54*: 237–244.

Simpson, K.L. and J.L. Fernandez. 1994. Mechanism of resistance of lactic acid bacteria to *trans*-isohumulone. J. Am. Soc. Brew. Chem. *52*: 9–11.

Simpson, K.L., B. Pettersson and F.G. Priest. 2001. Characterization of lactobacilli from Scotch malt whisky distilleries and description of *Lactobacillus ferintoshensis* sp. nov., a new species isolated from malt whisky fermentations. Microbiology *147*: 1007–1016.

Simpson, K.L., B. Pettersson and F.G. Priest. 2002a. *In* Validation of the publication of new names and new combinations previously effectively published outside the IJSEM. List no. 86. Int. J. Syst. Evol. Microbiol. *52*: 1075–1076.

Simpson, P.J., C. Stanton, G.F. Fitzgerald and R.P. Ross. 2002b. Genomic diversity within the genus *Pediococcus* as revealed by randomly amplified polymorphic DNA PCR and pulsed-field gel electrophoresis. Appl. Environ. Microbiol. *68*: 765–771.

Simpson, P.J., G.F. Fitzgerald, C. Stanton and R.P. Ross. 2006. Enumeration and identification of pediococci in powder-based products using selective media and rapid PFGE. J. Microbiol. Methods *64*: 120–125.

Simpson, W.J. and H. Taguchi. 1995. The genus *Pediococcus*, with notes on the genera *Tetragenococcus* and *Aerococcus*. *In* Wood and Holzapfel (Editors), The Genera of Lactic Acid Bacteria. Blackie Academic Press and Professional, London, pp. 125–172.

Sims, W. 1986. The isolation of pediococci from human saliva. Arch. Oral Biol. *11*: 967–972.

Skerman, V.B.D., V. McGowan and P.H.A. Sneath. 1980. Approved lists of bacterial names. Int. J. Syst. Bacteriol. *30*: 225–420.

Skerman, V.D.B., V. McGowan and P.H.A. Sneath. 1989. Approved Lists of Bacterial Names, Amended Edition. American Society for Microbiology, Washington, D.C.

Sneath, P.H.A., N.S. Mair, M.E. Sharpe and J.G. Holt (Editors). 1986. Bergey's Manual of Systematic Bacteriology. The Williams & Wilkins Co., Baltimore.

Song, Y., N. Kato, C. Liu, Y. Matsumiya, H. Kato and K. Watanabe. 2000. Rapid identification of 11 human intestinal *Lactobacillus* species by multiplex PCR assays using group- and species-specific primers derived from the 16S-23S rRNA intergenic spacer region and its flanking 23S rRNA. FEMS Microbiol. Lett. *187*: 167–173.

Sozzi, T. and M. Smiley. 1980. Antibiotic resistances of yogurt starter cultures *Streptococcus thermophilus* and *Lactobacillus bulgaricus*. Appl. Environ. Microbiol. *40*: 862–865.

Splittstoesser, D.F., L.L. Lienk, M. Wilkison and J.R. Stamer. 1975. Influence of wine composition on the heat resistance of potential spoilage organisms. Appl. Microbiol. *30*: 369–373.

Stamer, J.R. 1975. Recent developments in the fermentation of sauerkraut. *In* Carr, Cutting and Whiting (Editors), Lactic Acid Bacteria in Beverages and Food. Academic Press, London, pp. 267–280.

Stetter, K.O. and O. Kandler. 1973. Formation of DL-lactic acid by lactobacilli and characterization of a lactic acid racemase from several streptobacteria (author's transl). Arch. Mikrobiol. *94*: 221–247.

Stevenson, D.M., R.E. Muck, K.J. Shinners and P.J. Weimer. 2005. Use of real time PCR to determine population profiles of individual species of lactic acid bacteria in alfalfa silage and stored corn stover. Appl. Microbiol. Biotechnol. *71*: 329–338.

Stewart, C.S. 1992. Lactic acid bacteria in the rumen. *In* Wood (Editor), The Lactic Acid Bacteria in Health and Disease. Elsevier Applied Science Publishers, London, pp. 49–68.

Straub, B.W., M. Kicherer, S.M. Schilcher and W.P. Hammes. 1995. The formation of biogenic amines by fermentation organisms. Z. Lebensm. Unters. Forsch. *201*: 79–82.

Sugihara, T.F. and L. Kline. 1975. Further studies on a growth medium for *Lactobacillus sanfrancisco*. J. Milk Food Technol. *38*: 667–672.

Suhaimi, M., B. Bruyneel and W. Verstraete. 1987. Enrichment of an alkaline-adapted association of streptococci and lactobacilli. J. Appl. Bacteriol. *63*: 117–123.

Suzuki, K., W. Funahashi, M. Koyanagi and H. Yamashita. 2004. *Lactobacillus paracollinoides* sp. nov., isolated from brewery environments. Int. J. Syst. Evol. Microbiol. *54*: 115–117.

Swenson, J.M., R.R. Facklam and C. Thornsberry. 1990. Antimicrobial susceptibility of vancomycin-resistant *Leuconostoc, Pediococcus,* and *Lactobacillus* species. Antimicrob. Agents Chemother. *34*: 543–549.

Taguchi, H., M. Ohkochi, H. Uehara, K. Kojima and M. Mawatari. 1990. KOT medium, a new medium for the detection of beer spoilage lactic acid bacteria. J. Am. Soc. Brew. Chem. *48*: 72–75.

Takizawa, S., S. Kojima, S. Tamura, S. Fujinaga, Y. Benno and T. Nakase. 1994. *Lactobacillus kefirgranum* sp. nov. and *Lactobacillus parakefir* sp. nov., two new species from kefir grains. Int. J. Syst. Bacteriol. *44*: 435–439.

Talarico, T.L. and W.J. Dobrogosz. 1989. Chemical characterization of an antimicrobial substance produced by *Lactobacillus reuteri.* Antimicrob. Agents Chemother. *33*: 674–679.

Tamang, J.P., B. Tamang, U. Schillinger, C.M. Franz, M. Gores and W.H. Holzapfel. 2005. Identification of predominant lactic acid bacteria isolated from traditionally fermented vegetable products of the Eastern Himalayas. Int. J. Food Microbiol. *105*: 347–356.

Tamura, G. 1956. Hiochic acid, a new growth factor for *Lactobacillus homohiochii* and *Lactobacillus heterohiochii.* J. Gen. Appl. Microbiol. *2*: 431–434.

Tanasupawat, S., S. Okada and K. Komagata. 1998. Lactic acid bacteria found in fermented fish in Thailand. J. Gen. Appl. Microbiol. *44*: 193–200.

Tanasupawat, S., O. Shida, S. Okada and K. Komagata. 2000. *Lactobacillus acidipiscis* sp. nov. and *Weissella thailandensis* sp. nov., isolated from fermented fish in Thailand. Int. J. Syst. Evol. Microbiol. *50*: 1479–1485.

Tanasupawat, S., A. Pakdeeto, C. Thawai, P. Yukphan and S. Okada. 2007. *In* Validation of the publication of new names and new combinations previously effectively published outside the IJSEM. List no. 116. Int. J. Syst. Evol. Microbiol. *57*: 1371–1373.

Tankovic, J., R. Leclercq and J. Duval. 1993. Antimicrobial susceptibility of *Pediococcus* spp. and genetic basis of macrolide resistance in *Pediococcus acidilactici* HM3020. Antimicrob. Agents Chemother. *37*: 789–792.

Tannock, G.W. 1992. Lactic microflora of pigs, mice and rats. *In* Wood (Editor), The Lactic Acid Bacteria in Health and Disease. Elsevier Applied Science Publishers, London, pp. 21–48.

Tannock, G.W., A. Tilsala-Timisjarvi, S. Rodtong, J. Ng, K. Munro and T. Alatossava. 1999. Identification of *Lactobacillus* isolates from the gastrointestinal tract, silage, and yoghurt by 16S-23S rRNA gene intergenic spacer region sequence comparisons. Appl. Environ. Microbiol. *65*: 4264–4267.

Tannock, G.W., K. Munro, H.J. Harmsen, G.W. Welling, J. Smart and P.K. Gopal. 2000. Analysis of the fecal microflora of human subjects consuming a probiotic product containing *Lactobacillus rhamnosus* DR20. Appl. Environ. Microbiol. *66*: 2578–2588.

Tanous, C., A. Kieronczyk, S. Helinck, E. Chambellon and M. Yvon. 2002. Glutamate dehydrogenase activity: a major criterion for the selection of flavour-producing lactic acid bacteria strains. Antonie van Leeuwenhoek *82*: 271–278.

Taranto, M.P., J.L. Vera, J. Hugenholtz, G.F. De Valdez and F. Sesma. 2003. *Lactobacillus reuteri* CRL1098 produces cobalamin. J. Bacteriol. *185*: 5643–5647.

Teuber, M., L. Meile and F. Schwarz. 1999. Acquired antibiotic resistance in lactic acid bacteria from food. Antonie van Leeuwenhoek *76*: 115–137.

Teuber, M. 2000. Fermented Milk Products. *In* Lund, Baird-Parker and Gould (Editors), The Microbiological Safety and Quality of Food, Vol. I. Aspen Publishers, Gaitherburg, pp. 535–589.

Thomas, T.D., D.C. Ellwood and V.M. Longyear. 1979. Change from homo- to heterolactic fermentation by *Streptococcus lactis* resulting from glucose limitation in anaerobic chemostat cultures. J. Bacteriol. *138*: 109–117.

Tieking, M., M. Korakli, M.A. Ehrmann, M.G. Ganzle and R.F. Vogel. 2003. In situ production of exopolysaccharides during sourdough

fermentation by cereal and intestinal isolates of lactic acid bacteria. Appl. Environ. Microbiol. *69*: 945–952.

Tilsala-Timisjarvi, A. and T. Alatossava. 1997. Development of oligonucleotide primers from the 16S-23S rRNA intergenic sequences for identifying different dairy and probiotic lactic acid bacteria by PCR. Int. J. Food Microbiol. *35*: 49–56.

Titgemeyer, F. and W. Hillen. 2002. Global control of sugar metabolism: a Gram-positive solution. Antonie van Leeuwenhoek *82*: 59–71.

Todorov, S.D. and L.M.T. Dicks. 2005. Pediocin ST18, an anti-*Listerial* bacteriocin produced by *Pediococcus pentosaceus* ST18 isolated from boza, a traditional cereal beverage from Bulgaria. Proc. Biochem. *40*: 365–370.

Tong, H. and X. Dong. 2005a. *Lactobacillus concavus* sp. nov., isolated from the walls of a distilled spirit fermenting cellar in China. Int. J. Syst. Evol. Microbiol. *55*: 2199–2202.

Tong, H. and X. Dong. 2005b. *Lactobacillus concavus* sp. nov., isolated from the walls of a distilled spirit fermenting cellar in China. Int. J. Syst. Evol. Microbiol. *55*: 2199–2202.

Torriani, S., M. Vescovo and F. Dellaglio. 1987. Tracing *Pediococcus acidilactici* in ensiled maize by plasmid-encoded erythromycin resistance. J. Appl. Bacteriol. *63*: 543–553.

Torriani, S., C.A. VanReenen, G. Klein, G. Reuter, F. Dellaglio and L.M.T. Dicks. 1996. *Lactobacillus curvatus* subsp. *curvatus* subsp. nov. and *Lactobacillus curvatus* subsp. *melibiosus* subsp. nov.. and *Lactobacillus sake* subsp. *sake* subsp. nov. and *Lactobacillus sake* subsp. *carnosus* subsp. nov., new subspecies of *Lactobacillus curvatus* Abo-Elnaga and Kandler 1965 and *Lactobacillus sake* Katagiri, Kitahara, and Fukami 1934 (Klein et al. 1996, emended descriptions), respectively. Int. J. Syst. Bacteriol. *46*: 1158–1163.

Troili-Petersson, G. 1903. Studien über die Mikroorganismen des schwedischen Güterkäses. Zentbl. Bakteriol. II Abt. *11*: 120–143.

Trüper, H.G. and L. de Clari. 1997. Taxonomic note: necessary correction of specific epithets formed as substantives (nouns) 'in apposition'. Int. J. Syst. Bacteriol. *47*: 908–909.

Turner, D.L., L. Brennan, H.E. Meyer, C. Lohaus, C. Siethoff, H.S. Costa, B. Gonzalez, H. Santos and J.E. Suarez. 1999. Solution structure of plantaricin C, a novel lantibiotic. Eur. J. Biochem. *264*: 833–839.

Twomey, D., R.P. Ross, M. Ryan, B. Meaney and C. Hill. 2002. Lantibiotics produced by lactic acid bacteria: structure, function and applications. Antonie van Leeuwenhoek *82*: 165–185.

Tzanetakis, N. and E. Litopoulou-Tzanetaki. 1989. Biochemical activities of *Pediococcus pentosaceus* isolates of dairy origin. J. Dairy Sci. *72*: 859–863.

Valcheva, R., M. Korakli, B. Onno, H. Prevost, I. Ivanova, M.A. Ehrmann, X. Dousset, M.G. Ganzle and R.F. Vogel. 2005. *Lactobacillus hammesii* sp. nov., isolated from French sourdough. Int. J. Syst. Evol. Microbiol. *55*: 763–767.

Valcheva, R., M.F. Ferchichi, M. Korakli, I. Ivanova, M.G. Ganzle, R.F. Vogel, H. Prevost, B. Onno and X. Dousset. 2006. *Lactobacillus nantensis* sp. nov., isolated from French wheat sourdough. Int. J. Syst. Evol. Microbiol. *56*: 587–591.

Valerio, F., P. Lavermicocca, M. Pascale and A. Visconti. 2004. Production of phenyllactic acid by lactic acid bacteria: an approach to the selection of strains contributing to food quality and preservation. FEMS Microbiol. Lett. *233*: 289–295.

van Beek, S. and F.G. Priest. 2002. Evolution of the lactic acid bacterial community during malt whisky fermentation: a polyphasic study. Appl. Environ. Microbiol. *68*: 297–305.

Van Kerken, A.E. and O. Kandler. 1966. Die Laktobazillenflora des Tilsiterkäses. Milchwissenschaft *21*: 436–440.

Van Laer, H. 1892. Contribution á la histoire des ferments des hydrates de carbone. Acad. Roy. Belg. Cl. Sci. Collect. Octavo. Mem. *47*: 1–37.

Vancanneyt, M., J. Mengaud, I. Cleenwerck, K. Vanhonacker, B. Hoste, P. Dawyndt, M.C. Degivry, D. Ringuet, D. Janssens and J. Swings. 2004. Reclassification of *Lactobacillus kefirgranum* Takizawa *et al.* 1994 as *Lactobacillus kefiranofaciens* subsp. *kefirgranum* subsp. nov. and emended description of L. kefiranofaciens Fujisawa *et al.* 1988. Int. J. Syst. Evol. Microbiol. *54*: 551–556.

Vancanneyt, M., P. Neysens, M. De Wachter, K. Engelbeen, C. Snauwaert, I. Cleenwerck, R. Van der Meulen, B. Hoste, E. Tsakalidou, L. De Vuyst and J. Swings. 2005. *Lactobacillus acidifarinae* sp. nov. and *Lactobacillus zymae* sp. nov., from wheat sourdoughs. Int. J. Syst. Evol. Microbiol. *55*: 615–620.

Vancanneyt, M., S.M. Naser, K. Engelbeen, M. De Wachter, R. Van der Meulen, I. Cleenwerck, B. Hoste, L. De Vuyst and J. Swings. 2006. Reclassification of *Lactobacillus brevis* strains LMG 11494 and LMG 11984 as *Lactobacillus parabrevis* sp. nov. Int. J. Syst. Evol. Microbiol. *56*: 1553–1557.

Vandamme, P., B. Pot, M. Gillis, P. de Vos, K. Kersters and J. Swings. 1996. Polyphasic taxonomy, a consensus approach to bacterial systematics. Microbiol. Rev. *60*: 407–438.

Vasquez, A., T. Jakobsson, S. Ahrne, U. Forsum and G. Molin. 2002. Vaginal lactobacillus flora of healthy Swedish women. J. Clin. Microbiol. *40*: 2746–2749.

Vaughan, E.E., M.C. de Vries, E.G. Zoetendal, K. Ben-Amor, A.D. Akkermans and W.M. de Vos. 2002. The intestinal LABs. Antonie van Leeuwenhoek *82*: 341–352.

Vela, A.I., A. Fernandez, A. Espinosa de los Monteros, J. Goyache, P. Herraez, B. Tames, F. Cruz, L. Dominguez and J.F. Fernandez-Garayzabal. 2008. *Lactobacillus ceti* sp. nov., isolated from beaked whales (*Ziphius cavirostris*). Int. J. Syst. Evol. Microbiol. *58*: 891–894.

Ventura, M., I.A. Casas, L. Morelli and M.L. Callegari. 2000. Rapid amplified ribosomal DNA restriction analysis (ARDRA) identification of *Lactobacillus* spp. isolated from fecal and vaginal samples. Syst. Appl. Microbiol. *23*: 504–509.

Verachtert, H. and D. Iserentant. 1995. Properties of Belgian acid beers and their microflora I. The production of gueuze and related refreshing acid beers. Cerevisia *20*: 37–41.

Vogel, R.F., G. Bocker, P. Stolz, M. Ehrmann, D. Fanta, W. Ludwig, B. Pot, K. Kersters, K.H. Schleifer and W.P. Hammes. 1994. Identification of lactobacilli from sourdough and description of *Lactobacillus pontis* sp. nov. Int. J. Syst. Bacteriol. *44*: 223–229.

Walter, J., C. Hertel, G.W. Tannock, C.M. Lis, K. Munro and W.P. Hammes. 2001. Detection of *Lactobacillus, Pediococcus, Leuconostoc,* and *Weissella* species in human feces by using group-specific PCR primers and denaturing gradient gel electrophoresis. Appl. Environ. Microbiol. *67*: 2578–2585.

Waters, S.M., S. Doyle, R.A. Murphy and R.F. Power. 2005. Development of solution phase hybridisation PCR-ELISA for the detection and quantification of *Enterococcus faecalis* and *Pediococcus pentosaceus* in Nurmi-type cultures. J. Microbiol. Methods *63*: 264–275.

Weiss, N., U. Schillinger, M. Laternser and O. Kandler. 1981. *Lactobacillus sharpeae* sp. nov. and *Lactobacillus agilis* sp. nov., two new species of homofermentative, *meso*-diaminopimelic acid-containing lactobacilli isolated from sewage. Zentbl. Bakteriol. Mikrobiol. Hyg. I Abt. Orig. C *2*: 242–253.

Weiss, N., U. Schillinger, M. Laternser and O. Kandler. 1982. *In* Validation of the publication of new names and new combinations previously effectively published outside the IJSB. List no. 8. Int. J. Syst. Bacteriol. *32*: 266–268.

Weiss, N., U. Schillinger and O. Kandler. 1983a. *Lactobacillus trichodes,* and *Lactobacillus heterohiochii,* subjective synonyms of *Lactobacillus fructivorans.* Syst. Appl. Microbiol. *4*: 507–511.

Weiss, N., U. Schillinger and O. Kandler. 1983b. *Lactobacillus lactis, Lactobacillus leichmannii* and *Lactobacillus bulgaricus,* subjective synonyms of *Lactobacillus delbrueckii,* and description of *Lactobacillus delbrueckii* subsp. *lactis* comb. nov. and *Lactobacillus delbrueckii* subsp. *bulgaricus* comb. nov. Syst. Appl. Microbiol. *4*: 552–557.

Weiss, N. and U. Schillinger. 1984a. *In* Validation of the publication of new names and new combinations previously effectively published outside the IJSB. List no. 16. Int. J. Syst. Bacteriol. *34*: 503–504.

Weiss, N. and U. Schillinger. 1984b. *Lactobacillus sanfrancisco* sp. nov., nom. rev. Syst. Appl. Microbiol. *5*: 230–232.

Weiss, N., U. Schillinger and O. Kandler. 1984. *In* Validation of the publication of new names and new combinations previously effectively published outside the IJSB. List no. 14. Int. J. Syst. Bacteriol. *34*: 270–271.

Weiss, N. 1992. The genera *Pediococcus* and *Aerococcus. In* Balows, Trüper, Dworkin, Harder and Schleifer (Editors), The Prokaryotes, 2nd edn, Vol. 2. Springer-Verlag, New York.

Westby, A., L. Nuraida, J.D. Owens and P.A. Gibbs. 1993. Inability of *Lactobacillus plantarum* and other lactic acid bacteria to grow on D-ribose as sole source of fermentable carbohydrate. J. Appl. Bacteriol. *75*: 168–175.

Whiting, G.C. 1975. Some biochemical and flavour aspects of lactic acid bacteria in ciders and other alcoholic beverages. *In* Carr, Cutting and Whiting (Editors), Lactic Acid Bacteria in Beverages and Food. Academic Press, London, pp. 69–85.

Whittenbury, R. 1965. A study of some pediococci and their relationships to *Aerococcus viridans* and the enterococci. J. Gen. Microbiol. *40*: 97–106.

Wiese, B.G., W. Strohmar, F.A. Rainey and H. Diekmann. 1996. *Lactobacillus panis* sp. nov., from sourdough with a long fermentation period. Int. J. Syst. Bacteriol. *46*: 449–453.

Wilderdyke, M.R., D.A. Smith and M.M. Brashears. 2004. Isolation, identification, and selection of lactic acid bacteria from alfalfa sprouts for competitive inhibition of foodborne pathogens. J. Food Prot. *67*: 947–951.

Winslow, C.E., J. Broadhurst, R.E. Buchanan, C. Krumwiede, L.A. Rogers and G.H. Smith. 1917. The families and genera of the bacteria: preliminary report of the committee of the Society of American Bacteriologists on characterization and classification of bacterial types. J. Bacteriol. *2*: 505–566.

Wolf, G., E.K. Arendt, U. Pfahler and W.P. Hammes. 1990. Heme-dependent and heme-independent nitrite reduction by lactic acid bacteria results in different N-containing products. Int. J. Food Microbiol. *10*: 323–329.

Wood, B.J. and P.J. Warner. 2003. Genetics of Lactic Acid Bacteria. Springer-Verlag, New York.

Wu, C.W., L.J. Yin and S.T. Jiang. 2004. Purification and characterization of bacteriocin from *Pediococcus pentosaceus* ACCEL. J. Agric. Food Chem. *52*: 1146–1151.

Wylie, J.G. and A. Henderson. 1969. Identity and glycogen-fermenting ability of lactobacilli isolated from the vagina of pregnant women. J. Med. Microbiol. *2*: 363–366.

Yoon, J.H., S.S. Kang, T.I. Mheen, J.S. Ahn, H.J. Lee, T.K. Kim, C.S. Park, Y.H. Kho, K.H. Kang and Y.H. Park. 2000. *Lactobacillus kimchii* sp. nov., a new species from kimchi. Int. J. Syst. Evol. Microbiol. *50*: 1789–1795.

Yost, C.K. and F.M. Nattress. 2000. The use of multiplex PCR reactions to characterize populations of lactic acid bacteria associated with meat spoilage. Lett. Appl. Microbiol. *31*: 129–133.

Yuki, N., T. Shimazaki, A. Kushiro, K. Watanabe, K. Uchida, T. Yuyama and M. Morotomi. 2000. Colonization of the stratified squamous epithelium of the nonsecreting area of horse stomach by lactobacilli. Appl. Environ. Microbiol. *66*: 5030–5034.

Zanoni, P., J.A.E. Farrow, B.A. Phillips and M.D. Collins. 1987. *Lactobacillus pentosus* (Fred, Peterson, and Anderson) sp. nov., nom. rev. Int. J. Syst. Bacteriol. *37*: 339–341.

Zhang, B., H. Tong and X. Dong. 2005. *Pediococcus cellicola* sp. nov., a novel lactic acid coccus isolated from a distilled-spirit-fermenting cellar. Int. J. Syst. Evol. Microbiol. *55*: 2167–2170.

Zhou, X., S.J. Bent, M.G. Schneider, C.C. Davis, M.R. Islam and L.J. Forney. 2004. Characterization of vaginal microbial communities in adult healthy women using cultivation-independent methods. Microbiology *150*: 2565–2573.

Family II. **Aerococcaceae** fam. nov.

WOLFGANG LUDWIG, KARL-HEINZ SCHLEIFER AND WILLIAM B. WHITMAN

A.e.ro.coc.ca'ce.ae. N.L. masc. n. *Aerococcus* type genus of the family; suff. *-aceae* ending denoting family; N.L. fem. pl. n. *Aerococcaceae* the *Aerococcus* family.

The family *Aerococcaceae* is circumscribed for this volume on the basis of phylogenetic analyses of the 16S rRNA gene sequences and includes the genus *Aerococcus* and its close relatives. It is composed of nonmotile, Gram-positive ovoid cocci or coccobacilli. When it has been measured, cell walls have been found to contain the diamino acid lysine. Endospores are not formed. Facultatively anaerobic and catalase-negative. May grow in media containing 6.5 % NaCl.

Type genus: **Aerococcus** Williams, Hirch and Cowan 1953, 475[AL].

Genus I. **Aerococcus** Williams, Hirch and Cowan 1953, 475[AL]

MATTHEW D. COLLINS AND ENEVOLD FALSEN

A.ë.ro.coc'cus. Gr. masc. n. *aër* air, gas; Gr. n. *kokkos* a berry; N.L. masc. n. *Aerococcus* air coccus.

Cells are ovoid in shape (1–2 μm in diameter) and divide on two planes at right angles, giving rise to tetrad and cluster arrangements; some pairs and single cells may also be observed. α-Hemolytic. Gram-positive and nonmotile. Nonsporeforming. **Facultatively anaerobic. Catalase-negative. Oxidase-negative. Grows in 6.5% NaCl. Gas is not produced in MRS broth. Acid is produced from glucose and some other carbohydrates.** Hippurate is hydrolyzed by most strains. Leucine aminopeptidase and β-glucuronidase may or may not be produced. Arginine is not deaminated by most strains and urease is not produced. **Voges–Proskauer-negative. Vancomycin-sensitive.**

DNA G+C content (mol%): 35–44 (T_m).

Type species: **Aerococcus viridans** Williams, Hirch and Cowan 1953, 477.

Further descriptive information

Aerococci produce an α-hemolytic reaction on blood agar and form small (generally 1 mm or less), nonpigmented colonies (with the occasional exception of yellow pigment production by *Aerococcus viridans* strains) after 24 h incubation at 37°C. Aerococci generally consist of Gram-positive cocci arranged in clusters and tetrads except for *Aerococcus christensenii* which exhibits short chains. They are facultatively anaerobic; *Aerococcus christensenii* prefers anaerobic conditions. Aerococci are catalase-negative, but some strains may display a weak nonheme pseudocatalase activity. Cytochrome enzymes are absent. Aerococci are nutritionally fastidious. Pantothenic acid, nicotinic acid, and biotin are either required or stimulatory for the growth of *Aerococcus viridans*; guanine or another purine base is also required. Amino acids are needed for the growth of *Aerococcus viridans*, but the requirement is not specific or sharply defined (Miller and Evans, 1970). All aerococci grow in 6.5% NaCl, and *Aerococcus viridans* grows in 10% NaCl. Strains differ in their growth temperature characteristics. Most strains of *Aerococcus urinae* grow at 45°C whereas most strains of *Aerococcus viridans* do not. *Aerococcus urinae* and *Aerococcus viridans* do not grow at 10°C (Facklam and Elliott, 1995), although there are reports of the latter species (Evans, 1986) growing at this temperature. Most *Aerococcus viridans* strains and *Aerococcus sanguinicola* give a positive bile-esculin reaction, whereas *Aerococcus urinae* is bile-esculin negative. Aerococci are chemo-organotrophic; acid, but no gas, is produced from glucose and other sugars. Acid production from carbohydrates differs between species and also between the test systems employed (conventional tests or API kits). *Aerococcus urinae*, *Aerococcus urinaehominis*, *Aerococcus sanguinicola*, and a minority of strains of *Aerococcus viridans* produce β-glucuronidase. *Aerococcus urinae* produces leucine aminopeptidase but not pyrrolidonyl aminopeptidase, whereas *Aerococcus viridans* displays the opposite reactions (Christensen et al., 1997). By contrast, *Aerococcus sanguinicola* is positive for both leucine aminopeptidase and pyrrolidonyl aminopeptidase. None of the currently described *Aerococcus* species displays activity for alanyl-phenylalanine-proline arylamidase, glycyl-tryptophan arylamidase, β-mannosidase, *N*-acetyl-β-glucosaminidase, or urease. Aerococci do not hydrolyze gelatin and do not reduce nitrate to nitrite.

The cell-wall murein of *Aerococcus viridans* and *Aerococcus christensenii* are based on L-lysine (directly cross-linked, type A1α) (Collins et al., 1999d). Polyamines are not detected in *Aerococcus viridans* (Hamana, 1994). Aerococci lack respiratory menaquinones. The predominant long chain fatty acids of *Aerococcus viridans* are $C_{16:0}$, $C_{16:1\,\omega9c}$, $C_{18:0}$, $C_{18:1\,\omega9c}$, and $C_{20:0\,\omega9c}$, together with smaller amounts of $C_{14:0}$, $C_{16:1\,\omega9t}$, and $C_{18:1\,\omega7c}$ (Bosley et al., 1990). The major cellular fatty acids of *Aerococcus urinae* have been reported to be $C_{16:0}$, $C_{18:1\,\omega9c}$, and $C_{18:0}$; $C_{18:1\,\omega7t}$ has also been reported in this species (Zbinden et al., 1999). The reported G+C content of DNA of *Aerococcus viridans*, *Aerococcus urinae*, and *Aerococcus christensenii* is 35–40, 44.4 and 38.5 mol%, respectively. *Aerococcus viridans* is reported to have a relatively small genome ranging from 0.57×10^9 to 1×10^9 Da (Wiik et al., 1986).

Aerococcus urinae is susceptible to a wide range of antimicrobials including β-lactams but is resistant to sulfonamides and aminoglycosides. Christensen et al. (1996) found *Aerococcus urinae* was susceptible to penicillin, ampicillin, cephalosporins, tetracycline, doxycycline, chloramphenicol, furazolidone, erythromycin, clindamycin, vancomycin, teicoplanin, daptomycin, rifampin, clavulanate, and mupirocin but resistant to aminoglycosides, sulfonamides, trimethoprim, nalidixic acid, colistin, aztreonam, sulbactam, and bacitracin. *Aerococcus viridans* was not susceptible to bacitracin, clavulanate, colistin, furazolidone, penicillin, and tobramycin (Christensen et al., 1996). Penicillin resistance among strains of *Aerococcus viridans* has been reported, although there are reports describing isolates that are susceptible to penicillin (Kern and Vanek, 1987). Kern and Vanek (1987) reported that some strains of *Aerococcus viridans* were resistant to sulfatrimethyloprim-methoxazole and ofloxacin. Buu-Hoi et al. (1989) reported several strains of *Aerococcus viridans* that had acquired the *ermB* gene of *Enterococcus faecalis*. Other strains acquired the *tetM* gene to confirm tetracycline resistance. *Aerococcus viridans* is reported to produce a bacteriocin, designated viridicin (Ballester et al., 1980).

Aerococci have been recovered from a wide range of environments. *Aerococcus viridans* is generally saprophytic and is found in air, dust, vegetation, meat-curing brines, soil, and marine sources. The species can also be found in small numbers as part of the indigenous flora in the upper respiratory tract and on the skin of normal persons. *Aerococcus viridans* is a common airborne organism in hospital environments (Kerbaugh and Evans, 1968) and is associated, albeit infrequently, with human infections (Facklam and Elliott, 1995) such as endocarditis (Bru et al., 1986; Janosek et al., 1980; Untereker and Hanna, 1976), urinary tract infections (Colman, 1967), septic arthritis (Taylor and Trueblood, 1985), and acute childhood meningitis (Nathavitharana et al., 1983). *Aerococcus viridans* has been isolated from subclinical intramammary infections in dairy cows (Devriese et al., 1999) and is responsible for a fatal disease (gaffkemia) of lobsters (Fisher et al., 1978; Hitchner and Snieszko, 1947). *Aerococcus urinae* can cause infections in humans. This organism was originally characterized by Christensen et al. (1991, 1989) from urine samples of elderly persons with urinary tract infections. *Aerococcus urinae* infections are usually mild, but serious infections such as endocarditis, septicemia, urosepsis, and soft tissue infections have been reported (Christensen et al., 1991, 1995; Gritsch et al., 1999; Kristensen and Nielsen, 1995; Schuur et al., 1999; Zbinden et al., 1999). The three recently described species, *Aerococcus christensenii*, *Aerococcus sanguinicola*, and *Aerococcus urinaehominis*, have all been isolated from human sources including vagina, blood, and urine respectively (Collins et al., 1999d; Lawson et al., 2001a, 2001b).

Isolation procedures

Aerococci can be isolated on a variety of rich agar-containing media, such as heart infusion agar supplemented with 5% (v/v) sheep or horse blood at 37°C.

Maintenance procedures

Aerococci can be maintained on a variety of rich media generally used for other Gram-positive cocci, such as heart infusion or tryptic soy agar with or without animal blood (5% v/v). Strains can be maintained on cryogenic beads at −70°C or lyophilized for long-term preservation.

Taxonomic comments

The genus *Aerococcus* was created by Williams et al. (1953) to accommodate some Gram-positive, microaerophilic, catalase-negative, coccal-shaped organisms that differed from streptococci primarily by their characteristic tetrad cellular arrangement. Until recently, the genus contained only a single species, *Aerococcus viridans*, and historically, because

of its cellular morphology, this species was thought to resemble pediococci more than streptococci. Similarities were also apparent between *Aerococcus viridans* and the lobster pathogen "*Gaffkya homari*" (Deibel and Niven, 1960), and it is now evident that these represent a single species (Evans, 1986; Kelly and Evans, 1974; Wiik et al., 1986). *Pediococcus urinaeequi* has also been known for a long time to resemble *Aerococcus viridans* closely. These two species are biochemically and culturally indistinguishable, display similar antibiograms, possess almost identical 16S rRNA sequences (Christensen et al., 1996, 1997; Collins et al., 1990), and are considered to be the same species. DNA pairing investigations by Bosley et al. (1990) indicate that while most strains of *Aerococcus viridans* form a genetically highly related group, there is evidence of a second genomic species (designated genospecies 2) which is very closely related to *Aerococcus viridans*. During the past decade, four other *Aerococcus* species, *Aerococcus urinae*, *Aerococcus christensenii*, *Aerococcus urinaehominis*, and *Aerococcus sanguinicola*, have been described from human sources (Aguirre and Collins, 1992a; Collins et al., 1999d; Lawson et al., 2001a, 2001b). Comparative 16S rRNA gene sequencing studies have shown that this expanded *Aerococcus* genus forms a robust group among the catalase-negative, Gram-positive cocci (Collins et al., 1999d; Lawson et al., 2001a, 2001b) and is phylogenetically distinct from other genera such as *Streptococcus* and *Pediococcus* with which it has been historically associated. A tree depicting the phylogenetic interrelationships of aerococci and its close relatives is shown in Figure 94.

Differentiation of the genus *Aerococcus* from other genera

Although the genus *Aerococcus* is phylogenetically distinct, there are few phenotypic traits which serve to distinguish the genus from related genera. For example, when the genus was restricted to *Aerococcus viridans*, tests such as cellular arrangements, arginine dehydrolase (ADH), pyrrolidonyl arylamidase (PYRA), leucine aminopeptidase (LAP), vancomycin sensitivity, and growth at 45°C were useful aids in distinguishing aerococci from related genera (Bosley et al., 1990; Facklam and Elliott, 1995). However, with the possible exception of tetrad/cluster cellular arrangements (although *Aerococcus christensenii* displays short chains) and susceptibility to vancomycin, the newly defined *Aerococcus* species, in many instances, differ from *Aerococcus viridans* or from each other with respect to these and other traits, making simple identification of the genus somewhat difficult. For example, *Aerococcus urinae* produces LAP, is PYRA negative and grows at 45°C, whereas *Aerococcus viridans* displays the opposite results. The formation of tetrad/cluster cellular arrangements serves to distinguish aerococci from most other catalase-negative, Gram-positive cocci. Aerococci can be distinguished from other non-chain forming, Gram-positive cocci, such as *Pediococcus*, *Alloiococcus*, and *Gemella*, using conventional tests (Facklam and Elliott, 1995). For example, aerococci differ from pediococci in being sensitive to vancomycin. By contrast, pediococci are resistant to this antibiotic. Alloiococci differ from aerococci in not growing anaerobically (except for a few exceptional strains) and by failing to grow on a variety of commonly used media such as on sheep blood agar or in thioglycollate broth. Similarly, gemellae differ from most aerococcal species by failing to grow in 6.5% NaCl.

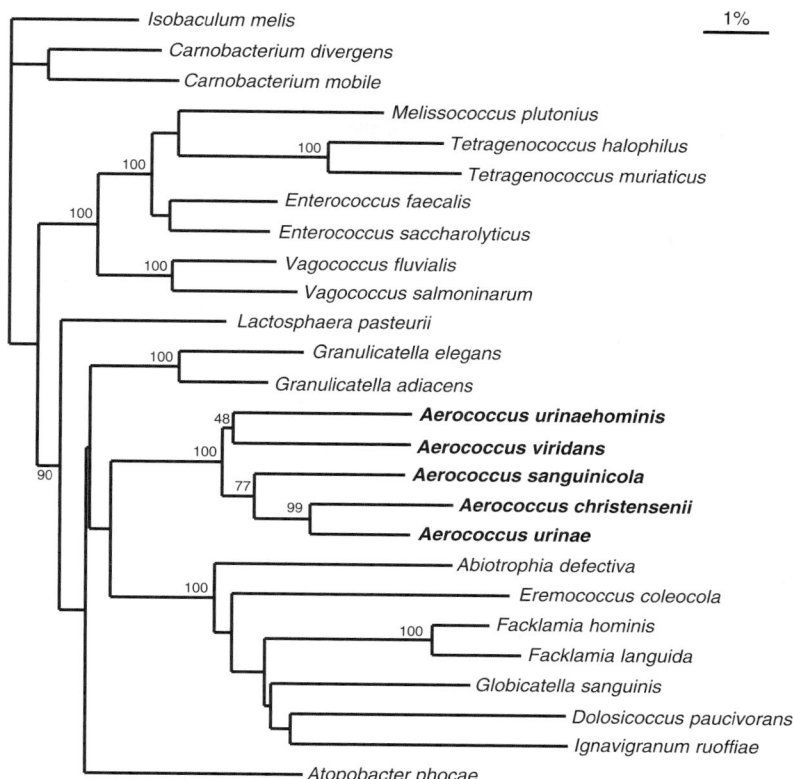

FIGURE 94. Neighbor-joining tree based on 16S rRNA sequences depicting the phylogenetic relationships of aerococci.

Differentiation of the species of the genus *Aerococcus*

In contrast to identification at the rank of genus, the phenotypic identification of aerococci at the species level is not problematic. All currently defined species can be easily recognized using conventional and miniaturized API test systems. Tests which are useful in distinguishing the various *Aerococcus* species from each other using conventional methods and the API Rapid ID32 Strep system are shown in Table 91 and Table 92, respectively.

List of species of the genus *Aerococcus**

1. **Aerococcus viridans** Williams, Hirch and Cowan 1953, 477[AL]
   vi.ri'dans. L. part. adj. *viridans* producing a green color.
   DNA G+C content (mol%): not reported.
   Type strain: ATCC 11563, CCM 1914, CCUG 4311, CIP 54.145, DSM 20340, HAMBI 1583, LMG 17931, NBRC 12219, NCAIM B.01070, NCTC 8251.
   GenBank accession number (16S rRNA gene): M58797.

2. **Aerococcus christensenii** Collins, Rodriguez Jovita, Hutson, Ohlén and Falsen 1999d, 1128[VP]
   christ.en.sen'i.i. N.L. gen. n. *christensenii* named after the Danish microbiologist Jens J. Christensen.
   DNA G+C content (mol%): 38.5 (T_m).
   Type strain: UW06, CCUG 28831, CIP 106115.
   GenBank accession number (16S rRNA gene): Y17005.

3. **Aerococcus sanguinicola** corrig. Lawson, Falsen, Truberg-Jensen and Collins 2001b, 478[VP]
   san.guin'i.co.la. L. n. *sanguis* blood; L. sulf -*cola* dweller; N.L. n. *sanguinicola* blood-dweller.
   DNA G+C content (mol%): not reported.
   Type strain: CCUG 43001, CIP 106533, JCM 11549.
   GenBank accession number (16S rRNA gene): AJ276512.

4. **Aerococcus urinae** Aguirre and Collins 1992b, 511[VP] (Effective publication: Aguirre and Collins 1992a, 403.)
   u.ri'nae. L. fem. n. *urinae* pertaining to urine.
   Two biotypes of *Aerococcus urinae* are recognized depending on their ability to hydrolyze esculin (Christensen et al., 1997). *DNA G+C content (mol%):* not known.
   Type strain: E2, ATCC 51268, CCUG 29291, CCUG 29564, CCUG 34223, CCUG 36881, CIP 104688, DSM 7446, NBRC 15544, NCFB 2893, NCIMB 702893, NCTC 12142.
   GenBank accession number (16S rRNA gene): M77819.

5. **Aerococcus urinaehominis** Lawson, Falsen, Ohlén and Coillins 2001a, 685[VP]

---

* *Editorial note.* Since this chapter was accepted for publication, two more *Aerococcus* species have been described, *Aerococcus suis* (Vela et al., 2007) and *Aerococcus urinaeequi* (Felis et al., 2005).

TABLE 91. Conventional tests which are useful in distinguishing *Aerococcus* species[a]

Test	1. *A. viridans*	2. *A. christensenii*	3. *A. sanguinicola*	4. *A. urinae*	5. *A. urinaehominis*
Maltose	d	–	+	–	+
Mannitol	d	–	–	+	–
Ribose	d	–	+	d	d
Sucrose	d	–	+	+	+
Trehalose	d	–	+	–	–
Esculin	+	–	+	d	+
PYRA	+	–	+	–	–
LAP	–	+	+	+	–
BE	d	–	d	–	–
NaCl 6.5%	+	–	+	+	+
Hippurate	d	+	+	+	+
VP	–	+	–	–	–

[a]Symbols: +, >85% positive; d, different strains give different reactions (16–84% positive); –, 0–15% positive. Abbreviations: PYRA, pyrrolidonylarylamidase; LAP, leucine amino peptidase; BE, Bile-esculin; VP, Voges–Proskauer; for conventional tests see Facklam and Elliott, (1995).

TABLE 92. Characteristics useful in distinguishing *Aerococcus* species using the API Rapid ID32 Strep system[a]

Test	1. *A. viridans*	2. *A. christensenii*	3. *A. sanguinicola*	4. *A. urinae*	5. *A. urinaehominis*
Acid from:					
D-Arabitol	–	–	–	d	–
Lactose	+	–	–	–	–
Mannitol	d	–	–	+	–
Maltose	+	–	+	–	+
MBDG	d	–	d	–	+
Ribose	d	–	–	d	+
Sorbitol	d	–	–	d	–
Sucrose	+	–	+	+	+
Trehalose	+	–	+	–	–
Production of:					
β-GLUR	–	–	+	+	+
PYR	d	–	+	–	–

[a]Symbols: +, >85% positive; d, different strains give different reactions (16–84% positive); –, 0–15% positive. Abbreviations: MBDG, methyl β-D-glucopyranoside; β-GLUR, β-glucuronidase; PYR, pyroglutamic acid arylamidase.

u.ri′nae.ho′mi.nis. L. fem. n. *urina* urine; L. n. *homo* man; L. gen. n. *hominis* of man. N.L. gen. n. *urinaehominis* pertaining to human urine, from which the organism was first isolated.

DNA G+C content (mol%): not reported.
Type strain: CCUG 42038B, CIP 106675.
GenBank accession number (16S rRNA gene): AJ278341.

Genus II. **Abiotrophia** (Bouvet, Grimonet and Grimont 1989) Kawamura, Hou, Sultana, Liu, Yamamoto and Ezaki 1995, 802[VP]

TAKAYUKI EZAKI AND YOSHIAKI KAWAMURA

A.bi.o.tro′phi.a. Gr. pref. *a-*, negative (un-); Gr. n. *bios* life; Gr. n. *trophe* nutrition; N.L. fem. n. *Abiotrophia* life-nutrition-deficiency.

Cells are mainly cocci, but pleomorphic ovoid cells, coccobacilli, and rod-shaped cells may occur. Pyridoxal hydrochloride (0.001%) or L-cysteine (0.01%) required for growth on blood agar. Nonsporeforming. Nonmotile. Gram-positive. Facultative anaerobes. Catalase- and oxidase-negative. **Lactic acid** is the major end product of glucose fermentation. Gas is not produced from glucose. Growth does not occur at 10 and 45°C or in the presence of 6.5% NaCl. Alpha-hemolytic on supplemented sheep blood agar. Pyrrolidonyl arylamidase and leucine arylamidase are positive. **Hippurate is not hydrolyzed. Argin-**ine dihydrolase is negative. Resistant to optochin and susceptible to vancomycin. The organisms have been isolated from human clinical specimens, such as blood with sepsis (Bouvet et al., 1989; Carey et al., 1975) and endocarditis (Bouvet, 1995; Bouvet et al., 1980, 1981).

DNA G+C content (mol%): 46.0–46.6.

Type species: **Abiotrophia defectiva** (Bouvet, Grimonet and Grimont 1989) Kawamura, Hou, Sultana, Liu, Yamamoto and Ezaki 1995, 802 (*Streptococcus defectives* Bouvet, Grimonet and Grimont 1989, 290.)

Further descriptive information

The phylogenetic position of the genus within the family *Aerococcaceae* is given in Figure 95. Only one species, *Abiotrophia defectiva*, is currentlyk included in the genus.

Lysine is the diamino acid at position 3 in the cell-wall peptidoglycan, and the peptide bridge consists of alanine or alanyl–alanine.

Abiotrophia is nutritionally fastidious: no growth or only slight growth occurs on a blood agar plate without supplements. Thiol compounds (0.01% L-cysteine is usually used) or vitamin B6 (0.001% pyridoxal hydrochloride is usually used) is required for growth. Growth occurs as satellite colonies adjacent to *Staphylococcus epidermidis* on blood agar.

Differentiation of the genus *Abiotrophia* from other genera

Characteristics differentiating *Abiotrophia defectiva* from other catalase-negative, Gram-positive cocci are listed in Table 93. *Abiotrophia defectiva* does not hydrolyze hippurate, and arginine dihydrolase-negative, whereas species in the genus *Granulicatella* (Collins and Lawson, 2000) are variable in these tests. *Abiotrophia defective* produces strong α-galactosidase and produces acid from sucrose, whereas species in the genus *Granulicatella* are variable in these biochemical tests. The mol% DNA G+C content of the genus *Granulicatella* is 36.6–37.4 mol%, which is lower than that for the genus *Abiotrophia* (46.0–46.6 mol%).

Taxonomic comments

Members of this genus were originally described as the nutritionally variant streptococci *Streptococcus defectiva* and *Streptococcus adjacens*, and the two species were transferred to different phylogenetic groups (family *Aerococaceae*. and family "*Carnobacteriaceae*". They were then transferred to a new genus, *Abiotrophia*, by Kawamura et al. (1995). Later, *Abiotrophia adiacens* was moved to a new genus, *Granulicatella* (Collins and Lawson, 2000) in the family "*Carnobacteriaceae*". The phylogenetic relationship between the two genera is shown in Figure 95.

Differentiation of the species of the genus *Abiotrophia*

Characteristics to differentiate *Abiotrophia defectiva* from closely related species of the genus *Granulicatella*, which also requires cysteine or pyridoxal hydrochloride to grow on a blood agar plate, are given in Table 94.

List of species of the genus *Abiotrophia*

1. **Abiotrophia defectiva** (Bouvet, Grimonet and Grimont 1989) Kawamura, Hou, Sultana, Liu, Yamamoto and Ezaki 1995, 802[VP] (*Streptococcus defectives* Bouvet, Grimonet and Grimont 1989, 290[VP])

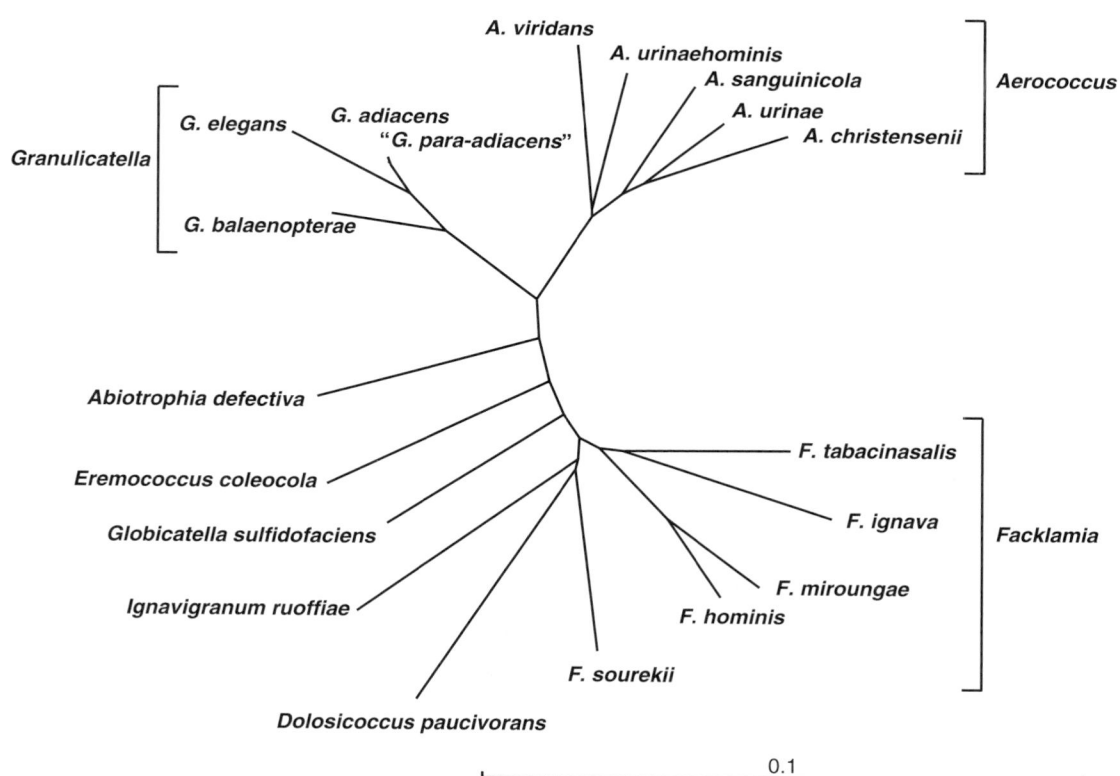

FIGURE 95. Phylogenetic position of genus *Abiotrophia* in the family *Aerococcaceae*.

TABLE 93. Phenotypic characteristics that differentiate the genus *Abiotrophia* from other catalase-negative, Gram-positive cocci[a]

Characteristic	*Abiotrophia*	*Aerococcus*	*Enterococcus*	*Gemella*	*Granulicatella*	*Lactococcus*	*Leuconostoc*	*Pediococcus*	*Streptococcus*	*Vagococcus*
Satellitism	+[b]	−	−	−	+[b]	−	−	−	−	−
Susceptibility to vancomycin (30 µg disk)[a]	Sensitive	Sensitive	Sensitive	Sensitive	Sensitive	Sensitive	Resistant	Resistant	Sensitive	Sensitive
Gas production from glucose	−	−	−	−	−	−	+	−	−	−
Leucine arylamidase	+	−	+	d	+	+	−	+	+	+
Pyrrolidonyl arylamidase	+	+	+	+	+	+	−	−	−[c]	+
Growth in presence of 6.5% NaCl	−	+	+	−	−	d	d	d	−[d]	+
Growth at 10°C	−	−	+	−	−	+	+	−	−	+
Growth at 45°C	−	+	+	−	−	−[e]	d	+	d	−
Motility	−	−	d	−	−	−	−	−	−	+

[a]Symbols: +, >85% positive; d, different strains give different reactions (16–84% positive); −, 0–15% positive; w, weak reaction; ND, not determined.
[b]More than 90% of the strains are positive.
[c]*Streptococcus pyogenes* strains exhibit activity.
[d]Some β-hemolytic streptococci growth in the presence of 6.5% NaCl.
[e]Some strains grow very slowly at 45°C.

TABLE 94. Characteristics that differentiate *Abiotrophia defectiva* from species of the genus *Granulicatella*[a]

Characteristic	*A. defectiva*	*G. adiacens*	*G. balaenopterae*	*G. elegans*
DNA G+C content (mol%)	46.0–46.6	36.6–37.4	37	37
Murein type	A1α	A4β	ND	A3α
Source:				
Human endocarditis, blood	+	+	−	+
Minke whale	−	−	+	−
Acid from:				
Sucrose	+	+	−	+
Trehalose	d	−	+	−
Tagatose	−	−	−	−
Pullulan	d	−	+	−
Hippurate hydrolysis	−	−	−	D
Arginine dihydrolase	−	−	+	+
α-Galactosidase	+	−	−	−
β-Glucuronidase	−	+	−	−
N-Acetyl-β-glucosaminidase	−	−	+	−

[a]Symbols: +, >85% positive; d, different strains give different reactions (16–84% positive); −, 0–15% positive; w, weak reaction; ND, not determined.

de.fec′ti.va. L. fem. adj. *defectiva* deficient.

The characteristics are as described for the genus and as listed in Table 93 and Table 94.

DNA G+C content (mol%): 46.0–46.6.

Type strain: SC10, ATCC 49176, CIP 103242, CCUG 27639, CCUG 27804, DSM 9849, LMG 14740.

GenBank accession number (16S rRNA gene): D50541.

Genus III. **Dolosicoccus** Collins, Rodriguez Jovita, Hutson, Falsen, Sjödén and Facklam 1999c, 1441[VP]

Matthew D. Collins

Do.lo′si.coc.cus. L. adj. *dolosus* crafty, deceptive; Gr. n. *kokkos* a grain; N.L. masc. n. *dolosicoccus* a deceptive coccus

Cells are ovoid, occurring singly, in pairs, or in or short chains. Gram-stain-positive and nonmotile. Nonspore-forming. Facultatively anaerobic and catalase and oxidase-negative. No growth at 10 or 45°C or in broth containing 6.5% NaCl. Weak acid production but no gas produced from glucose. Starch, esculin, gelatin, and urea are not hydrolyzed. Negative bile-esculin

reaction. Pyroglutamic acid arylamidase is produced but not leucine aminopeptidase. Voges–Proskauer-negative. Nitrate is not reduced to nitrite. **Vancomycin-sensitive.**

DNA G+C content (mol%): 40.5 (T_m).

Type species: **Dolosicoccus paucivorans** Collins, Rodriguez Jovita, Hutson, Falsen, Sjödén and Facklam 1999c, 1442^{VP}.

Further descriptive information

The genus contains only one species, *Dolosicoccus paucivorans*, and therefore the characteristics described refer to this species. Grows on 5% (v/v) horse or sheep blood agar producing a weak α-hemolytic reaction. Cells are nonpigmented. Pyruvate is not utilized and 0.04% tellurite is not tolerated. No dextrans or levans are formed on 5% sucrose. No acid or clot is formed in litmus milk. Using conventional heart infusion base medium, acid is produced from lactose (weak), maltose (weak), D-raffinose, ribose, and sucrose. Acid is not produced from L-arabinose, glycerol, melibiose, inulin, sorbitol, L-sorbose, or trehalose. Esculin is not hydrolyzed. Using conventional tests (Facklam and Elliott, 1995), pyroglutamic acid arylamidase is produced but leucine aminopeptidase is not. Using the API Rapid ID32 Strep system, acid may or may not be produced from lactose, maltose, mannitol, ribose, and sucrose. Acid is not produced from L-arabinose, D-arabitol, cyclodextrin, glycogen, melezitose, melibiose, methyl-β-D-glucopyranoside, pullulan, raffinose, sorbitol, tagatose, or trehalose. Alanine phenylalanine proline arylamidase, arginine dihydrolase, and pyroglutamic acid arylamidase may or may not be detected. Alkaline phosphatase, α-galactosidase, β-galactosidase, β-glucosidase, β-glucuronidase, glycyl tryptophan arylamidase, N-acetyl-β-glucosaminidase, β-mannosidase, and urease are not detected. Hippurate may or may not be hydrolyzed. Using the API ZYM system, esterase C4 (weak), ester lipase C8 (weak), and leucine arylamidase (weak) are detected. Alkaline phosphatase and acid phosphatase may or may not be detected. No activity is detected for cystine arylamidase, chymotrypsin, α-fucosidase, α-galactosidase, β-galactosidase, α-glucosidase, β-glucosidase, β-glucuronidase, N-acetyl-β-glucosaminidase, lipase C14, α-mannosidase, phosphoamidase, valine arylamidase, or trypsin.

Dolosicoccus paucivorans has been recovered from human blood (Collins et al., 1999c). Its habitat is not known.

Isolation procedures

Dolosicoccus paucivorans may be isolated at 37°C on rich agar-containing media containing 5% (v/v) animal blood. It can be

cultivated under aerobic or anaerobic conditions. There is no information on selective media for this species.

Maintenance procedures

Strains grow well on blood-based agars and in rich broths such as brain heart infusion or Todd–Hewitt at 37°C. For long-term preservation, strains can be stored on cryogenic beads at −70°C or lyophilized.

Taxonomic comments

The genus *Dolosicoccus* was proposed in 1999 for some Gram-stain-positive, catalase-negative, chain-forming, coccus-shaped organisms recovered from a human blood specimen (Collins et al., 1999c). Phylogenetically, the genus *Dolosicoccus* belongs to the order *Lactobacillales* and family *Aerococcaceae* of the phylum *Firmicutes*. Comparative 16S rRNA gene sequencing shows the genus *Dolosicoccus* represents a distinct subline among the Gram-stain-positive, catalase-negative cocci, with *Eremococcus coleocola*, *Globicatella sanguinis*, *Abiotrophia defectiva*, *Ignavigranum ruoffiae*, and *Facklamia* species as its nearest phylogenetic relatives (Collins et al., 1999c).

Differentiation of the genus *Dolosicoccus* from other genera

Using conventional phenotypic tests (Facklam and Elliott, 1995), the genus *Dolosicoccus* somewhat resembles *Dolosigranulum pigrum*, *Ignavigranum ruoffiae*, and *Facklamia* species in forming chains of cocci, being nonmotile, pyrrolidonyl-β-naphthylamide-positive, and catalase-negative, not growing at 10 or 45°C, having a negative bile-esculin reaction, and being susceptible to vancomycin. *Dolosicoccus paucivorans* differs from these taxa, however, in failing to produce leucine aminopeptidase and by not growing in 6.5% NaCl. *Dolosicoccus paucivorans* resembles *Globicatella sanguinis* in being leucine aminopeptidase-negative, but differs from this species by growing in 6.5% NaCl. In addition, *Globicatella sanguinis* differs markedly from *Dolosicoccus paucivorans* in fermenting a wide range of sugars (Collins et al., 1992). Classical phenotypic tests that are useful in distinguishing *Dolosicoccus paucivorans* from related Gram-stain-positive cocci are outlined in Table 95. *Dolosicoccus paucivorans* can also be readily identified using the commercially available API Rapid ID32 Strep system. Tests based on the API Rapid ID32 Strep system, which are useful in distinguishing *Dolosicoccus paucivorans* from related organisms, are shown in Table 96.

TABLE 95. Conventional phenotypic tests (Facklam and Elliott, 1995) which are useful in distinguishing *Dolosicoccus paucivorans* from related bacteria from human sources[a,b]

Test	Dolosicoccus paucivorans	Dolosigranulum pigrum	Facklamia hominis	Facklamia ignava	Facklamia languida	Facklamia sourekii	Globicatella sanguinis	Ignavigravum ruoffiae
LAP	−	+	+	+	+	+	−	+
Hippurate	d	−	+	+	−	+	+	−
Esculin	−	+	−	−	−	−	+	−
Sucrose	+	−	d	−	−	+	+	−
Sorbitol	−	−	−	−	−	+	d	−

[a]Symbols: +, >85% positive; d, different strains give different reactions (16–84% positive); −, 0–15% positive.
[b]Abbreviation: LAP, leucine aminopeptidase.

TABLE 96. API Rapid ID 32Strep system tests which are useful in distinguishing *Dolosicoccus paucivorans* from some related Gram-stain-positive, catalase-negative, coccus-shaped species from human and animal sources[a,b]

Test	*Dolosicoccus paucivorans*	*Eremococcus coleocola*	*Facklamia hominis*	*Facklamia ignava*	*Facklamia languida*	*Faclamia miroungae*	*Facklamia sourekii*	*Globicatella sanguinis*	*Globicatella sulfidifaciens*	*Ignavigranum ruoffiae*
Acid from:										
D-Arabitol	−	−	−	−	−	−	+	−	−	−
Glycogen	−	−	−	−	−	−	−	+	+	−
Lactose	d	−	−	−	−	−	−	d	−	−
Mannitol	d	−	−	−	−	−	+	+	−	d
Melibiose	−	−	−	−	−	−	−	+	+	−
MBDG	−	−	−	−	−	−	−	d	−	−
Raffinose	−	−	−	−	−	−	−	+	+	−
Ribose	d	−	−	−	−	−	−	+	−	−
Sorbitol	−	−	−	−	−	−	+	d	−	−
Sucrose	d	−	−	−	−	−	+	+	+	d
Trehalose	−	−	+	+	+	+	+	+	+	−
Activity for:										
ADH	d	+	+	−	−	+	−	−	−	+
APPA	d	−	+	+	−	+	−	+	+	−
α-Gal	−	−	+	−	−	−	−	+	+	−
β-Gal	−	−	+	−	−	−	−	d	d	−
NAG	−	+	−	−	−	w	−	−	−	−
URE	−	d	d	−	−	+	−	−	−	+
Hydrolysis of:										
Hippurate	d	+	+	+	−	−	+	+	d	−

[a]Symbols: +, >85% positive; d, different strains give different reactions (16–84% positive); −, 0–15% positive; w, weak or absent. Abbreviations: MBDG, methyl-β-D-glucoside; ADH, arginine dihydrolase; APPA, alanyl phenylalanine proline arylamidase; α-Gal, α-galactosidase; β-Gal, β-galactosidase; NAG, *N*-acetyl β-glucosaminidase; URE, urease.

List of species of the genus *Dolosicoccus*

1. **Dolosicoccus paucivorans** Collins, Rodriguez Jovita, Hutson, Falsen, Sjödén and Facklam 1995, 1442[VP]
   pau.ci'vo.rans. N.L. adj. *paucivorans* eating little, relating to the observation that the organism utilizes few carbohydrates.

   DNA G+C content (mol%): 40.5 (T_m).
   Type strain: 2992-95, ATCC BAA-56, CCUG 39307, CIP 106314.
   GenBank accession number (16S rRNA gene): AJ012666.

Genus IV. **Eremococcus** Collins, Rodriguez Jovita, Lawson, Falsen and Foster 1999f, 1383[VP]

MATTHEW D. COLLINS AND PAUL A. LAWSON

E.re.mo.coc'cus. Gr. adj. *eremos* lonely; Gr. n. *kokkos* a grain or berry; N.L. masc. n. *Eremococcus* a lonely or isolated coccus, referring to its distinct phylogenetic position.

Cells are cocci, some of which may elongated, and occur singly, in pairs, and in short chains. Gram-stain-positive and nonmotile. Nonsporeforming. **Facultatively anaerobic and catalase-negative. No growth at 10°C or in 10% NaCl. Gas is not produced. Acid is produced from D-glucose. Arginine dihydrolase and pyroglutamic acid arylamidase are produced. Hippurate is hydrolyzed but starch and gelatin are not hydrolyzed.** Nitrate is not reduced. Voges–Proskauer-negative. **The cell-wall murein is type L-lysine direct (A1α).**

   DNA G+C content (mol%): 40 (T_m).
   Type species: **Eremococcus coleocola** Collins, Rodriguez Jovita, Lawson, Falsen and Foster 1999f, 1383.

Further descriptive information

The genus contains only one species, *Eremococcus coleocola*, and therefore the characteristics provided below refer to this species. Grows on 5% (v/v) horse or sheep blood agar producing an α-hemolytic reaction. Cells are nonpigmented, nonmotile cocci which most commonly are arranged in short chains or pairs. Colonies are pinpoint, shiny, entire, circular, convex, and noncorroding on blood agar after 24 hours. Growth occurs at 42°C but not at 10°C; grows weakly in 6.5% NaCl but not in 10% NaCl. Acid is produced from glucose but from few other sugars. Acid may be produced from maltose and sometimes weakly from mannitol. Acid is not formed from L-arabinose, D-arabitol, cyclodextrin, glycogen, lactose, melibiose, melezitose, methyl-β-D-glucopyranoside, pullulan, raffinose, ribose, sorbitol, sucrose, tagatose, trehalose, or D-xylose. Activity is displayed for arginine dihydrolase, esterase C4, ester lipase C8, leucine arylamidase, *N*-acetyl-β-glucosaminidase, pyroglutamic acid arylamidase, and pyrrolydonyl arylamidase. Activity for alkaline phosphatase, acid phosphatase, urease, and valine arylamidase may or may not be detected. No activity is detected for alanine phenylalanine proline arylamidase, chymotrypsin, α-fucosidase, α-galactosidase, β-galactosidase, α-glucosidase, β-glucosidase, β-glucuronidase, glycine tryptophan arylamidase, α-mannosidase, β-mannosidase, lipase C14, pyrazinamidase, or trypsin. Hippurate is hydrolyzed, but esculin, starch, and gelatin are not hydrolyzed. Nitrate is not reduced and acetoin is not produced. The only known sources from which *Eremococcus coleocola* has been isolated are equine vaginal discharge and equine clitoral fossa (Collins et al., 1999f).

Isolation procedures

Eremococcus coleocola has been isolated from a vaginal discharge specimen and a clitoral fossa swab of horses using sheep blood agar at 37°C in air plus 5% CO_2. There is no information on enrichment or selective media for this species.

Maintenance procedures

Strains grow well on blood-based agar and in brain heart infusion and Todd–Hewitt broth at 37°C. For long- term preservation, strains can be maintained in media containing 15–20% glycerol at −70°C, or lyophilized.

Taxonomic comments

The genus *Eremococcus* was created in 1999 to accommodate a phylogenetically distinct catalase-negative, Gram-stain-positive, coccus-shaped bacterium originating from the reproductive tract of horses (Collins et al., 1999f). 16S rRNA gene sequencing shows that the genus belongs to the *Clostridium* subphylum of the *Firmicutes*. According to the revised roadmap for Volume 3 (Ludwig et al., this volume) *Eremococcus* has been placed in the new family *Aerococcaceae*, which includes *Aerococcus*, *Abiotrophia*, *Dolosicoccus*, *Facklamia*, *Ignavigranum*, and *Globicatella*. *Eremococcus coleocola* forms a distinct subline within this grouping and does not display a particularly close affinity with any of the aforementioned taxa.

Differentiation of the genus *Eremococcus* from other genera

Eremococcus coleocola can be readily distinguished from its nearest phylogenetic relatives using a combination of phenotypic tests.

Eremococcus coleocola can be readily distinguished from *Globicatella* species by its inability to ferment a broad range of sugars (Collins et al., 1999f). By contrast, globicatellae are relatively saccharolytic (Collins et al., 1992; Vandamme et al., 2001). In addition, globicatellae hydrolyze starch and are arginine dihydrolase-negative, whereas *Eremococcus coleocola* gives the opposite reactions. In terms of its restricted ability to ferment carbohydrates, *Eremococcus coleocola* more closely resembles some *Facklamia* species and *Ignavigranum ruoffiae*. However, *Eremococcus coleocola* can easily be distinguished from *Facklamia hominis* by failing to produce alanyl phenylalanine arylamidase and leucine arylamidase, whereas *Facklamia hominis* produces these enzymes. Similarly, *Facklamia ignava* differs from *Eremococcus coleocola* by producing alanyl phenylalanine arylamidase and being arginine dihydrolase-negative. In addition, unlike *Facklamia hominis* and *Facklamia ignava* which possess a type L-lysine-D-aspartic acid (variation A4α, for nomenclature see Schleifer and Kandler, 1972) cell-wall murein, *Eremococcus coleocola* contains an L-lysine-direct (variation A1α) type wall (Collins et al., 1997, 1998a). *Eremococcus coleocola* can be distinguished from *Facklamia languida* by hydrolyzing hippurate, producing arginine dihydrolase and failing to produce leucine arylamidase. *Facklamia languida* does not hydrolyze hippurate, is arginine dihydrolase-negative, and is leucine arylamidase-positive (Lawson et al., 1999a). *Facklamia sourekii* can be distinguished from *Eremococcus coleocola* by producing acid from a broader range of carbohydrates and by its positive leucine arylamidase and negative arginine dihydrolase reactions.

Eremococcus coleocola can be readily identified using API commercial kits. In particular, the API Rapid ID32 Strep code of 1, 0, 0, 0, 0, 3, 1, 0/1, 0, 0, 0/6 and the API Coryne code of 4, 1, 2, 0, 1, 3, 0 serve to distinguish *Eremococcus coleocola* from related species. Tests which are useful in distinguishing *Eremococcus coleocola* from its nearest phylogenetic relatives using the API Rapid ID32 Strep system are given in Table 96.

List of species of the genus *Eremococcus*

1. **Eremococcus coleocola** Collins, Rodriguez Jovita, Lawson, Falsen and Foster 1999f, 1383[VP]
co'le.o.co.la. Gr. n. *colea* vagina; L. masc. suff. *-cola* inhabitant; N.L. n. *coleocola* inhabitant of the vagina, referring to the isolation of the type strain.

DNA G+C content (mol%): 40 (T_m).
Type strain: M1832/95/2, ATCC BAA-57, CCUG 38207, CIP 106310.
GenBank accession number (16S rRNA gene): Y17780.

Genus V. **Facklamia** Collins, Falsen, Lemosy, Åkervall, Sjödén and Lawson 1997, 882[VP]

MATTHEW D. COLLINS AND PAUL A. LAWSON

Fack.lam′i.a. N.L. fem. n. *Facklamia* named after Richard R. Facklam, an American microbiologist.

Cells are ovoid, occurring in pairs, groups, or chains. Gram-stain-positive and nonmotile. Nonsporeforming. **Facultatively anaerobic and catalase-negative. No growth at 10 or 45°C. Gas is not produced. Acid is produced from D-glucose and some other sugars. Pyrrolidonylarylamidase and leucine aminopeptidase are produced.** Vancomycin-sensitive.
DNA G+C content (mol%): 40–42 (T_m).

Type species: **Facklamia hominis** Collins, Falsen, Lemosy, Åkervall, Sjödén and Lawson 1997, 882[VP].

Further descriptive information

Grows on 5% (v/v) horse or sheep blood agar producing a γ- or occasionally weak α-hemolytic reaction. Cells are nonpigmented, nonmotile cocci that most commonly are arranged in short chains or pairs; unlike other *Facklamia* species, cells of *Facklamia languida* are usually observed in clusters with little chaining. Acid is produced from glucose and other sugars, but reactions differ among species. *Facklamia sourekii* ferments a far greater range of sugars than the other species. None of the species produces acid from L-arabinose, cyclodextrin, glycogen, lactose, melibiose, melezitose, methyl-β-D-glucopyranoside, pullulan, raffinose, ribose, or tagatose. None of the species displays activity for alkaline phosphatase, β-glucosidase, β-glucuronidase, or β-mannosidase. Three species (*Facklamia hominis*, *Facklamia ignava*, and *Facklamia sourekii*) hydrolyze hippurate whereas the others do not. None of the *Facklamia* species hydrolyzes esculin or gelatin and none reduces nitrate to nitrite.

The cell-wall murein of *Facklamia hominis* and *Facklamia ignava* contains L-lysine as the dibasic amino acid, type L-lysine–D-aspartic acid (A4α, see Schleifer and Kandler, (1972) for nomenclature) (Collins et al., 1997, 1998a).

Facklamia species have been recovered from a variety of sources. Four of the six recognized species of the genus *Facklamia* have been isolated from human clinical sources (abscess, bone, cerebrospinal fluid, gall bladder, and vagina) (Collins et al., 1997, 1999b, 1998a; LaClaire and Facklam, 2000a; Lawson et al., 1999a). *Facklamia tabacinasalis* was originally isolated as a contaminant of powdered tobacco (snuff) (Collins et al., 1999a) and has also been isolated from sheep (Fernandez-Garazabal, Lawson, and Collins, unpublished). The only known strain of *Facklamia miroungae* was recovered from the nasal cavity of a juvenile southern elephant seal (*Mirounga leonina*). The habitats of *Facklamia* species are not known, but it has been speculated that the female genitourinary tract may be the natural habitat of human *Facklamia* species (LaClaire and Facklam, 2000a).

The antimicrobial susceptibilities of 18 strains of *Facklamia* species (*Facklamia hominis*, *Facklamia languida*, *Facklamia ignava*, *Facklamia sourekii*, and *Facklamia tabacinasalis*) were examined by LaClaire and Facklam (2000a). Seventeen per cent of strains were intermediate in resistance to penicillin, 44% were resistant to cefotaxime, and 33% presumptively resistant to cefuroxime. Twenty-two per cent were resistant to erythromycin and 33% to clindamycin. Twenty-eight per cent were presumptively resistant to trimethoprim-sulfamethoxazole and 17% to rifampin. There are appreciable differences in susceptibilities to antimicrobials among the various *Facklamia* species (LaClaire and Facklam, 2000a). Two out of five *Facklamia ignava* strains and one out of four *Facklamia hominis* strains were intermediate in resistance to penicillin, whereas all *Facklamia languida* and *Facklamia sourekii* strains were susceptible. One out of five *Facklamia ignava* strains, one out of three *Facklamia sourekii* strains, and all six *Facklamia languida* strains tested were resistant to cefotaxime whereas one out of four *Facklamia hominis* and five out of six *Facklamia languida* strains were presumptively resistant to cefuroxime. Some strains of *Facklamia ignava* (three out of five) and *Facklamia languida* (two out of six) were resistant to erythromycin whereas strains of *Facklamia hominis* (four strains) and *Facklamia sourekii* (three strains) were susceptible. Three out of four strains of *Facklamia hominis* were resistant to rifampin, but strains of *Facklamia ignava*, *Facklamia languida*, and *Facklamia sourekii* were susceptible.

Isolation procedures

Strains can be isolated on a variety of rich agar-containing media (such as heart infusion agar) supplemented with 5% (v/v) ani-

mal blood (sheep or horse). Strains grow at 37°C in ambient atmospheres or under anaerobic conditions.

Maintenance procedures

Strains can be maintained on agar media (such as heart infusion or Columbia agar) supplemented with blood (5% v/v). Strains grow in brain heart infusion broth. For long-term preservation, strains can either be stored at –70°C on cryogenic beads or lyophilized.

Taxonomic comments

The genus *Facklamia* was created in 1997 to accommodate a phylogenetically distinct group of catalase-negative, chain-forming, coccus-shaped organisms encountered in human clinical specimens. The genus was originally monospecific, containing the species *Facklamia hominis* (Collins et al., 1997). Subsequently, five other species have been assigned to the genus including *Facklamia ignava* (Collins et al., 1998a), *Facklamia languida* (Lawson et al., 1999a), and *Facklamia sourekii* (Collins et al., 1999b) from human clinical sources, *Facklamia miroungae* (Hoyles et al., 2001) from elephant seal, and *Facklamia tabacinasalis* (Collins et al., 1999a) from tobacco. According to the revised roadmap for Volume 3 (Ludwig et al., this volume) *Facklamia* has been placed in the new family *Aerococcaceae* in the order *Lactobacillaes* of the phylum *Firmicutes*. 16S rRNA gene sequencing shows that the nearest relative of *Facklamia* is *Ignavigranum ruoffiae*. Treeing analysis has shown that sometimes *Ignavigranum ruoffiae* branches close to the periphery of the genus *Facklamia*, whereas in other cases it phylogenetically intermingles with *Facklamia* species (Figure 96). *Facklamia languida* and *Facklamia miroungae* are phylogenetically closely related to the type species, *Facklamia*

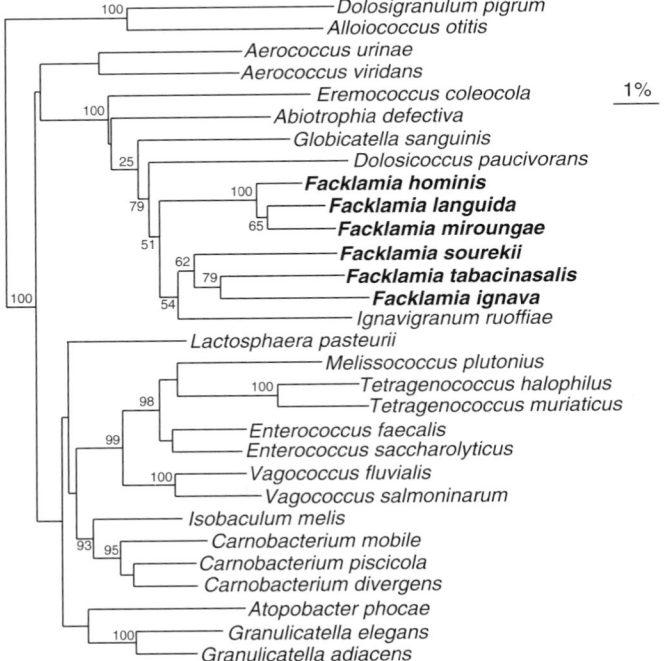

FIGURE 96. Neighbor-joining tree based on 16S rRNA sequences depicting the phylogenetic relationships of *Facklamia* species and close relatives. Bootstrap values determined from 500 replications.

hominis, and these three species form a robust phylogenetic cluster (Hoyles et al., 2001). It seems likely that in the future the genus *Facklamia* will be restricted to *Facklamia hominis* and its nearest relatives, with other *Facklamia* species possibly forming the nuclei of different genera/taxa.

Differentiation of the genus *Facklamia* from other genera

Identification of the genus *Facklamia* is not easy but can be aided using a combination of traditional tests (LaClaire and Facklam, 2000b). *Facklamia* species give positive pyrrolidonyl-β-naphthylamide (PYR) and leucine aminopeptidase reactions (LAP), grow in 6.5% NaCl broth, are susceptible to vancomycin, do not produce gas, give a negative bile-esculin reaction, are weak α- or γ-hemolytic on 5% (v/v) sheep blood agar, and fail to grow at 10°C and 45°C. This combination of reactions together with their cellular morphology is not, however, unique to the genus *Facklamia*. *Alloiococcus otitis*, *Dolosigranulum pigrum*, and *Ignavigranum ruoffiae* also share these characteristics (LaClaire

and Facklam, 2000b). *A. otitis* can be differentiated from the genus *Facklamia* by not growing anaerobically. *Facklamia* species can be differentiated from each other and from *Dolosigranulum pigrum* and *Ignavigranum ruoffiae* using additional traditional tests (Facklam and Elliott, 1995) including deamination of arginine, hydrolysis of hippurate and esculin, and acid production from sorbitol and sucrose broths (LaClaire and Facklam, 2000b) (see Table 97).

Differentiation of the species of the genus *Facklamia*

Species of the genus *Facklamia* can be distinguished from each other using the conventional biochemical tests listed in Table 97. *Facklamia* species can also be distinguished readily from each other using the commercially available API Rapid ID32 Strep system (Collins et al., 1999b, 1999a; Hoyles et al., 2001; Lawson et al., 1999a). A summary of the most useful tests for identifying *Facklamia* species using the API Rapid ID32 Strep system is shown in Table 98.

TABLE 97. Phenotypic characteristics of *Facklamia* species, *Ignavigranum ruoffiae*, and *Dolosigranulum pigrum* as determined by conventional biochemical tests[a,b]

Test	D. pigrum	F. hominis	F. ignava	F. miroungae	F. languida	F. sourekii	F. tabacinasalis	I. ruoffiae
Arginine	−	+	−	+	−	−	−	−
Hippurate	−	+	+	−	−	+	−	−
Esculin	+	−	−	−	−	−	−	−
Sucrose	−	d	−	−	−	+	+	−
Sorbitol	−	−	−	−	−	+	+	−

[a]Symbols: +, >85% positive; d, different strains give different reactions (16–84% positive); −, 0–15% positive.
[b]Results from LaClaire and Facklam (2000b).

TABLE 98. Tests distinguishing species of the genus *Facklamia* and *Ignavigranum ruoffiae* using the API Rapid ID32 Strep system[a]

Test	F. hominis	F. ignava	F. languida	F. miroungae	F. sourekii	F. tabacinasalis	I. ruoffiae
Acid from:							
D-Arabitol	−	−	−	−	+	−	−
Mannitol	−	−	−	−	+	−	−
Maltose	−	d	−	−	+	−	d
Sorbitol	−	−	−	−	+	−	−
Sucrose	−	−	−	−	+	−	D
Trehalose	−	+	+	+	+	−	−
Hydrolysis of:							
Hippurate	+	+	−	−	+	−	−
Activity for:							
ADH	+	−	−	+	−	−	+
APPA	+	+	−	+	−	−	−
α-Gal	+	−	−	−	−	+	−
β-Gal	+	−	−	−	−	v	−
GTA	d	d	+	+	−	−	−
NAG	−	−	−	+	−	−	−
PYRA	d	+	d	+	+	−	−
Urease	d	−	−	+	−	−	+

[a]Symbols: +, >85% positive; d, different strains give different reactions (16–84% positive); −, 0–15% positive. Abbreviations: ADH, arginine dihydrolase; APPA, alanine phenylalanine proline arylamidase; α-Gal, α-galactosidase; β-Gal, β-galactosidase; GTA, glycine tryptophan arylamidase; NAG, *N*-acetyl β-glucosaminidase; PYRA, pyroglutamic acid arylamidase; all species fail to produce acid from L-arabinose, cyclodextrin, glycogen, lactose, melibiose, melezitose, methyl-β-D-glucopyranoside, pullulan, raffinose, ribose, and tagatose. None display activity for alkaline phosphatase, β-glucosidase, β-glucuronidase, or β-mannosidase.

List of species of the genus *Facklamia*

1. **Facklamia hominis** Collins, Falsen, Lemosy, Åkervall, Sjödén and Lawson 1997, 882[VP]

   ho'mi.nis. L. gen. n. *hominis* of humans, from which the organisms were first isolated.

   DNA G+C content (mol%): 41.0.

   Type strain: ATCC 700628, CCUG 36813, CIP 105962, LMG 18980.

   GenBank accession number (16S rRNA gene): Y10772.

2. **Facklamia ignava** Collins, Lawson, Monasterio, Falsen, Sjödén and Facklam 1998a, 1083[VP] (Effective publication: Collins, Lawson, Monasterio, Falsen, Sjödén and Facklam 1998b, 2148.)

   ig.na.va. L. fem. adj. *ignava* lazy, unreactive.

   DNA G+C content (mol%): 42.0.

   Type strain: 164–97, ATCC 700631, CCUG 37419, CIP 105583, LMG 18981.

   GenBank accession number (16S rRNA gene): Y15716.

3. **Facklamia languida** Lawson, Collins, Falsen, Sjödén and Facklam 1999b, 935[VP] (Effective publication: Lawson, Collins, Falsen, Sjödén and Facklam 1999a, 1164.)

   lan.guida. L. adj. *languidus* languid; L. fem. adj. *languida* pertaining to the lack of activity of the organisms in biochemical tests.

   DNA G+C content (mol%): not reported.

   Type strain: 1144-97, CCUG 37842, CIP 105964.

   GenBank accession number (16S rRNA gene): Y18053.

4. **Facklamia miroungae** Hoyles, Foster, Falsen, Thomson and Collins 2001, 1403[VP]

   mi.roung'ae. N.L. gen. n. *miroungae* of *Mirounga*, named because the species was first isolated from the southern elephant seal, *Mirounga leonina.*

   DNA G+C content (mol%): not reported.

   Type strain: A/G13/99/2, CCUG 42728, CIP 106764.

   GenBank accession number (16S rRNA gene): AJ277381.

5. **Facklamia sourekii** Collins, Hutson, Falsen and Sjödén 1999b, 637[VP]

   schou.rek.i.i. (Czech pronunciation) N.L. gen. n. *sourekii* of Šourek, named after the Czech microbiologist Jiři Šourek.

   *DNA G+C content (mol%):*41.5.

   Type strain: SS 1019, ATCC 700629, de Moor M13582, CCUG 28783 A, CDC1, CIP 105940, CNCTC 2/84, LMG 18982.

   GenBank accession number (16S rRNA gene): Y17312.

6. **Facklamia tabacinasalis** Collins, Hutson, Falsen and Sjödén 1999a, 1249[VP]

   ta.ba.ci.na'sa.lis. N.L. n. *tabacum* tobacco; L. adj. *nasalis* pertaining to the nose; N.L. gen. fem. adj. *tabacinasalis* of snuff.

   DNA G+C content (mol%): 40.0 (T_m).

   Type strain: ATCC 700838, CCUG 30090, CIP 106117.

   GenBank accession number (16S rRNA gene): Y17820.

Genus VI. **Globicatella** Collins, Aguirre, Facklam, Shallcross and Williams 1995, 418[VP] (Effective publication: Collins, Aguirre, Facklam, Shallcross and Williams 1992, 436.)

MATTHEW D. COLLINS

Glo.bi.ca.tel'la L. n. *globus* round body, sphere; L. fem. n. *catella* small chain; N.L. fem. n. *Globicatella* a short chain made up of spheres.

Cells are ovoid, occurring singly, in pairs or short chains. Gram-positive, but sometimes cells stain Gram-negative. Nonmotile and nonsporeforming. **Facultatively anaerobic and catalase-negative. Grows in broth containing 6.5% NaCl. No growth at 10 or 45°C. Gas is not produced. Acid is produced from D-glucose and some other sugars. Pyroglutamic acid arylamidase is produced. Pyrrolidonyl arylamidase and leucine aminopeptidase may or may not be produced. Arginine dihydolase and urease are not produced.** Pyruvate is not utilized. **Voges–Proskauer-negative. Vancomycin-sensitive. Cell-wall murein is based on L-Lysine (type Lys-direct).**

DNA G+C content (mol%): 35.7–37 (T_m).

Type species: **Globicatella sanguinis** corrig. Collins, Aguirre, Facklam, Shallcross and Williams 1995, 418[VP] (Effective publication: Collins, Aguirre, Facklam, Shallcross and Williams 1992, 436.).

Further descriptive information

The genus currently contains two species, *Globicatella sanguinis* and *Globicatella sulfidifaciens*. Globicatellae grow on 5% (v/v) horse or sheep blood agar forming small viridans streptococcus-like colonies and producing a weak α-hemolytic reaction. *Globicatella sanguinis* does not tolerate tellurite (0.04%) or tetrazolium (0.25%). Globicatellae do not grow at 10°C and most do not grow at 45°C. Most strains grow in 6.5% NaCl, but occasional strains do not. Bile-esculin reaction for *Globicatella sanguinis* is variable (Facklam and Elliott, 1995). *Globicatella sulfidifaciens* does not grow on bile-esculin agar (Vandamme et al., 2001). Both *Globicatella* species produce acid from glucose, glycogen, maltose, melibiose (most strains), raffinose, trehalose, and sucrose. Most strains of *Globicatella sanguinis* form acid from mannitol and ribose whereas strains of *Globicatella sulfidifaciens* fail to produce acid from these substrates. *Globicatella sanguinis* is reported to form acid from maltotriose and methyl β-glucoside, whereas *Globicatella sulfidifaciens* does not (Vandamme et al., 2001). *Globicatella sulfidifaciens* ferments inulin, but *Globicatella sanguinis* does not (Vandamme et al., 2001). Acid production from lactose, methyl β-D-glucopyranoside and sorbitol is variable in *Globicatella sanguinis* but negative in *Globicatella sulfidifaciens*. Neither species produces acid from D-arabitol, cyclodextrin, melezitose, or D-tagatose. Starch and esculin are hydrolyzed by both species. *Globicatella sanguinis* hydrolyzes hippurate, but *Globicatella sulfidifaciens* gives variable results. Both *Globicatella* species produce alanine phenylalanine proline arylamidase, α-galactosidase, β-glucosidase, and pyroglutamic acid arylamidase. Activity for β-galactosidase and pyrrolidonyl arylamidase may or may not be detected. *Globicatella sulfidifaciens* displays activity for β-glucuronidase whereas most strains of *Globicatella sanguinis* do not show activity for this enzyme. Neither species shows activity for arginine dihydrolase, alkaline phosphatase, glycyl tryptophan arylamidase, N-acetyl-β-glucosaminidase, or urease. *Globicatella sulfidifaciens* produces hydrogen sulfide in Kligler's iron agar whereas *Globicatella sanguinis* does not (Vandamme et al., 2001).

Globicatella sulfidifaciens is resistant to neomycin and to the macrolide antibiotics erythromycin, clindamycin, and lincomycin (Vandamme et al., 2001).

Globicatella sanguinis has been isolated from a variety of human clinical specimens, including blood cultures (patients with sepsis, meningitis, and bacteremia), urine of patients with urinary tract infections, cerebrospinal fluid, and wounds (Collins et al., 1992; Facklam and Elliott, 1995). This species has been isolated in pure culture from blood and brain samples from lambs with meningoencephalitis (Vela et al., 2000). The habitat of *Globicatella sanguinis* is not known. *Globicatella sulfidifaciens* has been isolated from purulent joint and lung infections in calves, from a lung lesion of a sheep, and from a joint infection of a pig (Vandamme et al., 2001).

Isolation procedures

Globicatella species may be isolated on rich agar-containing media, such as brain heart infusion agar, with or without animal blood at 37°C. They can be cultivated under aerobic or anaerobic conditions. There is no information on selective media for these organisms.

Maintenance procedures

Strains can be maintained on media such as brain heart infusion agar or Columbia agar containing 5% (v/v) blood. Strains can be stored on cryogenic beads at –70°C or lyophilized for long-term preservation.

Taxonomic comments

The genus *Globicatella* was described to accommodate some coccus-shaped strains which somewhat resembled viridans-like streptococci and which also shared properties in common with aerococci (Collins et al., 1992). Until recently, *Globicatella* was monospecific containing the single species *Globicatella sanguinis*, but a second species, *Globicatella sulfidifaciens*, recovered from purulent infections in domestic animals, has been assigned to the genus by Vandamme et al. (2001). *Globicatella sanguinis* and *Globicatella sulfidifaciens* are genetically closely related exhibiting approximately 99.2% 16S rRNA sequence similarity (Vandamme et al., 2001). Very high levels of chromosomal DNA–DNA relatedness (mean DNA binding value of 68%) have been reported between the two species. The DNA-binding value is very close to the 70% guideline generally used for species delineation (Wayne et al., 1987). Although it is highly likely that *Globicatella sanguinis* and *Globicatella sulfidifaciens* are different species, the high DNA relatedness between these species and the method used (photobiotin-labeled probes in microplate wells as described by Ezaki et al., 1989) indicate that this finding should be re-examined using more high-fidelity techniques including ΔT_m determinations. Phylogenetically, the genus *Globicatella* forms a distinct subline among the catalase-negative, Gram-stain-positive cocci and is far removed from both streptococci and aerococci. The closest phylogenetic relatives of *Globicatella sanguinis* and *Globicatella sulfidifaciens* are *Facklamia* species, *Ignavigranum ruoffiae*, *Dolosicoccus paucivorans*, *Eremococcus coleocola*, and *Abiotrophia defectiva* (Collins et al., 1999c, 1999f).

Differentiation of the genus *Globicatella* from other genera

Globicatella sanguinis can be distinguished from viridans streptococci in being pyrrolidonyl-β-naphthylamide-positive and leucine aminopeptidase-negative and in growing in broth containing 6.5% NaCl (Facklam and Elliott, 1995). Aerococci also possess these characteristics but can be distinguished from *Globicatella* by cellular morphology. Aerococci form pairs and tetrad cellular arrangements, whereas *Globicatella sanguinis* and *Globicatella sulfidifaciens* form short chains of cocci. *Globicatella sanguinis* does not produce leucine aminopeptidase, which readily distinguishes it from *Ignavigranum ruoffiae* and *Facklamia* species. *Globicatella sanguinis* and *Globicatella sulfidifaciens* hydrolyze esculin and starch, and grow in broth containing 6.5% NaCl, which serves to distinguish them from *Dolosicoccus paucivorans* which is negative for these tests. Both *Globicatella* species also differ markedly from *Dolosicoccus paucivorans*, *Ignavigranum ruoffiae*, and *Facklamia* species in fermenting a wide range of carbohydrates.

Differentiation of the species of the genus *Globicatella*

Globicatella sulfidifaciens differs from *Globicatella sanguinis* in several biochemical reactions, such as fermenting inulin, failing to produce acid from mannitol, ribose, methyl β-glucoside, and maltotriose, and by producing β-glucuronidase. In addition, *Globicatella sulfidifaciens* is highly unusual in producing H_2S in Kligler's iron agar stabs (Vandamme et al., 2001). *Globicatella sanguinis* and *Globicatella sulfidifaciens* can readily be distinguished from each other by using the commercially available API Rapid ID32 Strep system. Useful distinguishing tests are listed in Table 99.

TABLE 99. Tests distinguishing *Globicatella sanguinis* and *Globicatella sulfidifaciens*[a]

Test	*G. sanguinis*	*G. sulfidifaciens*
Acid from:		
Mannitol	+	–
Ribose	+	–
Sorbitol	d	–
Lactose	d	–
Production of:		
H_2S[b]	–	+
Production of:		
β-Glucuronidase	–	+

[a]Symbols: +, >85% positive; d, different strains give different reactions (16–84% positive); –, 0–15% positive. Tests performed using the API Rapid ID32 Strep system.

[b]H_2S production in Kligler's iron agar.

List of species of the genus *Globicatella*

1. **Globicatella sanguinis** corrig. Collins, Aguirre, Facklam, Shallcross and Williams 1995, 418[VP] (Effective publication: Collins, Aguirre, Facklam, Shallcross and Williams 1992, 436.) san'gui.nis. L. gen. n. *sanguinis* of the blood.

DNA G+C content (mol%): 35.7 (HPLC).

Type strain: 1152-78, ATCC 51173, CCUG 32999, CIP 107044, DSM 7447, NBRC 15551, LMG 18987, NCIMB 702835.

GenBank accession number (16S rRNA gene): S50214.

2. **Globicatella sulfidifaciens** Vandamme, Hommez, Snauwaert, Hoste, Cleenwerck, Lefebvre, Vancanneyt, Swings, Devriese and Haesebrouck 2001, 1748[VP] sul'fi.di.fa'i.ens. L. n. *sulfidum* sulfide; L. v. *facere* to produce; N.L. part adj. *sulfidifaciens* sulfide-producing.

DNA G+C content (mol%): 35.8 (HPLC).

Type strain: GEM 604, CCUG 44365, CIP 107175, LMG 18844.

GenBank accession number (16S rRNA gene): AJ297627.

Genus VII. **Ignavigranum** Collins, Lawson, Monasterio, Falsen, Sjödén and Facklam 1999e, 100[VP]

MATTHEW D. COLLINS

Ig.na.vi.gra'num. L. adj. *ignavus* lazy, non-reacting; L. neut. n. *granum* grain, kernel; N.L. neut. n. *Ignavigranum* lazy grain.

Cells are ovoid, occurring in pairs, groups, or chains. Grampositive and nonmotile. No spores are formed. **Facultatively anaerobic and catalase-negative. Grows in 6.5% NaCl broth. No growth at 10°C; most strains do not grow at 45°C.** Gas is not produced. **Negative bile-esculin reaction.** Acid is produced from D-glucose and some other sugars. **Pyrrolidonylarylamidase and leucine aminopeptidase are produced. Vancomycin-sensitive. Cell-wall murein is based on L-lysine (type Lys-direct).**

DNA G+C content (mol%): 40 (T_m).

Type species: **Ignavigranum ruoffiae** Collins, Lawson, Monasterio, Falsen, Sjödén and Facklam 1999e, 100[VP]

Further descriptive information

The genus contains only one species, *Ignavigranum ruoffiae*, and therefore the characteristics provided here refer to this species. Grows on 5% (v/v) horse or sheep blood agar producing a weak α- or γ-hemolytic reaction. Cells are nonpigmented. Esculin and hippurate are not hydrolyzed. Voges–Proskauer negative. Acid is produced from very few carbohydrates.

Using the API Rapid ID32 Strep system, a few strains may produce acid from mannitol and sucrose. Acid is not produced from L-arabinose, D-arabitol, cyclodextrin, glycogen, lactose, maltose, melezitose, melibiose, pullulan, raffinose, ribose, sorbitol, tagatose, or trehalose. Hippurate is not hydrolyzed. Arginine dihydrolase, pyroglutamic acid arylamidase, and urease are detected, but alanine phenylalanine proline arylamidase, alkaline phosphatase, α-galactosidase, β-galactosidase, β-glucosidase, β-glucuronidase, glycyl tryptophan arylamidase, N-acetyl-β-glucosaminidase, and β-mannosidase are not detected. Using the API ZYM system, leucine arylamidase is detected. Activity for acid phosphatase, esterase C4, ester lipase C8, and phosphoamidase may or may not be detected. No activity is found for alkaline phosphatase, cystine arylamidase, chymotrypsin, α-fucosidase, α-galactosidase, β-galactosidase, α-glucosidase, β-glucosidase, β-glucuronidase, N-acetyl-β-glucosaminidase, lipase C14, α-mannosidase, valine arylamidase, or trypsin.

Ignavigranum has been isolated from human clinical specimens (blood, wound, ear abscess). Habitat is not known.

Isolation procedures

Strains can be isolated on a variety of rich agar-containing media (such as heart infusion agar) supplemented with 5% (v/v) animal blood (sheep or horse). Strains grow at 37°C in ambient atmospheres or under anaerobic conditions.

Maintenance procedures

Strains can be maintained on agar media (such as heart infusion or Columbia agar) supplemented with blood (5% v/v). Strains grow in brain heart infusion broth. For long-term preservation, strains can be stored at −70°C on cryogenic beads or lyophilized.

Taxonomic comments

The genus *Ignavigranum* was proposed in 1999 for some Gram-positive, catalase-negative, chain-forming cocci recovered from human clinical specimens (Collins et al., 1999e). According to the revised roadmap for Volume 3 (Ludwig et al., this volume) *Ignavigranum* has been placed in the new family *Aerococcaceae* in the phylum *Firmicutes*. Although *Ignavigranum ruoffiae* is a distinct species, comparative 16S rRNA gene sequencing shows it is closely related to the genus *Facklamia* (Hoyles et al., 2001). Treeing analysis shows the branching position of *Ignavigranum ruoffiae* is uncertain, with the species sometimes branching close to the periphery of the genus *Facklamia*, and in other cases phylogenetically intermingling with *Facklamia* species. Although *Ignavigranum ruoffiae* branches within the bounds of the genus *Facklamia* as presently defined, *Ignavigranum* probably merits separate genus status, whereas the genus *Facklamia* is in need of revision and should be restricted to the type species *Facklamia hominis* and its closest relatives (*Facklamia languida* and *Facklamia microungae*) (Hoyles et al., 2001).

Differentiation of the genus *Ignavigranum* from related genera

Ignavigranum resembles *Dolosigranulum pigrum* and *Facklamia* species in being nonmotile, pyrrolidonyl-β-naphthylamide, and leucine aminopeptidase-positive, growing in 6.5% NaCl broth, not growing at 10 or 45°C, having a negative bile-esculin reaction, being susceptible to vancomycin, and showing α- or γ-hemolytic reactions on sheep blood (LaClaire and Facklam, 2000b). It can, however, be distinguished from these taxa by using a combination of classical phenotypic tests (Facklam and Elliott, 1995) as outlined in Table 97. In addition, *Ignavigranum ruoffiae* can be identified readily by using the commercially available API Rapid ID32 Strep system. Miniaturized API Rapid ID32 Strep tests that are useful in distinguishing *Ignavigranum ruoffiae* from *Facklamia* species are shown in Table 98.

List of species of the genus *Ignavigranum*

1. **Ignavigranum ruoffiae** Collins, Lawson, Monasterio, Falsen, Sjödén and Facklam 1999e, 100[VP]
   ru.off′iae. N.L. gen. n. *ruoffiae* named after Kathryn L. Ruoff, an American microbiologist.

DNA G+C content (mol%): 40 (T_m).
Type strain: 1607-97, ATCC 700630, CCUG 37658, CIP 105896.
GenBank accession number (16S rRNA gene): Y16426.

References

Aguirre, M. and M.D. Collins. 1992a. Phylogenetic analysis of some *Aerococcus*-like organisms from urinary tract infections: description of *Aerococcus urinae* sp. nov. J. Gen. Microbiol. *138*: 401–405.

Aguirre, M. and M.D. Collins. 1992b. *In* Validation of the publication of new names and new combinations previously effectively published outside the IJSB. List no. 42. Int. J. Syst. Bacteriol. *42*: 511.

Ballester, J.M., M. Ballester and J.P. Belaich. 1980. Purification of the viridicin produced by *Aerococcus viridans*. Antimicrob. Agents Chemother. *17*: 784–788.

Bosley, G.S., P.L. Wallace, C.W. Moss, A.G. Steigerwalt, D.J. Brenner, J.M. Swenson, G.A. Hebert and R.R. Facklam. 1990. Phenotypic characterization, cellular fatty acid composition, and DNA relatedness of aerococci and comparison to related genera. J. Clin. Microbiol. *28*: 416–421.

Bouvet, A., A. Ryter, C. Frehel and J.F. Acar. 1980. Nutritionally deficient streptococci: electron microscopic study of 14 strains isolated in bacterial endocarditis. Ann. Microbiol. (Paris) *131B*: 101–120.

Bouvet, A., I. van de Rijn and M. McCarty. 1981. Nutritionally variant streptococci from patients with endocarditis: growth parameters in a semisynthetic medium and demonstration of a chromophore. J. Bacteriol. *146*: 1075–1082.

Bouvet, A., F. Grimont and P.A.D. Grimont. 1989. *Streptococcus defectivus* sp. nov. and *Streptococcus adjacens* sp. nov., nutritionally variant streptococci from human clinical specimens. Int. J. Syst. Bacteriol. *39*: 290–294.

Bouvet, A. 1995. Human endocarditis due to nutritionally variant streptococci: *Streptococcus adjacens* and *Streptococcus defectivus*. Eur. Heart. J. *16 Suppl B*: 24–27.

Bru, P., C. Manuel, C. Iacono, A. Vaillant, C. Malmejac and J. Houel. 1986. Indications and results of surgery in native valve infectious endocarditis. Apropos of 104 surgically-treated cases. Arch. Mal. Coeur Vaiss. *79*: 47–51.

Buu-Hoi, A., C. Le Bouguenec and T. Horaud. 1989. Genetic basis of antibiotic resistance in *Aerococcus viridans*. Antimicrob. Agents Chemother. *33*: 529–534.

Carey, R.B., K.C. Gross and R.B. Roberts. 1975. Vitamin B6-dependent *Streptococcus mitior* (*mitis*) isolated from patients with systemic infections. J. Infect. Dis. *131*: 722–726.

Christensen, J.J., B. Korner and H. Kjaergaard. 1989. *Aerococcus*-like organism: an unnoticed urinary tract pathogen. APMIS *97*: 539–546.

Christensen, J.J., E. Gutschik, A. Friis-Möller and B. Korner. 1991. Urosepticemia and fatal endocarditis caused by *Aerococcus*-like organisms. Scand. J. Infect. Dis. *23*: 717–721.

Christensen, J.J., B. Korner, J.B. Casals and N. Pringler. 1996. *Aerococcus*-like organisms: use of antibiograms for diagnostic and taxonomic purposes. J. Antimicrob. Chemother. *38*: 253–258.

Christensen, J.J., A.M. Whitney, L.M. Teixeira, A.G. Steigerwalt, R.R. Facklam, B. Korner and D.J. Brenner. 1997. *Aerococcus urinae*: intraspecies genetic and phenotypic relatedness. Int. J. Syst. Bacteriol. *47*: 28–32.

Collins, M.D., A.M. Williams and S. Wallbanks. 1990. The phylogeny of *Aerococcus* and *Pediococcus* as determined by 16S rRNA sequence analysis: description of *Tetragenococcus* gen. nov. FEMS Microbiol. Lett. *58*: 255–262.

Collins, M.D., M. Aguirre, R.R. Facklam, J. Shallcross and A.M. Williams. 1992. *Globicatella sanguis* gen. nov., sp. nov., a new Gram-positive catalase-negative bacterium from human sources. J. Appl. Bacteriol. *73*: 433–437.

Collins, M.D., M. Aguirre, R.R. Facklam, J. Shallcross and A.M. Williams. 1995. *In* Validation of the publication of new names and new combinations previously effectively published outisde the IJSB. List no. 53. Int. J. Syst. Bacteriol. *45*: 418.

Collins, M.D., E. Falsen, J. Lemozy, E. Äkervall, B. Sjödén and P.A. Lawson. 1997. Phenotypic and phylogenetic characterization of some *Globicatella*-like organisms from human sources: Description of *Facklamia hominis* gen. nov., sp. nov. Int. J. Syst. Bacteriol. *47*: 880–882.

Collins, M.D., P.A. Lawson, R. Monasterio, E. Falsen, B. Sjödén and R.R. Facklam. 1998a. *Facklamia ignava* sp. nov., isolated from human clinical specimens. J. Clin. Microbiol. *36*: 2146–2148.

Collins, M.D., P.A. Lawson, R. Monasterio, E. Falsen, B. Sjödén and R.R. Facklam. 1998b. In Validation of publication of new names and new combinations previously effectively published outside the IJSB. List no. 67. Int. J. Syst. Bacteriol. *48*: 1083–1084.

Collins, M.D., R.A. Hutson, E. Falsen and B. Sjödén. 1999a. *Facklamia tabacinasalis* sp. nov., from powdered tobacco. Int. J. Syst. Bacteriol. *49*: 1247–1250.

Collins, M.D., R.A. Hutson, E. Falsen and B. Sjödén. 1999b. *Facklamia sourekii* sp. nov., isolated from human sources. Int. J. Syst. Bacteriol. *49*: 635–638.

Collins, M.D., M.R. Jovita, R.A. Hutson, E. Falsen, B. Sjödén and R.R. Facklam. 1999c. *Dolosicoccus paucivorans* gen. nov., sp. nov., isolated from human blood. Int. J. Syst. Bacteriol. *49*: 1439–1442.

Collins, M.D., M.R. Jovita, R.A. Hutson, M. Ohlén and E. Falsen. 1999d. *Aerococcus christensenii* sp. nov., from the human vagina. Int. J. Syst. Bacteriol. *49*: 1125–1128.

Collins, M.D., P.A. Lawson, R. Monasterio, E. Falsen, B. Sjödén and R.R. Facklam. 1999e. *Ignavigranum ruoffiae* sp. nov., isolated from human clinical specimens. Int. J. Syst. Bacteriol. *49*: 97–101.

Collins, M.D., M. Rodriguez Jovita, P.A. Lawson, E. Falsen and G. Foster. 1999f. Characterization of a novel Gram-positive, catalase-negative coccus from horses: description of *Eremococcus coleocola* gen. nov., sp. nov. Int. J. Syst. Bacteriol. *49*: 1381–1385.

Collins, M.D. and P.A. Lawson. 2000. The genus *Abiotrophia* (Kawamura *et al.*) is not monophyletic: proposal of *Granulicatella* gen. nov., *Granulicatella adiacens* comb. nov., *Granulicatella elegans* comb. nov and *Granulicatella balaenopterae* comb. nov. Int. J. Syst. Evol. Microbiol. *50*: 365–369.

Colman, G. 1967. *Aerococcus*-like organisms isolated from human infections. J. Clin. Pathol. *20*: 294–297.

Deibel, R.H. and C.F. Niven, Jr. 1960. Comparative study of *Gaffkya homari*, *Aerococcus viridans*, tetrad-forming cocci from meat curing brines, and the genus *Pediococcus*. J. Bacteriol. *79*: 175–180.

Devriese, L.A., J. Hommez, H. Laevens, B. Pot, P. Vandamme and F. Haesebrouck. 1999. Identification of aesculin-hydrolyzing streptococci, lactococci, aerococci and enterococci from subclinical intramammary infections in dairy cows. Vet. Microbiol. *70*: 87–94.

Evans, J.B. 1986. Genus *Aerococcus* Williams, Hirch and Cowan 1953, 475. *In* Sneath, Mair, Sharpe and Holt (Editors), Bergey's Manual of Systematic Bacteriology, Vol. 2. The Williams & Wilkins Co., Baltimore, p. 1080.

Ezaki, T., Y. Hashimoto and E. Yabuuchi. 1989. Fluorometric DNA-DNA hybridization in microdilution wells as an alternative to membrane filter hybridization in which radioisotopes are used to determine genetic relatedness among bacterial strains. Int. J. Syst. Bacteriol. *39*: 224–229.

Facklam, R. and J.A. Elliott. 1995. Identification, classification, and clinical relevance of catalase-negative, gram-positive cocci, excluding the streptococci and enterococci. Clin. Microbiol. Rev. *8*: 479–495.

Felis, G.E., S. Torriani and F. Dellaglio. 2005. Reclassification of *Pediococcus urinaeequi* (*ex* Mees 1934) Garvie 1988 as *Aerococcus urinaeequi* comb. nov. Int. J. Syst. Evol. Microbiol. *55*: 1325–1327.

Fisher, W.S., E.H. Nilson, J.F. Steenbergen and D.V. Lightner. 1978. Microbial diseases of cultured lobsters. A review. Aquaculture *14*: 115–140.

Hamana, K. 1994. Polyamine distribution patterns in aerobic Gram positive cocci and some radio-resistant bacteria. J. Gen. Appl. Microbiol. *40*: 181–195.

Hitchner, E.R. and S.F. Snieszko. 1947. A study of a microorganism causing a bacterial disease of lobsters. J. Bacteriol. *54*: 48.

Hoyles, L., G. Foster, E. Falsen, L.F. Thomson and M.D. Collins. 2001. *Facklamia miroungae* sp. nov., from a juvenile southern elephant seal (*Mirounga leonina*). Int. J. Syst. Evol. Microbiol. *51*: 1401–1403.

Janosek, J., J. Eckert and A. Hudac. 1980. *Aerococcus viridans* as a causative agent of infectious endocarditis. J. Hyg. Epidemiol. Microbiol. Immunol. *24*: 92–96.

Kawamura, Y., X.G. Hou, F. Sultana, S.J. Liu, H. Yamamoto and T. Ezaki. 1995. Transfer of *Streptococcus adjacens* and *Streptococcus defectivus* to *Abiotrophia* gen. nov. as *Abiotrophia adiacens* comb. nov., and *Abiotrophia defectiva* comb. nov., respectively. Int. J. Syst. Bacteriol. *45*: 798–803.

Kelly, K.F. and J.B. Evans. 1974. Deoxyribonucleic acid homology among strains of the lobster pathogen "*Gaffkya homari*" and *Aerococcus viridans*. J. Gen. Microbiol. *81*: 257–260.

Kerbaugh, M.A. and J.B. Evans. 1968. *Aerococcus viridans* in the hospital environment. Appl. Microbiol. *16*: 519–523.

Kern, W. and E. Vanek. 1987. *Aerococcus* bacteremia associated with granulocytopenia. Eur. J. Clin. Microbiol. Infect. Dis. *6*: 670–673.

LaClaire, L. and R. Facklam. 2000a. Antimicrobial susceptibilities and clinical sources of *Facklamia* species. Antimicrob. Agents Chemother. *44*: 2130–2132.

LaClaire, L.L. and R.R. Facklam. 2000b. Comparison of three commercial rapid identification systems for the unusual gram-positive cocci *Dolosigranulum pigrum*, *Ignavigranum ruoffiae*, and *Facklamia* species. J. Clin. Microbiol. *38*: 2037–2042.

Lawson, P.A., M.D. Collins, E. Falsen, B. Sjoden and R.R. Facklam. 1999a. *Facklamia languida* sp. nov., isolated from human clinical specimens. J. Clin. Microbiol. *37*: 1161–1164.

Lawson, P.A., M.D. Collins, E. Falsen, B. Sjödén and R.R. Facklam. 1999b. In Validation of publication of new names and new combinations previously effectively published outside the IJSB. List no. 70. Int. J. Syst. Bacteriol. *49*: 935–936.

Lawson, P.A., E. Falsen, M. Ohlén and M.D. Collins. 2001a. *Aerococcus urinaehominis* sp. nov., isolated from human urine. Int. J. Syst. Evol. Microbiol. *51*: 683–686.

Lawson, P.A., E. Falsen, K. Truberg-Jensen and M.D. Collins. 2001b. *Aerococcus sanguicola* sp. nov., isolated from a human clinical source. Int. J. Syst. Evol. Microbiol. *51*: 475–479.

Miller, T.L. and J.B. Evans. 1970. Nutritional requirements for growth of *Aerococcus viridans*. J. Gen. Microbiol. *61*: 131–135.

Nathavitharana, K.A., S.N. Arseculeratne, H.A. Aponso, R. Vijeratnam, L. Jayasena and C. Navaratnam. 1983. Acute meningitis in early childhood caused by *Aerococcus viridans*. Br. Med. J. (Clin. Res. Ed.) *286*: 1248.

Schleifer, K.H. and O. Kandler. 1972. Peptidoglycan types of bacterial cell walls and their taxonomic implications. Bacteriol. Rev. *36*: 407–477.

Taylor, P.W. and M.C. Trueblood. 1985. Septic arthritis due to *Aerococcus viridans*. J. Rheumatol. *12*: 1004–1005.

Untereker, W.J. and B.A. Hanna. 1976. Endocarditis and osteomyelitis caused by *Aerococcus viridans*. Mt. Sinai J. Med. *43*: 248–252.

Vandamme, P., J. Hommez, C. Snauwaert, B. Hoste, I. Cleenwerck, K. Lefebvre, M. Vancanneyt, J. Swings, L.A. Devriese and F. Haesebrouck. 2001. *Globicatella sulfidifaciens* sp. nov., isolated from purulent infections in domestic animals. Int. J. Syst. Evol. Microbiol. *51*: 1745–1749.

Vela, A.I., E. Fernandez, A. las Heras, P.A. Lawson, L. Dominguez, M.D. Collins and J.F. Fernandez-Garayzabal. 2000. Meningoencephalitis associated with *Globicatella sanguinis* infection in lambs. J. Clin. Microbiol. *38*: 4254–4255.

Vela, A.I., N. Garcia, M.V. Latre, A. Casamayor, C. Sanchez-Porro, V. Briones, A. Ventosa, L. Dominguez and J.F. Fernandez-Garayzabal. 2007. *Aerococcus suis* sp. nov., isolated from clinical specimens from swine. Int. J. Syst. Evol. Microbiol. *57*: 1291–1294.

Wayne, L.G., D.J. Brenner, R.R. Colwell, P.A.D. Grimont, O. Kandler, M.I. Krichevsky, L.H. Moore, W.E.C. Moore, R.G.E. Murray, E. Stackebrandt, M.P. Starr and H.G. Trüper. 1987. Report of the ad hoc committee on the reconciliation of approaches to bacterial systematics. Int. J. Syst. Evol. Microbiol. *37*: 463–464.

Wiik, R., V. Torsvik and E. Egidius. 1986. Phenotypic and genotypic comparisons among strains of the lobster pathogen *Aerococcus viridans* and other marine *Aerococcus viridans*-like cocci. Int. J. Syst. Bacteriol. *36*: 431–434.

Williams, R.E.O., A. Hirch and S.T. Cowan. 1953. *Aerococcus*, a new bacterial genus. J. Gen. Microbiol. *8*: 475–480.

Zbinden, R., P. Santanam, L. Hunziker, B. Leuzinger and A. von Graevenitz. 1999. Endocarditis due to *Aerococcus urinae*: diagnostic tests, fatty acid composition and killing kinetics. Infection *27*: 122–124.

Family III. **Carnobacteriaceae** fam. nov.

WOLFGANG LUDWIG, KARL-HEINZ SCHLEIFER AND WILLIAM B. WHITMAN

Car.no.bac.te.ri.a′ce.ae. N.L. neut. n. *Carnobacterium* type genus of the family; suff. *-aceae* ending denoting family; N.L. fem. pl. n. *Carnobacteriaceae* the *Carnobacterium* family.

The family *Carnobacteriaceae* is circumscribed for this volume on the basis of phylogenetic analyzes of the 16S rRNA sequences and includes the genus *Carnobacterium* and its close relatives. It is composed of Gram-stain-positive rods or cocci that do not form endospores. Usually facultatively anaerobic, but some species grow aerobically or microaerophilically.

Usually catalase-negative. May be motile or nonmotile. The cell wall may contain the diamino acids lysine, ornithine or *meso*-diaminopimelate. Species may be psychrotolerant, halotolerant or alkaliphilic.

Type genus: **Carnobacterium** Collins, Farrow, Phillips, Feresu and Jones 1987, 314[VP].

Genus I. **Carnobacterium** Collins, Farrow, Phillips, Feresu and Jones 1987, 314[VP]

WALTER P. HAMMES AND CHRISTIAN HERTEL

Car.no.bac.te′ri.um. L. gen. n. *carnis* of flesh; N.L. neut. n. *bacterium* small rod; N.L. neut. n. *Carnobacterium* small rod from flesh.

Cells are **short to slender rods**, sometimes curved. Usually occurring singly or as pairs, sometimes short chains. **May or may not be motile. Nonsporeforming. Gram-stain-positive. Produces predominantly L(+)-lactic acid** from glucose. **One species** (*Carnobacterium pleistocenium*) **produces no lactate** but ethanol, acetic acid, and CO_2. **Gas production** from glucose **is variable** (dependent on substrate), frequently negative. **Facultatively anaerobic.**

Psychrotolerant. Most strains grow at 0 °C but not at 45 °C. No growth at 8% NaCl. **No growth on acetate (SL) agar or broth at pH 5.4,** good **growth at pH 9. Catalase- and oxidase-negative; some species exhibit catalase activity in the presence heme. Nitrate is not reduced** to nitrite. The **peptidoglycan is of the** *meso*-**diaminopimelic acid-direct type.** Major fatty acids are of the straight-chain saturated and monounsaturated types; cyclopropane ring containing derivatives may occur.

Found in vacuum packaged meat and related products, cheese, fish, meromictic antarctic lake, and pleistocenian ice in permafrost.

DNA G+C content (mol%): 33–42.

Type species: **Carnobacterium divergens** (Holzapfel and Gerber 1983) Collins, Farrow, Phillips, Feresu and Jones 1987, 315[VP] (*Lactobacillus divergens* Holzapfel and Gerber 1983, 530.).

Further descriptive information

Introductory remarks. The genus *Carnobacterium* comprises nine species as shown in Table 100. The genus is phylogenetically a member of the lactic acid bacteria group with a close relationship to the genera *Desemzia*, *Trichococcus*, and *Isobaculum* located in the *Carnobacterium* clade (Figure 97). Six species are associated with food of animal origin and/or living fish and three species have been isolated from cold environments with a low content of nutrients such as antarctic ice lakes and permafrost ice. Based on these quite different habitats, two ecological groups, I and II, respectively (Table 100) can be defined. The organisms within each group are adapted to their respective habitats in such a way that culturing of group I species requires rich media whereas rather simple media support the growth of group II species. This ecological grouping does not correlate with the phylogenetic relationship shown in Figure 98.

Cell morphology. The cell morphology varies depending on the species and age of the culture from coccobacilli (e.g., *Carnobacterium pleistocenium* and *Carnobacterium funditum*) to filaments of 13 µm and 20 µm in older cultures of *Carnobacterium funditum* and *Carnobacterium viridans*, respectively (Figure 99). Like lactobacilli, older cultures may looe their motility and Gram-stain-positive staining character. Otherwise, all carnobacteria stain clearly Gram-stain-positive. Motility by monotrichous polar to subpolar flagella occurs in *Carnobacterium mobile*, *Carnobacterium inhibens*, *Carnobacterium funditum*, and *Carnobacterium alterfunditum*. Nonmotile strains of *Carnobacterium mobile* have also been described (Laursen et al., 2005).

Culture characteristics. Colonies on agar are commonly white to creamy or buff, convex, and shiny. When grown on a semi-complex agar, *Carnobacterium pleistocenium* forms characteristic colonies that are conical in shape, with a denser consistency in the center and a darker color on the perimeter. The surface is granulated and rough with thinner, irregular, torn edges (Pikuta et al., 2005). The diameter of the colonies varies from 0.5–2 mm on optimal agar (e.g., MRS without acetate). *Carnobacterium funditum* and *Carnobacterium alterfunditum* produce a slightly yellow water-soluble pigment upon incubation for 2–4 weeks.

Clearing zones were described for *Lactobacillus maltaromicus* in assays for DNase and RNase activity (Miller et al., 1974) and for caseinolytic activity (Baya et al., 1991). Specific odors of cultures were described for *Carnobacterium alterfunditum* upon growth in Py-amygdalin medium (seeds of ripe plums), and for *Lactobacillus maltaromicus* in TSB medium and skim milk, a strong malty aroma is observed (Miller et al., 1974). By gas chromatography, 2-methylpropanal, 3-methylpropanal, 2-metylpropanol, and 3-methylpropanol were identified. These compounds are well known for their malty flavor. Carnobacteria are involved in the spoilage of food of animal origin, and numerous aroma active compounds are formed in addition therein, e.g., NH_3 from arginine catabolism (Leisner et al., 1994), acetic acid, and diacetyl (Joffraud et al., 2001).

Metabolism. Compared with the body of knowledge that has been accumulated in recent years for the metabolism of lactococci and lactobacilli, little is known about carnobacteria. Carnobacteria

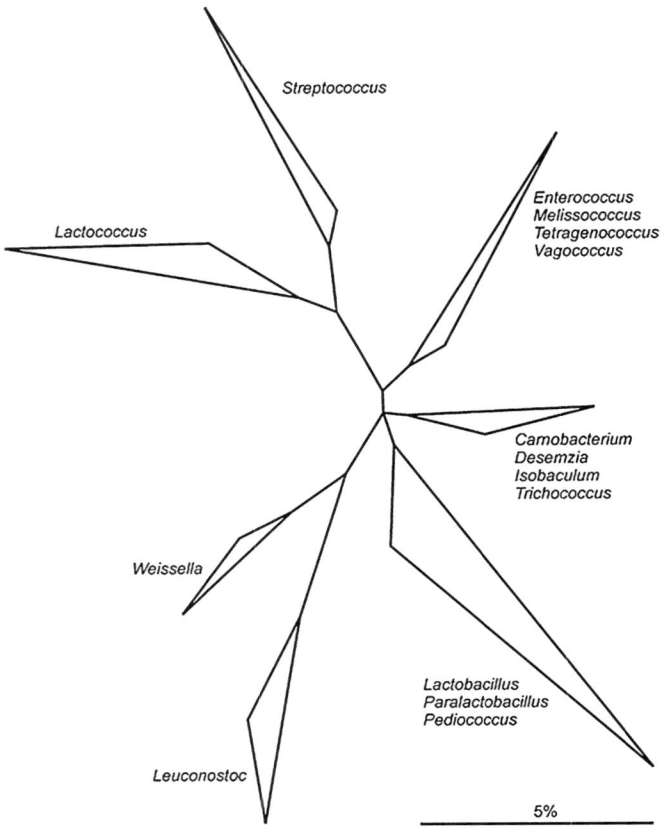

FIGURE 97. Phylogenetic tree depicting the position of carnobacteria in relation to lactic acid bacteria. The consensus tree is based on distance matrix analyzes of all available, at least 90% complete 16S rRNA sequences of Gram-stain-positive bacteria. The topology was evaluated and corrected according to the results of maximum-parsimony and maximum-likelihood analyzes with various datasets. Alignment positions sharing identical residues in at least 50% of all sequences of the depicted and closely related genera were considered. Multifurcations indicate that a common branching order could not be significantly determined or was not supported performing different alternative treeing approaches. The bar indicates 5% estimated sequence divergence.

TABLE 100. The species of the genus *Carnobacterium*

Species	Source of isolation[a]	Ecological group[b]
C. divergens	M, F, D	I
C. alterfunditum	A	II
C. funditum	A	II
C. gallinarum	M	I
C. inhibens	F	I
C. maltaromaticum	M, F, D	I
C. mobile	M	I
C. pleistocenium	P	II
C. viridans	M	I

[a]M, meat (meat products, poultry); F, fish; D, dairy products, including cheese; A, arctic ice lake; P, pleistocenian permafrost.
[b]I, associated with animals and products of animal origin; II, present in cold environment containing few nutrients.

are facultatively anaerobic and gain their energy by fermentation. However, they grow well in the presence of oxygen, and the growth yield of *Carnobacterium maltaromaticus* increases 10-fold when heme is added to the aerobically growing culture (Meisel et al., 1994). The authors showed that under these conditions, functional cytochromes of the *b* and *d* types can be detected. Similarly, cytochromes can be induced in lactococci and enterococci (Ritchey and Seely, 1976; Sijpesteijn, 1970). In *Lactococcus lactis*, *cydAB* genes encode a single cytochrome oxidase (bd) (Gaudu et al., 2002). Thus, when heme is available, *Carnobacterium maltaromaticus* has a potential for a respiratory metabolism and also exhibits catalase activity (Kloos and Wolfshohl, 1991). The latter property is common to all *Carnobacterium* species (Ringo et al., 2002) except *Carnobacterium pleistocenium* which was not investigated for this property. Respiration is connected with increased oxygen consumption, reduced lactate formation, and increased acetoin and CO_2 production.

The metabolism of carbohydrates by carnobacteria is facultatively heterofermentative as they ferment hexoses and pentoses. With the exception of *Carnobacterium pleistocenium*, hexoses are metabolized to L(+)-lactate, but CO_2, acetate, and ethanol are also formed. In addition, formate is produced anaerobically (Borch and Molin, 1989) and acetoin is produced aerobically. Pentoses are metabolized to L(+)-lactate, acetate, and ethanol (Holzapfel and Gerber, 1983). Glucose is metabolized via the glycolytic pathway (De Bruyn et al., 1988; De Bruyn et al., 1987). For the species of ecology group I, 75% of the lactate produced was shown to be derived from glucose and less than 10% was derived from formate and from acetate. The other products were assumed to originate from endogenous substrates in the rich growth medium. Differences exist for species of group II. For *Carnobacterium pleistocenium*, ethanol, acetate, and CO_2 (but no lactate) were found to be the products of carbohydrate fermentation. Thus, the physiological definition of lactic acid bacteria would not apply to this species and rather a metabolism similar to the Embden–Meyerhof pathway in yeast may be operative. Similarly, *Carnobacterium funditum* and *Carnobacterium alterfunditum* also do not produce lactate from glycerol; they produce only formate, ethanol, and acetate. Polysaccharides are hydrolyzed and fermented by some species or strains. Inulin is used by *Carnobacterium divergens*, *Carnobacterium gallinarum*, and certain strains of *Carnobacterium maltaromaticus*. Amylolytic activity is also found and is reported as utilization of starch for *Carnobacterium pleistocenium* or utilization of glycogen for *Carnobacterium mobile* and *Carnobacterium gallinarum*.

The presence of starch in plants and consequently in the intestinal tract of fish may be consistent with an adaption of carnobacteria to that habitat. Arginine hydrolysis produces an additional source of ATP generation and is a property of most carnobacteria, but is missing in *Carnobacterium viridans* and *Carnobacterium maltaromaticus*. The formation of biogenic amines creates a proton motive force in the bacteria (Konings, 2002). It is, however, of hygienic concern and is an essential contribution of carnobacteria to food spoilage (Laursen et al., 2005). *Carnobacterium divergens* and *Carnobacterium maltaromaticus* especially have a potential for tyramine formation (Straub et al., 1995).

Pathogenicity. *Lactobacillus piscicola* was the first *Lactobacillus* isolated from diseased trout (Cone, 1982) and later also from other farmed fish, fulfilling Koch's postulates. Isolates from

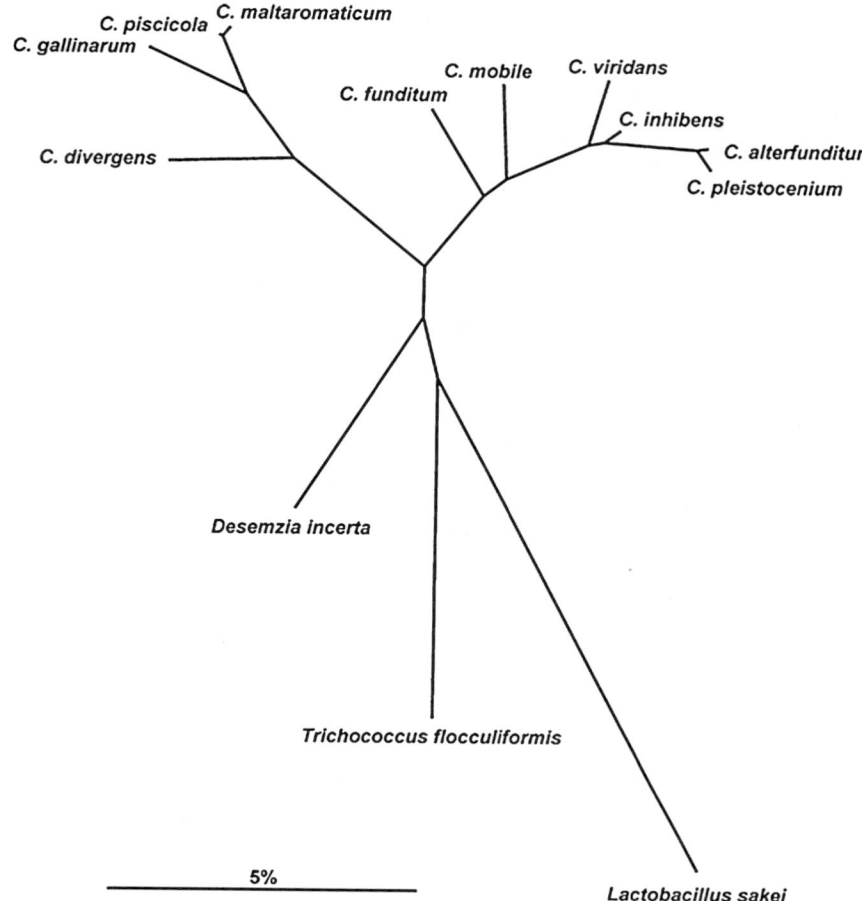

FIGURE 98. Phylogenetic tree reflecting the relationships of *Carnobacterium* species. The tree was reconstructed applying the maximum-parsimony analysis of all available at least 90% complete 16S rRNA sequences of *Lactobacillaceae*, carnobacteria, and close related species. Alignment positions sharing identical residues in at least 50% of all sequences of the depicted genera were considered. The bar indicates 5% estimated sequence divergence.

diseased striped bass (*Morone saxatilis*), but not from channel catfish (*Ictalurus punctatus*) or brown bullheads (*Ictalurus nebulosus*), killed fingerling rainbow trout (*Onchorynchus mykiss*) (LD$_{50}$ of two strains of *Lactobacillus piscicola*, 1.4×10^6 and 2.3×10^6 c.f.u./g, respectively), administered peritoneally (Baya et al., 1991). The dead or carrier state trout showed serious lesions of the inner organs. Striped bass were resistant, although the injected bacteria could be reisolated after two weeks from the kidneys, as was the case with moribund and dead trout. Thus, there appear to exist unknown strain specific virulence factors and specific, sensitive hosts. In general, diseased fish contain mixed infections in which carnobacteria are found together with *Aeromonas hydrophila* and *Pseudomonas fluorescens*. In addition, stress conditions, such as handling or spawning, cause a predisposition to infection by carnobacteria.

Remarkably, *Carnobacterium piscicola* has been isolated from human pus (Chmelar et al., 2002), and one of the strains investigated by Collins (1987) has been isolated from human blood plasma. Similar to the situation in lactobacilli, pathogenicity factors such as production of elastase, phospholipase, or hemolytic activity were not detected (Baya et al., 1991) with the exception of

hemolytic activity detected in *Carnobacterium viridans*. It can be concluded that true fish pathogenicity exists in certain strains of *Carnobacterium maltaromaticus* whereas for humans and warm blooded animals, carnobacteria are not pathogenic.

Ecology. Two ecological groups of *Carnobacterium* species have been introduced (see above). Group I species are associated with the man-made habitat of food of animal origin as well as with the natural habitats of intestines and gills of fish. Group II species occur in arctic ice lakes or pleistocenian ice. The group I species have been known since 1957 (Thornley, 1957) and were grouped as atypical lactobacilli ("Streptobacteria"; Reuter (1970, 1981); Hitchener et al., 1982; Holzapfel and Gerber,(1983); Shaw and Harding, 1984) mainly because they resemble lactobacilli in their morphology and especially in their phenotype, but do not grow on SL-agar (Rogosa et al., 1951). When associated with food, these carnobacteria frequently share their habitat with other lactic acid bacteria, especially with lactobacilli and leukonostocs.

Meat, meat products, and poultry. Meat, meat products, and poultry are substrates with a favorable pH (around neutral) and

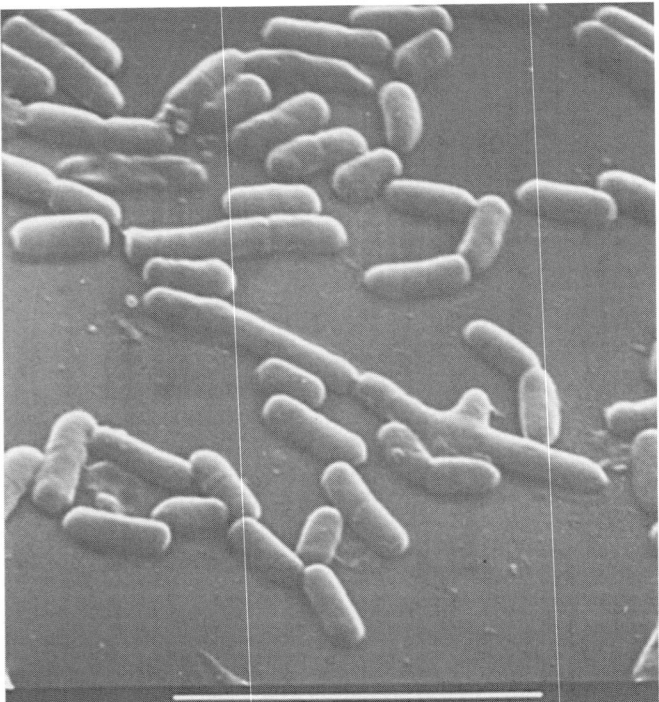

FIGURE 99. Electron micrograph of *Carnobacterium divergens* cells (bar = 5 μm).

high water activity, and they are rich in nutrients. They are, therefore, highly sensitive to spoilage and are stored refrigerated and often packaged in air-tight materials, sometimes under vacuum or controlled atmosphere. These conditions are selective for lactic acid bacteria and especially for carnobacteria. The preferred species isolated from meat and meat products are *Carnobacterium divergens* and *Carnobacterium maltaromaticus* (Laursen et al., 2005) which can grow to numbers of 10^6–10^8 c.f.u./g and cause spoilage. *Lactobacillus divergens* (now *Carnobacterium divergens*) was isolated from refrigerated beef (von Holy, 1983) and described as a heterofermentative *Lactobacillus* (*divergens*) species. Further isolates were originally allotted to *Lactobacillus carnis* and *Lactobacillus piscicola* and later became reclassified as *Carnobacterium maltaromaticus*. Spoiled products are characterized by discoloration, souring, and off-flavor, and include unprocessed meat, cooked ham, smoked pork loin, frankfurter sausage, and bologna type sausage. A green discolored bologna sausage was the source of isolation of *Carnobacterium viridans* (Hammes and Hertel, 2003; Schillinger and Holzapfel, 1995). During cold storage of meat, successions of lactic acid bacteria develop (Jones, 2004). Carnobacteria (*Carnobacterium divergens*) became the predominant element within four weeks, followed by other lactic acid bacteria. *Carnobacterium maltaromaticum* and *Carnobacterium divergens* were also isolated from poultry as were *Carnobacterium gallinarum* and *Carnobacterium mobile*, which were described as new species by Collins (1987). The strain employed in the species description was originally isolated by Thornley (1957) from poultry samples subjected to irradiation.

Fish and seafood. The intestines and gills of fish are natural habitats of carnobacteria. *Carnobacterium maltaromaticum* (*Lactobacillus piscicola*) was isolated from diseased salmonoid fish, striped bass, and catfish (Hammes and Hertel, 2003). This species, *Carnobacterium divergens*, and *Carnobacterium inhibens* (Ringo and Gatesoupe, 1998) are commonly associated with healthy sea and fresh water fish of various species. Their numbers depend on nutritional and environmental factors (e.g., polyunsaturated fatty acids in the feed and stress). The presence of carnobacteria (*Carnobacterium piscicola*-like) on gills of Atlantic charr at numbers of approximately 3×10^4 c.f.u./g has been shown (Ringo and Holzapfel, 2000). Some isolates exhibit a strong inhibitory activity against other fish pathogens. *Carnobacterium maltaromaticum*, *Carnobacterium divergens*, and *Carnobacterium mobile* are associated also with seafood, e.g., shrimp, gravid fish, cold smoked salmon, and seafood in vacuum packages or under modified atmosphere (Laursen et al., 2005; Leroi et al., 1998; Mauguin and Novel, 1994).

Dairy products including cheese. *Lactobacillus maltaromicus* was described by (Miller et al., 1974) as a species isolated from milk samples that had developed a distinct malty aroma. A greater reservoir of carnobacteria is cheese such as surface mold ripened soft cheese made from nonpasteurized milk. Brie cheese has a surface pH of 6.8–7.6 and enables the bacteria to grow to numbers of 10^8–10^9 c.f.u./g. *Carnobacterium divergens* and *Carnobacterium maltaromaticum* are also major components in the lactic acid bacteria association in the curd of mozzarella cheese made from nonpasteurized milk (Morea et al., 1999).

Ecological Group II carnobacteria. *Carnobacterium funditum* and *Carnobacterium alterfunditum* were isolated from the Antarctic Ace Lake. In that meromictic lake, the salinity increased from 0.6% at the surface to 4.3% at the bottom depth of 24 m where the sample was taken (Franzmann et al., 1991). The temperature varied little and was 1–2 °C. The organisms are adapted to that environment but grow best at 22–23 °C in a medium of 1.7 (*Carnobacterium funditum*) or 0.6% (*Carnobacterium alterfunditum*) salinity. It is assumed that the carnobacteria create a reduced environment and provide electron donors for exploitation of associated sulfate reducing bacteria. *Carnobacterium pleistocenium* was isolated from a pleistocenian ice sample from the permafrost tunnel in Fox, Alaska. The age was estimated to be 32,000 years. The organism grows optimally at 24 °C and has a unique carbohydrate metabolism that produces no lactic acid from glucose.

Sensitivity to chemotherapeutics. The pathogenicity of strains of *Carnobacterium maltaromaticus* to fish produced in aquaculture raised the interest in the sensitivity of carnobacteria to chemotherapeutics. The use of those compounds in aquaculture has been reviewed by Ringo and Gatesoupe (1998). Lai and Manchester (2000) determined the resistance of 97 isolates including all carnobacteria species except *Carnobacterium inhibens* and *Carnobacterium pleistocenium* and for numerous species found high resistance values (close to 100%) against ampicillin, cloxacillin, gentamicin, kanamycin, neomycin, and streptomycin. The results confirm the findings of Baya et al. (1991) who investigated the sensitivity of eight isolates of *Carnobacterium piscicola* from diseased fish and additionally found resistance to chlorotetracyclin, trimethoprim, furazolidon, nitrofurazone, and nitrofurantoin. All strains were sensitive to erythromycin, and the majority of strains were sensitive to penicillin and ampicillin. Similar results were reported by Ringo and Holzapfel,

(2002) who also included *Carnobacterium inhibens* in their study. A complete sensitivity to deferoxamin and resistance to vancomycin were also reported. It is notable that *Carnobacterium pleistocenium* that was isolated from a human-free environment established 32,000 years ago is sensitive to commonly used antibiotics such as ampicillin, kanamycin, gentamicin, tetracycline, rifampin, and chloramphenicol.

Antagonistic potential. A protective effect of carnobacteria used as probiotics against competing micro-organisms in fish has been observed by Robertson et al. (2000). Carnobacteria species isolated from Atlantic salmon exhibited *in vitro* activity against numerous Gram-stain-negative pathogens. The culture applied to rainbow trout (fry, fingerling) and small Atlantic salmon of 15 g improved the survival of the fish challenged with pathogens, e.g., for Atlantic salmon exposed to *Aeromonas salmoncida*, *Vibrio anguillarum*, *Vibrio ordalii*, and *Yersinia ruckerii*. Cultures were studied with the aim to protect food from growth of pathogens, especially *Listeria monocytogenes*, or to prevent spoilage (Brillet et al., 2004; Yamazaki et al., 2003). As carnobacteria do not acidify strongly, this undesired sensory effect is less pronounced. The strains used commonly form bacteriocins. These proteinaceous compounds were detected and attributed to lantibiotics (group I bacteriocins) and to group II compounds (Klaenhammer, 1993; Ouwehand, 1998; Tahiri et al., 2004). The bacteriocin formation is not always decisive for the protective effect, especially against *Listeria monocytogenes*, as on cold smoked salmon, a nonproducer exhibited the positive effect and producer strains enhanced it (Nilsson et al., 1999, 2004).

Enrichment, isolation, and cultivation procedures

The species of the ecological group I are similar to lactobacilli in their nutritional requirements. They are, however, not aciduric and either do not grow or do not readily grow on acetate rich media, e.g., SL-medium (Rogosa et al., 1951). The omission of acetate from MRS-medium (De Man et al., 1960) is commonly a favorable choice for isolation. Neutral to alkaline media in the range 6.8–9.0 favors their growth and pH 8.0–9.0 commonly prevents the growth of lactobacilli that are found in association with carnobacteria in food. For *Carnobacterium piscicola* isolates from diseased fish, a requirement for folic acid, riboflavin, pantothenate, and niacin but not for vitamin B_{12}, biotin, thiamine, or pyridoxal (Hiu et al., 1984) has been demonstrated. *Carnobacterium divergens* and *Carnobacterium piscicola* isolates from meat do not require thiamine (De Bruyn et al., 1987). The species of ecological group II are less fastidious. *Carnobacterium alterfunditum* and *Carnobacterium funditum* grow on mineral salts, a carbohydrate source, and yeast extract. These organisms do not grow on MRS-medium without acetate. *Carnobacterium pleistocenium* needs mineral salts, yeast extract, and peptone. Vitamins are also added to the medium, however, it is not known whether or not they are essential. To isolate ecological group I bacteria from environments dominated by carnobacteria, direct streaking or plating onto nonselective universal media is possible. For isolation from diseased fish, brain heart infusion or tryptic soy agar has been used (Hiu et al., 1984), and for isolation from vacuum packaged meat, standard-I-agar (Merck) (pH 7.2–7.5), caso-medium (Merck), or tryptic soy agar (Difco) with 0.3% added yeast extract and aerobic incubation at 25 °C for 3 d or 7 °C for 10 d is recommended (Hammes and Hertel, 2003).

When counting carnobacteria on the universal media, the pH should be adjusted to 8.0 to suppress growth of lactobacilli. The identity of typical catalase-negative bacteria needs verification by microscopy and the criteria in Table 101. The absence of heme has to be checked, as the group I species form the apoenzyme that becomes active in the presence of heme (see above). Selective cresol Red Thallium Acetate Sucrose (CTAS) as a selective medium for isolation and counting of *Carnobacterium divergens* and *Carnobacterium piscicola* was recommended by WPCM (1989). The preparation of that medium and the colony morphology seen on that medium have been described by Hammes and Hertel, (2003). For cultivation of the axenic cultures, APT medium (Evans and Niven, 1951) at pH 7.0, caso-medium with 0.3% yeast extract, or standard-I-medium can be used. In general and especially for more fastidious strains, D-MRS (pH 8.0–8.5) (Hammes et al., 1992) and incubation at 25 °C for 24–48 h, aerobically or in slightly reduced atmosphere, are recommended. D-MRS corresponds mainly with MRS-medium but does not contain acetate, and sucrose substitutes for glucose. According to Lai and Manchester, (2000), some strains of *Carnobacterium divergens* (cluster B) do not ferment sucrose and need, therefore, resubstitution of sucrose by glucose.

Cultivation of the group II bacteria *Carnobacterium funditum* and *Carnobacterium alterfunditum* is best performed under anaerobic conditions. The organisms grow poorly aerobically. Rich media such as SL-medium or MRS-medium without acetate do not support growth. Best growth is observed in medium 141 YF (Franzmann et al., 1991) consisting of mineral salts with added trace salts (141 salts) and vitamins, cysteine hydrochloride, Na_2S, 1% yeast extract, and 1% fructose. The optimum temperature and pH are 22–23 °C and 7.0–7.4, respectively. For optimal growth, *Carnobacterium funditum* and *Carnobacterium alterfunditum* require 1.7% and 0.6% NaCl, respectively (Franzmann et al., 1991). *Carnobacterium pleistocenium* (Pikuta et al., 2005) is cultured at optimum pH 7.3–7.5 and temperature 24 °C aerobically or anaerobically, in medium consisting of mineral salts, trace of vitamins and salts, 0.2 g/l yeast extract, and 3 g/l peptone. For determination of the utilization of carbohydrates, the mineral salt base medium with 0.1% yeast extract is supplemented with the respective carbohydrate at a concentration of 3 g/l.

Maintenance. The maintenance of cultures of carnobacteria can be performed as for lactobacilli. Care has to be taken that the pH does not drop markedly below 6.0. The organism grown in the above described media can be kept at 1–4 °C in stab culture for 2–3 weeks. The addition of 5% calcium carbonate can improve the vitality of the cultures. For long-term storage, lyophilization is the best choice. As cryoprotectives, skim milk, lactose, or horse serum are recommended. The dried cultures should be stored in sealed glass ampoules at 8–12 °C. This culture can be kept for years. Effective storage is also achieved by harvesting late logarithmic cells by centrifugation, resuspending in growth medium containing 20% glycerol, and storage in screw caps at –80 °C.

Procedures for testing special characteristics

The composition of the fatty acids has been used to differentiate groups or even species of carnobacteria. The composition of *Carnobacterium viridans* is, however, not known. Oleic acid is present in all species and is dominant in *Carnobacterium*

TABLE 101. Selected physiological characteristics of species of the genus *Carnobacterium*[a]

Characteristic	C. divergens	C. alterfunditum	C. funditum	C. gallinarum	C. inhibens	C. maltaromaticum	C. mobile	C. pleistocenium [c]	C. viridans
Growth at 0 °C	+	+	+	+	+	d	+	+	(2 °C+)
Growth at 30 °C	+	–	–	ND	+	+	+	–	+
Growth at 40 °C	+	– (+) [b]	– (+) [b]	ND	–	d	–	–	–
Motility	–	+	+	–	+	–	+	+	–
Arginine hydrolysis	+	+	–	+	+	–	+	ND	–
Voges–Proskauer test	+	ND	ND	+	ND	+	–	ND	–
Acid produced from:[c]									
Amygdalin	+	+	–	+	+	d	–	ND	–
Arabinose	–	–	–	–	–	–	–	+	–
Galactose	–	w	w	+	+	+	+	ND	+
Gluconate	+	– [d]	– [d]	ND	–	+	–	ND	–
Glycerol	ND	w	w	ND	–	+	+/–	–	–
Inulin	–	–	–	–	w	+	+	ND	–
Lactose	–	–	–	+	w	–	–	+	+
Mannitol	–	–	+	–	+	+	–	+	–
Melezitose	d	–	–	+	–	d	–	ND	–
Melibiose	–	–	–	–	–	+	–	ND	–
Methyl D-Glucoside	–	– [d]	– [d]	+	–	+	–	ND	–
Ribose	+	+	+	+	+	+	+	+	
D-Tagatose	–	ND	ND	+	–	–	d	ND	+
Trehalose	+	–	+	+	+	+	+	+	+
D-Turanose	–	– [d]	– [d]	+	–	d	–	ND	–
Xylose	–	–	–	+	–	–	–	ND	–
Hydrolysis of esculin	ND	+	–	+	+	+	+	–	ND

[a]Symbols: +, present; –, absent; d, 11–89% positive; w, weak; and ND, no data.

[b]According to Franzmann et al. (1991), sucrose fermentation by *Carnobacterium funditum* and *Carnobacterium alterfunditum* is + and weak, respectively, whereas Lai and Manchester (2000) report – for both species.

[c]All species produce acid from cellobiose, fructose, glucose, maltose, mannose, salicin and sucrose, but not from adonitol, dulcitol, glycogen, inositol, raffinose, rhamnose, and sorbitol.

[d]According to Lai and Manchester (2000); see text.

[e]Growth on: Pikuta et al. (2005); see text.

maltaromaticum, Carnobacterium divergens, Carnobacterium mobile, and *Carnobacterium gallinarum* (Collins et al., 1987; Ringo et al., 2002). *Carnobacterium divergens* also contains the biosynthetically related cyclopropane derivative 9,10 methylenoctadecanoic acid, and in summing up these two compounds, $C_{18:1}$ fatty acid constitute 40–50% of the total fatty acids. $C_{16:1}$ fatty acids are also present in all species in substantial amounts and were identified as ω9 isomers in *Carnobacterium divergens, Carnobacterium maltaromaticum, Carnobacterium mobile, Carnobacterium gallinarum*, and *Carnobacterium inhibens*, whereas in *Carnobacterium funditum, Carnobacterium alterfunditum*, and *Carnobacterium pleistocenium*, the ω7 isomer was detected. The group II species also contain eicosanoic acid ($C_{20:1}$) in the amount of 0.6–1.6%.

Except for the composition of fatty acids, the determination of classical biochemical and chemical characteristics are of limited value. Key criteria are compiled in Table 101. In a recent study (Laursen et al., 2005), 120 carnobacteria (mainly *Carnobacterium divergens* and *Carnobacterium maltaromaticum*) isolated from meat and seafood were subjected to numerical taxonomy relying on classical biochemical reactions as well as genotypic methods. These included carbohydrate fermentation and inhibition tests, SDS-PAGE electrophoresis of whole-cell proteins (protein profiling), plasmid profiling, intergenic space region (ISR) analysis, and examination of amplified-fragment length polymorphism (AFLP). The authors concluded that none of the phenotypic tests included in their study permitted a clear differentiation between *Carnobacterium divergens* and *Carnobacterium maltaromaticum*. Only the use of large numbers of phenotypic tests and data evaluation by numerical taxonomy would allow the identification of carnobacteria to the species level. The criteria listed in Table 101 can be obtained by using standard procedures such as the determination of the pattern of carbohydrate fermentation with the aid of automated systems, e.g., API 50 CH (bioMerieux).

For strain characterization, Laursen et al. (2005) used SDS-PAGE of whole-cell protein extracts as a nongenotypic identification method (Pot et al., 1994). For strain characterization, the authors had also included plasmid profiling according to Leisner et al. (2001) in their study. As about half of the strains of *Carnobacterium divergens* and *Carnobacterium maltaromaticum* do not contain plasmids, this identification method is applicable only to a limited number of strains.

Genotypic identification methods

As described in the chapter on the genus *Lactobacillus*, genotypic methods are increasingly applied for species identification. Most of these methods rely on the rRNA gene sequences, e.g., multiplex PCR for simultaneous identification of *Carnobacterium divergens*, *Carnobacterium gallinarum*, and *Carnobacterium maltaromicum* (Macian et al., 2004). Recently, it was demonstrated that the restriction fragment length polymorphism of the 16S–23S rRNA gene intergenic spacer region can be useful to identify carnobacteria at the species level (Laursen et al., 2005; Rachman et al., 2004). Nevertheless, the comparative 16S rRNA gene sequence analysis still remains a suitable tool for the rapid and reliable identification of *Carnobacterium* species. An exception is *Carnobacterium alterfunditum* and *Carnobacterium pleistocenium*, which show 99.7% sequence similarity.

Differentiation of the genus *Carnobacterium* from closely related taxa

Carnobacteria can be differentiated from coccoid lactic acid bacteria by microscopy. The reduced tolerance to low pH and the ability to grow at pH 9 differentiates the carnobacteria from the rod-shaped genera *Lactobacillus*, *Paralactobacillus*, and *Weissella*.

The latter is, in addition, obligately heterofermentative producing D- or DL-lactic acid and does not possess *meso*-diaminopimelic acid in the peptidoglycan. This characteristic amino acid is also absent in the other rod-shaped genera within the *Carnobacterium* clade (*Isobaculum* and *Desemzia*).

Taxonomic comments

Carnobacteria belong to the group of lactic acid bacteria and are closely related to species of the genera *Desemzia*, *Isobaculum*, and *Trichococcus*, forming together the *Carnobacterium* clade (Figure 97). Taken into consideration that phylogenetic distances vary with the dataset and methods used for tree construction, it is still an open question whether the closest relatives among the lactic acid bacteria are located in the *Lactobacillus* clade or *Enterococcus* clade. The phylogenetic relationship among the carnobacteria can only roughly be estimated because the positioning of species, especially of the genera *Desemzia* and *Trichococcus*, varies with the sequence positions selected and methods used for tree construction. As an example a tree constructed by applying the maximum-parsimony analysis is depicted in Figure 98. Interestingly, *Carnobacterium divergens*, *Carnobacterium gallinarum*, *Carnobacterium piscicola*, and *Carnobacterium maltaromaticum* remained in one group in most of the phylogenetic trees constructed by varying the sequence positions and treeing methods.

List of species of the genus *Carnobacterium*

1. **Carnobacterium divergens** (Holzapfel and Gerber 1983) Collins, Farrow, Phillips, Feresu and Jones 1987, 315[VP] (*Lactobacillus divergens* Holzapfel and Gerber 1983, 530.)

di.ver′gens. L. part. adj. *divergens* deviate, diverge.

Nonmotile, straight, slender and relatively short rods with rounded ends; generally 0.5–0.7 × 1.0–1.4 µm, occurring singly, in pairs, and short chains. Colonies cream colored to white, convex, shiny, varying from 0.5–1.5 mm on Standard I-agar and MRS-agar without acetate. Growth in acetate-containing media is suppressed in the presence of citrate and glucose, but is stimulated when ribose or fructose is added. Surface growth is generally affected by aerobiosis. Under certain conditions, heterofermentative, producing L(+)-lactic acid, CO_2, ethanol, and acetate from hexoses. In addition to L(+)-lactic acid, acetate and ethanol are produced from ribose. Grows in 10% NaCl. Final pH reached in MRS-broth (without acetate) between 5.0 and 5.3 after 4 d. No production of slime from sucrose or mannitol from fructose. No gas production from gluconate or malate. Produces catalase on heme-containing media. Gelatinase, indole, and H_2O_2 not produced. Additional physiological and biochemical properties are presented in Table 101.

Isolated from vacuum-packaged, refrigerated meat, fish.

DNA G+C content (mol%): 33.7–36.4 (T_m).

Type strain: 66, ATCC 35677, CCUG 30094, CIP 101029, DSM 20623, NBRC 15683, JCM 5816, JCM 9133, LMG 9199, NCIMB 11952, NRRL B-14830.

GenBank accession number (16S rRNA gene): M58816, X54270.

2. **Carnobacterium alterfunditum** Franzmann, Höpfl, Weiss and Tindall 1993, 188[VP] (Effective publication: Franzmann, Höpfl, Weiss and Tindall 1991, 261.)

al′ter.fun.di′tum. L. adj. *alter* another; L. adj. *funditus* from the bottom; N.L. neut. adj. *alterfunditum* another [carnobacterium] from the [lake] bottom.

Cells are rods (1.3 × 2.5–12.5 µm) occurring singly, in pairs, or short chains (typically of four cells), motile by a single subpolar flagellum. Old cells usually stain Gram-stain-stain-negative and are nonmotile. Anaerobic, with better growth at 20 °C. L(+)-Lactic acid is the major end product from D(+)-glucose with traces of ethanol, acetic acid, and formic acid. D(−)-Ribose is fermented to lactic acid and moderate amounts of ethanol, acetic acid, and formic acid. Glycerol is mainly fermented to acetic acid and formic acid, and traces of ethanol are produced. Gas is not produced. No growth in MRS or SL broths. Growth occurs in media containing 0.1% yeast extract without added sodium salts; the optimum concentration of NaCl is 0.1 M. Chopped meat medium and litmus milk remain unchanged. The optimum initial pH for growth is 7.0–7.4. Pyruvate, lactate, formate, acetate, methanol, betaine, trimethylammonium, chloride, glycine, and an atmosphere of H_2:CO_2 (2 bar, 80:20) do not stimulate growth. Gelatinase-negative. Additional physiological and biochemical characteristics are presented in Table 101. The cell wall contains *meso*-diaminopimelic acid. $C_{18:1\omega9c}$ is a major component of its fatty acids. Respiratory lipoquinones are not produced. The spent fermentation broth of PY-amygdalin-2% NaCl smells similar to the seeds of dried prunes.

Isolated from the anaerobic monimolimnion of an antarctic lake (with approximately the salinity of sea water) and rainbow trout.

DNA G+C content (mol%): 32–36 (T_m).

Type strain: pf4, ACAM 313, ATCC 49837, CCUG 34643, CIP 105796, DSM 5972, NBRC 15548, JCM 12498, LMG 13520.

GenBank accession number (16S rRNA gene): not available.

3. **Carnobacterium funditum** Franzmann, Höpfl, Weiss and Tindall 1993, 188[VP] (Effective publication: Franzmann, Höpfl, Weiss and Tindall 1991, 260.)

fun.di′tum. L. neut. adj. *funditum* from the bottom.

Cells are rods (0.8–1.3 ×1.7–0.8 μm) occurring singly, in pairs, or short chains (typically of four cells), motile by a single subpolar flagellum. Old cells usually stain Gram-stain-stain-negative and are nonmotile. Anaerobic, with better growth at 20 °C. (+)-Lactic acid is the major end product from D(+)-glucose with traces of ethanol, acetic acid, and formic acid. D(−)-Ribose is fermented to lactic acid and moderate amounts of ethanol, acetic acid, and formic acid. Glycerol is mainly fermented to acetic acid and formic acid, and traces of ethanol are produced. Gas is not produced. No growth in MRS or SL broths. At least 0.1% yeast extract is required for good growth. Chopped meat medium and litmus milk remain unchanged. Gelatinase-negative. The optimum initial pH for growth is between 7.0 and 7.4. Sodium is required for growth with optimal growth at 1.7%. Pyruvate, lactate, formate, acetate, methanol, betaine, trimethyl-ammonium, chloride, glycine, and an atmosphere of $H_2:CO_2$ (2 bar, 80:20) do not stimulate growth. Additional physiological and biochemical characteristics are presented in Table 101. $C_{18:1 \, \omega 9c}$ is a major component of its fatty acids. Respiratory lipoquinones are not produced.

Isolated from the anaerobic monimolimnion of an antarctic lake of about sea water salinity.

DNA G+C content (mol%): 32–35 (T_m).

Type strain: pf3, ACAM 312, ATCC 49836, CCUG 34644, CIP 106503, DSM 5970, NBRC15549, JCM 12499, LMG 14461.

GenBank accession number (16S rRNA gene): S86170.

4. **Carnobacterium gallinarum** Collins, Farrow, Phillips, Feresu and Jones 1987, 315[VP]

gal.li.na′rum. L. fem. n. *gallina* a hen; L. fem.gen.p.n. *gallinarum* of hens.

Gram-stain-positive, nonsporeforming, straight, slender rods which occur singly or in short chains. Cells are nonmotile. Colonies are white, convex, shiny, and circular. Facultatively anaerobic. L-(+)-Lactic acid is produced. No gas is produced from glucose. Lysine decarboxylase, ornithine decarboxylase, tryptophan desaminase, and urease-negative. Indole and H_2S are not produced. Additional physiological and biochemical characteristics are presented in Table 101. The cellular fatty acids are of the straight-chain saturated and monounsaturated types with tetradecanoic, hexadecanoic, and 9,10-octadecenoic acids predominating.

Isolated from ice slush from around chicken carcasses and various meat products.

DNA G+C content (mol%): 34.3–35.4 (T_m).

Type strain: MT44, ATCC 49517, CCUG 30095, CIP 103160, DSM 4847, JCM 12517, LMG 9841, NCIMB 12848, NRRL B-14832.

GenBank accession number (16S rRNA gene): AJ387905, X54269.

5. **Carnobacterium inhibens** Jöborn, Dorsch, Olsson, Westerdahl and Kjelleberg 1999, 1897[VP]

in.hi′bens. L. part. adj. *inhibens* inhibiting, referring to the growth-inhibitory activity that the bacterium shows.

Cells are motile (monotrichous), nonsporeforming rods occurring singly, in pairs, or as chains of four cells. No growth on MRS medium. Colonies (TSA, 20 °C) are 1–2 mm in diameter. Circular, entire, convex, and semitranslucent. The color of the colonies is whitish at aerobic growth conditions and buff at anaerobic growth conditions. pH range supporting growth 5.5–9.0. Catalase is produced on heme-containing media. H_2S not produced and nitrate not reduced. Hydrolyzes hippurate. Additional physiological and biochemical characteristics are presented in Table 101. The most abundant cellular fatty acids are $C_{16:0}$ (31.1%), $C_{16:1}$ (24.2%), and $C_{18:1 \, \omega 9}$c (23.4%). Other fatty acids are $C_{18:2 \, \omega 6,9c}$ or $C_{18:0}$ (10.8%), $C_{14:0}$ (5.4%), and $C_{18:0}$ (3.5%). The peptidoglycan type is not known.

Habitat is the intestines of healthy fish.

DNA G+C content (mol%): not determined.

Type strain: K1, CCUG 31728, CIP 106863, DSM 13024.

GenBank accession number (16S rRNA gene): Z773313.

6. **Carnobacterium maltaromaticum** (Miller, Morgan and Libbey 1974) Mora, Scarpellini, Franzetti, Colombo and Galli 2003, 677[VP] (*Lactobacillus maltaromicus* Miller, Morgan and Libbey 1974, 352; *Lactobacillus piscicola* Hiu, Holt, Sriranganathan, Seidler and Fryer 1984, 399; *Lactobacillus carnis* Shaw and Harding 1985, 296; *Carnobacterium piscicola* Collins, Farrow, Phillips, Feresu and Jones 1987, 315.)

malt.a.ro.mat.ic′um. N.L. neut. n. *maltum-i* malt; L. adj. *aromaticus -a -um* aromatic, fragrant; N.L. neut. adj. *maltaromaticum* possessing a malt-like aroma.

Cells are rods (0.5–0.7 × 3.0 μm) occurring singly or in chains. Nonmotile. Facultatively anaerobic. L-(+)-Lactic acid, ethanol, and acetate are produced heterofermentatively. Gas production is weak and frequently undetectable. Grows in MRS, TSBY, and brain heart infusion media. Additional physiological and biochemical characteristics are presented in Table 101. Major cellular fatty acids are straight-chain saturated and monounsaturated acids, with tetradecanoic, hexadecanoic, and 9- and 10-octadecenoic acids predominating.

Isolated from dairy products, meat, fish (healthy and diseased), and human plasma and pus.

DNA G+C content (mol%): 33.7–36.4 (T_m).

Type strain: ATCC 27865, CCUG 30142, CIP 103135, DSM 20342, JCM 1154, LMG 6903, NRRL B-14852.

GenBank accession number (16S rRNA gene): M58825, X54420.

7. **Carnobacterium mobile** Collins, Farrow, Phillips, Feresu and Jones 1987, 315[VP]

mo′bi.le. L. neut. adj. *mobile* movable or motile.

Gram-stain-positive, nonsporeforming, straight, slender rods which occur singly or in short chains. Cells are motile. Colonies are white, convex, shiny, and circular. Facultatively anaerobic. L-(+)-Lactic acid is produced heterofermentatively. Gas production from glucose is observed for most strains in arginine-MRS broth. All strains are lysine decarboxylase, ornithine decarboxylase, tryptophan desaminase, and urease-negative. Additional physiological and biochemical characteristics are presented in Table 101. The cellular fatty acids are of the straight-chain saturated and monounsaturated types, with hexadecanoic, hexadecenoic, and 9,10-octadecenoic acids predominating.

Isolated from irradiated chicken meat and from shrimp.

DNA G+C content (mol%): 35.5–37.2 (T_m).

Type strain: MT37L, ATCC 49516, CCUG 30096, CIP 103159, DSM 4848, JCM 12516, LMG 9842, NCIMB 12847, NRRL B-14831.

GenBank accession number (16S rRNA gene): AB083414, X54271.

8. **Carnobacterium pleistocenium** Pikuta, Marsic, Bej, Tang, Krader and Hoover 2005, 477[VP]

plei.sto.ce'ni.um. N.L. neut. adj. *pleistocenium* belonging to the Pleistocene, a geological epoch.

Cells are motile, small rods with rounded ends, 0.7–0.8 × 1.0–1.5 μm. Gram-stain-positive. Growth occurs at 22 °C in the pH range of 6.5–9.5. Range of NaCl for growth is 0–5% (w/v); optimum growth at 0.5% (w/v) NaCl. Facultatively anaerobic. Growth occurs with starch, peptone, Bacto tryptone, Casamino acids, and yeast extract. End products of growth are acetate, ethanol, and traces of carbon dioxide. Additional physiological and biochemical characteristics are presented in Table 101.

Isolated from a sample of permafrost from Fox Tunnel, Alaska.

DNA G+C content (mol%): 42 ± 1.5 (T_m).

Type strain: FTR1, ATCC BAA-754, CIP 108033, JCM 12174.

GenBank accession number (16S rRNA gene): AF450136.

9. **Carnobacterium viridans** Holley, Guan, Peirson and Yost 2002, 1884[VP]

vi'ri.dans. N.L. adj. *viridans* from L. v. *viridare* to make green, referring to the production of a green color in cured meat by the organism.

Cells are nonmotile, slightly curved rods, singly or in pairs, or as straight rods (0.8 × 3.6 ± 0.6 μm). Facultatively anaerobic. Grows satisfactorily in BHI, APT, M5, and CTSI media, but poorly on a variety of media including MRS and SL agar. Grows over a range of pH from 5.5–9.1. Produces predominantly L(+)-lactic acid from glucose and neither gas nor H_2S. Does not grow in 4% (w/v) NaCl but will tolerate 26.4% (w/v) NaCl (saturated brine) for long periods at 4 °C. Additional physiological and biochemical characteristics are presented in Table 101. The cell-wall peptidoglycan contains *meso*-DAP.

Isolated from green, discolored vacuum packaged bologna type sausage.

DNA G+C content (mol%): not determined.

Type strain: MPL-11, ATCC BAA-336, DSM 14451, JCM 12222.

GenBank accession number (16S rRNA gene): AF425608.

Genus II. **Alkalibacterium** Ntougias and Russell 2001, 1169[VP]

SPYRIDON NTOUGIAS AND NICHOLAS J. RUSSELL

Al.ka.li.bac'te.ri.um. N.L. n. *alkali* (from Ar. article *al* the; Ar. n. *qaliy* ashes of saltwort) alkali; L. neut. n. *bacterium* a small rod; N.L. neut. n. *Alkalibacterium* bacterium living under alkaline conditions.

Cells are **rods** up to 0.4–1.2 × 0.7–3.7 μm **in both liquid culture and on plates**. Cells occur singly, in pairs, or in clusters of up to five cells. **Gram-stain-positive. Motile by polar or peritrichous flagella. Endospores are** not formed. Facultatively anaerobic. Catalase- and oxidase-negative. Produce DL-lactate from glucose. On plates, forms round (0.5–2.5 mm diameter) **white colonies. Obligately alkaliphilic, growing only above pH 8.0 and up to pH 11 or higher. Halotolerant** with growth up to 15–17 % NaCl and often a broad optimum from 0 to 13 % NaCl. **Mesophilic (often psychrotolerant)** with growth from 4–15 to 35–45 °C and optimal growth at 20–37 °C. The major phospholipids are phosphatidyl-glycerol, diphosphatidylglycerol, phosphatidylserine, and an unknown phospholipid; glycolipids are also present. The cellular **fatty acids are saturated and unsaturated even-carbon chain**, with *n*-hexadecanoic, hexadecenoic, and octadecenoic acids as the major components. **Quinones are not present**.

DNA G+C content (mol%): 39.7–47.8.

Type species: **Alkalibacterium olivapovliticus** corrig. Ntougias and Russell 2001, 1169[VP].

Further descriptive information

Although colonies on plates are white, centrifuged biomass is yellow or orange. The peptidoglycan type is variable (A4β or A4α).

Enrichment and isolation procedures

Alkalibacterium species have been isolated from naturally alkaline habitats (e.g., soda lakes) as well as biotechnological processes (e.g., edible-olive wash-waters and indigo production). Both cultured and uncultured (clones) isolates have been reported.

Maintenance procedures

Cultures can be maintained on slants of agar (2%, w/v), containing 0.05 M L-sodium glutamate, 0.5% (w/v) yeast extract,

0.1 M sodium carbonate, 1 mM dipotassium hydrogen phosphate, 7.6 mM ammonium sulfate, and 0.1 mM magnesium sulfate for up to 2 months. Alternatively, reinforced clostridial broth (RCB) or agar (RCA) media containing 100 mM $NaHCO_3/Na_2CO_3$ buffer, pH 10 (alkali-RCB or alkali-RCA) can be used. $NaHCO_3/Na_2CO_3$ buffer should be sterilized separately by autoclaving. Cultures of *Alkalibacterium olivapovliticus* should be subcultured every 15 d. Viability of stored cultures on beads in glycerol media is poor. The best method of storing cultures is by freeze-drying; such cultures can be kept for more than 1 year.

Differentiation of the genus *Alkalibacterium* from other genera

Alkalibacterium is distinguishable from the closely related *Marinilactibacillus* on the basis of structure of the V6 region of 16S rRNA, the pH optimum and range of pH for growth, and minimum growth temperature (Ishikawa et al., 2003).

Alkalibacterium can be distinguished from the phylogenetically related genus *Alloiococcus* by differences in cellular morphology, motility, oxygen requirement, growth at pH 7, catalase reaction, and DNA base composition; from the genus *Dolosigranulum* by cellular morphology, pH tolerance, temperature and salt concentration ranges for growth, and peptidoglycan type; and from the genus *Carnobacterium* by pH and salt concentration ranges for growth, G+C content of DNA, and peptidoglycan type.

Taxonomic comments

The genus *Alkalibacterium* aligns with the lactic acid bacteria. Lactate and formate are the main end products of glucose fermentation, followed by acetate and ethanol. The amount of formate production increases relative to lactate at higher pH values (Ishikawa et al., 2003).

The peptidoglycan type of *Alkalibacterium psychrotolerans* and *Alkalibacterium iburiense* is A4α, instead of A4β, as recently re-examined (Yumoto et al., 2008).

The species *Alkalibacterium psychrotolerans* can be distinguished from the closely related species *Alkalibacterium olivapovliticus* on the basis of its flagellar position and strength of motility, pH range for growth, sugar fermentation pattern, and peptidoglycan type; and from *Marinilactibacillus psychrotolerans* on the basis of its optimum pH and range for growth, sugar fermentation pattern, colony color, peptidoglycan type and G+C content of DNA. The DNA–DNA similarity of *Alkalibacterium olivapovliticus* and *Alkalibacterium psychrotolerans* is 24.3 %.

The species *Alkalibacterium iburiense* can be distinguished from the closely related species *Alkalibacterium olivapovliticus* on the basis of its flagellar position and strength of motility, pH range for growth, temperature upper limit for growth, antibiotic sensitivity, peptidoglycan type and G+C content of DNA; from *Alkalibacterium psychrotolerans* on the basis of sugar fermentation pattern, peptidoglycan type and G+C content of DNA; and from *Marinilactibacillus psychrotolerans* on the basis of its optimum pH and range for growth, temperature lower limit for growth, sugar fermentation pattern, colony color, peptidoglycan type and G+C

content of DNA. The fatty acid composition of *Alkalibacterium iburiense* is similar to that of *Alkalibacterium olivapovliticus* rather than that of *Alkalibacterium psychrotolerans* and *Marinilactibacillus psychrotolerans*. The DNA–DNA similarities of *Alkalibacterium iburiense* with *Alkalibacterium psychrotolerans* and *Alkalibacterium olivapovliticus* are 14.1 % and 7.3 %, respectively.

The species *Alkalibacterium indicireducens* can be distinguished from the closely related species *Alkalibacterium olivapovliticus* on the basis of its flagellar position, pH range for growth, temperature range and optimum for growth, hydrolysis of cellulose, sugar fermentation pattern, antibiotic sensitivity, peptidoglycan type and G+C content of DNA; from *Alkalibacterium psychrotolerans* on the basis of its temperature range and optimum for growth, hydrolysis of cellulose, sugar fermentation pattern, sensitivity to trimethoprim and G+C content of DNA; from *Alkalibacterium iburiense* on the basis of its temperature range and optimum for growth, hydrolysis of cellulose, sugar fermentation pattern, peptidoglycan type and G+C content of DNA. The DNA–DNA similarities of *Alkalibacterium indicireducens* with *Alkalibacterium psychrotolerans*, *Alkalibacterium iburiense* and *Alkalibacterium olivapovliticus* are 40 %, 34 % and 21 %, respectively.

List of species of the genus *Alkalibacterium**

1. **Alkalibacterium olivapovliticus** corrig. Ntougias and Russell 2001, 1169[VP]

o.li.va.pov′lit.i.cus. L. n. *oliva* olives; Gr. n. *apovlito* waste disposal; N.L. gen. n. *olivapovliticus* from the waste of the olives.

Rods (0.4–0.6 ×1.3–2.9 μm) in liquid culture and on plates. Cells occur singly, in pairs, or in small clusters. Cells stain Gram-stain-negative but show Gram-stain-positive cell wall behavior in the KOH test and the aminopeptidase test. Weakly motile by polar flagella. Endospores are not formed. Colonies are small, round, and glistening, with diameter of 0.8–1.0 mm, appearing as pale white due to their small size, although the color of bulk biomass observed after centrifugation is yellow or orange. Obligately alkaliphilic, growing above pH 8–11 or higher and an optimum of pH 9–10. Halotolerant, growing in up to 15% NaCl with an optimum of 3–5% NaCl for the type strain and 0–10% NaCl for other strains. Psychrotolerant, growing over the range 4–37 °C with an optimum of 27–32 °C. The major phospholipids are phosphatidylglycerol, diphosphatidylglycerol, phosphatidylserine, and an unknown phospholipid; there are also four unidentified glycolipids. The cellular fatty acids are mainly saturated and unsaturated even-carbon chain, with hexadecanoic ($C_{16:0}$), hexadecenoic ($C_{16:1\ \omega9c}$), and octadecenoic ($C_{18:1\ \omega9c}$) acids as the major components. No quinones could be detected. Sugars which are metabolized include D(+)-glucose, D(+)-glucose-6-phosphate, maltose, D(+)-cellobiose, sucrose, mannose, and trehalose. Amino acids are not utilized as sole carbon and energy sources. Yeast extract could be utilized as the sole carbon and energy source. Cells are catalase- and oxidase-negative. Culture growth is inhibited by ampicillin, carbenicillin, chloramphenicol, penicillin G, kanamycin, streptomycin, and trimethoprim. Some strains are sensitive to amoxycillin or to miconazole and neo-

mycin. The peptidoglycan type is A4β, L-Orn–D-Asp. The natural habitat is the wash waters of edible olives. Reference strains include DSM 12937, NCIMB 13711 (strain WW2-SN4c; GenBank accession no. 16S rRNA gene, AF143512), DSM 12938, NCIMB 13712 (strain WW2-SN5; GenBank accession no. 16S rRNA gene, AF14513).

DNA G+C content (mol%): 39.7 (HPLC).
Type strain: WW2-SN4a, DSM 13175, NCIMB 13710.
GenBank accession number (16S rRNA gene): AF143511.

2. **Alkalibacterium psychrotolerans** Yumoto, Hirota, Nodasaka, Yokota, Hoshino and Nakajima 2004, 2382[VP]

psy.chro.to′le.rans. Gr. adj. *psychros* cold; L. part. adj. *tolerans* tolerating; N.L. neut. part. adj. *psychrotolerans* tolerating cold environments.

Cells are straight rods, 0.4–0.9 × 0.7–3.1 μm, stain Gram-stain-positive, and have peritrichous flagella. Endospores are not produced. Cells grow aerobically and anaerobically. Catalase- and oxidase-negative. Reduces indigo dye. Colonies are circular, convex, and pale white. Obligately alkaliphilic with growth at pH 9–12 but not pH 7–8 and an optimal pH of 9.5–10.5 at 27 °C. Halotolerant, growing at 0–17 % (w/v) NaCl with a broad optimum concentration of 2–12 %. Psychrotolerant, growing between 5 and 45 °C with an optimum growth temperature around 34 °C. The major cellular fatty acids are tetradecanoic ($C_{14:0}$), hexadecanoic ($C_{16:0}$), hexadecenoic ($C_{16:1\ \omega7c}$) and octadecenoic ($C_{18:1\ \omega9c}$) acids. Starch and gelatin are not hydrolyzed. Cells ferment D(+)-arabinose, D(+)-xylose, D(+)-glucose, and maltose, but do not ferment D(+)-melibiose, D(+)-raffinose, *myo*-inositol, mannitol, erythritol, sorbitol, xylitol, adonitol, dulcitol, sucrose, D(+)-galactose, inulin, and L(+)- or D(+)-rhamnose. Lactate is the major end product of glucose fermentation. No quinones can be detected. The peptidoglycan type is A4α, L-Lys (L-Orn)–D-Asp. Isolated from Polygonum Indigo (*Polygonum tinctorium* Lour.) produced in Date-city, Hokkaido, Japan.

---

*Four novel species, named *Alkalibacterium thalassium*, *Alkalibacterium pelagium*, *Alkalibacterium putridalgicola*, and *Alkalibacterium kapii*, which were isolated from marine organisms and salted foods collected in Japan and Thailand, have been accepted for publication in IJSEM in 2009 by M. Ishikawa et al.

DNA G+C content (mol%): 40.6 (HPLC).

Type strain: IDR2-2, JCM 12281, NCIMB 13981.

GenBank accession number (16S rRNA gene): AB125938.

3. **Alkalibacterium iburiense** Nakajima, Hirota, Nodasaka and Yumoto 2005, 1529[VP]

i.bu.ri.en'se. N.L. neut. adj. *iburiense* from Iburi, the place where the micro-organism was isolated.

Cells are Gram-stain-positive, peritrichously flagellated, straight rods (0.5–0.7 × 1.3–2.7 μm). Endospores are not produced. Cells grow aerobically and anaerobically. Aminopeptidase, catalase- and oxidase-negative. Reduces indigo dye. Colonies (2–2.5 mm in diameter) are circular, convex, and pale white. Obligately alkaliphilic with growth at pH 9–12 but not pH 7–8 and an optimal pH of 9.5–10.5 at 27 °C. Halotolerant, growing at 0–16 % (w/v) NaCl with a broad optimum concentration of 3–13 %. Psychrotolerant, growing between 5 and 45 °C with an optimum growth temperature of 30–37 °C. Gelatin and starch are not hydrolyzed. Cells ferment D-arabinose, glycogen and *N*-acetylglucosamine, but do not ferment arbutin, D-galactose, D-mannitol, D-sorbitol, melibiose, myo-inositol or raffinose. Fermentation of D-fructose, D-mannose, D-xylose, L-rhamnose, maltose, sucrose and trehalose is variable (all positive for the type strain). Lactate is the major end product of glucose fermentation. Culture growth is inhibited by amoxycillin (10 μg), ampicillin (10 μg) and penicillin G (1 IU), but is not inhibited by chloramphenicol (2 μg), kanamycin (10 μg), ketoconazole (25 μg), miconazole (25 μg), sulfamethoxazole (25 μg) and trimethoprim (25 μg). The major cellular fatty acids are hexadecanoic ($C_{16:0}$), hexadecenoic ($C_{16:1\ \omega9c}$) and octadecenoic ($C_{18:1\ \omega9c}$) acids. No quinones can be detected. The peptidoglycan type is A4α, L-Lys–D-Asp. Strain M3 was isolated from a contaminated culture in alkali broth, and reference strains 41A (GenBank accession no. 16S rRNA gene, AB188092) and 41C (GenBank accession no. 16S rRNA gene, AB188093) were isolated from polygonum indigo (*Polygonum tinctorium* Lour.) fermentation liquor produced in Date-city, Iburi, Hokkaido, Japan.

DNA G+C content (mol%): 42.6–43.2 (HPLC).

Type strain: M3, JCM 12662, NCIMB 14024.

GenBank accession number (16S rRNA gene): AB188091.

4. **Alkalibacterium indicireducens** Yumoto, Hirota, Nodasaka, Tokiwa and Nakajima 2008, 904[VP]

in.di.ci.re.du'cens. L. n. *indicum* indigo; L. part. adj. *reducens* bringing or leading back, reducing; N.L. part. adj. *indicireducens* indigo-reducing.

Cells are Gram-stain-positive, peritrichously flagellated, straight rods (0.4–1.2 × 1.7–3.7 μm). Endospores are not produced. Cells grow aerobically and anerobically. Aminopeptidase, catalase- and oxidase-negative. Reduces indigo dye. Colonies (0.5–2.0 mm in diameter) are circular, convex, and pale white. Obligately alkaliphilic with growth at pH 9–12.3 but not pH 7–8 and an optimal pH of 9.5–11.5. Halotolerant, growing at 0–15 % (w/v) NaCl with a broad optimum concentration of 1–11 %. Temperature range for growth is 15–40 °C with an optimum of 20–30 °C. Hydrolyzes starch, xylan and cellulose, but not casein or gelatin. Cells ferment D-glucose and sucrose, but do not ferment D-arabinose, D-galactose, D-mannose, D-xylose, inositol, L-rhamnose, maltose, melibiose, raffinose, sorbitol and trehalose. Fermentation of D-fructose is variable (positive for the type strain). Lactate is the major end product of glucose fermentation. Culture growth is inhibited by amoxycillin (10 μg), ampicillin (2 μg) and penicillin G (1 IU), but is not inhibited by chloramphenicol (10 μg), kanamycin (10 μg), ketoconazole (25 μg), miconazole (2 μg), sulfamethoxazole (25 μg) and trimethoprim (25 μg). The major cellular fatty acids are hexadecanoic ($C_{16:0}$), hexadecenoic ($C_{16:1\ \omega9c}$) and octadecenoic ($C_{18:1\ \omega\ 9c}$) acids. No quinones can be detected. The peptidoglycan type is A4α, L-Lys (L-Orn)–D-Asp. The type strain A11 and the reference strains F11 (GenBank accession no. 16S rRNA gene, AB268550) and F12 (GenBank accession no. 16S rRNA gene, AB268551) were isolated from Polygonum Indigo (*Polygonum tinctorium* Lour.) fermentation liquor samples obtained from Tokushima City and Aizumi-cho, Itanogun (Japan), respectively.

DNA G+C content (mol%): 47.0–47.8 (HPLC).

Type strain: A11, JCM 14232, NCIMB 14253.

GenBank accession number (16S rRNA gene): AB268549.

Genus III. **Allofustis** Collins, Higgins, Messier, Fortin, Hutson, Lawson and Falsen 2003, 813[VP]

PAUL A. LAWSON

Al.lo.fus'tis. Gr. prefix *allos* another, the other; L. masc. n. *fustis* stick; N.L. masc. n. *Allofustis* the other stick or rod.

Cells are **Gram-stain-positive, nonsporeforming and rod-shaped. Facultatively anaerobic. Catalase- and oxidase-negative**. Arginine dihydrolase, leucine arylamidase, and pyroglutamic acid arylamidase are produced. Indole-negative. Nitrate is not reduced. Voges–Proskauer-negative. The long-chain cellular fatty acids are of the straight-chain saturated and monounsaturated types. **The cell-wall murein is type L-lysine-direct (A1α).** The genus is classed in the family *Carnobacteriaceae*. Isolated from porcine semen.

DNA G+C content (mol%): 39.

Type species: **Allofustis seminis** Collins, Higgins, Messier, Fortin, Hutson, Lawson and Falsen 2003, 813[VP].

Isolation and maintenance procedures

Allofustis phocae has been isolated on sheep blood agar at 37 °C under anaerobic conditions. There is no information on enrichment or selective media for this species. The organism grows well on agar media (such as brain heart infusion or Columbia agar) supplemented with blood (5 % v/v). Grows well on chocolate agar in CO_2. The organism grows in brain heart infusion broth with serum (5 % v/v). Can be stored on cryogenic beads at −70 °C or lyophilized for long-term preservation.

Taxonomic comments

The genus *Allofustis* was created in 2003 to accommodate phylogenetically distinct catalase-negative, rod-shaped organisms encountered in porcine semen specimens (Collins et al., 2003). The genus is monospecific and phylogenetically belongs to the order *Lactobacillales* in the phylum *Firmicutes*, according to the revised roadmap to Volume 3 of the *Systematics* (Ludwig et al., this volume). Phylogenetic analysis using 16S rRNA gene sequences shows that *Allofustis seminis* clusters within a suprageneric grouping that includes *Alkalibacterium* species, *Alloiococcus otitis*,

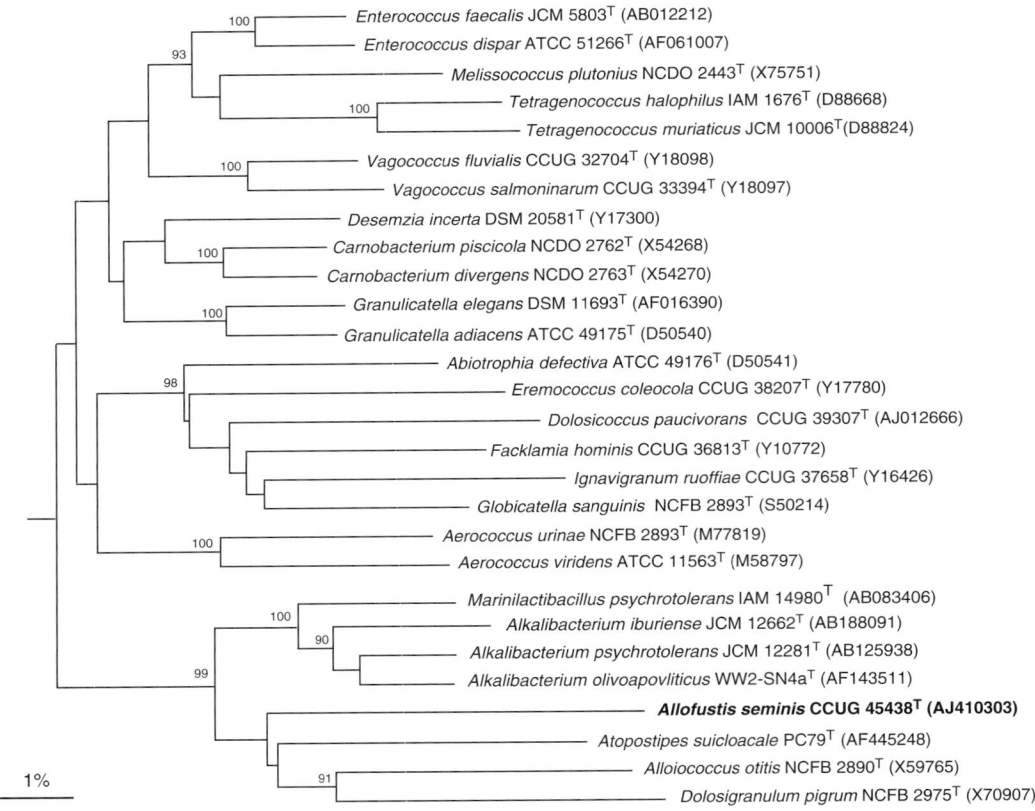

FIGURE 100. Unrooted neighbor-joining dendrogram depicting the estimated phylogenetic relationships of *Allofustis seminis* and its close relatives. The numbers on the branches refer to bootstrap values, determined from 1000 replications. Only values above 90% or higher are shown. Bar = 1% sequence divergence.

Atopostipes suicloacale, Dolosigranulum pigrum, and *Marinilactibacillus psychrotolerans. Allofustis seminis* forms a distinct subline within this grouping and does not display a particularly close affinity with any of the aforementioned taxa. A tree depicting the phylogenetic relationships of *Allofustis seminis* is shown in Figure 100. The predicted secondary structure of the V6 region of the 16S rRNA is also useful in the assignment of organisms within this suprageneric cluster of organisms (see Figure 101). In particular, nucleotides at positions 457–462 and the complementary nucleotide sequence at positions 471–476 appear to especially informative at the genus level. However, it remains

to be seen if these pairings found in *Allofustis seminis* are shown to be stable, as additional strains/species of the genus are isolated and described.

Differentiation of the genus *Allofustis* from other genera

In addition to differentiation by 16S rRNA gene sequence analysis, *Allofustis seminis* can be readily distinguished from its closest phylogenetic relatives *Alkalibacterium* species, *Alloiococcus otitis, Atopostipes suicloacale, Dolosigranulum pigrum,* and *Marinilactibacillus psychrotolerans* using a combination of morphological, biochemical, and chemotaxonomic criteria (see Table 102).

List of species of the genus *Allofustis*

1. **Allofustis seminis** Collins, Higgins, Messier, Fortin, Hutson, Lawson and Falsen 2003, 813[VP]

   sem.in′is. L. n. *semen* seed; L. gen. n. *seminis* of semen.

   Cells stain Gram-positive and are rod-shaped. Nonsporeforming. β-Hemolytic reaction on sheep blood. Facultatively anaerobic; grows well on chocolate agar anaerobically or in CO_2, at 37 °C. Catalase and oxidase-negative. Acid is not produced from L-arabinose, D-arabitol, cyclodextrin, glycogen, lactose, maltose, mannitol, mannose, melibiose, melezitose, pullulan, raffinose, D-ribose, sorbitol, sucrose, D-tagatose, or trehalose. Arginine dihydrolase, arginine arylamidase, acid phosphatase, alkaline phosphatase, alanine arylamidase, alanine

phenylalanine proline arylamidase, glycine arylamidase, glycyl tryptophan arylamidase, histidine arylamidase, leucyl glycine arylamidase, leucine arylamidase, phosphoamidase, proline arylamidase, phenyl alanine arylamidase, pyroglutamic acid arylamidase, β-mannosidase, serine arylamidase, tyrosine arylamidase, and valine arylamidase are produced. Activity for α-arabinosidase, esterase C4, ester lipase C8, α-galactosidase, β-galactosidase, β-galactosidase-6-phosphate, α-glucosidase, β-glucuronidase, glutamic acid decarboxylase, glutamyl glutamic acid arylamidase, lipase C14, α-mannosidase, chymotrypsin, trypsin, and urease is not detected. α-Fucosidase, β-glucosidase, and N-acetyl-β-glucosaminidase may or may not be detected.

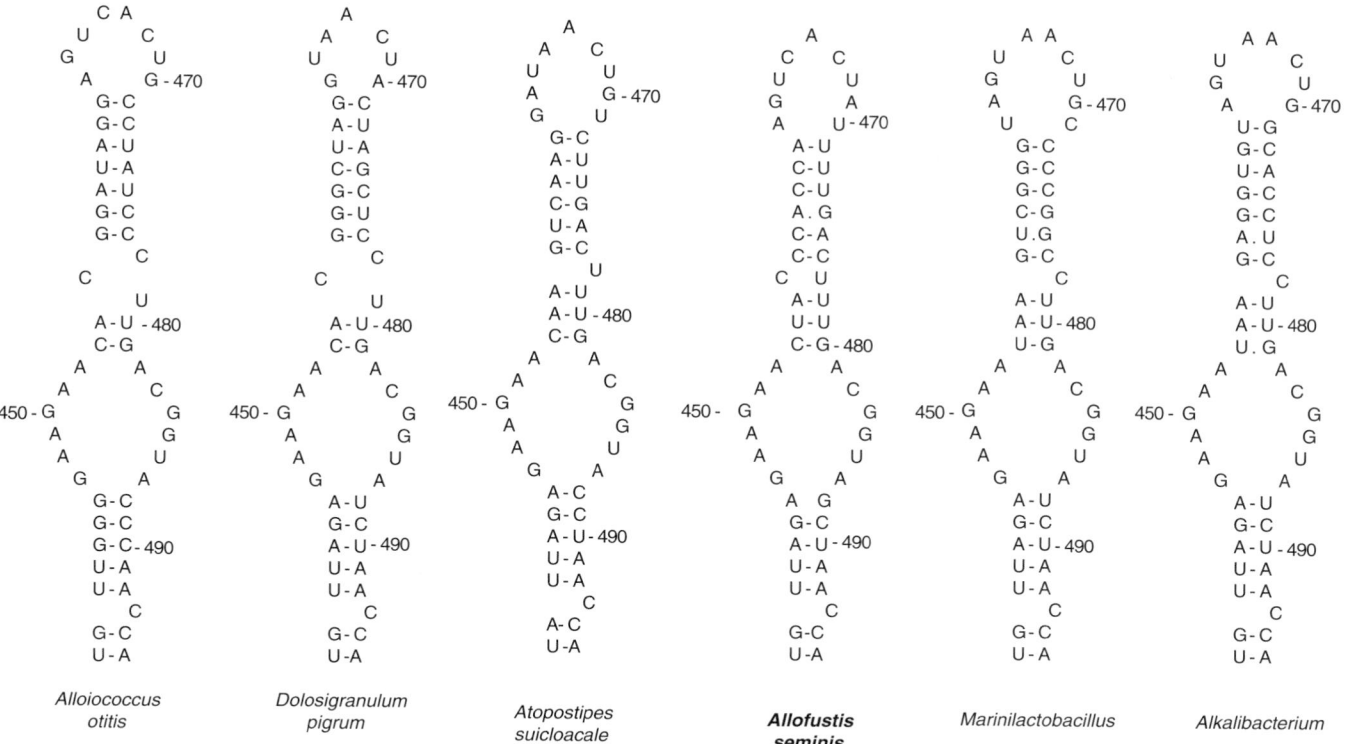

FIGURE 101. Nucleotide sequences and predicted secondary structure of the V6 region of the 16S rRNA of *Allofustis seminis* and its closest phylogenetic relatives. Numbers correspond to positions in the *Escherichia coli* sequence.

TABLE 102. Criteria that are useful in distinguishing the genus *Allofustis* from phylogenetically closely related taxa

Characteristics	*Allofustis*	*Alkalibacterium*[b]	*Alloiococcus*[c]	*Atopostipes*[d]	*Dolosigranulum*[e]	*Marilactibacillus*[f]
Cell morphology	Rod	Rod	Ovoid	Rod	Rod	Rod
Relationship to air	Facultatively anaerobic	Facultatively anaerobic	Aerobic	Facultatively anaerobic	Facultatively anaerobic	Facultatively anaerobic
Growth above 32 °C	+	+	+	−	+	+
Major cellular fatty acids	$C_{16:0}$, $C_{16:1}$, $C_{18:0}$, $C_{18:1w9c}$	$C_{16:0}$, $C_{16:1}$7c, $C_{16:1}$9c $C_{18:0}$,	$C_{16:0}$, $C_{18:1 w9c}$ $C_{18:2 w6}$, $C_{18:0 ante}$,	$C_{14:0}$, $C_{16:0}$, $C_{16:1}$, $C_{18:0}$, $C_{18:1}$	$C_{16:1w9c}$ $C_{16:0}$, $C_{16:1}$, $C_{18:1 w9c}$	$C_{16:0}$, $C_{16:1}$, $C_{18:0}$, $C_{18:1}$9c
Catalase	−	−	+	−	−	−
Acid from:						
Maltose	−	v	−	+	+	+
Sucrose	−	v	−	+	+	+
Production of:						
Arginine dihydrolase	+	ND	−	−	−	ND
β-Galactosidase	−	ND	+	+	+	ND
Murein type	L-Lys-direct	Orn–D-Glu, Orn–D-Asp	ND	L-Lys–D-Asp	Lys–D-Asp	Orn–D-Glu
DNAG+C content (mol%)	39	39.7–43.2	44–45	43.9	40.5	34.6–36.2

[a]Symbols: +, >85% positive; d, different strains give different reactions (16–84% positive); −, 0–15% positive; v, variable; w, weak reaction; ND, not determined.
[b]Data from Nakajima et al. (2005), Ntougias and Russell, (2001), Yumoto et al. (2004).
[c]Data from Aguirre and Collins, (1992a), CCUG Database http://www.ccug.gu.se.
[d]Data from Cotta et al. (2004b).
[e]Data from Aguirre et al. (1993), CCUG Database http://www.ccug.gu.se.
[f]Data from Ishikawa et al. (2003).

Hippurate and esculin are not hydrolyzed. Indole and acetoin are not produced, and nitrate is not reduced. The major long-chain cellular fatty acids are $C_{16:0}$, $C_{16:1}$, $C_{18:0}$, and $C_{18:1 \omega9c}$. The cell wall murein is based on L-lysine variation A1α (type: L-Lys-direct) (see Schleifer and Kandler, (1972) for nomenclature).

The only known source from which *Allofustis seminis* has been isolated is porcine semen. Habitat is not known.

DNA G+C content (mol%): 39 (HPLC).

Type strain: 01-570-1, CCUG 45438, CIP 107425, DSM 15817.

GenBank accession number (16S rRNA gene): AJ410303, AJ458446.

Genus IV. **Alloiococcus** Aguirre and Collins 1992a, 83[VP]

MATTHEW D. COLLINS

Al.loi.o.coc′cus. Gr. adj. *alloios* different; N.L. n. *coccus* coccus; N.L. masc. n. *Alloiococcus* different coccus, referring to the phylogenetic distinctiveness of the organism.

Cells are ovoid in shape and divide on an irregular plane giving rise to pairs, tetrads, and clusters. Gram-stain-positive and nonmotile. Nonsporeforming. **Nutritionally fastidious.** Grows in 6.5% NaCl (slowly) but not in 10% NaCl. **No growth at 10 or 45 °C. Aerobic.** Catalase may or may not be produced. Oxidase-negative. **Gas is not produced. Acid is not produced from glucose and other carbohydrates. Pyrrolidonyl arylamidase and leucine aminopeptidase-positive. Vancomycin sensitive.** Long-chain cellular fatty acids are of the straight-chain saturated and unsaturated types.

DNA G+C content (mol%): 44–45 (T_m).

Type species: **Alloiococcus otitis** Aguirre and Collins 1992a, 83[VP].

Further descriptive information

The genus contains only one species, *Alloiococcus otitis*, and therefore the characteristics provided below refer to this species. Colonies formed on BHIA with 5% (v/v) rabbit blood are small, moist, and slightly yellow at 3 d. After 4 d, colonies become darker yellow, dense, hard, involuted, and adherent to agar. Nonhemolytic, but cultures may become α-hemolytic after extended incubation. They grow on brucella agar supplemented with 5% (v/v) sheep blood but not without blood. Cultures grow in heart infusion and brain heart infusion broths very slowly, often requiring 2 d or more of incubation for growth. Alloiococci do not normally grow under anaerobic atmospheres, although some atypical strains have been reported to produce weak growth anaerobically (Miller et al., 1996). Catalase reaction is variable; cultures may give a negative or a weak catalase reaction, but they do not contain cytochrome enzymes. Acid is not formed from carbohydrate broth. Hippurate may or may not be hydrolyzed. Starch and esculin are not hydrolyzed. Most strains are capable of growing on bile-esculin medium but do not give positive reactions. Pyroglutamic acid arylamidase is produced and most strains are β-galactosidase-positive. Arginine dihydrolase, alanine phenylalanine proline arylamidase, glycyl tryptophan arylamidase, N-acetyl-β-glucosaminidase, pyrazinamidase, β-mannosidase, and urease are not produced. Voges–Proskauer-negative.

Polyamines are not detected in *Alloiococcus otitis* (Hamana, 1994). The long-chain fatty acids are of the straight-chain saturated and monounsaturated types, with $C_{16:0}$ (26–34%) and $C_{18:1}$ $_{\omega 9c}$ (14–15%) predominating (Bosley et al., 1995).

Alloiococci are β-lactamase-negative. All strains tested are resistant to trimethoprim-sulfamethoxazole, and, except for a single strain, to erythromycin. Alloiococci are susceptible or intermediately resistant to penicillin and ampicillin (range, 0.06 to 0.5 mg/ml). (Bosley et al., 1995).

Alloiococci have been isolated from human middle ear fluids (Bosley et al., 1995; Facklam and Elliott, 1995; Faden and Dryja, 1989). *Alloiococcus otitis* has been associated with chronic otitis media with effusion (Beswick et al., 1999; Faden and Dryja, 1989; Fraise et al., 2001; Hendolin et al., 1999). Several 16S rDNA PCR/probe based tests have been described for identifying and/or detecting *Alloiococcus otitis*

(Aguirre and Collins, 1992b; Beswick et al., 1999; Hendolin et al., 1999, 2000). PCR-based methods have detected alloiococci in culture negative specimens (Beswick et al., 1999; Hendolin et al., 1999).

Isolation procedures

Alloiococci do not grow on many commonly used microbiological media. They grow well in brain heart infusion broth supplemented with 0.07% lecithin and 0.5% Tween 80 (BHIS broth) and on brain heart infusion agar (BHIA) with 5% rabbit blood at 37 °C.

Maintenance procedures

Alloiococcus otitis grows in brain heart infusion broth with 0.07% lecithin and 0.5% Tween80 (BHIS broth) and on brain heart infusion agar (BHIA) or heart infusion agar (HIA) with 5% rabbit blood. Most strains do not grow on trypticase soy agar with 5% sheep blood or in thioglycolate or Todd–Hewitt broths. Strains can be stored on cryogenic beads at −70 °C or lyophilized for long-term preservation.

Taxonomic comments

The first strains of alloiococci were isolated from typanocentesis fluid collected from middle ears of children suffering from otitis media (Faden and Dryja, 1989). Based on the results of a phylogenetic study by Aguirre and Collins (1992a), a new genus and species *Alloiococcus otitis*, was proposed. Phylogenetically, *Alloiococcus otitis* is a member of the family *Carnobacteriaceae*, order *Lactobacillales* in the phylum *Firmicutes*, according to the revised roadmap to Volume 3 of the *Systematics* (Ludwig et al., this volume). *Dolosigranulum pigrum* is the nearest taxonomically named phylogenetic relative of *Alloiococcus*. However, several cloned 16S rDNA sequences derived from some uncultured bacteria associated with the sheep mite *Psoroptes ovis* display a closer phylogenetic affinity (approx. 4% 16S rRNA sequence divergence) with *Alloiococcus* than does *Dolosigranulum pigrum* (Hogg and Lehane, 1999). A tree depicting the phylogenetic relationships of *Alloiococcus otitis* is shown in Figure 102.

Differentiation of the genus *Alloiococcus* from other genera

Alloiococci can be readily identified in the routine laboratory. Using conventional tests (Facklam and Elliott, 1995), they resemble *Dolosigranulum*, *Ignavigranum*, and *Facklamia* species in being susceptible to vancomycin, nonmotile, pyrrolidonyl-β-naphthylamide and leucine aminopeptidase-positive, growing in 6.5% NaCl broth, not growing at 10 or 45 °C, having a negative bile-esculin reaction, and showing α- or γ-hemolytic reaction on sheep blood (LaClaire and Facklam, 2000b). Alloiococci, however, can be readily distinguished from these other taxa by not growing anaerobically and by their fastidious growth requirements. In addition, alloiococci do not produce acid from carbohydrates.

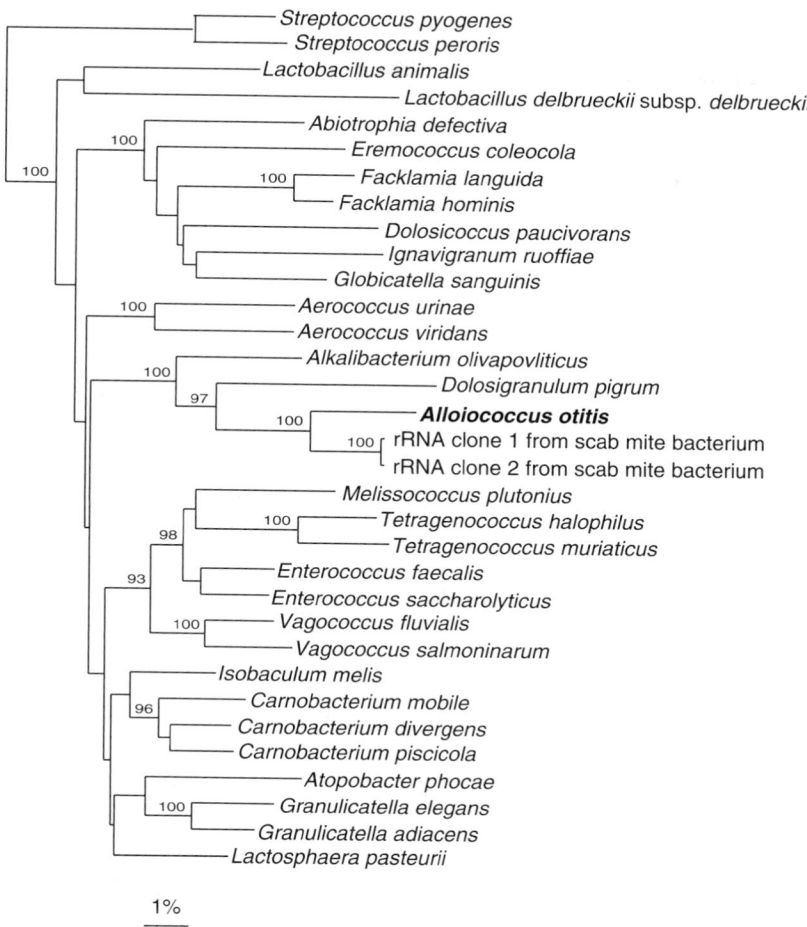

FIGURE 102. Neighbor-joining tree based on 16S rRNA sequences depicting the phylogenetic position of *Alloiococcus otitis* among its close relatives. Bootstrap values determined from 500 replicates.

List of species of the genus *Alloiococcus*

1. **Alloiococcus otitis**   Aguirre and Collins 1992a, 83[VP]

   o.ti′tis. N.L. gen. n. *otitis* of ear inflammation.

   The species epithet of *Alloiococcus otitis* is incorrect and should be *otitidis* (of the ear inflammation) (Von Graevenitz, 1993). This proposed change has, however, not been validated.

   DNA G+C content (mol%): 44–45 (T_m).
   Type strain: 7760, ATCC 51267, CCUG 32997, CIP 103508, NBRC 15545, NCIMB 702890.
   GenBank accession number (16S rRNA gene): X59765.

Genus V. **Atopobacter** Lawson, Foster, Falsen, Ohlén and Collins 2000, 1758[VP]

MATTHEW D. COLLINS AND PAUL A. LAWSON

A.to.po.bac′ter. Gr. adj. *atopos* having no place, strange; L. masc. n. *bacter* rod; M.L. masc. n. *Atopobacter* strange rod.

Cells are Gram-stain-positive, short, nonsporeforming, irregular rods. Facultatively anaerobic and catalase-negative. No growth at 45 °C. Gas is not produced. **Acid is produced from D-glucose and some other sugars. Arginine dihydrolase, pyroglutamic acid arylamidase, and pyrrolidonyl arylamidase are produced. Esculin, gelatin, hippurate, and urea are not hydrolyzed.** Nitrate is not reduced. Voges–Proskauer-negative. **The cell-wall murein is type L-ornithine–D-aspartic acid (A4β).**

Type species: **Atopobacter phocae** Lawson, Foster, Falsen, Ohlén and Collins 2000, 1759[VP].

Further descriptive information

The genus contains only one species, *Atopobacter phocae*, and therefore the characteristics provided below refer to this species. Cells consist of short, irregular rods. On Columbia agar supplemented with 5% horse blood, small (pin-sized), gray-colored, smooth colonies are formed after 24 h at 37 °C. Nonhemolytic. Growth occurs at 25 °C but not at 45 °C. Acid, but no gas, is produced from D-glucose. Acid is also formed from glycogen, maltose, pullulan, and D-ribose. Acid may or may not be formed from cyclodextrin, lactose, sucrose, and trehalose. Acid is not produced from D-arabitol, L-arabinose, mannitol, melibiose, melezitose, methyl-β-D-glucopyranoside, raffinose, sorbitol, tagatose, or D-xylose. Esculin, gelatin, hippurate, and urea are not hydrolyzed. Activity is detected for acid phosphatase, alkaline phosphatase, arginine dihydrolase, esterase C4, ester lipase C8, pyroglutamic acid arylamidase, pyrazinamidase, and pyrrolidonly arylamidase. Activity is not detected for chymotrypsin, cystine arylamidase, α-fucosidase, α-galactosidase, α-glucosidase, β-glucosidase, β-glucuronidase, glycine tryptophan arylamidase, α-mannosidase, β-mannosidase, lipase C14, N-acetyl β-glucosaminidase, trypsin, or valine arylamidase. Activity for alanyl phenylalanine proline arylamidase, β-galactosidase, and leucine arylamidase may or may not be detected. Nitrate is not reduced. The cell wall contains an L-ornithine–D-aspartic acid type murein (variation A4 β).

Atopobacter phocae was originally recovered from dead common seals (Lawson et al., 2000). The species has, however, been isolated subsequently in mixed culture from an otter head abscess (Foster, Lawson, and Collins, unpublished results). The habitat of *Atopobacter phocae* is not known.

Isolation

Atopobacter phocae has been isolated on sheep blood agar at 37 °C in air or in air plus 5% CO_2. There is no information on enrichment or selective media for this species.

Maintenance procedures

Strains grow well on blood-based agar at 37 °C; they also grow on suitable solid media without blood. Suitable liquid media for growth include brain heart infusion and Todd–Hewitt broth. Strains can be stored on cryogenic beads at −70 °C or lyophilized for long-term preservation.

Taxonomic comments

The genus *Atopobacter* was created by Lawson et al. (2000) to accommodate a novel Gram-stain-positive, facultatively anaerobic, catalase-negative, rod-shaped bacterium isolated from dead common seals. *Atopobacter* forms a distinct phylogenetic subline, within a radiation of taxa that includes *Carnobacterium*, *Desemzia*, *Enterococcus*, *Granulicatella*, *Isobaculum*, *Lactosphaera*, *Melissococcus*, *Tetragenococcus*, and *Vagococcus*. The closest phylogenetic relative of *Atopobacter* based on 16S rRNA gene sequencing appears to be the genus *Granulicatella*, but there is considerable uncertainty in the branching position of *Atopobacter* and bootstrap resampling analysis indicates its association with *Granulicatella* is not statistically significant (Lawson et al., 2000). *Atopobacter* displays 16S rRNA sequence divergence values of greater than 6% with its nearest relatives (i.e., *Carnobacterium*, *Desemzia*, *Enterococcus*, *Granulicatella*, *Isobaculum*, *Lactosphaera*, *Melissococcus*, *Tetragenococcus*, and *Vagococcus*). A tree depicting the phylogenetic relationships of *Atopobacter phocae* is shown in Figure 103.

Differentiation of the genus *Atopobacter* from other genera

Atopobacter phocae can be readily distinguished from its closest phylogenetic relatives using a combination of morphological,

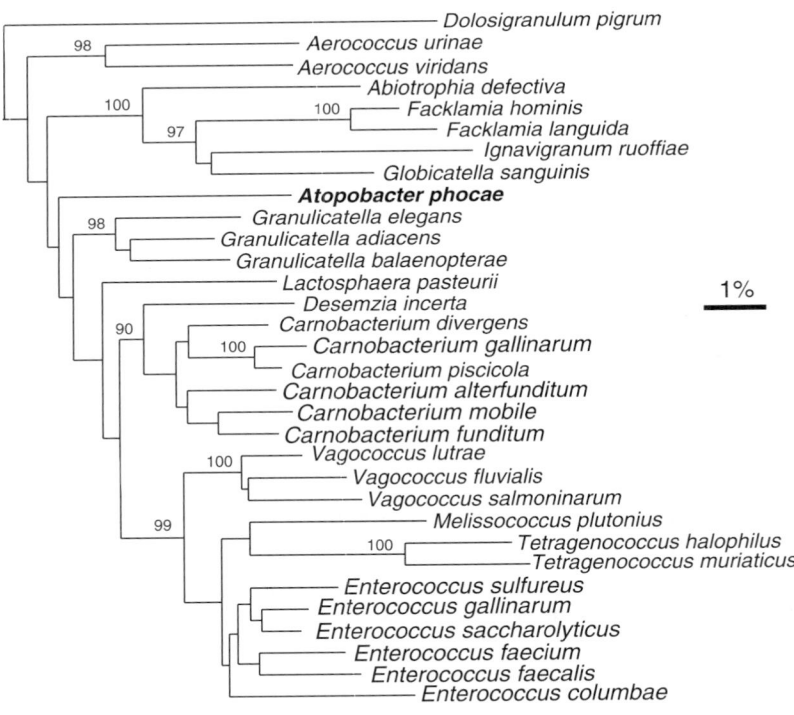

FIGURE 103. Unrooted tree showing the phylogenetic relationships of *Atopobacter phocae* and its closest relatives. The tree was constructed using the neighbor-joining method. Bootstrap values, expressed as a percentage of 500 replications, are given at the branching points.

biochemical, and chemotaxonomic criteria. *Atopobacter phocae* forms rod-shaped cells which distinguish it from *Granulicatella* species, which consist of cocci, occurring singly, in pairs, or in short chains. In addition, some granulicatellae are fastidious and require supplements such as L-cysteine and/or pyridoxal. *Atopobacter phocae* also differs from *Granulicatella* species in forming acid from a broader range of sugars (such as ribose and glycogen). *Atopobacter phocae* can be differentiated from *Lactosphaera pasteurii* as the latter consists of coccus-shaped cells. Also, *Atopobacter phocae* does not produce acid from L-arabinose, mannitol, melezitose, methyl-β-D-glucopyranoside, raffinose, or sorbitol, whereas *Lactosphaera pasteurii* does form acid from these substrates. In addition, *Atopobacter phocae* produces alkaline phosphatase and pyroglutamic acid arylamidase but not α-galactosidase, β-glucosidase, or β-glucuronidase, whereas *Lactosphaera pasteurii* produces the opposite reactions. The cell wall murein type of *Lactosphaera pasteurii* (variation A4

α, type L-lysine–D-aspartic acid) (Janssen et al., 1995) also differs from that of *Atopobacter phocae* (variation A4 β, type L-ornithine-D-aspartic acid) (Lawson et al., 2000). *Atopobacter phocae* can be distinguished from *Desemzia incerta*, another catalase-negative, rod-shaped species, by not hydrolyzing hippurate; producing acid from glycogen, pullulan, and ribose; and by failing to ferment methyl-β-D-glucopyranoside and raffinose. By contrast, *Desemzia incerta* hydrolyzes hippurate and produces acid from methyl-β-D-glucopyranoside and raffinose but not from glycogen, pullulan, or ribose. *Atopobacter phocae* also differs from *Desemzia incerta* by displaying activity for alkaline phosphatase but failing to produce α-galactosidase, β-glucosidase, β-glucuronidase, β-mannosidase, and N-acetyl β-glucosaminidase whereas *Desemzia incerta* produces the opposite reactions. *Desemzia incerta* also differs from *Atopobacter phocae* in possessing a murein based on L-lysine-D-glutamic acid (variation A4 α) (Stackebrandt et al., 1999).

List of species of the genus *Atopobacter*

1. **Atopobacter phocae** Lawson, Foster, Falsen, Ohlén and Collins 2000, 1759[VP]

pho'cae. L. gen. n. *phocae* of or pertaining to the common seal *Phoca vitulina*, from which the organism was first isolated.

Type strain: M1590/94/2, ATCC BAA-285, CCUG 42358, CIP 106392.
GenBank accession number (16S rRNA gene): Y16546.

Genus VI. **Atopococcus** VI. Collins, Wiernik, Falsen and Lawson 2005, 1695[VP]

PAUL A. LAWSON

A.to.po.coc'cus. Gr. adj. *atopos* having no place, strange; Gr. n. *coccus* a grain or berry; N.L. masc. n. *Atopococcus* a strange coccus.

Cells are cocci that occur in pairs or short chains. Gram-stain-positive. Nonmotile and spores are not formed. **Aerobic and catalase-negative.** Optimum growth temperature is 28–32 °C; **no growth above 32 °C. Acid is produced from D-glucose and some other carbohydrates.** Pyroglutamic acid arylamidase is produced. Arginine dihydrolase is not produced. Urease-negative. Nitrate is not reduced. The major long-chain fatty acids are the straight-chain and monounsaturated types. **The cell wall murein contains L-Lys, type A4α (L-Lys–D-Asp).** Isolated from moist, powdered tobacco.

DNA G+C content (mol%): 46.0.

Type species: **Atopococcus tabaci** Collins, Wiernik, Falsen, and Lawson 2005, 1695[VP].

Further descriptive information

The genus contains only one species, *Atopococcus tabaci*, and therefore the characteristics provided below refer to this species. Grows on Columbia blood agar base supplemented with 5% horse blood. α-Hemolytic on horse blood agar. Grows aerobically but not anaerobically. Optimal growth temperature on tryptic soy agar is 30 °C. Halotolerant, growing in 8–9% NaCl. Using API test kits, acid is produced from glucose, lactose, maltose, ribose, sucrose, and trehalose but not from L-arabinose, D-arabitol, cyclodextrin, glycogen, mannitol, melibiose, melezitose, methyl β-D-glucopyranoside, pullulan, raffinose, sorbitol, tagatose, or D-xylose. Esculin and hippurate are hydrolyzed but not gelatin. Activity is detected for acid phosphatase, esterase C4, β-glucosidase, β-galactosidase, leucine arylamidase, valine arylamidase, pyroglutamic acid arylamidase, and N-acetyl-β-

glucosaminidase. Activity may be detected for alkaline phosphatase, α-glucosidase, and urease. Alanine phenylalanine proline arylamidase, arginine dihydrolase, ester lipase C8, lipase C14, chymotrypsin, trypsin, phosphoamidase, cystine arylamidase, β-glucuronidase, glycyl trytophan arylamidase, α-fucosidase, α-mannosidase, and β-mannosidase are not detected. Nitrate is not reduced, and the Voges–Proskauer test is negative. The fatty acid content is: $C_{10:0}$ (0.7%), $C_{12:0}$ (2.6%), $C_{14:0}$ (8.3%), $C_{14:1}$ (4.5%), $C_{16:1\omega9c}$ (41.9%), $C_{16:0}$ (15.8%), *iso*-$C_{17:1}$ (5.3%), $C_{18:1\omega9c}$ (9.0%), and *iso*-$C_{19:1}$ (1.9%). No respiratory quinones were detected using TLC and HPLC as described by Collins et al. (1977). The cell wall contains A4α-type murein composed of L-Lys–L-Glu (for nomenclature, see Schleifer and Kandler, 1972). The amino acids lysine, alanine, and glutamic acid are present in molar ratios of 1.0 Lys:1.9 Ala:2.4 Glu. The partial hydrolysate contains the peptides L-Ala–D-Glu and L-Lys–D-Ala. Dinitrophenylation reveals that the N-terminus of the interpeptide bridge is glutamate. Since the peptide D-Ala–D-Glu was not detected, it is likely that the N-terminal glutamic acid is of the L-configuration. The only known source from which *Atopococcus tabaci* has been isolated is powdered tobacco (Collins et al., 2005).

Isolation procedures

Atopococcus tabaci was isolated from a sample of moist snuff tobacco from an undisclosed industrial source and was cultured on Columbia blood agar base supplemented with 5% horse blood at 37 °C under aerobic conditions. There is no information on enrichment or selective media for this species.

Maintenance procedures

Strains grow well on Columbia blood agar base supplemented with 5% horse blood at 37 °C under aerobic conditions. For long-term preservation, strains can be maintained in the same medium containing 15–20% glycerol at −70 °C or lyophilized.

Taxonomic comments

The genus *Atopococcus* was created to accommodate a phylogenetically distinct Gram-stain-positive, aerobic, catalase-negative, coccus originating from tobacco (Collins et al., 2005). Phylogenetic analysis using 16S rRNA gene sequences shows that *Atopococcus tabaci* is related to the *Lactobacillales* of the *Firmicutes*. In particular, the organism forms a loose cluster with *Alkalibacterium*, *Alloiococcus*, *Allofustis*, *Atopostipes*, *Dolosigranulum*, and *Marinilactibacillus*, within the family *Carnobacteriaceae* (Figure 3, Ludwig et al., this volume).

Differentiation of the genus *Atopococcus* from other genera

In addition to differentiation by 16S rRNA gene sequence analysis, a combination of morphology, biochemical, and chemotaxonomic criteria is used distinguish *Atopococcus tabaci* from its closest phylogenetic relatives, *Alkalibacterium* species,

Alloiococcus otitis, *Allofustis seminis*, *Atopostipes suicloacale*, *Dolosigranulum pigrum*, and *Marinilactibacillus* species. In particular, it can be distinguished from *Alkalibacterium olivoapovliticus* and *Marinilactobacillus psychrotolerans* because it is not alkaliphilic or rod-shaped. In addition, the cell walls of *Atopococcus tabaci* possess L-lysine (type L-Lys–L-Glu) whereas the walls of *Alkalibacterium olivoapovliticus* and *Marinilactobacillus psychrotolerans* contain ornithine as their dibasic amino acid (types Orn–D-Asp and Orn–D-Glu, respectively) (Ishikawa et al., 2003; Ntougias and Russell, 2001). Similarly, *Atopococcus tabaci* differs from *Alloiococcus otitis*, *Allofustis seminis*, and *Dolosigranulum pigrum* in having a growth temperature optimum of 30 °C. These other species all grow optimally at 37 °C (Aguirre and Collins, 1992a; Aguirre et al., 1993; Collins et al., 2003). In addition, *Alloiococcus otitis* is nutritionally very fastidious, and *Dolosigranulum pigrum* possesses the L-Lys–D-Asp type of cell wall murein (Aguirre and Collins, 1992a; Aguirre et al., 1993; Miller et al., 1996). *Allofustis seminis* is further differentiated from *Atopococcus tabaci* by its rod shape and cell-wall murein composed of directly cross-linked L-Lys (Collins et al., 2003). *Atopococcus tabaci* is also easily differentiated from *Atopostipes suicloacale*, which is a facultative anaerobe that requires rumen fluid for growth (Cotta et al., 2004b)

List of species of the genus *Atopococcus*

1. **Atopococcus tabaci**   Collins, Wiernik, Falsen and Lawson 2005, 1695[VP]

ta.ba′ci. N.L. gen. neut. n. *tabaci* of/from tobacco.

Cells stain Gram-positive, are coccoid in shape, and occur in pairs or short chains. Endospores are not formed, and cells are nonmotile. Aerobic. Catalase-negative. Acid is produced from glucose and some other sugars. Nitrate is not

reduced, and the Voges–Proskauer test is negative. The predominant long-chain fatty acids are $C_{14:0}$, $C_{16:1\omega9c}$, $C_{16:0}$, and $C_{18:1\omega9c}$. Other characteristics are as given for the genus. Isolated from powdered tobacco.

DNA G+C content (mol%): 46.0 (HPLC).

Type strain: CCUG 48253, CIP 108502, DSM 17538.

GenBank accession number (16S rRNA gene): AJ634917.

Genus VII. **Atopostipes** Cotta, Whitehead, Collins and Lawson 2004a, 1425[VP] (Effective publication: Cotta, Whitehead, Collins and Lawson 2004b, 193.)

Terence R. Whitehead, Paul A. Lawson and Michael A. Cotta

A.to.po.sti′pes. Gr. adj. *atopos* having no place, strange; L. masc. n. *stipes* rod; N.L. masc. n. *Atopostipes* a strange rod, referring to its distinct phylogenetic position.

Cells are short rods. Gram-stain-positive. Nonmotile and spores are not formed. **Facultatively anaerobic and catalase-negative.** Optimum growth temperature is 28–32 °C; **no growth occurs above 32 °C. Acid is produced from D-glucose and some other carbohydrates.** The end products of glucose metabolism are lactate, acetate, and formate. Urease-negative. Nitrate is not reduced. Indole is not formed. The cell-wall murein **contains L-Lys, type A4α (L-Lys–D-Asp).** Isolated from swine manure.

DNA G+C content (mol%): 43.9.

Type species: **Atopostipes suicloacalis** Cotta, Whitehead, Collins and Lawson 2004a, 1425[VP] (Effective publication: Cotta, Whitehead, Collins and Lawson 2004b, 194) (*Atopostipes suicloacale* [sic] Cotta, Whitehead, Collins and Lawson 2004b, 194).

Further descriptive information

The genus contains only one species, *Atopostipes suicloacalis*, and therefore the characteristics provided below refer to this species.

Grows on brain heart infusion agar supplemented with 10% rumen fluid. Cells are nonmotile rods. Colonies are pinhead to 0.5 mm in diameter, gray, smooth, circular, and entire after 48 h of incubation. Optimum growth temperature is 28–30 °C, with no growth observed above 32 °C. Acid is produced from glucose, and end products of glucose metabolism are lactate, acetate, and formate. Amygdalin, cellobiose, esculin, glucose, lactose, maltose, mannose, raffinose, and sucrose are utilized as carbon sources; grows weakly on lactate. Arabinose, cellulose, inulin, inositol, melibiose, rhamnose, sorbitol, trehalose, and xylose are not utilized as carbon sources. α-Arabinosidase, α-galactosidase, β-glucosidase, and *N*-acetyl-β-glucosaminidase activities are found in cells grown on glucose; in addition, β-galactosidase and 6-phospho-β-galactosidase are detected from cells grown on lactose. Arginine dihydrolase, arginine arylamidase, alkaline phosphatase, alanine arylamidase, α-fucosidase, α-glucosidase, β-glucuronidase, glycine arylamidase, glutamic acid decarboxylase, glutamyl glutamic acid arylamidase, histidine arylamidase, leucine arylamidase, leucyl glycine arylamidase, phenylalanine arylamidase, proline arylamidase, pyroglutamic acid arylamidase, serine arylamidase, tyrosine arylamidase, and urease are not produced. Nitrate is not reduced and indole is not produced. The only known source from which *Atopostipes suicloacalis* has been isolated is swine manure (Cotta et al., 2004b).

Isolation procedures

Atopostipes suicloacalis was isolated from a manure storage pit near Eureka, Illinois, USA. A manure sample was serially diluted and plated onto medium containing 40% clarified rumen fluid under anaerobic conditions (Cotta et al., 2003). There is no information on enrichment or selective media for this species.

Maintenance procedures

Strains grow well on routine growth medium (RGM; Cotta et al., 2003) containing 40% rumen fluid at 30 °C. For long-term preservation, strains can be maintained in medium containing 15–20% glycerol at –70 °C or lyophilized.

Taxonomic comments

The genus *Atopostipes* was created in 2004 to accommodate a phylogenetically distinct catalase-negative, Gram-stain-positive, rod-shaped bacterium originating from stored swine manure (Cotta et al., 2004b). Phylogenetic analysis using 16S rRNA gene sequences shows that *A. suicloacalis* is related to the *Lactobacillales* of the *Firmicutes*. In particular, the organism forms a loose cluster with *Alkalibacterium, Alloiococcus, Allofustis, Atopococcus, Dolosigranulum,* and *Marinilactibacillus,* within the family *Carnobacteriaceae* (Figure 104). It is clear from phylogenetic molecular profiling studies of environmental samples that, in addition to the diverse group of named organisms, some isolates and cloned sequences corresponding to as-yet uncultivated organisms are present in this phylogenetic grouping.

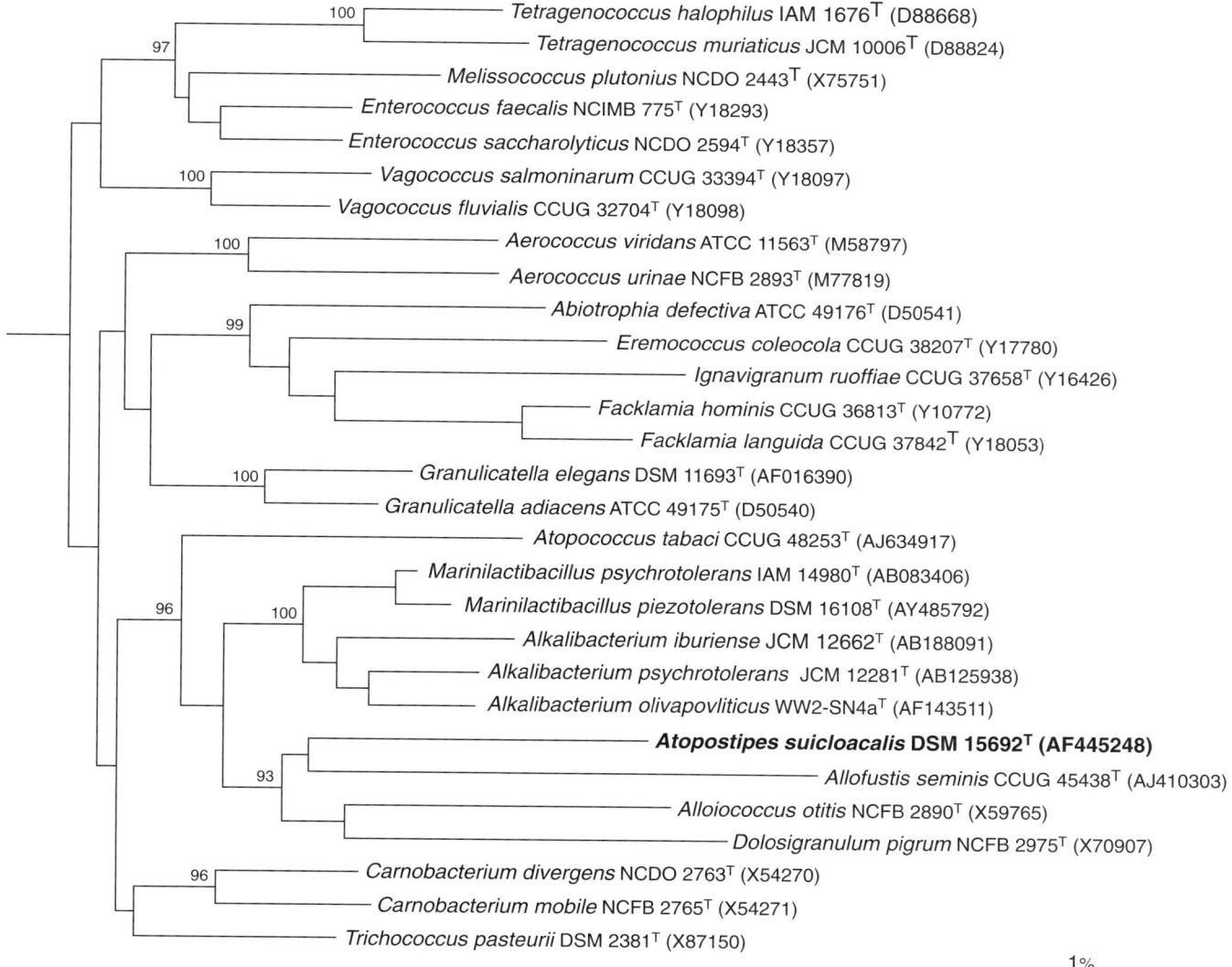

FIGURE 104. Unrooted phylogenetic tree of the 16S rRNA genes of *Atopostipes suicloacalis* and the lactic acid bacteria supra-generic cluster including the members of the families *Carnobacteriaceae, Aerococcaceae,* and *Enterococcaceae.* The phylogenetic tree was calculated by the neighbor-joining method. Bootstrap values greater than 90 are indicated. Bar = 1% sequence divergence.

Furthermore, many of the sequences cloned possess high similarity to those cloned from both animal and human sources (Leser et al., 2002; Paster et al., 2001). These sequences may represent novel taxa, including additional species of *Atopostipes*, that have not yet been described.

Differentiation of the genus *Atopostipes* from other genera

In addition to differentiation by 16S rRNA gene sequence analysis, *Atopostipes suicloacalis* can be readily distinguished from its closest phylogenetic relatives, *Alkalibacterium* species, *Alloiococcus otitis*, *Allofustis seminis*, *Atopococcus*, *Dolosigranulum pigrum*, and *Marinilactibacillus* species, using a combination of morphological, biochemical, and chemotaxonomic criteria.

Atopostipes suicloacalis differs from *Alkalibacterium* and *Marinilactibacillus* in cell-wall murein content: the former contains peptidoglycan type A4α (L-Lys–D-Asp), whereas the latter two contain A4β (Orn–D-Asp or Orn–D-Glu, Ishikawa et al., 2003; for nomenclature see Schleifer and Kandler, 1972). In addition, *Alkalibacterium* species are obligate alkalophiles, growing within a pH range of 8.5–10.8, with optimum growth at pH 9–10 (Ntougias and Russell, 2001). Similarly, *Alloiococcus otitis*, *Allofustis seminis*, and *Dolosigranulum pigrum* differ from *Atopostipes suicloacalis* in having higher growth temperature optima of around 37 °C and ovoid or coccoid morphologies (Aguirre and Collins, 1992a; Aguirre et al., 1993; Collins et al., 2003).

List of species of the genus *Atopostipes*

1. **Atopostipes suicloacalis** Cotta, Whitehead, Collins and Lawson 2004a, 1425[VP] (Effective publication: Cotta, Whitehead, Collins and Lawson 2004b, 194.) (*Atopostipes suicloacale* [sic] Cotta, Whitehead, Collins and Lawson 2004b, 194.)

su.i.clo.a.ca′lis. L. n. *sus* pig; L. adj. *cloacalis -e* pertaining to a sewer (manure canal); N.L. masc. n. *suicloacalis* pertaining to pig manure, referring to the isolation of the type strain.

Cells are short rods that stain Gram-stain-positive. No spores are observed and cells are nonmotile. Colonies are pinhead to 0.5 mm in diameter, gray, smooth, circular, and entire after 48 h incubation on brain heart infusion agar supplemented with 10% rumen fluid. Optimum growth temperature is 28–30 °C, with no growth above 32 °C. Other characteristics are as given for the genus. Isolated from swine manure.

DNA G+C content (mol%): 43.9 (T_m).
Type strain: PPC79, NRRL 23919, DSM 15692.
GenBank accession number (16S rRNA gene): AF445248.

Genus VIII. **Desemzia** Stackebrandt, Schumann, Swiderski and Weiss 1999, 187[VP]

ERKO STACKEBRANDT

De.sem′zi.a. N.L. fem. n. *Desemzia* arbitrary name, derived from the abbreviation DSMZ (Deutsche Sammlung von Mikroorganismen und Zellkulturen)

Short rods, occurring singly and occasionally in pairs. Gram-stain-positive. Cells from older cultures tend to lose the ability to retain the Gram stain. Microaerophilic. Nonsporeforming. Not acid-fast. Catalase-negative, oxidase-negative. Metabolism fermentative. **Peptidoglycan contains lysine as the diagnostic amino acid (peptidoglycan type L-lysine-D-glutamic acid; variation A4α).** Mycolic acids and isoprenoid quinones absent. **Straight-chain saturated and monounsaturated fatty acids, hexadecanoic ($C_{16:0}$), hexadecenoic ($C_{16:1}$) and *cis*-vaccenic ($C_{18:1\omega7}$) predominate.** The DNA G+C content is 40 mol%. Phylogenetically, a member of the family *Carnobacteriaceae*, order *Lactobacillales*, phylum *Firmicutes*, the genus *Carnobacterium* is the closest phylogenetic neighbor.

Type species: **Desemzia incerta** (Steinhaus 1941) Stackebrandt, Schumann, Swiderski and Weiss 1999, 187[VP] ("*Bacterium incertum*" Steinhaus 1941; *Brevibacterium incertum* Breed 1953, 14.).

Sources of data

Morphological, cultural and physiological properties described below were published by Stackebrandt et al. (1999) who combined new data with the original data of Steinhaus (1941), Breed (1953) and Jones and Keddie (1986). Some physiological data and the Riboprint™ patterns were generated for this chapter. The chemotaxonomic data have been published by Schleifer and Kandler (1972) and Collins and Kroppenstedt (1983) while the phylogenetic data originate from Stackebrandt et al. (1999), reanalyzed for this study.

Further taxonomic information

As the genus currently contains a single species only, the description of the genus is that of the species. Only the type strain DSM 20581[T] (=ATCC 8363[T], NCIB 9892[T]) has been analyzed in detail.

Phylogenetic analysis of the 16S rDNA indicates the genus *Desemzia* to be closely related to some genera of *Lactobacillus–Streptococcus* group of the *Firmicutes* (Stackebrandt et al., 1999); and in the revised roadmap for Volume 3 of the *Systematics* (Ludwig et al., this volume), it has been classified as a member of the family *Carnobacteriaceae*, order *Lactobacillales*, in the phylum *Firmicutes*. Comparison of the almost complete 16S rDNA of strain DSM 20581[T] (1510 bp; 98% of the *Escherichia coli* sequence, Brosius et al., 1978) to the homologous sequences of members of this group revealed that *Desemzia incerta* shows the highest similarity values to members of the genus *Carnobacterium* Collins et al. (1987) (95.1–96.2% sequence similarity), while those to members of *Enterococcus*, *Lactobacillus*, *Tetragenococcus*, *Trichococcus*, *Vagococcus*, *Weisella*, and related genera are distinctively lower (91–94.3% similarity). Phylogenetic dendrograms, reconstructed from evolutionary distance values (Jukes and Cantor, 1969), using the algorithm of De Soete (1983) and the neighbor-joining method (Felsenstein, 1993) shows *Desemzia incerta* to branch close to the deepest bifurcation point of *Carnobacterium* species. High bootstrap values found for the branching point of *Desemzia incerta* supports this phylogenetic affiliation (Figure 105, right dendrogram).

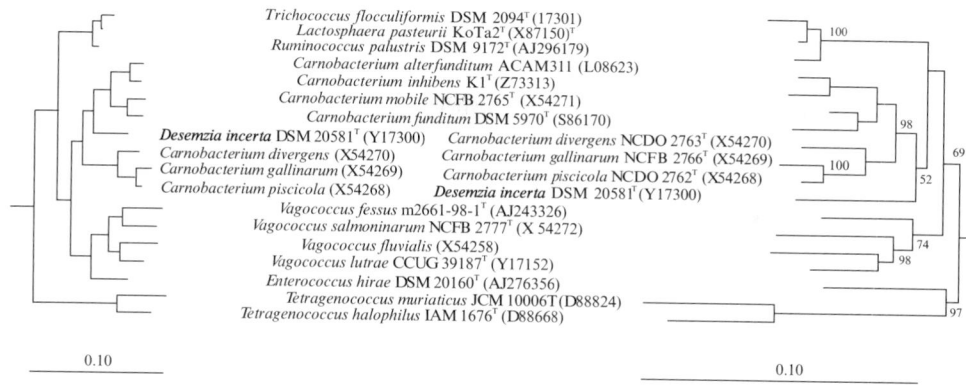

FIGURE 105. 16S rDNA sequence based phylogenetic dendrograms comparing the phylogenetic position of *Desemzia incerta* and related taxa by two different algorithms inferring phylogenetic relatedness. Phylogenetic analysis and 16S rDNA accession numbers of reference organisms have been described by Wallbanks et al. (1990), Franzmann et al. (1991), and Stackebrandt et al. (1999). Right dendrogram: Distance matrix algorithm of De Soete (1983). Bootstrap values (expressed as percentages of 100 replications) of 55% and higher are indicated at the branch points. Scale bar = 10 inferred nucleotide substitutions per 100 nucleotides. Left dendrogram: maximum-likelihood analysis (Felsenstein, 1993).

Using the same dataset but applying the maximum-likelihood analysis (Felsenstein, 1993) members of *Carnobacterium* form two groups, separated by the *Desemzia* lineage (Figure 105, left dendrogram). While the first group contains *Carnobacterium alterfundidum*, *Carnobacterium inhibens*, *Carnobacterium mobile* and *Carnobacterium fundidum*, the other group contains *Carnobacterium divergens*, *Carnobacterium gallinarum*, and *Carnobacterium piscicola*. Analysis of the 16S rDNA primary structures of the carnobacteria species indeed show the occurrence of a small set of mutually exclusive nucleotides which would support the phylogenetic separation (Table 103). On the other hand, sufficient nucleotides exist which demonstrates the phylogenetic uniqueness of *Desemzia incerta* (Table 104).

Riboprint™ analyses of *Desemzia inserta* and *Carnobacterium* species (Figure 106) reveals that the patterns of all type strains of the two genera are distinct. Whether they are species-specific can only be assessed when more strains of the species were included in Riboprint™ analyses.

Physiological reactions of strain DSM 20581[T] towards the substrate panel provided by the API ID32 Strep kit (API bioMérieux, Marcy l'Etoile, France) gave the following positve results: acid production from ribose, lactose, trehalose, raffinose, maltose, melibiose; sucrose; acids are not produced from mannitol, sorbitol, L-arabinose and melizitose. The reaction for fructose is different to that reported by Steinhaus (1941). Tests for β-glucosidase, β-glucoronidase, α-galactosidase, glycyl-tryptophan arylamidase, and β-mannosidase were positive, whereas tests for arginine dehydrolase, β-glucosidase, and β-galactosidase were negative. Acetoin production was negative. Using the API ZYM system, strain DSM 20581[T] hydrolyzed 2-naphtyl-butyrate, L-leucyl-2-naphtylamide and *N*-glutaryl-phenylalanine-2-naphtylamide. The Biolog identification system GP MicroPlate™ incubated under an atmosphere of 85% N_2 and 15% O_2, revealed that the following substrates were utilized: D-fructose, D-glucose, L-lactic acid, maltose, maltotriose, D-melibiose, D-mannose, D-trehalose, D-psicose, α-methyl D-glucoside and *N*-acetyl-D-glucosamine (reading after 48h of incubation). Substrates not utilized are listed in the species description. These data complement physiological and nutritional data given by Steinhaus (1941), Breed (1953) and Jones and Keddie (1986).

DNA–DNA reassocation values obtained for four *Carnobacterium* species that showed higher than 96.5% 16S rDNA sequence similarity (Collins et al., 1987) were lower than 45%. These low values clearly supported the validity of these species and without performing DNA-hybridization experiments between *Desemzia incerta* and the type strains of *Carnobacterium* species it was concluded on the basis of even lower 16S rDNA similarities that the species status of *Desemzia incerta* is confirmed.

Large amounts of inoculum of *Desemzia incerta* killed guinea pigs in 10–15 d but the organism was not pathogenic for rabbits (experiments performed with the original strain No. 41, originally classified as *Bacterium incertum* Steinhaus (1941). The pathogenic characteristics did not resemble those of *Listeria monocytogenes* and the ocular reaction indicative of *Listeria monocytogenes* was negative (Steinhaus, 1941).

Enrichment and isolation procedures

The following procedure has been indicated by Steinhaus (1941). The insect, *Tibicen linnei*, was etherized and the wings clipped off near the base. After washing the body for 1–2 min in a solution of 1:1000 mercuric chloride in 80 % alcohol, it was rinsed thoroughly in four washings of sterile saline. The sides of the abdomen were cut from near the posterior end up to the thorax with a pair of fine scissors which had been flamed. The top of the abdomen immediately back of the thorax was cut across, the resulting flap was then tilted back to leave the abdominal viscera exposed and the ovaries were examined. The ovaries were placed either directly on media in Petri dishes or into tubes of sterile saline to be streaked out on nutrient agar, glucose agar, North's gelatin chocolate agar (defibrinated blood added to medium at 80 °C) and blood agar and cultivated at 37 °C.

TABLE 103. Occurrence of 16S rDNA signatures demonstrating the phylogenetic heterogeneity of the genus *Carnobacterium* and the phylogenetic closeness of *Desemzia incerta* with members of *Carnobacterium* group

Nucleotide position[a]	*Carnobacterium* group 1	*Carnobacterium* group 2
	Desemzia incerta, C. alterfundidum, C. inhibens, C. mobile, C. fundidum	*C. divergens, C. gallinarum, C. piscicola*
70–98	U–A	A–G
77–92	Y[b]–R[c]	G–Y
81–88	U–A	A–U
1168	U	C
1436–1465	U–C	C–U
1453	G	A

[a]*Escherichia coli* nomenclature.
[b]Pyrimidine.
[c]Purine.

TABLE 104. Presence of unique nucleotides of *Desemzia incerta*, compared to the homologous composition in members of *Carnobacterium*

Nucleotide position	*Desemzia incerta*	*Carnobacterium* species
78–91	C–G	R[a]–Y
139–224	C–G	U–A
183–193	U–A	R–Y
458–474	A–G	G–G
615–625	U–A	C–G
669–737	A–U	G–C
771–808	A–U	G–C
929–1388	G–C	A–U

[a]R, purine.

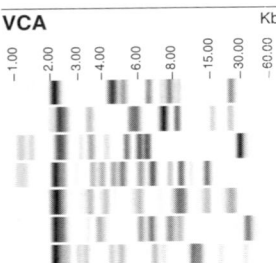

Desemzia incerta DSM 20581[T]
Carnobacterium piscicola DSM 20730[T]
Carnobacterium gallinarum DSM 4847[T]
Carnobacterium mobile DSM 4848[T]
Carnobacterium inhibens DSM 13024[T]
Carnobacterium funditum DSM 5970[T]
Carnobacterium alterfunditum DSM 5972[T]

FIGURE 106. Riboprint™ patterns of the DNA of *Desemzia incerta* and the type strains of *Carnobacterium* type strains. The pattern was generated by the Riboprint™ robot (Qualicon, Wilmington, Delaware, USA), in which DNA was cleaved by *Eco*RI, the resulting fragments were separated on a membrane and fragments of the rrn operons visualized by hybridization with a fluorescent rrn operon probe. The scale bar on top of the fragments indicate the length (in kb) of the fluorescence-positive fragments.

Maintenance procedures

Desemzia incerta is routinely maintained in the DSMZ-German Collection of Microorganisms and Cell Cultures in medium DSM 240 at 30°C (DSMZ, 1998). Medium 240 is listed as "*Corynebacterium* medium with blood", containing per liter distilled 10.0 g water casein peptone, tryptic digested, 5.0 g yeast extract, 5.0 g glucose, 5.0 g NaCl and agar 15.0 g, pH 7.2–7.4. Following sterilization and cooling to about 45°C, 5% defibrinated blood is added. The culture can be maintained long term under liquid nitrogen or freeze-dried.

The organism grows well on Columbia blood agar (Benson Dickenson, no. 4354005) within 24 h, while good growth is also observed in Caso medium (Oxoid, no. CM 129) and in PYG medium (DSM catalog of strains, medium no. 104) after 48 h. L-Lactate but no D-lactate (Hohorst, 1966) is produced from glucose in the latter two media, lowering the pH to 6.3. Growth temperature range on Columbia blood agar confirmed the data given by Jones and Keddie (1986).

Differentiation of *Desemzia* from other taxa

Desemzia incerta resembles *Carnobacterium* species in morphology, fermentative metabolism, base composition of DNA, and the composition of fatty acids, comprising predominantly straight-chain saturated and monounsaturated acids (Collins et al., 1987; Collins and Kroppenstedt, 1983). While neither phylogenetic analyses nor physiological and morphological properties of per se exclude the reclassification of *Desemzia incerta* as a member of *Carnobacterium* (Figure 105), the marked differences in the composition of the peptidoglycan and in the isomeric type of octadecenoic acid as predominant fatty acid are interpreted as being of sufficient high taxonomic significance to maintain the classification the genus *Desemzia* (Stackebrandt et al., 1999). *Desemzia incerta* and *Carnobacterium* species differ significantly in the amino acid composition of peptidoglycan. While *Desemzia incerta* possesses peptidoglycan of the L-lysine–D-glutamic acid type, variation A4α (Schleifer and Kandler, 1972), carnobacteria exhibit the meso-diaminopimelic acid direct cross-linked type, variation A1γ (Collins et al., 1987). Likewise, *Desemzia incerta* contains major amounts of vaccenic acid ($C_{18:1\omega7}$), while *Carnobacterium* species contain oleic acid ($C_{18:1\omega9}$) as the octadecenoic acid isomer. *Desemzia incerta* lacks isoprenoid quinones (Collins and Kroppenstedt, 1983) using high-performance liquid chromatography as described by Groth et al. (1996) and electron-impact mass spectrometry (Collins, 1994). None of the substances extracted by a method developed for lipoquinone analysis (Tindall, 1990) gave rise to chromatographic patterns of menaquinones or ubiquinones or to mass spectral nuclear fragments or molecular ions characteristic for isoprenoid quinones (Collins, 1994). No data on isoprenoid quinones are available for species of *Carnobacterium*.

The type strain of *Desemzia incerta* can be distinguished from the type strains of *Carnobacterium* species by several physiological reactions in which substrates are provided by the Biolog GP MicroPlate™ panel (Table 105).

Taxonomic comments

In the course of a study on bacteria isolated from seven orders of the class *Hexapoda*, a Gram-stain-positive bacterium was isolated by Steinhaus (1941) from the ovaries of the lyreman cicada, *Tibicen linnei*. Although the physiologic and cultural characteristics resembled those of the genus *Listeria* the isolate was tentatively classified as *Bacterium incertum* due to the taxonomic uncertainty which surrounded the *Listeria* group (Steinhaus, 1941) at that time. Breed (1953) transferred this species to the genus *Brevibacterium* as *Brevibacterium incertum* which over the years became a dumping ground for misclassified strains with superficial morphological and physiological similarities. Mainly on chemotaxonomic and phylogenetic grounds several species have been transferred to other genera such as *Arthrobacter* (*Brevibacterium protophormiae* (Stackebrandt et al., 1983), *Aureobacterium* (now *Microbacterium*, Takeuchi and Hatano, 1998) (*Brevibacterium saperdae*, *Brevibacterium testaceum* (Collins et al., 1983a), *Cellulomonas* (*Brevibacterium fermentans*,

Brevibacterium lyticum (Stackebrandt et al., 1982), *Curtobacterium* (*Brevibacterium albidum*, *Brevibacterium citreum*, *Brevibacterium luteum*, *Brevibacterium pusillum* (Yamada and Komagata, 1972), *Corynebacterium* (*Brevibacterium ammoniagenes* (Collins, 1987), *Brevibacterium divaricatum*, *Brevibacterium vitarumen* (Laneelle et al., 1980), *Exiguobacterium* (*Brevibacterium acetylicum* (Farrow et al., 1994), and *Microbacterium* (*Brevibacterium imperiale* (Collins et al., 1983b), *Brevibacterium oxydans* (Schumann et al., 1998). Analysis of the amino acid composition of peptidoglycan of *Brevibacterium incertum* (Schleifer and Kandler, 1972) revealed the presence of L-lysine while strains of the type species *Brevibacterium linens* contained meso-diaminopimelic acid. Subsequent studies on the composition of isoprenoid quinones and fatty acids (Collins and Kroppenstedt, 1983) confirmed the misclassification of *Brevibacterium incertum* as, in contrast to *Brevibacterium linens*, menaquinones were absent and fatty acids contained substantial amounts of monounsaturated fatty acids of the *cis*-vaccenic acid series. Collins and Kroppenstedt (1983) pointed out that the lack of isoprenoid quinones suggests a relatedness of *Brevibacterium incertum* to certain other Gram-stain-positive taxa, such as *Erysipelothrix*, *Gemella*, *Lactobacillus*, or *Streptococcus*

TABLE 105. Physiological reactions differentiating *Desemzia incerta* and type strains of *Carnobacterium* species[a,b]

Physiological test	*Desemzia incerta* (DSM 20581^T)	*C. alterfundidum* (DSM 5972^T)	*C. divergens* (DSM 20623^T)	*C. fundidum* (DSM 5970^T)	*C. gallinarum* (DSM 4847^T)	*C. mobile* (DSM 4848^T)	*C. piscicola* (DSM 20730^T)	*C. inhibens* (CCUG 31728^T)[c]
N-Acetyl-D-glucosamine	w	w	w	+	+	+	+	+
N-Acetyl-D-mannosamine	–	–	–	–	–	+	–	nd
Adenosine	–	–	+	–	+	w	+	nd
Amygdalin	–	–	–	–	–	–	+	+
Arbutin	–	–	–	–	+	–	+	+
Cellobiose	–	–	–	–	+	–	+	+
D-Fructose	+	+	+	+	+	+	+	+
Gentiobiose	–	–	–	–	–	–	+	+
α-D-Glucose	+	+	+	+	+	+	–	+
Glycerol	–	–	–	–	–	–	+	–
L-Lactic acid	+	–	+	–	–	+	–	nd
α-D-Lactose	–	–	–	–	+	–	–	+
Maltose	+	–	–	–	+	–	–	+
Maltotriose	+	–	–	–	–	–	+	+
D-Mannose	+	+	+	+	+	+	–	nd
D-Melibiose	w	–	–	–	+	+	+	+
3-Methyl glucose	–	–	–	–	–	–	+	–
β-Methyl-D-glucoside	–	+	–	–	–	–	–	nd
α-Methyl-D-glucoside	+	–	–	–	–	–	–	nd
Methyl pyruvate	–	+	+	+	+	+	+	nd
D-Psicose	+	w	w	–	+	+	–	nd
Pyruvic acid	–	–	–	+	+	w	+	nd
Salicin	–	–	–	–	+	–	–	+
Thymidine	–	–	+	–	–	–	+	nd
D-Trehalose	+	w	–	–	+	+	+	+
Uridine	–	–	+	–	+	+	+	+
Xylose	–	w	–	–	+	–	+	nd

[a]For symbols, see standard definitions; nd, not determined; w, weak reaction.
[b]Reactions carried out using the Biolog GP MicroPlate™ (24 h incubation).
[c]No positive reactions were obtained with *Carnobacterium inhibens* CCUG 31728^T. Data from Jöborn et al. (1999).

which lack respiratory quinones (Collins and Jones, 1981), while the type of fatty acids indicated a relationship to the lactic acid bacteria (see Kroppenstedt and Kutzner, 1978; Collins and Kroppenstedt, 1983). Subsequently, the species *Brevibacterium incertum* was listed as species *incertae sedis* in *Bergey's Manual of Systematic Bacteriology* (Jones and Keddie, 1986).

In order to determine the phylogenetic affiliation, the taxonomic position of *Brevibacterium incertum* was analyzed recently with respect to 16S rDNA nucleotide sequence (Rainey et al., 1996), base composition of DNA (Marmur, 1961; Mesbah et al., 1989), and isoprenoid quinines (Groth et al., 1996; Tindall, 1990). As a result of these studies *Brevibacterium incertum* was reclassified as *Desemzia incerta* by Stackebrandt et al. (1999).

Acknowledgements

I wish to thank the following DSMZ staff members for their support: Peter Schumann, riboprinting analysis; Anja Frühling, determination of physiological properties; and Jolantha Swiderski, 16S rRNA gene sequence analyses.

Further reading

Steinhaus, E. 1941. A study of the bacteria associated with thirty species of insects. J. Bacteriol. *42*, 757–790.

Stackebrandt, E., P. Schumann, J. Swiderski and N. Weiss. (1999). Reclassification of *Brevibacterium incertum* (Breed 1953) as *Desemzia incerta* gen. nov., comb. nov. Int. J. Syst. Bacteriol. *49*, 185–188.

List of species of the genus *Desemzia*

1. **Desemzia incerta** (Steinhaus 1941) Stackebrandt, Schumann, Swiderski and Weiss 1999, 187[VP] ("*Bacterium incertum*" Steinhaus 1941; *Brevibacterium incertum* Breed 1953, 14.)

in.cert'ae. L. fem. adj. *incerta* not firmly established, uncertain, undetermined, doubtful, dubious.

The description includes data compiled by Jones and Keddie (1986), Stackebrandt et al. (1999) and data generated in this study.

Short rods about 0.5–0.8 by 1.0–1.5 μm, with rounded ends and cocco-bacillary forms. Occur singly, in pairs or in chains. Gram-stain-positive; cells from older cultures tend to lose the ability to retain the Gram-stain. Not acid-fast. Endospores are not formed. Motile with one or two flagella. After 48h generally nonmotile. On PYE media colonies are tiny with no distinctive pigmentation. On chocolate agar colonies are filiform, thin, and transparent; the color of medium changes becoming yellowish-green. On blood agar alpha hemolysis at first; after 3d very strong beta hemolysis. Microaerophilic. Metabolism fermentative. L-lactic but no D-lactic acid is produced from glucose. Acid is produced from glucose, fructose, lactose, maltose, mannose, sucrose, raffinose, melibiose, ribose and trehalose. No acid is produced from galactose, rhamnose, mannitol, dulcitol, inositol, sorbitol, L-arabinose and melizitose. D-fructose, D-glucose, L-lactic acid, maltose, maltotriose, D-melebiose, D-mannose, D-trehalose, D-psicose, α-methyl-D-glucoside and N-acetyl-D-glucosamine are utilized under an atmosphere of 85% N_2 and 15% O_2. The following substrates were not utilized under an atmosphere of 85% N_2 and 15% O_2: dextrin, glycogen, inulin, mannan, Tween 40, Tween 60, N-acetyl-D-mannosamine, amygdalin, L-arabinose, D-arabitol, arbutin, cellobiose, L-fucose, D-galactose, D-galacturonic acid gentobiose, D-gluconic acid, *meso*-inositol, α-D-lactose, lactulose, D-mannitol, D-melezitose, D-raffinose, L-rhamnose, D-ribose, D-sorbitol, sucrose, xylose, glycerol, acetic acid, propionic acid, pyruvic acid, succinic acid, DL-alanine, L-serine, L-glutamic acid, adenosine, inosine, thymidine, and uridine.

Catalase-negative, oxidase-negative. Sodium hippurate, 2-naphtyl-butyrate, L-leucyl-2-naphtylamide and N-glutaryl-phenylalanine-2-naphtylamide are hydrolyzed. Gelatin, casein, and starch are not hydrolyzed. Cellulose is not degraded. Urease is not produced. DNase, arginine dehydrolase, β-glucosidase, and β-galactosidase-negative are not produced. Reactions for β-glucosidase, β-glucoronidase, α-galactosidase, glycyl-tryptophan arylamidase, and β-mannosidase are positive. Acetyl methyl carbinol, indol and H_2S are not produced. Nitrates are not reduced to nitrites. Acetoin production negative. The chemotaxonomic properties are as given for the genus.

Isolated from the ovaries of the lyreman cicada, *Tibicen linnei*.

DNA G+C content (mol%): 40 (HPLC).

Type strain: DSM 20581[T] (ATCC 8363, NCIB 9892).

GenBank accession number (16S rRNA gene): Y17300.

Genus IX. **Dolosigranulum** Aguirre, Morrison, Cookson, Gay and Collins 1994, 370[VP] (Effective publication: Aguirre, Morrison, Cookson, Gay and Collins 1993, 610.)

Mᴀᴛᴛʜᴇᴡ D. Cᴏʟʟɪɴs

Do.lo.si.gra'nu.lum. L. adj. *dolosus* crafty, deceitful; L neut. n. *granulum* a small grain; N.L. neut. n. *Dolosigranulum* a deceptive small grain.

Cells are ovoid, occurring in pairs, tetrads, or groups. Gram-stain-positive and nonmotile. Nonsporeforming. Facultatively anaerobic and catalase-negative. **Growth in 6.5% NaCl. No growth at 10 or 45 °C. Negative bile-esculin reaction. Gas is not produced in MRS broth. Acid is produced from D-glucose and some other sugars. Pyrrolidonylarylamidase and leucine aminopeptidase are produced.** Alanine phenylalanine proline arylamidase and urease are negative. **Does not deaminate arginine. Vancomycin-sensitive.** Pyruvate is not utilized. Voges–Proskauer-negative. **Cell-wall murein is based on L-lysine (type Lys–D-Asp).**

DNA G+C content (mol%): 40.5 (T_m).

Type species: **Dolosigranulum pigrum** Aguirre, Morrison, Cookson, Gay and Collins 1994, 370[VP] (Effective publication: Aguirre, Morrison, Cookson, Gay and Collins 1993, 610.)

Further descriptive information

The genus contains only one species, *Dolosigranulum pigrum*, and therefore the characteristics described below refer to this species. Grows on 5% (v/v) horse or sheep blood agar producing a weak α- or γ-hemolytic reaction. Cells are nonpigmented. Some strains grow better in aerobic conditions. Using the API Rapid ID32 Strep system, acid is produced from maltose, and most strains produce acid from sucrose. Acid is not formed from adonitol, D-arabitol, L-arabinose, cyclodextrin, glycogen, lactose, mannitol, melibiose, melezitose, pullulan, raffinose, sorbitol, tagatose, or trehalose. Acid production from ribose and methyl β-D-glucopyranoside is variable. Activity is detected for β-galactosidase and pyroglutamic acid arylamidase. Activity may or may not be detected for N-acetyl-β-glucosaminidase and glycyl tryptophan arylamidase. No activity is found for alkaline phosphatase, alanine phenylalanine proline arylamidase, arginine dihydrolase, α-galactosidase, β-glucosidase, β-glucuronidase, β-mannosidase, or urease. Hippurate is not hydrolyzed.

Antimicrobial susceptibilities of 27 strains of *Dolosigranumum pigrum* have been reported by LaClaire and Facklam (2000a). All were susceptible to amoxycillin, cefotaxime, cefuroxime, clindamycin, levofloxacin, meropenem, penicillin, quinupristin-dalfopristin, rifampin, tetracycline, and vancomycin. A single isolate was resistant to trimethoprim-sulfamethoxazole. Fifteen strains showed intermediate resistance to chloramphenicol. Two strains were susceptible, ten showed intermediate resistance, and 15 were resistant to erythromycin.

Strains of *Dolosigranulum pigrum* were originally recovered from spinal cord (removed at autopsy) of a person with acute multiple sclerosis and from eye and lens swabs (Aguirre et al., 1993). The species has also been recovered from other clinical sources such as eye, sputum, nose discharges, urine, and blood (LaClaire and Facklam, 2000a). The habitat of *Dolosigranulum pigrum* is not known.

Isolation procedures

Dolosigranulum pigrum may be isolated on rich agar-containing media such as heart infusion or Columbia agar containing 5% (v/v) animal blood at 37 °C, under aerobic or anaerobic conditions. There is no information on selective media for this species.

Maintenance procedures

Strains grow on blood-based agars such as Columbia agar containing 5% (v/v) horse blood, heart infusion agar supple-mented with 5% (v/v) rabbit blood, or brucella agar with 5% sheep blood. They also grow well in brain heart infusion and Todd–Hewitt broth at 37 °C. For long-term preservation, strains can be stored on cryogenic beads at –70 °C or lyophilized.

Taxonomic comments

The genus *Dolosigranulum* was described for some *Gemella*-like organisms recovered from human clinical specimens (Aguirre et al., 1993). Phylogenetic analyses presented in the revised road-map to Volume 3 of the *Systematics* (Ludwig et al., this volume) show that *Dolosigranulum* belongs in the family *Carnobacteriaceae*, order *Lactobacillales*, phylum *Firmicutes*. The nearest described phylogenetic relative of *Dolosigranulum* is *Alloiococcus otitis*. Cloned 16S rDNA sequences derived from some uncultured bacteria associated with the sheep mite *Psoroptes ovis* also show a close phylogenetic affinity with *Dolosigranulum* (Hogg and Lehane, 1999).

Differentiation of the genus *Dolosigranulum* from other genera

Dolosigranulum pigrum can be distinguished from other catalase-negative, Gram-stain-positive cocci using a combination of phenotypic tests. Using conventional biochemical tests (see Facklam and Elliott, 1995), *Dolosigranulum pigrum* is pyrrolidonyl-β-naphthylamide (PYR) and leucine aminopeptidase (LAP) positive, grows in 6.5% NaCl broth, does not grow at 10 or 45 °C, gives a negative bile-esculin reaction, and does not produce gas (LaClaire and Facklam, 2000b). This combination of characters is not unique to the *Dolosigranulum* genus but is shared by *Alloiococcus*, *Facklamia* species, and *Ignavigranum* (LaClaire and Facklam, 2000b). *Alloiococcus* can be distinguished from *Dolosigranulum* in being fastidious, by not growing anaerobically, and by not producing acid from sugars. *Dolosigranulum pigrum* differs from *Facklamia* species and *Ignavigranum* in conventional tests by hydrolyzing esculin (LaClaire and Facklam, 2000b). Biochemical tests which are useful in distinguishing *Dolosigranulum pigrum* from *Facklamia* species and *Ignavigranum ruoffiae* using the API Rapid ID32 Strep system are shown in Table 106.

TABLE 106. API Rapid ID32 Strep tests distinguishing *Dolosigranulum pigrum* from *Facklamia* species and *Ignavigranum ruoffiae*[a]

Test	D. pigrum	F. hominis	F. ignava	F. languida	F. miroungae	F. sourekii	F. tabacinasalis	I. ruoffiae
Acid from:								
D-Arabitol	–	–	–	–	–	+	–	–
Mannitol	–	–	–	–	–	+	–	d
Maltose	+	–	d	–	–	+	–	–
Sorbitol	–	–	–	–	–	+	–	–
Sucrose	+	–	–	–	–	+	–	d
Trehalose	–	–	+	+	+	+	–	–
Hydrolysis of:								
Hippurate	–	+	+	–	–	+	–	–
Activity for:								
ADH	–	+	–	–	+	–	–	+
APPA	–	+	+	–	+	–	–	
α-GAL	–	+	–	–	–	–	+	–
β-GAL	+	+	–	–	–	–	v	–
GTA	d	d	d	+	+	–	–	–
NAG	d	–	–	–	+	–	–	–
PYRA	+	d	+	d	+	+	–	+
URE	–	d	–	–	+	–	–	+

[a]Symbols: +, >85% positive; d, different strains give different reactions (16–84% positive); –, 0–15% positive; v, variable; w, weak reaction; ND, not determined. Abbreviations: ADH, arginine dihydrolase; APPA, alanine phenylalanine proline arylamidase; α-GAL., α-galactosidase; β-GAL, β-galactosidase; GTA, glycine tryptophan arylamidase; NAG, *N*-acetyl β-glucosaminidase; PYRA, pyroglutamic acid arylamidase; URE, urease.

List of species of the genus *Dolosigranulum*

1. **Dolosigranulum pigrum** Aguirre, Morrison, Cookson, Gay and Collins 1994, 370[VP] (Effective publication: Aguirre, Morrison, Cookson, Gay and Collins 1993, 610.)

pi'grum. L. n. adj. *pigrum* lazy.

DNA G+C content (mol%): 40.5 (T_m).

Type strain: R91/1468, ATCC 51524, CCUG 33392, CIP 104051, LMG 15126, NBRC 15550, NCIMB 702975.

GenBank accession number (16S rRNA gene): X70907.

Genus X. **Granulicatella** Collins and Lawson 2000, 367[VP]

PAUL A. LAWSON

Gra.nu.li.ca.tel'la. L. neut. dim. n. *granulum* small grain; L. fem. dim. n. *catella* small chain; N.L. fem. dim. n. *Granulicatella* small chain of small grains.

Cocci that occur as single cells, in pairs, or in short chains. Gram-stain-positive. Nonmotile. Nonsporeforming. Facultatively anaerobic. Catalase- and oxidase-negative. Acid but not gas is produced from glucose. No growth at 10 or 45 °C. Pyrrolidonyl arylamidase and leucine arylamidase are produced. Alkaline phosphatase, α-galactosidase, β-galactosidase, and urease are not produced. Acetoin is not produced. Found as part of the normal oral flora of the human pharynx and the human urogenital and intestinal tracts.

DNA G+C content (mol%): 36–37.5.

Type species: **Granulicatella adiacens** (Bouvet, Grimont and Grimont 1989) Collins and Lawson, 2000, 367[VP] (*Streptococcus adjacens* Bouvet, Grimont and Grimont 1989, 293; *Abiotrophia adiacens* Kawamura, Hou, Sultana, Liu, Yamamoto and Ezaki 1995, 802.).

Further descriptive information

The cell morphology is dependent on the culture conditions. Cocci in pairs or chains are observed under optimal nutritional conditions, but pleomorphic, elongated, swollen cells may be seen when growth conditions are suboptimal.

Some species are fastidious and grow poorly in media used routinely for streptococci, e.g., blood trypticase soy agar, and they require supplements such as L-cysteine and/or pyridoxal. Some grow as satellite colonies adjacent to other organisms such as *Staphylococcus epidermidis*. *Granulicatella elegans* requires L-cysteine and will not grow with only the addition of pyridoxal. *Granulicatella balaenopterae* does not require the addition of pyridoxal or L-cysteine to the blood agar or satellitism to support growth. Usually grow as small alpha-hemolytic colonies on chocolate agar but not on sheep blood agar unless the medium is supplemented or other bacteria are present to provide the compounds required for growth.

The cell-wall murein of *Granulicatella adiacens* is based on L-lysine variation A3α (type: L-Lys–L-Ala) (Kawamura et al., 1995), whereas the murein of *Granulicatella balaenopterae* is based on L-ornithine variation A4α (type: L-Orn–D-Asp) (Lawson et al., 1999). The murein type for *Granulicatella elegans* has not been determined (see Schleifer and Kandler, (1972) for nomenclature).

When discussing *Granulicatella* species, it is pertinent to include *Abiotrophia* due to their historical relationships, the habitats from which they are both isolated, and the similarity in phenotypic characteristics which may cause confusion in their initial identification. Both genera form part of the normal oral flora of the human pharynx and the human urogenital and intestinal tracts (Ruoff, 1991). A review of the literature shows

a growing number of case reports in which nutritionally variant streptococci (NVS) have been isolated from a range of sources and are responsible for a number of clinical and veterinary conditions (Christensen and Facklam, 2001). Like other viridans group streptococci, NVS cause sepsis and bacteremia and are responsible for a substantial proportion of cases of infectious endocarditis (Bouvet, 1995, 1980, 1981; Ruoff, 1991), including most of the so-called blood- culture negative cases of endocarditis (Casalta et al., 2002; Fournier and Raoult, 1999; Roberts et al., 1979). NVS have been implicated in a variety of other infections that are anatomically related to their natural habitats (Ruoff, 1991), for example, *Abiotrophia defectiva* and *Granulicatella adiacens* were isolated from two elderly patients with vitreous infections following cataract extraction (Namdari et al., 1999). In addition, NVS strains have been isolated from humans with infectious crystalline keratopathy (Ormerod et al., 1991) and from horses with corneal ulcers (Da Silva Curies et al., 1990). In a review of 30 cases of infective endocarditis caused by NVS isolates, Stein and Nelson (1987) noted that the clinical manifestations often were more severe than that in cases caused by enterococci or viridans streptococci. The infections were difficult to treat and had a relapse rate of 41% despite treatment with appropriate antibiotics.

In many laboratories, current methods of identification of organisms rely on phenotypic tests such as those developed in miniaturized biochemical kits including the API ID 32 Strep system (bioMerieux, France). However, a number of potential problems are inherent in the use of these phenotypic tests. Not all strains within a given species may be positive for a common trait (Beighton et al., 1991; Kilian et al., 1989) and a single strain may exhibit biochemical variability (Hillman et al., 1989; Teng et al., 2002). In addition, commercially available products are capable of identifying clinically isolated organisms, but the accuracy of these products for identifying the plethora of recently described genera and species of catalase-negative, Gram-stain-positive cocci is not established. Rapidly changing taxonomy within this group has made it difficult to keep the databases of identification products up-to-date. Tests for some of the phenotypic characteristics used to differentiate new genera and species may not be included in some of the commercially available products. As a result, routine identification based solely on phenotypic tests does not allow for an unequivocal identification of certain species. Consequently, molecular methods are increasingly being used in concert with phenotypic criteria as diagnostic tools in the identification of these organisms. Ohara-Nemoto et al. (1997) used PCR-RFLP using *Hae*III and *Msp*I to detect and discriminate 92 isolates with 11 strains of

Abiotrophia defectiva and 81 strains of *Granulicatella adiacens* from clinical specimens. However, use of this method did not allow the detection and identification of *Abiotrophia* species among bacteria in mixed cultures or the detection of atypical strains or unknown species. Primer sets for detection and identification by PCR have been described and tested by Roggenkamp et al. (1998). Four strains of *Granulicatella elegans*, eight strains of *Granulicatella adiacens*, and three strains of *Abiotrophia defectiva* strains in addition to 57 non-NVS strains were examined and the PCR strategy succeeded in separating NVS strains from non-NVS strains and correctly identifying the NVS strains. Furthermore, Kanamoto et al. (2000) investigated 45 *Abiotrophia* strains (including the type strains of *Abiotrophia defectiva*, *Granulicatella adiacens*, and *Granulicatella elegans*) from endocarditis patients by DNA–DNA hybridization, PCR of genomic DNA sequences, 16S rRNA gene PCR-restriction fragment length polymorphism analysis, 16S rRNA gene sequence homology, and phenotypic characteristics. The endocarditis isolates could be divided into four genetic groups representing the three type strains and a new group closely related to *Granulicatella adiacens*. These investigators proposed that this new group be named "*Abiotrophia para-adiacens*". The 45 endocarditis isolates were identified as 9 strains of *Abiotrophia defectiva*, 15 strains of *Granulicatella adiacens*, 13 strains of *Abiotrophia para-adiacens*, and 8 strains of *Granulicatella elegans*. Casalta et al. (2002) used PCR and 16S rRNA gene sequencing to identify *Granulicatella elegans* as the pathogen responsible for the damage caused to the cardiac valve of a patient with culture-negative endocarditis. Although sequence data, PCR primers, and probes have, for the most part, been derived from the 16S rRNA genes, alternative chronometers are used in the identification of Gram-stain-positive taxa. For example, *rpoB*, the gene encoding the highly conserved subunit of bacterial RNA polymerase, has previously been demonstrated to be a suitable target on which to base the identification of micro-organisms and has been used to identify enteric bacteria (Mollet et al., 1997). Drancourt et al. (2004) employed a single specific primer pair for PCR and sequencing method based on the sequence of the *rpoB* gene in the molecular identification of aerobic, Gram-stain-positive, catalase-negative species of the genera *Abiotrophia*, *Enterococcus*, *Gemella*, *Granulicatella*, and *Streptococcus*.

In addition to the authentication of the preliminary identification of both *Granulicatella* and *Abiotrophia* strains isolated in the laboratory, molecular tools are increasingly being used as the primary method of identifying these organisms. This is becoming especially relevant in the clinical environment where a high throughput of samples is encountered (Bosshard et al., 2004; Woo et al., 2001). Thus, these extremely powerful tools can unequivocally identify organisms with fastidious nutritional requirements and phenotypic similarities, which would previously have eluded detection using traditional cultivation techniques.

Isolation procedures

Strains can be isolated on blood agar at 37 °C anaerobically or in air enriched with CO_2. Growth occurs in Todd–Hewitt broth or brain heart infusion broth with the addition of 10 mg pyridoxal or 100 mg cystine per liter. *Granulicatella elegans* requires L-cysteine and will not grow with only the addition of pyridoxal. *Granulicatella balaenopterae* does not require the addition of pyridoxal or L-cysteine to the blood agar or satellitism for growth.

Maintenance procedures

Strains can be maintained on agar media (such as heart infusion or Columbia agar) supplemented with blood (5% v/v). Growth occurs in Todd–Hewitt broth or brain heart infusion broth with the addition of 10 mg pyridoxal or 100 mg cystine per liter. For long-term preservation, strains can either be stored at –70 °C on cryogenic beads or lyophilized.

Taxonomic comments

Nutritionally variant streptococci (NVS) were originally described by Frenkel and Hirsch (1961) as a new type of viridans group streptococci that exhibited satellitism around the colonies of other bacteria. Throughout the literature, these organisms have been referred to by a variety of terms, such as NVS (Bouvet et al., 1981; Cooksey et al., 1979), nutritionally deficient streptococci (Bouvet et al., 1980), satelliting streptococci (McCarthy and Bottone, 1974), vitamin B_6-dependent streptococci (Carey et al., 1975), and pyridoxal-dependent streptococci (Roberts et al., 1979) because of their fastidious nutritional requirements. The taxonomic status of these fastidious organisms was greatly clarified by Bouvet et al. (1989), who demonstrated the existence of two distinct species within the NVS by chromosomal DNA—DNA hybridizations; these species were named *Streptococcus adjacens* and *Streptococcus defectivus*. The studies of Bouvet et al. (1989) revealed *Streptococcus adjacens* and *Streptococcus defectivus* shared less than 10% DNA—DNA reassociation with each other and with other streptococcal species. Kawamura et al. (1995) used 16S rRNA gene sequencing to show that *Streptococcus adjacens* and *Streptococcus defectivus* formed a distinct clade that was phylogenetically far removed from the streptococci and proposed these species be transferred to the new genus *Abiotrophia* as *Abiotrophia adiacens* and *Abiotrophia defectiva*, respectively. The phylogenetic separateness of these NVS from authentic streptococci together with nutritional considerations, satellitism, and pyrrolidonyl-arylamidase production were the primary reasons for creating the genus *Abiotrophia*.

It is pertinent to note that in the 16S rRNA sequence analysis conducted by Kawamura et al. (1995), the exclusion of NVS organisms from the genus *Streptococcus* was justified on phylogenetic grounds (e.g., from tree topology considerations and from 16S rRNA sequence divergence values of generally >11% with streptococcal species). Although it was evident from this study (Kawamura et al., 1995) that NVS organisms shared higher 16S rRNA relatedness with certain other catalase-negative taxa (e.g., *Aerococcus*, *Carnobacterium*, and *Enterococcus*) than with streptococci, aside from a 7% 16S rRNA sequence divergence between *Abiotrophia adiacens* and *Abiotrophia defectiva*, there was little evidence to suggest that the newly delineated genus was not monophyletic. Since the description of *Abiotrophia* by Kawamura et al. (1995), however, two further species of this genus have been characterized. *Abiotrophia elegans* (Roggenkamp et al., 1998) was recovered from a patient with endocarditis, and *Abiotrophia balaenopterae* (Lawson et al., 1999) was isolated from a minke whale (*Balaenoptera acutorostrata*). Both new species display a closer phylogenetic affinity with *Abiotrophia adiacens* (16S rRNA sequence divergence approximately 3%) than with *Abiotrophia defectiva* (sequence divergence approximately 7%). With the description of a number of novel taxa, it has become increasingly evident that *Abiotrophia defectiva*, the type species of the genus, is phylogenetically closer to several non-*Abiotrophia* species including *Globicatella sanguinis*—a chain-forming coccus described by Collins et al. (1992) but not included in the study of Kawamura et al. (1995)—and species of the genera *Dolosicoccus* (Collins et al., 1999a), *Facklamia* (Collins et al., 1997), *Eremococcus* (Collins et al., 1999c), and *Ignavigranum* (Collins et al., 1999b), than to *Abiotrophia adiacens* and its close relatives. It is apparent that the genus is not monophyletic because there are two distinct lines within the *Abiotrophia* genus, namely *Abiotrophia defectiva* and a group comprising *Abiotrophia adiacens, Abiotrophia balaenopterae, Abiotrophia elegans*, and "*Abiotrophia para-adiacens*". To reflect these relationships in this particular clade of the phylum *Firmicutes*, Collins and Lawson (2000) created the genus *Granulicatella* to accommodate *Abiotrophia adiacens, Abiotrophia balaenopterae*, and *Abiotrophia elegans*, thus restricting the genus *Abiotrophia* to the type species, *Abiotrophia defectiva*. Phylogenetic

tree topologies show that *Granulicatella* belongs to a cluster of organisms that include *Atopobacter phocae* and, more loosely, with members of the genus *Trichococcus* (Figure 107).

Differentiation of the genus *Granulicatella* from closely related genera

Atopobacter phocae can be readily distinguished from its closest phylogenetic relatives using a combination of morphological, biochemical, and chemotaxonomic criteria. *Granulicatella* species form coccus-shaped cells that occur singly, in pairs, or in short chains. These features distinguish *Granulicatella* species from *Atopobacter* species, which consist of rod-shaped cells. In addition, *Granulicatella* also differs form *Atopobacter* in forming acid from a narrower range of sugars (such as ribose and glycogen) and by failing to produce α-galactosidase.

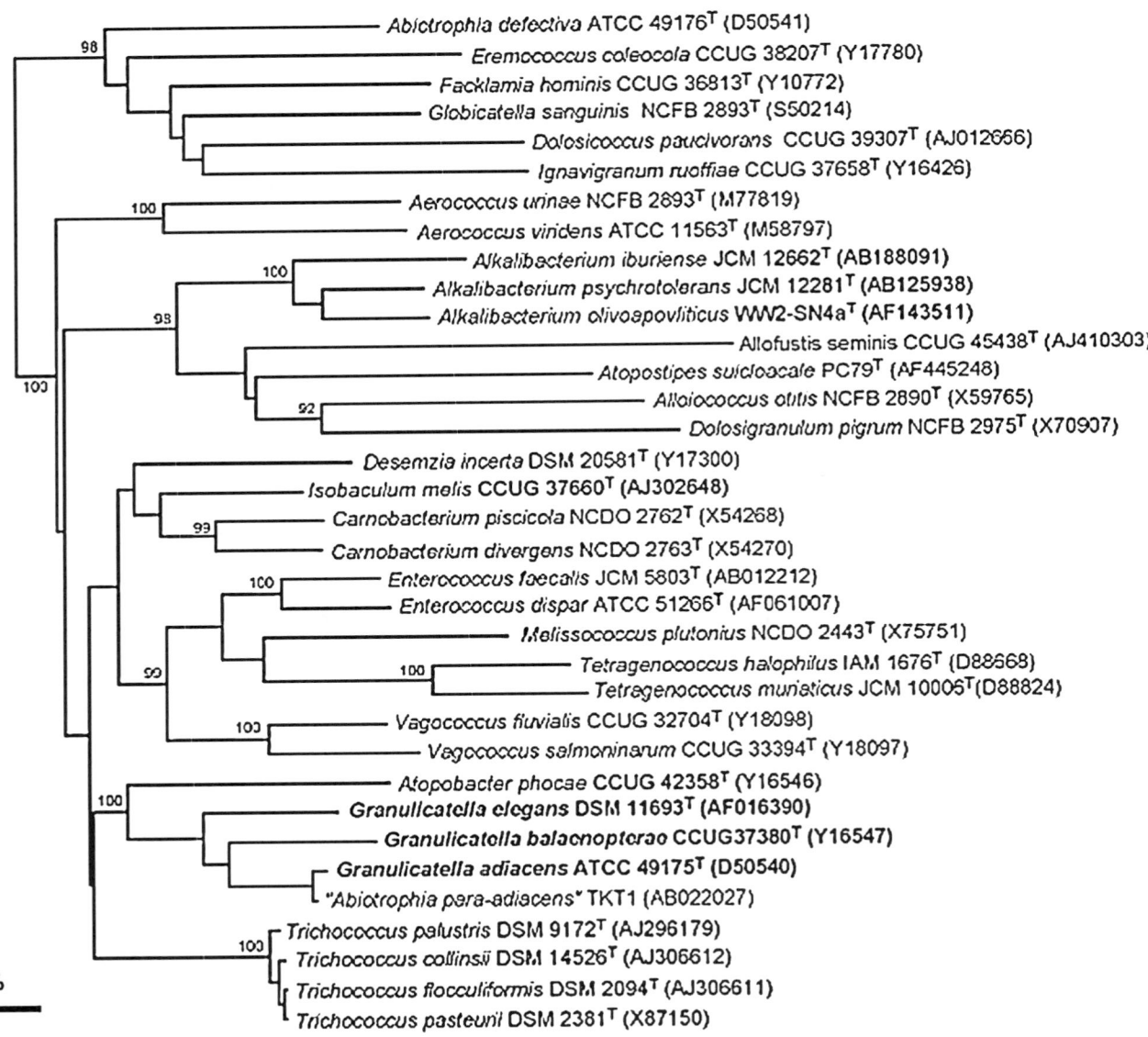

FIGURE 107. Neighbor-joining tree based on 16S rRNA sequences depicting the phylogenetic relationships of *Granulicatella* species and close relatives. Bootstrap values determined from 500 replications.

Granulicatella can be differentiated from *Trichococcus* by not producing acid from lactose and by their cell-wall murein structures. The former possesses either L-Orn–D-Asp (A4β) or L-Lys–L-Ala (A3β) whereas the latter contain L-Lys–D-Asp (A4α). *Granulicatella* also possesses a lower DNA G+C content, with values ranging between 36 and 37.5 mol%, compared to *Trichococcus*, which has values between 45 and 49 mol%. Some granulicatellae are fastidious and require supplements such as L-cysteine and/or pyridoxal, further distinguishing these organisms from *Atopobacter* and *Trichococcus* species. Criteria that are useful in distinguishing the genus *Granulicatella* from phylogenetically closely related taxa are

shown in Table 107. In addition, 16S rRNA gene sequencing serves to discriminate *Granulicatella* clearly from all closely related genera.

Differentiation of the species of the genus *Granulicatella* and *Abiotrophia*

When considering tests that differentiate species of *Granulicatella*, it is pertinent to include *Abiotrophia*, because these two genera may be confused when first isolated in the laboratory. Species of the genus *Granulicatella* and *Abiotrophia* can be distinguished from each other using a number of useful tests that are shown in Table 108.

List of species of the genus *Granulicatella*

1. **Granulicatella adiacens** (Bouvet, Grimont and Grimont 1989) Collins and Lawson 2000, 367[VP] (*Streptococcus adjacens* Bouvet, Grimont and Grimont 1989, 293; *Abiotrophia adiacens* Kawamura, Hou, Sultana, Liu, Yamamoto and Ezaki 1995, 802.)

ad′ia.cens. L. fem. adj. *adjacens* adjacent, indicating that this organism can grow as satellite colonies adjacent to other bacterial growth.

The characteristics are as given for the genus and as listed in Table 107 and Table 108 with the following additional information. Cellular morphology depends upon the conditions of growth; the organism is pleomorphic with chains including cocci, coccobacilli, and globular and rod-shaped cells when grown in cysteine- or pyridoxal-supplemented broth. Small ovoid cocci (diameter 0.4–0.6 μm) occur singly,

TABLE 107. Characteristics distinguishing the genus *Granulicatella* from phylogenetically closely related taxa[a,b]

Characteristic	*Granulicatella*	*Atopobacter*	*Trichococcus*
Cell morphology	Cocci	Rods	Cocci/ovoid
Relationship to air	Facultatively anaerobic	Facultatively anaerobic	Aerotolerant
Acid from:			
Cyclodextrin	D	+	–
Lactose	–	D	+
Ribose	–	+	D
Production of:			
α-Galactosidase	–	+	D
β-Galactosidase	–	D	–
β-Glucuronidase	D	–	–
Murein type	A4β/A3α	A4β	A4α
DNA G+C content (mol%)	36–37.5	ND	45–49

[a]Symbols: +, >85% positive; D, different taxa give different reactions; –, 0–15% positive; w, weak reaction; ND, not determined.
[b]Data obtained from Collins and Lawson (2000), Lawson et al. (2000), Liu et al. (2002).

TABLE 108. Characteristics differentiating *Granulicatella* species and *Abiotrophia defectiva*[a,b]

Characteristic	*Granulicatella adiacens*	*Granulicatella balaenopterae*	*Granulicatella elegans*	*Abiotrophia defectiva*
Production of acid from:				
Pullulan	–	+	–	d
Sucrose	+	–	+	+
Tagatose	+	–	–	–
Trehalose	–	+	–	d
Hydrolysis of:				
Hippurate	–	–	d	–
Arginine	–	+	+	–
Production of:				
Arginine dihydrolase	–	+	+	–
α-Galactosidase	–	–	–	+
β-Galactosidase	–	–	–	+

[a]Symbols: +, >85% positive; d, different strains give different reactions (16–84% positive); –, 0–15% positive; w, weak reaction; ND, not determined.
[b]Data obtained from Collins and Lawson (2000), Christensen and Facklam (2001), Lawson et al. (1999).

in pairs, or in chains of variable length in CDMT medium. Some tendency towards rod formation is observed in the stationary phase. Facultatively anaerobic with complex growth requirements. Grows as satellite colonies adjacent to *Staphylococcus epidermidis* on horse-blood trypticase soy agar and on stored sheep-blood agar. α-Hemolysis occurs on sheep-blood agar. Tiny colonies up to 0.2 mm in diameter are formed on fresh sheep-blood agar or on blood agar supplemented with pyridoxal or L-cysteine. Growth occurs in Todd–Hewitt broth (THB) enriched with 10 mg pyridoxal or 100 mg cysteine per liter. Growth occurs in CDMT semi-synthetic medium (Bouvet et al., 1981) and a red chromophore can be visualized by boiling the bacterium at pH 2 for 5 min. Lactic acid is the predominant acid produced. Resistant to optochin and susceptible to vancomycin. No production of exopolysaccharide occurs on 5% sucrose. Pyrrolidonyl arylamidase is produced, but alkaline phosphatase, alanine-phenylalanine-proline arylamidase, glycyl-tryptophan arylamidase, α-galactosidase, and β-galactosidase are not produced. Tagatose and sucrose are fermented. D-Ribose, L-arabinose, D-mannitol, melibiose, melezitose, pullulan, sorbitol, lactose, D-rhamnose, trehalose, starch, and glycogen are not fermented. Inulin is fermented by some strains. Some strains produce β-glucuronidase. Leucine aminopeptidase is produced. Arginine and hippurate are not hydrolyzed. Acetoin is not produced. The cell wall is of the A3α type. Isolated from the throat flora, urine, and blood of patients with endocarditis.

DNA G+C content (mol%): 36.6–37.4 (T_m).

Type strain: GaD, ATCC 49175, CCUG 27637 A, CCUG 27809, CIP 103243, DSM 9848, LMG 14496, NCTC 13000.

GenBank accession number (16S rRNA gene): D50540.

2. **Granulicatella balaenopterae** (Lawson, Foster, Falsen, Sjödén and Collins 1999) Collins and Lawson 2000, 368[VP] (*Abiotrophia balaenopterae* Lawson, Foster, Falsen, Sjödén and Collins 1999, 505.)

bal.aen.op′ter.ae. N.L. fem. n. *balaenopterae* pertaining to the minke whale, *Balaenoptera acutorostrata*, from which the organism was isolated.

The characteristics are as given for the genus and as listed in Table 107 and Table 108, with the following additional information. Tiny colonies up to 0.2 mm in diameter are formed on blood agar at 37 °C. Facultatively anaerobic. Acid is produced from glucose, maltose, pullulan, and trehalose but not from L-arabinose, D-arabitol, cyclodextrin, glycogen, lactose, mannitol, melibiose, melezitose, D-raffinose, sucrose, sorbitol, tagatose, and D-xylose. Arginine dihydrolase, pyroglutamic acid arylamidase (weak reaction), N-acetyl-glucosaminidase, ester lipase (C4), leucine arylamidase, and urease (weak reaction) activities are detected. Alkaline phosphatase, acid phosphatase, alanine-phenylalanine-proline arylamidase, α-galactosidase, β-galactosidase, β-galacturonidase, β-glucuronidase, glycyl-tryptophan arylamidase, α-mannosidase, chymotrypsin, trypsin, α-fucosidase, and pyrazinamidase activities are not detected. Esculin is hydrolyzed but hippurate and gelatin are not. Nitrate is not reduced. The cell wall contains an L-Orn–D-Asp directly cross-linked murein (type A4β). Isolated from the minke whale. Habitat is not known.

DNA G+C content (mol%): 37 (HPLC).

Type strain: M1975/96/1, ATCC 700813, CCUG 37380, CIP 105938.

GenBank accession number (16S rRNA gene): Y16547.

3. **Granulicatella elegans** (Roggenkamp, Abele-Horn, Trebesius, Tretter, Autenrieth and Heesemann 1998) Collins and Lawson 2000, 367[VP] (*Abiotrophia elegans* Roggenkamp, Abele-Horn, Trebesius, Tretter, Autenrieth and Heesemann 1998, 103.)

e′le.gans. L. adj. *elegans* choice, elegant.

The characteristics are as given for the genus and as listed in Table 107 and Table 108, with the following additional information. Cellular morphology is dependent on nutritional state. In sufficiently supplemented nutritional broth, the bacterium appears coccoid in short chains. Lack of appropriate growth factors results in pleomorphism and the appearance of elongated, swollen forms. The organism grows as a facultative anaerobe with complex growth requirements. It grows as satellite colonies adjacent to *Staphylococcus epidermidis* on trypticase soy sheep-blood agar plates, with α-hemolysis. Tiny colonies up to 0.2 mm in diameter are formed on Schaedler sheep-blood agar plates after 48 h, but only minimal growth is visible on chocolate agar plates. Growth occurs at 27 and 37 °C but not at 20 or 42 °C. Growth occurs in THB or casein-soy peptone bouillon when supplemented with 0.01% L-cysteine hydrochloride, but not when supplemented with 0.001% pyridoxal hydrochloride. A red chromophore can be visualized by boiling the microorganism at pH 2 for 5 min. Pyrrolidonyl arylamidase, arginine dihydrolase, and leucine aminopeptidase are produced but alkaline phosphatase, α-galactosidase, β-galactosidase, β-glucuronidase, β-mannosidase, β-glucosidase, glycyl-tryptophan arylamidase, β-mannosidase, and urease are not produced. Hippurate may or may not be hydrolyzed. Raffinose and sucrose may or may not be fermented. Trehalose, inulin, pullulan, tagatose, lactose, starch, glycogen, D-Arabitol, sorbitol, mannitol, melibiose, melezitose, L-arabinose, and ribose are not fermented. Isolated from a patient with endocarditis. Habitat is not known.

DNA G+C content (mol%): not determined.

Type strain: B1333, ATCC 700633, CCUG 38949, CIP 105513, DSM 11693.

GenBank accession number (16S rRNA gene): AF016390.

Other organisms

1. "**Abiotrophia para-adiacens**" Kanamoto, Sato and Inoue 2000, 497.

Kanamoto et al. (2000) described four groups of *Abiotrophia* strains, three pertaining to each of the three known species and a fourth group for which the name "*Abiotrophia para-adiacens*" was proposed. Since the latter name has not been effectively published or validly published, it lacks standing in the nomenclature. The organism requires pyridoxal for growth and produces a chromophore. Arginine dihydrolase, α-galactosidase, and β-galactosidase are not produced. β-Glucosidase and N-acetyl-β-glucosaminidase may or may not be produced. Trehalose and pullulan are not fermented. Tagatose and sucrose may or may not be fermented. Based upon the results, the authors described four groups: three pertained to each of the three known species, and a fourth group for which the name "*Abiotrophia para-adiacens*" was proposed.

GenBank accession number (16S rRNA gene): AB022027.

Genus XI. **Isobaculum** Collins, Hutson, Foster, Falsen and Weiss 2002, 209[VP]

MATTHEW D. COLLINS

Iso.bac′u.lum. Gr. adj. *isos* alike, similar; L. neut. n. *baculum* small rod; N.L. neut. n. *Isobaculum* the one like a stick or a rod.

Cells are nonsporeforming, nonmotile rods; filaments may be observed. Cells mainly stain Gram-positive, but some cells may appear Gram-negative. **Facultatively anaerobic and oxidase and catalase-negative. Grows at 10 °C but not at 45 °C or in broth containing 6.5% NaCl. Gas is not produced in MRS broth** (De Man et al., 1960). **Acid is produced from D-glucose; acetate and L(+) lactate are the end products of glucose metabolism. Arginine dihydrolase and pyrrolydonyl arylamidase are produced.** Pyruvate is not utilized. **Esculin is hydrolyzed but starch, gelatin, and urea are not hydrolyzed.** Nitrate is not reduced. Voges–Proskauer-negative. Sensitive to vancomycin. **The cell-wall murein is based on L-lysine (type A3α). Long-chain fatty acids are predominantly of the straight-chain saturated and monounsaturated types.** Respiratory menaquinones are absent.

DNA G+C content (mol%): 39 (T_m).

Type species: **Isobaculum melis** Collins, Hutson, Foster, Falsen and Weiss 2002, 209[VP].

Further descriptive information

The genus contains only one species, *Isobaculum melis*, and therefore the characteristics provided below refer to this species.

Cells are rod-shaped and stain Gram-positive, although some cells decolorize easily and appear Gram-negative. Nonpigmented and nonhemolytic. Produces a positive bile-esculin reaction. Pyrrolydonyl arylamidase-positive and leucine aminopeptidase-negative. Hippurate, starch, and urea are not hydrolyzed using conventional methods (Facklam and Elliott, 1995). Acid is produced in litmus milk, but clotting is not observed. In conventional heart infusion base medium (Facklam and Elliott, 1995), acid is produced from D-glucose, glycerol, D-ribose, and trehalose. Acid is not produced from arabinose, inulin, lactose, maltose, melezitose, melibiose, D-raffinose, sorbitol, sorbose, or sucrose. Using API systems, acid is formed from D-glucose, trehalose, and D-ribose but not from L-arabinose, D-arabitol, cyclodextrin, glycogen, lactose, melibiose, melezitose, maltose, mannitol, methyl-β-D-glucopyranoside, pullulan, D-raffinose, sorbitol, sucrose, D-tagatose, or D-xylose. Activity is displayed for acid phosphatase, arginine dihydrolase, esterase C-4, ester lipase C8, β-glucosidase, pyroglutamic acid arylamidase, pyrrolydonyl arylamidase, phosphoamidase, and β-mannosidase but not for alanine phenylalanine proline arylamidase, chymotrypsin, cystine arylamidase, α-fucosidase, α-galactosidase, β-galactosidase, α-glucosidase, β-glucuronidase, glycine tryptophan arylamidase, α-mannosidase, leucine arylamidase, lipase C14, pyrazinamidase, trypsin, urease, or valine arlyamidase. Activity for alkaline phosphatase and N-acetyl-β-glucosaminidase may or may not be detected. Nitrate is not reduced to nitrite. Sensitive to vancomycin (30 μm disc). The major long-chain cellular fatty acids are $C_{16:0}$,

$C_{18:0}$ and $C_{18:1\ \omega9c}$. The cell-wall murein is of the type L-Lys–L-Thr–Gly. The small intestine of a dead badger killed in a road accident is the only known source from which *Isobaculum melis* has been isolated (Collins et al., 2002).

Isolation

Isobaculum melis can be isolated on blood agar at 30 or 37 °C under aerobic or anaerobic conditions.

Maintenance procedures

Isobaculum melis can be maintained on brain heart infusion agar or Columbia blood (5%, v/v) agar. It grows well in brain heart infusion broth. For long-term preservation, organisms can be stored on cryogenic beads (Pro-Lab Diagnostics, UK) at −70 °C or lyophilized.

Taxonomic comments

The genus *Isobaculum* was proposed to accommodate a phylogenetically distinct, facultatively anaerobic, catalase-negative, Gram-stain-positive, rod-shaped bacterium originating from the intestine of a badger (Collins et al., 2002). According to the phylogenetic analysis used for the roadmap to this volume (Ludwig et al., above), the genus *Isobaculum* belongs to family *Carnobacteriaceae* in the order *Lactobacillales*, phylum *Firmicutes*. The nearest phylogenetic relatives of the *Isobaculum* are the genera *Carnobacterium* and *Desemzia*.

Differentiation of the genus *Isobaculum* from other genera

The closest phylogenetic relatives of the genus *Isobaculum* are carnobacteria and *Desemzia incerta*. The long-chain cellular fatty acids of *Isobaculum* closely resemble those of carnobacteria. Both *Isobaculum* and *Carnobacterium* species synthesize major amounts of $C_{18:1\ \omega9c}$ (oleic acid) whereas *Desemzia incerta* synthesizes major amounts of *cis*-vaccenic acid. The presence of an A3α murein type (L-Lys–L-Thr–Gly) in *Isobaculum melis* serves to distinguish it from members of the genus *Carnobacterium* which invariably contain walls based on *meso*-diaminopimelic acid, and from *Desemzia incerta* which contains an L-Lys–D-Glutamic acid murein type (Collins et al., 1987; Stackebrandt et al., 1999). *Isobaculum melis* possesses a distinct biochemical profile. The API Rapid ID32 Strep profile 30120310040 and API CORYNE profile 4140300 readily serve to distinguish *Isobaculum melis* from *Carnobacterium* species, *Desemzia incerta*, and other asporogenous rod-shaped Gram-stain-positive taxa. Tests which are useful in distinguishing *Isobaculum melis* from its nearest phylogenetic relatives using the API Rapid ID32 Strep system are given in Table 109.

List of species of the genus *Isobaculum*

1. **Isobaculum melis** Collins, Hutson, Foster, Falsen and Weiss 2002, 209[VP]

   me′lis. L. fem. n. *meles* badger; L. gen. fem. n. *melis* of the badger.

   DNA G+C content (mol%): 39 (T_m).
   Type strain: M577-94, CCUG 37660, DSM 13760.
   GenBank accession number (16S rRNA gene): AJ302648.

TABLE 109. Characteristics differentiating *Isobaculum melis* from its nearest phylogenetic relatives, *Desemzia* and *Carnobacterium*[a]

Test	*I. melis* CCUG 37660[T]	*D. incerta* CCUG 38799[T]	*C. alterfundium* CCUG 34643[T]	*C. divergens* NCDO 2763[T]	*C. funditum* CCUG 34644[T]	*C. gallinarum* CCUG 30095[T]	*C. mobile* CCUG 30096[T]	*C. piscicola* CCUG 34645[T]	*C. inhibens* CCUG 31728[T]
Production of acid from:									
Maltose	−	+	+	+	+	+	−	+	+
Mannitol	−	−	−	−	−	+	−	+	+
Ribose	+	+	−	+	−	+	+	+	+
Sucrose	−	+	+	+	+	+	+	+	+
Production of:									
ADH	+	−	−	+	−	+	+	+	−
GTA	−	+	−	+	+	+	−	+	−
PYRA	+	+	−	+	−	+	+	+	−
βNAG	+	+	−	+	−	+	−	+	−
Voges–Proskauer	−	−	−	+	+	+	−	+	−
Murein type	Lys–L-Thr–Gly	Lys–D-Glu	*m*-Dpm	*m*-Dpm	*m*-Dpm	*m*-Dpm	*m*-Dpm	*m*-Dpm	*m*-Dpm

[a]Abbreviations: ADH, arginine dihydrolase; GTA, glycine tryptophan arylamidase; PYRA, pyroglutamic acid arylamidase; βNAG, N-acetyl-β-glucosaminidase. Biochemical tests determined using API rapid ID32 Strep system. *m*-Dpm, *meso*-diaminopimelic acid.

Genus XII. **Marinilactibacillus** Ishikawa, Nakajima, Yanagi, Yamamoto and Yamasato 2003, 719[VP]

KAZUHIDE YAMASATO AND MORIO ISHIKAWA

Ma.ri.ni.lac.ti.ba.cil′lus. L. adj. *marinus* marine; L. n. *lac*, *lactis* milk; L. n. *bacillus* a small rod; N.L. masc. n. *Marinilactibacillus* marine lactic acid rodlet.

Cells are Gram positive, nonsporeforming, straight rods that occur singly, in pairs, or in short chains. **Motile with peritrichous flagella.** Facultative anaerobe. Catalase and oxidase negative. Negative for nitrate reduction and gelatin liquefaction. Hydrolyzes casein. Mesophilic and psychrotolerant. **Slightly halophilic and highly halotolerant. Alkaliphilic.** L(+)Lactic acid is the major end product from D(+)glucose; trace to small amounts of formate, acetate, and ethanol are produced with a molar ratio of approximately 2:1:1, without gas formation. The peptidoglycan is of the A4β, **Orn–D-Glu type.** Cellular fatty acids are primarily of the straight-chain saturated and monounsaturated even-carbon-numbered types. **The major fatty acids are $C_{16:0}$, $C_{16:1\,\omega7}$, $C_{18:0}$, and $C_{18:1\,\omega9}$ (oleic acid). Respiratory quinones and cytochromes are absent.**

DNA G + C content (mol%): 34.6–36.2.

Type species: **Marinilactibacillus psychrotolerans** Ishikawa, Nakajima, Yanagi, Yamamoto and Yamasato 2003, 719[VP].

Further descriptive information

Descriptive information is based on the characteristics of eight strains of *Marinilactibacillus psychrotolerans*. The genus *Marinilactibacillus* is a lactic acid bacterium, belonging to order *Lactobacillales*, class *Bacilli* in phylum *Firmicutes*.

Phylogenetic position of the genus *Marinilactibacillus* based on 16S rRNA gene sequence analysis is given in Figure 108. Phenotypic characteristics of *Marinilactibacillus psychrotolerans* are listed in Table 110, Table 111, and Table 112.

Lactate is the main end product of glucose fermentation under anaerobic cultivation. Lactate yields are 87–100% relative to the amount of consumed glucose. The ratio of the L(+) isomer to total lactate produced is 75–94% at optimum pH for growth. In addition to lactate, trace to small amounts of formate, acetate, and ethanol are produced with a molar ratio of approximately 2:1:1 (Table 111). The pH of the fermented medium has a large effect on the relative amount of lactate versus the other three products. As the initial acidity of media increases, the proportion of lactate also increases, while those of the other three products decrease in parallel, and vice versa on the alkaline side. The 2:1:1 molar ratio among the three products other than lactate is retained. It is assumed that *Marinilactibacillus psychrotolerans* has two pyruvate pathways mediated by lactate dehydrogenase and pyruvate formate-lyase, and the pathway used would be affected by the pH value of the fermentation medium (Gunzalus and Niven, 1942; Janssen et al., 1995; Rhee and Pack, 1980).

Marinilactibacillus psychrotolerans metabolizes glucose oxidatively. When grown aerobically with shaking, lactate yield decreases and acetate yield increases without production of formate and ethanol (Table 112).

Marinilactibacillus psychrotolerans is slightly halophilic (Kushner, 1978; Kusher and Kanekura, 1988). The optimum NaCl concentration for growth in GYPF broth, pH 8.5 (for composition, see *Maintenance procedures*) is between 2.0% (0.34M) and 3.75% (0.64M) (3.0–5.0% for strain IAM 14980[T]). For strain IAM 14980[T], the maximum specific growth rates, μ_{max} (h[-1]), are 0.38 in 0%; 0.36 in 0.5%; 0.40 in 1.0%; 0.42 in 1.5%, 2.0%, and 2.5%; 0.47 in 3.0%; 0.51 in 3.75%; 0.46 in 5.0%; and 0.36 in 7.5% NaCl. *Marinilactibacillus psychrotolerans* is highly halotolerant, able to grow at 17.0–20.5% (2.9–3.5M) NaCl. Strain IAM 14980[T] grows at 20% NaCl.

Marinilactibacillus psychrotolerans is alkaliphilic, having an optimum pH above 8.0 (Jones et al., 1994). For strain IAM 14980[T], the μ_{max} (h[-1]) values are 0.44 at pH 7.0; 0.48 at pH 7.5; 0.53 at pH 8.0;

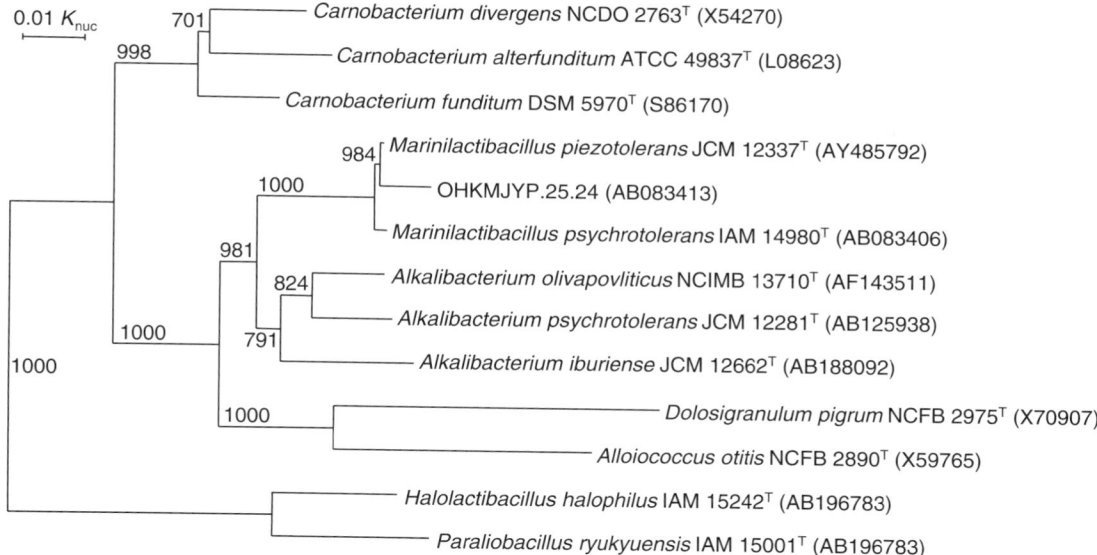

FIGURE 108. Phylogenetic relationships between *Marinilactibacillus* and some other related bacteria. The tree was constructed using the neighbor-joining method and is based on a comparison of approximately 1400 nt. Bootstrap values, based on 1000 replications, are given at the branching points.

TABLE 110. Phenotypic characteristics of *Marinilactibacillus psychrotolerans*[a,b]

Characteristic	*Marinilactibacillus psychrotolerans*
NaCl optima (%)	2–3.75
NaCl range (%)	0–17.5 to 20.5
pH optima	8.5–9.0
pH range	6.0–10
Temperature optima (°C)	37–40
Temperature range (°C)	−1.8–40 to 45
Gelatin liquefaction	−
Casein hydrolysis	−
Nitrate reduction	−
Arginine hydrolysis	w
Fermentation of:	
D-Ribose, D-xylose, D-glucose, D-fructose, D-mannose, maltose, sucrose, D-cellobiose, D-trehalose, D-mannitol, α-methyl-D-glucoside, D-salicin, gluconate	+
L-Arabinose, D-galactose, lactose, D-rhamnose	(+)
Glycerol	(w)
D-Arabinose	v
Melibiose, D-raffinose, D-melezitose, adonitol, dulcitol, D-sorbitol, inulin, starch	(−)
myo-Inositol	−
Gas from gluconate	−
Major fatty acid composition (% of total):[c]	
$C_{12:0}$	0.4
$C_{14:0}$	5.2
$C_{15:0}$	0.4
$C_{16:0}$	32.0
$C_{16:1}$	3.1
$C_{16:1\ \omega7}$	19.3
$C_{18:0}$	6.9
$C_{18:1\ \omega9}$ (oleic acid)	30.2
$C_{18:2}$	1.1
$C_{20:1}$	1.4

[a]Symbols: +, all strains positive; (+), most strains positive; (w), most strains weakly positive; v, strain-to-strain variation; (−), most strains negative; −, all strains negative.
[b]Ishikawa et al. (2003).
[c]Data for strain IAM 14980[T].

TABLE 111. Effect of pH on the balance of glucose fermentation by *Marinilactibacilus pyschrotolerans* IAM 14980[Ta]

Initial pH	End products [mol/(mol glucose)]				Lactate yield per consumed glucose (%)	Carbon recovery (%)
	Lactate	Formate	Acetate	Ethanol		
7	2.02	0.15	0.04	0.05	101	107
8	1.5	0.52	0.2	0.19	75	97
9	1.29	0.81	0.35	0.2	66	98

[a]Data from Ishkawa et al. (2003).

TABLE 112. Products from glucose under aerobic and anaerobic cultivation conditions by *Marinilactibacilus psychrotolerans* IAM 14980[Ta,b]

Cultivation condition	Glucose consumed (mM)	Products (mM)					Carbon recovery (%)
		Pyruvate	Lactate	Formate	Acetate	Ethanol	
Aerobic	33.2	ND	21.5	ND	40	ND	73
Anaerobic	38.6	ND	63.9	7.5	5.1	4.1	94

[a]ND, not detected.
[b]Data from Ishikawa et al. (2005).

0.50 at pH 8.5; 0.51 at pH 9.0; 0.50 at pH 9.5; and 0.34 at pH 10.0. When the isolates were grown in 2.5% NaCl GYPF broth, pH 8.5, the final pH could be as low as 4.7–5.2, which is 0.8–1.3 pH units lower than the minimum pH required to initiate growth.

Marinilactibacillus psychrotolerans is mesophilic. The μ_{max} (h^{-1}) values for *Marinilactibacillus psychrotolerans* IAM 14980^T are 0.36 at 25 °C; 0.52 at 30 °C; 0.60 at 37 °C; 0.42 at 40 °C; and 0.02 at 42.5 °C. It grows very well at lower temperatures with respect to the extent of maximum growth, although growth rates are low. For strain IAM 14980^T, OD$_{660}$ values are 0.78 at −1.8 °C (the freezing point of seawater; 21 d incubation), 0.70 at 0 °C (21 d), 0.89 at 5 °C (9 d), and 1.14 at 37 °C (the optimum temperature; 15 h).

Marinilactibacillus psychrotolerans ferments a fairly wide range of carbohydrates (pentoses, hexoses, and disaccharides) and related carbon compounds. Fermentation of sugar alcohols and polysaccharides are negative for most of strains except for glycerol (weak) and D-mannitol (Table 110).

For eight strains of *Marinilactibacillus psychrotolerans*, the compared sequences of 16S rRNA genes (1458–1479 nt) are identical and the DNA–DNA hybridization values are 74–100%. The sequence similarity between *Marinilactibacillus psychrotolerans* IAM 14980^T and *Marinilactibacillus piezotolerans* (Toffin et al., 2005) JCM 12337^T is 99.6%. The phylogenetically closest genus is *Alkalibacterium* (Ntougias and Russell, 2001): *Marinilactibacillus psychrotolerans* IAM 14980^T has 96.2% similarity to *Alkalibacterium olivapovlyticus* NCIMB 13710^T, 94.6% similarity to *Alkalibacterium iburiense* JCM 12662^T (Nakajima et al., 2005), and 95.1% similarity to *Alkalibacterium psychrotolerans* JCM 12281^T (Yumoto et al., 2004). The similarities of *Marinilactibacillus psychrotolerans* to *Carnobacteriun funditum*, *Carnobacterium alterfunditum*, *Alloiococcus otitis*, *Dolosigranulum pigrum*, and *Desemzia incerta* are 93.2%, 92.8%, 91.1%, 90.9%, and 91.9%, respectively.

Marinilactibacillus psychrotolerans was isolated from a living sponge collected from the Oura beach, Miura Peninsula, Japan, and a raw Japanese ivory shell and decomposing alga collected from Okinawa, Japan. *Marinilactibacillus piezotolerans* was isolated from a sediment core collected at 4.15 m below the sea floor from a water depth of 4790.4 m in the Nankai Trough, off the coast of Japan in the Pacific Ocean (Toffin et al., 2004, 2005).

Halolactibacillus species (*Halolactibacillus halophilus* and *Halolactibacillus miurensis*; Ishikawa et al., 2005) were isolated from living and decaying marine organisms. *Halolactibacillus* is slightly halophilic with an optimum of 2.0–3.0%, and highly halotolerant with a range of 0–25.5%. *Halolactibacillus* is alkaliphilic with an optimum of pH 8.0–9.5 and a range of pH 6.0–10.0.

Carnobacterium funditum and *Carnobacterium alterfunditum* were isolated from water of possible seawater origin in Ace Lake in Antarctica (Franzmann et al., 1991). These two species grow preferably under conditions of low salinity. These members of *Marinilactibacillus*, *Halolactibacillus*, and *Carnobacterium* have adapted to marine environments characterized by salinity and slightly alkaline reaction and can be called marine lactic acid bacteria.

Enrichment and isolation procedures

Marinilactibacillus psychrotolerans was isolated by enrichment culture from marine organisms. For isolation, cultures are enriched and subcultured in 7% NaCl GYPF or GYPB isolation medium containing peptone, yeast extract, fish or beef extract, pH 9.5 or 10.0, at 30 °C under anaerobic conditions. The first enrichment culture, whose pH has decreased below 7.0, is subcultured. The second enrichment culture is incubated anaerobically; it is pour-plated with an agar medium supplemented with CaCO$_3$ and overlaid with an agar medium containing 0.1% sodium thioglycolate. Prolonged incubation in enrichment culture should be avoided, as cells tend to autolyse. Another enrichment medium used is GYPFSK isolation broth of similar composition containing 12% NaCl, pH 7.5. For the second enrichment, the same medium containing 18% NaCl is used. Incubation is conducted at 30 °C in standing culture, and the culture is pour-plated with 12% NaCl GYPFSK agar supplemented with CaCO$_3$. The compositions and preparation methods of media were described by Ishikawa et al. (2003).

Maintenance procedures

Marinilactibacillus psychrotolerans is maintained by serial transfer in a stab culture stored at 5–10 °C at 1–2 month intervals. The medium is 7% NaCl GYPF or GYPB agar supplemented with 12 g Na_2CO_3, 3 g $NaHCO_3$, and 5 g $CaCO_3$ per liter. The 7% NaCl GYPF or GYPB agar is composed of (per liter) 10 g glucose, 5 g yeast extract, 5 g peptone, 5 g fish or beef extract, 1 g K_2HPO_4, 70 g NaCl, 1 g sodium thioglycolate, 5 ml salt solution (per ml, 40 mg $MgSO_4 \cdot 7H_2O$, 2 mg $MnSO_4 \cdot 4H_2O$, 2 mg $FeSO_4 \cdot 7H_2O$), and 13 g agar. The final pH is 9.0. A solution of the main components, carbonate buffer compounds, and $CaCO_3$ are sterilized separately.

Marinilactibacillus psychrotolerans can be maintained in 2.5% NaCl GYPF or GYPB agar, pH 8.5, supplemented with 5 g $CaCO_3$ per liter. To prepare this medium, a double-strength solution of the main components is adjusted to pH 8.5, sterilized by filtration, and aseptically mixed with an equal volume of autoclaved 2.6% agar solution. Then, autoclaved $CaCO_3$ (as a slurry with a small amount of water) is added.

Strains are maintained by freezing at –80 °C or below in 2.5% GYPFK broth (i.e., GYPF broth in which the concentration of K_2HPO_4 is increased to 1%) supplemented with 10% (w/v) glycerol. Strains are kept by L-drying in an adjuvant solution composed of (per liter) 3 g sodium glutamate, 1.5 g adonitol, and 0.05 g cysteine hydrochloride in 0.1 M phosphate buffer (KH_2PO_4/K_2HPO_4), pH 7.0 (Sakane and Imai, 1986). Strains can be kept by freeze-drying with a standard suspending fluid containing an appropriate concentration of NaCl.

Differentiation of *Marinilactibacillus* from other related genera and species

Characteristics used to differentiate *Marinilactibacillus psychrotolerans* from other marine lactic acid species isolated from saline/alkaline environments are listed in Table 113. For comparison, the catalase-positive, lactic-acid-producing *Paraliobacillus ryukyuensis* is included in the table. *Marinilactibacillus psychrotolerans*, *Halolactibacillus halophilus*, and *Halolactibacillus miurensis* are marine lactic acid bacteria (Ishikawa et al., 2003, 2005), and *Paraliobacillus ryukyuensis* is of marine origin and produces lactic acid as a main product from glucose under anaerobic conditions (Ishikawa et al., 2002). In addition to the phenotypic similarities (Table 113), these four species produce formate, acetate, and ethanol with a molar ratio of 2:1:1 as well as lactate. The amounts of these three products produced relative to the amount of lactate increase as acidity in fermentation medium decreases (Table 111; see chapter on *Halolactibacillus*, Table 24). Although they are catalase-negative, *Marinilactibacillus psychrotolerans* and *Halolactibacillus* species are similar in their ability

TABLE 113. Characteristics that distinguish *Marinilactibacillus psychrotolerans* from other lactic acid rods and *Paraliobacillus ryukyuensis*, which are phenotypically similar and were isolated from saline/alkaline environments[a]

Characteristic	Marinilactibacillus psychrotolerans[b]	Alkalibacterium iburiense[c]	Alkalibacterium olivapovliticus[d]	Alkalibacterium psychrotolerans[c]	Carnobacterium alterfunditum[f]	Carnobacterium funditum[f]	Halolactibacillus halophilus[g]	Paraliobacillus ryukyuensis[h]
Spore formation	–	–	–	–	–	–	–	+
Motility (flagellation)	+ (Pe)	+ (Pe)	+ (Po)	+ (Pe)	+ (Pe)	+ (Pe)	+ (Pe)	+ (Pe)
Catalase	–	–	–	–	–	–	–	+
NaCl range (%)	0–20	0–14 to 16	3–15	0–17	≥0, <8.8	>0, <8.8	0–25.5	0–22
NaCl optima (%)	2.0–3.75	3–13	3–5	2–12	0.6	1.7	2–3	0.75–3
pH range	6.0–10.0	9–12	8.5–10.8	9–12	7–7.3	7–7.4	6–10	5.5–9.5
pH optima	8.5–9.0	9.5–10.5	9.0–10.2	9.5–10.5	5.9	5.9	8–9.5	7–8.5
Growth at –1.8 °C	+	–	–	–	–	–	–	–
Major isoprenoid quinones	None	None	None	None	ND	ND	None	MK-7
Peptidoglycan type	Orn–D-Glu	Lys–D-Asp	Orn–D-Asp	Lys(Orn)–D-Glu	*m*-Dpm	*m*-Dpm	*m*-Dpm	*m*-Dpm
G + C content (mol%)	34.6–36.2	42.6–43.2	39.7±1.0	40.6	33.3–33.4	31.6–34.0	38.5–40.7	35.6
Major cellular fatty acids	$C_{16:0}$, $C_{16:1\omega7}$, $C_{18:1\omega9}$	$C_{16:1\omega9}$, $C_{18:1\omega9}$	$C_{16:0}$, $C_{16:1\omega9}$	$C_{14:0}$, $C_{16:0}$, $C_{16:1\omega7}$, $C_{18:1\omega9}$	$C_{16:1c7}$, $C_{16:0}$, $C_{18:1c9}$	$C_{16:1c7}$, $C_{16:0}$, $C_{18:1c9}$	$C_{13:0\ ante}$, $C_{16:0}$	ND
Isolation source	Decaying marine algae, living sponge, raw Japanese ivory shell	Fermented polygram indigo	Wash-waters of edible olives	Fermented polygram indigo	Antarctic lake of possible seawater origin	Antarctic lake of possible seawater origin	Decaying marine algae, living sponge	Decaying marine alga

[a]Symbols: +, positive; –, negative; Pe, peritrichous; Po, polar; ND, no data; *m*-Dpm, *meso*-diaminopimelic acid; Orn, ornithine; Asp, aspartic acid; Glu, glutamic acid.
[b]Data from Ishikawa et al. (2003).
[c]Data from Nakajima et al. (2005).
[d]Data from Ntougias and Russell (2001).
[e]Data from Yumoto et al. (2004).
[f]Data from Franzmann et al. (1991).
[g]Data from Ishikawa et al. (2005).
[h]Data from Ishikawa et al. (2002).

to metabolize glucose oxidatively under aerobic conditions (Table 112; see Table 25 in the chapter on *Halolactibacillus*). In spite of physiological and biochemical similarities, *Marinilactibacillus* is phylogenetically distinguished from *Halolactibacillus* and *Paraliobacillus*. Both genera belong to the discrete cluster within *Bacillus* rRNA group 1 in the family *Bacillaceae*. *Marinilactibacillus psychrotolerans* is chemotaxonomically distinguished from *Halolactibacillus* and *Paraliobacillus* by its peptidoglycan type and cellular fatty acid composition and from *Paraliobacillus*, in addition, by sporeformation and lack of catalase production (Table 113).

Taxonomic comments

The description of the genus *Marinilactibacillus* is based on *Marinilactibacillus psychrotolerans*. Another validly published species of this genus is *Marinilactibacillus piezotolerans* (Toffin et al., 2005). The sequence similarity of the 16S rRNA gene between the two species is 99.6% (1287 nt) when calculated with the sequence for strain LT20ᵀ (accession no. AY485792; Toffin et al., 2005) and is 99.5% (1456 nt) when calculated with

the sequence for strain JCM 12337ᵀ (accession no. AB247277). The similarity between the two sequences of *Marinilactibacillus piezotolerans* is 99.2% (1338 nt). Several characteristics described for *Marinilactibacillus piezotolerans* differ from some experimental results obtained for strain JCM 12337ᵀ (Yamasato and Ishikawa, unpublished data). Catalase activity is negative (described as positive); G + C content of the DNA is 35.5 mol% rather than 42 mol%; major cellular fatty acids are C_{16} and C_{18} acids rather than C_{14} and C_{16} acids; and fermentation of maltose, sucrose, and D-galactose is positive rather than negative. It may be the case that *Marinilactibacillus piezotolerans* is inadequately described, and further critical characterization is needed.

Bacterial strains identified as *Marinilactibacillus psychrotolerans* by molecular method have been isolated from deep-sea sediments (Inagaki et al., 2003), soft cheeses (Maoz et al., 2003; Feurer et al., 2004; Ishikawa et al., 2007), and spoiled dry-cured hams (Rastelli et al., 2005). In spoiled dry-cured hams and mold-ripened soft cheeses (Ishikawa et al., 2005), *Marinilactibacillus psychrotolerans* constituted a predominant flora.

List of species of the genus *Marinilactibacillus*

1. **Marinilactibacillus psychrotolerans** Ishikawa, Nakajima, Yanagi, Yamamoto and Yamasato 2003, 719ᵛᴾ

   psy.chro.to′le.rans. Gr. adj. *psychros* cold; L. part. adj. *tolerans* tolerating; N.L. adj. *psychrotolerans* tolerating cold temperature.

   The characteristics are as described for the genus and as listed in Table 110, Table 111, and Table 112. The morphology is shown in Figure 109. Deep colonies in agar medium are pale yellow, opaque, and lenticular, with diameters of 2–4 mm. The cells are typically 0.4–0.5 × 2.3–4.5 μm and elongated in older cultures. Grows evenly throughout a column of semisolid agar medium.

   Source: Living sponge, decaying marine algae, raw Japanese ivory shell.

   DNA G + C content (mol%): 34.6–36.2 (HPLC).

   Type strain: M13-2, IAM 14980, JCM 21451, NBRC 100002, NCIMB 13873, NRIC 0150.

   GenBank accession number (16S rRNA): AB083406.

2. **Marinilactibacillus piezotolerans** Toffin, Zink, Kato, Pignet, Bidault, Bienvenu, Birrien and Prieur 2005, 349ᵛᴾ

   pie.zo.to′le.rans. Gr. v. *piezo* to press; L. part. adj. *tolerans* tolerating, N.L. part. adj. *piezotolerans* tolerating high hydrostatic pressure.

   Source: Deep subseafloor sediment.

   DNA G + C content (mol%): 35.5 (HPLC).

   Type strain: LT20, DSM 16108, JCM 12337.

   GenBank accession number (16S rRNA): AY485792, AB247277.

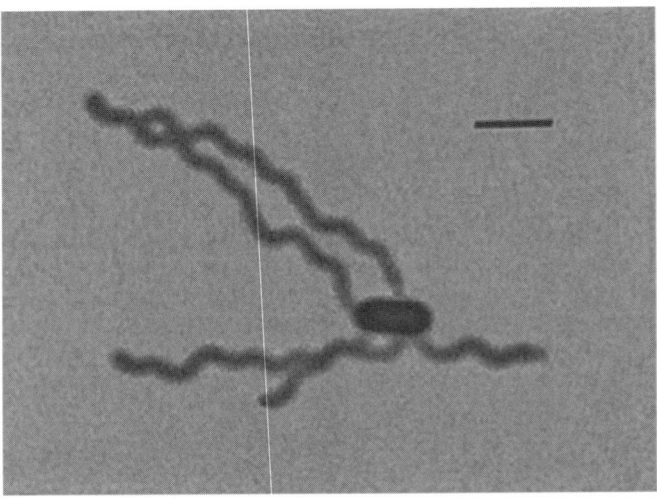

FIGURE 109. Photomicrograph of a cell and peritrichous flagella of *Marinilactibacillus psychrotolerans* IAM 14980ᵀ grown anaerobically at 30 °C for 1 d on 2.5% NaCl GYPFK agar. Bar = 2 μm.

Genus XIII. **Trichococcus** Scheff, Salcher and Lingens 1984b, 356[VP] (Effective publication: Scheff, Salcher and Lingens 1984a, 118.) emend. Liu, Tanner, Schumann, Weiss, McKenzie, Janssen, Seviour, Lawson, Allen and Seviour 2002, 1124

FRED A. RAINEY

Tri.cho.coc.cus. Gr. n. *thrix* hair; L. masc. n. *coccus* a grain or berry; N.L. masc. n. *Trichococcus* a hair of cocci.

Pleiomorphic cells, can be spherical to ovoid, olive shaped with tapered ends or rod shaped. Cells are $0.75–2.5\,\mu m \times 0.7–1.5\,\mu m$. Cells occur singly, in pairs, or in short or long chains. Motility observed in some species. **Gram-stain-positive.** Nonsporeforming. Peptidoglycan type is A4α, L-Lys-D-Asp. **Aerotolerant. Fermentative metabolism.** Catalase and oxidase negative. **Psychroactive mesophile**; optimum temperature for growth is 23–30 °C. Growth occurs in the range −5 to 40 °C and pH 5.5–9.0. Indole is not produced. Under aerobic conditions lactate and acetate produced from glucose; under anerobic conditions lactate, formate, acetate, and ethanol produced from glucose. Isolated from activated sludge, swamps, soil, hydrocarbon spill site, and penguin guano.

DNA G+C content (mol%): 45–49.

Type species: **Trichococcus flocculiformis** Scheff, Salcher and Lingens 1984b, 356[VP] (Effective publication: Scheff, Salcher and Lingens 1984a, 118.) emend. Liu, Tanner, Schumann, Weiss, McKenzie, Janssen, Seviour, Lawson, Allen and Seviour 2002, 1124.

Further descriptive information

The species and strains assigned to this genus do not demonstrate a uniform morphology and the morphotype varies depending on the culture media used for growth (Liu et al., 2002). The morphologies of *Trichococcus flocculiformis* strains Echt[T], NDP, and Ben 200, *Trichococcus pasteurii* strain KoTa2[T] and *Trichococcus collinsii* strains 37AN3*[T] and 45AN2 were observed by scanning electron microscopy after growth on R2A and SR2A agar (Liu et al., 2001) and after growth in R2A broth with and without shaking (Liu et al., 2002). It was found that these strains are pleomorphic under different conditions of growth, and morphologies can differ from those presented in the original species descriptions. The type strain of the *Trichococcus flocculiformis* Echt[T] was found to grow as regular cocci in chains on most culture media including SR2A, but it grew as larger tapered rod-shaped cells when cultured on R2A agar. These morphological differences were also observed for *Trichococcus*

pasteurii KoTa2[T] when it was grown on R2A agar. *Trichococcus flocculiformis* strain NDP displayed cells with tapered ends in pairs when grown on both SR2A and R2A agar. These morphological differences were also observed in strains of the same species, *Trichococcus collinsii*; in the case of strains 37AN3*[T] and 45AN2 in that cells of strain 37AN3*[T] appeared swollen, often tapered, paired, and irregular on both R2A and SR2A while strain 45AN2 grew as regular single or paired cocci on SR2A and as larger cocci with tapered ends, often in pairs on R2A (Liu et al., 2002). Strain Ben 200 (*Trichococcus flocculiformis*) grew as regular cocci in chains on SR2A agar and pleiomorphic irregular cells in chains on R2A agar. Changes in morphology between strains grown in static and shaken broth cultures were also observed for strains Ben 77, Ben 200, Ben 201, KoTa2[T], 37AN3*[T], 45AN2, and NDP (Liu et al., 2002). No differences in cell morphology were observed between static or shaken cultures of *Trichococcus flocculiformis* Echt[T]. No information on changes in morphology due to conditions of growth has been reported for *Trichococcus patagoniensis* (Pikuta et al., 2006).

The cell wall composition has been determined for all species of the genus with the exception of *Trichococcus patagoniensis* and found to be A4 with an L-Lys–D-Asp interpeptide bridge. Fatty acid profiles of 8 strains including the type strains of the species *Trichococcus flocculiformis*, *Trichococcus pasteurii*, and *Trichococcus collinsii* were determined for cells grown on Tryptic Soy agar at room temperature for 3 d (Liu et al., 2002). The fatty acid composition of *Trichococcus palustris* grown at both 6 and 25 °C was determined by Zhilina et al. (1995). In the study of Pikuta et al. (2006) the fatty acid profiles were determined for *Trichococcus patagoniensis* PmagG1[T] and *Trichococcus pasteurii* ATCC 35945[T] after growth on liquid medium incubated at 22 °C for 4 d. The fatty acid data of the type strains of the *Trichococcus* species from these studies is assembled and shown in Table 114.

Although different culture media and conditions were used in the cultivation of the strains compared in Table 114, the major fatty acids of the species of the genus *Trichococcus* are consistently $C_{14:0}$, $C_{16:0}$, and $C_{16:1}$. Interestingly, in the study of Lui

TABLE 114. Cellular fatty acids composition (% of total) of type strains of the genus *Trichococcus*[a]

Fatty acids	*T. flocculiformis* Echt[b]	*T. collinsii* 37AN3*[b]	*T. palustris* Z-7189[c]	*T. pasteurii* KoTa2[b]	*T. patagoniensis* PmagG1[d]
$C_{12:0}$	ND	6	ND	1	1
$C_{14:0}$	14	46	21	28 (16)	11
$C_{16:0}$	15	18	15	16 (15)	16
$C_{16:1}$	46	27	20	42	NR
$C_{16:1\ 7c}$	NR	NR	NR	NR (48)	43
$C_{18:0}$	2	2	4	2	3
$C_{18:1\ \omega9c}$	18	2	22	6 (17)	22
$C_{18:1\ \omega7c}$	ND	ND	2	ND	1

[a]Symbols: ND, not detected; NR, not reported.
[b]Data from Liu et al. (2002); values in parentheses are from Pikuta et al. (2006).
[c]Data from Zhilina et al. (1995).
[d]Data from Pikuta et al. (2006).

et al. (2002), one of the major fatty acids in all strains was $C_{16:1}$, while in the study of Pikuta et al. (2006), $C_{16:1}$ was not detected in *Trichococcus patagoniensis* and $C_{16:1\ 7c}$ was detected in both *Trichococcus patagoniensis* and *Trichococcus pasteurii*.

Species of the genus *Trichococcus* have been found to utilize a variety of sugars, sugar alcohols, and polysaccharides. These utilizations have been tested using a number of approaches including determination of growth, production of acid, and the metabolism of BIOLOG compounds. In the emended description of the genus *Trichococcus*, Liu et al. (2002) included a large number of BIOLOG system metabolism results as genus characteristics. This resulted in a very wide ranging genus description with respect to substrate utilization tests, and that may result in many genus description emendations as new species are added to the genus. With the addition of *Trichococcus patagoniensis*, for which no BIOLOG substrates have been tested, the genus description as presented by Liu et al. (2002) may require further emendation. Pikuta et al. (2006) demonstrated that the utilization patterns for a number of substrates (D-trehalose, lactate, citrate, D-ribose, maltose, and α- D-lactose) could be used to differentiate between the type strains of the five species of the genus *Trichococcus*. The substrates used or not used by the species of this genus are provided in the species descriptions below when available.

Fermentation of glucose by strains of this genus results in the production of lactate, formate, acetate, ethanol, and CO_2. In addition, traces of isobutyrate were detected as end products of the fermentation of glucose by *Trichococcus palustris* (Zhilina et al., 1995). Growth of strains under oxic conditions resulted in the production of lactate and acetate (Liu et al., 2002).

Enrichment, isolation and growth conditions

The strains assigned to the genus *Trichococcus* were isolated in the study of the microbial composition of sewage sludge and duck pond sediment (*Trichococcus flocculiformis* and *Trichococcus pasteurii*), or hydrocarbon contaminated sites (*Trichococcus collinsii*), or from studies aimed at the isolation of psychrophiles (*Trichococcus palustris* and *Trichococcus patagoniensis*). There is no clear selective isolation approach for the isolation of species of the genus *Trichococcus*, but considering all strains grow at 4 °C or below, it could be considered that incubation of enrichment cultures at temperatures less than 4 °C could select for these organisms. *Trichococcus flocculiformis* Echt[T] was isolated by streaking a single drop of bulking sludge on UA medium and incubating the plates at 25 °C for several weeks. UA medium contained (g/l): peptone (Oxoid), 1 g; glucose, 1 g; urea, 20 g; KH_2PO_4, 2 g; phenol red, 0.012 g; agar (Merck), 12 g. Although isolated under aerobic conditions, this species grew under anaerobic conditions facilitated by the Anaerobic Generating System (Oxoid) as well as in agar deeps and in liquid culture under a layer of paraffin (Scheff et al., 1984b). Strains 37AN3[T] and 45AN2, assigned to the species *Trichococcus collinsii*, were isolated from gas condensate-contaminated groundwater sediments (Gieg et al., 1999). Both aerobic and anaerobic heterotrophic organisms were isolated on tryptic soy broth with mineral salts and vitamins added (Gieg et al., 1999). The paper of Gieg et al. (1999) does not provide details on the temperatures or length of incubation of these enrichments. *Trichococcus pasteurii* was isolated from anoxic digestor sludge from the municipal sewage plant of the city of Konstanz, Germany and described

as *Ruminococcus pasteurii* (Schink, 1984). The strain KoTa2[T], described as *Ruminococcus pasteurii*, was isolated in a study to enumerate tartrate fermenting bacteria. The anaerobic tartrate fermenting bacteria were enumerated by MPN methods in mineral medium containing 10 mM L-, D-, and m-tartrate (Schink, 1984). Pure cultures were obtained using serial dilution in agar shakes (Schink, 1984). In a study to isolate psychroactive bacteria from a swamp ecosystem in which organic matter is decomposed at −2 to −4 °C in winter but is exposed to much higher temperatures during the summer months, a coccoid organism strain Z-7189[T] was isolated (Zhilina et al., 1995). This organism was initially assigned to the genus *Ruminococcus* as *Ruminococcus palustris* (Zhilina et al., 1995) and later transferred to the genus *Trichococcus* as *Trichococcus palustris* (Liu et al., 2002). Strain Z-7189[T] was isolated from swamp samples collected in winter and maintained in a frozen state until used for inoculation (Zhilina et al., 1995). Enrichment cultures on glucose were established with 10% inoculum in bicarbonate Pfennig medium (Zhilina, 1978) supplemented with yeast extract, 2 g/l; trace element solution, 2 ml/l (Liu et al., 2002); vitamin solution, 10 ml/l (Wolin et al., 1963); $Na_2S·9H_2O$, 0.5 g/l; resazurin, 0.001 g/l and glucose, 0.5 g/l. The enrichment culture was incubated at 6 °C, and cultures were subsequently selected that had good growth rates at 6 °C but grew better at 20–30 °C (Zhilina et al., 1995). These organisms were considered to be psychroactive organisms and a coccoid organism designated Z-7189 was isolated using the serial dilution technique in liquid medium followed by colony isolation using agar roll-tubes (Zhilina et al., 1995). Penguin guano was the source of *Trichococcus patagoniensis* PmagG1[T] (Pikuta et al., 2006). The guano was collected from a shallow marine tidal pool and stored at 4 °C until homogenized and used to inoculate a glucose-containing enrichment culture. The culture medium contained (g/l): NaCl, 30.0; KCl, 0.3; KH_2PO4, 0.3; $MgSO_4·7H_2O$, 0.1; NH_4Cl, 1.0; $CaSO_4·7H_2O$, 0.0125; $NaHCO_3$, 0.4; $Na_2S·9H_2O$, 0.4; resazurin, 0.0001; yeast extract, 0.1; D-glucose, 5.0; vitamin solution (Wolin et al., 1963), 2 ml; and trace mineral solution (Whitman et al., 1982), 1 ml. The pH was 7.8 and the headspace gas N_2 (Pikuta et al., 2006). The culture was incubated at 4 °C and serially diluted to obtain a pure culture using the roll tube method.

Taxonomic comments

The genus *Trichococcus* currently comprises five species which form a phylogentically coherent group within the *Firmicutes*. It was the lack of a 16S rRNA gene sequence for *Trichococcus flocculiformis* and the morphological differences between strains Echt[T] and KoTa2[T] that led to the description of KoTa2[T] initially as a *Ruminococcus* species (Schink, 1984) and subsequently as the type species of the genus *Lactosphaera* (Janssen et al., 1995). With the determination of the 16S rRNA gene sequence of *Trichococcus flocculiformis* Echt[T], it became clear that *Lactosphaera pasteurii* KoTa2[T] and *Ruminococcus palustris* Z-7189[T] were members of the genus *Trichococcus* (Liu et al., 2002; Stackebrandt et al., 1999). In addition, the strains Ben 200, 37AN3*[T], 45AN2, and NDP were shown to group within the *Trichococcus* group (Liu et al., 2002). All of the 16S rRNA gene sequences available for *Trichococcus* strains share greater than 99% sequence similarity (Table 115). The closest relatives to species of the genus *Trichococcus* that are related at the 94–95% similarity level are *Isobaculum melis*

TABLE 115. 16S rRNA gene sequence similarities of the type strains of the five species of the genus *Trichococcus*

Sequence	*T. flocculiformis* Echt[T]	*T. collinsii* 37AN3*[T]	*T. palustris* Z-7189[T]	*T. pasteurii* KoTa2[T]	*T. patagoniensis* PmagG1[T]
AJ306611	–				
AJ306612	99.6	–			
AJ296179	99.0	99.3	–		
X87150	99.8	99.9	99.1	–	
AF394926	99.6	100	99.3	99.9	–

(Collins et al., 2002), *Desemzia incerta* (Stackebrandt et al., 1999), and the species of the genus *Carnobacterium*. Although these 16S rRNA gene sequence similarity values are high and in many situations would be considered to represent species of the same genus, there are distinct cell wall peptidoglycan type differences between the species of the genus *Trichococcus* which have an A4α, L-Lys–D-Asp peptidoglycan and those of the genera *Carnobacterium* (*meso*-diaminopimelic acid), *Isobaculum* (A3α, Lys–L-Thr–Gly), and *Desemzia* (A4α, Lys–D-Glu) (Collins et al., 2002; Stackebrandt et al., 1999).

With such high 16S rRNA gene sequence similarities between all of the species of the genus *Trichococcus* and considering the 100% similarity between *Trichococcus collinsii* 37AN3*[T] and *Trichococcus patagoniensis* PmagG1[T], DNA–DNA studies were carried out by Liu et al. (2002) and Pikuta et al. (2006). The results of these DNA–DNA hybridizations demonstrated that strain 37AN3*[T] was a distinct species with DNA–DNA reassociation of 34, 58, and 45% to *Trichococcus palustris* Z-7189[T], *Trichococcus pasteurii* KoTa2[T], and *Trichococcus patagoniensis* PmagG1[T], respectively (Liu et al., 2002; Pikuta et al., 2006). Reassocia-tion values of 40 and 47% were determined for *Trichococcus palustris* vs *Trichococcus pasteurii* and *Trichococcus patagoniensis*, repectively (Liu et al., 2002; Pikuta et al., 2006). *Trichococcus flocculiformis* Echt[T] has not been hybridized with the type strain of any of the other species of the genus *Trichococcus*. However, *Trichococcus flocculiformis* Echt[T] has been hybridized with strains Ben 77, Ben 200, Ben 201, and NDP demonstrating that those strains had DNA–DNA reassociation values greater than 70% and all belong to the species *Trichococcus flocculiformis* (Liu et al., 2002). Strain Ben 201 was hybridized with the type strains of the species *Trichococcus palustris*, *Trichococcus pasteurii*, and *Trichococcus collinsii*, and the result demonstrates that these strains (Ben 77, Ben 200, Ben 201, and NDP) all belong to the same species, i.e. *Trichococcus flocculiformis* (Liu et al., 2002). Phenotypic differentiation of the species of the genus *Trichococcus* is based on substrate utilization tests. In describing *Trichococcus patagoniensis*, Pikuta et al. (2006) demonstrated that the utilization patterns for the carbon sources D-trehalose, lactate, citrate, D- ribose, maltose, and α-D-lactose can be used to differentiate the species.

List of the species of the genus *Trichococcus*

The species descriptions presented here are a combination of characters presented in the original description of these species as well as data presented in subsequent publications.

1. **Trichococcus flocculiformis** Scheff, Salcher and Lingens 1984b, 356[VP] (Effective publication: Scheff, Salcher and Lingens 1984a, 118.) emend. Liu, Tanner, Schumann, Weiss, McKenzie, Janssen, Seviour, Lawson, Allen and Seviour 2002, 1124.

floc.cu.li′form.is. L. n. *floccus* a flock of wool; N.L. dim. adj. *flocculus* like a small floc of wool; L. n. *forma* shape; L. adj. *flocculiformis* small-floc-shaped.

Cells are spherical to ovoid or rod shaped, 1.0–2.5 μm × 1.0–1.5 μm. Occur in filaments of 20 to several hundred cells, as single cells, as pairs, or are rod shaped. Gram-stain-positive. Nonsporeforming. Colonies are white, opaque, circular, convex with lobate or erose margins on Nutrient Agar. The temperature range for growth is 4–39 °C with optimum growth at 25–30 °C. The pH range for growth is 5.8–9.0, with optimum growth at pH 8.0. Facultative anaerobe. Aerotolerant. Fermentative metabolism. Acid produced from fructose, galactose, glucose, mannose, ribose, xylose, cellobiose, lactose, maltose, sucrose, and trehalose. Acid is not produced from DL-arabinose, erythritol, galactitol, ribitol, starch, agar, esculin, casein, cellulose, chitin, DNA, gelatin, pectin, tributyrin, Tween 80, and urea. Some strains produce acid from inositol, sorbitol, mannitol, raffinose, and adonitol. Nitrate reduction variable. Urease variable. Hydrogen sulfide not produced. Acetylmethylcarbinol not produced. Methyl red test positive. Fatty acids include $C_{14:0}$ (8–14%), $C_{16:0}$ (12–24%), $C_{16:1}$ (39–50%), $C_{17:1iso}$ (1–4%), $C_{18:0}$ (1–3%), and $C_{18:1 ω9c}$ (14–23%).

Isolated from bulking sludge and duck pond sediment.

DNA G+C content (mol%): 46.8 (T_m).

Type strain: Echt, ATCC 51221, DSM 2094. Other strains include Ben 77, Ben 200, Ben 201, and NDP (Liu et al., 2002).

GenBank accession number (16S rRNA gene): AJ306611, Y17301.

2. **Trichococcus collinsii** Liu, Tanner, Schumann, Weiss, McKenzie, Janssen, Seviour, Lawson, Allen and Seviour 2002, 1124[VP] emend. Pikuta, Hoover, Bej, Marsic, Whitman, Krader and Tang 2006, 2060.

coll.ins′i.i. N.L. gen. n. *collinsii* referring to Matthew D. Collins, a contemporary English microbiologist who contributed significantly to our understanding of the lactic acid bacteria.

Cells are single or paired cocci or tapered cells. Small chains of cocci form in R2A broth. The temperature range for growth is –5 to 36 °C with optimum growth at 25–30 °C. The pH range for growth is 6.0–9.0, with optimum growth at pH 7.5. Facultative anaerobe. Aerotolerant. Fermentative metabolism. Grow on citrate, D-malate, allantoin, and L-tartrate, and produce acid from mannitol. The type strain produces acid from inositol, sorbitol, raffinose, and adonitol. Nitrate reduction negative. Urease-negative.

Isolated from hydrocarbon spill site.

DNA G+C content (mol%): 47.0 (HPLC).

Type strain: 37AN3*, ATCC BAA-296, DSM 14526. Another strain is 45AN2 (Liu et al., 2002).

GenBank accession number (16S rRNA gene): AJ306612.

3. **Trichococcus palustris** (Zhilina, Kotsyurbenko, Osipov, Kostrinka and Zavarzin 1995) Liu, Tanner, Schumann, Weiss, McKenzie, Janssen, Seviour, Lawson, Allen and Seviour 2002b, 1125[VP] (*Ruminococcus palustris* Zhilina, Kotsyurbenko, Osipov, Kostrinka and Zavarzin 1995, 577.)

pa.lu'stris. L. adj. *palustris* swamp-inhabiting ruminococcus.

Cells are coccoid or elongated with slightly tapered ends, 0.75–1.3 μm × 0.7–1.0 μm. Occur as single cells, or in pairs, short chains joined by a mucous capsule. Gram-stain-positive. Nonsporeforming. The temperature range for growth is 0–33 °C with optimum growth at 30 °C. The pH range for growth is 6.2–8.4, with optimum growth at pH 7.5. Facultative anaerobe. Aerotolerant. Fermentative metabolism. Ferments glucose, sucrose, fructose, mannose, maltose, raffinose, lactose, cellobiose, mannitol, sorbitol, pyruvate, and *N,N*-acetyl-D-glucosamine. Fucose, arabinose, rhamnose, xylose, sorbose, galactose, ribose, trehalose, melibiose, erythritol, adonite, ducitol, inositol, betaine, choline chloride, acetate, malate, formate, butyrate, propionate, lactate, succinate, fumarate, mono-, trimethylamine, L-histine HCl, DL-methionine, DL-serine, L-glutamate, L-glutamine, Casamino acids, glycine, D-glucosamine HCl, glycogen, starch, microcrystalline cellulose, peptone, ethanol, butanol, propanol, methanol, glycerol, H₂ + CO₂ are not utilized. Nitrate not reduced. Urease-negative.

Isolated from a swamp in floodplain of the Yasnushka River near the Abramtsev settlement, Moscow Region, Russia.

DNA G+C content (mol%): 47.5 (T_m).

Type strain: Z-7189, CIP 105359, DSM 9172.

GenBank accession number (16S rRNA gene): AF296179.

4. **Trichococcus pasteurii** (Schink 1984) Liu, Tanner, Schumann, Weiss, McKenzie, Janssen, Seviour, Lawson, Allen and Seviour 2002b, 1124[VP] (*Ruminococcus pasteurii* Schink, 1984, 413; *Lactosphaera pasteurii* Janssen, Evers, Rainey, Weiss, Ludwig, Harfoot and Schink 1995, 570.)

pas.teu'ri.i. N.L. gen. n. *pasteurii* referring to Louis Pasteur, who probably first enriched and observed this bacterium during studies on tartrate fermentation.

Cells are nonmotile cocci that form pairs or small irregular packets of cells, 1.0–1.5 μm in diameter. Spherical, ovoid, or olive shaped and occur as single cells, in pairs, in chains, or as irregular conglomerates. Nonsporeforming. The temperature range for growth is 0–42 °C with

optimum growth at 25–30 °C. The pH range for growth is 5.5–9.0. Growth occurs up to 2% NaCl (w/v). Biotin is required as a growth factor. Aerotolerant. Fermentative metabolism. Nitrate, sulfate, sulfite, thiosulfate, sulfur, and fumarate are not reduced. L-Tartrate, pyruvate, oxaloacetate, malate, citrate, mannitol, sorbitol, glucose, galactose, mannose, L-rhamnose, fructose, maltose, lactose, sucrose, cellobiose, raffinose, trehalose, sorbose, starch, and oat spelt xylan are utilized. Laminarin is weakly used. D-Tartrate, *meso*-tartrate, xylose, ribose, arabinose, malonate, succinate, DL-3-hydroxybutyrate, lactate, amino acids, alcohols, chitin, gumkaraya, carboxymethyl cellulose, amorphous cellulose, mannan, lichenan, gum locust bean, pullulan, arabinogalactan, and glycogen are not utilized. The end products of L-tartrate, pyruvate, and citrate are acetate, formate, and CO₂. Glucose and other C-sources are fermented to L-lactate, acetate, formate, and ethanol. Urease and gelatin hydrolysis negative.

Isolated from anoxic digestor sludge.

DNA G+C content (mol%): 45.0 (T_m).

Type strain: KoTa2, ATCC 35945, CCUG 37395, CIP 104580, DSM 2381.

GenBank accession number (16S rRNA gene): L76599, X87150.

5. **Trichococcus patagoniensis** Pikuta, Hoover, Bej, Marsic, Whitman, Krader and Tang 2006, 2060[VP]

pa.ta.go.ni.en'sis. N.L. masc. adj. *patagoniensis* pertaining to Patagonia, the region of South America where the sample for the type strain was collected.

Cells are motile cocci with a diameter of 1.3–2.0 μm. Cells are spherical, ovoid, or olive shaped and occur as single cells, in pairs, in chains, or as irregular conglomerates. Gram-variable. After 14 d incubation at 4 °C, surface colonies are circular, slimy/mucoid, white, with entire margins, and 0.5–4.0 mm in diameter. The temperature range for growth is –5 to 35 °C with optimum growth in the range 28–30 °C. Slightly alkalophilic; pH range for growth is 6.0–10.0, with optimum growth at pH 8.5. Growth occurs at NaCl concentrations of 0–6.5% (w/v), with optimum growth at 0.5% NaCl. Facultative anaerobe. Heterotrophic growth occurs on D-glucose, D-fructose, maltose, D-mannitol, D-mannose, D-ribose, D-arabinose, sucrose, starch, pyruvate, and citrate. No growth occurs on formate, acetate, lactate, propionate, butyrate, ethanol, methanol, glycerol, acetone, betaine, trimethylamine, triethylamine, peptone, Bacto tryptone, Casamino acids, yeast extract, pectin, and D-trehalose. The metabolic end products are lactate, formate, acetate, ethanol, and CO₂. Sensitive to ampicillin, kanamycin, gentamicin, tetracycline, rifampicin (all at 250 μg/ml), and chloramphenicol (125 μg/ml).

Isolated from guano of Magellanic penguins (*Spheniscus magellanicus*) inhabiting the southern region of Chilean Patagonia.

DNA G+C content (mol%): 45.8 (HPLC).

Type strain: PmagG1, ATCC BAA-756, CIP 108035, JCM 12176.

GenBank accession number (16S rRNA gene): AF394926.

References

Aguirre, M. and M.D. Collins. 1992a. Phylogenetic analysis of *Alloiococcus otitis* gen. nov., sp. nov., an organism from human middle ear fluid. Int. J. Syst. Bacteriol. *42*: 79–83.

Aguirre, M. and M.D. Collins. 1992b. Development of a polymerase chain reaction-probe test for identification of *Alloiococcus otitis*. J. Clin. Microbiol. *30*: 2177–2180.

Aguirre, M., D. Morrison, B.D. Cookson, F.W. Gay and M.D. Collins. 1993. Phenotypic and phylogenetic characterization of some *Gemella*-like organisms from human infections: description of *Dolosigranulum pigrum* gen. nov., sp. nov. J. Appl. Bacteriol. *75*: 608–612.

Aguirre, M., D. Morrison, B.D. Cookson, F.W. Gay and M.D. Collins. 1994. Phenotypic and phylogenetic characterization of some *Gemella*-like organisms from human infections: description of *Dolosigranulum pigrum* gen. nov., sp. nov. Int. J. Syst. Bacteriol. *44*: 370.

Baya, A.M., A.E. Toranzo, B. Lupiani, T. Li, B.S. Roberson and F.M. Hetrick. 1991. Biochemical and serological characterization of *Carnobacterium* spp. isolated from farmed and natural populations of striped bass and catfish. Appl. Environ. Microbiol. *57*: 3114–3120.

Beighton, D., J.M. Hardie and R.A. Whiley. 1991. A scheme for the identification of viridans streptococci. J. Med. Microbiol. *35*: 367–372.

Beswick, A.J., B. Lawley, A.P. Fraise, A.L. Pahor and N.L. Brown. 1999. Detection of *Alloiococcus otitis* in mixed bacterial populations from middle-ear effusions of patients with otitis media. Lancet *354*: 386–389.

Borch, E. and G. Molin. 1989. The aerobic growth and product formation of *Lactobacillus*, *Leuconostoc*, *Brochothrix*, and *Carnobacterium* in batch cultures. Appl. Microbiol. Biotechnol. *30*: 81–88.

Bosley, G.S., A.M. Whitney, J.M. Pruckler, C.W. Moss, M. Daneshvar, T. Sih and D.F. Talkington. 1995. Characterization of ear fluid isolates of *Alloiococcus otitidis* from patients with recurrent otitis media. J. Clin. Microbiol. *33*: 2876–2880.

Bosshard, P.P., S. Abels, M. Altwegg, E.C. Böttger and R. Zbinden. 2004. Comparison of conventional and molecular methods for identification of aerobic catalase-negative Gram-positive cocci in the clinical laboratory. J. Clin. Microbiol. *42*: 2065–2073.

Bouvet, A., A. Ryter, C. Frehel and J.F. Acar. 1980. Nutritionally deficient streptococci: electron microscopic study of 14 strains isolated in bacterial endocarditis. Ann. Microbiol. (Paris) *131B*: 101–120.

Bouvet, A., I. van de Rijn and M. McCarty. 1981. Nutritionally variant streptococci from patients with endocarditis: growth parameters in a semisynthetic medium and demonstration of a chromophore. J. Bacteriol. *146*: 1075–1082.

Bouvet, A., F. Grimont and P.A.D. Grimont. 1989. *Streptococcus defectivus* sp. nov. and *Streptococcus adjacens* sp. nov., nutritionally variant streptococci from human clinical specimens. Int. J. Syst. Bacteriol. *39*: 290–294.

Bouvet, A. 1995. Human endocarditis due to nutritionally variant streptococci: *Streptococcus adjacens* and *Streptococcus defectivus*. Eur. Heart. J. *16 Suppl B*: 24–27.

Breed, R.S. 1953. The families developed from *Bacteriaceae* Cohn with a description of the family *Brevibacteriaceae*. Riass. Commun. VI Congr. Int. Microbiol. Roma *1*: 10–15.

Brillet, A., M.F. Pilet, H. Prevost, A. Bouttefroy and F. Leroi. 2004. Biodiversity of *Listeria monocytogenes* sensitivity to bacteriocin-producing *Carnobacterium* strains and application in sterile cold-smoked salmon. J. Appl. Microbiol. *97*: 1029–1037.

Brosius, J., M.L. Palmer, P.J. Kennedy and H.F. Noller. 1978. The complete nucleotide sequence of a 16S ribosomal RNA gene from *Escherichia coli*. Proc. Natl. Acad. Sci. U.S.A. *75*: 4801–4805.

Carey, R.B., K.C. Gross and R.B. Roberts. 1975. Vitamin B6-dependent *Streptococcus mitior* (*mitis*) isolated from patients with systemic infections. J. Infect. Dis. *131*: 722–726.

Casalta, J.P., G. Habib, B. La Scola, M. Drancourt, T. Caus and D. Raoult. 2002. Molecular diagnosis of *Granulicatella elegans* on the cardiac valve of a patient with culture-negative endocarditis. J. Clin. Microbiol. *40*: 1845–1847.

Chmelar, D., A. Matusek, J. Korger, E. Durnova, M. Steffen and E. Chmelarova. 2002. Isolation of *Carnobacterium piscicola* from human pus: case report. Folia Microbiol. (Praha) *47*: 455–457.

Christensen, J.J. and R.R. Facklam. 2001. *Granulicatella* and *Abiotrophia* species from human clinical specimens. J. Clin. Microbiol. *39*: 3520–3523.

Collins, M.D., T. Pirouz, M. Goodfellow and D.E. Minnikin. 1977. Distribution of menaquinones in actinomycetes and corynebacteria. J. Gen. Microbiol. *100*: 221–230.

Collins, M.D. and D. Jones. 1981. Distribution of isoprenoid quinone structural types in bacteria and their taxonomic implication. Microbiol. Rev. *45*: 316–354.

Collins, M.D., D. Jones, R.M. Keddie, R.M. Kroppenstedt and K.H. Schleifer. 1983a. Classification of some coryneform bacteria in a new genus *Aureobacterium*. Syst. Appl. Microbiol. *4*: 236–252.

Collins, M.D., D. Jones and R.M. Kroppenstedt. 1983b. Reclassification of *Brevibacterium imperiale* (Steinhaus) and *Corynebacterium laevaniformans* (Dias and Bhat) in a redefined genus *Microbacterium* (Orla-Jensen), as *Microbacterium imperiale* comb. nov. and *Microbacterium laevaniformans* nom. rev. comb. nov. Syst. Appl. Microbiol. *4*: 65–78.

Collins, M.D. and R.M. Kroppenstedt. 1983. Lipid composition as a guide to the classification of some coryneform bacteria containing an A-4-alpha type peptidoglycan. Syst. Appl. Microbiol. *4*: 95–104.

Collins, M.D. 1987. Transfer of *Brevibacterium ammoniagenes* (Cooke and Keith) to the genus *Corynebacterium* as *Corynebacterium ammoniagenes* comb. nov. Int. J. Syst. Bacteriol. *37*: 442–443.

Collins, M.D., J.A.E. Farrow, B.A. Phillips, S. Feresu and D. Jones. 1987. Classification of *Lactobacillus divergens*, *Lactobacillus piscicola*, and some catalase-negative, asporogenous, rod-shaped bacteria from poultry in a new genus, *Carnobacterium*. Int. J. Syst. Bacteriol. *37*: 310–316.

Collins, M.D., M. Aguirre, R.R. Facklam, J. Shallcross and A.M. Williams. 1992. *Globicatella* sanguis gen. nov., sp. nov., a new Gram-positive catalase-negative bacterium from human sources. J. Appl. Bacteriol. *73*: 433–437.

Collins, M.D. 1994. Isoprenoid quinones. *In* Goodfellow and O'Donnell (Editors), Chemical Methods in Prokaryotic Systematics. John Wiley & Sons, New York, pp. pp. 265–309.

Collins, M.D., E. Falsen, J. Lemozy, E. Akervall, B. Sjödén and P.A. Lawson. 1997. Phenotypic and phylogenetic characterization of some *Globicatella*-like organisms from human sources: description of *Facklamia hominis* gen. nov., sp. nov. Int. J. Syst. Bacteriol. *47*: 880–882.

Collins, M.D., M.R. Jovita, R.A. Hutson, E. Falsen, B. Sjödén and R.R. Facklam. 1999a. *Dolosicoccus paucivorans* gen. nov., sp. nov., isolated from human blood. Int. J. Syst. Bacteriol. *49*: 1439–1442.

Collins, M.D., P.A. Lawson, R. Monasterio, E. Falsen, B. Sjödén and R.R. Facklam. 1999b. *Ignavigranum ruoffiae* sp. nov., isolated from human clinical specimens. Int. J. Syst. Bacteriol. *49*: 97–101.

Collins, M.D., M. Rodriguez Jovita, P.A. Lawson, E. Falsen and G. Foster. 1999c. Characterization of a novel Gram-positive, catalase-negative coccus from horses: description of *Eremococcus coleocola* gen. nov., sp. nov. Int. J. Syst. Bacteriol. *49*: 1381–1385.

Collins, M.D. and P.A. Lawson. 2000. The genus *Abiotrophia* (Kawamura *et al.*) is not monophyletic: proposal of *Granulicatella* gen. nov., *Granulicatella adiacens* comb. nov., *Granulicatella elegans* comb. nov and *Granulicatella balaenopterae* comb. nov. Int. J. Syst. Evol. Microbiol. *50*: 365–369.

Collins, M.D., R.A. Hutson, G. Foster, E. Falsen and N. Weiss. 2002. *Isobaculum melis* gen. nov., sp. nov., a *Carnobacterium*-like organism isolated from the intestine of a badger. Int. J. Syst. Evol. Microbiol. *52*: 207–210.

Collins, M.D., R. Higgins, S. Messier, M. Fortin, R.A. Hutson, P.A. Lawson and E. Falsen. 2003. *Allofustis seminis* gen. nov., sp. nov., a novel Gram-positive, catalase-negative, rod-shaped bacterium from pig semen. Int. J. Syst. Evol. Microbiol. *53*: 811–814.

Collins, M.D., A. Wiernik, E. Falsen and P.A. Lawson. 2005. *Atopococcus tabaci* gen. nov., sp. nov., a novel Gram-positive, catalase-negative, coccus-shaped bacterium isolated from tobacco. Int. J. Syst. Evol. Microbiol. *55*: 1693–1696.

Cone, D.K. 1982. A *Lactobacillus* sp. from diseased female rainbow trout, *Salmo gairderni* Richardson, in Newfoundland, Canada. J. Fish Dis. *5*: 479–485.

Cooksey, R.C., F.S. Thompson and R.R. Facklam. 1979. Physiological characterization of nutritionally variant streptococci. J. Clin. Microbiol. *10*: 326–330.

Cotta, M.A., T.R. Whitehead and R.L. Zeltwanger. 2003. Isolation, characterization and comparison of bacteria from swine faeces and manure storage pits. Environ. Microbiol. *5*: 737–745.

Cotta, M.A., T.R. Whitehead, M.D. Collins and P.A. Lawson. 2004a. *In* Validation of publication of new names and new combinations previously effectively published outside the IJSEM. List no. 99 Int. J. Syst. Evol. Microbiol. *54*: 1425–1426.

Cotta, M.A., T.R. Whitehead, M.D. Collins and P.A. Lawson. 2004b. *Atopostipes* suicloacale gen. nov., sp. nov., isolated from an underground swine manure storage pit. Anaerobe *10*: 191–195.

Da Silva Curies, C.J.M. J.M., S.S. Lang and R.W. Bellhorn. 1990. Nutritionally variant streptococci associated with corneal ulcers in horses: 35 cases (1982–1988). J. Am. Vet. Med. Assoc. *197*: 624–626.

De Bruyn, I.N., A.I. Louw, L. Visser and W.H. Holzapfel. 1987. *Lactobacillus divergens* is a homofermentative organism. Syst. Appl. Microbiol. *9*: 173–175.

De Bruyn, I.N., W.H. Holzapfel, L. Visser and A.I. Louw. 1988. Glucose metabolism by *Lactobacillus divergens*. J. Gen. Microbiol. *134*: 2103–2109.

De Man, J.C., M. Rogosa and M.E. Sharpe. 1960. A medium for the cultivation of lactobacilli. J. Appl. Bacteriol. *23*: 130–135.

DeSoete, G. 1983. A least square algorithm for fitting additive trees to proximity data. Psychometrika *48*: 621–626.

Drancourt, M., V. Roux, P.E. Fournier and D. Raoult. 2004. *rpoB* gene sequence-based identification of aerobic Gram-positive cocci of the genera *Streptococcus, Enterococcus, Gemella, Abiotrophia*, and *Granulicatella*. J. Clin. Microbiol. *42*: 497–504.

DSMZ. 1998. Catalog of Strains. DSMZ, 6th edn, Mascheroder Weg 1b, 38124 Braunschweig, Germany.

Evans, J.B. and C.F. Niven, Jr. 1951. Nutrition of the heterofermentative lactobacilli that cause greening of cured meat proucts. J. Bacteriol. *62*: 599–603.

Facklam, R. and J.A. Elliott. 1995. Identification, classification, and clinical relevance of catalase-negative, gram-positive cocci, excluding the streptococci and enterococci. Clin. Microbiol. Rev. *8*: 479–495.

Faden, H. and D. Dryja. 1989. Recovery of a unique bacterial organism in human middle ear fluid and its possible role in chronic otitis media. J. Clin. Microbiol. *27*: 2488–2491.

Farrow, J.A., S. Wallbanks and M.D. Collins. 1994. Phylogenetic interrelationships of round-spore-forming bacilli containing cell walls based on lysine and the non-spore-forming genera *Caryophanon, Exiguobacterium, Kurthia*, and *Planococcus*. Int. J. Syst. Bacteriol *44*: 74–82.

Felsenstein, J. 1993. PHYLIP (Phylogeny Interference Package), 3.5.1 edn. Department of Genetics, University of Washington, Seattle.

Feurer, C., F. Irlinger, H.E. Spinnler, P. Glaser and T. Vallaeys. 2004. Assessment of the rind microbial diversity in a farmhouse-produced vs a pasteurized industrially produced soft red-smear cheese using

both cultivation and rDNA-based methods. J. Appl. Microbiol. *97*: 546–556.

Fournier, P.E. and D. Raoult. 1999. Nonculture laboratory methods for the diagnosis of infectious endocarditis. Curr. Infect. Dis. Rep. *1*: 136–141.

Fraise, A.P., A.L. Pahor and A.J. Beswick. 2001. Otitis media with effusion: the role of *Alloiococcus otitidis*. Ann. Med. *33*: 1–3.

Franzmann, P.D., P. Hopfl, N. Weiss and B.J. Tindall. 1991. Psychotrophic, lactic acid-producing bacteria from anoxic waters in Ace Lake, Antarctica: *Carnobacterium funditum* sp. nov. and *Carnobacterium alterfunditum* sp. nov. Arch. Microbiol. *156*: 255–262.

Franzmann, P.D., P. Hopfl, N. Weiss and B.J. Tindall. 1993. *In* Validation of the publication of new names and new combinations previously effectively published outside the IJSB. List no. 44. Int. J. Syst. Bacteriol. *43*: 188–189.

Frenkel, A. and W. Hirsch. 1961. Spontaneous development of L forms of streptococci requiring secretions of other bacteria or sulphydryl compounds for normal growth. Nature *191*: 728–730.

Gaudu, P., K. Vido, B. Cesselin, S. Kulakauskas, J. Tremblay, L. Rezaiki, G. Lamberret, S. Sourice, P. Duwat and A. Gruss. 2002. Respiration capacity and consequences in *Lactococcus lactis*. Antonie van Leeuwenhoek *82*: 263–269.

Gieg, L.M., R.V. Kolhatkar, M.J. McInerney, R.S. Tanner, S.H. Harris, K.L. Sublette and J.M. Suflita. 1999. Intrinsic bioremediation of petroleum hydrocarbons in a gas condensatecontaminated aquifer. Environ. Sci. Technol. *33*: 2550–2560.

Groth, I., P. Schumann, N. Weiss, K. Martin and F.A. Rainey. 1996. *Agrococcus jenensis* gen. nov., sp. nov., a new genus of actinomycetes with diaminobutyric acid in the cell wall. Int. J. Syst. Bacteriol. *46*: 234–239.

Gunzalus, I.C. and C.F. Niven. 1942. The effect of pH on the lactic acid fermentation. J. Biol. Chem. *145*: 131–136.

Hamana, K. 1994. Polyamine distribution patterns in aerobic Gram positive cocci and some radio-resistant bacteria. J. Gen. Appl. Microbiol. *40*: 181–195.

Hammes, W.P., N. Weiss and W. Holzapfel. 1992. The genera *Lactobacillus* and *Carnobacterium*. *In* Balows, Trüper, Dworkin, Harder and Schleifer (Editors), The Prokaryotes. A Handbook on the Biology of Bacteria: Ecophysiology, Isolation, Identification, Applications, 2nd edn. Springer-Verlag, New York, pp. 1535–1594.

Hammes, W.P. and C. Hertel. 2003. The genera *Lactobacillus* and *Carnobacterium*. *In* Dworkin (Editor), The Prokaroyes, 3rd edn. Springer-Verlag, New York, pp. release 3.15.

Hendolin, P.H., U. Karkkainen, T. Himi, A. Markkanen and J. Ylikoski. 1999. High incidence of *Alloiococcus otitis* in otitis media with effusion. Pediatr. Infect. Dis. J. *18*: 860–865.

Hendolin, P.H., L. Paulin and J. Ylikoski. 2000. Clinically applicable multiplex PCR for four middle ear pathogens. J. Clin. Microbiol. *38*: 125–132.

Hillman, J.D., S.W. Andrew, S. Palner and P. Strashenko. 1989. Adaptive changes in a strain of *Streptococcus mutans* during colonization of the human oral cavity Microb. Ecol. Health Dis. *2*: 231–239.

Hitchener, B.J., F.A. Egan and P.J. Rogers. 1982. Characteristics of lactic acid bacteria isolated from vacuum-packaged beef. J. Appl. Bacteriol. *52*: 31–37.

Hiu, S.F., R.A. Holt, N. Sriranganathan, R.J. Seidler and J.L. Fryer. 1984. *Lactobacillus piscicola*, a new species from salmonid fish. Int. J. Syst. Bacteriol. *34*: 393–400.

Hogg, J.C. and M.J. Lehane. 1999. Identification of bacterial species associated with the sheep scab mite (*Psoroptes ovis*) by using amplified genes coding for 16S rRNA. Appl. Environ. Microbiol. *65*: 4227–4229.

Hohorst, H.J. 1966. L-Laktat Bestimmung. *In* Bergmeyer (Editor), Methoden der Enzymatischen Analyse. Verlag Chemie, Weinheim, pp. 266–270.

Holley, R.A., T.Y. Guan, M. Peirson and C.K. Yost. 2002. *Carnobacterium viridans* sp. nov., an alkaliphilic, facultative anaerobe isolated from refrigerated, vacuum-packed bologna sausage. Int. J. Syst. Evol. Microbiol. *52*: 1881–1885.

Holzapfel, W.H. and E.S. Gerber. 1983. *Lactobacillus divergens* sp. nov., a new heterofermentative *Lactobacillus* species producing L.(+)-lactate. Syst. Appl. Microbiol. *4*: 522–534.

Inagaki, F., M. Suzuki, K. Takai, H. Oida, T. Sakamoto, K. Aoki, K.H. Nealson and K. Horikoshi. 2003. Microbial communities associated with geological horizons in coastal subseafloor sediments from the sea of okhotsk. Appl. Environ. Microbiol. *69*: 7224–7235.

Ishikawa, M., S. Ishizaki, Y. Yamamoto and K. Yamasato. 2002. *Paraliobacillus ryukyuensis* gen. nov., sp. nov., a new Gram-positive, slightly halophilic, extremely halotolerant, facultative anaerobe isolated from a decomposing marine alga. J. Gen. Appl. Microbiol. *48*: 269–279.

Ishikawa, M., K. Nakajima, M. Yanagi, Y. Yamamoto and K. Yamasato. 2003. *Marinilactibacillus psychrotolerans* gen. nov., sp. nov., a halophilic and alkaliphilic marine lactic acid bacterium isolated from marine organisms in temperate and subtropical areas of Japan. Int. J. Syst. Evol. Microbiol. *53*: 711–720.

Ishikawa, M., K. Nakajima, Y. Itamiya, S. Furukawa, Y. Yamamoto and K. Yamasato. 2005. *Halolactibacillus halophilus* gen. nov., sp. nov. and *Halolactibacillus miurensis* sp. nov., halophilic and alkaliphilic marine lactic acid bacteria constituting a phylogenetic lineage in *Bacillus* rRNA group 1. Int. J. Syst. Evol. Microbiol. *55*: 2427–2439.

Ishikawa, M., K. Kodama, H. Yasuda, A. Okamoto-Kainuma, K. Koizumi and K. Yamasato. 2007. Presence of halophilic and alkaliphilic lactic acid bacteria in various cheeses. Lett. Appl. Microbiol. *44*: 308–313.

Janssen, P.H., S. Evers, F.A. Rainey, N. Weiss, W. Ludwig, C.G. Harfoot and B. Schink. 1995. *Lactosphaera* gen. nov., a new genus of lactic-acid bacteria, and transfer of *Ruminococcus pasteurii* Schink 1984 to *Lactosphaera pasteurii* comb. nov. Int. J. Syst. Bacteriol. *45*: 565–571.

Joborn, A., M. Dorsch, J.C. Olsson, A. Westerdahl and S. Kjelleberg. 1999. *Carnobacterium inhibens* sp. nov., isolated from the intestine of Atlantic salmon (*Salmo salar*). Int. J. Syst. Bacteriol. *49*: 1891–1898.

Joffraud, J.J., F. Leroi, C. Roy and J.L. Berdague. 2001. Characterization of volatile compounds produced by bacteria from the spoilage flora of cold-smoked salmon. Int. J. Food Microbiol. *66*: 175–184.

Jones, B.E., W.D. Grant, N.C. Collins and W.E. Mwatha. 1994. Alkaliphiles: diversity and identification. *In* Priest, Ramos-Cormenzana and Tindall (Editors), Bacterial Diversity and Systematics. Plenum, New York, pp. 195–230.

Jones, D. and R.M. Keddie. 1986. *Brevibacterium. In* Sneath, Mair, Sharp and Holt (Editors), Bergey's Manual of Systematic Bacteriology, Vol. 2. The Williams & Wilkins Co., Baltimore, pp. pp 1301–1313.

Jones, R.J. 2004. Observations on the succession dynamics of lactic acid bacteria populations in chill-stored vacuum-packaged beef. Int. J. Food Microbiol. *90*: 273–282.

Jukes, T.H. and C. Cantor. 1969. Evolution of protein molecules. *In* Murano (Editor), Mammalian Protein Metabolism. Academic Press, New York pp. 21–132.

Kanamoto, T., S. Sato and M. Inoue. 2000. Genetic heterogeneities and phenotypic characteristics of strains of the genus *Abiotrophia* and proposal of *Abiotrophia* para-adiacens sp. nov. J. Clin. Microbiol. *38*: 492–498.

Kawamura, Y., X.G. Hou, F. Sultana, S.J. Liu, H. Yamamoto and T. Ezaki. 1995. Transfer of *Streptococcus adjacens* and *Streptococcus defectivus* to *Abiotrophia* gen. nov. as *Abiotrophia adiacens* comb. nov., and *Abiotrophia defectiva* comb. nov., respectively. Int. J. Syst. Bacteriol. *45*: 798–803.

Kilian, M., L. Mikkelsen and J. Henrichsen. 1989. Taxonomic study of viridans streptococci: description of *Streptococcus gordonii* sp. nov. and emended descriptions of *Streptococcus sanguis* (White and Niven 1946), *Streptococcus oralis* (Bridge and Sneath 1982), and *Streptococcus mitis* (Andrewes and Horder 1906). Int. J. Syst. Bacteriol. *39*: 471–484.

Klaenhammer, T.R. 1993. Genetics of bacteriocins produced by lactic acid bacteria. FEMS Microbiol. Rev. *12*: 39–85.

Kloos, W.E. and J.F. Wolfshohl. 1991. *Staphylococcus cohnii* subspecies: *Staphylococcus cohnii* subsp. *cohnii* subsp. nov. and *Staphylococcus cohnii* subsp. *urealyticum* subsp. nov. Int. J. Syst. Bacteriol. *41*: 284–289.

Konings, W.N. 2002. The cell membrane and the struggle for life of lactic acid bacteria. Antonie van Leeuwenhoek *82*: 3–27.

Kroppenstedt, R.M. and H.J. Kutzner. 1978. Biochemical taxonomy of some problem actinomycetes. Zentralbl. Bakteriol. Parasitenkd. Infektionskr. Hyg. Abt. I *Suppl. 6*: 125–133.

Kushner, D.J. 1978. Life in high salt and solute concentrations: halophilic bacteria. *In* Kushner (Editor), Microbial Life in Extreme Environments. Academic Press, London, pp. 318–346.

Kushner, D.J. and K. Kamekura. 1988. Physiology of halophilic eubacteria. *In* Rodriguez-Valera (Editor), Halophilic Bacteria, Vol. 1. CRC Press, Boca Raton, FL, pp. 109–140.

LaClaire, L. and R. Facklam. 2000a. Antimicrobial susceptibility and clinical sources of *Dolosigranulum pigrum* cultures. Antimicrob. Agents Chemother. *44*: 2001–2003.

LaClaire, L.L. and R.R. Facklam. 2000b. Comparison of three commercial rapid identification systems for the unusual gram-positive cocci *Dolosigranulum pigrum, Ignavigranum ruoffiae*, and *Facklamia* species. J. Clin. Microbiol. *38*: 2037–2042.

Lai, S. and L.N. Manchester. 2000. Numerical phenetic study of the genus *Carnobacterium*. Antonie van Leeuwenhoek *78*: 73–85.

Laneelle, M.A., J. Asselineau, M. Welby, M.V. Norgard, T. Imaeda, M.C. Pollice and L. Barksdale. 1980. Biological and chemical bases for the reclassification of *Brevibacterium vitarumen* (Bechdel et al.) Breed (Approved Lists, 1980) as *Corynebacterium* vitarumen (Bechdel et al.) comb. nov. and *Brevibacterium liquefaciens* Okabayashi and Masuo (Approved Lists, 1980) as *Corynebacterium liquefaciens* (Okabayashi and Masuo) comb. nov. Int. J. Syst. Bacteriol. *30*: 539–546.

Laursen, B.G., L. Bay, I. Cleenwerck, M. Vancanneyt, J. Swings, P. Dalgaard and J.J. Leisner. 2005. *Carnobacterium divergens* and *Carnobacterium maltaromaticum* as spoilers or protective cultures in meat and seafood: phenotypic and genotypic characterization. Syst. Appl. Microbiol. *28*: 151–164.

Lawson, P.A., G. Foster, E. Falsen, B. Sjoden and M.D. Collins. 1999. *Abiotrophia balaenopterae* sp. nov., isolated from the minke whale (*Balaenoptera acutorostrata*). Int. J. Syst. Bacteriol. *49*: 503–506.

Lawson, P.A., G. Foster, E. Falsen, M. Ohlen and M.D. Collins. 2000. *Atopobacter phocae* gen. nov., sp. nov., a novel bacterium isolated from common seals. Int. J. Syst. Evol. Microbiol. *50*: 1755–1760.

Leisner, J.J., J.C. Millan, H.H. Huss and L.M. Larsen. 1994. Production of histamine and tyramine by lactic acid bacteria isolated from vacuum-packed sugar-salted fish. J. Appl. Bacteriol. *76*: 417–423.

Leisner, J.J., M. Vancanneyt, G. Rusul, B. Pot, K. Lefebvre, A. Fresi and L.K. Tee. 2001. Identification of lactic acid bacteria constituting the predominating microflora in an acid-fermented condiment (tempoyak) popular in Malaysia. Int. J. Food Microbiol. *63*: 149–157.

Leroi, F., J.J. Joffraud, F. Chevalier and M. Cardinal. 1998. Study of the microbial ecology of cold-smoked salmon during storage at 8 degrees C. Int. J. Food Microbiol. *39*: 111–121.

Leser, T.D., J.Z. Amenuvor, T.K. Jensen, R.H. Lindecrona, M. Boye and K. Moller. 2002. Culture-independent analysis of gut bacteria: the pig gastrointestinal tract microbiota revisited. Appl. Environ. Microbiol. *68*: 673–690.

Liu, J.R., C.A. McKenzie, E.M. Seviour, R.I. Webb, L.L. Blackall, C.P. Saint and R.J. Seviour. 2001. Phylogeny of the filamentous bacterium 'Nostocoida limicola' III from activated sludge. Int. J. Syst. Evol. Microbiol. 51: 195–202.

Liu, J.R., R.S. Tanner, P. Schumann, N. Weiss, C.A. McKenzie, P.H. Janssen, E.M. Seviour, P.A. Lawson, T.D. Allen and R.J. Seviour. 2002. Emended description of the genus Trichococcus, description of Trichococcus collinsii sp. nov., and reclassification of Lactosphaera pasteurii as Trichococcus pasteurii comb. nov. and of Ruminococcus palustris as Trichococcus palustris comb. nov. in the low-G+C Gram-positive bacteria. Int. J. Syst. Evol. Microbiol. 52: 1113–1126.

Macian, M.C., E. Chenoll and R. Aznar. 2004. Simultaneous detection of Carnobacterium and Leuconostoc in meat products by multiplex PCR. J. Appl. Microbiol. 97: 384–394.

Maoz, A., R. Mayr and S. Scherer. 2003. Temporal stability and biodiversity of two complex antilisterial cheese-ripening microbial consortia. Appl. Environ. Microbiol. 69: 4012–4018.

Marmur, J. 1961. A procedure for the isolation of deoxyribonucleic acid from microorganisms. J. Mol. Biol. 3: 208–218.

Mauguin, S. and G. Novel. 1994. Characterization of lactic acid bacteria isolated from seafood. J. Appl. Bacteriol. 76: 616–625.

McCarthy, L.R. and E.J. Bottone. 1974. Bacteremia and endocarditis caused by satelliting streptococci. Am. J. Clin. Pathol. 61: 585–591.

Meisel, J., G. Wolf and W.P. Hammes. 1994. Heme-dependent cytochrome formation in Lactobacillus maltaromicus. Syst. Appl. Microbiol. 17: 20–23.

Mesbah, M., U. Premachandran and W.B. Whitman. 1989. Precise measurement of the G + C content of deoxyribonucleic acid by high-performance liquid chromatography. Int. J. Syst. Bacteriol. 39: 159–167.

Miller, A., M.E. Morgan and L.M. Libbey. 1974. Lactobacillus maltaromicus, a new species producing a malty aroma. Int. J. Syst. Bacteriol. 24: 346–354.

Miller, P.H., R.R. Facklam and J.M. Miller. 1996. Atmospheric growth requirements for Alloiococcus species and related gram-positive cocci. J. Clin. Microbiol. 34: 1027–1028.

Mollet, C., M. Drancourt and D. Raoult. 1997. rpoB sequence analysis as a novel basis for bacterial identification. Mol. Microbiol. 26: 1005–1011.

Mora, D., M. Scarpellini, L. Franzetti, S. Colombo and A. Galli. 2003. Reclassification of Lactobacillus maltaromicus (Miller et al. 1974) DSM 20342^T and DSM 20344 and Carnobacterium piscicola (Collins et al. 1987) DSM 20730^T and DSM 20722 as Carnobacterium maltaromaticum comb. nov. Int. J. Syst. Bacteriol. 53: 675–678.

Morea, M., F. Baruzzi and P.S. Cocconcelli. 1999. Molecular and physiological characterization of dominant bacterial populations in traditional mozzarella cheese processing. J. Appl. Microbiol. 87: 574–582.

Nakajima, K., K. Hirota, Y. Nodasaka and I. Yumoto. 2005. Alkalibacterium iburiense sp. nov., an obligate alkaliphile that reduces an indigo dye. Int. J. Syst. Evol. Microbiol. 55: 1525–1530.

Namdari, H., K. Kintner, B.A. Jackson, S. Namdari, J.L. Hughes, R.R. Peairs and D.J. Savage. 1999. Abiotrophia species as a cause of endophthalmitis following cataract extraction. J. Clin. Microbiol. 37: 1564–1566.

Nilsson, L., L. Gram and H.H. Huss. 1999. Growth control of Listeria monocytogenes on cold-smoked salmon using a competitive lactic acid bacteria flora. J. Food Prot. 62: 336–342.

Nilsson, L., Y.Y. Ng, J.N. Christiansen, B.L. Jorgensen, D. Grotinum and L. Gram. 2004. The contribution of bacteriocin to inhibition of Listeria monocytogenes by Carnobacterium piscicola strains in cold-smoked salmon systems. J. Appl. Microbiol. 96: 133–143.

Ntougias, S. and N.J. Russell. 2001. Alkalibacterium olivoapovliticus gen. nov., sp. nov., a new obligately alkaliphilic bacterium isolated from edible-olive wash-waters. Int. J. Syst. Evol. Microbiol. 51: 1161–1170.

Ohara-Nemoto, Y., S. Tajika, M. Sasaki and M. Kaneko. 1997. Identification of Abiotrophia adiacens and Abiotrophia defectiva by 16S rRNA gene PCR and restriction fragment length polymorphism analysis. J. Clin. Microbiol. 35: 2458–2463.

Ormerod, L.D., K.L. Ruoff, D.M. Meisler, P.J. Wasson, J.C. Kintner, S.P. Dunn, J.H. Lass and I. van de Rijn. 1991. Infectious crystalline keratopathy. Role of nutritionally variant streptococci and other bacterial factors. Ophthalmology 98: 159–169.

Ouwehand, A.C. 1998. Antimicrobial components from lactic acid bacteria. In Salminen and von Wright (Editors), Lactic Acid Bacteria. Marcel Dekker, Inc., New York, pp. 139–159.

Paster, B.J., S.K. Boches, J.L. Galvin, R.E. Ericson, C.N. Lau, V.A. Levanos, A. Sahasrabudhe and F.E. Dewhirst. 2001. Bacterial diversity in human subgingival plaque. J. Bacteriol. 183: 3770–3783.

Pikuta, E.V., D. Marsic, A. Bej, J. Tang, P. Krader and R.B. Hoover. 2005. Carnobacterium pleistocenium sp. nov., a novel psychrotolerant, facultative anaerobe isolated from permafrost of the Fox Tunnel in Alaska. Int. J. Syst. Evol. Microbiol. 55: 473–478.

Pikuta, E.V., R.B. Hoover, A.K. Bej, D. Marsic, W.B. Whitman, P.E. Krader and J. Tang. 2006. Trichococcus patagoniensis sp. nov., a facultative anaerobe that grows at −5 °C, isolated from penguin guano in Chilean Patagonia. Int. J. Syst. Evol. Microbiol. 56: 2055–2062.

Pot, B., P. Vandamme and K. Kersters. 1994. Analysis of electrophoretic whole-organisms protein fingerprints. In Goodfellow and O'Donnel (Editors), Modern Microbial Methods: Chemical Methods in Procaryotic Systematics. Wiley, Chichester, England, pp. 493–521.

Rachman, C., P. Kabadjova, R. Valcheva, H. Prevost and X. Dousset. 2004. Identification of Carnobacterium species by restriction fragment length polymorphism of the 16S-23S rRNA gene intergenic spacer region and species-specific PCR. Appl. Environ. Microbiol. 70: 4468–4477.

Rainey, F.A., N. Ward-Rainey, R.M. Kroppenstedt and E. Stackebrandt. 1996. The genus Nocardiopsis represents a phylogenetically coherent taxon and a distinct actinomycete lineage: Proposal of Nocardiopsaceae fam. nov. Int. J. Syst. Bacteriol. 46: 1088–1092.

Rastelli, E., G. Giraffa, D. Carminati, G. Parolari and S. Barbuti. 2005. Identification and characterization of halotolerant bacteria in spoiled dry-cured hams. Meat Sci. 70: 241–246.

Reuter, G. 1970. Laktobazillen undeng verwandte Mikrooganismen in Fleisch und Fleischerzeugnissen. 2. Mitteilung: Die Charakterisierung der isolierten Laktobazillenstämme. Fleischwirtschaft 50: 954–932.

Reuter, G. 1981. Psychrotrophic lactobacilli in meat products. In Roberts, Hobbs, Christian and Skovgaard (Editors), Psychrotrophic Microorganisms in Spoilage and Pathogenicity. Academic Press, London, pp. 253–258.

Rhee, S.K. and M.Y. Pack. 1980. Effect of environmental pH on fermentation balance of Lactobacillus bulgaricus. J. Bacteriol 144: 865–868.

Ringo, E. and F.J. Gatesoupe. 1998. Lactic acid bacteria in fish: a review. Aquaculture 160: 177–203.

Ringo, E. and W.H. Holzapfel. 2000. Identification and characterization of carnobacteria associated with the gills of Atlantic salmon (Salmo salar L.). Syst. Appl. Microbiol. 23: 523–527.

Ringo, E., M. Seppola, A. Berg, R.E. Olsen, U. Schillinger and W. Holzapfel. 2002. Characterization of Carnobacterium divergens strain 6251 isolated from intestine of Arctic charr (Salvelinus alpinus L.). Syst. Appl. Microbiol. 25: 120–129.

Ritchey, T.W. and H.W. Seely, Jr. 1976. Distribution of cytochrome-like respiration in streptococci. J. Gen. Microbiol. 93: 195–203.

Roberts, R.B., A.G. Kreiger, N.L. Schiller and K.C. Gross. 1979. Viridans streptococcal endocarditis: the role of various species, including pyridoxal-dependant streptococci. Rev. Infect. Dis. 1: 955–966.

Robertson, P.A.W., C. O'Dowd, C. Burrells, P. Williams and B. Austin. 2000. Use of Carnobacterium sp. as a probiotic for Atlantic salmon (Salmo salar L.) and rainbow trout (Oncorhynchus mykiss, Walbaum). Aquaculture 185: 235–243.

Roggenkamp, A., M. Abele-Horn, K.H. Trebesius, U. Tretter, I.B. Autenrieth and J. Heesemann. 1998. *Abiotrophia elegans* sp. nov., a possible pathogen in patients with culture-negative endocarditis. J. Clin. Microbiol. *36*: 100–104.

Rogosa, J., J.A. Mitchell and R.F. Wiseman. 1951. A selective medium for isolation and enumeration of oral and fecal lactobacilli. J. Bacteriol. *62*: 132–133.

Ruoff, K.L. 1991. Nutritionally variant streptococci. Clin. Microbiol. Rev. *4*: 184–190.

Sakane, K. and K. Imai. 1986. Preservation of various bacteria by L-drying, Part II. Jpn. J. Freez. Dry. *32*: 47–53.

Scheff, G., O. Salcher and F. Lingens. 1984a. *Trichococcus flocculiformis* gen. nov. sp. nov. A new Gram-positive filamentous bacterium isolated from bulking sludge. Appl. Microbiol. Biotechnol. *19*: 114–119.

Scheff, G., O. Salcher and F. Lingens. 1984b. *In* Validation of the publication of new names and new combinations previously effectively published outside the IJSB. List no. 15. Int. J. Syst. Bacteriol. *34*: 355–357.

Schillinger, U. and W.H. Holzapfel. 1995. The genus *Carnobacterium*. *In* Wood and Holzapfel (Editors), The Genera of Lactic Acid Bacteria. Blackie Academic and Professional, London, pp. 307–326.

Schink, B. 1984. Fermentation of tartrate enantiomers by anaerobic bacteria, and description of two new species of strict anaerobes, *Ruminococcus pasteurii* and *Ilyobacter tartaricus*. Arch. Microbiol. *139*: 409–414.

Schleifer, K.H. and O. Kandler. 1972. Peptidoglycan types of bacterial cell walls and their taxonomic implications. Bacteriol Rev *36*: 407–477.

Schumann, P., F. A. Rainey, J. Burkhardt, E. Stackebrandt and N. Weiss. 1998. Reclassification of *Brevibacterium oxydans* (Chatelain and Second 1966) as *Microbacterium oxydans* comb. nov.. Int. J. Syst. Bacteriol.: 175–177.

Shaw, B.G. and C.D. Harding. 1984. A numerical taxonomic study of lactic acid bacteria from vacuum-packed beef, pork, lamb and bacon. J. Appl. Bacteriol. *56*: 25–40.

Shaw, B.G. and C.D. Harding. 1985. Atypical lactobacilli from vacuum-packaged meats: comparison by DNA hybridization, cell composition and biochemical tests with a description of *Lactobacillus carnis* sp. nov. Syst. Appl. Microbiol. *6*: 291–297.

Sijpesteijn, A.K. 1970. Induction of cytochrome formation and stimulation of oxidative dissimilation by hemin in *Streptococcus lactis* and *Leuconostoc mesenteroides*. Antonie van Leeuwenhoek *36*: 335–348.

Stackebrandt, E., H. Seiler and K.H. Schleifer. 1982. Union of the genera *Cellulomonas* Bergey et al. and *Oerskovia* Prauser et al. in a redefined genus *Cellulomonas*. Zentralbl. Bakteriol. Parasitenkd. Infektionskr. Hyg. Abt. 1 Orig. *C3*: 401–409.

Stackebrandt, E., V.J. Fowler, F. Fiedler and H. Seiler. 1983. Taxonomic studies on *Arthrobacter nicotianae* and related taxa: description of *Arthrobacter uratoxydans* sp. nov. and *Arthrobacter sulfureus* sp. nov. and reclassification of *Brevibacterium protophormiae* as *Arthrobacter protophormiae* comb. nov. Syst. Appl. Microbiol. *4*: 470–486.

Stackebrandt, E., P. Schumann, J. Swiderski and N. Weiss. 1999. Reclassification of *Brevibacterium incertum* (Breed 1953) as *Desemzia incerta* gen. nov., comb. nov. Int. J. Syst. Bacteriol. *49*: 185–188.

Stein, D.S. and K.E. Nelson. 1987. Endocarditis due to nutritionally deficient streptococci: therapeutic dilemma. Rev Infect Dis *9*: 908–916.

Steinhaus, E. 1941. A study of the bacteria associated with thirty species of insects. J. Bacteriol. *42*: 757–790.

Straub, B.W., M. Kicherer, S.M. Schilcher and W.P. Hammes. 1995. The formation of biogenic amines by fermentation organisms. Z. Lebensm. Unters. Forsch. *201*: 79–82.

Tahiri, I., M. Desbiens, R. Benech, E. Kheadr, C. Lacroix, S. Thibault, D. Ouellet and I. Fliss. 2004. Purification, characterization and amino acid sequencing of divergicin M35: a novel class IIa bacteriocin produced by *Carnobacterium divergens* M35. Int. J. Food Microbiol. *97*: 123–136.

Takeuchi, M. and K. Hatano. 1998. Union of the genera *Microbacterium* Orla-Jensen and *Aureobacterium* Collins *et al*. in a redefined genus *Microbacterium*. Int. J. Syst. Bacteriol. *48*: 739–747.

Teng, L.J., P.R. Hsueh, J.C. Tsai, P.W. Chen, J.C. Hsu, H.C. Lai, C.N. Lee and S.W. Ho. 2002. *groESL* sequence determination, phylogenetic analysis, and species differentiation for viridans group streptococci. J. Clin. Microbiol. *40*: 3172–3178.

Thornley, M.J. 1957. Observations on the microflora of minced chicken meat irradiated with 4 MeV cathode rays. J. Appl. Bacteriol. *20*: 286–298.

Tindall, B.J. 1990. Lipid composition of *Halobacterium lacusprofundi*. FEMS Microbiol. Lett. *66*: 199–202.

Toffin, L., G. Webster, A.J. Weightman, J.C. Fry and D. Prieur. 2004. Molecular monitoring of culturable bacteria from deep-sea sediment of the Nankai Trough, Leg 190 Ocean Drilling Program. FEMS Microbiol. Ecol. *48*: 357–367.

Toffin, L., K. Zink, C. Kato, P. Pignet, A. Bidault, N. Bienvenu, J.L. Birrien and D. Prieur. 2005. *Marinilactibacillus piezotolerans* sp. nov., a novel marine lactic acid bacterium isolated from deep sub-seafloor sediment of the Nankai Trough. Int. J. Syst. Evol. Microbiol. *55*: 345–351.

Von Graevenitz, A. 1993. Revised nomenclature of *Alloiococcus otitis*. J. Clin. Microbiol. *31*: 472.

von Holy, A. 1983. Bacteriological Studies on the Extension of the Shelf Life of Raw Minced Beef. MSc thesis, University of Pretoria, Pretoria, South Africa.

Wallbanks, S., A.J. Martinez-Murcia, J.L. Fryer, B.A. Phillips and M.D. Collins. 1990. 16S ribosomal RNA sequence determination for members of the genus *Carnobacterium* and related lactic acid bacteria and description of *Vagococcus salmoninarum* sp. nov. Int. J. Syst. Bacteriol. *40*: 224–230.

Whitman, W.B., E. Ankwanda and R.S. Wolfe. 1982. Nutrition and carbon metabolism of *Methanococcus voltae*. J. Bacteriol. *149*: 852–863.

Wolin, E.A., M.G. Wolin and R.S. Wolfe. 1963. Formation of methane by bacterial extracts. J. Biol. Chem. *238*: 2882–2886.

Woo, P.C.Y., K.H.L. Ng, S.K.P. Lau, K-T Yip, A.M.Y. Fung, K-W Leung, D.M. W. Tam, T.-L. Que and K.-Y. Yuen. 2001. Usefulness of the MicroSeq 500 16S ribosomal DNA-based bacterial identification system for identification of clinically significant bacterial isolates with ambiguous biochemical profiles. J. Clin. Microbiol. *41*: 1996–2001.

WPCM. 1989. (IUMS/JCFMH Working Party for Cultural Media). Cresol red Thallium Acetate Sucrose (CTAS) Agar. *In* Baird, Corry, Curtis, Mossel and Skovgaard (Editors), Pharmacopoeia of Cultural Media for Food Microbiology: Additional Monographs. Elsevier, Amsterdam, The Netherlands., pp. 129 – 131.

Yamada, K. and K. Komagata. 1972. Taxonomic studies on coryneform bacteria. V. Classification of coryneform bacteria. J. Gen. Appl. Microbiol. *18*: 417–431.

Yamazaki, K., M. Suzuki, Y. Kawai, N. Inoue and T.J. Montville. 2003. Inhibition of *Listeria monocytogenes* in cold-smoked salmon by *Carnobacterium piscicola* CS526 isolated from frozen surimi. J. Food Prot. *66*: 1420–1425.

Yumoto, I., K. Hirota, Y. Nodasaka, Y. Yokota, T. Hoshino and K. Nakajima. 2004. *Alkalibacterium psychrotolerans* sp. nov., a psychrotolerant obligate alkaliphile that reduces an indigo dye. Int. J. Syst. Evol. Microbiol. *54*: 2379–2383.

Yumoto, I., K. Hirota, Y. Nodasaka, Y. Tokiwa and K. Nakajima. 2008. *Alkalibacterium indicireducens* sp. nov., an obligate alkaliphile that reduces indigo dye. Int. J. Syst. Evol. Microbiol. *58*: 901–905.

Zhilina, T.N. 1978. Growth of a pure *Methanosarcina* culture, biotype 2, on acetate. (En. translation). Mikrobiologiya *47*: 321–323.

Zhilina, T.N., O. R. Kotsyurbenko, G. A. Osipov, N. A. Kostrinka and G.A. Zavarzin. 1995. *Ruminococcus palustris* sp. nov. - a psychroactive anaerobic organism from a swamp. Microbiology (En. translation of *Mikrobiologiya*) *64*: 674–680.

Family IV. **Enterococcaceae** fam. nov.

WOLFGANG LUDWIG, KARL-HEINZ SCHLEIFER AND WILLIAM B. WHITMAN

En.te.ro.coc.ca'ce.ae. N.L. masc. n. *Enterococcus* type genus of the family; suff. *-aceae* ending denoting family; N.L. fem. pl. n. *Enterococcaceae* the *Enterococcus* family.

The family *Enterococcaceae* is circumscribed for this volume on the basis of phylogenetic analyses of the 16S rRNA sequences and includes the genus *Enterococcus, Melissococcus, Tetragenococcus,* and *Vagococcus* (Figure 3). It is composed of ovoid cocci that contain a Gram-stain-positive cell wall. Endospores are not formed. Facultative anaerobic, anaerobic or microaerophilic chemo-organotrophic growth. Catalase-negative. Some species are carboxyophilic or halophilic. May be resistant to bile.

Type genus: **Enterococcus** (*ex* Thiercelin and Jouhaud 1903) Schleifer and Kilpper-Bälz 1984, 32.

Genus I. **Enterococcus** (*ex* Thiercelin and Jouhaud 1903) Schleifer and Kilpper-Bälz 1984, 32[VP]

PAVEL ŠVEC AND LUC A. DEVRIESE

En.te.ro.coc'cus. Gr. n. *enteron* intestine; L. masc. n. *coccus* a grain, berry; N.L. masc. n. *Enterococcus* intestinal coccus

Gram-positive. Cells are ovoid, occur singly, in pairs, or in short chains, and are frequently elongated in the direction of the chain. Nonsporeforming. Strains of some species may be motile by scanty flagella. Some species are yellow pigmented. Facultatively anaerobic. Certain species are carboxyphilic (CO_2-dependent). Catalase-negative, but some strains reveal pseudocatalase activity when cultivated on blood-containing agar media. Hemolytic activity is variable and largely species-dependent.

Optimal growth of most species at 35–37 °C. Many, but not all, species are able to grow at 42 and even at 45 °C, and (slowly) at 10 °C. **Very resistant to drying**.

Chemo-organotrophic growth. Generally **complex nutrient requirements**. Fermentative metabolism. Homofermentative lactic acid fermentation. Predominant end product of glucose fermentation is L(+)-lactic acid. Certain characteristics are common to all described species, although rare exceptions may occur and certain test results have not yet been reported in the lesser known species: **resistance to 40% (v/v) bile**, production of β-glucosidase, leucine arylamidase, hydrolysis of esculin, production of acid from *N*-acetylglucosamine, amygdalin, arbutin, cellobiose, D-fructose, galactose, β-gentiobiose, glucose, lactose, maltose, D-mannose, methyl β-D-glucoside, ribose, salicin and trehalose. The following tests are mostly negative: urease, production of acid from D-arabinose, **erythritol, D- and L-fucose**, methyl α-D-xyloside and L-xylose. It should be noted that **certain characteristics traditionally considered to be typical for the genus do not apply to several of the more recently described species**: Lancefield group D antigen, resistance to 0.4% sodium azide or 6.5% NaCl, growth at 10 °C and 45 °C, production of pyrrolidonyl arylamidase and acetoin.

Most species are part of the intestinal flora of mammals and birds, and (much less well-known) other animals as well. Other species are associated with plants, or have been isolated from water.

DNA G+C content (mol%): 35.1–44.9.

Type species: **Enterococcus faecalis** (Andrewes and Horder 1906) Schleifer and Kilpper-Bälz 1984, 33[VP] (*Streptococcus faecalis* Andrewes and Horder 1906, 713.).

Further descriptive information

The genus *Enterococcus* has been separated from the genus *Streptococcus* on the basis of results of DNA–DNA and DNA–rDNA hybridization studies (Schleifer and Kilpper-Bälz, 1984).

Subsequent 16S rDNA studies (Ludwig et al., 1985; Schleifer and Kilpper-Bälz, 1987) not only confirmed this separation, but also demonstrated that the enterococci also differed from the genus *Lactococcus* and certain other Gram-positive cocci. The genera *Streptococcus* and *Lactococcus* are more distantly related to the genus *Enterococcus* than are the coccus- or rod-shaped bacteria of the genera *Vagococcus, Carnobacterium, Tetragenococcus, Aerococcus, Alloiococcus, Dolosigranulum, Facklamia, Globicatella, Granulicatella, Melissococcus, Eremococcus, Ignavigranum,* and *Abiotrophia.* Indeed, according to the 16S rRNA phylogenetic analysis in the revised roadmap to Volume 3 (Figure 1 and Figure 3; Ludwig et al., this volume), *Enterococcus, Melissococcus, Tetragenococcus,* and *Vagococcus* are in a new family, *Enterococcaceae,* within the order *Lactobacillales.*

The cell-wall peptidoglycan type is lysine-D-asparagine (with D-isoasparagine as cross-bridge) as described for *Enterococcus avium, Enterococcus casseliflavus, Enterococcus dispar, Enterococcus durans, Enterococcus faecium, Enterococcus gallinarum, Enterococcus hirae, Enterococcus malodoratus, Enterococcus mundtii, Enterococcus pseudoavium* and *Enterococcus raffinosus,* except *Enterococcus faecalis,* which has a lysine–alanine$_{2-3}$-type peptidoglycan (Collins et al., 1989b, 1986, 1984, 1991a; Farrow and Collins, 1985; Schleifer and Kilpper-Bälz, 1984). Position 2 of the glycerol molecules that constitute the backbone of the teichoic acids is esterified either with the glucose disaccharide kojibiose (as in *Enterococcus faecalis*) or the glucose trisaccharide kojitriose (as in *Enterococcus faecium*). Long-chain fatty acids are predominantly of the straight-chain saturated or monounsaturated types. Several species produce cyclopropane ring acids (Collins et al., 1989b, 1986; Schleifer and Kilpper-Bälz, 1984). The type group D antigen is located between the cell wall and the cell membrane (Smith and Shattock, 1964). It is a teichoic acid polymer of glycerol phosphate containing a high proportion of glucose (Wicken et al., 1963). Its lipid component is a 1-kojibiosyl diglyceride, with the sugars linked to the diglyceride by a phosphatidyl substituent (Toon et al., 1972). This antigen has been traditionally considered to be a typical trait of the genus *Enterococcus* but not all species possess it and its production was not determined for a few recently described species. On the other hand it can be found in several streptococci such as *Streptococcus bovis, Streptococcus suis* and *Streptococcus alactolyticus,* certain leuconostocs and pediococci (Devriese et al., 1993b).

Although enterococci have complex nutritional requirements, their growth on commonly used bacteriological media is usually profuse. They require several amino acids, B vitamins and purine and pyrimidine bases (Garg and Mital, 1991). Colonies are always regular and circular with a smooth surface, up to 5 mm in diameter. Certain species produce a yellow carotenoid pigment on agar media. An unusual C_{32} carotenoid aldehyde has been detected in the *Enterococcus casseliflavus* pigment by Taylor et al. (1971).

The enterococci are facultatively anaerobic with a preference for anaerobic conditions. They are not able to synthesize porphyrins and therefore lack cytochrome pigments, but they may have flavin-containing NADH peroxidases (Miller et al., 1990). Superoxide dismutase induced by molecular oxygen enables survival of *Enterococcus faecalis* in aerobic conditions (Gregory and Fridovich, 1973). Demethylmenaquinones with nine isoprenoid subunits may function as noncytochrome electron carriers in *Enterococcus faecalis*, while menaquinones with eight isoprenoid subunits have been detected in *Enterococcus casseliflavus* (Collins and Jones, 1979). Membrane-associated demethylmenaquinones of *Enterococcus faecalis* are involved in substantial production of cellular superoxide (O_2^-) and reactive oxygen compounds (H_2O_2, hydroxyl radical). This reaction is inhibited by exogenous fumarate and hematin (Huycke et al., 2001). Enterococci can ferment a wide variety of substrates. The main route of energy production is the homofermentative formation of mainly L-lactic acid from glucose through the Embden–Meyerhof–Parnas pathway. Under the aerobic conditions glucose is metabolized to acetic acid, acetoin and CO_2. Pyruvate is stoichiometrically converted to lactate by *Enterococcus faecalis* at pH 5.0–6.0, but at neutral or slightly alkaline pH it is converted to formate, ethanol, and acetate at the ratio 2:1:1 (Gunsalus et al., 1955). In nutrient deficiency, pyruvate is converted to ethanol and acetate (Garg and Mital, 1991). *Enterococcus faecalis* metabolizes malate by an inducible NAD-specific malic enzyme and malate permease is formed in the presence of substrate (London and Meyer, 1970). Gluconate is phosphorylated to 6-phosphogluconate followed by its dissimilation to lactate and CO_2 as described for *Enterococcus faecalis* (Garg and Mital, 1991). In aerated glycerol media *Enterococcus faecalis* produces mainly acetic acid and CO_2 with trace amounts of acetylmethylcarbinol. *Enterococcus faecium* can oxidize glycerol to acetic acid, CO_2 and low amounts of hydrogen peroxide (London and Appleman, 1962). Hippurate is hydrolyzed to glycine and benzoic acid. Energy can be obtained by degradation of some amino acids, as are arginine, tyrosine, serine, agmatine, phenylalanine, and canavanine (Deibel, 1964).

The enterococcal chromosome represents a single circular DNA of a size ranging from 3000–3250 kb for *Enterococcus faecalis*, 2550–2995 kb for *Enterococcus faecium*, 3445 kb for *Enterococcus avium* and 3070 kb for *Enterococcus durans* as revealed by physical mapping. *Enterococcus faecium*, *Enterococcus avium*, and *Enterococcus durans* contains six *rrn* operons; *Enterococcus faecalis* four *rrn* operons (Oana et al., 2002). The full genome sequence of *Enterococcus faecalis* V583 is available (Paulsen et al., 2003) and genome sequencing of *Enterococcus faecium* ATCC BAA-472 is in progress (information available at http://hgsc.bcm.tmc.edu/microbial/efaecium/and http://genome.ornl.gov/microbial/efae/). The *Enterococcus faecalis* V583 chromosome contains 3,218,031 bp consisting of 87.9% coding sequences. A total of

3182 protein-coding genes were detected with a mean size of 889 bp. Among these, 1760 were similar to known protein-coding genes. Furthermore, 495 showed similarity with conserved hypothetical functions, 221 were of unknown function and 706 protein coding genes did not match to any known protein. Twelve rRNA- and 68 tRNA-coding genes were detected. A total of 35 probable PTS-type sugar transporters involved in acquisition and fermentation of substrates were found. A total of 519 proteins were found which are conserved in low-G+C Gram-positive bacteria and are known to be involved in transcription, translation, and protein synthesis. Proteins from large paralogous families such as the PTS and ABC transporter families were also detected (Paulsen et al., 2003; Tendolkar et al., 2003).

Enterococci harbor a wide variety of plasmids and transposons that are involved in transfer of antibiotic resistance and virulence factor determinants. Pheromone-responding plasmids (e.g., pAD1, pAM322, pCF10 or pPD1) are typical for *Enterococcus faecalis*. They encode a surface adhesin called "aggregation substance" which binds donor as well as recipient cells. This leads to formation of cellular aggregates, enabling direct cell contact and efficient plasmid transfer. Antibiotic- or UV-resistance genes or cytolysin-encoding genes may be carried by these plasmids. Inc18 are low-copy-number plasmids (less than 10 copies per cell) of size 25–30 kb. Rolling Circle Replicating (RCR) plasmids are high-copy-number plasmids smaller than 10 kb that replicate by a rolling-circle mechanism. Inc18 and RCR plasmids can be replicated in many other Gram-positive bacteria (Devriese et al., 1992a; Weaver, 2000, 2002).

Besides plasmids, a wide variety of transposons have been isolated in enterococci. Tn3-family transposons, composite transposons and conjugative transposons classes are generally recognized. Tn917 and Tn1546 transposons represent typical members of the Tn3 group. Tn917 carries macrolides-lincosamide-streptogramin resistance genes and Tn1546 encodes *vanA* vancomycin-resistance genes. Conjugative transposons encoding resistance at least to one antibiotic reveal a broad host range. They integrate into the host DNA by transposition, but their transfer to other cells is enabled by conjugation. This common group of transposons contains, e.g., Tn916, Tn918, or Tn1549-encoding VanB phenotype vancomycin-resistance genes. Composite transposons are DNA sequences mobilized by terminal insertion sequences. They may encode sequences involved in antibiotic resistance (aminoglycosides, vancomycin) or mercury resistance or they can bear parts of plasmids or other transposons (Devriese et al., 1992a; Tendolkar et al., 2003; Weaver, 2000; Weaver et al., 2002). About a quarter of the fully sequenced genome of *Enterococcus faecalis* V583 consists of mobile and exogenous DNA including conjugative and composite transposons, a pathogenicity island, phage regions, integrated plasmid remnants and 38 insertion sequence elements. This high number of mobile DNA elements probably significantly contributes to acquisition and accumulation of virulence and resistance factors (Paulsen et al., 2003).

Enterococci produce bacteriocins called "enterocins" that are active against other enterococci as well as other bacterial groups including *Listeria* spp. Genes responsible for their production are located both on chromosome as well as on plasmids. A few enterocins from *Enterococcus faecalis* (AS-48, cytolysins $CylL_L$ and $CylL_S$) and *Enterococcus faecium* (Bacteriocin 31 and Enterocins A, B, P, and 50) have been studied in detail. They generally belong to class II bacteriocins characterized as small, heat-stable proteins that are synthesized on ribosomes but do not undergo post-translational modifications. Enterocin-producing enterococci have been isolated from dairy products, fermented meat, as well as vegetables and plants (Franz et al., 1999). Ott et al. (2001) found antagonistic activity in *Enterococcus faecalis*, *Enterococcus mundtii*, *Enterococcus casseliflavus*, and *Enterococcus faecium* isolated from grasses. Partially purified enterocins from these plant-associated strains were active against a wide range of lactic acid bacteria, clostridia, and *Listeria* strains. Pompei et al. (1992b) and Berlutti et al. (1993) detected bacteriolytic activity in most human clinical strains tested. Different strains produced bacteriolytic enzymes of different properties that enable, together with biochemical tests, identification to the species level.

Enterococcal bacteriophages isolated from small intestine of white rats were characterized by Rogers and Sarles (1963). Bachrach et al. (2003) isolated enterococcal bacteriophages from human saliva and all enterococcal strains tested by Tzannetis et al. (1970) were lysogenic. A lytic enzyme obtained from a phage-infected *Enterococcus faecalis* culture has been used to induce *Enterococcus faecalis* protoplast formation (Bleiweis and Zimmerman, 1961). Phage typing has been used for characterization and typing of enterococci (e.g., Smyth et al., 1987) but its application as described in literature is scanty.

Enterococci are intrinsically resistant to many β-lactams, fluoroquinolones, lincosamides, trimethoprim-sulfamethoxazole and low concentrations of aminoglycosides. Acquired antibiotic resistance mediated by extrachromosomal DNA elements (plasmids, transposons) includes resistance to chloramphenicol, tetracyclines, macrolides, lincosamides, streptogramins and quinolones as well as high-level resistance to aminoglycosides, β-lactams and glycopeptides (Cetinkaya et al., 2000; Devriese et al., 1992a; Teixeira and Facklam, 2003). Vancomycin-resistant strains are a serious problem. Six vancomycin-resistance phenotypes have been described and characterized as VanA, VanB, VanD, VanE, and VanG. They are determined by transposable elements or can be chromosomally encoded and nontransferable as in case of the VanC phenotype. The VanA phenotype displays high-level inducible resistance to vancomycin and teicoplanin; VanB is characterized by moderate inducible resistance to vancomycin. They occur most frequently in *Enterococcus faecalis* and *Enterococcus faecium*. The VanC phenotype genes encode intrinsic low-level vancomycin resistance typical of *Enterococcus gallinarum* and *Enterococcus casseliflavus*. The VanD phenotype characterized by moderate resistance levels to vancomycin and low-level resistance to teicoplanin has been found in *Enterococcus faecium*. VanE and VanG phenotypes exhibit low-level and moderate-level resistance to vancomycin, respectively, but both these phenotypes characterized in *Enterococcus faecalis* reveal full sensitivity to teicoplanin (Cetinkaya et al., 2000; McKessar et al., 2000; Tendolkar et al., 2003). Resistance to β-lactam antibiotics found in *Enterococcus faecalis* and *Enterococcus faecium* clinical strains is mediated mostly by production of penicillin-binding protein (PBP5). Production of β-lactamase by these two species is very rare. High-level aminoglycoside resistance is mediated by a single mutation within the 30S ribosomal subunit protein as in the case of streptomycin or through the production of aminoglycoside-modifying enzymes (Shepard and Gilmore, 2002). In general, *Enterococcus faecium* shows higher prevalence of resistance than *Enterococcus faecalis* and other enterococcus species.

Although enterococci are commensal inhabitants of humans, they are increasingly isolated from a variety of nosocomial and other infections. *Enterococcus faecalis* represents the most commonly isolated enterococcal species from human clinical material (80–90%) followed by *Enterococcus faecium* (8–16%). *Enterococcus avium*, *Enterococcus casseliflavus*, *Enterococcus cecorum*, *Enterococcus dispar*, *Enterococcus durans*, *Enterococcus gallinarum*, *Enterococcus gilvus*, *Enterococcus hirae*, *Enterococcus mundtii*, *Enterococcus pallens*, and *Enterococcus raffinosus* are rarely isolated (Teixeira and Facklam, 2003). Enterococci are found mainly in urinary tract and wound infections, bacteremias, and endocarditis. They are less frequently involved in respiratory, biliary tract, and central nervous system infections, otitis, sinusitis, septic arthritis, endophthalmitis and burn wounds (Jett et al., 1994; Teixeira and Facklam, 2003). Adherence of enterococcal cells to human cells and subsequent invasion of host tissues is supported by aggregation substance, surface carbohydrates, cytolysins, protease, hyaluronidase, lipase, gelatinase, and superoxide production (Elsner et al., 2000; Jett et al., 1994; Mundy et al., 2000). All these traits have been extensively characterized in *Enterococcus faecalis* but they are less well-known in *Enterococcus faecium* or other species.

Enterococci are important in septicemia, osteomyelitis, and endocarditis in poultry and ducks and they may cause sporadic infections in other animal species (Devriese et al., 1992a; Vancanneyt et al., 2001; Wood et al., 2002).

Enterococcus faecalis represents the prevailing enterococcal species colonizing the human digestive tract, although *Enterococcus faecium* may prevail in some persons or localities depending probably on age, diet, or physiological conditions of organisms (Devriese et al., 1992a; Devriese et al., 1995). One gram of human feces contains about 10^5–10^7 enterococcal cells (Murray and Weinstock, 1999). *Enterococcus gallinarum*, *Enterococcus casseliflavus*, and *Enterococcus mundtii* have been found in stool samples by Van Horn and Rodney (1998). Enterococci are less commonly isolated from the vagina and from oral cavity (Hardie and Whiley, 1997; Murray, 1990). More species are found in animals. Poultry is commonly inhabited by *Enterococcus faecium*, *Enterococcus cecorum*, *Enterococcus faecalis*, *Enterococcus hirae*, and *Enterococcus durans*; in contrast *Enterococcus gallinarum* and *Enterococcus avium* are rare (Devriese et al., 1991, 1987). *Enterococcus faecalis* is a prevailing species in rectum and tonsils of dogs and cats (Devriese et al., 1992b). Baele et al. (2002) found *Enterococcus columbae* to be a common inhabitant of pigeon intestines followed by *Enterococcus cecorum*. Mundt (1963a) and Mallon et al. (2002) found enterococci in feces of wild mammals, reptiles and birds. *Enterococcus casseliflavus* has been found to be a common inhabitant of the gut of garden snails (Charrier et al., 1998; Švec et al., 2002). Enterococci are often isolated from nectar-feeding insects (Martin and Mundt, 1972) and have been found in gut of termites (Bauer et al., 2000; Brune and Friedrich, 2000; Švec et al., 2006). The recently described *Enterococcus canis* (De Graef et al., 2003) and *Enterococcus hermanniensis* (Koort et al., 2004) have been isolated from canine tonsils, *Enterococcus canintestini* from dog feces (Naser et al., 2005c), *Enterococcus phoeniculicola* from a Red-billed Woodhoopoe (Law-Brown and Meyers, 2003) and *Enterococcus devriesei* from bovine materials and from the air of a poultry slaughter by-product processing plant (Švec et al., 2005b).

Enterococci can also be isolated from food, plants, soil, and water. Although these bacteria are considered to be only a temporary part of the microflora of plants, in good conditions they can propagate on their surface (Mundt, 1961; Stirling and Whittenbury, 1963). *Enterococcus faecalis*, *Enterococcus faecium*, *Enterococcus hirae*, *Enterococcus mundtii*, *Enterococcus casseliflavus*, and *Enterococcus sulfureus* were isolated from plants (Müller et al., 2001; Ott et al., 2001). They are generally isolated more often from flowers than from buds or leaves. This indicates that insects are probably involved in the dissemination of enterococci. Soil can be contaminated by enterococci from animals and/or plants thanks to wind or rain (Mundt, 1961). It is not naturally inhabited by enterococci as shown by Medrek and Lidsky (1960) who isolated enterococci in only 8 out of 369 samples of soils that were unaffected by human activity. Occurrence of enterococci in waters is generally considered to be the result of fecal contamination. Typically, *Enterococcus faecalis*, *Enterococcus faecium*, *Enterococcus durans*, and *Enterococcus hirae* can be commonly found in contaminated waters; less often *Enterococcus avium*, *Enterococcus gallinarum*, *Enterococcus cecorum*, and *Enterococcus columbae* are isolated (Godfree et al., 1997; Pourcher et al., 1991). Pristine waters in Finland mainly contained *Enterococcus casseliflavus* (Niemi et al., 1993). Because of their long survival capacities, enterococci are used as indicators of distant contamination in water supplies. Environment-associated strains are often biochemically different from strains isolated from animals and many strains cannot be unequivocally assigned to known species (Devriese et al., 1992a; Martin and Mundt, 1972; Müller et al., 2001; Mundt, 1961, 1963b; Niemi et al., 1993; Ulrich and Müller, 1998).

Enterococci are a common part of many kinds of food – especially those of animal origin such as milk and milk products, meat, and fermented sausages. They are generally considered to be secondary contaminants of food often playing a role in its spoilage, although certain strains positively influence ripening and aroma development of some types of cheeses. They are even used as probiotic cultures in some products (Franz et al., 1999; Giraffa, 2002). On the other hand, presence of virulence factors in starter and food enterococcal strains as well as transfer of virulence determinants to starter strains has been demonstrated (Eaton and Gasson, 2001). Devriese et al. (1995) found *Enterococcus faecium* to prevail in cheese and cheese–meat combined foods. *Enterococcus faecalis* was common in crustaceans. Meat products contained mostly *Enterococcus faecium* followed by *Enterococcus faecalis*, and *Enterococcus durans/Enterococcus hirae*. Products containing turkey meat contained *Enterococcus gallinarum*. *Enterococcus hermanniensis* has been isolated from broiler meat (Koort et al., 2004) and *Enterococcus devriesei* from vacuum-packaged charcoal-broiled river lampreys (Švec et al., 2005b). Kirk et al. (1997) isolated *Enterococcus faecium* and *Enterococcus faecalis* as prevailing species from frozen chicken carcasses. Enterococci can contaminate milk and dairy products during various processing steps, where *Enterococcus faecalis* and *Enterococcus faecium* prevail (Garg and Mital, 1991; Mannu et al., 1999).

Enrichment and isolation procedures

As enterococci have complex nutritional requirements it is not possible to cultivate them in defined media, but they grow well on commonly used rich complex bacteriological media (e.g., Todd–Hewitt, trypticase soy, brain heart infusion, starch agar, blood-containing agars). Some species grow very poorly on MRS broth and agar commonly used for cultivation of lactic acid bacteria. More than 60 different selective media have been described for isolation of enterococci. Unfortunately, they all allow growth of some other bacteria while many enterococcal species are completely or partially inhibited. The most frequently used media for isolation and enumeration of enterococci from various samples contain sodium azide as a selective agent.

Maintenance procedures

Lyophilization, coating of glass beads in −70 °C, or storage in liquid nitrogen can be used for long-term preservation of enterococci. They can also be kept for 1–2 years at −20 °C in lyophilization medium or tubes of stab-inoculated agar (YGLP, BHI) stored at 4 °C. Maintenance in Litmus Milk plus chalk at 4 °C is generally suitable for short-term preservation (3 months), although some strains can survive in these conditions for up to 5 years.

Differentiation of the genus *Enterococcus* from other genera

Classical biochemical identification of strains as belonging to the genus *Enterococcus* and their differentiation from other related genera is achieved primarily via the species identification. For practical purposes the following approach can be used: catalase-negative, Gram-positive cocci, showing good growth on enterococcus-selective media containing 0.4% sodium azide, and able to grow in 6.5% (w/v) NaCl broth, can be identified presumptively as belonging to the genus *Enterococcus*. Typically, only strains of the *Streptococcus bovis* species group show colony characteristics similar to those of the "classical" enterococci on these selective media, but these strains do not grow in 6.5% NaCl broth. However, it should be kept in mind that this procedure excludes several enterococcal species. Molecular identification methods generally obviate this type of difficulties.

Taxonomic comments

A most important phenomenon is the existence of species groups within the genus *Enterococcus* (Table 116, Figure 110). Certain species appear to form distinct lineages, but most others can be allotted to these groups (Švec et al., 2001; Tyrrell et al., 2002a; Vancanneyt et al., 2001; Williams et al., 1991). Members of such groups exhibit similar phenotypic characteristics, and species separation can be problematic. 16S rRNA gene sequence similarities between species can be as high as 99.8% within some groups. Despite these close relationships and similarities the species are well separated by DNA–DNA similarity determinations and certain molecular identification techniques. The *Enterococcus faecium* group contains *Enterococcus faecium*, *Enterococcus durans*, *Enterococcus canis*, *Enterococcus hirae*, *Enterococcus mundtii*, *Enterococcus ratti*, and *Enterococcus villorum*. These species mostly show identical growth and physiological characteristics. Discrimination of individual species of this group by biochemical tests is often unreliable. *Enterococcus durans*, *Enterococcus hirae*, and *Enterococcus villorum* are especially difficult to differentiate, although they can be clearly distinguished by whole-cell protein profile analysis using SDS-PAGE, tDNA-PCR and arbitrarily primed PCR analysis (Devriese et al., 2002). The *Enterococcus avium* group contains *Enterococcus*

TABLE 116. Tests differentiating species groups and species of the genus *Enterococcus*[a]

Test	E. faecalis	E. caccae	E. haemoperoxidus	E. moraviensis	E. silesiacus	E. termitis	E. avium	E. devriesei	E. gilvus	E. malodoratus	E. pseudoavium	E. raffinosus	E. faecium	E. canis	E. durans	E. hirae	E. mundtii	E. ratti	E. villorum	E. cecorum	E. columbae	E. gallinarum	E. casseliflavus	E. italicus	E. camelliae	E. aquimarinus	E. asini	E. canintestini	E. dispar	E. hermanniensis	E. pallens	E. phoeniculicola	E. saccharolyticus	E. sulfureus
Motility	−	−	−	−	−	−	−		−	−	−	−	−	−	−	−	−	−	−	−	d[b]	+[b]	+[b]	−	−	−	−	−	−	−	−	−	−	−
Pyruvate utilization	+	+	+	+	−	−	+		+	+	+	+	+	+	−	−	−	+	−	+	+	−	d	−	−	−	−	−	+	−	+[b]	−	−	−
Production of:																																		
Alkaline phosphatase	−	−	−	−	−	−	−	−	−	−	−	−	−	d	−	−	−	−	−	d[b]	d[b]	−	−	−	−	−	+	+	−	−	−	−	−	−
Arginine dehydrolase	+	+	+	+	+	+	−	−	−	−	−	−	+	d	+	+	+	+	+	−	−	+	+	−	−	−	+	d−	+	−	−	−	−	−
Pyrrolidonyl arylamidase	+	+	+	d	+	−	+	d	+	+	−	+	+	+	+	+	+	+	+	−	−	+	+	+		+	+	d−	d[b]	+	−	−	+	+
Acid from:																																		
Adonitol	−	−	−	−	−	−	+	d	−	+	+	d	−	−	−	−	−	−	−	d	d	−	−	−	−	−	−	−	−	w	−	−	−	−
D-Arabitol	−	−	−	+	−	−	+	+	+	+	+	+	−	+	−	−	−	−	−	+	+	d	−	−	−	+	+	+	+	+	+	+	+	+
Inulin	d	−	+	+	−	+	−	d	+	+	−	+	+	+	−	−	+	−	−	+	+	−	−	+	+	+	+	+	−	+	+	+	+	+
Melezitose	+	+	+	+	+	+	+	d	+	+	+	+	−	+	−	d	−	+	+	d	d	d	+	−	−	+	+	+	+	−	+	+	+	+
Ribose	+	+	+	+	−	+	+	−	+	+	+	+	+	+	+	+	+	+	+	+	−	+	+			+	+	+	+	+	+	+	+	+
L-Sorbose	−	−	−	−	−	+	−	d	+	−	−	+	−	−	−	−	+	−	+	−	−	−	−	−	−	−	−	−	−	−	−	−	−	−
Pigment production	−	−	−	−	−	−	−	−	+	−	−	−	−	−	−	−	+	−	−	−	−	−	+[b]	−	−	−	−	−	−	−	−	−	−	−
Growth:																																		
at 10°C	+	+	+	+	+	+	+	+	−	+	+	+	+	d	+	+	+	−	+	−		+	+	w	+	−	w	+	+	+	+	−	+	+
at 45°C	+	d[b]	−	−	−	−	+	−	+	+	+	+	+	−	+	+	+	+	+	d	d[b]	+	+	d	+	+	w	−	+	−	+	−	+	+
in 6.5% NaCl	+	+	w	w	+	+	d	+	−	+	+	+	+	+	+	d	+	−	+	−	d[b]	+	+	−	−	+	−	+	+	d	+	−	w	+
Production of:																																		
Acetoin	+	+	+	+	+	−	+	+	−	+	+	+	+	d	+	+	+	w	+	+	+	+	+	+		+	+	+	+	+	+	+	+	+
D group antigen	+	+	d	+	+	+	d	d	+	+	+	d	+	d	+	+	+	+	+	d	−	+	+	−	−	+	+	−	d[b]	+	+	+	+	+
Leucine arylamidase	+	+	−	d	+	+	+	+	+	+	+	+	+	+	+	+	+	+	+	+	d[b]	+	+			+	+	d−	d[b]	+	+	+	+	+
α-Galactosidase	−	−	d	d	d	−	−	−	+	−	−	+	+	d	−	d	d	−	+	d	d	+	+	+		+	+	d−	d[b]	+	+	+	+	+
β-Galactosidase	−	−	−	d	+	−	+	−	+	−	+	−	−	+	−	−	+	+	+	d	d	+	+			−	+	d−	d[b]	−	+	−	+	+
β-Glucuronidase	−	−	−	−	−	−	−	−	−	−	−	−	−	−	−	−	+	−	−	d[b]	−	d	−	−		−	−	−	−	−	−	−	−	−
Hydrolysis of:																																		
Esculin	+	+	+	+	+	+	+	+	+	+	+	+	+	+	+	+	+	+	+	+	+	+	+	+	+	+	+	+	+	w	+	−	+	+
Hippurate	+	d[b]	+	+	−	+	d	+	−	d	+	−	d	−	d	d	−	d	−	d	+	+	−	−	+	−	+	−	d[b]	−	+	−	+	−
Starch	−	−	w	w	+	+	−	−	+	+	−	+	−	−	−	d	−	+	−	d	+	−	+	w	w	−	+	+	−	d	+	−	+	+
Acid from:																																		
N-Acetylglucosamine	+	+	+	+	+	+	+	+	+	+	+	+	+	+	+	+	+	+	+	+	+	+	+	−	−	+	+	+	+	+	+	−	+	+
Amygdalin	+	+	+	+	+	+	+	+	+	+	+	+	+	+	+	+	+	+	+	d	d	+	+	d		+	+	+	+	d	+	−	+	+
D-Arabinose	−	−	d	+	d	−	−	d	−	+	+	+	+	d	+	+	+	+	+	d	d	+	+	−		+	+	+	+	d	+[bc]	−	+	−
L-Arabinose	−	−	−	−	−	−	+	d	−	−	−	+	+	d	+	+	+	+	+	d	d	+	+	−		+	+	+	+	d	−	−	+	+
L-Arabitol	−	−	−	−	−	−	+	d	−	+	+	+	+	+	−	−	−	−	−	d	d	+	+	−		+	+	+	−	−	−	−	+	+
Arbutin	+	+	+	+	+	+	+	+	+	+	+	+	+	+	+	+	+	+	+	+	+	+	+	d		+	+	+	+	+	+	−	+	+
Cellobiose	+	+	+	+	+	+	+	+	+	+	+	+	+	+	+	+	+	+	+	+	+	+	+	d		+	+	+	+	+	+	+	+	+
Dulcitol	−	−	−	−	−	−	d	−	−	d	−	−	−	−	−	−	−	−	−	d	+	−	−	−		+	−	−	−	−	−	+	+	−

(continued)

TABLE 116. (continued)

	E. faecalis group						E. avium group						E. faecium group							E. cecorum group		E. gallinarum group		E. italicus group		Phylogenetically distinct species								
	E. faecalis	*E. caccae*	*E. haemoperoxidus*	*E. moraviensis*	*E. silesiacus*	*E. termitis*	*E. avium*	*E. devriesei*	*E. gilvus*	*E. malodoratus*	*E. pseudoavium*	*E. raffinosus*	*E. faecium*	*E. canis*	*E. durans*	*E. hirae*	*E. mundtii*	*E. ratti*	*E. villorum*	*E. cecorum*	*E. columbae*	*E. gallinarum*	*E. casseliflavus*	*E. italicus*	*E. camelliae*	*E. aquimarinus*	*E. asini*	*E. canintestini*	*E. dispar*	*E. hermanniensis*	*E. pallens*	*E. phoeniculicola*	*E. saccharolyticus*	*E. sulfureus*
Erythritol	–	–	–	–	–	–	–	–	–	–	–	–	–	–	–	–	–	–	–	–	–	–	–	–	–	–	–	–	–	–	–	–	–	–
D-Fructose	+	+	+	+	+	+	+	+	+	+	+	+	+	+	+	+	+	+	+	+	+	+	+	+	+	+	+	+	+	+	+	+	+	+
D-Fucose	–	–	–	–	–	–	–	–	–	–	–	–	–	–	–	–	–	–	–	–	–	–	–	–	–	–	–	–	–	–	–	–	–	–
L-Fucose	–	–	–	–	–	–	–	–	–	–	–	–	–	–	–	–	–	–	–	–	–	–	–	–	–	–	–	–	–	–	–	–	–	–
Galactose	+	+	+	+	+	+	+	+	+	+	+	+	+	+	+	+	+	+	+	d	d	+	+	+	–	+	+	+	+	d	+	+	+	+
β-Gentiobiose	+	–	+	+	d	+	+	d	+	+	+	d	d	+	+	+	+	+	d	–	–	+	+	d	–	–	+	+	d	w	+	+	d	+
Gluconate	d	–	–	+	–	+	+	d	+	+	+	–	d	d	d	d	–	+	w	d	d	d	d	–	–	+	+	+	d	–	+	d	+	+
D-Glucose	+	+	+	+	+	+	+	+	+	+	+	+	+	+	+	+	+	+	+	+	+	+	+	+	+	+	+	+	+	+	+	+	+	+
Glycerol	+[b]	+	+	+	+	+	d	d	+	d	–	+	d	+	d	d	+	+	w	d[b]	d	d[b]	d	–	–	+	–	d–	d[b]	+	+	d[b]	d	d
Glycogen	–	–	d	d	–	–	+	–	+	d	–	–	–	–	–	–	–	–	–	–	–	d[b]	d	–	–	+	–	d	–	–	–	+	–	–
Inositol	–	–	–	–	–	–	d	–	+	–	+	+	–	–	–	–	–	–	–	–	–	–	–	–	–	–	–	d+	–	–	–	+	–	–
2-Keto-gluconate	d	+	–	–	–	–	+	+	–	+	+	+	+	+	–	–	–	–	–	d	d	d	+	–	–	+	+	d+	–	–	+	–	+	+
5-Keto-gluconate	–[b]	–	–	–	–	–	d	–	+	d	–	–	–	–	–	–	–	–	–	d	d	d	d	–	–	–	+	d+	–	–	+	+	–	–
Lactose	+[b]	+	+	+	+	+	+	+	+	+	+	+	+	+	+	+	+	+	d[b]	d[b]	d	+	+	+	–	+	+	d	d[b]	w	+	+	+	+
D-Lyxose	–	–	–	–	–	–	–	d	–	–	–	–	–	–	–	–	–	+	–	–	–	–	–	–	–	–	–	–	–	–	–	–	–	–
Maltose	+	+	+	+	+	+	+	+	+	+	+	+	d	+	+	+	+	+	d	+	+	+	d	+	+	+	+	+	+	+	+	+	+	+
Mannitol	+[b]	–	d	d	–	+	+	d	+	+	+	+	d	–	d	d[b]	d	d	d[b]	d[b]	d	d	d	d	–	+	+	d–	d[b]	d	+	d+	+	+
D-Mannose	+	+	+	+	+	+	+	+	+	+	+	+	d	+	+	d[b]	d	–	d[b]	d[b]	d	d	d	+	–	+	+	+	+	+	+	+	+	+
Melibiose	–	–	–	–	–	–	d	–	+	+	+	+	d	–	d[b]	d[b]	–	–	d[b]	–	d	+	+	–	–	+	–	–	d[b]	–	–	+	–	+
α-Methyl-D-glucoside	–	–	–	–	–	–	–	–	+	–	+	d	d	–	d	d[b]	–	–	–	d	d	+	+	–	–	+	+	d–	d[b]	–	+	d–	+	+
α-Methyl-D-mannoside	–	d	–	d	–	–	d	–	–	–	–	–	d	–	–	d	–	–	–	d	d	d	d	–	–	+	d	+	+	–	+	+	+	+
D-Raffinose	–	–	–	–	–	–	–	d	–	+	–	+	d[b]	d	d	–	d	–	d[b]	d	d	d	d[b]	–	d[b]	+	+	–	d[b]	d	+	–	+	–
Rhamnose	d	–	–	d	–	–	d	d	d	+	+	+	d	–	d	d	d	–	–	d	d	d	d	–	–	+	w	+	d[b]	d	+	–	+	+
Sorbitol	d	–	d	–	+	–	d	d	+	+	+	+	d[b]	–	d	d	d	–	d[b]	d[b]	d	d	d	d	–	+	–	+	d[b]	–	+	d[b]	+	–
Starch (Amidon)	d	–	–	–	–	–	–	–	+	+	+	+	+	d	d[b]	d[b]	d	d	+	+	d	d	–	–	–	+	–	d+	–	–	+	+	–	+
Sucrose	+[b]	+	+	+	+	+	+	+	+	+	+	+	+	+	+	+	+	+	+	+	+	+	+	–	+	+	+	+	–	+	+	–	+	+
D-Tagatose	+	+	+	+	+	+	d	d	d	+	+	+	d	–	d	d	d	–	d[b]	d	d	d	d	–	–	+	w	d	+	+	+	+	+	+
Trehalose	+[b]	+	+	+	+	+	+	+	+	+	+	+	+	+	+	+	+	+	+	d	–	+	+	+	–	+	+	+	+	w	+	w	+	+
D-Turanose	–	d	–	+	–	–	+	d	–	d	+	–	–	d	–	–	+	d	–	d	d	–	–	–	–	+	–	d+	–	–	–	d+	+	–
Xylitol	–	–	–	–	+	–	+	+	+	+	+	–	+	+	–	–	+	–	–	–	–	+	+	–	–	–	–	–	–	–	–	+	+	–
D-Xylose	–	–	–	+	+	+	–	–	–	d	d	d[b]	d[b]	d	–	–	+	d	d	–	d	+	+	+	+	–	+	–	–	–	+	+	+	–
L-Xylose	–	–	–	–	–	–	–[c]	–[c]	–	–	–	–	d[b]	d	–	–	+	–	–	–	d	+	–	–	–	–	–	–[c]	–	–	–[c]	+	+	–

Data obtained from Devriese et al., 1992a; Pompei et al., 1992b; Devriese et al., 1993b; Devriese and Pot, 1995; Teixeira et al., 1995; Manero and Blanch, 1999; Baele et al., 2002; Devriese et al., 2002; Teixeira and Facklam, 2003; Carvalho et al., 2004 as well as from the original descriptions of enterococcal species (for references see "List of species of the genus *Enterococcus*").

[a] For symbols see standard definitions; w, weak reaction.

[b] Refers to discordant results appearing in literature; see the species description in the "List of species of the genus *Enterococcus*" for more details

[c] Enantiomers are not stated in literature

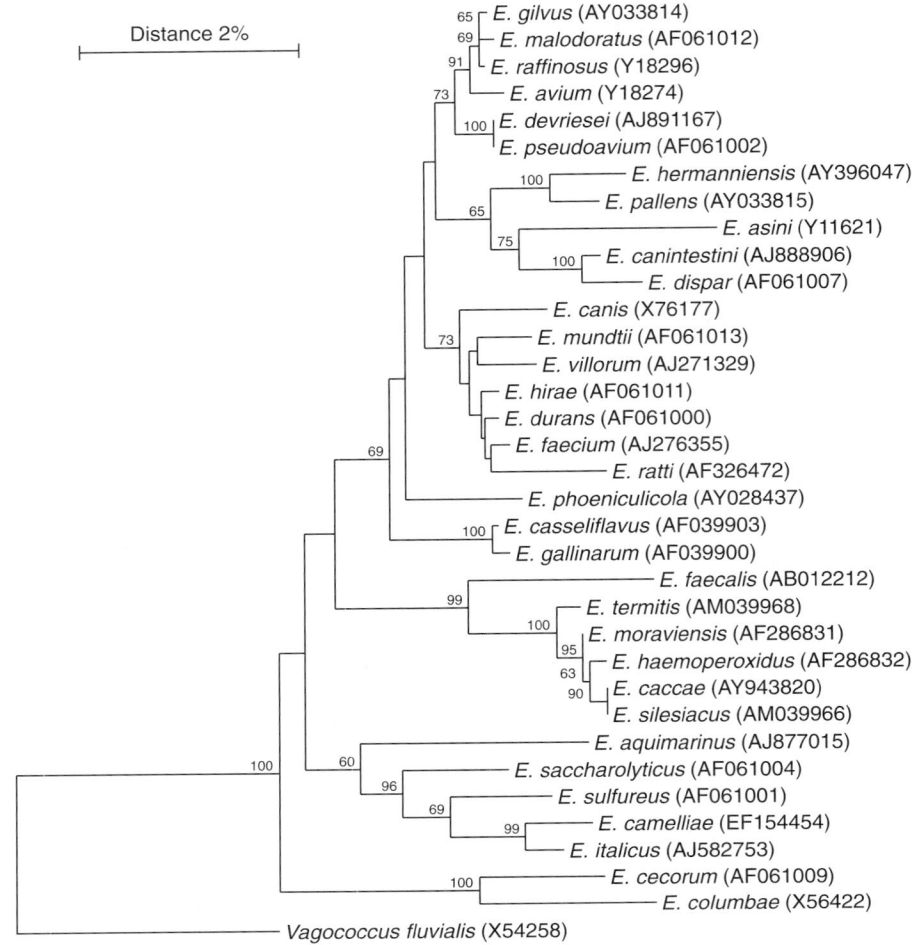

FIGURE 110. Distance matrix tree based on 16S RNA gene sequence comparisons showing the phylogenetic relationships of *Enterococcus* species. *Vagococcus fluvialis* (X54258) sequence was used as the outgroup. Bootstrap percentage values (500 tree replications) higher than 50% are indicated at the branch points. The tree was constructed by the neighbor-joining method.

avium, Enterococcus devriesei, Enterococcus gilvus, Enterococcus malodoratus, Enterococcus pseudoavium, and *Enterococcus raffinosus.* These species form mostly small colonies with strong greening hemolysis and they often grow but only weakly on enterococcus selective media. Growth at 10 °C, 45 °C and in 6.5% NaCl as well as D antigen production may be negative. Members of this group are typically adonitol- and L-sorbose-positive. The *Enterococcus faecalis* group contains *Enterococcus faecalis, Enterococcus caccae, Enterococcus hemoperoxidus, Enterococcus moraviensis, Enterococcus silesiacus* and *Enterococcus termitis.* Although phylogenetic distances are larger in this group the six species share many phenotypic traits. They form similar dark red colonies with a metallic sheen on Slanetz–Bartley agar. Growth at 10 °C, in 6.5% NaCl as well as production of D antigen is positive. Nevertheless, these species can be differentiated from each other by a few biochemical tests. The *Enterococcus gallinarum* group consists of *Enterococcus gallinarum* and *Enterococcus casseliflavus.* Intrinsic low-level vancomycin resistance and motility are typical for this group, although nonmotile strains may be rarely isolated (Clark et al., 1998; Patel et al., 1998; Vincent et al., 1991). *Enterococcus cecorum* and *Enterococcus columbae* species are more distantly phylogenetically related, but they show,

nevertheless, remarkable phenotypic similarities. They grow poorly on enterococcus-selective media and they are carboxyphilic: their growth is strongly enhanced by cultivation in a CO_2 atmosphere. The alkaline phosphatase that may be produced by these two species is a unique trait in the genus *Enterococcus.* The *Enterococcus italicus* group contains *Enterococcus italicus* and *Enterococcus camelliae.* These two recently described species reveal low biochemical activity in comparison with the other enterococcal species

Five species have been reclassified from the genus *Enterococcus* since it was established by Schleifer and Kilpper-Bälz (1984). *Enterococcus seriolicida* (Kusuda et al., 1991) was found to be synonymous with *Lactococcus garvieae* and reclassified by Teixeira et al. (1996). The species *Enterococcus porcinus* (Teixeira et al., 2001) was found to be a junior synonym of *Enterococcus villorum* and validly reclassified by De Graef et al. (2003). Similarly, *Enterococcus solitarius* (Collins et al., 1989b) was reclassified to the genus *Tetragenococcus* as *Tetragenococcus solitarius* (Ennahar and Cai, 2005). Recently, Naser et al. (2006) reclassified *Enterococcus flavescens* (Pompei et al., 1992a) as *Enterococcus casseliflavus* and *Enterococcus saccharominimus* (Vancanneyt et al., 2004) as *Enterococcus italicus.*

Miscellaneous comments

A number of schemes based on biochemical and physiological characteristics have been proposed for the identification of enterococci (e.g., Day et al., 2001; Devriese et al., 1993b; Facklam and Collins, 1989; Manero and Blanch, 1999). However, identification of enterococci based on biochemical tests only is difficult and often unreliable. The situation is complicated by different methodological approaches that provide different results for the same test, and discrepancies can be found between the conventional methods and various commercial kits (e.g., Hudson et al., 2003; Bosshard et al., 2004; Winston et al., 2004). Moreover, strains of the same species isolated from different sources may reveal different biochemical characteristics (Devriese et al., 1995; Švec et al., 2002). Generally speaking, it can be concluded that only *Enterococcus faecalis* is relatively easily and reliably identified using growth characteristics and biochemical testing. In sample types in which *Enterococcus faecium* is the only frequent representative of the *Enterococcus faecium* species group, identification is possible in this way. But in all other instances it is advisable to confirm identification of this important species by other means.

Full as well as partial 16S rRNA gene sequencing has been used for identification of enterococci (Monstein et al., 2001; Patel et al., 1998). This method is valuable for identification at the genus and species-group level but it should be kept in mind that 16S rRNA gene sequence similarities between species can be as high as 99.8% as in the case of *Enterococcus casseliflavus*–*Enterococcus gallinarum* or 99.7% for *Enterococcus faecium*–*Enterococcus durans* (Williams et al., 1991). However, the DNA–DNA similarity values of these species are low and clearly indicate their separate species status. The DNA–DNA hybridization method can be considered as a gold standard for description of novel enterococcus species, but this time-consuming technique is yet not applicable for routine identification. PCR methods based on detection of D-Ala:D-Ala ligase gene-targeted primers ($ddl_{\text{E.faecalis}}$ and $ddl_{\text{E.faecium}}$) and *vanA*, *vanB*, and *vanC* genes are applied for differentiation of *Enterococcus faecalis*, *Enterococcus faecium*, *Enterococcus gallinarum*, and *Enterococcus casseliflavus* (Dudka-Malen et al., 1995). Differentiation of *Enterococcus durans* and *Enterococcus hirae* by the *ddl*-PCR was described by Knijff et al. (2001). Another method is based on partial sequencing of the D-Ala:D-Ala ligase gene (Ozawa et al., 2000).

tRNA intergenic spacer PCR (tDNA-PCR) followed by capillary electrophoresis of the PCR products proved to be a good tool for identification of all enterococcus species as shown by Baele et al. (2000). A related method uses PCR amplification of 16S–23S rRNA intergenic spacer (Tyrrell et al., 1997). Evaluation of repetitive sequence based PCR (rep-PCR) fingerprinting with the $(\text{GTG})_5$ primer for identification of enterococci was described by Švec et al. (2005c). Arbitrarily primed PCR with a D11344 primer has been shown to be a useful method for species-specific identification of enterococci by Descheemaeker et al. (1997) and Devriese et al. (2002). Broad-range PCRs using 16S rDNA-targeted primers differentiate enterococci into species groups (Monstein et al., 1998). Hybridization of rRNA gene sequence-targeting probes was evaluated by Behr et al. (2000) and Manero and Blanch (2002). Application of multilocus sequence analysis (MLSA) using *rpoA*, *pheS* and *atpA* genes was evaluated for identification of enterococci by Naser et al. (2005a, 2005b).

Analysis of whole-cell protein profiles obtained by SDS-PAGE is a suitable method corresponding to DNA–DNA hybridization results (De Graef et al., 2003; Devriese et al., 2002; Koort et al., 2004; Merquior et al., 1994; Müller et al., 2001; Tyrrell et al., 2002a; Vancanneyt et al., 2001). Other biochemical and physical identification methods applied in enterococcus identification include long-chain fatty acid analysis (Fortina et al., 2004; Tyrrell et al., 2002a), cell membrane fatty acid methyl ester analysis (Lang et al., 2001), temperature-gradient gel electrophoresis (TGGE) and denaturing-gradient gel electrophoresis (DGGE) of 16S rDNA variable regions (Ercolini et al., 2001; Monstein et al., 2001) and vibrational spectroscopy (Kirschner et al., 2001).

Acknowledgements

Special thanks to M. Vancanneyt for his comments and suggestions during the preparation of this chapter. P. Švec is indebted to the Ministry of Education, Youth and Sports of the Czech Republic (project MSM0021622416).

Further reading

Gilmore, M.S., D.B. Clewell, P.M. Courvalin, G.M. Dunny, B.E. Murray and L.B. Rice. 2002. The Enterococci: Pathogenesis, Molecular Biology and Antibiotic Resistance, ASM Press, Washington, DC. (Gilmore et al., 2002)

List of species of the genus Enterococcus

1. **Enterococcus faecalis** (Andrewes and Horder 1906) Schleifer and Kilpper-Bälz 1984, 33^VP (*Streptococcus faecalis* Andrewes and Horder 1906, 713)

   fae.cal'is. L. n. *faex* dregs; N.L. adj. *faecalis* relating to feces.

   See Table 116 for most of the biochemical and physiological characteristics. The other traits are as follows: Usually nonhemolytic. Pseudocatalase may be produced when cultivated on blood-containing agar media. Strains survive heating at 60 °C for 30 min. Tetrazolium is reduced to formazan and typical red colonies are formed when grown on tetrazolium-containing selective media. Grows on media containing sodium azide. The cell-wall peptidoglycan is Lys–Ala$_{2-3}$ type. The major fatty acids are hexadecanoic, octadecenoic and *cis*-11,12-methylenoctadecanoic. Most strains contain demethylmenaquinones with nine isoprene units. Malate, serine, citrate, gluconate and arginine are utilized as an energy source. Tyrosine is decarboxylated to tyramine.

   Aberrant results may be seen in asaccharolytic variant strains from human clinical material, which may be glycerol-, lactose-, mannitol-, sucrose- and trehalose-negative (Facklam and Collins, 1989; Teixeira et al., 2001).

   Isolated from human and veterinary clinical materials, from food, and the environment. Typically associated with intestines of humans and animals.

   See generic description for further features.

   DNA G+C content (mol%): 37.0–40.0 (T_m).

Type strain: ATCC 19433, ATCC 19433-U, CCM 7000, CCUG 19916, CIP 103015, DSM 20478, HAMBI 1711, JCM 5803, JCM 8726, LMG 7937, NBRC 100480, NBRC 100481, NCAIM B.01312, NCIMB 775, NCTC 775.

GenBank accession number (16S rRNA gene): AB012212, AJ301831.

2. **Enterococcus aquimarinus** Švec, Vancanneyt, Devriese, Naser, Snauwaert, Lefebvre, Hoste and Swings 2005a, 2186[VP]

a.qui.ma.ri′nus. L. fem. n. *aqua* water; L. adj. *marinus* of the sea; N.L. masc. adj. *aquimarinus* pertaining to sea water

See Table 116 for most of the biochemical and physiological characteristics. The other traits are as follows: Growth at 42 °C and in 6.5% NaCl is positive. Weak growth on Slanetz–Bartley medium.

Isolated from sea water.

DNA G+C content (mol%): 38.7 (HPLC).

Type strain: API 8407116, CCM 7283, LMG 16607.

GenBank accession number (16S rRNA gene): AJ877015.

3. **Enterococcus asini** de Vaux, Laguerre, Diviès and Prévost 1998, 386[VP]

a.si′ni. L. gen. sing. *asini* of donkeys, *Equus asinus.*

See Table 116 for most of the biochemical and physiological characteristics. The other traits are as follows: Grows in 4% NaCl but not in 6.5% NaCl. Strains survive heating at 60 °C for 30 min.

Isolated from cecal contents of donkeys (*Equus asinus*). See generic description for further features.

DNA G+C content (mol%): 39.4 (UV spectroscopy).

Type strain: AS2, ATCC 700915, CCM 4895, CCUG 44928, DSM 11492, LMG 18727, NBRC 100681.

GenBank accession number (16S rRNA gene): Y11621.

4. **Enterococcus avium** (*ex* Nowlan and Deibel 1967) Collins, Jones, Farrow, Kilpper-Bälz and Schleifer 1984, 220[VP] ("*Streptococcus avium*" Nowlan and Deibel 1967, 295)

av.i′um. L. n. *avis* bird; L. gen. pl. n. *avium* of birds.

See Table 116 for most of the biochemical and physiological characteristics. The other traits are as follows: Mostly α-hemolytic. Folinic acid is required for growth. H_2S production is positive.

Isolated from human and veterinary clinical materials, from food and the environment.

See generic description for further features.

DNA G+C content (mol%): 39.0–40.0 (T_m).

Type strain: Guthof E6844, ATCC 14025, CCM 4049, CCUG 44928, CIP 103019, DSM 20679, JCM 8722, LMG 10744, NBRC 100477, NCIMB 702369, NCTC 9938, VKM B-1673.

GenBank accession number (16S rRNA gene): Y18274, AJ301825, AF133535.

5. **Enterococcus caccae** Carvalho, Shewmaker, Steigerwalt, Morey, Sampson, Joyce, Barrett, Teixeira and Facklam 2006, 1507[VP]

cac′cae. Gr. n. *kakke* feces; N.L. gen. n. *caccae* of feces.

See Table 116 for most of the biochemical and physiological characteristics. The other traits are as follows: Small colonies up to 0.5 mm in diameter are formed on sheep-blood agar at 37 °C. Growth at 45 °C and in 6.5% NaCl is positive.

Positive production of β-glucosidase, glycyl-tryptophan arylamidase, β-mannosidase, pyroglutamic acid arylamidase and N-acetyl-β-glucosaminidase. Negative production of alanine-phenyl-alanine-proline arylamidase, urease and acidification of cyclodextrin and pullulan. Hippurate hydrolysis negative in API Rapid ID 32 kit but positive using conventional testing.

Isolated from human stool samples.

DNA G+C content (mol%): 32.5 (T_m).

Type strain: 2215–02, SS-1777, ATCC BAA-1240, CCUG 51564, DSM 19114.

GenBank accession number (16S rRNA gene): AY943820.

6. **Enterococcus camelliae** Sukontasing, Tanasupawat, Moonmangmee, Lee and Suzuki 2007, 2153[VP]

ca.mel.li′ae. N.L. gen. n. *camelliae* of Camellia, isolated from fermented tea (*Camellia sinensis*) leaves.

See Table 116 for most of the biochemical and physiological characteristics. The other traits are as follows: The species contains DMK-7 (94.7%), DMK-8 (4.2%) and DMK-9 (1.1%) demethylmenaquinones. $C_{16:1}$ (30.5%) and $C_{18:1}$ (20.9%) are predominant straight-chain fatty acids. The species grows in pH 5–9.0, at 15–45 °C and in 2–6% NaCl. Riboflavin, niacin and calcium-pantothenate are required for growth. See generic description for further features.

Isolated from fermented tea (*Camellia sinensis*) leaves.

DNA G+C content (mol%): 37.8.

Type strain: FP15-1, KCTC 13133, NBRC 101868, NRIC 0105, TISTR 932, PCU 277.

GenBank accession number (16S rRNA gene): EF154454.

7. **Enterococcus canintestini** Naser, Vancanneyt, De Graef, Devriese, Snauwaert, Lefebvre, Hoste, Švec, Decostere, Haesebrouck and Swings 2005c, 2181[VP]

can.in.tes′ti.ni. L. gen. n. *canis* of a dog; L. neut. n. *intestinum* gut; N.L. gen. n. *canintestini* of the gut of dog.

See Table 116 for most of the biochemical and physiological characteristics. The other traits are as follows: Growth at 42 °C is positive. Tetrazolium reduction on Slanetz–Bartley agar is positive. Positive for salicin and negative for methyl β-D-xylopyranoside acidification.

Isolated from feces of healthy dogs.

DNA G+C content (mol%): 36.0–37.0 (HPLC)

Type strain: CCM 7285, LMG 13590

GenBank accession number (16S rRNA gene): AJ888906

8. **Enterococcus canis** De Graef, Devriese, Vancanneyt, Baele, Collins, Lefebvre, Swings and Haesebrouck 2003, 1072[VP]

ca′nis. L. gen. n. *canis* of a dog.

See Table 116 for most of the biochemical and physiological characteristics. The other traits are as follows: α-hemolytic. Growth at 42 °C is positive. Pinpoint-sized colonies are formed on Slanetz–Bartley agar after 48 h without tetrazolium reduction. Urease may be produced.

Isolated from dog feces.

See generic description for further features.

DNA G+C content (mol%): 41.7–43.0 (HPLC).

Type strain: CCUG 46666, LMG 12316, CCM 7125, NBRC 100695.

GenBank accession number (16S rRNA gene): X76177.

9. **Enterococcus casseliflavus** (*ex* Vaughan, Riggsby and Mundt 1979) Collins, Jones, Farrow, Kilpper-Bälz and Schleifer 1984, 221[VP] (*Streptococcus faecium* subsp. *casseliflavus* Mundt and Graham 1968, 2007; *Streptococcus casseliflavus* Vaughan, Riggsby and Mundt 1979, 212)

cass.el.i.fla′vus. N.L. n. *casseli* of Cassel (Cassel yellow); L. adj. *flavus* yellow; *casseliflavus* yellow-colored.

See Table 116 for most of the biochemical and physiological characteristics. The other traits are as follows: α-hemolytic on horse blood. Growth at pH 9.6 is positive. Strains survive heating at 60 °C for 30 min. Tyrosine decarboxylation-negative. The species contains menaquinones; MK-7 and MK-8 are the major isoprenologs. Hexadecanoic and octadecenoic acids are the major nonhydroxylated long-chain fatty acids.

Originally described as D-raffinose acidification-negative but stated to be positive by Devriese et al. (1992a) and Teixeira and Facklam (2003). The species is generally considered to be motile and pigmented although nonmotile and nonpigmented strains have been reported from human clinical material (Clark et al., 1998; Vincent et al., 1991).

Isolated from human and veterinary clinical materials, from food, and the environment. It is considered to be typically plant-associated.

See generic description for further features.

DNA G+C content (mol%): 40.5–44.9 (T_m).

Type strain: ATCC 25788, CCUG 18657, CIP 103018, CCM 2478, DSM 20680, JCM 8723, LMG 10745, NBRC 100478, NCIMB 11449, NCTC 12361, NRRL B-3502, MUTK 20.

GenBank accession number (16S rRNA gene): AJ301826, Y18161.

10. **Enterococcus cecorum** (Devriese, Dutta, Farrow, Van de Kerckhove and Phillips 1983) Williams, Farrow and Collins 1989a, 495[VP] (Effective publication: Williams, Farrow and Collins 1989b, 188.) (*Streptococcus cecorum* Devriese, Dutta, Farrow, Van de Kerckhove and Phillips 1983, 774)

ce.co′rum. L. n. *cecum* sacculated diverticulum of the large intestine (cecum); N.L. gen. pl. n. *cecorum* of ceca.

See Table 116 for most of the biochemical and physiological characteristics. Other traits are as follows: hemolytic on sheep-blood agar. Weak growth at pH 9.6. Strains do not survive 60 °C for 30 min. The species prefers an increased CO_2 content in atmosphere (carboxyphilic). No growth on Slanetz–Bartley medium and kanamycin esculin azide agar. Gelatinase- and tyrosinase-negative. Hexadecanoic and octadecenoic acids are the major nonhydroxylated long-chain fatty acids.

Originally described as alkaline phosphatase- and β-glucuronidase production-positive and mannitol and sorbitol acidification-negative, but strains revealing opposite results were described by Devriese et al. (1992b).

Isolated from animals, rarely from human clinical material and from water.

See generic description for further features.

DNA G+C content (mol%): 37.0–38.0 (T_m).

Type strain: A60, ATCC 43198, CCM 3659, CCUG 27299, CIP 103676, DSM 20682, JCM 8724, LMG 12902, NBRC 100674, NCIMB 702674, NCTC 12421.

GenBank accession number (16S rRNA gene): AF061009, AJ301827, X54290.

11. **Enterococcus columbae** Devriese, Ceyssens, Rodrigues and Collins 1993a, 188[VP] (Effective publication: Devriese, Ceyssens, Rodrigues and Collins 1990, 251.)

co.lumb′ae. L. n. *columba* pigeon; L. gen. n. *columbae* of a pigeon.

See Table 116 for most of the biochemical and physiological characteristics. The other traits are as follows: The species prefers an increased CO_2 content in atmosphere (carboxyphilic). No growth on Slanetz–Bartley medium. Grows on bile esculin agar in 3–10% CO_2 in air but not in the normal atmosphere.

Motility and growth in 6.5% NaCl are stated to be negative and alkaline phosphatase and leucine arylamidase to be positive in the original species description, but later tests revealed that the species is variably motile, grows in 6.5% NaCl and is mostly negative for alkaline phosphatase and leucine arylamidase production (Baele et al., 2002).

Dominant bacterium in the small intestines of pigeons; rarely isolated from water.

See generic description for further features.

DNA G+C content (mol%): 38.2 (T_m).

Type strain: STR 345, ATCC 51263, CCM 4376, DSM 7374, LMG 11740, NCIMB 13013.

GenBank accession number (16S rRNA gene): AF061006, AJ301828, X56422, Y18275.

12. **Enterococcus devriesei** Švec, Vancanneyt, Koort, Naser, Hoste, Vihavainen, Vandamme, Swings and Björkroth 2005b, 2482[VP]

de′vrie.se.i. N.L. gen. n. *devriesei* of Devriese, in honor of the Belgian microbiologist Luc A. Devriese.

See Table 116 for most of the biochemical and physiological characteristics. The other traits are as follows: α-hemolysis on bovine blood agar. Positive growth on azide-containing selective media and bile-esculin agar. Slow growth at 4 °C. Positive salicin and negative methyl β-xyloside acidification.

Isolated from bovine materials, vacuum-packaged charcoal-broiled river lampreys and the air of a poultry slaughter by-product processing plant

DNA G+C content (mol%): 40.0 (HPLC).

Type strain: CCM 7299, LMG 14595.

GenBank accession number (16S rRNA gene): AJ891167.

13. **Enterococcus dispar** Collins, Rodrigues, Piggot and Facklam 1991b, 456[VP] (Effective publication: Collins, Rodrigues, Piggot and Facklam 1991a, 97.)

dis′par. L. adj. *dispar* dissimilar, different.

See Table 116 for most of the biochemical and physiological characteristics. The other traits are as follows: Strains survive heating at 60 °C for 30 min.

Originally described as melibiose acidification-positive but stated as negative according to Devriese et al. (1993b).

Isolated from human clinical material and from dog feces.

See generic description for further features.

DNA G+C content (mol%): 39.0 (T_m).

Type strain: E18–1, ATCC 51266, CCM 4282, CCUG 33309, CIP 103646, DSM 6630, HAMBI 2231, LMG 13521, NBRC 100678, NCFB 2821, NCIMB 13000.

GenBank accession number (16S rRNA gene): AF061007, AJ301829, Y18358.

14. **Enterococcus durans** (*ex* Sherman and Wing 1937) Collins, Jones, Farrow, Kilpper-Bälz and Schleifer 1984, 222^VP (*"Streptococcus durans"* Sherman and Wing 1937)

du'rans. L. part. adj. *durans* hardening, resisting.

See Table 116 for most of the biochemical and physiological characteristics. The other traits are as follows: Strains may be α-hemolytic and rarely β-hemolytic. Respiratory quinones absent. Hexadecanoic and octadecenoic acids are the major nonhydroxylated long-chain fatty acids; substantial amounts of *cis*-11,12-methylenoctadecanoic are also present (approx. 10%).

Originally described as negative for melibiose and sucrose acidification, but strains positive for sucrose as well as melibiose acidification were described by Devriese et al. (2002). Melibiose-positive strains were isolated from human clinical material by Facklam and Collins (1989).

Isolated from human and veterinary clinical materials, from food and the environment.

See generic description for further features.

DNA G+C content (mol%): 38.0–40.0 (T_m).

Type strain: 98D, ATCC 19432, CCM 5612, CCUG 7972, CIP 55.125, DSM 20633, JCM 8725, LMG 10746, NBRC 100479, NCIMB 700596, NCTC 8307.

GenBank accession number (16S rRNA gene): AJ276354.

15. **Enterococcus faecium** (Orla-Jensen 1919) Schleifer and Kilpper-Bälz 1984, 33^VP (*Streptococcus faecium* Orla-Jensen 1919, 139)

fae'ci.um. L. n. *faex* dregs; L. gen. pl. n. *faecium* of the dregs, of feces.

See Table 116 for most of the biochemical and physiological characteristics. The other traits are as follows: Some strains may be α-hemolytic. Growth at pH 9.6 is positive. Survives heating at 60 °C for 30 min. Negative for citrate, malate and serine utilization and gelatin hydrolysis. Reduction of tellurite and tetrazolium-negative. The major fatty acids are hexadecanoic, octadecenoic and *cis*-11,12-methylenoctadecanoic. Cells do not contain menaquinones. The cell-wall peptidoglycan is Lys–D–Asp type.

Originally described as negative for D-xylose acidification, but most of the canine and bovine strains are positive (Devriese et al., 1987). Originally described as raffinose and sorbitol acidification-negative, but strains from poultry are usually raffinose-positive and strains from dogs as well as rare strains from humans are sorbitol-positive (Devriese et al., 1992b; Devriese et al., 1995).

Isolated from human and veterinary clinical materials, from food, and the environment.

See generic description for further features.

DNA G+C content (mol%): 37.0–40.0 (T_m).

Type strain: ATCC 19434, CCM 7167, CCUG 542, CIP 103014, CFBP 4248, DSM 20477, LMG 11423, HAMBI 1710, JCM 5804, JCM 8727, NBRC 100485, NCIMB 11508, NCTC 7171.

GenBank accession number (16S rRNA gene): AB012213, AJ276355, AJ301830, Y18294.

16. **Enterococcus gallinarum** (Bridge and Sneath 1982) Collins, Jones, Farrow, Kilpper-Bälz and Schleifer 1984, 222^VP (*Streptococcus gallinarum* Bridge and Sneath 1982, 414.)

gall.in.ar'um. L. fem. gen. pl. n. *gallinarum* of hens.

See Table 116 for most of the biochemical and physiological characteristics. The other traits are as follows: β-hemolytic on horse-blood agar. Most strains survive heating at 60 °C for 15 min but not for 30 min. Gelatinase production-negative. Low levels of menaquinones produced, with MK-8 predominant. Hexadecanoic and octadecenoic acids are the major nonhydroxylated long-chain fatty acids; *cis*-11,12-methylenoctadecanoic is present in small amounts.

Originally described as negative for glycerol acidification, but stated to be positive by Devriese and Pot (1995) and variable by Devriese et al. (1983) and Collins et al. (1986). Originally described as positive for sorbitol acidification, but found to be negative by Devriese et al. (1992a) and Teixeira and Facklam (2003). Devriese and Pot (1995) concluded that this characteristic may be variable. Generally considered to be motile although it was originally described as nonmotile and nonmotile strains were isolated from human clinical material (Clark et al., 1998; Patel et al., 1998).

Isolated from human and veterinary clinical materials, from food and the environment.

See generic description for further features.

DNA G+C content (mol%): 39.0–40.0 (T_m).

Type strain: F87/276, ATCC 49573, CCM 4054, CCUG 18658, CIP 103013, DSM 20628, HAMBI 1717, JCM 8728, LMG 13129, NBRC 100675 = NCIMB 702313, NCTC 1142, NCTC 11428, NCTC 12359, PB218.

GenBank accession number (16S rRNA gene): AJ301833.

17. **Enterococcus gilvus** Tyrrell, Turnbull, Teixeira, Lefebvre, Carvalho, Facklam and Lovgren 2002b, 1075^VP (Effective publication: Tyrrell, Turnbull, Teixeira, Lefebvre, Carvalho, Facklam and Lovgren 2002a, 1143.)

gil.vus. L. adj. *gilvus* pale yellow, referring to the pale yellow pigmentation of the bacterium.

See Table 116 for most of the biochemical and physiological characteristics. The other traits are as follows: Tetrazolium reduction is positive. Black colonies are not produced on tellurite-containing media.

Isolated from human clinical material.

See generic description for further features.

DNA G+C content (mol%): not determined.

Type strain: PQ1, ATCC BAA-350, CCM 7168, CCUG 45553, DSM 15689, LMG 21841, NBRC 100696.

GenBank accession number (16S rRNA gene): AY033814.

18. **Enterococcus haemoperoxidus** Švec, Devriese, Sedláček, Baele, Vancanneyt, Haesebrouck, Swings and Doškař 2001, 1571^VP

hae.mo.per.o'xi.dus. Gr. n. *haema* blood; Gr. pref. *per* intensification; Gr. adj. *oxys* sour; N.L. adj. *haemoperoxidus* blood peroxide, derived from the ability of the species to decompose hydrogen peroxide into oxygen and water when cultivated on blood-agar media.

See Table 116 for most of the biochemical and physiological characteristics. The other traits are as follows: Growth is strongly inhibited at 42 °C. Positive catalase reaction when cultivated on blood-agar but negative when grown on non-blood-containing media.

Isolated from surface waters.

See generic description for further features.

DNA G+C content (mol%): 35.3–35.5 (HPLC).

Type strain: 440, ATCC BAA-382, CCM 4851, CIP 107129, LMG 19487, NBRC 100709.

GenBank accession number (16S rRNA gene): AF286832.

19. **Enterococcus hermanniensis** Koort, Coenye, Vandamme, Sukura and Björkroth 2004, 1826[VP]

her.man.ni.en′sis. N.L. adj. *hermanniensis* pertaining to Hermanni, a locality in Helsinki, Finland.

See Table 116 for most of the biochemical and physiological characteristics. The other traits are as follows: α-hemolytic on bovine blood agar. Slow growth as maroon colonies on azide-containing media.

Isolated from broiler meat and canine tonsils.

See generic description for further features.

DNA G+C content (mol%): 36.6–37.1 (HPLC).

Type strain: CCUG 48100, CCM 7222, LMG 12317.

GenBank accession number (16S rRNA gene): AY396047.

20. **Enterococcus hirae** Farrow and Collins 1985, 74[VP]

hir.ae. L. gen. sing. n. *hirae* of the intestine or gut.

See Table 116 for most of the biochemical and physiological characteristics. The other traits are as follows: Growth at pH 9.6 is positive. Hexadecanoic, octadecenoic, and *cis*-11,12-methylenoctadecanoic acids are the major nonhydroxylated long-chain fatty acids. Menaquinones are absent.

Originally described as positive for melibiose and sucrose acidification, but negative strains were described by Facklam and Collins (1989) and Devriese et al. (2002).

Isolated from human and veterinary clinical materials, from food and the environment.

See generic description for further features.

DNA G+C content (mol%): 37.0–38.0 (T_m).

Type strain: E.E. Snell strain R, ATCC 8043, ATCC 9790, CCM 2423, CCUG 1332, CCUG 18659, CCUG 19917, CIP 53.48, CFBP 4250, DSM 20160, HAMBI 644, HAMBI 1709, NBRC 3181, JCM 8729, LMG 6399, NCCB 46070, NCCB 58005, NCIMB 6459, NCTC 12367.

GenBank accession number (16S rRNA gene): AF061011, AJ276356, AJ301834, Y17302.

21. **Enterococcus italicus** Fortina, Ricci, Mora and Manachini 2004, 1720[VP]

i.ta′li.cus. L. adj. *italicus* from Italy, where the bacterium was first isolated.

See Table 116 for most of the biochemical and physiological characteristics. The other traits are as follows: Positive α-hemolysis on blood-containing agar media. Growth is positive at 5% NaCl, pH 9.6 and 42 °C, variable in 6.0% NaCl but negative in 6.5% NaCl. Does not grow on kanamycin esculin azide agar.

Isolated from dairy products.

See generic description for further features.

DNA G+C content (mol%): 39.9–41.1 (T_m).

Type strain: DSM 15952, LMG 22039, TP1.5.

GenBank accession number (16S rRNA gene): AJ582753.

22. **Enterococcus malodoratus** (*ex* Pette 1955) Collins, Jones, Farrow, Kilpper-Bälz and Schleifer 1984, 222[VP] ("*Streptococcus faecalis* subsp. *malodoratus*" Pette 1955)

mal.od.or.a′tus. L. n. *malus* ill; N.L. part. adj. *odoratus* odoous; N.L. adj. *malodoratus* ill-smelling.

See Table 116 for most of the biochemical and physiological characteristics. The other traits are as follows: Some strains produce slime. Survival at 60 °C for 30 min is mostly negative. H_2S is produced. Respiratory quinones absent. Hexadecanoic and tetradecanoic acids are the major non-hydroxylated long-chain fatty acids.

Isolated from dairy products.

See generic description for further features.

DNA G+C content (mol%): 40.0–41.0 (T_m).

Type strain: ATCC 43197, CCM 4056, CCUG 30572, CIP 103012, DSM 20681, HAMBI 1569, JCM 8730, LMG 10747, NBRC 100489, NCIMB 700846, NCTC 12365.

GenBank accession number (16S rRNA gene): AF061012, AJ301835, Y18339.

23. **Enterococcus moraviensis** Švec, Devriese, Sedláček, Baele, Vancanneyt, Haesebrouck, Swings and Doškař 2001, 1572[VP]

mo.ra.vi.en′sis. N.L. adj. *moraviensis* pertaining to Moravia, the region in the Czech Republic.

See Table 116 for most of the biochemical and physiological characteristics. The other traits are as follows: Growth is strongly inhibited at 42 °C.

Isolated from surface waters.

See generic description for further features.

DNA G+C content (mol%): 35.6–36.3 (HPLC).

Type strain: 330, ATCC BAA-383, CCM 4856, CIP 107130, LMG 19486, NBRC 100710.

GenBank accession number (16S rRNA gene): AF286831.

24. **Enterococcus mundtii** Collins, Farrow and Jones 1986, 10[VP]

mund′ti.i. N.L. gen. n. *mundtii* of Mundt; named after the late J.O. Mundt, an American microbiologist.

See Table 116 for most of the biochemical and physiological characteristics. The other traits are as follows: Growth at pH 9.6 is positive. The long-chain fatty acids are of the straight-chain saturated, monounsaturated and cyclopropane ring types. The major fatty acids are hexadecanoic, octadecenoic, and *cis*-11,12-methylenoctadecanoic acids.

The species is typically associated with plants but it was rarely isolated from human clinical material and from animals.

See generic description for further features.

DNA G+C content (mol%): 38.0–39.0 (T_m).

Type strain: ATCC 43186, CCM 4058, CCUG 18656, CFBP 4251, CIP 103010, DSM 4838, HAMBI 1570, JCM 8731, LMG 10748, MUTK 559, NBRC 100490, NCIMB 702375, NCTC 12363.

GenBank accession number (16S rRNA gene): AF061013, AJ301836, Y18340.

25. **Enterococcus pallens** Tyrrell, Turnbull, Teixeira, Lefebvre, Carvalho, Facklam and Lovgren 2002b, 1075[VP] (Effective publication: Tyrrell, Turnbull, Teixeira, Lefebvre, Carvalho, Facklam and Lovgren 2002a, 1144.)

pall'ens. L. adj. *pallens* yellowish, referring to the yellow pigmentation of the bacterium.

See Table 116 for most of the biochemical and physiological characteristics. Originally described as positive for pyruvate utilization and arabinose acidification but stated as negative for these tests by Teixeira and Facklam (2003).

Isolated from human clinical material.

See generic description for further features.

DNA G+C content (mol%): not determined.

Type strain: PQ2, ATCC BAA-351, CCM 7169, DSM 15690, LMG 21842, CCUG 45554.

GenBank accession number (16S rRNA gene): AY033815.

26. **Enterococcus phoeniculicola** Law-Brown and Meyers 2003, 684[VP]

phoe.ni.cu.li'co.la. N.L. n. *Phoeniculus* the genus of the woodhoopoe; L. suff. *-cola* inhabitant; N.L. masc. n. *phoeniculicola* growing in *Phoeniculus*.

See Table 116 for most of the biochemical and physiological characteristics. The other traits are as follows: Nonhemolytic. Does not grow in 6% NaCl and in the presence 40% bile.

Described as delayed positive (48 h) for mannitol, sorbitol and L-sorbose acidification and negative for D-raffinose acidification but opposite results of these tests were revealed in the type strain by Carvalho et al. (2004).

Isolated from the uropygial gland of a Red-billed Woodhoopoe (*Phoeniculus purpureus*).

See generic description for further features.

DNA G+C content (mol%): not determined.

Type strain: JLB-1, ATCC BAA-412, CCM 7236, CCUG 48923, DSM 14726, LMG 22471, KCTC 3818, NBRC 100711.

GenBank accession number (16S rRNA gene): AY028437.

27. **Enterococcus pseudoavium** Collins, Facklam, Farrow and Williamson 1989c, 371[VP] (Effective publication: Collins, Facklam, Farrow and Williamson 1989b, 287.)

pseu.do. av.i'um. Gr. adj. *pseudes* false; L. n. *avis* bird; L. gen. plur. *avium* of birds; N.L. neut. pl. n. *pseudoavium* false [*Enterococcus*] *avium*, owing to its similarity to this species.

See Table 116 for most of the biochemical and physiological characteristics. The other traits are as follows: α-hemolytic. The long-chain cellular fatty acids are of the straight-chain saturated and mono-unsaturated types. The major types are tetradecanoic and hexadecanoic acids.

Habitat unknown. The type strain was isolated from bovine mastitis, however no other strains have been found despite intensive investigation of this habitat by several groups.

See generic description for further features.

DNA G+C content (mol%): 40.0 (T_m).

Type strain: ATCC 49372, CCUG 33310, CIP 103647, DSM 5632, CCM 4215, JCM 8732, LMG 11426, NCIMB 13084.

GenBank accession number (16S rRNA gene): AF061002, AJ301837, Y18356.

28. **Enterococcus raffinosus** Collins, Facklam, Farrow and Williamson 1989c, 371[VP] (Effective publication: Collins, Facklam, Farrow and Williamson 1989b, 286.)

raf'fi.no.sus. N.L. adj. *raffinosus* of raffinose, referring to ability to metabolize raffinose.

See Table 116 for most of the biochemical and physiological characteristics. The other traits are as follows: The long-chain cellular fatty acids are of the straight-chain saturated, monounsaturated and cyclopropane ring types. The major types are tetradecanoic and hexadecanoic acids.

Isolated from human clinical material and rarely from animal sources or from environmental samples.

See generic description for further features.

DNA G+C content (mol%): 39.0–40.0 (T_m).

Type strain: 1789/79, ATCC 49427, CCM 4216, CCUG 29292, CIP 103329, DSM 5633, JCM 8733, LMG 12888, NBRC 100492, NCTC 12192.

GenBank accession number (16S rRNA gene): Y18296.

29. **Enterococcus ratti** Teixeira, Carvalho, Espinola, Steigerwalt, Douglas, Brenner and Facklam 2001, 1742[VP]

rat'ti. N.L. gen. masc. n. *ratti* of the rat.

See Table 116 for most of the biochemical and physiological characteristics. The other traits are as follows: Some strains may be α-hemolytic.

Isolated from intestines and feces of infant rats with diarrhea.

See generic description for further features.

DNA G+C content (mol%): not determined.

Type strain: DS 2705-87, ATCC 700914, CCUG 43228, CCM 7235, CIP 107173, DSM 15687, LMG 21828, NBRC 100698, NCIMB 13635.

GenBank accession number (16S rRNA gene): AF539705.

30. **Enterococcus saccharolyticus** (Farrow, Kruze, Phillips, Bramley and Collins 1984) Rodrigues and Collins 1991, 178[VP] (Effective publication: Rodrigues and Collins 1990, 233.) (*Streptococcus saccharolyticus* Farrow, Kruze, Phillips, Bramley and Collins 1984, 480)

sac.cha.ro.ly'ti.cus. Gr. n. *sakchar* sugar; Gr. adj. *lyticus* able to loosen; M.L. adj. *saccharolyticus* sugar-digesting.

See Table 116 for most of the biochemical and physiological characteristics. The other traits are as follows: Nonhemolytic. Strains do not survive at 60 °C for 30 min. Gelatin hydrolysis is negative.

Isolated from cow feces and from straw bedding.

See generic description for further features.

DNA G+C content (mol%): 37.6–38.3 (T_m).

Type strain:, ATCC 43076, CCM 4377, CCUG 27643, CCUG 33311, CIP 103246, DSM 20726, HF 62, JCM 8734, LMG 11427, NBRC 100493, NCIMB 702594.

GenBank accession number (16S rRNA gene): AF061004, AJ 301839, X55767, Y18357.

31. **Enterococcus silesiacus** Švec, Vancanneyt, Sedláček, Naser, Snauwaert, Lefebvre, Hoste and Swings 2006, 580[VP]

si.le'si.a.cus. N.L. masc. adj. *silesiacus* pertaining to Silesia, the region in the Czech Republic from which the type strain originates.

See Table 116 for most of the biochemical and physiological characteristics. The other traits are as follows: The species grows on kanamycin esculin azide agar and bile esculin agar with positive esculin reaction. Weak growth on Slanetz and Bartley agar in small dark-red colonies. Positive catalase reaction when cultivated on blood-agar but negative when grown on non-blood-containing media. Salicin positive and methyl β-xylopyranoside negative acidification.

> Isolated from surface waters.
>
> *DNA G+C content (mol%): 35.6–36.7 (HPLC).*
>
> *Type strain: W442, CCM 7319, LMG 23085.*
>
> *GenBank accession number (16S rRNA gene): AM039966.*

32. **Enterococcus sulfureus** Martinez-Murcia and Collins 1991b, 580[VP] (Effective publication: Martinez-Murcia and Collins 1991a, 72.)

sulfu're.us. L. adj. *sulfureus* of sulfur.

> See Table 116 for most of the biochemical and physiological characteristics. The other traits are as follows: Growth is positive at 40 °C, but not at 45 °C.
>
> Isolated from plants.
>
> See generic description for further features.
>
> *DNA G+C content (mol%): 38.0 (T_m).*
>
> *Type strain: ATCC 49903, CCM 4283, CCUG 30571, CCUG 33313, CIP 104373, DSM 6905, HAMBI 2232, LMG 13084, MUTK 31, NBRC 100680, NCIMB 13117.*
>
> *GenBank accession number (16S rRNA gene): AF061001, AJ301841, X55133, Y18341.*

33. **Enterococcus termitis** Švec, Vancanneyt, Sedláček, Naser, Snauwaert, Lefebvre, Hoste and Swings 2006, 580[VP]

ter.mi'tis. L. n. *termes -itis* a worm that eats wood, a woodworm, and in zoology the name of a scientific genus; L. gen. n. *termitis* of a termite.

> See Table 116 for most of the biochemical and physiological characteristics. The other traits are as follows: The species grows on kanamycin esculin azide agar and bile esculin

agar with positive esculin reaction. Weak growth on Slanetz and Bartley agar in small dark-red colonies. Salicin positive and methyl β-xylopyranoside negative acidification.

> Isolated from the gut of a termite.
>
> *DNA G+C content (mol%): 37.1 (HPLC).*
>
> *Type strain: CCM 7300, LMG 8895.*
>
> *GenBank accession number (16S rRNA gene): AM039968.*

34. **Enterococcus villorum** Vancanneyt, Snauwaert, Cleenwerck, Baele, Descheemaeker, Goossens, Pot, Vandamme, Swings, Haesebrouck and Devriese 2001, 398[VP] (*Enterococcus porcinus* Teixeira, Carvalho, Espinola, Steigerwalt, Douglas, Brenner and Facklam 2001, 1742).

vil.lo'rum. L.n. *villus* rough hair, anatomical term for flocculate structures in the small intestine; L. gen. pl. *villorum.*

> See Table 116 for most of the biochemical and physiological characteristics. The other traits are as follows: α-hemolytic. Grows on azide-containing media but slower than related enterococci. Colonies on Slanetz–Bartley agar are pink.
>
> Originally described as D-raffinose, sucrose and α-methyl-D-glucoside acidification-negative and melibiose acidification-positive but strains revealing opposite results were described by Devriese et al. (2002).
>
> Isolated from pigs and from birds.
>
> See generic description for further features.
>
> *DNA G+C content (mol%): 35.1–35.3 (HPLC).*
>
> *Type strain: 88-5474, ATCC 700913, CCM 4887, CCUG 45025, DSM 15688, JCM 11557, LMG 12287, NBRC 100699.*
>
> *GenBank accession number (16S rRNA gene): AF335596, AJ271329.*

Genus II. **Melissococcus** Bailey and Collins 1983, 672[VP] (Effective publication: Bailey and Collins 1982b, 216.)

LEON M. T. DICKS AND WILHELM H. HOLZAPFEL

Me.lis'so.coc'cus. Gr. n. *melissa* bee; Gr. n. *coccus* berry; N.L. masc. n. *Melissococcus* coccus of the (honey) bee.

Cells are **lanceolate** cocci and occur **singly, in pairs, or in chains of varying lengths** (Figure 111). **Pleomorphic** and rod-like forms have also been described. Individual cells measure $0.5–0.7 \times 1.0\,\mu m$. Gram-positive, but destain easily. Stain negative with nigrosin. Nonmotile, catalase-negative, asporogenic, and non-acid-fast. **No aerobic growth.** Anaerobic to microaerophilic, with **best growth in the presence of 1–5% (v/v) CO_2.** CO_2 levels higher than 5% inhibit some strains, but inhibition may be overcome by inclusion of self-prepared autolyzed yeast extract.

Cell-wall peptidoglycan contains lysine and is of the Lys–Ala type. Respiratory quinones are absent. The long-chain fatty acids are primarily straight-chain, monounsaturated (cis-vaccenic acid type), and cyclopropane ring acids. There is only one species in the genus.

Type species: **Melissococcus plutonius** corrig. (*ex* White 1912) Bailey and Collins 1983, 672 (Effective publication: Bailey and Collins 1982b, 216.)

Further descriptive information

Based on 16S rRNA sequence analysis, *Melissococcus plutonius* is only remotely related (approx. 87–89% sequence similarity) to

the genus *Streptococcus* (Cai and Collins, 1994). The species is, however, phylogenetically closely related to the genus *Enterococcus* (94–96%, Figure 112) and is considered to be on the periphery of that genus. The close affinity between the two genera is further supported by the fact that they both belong to the Lancefield D antigen group (Glinski, 1972a). Close, albeit more distant, relationships were recorded with *Carnobacterium* species (92%), *Vagococcus* species (93%), and *Tetragenococcus halophilus* (approx. 94%). According to the 16S rRNA phylogenetic analysis in the revised roadmap to Volume 3 (Figure 1 and Figure 3; Ludwig et al., this volume), *Melissococcus* is classified in the new family *Enterococcaceae* along with the genera *Enterococcus*, *Tetragenococcus*, and *Vagococcus*, within the order *Lactobacillales*.

Stringent growth requirements, competition from other bacteria during isolation, and the small number of phenotypic characteristics whereby *Melissococcus plutonius* can be differentiated from other bacterial species, led to the developing of more reliable identification methods. Further differentiation of *Melissococcus plutonius* from *Streptococcus faecalis* and related enterococci is obtained by serological (Bailey and Gibbs, 1962) and chemical (Bailey and Collins, 1982a) methods. Enzyme-linked immunosorbent assay (ELISA) has proven a rapid and

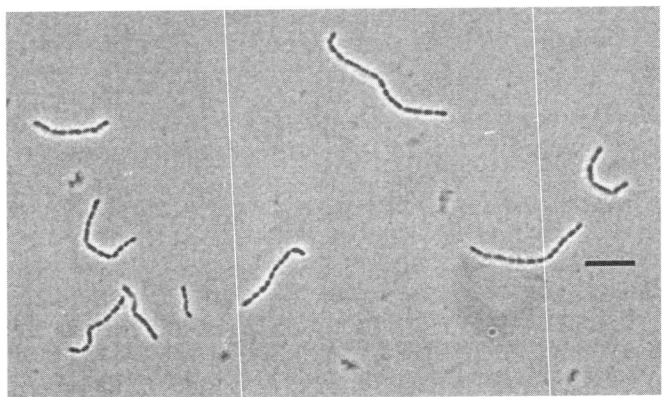

FIGURE 111. Morphology of cells of *Melissococcus plutonius* under phase-contrast microscopy, after anaerobic growth at 30 °C in a special medium (0.25% Peptone, Oxoid L37, 1% glucose, 0.2% soluble starch, 0.25% yeast extract, Oxoid L21, 0.5% Neopeptone, Difco 0119, 0.2% Trypticase BBL 11921, 50 ml 1 M phosphate buffer, pH 7.6, 0.025% Cysteine·HCl·H₂O and 2% agar). Bar = 10 µm.

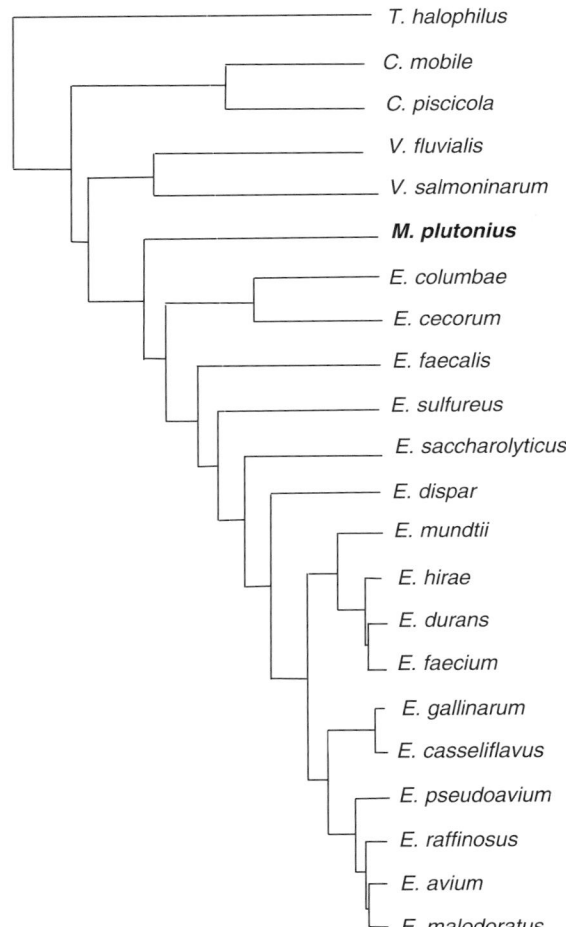

FIGURE 112. Dendrogram showing the phylogenetic relationships based on 16S rRNA gene, of *Melissococcus plutonius*, enterococci, and other members of the *Firmicutes*. Clustering was by the unweighted pair group method. *T = Tetragenococcus, C = Carnobacterium, V = Vagococcus, E = Enterococcus.* (Modified and reprinted with permission from Cai and Collins, *International Journal of Systematic Bacteriology 44:* 365–367, 1994.)

extremely sensitive method to identify strains of *Melissococcus plutonius* (Pinncock and Featherstone, 1984). Serological tests with polyclonal antisera have also been used to some extent (Allen and Ball, 1993).

A rapid PCR detection method, with primers designed from the 16S rRNA sequence of the type strain of *Melissococcus plutonius* (LMG 9003, NCDO 2443), was developed by Govan et al. (1998). The primers selected were 5′ GAAGAGGAGT-TAAAAGGCGC 3′ (primer 1) and 5′ TTATCTCTAAGGCGT-TCAAAGG 3′ (primer 2). An 812 bp fragment was generated by PCR. Detection was highly specific and allowed for the identification of *Melissococcus plutonius* among other bacterial strains present in bee larvae.

A hemi-nested PCR assay designed to detect *Melissococcus plutonius* was developed by Djordjevic et al. (1998). Three oligonucleotide primers were used. The first pair (MP1 and MP2) flanked the 16S rRNA gene from positions 893–1377, yielding a 486 bp fragment. The third primer amplified 25 nucleotides within the latter region (from 1144 to 1168). The resultant 276 bp fragment amplified was species specific with the ability to detect 1–10 cells per ml. The method proved reliable in the detection of *Melissococcus plutonius* in infected brood. The annealing temperatures used in the PCR reactions seem to be critical. PCR fragments of the same length were amplified from *Melissococcus plutonius* and *Enterococcus faecalis* when the annealing temperature was lowered by 5 °C (Djordjevic et al., 1998). However, digestion of the DNA fragments with *Hinf*I, clearly differentiated between DNA from *Melissococcus plutonius* and *Enterococcus faecalis*.

Behr et al. (2000) designed a comprehensive set of rRNA targeted species-specific oligonucleotide hybridization probes for the detection and identification of *Enterococcus* species. The array of probes used was designed from 16S and 23S rRNA sequences according to the multiple probe concept. Probe Mplu464 (5′ GTCACGAGGAAAACAGTT 3′) which spans position 465–483 of the 16S rRNA gene, and probe Enc01cV (5′ AGGTTAAGT-GAACAAGGG 3′), position 1–17 of the 23S rRNA, are specific for *Melissococcus plutonius* (Behr et al., 2000).

It is generally assumed that European foulbrood is caused by infection from *Melissococcus plutonius* via food intake. However, more recent studies (Kanbar et al., 2004) indicated that the mite *Varroa destructor* may also be a vector. The latter authors have identified strains of *Melissococcus plutonius* with strong tyramine production, of which the toxicity has been demonstrated by using the protozoon *Stylonychia lemnae*.

The cultural characteristics and serological relationships of strains of *Melissococcus plutonius* isolated from *A. mellifera* larvae from Brazil, China and the UK are discussed by Allen and Ball (1993). Nigrosin smears of cells clearly show the variation in morphology depicted for *Melissococcus plutonius*.

Enrichment and isolation procedures

Melissococcus plutonius only grows under anaerobic to microaerophilic conditions, with best growth in the presence of 1–5% (v/v) CO_2. Growth is greatly stimulated in the presence of self-prepared autolyzed yeast extract. Strains of this species are usually cultivated on complex growth media only. Two chemically defined growth media were developed by Bailey (1984). They differ slightly in amino acid composition, with the major differences in salts, carbohydrates (starch and glucose), and bases. The carbohydrates and salts, at double strength, are sterilized

Dendrogram labels:
- T. halophilus
- C. mobile
- C. piscicola
- V. fluvialis
- V. salmoninarum
- **M. plutonius**
- E. columbae
- E. cecorum
- E. faecalis
- E. sulfureus
- E. saccharolyticus
- E. dispar
- E. mundtii
- E. hirae
- E. durans
- E. faecium
- E. gallinarum
- E. casseliflavus
- E. pseudoavium
- E. raffinosus
- E. avium
- E. malodoratus

separately at 116 °C for 20 min. The rest of the constituents, also at double strength, are filter-sterilized and then added to the melted agar at a 1:1 ratio. The medium is freshly prepared and the plates immediately used. Incubation is at 35 °C in anaerobic jars in the presence of 5% (v/v) CO_2.

Methionine, glucose, thymine, xanthine, pyridoxal HCl, and a number of vitamins (Bailey, 1984) are essential. Methionine cannot be substituted for by cysteine or cystine. The presence of yeast extract, Difco (1–3 g/l), or peptone, Oxoid L37 (10 g/l), maintains the growth of most strains (Bailey, 1984). Starch is not required by all strains. Glucose can be replaced by sucrose or melezitose for some strains, although glucose is preferred.

Isolation of *Melissococcus plutonius* from infected bee larvae is usually done by crushing the larvae in sterile distilled water, followed by streaking onto the following medium (g/100 ml): yeast extract (1.0), glucose (1.0%), soluble starch (1.0%), KH_2PO_4 (1.36%), and agar (2.0%) in distilled water, adjusted to pH 6.6 with 5 M KOH (Anderson, 1990; Bailey, 1981). The plates are incubated anaerobically at 37 °C for 9 d and then examined for the presence of white colonies.

A variation of the above medium, with filter-sterilized nalidixic acid added to a final concentration of 3 µg/ml, proved very selective for the isolation of *Melissococcus plutonius* from diseased material and honey contaminated with *Paenibacillus larvae* (Alippi, 1991; Hornitzky and Smith, 1998).

Maintenance procedures

A maintenance medium consisting of peptone (Neopeptone, Difco, or Oxoid Peptone L37), cysteine (0.1%), glucose (1%), soluble starch (1%), and KH_2PO_4 (1 mol/l), pH 6.6, was developed by Bailey and Collins (1982a), in which the yeast extract was exchanged for peptone (Bailey, 1984). Although some strains grow in the presence of Na_2HPO_4, it is not preferred. The medium is autoclaved in screw-capped bottles at 116 °C for 20 min, sealed, and used immediately. Cultures are streaked onto the surface of agar slopes (2% w/v) and incubated at 35 °C in the presence of hydrogen and 5% CO_2 (v/v).

Taxonomic comments

Melissococcus plutonius was first isolated from diseased honeybee larvae (*Apis mellifera*) with symptoms of European foulbrood. White (1912, 1920) classified the bacterium as *Bacillus pluton*, based on microscopic examinations. Upon a more detailed description of the organism, it was reclassified as *Streptococcus pluton* (Bailey, 1957). Strains closely related to *Streptococcus pluton* were subsequently also isolated from diseased larvae of *Apis indica* (Diwan et al., 1971), *Apis cerana* (Bailey, 1974), and *Apis laboriosa* (Allen et al., 1990). The species name "*Streptococcus pluton*" was never included in the Approved Lists of Bacterial Names (Skerman et al., 1980). Bailey and Collins (1982b) assigned "*Streptococcus pluton*" to a new genus, *Melissococcus*, based on phenotypic characteristics (Bailey and Collins, 1982a). The genus was subsequently listed in Validation List no. 11 (Bailey and Collins, 1983) and was treated as *Genus Incertae Sedis* in *Bergey's Manual of Systematic Bacteriology* (Hardie, 1986). The specific epithet "*pluton*" (Greek god of the underworld) was corrected to *plutonius*, pertaining to Pluto or the underworld (Trüper and de Clari, 1998).

The close phylogenetic relatedness recorded between the genera *Melissococcus* and *Enterococcus* (Behr et al., 2000; Cai and Collins, 1994; Djordjevic et al., 1998; Lawson et al., 2000) suggests that they should be included in a single genus. According to the International Code of Nomenclature, Bailey and Collins (1982b) consider the genus name *Melissococcus* to have preference over the genus name *Enterococcus* (Schleifer and Kilpper-Bälz, 1984). However, Cai and Collins (1994) argued that a change in genus name would only cause confusion. Still, it should be noted that the genus name *Enterococcus* was originally proposed by Thiercelin and Jouhaud (1903) for Gram-positive diplococci from intestinal origin, thus confirming the priority of the genus name *Enterococcus*. Today, the genus *Enterococcus* is well established, with more than 25 species, and is widely accepted by the scientific community. The fastidious cultural requirements of *Melissococcus plutonius*, its low G+C content (29–30 mol%), and its branching at the periphery of the *Enterococcus* cluster (Figure 112) warrant separate genus status. Confirmation of the latter was obtained from separate 16S rRNA and 23S rRNA sequence analyses (Figure 113).

Bailey (1974) noted that strains he isolated from *Apis cerana* differed distinctly from all known strains of "*Streptococcus pluton*" previously isolated from *Apis mellifera*. Based on the limited phenotypic data available at the time, he did, however, consider the strains to be more closely related to each other than to any other known bacterial species and therefore grouped them into a single species. The question of whether strains of *Streptococcus pluton* isolated from *Apis cerana* could be a variety of *Streptococcus pluton* from *Apis mellifera* was raised, but left unanswered. In a later study, Bailey (1984) noted that a strain isolated from bees in Brazil were slightly different from other strains of *Melissococcus plutonius* (then *Melissococcus pluton*), but only regarding their serological reactions and higher DNA base composition, recorded as 31.4 mol% G+C. The Brazil strains, however, were well separated from the genus *Streptococcus* (Bailey, 1984).

To date, *Melissococcus* is considered a monospecific genus. Some proof as to the genetic homogeneity of the genus came from an elaborate study conducted on strains of *Melissococcus plutonius* isolated from *Apis mellifera* collected from various states of Australia (Djordjevic et al., 1999). Although minor differences were observed in the restriction endonuclease (*Alu*I, *Cfo*I, *Rsa*I, and *Dra*I) profiles for a few strains, neither SDS-PAGE of whole-cell proteins nor immunoblotting revealed strain-specific patterns. This high degree of genetic (and phenotypic) homogeneity among *Melissococcus plutonius* isolates led the authors to hypothesize that the host (*Apis* species) exerts little selective pressure from its immune response to stimulate immune evasion mechanisms in the bacteria. This may well be the case, since *Melissococcus plutonius* only infects the larvae and pupal stages of the host, leaving the insect with little opportunity to develop an immune response (Djordjevic et al., 1999). The seemingly limited selective pressure from the environment may also contribute to the high degree of genetic homogeneity observed among strains of *Melissococcus plutonius*. It is tempting to assume that the homogeneous status of the species is also indicative of a low profile of genetic recombination events among strains. Furthermore, since no plasmids (at least none larger than 15 kb) could be isolated from any of the 49 strains studied (Djordjevic et al., 1999), it can be assumed that the differences recorded in DNA profiles, albeit small, are located on the genome. Alternative plasmid isolation methods may, however, prove this assumption wrong.

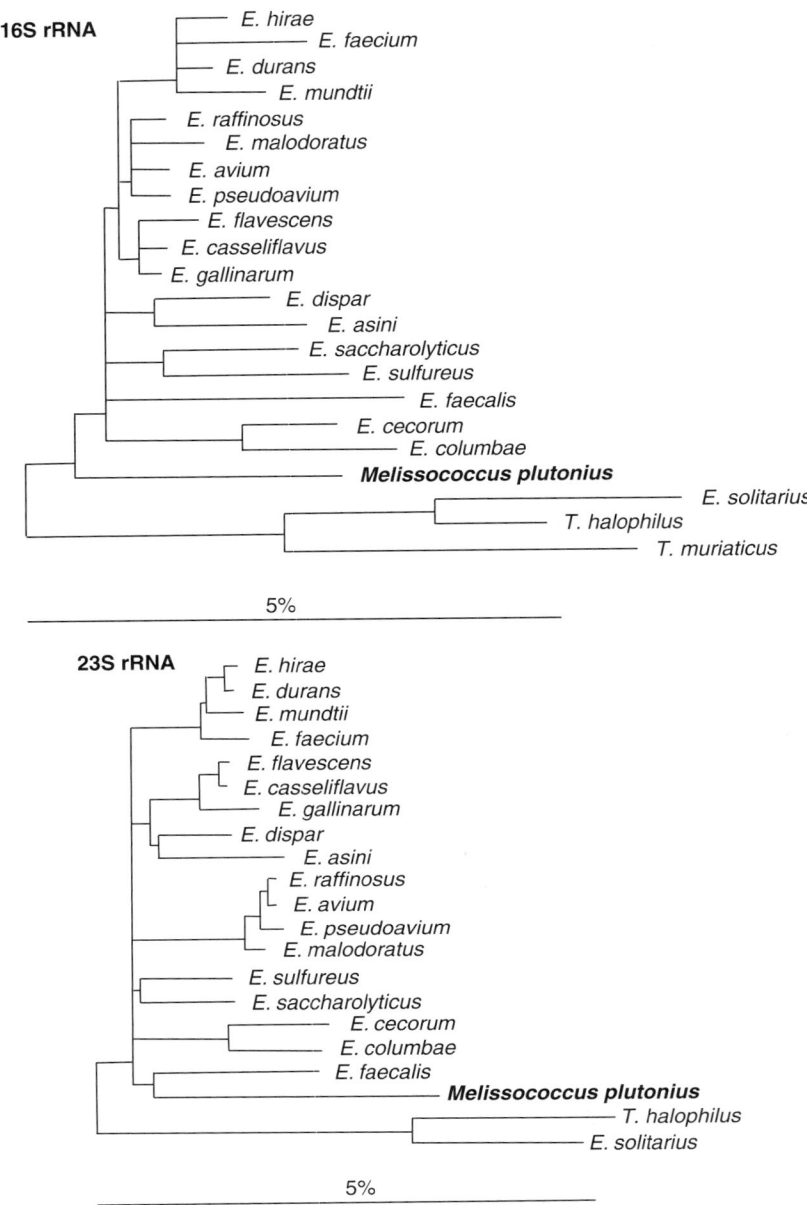

FIGURE 113. Phylogenetic relationships of *Melissococcus plutonius* and the type strains of *Enterococcus* species and *Tetragenococcus* species based on 16S rRNA and 23S rRNA. (Modified and reprinted with permission from Behr et al., *Systematic and Applied Microbiology 23:* 563–572, 2000.)

List of species of the genus *Melissococcus*

1. **Melissococcus plutonius** corrig. (*ex* White 1912) Bailey and Collins 1983, 672^{VP} (Effective publication: Bailey and Collins 1982b, 216.)

plu'ton.ius. L. masc. adj. *plutonius* of or belonging to Pluto, Roman god of the underworld.

The specific epithet "*pluton*" (Pluton, Greek god of the underworld) has been corrected to *plutonius* (Trüper and de Clari, 1998). The following is based on previous descriptions of "*Bacillus pluton*" (White, 1912, 1920), "*Streptococcus pluton*" (Bailey, 1963a, 1963b, 1974; Bailey and Collins, 1982a; Bailey and Gibbs, 1962; Glinski, 1972b) and *Melissococcus pluton* (Bailey and Collins, 1982b).

Colony morphology varies, depending on the source of isolation. Colonies are usually easily visible (up to 1 mm in diameter), dense (opaque), white or granular, and dome-shaped. However, small, flat, transparent, umbonate colonies with clear centers and granular peripheries have also been described. Growth is enhanced if autolyzed yeast extract instead of standard Difco yeast extract is included in the medium, to the extent that small, flat, transparent colonies divert to dome-shaped, opaque, white colonies. Isolation and cultivation is difficult due to the extremely fastidious nature of the organism. Apart from yeast extract, the growth medium usually contains a fermentable sugar, starch, peptone, cysteine or cystine, and potassium. The Na:K ratio required for growth is 1 or less. Glucose and fructose are the only sugars usually fermented. Some strains metabolize sucrose, melezitose, maltose, and salicin. Citrate is not metabolized. Lactic acid is the major organic acid produced, with small amounts of acetic, isobutyric, and succinic acids. Final pH in the growth medium is 5.3. Optimum growth at 35 °C, with some strains growing between 20 and 45 °C. Optimum pH is 6.5–6.6. Possess Lancefield group D antigen. Peptidoglycan type: Lys–Ala. The major fatty acids are hexadecanoic and lactobacillic acids.

Causative agent of European foulbrood of the honeybee. Isolated from larvae of *Apis mellifera* and *Apis cerana* with symptoms of European foulbrood. More recently, tyramine-producing strains of *Melissococcus plutonius* were isolated from the *Varroa destructor* mite (Kanbar et al., 2004) which is an indication that European foulbrood may also be transmitted by mites.

DNA G+C content (mol%): 29–30 (T_m).

Type strain: ATCC 35311, CIP 104052, LMG 20360, NCIMB 702443.

GenBank accession number (16S rRNA gene): AY862507, X75751.

Genus III. **Tetragenococcus** Collins, Williams and Wallbanks 1993, 188[VP] (Effective publication: Collins, Williams and Wallbanks 1990b, 261.)

LEON M.T. DICKS, WILHELM. H. HOLZAPFEL, MASATAKA SATOMI, BON KIMURA AND TATEO FUJII

Tet.ra.geno.coccus. Gr. pref. *tetra* four; G. v. suff. *genes* producing or forming; Gr. n. *kokkos* a grain or berry; N.L. masc. n. *Tetragenococcus* tetrad arrangement of cells.

The cells are **nonmotile, spherical, and occasionally ovoid** and divide in two planes at right angles to **form tetrads**. Cells **may also form pairs or occur singly, especially during early or mid-exponential growth**. Individual cells measure 0.5–1.0 μm. The cells are Gram-positive, asporogenic, and catalase- and oxidase-negative. No cytochromes. **Homofermentative with no CO₂** production from glucose. **Chemo-organotroph. Facultatively anaerobic. Moderately halophilic. Growth in the presence of 4–18% NaCl**, with optimum concentration of 5–10% NaCl; most of the strains will grow at 1% and 25% NaCl. **No growth at pH 4.5** and optimum pH is 7.0–8.0. **No growth at 10 and 45 °C**, optimum temperature of growth is 25–35 °C. Produces ʟ(+)-lactic acid from glucose. Traces of ᴅ(–)-lactic acid may be produced and is species-specific. **Arginine is not hydrolyzed** and nitrate not reduced. *Tetragenococcus halophilus* requires the growth factors nicotinic acid, pantothenic acid, and biotin. Cellular fatty acids are long-chain saturated, monounsaturated, and cyclopropane-ring types, predominantly of the ω7 isomer (*cis*-vaccenic acid) type (Collins et al., 1990b). Peptidoglycan type is Lys–ᴅ-Asp (Satomi et al., 1997). There are four species in the genus: *Tetragenococcus halophilus* (type = ATCC 33315[T]), *Tetragenococcus muriaticus* (type = JCM 10006[T]), *Tetragenococcus koreensis* (type = DSM 16501[T]; KCTC 3924[T]), and *Tetragenococcus solitarius* (type = DSM 5634[T]).

DNA G+C content (mol%): 34–36 mol% (T_m), 36–38.3 (HPLC).

Type species: **Tetragenococcus halophilus** (Mees 1934) Collins, Williams and Wallbanks 1993, 188[VP] (Effective publication: Collins, Williams and Wallbanks 1990b, 261.) (*Pediococcus halophilus* Mees 1934, 96.).

Further descriptive information

The genus contains four species, *Tetragenococcus halophilus* (Collins et al., 1990b), *Tetragenococcus muriaticus* (Röling et al., 1999; Satomi et al., 1997), *Tetragenococcus koreensis* (Lee et al., 2005) and *Tetragenococcus solitarius* (Ennahar and Cai, 2005). The species *Tetragenococcus halophilus* was renamed from *Pediococcus halophilus* (Collins et al., 1990b) and is synonymous to "*Pediococcus soyae*" (Weiss, 1992). Strains from both species are found in the secondary fermentation of soy sauce, preceded by a fungal fermentation, and in pickling brines (Back and Stackebrandt, 1978; Garvie, 1986; Kobayashi et al., 2000a; Satomi et al., 1997). The name *Tetragenocoocus halophila*, as recorded in the Validation List no. 49 (1994) was later rejected and corrected to *Tetragenococcus halophilus* (Euzéby and Kudo, 2001). *Tetragenococcus koreensis* sp. nov. was isolated from kimchi, a traditional Korean dish (Lee et al., 2005). *Tetragenococcus solitarius* is a basonym of *Enterococcus solitarius*, isolated from humans and first described by Collins et al. (1989b). The description of the genus is based on the description of *Pediococcus halophilus* (Garvie, 1986) and *Tetragenococcus* gen. nov. (Collins et al., 1990b).

The cell morphology of *Tetragenococcus halophilus* and *Tetragenococcus muriaticus* is similar to that of *Pediococcus* species (Holzapfel et al., 2006). The cell morphology of *Tetragenococcus koreensis* and *Tetragenococcus solitarius* are similar and characteristic of the genus (Ennahar and Cai, 2005; Lee et al., 2005).

Tetragenococcus halophilus exhibits a low 16S rRNA sequence similarity with the "true" pediococci, i.e., *Pediococcus acidilactici* (89.7%), *Pediococcus pentosaceus* (88.3%), *Pediococcus damnosus* (88.7%), *Pediococcus parvulus* (87.4%), *Pediococcus dextrinicus* (88.6%), and *Pediococcus urinaeequi* (90.4%), indicating that it forms a distinct line of descent (Collins et al., 1990b). The phylogenetic relatedness of *Tetragenococcus halophilus* (*Pediococcus halophilus*) and *Tetragenococcus muriaticus* with other members of the lactic acid bacteria is indicated in Figure 114. This relatedness, particularly towards *Pediococcus* species, is also illustrated in *The Prokaryotes* chapter on the genus *Pediococcus* (Holzapfel et al., 2006). The 16S rDNA sequence of the type strain of *Tetragenococcus halophilus* (IAM 1676[T]) is 93.4% homologous to the sequence of the type strain of *Tetragenococcus muriaticus* (JCM 10006[T]), indicating that the two species belong to the same phylogenetic group (Satomi et al., 1997). The phylogenetic relatedness of *Tetragenococcus halophilus* and *Tetragenococcus muriaticus* with members of *Aerococcus*, *Enterococcus*, *Vagococcus*, *Carnobacterium*, *Pediococcus*, and *Lactococcus* is indicated in the dendrogam in Figure 115. A much closer relationship exists with enterococci and carnobacteria than with pediococci (Figure 114 and Figure 115). A more recent 16S rRNA phylogenetic analysis, presented in the revised roadmap to Volume 3 (Figure 1 and Figure 3;

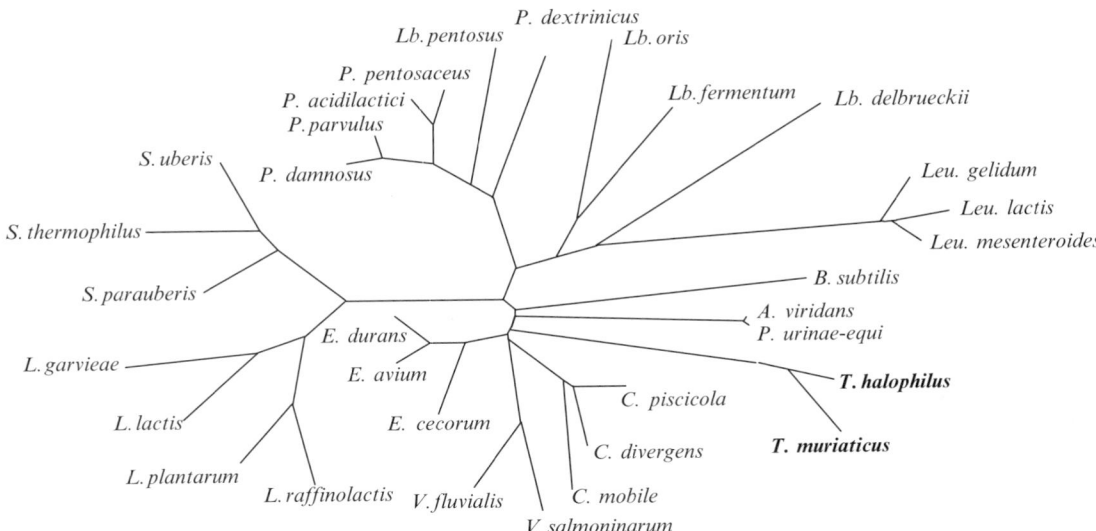

FIGURE 114. Unrooted tree showing the phylogenetic relationships based on 16S rRNA, among *Tetragenococcus halophilus (Pediococcus halophilus)*, *Tetragenococcus muriaticus*, pediococci, aerococci, and other lactic acid bacteria. A total of 1340 nucleotides, positions 107 (G) to 1433 (A), based on the *Escherichia coli* numbering system, was compared. *T = Tetragenococcus, P = Pediococcus, A = Aerococcus, C = Carnobacterium, V = Vagococcus, L = Lactococcus, S = Streptococcus, Lb = Lactobacillus, Leu = Leuconostoc,* and *E = Enterococcus*. (Modified and reprinted with permission from Collins et al., FEMS Microbiology Letters *70:* 255–262, 1990b.)

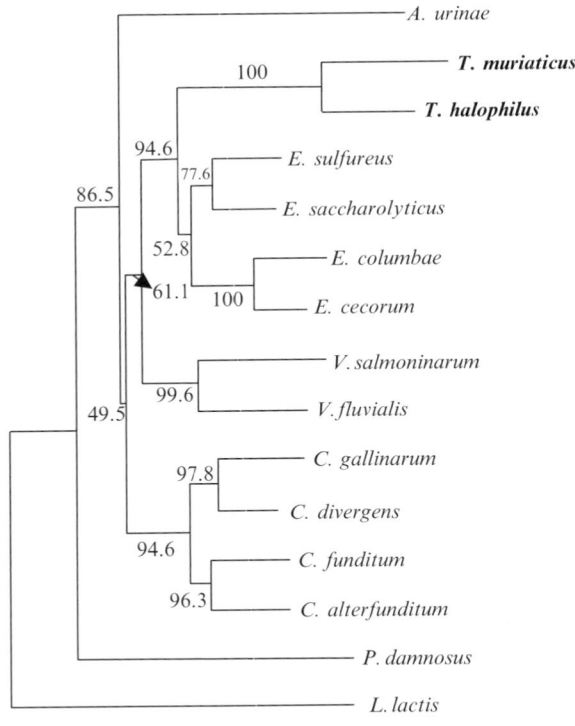

FIGURE 115. Phylogenetic tree based on 16S rDNA sequencing, showing the relationships among *Tetragenococcus halophilus, Tetragenococcus muriaticus,* and other lactic acid bacteria. *T = Tetragenococcus, A = Aerococcus, E = Enterococcus, V = Vagococcus, C = Carnobacterium, P = Pediococcus,* and *L = Lactococcus*. The numbers at the nodes indicate the percentages of occurrence in 1000 bootstrapped trees. Only values greater than 40% are shown. Bar = 0.0099 K_{nuc}. (Modified and reprinted with permission from Satomi et al., International Journal of Systematic Bacteriology *47:* 832–836, 1997.)

Ludwig et al., this volume), classifies *Tetragenococcus* in the new family *Enterococcaceae* along with the genera *Enterococcus, Melissococcus,* and *Vagococcus,* within the order *Lactobacillales.*

Tetragenococcus koreensis shares 98.1% 16S rRNA sequence homology with the type strain of *Tetragenococcus halophilus* (IAM 1676, ATCC 33315, DSM 20339). However, the DNA–DNA relatedness of the two species is only 9.7% (Lee et al., 2005), indicating that they are genetically diverse. *Enterococcus solitarius* described by Collins et al. (1989b) is genetically closer related to the genus *Tetragenoccus* as the species revealed approximately 94% 16S rRNA sequence homology with *Tetragenococcus halophilus* and *Tetragenococcus muriaticus* (Ennahar and Cai, 2005). *Tetragenococcus solitarius* shares low DNA homology with *Tetragenococcus halophilus* (23%) and *Tetragenococcus muriaticus* (54%) and is thus a genetically distinct species (Table 117). On a phenotypic level *Tetragenococcus solitarius* is distinguished from *Enterococcus* spp. by its inability to ferment lactose (Ennahar and Cai, 2005). Strains of *Tetragenococcus halophilus* and *Tetragenococcus muriaticus* share a DNA similarity of less than 50% (Table 117). The phenotypic properties of the four species are listed in Table 118 and Table 119. The genetic relatedness of the four *Tetragenococcus* spp, based on 16S rRNA sequence analyses, is shown in Figure 116.

Glucose is metabolized homofermentatively to L(+)-lactate, probably by the Embden–Meyerhof pathway. At low growth rates, under glucose limiting conditions, and in the presence of 10% NaCl, *Tetragenococcus halophilus* shows a mixed acid fermentation with the production of two formates, one acetate, and one ethanol per glucose (Röling and van Verseveld, 1996, 1997).

A cryptic plasmid (pUCL287), 8.7 kb in size and with a novel theta-type mechanism of replication, was isolated from *Tetragenococcus halophilus* ATCC 33315 (Benachour et al., 1995). The replicon, *Rep*287, has no homology with plasmids pAMβ1, pIP501, and pUCL22, representatives of the most common theta-type replicon groups in Gram-positive bacteria. Plasmid pD1, isolated

TABLE 117. DNA similarity values recorded between *Tetragenococcus* spp.[a,b]

	Percentage relatedness to labeled DNA from:		
Strain	*T. halophilus* IAM 1676[T]	*T. halophilus* IAM 1678	*T. muriaticus* JCM 10006[T]
T. halophilus:			
IAM 1676[T]	100	ND	49
IAM 1678	92	100	21
IAM 1673	79	ND	ND
IAM 1681	100	ND	ND
T. muriaticus:			
JCM 10006[T]	47	36	100
X-2	45	ND	97
T. koreensis			
DSM 16501[T]	9.7	ND	ND
T. solitarius			
DSM 5634[T]	23	ND	54

[a]ND, Not determined.
[b]Data from Ennahar and Cai (2005), Lee et al. (2005), Satomi et al. (1997).

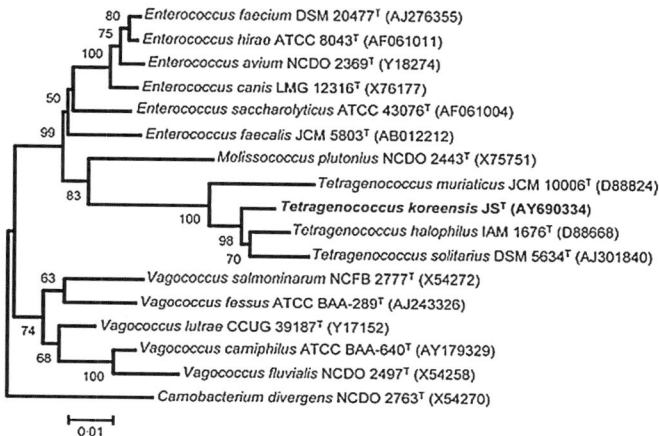

FIGURE 116. Neighbor-joining tree, based on 16S rRNA gene sequences, showing the phylogenetic relatedness of *Tetragenococcus koreensis* and *Tetragenococcus solitarius* with *Tetragenococcus muriaticus*, *Tetragenococcus halophilus* and other related lactic acid bacteria (Taken from Lee et al., 2005).

from *Tetragenococcus halophilus* strain D10, encodes the decarboxylation of aspartate to alanine (Higuchi et al., 1998).

Infection by bacteriophages has been reported (Hanagata et al., 2003). Detailed information on sensitivity to drugs and antibiotics is not available. No strains are pathogenic to plants or animals.

Tetragenococci have been isolated from salted food such as anchovies, soy sauce, pickling brines, and fish sauce, indicating that tetragenococci are widely and commonly distributed in the environment, including in high concentrations of salt with carbohydrate, but with less relationship to a milk and dairy environment. *Tetragenococcus halophilus* is used as a starter in soy sauce.

Enrichment and isolation procedures

Members of the genus *Tetragenococcus* are halophilic and slightly alkaliphilic. For isolation, a medium is employed with high levels of NaCl, with glucose as carbon source, and with the pH adjusted to 7.0–8.0. Growth media generally described for pedio-

cocci, e.g., MRS (De Man et al., 1960) or Selective SL Medium (Rogosa et al., 1951), with the addition of 4–6% NaCl and pH adjusted to 7.0, may be used (Weiss, 1992). GYP medium (glucose yeast extract phosphate) supplemented with 10% NaCl, 1% $MgSO_4$, 0.1% KCl, and 0.3% $CaCO_3$, and adjusted to pH 7.0 with 1.0 N NaOH (Satomi et al., 1997) may also be used. Colonies on these media supplemented with 0.3% $CaCO_3$ and surrounded by a clear zone resulting from lactic acid production, are tentatively identified as *Tetragenococcus*. Growth at 30 °C may occur either aerobically or anaerobically. Cycloheximide may be added to suppress yeast growth. All members of the genus *Tetragenococcus* grow well in GYP broth (Garvie, 1978), supplemented with 5% NaCl (Collins et al., 1990b).

Histamine production is tested by inoculating the strains in modified GYP medium ((Kimura et al., 2001) composed of 0.5% (w/v) Bacto Peptone, 0.2% (w/v) Lab-lemco powder, 1% (w/v) yeast extract, 1% (w/v) glucose, 0.2% (w/v) $NaCOOH_3 \cdot 3H_2O$, 0.05% (w/v) Tween 80, 0.02% (w/v) $MgSO_4 \cdot 7H_2O$, 0.001% (w/v) $MgSO_4 \cdot 3H_2O$, 0.001% (w/v) $FeSO_4 \cdot 7H_2O$, 7% (w/v) NaCl, 1% (w/v) histidine monohydrochloride, and 500 mM 2-morpholinoethansulfonic acid monohydrate (MES), adjusted to pH 6.5 before autoclaving. Incubation at 30 °C should be in an atmosphere of 90% N_2 and 10% CO_2.

Maintenance procedures

For short-term storage, cultures may be kept at 4 °C as stabs in appropriate growth media supplemented with 0.3% $CaCl_2$ as pH neutralizer, with monthly transfers to freshly prepared medium. For long-term storage, exponential-phase cultures may be mixed with an equal volume of 80% (v/v) sterile glycerol and stored in cryotubes at –20 or –80 °C. Lyophilization with sterile skim milk or storage in liquid nitrogen is recommended for longer-term storage.

Differentiation from closely related taxa

The main characteristics distinguishing *Tetragenococcus* species from the "true" pediococci, *Pediococcus acidilactici*, *Pediococcus claussenii*, *Pediococcus damnosus*, *Pediococcus inopinatus*, *Pediococcus parvulus*, and *Pediococcus pentosaceus* are their halophilic, non-aciduric, and facultative anaerobic growth (Holzapfel et al., 2006; Weiss, 1992). They are differentiated from the L(+)-lactic acid producing species *Pediococcus claussenii* and the atypical "pediococci" *Pediococcus dextrinicus* and *Pediococcus urinaeequi* by growth in the presence of 10% NaCl and the fermentation of melezitose, and from the tetrad-forming *Aerococcus viridans* and *Pediococcus urinaeequi* by growth at pH 5.0, no growth at pH 9.0, and being facultatively anaerobic and not microaerophilic (Holzapfel et al., 2006; Weiss, 1992). Resistance to vancomycin (30 μg, disk method), the production of pyrrolidonyl-arylase, and lack of leucine aminopeptidase, and the presence of the fatty acids $C_{16:1\ \omega9c}$ and $C_{18:1\ \omega9c}$ are further characteristics which differentiate *Tetragenococcus* species from *Aerococcus* species (Uchida and Mogi, 1972; Weiss, 1992). According to DNA similarity determination, *Tetragenococcus halophilus* is less than 20% related to the type strain of *Aerococcus viridans* (Bosley et al., 1990).

Taxonomic comments

Digestion of PCR amplified 16S rDNA with *Alu*I, *Mbo*I (Satomi et al., 1997), *Afa*I, and *Mbo*I (Kobayashi et al., 2000a) clearly differentiated strains of *Tetragenococcus halophilus* from *Tetragenococcus muriaticus*. The two species may also be differentiated by

TABLE 118. Phenotypic characteristics of *Tetragenococcus* spp.[a,b]

Characteristic	T. halophilus	T. muriaticus	T. koreensis	T. solitarius
Growth at 15 and 40 °C	+	+	−	+
Growth at 45 °C	−	−	−	−
Growth at pH 4.2	−	−	−	−
Growth at pH 4.5	−	−	−	−
Growth at pH 7.0	+	+	+	+
Growth at pH 8.5	+	+	+	+
Optimal growth pH	7.5–8.0	7.5–8.0	7.0	7.5–8.0
Optimal NaCl concentration	7–10%	7–10%	2–5%	7–10%
Growth in absence of NaCl	+	−	+	+
Acid production from:				
Amygdalin	+	−	−	+
Arabinose	+	−	−	−
Arabitol	+	−	+	−
Arbutin	+	−	+	+
Cellobiose	+	−	−	+
Dextrin	−	−	ND	ND
Dulcitol	−	−	−	+
Fructose	+	+	+	+
Galactose	+	−	+	+
Gluconate	+	−	+	+
Glucose	+	+	+	+
Glycerol	−	−	−	−
Inulin	−	−	−	−
Lactose	−	−	−	−
Maltose	+	−	+	+
Mannitol	−	+	+	+
Maltotriose	D	−	ND	ND
Mannose	+	+	+	+
Mannitol	D	+	+	+
Melezitose	−	−	+	+
Melibiose	−	−	−	−
Raffinose	−	−	−	−
Rhamnose	−	−	−	−
Ribose	+	+	+	−
Sorbitol	−	−	−	−
Starch	−	−	−	
Sucrose	+	−	+	+
Tagatose	−	−	−	+
Trehalose	D	+	+	+
Turanose	+	−	+	+
Xylose	+	−	+	−
DNA G+C content (mol%)	34–36 (T_m)	36.5 (HPLC)	38.8 (HPLC)	38.0 (HPLC)

[a]Symbols: +, positive reaction or fermented; −, negative reaction or not fermented; ND, not determined.
[b]Data from Garvie (1986), Collins et al. (1990b), Röling and van Verseveld (1996), Satomi et al. (1997), Weiss (1992), Kobayashi et al. (2000b), Ennahar and Cai (2005), Lee et al. (2005). Sugar fermentation reactions are as indicated for the type strains.

their ability to ferment the carbohydrates amygdalin, arabinose, arabitol, arbutin, cellobiose, galactose, gluconate, maltose, sucrose, turanose and xylose (Table 118). Generally, the major difference in the sugar fermentation pattern of *Tetragenococcus muriaticus* from *Tetragenococcus halophilus* is the inability of *Tetragenococcus muriaticus* to ferment L-arabinose or sucrose and its ability to ferment D-mannitol (Kobayashi et al., 2000b; Satomi et al., 1997). However, variations of sugar fermentation patterns of *Tetragenococcus halophilus* are significant among strains (Röling and van Verseveld, 1996; Uchida, 1982). It is difficult to distinguish between *Tetragenococcus halophilus* and *Tetragenococcus muriaticus* by using sugar fermentation patterns alone.

Histamine formation is characteristic for most strains of *Tetragenococcus muriaticus* with optimal production at pH 5.8, in the presence of 5–7% (w/v) NaCl, and glucose concentrations exceeding 1% (w/v) (Kimura et al., 2001), though the ability to produce histamine is strain specific (Kobayashi et al., 2000b). Histamine is also produced in the presence of 20% (w/v) NaCl, with maximum accumulation during late stationary growth, independent of NaCl concentrations.

List of species of the genus *Tetragenococcus*

1. **Tetragenococcus halophilus** (Mees 1934) Collins, Williams and Wallbanks 1993, 188[VP] (Effective publication: Collins, Williams and Wallbanks 1990b, 261.) (*Pediococcus halophilus* Mees 1934, 96)

hal.o.phi'lus. Gr. n. hals, *halos* salt; Gr. adj. *philos* loving; N.L. adj. *halophilus* salt-loving.

The description of this species is based on the description of *Pediococcus halophilus* as by Garvie (1986) and Weiss (1992).

TABLE 119. Fatty acid composition (as % of total fatty acids) of *Tetragenococcus* spp.[a,b]

Fatty acid	*T. halophilus* IAM 1676[T]	*T. muriaticus* JCM 10006[T]	*T. koreensis* DSM 16501[T]	*T. solitarius* DSM 5634[T]
Saturated:				
12:0	–	0.9	–	2.7
14:0	8.8	8.0	7.4	18.5
16:0	28.9	21.9	22.4	23.3
17:0 cyclo	1.5	–	–	2.8
18:0	–	1.1	–	2.7
19:0 cyclo ω8*c*	14.6	6.3	26.8	12.0
19:0 cyclo ω10*c*	–	–	–	3.7
Unsaturated:				
16:1ω7*c*	16.1	18.9	9.6	13.4
18:1ω7*c*/9*t*/12*t*	25.4	36.6	33.8	11.9
18:1ω9*c*	2.1	2.0	–	4.0
18:2ω6,9*c*	–	1.3	–	2.8
Branched:				
15:0 iso 2-OH	2.5	2.0	–	2.5
16:0 2-OH	–	1.0	–	–

[a]–, Not detected.
[b]Data from Lee et al. (2005).

The morphology is as described for the genus *Pediococcus* (Garvie, 1986) and for the genus *Tetragenococcus* (Collins et al., 1990b). Cells are spherical (0.6–1.0 μm in diameter) and occasionally ovoid. Cell division occurs in two planes at right angles to form tetrads. Cells may from pairs. L(+)-lactate is the major end product of glucose metabolism: traces of D(−)-lactate is formed (3–25%). Slow growth on agar, especially under aerobic conditions. Facultatively anaerobic, with slow growth in broth (typically 4–5 d).

Glucose, fructose, ribose, maltose, mannose, melezitose, ribose, and sucrose are usually fermented. The fermentation of arabinose, glycerol, maltotriose, mannitol, melibiose, raffinose, and trehalose varies among strains. Dextrin, lactose, sorbitol, starch, and xylose are usually not fermented.

Growth at 15 and 40 °C, but not at 45 °C. Optimal growth between 30 and 35 °C. Growth at pH 5.0–9.6, with optimal range between pH 7.0 and 8.0. The final growth pH is about 5.0. Growth occurs in 4–18% NaCl, with optimum concentration of 5–10% NaCl; 20–26% NaCl may be tolerated. Most strains are able to grow in the absence of NaCl.

Susceptible to vancomycin and pyrrolidonyl-arylamidase-positive (Bosley et al., 1990). The carbohydrates fermented by the type strain are listed in Table 118. Peptidoglycan type is Lys–D-Asp. The fatty acid profile is listed in Table 119. Isolated from soy sauce and pickling brines. Type species of the genus.

DNA G+C content (mol%): 34–36 (T_m), 36–37 (HPLC).

Type strain: ATCC 33315, CCUG 32204, CIP 102263, DSM 20339, IAM 1676, JCM 5888, LMG 11490, NBRC 10049, NCIMB 12011.

GenBank accession number (16S rRNA gene): D88668.

2. **Tetragenococcus muriaticus** Satomi Kimura, Mizoi, Sato and Fujii 1997, 835[VP]

mu.ri.a'ti.cus. N.L. adj. *muriaticus* briny, pickled.

This description is based on Satomi et al. (1997). Cells are Gram-positive cocci, 0.5–0.8 μm in diameter and form tetrads or pairs. Colonies on 10% NaCl-GYP agar plates appear smooth, entire, white, convex, and 1–1.5 mm in diameter. Arginine is not hydrolyzed and nitrate not reduced to nitrite. Growth occurs at 15 to 40 °C, but not at 10 or 45 °C. Optimal growth between 25 and 30 °C. Growth at pH 5.0 to 9.6, but not at 4.2; optimal growth pH is between pH 7.5 and 8.0. Growth in the presence of 1–25% NaCl, with optimal growth between 7 and 10%. Requires Na[+] for growth (no growth in the absence of NaCl). L(+)-lactate is the major end product of glucose metabolism.

Glucose, fructose, mannose, mannitol, ribose, and trehalose are usually fermented. The fermentation of maltose seems to be strain specific. Arabinose, dextrin, gluconate, glycerol, lactose, maltotriose, melezitose, raffinose, sorbitol, starch, sucrose, and xylose are usually not fermented. The carbohydrates fermented by the type strain are listed in Table 118. The fatty acid profiles are listed in Table 119. Isolated from fermented squid liver sauce.

DNA G+C content (mol%): 36–37 (HPLC).

Type strain: X-1, CIP 105747, JCM 10006, LMG 18498, NBRC 100499.

GenBank accession number (16S rRNA gene): AJ301843, D88824, D87680.

3. **Tetragenococcus koreensis** Lee, Kim, Vancanneyt, Swings, Kim, Kang and Lee 2005, 1412[VP]

ko.re.en'sis. N.L. masc. adj. *koreensis* pertaining to Korea.

The description is according to Lee et al. (2005). Non-motile, nonsporeforming cocci (approx. 1 μm in length). Catalase and oxidase-negative. Grows on glucose yeast peptone (GYP) sodium acetate mineral salts agar (pH 7.0) and trypticase soy agar at 30 °C. Colonies develop aerobically and anaerobically on plates incubated in an anaerobic jar. No growth on R2A agar, nutrient agar or MacConkey agar at 30 °C under either aerobic or anaerobic conditions. Fatty acid profile: $C_{18:1}$ ω7*c* (33.8%), $C_{19:0}$ cyclo ω8*c* (26.8%) and $C_{16:0}$ (22.4%). Growth occurs at 15–30 °C on GYP/sodium acetate/mineral salts agar but not at 4 °C or above 37 °C. Optimum growth temperature is 30 °C. Optimal pH is 9.0. Growth in the presence of 8% (w/v) NaCl, but 2–5% (w/v) is optimal. Facultatively aerobic. Homofermentative and produces mainly lactic acid. The peptidoglycan type is A4 αL-lys–D-Asp. Nitrate is not reduced to nitrite or nitrogen gas. β-Glucosidase- and β-galactosidase-positive, but protein is not hydrolysed. Characterized by the production of rhamnolipid biosurfactant.

Glucose, fructose, salicin and *N*-acetylglucosamine are usually fermented. L-fucose, 5-ketogluconate, lactose, D-lyxose, D-melibiose, methyl β-D-xyloside, D-raffinose, L-rhamnose, L-xylose, glycogen, starch, adonitol, L-arabitol, erythritol, glycerol, sorbitol and inulin are not fermented. Carbohydrates fermented by the type strain are listed in Table 118. The fatty acid profiles are listed in Table 119. Isolated from the traditional Korean food kimchi in Daejeon, South Korea.

DNA G+C content (mol%): 38.3 (HPLC).

Type strain: JS[T] (=KCTC 3924[T]=DSM 16501[T]=LMG 22864[T]).

GenBank accession number (16S rRNA gene): AY690334.

4. **Tetragenococcus solitarius** Ennahar and Cai 2005, 592[VP]

sol.i.tar.i'us. L. adj. solitarius alone, lonely.

Basonym: *Enterococcus solitarius* Collins et al., (1989b).

The description for the species *Tetragenococcus solitarius* comb. nov. is as described for *Enterococcus solitarius* by Collins et al. (1989b). A few additions and corrections to the description of the species was proposed by Ennahar and Cai (2005).

Glucose, fructose, mannose, salicin and *N*-acetylglucosamine are usually fermented. Adonitol, L-arabitol, erythritol, D-fucose, L-fucose, glycogen, 5-ketogluconate, D-lyxose, methyl β-D-xyloside, sorbose, L-xylose and xylitol are not fermented. Carbohydrates fermented by the type strain are listed in Table 118. *Enterococcus solitarius* was first isolated from humans (Collins et al., 1989b).

DNA G+C content (mol%): 38.0 (HPLC).

Type strain: 885/78[T] (=ATCC 49428[T]=CCUG 29293[T]=CIP 103330[T]=DSM 5634[T]=JCM 8736[T]=LMG 12890[T]=NCTC 12193[T]).

GenBank accession number (16S rRNA gene): AJ301840.

Genus IV. **Vagococcus** Collins, Ash, Farrow, Wallbanks and Williams 1990a, 212[VP] (Effective publication: Collins, Ash, Farrow, Wallbanks and Williams 1989a, 459.)

MATTHEW D. COLLINS

Va.go.coc'cus. L. adj. *vagus* wandering; Gr. n. *kokkos* a grain, berry; N.L. masc. n. *Vagococcus* wandering coccus, referring to motility.

Cells are ovoid, occurring singly, in pairs, or chains. Grampositive. May or may not be motile. Nonsporeforming. **Facultatively anaerobic and catalase-negative. Grows at 10 °C but not at 45 °C. Gas is not produced in MRS broth. Acid is produced from D-glucose and some other sugars. Pyrrolidonyl arylamidase and leucine arylamidase are produced. Arginine dihydrolase is not produced. Hippurate is not hydrolyzed. Susceptible to vancomycin.**

DNA G+C content (mol%): 34–40 (T_m).

Type species: **Vagococcus fluvialis** Collins, Ash, Farrow, Wallbanks and Williams 1990a, 212[VP] (Effective publication: Collins, Ash, Farrow, Wallbanks and Williams 1989a, 459.).

Further descriptive information

Vagococci grow on 5% (v/v) horse or sheep blood agar producing an α- hemolytic reaction. Cells are ovoid in shape and are most commonly arranged in short chains or pairs. They are nonpigmented. Most strains of *Vagococcus fluvialis* are motile by means of peritrichous flagella; *Vagococcus salmoninarum* is nonmotile. Most strains of vagococci grow at 10 °C but not at 45 °C; strains of *Vagococcus salmoninarum* do not grow at 40 °C whereas *Vagococcus fluvialis* may or may not grow at this temperature. Vagococci generally do not grow in 6.5% NaCl but reports vary. *Vagococcus salmoninarum* is considered not to grow in 6.5% NaCl (Schmidtke and Carson, 1994), but Teixeira et al. (1997) reported growth at this salt concentration. *Vagococcus fluvialis* strains do not grow or occasionally produce very weak growth in 6.5% NaCl (Pot et al., 1994). Some strains of vagococci grow at pH 9.6, but others do not. Vagococci produce acid, but not gas, from glucose and some other sugars; acid production from sugars varies among species. L-Lactic acid is produced from glucose. None produce acid from D-arabitol, L-arabinose, glycogen, melibiose, melezitose, pullulan, or raffinose. Some strains of *Vagococcus salmoninarum* and *Vagococcus fluvialis* produce acid from starch; there are no published data on acid production from starch for *Vagococcus fessus* and *Vagococcus lutrae*. Vagococci do not produce alanine phenylalanine proline arylamidase or β-glucuronidase. *Vagococcus fessus*, *Vagococcus lutrae*, *Vagococcus salmoninarum*, and most strains of *Vagococcus fluvialis* produce leucine arylamidase. Some vagococci produce acetoin (positive Voges–Proskauer test), whereas others do not. *Vagococcus salmoninarum*, *Vagococcus fessus*, and *Vagococcus lutrae* are Voges–Proskauer-negative, but some strains of *Vagococcus fluvialis* give a positive reaction. Vagococci do not hydrolyze gelatin and do not reduce nitrate to nitrite. Most strains of *Vagococcus fluvialis* and *Vagococcus salmoninarum* produce a positive bile-esculin reaction; there is no published data on the bile-esculin reactions of *Vagococcus fessus* and *Vagococcus lutrae*.

Vagococcus fluvialis is reported to react with Lancefield group N antisera. *Vagococcus salmoninarum* does not possess a Lancefield group N antigen (Schmidtke and Carson, 1994). Some isolates (including the type strain) of *Vagococcus fluvialis* and the type strain of *Vagococcus salmoninarum* have been reported to give a weak reaction with Lancefield group D antiserum (Teixeira et al., 1997). The cell-wall murein of *Vagococcus fluvialis* is type L-Lysine–D-Aspartic acid (Collins et al., 1989a). *Vagococcus fluvialis* contains poly (glycerophosphate) lipoteichoic acids solely substituted with D-alanine ester. The long-chain cellular fatty acids of *Vagococcus fluvialis* and *Vagococcus salmoninarum* are of the straight-chain saturated and monounsaturated types with $C_{16:0}$, $C_{16:1\,\omega9}$, and $C_{18:1\,\omega9}$ predominating (Schleifer et al., 1985; Wallbanks et al., 1990).

The antimicrobial susceptibilities of only *Vagococcus fluvialis* and *Vagococcus salmoninarum* have been investigated. Teixeira et al. (1997) reported *Vagococcus fluvialis* (type strain plus seven other isolates) and *Vagococcus salmoninarum* (type strain) were susceptible to ampicillin, cefotaxime, trimethylprim-sulfamethoxazole, and vancomycin but resistant to clindamycin, lomefloxacin, and ofloxacin. Strain to strain variation was observed in relation to susceptibilities to 18 other antimicrobial agents (Teixeira et al., 1997). Schmidtke and Carson (1994) reported strains of *Vagococcus salmoninarum* were resistant to streptomycin but susceptible to ampicillin, co-trimoxazole, penicillin G, and vancomycin.

Vagococcus fluvialis was originally isolated from chicken feces and river water (Hashimoto et al., 1979, 1974). Subsequently, it has been recovered from diverse sources and may be an opportunistic pathogen of humans and animals. The species has been isolated from a variety of human clinical specimens including blood, peritoneal fluid, cerebrospinal fluid, and wounds (Teixeira et al., 1997), and from various domestic animals, i.e., various lesions of pigs, lesions and tonsils of cattle and cats, and tonsils of horses (Pot et al., 1994; Teixeira et al., 1997). *Vagococcus salmoninarum* was originally isolated in the USA from diseased rainbow trout (*Oncorhynchus mykiss*). The diseased fish had a condition similar to post-spawning stress syndrome caused by *Carnobacterium piscicola*. *Vagococcus salmoninarum* has since been recovered from Atlantic salmon (*Salmo salar*) and rainbow trout in Tasmania and Australia and brown trout (*Salmo trutta*) in Norway, with peritonitis (Schmidtke and Carson, 1994). Little is known about the natural occurrence and distribution of *Vagococcus lutrae* and *Vagococcus fessus*. *Vagococcus lutrae* has, so far, only been reported to be isolated from a common otter (*Lutra lutra*) (Lawson et al., 1999), whereas *Vagococcus fessus* has been isolated from a seal and harbor porpoise (Hoyles et al., 2000).

Isolation

Vagococcus species can be isolated on blood-supplemented agars, such as brain heart infusion or Columbia agar containing 5% (v/v) bovine, horse, or sheep blood incubated at 37 °C in air or air with enhanced CO_2 (5–10%). For the isolation of *Vagococcus salmoninarum*, blood agar base no. 2 (Oxoid) supplemented with 7% (v/v) sheep blood and incubated in air at 25 °C, has been used (Schmidtke and Carson, 1994).

Maintenance procedures

Strains can be maintained on media such as brain heart infusion agar with or without animal blood (5% v/v). They can be stored on cryogenic beads at −70 °C or lyophilized for long-term preservation.

Taxonomic comments

The genus *Vagococcus* was proposed in 1989 to accommodate some motile, Lancefield group N cocci which were first described by Hashimoto et al. (1974) and which were phylogenetically distinct from lactococci (Collins et al., 1989a). The genus was originally monospecific, but subsequently three other species have been described. Phylogenetically, *Vagococcus* is a member of the family *Enterococcaceae* in the phylum *Firmicutes* and is closely related to, but distinct from, *Enterococcus*, *Melissococcus plutonius*, and *Tetragenococcus* (Hoyles et al., 2000).

Differentiation of the genus *Vagococcus* from other genera

When the genus *Vagococcus* was originally proposed, it was relatively easy to identify, but the expanded genus is difficult to differentiate from other related genera solely on the basis of phenotypic properties. *Vagococcus fluvialis* can be distinguished from lactococci by being motile (most strains), producing acid from glycerol, sorbitol, and starch, by failing to hydrolyze arginine, and by failing to grow (or producing sparse growth) in 6.5% NaCl. The differentiation of *Vagococcus fluvialis* strains from some enterococci can be problematic. *Vagococcus fluvialis* may be distinguished from motile enterococcal species, *Enterococcus casseliflavus* and *Enterococcus gallinarum*, in failing to produce acid from L-arabinose, melibiose, raffinose, or D-xylose (Pot et al., 1994; Teixeira et al., 1997). Similarly, most *Enterococcus* species show growth in 6.5% NaCl and at 45° C, while most strains of *Vagococcus fluvialis* do not grow or are strongly inhibited in these conditions.

Differentiation of the species of the genus Vagococcus

The four described *Vagococcus* species can be distinguished readily from each other using commercially available API test systems. A summary of the most useful tests for differentiating *Vagococcus* species using the API Rapid ID32 Strep system is shown in Table 120.

List of species of the genus Vagococcus*

1. **Vagococcus fluvialis** Collins, Ash, Farrow, Wallbanks and Williams 1990a, 212[VP] (Effective publication: Collins, Ash, Farrow, Wallbanks and Williams 1989a, 459.)

   flu.vi.al′is. L. adj. *fluvialis* belonging to a river.

   Type strain: M-29C, ATCC 49515, CCUG 32704, CIP 102976, DSM 5731, LMG 9464, NCIMB 13038.

   GenBank accession number (16S rRNA gene): X54258.

2. **Vagococcus fessus** Hoyles, Lawson, Foster, Falsen, Ohlén, Grainger and Collins 2000, 1154[VP]

   fes.sus. L. adj. *fessus* weary, pertaining to the general biochemical unreactivity of the organism.

   Type strain: M2661/98/1, ATCC BAA-289, CCUG 41755, CIP 106499.

   GenBank accession number (16S rRNA gene): AJ243326.

3. **Vagococcus lutrae** Lawson, Foster, Falsen, Ohlén and Collins 1999, 1254[VP]

   lu′trae. L. fem. gen. n. *lutrae* of an otter, pertaining to the common otter, *Lutra lutra*, the mammal from which the bacterium was isolated.

   Type strain: M1134/97/1, ATCC 700839, CCUG 39187, CIP 106118.

   GenBank accession number (16S rRNA gene): Y17152.

4. **Vagococcus salmoninarum** Wallbanks, Martinez-Murcia, Fryer, Phillips and Collins 1990, 229[VP]

   sal.mo.ni.na′rum. M.L. *salmoninae* subfamily of the *Salmonidae*; M.L. gen. n. *salmoninarum* of the *Salmonidae*.

   Type strain: OS1-68, ATCC 51200, CCUG 33394, CIP 104684, DSM 6633, LMG 11491, NCFB 2777, NCIMB 13133.

   GenBank accession number (16S rRNA gene): X54272.

---

* Editorial note: Since the acceptance of this chapter, two more *Vagococcus* species have been described, *Vagococcus carniphilus* (Shewmaker et al., 2004) and *Vagococcus carniphilus* (Lawson et al., 2007).

TABLE 120. Tests based on the API Rapid ID32 Strep system which is useful in distinguishing species of the genus *Vagococcus*[a,b]

Test	V. fluvialis	V. fessus	V. lutrae	V. salmoninarum
Acid from:				
Cyclodextrin	+	d	+	−
Lactose	d	−	−	−
MBDG	d	−	+	−
Maltose	+	d	+	d
Mannitol	+	−	−	−
Sorbitol	+	−	+	−
Sucrose	d	−	+	d
D-Tagatose	d	−	−	−
Trehalose	+	−	+	+
Production of:				
α-GAL	−	−	+	−
β-GAL	−	−	+	−
GTA	+	−	+	+
β-NAG	d	−	+	−
PAL	d	−	+	−
URE	d	d	−	−

[a]Symbols: +, >85% positive; d, different strains give different reactions (16–84% positive); −, 0–15% positive. Abbreviations: PAL, alkaline phosphatase; α-GAL, α-galactosiodase; β-GAL, β-galactosidase; GTA, glycyl tryptophan arylamidase; β-NAG, N-acetyl-β-glucosaminidase; URE, urease.

[b]All species fail to produce acid from D-arabitol, L-arabinose, glycogen, melibiose, melezitose, pullulan, or raffinose; produce pyroglutamic acid arylamidase; do not produce arginine dihydrolase, alanine phenylalanine proline arylamidase, or β-glucuronidase; and do not hydrolyze hippurate.

References

Alippi, A.M. 1991. A comparison of laboratory techniques for the detection of significant bacteria of the honey bees, *Apis mellifera*. J. Apic. Res. *30:* 75–80.

Allen, M.F., B.V. Ball and B.A. Underwood. 1990. An isolate of *Melissococcus pluton* from *Apis laboriosa*. J. Invertebr. Pathol. *55:* 439–440.

Allen, M.F. and B.V. Ball. 1993. The cultural characteristics and serological relationships of isolates of *Melissococcus pluton*. J. Apic. Res. *32:* 80–88.

Anderson, D.L. 1990. Pests and pathogens of the honeybee (*Apis mellifera*) in Fiji. J. Apic. Res. *29:* 53–59.

Andrewes, F.W. and T.J. Horder. 1906. A study of the streptococci pathogenic for man. Lancet *ii:* 708–713.

Bachrach, G., M. Leizerovici-Zigmond, A. Zlotkin, R. Naor and D. Steinberg. 2003. Bacteriophage isolation from human saliva. Lett. Appl. Microbiol. *36:* 50–53.

Back, W. and E. Stackebrandt. 1978. DNA–DNA homology studies on genus *Pediococcus*. Arch. Microbiol. *118:* 79–85.

Baele, M., P. Baele, M. Vaneechoutte, V. Storms, P. Butaye, L.A. Devriese, G. Verschraegen, M. Gillis and F. Haesebrouck. 2000. Application of tRNA intergenic spacer PCR for identification of *Enterococcus* species. J. Clin. Microbiol. *38:* 4201–4207.

Baele, M., L.A. Devriese, P. Butaye and F. Haesebrouck. 2002. Composition of enterococcal and streptococcal flora from pigeon intestines. J. Appl. Microbiol. *92:* 348–351.

Bailey, L. 1957. The isolation and cultural characteristics of *Streptococcus pluton* and further observations on "Bacterium eurydice". J. Gen. Microbiol. *17:* 39–48.

Bailey, L. and A.J. Gibbs. 1962. Cultural characters of *Streptococcus pluton* and its differentiation from associated enterococci. J. Gen. Microbiol. *28:* 385–391.

Bailey, L. 1963a. The habitat of "*Bacterium eurydice*". J. Gen. Microbiol. *31:* 147–150.

Bailey, L. 1963b. The pathogenicity for honey-bee larvae of microorganisms associated with European foulbrood. J. Insect Pathol. *5:* 198–205.

Bailey, L. 1974. An unusual type of *Streptococcus pluton* from the eastern hive bee. J. Invertebr. Pathol. *23:* 246–247.

Bailey, L. 1981. Honey Bee Pathology. Academic Press, London and New York.

Bailey, L. and M.D. Collins. 1982a. Taxonomic studies on *Streptococcus pluton*. J. Appl. Bacteriol. *53:* 209–214.

Bailey, L. and M.D. Collins. 1982b. Reclassification of *Streptococcus pluton* (White) in a new genus *Melissococcus*, as *Melissococcus pluton* nom. rev., comb. nov. J. Appl. Bacteriol. *53:* 215–217.

Bailey, L. and M.D. Collins. 1983. *In* Validation of the publication of new names and new combinations previously effectively published outside the IJSB. List no. 11. Int. J. Syst. Bacteriol. *33:* 672–674.

Bailey, L. 1984. A strain of *Melissococcus pluton* cultivable on chemically defined media. FEMS Microbiol. Lett. *25:* 139–141.

Bauer, S., A. Tholen, J. Overmann and A. Brune. 2000. Characterization of abundance and diversity of lactic acid bacteria in the hindgut of wood- and soil-feeding termites by molecular and culture-dependent techniques. Arch. Microbiol. *173:* 126–137.

Behr, T., C. Koob, M. Schedl, A. Mehlen, H. Meier, D. Knopp, E. Frahm, U. Obst, K. Schleifer, R. Niessner and W. Ludwig. 2000. A nested array of rRNA targeted probes for the detection and identification of enterococci by reverse hybridization. Syst. Appl. Microbiol. *23:* 563–572.

Benachour, A., J. Frere and G. Novel. 1995. pUCL287 plasmid from *Tetragenococcus halophila* (*Pediococcus halophilus*) ATCC 33315 represents a new theta-type replicon family of lactic acid bacteria. FEMS Microbiol. Lett. *128:* 167–175.

Berlutti, F., M.C. Thaller, S. Schippa, F. Pantanella and R. Pompei. 1993. A new approach to use of bacteriolytic enzymes as a tool for species identification: selection of species-specific indicator strains with bacteriolytic activity towards *Enterococcus* strains. Int. J. Syst. Bacteriol. *43:* 63–68.

Bleiweis, A.S. and L.N. Zimmerman. 1961. Formation of two types of osmotically fragile bodies from *Streptococcus faecalis* var. liquifaciens. Can. J. Microbiol. *7:* 363–373.

Bosley, G.S., P.L. Wallace, C.W. Moss, A.G. Steigerwalt, D.J. Brenner, J.M. Swenson, G.A. Hebert and R.R. Facklam. 1990. Phenotypic characterization, cellular fatty acid composition, and DNA related-ness of aerococci and comparison to related genera. J. Clin. Microbiol. *28:* 416–421.

Bosshard, P.P., S. Abels, M. Altwegg, E.C. Böttger and R. Zbinden. 2004. Comparison of conventional and molecular methods for identification of aerobic catalase-negative Gram-positive cocci in the clinical laboratory. J. Clin. Microbiol. *42:* 2065–2073.

Bridge, P.D. and P.H.A. Sneath. 1982. *Streptococcus gallinarum* sp. nov. and *Streptococcus oralis* sp. nov. Int. J. Syst. Bacteriol. *32:* 410–415.

Brune, A. and M. Friedrich. 2000. Microecology of the termite gut: structure and function on a microscale. Curr. Opin. Microbiol. *3:* 263–269.

Cai, J. and M.D. Collins. 1994. Evidence for a close phylogenetic relationship between *Melissococcus pluton*, the causative agent of European foulbrood disease, and the genus *Enterococcus*. Int. J. Syst. Bacteriol. *44:* 365–367.

Carvalho, M.D.S., A.G. Steigerwalt, R.E. Morey, P.L. Shewmaker, L.M. Teixeira and R.R. Facklam. 2004. Characterization of three new enterococcal species, *Enterococcus* sp. nov. CDC PNS-E2, *Enterococcus* sp. nov. CDC PNS-E3, and *Enterococcus* sp. nov. CDC PNS-E3, isolated from human clinical specimens. J. Clin. Microbiol. *42:* 1192–1198.

Carvalho, M.D.S. G., P.L. Shewmaker, A.G. Steigerwalt, R.E. Morey, A.J. Sampson, K. Joyce, T.J. Barrett, L.M. Teixeira and R.R. Facklam. 2006. *Enterococcus caccae* sp. nov., isolated from human stools. Int. J. Syst. Evol. Microbiol. *56:* 1505–1508.

Cetinkaya, Y., P. Falk and C.G. Mayhall. 2000. Vancomycin-resistant enterococci. Clin. Microbiol. Rev. *13:* 686–707.

Charrier, M., Y. Combet-Blanc and B. Ollivier. 1998. Bacterial flora in the gut of *Helix aspersa* (Gastropda Pulmonata): evidence for a permanent population with a dominant homolactic intestinal bacterium, *Enterococcus casseliflavus*. Can. J. Microbiol. *44:* 20–27.

Clark, N.C., L.M. Teixeira, R.R. Facklam and F.C. Tenover. 1998. Detection and differentiation of *vanC-1*, *vanC-2*, and *vanC-3* glycopeptide resistance genes in enterococci. J. Clin. Microbiol. *36:* 2294–2297.

Collins, M.D. and D. Jones. 1979. Distribution of isoprenoid quinones in streptococci of serological group D and group N. J. Gen. Microbiol. *114:* 27–33.

Collins, M.D., D. Jones, J.A.E. Farrow, R. Kilpper-Balz and K.H. Schleifer. 1984. *Enterococcus avium* nom. rev., comb. nov., *Enterococcus casseliflavus* nom. rev., comb. nov., *Enterococcus durans* nom. rev., comb. nov., *Enterococcus gallinarum* comb. nov., and *Enterococcus malodoratus* sp. nov. Int. J. Syst. Bacteriol. *34:* 220–223.

Collins, M.D., J.A.E. Farrow and D. Jones. 1986. *Enterococcus mundtii* sp. nov. Int. J. Syst. Bacteriol. *36:* 8–12.

Collins, M.D., C. Ash, J.A.E. Farrow, S. Wallbanks and A.M. Williams. 1989a. 16S ribosomal ribonucleic acid sequence analyses of lactococci and related taxa: description of *Vagococcus fluvialis* gen. nov., sp. nov. J. Appl. Bacteriol. *67:* 453–460.

Collins, M.D., R.R. Facklam, J.A. Farrow and R. Williamson. 1989b. *Enterococcus raffinosus* sp. nov., *Enterococcus solitarius* sp. nov. and *Enterococcus pseudoavium* sp. nov. FEMS Microbiol. Lett. *48:* 283–288.

Collins, M.D., R.R. Facklam, J.A.E. Farrow and R. Williamson. 1989c. *In* Validation of the publication of new names and new combincations previously effectively published outside the IJSB. List no. 30. Int. J. Syst. Bacteriol. *39:* 371.

Collins, M.D., C. Ash, J.A.E. Farrow, S. Wallbanks and A.M. Williams. 1990a. *In* Validation of the publication of new names and new combinations previously effectively published outside the IJSB. List no. 33. Int. J. Syst. Bacteriol. *40:* 212.

Collins, M.D., A.M. Williams and S. Wallbanks. 1990b. The phylogeny of *Aerococcus* and *Pediococcus* as determined by 16S rRNA sequence analysis: description of *Tetragenococcus* gen. nov. FEMS Microbiol. Lett. *58:* 255–262.

Collins, M.D., U.M. Rodrigues, N.E. Pigott and R.R. Facklam. 1991a. *Enterococcus dispar* sp. nov. a new *Enterococcus* species from human sources. Lett. Appl. Microbiol. *12:* 95–98.

Collins, M.D., U.M. Rodrigues, N.E. Pigott and R.R. Facklam. 1991b. *In* Validation of the publication of new names and new combinations previously effectively published outside the IJSB. List no. 38. Int. J. Syst. Bacteriol. *41:* 456–458.

Collins, M.D., A.M. Williams and S. Wallbanks. 1993. *In* Validation of the publication of new names and new combinations previously effectively published outside the IJSB. List no. 44. Int. J. Syst. Bacteriol. *43:* 188–189.

Day, A.M., J.A. Sandoe, J.H. Cove and M.K. Phillips-Jones. 2001. Evaluation of a biochemical test scheme for identifying clinical isolates of *Enterococcus faecalis* and *Enterococcus faecium*. Lett. Appl. Microbiol. *33:* 392–396.

De Graef, E.M., L.A. Devriese, M. Vancanneyt, M. Baele, M.D. Collins, K. Lefebvre, J. Swings and F. Haesebrouck. 2003. Description of *Enterococcus canis* sp. nov. from dogs and reclassification of *Enterococcus porcinus* Teixeira *et al.* 2001 as a junior synonym of *Enterococcus villorum* Vancanneyt *et al.* 2001. Int. J. Syst. Evol. Microbiol. *53:* 1069–1074.

De Man, J.C., M. Rogosa and M.E. Sharpe. 1960. A medium for the cultivation of lactobacilli. J. Appl. Bacteriol. *23:* 130–135.

de Vaux, A., G. Laguerre, C. Divies and H. Prevost. 1998. *Enterococcus asini* sp. nov. isolated from the caecum of donkeys (*Equus asinus*). Int. J. Syst. Bacteriol. *48:* 383–387.

Deibel, R.H. 1964. The Group D streptococci. Bacteriol. Rev. *28:* 330–366.

Descheemaeker, P., C. Lammens, B. Pot, P. Vandamme and H. Goossens. 1997. Evaluation of arbitrarily primed PCR analysis and pulsed-field gel electrophoresis of large genomic DNA fragments for identification of enterococci important in human medicine. Int. J. Syst. Bacteriol. *47:* 555–561.

Devriese, L.A. and B. Pot. 1995. The genus *Enterococcus*. *In* Wood and Holzapfel (Editors), The Genera of Lactic Acid Bacteria. Blackie Academic & Professional, London, pp. 327–367.

Devriese, L.A., G.N. Dutta, J.A.E. Farrow, A. van de Kerckhove and B.A. Phillips. 1983. *Streptococcus cecorum*, a new species isolated from chickens. Int. J. Syst. Bacteriol. *33:* 772–776.

Devriese, L.A., A. Van De Kerckhove, R. Kilpper-Bälz and K.H. Schleifer. 1987. Characterization and identification of *Enterococcus* species isolated from the intestines of animals. Int. J. Syst. Bacteriol. *37:* 257–259.

Devriese, L.A., K. Ceyssens, U.M. Rodrigues and M.D. Collins. 1990. *Enterococcus columbae*, a species from pigeon intestines. FEMS Microbiol. Lett. *59:* 247–251.

Devriese, L.A., J. Hommez, R. Wijfels and F. Haesebrouck. 1991. Composition of the enterococcal and streptococcal intestinal flora of poultry. J. Appl. Bacteriol. *71:* 46–50.

Devriese, L.A., M.D. Collins and R. Wirth. 1992a. The genus *Enterococcus*. *In* Balows, Trüper, Dworkin, Harder and Schleifer (Editors), The Prokaryotes. A Handbook on the Biology of Bacteria: Ecophysiology, Isolation, Identification, Applications, 2nd edn, Vol. 2. Springer-Verlag, New York, pp. 1465–1481.

Devriese, L.A., J.I.C. Colque, P. Deherdt and F. Haesebrouck. 1992b. Identification and composition of the tonsillar and anal enterococcal and streptococcal flora of dogs and cats. J. Appl. Bacteriol. *73:* 421–425.

Devriese, L.A., K. Ceyssens, U.M. Rodrigues and M.D. Collins. 1993a. *In* Validation of the publication of new names and new combinations previously effectively published outside the IJSB. List no. 44. Int. J. Syst. Bacteriol. *43:* 188–189.

Devriese, L.A., B. Pot and M.D. Collins. 1993b. Phenotypic identification of the genus *Enterococcus* and differentiation of phylogenetically distinct enterococcal species and species groups. J. Appl. Bacteriol. *75:* 399–408.

Devriese, L.A., B. Pot, L. Van Damme, K. Kersters and F. Haesebrouck. 1995. Identification of *Enterococcus* species isolated from foods of animal origin. Int. J. Food Microbiol. *26:* 187–197.

Devriese, L.A., M. Vancanneyt, P. Descheemaeker, M. Baele, H.W. Van Landuyt, B. Gordts, P. Butaye, J. Swings and F. Haesebrouck. 2002. Differentiation and identification of *Enterococcus durans, E. hirae* and *E. villorum.* J. Appl. Microbiol. *92:* 821–827.

Diwan, V.V., K.K. Kshirsagar, A.V. Ramana Rao, D. Raghunath, C.S. Bhambure and S.H. Godbole. 1971. Occurrence of a new bacterial disease of Indian honey-bees *Apis indica* F. Curr. Sci. *40:* 196–197.

Djordjevic, S.P., K. Noone, L. Smith and M.A.Z. Hornitzky. 1998. Development of a hemi-nested PCR assay for the specific detection of *Melissococcus pluton.* J. Apic. Res. *37:* 165–174.

Djordjevic, S.P., L.A. Smith, W.A. Forbes and M.A. Hornitzky. 1999. Geographically diverse Australian isolates of *Melissococcus pluton* exhibit minimal genotypic diversity by restriction endonuclease analysis. FEMS Microbiol. Lett. *173:* 311–318.

Dudka-Malen, S., S. Evers and P. Courvalin. 1995. Detection of glycopeptide resistance genotypes and identification to the species level of clinically relevant enterococci by PCR. J. Clin. Microbiol. *33:* 24–27.

Eaton, T.J. and M.J. Gasson. 2001. Molecular screening of *Enterococcus* virulence determinants and potential for genetic exchange between food and medical isolates. Appl. Environ. Microbiol. *67:* 1628–1635.

Elsner, H.A., I. Sobottka, D. Mack, M. Claussen, R. Laufs and R. Wirth. 2000. Virulence factors of *Enterococcus faecalis* and *Enterococcus faecium* blood culture isolates. Eur. J. Clin. Microbiol. Infect. Dis. *19:* 39–42.

Ennahar, S. and Y. Cai. 2005. Biochemical and genetic evidence for the transfer of *Enterococcus solitarius* Collins *et al.* 1989 to the genus *Tetragenococcus* as *Tetragenococcus solitarius* comb. nov. Int. J. Syst. Evol. Microbiol. *55:* 589–592.

Ercolini, D., G. Moschetti, G. Blaiotta and S. Coppola. 2001. Behavior of variable V3 region from 16S rDNA of lactic acid bacteria in denaturing gradient gel electrophoresis. Curr. Microbiol. *42:* 199–202.

Euzéby, J.P. and T. Kudo. 2001. Corrigenda to the validation lists. Int. J. Syst. Evol. Microbiol. *51:* 1933–1938.

Facklam, R.R. and M.D. Collins. 1989. Identification of *Enterococcus* species isolated from human infections by a conventional test scheme. J. Clin. Microbiol. *27:* 731–734.

Farrow, J.A.E., J. Kruze, B.A. Phillips, A.J. Bramley and M.D. Collins. 1984. Taxonomic studies on *Streptococcus bovis* and *Streptococcus equinus:* description of *Streptococcus alactolyticus* sp. nov. and *Streptococcus saccharolyticus* sp. nov. Syst. Appl. Microbiol. *5:* 467–482.

Farrow, J.A.E. and M.D. Collins. 1985. *Enterococcus hirae*, a new species that includes amino acid assay strain NCDO 1258 and strains causing growth depression in young chickens. Int. J. Syst. Bacteriol. *35:* 73–75.

Fortina, M.G., G. Ricci, D. Mora and P.L. Manachini. 2004. Molecular analysis of artisanal Italian cheeses reveals *Enterococcus italicus* sp. nov. Int. J. Syst. Evol. Microbiol. *54:* 1717–1721.

Franz, C.M., W.H. Holzapfel and M.E. Stiles. 1999. Enterococci at the crossroads of food safety? Int. J. Food Microbiol. *47:* 1–24.

Garg, S.K. and B.K. Mital. 1991. Enterococci in milk and milk products. Crit. Rev. Microbiol. *18:* 15–45.

Garvie, E.I. 1978. *Streptococcus raffinolactis* Orla Jensen and Hansen, a group N *Streptococcus* found in raw milk. Int. J. Syst. Bacteriol. *28:* 190–193.

Garvie, E.I. 1986. Genus *Pediococcus. In* Sneath, Mair, Sharpe and Holt (Editors), Bergey's Manual of Systematic Bacteriology, Vol. 2. The Williams & Wilkins Co., Baltimore, pp. 1075–1079.

Gilmore, M.S., D.B. Clewell, P.M. Courvalin, G.M. Dunny, B.E. Murray and L.B. Rice. 2002. The Enterococci: Pathogenesis, Molecular Biology and Antibiotic Resistance. ASM Press, Washington, D.C.

Giraffa, G. 2002. Enterococci from foods. FEMS Microbiol. Rev. *26:* 163–171.

Glinski, Z. 1972a. Investigations on the properties and antigenic structure of *Streptococcus pluton.* IV. Antigenic structure of *Streptococcus pluton.* Med. Weter. *28:* 603–611.

Glinski, Z. 1972b. Investigations on the properties and antigenic structure of *Streptococcus pluton.* L. Morphological and cultural characteristics. Med. Weter. *28:* 399–405.

Godfree, A.F., D. Kay and M.D. Wyer. 1997. Faecal streptococci as indicators of faecal contamination in water. J. Appl. Microbiol. Symp. Suppl. *83:* 110S–119S.

Govan, V.A., V. Brözel, M.H. Allsopp and S. Davison. 1998. A PCR detection method for rapid identification of *Melissococcus pluton* in honeybee larvae. Appl. Environ. Microbiol. *64:* 1983–1985.

Gregory, E.M. and I. Fridovich. 1973. Induction of superoxide dismutase by molecular oxygen. J. Bacteriol. *114:* 543–548.

Gunsalus, I.C., B.L. Horecker and W.A. Wood. 1955. Pathways of carbohydrate metabolism in microorganisms. Bacteriol. Rev. *19:* 79–128.

Hanagata, H., O. Shida and H. Takagi. 2003. Taxonomic homogeneity of a salt-tolerant lactic acid bacteria isolated from shoyu mash. J. Gen. Appl. Microbiol. *49:* 95–100.

Hardie, J.M. 1986. Genus *Streptococcus* Rosenbach 1884, 22. *In* Sneath, Mair, Sharpe and Holt (Editors), Bergey's Manual of Systematic Bacteriology, Vol. 2. The Williams & Wilkins Co., Baltimore, pp. 1043–1071.

Hardie, J.M. and R.A. Whiley. 1997. Classification and overview of the genera *Streptococcus* and *Enterococcus.* J. Appl. Microbiol. Symp. Suppl. *83:* S1–S11.

Hashimoto, H., R. Noborisaka and R. Yanagawa. 1974. Distribution of motile streptococci in faeces of man and animals and in river and sea water. Japan J. Bacteriol. *29:* 387–393.

Hashimoto, H., H. Kawakami, K. Tomokane, Z. Yoshii, G. Hahn and A. Tolle. 1979. Isolation and characterisation of motile group N streptococci. J. Faculty Appl. Biol. Sc. Hiroshima University *18:* 207–216.

Higuchi, T., K. Uchida and K. Abe. 1998. Aspartate decarboxylation encoded on the plasmid in the soy sauce lactic acid bacterium, *Tetragenococcus halophila* D10. Biosci. Biotechnol. Biochem. *62:* 1601–1603.

Holzapfel, W.H., C.M.A.P. Franz, W. Ludwig, W. Back and L.M.T. Dicks. 2006. The genera *Pediococcus* and *Tetragenococcus. In* Dworkin, Falkow, Rosenberg, Schleifer and Stackebrandt (Editors), The Prokaryotes 3rd Edn. Vol. 4. Springer Verlag, New York, pp. 229–266.

Hornitzky, M.A.Z. and L. Smith. 1998. Procedures for the culture of *Melissococcus pluton* from diseased brood and bulked honey samples. J. Apic. Res. *37:* 293–294.

Hoyles, L., P.A. Lawson, C. Foster, E. Falsen, M. Ohlén, J.M. Grainger and M.D. Collins. 2000. *Vagococcus fessus* sp. nov., isolated from a seal and a harbour porpoise. Int. J. Syst. Evol. Microbiol. *50:* 1151–1154.

Hudson, C.R., P.J. Fedorka-Cray, M.C. Jackson-Hall and L.M. Hiott. 2003. Anomalies in species identification of enterococci from veterinary sources using a commercial biochemical identification system. Lett. Appl. Microbiol. *36:* 245–250.

Huycke, M.M., D. Moore, W. Joyce, P. Wise, L. Shepard, Y. Kotake and M.S. Gilmore. 2001. Extracellular superoxide production by *Enterococcus faecalis* requires demethylmenaquinone and is attenuated by functional terminal quinol oxidases. Mol. Microbiol. *42:* 729–740.

Jett, B.D., M.M. Huycke and M.S. Gilmore. 1994. Virulence of enterococci. Clin. Microbiol. Rev. *7:* 462–478.

Kanbar, G., W. Engels, G.J. Nicholson, R. Hertle and G. Winkelmann. 2004. Tyramine functions as a toxin in honey bee larvae during *Varroa*-transmitted infection by *Melissococcus pluton*. FEMS Microbiol. Lett. *234:* 149–154.

Kimura, B., Y. Konagaya and T. Fujii. 2001. Histamine formation by *Tetragenococcus muriaticus*, a halophilic lactic acid bacterium isolated from fish sauce. Int. J. Food Microbiol. *70:* 71–77.

Kirk, M., R.L.R. Hill, M.W. Casewell and D. Beighton. 1997. Isolation of vancomycin-resistant enterococci from supermarket poultry. *In* Horaud, Bouvet, Leclercq, Montclos and Sicard (Editors), Streptococci and the Host, Proceedings of the XIIIth Lancefield International Symposium on Streptococci and Streptococcal Diseases. Plenum Press, New York, pp. 289–291.

Kirschner, C., K. Maquelin, P. Pina, N.A. Ngo Thi, L.P. Choo-Smith, G.D. Sockalingum, C. Sandt, D. Ami, F. Orsini, S.M. Doglia, P. Allouch, M. Mainfait, G.J. Puppels and D. Naumann. 2001. Classification and identification of enterococci: a comparative phenotypic, genotypic, and vibrational spectroscopic study. J. Clin. Microbiol. *39:* 1763–1770.

Knijff, E., F. Dellaglio, A. Lombardi, C. Andrighetto and S. Torriani. 2001. Rapid identification of *Enterococcus durans* and *Enterococcus hirae* by PCR with primers targeted to the *ddl* genes. J. Microbiol. Methods *47:* 35–40.

Kobayashi, T., B. Kimura and T. Fujii. 2000a. Strictly anaerobic halophiles isolated from canned Swedish fermented herrings (Surströmming). Int. J. Food Microbiol. *54:* 81–89.

Kobayashi, T., B. Kimura and T. Fujii. 2000b. Differentiation of *Tetragenococcus* populations occurring in products and manufacturing processes of puffer fish ovaries fermented with rice-bran. Int. J. Food Microbiol. *56:* 211–218.

Koort, J., T. Coenye, P. Vandamme, A. Sukura and J. Bjorkroth. 2004. *Enterococcus hermanniensis* sp. nov., from modified-atmosphere-packaged broiler meat and canine tonsils. Int. J. Syst. Evol. Microbiol. *54:* 1823–1827.

Kusuda, R., K. Kawai, F. Salati, C.R. Banner and J.L. Fryer. 1991. *Enterococcus seriolicida* sp. nov., a fish pathogen. Int. J. Syst. Bacteriol. *41:* 406–409.

Lang, M.M., S.C. Ingham and B.H. Ingham. 2001. Differentiation of *Enterococcus* spp. by cell membrane fatty acid methyl ester profiling, biotyping and ribotyping. Lett. Appl. Microbiol. *33:* 65–70.

Law-Brown, J. and P.R. Meyers. 2003. *Enterococcus phoeniculicola* sp. nov., a novel member of the enterococci isolated from the uropygial gland of the Red-billed Woodhoopoe, *Phoeniculus purpureus*. Int. J. Syst. Evol. Microbiol. *53:* 683–685.

Lawson, P.A., G. Foster, E. Falsen, M. Ohlén and M.D. Collins. 1999. *Vagococcus lutrae* sp. nov., isolated from the common otter (*Lutra lutra*). Int. J. Syst. Bacteriol. *49:* 1251–1254.

Lawson, P.A., G. Foster, E. Falsen, M. Ohlén and M.D. Collins. 2000. *Atopobacter phocae* gen. nov., sp. nov., a novel bacterium isolated from common seals. Int. J. Syst. Evol. Microbiol. *50:* 1755–1760.

Lawson, P.A., E. Falsen, M.A. Cotta and T.R. Whitehead. 2007. *Vagococcus elongatus* sp. nov., isolated from a swine-manure storage pit. Int. J. Syst. Evol. Microbiol. *57:* 751–754.

Lee, M., M.K. Kim, M. Vancanneyt, J. Swings, S.H. Kim, M.S. Kang and S.T. Lee. 2005. *Tetragenococcus koreensis* sp. nov., a novel rhamnolipid-producing bacterium. Int. J. Syst. Evol. Microbiol. *55:* 1409–1413.

London, J. and M.D. Appleman. 1962. Oxidative and glycerol metabolism of two species of enterococci. J. Bacteriol. *84:* 597–598.

London, J. and E.Y. Meyer. 1970. Malate utilization by a group D *Streptococcus*: regulation of malic enzyme synthesis by an inducible malate permease. J. Bacteriol. *102:* 130–137.

Ludwig, W., E. Seewaldt, R. Kilpper-Bälz, K.H. Schleifer, L. Magrum, C.R. Woese, G.E. Fox and E. Stackebrandt. 1985. The phylogenetic position of *Streptococcus* and *Enterococcus*. J. Gen. Microbiol. *131:* 543–551.

Mallon, D.J., J.E. Corkill, S.M. Hazel, J.S. Wilson, N.P. French, M. Bennett and C.A. Hart. 2002. Excretion of vancomycin-resistant enterococci by wild mammals. Emerg. Infect. Dis. *8:* 636–638.

Manero, A. and A.R. Blanch. 1999. Identification of *Enterococcus* spp. with a biochemical key. Appl. Environ. Microbiol. *65:* 4425–4430.

Manero, A. and A.R. Blanch. 2002. Identification of *Enterococcus* spp. based on specific hybridization with 16S rDNA probes. J. Microbiol. Methods *50:* 115–121.

Mannu, L., A. Paba, M. Pes, R. Floris, M.F. Scintu and L. Morelli. 1999. Strain typing among enterococci isolated from home-made Pecorino Sardo cheese. FEMS Microbiol. Lett. *170:* 25–30.

Martin, J.D. and J.O. Mundt. 1972. Enterococci in insects. Appl. Microbiol. *24:* 575–580.

Martinez-Murcia, A.J. and M.D. Collins. 1991a. *Enterococcus sulfureus*, a new yellow-pigmented *Enterococcus* species. FEMS Microbiol. Lett. *64:* 69–74.

Martinez-Murcia, A.J. and M.D. Collins. 1991b. *In* Validation of the publication of new names and new combinations previously effectively published outside the IJSB. List no. 39. Int. J. Syst. Bacteriol. *41:* 580–581.

McKessar, S.J., A.M. Berry, J.M. Bell, J.D. Turnidge and J.C. Paton. 2000. Genetic characterization of *vanG*, a novel vancomycin resistance locus of *Enterococcus faecalis*. Antimicrob. Agents Chemother. *44:* 3224–3228.

Medrek, T.F. and W. Litsky. 1960. Comparative incidence of coliform bacteria and enterococci in undisturbed soil. Appl. Microbiol. *8:* 60–63.

Mees, R.H. 1934. Onderzoekingen over de Biersarcina. Technical University, Delft, Holland.

Merquior, V.L.C., J.M. Peralta, R.R. Facklam and L.M. Teixeira. 1994. Analysis of electrophoretic whole-cell protein profiles as a tool for characterization of *Enterococcus* species. Curr. Microbiol. *28:* 149–153.

Miller, H., L.B. Poole and A. Claiborne. 1990. Heterogeneity among the flavin-containing NADH peroxidases of group D streptococci. Analysis of the enzyme from *Streptococcus faecalis* ATCC 9790. J. Biol. Chem. *265:* 9857–9863.

Monstein, H.J., M. Quednau, A. Samuelsson, S. Ahrné, B. Isaksson and J. Jonasson. 1998. Division of the genus *Enterococcus* into species groups using PCR-based molecular typing methods. Microbiology *144:* 1171–1179.

Monstein, H.J., S. Ahrné, G. Molin, S. Nikpour-Badr and J. Jonasson. 2001. Identification of enterococcal isolates by temperature gradient gel electrophoresis and partial sequence analysis of PCR-amplified 16S rDNA variable V6 regions. APMIS *109:* 209–216.

Müller, T., A. Ulrich, E.M. Ott and M. Müller. 2001. Identification of plant-associated enterococci. J. Appl. Microbiol. *91:* 268–278.

Mundt, J.O. 1961. Occurrence of enterococci: bud, blossom, and soil studies. Appl. Microbiol. *9:* 541–544.

Mundt, J.O. 1963a. Occurrence of enterococci in animals in a wild environment. Appl. Microbiol. *11:* 136–140.

Mundt, J.O. 1963b. Occurrence of enterococci on plants in a wild environment. Appl. Microbiol. *11:* 141–144.

Mundt, J.O. and W.F. Graham. 1968. *Streptococcus faecium* var. casselifavus, nov. var. J. Bacteriol. *95:* 2005–2009.

Mundy, L.M., D.F. Sahm and M. Gilmore. 2000. Relationships between enterococcal virulence and antimicrobial resistance. Clin. Microbiol. Rev. *13:* 513–522.

Murray, B.E. 1990. The life and times of the *Enterococcus*. Clin. Microbiol. Rev. *3:* 46–65.

Murray, B.E. and G.M. Weinstock. 1999. Enterococci: new aspects of an old organism. Proc. Assoc. Am. Physic. *111:* 328–334.

Naser, S., F.L. Thompson, B. Hoste, D. Gevers, K. Vandemeulebroecke, I. Cleenwerck, C.C. Thompson, M. Vancanneyt and J. Swings. 2005a. Phylogeny and identification of enterococci by *atpA* gene sequence analysis. J. Clin. Microbiol. *43:* 2224–2230.

Naser, S.M., F.L. Thompson, B. Hoste, D. Gevers, P. Dawyndt, M. Vancanneyt and J. Swings. 2005b. Application of multilocus sequence analysis (MLSA) for rapid identification of *Enterococcus* species based on *rpoA* and *pheS* genes. Microbiology *151:* 2141–2150.

Naser, S.M., M. Vancanneyt, E. De Graef, L.A. Devriese, C. Snauwaert, K. Lefebvre, B. Hoste, P. Švec, A. Decostere, F. Haesebrouck and J. Swings. 2005c. *Enterococcus canintestini* sp. nov., from faecal samples of healthy dogs. Int. J. Syst. Evol. Microbiol. *55:* 2177–2182.

Naser, S.M., M. Vancanneyt, B. Hoste, C. Snauwaert, K. Vandemeulebroecke and J. Swings. 2006. Reclassification of *Enterococcus flavescens* Pompei *et al.* 1992 as a later synonym of *Enterococcus casseliflavus* (ex Vaughan *et al.* 1979) Collins *et al.* 1984 and *Enterococcus saccharominimus* Vancanneyt *et al.* 2004 as a later synonym of *Enterococcus italicus* Fortina *et al.* 2004. Int. J. Syst. Evol. Microbiol. *56:* 413–416.

Niemi, R.M., S.I. Niemelä, D.H. Bamford, J. Hantula, T. Hyvärinen, T. Forsten and A. Raateland. 1993. Presumptive fecal streptococci in environmental samples characterized by one-dimensional sodium dodecyl sulfate-polyacrylamide gel electrophoresis. Appl. Environ. Microbiol. *59:* 2190–2196.

Nowlan, S.P. and R.H. Deibel. 1967. Group Q streptococci. I. Ecology, serology, physiology, and relationship to established enterococci. J. Bacteriol. *94:* 291–296.

Oana, K., Y. Okimura, Y. Kawakami, N. Hayashida, M. Shimosaka, M. Okazaki, T. Hayashi and M. Ohnishi. 2002. Physical and genetic map of *Enterococcus faecium* ATCC19434 and demonstration of intra- and interspecific genomic diversity in enterococci. FEMS Microbiol. Lett. *207:* 133–139.

Orla-Jensen, S. 1919. The lactic acid bacteria. Memoirs of the Academy of the Royal Society of Denmark. Section of Sciences Series *8:* 81–197.

Ott, E.M., T. Müller, M. Müller, C.M. Franz, A. Ulrich, M. Gabel and W. Seyfarth. 2001. Population dynamics and antagonistic potential of enterococci colonizing the phyllosphere of grasses. J. Appl. Microbiol. *91:* 54–66.

Ozawa, Y., P. Courvalin and M. Galimand. 2000. Identification of enterococci at the species level by sequencing of the genes for D-alanine:D-alanine ligases. Syst. Appl. Microbiol. *23:* 230–237.

Patel, R., K.E. Piper, M.S. Rouse, J.M. Steckelberg, J.R. Uhl, P. Kohner, M.K. Hopkins, F.R. Cockerill, 3rd and B.C. Kline. 1998. Determination of 16S rRNA sequences of enterococci and application to species identification of nonmotile *Enterococcus gallinarum* isolates. J. Clin. Microbiol. *36:* 3399–3407.

Paulsen, I.T., L. Banerjei, G.S. Myers, K.E. Nelson, R. Seshadri, T.D. Read, D.E. Fouts, J.A. Eisen, S.R. Gill, J.F. Heidelberg, H. Tettelin, R.J. Dodson, L. Umayam, L. Brinkac, M. Beanan, S. Daugherty, R.T. DeBoy, S. Durkin, J. Kolonay, R. Madupu, W. Nelson, J. Vamathevan, B. Tran, J. Upton, T. Hansen, J. Shetty, H. Khouri, T. Utterback, D. Radune, K.A. Ketchum, B.A. Dougherty and C.M. Fraser. 2003. Role of mobile DNA in the evolution of vancomycin-resistant *Enterococcus faecalis.* Science *299:* 2071–2074.

Pette, J.W. 1955. De vorming van zwavelwaterstof in goudse kaas, veroorzaakt door melkzuurbacterien. Nether. Milk Dairy J. *10:* 291–302.

Pinncock, D.E. and N.E. Featherstone. 1984. Detection and quantification of *Melissococcus pluton* infection in honey bee colonies by means of enzyme-linked immunosorbent assay. J. Apic. Res. *23:* 168–170.

Pompei, R., F. Berlutti, M.C. Thaller, A. Ingianni, G. Cortis and B. Dainelli. 1992a. *Enterococcus flavescens* sp. nov., a new species of enterococci of clinical origin. Int. J. Syst. Bacteriol. *42:* 365–369.

Pompei, R., M.C. Thaller, F. Pittaluga, O. Flore and G. Satta. 1992b. Analysis of bacteriolytic activity patterns, a novel approach to the taxonomy of enterococci. Int. J. Syst. Bacteriol. *42:* 37–43.

Pot, B., L.A. Devriese, J. Hommez, C. Miry, K. Vandemeulebroecke, K. Kersters and F. Haesebrouck. 1994. Characterization and identification of *Vagococcus fluvialis* strains isolated from domestic animals. J. Appl. Bacteriol. *77:* 362–369.

Pourcher, A.M., L.A. Devriese, J.F. Hernandez and J.M. Delattre. 1991. Enumeration by a miniaturized method of *Escherichia coli, Streptococcus bovis* and enterococci as indicators of the origin of faecal pollution of waters. J. Appl. Bacteriol. *70:* 525–530.

Rodrigues, U. and M.D. Collins. 1990. Phylogenetic analysis of *Streptococcus saccharolyticus* based on 16S rRNA sequencing. FEMS Microbiol. Lett. *59:* 231–234.

Rodrigues, U. and M.D. Collins. 1991. *In* Validation of the publication of new names and new combinations previously effectively published outside the IJSB. List no. 36. Int. J. Syst. Bacteriol. *41:* 178–179.

Rogers, C.G. and W.B. Sarles. 1963. Characterization of *Enterococcus* bacteriophages from the small intestine of the rat. J. Bacteriol. *85:* 1378–1385.

Rogosa, J., J.A. Mitchell and R.F. Wiseman. 1951. A selective medium for isolation and enumeration of oral and fecal lactobacilli. J. Bacteriol. *62:* 132–133.

Röling, W.F.M. and H.W. van Verseveld. 1996. Characterization of *Tetragenococcus halophila* populations in Indonesian soy mash (kecap) fermentation. Appl. Environ. Microbiol. *62:* 1203–1207.

Röling, W.F.M. and H.W. van Verseveld. 1997. Growth, maintenance and fermentation pattern of the salt-tolerant lactic acid bacterium *Tetragenococcus halophila* in anaerobic glucose limited retention cultures. Antonie van Leeuwenhoek *72:* 239–243.

Röling, W.F.M., A.B. Prasetyo, A.H. Stouthamer and H.W. van Verseveld. 1999. Note: Physiological aspects of the growth of the lactic acid bacterium *Tetragenococcus halophila* during Indonesian soy sauce (kecap) production. J. Appl. Microbiol. *86:* 348–352.

Satomi, M., B. Kimura, M. Mizoi, T. Sato and T. Fujii. 1997. *Tetragenococcus muriaticus* sp. nov., a new moderately halophilic lactic acid bacterium isolated from fermented squid liver sauce. Int. J. Syst. Bacteriol. *47:* 832–836.

Schleifer, K.H. and R. Kilpper-Bälz. 1984. Transfer of *Streptococcus faecalis* and *Streptococcus faecium* to the genus *Enterococcus* nom. rev. as *Enterococcus faecalis* comb. nov. and *Enterococcus faecium* comb. nov. Int. J. Syst. Bacteriol. *34:* 31–34.

Schleifer, K.H., J. Kraus, C. Dvorak, R. Kilpper-Bälz, M.D. Collins and W. Fischer. 1985. Transfer of *Streptococcus lactis* and related streptococci to the genus *Lactococcus* gen. nov. Syst. Appl. Microbiol. *6:* 183–195.

Schleifer, K.H. and R. Kilpper-Bälz. 1987. Molecular and chemotaxonomic approaches to the classification of streptococci, enterococci and lactococci: a review. Syst. Appl. Microbiol. *10:* 1–19.

Schmidtke, L.M. and J. Carson. 1994. Characteristics of *Vagococcus salmoninarum* isolated from diseased salmonid fish. J. Appl. Bacteriol. *77:* 229–236.

Shepard, B.D. and M.S. Gilmore. 2002. Antibiotic-resistant enterococci: the mechanisms and dynamics of drug introduction and resistance. Microb. Infect. *4:* 215–224.

Sherman, J.M. and H.U. Wing. 1937. *Streptococcus durans.* J. Dairy Sci. *28:* 165–167.

Shewmaker, P.L., A.G. Steigerwalt, R.E. Morey, G. Carvalho Mda, J.A. Elliott, K. Joyce, T.J. Barrett, L.M. Teixeira and R.R. Facklam. 2004. *Vagococcus carniphilus* sp. nov., isolated from ground beef. Int. J. Syst. Evol. Microbiol. *54:* 1505–1510.

Skerman, V.B.D., V. McGowan and P.H.A. Sneath. 1980. Approved lists of bacterial names. Int. J. Syst. Bacteriol. *30:* 225–420.

Smith, D.G. and P.M.F. Shattock. 1964. The cellular location of antigens in streptococci of groups D, N, and Q. J. Gen. Microbiol. *34:* 165–175.

Smyth, C.J., H. Matthews, M.K. Halpenny, H. Brandis and G. Colman. 1987. Biotyping, serotyping and phage typing of *Streptococcus faecalis* isolated from dental plaque in the human mouth. J. Med. Microbiol. *23:* 45–54.

Stirling, A.C. and R. Whittenbury. 1963. Sources of the lactic acid bacteria occurring in silage. J. Appl. Bacteriol. *26:* 86–90.

Sukontasing, S., S. Tanasupawat, S. Moonmangmee, J.S. Lee and K. Suzuki. 2007. *Enterococcus camelliae* sp. nov., isolated from fermented tea leaves in Thailand. Int. J. Syst. Evol. Microbiol. *57:* 2151–2154.

Švec, P., L.A. Devriese, I. Sedláček, M. Baele, M. Vancanneyt, F. Haesebrouck, J. Swings and J. Doškař. 2001. *Enterococcus haemoperoxidus* sp. nov. and *Enterococcus moraviensis* sp. nov., isolated from water. Int. J. Syst. Evol. Microbiol. *51:* 1567–1574.

Švec, P., L.A. Devriese, I. Sedláček, M. Baele, M. Vancanneyt, F. Haesebrouck, J. Swings and J. Doškař. 2002. Characterization of yellow-pigmented and motile enterococci isolated from intestines of the garden snail *Helix aspersa.* J. Appl. Microbiol. *92:* 951–957.

Švec, P., M. Vancanneyt, L.A. Devriese, S.M. Naser, C. Snauwaert, K. Lefebvre, B. Hoste and J. Swings. 2005a. *Enterococcus aquimarinus* sp. nov., isolated from sea water. Int. J. Syst. Evol. Microbiol. *55:* 2183–2187.

Švec, P., M. Vancanneyt, J. Koort, S.M. Naser, B. Hoste, E. Vihavainen, P. Vandamme, J. Swings and J. Bjorkroth. 2005b. *Enterococcus devriesei* sp. nov., associated with animal sources. Int. J. Syst. Evol. Microbiol. *55:* 2479–2484.

Švec, P., M. Vancanneyt, M. Seman, C. Snauwaert, K. Lefebvre, I. Sedláček and J. Swings. 2005c. Evaluation of (GTG)$_5$-PCR for identification of *Enterococcus* spp. FEMS Microbiol. Lett. *247:* 59–63.

Švec, P., M. Vancanneyt, I. Sedláček, S.M. Naser, C. Snauwaert, K. Lefebvre, B. Hoste and J. Swings. 2006. *Enterococcus silesiacus* sp. nov. and *Enterococcus termitis* sp. nov. Int. J. Syst. Evol. Microbiol. *56:* 577–581.

Taylor, R.F., M. Ikawa and W. Chesbro. 1971. Carotenoids in yellow-pigmented enterococci. J. Bacteriol. *105:* 676–678.

Teixeira, L.M., R.R. Facklam, A.G. Steigerwalt, N.E. Pigott, V.L. Merquior and D.J. Brenner. 1995. Correlation between phenotypic characteristics and DNA relatedness within *Enterococcus faecium* strains. J. Clin. Microbiol. *33:* 1520–1523.

Teixeira, L.M., V.L.C. Merquior, M.D.E. Vianni, M.D.S. Carvalho, S.E.L. Fracalanzza, A.G. Steigerwalt, D.J. Brenner and R.R. Facklam. 1996. Phenotypic and genotypic characterization of atypical *Lactococcus garvieae* strains isolated from water buffalos with subclinical mastitis and confirmation of *L. garvieae* as a senior subjective synonym of *Enterococcus seriolicida*. Int. J. Syst. Bacteriol. *46:* 664–668.

Teixeira, L.M., M.D.G. Carvalho, V.L.C. Merquior, A.G. Steigerwalt, D.J. Brenner and R.R. Facklam. 1997. Phenotypic and genotypic characterization of *Vagococcus fluvialis*, including strains isolated from human sources. J. Clin. Microbiol. *35:* 2778–2781.

Teixeira, L.M., M.G. Carvalho, M.M. Espinola, A.G. Steigerwalt, M.P. Douglas, D.J. Brenner and R.R. Facklam. 2001. *Enterococcus porcinus* sp. nov. and *Enterococcus ratti* sp. nov., associated with enteric disorders in animals. Int. J. Syst. Evol. Microbiol. *51:* 1737–1743.

Teixeira, L.M. and R.R. Facklam. 2003. *Enterococcus. In* Murray, Baron, Jorgensen, Pfaller and Yolken (Editors), Manual of Clinical Microbiology, 8th edn. ASM Press, Washington, D.C., pp. 422–433.

Tendolkar, P.M., A.S. Baghdayan and N. Shankar. 2003. Pathogenic enterococci: new developments in the 21st century. Cell. Mol. Life Sci. *60:* 2622–2636.

Thiercelin, E. and L. Jouhaud. 1903. Reproduction de l'entérocoque; taches centrales; granulations péripheriques et microblastes. C. R. Séances Soc. Biol. Paris *55:* 686–688.

Toon, P., P.E. Brown and J. Baddiley. 1972. The lipid-teichoic acid complex in the cytoplasmic membrane of *Streptococcus faecalis* N.C.I.B. 8191. Biochem. J. *127:* 399–409.

Trüper, H.G. and L. de Clari. 1998. Taxonomic note: erratum and correction of further specific epithets formed as substantives (nouns) 'in apposition'. Int. J. Syst. Bacteriol. *48:* 615.

Tyrrell, G.J., R.N. Bethune, B. Willey and D.E. Low. 1997. Species identification of enterococci via intergenic ribosomal PCR. J. Clin. Microbiol. *35:* 1054–1060.

Tyrrell, G.J., L. Turnbull, L.M. Teixeira, J. Lefebvre, M.D.S. Carvalho, R.R. Facklam and M. Lovgren. 2002a. *Enterococcus gilvus* sp. nov. and *Enterococcus pallens* sp. nov. isolated from human clinical specimens. J. Clin. Microbiol. *40:* 1140–1145.

Tyrrell, G.J., L.A. Turnbull, L.M. Teixeira, J. Lefebvre, M.G.S. Carvalho, R.R. Facklam and M. Lovgren. 2002b. *In* Validation of publication of new names and new combinations previously effectively published outside the IJSEM. List no. 86. Int. J. Syst. Evol. Microbiol. *52:* 1075–1076.

Tzannetis, S., J. Leonardopoulos and J. Papavassiliou. 1970. Enterocinogeny and lysogeny in enterococci. J. Appl. Bacteriol. *33:* 358–362.

Uchida, K. and K. Mogi. 1972. Cellular fatty acid spectra of *Pediococcus* species in relation to their taxonomy. J. Gen. Appl. Microbiol. *18:* 109–129.

Uchida, K. 1982. Multiplicity in soy pediococci carbohydrate fermentation and its application for analysis of their flora. J. Gen. Appl. Microbiol. *28:* 215–223.

Ulrich, A. and T. Müller. 1998. Heterogeneity of plant-associated streptococci as characterized by phenotypic features and restriction analysis of PCR-amplified 16S rDNA. J. Appl. Microbiol. *84:* 293–303.

Van Horn, K.G. and K.M. Rodney. 1998. Colonization and microbiology of the motile enterococci in a patient population. Diagn. Microbiol. Infect. Dis. *31:* 525–530.

Vancanneyt, M., C. Snauwaert, I. Cleenwerck, M. Baele, P. Descheemaeker, H. Goossens, B. Pot, P. Vandamme, J. Swings, F. Haesebrouck and L.A. Devriese. 2001. *Enterococcus villorum* sp. nov., an enteroadherent bacterium associated with diarrhoea in piglets. Int. J. Syst. Evol. Microbiol. *51:* 393–400.

Vancanneyt, M., M. Zamfir, L.A. Devriese, K. Lefebvre, K. Engelbeen, K. Vandemeulebroecke, M. Amar, L. De Vuyst, F. Haesebrouck and J. Swings. 2004. *Enterococcus saccharominimus* sp. nov., from dairy products. Int. J. Syst. Evol. Microbiol. *54:* 2175–2179.

Vaughan, D.H., W.S. Riggsby and J.O. Mundt. 1979. Deoxyribonucleic acid relatedness of strains of yellow-pigmented group D streptococci. Int. J. Syst. Bacteriol. *29:* 204–212.

Vincent, S., R.G. Knight, M. Green, D.F. Sahm and D.M. Shlaes. 1991. Vancomycin susceptibility and identification of motile enterococci. J. Clin. Microbiol. *29:* 2335–2337.

Wallbanks, S., A.J. Martinez-Murcia, J.L. Fryer, B.A. Phillips and M.D. Collins. 1990. 16S ribosomal RNA sequence determination for members of the genus *Carnobacterium* and related lactic acid bacteria and description of *Vagococcus salmoninarum* sp. nov. Int. J. Syst. Bacteriol. *40:* 224–230.

Weaver, K.E. 2000. Enterococcal genetics. *In* Fischetti, Novick, Ferretti, Portnoy and Rood (Editors), Gram-Positive Pathogens. ASM Press, Washington, D.C., pp. 259–271.

Weaver, K.E., L.B. Rice and G. Churchward. 2002. Plasmids and transposons. *In* M. Gilmore, Clewell, P. Courvalin, Dunny, Murray and Rice (Editors), The Enterococci: Pathogenesis, Molecular Biology and Antibiotic Resistance. ASM Press, Washington, D.C., pp. 219–263.

Weiss, N. 1992. The genera *Pediococcus* and *Aerococcus. In* Balows, Trüper, Dworkin, Harder and Schleifer (Editors), The Prokaryotes, 2nd edn, Vol. 2. Springer-Verlag, New York.

White, G.F. 1912. The Cause of European Foulbrood. U.S. Department of Agriculture Bureau of Entomology Circular No. 157. U.S. Department of Agriculture, Washington, D.C.

White, G.F. 1920. European foulbrood; Bureau of Entomology Bulletin 810. U.S. Department of Agriculture, Washington, D.C.

Wicken, A.J., S.D. Elliott and J. Baddiley. 1963. The identity of streptococcal group D antigen with teichoic acid. J. Gen. Microbiol. *31:* 231–239.

Williams, A.M., J.A.E. Farrow and M.D. Collins. 1989a. *In* Validation of the publication of new names and new combinations previously effectively published outside the IJSB. List no. 31. Int. J. Syst. Bacteriol. *39:* 495–497.

Williams, A.M., J.A.E. Farrow and M.D. Collins. 1989b. Reverse transcriptase sequencing of 16S ribosomal RNA from *Streptococcus cecorum.* Lett. Appl. Microbiol. *8:* 185–189.

Williams, A.M., U.M. Rodrigues and M.D. Collins. 1991. Intrageneric relationships of enterococci as determined by reverse transcriptase sequencing of small-subunit rRNA. Res. Microbiol. *142:* 67–74.

Winston, L.G., S. Pang, B.L. Haller, M. Wong, H.F. Chambers, 3rd and F. Perdreau-Remington. 2004. API 20 strep identification system may incorrectly speciate enterococci with low level resistance to vancomycin. Diagn. Microbiol. Infect. Dis. *48:* 287–288.

Wood, A.M., G. MacKenzie, N.C. McGiliveray, L. Brown, L.A. Devriese and M. Baele. 2002. Isolation of *Enterococcus cecorum* from bone lesions in broiler chickens. Vet. Rec. *150:* 27.

Family V. **Leuconostocaceae** fam. nov.

Leu.co.nos.to.ca'ce.ae. N.L. neut. n. *Leuconostoc* type genus of the family; suff. -*aceae* ending denoting family; N.L. fem. pl. n. *Leuconostocaceae* the *Leuconostoc* family.

The family *Leucostocaceae* is circumscribed in this volume on the basis of phylogenetic analyses of 16S rRNA sequences and includes the genera *Leuconostoc*, *Oenococcus*, and *Weissella*.

Cells are Gram-stain-positive, nonmotile and asporogenous. Usually ellipsoid to spherical cells, often elongated. Cells of *Weissella* are either short rods or ovoid. Occur in pairs or short chains. Facultatively anaerobic, catalase-negative, devoid of cytochromes. Glucose is fermented heterofermentatively to lactate, CO_2, acetate and ethanol. With the exception of some DL-lactate producing *Weissella* D-lactate is a typical end product of glucose fermentation.

Type genus: **Leuconostoc** Van Tieghem 1878, 198[AL].

Genus I. **Leuconostoc** van Tieghem 1878, 198[AL] emend. mut. char. (Hucker and Pederson 1930), 66[AL]

WILHELM. H. HOLZAPFEL, JOHANN A. BJÖRKROTH AND LEON M. T. DICKS

Leu.co.nos'toc. Gr. adj. *leucus* clear, light; N.L. neut. n. *Nostoc* algal generic name; N.L. neut. n. *Leuconostoc* colorless nostoc.

The genus *Leuconostoc* comprises 10 species, with three subspecies for *Leuconostoc mesenteroides*. *Leuconostoc paramesenteroides* has been reclassified as *Weissella paramesenteroides* (Collins et al., 1993) and *Leuconostoc oenos* has been assigned to the genus *Oenococcus* as *Oenococcus oeni* sp. nov. (Dicks et al., 1995a).

The description for the genus is as published by Garvie (1986).

The cells are Gram-stain-positive, nonmotile and asporogenous. **Ellipsoidal to spherical cells, often elongated**, usually in pairs or chains. Cells grown in a glucose medium are elongated and appear morphologically closer to lactobacilli than to streptococci. Most strains form coccoid cells when cultured in milk. Cells may occur singularly or in pairs, and form short to medium length chains. When grown on solid medium, cells are elongated and can be mistaken for rods. True cellular capsules are not formed. Certain strains of *Leuconostoc mesenteroides* produce extracellular dextran, which forms an electron-dense coat on the cell surface.

Chemo-organotrophic, facultatively anaerobic, catalase-negative, no cytochromes present. Nonproteolytic. Nitrate is not reduced. Nonhemolytic. Indole is not formed. Arginine is not hydrolyzed and milk is usually not acidified and curdled. Although growth may occur at pH 4.5, the species are **nonacidophilic** and prefer an initial medium pH of 6.5. The optimal growth temperature is between 20 and 30 °C, but growth may occur at 5 °C. Grows in rich media supplemented with growth factors and amino acids. All species require nicotinic acid, thiamine, biotin and either pantothenic acid or its derivative 4'O-(α-glucopyranosyl)-D-pantothenic acid. Cobalamin and ρ-aminobenzoic acid are not required. Growth in broth is stimulated by the addition of 0.05% cysteine·HCl. Growth is uniform, except when cells in long chains sediment. In stab cultures growth is more concentrated in the lower two thirds. **Growth on surface plates is poor, but is stimulated when incubated in the presence of a gas mixture of 19.8% CO_2, 11.4% H_2, and nitrogen as balance.** Colonies develop usually only after 3–5 d, are smooth, round grayish white and less than 1 mm in diameter. Glucose is fermented to equimolar amounts of D(-)-lactic acid, CO_2, and ethanol or acetate via a combination of the hexose-monophosphate and phosphoketolase pathways. Fructose 1,6-diphosphate aldolase is absent. All species contain an active glucose-6-phosphate dehydrogenase (G-6-P-DH). Carbon dioxide and D-ribulose-5-P are produced from glucose. Xylulose 5-P phosphoketolase converts fermentable sugars to ethanol and D(-)-lactic acid. Some strains may produce additional acetate instead of ethanol when grown in the presence of oxygen. **L-malate is decarboxylated to L(+)-lactate by some strains and only in the presence of a fermentable carbohydrate.** Polysaccharides and alcohols, except mannitol, are not metabolized. **Nicotinamide adenine dinucleotide (NAD)-dependent G-6-P-DH is present.** Nonpathogenic to humans, animals and plants. The phenotypic characteristics used to differentiate the genus *Leuconostoc* from other genera of the LAB are listed in Table 121. It is evident that the genera *Leuconostoc* and *Weissella* cannot be separated by physiological means only.

The amino acids in the cross-linking peptide of the cell-wall peptidoglycan are a combination of alanine, serine and lysine; information on the peptidoglycan type is given under the description of each species.

DNA G+C content (mol%): 38–44 (T_m and Bd).

Type species: **Leuconostoc mesenteroides** (Tsenkovskii 1878) van Tieghem 1878, 191[AL] (*Ascococcus mesenteriodes* Tsenkovskii 1878, 159.).

Further descriptive information

Leuconostocs require a rich and complex growth medium (Garvie, 1967a, 1986; Reiter and Oram, 1962). Best growth is obtained under facultatively anaerobic conditions, in the presence of 0.05% (w/v) cysteine·HCl and 19.8% CO_2, 11.4% H_2 and N_2 as balance (Dellaglio et al., 1995). Optimal growth is achieved at pH 6–7, depending on the medium used (Dellaglio et al., 1995). Growth of *Leuconostoc mesenteroides* stops when the internal pH decreases to 5.4–5.7 (McDonald et al., 1990). Optimal growth is obtained between 20 and 30 °C, with minimum growth for most species at 5 °C (Garvie, 1986). However, growth at 1 °C has been recorded for strains of *Leuconostoc gelidum* and *Leuconostoc carnosum* (Shaw and Harding, 1989) and at 4 °C for strains of *Leuconostoc gasicomitatum* (Björkroth et al., 2000). All three species originate from cold-stored modified atmosphere packaged meat products (Björkroth et al., 2000; Shaw and Harding, 1989). Strains of *Leuconostoc mesenteroides* subsp. *mesenteroides* have a relatively short generation time and good

TABLE 121. Major phenotypic characteristics differentiating members of the genus *Leuconostoc* from other genera of the lactic acid bacteria (LAB) (Björkroth and Holzapfel, 2006; modified)[a]

Characteristic	*Leuconostoc*	*Weissella*	Homofermentative ovoid-shaped LAB (*Enterococcus, Lactococcus, Streptococcus*)	Homofermentative *Lactobacillus*	Heterofermentative *Lactobacillus*	*Carnobacterium*	*Pediococcus*
Morphology	Ovoid to short rods (pairs and chains)	Ovoid or rods (pairs and chains)	Ovoid (pairs and chains)	Rods	Rods	Rods	Round cocci (tetrads and pairs)
CO_2 from glucose	+	+	–	–	+	– (+)	
Hydrolysis of arginine	–	+ or –	– or +	+ or –	– or +	+	– or +
Dextran from sucrose	–/+	–/+	– or +	– or +	– or +	–	–
Lactic acid isomer from glucose	D(–)	D(–) or DL	L(+)	D(–), DL, or L(+)	DL	L(+)	DL or L(+)
Peptidoglycan[b]	LysAla/Ser	LysAla/Ser[c]	Lys–Asp	Mainly Lys–Asp and *m*-A2pm	Lys–Asp and others[d]	*m*-A2pm	Lys–Asp

[a]Symbols: +, positive reaction; –, negative reaction; +/–, mostly positive; –/+, mostly negative.

[b]Abbreviations: Lys, lysine; Ala, alanine; Ser, serine; Asp, aspartate; *m*-A2pm, 2,6-diaminopimelic acid (2,6-diaminoheptanedioic acid).

[c]No Lys–Asp-types, but differentiations of Lys–Ala/Ser types, with the exception of *Weissella kandleri*.

[d]Related to those of *Leuconostoc* and *Weissella*.

growth is obtained within 24 h of incubation at 30 °C. *Leuconostoc mesenteroides* subsp. *cremoris* may require 48 h of incubation, and grows preferably at 22–30 °C.

Carbohydrate fermentation patterns vary considerably among *Leuconostoc* species (Table 122), are often strain dependent and thus of limited value in species differentiation. Furthermore, the fermentation of sugars also depends on culture conditions. Certain strains of *Leuconostoc*, when grown under acidic conditions, ferment sucrose (Garvie, 1967b). Nonpolysaccharide producing strains of *Leuconostoc mesenteroides* form dextran in media containing tomato juice or orange juice (Langston and Bouma, 1959; Pederson and Albury, 1955). *Leuconostoc mesenteroides* subsp. *cremoris* is easily distinguished from the other species by the fermentation of only glucose, galactose, and lactose.

The electrophoretic banding patterns of the NAD-dependent D-LDHs are quite uniform (Garvie, 1969, 1980; Gasser, 1970; London, 1976), with little variation between *Leuconostoc mesenteroides* and *Leuconostoc lactis*. *Leuconostoc* species have a single NAD-dependent lactate dehydrogenase (D(–)nLDH). Several strains have NAD-independent L(+)- and D(–)-LDHs (L(+)-iLDH and D(–)-iLDH). However, the D(–)-iLDH seems to be more active *in vitro* (Garvie, 1980).

Leuconostoc mesenteroides is classified into five immunological groups with antisera prepared from G-6-P-DH of *Leuconostoc mesenteroides* and *Leuconostoc lactis* (Gasser and Hontebeyrie, 1977; Hontebeyrie and Gasser, 1975). *Leuconostoc mesenteroides* subsp. *dextranicum* and *Leuconostoc mesenteroides* subsp. *mesenteroides* belong to a single G-6-P-DH immunological group.

Leuconostoc species are clearly differentiated based on their soluble cell protein profiles. Groupings obtained by numerical analysis of the banding patterns corresponded well with results obtained by DNA–DNA hybridization (Dicks et al., 1993, 1990). The method proved valuable in the differentiation of *Leuconostoc* species from *Weissella* species (Dicks, 1995; Tsakalidou et al., 1994) and *Oenococcus oeni* (Dicks et al., 1995a). Numerical analysis of 16 and 23S rRNA gene RFLP patterns have also proved to be a good tool for differentiation of *Leuconostoc* species. This approach with *Hin*dIII as the restriction endonuclease was found to provide species-specific patterns for leuconostocs (Björkroth et al., 2000).

The major fatty acids recorded for *Leuconostoc* species are myristic (tetradecanoic) acid ($C_{14:0}$), palmitic (hexadecanoic) acid ($C_{16:0}$), palmitoleic (9-hexadecanoic) acid ($C_{16:1\ \omega7}$), oleic (*cis*-9-octadecanoic) acid ($C_{18:1\ \omega9}$) and dihydrosterculic (*cis*-9,10-methyleneoctadecanoic) acid ($C_{19\ cyclo9}$) (Schmitt et al., 1989; Shaw and Harding, 1989; Tracey and Britz, 1989). *Leuconostoc* species differ from the other lactic acid bacteria in containing oleic acid, and not vaccenic acid, as the dominant $C_{18:1}$ fatty acid (Tracey and Britz, 1989). *Leuconostoc carnosum* and *Leuconostoc gelidum* are clearly differentiated based on their fatty acid profiles (Shaw and Harding, 1989).

The 16S rRNA relatedness among *Leuconostoc* species is shown in Table 123. DNA–DNA hybridization results obtained with the different reassociation techniques, i.e., Brenner's hydroxylapatite method (Hontebeyrie and Gasser, 1977), Denhardt's membrane filter method (Farrow et al., 1989; Garvie, 1976; Vescovo et al., 1979), De Ley's spectrophotometric method (Schillinger et al., 1989), and Crosa's S1 nuclease procedure (Shaw and Harding, 1989) corresponded well.

Metabolism

Leuconostocs have complex nutritional requirements (Garvie, 1967a, 1967b, 1986; Holzapfel and Schillinger, 1992; Reiter and Oram, 1962; Tracey and Britz, 1989; Weiler and Radler, 1970, 1972). Nicotinic acid, thiamine, biotin, and pantothenic acid (or its derivatives) are generally required for growth (Garvie, 1986). Leuconostocs do not require cobalamin or p-aminobenzoic acid (Garvie, 1986). *Leuconostoc lactis* does not require folic acid, whereas *Leuconostoc mesenteroides* subsp. *mesenteroides* requires glutamic acid and valine for growth (Garvie, 1967a). *Leuconostoc* species can be divided in two groups according to their amino acid requirements, namely the dextran forming strains of *Leuconostoc mesenteroides* subsp. *dextranicum* and *Leuconostoc mesenteroides* subsp. *mesenteroides* which require up to eight amino acids, and the non-dextran-forming strains which require more than eight amino acids. Methionine, valine, and glutamic acid are required by all strains. Riboflavin and folic acid are often required and some strains do not grow in the absence of adenine, guanine, xanthine, and uracil. None of the leuconostocs require alanine. All strains require thiamine, panthothenic acid, and biotin (Garvie, 1967a).

TABLE 122. Phenotypic characteristics differentiating *Leuconostoc* species[a,b]

Characteristic	*L. mesenteroides* subsp. *mesenteroides*	subsp. *dextranicum*	subsp. *cremoris*	*L. argentinum*	*L. carnosum*	*L. citreum*	*L. durionis*	*L. fallax*	*L. ficuleum*	*L. fructosum*	*L. gasicomitatum*	*L. gelidum*	*L. inhae*	*L. kimchii*	*L. lactis*	*L. pseudomesenteroides*
Cell morphology	Coccoid to elongated cocci	Coccoid to elongated cocci	Coccoid to elongated cocci	Coccoid to elongated cocci	Coccoid to elongated cocci	Coccoid to elongated cocci	Rods in pairs or chains	Coccoid to elongated cocci	Short rods	Short rods	Coccoid to elongated cocci	Coccoid to elongated cocci	Coccoid to elongated cocci	Coccoid or ovoid	Coccoid to elongated cocci	Coccoid to elongated cocci
Acid from:																
Amygdalin	ND	ND	ND	ND	ND	ND	−	ND	ND	ND	−	ND	d	+	ND	ND
Arabinose	+	−	−	d	−	+	−	−	−	−	+	+	d	+	−	d
Arbutin	d	d	−	−	−	+	−	−	ND	ND	−	+	−	ND	−	d
Cellulose	d	d	−	ND	ND	ND	ND	ND	ND	ND	ND	ND	ND	ND	−	ND
Cellobiose	d	−	−	d	d	d	+	+	−	−	+	+	+	+	+	d
Fructose	+	+	−	d	+	+	+	+	+	+	+	+	+	+	+	+
Galactose	+	d	+	+	−	−	+	−	−	−	d	−	d	+	+	d
Lactose	d	+	d	+	−	−	−	−	+	−	−	−	+	−	+	d
Maltose	+	+	−	+	−	+	−	+	−	−	+	d	+	+	+	+
Mannitol	d	−	−	d	−	d	+	(d)	+	+	−	−	+	+	−	−
Mannose	d	d	−	+	d	+	−	+	−	−	+	+	−	+	d	+
Melibiose	d	d	−	+	d	−	−	−	−	−	−	+	−	−	d	d
Raffinose	d	d	−	+	−	−	−	−	−	−	−	−	ND	+	−	d
Ribose	d	−	−	−	d	−	+	+	ND	ND	+	d	d	+	d	+
Salicin	d	−	−	+	d	+	+	−	(+)	ND	+	+	+	+	+	d
Sucrose	+	+	−	+	+	+	+	+	+	+	−	+	+	+	+	+
Trehalose	+	+	−	d	+	+	+	(d)	+	−	+	+	−	+	−	d
Xylose	d	d	−	d	−	−	−	−	−	−	−	−	−	−	−	+
Ammonia from arginine	−	−	−	−	−	−	+	+	−	−	−	−	−	−	−	−
Lactic acid configuration	D	D	D	D	D	D	D	D	D	D	D	D	D	D	D	D
Hydrolysis of esculin	+	+	−	−	d	+	−	ND	+	+	+	+	+	+	−	d
Dextran production	+	+	−	−	+	ND	ND	ND	−	−	+	+	d	+	−	ND
Growth at pH 4.8	−	−	−	ND	ND	ND	+	+	ND	ND	ND	ND	+	ND	ND	ND
Requirement of TJF	−	−	−	−	−	−	−	−	ND	ND	ND	−	−	−	−	−
Growth in 10% ethanol	−	−	+	ND	ND	ND	ND	ND	ND	ND	ND	ND	ND	ND	ND	ND
NAD-dependent G6PDH present	−	−	−	ND	ND	ND	ND	ND	ND	+	ND	ND	ND	ND	+	ND
Growth at 37°C	d	+	−	+	−	d	+	+	+	+	−	MD	−	+	+	+
Peptidoglycan type	Lys–Ser–Ala$_2$	Lys–Ser–Ala$_2$	Lys–Ser–Ala$_2$	Lys–Ala$_2$	Lys–Ala$_2$	Lys–Ala$_2$	Lys–Ala$_2$	Lys–Ala$_2$	Lys–Ala$_2$	Lys–Ala$_2$	Lys–Ala$_2$	Lys–Ala$_2$	ND	ND	Lys–Ala$_2$	Lys–Ser–Ala$_2$

[a]+, 90% or more of strains positive; −, 90% or more of strains negative; d, 11–98% of strains positive; (+), delayed reaction; ND, no data. D, 90% or more of the lactic acid is D(−).

[b]Data for *Leuconostoc mesenteroides* are from Garvie (1986), for *Leuconostoc argentinum* from Dicks et al. (1993), for *Leuconostoc carnosum* from Shaw and Harding (1989) for *Leuconostoc citreum* from Farrow et al. (1989) and Schillinger et al. (1989), for *Leuconostoc durionis* from Leisner et al. (2005), for *Leuconostoc fallax* from Martinez-Murcia and Collins (1991) and Barrangou et al. (2002), for *Leuconostoc ficuleum* and *Leuconostoc fructosum* from Antunes et al. (2002), for *Leuconostoc gasicomitatum* from Björkroth et al. (2000), for *Leuconostoc gelidum* from Shaw and Harding (1989), for *Leuconostoc inhae* from Kim et al. (2003), for *Leuconostoc kimchii* from Kim et al. (2000b), for *Leuconostoc lactis* from Garvie (1986), and for *Leuconostoc pseudomesenteroides* from Farrow et al. (1989). The peptidoglycan type of *Leuconostoc carnosum*, *Leuconostoc gelidum*, and *Leuconostoc argentinum* was obtained from the German Collection of Microorganisms and Cell Cultures (DSMZ), Braunschweig, Germany.

TABLE 123. Percentage sequence similarities for a 1491-nucleotide region of 16S rRNA of some species of the genus *Leuconostoc*, compared to *Weissella paramesenteroides* (compiled from Björkroth and Holzapfel, 2006)[a]

Species	*L. argentinum*	*L. carnosum*	*L. citreum*	*L. fallax*	*L. gelidum*	*L. lactis*	*L. mesenteroides* subsp. *cremoris*	*L. mesenteroides* subsp. *mesenteroides*	*L. pseudomesenteroides*	*Weissella paramesenteroides*
L. mesenteroides subsp. *mesenteroides*	97.5	97.8	97.7	94.6	97.9	98.2	100	100	99.5	91.8
L. mesenteroides subsp. *cremoris*	97.7	97.8	97.7	94.6	97.9	98.2	100	100	99.5	91.8
L. argentinum	100	97.5	98.7	92.4	97.1	99.3	97.7	97.5	97.8	90.3
L. carnosum	97.5	100	97.4	94.3	98.3	97.3	97.8	97.8	97.8	91.0
L. citreum	98.7	97.4	100	94.5	97.4	98.5	97.7	97.7	97.5	91.2
L. fallax	92.4	94.3	94.5	100	93.5	93.6	94.6	94.6	94.6	92.4
L. gelidum	97.1	98.3	97.4	93.5	100	97.6	97.9	97.9	98.0	90.9
L. lactis	99.3	97.3	98.5	93.6	97.6	100	98.2	98.2	98.0	91.3
L. pseudomesenteroides	97.8	97.8	97.5	94.6	98.0	98.0	99.5	99.5	100	91.7
Weissella paramesenteroides	90.3	91.0	91.2	92.4	90.9	91.3	91.8	91.8	91.7	100

[a]The sequence data used for constructing this table originates from the National Centre for Biotechnology Information Entrez database (http://www.ncbi.nlm.nih.gov/entrez/query.fcgi?db=Nucleotide).
[b]Strains used are as follows: *Leuconostoc argentinum*, LMG 18534[T]; *Leuconostoc carnosum*, LMG 11498[T]; *Leuconostoc citreum*, LMG 9824[T]; *Leuconostoc fallax*, LMG 13177[T]; *Leuconostoc gelidum*, LMG 9850[T]; *Leuconostoc lactis*, LMG 8894[T]; *Leuconostoc mesenteroides* subsp. *cremoris*, LMG 6909[T]; *Leuconostoc mesenteroides* subsp. *mesenteroides*, LMG 6893[T]; *Leuconostoc pseudomesenteroides*, LMG 11482[T]; *Weissella paramesenteroides* LMG 9852[T]

Growth is dependent on the metabolism of a fermentable carbohydrate (Garvie, 1986). Under microaerophilic conditions, glucose is converted to equimolar amounts of $D(-)$-lactate, ethanol and CO_2 via a combination of the hexose-monophosphate (6-P-gluconate) and pentose phosphate pathways, with glucose-6-phosphate dehydrogenase and xylulose-5-phospoketolase as the key enzymes (Cogan, 1987; De Moss et al., 1951; Garvie, 1986; Schmitt et al., 1992). Fructose 1,6-diphosphate aldolase is absent. Reduced NAD (NADH) is regenerated to NAD by lactic dehydrogenase (LDH), acetaldehyde dehydrogenase, and alcohol dehydrogenase (Condon, 1987). However, in the presence of oxygen, strains of *Leuconostoc mesenteroides* use NADH oxidases and NADH peroxidases as alternative mechanisms to regenerate NAD (Condon, 1987). Acetate, instead of ethanol, and double the amount of ATP are produced (Condon, 1987; Ito et al., 1983; Johnson and McCleskey, 1957; Keenan, 1968).

Fructose is fermented by all *Leuconostoc* species except *Leuconostoc mesenteroides* subsp. *cremoris* and some strains of *Leuconostoc argentinum* (Table 122). If fructose serves as hydrogen acceptor, mannitol is formed instead of ethanol (Kandler et al., 1983).

Lactose is fermented by all strains of *Leuconostoc mesenteroides* subsp. *dextranicum*, *Leuconostoc lactis* (Garvie, 1986), and *Leuconostoc argentinum* (Dicks et al., 1993). Variable results were recorded for *Leuconostoc mesenteroides* subsp. *mesenteroides* and subsp. *cremoris* and for *Leuconostoc pseudomesenteroides* (Garvie, 1986). *Leuconostoc gelidum*, *Leuconostoc carnosum*, *Leuconostoc citreum*, and *Leuconostoc gasicomitatum* do not ferment lactose (Garvie, 1986). Most lactobacilli transport lactose and galactose into the cell via specific permeases (Lawrence and Thomas, 1979). It is not yet known if *Leuconostoc* species contain such a transport system. It is also not known if the Leloir pathway exists in *Leuconostoc* species.

Since phosphoketolase is one of the key enzymes in the heterolactic fermentation pathway (Kandler et al., 1983), it is assumed that pentoses are, as a rule, fermented by all *Leuconostoc* species. However, the fermentation profile of arabinose, ribose, and xylose were different for each of the *Leuconostoc* species (Table 122). It might be that the genes encoding pentose fermentation are plasmid linked.

The formation of dextran from sucrose has been recorded for *Leuconostoc mesenteroides* subsp. *mesenteroides* and subsp. *dextranicum* (Garvie, 1986), *Leuconostoc carnosum*, *Leuconostoc citreum*, *Leuconostoc gelidum* (Shaw and Harding, 1989), and *Leuconostoc gasicomitatum* (Björkroth et al., 2000) but not *Leuconostoc mesenteroides* subsp. *cremoris*, *Leuconostoc lactis* (Garvie, 1986), and *Leuconostoc argentinum* (Dicks et al., 1993). The ability to form dextran is often lost when serial transfers are made in media of increasing salt concentrations (Pederson and Albury, 1955). Non-dextran-producing leuconostocs may be reverted to producing dextran when inoculated in media containing tomato or orange juice (Pederson and Albury, 1955).

Citrate and malate are the most frequently fermented organic acids. All strains of *Leuconostoc mesenteroides* subsp. *cremoris* and most of the dairy leuconostocs (*Leuconostoc mesenteroides* subsp. *mesenteroides*, *Leuconostoc mesenteroides* subsp. *dextranicum* and *Leuconostoc lactis*) ferment citrate in the presence of a fermentable carbohydrate (Garvie, 1986; Schmitt et al., 1997). It is not known if *Leuconostoc gelidum*, *Leuconostoc carnosum*, (Shaw and Harding, 1989), *Leuconostoc gasicomitatum* (Björkroth et al., 2000), *Leuconostoc citreum*, *Leuconostoc pseudomesenteroides* (Farrow et al., 1989), and *Leuconostoc argentinum* (Dicks et al., 1993) are able to ferment citrate and malate. Malate is converted into $L(+)$-lactate and CO_2 by strains of *Leuconostoc mesenteroides* subsp. *mesenteroides*. Citrate is metabolized to acetate and lactate. At low pH, however, diacetyl and acetoin may be produced from citrate (Cogan et al., 1981). All strains of *Leuconostoc mesenteroides* subsp. *cremoris* utilized citrate (Garvie, 1984). By contrast, acetate and tartrate are not utilized. Increased flavor production occurs at reduced pH, and is especially associated with citrate-lyase-positive strains in the presence of citrate (Cogan, 1975; Collins and Speckman, 1974; Speckman and Collins, 1968), although considerable variations were reported in the amount of diacetyl produced among strains of *Leuconostoc mesenteroides* subsp. *cremoris* (Walker and Gilliland, 1987).

Additional pyruvic acid formed during the co-metabolism of citrate and glucose is used to reoxidize NADH and NADPH (Cogan, 1987). Because of this, acetaldehyde is not reduced to ethanol, and acetyl-phosphate is converted to acetate and ATP by an active acetate kinase. The increased level of ATP results in more rapid growth compared with cells grown on glucose without citrate. Apart from generating more ATP, citrate could serve as a carbon source for the synthesis of lipids and possibly other essential cell constituents (Schmitt et al., 1992).

Little is known about the production of biogenic amines by leuconostocs. No tyramine formation was detected in strains of *Leuconostoc* species isolated from fresh and vacuum-packaged meat (Edwards et al., 1987). Some strains of *Leuconostoc mesenteroides* subsp. *mesenteroides* produce tyramine and tryptamine (Bover-Cid and Holzapfel, 1999, 2000; Bover-Cid et al., 1999).

Certain strains isolated from sourdough degrade phytic acid. In one study, *Leuconostoc mesenteroides* strain 38 improved the Ca^{2+} and Mg^{2+} solubility during a 9 h fermentation of phytate-rich whole wheat flour (Lopez et al., 2000).

Habitat

Leuconostocs are obligately fermentative chemo-organotrophs (Garvie, 1967a, 1986). They are fastidious in their nutrient requirements and share numerous natural and artificial habitats with other LAB, and particularly with the lactobacilli, pediococci and carnobacteria (Holzapfel and Schillinger, 1992).

Reports on the association of leuconostocs and weissellas with humans and animals are relatively rare and mainly refer to the digestive tract. In the small and large intestines, the numbers of leuconostocs may range around $10^5/g$. Several workers have reported the isolation of vancomycin-resistant leuconostocs from clinical sources (Coovadia et al., 1987, 1988; Hardy et al., 1988; Horowitz et al., 1987; Isenberg et al., 1988; Luetticken and Kunstmann, 1988; Rubin et al., 1988; Ruoff et al., 1988; Wenocur et al., 1988).

Leuconostoc species are associated with a wide variety of meat products, including fresh and vacuum packaged meat, poultry and processed and fermented meat (Holzapfel, 1998; Holzapfel and Schillinger, 1992; Reuter, 1975; Von Holy and Holzapfel, 1989). *Leuconostoc carnosum*, *Leuconostoc gasicomitatum*, and *Leuconostoc gelidum* are known to cause spoilage in certain meat products (Björkroth et al., 2000, 1998; Dykes et al., 1994). Spoilage of cold-smoked salmon or trout by *Leuconostoc mesenteroides* subsp. *mesenteroides* and *Leuconostoc citreum* has been reported (Hansen and Huss, 1998; Lyhs et al., 1999), although only on a few occasions. *Leuconostoc citreum* has been isolated from low-salted fermented fish products (Paludan-Muller et al., 1999).

Leuconostoc mesenteroides and *Leuconostoc lactis* are used as starter cultures in the production of buttermilk, butter, quarg (cream cheese), Gouda and Edam (Collins, 1972; Collins and Speckman, 1974; Garvie, 1984; Quist et al., 1987). Their use in formulated starter cultures for kefir production has also been reported (Duitschaever et al., 1987; Marshall and Cole, 1985). *Leuconostoc argentinum* has been isolated from Argentine raw milk (Dicks et al., 1993).

Leuconostoc mesenteroides subsp. *mesenteroides* plays an important role in the fermentation of vegetables such as sauerkraut and cucumbers. Although not the dominant species, *Leuconostoc mesenteroides* subsp. *mesenteroides* initiates the fermentation of sauerkraut and is then succeeded by the more acid-tolerant lactobacilli (Pederson, 1930; Stamer, 1975). Meanwhile, it has been shown that, in addition to *Leuconostoc mesenteroides* subsp. *mesenteroides*, *Leuconostoc fallax* is also involved in the early stages of sauerkraut fermentation, with growth and fermentation patterns of *Leuconostoc fallax* strains highly similar to those of *Leuconostoc mesenteroides* (Barrangou et al., 2002). A final pH of 3.5 was reached during cabbage juice fermentation, with lactic acid, acetic acid, and mannitol as the main fermentation end products for both species. None of the *Leuconostoc fallax* strains exhibited the malolactic reaction, a characteristic of most *Leuconostoc mesenteroides* strains. Thus, the microbial ecology of sauerkraut fermentation may be more complex than previously indicated (Barrangou et al., 2002). *Leuconostoc citreum*, *Leuconostoc gelidum*, *Leuconostoc kimchi*, and *Leuconostoc mesenteroides* dominate the early stage of fermentation in kimchi, a traditional Korean food produced from Chinese cabbage, radishes and cucumbers (Choi et al., 2003; Kim et al., 2000a, 2000b).

Leuconostoc mesenteroides subsp. *mesenteroides* and subsp. *dextranicum* are associated with tapai, a sweet, fermented glutinous rice or cassava and chili bo, a nonfermented chili and cornstarch (Björkroth et al., 2002). *Leuconostoc mesenteroides* subsp. *dextranicum* plays a role in sour-dough fermentation (Lönner and Prove-Akesson, 1989). *Leuconostoc mesenteroides* subsp. *mesenteroides* is found in acidic leavened breads and pancake-like products such as idli or dosa and tef (Gashe, 1985; Mukherjee et al., 1965; Soni et al., 1986; Steinkraus, 1983).

Leuconostoc mesenteroides subsp. *mesenteroides* is present in high numbers in vegetable products such as Sayur-Asin prepared from mustard cabbage (Puspito and Fleet, 1985) and in starchy products such as cassava (Okafor, 1977) or kocho, produced from *Ensete ventricos* (Gashe, 1987). *Leuconostoc mesenteroides* subsp. *mesenteroides* is also involved in the submerged fermentation of coffee beans (Frank and Dela Cruz, 1964; Jones and Jones, 1984; Müller, 1996).

Leuconostoc mesenteroides subsp. *mesenteroides* is also found in sugarcane, where it produces polysaccharides and causes souring (Tilbury, 1975; van Tieghem, 1878).

Enrichment and isolation procedures

Most leuconostocs are relatively insensitive to oxygen, although more prolific growth is often observed under reduced atmospheres. Microaerophilic to anaerobic conditions are recommended. Incubation in the presence of N_2, H_2 or CO_2, or with anaerobic gas-generating kits such as GasPak (Oxoid) or Anaerocult A (Merck) is advisable. Cysteine hydrochloride (0.05–0.1%, m/v) may be added to liquid media.

Various enrichment and growth media have been described, of which most are listed by Holzapfel and Schillinger (1992) and Schillinger and Holzapfel (2002).

Some of the more general growth media include MRS (De Man et al., 1960), Rogosa SL-medium (Rogosa et al., 1951), glucose-yeast extract (Whittenbury, 1965b), acetate medium (Whittenbury, 1965a), yeast extract-glucose-citrate (YGC) broth (Garvie, 1967b), and HP Medium (Pearce and Halligan, 1978).

Tetrazolium-sucrose (TS) medium (Cavett et al., 1965), thallous-acetate-tetrazolium-sucrose (TTS) medium (Cavett et al., 1965), and HHᴅ-medium (McDonald et al., 1987) have been used in the isolation of leuconostocs from plant material. For isolations from meat, MRS medium, adjusted to pH 5.7 and supplemented with 0.2% of potassium sorbate (Reuter, 1970, 1985), and thallium acetate medium (Barnes, 1956) have been used. For the collective enumeration of leuconostocs and *Lactococcus lactis* subsp. *diacetylactis* in fermented dairy products, a whey-containing calcium lactate and Casamino acids (WACCA) medium has been suggested (Galesloot et al., 1961). For the selective isolation and enumeration of *Leuconostoc* strains from mixed starter cultures, a chemically defined medium was developed by Foucaud et al. (1997). The medium contains lactose, Mn^{2+}, Mg^{2+}, 12 amino acids, 8 vitamins, adenine, uracil, and Tween 80.

Maintenance procedures

Strains may be stored as stab cultures in MRS medium (1%, w/v agar) for 1–2 weeks at 4 °C. Alternatively, cultures may be stored for several months in yeast glucose litmus milk supplemented with calcium carbonate (Sharpe, 1981). Long-term preservation (6–24 months) may be obtained by storing dehydrated cultures on granular pumice stone at room temperature (Juven, 1979). *Leuconostoc lactis* and *Leuconostoc mesenteroides* subsp. *cremoris* were stored at –30 °C for 3 months without loss of viability (Oberman et al., 1986). Initial freezing was at –70 °C. Sterile milk or cream (18%) yielded equally good results as protecting agents. Most strains can be kept at –80 °C. Best results are obtained when the strains are grown in optimal growth medium, harvested by centrifugation, washed in a phosphate buffer (e.g., quarter-strength Ringer's solution), resuspended in the fresh growth medium containing 10–15% sterile glycerol, and distributed in small cryotubes before freezing. Strains lyophilized in the presence of 40% (w/v) sterile skim milk and stored at 4–8 °C remain viable for several years.

Taxonomic comments

According to the 16S rRNA phylogenetic analysis in the revised roadmap to Volume 3 (Figure 1 and Figure 3; Ludwig et al., this volume), *Leuconostoc*, *Oenococcus*, and *Weissella* are in a new family, *Leuconostocaceae*, within the order *Lactobacillales*. Earlier studies on the 23S rRNA elongation factor Tu and the β-subunit of ATPase showed that the genus *Leuconostoc* groups with the genera *Weissella*, *Lactobacillus*, *Lactococcus*, *Enterococcus*, and *Carnobacterium* (Ludwig et al., 1993; Ludwig and Schleifer, 1994), consistent with the current roadmap for the *Firmicutes*. *Lactobacillus*, *Paralactobacillus* and *Pediococcus* species are phylogenetically closely related to each other, forming the family *Lactobacillaceae*, and also to the *Leuconostocaceae* (Figure 3).

Based on 16S rRNA analyses (Collins et al., 1993; Martinez-Murcia and Collins, 1990, 1991) and 23S rRNA sequencing (Martinez-Murcia et al., 1993), the *Leuconostoc* group is divided into the three distinct evolutionary lines *Leuconostoc sensu stricto*, *Leuconostoc paramesenteroides*, and *Leuconostoc oenos*. *Leuconostoc paramesenteroides* has been reclassified as *Weissella paramesenteroides*

(Collins et al., 1993) and *Leuconostoc oenos* as *Oenococcus oeni* (Dicks et al., 1995a).

The *Leuconostoc sensu stricto* species *Leuconostoc argentinum*, *Leuconostoc carnosum*, *Leuconostoc citreum*, *Leuconostoc gasicomitatum*, *Leuconostoc gelidum*, *Leuconostoc kimchii*, *Leuconostoc lactis*, *Leuconostoc mesenteroides* (subsp. *cremoris*, subsp. *dextranicum* and subsp. *mesenteroides*), and *Leuconostoc pseudomesenteroides*) display 97.1–99.5% 16S rRNA sequence similarity. *Leuconostoc fallax* possesses 94–95% 16S rRNA similarity with the latter species (Martinez-Murcia and Collins, 1991). The *Leuconostoc sensu stricto* line is further divided into three evolutionary branches including *Leuconostoc carnosum*, *Leuconostoc gasicomitatum*, and *Leuconostoc gelidum* in the first branch, *Leuconostoc citreum* and *Leuconostoc lactis* in the second, and *Leuconostoc mesenteroides* and *Leuconostoc pseudomesenteroides* in the third branch. *Leuconostoc argentinum* joins the *Leuconostoc citreum - Leuconostoc lactis* branch.

Phenotypically, the leuconostocs, lactobacilli, and pediococci share many characteristics and are often isolated from the same habitat (Garvie, 1976; Sharpe et al., 1972). Based on 16S rRNA sequences, *Leuconostoc mesenteroides* subsp. *cremoris*, *Leuconostoc lactis*, *Leuconostoc mesenteroides*, *Lactobacillus confusus* (now *Weissella confusa*), *Lactobacillus viridescens* (*Weissella viridescens*), *Lactobacillus halotolerans* (*Weissella halotolerans*), *Lactobacillus kandleri* (*Weissella kandleri*), *Lactobacillus minor* (*Weissella minor*), *Oenococcus oeni*, and *Leuconostoc paramesenteroides* (*Weissella paramesenteroides*) form a natural phylogenetic group, referred to as the "leuconostoc branch" of the lactobacilli (Yang and Woese, 1989). Included in the latter phylogenetic group are some of the other *Leuconostoc* species more recently described, including *Leuconostoc carnosum*, *Leuconostoc citreum*, *Leuconostoc gasicomitatum*, *Leuconostoc gelidum*, and *Leuconostoc pseudomesenteroides*. Information on 16S rRNA gene sequences for most *Leuconostoc* species is given in Table 124.

TABLE 124. 16S rRNA gene sequence information for *Leuconostoc* spp. (modified from Barrangou et al., 2002; Antunes et al., 2002; Kim et al., 2003; Leisner et al., 2005)

Species/subspecies	Accession no.[a]	Total length (nt)
Leuconostoc amelibiosum	S78390	1490
Leuconostoc argentinum	AF175403	1471
Leuconostoc carnosum	AB022925	1450
Leuconostoc citreum	AB022923	1448
Leuconostoc durionis	AJ780981	1502
Leuconostoc fallax	S63851	1504
Leuconostoc ficulneum	AF360736	
Leuconostoc fructosum	AF360737	
Leuconostoc gasicomitatum	AF231131	1500
Leuconostoc gelidum	AB022921	1445
Leuconostoc inhae	AF439560/AY117686	1474/1505
Leuconostoc kimchii	AF173986	1505
Leuconostoc lactis	AB023968	1451
Leuconostoc mesenteroides subsp. *cremoris*	M23034	1493
Leuconostoc mesenteroides subsp. *mesenteroides*	AB023243	1440
Leuconostoc pseudomesenteroides	AB023237	1448

[a]GenBank accession numbers do not necessarily refer to the ATCC type strain. Accession numbers and corresponding nucleotide sequences are from the National Center for Biotechnology Information.

Leuconostoc mesenteroides, *Leuconostoc dextranicum*, and *Leuconostoc cremoris* are genetically closely related and were reclassified as subspecies of *Leuconostoc mesenteroides* (Garvie, 1986). DNA–DNA hybridizations indicated that *Leuconostoc lactis* and *Leuconostoc paramesenteroides* are genetically not closely related to the other *Leuconostoc* species (Dicks et al., 1990; Fantuzzi et al., 1992; Farrow et al., 1989; Garvie, 1976; Hontebeyrie and Gasser, 1977; Schillinger et al., 1989; Shaw and Harding, 1989). In DNA-rRNA hybridizations, *Leuconostoc lactis* clustered with *Leuconostoc mesenteroides*, *Leuconostoc amelibiosum*, and *Leuconostoc fructosus* (Garvie, 1981; Schillinger et al., 1989). These results are in good agreement with the 16S rRNA sequence analyses which indicated that the leuconostocs are phylogenetically related to *Weissella confusa/Weissella viridescens*, the former so-called "leuconostoc branch" of the lactobacilli (Yang and Woese, 1989).

According to the DNA similarity values obtained by Farrow et al. (1989), *Leuconostoc pseudomesenteroides* is genetically not closely related to *Leuconostoc citreum*, *Leuconostoc lactis*, *Leuconostoc mesenteroides* subsp. *mesenteroides*, and *Leuconostoc mesenteroides* subsp. *dextranicum*. However, Dicks et al. (1990) and Fantuzzi et al. (1992) recorded DNA similarity values of 78 and 82% between *Leuconostoc pseudomesenteroides*, and *Leuconostoc mesenteroides* subsp. *mesenteroides* and *Leuconostoc mesenteroides* subsp. *dextranicum*, respectively. The taxonomic status of *Leuconostoc pseudomesenteroides* is, therefore, uncertain.

Specific DNA probes, some designed from variable regions V1 and V3 of 16S rRNA (Klijn et al., 1991) have been useful in the identification of leuconostocs, especially isolates from meat (Nissen et al., 1994) and dairy products (Ward et al., 1995).

Ribotyping proved valuable in the identification of *Leuconostoc carnosum* strains isolated from ham and processed meat and the description of *Leuconostoc gasicomitatum* (Björkroth et al., 2000) and *Weissella cibaria* (Björkroth et al., 2002). Villiani et al. (1997), on the other hand, found the technique less reliable in the identification of *Leuconostoc mesenteroides*. However, they did not utilize a numerical approach in interpreting the patterns. They also found that *Leuconostoc lactis* shares same pattern with *Leuconostoc mesenteroides* subsp. *dextranicum* and *mesenteroides* and that the pattern of *Leuconostoc mesenteroides* subsp. *cremoris* is very different from the patterns of other leuconostocs, all findings are not supported by other studies (Björkroth et al., 2000, 1998). Barrangou et al. (2002) used *Rsa*I digestion of ITS-PCR products and sequencing of a 350-bp variable region of the 16S rRNA gene for identification of *Leuconostoc fallax* strains from sauerkraut fermentation at the genus and species levels; RAPD typing could differentiate isolates at the strain level. Barrangou et al. (2002)

indeed suggested that the lack of molecular identification methods for *Leuconostoc fallax* could be responsible for earlier failure to distinguish *Leuconostoc fallax* from *Leuconostoc mesenteroides*. This may also be exemplified by earlier reports indicating *Leuconostoc mesenteroides* to predominate in fermented rice cake (Cooke et al., 1987), whereas later *Leuconostoc fallax* was shown to be the prevalent *Leuconostoc* species (Kelly et al., 1995).

Macrorestriction patterns separated by using pulsed field gel electrophoresis (PFGE) were used to differentiate among *Leuconostoc mesenteroides* subsp. *mesenteroides*, *Leuconostoc pseudomesenteroides*, *Leuconostoc citreum*, and *Leuconostoc fallax* (Kelly et al., 1995) and in the characterization of *Leuconostoc carnosum* (Björkroth et al., 1998). Good results were obtained with *Apa*I and *Sma*I digests. According to the current knowledge, the strength of this method lies in distinguishing different clonal lineages.

Lee et al. (2000) developed a multiplex polymerase chain reaction (PCR) assay for the identification of *Leuconostoc* species, by using species-specific primers targeted to the genes encoding 16S rRNA. Strains are differentiated within 3h. The technique also proved reliable in the differentiation of leuconostocs from *Carnobacterium* species, *Lactobacillus curvatus*, and *Lactobacillus sakei* (Yost and Nattress, 2001).

Further reading

Cavin, J.F., P. Schmitt, A. Arias, J. Lin and C. Divies. 1988. Plasmid profiles in *Leuconostoc* species. Microbiol. Alim. Nutr. *6:* 55–62.

Coffey, A., A. Harrington, K. Kerney, C. Daly and G. Fitzgerald. 1994. Nucleotide sequence and structural organization of the small, broad-host-range plasmid pC1411 from *Leuconostoc lactis* 533. Microbiol. (Reading) *140:* 2263–2269.

Daba, H., S. Pandian, J.F. Gosselin, R.E. Simard, J. Huang and C. Lacroix. 1991. Detection and activity of a bacteriocin produced by *Leuconostoc mesenteroides*. Appl. Environ. Microbiol. *57:* 3450–3455.

Hastings, J.W. and M.E. Stiles. 1991. Antibiosis of *Leuconostoc gelidum* isolated from meat. J. Appl. Bacteriol. *70:* 127–134.

Jarvis, A.W. 1989. Bacteriophages of lactic acid bacteria. J. Dairy Sci. *72:* 3406–3428.

Orberg, P.K. and W.E. Sandine. 1984. Common occurrence of plasmid DNA and vancomycin resistance in *Leuconostoc* spp. Appl. Environ. Microbiol. *48:* 1129–1133.

O'Sullivan, T. and C. Daly. 1982. Plasmid DNA in *Leuconostoc* species. Ir. J. Food Science Technol. *6:* 206.

Saxelin, M. L., E.L. Nurmiaho-Lassila, V.T. Merilainen and R.I. Forsen. 1986. Ultrastructure and host specificity of bacteriophages of *Streptococcus cremoris*, *Streptococcus lactis* ssp. *diacetylactis* and *Leuconostoc cremoris* from Finnish fermented milk viili. Appl. Environ. Microbiol. *52:* 771–777.

List of species of the genus *Leuconostoc*

1. **Leuconostoc mesenteroides**   (Tsenkovskii 1878) van Tieghem 1878, 198[AL] (*Ascococcus mesenteroides* Tsenkovskii 1878, 159)

me.sen.ter.oi′des. Gr. n. *mesenterium* the mesentery; L. suff.-*oides* resembling, similar; N.L. adj. *mesenteroides* mesentery-like.

Gram-stain-positive, catalase-negative, nonmotile, asporogenous, spherical or lenticular cells (especially when grown on agar), arranged in pairs or chains. Colonies are small (less than 1.0mm in diameter), smooth and grayish white. Growth is facultatively anaerobic, with sedimentation when long chains of cells are formed. Growth occurs between 5 °C

and 30 °C, with optimum growth between 20 °C and 30 °C. Arginine is not hydrolyzed and milk is seldom acidified and curdled. Nonproteolytic, indole is not formed and nitrates are not reduced. Acetate and tartrate are not fermented. Growth at pH 6.5, but not at pH 4.8 and not in the presence of 10% (v/v) ethanol. Nonhemolytic.

Three subspecies have been proposed including *Leuconostoc mesenteroides* subsp. *mesenteroides*, *Leuconostoc mesenteroides* subsp. *dextranicum*, and *Leuconsotoc mesenteroides* subsp. *cremoris* (Garvie, 1983).

1a. **Leuconostoc mesenteroides subsp. mesenteroides** (Tsenkovskii 1878) van Tieghem 1878, 198[AL] (*Ascococcus mesenteroides* Tsenkovskii 1878, 159)

Morphology as described for the species. Dextran is produced from the fermentation of sucrose, especially when incubated between 20 °C and 25 °C. Colonies are not uniform on sucrose agar and depend on the chemical structure of the dominant type of dextran formed. These differences do not have taxonomic value. Some strains produce a heme-requiring catalase (Whittenbury, 1964). Cells cultured in a glucose-containing broth are killed after 30 min at 55 °C but may withstand heating to 80 °C and 85 °C in a polysaccharide-containing medium. Growth occurs from 10 °C to 37 °C (optimum, 20–30 °C). Growth in the presence of 3.0% (m/v) NaCl. Some strains grow in the presence of 6.5% (m/v) NaCl. Milk supplemented with yeast extract and glucose is acidified and curdled, but not by all strains. Gas production in yeast glucose litmus milk has been recorded for some strains. Glutamic acid and valine are required. Some strains convert L-malate to L-lactate. Acid is produced from arabinose, fructose, galactose, maltose, mannose, ribose, sucrose, and trehalose. Variable reactions have been recorded for the fermentation of amygdalin, arbutin, cellobiose, lactose, mannitol, melibiose, raffinose, salicin, xylose, and esculin. Some strains ferment citrate. The requirements for riboflavin, pyridoxal, and folic acid seem to be strain-specific. Tween 80, uracil, and a combination of uracil, guanine, adenine, and xanthine are not required for growth. The murein type is L-Lys–L-Ser–L-Ala$_2$ and L-Lys–L-Ala$_2$, respectively.

DNA G+C content (mol%): 37.0–39.0, but values of 40.0 and 41.0 have been recorded for some strains (T_m).

Type strain: 12954, ATCC 8293, DSM 20343, CCUG 30066, CIP 102305, HAMBI 2347, JCM 6124, LMG 6893, NBRC 100496, NRRL B-1118, NRRL B-3470, VKM B-1601.

GenBank accession number (16S rRNA gene): M23035.

1b. **Leuconostoc mesenteroides subsp. cremoris** (Knudsen and Sørensen 1929) Garvie 1983, 118[VP] (*Leuconostoc cremoris* (Knudson and Sørensen 1929) Garvie 1960, 288; *Betacoccus cremoris* Knudsen and Sørensen 1929, 81)

cre.mor'is. L. n.. *cremor* cream; L. gen. n. *cremoris* of cream.

Morphology as described for the species. Cells may form long chains with resultant flocculent growth in broth. Dextran is not produced from the fermentation of sucrose. Citrate is metabolized to acetoin and diacetyl (Speckman and Collins, 1968). The latter end products are not always detected, because pyruvate is an electron donor in the regeneration of NAD and the formation of D-(−)-lactate. Most strains do not ferment sucrose, but mutants have been reported (Whittenbury, 1966). Compared to the other two subspecies, a limited number of carbohydrates are fermented. All strains ferment glucose and lactose. Amygdalin, arabinose, arbutin, cellobiose, fructose, mannitol, mannose, melibiose, raffinose, salicin, sucrose, trehalose, and xylose are not fermented. Esculin is not hydrolyzed. The fermentation of galactose and maltose is strain-specific. A large number of vitamins and amino acids are required. Riboflavin, pyridoxal, folic acid, uracil, and a combination of uracil,

guanine, adenine, and xanthine are required by all strains. Glucose litmus milk is acidified and curdled by some strains. Gas production in litmus milk has not been described. Growth is facultatively anaerobic, and optimal between 18 °C and 25 °C. All known strains have been isolated from dairy products and it would appear that the strains have adapted from *Leuconostoc mesenteroides* subsp. *mesenteroides*. It is not always possible to differentiate between *Leuconostoc mesenteroides* subsp. *cremoris* and *Leuconostoc mesenteroides* subsp. *dextranicum* on a phenotypic level. The murein type is L-Lys–L-Ser–L-Ala$_2$.

DNA G+C content (mol%): 38.0–40.0 (T_m).

Type strain: ATCC 19254, CCUG 21965, CIP 103009, DSM 20346, LMG 6909, NCIMB 12008, NRRL B-3252, VKM B-1420.

GenBank accession number (16S rRNA gene): M23034.

1c. **Leuconostoc mesenteroides subsp. dextranicum** (Beijerinck 1912) Garvie 1983, 118[VP] (*Leuconostoc dextranicum* (Beijerinck 1912) Hucker and Pederson 1930, 67; *Lactococcus dextranicus* Beijerinck 1912, 27)

dex.tra'ni.cum. N.L. n. *dextranum* dextran; N.L. neut. adj. *dextranicum* relating to dextran.

Morphology as described for the species. Dextran is produced from sucrose, but less actively than by *Leuconostoc mesenteroides* subsp. *mesenteroides*. Fewer carbohydrates are fermented than by *Leuconostoc mesenteroides* subsp. *mesenteroides*. More amino acids and vitamins are required for growth than are required by *Leuconostoc mesenteroides* subsp. *mesenteroides*. Growth occurs from 10 °C to 37 °C (optimum, 20–30 °C). Glucose litmus milk is acidified and curdled and gas production may be formed but not by all strains. Glucose, fructose, lactose, maltose, sucrose and trehalose are fermented by all strains. Variable reactions have been recorded for the fermentation of amygdalin, cellobiose, galactose, mannitol, mannose, melibiose, raffinose, salicin, xylose, and esculin. Arabinose and arbutin are not fermented. L-malate is not metabolized. The requirements for riboflavin, pyridoxal, and folic acid seems to be strain-specific. Uracil is not required for growth, however, a combination of uracil, guanine, adenine, and xanthine are required by some strains. The murein type is L-Lys–L-Ser–L-Ala$_2$.

DNA G+C content (mol%): 37.0–40.0 (T_m).

The type strain is ATCC 19255, CCUG 21966, CCUG 30065, CIP 102423, DSM 20484, JCM 9700, LMG 6908, NCIMB 12007, NBRC 100495, NCAIM B.01658, NCIMB 12007, NRRL B-3469, VKM B-1225.

GenBank accession number (16S rRNA gene): not available.

2. **Leuconostoc argentinum** Dicks, Fantuzzi, Gonzalez, Du Toit and Dellaglio 1993, 348[VP]

ar.gen.ti'num. N.L. adj. *argentinum* pertaining to Argentina, where the bacterium was isolated.

According to Vancanneyt et al. (2006), *Leuconostoc argentinum* is a later synonym of *Leuconostoc lactis*.

Gram-stain-positive coccus, nonmotile, nonsporulating, catalase-negative and oxidase-negative. Spherical and sometimes lenticular cells, usually in pairs and short chains. Colonies on MRS agar are round, smooth, grayish white and usually small (less than 1.0 mm in diameter). Growth occurs between 10 and 39 °C. Optimum growth temperature is between 27 °C and 30 °C. Facultative anaerobic. Heterofermentative, with production of carbon dioxide from D-glucose. More than 97% of the lactate is produced as the D-(−)-isomer. Arginine and esculin are not hydrolyzed. Voges–Proskauer-negative. Indole is not formed and nitrate is not reduced. Polysaccharides are not produced from sucrose. Milk is coagulated without added yeast extract, but only after extended incubation (48 h) at optimal growth temperature. All strains produce acid from galactose, D-glucose, D-mannose, maltose, N-acetylglucosamine, lactose, melibiose, and sucrose. Most strains produce acid from D-fructose and D-raffinose. Some strains produce acid from L-arabinose, D-xylose, mannitol, α-methyl-D-glucoside, cellobiose, trehalose, α-gentiobiose, and D-turanose. Acid is not produced from ribose, amygdalin, arbutin, salicin, D-tagatose, glycerol, erythritol, D-arabinose, L-xylose, adonitol, α-methyl-xyloside, sorbose, rhamnose, dulcitol, inositol, sorbitol, α-methyl-D-mannoside, inulin, melezitose, starch, glycogen, xylitol, D-lyxose, D-fucose, L-fucose, D-arabitol, gluconate, 2-ketogluconate, and 5-ketogluconate.

DNA G+C content (mol%): 40.5 (T_m).

Isolated from raw milk in Argentina. Represents 7–10% of the *Leuconostoc* species isolated.

Type strain: LL76, CCUG 38887, CIP 103889, DSM 8581, JCM 11052.

GenBank accession number (16S rRNA gene): AF175403.

Note: The description of the type strain corresponds to that of the species, except that acid is typically produced from α-methyl-D-glucoside, trehalose, and D-turanose and acid is not produced from L-arabinose, D-xylose, cellobiose, and β-gentiobiose.

3. **Leuconostoc carnosum** Shaw and Harding 1989, 222[VP]

car.no′sum. L. neut. adj. *carnosum* pertaining to flesh.

Gram-stain-positive, nonmotile, nonspore-forming, spherical cells arranged in pairs or chains. Some cells are lenticular. Colonies are small, smooth, round, and grayish-white. Growth occurs at 10 °C. Most strains do not grow at 37 °C. Carbon dioxide and D-(−)-lactate are produced from D-glucose. Catalase is not produced. Arginine is not hydrolyzed. Some strains hydrolyze esculin. All strains are β-galactosidase and Voges–Proskauer-negative. Deoxyribonuclease-positive. Tween 20 and Tween 80 are not hydrolyzed, but some strains hydrolyze Tween 40 and Tween 60. Most strains produce dextran from sucrose. Acid is produced from D-fructose, D-glucose, α-methyl-D-glucoside, sucrose, and trehalose. Acid is not produced from amygdalin, D-arabinose, L-arabinose, arbutin, dulcitol, erythritol, galactose, gluconate, glycerol, lactose, mannitol, D-melezitose, raffinose, rhamnose, sorbitol, L-sorbose, or D-xylose. Some strains produce acid from cellobiose, maltose, D-mannose, melibiose, ribose, and salicin. The cellular fatty acid are of the straight-chain saturated, monounsaturated, and cyclopropane ring types, with tetradecanoic, hexadecanoic, *cis*-9,10-hexadecenoic, *cis*-11,12-octadecenoic, and *cis*-11,12-methyleneoctadecanoic acids predominating. Some strains produce methylenehexadecanoic acid.

DNA G+C content (mol%): 39.0 (T_m).

Isolated from vacuum-packaged meat stored at low temperatures.

Type strain: ATCC 49367, CCUG 30059, CIP 103319, DSM 5576, JCM 9695, LMG 11498, NCIMB 12898 (formerly NCFB 2776), SML40.

GenBank accession number (16S rRNA gene): X95977.

Note: The description of the type strain corresponds to that of the species, except that no growth occurs at 37 °C, dextran is produced from sucrose, esculin is not hydrolyzed, Tween 40 and Tween 60 are not hydrolyzed, acid is produced from D-mannose and ribose, and acid is not produced from cellobiose, maltose, melibiose, or salicin.

4. **Leuconostoc citreum** Farrow, Facklam and Collins 1989, 280[VP]

cit′re.um. N.L. adj. *citreum* lemon-colored.

Gram-stain-positive, nonmotile, spherical or lenticular cells (approx. 0.5–0.7 × 0.7–1.2 μm), arranged in pairs or short chains. Colonies produce a lemon yellow pigment. Grows occurs at 10 ° C and 30 ° C. Some strains grow at 37 °C, but not at 40 °C. Acid is produced from N-acetylglucosamine, esculin, D-fructose, glucose, D-mannose, maltose, α-methyl-D-glucoside, sucrose, trehalose, and D-turanose. The majority of strains produce acid from amygdalin, arbutin, L-arabinose, cellobiose, β-gentiobose, gluconate, D-mannitol, and salicin. A few strains produce acid from 2-ketogluconate, galactose, and D-xylose. Acid is not produced from adonitol, D-arabitol, L-arabitol, D-arabinose, dulcitol, erythritol, D-fucose, L-fucose, glycerol, glycogen, inositol, inulin, 5-ketogluconate, lactose, D-xylose, melezitose, melibiose, α-methyl-D-mannoside, β-methyl-xyloside, raffinose, ribose, L-rhamnose, sorbitol, L-sorbose, starch, D-tagatose, xylitol, or L-xylose. Arginine dihydrolase, β-D-xylosidase, and urease-negative.

Isolated from food and clinical sources.

DNA G+C content (mol%): 38.0–40.3 (T_m).

The type strain is ATCC 49370, CCUG 30060, CIP 103315, DSM 5577, JCM 9698, LMG 9849, NCIMB 13121.

GenBank accession number (16S rRNA gene): AF111948.

Note: In most respects the description of the type strain corresponds to the description of the species. The type strain produces acid from amygdalin, L-arabinose, arbutin, cellobiose, β-gentiobiose, gluconate, mannitol, and salicin. Acid is not produced from melibiose, raffinose, or D-xylose.

Leuconostoc amelibiosum is a later heterotypic synonym of *Leuconostoc citreum* (Takahashi et al., 1992).

5. **Leuconostoc amelibiosum** Schillinger, Holzapfel and Kandler 1989, 495[VP] (Effective publication: Schillinger, Holzapfel and Kandler 1989, 54.)

a′me.li.bi.o′sum, Gr. pref. *a* not; N.L. adj. *melibiosum* referring to the disaccharide melibiose; *amelibiosum* not fermenting melibiose.

Gram-stain-positive, nonmotile, catalase-negative, nonspore-forming, elongated cocci (0.8–1.0 × 1.0–1.2 μm), arranged in pairs or short chains. Facultatively anaerobic. Growth at 15 °C but not at 5 °C or 45 °C. Heterofermentative, producing CO_2, acetic acid (ethanol) and D-(−)-lactate from hexoses and gluconate. Arabinose, cellobiose, fructose, glucose, maltose, sucrose, salicin, and trehalose are fermented. Lactose, mannitol, melibiose, raffinose, rhamnose, ribose, sorbitol, and xylose are not fermented. Dextran is produced from sucrose. Peroxide is not produced. Arginine is not hydrolyzed and nitrate is not reduced. The murein type is Lys–Ala$_2$.

Isolated from sugar refineries.

DNA G+C content (mol%): 41.6 (T_m).

Type strain: CCUG 30058, DSM 20188, LMG 9824.

GenBank accession number (16S rRNA gene): X53963.

5. **Leuconostoc durionis** Leisner, Vancanneyt, Van der Meulen, Lefebvre, Engelbeen, Hoste, Laursen, Bay, Rusul, De Vuyst and Swings 2005, 1269[VP]

du.ri.o'nis. N.L. gen. n. *durionis* of *Durio*, the generic name of *Durio zibenthinus*, the durian fruit).

Gram-stain-positive, catalase-negative, nonspore-forming rods. Single cells, as pairs or in chains. Nonmotile in APT broth. Colonies on APT agar are off-white, round with smooth surfaces and 1–2 mm in diameter. Colonies develop on Rogosa agar but not on STA agar. Optimal growth at 30 °C. Growth occurs at 5 °C and 35 °C, but not at 45 °C. Growth usually occurs at pH 3.9 in APT broth, but not below pH 4.15 in La broth. Growth in the presence of 8.0% NaCl. Nitrate is not reduced and ammonia is not produced from arginine. D(–)-lactic acid and acetic acid are produced from D-glucose. Carbon dioxide is produced from the fermentation of fructose, but not from D-glucose. Cell growth is not affected by nisin produced by *Lactococcus lactis* ATCC 11454 or pediocin PA-1 produced by *Pediococcus acidilactici* PAC-1.0. Ribose, D-glucose, D-fructose, mannitol, sucrose, trehalose, D-turanose, and gluconate are fermented, as recorded after 5 d of incubation at 30 °C. The following carbohydrates are not fermented: glycerol, erythritol, D-arabinose, L-arabinose, D-xylose, L-xylose, adonitol, methyl-β-xyloside, galactose, D-mannose, L-sorbose, rhamnose, dulcitol, inositol, sorbitol, methyl α-D-mannoside, methyl-α-D-glucoside, N-acetylglucosamine, amygdalin, arbutin, esculin, salicin, cellobiose, maltose, lactose, melibiose, inulin, melezitose, D-raffinose, starch, glycogen, xylitol, β-gentiobiose, D-lyxose, D-tagatose, D-fucose, L-fucose, D-arabitol, L-arabitol, 2-ketogluconate, and 5-ketogluconate. Strain LMG 22557 ferments methyl-α-D-glucoside and maltose, but only weakly.

DNA G+C content (mol%): 44 (HPLC).

Isolated from tempoyak, a Malaysian acid-fermented condiment made from pulp of the durian fruit.

Type strain: D-24, CCUG 49949, LAB 1679, LMG 22556.

GenBank accession number (16S rRNA gene): AJ780981.

Note: The type strain and strains LMG 22557 (LAB1663, D-6) and LMG 22558 (LAB 1674, D-18) were all isolated from pulp of the durian fruit.

6. **Leuconostoc fallax** Martinez-Murcia and Collins 1992, 191[VP] (Effective publication: Martinez-Murcia and Collins 1991, 59.)

fal'lax. L. adj. *fallax* deceptive.

Gram-stain-positive, nonmotile, spherical or lenticular cells, 1.0–1.2 × 2.0–2.4 μm, arranged in pairs and short chains. Facultatively anaerobic. Growth at 10 °C and 40 °C but not at 45 °C. In glucose broth, *Leuconostoc fallax* grows and lowers the pH to 3.9. It is resistant to 9% (v/v) ethanol, and tolerates 5% (w/v) salt. No malolactic fermentation (Barrangou et al., 2002; Middelhoven and Klijn, 1997). The cell-wall murein is type A3α, L-Lys–L-Ala–L-Ala. Arginine, is not hydrolyzed. D(–)-lactate is produced from D-glucose. Acid is produced from N-acetylglucosamine, D-fructose, D-glucose, D-mannose, α-methyl-D-glucoside, maltose, ribose, sucrose, and D-turanose. Delayed acid production from mannitol and trehalose. Acid is not produced from adonitol, arabinose, arabitol, cellobiose, arbutin, amygdalin, cellobiose, dulcitol,

erythritol, fucose, galactose, glycerol, lactose, melezitose, melibiose, D-raffinose, rhamnose, salicin, sorbitol, L-sorbose, D-tagatose, and xylose.

Originally isolated from sauerkraut.

DNA G+C content (mol%): 40 (T_m) for type strain DSM 20189.

Type strain: ATCC 700006, CCUG 30061, CIP 104855, DSM 20189, JCM 9694, LMG 13177.

GenBank accession number (16S rRNA gene): S63851.

7. **Leuconostoc ficulneum** Antunes, Rainey, Nobre, Schumann, Ferreira, Ramos, Santos and da Costa 2002, 653[VP]

fi.cul'ne.um. L. adj. *ficulneum* pertaining to figs and fig trees.

Cells are Gram-stain-positive, short rods 0.4–1.0 μm by 1.8–2.9 μm, nonmotile and asporogenous. Catalase is produced, but cytochrome oxidase is absent. Colonies are small, smooth, round and grayish-white. Optimum growth is at 30 °C. No growth at 6 °C and 40 °C. The optimum pH ranges between 6.5 and 7.0. No growth at pH 4.5. Growth occurs in the presence of 7.0% (m/v) NaCl and 40% (m/v) glucose. The cell-wall murein is type A3α, variation L-Lys←L-Ala. The predominant cellular fatty acids are hexadecanoic acid ($C_{16:0}$) and 11,12-octadecenoic acid ($C_{18:1\,\omega7c}$). Arginine, elastin, fibrin, gelatin, casein, starch, and esculin are not hydrolyzed. Dextran is produced from the fermentation of sucrose. Fructose (5%, m/v) is fermented to lactate via the Embden–Meyerhof–Parnas pathway to lactate. Other fermentation products from fructose are mannitol and acetate. D(–)-lactate is produced from D-glucose. Gluconate, mannitol, and trehalose are fermented, albeit slowly. Sucrose and D-turanose are fermented, but not actively. Acid is not produced from D-arabinose, L-arabinose, D-arabitol, L-arabitol, cellobiose, erythritol, D-fucose, L-fucose, galactitol, galactose, gluconate, N-acetylglucosamine, methyl-α-D-glucoside, β-gentiobiose, 2-ketogluconate, 5-ketogluconate, glycerol, glycogen, inositol, inulin, lactose, D-lyxose, maltose, D-mannose, methyl-D-mannoside, melezitose, melibiose, D-raffinose, rhamnose, ribitol, ribose, L-sorbose, starch, sucrose, D-tagatose, trehalose, D-turanose, D-xylose, methyl-β-xyloside, L-xylose, and xylitol.

Isolated from a ripe fig bought in Portugal.

DNA G+C content (mol%): 42.6 (HPLC).

Type strain: FS-1, DSM 13613, NRRL B-23447, JCM 12225.

GenBank accession number (16S rRNA gene): AF360736.

8. **Leuconostoc fructosum** (Kodama 1956) Antunes, Rainey, Nobre, Schumann, Ferreira, Ramos, Santos and da Costa 2002, 654[VP] (Basonym: *Lactobacillus fructosus* Kodama 1956.)

fruc.to'sum. N.L. adj. *fructosum* of fructose, pertaining to fructose.

The description is based on the results of Kodama (1956, 1963), Schillinger et al. (1989), and Antunes et al. (2002).

Gram-stain-positive, short, rod-shaped (0.5–0.8 μm by 2.0–4.0 μm), nonmotile, asporogenous cells. Colonies are small, smooth, round and grayish-white. The optimum growth temperature is 30 °C. Growth at 6 °C and 40 °C, but not at 45 °C. Growth occurs in the presence of 8.0% (m/v) NaCl and 40% (m/v) glucose. The cell-wall murein is type A3α, variation L-Lys←L-Ala. The predominant cellular fatty acids are 9,10-octadecanoic acid ($C_{18:1\,\omega9c}$) and 11,12-octadecanoic acid ($C_{18:1\,\omega7c}$). Arginine, elastin, fibrin, gelatin, casein, starch, and esculin are not hydrolyzed. Dextran is not produced from the fermentation of sucrose. Fructose is metabolized via the Embden–Meyerhof–Parnas pathway. D(−)-lactate is produced from D-glucose. Mannitol is fermented. Acid is not produced from D-arabinose, L-arabinose, D-arabitol, L-arabitol, cellobiose, erythritol, D-fucose, L-fucose, galactitol, galactose, gluconate, N-acetylglucosamine, methyl-α-D-glucoside, β-gentiobiose, 2-ketogluconate, 5-ketogluconate, glycerol, glycogen, inositol, inulin, lactose, D-lyxose, maltose, D-mannose, methyl-D-mannoside, melezitose, melibiose, D-raffinose, rhamnose, ribitol, ribose, L-sorbose, starch, sucrose, D-tagatose, trehalose, D-turanose, D-xylose, methyl-β-xyloside, L-xylose, and xylitol.

Isolated from flowers in Japan.

DNA G+C content (mol%): 43.4 (HPLC).

The type strain is ATCC 13162, CCUG 32246, CIP 102985, DSM 20349, JCM 1119, LMG 9498, NBRC 3516, NCIB 10784, NRRL B-2041.

GenBank accession number (16S rRNA gene): AF360737, X61140.

9. **Leuconostoc gasicomitatum** Björkroth, Geisen, Schillinger, Weiss, De Vos, Holzapfel, Korkeala and Vandamme 2001, 264[VP] (Effective publication: Björkroth, Geisen, Schillinger, Weiss, De Vos, Holzapfel, Korkeala and Vandamme 2000, 3771.)

ga.si.co.mi.ta′tum. N.L. neut. n. *gasum* gas, L. neut. adj. *comitatum* accompanied, N.L. adj. *gasicomitatum* accompanied by gas, referring to the association with gaseous spoilage.

Gram-stain-positive, nonmotile, nonspore-forming spherical or oval cells (0.5–1.0 μm in diameter). Colonies are small and grayish-white. Catalase is not produced. Growth occurs at 4 °C and 15 °C, is slow at 30 °C, and does not occur at 37 °C. Metabolism is heterofermentative. Carbon dioxide and D(−)-lactate are produced from D-glucose. Arginine is not hydrolyzed. Exopolysaccharides are produced from the fermentation of sucrose. No growth in the presence of 6.5–12.0% (m/v) NaCl. The murein type is A3α, L-Lys–L-Ala–L-Ala. L-arabinose, ribose, D-xylose, glucose, fructose, mannose, methyl-α-D-glucoside, N-acetylglucosamine, esculin, cellobiose, maltose, melibiose, sucrose, trehalose, raffinose, gentiobiose, turanose, and 5-ketogluconate are fermented. Some isolates ferment galactose and gluconate. Glycerol, erythritol, D-arabinose, L-xylose, adonitol, β-methyl-D-xyloside, galactose, sorbose, rhamnose, dulcitol, inositol, mannitol, sorbitol, α-methyl-D-mannoside, amygdalin, arbutin, salicin, lactose, inulin, melezitose, starch, glycogen, xylitol, D-lyxose, D-tagatose, fucose, arabitol, gluconate, and 2-ketogluconate are not fermented. β-Galactosidase is produced.

Isolated from MA-packaged, tomato-marinated broiler meat strips showing extreme gaseous spoilage.

DNA G+C content (mol%): 37 (T_m).

Type strain: TB 1-10, JCM 12535, LMG 18811.

GenBank accession number (16S rRNA gene): AF231131.

Note: The description of the type strain corresponds to that of the species with the exception that esterase (C_4), esterase lipase (C_8), lipase (C_{14}), acid phosphatase and naphthol-AS-BI-phosphohydrolase activities are also present. Galactose and gluconate are fermented. The type strain and strains LMG 18812, LMG 18813, and LMG 18889 have been deposited in the BCCM/LMG Bacteria Collection.

10. **Leuconostoc gelidum** Shaw and Harding 1989, 222[VP]

ge′li.dum. L. neut. adj. *gelidum* cold (referring to the ability to grow on chill-stored meat.

Gram-stain-positive, nonmotile, asporogenous spherical cells (may be lenticellular), arranged in pairs or chains. Colonies are small, smooth, round and grayish-white. Growth occurs at 1 °C. Most strains do not grow at 37 °C. Carbon dioxide and D(−) lactate are produced from glucose. Catalase-negative. Arginine is not hydrolyzed. All strains hydrolyze esculin and are β-galactosidase-positive. Voges–Proskauer-negative. Deoxyribonuclease-positive. Tweens 20, 40, 60, and 80 are not hydrolyzed. Most strains produce dextran from sucrose. Acid is produced from amygdalin, L-arabinose, arbutin, cellobiose, D-fructose, D-glucose, D-mannose, melibiose, α-methyl-D-glucoside, raffinose, salicin, sucrose, trehalose and D-xylose. Acid is not produced from D-arabinose, dulcitol, erythritol, galactose, gluconate, glycerol, inositol, lactose, mannitol, D-melezitose, rhamnose, sorbitol, and L-sorbose. Some strains produce acid from maltose and ribose. The cellular fatty acids are of the straight-chain saturated, monounsaturated, and cyclopropane ring types, with tetradecanoic, hexadecanoic, cis-9,10-hexadecenoic, cis-11,12-octadecenoic, and cis-11,12-methyleneoctadecanoic acids predominating. Small quantities of methylenehexadecanoic acid are also present.

Isolated from vacuum-packaged meat stored at low temperatures.

DNA G+C content (mol%): 37 (T_m).

Type strain: SML9, ATCC 49366, CCUG 30198, CCUG 32249, CCUG 32696, CIP 103318, DSM 5578, JCM 9697, LMG 18297, NCIMB 12897.

GenBank accession number (16S rRNA gene): AF175402.

Note: The description of the type strain corresponds to that of the species, except that no growth occurs at 37 °C, dextran is produced from sucrose, and ribose is fermented, but maltose is not.

11. **Leuconostoc inhae** Kim, Lee, Jang, Kim and Han 2003, 1125[VP]

in′ha.e. N.L. gen. n. *inhae* of Inha, from the Inha University, Republic of Korea.

Gram-stain-positive, spherical or lenticular cells (0.8–1.0 by 0.8–1.8 μm), single or arranged in pairs. Colonies on MRS agar are small (1.0–1.5 mm in diameter), round, smooth, convex, opaque and grayish-white. Some strains produce exopolysaccharides. Cells are nonmotile and do not produce endospores. Growth is facultatively anaerobic and metabolism obligately heterofermentative. Catalase is not produced. Arginine is not hydrolyzed. Growth occurs in the presence of 3% (m/v) NaCl but not in the presence of 7% (m/v) NaCl. Growth is recorded at pH 4.8 but not at pH 3.8. All strains grow at 25 °C and 30 °C but not at 37 °C. Good growth is obtained at 1–5 °C. D-Lactic acid is produced from D-glucose. Acid is produced from L-arabinose, cellobiose, esculin, D-fructose, β-gentiobiose, D-glucose, D-mannose, mannitol, maltose, methyl-α-D-glucoside, N-acetylglucosamine, ribose, sucrose, and trehalose. Acid is not produced from adonitol, D-arabinose, D-arabitol, arbutin, 2-ketogluconate, 5-ketogluconate, dulcitol, erythritol, D-fucose, L-fucose, glycerol, glycogen, inositol, inulin, lactose, D-lyxose, melezitose, melibiose, methyl α-D-mannoside, methyl-β-xyloside, D-raffinose, rhamnose, sorbitol, L-sorbose, starch, D-tagatose, xylitol, D-xylose, and L-xylose. Some strains produce acid from amygdalin, galactose, gluconate, salicin, and D-turanose.

Isolated from kimchi.

DNA G+C content (mol%): 39.99 ± 0.5 (HPLC).

Type strain: IH003, DSM 15101, KCTC 3774.

GenBank accession number (16S rRNA gene): AF439560.

Note: The description of the type strain corresponds to the description of five other isolates, except that no growth occurs at pH 3.8 and exopolysaccharides are not produced from sucrose. Amygdalin, galactose, gluconate, salicin, and D-turanose are also not fermented.

12. **Leuconostoc kimchii** Kim, Chun and Han 2000b, 1918[VP]

kim'chi.i. N.L. gen. n. *kimchii* of kimchi, a traditional Korean food made by fermentation of Chinese cabbage.

Gram-stain-positive, spherical or ovoid cells, single or in pairs. Nonmotile and asporogenous. Catalase is not produced. Growth is facultatively anaerobic at 15 °C and 37 °C but not at 45 °C. Growth occurs in the presence of 7% (m/v) NaCl but not in the presence of 10% (m/v) NaCl. Dextran is produced from the fermentation of sucrose. Carbon dioxide and D-lactate are produced from D-glucose. Arginine is not hydrolyzed. Amygdalin, arabinose, cellobiose, galactose, gluconate, lactose, mannitol, ribose, salicin, trehalose, turanose, fructose, maltose, mannose and sucrose are fermented. Melezitose, sorbitol, starch, rhamnose, inulin, melibiose, raffinose, and xylose are not fermented. The majority of cellular fatty acids are straight-chain saturated, mono-unsaturated, and cyclopropane rings.

Isolated from kimchi.

DNA G+C content (mol%): 37 (T_m) (HPLC).

Type strain: IH25, KCTC 2386, IMSNU 11154.

GenBank accession number (16S rRNA gene): AF173986.

Note: The type strain is the only strain available.

13. **Leuconostoc lactis** Garvie 1960, 290[AL]

lac'tis. L. n. *lac* milk; L. gen.n. *lactis* of milk.

Gram-stain-positive, catalase-negative, nonmotile, asporogenous, spherical or lenticular cells (especially when grown on agar), arranged in pairs or chains. Colonies are small (less than 1.0 mm in diameter), smooth and grayish white. Dextran is not produced from sucrose. The amino acid requirements are complex. Lactose is fermented more readily than by other species and strains may acidify and even clot unsupplemented milk. Citrate may be dissimulated and acetoin and diacetyl formed (Cogan et al., 1981). Fructose, galactose, glucose, lactose, maltose, and sucrose are fermented. Amygdalin, arabinose, arbutin, cellobiose, mannitol, trehalose, and xylose are not fermented. Esculin is not hydrolyzed. Guanine, ade-

nine, xanthine, pyridoxal, and folic acid are not required for growth. Glucose litmus milk is acidified and curdled by some strains, albeit weakly. Gas production in litmus milk has not been described. Growth occurs in the presence of 3.0% (m/v) NaCl but not in the presence of 6.5% (m/v) NaCl. Growth occurs at pH 6.5 and 4.8. Heat resistance is higher than in other species and cells may survive 60 °C for 30 min. The species may not be widely distributed as recorded isolations are few and are mostly from dairy sources.

DNA G+C content (mol%): 43–45 (T_m) (Garvie, 1986), or 43 (HPLC) (Vancanneyt et al., 2006).

Type strain: ATCC 19256, DSM 20202, CCUG 30064, CIP 102422, HAMBI 2346, JCM 6123, LMG 8894, NCIMB 13091, NRRL B-3468.

GenBank accession number (16S rRNA gene): AB023968, M23031.

14. **Leuconostoc pseudomesenteroides** Farrow, Facklam and Collins 1989, 283[VP]

pseu.do.me.sen.ter.oi'des. Gr. adj. *pseudes* false; Gr. n. *mesenterium* the mesentery; Gr. suff. *eides* form, shape; N.L. adj. *pseudomensenteroides* not the true mesenteroides.

Gram-stain-positive, nonmotile, spherical or lenticular cells (*ca.* 0.5–0.7 × 0.7–1.2 μm), arranged in pairs or short chains. Most strains are nonpigmented. Growth occurs at 10 °C and 37 °C. Acid is produced from N-acetylglucosamine, D-fructose, D-glucose, maltose, D-mannose, melibiose, α-methyl-D-glucoside, raffinose, ribose, trehalose, and D-xylose. Most strains produce acid from L-arabinose, cellobiose, esculin, galactose, β-gentiobiose, lactose, sucrose, and D-turanose. A few strains produce acid from amygdalin, arbutin, 2-ketogluconate, 5-ketogluconate, and mannitol (after 7 d). Acid is not produced from adonitol, D-arabinose, D-arabitol, L-arabitol, dulcitol, erythritol, D-fucose, L-fucose, glycerol, inositol, inulin, D-lyxose, melezitose, α-methyl-D-mannoside, L-rhamnose, L-sorbose, sorbitol, D-tagatose, L-xylose, and xylitol. Arginine is not hydrolyzed. Urease, L-isoleucine arylamidase, and L-proline aralymidase-negative.

Isolated from dairy sources, food, and clinical sources.

DNA G+C content (mol%): 38.1–40.8 (T_m).

Type strain: ATCC 12291, DSM 20193, CCUG 30063, CIP 103316, JCM 9696, LMG 11482, NCCB 83005, NCIMB 8699.

GenBank accession number (16S rRNA gene): AB023237, X95979.

Note: In most respects the description of the type strain resembles the description of the species. The type strain produces acid from L-arabinose, cellobiose, 2-ketogluconate, 5-ketogluconate, galactose, β-gentiobiose, gluconate, and lactose. Acid is not produced from amygdalin arbutin, mannitol (after 7 d), salicin, or starch.

Genus II. **Oenococcus** Dicks, Dellaglio and Collins 1995a, 396[VP]

LEON M.T. DICKS AND WILHELM H. HOLZAPFEL

oe.no.coc'cus. Gr. n. *oinos* wine; Gr. masc. n. *kokkos* berry, little round thing; N.L. masc. n. *oenococcus* little round thing from wine.

Cells are Gram-stain-positive, nonmotile, and asporogenous. **Ellipsoidal to spherical in shape**, usually arranged in pairs or chains. Cell morphology varies with strain and is influenced by

growth medium and age. Chemo-organotrophic, facultatively anaerobic, catalase-negative, no cytochromes present. Usually nonproteolytic. Exoprotease activity has been reported for a strain isolated from a red wine produced in Argentina (Manca de Nadra et al., 2005). Nitrate is not reduced. Nonhemolytic. Indole is not formed. **Acidophilic (prefers an initial growth pH of 4.8) or non-acidophilic (growth at pH 5.0–7.5; pH 6.0–6.8 as optimum), depending on the species.** Growth in media supplemented with 5% (v/v) ethanol is usual, but growth in the presence of 10% (v/v) ethanol depends on the species. Growth in broth is slow and usually uniform. **Surface growth is enhanced by incubation in a 10% CO_2 atmosphere.** Colonies usually develop only after 5 d and are less than 1 mm in diameter. Grows between 20 and 30 °C. Requires a rich medium with complex growth factors and amino acids. **Tomato juice, grape juice, pantothenic acid, or 4′-O-(β-glucopyranosyl)-D-pantothenic acid may be required for growth and depends on the species.** Glucose is fermented to equal amounts of D(−)-lactic acid, CO_2, and ethanol or acetate via a pathway not yet fully elucidated. **L-malate is decarboxylated to L(+)-lactate in the presence of a fermentable carbohydrate. Polysaccharides and most alcohols are not metabolized. Nicotinamide adenine dinucleotide (NAD)-dependent glucose-6-phosphate dehydrogenase (G-6-P-DH) not present.** Habitat: wine, cider and distilled shochu residue.

DNA G+C content (mol%): 37–43 (T_m and HPLC).

Type species: **Oenococcus oeni** (Garvie 1967b) Dicks, Dellaglio and Collins 1995a, 397[VP] (*Leuconostoc oeni* corrig. Garvie 1967b, 431).

Further descriptive information

Oenococcus oeni is distinguished from *Leuconostoc* species by growth at an initial pH of 4.8 and in media containing 10% (v/v) ethanol, requirement of a tomato juice growth factor

(most strains), and the lack of NAD-dependent G-6-P-DH (Garvie, 1986). The electrophoretic mobilities of D(−)-lactate dehydrogenase (D-LDH), 6-P-G-DH, and alcohol dehydrogenase (ADH) also clearly distinguish *Oenococcus oeni* from *Leuconostoc* species (Garvie, 1986) Antisera prepared from *Oenococcus oeni* show no cross-reactivity with anti-G-6-P-DH and anti-D(−)-LDH of *Leuconostoc lactis* and anti-G-6-P-DH of *Leuconostoc mesenteroides* subsp. *mesenteroides* (Gasser and Hontebeyrie, 1977). *Oenococcus oeni* is further distinguished from *Leuconostoc* species by total soluble cell protein patterns (Dicks, 1989, 1995; Dicks et al., 1990).

Evidence for separate genotypic status of *Oenococcus oeni* comes from DNA–DNA hybridization studies (Dicks et al., 1990; Farrow et al., 1989; Garvie, 1976; Hontebeyrie and Gasser, 1977; Schillinger et al., 1989), confirmed by DNA-rRNA hybridization (Garvie, 1984; Schillinger et al., 1989) and rRNA sequence analyses (Martinez-Murcia and Collins, 1990; Yang and Woese, 1989). Comparative 16S rRNA (Martinez-Murcia and Collins, 1990; Yang and Woese, 1989) and 23S rRNA sequencing analyses (Martinez-Murcia et al., 1993) have clearly shown that *Oenococcus oeni* forms a distinct line of descent, separate from the *Leuconostoc sensu stricto* (including *Leuconostoc mesenteroides*) and the *Leuconostoc paramesenteroides* (now *Weissella paramesenteroides*) groups of species (Figure 117).

Considerable variation in carbohydrate fermentation profiles has been recorded for strains of *Oenococcus oeni*, despite the genetically homogeneous nature of the species. This variation in phenotype is further emphasized by the differences recorded in fatty acid profiles (Tracey and Britz, 1989).

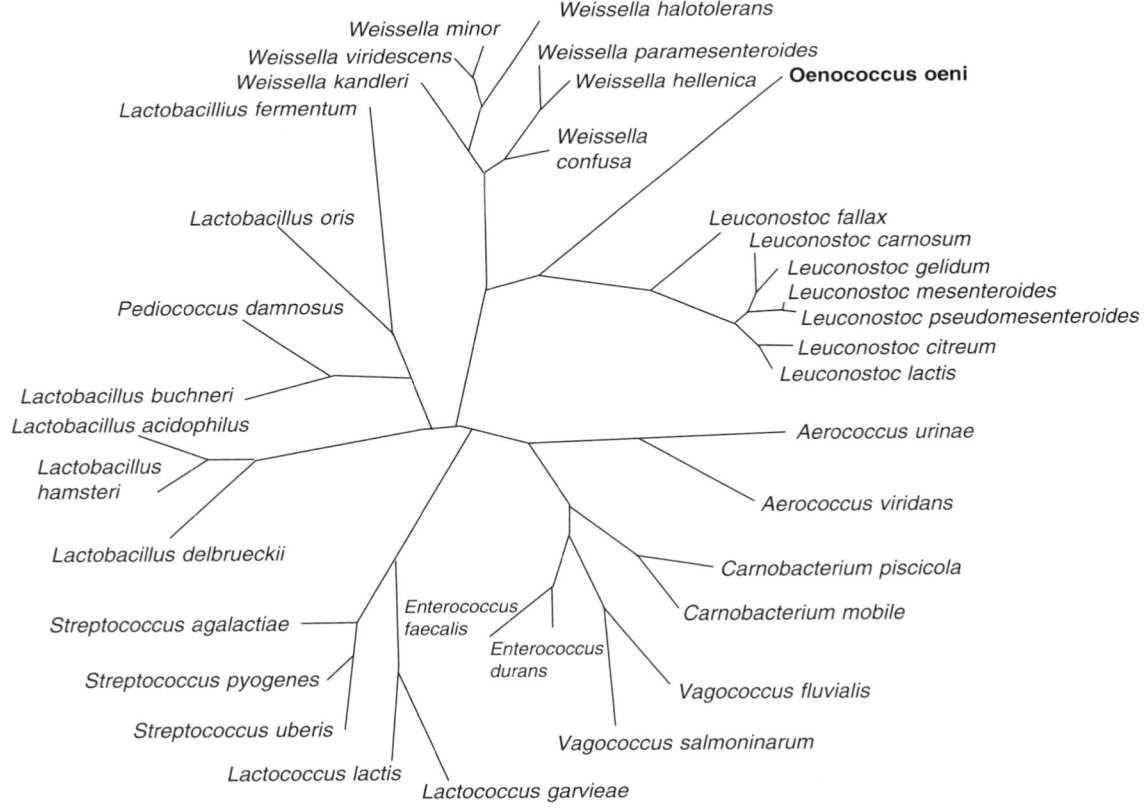

FIGURE 117. Unrooted phylogenetic tree showing the relationship of *Oenococcus oeni* and lactic acid bacteria based on 16S rRNA sequences. (Modified and reprinted from Dicks et al., 1995. International Journal of Systematic Bacteriology *45:* 395–397)

Numerical analysis of these profiles have grouped the strains into four clusters defined at $r = 0.92$, with five strains unassigned. The phenotypic characteristics which distinguish *Oenococcus oeni* from *Leuconostoc* species and *Weissella paramesenteroides* are listed in Table 125.

The development of ethanol resistance in *Oenococcus oeni* is a complex and multi-layered phenomenon. It depends on the severity and duration of the ethanol shock and on culture conditions such as medium composition, pH, and temperature (Bauer and Dicks, 2004; Desroche et al., 2005). The synthesis of low-molecular-mass stress proteins is induced and may be involved in cell adaption (Bourdineaud et al., 2004; Guzzo et al., 1997, 2000; Teixeira et al., 2002; Tourdot-Marechal et al., 2000).

Proteolytic activities and the resulting effects in wine fermentation have been studied by Manca de Nadra et al. (1999, 1997a, 1997b). The ability of *Oenococcus oeni* strains to metabolize methionine in wine with the formation of volatile sulfur compounds was reported by Pripis-Nicolau et al. (2004).

Oenococcus kitaharae is distinguished from *Oenococcus oeni* by the following characteristics: L-malate is not decarboxylated to L(+)-lactate in the presence of a fermentable carbohydrate, acid is produced from the fermentation of maltose, optimal growth at pH 6.0–6.8 and at 30 °C, and no growth in the presence of 10% (v/v) ethanol (Endo and Okada, 2006). Growth is not stimulated in the presence of tomato juice and cells can be cultured in MRS or BHI medium (Endo and Okada, 2006).

Metabolism

The metabolic pathway has not been fully elucidated and the following pertain to *Oenococcus oeni*, unless stated otherwise. Glucose-6-phosphate dehydrogenase and xylulose-5-phosphoketolase are the key metabolic enzymes (Garvie, 1986). NAD and NADP serve as coenzymes of the glucose-6-phosphate dehydrogenase, but only NADP is required (Garvie, 1975). Glucose is phosphorylated and then oxidized to 6-phosphogluconate, followed by decarboxylation. The resulting pentose phosphate is converted to lactic acid and ethanol and/or acetic acid (Garvie, 1986).

Oenococcus oeni and *Oenococcus kitaharae* prefer a complex growth medium, which contains vitamins, nucleotide bases, and amino acids. Nicotinic acid, thiamine, biotin, and pantothenic acid may be required and depends on the species. Most strains of *Oenococcus oeni* require a gluco-derivative of pantothenic acid (Amachi et al., 1971), also known as tomato juice factor (TJF) (Garvie and Mabbitt, 1967). All strains of *Oenococcus oeni* grow at a pH of 3.9 (optimal pH 4.8) and are resistant to ethanol levels of 10% (v/v) (Garvie, 1986). *Oenococcus kitaharae* is less acidophilic and grows at pH 5.0–7.5 (pH 6.0–6.8 as optimum) and is less resistant to ethanol (growth in the presence of 5%, v/v, ethanol but not 10%, v/v, ethanol). Growth of *Oenococcus kitaharae* is not stimulated in the presence of tomato juice and cells can be cultured in MRS or BHI medium (Endo and Okada, 2006). Glucose is fermented by all species, but *Oenococcus oeni* prefers fructose. In the absence of FDP-aldolase, hexoses are fermented by a combination of the hexose monophosphate system and the phosphoketolase pathway, yielding equimolar amounts of D(−)-lactate, ethanol/acetate, and CO_2 (De Moss et al., 1951; Gunsalus and Gibbs, 1952). All strains ferment lactose, maltose, and sucrose (Garvie, 1967a, 1967b, 1986).

Malic acid is actively metabolized and converted to L(+)-lactic acid and CO_2 by *Oenococcus oeni*, a process which seems to be regulated by glucose. As little as 2 mM glucose inhibits malolactic fermentation by 50%, with 5 mM or higher leading to a 70% inhibition (Miranda et al., 1997). Similar results were obtained in the presence of galactose, trehalose, and maltose, but ribose and 2-deoxyglucose had no effect on the conversion of L-malic acid (Miranda et al., 1997). The inhibition exerted by glucose is reversed by the addition of fructose or citrate, which leads to an increase in intracellular concentrations of glucose 6-phosphate, 6-phosphogluconate, and glycerol 3-phosphate. The effect of glucose 6-phosphate, 6-phosphogluconate, NAD(P)H, and NADP$^+$ on the malolactic enzyme is discussed by Miranda et al. (1997). Gallic acid (3,4,5-trihydroxybenzoic acid - $C_6H_2(OH)_3CO_2H$ - a colorless crystalline organic acid found in gallnuts, sumac, tea leaves, oak bark, and many other plants, both in its free state and as part of the tannin molecule) and free anthocyanins activate cell growth and the rate of malic acid degradation (Vivas et al., 1997). Phenolic compounds, typically associated with red wine fermentation, reduce the rate of sugar consumption by *Oenococcus oeni* but enhance citric acid consumption, thereby resulting in an increased yield of acetic acid (Rozès et al., 2003). The genes encoding the malolactic enzyme and malate permease (*mleA, mleP*) have been sequenced (Labarre et al., 1996). *Oenococcus kitaharae* does not decarboxylate L-malate to L(+)-lactate in the presence of a fermentable carbohydrate (Endo and Okada, 2006).

Little is known about the production of biogenic amines. Tyramine formation could not be detected for most *Oenococcus oeni* isolated from wine (Cilliers and van Wyk, 1985). Moreno-Arribas et al. (2003) could also not detect any potential among *Oenococcus oeni* strains to form biogenic amines. However, Gardini et al. (2005) reported tyramine formation by a strain of *Oenococcus oeni* isolated from Italian red wine. Coton et al. (1998) reported that approximately 50% of the *Oenococcus oeni* strains they studied contain the histidine decarboxylase gene, thus are capable of histamine production in wine. Guerini et al. (2002) found that >60% out of 44 *Oenococcus oeni* strains, isolated from different Italian wines, were able to produce histamine at concentrations ranging from 1.0 to 33 mg/l, while about 16% showed an additional capability to form both putrescine and cadaverine. The formation of putrescine from arginine by some strains could be demonstrated. Marcobal et al. (2004) identified the ornithine decarboxylase gene in the putrescine producer *Oenococcus oeni* BIFI-83. The gene was found to contain a 2235-nucleotide open-reading frame encoding a protein with 745 amino acid residues and a deduced molecular mass of 81 kDa.

Practical importance of malolactic fermentation by *Oenococcus oeni*. Malolactic fermentation (MLF) is important in wines with a low pH, as is often found in cooler viticultural regions. This secondary fermentation not only renders the wine microbiologically stable (Davis et al., 1985b; Wibowo et al., 1985), but also decreases the acidity and often enhances the organoleptic properties of the wine (Rodriguez et al., 1990; Rossi et al., 1978, 1993). However, in wines with a low acidity, a further increase in pH may render the wine unpalatable. Under these conditions, and also after bottling, *Oenococcus oeni* may be considered a spoilage organism. The increase in pH may also lead to spoilage by

TABLE 125. Phenotypic characteristics of *Oenococcus oeni*, *Oenococcus kitaharae* and *Leuconostoc* species[a,b]

Characteristic	*O. oeni*	*O. kitaharae*	*L. mesenteroides* subsp. *mesenteroides*	*L. mesenteroides* subsp. *dextranicum*	*L. mesenteroides* subsp. *cremoris*	*L. pseudomesenteroides*	*L. citreum*	*L. carnosum*	*L. gasicomitatum*	*L. gelidum*	*L. lactis*	*L. argentinum*	*L. fallax*
Acid from:													
Arabinose	d	−	+	−	−	d	+	−	+	+	−	d	−
Arbutin	ND	ND	d	−	−	d	+	−	+	+	−	d	−
Cellobiose	d	d	d	−	−	d	d	d	+	+	−	d	−
Cellulose	d	ND	d	d	−	ND	ND	ND	ND	ND	−	ND	ND
Fructose	+	+	+	+	−	+	+	+	+	+	+	d	+
Galactose	d	+	+	d	+	d	−	−	d	−	+	+	−
Lactose	−	−	d	+	d	d	−	−	−	−	+	+	+
Maltose	−	+	+	+	−	+	d	−	+	d	+	d	(d)
Mannitol	−	−	d	d	−	−	+	d	−	−	−	+	+
Mannose	d	+	d	d	−	+	+	d	+	+	d	+	+
Melibiose	d	−	d	d	−	d	−	d	+	+	d	+	−
Raffinose	−	(+)	d	d	−	d	−	−	+	+	d	−	+
Ribose	d	(+)	d	+	−	+	−	d	+	d	−	−	−
Salicin	d	d	d	−	−	d	+	d	−	−	d	d	(d)
Sucrose	−	−	+	+	−	d	+	+	−	+	+	+	+
Trehalose	+	+	+	+	−	+	+	+	+	+	−	d	−
Xylose	d	−	d	d	−	d	−	−	−	+	−	d	D
Lactic acid configuration	D	D	D	D	D	D	D	D	D	D	D	D	D
Hydrolysis of esculin	+	ND	+	+	−	+	+	d	+	+	−	−	ND
Dextran production	−	−	+	+	−	ND	ND	+	+	+	−		ND
Growth at pH 4.8	+	d	−	−	−	ND	ND	ND	ND	ND	−	ND	ND
Requirement of TJF	d	−	−	−	−	−	−	−	−	−	−	−	−
Growth in 10% ethanol	+	−	−	−	−	ND	ND	ND	ND	ND	−	ND	ND
NAD-dependent G6PDH present[c]	−	ND	+	+	+	ND	ND	ND	ND	MD	+	ND	ND
Growth at 37°C	d	−	d	+	−	+	d	−			+	+	+
Peptidoglycan type	Lys–Ser2 or Lys–Ala–Ser	Lys–Ser–Ala2	Lys–Ser–Ala2	Lys–Ser–Ala2	Lys–Ser–Ala2	Lys–Ser–Ala2	Lys–Ala2	Lys–Ala2	Lys–Ala2	Lys–Ala2	Lys–Ala2	Lys–Ala2	Lys–Ala2
Cell morphology	Coccoid to elongated cocci	Coccoid to elongated cocci	Coccoid to elongated cocci	Coccoid to elongated cocci	Coccoid to elongated cocci	Coccoid to elongated cocci	Coccoid to elongated cocci	Coccoid to elongated cocci	Coccoid to elongated cocci	Coccoid to elongated cocci	Coccoid to elongated cocci	Coccoid to elongated cocci	Coccoid to elongated cocci

[a] +, 90% or more of strains positive; −, 90% or more of strains negative; d, 11–98% of strains positive; (), delayed reaction; ND, no data; D, 90% or more of the lactic acid is D(−). Data for *Leuconostoc mesenteroides*, *Leuconostoc lactis*, and *Oenococcus oeni* is from Garvie (1986) and *Oenococcus kitaharae* from Endo and Okada (2006). Data for *Leuconostoc carnosum* and *Leuconostoc gelidum* is from Shaw and Harding (1989), *Leuconostoc pseudomesenteroides* from Farrow et al. (1989), for *Leuconostoc citreum* from Farrow et al. (1989) and Schillinger et al. (1989), for *Leuconostoc argentinum* from Dicks et al. (1993), for *Leuconostoc fallax* from Martinez-Murcia and Collins (1991) and for *Leuconostoc gasicomitatum* from Björkroth et al. (2000). The peptidoglycan type of *Leuconostoc carnosum*, *Leuconostoc gelidum*, and *Leuconostoc argentinum* was obtained from the German Collection of Microorganisms and Cell Cultures (DSMZ, Braunschweig).

[b] Hydrolysis of arginine is negative for all species. All representatives are heterofermentative and produce D(−)lactic acid from glucose, in addition to CO₂, and ethanol and/or acetic acid.

[c] Glucose-6-phosphate dehydrogenase.

various other *Lactobacillus* and *Leuconostoc* species (Dittrich, 1993). Selected strains of *Oenococcus oeni* are used as starter cultures for improved control of the MLF, and are commercially available in the major wine-growing areas of industrialized countries.

Habitat

Oenococcus oeni is almost exclusively found in grape must, wine, and fortified wine (Davis et al., 1985a; Dicks et al., 1995b; Wibowo et al., 1985), where it plays an important role in the decarboxylation of L-malic acid to L(+)-lactic acid. *Oenococcus oeni* was also found to be the predominant species among lactic acid bacteria (LAB) in South African brandy base wines produced without sulfur dioxide (du Plessis et al., 2004). *Oenococcus oeni* strains were found to exert a more favorable influence on the quality of base wine and distillates than other LAB. Bacteriophages specific to *Oenococcus oeni* have been isolated from sugar cane (Nel et al., 1987), which would indicate that the species may also be associated with habitats rich in sugar. However, isolation of strains of *Oenococcus oeni* from habitats other than wine and cider has not been reported. All known strains of *Oenococcus kitaharae* have been isolated from a composting distilled shochu residue collected at a shochu distillery in Miyazaki Prefecture in the Southern Kyushu area of Japan.

Enrichment and isolation procedures

Growth is supported by a complex combination of amino acids, peptides, fermentable carbohydrates, fatty acids, nucleic acids, and vitamins. Biotin, nicotine, thiamine, and pantothenic acids, or derivatives thereof, are required by most strains.

Growth on plates is greatly enhanced by semi-anaerobic or microaerophilic incubation, i.e., in the presence of 5% CO_2, N_2, and H_2, or in an anaerobic jar with GasPak (Oxoid) or Anaerocult A (Merck). Cysteine hydrochloride (0.05–0.1%) is often added to broth media. Incubation is usually between 20 and 30 °C. Depending on the strain and other growth factors, an incubation period from 48 h up to 10 d may be necessary.

Several growth media have been described for *Oenococcus oeni* (summarized in Table 126), all of which are acidic with an initial pH between 4.8 and 5.5. The acidic tomato broth (ATB) medium described by Garvie (1967b) contains 25% (v/v) tomato juice. A modification of the latter medium, described by Dicks et al. (1990) contains 25% (v/v) grape juice (acidic grape medium). Weiler

and Radler (1970) described a more complex medium which, in addition to glucose, contains yeast extract and peptone, Tween 80 (0.1%, v/v), diammonium hydrogen citrate (0.2%, w/w), and sodium acetate (0.5%, w/v). The pH of the latter medium is 5.4 and contains sorbic acid (0.05%, w/v, final concentration) to suppress the growth of yeast. Davis et al. (1985b) used MRS (De Man et al., 1960) or tomato juice medium (Ingraham et al., 1960), adjusted to pH 5.5 and supplemented with cycloheximide (50 µg/ml). The same medium, supplemented with wine (40–80%, v/v), enhanced growth (Davis et al., 1985a). Cavin et al. (1988) incorporated Tween 80, D-fructose (3.5%, w/v), and L-malic acid (1%, w/v) to their medium (FT medium), which they buffered by adding $CaCl_2$ (0.013%, w/v) and KH_2PO_4 (0.06%, w/v). A modification of the FT-medium, which contains lower concentrations of fructose (0.04%, w/v) and glucose (0.01%, w/v), cellulose MN 300 (10%, w/v), and bromocresol green (0.01%, w/v) was used to differentiate malolactic strains from strains unable to conduct the fermentation (Cavin et al., 1989). Sugar fermentation reactions may be studied by using the "sugar basal broth" described by Garvie (1984), modified by adjusting the pH to 5.2, and substituting bromocresol purple with bromocresol green (0.004%, w/v).

Oenococcus kitaharae is less fastidious and grows in MRS and BHI (brain heart infusion) medium (Endo and Okada, 2006).

Maintenance procedures

For short-term maintenance stab cultures in ATB, pH 4.8 (Garvie and Mabbitt, 1967), may be used. The addition of $CaCO_3$ neutralizes acid formation. Most strains survive storage at –80 °C in the presence of 40% (v/v) glycerol (final concentration) for at least a year. Cells may also be lyophilized. Since surface growth is often poor, best results are obtained by growing the cells in 10 ml liquid medium to mid-exponential phase. The cells are then harvested by centrifugation, resuspended in 1 ml sterile saline and an equal volume of sterile 40% (w/v) skim milk, frozen, and then lyophilized.

Taxonomic and nomenclatural comments

The genus *Oenococcus* contains two species, *Oenococcus oeni*, which was formerly considered to belong to the genus *Leuconostoc* as *Leuconostoc oenos*, and *Oenococcus kitaharae*. When the *Leuconostoc oenos* taxon was assigned to the genus *Leuconostoc* by

TABLE 126. Summary of media used for the isolation of *Oenococcus oeni* from wine[a]

Medium	pH	Selectivity	Most important components	Reference
Acidic Tomato Broth (ATB)	4.8	Semi-selective	Tomato juice (25%)	Garvie (1967a)
Weiler and Radler medium	5.3–5.4	Semi-selective	Sodium acetate (0.5%), Tween 80 (0.1%), sorbic acid (0.05%)	Weiler and Radler (1970)
M104 medium	4.5	Selective	Tomato juice (25%), DL-malic acid (1%), Tween 80 (0.1%)	Barillière (1981)
Tomato juice-Glucose-Fructose-Malate Broth (TGFMB)	5.5	Selective	Tomato juice (diluted), fructose (0.3 %), malate (0.2 %), Tween 80, cycloheximide (0.005%)	Izuagbe et al. (1985)
Fructose and Tween 80 (FT 80) medium	5.2	Semi-selective	Fructose (3.5 %), L-malic acid (1%), Tween 80 (0.1%)	Cavin et al. (1988)
Enriched tomato juice broth (ETJB)	4.8	Semi-selective	Tomato juice broth (Difco) (2%), Tween 80 (0.1%), cysteine (1%)	Britz and Tracey (1990)
UBA-agar		Selective	Apple juice (20%), cycloheximide (0.005%)	Rodriguez et al. (1990)

[a]Data from Schillinger and Holzapfel (2002). *Oenococcus kitaharae* does not require tomato juice in the growth medium and can be cultured in MRS (De Man et al., 1960) and BHI (brain heart infusion) medium (Endo and Okada, 2006).

Garvie (1967b), the specific epithet was incorrectly presented as *oenos*, which is the nominative form of this noun where the genitive is required. It is also incorrectly Latinized as a nominative. The correct latinization is *oinus*. The specific epithet *oenos* was changed to the correct Latin genitive form *oeni* by Dicks et al. (1995a). The description for the genus is as published by Garvie (1986) and amended by Dicks et al. (1995a).

Various DNA reassociation techniques have been used to clarify the taxonomic status of *Oenococcus oeni*, *Leuconostoc* species, and *Weissella* species (Farrow et al., 1989; Garvie, 1976; Hontebeyrie and Gasser, 1977; Schillinger et al., 1989; Shaw and Harding, 1989; Vescovo et al., 1979). All DNA–DNA similarity studies indicate that *Oenococcus oeni* is a genetically distinct and homogeneous species (Table 127).

Phylogenetically, *Oenococcus oeni* is well separated from the leuconostocs (the *Leuconostoc sensu stricto* group), the *Weissella* ("*Leuconostoc*") *paramesenteroides* group, and all other LAB. *Oenococcus oeni*, *Leuconostoc*, and *Weissella* species group into three distinct evolutionary lines, as shown by 16S rRNA (Collins et al., 1993; Martinez-Murcia and Collins, 1990, 1991) and 23S rRNA (Martinez-Murcia et al., 1993) analyses. Furthermore, results obtained by 16S and 23S rRNA sequence analyses (Martinez-Murcia et al., 1993), DNA–DNA hybridizations (Dicks et al., 1995b; Garvie, 1981; Schillinger et al., 1989), and total soluble-cell-protein profiles (Dicks et al., 1995b), clearly distinguish strains of *Oenococcus oeni* from all leuconostocs. In the 16S rRNA phylogenetic analysis presented in the revised roadmap to Volume 3 (Figure 1 and Figure 3; Ludwig et al., this volume), *Oenococcus*, *Weissella*, and *Leuconostoc* form a new family, *Leuconostocaceae*, within the order *Lactobacillales*.

Applying molecular and chemotaxonomic techniques, Sato et al. (2001) studied the intraspecific diversity of *Oenococcus oeni* strains isolated during red wine-making in Japan. They found high DNA–DNA relatedness and a strong similarity among 16S rDNA sequences of the studied isolates with the type strain of *Oenococcus oeni*, thereby confirming that *Oenococcus oeni* comprises one species. Using pulsed-field gel electrophoresis (PFGE), four patterns could be identified among the strains, while three different patterns of lactate dehydrogenase mobility were seen with a strong correlation between PFGE pattern and mobility.

The hypothesis that *Oenococcus oeni* is a "fast-evolving" or "tachytelic" species (Yang and Woese, 1989) was proven wrong by sequencing of the *rpoC* gene (Morse et al., 1996). The latter is shown in Figure 118.

A number of molecular taxonomic methods have been applied to differentiate *Oenococcus oeni* from other LAB. This includes *in situ* hybridization with DNA probes derived from genomic sequences (Sohier and Lonvaud-Funel, 1998) and spacer regions of 16S/23S sequences (Le Jeune and Lonvaud-Funel, 1997; Zavaleta et al., 1997a), ribotyping (Viti et al., 1996; Zavaleta et al., 1997b), 16S rDNA RFLP (Sato et al., 2000), restriction endonuclease analysis of genomic DNA (Daniel et al., 1993; Kelly et al., 1993; Lamoureux et al., 1993; Tenreiro et al., 1994; Viti et al., 1995), RAPD-PCR analyses (Zapparoli et al., 2000; Zavaleta et al., 1997b), and plasmid profiling (Prévost et al., 1995). Rodas et al. (2003) developed a 16S-ARDRA protocol for identification of LAB from grape must and wine. The protocol is based on the direct amplification of 16S rDNA followed by digestion with the restriction enzymes *Bfa*I, *Mse*I, and *Alu*I, in sequence, and could discriminate 32 of the 36 LAB reference species tested, while allowing the identification of 342 isolates from musts and wines. Pinzani et al. (2004) developed a real-time polymerase chain reaction (PCR) method for rapid detection and quantification of *Oenococcus oeni* in wine. It stated that this rapid quantification of *Oenococcus oeni* will allow prompt corrective measures to regulate the bacterial growth, and thus contributes to improving the malolactic fermentation in wines. Reguant and Bordons (2003) applied a multiplex RAPD-PCR for the typification of *Oenococcus oeni* and the study of population dynamics during malolactic fermentation. The method is based on the combination of one random 10-mer and one specific 23-mer oligonucleotide in a single PCR. Fröhlich (2002) and Fröhlich et al. (2002) developed gene probes for the specific detection and identification of *Oenococcus oeni* in wine production, and also used FISH and micromanipulation. The gene probe invention has been patented (Fröhlich et al., 2003). The construction of the probes provides a suitable signal for detection, e.g., by a good fluorescent yield.

Oenococcus kitaharae has only recently been described (Endo and Okada, 2006) and differ from *Oenococcus oeni* by being less

TABLE 127. Levels of DNA–DNA similarity (%) for *Oenococcus oeni*, *Oenococcus kitaharae*, *Leuconostoc* species, and other related species[a,b]

Species or subspecies	Oenococcus oeni	Lactobacillus kimchi	Leuconostoc carnosum	Leuconostoc lactis	Leuconostoc mesenteroides subsp. mesenteroides	Weissella confusa	Weissella paramesenteroides	Weissella viridescens
O. oeni	100	ND	ND	3–11	8–15	ND	5–7	ND
O. kitaharae	25–30	ND	ND	ND	ND	ND	ND	ND
L. argentinum	ND	7	ND	35–39	9	ND	ND	ND
L. carnosum	ND	7	78–16	0–25	19–32	ND	0–6	ND
L. citreum	ND	2	ND	23–32	21–39	ND	10–22	ND
L. fallax	ND	1	ND	ND	28–41	ND	25	ND
L. gelidum	ND	17	3–21	0–10	9–31	ND	0–3	ND
L. lactis	3–11	2	7–22	74–100	16–49	ND	0–25	ND
L. mesenteroides subsp. *mesenteroides*	8–15	2	1–17	7–38	73–108	ND	7–18	ND
L. mesenteroides subsp. *cremoris*	ND	24	4–10	5–35	66–106	ND	5–10	ND
L. mesenteroides subsp. *dextranicum*	ND	2	5–8	6–35	84–110	ND	5–19	ND
L. pseudomesenteroides	ND	2	ND	13–36	18–48	ND	9–22	ND

[a]Data compiled from Dicks et al. (1990), Endo and Okada (2006), Farrow et al. (1989), Garvie (Bfa1976), Hontebeyrie and Gasser (1977), Schillinger et al. (1989), Shaw and Harding (1989), Vescovo et al. (1979), and Kim et al. (2000b).

[b]ND, Not determined.

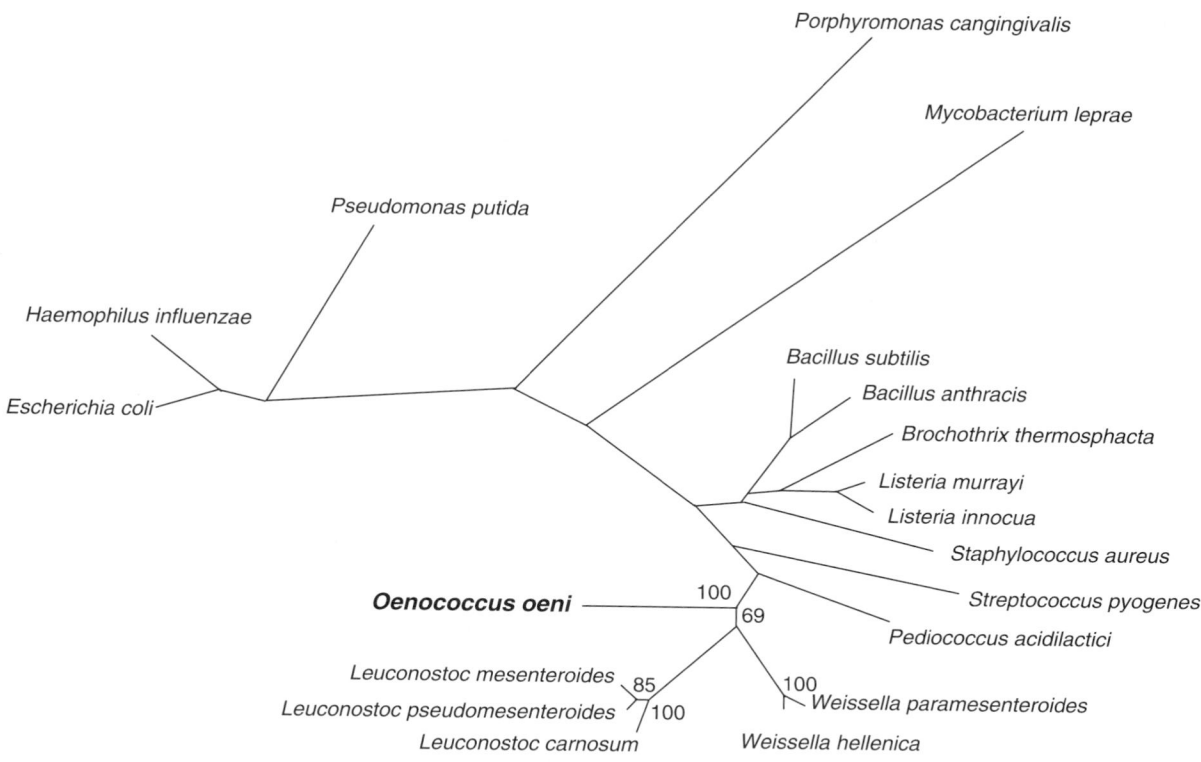

FIGURE 118. A phylogenetic tree constructed from the amino acid sequence of the second codon position of the *rpoC* gene. (Reprinted with permission from Morse et al., 1996. International Journal of Systematic Bacteriology *46:* 1004–1009)

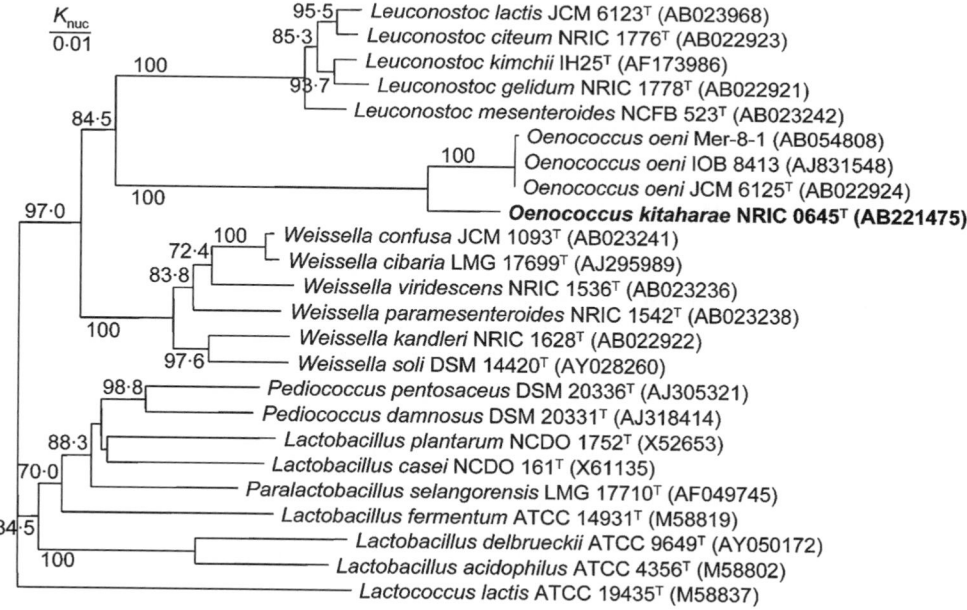

FIGURE 119. Neighbor-joining phylogenetic tree showing the relationship between *Oenococcus kitaharae, Oenococcus oeni* and related taxa (Reprinted with permission from Endo and Okada, 2006. International Journal of Systematic and Evolutionary Microbiology *56:* 2345–2348)

acidophilic, less resistant to ethanol and not able to convert L-malate to L(+)-lactate. The species is also less fastidious and grows in MRS and BHI medium not supplemented with addi-tional growth factors such as tomato juice (Endo and Okada, 2006). *Oenococcus kitaharae* is 96.0% related to *Oenococcus oeni,* based on 16S rRNA sequence analyses (Figure 119).

Further reading

Arendt, E.K., A. Lonvaud and W.P. Hammes. 1991. Lysogeny in *Leuconostoc oenos*. J. Gen. Microbiol. *137*: 2135–2139.

Fremaux, C., M. Aigele and A. Lonvaud-Funel. 1993. Sequence analysis of *Leuconostoc oenos* DNA: organization of pLo13, a cryptic plasmid. Plasmid. *25*: 212–223

Gindreau, E., S. Torolois and A. Lonvaud-Funel. 1997. Identification and sequence analysis of the region encoding the site-specific integration system from *Leuconostoc oenos (Oenococcus oeni)* temperate bacteriophage φ10MC. FEMS Microbiol. Lett. *147*: 279–285

Henick-Kling, T., T.H. Lee and D.J.D. Nicholas. 1986. Inhibition of bacterial growth and malolactic fermentation in wine by bacteriophage. J. Appl. Bacteriol. *61*: 287–293.

Janse, B.J.H., B.D. Wingfield, I.S. Pretorius and H.J.J. van Vuuren. 1987. Plasmids in *Leuconostoc oenos*. Plasmid. *17*: 173–175

Novick, C.P. 1989. Staphylococcal plasmids and their replication. Annu. Rev. Microbiol. 43: 537–565

Orberg, P.K. and W.E. Sandine. 1984.Common occurrence of plasmid DNA and vancomycin resistance in *Leuconostoc* spp. Appl. Environ. Microbiol. *48*: 1129–1133.

Santos, R., G. Vieira, M.A. Santos and H. Paveia. 1996. Characterization of temperate bacteriophages *Leuconostoc oenos* of and evidence of two prophage attachment sites in the genome of starter strain PSU-1. J. Appl. Bacteriol. *81*: 83–392.

Sozzi, T., R. Maret and J.M. Poulin. 1976. Mise en évidence de bactériophages dans le vin. Experentia. *32*: 568–569.

Tenreiro, R., M.A. Santos, L. Brito, H. Paveia and G. Vieira. 1993. Bacteriophages induced by mitomycin C treatment of *Leuconostoc oenos* strains from Portuguese wines. Lett. Appl. Microbiol. *16*: 207–209.

Zé-Zé, L., R.Tenreiro, L. Brito, M.A. Santos and H. Paveia. 1998. Physical map of the genome of *Oenococcus oeni* PSU-1 and localization of genetic markers. Microbiol.*144*: 1145–1156.

Zé-Zé, L., R. Tenreiro and H. Paveia. 2000. The *Oenococcus oeni* genome: physical and genetic mapping of strain GM and comparison with the genome of a "divergent" strain, PSU-1. Microbiol. *146*: 195–3204.

Zúñiga, M., I. Pardo and S. Ferrer. 1996. Nucleotide sequence of plasmid p4028, a cryptic plasmid from *Leuconostoc oenos*. Plasmid. *35*: 67–74.

List of species of the genus *Oenococcus*

1. **Oenococcus oeni**   (Garvie 1967b) Dicks, Dellaglio and Collins 1995a, 397[VP] (*Leuconostoc oeni* corrig. Garvie 1967b, 431)

oe'ni. Gr. gen. n. *oinou* of wine; N.L. gen. n. *oeni* of wine.

Acidophilic and grows in grape must and wine at pH 3.5–3.8. Prefers an initial growth pH of 4.8. Growth is not inhibited in the presence of 10% (v/v) ethanol. Optimum growth temperature is 22 °C. Growth at 15 °C is slow. Requires a rich growth medium supplemented with grape juice, tomato juice, pantothenic acid, or a glucose derivative of pantothenate. Converts L-malate to L(+)-lactate and CO_2 in the presence of a fermentable carbohydrate. Glucose is fermented to D(−)-lactic acid, CO_2, and ethanol or acetate. Metabolically inactive and ferments only a few carbohydrates. Prefers fructose and usually ferments trehalose. In grape must or wine, pentose sugars (arabinose or xylose) are fermented before glucose, resulting in diauxic growth. Variable reactions have been recorded for arabinose, cellobiose, galactose, mannose, melibiose, salicin, and xylose fermentations. Sucrose, lactose, maltose, mannitol, and raffinose are not fermented. Esculin is hydrolyzed. Arginine is hydrolyzed by certain strains, but only in wine or related habitats. Some strains ferment citrate in the presence of a fermentable carbohydrate, but this characteristic is often lost when strains are subcultured. Glucose-6-phosphate dehydrogenase activity is only obtained in the presence of NADP. D(−)-LDH migrates slowly and at least nine electrophoretically different D(−)-LDH bands have been obtained. Growth in yeast glucose milk is unusual, but, if it occurs, acid and gas are not produced. Certain strains produce extracellular polysaccharides in medium supplemented with grape juice.

Oenococcus oeni is a phylogenetically homogeneous species based on 16S–23S-rDNA intergenic sequence analysis (Zavaleta et al., 1997a). Isolated from wine. The amino acid sequence of the interpeptide bridge of cell-wall peptidoglycan is L-Lys–L-Ala–L-Ser or L-Lys–L-Ser–L-Ser.

DNA G+C content (mol%): 38–42 (T_m).

Type strain: ATCC 23279, CCUG 30199, CCUG 32250, CIP 106144, DSM 20252, JCM 6125, LMG 9851, NBRC 100497, NCIMB 11648, NRRL B-3472.

GenBank accession number (16S rRNA gene): AB022924, M35820, X95980.

2. **Oenococcus kitaharae**   (Endo and Okada 2006)

ki.ta.ha'rae. N.L. gen. n. *kitaharae* of Kitahara, in honor of the Japanese microbiologist Kakuo Kitahara.

Cells are Gram-stain-positive, nonmotile and small ellipsoidal cocci (0.2–0.4 by 0.5–0.8 μm), usually arranged in pairs. Growth on agar medium is enhanced when incubated anaerobically. Colonies are white, smooth and approximately 1 mm in diameter on MRS and BHI (brain heart infusion) agar and less than 1 mm in diameter on MRS agar under aerobic conditions after 7 d at 30 °C. Incubation under anaerobic conditions yield colonies of approximately 2 mm in diameter after 5 d at 30 °C. Heterofermentative and produce lactic acid, carbon dioxide and ethanol or acetic acid from D-glucose. D(−)- and L(+)-lactate are produced (9:1). Nitrate is not reduced. Acid is produced from D-glucose, D-fructose, D-galactose, D-mannose, maltose, melibiose and D-trehalose. D-ribose, D-gluconate and raffinose are weakly fermented. L-arabinose, D-xylose, L-rhamnose, lactose, sucrose, D-melezitose, D-mannitol, D-sorbitol and starch are nor fermented. Fermentation of cellobiose and D-salicin is strain-dependent. Dextran is not formed from sucrose. L-malate is not decarboxylated to L-lactate in the presence of a fermentable carbohydrate. Growth between 20 and 30 °C, but not at 15 or 37 °C. Optimum temperature for growth is 30 °C. All strains grow at pH values ranging from 5.0 to 7.5; some strains grow at pH 4.5. Optimum pH for growth is between 6.0 and 6.8. Growth is not stimulated in the presence of tomato juice. Growth is observed in MRS and BHI broth containing 1 % (w/v) NaCl but not 2.5 % (w/v) NaCl. Growth is observed in broth containing 5 % (v/v) ethanol but not 10 % (v/v) ethanol. All known strains were isolated from a composting distilled shochu residue collected at a shochu distillery in Miyazaki Prefecture in the Southern Kyushu area of Japan.

DNA G+C content (mol%): 41–43 mol% (type strain, 41 mol%) (HPLC).

Type strain: NRIC 0645[T] (=JCM 13282[T]=DSM 17330[T]).

Genus III. **Weissella** Collins, Samelis, Metaxopoulos and Wallbanks 1994, 370[VP]
(Effective publication: Collins, Samelis, Metaxopoulos and Wallbanks 1993, 597.)

JOHANNA BJÖRKROTH, LEON M.T. DICKS AND WILHELM. H. HOLZAPFEL

Weiss.el'la. N.L. dim. fem. n. named after Norbert Weiss, a German microbiologist known for his many research contributions to the taxonomy of the lactic acid bacteria.

The description for the genus is as published by Collins et al. (1993). The cells are **either short rods with rounded tapered ends** (see Figure 120 for *Weissella cibaria*) **or ovoid** and occur in pairs or in short chains. A tendency toward pleomorphism may be found for strains in species such as *Weissella minor* (see Figure 121). **Gram-stain-positive, nonmotile, and asporogenous.** Chemo-organotrophic, facultatively anaerobic, catalase-negative, and devoid of cytochromes. Obligately fermentative chemo-organotrophs; **ferment glucose heterofermentatively and have complex nutritional requirements.** The genus *Weissella*

belongs to lactic acid bacteria (LAB) because of their metabolism and phylogenetic position. Carbohydrate fermentation via the hexose-monophosphate and the phosphoketolase pathways. End products of glucose fermentation are CO_2, ethanol, and/or acetate. The configuration of lactic acid is either DL- or D(−), depending on the species. Amino acids, peptides, fermentable carbohydrates, fatty acids, nucleic acids, and vitamins are generally required for growth. Biotin, nicotine, thiamine, and pantothenic acid or its derivatives are required by all strains. Arginine is not hydrolyzed by all species. Growth occurs at 15 °C, with some species growing at 42–45 °C. Cell-wall peptidoglycan is based on lysine as diamino acid, and, with the exception of *Weissella kandleri*, all contain alanine or alanine and serine in the interpeptide bridge. In addition, the interpeptide bridge of *Weissella kandleri* (Lys–L-Ala–Gly–L-Ala$_2$) contains glycine (Holzapfel and van Wyk, 1982). The phenotypic characteristics of *Weissella* spp. are listed in Table 128.

DNA G+C content (mol%): 37–47 (T_m and Bd).

Type species: **Weissella viridescens** (Niven and Evans, 1957) Collins, Samelis, Metaxopoulos and Wallbanks 1994, 370[VP] (Effective publication: Collins, Samelis, Metaxopoulos and Wallbanks 1993, 601.) (*Lactobacillus viridescens* Niven and Evans 1957, 758; *Lactobacillus corynoides* subsp. *corynoides* Kandler and Abo-Elnaga 1966, 753)

Further descriptive information

According to the 16S rRNA phylogenetic analysis in the revised roadmap to Volume 3 (Figure 1 and Figure 3; Ludwig et al., this volume), *Weissella*, *Leuconostoc*, and *Oenococcus* are in a new family, *Leuconostocaceae*, within the order *Lactobacillales* in the phylum *Firmicutes*, i.e., Gram-positive bacteria with a DNA base composition of less than 50 mol% G+C. It has been concluded from the 16S and 23S rRNA sequence analyses (Martinez-Murcia and Collins, 1990) that *Leuconostoc paramesenteroides* is phylogenetically distinct from *Leuconostoc mesenteroides* and forms a natural grouping with the heterofermentative former *Lactobacillus* species *Lactobacillus confusus*, *Lactobacillus kandleri*, *Lactobacillus minor*, and *Lactobacillus viridescens*.

Table 129 shows 16S rRNA-encoding gene sequence similarity levels within *Weissella*. *Weissella confusa*, *Weissella cibaria*, *Weissella halotolerans*, *Weissella hellenica*, *Weissella kandleri*, *Weissella koreensis*, *Weissella minor*, *Weissella paramesenteroides*, *Weissella soli*, *Weissella thailandensis* and *Weissella viridescens* have 93.9–99.2% 16S rRNA-encoding gene sequence similarity over the length of 1400–1500 bp determined. Based on the sequences encoding the 16S rRNA genes, the genus *Weissella* is differentiated into four main phylogenetic branches. The first branch contains *Weissella hellenica*, *Weissella paramesenteroides*, and *Weissella thailandensis*. The second branch consists of *Weissella confusa* and *Weissella cibaria* (Björkroth et al., 2002; Choi et al., 2002). *Weissella minor*, *Weissella viridescens*, and *Weissella halotolerans* form a third branch with a 16S rDNA similarity of 96.8–99.1%. The fourth branch comprises *Weissella kandleri* and the newly described

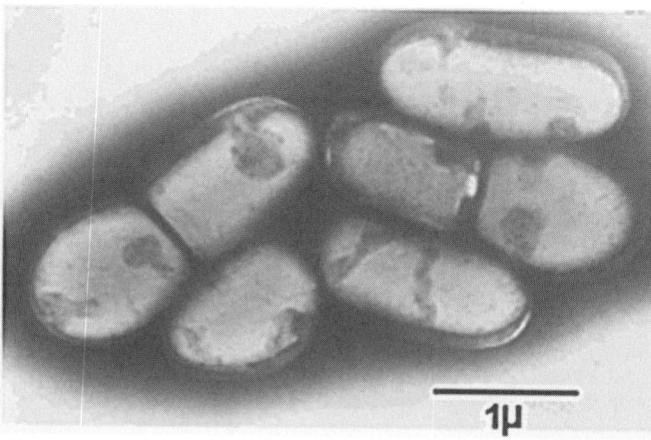

FIGURE 120. Transmission electron micrograph of negatively stained (1% of phosphotungstic acid) cells of *Weissella cibaria* LMG 17699[T].

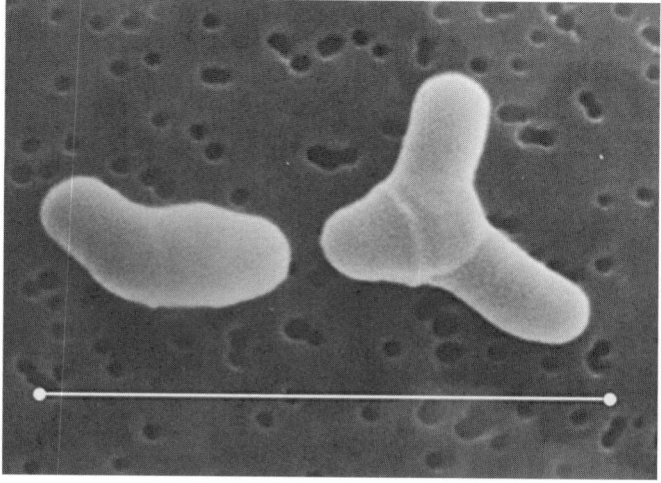

FIGURE 121. Scanning electron micrograph of a strain of *Weissella minor*, showing irregular and pleomorphic morphology. Bar = 5 μm.

TABLE 128. Characteristics differentiating the species of the genus *Weissella*[a,b]

Characteristic	W. viridescens	W. cibaria	W. confusa	W. halotolerans	W. hellenica	W. kandleri	W. korensis	W. minor	W. paramesenteroides	W. soli	W. thailandensis
Growth at 37°C	ND	+	+	ND	ND	ND	+	ND	ND	+	+
Peptidoglycan type	Lys–Ala–Ser	L-Lys–L-Ala (L-Ser)–L-Ala	Lys–Ala	Lys–Ala–Ser	Lys–Ala–Ser(L-Ala)	Lys–Ala–Gly–Ala₂	Lys–Ala–Ser	Lys–Ser–Ala₂	Lys–Ala₂ or Lys–Ser–Ala₂	ND	L-Lys–Ala₂
Cell morphology	Small irregular rods	Short rods growing in pairs	Short rods thickened at one end	Irregular short or coccoid rods	Large spherical or lenticular cells	Irregular rods	Irregular short rods or coccoid	Irregular short coccoid rods with rounded to tapered ends	Spherical or lenticular	Short rods often thickened at one end	Cocci in pairs or chains
Acid from:											
Amygdalin	ND	+	ND	ND	ND	ND	ND	ND	ND	−	−
L-Arabinose	−	+	−	−	+	−	+	−	d	+	+
Arbutin	ND	+	ND	ND	ND	ND	ND	ND	ND	−	ND
Cellobiose	−	+	+	ND	−	−	−	+	(d)	+	−
Cellulose	ND	ND	ND	ND	ND	ND	ND	ND	ND	ND	ND
Fructose	ND	+	ND	ND	+	ND	ND	ND	ND	−	+
Galactose	−	−	+	−	−	+	−	−	+	−	+
Lactose	ND	−	ND	ND	ND	ND	ND	ND	ND	−	ND
Maltose	+	+	+	+	+	−	−	+	+	+	−
Mannitol	ND	−	ND	ND	ND	ND	ND	+	+	−	ND
Mannose	ND	+	ND	ND	+	ND	ND	−	ND	+	ND
Melibiose	−	−	−	−	−	−	−	−	+	+	+
Raffinose	−	−	−	+	−	+	+	+	d	+	+
Ribose	−	+	+	+	ND	+	ND	+	ND	+	+
Salicin	ND	+	ND	ND	ND	ND	ND	ND	ND	+	+
Sucrose	d	+	+	−	+	−	−	+	+	+	+
Trehalose	d	−	−	−	+	−	−	+	+	+	+
Xylose	−	+	+	−	−	−	+	−	d	+	−
Ammonia from arginine	−	+	+	+	−	+	+	+	−	+	−
Lactic acid configuration	DL	DL	DL	DL	D	DL	D	DL	D	D	D
Hydrolysis of esculin	ND	+	+	−	ND	ND	ND	ND	ND	+	+
Dextran production	ND	+	+	ND	ND	ND	ND	ND	ND	−	−

a +, 90% or more of strains positive; −, 90% or more of strains negative; d, 11–98% of strains positive; (), delayed reaction; ND, no data., D, 90% or more of the lactic acid is D(−); DL, more than 25% of the total lactic acid is L(+).

b Data for *Weissella confusus, Weissella halotolerans, Weissella hellenica, Weissella kandleri, Weissella minor, Weissella paramesenteroides,* and *Weissella viridescens* are from Collins et al. (1993). Data for *Weissella thailandensis* are from Tanasupawat et al. (2000). Data for *Weissella cibaria* are from Björkroth et al. (2002). Data for *Weissella soli* are from Magnusson et al. (2002) and data for *Weissella korensis* are from Lee et al. (2002).

TABLE 129. 16S rRNA encoding gene sequence similarities between the different *Weissella* species[a,b]

Species	*W. confusa*	*W. cibaria*	*W. halotolerans*	*W. hellenica*	*W. kandleri*	*W. koreensis*	*W. minor*	*W. paramesenteroides*	*W. soli*	*W. viridescens*	*L. mesenteroides* subsp. *mesenteroides*
W. confusa NCIMB 9311[T]	100										
W. cibaria LMG 17699[T]	99.2	100									
W. halotolerans DSM 20190[T]	94.9	NG	100								
W. hellenica NCFB 2973[T]	97.2	NG	94.3	100							
W. kandleri NCFB 2753[T]	96.8	NG	95.6	95.5	100						
W. koreensis KCTC 3621[T]	NG	NG	NG	NG	97.2	100					
W. minor NCFB 1973[T]	96.8	NG	96.6	96.3	96.5	NG	100				
W. paramesenteroides DSM 20288[T]	96.8	NG	93.9	98.6	95.3	NG	95.8	100			
W. soli LMG 20113[T]	95.3	NG	NG	NG	95.5	NG	NG	NG	100		
W. viridescens NCIMB 8965[T]	97.0	NG	96.6	96.4	96.7	NG	99.1	95.8	NG	100	
L. mesenteroides subsp. *mesenteroides* NCIMB 8023[T]	91.5	NG	88.7	90.2	90.5	NG	90.2	91.1	NG	89.9	100

[a]NG, Not given.

[b]Data for *Weissella confusus*, *Weissella halotolerans*, *Weissella hellenica*, *Weissella kandleri*, *Weissella minor*, *Weissella paramesenteroides*, and *Weissella viridescens* are from Collins et al. (1993); for *Weissella thailandensis* from Tanasupawat et al. (2000); for *Weissella cibaria* from Björkroth et al. (2002); for *Weissella soli* from Magnusson et al. (2002); and data for *Weissella korensis* from Lee et al. (2002).

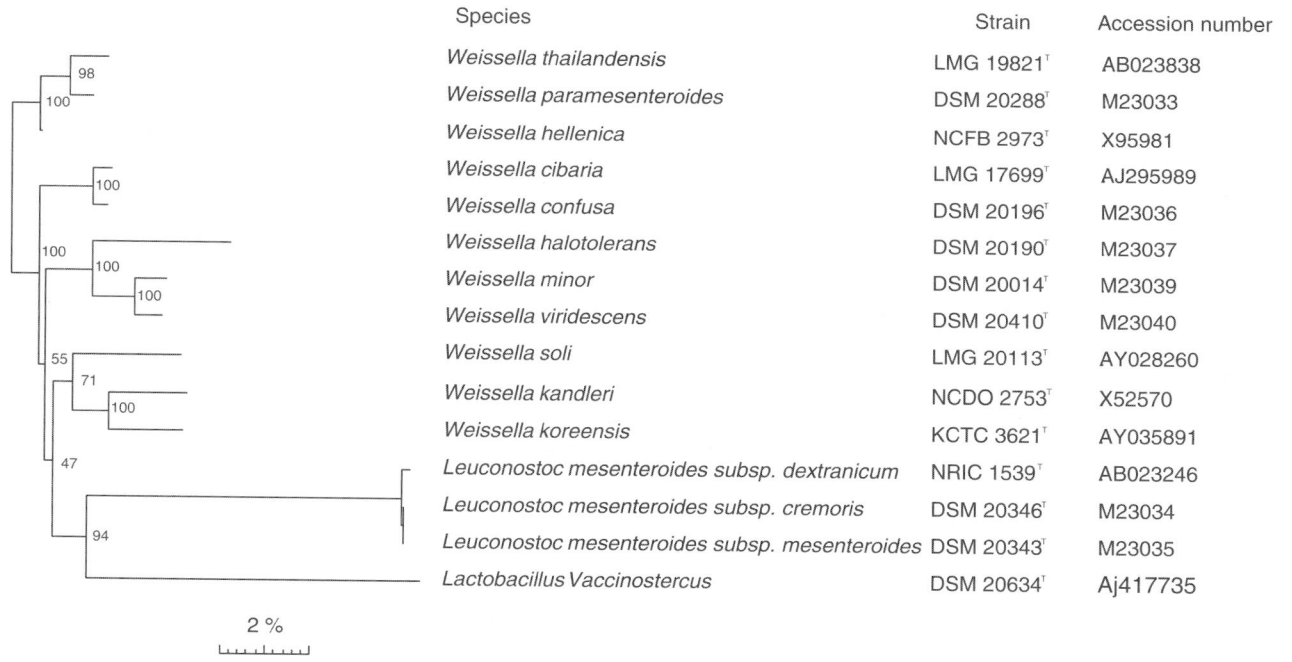

Species	Strain	Accession number
Weissella thailandensis	LMG 19821[T]	AB023838
Weissella paramesenteroides	DSM 20288[T]	M23033
Weissella hellenica	NCFB 2973[T]	X95981
Weissella cibaria	LMG 17699[T]	AJ295989
Weissella confusa	DSM 20196[T]	M23036
Weissella halotolerans	DSM 20190[T]	M23037
Weissella minor	DSM 20014[T]	M23039
Weissella viridescens	DSM 20410[T]	M23040
Weissella soli	LMG 20113[T]	AY028260
Weissella kandleri	NCDO 2753[T]	X52570
Weissella koreensis	KCTC 3621[T]	AY035891
Leuconostoc mesenteroides subsp. *dextranicum*	NRIC 1539[T]	AB023246
Leuconostoc mesenteroides subsp. *cremoris*	DSM 20346[T]	M23034
Leuconostoc mesenteroides subsp. *mesenteroides*	DSM 20343[T]	M23035
Lactobacillus Vaccinostercus	DSM 20634[T]	Aj417735

2 %

FIGURE 122. Phylogenetic tree based on almost entire 16S rRNA gene sequences of *Weissella* and related species. Bootstrap probability values from 1000 trees resampled are given at the branch points.

species *Weissella soli* and *Weissella koreensis*. A phylogenetic tree for *Weissella*, based on almost entire 16S rRNA sequences, is shown in Figure 122.

For some *Weissella* species, species-specific sequences have been located in helix 1007/1022 of the variable region V6 in the 16S rRNA gene (Collins et al., 1993). Rapid identification of *Weissella* species is possible by comparison of DNA patterns generated after restriction enzyme digests (*Mn*/I, *Mse*I, and *Bce*AI) of a 725-bp 16S rDNA fragment (Jang et al., 2002).

Numerical analyses of macromolecule patterns have proved useful in the identification of *Weissella* species. The relatedness between

Weissella species and *Leuconostoc* species has been studied by comparison of total soluble cell protein patterns (Dicks, 1995; Tsakalidou et al., 1997). This method and rRNA gene restriction patterns (ribotypes) differentiate *Weissella confusa* from *Weissella cibaria* and group the strains into species-specific clusters (Björkroth et al., 2002). Comparison of cellular fatty acid profiles of *Weissella* species (Samelis et al., 1998) correspond well with results recorded in other taxonomic studies and prove valuable in the differentiation of *Weissella viridescens*, *Weissella paramesenteroides*, *Weissella hellenica*, and typical arginine-negative *Weissella* strains isolated from meat. *Weissella viridescens* is identified based on the presence of eicosenoic ($C_{20:1}$) acid. Unlike *Weissella paramesenteroides*, *Weissella hellenica* does not contain cyclopropane fatty acids with 19 carbon atoms (i.e., dihydrosterculic or lactobacillic acid), and *Weissella viridescens* does not contain these fatty acids or has very low quantities. Meat isolates identified as *Weissella viridescens* or *Weissella hellenica* share similar fatty acid profiles with the respective type strains, whereas the "wild" atypical *Weissella* isolates more closely resemble *Weissella paramesenteroides* and *Leuconostoc mesenteroides* subsp. *mesenteroides*. Major cellular fatty acids associated with *Weissella thailandensis* and *Weissella koreensis* include straight-chained $C_{16:0}$ and $C_{20:1}$ (Tanasupawat et al., 1998, 2000) and $C_{18:1:\omega c}$ and $C_{16:0}$, respectively.

Levels of DNA–DNA reassociation among some type strains of *Weissella* species, *Leuconostoc* species, and *Oenococcus oeni* are shown in Table 130. Five strains of *Weissella thailandensis* possess DNA–DNA reassociation levels of 84.6–106.1% (Tanasupawat et al., 2000). The type strain of *Weissella thailandensis* has DNA–DNA similarity levels of 16.7–21.5% with the type strains of *Weissella paramesenteroides*, *Weissella hellenica*, and *Weissella confusa* (Tanasupawat et al., 2000). *Weissella cibaria* has 76–100% DNA–DNA similarity levels within the species and 22–49% with *Weissella confusa* (Björkroth et al., 2002), its closest phylogenetic neighbor. In the study by Vescovo et al. (1979), *Weissella confusa* LMG 18507 (NCFB 1937) was reported to possess a DNA–DNA reassociation level of 93% with the type strain of *Weissella confusa* (ATCC 10881). However, Björkroth et al. (2002) detected a DNA–DNA reassociation level of 100% between this strain and the type strain of *Weissella cibaria*, suggesting that this strain belongs to *Weissella cibaria*. This was also warranted by the other results of this polyphasic taxonomic study.

Weissella viridescens FemX (FemX(Wv)) belongs to the Fem family of nonribosomal peptidyl transferases that use aminoacyl-tRNA as the amino acid donor to synthesize the peptide cross-bridge found in the peptidoglycan of many species of pathogenic Gram-stain-positive bacteria. The crystal structure of FemX(Wv) in complex with the peptidoglycan precursor UDP-MurNAc-pentapeptide has recently been determined and the site-directed mutagenesis of nine residues located in the binding cavity for this substrate has been reported (Maillard et al., 2005). Two substitutions, Lys36Met and Arg211Met, depressed FemX(Wv) transferase activity below detectable levels without affecting protein folding. Analogs of UDP-MurNAc-pentapeptide lacking the phosphate groups or the C-terminal ᴅ-alanyl residues are not substrates of the enzyme. These results indicate that Lys36 and Arg211 participate in a complex hydrogen bond network that connects the C-terminal ᴅ-Ala residues to the phosphate groups of UDP-MurNAc-pentapeptide and constrains the substrate in a conformation that is essential for transferase activity.

A physical and genetic map of *Weissella paramesenteroides* DSM 20288[T] chromosome has been constructed. A total of 21 recognition sites of the restriction enzymes *Asc*I, I-*Ceu*I, *Not*I, and *Sfi*I were mapped on the chromosome which was found to be circular with an estimated size of 2026 kb. The localization of important chromosomal regions such as *oriC* and *terC* have been addressed and a total of 23 genetic markers have been mapped, including eight *rrn* operons that were precisely assigned in 37% of the *Weissella paramesenteroides* chromosome. The transcription direction of *rrn* loci was determined and three different *rrn* clusters were recognized regarding the presence/absence of tRNA genes in ITS regions.

Isolation and maintenance of *Weissella* spp.

Weissella are routinely cultured using the general growth media for lactic acid bacteria, such as MRS (De Man et al., 1960) and Rogosa SL-medium (Rogosa et al., 1951). There is no specific selective medium or enrichment method for *Weissella* species. The intrinsic resistance to vancomycin may be useful in certain approaches (Björkroth and Holzapfel, 2003). Strains can be maintained as described in the chapter for *Leuconostoc*. For more information, see also Schillinger and Holzapfel (2002).

Distinguishing the genus *Weissella* from *Leuconostoc* spp. and other lactic acid bacteria

Table 121 in the genus *Leuconostoc* chapter, above, shows the main criteria for distinguishing *Weissella* from other LAB. Clearly, it is most difficult to distinguish *Weissella* from *Leuconostoc* and heterofermentative *Lactobacillus* by phenotypic characteristics; this poses a great challenge for the development of genus-specific molecular methods for discrimination among these three genera. For species identification, classical phenotypic criteria may not be sufficient, warranting the use of molecular

TABLE 130. Levels (%) of DNA–DNA reassociation between type strains of some species of the genera *Weissella*, *Leuconostoc*, and *Oenococcus*[a,b]

Species	W. viridescens	W. cibaria	W. confusa	W. koreensis	W. paramesenteroides	W. soli	W. thailandensis
W. viridescens	83–92	ND	ND	13–16	ND	ND	ND
W. confusa	ND	42–49	93–105	ND	ND	17	15.7–16.7
W. hellenica							19.3–28.2
W. kandleri	ND	ND		20–21		16	
W. paramesenteroides	ND	ND	ND		82–100	ND	21.5–36.0
L. argentinum	ND	ND	ND	ND	ND	ND	ND
L. carnosum	ND	ND	ND	ND	0–6	ND	ND
L. citreum	ND	ND	ND	ND	10–22	ND	ND
L. fallax	ND	ND	ND	ND	25	ND	ND
L. gelidum	ND	ND	ND	ND	0–3	ND	ND
L. lactis	ND	ND	ND	ND	0–25	ND	ND
L. mesenteroides subsp. *mesenteroides*	ND	ND	ND	ND	7–18	ND	ND
L. mesenteroides subsp. *cremoris*	ND	ND	ND	ND	5–10	ND	ND
L. mesenteroides subsp. *dextranicum*	ND	ND	ND	ND	5–19	ND	ND
L. pseudomesenteroides	ND	ND	ND	ND	9–22	ND	ND
O. oeni	ND	ND	ND	ND	5–7	ND	ND

[a]Data compiled from Garvie (1976); Hontebeyrie and Gasser (1977); Vescovo et al. (1979); Farrow et al. (1989); Schillinger et al. (1989); Shaw and Harding (1989); Dicks et al. (1990); Tanasupawat et al. (2000); Björkroth et al. (2002), Lee et al. (2002); Magnusson et al. (2002).

[b]ND, not determined.

approaches such as the numerical analysis of macromolecule patterns and sequencing and employment of specific signature bases in the rRNA-encoding genes.

Metabolism

All species are heterofermentative and produce CO_2, ethanol, and/or acetate while fermenting glucose. Carbohydrate fermentation is via the hexose-monophosphate and the phosphoketolase pathways. The configuration of lactic acid is either DL or D(−), depending on the species. There is, however, variation between the species in the production of D(−)- or DL-lactic acid isomers (Table 128).

Habitat

The habitats of *Weissella* species are variable. *Weissella viridescens*, *Weissella halotolerans*, and *Weissella hellenica* have been associated with meat and meat products. *Weissella viridescens* may cause spoilage of cured meat due to green discoloration (Niven and Evans, 1957). This species is also considered as heat resistant (Niven et al., 1954), a property not very common for lactic acid bacteria (LAB). *Weissella cibaria*, *Weissella confusa*, and *Weissella koreensis* have been described for fermented foods of vegetable origin (Björkroth et al., 2002; Lee et al., 2002), whereas *Weissella confusa* is also associated with Greek cheese (Samelis et al., 1994), Mexican pozol (Ampe et al., 1999), and Malaysian chili bo (Leisner et al., 1999). *Weissella soli* (Magnusson et al., 2002) is the only species known to originate in soil.

Weissella confusa forms part of the normal microbiota in human intestines (Stiles and Holzapfel, 1997; Tannock et al., 1999; Walter et al., 2001). *Weissella cibaria* and *Weissella confusa* have also been associated with clinical samples of humans and animals (Björkroth et al., 2002). In addition, *Weissella confusa* has been documented as a cause for a systemic infection in a healthy primate (*Cercopitheus mona*; Vela et al., 2003). A vancomycin-resistant strain of *Weissella confusa* has also been reported to have caused an abscess (Bantar et al., 1991).

Taxonomic comments

The genus *Weissella* was proposed in 1993 (Collins et al., 1993), mainly as a result of phylogenetic studies targeting both 16S (Collins et al., 1993; Martinez-Murcia and Collins, 1990; Martinez-Murcia et al., 1993) and 23S (Martinez-Murcia et al., 1993) rRNA sequences. Based on these results, *Leuconostoc paramesenteroides* (Garvie, 1967a), *Lactobacillus viridescens* (Kandler and Abo-Elnaga, 1966; Niven and Evans, 1957), *Lactobacillus confusus* (Holzapfel and Kandler, 1969), *Lactobacillus kandleri* (Holzapfel and van Wyk, 1982), *Lactobacillus minor* (Kandler et al., 1983), and *Lactobacillus halotolerans* (Kandler et al., 1983) were reclassified as *Weissella paramesenteroides*, *Weissella viridescens*, *Weissella confusa*, *Weissella kandleri*, *Weissella minor*, and *Weissella halotolerans*, respectively. Subsequent designations to the genus include *Weissella hellenica* (Collins et al., 1993), *Weissella thailandensis* (Tanasupawat et al., 2000), *Weissella cibaria* (Björkroth et al., 2002), *Weissella soli* (Magnusson et al., 2002), and *Weissella koreensis* (Lee et al., 2002). *Weissella kimchii* proposed by Choi et al. (2002) has been found to be a later heterotypic synonym of *Weissella cibaria* (Ennahar and Cai, 2004).

List of species of the genus *Weissella*

1. **Weissella viridescens** (Niven and Evans 1957) Collins, Samelis, Metaxopoulos and Wallbanks 1994, 370[VP] (Effective publication: Collins, Samelis, Metaxopoulos and Wallbanks 1993, 601.) (*Lactobacillus viridescens* Niven and Evans 1957, 758; *Lactobacillus corynoides* subsp. *corynoides* Kandler and Abo-Elnaga 1966, 753)

vi.ri.des′cens. L. pres. part. *viridescens* growing green, greening.

Small, often slightly irregular rods, generally 0.7–0.9 × 2.0–5.0 μm, with rounded to tapered ends occurring singly or in pairs. Gram-stain-positive, nonmotile, and nonspore-forming. The interpeptide bridge of the peptidoglycan structure is Lys–Ala Ser. The configuration of lactic acid produced is DL. Growth factor requirements: pantothenic acid, niacin, thiamine, riboflavin, and biotin are essential; folic acid and pyridoxal may be stimulatory. Does not produce ammonia from arginine; dextran production from sucrose has not been studied. Detected in discolored cured meat products and pasteurized milk.

DNA G+C content (mol%): 41–44 (Bd, T_m).

Type strain: ATCC 12706, CCM 56, BCRC 11650, CCUG 21533, CCUG 30502, CECT 283, CIP 102810, DSM 20410, IAM 13546, JCM 1174, LMG 3507, NCCB 71015, NCFB 1655, NCIMB 8965, NRIC 1536, NRRL B-1951, VKM B-1528.

GenBank accession number (16S rRNA gene): AB023236, M23040, X52568.

Additional remarks: Lactobacillus viridescens is incorrectly cited on the Approved Lists of Bacterial Names as *Lactobacillus viridescens* Kandler and Abo-Elnaga 1966, 573.

2. **Weissella cibaria** Björkroth, Schillinger, Geisen, Weiss, Hoste, Holzapfel, Korkeala and Vandamme 2002, 147[VP]

ci.ba′ri.a. L. adj. *cibarius* pertaining to food.

Cells are short rods growing in pairs, 0.8–1.2 × 1.5–2 μm. Gram-stain-positive, nonmotile, and nonspore-forming. The interpeptide bridge of the peptidoglycan structure is Lys–Ala–(Ser)-Ala. The configuration of lactic acid produced is DL. Growth occurs at 45 °C and at 15 °C but not at 4 °C. Dextran is produced from sucrose and ammonia from arginine.

Isolated from Malaysian fermented foods and other food sources but also from clinical samples including human gall, otitis smear (dog), and human feces.

DNA G+C content (mol%): 44 (T_m).

Type strain: LMG 17699, CCUG 41967, JCM 12495.

GenBank accession number (16S rRNA gene): AJ295989.

3. **Weissella confusa** (Holzapfel and Kandler 1969) Collins, Samelis, Metaxopoulos and Wallbanks 1994, 370[VP] (Effective publication: Collins, Samelis, Metaxopoulos and Wallbanks 1993, 599.) (*Lactobacillus coprophilus* subsp. *confusus* Holzapfel and Kandler 1969, 665; *Lactobacillus confusus* Sharpe, Garvie and Tilbury 1972, 396)

con.fu′sus. L. v. *confundere* to confuse; L. past part. *confusus* confused, an allusion to its original confusion with *Leuconostoc*.

Short rods, 0.8–1.0 × 1.5–3.0 μm, with tendency to thicken at one end; occurring singly, rarely in short chains. Gram-stain-positive, nonmotile, and nonspore-forming. The interpeptide bridge of the peptidoglycan structure is Lys–Ala. The configuration of lactic acid produced is DL.

Growth at 45 °C is variable, some strains grow well. Dextran is produced from sucrose and ammonia from arginine.

Isolated from sugarcane, carrot juice, and fermented foods, occasionally found in raw milk, saliva, sewage, and clinical samples.

DNA G+C content (mol%): 45–47 (T_m).

Type strain: TCC 10881, ATCC 10881, BCRC 14002, CCUG 30113, CIP 103172, CIP 54.169, DSM 20196, JCM 1093, LMG 9497, NCIMB 9311, NRRL B-1064.

GenBank accession number (16S rRNA gene): AB023241, M23036.

4. **Weissella halotolerans** (Kandler, Schillinger and Weiss 1983) Collins, Samelis, Metaxopoulos and Wallbanks 1994, 370[VP] (Effective publication: Collins, Samelis, Metaxopoulos and Wallbanks 1993, 599.) (*Lactobacillus halotolerans* Kandler, Schillinger and Weiss, 1983, 283)

ha.lo.to'le.rans. Gr. n. *hals* salt; L. pres. part. *tolerans* tolerating, enduring; N.L. part. adj. *halotolerans* salt-tolerating.

Irregular, short, or coccoid rods with rounded to tapered ends, generally 0.5–0.7 × 1.0–3 μm, sometimes longer, with tendency to form coiling chains, clumping together. Gram-stain-positive, nonmotile, and nonspore-forming. The interpeptide bridge of the peptidoglycan structure is Lys–Ala–Ser. The configuration of lactic acid produced is DL. No growth at 45 °C. Good growth in the presence of 12% NaCl and very weak growth in the presence of 14% NaCl. Does not produce ammonia from arginine; dextran production from sucrose has not been studied.

Isolated from meat products.

DNA G+C content (mol%): 45 (T_m).

Type strain: R61, ATCC 35410, BCRC 14050, CCUG 33457, CECT 573, CIP 103005, DSM 20190, JCM 1114, LMG 9469, NRIC 1627.

GenBank accession number (16S rRNA gene): AB022926, M23037.

5. **Weissella hellenica** Collins, Samelis, Metaxopoulos and Wallbanks 1994, 370[VP] (Effective publication: Collins, Samelis, Metaxopoulos and Wallbanks 1993, 601.)

hel. len'ic.a. Gr. adj. *Hellenikos* Greek; N.L. fem. adj. *hellenica* Greece, where the bacterium was isolated.

Spherical but sometimes lenticular cells occurring in pairs or short chains, with a tendency to form clusters. Gram-stain-positive, nonmotile, and nonspore-forming. The interpeptide bridge of the peptidoglycan structure is Lys–Ala–Ser. The configuration of lactic acid produced is D(−). No growth occurs at 37 °C; all strains grow at 10 °C and 4 °C (delayed). Dextran is not formed from sucrose; ammonia is not formed from arginine.

Isolated from fermented sausages.

DNA G+C content (mol%): 39–40 (T_m).

Type strain: LV346 (= ATCC 51523 = CCUG 33494 = DSM 7378 = NBRC 15553 = JCM 10103 = LMG 15125 = NCIMB 702973.

GenBank accession number (16S rRNA gene): X95981.

6. **Weissella kandleri** (Holzapfel and van Wyk 1982) Collins, Samelis, Metaxopoulos and Wallbanks 1994, 370[VP] (Effective publication: Collins, Samelis, Metaxopoulos and Wallbanks 1993, 599.) (*Lactobacillus kandleri* Holzapfel and van Wyk, 1982, 501)

kand'le.ri. N.L. gen. n. *kandleri* of Kandler; named for Otto Kandler, a German microbiologist.

Partly irregular rods, generally 0.6–0.8 × 3.0–15 μm, occurring singly or in pairs, seldom in short chains. Gram-stain-positive, nonmotile, and nonspore-forming. The interpeptide bridge of the peptidoglycan structure is Lys–Ala–Gly–Ala₂. The configuration of lactic acid produced is DL. No growth at 45 °C. Dextran is produced from sucrose and ammonia is produced from arginine.

Isolated from a desert spring.

DNA G+C content (mol%): 39 (T_m)

Type strain: L250, ATCC 51149, BCRC 14624, CCUG 32237, CIP 102809, DSM 20593, JCM 5817, LMG 18979, NCIMB 702753, NRIC 1628.

GenBank accession number (16S rRNA gene): AB022922, M23038.

7. **Weissella koreensis** Lee, Lee, Ahn, Mheen, Pyun and Park 2002, 1260[VP]

ko.re.en'sis. N.L. adj. *koreensis* of Korea, where the novel organisms were isolated.

Cells are irregular short rod-shaped or coccoid. Gram-stain-positive, nonmotile, and nonspore-forming. The interpeptide bridge of the peptidoglycan structure is Lys–Ala–Ser. The configuration of lactic acid produced is mainly D(−). Growth occurs between 10 and 37 °C but not at 42 °C. Ammonia is produced from arginine and dextran is produced from sucrose.

Detected in kimchi, a Korean fermented food product.

DNA G+C content (mol%): 37 (HPLC).

Type strain: S-5623, JCM 11263, KCCM 41516, KCTC 3621.

GenBank accession number (16S rRNA gene): AY035891.

8. **Weissella minor** (Kandler, Schillinger and Weiss 1983) Collins, Samelis, Metaxopoulos and Wallbanks 1994, 370[VP] (Effective publication: Collins, Samelis, Metaxopoulos and Wallbanks 1993, 599.) (*Lactobacillus minor* Kandler, Schillinger and Weiss 1983, 284; *Lactobacillus corynoides* subsp. *minor* Abo-Elnaga and Kandler 1965, 128; *Lactobacillus viridescens* subsp. *minor* Kandler and Abo-Elnaga 1966, 754)

mi'nor. L. comp. adj. *minor* smaller.

Irregular short rods with rounded to tapered ends (see Figure 121), generally 0.6–0.8 × 1.5–2.0 μm, sometimes longer, often bent with unilateral swellings, occurring in pairs or short chains, with a tendency to form loose clusters. Gram-stain-positive, nonmotile, and nonspore-forming. The interpeptide bridge of the peptidoglycan structure is Lys–Ser–Ala₂. The configuration of lactic acid produced is DL. Does not produce dextran from sucrose, but produces ammonia from arginine. No growth at 45 °C.

Isolated from the sludge of milking machines.

DNA G+C content (mol%): 44 (T_m).

Type strain: 3, ATCC 35412, BCRC 14049, CCUG 30668, CECT 572, CIP 102978, DSM 20014, JCM 1168, LMG 9847, NCIMB 701973, NRIC 1625.

GenBank accession number (16S rRNA gene): M23039.

9. **Weissella paramesenteroides** (Garvie 1967a) Collins, Samelis, Metaxopoulos and Wallbanks 1994, 370 (Effective publication: Collins, Samelis, Metaxopoulos and Wallbanks 1993, 601.) (*Leuconostoc paramesenteroides* Garvie 1967a, 446)

pa.ra.me.sen.ter.oi'des. Gr. prep. *para* resembling; N.L. *mesenteroides* a specific epithet, mesentery-like; N.L. adj. *paramesenteroides* resembling *Leuconostoc mesenteroides*.

Spherical but often lenticular cells usually occurring in pairs and chains. Gram-stain-positive, nonmotile, and nonspore-forming. The interpeptide bridge of the peptidoglycan structure is Lys–Ala$_2$; Lys–Ser–Ala$_2$. The configuration of lactic acid produced is D(–). Does not produce ammonia from arginine. Dextran is not formed from sucrose, and amino acid requirements are complex and variable. Many cells grow well at 30 °C but some prefer reducing conditions and a temperature of 18–24 °C (Garvie, 1967a). Pseudocatalase may be present if organisms are grown in a medium with a low glucose content (Whittenbury, 1964). Tolerates NaCl well, and some strains may be detected in foods possessing high levels of salt. Tolerates acidic pH and may grow in media with an initial pH below 5.0.

Before reclassification as *Weissella*, *Weissella paramesenteroides* strains were considered to be nondextran-forming *Leuconostoc mesenteroides* strains. Phenotypic differentiation of *Weissella paramesenteroides* and nondextran-forming strains of *Leuconostoc mesenteroides* is difficult.

DNA G+C content (mol%): 37–38 (T_m).

Type strain: ATCC 33313, CCUG 30068, CIP 102421, DSM 20288, JCM 9890, LMG 9852, NCIMB 13092, NRIC 1542, NRRL B-1186, NRRL B-3471.

GenBank accession number (16S rRNA gene): AB023238, M23033, X95982.

10. **Weissella soli** Magnusson, Jonsson, Schnürer and Roos 2002, 833[VP]

so′li. L. n. *solum* soil; L. gen. n. *soli* of the soil.

Cells are rods, often thickened at one end, occurring singly or in pairs, 0.9 × 1.2–3.0 µm. The cells in a pair are often different size. Gram-stain-positive, nonmotile, and nonspore-forming. The interpeptide bridge of the peptidoglycan structure has not been determined. The configuration of lactic acid produced is mainly D(–). Growth occurs between 4 and 40 °C but not at 45 °C. Ammonia is produced from arginine, but no dextran is produced from sucrose.

Detected in soil.

DNA G+C content (mol%): 43 (HPLC).

Type strain: Mi268, DSM 14420, JCM 12536, LMG 20113.

GenBank accession number (16S rRNA gene): AY028260.

11. **Weissella thailandensis** Tanasupawat, Shida, Okada and Komagata 2000, 1484[VP]

thai.lan′den.sis. N.L. fem. adj. *thailandensis* pertaining to Thailand where the strains were first isolated.

Cells are spherical, 0.5 × 0.7 µm. Gram-stain-positive, nonmotile, and nonspore-forming The interpeptide bridge of the peptidoglycan structure is Lys–Ala$_2$. The configuration of lactic acid produced is D(–). Grows between 25 and 37 °C but not at 42 °C. Does not produce dextran from sucrose or ammonia from arginine.

Detected in pla-ra, a fermented Thai fish product.

DNA G+C content (mol%): 38–41 (method not disclosed).

Type strain: FS61-1, PCU 210, HSCC 1412, CIP 106751, JCM 10695, KCTC 3751, NRIC 0298, TISTR 1384.

GenBank accession number (16S rRNA gene): AB023838.

References

Abo-Elnaga, I.G. and O. Kandler. 1965. On the taxonomy of the species *Lactobacillus* Beijerinck. III. The need for vitamins. Zentbl. Bakteriol. Parasitenkd. Infektionskr. Hyg. *119:* 661–672.

Amachi, T., S. Imamoto and H. Yoshizumi. 1971. A growth factor for malo-lactic fermentation bacteria. II. Structure and synthesis of a novel pantothenic acid derivative isolated from tomato juice. Agric. Biol. Chem. *35:* 1222–1230.

Ampe, F., N. ben Omar, C. Moizan, C. Wacher and J.P. Guyot. 1999. Polyphasic study of the spatial distribution of microorganisms in Mexican pozol, a fermented maize dough, demonstrates the need for cultivation-independent methods to investigate traditional fermentations. Appl. Environ. Microbiol. *65:* 5464–5473.

Antunes, A., F.A. Rainey, M.F. Nobre, P. Schumann, A.M. Ferreira, A. Ramos, H. Santos and M.S. da Costa 2002. *Leuconostoc ficulneum* sp. nov., a novel lactic acid bacterium isolated from a ripe fig, and reclassification of *Lactobacillus fructosus* as *Leuconostoc fructosum* comb. nov. Int. J. Syst. Evol. Microbiol. *52:* 647–655.

Bantar, C.E., S. Relloso, F.R. Castell, J. Smayevsky and H.M. Bianchini. 1991. Abscess caused by vancomycin-resistant *Lactobacillus confusus*. J. Clin. Microbiol. *29:* 2063–2064.

Barillière, J.M. 1981. Etude de la thermorésistance de souches de levure et de bactéries lactiques isoleés de vin: influence de degré alcoolique et de la teneur en SO$_2$. Ecole Nationale Supérieure d'Agronomie, Montpellier.

Barnes, E.M. 1956. Methods for the isolation of faecal streptococci (Lancefield Group D) from bacon factories. J. Appl. Bacteriol. *19:* 193–203.

Barrangou, R., S.S. Yoon, F. Breidt, H.P. Fleming and T.R. Klaenhammer. 2002. Identification and characterization of *Leuconostoc fallax* strains isolated from an industrial sauerkraut fermentation. Appl. Environ. Microbiol. *68:* 2877–2884.

Bauer, R. and L.M.T. Dicks. 2004. Control of malolactic fermentation in wine. A review. S. Afr. J. Enol. Vitic. *25:* 1–8.

Beijerinck, M.W. 1912. Mutation bei Mikroben. Folia Mikrobiologiya (Delft) *1:* 4–100.

Björkroth, J. and W. Holzapfel. 2003. Genera *Leuconostoc, Oenococcus,* and *Weisella*. In Dworkin, Falkow, Rosenberg, Schleifer and Stackbrandt (Editors), The Prokaryotes: An Evolving Electronic Resource for the Microbiological Community, 3rd edn. Springer, New York.

Björkroth, J. and W.H. Holzapfel. 2006. Genera *Leuconostoc, Oenococcus* and *Weissella*. In Dworkin (Editor), The Prokaryotes, 3 edn, Vol. 4. Springer, Springer, pp. 267–319.

Björkroth, K.J., P. Vandamme and H.J. Korkeala. 1998. Identification and characterization of *Leuconostoc carnosum*, associated with production and spoilage of vacuum-packaged, sliced, cooked ham. Appl. Environ. Microbiol. *64:* 3313–3319.

Björkroth, K.J., R. Geisen, U. Schillinger, N. Weiss, P. De Vos, W.H. Holzapfel, H.J. Korkeala and P. Vandamme. 2000. Characterization of *Leuconostoc gasicomitatum* sp. nov., associated with spoiled raw tomato-marinated broiler meat strips packaged under modified-atmosphere conditions. Appl. Environ. Microbiol. *66:* 3764–3772.

Björkroth, K.J., R. Geisen, U. Schillinger, N. Weiss, P. De Vos, W.H. Holzapfel, H.J. Korkeala and P. Vandamme. 2001. In Validation of publication of new names and new combinations previously effectively published outside the IJSEM. List no. 79. Int. J. Syst. Evol. Microbiol. *51:* 263–265.

Björkroth, K.J., U. Schillinger, R. Geisen, N. Weiss, B. Hoste, W.H. Holzapfel, H.J. Korkeala and P. Vandamme. 2002. Taxonomic study of *Weissella confusa* and description of *Weissella cibaria* sp. nov., detected in food and clinical samples. Int. J. Syst. Evol. Microbiol. *52:* 141–148.

Bourdineaud, J.P., B. Nehme, S. Tesse and A. Lonvaud-Funel. 2004. A bacterial gene homologous to ABC transporters protect *Oenococcus oeni* from ethanol and other stress factors in wine. Int. J. Food Microbiol. *92:* 1–14.

Bover-Cid, S. and W.H. Holzapfel. 1999. Improved screening procedure for biogenic amine production by lactic acid bacteria. Int. J. Food Microbiol. *53:* 33–41.

Bover-Cid, S., M. Izuierdo-Pulido, M. Carmen Vidal-Carou and W.H. Holzapfel. 1999. Determination of amino acid decarboxylase activity of lactic acid bacteria. Food Microbiology and Food Safety into the Next Millenium, Proceedings of the 17th International Symposium on Food Microbiology and Hygiene (ICFMH), pp. 634–637, Veldhoven, The Netherlands.

Bover-Cid, S. and W.H. Holzapfel. 2000. Biogenic amine production by bacteria. *In* Morgan, White, Sánchez-Jiménez and S. Bardócz (Editors), Biogenic Active Amines in Food, Vol. IV. First General Workshop, COST 917, European Communities 2000, Luxembourg.

Britz, T.J. and R.P. Tracey. 1990. The combination effect of pH, SO2, ethanol and temperature on the growth of *Leuconostoc oenos*. J. Appl. Bacteriol. *68:* 23–31.

Cavett, J.J., G.J. Dring and A.W. Knight. 1965. Bacterial spoilage of thawed frozen peas. J. Appl. Bacteriol. *28:* 241–251.

Cavin, J.F., P. Schmitt, A. Arias, J. Lin and C. Diviès. 1988. Plasmid profiles in *Leuconostoc* species. Microbiol. Alim. Nutr. *6:* 55–62.

Cavin, J.F., H. Prevost, J. Lin, P. Schmitt and C. Divies. 1989. Medium for screening *Leuconostoc oenos* strains defective in malolactic fermentation. Appl. Environ. Microbiol. *55:* 751–753.

Choi, H.J., C.I. Cheigh, S.B. Kim, J.C. Lee, D.W. Lee, S.W. Choi, J.M. Park and Y.R. Pyun. 2002. *Weissella kimchii* sp. nov., a novel lactic acid bacterium from kimchi. Int. J. Syst. Evol. Microbiol. *52:* 507–511.

Choi, I.-K., S.-H. Jung, B.-J. Kim, S.-Y. Park, J. Kim and H.-U. Han. 2003. Novel *Leuconostoc citreum* starter culture system for fermentation of kimchi, a fermented cabbage product. Antonie van Leeuwenhoek *84:* 247–253.

Cilliers, J.D. and C.J. van Wyk. 1985. Histamine and tyramine content of South African wine. S. Afr. J. Enol. Vitic. *6:* 35–40.

Cogan, T.M. 1975. Citrate utilization in milk by *Leuconostoc cremoris* and *Streptococcus diacetilactis*. J. Dairy Res. *42:* 139–146.

Cogan, T.M., M. O'Dowd and D. Mellerick. 1981. Effects of pH and sugar on acetoin production from citrate by *Leuconostoc lactis*. Appl. Environ. Microbiol. *41:* 1–8.

Cogan, T.M. 1987. Co-metabolism of citrate and glucose by *Leuconostoc* spp.: effects on growth, substrates and products. J. Appl. Bacteriol. *63:* 551–558.

Collins, E.B. 1972. Biosynthesis of flavor compounds by microorganisms. J. Dairy Sci. *55:* 1022–1028.

Collins, E.B. and R.A. Speckman. 1974. Influence of acetaldehyde on growth and acetoin production by *Leuconostoc citrovorum*. J. Dairy Sci. *57:* 1428–1431.

Collins, M.D., J. Samelis, J. Metaxopoulos and S. Wallbanks. 1993. Taxonomic studies on some *Leuconostoc*-like organisms from fermented sausages: description of a new genus *Weissella* for the *Leuconostoc paramesenteroides* group of species. J. Appl. Bacteriol. *75:* 595–603.

Collins, M.D., J. Samelis, J. Metaxopoulous and S. Wallbanks. 1994. *In* Validation of the publication of new names and new combinations previously effectively published outside the IJSB. List no. 49. Int. J. Syst. Bacteriol. *44:* 370–371.

Condon, S. 1987. Responses of lactic acid bacteria to oxygen. FEMS Microbiol. Rev. *46:* 269–280.

Cooke, R.D., D.R. Twiddy and P.J.A. Reilly. 1987. Lactic acid fermentation as a low-cost means of food preservation in tropical countries. FEMS Microbiol. Lett. *46:* 369–379.

Coovadia, Y.M., Z. Solwa and J. van den Ende. 1987. Meningitis caused by vancomycin-resistant *Leuconostoc* sp. J. Clin. Microbiol. *25:* 1784–1785.

Coovadia, Y.M., Z. Solwa and J. van Den Ende. 1988. Potential pathogenicity of *Leuconostoc*. Lancet *1:* 306.

Coton, E., G.C. Rollan and A. Lonvaud-Funel. 1998. Histidine carboxylase of *Leuconostoc oenos* 9204: Purification, kinetic properties, cloning and nucleotide sequence of the *hdc* gene. J. Appl. Microbiol. *84:* 143–151.

Daniel, P., E. Dewaele and J.N. Hallet. 1993. Optimization of transverse alternating field electrophoresis for strain identification of *Leuconostoc oenos*. Appl. Microbiol. Biotechnol. *38:* 638–641.

Davis, C.R., D. Wibowo, R. Eschenbruch, T.H. Lee and G.H. Fleet. 1985a. Practical implications of malo-lactic fementation: a review. Am. J. Enol. Vitic. *36:* 290–301.

Davis, G., N.F.A. Silveira and G.H. Fleet. 1985b. Occurrence and properties of bacteriophages of *Leuconostoc oenos* in Australian wines. Appl. Environ. Microbiol. *50:* 872–876.

De Man, J.C., M. Rogosa and M.E. Sharpe. 1960. A medium for the cultivation of lactobacilli. J. Appl. Bacteriol. *23:* 130–135.

De Moss, R.D., R.C. Bard and I.C. Gunsalus. 1951. The mechanism of the hetero-lactic fermentation: a new route of ethanol formation. J. Bacteriol. *62:* 499–511.

Dellaglio, F., L.M.T. Dicks and S. Torriani. 1995. The genus *Leuconostoc*. *In* Wood and Holzapfel (Editors), The Genera of Lactic Acid Bacteria. Blackie Academic & Professional, Glasgow, pp. 233–278.

Desroche, N., C. Beltramo and J. Guzzo. 2005. Determination of an internal control to apply reverse transcription quantitative PCR to study stress response in the lactic acid bacterium *Oenococcus oeni*. J. Microbiol. Methods *60:* 325–333.

Dicks, L.M.T. 1989. A taxonomic study of *Leuconostoc oenos*. University of Stellenbosch, South Africa.

Dicks, L.M.T., H.J.J. van Vuuren and F. Dellaglio. 1990. Taxonomy of *Leuconostoc* species, particularly *Leuconostoc oenos*, as revealed by numerical analysis of total soluble cell protein patterns, DNA base compositions, and DNA-DNA hybridizations. Int. J. Syst. Bacteriol. *40:* 83–91.

Dicks, L.M.T., L. Fantuzzi, F.C. Gonzalez, M. du Toit and F. Dellaglio. 1993. *Leuconostoc argentinum* sp. nov, isolated from Argentine raw milk. Int. J. Syst. Bacteriol. *43:* 347–351.

Dicks, L.M.T. 1995. Relatedness of *Leuconostoc* species of the *Leuconostoc* sensu stricto line of descent, *Leuconostoc oenos* and *Weissella paramesenteroides* revealed by numerical analysis of total soluble cell protein patterns. Syst. Appl. Microbiol. *18:* 99–102.

Dicks, L.M.T., F. Dellaglio and M.D. Collins. 1995a. Proposal to reclassify *Leuconostoc oenos* as *Oenococcus oeni* corrig. gen. nov., comb. nov. Int. J. Syst. Bacteriol. *45:* 395–397.

Dicks, L.M.T., P.A. Loubser and O.P.H. Augustyn. 1995b. Identification of *Leuconostoc oenos* from South African fortified wines by numerical analysis of total soluble cell protein patterns and DNA-DNA hybridizations. J. Appl. Bacteriol. *79:* 43–48.

Dittrich, W. 1993. Mikrobiologie des weines und schaumweines. *In* Dittrich (Editor), Mikrobiologie der Lebensmittel: Getränke. Behr's Verlag, Jamburg, pp. 183–259.

du Plessis, H.W., L.M.T. Dicks, I.S. Pretorius, M.G. Lambrechts and M. du Toit. 2004. Identification of lactic acid bacteria isolated from South African brandy base wines. Int. J. Food Microbiol. *91:* 19–29.

Duitschaever, C.L., N. Kemp and D. Emmons. 1987. Pure culture formulation and procedure for the production of kefir. Milchwissenschaft-Milk Sci. Int. *42:* 80–82.

Dykes, G.A., T.E. Cloete and A. von Holy. 1994. Identification of *Leuconostoc* species associated with the spoilage of vacuum-packaged Vienna sausages by DNA-DNA hybridization. Food Microbiol. *11:* 271–274.

Edwards, R.A., R.H. Dainty, C.M. Hibbard and S.V. Ramantanis. 1987. Amines in fresh beef of normal pH and the role of bacteria in changes in concentration observed during storage in vacuum packs at chill temperatures. J. Appl. Bacteriol. *63:* 427–434.

Endo, A. and S. Okada. 2006. *Oenococcus kitaharae* sp. nov., a non-acidophilic and non-malolactic-fermenting oenococcus isolated from a composting distilled shochu residue. Int. J. Syst. Evol. Microbiol. *56:* 2345–2348.

Ennahar, S. and Y. Cai. 2004. Genetic evidence that *Weissella kimchii* Choi *et al.* 2002 is a later heterotypic synonym of *Weissella cibaria* Björkroth *et al.* 2002. Int. J. Syst. Evol. Microbiol. *54:* 463–465.

Fantuzzi, L., L.M.T. Dicks, M. Dutoit, R. Reniero, V. Bottazzi and F. Dellaglio. 1992. Identification of *Leuconostoc* strains isolated from Argentine raw milk. Syst. Appl. Microbiol. *15:* 229–234.

Farrow, J.A.E., R.R. Facklam and M.D. Collins. 1989. Nucleic acid homologies of some vancomycin-resistant leuconostocs and description of *Leuconostoc citreum* sp. nov. and *Leuconostoc pseudomesenteroides* sp. nov. Int. J. Syst. Bacteriol. *39:* 279–283.

Foucaud, C., A. Francois and J. Richard. 1997. Development of a chemically defined medium for the growth of *Leuconostoc mesenteroides*. Appl. Environ. Microbiol. *63:* 301–304.

Frank, H.A. and A.S. Dela Cruz. 1964. Role of incidental microflora in natural decomposition of mucilage layer in Kona coffee. Food Sci. *29:* 850–853.

Fröhlich, J. 2002. Fluorescence in situ hybridization (FISH) and single cell micromanipulation as novel applications for identification and isolation of new *Oenococcus* strains. Lallemand, Langenlois.

Fröhlich, J., S. Hirschauer, Salzbrunn and H. König. 2002. Fluorezenz in situ-hybridisierung (FISH) und einzelzell-mikromanipulation als neue anwendung bei der identifizierung und isolierug nuer *Oenococcus*- Stämme. Deutsches Weinbau-Jahrbuch *54:* 241–246.

Fröhlich, J., H. König and B. Bandenburg. 2003. Gene probes for the detection of *Oenococcus* species. Patent WO/2003/066894.

Galesloot, T.E., E. Hassing and J. Stadhouders. 1961. Agar media for the isolation and enumeration of aromabacteria in starters. Neth. Milk Dairy J. *15:* 127–150.

Gardini, F., A. Zaccarelli, N. Belletti, F. Faustini, A. Cavazza, M. Martuscelli, D. Mastrocola and G. Suzzi. 2005. Factors influencing biogenic amine production by a strain of *Oenococcus oeni* in a model system. Food Control *16:* 609–616.

Garvie, E.I. 1960. The genus *Leuconostoc* and its nomenclature. J. Dairy Res. *27:* 283–292.

Garvie, E.I. 1967a. The growth factor and amino acid requirements of species of the genus *Leuconostoc*, including *Leuconostoc paramesenteroides* (sp. nov.) and *Leuconostoc oenos*. J. Gen. Microbiol. *48:* 439–447.

Garvie, E.I. 1967b. *Leuconostoc oenos* sp. nov. J. Gen. Microbiol. *48:* 431–438.

Garvie, E.I. and L.A. Mabbitt. 1967. Stimulation of the growth of *Leuconostoc oenos* by tomato juice. Arch. Microbiol. *55:* 398–407.

Garvie, E.I. 1969. Lactic dehydrogenases of strains of the genus *Leuconostoc*. J. Gen. Microbiol. *58:* 85–94.

Garvie, E.I. and R. Tilbury. 1972. Some slime-forming heterofermentative species of the genus *Lactobacillus*. Appl. Microbiol. *23:* 389–397.

Garvie, E.I. 1975. Some properties of gas forming lactic acid bacteria and their significance in classification. *In* Carr, Cutting and Whiting (Editors), Lactic Acid Bacteria in Beverages and Food. Academic Press, London.

Garvie, E.I. 1976. Hybridization between deoxyribonucleic acids of some strains of heterofermentative lactic acid bacteria. Int. J. Syst. Bacteriol. *26:* 116–122.

Garvie, E.I. 1980. Bacterial lactate dehydrogenases. Microbiol. Rev. *44:* 106–139.

Garvie, E.I. 1981. Subdivisions within the genus *Leuconostoc* as shown by RNA-DNA hybridization. J. Gen. Microbiol. *127:* 209–212.

Garvie, E.I. 1983. *Leuconostoc mesenteroides* subsp. *cremoris* (Knudsen and Sørensen) comb. nov. and *Leuconostoc mesenteroides* subsp. *dextranicum* (Beijerinck) comb. nov. Int. J. Syst. Bacteriol. *33:* 118–119.

Garvie, E.I. 1984. Separation of species of the genus *Leuconostoc* and differentiation of the leuconostocs from the other lactic acid bacteria. *In* Bergan (Editor), Methods in Microbiology, Vol. 16. Academic Press, London, pp. 147–178.

Garvie, E.I. 1986. Genus *Leuconostoc* van Tieghem 1878, 198[AL] emended mut. char. Hucker and Pederson 1930, 66[AL]. *In* Sneath, Mair, Sharpe and Holt (Editors), Bergey's Manual of Systematic Bacteriology. The Williams & Wilkins Co., Baltimore, pp. 1071–1075.

Gashe, B.A. 1985. Involvement of lactic-acid bacteria in the fermentation of Tef eragrostis-Tef, an Ethiopian fermented food. J. Food Sci. *50:* 800–801.

Gashe, B.A. 1987. Kocho fermentation. J. Appl. Bacteriol. *62:* 473–478.

Gasser, F. 1970. Electrophoretic characterization of lactic dehydrogenases in the genus *Lactobacillus*. J. Gen. Microbiol. *62:* 223–239.

Gasser, F. and M. Hontebeyrie. 1977. Immunological relationships of glucose-6-phosphate-dehydrogenase of *Leuconostoc mesenteroides* NCDO 768 (=ATCC 12291). Int. J. Syst. Bacteriol. *27:* 6–8.

Guerini, S., S. Mangani, L. Granchi and M. Vincenzini. 2002. Biogenic amine production by *Oenococcus oeni*. Curr. Microbiol. *44:* 374–378.

Gunsalus, I.C. and M. Gibbs. 1952. The heterolactic fermentation. II. Position of C[14] in the products of glucose dissimilation by *Leuconostoc mesenteroides*. J. Biol. Chem. *194:* 871–875.

Guzzo, J., F. Delmas, F. Pierre, M.P. Jobin, B. Samyn, J. Van Beeumen, J.F. Cavin and C. Divies. 1997. A small heat shock protein from *Leuconostoc oenos* induced by multiple stresses and during stationary growth phase. Lett. Appl. Microbiol. *24:* 393–396.

Guzzo, J., M.P. Jobin, F. Delmas, L.C. Fortier, D. Garmyn, R. Tourdot-Marechal, B. Lee and C. Divies. 2000. Regulation of stress response in *Oenococcus oeni* as a function of environmental changes and growth phase. Int. J. Food Microbiol. *55:* 27–31.

Hansen, L.T. and H.H. Huss. 1998. Comparison of the microflora isolated from spoiled cold-smoked salmon from three smokehouses. Food Res. Int. *31:* 703–711.

Hardy, S., K.L. Ruoff, E.A. Catlin and J.J. Santos. 1988. Catheter-associated infection with a vancomycin-resistant Gram-positive coccus of the *Leuconostoc* spp. Pediatr. Infect. Dis. J. *7:* 519–520.

Holzapfel, W. and O. Kandler. 1969. Taxonomy of the species *Lactobacillus* Beijerinck. VI. *Lactobacillus coprophilus* subsp. *confusus* nov. subsp., a new variety of the subspecies *Betabacterium*. Zentbl. Bakteriol. Parasitenkd. Infektionskr. Hyg. *123:* 657–666.

Holzapfel, W. 1998. The Gram-positive bacteria associated with meat and meat products. *In* Davies and Board (Editors), The Microbiology of Meat and Poultry. Blackie Academic & Professional, London, pp. 35–74.

Holzapfel, W.H. and E.P. van Wyk. 1982. *Lactobacillus kandleri* sp. nov., a new species of the subgenus *Betabacterium* with glycine in the peptidoglycan. Zentbl. Bakteriol. Parasitenkd. Infektionskr. Hyg. *C3:* 495–502.

Holzapfel, W.H. and U. Schillinger. 1992. The genus Leoconostoc. *In* Balows, Trüper, Dworkin, Harder and Schleifer (Editors), The Prokaryotes, 2nd edn. Springer Verlag, New York, pp. 1508–1534.

Hontebeyrie, M. and F. Gasser. 1975. Comparative immunological relationships of two distinct sets of isofunctional dehydrogenases in genus *Leuconostoc*. Int. J. Syst. Bacteriol. *25:* 1–6.

Hontebeyrie, M. and F. Gasser. 1977. Deoxyribonucleic acid homologies in genus *Leuconostoc*. Int. J. Syst. Bacteriol. *27:* 9–14.

Horowitz, H.W., S. Handwerger, K.G. van Horn and G.P. Wormser. 1987. *Leuconostoc*, an emerging vancomycin-resistant pathogen. Lancet *2:* 1329–1330.

Hucker, G.J. and C.S. Pederson. 1930. Studies on the Coccoceae XVI. The genus *Leuconostoc*. N.Y. Agric. Exp. Sta. Bull. *167:* 3–80.

Ingraham, J.L., R.H. Vaughn and G.M. Cooke. 1960. Studies of the malolactic organisms of Californian wines. Am. J. Enol. Vitic. *11:* 1–4.

Isenberg, H.D., E.M. Vellozzi, J. Shapiro and L.G. Rubin. 1988. Clinical laboratory challenges in the recognition of *Leuconostoc* spp. J. Clin. Microbiol. *26:* 479–483.

Ito, S., T. Kobayashi, Y. Ohta and Y. Akiyama. 1983. Inhibition of glucose catabolism by aeration in *Leuconostoc mesenteroides*. J. Ferm. Technol. *61:* 353–358.

Izuagbe, Y.S., T.P. Dohman, W.E. Sandine and D.A. Heatherbell. 1985. Characterization of *Leuconostoc oenos* isolated from Oregon wines. Appl. Environ. Microbiol. *50:* 680–684.

Jang, J., B. Kim, J. Lee, J. Kim, G. Jeong and H. Han. 2002. Identification of *Weissella* species by the genus-specific amplified ribosomal DNA restriction analysis. FEMS Microbiol. Lett. *212:* 29–34.

Johnson, M.K. and C.S. McCleskey. 1957. Studies on the aerobic carbohydrate metabolism of *Leuconostoc mesenteroides*. J. Bacteriol. *74:* 22–25.

Jones, K.L. and S.E. Jones. 1984. Fermentations involved in the production of cocoa, coffee and tea. Progr. Ind. Microbiol. *19:* 411–456.

Juven, B.J. 1979. A simple method for long-term preservation of stock cultures of lactic acid bacteria. J. Appl. Bacteriol. *47:* 379–382.

Kandler, O. and I.G. Abo-Elnaga. 1966. [On the taxonomy of the genus *Lactobacillus* Beijerinck. V. *Lactobacillus coprophilus* nov. spec., a new species of the subgenus *Betabacterium*]. Zentbl. Bakteriol. Parasitenkd. Infektionskr. Hyg. *120:* 755–759.

Kandler, O., U. Schillinger and N. Weiss. 1983. *Lactobacillus halotolerans* sp. nov., nom. rev. and *Lactobacillus minor* sp. nov., nom. rev. Syst. Appl. Microbiol. *4:* 280–285.

Keenan, T.W. 1968. Production of acetic acid and other volatile compounds by *Leuconostoc citrovorum* and *Leuconostoc dextranicum*. Appl. Microbiol. *16:* 1881–1885.

Kelly, W.J., C.M. Huang and R.V. Asmundson. 1993. Comparison of *Leuconostoc oenos* strains by pulsed field gel electrophoresis. Appl. Environ. Microbiol. *59:* 3969–3972.

Kelly, W.J., R.V. Asmundson, G.L. Harrison and C.M. Huang. 1995. Differentiation of dextran-producing *Leuconostoc* strains from fermented rice cake (puto) using pulsed-field gel electrophoresis. Int. J. Food Microbiol. *26:* 345–352.

Kim, B., J. Lee, J. Jang, J. Kim and H. Han. 2003. *Leuconostoc inhae* sp. nov., a lactic acid bacterium isolated from kimchi. Int. J. Syst. Evol. Microbiol. *53:* 1123–1126.

Kim, B.J., H.J. Lee, S.Y. Park, J. Kim and H.U. Han. 2000a. Identification and characterization of *Leuconostoc gelidum* isolated from kimchi, a fermented cabbage product. J. Microbiol. *38:* 132–135.

Kim, J., J. Chun and H.U. Han. 2000b. *Leuconostoc kimchii* sp. nov., a new species from kimchi. Int. J. Syst. Evol. Microbiol. *50:* 1915–1919.

Klijn, N., A.H. Weerkamp and W.M. Devos. 1991. Identification of mesophilic lactic acid bacteria by using polymerase chain reaction amplified variable regions of 16S ribosomal RNA and specific DNA probes. Appl. Environ. Microbiol. *57:* 3390–3393.

Knudsen, S. and A. Sørensen. 1929. Contributions to the bacteriology of starters. Aarsskrift K. Veterinagr-Og Landbohojskole Copenhagen: 131–138.

Kodama, R. 1956. Studies on the nutrition of lactic acid bacteria. Part IV. *Lactobacillus fructosus* nov. sp., a new species of lactic acid bacteria. J. Agric. Chem. Soc. Jap. *30:* 705–708.

Kodama, R. 1963. The nutrition of *Lactobacillus fructosus* and its application to microbiological determination of nicotinamide and fructose. Ann. Report Inst. Ferm. Osaka *1:* 11–24.

Labarre, C., J. Guzzo, J.F. Cavin and C. Divies. 1996. Cloning and characterization of the genes encoding the malolactic enzyme and the malate permease of *Leuconostoc oenos*. Appl. Environ. Microbiol. *62:* 1274–1282.

Lamoureux, M., H. Prevost, J.F. Cavin and C. Divies. 1993. Recognition of *Leuconostoc oenos* strains by the use of DNA restriction profiles. Appl. Microbiol. Biotechnol. *39:* 547–552.

Langston, C.W. and C. Bouma. 1959. A study of the microorganisms from grass silage. I. The cocci. Appl. Microbiol. *8:* 212–222.

Lawrence, R.C. and T.D. Thomas. 1979. The fermentation of milk by lactic acid bacteria. *In* Bull, Ellwood and Ratledge (Editors), Microbial Technology: Current State, Future Prospects. University Press, Cambridge, pp. 187–219.

Le Jeune, C. and A. Lonvaud-Funel. 1997. Sequence of DNA 16S/23S spacer region of *Leuconostoc oenos* (*Oenococcus oeni*): application to strain differentiation. Res. Microbiol. *148:* 79–86.

Lee, H.J., S.Y. Park and J. Kim. 2000. Multiplex PCR-based detection and identification of *Leuconostoc* species. FEMS Microbiol. Lett. *193:* 243–247.

Lee, J.S., K.C. Lee, J.S. Ahn, T.I. Mheen, Y.R. Pyun and Y.H. Park. 2002. *Weissella koreensis* sp. nov., isolated from kimchi. Int. J. Syst. Evol. Microbiol. *52:* 1257–1261.

Leisner, J.J., B. Pot, H. Christensen, G. Rusul, J.E. Olsen, B.W. Wee, K. Muhamad and H.M. Ghali. 1999. Identification of lactic acid bacteria from chili bo, a Malaysian food ingredient. Appl. Environ. Microbiol. *65:* 599–605.

Leisner, J.J., M. Vancanneyt, R. Van der Meulen, K. Lefebvre, K. Engelbeen, B. Hoste, B.G. Laursen, L. Bay, G. Rusul, L. De Vuyst and J. Swings. 2005. *Leuconostoc durionis* sp. nov., a heterofermenter with no detectable gas production from glucose. Int. J. Syst. Evol. Microbiol. *55:* 1267–1270.

London, J. 1976. The ecology and taxonomic status of the lactobacilli. Annu. Rev. Microbiol. *30:* 279–301.

Lönner, C. and K. Prove-Akesson. 1989. Effects of lactic acid bacteria on the properties of sour dough bread. Food Microbiol. *6:* 19–35.

Lopez, H.W., A. Ouvry, E. Bervas, C. Guy, A. Messager, C. Demigne and C. Remesy. 2000. Strains of lactic acid bacteria isolated from sour doughs degrade phytic acid and improve calcium and magnesium solubility from whole wheat flour. J. Agric. Food Chem. *48:* 2281–2285.

Ludwig, W., J. Neumaier, N. Klugbauer, E. Brockmann, C. Roller, S. Jilg, K. Reetz, I. Schachtner, A. Ludvigsen, M. Bachleitner, G. Wallner, U. Fischer and K.H. Schleifer. 1993. Phylogenetic relationships of bacteria based on comparative sequence analysis of elongation factor Tu and ATP-synthase beta-subunit genes. Antonie van Leeuwenhoek *64:* 285–305.

Ludwig, W. and K.H. Schleifer. 1994. Bacterial phylogeny based on 16S and 23S rRNA sequence analysis. FEMS Microbiol. Rev. *15:* 155–173.

Luetticken, R. and G. Kunstmann. 1988. Vancomycin-resistant *Streptococcaceae* from clinical material. Zentbl. Bakteriol. Mikrobiol. Hyg. Ser. A *267:* 379–382.

Lyhs, U., J. Björkroth and H. Korkeala. 1999. Characterisation of lactic acid bacteria from spoiled, vacuum-packaged, cold-smoked rainbow trout using ribotyping. Int. J. Food Microbiol. *52:* 77–84.

Magnusson, J., H. Jonsson, J. Schnurer and S. Roos. 2002. *Weissella soli* sp. nov., a lactic acid bacterium isolated from soil. Int. J. Syst. Evol. Microbiol. *52:* 831–834.

Maillard, A.P., S. Biarrotte-Sorin, R. Villet, S. Mesnage, A. Bouhss, W. Sougakoff, C. Mayer and M. Arthur. 2005. Structure-based site-directed mutagenesis of the UDP-MurNAc-pentapeptide-binding cavity of the FemX alanyl transferase from *Weissella viridescens*. J. Bacteriol. *187:* 3833–3838.

Manca de Nadra, M.C., M.E. Farías, M.V. Moreno-Arribas, E. Pueyo and M.C. Polo. 1997a. Proteolytic activity of *Leuconostoc oenos*. Effect on proteins and polypeptides from white wine. FEMS Microbiol. Lett. *150:* 135–139.

Manca de Nadra, M.C., M.E. Farías, V. Moreno-Arribas, E. Pueyo and M.C. Polo. 1997b. A proteolytic effect of *Oenococcus oeni* on the nitrogenous macromolecular fraction of red wine. FEMS Microbiol. Lett. *174:* 135–139.

Manca de Nadra, M.C., M. E. Farías, V. Moreno-Arribas, E. Pueyo and M. C. Polo. 1999. A proteolytic effect of *Oenococcus oeni* on the nitrogenous macromolecular fraction of red wine. FEMS Microbiol. Lett. *174:* 41–47.

Manca de Nadra, M.C., M. E. Farías, E. Pueyo and M. C. Polo. 2005. Protease activity of *Oenococcus oeni* viable cells on red wine nitrogenous macromolecular fraction in presence of SO_2 and ethanol. Food Control *16:* 851–854.

Marcobal, A., B. de las Rivas, M.V. Moreno-Arribas and R. Muñoz. 2004. Identification of the ornithine decarboxylase gene in the putrescine-producer *Oenococcus oeni* BIFI-83. FEMS Microbiol. Lett. *239:* 213–220.

Marshall, V.M. and W.M. Cole. 1985. Methods for making kefir and fermented milks based on kefir. J. Dairy Res. *52:* 451–456.

Martinez-Murcia, A.J. and M.D. Collins. 1990. A phylogenetic analysis of the genus *Leuconostoc* based on reverse transcriptase sequencing of 16 S rRNA. FEMS Microbiol. Lett. *58:* 73–83.

Martinez-Murcia, A.J. and M.D. Collins. 1991. A phylogenetic analysis of an atypical *Leuconostoc*: description of *Leuconostoc fallax* sp. nov. FEMS Microbiol. Lett. *66:* 55–59.

Martinez-Murcia, A.J. and M.D. Collins. 1992. *In* Validation of the publication of new names and new combinations previously effectively published outside the IJSB. List no. 40. Int. J. Syst. Bacteriol. *42:* 191–192.

Martinez-Murcia, A.J., N.M. Harland and M.D. Collins. 1993. Phylogenetic analysis of some leuconostocs and related organisms as determined from large-subunit rRNA gene sequences: assessment of congruence of small- and large-subunit rRNA derived trees. J. Appl. Bacteriol. *74:* 532–541.

McDonald, L.C., R.F. McFeeters, M.A. Daeschel and H.P. Fleming. 1987. A differential medium for the enumeration of homofermentative and heterofermentatinve lactic acid bacteria. Appl. Environ. Microbiol. *53:* 1382–1384.

McDonald, L.C., L.C. Flemming and H.M. Hanssen. 1990. Acid tolerance of *Leuconostoc mesenteroides* and *Lactobacillus plantarum*. Appl. Environ. Microbiol. *56:* 2120–2124.

Middelhoven, W.J. and N. Klijn. 1997. *Leuconostoc fallax*, an acid and ethanol tolerant lactic acid bacterium. J. Sci. Food Agric. *75:* 57–60.

Miranda, M., A. Ramos, M. Veiga-da-Cunha, M.C. Loureiro-Dias and H. Santos. 1997. Biochemical basis for glucose-induced inhibition of malolactic fermentation in *Leuconostoc oenos*. J. Bacteriol. *179:* 5347–5354.

Moreno-Arribas, M.V., M.C. Polo, F. Jorganes and R. Munoz. 2003. Screening of biogenic amine production by lactic acid bacteria isolated from grape must and wine. Int. J. Food Microbiol. *84:* 117–123.

Morse, R., M.D. Collins, K. O'Hanlon, S. Wallbanks and P.T. Richardson. 1996. Analysis of the β' subunit of DNA-dependent RNA polymerase does not support the hypothesis inferred from 16S rRNA analysis that *Oenococcus oeni* (formerly *Leuconostoc oenos*) is a tachytelic (fast-evolving) bacterium. Int. J. Syst. Bacteriol. *46:* 1004–1009.

Mukherjee, S.K., M. N. Albury, C. S. Pederson, A. G. Van Veen and K.H. Steinkraus. 1965. Role of *Leuconostoc mesenteroides* in leavening the batter of idli, a fermented food of India. Appl. Microbiol. *13:* 227–231.

Müller, G. 1996. Kaffee, kakao, tee, vanile, tabak. *In* Müller, Holzapfel and Weber (Editors), Mikrobiologie der Lebensmittel pflanzlicher Herkunft. Behr's Verlag, Hamburg, pp. 431–450.

Nel, L., B.D. Wingfield, L.J. Vandermeer and H.J.J. Vanvuuren. 1987. Isolation and characterization of *Leuconostoc oenos* bacteriophages from wine and sugarcane. FEMS Microbiol. Lett. *44:* 63–67.

Nissen, H., A. Holck and R.H. Dainty. 1994. Identification of *Carnobacterium* spp. and *Leuconostoc* spp. in meat by genus-specific 16S rRNA probes. Lett. Appl. Microbiol. *19:* 165–168.

Niven, C.F., Jr., L.G. Buettner and J.B. Evans. 1954. Thermal tolerance studies on the heterofermentative lactobacilli that cause greening of cured meat products. Appl. Microbiol. *2:* 26–29.

Niven, C.F., Jr. and J.B. Evans. 1957. *Lactobacillus viridescens* nov. spec., a heterofermentative species that produces a green discoloration of cured meat pigments. J. Bacteriol. *73:* 758–759.

Oberman, H., Z. Libudzisz and A. Piatkiewicz. 1986. Physiological activity of deep-frozen concentrates of *Leuconostoc* strains. Nahrung-Food *30:* 147–154.

Okafor, N. 1977. Microorganisms associated with cassava fermentation for garri production. J. Appl. Bacteriol. *42:* 279–284.

Paludan-Muller, C., H.H. Huss and L. Gram. 1999. Characterization of lactic acid bacteria isolated from a Thai low-salt fermented fish product and the role of garlic as substrate for fermentation. Int. J. Food Microbiol. *46:* 219–229.

Pearce, L.E. and A.C. Halligan. 1978. Cultural characteristics of *Leuconostoc* strains from cheese starters. Congrilait 20th International Dairy Congress, Paris. pp. 520–521.

Pederson, C.S. 1930. Floral changes in the fermentation of sauerkraut. N.Y.S. Agric. Exp. Sta. Techn. Bull. *168:* 137.

Pederson, C.S. and M.N. Albury. 1955. Variation among the heterofermentative lactic acid bacteria. J. Bacteriol. *70:* 702–708.

Pinzani, P., L. Bonciani, M. Pazzagli, C. Orlando, S. Guerrini and L. Granchi. 2004. Rapid detection of *Oenococcus oeni* in wine by real-time quantitative PCR. Lett. Appl. Microbiol. *38:* 118–124.

Prévost, H., J.F. Cavin, M. Lamoureux and C. Diviès. 1995. Plasmid and chromosome characterization of *Leuconostoc oenos* strains. Am. J. Enol. Vitic. *46:* 43–48.

Pripis-Nicolau, L., G. de Revel, A. Bertrand and A. Lonvaud-Funel. 2004. Methionine catabolism and production of volatile sulphur compounds by *Oenococcus oeni*. J. Appl. Microbiol. *96:* 1176–1184.

Puspito, H. and G.H. Fleet. 1985. Microbiology of Sayur-Asin fermentation. Appl. Microbiol. Biotechnol. *22:* 442–445.

Quist, K., B.D. Thomsen and E. Hoier. 1987. Effect of ultrafiltered milk and use of different starters on the manufacture, fermentation and ripening of Havarti cheese. J. Dairy Res. *54:* 437–446.

Reguant, C. and A. Bordons. 2003. Typification of *Oenococcus oeni* strains by multiplex RAPD-PCR and study of population dynamics during malolactic fermentation. J. Appl. Microbiol. *95:* 344–353.

Reiter, B. and J.D. Oram. 1962. Nutritional studies on cheese starters. 1. Vitamin and amino acid requirements of single strain starters. J. Dairy Res. *29:* 63.

Reuter, G. 1970. Mikrobiologische Analyse von Lebensmitteln mit selektiven Medien. Lebensmittelhyg. *21:* 30–35.

Reuter, G. 1975. Classification problems, ecology and some biochemical activities of lactobacilli in meat products. *In* Carr, Cutting and Whiting (Editors), Lactic Acid Bacteria in Beverages and Food. Academic Press, London, New York, pp. 221–229.

Reuter, G. 1985. Elective and selective media for lactic acid bacteria. Int. J. Food Microbiol. *2:* 55–68.

Rodas, A.M., S. Ferrer and I. Pardo. 2003. 16S-ARDRA, a tool for identification of lactic acid bacteria isolated from grape must and wine. Syst. Appl. Microbiol. *26:* 412–422.

Rodriguez, S.B., E. Amberg, R.J. Thornton and M.R. Mclellan. 1990. Malolactic fermentation in Chardonnay: growth and sensory effects of commercial strains of *Leuconostoc oenos*. J. Appl. Bacteriol. *68:* 139–144.

Rogosa, J., J.A. Mitchell and R.F. Wiseman. 1951. A selective medium for isolation and enumeration of oral and fecal lactobacilli. J. Bacteriol. *62:* 132–133.

Rossi, I., L. Costamagna and F. Cleventi. 1978. La flora malolactica in alcuni vini dell' Italia centrole. Ann. Facolta di Agraria dell' University of Perugia *33:* 187–196.

Rossi, I., E. Dinardo, H. Salicone and H. Bertolucci. 1993. Characteristics of lactic acid bacteria and wine quality. Proceedings of the Symposium on Biotechnology and Molecular Biology of Lactic Acid Bacteria for the improvement of Foods and Feeds Quality. *In* Zamorani, Manachini, Bottazzi, and Coppola (Editors), Istituto Poligrafico e Zecca dello Stato, Rome, Italy, pp. 264–273.

Rozès, N., L. Arola and A. Bordons. 2003. Effect of phenolic compounds on the co-metabolism of citric acid and sugars by *Oenococcus oeni* from wine. Lett. Appl. Microbiol. *36:* 337–341.

Rubin, L.G., E. Vellozzi, J. Shapiro and H.D. Isenberg. 1988. Infection with vancomycin-resistant "streptococci" due to *Leuconostoc* species. J. Infect. Dis. *157:* 216.

Ruoff, K.L., D.R. Kuritzkes, J.S. Wolfson and M.J. Ferraro. 1988. Vancomycin-resistant Gram-positive bacteria isolated from human sources. J. Clin. Microbiol. *26:* 2064–2068.

Samelis, J., F. Maurogenakis and J. Metaxopoulos. 1994. Characterisation of lactic acid bacteria isolated from naturally fermented Greek dry salami. Int. J. Food Microbiol. *23:* 179–196.

Samelis, J., J. Rementzis, E. Tsakalidou and J. Metaxopoulos. 1998. Usefulness of rapid GC analysis of cellular fatty acids for distinguishing *Weissella viridescens*, *Weissella paramesenteroides*, *Weissella hellenica* and some non-identifiable, arginine-negative *Weissella* strains of meat origin. Syst. Appl. Microbiol. *21:* 260–265.

Sato, H., F. Yanagida, T. Shinohara and K. Yokotsuka. 2000. Restriction fragment length polymorphism analysis of 16S rRNA genes in lactic acid bacteria isolated from red wine. J. Biosci. Bioeng. *90:* 335–337.

Sato, H., F. Yanagida, T. Shinohara, M. Suzuki, K. Suzuki and K. Yokotsuka. 2001. Intraspecific diversity of *Oenococcus oeni* isolated during red wine-making in Japan. FEMS Microbiol. Lett. *202:* 109–114.

Schillinger, U., W. Holzapfel and O. Kandler. 1989. Nucleic acid hybridization studies on *Leuconostoc* and heterofermentative lactobacilli and description of *Leuconostoc amelibiosum* sp. nov. Syst. Appl. Microbiol. *12:* 48–55.

Schillinger, U. and W.H. Holzapfel. 2002. Culture media for lactic acid bacteria. *In* Corry, Curtis and Baird (Editors), Culture Media for Food Microbiology: Progress in Industrial Microbiology, 2nd edn. Elsevier, Amsterdam, pp. 34.

Schmitt, P., A.G. Mathot and C. Divies. 1989. Fatty acid composition of the genus *Leuconostoc*. Milchwissenschaft-Milk Sci. Int. *44:* 556–559.

Schmitt, P., C. Davies and A. Cardona. 1992. Origin of end-products from the co-metabolism of glucose and citrate by *Leuconostoc mesenteroides* subsp. *cremoris*. Appl. Microbiol. Biotechnol. *36:* 679–683.

Schmitt, P., C. Vasseur, V. Phalip, D.Q. Huang, C. Diviés and H. Prevost. 1997. Diacetyl and acetoin production from the co-metabolism of citrate and xylose by *Leuconostoc mesenteroides* subsp. *mesenteroides*. Appl. Microbiol and Biotechnol *47:* 715–718.

Sharpe, M.E., E.I. Garvie and R.H. Tilbury. 1972. Some slime-forming heterofermentative species of the genus *Lactobacillus*. Appl. Microbiol. *23:* 389–397.

Sharpe, M.E. 1981. The genus *Lactobacillus*. *In* Starr, Stolp, Trüper, Balows and Schlegel (Editors), The Prokaryotes. A Handbook on Habits, Isolation and Identification of Bacteria. Springer-Verlag, Berlin, pp. 1653–1679.

Shaw, B.G. and C.D. Harding. 1989. *Leuconostoc gelidum* sp. nov. and *Leuconostoc carnosum* sp. nov. from chill-stored meats. Int. J. Syst. Bacteriol. *39:* 217–223.

Sohier, D. and A. Lonvaud-Funel. 1998. Rapid and sensitive in situ hybridization method for detecting and identifying lactic acid bacteria in wine. Food Microbiol. *15:* 391–397.

Soni, S.K., D.K. Sandhu, K.S. Vilkhu and N. Kamra. 1986. Microbiological studies on Dosa fermentation. Food Microbiol. *3*: 45–54.

Speckman, R.A. and E.B. Collins. 1968. Diacetyl biosynthesis in *Streptococcus diacetilactis* and *Leuconostoc citrovorum*. J. Bacteriol. *95*: 174–180.

Stamer, J.R. 1975. Recent developments in the fermentation of sauerkraut. *In* Carr, Cutting and Whiting (Editors), Lactic Acid Bacteria in Beverages and Food. Academic Press, London, pp. 267–280.

Steinkraus, K.H. 1983. Lactic acid fermentation in the production of foods from vegetables, cereals and legumes. Antonie van Leeuwenhoek *49*: 337–348.

Stiles, M.E. and W.H. Holzapfel. 1997. Lactic acid bacteria of foods and their current taxonomy. Int. J. Food Microbiol. *36*: 1–29.

Takahashi, M., S. Okada, T. Uchimura and M. Kozaki. 1992. *Leuconostoc amelibiosum* Schillinger, Holzapfel, and Kandler 1989 is a later subjective synonym of *Leuconostoc citreum* Farrow, Facklam, and Collins 1989. Int. J. Syst. Bacteriol. *42*: 649–651.

Tanasupawat, S., S. Okada and K. Komagata. 1998. Lactic acid bacteria found in fermented fish in Thailand. J. Gen. Appl. Microbiol. *44*: 193–200.

Tanasupawat, S., O. Shida, S. Okada and K. Komagata. 2000. *Lactobacillus acidipiscis* sp. nov. and *Weissella thailandensis* sp. nov., isolated from fermented fish in Thailand. Int. J. Syst. Evol. Microbiol. *50*: 1479–1485.

Tannock, G.W., A. Tilsala-Timisjarvi, S. Rodtong, J. Ng, K. Munro and T. Alatossava. 1999. Identification of *Lactobacillus* isolates from the gastrointestinal tract, silage, and yoghurt by 16S-23S rRNA gene intergenic spacer region sequence comparisons. Appl. Environ. Microbiol. *65*: 4264–4267.

Teixeira, H., M.G. Goncalves, N. Rozes, A. Ramos and M.V. San Romao. 2002. Lactobacillic acid accumulation in the plasma membrane of *Oenococcus oeni*: a response to ethanol stress? Microb. Ecol. *43*: 146–153.

Tenreiro, R., M.A. Santos, H. Paveia and G. Vieira. 1994. Inter-strain relationships among wine leuconostocs and their divergence from other *Leuconostoc* species, as revealed by low-frequency restriction fragment analysis of genomic DNA. J. Appl. Bacteriol. *77*: 271–280.

Tilbury, R.H. 1975. Occurrence and effects of lactic acid bacteria in the sugar industry. *In* Carr, Cutting and Whiting (Editors), Lactic Acid Bacteria in Beverages and Food. Academic Press, London, pp. 177–191.

Tourdot-Marechal, R., D. Gaboriau, L. Beney and C. Divièes. 2000. Membrane fluidity of stressed cells of *Oenococcus oeni*. Int. J. Food Microbiol. *55*: 269–273.

Tracey, R.P. and T.J. Britz. 1989. Cellular fatty acid composition of *Leuconostoc oenos*. J. Appl. Bacteriol. *66*: 445–456.

Tsakalidou, E., E. Manolopoulou, E. Kabaraki, E. Zoidou, B. Pot, K. Kersters and G. Kalatzopoulos. 1994. The combined use of whole-cell protein extracts for the identification of (SDS-PAGE) and enzyme activity screening of lactic acid bacteria isolated from traditional Greek dairy products. Syst. Appl. Microbiol. *17*: 444–458.

Tsakalidou, E., J. Samelis, J. Metaxopoulos and G. Kalantzopoulos. 1997. Atypical *Leuconostoc*-like *Weissella* strains isolated from meat, shaving low phenotypic relatedness with the so far recognized arginine-negative *Weissella* spp. as revealed by SDS-PAGE of whole cell proteins. Syst. Appl. Microbiol. *20*: 659–664.

Tsenkovskii, L. 1878. Gel formation in sugar beet solutions. Proc. Soc. Sci. Nat. Imper. Univ. Kharkov *12*: 137–167.

van Tieghem, P.E.L. 1878. Sur la gomme de sucrerie. Ann. Sci. Nat. Bot. 6eSer.*67*: 180–202.

Vancanneyt, M., M. Zamfir, M. De Wachter, I. Cleenwerck, B. Hoste, F. Rossi, F. Dellaglio, L. De Vuyst and J. Swings. 2006. Reclassification of *Leuconostoc argentinum* as a later synonym of *Leuconostoc lactis*. Int. J. Syst. Evol. Microbiol. *56*: 213–216.

Vela, A.I., C. Porrero, J. Goyache, A. Nieto, B. Sanchez, V. Briones, M.A. Moreno, L. Dominguez and J.F. Fernández-Garayzábal. 2003. *Weissella confusa* infection in primate (*Cercopithecus mona*). Emerg. Infect. Dis. *9*: 1307–1309.

Vescovo, M., F. Dellaglio, V. Bottazzi and P.G. Sarra. 1979. Deoxyribonucleic acid homology among *Lactobacillus* species of the subgenus *Betabacterium* Orla-Jensen. Microbiologica *2*: 317–330.

Villani, F., G. Moschetti, G. Blaiotta and S. Coppola. 1997. Characterization of strains of *Leuconostoc mesenteroides* by analysis of soluble whole-cell protein pattern, DNA fingerprinting and restriction of ribosomal DNA. J. Appl. Microbiol. *82*: 578–588.

Viti, C., S. Ventura, L. Granchi and L. Giovannetti. 1995. Genome diversity in several strains of *Leuconostoc oenos* from the Chianti region. Ann. Microbiol. Enzimol. *45*: 151–158.

Viti, C., L. Giovannetti, L. Granchi and S. Ventura. 1996. Species attribution and strain typing of *Oenococcus oeni* (formerly *Leuconostoc oenos*) with restriction endonuclease fingerprints. Res. Microbiol. *147*: 651–660.

Vivas, N., A. Lonvaud-Funel and Y. Glories. 1997. Effect of phenolic acids and anthocyanins on growth, viability and malolactic activity of a lactic acid bacterium. Food Microbiol. *14*: 291–299.

Von Holy, A. and W.H. Holzapfel. 1989. Spoilage of vacuum packaged processed meats by lactic acid bacteria, and economic consequences. Xth WAFVH International Symposium, Stockholm, Sweden, 6–9 July 1989.

Walker, D.K. and S.E. Gilliland. 1987. Buttermilk manufacture using a combination of direct acidification and citrate fermentation by *Leuconostoc cremoris*. J. Dairy Sci. *70*: 2055–2062.

Walter, J., C. Hertel, G.W. Tannock, C.M. Lis, K. Munro and W.P. Hammes. 2001. Detection of *Lactobacillus*, *Pediococcus*, *Leuconostoc*, and *Weissella* species in human feces by using group-specific PCR primers and denaturing gradient gel electrophoresis. Appl. Environ. Microbiol. *67*: 2578–2585.

Ward, L.J., J.C. Brown and G.P. Davey. 1995. Detection of dairy *Leuconostoc* strains using the polymerase chain reaction. Lett. Appl. Microbiol. *20*: 204–208.

Weiler, H.G. and F. Radler. 1970. Milchsäurebakterien aus wein und von rebenblättern. Zentbl. Bakteriol. Parasitenkd. Infektkrankh. Hyg. Abt. 2 Orig. *124*: 707–732.

Weiler, H.G. and F. Radler. 1972. Vitamin-und Aminosäurebedarf von Milchsäurebakterien aus Wein und von Rebenblättern. Mitteilungen der Höheren Bundeslehr- und Versuchsanstalt für Wein und Obstbau Klosterneuburg Serie B Obst und Garten *22*: 4–18.

Wenocur, H.S., M.A. Smith, E.M. Vellozzi, J. Shapiro and H.D. Isenberg. 1988. Odontogenic infection secondary to *Leuconostoc* species. J. Clin. Microbiol. *26*: 1893–1894.

Whittenbury, R. 1964. Hydrogen peroxide formation and catalase activity in the lactic acid bacteria. J. Gen. Microbiol. *35*: 13–26.

Whittenbury, R. 1965a. A study of some pediococci and their relationships to *Aerococcus viridans* and the enterococci. J. Gen. Microbiol. *40*: 97–106.

Whittenbury, R. 1965b. The enrichment and isolation of lactic acid bacteria from plant material. Zentbl. Parasitenkd. Infektionskr. Hyg. Abt. 1 Suppl. *1*: 395–398.

Whittenbury, R. 1966. A study of the genus *Leuconostoc*. Arch. Mikrobiol. *53*: 317–327.

Wibowo, D., R. Eschenbruch, C.R. Davis, G.H. Fleet and T.H. Lee. 1985. Occurrence and growth of lactic acid bacteria in wine: a review. Am. J. Enol. Vitic. *36*: 302–313.

Yang, D. and C.R. Woese. 1989. Phylogenetic structure of the leuconostocs: an interesting case of a rapidly evolving organism. Syst. Appl. Microbiol. *12*: 145–149.

Yost, C.K. and F.M. Nattress. 2001. The use of multiplex PCR reactions to characterize populations of lactic acid bacteria assoicated with meat spoilage. Lett. Appl. Microbiol. *32*: 368.

Zapparoli, G., C. Reguant, A. Bordons, S. Torriani and F. Dellaglio. 2000. Genomic DNA fingerprinting of *Oenococcus oeni* strains by pulsed-field gel electrophoresis and randomly amplified polymorphic DNA-PCR. Curr. Microbiol. *40*: 351–355.

Zavaleta, A.I., A.J. Martinez-Murcia and F. Rodriguez-Valera. 1997a. 16-23S rDNA intergenic sequences indicate that *Leuconostoc oenos* is phylogenetically homogenous. Microbiology *142*: 2105–2114.

Zavaleta, A.I., A.J. Martinez-Murcia and F. Rodriguez-Valera. 1997b. Intraspecific genetic diversity of *Oenococcus oeni* as derived from DNA fingerprinting and sequence analyses. Appl. Environ. Microbiol. *63*: 1261–1267.

Family VI. **Streptococcaceae** Deibel and Seeley 1974, 490[AL]

THE EDITORIAL BOARD

Strep.to.coc.ca′ce.ae. N.L. masc. n. *Streptococcus* type genus of the family; -*aceae* ending to denote family; N.L. fem. pl. n. *Streptococcaceae* the *Streptococcus* family.

The family *Streptococcaceae* is circumscribed on the basis of 16S rDNA sequence analysis; and contains the genera *Streptococcus*, *Lactococcus*, and *Lactovum* (see Figure 1 and Figure 3). It is composed of Gram-positive, ovoid or spherical cocci. When it has been determined, cell walls have been found to contain the diamino acid lysine. Endospores are not formed. Facultative anaerobes that may require CO_2 for growth and are catalase-negative.

DNA G+C content (mol%): 33–46 mol%.

Type genus: **Streptococcus** Rosenbach 1884, 22.

Genus I. **Streptococcus** Rosenbach 1884, 22[AL]

ROBERT A. WHILEY AND JEREMY M. HARDIE

Strep.to.coc′cus. Gr. adj. *streptus* pliant; Gr. n. *kokkos* a grain, berry; N.L. masc. n. *Streptococcus* pliant coccus.

Streptococcus strains are normally **spherical or ovoid**, less than 2 μm in diameter, **occurring in chains or in pairs** when grown in liquid media. Cells are **nonmotile. Endospores are not formed. Gram-positive.** Virtually all species are **facultatively anaerobic**, some requiring additional CO_2 for growth. **Chemo-organotrophic with fermentative metabolism.** Carbohydrates are fermented to **produce mainly lactic acid** but no gas. **Catalase-negative.** Nutritional requirements are complex and variable. **The peptidoglycan is of group A** (Schleifer and Kandler, 1972) with L-lysine as the diamino acid in position 3 of the peptide subunit. Menaquinones are not present. Cell-wall polysaccharides form the basis of the Lancefield serological grouping scheme (Lancefield, 1933). Rhamnose is a common constituent of almost all streptococcal cell walls and its notable absence among members of the Mitis species group which includes *Streptococcus pneumoniae*, together with the presence of significant amounts of ribitol, are valuable chemotaxonomic markers for these taxa.

Temperature optimum usually about 37°C, but maximum and minimum temperatures vary somewhat among species. Many species exist as commensals or parasites on man and animals, some are highly pathogenic.

DNA G+C content (mol%): 33–46.

Type species: **Streptococcus pyogenes** Rosenbach 1884, 23[AL].

Further descriptive information

Phylogenetic position. Based on 16S rRNA gene sequence analysis the genus *Streptococcus* belongs within the low (< 50 mol%) G+C (*Clostridium–Bacillus*) branch of the Gram-positive eubacteria (Ludwig et al., 1985; Schleifer and Ludwig, 1995), and is a member (type genus) of the family *Streptococcaceae*. Following extensive taxonomic revision of *Streptococcus* (Collins et al., 1984b; Schleifer and Kilpper-Bälz, 1984, 1987), the genus currently consists of over 50 recognized species which, for the most part, fall within "species groups" named "Pyogenic", "Bovis", "Mutans", "Mitis", "Anginosus", and "Salivarius" (Bentley et al., 1991; Kawamura et al., 1995b).

Cell morphology. *Streptococcus* cells are usually spherical or ovoid in shape and are arranged in chains or pairs. Chain formation is best seen in liquid cultures. Some species (e.g., *Streptococcus mutans*) grow as short rods under certain cultural conditions, and several of the oral streptococci appear to be pleomorphic on primary isolation. Growth is by elongation in the axis of the chain with cell division in one plane, at right angles to the long axis. Chain length varies considerably between species and strains and is also dependent on medium composition; long chains of over 50 cells may be produced. Cells of *Streptococcus pneumoniae* characteristically form pairs (diplococci).

A recently described species, *Streptococcus minor* (Vancanneyt et al., 2004), has cells arranged predominantly in small groups.

Cell-wall composition. The cell-wall composition is characteristic of Gram-positive bacteria, consisting mainly of peptidglycan to which are attached a variety of carbohydrate, teichoic acid, and surface-protein antigens. The peptidoglycan is of group A (Schleifer and Kandler, 1972) with L-lysine as the diamino acid in position 3 of the peptide subunit. The most common peptidoglycan type is Lys–Ala$_n$, where an interpeptide bridge consisting of one up to four L-alanine residues is present. In some species the alanine residues can be partly replaced by L-threonine and/or L-serine. A few species, e.g., *Streptococcus acidominimus*, *Streptococcus canis* and some strains of *Streptococcus salivarius* and *Streptococcus bovis* have no alanine residues in their interpeptide bridge. Several species, e.g., *Streptococcus oralis* and *Streptococcus suis*, lack an interpeptide bridge and have a directly cross-linked peptidoglycan.

The qualitative composition of the cell-wall polysaccharides is known for many of the streptococcal species and these form the basis of Lancefield serological grouping (Schleifer and Kilpper-Bälz, 1987), in which strains are designated by a letter of the alphabet according to which cell-wall associated group antigen they possess (Lancefield, 1933). The amino sugars glucosamine and muramic acid are always present, while galactosamine is a variable component. The common reducing sugars are glucose, galactose, and rhamnose, which occur in various combinations in different species (Colman and Williams, 1965). The absence of rhamnose together with the presence of ribitol in *Streptococcus pneumoniae*, *Streptococcus oralis*, and *Streptococcus mitis* are chemotaxonomic markers for these species (Kilpper-Bälz et al., 1985; Price et al., 1986), while the polyols glucitol (sorbitol) and glycerol are present in *Streptococcus agalactiae* and *Streptococcus ratti*, respectively (Pritchard et al., 1981; Schleifer et al., 1984). Serological groups A, A-variant, and C all contain a backbone of α-1,3- and α-1,2-glycosidically linked rhamnose residues in which the group A-variant polysaccharide is not substituted, the group A carries *N*-acetylglucosamine residues, and group C contains *N*-acetylgalactosamine residues as immunodominant substitutions (Braun, 1983; Coligan et al., 1978). The group-B-specific polysaccharide found in *Streptococcus agalactiae* consists of a rhamnose, glucitol, and a phosphate-containing backbone with trisaccharide side chains of rhamnose, galactose, and *N*-acetylglucosamine linked to the 4-position of a rhamnose in the backbone of the polysaccharide (Pritchard et al., 1984). Rhamnose has been found to contribute to the immunodominant structure of group B and G polysaccharides, possibly explaining occasional cross-reactions between groups B and G (Schleifer and Kilpper-Bälz, 1987).

Serological classification has formed a major part of streptococcal taxonomy since the grouping scheme of Lancefield (1933) based on group-specific polysaccharide antigens (also referred to as C-substances) associated with the cell wall (originally groups D and N streptococci, now classified as *Enterococcus* and *Lactococcus* spp. respectively in which the group-specific antigens are teichoic acids, were included in *Streptococcus*). The Lancefield grouping scheme, where strains are designated by an upper-case letter of the alphabet (A, B, C, E, F, G, etc.), has been, and still is, extremely useful for differentiating between β-hemolytic streptococci from human and animal infections. The system is not comprehensive in that non-hemolytic and α-hemolytic species in particular may not possess recognized Lancefield grouping antigens or may be heterogeneous with respect to the group antigens possessed by different strains of a species.

Not all species form capsules. Some may form hyaluronic acid capsules during the early phase of growth, while an antigenically distinct polysaccharide capsule type has been identified on the surface of 90 different pneumococci (Henrichsen, 1995). Several streptococcal species produce soluble and insoluble extracellular polysaccharides when grown in the presence of sucrose, but these do not necessarily form morphologically distinctive capsules.

Important to the success of streptococci as commensals and pathogens is the expression of arrays of adhesins on their surface that mediate binding to a wide range of substrates available within the mammalian host. Several species, particularly those found within the oral cavity, can bind salivary glycoproteins and bacteria-derived salivary components. Examples of the latter include the antigen I/II family of adhesions expressed by *Streptococcus mutans*, *Streptococcus sobrinus*, *Streptococcus gordonii*, *Streptococcus oralis*, and *Streptococcus intermedius* and the Lral family expressed by *Streptococcus sanguis* and *Streptococcus parasanguis* (Jenkinson and Demuth, 1997; Whittaker et al., 1996). Among pyogenic and oral streptococcal species adhesins are expressed that recognize extracellular matrix and serum components, particularly fibronectin and plasminogen, as well as host and other microbial cells. The literature on streptococcal binding and colonization and the relevance to pathogenicity has been extensively reviewed by Jenkinson and Lamont (1997). Cell-wall-associated protein adhesins often contain repeating blocks of amino acids and binding can take place both within the repeating blocks and within non-repeating regions (Fischetti, 1989; McNab et al., 1994). Sequences and functions of these proteins have become assorted through gene-duplication and horizontal transfer between populations (Jenkinson and Lamont, 1997).

Nutrition and growth. Streptococci are facultatively anaerobic with some species such as *Streptococcus mutans*, members of the Anginosus species group, *Streptococcus pneumoniae*, and many others requiring the addition of 5% CO_2 to the atmosphere for growth. Streptococci are characteristically susceptible to vancomycin, unable to produce gas from glucose in MRS broth, mainly unable to produce pyrrolidonyl arylamidase (PYR), able to produce leucine arylaminopeptidase with only occasional exceptions, and exhibit variable reactions for growth in 6.5% NaCl-containing broth and the hydrolysis of esculin in the presence of 40% bile. These tests, together with the absence of motility and chain formation, serve to distinguish streptococci from other genera of facultatively anaerobic, catalase-negative, Gram-positive cocci (Facklam and Elliott, 1995, 2002; Ruoff, 2003).

All streptococci ferment carbohydrates, producing predominantly lactic acid; minor amounts of acetic and formic acids, ethanol, and CO_2 may also be produced. Although all species ferment glucose, a wide range of other carbohydrates is utilized and variations in fermentation patterns between species can be used for identification purposes. Streptococci are catalase-negative with the exception of the recently described species *Streptococcus didelphis* which, on initial isolation on blood agar, gives vigorous catalase activity that is lost on subsequent subculturing (Rurangirwa et al., 2000).

The complex nutritional requirements, generally including amino acids, peptides, purines, pyrimidines, and vitamins, are normally provided by using complex media. For optimal growth in liquid media, addition of a fermentable carbohydrate is also necessary.

Colonial and cultural characteristics. Growth on solid media is often enhanced by the addition of blood, serum, or glucose. Colonies on glucose are usually 0.5–1.0 mm in diameter after 24 h at 37°C and show little or no increase after prolonged incubation. Virtually all species are non-pigmented with the exception of some strains of *Streptococcus agalactiae*, which may have a yellow, orange, or brick-red pigment. Growth in liquid media is increased greatly by the addition of glucose, but the rapid fall in pH quickly inhibits growth unless the medium is highly buffered, as in Todd–Hewitt broth (Todd and Hewitt, 1932), or the medium is pH-controlled by the continuous addition of alkali. In batch culture, some strains produce a granular type of growth in broth with a clear supernatant fluid; these are usually found to have produced long chains. Other strains tend to produce a more diffuse growth, even turbidity. In continuous culture, both the macroscopic and microscopic appearance of the growth may vary according to the dilution rate and the limiting nutrient (Ellwood et al., 1974). Several types of reaction on blood agar are produced by different streptococci. These are visualized from surface growth on layered blood agar plates (Parker, 1983); sheep or horse blood is generally used. Reducing sugars may inhibit hemolysis by *Streptococcus pyogenes* (Facklam and Wilkinson, 1981).

β-Hemolysis is characterized by a sharply defined zone of clearing around the colonies with the zone size varying from strain to strain. With α-hemolysis, a zone of greenish discoloration occurs around the colony, usually 1–3 mm in width, and the margin is indistinct. A third type of hemolysis, α-prime, has also been described and resembles α-hemolysis with an obvious outer ring of clearing around the zone of discolored (green) erythrocytes (Parker, 1983). Non-hemolytic (or γ-hemolytic) streptococci have no effect on blood agar. Both the type and extent of hemolysis are influenced by the composition of the basal medium, the type and concentration of blood used and the cultural conditions. In the United States, 5% sheep blood agar is generally used for recognition of hemolytic streptococci, especially as the medium does not support the growth of hemolytic *Hemophilus* species. Most laboratories in the United Kingdom prefer 5% horse blood agar, on which good hemolytic zones may be seen around the surface colonies.

Genomes. The size range for completed streptococcal genomes is 1.85 (*Streptococcus pyogenes*) to 2.21 (*Streptococcus agalactiae*) Mbp. *Streptococcus mutans* (2.03 Mbp) and *Streptococcus pneumoniae* (2.04–2.16 Mbp) have also been sequenced. Sequences of the genomes of representative strains of several more species are due for completion; these species include *Streptococcus equi*, *Streptococcus gordonii*, *Streptococcus mitis*, *Streptococcus suis*, *Streptococcus thermophilus*, and *Streptococcus uberis*.

16S rRNA. The genus as a whole is very diverse and 16S rRNA gene sequencing has shown it to be divided into "species-groups" that encompass the majority of the total of 55 species at the time of writing (Bentley et al., 1991; Kawamura et al., 1995b; Tapp et al., 2003); see Figure 123.

Horizontal gene transfer. Horizontal gene transfer and recombination of genes have played a major role in generating streptococcal diversity both in the oral streptococci, particularly within the Mitis and Anginosus species groups, and among the pyogenic, β-hemolytic streptococci (Dowson et al., 1989; Kalia et al., 2001; Kehoe et al., 1996; Lunsford, 1998). Competence for transformation is not constitutive in streptococci but is regulated by genes *comA–comE* from two *com* operons: these operons include i) *comC*, encoding a competence-stimulating peptide (or pheromone) (CSP) together with *comD* and *comE*, encoding a two-component regulatory system, and ii) *comA* and *comB*, encoding the CSP-secretion apparatus (Håvarstein et al., 1996; Morrison, 1997; Whatmore et al., 1999). Recently, a strain of *Streptococcus infantis* (Atu-4) has been shown to be competent in the absence of the competence-pheromone gene *comC* (Ween et al., 2002). Competence in this strain was not regulated by cell density; loss of *comE* gene function resulted in loss of competence, indicating that mutations in *comD* or *comE* have given rise to a phenotype where competence is achieved without production of CSP. *Streptococcus mutans* has also been shown to contain a peptide-secretion-like apparatus (Petersen and Scheie, 2000) and to possess the *comCDE* operon (Ajdic et al., 2002). This species, like *Streptococcus gordonii*, has been shown to possess genes for DNA uptake (*comYA* and *comYB*) (Ajdic et al., 2002; Lunsford and Roble, 1997). *Streptococcus mutans* strain UA159, like sequenced *Streptococcus pneumoniae* strains TIGR4 and R6, both naturally transformable species, contain no temperate bacteriophage genomes. This is in contrast to *Streptococcus pyogenes* in which transformation via a competence pathway has not been described and in which approximately 10% of the genome is accounted for by bacteriophage and transposon genes (Ferretti et al., 2001).

Bacteriophages. Active virulent and temperate phages have been described. Phages are present within a high proportion of strains of group A streptococci and contribute to the pathogenic potential of this species (Ferretti et al., 2001; Hynes et al., 1995). The pyrogenic exotoxins SpeA and SpeC are bacteriophage-encoded (Cunningham, 2000). Sequencing of the *Streptococcus pyogenes* entire genome has revealed the presence of complete or partial sequences of four bacteriophage genomes containing genes for one or more previously undiscovered superantigen-like proteins (Desiere et al., 2001; Ferretti et al., 2001): at least six potential virulence factors are encoded and the importance of bacteriophages in horizontal gene transfer and the possible contribution of these events to the generation of new strains with increased pathogenic potential is recognized. Genes encoding proteins Pb1A and Pb1B involved in the binding of *Streptococcus mitis* to human platelets, with obvious relevance to the pathogenesis of infective endocarditis, are encoded by a lysogenic bacteriophage (Bensing et al., 2001).

Bacteriocins. Several species of streptococcus have been shown to be able to inhibit closely related strains through the production of bacteriocins (Tagg et al., 1976). Bacteriocin typing systems for *Streptococcus mutans* have been used to study the distribution of particular types within individuals and among family groups, and also to investigate transmission between mothers and fathers and their children (Berkovitz and Jordan, 1975; Kelstrup et al., 1970; Rogers, 1976; van Loveren et al., 2000). The ability of a particular strain of *Streptococcus mutans* to inhibit other bacteria is thought to play an important role in its successful transmission and subsequent colonization of a new host (Gronroos et al., 1998), and production of bacteriocin-like inhibitory substances (BLISs) by members of the normal oral flora have suggested an important role for these in protection against *Streptococcus pyogenes* infection (Dierksen et al., 2000).

Antibiotic sensitivity. On the whole, streptococci remain susceptible to most prescribed antibiotics although there are important exceptions. The MIC of penicillin to group A streptococci has not significantly changed over the past 70 or 80 years (Macris et al., 1998), whereas macrolide resistance has been reported to be increasing in Europe (Kaplan et al., 1999). A major cause for concern is the emergence and steady increase of penicillin-resistance within *Streptococcus pneumoniae* since the late 1970s (Casal, 1982). This has been due to the emergence of altered forms of penicillin-binding proteins (PBPs) PBP1a, PBP2x, and PBP2b, having decreased affinity for the antibiotic due to interspecies recombinational events involving viridans streptococcal species including *Streptococcus mitis* and *Streptococcus oralis* (Coffey et al., 1996; Dowson et al., 1994; Sibold et al., 1994). PBP-derived penicillin resistance in *Streptococcus pneumoniae* is associated with the emergence of extremely diverse resistant isolates as well as the geographic spread of resistant clones, notably capsular serotypes 23F 6B, 9V, and 19F clones (Coffey et al., 1995; Dowson et al., 1994).

Ecology and pathogenicity. Streptococci are associated with warm-blooded animals and birds. Most species can be regarded as commensal, being usually found on mucosal surfaces in the oral cavity, upper respiratory tract, and gastrointestinal tract, and under appropriate conditions can cause localized and systemic infections. The success of streptococci as commensals and potential pathogens has been attributed to i) "the ability

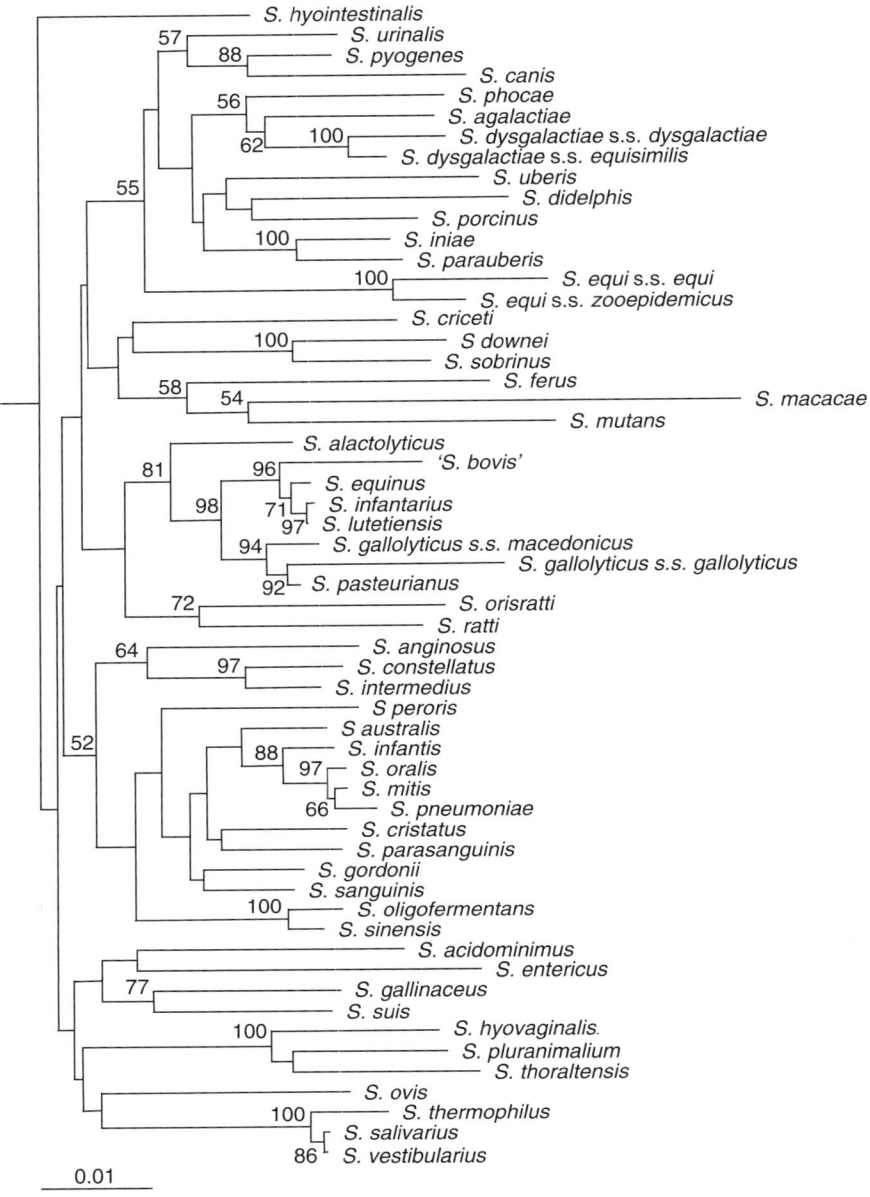

FIGURE 123. Phylogenetic tree of the streptococci based on 16S rRNA gene sequence comparisons over 1239 aligned bases. The tree was constructed using the neighbor-joining method after distance analysis of aligned sequences. Numbers represent bootstrap values for each branch, based on data for 100 trees. The scale bar shows the number of nucleotide substitutions per site.

to adhere to almost any surface present within their natural environment", ii) "the ability to rapidly utilize available nutrients under fluctuating environmental conditions", and iii) "the abilty to tolerate, resist, or even destroy host immune defences" (Jenkinson and Lamont, 1997). The pyogenic streptococci together with *Streptococcus pneumoniae* are the major pathogens. *Streptococcus pyogenes* can colonize the throat or skin and can cause a number of suppurative infections and non-suppurative sequelae. This species is the most common cause of bacterial pharyngitis, impetigo, and scarlet fever as well as erysipelas and other spreading infections. Group A streptococci have been found to be responsible for streptococcal-associated necro-

tizing fasciitis and streptococcal toxic-shock syndrome. Post-streptococcal infection sequelae associated with this organism include acute rheumatic fever, glomerulonephritis, and reactive arthritis. Group A streptococci have also been associated with Tourette's syndrome, tics, and movement and attention-deficit disorders (Cunningham, 2000). Studies have suggested that group A streptococci are able to invade as well as adhere to epithelial cells (LaPenta et al., 1994). Other pyogenic streptococci including those of Lancefield groups C and G are widely distributed in animals and man, where they exist as commensals and can cause serious infections including septicemia, endocarditis, septic arthritis, pneumonia, meningitis, pharyngitis, otitis media, and cellulitis. These streptococci share several features with group A streptococci, including the possession of M protein antigens, C5a peptidase, hyaluronic acid capsule, streptokinase, and other virulence factors (Efstratiou, 1997). *Streptococcus pneumoniae* causes pneumonia, meningitis, otitis media, and septicemia and is a serious pathogen among patients with underlying illness and compromised immunity (Ross, 1996). This species is the commonest cause of pneumonia, accounting for approximately 30% of community acquisitions (Brown and Lerner, 1998; Fedson, 1997).

Enrichment and isolation procedures

Streptococci are isolated from a wide range of clinical specimens. Viability is maintained during the transport of clinical specimens to the laboratory if recommended procedures are followed (Isenberg et al., 1985; Ruoff et al., 1999). It has been recommended that throat swabs for culturing group A streptococci (*Streptococcus pyogenes*) be transported directly to the laboratory, but a suitable transport medium should be employed where a delay of more than 2 h is anticipated before processing (Ruoff et al., 1999). Alternative, commercially available, systems for extended transport times are recommended (Facklam and Carey, 1985). Although nutritionally fastidious, streptococci can be cultured on most commonly used complex agar media. Incorporation of 5% defibrinated animal blood (horse or sheep) into agar media allows determination of hemolysis and hemolytic colonies can readily be picked off for subculturing although their presence may be masked if large numbers of other organisms are present. Detailed isolation procedures from clinical specimens have been described by Ruoff et al. (1999). Growth of streptococci is enhanced by incubation under anaerobic conditions at 37°C and an anaerobic atmosphere is essential for detection of hemolysis from *Streptococcus pyogenes* as air and CO_2 will interfere with the hemolytic reaction. Incubation in elevated levels of CO_2 is optimal for the recovery of capnophilic streptococci.

Selective media. Selective media have been described for the isolation of streptococci and media for specific groups of streptococci are also available. The oral streptococci can be isolated and partially characterized through the production of extracellular polysaccharide by plating on sucrose-containing agar media such as Trypticase-Yeast Extract Cystine (TYC) 5% sucrose agar (de Stoppelaar et al., 1967) or Mitis Salivarius agar containing the selective agents trypan blue (75 mg/l), crystal violet (0.8 mg/l), and potassium tellurite (3.5 mg/l) in addition to 5% sucrose. Selective media for isolating the cariogenic streptococcus *Streptococcus mutans* have been developed based on TYC and Mitis Salivarius media. These employ the use of increased levels of sucrose and the addition of bacitracin (Gold et al., 1973; Ikeda and Sandham, 1972; Linke, 1977; Tanzer et al., 1983; Van Palenstein Helderman et al., 1983; Wade et al., 1986). Members of the Anginosus species group can be isolated on a semi-selective agar medium (NAS) containing 40 g/l Sensitivity Agar (Lab M) + 30 μg/ml nalidixic acid, 1 mg/ml sulfamethazine, and 5% (v/v) defibrinated horse blood (Whiley et al., 1993). NAS will also select for *Streptococcus mutans* and *Streptococcus sobrinus* although recovery of the latter is strain-dependent. Isolation of *Streptococcus pyogenes* on sheep-blood agar plus trimethoprim-sulfamethoxazole incubated anaerobically and other media have been compared (Kellogg, 1990). Enhanced

recovery of β-hemolytic streptococci on an alkaline pH-adjusted agar medium that interferes with the production and/or release of bacteriocin-like inhibitory molecules from *Streptococcus salivarius* has been described (Dierksen et al., 2000).

Maintenance procedures

Most strains can be maintained for short periods by weekly subculturing on appropriate media. For long-term maintenance, strains can be preserved either deep frozen at −70 to −80°C or in liquid nitrogen. Freezing cultures on glass beads is particularly useful since one or two beads can be removed whenever the culture needs to be regenerated. Cells for freezing are harvested from blood agar and suspended in a diluent containing 1% tryptone, 0.5% yeast extract, 0.1% glucose, 0.1% cysteine-HCl, and 2% bovine serum (Bowden and Hardie, 1971). Commercially available deep-freeze storage vial systems are also well suited for storing streptococcal strains. The most reliable method for long-term preservation is lyophilization using standard methods (Kirsop and Snell, 1984).

Procedures for testing special characteristics

Hemolysis. Determined on a blood agar base medium such as Todd–Hewitt, brain heart infusion, proteose peptone, etc., plus 5% defibrinated horse or sheep blood. Streak, stab, or pour plates are used. For streak-plating, layered plates are prepared with a thin layer of blood agar poured on top of a nutrient agar base. Inoculation is by streaking with the loop and deep growth can be obtained by stabbing the loop into the medium. Anaerobic incubation is recommended but not with a candle jar or CO_2 incubator (Facklam and Wilkinson, 1981). Pour plates are prepared by adding the inoculum to 15 ml melted agar base cooled to 50–55°C, and then adding 0.6 ml defibrinated blood, mixing, and pouring into a Petri dish. It has been recommended that plates be examined microscopically (60× magnification) so that unlysed erythrocytes can be recognized within zones. This is especially important for recognizing α-prime hemolysis (a zone of partial hemolysis surrounded by a zone of complete hemolysis). In practise, plates are usually examined macroscopically in the clinical laboratory.

Streptococcus agalactiae strains show a marked enlargement of the zone of β-hemolysis when, after 18 h anaerobic incubation at 37°C, they are removed to 4°C for 6 h and then reincubated anaerobically at 37°C (Okazaki et al., 2003).

Serological determination. Established by Lancefield (1933) as one of the most important characteristics in identifying β-hemolytic streptococci, serological determination of group-specific carbohydrate antigens present in the cell walls of isolates has remained an important routine identification technique. Traditional antigen extraction includes growth of isolates in Todd–Hewitt or other suitable broth followed by antigen extraction with agents that have variously included hot hydrochloric acid, hot formamide, heat (autoclave), Pronase B, nitrous acid, or *Streptomyces albus* enzymes with or without lysozyme (Ruoff et al., 1999). Antigen identification has been achieved using ring precipitation in small capillary tubes, double diffusion in gels (Rotta et al., 1971), countercurrent electrophoresis (Dajani, 1973), coagglutination, or immunofluorescence (Cars et al., 1975; Christensen et al., 1973). More recently commercial streptococcal grouping kits have become available and are widely used especially for routine diagnostic purposes. These are based on latex agglutination where latex particles are coated with a Lancefield-group-specific antiserum and mixed with extracted antigens. Positive reactions are visualized by clumping of the particles.

Streptococcus pyogenes (Lancefield group A streptococcus, GAS) is divided into serotypes on the basis of streptococcal surface M proteins with more than 80 types currently known of which types 1, 3, 11, 12, and 28 are most common in invasive and toxic infections (Cunningham, 2000; Lancefield, 1928). The difficulties associated with M protein serotyping (difficulty reading results, the discovery of new serotypes and the high cost of antiserum production) have largely been overcome by molecular approaches, including M protein gene (*emm*) typing and multilocus sequence typing (Beall et al., 1996; Enright et al., 2001). Streptococci of Lancefield group C and G have been shown to express M proteins with high sequence homology to those of *Streptococcus pyogenes* (for a review of the literature, see Cunningham, 2000), where they are presumed to have the same function (M protein inhibits activation of the alternative complement pathway and phagocytosis by polymorphonuclear leukocytes).

T and R proteins are also present on the surface of *Streptococcus pyogenes* (see the description of *Streptococcus pyogenes*, this chapter). T typing is an important adjunct to M typing and is particularly useful when the M type is not known. Associations between certain T and M types in particular strains have been useful in speeding up M or *emm* typing (Beall et al., 1998). In contrast to the M protein, the T protein molecule has its most conserved region towards the amino-terminal and T proteins are not present in Lancefield group C or G streptococci (Jones et al., 1991).

Serological typing of *Streptococcus pneumoniae* depends on recognition of different capsular polysaccharide antigens by means of the Quellung reaction. Culture suspension and a capsule-specific antiserum are mixed on a slide with saturated aqueous methylene blue. After 10 min under a cover slip the slides are examined under the microscope for swelling and increased visibility of the capsule around the cocci (Austrian, 1976).

Biochemical and physiological tests. A wide range of tests has been used for characterizing streptococci, including carbohydrate fermentations, production of acetyl methyl carbinol (acetoin) from glucose in the Voges–Proskauer reaction, production of ammonia from arginine, hydrolysis of esculin, hippurate, and starch, reduction of litmus milk, production of H_2O_2, tolerance to NaCl and bile, and formation of extracellular polysaccharide from sucrose (Facklam and Wilkinson, 1981; Parker, 1983). Conventionally these tests have been carried out in meat extract or serum water broths with appropriate additives (Cowan and Steel, 1974). Nowadays, identification schemes for the streptococci frequently rely on the inclusion of tests designed to detect preformed enzyme activities after a few hours of incubation by using chromogenic or fluorogenic substrates including glycosides, aminoglycosides, and peptidases. Commercially available test identification kits designed specifically for streptococci have been developed and include preformed enzyme tests as well as miniaturized formats for carbohydrate fermentation and other traditional biochemical tests. These test systems are particularly useful in the clinical laboratory although they may lag behind in facilitating the identification of more recently recognized species. Commer-

cial test kits arguably also have the advantage of enabling standardization of test methods used in different laboratories and thus can help avoid interlaboratory discrepancies in reported test reactions for a given species. However, care must be taken to specify exactly which test format and substrate type is being used as results may vary for a given enzyme (Ahmet et al., 1995; Kilian et al., 1989b).

Several key tests are particularly useful to presumptively identify streptococci, particularly in the clinical laboratory. Bacitracin (0.04 U disks) and sulfamethoxazole (SXT) (23.75 µg SXT plus 1.25 µg trimethoprim, disks) can be placed on a blood agar plate with a β-hemolytic streptococcus. *Streptococcus pyogenes* is usually bacitracin-sensitive and SXT-resistant, group B streptococci (*Streptococcus agalactiae*) are resistant to both of these agents, and group C and G large-colony forming streptococci are usually bacitracin-resistant and SXT-susceptible (Facklam and Carey, 1985). Group B streptococci can also be distinguished by the CAMP test for detecting a diffusible extracellular protein referred to as the CAMP factor: a single streak of the streptococcus is made perpendicular to, but not touching, a streak of β-lysin-producing *Staphylococcus aureus*, either on sheep or bovine blood agar. After incubation aerobically or in a candle jar, CAMP-positive strains produce an arrowhead-shaped enlargement of the zone of lysis around the staphylococcus at the junction between the two streaks of growth. Detection of the enzyme pyrrolidonyl arylamidase (PYR) has been found to be more specific than bacitracin sensitivity for *Streptococcus pyogenes* identification (Facklam et al., 1982). This test is carried out in PYR broth (Todd–Hewitt broth supplemented with 0.01% L-pyrrolidonyl-β-naphthylamide) inoculated and incubated at 36°C for 4h. On addition of PYR reagent (*N,N*-dimethylaminocinnamaldehyde), a positive reaction is indicated by the formation of a deep red color due to reaction of liberated β-naphthylamine with the PYR reagent.

Streptococcus pneumoniae can be distinguished from other α-hemolytic streptococci on the basis of its sensitivity to a 5 µg optochin (ethylhydrocupreine) disk on a blood agar plate. Sensitivity is indicated by an inhibition zone of >14 mm with a 6 mm disk or >16 mm with a 10 mm disk (Facklam and Carey, 1985). The bile solubility test in which colonies are covered with 10% deoxycholate also differentiates this species by the lysis of pneumococcal colonies, but is considered by some to be more complicated and less reliable than the optochin test (Lund, 1959).

The hydrolysis of sodium hippurate is a character shared by several species that include *Streptococcus acidominimus*, *Streptococcus agalactiae*, *Streptococcus devriesei*, *Streptococcus didelphis*, *Streptococcus hyovaginalis*, *Streptococcus oligofermentans*, *Streptococcus pluranimalium*, *Streptococcus uberis*, and some strains of *Streptococcus thoraltensis*. A rapid test (Hwang and Ederer, 1975) has been described in which a loopful of an overnight culture of streptococcus grown on blood agar is suspended in 0.4 ml 1% aqueous sodium hippurate. After 2h incubation at 37°C, 0.2 ml 3.5% ninhydrin made up in a 1:1 mixture of butanol and acetone is added. After further incubation for 10 min a deep purple color, due to the reaction of ninhydrin with glycine, indicates a positive reaction.

Production of extracellular polysaccharides (ECPs), i.e., glucans (dextrans) and fructans (levans) from sucrose is an important characteristic of several of the oral streptococci, including *Streptococcus mutans*, *Streptococcus salivarius*, *Streptococcus sanguis*, and *Streptococcus oralis*, together with some slime-producing strains designated *Streptococcus bovis*. Such polysaccharides can be detected in the supernatant of cultures in buffered 5% sucrose broth, either by observing gel formation in some cases or by differential precipitation with ethanol. A precipitate after addition of 1.2 volumes ethanol indicates glucan formation, whereas fructans are precipitated by 2.5 volumes (Hehre and Neill, 1946). Standardization of the tests for polysaccharide production can prove to be difficult in practice. A useful indication of the presence of these products can be obtained more simply by observation of colonial morphology on 5% sucrose agar plates (Colman and Ball, 1984; Parker, 1983).

Streptococci of group D are differentiated from *Enterococcus* species by the ability of the latter to hydrolyze esculin in the presence of 40% bile and to grow in the presence of 6.5% NaCl.

Taxonomic comments

Since the publication of Volume 2 of the First Edition of *Bergey's Manual of Systematic Bacteriology*, in which not only facultatively anaerobic but also several strictly anaerobic *Streptococcus* species were included (Hardie, 1986c), the genus *Streptococcus* has undergone considerable taxonomic revision. The period since has seen the application of chemotaxonomic techniques, whole-genomic DNA hybridization and 16S rRNA sequencing (Bentley et al., 1991; Kawamura et al., 1995b; Schleifer and Kilpper-Bälz, 1987). These methods have given insight into the natural (phylogenetic) relationships between *Streptococcus* and other genera, revealed the natural grouping of species into "species groups" and form the basis of species delineation (Stackebrandt et al., 2002). These approaches have surmounted the difficulties faced by previous workers who were, in the main, reliant on the biochemical, physiological, and serological characterization of strains, but who had often been well aware of the limitations of the approaches then available and of the undue emphasis sometimes given to them in streptococcal classification and identification (Jones, 1978). *Streptococcus* was divided into three genera that included *Streptococcus* (*sensu stricto*), *Enterococcus*, and *Lactococcus* (Schleifer and Kilpper-Bälz, 1984; Schleifer et al., 1985). *Streptococcus*, comprising the pyogenic and oral species included both human and animal pathogenic and saprophytic streptococci, *Lactococcus* was formed from the lactic streptococci (Kilpper-Bälz et al., 1982; Schleifer et al., 1985), and *Enterococcus* from the enterococci (Collins et al., 1984b; Schleifer and Kilpper-Bälz, 1984).

The anaerobic species previously included within *Streptococcus* (*Streptococcus parvulus*, *Streptococcus morbillorum*, *Streptococcus hansenii*, and *Streptococcus pleomorphus*) have, for the most part, been excluded and assigned to other genera. *Streptococcus parvulus* was removed to the genus *Atopobium* as *Atopobium parvulus* (comb. nov.) (Collins and Wallbanks, 1992), *Streptococcus morbillorum* was reclassified as *Gemella morbillorum* comb. nov. (Kilpper-Bälz and Schleifer, 1988), and *Streptococcus hansenii* has been proposed as a member of the genus *Ruminococcus* (Ezaki et al., 1994). The latter proposal has been argued against on the grounds that the *Ruminococcus* type species, *Ruminococcus flavefaciens*, belongs to a separate phyletic group to that containing "*Ruminococcus hansenii*" (Rainey and Janssen, 1995; Willems and Collins, 1995). In the case of *Streptococcus pleomorphus* (Barnes et al., 1977), 16S rRNA gene sequence cataloging (Ludwig et al.,

1988) and 16S rRNA sequencing (Kawamura et al., 1995b) have placed this species outside the genus *Streptococcus* and indicated that *Streptococcus pleomorphus* is most closely related to *Clostridium innocuum*. Recent analysis has placed this species within the family *Erysipelotrichaceae* together with *Eubacterium biforme* and *Eubacterium cylindroides* (W. Wade, personal communication). See the *Eubacterium* section of this volume).

Nutitionally variant streptococci (Cooksey et al., 1979; Freney et al., 1992), originally referred to as "thiol-dependent", "vitamin B6-dependent" and "symbiotic", streptococci have been excluded from *Streptococcus* and reclassified as *Abiotrophia defectiva* (Kawamura et al., 1995a), *Granulicatella adiacens*, *Granulicatella balaenopterae*, and *Granulicatella elegans* (Collins and Lawson, 2000). These organisms have been isolated from clinical specimens including blood, abscesses, oral ulcers, and urethral samples (Christensen and Facklam, 2001).

By 16S rRNA gene sequence analysis and reassociation data the species within *Streptococcus sensu stricto* comprise distinct "species groups" along with several species remaining ungrouped (Bentley et al., 1991; Kawamura et al., 1995b; Schleifer and Kilpper-Bälz, 1987). The species groups have been designated "Pyogenic", "Bovis", "Mutans", "Mitis", "Anginosus", and "Salivarius" and comprise the majority of described species. The oral streptococci include the "Salivarius", "Mutans", "Anginosus", and "Mitis" species groups of Bentley et al. (1991) and Kawamura et al. (1995b) of which the latter two groups formed subgroups within a single "Oralis group" according to the classification of Schleifer and Kilpper-Bälz (1987).

Several streptococcal species and species groups have confused taxonomic and nomenclatural histories that require clarification if the present classification and nomenclature of the genus are to be understood.

Nomenclatural note. Several nomenclatural changes to established streptococcal species epithets have recently been made (Trüper and de Clari, 1997, 1998) in the interests of grammatical correctness and in accordance with the International Code of Nomenclature of Bacteria (Lapage et al., 1992). These include *Streptococcus cricetus*, *Streptococcus crista*, *Streptococcus rattus*, *Streptococcus sanguis*, and *Streptococcus parasanguis* which have been changed to *Streptococcus criceti*, *Streptococcus cristatus*, *Streptococcus ratti*, *Streptococcus sanguinis*, and *Streptococcus parasanguinis*, respectively. Argument against this development has, quite correctly in the opinion of these authors, been put forward on the basis that changes in names of species long-standing in the literature causes unnecessary confusion and therefore cannot be justified (Kilian, 2001). The Judicial Commission of the International Committee on Systematic Bacteriology ruled recently that priority be given to the stabilization of nomenclature over orthographic correctness and that names on the Approved Lists of Bacterial Names, the Validation Lists and the Notification Lists should not be changed on grammatical grounds (Amendment to Rule 61 of the Bacteriological Code, Minute 7. Session 1 of the meeting of the Judicial Commission held 14, 15, and 18 August 1999, Sydney, Australia, published in the *International Journal of Systematic* and *Evolutionary Microbiology* [2000], vol. 50, pp. 2239–2244). Despite the present authors' misgivings concerning these changes, the corrected specific epithets are becoming more frequently used in the literature and therefore are employed in this chapter.

Streptococcal species groups

Pyogenic group. In the previous edition of the *Manual* six named species were included within the "pyogenic hemolytic streptococci": *Streptococcus pyogenes* (Lancefield group A and "group A-variant), *Streptococcus agalactiae* (Lancefield group B), *Streptococcus equi* (Lancefield group C), *Streptococcus iniae*, *Streptococcus dysgalactiae* (Lancefield group C), and *Streptococcus pneumoniae* together with other streptococci of Lancefield groups C, G, and L, and groups E, P, U, and V (Rotta, 1986). These streptococci largely comprised β-hemolytic species usually containing a polysaccharide Lancefield group antigen in the cell wall and producing pyogenic infections in man and animals. Exceptions to these characteristics were acknowledged at the time, particularly the inclusion of *Streptococcus pneumoniae* which previous authors had listed either as a separate group (Jones, 1978) or within the viridans streptococci (Bridge and Sneath, 1983). With the exception of *Streptococcus pneumoniae*, all the taxa listed previously in the "pyogenic hemolytic streptococci" are included within the Pyogenic group here. All are associated with pyogenic infections in man and/or animals. The species within the present Pyogenic species group include *Streptococcus pyogenes*, *Streptococcus agalactiae*, *Streptococcus canis*, *Streptococcus equi*, *Streptococcus iniae*, *Streptococcus uberis*, *Streptococcus parauberis*, *Streptococcus phocae*, *Streptococcus urinalis*, and *Streptococcus didelphis* with strains of Lancefield groups C, G, and L, and groups E,P,U, and V comprising *Streptococcus dysgalactiae* and *Streptococcus porcinus*, respectively. *Streptococcus uberis* and *Streptococcus parauberis* strains were previously both included as distinct genotypes within *Streptococcus uberis* by DNA–DNA hybridization and designated *Streptococcus uberis* types I and II, respectively (Collins et al., 1984a; Fuller et al., 2001, 1979a). 16S rRNA gene sequence analysis later demonstrated that type I and II strains were sufficiently dissimilar to warrant separate species status (Williams and Collins, 1990a).

Within the Pyogenic group, the β-hemolytic streptococci of Lancefield group C, large-colony type of Lancefield group G and related strains, previously assigned to *Streptococcus dysgalactiae*, "*Streptococcus equisimilis*", *Streptococcus equi*, and *Streptococcus zooepidemicus*, have perhaps the most confused taxonomic background. The β-hemolytic streptococci of Lancefield group C isolated from human respiratory tracts, vaginas, and skin were originally called *Streptococcus equisimilis* (Frost and Engelbrecht, 1936) and the name *Streptococcus dysgalactiae* was given to non-β-hemolytic strains of bovine origin that were otherwise identical to *Streptococcus equisimilis* (Breed et al., 1948). Neither species was included in the Approved Lists of Bacterial names (Skerman et al., 1980), but later the name *Streptococcus dysgalactiae* was revived for α-hemolytic, Lancefield group C strains (Garvie et al., 1983). Subsequently DNA–DNA hybridization studies demonstrated that α-, and β-hemolytic Lancefield group C strains, large-colony type strains of Lancefield group G and β-hemolytic Lancefield group L streptococci together with isolates previously classified as *Streptococcus dysgalactiae* and "*Streptococcus equisimilis*" belonged within a single species (Farrow and Collins, 1984a; Kilpper-Bälz and Schleifer, 1984) for which the name *Streptococcus dysgalactiae* was proposed and an emended description given (Farrow and Collins, 1984a). Vandamme et al. (1996) divided *Streptococcus dysgalactiae* strains into two clusters (named clusters I and III in that study) based on whole-cell-derived polypeptide patterns by SDS-PAGE. Cluster I strains, isolated from animal sources, were

of Lancefield groups C and L, whereas cluster III strains were of human origin, belonged to Lancefield C and G and, unlike cluster I strains, produced streptokinase activity on human plasminogen. Clusters I and III were named *Streptococcus dysgalactiae* subsp. *dysgalactiae* and *Streptococcus dysgalactiae* subsp. *equisimilis* (cluster III strains, i.e., of human origin), respectively. Vieira et al. (1998) used DNA–DNA reassociation, multilocus enzyme electrophoresis (MLEE), and physiological characterization to also recognize these two subspecies within *Streptococcus dysgalactiae* in close agreement with Vandamme et al. (1996). Despite the evidence from DNA–DNA hybridization experiments and overall close phenotypic similarity some authors prefer to differentiate human group C and group G strains as separate species and to subdivide strains primarily according to habitat and pathogenicity (Efstratiou, 1997, 1994). The pivotal DNA–DNA hybridization studies of Kilpper-Bälz and Schleifer (1984) and Farrow and Collins (1984a) also demonstrated that Lancefield-group-C-reacting strains, *Streptococcus equi* (Sand and Jensen, 1888) and "*Streptococcus zooepidemicus*" (Frost and Engelbrecht, 1936) belonged to a single species. Farrow and Collins (1984a) argued that the separate identity of "*Streptococcus zooepidemicus*" be maintained on phenotypic grounds, specifically the latter's ability to produce acid from sorbitol but not from trehalose and to hydrolyze esculin, and they therefore proposed the names *Streptococcus equi* subsp. *equi* and *Streptococcus equi* subsp. *zooepidemicus*. A high level of similarity (98.9%) observed between fragments of manganese-dependent superoxide dismutase genes from *Streptococcus equi* subsp. *equi* and subsp. *zooepidemicus* has also been taken to indicate that these do indeed belong to a single species (Poyart et al., 1998).

Biochemical, physiological, and antigenic characteristics of species within the Pyogenic species group are given in Table 131 and Table 132.

Mutans group. The mutans streptococci are associated with dental plaque in man and several animal species and includes the species *Streptococcus mutans* (Clarke, 1924), generally implicated as the primary pathogen in the etiology of human dental caries (Hamada and Slade, 1980). 16S rRNA data place *Streptococcus mutans*, *Streptococcus sobrinus*, *Streptococcus criceti*, *Streptococcus ratti*, *Streptococcus downei*, and *Streptococcus macacae* together to form the loose conglomeration of species referred to as the "Mutans group" (Bentley et al., 1991; Kawamura et al., 1995b; Schleifer et al., 1984). *Streptococcus ferus* is a peripheral member of this species group by DNA–DNA hybridization (Schleifer et al., 1984) but has been found to cluster closest to *Streptococcus sanguinis* by MLEE (Gilmour et al., 1987). In addition, there is no strong evidence to support the inclusion of *Streptococcus ferus* within the Mutans group by 16S rRNA or *sodA* sequence comparisons alone (Whatmore and Whiley, 2002), although combined analysis of 16S rRNA and the RNase P RNA gene (*rnpB*) supported inclusion of this species (Täpp et al., 2003). The organism designated *Streptococcus mutans* which was first isolated from carious human teeth by Clarke (1924), was not included in the Eighth Edition of the *Manual of Determinative Bacteriology* but was later added to the Approved Lists of Bacterial Names. Studies on the etiological agent of dental caries prompted renewed interest in *Streptococcus mutans* in the 1960s. Evidence of biochemical, serological, and genetic (G+C determinations and DNA–DNA reassociations) heterogeneity among strains resulted initially in the proposal of four subspecies of *Streptococcus* *mutans*, namely subsp. *mutans*, subsp. *sobrinus*, subsp. *rattus*, and subsp. *cricetus* (Coykendall, 1974). This was followed by a proposal for the recognition of *Streptococcus mutans*, *Streptococcus sobrinus*, *Streptococcus cricetus*, *Streptococcus rattus*, and *Streptococcus ferus* as distinct species (Coykendall, 1977, 1983) to which were later added two more "mutans-like" species, *Streptococcus macacae* and *Streptococcus downei*, from the dental plaque of monkeys (Beighton et al., 1984; Whiley and Hardie, 1988) (for reviews of the literature, see Hamada and Slade, (1980), and Whiley and Beighton, 1998). Measured by molecular methods the Mutans group overall is a relatively loose one with the member species having deep lines of descent.

Another species named *Streptococcus orisratti* (Zhu et al., 2000), isolated from the teeth of laboratory rats, had the highest overall 16S rRNA sequence similarity (1265 bases) (94.1%) with *Streptococcus ratti*, although the combined phenotypic and molecular data on this species does not whole-heartedly support its inclusion within this relatively loose species-group (see Figure 123).

At the time of writing, a further member of this species group has been described, *Streptococcus devriesei*, isolated from equine teeth (Collins et al., 2004).

Recent nomenclatural changes have taken effect within the Mutans group, with the names *Streptococcus rattus* and *Streptococcus cricetus* being changed to *Streptococcus ratti* and *Streptococcus criceti*, respectively, on grammatical grounds (Trüper and de Clari, 1997, 1998).

Biochemical, physiological, and antigenic characteristics of species within the Mutans group are given in Table 131 and Table 133.

Anginosus group. The Anginosus group includes streptococci isolated from the oral cavity, upper respiratory, intestinal, and urogenital tracts, that are associated with purulent infections at oral and non-oral sites (Gossling, 1988). The three species currently included here are *Streptococcus intermedius*, *Streptococcus anginosus*, and *Streptococcus constellatus* (subsp. *constellatus* and subsp. *pharyngis*) together with additional centers of variation for which species epithets have not yet been proposed. The classification of the streptococci listed here has a convoluted history involving several nomenclatural changes: the group includes the non-hemolytic strains isolated from dental abscesses, originally given the name "*Streptococcus milleri*" (Guthof, 1956), together with the non-hemolytic streptococci from the human respiratory tract called "*Streptococcus MG*" (Mirick et al., 1944), the minute-colony-forming streptococci of Lancefield groups F and G (Bliss, 1937; Long and Bliss, 1934), hemolytic and non-hemolytic isolates possessing the "type antigens" of Lancefield group F (Ottens and Winkler, 1962), *Streptococcus intermedius* (Holdeman and Moore, 1974), *Streptococcus anginosus* (Andrewes and Horder, 1906; Deibel and Seeley, 1974), *Streptococcus constellatus* (Holdeman and Moore, 1974), "*Streptococcus anginosus-constellatus*", and "*Streptococcus MG-intermedius*" (Facklam, 1977), the "*Streptococcus milleri*-group" (Gossling, 1988), and *Streptococcus anginosus*, *Streptococcus constellatus*, and *Streptococcus intermedius*, according to the emended descriptions given for these species by Facklam (1984, 1985). These earlier attempts at classification relied heavily on phenotypic characteristics with considerable emphasis given to hemolytic reactions on blood agar, possession of Lancefield group antigens or type antigens, and fermentation of lactose (Whiley and

Beighton, 1998; Whiley and Hardie, 1989). Overall these studies produced a conflicting picture with evidence reported both for and against the separation of "*Streptococcus milleri*" strains into more than one species (see Whiley and Beighton, 1998). Later studies based on DNA–DNA reassociation data were also somewhat contradictory, with some authors concluding that strains resembling "*Streptococcus milleri*" be included within a single species (Coykendall et al., 1987; Ezaki et al., 1986; Farrow and Collins, 1984a; Welborn et al., 1983) and others proposing that more than one species be recognized (Kilpper-Bälz and Schleifer, 1984; Knight and Shlaes, 1988; Whiley and Hardie, 1989). The situation may not have been helped by differences in the stringencies of the experimental methods employed and the selection of collections of strains not representative of the "*Streptococcus milleri*-group" as a whole (Ezaki et al., 1986; Knight and Shlaes, 1988). Studies using the S1-nuclease DNA–DNA hybridization method under optimum and stringent conditions have demonstrated that three closely related but distinct species are present within these streptococci, namely *Streptococcus anginosus*, *Streptococcus constellatus*, and *Streptococcus intermedius*, each with an emended description (Whiley and Beighton, 1991), and have also shown that the phenotypic characteristics emphasized in earlier studies, namely hemolytic reaction, lactose fermentation, and Lancefield grouping are not discriminatory at the species level (Whiley et al., 1990a). Similar studies have revealed further centres of taxonomic variation within *Streptococcus anginosus* and *Streptococcus constellatus*, particularly β-hemolytic strains of Lancefield group C (Whiley et al., 1997, 1999), and subsequently it has been proposed that *Streptococcus constellatus* constitutes two subspecies, *Streptococcus constellatus* subsp. *constellatus*, isolated from a relatively broad clinical background, and *Streptococcus constellatus* subsp. *pharyngis*, mainly from the human throat and cases of pharyngitis (Whiley et al., 1999). Additionally, two DNA similarity groups were demonstrated within *Streptococcus anginosus* and were also centered on β-hemolytic strains of Lancefield group C, although lack of discriminating biochemical characteristics for the identification of fresh isolates prevented a formal taxonomic proposal from being made. Support for the present division of the Anginosus group into the three species *Streptococcus anginosus*, *Streptococcus constellatus*, and *Streptococcus intermedius* has also been provided by 16S rRNA gene sequence data (Bentley et al., 1991; Kawamura et al., 1995b; Poyart et al., 1998) and further heterogeneity has been demonstrated by ribotyping, serotyping, and macrorestriction fingerprinting by pulsed-field gel electrophoresis (PFGE) (Bartie et al., 2000; Doit et al., 1994; Inoue et al., 1998; Jacobs et al., 2000a). However, in a recent study based solely on whole-cell-derived polypeptide patterns the case for inclusion of all these strains into a single species has been revived (Vandamme et al., 1998).

An unusual population of strains reacting to oligonucleotide probes to the species-specific, 213–231-bp region of the 16S rRNA gene from both *Streptococcus intermedius* and *Streptococcus constellatus* (Jacobs et al., 1996) was recently shown by DNA–DNA hybridization to belong to *Streptococcus constellatus* (Jacobs et al., 2000b). Further support for the recognition of the streptococci comprising the Anginosus group into distinct species has been the demonstration of a human-specific cytotoxin, intermedilysin, the gene for which is only present in strains of *Streptococcus intermedius* (Nagamune et al., 1996, 2000).

TABLE 131. Characteristics of *Streptococcus* species[a]

Characteristic	S. pyogenes	S. acidominimus[c]	S. agalactiae	S. alactolyticus	S. anginosus	S. australis	S. canis	S. constellatus subsp. constellatus	S. constellatus subsp. pharyngis	S. criceti	S. cristatus	S. devriesei	S. dulcphis
Cell size (μm)	0.5–1.0	ND	0.6–1.2	0.5–1.0	0.5–1.0	ND	ND	0.5–1.0	0.5–1.0	0.5	~1.0	ND	ND
Catalase	-	-	-	-	-	-	-	-	-	-	-	-	c
Acid production from:													
N-Acetyl-glucosamine	+	ND	+	+	d	-	+	-	+	+	+	-	ND
Adonitol	-	ND	-	-	ND	ND	+	+	ND	ND	ND	-	ND
Amygdalin	-	ND	-	d	+	ND	-	d	+	+	ND	+	ND
Arabinose	-	-	-	d	-	-	+	d	+	+	+	+	-
Arbutin	-	ND	-	d	+	ND	+	ND	ND	+	+	+	+
Arabitol	p	-	-	+	d	ND	ND	d	ND	+	-	-	ND
Cellobiose	p	-	p	+	-	ND	+	p	+	+	+	+	+
Cyclodextrin	ND	ND	ND	ND	ND	ND	ND	ND	ND	ND	ND	-	ND
Dextran	ND	ND	ND	ND	ND	ND	ND	ND	ND	ND	ND	p	-
Dulcitol	-	-	-	-	ND	ND	ND	ND	ND	ND	ND	+	ND
Erythritol	-	-	-	-	ND	ND	ND	ND	ND	ND	ND	+	ND
Esculin	ND	ND	+	-	ND	ND	ND	ND	ND	+	+	+	ND
Fructose	+	+	+	+	ND	ND	+	+	+	+	+	+	+
Fucose	-	-	ND	-	ND	ND	ND	ND	ND	ND	ND	+	ND
Galactose	+	ND	+	+	ND	ND	+	ND	+	+	+	+	ND
Gentiobiose	-	ND	+	p	ND	ND	ND	ND	ND	+	+	+	ND
Gluconate	ND	ND	ND	-	ND	ND	ND	ND	ND	+	ND	+	ND
2-Ketogluconate	ND	ND	ND	-	ND	ND	ND	ND	ND	ND	ND	+	ND
5-Ketogluconate	ND	+	+	p	ND	ND	ND	ND	ND	+	ND	+	ND
Glucose	+	+	+	+	+	-	+	+	+	+	+	+	+
Glycerol	p	+	+	p	-	-	+	+	+	+	+	-	ND
Glycogen	p	ND	-	-	-	ND	c	-	-	-	ND	-	-
Inositol	ND	-	ND	-	ND	ND	ND	ND	ND	ND	ND	p	ND
Inulin	ND	+	ND	-	+	+	ND	p	-	p	p	+	p
Lactose	+	+	p	-	+	+	+	+	+	+	+	+	+
Lyxose	-	-	p	-	ND	ND	ND	+	ND	+	+	+	ND
Maltose	+	+	+	+	+	+	+	+	+	+	+	+	+
Mannose	ND	ND	ND	+	ND	-	ND	-	ND	+	ND	+	+
Mannitol	-	+	-	p(−)	d(−)	-	-	-	-	+	ND	-	-
Melibiose	-	ND	-	d	p	-	-	ND	-	-	ND	+	ND
Melezitose	-	ND	+	p	ND	ND	p	ND	ND	-	ND	+	+
Methyl-D-mannoside	+	-	+	p	+	-	p	d	-	+	ND	+	+
Methyl-D-glucoside	-	-	p	p	+	-	-	p	-	+	ND	+	+
β-Methyl-xyloside	ND	-	p	ND	p	+	-	ND	ND	+	p	+	p
Methyl-D-xyloside	-	-	+	+	p	-	ND	-	-	+	ND	-	+
Pullulan	p	-	-	-	p	-	-	-	-	-	-	-	+
Raffinose	-	ND	+	+	p	-	-	ND	-	+	ND	-	+
Rhamnose	ND	-	p	+	+	+	-	-	-	+	ND	-	+
Ribose	-	ND	-	-	+	-	-	-	-	+	ND	+	-
Salicin	+	-	+	P	-	-	-	-	-	+	+	+	-
Sorbitol	-	ND	ND	P	ND	ND	ND	ND	ND	ND	ND	p	ND
Sorbose	ND	-	-	+	+	ND	ND	ND	ND	ND	ND	ND	ND
Starch	ND	+	p	+	+	+	+	+	+	+	+	+	+
Sucrose	+	+	+	+	+	+	+	+	+	+	+	+	+
Tagatose	-	ND	p	p	-	-	p	p	+	+	-	-	-
Trehalose	+	-	+	+	ND	-	-	p	-	+	+	-	+
D-Turanose	ND	ND	ND	-	ND	ND	-	ND	ND	ND	ND	ND	ND
Xylose	-	-	-	-	+	ND	-	-	-	-	-	-	+
Xylitol	ND	-	ND	ND	ND	ND	ND	ND	ND	ND	ND	ND	ND
Hydrolysis of:													
Arginine	+	-	+	+	+	+	+	+	+	d[b](+)	d[b](+)	-	d[b](+)
Esculin	p	+	-	-	+	-	ND	ND	ND	-	-	-	-
Gelatin	ND	ND	ND	ND	ND	ND	ND	ND	ND	ND	ND	ND	ND

Characteristic	1	2	3	4	5	6	7	8	9	10	11	12
Hippurate	–	+[h,m] (–)	+	–[b] (+)	–	–	–	–	–	–	+	+[b] (–)
Starch	–	ND	–	ND	–	ND	–	–	–	–	ND	–
Production of:												
Acid phosphatase	ND	ND	ND	ND	ND	ND	ND	ND	ND	ND	ND	ND
Alkaline phosphatase	+	+	+	+	+	+	+	+	+	–	+	+
α-L-Arabinosidase	ND	ND	ND	ND	ND	ND	ND	ND	ND	ND	–	ND
α-D-Galactosidase	–	d	+	–	–	d	d	+	+	–	–	+
β-D-Galactosidase	ND	ND	d	+	+	+	+	+	+	–	+	d
Phospho-β-galactosidase	ND	ND	ND	ND	ND	ND	ND	ND	ND	ND	ND	–
α-D-Glucosidase	ND	ND	d	ND	d	ND	ND	+	ND	ND	ND	ND
β-D-Glucosidase	–	ND	p	d	+	p	+	+	+	–	–	–
α-L-Fucosidase	ND	ND	–	–	–	–	–	–	–	–	+	–
β-D-Fucosidase	ND	ND	ND	ND	ND	ND	ND	ND	ND	ND	ND	–
β-L-Fucosidase	ND	ND	ND	ND	ND	ND	ND	ND	ND	ND	ND	–
β-D-Glucuronidase	d	d	–	–	–	d	–	–	–	+	–	+
β-Glucosaminidase	ND	ND	ND	ND	ND	ND	ND	ND	ND	ND	ND	+
N-Acetyl-β-D-galactosaminidase	ND	ND	ND	ND	ND	ND	ND	ND	ND	ND	ND	ND
N-Acetyl-α-D-glucosaminidase	–	–	–	–	–	–	–	–	–	–	–	–
N-Acetyl-β-D-glucosaminidase	ND	ND	ND	ND	ND	ND	ND	ND	ND	ND	ND	ND
α-Maltosidase	ND	ND	–	ND	ND	ND	ND	+	ND	ND	ND	ND
β-Maltosidase	ND	ND	ND	ND	ND	ND	ND	ND	ND	ND	ND	ND
Glycyl-tryptophan arylamidase	–	–	+	ND	ND	+	ND	+	+	ND	ND	–
β-D-Lactosidase	ND	ND	ND	ND	ND	ND	ND	ND	ND	ND	ND	+
Leucine arylamidase	+	+	+	+	+	+	+	+	+	ND	+	+
Leucine aminopeptidase	ND	ND	ND	ND	ND	ND	ND	ND	ND	ND	ND	ND
Lysine decarboxylase	ND	ND	ND	ND	ND	ND	ND	ND	ND	ND	ND	ND
α-D-Mannosidase	–	–	–	–	–	–	–	–	–	–	–	–
β-Mannosidase	ND	ND	ND	ND	ND	ND	ND	ND	ND	ND	ND	ND
Ornithine decarboxylase	+	ND	ND	ND	ND	ND	ND	ND	ND	ND	ND	ND
Phosphoamidase	+	ND	+	ND	ND	+	ND	+	+	ND	ND	+
Pyroglutamic acid arylamidase	ND	ND	ND	ND	ND	ND	ND	ND	ND	ND	ND	ND
Pyrrolidonylarylamidase	+	ND	+	ND	ND	+	ND	+	+	ND	ND	ND
Sialidase (neuraminidase)	–	–	–	–	–	–	–	–	–	ND	ND	ND
Ala-phe-pro arylamidase	d	ND	+	+	+	+	+	+	+	–	+	+
α-Xylosidase	ND	ND	ND	ND	ND	ND	ND	ND	ND	ND	ND	ND
β-D-Xylosidase	ND	ND	ND	ND	ND	ND	ND	ND	ND	ND	ND	ND
Chymotrypsin	+	ND	ND	ND	ND	–	ND	–	–	–	ND	ND
Hyaluronidase	+	ND	d	d	d	d	d	d	d	–	+	d
Urease	–	ND	+[b] (–)	+[b] (–)	–[b] (+)	–[b]	ND	d	d	d	ND	+[b] (–)
Valine aminopeptidase	–	–	ND	ND	ND	ND	ND	ND	ND	–	ND	ND
Acetoin (V–P)	–	–	+[b,u] (–)	+	+	–	+	+	+	–	–	+
H₂O₂	ND	ND	ND	d	d	ND	ND	ND	ND	ND	ND	ND
IgA1 protease	ND	ND	–	ND	ND	ND	ND	+	+	–	ND	ND
Extracellular polysaccharide	ND	ND	–	–	–	–	–	–	–	+	d	ND
Amylase binding	ND	ND	+	ND	ND	ND	ND	–	–	+	+	ND
CAMP reaction	–	–	d	+[b] (–)	d	d	–[b] (+)	–	–	–	ND	ND
Growth at:												
10°C	–	–	d	+	d	ND	ND	ND	ND	ND	+	ND
45°C	–	–	–	–	–	–	–	–	–	–	+	ND
Growth in presence of:												
6.5% w/w NaCl	–	–	d	–	–	ND	ND	ND	ND	ND	ND	ND
40% bile	–	ND	d	–	–	ND	ND	ND	ND	ND	d	ND
pH 9.6	–	–	–	–	–	–	+	+	+	+	+	ND
Hemolytic Reaction	β	α,NG	β(α,–)	α,NH	NH,α,β	α	β	β,α,NH	β	NH,α	α	β
Lancefield group antigen	A		B	D (G)	NG, F, C, G, A	NG	G	NG, F, C, G, A	C	NG		NG
Peptidoglycan type	Lys–Ala₁₋₃	Lys–Ser–Gly	Lys–Ala₁₋₃ (Ser)	ND	Lys–Ala₁₋₃	Lys–Ala–Gly	Lys–Thr–Gly	Lys–Ala₁₋₃	ND	Lys–Thr–Ala		
Major cell-wall carbohydrate constituents	Rha, GlcNAc	Rha, Gal, (GlcNAc)	Rha, Gal, GlcNAc, Glucitol	ND	Rha, Glc, Gal, GalNAc	ND	ND	Rha, Glc, Gal, GlcNAc	ND	Rha, Glc, Gal		
Host	Man	Cattle, milk	Man, animals	Animals	Man	Man	Animals, man	Man	Man	Animals (man)	Man	Animals
Source of data	19, 26, 27, 40	16, 19, 40	19, 26	23, 35, 47	1, 27, 38, 51, 52, 55	58	13, 19	27, 51, 54, 55	27, 51, 54, 55	4, 12, 27, 33	25, 27	38

(continued)

TABLE 131. (continued)

Characteristic	S. downei	S. dysgalactiae subsp. dysgalactiae	S. dysgalactiae subsp. equisimilis	S. entericus	S. equi subsp. equi	S. equi subsp. zooepidemicus	S. equinus	S. ferus	S. gallinaceus	S. gallolyticus subsp. gallolyticus	S. gallolyticus subsp. macedonicus	S. gordonii	S. hyointestinalis	S. hyovaginalis	S. infantarius
Cell size (µm)	ND	ND	ND	ND	ND	ND	ND	0.5	ND	ND	ND	ND	ND	ND	ND
Catalase	–	–	–	–	–	–	–	–	–	–	–	–	–	–	–
Acid production from:															
N-Acetyl-glucosamine	ND	ND	ND	ND	ND	ND	+	+	ND	ND	+	+	+	+	ND
Adonitol	–	ND	ND	ND	ND	ND	+	+	ND	ND	–	+	+	ND	ND
Amygdalin	ND	ND	ND	ND	ND	ND	+	+	ND	+	–	+	d	ND	ND
Arabinose	–	ND	ND	–	–	–	+	–	–	+	+	–	–	–	–
Arbutin	ND	ND	ND	–	ND	ND	+	d	–	+	d	–	d	+	–
Arabitol	ND	ND	ND	+	ND	ND	+	+	–	ND	+	–	ND	–	–
Cellobiose	–	ND	ND	ND	ND	ND	–	ND	–	+	+	+	ND	+	–
Cyclodextrin	ND	ND	ND	ND	ND	ND	P	ND	ND	–	ND	ND	ND	–	–
Dextran	ND	ND	ND	ND	ND	ND	P	ND	ND	–	–	ND	ND	ND	–
Dulcitol	ND	ND	ND	ND	ND	ND	–	–	ND	ND	ND	ND	ND	–	–
Erythritol	ND	ND	ND	ND	ND	ND	ND	ND	ND	ND	ND	ND	–	ND	–
Esculin	ND	ND	ND	ND	ND	ND	+	ND	ND	ND	+	+	+	+	–
Fructose	+	ND	ND	ND	ND	ND	+	+	ND	ND	+	+	+	+	+
Fucose	–	ND	ND	ND	ND	ND	–	–	ND	ND	–	+	–	–	–
Galactose	+	ND	ND	ND	ND	ND	+	+	ND	ND	+	+	+	+	+
Gentiobiose	ND	ND	ND	ND	ND	ND	P	ND	ND	ND	+	ND	P	+	–
Gluconate	ND	ND	ND	ND	ND	ND	–	–	ND	ND	–	–	–	–	–
2-Ketogluconate	ND	ND	ND	ND	ND	ND	ND	ND	ND	ND	ND	ND	–	–	–
5-Ketogluconate	ND	ND	ND	ND	ND	ND	P	–	ND	ND	–	+	–	+	–
Glucose	+	+	+	+	+	+	+	+	+	+	+	+	+	+	+
Glycerol	–	P	P	ND	ND	+	P	d	d	–	P	d	P	–	d
Glycogen	–	–	–	+	–	+	+	P	–	+	–	P	–	–	ND
Inositol	+	ND	–	–	ND	ND	P	P	ND	P	–	ND	–	–	–
Inulin	+	P	P	ND	ND	ND	P	d	ND	d	+	d	P	+	+
Lactose	ND	ND	ND	ND	ND	ND	P	+	ND	+	+	+	+	+	ND
Lyxose	+	+	+	+	+	+	+	+	+	+	+	+	+	+	+
Maltose	+	+	+	+	+	+	+	+	+	+	+	+	+	+	+
Mannose	+	+	+	+	+	+	+	+	+	+	+	+	+	+	d^a(+)
Mannitol	–	d	d	–	–	–	d^b(–)	d^a P	–	–^a(+)	–	–^b(d)	P	P	d^b(+)
Melibiose	ND	ND	ND	ND	ND	ND	ND	ND	ND	+	ND	ND	+	ND	–
Melezitose	ND	ND	ND	ND	ND	ND	ND	ND	ND	+	ND	ND	ND	ND	ND
Methyl-D-mannoside	ND	ND	ND	+	+	+	ND	d	+	+	–	P	–	+	+
Methyl-D-glucoside	ND	ND	ND	+	+	+	P	–	+	+	P	P	+	+	P
β-Methylxyloside	ND	ND	ND	ND	ND	ND	ND	ND	ND	P	–	ND	ND	–	+
Methyl-D-xyloside	ND	ND	ND	ND	ND	ND	–	ND	ND	P	+	ND	–	–	–
Pullulan	–	–	–	+	–	+	P	P	ND	P	–	P	+	P	d^b(+)
Raffinose	ND	ND	ND	ND	ND	ND	–	ND	ND	–	+	P	–	+	–
Rhamnose	ND	+	+	–	+	+	+	+	+	ND	–	–	+	+	–
Ribose	–^a(+)	P	d	–	+	+	P	d^b(+)	+	–^a(–)	P	–^b(d)	+	P	d^b(+)
Salicin	+	+	+	+	+	+	+	+	+	+	+	+	+	+	+
Sorbitol	–	d	d	–	–	–	P	+	–	+	P	–	–	–	–
Sorbose	ND	P	P	ND	ND	ND	P	+	ND	–	–	–	–	–	–
Starch	+	+	+	+	+	+	+	+	+	+	+	+	+	+	+
Sucrose	+	+	+	+	+	+	+	+	+	+	+	+	+	+	–
Tagatose	ND	ND	ND	ND	ND	ND	P	P	ND	ND	ND	ND	ND	–	–
Trehalose	ND	ND	+	ND	ND	P	P	+	ND	ND	ND	ND	ND	ND	ND
D-Turanose	ND	ND	+	ND	ND	ND	–	P	ND	ND	ND	ND	+	–	ND
Xylose	–	ND	ND	ND	ND	ND	–	+	ND	ND	ND	ND	+	–	–
Xylitol	–	ND	ND	ND	ND	ND	–	–	ND	ND	–	–	–	–	–
Hydrolysis of:															
Arginine	–	+	d^b(+)	+	+	+	+	–	+	–	–	+	+^a(–)	–	P
Esculin	ND	P	ND	ND	P	P	+	ND	+	+	–	+	–	–	ND
Gelatin	ND	ND	ND	ND	–	–	–	ND	–	ND	ND	ND	–	ND	ND

Characteristic	1	2	3	4	5	6	7	8	9	10	11	12	13	14	15
Hippurate	−	−	d[b](−)	−	−	−	−	−	−	−	−	−	−	+	−
Starch	−	−	−	ND	−	+	+	ND	ND	+	+	d	+	ND	+
Production of:															
Acid phosphatase	ND	ND	ND	ND	ND	ND	ND	ND	ND	ND	ND	+	ND	ND	ND
Alkaline phosphatase	−	+	+	+	+	+	d	d	d	−	+	+	+	−	+
α-L-Arabinosidase	ND	ND	ND	ND	ND	ND	ND	ND	ND	ND	ND	−	ND	+	ND
α-D-Galactosidase	ND	−	+	ND	+	−	+	+	+	d	+	d	d	+	+
β-D-Galactosidase	+	+	+	+	+	+	+	−	+	d	+	+	−	+	−
Phospho-β-galactosidase	ND	ND	ND	ND	ND	ND	ND	ND	ND	ND	ND	+	ND	ND	ND
α-D-Glucosidase	ND	ND	−	+	−	+	+	+	+	+	+	d	ND	ND	d
β-D-Glucosidase	ND	ND	ND	ND	ND	ND	ND	ND	ND	−	+	d	d	d	d
α-L-Fucosidase	ND	ND	ND	ND	ND	−	ND	ND	ND	ND	ND	+	ND	ND	ND
β-D-Fucosidase	ND	ND	ND	ND	ND	ND	ND	ND	ND	ND	ND	−	ND	ND	ND
β-L-Fucosidase	ND	ND	ND	ND	ND	ND	ND	ND	ND	ND	ND	−	ND	ND	ND
β-D-Glucuronidase	+	+	+	−	+	+	+	+	−	+	+	+	+	+	−
β-Glucosaminidase	ND	ND	ND	ND	ND	ND	ND	ND	ND	ND	ND	d	ND	ND	ND
N-Acetyl-β-D-galactosaminidase	ND	ND	ND	ND	ND	ND	ND	ND	ND	ND	ND	ND	ND	ND	ND
N-Acetyl-α-D-glucosaminidase	ND	ND	ND	ND	ND	ND	ND	ND	ND	ND	ND	+	ND	ND	ND
N-Acetyl-β-D-glucosaminidase	ND	ND	−	−	ND	−	−	−	−	−	+	−	ND	ND	−
α-Maltosidase	ND	ND	ND	ND	ND	+	ND	+	−	ND	+	+	ND	ND	+
β-Maltosidase	ND	ND	ND	ND	ND	ND	ND	ND	ND	ND	ND	ND	ND	ND	ND
Glycyl-tryptophan arylamidase	ND	ND	ND	ND	ND	ND	ND	ND	+	ND	ND	−	ND	ND	ND
β-D-Lactosidase	ND	ND	ND	ND	ND	ND	ND	ND	ND	ND	ND	d	ND	ND	−
Leucine arylamidase	+	+	+	+	+	+	+	+	+	+	+	+	+	−	+
Leucine aminopeptidase	ND	ND	ND	ND	ND	ND	ND	ND	ND	ND	ND	+	ND	ND	+
Lysine decarboxylase	ND	ND	−	−	ND	−	ND	−	−	ND	−	+	ND	ND	ND
α-D-Mannosidase	ND	ND	ND	ND	ND	ND	ND	ND	ND	ND	ND	ND	ND	ND	ND
β-Mannosidase	ND	ND	ND	ND	ND	ND	ND	ND	ND	ND	ND	+	ND	ND	ND
Ornithine decarboxylase	−	−	ND	+	ND	ND	ND	d	−	d	d	+	ND	ND	−
Phosphoamidase	ND	ND	ND	ND	ND	ND	ND	ND	ND	ND	ND	d	ND	ND	ND
Pyroglutamic acid arylamidase	ND	ND	ND	ND	ND	ND	ND	ND	ND	ND	ND	ND	ND	ND	ND
Pyrolidonylarylamidase	−	−	−	−	−	−	−	−	−	−	−	−	−	−	−
Sialidase (neuraminidase)	ND	ND	ND	ND	ND	ND	ND	ND	ND	ND	ND	−	ND	ND	ND
Ala-phe-pro arylamidase	ND	ND	ND	ND	ND	ND	ND	+	+	+	+	+	+	+	+
α-Xylosidase	ND	ND	ND	ND	ND	ND	ND	ND	−	ND	ND	−	ND	ND	−
β-D-Xylosidase	ND	ND	ND	ND	ND	ND	ND	ND	−	ND	ND	+	ND	ND	−
Chymotrypsin	ND	ND	ND	ND	ND	ND	ND	ND	−	ND	ND	−	ND	ND	ND
Hyaluronidase	+	+	+	+	+	+	+	+	−	+	+	+	+	+	+
Urease	−	−	−	−	−	−	−	−	−	−	−	−	−	−	−
Valine aminopeptidase	ND	ND	ND	ND	ND	−	+	+	ND	+	+	−	+	−	+
Acetoin (V-P)	+	−	+	−	+	+	+	+	+	+	+	+	+	+	+
H₂O₂	−	ND	ND	+	ND	−	−	−	−	−	−	+	−	ND	ND
IgA1 protease	+	+	+	+	+	−	+	−	ND	+	+	+	+	ND	+
Extracellular polysaccharide	ND	ND	ND	ND	ND	ND	ND	+	ND	+	+	−	−	ND	−
Amylase binding	ND	ND	ND	ND	ND	ND	ND	ND	ND	ND	ND	+	ND	ND	ND
CAMP reaction	−	−	−	−	−	−	−	−	ND	ND	ND	ND	ND	ND	ND
Growth at:															
10°C	ND	−	−	−	−	−	−	−	ND	+	+	ND	ND	ND	ND
45°C	−	−	−	−	−	+	+	+	ND	+	+	ND	ND	ND	ND
Growth in presence of:															
6.5% w/w NaCl	d	−	−	−	−	−	+	+	ND	−	−	−	−	−	−
40% bile		ND	ND	+	ND	−	+	+	ND	ND	ND	d	ND	ND	ND
pH 9.6	−	−	−	+	−	−	−	−	ND	ND	ND	ND	ND	ND	ND
Hemolytic Reaction	ND	α, NH	β	α	β	β	α	α	α	α, NH	α, NH	α	α	α	α
Lancefield group antigen	NG	C	C, G, L, A	D	C	C	D	NG	D, NG	D, NG	D, NG	NG, H	NG	NG	NG, D
Peptidoglycan type	Lys–Thr–Ala	Lys–Ala$_{1-3}$	Lys–Ala$_{1-3}$	ND	Lys–Ala$_{1-3}$	Lys–Ala$_{1-3}$	Lys–Thr–Ala	Lys–Ala$_{1-3}$	Rha, Glc	Rha, Glc	Rha, Glc	Lys–Ala	Lys–Ala (Ser)	ND	ND
Major cell-wall carbohydrate constituents	ND	Rha, GalNAc	ND	ND	Rha, GalNAc	ND	ND	Rha, Glc	ND	ND	ND	Rha, Glc, Gal, GlcNAc, (GalNAc)	ND	ND	Man
Host	Animals	Animals	Man, animals	Animals	Animals	Animals, man	Animals	Animals	Birds	Animals, man	Dairy	Man	Animals	Animals	Man
Source of data	5, 57	19, 22, 48, 50	19, 22, 48, 50	49	9, 19, 22, 27	9, 19, 22, 27	21, 41	2, 12	11	7, 24, 34, 36, 41, 45	7, 24, 34, 36, 41, 45	3, 30	14	15	36, 42

(continued)

TABLE 131. (continued)

Characteristic	S. infantis	S. iniae	S. intermedius	S. luetkiensis	S. macacae	S. minor	S. mitis	S. mutans	S. oligofermentans	S. oralis	S. orisratti	S. ovis	S. parasanguinis	S. parauberis	S. pasteurianus	S. peroris	S. phocae
Cell size (μm)	0.6–1.0	~1.5	0.5–1.0	ND	ND	□1.0	ND	0.5–0.75	0.7	ND	ND	ND	0.8–1.0	ND	ND	0.6–0.8	ND
Catalase	−	−	−	−	−	ND	−	−	ND	−	ND	−	−	−	−	ND	−
Acid production from:																	
N-Acetyl-l-glucosamine	ND	+	+	ND	+	+	+	+	ND	+	ND	−	+	+	+	ND	+
Adonitol	ND	−	d	ND	−	d	ND	d	ND	ND	+	ND	d	d	ND	ND	ND
Amygdalin	−	−	d	ND	+	d	−	+	−	−	+	ND	d	+	−	−	ND
Arabinose	−	+	+	−	−	d	−	+	−	−	+	−	d	−	−	−	−
Arbutin	−	+	d	ND	ND	+	d	+	ND	−	+	ND	d	+	ND	ND	−
Arabitol	−	ND	d	ND	ND	ND	ND	ND	ND	ND	ND	ND	ND	ND	−	ND	ND
Cellobiose	ND	+	+	ND	+	+	+	+	+	+	+	d	+	ND	+	+	+
Cyclodextrin	−	ND	−	ND	−	ND	−	+	+	−	+	−	+	d	+	+	+
Dextran	ND	+	ND	ND	ND	ND	ND	ND	ND	ND	ND	ND	ND	d	ND	ND	ND
Dulcitol	ND	−	ND	ND	ND	ND	ND	ND	ND	ND	ND	ND	ND	−	ND	ND	−
Erythritol	ND	ND	ND	ND	ND	−	ND	−	ND	−	ND	ND	ND	−	ND	ND	ND
Esculin	ND	+	ND	ND	ND	+	+	+	+	+	+	+	+	+	+	+	+
Fructose	ND	+	ND	ND	+	+	ND	+	+	+	+	+	+	ND	+	+	+
Fucose	ND	ND	ND	ND	−	−	ND	−	ND	ND	ND	+	ND	−	ND	−	−
Galactose	ND	d	ND	+	+	+	+	+	+	+	+	+	+	+	+	+	+
Gentiobiose	ND	+	ND	+	ND	+	ND	+	ND	ND	ND	+	ND	+	ND	ND	−
Gluconate	ND	−	ND	ND	ND	−	ND	+	ND	ND	ND	ND	ND	+	ND	ND	ND
2-Ketogluconate	ND	−	ND	ND	ND	+	ND	+	ND	ND	ND	ND	ND	+	ND	ND	ND
5-Ketogluconate	ND	ND	ND	ND	+	−	ND	+	ND	ND	ND	ND	ND	−	ND	ND	ND
Glucose	ND	+	+	ND	+	+	+	+	+	+	+	+	+	+	+	+	+
Glycerol	ND	+	ND	ND	ND	+	ND	−	+	d	+	+	+	+	+	+	d
Glycogen	ND	+	−	ND	ND	+	ND	+	ND	ND	ND	+	ND	+	ND	ND	−
Inositol	ND	−	ND	ND	ND	−	ND	−	ND	ND	+	+	ND	−	ND	ND	ND
Inulin	d	−	ND	ND	−	−	ND	+	d	−	+	+	+	−	ND	+	−
Lactose	+	+	+	+	+	+	+	+	+	+	+	+	+	d (+)	+	−	−
Lyxose	+	+	+	ND	ND	−	ND	+	+	+	+	+	+	+	+	+	ND
Maltose	ND	+	ND	ND	+	+	+	+	+	+	+	+	+	d (+)	+	ND	−
Mannitol	−	+	−	ND	+	−	−	−	−	−	+	−	−	+	−	−	−
Mannose	ND	+	+	ND	+	+	+	+	ND	+	+	+	+	+	+	−	−
Melibiose	−	−	−	ND	ND	−	ND	−	ND	−	+	−	ND	d	ND	−	−
Melezitose	ND	+	ND	ND	ND	−	ND	+	ND	ND	ND	+	ND	−	ND	ND	ND
Methyl-D-mannoside	ND	+	d	ND	ND	d	ND	−	ND	−	ND	−	ND	−	ND	ND	−
Methyl-D-glucoside	ND	−	d	ND	ND	d	ND	+	ND	ND	ND	+	ND	+	+	ND	−
β-Methylyloside	ND	ND	ND	ND	ND	ND	ND	ND	ND	ND	ND	ND	ND	ND	ND	ND	ND
Methyl-D-xyloside	d	−	ND	ND	ND	−	ND	−	ND	ND	ND	+	ND	−	+	ND	−
Pullulan	−	ND	−	ND	ND	−	−	−	+	+	+	−	+	ND	+	−	−
Raffinose	−	+	+	ND	ND	d	d	+	ND	−	ND	−	ND	−	+	ND	−
Rhamnose	ND	+	+	ND	+	−	−	−	ND	−	ND	−	+	d	+	+	−
Ribose	−	+	d	ND	+	d	d	+	d	d	+	+	+	d	+	−	−
Salicin	ND	+	−	ND	−	+	d	+	ND	d	+	−	d (−)	d	ND	ND	ND
Sorbitol	−	−	ND	d	+	d	d	+	+	+	+	+	+	d	+	+	−
Sorbose	ND	+	ND	+	+	−	+	−	−	−	+	+	+	d	+	+	−
Starch	ND	+	+	ND	+	−	−	−	+	d	+	+	+	+	+	+	−
Sucrose	+	−	+	ND	−	+	+	+	+	+	+	+	+	+	+	−	−
Tagatose	+	+	+	ND	+	+	−	−	d	d	+	d	d	d	d	+	d
Trehalose	−	−	ND	ND	ND	−	ND	+	ND	+	+	+	+	+	+	−	−
D-Turanose	ND	−	ND	ND	ND	−	ND	ND	ND	ND	ND	ND	−	−	ND	ND	−
Xylose	ND	−	ND	ND	ND	−	ND	ND	ND	ND	ND	ND	ND	ND	ND	ND	−
Xylitol	ND	ND	ND	ND	ND	ND	ND	ND	ND	ND	ND	ND	ND	ND	ND	ND	ND
Hydrolysis of:																	
Arginine	−	+[a] (−)	+	−	−	+	d[a] (−)	−	−	−[a] (−)	−	d	+	+[a] (d)	−	−	−
Esculin	−	+	+	+	+	+	d[a] (−)	+	−	d	+	+	d	+	+	−	−
Gelatin	ND	ND	ND	ND	ND	ND	ND	ND	ND	ND	ND	ND	ND	ND	ND	ND	ND

	1	2	3	4	5	6	7	8	9	10	11	12	13	14	15	16
Hippurate	−	+	−	d	−	+	−	−	+	−	+	−	d[a](+)	+	−	−
Starch	ND	+	ND	+	+	−	ND	+	ND	+	+	ND	ND	ND	ND	−
Production of:[b]																
Acid phosphatase	ND	ND	ND	ND	ND	ND	ND	ND	ND	+	ND	ND	ND	ND	ND	ND
Alkaline phosphatase	−	+	+	d	d	−	d	d	d	+	+	d	+	+	d	+
α-L-Arabinosidase	ND	ND	ND	−	−	−	ND	−	ND	+	ND	ND	−	−	ND	ND
α-D-Galactosidase	+	−	+	+	+	+	+	+	+	+	−	ND	+	+	+	ND
β-D-Galactosidase	+	+	−	−	−	−	+	−	d	d	−	d	−	−	−	ND
Phospho-β-galactosidase	ND	ND	ND	ND	d	−	−	+	−	+	+	−	+	+	−	ND
α-D-Glucosidase	ND	+	+	+	+	+	+	+	ND	+	ND	+	+	+	+	ND
β-D-Glucosidase	−	d	d	+	−	−	−	+	ND	+	ND	+	−	+	−	ND
α-L-Fucosidase	ND	−	−	−	−	−	d	−	ND	−	ND	d	−	−	−	ND
β-D-Fucosidase	d	+	ND	ND	ND	p	−	−	ND	−	ND	−	−	−	ND	ND
β-L-Fucosidase	ND	ND	ND	ND	ND	−	ND	−	ND	−	ND	−	−	−	−	ND
β-D-Glucuronidase	−	−	−	−	−	−	−	−	ND	−	ND	−	−	−	−	−
β-D-Glucosaminidase	ND	ND	ND	ND	ND	−	ND	−	ND	+	ND	−	ND	ND	ND	ND
N-Acetyl-β-D-galactosaminidase	ND	+	+	+	+	+	+	+	ND	+	ND	+	ND	+	ND	ND
N-Acetyl-α-D-glucosaminidase	ND	−	ND	ND	ND	ND	ND	−	ND	+	ND	ND	ND	ND	ND	ND
N-Acetyl-β-D-glucosaminidase	d	+	−	+	+	+	+	+	ND	+	ND	+	+	+	+	ND
α-Maltosidase	ND	ND	ND	ND	ND	ND	−	−	ND	−	ND	ND	ND	ND	ND	ND
β-Maltosidase	ND	ND	ND	ND	ND	ND	−	d	ND	+	ND	ND	ND	ND	ND	ND
Glycyl-tryptophan arylamidase	−	−	ND	ND	ND	ND	d	d	ND	−	ND	ND	ND	ND	ND	ND
β-D-Lactosidase	ND	ND	ND	ND	ND	ND	−	+	ND	+	ND	p	+	+	−	ND
Leucine arylamidase	+	+	+	+	+	+	−	+	ND	+	+	+	+	+	+	+
Leucine aminopeptidase	ND	ND	ND	+	+	+	ND	+	ND	+	ND	+	+	+	+	+
Lysine decarboxylase	ND	ND	ND	ND	ND	−	−	−	ND	ND	ND	−	ND	ND	ND	ND
α-D-Mannosidase	−	−	−	−	d	−	ND	−	ND	−	−	−	−	−	−	−
β-D-Mannosidase	ND	ND	ND	ND	ND	ND	ND	−	ND	ND	ND	ND	ND	ND	ND	ND
Ornithine decarboxylase	ND	ND	ND	ND	ND	−	ND	−	ND	+	ND	+	ND	ND	ND	ND
Phosphoamidase	ND	ND	ND	ND	ND	+	ND	+	ND	d	ND	+	ND	ND	ND	ND
Pyroglutamic acid arylamidase	d	+	ND	ND	ND	ND	d	−	ND	−	ND	d	ND	+	+	ND
Pyrrolidonylarylamidase	+	ND	ND	ND	ND	−	ND	+	ND	−	+	−	ND	+	+	ND
Sialidase (neuraminidase)	ND	ND	ND	ND	ND	−	−	−	ND	−	ND	d	ND	ND	ND	ND
Ala-phe-pro arylamidase[c]	ND	ND	ND	ND	ND	ND	ND	ND	ND	+	ND	+	ND	+	ND	ND
α-Xylosidase	ND	ND	ND	ND	ND	ND	ND	−	ND	−	ND	−	ND	−	ND	ND
β-D-Xylosidase	−	+	−	+	+	+	+	+	ND	+	−	+	ND	+	+	ND
Chymotrypsin	ND	+	ND	ND	ND	−	−	+	ND	−	ND	+	ND	−	+	ND
Hyaluronidase	ND	ND	ND	ND	ND	ND	ND	−	ND	−	−	−	ND	−	−	−
Urease	−	+	−	−	−	−	−	−	ND	−	−	−	−	−	−	−
Valine aminopeptidase	ND	ND	ND	ND	ND	+	ND	+	ND	−	ND	−	ND	+	ND	ND
Acetoin (V-P)	−	−	+	+	+	−	+	+	+	+	+	−	+[b](d)	d	+	−
H₂O₂	ND	ND	ND	ND	ND	−	+	d	+	−	−	+	ND	+	+	ND
IgA1 protease	ND	ND	+	−	−	+	+	+	+	+	ND	−	ND	+	ND	ND
Extracellular polysaccharide	ND	−	−	−	−	−	−	−	+	−	ND	−	−[b](+)	−	−	ND
Amylase binding	ND	−	−	−	−	−	p	d	d	d	ND	ND	+[b](d)	+	ND	ND
CAMP reaction	−	+	−	−	−	−	−	−	ND	ND	ND	−	α, NH	−	−	ND
Growth at:																
10°C	ND	+	ND	ND	ND	ND	ND	−	ND	ND	+	ND	ND	ND	ND	ND
45°C	ND	−	ND	ND	ND	−	ND	d	−	ND	+	−	−[b](+)	ND	ND	ND
Growth in presence of:																
6.5% w/w NaCl	ND	d	ND	−	−	−	−	d	−	−	−	ND	−[b](+)	ND	−	−
40% bile	ND	−	ND	+	+	+	+	ND	+	+	+	ND	+	ND	+	−
pH 9.6	ND	−	ND	−	−	−	d	ND	d	−	−	ND	ND	ND	ND	ND
Hemolytic Reaction	α	β	NHα,β	α	α	α	α	NHα(β)	α	α	α	α	α, NH	α	α	β
Lancefield group antigen[d]	ND	NG, F, C	NG, F, C	NG, D	NG	NG	NG, O*, K	NG	NG	NG	A	NG	NG	NG(F, G, C, B)	NG	NG, F, C
Peptidoglycan type[e]	Lys-Ala₂₋₃	Lys-Ala₂₋₃	Lys-Ala₂₋₃	ND	ND	Lys-Ala₂₋₃	Lys-direct	Lys-Ala₂₋₃	ND	Lys-direct	ND	ND	ND	ND	D	ND
Major cell-wall carbohydrate constituents[g]	Rha, Glc, Gal, GlcNAc	Rha, Glc	Rha, Glc, Gal, GlcNAc	ND	ND	Rha, Glc	Gal, Glu, Rtf	Rha, Glc	ND	Glc, Gal, GlcNAc, GalNAc, (Rha), Rtf	ND	ND	ND	Rha, Glc	ND	ND
Host	Man	Fish, man	Man	Man	Animals	Animals	Man	Man	Man	Man	Animals	Animals	Animals	Man	Man	Animals (seals)
Source of data[ab]	28	17, 18, 35	1, 27, 40, 51, 52, 55	36, 42	4	46	3, 30	6, 12, 27, 30	44	3, 30, 32	61	10	9, 60	35, 42	29	9, 43

(continued)

TABLE 131. (continued)

Characteristic	S. pluranimalium	S. pneumoniae	S. porcinus	S. ratti	S. salivarius	S. sanguinis	S. sinensis	S. sobrinus	S. suis	S. thermophilus	S. thoraltensis	S. uberis	S. urinalis	S. vestibularis
Cell size (μm)	ND	0.5–1.25	ND	0.5	0.8–1.0	ND	0.82–0.98	0.5	<2.0	0.7–1.0	ND	ND	ND	~1.0
Catalase	−	−	−	−	−	−	−	−	−	−	−	−	ND	−
Acid production from:														
N-Acetyl-glucosamine	d	d	+	+	+	+	ND	−	ND	−	+	+	−	d
Adonitol	d^d	ND	d	+	+	d	ND	−	ND	−	d	−	ND	−
Amygdalin	−	+	d	+	+	d	ND	−	ND	d	d	−	ND	d
Arabinose	−	−	−	−	−	−	−	−	−	−	+	−	ND	d
Arbutin	d^d	+	d	+	+	d	ND	−	ND	−	+	+	−	−
Arabitol	d^d	ND	d	ND	+	−	ND	−	ND	−	−	+	ND	−
Cellobiose	d^d	ND	d	+	+	−	+	+	−	d	+	+	−	d
Cyclodextrin	−	−	−	−	−	ND	−	ND	−	−	+	ND	ND	ND
Dextran	−	−	ND	ND	ND	ND	ND	ND	ND	ND	ND	+	ND	+
Dulcitol	−	−	−	−	−	−	−	−	−	−	−	−	−	−
Esculin	ND	ND	ND	ND	ND	d	ND	−	ND	ND	ND	+	ND	+
Fructose	+	+	+	+	+	+	+	+	+	+	+	+	+	+
Fucose	−	+	+	−	−	+	ND	−	ND	−	+	−	−	+
Galactose	d	ND	+	ND	+	d	ND	d	ND	+	+	−	+	ND
Gentiobiose	d^d	ND	P	ND	P	ND	ND	d	ND	d	+	ND	+	ND
Gluconate	−	ND	−	ND	ND	+	ND	ND	ND	ND	+	−	ND	ND
2-Ketogluconate	−	ND	−	ND	ND	ND	ND	ND	ND	ND	+	−	−	ND
5-Ketogluconate	+	ND	−	ND	−	ND	ND	ND	ND	ND	+	+	ND	ND
Glucose	+	+	+	+	+	+	+	+	+	+	+	+	+	+
Glycerol	ND	+	P	+	−	−	−	−	−	−	d	d	−	−
Glycogen	+	+	−	−	−	ND	ND	+	+	ND	+	−	+	ND
Inositol	−	ND	−	−	−	ND	ND	−	−	ND	+	−	−	−
Inulin	P	+	+	P	P	P	−	P	+	−	−	−	+	P
Lactose	d^d	−	P	+	P	d	ND	d	+	+	d	+	+	+
Lyxose	d	+	+	+	+	+	+	+	+	+	+	+	+	+
Maltose	d^d	+	+	+	+	+	−	+	+	+	+	+	+	+
Mannose	d^{a,b} (−)	d^{a,b} (−)	P	+	−	d	ND	−	ND	d	+	−	ND	−
Mannitol	d^d	d	P	P	+	+	+	d^b (−)	d^b (−)	P	d	−	+	ND
Melibiose	d^d	d	−	−	−	−	−	−	−	d	d	−	−	−
Melezitose	+	d^a (+)	ND	ND	ND	ND	ND	ND	ND	ND	d	ND	ND	ND
Methyl-D-mannoside	−	ND	ND	ND	+	ND	+	ND	ND	ND	d	−	ND	+
Methyl-D-glucoside	−	ND	−	ND	−	d	−	−	d	d	d	ND	−	+
β-Methylxyloside	−	ND	ND	ND	−	−	−	−	−	−	−	−	−	−
Methyl-D-xyloside	−	ND	ND	ND	+	ND	ND	ND	d	ND	+	ND	ND	+
Pullulan	−	+	+	−	−	d	−	−	d	+	+	ND	+	+
Raffinose	P	−	+	−	−	d	ND	−	d	ND	d	+	−	−
Rhamnose	d	d	+	+	+	d	ND	−	d	d	d	−	+	d
Ribose	d^d	−	P	+	+	−	+	+	+	−	−	+	ND	−
Salicin	d^{a,b} (−)	−	P	−	+	+	−	d^b (+)	+	−	+	+	−	+
Sorbitol	−	ND	+	+	+	−	ND	−	−	ND	d	−	ND	+
Sorbose	d	d	P	+	−	P	+	+	ND	−	+	d	ND	−
Starch	d	P	P	+	+	d	+	+	−	+	+	+	+	+
Sucrose	+	−	+	+	+	+	+	+	+	+	+	+	+	+
Tagatose	−	d^a (d)	−	−	−	P	−	−	−	ND	d^a (+)	−	−	−
D-Turanose	−	ND	−	+	P	ND	+	+	+	ND	+	P	+	+
Xylose	−	+	ND	ND	ND	−	ND	−	ND	ND	d	d	ND	ND
Xylitol	−	ND	ND	ND	ND	ND	ND	ND	ND	ND	−	−	−	−
Hydrolysis of:														
Arginine	d^b (−)	+	d^b (+)	+	−	+	+	−^{b} (−)	+	−	+	+	−^{b} (−)	+^{b} (d)
Esculin	d^b (+)	d	d	+	+	d^b (+)	+	+	+	−	+	+	+	+^{b} (+)
Gelatin	ND	ND	d	ND	ND	ND	ND	ND	ND	−	ND	ND	ND	ND

	1	2	3	4	5	6	7	8	9	10	11	12	13
Hippurate	+ᵇ (−)	−	−	−	+	−	−	−	+	−	−	−	−
Starch	ND	−	−	ND	+	ND	ND	ND	ND	ND	ND	ND	+
Production of:													
Acid phosphatase	ND	ND	ND	ND	+	ND	−	−	ND	+	+	+	−
Alkaline phosphatase	d	−	+	+	+	+	d	d	−	ND	d	+	+
α-L-Arabinosidase	d	d	−	d	−	−	d	d	−	−	ND	+	ND
α-D-Galactosidase	d	+	ND	−	+	+	d	−	ND	ND	+	ND	ND
β-D-Galactosidase	d	+	ND	d	p	p	d	d	ND	+	d	d	ND
Phospho-β-galactosidase	ND	ND	ND	ND	+	ND	−	ND	+	ND	ND	+	ND
α-D-Glucosidase	+	+	ND	+	+	+	+	+	+	+	+	+	ND
β-D-Glucosidase	dᵈ	+	ND	−	−	−	−	d	−	−	−	−	−
α-L-Fucosidase	d	d	ND	−	+	−	−	−	+	−	−	−	ND
β-D-Fucosidase	d	d	ND	−	−	−	d	−	ND	ND	−	ND	−
β-L-Fucosidase	ND	ND	ND	−	−	−	−	−	ND	ND	ND	ND	ND
β-D-Glucuronidase	dᵈ	−	ND	+	+	+	−	+	ND	ND	+	+	+
β-Glucosaminidase	ND	+	ND	p	+	−	−	+	+	ND	+	ND	−
N-Acetyl-β-D-galactosaminidase	ND	d	ND	p	−	ND	+	+	ND	ND	+	ND	ND
N-Acetyl-α-D-glucosaminidase	+	+	ND	+	+	ND	+	+	+	ND	+	ND	+
N-Acetyl-β-D-glucosaminidase	d	−	ND	+	+	ND	+	+	+	ND	+	ND	ND
α-Maltosidase	−	+	ND	−	−	−	−	−	−	ND	+	ND	−
β-Maltosidase	ND	+	ND	d	−	ND	d	−	ND	ND	ND	ND	−
Glycyl-tryptophan arylamidase	+	+	ND	−	p	ND	+	p	ND	ND	+	ND	−
β-D-Lactosidase	ND	−	ND	p	p	ND	−	−	+	ND	−	ND	−
Leucine arylamidase	+	d	ND	+	+	ND	+	−	+	ND	+	ND	+
Leucine aminopeptidase	d	+	ND	+	+	+	+	+	+	ND	+	+	+
Lysine decarboxylase	ND	ND	ND	ND	ND	ND	ND	ND	ND	ND	ND	ND	−
α-D-Mannosidase	ND	ND	ND	+	+	ND	−	+	ND	ND	−	ND	−
β-D-Mannosidase	−	+	ND	−	−	−	−	−	ND	ND	−	ND	−
Ornithine decarboxylase	ND	ND	ND	ND	ND	ND	ND	ND	ND	ND	ND	ND	−
Phosphoamidase	ND	ND	ND	+	+	ND	d	+	ND	ND	+	ND	+
Pyroglutamic acid arylamidase	d	−	ND	d	+	−	+	+	ND	ND	d	+	+
Pyrolidonylarylamidase	dᵇ(−)	d⁵(+)	−ᵇ(+)	−	−	−	−	−	ND	ND	−	−	ND
Sialidase (neuraminidase)	ND	+	ND	+	+	+	−	ND	ND	ND	ND	+	ND
Ala-phe-pro arylamidase	+	+	+	+	+	+	+	+	+	ND	+	+	+
α-Xylosidase	ND	ND	ND	+	d	ND	d	ND	ND	ND	ND	ND	ND
β-D-Xylosidase	ND	ND	ND	d	−	ND	−	ND	ND	ND	ND	ND	ND
Chymotrypsin	ND	+	ND	+	+	ND	+	−	ND	ND	ND	ND	ND
Hyaluronidase	−	−	ND	d	d	ND	d	ND	ND	ND	ND	ND	ND
Urease	−	+	−	−	−	−	−	−	−	ND	−	ND	−
Valine aminopeptidase	d	ND	ND	d	d	ND	+	ND	ND	ND	+	ND	−
Acetoin (V-P)	−	dᵇ(+)	dᵇ(+)	d	d⁵(+)	+	+	−	+	+	+ᵇ(d)	−	+
H₂O₂	ND	ND	ND	+	+	ND	+	−	+	ND	+ᵇ(d)	ND	−
IgA1 protease	ND	ND	ND	−	+	ND	+	d	+	ND	ND	ND	ND
Extracellular polysaccharide	ND	−	ND	+	+	+	+	−	−	−	−	ND	−
Amylase binding	ND	ND	ND	d	d	ND	ND	ND	ND	ND	ND	ND	−
CAMP reaction	ND	+	ND	ND	ND	ND	ND	ND	ND	−	−	ND	ND
Growth at:													
10°C	ND	−	−	−	ND	ND	ND	−	−	ND	−	ND	−
45°C	ND	−	−	d	d	ND	d	−	−	+	−	ND	−
Growth in presence of:													
6.5% w/w NaCl	dᵈᵇ(+)	−	ND	ND	d	−	d	−	−	−	−ᵃᵇ(d)	+	−
40% bile	ND	−	ND	d	d	+	d	d	d	ND	d	ND	+
pH 9.6	ND	−	ND	−	d	+	−	−	−	ND	d	ND	ND
Hemolytic Reaction	α	α	β	NH	NH,α,(β)	α	NH,α	α (β)ᶜ	α,NH	α	α	NH	α
Lancefield group antigen	ND	NG	E,P,U,V	NG	NG,K	NG	NG,H	R,S,T,NG	NG	NG	NG	NG	NG
Peptidoglycan typeˣ	ND	Lys-Ala₂ (Ser)	Lys-Ala₂₋₄	Lys-Ala₃₋₃	Lys-Ala₂₋₄→Lys-Thr-Gly	Lys-Thr-Ala	Lys-Ala₂₋₃	Lys-direct (Lys-Ala₁₋₂)	Lys-Ala₂₋₃	ND	Lys-Ala₁₋₃, Glc, GlcNAc	ND	ND
Major cell-wall carbohydrate constituentsᶻ	Glc, (Gal), GalNAc, Rh	—	ND	Rha, Gal, Gro	Rha, Glc, Gal, GalNAc	Rha, Glc, Gal	Rha, Glc, Gal	Rha, Glc, (Gal), (GalNAc), man	ND	Rha, Glc, Gal	Rha, Glc, GlcNAc	ND	ND
Host	Animals	Man	Animals	Man, animals	Man	Man	Man, animals	Animals	(Dairy products)	Animals	Animals	Man	Man
Source of dataᵃᵇ	16	19, 27, 40	8, 9	12, 33	3, 19, 30, 40	60	6, 12, 33	19, 31	20, 39, 55	15	8, 9, 19, 26, 59	9	56

(continued)

TABLE 131. (continued)

[a]Symbols: +, >85% positive; d, different strains give different reactions (16–84% positive); −, 0–15% positive; w, weak reaction; ND, not determined.

[b]Acid production from carbohydrates often weak and difficult to interpret.

[c]Isolates become catalase-negative on subculturing.

[d]Strains from bovine genital tract differ from those from other sources.

[e]Reaction varies according to test kit used.

[f]Reports vary.

[g]Some strains give weak reactions.

[h]Characteristic score differs from Facklam (2002), with the latter's score given in parentheses.

[i]Acid production delayed.

[j]Acid production slow.

[k]Weak reaction.

[l]Some reports negative.

[m]Slow reaction.

[n]Positive or weak reaction on Müller–Hinton agar.

[o]Preformed enzyme activities detected are dependent on the test format (substrate) used and may be affected by culture conditions. Individual enzyme activities reported in the literature may therefore vary.

[p]Substrate-dependent.

[q]Results may vary.

[r]Reactions frequently weak.

[s]Conflicting results reported.

[t]Ala-phe-pro, alanyl-phenylalanyl-proline [arylamidase].

[u]Reported results differ.

[v]β-Hemolysis may be obtained on horse blood-containing agar.

[w]O antigen present in biovar 1 strains.

[x]Abbreviation of peptidoglycan type according to Schleifer and Kandler (1972).

[y]Ribitol teichoic acid and phosphocholine present.

[z]Gal (galactose); Glu (glucose); GalNAc (N-acetylgalactosamine); GlcNAc (N-acetylglucosamine; Gro (glycerol); Rha (rhamnose); Rtl (ribitol).

[aa]Occasionally man.

[ab]Data obtained from: 1. Ahmet et al. (1995); 2. Baele et al. (2003); 3. Beighton et al. (1991); 4. Beighton et al. (1984); 5. Beighton et al. (1981); 6. Beighton et al. (1991); 7. Brooker et al. (1994); 8. Collins et al. (1984a); 9. Collins et al. (2000); 10. Collins et al. (2001); 11. Collins et al. (2002); 12. Coykendall (1977); 13. Devriese e al. (1986); 14. Devriese et al. (1988); 15. Devriese et al. (1997); 16. Devriese et al. (1999); 17. Eldar et al. (1994); 18. Eldar et al. (1995); 19. Facklam (2002); 20. Farrow and Collins (1984b, 1984a); 21. Farrow et al. (1984); 22. Flint et al. (1999); 23. Handley et al. (1991); 24. Hardie (1986a, 1986b); 25. Hardie and Whiley (1995); 26. Kawamura et al. (1998); 27. Kawamura et al. (1998); 28. Kilian et al. (1989b); 29. Kilpper-Bälz and Schleifer (1987); 30. Kilpper-Bälz et al. (1985); 31. Kral and Daneo-Moore (1981); 32. Osawa et al. (1995); 33. Pier and Madin (1976); 34. Poyart et al. (2002); 35. Robinson et al. (1988); 36. Rurangirwa et al. (2000); 37. Schleifer et al. (1991); 38. Schleifer and Kilpper-Bälz (1987); 39. Schlegel et al. (2003); 40. Schlegel et al. (2000); 41. Skaar et al. (1994); 42. Tong et al. (2003); 43. Tsakalidou et al. (1998); 44. Vancanney et al. (2004); 45. Vandamme et al. (1999); 46. Vandamme et al. (1999); 47. Vela et al. (2002); 48. Vieira et al. (1998); 49. Whiley and Beighton (1991); 50. Whiley and Beighton (1998); 51. Whiley et al. (1990a, 1990b); 52. Whiley et al. (1999); 53. Whiley and Hardie (1988); 54. Whiley et al. (1988); 55. Willcox et al. (2001); 56. Williams and Collins (1990a); 57. Woo et al. (2002); 58. Zhu et al. (2000); 59. Collins et al. (2004).

TABLE 132. Biochemical tests differentiating β-hemolytic streptococcal species (Pyogenic group plus Anginosus group)[a]

Characteristic	*S. pyogenes*	*S. agalactiae*	*S. canis*	*S. didelphis*	*S. dysgalactiae* subsp. *dysgalactiae*	*S. dysgalactiae* subsp. *equisimilis*	*S. equi* subsp. *equi*	*S. equi* subsp. *zooepidemicus*	*S. iniae*	*S. phocae*	*S. porcinus*	*S. anginosus*	*S. constellatus* subsp. *constellatus*	*S. constellatus* subsp. *pharyngis*	*S. intermedius*
Catalase	−	−	−	+[b]	−	−	−	−	−	−	−	−	−	−	−
Acid production from:															
Glycogen	d	−	−[c]	−	−	d	−	+	+	d	−	−	−	ND	−
Mannitol	−	−	−	−	d	d	−	−	+	−	d	d[d] (−)	−	ND	−
Raffinose	−	−	−	d	−	−	−	−	−	+	+	d	−	−	−
Ribose	−	+	+	+	+	+	−	+	+	+	+	−	−	ND	−
Sorbitol	−	−	−	−	d	−	−	+	−	−	d	−	−	ND	−
Sucrose	+	+	+	ND	+	+	+	+	+	−	+	+	ND	ND	+
Trehalose	+	+	d	+	+	+	−	d	+	−	+	+	d	+	+
Hydrolysis of:															
Arginine	+	+	+	d[d] (+)	+	+	+	+	+[d] (−)	−	d[d] (+)	+	+	+	+
Esculin	d	−	+	d	d	d	d	d	+	−	d[d] (+)	+	+	+	+
Hippurate	−	+	−	+	−	d[d] (−)	−	−	−	−	−[d,f]	−	−	−	−
Starch	−	−	−	−	−	−	+	+	+	−	−	−	−	−	ND
Production of:[e]															
α-D-galactosidase	−	d	d	d	−	−	−	−	−	ND	−	−	−	−	+
β-D-galactosidase	d	−	+	−	−	+	−	−	−	ND	−	d	−	+	+
β-D-glucuronidase	d	d	d	+	+	+	+	+	ND	−	d	−	−	−	+
Pyrolidonylarylamidase	+	−	−	−	−	−	−	−	+	−	−[d] (+)	−	−	−	−
Acetoin (V–P)	−	+[d,f] (−)	−[d] (+)	−	−	−	−	−	−	−	d[d] (+)	+	+	+	+
CAMP reaction	−	+	−	−	−	−	−	−	−	−	+	−	−	−	−
Lancefield group antigen	A	B	G	NG	C	C, G, L, A	C	C	NG	NG, F, C	E, P, U, V	NG,F,C,G,A	NG,F,C,G,A	C	NG,F,C
Host	Man	Man, animals	Animals, man	Animals	Animals	Man, animals	Animals	Animals, Man	Fish, animals,	Animals	Animals	Man	Man	Man	Man

[a]Symbols: +, >85% positive; d, different strains give different reactions (16–84% positive); −, 0–15% positive; w, weak reaction; ND, not determined.

[b]Isolates become catalase-negative on subculturing.

[c]Some strains give weak reactions.

[d]Character score differs from Facklam (2002) with the latter's score also following in parentheses. An additional, useful test according to Facklam (2002) is sensitivity to bacitracin for *S. pyogenes* and *S. phocae* only

[e]Preformed enzyme activities detected are dependent on the test format (substrate) used and may be affected by culture conditions. Individual enzyme activities reported in the literature may therefore vary.

[f]Reported results differ.

TABLE 133. Characteristics of *Streptococcus* species: oral streptococcal species groups[a,b]

	Anginosus group				Mutans group								
Characteristic	*S. anginosus*	*S. constellatus* subsp. *constellatus*	*S. constellatus* subsp. *pharyngis*	*S. intermedius*	*S. criceti*	*S. devriesei*	*S. downei*	*S. ferus*	*S. macacae*	*S. mutans*	*S. orisratti*	*S. ratti*	*S. sobrinus*
Acid production from:													
Amygdalin	+	d	+	d	+	+	ND	+	+	d	+	+	−
Arbutin	+	d	+	+	+	+	ND	d	ND	+	+	+	−
Glycogen	−	−	ND	−	−	−	−	+	−	−	+	−	−
Inulin	−	−	−	−	+	d	+	d	−	+	+	d	d
Lactose	+	d	+	+	d	+	+	+	ND	+	+	d	d
Mannitol	dᶜ (−)	−	−	−	+	+	+	dᶜ (+)	+	+	−	+	+
Melibiose	d	−	−	−	+	+	−	d	−	d	+	d	−
Raffinose	d	−	−	−	+	+	−	d	+	+	+	+	−
Ribose	−	−	ND	−	−	−	ND	dᶜ (+)	−	−	ND	−	−
Sorbitol	−	−	−	−	+	+	−ᶜ (+)	dᶜ (+)	+ᵇ	+	−	+	dᶜ (+)
Trehalose	+	d	+	+	d	+	+	+	+	+	+	+	+
Hydrolysis of:													
Arginine	+	+	+	+	−	−	−	−	−	−	−	+	−
Esculin	+	+	+	+	dᶜ (+)	ND	−	+	+	+	+	+	dᶜ (+)
Hippurate	−	−	−	−		+	−	−	ND	−	ND	−	−
Production of:ᵉ													
Acetoin (V–P)	+	+	+	+	+	+	+	+	+	+	−	+	+
N-Acetyl-β-D-glucosaminidase	−	−	d	+	ND	ND	ND	−	ND	−	ND	ND	−
Alkaline phosphatase	+	+	+	+	ND	−	-	d	ND	−	−	ND	
Extracellular polysaccharide	−	−	−	−	+	ND	+	+	+	+	−	+	+
β-D-Galactosidase	d	−	+	+	ND	−	ND	−	ND	−	−	ND	−
α-D-Glucosidase	d	+	+	+	ND	ND	ND	+	ND	+	ND	ND	+
β-D-Glucosidase	+	−	+	d	ND	+	ND	+	ND	+	ND	ND	−
β-D-Fucosidase	−	−	+	+	ND	ND	ND	ND	ND	−	ND	ND	−
Urease	−	−	−	−	−	−	−	−	−	−	−	−	−
Lancefield group antigen	NG, F, C, G, A	NG, F, C, G, A	C	NG, F, C	NG	ND	NG	NG	NG	NG	A	NG	NG
Host	Man	Man	Man	Man	Animals (man)ʲ	Animals	Animals	Animals	Animals	Man	Animals	Man, animals	Man, animals

[a]Symbols: +, >85% positive; d, different strains give different reactions (16–84% positive); −, 0–15% positive; w, weak reaction; ND, not determined.

[b]Acid production is slow.

[c]Character score differs from Facklam (2002), with the latter's score given in superscript.

[d]Some reports negative.

[e]Preformed enzyme activities detected are dependent on the test format (substrate) used and may be affected by culture conditions. Individual enzyme activities reported in the literature may therefore vary.

[f]Substrate-dependent.

[g]Substrate-dependent.

[h]Substrate-dependent.

[i]O antigen present in biovar 1 strains.

S. australis	*S. cristatus*	*S. gordonii*	*S. infantis*	*S. mitis*	*S. oligofermentans*	*S. oralis*	*S. parasanguinis*	*S. peroris*	*S. pneumoniae*	*S. sanguinis*	*S. sinensis*	*S. salivarius*	*S. thermophilus*	*S. vestibularis*
	Mitis group											Salivarius group		
ND	–	+	–	–	–	–	d	–	–	d	ND	+	–	d
ND	+	+	–	–	–	–	d	–	–	d	ND	+	d	d
–	ND	–	–	–	ND	–	–	–	+	–	–	–	–	–
ND	–	d	d	–	–	–	–	–	+	d	–	d	–	–
+	d	+	+	+	d	+	+	+	+	+	+	d	+	d
–	–	–	–	–	–	–	–	–	d[c] (–)	–	–	–	–	–
–	–	d	–	d	–	d	d	–	+	d	–	–	d	–
–	–	d	–	d	d	d	d	–	+	d	–	d	d	–
–	–	–	–	d	–	d	ND	–	–	–	–	–	d	–
–	–	–[c,f]	–	–[c,f]	–	–	d[c] (–)	–	–	d	–	–	–	–
–	+	+	–	d	d	d	d	–	+[c,f]	+	+	d	–	d
+	d[c] (+)	+	–	d[c] (–)		–	+	–	+	+	+	–	–	–
ND	–	+	–	–	–	d[d]	d	–	d	d[c] (+)	+	+	–	+[c,f]
–	ND	–	–	–	+	–	ND	–	–	–	–	–	–	–
–	–	–	–	–	–	–	–	–	–	–	+	d[c] (+)	+	+[c,f]
–[h]	+	+	d	+	ND	+	+	–	d	d	–	–	–	–
+	–	+	–	d	d	+	+	d	–	–	–	+	–	–
–	d	+	ND	–	ND	d	–	ND	–	+	ND	+	ND	–
–[f]	d	d	+	d	–	+	+	–[g]	+	d	–	d	+	+
ND	–	d	ND	+	ND	+	+	ND	+	–	ND	+	–	d
–	–	+	–	–	ND	–	d	–	d	d	+	+	–	–
ND	–	–	d	d	ND	–	d	–	d	d	ND	d	ND	–
–	–	–	–	–	ND	–	–	–	–	–	–	d	–	+
NG	ND	NG, H	ND	NG, O^i, K	ND	NG	NG (F, G, C, B)	ND	NG	NG, H	NG	NG, K	NG	NG
Man	Man	Man	Man	Man	Man	Man	Man	Man	Man	Man	Man	Man	(Dairy products)	Man

Biochemical, physiological, and antigenic characteristics of species within the Anginosus group are given in Table 131, Table 132, and Table 133.

Salivarius group. The close relationships between two oral species, *Streptococcus salivarius* and *Streptococcus vestibularis*, and *Streptococcus thermophilus*, a species isolated from dairy sources but of unknown habitat, have been demonstrated in several studies using DNA–DNA reassociation and by the presence of significant amounts of eicosenoic acid (12–17%) (Farrow and Collins, 1984b; Kilpper-Bälz et al., 1982; Whiley and Hardie, 1988). Initially a proposal was made for recognition of *Streptococcus salivarius* and *Streptococcus thermophilus* as subsp. *salivarius* and subsp. *thermophilus*, respectively (Farrow and Collins, 1984b). Subsequent data supported the recognition of *Streptococcus salivarius* and *Streptococcus thermophilus* as distinct species (Schleifer et al., 1991, 1987), and was followed by the addition of *Streptococcus vestibularis* as the third member of this group (Whiley and Hardie, 1988).

Phenotypic similarities between *Streptococcus salivarius* and *Streptococcus bovis* have been noted as a possible source of confusion in routine identification. Similarities with some biochemical reactions, reaction of some strains of *Streptococcus salivarius* with Lancefield group D antiserum and growth of both species on bile-esculin agar (Coykendall and Gustafson, 1985) have been highlighted, although extended physiological characterization has been shown to differentiate them (Ruoff et al., 1989).

Biochemical, physiological, and antigenic characteristics of species within the Salivarius group are given in Table 131 and Table 133.

Mitis group. This species group currently includes *Streptococcus mitis*, *Streptococcus oralis*, *Streptococcus sanguinis*, *Streptococcus gordonii*, *Streptococcus parasanguinis*, *Streptococcus cristatus*, *Streptococcus peroris*, *Streptococcus infantis*, *Streptococcus australis*, *Streptococcus sinensis*, and *Streptococcus oligofermentans*, mainly isolated from the normal oral and pharyngeal flora in man, together with the potentially highly pathogenic species *Streptococcus pneumoniae* that may also be found resident in the upper respiratory tract of healthy humans. Further centres of variation have been described within these streptococci (Kawamura et al., 2000). *Streptococcus mitis* (biovar 1) (Kilian et al., 1989b) and *Streptococcus oralis* comprise the majority of the earliest, "pioneer" species to colonize the mouths of healthy human neonates (Pearce et al., 1995), while *Streptococcus sanguis*, *Streptococcus oralis*, and *Streptococcus gordonii* were found to be the most common streptococci isolated from bacterial endocarditis patients (Douglas et al., 1993), and *Streptococcus oralis* and *Streptococcus mitis*, together with *Streptococcus salivarius*, have been found to be a major cause of infections in neutropenic patients (Beighton et al., 1994; Elting et al., 1992; Jacobs et al., 1995). The taxonomic histories of several of the species included here, particularly those referred to in the past as "*Streptococcus viridans*", "*Streptococcus sanguis* type II", "*Streptococcus mitior*" and *Streptococcus mitis*, have frequently been sources of confusion involving nomenclatural changes and accompanying definitions (see Whiley and Beighton, 1998). Despite the inclusion of the name *Streptococcus mitis* and exclusion of "*Streptococcus mitior*" (Colman and Williams, 1972) from the Approved Lists of Bacterial Names (1980), the previous edition of the *Manual* did not

list *Streptococcus mitis* on the grounds that this species had always been ill-defined (Andrewes and Horder, 1906; Sherman et al., 1943) with the name "*Streptococcus mitior*" instead preferred "for a group of α-hemolytic streptococci that can be defined with a reasonable degree of precision" (Hardie, 1986a). The history of these streptococci is best traced commencing with *Streptococcus sanguis* (White and Niven, 1946) a species isolated from blood and endocarditis vegetations, able to hydrolyze arginine and esculin and form dextran from sucrose, which resembled streptococci also referred to as "*Streptococcus s.b.e*" (Loewe et al., 1946; Washburn et al., 1946). Serological analyses resulted in the recognition of *Streptococcus sanguis* types I, I-II, and II (Washburn et al., 1946), and the description of dextran-forming strains belonging to Lancefield group H from bacterial endocarditis (Hehre and Neill, 1946). Confusion surrounding the precise strain used in producing the original Lancefield group H antiserum understandably led to disagreement regarding the relationship between strains called *Streptococcus sanguis* and possession of a Lancefield group H antigen (Dodd, 1949; Farmer, 1954; Porterfield, 1950), with *Streptococcus sanguis* recognized as including both dextran-producing and non-producing strains by these authors. Evidence of heterogeneity within *Streptococcus sanguis* came from the demonstration of a lack of rhamnose but presence of significant amounts of anhydroribitol in the cell walls of the type II strains (Colman and Williams, 1965) and the separation of the latter from the remainder of *Streptococcus sanguis* strains by numerical taxonomy (Carlsson, 1968; Colman, 1968). Dextran formation was a variable characteristic of all clusters. Colman and Williams (1972) used the name "*Streptococcus mitior*" for those strains unable to hydrolyze arginine or esculin and without rhamnose in their cell walls. Subsequently, confusion arose through the adoption of an alternative nomenclature in the USA, in which strains referred to as *Streptococcus sanguis* by Colman and Williams (1972) were called *Streptococcus sanguis* biotype I and "*Streptococcus mitior*" strains were divided into *Streptococcus sanguis* biotype II and *Streptococcus mitis* on the basis of fermentation and non-fermentation of raffinose, respectively (Facklam, 1977). The data from DNA–DNA hybridizations continued to provide unremitting evidence of heterogeneity amongst these streptococci, with those similarity groups obtained that corresponded to *Streptococcus sanguis* according to Colman and Williams' definition (1972) remaining distinct from those corresponding to "*Streptococcus mitior*" (Coykendall and Munzenmaier, 1978; Coykendall and Specht, 1975; Schmidhuber et al., 1987; Welborn et al., 1983). The fact that more than one similarity group was present within strains of both of these "species" provided irrefutable evidence that several taxa existed. Differentiation at the species level on the basis of raffinose fermentation was not upheld by these studies. *Streptococcus sanguis* and *Streptococcus mitis* both appeared in the Approved Lists of Bacterial Names according to descriptions from the Eighth Edition of *Bergey's Manual of Determinative Bacteriology* (Deibel and Seeley, 1974; Skerman et al., 1980). The type strains assigned were NCTC 7863 (ATCC 10556) for *Streptococcus sanguis* and NCTC 3165 (ATCC 33399) for *Streptococcus mitis* although the latter strain was phenotypically more typical of *Streptococcus sanguis* in being able to hydrolyze arginine and esculin and with only 40% overall DNA similarity demonstrated between it and other, typical *Streptococcus mitis* strains (Kilpper-Bälz et al., 1985; Welborn et al., 1983). Strain NCTC 3165 was eventually pronounced inappropriate as the *Streptococcus mitis* type strain, reassigned to the species *Streptococcus gordonii* and replaced with strain NS51 (NCTC 12261) (Kilian et al., 1989a).

Further confusion of the overall taxonomic picture of this group was caused by the description of *Streptococcus oralis*, based on the results of a numerical taxonomic study (Bridge and Sneath, 1982). This species was later redefined by Kilpper-Bälz et al. (1985) and restricted to strains with cell walls containing directly cross-linked peptidoglycan with lysine as the diamino acid (Lys-direct), with ribitol and choline as characteristic cell-wall constituents but with rhamnose absent. Several strains previously referred to as "*Streptococcus viridans*" were included within *Streptococcus oralis* (Kilpper-Bälz et al., 1985). The omission of the name "*Streptococcus mitior*" from the Approved Lists of Bacterial Names (Skerman et al., 1980) and the subsequent valid publication of the species epithet *oralis*, albeit originally for a heterogeneous collection of strains encompassing more than one species, meant that the name "*Streptococcus mitior*" had no official standing in the literature. Kilian et al. (1989b) studied 151 viridans streptococci that included reference and type strains from earlier studies together with clinical isolates. These were examined using biochemical and physiological tests and serological reactions including the presence of the Lancefield group H antigen derived from several candidate strains. This study provided much of the current view of the classification of the Mitis group of streptococci and resulted in emended descriptions of *Streptococcus sanguis*, *Streptococcus oralis*, and *Streptococcus mitis*, together with the proposal of a new species, *Streptococcus gordonii*, and incorporated strain NCTC 3165, which had previously been rejected as the type strain of *Streptococcus mitis*.

It needs to be noted here that in the case of *Streptococcus mitis*, efforts to delineate the species more clearly have continued with its division into biotypes (Kilian et al., 1989b), by comparison of whole cell-derived polypeptide patterns by SDS-PAGE (Vandamme et al., 1998), and by sequence analysis of housekeeping genes (Whatmore et al., 2000). These studies and others have demonstrated heterogeneity within *Streptococcus mitis*, particularly biovar 2 strains (Kilian et al., 1989b) which the data indicate are of disparate taxonomic affiliation (De Gheldre et al., 1999; Kikuchi et al., 1995; Vandamme et al., 1998).

Further additions to the Mitis species group have included *Streptococcus parasanguinis* (*Streptococcus parasanguis*, Whiley et al., 1990b) isolated originally from human clinical specimens, *Streptococcus cristatus* (*Streptococcus crista*; Handley et al., 1991), *Streptococcus peroris*, *Streptococcus infantis* (Kawamura et al., 1998), *Streptococcus australis* (Willcox et al., 2001), and *Streptococcus oligofermentans* (Tong et al., 2003), all isolated from the human oral cavity, together with *Streptococcus sinensis* from the blood of an endocarditis patient (Woo et al., 2002). The DNA G+C content for the latter species was determined to be 53 ± 2.9 mol% which would appear to be high for a streptococcus.

An example of further taxa within the group is the "tufted mitior" strains originally described by Handley et al. (1985) on the basis of unusual tufted surface fibrils seen by electron microscopy. "Tufted mitior" strains were shown to constitute a distinct taxon at the species level by DNA–DNA hybridization and 16S rRNA gene sequence comparisons, but the lack of suitable biochemical tests for differentiation from other streptococci, in particular *Streptococcus mitis*, so far precludes any formal taxonomic proposals (Kawamura et al., 2000).

Despite listing *Streptococcus pneumoniae* within the "pyogenic hemolytic streptococci" in the previous edition of the *Manual*, the close relationships between *Streptococcus pneumoniae* and members of the Mitis group, in particular *Streptococcus oralis* and *Streptococcus mitis*, have been recognized for some time (Bentley et al., 1991; Kawamura et al., 1995b, 1999; Kilpper-Bälz et al., 1985). These three species share over 99% 16S rRNA gene sequence similarity although the estimated level of sequence similarity over their entire genomes is < 60%. Interspecies recombination events have been shown to have taken place involving *Streptococcus pneumoniae*, *Streptococcus oralis*, and *Streptococcus mitis*, a process particularly evident in genes under relatively intense selective pressure, such as the PBPs (Dowson et al., 1997, 1993, 1994).

The names *Streptococcus sanguis*, *Streptococcus parasanguis*, and *Streptococcus crista* have been changed on grammatical grounds to *Streptococcus sanguinis*, *Streptococcus parasanguinis*, and *Streptococcus cristatus*, respectively (Trüper and de Clari, 1997).

Biochemical, physiological, and antigenic characteristics of species within the Mitis group are given in Table 131 and Table 133.

Bovis group. The species within this group represent a collection of streptococci of human and animal origin whose classification has long been problematic and is currently undergoing considerable revision in the light of data from molecular methods. The two earliest recognized species names here that are still in use today are *Streptococcus bovis* (Andrewes and Horder, 1906) and *Streptococcus equinus* (Orla-Jensen, 1919; Sherman and Wing, 1937). Both of these species were listed in the Approved Lists of Bacterial Names (Skerman et al., 1980) and in the previous edition of *Bergey's Manual of Systematic Bacteriology* (Hardie, 1986b). Differentiation of *Streptococcus equinus* and *Streptococcus bovis* by acid from lactose and raffinose and production of α-galactosidase by *Streptococcus bovis* correlated with the isolation of *Streptococcus equinus* from horse feces and *Streptococcus bovis* from bovine feces, mastitis, human feces, and human clinical sources (Colman, 1990; Schlegel et al., 2000). *Streptococcus bovis* strains of human origin were described as "typical" (Facklam, 1972) or "biotype I" (Parker and Ball, 1976) when able to produce acid from mannitol and inulin but rarely from sorbitol or arabinose, and to produce copious amounts of extracellular polysaccharide (glucan) from sucrose (see Coykendall (1989a) for review of the literature). Strains unable to ferment mannitol or produce glucans were called "variant" (Facklam, 1972) or "biotype II" (Parker and Ball, 1976). Coykendall and Gustafsson (1985) subdivided biotype II strains into biotypes II/1 and II/2 on the ability of biotype II/2 strains to produce acid from trehalose, to produce β-glucuronidase and β-galactosidase more frequently and to less frequently split starch.

Successive classification studies, initially based on phenotypic characteristics and more recently using nucleic acid based approaches (Coykendall and Gustafson, 1985; Farrow et al., 1984; Knight and Shlaes, 1985; Osawa et al., 1995; Poyart et al., 2002; Schlegel et al., 2000) have resulted in the gradual recognition of several additional species. Consequently the name "bovis", long familiar in the streptococcal literature, is increasingly associated with a group of species sometimes variously referred to as the "Bovis group" (Kawamura et al., 1995b), the "*Streptococcus bovis/Streptococcus equinus* complex" (Schlegel et al., 2000) and the "*Streptococcus bovis/Streptococcus equinus* group"

(Poyart et al., 2002). In addition, DNA–DNA hybridization studies have demonstrated that the type strains of *Streptococcus bovis* and *Streptococcus equinus* belong to the same similarity group, leading to the recognition of the names *Streptococcus equinus* and *Streptococcus bovis* as subjective synonyms with the specific epithet *Streptococcus equinus* having priority under Rule 19 of the International Code of Nomenclature of Bacteria (Farrow et al., 1984; Kilpper-Bälz et al., 1982; Nelms et al., 1995). This latter development may not be welcomed by many clinical bacteriologists who usually associate the names "bovis" and "equinus" with human and animal (equine) sources, respectively.

Currently the species included within the "bovis/equinus" complex are i) *Streptococcus equinus* (Farrow et al., 1984), which encompasses mainly mannitol-negative strains that previously would have been described as *Streptococcus bovis* biotype II/1, ii) *Streptococcus gallolyticus* (Osawa et al., 1995), originally comprising strains previously identified as *Streptococcus bovis* biotype I from human clinical and animal sources and currently also including strains previously assigned to *Streptococcus caprinus* by Brooker et al. (1994), *Streptococcus macedonicus* (Tsakalidou et al., 1998), and strains previously designated *Streptococcus waius* (Flint et al., 1999; Manachini et al., 2002; Poyart et al., 2002), iii) *Streptococcus infantarius* (Bouvet et al., 1997; Poyart et al., 2002) which currently includes *Streptococcus bovis* biotype II/1 strains previously given subspecific status as *Streptococcus infantarius* subsp. *infantarius* (Schlegel et al., 2000), iv) *Streptococcus alactolyticus* (Farrow et al., 1984), v) *Streptococcus lutetiensis* which includes strains previously described as *Streptococcus infantarius* subsp. *coli* (Poyart et al., 2002; Schlegel et al., 2000), and vi) *Streptococcus pasteurianus* (Poyart et al., 2002), comprising strains previously identified as *Streptococcus bovis* biotype II/2.

Recent studies employing DNA–DNA reassociation and gene sequence data support the inclusion of *Streptococcus macedonicus* and *Streptococcus waius* within *Streptococcus gallolyticus*, with the latter having nomenclatural priority (Poyart et al., 2002; Schlegel et al., 2003). The present authors agree with this proposal although the levels of DNA reassociation, occasionally obtained marginally below the generally accepted 70% cut-off for inclusion within a single species, will no doubt continue to fuel debate. Further, Schlegel et al. (2003) have also proposed that strains previously classified as *Streptococcus pasteurianus* be included within *Streptococcus gallolyticus* with the proposed recognition of *Streptococcus gallolyticus* subsp. *gallolyticus*, subsp. *macedonicus*, and subsp. *pasteurianus*. On balance, the present authors would agree with two of the latter proposals with the exception of the case for *Streptococcus gallolyticus* subsp. *pasteurianus* where the evidence for inclusion is less clear overall. *Streptococcus pasteurianus* is recognized as a separate species here.

In the often-quoted DNA–DNA hybridization study of bovis/equinus strains by Farrow et al. (1984), six similarity groups were delineated of which their group 3 has yet to be assigned to a named species. A representative strain of group 3 remained ungrouped in a study using ribotyping and DNA–DNA hybridization (Schlegel et al., 2000). Group 3 strains were isolated from raw milk and bovine mastitis cases and included strain NCDO 2127, previously described as an anomalous strain of *Streptococcus bovis* (Garvie and Bramley, 1979a).

Biochemical, physiological, and antigenic characteristics of species within the Bovis group are given in Table 131 and Table 134.

"**Hyovaginalis group**". *Streptococcus hyovaginalis*, *Streptococcus thoraltensis* (Devriese et al., 1997), and *Streptococcus pluranimalium* (Devriese et al., 1999) are isolated from the genital and respiratory tracts of domestic animals and birds and together comprise another species group by 16S rRNA gene sequence analysis.

Biochemical, physiological, and antigenic characteristics of species within the Hyovaginalis group are given in Table 131 and Table 135.

Species not assigned to recognized species groups

Included here are the species that i) remain ungrouped, forming separate lines of descent, ii) are only weakly affiliated to one of the main species groups already listed listed above, or iii) are associated phylogenetically with each other and may be regarded as members of additional, previously unnamed, species groups.

Biochemical, physiological, and antigenic characteristics of species included here are given in Table 131 and Table 135.

Streptococcus ovis (Collins et al., 2001), isolated from abscesses and other clinical conditions in sheep, and *Streptococcus hyointestinalis* (Devriese et al., 1988), from the intestines of pigs, both remain ungrouped by 16S rRNA gene sequence analysis. *Streptococcus minor* (Vancanneyt et al., 2004), a species from dogs, cats, and cattle, shares the highest 16S rRNA gene sequence homology with *Streptococcus ovis*.

Streptococcus acidominimus a generally non-pathogenic species associated with cattle, the pathogenic species from pig infections *Streptococcus suis*, together with a recently described species *Streptococcus entericus* isolated from cows with catarrhal enteritis (Vela et al., 2002) are closest neighbors to each other by 16S rRNA gene sequence analysis and form a loose species group. It should be noted however that in one study *Streptococcus suis* clustered together with the Pyogenic group by 16S rRNA gene sequence comparisons (Flint et al., 1999) and in Figure 123 of this chapter *Streptococcus acidominimus*, *Streptococcus suis*, *Streptococcus entericus*, and *Streptococcus gallinaceus* (Collins et al., 2002) are loosely associated.

Streptococcus ferus, although discussed previously within the context of the Mutans group, was originally isolated from the teeth of rats in sugar cane fields and has also been isolated from the tonsils and nasal conchae of piglets (Baele et al., 2003). In several studies this species appears peripheral to the Mutans species group (Gilmour et al., 1987; Schleifer et al., 1984; Whatmore and Whiley, 2002).

Identification of streptococci

Routine identification of isolates is mainly carried out by biochemical testing using commercially available test kits. However, the recent addition of new species and reporting of as-yet-unnamed genospecies means that problematic areas persist, notably between certain members of the Mitis species group (Kawamura et al., 1999; Kikuchi et al., 1995) and the Anginosus group (Whiley et al., 1999). Incorporation of chromogenic and fluorogenic substrates has greatly improved the resolving power of biochemical test schemes (Beighton et al., 1991; Freney et al., 1992; Kikuchi et al., 1995; Kilian et al., 1989b) and has led to the description of biovars or biotypes within several species which need to be accounted for in any proposed identification scheme (Beighton et al., 1991; Kilian et al., 1989b).

TABLE 134. Differential characteristics of the Bovis group[a,b]

Characteristic	S. alactolyticus	S. equinus	S. gallolyticus subsp. gallolyticus	S. gallolyticus subsp. macedonicus	S. infantarius	S. lutetiensis	S. pasteurianus
Acid production from:							
Glycogen	–	d	+	–	d	–	–
Inulin	–	d	d	–	–	ND	–
Lactose	–	d	+	+	+	+	+
Mannitol	d[c] (–)	–	+	–	–	–	–
Melibiose	d	d[c] (–)	d[c] (–)	ND	d	–	+
Melezitose	d	–	d	–	–	–	–
Pullulan	–	–	d	–	d	–	+
Raffinose	+	d	d	+	d	+	+
Starch	d	d	+	d	d	d	–
Tagatose	–	–	–	–	–	–	d
Trehalose	d	d	+	–	–	–	+
Hydrolysis of:							
Esculin	+	+	+	–	d	+	+
Starch	ND	–	+	–	+	d	–
Production of:[a]							
Acetoin (V-P)	+	+	+	+	+	+	+
N-Acetyl-β-D-glucosaminidase	–	–	ND	–	–	–	–
Extracellular polysaccharide	–	–	+	–	–	–	–
α-D-Galactosidase	+	d	d	d	+	+	+
β-D-Galactosidase	–	–	d	d	–	–	+
β-D-Glucosidase	d	+	+	–	–	+	+
β-D-Glucuronidase	–	–	–	–	d	–	+
β-Mannosidase	–	d	d	–	–	–	+
Pyrolidonylarylamidase	–	–	–	–	–	–	–
Lancefield group antigen	D(G)	D	D, NG	D, NG	NG, D	NG, D	D
Host	Animals	Animals	Animals, man	Dairy	Man	Man	Man

[a]Symbols: +, >85% positive; d, different strains give different reactions (16–84% positive); –, 0–15% positive; w, weak reaction; ND, not determined.

[b]Preformed enzyme activities detected are dependent on the test format (substrate) used and may be affected by culture conditions. Individual enzyme activities reported in the literature may therefore vary.

[c]Character score differs from Facklam (2002), with the latter's score given in parentheses.

TABLE 135. Characteristics differentiating unusual streptococci including the Hyovaginalis group[a]

Characteristic	Species groups						Hyovaginalis group			Unusual (ungrouped) streptococcal species						
	Anginosus group	Bovis group	Mitis group	Mutans group	Pyogenic group	Salivarius group	S. hyovaginalis	S. pluranimalium	S. thoraltensis	S. acidominimus[b]	S. entericus	S. gallinaceus	S. hyointestinalis	S. minor	S. ovis	S. suis
Acid production from:																
Glycogen	−	d	−	d	d	−	−	−	−	−	+	+	−	+	+	+
Inulin	−	d	−	d	d	d	−	−	+	+	+	ND	−	d	+	+
Mannitol	−	d	−	d	d	−	+	d[c]	+	+	−	+	−	+[b]	+	−
Pullulan	d	d	d	−	ND	d	−	d[c]	+	ND	−	+	d	ND	+	d
Raffinose	−	d	d	d	−	d	d	−	d	ND	−	+	d	d	+	d
Ribose	−	−	d	d	+	d	d	d[c]	d[d] (+)	d	−	+	−	d	+	−
Sorbitol	−	−	−	d	d	−	+	d	d	ND	−	−	−	−	+	−
Tagatose	d	−	d	d	d	d	+	d	d	−	−	−	−	d	d	+
Trehalose	d	d	d	+	d	d	+	+	+	−	+	+	+	+	+	+
Hydrolysis of:																
Arginine	+	−	d	−	d	−	−	d[d] (−)	+	+	+	+	+[d] (−)	+	d	+
Esculin	+	d	d	d	d	d	+	d[d] (+)	+	+	+	+	−	+	+	+
Hippurate	−	−	−	−	d	−	+	+[d] (−)	d	+[d,e] (−)	−	+	−	+	−	−
Production of:[b]																
Acetoin (V–P)	+	+	−	+	d	d	+[i]	−	+	−	−	−	+	ND	−	−
N-Acetyl-β-D-glucosaminidase	d	−	d	−	ND	−	+	d	+	−	−	−	ND	ND	−	+
Alkaline phosphatase	+	−	d	−	+	d	+	d	d[f]	ND	−	−	+	d	d	−
α-D-Galactosidase	−	d	d	d	d	−	−	d	d	ND	−	+	d	d[g]	d	+
β-D-Galactosidase	d	d	d	−	d	d	+	d	−[h]	ND	−	+	−	−	d	d
β-D-Glucuronidase	−	d	d	+	d	−	+[j]	d[c]	+	+	−	−	−	−	−	+
Leucine arylamidase	+	+	ND	+	d	+	+[j]	+	+	ND	d	ND	+	+	+	+
β-Mannosidase	ND	d	d	−	ND	−	−	−	−	−	−	−	ND	ND	−	d
Pyrolidonylarylamidase	−	−	d	−	d	−	d[d] (−)	d[d] (−)	−	+	d	ND	−	−	ND	
Hemolytic reaction	α, NH	α, NH	α	α, NH	β, (α, NH)	α, NH	α	α	α	α	α	α	α	α	α	α(β)[k]
Lancefield group antigen	NG, F, C, G, A	D, NG	NG, H, K, O	NG, (A)	A, B, C, G, F, L, E, P, U, V, NG	NG, K	NG	ND	NG	NG	D	D, NG	NG	NG	ND	R, S, T, NG
Host	Man	Man, animals, dairy	Man	Man, animals	Man, animals	Man, dairy products	Animals	Animals	Animals	Cattle, milk	Animals	Birds	Animals	Animals	Animals	Animals, man

[a]Symbols: +, >85% positive; d, different strains give different reactions (16–84% positive); −, 0–15% positive; w, weak reaction; ND, not determined.

[b]Acid production delayed.

[c]Strains from bovine genital tract differ from those from other sources.

[d]Character score differs from Facklam (2002), with the latter's score given in parentheses.

[e]Slow reaction.

[f]Weak reaction.

[g]Substrate-dependent.

[h]Results may vary.

[i]Reactions frequently weak.

[k]β-Hemolysis may be obtained on horse blood-containing agar.

Alternative approaches have been used although none have been successfully applied to all the currently recognized species within the genus and are often useful within particular groups of species. Examples include whole cell polypeptide signatures by SDS-PAGE (Vandamme et al., 1998), pyrolysis-mass spectroscopy (Magee et al., 1997), Fourier transform infra-red spectroscopy (van der Mei et al., 1993), monoclonal antibodies (De Soet and De Graaff, 1990), RFLP patterns, ribotyping (Rudney and Larson, 1993, 1994; Rudney et al., 1992) and PCR-based approaches (Alam et al., 1999; Li et al., 2001; Rudney and Larson, 1999; Truong et al., 2000). Sequence analysis of internal fragments of the manganese-dependent superoxide dismutase gene ($sodA_{int}$) has considerable potential as a gene-based approach to identification (Hoshino et al., 2005; Kawamura et al., 1999; Poyart et al., 1998; Whatmore and Whiley, 2002). Comparison between sequence similarities of 16S rRNA and $sodA_{int}$ have shown the latter to constitute a more discriminating target, particularly for differentiating closely related species such as *Streptococcus pneumoniae*, *Streptococcus oralis*, and *Streptococcus mitis* within the Mitis group.

Recent analysis of the gene encoding the RNA subunit of endoribonuclease P, *rnpB*, has shown this to be potentially extremely useful for phylogenetic analysis and species discrimination (Tapp et al., 2003). A similar tree topology to that derived from 16S rRNA was obtained for *rnpB*. From this study it was concluded that a combined approach with both *rnpB* and 16S rRNA gene sequence data included in the analysis would result in a better phylogenetic hypothesis for the genus *Streptococcus*. More recently still, evidence has been published of two other genes (*dnaJ*, a member of the Hsp70 family and *gyrB*, which encodes the B-subunit protein of DNA gyrase (topoisomerase type II)) showing sufficient intra species divergence to be useful for identification and for discriminating between closely related streptococcal strains (Itoh et al., 2006). However, studies indicate that reliance on single gene sequence data for discrimination may result in a degree of misidentification of some strains due to previous horizontal gene transfer events,

notably within *Streptococcus sanguinis*, members of the anginosus species group, *Streptococcus mitis*, *Streptococcus parasanguinis*, *Streptococcus thoraltensis* and *Streptococcus criceti* (Hoshino et al., 2005; Itoh et al., 2006) with some authors recommending multilocus sequence analysis for optimal results (Hoshino et al., 2005).

The extent to which these transfer events have occurred remains to be elucidated and no doubt will be found to be much more frequent within some species than in others.

High-speed, microarray-based technology may provide the basis for streptococcal species identification in the not too distant future; the study by Tung et al. (2006) used a panel of immobilized oligonucleotide probes designed from rDNA intergenic spacer (ITS) regions to hybridize with PCR amplified ITS regions from strains with promising results.

Differential characteristics tables in this chapter. Differential biochemical tests for the streptococcal species are given in Table 132, Table 133, Table 134, and Table 135. The tests listed in each table have been selected from Table 131, i.e., from the collected data obtained from the literature, and do not form a working laboratory test scheme currently in use. Facklam (2002) has recently published a comprehensive set of tables for identifying the majority of the species described in this chapter. These are compiled from empirical data on relatively large numbers of human isolates examined (i.e., excluding animal and environmental isolates) at the Centers for Disease Control and Prevention (CDC), Atlanta, GA, USA, and as such may more accurately represent the properties of the human pathogens.

Further reading

Cleary, P. and Q. Cheng. 2002. Medically important beta-hemolytic streptococci. *In* Dworkin (Editor-in-Chief), The Prokaryotes: An evolving electronic resource for the Microbial Community. Springer www.springer-ny.com.

Hardie, J.M. and R.A. Whiley. 2002. The genus *Streptococcus* – oral. *In* Dworkin (Editor-in-Chief), The Prokaryotes: An evolving electronic resource for the Microbial Community. Springer www.springer-ny.com.

List of species of the genus *Streptococcus*

1. **Streptococcus pyogenes** Rosenbach 1884, 23[AL]

py.og′en.es. Gr. n. *pyum* pus; Gr. v. *gennaio* beget; N.L. adj. *pyogenes* pus-producing.

Constitutes Lancefield's (1933) group A streptococci (GAS). Cells are spherical and 0.5–1.0 μm in diameter; ovoid forms may occur usually in older cultures. Growth in chains of short or moderate length, in clinical material may also occur in pairs with long chains frequently observed in broth cultures. After overnight growth on blood agar three major visual colony types exhibiting β- (complete) hemolysis may form: mucoid, matte (dehydrated mucoid) or glossy. The colony type depends largely on production of hyaluronic acid and on growth conditions. Growth does not occur at 10°C, 45°C, in the presence of 6.5% NaCl, at pH 9.6, or in the presence of 40% bile. The optimum temperature for growth is 37°C. Growth is enhanced by supplementation of broth with blood or serum. A chemically defined medium has been described (Ginsburg and Grossowicz, 1957).

Energy metabolism is fermentative and the final pH in glucose broth is 4.8–6.0. Acid is produced from *N*-acetylglucosamine, fructose, glucose, galactose, lactose, maltose, salicin, sucrose, methyl D-glucoside, and trehalose. Acid is not produced from adonitol, amygdalin, arbutin, arabinose, arabitol, dulcitol, cyclodextrin, erythritol, gluconate, melibiose, methyl D-mannoside, sorbose, xylose, inulin, mannitol, raffinose, ribose, sorbitol, tagatose, melezitose, methyl D-xyloside.

Strains produce alkaline phosphatase and pyrrolidonyl arylamidase. Few strains produce β-glucuronidase. Strains do not produce α-galactosidase, β-galactosidase, β-glucosidase, *N*-acetyl-β-glucosaminidase, or glycyl-tryptophan arylamidase.

Arginine is hydrolyzed. Hippurate is not hydrolyzed. Esculin hydrolysis is a variable characteristic. Acetoin (Voges–Proskauer reaction) is not produced. Urease is not produced.

Chemically the cell wall is made up of four constituents: protein, polysaccharide, and peptidoglycan are interwoven rather than arranged in separate concentric layers as assumed previously; the fourth component is teichoic acid. Cell-wall thickness is about 20 nm. On the outer layer of the cell wall is a fringe of fimbriae, earlier referred to as a microcapsule. In some strains the cell wall with fimbriae can be enveloped by a capsule of hyaluronic acid. The rigidity of the cell wall is provided by peptidoglycan. The peptidoglycan molecule is of the A3 α-type, Lys–Ala$_{1-3}$ (Schleifer and Kandler, 1972), containing a polysaccharide polymer with tetrapeptide subunits that consist of L-Ala, D-iso-Gln, L-Lys, and D-Ala cross-linked through a tetrapeptide bridge L-Ala–L-Ala located between the terminal D-Ala and the subterminal L-Lys. Some of the peptide subunits are not cross-linked, possess one more D-Ala and thus form pentapeptides. The cell-wall polysaccharide ("C" polysaccharide), which is specific to the species (group A-specific), is attached to the peptidoglycan by phosphate-containing bridges composed of one or more units having glycerol or glyceryl-rhamnoside as an organic moiety. The antigens of the protein layer are linked to the peptidoglycan by covalent bonds.

Cell-wall antigenic components. Antigenic components essential in classification and diagnosis are located mainly in the cell wall. The group A specificity of *Streptococcus pyogenes* is due to the presence of a polysaccharide, a multibranch structure and composed of *N*-acetylglucosamine and rhamnose in a molar ratio of 1:2.47, formula weight 8000. The immunodominant component of the polysaccharide is *N*-acetylglucosamine, located in the terminal positions of the molecule. Enzymic or chemical degradation of this compound unmasks the subterminal rhamnose–rhamnose structure, which is also antigenic and represents the immunodeterminant structure of the "group A-variant" polysaccharide.

M-protein type antigen. These are situated at the cell surface and constitute heat-stable, trypsin-sensitive protein fibrils extending some 50–60 nm from the cell wall at the N terminus, mainly of an alpha-helical coiled-coil structure with heptad repeats (Fischetti, 1989). From the end of the molecule distal to the cell surface the M protein consists of a non-helical region followed by a coiled-coil region extending for some 50 nm to the cell wall. A proline–glycine-rich region is located within the peptidoglycan with a short section of the coiled-coil rod within the group carbohydrate portion of the wall. The anchor region or membrane anchor segment extends through the cell membrane projecting a charged tail of six amino acid residues into the cytoplasm at the C terminus of the molecule. Those regions concerned with anchorage are highly conserved and non-immunogenic. Domains in the middle of the M protein sequence are weakly immunogenic, semiconserved tandemly repeated linear sequences. The region distal to the cell surface is hypervariable, immunogenic, and responsible for the type specificity of these organisms. Lysin-extracted M protein from GAS strains examined by Western blotting using a broadly cross-reactive monoclonal antibody demonstrated that different serotypes varied in size within the range 41–80 kDa with M6 strains isolated over a period of 40 years exhibiting a similar size range. This has been explained on the basis of the presence of extensive DNA sequence repeats leading to the likelihood of recombinational events or replicative slippage which

generate deletions and duplications within the M-protein gene (*emm*) leading to the production of different sized M proteins. More than 80 different serological M protein types within GAS are recognized, each of which is associated with specific T-antigen patterns, and together typing based on these antigens has formed a basis for epidemiological studies of GAS. M protein serotype identification was originally described by Lancefield (1928) in a tube precipitin test using standardized typing antisera. The approach is prone to difficulties and impracticalities including the cost of antiserum preparation for all known serotypes, raising antisera with sufficiently high titer against opacity-factor-positive strains, ambiguities in results, the expression of new and therefore untypable M proteins by strains, and untypability due to lack of reactivity with available antisera (Beall et al., 1997, 1996; Facklam, 1977). Rapid sequencing *emm*-specific PCR products (cycle sequencing) of the variable 5′ *emm* gene regions in order to accurately deduce *emm* types corresponding to GAS M serotypes has proven to be a practical alternative approach. This has been shown to be effective in surveying the sequence variability of the M virulence protein and combined with T typing to be a useful tool for monitoring GAS diversity (Beall et al., 1997, 1996).

T-protein antigens. The T protein antigen is a relatively stable, trypsin-resistant surface protein of unknown function but is well defined in >95% of GAS, which provides a valuable additional strain characteristic for epidemiological studies. It is detected by the agglutination reaction using T-specific antisera. Although expressed independently of M proteins certain T proteins occur only in association with specific M serotypes (Beall et al., 1997, 1998) and in some instances more than one T protein is to be found on a single strain (Griffith, 1934). Strains producing untypable or no M protein may be typed by T protein typing. The T6 protein structural and sequence characteristics have been described in detail (Schneewind et al., 1990) and have revealed a protein consisting of 537 amino acids (M_r 57,675) with an N terminus exhibiting features of a typical signal peptide and a C terminus having a high degree of homology with the membrane anchor region of surface proteins found in other Gram-positive bacteria, including streptococcal M protein, wall-associated protein precursor wapA from *Streptococcus mutans* and staphylococcal protein A. The T6 protein appears to be a compact molecule, unlike the fibrillar M protein and despite the presence of typical cell-surface-associated C-terminal domains. It also contains an unusually high proportion (21%) of serine and threonine residues.

R protein antigens are also trypsin-resistant and occur in some strains of some types only. They have been found in some strains of Lancefield groups other than A (i.e., groups B, C, and G) and have been identified in four antigenic forms, the most common being R28 and R3 (Wilkinson, 1972). They have no known relation to virulence.

Pathogenicity. *Streptococcus pyogenes* (GAS) is an important human pathogen which colonizes the throat or skin, causing suppurative infections and non-suppurative sequelae. GAS cause pharyngitis, scarlet fever, impetigo, cellulitis including erysipelas septicemia, and invasive diseases including streptococcal "toxic-shock" syndrome and necrotizing fasciitis. Post-infection conditions include glomerulonephritis (a seasonal condition associated with skin strains), acute rheumatic fever and rheumatic heart disease, and streptococcal reactive arthritis. Recent associations with GAS include Tourette's syndrome, tics, and movement and attention-deficit disorders (Cunningham, 2000). Patterns of association between M serotypes and clinical conditions are discernable: M serotypes including M types 1, 3, 5, 6, 14, 18, 19, and 24 are associated with throat infection and rheumatic fever, M types 2, 49, 57, 59, 60, and 61 with pyoderma and acute glomerulonephritis (Bisno, 1980, 1995; Stollerman, 1975, 1997) and several M serotypes are associated with invasive and toxic disease including M types 1, 3, 11, 12, and 28.

Extracellular surface molecules and virulence factors. GAS are surrounded with a hyaluronic acid capsule. This consists of glucuronic acid and N-acetylglucosamine repeating units, is weakly immunogenic due to antigenic similarity with the host and serves to make the bacterium resistant to phagocytosis. Experiments have shown non-encapsulated GAS mutants to be important in adherence in the pharynx through binding to CD44 on epithelial cells and also to be less virulent in a mouse skin model (Schrager et al., 1998, 1996).

M proteins function in the resistance of GAS to phagocytosis and are a primary virulence mechanism for survival in host tissues. This is at least in part due to the ability to bind plasma fibrinogen at the N terminal of the M protein which in turn effectively masks complement receptors on the streptococcal cell, thus preventing opsonization via the alternative complement pathway (Jacks-Weis et al., 1982). M proteins are divided into Class I and Class II based on their reaction with antibodies against the M protein C repeat region. Class I M protein serotypes have been reported to contain the surface-exposed epitope that reacts with antibodies against the C repeat region and are opacity factor negative while the opposite is true for Class II M proteins. Studies have shown an association between the presence of the Class I epitope and the production of rheumatic fever by strains (Bessen et al., 1989, 1990). Antibodies to the M protein confer immunity to infection but persons may be repeatedly infected with GAS of different M protein serotypes. Multilocus sequence typing (MLST) based on housekeeping genes has indicated that *Streptococcus pyogenes emm* types define clones or clonal complexes (Enright et al., 2001)

Lipoteichoic acid (LTA) an amphipathic molecule associated with M proteins is thought to account for around 60% of the adhesion of GAS to epithelial cells through interaction with fatty-acid binding sites on fibronectin and the epithelial cell surface. At least 11 adhesins have been proposed as contributors to GAS adhesion (Cunningham, 2000).

Plasminogen-binding proteins. These are thought to enhance the invasive potential of GAS by enabling movement through normal tissue barriers. Plasminogen-plasmin-binding proteins include extracellular streptokinase together with glyceraldehyde-3-phosphate dehydrogenase and enolase on the cell surface (Lottenberg et al., 1994; Pancholi and Fischetti, 1998; Winram and Lottenberg, 1996). Streptokinase has been associated with the pathogenesis of acute post-streptococcal glomerulonephritis (Holm et al., 1992).

Exotoxins and superantigens. Factors associated with invasive disease include pyrogenic exotoxins A, B, C, and F (Norrby-Teglund et al., 1994a), and streptococcal superantigen (Mollick et al., 1993; Norrby-Teglund et al., 1994a, 1994b). The latter include several recently described superantigens with strong mitogenic activity (Kamezawa et al., 1997; Proft et al., 1999). The genes for pyrogenic exotoxins A and C are bacteriophage-encoded (Weeks and Ferretti, 1984) while exotoxin B (streptococcal proteinase) is chromosomally encoded. Exotoxins A, B, and C are probably important in the clinical manifestations of streptococcal toxic-shock syndrome. They are also responsible for the signs seen in cases of scarlet fever which include rash, strawberry tongue, and desquamation of the skin. Exotoxin B may contribute to survival of GAS in the host through increased resistance to phagocytosis and dissemination to organs (Lukomski et al., 1998).

C5a peptidase inhibits the recruitment of phagocytes to infection sites by cleavage and inactivation of the complement-derived chemotaxin C5a at its polymorphonucleargranulocyte (PMN)-binding site.

Streptolysin O (SLO) and streptolysin S (SLS) are among the GAS proteins involved in the breakdown of host tissues and cells and are able to lyse red blood cells, inhibit normal cell function, and destroy cells and tissues. Streptolysin O (SLO) is a secreted, oxygen-sensitive, thiol-activated, multi-subunit pore-forming toxin that interacts with cholesterol in the target cell. SLO is related to other thiol-activated cytotoxins such as pneumolysin, listeriolysin, and perfringolysin. Streptolysin S (SLS), like SLO, is produced by the majority of *Streptococcus pyogenes* strains, but in contrast is oxygen-stable. SLS is responsible for the β-hemolytic reaction around colonies on blood agar and has recently been shown to be encoded by a contiguous nine-gene locus (*sagA–sagI*) and sequence homology suggests SLS to be related to the bacteriocin family of microbial toxins (Nizet et al., 2000).

Genome analysis. The complete genome sequence for M1 strain SF370 (GenBank accession no.AE004092) has been published (Ferretti et al., 2001) and can be accessed at http://www. genome.ou.edu/strep.html. The 1.85-Mbp sequence contains 1752 predicted protein-encoding genes of which approximately one-third were reported to have no identifiable function. The remaining predicted genes fell into known function categories. Forty-six encoded virulence factor associated genes were reported reflecting the versatile pathogenic potential of the organism. Virulence-factor-encoding genes were scattered throughout the genome with the exception of those in the *emm* region although this region does not appear to have the organization typical of a pathogenicity island. A minimum of six new superantigen-like protein-encoding genes were found associated in many cases with mobile genetic elements. These proteins have at least one related protein in other Gram-positive species and thus suggest that they may have spread through horizontal gene transfer.

16S rRNA gene sequence analysis places this species within the Pyogenic species group.

DNA G+C content (mol%): 34.5–38.5 (method not stated); genome sequencing of M1 strain SF370 gave a mean G+C content of 38.5mol%.

Type strain: SF 130, ATCC 12344, NCTC 8198, DSM 20565, IMET 3002, T1.

GenBank accession number (16S rRNA gene): AB002521.

2. **Streptococcus acidominimus** Ayers and Mudge 1922, 49[AL] a.ci.do.mi′ni.mus. L. adj. *acidus* sour, acid; L. sup. adj. *minimus* very least; N.L. masc. adj. *acidominimus* probably intended to mean that this organism produces the least amount of acid.

Cells are coccoid and occur in short chains. Growth on blood agar produces α-hemolysis. Grows only on complex media; minimal nutritional requirements are unknown. Growth does not occur in 6.5% NaCl-containing broth.

Strains are weakly fermentative with, in most cases, the final pH of carbohydrate-containing broths not falling below 6.0 making the determination of fermentation reactions difficult. Acid is produced from fructose, glucose, inulin, lactose, mannitol, and sucrose. Inulin and mannitol fermentation are both delayed. Adonitol, arabitol, arabinose, D-cyclodextrin, dulcitol, erythritol, D- and L-fucose, methyl α-D-glucoside, 2- and 5-ketogluconates, gluconate, glycogen, inositol, lyxose, methyl α-D-mannoside, pullulan, rhamnose, ribose, L-sorbose, sorbitol, starch, trehalose, D-turanose, D- and L-xylose, xylitol, and methyl β-xyloside are not fermented. Leucine arylamidase, β-glucuronidase, pyrrolidonyl arylamidase, and alanine-phenylalanine-proline arylamidase are produced and 4-methyl-umbelliferyl (4MU)-phosphate, L-phenylalanine 7 amido-4-coumarin (AMC), and L-tryptophan AMC are hydrolyzed. Esculin is hydrolyzed but arginine and hippurate are not. Acetoin is not produced (Voges–Proskauer reaction). Alkaline phosphatase, β-mannosidase, and urease are not produced. No group-specific antigen has been demonstrated although reaction with group E antiserum has been reported previously (Wilson and Miles, 1975). The peptidoglycan type is Lys–Ser–Gly.

Strains are isolated from the bovine vagina, occasionally found on the skin of calves and in raw milk.

Similarities between the biochemical characteristics of this species and another member of the bovine genital flora, *Streptococcus pluranimalium*, have been investigated further by Devriese et al. (1999) who described improved identification criteria. Human isolates previously identified as *Streptococcus acidominimus* have been identified as *Facklamia sourekii* (Facklam, 2002; LaClaire and Facklam, 2000). On the basis of whole-cell-derived polypeptide patterns by SDS-PAGE and 16S rRNA gene sequence comparisons, only a distant relationship was found between these two species. 16S rRNA analysis shows a loose association between *Streptococcus acidominimus*, *Streptococcus suis*, and *Streptococcus entericus* (Vela et al., 2002), although insufficient for them to be regarded as forming a species group.

DNA G+C content (mol%): 39.7 (T_m).

Type strain: ATCC 51725, DSM 20622, CCUG 27296, CIP 82.4, LMG 17755, NCIMB 702025, NCTC 12957.

GenBank accession number (16S rRNA gene): X58301.

3. **Streptococcus agalactiae** Lehmann and Neumann 1896, 126[AL] nom. cons. Opin. 8 Jud. Comm. 1954, 152.

a.ga.lac′ti.ae. Gr. n. *agalactia* want of milk; N.L. gen. n. *agalactiae* of agalactia.

Constitutes Lancefield's (1933) group B streptococci. Cells are spherical or ovoid, 0.6–1.2 μm in diameter, occurring in chains of seldom less than four cells and frequently very long. Grows readily on blood agar and exhibits various types of hemolysis viz. typical β-hemolysis, but with a narrow zone, α-double zone, or no hemolysis. Some strains produce a characteristic opaque β-hemolysis, probably due to a soluble hemolysin with low hemolytic activity. This hemolysin is different from both hemolysin O and S. The CAMP factor (Christie et al., 1944) produced by most group B streptococci binds to erythrocyte membrane altered by *Staphylococcus aureus* sphingomyelinase C and this results in enhanced lysis of erythrocytes. The CAMP factor is not totally specific for group B streptococci since it has also been found in group C, F, and G streptococci. Many strains can grow in media containing 40% bile. Some strains produce a yellow, orange, or brick-red pigment and production may be enhanced by addition of starch to the medium or by anaerobic incubation. Energy-yielding metabolism is fermentative with lactic acid constituting the major end product. The final pH in glucose broth is 4.2–4.8. Acid is produce from glucose, maltose, ribose, sucrose, and trehalose. Glycerol is fermented only aerobically. Acid from lactose and salicin are variable characteristics of this species. Acid is not produced from xylose, arabinose, raffinose, inulin, mannitol, or sorbitol. Strains produce alkaline phosphatase. Production of α-galactosidase and β-glucuronidase are variable characteristics. Pyrrolidonyl arylamidase is not produced. Acetoin (Voges–Proskauer reaction) has been reported both as a positive (Hardie, 1986a) and as a negative (Facklam, 2002) characteristic. Strains hydrolyze arginine and hippurate but not esculin.

The group B-specific polysaccharide antigen is composed of rhamnose, *N*-acetylglucosamine, and galactose (Curtis and Krause, 1964). Serological cross-reaction may be encountered with group G strains of streptococci due to the presence of rhamnose in the immunodominant portion of the polysaccharides. The peptidoglycan type is Lys–Ala$_{1-3}$(Ser).

Serotypes Ia, Ib, II, III, IV, V, and VI are recognized based on the capsular polysaccharide antigen and surface protein antigen. The serotype-specific capsular polysaccharides are essential for pathogenesis (Rubens et al., 1987). Types I, II, III, and V cause the majority of neonatal disease in humans (Baker, 2000; Blumberg et al., 1996). Serotype III strains are important in that they cause approximately 60% of invasive neonatal infections although they represent only approximately one-third of strains isolated from asymptomatically colonized individuals (Baker and Barrett, 1974; Dillon et al., 1987).

Pathogenicity and virulence factors. Initially recognized as a cause of puerperal sepsis, in man group B streptococci can produce a variety of clinical conditions, some of them very serious. Disease can be invasive (normally defined as occurring when group B streptococci can be isolated from usually sterile site such as blood or cerebrospinal fluid) or non-invasive. Among pregnant women, infection ranges from mild urinary tract infection, amnionitis, and endometritis to sepsis and meningitis. Wound infections, cellulitis, fasciitis, endocarditis, and osteomyelitis may occur. In neonates and children may cause sepsis, pneumonia, meningitis, and pyelonephritis. Neonatal infections are divided into early and late-onset. In the former, occurrence is usually within the first 7 d of life where prolonged rupture of the membranes and chorioamnionitis features pneumonia, septicemia, or meningitis. The organism may be cultured from the mother's genital tract. Late-onset disease occurs in the first 3 months of life in otherwise healthy individuals and features septicemia, meningitis, and other foci of infection.

Fatality rates in both early and late-onset disease is around 5.8% of cases in the USA (Schuchat, 1998; Zangwill et al., 1992). In animals these organisms represent a highly contagious, economically important, obligate pathogen of the mammary gland and are one of the main causes of bovine mastitis. Sources of infection in man are the vaginal mucosa, the upper respiratory tract, urine, stool, and, in animals, milk and udder tissue. Virulence factors include the serotype-specific polysaccharide capsule, secreted enzymes hyaluronidase and C5a-peptidase, and a surface-associated beta-antigen protein able to bind human immunoglobulin A via the Fc portion. Experiments involving a gene knock-out mutant of the superoxide dismutase gene (*sodA*) in *Streptococcus agalactiae* demonstrated increased susceptibility to killing by macrophages and therefore indicated production of superoxide dismutase to be a contributing factor in virulence (Poyart et al., 2001). Restriction digest pattern (RDP) analysis of serotype III strains divided these into three separate lineages. RDP type III-3 strains differed in relatively circumscribed areas of the genome containing virulence genes *scpB* (C5A-ase), *lmb* (laminin-binding protein), *hylA* (hyaluronate lyase), and *cps* (putative cell-surface protein) (Bohnsack et al., 2002).

Taxonomic note and further comment: A non-hemolytic streptococcus pathogenic for fish, causing meningoencephalitis, was described and named *Streptococcus difficile* (Eldar et al., 1994). Despite relative binding ratios as low as 30.8 % with *Streptococcus agalactiae* in DNA–DNA hybridization experiments, later studies have placed these strains within *Streptococcus agalactiae*, belonging to the capsular type Ib (Berridge et al., 2001; Vandamme et al., 1997). RFLP analysis of whole-cell DNA probed with rRNA, fragments of genes encoding hyaluronidase, C5a peptidase, α-antigen, β-antigen, and two randomly selected genomic DNA fragments combined with data from MLEE studies and serotyping demonstrated *Streptococcus agalactiae* to be predominantly clonal in population structure, consisting of six lineages. Association was found between specific lineages and putative virulence factors but not with pathogenic potential (Hauge et al., 1996). It was notable that capsular serotype III strains divided into two distant evolutionary lineages, one of which lacked expression of hyaluronidase activity. MLEE analysis has also revealed two lineages within group III strains (Quentin et al., 1995).

Kilpper-Bälz and Schleifer (1984) demonstrated by DNA–DNA hybridization that a strain of Lancefield group M belonged to *Streptococcus agalactiae*.

16S rRNA sequencing places this species within the Pyogenic species group.

Genome analysis. The genomes of serotype V strain 2603 V/R (2.16 Mbp) (Tettelin et al., 2002) and serotype III strain NEM316 (ATCC 12403) (2.21 Mbp) (Glaser et al., 2002) have been sequenced, 2175 and 2118 predicted genes were identified, respectively, of which 61.3% (2603 V/R) and 62% (NEM 316) were assigned predicted functions, 28.6% (2603 V/R) and 25% (NEM 316) matched proteins of unknown function, and 10% (2603 V/R) and 9% (NEM 316) were unique to *Streptococcus agalactiae*.

These studies revealed extensive similarity with the human pathogenic streptococci *Streptococcus pyogenes* and *Streptococcus pneumoniae*. Many of the genes unique to *Streptococcus agalactiae* were associated with islands containing putative virulence genes and mobility elements supporting the acquisition of virulence traits via horizontal gene transfer.

Comparative genome hybridizations using microarray technology revealed considerable heterogeneity among strains, even of the same serotype, providing evidence that genetic acquisition, duplication, and reassortment events have provided this species with the diversity required for it to adapt to new environments and become a successful human pathogen (Tettelin et al., 2002).

DNA G+C content (mol%): 34 (T_m).

Type strain: G19, ATCC 13813, CCUG 4208, CIP 103227, DSM 2134, JCM 5671, LMG 14694 NCTC 8181.

GenBank accession number (16S rRNA gene): AB002479, AB175037, X59032.

4. **Streptococcus alactolyticus** Farrow, Kruze, Phillips, Bramley and Collins 1985, 224[VP] (Effective publication: Farrow, Kruze, Phillips, Bramley and Collins 1984, 480.)

a.lac.to.ly'ti.cus. Gr. pref. *a* not; L. n. *lac lactis* milk; Gr. adj. *lyticus* dissolving; N.L. masc. adj. *alactolyticus* not milk-digesting.

Cells are cocci, occurring mostly in pairs or short chains. Colonies on blood agar are circular, smooth and entire, α-hemolytic, or non-hemolytic. Growth occurs at 45°C but not at 50°C or in 6.5% NaCl.

Chemo-organotroph: metabolism fermentative. Acid is produced from cellobiose, fructose, galactose, D-glucose, D-mannose, maltose, *N*-acetylglucosamine, D-raffinose, and sucrose. The majority of strains produce acid from melibiose and salicin. Acid is not produced from D-arabitol, L-arabitol, D-arabinose, L-arabinose, adonitol, cyclodextrin, 2-ketogluconate, 5-ketogluconate, dulcitol, erythritol, D-fucose, L-fucose, gluconate, glycerol, glycogen, inulin, lactose, D-lyxose, methyl α-D-mannoside, methyl β-xyloside, melezitose, pullulan, rhamnose, ribose, L-sorbose, sorbitol, D-tagatose, D-xylose, L-xylose, and xylitol. Acid production from amygdalin, arbutin, mannitol, methyl α-D-glucoside, starch, trehalose, and D-turanose are variable characteristics of this species. Strains produce urease, α-galactosidase, and leucine arylamidase. Alkaline phosphatase, arginine dehydrolase, β-galactosidase, β-glucuronidase, glycyl tryptophan arylamidase, *N*-acetyl-β-glucosamine, and pyrrolidonyl arylamidase are not produced. Esculin is hydrolyzed but hippurate is not. Acetoin is produced (Voges–Proskauer reaction). Strains contain the Lancefield group D antigen. Occasionally strains belong to Lancefield group G (Robinson et al., 1988; Vandamme et al., 1999). Strains are isolated from the intestines of pigs and feces of chickens.

DNA G+C content (mol%): 39.9–41.3 (T_m).

Type strain: GP2, ATCC 43077, CCUG 27297, CIP 103244, DSM 20728, HAMBI 1719, LMG 14808, NCIMB 701091.

GenBank accession number (16S rRNA gene): AF201899, X58319.

Additional comments: Ureolytic streptococci isolated from pig feces and colons and with broadly similar biochemical characteristics were proposed as *Streptococcus intestinalis* (Robinson et al., 1988). Differences in phenotype between *Streptococcus intestinalis* and *Streptococcus alactolyticus* (Farrow and Collins, 1984a) included β-hemolysis on blood agar, possession of Lancefield group G antigen or ungroupable against Lancefield grouping antisera, together with the inability to produce acid from mannitol and raffinose, for the former. In a subsequent study (Vandamme et al., 1999), strains representing both species were studied by biochemical testing and by comparison of whole-cell-derived polypeptide patterns by SDS-PAGE. These authors found strains of both species to be "phenotypically indistinguishable" and were unable to confirm the reported β-hemolysis given on blood agar or reaction of some strains to Lancefield group G antiserum and therefore concluded that the name *Streptococcus intestinalis* (Robinson et al., 1988) is a junior synonym of *Streptococcus alactolyticus*.

16S rRNA gene sequence analysis places *Streptococcus alactolyticus* within the Bovis species group.

5. **Streptococcus anginosus** (Andrewes and Horder 1906) Smith and Sherman 1938, 189[AL] emend. Whiley and Beighton 1991, 4 (Andrewes and Horder 1906, 367.)

an.gi.no'sus. N.L. adj. *anginosus* pertaining to angina.

Cells are cocci, small (0.5–1.0 μm diameter), forming short chains. Colonies on blood (horse) agar are 0.5–2.0 mm in diameter, white or translucent, convex, and entire; some strains produce colonies that are 0.5–1.0 mm in diameter, white, and matte. Growth is reduced under aerobic conditions, is frequently enhanced by the addition of CO_2 with some strains requiring anaerobic incubation. Most strains give α-hemolysis or no hemolysis on blood agar with some strains giving β-hemolysis. No extra-cellular polysaccharides are produced on sucrose-containing medium. Strains produce alkaline phosphatase and leucine arylamidase. Virtually all strains produce β-glucosidase. α-Galactosidase and α-glucosidase are sometimes produced. Strains do not produce α-fucosidase, β-fucosidase, β-glucuronidase, β-*N*-acetylglucosaminidase, β-*N*-acetylgalactosaminidase, pyrrolidonyl arylamidase, or sialidase. Over 50% of strains have been reported to produce β-galactosidase when incubated in CO_2 (Ahmet et al., 1995). Acid is produced from amygdalin, arbutin, maltose, and glucose, and is frequently produced from cellobiose, lactose, salicin, and trehalose. Strains of this species that are able to produce acid from mannitol, raffinose, and melibiose are frequently associated with the human female genital tract. Strains do not ferment arabinose, glycerol, glycogen inulin, ribose, sorbitol, or tagatose. Arginine and esculin are hydrolyzed but urea and hippurate are not. Acetoin is produced (Voges–Proskauer reaction). Hydrogen peroxide production is sometimes detected. Hyaluronidase is produced by a few strains only. The majority of strains are either serologically ungroupable or belong to Lancefield group F although strains can possess the Lancefield group A, C, or G antigen. The cell-wall peptidoglycan type is Lys–Ala$_{1–3}$. Strains are isolated from the human oral cavity, upper respiratory tract, and vagina. Strains of this species are isolated from a variety of human purulent infections.

DNA G+C content (mol%): 38–40 (T_m).

Type strain: Havill III, ATCC 33397, CCUG 27298, CCUG 35776, CIP 102921, DSM 20563, HAMBI 1525, LMG 14502, NCTC 10713.

GenBank accession number (16S rRNA gene): AF104678, X58309.

Additional comments: The often confusing history of the classification of strains currently designated *Streptococcus anginosus* together with the collection of closely related strains variously referred to in the past as "*Streptococcus milleri*" (Guthof, 1956), "*Streptococcus MG*" (Mirick et al., 1944), the minute-colony-forming streptococci of Lancefield groups F and G (Bliss, 1937; Long and Bliss, 1934), hemolytic and non-hemolytic isolates possessing the "type antigens" of Lancefield group F (Ottens and Winkler, 1962), *Streptococcus intermedius* (Holdeman and Moore, 1974), *Streptococcus anginosus* (Andrewes and Horder, 1906; Deibel and Seeley, 1974), *Streptococcus constellatus* (Holdeman and Moore, 1974), "*Streptococcus anginosus–constellatus*", "*Streptococcus MG–intermedius*" (Facklam, 1977), the "*Streptococcus milleri*-group" (Gossling, 1988), and *Streptococcus anginosus*, *Streptococcus constellatus*, and *Streptococcus intermedius* (Facklam, 1984, 1985), is detailed in the previous section on the Anginosus species group.

DNA–DNA hybridization studies have demonstrated centers of variation within strains designated *Streptococcus anginosus*, requiring further characterization before formal taxonomic proposals can be made (Whiley et al., 1999).

16S rRNA gene sequence analysis places *Streptococcus anginosus* within the Anginosus species group.

6. **Streptococcus australis** Willcox, Zhu and Knox 2001, 1281[VP]

aus.tra'lis. L. masc. adj. *australis* of the south, southern, relating to the region in which the organism was isolated.

Cells are cocci, forming short chains. Colonies are approximately 0.5–1.0 mm on blood agar plates and are α-hemolytic on Columbia blood (sheep) agar. Facultatively anaerobic. The following biochemical characteristics were obtained using Rapid ID32 Strep kits (bioMerieux). All strains produce alkaline phosphatase, alanine-phenylalanine-proline arylamidase, and glycyl-tryptophan arylamidase, produced acid from lactose, maltose, pullulan, and sucrose, and hydrolyzed arginine. No strains are able to produce acetoin (Voges–Proskauer reaction), α-galactosidase, β-glucosidase, β-glucuronidase, methyl β-D-glucopyranoside, pyroglutamic acid arylamidase, *N*-acetyl-β-glucosaminidase, β-mannosidase, or urease. β-Galactosidase production is dependent on the substrate used: 2-naphthyl β-D-galactopyranoside (−), or *p*-nitrophenyl β-D-galactopyranoside (+). Strains are unable to produce acid from L-arabinose, D-arabinose, glycogen, mannitol, melibiose, melezitose, methyl β-D-glucopyranoside, raffinose, ribose, sorbitol, tagatose, or trehalose. Hippurate is not hydrolyzed. Neuraminidase is not produced. Strains have been isolated from the saliva of healthy children between 8 and 12 years but not from adults. The peptidoglycan type is Lys–Ala–Gly. Strains do not react with Lancefield grouping antisera.

16S rRNA gene sequence analysis places this species within the Mitis species group.

DNA G+C content (mol%): 43.5 (T_m) (for the type strain AI-1[T]).

Type strain: AI-1, ATCC 700641, CIP 107167, NCTC 13166.

GenBank accession number (16S rRNA gene): AF184974, AY485604.

7. **Streptococcus canis** Devriese, Hommez, Kilpper-Bälz and Schleifer 1986, 424[VP]

ca′nis. L. gen. n. *canis* of a dog.

Cells are cocci, occurring in pairs or short chains. Wide zones of β-hemolysis produced on blood (sheep) agar and colonies are large, circular with entire edges. A precipitate and clear supernatant are formed in broth cultures. CAMP factor-negative. Growth does not occur in the presence of 6.5% NaCl or 40% bile. Strains produce acid from *N*-acetylglucosamine, D-fructose, galactose, D-glucose, lactose, maltose, D-mannose, ribose, salicin, starch, and sucrose. Acid is not produced from adonitol, amygdalin, L-arabinose, D-arabinose, D-arabitol, dulcitol, erythritol, D-fucose, L-fucose, β-gentiobiose, gluconate, 2-ketogluconate, inositol, inulin, D-lyxose, mannitol, methyl α-D-mannoside, melezitose, melibiose, D-raffinose, rhamnose, sorbitol, L-sorbose, D-tagatose, D-turanose, xylitol, D-xylose, L-xylose, or methyl β-xyloside. Acid production from methyl D-glucoside is a variable characteristic of this species. A few strains produce acid from trehalose and some give weak acid from glycogen. Strains produce L-aminopeptidase, alkaline phosphatase, leucine arylamidase, and arginine dehydrolase. Most strains produce β-galactosidase, a few produce hyaluronidase and β-glucuronidase. α-Galactosidase is a variable characteristic of this species. Strains do not produce pyrrolidonyl arylamidase or acetoin (Voges–Proskauer reaction) and do not hydrolyze hippurate, fibrinolysin, tyrosine, or starch. Esculin and arginine are hydrolyzed.

The peptidoglycan type is Lys–Thr–Gly. Strains belong to Lancefield group G. Strains are isolated from the skin, upper respiratory tract, anus, and genitals of dogs, cow udders, and the genital tracts of female cats. Responsible for infections in animals including toxic shock and necrotizing fasciitis in dogs. *Streptococcus pyogenes* has been isolated from human infections (wound infection and bacteremia) and it has been speculated that the incidence of this streptococcus in the human population may be underestimated (Whatmore et al., 2001).

16S rRNA gene sequence analysis places this species within the Pyogenic species group.

DNA G+C content (mol%): 39–40 (T_m).

Type strain: STR-T1, ATCC 43496, CCUG 27661, CIP 103223, DSM 20715, LMG 15890.

GenBank accession number (16S rRNA gene): AB002483, X59061.

8. **Streptococcus constellatus** (Prévot 1924) Holdeman and Moore 1974, 266[AL] emend. Whiley, Hall, Hardie and Beighton 1999, 1448 ("*Diplococcus constellatus*" Prévot (1924).)

con.stel.la′tus. L. adj. *constellatus* studded with stars.

Cells are cocci, 0.5–1.0 µm in diameter, forming short chains. Colonies on blood agar are 0.5–2.0 mm in diameter, white or translucent, convex, and entire. Some strains produce colonies that are 0.5–1.0 mm diameter, white, and matte. Growth is reduced under aerobic conditions and is frequently enhanced by the addition of CO_2. Some strains require anaerobic conditions for growth. Alkaline phos-

phatase and leucine arylamidase are produced. Most strains produce α-glucosidase. A few produce β-galactosidase, β-glucosidase, β-fucosidase, β-*N*-acetylgalactosaminidase, and β-*N*-acetylglucosaminidase. Pyrrolidonyl arylamidase, α-galactosidase, β-glucuronidase, α-L-fucosidase, and sialidase are not produced. Arginine and esculin are hydrolyzed but urea and hippurate are not. Few strains produce hydrogen peroxide. Most produce hyaluronidase, but neuraminidase (sialidase) is not produced. Acetoin is produced (Voges–Proskauer reaction). Acid is produced from glucose and *N*-acetylglucosamine and is frequently produced from salicin and trehalose. Acid production from lactose and amygdalin are variable characteristics of this species. A few strains produce acid from arbutin, cellobiose, mannitol, melibiose, and raffinose. Strains do not produce acid from arabinose, glycerol, inulin, or sorbitol.

Strains are frequently either β-hemolytic and of Lancefield group F or non-hemolytic (non-β) and serologically ungroupable using Lancefield grouping antisera. Strains may also possess the Lancefield group A, C, or G antigens. Habitats are the oral cavity and upper respiratory tract in man. Strains are isolated from purulent infections in man.

For a description of the often confusing history of the classification of strains currently designated *Streptococcus constellatus* together with closely related strains, see section on the Anginosus species group.

Recently *Streptococcus constellatus* was divided into subsp. *pharyngis* and subsp. *constellatus* based on DNA–DNA hybridization and which display different clinical associations (Whiley et al., 1999).

8a. **Streptococcus constellatus subsp. constellatus** Whiley, Hall, Hardie and Beighton 1999, 1448[VP]

The description of this subspecies corresponds to that given above for *Streptococcus constellatus* with the exception that β-*N*-acetylglucosaminidase, β-*N*-acetylgalactosaminidase, β-D-glucosidase, β-galactosidase, and β-D-fucosidase are not produced. Lactose is sometimes fermented.

Strains of *Streptococcus constellatus* subsp. *constellatus* are recovered from a wide range of anatomical and infection sites compared to subsp. *pharyngis*.

DNA G+C content (mol%): 37–38 (T_m).

Type strain: ATCC 27823, CCUG 24889, CIP 103247, DSM 20575, LMG 14507, NCIMB 702226, NCTC 11325.

GenBank accession number (16S rRNA gene): AF104676, X58310.

8b. **Streptococcus constellatus subsp. pharyngis** Whiley, Hall, Hardie and Beighton 1999, 1448[VP]

pha. ryn′gis. Gr. n. *pharynx* throat; N.L. gen. n. *pharyngis* of the throat.

Strains are β-hemolytic and belong to Lancefield group C. β-*N*-acetylgalactosaminidase, β-D-fucosidase, β-galactosidase, and α- and β-glucosidases are produced. The majority of strains produce β-*N*-acetylglucosaminidase. Strains do not produce α-galactosidase, β-glucuronidase, pyrrolidonyl arylamidase, or sialidase (neuraminidase). Acid is produced from amygdalin, arbutin, lactose, glucose, and β-D-*N*-acetylglucosamine. Acid is not produced from inulin, mannitol, melibiose, raffinose, or sorbitol. The majority of strains produce hyaluronidase. Strains exhibit a predilection for sites within the throat and have mainly been isolated from human throat infections (pharyngitis) with one strain from an abdominal mass.

DNA G+C content (mol%): 35–37 (T_m).

Type strain: MM9889a, NCTC 13122.

GenBank accession number (16S rRNA gene): AY309095.

9. **Streptococcus criceti** corrig. Coykendall 1977, 28[AL]

cri′ce.ti. N.L. gen. n. *criceti* of the hamster.

Cells are coccoid, 0.5 μm in diameter, occurring in pairs or chains. Colonies on sucrose agar are approximately 1 mm in diameter, rough, heaped, often glossy and may be surrounded by liquid containing soluble extracellular glucan. Colonies on blood agar are 2–3 mm in diameter, smooth, and round with the majority of strains producing no hemolysis. Some strains produce α-hemolysis or are non-hemolytic. Strains are facultatively anaerobic, growing best in an atmosphere with reduced O_2 and added CO_2. Human isolates are susceptible to 2 U/ml bacitracin. Acid is produced from *N*-acetylglucosamine, amygdalin, arbutin, cellobiose, galactose, glucose, inulin, lactose, maltose, mannitol, mannose, melibiose, raffinose, salicin, sorbitol, and sucrose. Acid is not produced from arabinose, glycerol, glycogen, melezitose, methyl D-glucoside, rhamnose, ribose, starch, or xylose. Acid from lactose and trehalose are variable characteristics of this species. The final pH in glucose broth is 4.1–4.2. Ammonia is not produced from arginine. Acetoin is produced (Voges–Proskauer reaction). Esculin is usually hydrolyzed but hippurate and starch are not. Extracellular polysaccharide is produced from sucrose-containing medium. Hydrogen peroxide is not produced. Most strains possess the *a* antigen of the serological scheme of Bratthall (1970), but some rat isolates lack it (Coykendall, 1977). The peptidoglycan type is Lys–Thr–Ala.

Strains have been isolated from the mouth of hamsters, wild rats, and, occasionally, man.

16S rRNA gene sequence analysis places this species within the Mutans species group of cariogenic streptococci.

DNA G+C content (mol%): 42–44 (T_m).

Type strain: HS6, NCTC 12277, ATCC 19642, CCUG 27300, CIP 102510, DSM 20562, HAMBI 1517, LMG 14508.

GenBank accession number (16S rRNA gene): AJ420198, X58305.

Additional comments: Streptococcus criceti is the corrected name for strains previously called *Streptococcus cricetus* by Coykendall (1977) (Trüper and de Clari, 1998).

10. **Streptococcus cristatus** corrig. Handley, Coykendall, Beighton, Hardie and Whiley 1991, 546[VP]

cris′ta.tus. L. adj. *cristatus* ornamented by a crest.

Cells are coccoid, approximately 1 μm in diameter, and grow in chains. Cells have tufts of fibrils in a lateral position. Colonies grown for 2 d at 37°C on blood agar are grayish, white, glossy, entire, 1–2 mm in diameter, and are α-hemolytic. Glucan production on sucrose-containing medium is a variable characteristic. Acid is produced from arbutin, maltose, trehalose, and *N*-acetylglucosamine. The majority of strains produce acid from galactose and lactose but not from lactulose, or pullulan. Acid is not produced from amygdalin, arabinose, inulin, mannitol, melibiose, melezitose, raffinose, rhamnose, ribose, sorbose, sorbitol, tagatose, turanose, or xylitol. Strains are able to produce β-*N*-acetylglucosaminidase, β-*N*-acetylgalactosaminidase, and α-fucosidase. Strains do not produce α-arabinosidase, β-D-fucosidase, α-galactosidase, α-glucosidase, β-glucosidase, sialidase (neuraminidase), or alkaline phosphatase. Some strains produce β-galactosidase. Arginine is hydrolyzed by the majority of strains but esculin and urea are not. Acetoin is not produced (Voges–Proskauer reaction). Hydrogen peroxide is not produced. Salivary amylase is bound.

16S rRNA gene sequence analysis places this species within the Mitis species group.

DNA G+C content (mol%): 42.6–43.2 (T_m).

Type strain: CR311, ATCC 51100, CCUG 33481, CIP 105954, DSM 8249, LMG 16320, NCTC 12479.

GenBank accession number (16S rRNA gene): AB008313, AY188347, AY584476.

Additional comments: Streptococcus cristatus is naturally transformable, is involved in "corncob" formation within dental plaque in association with *Fusobacterium nucleatum* (Correia et al., 1996), and may play a significant role *in vivo* in the establishment of periodontopathic plaque through its ability to adhere to several periodontal pathogens (Yao et al., 1996).

Isolated from human throats and oral cavities.

Streptococcus cristatus is the corrected name for strains previously called *Streptococcus crista* by Handley et al. (1991) (Trüper and de Clari, 1997).

11. **Streptococcus devriesei** Collins, Lundström, Welinder-Olsson, Hansson, Wattle, Hudson and Falsen 2004, 631[VP] (Effective publication: Collins, Lundström, Welinder-Olsson, Hansson, Wattle, Hudson and Falsen 2004, 150.)

de.vries.e.i. N.L. gen. n. *devriesei* of Devriese, to honor Luc A. Devriese, a contemporary Belgian microbiologist, in recognition of his outstanding contributions to the taxonomy of streptococci and related organisms.

Strains are cocci, occurring singly, in pairs or short chains. α-hemolysis is formed on blood agar. Acid is produced (NVI Streptococci kit) from esculin, inulin, lactose, mannitol, raffinose, sucrose, salicin, and trehalose; sorbitol is a variable characteristic; glycerine is not fermented. Acid is produced (API 50 CH kit) from amygdalin, D-arabitol, arbutin, cellobiose, esculin, fructose, galactose, β-gentiobiose, glucose, gluconate, lactose, maltose mannose, melibiose, D-raffinose, salicin, sorbitol, sucrose, D-tagatose, and trehalose; no acid is formed from adonitol, D- or L-arabinose, 2-ketogluconate, 5-ketogluconate, erythritol, D- or L-fucose, glycerol, glycogen, inositol, D-lyxose, melezitose, methyl α-D-glucoside, methyl α-D-mannoside, methyl β-xyloside, *N*-acetylglucoasamine, D-ribose, rhamnose, starch, D-turanose, xylitol, or D- or L-xylose; some strains ferment dulcitol, inulin, and L-sorbose. Acid is produced (API Rapid ID 32 Strep system) from lactose, maltose, mannitol, melibiose, methyl β-D-glucopyranoside, D-raffinose, sorbitol, tagatose, trehalose, and sucrose; acid is not produced from D-arabitol, L-arabinose, cyclodextrin, glycogen, melezitose, *N*-acetylglucosamine, pullulan, or D-ribose. Alanyl-phenylalanine-proline arylamidase, α-galactosidase, β-glucosidase, β-mannosidase (weak) are produced. Strains do not produce alkaline phosphatase, arginine dihydrolase, β-galactosidase, β-glucuronidase, glycyl-tryptophan arylamidase, pyroglutamic acid arylamidase, or urease. Acetoin is produced (Voges–Proskauer reaction). Nitrate is not reduced.

Isolated from the teeth of horses.

16S rRNA gene sequence analysis places this species within the Mutans group of streptococci.

Type strain: CCUG 47155, CIP 107809.

DNA G+C content (mol%): 42 (HPLC).

GenBank accession number (16S rRNA gene): AJ564067.

12. **Streptococcus didelphis** Rurangirwa, Teitzel, Cui, French, McDonough and Besser 2000, 765[VP]

di.del′phis. N.L. Gr.-derived n. *Didelphis* taxonomic genus name of the American opossum; N.L. gen. n. *didelphis* of the opossum.

Strains are coccoid, forming chains in broth culture. Growth on Columbia blood agar (sheep blood) produces small, translucent colonies giving β-hemolysis. Does not grow on MacConkey agar. On isolation on blood agar and initial passages on non-blood-containing media, vigorous catalase activity is detected but is lost after several passages. All strains produce acid from ribose and trehalose. The majority are unable to produce acid from lactose and raffinose. Strains are unable to produce acid from arabinose, glycogen, inulin, mannitol, salicin, or sorbitol. Alkaline phosphatase, β-glucuronidase, and leucine arylamidase are produced. α-Galactosidase production is a variable characteristic of this species. β-Galactosidase and pyrrolidonyl arylamidase are not produced. Hippurate is hydrolyzed but esculin is not. The majority of strains hydrolyze arginine. Acetoin is not produced (Voges–Proskauer reaction). Strains are non-groupable with Lancefield antisera. Long-chain fatty acid methyl ester analysis has demonstrated the presence of 12:0, 14:0, 15:0, 16:1ω9c, 16:1ω5c, 16:0, 17:0 anteiso, 17:1ω8c, 17:0, 18:1ω9c, 18:1ω7c, 18:0, 20:4ω6,9,12, 15c, summed feature 3 (16:1ω7c/15 iso 2-OH and 15:0 iso 2-OH/16:1ω7c) and summed feature 5 (18:2ω6, 9c/18:0 anteiso and 18:0 anteiso/18:2ω6,9c), where summed feature are groups of FAMEs that cannot be resolved by G+C analysis due to peak overlap.

Streptococcus didelphis is isolated from the tissues of opossums with suppurative dermatitis and hepatic fibrosis.

16S rRNA gene sequence analysis demonstrated that the closest relative of *Streptococcus didelphis* is *Streptococcus dysgalactiae* within the Pyogenic species group.

Type strain: WADDL 94-11374-1, ATCC 700828, CCUG 45419, CIP 106980.

DNA G+C content (mol%): not determined.

GenBank accession number (16S rRNA gene): AF176103.

13. **Streptococcus downei** Whiley, Russell, Hardie and Beighton 1988, 27[VP]

down.e.i. N.L. gen. n. *downei* Downe, the village in Kent, United Kingdom, where the type strain was isolated.

Cells are coccoid, forming chains. Colonies on 5% sucrose agar (TYC) are large, 2–3 mm in diameter, white, and conical with an erose edge surrounded by a distinct halo. Colonies on Mitis-Salivarius agar are small, dark blue,

crinkled, up to 1 mm in diameter, with an erose edge that slightly pits the agar, but are easily dislodged though difficult to disperse. No growth occurs at 45°C, at pH 9.6, or in the presence of 6.5% NaCl. Variable growth is obtained on 10% and 40% bile agar. Acid is produced from glucose, sucrose, fructose, galactose, mannose, mannitol, lactose, maltose, salicin, trehalose, and inulin. Acid is not produced from adonitol, melezitose, melibiose, sorbose, cellobiose, glycogen, soluble starch, inositol, xylitol, sorbitol, glycerol, arabinose, pullulan, or raffinose. Starch, esculin, and hippurate are not hydrolyzed. Ammonia is not produced from arginine. Hydrogen peroxide and intracellular polysaccharides are not produced. Extracellular polysaccharide is produced on sucrose agar and in sucrose broth although could not be demonstrated by ethanol precipitation in the original description (Beighton et al., 1981). Acetoin is produced (Voges–Proskauer reaction). Strains possess a serologically distinct polysaccharide antigen designated h (Beighton et al., 1981). The peptidoglycan type is Lys–Thr–Ala. Cellular long-chain fatty acid composition consists of major amounts of hexadecanoic ($C_{16:0}$, palmitic), octadecanoic ($C_{18:0}$, stearic), octadecenoic ($C_{18:1}$, vaccenic), and eicosenoic ($C_{20:1}$) acids together with minor amounts of tetradecenoic ($C_{14:0}$, myristic), hexadeneoic ($C_{16:1}$, palmitoleic), octadecenoic ($C_{18:1}$, oleic), eicosanoic ($C_{20:0}$, arachidic), and cyclopropane (*cis*-9,10-methyleneoctadecanoic) acids. Strains have been isolated from the dental plaque of monkeys (*Macaca fascicularis*) and are cariogenic when monoassociated with germ-free rats.

16S rRNA gene sequence analysis places this specis within the Mutans species group.

DNA G+C content (mol%): 41–42 (T_m).

Type strain: MFe28, ATCC 33748, CCUG 24890, CIP 103222, DSM 5635, LMG 14514, NCTC 11391.

GenBank accession number (16S rRNA gene): AJ, 420200, AY188350, X58306.

14. **Streptococcus dysgalactiae** (*ex* Diernhofer 1932) Garvie, Farrow and Bramley 1983, 404[VP]

dys.ga.lac′ti.ae. Gr. pref. *dys* ill, hard; Gr. n. *galactia* pertaining to milk; N.L. n. *dysgalactia* loss or impairment of milk secretion; N.L. gen. n. *dysgalactiae* of dysgalactia.

Cells are coccoid or ovoid occurring in pairs or chains. Strains may produce α, β, or no hemolysis on blood agar. The optimum temperature for growth is approximately 37°C. Growth does not occur at 10°C or at 45°C, in 6.5% NaCl, or 0.1% methylene blue milk, or at pH 9.6. Does not survive heating at 60°C for 30 min. Chemo-organotroph: metabolism fermentative. Growth occurs only on complex media. Acid is produced from glucose, maltose, ribose, starch, sucrose, and trehalose. Most strains produce acid from glycerol when incubated aerobically. A few strains produce acid from sorbitol. Variable results are obtained for acid production from glycogen, lactose, and salicin. Acid is not produced from arabinose, inulin, mannitol, or raffinose. The majority of strains do not hydrolyze hippurate. Esculin hydrolysis is a variable characteristic. Most strains produce hyaluronidase. Arginine is hydrolyzed. Acetoin is not produced (Voges–Proskauer reaction). Leucine arylamidase and β-D-glucuronidase are produced. Alkaline phosphatase, α-galactosidase, and pyrrolidonyl arylamidase are not produced. β-Galactosidase is a variable characteristic of the species. Major non-hydroxylated long-chain fatty acids are hexadecanoic ($C_{16:0}$), and octadecenoic ($C_{18:1}$) acids; cyclopropane-ring-containing fatty acids are absent. Menaquinones are not produced. Peptidoglycan contains lysine as diamino acid (type Lys–Ala$_{1-3}$). Strains may react with Lancefield group C, G, or L (and occasionally A) antisera.

Habitats include the respiratory and genital tracts of various animals, miscellaneous infections of humans and domestic animals. Strains isolated from humans share several virulence determinants in common with *Streptococcus pyogenes* (Efstratiou, 1997), including M protein antigens (Efstratiou, 1997; Lawal et al., 1982; Maxted and Potter, 1967), C5a peptidase (Chen and Cleary, 1990), a hyaluronic acid capsule (Balke et al., 1985), production of streptokinase and streptolysin O (Okumura et al., 1994) and the ability to bind mammalian proteins (Tewodoros and Kronvall, 1993).

16S rRNA gene sequence analysis places *Streptococcus dysgalactiae* within the Pyogenic species group.

DNA G+C content (mol%): 38.1–40.2 (T_m).

Additional comments: The confused taxonomic history of strains included here is summarized in the section describing the Pyogenic species group. The weight of evidence supports the inclusion of strains previously designated *Streptococcus dysgalactiae* and *Streptococcus equisimilis* into a single species, each with subspecies status, i.e., *Streptococcus dysgalactiae* subsp. *dysgalactiae* and *Streptococcus dysgalactiae* subsp. *equisimilis*, respectively (Vandamme et al., 1996).

14a. **Streptococcus dysgalactiae subsp. dysgalactiae** Vandamme, Pot, Falsen, Kersters and Devriese 1996, 780^{VP}

The description for *Streptococcus dysgalactiae* subsp. *dysgalactiae* is as given above for the species with the exception that strains are α-hemolytic and non-hemolytic, isolated commonly from bovine sources. Strains may react with Lancefield group C antiserum. Streptokinase activity on human plasminogen and proteolytic activity do not occur. Hippurate is not hydrolyzed. Glycogen is not fermented. β-D-Galactosidase and α-L-glutamate aminopeptidase are not produced. Sorbitol fermentation is a variable characteristic of this subspecies. Bacitracin sensitive. Strains are commonly associated with bovine mastitis and are isolated from the bovine vagina.

DNA G+C content (mol%): 38.5–39.8 (T_m).

Type strain: 134, ATCC 43078, CCUG 27301, CIP 102914, DSM 20662, LMG 15885, LMG 16023, NCIMB 702023.

GenBank accession number (16S rRNA gene): AB002485, AY584478, X59030.

14b. **Streptococcus dysgalactiae subsp. equisimilis** Vandamme, Pot, Falsen, Kersters and Devriese 1996, 780^{VP}

The description for *Streptococcus dysgalactiae* subsp. *equisimilis* is as given above for the species with the exception that strains are β-hemolytic and may react with Lancefield group A, C, G, or L antisera. Streptokinase activity occurs on human plasminogen and proteolytic activity on human fibrin is a variable characteristic of this subspecies. Hippurate hydrolysis is a variable characteristic (see below). Acid is not produced from sorbitol. Glycogen fermentation is a variable characteristic. α-L-glutamate aminopeptidase and β-D-galactosidase are produced. Bacitracin sensitivity is variable.

Vieira et al. (1998) characterized *Streptococcus dysgalactiae* subsp. *equisimilis* strains as follows: Human groups C and G (glycogen fermentation and hippurate hydrolysis both negative, proteolytic activity observed on human fibrin, bacitracin-resistant); Animal group C (differing from human group C strains in being able to ferment glycogen, but unable to produce proteolytic activity on human fibrin); group L (as animal group C strains, but with hippurate hydrolysis a variable characteristic and strains sensitive to bacitracin).

Recently, workers have isolated Lancefield group A-antigen-carrying strains of *Streptococcus dysgalactiae* subsp. *equisimilis* from human blood (Bert and Lambert-Zechovsky, 1997; Brandt et al., 1999).

DNA G+C content (mol%): 38.1–40.2 (T_m).

Type strain: CCUG 36637, CCUG 36913, CIP 105120, LMG 16026, NCIMB 701356, NCTC 4540.

GenBank accession number (16S rRNA gene): AB008926.

15. **Streptococcus entericus** Vela, Fernández, Lawson, Latre, Falsen, Domínguez, Collins and Fernández-Garayzábal 2002, 668^{VP}

en.te′ri.cus. N.L. adj. *entericus* from Gr. n. *enteron* gut, pertaining to the gut.

Cells are coccoid, arranged in short chains. Colonies after 24 h incubation on blood agar are circular and 1 mm in diameter, and produce α-hemolysis. Growth occurs at 30°C, at 37°C, and at pH 9.6, but not at 10°C, 45°C, or in broth containing 6.5% NaCl. Bile-esculin test is negative.

Acid is produced from starch, cyclodextrin, glycogen, lactose, maltose, trehalose, sucrose, and methyl β-D-glucopyranoside. Acid is not produced from D-arabitol, L-arabinose, inulin, mannitol, melibiose, melezitose, pullulan, ribose, raffinose, sorbitol, or tagatose. Produces β-glucosidase, leucine arylamidase, β-galactosidase, alanine-phenylalanine-proline arylamidase, and glycyl-tryptophan arylamidase. Does not produce β-glucuronidase, α-galactosidase, alkaline phosphatase, pyroglutamic acid arylamidase, N-acetyl-β-glucosaminidase, or β-mannosidase.

Esculin is hydrolyzed but arginine, hippurate, and urea are not. Acetoin is not produced (Voges–Proskauer reaction).

The single strain examined reacts with Lancefield group D antiserum. Vancomycin-sensitive.

At the time of writing only two isolates, presumed to be of the same clinical strain, from a calf have been studied.

Isolated during post-mortem examination from feces and jejunum of a Holstein–Friesian calf diagnosed with catarrhal enteritis. Habitat is unknown.

16S rRNA gene sequence analysis does not place *Streptococcus entericus* within an established species group. *Streptococcus acidominimus* is the nearest described relative.

DNA G+C content (mol%): not reported.

Type strain: CECT 5353, CCUG 44616, JCM 12180.

GenBank accession number (16S rRNA gene): AJ409287.

16. **Streptococcus equi**   Sand and Jensen 1888, 436[AL]

e′qui. L. n. *equus* horse; L. gen. n. *equi* of a horse.

Cells are spherical or ovoid, occurring in pairs or chains. Growth on blood agar plates produces colonies that are surrounded by a wide zone of β-hemolysis. The optimum temperature for growth is around 37°C. Growth does not occur at 10°C or at 45°C, in 6.5% NaCl, in 10% bile, in 0.1% methylene blue, or at pH 9.6. Cells do not survive heating at 60°C for 30 min. Growth only occurs on complex media; minimal nutritional requirements are unknown. Chemo-organotroph: metabolism fermentative. Final pH in glucose broth is around 4.6–5.0. Acid is produced from glucose, glycogen, maltose, pullulan, salicin, sucrose, and starch. Acid production from lactose, trehalose, and sorbitol is a characteristic of some strains (see below). Acid is not produced from arabinose, glycerol, inulin, mannitol, melezitose, raffinose, or xylose. Reports of acid from ribose vary in the literature. Strains produce alkaline phosphatase, leucine arylamidase, and β-glucuronidase. Both α- and β-galactosidase, *N*-acetyl-β-glucosaminidase, and pyrrolidonyl arylamidase are not produced. Most strains do not produce β-galactosidase. Arginine is hydrolyzed. Esculin hydrolysis is a variable characteristic. Hippurate is not hydrolyzed. Gelatin is not liquefied. Acetoin is not produced (Voges–Proskauer reaction).

Strains react with Lancefield group C antiserum. The major non-hydroxylated long-chain fatty acids are hexadecanoic ($C_{16:0}$) and octadecenoic ($C_{18:1}$) acids; cyclopropane-ring-containing fatty acids are absent. Menaquinones are absent. Capsules may be produced.

Causes infections in animals and humans. Animal infections include equine strangles with the organisms being isolated from submaxillary glands and mucopurulent discharges of the upper respiratory system of horses and from their immediate environment. Causes bovine mastitis.

16S rRNA gene sequence analysis places *Streptococcus equi* within the Pyogenic species group.

The close relationship between strains previously designated *Streptococcus equi* and *Streptococcus zooepidemicus* has resulted in the proposal that they should be included within a single species but be given subspecies status (Farrow and Collins, 1984a). Details are given in the section on the Pyogenic species group.

The genome currently being sequenced is approximately 2.3 Mb, with a G+C content of around 41 mol%.

16a. **Streptococcus equi subsp. equi** (Sand and Jensen 1888) Howey, Lock and Moore 1990, 318[VP]

The description for this subspecies is as above with the exception that strains are not able to produce acid from lactose, ribose, trehalose, or sorbitol. The cell-wall peptidoglycan is type Lys–Ala$_{1-3}$. Isolated from horses with equine strangles.

DNA G+C content (mol%): 40.7 (type strain) (T_m).

Type strain: C 15, ATCC 33398, CCUG 23255, CIP 102910, DSM 20561, LMG 15886, NCTC 9682.

GenBank accession number (16S rRNA gene): AB002515, X58314.

16b. **Streptococcus equi subsp. zooepidemicus** (*ex* Frost and Engelbrecht 1936) Farrow and Collins 1985, 224[VP] (Effective publication: Farrow and Collins 1984a, 491.)

zo.ö.e.pi.démi.cus. Gr. n. *zoon* an animal; Gr. adj. *epidemius* among people; prevalent, epidemic; M.L. adj. *zooepidemicus* prevalent among animals.

The description for this subspecies is as above with the exception that strains are able to produce acid from lactose and sorbitol. Reports of acid from ribose vary in the literature. Acid from trehalose is a variable characteristic. The cell-wall peptidoglycan is type Lys–Ala$_{2-3}$.

Strains are isolated from the blood stream, inflammatory exudates, and lesions of diseased animals. Frequent cause of bovine mastitis. Isolated from human infections usually traceable to the consumption of contaminated dairy products (Facklam, 2002).

DNA G+C content (mol%): 41.3–42.7 (T_m).

Type strain: S 34, ATCC 43079, CCUG 23256, CIP 103228, DSM 20727, LMG 16030, NCIMB 701358, NCTC 4676,.

GenBank accession number (16S rRNA gene): AB002516.

Additional comments: The type strain is β-galactosidase-negative, and does not hydrolyze esculin or hippurate.

17. **Streptococcus equinus**   Andrewes and Horder 1906, 712[AL]

e.qui′nus. L. adj. *equinus* pertaining to a horse.

Cells are spherical or ovoid, occurring in pairs to moderately long chains. Colonies on blood agar are small, circular, and entire with the majority of strains producing an α-hemolytic reaction of varying intensity. Growth occurs at 45°C but not at 10°C or at 50°C. Some strains survive heating at 60°C for 30 min. No growth occurs in 6.5% NaCl, 0.1% methylene blue milk, or at pH 9.6. Growth occurs in 40% bile. Chemo-organotroph: metabolism fermentative. Acid is produced from glucose, amygdalin, arbutin, cellobiose, fructose, galactose, D-mannose, maltose, *N*-acetylglucosamine, salicin, and sucrose. The majority of strains produce acid from β-gentiobiose. Variable results are obtained for acid production from glycogen, inulin, lactose, melibiose, D-raffinose, starch and trehalose. Strains do not produce acid from D- or L-arabinose, D- or L-arabitol, 2-ketogluconate, 5-ketogluconate, dulcitol, erythritol, D- or L-fucose, glycerol, gluconate, inositol, D-lyxose, mannitol, methyl β-xyloside, melezitose, ribose, rhamnose, L-sorbose, sorbitol, D-turanose, D-tagatose, D- or L-xylose, or xylitol. Esculin is hydrolyzed but arginine, hippurate, and gelatin are not. Acetoin is produced (Voges–Proskauer reaction). Strains produce leucine arylamidase. Strains do not produce alkaline phosphatase, pyrrolidonyl arylamidase, β-galactosidase, or β-glucuronidase. Strains contain the Lancefield group D antigen.

Isolated mainly from animals, including the alimentary tract of cows, horse, sheep, and other ruminants. Rarely from human clinical specimens.

16S rRNA gene sequence analysis places *Streptococcus equinus* within the Bovis species group.

DNA G+C content (mol%): 36.2–38.6 (T_m).

Type strain: Hl2B, ATCC 9812, CCUG 27302, CIP 102504, DSM 20558, HAMBI 1572, JCM 7879, LMG 14897, NBRC 12553 NCTC 12969.

GenBank accession number (16S rRNA gene): AF429765, AJ301607, X58318.

Additional comments: The description given above encompasses Lancefield group D strains that would previously have been identified as *Streptococcus equinus* (Andrewes and Horder, 1906) and *Streptococcus bovis* (Orla-Jensen 1919, 137; *emend. mut. char.* Sherman and Wing 1937, 57[AL]). DNA–DNA hybridization has led to the recognition that the names *Streptococcus equinus* and *Streptococcus bovis* are subjective synonyms with the specific epithet *Streptococcus equinus* having priority.

The "bovis/equinus" complex currently includes the following species: (i) *Streptococcus equinus* (Farrow et al., 1984), (ii) *Streptococcus gallolyticus* (Osawa et al., 1995), (iii) *Streptococcus infantarius* (Bouvet et al., 1997; Poyart et al., 2002), (iv) *Streptococcus alactolyticus* (Farrow et al., 1984), (v) *Streptococcus lutetiensis*, and (vi) *Streptococcus pasteurianus* (Poyart et al., 2002). The classification of the streptococci included here is detailed in the section on the Bovis group.

18. **Streptococcus ferus** Coykendall 1983, 883[VP] emend. Baele, Devriese, Vancanneyt, Vaneechoutte, Snauwaert, Swings and Haesebrouck 2003, 145.

fe'rus. L. adj. *ferus* wild, referring to wild rats from which the organism was isolated.

Cells are relatively small, lanceolate coccobacilli that occur singly or in pairs or forming short chains. Colonies on blood agar are small, dry, white, and adhering to, pitting, and corroding the agar surface. Growth is slightly enhanced in the presence of 5% CO_2, is strongly inhibited at 25°C and to a lesser extent at 30°C. Growth is equally good at 42°C and 37°C. Cells precipitate in brain heart infusion broth and strains are unable to grow in the presence of 6.5% NaCl or 20 μg/ml bacitracin. Growth occurs in the presence of 0.0013 g/l crystal violet plus 0.33 g/l thallous sulfate (Edwards medium) and in the presence of 0.4 g/l sodium azide plus 0.1 g/l tetrazolium chloride (Slanetz–Bartley medium) where white colonies are formed. Strains are resistant to bile (bile–esculin agar). Acid is produced from *N*-acetylglucosamine, esculin, amygdalin, cellobiose, D-fructose, galactose, D-glucose, glycogen, β-gentiobiose, lactose, D-mannose, maltose, maltotriose, sucrose, salicin, starch, and trehalose. Variable results are obtained for acid from D-tagatose, arbutin, inulin, mannitol, melibiose, D-raffinose, and sorbitol. Raffinose fermentation was previously recorded as a negative characteristic for this species (Coykendall, 1977, 1983), but strains isolated from pigs are able to produce acid from this carbohydrate (Baele et al., 2003). No acid is produced from adonitol, D- or L-arabinose, D- or L-arabitol, dulcitol, erythritol, D- or L-fucose, gluconate, glycerol, inositol, 2- or 5-ketogluconate, D-lyxose, melezitose, methyl β-glycoside, methyl α-D-glycoside, rhamnose, ribose, L-sorbose, D-turanose, xylitol, or D- or L-xylose. Strains produce leucine arylamidase and β-D-glucosidase and give positive tests for enzymic hydrolysis of L-valine 7-amido-4-methylcoumarin (AMC), L-phenylalanine AMC, L-tryptophan AMC, methyl α- and β-glucoside, *p*-nitrophenyl (pNP) β-D-glucoside, pNP β-D-cellobioside, proline and leucine *p*-nitroanilide, pNP phosphate, and pNP α-D-maltoside. The majority of strains are positive for the production of acetoin (Voges–Proskauer reaction), L-arginine AMC, and L-isoleucine AMC. Variable reactions are obtained for alkaline phosphatase and α-galactosidase. Negative test results are obtained for the hydrolysis of hippurate, the production of pyrrolidonyl arylamidase, β-glucuronidase, and β-galactosidase, hydrolysis of arginine, L-pyroglutamic acid AMC, 4MU *N*-acetyl-β-D-glucosaminide, 4MU phosphate, and 4MU β-D-glucuronide.

The cell-wall peptidoglycan type is Lys–Ala$_{2-3}$.

Strains have been isolated from the oral cavities of wild rodents and from the tonsils and nasal conchae of piglets.

DNA G+C content (mol%): 43–45 (T_m).

Type strain: 8S1, ATCC 33477, CCUG 34784, CCUG 34834, CIP 103225, DSM 20646, HAMBI 1522, LMG 16520, NCTC 12278.

GenBank accession number (16S rRNA gene): AJ420197, AY058218, AY584479.

Additional comments: Strains of this species were originally isolated from the oral cavities of wild rats living in sugar cane fields and eating a sucrose-rich diet (Coykendall, 1974). Strains gave colonies on sucrose agar that were approximately 1 mm in diameter, raised, somewhat adherent, but not showing drops of glucan-containing liquid on or around the colony. Strains were inhibited by bacitracin (20 U/ml). Growth did not occur at 45°C or in the presence of 6.5% NaCl. Acid was produced from mannitol and sorbitol but not from raffinose. The final pH in glucose broth was 4.2–4.5. Arginine is not hydrolyzed to produce ammonia. Both extracellular and intracellular polysaccharides were produced form sucrose. Strains possessed the c antigen of the serological scheme of Bratthall (1970) for mutans group streptococci.

Although originally proposed as a species by Coykendall (1977), the name *Streptococcus ferus* did not appear in the Approved Lists of Bacterial Names (Skerman et al., 1980) but was later accorded species status (Coykendall, 1983). The biochemical and physiological resemblance of *Streptococcus ferus* to other member species of the Mutans group of streptococci has given rise to the view that this constitutes a further member of the group. However, the demonstration that *Streptococcus ferus* was most closely related to *Streptococcus sanguinis* (*Streptococcus sanguis*), according to genetic distance measured by MLEE, casts doubt on this view. In addition 16S rRNA gene sequence analysis has not shown *Streptococcus ferus* clearly to belong to any species group (Whatmore and Whiley, 2002). However, analysis of combined sequence data from 16S rRNA and endoribonuclease P (RNase P) genes does place *Streptococcus ferus* within the Mutans group (Tapp et al., 2003).

Recently an emended description of *Streptococcus ferus* was published to take into account isolates obtained from the tonsils and nasal conchae of piglets (Baele et al., 2003).

19. **Streptococcus gallinaceus** Collins, Hutson, Falsen, Inganäs and Bisgaard 2002, 1163[VP]

gal.li.na'ce.us. L. n. *gallus* rooster, genus name of the chicken; L. masc. adj. *gallinaceus* pertaining to a domestic fowl.

Cells are coccoid and occur singly, in pairs, or in short chains. Colonies are 0.5–1.0 mm in diameter after 24 h on sheep and horse blood agar and α-hemolysis is produced. Using commercial API systems strains produce acid from D-glucose, lactose, maltose, mannitol, melibiose, methyl β-D-glucopyranoside, pullulan, D-raffinose, D-ribose, trehalose, and sucrose. Acid is not produced from D-arabitol, L-arabinose, cyclodextrin, glycogen, melezitose, sorbitol,

D-tagatose, or xylose. Strains produce alanyl-phenylalanine-proline arylamidase, arginine dihydrolase, α-galactosidase, β-galactosidase, α-glucosidase, β-glucosidase, and glycyl-tryptophan arylamidase. Strains do not produce alkaline phosphatase, β-glucuronidase, β-mannosidase, pyroglutamic acid arylamidase, or urease. N-Acetyl-β-glucosaminidase and pyrazinamidase are either weakly positive or negative. Esculin is hydrolyzed. Gelatin and hippurate are not hydrolyzed. Acetoin is not produced (Voges–Proskauer reaction). Nitrate is not reduced.

Strains are isolated from chickens with sepsis. Habitat is unknown.

16S rRNA gene sequence analysis places this species outside the recognized species groups, representing a new subline within *Streptococcus*, in loose association with *Streptococcus acidominimus*.

DNA G+C content (mol%): 40.5.

Type strain: CCUG 42692, CIP 107087, DSM 15349, JCM 12181.

GenBank accession number (16S rRNA gene): AJ307888.

20. **Streptococcus gallolyticus** Osawa, Fujisawa and Sly 1995, 362[VP] (Effective publication: Osawa, Fujisawa and Sly 1995, 78.)

gal.lo.ly′ti.cus. Gr. n. *gallo* gallate; Gr. adj. *lyticus* able to loosen; N.L. adj. *gallolyticus* gallate-digesting.

Cells are coccoid, mostly in pairs or short chains. Some strains require CO_2. Colonies on blood (horse) agar or brain heart infusion agar are 1 mm diameter at 24 h, α- or non-hemolytic, circular, smooth, unpigmented, and entire. Some strains give growth at 10°C and 45°C. The optimum temperature for growth is around 37°C. Growth in the presence of 6.5% NaCl is a variable characteristic. Chemo-organotroph. Some strains can decarboxylate gallic acid or produce tannase activity. Acid is produced from lactose, maltose, and sucrose. A few strains produce acid from arabinose. Acid is not produced from arabitol, cyclodextrin, melezitose, tagatose, ribose, or sorbitol. Acid production from amygdalin, glycogen, inulin, mannitol, melibiose, methyl D-glucopyranoside, pullulan, raffinose, starch, and trehalose are variable characteristics of this species as is esculin hydrolysis. Hippurate, urea, and arginine are not hydrolyzed. Acetoin is produced (Voges–Proskauer reaction). Leucine arylamidase and alanyl-phenylalanyl-proline arylamidase are produced. Most strains produce α-galactosidase. β-Galactosidase, β-glucosidase, β-glucuronidase, and β-mannosidase activities are variable characteristics. Alkaline phosphatase and pyrrolidonyl arylamidase are not produced. Some strains carry the Lancefield group D antigen.

Strains are isolated frequently from the alimentary tracts of various animals, dairy products, and from human clinical sources, including blood, feces, and endocarditis associated with colonic cancer.

DNA G+C content (mol%): 34.6–38 (T_m).

Additional comments: Strains phenotypically resembling *Streptococcus gallolyticus* that had been isolated from feral goats grazing tannin-rich *Acacia* species were proposed as *Streptococcus caprinus* (Brooker et al., 1994). Subsequent comparison of type strains by phenotype, 16S rRNA gene sequence similarity and DNA–DNA hybridization demonstrated that *Streptococcus gallolyticus* and *Streptococcus caprinus* were subjective synonyms and that under Rule 24b(2) of the International Code of Nomenclature of Bacteria, according to the order of receipt of their respective effective publications for validation, higher priority be given to *Streptococcus gallolyticus* (Sly et al., 1997). Poyart et al. (2002) used DNA–DNA reassociation, 16S rRNA and *sodA* gene sequence analyzes to demonstrate that a single cluster is formed by strains assigned to *Streptococcus gallolyticus* and *Streptococcus macedonicus*. In the opinion of those authors strains originally classified as *Streptococcus macedonicus* constitute an esculin-negative "variant" of *Streptococcus gallolyticus*. Strains previously described as *Streptococcus waius* (Flint et al., 1999) are also included here as these were recently reclassified as *Streptococcus macedonicus* (Manachini et al., 2002). Schlegel *et al.* (2003) have proposed that the strains thus grouped together within *Streptococcus macedonicus* were sufficiently closely related to *Streptococcus gallolyticus* to deserve subspecies status and have proposed the recognition of *Streptococcus gallolyticus* subsp. *macedonicus* alongside subsp. *gallolyticus*. These authors have also proposed that "*Streptococcus bovis*" biotype II/2 strains be recognized as subsp. *pasteurianus*, but the evidence for this is less convincing.

The classification of species comprising the Bovis group or complex is detailed above.

20a. **Streptococcus gallolyticus subsp. gallolyticus** (Osawa, Fujisawa and Sly 1996) Schlegel, Grimont, Ageron, Grimont and Bouvet 2003, 643[VP] (*Streptococcus gallolyticus* Osawa, Fujisawa and Sly 1996, 362)

Strains are non-hemolytic and do not grow in the presence of 6.5% NaCl. Strains produce tannase activity and can decarboxylate gallate. Acid is produced from glycogen, mannitol, methyl D-glucoside, trehalose, and starch. Acid production from inulin, pullulan, and raffinose are variable characteristics. Esculin is hydrolyzed. α-Galactosidase and β-glucosidase are produced. β-Galactosidase and β-mannosidase are variable characteristics. Extracellular polysaccharide is produced in 5% sucrose-containing medium.

Strains are isolated from the alimentary tracts of various animals including marsupials (koala bear, kangaroos, brushtails, and possums) and mammals, septicemia of pigeons, and human clinical cases, including blood from endocarditis following colonic cancer and feces.

DNA G+C content (mol%): 37–38 (T_m).

Type strain: ACM 3611, CCUG 35224, CIP 105428, HDP 98035, JCM 10005, LMG 16802.

GenBank accession number (16S rRNA gene): X94337.

20b. **Streptococcus gallolyticus subsp. macedonicus** (Tsakalidou, Zoidou, Pot, Wassill, Ludwig, Devriese, Kalantzopoulos, Schleifer and Kersters 1998) Schlegel, Grimont, Ageron, Grimont and Bouvet 2003, 643[VP] (*Streptococcus macedonicus* Tsakalidou, Zoidou, Pot, Wassill, Ludwig, Devriese, Kalantzopoulos, Schleifer and Kersters 1998, 525.)

ma.ce.do′ni.cus. L. adj. *macedonicus* of Macedonia, northern Greece, where the bacterium was first isolated.

Strains are mostly α-hemolytic with a minority of non-hemolytic reported. Growth in the presence of 6.5% NaCl is a variable characteristic. Strains do not produce tannase activity and cannot decarboxylate gallate. Acid is not produced from glycogen, inulin, mannitol, pullulan, raffinose, trehalose, and starch. Acid production from methyl D-glucoside and starch and are variable characteristics. Esculin is not hydrolyzed. β-Galactosidase is produced (β-GAR test) and α-galactosidase is produced by most strains. β-Glucosidase and β-mannosidase are not produced and esculin is not hydrolyzed.

Strains originally called *Streptococcus macedonicus* have been isolated from dairy products, including Greek Kasseri cheese and Italian cheese, and from sour mash, while strains originally named *Streptococcus waius* were isolated from biofilms on stainless steel samples exposed to pasteurized skim milk and dairy products.

DNA G+C content (mol%): 34.6–38 (T_m).

Type strain: ACA-DC 206, ATCC BAA-249, CCUG 39970, CIP 105683, HDP 98362, JCM 11119, LAB 617, LMG 18488.

GenBank accession number (16S rRNA gene): Z94012.

21. **Streptococcus gordonii** Kilian, Mikkelsen and Henrichsen 1989b, 481[VP]

gor.don'i.i. N.L. gen. n. *gordonii* of Gordon, in honor of British microbiologist Mervyn H. Gordon, who pioneered classification of viridans streptococci by fermentation tests.

Cells are coccoid and grow in short chains in serum broth. Growth on blood (horse) agar produces α-hemolysis and pronounced greening on chocolate agar.

Strains are not able to grow in 4% (w/v) NaCl. Metabolism is fermentative: acid is produced from galactose, N-acetylglucosamine, amygdalin, arbutin, fructose, glucose, lactose, salicin, esculin, maltose, mannose, D-cellobiose, sucrose, and trehalose. Inulin is fermented by the majority of strains. Some strains produce acid from inulin, methyl D-glucoside, raffinose, melibiose, and tagatose. Acid is not produced from arabinose, arabitol, adonitol, cyclodextrin, dulcitol, gluconate, glycerol, glycogen, mannitol, sorbitol, D-melezitose, pullulan, rhamnose, ribose, sorbose, xylose, or xylitol. Reported acid production from starch varies in the literature. Strains produce acid phosphatase, alkaline phosphatase, alanyl-phenylalanyl-proline arylamidase, chymotrypsin, leucine aminopeptidase, β-mannosidase, α-D-fucosidase, β-D-lactosidase, α-maltosidase, phospho-β-galactosidase, valine aminopeptidase, β-D-glucosaminidase, and N-acetyl-β-D-glucosaminidase. Strains do not produce α-mannosidase, β-maltosidase, α-L-fucosidase, β-D-fucosidase, β-D-glucuronidase, α-L-arabinosidase, N-acetyl-α-D-glucosaminidase, α- or β-xylosidase, or neuraminidase (sialidase). Glycyl-tryptophan arylamidase, methyl β-D-glucopyranoside, and N-acetylgalactosaminidase are variable characteristics. The percentages of strains testing positive for the following glycoside hydrolase activities is variable and dependent on the substrates used (i.e., the chromogenic naphthol- or nitrophenol-derivatives, or the fluorogenic derivative 4-methyl-umbelliferyl-) – therefore caution should be observed when incorporating these for characterizing or identifying isolates: α-galactosidase, β-D-galactosidase, α-glucosidase, β-D-glucosidase (see Kilian

et al., (1989b), Beighton et al., (1991). Arginine and esculin are hydrolyzed. Hippurate and urea are not hydrolyzed. The majority of strains hydrolyze starch. Acetoin is not produced (Voges–Proskauer reaction). H_2O_2 is produced. Virtually all strains produce extracellular polysaccharide (dextran) from sucrose and all bind salivary amylase. Strains do not produce IgA1 protease or hyaluronidase.

The cell walls contain glycerol teichoic acid and rhamnose. The peptidoglycan type is Lys–Ala$_{1-3}$. Some strains react with Lancefield group H antiserum or are ungroupable. Detection of the group H antigen is dependent on the source of the group H antiserum due to differences in immunizing strains used in antiserum production (Kilian et al., 1989b). Strains are inhibited by nitrofurazon but not by bacitracin or sulfafurazole.

Kilian et al. (1989b) described three biovars (1–3) within *Streptococcus gordonii* with the following differential characteristics: biovar 1 was able to produce acid from melibiose, inulin, and raffinose, and produce extracellular polysaccharide (dextran) from sucrose; biovars 2 and 3 were unable to ferment melibiose, or raffinose; the fermentation of inulin and the production of extracellular polysaccharide were variable characteristics of biovar 2 and 3, respectively. Strains are isolated from the human oral cavity and pharynges, and constitute one of the early tooth colonizing (pioneer) species.

16S rRNA gene sequence analysis has placed *Streptococcus gordonii* within the Mitis species group.

The genome (strain Challis) is currently being sequenced.

DNA G+C content (mol%): 40–43 (T_m).

Type strain: SK3, ATCC 10558, CCUG 25608, CCUG 33482, CCUG 35801, CIP 105258, DSM 6777, LMG 14518, NCTC 7865.

GenBank accession number (16S rRNA gene): AF003931, AY485606, D38483.

22. **Streptococcus hyointestinalis** Devriese, Kilpper-Balz and Schleifer 1988, 440[VP]

hy.o.in.tes.ti.na'lis. Gr. n. *hyos* pig; L. adj. *intestinalis* associated with intestine; N.L. adj. *hyointestinalis* associated with pig intestines.

Cells are coccoid, occurring in chains and forming a sediment with a clear supernatant in broth culture. Colonies on blood (ox) agar are transparent, 1–1.5 mm diameter after 1 d incubation, 1.5–2.5 mm diameter after 3 d incubation, circular, with entire edges and with narrow zones of incomplete hemolysis. The optimum growth temperature is 37°C. Strains do not grow in the presence of 6.5% NaCl of 40% bile. Acid is formed from glucose, fructose, mannose, galactose, lactose, N-acetylglucosamine, arbutin, salicin, maltose, sucrose, trehalose, and starch. Acid is not produced from adonitol, D- and L-arabinose, D- and L-arabitol, dulcitol, erythritol, L-fucose, gluconate, 2- and 5-ketogluconate, glycerol, glycogen, inositol, L-lyxose, mannitol, methyl α-D-glucoside, methyl α-D-mannoside, pullulan, ribose, rhamnose, sorbitol, L-sorbose, D-tagatose, D-turanose, D- and L-xylose, methyl α-xyloside, or xylitol. Variable reactions are obtained for the fermentation of raffinose, amygdalin, and cellobiose. Strains produce amylase, alkaline phosphatase, and leucine arylamidase. β-Glucuronidase, β-galactosidase, pyrrolidonyl arylamidase, arginine dehydrolase, and hyaluronidase are not produced. Variable reactions are obtained with α-galactosidase and β-gentiobiose activity tests. Acetoin is produced (Voges–Proskauer reaction). Esculin is hydrolyzed but hippurate and gelatin are not. Hydrolysis of starch is positive or weakly positive (on Mueller–Hinton medium). No dextran or levan are produced from sucrose. No reaction is obtained with streptococcal group A, B, C, D, F, and G antisera. The peptidoglycan type is Lys–Ala (Ser).

Habitat: the intestines of pigs.

16S rRNA sequence analysis places this species outside the recognized species groups.

DNA G+C content (mol%): 42–43 (T_m).

Type strain: S93, ATCC 49169, CCUG 27888, CIP 103372, DSM 20770, LMG 14579, NCTC 12224.

GenBank accession number (16S rRNA gene): AB002518, AF201898, X58313.

23. **Streptococcus hyovaginalis** Devriese, Pot, Vandamme, Kersters, Collins, Alvarez, Haesebrouck and Hommez 1997, 1077^{VP}

hy.o.va.gi.na'lis. Gr. n. *hyos* pig; L. n. *vagina* sheath, vagina; N.L. masc. adj. *hyovaginalis* associated with pig vaginas.

Cells are coccoid, occurring as long chains and forming sediment with a clear supernatant in broth culture. Growth on blood (bovine) agar produces colonies that are regular, ≤0.5 mm diameter, non-pigmented, translucent, and surrounded by a sharply demarcated zone of α-hemolysis. Growth at 37°C is better than at 25°C, 30°C, or 42°C, and is enhanced somewhat by the presence of 5% CO_2. Growth does not occur in the presence of 6.5% NaCl, bile–esculin, or on Slanetz–Bartley agar. Acid is produced from cellobiose, galactose, D-glucose, N-acetylglucosamine, D-fructose, lactose, maltose, D-mannose, mannitol, salicin, sucrose, sorbitol, and trehalose. Strains do not produce acid from D- and L-arabitol, amygdalin, D- and L-arabinose, dulcitol, erythritol, D- and L- fucose, gluconate, glycerol, glycogen, β-gentiobiose, inositol, inulin, 2- and 5-ketogluconate, D-lyxose, methyl α-D-mannoside, methyl α-D-glucoside, melibiose, melezitose, D-raffinose, rhamnose, L-sorbose, starch, D-tagatose, D-turanose, pullulan, xylitol, D- and L-xylose, and methyl β-xyloside. Variable reactions for acid production from β-glucosidase, methyl β-glucopyranoside, and ribose. Ungroupable in Lancefield serological testing. Strains produce alkaline phosphatase, alanine phenylalanine proline arylamidase, N-acetyl-β-glucosaminidase, β-galactosidase, and β-glucuronidase. Leucine arylamidase (often weak) is produced. Pyrrolidonyl arylamidase, α-galactosidase, arginine dihydrolase, glycyltryptophan arylamidase, β-mannosidase, and urease are not produced. Acetoin is produced (Voges–Proskauer reaction). Hippurate is hydrolyzed.

Strains are associated with pigs, in particular the vaginal fluids. Involvement of this species with pathological processes is unclear.

16S rRNA gene sequence analysis has shown *Streptococcus hyovaginalis* to form a distinct subline within the streptococci in association with but distinct from *Streptococcus thoraltensis* and peripherally associated with the Pyogenic species group.

DNA G+C content (mol%): 40 (type strain LMG 14710) (T_m).

Type strain: SHV515, ATCC 70086, CCUG 37866, CIP 105517, DSM 12219, LMG 14710.

GenBank accession number (16S rRNA gene): Y07601.

24. **Streptococcus infantarius** Schlegel, Grimont, Collins, Régnault, Grimont and Bouvet 2000, 1432^{VP}

in.fan.ta'ri.us. L. adj. *infantarius* relating to infants, the source of the type strain.

Cells are coccoid occurring in pairs or short chains. Colonies on blood agar are circular, 1 mm diameter after 24h incubation, unpigmented, and α-hemolytic. Growth is enhanced in a 5% CO_2 atmosphere. Homogeneous growth occurs in buffer glucose and brain heart infusion broths. Growth occurs in MRS broth without gas production. No growth occurs in 6.5% NaCl broth. All strains produce acid from glycogen, lactose, maltose, melibiose, starch, and sucrose. Acid is not produced from arabinose, arabitol, cyclodextrin, inulin, D-mannitol, melezitose, ribose, sorbitol, D-tagatose, or trehalose. Acid production from glycogen, melibiose, methyl β-D-glucopyranoside, pullulan, D-raffinose, and starch are variable characteristics of this species. Most strains produce acid from lactose and pullulan. Strains produce leucine aminopeptidase arylamidase and alanyl-phenylalanyl-proline arylamidase. Virtually all strains are α-galactosidase-positive. Strains are negative for N-acetyl-β-glucosaminidase, β-galactosidase, β-glucuronidase, glycyl-tryptophan arylamidase, β-mannosidase, alkaline phosphatase, and pyrrolidonyl arylamidase activities. Most strains are negative for β-glucosidase and methyl β-D-glucopyranoside. Arginine, hippurate, and urea are not hydrolyzed. Most strains do not hydrolyze esculin. Acetoin is produced (Voges–Proskauer reaction). Approximately 40% of strains carry the Lancefield group D antigen. Extracellular polysaccharide is not produced on 5% sucrose containing media. The type strain is esculin- and β-glucosidase-negative, produces acid from glycogen, pullulan, D-raffinose, and starch but not from melibiose or methyl β-D-pyranoside.

Strains are isolated from clinical specimens, including blood and a case of endocarditis, or from food products (dairy and frozen peas). The type strain was isolated from the feces of an infant human.

16S rRNA gene sequence analysis places *Streptococcus infantarius* within the Bovis species group.

DNA G+C content (mol%): not reported.

Type strain: HDP 90056, ATCC BAA-102, CCUG 43820, CIP 103233 NCIMB 700599.

GenBank accession number (16S rRNA gene): AF429762.

Additional comments: Strains of *Streptococcus infantarius* that produced β-glucosidase, hydrolyzed esculin and did not ferment glycogen, starch, melibiose, or pullulan were designated a separate subspecies, *Streptococcus infantarius* subsp. *coli* (Schlegel et al., 2000). These have since been proposed as a distinct species, *Streptococcus lutetiensis*, and the evidence from DNA–DNA reassociation supports this (Poyart et al., 2002).

25. **Streptococcus infantis** Kawamura, Hou, Todome, Sultana, Hirose, Shu, Ezaki and Ohkuni 1998, 926^{VP}

in.fan.tis. L. n. *infans* infant; L. gen. n. *infantis* of a human infant, from whom the organism was isolated.

Cells are coccoid, approximately 0.6–1.0 nm in diameter and grow singly or in short chains. Facultatively anaerobic. α-hemolysis is produced on Columbia blood (sheep) agar with pin-point colonies being formed at 37°C under aerobic conditions for 1 d and approximately 0.3–0.8 mm diameter colonies after 2 d incubation. Strains produce acid from lactose, maltose, sucrose, and tagatose. Acid is not produced from amygdalin, L-arabinose, D-arabinose, arabitol, arbutin, cyclodextrin, glycogen, mannitol, methyl β-D-glucopyranoside, melibiose, melezitose, pullulane, raffinose, ribose, sorbitol, or trehalose. Acid production from inulin is a variable characteristics of this species (the type strain is positive for these tests). Strains produce β-galactosidase (with both 2-naphthyl β-D-galactopyranoside and *p*-nitrophenyl β-D-galactopyranoside substrates) and alanyl-phenylalanyl-proline arylamidase, but do not produce alkaline phosphatase, α-galactosidase, β-glucosidase, β-mannosidase, β-glucuronidase, glycine-tryptophan arylamidase, or urease. β-D-Fucosidase, *N*-acetyl-β-glucosaminidase, and pyrrolidonyl arylamidase are variable characteristics of this species. Arginine, esculin, and hippurate are not hydrolyzed. Acetoin is not produced (Voges–Proskauer reaction).

Strains have been isolated from the human tooth surface and pharynx.

16S rRNA gene sequence analysis places *Streptococcus infantis* within the Mitis species group.

DNA G+C content (mol%): 39.9–40.4 (HPLC).

Type strain: 0-122, ATCC 700779, CCUG 39817, CIP 105949, DSM 12492, GTC 849, JCM 10157, LMG 18720.

GenBank accession number (16S rRNA gene): AB008315, AY485603.

Additional comments: Competence for genetic transformation has been observed in the absence of the *comC* gene which encodes streptococcal competence pheromone (Ween et al., 2002).

26. **Streptococcus iniae** Pier and Madin 1976, 547[AL]

in'i.ae. N.L. gen. n. of the dolphin, *Inia*.

Cells are coccoid, encapsulated, up to 1.5 μm in diameter and are arranged in long chains when cultured in broth. Colonies on blood (sheep) agar are small (up to 1 mm diameter), white, umbonate with an entire opaque border, an opaque center spot and a translucent ring of growth separating the border and the center. A small to moderate area of β-hemolysis is produced on sheep blood agar with a diffuse zone of α-hemolysis beyond while α-hemolysis is produced on human and bovine blood agar. Characteristic growth in Todd–Hewitt broth produces coarse, white, granular sediment with clear supernatant. Acid is produced from dextran, fructose, galactose, glucose, maltose, mannitol, mannose, salicin, sucrose, and trehalose. Acid is not produced from arabinose, dulcitol, glycerol, inositol, inulin, lactose, melibiose, raffinose, rhamnose, sorbitol, or xylose. The original description of *Streptococcus iniae*, above, was taken from the reactions of the type strain (ATCC 29178). From the description of multiple strains given by Eldar et al. (1995), *Streptococcus iniae* is able to produce acid from ribose, galactose, D-glucose, D-fructose, D-mannose, man-

nitol, methyl D-mannoside, *N*-acetylglucosamine, arbutin, esculin, salicin, cellobiose, maltose, sucrose, melezitose, starch, glycogen, and gentiobiose. Strains are unable to produce acid from adonitol, D- and L-arabinose, galactose, methyl D-glucoside, L-sorbose, D- and L-xylose, methylxyloside, rhamnose, lactose, D-raffinose, D- and L-fucose, glycerol, sorbitol, inositol, dulcitol, erythritol, adonitol, xylitol, D- and L-arabitol, amygdalin, inulin, lyxose, gluconate, 2-ketogluconate, 5-ketogluconate, turanose, or tagatose. Strains produce pyrrolidonyl arylamidase, alkaline phosphatase, leucine arylamidase, and arginine dihydrolase. α- and β-Galactosidases are not produced. Esculin and starch are hydrolyzed but hippurate is not. Acetoin is not produced (Voges–Proskauer reaction). Strains are nongroupable with Lancefield antisera.

The peptidoglycan type is Lys–Ala$_{1-3}$.

The species was initially isolated from abscess foci in an Amazonian freshwater dolphin (*Inia geoffrensis*) and also causes meningoencephalitis in a variety of fish species, is associated with disease outbreaks in aquaculture farms and has been linked with transmission from fish to humans, causing severe infection in the latter host (cellulitis). Virulence in this species has been shown to be associated with a distinct genetic profile as shown by PGFE (Fuller et al., 2001).

16S rRNA gene sequence analysis places *Streptococcus iniae* within the Pyogenic species group.

DNA G+C content (mol%): 32.9 (T_m).

Type strain: PW, ATCC 29178, CCUG 27303, CIP 102508, DSM 20576, LMG 14520.

GenBank accession number (16S rRNA gene): AF335572.

Additional comments: A streptococcus that causes meningoencephalitis in fish in Israeli farms was given the name *Streptococcus shiloi* which was subsequently shown to be a junior synonym of *Streptococcus iniae* (Eldar et al., 1994, 1995).

27. **Streptococcus intermedius** Prévot 1925,417[AL] emend. Whiley and Beighton 1991, 2.

in.ter.me'di.us. L. adj. *intermedius* intermediate.

Cells are small (0.5–1.0 μm in diameter), coccoid, and grow in short chains. Colonies on blood agar are 0.5–2.0 mm in diameter, white or translucent, convex and entire. Some strains also produce colonies that are 0.5–1.0 mm diameter, white and matte. The majority of strains are non-hemolytic or α-hemolytic on blood (horse or sheep) agar with occasional strains giving β-hemolysis. β-Hemolysis is always observed when *Streptococcus intermedius* is grown on agar plates containing human blood mainly due to the production of the human-specific cytotoxin intermedilysin (Nagamune et al., 1996, 2000). Growth is reduced under aerobic conditions and is frequently enhanced by the addition of CO_2. Some strains require anaerobic conditions for growth. Extracellular polysaccharide is not produced on sucrose-containing media. Strains produce alkaline phosphatase, β-fucosidase, β-galactosidase, α-glucosidase, β-*N*-acetylgalactosaminidase, β-*N*-acetylglucosaminidase, leucine arylamidase, and neuraminidase (sialidase). β-Glucosidase production is a variable characteristic of this species. α-Galactosidase, β-glucuronidase, and pyrrolidonyl arylamidase are not produced. Acid is produced from glucose and trehalose, virtually all strains produce acid from lactose. The majority of strains produce acid from amygdalin, cellobiose, and salicin. Acid is not produced from arabinose, glycerol, inulin, or sorbitol. Acetoin is produced (Voges–Proskauer reaction) and arginine and esculin are hydrolyzed but urea and hippurate are not. Virtually all strains produce hyaluronidase. Hydrogen peroxide is not produced.

Most strains are serologically ungroupable with Lancefield grouping antisera, but strains are isolated that react with Lancefield groups F and C.

Habitats are the human oral cavity and upper respiratory tract, and this species has been reported to occur in feces. Strains are isolated from human purulent infections, notably liver and brain abscesses.

16S rRNA gene sequence analysis places *Streptococcus intermedius* within the Anginosus species group.

DNA G+C content (mol%) of the DNA: 37–38 (T_m).

Type strain: ATCC 27335, CCUG 17827, CCUG 32759, CIP 103248, DSM 20573, HAMBI 1571, LMG 17840, NCTC 11324.

GenBank accession number (16S rRNA gene): AF104671, X58311.

Additional comments: For a description of the often confusing history of the classification of strains currently designated *Streptococcus intermedius* together with closely related strains see the section on the Anginosus species group.

28. **Streptococcus lutetiensis** Poyart, Quesne and Trieu-Cuot 2002, 1253[VP]

lu.te′ti.en.sis. L. fem. n. *lutetia*, Lutèce, Paris; N.L. masc. adj. *lutetiensis* pertaining to Paris, where the species was characterized

Cells are coccoid, occurring in pairs or short chains. Colonies on blood agar or nutrient agar are circular, smooth, entire, non-pigmented, and produce α-hemolysis. Homogeneous growth is obtained in brain heart infusion and in glucose broths after 18 h incubation at 37°C. Growth occurs in MRS broth without gas production. No growth occurs in the presence of 6.5% NaCl.

Acid is produced from lactose, maltose, methyl β-D-glucopyranoside, raffinose, and sucrose. Acid is not produced from L-arabinose, D-arabitol, cyclodextrin, glycogen, inulin, D-mannitol, melezitose, melibiose, pullulan, ribose, sorbitol, tagatose, or trehalose.

Acid production from starch is a variable characteristic. Strains produce alanyl-phenylalanyl-proline arylamidase, leucine aminopeptidase, α-galactosidase, and β-glucosidase. Strains do not produce arginine dihydrolase, alkaline phosphatase, pyrrolidonyl arylamidase, β-galactosidase, β-glucuronidase, β-mannosidase, N-acetyl-β-glucosaminidase, or glycyltryptophan arylamidase.

Esculin is hydrolyzed but hippurate and urea are not. Acetoin is produced (Voges–Proskauer reaction). Some strains react with Lancefield group D antiserum. No extracellular polysaccharide is produced on 5% sucrose medium.

Strains have been isolated from human specimens, including feces, blood (including infective endocarditis), urine, and cerebrospinal fluid. The type strain is of unknown human origin.

16S rRNA gene sequence analysis places *Streptococcus lutetiensis* within the Bovis species group.

DNA G+C content (mol%) of the DNA: has not been determined.

Type strain: NEM 782, CIP 106849, DSM 15350.

GenBank accession number (16S rRNA gene): AJ297215.

Additional comments: Strains of *Streptococcus lutetiensis* were previously designated *Streptococcus infantarius* subsp. *coli* (Schlegel et al., 2000). However, despite the close relationship between these and strains designated *Streptococcus infantarius* subsp. *infantarius* the level of DNA similarity between the two groups has consistently fallen below 70% in different studies (Poyart et al., 2002; Schlegel et al., 2000), indicating that separate species status is appropriate.

The classification of *Streptococcus lutetiensis* and close relatives within the Bovis group is detailed above.

29. **Streptococcus macacae** Beighton, Hayday, Russell and Whiley 1984, 333[VP]

ma.ca.cae. N.L. fem. n. *Macaca* genus name of the macaque; N.L. gen. n. *macacae* of macaque.

Cells are coccoid and grow in chains. Strains grow poorly in air, but growth is stimulated by CO_2; hence growth occurs in candle jars and anaerobic jars in an atmosphere of 90% H_2, and 10% (v/v) CO_2. Strains produce greening of (horse) blood agar plates. On sucrose-containing medium, coherent colonies are produced that are easily detachable from the agar surface, transparent, and 1–2 mm in diameter after 3 d incubation. In addition, on sucrose agar a less frequent colonial form is observed which is vivid white, <1 mm diameter, erose, crumbly, and detachable. The transparent colonial type gives rise to the two observed colony forms but the vivid white type retains the original form. Growth occurs on media containing 10% or 40% (w/v) bile but not in the presence of 6.5% NaCl, at 45°C, or at pH 9.6. Acid is produced from N-acetylglucosamine, amygdalin, cellobiose, fructose, glucose, galactose, maltose, mannitol, raffinose, sucrose, and trehalose. Sorbitol is fermented slowly. Acid is not produced from adonitol, arabinose, dextrin, glycerol, glycogen, inositol, inulin, melizitose, melibiose, ribose, starch, sorbose, xylitol, or xylose. Esculin is hydrolyzed, starch and blue dextran are weakly hydrolyzed, but arginine is not hydrolyzed. H_2O_2 is not produced. Strains do not grow in the presence of 20 μg/ml bacitracin. *Streptococcus macacae* strains possess the serotype c antigen described by Bratthall (1970). Cell-wall carbohydrates include glucose and rhamnose. Extracellular polysaccharide is produced from sucrose.

Strains are isolated from oral samples, primarily dental plaque, of monkeys (*Macaca fascicularis*)

16S rRNA gene sequence analysis places *Streptococcus macacae* within the Mutans group.

DNA G+C content (mol%): 35–36 (T_m).

Type strain: 25-1, ATCC 35911, CCUG 27653, CIP 102912, DSM 20724, LMG 15097, NCTC 11558.

GenBank accession number (16S rRNA gene): AJ420199 AY188351, X58302.

30. **Streptococcus minor** Vancanneyt, Devriese, Graef, Baele, Lefebvre, Snauwaert, Vandamme, Swings and Haesebrouck 2004, 451[VP]

mi′nor. L. masc. adj. *minor* smaller.

Cells are ovoid, very small (<1 μm in diameter), and arranged predominantly in small groups. On Columbia sheep blood agar colonies are nonpigmented, regular, translucent, approximately 0.5 mm in diameter, producing α-hemolysis in a narrow zone around the colonies. Incubation in 5% CO_2 slightly enhances growth. Growth at 37°C is better than at 42°C with growth at 25°C almost equal to that observed at 30°C or 37°C. Homogeneous growth is obtained in brain heart infusion broth (Oxoid). Growth is obtained on Edwards medium (Oxoid) but not on Slanetz–Bartley agar (Oxoid). Blackening is observed on bile–esculin agar (Oxoid).

Acid is produced from N-acetylglucosamine, cellobiose, D-fructose, galactose, β-gentiobiose (often delayed), D-glucose, glycogen, lactose, D-mannose, mannitol (often delayed), maltose, salicin, sucrose, and trehalose. Acid is not produced from adonitol, DL-arabinose, DL-arabitol, dulcitol, erythritol, DL-fucose, gluconate, glycerol, inositol, 2- and 5-ketogluconate, D-lyxose, methyl α-D-mannoside, methyl α-glucoside, methyl β-xyloside, melibiose, melezitose, rhamnose, ribose, L-sorbose, D-turanose, DL-xylose, or xylitol. Acid from amygdalin, arbutin, inulin, D-raffinose, starch, sorbitol, and D-tagatose is strain-dependent.

Strains produce leucine arylamidase and are able to hydrolyze L-valine-AMC, L-phenylalanine-AMC, 4MU-α-D-glucoside, L-tryptophan-AMC, L-arginine-AMC, L-isoleucine-AMC, and esculin. Strains do not produce pyrrolidonyl arylamidase, β-glucuronidase, or β-galactosidase. Hippurate is not hydrolyzed. Production of α-galactosidase (often weak) and alkaline phosphatase and hydrolysis of p-nitrophenyl phosphate are variable characteristics.

Strains do not react with Lancefield groups A, B, C, D, G, or F antisera.

Strains are isolated from the tonsils and intestinal tract of dogs and tonsils of cats and cattle.

The highest 16S rRNA gene sequence similarity is with *Streptococcus ovis* (95.9%).

DNA G+C content (mol%): 40.6–41.5 (HPLC).

Type strain: ON59, LMG 21734, CCUG 47487.

GenBank accession number (16S rRNA gene): AY232832.

31. **Streptococcus mitis** Andrewes and Horder 1906, 712[AL] emend. Kilian, Mikkelsen and Henrichsen 1989b, 483.

mi'tis. L. adj. *mitis* mild.

Cells are coccoid, growing in pairs or short chains in serum broth. Growth on blood (horse) agar produces α-hemolysis and pronounced greening on chocolate agar. Strains unable to grow in the presence of 6.5% NaCl. A few strains may grow in 4% NaCl. Metabolism is fermentative. Acid is produced from galactose, glucose, N-acetylglucosamine, fructose, lactose, mannose, and sucrose. Acid production from sorbitol, salicin, D-cellobiose, maltose, D-melibiose, pullulan, D-trehalose, D-raffinose, and ribose are variable characteristics of this species. Strains do not produce acid from amygdalin, arabinose, arabitol, arbutin, cyclodextrin, dulcitol, esculin, glycerol, glycogen, mannitol, methyl D-glucoside, D-melezitose, rhamnose, L-sorbose, starch, tagatose, or xylose. Reports of acid from inulin and melibiose vary in the literature. Strains produce leucine aminopeptidase, alanyl-phenylalanyl-proline arylamidase, valine aminopeptidase, α-D-glucosidase,

N-acetyl-β-D-glucosaminidase, and chymotrypsin. Production of alkaline phosphatase, acid phosphatase, phosphoamidase, α-glucosidase, α-maltosidase, β-D-fucosidase, β-glucosaminidase, glycyl-tryptophan-arylamidase, α- and β-D-galactosidases, lactosidase, phospho-β-galactosidase, and sialidase are variable characteristics. Strains do not produce N-acetyl-α-D-glucosaminidase, N-acetylgalactosaminidase, α- or β-mannosidases, β-maltosidase, β-D-glucosidase, β-D-glucuronidase, α-L-arabinosidase, pyrrolidonyl arylamidase, hyaluronidase or α-L-fucosidase, or α- or β-xylosidases.

The percentage of strains testing positive for the following glycoside hydrolase activities is variable and dependent on the substrates used (i.e., the chromogenic naphthol- or nitrophenol-derivatives, or the fluorogenic derivative 4-methyl-umbelliferyl-) – therefore caution should be observed when incorporating these for characterizing or identifying isolates: α-galactosidase, β-D-galactosidase, α-glucosidase, β-D-glucosidase, and β-D-fucosidase (see Kilian et al., 1989b; Beighton et. al., 1991). Acetoin is not produced (Voges–Proskauer reaction). Extracellular polysaccharide is not produced. Hydrogen peroxide is produced. Strains do not hydrolyze esculin, urea, or hippurate, and do not bind salivary amylase. Arginine and starch hydrolysis and the production of IgA_1 protease are variable characteristics. The cell wall contains ribitol teichoic acid and lacks significant amounts of rhamnose. The peptidoglycan type is Lys-direct. Some strains react with Lancefield group K and O antisera.

Kilian et al. (1989b) described two biovars (1 and 2) within *Streptococcus mitis* with the following differential characteristics: biovar 1 was able to ferment trehalose more often, unable to hydrolyze arginine or produce β-glucosaminidase (positive characteristics of biovar 2), unable to ferment sorbitol or produce phospho-β-galactosidase or β-lactosidase (a variable characteristic of biovar 2), was less often able to ferment maltose, melibiose, and raffinose, or to hydrolyze starch (positive characteristics of biovar 2), and was sometimes able to produce IgA_1 protease (negative characteristic of biovar 2). Biovar 2 was positive for the production of α- and β-galactosidase (variable characteristics of biovar 1) and included certain strains able to produce β-glucosidase (a negative characteristic of biovar 1).

Strains are isolated from the human oral cavity (dental plaque and mucosal surfaces) and pharynges. *Streptococcus mitis* biovar 1 constitutes one of the earliest colonizers of the tooth surface in the development of dental plaque (Fransen et al., 1991). This species is a significant cause of life-threatening infection in immunocompromised patients (Beighton et al., 1994). Evidence from MLEE suggests the existence of both transient and persistent clones of *Streptococcus mitis* biovar 1 in adults with persistence maintained by successive clones rather than stable strains (Hohwy et al., 2001).

16S rRNA gene sequence analysis places *Streptococcus mitis* within the Mitis group.

It is anticipated that genome sequencing will be completed for strain NCTC 12261 in 2005.

DNA G+C content (mol%): 39–41 (T_m).

Type strain: NS51, ATCC 49456, CCUG 35790, CCUG 31611, CIP 103335, DSM 12643, NCTC 12261.

GenBank accession number (16S rRNA gene): AB002520, AF003929, D38482.

Additional comments: Data from several sources have shown *Streptococcus mitis* biovar 2 to be heterogeneous species comprising strains of disparate taxonomic affiliation requiring further study (De Gheldre et al., 1999; Kikuchi et al., 1995; Vandamme et al., 1998; Whatmore et al., 2000).

The classification of *Streptococcus mitis* and relatives is detailed in the section on the Mitis species group.

The present type strain, NS51^T, replaced strain NCTC 3165, which was shown to belong to *Streptococcus gordonii* (Coykendall, 1989b; Kilian et al., 1989a).

32. **Streptococcus mutans** Clarke 1924, 144^AL

mu'tans. L. part. adj. *mutans* changing.

Cells are coccoid, approximately 0.5–0.75 µm in diameter, occurring in pairs or as short- to medium-length chains without capsules. In acid conditions in broth and on some solid media, *Streptococcus mutans* may form short rods (1.5–3.0 µm in length) and rod-shaped morphology may be evident on primary isolation from oral specimens. Growth on blood agar after incubation anaerobically for 2 d produces colonies that are white or gray, circular or irregular, 0.5–1.0 mm in diameter, sometimes rather hard and tending to adhere to the surface of the agar and slightly pitting into the agar surface. Hemolytic reaction on blood agar is usually α-hemolytic or non-hemolytic with very occasionally strains giving β-hemolysis. Growth on sucrose-containing agar typically produces rough, heaped colonies, about 1 mm in diameter with soluble extracellular polysaccharide visible as beads, droplets, or puddles of liquid on or surrounding the colonies. Some strains may form smooth or mucoid colonies (Edwardsson, 1968, 1970). Facultatively anaerobic; while most strains grow in air growth is optimum at 37°C under anaerobic condition with some strains CO_2-dependent. A few strains have been reported to grow at 45°C, but no growth occurs at 10°C. Acid is produced from arbutin, D-cellobiose, esculin, galactose, glucose, inulin, lactose, maltose, mannitol, mannose, N-acetylglucosamine, salicin, sorbitol, sucrose, trehalose, and D-raffinose. Melibiose and amygdalin fermentation are variable characteristics of *Streptococcus mutans* with strains that cannot ferment melibiose being less frequently able to produce acid from raffinose or amygdalin. Acid is not produced from adonitol, arabinose, cyclodextrin, dulcitol, erythritol, glycerol, glycogen, gluconate, inositol, methyl D-glucoside, methyl D-mannoside, methyl D-xyloside, melezitose, rhamnose, ribose, sorbose, starch, or xylose. Strains hydrolyze esculin but not arginine. Starch is hydrolyzed by the majority of strains but hippurate and urea are not. Strains produce leucine aminopeptidase, valine aminopeptidase, chymotrypsin, phosphoamidase, α- and β-glucosidase, α-galactosidase, and β-lactosidase. Strains do not produce alkaline phosphatase, β-glucosaminidase, glycyl tryptophan arylamidase, N-acetyl-β-D-galactosaminidase, N-acetyl-β-D-glucosaminidase, β-D-glucuronidase, α-L-fucosidase, β-D-fucosidase, β-mannosidase, β-maltosidase, α-arabinosidase, pyrrolidonyl arylamidase, urease, hyaluronidase, orneuraminidase (sialidase). α-maltosidase is a variable characteristic of this species. Melibiose non-

fermenting strains do not produce α-galactosidase or α-glucosidase. Test results for β-D-galactosidase production have been shown to vary depending on the substrate derivative employed (Kilian et al., 1989b). Acetoin is produced by the majority of strains. Hydrogen peroxide is not produced. Strains are non-groupable with Lancefield grouping antisera. Strains contain the cell-wall polysaccharide-typing antigens c, e, or f, according to the serological scheme of Bratthall (1970) and Perch et al. (1974). Extracellular polysaccharide (dextran) is produced from sucrose. The majority of strains are resistant to bacitracin.

Strains are isolated from the oral cavity where the primary habitat is the tooth surface (dental plaque) and colonization is favored by high levels of dietary sucrose. Strains have been isolated from feces.

16S rRNA gene sequence analysis places this species within the Mutans species group.

DNA G+C content (mol%): 36–38 (T_m and buoyant density).

Type strain: ATCC 25175, CCUG 6519, CCUG 11877, CCUG 17824, CIP 103220, DSM 20523, HAMBI 1519, JCM 5705, LMG 14558, NBRC 13955, NCTC 10449.

GenBank accession number (16S rRNA gene): AJ243965, AY188348, X58303.

Additional comments: The complete genome sequence has been determined for strain UA159 (GenBank accession no. AE0141333) composed of 2,030,936 bp with a mean G+C content of 36.82 mol%. The genome contains 1963 open reading frames (ORFs) of which 63% were assigned putative functions at the time of publication (Ajdic et al., 2002). Analysis has shown that almost 15% of the genome is accounted for by non-oxidative pathways for carbohydrate metabolism and associated transport systems and reinforces the suggestion that *Streptococcus mutans* is able to metabolize a wider range of carbohydrates than any other Gram-positive bacterium sequenced to date.

A recent publication has described an additional serotype (k) within this species. Serotype k strains were isolated from the dental plaque of children and were less susceptible to phagocytosis than other serotypes (Nakano et al., 2004).

33. **Streptococcus oligofermentans** Tong, Gao and Dong 2003, 1103^VP

o.li.go.fer.men'tans. Gr. adj. *oligos* little, scanty; L. part. adj. *fermentans* fermenting; N.L. part. adj. *oligofermentans* fermenting few compounds.

Cells are coccoid, approximately 0.7 µm in diameter after 24 h growth in brain heart infusion medium at 37°C, arranged in short chains. Colonies on BHI blood agar are approximately 0.5–1.0 mm in diameter after 24 h, even, locally rough, dark yellow, and α-hemolytic. Optimum growth temperature is 37°C with a temperature range for growth of 25–41°C. Optimum pH for growth is 7.0 with a pH range for growth of 5.3–8.95. Acid is produced from sucrose, D-glucose, mannose, and maltose. Fermentation of lactose, trehalose, and raffinose is variable. Acid is not produced from mannitol, salicin, sorbitol, arabinose, inulin, melibiose, cellobiose, arbutin, amygdalin, ribose, starch, or glycogen. Hippurate is hydrolyzed. Arginine and esculin are not hydrolyzed. Acetoin is not produced (Voges–Proskauer reaction). Differential biochemical test data from Tong et al. (2003) indicates that the minority (11–49%) of strains produce alkaline phosphatase and α-galactosidase with no strain producing β-galactosidase. Strains produce leucine arylamidase but not β-glucuronidase or pyrrolidonyl arylamidase.

Habitat is dental plaque and saliva of humans.

16S rRNA gene sequence data indicates that *Streptococcus oligofermentans* is a member of the Mitis group of species and is related most closely to *Streptococcus sinensis* (Woo et al., 2002). DNA–DNA relatedness between type strain LMG 21535 and other closely related, Mitis group species was 7.1–16.4%.

DNA G+C content (mol%): 39.5 ± 0.8 (39.9 mol% for the type strain) (T_m).

Type strain: CGMCC (China General Microbiological Culture Collection Center, Beijing) AS 1.3089, LMG 21535.

GenBank accession number (16S rRNA gene): AY099095.

34. **Streptococcus oralis** Bridge and Sneath 1982, 414[VP] emend. Kilpper-Bälz, Wenzig and Schleifer 1985, 487 emend. Kilian, Mikkelsen and Henrichsen 1989b, 482.

o.ra'lis. N.L. adj. *oralis* of the mouth.

Cells are coccoid, non-encapsulated, and arranged in short chains. α-hemolysis is produced on blood (horse) agar often with pronounced greening. Growth occurs in the presence of 0.0004% crystal violet but not with 3% NaCl. Tetrazolium (0.1%) and nitrite are reduced, and lysine is not decarboxylated. Gluconate is usually not oxidized. Tetrathionate reductase is usually produced, lipase production is variable, and deoxyribonuclease production is negative. Metabolism is fermentative. Acid is produced from fructose, galactose, glucose, N-acetylglucosamine, lactose, maltose, pullulan, and sucrose. Acid production from raffinose, tagatose, and erythritol are variable characteristics of this species as are acid from glycerol, melibiose, ribose, salicin, and starch, with the percentages of strains giving positive results in the latter tests varying between reports in the literature. A minority of strains produce acid from D-trehalose and melezitose. Acid is not produced from amygdalin, arabinose, arabitol, arbutin, cellobiose, cyclodextrin, dulcitol, esculin, glycogen, inulin, mannitol, melezitose, methyl D-glucoside, rhamnose, sorbitol, or sorbose. Melibiose has been reported as being fermented weakly with terminal pH between 5.6 and 6.2 (Kilian et al., 1989b).

Strains produce alkaline phosphatase, acid phosphatase, leucine aminopeptidase, valine aminopeptidase, chymotrypsin, α-D-glucosidase, sialidase (neuraminidase), N-acetyl-β-D-glucosaminidase, N-acetyl-β-D-galactosaminidase, glycyl tryptophan arylamidase, and β-glucosaminidase. The production of α-D-galactosidase and β-D-galactosidase are variable characteristics of this species with the proportion of positive strains varying between reports and dependent upon the test method employed (Beighton et al., 1991; Kilian et al., 1989b). Phospho-β-galactosidase, α-arabinosidase, alanine-phenylalanyl-proline arylamidase, α-L-fucosidase, β-D-fucosidase, β-D-glucosidase, β-maltosidase, β-mannosidase, β-glucuronidase, β-lactosidase, α- and β-xylosidase, and hyaluronidase are not produced.

Arginine and esculin are not usually hydrolyzed although a minority of strains has been reported as positive for esculin (Beighton et al., 1991; Facklam, 2002). Urea and hippurate are not hydrolyzed. Acetoin production is usually reported as negative although this was not the case in the emended description given for this species by Kilpper-Bälz et al. (1985). Hydrogen peroxide is produced. The methyl red reaction, growth with sodium azide, and dextran production are variable. IgA1 protease is produced. Strains do not bind salivary amylase.

The cell walls contain ribitol teichoic acid and choline. Galactose or glucose or both and galactosamine or glucosamine or both are present as major sugar constituents. Rhamnose is absent or present only in minor amounts. The peptidoglycan type is the Lys-direct type.

Strains are isolated from the human mouth. This species is a significant cause of life-threatening infections in immunocompromised patients (Beighton et al., 1994), is one of the commonest streptococci isolated from infective endocarditis (Douglas et al., 1993) and is a primary colonizer of dental plaque (Fransen et al., 1991). The demonstration of a distinct acidoduric population of *Streptococcus oralis* strains means that a more significant role may eventually be recognized for this species in the etiology of dental caries (Alam et al., 2000).

16S rRNA gene sequence analysis places *Streptococcus oralis* within the Mitis species group. The classification of this species is detailed in the section on the Mitis group.

DNA G+C content (mol%): 38–42 (T_m).

Type strain: LVG1, SK23, PB 182, ATCC 35037, CCUG 13229, CCUG 24891, CIP 102922, DSM 20627, LMG 14532, NCTC 11427.

GenBank accession number (16S rRNA gene): AF003932, AY485602, X58308.

35. **Streptococcus orisratti** Zhu, Willcox and Knox 2000, 60[VP] o.ris.rat'ti. L. gen. n. *oris* of the mouth; N.L. gen. masc. n. *ratti* of the rat; N.L. gen. n. *orisratti* of the mouth of the rat.

Cells are coccoid, occurring in pairs or short chains. Cells are Gram-positive and catalase-negative. Colonies on sheep-blood agar are 0.5–1.0 mm diameter, white, circular or irregular, and are α-hemolytic. Strains grow at 45°C, in 40% bile, and at pH 9.6, but not at 50°C or in the presence of 6.5% NaCl. Growth in glucose-containing culture gives a terminal pH of 4.3–4.4. Acid is produced from amygdalin, arbutin, cellobiose, fructose, galactose, glucosamine, glucose, glycogen, inulin, lactose, maltose, mannose, melibiose, raffinose, salicin, starch, sucrose, and trehalose but not from arabinose, mannitol, melezitose, or sorbitol. Esculin and starch are hydrolyzed but arginine is not. Glucosyl-transferase and fructosyl-transferase are not produced. *Streptococcus orisratti* belongs to Lancefield group A.

Despite an overall phenotypic resemblance to previous descriptions of *Streptococcus bovis*, and the inability to ferment mannitol or sorbitol, and lack of glucosyl- and fructosyltransferase activities, 16S rRNA gene sequence analysis places *Streptococcus orisratti* as a phylogenetically distinct taxon within the relatively loose cluster that is the "Mutans group" of species (Tapp et al., 2003; Zhu et al., 2000).

Strains have been isolated from the tooth surface of laboratory rats.

DNA G+C content (mol%): 39.6–43.5 (T_m).

Type strain: A63, ATCC 700640, CIP 106965.

GenBank accession number (16S rRNA gene): AF124350.

36. **Streptococcus ovis** Collins, Hutson, Hoyles, Falsen, Nikolaitchouk and Foster 2001, 1149^{VP}

o'vis. L. gen. n. *ovis* of the sheep.

Cells are coccoid, occurring singly, in pairs or short chains. Colonies are approximately 0.5–1.0 mm diameter on sheep and horse blood agar plates and are α-hemolytic. Esculin-positive colonies are produced on Edwards medium. Biochemical characteristics were obtained using API commercial test systems. Strains produce alanyl-phenylalanine-proline arylamidase, α-glucosidase, β-glucosidase, esterase C-4 (weak reaction), leucine arylamidase, and phosphoamidase (weak reaction). Strains do not produce alkaline phosphatase, chymotrypsin, ester lipase C8, lipase C14, α-fucosidase, β-galactosidase, β-glucuronidase, α-mannosidase, β-mannosidase, *N*-acetyl-β-glucosaminidase, pyroglutamic acid arylamidase, trypsin, or urease. Variable reactions are obtained for acid phosphatase, arginine dihydrolase, α- galactosidase, cystine arylamidase, valine arylamidase, and glycyl-tryptophan arylamidase. Strains do not hydrolyze hippurate or produce acetoin (Voges–Proskauer reaction). Strains produce acid from glucose, glycogen, lactose, maltose, mannitol, raffinose, trehalose, sorbitol, and sucrose. The majority of strains produce acid from D-tagatose but not from cyclodextrin. Acid is not produced from D-arabitol, L-arabinose, melibiose, melezitose, methyl β-D-glucopyranoside, pullulan, or ribose.

Strains have been isolated from clinical specimens of sheep but are of unknown habitat.

16S rRNA gene sequence analysis has shown that *Streptococcus ovis* forms a distinct subline within the genus and does not display a close association with any other described species.

DNA G+C content (mol%): 38 (T_m).

Type strain: S 369/98/1, CCUG 39485, LMG 19174, CIP 107097.

GenBank accession number (16S rRNA gene): Y17358.

37. **Streptococcus parasanguinis** corrig. Whiley, Fraser, Douglas, Hardie, Williams and Collins 1990c, 321^{VP} (Effective publication: Whiley, Fraser, Douglas, Hardie, Williams and Collins 1990b, 120.)

pa.ra.san'guin.is. L. gen. n. *sanguinis* of the blood, specific epithet of *Streptococcus sanguinis;* N.L. gen. n. *parasanguinis* alongside of *Streptococcus sanguinis,* indicating the close similarity between the two species.

Cells are coccoid, approximately 0.8–1.0 μm in diameter, occurring in chains. α-Hemolysis is produced on blood agar. Most strains grow at 45°C and in the presence of 40% bile but not in the presence of 4% NaCl. Acid is produced from *N*-acetylglucosamine, fructose, galactose, glucose, lactose, maltose, and sucrose. Virtually all strains produce acid from raffinose, melibiose, and salicin. A few strains produce acid from amygdalin, sorbitol, or starch. Acid is not produced from adonitol, arabinose, glycerol, glycogen, inulin, mannitol, melezitose, or xylose. Acid production from cellobiose, arbutin, and trehalose are variable characteristics of this species. Strains produce leucine arylamidase, β-*N*-acetylgalactosaminidase,

β-*N*-acetylglucosaminidase, β-galactosidase, and α-glucosidase. α-Galactosidase and alkaline phosphatase are produced by most strains. β-D-Fucosidase, α-L-fucosidase, β-glucosidase, and arabinosidase are variable characteristics of *Streptococcus parasanguinis.* Pyrrolidonyl arylamidase and sialidase are not produced. Arginine is hydrolyzed but urea is not. Some strains hydrolyze esculin. Acetoin is not produced (Voges–Proskauer reaction). Hydrogen peroxide is produced. Extracellular polysaccharide is not produced. Cells bind salivary amylase.

Strains have been isolated from human clinical sources that include throat, blood, and urine.

16S rRNA gene sequence analysis places *Streptococcus parasanguinis* within the Mitis species group.

DNA G+C content (mol%): 40.6–42.7 (T_m).

Type strain: 55898, ATCC 15912, CCUG 30417, CIP 104372, DSM 6778, LMG 14537.

GenBank accession number (16S rRNA gene): AF003933, AY485605.

Additional comments: Immunization of rats with lipoprotein receptor protein (FimA) from *Streptococcus parasanguinis* confers protection against endocarditis infection indicating that FimA may represent a promising vaccinogen to control infective endocarditis (Kitten et al., 2002).

Streptococcus parasanguinis is the corrected name for strains previously called *Streptococcus parasanguis* (Trüper and de Clari, 1997; Whiley et al., 1990b).

38. **Streptococcus parauberis** Williams and Collins 1990b, 470^{VP} (Effective publication: Williams and Collins 1990a, 486.)

para.u'ber.is. Gr. pref. *para* alongside of or near; L. gen. n. *uberis* of an udder, specific epithet of *Streptococcus uberis;* N.L. gen. n. *parauberis* alongside of *Streptococcus uberis,* indicating the close similarity between the two species.

Cells are coccoid, occurring as pairs or chains of moderate length. Growth on blood agar produces weak α-hemolysis or no hemolysis. The optimum temperature for growth is around 35–37°C. Strains may or may not survive heating at 60°C for 30 min. Growth occurs in the presence of 4% NaCl but not 6.5% NaCl, nor at pH 9.6. Acid is produced from arbutin, fructose, galactose, glucose, lactose, maltose, mannose, *N*-acetylglucosamine, salicin, sucrose, and trehalose. Some strains produce acid from dulcitol, inulin, melezitose, raffinose, starch and D-tagatose. Acid is not produced from adonitol, D- or L-arabinose, erythritol, D- or L-fucose, glycerol, glycogen, gluconate, 2-ketogluconate, 5-ketogluconate, inositol, lyxose, melibiose, methyl α-D-glucoside, methyl α-D-mannoside, methyl α-xyloside, rhamnose, sorbose, turanose, xylitol, D- or L-xylose. Acid production from amygdalin, cellobiose, β-gentiobiose, mannitol, ribose, and sorbitol are variable characteristics of this species. Alkaline phosphatase, leucine arylamidase, and pyrrolidonyl arylamidase are produced. Some strains produce α-galactosidase and β-glucuronidase. β-Galactosidase is not produced. Arginine and esculin are hydrolyzed, hippurate hydrolysis is a variable characteristic of this species, and tyrosine is not decarboxylated. Acetoin is produced (Voges–Proskauer reaction). The cell-wall peptidoglycan type is Lys–Ala$_{1-3}$.

Serologically heterogeneous, including strains reacting with Lancefield E, P, and U antisera. The V2 region of 16SrRNA of this species contains the sequence UAA GUA CAC AUG UAC UNA AUU UAA AAG GAG CAA U (positions 187–220) which serves to distinguish it from *Streptococcus uberis*. *Streptococcus parauberis* is found on the lips and skin of cows, in raw milk, and on udder tissue. It causes mastitis in cows.

16S rRNA gene sequence analysis places *Streptococcus parauberis* within the Pyogenic species group.

DNA G+C content (mol%): 34.8–36.5 (method not reported).

Type strain: BC45 RH, CCUG 39954, CIP 103956, DSM 6631, LMG 12174, NCIMB 702020.

GenBank accession number (16S rRNA gene): AY584477; X89967 (isolated from turbot, *Scophthalmus maximus*).

Additional comments: Strains of *Streptococcus uberis* and *Streptococcus parauberis* were previously referred to as *Streptococcus uberis* types I and II, respectively on the basis of DNA–DNA similarity data (Collins et al., 1984a; Garvie and Bramley, 1979b). These two species are physiologically and biochemically indistinguishable and species-specific probes have been designed to differentiate them (Bentley et al., 1993).

39. **Streptococcus pasteurianus** Poyart, Quesne and Trieu-Cuot 2002, 1253[VP]

pas.teu'ri.an.us. N.L. masc. adj. *pasteurianus* pertaining to the Pasteur Institute, where the species was characterized.

Cells are coccoid, occurring in pairs or short chains. Colonies on blood agar or nutrient agar are circular, smooth, entire, non-pigmented, and produce α-hemolysis. Homogeneous growth is obtained in brain heart infusion and in glucose broths after 18 h incubation at 37°C. Growth occurs in MRS broth without gas production. No growth occurs in the presence of 6.5% NaCl.

Acid is produced from lactose, maltose, melibiose, methyl D-glucoside, raffinose, sucrose, and trehalose. Acid from tagatose is a variable characteristic of this species. Acid is not produced from L-arabinose, D-arabitol, cyclodextrine, glycogen, inulin, D-mannitol, melezitose, pullulan, ribose, sorbitol, or starch.

Strains produce alanyl-phenylalanyl-proline arylamidase, leucine aminopeptidase, α-galactosidase, β-galactosidase, β-glucosidase, β-glucuronidase, and β-mannosidase. Strains do not produce arginine dihydrolase, alkaline phosphatase, glycyltryptophan arylamidase, pyrrolidonyl arylamidase, or N-acetyl-β-glucoaminidase.

Acetoin is produced (Voges–Proskauer reaction). Esculin is hydrolyzed but hippurate and urea are not.

Isolated from human clinical specimens including cerebrospinal fluid, blood, and urine.

16S rRNA gene sequence analysis places this species within the Bovis species group.

Streptococcus pasteurianus comprises strains previously assigned to *Streptococcus bovis* biotype II/2 and also proposed as *Streptococcus gallolyticus* subsp. *pasteurianus* (Schlegel et al., 2003). The classification of this species and its close relatives is detailed in the section on the Bovis species group.

DNA G+C content (mol%) of the DNA: has not been determined.

Type strain: NEM 1202, CIP 107122, DSM 15351, JCM 12261.

GenBank accession number (16S rRNA gene): AJ297216.

40. **Streptococcus peroris** Kawamura, Hou, Todome, Sultana, Hirose, Shu, Ezaki and Ohkuni 1998, 926[VP]

per.or.is. L. adj. *per* through; L. n. *os oris* oral cavity; L. adj. *peroris* pertaining to the oral cavity, from where the organism was isolated.

Cells are coccoid, approximately 0.6–0.8 μm in diameter, and grow in short chains. Facultatively anaerobic. α-hemolysis is produced on Columbia blood (sheep) agar with pin-point colonies being formed at 37°C under aerobic conditions for 1 d and approximately 0.3–0.8 mm diameter colonies after 3 d incubation. Strains produce acid from lactose, maltose, and sucrose. Acid is not produced from amygdalin, L-arabinose, D-arabinose, arabitol, arbutin, cyclodextrine, glycogen, inulin, mannitol, methyl β-D-glucopyranoside, melibiose, melezitose, pullulane, raffinose, ribose, sorbitol, tagatose, or trehalose. Strains produce β-galactosidase (*p*-nitrophenyl β-D-galactopyranoside substrate) and alanyl-phenylalanyl-proline arylamidase, but do not produce α-galactosidase, β-galactosidase (2-naphthyl β-D-galactopyranoside substrate), β-glucosidase, β-mannosidase, β-D-fucosidase, β-glucuronidase, glycine-tryptophan arylamidase, N-acetyl-β-glucosaminidase, pyrrolidonyl arylamidase, or urease. Alkaline phosphatase is a variable characteristic of this species. Arginine, esculin, and hippurate are not hydrolyzed. Acetoin is not produced (Voges–Proskauer reaction).

Strains have been isolated from the human tooth surface and pharynx.

16S rRNA gene sequence analysis places this species within the Mitis species group.

DNA G+C content (mol%): 39.8–40.5 (HPLC).

Type strain: O-66, ATCC 700780, CCUG 39814, CIP 105950, DSM 12493, GTC 848, JCM 10158, LMG 18719.

GenBank accession number (16S rRNA gene): AB008314.

41. **Streptococcus phocae** Skaar, Gaustad, Tønjum, Holm and Stenwig 1994, 649[VP]

pho'cae. L. gen. n. *phocae* of a seal.

Cells are coccoid, 1 μm in diameter, growing singly, in pairs, or chains. Colonies range from pinpoint up to 0.8 mm in diameter after aerobic incubation for 24 h at 37°C. β-Hemolysis is produced on blood agar plates. No growth is obtained at 10°C, at 45°C or in the presence of 40% bile or 6.5% NaCl. Acid is produced from N-acetylglucosamine, D-fructose, glucose, maltose, D-mannose, and ribose. Acid is not produced from inulin, lactose, mannitol, D-raffinose, salicin, sorbitol, or trehalose. Acid from glycogen and starch are variable characteristics of this species. A minority of strains (i.e., < 10%) produce acid from galactose, glycerol, melezitose, sucrose, and D-turanose. Alkaline phosphatase is produced by the majority of strains. Pyrrolidonyl arylamidase is not produced in the majority of strains. Acetoin is not produced (Voges–Proskauer reaction). Hyaluronidase is not produced. Arginine, esculin, and hippurate are not hydrolyzed. Starch is hydrolyzed by < 10% of strains. Strains are sensitive to bacitracin (0.04 IU) but not to optochin. Lancefield group C, F, or non-groupable. Isolated from the organs (lungs, liver, spleen, and kidneys) of septicemic seals.

16S rRNA gene sequence analysis places this species within the Pyogenic group.

DNA G+C content (mol%): 38.6 (type strain) (T_m).

Type strain: 8399 H1, ATCC 51973, CCUG 35103, CIP 104665, LMG 16735, NCTC 12719.

GenBank accession number (16S rRNA gene): AF235052, AJ621053.

42. **Streptococcus pluranimalium** Devriese, Vandamme, Collins, Alvarez, Pot, Hommez, Butaye and Haesebrouck 1997, 1225[VP]

plur.an.im.al'i.um. L. adj. *pluris* many; L. gen. pl. n. *animalium* from animals; N.L. gen. n. *pluranimalium* from many animals.

Cells are coccoid, forming chains or arranged in groups. Growth occurs equally well at 37°C and 42°C. Growth in broth culture produces precipitated cells with a clear supernatant. Colonies produce greening (α-hemolysis) on blood agar and are less than 1 mm in diameter. May grow in 6.5% NaCl broth. Acetoin is not produced (Voges–Proskauer reaction). Acid is produced from glucose, fructose, and trehalose, while other carbohydrate reactions may be weak and slow to develop in some strains. Acid is not produced from adonitol, D- and L-arabitol, D- and L-arabinose, D-cyclodextrin, dulcitol, erythritol, D- and L-fucose, gluconate, glycogen, inositol, inulin, 2- and 5-ketogluconate, D-lyxose, methyl α-D-glucoside, methyl α-D-mannoside, pullulan, rhamnose, starch, L-sorbose, D-turanose, D- and L-xylose, xylitol, and methyl β-xyloside. Variable carbohydrate reactions are obtained for *N*-acetylglucosamine, amygdalin, arbutin, cellobiose, galactose, β-gentiobiose, lactose, D-mannose, mannitol, maltose, melezitose, melibiose, methyl β-glucopyranoside, raffinose, ribose, sucrose, salicin, sorbitol, and D-tagatose. All strains produce leucine arylamidase, alanine-phenylalanine-proline arlyamidase, and are able to hydrolyze 4-methylumbelliferyl (MU)-phosphate, L-phenylalanine 7-amido-4-methyl-coumarin (AMC), and L-tryptophan AMC. The following are variable characteristics of this species: arginine dihydrolase, pyrrolidonyl arylamidase, α-galactosidase, β-galactosidase, β-glucuronidase, β-glucosidase, alkaline phosphatase, and *N*-acetyl-β-glucosaminidase activities, hydrolysis of 4-MU β-glucoside, L-valine AMC, 4-MU α-D-glucoside, L-pyroglutamic acid AMC, L-arginine AMC, 4-MU *N*-acetyl-β-D-glucosaminide, 4-MU β-D-glucuronide, L-isoleucine AMC, *p*-nitrophenyl β-D-glucoside, *p*-nitrophenyl β-D-cellobioside, proline, and leucine *p*-nitroanilide. Strains do not have β-mannosidase or urease activities and are unable to hydrolyze *p*-nitrophenyl α-D-maltoside. Hippurate is hydrolyzed. Esculin hydrolysis is a variable reaction of this species with the majority of strains hydrolyzing this substrate on Edwards agar and giving a brown discoloration of this medium. Amongst strains of bovine origin those from the genital tract differ from other strains in being unable to grow in 6.5% NaCl broth, produce acid from amygdalin, arbutin, cellobiose, β-gentiobiose, melezitose, raffinose, ribose, or

sorbitol, in less frequently producing acid from mannitol or having β-glucosidase activity and in more frequently having β-glucuronidase activity. Strains are isolated from a relatively wide range of animal hosts and sites including the genital tract, udder, and tonsils of cattle, tonsils of the cat and goat, and the lung, crop, and a pox lesion of the canary.

16S rRNA gene sequence analysis has not shown this species to be associated with the main recognized species groups within the genus but is associated with *Streptococcus hyovaginalis* and *Streptococcus thoraltensis*.

Strains of *Streptococcus pluranimalium* resemble *Streptococcus acidominimus* and some reference strains of the latter have been re-identified as *Streptococcus pluranimalium* (Devriese et al., 1999; Facklam, 2002).

DNA G+C content (mol%): 38.5 (HPLC).

Type strain: T70, ATCC 700864, LMG 14177, CCUG 43803, CIP 106120.

GenBank accession number (16S rRNA gene): not available from type strain; Y18026 for strain LMG 14257.

43. **Streptococcus pneumoniae** (Klein 1884) Chester 1901, 63[AL] (*Micrococcus pneumoniae* Klein 1884, 329)

pneu.mo'ni.ae. Gr. n. *pneumon* the lungs; M.L. fem. n. *pneumonia* pneumonia. M.L. gen. n. *pneumoniae* of pneumonia.

Cells are oval or spherical, coccus-like forms, 0.5–1.25 μm in diameter, typically in pairs, occasionally singly or in short chains. The distal end of each pair of cells tends to be pointed or lance-shaped. On primary isolation, generally heavily encapsulated with polysaccharide (termed SSS, specific soluble substance). Continued growth in laboratory media promotes chain formation. Gram-positive reaction of young cells may be lost as culture ages and subsequently stains Gram-negative.

Strong α-hemolysis on blood agar when cultures are incubated aerobically. Anaerobic incubation results in β-hemolysis due to pneumolysin O (identical to streptolysin O) activity. The addition of blood, serum, or ascitic fluid to media enhances growth especially on primary isolation. Mucoid colonies result from copious capsular polysaccharide synthesis. Smooth colonies are glistening and dome-shaped, and reflect decreased capsular polysaccharide. Rough colonies occur rarely and have a wrinkled, mycelium-like appearance. "Phantom" colonies reflect early and rapid partial autolysis of a mucoid colony which is suppressed by incubation under increased CO_2 tension.

Metabolism is fermentative, yielding primarily low levels of lactic acid. Final pH in glucose broth is approximately 5.0. Aerobically a significant quantity of H_2O_2 accumulates as well as acetic and formic acids. Acid is produced from glucose, galactose, fructose, sucrose, lactose, maltose, raffinose, glycogen, trehalose, and inulin. Strains frequently ferment *N*-acetylglucosamine and salicin. Slow acid production occurs from glycerol (aerobic incubation), xylose, arabinose, and erythritol. Some strains may ferment mannitol. Acid is not produced from amygdalin, arbutin, ribose, tagatose, dulcitol, or sorbitol.

Strains produce α-galactosidase, β-galactosidase, glycyl-tryptophan arylamidase, α-glucosidase, *N*-acetyl-β-galactosaminidase, sialidase (neuraminidase), and hyaluronidase. A minority of strains produce leucine arylamidase, pyrrolidonyl arylamidase, α-fucosidase, β-fucosidase, and β-glucosidase.

Strains do not produce alkaline phosphatase, urease, or β-glucuronidase.

Hydrogen peroxide is produced. Acetoin is not produced (Voges–Proskauer reaction).

Esculin and arginine hydrolysis have been reported variously in the literature with the majority listing esculin hydrolysis as a variable characteristic and arginine hydrolysis as a positive characteristic. These tests are listed as "+" for arginine and "d" for esculin as in the previous edition of the *Manual*.

Pneumococci are inhibited by approximately 1:400 ethylhydrocupreine HCl (optochin); this sensitivity test and the bile-solubility test can be used to differentiate pneumococci from other streptococci (Bowers and Jeffries, 1955).

The cell-wall peptidoglycan type is Lys–Ala$_2$ (Ser). Major constituents of the cell-wall polysaccharides are glucose, *N*-acetylgalactosamine, and ribitol with trace amounts of galactose. The C-polysaccharide antigen (phosphorylcholine-containing teichoic acid) is covalently attached to the *N*-acetylmuramic acid of the peptidoglycan backbone and is uniformly distributed over the cell surface. C-Polysaccharide is responsible for pneumococcal reaction with C-reactive protein, an acute-phase human serum protein, and also for the bacterium's susceptibility to pneumococcal autolysin (an *N*-acetylmuramic acid-L-alanine amidase) where autolysin release from C-polysaccharide linked to a lipid moiety (i.e., the Forssman antigen) during cessation of growth due to starvation or treatment with detergents or penicillin, results in cellular autolysis.

Pneumococcal capsules. Ninety antigenically distinct capsular serotypes are recognized differing in component sugars and/or linkages (Henrichsen, 1995) and the structures of approximately half have been determined (Henrichsen, 1995; van Dam et al., 1990). Together these serotypes form the basis of pneumococcal serological typing (Kauffmann et al., 1960) in which capsular serotypes are designated numerically (serotype 1, etc.), and antigenically related serotypes form serogroups designated a common serogroup number within which all member serotypes are designated a different letter e.g., 7A, 7B, etc. Pneumococcal capsules are generally composed of repeating oligosaccharide units consisting of 2–10 monosaccharides which may be substituted with other organic and inorganic molecules (van Dam et al., 1990). Genes encoding the enzymes essential for the synthesis of a specific capsular polysaccharide are linked within the chromosome and are flanked by genes, common to all capsular serotypes, which are involved in regulation and polysaccharide transport (Hardy et al., 2000).

Virulence factors. Several factors, most found at the cell surface, contribute to the pathogenicity of *Streptococcus pneumoniae*. These include factors that aid the pathogen in escaping the host immune system and involved in the inflammation resulting from infection (Alonso DeVelasco et al., 1995).

Capsules. Capsules of *Streptococcus pneumoniae* are regarded as being essential for virulence and serve to protect from phagocytosis (Alonso DeVelasco et al., 1995). Encapsulated strains have been found to be at least five times as virulent as non-encapsulated strains (Avery and Dubos, 1931; Watson and Musher, 1990). Of the 90 known capsular serotypes, those associated more often with invasive disease in children include serotypes 1, 5, 6B, 7F, 9V, 14, 18C, 19F, and 23F, with the spectrum of serotypes being less restricted in adult disease (Butler, 1997; Hausdorff et al., 2000; Sniadack et al., 1995).

Cell-wall polysaccharide. This induces inflammatory responses in the host by activation of the alternative complement pathway and stimulation of interleukin-1 production by monocytes (Riesenfeld-Orn et al., 1989; Winkelstein and Tomasz, 1977, 1978).

IgA$_1$ protease. This is produced and probably acts as a defence against the host immune system at mucosal surfaces.

Pneumolysin. This thiol-activated cytotoxin is released under the influence of autolysin. At high concentrations pneumolysin causes transmembrane ring formation on host cells resulting in lysis and at lower concentrations has several effects, including stimulating inflammatory cytokine (interleukin-1 and tumour necrosis factor) production by monocytes, slowing the beating of cilia and disrupting the integrity of human respiratory epithelium in culture, decreasing the migration and bactericidal activity of neutrophils, and inhibiting lymphocyte proliferation and antibody synthesis. Pneumolysin is able to activate complement via antibody binding of the Fc region. Pneumolysin has some sequence homology with C-reactive protein (CRP), an acute-phase protein that protects mice against pneumococci.

Autolysin. This is a cell-wall-degrading enzyme which, when activated during cell starvation or penicillin treatment, causes autolysis of the pneumococcus and release of inflammatory mediators such as peptidoglycan and teichoic acid and the release of pneumolysin from the cytoplasm.

Pneumococcal proteins studied in relation to virulence include neuraminidase (Berry et al., 1988), hyaluronidase (Berry et al., 1994), pneumococcal surface proteins PspA (Briles et al., 1988; Yother and Briles, 1992), PsaA (Sampson et al., 1994), SpsA (Hammerschmidt et al., 1997), and IgA$_1$ protease (Male, 1979; Poulsen et al., 1996).

Pathogenicity. The clinical patterns of pneumococcal infections in man are numerous. They include pneumonia, meningitis, otitis media, and some less frequent conditions such as abscesses, conjunctivitis, pericarditis, and arthritis. In animals pneumococci may occasionally cause mastitis and septicemia in cows, sheep and goats, and respiratory tract infections in monkeys.

Strains are isolated from the upper respiratory tract, inflammatory exudates and various body fluids of diseased humans and, rarely, domestic animals. The normal habitat in humans is the nasopharynx with an estimate of as high as 60% of the population colonized at any one time (Austrian, 1986).

DNA G+C content (mol%): 38.5 (chemical analysis) to 30 (T_m) and 42 (Bd). Genome sequencing of strain R6 gave an overall G+C content of 40.0 mol%.

Type strain: SV 1, ATCC 33400, CCUG 28588, CIP 102911, DSM 20566, LMG 14545, NCTC 7465.

GenBank accession number (16S rRNA gene): AF003930, AY485600, X58312.

Additional comments: Streptococcus pneumoniae is naturally competent for transformation with competence under the regulation of the *com* operon (Pestova et al., 1996). The process of horizontal gene transfer, often between *Streptococcus pneumoniae* and commensal oral streptococci including *Streptococcus mitis* and *Streptococcus oralis*, has been shown to be important in the evolution of penicillin resistance and virulence determinants (Dowson et al., 1997). Low-affinity PBPs arising in resistant strains are encoded by mosaic genes, with oral commensal species *Streptococcus mitis* and *Streptococcus oralis* as gene donors for PBP2B (Dowson et al., 1993) and PBP2X (Sibold et al., 1994), respectively. Recombination events at the capsular polysaccharide synthesis locus give rise to serotype changes among natural populations (Coffey et al., 1998). Other virulence determinants displaying evidence for gene transfer include autolysin (*lytA*), neuraminidase (*nanA*) and IgA$_1$ protease (*iga*). It has been postulated that atypical *Streptococcus pneumoniae* strains lacking one or more of the defining phenotypic characteristics of the species, i.e., optochin susceptibility, bile solubility, and agglutination with antipneumococcal polysaccharide capsule antibodies, and "*Streptococcus mitis*-like" strains harboring pneumococcal virulence factor genes may act as reservoirs of DNA for recombination events (Whatmore et al., 2000).

The genomes of three strains of *Streptococcus pneumoniae* have been sequenced. These include serotype 4 strain TIGR4 (Tettelin et al., 2001), avirulent (non-encapsulated) strain R6 (Hoskins et al., 2001), and >90% of the genome of serotype 19 clinical isolate G54 (Dopazo et al., 2001).

16S rRNA gene sequence analysis places *Streptococcus pneumoniae* within the Mitis species group.

44. **Streptococcus porcinus** Collins, Farrow, Katic and Kandler 1985, 224[VP] (Effective publication: Collins, Farrow, Katic and Kandler 1984a, 409.)

por.ci′nus. L. adj. *porcinus* of pigs, pertaining to pigs.

Cells are spherical to ovoid cocci in chains of small to medium length. Colonies on blood agar are normally small, elevated, and entire. β-Hemolysis is produced on blood agar. Growth does not occur at 10°C or at 45°C. Does not survive heating at 60°C for 30 min. Chemo-organotroph: metabolism fermentative. Acid is produced from glucose, *N*-acetylglucosamine, fructose, galactose, mannose, ribose, sorbitol, and trehalose. Most strains produce acid from cellobiose, glycerol, mannitol, maltose, salicin, and sucrose. Acid from amygdalin, lactose, and starch are variable characteristics of this species. Acid is not produced from D- and L-arabinose, D- and L-arabitol, adonitol, dulcitol, erythritol, D- and L-fucose, gluconate, glycogen, inositol, inulin, 2-ketogluconate, 5-ketogluconate, lyxose, melezitose, melibiose, methyl α-D-glucoside, methyl α-D-mannoside, raffinose, rhamnose, L-sorbose, tagatose, D-turanose, xylitol, D-xylose, and L-xylose. Acid production from arbutin, gentiobiose and startch are variable characteristics.

Strains produce alkaline phosphatase and leucine arylamidase. α-Galactosidase, β-galactosidase, and pyrrolidonyl arylamidase are not produced. β-Glucuronidase production is a variable characteristic. Arginine is hydrolyzed, some strains hydrolyze esculin and most hydrolyze gelatin if incubated anaerobically. Hippurate is not hydrolyzed. Acetoin production (Voges–Proskauer reaction) is a variable characteristic of this species. Strains may react with Lancefield group E,P,U, or V antiserum. The peptidoglycan contains lysine as the diamino acid (type: Lys–Ala$_{2-4}$). Menaquinones are absent. Major nonhydroxylated long-chain fatty acids are hexadecanoic ($C_{16:0}$) and octadecenoic (*cis*-vaccenic acid); cyclopropane fatty acids are absent.

Collins et al. (1984a) describe the type strain, NCTC 10999, as producing acetoin and β-glucuronidase, reacting with Lancefield group V antisera, hydrolyzing esculin, and producing acid from arbutin, amygdalin, cellobiose, glycerol, lactose, salicin, and sucrose, with all other characteristics being as given above.

Strains are commonly associated with abscesses in the cervical lymph nodes of pigs; also isolated from other diseases of pigs (pneumonia and septicemia) and from milk.

16S rRNA gene sequence analysis places this species within the Pyogenic species group.

DNA G+C content (mol%): 37.1–37.7 (T_m).

Type strain: ATCC 43138, CCUG 27628, CIP 103218, DSM 20725, LMG 15980, NCTC 10999.

GenBank accession number (16S rRNA gene): AB002523, X58315.

45. **Streptococcus ratti** corrig. Coykendall 1977, 28[VP]

rat′ti. L. gen. n. *ratti* of a rat.

Cells are coccoid, 0.5 μm in diameter and are in pairs or form chains. Growth occurs in air but is generally improved by an atmosphere of reduced O_2 content, with added CO_2. An adhesive extracellular glucan is produced from sucrose. Colonies on sucrose agar are often rough, heaped, and with beads or puddles of liquid (containing glucan) or some in the case of some strains, firm, "rubbery" colonies are formed. Strains are non-hemolytic on blood agar. The terminal pH in glucose broths is 4.2–4.4. Acid is produced from amygdalin, arbutin, cellobiose, methyl D-glucoside, *N*-acetylglucosamine, galactose, glucose, maltose, mannitol, raffinose, salicin, sorbitol, trehalose, sucrose, lactose, and inulin. The majority of strains produce acid from inulin, lactose, and melibiose. Acid is not produced from adonitol, arabinose, dulcitol, erythritol, glycerol, inositol, melezitose, methyl D-mannoside, methyl D-xyloside, rhamnose, ribose, sorbose, or xylose. Arginine and esculin are hydrolyzed but starch is not. Acetoin is produced (Voges–Proskauer reaction). Hydrogen peroxide is not produced. Strains can grow in the presence of 20 μg/ml bacitracin.

Strains of *Streptococcus ratti* were originally described (Coykendall, 1977) as a distinct serovar of *Streptococcus mutans*, designated *type b* (Bratthall, 1970). Two polysaccharide antigens, apparently with identical immunodeterminants, have been purified from this serovar (Mukasa and Slade, 1973) and a glycerol teichoic acid substituted with a galactosyl moiety has also been described (Vaught and Bleiweis, 1974). *Streptococcus ratti* was first obtained from a laboratory rat, but has also been isolated from the human mouth. However, this species is less commonly isolated from humans than other species belonging to the "mutans group" of streptococci.

16S rRNA gene sequence analysis places this species within the Mutans species group.

DNA G+C content (mol%): 41–43 (T_m).

Type strain: FAI, ATCC 19645 CCUG 27502, CCUG 27642, CIP 102509, DSM 20564, HAMBI 1518, LMG 14650.

GenBank accession number (16S rRNA gene): AJ420201.

Additional comments: Streptococcus ratti is the corrected name for strains previously called *Streptococcus rattus* by Coykendall (1977) (Trüper and de Clari, 1997).

46. **Streptococcus salivarius** Andrewes and Horder 1906, 712[AL]

sa.li.va′ri.us. L. adj. *salivarius* salivary, slimy.

Cells are spherical to ovoid, 0.8–1.0 μm in diameter, forming chains of varying length from short to very long. Grows readily on suitable media in the presence of O$_2$. Growth occurs at 45°C but not at 10°C. The majority of strains are non-hemolytic on blood agar with occasional α-hemolytic and β-hemolytic strains found. Smooth and rough variants occur, with the rough variant often reverting after subculture in broth. On sucrose-containing agar most isolates produce soluble fructan (levan) which results in the development of large mucoid colonies. Some strains also produce insoluble glucans (dextrans). Colonies on sucrose agar vary from smooth to rough depending upon the relative proportions of the different polysaccharides synthesized. Acid is produced from amygdalin, arbutin, cellobiose, esculin, fructose, glucose, sucrose, maltose, mannose, *N*-acetylglucosamine, pullulan, and salicin. Acid is not produced from arabinose, arabitol, adonitol, cyclodextrin, dulcitol, gluconate, glycogen, glycerol, mannitol, melibiose, rhamnose, ribose, sorbitol, L (−) sorbose, starch, tagatose, or xylose. The production of acid from galactose, lactose, methyl D-glucoside, inulin, raffinose, and trehalose are variable characteristics. The final pH produced in glucose broth is 4.0–4.4. Strains produce α-arabinosidase, acid- and alkaline phosphatases, alanyl-phenylalanyl-proline arylamidase, chymotrypsin, α-D- and glucosidases, leucine aminopeptidase, leucine arylamidase, α-maltosidase phosphoamidase, and valine aminopeptidase but not α-L- and β-D-fucosidase, α-D-galactosidase, glucosaminidase, *N*-acetyl-α-D-glucosaminidase, β-*N*-acetylgalactosidase, β-*N*-acetylglucosaminidase, β-D-glucuronidase, glycyl-tryptophan arylamidase, hyaluronidase, α- and β-mannosidase, pyroglutamic acid arylamidase, pyrrolidonyl arylamidase, phosphor-β-galactosidase, β-sialidase (neuraminidase), or α- and β-xylosidase. β-D-Fucosidase, β-D-galactosidase, β-maltosidase, β-lactosidase, and urease are variable characteristics of this species. However, it should be noted that the proportion of strains reported to be positive in these tests varies within the literature. Esculin and starch are hydrolyzed. Urea is hydrolyzed by some strains. Arginine and hippurate are not hydrolyzed. Acetoin is produced (Voges–Proskauer reaction) by the majority of strains. Hydrogen peroxide and IgA$_1$ protease are not produced. Approximately 50% of strains react with Lancefield group K antiserum. Two types of peptidoglycan have been found in *Streptococcus salivarius* (Lys–Ala$_{2-3}$ and Lys–Thr–Gly) (Schleifer and Kilpper-Bälz, 1987).

Strains are isolated from the oral cavities of man and animals, being associated particularly with the tongue and saliva, and in feces. It is occasionally isolated from the blood of patients with infective endocarditis. Some strains have been shown to be cariogenic in gnotobiotic rats.

16S rRNA gene sequence analysis places this species within the Salivarius species group.

DNA G+C content (mol%): 39–42 (T_m).
Type strain: ATCC 7073, CCUG 11878, CCUG 17825, CIP 102503, DSM 20560, HAMBI 1716, JCM 5707, LMG 11489, NCIMB 701779, NCTC 8618.
GenBank accession number (16S rRNA gene): AY188352.

47. **Streptococcus sanguinis** corrig. White and Niven 1946, 722[AL] emend. Kilian, Mikkelsen and Henrichsen 1989b, 482

san′gui.nis. L. n. *sanguis* blood; L. gen. n. *sanguinis* of the blood.

Cells are coccoid and occur in short chains in serum broth. α-Hemolysis is produced on blood (horse) agar and pronounced greening is seen on chocolate agar. The majority of strains are inhibited in the presence of 4% (w/v) NaCl and no growth occurs in 6.5% NaCl. Acid is produced from galactose, *N*-acetylglucosamine, fructose, glucose, maltose, mannose, lactose, salicin, sucrose, and trehalose. Acid is not produced from adonitol, arabinose, arabitol, cyclodextrin, dulcitol, glycerol, gluconate, glycogen, mannitol, D-melezitose, rhamnose, ribose, sorbose, tagatose, or xylose. Acid from amygdalin, arbutin, D-cellobiose, esculin, inulin, melibiose, methyl D-glucoside, pullulan, raffinose, sorbitol, and starch. Starch is weakly fermented producing a final pH of around 5.6–6.2. Reports of the proportion of strains able to ferment inulin vary in the literature.

Strains produce leucine aminopeptidase, α-maltosidase, and alanyl-phenylalanyl-proline arylamidase. Production of acid phosphatase, chymotrypsin, β-D-fucosidase, glycyl-tryptophan arylamidase, β-D-lactosidase, *N*-acetyl-β-D-glucosaminidase, phospho-β-galactosidase, phosphoamidase, and valine aminopeptidase are variable characteristics of this species. Strains do not produce alkaline phosphatase, α-L-arabinosidase, α-L-fucosidase, α-D-glucosidase, β-maltosidase, α- and β-mannosidases, β-L-fucosidase, β-glucuronidase, β-glucosaminidase, *N*-acetyl-α-D-glucosaminidase, *N*-acetyl-β-D-galactosaminidase, ornithine and lysine decarboxylases, sialidase (neuraminidase), urease, or α- and β-xylosidases. Detection of β-D-glucosidase, α-D-galactosidase, β-D-galactosidase, and *N*-acetyl-β-D-glucosaminidase activities are dependent on the substrate used and the percentage of strains reported as positive for these properties may vary considerably between studies.

Arginine is hydrolyzed. Esculin and starch are variable characteristics, being hydrolyzed by the majority of strains. Urea and hippurate are not hydrolyzed. Hydrogen peroxide and IgA1 protease are produced. Acetoin is not produced (Voges–Proskauer reaction). Dextran is produced from sucrose-containing media by the majority of strains. Hyaluronidase is not produced and strains do not bind salivary amylase. Some strains react to Lancefield group H antiserum raised against strain "Blackburn" (Kilian et al., 1989b).

The cell wall contains glycerol teichoic acid and rhamnose, and the peptidoglycan type is Lys–Ala$_{1-3}$.

Streptococcus sanguinis has been separated into biovars (Kilian et al., 1989b), or biotypes (Beighton et al., 1991), on the basis of several biochemical and serological characteristics.

Strains are isolated from the human mouth. *Streptococcus sanguinis* is one of the commonest streptococci isolated from infective endocarditis (Douglas et al., 1993) and is a primary colonizer of dental plaque (Fransen et al., 1991).

16S rRNA gene sequence analysis places this species within the Mitis species group.

DNA G+C content (mol%): 43–46% (T_m).

Type strain: SK 1, ATCC 10556, CCUG 17826, CCUG 35770, CIP 55.128, DSM 20567, JCM 5708, LMG 14702, NCTC 7863.

GenBank accession number (16S rRNA gene): AB002524, AF003928, X53653.

Additional comments: Streptococcus sanguinis is the corrected name for strains previously called *Streptococcus sanguis* (Trüper and de Clari, 1997).

48. **Streptococcus sinensis** Woo, Tan, Leung, Lau, Teng, Wong and Yuen 2002, 1438[VP] (Effective publication: Woo, Tan, Leung, Lau, Teng, Wong and Yuen 2002, 810.)

si.nen.sis. L. gen. n. *sinae* of China; N.L. masc. adj. *sinensis* pertaining to China, the country where the bacterium was isolated.

Cells are coccoid, 0.82–0.98 μm in diameter, occurring in chains. Facultatively anaerobic. Growth for 24h incubation at 37°C in ambient air sheep blood agar produces α-hemolytic, gray colonies of 0.5–1mm diameter. No enhancement of growth is seen in 5% CO_2. Growth occurs in 10 and 40% bile and on bile-esculin agar but not in 6% NaCl.

Acid is produced from glucose, lactose, salicin, sucrose, pullulan, trehalose, cellobiose, hemicellulase, mannose, maltose, and starch. Acid is not produced from inulin, mannitol, ribose, raffinose, or sorbitol. Strain HKU4 produces leucine arylamidase and β-glucosidase but not urease, lysine decarboxylase or ornithine decarboxylase. Esculin and arginine are hydrolyzed. Acetoin is produced (Voges–Proskauer reaction).

Non-groupable with Lancefield groups A, B, C, D, F, and G antisera. Resistant to optochin and bacitracin.

Strain HKU4, on which the species description is based, was isolated from the blood culture of a patient with infective endocarditis.

16S rRNA gene sequence analysis indicates that this species is affiliated with the Mitis and Anginosus species groups with 3.6, 3.7, 4.3, 4.7, and 5.9% base differences between it and *Streptococcus gordonii, Streptococcus intermedius, Streptococcus constellatus, Streptococcus sanguinis,* and *Streptococcus anginosus,* respectively. It is listed within the Mitis group in Table 133.

DNA G+C content (mol%): 53±2.9 (T_m).

Type strain: HKU4, DSM 14990, LMG 21517.

GenBank accession number (16S rRNA gene): AF432856.

49. **Streptococcus sobrinus** Coykendall 1983, 883[VP]

so.bri'nus. L. masc. n. *sobrinus* male cousin on mother's side (referring to the "distant relationship" between this species and *Streptococcus mutans*).

Cells are coccoid, 0.5 μm in diameter, and occur in pairs or chains, often long chains. Colonies on sucrose agar are about 1mm in diameter, rough, heaped, often showing

drops of glucan-containing liquid on or around the colony. Some strains are α-hemolytic on blood agar, others nonhemolytic. Acid is produced from glucose, maltose, mannose, sucrose, tagatose, and trehalose. Acid production is detectable from cellobiose and mannitol after extended incubation (96 and 48 h, respectively). Acid is not produced from adonitol, amygdalin, arabinose, arbutin, dulcitol, erythritol, esculin, gluconate, glycerol, glycogen, inositol, melibiose, methyl D-glucoside, methyl D-mannoside, methyl D-xyloside, *N*-acetylglucosamine, melibiose, melizitose, pullulan, raffinose, rhamnose, salicin, sorbose, starch, or xylose. Acid production from galactose, lactose, and sorbitol are variable characteristics of this species. Strains produce α-glucosidase. Alkaline phosphatase, α-arabinosidase, α-galactosidase, β-galactosidase, α-fucosidase, β-fucosidase, β-glucosidase, β-glucuronidase, *N*-acetylgalactosaminidase, *N*-acetylglucosaminidase, pyrrolidonyl arylamidase, glycyl tryptophan arylamidase, hyaluronidase, sialidase, and urease are not produced. Ammonia is not produced from arginine. Esculin is not hydrolyzed. Acetoin is produced (Voges–Proskauer reaction). Most strains produce hydrogen peroxide. Significant amounts of intra-cellular polysaccharide (IPS) are not produced. The peptidoglycan type is Lys–Thr–Ala. Strains usually belong to the serological type d or g of Bratthall (1970). However, the type strain SL-1 (ATCC 33478) does not belong to these serological types.

The habitat of *Streptococcus sobrinus* is the human tooth surface. Strains have been shown to be cariogenic in experimental animals and may be associated with human dental caries.

16S rRNA gene sequence analysis places this species within the Mutans species group.

The genome of strain 6715 is currently being sequenced (http://www.tigr.org/tdb/mdb/mdbinprogress.html).

DNA G+C content (mol%): 44–46 (T_m).

Type strain: SL1, NCTC 12279, ATCC 33478, CCUG 25735, CIP 103230, DSM 20742, HAMBI 1516, LMG 14641.

GenBank accession number (16S rRNA gene): AJ243966, AY188349, X58307.

50. **Streptococcus suis** (*ex* Elliott 1966) Kilpper-Bälz and Schleifer 1987, 160[VP]

su'is. L. n. *sus* a pig; L. gen. n. *suis* of a pig.

Cells are coccoid, < 2 μm in diameter and occur singly, in pairs, or (rarely) in short chains. There is some tendency towards rod formation. Many strains produce β-hemolysis on horse blood agar and all are α-hemolytic on sheep blood agar. No growth occurs at 10 or 45°C, in 6.5% NaCl or 0.04% tellurite. Some strains are resistant to bile. All are resistant to optochin. Chemo-organotroph with fermentative metabolism. Acid is produced from D-glucose, sucrose, lactose, maltose, salicin, trehalose, and inulin. Acid production from raffinose and melibiose are variable characteristics of this species. Acid is not produced from L-arabinose, D-mannitol, D-sorbitol, glycerol, melizitose, or D-ribose. Strains produce L-ornithine decarboxylase, β-acetylglucosaminidase, α-galactosidase, β-glucuronidase, and leucine arylamidase. β-Galactosidase and hyaluronidase production are variable characteristics. Strains do not produce acid or alkaline phosphatases or pyrrolidonyl arylamidase. Strains hydrolyze L-arginine, esculin, salicin, starch, and glycogen but not hippurate. Acetoin is not produced. Strains do not produce extracellular polysaccharide on sucrose-containing medium.

Type of peptidoglycan is usually lysine direct and, in a few cases, Lys–Ala$_{1-2}$. Cell-wall polysaccharides contain rhamnose, glucose, galactose, and glucosamine.

Strains assigned to *Streptococcus suis* by Kilpper-Bälz and Schleifer (1987) belonged to Lancefield groups R, S, RS, and T, or were ungroupable, and also included morphologically and biochemically similar strains described previously as serovars (1–8) of "*Streptococcus suis*" (Perch et al., 1983; Windsor and Elliott, 1975). Additional serotypes have been described (Gottschalk et al., 1991) and 35 different antigenic carbohydrate types of this species are described. The most common strains identified are Lancefield group R (type 2) and these are the only serotype identified from humans (Facklam, 2002). Lancefield group S strains are type 1.

Streptococcus suis is an important pig pathogen and has been isolated from cases of bacteremias and meningitis in piglets and from respiratory disease. Human infections are often associated with patients having worked in contact with pigs (Luttiken et al., 1986). The great extent of genetic diversity within this species has been noted by the demonstration of 60 pulsotypes by PFGE from a collection of 302 isolates from swine (Vela et al., 2003). 16S rRNA sequencing has shown a higher degree of diversity for serotypes 32, 33, and 34 (Chatellier et al., 1998).

16S rRNA gene sequence analysis places this species outside the main recognized species groups in most studies although it joined the Pyogenic species group in one study (Flint et al., 1999).

The genome of strain P1/7 is currently being sequenced and is approximately 1.7 Mb in size with a mol% G+C content of around 40.

DNA G+C content (mol%): 38–42 (T_m).

Type strain: Henrichsen S735, ATCC 43765, CCUG 7984, CIP 103217, DSM 9682, LMG 14181, NCTC 10234.

GenBank accession number (16S rRNA gene): AB002525 AF009477.

51. **Streptococcus thermophilus** (*ex* Orla-Jensen 1919) Schleifer, Ehrmann, Krusch and Neve 1995, 619[VP] (Effective publication: Schleifer, Ehrmann, Krusch and Neve 1991, 387.)

ther.mo'phil.us. Gr. n. *therme* heat; Gr. adj. *philos* loving. N.L. adj. *thermophilus* heat-loving.

Cells are spherical or ovoid, 0.7–1.0 μm in diameter, occurring in pairs to long chains. Irregular segments and cells can occur at 45°C. α-Hemolysis is formed on blood agar. Chemo-organotrophic, facultatively anaerobic. Able to grow at 45°C and most strains will survive heating to 65°C for 30 min. No growth at 15°C. No growth at pH 9.6 or in the presence of 0.1% methylene blue. Variable growth is observed in broth containing 2% NaCl, no growth at 3% NaCl. B-Vitamins and some amino acids are required. Acid is produced from fructose, glucose, lactose, mannose, and sucrose. Acid is not produced from *N*-acetylgalactosamine, *N*-acetylglucosamine, adonitol, amygdalin, arabinose, cellobiose, cyclodextrin, dulcitol, erythritol, gluconate, glycerol, glycogen, inulin, maltose, mannitol, methyl D-glucoside, methyl D-mannoside, methyl D-xyloside, rhamnose, salicin, sorbitol, tagatose, trehalose, and xylose. Variable reactions are observed for arbutin, galactose, melizitose, melibiose, raffinose, and ribose. Acetoin is produced (Voges–Proskauer reaction). Strains do not hydrolyze esculin, casein, gelatin, or hippurate. Ammonia is not produced from arginine, which is in accordance with Teuber and Geis (1981) but in contrast to Hardie (1986a). Starch hydrolysis is a variable characteristic of this species. β-Galactosidase and leucine arylamidase are produced. Acid- and alkaline-phosphatase, α-fucosidase, α-galactosidase, α-glucosidase, β-glucosidase, β-glucuronidase, cysteine arylamidase, *N*-acetyl-β-glucoaminidase, pyrrolidonyl arylamidase, and valine arylamidase are not produced. DNase and urease are not produced. The peptidoglycan type is Lys–Ala$_{2-3}$. No group-specific antigen has been demonstrated.

Streptococcus thermophilus is found in dairy sources including heated and pasteurized milk.

16S rRNA gene sequence analysis places this species within the Salivarius group.

The genome sequencing of *Streptococcus thermophilus* strain LMG 18311 is currently under way.

DNA G+C content (mol%): 37–40 (T_m).

Type strain: ATCC 19258, CCUG 21957, CIP 102303, DSM 20617, LMG 6896, NCIMB 8510.

GenBank accession number (16S rRNA gene): AY188354, X68418.

52. **Streptococcus thoraltensis** Devriese, Pot, Vandamme, Kersters, Collins, Alvarez, Haesebrouck and Hommez 1997, 1077[VP]

thor.al.ten'sis. L. masc. adj. *thoraltensis* from Thoraltum, L. name of Torhout, the town where the strains were isolated.

Cells are coccoid, arranged as short chains, pairs or groups. Colonies on blood (bovine) agar incubated in 5% CO$_2$ are approximately 1 mm in diameter, regular, opaque, and are α-hemolytic. Aerobic growth produces pinpoint colonies. Growth occurs at 25–37°C but is strongly inhibited at 42°C. Culture in broth produces sediment with clear supernatant. Growth occurs in the presence of 6.5% NaCl after 2 d. Strains will grow on Slanetz–Bartley medium without the reduction of triphenyltetrazolium chloride when incubated in the presence of 5% CO$_2$. Growth is strongly inhibited on bile–esculin medium. Strains produce acid from L-arabinose, arbutin, cellobiose, fructose, galactose, β-gentiobiose, methyl β-D-glucopyranoside, glucose, *N*-acetylglucosamine, inulin, lactose, maltose, mannose, mannitol, ribose, salicin, starch, sucrose, trehalose, and pullulan. Acid is not produced from adonitol, D-arabinose, D- and L-arabitol, cyclodextrin, dulcitol, erythritol, D-and L-fucose, glycerol, glycogen, gluconate, 2- and 5-ketogluconate, inositol, D-lyxose, methyl α-D-mannoside, rhamnose, L-sorbose, xylitol, L-xylose, or methyl β-xyloside. Strains produce leucine arylamidase, arginine dihydrolase, alanine phenylalanine proline arylamidase, β-glucuronidase, alkaline phosphatase (often weak), and β-glucosidase. Acetoin is produced (Voges–Proskauer reaction). Strains do not produce pyrrolidonyl arylamidase, β-mannosidase, urease, and *N*-acetyl-β-D-glucosaminidase. α-Galactosidase, alkaline phosphatase, and hippurate hydrolysis are variable characteristics of this species. β-Galactosidase test results may vary depending on the substrate used. Sorbitol, amygdalin, melibiose, melezitose, raffinose, D-turanose, tagatose, D-xylose, and methyl α-D-glucoside are variable characteristics of this species. Strains are ungroupable by Lancefield grouping reaction.

16S rRNA gene sequence analysis has shown *Streptococcus thoraltensis* to form a distinct subline within the streptococci in association with, but distinct from, *Streptococcus hyovaginalis*.

Strains are associated with pigs, occurring most frequently in vaginal fluids and in the intestine. Involvement with pathological processes in these animals is unclear.

DNA G+C content (mol%): 40 (T_m).

Type strain: S69, ATCC 700865, CCUG 32906, CCUG 37868, CIP 105518, LMG 13593.

GenBank accession number (16S rRNA gene): Y09007.

53. **Streptococcus uberis** Diernhofer 1932, 370[AL]

u'ber.is. L. n. *uber* udder, teat; L. gen. n. *uberis* of an udder.

Cells are coccoid, occurring as pairs to chains of moderate length. α-Hemolysis is produced on blood agar but β-hemolysis is not. Growth occurs in the presence of 4% but not 6.5% NaCl. Growth does not occur at pH 9.6. Growth is slow or does not occur at 10°C and does not occur at 45°C. Does not survive heating at 65°C for 30 min.

Acid is produced from amygdalin, arbutin, cellobiose, β-gentiobiose, inulin, lactose, maltose, mannitol, ribose, salicin, sorbitol, sucrose, *N*-acetylglucosamine, fructose, galactose, glucose, mannose, and trehalose. The majority of strains produce acid from tagatose. A minority of strains produce acid from glycogen, methyl α-D-glucoside and starch. Acid is not produced from adonitol, D-arabinose, L-arabinose, D-arabitol, L-arabitol, erythritol, D-fucose, L-fucose, dulcitol, glycerol, gluconate, 2-ketogluconate, 5-ketogluconate, inositol, lyxose, melibiose, melizitose, methyl α-mannoside, methyl α-xyloside, raffinose, rhamnose, sorbose, turanose, xylitol, D-xylose, or L-xylose. Strains produce leucine arylamidase, pyrrolidonyl arylamidase, and β-glucuronidase. A few strains produce alkaline phosphatase. Strains do not produce α- or β-galactosidase.

Arginine, esculin, and hippurate are hydrolyzed. Acetoin is produced (Voges–Proskauer reaction). Strains reduce tetrazolium, but do not decarboxylate tyrosine or produce DNase. The peptidoglycan type is Lys–Ala$_{1-3}$.

Strains are isolated from lips and skin of cows, raw milk, and udder tissue. All isolates previously reported from humans have been reidentified as *Globicatella sanguinis* (Facklam, 1977, 2002)

16S rRNA gene sequence analysis places this species within the Pyogenic species group.

DNA G+C content (mol%): 36–37.5 (T_m).

Type strain: ATCC 19436, CCUG 17930, CCUG 27579, CIP 103219, DSM 20569, JCM 5709, LMG 9465, NCTC 3858.

GenBank accession number (16S rRNA gene): AB002526, AB023573.

Additional comments: Strains of *Streptococcus uberis* and *Streptococcus parauberis* were previously referred to as *Streptococcus uberis* types I and II, respectively on the basis of DNA–DNA similarity data (Collins et al., 1984a; Garvie and Bramley, 1979b). These two species are physiologically and biochemically indistinguishable, and species-specific probes have been designed to differentiate them (Bentley et al., 1993).

The genome of *Streptococcus uberis* strain 0140J is currently being completed. The genome is approximately 1.7 Mb in size with a mol% G+C content of around 40.

54. **Streptococcus urinalis** Collins, Hutson, Falsen, Nikolaitchouk, LaClaire and Facklam 2000, 1177[VP]

u.ri.na'lis. M.L. adj. *urinalis* pertaining to urine.

Cells are ovoid and occur singly, in pairs or short chains. Facultatively anaerobic. Non-hemolytic on blood agar. Growth occurs in 6.5% NaCl but not at 10°C. Gas is not produced in MRS broth. Acid and clot are formed in litmus milk. Strains produce acid from glucose, lactose, ribose, sucrose, maltose, and trehalose, although lactose fermentation has been reported as a negative characteristic when testing strains using a commercial test system (API) (Collins et al., 2000). Acid is not produced from L-arabinose, D-arabitol, cyclodextrin, glycogen, glycerol, inulin, mannitol, melibiose, melezitose, methyl β-D-glucopyranoside, *N*-acetylglucosamine, pullulan, sorbitol, sorbose, raffinose, or tagatose. Strains are positive for leucine aminopeptidase, pyrrolidonyl arylamidase, acid phosphatase, alkaline phosphatase, α-glucosidase, β-glucosidase, pyroglutamic acid arylamidase, and leucine arylamidase. Strains are negative for alanine-phenylalanine-proline arylamidase, chymotrypsin, esterase C4, ester lipase C8, α-fucosidase, α-galactosidase, β-galactosidase, β-galacturonidase, β-glucuronidase, glycyl-tryptophan arylamidase, lipase C14, α-mannosidase, β-mannosidase, trypsin, and valine arylamidase. Hippurate, urea, starch, and arginine are not hydrolyzed. Pyruvate is not utilized. Extracellular polysaccharide is not produced. Acetoin is produced (Voges–Proskauer reaction). Strains are vancomycin-sensitive and bacitracin-resistant. Bile-esculin-positive. Strains are not grouped with Lancefield antisera for groups A, B, C, D, E, F, and G.

Strains have been isolated from the urine of patients suffering from cystitis.

16S rRNA gene sequence analysis places this species within the Pyogenic species group.

DNA G+C content (mol%): 39 (T_m).

Type strain: 2285-97, CCUG 41590, CIP 106463.

GenBank accession number (16S rRNA gene): AJ131965.

55. **Streptococcus vestibularis** Whiley and Hardie 1988, 338[VP]

ves.tib.u.lar'is. L. n. *vestibulum* entrance hall or forecourt; N.L. adj. *vestibularis* pertaining to the vestibule of the mouth where the organism was originally isolated.

Cells are coccoid, approximately 1 μm in diameter growing in chains. α-hemolysis is produced on blood (horse) agar. Colonies growing anaerobically for 3 d on sucrose-containing (Mitis Salivarius) agar are 1–2 mm in diameter, dark blue, matte, and umbonate with undulate edges. Colonies grown aerobically on Mitis Salivarius agar are

1–2 mm in diameter, dark blue, convex, glossy with entire edges. Both aerobic and anaerobic growth on 5% sucrose containing agar (TYC) produce 1–2 mm diameter colonies, white, convex and glossy with entire edges. Growth does not occur at 10°C or at 45°C, in the presence of 4% NaCl or 0.0004% (w/v) crystal violet. The majority of strains grow in the presence of 10% (w/v) bile but not 40% (w/v). The type strain does not grow in 10% bile. Extracellular polysaccharide is not produced on sucrose agar. Acid is produced from N-acetylglucosamine, fructose, galactose, glucose, lactose, maltose, mannose, salicin and sucrose. Acid is not produced from adonitol, arabinose, dextrin, dulcitol, fucose, glycerol, glycogen, inositol, inulin, mannitol, melezitose, melibiose, raffinose, ribitol, ribose, sorbitol, starch, or xylose. The majority of strains produce acid from cellobiose and amygdalin. Few strains produce acid from trehalose or D-glucosamine hydrochloride. Arbutin fermentation is a variable characteristic of *Streptococcus vestibularis*. Strains which ferment trehalose but not cellobiose and amygdalin constitute a separate biotype from the majority of the strains.

β-Galactosidase is produced. Alkaline phosphatase, α-galactosidase, α-fucosidase, β-glucosidase, glucuronidase, β-fucosidase, N-acetylglucosaminidase, N-acetylgalactosaminidase, hyaluronidase, neuraminidase, and pyrrolidonyl arylamidase are not produced. Urease and hydrogen peroxide are produced. α-Glucosidase is a variable characteristic of this species. Esculin and starch are hydrolyzed. Acetoin is produced (Voges–Proskauer reaction). Arginine is not hydrolyzed to produce ammonia. Cellular long-chain fatty acids consist of major amounts of hexadecanoic ($C_{16:0}$, palmitoleic), octadecenoic ($C_{18:1\omega7}$, cis-vaccenic) acids together with tetradecanoic ($C_{14:0}$, myristic), hexadecenoic ($C_{16:1}$, palmitoleic), octadecanoic ($C_{18:0}$, stearic), octadecenoic ($C_{18:1\omega9}$, oleic), and eicosenoic ($C_{20:1}$) acids.

Strains are isolated mainly from the oral vestibular mucosa of the human oral cavity.

16S rRNA gene sequence analysis places this species within the Salivarius species group.

DNA G+C content (mol%): 38–40 (T_m).

Type strain: MM1, ATCC 49124, CCUG 24893, CIP 103363, DSM 5636, LMG 13516, NCTC 12166.

GenBank accession number (16S rRNA gene): AY188353, X58321.

Species *Incertae Sedis*

56. **Streptococcus pleomorphus** Barnes, Impey, Stevens and Peel 1977, 52[AL].

ple.o.mor'phus. N.L. adj. *pleomorphus* many forms.

Obligatory anaerobic. Cells are pleomorphic and occur singly, in pairs or short chains. Variations in size and shape of cells occur depending on the media and growth conditions. Cells may stain Gram-negative within 24 h of culture. Colonies after 3 d of incubation on growth media such as RCM, supplemented BGP agar (Barnes et al., 1978), or VL agar (Barnes and Impey, 1970; Beerens et al., 1963) are 2–3 mm in diameter, circular, convex with irregular edge. Some strains are weakly β-hemolytic on VL agar. Growth in broth is flocculent. Good growth occurs at 37°C and at 45°C but none occurs at 20°C. No growth occurs in air or in air plus 10% CO_2. Carbohydrates are required for growth. Glucose is fermented to L-lactic acid; no gas production occurs. Traces of butyric, formic, and sometimes acetic and succinic acids are also found. The terminal pH is 4.4–5.0. Acid is produced from glucose, fructose, and, usually, mannose. No acid is produced from arabinose, cellobiose, dextrin, galactose, inositol, lactose, maltose, mannitol, salicin, starch, sucrose, or xylose. No change in cysteine milk, gelatin is not liquefied, indole is not produced and nitrates are not reduced. Small amounts of H_2S are detected in media containing ferrous sulfate and sodium thiosulfate. Growth occurs in the presence of polymyxin B (10 μg/ml), neomycin (100 μg/ml), and kanamycin (100 μg/ml), and most strains are resistant to brilliant green (1/100,000).

The peptidoglycan type is Lys–Thr–Ala (Ser).

Strains have been isolated from the intestines of chickens and occasionally from human feces.

DNA G+C content (mol%): 39 (type strain) (T_m).

Type strain: EBF 61/60B, ATCC 29734, CCUG 11733, DSM 20574, JCM 10414, NCTC 11087.

GenBank accession number (16S rRNA gene): M23730.

Additional comments: Previously listed in *Bergey's Manual of Systematic Bacteriology* within the "anaerobic streptococci" (Hardie, 1986c), subsequent 16S rRNA gene sequence cataloging (Ludwig et al., 1988) and 16S rRNA sequencing (Kawamura et al., 1995a) place this species outside the genus *Streptococcus* and most closely related to *Clostridium innocuum*. Recent analysis has placed this species within the family *Erysipelotrichaceae* together with *Eubacterium biforme* and *Eubacterium cylindroides* (W. Wade, personal communication). See *Eubacterium* section of this volume).

"Lancefield group M Streptococcus"

In the previous edition of the *Manual* (Rotta, 1986) streptococci of Lancefield group M were also listed under *Species Incertae Sedis*. The description given included three biovars:

All three biovars produce acid from glucose, lactose, maltose, and sucrose but not from adonitol, dulcitol, melibiose, melezitose, inositol, rhamnose, or xylitol. Strains are unable to grow in the presence of 40% bile, 6.5% NaCl, or 0.04% tellurite.

Biovar I. α-Hemolytic, from humans, do not hydrolyze arginine, give a final pH in glucose broth of 4.6–5.2. Strains of this biovar also contain a heat-stable "group" antigen that is resistant to pepsin. Isolated from cases of endocarditis, abscesses, and the nasopharynx and vagina.

Biovar II. β-Hemolytic, from animals, are able to hydrolyze arginine, attain a final pH of 6.3–7.2 in glucose broth. Strains also contain both of the additional "group" antigens found in biovar I and III strains. Isolated from the urethra, vagina, and tonsillar area of dogs.

Biovar III. β-Hemolytic, also from animals, hydrolyze arginine, produce a lower final pH of 5.9–6.7 in glucose broth. Strains also contain a heat-labile "group" antigen that is pepsin-sensitive.

The study by Colman (1968) demonstrated that streptococci of Lancefield group M clustered on the one hand with strains designated "*Streptococcus mitior*" as well as including distinct, β-hemolytic strains, on the other. During the DNA–DNA reassociation studies of Kilpper-Bälz and Schleifer (1984), a Lancefield group M strain was shown to belong to *Streptococcus agalactiae*. These data indicate that Lancefield group M streptococci belong to disparate taxa where possession of this antigen can be considered a variable characteristic.

Recently described species

Arbique et al. (2004) have recently described the new species *Streptococcus pseudopneumoniae* (type strain ATCC BAA-960), to which they have assigned unusual viridans streptococci resembling *Streptococcus pneumoniae* (R.R. Facklam, personal communication). *Streptococcus pseudopneumoniae* strains were characterized as resistant to optochin under increased CO_2 content, but susceptible when under an "ambient" atmosphere, not soluble in bile, and as being indistinguishable from *Streptococcus pneumoniae* when tested using the AccuProbe Pneumococcus test kit (based on 16S rRNA gene sequence), on the basis of PCR detection of the pneumolysin (*ply*) gene and by sequencing the superoxide dismutase (*sodA*) gene. However these strains formed a DNA similarity group distinct from all other streptococcal species tested, including *Streptococcus pneumoniae*, by DNA–DNA reassociation.

Between completion of this chapter and publication several new species and a subspecies have been described and are listed below.

Streptococcus caballi, isolated from the hindgut of horses with oligofructose-induced laminitis (Milinovich et al., 2008).

Streptococcus castoreus, isolated from a beaver (*Castor fiber*) (Lawson et al., 2005b).

Streptococcus dentirousetti, isolated from the oral cavities of bats (Takada and Hirasawa, 2008).

Streptococcus equi subsp. *ruminatorum*, isolated from mastitis in small ruminants (Fernandez et al., 2004)

Streptococcus halichoeri, isolated from gray seals (*Halichoerus grypus*) (Lawson et al., 2004).

Streptococcus henryi, isolated from the hindgut of horses with oligofructose-induced laminitis (Milinovich et al., 2008).

Streptococcus ictaluri, isolated from channel catfish (*Ictalurus punctatus*) (Shewmaker et al., 2007).

Streptococcus marimammalium, isolated from seals (Lawson et al., 2005a).

Streptococcus massiliensis, isolated from a human blood culture (Glazunova et al., 2006).

Streptococcus orisuis, isolated from the oral cavity of a pig (Takada and Hirasawa, 2007).

Streptococcus pseudoporcinus, isolated from the human female genitourinary tract (Bekal et al., 2006).

Acknowledgements

The authors would like to thank Professor William Wade for help and guidance with the construction of the phylogenetic tree of the genus *Streptococcus* (Figure 123).

Genus II. **Lactococcus** Schleifer, Kraus, Dvorak, Kilpper-Bälz, Collins and Fischer 1986, 354 [VP] (Effective publication: Schleifer, Kraus, Dvorak, Kilpper-Bälz, Collins and Fischer 1985, 189.)

MICHAEL TEUBER

Lac.to.coc′cus. L. n. *lac* milk, L. gen. n. *lactis* of milk, Gr. n. *coccus* (*kokkos*) a grain or berry; N.L. masc. n. *Lactococcus* milk coccus

Spheres or ovoid cells occur singly, in pairs, or in chains, and are often elongated in the direction of the chain. Gram-stain-positive. Endospores are not formed. Nonmotile. Not β-hemolytic. Facultatively anaerobic; catalase-negative. Growth at 10°C but not at 45°C. Usually grows in 4% (w/v) NaCl with the exception of *Lactococcus lactis* subsp. *cremoris* which only tolerates 2% (w/v) NaCl. Chemo-organotroph. Metabolism fermentative. The predominant end product of glucose fermentation is L(+)-lactic acid. May contain menaquinones. Most strains react with group N antisera (Lancefield, 1933). Nutritional requirements are complex and variable. The peptidoglycan is of group A with L-lysine as diamino acid in position 3 of the peptide subunit. Typical inhabitants of plants, animals, and their products.

DNA G+C content (mol%): 34–43 (T_m).

Type species: **Lactococcus lactis** (Lister 1873) Schleifer, Kraus, Dvorak, Kilpper-Bälz, Collins and Fischer 1986, 354 [VP] (Effective publication: *Lactococcus lactis* Schleifer, Kraus, Dvorak, Kilpper-Bälz, Collins and Fischer 1985, 189.) (*Bacterium lactis* Lister 1873, 408; *Streptococcus lactis* Löhnis 1909, 554.)

Further descriptive information

Phylogenetic treatment. The phylogenetic position of the lactococci within the *Firmicutes* was established by comparison of the 16S rRNA sequences (Schleifer and Ludwig, 1995) and in the current analysis used in the roadmap to Volume 3 (Ludwig et al., this volume) *Lactococcus* is a member of the family *Streptococcaceae*, along with *Streptococcus* and *Lactovum*, in the order *Lactobacillales* (Figure 1 and Figure 3). The lactococci are clearly separated from pathogenic genera of streptococci (Stackebrandt and Teuber, 1988). A phylogenetic tree reflecting the close relationships within the genus *Lactococcus* is shown in Figure 124. The five accepted species *Lactococcus garvieae*, *Lactococcus lactis*, *Lactococcus raffinolactis*, *Lactococcus piscium*, and *Lactococcus plantarum* are clearly separated.

Based on this phylogeny, Tailliez (2001) proposed that the lactic acid bacteria including the lactococci appeared before the photosynthetic cyanobacteria. As these were found in sediments dated 2.75 billion years ago, lactic acid bacteria may have emerged 3 billion years ago, before the atmosphere contained appreciable amounts of oxygen, which is consistent with their poor adaptation to aerobic environments.

Cell morphology. A typical scanning electron micrograph of *Lactococcus lactis* is presented in Figure 125. The elongated shape of the ovoid cells is evident.

Cell-wall composition. The peptidoglycan type (Schleifer and Kandler, 1972) was an important early feature that enabled differentiation of the five species: Lys–D-Asp in *Lactococcus lactis*

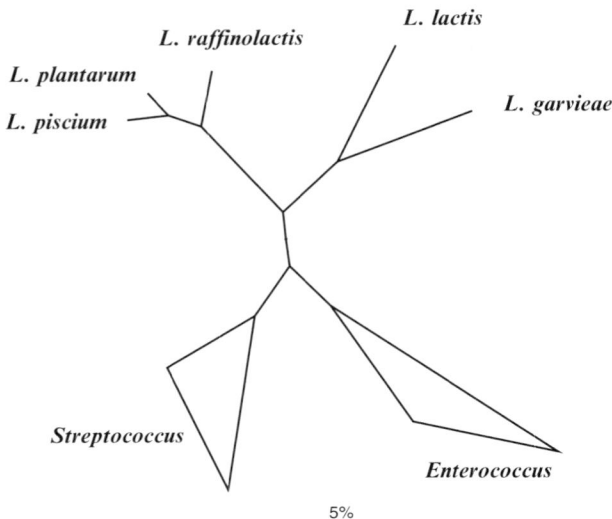

FIGURE 124. Phylogenetic tree of the species of the genus *Lactococcus*. The tree was constructed by a maximum-likelihood analysis of more than 50,000 full 16S rRNA sequences. The bar indicates 5% estimated sequence differences (courtesy of Wolfgang Ludwig, TU Munich, Germany).

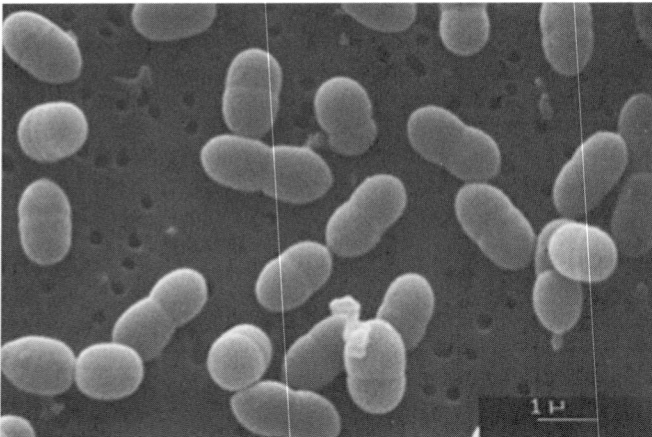

FIGURE 125. Scanning electronmicrograph of *Lactococcus lactis* subsp. *lactis* Bu2 (Neve et al., 1987).

and its subspecies, Lys–Ala–Gly–Ala in *Lactococcus garvieae,* Lys–Ser–Ala in *Lactococcus plantarum,* and Lys–Thr–Ala in *Lactococcus raffinolactis* (Schleifer et al., 1985); the peptidoglycan type of *Lactococcus piscium* has not been reported.

The group N antigen previously important for classification, differentiation, and identification within the large group of morphologically very similar streptococci (Lancefield, 1933) is a cell-wall-associated glycerol teichoic acid containing galactose phosphate (Smith and Shattock, 1964). Commercially available antibodies for its determination are no longer on the market.

Colonial and cultural characteristics. Lactococci produce small, translucent to whitish colonies that are circular, smooth, and entire within 1–2 d of incubation on semisolid complex media like blood agar, Elliker agar (Elliker et al., 1956), or

M17 medium (Terzaghi and Sandine, 1975). Incubation under reduced oxygen pressure is not necessary.

Nutrition and growth conditions. Lactococci, like lactobacilli, are typical inhabitants of plants and animals and products derived from these organisms (see *Ecology* section). As a consequence they are nutritionally fastidious because they have adapted to being supplied by their host environment with carbohydrates, amino acids, vitamins, nucleic acid derivatives, fatty acids, and other compounds. The sum of these requirements is obvious from the composition of a defined medium that was developed to achieve exponential growth of *Lactococcus lactis* (Jensen and Hammer, 1993): the amino acids L-alanine, L-arginine, L-asparagine, L-cysteine, L-glutamate, L-glutamine, glycine, L-histidine, L-isoleucine, L-leucine, L-lysine, L-methionine, L-phenylalanine, L-proline, L-serine, L-threonine, L-tryptophan, L-tyrosine, and L-valine, glucose as carbohydrate, the short-chain fatty acid acetate, the vitamins biotin, pyridoxal-HCl, folic acid, riboflavin, niacinamide, thiamine, and pantothenate, and the micronutrients Mo, B, Co, Cu, Mn, and Zn, in addition to Mg and Ca.

Lactococci are mesophilic (grow in the range of 10–40°C) but some may grow at as low as at 7°C upon prolonged incubation for 10–14 d (Sakala et al., 2002b).

Lactococci are microaerophilic and growth is not severely affected by aeration. Under aerobic growth conditions, however, highly toxic oxygen compounds such as superoxide, hydrogen peroxide, and hydroxyl radicals are generated by *Lactococcus lactis.* Although lactococci are catalase-negative, they possess NADH oxidases/peroxidases, and superoxide dismutases which are up-regulated under aerobic stress conditions (Hansson and Häggström, 1984). In the presence of a heme source in a complex medium, *Lactococcus lactis* has been demonstrated to exhibit respiration due to the presence of a terminal cytochrome oxidase (*bd*) (Gaudu et al., 2002). This leads to increased survival and growth yield, decreased lactate production, and increased acetate and diacetyl in the medium. So lactococci may be on their way from a fermentative to an oxidative route of living.

Lactococci grow best at near neutral pH values in buffered media but cease to grow at about pH 4.5.

Metabolism, metabolic pathways, and genetics. The metabolism of lactococci in an unstirred culture such as in milk fermentations is fermentative; it means energy for growth is produced by substrate-level phosphorylation. The study of metabolism and metabolic pathways is tremendously advanced due to the application of *Lactococcus lactis* strains of the subspecies *cremoris* and *lactis* as starter cultures for the manufacture of fermented dairy products like hard, semihard, soft, and fresh cheeses, sour milk, and sour cream (Teuber, 2000) as compiled in Table 136.

Lactose metabolism. Lactococci are homofermentative lactic acid bacteria. During growth in milk, lactose is converted to lactic acid. Lactose is taken up by a PEP-dependent phosphotransferase system. During transport, lactose is phosphorylated to lactose 6-phosphate which is subsequently hydrolyzed by a phospho-β-galactosidase into glucose and galactose 6-phosphate (gal-6-P). Glucose is phosphorylated by glucokinase and metabolized to lactate by the glycolytic pathway. Gal-6-P is converted to the glycolytic intermediates glyceraldehyde 3-phos-

TABLE 136. Lactococci as components of starter cultures for fermented dairy products[a]

Type of product	Composition of starter culture
Cheese type without eye formation (Cheddar, Camembert, Tilsit)	*Lactococcus lactis* subsp. *cremoris*, 95–98%; *Lactococcus lactis* subsp. *lactis*, 2–5%
Cottage cheese, quarg, fermented milk, cheese types with few or small eyes (e.g., Edam)	*Lactococcus lactis* subsp. *cremoris*, 95% and *Leuconostoc mesenteroides* subsp. *cremoris*, 5% or *Lactococcus lactis* subsp. *cremoris*, 85–90%, *Lactococcus lactis* subsp. *lactis*, 3%, *Leuconostoc mesenteroides* subsp. *cremoris*, 5%
Cultured butter, fermented milk, buttermilk, cheese types with round eyes (e.g., Gouda)	*Lactococcus lactis* subsp. *cremoris*, 70–75%, *Lactococcus lactis* subsp. *lactis* biovar diacetylactis, 15–20%, and *Leuconostoc mesneteroides* subsp. *cremoris*, 2–5%
Taette (Scandinavian ropy milk)	*Lactococcus lactis* subsp. *cremoris* (ropy strain producing extracellular polysaccharides)
Viili (Finnish ropy milk)	*Oidium lactis* (yeast covering surface); *Lactococcus lactis* subsp. *cremoris* (ropy strain)
Casein	*Lactococcus lactis* subsp. *cremoris*
Kefir	Kefir grains containing lactose-fermenting yeasts (e.g., *Candida kefir*), *Lactobacillus kefir*, *Lactobacilus kefiranofacians*, *Lactococcus lactis* subsp. *lactis*

[a]The quantitative composition has been taken from the culture catalogue of a major worldwide supplier.

phate and dihydroxyacetone-phosphate by enzymes of the tagatose pathway including galactose-6-P isomerase, tagatose-6-phosphate kinase, and tagatose-1,6-diphosphate aldolase. So far the *lac*-phenotype has always been found to be correlated to the presence of specific plasmids in different strains. In *Lactococcus lactis* subsp. *cremoris* NCDO712, the structural genes for the *lac*-specific PTS-system, the phospho-β-galactosidase, and the enzymes of the tagatose pathway are organized in a 7.8 kb lac-operon located on a 56.6 kb conjugative plasmids. A specific repressor (lacR) is located at the start of the operon. This gene is expressed in the presence of glucose and repressed during growth in milk. Transcription of the lac-operon into two polycistronic transcripts (*lacA–E*, *lacA–X*) is regulated by the repressor LacR and tagatose 6-phosphate as the inducer (van Rooijen and de Vos, 1990).

Exopolysaccharides. The production of significant amounts of exopolysaccharides (EPS) is quite common among lactococci and an important technological trait for the produced sour milk or cheese (von Wight and Tynkkynen, 1987). The strains can be assigned to three major groups: group I contains strains that produce EPS containing galactose, glucose, and rhamnose; the EPSs of group II strains consist only of galactose, and those of group III contain galactose and glucose. Additional strains show unique EPS sugar composition. The genes for EPS synthesis are found on plasmids larger than 20 kb and clustered in large operons of up to fourteen coordinately expressed genes; one of these plasmids, the 42,180-bp pNZ4000, has been completely sequenced (van Kranenburg et al., 2000).

Amino acid and protein metabolism. Dairy lactococci are dependent upon the presence of some essential amino acids for growth (Anderson and Elliker, 1953; Jensen and Hammer, 1993), whereas some *Lactococcus lactis* subsp. *lactis* strains isolated from nondairy environments are prototrophic for amino acids. The analysis of the genome sequence of *Lactococcus lactis* IL1403 revealed the genetic potential (95 putative genes) to synthesize all 20 standard amino acids. Mutational inactivation of some amino acid biosynthetic pathways, a possible consequence of the adaption to growth in milk, can explain the auxotrophic phenotype of dairy lactococci. Several operons involved in the synthesis of amino acids have been characterized in detail in lactococci from dairy and nondairy origins. These include those for the biosynthesis of aromatic amino acids, histidine, threonine, and the branched-chain amino acids.

Due to this amino acid auxotrophy and the lack of sufficient amounts of free amino acids in milk, dairy lactococci need a proteolytic system for growth in milk. This system consists of a cell-wall-bound protease (PrtP), three peptide transporter (Opp, DtpT, DtpP), and a variety of peptidases (Pep). The genes for the majority, if not all, of the enzymes necessary for casein degradation and transport of the degradation products were cloned, sequenced, and analyzed in detail. From these data, the following pictures emerges. Caseins of the milk are partially degraded by the cell-wall-bound protease into a large number of oligopeptides; some of these oligopeptides (≤10 amino acid residues) are taken up by the oligo- and the di/tripeptide transport systems and subsequently hydrolyzed to amino acids by a plethora of peptidases. Mutants missing transporter or/and peptidase genes have been constructed by targeting deletion or disruption of the corresponding genes. Those missing the Opp system but still having di/tripeptide transport activity are unable to grow in milk. Mutants with an increasing number of peptidase mutations showed decreasing growth rates in milk; a five-fold peptidase mutant grew 10 times slower than the wild-type. The genes for the peptide transport systems and the peptidase are located on the chromosome; those for the proteases are located exclusively on plasmids.

Genetics. Sophisticated and efficient transformation and special vector systems have been developed which aid analysis of gene regulation and the cloning and expression of homologous and heterologous genes. Transformation of lactococci is best performed by electroporation (Harlander, 1987). Several protocols allow successful transformation even of strains that were resistant to the formerly used method of polyethylene glycol-induced protoplast transformation. Cloning and expression of genes from different sources, random and targeted inactivation of genes, selection for promoter-, terminator-, and signal-sequences, and anchoring of proteins to the surface of lactococci are state of the art techniques.

A homologous *Lactococcus* system makes use of the autoregulatory properties of the nisin gene cluster. Nisin, at concentrations far below the minimal inhibitory concentration, acts as an inducer from outside via a two component signal transduction system consisting of a histidine protein kinase (NisK), and a response repressor (NisR). The two genes, *nisK* and *nisR* (under the control of a constitutive promoter), were delivered by recombinant plasmids or are integrated into the chromosome of the expression strain. The gene to be expressed is fused to the *nisA* promoter that is part of the expression vectors. Vectors for transcriptional and translational fusions have been constructed. Homologous as well as heterologous proteins can make up to 47% of the total cell protein with increasing amounts of the inducer.

Several systems to anchor and display heterologous (poly) peptides at the cell surface of lactococci have been developed (Leenhouts et al., 1999). This allowed changes of the outside composition of the cells which may help in the understanding of the mechanisms of protein targeting. Changes in the surface composition may also influence interactions between the bacteria and the environment and allow potentially important biotechnological applications such as immobilization of enzymes at the bacterial surface, fixing of cells to special carrier surfaces, and display of entire peptide libraries. *Lactococcus lactis* presenting proper antigens may in the future be used as live bacterial vaccine delivery systems (Steidler et al., 1998).

A variety of peptides and proteins could be anchored and at least partially presented at the cell surface of lactococci (Wells and Mercenier, 2003). These includes enzymes like β-lactamase, α-amylase, and nucleases, and epitopes of the human cytomegalovirus of the human immunodeficiency virus, the *Plasmodium falciparum* merocoite stage surface antigen, and the *Clostridium tetani* toxin C fragment.

Genome analysis. Discovery of extrachromosomal elements (plasmids) in dairy lactococci (Cords et al., 1974) and the elucidation of the complete genome sequence of *Lactococcus lactis* subsp. *lactis* IL1403 (Bolotin et al., 1999, 2001) are two milestones in the genetics of these bacteria.

Lactococcus lactis is an AT-rich Gram-positive bacterium; its genome consists of 2345 ± 5 kb with a mean G+C content of 35.4 mol%. The sequencing data allowed the prediction of some very interesting features of the lactococcal genome. The presence of 42 copies of the five different IS-elements – two of the five were previously unknown – indicate the importance of these elements for genetic exchange. The genome of IL1403 also carries six potential or rudimentary prophages (Chopin et al., 2001). The genome sequencing data confirmed the presence of six rRNA operons.

A total of 62 tRNA genes were detected, the majority organized in four large operons, and 1495 protein-encoding genes were identified, mostly oriented collinear to the direction of replication. The total number of lactococcal genes is estimated to be about 2300, approximately one gene per kilobase of DNA.

16S rDNA sequence analysis. This technique is now the gold standard for the assignment of an unknown, newly isolated strain to the genus *Lactococcus* and its species. It is also the basis for the phylogenetic positioning as presented in Figure 124. In addition, appropriate 16S rDNA and 16S rRNA sequences have been used to construct genus- and species-specific nucleotide probes (fluorescent or radioactively labeled) for the identification and differentiation of lactococci *in situ* and in single colonies (Bauer et al., 2000; Betzl et al., 1990).

DNA–DNA similarity. DNA–DNA hybridization studies were an essential milestone for the clustering of the genus *Lactococcus* and its species and subspecies (Schleifer et al., 1985). DNA/DNA similarities between the different *Lactococcus* species are about 15–20%, whereas within the *Lactococcus lactis* subspecies (*Lactococcus lactis* subsp. *cremoris*, *Lactococcus lactis* subsp. *lactis*, and *Lactococcus lactis* subsp. *hordniae*) the similarity is more than 80%.

Plasmids, phages, and bacteriocins

Plasmids. Dairy lactococci possess an unusually large complement of plasmids, ranging in molecular size from about 2 to more than 100 kb (Gasson and Shearman, 2003; Teuber, 1995). The number of plasmids from different strains isolated from dairy cultures varies from 1 to 12. Up to 10% or more of the coding capacity of lactococci consists of extrachromosomal DNA. In the natural habitat, plasmids can be transferred among lactococci and even into other bacterial species of other genera like *Lactobacillus*, *Bacillus*, *Leuconostoc*, *Pediococcus*, *Listeria*, and *Streptococcus* by conjugation and mobilization (Gasson and Davies, 1980; McKay et al., 1973; Neve et al., 1984, 1987; Walsh and McKay, 1981). Transfer of plasmid DNA by transduction is also possible in experimental systems.

Plasmid curing performed by treatment with acridine dyes, ethidium bromide, nalidixic acid, growth at elevated temperature, or protoplast regeneration provided evidence for plasmid linkage of a variety of functions important in fermentation technology (Gasson, 1983). These include lactose transport and metabolism, casein degradation by cell-wall protease, citrate and oligopeptide transport (permeases), bacteriophage protection by restriction/modification and abortive infection, formation of extracellular polysaccharides, bacteriocin production and immunity, insertion (IS) element-dependent recombination and cointegrate formation, group II-intron based lactococcal sex factor, conjugal transfer and mobilization, antibiotic resistance, UV- and metal-resistance, and plasmid replication (Teuber, 1995).

Antibiotic resistance. The molecular analysis of an antibiotic resistant strain of *Lactococcus lactis* subsp. *lactis* isolated from a raw milk soft cheese revealed a 29,815-bp multiresistance plasmid pK214 (Perreten et al., 1997). It assembles resistance genes like *tetS*, *str* (streptomycin adenylase), *cat* (chloramphenicol acetyltransferase), and a new multiple drug transporter (*mdtA*) which had previously been detected in *Listeria monocytogenes* and *Staphylococcus aureus*. *mdtA* encodes a new efflux protein. This observation proves that lactococci are able to pick up multiple antibiotic resistance elements in an antibiotic-challenged habitat (Teuber et al., 1999) from other members of the microbial gene exchange community (Gram-positive and Gram-negative) that share genetic pieces (Baquero, 2004).

Bacteriophages in dairy fermentations. Because sour milk and cheese are normally not made from sterilized milk and because most cheese fermentations are open systems (open vats, control of fermentation by direct and open inspection with hands, tools, etc.), bacteriophages are a common threat. Even pasteurized milk may still contain virulent bacteriophages and residual lactic acid bacteria (Neve and Teuber, 1991).

The phages of the mesophilic dairy lactococci have been investigated in detail. At least 11 genetically distinct phage types have been described (Jarvis et al., 1991). They may be lytic or temperate. Many starter culture strains of lactococci carry prophages that may be released during sour milk and cheese fermentations. The microbiology and genetics of these phages have been investigated thoroughly, leading to clear information on phage adsorption, injection, replication, and release. The host specificities have been found to vary substantially. Bacteriophage-resistant mechanisms have been characterized as being due to inhibition of adsorption, DNA injection, restriction/modification, or abortive infection. Many of these traits are coded for on plasmids and are therefore amenable to genetic engineering and transfers from cell to cell. In cheese factories, high bacteriophage levels in whey and curds are found under conditions that may also lead to contamination with undesirable bacteria.

In addition to the use of phage-resistant strains developed by genetic engineering (Walker and Klaenhammer, 2003) or selected by challenge with bacteriophage cocktails (Möller and Teuber, 1988), there are many generally accepted technological measures that reduce or avoid the bacteriophage problem. These include use of starters containing phage-unrelated or phage-insensitive strains, production of phage-free bulk starter, aseptic propagation systems, use of phage-inhibitory media, reduction of phage contamination of processing plants, starter culture rotations, air conditioning and free air flow, avoidance of aerosol generation, cleaning/chlorination of vats between refills, use of cleaning-in-place (CIP) systems, segregation of starter room and cheese process equipment, removal of deposits on bulk starter vessels, good factory design (suitable location of whey storage tanks, whey-handling systems and wastewater treatment plants), regular use of an aerosol containing an agent such as chlorine to reduce contamination in the atmosphere, and other measures such as use of cultures for direct setting of the milk, inspection of jacket/agitators for pin-holing, and early renneting.

Slow acidification due to bacteriophage attack of the starter culture may lead to good proliferation conditions for undesirable contaminant bacteria such as staphylococci, enterococci, listeria, enteric bacteria, and others. The use of strains resistant to specific phages may induce the selection of new, less virulent phages not encountered previously (Moineau et al., 1993). The proper handling of the bacteriophage issue still remains a major challenge for proper dairy fermentations.

Bacteriocins. Bacteriocins, proteinaceous compounds that kill closely related bacteria, are produced by a variety of lactococcal strains (Skaugen et al., 2003). In a survey of 280 strains isolated from dairy environments, 5% were found to produce such substances. On the basis of biochemical and physical properties, host range, and cross-reactivity, eight bacteriocin types were predicted (Geis et al., 1983; Klaenhammer, 1993; Skaugen et al., 2003). Many of these bacteriocins have been characterized by genetic and biochemical methods. These belong mainly to two defined classes: lantibiotics class I, and small, heat-stable nonlantibiotics class II (Nes et al., 1996).

Nisin, a class I bacteriocin, is a small peptide (34 amino acid residues) produced by several *Lactococcus lactis* subsp. *lactis* strains (Hirsch, 1953). It strongly inhibits the growth of a wide range of Gram-positive bacteria. Two natural variants, nisin A and nisin Z, were found. Nisin is ribosomally synthesized as a 57-amino acid precursor peptide which is subjected to various modifications. The mature peptide (34 amino acids) shows some unusual features including the dehydrated amino acids dehydro-alanine and -butyrine and lanthionine and β-methyllanthionine residues which form five intracellular thioester bridges. Nisin is able to form pores in the membrane of Gram-positive bacteria mediating the efflux of ions, amino acids, and ATP from cells

(Sahl et al., 1987). The structural genes for nisin A and Z were cloned and sequenced. The structural genes are part of a large gene cluster of eleven genes that are involved in all aspects of the nisin biosynthesis. An autoregulated pathway for nisin has been proposed; via a two-component signal system nisin activates the *nisA* promoter. This results in the production of pre-nisin, which is modified by a membrane-bound enzyme system. Subsequently, the precursor nisin is translocated by an ABC transporter and activated by proteolytic cleavage by an extracellular proteinase.

Nisin production is always linked to the ability to ferment sucrose (Rauch and De Vos, 1992). This linkage was confirmed by curing experiments and conjugal transfer. It was shown that the nisin-sucrose element is part of a conjugative transposon that integrates in several different sites of the lactococcal chromosome by a mechanism of transposition similar to that of the *Tn*916 family. Nisin is a legal food preservative/additive in many countries and is used to protect high-moisture food commodities against the pathogenic *Listeria monocytogenes* or *Clostridium botulinum*, but also against spoilage by clostridia and other Gram-positive bacteria.

Pathogenicity

Lactococcus garvieae must be regarded as a fish pathogen and responsible for mastitis in cows and buffalos. It has been denoted as an emerging pathogen of increased clinical significance in both veterinary and human medicine (Facklam and Elliott, 1995; Texeira et al., 1996; Vela et al., 2000). Although not very common, *Lactococcus garvieae* was recently isolated in Switzerland from 4.5% of 150 milk samples collected from cows with either chronic or subclinical mastitis (Thomann and Perreten, Institute of Veterinary Bacteriology, University of Berne, Switzerland; personal communication, 2003/2004). *Lactococcus garvieae* is the etiologic agent of hemorrhagic septicemia in farmed trout and is characterized by bilateral exophthalmos, darkening of the skin, congestion of the intestine, liver, kidney, spleen, and brain, and a characteristic hemorrhagic enteritis (Domenech et al., 1993). The disease has been termed lactococcosis to distinguish it from streptococcosis. It is a worldwide bacterial disease affecting different fish species such as eels, yellowtails, farmed trout, and prawns. *Enterococcus seriolicida* is a junior synonym of *Lactococcus garvieae* (Eldar et al., 1999). Pathogenicity and virulence factors have not yet been identified.

In contrast, *Lactococcus piscium* seems to be a meat spoilage bacterium encountered in vacuum-packed, chilled meat (Sakala et al., 2002b). The isolation of the type strain from a diseased rainbow trout (Williams et al., 1990b) cannot be taken as evidence for its pathogenicity because no other cases have been reported since.

In rare instances, *Lactococcus lactis* has been isolated from human cases of urinary tract and wound infections, and from patients with endocarditis (Aguirre and Collins, 1993). In this behavior it resembles other lactic acid bacteria.

Ecology

The most important habitats for lactococci are in the dairy industry and are shown in Table 136.

The lactococci comprise the species *Lactococcus lactis*, *Lactococcus garvieae*, *Lactococcus plantarum*, *Lactococcus piscium*, and *Lactococcus raffinolactis* (Table 137).

TABLE 137. *Lactococcus* species isolated during environmental screening (modified from Klijn et al., 1995)[a]

Sample	L. lactis subsp. lactis	L. lactis subsp. cremoris	L. garviae	L. raffinolactis	New, undefined Lactococcus sp.
Cheese plant:					
Cheese milk	+	+			
Cheese whey	+	+			
Waste whey	+	+	(+)		
Waste water tank	+	+	+	+	+
Waste water disposal site soil	(+)	(+)		(+)	
Grass	(+)	(+)	(+)	(+)	
Farm samples:					
Raw milk	+	+	+	+	+
Milk machine	+	+			
Udder	+	+			
Saliva, cow	(+)	(+)			
Saliva, bull	(+)				
Skin, cow	(+)		(+)		
Skin, bull	(+)		(+)		
Grass		(+)		(+)	
Soil	(+)		(+)	(+)	
Silage	(+)				

[a]+, Detected by direct plating; (+), detected by plating after enrichment.

Lactococcus lactis subsp. *lactis* and *Lactococcus lactis* biovar "diacetylactis" (according to Schleifer et al., (1985), the subspecific status *Lactococcus lactis* subsp. "diacetylactis" is no longer valid; because of the importance of diacetyl-forming lactococci, especially in dairy fermentations, we still distinguish, for practical reasons, between *Lactococcus lactis* subsp. *lactis* and *Lactococcus lactis* biovar "diacetylactis", instead) have commonly been detected directly or following enrichment in plant material, including fresh and frozen corn, corn silks, navy beans, cabbage, lettuce, peas, wheat middlings, grass, clover, potatoes, cucumbers, and cantaloupe (Sandine et al., 1972). Lactococci are usually not found in fecal material or soil. Survival of *Lactococcus lactis* MG1363 ingested by human volunteers disappeared within 4 h from their ileal fluids (Vesa et al., 2000). Only small numbers occur on the surface of the cow and in its saliva. This was ascertained by an environmental screening. *Lactococcus lactis* and its subspecies *Lactococcus lactis* subsp. *lactis* and *Lactococcus lactis* subsp. *cremoris* were detected without enrichment in raw milk, milking machines, and the udders at 10^3–10^4 c.f.u. per gram or cm², respectively, after enrichment in addition in the saliva and on the skin of cows and bulls, on grass and in soil and silage (Klijn et al., 1995). In a dairy environment, lactococci were abundant in cheese milk, cheese whey, waste whey, and the waste water tank. After enrichment, *Lactococcus lactis* was also found in the waste water disposal site soil and on grass. *Lactococcus garvieae* and *Lactococcus raffinolactis* were also consistently detected in raw milk, the skin and saliva of cows, and on grass (see Table 137). Since raw cow's milk consistently contains *Lactococcus lactis* subsp. *lactis*, *Lactococcus lactis* subsp. *lactis* biovar "diacetylactis" and *Lactococcus lactis* subsp. *cremoris*, it is tempting to suggest that lactococci enter the milk from the exterior of the udder during milking and from the feed, which may be the primary source of inoculation. *Lactococcus lactis* subsp. *cremoris* seems to have its main habitat in milk, fermented milk, cheese, and starter cultures.

A study of lactococci isolated from animal material in Belgium revealed the following habitats (Pot et al., 1996): *Lactococcus lactis* subsp. *lactis*, from bovine intestine, milk from healthy cows and cows with mastitis, tonsils of cats, dogs and goats; *Lactococcus garvieae* from bovine milk and tonsils, tonsils of dogs, feces of cats and horses, conjunctiva of a turtle; *Lactococcus raffinolactis*, from bovine tonsils, goat intestine; and *Lactococcus piscium* was consistently detected in vacuum-packed, refrigerated beef in Japan, when incubation of the plated samples was at 7°C for 10–14 d (Sakala et al., 2002b). The type strain of *Lactococcus piscium* was isolated from a diseased rainbow trout (Williams et al., 1990b).

New, undefined *Lactococcus* species. The results of screening of a dairy environment have shown that new, not yet differentiated *Lactococcus* species can be shown to exist in nature if modern molecular methods such as 16S RNA gene sequencing are employed (Klijn et al., 1995). Another, interesting habitat seems to be the microflora of the gut of wood-feeding termites like *Reticulitermes flaviceps*. The Gram-postive cocci comprising about 50% of the culturable microflora contained new species closely related to *Enterococcus faecalis*. Three other studied strains cluster closely with *Lactococcus garvieae*, but definitely represent a new species (Bauer et al., 2000).

In summary, animal and plant environments are the main habitats of the investigated lactococci. To what extent the lactococci used by the dairy industry have gone through recent phases of evolutionary changes due to their deliberate use as starter cultures, remains to be established. Their original habitat is and was the dairy cow and the raw milk extracted from it. In a recent attempt to reisolate *Lactococcus lactis* subsp. *cremoris*, it was found in samples of Chinese and Moroccan raw milk, and one corn (*Zea mays*) sample, but not in *Rubus discolor* (Himalayan blackberry), *Hypericum calycinum* (gold flower), *Prunus* sp. (plum), *Hedera helix* (English ivy), *Acer platanoides* (green lace tree), and *Gingko biloba* (maidenhair tree) (Salama et al., 1993).

Industrial application. The present-day significance is in the large-scale use of the lactococci for industrial fermentations, especially dairy products (Cogan and Accolas, 1996; Teuber, 2000). The species *Lactococcus lactis* and its subspecies used on a large scale by the dairy industry are generally recognized as safe (GRAS) for human consumption. The world cheese production based on starter cultures of lactic acid bacteria was more than 16 million metric tonnes in 2000 (Food and Agriculture Organization of the United Nations, 2001), most of it made with lactococci. No similar data are available for the other lactococcal products such as cultured butter, sour milk, and casein. The deliberate use of the lactococci in the dairy industry as starter cultures for many different products is documented in Table 136. This development was initiated at the turn of the century when Weigmann, Storch, and Conn identified lactococci as the essential components of the mesophilic microflora in spontaneously fermented cream and milk. This finding led to the introduction of pure starter cultures of lactic acid bacteria to the dairy field for use in the fermentation and ripening of milk, cream, and cheese (Weigmann, 1905-1908).

In recent years, the physiology, biochemistry, genetics, and molecular biology of the lactococci have gained much attention due to the great economic importance of these organisms (see above). Lactococci are employed in single and mixed cultures for the production of different kinds of cheeses, fermented milks, cultured butter, and casein (see Table 136).

The biochemical and technological functions of lactococci necessary for milk fermentation and cheese production can be summarized as follows:

1. Formation of lactic acid from lactose (early function in fermentation). The resulting lowered pH values (4.0–5.6) as compared with milk (6.6–6.7) prevent or retard growth of spoilage bacteria, especially clostridia, staphylococci, *Enterobacteriaceae*, and psychrotolerant proteobacteria like *Pseudomonas*. If the isoelectric point of casein at pH 4.6–4.8 is approached, casein is precipitated and the milk curdles. This effect is used in the production of cottage cheese, quarg, sour milk, yogurt, and casein. Starter bacteria for this purpose are *Lactococcus lactis* subsp. *lactis* and *Lactococcus lactis* subsp. *cremoris* (besides lactobacilli and *Streptococcus thermophilus*).

2. Formation of diacetyl from citrate (early to medium function in fermentation). Diacetyl is the most characteristic aroma compound provided by *Lactococcus lactis* subsp. *lactis* biovar "diacetylactis". It is derived from the citrate of milk (about 0.1%) that is present in solution and in the casein micelles as a casein-citrate-calcium-phosphate complex. The pathway goes through oxaloacetate, pyruvate, and α-acetolactate with a coupled release of carbon dioxide, which induces eye formation in cheese but also unwanted floating of the curd in the manufacture of cottage cheese or quark if an unbalanced mixed starter is used (see Table 136).

3. Limited proteolysis during cheese ripening (medium to late function in cheese fermentation). Many strains of all lactococcal species possess a β-casein-specific, cell-wall-associated protease, together with a complement of peptidases including an aminopeptidase, an X-prolyl-dipeptidyl aminopeptidase, and a dipeptidase, which are necessary for growth in milk but also function during cheese ripening.

Propagation and preservation of starter cultures. Starter cultures are needed and applied by the industry producing food and feed as indicated in Table 136. Dairy products are almost always made with commercial cultures. Liquid, dried, and frozen starter cultures are in use. Starter cultures must have a high survival rate of micro-organisms coupled with optimum activity for the desired technological performance, e.g., the fermentation of lactose to lactate, controlled proteolysis of casein, and production of aroma compounds like diacetyl. Because the genes for lactose and citrate fermentation and those for certain proteases are located on plasmids, continuous culture has not been successful because fermentation-defective variants develop easily. In most instances, pasteurized or sterilized skim milk is the basic nutrient medium for the large-scale production of starter cultures because it ensures that only lactococci fully adapted to the complex substrate, milk, will develop. For liquid starter cultures, the basic milk medium may be supplemented with yeast extract, glucose, lactose, and calcium carbonate. To obtain optimum activity and survival, it may be necessary to neutralize the lactic acid that is produced by addition of sodium or ammonium hydroxide. Because many strains of dairy lactococci produce hydrogen peroxide during growth under microaerophilic conditions, it has been beneficial to add catalase to the growth medium, thus leading to cell densities of more than 10^{10} viable units per ml of culture.

For the preparation of concentrated starters, the media are clarified by proteolytic digestion of skim milk with papain or bacterial enzymes to avoid precipitation of casein in the separator used to collect the lactococcal biomass. The available self-cleaning clarifiers, e.g., bactofuges, concentrate the fermentation broth to cell concentrations of about 10^{12} c.f.u./ml. These concentrates can either be lyophilized or are preferentially transferred drop-wise into liquid nitrogen. The formed pellets are packed in metal cans or cartons and are kept and shipped at –70°C (dry ice). These modern starter preparations allow the direct inoculation of cheese vats because the bacteria immediately resume exponential growth if harvested in the late exponential growth phase. This modern technology shifts most of the microbiological work and responsibility from the cheese factory to the starter producer.

The classical liquid starter culture with about 10^9 c.f.u./ml and traditional lyophilized cultures containing about 10^{11} c.f.u./gram must be further propagated in the factory. However, lyophilized cultures can be transported easily and kept at ambient temperature for several months. Concentrated lyophilized starters are also suited for direct vat inoculation having a short lag phase, however, before growth is resumed. Direct vat inoculation may have a certain advantage regarding protection against bacteriophages which are common in the open dairy fermentation systems (see above).

Enrichment and isolation procedures

Enrichment and isolation. Lactococci are nutritionally fastidious. They all require complex media for optimal growth. In synthetic media, all strains require amino acids such as isoleucine, valine, leucine, histidine, methionine, arginine, and proline, and the vitamins niacin, Ca-pantothenate, and biotin (Anderson and Elliker, 1953; Jensen and Hammer, 1993).

Isolation from plant material. Plant material such as grass and herbages is the natural source of lactic acid bacteria. Ensilage allows the enrichment of lactococci, leuconostocs, and pediococci. The isolation procedure takes advantage of the sequential growth of the above-mentioned genera. For this purpose, the procedure of Whittenbury (1965) is recommended.

Isolation from cheese. The main problem in isolating lactic acid bacteria including lactococci from dairy products is a proper dissolution or dispersion of the solid or semisolid fat containing material. An accepted method is the one by Olson et al. (1978), as follows.

Using aseptic technique, thoroughly comminute or mix each sample until representative portions can be removed. Heat 99-ml dilution blanks of sterile, freshly prepared (less than 7 d old), aqueous 2% sodium citrate to 40°C. Aseptically transfer 11 g of cheese to a sterile blender container previously warmed to 40°C and add the warmed sodium citrate blank. Mix for 2 min at a speed sufficient to emulsify the sample properly, invert the container to rinse particles from the interior walls, and remix for approximately 10 s. Inadvertent heating to temperatures in excess of 40°C from friction in agitation may occur with some mechanical blenders. This should be determined before use of the blender so corrective action can be taken if needed. If heating is unavoidable with equipment available, mixing periods of less than 2 min should be used provided that complete emulsification is obtained. The 1:10 dilution should be plated or further diluted immediately with great care being taken to avoid air bubbles or foam.

Isolation of lactic acid bacteria and lactococci from cultured milk, cultured cream, yogurt, acidophilus milk, Bulgarian buttermilk, and similar cultured or acidified semifluid products follows a similar technique (Olson et al., 1978).

No satisfactory selective medium is available for the isolation of lactococci. Two media, both commercially available, are generally accepted to give reliable growth of these organisms. The medium proposed by Elliker et al. (1956) is widely used for the isolation and enumeration of lactococci. Elliker Agar Medium for Isolation of Lactococci contains (per l): tryptone, 20.0 g; yeast extract, 5.0 g; gelatin, 2.5 g; glucose, 5.0 g; lactose, 5.0 g; sucrose, 5.0 g; sodium chloride, 4.0 g; sodium acetate, 1.5 g; ascorbic acid, 0.5 g; agar, 15.0 g. The medium has a pH of 6.8 before autoclaving. This medium is probably the most cited for the isolation and growth of lactococci, although it is unbuffered. This disadvantage can be overcome by the addition of suitable buffer substances. Addition of 0.4% (w/v) of diammonium phosphate improves the enumeration of lactic streptococci on Elliker agar. Colony counts were up to about eight times greater due to improved buffering capacity (Barach, 1979). M17 medium (Terzaghi and Sandine, 1975), a complex medium supplemented by 1.9% β-disodium glycerophosphate, results in improved growth of lactococci. M17 Medium for Isolation of Lactococci contains (per l): phytone peptone, 5.0 g; polypeptone, 5.0 g; yeast extract, 5.0 g; beef extract, 2.5 g; lactose, 5.0 g; ascorbic acid, 0.5 g; β-disodium glycerophosphate, 19 g; 1.0 M $MgSO_4 \cdot 7H_2O$, 1.0 ml; glass-distilled water, 1 l. The medium is sterilized at 121°C for 15 min. The pH of the broth is 7.1.

Solid medium contains 10 g agar/l of medium. This medium is useful for the isolation of all strains of *Lactococcus lactis* subsp. *cremoris*, *Lactococcus lactis* subsp. *lactis*, *Lactococcus lactis* biovar "diacetylactis", and *Streptococcus thermophilus* and mutants of those strains lacking the ability to ferment lactose. Addition of a pH indicator dye (bromcresol purple) and reduction of β-disodium glycerophosphate (5 g/l) allows an easy differentiation between lactose-fermenting (large yellow colonies) and nonfermenting (small white colonies) strains (Kondo and McKay, 1984). This medium has become the standard for genetic investigations of lactococci. Lactococcal bacteriophages can be efficiently demonstrated and distinguished on M17 agar. Plaques larger than 6 mm in diameter could be observed as could turbid plaques, indicating lysogeny (Terzaghi and Sandine, 1975).

Enumeration of citrate-fermenting bacteria in lactic starter cultures and dairy products. To control gas and aroma (diacetyl) production in the fermentation of various dairy products, it is important to know the quantitative composition of the starter cultures used. *Leuconostoc* species and *Lactococcus lactis* biovar "diacetylactis" are components of many mesophilic starter cultures (Table 136). These organisms are able to ferment citrate with concomitant production of CO_2 and diacetyl. For the collective enumeration of leuconostocs and *Lactococcus lactis* biovar "diacetylactis" in starters and fermented dairy products, a whey agar containing calcium lactate and Casamino acids (WACCA) has been introduced by Galesloot et al. (1961). A modified medium based on the different action on the lactose analog 5-bromo-4-chloro-3-indolyl-β-D-galactopyranoside (Xgal) has been suggested (Vogensen et al., 1987).

Maintenance procedures

Lactococci can conveniently be kept for several weeks at refrigeration temperature in the buffered M17 medium in stab cultures. Freeze drying in neutralized skimmed milk yields cultures that can be kept for several years when sealed *in vacuo*. Starter cultures keep well in freeze-dried preparations or when deep frozen in liquid nitrogen and stored at −70°C (Sandine, 1996).

Differentiation from closely related taxa

When isolating lactic acid bacteria and lactococci from plant and food sources, it is necessary to distinguish lactococcal colonies from pediococci and leuconostocs which often grow on the same media along with lactococci. The method of choice in the microbiological laboratory is to analyze the fermentation products from glucose: L(+)-lactic acid in *Lactococcus*, DL-lactic acid in *Pediococccus*, and D(−)-lactic acid, CO_2, acetic acid, and ethanol in *Leuconostoc*.

Taxonomic comments

The first studies of the lactococci were those by Joseph Lister (1873), who was attempting to prove Pasteur's germ theory of fermentative changes. In experiments with boiled milk as a nutrient medium, he obtained, by chance, the first pure bacterial culture. It is worthwhile to recall his original discussion of this discovery, marking the start of bacterial taxonomy:

"Admitting then that we had here to deal with only one bacterium, it presents such peculiarities both morphologically and physiologically as to justify us, I think, in regarding it a definite and recognizable species for which I venture to suggest the name *Bacterium lactis*. This I do with diffidence, believing that up to this time no bacterium has been defined by reliable characters. Whether this is the only bacterium that can occasion the lactic acid fermentation, I am not prepared to say."

This bacterium was later renamed *Streptococcus lactis* (Löhnis, 1909). On the basis of exhaustive reinvestigations including cell-wall analysis, long-chain fatty acid composition, unsaturated fatty acid patterns, menaquinone composition, immunological cross-reactions of purified superoxide dismutases, biochemical properties, and DNA/DNA hybridization, Schleifer et al. (1985) proposed that the N streptococci (Lancefield, 1933) be separated from the oral streptococci, the enterococci, and the hemolytic streptococci. They suggested the new genus name *Lactococcus*. The molecular taxonomy of the lactococci has yielded the detection and differentiation of new species besides *Lactococcus lactis: Lactococcus garvieae, Lactococcus raffinolactis, Lactococcus plantarum,* and *Lactococcus piscium.*

Further reading

Hardie, Jeremy, M. 1986. Genus Streptococcus Rosenbach 1884, 22. *In* Sneath, Mair, Sharpe and Holt (Editors), *Bergey's Manual of Systematic Bacteriology,* Vol. 2, The Williams & Wilkins Co., Baltimore. pp.1043–1071.

Teuber, M. and Geis, A. 2002. The genus *Lactococcus.* The Prokaryotes. 3rd electronic edition. Springer, New York.

Wood, B.J.B. and Warner, P.J. (Editors). 2003. Genetics of Lactic Acid Bacteria. *In* The Lactic Acid Bacteria Vol. 3, Kluwer Academic/Plenum Publishers, New York.

List of species of the genus *Lactococcus*

Characteristics differentiating the species and subspecies of the genus *Lactococcus* are listed in Table 138 and Table 139.

1a. **Lactococcus lactis subsp. lactis** (Lister 1873) Schleifer, Kraus, Dvorak, Kilpper-Bälz, Collins and Fischer 1986, 354[VP]

(Effective publication: Schleifer, Kraus, Dvorak, Kilpper-Bälz, Collins and Fischer 1985, 190.) (*Bacterium lactis* Lister 1873, 408; *Streptococcus lactis* Löhnis 1909, 554) lac'tis. L. n. *lac* milk; L. gen. n. *lactis* of milk.

TABLE 138. Characteristics differentiating species and subspecies of the genus *Lactococcus*[a,b]

Characteristic	*L. lactis* subsp. lactis	*L. lactis* subsp. cremoris	*L. lactis* subsp. hordniae	*L. garviae*	*L. piscium*	*L. plantarum*	*L. raffinolactis*
Peptidoglycan type[c]	Lys–D-Asp	Lys–D-Asp	Lys–D-Asp	Lys–Ala–Gly–Ala	ND	Lys–Ser–Ala	Lys–Thr–Ala
Major menaquinones[d]	MK-9, MK-8	MK-9, MK-8	MK-8, MK-9	MK-9, MK-8	ND	–	–
Acid production from:							
Galactose	+	+	–	+	+	–	+
Lactose	+	+	–	+	+	–	+
Maltose	+	–	–	V	+	–	+
Melibiose	–	–	–	V	+	+	+
Melizitose	–	–	–	–	+	–+	+
Raffinose	–	–	–	–	+	+	V
Ribose	+	–	–	+	–	–	+
Hydrolysis of arginine	+	–	+	+	–	–	V

[a]Adapted from Schleifer (1987) and Sakala et al. (2002a).
[b]+, Positive, –, negative, V, variable, ND, not determined.
[c]Abbreviations according to Schleifer and Kandler (1972): Asp, aspartic acid; Gly, glycine; Lys, lysine; Ser, serine; Thr, threonine; Ala, alanine.
[d]Abbreviations according to Collins and Jones (1979): MK-8, menaquinone with n = 8 isoprene units; MK-9, menaquinone with n = 9 isoprene units.

TABLE 139. Physiological and other properties of dairy lactococci used for identification and differentiation[a]

Properties	*L. lactis* subsp. *lactis*	*L. lactis* subsp. *lactis* biovar diacetylactis	*L. lactis* subsp. *cremoris*
Growth at 4°C	+	+	–
Growth at 10°C	+	+	+
Growth at 45°C	–	–	–
Growth in 4% NaCl	+	+	–
Growth in 6.5% NaCl	–	–	–
Growth at pH 9.2	+	+	–
Growth with methylene blue (0.1% milk)	+	+	–
Growth in presence of bile (40%)	+	+	+
NH₃ from arginine	+	+	–
CO₂ from citrate	–	+	–
Diacetyl and acetoin	–	+	–
Fermentation of maltose	+	+	Rarely
Hydrolysis of starch	–	–	–
Heat resistance (30 min at >60°C)	V	V	V
Serological group[b]	N	N	N
DNA G+C content (mol%)	33.8–36.8	33.6–34.7	35.0–36.1

[a]+, Positive; –, negative, V, variable.
[b]Lancefield (1933).

Colonies on blood agar or nutrient agar are circular, smooth, and entire. Nonpigmented. Nonhemolytic (some strains may produce a weak α-reaction). Ovoid cells elongated in the direction of the chain. Mostly in pairs or short chains. Gram-positive. Nonmotile. Facultatively anaerobic. Catalase-negative. Growth at 10°C but not at 45°C. Grows in 4% (w/v) NaCl and 0.1% methylene blue milk. Chemoorganotrophic; fermentative metabolism. All strains produced acid from galactose, glucose, fructose, lactose, maltose, mannose, *N*-acetylglusosamine, ribose, and trehalose. Most strains produced acid from arbutin, cellobiose, β-gentiobiose, and salicin. Acid not produced from D-arabinose, L-arabinose, D-arabitol, L-arabitol, adonitol, 2-ketogluconate, 5-keto-gluconate, dulcitol, erythritol, D-fucose, L-fucose, gluconate, glycerol, glycogen, inositol, melibiose, melizitose, α-methyl-D-glucoside, α-methyl-D-mannoside, D-lyxose, raffinose, rhamnose, L-sorbose, sorbitol, D-tagatose, D-turanose, xylitol, and L-xylose. Variable results may be obtained from amygdalin, inulin, mannitol, sucrose, starch, and D-xylose. Esculin and hippurate hydrolyzed. Arginine dehydrolase and leucine arylamidase-positive. Alkaline phosphatase, α-galactosidase, β-galactosidase, and β-glucuronidase-negative. A few strains are pyrrolidonyl arylamidase-positive. Some strains can utilize citrate (in conjunction with a fermentable carbohydrate) with the production of CO_2, acetoin, and diacetyl (if they possess a proper plasmid encoding a citrate permease, CitP). Reacts with Lancefield serological group N antiserum.

The petidoglycan type is Lys–D-Asp. Low levels of menaquinones produced, with MK-9 predominating. Contains poly(glycerophosphate)-lipoteichoic acid partially substituted with α-galactosyl residues and D-alanine esters. The major glycolipid is Glc(α1–2)Glc(α1–3)acyl$_2$Gro. Aminophospholipids are not present. Major nonhydroxylated long-chain fatty acids are hexadecanoic, *cis*-11,12-octadecenoic, and *cis*-11,12-methylenoctadecanoic acids.

The level of 16S rRNA similarity between the type strain of *Lactococcus lactis* subsp. *lactis* and the type strains of the other *Lactococcus lactis* subspecies is 98–99%. The level of 16S rRNA sequence similarity between *Lactococcus lactis* subsp. *lactis* and the type strains of the other *Lactococcus* species is 90–93%, and it is 98–99% between *Lactococcus lactis* subsp. *lactis* and other strains of the other subspecies of *Lactococcus lactis*.

DNA G+C content (mol%): 34.4–36.3 (T_m).

Type strain: ATCC 19435, CCUG 7980, CIP 70.56, DSM 20481, HAMBI 1591, JCM 5805, LMG 6890, NCIMB 6681, NCTC 6681, VKM B-1662.

GenBank accession number (16S rRNA gene): AB100803.

1b. **Lactococcus lactis subsp. cremoris** (Orla-Jensen 1919) Schleifer, Kraus, Dvorak, Kilpper-Bälz, Collins and Fischer 1986, 354 [VP] (Effective publication: Schleifer, Kraus, Dvorak, Kilpper-Bälz, Collins and Fischer 1985, 192.) (*Streptococcus cremoris* Orla-Jensen 1919)

cre′moris. L. n. *cremor* cream; L. gen. n. *cremoris* of cream.

In most respects, the description of *Lactococcus lactis* subsp. *cremoris* corresponds to the description of *Lactococcus lactis* subsp. *lactis*. It differs in the following character-

istics. Grows in 2% but not 4% (w/v) NaCl. No growth at 40°C. Acid not produced from maltose and ribose. Most strains do not produce acid from β-gentiobiose, salicin, and trehalose. Arginine dehydrolase-negative. Produces detectable quantity of CO_2.

DNA G+C content (mol%): 34.8–36.3 (T_m).

Type strain: ATCC 19257, CCUG 21953, CIP 102301, DSM 20069, HAMBI 1588, BBRC 3427, LMG 6897, NCIMB 8662.

GenBank accession number (16S rRNA gene): AB100802, M58836.

1c. **Lactococcus lactis subsp. hordniae** (*ex* Latorre-Guzman, Kado and Kunkee 1977) Schleifer, Kraus, Dvorak, Kilpper-Bälz, Collins and Fischer 1986, 354 [VP] (Effective publication: Schleifer, Kraus, Dvorak, Kilpper-Bälz, Collins and Fischer 1985, 193.) (*Lactobacillus hordniae* Latorre-Guzman, Kado and Kunkee 1977, 365)

hord′ni.ae. N.L. fem. n. *Hordnia* generic name; N.L. gen. n. *hordniae* of *Hordnia circellata* (name of the leaf hopper from which the organism was isolated).

In most respects, the description of *Lactococcus lactis* subsp. *hordniae* corresponds to the description of *Lactococcus lactis* subsp. *lactis*. It differs in the following characteristics: Grows in 2% but not 4% (w/v) NaCl. No growth at 40°C. Acid not produced from galactose, lactose, maltose, or ribose. Hippurate is not hydrolyzed. *cis*-11,12-Methylenoctanoic acid is absent.

DNA G+C content (mol%): 35.2 (T_m).

Type strain: ATCC 29071, CCUG 32210, CIP 102973, DSM 20450, HAMBI 1590, JCM 1180, LMG 8520, NCIMB 702181.

GenBank accession number (16S rRNA gene): AB100804.

2. **Lactococcus garvieae** (Collins, Farrow, Phillips and Kandler 1983) Schleifer, Kraus, Dvorak, Kilpper-Bälz, Collins and Fischer 1986, 354[VP] (Effective publication: Schleifer, Kraus, Dvorak, Kilpper-Bälz, Collins and Fischer 1985, 183.) (*Streptococcus garvieae* Collins, Farrow, Phillips and Kandler 1983, 3430)

gar′vie.ae. N.L. gen. n. *garvieae* of Garvie, named for E.I. Garvie, a British microbiologist.

Colonies on blood agar or nutrient agar are circular, smooth, and entire. Nonpigmented. Not β-hemolytic. Ovoid cells elongated in the direction of the chain; mostly in pairs or short chains. Gram-positive. Nonmotile. Facultatively anaerobic. Catalase-negative. Grows at 10°C and 40°C but not at 45°C. Grows in 4% (w/v) NaCl. Grows in and reduces 0.1% methylene blue milk. Chemo-organotrophic; fermentative metabolism. Acid produced from galactose, glucose, fructose, cellobiose, amygdalin, arbutin, mannose, ribose, trehalose, salicin, β-gentiobiose, and *N*-acetylglucosamine. Acid not produced from D-arabinose, L-arabinose, D-arabitol, L-arabitol, adonitol, 2-ketogluconate, 5-ketogluconate, dulcitol, erythritol, D-fucose, L-fucose, glycogen, inulin, inositol, melibiose, melizitose, β-methylxyloside, α-methyl-D-glucoside, α-methyl-D-mannoside, D-lyxose, raffinose, rhamnose, L-sorbose, sorbitol, D-turanose, xylitol, D-xylose, and L-xylose. Variable results may be obtained from maltose, mannitol, sucrose, and D-tagatose. Esculin hydrolyzed. Hippurate hydrolysis variable. Starch hydrolysis negative. Arginine dehydrolase, leucine arylamidase, and pyrrolidonyl arylamidase-positive. α-Galactosidase, β-galactosidase, and alkaline phosphatase-negative. Some strains react with Lancefield group N antisera.

The petidoglycan type is Lys–Ala–Gly–Ala. Low levels of menaquinones produced with MK-9 predominating. Contains poly[Gal(α1–6)Gal(α1–3)Gro-1-P(2←1αGal)] lipoteichoic acid. The major glycolipid is Glc(α1–2)Glc(α1–3) acyl$_2$Gro. Aminophospholipids are not detected. Major nonhydoxylated long-chain fatty acids are hexadecanoic, cis-11,12-octadecenoic and cis-11,12-methylene-octadecanoic acids.

The level of 16S rRNA sequence similarity between $Lactococcus\ garvieae$ and the type strains of the other $Lactococcus$ species is 91–93%.

DNA G+C content (mol%): 38.3–38.7 (T_m).

Type strain: ATCC 43921, CCUG 32208, CIP 102507, DSM 20684, HAMBI 1592, JCM 10343, JCM 12256, LMG 8893, NCIMB 702155.

GenBank accession number (16S rRNA gene): X54262.

3. **Lactococcus piscium** Williams, Fryer and Collins 1990a, 320[VP] (Effective publication: Williams, Fryer and Collins 1990b, 113.)

pis′cium. L. n. *piscis* fish; L. gen. pl. n. *piscium* of fishes.

Cells are short rods to ovoid in shape, mostly in pairs or short chains. Gram-positive. Nonmotile. Facultatively anaerobic. Catalase-negative. Grows at 5°C and 30°C; no growth at 40°C. Acid is produced from amygdalin, L-arabinose, arbutin, N-acetylglucosamine, cellobiose, D-fructose, galactose, β-gentiobiose, gluconate, glucose, lactose, maltose, D-mannose, mannitol, melibiose, melizitose, D-raffinose, ribose, salicin, sucrose, trehalose, D-turanose, and D-xylose. Acid is not produced from adonitol, D-arabinose, D-arabitol, L-arabitol, dulcitol, erythritol, D-fucose, L-fucose, glycogen, glycerol, inositol, inulin, 2-ketogluconate, 5-ketogluconate, D-lyxose, α-methyl-xyloside, rhamnose, L-sorbose, sorbitol, D-tagatose, xylitol, and L-xylose. Esculin is hydrolyzed. Starch hydrolysis is slow and weak. Arginine hydrolysis and urease-negative. H$_2$S is not produced.

The cell-wall type has not been determined. The long-chain fatty acids are of the straight-chain saturated, mono-unsaturated, and cyclopropanoic types. The major acids correspond to hexadecanoic acid, ω7-octadecenoic acid, and ω7-methylene-octadecenoic acid.

The level of 16S rRNA sequence similarity between $Lactococcus\ piscium$ and the type strains of the other $Lactococcus$ species is 92–93%.

DNA G+C content (mol%): 38.5 (T_m).

Type strain: HRIA 68, ATCC 700018, CCUG 32207, CCUG 32732, CIP 104371, DSM 6634, JCM 11055, NCIMB 13196.

GenBank accession number (16S rRNA gene): X53905.

4. **Lactococcus plantarum** (Collins, Farrow, Phillips and Kandler 1983) Schleifer, Kraus, Dvorak, Kilpper-Bälz, Collins and Fischer 1986, 354 [VP] (Effective publication: Schleifer, Kraus, Dvorak, Kilpper-Bälz, Collins and Fischer 1985, 193.)

(*Streptococcus plantarum* Collins, Farrow, Phillips and Kandler 1983, 3430)

plan.ta′rum. L. n. *planta* plant; N.L. gen. pl. n. *plantarum* of plants.

Colonies on blood agar or nutrient agar are circular, smooth, and entire. Nonpigmented. Not β-hemolytic. Spheres or ovoid cells elongated in the direction of the chain; mostly in pairs or short chains. Gram-positive. Nonmotile. Facultatively anaerobic. Catalase-negative. Grows at 10°C but not at 45°C. Grows in 4% (w/v) NaCl. Does not grow in 0.1% methylene blue milk. Chemo-organotrophic; fermentative metabolism. Acid produced from amygdalin, arbutin, cellobiose, dextrin, fructose, glucose, maltose, D-mannose, mannitol, melizitose, N-acetylglucosamine, salicin, sorbitol, sucrose, and trehalose. Acid not produced from adonitol, D-arabinose, L-arabinose, arabitol, 2-ketogluconate, 5-ketogluconate, dulcitol, erythritol, D-fucose, L-fucose, galactose, gluconate, glycogen, glycerol, inositol, inulin, lactose, D-xylose, melibiose, α-methyl-D-glucoside, α-methyl-D-mannoside, β-methylxyloside, raffinose, ribose, rhamnose, L-sorbose, D-tagatose, D-xylose, L-xylose, and xylitol. Variable results may be obtained from β-gentiobiose and turanose. Esculin hydrolyzed. Starch, hippurate, and gelatin not hydrolyzed. Leucine arylamidase-positive. β-Galactosidase, β-glucuronidase, alkaline phosphatase, arginine dehydrolase, and pyrrolidonyl arylamidase-negative. Some strains are α-galactosidase-positive. Reacts with Lancefield group N antisera.

The peptidoglycan type is Lys–Ser–Ala. Menaquinones are absent. Contains poly(glycerophosphate)-lipoteichoic acid substituted solely with D-alanine esters. The major glycolipid is Glc(α1–2)Glc(α1–3)acyl$_2$Gro. Contains D-alanyl- and L-lysylphosphatidylglycerol. Major nonhydroxylated long-chain fatty acids are hexadecanoic, cis-11,12-octadecenoic acids, and cis-11,12-methyleneoctadecanoic acid.

The level of 16S rRNA sequence similarity between $Lactococcus\ plantarum$ and the type strains of the other $Lactococcus$ species is 90–93%.

DNA G+C content (mol%): 36.9–38.1 (T_m).

Type strain: ATCC 43199, CCUG 39180, CIP 102506, DSM 20686, HAMBI 1593, JCM 11056, LMG 8517, NCIMB 11945.

GenBank accession number (16S rRNA gene): X54259.

5. **Lactococcus raffinolactis** (Orla-Jensen 1932) Schleifer, Kraus, Dvorak, Kilpper-Bälz, Collins and Fischer 1988, 220[VP] (Effective publication: Schleifer, Kraus, Dvorak, Kilpper-Bälz, Collins and Fischer 1985, 194.) (*Streptococcus raffinolactis* Orla-Jensen and Hansen 1932, 6)

raf.fi.no.lac′tis. N.L. adj. *raffinosum* raffinose; L. gen. n. *lactis* of milk; L. gen. n. *raffinolactis* raffinose fermenting bacterium from milk.

Colonies on blood agar are circular, smooth, and entire. Nonpigmented. Not β-hemolytic. Spheres or ovoid cells elongated in the direction of the chain; mostly in pairs or short chains. Gram-positive. Nonmotile. Facultatively anaerobic. Catalase-negative. Growth at 10°C, but not at 40°C. Does not grow in 4% (w/v) NaCl or 0.1% methylene blue milk. Chemo-organotrophic; fermentative metabolism.

Acid is usually produced from arbutin, D-fructose, galactose, lactose, maltose, D-mannose, melizitose, melibiose, N-acetyglucosamine, raffinose, salicin, starch, sucrose, and trehalose. Most strains produce acid from D-xylose. Acid not produced from adonitol, arabitol, D-arabinose, dulcitol, erythritol, D-fucose, gluconate, glycerol, glycogen, inositol, 2-keto-gluconate, D-lyxose, α-methyl-D-mannoside, β-methylxyloside, rhamnose, sorbitol, D-tagatose, L-xylose, and xylitol. Variable results may be obtained with amygdalin, L-arabinose, dextrin, β-gentiobiose, inulin, mannitol, melizitose, α-methyl-D-glucoside, ribose, L-sorbose, and turanose. Esculin hydrolyzed. Casein, hippurate, and gelatin not hydrolyzed. α-Galactosidase and leucine arylamidase-positive. Alkaline phosphatase, β-galactosidase, β-glucuronidase, and pyrrolidonyl arylamidase-negative. Most strains are arginine dehydrolase-negative. Reacts with Lancefield group N antisera.

The peptidoglycan type is Lys–Thr–Ala. Menaquinones are absent. Contains poly(glycerophosphate)-lipoteichoic acid substituted solely with D-alanine ester. The major glycolipid is $Glc(\alpha1\text{–}2)Glc(\alpha1\text{–}3)acyl_2Gro$. Contains D-alanyl- and L-lysylphosphatidylglycerol. Major nonhydoxylated long-chain fatty acids are hexadecanoic and cis-11,12-octadecenoic acids. cis-11,12-Methylenoctadecenoic acid is absent.

The level of 16S rRNA sequence similarity between Lactococcus raffinolactis and the type strains of the other Lactococcus species is 92–93%.

DNA G+C content (mol%): 40–43 (T_m).

Type strain: ATCC 43920, CCUG 32209, CIP 102300, DSM 20443, HAMBI 1589, JCM 5706, LMG 13095, NCIMB 13197.

GenBank accession number (16S rRNA gene): not available for type strain.

Additional remarks: X54261 is the GenBank accession number (16S rRNA) for strain NCIMB 702156.

Genus III. **Lactovum** Matthies, Gössner, Acker, Schramm and Drake 2005, 547^{VP} (Effective publication: Matthies, Gössner, Acker, Schramm and Drake 2004, 853.)

HAROLD L. DRAKE

Lact.o'vum. L. n. lac (gen. lactis) milk. L. neut. n. ovum egg; N.L. neut. n. Lactovum egg from milk (to indicate that the capacity to form lactate yields egg-shaped cells).

Gram-stain-positive, **mesophilic**, **anaerobic** with an **aerotolerant**, **chemo-organotrophic**, fermentative metabolism. Cells are ovoid, do not form spores, and lack flagella (Figure 126).

DNA G+C content (mol%): 37.6 (HPLC).

Type species: **Lactovum miscens** Matthies, Gössner, Acker, Schramm and Drake 2005, 547^{VP} (Effective publication: Matthies, Gössner, Acker, Schramm and Drake 2004, 853.)

Further descriptive information

Members of the genus Lactovum are currently represented by a single species, i.e., the type species Lactovum miscens. Lactovum miscens is phylogenetically distantly related to the genera Lactococcus and Streptococcus. Sequence similarity of the 16S rRNA gene sequence of Lactovum miscens to that of its closest relatives, Lactococcus garvieae and Lactococcus piscium, is 89.3%. Sequence similarity of the 16S rRNA gene sequence of Lactovum miscens to that of its closest streptococcal relative, Streptococcus pyogenes, is 88.2% (Matthies et al., 2004).

Enrichment and isolation procedures

The type species of Lactovum was isolated from an acidic forest floor (L horizon) solution obtained from a beech forest in east-central Germany (Matthies et al., 2004). Forest floor solution was supplemented with N-acetylglucosamine and incubated at 15°C. An N-acetylglucosamine enrichment was streaked onto solidified anoxic medium and incubated at 15°C in O_2-free gas jars. The type species was derived from an isolated colony.

Maintenance procedures

The type species is easily maintained in the phosphate-buffered, yeast extract medium described by Matthies et al. (2004). Stability under long-term storage conditions has not been determined.

List of species of the genus Lactovum

1. **Lactovum miscens** Matthies, Gössner, Acker, Schramm and Drake 2005, 547^{VP} (Effective publication: Matthies, Gössner, Acker, Schramm and Drake 2004, 853.)

    mis.cens. L. part. adj. miscens mixing (to indicate a mixed fermentative metabolism).

    Cells are 0.7 × 1.0 μm, contain multiple cell-wall layers and intracellular membranes. Cells often occur in pairs. Grows at 0–35°C and pH 3.5–7.5. Doubling time is 4.7 h at pH 6.3 and 25°C. Cell lysis is minimal in broth cultures. Grows on glucose, galactose, fructose, mannitol, glucosamine, N-acetylglucosamine, cellobiose, and maltose.

    Displays a mixed fermentative metabolism (Figure 127). Certain sugars (e.g., glucose) are metabolized to lactate via homolactate fermentation, while other sugars (e.g., galactose) are metabolized to ethanol, formate, and acetate. O_2, nitrate, sulfate, and Fe(III) are not utilized as terminal electron acceptors. Cells are aerotolerant, do not contain membranous or cytoplasmic cytochromes, and lack catalase and NADH oxidase.

    DNA G+C content (mol%): 37.6 (HPLC).

    Type strain: anNAG3, ATCC BAA-490, DSM 14925.

    GenBank accession number (16S rRNA gene): AJ439543.

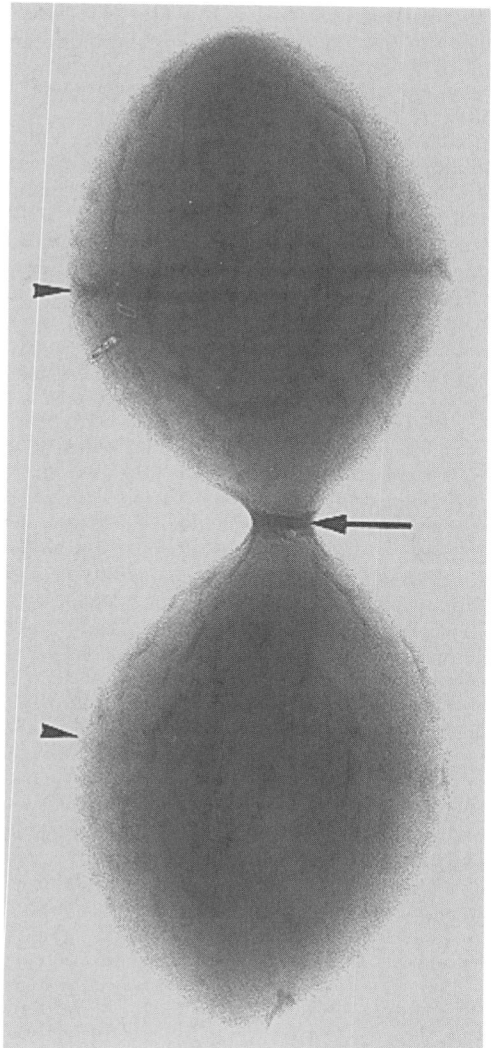

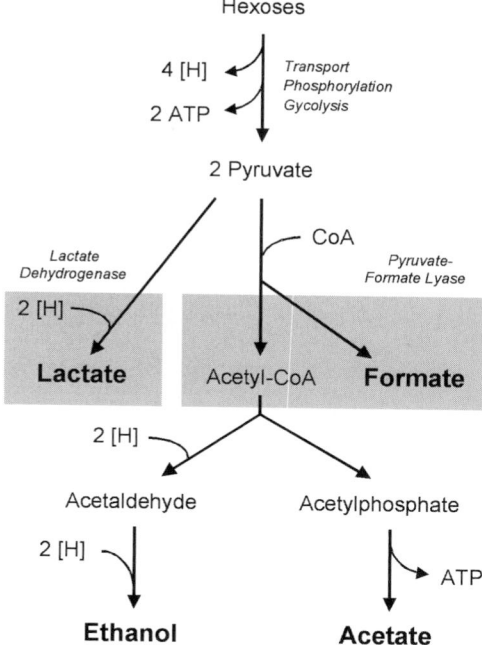

FIGURE 127. Metabolic transformations of *Lactovum miscens*. The route by which pyruvate is metabolized is dependent upon the sugar that is fermented. Abbreviations: [H], reductant; ATP, adenosine triphosphate; CoA, coenzyme A. (Taken from Matthies et al., 2004; used with permisssion.)

FIGURE 126. Transmission electron micrograph of *Lactovum miscens* (DSM 14925). Arrow points identify regions of cells forming septa prior to division. Cells often occur in pairs; an amorphous cellular bridge (arrow) occurs between paired cells. Bar = 0.5 μm. (Taken from Matthies et al., 2004; used with permisssion.)

References

Aguirre, M. and M.D. Collins. 1993. Lactic acid bacteria and human clinical infection. J. Appl. Bacteriol. *75*: 95–107.

Ahmet, Z., M. Warren and E.T. Houang. 1995. Species identification of members of the *Streptococcus milleri* group isolated from the vagina by ID 32 Strep system and differential phenotypic characteristics. J. Clin. Microbiol. *33*: 1592–1595.

Ajdic, D., W.M. McShan, R.E. McLaughlin, G. Savic, J. Chang, M.B. Carson, C. Primeaux, R. Tian, S. Kenton, H. Jia, S. Lin, Y. Qian, S. Li, H. Zhu, F. Najar, H. Lai, J. White, B.A. Roe and J.J. Ferretti. 2002. Genome sequence of *Streptococcus mutans* UA159, a cariogenic dental pathogen. Proc. Natl. Acad. Sci. USA *99*: 14434–14439.

Alam, S., S.R. Brailsford, R.A. Whiley and D. Beighton. 1999. PCR-Based methods for genotyping viridans group streptococci. J. Clin. Microbiol. *37*: 2772–2776.

Alam, S., S.R. Brailsford, S. Adams, C. Allison, E. Sheehy, L. Zoitopoulos, E.A. Kidd and D. Beighton. 2000. Genotypic heterogeneity of

Streptococcus oralis and distinct aciduric subpopulations in human dental plaque. Appl. Environ. Microbiol. *66*: 3330–3336.

Alonso DeVelasco, E., A.F. Verheul, J. Verhoef and H. Snippe. 1995. *Streptococcus pneumoniae*: virulence factors, pathogenesis, and vaccines. Microbiol. Rev. *59*: 591–603.

Anderson, A.W. and P.R. Elliker. 1953. The nutritional requirements of lactic sreptococci isolated from starter cultures. I. Growth in a synthetic medium. J. Dairy Sci. *36*: 161–167.

Andrewes, F.W. and T.J. Horder. 1906. A study of the streptococci pathogenic for man. Lancet *ii*: 708–713.

Arbique, J.C., C. Poyart, P. Trieu-Cuot, G. Quesne, M.D.S. Carvalho, A.G. Steigerwalt, R.E. Morey, D. Jackson, R.J. Davidson and R.R. Facklam. 2004. Accuracy of phenotypic and genotypic testing for identification of *Streptococcus pneumoniae* and description of *Streptococcus pseudopneumoniae* sp. nov. J. Clin. Microbiol. *42*: 4686–4696.

Austrian, R. 1976. The Quellung reaction, a neglected microbiologic technique. Mt. Sinai J. Med. *43*: 699–709.

Austrian, R. 1986. Some aspects of the pneumococcal carrier state. J. Antimicrob. Chemother. *18 Suppl A*: 35–45.

Avery, O.T. and R. Dubos. 1931. The protective action of a specific enzyme against type III pneumococcus infection in mice. J. Exp. Med. *54*: 73–89.

Ayers, S.H. and C.S. Mudge. 1922. The streptococci of the bovine udder. J. Infect. Dis. *31*: 40–50.

Baele, M., L.A. Devriese, M. Vancanneyt, M. Vaneechoutte, C. Snauwaert, J. Swings and F. Haesebrouck. 2003. Emended description of *Streptococcus ferus* isolated from pigs and rats. Int. J. Syst. Evol. Microbiol. *53*: 143–146.

Baker, C.J. and F.F. Barrett. 1974. Group B streptococcal infections in infants: the importance of the various serotypes. J. Am. Med. Assoc. *230*: 1158–1160.

Baker, C.J. 2000. Group B streptococcal infections. *In* Stevens and Kaplan (Editors), Streptococcal Infections: Clinical Aspects, Microbiology, and Molecular Pathogenesis. Oxford University Press, New York, pp. 222–237.

Balke, E., R. Weiss and A. Seipp. 1985. Hyalurodinase activity of beta-hemolytic streptococci of the Lancefield group C. Zentbl. Bakteriol. Mikrobiol. Hyg. A *259*: 194–200.

Baquero, F. 2004. From pieces to patterns: Evolutionary engineering in bacterial pathogens. Nat. Rev. Microbiol. *2*: 510–518.

Barach, J.T. 1979. Improved enumeration of lactic acid streptococci on Elliker agar containing phosphate. Appl. Environ. Microbiol. *38*: 173–174.

Barnes, E.M. and C.S. Impey. 1970. The isolation and properties of the predominant anaerobic bacteria in the caeca of chickens and turkeys. Br. Poult. Sci. *11*: 467–481.

Barnes, E.M., C.S. Impey, B.J.H. Stevens and J.L. Peel. 1977. *Streptococcus pleomorphus* sp. nov., anaerobic *Streptococcus* isolated mainly from ceca of birds. J. Gen. Microbiol. *102*: 45–53.

Barnes, E.M., P.W. Ross, C.D. Wilson, J.M. Hardie, P.D. Marsh, G.C. Mead, M.E. Sharpe, D.A.A. Mossel, N.P. Burman, A.W. Evans and M. Ingram. 1978. Isolation media for streptococci: proceedings of a discussion meeting, Streptococci, Society for Applied Bacteriology Symposium Series No. 7. Academic Press, London, New York, and San Francisco, pp. 1–49, 371–395.

Bartie, K.L., M.J. Wilson, D.W. Williams and M.A. Lewis. 2000. Macrorestriction fingerprinting of "*Streptococcus milleri*" group bacteria by pulsed-field gel electrophoresis. J. Clin. Microbiol. *38*: 2141–2149.

Bauer, S., A. Tholen, J. Overmann and A. Brune. 2000. Characterization of abundance and diversity of lactic acid bacteria in the hindgut of wood- and soil-feeding termites by molecular and culture-dependent techniques. Arch. Microbiol. *173*: 126–137.

Beall, B., R. Facklam and T. Thompson. 1996. Sequencing *emm*-specific PCR products for routine and accurate typing of group A streptococci. J. Clin. Microbiol. *34*: 953–958.

Beall, B., R. Facklam, T. Hoenes and B. Schwartz. 1997. Survey of *emm* gene sequences and T-antigen types from systemic *Streptococcus pyogenes* infection isolates collected in San Francisco, California; Atlanta, Georgia; and Connecticut in 1994 and 1995. J. Clin. Microbiol. *35*: 1231–1235.

Beall, B., R.R. Facklam, J.A. Elliott, A.R. Franklin, T. Hoenes, D. Jackson, L. Laclaire, T. Thompson and R. Viswanathan. 1998. Streptococcal emm types associated with T-agglutination types and the use of conserved *emm* gene restriction fragment patterns for subtyping group A streptococci. J. Med. Microbiol. *47*: 893–898.

Beerens, H., C. Romond and C. Neut. 1963. Les bacilles anaerobies non sporules a Gram-negatif favorises par la bile. Ann. Inst. Pasteur Lille *14*: 5.

Beighton, D., R.R. Russell and H. Hayday. 1981. The isolation of characterization of *Streptococcus mutans* serotype h from dental plaque of monkeys (*Macaca fascicularis*). J. Gen. Microbiol. *124*: 271–279.

Beighton, D., H. Hayday, R.R.B. Russell and R.A. Whiley. 1984. *Streptococcus macacae* sp. nov. from dental plaque of monkeys (*Macaca fascicularis*). Int. J. Syst. Bacteriol. *34*: 332–335.

Beighton, D., J.M. Hardie and R.A. Whiley. 1991. A scheme for the identification of viridans streptococci. J. Med. Microbiol. *35*: 367–372.

Beighton, D., A.D. Carr and B.A. Oppenheim. 1994. Identification of viridans streptococci associated with bacteraemia in neutropenic cancer patients. J. Med. Microbiol. *40*: 202–204.

Bekal, S., C. Gaudreau, R.A. Laurence, E. Simoneau and L. Raynal. 2006. *Streptococcus pseudoporcinus* sp. nov., a novel species isolated from the genitourinary tract of women. J. Clin. Microbiol. *44*: 2584–2586.

Bensing, B.A., I.R. Siboo and P.M. Sullam. 2001. Proteins PblA and PblB of *Streptococcus mitis*, which promote binding to human platelets, are encoded within a lysogenic bacteriophage. Infect. Immun. *69*: 6186–6192.

Bentley, R.W., J.A. Leigh and M.D. Collins. 1991. Intrageneric structure of *Streptococcus* based on comparative-analysis of small-subunit ribosomal-RNA sequences. Int. J. Syst. Bacteriol. *41*: 487–494.

Bentley, R.W., J.A. Leigh and M.D. Collins. 1993. Development and use of species-specific oligonucleotide probes for differentiation of *Streptococcus uberis* and *Streptococcus parauberis*. J. Clin Microbiol. *31*: 57–60.

Berkovitz, R.J. and H.V. Jordan. 1975. Similarity of bacteriocins of *Streptococcus mutans* from mother and infant. Arch. Oral Biol. *20*: 725–730.

Berridge, B.R., H. Bercovier and P.F. Frelier. 2001. *Streptococcus agalactiae* and *Streptococcus difficile* 16S-23S intergenic rDNA: genetic homogeneity and species-specific PCR. Vet. Microbiol. *78*: 165–173.

Berry, A.M., J.C. Paton, E.M. Glare, D. Hansman and D.E. Catcheside. 1988. Cloning and expression of the pneumococcal neuraminidase gene in *Escherichia coli*. Gene *71*: 299–305.

Berry, A.M., R.A. Lock, S.M. Thomas, D.P. Rajan, D. Hansman and J.C. Paton. 1994. Cloning and nucleotide sequence of the *Streptococcus pneumoniae* hyaluronidase gene and purification of the enzyme from recombinant *Escherichia coli*. Infect. Immun. *62*: 1101–1108.

Bert, F. and N. Lambert-Zechovsky. 1997. Analysis of a case of recurrent bacteraemia due to group A *Streptococcus equisimilis* by pulsed-field gel electrophoresis. Infection *25*: 250–251.

Bessen, D., K.F. Jones and V.A. Fischetti. 1989. Evidence for two distinct classes of streptococcal M protein and their relationship to rheumatic fever. J. Exp. Med. *169*: 269–283.

Bessen, D.E. and V.A. Fischetti. 1990. Differentiation between two biologically distinct classes of group A streptococci by limited substitutions of amino acids within the shared region of M protein-like molecules. J. Exp. Med. *172*: 1757–1764.

Betzl, D., W. Ludwig and K.H. Schleifer. 1990. Identification of lactococci and enterococci by colony hybridization with 23S rRNA-targeted oligonucleotide probes. Appl. Environ. Microbiol. *56*: 2927–2929.

Bisno, A.L. 1980. The concept of rheumatogenic and non-rheumatogenic group A streptococci. *In* Read and Zabriskie (Editors), Streptococcal Diseases and the Immune Response. Academic Press, New York, pp. 789–803.

Bisno, A.L. 1995. Steptococcus pyogenes. *In* Mandell, Bennett and Dolin (Editors), Principles and Practice of Infectious Diseases, Vol. 2. Churchill Livingstone, New York, pp. 1786–1799.

Bliss, E.A. 1937. Studies on minute haemolytic streptococci: serological differentiation. J. Bacteriol. *33*: 625–642.

Blumberg, H.M., D.S. Stephens, M. Modansky, M. Erwin, J. Elliot, R.R. Facklam, A. Schuchat, W. Baughman and M.M. Farley. 1996. Invasive group B streptococcal disease: the emergence of serotype V. J. Infect. Dis. *173*: 365–373.

Bohnsack, J.F., A.A. Whiting, R.D. Bradford, B.K. Van Frank, S. Takahashi and E.E. Adderson. 2002. Long-range mapping of the *Streptococcus agalactiae* phylogenetic lineage restriction digest pattern type III-3 reveals clustering of virulence genes. Infect. Immun. *70*: 134–139.

Bolotin, A., S. Mauger, K. Malarme, S.D. Ehrlich and A. Sorokin. 1999. Low-redundancy sequencing of the entire *Lactococcus lactis* IL1403 genome. Antonie van Leeuwenhoek *76*: 27–76.

Bolotin, A., P. Wincker, S. Mauger, O. Jaillon, K. Malarme, J. Weissenbach, S.D. Ehrlich and A. Sorokin. 2001. The complete genome sequence of the lactic acid bacterium *Lactococcus lactis* ssp. lactis IL1403. Genome Res. *11*: 731–753.

Bouvet, A., F. Grimont, M.D. Collins, F. Benaoudia, C. Devine, B. Regnault and P.A. Grimont. 1997. *Streptococcus infantarius* sp. nov. related to *Streptococcus bovis* and *Streptococcus equinus*. Adv. Exp. Med. Biol. *418*: 393–395.

Bowden, G.H. and J.M. Hardie. 1971. Anaerobic organisms from the human mouth. *In* Shapton and Board (Editors), Isolation of Anaerobes, Society for Applied Bacteriology, Technical Series No.5. Academic Press, London and New York, pp. 177–205.

Bowers, E.F. and L.R. Jeffries. 1955. Optochin in the identification of *Streptococcus pneumoniae*. J. Clin. Pathol. London *8*: 58–60.

Brandt, C.M., G. Haase, N. Schnitzler, R. Zbinden and R. Lutticken. 1999. Characterization of blood culture isolates of *Streptococcus dysgalactiae* subsp. *equisimilis* possessing Lancefield's group A antigen. J. Clin. Microbiol. *37*: 4194–4197.

Bratthall, D. 1970. Demonstration of five serological groups of streptococcal strains resembling *Streptococcus mutans*. Odontol. Revy. *21*: 143–152.

Braun, D.G. 1983. The use of streptococcal antigens to probe the mechanisms of immunity. Microbiol. Immunol. *27*: 823–836.

Breed, R.S., E.G.D. Murray and A.P. Hitchens. 1948. Bergey's Manual of Determinative Bacteriology, 6th edn. The Williams & Wilkins Co., Baltimore.

Bridge, P.D. and P.H.A. Sneath. 1982. *Streptococcus gallinarum* sp. nov. and *Streptococcus oralis* sp. nov. Int. J. Syst. Bacteriol. *32*: 410–415.

Bridge, P.D. and P.H. Sneath. 1983. Numerical taxonomy of *Streptococcus*. J. Gen. Microbiol. *129*: 565–597.

Briles, D.E., J. Yother and L.S. McDaniel. 1988. Role of pneumococcal surface protein A in the virulence of *Streptococcus pneumoniae*. Rev. Infect. Dis. *10 Suppl 2*: 372–374.

Brooker, J.D., L.A. O'Donovan, L.A. Skene, K. Clarke, L. Blackall and P. Muslera. 1994. *Streptococcus caprinus* sp. nov., a tannin-resistant ruminal bacterium from feral goats. Lett. Appl. Microbiol. *18*: 313–318.

Brown, P.D. and S.A. Lerner. 1998. Community-acquired pneumonia. Lancet *352*: 1295–1302.

Butler, J.C. 1997. Epidemiology of pneumococcal serotypes and conjugate vaccine formulations. Microb. Drug Resist. *3*: 125–129.

Carlsson, J. 1968. A numerical taxonomic study of human oral streptococci. Odontol. Revy *19*: 137–160.

Cars, O., U. Forsum and E. Hjelm. 1975. New immunofluorescence method for the identification of group A, B, C, E, and G streptococci. Acta Pathol. Microbiol. Scand. B *83*: 145–152.

Casal, J. 1982. Antimicrobial susceptibility of *Streptococcus pneumoniae*: serotype distribution of penicillin-resistant strains in Spain. Antimicrob. Agents Chemother. *22*: 222–225.

Chatellier, S., J. Harel, Y. Zhang, M. Gottschalk, R. Higgins, L.A. Devriese and R. Brousseau. 1998. Phylogenetic diversity of *Streptococcus suis* strains of various serotypes as revealed by 16S rRNA gene sequence comparison. Int. J. Syst. Bacteriol. *48*: 581–589.

Chen, C.C. and P.P. Cleary. 1990. Complete nucleotide sequence of the streptococcal C5a peptidase gene of *Streptococcus pyogenes*. J. Biol. Chem. *265*: 3161–3167.

Chester, F.D. 1901. A Manual of Determinative Bacteriology. The Macmillan Co., New York.

Chopin, A., A. Bolotin, A. Sorokin, S.D. Ehrlich and M. Chopin. 2001. Analysis of six prophages in *Lactococcus lactis* IL1403: Different genetic structure of temperate and virulent phage populations. Nucleic Acids Res. *29*: 644–651.

Christensen, J.J. and R.R. Facklam. 2001. *Granulicatella* and *Abiotrophia* species from human clinical specimens. J. Clin. Microbiol. *39*: 3520–3523.

Christensen, P., G. Kahlmeter, S. Jonsson and G. Kronvall. 1973. New method for the serological grouping of streptococci with specific antibodies adsorbed to protein A-containing staphylococci. Infect. Immun. *7*: 881–885.

Christie, R., N.E. Atkins and E. Munch-Peterson. 1944. A note on a lytic phenomenon shown by group B streptococci. Aust. J. Exp. Biol. Med. Sci. *22*: 197–200.

Clarke, J.K. 1924. On the bacterial factor in the aetiology of dental caries. Br. J. Exp. Pathol. *5*: 141–147.

Coffey, T.J., C.G. Dowson, M. Daniels and B.G. Spratt. 1995. Genetics and molecular biology of beta-lactam-resistant pneumococci. Microb. Drug Resist. *1*: 29–34.

Coffey, T.J., S. Berrón, M. Daniels, M.E. Garcia-Leoni, E. Cercenado, E. Bouza, A. Fenoll and B.G. Spratt. 1996. Multiply antibiotic-resistant *Streptococcus pneumoniae* recovered from Spanish hospitals (1988–1994): novel major clones of serotypes 14, 19F and 15F. Microbiology *142*: 2747–2757.

Coffey, T.J., M.C. Enright, M. Daniels, J.K. Morona, R. Morona, W. Hryniewicz, J.C. Paton and B.G. Spratt. 1998. Recombinational exchanges at the capsular polysaccharide biosynthetic locus lead to frequent serotype changes among natural isolates of *Streptococcus pneumoniae*. Mol. Microbiol. *27*: 73–83.

Cogan, T.M. and J.P. Accolas (Editors). 1996. Dairy Starter Cultures. Verlag Chemie Publishers, New York.

Coligan, J.E., T.J. Kindt and R.M. Krause. 1978. Structure of the streptococcal groups A, A-variant and C carbohydrates. Immunochemistry *15*: 755–760.

Collins, M.D. and D. Jones. 1979. Distribution of isoprenoid quinones in streptococci of serological group D and group N. J. Gen. Microbiol. *114*: 27–33.

Collins, M.D., J.A.E. Farrow, B.A. Phillips and O. Kandler. 1983. *Streptococcus garvieae* sp. nov. and *Streptococcus plantarum* sp. nov. J. Gen. Microbiol. *129*: 3427–3431.

Collins, M.D., J.A.E. Farrow, V. Katic and O. Kandler. 1984a. Taxonomic studies on streptococci of serological Group E, Group P, Group U, and Group V, description of *Streptococcus porcinus* sp. nov. Syst. Appl. Microbiol. *5*: 402–413.

Collins, M.D., D. Jones, J.A.E. Farrow, R. Kilpper-Bälz and K.H. Schleifer. 1984b. *Enterococcus avium* nom. rev., comb. nov., *Enterococcus casseliflavus* nom. rev., comb. nov., *Enterococcus durans* nom. rev., comb. nov., *Enterococcus gallinarum* comb. nov., and *Enterococcus malodoratus* sp. nov. Int. J. Syst. Bacteriol. *34*: 220–223.

Collins, M.D., J.A.E. Farrow, V. Katic and O. Kandler. 1985. *In* Validation of the publication of new names and new combinations previously effectively published outside the IJSB. List no. 17. Int. J. Syst. Bacteriol. *35*: 223–225.

Collins, M.D. and S. Wallbanks. 1992. Comparative sequence analyses of the 16S rRNA genes of *Lactobacillus minutus*, *Lactobacillus rimae* and *Streptococcus parvulus*: proposal for the creation of a new genus *Atopobium*. FEMS Microbiol. Lett. *74*: 235–240.

Collins, M.D., R.A. Hutson, E. Falsen, N. Nikolaitchouk, L. LaClaire and R.R. Facklam. 2000. An unusual *Streptococcus* from human urine, *Streptococcus urinalis* sp. nov. Int. J. Syst. Evol. Microbiol. *50*: 1173–1178.

Collins, M.D. and P.A. Lawson. 2000. The genus *Abiotrophia* (Kawamura *et al.*) is not monophyletic: proposal of *Granulicatella* gen. nov., *Granulicatella adiacens* comb. nov., *Granulicatella elegans* comb. nov and *Granulicatella balaenopterae* comb. nov. Int. J. Syst. Evol. Microbiol. *50*: 365–369.

Collins, M.D., R.A. Hutson, L. Hoyles, E. Falsen, N. Nikolaitchouk and G. Foster. 2001. *Streptococcus ovis* sp. nov., isolated from sheep. Int. J. Syst. Evol. Microbiol. *51*: 1147–1150.

Collins, M.D., R.A. Hutson, E. Falsen, E. Inganas and M. Bisgaard. 2002. *Streptococcus gallinaceus* sp. nov., from chickens. Int. J. Syst. Evol. Microbiol. *52*: 1161–1164.

Collins, M.D., T. Lundstrom, C. Welinder-Olsson, I. Hansson, O. Wattle, R.A. Hudson and E. Falsen. 2004. *Streptococcus devriesei* sp. nov., from equine teeth. Syst. Appl. Microbiol. *27*: 146–150.

Colman, G. and R.E. Williams. 1965. The cell walls of streptococci. J. Gen. Microbiol. *41*: 375–387.

Colman, G. 1968. The application of computers to the classification of streptococci. J. Gen. Microbiol. *50*: 149–158.

Colman, G. and R.E.O. Williams. 1972. Taxonomy of some human viridans streptococci. Academic Press, London and New York.

Colman, G. and L.C. Ball. 1984. Identification of streptococci in a medical laboratory. J. Appl. Bacteriol. *57*: 1–14.

Colman, G. 1990. *Streptococcus* and *Lactococcus*. Chapter 2.7. *In* Parker and Collier (Editors), Topley and Wilson's Principles of Bacteriology, Virology, and Immunity, 8th edn, Vol. 2: Systematic Bacteriology. Edward Arnold, London, Melbourne, Aukland, pp. 119–159.

Cooksey, R.C., F.S. Thompson and R.R. Facklam. 1979. Physiological characterization of nutritionally variant streptococci. J. Clin. Microbiol. *10*: 326–330.

Cords, B.R., L.L. McKay and P. Guerry. 1974. Extrachromosomal elements in group N streptococci. J. Bacteriol. *117*: 1149–1152.

Correia, F.F., J.M. DiRienzo, T.L. McKay and B. Rosan. 1996. scbA from *Streptococcus crista* CC5A: an atypical member of the *lraI* gene family. Infect. Immun. *64*: 2114–2121.

Cowan, S.T. and K.J. Steel. 1974. Cowan and Steel's Manual for Identification of Medical Bacteria, 2nd edn. Cambridge University Press, London.

Coykendall, A.L. 1974. Four types of *Streptococcus mutans* based on their genetic, antigenic and biochemical characteristics. J. Gen. Microbiol. *83*: 327–338.

Coykendall, A.L. and P.A. Specht. 1975. DNA base sequence homologies among strains of *Streptococcus sanguis*. J. Gen. Microbiol. *91*: 92–98.

Coykendall, A.L. 1977. Proposal to elevate the subspecies of *Streptococcus mutans* to species status, based on their molecular composition. Int. J. Syst. Bacteriol. *27*: 26–30.

Coykendall, A.L. and A.J. Munzenmaier. 1978. Deoxyribonucleic acid base sequence studies on glucan-producing and glucan-negative strains of *Streptococcus mitior*. Int. J. Syst. Bacteriol. *28*: 511–515.

Coykendall, A.L. 1983. *Streptococcus sobrinus* nom. rev. and *Streptococcus ferus* nom. rev.: habitat of these and other mutans streptococci. Int. J. Syst. Bacteriol. *33*: 883–885.

Coykendall, A.L. and K.B. Gustafson. 1985. Deoxyribonucleic acid hybridizations among strains of *Streptococcus salivarius* and *Streptococcus bovis*. Int. J. Syst. Bacteriol. *35*: 274–280.

Coykendall, A.L., P.M. Wesbecher and K.B. Gustafson. 1987. *Streptococcus milleri*, *Streptococcus constellatus*, and *Streptococcus intermedius* are later synonyms of *Streptococcus anginosus*. Int. J. Syst. Bacteriol. *37*: 222–228.

Coykendall, A.L. 1989a. Classification and identification of the viridans streptococci. Clin. Microbiol. Rev. *2*: 315–328.

Coykendall, A.L. 1989b. Rejection of the type strain of *Streptococcus mitis* (Andrewes and Horder 1906). Request for an Opinion. Int. J. Syst. Bacteriol. *39*: 207–209.

Cunningham, M.W. 2000. Pathogenesis of group A streptococcal infections. Clin. Microbiol. Rev. *13*: 470–511.

Curtis, S.N. and R.M. Krause. 1964. Antigenic relationships between groups B and G streptococci. J. Exp. Med. *120*: 629–637.

Dajani, A.S. 1973. Rapid identification of beta hemolytic streptococci by counterimmunoelectrophoresis. J. Immunol. *110*: 1702–1705.

De Gheldre, Y., P. Vandamme, H. Goossens and M.J. Struelens. 1999. Identification of clinically relevant viridans streptococci by analysis of transfer DNA intergenic spacer length polymorphism. Int. J. Syst. Bacteriol. *49*: 1591–1598.

De Soet, J.J. and J. De Graaff. 1990. Monoclonal antibodies for enumeration and identification of mutans streptococci in epidemiological studies. Arch. Oral Biol. *35*: 165S–168S.

de Stoppelaar, J.D., J. van Houte and C.E. de Moor. 1967. The presence of dextran-forming bacteria, resembling *Streptococcus bovis* and *Streptococcus sanguis*, in human dental plaque. Arch. Oral Biol. *12*: 1199–1202.

Deibel, R.M. and H.W. Seeley. 1974. *Streptococcaceae*. *In* Buchanan and Gibbons (Editors), Bergey's Manual of Determinative Bacteriology, 8th edn. The Williams & Wilkins Co., Baltimore, pp. 490–509.

Desiere, F., W.M. McShan, D. van Sinderen, J.J. Ferretti and H. Brussow. 2001. Comparative genomics reveals close genetic relationships between phages from dairy bacteria and pathogenic streptococci: evolutionary implications for prophage-host interactions. Virology *288*: 325–341.

Devriese, L.A., J. Hommez, R. Kilpper-Bälz and K.H. Schleifer. 1986. *Streptococcus canis* sp. nov. - a species of Group G streptococci from animals. Int. J. Syst. Bacteriol. *36*: 422–425.

Devriese, L.A., R. Kilpper-Bälz and K.H. Schleifer. 1988. *Streptococcus hyointestinalis* sp. nov. from the gut of swine. Int. J. Syst. Bacteriol. *38*: 440–441.

Devriese, L.A., B. Pot, P. Vandamme, K. Kersters, M.D. Collins, N. Alvarez, F. Haesebrouck and J. Hommez. 1997. *Streptococcus hyovaginalis* sp. nov. and *Streptococcus thoraltensis* sp. nov., from the genital tract of sows. Int. J. Syst. Bacteriol. *47*: 1073–1077.

Devriese, L.A., P. Vandamme, M.D. Collins, N. Alvarez, B. Pot, J. Hommez, P. Butaye and F. Haesebrouck. 1999. *Streptococcus pluranimalium* sp. nov., from cattle and other animals. Int. J. Syst. Bacteriol. *49*: 1221–1226.

Dierksen, K.P., N.L. Ragland and J.R. Tagg. 2000. A new alkaline pH-adjusted medium enhances detection of beta-hemolytic streptococci by minimizing bacterial interference due to *Streptococcus salivarius*. J. Clin. Microbiol. *38*: 643–650.

Diernhofer, K. 1932. Asculinbouillon als hilfsmittel für die differenzierung von euter- und milchstreptokokken bei massenuntersuchungen. Milchw. Forsch. *13*: 368–374.

Dillon, H.C., Jr., S. Khare and B.M. Gray. 1987. Group B streptococcal carriage and disease: a 6-year prospective study. J. Pediatr. *110*: 31–36.

Dodd, R. 1949. Serologic relationship between *Streptococcus* group H and *Streptococcus sanguis*. Proc. Soc. Exp. Biol. Med. *70*: 598–599.

Doit, C., F. Grimont, R.A. Whiley, B. Regnault, P.A.D. Grimont, J.M. Hardie and A. Bouvet. 1994. Ribotypes of the 'Streptococcus milleri-group' allow discrimination between strains of *Streptococcus constellatus*, *Streptococcus intermedius*, and *Streptococcus anginosus*. *In* Totolian (Editor), Pathogenic Streptococci Present and Future. Lancer Publications, St. Petersburg, Russian Federation, pp. 531–532.

Domenech, A., J. Prieta, J.F. Fernandez-Garayzabal, M.D. Collins, D. Jones and L. Dominguez. 1993. Phenotypic and phylogenetic evidence for a close relationship between *Lactococcus garvieae* and *Enterococcus seriolicida*. Microbiologia *9*: 63–68.

Dopazo, J., A. Mendoza, J. Herrero, F. Caldara, Y. Humbert, L. Friedli, M. Guerrier, E. Grand-Schenk, C. Gandin, M. de Francesco, A. Polissi, G. Buell, G. Feger, E. García, M. Peitsch and J.F. García-Bustos. 2001. Annotated draft genomic sequence from a *Streptococcus pneumoniae* type 19F clinical isolate. Microb. Drug Resist. *7*: 99–125.

Douglas, C.W., J. Heath, K.K. Hampton and F.E. Preston. 1993. Identity of viridans streptococci isolated from cases of infective endocarditis. J. Med. Microbiol. *39*: 179–182.

Dowson, C.G., A. Hutchison, J.A. Brannigan, R.C. George, D. Hansman, J. Linares, A. Tomasz, J.M. Smith and B.G. Spratt. 1989. Horizontal transfer of penicillin-binding protein genes in penicillin-resistant clinical isolates of *Streptococcus pneumoniae*. Proc. Natl. Acad. Sci. U.S.A *86*: 8842–8846.

Dowson, C.G., T.J. Coffey, C. Kell and R.A. Whiley. 1993. Evolution of penicillin resistance in *Streptococcus pneumoniae*; the role of *Streptococcus mitis* in the formation of a low affinity PBP2B in *S. pneumoniae*. Mol. Microbiol. *9*: 635–643.

Dowson, C.G., T.J. Coffey and B.G. Spratt. 1994. Origin and molecular epidemiology of penicillin-binding-protein-mediated resistance to beta-lactam antibiotics. Trends Microbiol. *2*: 361–366.

Dowson, C.G., V. Barcus, S. King, P. Pickerill, A. Whatmore and M. Yeo. 1997. Horizontal gene transfer and the evolution of resistance and virulence determinants in *Streptococcus*. J. Appl. Microbiol. *83*: 42S-51S.

Edwardsson, S. 1968. Characteristics of caries-inducing human streptococci resembling *Streptococcus mutans*. Arch. Oral Biol. *13*: 637–646.

Edwardsson, S. 1970. The caries-inducing property of variants of *Streptococcus mutans*. Odontol. Revy *21*: 153–157.

Efstratiou, A., G. Colman, G. Hahn, J.F. Timoney, J.M. Boeufgras and D. Monget. 1994. Biochemical differences among human and animal streptococci of Lancefield group C or group G. J. Med. Microbiol. *41*: 145–148.

Efstratiou, A. 1997. Pyogenic streptococci of Lancefield groups C and G as pathogens in man. J. Appl. Bacteriol. *83*: 72S-79S.

Eldar, A., Y. Bejerano and H. Bercovier. 1994. *Streptococcus shiloi* and *Streptococcus difficile*: two new streptococcal species causing a meningoencephalitis in fish. Curr. Microbiol. *28*: 139–143.

Eldar, A., P.F. Frelier, L. Assenta, P.W. Varner, S. Lawhon and H. Bercovier. 1995. *Streptococcus shiloi*, the name for an agent causing septicemic infection in fish, is a junior synonym of *Streptococcus iniae*. Int. J. Syst. Bacteriol. *45*: 840–842.

Eldar, A., M. Goria, C. Ghittino, A. Zlotkin and H. Bercovier. 1999. Biodiversity of *Lactococcus garvieae* strains isolated from fish in Europe, Asia, and Australia. Appl. Environ. Microbiol. *65*: 1005–1008.

Elliker, P.R., A.W. Anderson and G. Hannesson. 1956. An agar medium for lactic acid streptococci and lactobacilli. J. Dairy Sci. *39*: 1611–1612.

Elliott, S.D. 1966. Streptococcal infection in young pigs. I. An immunochemical study of the causative agent (PM streptococcus). J. Hyg. Lond. *64*: 205–212.

Ellwood, D.C., J.R. Hunter and V.M. Longyear. 1974. Growth of *Streptococcus mutans* in a chemostat. Arch. Oral Biol. *19*: 659–664.

Elting, L.S., G.P. Bodey and B.H. Keefe. 1992. Septicemia and shock syndrome due to viridans streptococci: a case-control study of predisposing factors. Clin. Infect. Dis. *14*: 1201–1207.

Enright, M.C., B.G. Spratt, A. Kalia, J.H. Cross and D.E. Bessen. 2001. Multilocus sequence typing of *Streptococcus pyogenes* and the relationships between emm type and clone. Infect. Immun. *69*: 2416–2427.

Ezaki, T., R. Facklam, N. Takeuchi and E. Yabuuchi. 1986. Genetic relatedness between the type strain of *Streptococcus anginosus* and minute-colony-forming beta-hemolytic streptococci carrying different Lancefield grouping antigens. Int. J. Syst. Bacteriol. *36*: 345–347.

Ezaki, T., N. Li, Y. Hashimoto, H. Miura and H. Yamamoto. 1994. 16s Ribosomal RNA sequences of anaerobic cocci and proposal of *Ruminococcus hansenii* comb. nov. and *Ruminococcus productus* comb. nov. Int. J. Syst. Bacteriol. *44*: 130–136.

Facklam, R. and J.A. Elliott. 1995. Identification, classification, and clinical relevance of catalase-negative, gram-positive cocci, excluding the streptococci and enterococci. Clin. Microbiol. Rev. *8*: 479–495.

Facklam, R.R. 1972. Recognition of group D streptococcal species of human origin by biochemical and physiological tests. Appl. Microbiol. *23*: 1131–1139.

Facklam, R.R. 1977. Physiological differentiation of viridans streptococci. J. Clin. Microbiol. *5*: 184–201.

Facklam, R.R. and H.W. Wilkinson. 1981. The family *Streptococcaceae* (medical aspects). *In* Starr, Stolp, Trüper, Balows and Schlegel (Editors), The Prokaryotes: A Handbook on Habitats, Isolation, and Identification of Bacteria, Vol. 2. Springer Verlag, Berlin, pp. pp. 1572–1597.

Facklam, R.R., L.G. Thacker, B. Fox and L. Eriquez. 1982. Presumptive identification of streptococci with a new test system. J. Clin. Microbiol. *15*: 987–990.

Facklam, R.R. 1984. The major differences in the American and British *Streptococcus* taxonomy schemes with special reference to *Streptococcus milleri*. Eur. J. Clin. Microbiol. Infect. Dis. *3*: 91–93.

Facklam, R.R. 1985. *Streptococcus milleri*: reply to D.M. Yajko and W.K. Hadley. Eur. J. Clin. Microbiol. Infect. Dis. *4*: 356.

Facklam, R.R. and R.B. Carey. 1985. Streptococci and Aerococci. *In* Lennette, Balows, Hausler and Shadomy (Editors), Manual of Clinical Microbiology, 4th edn. American Society for Microbiology, Washington, D.C., pp. 154–175.

Facklam, R.R. 2002. What happened to the streptococci: overview of taxonomic and nomenclature changes. Clin. Microbiol. Revs. *15*: 613–630.

Farmer, E.D. 1954. Serological subdivisions among the Lancefield group H streptococci. J. Gen. Microbiol. *11*: 131–138.

Farrow, J.A.E. and M.D. Collins. 1984a. Taxonomic studies on streptococci of serological groups C, G, and L and possibly related taxa. Syst. Appl. Microbiol. *5*: 483–493.

Farrow, J.A.E. and M.D. Collins. 1984b. DNA base composition, DNA-DNA homology and long-chain fatty acid studies on *Streptococcus thermophilus* and *Streptococcus salivarius*. J. Gen. Microbiol. *130*: 357–362.

Farrow, J.A.E., J. Kruze, B.A. Phillips, A.J. Bramley and M.D. Collins. 1984. Taxonomic studies on *Streptococcus bovis* and *Streptococcus equinus*: description of *Streptococcus alactolyticus* sp. nov. and *Streptococcus saccharolyticus* sp. nov. Syst. Appl. Microbiol. *5*: 467–482.

Farrow, J.A.E., J. Kruze, B.A. Phillips, A.J. Bramley and M.D. Collins. 1985. *In* Validation of the publication of new names and new combinations previously effectively published outside the IJSB Validation List no17. Int. J. Syst. Bacteriol. *35*: 223–225.

Fedson, D.S. 1997. Pneumococcal vaccination: four issues for western Europe. Biologicals *25*: 215–219.

Fernandez, E., V. Blume, P. Garrido, M.D. Collins, A. Mateos, L. Dominguez and J.F. Fernandez-Garayzabal. 2004. *Streptococcus equi* subsp. *ruminatorum* subsp. nov., isolated from mastitis in small ruminants. Int. J. Syst. Evol. Microbiol. *54*: 2291–2296.

Ferretti, J.J., W.M. McShan, D. Ajdic, D.J. Savic, G. Savic, K. Lyon, C. Primeaux, S. Sezate, A.N. Suvorov, S. Kenton, H.S. Lai, S.P. Lin, Y. Qian, H.G. Jia, F.Z. Najar, Q. Ren, H. Zhu, L. Song, J. White, X. Yuan, S.W. Clifton, B.A. Roe and R. McLaughlin. 2001. Complete genome sequence of an M1 strain of *Streptococcus pyogenes*. Proc. Natl. Acad. Sci. U.S.A. *98*: 4658–4663.

Fischetti, V.A. 1989. Streptococcal M protein: molecular design and biological behavior. Clin. Microbiol. Rev. *2*: 285–314.

Flint, S.H., L.J.H. Ward and J.D. Brooks. 1999. *Streptococcus waius* sp. nov., a thermophilic streptococcus from a biofilm. Int. J. Syst. Bacteriol. *49*: 759–767.

Food and Agriculture Organization of the United Nations. 2001. 2000 production.

Fransen, E.V.G., V. Pedrazzoli and M. Kilian. 1991. Ecology of viridans streptococci in the oral cavity and pharynx. Oral Microbiol. Immunol. *6*: 129–133.

Freney, J., S. Bland, J. Etienne, M. Desmonceaux, J.M. Boeufgras and J. Fleurette. 1992. Description and evaluation of the semiautomated 4-hour rapid ID32 strep method for identification of Streptococci and members of related genera. J. Clin. Microbiol. *30*: 2657–2661.

Frost, W.D. and M.A. Engelbrecht. 1936. A revision of the genus *Streptococcus*. Dept. Agr. Bacteriol. Univ. Wisconsin, Madison: 1–4.

Fuller, J.D., D.J. Bast, V. Nizet, D.E. Low and J.C. de Azavedo. 2001. *Streptococcus iniae* virulence is associated with a distinct genetic profile. Infect. Immun. *69*: 1994–2000.

Galesloot, T.E., E. Hassing and J. Stadhouders. 1961. Agar media for the isolation and enumeration of aromabacteria in starters. Neth. Milk Dairy J. *15*: 127–150.

Garvie, E.I. and A.J. Bramley. 1979a. *Streptococcus bovis*–an approach to its classification and its importance as a cause of bovine mastitis. J. Appl. Bacteriol. *46*: 557–566.

Garvie, E.I. and A.J. Bramley. 1979b. *Streptococcus uberis*: approach to its classification. J. Appl. Bacteriol. *46*: 295–304.

Garvie, E.I., J.A.E. Farrow and A.J. Bramley. 1983. *Streptococcus dysgalactiae* (Diernhofer) nom. rev. Int. J. Syst. Bacteriol. *33*: 404–405.

Gasson, M.J. and F.L. Davies. 1980. High-frequency conjugation associated with *Streptococcus lactis* donor cell aggregation. J. Bacteriol. *143*: 1260–1264.

Gasson, M.J. 1983. Plasmid complements of *Streptococcus lactis* NCDO 712 and other lactic streptococci after protoplast-induced curing. J. Bacteriol. *154*: 1–9.

Gasson, M.J. and C.A. Shearman. 2003. Plasmid biology, conjugation, and transposition. *In* Wood and Wamer (Editors), Genetics of Lactic Acid Bacteria. Kluwer Academic/Plenum Publishers, New York, pp. 25–44.

Gaudu, P., K. Vido, B. Cesselin, S. Kulakauskas, J. Tremblay, L. Rezaiki, G. Lamberret, S. Sourice, P. Duwat and A. Gruss. 2002. Respiration capacity and consequences in *Lactococcus lactis*. Antonie van Leeuwenhoek *82*: 263–269.

Geis, A., J. Singh and M. Teuber. 1983. Potential of lactic streptococci to produce bacteriocin. Appl. Environ. Microbiol. *45*: 205–211.

Gilmour, M.N., T.S. Whittam, M. Kilian and R.K. Selander. 1987. Genetic relationships among the oral streptococci. J. Bacteriol. *169*: 5247–5257.

Ginsburg, I. and N. Grossowicz. 1957. Group A hemolytic streptococci. I. A chemically defined medium for growth from small inocula. Proc. Soc. Exp. Biol. Med. *96*: 108–112.

Glaser, P., C. Rusniok, C. Buchrieser, F. Chevalier, L. Frangeul, T. Msadek, M. Zouine, E. Couvé, L. Lalioui, C. Poyart, P. Trieu-Cuot and F. Kunst. 2002. Genome sequence of *Streptococcus agalactiae*, a pathogen causing invasive neonatal disease. Mol. Microbiol. *45*: 1499–1513.

Glazunova, O.O., D. Raoult and V. Roux. 2006. *Streptococcus massiliensis* sp. nov., isolated from a patient blood culture. Int. J. Syst. Evol. Microbiol. *56*: 1127–1131.

Gold, O.G., H.V. Jordan and J. Van Houte. 1973. A selective medium for *Streptococcus mutans*. Arch. Oral Biol. *18*: 1357–1364.

Gossling, J. 1988. Occurrence and pathogenicity of the *Streptococcus milleri* group. Rev. Infect. Dis. *10*: 257–285.

Gottschalk, M., R. Higgins, M. Jacques, M. Beaudoin and J. Henrichsen. 1991. Characterization of six new capsular types (23–28) of *Streptococcus suis*. J. Clin. Microbiol. *29*: 2590–2594.

Griffith, F. 1934. The serological classification of *Streptococcus pyogenes*. J. Hyg. Camb. *34*: 542–584.

Gronroos, L., M. Saarela, J. Matto, U. Tanner-Salo, A. Vuorela and S. Alaluusua. 1998. Mutacin production by *Streptococcus mutans* may promote transmission of bacteria from mother to child. Infect. Immun. *66*: 2595–2600.

Guthof, O. 1956. Pathogenic strains of *Streptococcus viridans*; streptocci found in dental abscesses and infiltrates in the region of the oral cavity. Zentbl. Bakteriol. Parasitenkd. Infektkrankh. Hyg. Abt. I *166*: 553–564.

Hamada, S. and H.D. Slade. 1980. Biology, immunology, and cariogenicity of *Streptococcus mutans*. Microbiol. Rev. *44*: 331–384.

Hammerschmidt, S., S.R. Talay, P. Brandtzaeg and G.S. Chhatwal. 1997. SpsA, a novel pneumococcal surface protein with specific binding to secretory immunoglobulin A and secretory component. Mol. Microbiol. *25*: 1113–1124.

Handley, P., A. Coykendall, D. Beighton, J.M. Hardie and R.A. Whiley. 1991. *Streptococcus crista* sp. nov., a viridans *Streptococcus* with tufted fibrils, isolated from the human oral cavity and throat. Int. J. Syst. Bacteriol. *41*: 543–547.

Handley, P.S., P.L. Carter, J.E. Wyatt and L.M. Hesketh. 1985. Surface structures (peritrichous fibrils and tufts of fibrils) found on *Streptococcus sanguis* strains may be related to their ability to coaggregate with other oral genera. Infect. Immun. *47*: 217–227.

Hansson, L. and M.H. Häggström. 1984. Effect of growth conditions on the activities of superoxide dismutase and NADH-oxidase/NADH-peroxidase in *Streptococcus lactis*. Curr. Microbiol. *10*: 345–352.

Hardie, J.M. 1986a. Anaerobic streptococci. *In* Sneath, Mair, Sharpe and Holt (Editors), Bergey's Manual of Systematic Bacteriology, 1st edn, Vol. 2. The Williams & Wilkins Co., Baltimore, pp. 1043–1071.

Hardie, J.M. 1986b. Other streptococci. *In* Sneath, Mair, Sharpe and Holt (Editors), Bergey's Manual of Systematic Bacteriology, 1st edn, Vol. 2. The Williams & Wilkins Co., Baltimore, pp. 1068–1071.

Hardie, J.M. 1986c. Anaerobic Streptococci. *In* Sneath, Mair, Sharpe and Holt (Editors), Bergey's Manual of Systematic Bacteriology, 1st edn, Vol. 2. The Williams & Wilkins Co., Baltimore, pp. 1066–1068.

Hardie, J.M. and R.A. Whiley. 1995. The genus *Streptococcus*. *In* Wood and Holzapfel (Editors), The Lactic Acid Bacteria Volume 2: The Genera of Lactic Acid Bacteria, Vol. 2. Blackie Academic & Professional, London, Glasgow, Weinhein, New York, Tokyo, Melbourne, Madras, pp. 55–124.

Hardy, G.G., M.J. Caimano and J. Yother. 2000. Capsule biosynthesis and basic metabolism in *Streptococcus pneumoniae* are linked through the cellular phosphoglucomutase. J. Bacteriol. *182*: 1854–1863.

Harlander, S.K. 1987. Transformation of *Streptococcus lactis* by electroporation. *In* Ferretti and Curtiss (Editors), Streptococcal Genetics. American Society for Microbiology, Washington, D.C., pp. 229–233.

Hauge, M., C. Jespersgaard, K. Poulsen and M. Kilian. 1996. Population structure of *Streptococcus agalactiae* reveals an association between specific evolutionary lineages and putative virulence factors but not disease. Infect. Immun. *64*: 919–925.

Hausdorff, W.P., J. Bryant, P.R. Paradiso and G.R. Siber. 2000. Which pneumococcal serogroups cause the most invasive disease: implications for conjugate vaccine formulation and use, part I. Clin. Infect. Dis. *30*: 100–121.

Håvarstein, L.S., P. Gaustad, I.F. Nes and D.A. Morrison. 1996. Identification of the streptococcal competence-pheromone receptor. Mol. Microbiol. *21*: 863–869.

Hehre, E.J. and J.M. Neill. 1946. Formation of serologically reactive dextrans by streptococci from subacute bacterial endocarditis. J. Exp. Med. *83*: 147–162.

Henrichsen, J. 1995. Six newly recognized types of *Streptococcus pneumoniae*. J. Clin. Microbiol. *33*: 2759–2762.

Hirsch, A. 1953. The evolution of lactic streptococci. J. Dairy Res. *20*: 290–293.

Hohwy, J., J. Reinholdt and M. Kilian. 2001. Population dynamics of *Streptococcus mitis* in its natural habitat. Infect. Immun. *69*: 6055–6063.

Holdeman, L.V. and W.E.C. Moore. 1974. New genus, *Coprococcus*, twelve new species, and emended descriptions of four previously described species of bacteria from human feces. Int. J. Syst. Bacteriol. *24*: 260–277.

Holm, S.E., J. Ferretti, D. Simon and K. Johnston. 1992. Deletion of a streptokinase gene eliminates the nephritogenic capacity of a type 49 strain. *In* Orifici (Editor), New Perspectives on Streptococci and Streptococcal Infections: Proceedings of the XI Lancefield International Symposium. Zentbl. Bacteriol. Suppl. 22, New York, pp. 261–263.

Hoshino, T., T. Fujiwara and M. Kilian. 2005. Use of phylogenetic and phenotypic analyses to identify nonhemolytic streptococci isolated from bacteremic patients. J. Clin Microbiol. *43*: 6073–6085.

Hoskins, J., W.E. Alborn, Jr., J. Arnold, L.C. Blaszczak, S. Burgett, B.S. DeHoff, S.T. Estrem, L. Fritz, D.J. Fu, W. Fuller, C. Geringer, R. Gilmour, J.S. Glass, H. Khoja, A.R. Kraft, R.E. Lagace, D.J. LeBlanc, L.N. Lee, E.J. Lefkowitz, J. Lu, P. Matsushima, S.M. McAhren, M. McHenney, K. McLeaster, C.W. Mundy, T.I. Nicas, F.H. Norris, M. O'Gara, R.B. Peery, G.T. Robertson, P. Rockey, P.M. Sun, M.E. Winkler, Y. Yang, M. Young-Bellido, G. Zhao, C.A. Zook, R.H. Baltz, S.R. Jaskunas, P.R. Rosteck, Jr., P.L. Skatrud and J.I. Glass. 2001. Genome of the bacterium *Streptococcus pneumoniae* strain R6. J. Bacteriol. *183*: 5709–5717.

Howey, R.T., C.M. Lock and L.V.H. Moore. 1990. Subspecies names automatically created by Rule 46. Int. J. Syst. Bacteriol. *40*: 317–319.

Hwang, M.N. and G.M. Ederer. 1975. Rapid hippurate hydrolysis method for presumptive identification of group B streptococci. J. Clin. Microbiol. *1*: 114–115.

Hynes, W.L., L. Hancock and J.J. Ferretti. 1995. Analysis of a second bacteriophage hyaluronidase gene from *Streptococcus pyogenes*: evidence for a third hyaluronidase involved in extracellular enzymatic activity. Infect. Immun. *63*: 3015–3020.

Ikeda, T. and H.J. Sandham. 1972. A medium for the recognition and enumeration of *Streptococcus mutans*. Arch. Oral Biol. *17*: 601–604.

Inoue, M., H. Eifuku-Koreeda, K. Kitada, N. Takamatsu-Matsushita, Y. Okada and E. Osano. 1998. Serotype variation in *Streptococcus anginosus*, *S. constellatus* and *S. intermedius*. J. Med. Microbiol. *47*: 435–439.

Isenberg, H.D., J.A. Washington, II, A. Balows and A. Sonnenwirth, C,. 1985. Collection, handling, and processing of specimens. *In* Lennette, Balows, Hausler and Shadomy (Editors), Manual of Clinical Microbiology, 4th edn. American Society for Microbiology, Washington, D.C., pp. 73–98.

Itoh, Y., Y. Kawamura, H. Kasai, M.M. Shah, P.H. Nhung, M. Yamada, X. Sun, T. Koyana, M. Hayashi, K. Ohkusu and T. Ezaki. 2006. *dnaJ* and *gyrB* gene sequence relationship among species and strains of genus *Streptococcus*. Syst. Appl. Microbiol. *29*: 368–374.

Jacks-Weis, J., Y. Kim and P.P. Cleary. 1982. Restricted deposition of C3 on M⁺ group A streptococci: correlation with resistance to phagocytosis. J. Immunol. *128*: 1897–1902.

Jacobs, J.A., H.C. Schouten, E.E. Stobberingh and P.B. Soeters. 1995. Viridans streptococci isolated from the bloodstream. Relevance of species identification. Diagn. Microbiol. Infect. Dis. *22*: 267–273.

Jacobs, J.A., C.S. Schot, A.E. Bunschoten and L.M. Schouls. 1996. Rapid species identification of '*Streptococcus milleri*' strains by line blot hybridization: identification of a distinct 16S rRNA population closely related to *Streptococcus constellatus*. J. Clin. Microbiol. *34*: 1717–1721.

Jacobs, J.A., C.S. Schot and L.M. Schouls. 2000a. The *Streptococcus anginosus* species comprises five 16S rRNA ribogroups with different phenotypic characteristics and clinical relevance. Int. J. Syst. Evol. Microbiol. *50*: 1073–1079.

Jacobs, J.A., L.M. Schouls and R.A. Whiley. 2000b. DNA-DNA reassociation studies of *Streptococcus constellatus* with unusual 16S rRNA sequences. Int. J. Syst. Evol. Microbiol. *50*: 247–249.

Jarvis, A.W., G.F. Fitzgerald, M. Mata, A. Mercenier, H. Neve, I.B. Powell, C. Ronda, M. Saxelin and M. Teuber. 1991. Species and type phages of lactococcal bacteriophages. Intervirology *32*: 2–9.

Jenkinson, H.F. and D.R. Demuth. 1997. Structure, function and immunogenicity of streptococcal antigen I/II polypeptides. Mol. Microbiol. *23*: 183–190.

Jenkinson, H.F. and R.J. Lamont. 1997. Streptococcal adhesion and colonization. Crit. Rev. Oral Biol. Med. *8*: 175–200.

Jensen, P.R. and K. Hammer. 1993. Minimal requirements for exponential growth of *Lactococcus lactis*. Appl. Environ. Microbiol. *59*: 4363–4366.

Jones, D. 1978. Composition and differentiation of the genus *Streptococcus*. *In* Skinner and Quesnel (Editors), Streptococci, Society for Applied Bacteriology Symposium Series No. 7. Academic Press, London, New York, and San Fransisco, pp. 1–49.

Jones, K.F., O. Schneewind, J.M. Koomey and V.A. Fischetti. 1991. Genetic diversity among the T-protein genes of group A streptococci. Mol. Microbiol. *5*: 2947–2952.

Judical Commission. 1954. Opinion 8. Int. Bull. Bacteriol. Nomencl. Taxon. *4*: 145–146.

Kalia, A., M.C. Enright, B.G. Spratt and D.E. Bessen. 2001. Directional gene movement from human-pathogenic to commensal-like streptococci. Infect. Immun. *69*: 4858–4869.

Kamezawa, Y., T. Nakahara, S. Nakano, Y. Abe, J. Nozaki-Renard and T. Isono. 1997. Streptococcal mitogenic exotoxin Z, a novel acidic superantigenic toxin produced by a T1 strain of *Streptococcus pyogenes*. Infect. Immun. *65*: 3828–3833.

Kaplan, E.L., D.R. Johnson, M.C. Del Rosario and D.L. Horn. 1999. Susceptibility of group A beta-hemolytic streptococci to thirteen antibiotics: examination of 301 strains isolated in the United States between 1994 and 1997. Pediatr. Infect. Dis. J. *18*: 1069–1072.

Kauffmann, F., E. Lund and B.E. Eddy. 1960. Proposals for a change in the nomenclature of Diplococcus pneumoniae and a comparison of the Danish and American type designations. Int. Bull. Bacteriol. Nomencl. Taxon. *10*: 31–40.

Kawamura, Y., X.G. Hou, F. Sultana, S.J. Liu, H. Yamamoto and T. Ezaki. 1995a. Transfer of *Streptococcus adjacens* and *Streptococcus defectivus* to *Abiotrophia* gen. nov. as *Abiotrophia adiacens* comb. nov., and *Abiotrophia defectiva* comb. nov., respectively. Int. J. Syst. Bacteriol. *45*: 798–803.

Kawamura, Y., X.G. Hou, F. Sultana, H. Miura and T. Ezaki. 1995b. Determination of 16S ribosomal RNA sequences of *Streptococcus mitis* and *Streptococcus gordonii* and phylogenetic relationships among members of the genus *Streptococcus*. Int. J. Syst. Bacteriol. *45*: 406–408.

Kawamura, Y., X.G. Hou, Y. Todome, F. Sultana, K. Hirose, S.E. Shu, T. Ezaki and H. Ohkuni. 1998. *Streptococcus peroris* sp. nov. and *Streptococcus infantis* sp. nov., new members of the *Streptococcus mitis* group, isolated from human clinical specimens. Int. J. Syst. Bacteriol. *48*: 921–927.

Kawamura, Y., R.A. Whiley, S.E. Shu, T. Ezaki and J.M. Hardie. 1999. Genetic approaches to the identification of the mitis group within the genus *Streptococcus*. Microbiology *145*: 2605–2613.

Kawamura, Y., R.A. Whiley, L.C. Zhao, T. Ezaki and J.M. Hardie. 2000. Taxonomic study of "tufted mitior" strains of streptococci (*Streptococcus sanguinis* biotype II): recognition of a new genospecies. Syst. Appl. Microbiol. *23*: 245–250.

Kehoe, M.A., V. Kapur, A.M. Whatmore and J.M. Musser. 1996. Horizontal gene transfer among group A streptococci: implications for pathogenesis and epidemiology. Trends Microbiol. *4*: 436–443.

Kellogg, J.A. 1990. Suitability of throat culture procedures for detection of group A streptococci and as reference standards for evaluation of streptococcal antigen detection kits. J. Clin. Microbiol. *28*: 165–169.

Kelstrup, J., S. Richmond, C. West and R.J. Gibbons. 1970. Fingerprinting human oral streptococci by bacteriocin production and sensitivity. Arch. Oral Biol. *15*: 1109–1116.

Kikuchi, K., T. Enari, K.I. Totsuka and K. Shimizu. 1995. Comparison of phenotypic characteristics, DNA-DNA hybridization results, and results with a commercial rapid biochemical and enzymatic-reaction system for Identification of viridans group streptococci. J. Clin. Microbiol. *33*: 1215–1222.

Kilian, M., L. Mikkelsen and J. Henrichsen. 1989a. Replacement of the type strain of *Streptococcus mitis*: request for an opinion. Int. J. Syst. Bacteriol. *39*: 498–499.

Kilian, M., L. Mikkelsen and J. Henrichsen. 1989b. Taxonomic study of viridans Streptococci: description of *Streptococcus gordonii* sp. nov. and emended descriptions of *Streptococcus sanguis* (White and Niven 1946), *Streptococcus oralis* (Bridge and Sneath 1982), and *Streptococcus mitis* (Andrewes and Horder 1906). Int. J. Syst. Bacteriol. *39*: 471–484.

Kilian, M. 2001. Recommended conservation of the names *Streptococcus sanguis*, *Streptococcus rattus*, *Streptococcus cricetus*, and seven other names included in the Approved Lists of Bacterial Names. Request for an Opinion. Int. J. Syst. Evol. Microbiol. *51*: 723–724.

Kilpper-Bälz, R., G. Fischer and K.H. Schleifer. 1982. Nucleic acid hybridization of group N and group D streptococci. Curr. Microbiol. *7*: 245–250.

Kilpper-Bälz, R. and K.H. Schleifer. 1984. Nucelic acid hybridization and cell wall composition studies of pyogenic streptococci. FEMS Microbiol. Lett. *24*: 355–364.

Kilpper-Bälz, R., P. Wenzig and K.H. Schleifer. 1985. Molecular relationships and classification of some viridans Streptococci as *Streptococcus oralis* and emended description of *Streptococcus oralis* (Bridge and Sneath 1982). Int. J. Syst. Bacteriol. *35*: 482–488.

Kilpper-Bälz, R. and K.H. Schleifer. 1987. *Streptococcus suis* sp. nov., nom. rev. Int. J. Syst. Bacteriol. *37*: 160–162.

Kilpper-Bälz, R. and K.H. Schleifer. 1988. Transfer of *Streptococcus morbillorum* to the genus *Gemella* as *Gemella morbillorum* comb. nov. Int. J. Syst. Bacteriol. *38*: 442–443.

Kirsop, B.E. and J.J.S. Snell. 1984. Maintenance of Microorganisms: A Manual of Laboratory Methods. Academic Press, London.

Kitten, T., C.L. Munro, A. Wang and F.L. Macrina. 2002. Vaccination with FimA from *Streptococcus parasanguis* protects rats from endocarditis caused by other viridans streptococci. Infect. Immun. *70*: 422–425.

Klaenhammer, T.R. 1993. Genetics of bacteriocins produced by lactic acid bacteria. FEMS Microbiol. Rev. *12*: 39–85.

Klein, E. 1884. Micro-organisms and disease. Practitioner *33*: 21–40.

Klijn, N., A.H. Weerkamp and W.M. Devos. 1995. Detection and characterization of lactose-utilizing *Lactococcus* spp. in natural ecosystems. Appl. Environ. Microbiol. *61*: 788–792.

Knight, R.G. and D.M. Shlaes. 1985. Physiological characteristics and deoxyribonucleic acid relatedness of human isolates of *Streptococcus bovis* and *Streptococcus bovis* (var.). Int. J. Syst. Bacteriol. *35*: 357–361.

Knight, R.G. and D.M. Shlaes. 1988. Physiological characteristics and deoxyribonucleic acid relatedness of *Streptococcus intermedius* strains. Int. J. Syst. Bacteriol. *38*: 19–24.

Kondo, J.K. and L.L. McKay. 1984. Plasmid transformation of *Streptococcus lactis* protoplasts: optimization and use in molecular cloning. Appl. Environ. Microbiol. *48*: 252–259.

Kral, T.A. and L. Daneo-Moore. 1981. Biochemical differentiation of certain oral streptococci. J. Dent. Res. *60*: 1713–1718.

LaClaire, L.L. and R.R. Facklam. 2000. Comparison of three commercial rapid identification systems for the unusual gram-positive cocci *Dolosigranulum pigrum*, *Ignavigranum ruoffiae*, and *Facklamia* species. J. Clin. Microbiol. *38*: 2037–2042.

Lancefield, R.C. 1928. The antigenic complex of *Streptococcus hemolyticus*. I. Demonstration of a type-specific substance in extracts of *Streptococcus hemolyticus*. J. Exp. Med. *47*: 9–10.

Lancefield, R.C. 1933. A serological differentiation of human and other groups of hemolytic streptococci. J. Exp. Med. *57*: 571–595.

Lapage, S.P., P.H.A. Sneath, E.F. Lessel, V.B.D. Skerman, H.P.R. Seeliger and W.A. Clark. 1992. International Code of Nomenclature of Bacteria (1990 Revision). Bacteriological Code. Published for IUMS by the American Society for Microbiology, Washington, D.C.

LaPenta, D., C. Rubens, E. Chi and P.P. Cleary. 1994. Group A streptococci efficiently invade human respiratory epithelial cells. Proc. Natl. Acad. Sci. U.S.A. *91*: 12115–12119.

Latorre-Guzman, B.A., C.I. Kado and R.E. Kunkee. 1977. *Lactobacillus hordniae*, a new species from leafhopper (*Hordnia circellata*). Int. J. Syst. Bacteriol. *27*: 362–370.

Lawal, S.F., A.O. Coker, E.O. Solanke and O. Ogunbi. 1982. Serotypes among Lancefield group G streptococci isolated in Nigeria. J. Med. Microbiol. *15*: 123–125.

Lawson, P.A., G. Foster, E. Falsen, N. Davison and M.D. Collins. 2004. *Streptococcus halichoeri* sp. nov., isolated from grey seals (*Halichoerus grypus*). Int. J. Syst. Evol. Microbiol. *54*: 1753–1756.

Lawson, P.A., G. Foster, E. Falsen and M.D. Collins. 2005a. *Streptococcus marimammalium* sp. nov., isolated from seals. Int. J. Syst. Evol. Microbiol. *55*: 271–274.

Lawson, P.A., G. Foster, E. Falsen, S.J. Markopoulos and M.D. Collins. 2005b. *Streptococcus castoreus* sp. nov., isolated from a beaver (*Castor fiber*). Int. J. Syst. Evol. Microbiol. *55*: 843–846.

Leenhouts, K., G. Buist and J. Kok. 1999. Anchoring of proteins to lactic acid bacteria. Antonie van Leeuwenhoek *76*: 367–376.

Lehmann, K.B. and R. Neumann. 1896. Atlas und Grundriss der Bakteriologie und Lehrbuch der speciellen backteriologischen *In* Lehmann (Editor), Diagnostik, 1st edn, München, pp. 1–448.

Li, Y., P.W. Caulfield, I.R. Emanuelsson and E. Thornqvist. 2001. Differentiation of *Streptococcus mutans* and *Streptococcus sobrinus* via genotypic and phenotypic profiles from three different populations. Oral Microbiol. Immunol. *16*: 16–23.

Linke, H.A.B. 1977. A new medium for the isolation of *Streptococcus mutans* and its differentiation from other oral streptococci. J. Clin. Microbiol. *5*: 604–609.

Lister, J. 1873. A further contribution to the natural history of bacteria and the germ theory of fermentative changes. Quart. Microbiol. Sci. *13*: 380–408.

Loewe, L., N. Plummer, C.F. Niven, Jr. and J.M. Sherman. 1946. *Streptococcus s.b.e.* in subacute bacterial endocarditis. J. Am. Med. Assoc. *130*: 257.

Löhnis, F. 1909. Die Benennung der Milchsaurebakterien. Aentbl. Bakteriol. Parasitenk. Infektionskr. Hyg. Abt. B *22*: 553–555.

Long, P.H. and E.A. Bliss. 1934. Studies on minute haemolytic streptococci: isolation and cultural characteristics of minute beta-haemolytic streptococci. J. Exp. Med. *60*: 619–631.

Lottenberg, R., D. Minning-Wenz and M.D. Boyle. 1994. Capturing host plasmin(ogen): a common mechanism for invasive pathogens? Trends Microbiol. *2*: 20–24.

Ludwig, W., E. Seewaldt, R. Kilpper-Bälz, K.H. Schleifer, L. Magrum, C.R. Woese, G.E. Fox and E. Stackebrandt. 1985. The phylogenetic position of *Streptococcus* and *Enterococcus*. J. Gen. Microbiol. *131*: 543–551.

Ludwig, W., M. Weizenegger, R. Kilpper-Bälz and K.H. Schleifer. 1988. Phylogenetic relationships of anaerobic streptococci. Int. J. Syst. Bacteriol. *38*: 15–18.

Lukomski, S., E.H. Burns, Jr., P.R. Wyde, A. Podbielski, J. Rurangirwa, D.K. Moore-Poveda and J.M. Musser. 1998. Genetic inactivation of an extracellular cysteine protease (SpeB) expressed by *Streptococcus pyogenes* decreases resistance to phagocytosis and dissemination to organs. Infect. Immun. *66*: 771–776.

Lund, E. 1959. Diagnosis of pneumococci by the optochin and bile tests. Acta Pathol. Microbiol. Scand. *47*: 308–315.

Lunsford, R.D. and A.G. Roble. 1997. *comYA*, a gene similar to *comGA* of *Bacillus subtilis*, is essential for competence-factor-dependent DNA transformation in *Streptococcus gordonii*. J. Bacteriol. *179*: 3122–3126.

Lunsford, R.D. 1998. Streptococcal transformation: essential features and applications of a natural gene exchange system. Plasmid *39*: 10–20.

Luttiken, R., N. Temme, G. Hahn and E.W. Bartelheimer. 1986. Meningitis caused by *Streptococcus suis*: case report and review of the literature. Infection *14*: 181–185.

Macris, M.H., N. Hartman, B. Murray, R.F. Klein, R.B. Roberts, E.L. Kaplan, D. Horn and J.B. Zabriskie. 1998. Studies of the continuing susceptibility of group A streptococcal strains to penicillin during eight decades. Pediatr. Infect. Dis. *17*: 377–381.

Magee, J.T., J.M. Hindmarch and C.W. Douglas. 1997. A numerical taxonomic study of *Streptococcus sanguis*, *S. mitis* and similar organisms using conventional tests and pyrolysis mass spectrometry. Zentbl. Bakteriol. *285*: 195–203.

Male, C.J. 1979. Immunoglobulin A1 protease production by *Haemophilus influenzae* and *Streptococcus pneumoniae*. Infect. Immun. *26*: 254–261.

Manachini, P.L., S.H. Flint, L.J.H. Ward, W. Kelly, M.G. Fortina, C. Parini and D. Mora. 2002. Comparison between *Streptococcus macedonicus* and *Streptococcus waius* strains and reclassification of *Streptococcus waius* (Flint *et al.* 1999) as *Streptococcus macedonicus* (Tsakalidou *et al.* 1998). Int. J. Syst. Evol. Microbiol. *52*: 945–951.

Matthies, C., A. Gössner, G. Acker, A. Schramm and H.L. Drake. 2004. *Lactovum miscens* gen. nov., sp. nov., an aerotolerant, psychrotolerant, mixed-fermentative anaerobe from acidic forest soil. Res. Microbiol. *155*: 847–854.

Matthies, C., A. Gössner, G. Acker, A. Schramm and H.L. Drake. 2005. *In* Validation of publication of new names and new combinations previously effectively published outside the IJSEM. List no. 102. Int. J. Syst. Evol. Microbiol. *55*: 547–549.

Maxted, W.R. and E.V. Potter. 1967. The presence of type 12 M-protein antigen in group G streptococci. J. Gen. Microbiol. *49*: 119–125.

McKay, L.L., B.R. Cords and K.A. Baldwin. 1973. Transduction of lactose metabolism in *Streptococcus lactis* C2. J. Bacteriol. *115*: 810–815.

McNab, R., H.F. Jenkinson, D.M. Loach and G.W. Tannock. 1994. Cell-surface-associated polypeptides CshA and CshB of high molecular mass are colonization determinants in the oral bacterium *Streptococcus gordonii*. Mol. Microbiol. *14*: 743–754.

Milinovich, G.J., P.C. Burrell, C.C. Pollitt, A. Bouvet and D.J. Trott. 2008. *Streptococcus henryi* sp. nov. and *Streptococcus caballi* sp. nov., isolated from the hindgut of horses with oligofructose-induced laminitis. Int. J. Syst. Evol. Microbiol. *58*: 262–266.

Mirick, G.S., L. Thomas, E.C. Curnen and F.L. Horsfall. 1944. Studies on a nonhemolytic streptococcus isolated from the respiratory tract of human beings: I. biological characteristics of *Streptococcus MG*. J. Exp. Med. *80*: 391–406.

Moineau, S., S. Pandian and T.R. Klaenhammer. 1993. Restriction modification systems and restriction endonucleases are more effective on lactococcal bacteriophages that have emerged recently in the dairy industry. Appl. Environ. Microbiol. *59*: 197–202.

Möller, V. and M. Teuber. 1988. Selection and characterization of phage-resistant mesophilic lactococci from mixed-strain dairy starter cultures. Milchwissenschaft-Milk Sci. Int. *43*: 482–486.

Mollick, J.A., G.G. Miller, J.M. Musser, R.G. Cook, D. Grossman and R.R. Rich. 1993. A novel superantigen isolated from pathogenic strains of *Streptococcus pyogenes* with aminoterminal homology to staphylococcal enterotoxins B and C. J. Clin. Invest. *92*: 710–719.

Morrison, D.A. 1997. Streptococcal competence for genetic transformation: regulation by peptide pheromones. Microb. Drug Resist. *3*: 27–37.

Mukasa, H. and H.D. Slade. 1973. Structure and immunological specificity of the *Streptococcus mutans* group b cell wall antigen. Infect. Immun. *7*: 578–585.

Nagamune, H., C. Ohnishi, A. Katsuura, K. Fushitani, R.A. Whiley, A. Tsuji and Y. Matsuda. 1996. Intermedilysin, a novel cytotoxin specific for human cells secreted by *Streptococcus intermedius* UNS46 isolated from a human liver abscess. Infect. Immun. *64*: 3093–3100.

Nagamune, H., R.A. Whiley, T. Goto, Y. Inai, T. Maeda, J.M. Hardie and H. Kourai. 2000. Distribution of the intermedilysin gene among the anginosus group streptococci and correlation between intermedilysin production and deep-seated infection with *Streptococcus intermedius*. J. Clin. Microbiol. *38*: 220–226.

Nakano, K., R. Nomura, I. Nakagawa, S. Hamada and T. Ooshima. 2004. Demonstration of *Streptococcus mutans* with a cell wall polysaccharide specific to a new serotype, k, in the human oral cavity. J. Clin. Microbiol. *42*: 198–202.

Nelms, L.F., D.A. Odelson, T.R. Whitehead and R.B. Hespell. 1995. Differentiation of ruminal and human *Streptococcus bovis* strains by DNA homology and 16s rRNA probes. Curr. Microbiol. *31*: 294–300.

Nes, I.F., D.B. Diep, L.S. Håvarstein, M.B. Brurberg, V. Eijsink and H. Holo. 1996. Biosynthesis of bacteriocin in lactic acid bacteria. Antonie van Leeuwenhoek *70*: 113–128.

Neve, H., A. Geis and M. Teuber. 1984. Conjugal transfer and characterization of bacteriocin plasmids in group N (lactic acid) streptococci. J. Bacteriol. *157*: 833–838.

Neve, H., A. Geis and M. Teuber. 1987. Conjugation, a common plasmid transfer mechanism in lactic acid streptococci of dairy starter cultures. Syst. Appl. Microbiol. *9*: 151–157.

Neve, H. and M. Teuber. 1991. Basic microbiology and molecular biology of bacteriophage of lactic acid bacteria in dairies. Bull. Int. Dairy, Fed. *263*: 3–15.

Nizet, V., B. Beall, D.J. Bast, V. Datta, L. Kilburn, D.E. Low and J.C. De Azavedo. 2000. Genetic locus for streptolysin S production by group A streptococcus. Infect. Immun. *68*: 4245–4254.

Norrby-Teglund, A., D. Newton, M. Kotb, S.E. Holm and M. Norgren. 1994a. Superantigenic properties of the group A streptococcal exotoxin SpeF (MF). Infect. Immun. *62*: 5227–5233.

Norrby-Teglund, A., M. Norgren, S.E. Holm, U. Andersson and J. Andersson. 1994b. Similar cytokine induction profiles of a novel streptococcal exotoxin, MF, and pyrogenic exotoxins A and B. Infect. Immun. *62*: 3731–3738.

Okazaki, N., R. Osawa, R. Suzuki, T. Nikkawa and R.A. Whiley. 2003. Novel observation of hot-cold-hot hemolysis exhibited by group B streptococci. J. Clin Microbiol. *41*: 877–879.

Okumura, K., A. Hara, T. Tanaka, I. Nishiguchi, W. Minamide, H. Igarashi and T. Yutsudo. 1994. Cloning and sequencing the streptolysin O genes of group C and group G streptococci. DNA Seq. *4*: 325–328.

Olson, N.F., R.E. Anderson and R. Sellars. 1978. Microbiological methods for cheese and other cultured products. *In* Marth (Editor), Standard Methods for the Examination of Dairy Products, 14th edn. American Public Health Association, Washington, D.C., pp. 161–164.

Orla-Jensen, A.D. 1932. The bacteriological flora of spontaneously soured milk and of commercial starters for butter making. Zentbl. Bakteriol. II Abt. *86*: 6–29.

Orla-Jensen, S. 1919. The Lactic Acid Bacteria. Host & Son, Copenhagen.

Osawa, R., T. Fujisawa and L.I. Sly. 1995. *Streptococcus gallolyticus* sp. nov., gallate degrading organisms formerly assigned to *Streptococcus bovis*. Syst. Appl. Microbiol. *18*: 74–78.

Osawa, R., T. Fujisawa and I. Sly. 1996. *In* Validation of the publication of new names and new combinations previously effectively published outside the IJSB. List no. 56. Int. J. Syst. Bacteriol. *46*: 362–363.

Ottens, H. and K.C. Winkler. 1962. Indifferent and haemolytic streptococci possessing group-antigen F. J. Gen. Microbiol. *28*: 181–191.

Pancholi, V. and V.A. Fischetti. 1998. Alpha-enolase, a novel strong plasmin(ogen) binding protein on the surface of pathogenic streptococci. J. Biol. Chem. *273*: 14503–14515.

Parker, M.T. and L.C. Ball. 1976. Streptococci and aerococci associated with systemic infection in man. J. Med. Microbiol. *9*: 275–302.

Parker, M.T. 1983. *Streptococcus* and *Lactobacillus*. *In* Wilson, Miles and Parker (Editors), Topley and Wilson's Principles of Bacteriology, Virology, and Immunity, Vol. 2. Edward Arnold, London, pp. 173–217.

Pearce, C., G.H. Bowden, M. Evans, S.P. Fitzsimmons, J. Johnson, M.J. Sheridan, R. Wientzen and M.F. Cole. 1995. Identification of pioneer viridans streptococci in the oral cavity of human neonates. J. Med. Microbiol. *42*: 67–72.

Perch, B., E. Kjems and T. Ravn. 1974. Biochemical and serological properties of *Streptococcus mutans* from various human and animal sources. Acta Pathol. Microbiol. Scand. B Microbiol. Immunol. *82*: 357–370.

Perch, B., K.B. Pedersen and J. Henrichsen. 1983. Serology of capsulated streptococci pathogenic for pigs: six new serotypes of *Streptococcus suis*. J. Clin. Microbiol. *17*: 993–996.

Perreten, V., F. Schwarz, L. Cresta, M. Boeglin, G. Dasen and M. Teuber. 1997. Antibiotic resistance spread in food. Nature *389*: 801–802.

Pestova, E.V., L.S. Havarstein and D.A. Morrison. 1996. Regulation of competence for genetic transformation in *Streptococcus pneumoniae* by an auto-induced peptide pheromone and a two-component regulatory system. Mol. Microbiol. *21*: 853–862.

Petersen, F.C. and A.A. Scheie. 2000. Genetic transformation in *Streptococcus mutans* requires a peptide secretion-like apparatus. Oral Microbiol. Immunol. *15*: 329–334.

Pier, G.B. and S.H. Madin. 1976. *Streptococcus iniae* sp. nov., a beta-hemolytic streptococcus isolated from an Amazon freshwater dolphin, *Inia geoffrensis*. Int. J. Syst. Bacteriol. *26*: 545–553.

Porterfield, J.S. 1950. Classification of the streptococci of subacute bacterial endocarditis. J. Gen. Microbiol. *4*: 92–101.

Pot, B., L.A. Devriese, D. Ursi, P. Vandamme, F. Haesebrouck and K. Kersters. 1996. Phenotypic identification and differentiation of *Lactococcus* strains isolated from animals. Syst. Appl. Microbiol. *19*: 213–222.

Poulsen, K., J. Reinholdt and M. Kilian. 1996. Characterization of the *Streptococcus pneumoniae* immunoglobulin A1 protease gene (*iga*) and its translation product. Infect. Immun. *64*: 3957–3966.

Poyart, C., G. Quesne, S. Coulon, P. Berche and P. Trieu-Cuot. 1998. Identification of streptococci to species level by sequencing the gene encoding the manganese-dependent superoxide dismutase. J. Clin. Microbiol. *36*: 41–47.

Poyart, C., E. Pellegrini, O. Gaillot, C. Boumaila, M. Baptista and P. Trieu-Cuot. 2001. Contribution of Mn-cofactored superoxide dismutase (SodA) to the virulence of *Streptococcus agalactiae*. Infect. Immun. *69*: 5098–5106.

Poyart, C., G. Quesne and P. Trieu-Cuot. 2002. Taxonomic dissection of the *Streptococcus bovis* group by analysis of manganese-dependent superoxide dismutase gene (*sodA*) sequences: reclassification of 'Streptococcus infantarius subsp. coli' as *Streptococcus lutetiensis* sp. nov. and of *Streptococcus bovis* biotype II.2 as *Streptococcus pasteurianus* sp. nov. Int. J. Syst. Evol. Microbiol. *52*: 1247–1255.

Prévot, A.R. 1924. *Diplococcus constellatus* (n. sp.). C. R. Séances Soc. Biol. (Paris) *91*: 426–428.

Prévot, A.R. 1925. Les streptocoques anaérobies. Ann. Inst. Pasteur (Paris) *39*: 417–447.

Price, T., G.L. French, H. Talsania and I. Phillips. 1986. Differentiation of *Streptococcus sanguis* and *S. mitior* by whole-cell rhamnose content and possession of arginine dihydrolase. J. Med. Microbiol. *21*: 189–197.

Pritchard, D.G., G.B. Brown, B.M. Gray and J.E. Coligan. 1981. Glucitol is present in the group-specific polysaccharide of group B *Streptococcus*. Curr. Microbiol. *5*: 283–287.

Pritchard, D.G., B.M. Gray and H.C. Dillon, Jr. 1984. Characterization of the group-specific polysaccharide of group B *Streptococcus*. Arch. Biochem. Biophys. *235*: 385–392.

Proft, T., S.L. Moffatt, C.J. Berkahn and J.D. Fraser. 1999. Identification and characterization of novel superantigens from *Streptococcus pyogenes*. J. Exp. Med. *189*: 89–102.

Quentin, R., H. Huet, F.S. Wang, P. Geslin, A. Goudeau and R.K. Selander. 1995. Characterization of *Streptococcus agalactiae* strains by multilocus enzyme genotype and serotype: identification of multiple virulent clone families that cause invasive neonatal disease. J. Clin. Microbiol. *33*: 2576–2581.

Rainey, F.A. and P.H. Janssen. 1995. Phylogenetic analysis by 16S ribosomal DNA sequence comparison reveals two unrelated groups of species within the genus *Ruminococcus*. FEMS Microbiol. Lett. *129*: 69–73.

Rauch, P.J. and W.M. De Vos. 1992. Characterization of the novel nisin-sucrose conjugative transposon Tn*5276* and its insertion in *Lactococcus lactis*. J. Bacteriol. *174*: 1280–1287.

Riesenfeld-Orn, I., S. Wolpe, J.F. Garcia-Bustos, M.K. Hoffmann and E. Tuomanen. 1989. Production of interleukin-1 but not tumor necrosis factor by human monocytes stimulated with pneumococcal cell surface components. Infect. Immun. *57*: 1890–1893.

Robinson, I.M., J.M. Stromley, V.H. Varel and E.P. Cato. 1988. *Streptococcus intestinalis*, a new species from the colons and feces of pigs. Int. J. Syst. Bacteriol. *38*: 245–248.

Rogers, A.H. 1976. Bacteriocinogeny and properties of some bacteriocins of *Streptococcus mutans*. Arch. Oral Biol. *21*: 99–104.

Rosenbach, F.J. 1884. Micro-organismen bei den Wund-Infections-Krankheiten des Menschen. J.F.Bergmann, Weisbaden.

Ross, P.W. 1996. *Streptococcus pneumoniae*. In Collee, Fraser, Marmion and Simmons (Editors), Practical Medical Microbiology. Churchill Livingstone, New York.

Rotta, J., R.M. Krause, R.C. Lancefield, W. Everly and H. Lackland. 1971. New approaches for laboratory recognition of M types of group A streptococci. J. Exp. Med. *134*: 1298–1315.

Rotta, J. 1986. Pyogenic haemolytic streptococci. In Sneath, Mair, Sharpe and Holt (Editors), Bergey's Manual of Systematic Bacteriology, 1st edn, Vol. 2. The Williams & Wilkins Co., Baltimore, pp. 1047–1055.

Rubens, C.E., M.R. Wessels, L.M. Heggen and D.L. Kasper. 1987. Transposon mutagenesis of type III group B *Streptococcus*: correlation of capsule expression with virulence. Proc. Natl. Acad. Sci. U.S.A. *84*: 7208–7212.

Rudney, J.D., E.K. Neuvar and A.H. Soberay. 1992. Restriction endonuclease-fragment polymorphisms of oral viridans streptococci, compared by conventional and field-inversion gel electrophoresis. J. Dent. Res. *71*: 1182–1188.

Rudney, J.D. and C.J. Larson. 1993. Species identification of oral viridans streptococci by restriction fragment polymorphism analysis of ribosomal RNA genes. J. Clin. Microbiol. *31*: 2467–2473.

Rudney, J.D. and C.J. Larson. 1994. Use of restriction fragment polymorphism analysis of ribosomal RNA genes to assign species to

unknown clinical isolates of oral viridans streptococci. J. Clin. Microbiol. *32*: 437–443.

Rudney, J.D. and C.J. Larson. 1999. Identification of oral mitis group streptococci by arbitrarily primed polymerase chain reaction. Oral Microbiol. Immunol. *14*: 33–42.

Ruoff, K.L., S.I. Miller, C.V. Garner, M.J. Ferraro and S.B. Calderwood. 1989. Bacteremia with *Streptococcus bovis* and *Streptococcus salivarius*: clinical correlates of more accurate identification of isolates. J. Clin. Microbiol. *27*: 305–308.

Ruoff, K.L., R.A. Whiley and D. Beighton. 1999. *Streptococcus*. In Murray, Baron, Pfaller, Tenover and Yolken (Editors), Manual of Clinical Microbiology, 7th edn. ASM Press, Washington, D.C., pp. 283–296.

Ruoff, K.L. 2003. *Aerococcus, Abiotrophia*, and other infrequently isolated aerobic catalase-negative, gram-positive cocci. In Murray, Baron, Jorgensen, Pfaller and Yolken (Editors), Manual of Clinical Microbiology, 8th edn. ASM Press, Washington, D.C., pp. 434–444.

Rurangirwa, F.R., C.A. Teitzel, J. Cui, D.M. French, P.L. McDonough and T. Besser. 2000. *Streptococcus didelphis* sp. nov., a streptococcus with marked catalase activity isolated from opossums (*Didelphis virginiana*) with suppurative dermatitis and liver fibrosis. Int. J. Syst. Evol. Microbiol. *50*: 759–765.

Sahl, H.G., M. Kordel and R. Benz. 1987. Voltage-dependent depolarization of bacterial membranes and artificial lipid bilayers by the peptide antibiotic nisin. Arch. Microbiol. *149*: 120–124.

Sakala R.M., H. Hayashidani, Y. Kato, C. Kaneuchi and M. Ogawa. 2002a. Isolation and characterization of *Lactococcus piscium* strains from vacuum-packaged refrigerated beef. J. Appl Microbiol. *92*: 173–179.

Sakala, R.M., H. Hayashidani, Y. Kato, T. Hirata, Y. Makino, A. Fukushima, T. Yamada, C. Kaneuchi and M. Ogawa. 2002b. Change in the composition of the microflora on vacuum-packaged beef during chiller storage. Int. J. Food Microbiol. *74*: 87–99.

Salama, M.S., W.E. Sandine and S.J. Giovannoni. 1993. Isolation of *Lactococcus lactis* subsp. *cremoris* from nature by colony hybridization with rRNA probes. Appl. Environ. Microbiol. *59*: 3941–3945.

Sampson, J.S., S.P. O'Connor, A.R. Stinson, J.A. Tharpe and H. Russell. 1994. Cloning and nucleotide sequence analysis of *psaA*, the *Streptococcus pneumoniae* gene encoding a 37-kilodalton protein homologous to previously reported *Streptococcus* sp. adhesins. Infect. Immun. *62*: 319–324.

Sand, G. and C.O. Jensen. 1888. Die aetiologie der druise. Dtsch. Z. Tiermed. Verg. Pathol. *13*: 437–464.

Sandine, W.E., C. Radich and P.R. Elliker. 1972. Ecology of lactic streptococci, a review. J. Milk Food Technol. *35*: 176–184.

Sandine, W.E. 1996. Commercial Production of Dairy Starter Cultures. In Cogan and Accolas (Editors), Dairy Starter Cultures. VCH Publishers, Inc., New York, pp. 191–206.

Schlegel, L., F. Grimont, M.D. Collins, B. Regnault, P.A.D. Grimont and A. Bouvet. 2000. *Streptococcus infantarius* sp. nov., *Streptococcus infantarius* subsp. *infantarius* subsp. nov. and *Streptococcus infantarius* subsp. *coli* subsp. nov., isolated from humans and food. Int. J. Syst. Evol. Microbiol. *50*: 1425–1434.

Schlegel, L., F. Grimont, E. Ageron, P.A.D. Grimont and A. Bouvet. 2003. Reappraisal of the taxonomy of the *Streptococcus bovis Streptococcus equinus* complex and related species: description of *Streptococcus gallolyticus* subsp. *gallolyticus* subsp. nov., *S. gallolyticus* subsp. *macedonicus* subsp. nov. and *S. gallolyticus* subsp. *pasteurianus* subsp. nov. Int. J. Syst. Evol. Microbiol. *53*: 631–645.

Schleifer, K.H. and O. Kandler. 1972. Peptidoglycan types of bacterial cell walls and their taxonomic implications. Bacteriol. Rev. *36*: 407–477.

Schleifer, K.H. and R. Kilpper-Bälz. 1984. Transfer of *Streptococcus faecalis* and *Streptococcus faecium* to the genus *Enterococcus* nom. rev. as *Enterococcus faecalis* comb. nov. and *Enterococcus faecium* comb. nov. Int. J. Syst. Bacteriol. *34*: 31–34.

Schleifer, K.H., R. Kilpper-Bälz, J. Kraus and F. Gehring. 1984. Relatedness and classification of *Streptococcus mutans* and "mutans-like" streptococci. J. Dent. Res. *63*: 1047–1050.

Schleifer, K.H., J. Kraus, C. Dvorak, R. Kilpper-Bälz, M.D. Collins and W. Fischer. 1985. Transfer of *Streptococcus lactis* and related streptococci to the genus *Lactococcus* gen. nov. Syst. Appl. Microbiol. *6*: 183–195.

Schleifer, K.H., J. Kraus, G. Dvorak, R. Kilpper-Bälz, M.D. Collins and W. Fischer. 1986. *In* Validation of the publication of new names and new combinations previously effectively published outside the IJSB. List no. 20. Int. J. Syst. Bacteriol. *36*: 354–356.

Schleifer, K.H. 1987. Recent changes in the taxonomy of lactic acid bacteria. FEMS Microbiol. Rev. *46*: 201–203.

Schleifer, K.H. and R. Kilpper-Bälz. 1987. Molecular and chemotaxonomic approaches to the classification of streptococci, enterococci and lactococci: a review. Syst. Appl. Microbiol. *10*: 1–19.

Schleifer, K.H., J. Kraus, C. Dvorak, R. Kilpper-Bälz, M.D. Collins and W. Fischer. 1988. *In* Validation of the publication of new names and new combinations previously effectively published outside the IJSB. List no. 25. Int. J. Syst. Bacteriol. *38*: 220–222.

Schleifer, K.H., M. Ehrmann, U. Krusch and H. Neve. 1991. Revival of the species *Streptococcus thermophilus* (*ex* Orla-Jensen, 1919) nom. rev. Syst. Appl. Microbiol. *14*: 386–388.

Schleifer, K.H. and W. Ludwig. 1995. Phylogenetic relationships of lactic acid bacteria. *In* Wood and Holzapfel (Editors), The Genera of Lactic Acid Bacteria. Blackie Academic & Professional, London, pp. 7–18.

Schmidhuber, S., R. Kilpper-Bälz and K.H. Schleifer. 1987. A taxonomic study of *Streptococcus mitis*, *S. oralis*, and *S. sanguis*. Syst. Appl. Microbiol. *10*: 74–77.

Schneewind, O., K.F. Jones and V.A. Fischetti. 1990. Sequence and structural characteristics of the trypsin-resistant T6 surface protein of group A streptococci. J. Bacteriol. *172*: 3310–3317.

Schrager, H.M., J.G. Rheinwald and M.R. Wessels. 1996. Hyaluronic acid capsule and the role of streptococcal entry into keratinocytes in invasive skin infection. J. Clin. Invest. *98*: 1954–1958.

Schrager, H.M., S. Alberti, C. Cywes, G.J. Dougherty and M.R. Wessels. 1998. Hyaluronic acid capsule modulates M protein-mediated adherence and acts as a ligand for attachment of group A *Streptococcus* to CD44 on human keratinocytes. J. Clin. Invest. *101*: 1708–1716.

Schuchat, A. 1998. Epidemiology of group B streptococcal disease in the United States: shifting paradigms. Clin. Microbiol. Rev. *11*: 497–513.

Sherman, J.M. and H.U. Wing. 1937. *Streptococcus durans*. J. Dairy Sci. *28*: 165–167.

Sherman, J.M., C.F. Niven and K.L. Smiley. 1943. *Streptococcus salivarius* and other non-hemolytic streptococci of the human throat. J. Bacteriol. *45*: 249–263.

Shewmaker, P.L., A.C. Camus, T. Bailiff, A.G. Steigerwalt, R.E. Morey and G. Carvalho Mda. 2007. *Streptococcus ictaluri* sp. nov., isolated from Channel Catfish *Ictalurus punctatus* broodstock. Int. J. Syst. Evol. Microbiol. *57*: 1603–1606.

Sibold, C., J. Henrichsen, A. Konig, C. Martin, L. Chalkley and R. Hakenbeck. 1994. Mosaic *pbpX* genes of major clones of penicillin-resistant *Streptococcus pneumoniae* have evolved from *pbpX* genes of a penicillin-sensitive *Streptococcus oralis*. Mol. Microbiol. *12*: 1013–1023.

Skaar, I., P. Gaustad, T. Tønjum, B. Holm and H. Stenwig. 1994. *Streptococcus phocae* sp. nov., a new species isolated from clinical specimens from seals. Int. J. Syst. Bacteriol. *44*: 646–650.

Skaugen, M., L.M. Cintas and I.F. Nes. 2003. Genetics of bacteriocin production in lactic acid bacteria. *In* Wood and Warner (Editors), Genetics of Lactic Acid Bacteria. The Lactic Acid Bacteria, Vol. 3. Kluwer Academic/Plenum Publishers, New York, pp. 225–260.

Skerman, V.B.D., V. McGowan and P.H.A. Sneath. 1980. Approved lists of bacterial names. Int. J. Syst. Bacteriol. *30*: 225–420.

Sly, L.I., M.M. Cahill, R. Osawa and T. Fujisawa. 1997. The tannin-degrading species *Streptococcus gallolyticus* and *Streptococcus caprinus* are subjective synonyms. Int. J. Syst. Bacteriol. *47*: 893–894.

Smith, D.G. and P.M.F. Shattock. 1964. The cellular location of antigens in streptococci of groups D, N, and Q. J. Gen. Microbiol. *34*: 165–175.

Smith, F.R. and J.M. Sherman. 1938. The hemolytic streptococci of human feces. J. Infect. Dis. *62*: 186–189.

Sniadack, D.H., B. Schwartz, H. Lipman, J. Bogaerts, J.C. Butler, R. Dagan, G. Echaniz-Aviles, N. Lloyd-Evans, A. Fenoll, N.I. Girgis and e. al. 1995. Potential interventions for the prevention of childhood pneumonia: geographic and temporal differences in serotype and serogroup distribution of sterile site pneumococcal isolates from children-implications for vaccine strategies. Pediatr. Infect. Dis. J. *14*: 503–510.

Stackebrandt, E. and M. Teuber. 1988. Molecular taxonomy and phylogenetic position of lactic acid bacteria. Biochimie *70*: 317–324.

Stackebrandt, E., W. Frederiksen, G.M. Garrity, P.A. Grimont, P. Kämpfer, M.C. Maiden, X. Nesme, R. Rosselló-Mora, J. Swings, H.G. Truper, L. Vauterin, A.C. Ward and W.B. Whitman. 2002. Report of the *ad hoc* committee for the re-evaluation of the species definition in bacteriology. Int. J. Syst. Evol. Microbiol. *52*: 1043–1047.

Steidler, L., J. Viaene, W. Fiers and E. Remaut. 1998. Functional display of a heterologous protein on the surface of *Lactococcus lactis* by means of the cell wall anchor of *Staphylococcus aureus* protein A. Appl. Environ. Microbiol. *64*: 342–345.

Stollerman, G.H. 1975. Rheumatic fever and streptococcal infection. *In* Stollerman (Editors), Clinical Cardiology Monographs. Grune and Stratton, New York, pp. 1–303.

Stollerman, G.H. 1997. Rheumatic fever. Lancet *349*: 935–942.

Tagg, J.R., A.S. Dajani and L.W. Wannamaker. 1976. Bacteriocins of gram-positive bacteria. Bacteriol. Rev. *40*: 722–756.

Tailliez, P. 2001. Mini-revue: Les bactéries lactiques, ces etre vivants apparus il y a prè de 3 milliards d'annés. Lait *81*: 1–11.

Takada, K. and M. Hirasawa. 2007. *Streptococcus orisuis* sp. nov., isolated from the pig oral cavity. Int. J. Syst. Evol. Microbiol. *57*: 1272–1275.

Takada, K. and M. Hirasawa. 2008. *Streptococcus dentirousetti* sp. nov., isolated from the oral cavities of bats. Int. J. Syst. Evol. Microbiol. *58*: 160–163.

Tanzer, J.M., A.C. Borjesson, A. Kurasz, L. Laskowski, M. Testa and B. Krasse. 1983. GSBT, an alternative to MSB agar. J. Dent. Res. *62*: 241.

Tapp, J., M. Thollesson and B. Herrmann. 2003. Phylogenetic relationships and genotyping of the genus *Streptococcus* by sequence determination of the RNase P RNA gene, *rnpB*. Int. J. Syst. Evol. Microbiol. *53*: 1861–1871.

Terzaghi, B.E. and W.E. Sandine. 1975. Improved medium for lactic streptococci and their bacteriophages. Appl. Microbiol. *29*: 807–813.

Tettelin, H., K.E. Nelson, I.T. Paulsen, J.A. Eisen, T.D. Read, S. Peterson, J. Heidelberg, R.T. DeBoy, D.H. Haft, R.J. Dodson, A.S. Durkin, M. Gwinn, J.F. Kolonay, W.C. Nelson, J.D. Peterson, L.A. Umayam, O. White, S.L. Salzberg, M.R. Lewis, D. Radune, E. Holtzapple, H. Khouri, A.M. Wolf, T.R. Utterback, C.L. Hansen, L.A. McDonald, T.V. Feldblyum, S. Angiuoli, T. Dickinson, E.K. Hickey, I.E. Holt, B.J. Loftus, F. Yang, H.O. Smith, J.C. Venter, B.A. Dougherty, D.A. Morrison, S.K. Hollingshead and C.M. Fraser. 2001. Complete genome sequence of a virulent isolate of *Streptococcus pneumoniae*. Science *293*: 498–506.

Tettelin, H., V. Masignani, M.J. Cieslewicz, J.A. Eisen, S. Peterson, M.R. Wessels, I.T. Paulsen, K.E. Nelson, I. Margarit, T.D. Read, L.C. Madoff, A.M. Wolf, M.J. Beanan, L.M. Brinkac, S.C. Daugherty, R.T. DeBoy, A.S. Durkin, J.F. Kolonay, R. Madupu, M.R. Lewis, D. Radune, N.B. Fedorova, D. Scanlan, H. Khouri, S. Mulligan, H.A. Carty, R.T. Cline, S.E. Van Aken, J. Gill, M. Scarselli, M. Mora, E.T. Iacobini, C. Brettoni, G. Galli, M. Mariani, F. Vegni, D. Maione, D. Rinaudo, R. Rappuoli, J.L. Telford, D.L. Kasper, G. Grandi and C.M. Fraser. 2002. Complete genome sequence and comparative genomic analysis of an emerging human pathogen, serotype V *Streptococcus agalactiae*. Proc. Natl. Acad. Sci. U.S.A. *99*: 12391–12396.

Teuber, M. and A. Geis. 1981. The family *Streptococcaceae* (non-medical aspects). *In* Starr, Stolp, Trüper, Balows and Schlegel (Editors), The Prokaryotes. Springer-Verlag, New York, pp. 1614–1630.

Teuber, M. 1995. The Genus *Lactococcus*. *In* Wood and Holzapfel (Editors), The Genera of Lactic Acid Bacteria. Blackie Academic & Professional, London, pp. 173–234.

Teuber, M., L. Meile and F. Schwarz. 1999. Acquired antibiotic resistance in lactic acid bacteria from food. Antonie van Leeuwenhoek *76*: 115–137.

Teuber, M. 2000. Fermented Milk Products. *In* Lund, Baird-Parker and Gould (Editors), The Microbiological Safety and Quality of Food, Vol. I. Aspen Publishers, Gaitherburg, pp. 535–589.

Tewodoros, W. and G. Kronvall. 1993. Distribution of presumptive pathogenicity factors among beta-hemolytic streptococci isolated from Ethiopia. APMIS *101*: 295–305.

Texeira, L.M., V.L.C. Merquior, M.D.E. Vianni, M.D.S. Carvalho, S.E.L. Fracalanzza, A.G. Steigerwalt, D.J. Brenner and R.R. Facklam. 1996. Phenotypic and genotypic characterization of atypical *Lactococcus garvieae* strains isolated from water buffalo with subclinical mastitis and conformation of *L. garvieae* as a senior subjective synonym of *Enterococcus seriolicida*. Int. J. Syst. Bacteriol. *46*: 664–668.

Todd, E.W. and L.F. Hewitt. 1932. A new culture medium for the production of antigenic streptococcal haemolysin. J. Pathol. Bacteriol. *35*: 973–974.

Tong, H., X. Gao and X. Dong. 2003. *Streptococcus oligofermentans* sp. nov., a novel oral isolate from caries-free humans. Int. J. Syst. Evol. Microbiol. *53*: 1101–1104.

Truong, T.L., C. Ménard, C. Mouton and L. Trahan. 2000. Identification of mutans and other oral streptococci by random amplified polymorphic DNA analysis. J. Med. Microbiol. *49*: 63–71.

Trüper, H.G. and L. de Clari. 1997. Taxonomic note: necessary correction of specific epithets formed as substantives (nouns) 'in apposition'. Int. J. Syst. Bacteriol. *47*: 908–909.

Trüper, H.G. and L. de Clari. 1998. Taxonomic note: erratum and correction of further specific epithets formed as substantives (nouns) 'in apposition'. Int. J. Syst. Bacteriol. *48*: 615.

Tsakalidou, E., E. Zoidou, B. Pot, L. Wassill, W. Ludwig, L.A. Devriese, G. Kalantzopoulos, K.H. Schleifer and K. Kersters. 1998. Identification of streptococci from Greek Kasseri cheese and description of *Streptococcus macedonicus* sp. nov. Int. J. Syst. Bacteriol. *48*: 519–527.

Tung, S.K., L.J. Teng, M. Vaneechoutte, H.M. Chen and T.C. Chang. 2006. Array-based identification of species of the genera *Abiotrophia*, *Enterococcus*, *Granulicatella*, and *Streptococcus*. J. Clin. Microbiol. *44*: 4414–4424.

van Dam, J.E., A. Fleer and H. Snippe. 1990. Immunogenicity and immunochemistry of *Streptococcus pneumoniae* capsular polysaccharides. Antonie. van Leeuwenhoek *58*: 1–47.

van der Mei, H.C., D. Naumann and H.J. Busscher. 1993. Grouping of oral streptococcal species using Fourier-transform infrared spectroscopy in comparison with classical microbiological identification. Arch. Oral Biol. *38*: 1013–1019.

van Kranenburg, R., M. Kleerebezem and W.M. de Vos. 2000. Nucleotide sequence analysis of the lactococcal EPS plasmid pNZ4000. Plasmid *43*: 130–136.

van Loveren, C., J.F. Buijs and J.M. ten Cate. 2000. Similarity of bacteriocin activity profiles of mutans streptococci within the family when the children acquire the strains after the age of 5. Caries Res. *34*: 481–485.

Van Palenstein Helderman, W.H., M. Ijsseldijk and J.H. Huis in 't Veld. 1983. A selective medium for the two major subgroups of the bacterium *Streptococcus mutans* isolated from human dental plaque and saliva. Arch. Oral Biol. *28*: 599–603.

van Rooijen, R.J. and W.M. de Vos. 1990. Molecular cloning, transcriptional analysis and nucleotide sequence of *lacR*, a gene encoding the repressor of the lactose phosphotransferase system in *Lactococcus lactis*. J. Biol. Chem. *265*: 8499–8503.

Vancanneyt, M., L.A. Devriese, E.M. De Graef, M. Baele, K. Lefebvre, C. Snauwaert, P. Vandamme, J. Swings and F. Haesebrouck. 2004. *Strep-*

tococcus minor sp. nov., from faecal samples and tonsils of domestic animals. Int. J. Syst. Evol. Microbiol. *54*: 449–452.

Vandamme, P., B. Pot, E. Falsen, K. Kersters and L.A. Devriese. 1996. Taxonomic study of Lancefield streptococcal groups C, G, and L (*Streptococcus dysgalactiae*) and proposal of *S. dysgalactiae* subsp. *equisimilis* subsp. nov. Int. J. Syst. Bacteriol. *46*: 774–781.

Vandamme, P., L.A. Devriese, B. Pot, K. Kersters and P. Melin. 1997. *Streptococcus difficile* is a nonhemolytic group B, type Ib *Streptococcus*. Int. J. Syst. Bacteriol. *47*: 81–85.

Vandamme, P., U. Torck, E. Falsen, B. Pot, H. Goossens and K. Kersters. 1998. Whole-cell protein electrophoretic analysis of viridans streptococci: evidence for heterogeneity among *Streptococcus mitis* biovars. Int. J. Syst. Bacteriol. *48*: 117–125.

Vandamme, P., L.A. Devriese, F. Haesebrouck and K. Kersters. 1999. *Streptococcus intestinalis* Robinson *et al.* 1988 and *Streptococcus alactolyticus* Farrow *et al.* 1984 are phenotypically indistinguishable. Int. J. Syst. Bacteriol. *49*: 737–741.

Vaught, R.M. and A.S. Bleiweis. 1974. Antigens of *Streptococcus mutans*. II. Characterization of an antigen resembling a glycerol teichoic acid in walls of strain BHT. Infect. Immun. *9*: 60–67.

Vela, A.I., J. Vazquez, A. Gibello, M.M. Blanco, M.A. Moreno, P. Liebana, C. Albendea, B. Alcala, A. Mendez, L. Domínguez and J.F. Fernández-Garayzábal. 2000. Phenotypic and genetic characterization of *Lactococcus garvieae* isolated in Spain from lactococcosis outbreaks and comparison with isolates of other countries and sources. J. Clin. Microbiol. *38*: 3791–3795.

Vela, A.I., E. Fernández, P.A. Lawson, M.V. Latre, E. Falsen, L. Domínguez, M.D. Collins and J.F. Fernández-Garayzábal. 2002. *Streptococcus entericus* sp. nov., isolated from cattle intestine. Int. J. Syst. Evol. Microbiol. *52*: 665–669.

Vela, A.I., J. Goyache, C. Tarradas, I. Luque, A. Mateos, M.A. Moreno, C. Borge, J.A. Perea, L. Domínguez and J.F. Fernández-Garayzábal. 2003. Analysis of genetic diversity of *Streptococcus suis* clinical isolates from pigs in Spain by pulsed-field gel electrophoresis. J. Clin. Microbiol. *41*: 2498–2502.

Vesa, T., P. Pochart and P. Marteau. 2000. Pharmacokinetics of *Lactobacillus plantarum* NCIMB 8826, *Lactobacillus fermentum* KLD, and *Lactococcus lactis* MG 1363 in the human gastrointestinal tract. Aliment. Pharmacol. Ther. *14*: 823–828.

Vieira, V.V., L.M. Teixeira, V. Zahner, H. Momen, R.R. Facklam, A.G. Steigerwalt, D.J. Brenner and A.C.D. Castro. 1998. Genetic relationships among the different phenotypes of *Streptococcus dysgalactiae* strains. Int. J. Syst. Bacteriol. *48*: 1231–1243.

Vogensen, E.F., T. Karst, J.J. Larsen, B. Kringelum, D. Ellekjaer and E. Waagner Nielsen. 1987. Improved direct differentiation between *Leuconostoc cremoris*, *Streptococcus lactis*, and *Streptococcus cremoris/Streptococcus lactis* on agar. Milchwissenschaft *42*: 646–648.

von Wight, A. and S. Tynkkynen. 1987. Construction of *Streptococcus lactis* sub. lactis strains with a single plasmid associated with mucoid phenotype. Appl. Environ. Microbiol. *53*: 1385–1386.

Wade, W.G., M.J. Aldred and D.M. Walker. 1986. An improved medium for isolation of *Streptococcus mutans*. J. Med. Microbiol. *22*: 319–323.

Walker, S.A. and T.R. Klaenhammer. 2003. The genetics of phage resistance in *Lactococcus lactis*. *In* Wood and Warner (Editors), Genetics of Lactic Acid Bacteria. Kluwer Academic/ Plenum Publishers, New York pp. 291–315.

Walsh, P.M. and L.L. McKay. 1981. Recombinant plasmid associated cell aggregation and high-frequency conjugation of *Streptococcus lactis* ML3. J. Bacteriol. *146*: 937–944.

Washburn, M.R., J.C. White and C.F. Niven, Jr. 1946. *Streptococcus sbe*: immunological characteristics. J. Bacteriol. *51*: 723–729.

Watson, D.A. and D.M. Musher. 1990. Interruption of capsule production in *Streptococcus pneumoniae* serotype 3 by insertion of transposon Tn*916*. Infect. Immun. *58*: 3135–3138.

Weeks, C.R. and J.J. Ferretti. 1984. The gene for type A streptococcal exotoxin (erythrogenic toxin) is located in bacteriophage T12. Infect. Immun. *46*: 531–536.

Ween, O., S. Teigen, P. Gaustad, M. Kilian and L.S. Håvarstein. 2002. Competence without a competence pheromone in a natural isolate of *Streptococcus infantis.* J. Bacteriol. *184*: 3426–3432.

Weigmann, H. 1905–1908. Das Reinzuchtsystem in der Butterbereitung und in der Käserei. *In* Lafarge (Editors), Handbuch der Technischen Mykologie, Vol. 2. Gustav Fischer Verlag, Jena, Germany, pp. 293–309.

Welborn, P.R., W.K. Hadley, E. Newborn and D.M. Yajko. 1983. Characterization of strains of viridans streptococci by deoxyribonucleic acid hybridization and physiological tests. Int. J. Syst. Bacteriol. *33*: 293–299.

Wells, J.M. and A. Mercenier. 2003. Lactic acid bacteria as mucosal delivery vehicles. *In* Wood and Warner (Editors), Genetics of Lactic Acid Bacteria. Kluwer Academic/Plenum Publishers, New York, pp. 261–290.

Whatmore, A.M., V.A. Barcus and C.G. Dowson. 1999. Genetic diversity of the streptococcal competence (*com*) gene locus. J. Bacteriol. *181*: 3144–3154.

Whatmore, A.M., A. Efstratiou, A.P. Pickerill, K. Broughton, G. Woodard, D. Sturgeon, R. George and C.G. Dowson. 2000. Genetic relationships between clinical isolates of *Streptococcus pneumoniae, Streptococcus oralis,* and *Streptococcus mitis*: characterization of "Atypical" pneumococci and organisms allied to *S. mitis* harboring *S. pneumoniae* virulence factor-encoding genes. Infect. Immun. *68*: 1374–1382.

Whatmore, A.M., K.H. Engler, G. Gudmundsdottir and A. Efstratiou. 2001. Identification of isolates of *Streptococcus canis* infecting humans. J. Clin. Microbiol. *39*: 4196–4199.

Whatmore, A.M. and R.A. Whiley. 2002. Re-evaluation of the taxonomic position of *Streptococcus ferus.* Int. J. Syst. Evol. Microbiol. *52*: 1783–1787.

Whiley, R.A. and J.M. Hardie. 1988. *Streptococcus vestibularis* sp. nov. from the human oral cavity. Int. J. Syst. Bacteriol. *38*: 335–339.

Whiley, R.A., R.R.B. Russell, J.M. Hardie and D. Beighton. 1988. *Streptococcus downei* sp. nov. for strains previously described as *Streptococcus mutans* serotype H. Int. J. Syst. Bacteriol. *38*: 25–29.

Whiley, R.A. and J.M. Hardie. 1989. DNA-DNA hybridization studies and phenotypic characteristics of strains within the '*Streptococcus milleri* group'. J. Gen. Microbiol. *135*: 2623–2633.

Whiley, R.A., H. Fraser, J.M. Hardie and D. Beighton. 1990a. Phenotypic differentiation of *Streptococcus intermedius, Streptococcus constellatus,* and *Streptococcus anginosus* strains within the "*Streptococcus milleri* group". J. Clin. Microbiol. *28*: 1497–1501.

Whiley, R.A., H.Y. Fraser, C.W. Douglas, J.M. Hardie, A.M. Williams and M.D. Collins. 1990b. *Streptococcus parasanguis* sp. nov., an atypical viridans *Streptococcus* from human clinical specimens. FEMS Microbiol. Lett. *56*: 115–121.

Whiley, R.A., H.Y. Fraser, C.W.I. Duouglas, J.M. Hardie, A.M. Williams and M.D. Collins. 1990c. *In* Validation of the publication of new names and new combinations previously effectively published outside the IJSB. List no. 34. Int. J. Syst. Bacteriol. *40*: 320–321.

Whiley, R.A. and D. Beighton. 1991. Emended descriptions and recognition of *Streptococcus constellatus, Streptococcus intermedius,* and *Streptococcus anginosus* as distinct species. Int. J. Syst. Bacteriol. *41*: 1–5.

Whiley, R.A., L. Freemantle, D. Beighton, J.R. Radford, J.M. Hardie and G. Tillotsen. 1993. Isolation, identification and prevalence of *Streptococcus anginosus, S. intermedius,* and *S. constellatus* from the human mouth. Microb. Ecol. Health Dis. *6*: 285–291.

Whiley, R.A., L.M.C. Hall, J.M. Hardie and D. Beighton. 1997. Genotypic and phenotypic diversity within *Streptococcus anginosus.* Int. J. Syst. Bacteriol. *47*: 645–650.

Whiley, R.A. and D. Beighton. 1998. Current classification of the oral streptococci. Oral Microbiol. Immunol. *13*: 195–216.

Whiley, R.A., L.M.C. Hall, J.M. Hardie and D. Beighton. 1999. A study of small-colony, beta-haemolytic, Lancefield group C streptococci within the anginosus group: description of *Streptococcus constellatus* subsp. *pharyngis* subsp. nov., associated with the human throat and pharyngitis. Int. J. Syst. Bacteriol. *49*: 1443–1449.

White, J.C. and C.F. Niven, Jr. 1946. *Streptococcus* S.B.E.: a streptococcus associated with subacute bacterial endocarditis. J. Bacteriol. *51*: 711–722.

Whittaker, C.J., C.M. Klier and P.E. Kolenbrander. 1996. Mechanisms of adhesion by oral bacteria. Annu. Rev. Microbiol. *50*: 513–552.

Whittenbury, R. 1965. The enrichment and isolation of lactic acid bacteria from plant material. Zentbl. Parasitenkd. Infektionskr. Hyg. Abt. 1 Suppl. *1*: 395–398.

Wilkinson, H.W. 1972. Comparison of streptococcal R antigens. Appl. Microbiol. *24*: 669–670.

Willcox, M.D., H. Zhu and K.W. Knox. 2001. *Streptococcus australis* sp. nov., a novel oral streptococcus. Int. J. Syst. Evol. Microbiol. *51*: 1277–1281.

Willems, A. and M.D. Collins. 1995. Phylogenetic analysis of *Ruminococcus flavefaciens,* the type species of the genus *Ruminococcus,* does not support the reclassification of *Streptococcus hansenii* and *Peptostreptococcus productus* as ruminococci. Int. J. Syst. Bacteriol. *45*: 572–575.

Williams, A.M. and M.D. Collins. 1990a. Molecular taxonomic studies on *Streptococcus uberis* type I and type II: description of *Streptococcus parauberis* sp. nov. J. Appl. Bacteriol. *68*: 485–490.

Williams, A.M. and M.D. Collins. 1990b. *In* Validation of the publication of new names and new combinations previously effectively published outside the IJSB. List no. 35. Int. J. Syst. Bacteriol. *40*: 470–471.

Williams, A.M., J.L. Fryer and M.D. Collins. 1990a. *In* Validation of the publication of new names and new combinations previously effectively published outside the IJSB. List no. 34. Int. J. Syst. Bacteriol. *40*: 320–321.

Williams, A.M., J.L. Fryer and M.D. Collins. 1990b. *Lactococcus piscium* sp. nov. a new *Lactococcus* species from salmonid fish. FEMS Microbiol. Lett. *56*: 109–113.

Wilson, G.S. and A.A. Miles (Editors). 1975. Topley and Wilson's Principles of Bacteriology and Immunity, 6th edn, vol. 1. Arnold, London.

Windsor, R.S. and S.D. Elliott. 1975. Streptococcal infection in young pigs. IV. Outbreak of streptococcal meningitis in weaned pigs. J. Hyg. *75*: 69–78.

Winkelstein, J.A. and A. Tomasz. 1977. Activation of alternative pathway by pneumococcal cell walls. J. Immunol. *118*: 451–454.

Winkelstein, J.A. and A. Tomasz. 1978. Activation of alternative complement pathway by pneumococcal cell wall teichoic acid. J. Immunol. *120*: 174–178.

Winram, S.B. and R. Lottenberg. 1996. The plasmin-binding protein Plr of group A streptococci is identified as glyceraldehyde-3-phosphate dehydrogenase. Microbiology *142*: 2311–2320.

Woo, P.C.Y., D.M.W. Tam, K.W. Leung, S.K.P. Lau, J.L.L. Teng, M.K.M. Wong and K.Y. Yuen. 2002. *Streptococcus sinensis* sp. nov., a novel species isolated from a patient with infective endocarditis. J. Clin. Microbiol. *40*: 805–810.

Yao, E.S., R.J. Lamont, S.P. Leu and A. Weinberg. 1996. Interbacterial binding among strains of pathogenic and commensal oral bacterial species. Oral Microbiol. Immunol. *11*: 35–41.

Yother, J. and D.E. Briles. 1992. Structural properties and evolutionary relationships of PspA, a surface protein of *Streptococcus pneumoniae,* as revealed by sequence analysis. J. Bacteriol. *174*: 601–609.

Zangwill, K.M., A. Schuchat and J.D. Wenger. 1992. Group B streptococcal disease in the United States, 1990: report from a multistate active surveillance system. CDC Surveillence Summaries. Morb. Mortal. Wkly. Rep. *41*: 25–32.

Zhu, H., M.D.P. Willcox and K.W. Knox. 2000. A new species of oral *Streptococcus* isolated from Sprague–Dawley rats, *Streptococcus orisratti* sp. nov. Int. J. Syst. Evol. Microbiol. *50*: 55–61.

Class II. **Clostridia** class. nov.

FRED A. RAINEY

Clos.tri′di.a. N.L. fem. pl. n. *Clostridiales* type order of the class; dropping the ending to denote a class; N.L. neut. pl. n. *Clostridia* the class of the *Clostridiales*.

The class is phenotypically, chemotaxonomically, physiologically, and ecologically diverse. Contains both Gram-stain-positive and Gram-stain-negative organisms. The class forms a phylogenetically coherent cluster based on the analysis of 16S rRNA gene sequences.

The class contains the orders *Clostridiales*, *Halanaerobiales*, and *Thermoanaerobacterales*.

Type order: **Clostridiales** Prévot 1953.

Reference

Prévot, A.R. 1953. *In* Hauduroy, Ehringer, Guillot, Magrou, Prévot, Rosset and Urbain (Editors), Dictionnaire des Bactéries Pathogènes, 2nd edn. Masson, Paris.

Order I. **Clostridiales** Prévot 1953

FRED A. RAINEY

Clos.tri′di.a′les. N.L. neut. n. *Clostridium* type genus of the order; suff. -*ales* ending denoting an order; N.L. fem. pl. n. *Clostridiales* the order *Clostridium*.

The order is phenotypically, chemotaxonomically, physiologically and ecologically diverse. Contains both Gram-stain-positive and Gram-stain-negative organisms. The order forms a phylogenetically coherent cluster based on the analysis of 16S rRNA gene sequences. The order contains the families *Clostridiaceae*, *Eubacteriaceae*, *Gracilibacteraceae*, *Heliobacteriaceae*, *Lachnospiraceae*, *Peptococcaceae*, *Peptostreptococcaceae*, *Ruminococcaceae*, *Syntrophomonadaceae*, and *Veillonellaceae*, along with nine families *incertae sedis* (see Figure 1, Figure 4, Figure 5 and Figure 6).

Type genus: **Clostridium** Prazmowski 1880, 23[AL].

References

Prazmowski, A. 1880. Untersuchung über die Entwickelungsgeschichte und Fermentwirking einiger Bacterien-Arten Hugo Voigt, Leipzig, pp. 1–58.

Prévot, A.R. 1953. *In* Hauduroy, Ehringer, Guillot, Magrou, Prévot, Rosset and Urbain (Editors), Dictionnaire des Bactéries Pathogènes, 2nd edn. Masson, Paris.

Family I. **Clostridiaceae** Pribram 1933, 90[AL]

JUERGEN WIEGEL

Clos.tri.di.a′ce.ae. N.L. neut. n. *Clostridium* type genus of the family; suff. -*aceae* ending to denote a family; N.L. fem. pl. n. *Clostridiaceae* the *Clostridium* family.

Phylogenetic and taxonomic comments

The family *Clostridiaceae* is the first of 19 families within the order *Clostridiales* of the class "*Clostridia*"of the phylum *Firmicutes*. In the current volume, the family contains 13 recognized genera. However, in this volume the only species of the genus *Anaerobacter*, *Anaerobacter polyendosporus*, is reassigned to *Clostridium* as *Clostridium polyendosporum*, and the only species of the genus *Thermobrachium*, *Thermobrachium celere*, is reassigned in this volume to the genus *Calaromator* as *Calaromator celer*, thus reducing the number of genera to 11. The type genus of the *Clostridiaceae*, *Clostridium*, presently contains about 190 recognized species (Euzéby, 1997). However, some of these are not validly published, either because they lack a designated type strain or, if published after 2000, have not been deposited in two culture collections in two different countries. Furthermore, as indicated by the analysis of Collins et al. (1994) and discussed by Wiegel et al. (2006), probably more than half of the validly named clostridial species do not belong to the genus *Clostridium sensu stricto* (Collins' Group I) and should be reassigned to new genera and families. Thus, extensive modification of the systematics of *Clostridiaceae* is to be expected, and the reader should consult Jean Euzéby's List of Prokaryotic Names with Standing in Nomeclature (http://www.bacterio.cict.fr/), which is updated monthly on publication of the *International Journal of Systematic and Evolutionary Microbiology*.

The systematics of *Clostridiaceae* used for this volume differs significantly from what has been published previously (Hippe et al., 1992). The new system relies heavily on phylogenetic analyses of the 16S rRNA gene and attempts to circumvent some of the problems of the previously ill-defined genus *Clostridium*. These problems have been evident for decades prior to 16S rRNA gene sequence analyzes, e.g., from the wide range of values for the G+C content of DNA (22–55 mol%) and the variability of sporulation and Gram staining. Thus, several of the species which were not in the radiation of the genus *Clostridium sensu stricto* have been transferred to novel genera and other families, in some cases even in different orders. Examples of well known species being transferred in this way include the transfer of *Clostridium thermosaccharolyticum* to *Thermaoanaerobacterium* (Lee et al., 1993) and *Clostridium thermoaceticum* to *Moorella*, both in the family "*Thermoanaerobacteraceae*" of the order "*Thermoanaerobacterales*" (Collins et al., 1994; Ludwig et al., 2009; Stackebrandt et al., 1999a; Stackebrandt and Rainey, 1997). Even though many of the "clostridia-like" thermophiles have been transferred to the *Thermoanaerobacterales*, the family *Clostridiaceae sensu stricto* and genus *Clostridium sensu stricto* still contains many thermophiles, such as *Caloramator* and *Thermohalobacter* spp. and *Clostridium thermobutyricum*, respectively. Thus, thermophily in itself is not a reliable phylogenetic marker.

All species of this genus *Clostridium sensu stricto* (previously called Group I of Collins et al., 1994) are obligately anaerobic, sporulating rod-shaped bacteria with a G+C content of their chromosomal DNA of around 22–35 mol%. Most, especially the mesophilic species, stain Gram-positive and contain *meso*-diaminopimelic acid in their cell walls. An exception is *Clostridium thermobutyricum*, which stains Gram-negative (Wiegel et al., 1989b). They are not capable of dissimilatory sulfate reduction. Thus, most properties are still consistent with the definition of *Clostridium* as given by Hippe et al. (1992). However, several of the other genera within the family *Clostridiaceae* have different properties. Some genera contain species that are asporogenic (but contain sporulation genes) or are even nonsporogenic or lack sporulation genes (Brill and Wiegel, 1997; Onyenwoke et al., 2004). Some species, especially many of the thermophilic taxa such as *Clostridium thermocellum*, stain Gram-negative at all growth phases even though they possess a Gram-type-positive cell wall without lipopolysaccharide (Wiegel, 1981).

Since the organization of the family *Clostridiaceae* is still in flux, it is expected to change in the near future. As discussed by Rainey et al. (2006), some of the genera included in this family were placed there in part because some members had been previously published as species of *Clostridium*. Likewise, many of the 13 genera in the present family *Clostridiaceae* are only represented by a single species. Thus, as further species are described, their systematic position will have to be adjusted accordingly. Current evidence suggests that this family contains three phylogenetic groups in addition to close relatives of *Clostridium sensu stricto*. These include the clusters of *Alkaliphilus, Anoxynatronum, Natronincola*, and *Tindallia; Caloranaerobacter* and *Thermohalobacter*; and *Caminicella*. If future investigations warrant it, some or all of these groups may be moved to separate families.

Morphological and physiological comments

The taxa presently placed into the *Clostridiaceae* are generally obligately anaerobic rods. Their cells usually stain Gram-positive, although a significant number, especially the thermophilic species, stain Gram-negative at all growth phases. The typical cell wall, especially for the species of the type genus *Clostridium sensu stricto*, contains *meso*-diaminopimelic acid and is of the A1γ peptidoglycan type. Species such as *Clostridium thermoalkaliphilum*, which is not a member of the *Clostridium sensu stricto* and has the A4β type of cell wall type that contains a L-ornithine–D-asparagine interpeptide bridge, will probably be reclassified into other families. Aside from species with rod-shaped morphologies, the family *Clostridiaceae sensu stricto* includes species or genera with coccoid or polymorphic morphologies. For instance, *Sarcina* species are characterized by their very distinct morphology, forming symmetrical packages of multiples of four cells and producing cellulose as a "glue-like" material outside of their cells. Most of the proposed species belonging to the *Clostridiaceae* form endospores or have been shown to contain sporulation-specific genes (Onyenwoke et al., 2004). Although *Anaerobacter* falls within the radiation of *Clostridiaceae sensu stricto*

(Stackebrandt et al., 1999a), it produces many spores within one cell instead of only one spore per cell, as most other clostridia.

Other physiological characteristics of the *Clostridiaceae* are highly variable. Nearly all members of the type genus *Clostridium sensu stricto* form butyrate as a major fermentation product. Excepts include *Clostridium ljungdahlii*, *Tindallia*, and *Natronincola*, which produce acetate using the Wood–Ljungdahl homoacetogenic pathway. These bacteria produce acetate as the sole acid from fructose as well as autotrophically from H_2/CO_2 or CO. Most clostridial species are described as heterotrophs that depend upon substrate level phosphorylation for ATP generation, but exceptions are observed. *Clostridium ljungdahlii* and *Natronincola histidinovorans* use electron transport phosphorylation involving the Wood–Ljungdahl pathway. *Natronincola histidinovorans* also couples the oxidation of histidine to fumarate reduction via an electron transport chain and gains ATP by an anaerobic respiration. Thiosulfate can also function as an electron acceptor. Members of the *Clostridiaceae* are glycolytic, saccharolytic, peptolytic (including amino acid utilization), and/or chemolithoautotrophic. The fermentation products include various organic acids and alcohols, representing various pathways such as mixed acid, solvent, and homoacetogenic fermentation.

Although most species are neutrophiles, several alkaliphilic, alkalithermophilic, moderate halophilic, haloalkaliphilic, and slightly acidophilic species have been described.

Comments on habitats

Habitats from which members of the *Clostridiaceae* have been isolated are very diverse, various types of feces and manure piles, sewage sludge, freshwater and marine sediments, salt lakes, and various parts of the human body. Habitats also vary greatly with respect to temperature, i.e., from permafrost, common mesobiotic habitats, and geothermally and anthropogenically heated sources, Since many species of the *Clostridiaceae* form desiccation and heat-resistant spores, it is not surprising that members of the family can be generally regarded as ubiquitous. However, some species are more restricted in their habitats.

The family also includes clinically important and highly pathogenic species, many within the genus *Clostridium sensu stricto*, as well as biotechnologically important species such as the solvent-producer *Clostridium acetobutylicum* and related species.

Genera and species belonging to Family I. *Clostridiaceae*

Order I. *Clostridiales* (Prévot, 1953)
    Family I. *Clostridiaceae*[AL] (Pribram, 1933)
        Genus I. *Clostridium*[AL(T)] (Prazmowski, 1880)
            Clostridium butyricum[AL(T)] (genus presently contains 189 recognized species, but less than half are expected to be verified as clostridia *sensu stricto*)
        Genus II. *Alkaliphilus*[VP] (Takai et al., 2001)
            Alkaliphilus transvaalensis[VP(T)] (Takai et al., 2001)
            Alkaliphilus crotonatoxidans[VP] (Cao et al., 2003)
        Genus III. *Anaerobacter*[VP] (Duda et al., 1987)
            Anaerobacter polyendosporus[VP(T)] (type and only species, reassigned to *Clostridium*)
        Genus IV. *Anoxynatronum*[VP] (Garnova et al., 2003b)
            Anoxynatronum sibiricum[VP(T)] (Garnova et al., 2003b)
        Genus V. *Caloramator*[VP] (Patel et al., 1987; (Collins et al., 1994)

Caloramator fervidus[VP(T)] (Patel et al., 1987; (Collins et al., 1994)

"*Caloramator celer*" (bas. *Thermobrachium*) Wiegel, 2009

Caloramator coolhaasii[VP] (Plugge et al., 2000)

Caloramator indicus[VP] (Chrisostomos et al., 1996),

Caloramator proteoclasticus[VP] (Tarlera et al., 1997)

Caloramator viterbiensis corrig.[VP] (Seyfried et al., 2002)

Genus VI. *Caloranaerobacter*[VP] (Wery et al., 2001)

    Caloranaerobacter azorensis[VP(T)] (Wery et al., 2001)

Genus VII. *Caminicella*[VP] (Alain et al., 2002)

    Caminicella sporogenes[VP(T)] (Alain et al., 2002)

Genus VIII. *Natronincola*[VP] (Zhilina et al., 1998)

    Natronincola histidinovorans[VP(T)] (Zhilina et al., 1998)

Genus IX. *Oxobacter*[VP] (Collins et al., 1994)

Oxobacter pfennigii[VP(T)] (Collins et al., 1994)

Genus X. *Sarcina*[AL] (Goodsir, 1842)

    Sarcina ventriculi[AL(T)] (Goodsir, 1842)

    Sarcina maxima[AL] (Lindner, 1888)

Genus XI. *Thermobrachium*[VP] (Engle et al., 1996)

    Thermobrachium celere[VP(T)] (type and only species, reassigned to *Caloramator*) Wiegel, 2009

Genus XII. *Thermohalobacter*[VP] (Cayol et al., 2000a)

    Thermohalobacter berrensis[VP(T)] (presently not freely available; Euzéby, 2007)

Genus XIII. *Tindallia*[VP] (Kevbrin et al., 1998)

    Tindallia magadiensis[VP(T)] (Kevbrin et al., 1998)

    Tindallia californiensis[VP] (Pikuta et al., 2003a)

    Tindallia texcoconensis[VP] (Alazard et al., 2009)

Genus I. **Clostridium** Prazmowski 1880, 23[AL]

FRED A. RAINEY, BECKY JO HOLLEN AND ALANNA SMALL

Clos.tri'di.um. Gr. n. *closter* a spindle; N.L. neut. dim. n. *Clostridium* a small spindle.

Rods, usually stain Gram-positive at least in very early stages of growth, although in some species Gram-stain-positive cells have not been seen. Motile or nonmotile. When motile, cells usually are peritrichous. The majority of species form oval or spherical **endospores** that usually distend the cell.

Usually chemoorganotrophic; some species are chemoautotrophic or chemolithotrophic as well. Usually produce mixtures of organic acids and alcohols from carbohydrates or peptones. **May be saccharolytic, proteolytic, neither, or both.** May metabolize carbohydrates, alcohols, amino acids, purines, steroids, or other organic compounds. Some species fix atmospheric nitrogen. **Do not carry out a dissimilatory sulfate reduction.** Usually catalase-negative, although trace amounts of catalase may be detected in some strains.

The cell wall usually contains *meso*-diaminopimelic acid (*meso*-DAP).

Most species are obligately **anaerobic**, although tolerance to oxygen varies widely; some species will grow but not sporulate in the presence of air at atmospheric pressure.

For most species, growth is most rapid at pH 6.5–7 and at temperatures between 30 and 37°C.

DNA G+C content (mol%): 22–53.

Type species: **Clostridium butyricum** Prazmowski 1880, 24[AL].

Taxonomic comments

There are currently 168 validly published species of the genus *Clostridium*. Of these, 77 fall within the 16S rRNA gene sequence cluster I of the clostridia as defined by Collins et al. (1994) and are listed in Table 140. The remaining 81 species fall outside of cluster I (*Clostridium sensu stricto*) and so, although currently assigned to the genus *Clostridium*, their genus assignment is in many cases uncertain (Table 141). Many of these species fall within the radiation of other established genera. The phylogenetic relationships of the species falling within clusters I and II of the clostridia are shown in Figure 128. The characteristics of the species of clusters I and II falling with the family *Clostridiaceae* are given in the descriptions below and in Table 142. The polyphyletic nature of the genus *Clostridium* and possible solutions to the restructuring of the genus has been described and discussed at length previously (see Collins et al., 1994; Rainey et al., 1993; Stackebrandt et al., 1999a; Wiegel et al., 2005).

Since Collins et al. (1994) demonstrated the extensive phylogenetic diversity of the species then assigned to the genus *Clostridium*, a number of species have been reclassified to novel or existing genera. However, species (27 since 1995) continue to be added to the genus *Clostridium* even though they are not members of cluster I, the *Clostridium sensu stricto*, or the family *Clostridiaceae*. The addition of these new species to the genus *Clostridium* is adding to the taxonomic confusion associated with this taxon.

List of species of the genus *Clostridium*

Species designated with an asterisk (*) are descriptions taken from the previous edition of *Bergey's Manual* (Cato et al., 1986). Additional information on the phylogenetic position of these species has been added when available. In the previous edition of this volume (Cato et al., 1986), species were grouped together and compared on the basis of phenotypic characteristics. The main criteria for clustering the species for comparative purposes included the production of acid from glucose, gelatin hydrolysis, meat digestion, starch hyrolysis, and combinations thereof.

The characteristics of the species of the genus *Clostridium sensu stricto* and those within the family *Clostridiaceae* are given in the species decription as well as in Table 142. The species are listed in Table 142 in the order in which they cluster in the 16S rRNA gene sequence based dendrogram Figure 128. The characteristics of other species that fall outside of the genus *Clostridium sensu stricto* for which Tables of characteristics were provided in the previous edition of this volume (Cato et al., 1986) are given in Table 143, Table 144, and Table 145.

TABLE 140. Species of the genus *Clostridium* falling within the radiation of cluster I and considered to represent the genus *Clostridium sensu stricto*

Clostridium butyricum
Clostridium acetireducans
Clostridium acetobutylicum
Clostridium acidisoli
Clostridium aciditolerans
Clostridium aestuarii
Clostridium akagii
Clostridium algidicarnis
Clostridium argentinense
Clostridium aurantibutyricum
Clostridium baratii
Clostridium beijerinckii
Clostridium botulinum
Clostridium bowmanii
Clostridium cadaveris
Clostridium carboxidivorans
Clostridium carnis
Clostridium celatum
Clostridium cellulovorans
Clostridium chartatabidum
Clostridium chauvoei
Clostridium cochlearium
Clostridium colicanis
Clostridium collagenovorans
Clostridium diolis
Clostridium disporicum
Clostridium drakei
Clostridium estertheticum subsp. *estertheticum*
Clostridium estertheticum subsp. *laramiense*
Clostridium fallax
Clostridium frigidicarnis
Clostridium frigoris
Clostridium ganghwense
Clostridium gasigenes
Clostridium grantii
Clostridium hemolyticum
Clostridium homopropionicum
Clostridium intestinale

(continued)

TABLE 140. (continued)

Clostridium isatidis
Clostridium kluyveri
Clostridium lacusfryxellense
Clostridium ljungdahlii
Clostridium lundense
Clostridium magnum
Clostridium malenominatum
Clostridium nitrophenolicum
Clostridium novyi
Clostridium oceanicum
Clostridium paraputrificum
Clostridium pascui
Clostridium pasteurianum
Clostridium peptidivorans
Clostridium perfringens
Clostridium polyendosporum
Clostridium psychrophilum
Clostridium puniceum
Clostridium putrefaciens
Clostridium quinii
Clostridium roseum
Clostridium saccharobutylicum
Clostridium saccharoperbutylacetonicum
Clostridium sardinense
Clostridium sartagoforme
Clostridium scatologenes
Clostridium schirmacherense
Clostridium septicum
Clostridium sporogenes
Clostridium subterminale
Clostridium tepidiprofundi
Clostridium tertium
Clostridium tetani
Clostridium tetanomorphum
Clostridium thermobutyricum
Clostridium thermopalmarium
Clostridium thiosulfatireducens
Clostridium tyrobutyricum
Clostridium uliginosum
Clostridium vincentii

1. **Clostridium butyricum**[*] Prazmowski 1880, 24[AL]

bu.ty ri.cum. Gr. n. *boutyron* butter; N.L. neut. adj. *butyricum* related to butter, butyric.

Type species of the genus *Clostridium*.

Cells in PYG broth are Gram-stain-positive straight rods with rounded ends, motile and peritrichous, 0.5–1.7 × 2.4–7.6 μm. Occur singly, in pairs, or in short chains and occasionally as long filaments. Cells are often granulose-positive. Spores are oval, central to subterminal, and usually do not swell the cell. No exosporium or appendages are present. Sporulation occurs readily both in broth and on solid media. Cell walls contain *meso*-DAP and glucose. Glutamic acid and alanine are present.

Surface colonies on blood agar plates are 1–6 mm in diameter, circular to irregular, lobate or slightly scalloped, raised to convex, translucent, gray-white, shiny or dull, smooth, with a granular or mottled internal structure.

Cultures in PYG broth are turbid with a smooth or flocculent sediment and have a pH of 4.6–5.0 after incubation for 5 d.

The optimum temperature range for growth is 30–37 °C; many strains grow equally well at 25 °C and growth can occur at 10 °C. Growth is stimulated by a fermentable carbohydrate and inhibited by 6.5% NaCl. Strains of *Clostridium butyricum* grow readily in glucose-mineral salts medium with biotin as the only required vitamin.

Pectin is strongly fermented by all strains tested. Pectin degradation has been implicated in the formation of wetwood in living hardwood trees and in the anaerobic digestion of fruit and vegetable wastes, and strains of *Clostridium butyricum* which carry out these processes have been isolated from these sources. Chlorinated hydrocarbon pesticides can be degraded in the presence of glucose by strains of this species. Atmospheric N_2 is fixed.

Strains contain an iron-sulfur-thiamin pyrophosphate enzyme involved in the reduction of ferredoxin by pyruvate. DNase activity has been detected. Neutral red and resazurin are reduced.

Products in PYG broth are butyric, acetic, and formic acids; lactic and succinic acids; butanol and ethanol are

TABLE 141. Species of the genus *Clostridium* not falling within the radiation of cluster I (genus *Clostridium sensu stricto*)

Cluster II
Clostridium histolyticum
Clostridium limosum
Clostridium proteolyticum
Cluster III
Clostridium aldrichii
Clostridium alkalicellulosi
Clostridium cellobioparum
Clostridium cellulolyticum
Clostridium hungatei
Clostridium josui
Clostridium papyrosolvens
Clostridium stercorarium subsp. *stercorarium*
Clostridium stercorarium subsp. *leptospartum*
Clostridium stercorarium subsp. *thermolacticum*
Clostridium straminisolvens
Clostridium termitidis
Clostridium thermocellum
Clostridium thermosuccinogenes
Cluster IV
Clostridium cellulosi
Clostridium leptum
Clostridium methylpentosum
Clostridium orbiscindens
Clostridium sporosphaeroides
Clostridium viride
Cluster XI
Clostridium aceticum
Clostridium bartlettii
Clostridium bifermentans
Clostridium caminithermale
Clostridium difficile
Clostridium felsineum
Clostridium formicaceticum
Clostridium ghonii
Clostridium glycolicum
Clostridium halophilum
Clostridium hiranonis
Clostridium irregulare
Clostridium litorale
Clostridium lituseburense
Clostridium mangenotii
Clostridium mayombei
Clostridium parodoxum
Clostridium sordellii
Clostridium sticklandii
Clostridium thermoalcaliphilum
Cluster XII
Clostridium acidurici
Clostridium purinilyticum
Clostridium ultunense
Cluster XIVa
Clostridium aerotolerans
Clostridium aldenense
Clostridium agidixylanolyticum
Clostridium aminophilum
Clostridium aminovalericum
Clostridium amygdalinum
Clostridium asparagiforme
Clostridium bolteae
Caloramator celerecresens
Clostridium citroniae
Clostridium clostridioforme

(continued)

TABLE 141. (continued)

Clostridium coccoides
Clostridium fimetarium
Clostridium glycyrrhizinilyticum
Clostridium hathewayi
Clostridium herivorans
Clostridium hylemonae
Clostridium indolis
Clostridium innocum
Clostridium jejuense
Clostridium methoxybenzovorans
Clostridium nexile
Clostridium oroticum
Clostridium phytofermentans
Clostridium polysaccharolyticum
Clostridium populeti
Clostridium propionicum
Clostridium proteoclasticum
Clostridium saccharolyticum
Clostridium scindens
Clostridium sphenoides
Clostridium symbiosum
Clostridium xylanolyticum
Clostridium xylanovorans
Cluster XIVb
Clostridium colinum
Clostridium lactatifermentans
Clostridium lentocellum
Clostridium neopropionicum
Cluster XVIII
Clostridium cocleatum
Clostridium ramosum
Clostridium saccharogumia
Clostridium spiroforme
Cluster XIX
Clostridium rectum
Additional *Clostridium* species
Clostridium cylindrosporum

sometimes produced. Pyruvate is converted to acetate, butyrate, and sometimes formate. Neither lactate nor threonine is utilized. Products from the fermentation of pectin are large amounts of methanol, acetate, H_2, and CO_2, and moderate amounts of butyrate and ethanol.

All strains tested are susceptible to clindamycin, chloramphenicol, erythromycin, and tetracycline; 9 of 37 strains are resistant to penicillin. Three clinical isolates, resistant to penicillin, produce a β-lactamase. Two clinical isolates tested are susceptible to cefoxitin (4 μg/ml), clindamycin (1 μg/ml), and metronidazole (1 μg/ml). Culture supernatants are not toxic to mice. A bacteriocin produced by the type strain is active against cell membrane functions of several species of clostridia, particularly those of *Clostridium pasteurianum*.

Other characteristics of the species are given in Table 142.

Source: soil, freshwater and marine sediments, cheese, rumen of healthy calves, animal and human feces (including feces of healthy infants), snake venom, and, although seldom in pure culture, from a wide variety of human and animal clinical specimens including those from blood, urine, lower respiratory tract, pleural cavity, abdomen, wounds, and abscesses.

DNA G+C content (mol%): 27–28 (T_m).

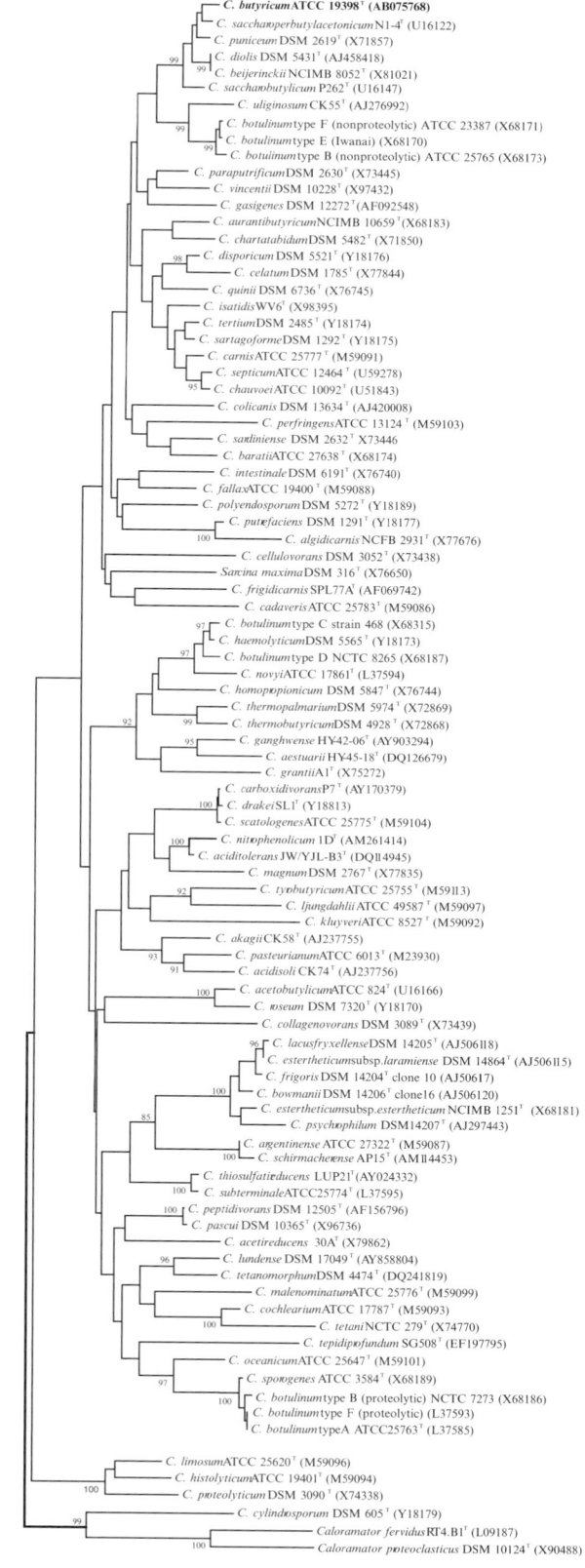

C. butyricum ATCC 19398ᵀ (AB075768)
C. saccharoperbutylacetonicum N1-4ᵀ (U16122)
C. puniceum DSM 2619ᵀ (X71857)
C. diolis DSM 5431ᵀ (AJ458418)
C. beijerinckii NCIMB 8052ᵀ (X81021)
C. saccharobutylicum P262ᵀ (U16147)
C. uliginosum CK55ᵀ (AJ276992)
C. botulinum type F (nonproteolytic) ATCC 23387 (X68171)
C. botulinum type E (Iwanai) (X68170)
C. botulinum type B (nonproteolytic) ATCC 25765 (X68173)
C. paraputrificum DSM 2630ᵀ (X73445)
C. vincentii DSM 10228ᵀ (X97432)
C. gasigenes DSM 12272ᵀ (AF092548)
C. aurantibutyricum NCIMB 10659ᵀ (X68183)
C. chartatabidum DSM 5482ᵀ (X71850)
C. disporicum DSM 5521ᵀ (Y18176)
C. celatum DSM 1785ᵀ (X77844)
C. quinii DSM 6736ᵀ (X76745)
C. isatidis WV6ᵀ (X98395)
C. tertium DSM 2485ᵀ (Y18174)
C. sartagoforme DSM 1292ᵀ (Y18175)
C. carnis ATCC 25777ᵀ (M59091)
C. septicum ATCC 12464ᵀ (U59278)
C. chauvoei ATCC 10092ᵀ (U51843)
C. colicanis DSM 13634ᵀ (AJ420008)
C. perfringens ATCC 13124ᵀ (M59103)
C. sardiniense DSM 2632ᵀ X73446
C. baratii ATCC 27638ᵀ (X68174)
C. intestinale DSM 6191ᵀ (X76740)
C. fallax ATCC 19400ᵀ (M59088)
C. polyendosporum DSM 5272ᵀ (Y18189)
C. putrefaciens DSM 1291ᵀ (Y18177)
C. algidicarnis NCFB 2931ᵀ (X77676)
C. cellulovorans DSM 3052ᵀ (X73438)
Sarcina maxima DSM 316ᵀ (X76650)
C. frigidicarnis SPL77Aᵀ (AF069742)
C. cadaveris ATCC 25783ᵀ (M59086)
C. botulinum type C strain 468 (X68315)
C. haemolyticum DSM 5565ᵀ (Y18173)
C. botulinum type D NCTC 8265 (X68187)
C. novyi ATCC 17861ᵀ (L37594)
C. homopropionicum DSM 5847ᵀ (X76744)
C. thermopalmarium DSM 5974ᵀ (X72869)
C. thermobutyricum DSM 4928ᵀ (X72868)
C. ganghwense HY-42-06ᵀ (AY903294)
C. aestuarii HY-45-18ᵀ (DQ126679)
C. grantii A1ᵀ (X75272)
C. carboxidivorans P7ᵀ (AY170379)
C. drakei SL1ᵀ (Y18813)
C. scatologenes ATCC 25775ᵀ (M59104)
C. nitrophenolicum 1Dᵀ (AM261414)
C. aciditolerans JW/YJL-B3ᵀ (DQ114945)
C. magnum DSM 2767ᵀ (X77835)
C. tyrobutyricum ATCC 25755ᵀ (M59113)
C. ljungdahlii ATCC 49587ᵀ (M59097)
C. kluyveri ATCC 8527ᵀ (M59092)
C. akagii CK58ᵀ (AJ237755)
C. pasteurianum ATCC 6013ᵀ (M23930)
C. acidisoli CK74ᵀ (AJ237756)
C. acetobutylicum ATCC 824ᵀ (U16166)
C. roseum DSM 7320ᵀ (Y18170)
C. collagenovorans DSM 3089ᵀ (X73439)
C. lacusfryxellense DSM 14205ᵀ (AJ506118)
C. estertheticum subsp. laramiense DSM 14864ᵀ (AJ506115)
C. frigoris DSM 14204ᵀ clone 10 (AJ506117)
C. bowmanii DSM 14206ᵀ clone16 (AJ506120)
C. estertheticum subsp. estertheticum NCIMB 1251ᵀ (X68181)
C. psychrophilum DSM14207ᵀ (AJ297443)
C. argentinense ATCC 27322ᵀ (M59087)
C. schirmacherense AP15ᵀ (AM114453)
C. thiosulfatireducens LUP21ᵀ (AY024332)
C. subterminale ATCC25774ᵀ (L37595)
C. peptidivorans DSM 12505ᵀ (AF156796)
C. pascui DSM 10365ᵀ (X96736)
C. acetireducens 30Aᵀ (X79862)
C. lundense DSM 17049ᵀ (AY858804)
C. tetanomorphum DSM 4474ᵀ (DQ241819)
C. malenominatum ATCC 25776ᵀ (M59099)
C. cochlearium ATCC 17787ᵀ (M59093)
C. tetani NCTC 279ᵀ (X74770)
C. tepidiprofundum SG508ᵀ (EF197795)
C. oceanicum ATCC 25647ᵀ (M59101)
C. sporogenes ATCC 3584ᵀ (X68189)
C. botulinum type B (proteolytic) NCTC 7273 (X68186)
C. botulinum type F (proteolytic) (L37593)
C. botulinum type A ATCC25763ᵀ (L37585)
C. limosum ATCC 25620ᵀ (M59096)
C. histolyticum ATCC 19401ᵀ (M59094)
C. proteolyticum DSM 3090ᵀ (X74338)
C. cylindrosporum DSM 605ᵀ (Y18179)
Caloramator fervidus RT4.B1ᵀ (L09187)
Caloramator proteoclasticus DSM 10124ᵀ (X90488)

0.01

FIGURE 128. 16S rRNA gene sequence based phylogeny of clusters I and II of the genus *Clostridium* as defined by Collins et al. (1994).

Type strain: ATCC 19398, CCUG 4217, CIP 103309, DSM 10702, HAMBI 482, IAM 14194, NBRC 13949, JCM 1391, KCTC 1786, KCTC 1871, LMG 1217, NCCB 89156, NCIMB 7423, NCTC 7423, VKM B-1773.

GenBank accession number (16S rRNA gene): AB075768, M59085.

Further comments: 16S rRNA gene studies have established the phylogenetic position of *Clostridium butyricum* in relationship to other *Clostridium* species. As the type species of the genus, *Clostridium butyricum* is the core of the true genus *Clostridium* in cluster I of the clostridia as defined by Collins et al. (1994). The closest relatives to the type species of the genus based on sequence similarity values are *Clostridium saccharoperbutylacetonicum* (97.6%), *Clostridium beijerinckii* (97.5%), *Clostridium diolis* (97.5%), *Clostridium puniceum* (97.3%), and *Clostridium saccharobutylicum* (97.1%).

DNA/DNA homology experiments established that *Clostridium butyricum* is a distinct species that includes strains previously identified as other species. In addition, many strains previously labeled *Clostridium butyricum* are now recognized as *Clostridium beijerinckii*. The two species are most easily differentiated by distinct electrophoretic patterns of cellular proteins and by nutritional requirements. *Clostridium butyricum* will grow well even in a third serial transfer in a medium containing only glucose, mineral salts, and biotin, while *Clostridium beijerinckii* requires other vitamins and growth factors such as are provided by yeast extract. The two species also differ in phospholipid composition. *Clostridium butyricum* has ethanolamine as its major nitrogenous phospholipid base, while *Clostridium beijerinckii* has *N*-methylethanolamine and ethanolamine.

Historically, "*Clostridium pseudotetanicum*" has been differentiated from *Clostridium butyricum* by spore location. However, both terminal and subterminal spores are seen in cultures of strains of the two species. The reference strain of "*Clostridium pseudotetanicum*" (Prévot, 1938) Smith and Hobbs 1974, 567 (ATCC 25779; NCIB 10630) is phenotypically indistinguishable from strains of *Clostridium butyricum*. Bulk RNA preparations from ATCC 25779 are 100% homologous with the 23S RNA from ATCC 19398, the type strain of *Clostridium butyricum* (Johnson and Francis, 1975). In addition, the electrophoretic patterns of soluble cellular proteins of the two strains are identical. Thus, "*Clostridium pseudotetanicum*" appears to be a later synonym of *Clostridium butyricum* pending confirmation by DNA–DNA homology determinations.

2. **Clostridium aceticum*** (*ex* Wieringa 1940) Gottschalk and Braun 1981, 476ⱽᴾ

a.ce'ti.cum. L. n. *acetum* vinegar; L. adj. suff. -*icus* belonging to; N.L. neut. adj. *aceticum* related to acetic acid, which it produces.

Cells in broth cultures in an atmosphere of 67% H_2 and 33% CO_2 are Gram-stain-negative rods, motile and peritrichous, 0.3–1.0 μm × 4.0–8.0 μm. With fructose as the substrate, cells may be up to 40 μm in length. Spores are round, terminal, and swell the cell. Sporulation occurs most readily on fructose agar medium after 2 d incubation. Cell walls contain *meso*-DAP and are composed of at least two layers. Colonies are not readily formed. After prolonged incubation

TABLE 142. Characteristics of the species of the genus *Clostridium*[a]

Characteristic	C. butyricum	C. saccharoperbutylacetonicum	C. puniceum	C. beijerinckii	C. diolis	C. saccharobutylicum	C. uliginosum	C. gasigenes	C. parapurificum	C. vincentii	C. aurantibutyricum	C. chartatabidum	C. celatum	C. disporicum	C. quinii	C. isatidis
Products from PYG[b]	BAF(ls2,4)	1,4 (a,b)	AB4(lf)	BA (Fpls,2,4)	BA	1, 2, 4(a,b)	A (b,l)	ABL 2,4	BAL(sf)	A (b,f)	AL (bps)	ABE	AFb2(s)	AL(bs2)	FAL2 (b)	ALF2
Motility	±	+	+	+	+	+	+	+	±	+	+	+	−	−	+	+
H₂ produced	4	+	4	4	NT	+	+	+	4	+	4	+	4	NT	+	+
Indole produced	−	+	−	−	−	−	NT	NT	−	NT	−	NT	−	−	−	−
Lecithinase produced	−	NT	−	−	NT	NT	NT	NT	−	NT	−	NT	−	−	−	NT
Lipase produced	−	NT	−	−	NT	NT	NT	NT	−	NT	+	NT	−	−	−	NT
Esculin hydrolyzed	+	+	+	+	+	+	NT	+	+	NT	+	+	+	+	+	+
Starch hydrolyzed	+	NT	+	±	−	+	NT	+	+	−	+	−	−	+	−	−
Nitrate reduced	−	NT	−	−	−	NT	−	NT	−/+	−	+	NT	d	NT	−	−
Substrate utilized and/or acid produced from:																
Amygdalin	±	+	−w	−/+	+	+	NT	NT	d	NT	−	−	+	NT	NT	+
Arabinose	±	+	−w	±	w	+	−	−	−	NT	w	NT	−	−	NT	−
Cellobiose	+	+	d	+	+	+	+	+	+	+	+	+	+	+	+	+
Fructose	+	NT	w+	+	NT	NT	NT	+	+	+	+	+	+	+	+	+
Galactose	+	NT	+	+	+	NT	NT	NT	+	NT	w	−	+	+	+	+
Glycogen	+	−	−	d	+	+	NT	NT	±	NT	w	NT	−	NT	NT	−
Inositol		NT	−		−	+	NT	+	NT	NT	−	NT	NT	−	NT	NT
Inulin	−/+	+/−	NT	d	NT	−/+	NT	NT	−	NT	−	NT	−	NT	NT	−
Lactose	+	+	−/+	+	w	+	−	−	+	+	+	−/+	+	+	−	−
Maltose	+	+	+w	+	NT	+	−	+	+	+	+	−	+	+	+	+
Mannitol	−/+	−/+	−	d	+	−/+	+	−	−	NT	−	−	−	+	+	
Mannose	+	NT	d	+	w	+	+	+	+	+	w	−	+	+	+	+
Melezitose	−/+	−	−	−/+	w	−	NT	NT	−	NT	−	NT	−	−	NT	−
Melibiose	+	+	−w	d	+	+	NT	NT	−	NT	w	−	−	+	NT	NT
Raffinose	+	+	−w	±	−	+	−	−	−	NT	+	NT	−	+	+	−
Rhamnose	−	−	−	−/+	NT	−	NT	NT	−	−	−	−	−	−	−	−
Ribose	+	w	−w	−/+	w	w	NT	NT	w−	NT	−	NT	d	+	−	−
Salicin	+	+	±	−/+	+	+	NT	+	+	NT	w	+	+	+	+	+
Sorbitol	−	−		−/+	w	−	NT	−	−	NT	−	NT	−	−	−	+
Starch	+	+	d	±	−	+	NT	+	+	−	w	−	−	+	+	+
Sucrose	+	NT	+	+	NT	NT	NT	NT	+	+	+	+	+	+	+	+
Trehalose	+	+	−/+	±	+	+	NT	+	−/+	NT	−	−	+	+	NT	+
Xylose	+	NT	d	+	+	NT	+	−	−	+	−	+	−	+	+	−
Milk reaction	c	+	c	c	−	c	NT	NT	c	NT	c	NT	c	−	NT	NT
Meat digested	−	NT	−	−	NT	NT	NT	NT	NT	NT	−	NT	NT	−	NT	−

(continued)

TABLE 142. (continued)

Characteristic	C. sartagoforme	C. tertium	C. carnis	C. chauvoei	C. septicum	C. colicanis	C. perfringens	C. sardinense	C. baratii	C. fallax	C. intesinale	C. polyendosporum	C. algidicarnis	C. putrefaciens	C. cadaveris	C. frigidicarnis	C. cellulovorans
Products from PYG[b]	BAF(L)	ABL(fs2)	BALf(s)	ABF(ls)4	BA(Fpl2)	NT	ABL (pfs)	BAL(Fp)	BAL (fps)	ABL(s)	AB(lfs)	A (lb,2,4)	AB	afl(s)	A2,4 (fpls)	B,ib,iv,1,o,2,4	Baf
Motility	d	+	±	d	±	−	−	+	−	−/+	+	−	−	−	±	+	−
H₂ produced	4	4	4	4	4	NT	4	4	4	4	+	+	NT	−	4	+	+
Indole produced	−	−	−	−	−	−	−	−	−	−	−	−	−	−	+	−	NT
Lecithinase produced	−	−	−	−	−	−	+	+	+	−	−	−	−	−	−	+	NT
Lipase produced	−	−	−	−	−	−	−	−	−	−	−	−	NT	−	−	−	NT
Esculin hydrolyzed	+	+	+	+	+	+	d	+	+	+	+	NT	−	−	−	−	NT
Starch hydrolyzed	−/+	d	−	−	−	w	±	d	d	−	−	+	−	−	−	−	−
Nitrate reduced	−/+	±	−	±	d	−	±	±	d	−/+	−	−	−	−	−	NT	NT
Substrate utlized and/or acid produced from:																	
Amygdalin	±	d	w	−	−	−	−w	−	−	−	+		NT	−	−	NT	NT
Arabinose	−	−	−	−	−		−	−	−	−	−	−	−	−	−	NT	−
Cellobiose	+	+	w+	−	+w	+	−/+	+w	+	−w	+	+	−	−	−	−	+
Fructose	+	+	d	−w	+	+	+	+	+	+	+	+	+	−	d	+	+
Galactose	+	+	d	w+	+w	+	+w	+	+w	w+	+	+	NT	−	−	−	+
Glycogen	±	+w	−	−	−	−	d	−	−/+	−	−	−	NT	−	−	NT	NT
Inositol	−	NT		−	−	−	±	−	NT	NT	−	−	NT	−	−	NT	NT
Inulin	−	−w	−	−	−	NT	−w	−	−	−	−	+	NT	−	−	NT	NT
Lactose	+	+	d	+w	+	+	+	+w	+w	w−	+	−	−	−	−	−	+
Maltose	±	+	w+	+w	+	+	+	+w	+w	+	−	+	−	−	−	+	+
Mannitol	+	+w	−	−	−	−	−	−	−	−	+	+	−	−	−	−	NT
Mannose	+	+	+w	+w	+	+	+	+	+	+w	+	+	+ slow	−	−w	+	+
Melezitose	−/+	d	−	−	−	−	−	−	−	−	−	NT	−	−	−	NT	−
Melibiose	+	+w	−	−	−	−	−w	−	−w	−	−	NT	−	−	−	NT	NT
Raffinose	−/+		−	−	−	−	d	−	−	−	−	−	+	−	−	NT	NT
Rhamnose	−/+	−w	−	−	−	−	−	−	−	−	−	−	−	−	−	NT	−
Ribose	d	+w	−	d	d	+	d	d	−w	w+	−	NT	+	−	−	NT	NT
Salicin	+	+	w	−	d	w	−	+w	+w	−	+	+	−	−	−	NT	NT
Sorbitol	−	−	−	−	−	−	−/+	−	−	−	+	+	−	−	−	+	−
Starch	±	+w	−w	−	−	w	d	−w	d	w+	−	−	−	−	−	NT	−
Sucrose	+	+	+w	+w	−	+	+	+w	+	−	+	+	−	−	−	NT	+
Trehalose	+	±	−	−	+w	−	d	d	−	−	+	+	−	−	−	+	−
Xylose	−/+	±	−	−	−	−	−	−	−	−	+	+	−	−	−	−	−
Milk reaction	−c	c	−	c	cd	NT	dc	c	c	c	NT	NT	−	−	cd	+	NT
Meat digested	−	−	−	−	−	NT	±	−	−	−	NT	NT	−	+	+	+	NT

(continued)

TABLE 142. (continued)

Characteristic	C. aestuarii	C. ganghwense	C. grantii	C. haemolyticum	C. homopropionicum	C. thermobutyricum	C. thermopalmarium	C. aciditolerans	C. nitrophenolicum	C. magnum	C. scatologenes	C. drakei	C. carboxidivorans	C. kluyveri	C. ljungdahlii	C. tyrobutyricum	C. akagii	C. acidisoli
Products from PYG[b]	Bpg	G2	A(f2)	APB(s)	Ab4 from fructose	B(al)	B(al2)	AB2	A(fp)	A	AC (cfpibivvls)	A(2b4)	A(2b4)	Cba	A (2)	BA(sflp)	ABL(f)	ABL(f)
Motility	+	+	+	+	+	+	+	+	−	+	+	+	+	+	+	±	+	+
H₂ produced	+	−	−	4	+	+	+	+*	NT	+	4	+	+	2	+	4	+	+
Indole produced	−	+	NT	+	NT	−	NT	+	+	NT	−/+	−	−	NG	NT	−	NT	NT
Lecithinase produced	−	NT	NT	+	NT	NT	NT	+	NT	NT	−	NT	NT	NG	NT	−	NT	NT
Lipase produced	NT	NT	NT	−	NT	NT	NT	−	NT	NT	−	NT	NT	NG	NT	−	NT	NT
Esculin hydrolyzed	−	+	NT	−	NT	NT	NT	NT	NT	NT	−/+	−	−	−	NT	−	NT	NT
Starch hydrolyzed	NT	NT	NT	−	NT	NT	NT	−	NT	NT	−	NT	NT	−	NT	−	NT	NT
Nitrate reduced	−	NT	−	−	−	−	NT	−	−	−	−	−	−	−	NT	−/+	−	−
Substrate utilized and/or acid produced from:																		
Amygdalin	NT	NT	NT	−	NT	NT	NT	NT	NT	NT	−	−	−	−	NT	−	NT	NT
Arabinose	−	−	NT	−	−	−	−	−	−	−	−w	+(L−)	+(L−)	NT	+	−	+	+
Cellobiose	−	+	+	−	NT	+	+	+	−	NT	−w	+	+	−	+	+	NT	NT
Fructose	−	NT	+	d	+	+	+	+	+	+	+	+	+	NT	−	+	NT	NT
Galactose	−	NT	−	−w	NT	−	−	+	+	NT	−	NT	NT	−	NT	−	NT	NT
Glycogen	NT	NT	NT	−	NT	NT	NT	−	NT	NT	NT	+	+	NT	NT	NT	NT	NT
Inositol	NT	NT	NT	d	NT	NT	NT	−	NT	NT	NT	NT	NT	NT	NT	−	NT	NT
Inulin	−	−	−	−	NT	NT	−	+	NT	NT	−	−	−	−	−	−	−	+
Lactose	−	+	+	−w	NT	+	+	+	+	−	−	−	−	−	−	−	+	+
Maltose	+	+	+	−w	NT	NT	+	−	−	NT	−	+	+	NT	NT	−w	−	+
Mannitol	−	+	+	d	NT	−	NT	+	+	NT	+w	+	+	NT	−	+w	+	+
Mannose	−	−	NT	−w	NT	NT	NT	NT	NT	NT	−	+	+	NT	−	−	−	+
Melezitose	NT	NT	NT	w−	NT	NT	−	NT	−	NT	−	−	−	NT	−	−	NT	NT
Melibiose	NT	−	NT	−w	NT	−	NT	+	NT	NT	−	−	−	NT	NT	−	−	+
Raffinose	−	−	NT	−w	NT	−	+	NT	NT	NT	d	+	+	NT	NT	−	−	+
Rhamnose	−	NT	NT	d	NT	+	+	+	NT	−	−w	+	+	−	+	−	NT	NT
Ribose	−	+	NT	−	NT	NT	−	NT	−	NT	−w	NT	NT	NT	NT	−	+	+
Salicin	−	−	NT	−	NT	NT	NT	−	−	NT	−	−	−	NT	−	−	−	−
Sorbitol	NT	NT	NT	−	NT	−	−	−	−	−	+	+	−	−	NT	NT
Starch	+	−	−	−	NT	NT	+	+	+	+	−	+	+	−	−	−	−	+
Sucrose	−	−	NT	−w	NT	NT	NT	−	+	NT	−	−	+	−	NT	−	−	−
Trehalose	−	−	−	−	−	+	+	+	NT	+	±	+	+	−	+	−/+	+	+
Xylose	NT	NT	NT	cd	NT	NT	NT	NT	NT	NT	NT	NT	NT	NT	NT	−	NT	NT
Milk reaction	NT	NT	NT	cd	NT	NT	NT	NT	NT	NT	−	NT	NT	NT	NT	−	NT	NT
Meat digested	NT	NT	NT	−/+	NT	−	NT	NT	NT	NT	−	NT	NT	NT	NT	−	NT	NT

(continued)

TABLE 142. (continued)

Characteristic	C. pasteurianum	C. acetobutylicum	C. roseum	C. cochlearium	C. tetani	C. malenominatum	C. lundense	C. tetanomorphum	C. acetireducans	C. pascui	C. peptidivorans	C. subterminale	C. thiosulfatireducens	C. argentinense	C. schirmacherense	C. bowmanii	C. psychrophilum	C. estertheticum subsp. estertheticum
Products from PYG[b]	ABf(1s)	BAl4(s)	BAs4	Bap(fls4)	ABp4(2ls)	AB (fpls)	NT	AB2,4	B (from pyruvate)	AB2 (from glutamate)	AF (from histidine)	ABiVib (fpicls2)	ABibiv (from peptone)	Abivib (l2,4)	FAPIV	ABfl2,4	L2,4 b	BAfl4,2
Motility	−/+	±	−/+	±	−/+	±	−		−	+	−	+/−	+	+	NT	+	+	+
H₂ produced	4	4	4	4	4	4	NT	+	+	+	+	4	NT	4	1	+	+	+
Indole produced	−	−	−	−/+	d	+	+	+/−	NT	+	NT	−	NT	−	−	+	+	+
Lecithinase produced	−	−	−	−/+	−	−	NT	−	NT	−	NT	−	NT	−	−	NT	NT	NT
Lipase produced	−	−	−	NT	−	−	+	NT	NT	−	NT	−/+	NT	−		NT	NT	NT
Esculin hydrolyzed	−	+	+	NT	−	−	+	NT	NT	NT	NT	−	NT	−	+	NT	NT	NT
Starch hydrolyzed	−	+	−	−	−	−	NT	NT	NT	NT	NT	−	NT	NT		NT	NT	NT
Nitrate reduced	−	−	−	−	−	−/+	NT	−/+	−	−	−	−	−	−	−	NT	NT	w(+)
Substrate utilized and/or acid produced from:																		
Amygdalin	−	−	−	−	−	−	NT	NT	−	−	NT	−	NT	−	−	−	−	−
Arabinose	w	d	+	NT	−	−	−	−	−	−	−	−	−	−	−	−	+	+
Cellobiose	−	+	+	−	−	−	NT	+	−	−	−	−	NT	−	NT	−	+	w(+)
Fructose	+	+	+	−	−	−	NT	+	−	−	−	−	−	−	NT	NT	NT	NT
Galactose	w−	+	+	NT	−	−	NT	−	−	−	−	−	−	−	−	+	−	+
Glycogen	−	d	−	−	−	−	NT	−/+	−	−	NT	−	NT	−	−	−	−	−
Inositol	NT	NT	−	NT	−	−	NT	+	−	−	−	−	NT	−	−	−	−	+
Inulin	−	−	−	NT	−	−	NT	NT	−	−	NT	−	NT	−	−	+	+	+
Lactose	−w	d	+	−	−	−	−	−	−	−	−	−	NT	−	−	+	+	+
Maltose	+	+	−	−	−	−	−	+	−	−	NT	−	−	−	−	+	+	+
Mannitol	+	−/+	−	NT	−	−	−	NT	−	−	−	−	−	−	−	−	−	+
Mannose	+	+	+	NT	−	−	−	+/−	−	−	−	−	NT	−	−	+	+	+
Melezitose	+	−	−	NT	−	−	−	NT	−	−	NT	−	NT	−	NT	−	−	−
Melibiose	+	−	−	NT	−	−	NT	NT	−	−	−	−	NT	−		−	−	+
Raffinose	+	−	−	NT	−	−	−	NT	−	−	−	−	NT	−	+	−	−	+
Rhamnose	−	−	+	NT	−	−	−	−	−	−	NT	−	NT	−	+	−	−	+
Ribose	−	−	−	−	−	−	NT	+	−	−	−	−	−	−	+	w	−	−
Salicin	−	+	+	NT	−	−	−	+	−	−	NT	−	NT	−	NT	+	−	+
Sorbitol	+	-	-	NT	−	−	−	+	−	−	−	−	NT	−	−	−	−	+
Starch	−w	+	+	−	−	−	NT	NT	−	−	NT	−	NT	−	NT	−	−	w(+)
Sucrose	+	+	+	−	−	−	−	NT	−	−	−	−	NT	−	−	+	+	+
Trehalose	+	−/+	−	−	−	−	NT	−	−	−	NT	−	NT	−	−	+	"+	−
Xylose	w	±	+	−	−	−	−	+	−	−	−	−	−	−	−	+	+	+
Milk reaction	−	c	cd	−	d−	−	NT	−	NT	NT	NT	dc	NT	d	NT	NT	NT	NT
Meat digested	−	−	−	−	−/+	−	NT	−	NT	NT	NT	+/−	NT	+/−	NT	NT	NT	NT

(continued)

TABLE 142. (continued)

Characteristic	*C. estertheticum* subsp. *laramiense*	*C. frigoris*	*C. lacusfryxellense*	*C. collagenovorans*	*C. tepidiprofundi*	*C. oceanicum*	*C. novyi* type A	*C. novyi* type B	*C. novyi* type C	*C. sporogenes*	*C. botulinum* types C, D	*C. botulinum* types B, E, F (saccharolytic)	*C. botulinum* types A, B, F (proteolytic)	*C. botulinum* type G	*C. proteolyticum*	*C. histolyticum*	*C. limosum*
Products from PYG[b]	B4laf2	Bfla2	BFal2	Aibpiv	A2f	ALb (ficspibiv2,3,4)	ABP	PBA	PBaf	ABibiv2 (picvls4)	BPA (vls)	BA(l)	ABiVib (icvp2,3,4)	Abivib(l2,4)	A2biv	A(fls)	A(fls)
Motility	+	+	+	+	–	±	±	±	+	±	±	+	±	+	–	+/–	d
H₂ produced	+	+	+	+	+	4	4	d	1	4	4	4	4	4	+	2	–1
Indole produced	–	NT	NT	NT	NT	–	–	±	+	–	–/+	–	–	–	NT	–	–
Lecithinase produced	–	NT	NT	NT	NT	–/+	+	+	–	–	–/+	–	–	–	NT	–	+
Lipase produced	+	NT	NT	NT	NT	–	+	–	–	+	+	+	+	+	NT	–	–
Esculin hydrolyzed	–	NT	NT	NT	NT	+	–	–	–	+	–	–	–	+	NT	–	–
Starch hydrolyzed	+	–	–	–	NT	d	–	–	–	–	–	–	±	–	NT	NT	NT
Nitrate reduced	+	NT	NT	NT	–	–	–	–	–	–	–	–	–	–	NT	–	–
Substrate utilized and/or acid produced from:																	
Amygdalin	–	+	+	NT	NT	–	–	–	–	–	–	–w	–	–	–	–	–
Arabinose	–/(–)	+	–	NT	NT	–	–	–	–	–	–	–	–	–	–	–	–
Cellobiose	–/(–)	+	+	–	–	d	–	–	–	–	–	–	–	–	–	–	–
Fructose	+	NT	NT	–	–	w+	–w	d	–	–w	w–	+w	–w	–	–	–	–
Galactose	+	+	+	–	NT	d	d	–	–	–	–w	–/+	–	–	–	–	–
Glycogen	+	+	+	NT	NT	–	–	–	–	–	–	–w	–	–	–	–	–
Inositol	+	w	+	NT	NT	–	±	±	w	–	–	±	–w	–	–	–	–
Inulin	–/(–)	+	+	NT	NT	–	–	–	–	–	–	–w	–	–	–	–	–
Lactose	–	+	+	NT	NT	–	–	–	–	–	–	–	–	–	–	–	–
Maltose	+	+	–	NT	+	+w	d	d	–	–w	d	+w	–w	–	–	–	–
Mannitol	+	–	+	–	NT	–	–	–	–	–	–	–	–	–	–	–	–
Mannose	+	+	–	–	NT	+w	–	w+	w	–	d	+w	–	–	–	–	–
Melezitose	–	–	+	NT	NT	–	–	–	–	–	–	d	–	–	–	–	–
Melibiose	+	+	+	NT	NT	–	–	–	–	–	–w	–	–	–	–	–	–
Raffinose	+	+	+	NT	NT	–	–	–	–	–	–	–	–	–	–	–	–
Rhamnose	+	+	–	NT	NT	–	–	–	–	–	–	–	–	–	–	–	–
Ribose	–	w	+	–	NT	–	d	–w	–	–	d	d	–	–	–	–	–
Salicin	+	+	+	NT	NT	–w	–/+	–	–	–	–	–	–	–	–	–	–
Sorbitol	+	–	–	NT	NT	–	–	–	–	–	–	±	–w	–	–	–	–
Starch	+	+	+	–	+	–	–	–	–	–	–	d	–	–	–	–	–
Sucrose	+	+	+	–	–	–	–	–	–	–	–	+w	–	–	–	–	–
Trehalose	–	+	+	NT	NT	–w	–	–	–	–	–	w+	–	–	–	–	–
Xylose	–/(–)	+	+	–	–	–	–	–	–	–	–	–	–	–	–	–	–
Milk reaction	–	NT	NT	c	NT	d–	c	d	–	d	dc	c–	d	d	NT	d	d
Meat digested	+	NT	NT	NT	NT	+	–	+	+	+	±	–	+	+/–	NT	+	+

TABLE 143. Characteristics of *Clostridium* species that fall outside cluster I of the clostridia and indicated as other characteristics in the species descriptions[a]

Characteristics	C. aceticum	C. acidurici	C. aminovalericum	C. formicaceticum	C. leptum	C. polysaccharolyticum	C. propionicum	C. purinilyticum	C. sporosphaeroides	C. sticklandii
Products from PYG[b]	A	Af	Af	A(Fs)	A(2)	Fabp2	PiVibbas (1)	FA	ABp	Aivbpib
Motility	+	+	0	+	−	+	+	+	−	+
H_2 produced	−	−	4	−	4		4	−	4	1
Indole produced	NT	−	−	−	−	−		−	−	−
Esculin hydrolyzed	−	−	+	−	+/−	+	−	−	−	−
Starch hydrolyzed	NT	−	+	−	−	+	−	−	−	−
Nitrate reduced	NT	−	−	−	−	−	−	−	−	−
Acid produced from:										
Amygdalin	−	−	−	−	−w	−	−	−	−	−
Cellobiose	NT	−	−	−	−	w	−	−	−	−
Fructose	+	−	−	w	−w	−	−	−	−	−
Glycogen	NT	−	−	−	−w	w	−	−	−	−
Lactose	−	−	−	−	−/+	−	−	−	−	−
Maltose	−	−	−	−	+	−	−	−	−	−
Ribose	+	−	−	−	w−	−	−	−	−	−
Starch	−	−	−	−	−	w	−	−	−	−
Sucrose	−	−	−	−	+/−	−	−	−	−	−
Trehalose	NT	−	−	−	w−	−	−	−	−	−
Xylose	−	−	−	−	w−	−	−	−	−	−

[a]Symbols: +, reaction positive for 90–100% of strains; −, reaction negative for 90–100% of strains; ±, 61–89% of strains positive; −/+, 11–39% of strains positive; d, 40–60% of strains positive; w, weak; numbers (hydrogen) represent abundant (4) to negative on a "−" to "4" scale; c (milk), curd; a (milk), acid; tr, trace; NT, not tested. Where two reactions are listed, the first is more usual and occurs in 60–90% of strains.

[b]Products (listed in the order of amounts usually detected): a, acetic acid; b, butyric acid; l, lactic acid; s, succinic acid; p, propionic acid; f, formic acid; iv, isovaleric acid; ib, isobutyric acid; v, valeric acid; c, caproic acid; ic, isocaproic acid; 2, ethanol; 3, propanol; 4, butanol; i4, isobutanol; i5, isopentanol. Capital letters indicate at least 1 meq/1000 ml of culture; small letters indicate less than 1 meq/1000 ml. Products in parentheses are not detected uniformly.

in roll tubes in an agar medium containing mud extract and a CO_2-H_2 (1:2) atmosphere, "a barely visible, light-brown tuft of cellular material" can be seen.

Growth occurs in the temperature range 25–37 °C with an optimum temperature for growth of 30 °C; growth is poor at 45 °C. Growth occurs in the pH range pH 7.5–9.5 with an optimum for autotrophic growth at 8.3. Strains of this species grow chemolithotropically in an atmosphere of CO_2 and H_2 and converting these substrates to acetate. They also utilize the organic substrates fructose, ribose, glutamate, fumarate, malate, pyruvate, serine, formate, ethylene glycol, and ethanol, but, in the presence of organic substrates CO_2 and H_2, are not converted to acetate. Amygdalin, dulcitol, adonitol, citrate, succinate, glycine, threonine, lactate, maltose, sucrose, xylose, methanol, 2-propanol, and glycerol are not utilized. H_2 is produced only in the stationary growth phase and inhibits growth in fructose medium at pH 8.5 if the bicarbonate concentration is very low. Esculin, gelatin, and starch are not hydrolyzed. Atmospheric N_2 is fixed. Gluconate is fermented to pyruvate and glyceraldehyde-3-phosphate by a modified Entner–Doudoroff pathway.

Source: soil, lake sediment, and sewage sludge.

DNA G+C content (mol%): 33 (T_m).

Type strain: ATCC 35044, DSM 1496.

GenBank accession number (16S rRNA gene): Y18183.

Further comments: Clostridium aceticum is not a member of the genus *Clostridium sensu stricto* and falls in the cluster XI as defined by Collins et al. (1994). The 16S rRNA gene sequence of the type strain DSM 1496[T] shows highest similarity to the species *Clostridium felsenium* (97.7%) and *Clostridium formicaceticum* (97.6%). The closest non-clostridial species is *Anerovirgula multivorans* (96.8%) (Pikuta et al., 2006).

Clostridium aceticum can be differentiated from *Clostridium felsenium* in that *Clostridium felsenium* produces acid from glucose, sucrose, and xylose; produces acetate, butyrate and H_2 as end products of fermentation; hydrolyzes esculin and gelatin (see description of *Clostridium felsenium* below).

The species is most easily differentiated from *Clostridium formicaeceticum*, which it most closely resembles phenotypically, by its ability to form acetate from CO_2 and H_2 and to utilize formate, serine, or ethylene glycol, but not methanol, as substrates.

3. **Clostridium acetireducens** Örlygsson, Krooneman, Collins, Pascual and Gottschal 1996, 458[VP]

a.ce.ti.re.du′cens. L. neut. n *acetum* vinegar; L. pres. part. *reducens* reducing; L. neut. adj. *acetireducens* vinegar- or acetic acid-reducing.

TABLE 144. Characteristics of *Clostridium* species that fall outside cluster 1 of the clostridia and indicated as other characteristics in the species descriptions[a]

Characteristic / Products from PYG[b]	C. articum PA(b)	C. bifermentans AF(ivic pibbls2)	C. cellobioparum AlI2	C. clostridioforme A(Fls2)	C. coccoides AS	C. cocleatum AF(Ls)	C. cotinum FAp(1)	C. difficile BAicivib (fv|2,4)	C. felsineum AB4 (lsf)	C. innocuum BLa (ls)	C. litusburense BAiVp-fib (2,3,i4)	C. nexile AF2 (ls)	C. oroticum AF2(ls)	C. papyrosolvens AL2	C. ramosum FAl(s2)	C. rectum Bapv	C. saccharolyticum Afl2	C. sordellii A (FiCpibivl)	C. sphenoides AF(ls2)	C. spiroforme AFl(s2)	C. symbiosum ABL(f2,4)
Motility	+	+	+	-/+	-	-	+	±	d	-	+	-	-	+	-	-	-	±	+	-	±
H₂ produced	tr	4	4	4	4	4	4	4	4	4	4	4	4	4	d	4	4	4	4	4	4
Indole produced	+	+	-	-/+	-	-	-	-	-	-	-	-	-	-	-	-	+	+	+	-	-
Lecithinase produced	NT	+	-	-	-	-	-	-	-	-	+	-	-	-	-	-	-	+	+	-	-
Lipase produced	NT	+	-	-	-	-	-	-	-	-	-	-	-	-	-	-	-	-	-	-	-
Esculin hydrolyzed	+	±	+	+	+	+	+	+	+	+	-	+	+	+	+	+	+	-/+	+	-/+	+
Starch hydrolyzed	-	-	-	-	-	-	-	-	-/+	-	-	-	-	+	-	-	+	-	d	-	-
Nitrate reduced	-	-	+	+	-	-	-	-	-	-	-	-	±	-	-	-	+	-	±	-	-
Substrate utilized and/or acid produced from:																					
Amygdalin	-	-	-	-w	+	±	-w	-	-	-	-	-w	-	-	-	-	-	-	-	-w	-
Arabinose	-	-	+	d	+	+w	-	-	+w	-	-	-	+	+	+	-	+	-	-w	-	d
Cellobiose	w	+	+	±w	+	+	d	+w	+w	+	-	+	+	+	+	-	w	-	+	-/+	+
Fructose	+	d	+	+	+	+	+	+	+	+	+	w+	+	±	+	-	+	-	w+	+	+
Galactose	NT	-	+	w+	+	+	w	±	+	+	+	+w	+	+	+	w	-	-	w+	-	+w
Glycogen	-	-	-	-	+	-	-	-	-	-	-	-	-	-	-	-	-	-	-w	-	-
Inositol	NT	-	NT	NT	NT	NT	NT	-	-	NT	-	NT	NT	NT	NT	NT	NT	-	NT	NT	NT
Inulin	NT	-	-	-	-	+w	d	-	d	+	-	d	+w	-	-	-	-	-	-	+	-
Lactose	-	-	w	±	+	+	+	-	+w	+	-	+w	+	+	+	w	w	-	w+	+w	-/+
Maltose	-	w-	+	+w	+	-/+	d	+w	-/+	+	-	-w	+	+	+	-	+w	w+	+	-/+	d
Mannitol	w	-	-	-	+	-	d	±	+	+	-	-	±	-	±	-	w	w+	w+	+	d
Mannose	+	-w	+	+w	+	+	+	±	+	+	+w	-w	+	+	+	-	+	-w	w+	+	d
Melezitose	-	-	-	-/+	+	-	-	d	-	-	-	-	±	NT	-	NT	-	-	-w	-	NT
Melibiose	-	-	+	d	+	-	+w	-	-	-	-	-w	-	-	-w	-	w	-	d	-	-
Raffinose	-	-	+	±	+	-/+	+	-	-/+	-w	-	d	+	+	+	-	+	-	+w	+	-
Rhamnose	-	-	±	±	+	-	-	-	+w	+	-	-	+	-	d	-	+	-	+w	-	+w
Ribose	-	-	+	d	+	-	-w	-	+	±	-w	±	+	+	d	-	-	-w	-w	-/+	-
Salicin	w	-	w	±	+	d	w+	-w	+	+	-	±	+	+	+	w	w	+	w+	-	-
Sorbitol	-	-w	-	-	+	-	-	-w	-/+	+	-	-	-w	-	-/+	+	-	-	-	+	-
Starch	NT	-	-	-w	-	-	-w	-w	-/+	+	-w	-w	-	+	+	w	w	-w	w-	+	-
Sucrose	-	-	+	+	+	+	+	-w	+	+	+w	+w	+	+	+	w	w	-	w-	+	-
Trehalose	-	-	+	±	+	+	+	-w	+	-w	+	-w	±	+	+	+	w	+	d	+	-
Xylose	+	-	+	-/+	+	+	-	-w	+	-w	w+	w+	+	+	-w	+	w	-	d	-	-w
Milk reaction	a	d	-	c	c	c	c	-	c	-	cd	-c	c	c	c	c	c	d	c	c	-c
Meat digestion	+	+	-	-	-	-	-	-	-	-	+	-	-	-	-	-	-	+	-	-	-

[a]Symbols: +, reaction positive for 90–100% of strains; -, reaction negative for 90–100% of strains; ±, 61–89% of strains positive; d, 40–60% of strains positive; w, weak; numbers (hydrogen) represent abundant (4) to negative on a "-" to "4" scale; c (milk), curd; a (milk), acid; tr, trace; NT, not tested. Where two reactions are listed, the first is more usual and occurs in 60–90% of strains.

[b]Products (listed in the order of amounts usually detected): a, acetic acid; b, butyric acid; p, propionic acid; f, formic acid; iv, isovaleric acid; ib, isobutyric acid; v, valeric acid; c, caproic acid; ic, isocaproic acid; 2, ethanol; 3, propanol; 4, butanol; i4, isobutanol; i5, isopentanol. Capital letters indicate at least 1 meq/1000ml of culture; small letters indicate less than 1 meq/1000ml. Products in parentheses are not detected uniformly.

TABLE 145. Characteristics of *Clostridium* species that fall outside cluster I of the clostridia and indicated as other characteristics in the species descriptions[a]

Characteristics	*C. ghonii*	*C. irregulare*	*C. mangenotii*	*C. thermocellum*[b]
Products from PYG[c]	Abivicib4i4 (fpls2,3)	Aiv (fpibl)	Afpibivic	A2l
Motility	+	+	−	−
H₂ produced	1–3	1	3	4
Indole produced	+	−	+	−
Lecithinase produced	+	−	−	−
Lipase produced	+	−	−	−
Esculin hydrolyzed	+	−	−	−
Nitrate reduced	−	−	−	+
Milk reaction	D	−	d	−
Meat digested	+	−	+	−

[a]Symbols: +, reaction positive for 90–100% of strains; −, reaction negative for 90–100% of strains; ±, 61–89% of strains positive; −/+, 11–39% of strains positive; d, 40–60% of strains positive; D, different reactions occur; w, weak; numbers (hydrogen) represent abundant (4) to negative on a "−" to "4" scale; c (milk), curd; a (milk), acid; tr, trace; NT, not tested. Where two reactions are listed, the first is more usual and occurs in 60–90% of strains.
[b]Produces weak acid from cellobiose and cellulose.
[c]Products (listed in the order of amounts usually detected): a, acetic acid; b, butyric acid; l, lactic acid; s, succinic acid; p, propionic acid; f, formic acid; iv, isovaleric acid; ib, isobutyric acid; v, valeric acid; c, caproic acid; ic, isocaproic acid; 2, ethanol; 3, propanol; 4, butanol; i4, isobutanol; i5, isopentanol. Capital letters indicate at least 1 meq/1000 ml of culture; small letters indicate less than 1 meq/1000 ml. Products in parentheses are not detected uniformly.

Straight rods, $1.0 \times 6 \mu m$ in length. Gram-stain-positive cell wall with thin S-layer. Endospores and flagella are not observed. Nonmotile. Strict anaerobe. Growth occurs in the temperature range 30–45 °C and optimally at 39–43 °C. The pH range for growth is 6.0–8.2, with optimal growth at pH 6.4–7.6. Acetate is used as an electron acceptor when growing on leucine, valine, isoleucine, and alanine. The reduced end product of growth is butyrate. Growth on leucine, valine, isoleucine, and alanine only occurs when acetate is available. Yeast extract required for growth. Cannot degrade sugars. Peptone (pancreatic digest of casein), serine, threonine, α-ketobutyrate, α-ketoisocaproate, and α-keto-3-methylvalerate are utilized as growth substrates. Unable to grow on saccharides, citrate, succinate, propionate, H₂-acetate, ethanol, and other amino acids. Fumarate, sulfate, sulfite, nitrate, nitrate, and oxygen cannot be used as electron acceptors.

Source: samples of an anaerobic bioreactor at the AVEBE potato starch factory in De Krim, Netherlands.

DNA G+C content (mol%): 28.5 (T_m).

Type strain: 30A, DSM 10703.

GenBank accession number (16S rRNA gene): X79862.

Further comments: 16S rRNA gene sequence comparisons show *Clostridium acetireducans* to fall within the radiation of cluster I of the clostridia as defined by Collins et al. (1994). The 16S rRNA gene sequence of *Clostridium acetireducans* shows ~ 94% similarity to those of the species *Clostridium pascui*, *Clostridium cochlearum*, and *Clostridium tetanomorphum*. In the original description of *Clostridium acetireducens* (Örlygsson et al., 1996), the type strain is incorrectly cited as DSM 7310 when it is, in fact, DSM 10703.

4. **Clostridium acetobutylicum**[*] McCoy, Fred, Peterson and Hastings 1926, 483[AL] emend. Keis, Shaheen and Jones 2001, 2100.

a.ce.to.bu.ty′li.cum. En. n. *acetone*; N.L. adj. *butylicum* butylic; N.L. neut. adj. *acetobutylicum* referring to production of acetone and butyl alcohol.

This description includes that provided by Cato et al. (1986) and Keis et al. (2001).

Cells in PYG broth cultures are straight rods, motile and peritrichous, $0.5–0.9 \times 1.6–6.4 \mu m$. Granulose (a starch-like polymer) is often present. Gram-stain-positive, becoming Gram-stain-negative in older cultures. Spores are oval and subterminal, slightly swelling the cell.

Cell walls contain *meso*-DAP, glucose, rhamnose, galactose, and mannose. The wall is triple-layered. Surface colonies on blood agar plates are 1–5 mm, flat to raised, granular, translucent to semiopaque with irregular margins and occasionally with a mosaic internal structure.

Cultures in PYG broth are turbid with a smooth sediment, and have a pH of 4.5–5.0 after incubation for 5 d.

The optimum temperature for growth is 37 °C. A fermentable carbohydrate, biotin, and *p*-aminobenzoic acid are required. No growth in the presence of 6.5% NaCl or 20% bile. Acetyl methyl carbinol is produced. Neutral red is reduced. Abundant gas is produced in glucose agar deep cultures. H₂S is produced by one of nine strains tested. Fixes atmospheric N₂.

Strains produce an inducible carboxymethyl cellulase and cellobiase. NADH and NADPH-ferredoxin and rubredoxin oxidoreductases also are present. Superoxide dismutase and deoxyribonuclease are produced.

Fermentation products include acetic, butyric, and lactic acids; butanol; acetone; CO₂; and large amounts of H₂. Small amounts of succinic acid may be formed. Ethanol was detected with high pressure liquid chromatography. During exponential growth, products are acetate and butyrate. Production of butanol and acetone is highest after 18 h when the organisms are in their stationary growth phase and is associated with morphological changes in the cells. Pyruvate is converted to acetate, butyrate, and butanol. Neither lactate nor threonine is utilized.

The type strain produces the amino acids lysine, arginine, aspartic acid, threonine, serine, glutamic acid, alanine, valine, isoleucine, leucine, and tyrosine in broth.

Strains are susceptible to chloramphenicol, clindamycin, erythromycin, penicillin G, and tetracycline. Culture supernatants are nontoxic to mice. Other characteristics of the species are listed in Table 142.

Characteristics added in the emendation of the species by Keis et al. (2001) include fermentation of amygdalin and raffinose.

Source: soil, lake sediment, well water, clam gut, bovine feces, canine feces, human feces.

DNA G+C content (mol%): 28–29 (T_m).

Type strain: ATCC 824, BCRC 10639, CCUG 42182C, CECT 508, DSM 792, IFO (now NBRC) 13948, JCM 1419, KCTC 1790, LMG 5710, NCCB 84048, NCCB 29024, NCIMB 13357, NRRL B-527, VKM B-1787.

GenBank accession number (16S rRNA gene): U16166, X78070.

Further comments: 16S rRNA gene sequence comparison show *Clostridium acetobutylicum* to fall within the radiation of cluster I of the clostridia as defined by Collins et al. (1994). The closest relatives of *Clostridium acetobutylicum* based on sequence similarity values are *Clostridium sardiniense* (94.8%) and *Clostridium collagenovorans* (94.1%).

5. **Clostridium acidisoli** Kuhner, Matthies, Acker, Schmittroth, Gößner and Drake 2000, 880[VP]

a.ci.di.so'li. L. adj. *acidus* acidic; L. neut. n. *solum* soil; N.L. gen. neut. n. *acidisoli* of acidic soil.

Rod-shaped cells, 1.0×3–7 μm, occur singly or in chains. Cells in chains do not separate after septum formation. Motile with 8–12 peritrichous flagella. Gram-stain-positive. Sporeforming. Anaerobe. Catalase-, oxidase-, tryptophanase-, and urease-negative. No cytochromes. Chemoorganotroph with fermentative metabolism. Growth occurs in the temperature range 5–37 °C and optimally at 25–30 °C. Acid tolerant. The pH range for growth is 3.6–6.9, with no defined pH optimal for growth between pH 3.6 and 6.6. Doubling time on glucose at 30 °C and pH 4.0 is ~3.5 h. Glucose, cellobiose, xylose, arabinose, maltose, mannose, salicin, mannitol, lactose, sucrose, glycerol, melezitose, raffinose, and rhamnose support growth. Does not grow on sorbitol, trehalose, H_2/CO_2, CO/CO_2, vanillate, Casamino acids, peptone, or various purines and pyrimidines. Glucose is fermented to acetate, butyrate, lactate, H_2, and CO_2. Under certain conditions, formate is formed. Nitrate and sulfate are not reduced. Nitrogen is fixed at pH 3.7.

Source: an acidic peat bog.

DNA G+C content (mol%): 30.7 (HPLC).

Type strain: CK74, ATCC BAA-167, DSM 12555.

GenBank accession number (16S rRNA gene): AJ237756.

Further comments: 16S rRNA gene sequence comparison show *Clostridium acidisoli* to fall within the radiation of cluster I of the clostridia as defined by Collins et al. (1994).The 16S rRNA gene sequence of *Clostridium acidisoli* shows ~ 97% similarity to those of the species *Clostridium akagii* and *Clostridium pasteurianum.* *Clostridium acidisoli* differs from *Clostridium pasteurianum* in substrate utilization range, forming lactate but not ethanol, lacking a carbon monoxide dehydrogenase, growing at pH 3.6 and fixing nitrogen at low pH. *Clostridium acidisoli* shares the ability to fix nitrogen at pH 3.7 with its other close relative *Clostridium akagii* (see description below) but differs in range of substrates utilized and cellular morphology.

6. **Clostridium aciditolerans** Lee, Romanek and Wiegel 2007, 314[VP]

a.ci.di.to'le.rans. N.L. n. *acidum* an acid; L. part. adj. *tolerans* tolerating; N.L. part. adj. *aciditolerans* acid-tolerating.

Straight to slightly curved rods, 0.5–1×3.0–9.0 μm. Type strain stains Gram-stain-negative at all growth phases, but have Gram positive wall structure. Retarded peritrichous flagellation. Forms subterminal spores that are oval in shape and do not swell the cell. After 1–2 d, colonies are irregular, mostly translucent, and less than 1.5 mm in diameter. Anaerobic growth. Growth occurs in the temperature range 20–45 °C with optimal growth around 35 °C. No growth at or below 18 °C or at or above 47 °C. The $pH^{25°C}$ range for growth is pH 3.8–8.9 with optimal growth at pH 7.0–7.5. Growth at pH 4.5–5.0 takes place at 52% of the optimal growth rate; no growth is observed at or below pH 3.5 or at or above pH 9.2. The salinity range for growth is from 0–1.5% NaCl (w/v). Methyl red test, indole production, and lecithinase are positive, but Voges–Proskauer reaction and lipase are negative. Gelatin is hydrolyzed, but casein is not. Heamolysis occurs on blood agar. Grows on peptone–yeast extract (PY), peptone–yeast extract–glucose (PYG), reinforced clostridial medium (RCM, Difco), and thioglycolate broth (Difco). The following substrates serve as carbon and energy source in the presence of 0.02%yeast extract: beef extract, Casamino acids, peptone, tryptone, cellobiose, fructose, galactose, glucose, lactose, maltose, mannose, raffinose, ribose, sucrose, xylose, pyruvate, glutamate, and inulin. Acetate, lactate, arabinose, trehalose, inositol, mannitol, sorbitol, xylitol, and cellulose do not support growth. No autotrophic growth on CO_2 and H_2 (80:20). Fe(III), nitrate, thiosulfate, elemental sulfur, sulfate, sulfite, MnO_4, and fumarate are not used as electron acceptors. The main end products from glucose fermentation are acetate, butyrate, and ethanol. Resistant to tetracycline (10 μM) and sensitive to ampicillin (10 μM), chloramphenicol (10 μM), erythromycin (10 μM), rifampin (10 μM), and streptomycin (10 μM).

Source: a constructed wetland system receiving acid sulfate water.

DNA G+C content (mol%): 30.8 (HPLC).

Type strain: JW/YJL-B3, ATCC BAA-1220, DSM 17425.

GenBank accession number (16S rRNA gene): DQ114945.

Further comments: 16S rRNA gene sequence comparison shows *Clostridium aciditolerans* to fall within the radiation of cluster I of the clostridia as defined by Collins et al. (1994). The 16S rRNA gene sequence of *Clostridium aciditolerans* shows 96.2% similarity to *Clostridium drakei. Clostridium aciditolerans* can be differentiated from *Clostridium drakei* (see description below) in its ability to grow at pH 3.8; its ability to utilize lactose, maltose, and raffinose; and lack of growth on cellulose, inositol, and mannitol.

7. **Clostridium acidurici*** (corrig. Liebert 1909) Barker 1938, 323[AL]

a.ci.du'ri.ci. N.L. n. *acidum uricum* uric acid; N.L. gen. n. *acidurici* of uric acid.

Cells in PY-0.3% uric acid broth stain Gram-variable to Gram-stain-negative, motile and peritrichous, 0.5–0.7×2.5–4.0 μm, occurring singly. Spores are oval, terminal and subterminal, and swell the cell. Sporulation occurs most reliably on chopped-meat uric-acid agar slants incubated at

30 °C for 1 week in an atmosphere of N_2. Cell walls contain *meso*-DAP. Surface colonies on uric acid agar are spreading, rhizoid, transparent, colorless, and flat, clearing the agar. No growth in gelatin or milk, or on egg-yolk or blood agar.

Broth cultures supplemented with 0.3% uric acid have a smooth sediment with no turbidity and a pH of 7.4–7.7 after incubation under N_2 for 6 d. Most rapid growth occurs in media at an initial pH of 7.6–8.1; there is poor growth below pH 6.5 or above 9.0. Growth occurs between 19 °C and 37 °C. Uric acid, xanthine, guanine, or hypoxanthine is required as a carbon and energy source. Selenite and tungstate stimulate xanthine dehydrogenase and formate dehydrogenase activity.

Products in PY-urate broth are acetate, NH_3, and CO_2. No H_2 is produced. No carbohydrates are fermented. The type strain is resistant to erythromycin, penicillin, and tetracycline. It is moderately sensitive to chloramphenicol and clindamycin. Culture supernatants of the type strain are nontoxic to mice.

Other characteristics of the species are given in Table 143.

Source: soil, chicken droppings, wild birds.

DNA G+C content (mol%): 28 (T_m).

Type strain: ATCC 7906, CIP 104303, DSM 604, NCCB 46094.

GenBank accession number (16S rRNA gene): M59084.

Further comments: 16S rRNA gene sequence comparisons show *Clostridium acidurici* to fall within the radiation of the cluster XII of the clostridia as defined by Collins et al. (1994). The closest relatives based on sequence similarity values are *Clostridium purinolyticum* (93.8%) and *Eubacterium angustum* (92.1%).

8. **Clostridium aerotolerans** van Gylswyk and van der Toorn 1987, 104[VP] emend. Chamkha, Garcia and Labat 2001a, 2109.

ae.ro.to'le.rans. Gr. masc. n. *aer* air, gas; L. v. *tolerare* to endure, to'put up with, tolerate; L. neut. part. adj. *aerotolerans* air-tolerating.

Straight rods, 0.4–0.6 × 1.5–3.0 µm. Grow under strictly anaerobic conditions or in unreduced media. Gram-stain-negative. Motile by peritrichous flagella. Produces single, terminal endospores. Spores are slightly oval and measure 0.6–0.8 µm. Colonies are round, have smooth edges, convex, and 2.5 mm in diameter. Appear light brown at the center but become translucent around the edges. The temperature range for growth is 15–45 °C with an optimum at 38–39 °C. The pH range for growth is 5.2–7.0 with an optimum at pH 6.6–6.9.

Ferments xylan to produce formic, acetic, and lactic acid; ethanol; carbon dioxide; and hydrogen. Maximum growth occurs on glucose, yeast extract, cysteine hydrochloride·H_2O, and Na_2S·9H_2O without reducing agent. Reduces various cinnamic acids. Digests xylan. Does not reduce sulfate. Gelatin is not hydrolyzed. Strains cannot digest casein. Indole is not produced. Urease, catalase, and nitrate reducing enzymic activity not observed. Various cinnamic acids including cinnamic; *o-, m-, p*-coumaric; *o-, m-, p*-methoxycinnamic; *p*-methylcinnamic; caffeic; ferulic; isoferulic; and 3,4,5-trimethoxycinnamic acids are reduced to their corresponding phenylpropionic acids, i.e.,3-phenyl,

3-(2-hydroxyphenyl), 3-(3-hydroxyphenyl), 3-(4-hydroxyphenyl), 3-(2-methoxyphenyl), 3-(3-methoxyphenyl), 3-(4-methoxyphenyl), 3-(4-methylphenyl), 3-(3,4-dihydroxyphenyl), 3-(4-hydroxy-3-methoxyphenyl), 3-(3-hydroxy-4- methoxyphenyl), and 3-(3,4,5-trimethoxyphenyl) propionic acids. In addition to the reduction of *p*-coumaric acid to 3-(4-hydroxyphenyl) propionic acid, *Clostridium aerotolerans* involves another pathway, i.e., *p*-coumaric acid can be decarboxylated to yield 4-vinylphenol which is reduced to 4-ethylphenol. Supplementation of glucose markedly accelerates all these conversions. Cinnamyl alcohol and phenylpropionic acids, including 3-phenylpropionic acid and 3,4-dihydroxyphenylpropionic acid, are not transformed.

Source: rumen of a sheep.

DNA G+C content (mol%): 40 (T_m).

Type strain: X8A62 = ATCC 43524, DSM 5434.

GenBank accession number (16S rRNA gene): X76163.

Further comments: 16S rRNA gene sequence comparisons show *Clostridium aerotolerans* to fall within the radiation of cluster XIV of the clostridia as defined by Collins et al. (1994). The closest relatives based on sequence similarity are *Clostridium xylanolyticum* (99.4%), *Clostridium saccharolyticum* (98.4%), *Clostridium algidixylanolyticum* (98.1%), and *Clostridium amygdalinum* (98.0%).

9. **Clostridium aestuarii** Kim, Jeong and Chun 2007, 1317[VP]

aes.tu.a'ri.i. L. gen. n. *aestuarii* of the tidal flat.

Rod-shaped cells, 0.7–0.8 × 2–4 µm. Motile with peritrichous flagella. Forms circular and yellowish colonies on MRCM. Gram-stain-negative. KOH-negative. Forms oval, terminal spores. Strictly anaerobic and chemoheterotrophic. Temperature range for growth is 15–30 °C with optimal growth at 30 °C. The optimum pH for growth on MRCM is 7.0 with a growth occurring in the range pH 5.5–8.5. Requires 1–10% (w/v) artificial sea salts (optimum 4%) for growth. Does not grow on reinforced clostridial medium containing 0–5% (w/v) NaCl alone. Glucose, maltose, and sucrose are utilized. Arabinose, cellobiose, fructose, galactose, glycerol, lactose, mannitol, mannose, melezitose, raffinose, rhamnose, ribose, salicin, sorbitol, trehalose, and xylose are not utilized. The fermentation end products from glucose are butyric acid, propionic acid, glycerol, and H_2. Nitrate is not reduce. Produces alkaline phosphatase, esterase (C4), esterase lipase (C8), valine arylamidase, acid phosphatase, and naphthol-ASBI-phosphohydrolase but not lipase (C14), leucine arylamidase, cystine arylamidase, trypsin, α-chymotrypsin, α-galactosidase, β-galactosidase, β-glucuronidase, α-glucosidase, β-glucosidase, *N*-acetyl-β-glucosaminidase, α-mannosidase, or α-fucosidase. Indole is not produced and urease-negative. Cells are catalase-negative and lecithinase-negative. Esculin is hydrolyzed, but gelatin is not.

Source: a tidal flat sediment on Ganghwa Island, South Korea.

DNA G+C content (mol%): not determined.

Type strain: HY-45-18, IMSNU 40129, JCM 13194, KCTC 5147.

GenBank accession number (16S rRNA gene): DQ126679.

Further comments: 16S rRNA gene sequence comparisons show *Clostridium aestuarii* to fall within the radiation of cluster I of the clostridia as defined by Collins et al. (1994). The 16S rRNA gene sequence of *Clostridium aestuarii* (DQ126679) shows 96.5% 16S rRNA gene sequence similarity to that of *Clostridium ganghwense*. Other validly published species share less than 94% similarity in their 16S rRNA gene sequences with that of *Clostridium aestuarii*. Both *Clostridium aestuarii* and *Clostridium ganghwense* (see description below) were isolated from tidal flat sediments on Ganghwa Isand, South Korea and require sea salts for growth. *Clostridium aestuarii* can be differentiated from *Clostridium ganghwense* by its ability to utilize sucrose but not mannose, fructose, salicin, gelatin, and cellobiose. Differences are also observed in the end products of glucose fermentation with *Clostridium aestuarii* forming butyrate, propionate, glycerol, and H_2 while *Clostridium ganghwense* forms glycerol, ethanol, and CO_2.

10. **Clostridium akagii** Kuhner, Matthies, Acker, Schmittroth, Gößner and Drake 2000, 879[VP]

a.ka′gi.i. L. gen. n. *akagii* named after James M. Akagi, who worked for more than three decades at the University of Kansas on the physiology of various anaerobic bacteria.

Rod-shaped cells, 1.0×2–$11\,\mu m$, occur singly or in chains and linked by a connecting filament. Motile with 10–15 peritrichous flagella. Sporeforming. Gram-stain-negative and lack an outer membrane. Anaerobic, chemoorganotrophic, fermentative growth. Growth occurs in the temperature range 5–30 °C and optimally at 20–25 °C. The pH range for growth is 3.7–7.1 with no distinct optimum between 4.2 and 6.8. Ferments glucose, cellobiose, xylose, arabinose, maltose, mannose, and salicin. Does not utilize mannitol, lactose, sucrose, glycerol, melezitose, raffinose, rhamnose, sorbitol, trehalose, H_2/CO_2, CO/CO_2, vanillate, Casamino acids, peptone, purines, and pyrimidines. End products of glucose fermentation are acetate, butyrate, lactate, H_2, and CO_2. Under certain conditions, formate is also produced. Doubling time of ~11 h at pH 4.0, 25 °C growing on glucose. Nitrate and sulfate are not reduced. Nitrogen is fixed at pH 3.7. Does not contain cytochromes. Negative for catalase, oxidase, tryptophanase, and urease activity.

Source: acidic beech liter.
DNA G+C content (mol%): 31.4 (HPLC).
Type strain: CK 58, ATCC BAA-166, DSM 12554.
GenBank accession number (16S rRNA gene): AJ237755.

Further comments: The 16S rRNA gene sequence of *Clostridium akagii* shows ~97% similarity to those of the species *Clostridium acidisoli* and *Clostridium pasteurianum*, falling within cluster I of the clostridia as defined by Collins et al. (1994). These species are differentiated on the basis of morphology and substrate utilization. *Clostridium akagii* differs from *Clostridium pasteurianum* in morphology, substrate range, the formation of lactate but not ethanol, growth and N_2 fixation at pH 3.7, and the lack of carbon monoxide dehydrogenase.

11. **Clostridium aldenense** Warren, Tyrrell, Citron and Goldstein 2007, 893[VP] (Effective publication: Warren, Tyrrell, Citron and Goldstein 2006, 2420.)

al.de.nen′se. N.L. neut. adj. *aldenense* pertaining to R. M. Alden Research Laboratory and its first patron, Rose M. Alden Goldstein.

Rod-shaped cells 2–5 × 0.8–1.1 µm. Colonies that are 1–2 mm in diameter, flat, opaque to white, and nonhemolytic on Brucella blood agar plates after 48 h incubation at 37 °C. Gram-stain-negative. Spores only rarely formed.

Positive indole reaction. Acid from glucose, maltose, mannose, raffinose, sucrose, and xylose but not from cellobiose, esculin, mannitol, melezitose, rhamnose, sorbitol, or starch. Urea, starch, and gelatin are not hydrolyzed, and nitrate is not reduced. Fermentation of arabinose, lactose, salicin, and trehalose, and esculin hydrolysis are variable. Enzymically, α-galactosidase and β-galactosidase are positive, but arginine dihydrolase, β-galactosidase-6-phosphate, α-glucosidase, β-glucosidase, β-glucuronidase, N-acetyl-β-glucosaminidase, arginine arylamidase, praline arylamidase, leucyl glycine arylamidase, phenylalanine arylamidase, leucine arylamidase, pyroglutamic acid arylamidase, tyrosine arylamidase, alanine arylamidase, and glycine arylamidase are negative. Variable reactions are produced for α-arabinosidase and alkaline phosphatase.

Source: human clinical infections.
DNA G+C content (mol%): not reported.
Type strain: RMA 9741, ATCC BAA-1318, CCUG 52204.
GenBank accession number (16S rRNA gene): DQ279736.

Further comments: Comparison of the 16S rRNA gene sequence of *Clostridium aldenense* shows it to cluster with the species of the *Clostridium clostridiforme* group. This group was designated cluster XIVa by Collins et al. (1994). 16S rRNA gene sequence similarities are in the range 95.8–96.7% to the species *Clostridium citroniae* (96.7%), *Clostridium clostridiforme* (96.2%), *Clostridium asparagiforme* (95.9%), and *Clostridium bolteae* (95.8%). This species differs from the other species of the *Clostridium clostridiforme* group in that it is indole-positive. *Clostridium aldenense* and *Clostridium citroniae* can be differentiated by their profile numbers obtained by Rapid ID 32A and individual phenotypic tests such as raffinose, rhamnose, α-galactosidase, and β-galactosidase (Warren et al., 2006).

12. **Clostridium aldrichii** Yang, Chynoweth, Williams and Li 1990, 270[VP]

al.dri′chi.i. N.L. gen. n. *aldrichii* of Aldrich, named for Henry C. Aldrich, a professor of the University of Florida, Gainesville, for his contributions to ultrastructural research on the strictly anaerobic bacteria.

Cells are rods, 3–5 × 0.5–1.0 µm. Slightly rounded ends, pleomorphic, and occasionally swollen at the ends. Cells are found singly, in pairs, or short chains. Strict anaerobe. Gram-stain-positive. Motile with a bundle of flagella at one end. Oblong spores occur subterminally after 2 weeks, measuring 1–2 × 0.5–1.0 µm. Colonies are round, smooth, semitransparent to grayish white, and 1–2 mm in diameter. Cellulolytic activity creates clear zones around colonies 4–6 mm in diameter and 2–3 mm in width. Growth occurs in the temperature range 20–45 °C with an optimum for growth 35 °C. Growth occurs in the pH range 6.2–7.8 with optimal growth at pH 7.0.

Trypticase, yeast extract, and trace vitamins aid growth. Does not require rumen fluid. Grows on cellulose, xylan, and cellobiose. Ferments cellobiose to acetic, propionic, isobutyric, butyric, isovaleric, lactic, and succinic acids, hydrogen, and carbon dioxide. Hydrolyzes esculin. Fructose, glucose, glycogen, inositol, lactose, maltose, mannitol, mannose, starch, sucrose, and xylose are not utilized. Milk is not clotted, and gelatin is not liquified. Does not produce indole or hydrogen sulfide. Tests negative for nitrate reduction and lecithinase A.

Source: a 3-month-old poplar wood-fed continuously stirred anaerobic digester.

DNA G+C content (mol%): 40 (T_m).

Type strain: P-1, ATCC 49358, DSM 6159, OGI (now OCM) 122.

GenBank accession number (16S rRNA gene): X71846.

Further comments: Clostridium aldrichii is not a member of the genus *Clostridium sensu stricto* and falls in the cluster III as defined by Collins et al. (1994). The 16S rRNA gene sequence of the type strain DSM 6159 shows highest similarity to the species *Acetivibrio cellulolyticus* (ATCC 33288) (98.8%), *Clostridium straminisolvens* (95.9%), *Clostridium alkalicellulosi* (95.5%), and *Clostridium thermocellum* (94.3%). *Clostridium aldrichii* can be differentiated from its closest relatives by its lower temperature range for growth as well as its extensive range of end products of fermentation including propionic, isobutyric, butyric, isovaleric, and succinic aicds.

13. **Clostridium algidicarnis** Lawson, Dainty, Kristiansen, Berg and Collins 1995, 197[VP] (Effective publication: Lawson, Dainty, Kristiansen, Berg and Collins 1994, 156.)

al.gid.i.carn′is. L. adj. *algidus* cold; L. gen. n. *carnis* of meat; N.L. gen. n. *algidicarnis* of cold meat

Rod-shaped cells, 0.5–1.0 × 2.5 μm, occur singly, in pairs, or in chains of 3–4 cells. Anaerobe. Gram-stain-positive. Nonmotile. Colonies are creamy gray in color, raised, convex, and 2–3 mm in diameter. Cells form ovoid terminal spores. Growth occurs in the temperature range 4–40 °C with an optimum at 37 °C.

Ferments fructose, glucose, mannose, ribose, and *N*-acetylglucosamine to produce acetic acid, butyric acid, propionic acid, and gas. Arabinose, cellobiose, glycerol, lactose, maltose, mannitol, melezitose, melibiose, raffinose, rhamnose, salicin, sorbitol, starch, sucrose, and trehalose are not utilized. Cells do not hydrolyze esculin. Sulfite is reduced, but nitrate is not. Cells test negative for urease and lecithinase activity. Cells contain arylamidase acitivity. Cells digest casein weakly, but not gelatin, milk, or cooked meat. Indole is not produced

Source: spoiled, vacuum-packed cooked pork.

DNA G+C content (mol%): not reported.

Type strain: DSM 15099, NCIMB 702931(formerly NCFB 2931).

GenBank accession number (16S rRNA gene): AF127023, X77676.

Further comments: Comparison of the 16S rRNA gene sequence of *Clostridium algidicarnis* shows it to share high similarity with that of *Clostridium putrefaciens* at ~99.0% within cluster I of the clostridia as defined by Collins et al. (1994). At the time of the original description of *Clostridium algidicarnis*, the 16S rRNA gene sequence of *Clostridium putrefaciens* was either not available or not included in the analysis. The comparisons made at the time were between *Clostridium algidicarnis* and *Clostridium tryobutyricum* (Lawson et al., 1994). Further phenotypic and genetic comparisons should

be made between the type strains of the species *Clostridium algidicarnis* and *Clostridium putrefaciens*.

14. **Clostridium algidixylanolyticum** Broda, Saul, Bell and Musgrave 2000a, 629[VP]

al.gi.di.xy.la.no.ly′ti.cum. L. adj. *algidus* cold; Gr. derived N.L. n. *xylanum* xylan; Gr. adj. *lyticus* dissolving; N.L. gen. n. *algidixylanolyticum* cold xylan-dissolving.

Tapered rods, 0.5–0.8 × 1.8–2.8 μm, occurring singly. Motile. Colonies measure 0.8–2.5 mm in diameter. On sheep-blood agar colonies are circular with a grayish white to translucent color. β-Hemolytic. Spores are produced in the late stationary phase of growth and do not cause swelling of the paternal cells. Growth occurs in the temperature range 2.5–32.2 °C with an optimum at 25.5–30 °C. Growth occurs in the pH range 4.7–9.1 with an optimum at 6.8–7.0. Saccharoclastic. Ferments arabinose, cellobiose, fructose, galactose, glucose, inulin, lactose, maltose, mannose, raffinose, rhamnose, salicin, sucrose, and xylose. Fermentation products from growth on PYGS broth include acetate, formate, lactate, ethanol, butyrate, butanol, hydrogen, and carbon dioxide. Starch can be hydrolyzed and xylan degraded.

Source: vacuum-packed, temperature abused raw lamb.

DNA G+C content (mol%): 38.4 (T_m).

Type strain: SPL73, ATCC BAA-156, DSM 12273.

GenBank accession number (16S rRNA gene): AF092549.

Further comments: 16S rRNA gene sequence comparisons place *Clostridium agidixylanolyticum* within cluster XIVa as defined by Collins et al. (1994). It groups with the species *Clostridium aerotolerans* (98.1%), *Clostridium xylanolyticum* (97.8%), *Clostridium saccharolyticum* (97.5%), *Clostridium amygdalinum* (97.4%), and *Clostridium celerecrescens* (97.1%). In the paper describing this species, comparisons were made both to psychrotolerant species as well as these phylogenetically related xylanolytic species (Broda et al., 2000a). Differences in phenotype and fatty acid patterns were used to delineate this species from related species.

15. **Clostridium alkalicellulosi** corrig. Zhilina, Kevbrin, Tourova, Lysenko, Kostrikina and Zavarzin 2006, 925[VP] (Effective publication: Zhilina, Kevbrin, Tourova, Lysenko, Kostrikina and Zavarzin 2005, 564.)

al.kal.i.cell.ulosi. Arabic article *al* the; Arabic n. *qaliy* ashes of saltwort, soda; N.L. n. *cellulosum* cellulose; N.L. gen. n. *alkalicellulosi* of alkaline cellulose, intended to mean that the bacterium utilizes cellulose under alkaline conditions.

Cells are straight or slightly curved rods 0.5–0.7 × 1.1–2.5 μm. Gram-stain-positive, sporeforming and nonmotile. Some strains are motile due to presence of flagella. Spores are 0.7–1.0 μm in diameter, spherical and terminal; resistant to drying and moderately heat-resistant. Growth occurs after heating for 30 min at 80 °C but not 100 °C. On medium containing microcrystalline cellulose colonies are 0.5–1.0 mm in diameter, circular, and flat with a yellowish center and a denser granular edge. On cellobiose, colonies are 1–2 mm in diameter, white, dense, and flat with smooth edges. Mesophile, growing at 18–42 °C with an optimum temperature for growth at 35–45 °C. Obligate extreme alkaliphile growing at pH 8.0–10.2 with optimum pH for growth of 9.0. Sodium ions are obligately required. Weakly halophilic growing at 0.017–0.4M Na$^+$ with an optimum at 0.15–0.3 M Na$^+$.

Strict anaerobe with fermentative metabolism. Catalase-negative. Dinitrogen fixer. Chemoorganoheterotroph. Cellulolytic, utilizing a narrow range of polymers and sugars. Cellulose, cellobiose, and xylan are used as carbon and energy sources. No other mono-, di-, or poly saccharides are utilized. Capable of decomposing plant and algal debris. Yeast extract or vitamins are required for anabolism. Organic acids, alcohols, and protein substrates are not utilized. End products from fermentation of cellobiose and or cellulose are lactate, ethanol, acetate, hydrogen, and traces of formate. Tolerant of up to 48mM Na$_2$S in culture medium. Grows in the presence of kanamycin and neomycin. Growth is suppressed by chloramphenicol, streptomycin, penicillin, ampicillin, ampiox, bacillin, novobiocin, and bacitracin. Inhabits steppe soda lakes with a moderate salt content and usually with coastal vegetation.

Source: the anaerobic cellulose degrading community of Verkhnee Beloe Lake (Buryatiya, Russia).

DNA G+C content (mol%): 29.9–30.2 (T_m).

Type strain: Z-7026, DSM 17461, VKM 2349.

GenBank accession number (16S rRNA gene): AY959944.

Further comments: Strain Z-7026^T was originally described as *Clostridium alkalicellum* (sic) (Zhilina et al., 2005). The original spelling of the specific epithet, *alkalicellum* (sic), has been corrected on validation according to Rule 61. *Clostridium alkalicellulosi* falls within the cluster III of the clostridia as defined by Collins et al. (1994). 16S rRNA gene sequence comparisons show the closest phylogenetic relatives of *Clostridium alkalicellulosi* to be *Clostridium straminisolvens* (96.2%), *Acetivibrio cellulolyticus* (95.7%), *Clostridium thermocellum* (95.4%), and *Clostridium aldrichii* (95.45). *Clostridium alkalicellulosi* can be further differentiated from these related species on the basis of its temperature and pH ranges and optima for growth, production of trace amounts of formate and the relatively low G+C content of its DNA at 29.9 mol%. In contrast to *Clostridium straminisolvens* which can tolerate oxygen, *Clostridium alkalicellulosi* is a strict anaerobe (Kato et al., 2004).

16. **Clostridium aminophilum** Paster, Russell, Yang, Chow, Woese and Tanner 1993, 109^{VP}

a.mi.no'phi.lum. chem. term *amino* amino; G. adj. *philos* loving; N.L. neut. adj. *aminopilum* loving amino acids.

Irregular Gram-stain-positive rods, 1.0 × 1.5 μm. Nonmotile with occasional central or subterminal spores. Cells survive heating to 80°C for 10mins. Obligate anaerobe. Growth occurs in the temperature range 25–45°C and at pH 5.3. Asaccharolytic. Utilizes amino acids or peptides. Preferred carbon sources are glutamine, glutamate, serine, and histidine. Some strains deaminate pyroglutamate. Ferments Casamino acids to ammonia, acetate, butyrate, and traces of lactate and succinate. Hydrogen is not produced. Sodium required for growth. Indole is produced. Sulfate is reduced. Gelatin is not hydrolyzed. Casein is not digested. Tests negative for urease, lipase, lecithinase, and esculinase activity.

Source: fistulated cattle.

DNA G+C content (mol%): 52.5 (T_m).

Type strain: F, ATCC 49906, DSM 10710.

GenBank accession number (16S rRNA gene): L04165.

Further comments: Phylogenetic analysis of the 16S rRNA gene shows *Clostridium aminophilum* to fall within the radiation of cluster XIVa of the clostridia as defined by Collins et al. (1994). *Clostridium aminophilum* repesents a distinct lineage within this cluster and is not closely related to other taxa. The closest relatives are the species *Clostridium sybiosum* (92.1%), *Clostridium celerecrescens* (92.0%), *Clostridium spenoides* (91.9%), and *Clostridium amygdalinum* (91.9%).

17. **Clostridium aminovalericum*** Hardman and Stadtman 1960, 552^{AL}

a.mi′no.va.ler′i.cum. N.L. n. *acidum aminovalericum* aminovaleric acid; L. adj. suff. *-icus* related to; N.L. neut. adj. *aminovalericum* referring to ability to ferment aminovaleric acid strongly.

Cells in PYG broth cultures are straight rods, motile and peritrichous, 0.3–0.5 × 1.5–5.2 μm, occurring singly and in pairs. Cells stain Gram-positive but rapidly become Gram-stain-negative as cultures reach maximum stationary phase. Spores are small, spherical, and terminal, swelling the cell. Sporulation occurs most reliably on chopped-meat agar slants incubated at 30°C for 5d. Cell walls contain *meso*-DAP.

Surface colonies on blood agar plates are 0.5–1mm, circular, entire, flat to convex, translucent to opaque, granular, gray, dull, smooth, and weakly hemolytic.

Cultures in PYG broth are turbid with a smooth sediment and have a pH of 5.9–6.2 after incubation for 1 week. Optimum temperature for growth is 37°C; grows at 25°C and 30°C; poor growth at 45°C. Growth is inhibited by 6.5% NaCl and by 20% bile. Hippurate is hydrolyzed by the type strain. Neutral red and resazurin are reduced. Abundant gas is produced in PYG agar deep cultures. Deoxyribonuclease is present.

Products in PYG broth at a pH of 6.1 include major amounts of acetic acid and abundant H$_2$. At a pH of 7.4–7.7, acetate, ammonia, propionate, and valerate are produced from aminovalerate as the sole energy source. Phenylacetic acid has been detected in the one strain tested. Culture supernatants of the type strain are nontoxic to mice. The type strain is sensitive to chloramphenicol, erythromycin, penicillin G, and tetracycline. Resistance to clindamycin is variable. Other characteristics of the species are given in Table 143.

Source: sewage sludge, rumen contents of bloating calves, urine specimens from pregnant women with bacteriuria, hamster feces, and human feces.

DNA G+C content (mol%): 33 (T_m).

Type strain: ATCC 13725, CIP 104304, DSM 1283, JCM 11016, NCIB 10631.

GenBank accession number (16S rRNA gene): X73436.

Further comments: 16S rRNA gene sequence comparisons show *Clostridium aminovalericum* to fall within the radiation of the cluster XVIa of the clostridia as defined by Collins et al. (1994). The closest relatives based on sequence similarity values include *Clostridium jejuense* (95.2%), *Clostridium xylanolyticum* (93.5%), and *Clostridium populeti* (93.2%). *Clostridium aminovalericum* is distinguished by its ability to grow with aminovalerate as its sole source of energy.

18. **Clostridium amygdalinum** Parshina, Kleerebezem, Sanz, Lettinga, Nozhevnikova, Kostrikina, Lysenko and Stama 2003, 1797^{VP}

a.myg.da.li′num. L. neut. adj. *amygdalinum* made from almonds, referring to the smell of benzaldehyde, which is reduced by the type strain.

Oval or straight rod-shaped cells, 0.5–1.0 × 0.5–10 μm, occurring singly, in pairs, or in chains. Gram-stain-positive. Cells are motile by means of one terminal flagellum, however, older cells lose their motility. Cell chains, as well as swelling cells, are formed in the late-stationary phase of growth. Round, free spores are formed only in nitrogen-free medium. Colonies on agar are circular, approximately 1 mm in diameter, cream with a yellowish elevated center with a slightly undulated margin. Anaerobic, but aerotolerant up to 50% air in gas phase. N_2 not fixed. Moderately thermophilic. Temperature range for growth is 20–60 °C with an optimum at 45 °C. Growth occurs in the pH range 6.5–8.0 with an optimum at pH 7.0–7.5. Substrates utilized as carbon and energy sources include yeast extract, glucose, sucrose, fructose, ribose, arabinose, xylose, melibiose, maltose, cellobiose, crotonate, casitone, pyruvate, lactate, ethanol, inositol, glycerol, mannitol, xylan, betaine, starch, casein, cysteine, serine, and threonine. Galactose, rhamnose, lactose, mannose, gelatin, cellulose, methanol, and $CO_2{:}H_2$ are not utilized. Obligate requirement for yeast extract (1–2 g/l). Major end products formed from yeast extract are H_2, CO_2, and acetate. Minor amounts of propionate, butyrate, and valerate are formed. End products from glucose fermentation are ethanol, acetate, H_2, and CO_2. Benzaldehyde, sulfite, and thiosulfite are used as electron acceptors. Sulfate, dithionite, disulfite, sulfur, and nitrate are not used as electron acceptors. Catalase-negative. Produces indole. Does not liquefy gelatin.

Source: anaerobic-digester sludge.

DNA G+C content (mol%): 32 (HPLC).

Type strain: BR-10, ATCC BAA- 501, DSM 12857.

GenBank accession number (16S rRNA gene): AY353957.

Further comments: Clostridium amygdalinum falls within the radiation of cluster XIVa (Collins et al., 1994) and shows highest 16S rRNA gene sequence similarity to the *Clostridum* species *Clostridium saccharolyticum* (98.9%), *Clostridium indolis* (98.4%), *Clostridium celerecrescens* (98.0%), and *Clostridium aerotolerans* (97.9%). Strain BR-10^T differs from *Clostridium saccharolyticum* in its inability to utilize galactose, rhamnose, lactose, and mannose. *Clostridium amygdalinum* degrades starch while *Clostridium saccharolyticum* does not. The species to which strain BR-10^T is most closely related are mesophiles while *Clostridium amygdalinum* is a moderate thermophile. Strain BR-10^T is also capable of using sulfite and thiosulfate as electron acceptors.

19. **Clostridium arcticum**[*] (*ex* Jordan and McNicol (1979) Cato, George and Finegold 1988, 220^{VP} (Effective publication: Cato, George and Finegold 1986, 1154.)

arc′ti.cum. L. neut. adj. *arcticum* related to the Arctic.

This description is based on the description by Jordan and McNicol (1979) and on the study of Jordan and McNicol strain III, one of the strains included in the original description.

Cells in PYG broth are straight or slightly curved motile rods, Gram-stain-negative, and 0.5–0.7 × 3.2–4.6 μm. They

occur singly or in pairs. Spores are round, terminal, and swell the cell. Cell wall content has not been determined.

There is no growth on blood agar or egg-yolk agar plates. On trypticase soy agar, colonies are pinpoint, circular, convex, and creamy or yellowish. On Jensen's N_2-free medium, colonies are yellow. In PYG deep agar cultures, colonies are white balls with slime that adheres to the sides of the tube.

Cultures grow slowly in PYG broth, which becomes turbid with little sediment; the pH after 7 d is 5.4. The optimum temperature for growth is 22–25 °C. Strains grow at 5 °C and 37 °C but more slowly and not as well. Growth is stimulated by fermentable carbohydrate. Resazurin is reduced. Atmospheric N_2 is fixed. Products in PYG broth culture are propionic and acetic acids. Only traces of H_2 are detected. Lactate in chopped-meat carbohydrate is converted to propionate. Culture supernatants are nontoxic for mice.

Other characteristics of the species are given in Table 144.

Source: arctic soil where the species represented 19% of the anaerobic N_2-fixing strains isolated.

DNA G+C content (mol%): not reported.

Type strain: Jordan and McNicol no. III.

GenBank accession number (16S rRNA gene): not reported.

20. **Clostridium argentinense** Suen, Hatheway, Steigerwalt and Brenner 1988, 380^{VP}

ar.gen.tin.en′se. N.L. neut. adj. *argentinense* coming from Argentina.

Straight to slightly curved rods, 0.5–2.0 × 1.6–9.4 μm. Anaerobic. Motile with peritrichous flagella. Gram-stain-positive. Form spores. Both small, smooth colonies and large, rough colonies form on blood agar. Cells are β-hemolytic on rabbit blood and weakly to not hemolytic on sheep blood. Optimal growth occurs at 35–37 °C and at a pH of 7.0.

Does not ferment sugars. Produces acetic, isobutyric, butyric, isovaleric, and phenylacetic acids. Tests negative for catalase, urease, lipase, deoxyribonuclease, and indole production. Cells are proteolytic and gelatinolytic. Strains may or may not produce a neuroparalytic toxin that causes botulism in laboratory animals. Toxin is neutralized with type G botulinal antitoxin.

Source: soil, amniotic fluid, autopsy specimens, blood, and wounds in Argentina, Switzerland, and the USA.

DNA G+C content (mol%): 28–30 (T_m).

Type strain: ATCC 27322.

GenBank accession number (16S rRNA gene): X68316.

Further comments: The 16S rRNA gene sequence of *Clostridium argentinense* strain ATCC 27322 shows highest similarity to *Clostridium schirmacherense* (99.4%). *Clostridium argentinense* differs from *Clostridium schirmacherense* in its more mesophilic temperature range for growth; *Clostridium schirmacherense* grows at 5–35 °C. In addition, the end products on PYG medium differ in that *Clostridium argentinense* does not produce formate or propionate. *Clostridium schirmacherense* does hydrolyze esculin but is not β-hemolytic. *Clostridium argentinense* also has greater than 99% 16S rRNA gene sequence similarity with type G *Clostridium botulinum* strain ATCC 27322 (M59087).

21. **Clostridium asparagiforme** Mohan, Namsolleck, Lawson, Osterhoff, Collins, Alpert and Blaut 2007, 1933^{VP} (Effective

publication: Mohan, Namsolleck, Lawson, Osterhoff, Collins, Alpert and Blaut 2006, 297.)

as.pa.ra.gi.for′me. L. masc. n. *asparagus* asparagus, L. neut. suffix *-forme* having the shape of, N.L. neut. adj. *asparagiforme* having the shape of asparagus stems.

Rod-shaped cells with tapered ends, $0.6–0.8 \times 2.3–5.0\,\mu m$. Gram-stain-positive. Sporesforming. On Columbia blood agar, nonhemolytic, slightly raised, white, and irregular colonies with a diameter of 2–2.5 mm are formed after 48 h at 37 °C.

Anaerobe. Hydrolyzes glucose, L-tryptophan, and the *p*-nitrophenyl derivatives of β-D-galactopyranoside, α-D-galactopyranoside, β-D-glucouronide, α-D-glucopyranoside, and α-L-fucopyranoside but not D-mannose, arabinose, xylose, and D-raffinose. The major end products of glucose fermentation are acetate, lactate, and ethanol with minor amounts of hydrogen and formate. Using the API ZYM system, activity was detected for esterase, acid phosphatase, esterase lipase, naphthol-AS-BI-phosphohydrolase, β-glucuronidase, and α-glucuronidase. The major cellular fatty acids are $C_{18:1\ DMA\ 9c}$ (15.23%), $C_{16:0}$ (14.84%), $C_{16:1\ DMA\ 9c}$ (13.76%), $C_{16:0\ DMA}$ (8.14%), $C_{17:1\ 8c}$ (6.92%). Other minor components include: $C_{14:0}$ (5.84%), $C_{15:1\ 7c}$ (5.77%) $C_{18:1\ 9c}$ (5.77%), $C_{18:1\ DMA\ 9c}$ (5.42%), $C_{16:0\ aldehyde}$ (3.46%), $C_{16:1\ 9c}$ (3.04%), $C_{18:1\ 11c}$ (2.69%), $C_{17:1\ 8c}$ (2.11%), $C_{14:0\ DMA}$ (1.29%), and $C_{16:1\ DMA\ 7c}$ (1.2%).

Source: human fecal sample.

DNA G+C content (mol%): 53.0 (HPLC).

Type strain: N6, CCUG 48471, DSM 15981.

GenBank accession number (16S rRNA gene): AJ582080.

Further comments: 16S rRNA gene sequence comparisons show *Clostridium asparagiforme* to fall within the radiation of cluster XIVa of the clostridia as defined by Collins et al. (1994). The closest relatives of *Clostridium asparagiforme* based on 16S rRNA gene sequence similarity values are *Clostridium bolteae* (98.5%) and *Clostridium clostridiforme* (97.2%).

22. **Clostridium aurantibutyricum**[*] Hellinger 1944, 46[AL]

au.ran.ti.bu.ty′ri.cum. N.L. adj. *aurantium* orange; N.L. n. *acidum butyricum* butyric acid; N.L. neut. adj. *aurantibutyricum* probably intended to mean the orange-colored organism producing butyric acid.

Cells in PYG broth cultures are motile and peritrichous, straight rods $0.5–0.8 \times 2.8$ to $6.3\,\mu m$, occurring singly and in pairs, Gram-stain-positive, rapidly becoming Gram-stain-negative in older cultures; often granulose positive. Spores are oval and subterminal, swelling the cell. Sporulation occurs most readily on chopped-meat agar slants incubated at 30 °C for 1 week. Cell walls contain *meso*-DAP; cell-wall sugars are rhamnose and traces of glucose, galactose, and mannose.

Surface colonies on blood agar plates are 1–2 mm in diameter, circular to slightly irregular, entire, raised to low convex, translucent, gray to pink-orange, dull, smooth, with a mosaic internal structure. PYG broth cultures are turbid with a heavy, ropy, or viscous sediment and have a pH of 5.4 after incubation for 6 d. The optimum temperature for growth is 37 °C. Moderate growth occurs at 30 °C but not at 25 °C or 45 °C. Growth is inhibited by 6.5% NaCl and by 20% bile. Abundant gas is detected in PYG deep agar cultures.

Neutral red and resazurin are reduced. On egg-yolk agar, lipase is produced with a zone of opacity extending beyond the area of lipase production. There is stormy fermentation in milk; a solid curd is formed with 50% digestion in 3 weeks.

Products of fermentation in PYG broth are acetate and lactate with small amounts of butyrate, propionate, and succinate. Large amounts of butyrate are produced in chopped-meat carbohydrate broth. Large amounts of butanol as well as acetone and 2-propanol are formed. Pectic lyase enzymes and pectinesterase, but no pectic hydrolase, are formed.

The type strain is susceptible to chloramphenicol, clindamycin, erythromycin, penicillin G, and tetracycline. Culture supernatants of the type strain are nontoxic to mice.

Other characteristics of the species are given in Table 142.

Source: rotting hibiscus stumps, flax, rotting potatoes, soil, sewage sludge; bovine, human infant, and adult feces.

DNA G+C content (mol%): 27 (T_m).

Type strain: ATCC 17777, CIP 104305, DSM 793, NCIB 10659.

GenBank accession number (16S rRNA gene): X68183.

Further comments: 16S rRNA gene sequence comparisons show *Clostridium aurantibutyricum* to fall within the radiation of the cluster I of the clostridia as defined by Collins et al. (1994). 16S rRNA gene sequence comparison of the type strain of these species shows it to have highest similarity to the species *Clostridium chartatabidum* (97.9%), *Clostridium paraputrificum* (97.0%), and *Clostridium beijerinckii* (96.5%).

23. **Clostridium baratii**[*] corrig. (Prévot 1938) Holdeman and Moore 1970, 60[AL] (*Inflabilis barati* Prévot 1938, 77; *Clostridium perenne* (Prévot 1940), McClung and McCoy 1957, 673; *Clostridium paraperfringens* Nakamura, Tamai and Nishida 1970, 137.)

ba.ra′ti.i. N.L. gen. n. *baratii* in honor of Barat, French bacteriologist.

Cells in PYG broth are Gram-stain-positive, often granulose positive, nonmotile, straight rods, $0.5–1.9 \times 1.6–10.2\,\mu m$ and usually occur singly, occasionally in pairs. Spores are round to oval, subterminal to terminal, and swell the cell. Strains sporulate poorly and spores may be found more readily in chopped-meat carbohydrate broth or PY broth cultures than in cultures on agar slants or plates. Cell walls contain *meso*-DAP. Surface colonies on blood agar plates are 0.5–2 mm in diameter, circular to irregular, entire to lobate, flat to low convex, granular to mosaic, translucent to opaque, with a smooth, shiny surface. Most strains are β-hemolytic, but some show α- or no hemolysis.

Cultures in PYG broth are turbid with a heavy, sometimes ropy sediment, and have a pH of 4.5–4.8 after incubation for 5 d. Growth is equally abundant at 30 °C, 37 °C, and 45 °C, but less at 25 °C. Growth is inhibited by 6.5% NaCl and by 20% bile.

Abundant gas is produced in PYG deep agar cultures. Ammonia is produced. Neutral red and resazurin are reduced. Acetyl methyl carbinol is formed by four of eight strains tested.

Products of fermentation in PYG broth are butyric, acetic, and lactic acids; smaller amounts of formic, propionic, and succinic acids are sometimes produced. Butanol is not detected. Abundant H_2 is produced. Pyruvate is converted to acetate and butyrate; most strains convert threonine to propionate. Lactate is not utilized. All strains tested are sensitive to chloramphenicol, penicillin G, and tetracycline; nearly all strains are resistant to clindamycin and erythromycin. Culture supernatants are nontoxic to mice.

Other characteristics of the species are given in Table 142.

Source: normal human and rat feces; war wounds; peritoneal fluid, infections of the eye, ear, and prostate; soil; sediments in Puget Sound; soil from Antarctica.

DNA G+C content (mol%): 28 (T_m).

Type strain: ATCC 27638, CCUG 24033, CIP 104306, BCRC 14541, DSM 601, JCM 1385.

GenBank accession number (16S rRNA gene): X68174.

Further comments: 16S rRNA gene sequence comparisons show *Clostridium baratii* to fall within the radiation of cluster I of the clostridia as defined by Collins et al. (1994) and to be most closely related to a number of species of the genus *Eubacterium*, namely *Eubacterium budayi* (99.5%), *Eubacterium nitritogenes* (99.0%), *Eubacterium moniliforme* (98.2%), and *Eubacterium multiforme* (97.7%). These *Eubacterium* species are unrelated to the type species of the genus *Eubacterium*, and further phenotypic characterizations and comparisons of these species and *Clostridium baratii* are required.

24. **Clostridium bartlettii** Song, Liu, McTeague, Summanen and Finegold 2004a, 1425[VP] (Effective publication: Song, Liu, McTeague, Summanen and Finegold 2004b, 182.)

bart.let′ti. i. N.L. gen. n. *bartlettii* to honor John G. Bartlett, for his contributions to the role of intestinal flora in disease and to our knowledge of infectious diseases in general.

Rod-shaped cells, 1.0–1.5 × 5.0–50 μm, forming yellowish, circular, umbonate, entire, dull, opaque colonies on Brucella blood agar plates which have a small zone of β-hemolysis and attain a diameter of 2–3 mm after 48 hours. Gram-stain-positive. Occasional free spores. Obligately anaerobic. In PY broth and PYG broth, major amounts of acetic acid, isovaleric acid, and isobutyric acid are produced by all isolates. The major nonvolatile fatty acid produced is phenylacetic acid; occasionally, lactic acid or methylmalonic acid are also detected. Esculin, gelatin, and urea are not hydrolyzed. Indole is not produced and nitrate is not reduced. Lecithinase, lipase, and catalase are absent. Sensitive to kanamycin (1000 μg) and vancomycin (5 μg) and resistant to colistin (10 μg) special potency identification disks. Resistant to 20% bile in PYG broth. Acid is produced from fructose, glucose, maltose, mannitol, ribose, and sucrose when grown in PRAS carbohydrate broths. Acid is not produced from lactose, mannose, melibiose, melezitose, salicin, or xylose. With the API ZYM system, acid phosphatase, galactosidase, and β-glucuronidase are detected. In the BIOLOG system, the type strain utilizes D-cellobiose, dextrin, D-fructose, gentiobiose, α-D-glucose, glucose-6-phosphate, maltose, maltotriose, D-mannitol, D-mannose, 3-methyl-D-glucose, β-methyl-D-glucoside, palatinose, sucrose, D-trehalose, and turanose. The type strain also utilizes glyoxylic acid, α-keto-butyric acid, pyruvic acid, pyruvic acid methy ester, alaninamide, L-alanine, L-alanyl–L-glutamine, L-alanyl–L-histidine, L-alanyl–L-threonine,

glycyl–L-methionine, L-methionine, L-phenylalanine, L-serine, and L-threonine for growth. The predominant cellular fatty acid produced is $C_{18:1\ 9c}$. Intermediate susceptibility to ampicillin (MIC≤1 μg/ml); susceptible to bacitracin (MIC≤2 μg/ml), cefoxitin (MIC≤2 μg/ml), clindamycin (MIC 0.25 μg/ml), imipenem (MIC≤4 μg/ml), metronidazole (MIC≤1 μg/ml), and vancomycin (MIC 4 μg/ml). Resistant to trimethoprim/sulfamethoxazole 2/1 (MIC≥64 μg/ml trimethoprim).

Source: human fecal material, habitat most likely the human gut.

DNA G+C content (mol%): 29.8 (T_m)

Type strain: WAL 16138, ATCC BAA-827, CCUG 48940.

GenBank accession number (16S rRNA gene): AY438672.

Further comments: 16S rRNA gene sequence shows *Clostridium bartlettii* to fall within the radiation of the cluster XI as defined by Collins et al. (1994). The closest relatives are the *Clostridium* species *Clostridium mayombei* (96.9%), *Clostridium glycolicum* (96.7%), *Clostridium lituseburense* (96.5%), *Clostridium irregulare* (96.2%), and *Clostridium sordellii* (96.0%).

25. **Clostridium beijerinckii**[*] Donker 1926, 145[AL] emend. Keis, Shaheen and Jones 2001, 2100.

beijer.inck′i.i. N.L. gen. n. *beijerinckii* named for M.W. Beijerinck, Dutch bacteriologist.

Cells in PYG broth culture are straight rods with rounded ends, motile and peritrichous, 0.5–1.7 × 1.7–8.0 μm, occurring singly, in pairs, or in short chains. They are Gram-stain-positive, becoming Gram-stain-negative in older cultures. Spores are oval, eccentric to subterminal, and swell the cell, with no exosporium or appendages. Sporulation occurs readily on chopped-meat agar slants incubated at 30 °C. Cell walls contain *meso*-DAP; cell-wall sugars are glucose and galactose. Surface colonies on blood agar plates are 1–5 mm, circular to irregular, entire to scalloped, flat to raised, translucent, gray, shiny, and smooth. Strains may be either α, β, or nonhemolytic.

Cultures in PYG broth are turbid with a smooth to flocculent sediment and have a pH of 4.6–5.4 after incubation for 5 d.

Optimum temperature for growth is 37 °C. Cultures grow well at 30 °C but poorly if at all at 25 °C or 45 °C. Growth is stimulated by a fermentable carbohydrate, inhibited by 6.5% NaCl or 20% bile. Strains are nutritionally fastidious, requiring a complex mixture of growth factors such as are supplied by yeast extract. Abundant gas is detected in deep cultures in PYG agar. Ammonia is produced by 10 of 40 strains tested. Neutral red is reduced; most strains reduce resazurin. Atmospheric N_2 is fixed. Ammonium salts are utilized as the sole N_2 source to produce a wide variety of amino acids including alanine, valine, aspartic acid, and threonine. A neuraminidase is produced by strains of this species. Two of three strains tested produce an extracellular β-glucuronidase. The activity of a ferro-flavoprotein hydrogenase isolated from one strain has been investigated. The emendation of Keis et al. (2001) included the ability to ferment sucrose and the ability of the majority of the strains to utilize the alcohol sugars D- and L- arabitol, dulcitol, and inositol, but glycerol only weakly. All of the strains of *Clostridium beijerinckii* were also able to ferment methylglucopyranoside, turanose, dextrin, and pectin.

Products in PYG broth are butyric and acetic, moderate amounts of succinic, lactic, and formic acids; traces of propionic acid may also be detected. Although not a stable trait, most strains produce substantial amounts of n-butanol, and some produce moderate amounts of acetone or 2-propanol. Pyruvate is converted to butyrate and acetate; neither threonine nor lactate is utilized. All strains tested are sensitive to erythromycin and tetracycline. Of 21 strains, one is resistant to clindamycin, one to chloramphenicol, and one to penicillin G. Culture supernatants are nontoxic to mice.

Other characteristics of the species are given in Table 142.

Source: soil, infected wounds, fermenting olives, spoiled candy, human feces.

DNA G+C content (mol%): 26–28 (T_m).

Type strain: ATCC 25752, CIP 104308, DSM 791, JCM 1390, LMG 5716, NCTC 13035.

GenBank accession number (16S rRNA gene): X68179.

Further comments: 16S rRNA gene sequence comparisons show *Clostridium beijerinckii* to fall within the radiation of cluster I of the clostridia as defined by Collins et al. (1994). The closest relatives based on sequences similarities include *Clostridium diolis* (99.9%), *Clostridium saccharoperbutylacetonicum* (99.1%), *Clostridium puniceum* (98.6%), *Clostridium saccharobutylicum* (98.6%), and *Clostridium butryicum* (97.7%). This species is most easily differentiated from *Clostridium butyricum* which it resembles most closely phenotypically by its requirement for growth factors present in yeast extract. Patterns of soluble cellular proteins of these two species, as determined by polyacrylamide gel electrophoresis, are distinct. Differential fermentation patterns are helpful but not absolute. *Clostridium butyricum* usually ferments ribose and glycerol but not inositol; *Clostridium beijerinckii* usually ferments inositol but not ribose or glycerol.

26. **Clostridium bifermentans**[*] (Weinberg and Séguin 1918) Bergey, Harrison, Breed, Hammer and Huntoon 1923, 323[AL] emend. Chamkha, Bharat, Garcia and Labat 2001c, 195 (*Bacillus bifermentans* Weinberg and Séguin 1918, 128.)

bi.fer.men′tans. L. pref. *bis* twice; L. part. adj. *fermentans* leavening; N.L. adj. *bifermentans* fermenting both carbohydrates and amino acids.

Cells in PYG broth are Gram-stain-positive straight rods, 0.6–1.9 × 1.6–11.0 µm, motile and peritrichous, occurring singly, in pairs, or in short chains. Spores are oval, central to subterminal, and usually do not swell the cell. Sporulation occurs readily both in PY broth and on chopped-meat agar slants. The composition of sporulation medium can affect the chemical content of spores, their germination rate, and their resistance to heat and to chemical agents. Spores have an exosporium. Six different types of spores have been identified depending on the presence, type, or absence of spore appendages. The significance or functional role of these appendages is unknown.

Cell walls of most strains contain *meso*-DAP; in 9 of 32 strains tested, DAP was not detected. Cell-wall sugars of most strains are glucose, rhamnose, and mannose; walls of some strains contain galactose rather than mannose; in some strains only glucose was detected.

Surface colonies on blood agar plates are 0.5–4mm, circular with irregular margins, flat or raised, lobate or scalloped,

translucent or opaque, granular or slightly mottled, gray, shiny, and smooth. Individual colonies can often be seen best on 4% agar plates. Most strains are β-hemolytic. Addition of beef liver catalase to plating media increases the percentage recovery of the species. Cultures in PYG broth are turbid with a heavy, often ropy, sediment. The optimum temperature for growth is 30–37 °C. Most strains grow nearly as well at 25 °C and 45 °C. Growth is inhibited by 6.5% NaCl and by 20% bile. Abundant gas is detected in deep PYG agar cultures. Ammonia is produced. Neutral red is reduced; reduction of resazurin is variable. Fermentation products from PYG broth include large amounts of acetic and formic acids, smaller amounts of isobutyric, isovaleric, isocaproic, hydrocinnamic, benzoic, and propionic acids, and ethyl alcohol. Trace amounts of butyric and phenylacetic acids and propyl and isobutyl alcohols are produced by some strains. In young cultures only acetic and formic acids may be detected. Abundant H_2 is produced. The glucose analog 1,2-*O*-isopropylidene-β-glucofuranose ("monoacetone glucose") is utilized as a carbon source with different proportions of volatile fatty acids produced from those detected from glucose; a larger percentage of propionic, butyric, isovaleric, and valeric acids, and a smaller percentage of isobutyric acid. Valine is converted to isobutyrate, leucine to isovalerate and isocaproate, isoleucine to isovalerate, and threonine to propionate. Pyruvate is converted by most strains to acetate and formate; excess isobutyrate, isovalerate, and isocaproate may be produced. Lactate is not converted to propionate. Proline, serine, threonine, arginine, and aspartate are utilized; δ-aminovalerate and α-aminobutyrate are produced. DNase is produced by 10 strains tested.

All strains tested are susceptible to chloramphenicol, erythromycin, and penicillin G. One of 72 strains is resistant to clindamycin; one strain (a different one) is resistant to tetracycline. Three strains tested are sensitive to 8µg of nalidixic acid/ml. Culture supernatants are nontoxic to mice.

Other characteristics of the species are given in Table 144.

Source: soil; fresh water; marine sediments; human feces; normal cervical flora; snake venom; a goat stomach ulcer; wounds in horses and sheep; clinical specimens including wounds, abscesses, and blood; clam gut; cheese fondue; canned tomatoes; vacuum packed smoked fish.

DNA G+C content (mol%): 27 (T_m).

Type strain: ATCC 638, CCUG 36626, BCRC 14542, CIP 104309, DSM 14991, JCM 1386, NCIMB 10716, NCTC 13019.

GenBank accession number (16S rRNA gene): AB075769, X75906.

Further comments: 16S rRNA gene sequence shows *Clostridium bifermentans* to fall within the radiation of the cluster XI as defined by Collins et al. (1994). The closest relatives are *Clostridium ghonii*, *Eubacterium tenue*, and *Clostridium sordellii*, all sharing 97.7% 16S rRNA gene sequence similarity.

27. **Clostridium bolteae** Song, Liu, Molitoris, Tomzynski, Lawson, Collins and Finegold 2003a, 935[VP] (Effective publication: Song, Liu, Molitoris, Tomzynski, Lawson, Collins and Finegold 2003b, 88.)

bolt.e.ae. N.L. gen. n. *bolteae* to honor the American Ellen Bolte, who first proposed a bacterial role in late-onset autism and stimulated work in this area.

Rod-shaped cells, 1.0–1.2 × 2.0–5.0 μm, that form occasional subterminal spores. Gram-stain-positive. Obligately anaerobic. Colonies are 2–3 mm in diameter, gray, waxy, circular with slightly scalloped edges, slightly raised, smooth, dull, and opaque after anaerobic incubation for 48 hours at 37 °C when grown on Brucella blood agar.

Saccharolytic. Glucose metabolism produces acetate and lactate as end products. Acid is produced from arabinose, fructose, glucose, glycerol, maltose, mannose, melezitose, sorbitol, sucrose, trehalose, and xylose. Acid is not produced from erythritol, inulin, lactose, mannitol, melibiose, ribose, or salicin. Esculin, gelatin, and urea are not hydrolyzed. Lecithinase- and lipase-negative. Indole-negative. Nitrate is not reduced. Using the API ZYM system, acid phosphatase, alkaline phosphatase (weak reaction), α-galactosidase, β-galactosidase, and α-glucosidase are detected. The predominant cellular fatty acids produced are $C_{16:0}$, $C_{14:0}$, and $C_{16:1\ DMA\ 9c}$. Susceptible to kanamycin (1000 μg) and colistin sulfate (10 μg) identification disks; susceptible or had intermediate susceptibility to bacitracin (MIC ≤ 2 μg/ml), cefoxitin (MIC 32 μg/ml), clindamycin (MIC ≤ 4 μg/ml), imipenem (MIC ≤ 4 μg/ml), metronidazole (MIC 0.25 μg/ml), ramoplanin (MIC ≤ 16 μg/ml), trimethoprim/sulfamethoxazole 2/1 (MIC < 16 μg/ml trimethoprim), and vancomycin (MIC 2 μg/ml). Resistant to ampicillin (MIC > 2 μg/ml) and usually resistant to piperacillin and ticarcillin and to the vancomycin (5 μg) identification disk.

Source: human gut, human fecal material, blood, and intra-abdominal abscess.

DNA G+C content (mol%): 50.5 (HPLC).

Type strain: WAL 16351, ATCC BAA-613, CCUG 46953.

GenBank accession number (16S rRNA gene): AJ508452.

Further comments: 16S rRNA gene sequence shows *Clostridium bolteae* to fall within the radiation of the cluster XIVa as defined by Collins et al. (1994). *Clostridium bolteae* is most closely related to the species *Clostridium asparagiforme* (98.5%), *Clostridium clostridioforme* (98.2%), and *Clostridium citroniae* (96.9%). *Clostridium bolteae* was differentiated from *Clostridium clostridioforme* on the basis of its lack of acid production from lactose and salicin, its acid production from melezitose and sorbitol, and its lack of esculin hydrolysis and β-glucouronidase production.

28. **Clostridium botulinum**[*] (van Ermengem 1896) Bergey, Harrison, Breed, Hammer and Huntoon 1923, 328[AL] (*Bacillus botulinus* van Ermengem 1896, 443.)

bo.tu.li′num. L. n. *botulus* sausage; N.L. adj. *botulinum* pertaining to sausage.

The species includes seven toxin types, A, B, C, D, E, F, and G, differentiated by the antigenic specificity of their individual toxins. All strains of the species produce neurotoxins with similar effects on an affected host, but the toxins of the different types are serologically distinct. These distinctions do not necessarily correlate with observed phenotypic differences. The species was divided into three metabolic groups by Holdeman and Brooks (1970). Strains of *Clostridium botulinum* type G, described by Gimenez and Ciccarelli (1970), are metabolically distinct and were placed in a separate group by Smith and Hobbs (1974). These groups and other species that are phenotypically similar are: (a) strains

of type A, proteolytic strains of types B and F, and *Clostridium sporogenes*; (b) strains of type E and saccharolytic strains of types B and F; (c) strains of types C and D and *Clostridium novyi* type A; and (d) strains of type G and *Clostridium subterminale*. The validity of this grouping has been confirmed by the data of Johnson and Francis (1975) in that the metabolic types correlate well with rRNA homology groups I-F, I-A, I-H, and I-K, respectively. The toxins of all types are pathogenic to laboratory animals through the action of a neurotoxin. Some toxins, particularly those of the nonproteolytic strains, require trypsin activation for effectiveness in laboratory toxin testing. Human disease (botulism) with similar symptoms is caused by toxins elaborated under anaerobic conditions usually by colonization of food, less often by colonization of wounds, or colonization of the intestinal tract as in infant botulism (Wilcke et al., 1980).

This description is based on those by Smith and Hobbs (1974), Holdeman et al. (1977a), and on the study of the type and 15 other strains of *Clostridium botulinum* type A, 14 proteolytic strains of type B, 2 saccharolytic strains of type B, 24 strains of type C, 5 strains of type D, 9 strains of type E, 3 proteolytic strains of type F, 4 saccharolytic strains of type F, and 6 strains of type G.

(a) Type A and proteolytic strains of types B and F. Cells in PYG broth are usually motile and peritrichous, straight to slightly curved rods, 0.6–1.4 × 3.0–20.2 μm. Spores are oval and subterminal and swell the cell. Sporulation occurs most readily on egg-yolk agar plates incubated for 2 d or on chopped-meat agar slants incubated at 30 °C for 1 week.

Cell walls contain *meso*-DAP and glucose. A cell-wall protein with a common antigenic specificity has been isolated from each of these types. Surface colonies on blood agar plates are 2–6 mm in diameter, circular to irregular with a scalloped or rhizoid margin, flat to raised, translucent to semiopaque, gray, often with a mottled or crystalline internal structure, and are β-hemolytic.

Cultures in PYG broth are turbid with a smooth or flocculent sediment and have a pH of 5.6–6.2 after incubation for 1 week. Ammonia produced from the deamination of amino acids often masks acid production from carbohydrates.

Optimum temperature for growth is 30–40 °C. Some strains grow well at 25 °C and a few at 45 °C. Growth is inhibited by 6.5% NaCl, 20% bile, and at a pH of 8.5. The bile acids litho-cholic and chenodeoxycholic are the most inhibitory. Strains of *Clostridium perfringens* and *Clostridium sporogenes* isolated from soil also can inhibit growth.

Toxin production is delayed in an atmosphere of 100% CO_2 and pressurized CO_2 is lethal to strains, depending on the amount of pressure and the length of exposure.

Gelatin, milk, and meat are digested. Ammonia and H_2S are produced.

Strains of *Clostridium botulinum* type A and type B reduce proline to δ-aminovalerate. Arginine, glycine, phenylalanine, serine, tyrosine, and tryptophan are also utilized for growth; valine, α-aminobutyrate, and γ-aminobutyrate are produced. The aromatic amino acids, phenylalanine, tyrosine, and tryptophan are reduced to phenylpropionic, *p*-hydroxyphenylpropionic, and indolepropionic acids, respectively. *Clostridium botulinum* type A and proteolytic types B and F convert valine to isobutyrate, and isoleucine and leucine to isovalerate; types A and B, but not type F, produce isocaproate from leucine.

Fermentation products in PYG broth include large amounts of butyric and acetic acids with moderate amounts of isobutyric and isovaleric acids. Isocaproic, propionic, and valeric acids, and ethanol, propanol, and butanol may also be detected. Hydrocinnamic acid is produced in trypticase soy broth cultures by all strains of types A and (presumably proteolytic) strains of types B and F tested; abundant H_2 gas is produced. Pyruvate is converted to acetate, butyrate, ethanol, and butanol; excess amounts of isovalerate from those produced in basal medium may also be detected. Most strains convert threonine to propionate; lactate is not utilized.

All strains are sensitive to chloramphenicol, penicillin G, and tetracycline. One strain of proteolytic type B is resistant to clindamycin and erythromycin; all other strains are sensitive to these antibiotics. Types A and B are resistant to cycloserine, sulfamethoxazole, and trimethoprim. Sensitive *in vitro* to tetracycline (0.5 μg/ml), metronidazole (1 mg/ml), penicillin (4 μg/ml), rifampin (2 μg/ml), and erythromycin (4 μg/ml); more than 90% are sensitive to chloramphenicol (4 ug/ml), clindamycin (4 μg/ml), cefoxitin (1 μg/ml), and vancomycin (8 μg/ml); they are resistant to nalidixic acid and gentamicin.

Culture supernatants are toxic to mice. Culture supernatants of type A are toxic to chickens, turkeys, pheasants, and peafowl. Although bacteriophages have been demonstrated in all types and two phages have been isolated from a strain of type A, there has been no report of the mediation of toxin production by phage in these strains.

Other characteristics are given in Table 142.

Source: soil, marine and lake sediments; animal, bird, and fish intestines; food (particularly improperly preserved vegetables, meat, and fish). (The types isolated reflect those present in the soil or sediment of the area. *Clostridium botulinum* toxin type F has been implicated in one case of infant botulism, but otherwise only types A and B have been reported. These two types also are most frequently isolated in outbreaks of food poisoning and cases of wound botulism.)

DNA G+C content (mol%): 26–28 (T_m).

Type strain: Type A, ATCC 25763, CIP 104310, NCIB 10640.

GenBank accession number (16S rRNA gene): L37585.

Reference strains: Proteolytic type B, ATCC 7949 (NCIB 10657); proteolytic type F, ATCC 25764 (NCIB 10658) (GenBank accession number L37593).

Further comments: Clostridium botulinum type A and proteolytic strains of types B and F have high DNA–DNA homology with *Clostridium sporogenes* and cannot be distinguished metabolically or biochemically. Strains of these types are identified by toxin neutralization tests in mice. Although cellular protein electrophoretic patterns of the proteolytic *Clostridium botulinum* strains cannot be used to identify the toxin type, they are distinct from patterns given by strains of nontoxic *Clostridium sporogenes* as are gas chromatographic patterns of trimethylsilyl derivatives of whole-cell hydrolysates of strains of the two species. 16S rRNA gene sequence comparisons show these strains to cluster with the type strain of *Clostridium sporogenes* with greater than 99.5% sequence similarity (Hutson et al., 1993; Figure 128).

(b) Type E and saccharolytic strains of types B and F. Cells in PYG broth cultures are motile and peritrichous, straight rods, 0.8–1.6 × 1.7–15.7 μm, occurring singly and in pairs. Spores are oval, eccentric to subterminal, and usually swell the cell. Sporulation occurs readily in broth and on solid media. Cell walls contain *meso*-DAP.

Surface colonies on blood agar plates are β-hemolytic, 1–5 mm in diameter, irregular with lobate or scalloped margins, raised, translucent to opaque, gray-white, with a mottled or mosaic internal structure.

Cultures in PYG broth are turbid with a smooth sediment, and have a pH of 5.2–5.5 after incubation for 1–2 d.

Optimum temperature for growth ranges from 25–37 °C. Little or no growth occurs at 45 °C. Growth is stimulated by a fermentable carbohydrate and is inhibited by 6.5% NaCl, 20% bile, or a pH of 8.5. Nonproteolytic strains of types B and F and toxic strains of type E are inhibited by soil strains of *Clostridium perfringens*, while type E strains that have lost toxicity are unaffected. A boticin isolated from a nontoxigenic strain of *Clostridium botulinum* type E inhibited both vegetative growth and spore germination of 10 out of 12 toxic and 3 out of 6 nontoxic strains of *Clostridium botulinum* E, as well as 2 out of 2 strains of nonproteolytic *Clostridium botulinum* type B. Plasmids that may be related to production of "boticin E" have been demonstrated in both toxic type E and nontoxic type E-like strains. Gelatin is digested but milk and meat are not.

Fermentation products in PYG broth cultures are butyric and acetic acids. Large amounts of H_2 are detected in headspace gas. Most strains convert pyruvate to acetate and butyrate. Neither lactate nor threonine is utilized.

All strains tested are susceptible to chloramphenicol, clindamycin, erythromycin, penicillin, and tetracycline. They also are susceptible to metronidazole, rifampin, cefoxitin, and vancomycin, but are resistant to nalidixic acid and gentamicin (Swenson et al., 1980).

Culture supernatants are toxic to mice. Supernatants of strains of type E are toxic to gallinaceous birds. Other characteristics are given in Table 142.

Source: soil, marine and lake sediments, food, fish, birds, mammals.

DNA G+C content (mol%): 27–29 (T_m).

Reference strains: Nonproteolytic type B, ATCC 25765 (NCIB 10642) (GenBank accession number 16S rRNA: X68173); nonproteolytic type F, ATCC 27321 (NCIB 10641); and type E, ATCC 9564 (NCIB 10660).

Further comments: 16S rRNA gene sequence comparisons show these strains to form a distinct cluster unrelated to the type strain of this species (Hutson et al., 1993) and clustering with *Clostridium uliginosum* (Figure 128).

(c) Type C and type D. Cells in PYG broth are straight rods, motile and peritrichous, 0.5–2.4 × 3.0–22.0 μm, and occur singly or in pairs. Spores are oval, subterminal, and swell the cell. Sporulation of most strains occurs most readily on chopped-meat agar slants incubated at 30 °C for 1 week. Cell walls contain *meso*-DAP.

Surface colonies on blood agar plates are β-hemolytic, 1–5 mm in diameter, circular to slightly irregular, slightly scalloped or lobate, flat to raised, translucent, gray-white, with a mottled or mosaic internal structure.

Cultures in PYG broth are turbid with a smooth or flocculent sediment and have a pH of 5.2–5.7 after incubation for 1–2 d.

Optimum temperature for growth is 30–37 °C; most strains grow well at 45 °C and poorly, if at all, at 25 °C. Growth is stimulated by a fermentable carbohydrate but is inhibited by 6.5% NaCl, 20% bile, or a pH of 8.5. There was no inhibition of growth by strains of *Clostridium perfringens* isolated from soil samples from the USA, but several species of *Bacillus* isolated from samples of mud from England, France, and Spain were inhibitory to growth of a strain of *Clostridium botulinum* type C.

Gelatin is digested; milk is acidified, curdled, and digested by 20 of 29 strains tested; meat is digested by 20 of 28 strains tested. Production of ammonia and H$_2$S varies among strains.

Strains of type C utilize glutamic acid, serine, glycine, arginine, and aspartic acid; δ-aminovalerate is not produced.

Products in PYG broth are butyric, propionic, and acetic acids; traces of valeric, succinic, and lactic acids may be detected. Abundant H$_2$ is formed. Lactate is converted to propionate. Pyruvate is converted to acetate and butyrate; propionate is formed from pyruvate by some strains. Five of 24 strains convert threonine to propionate.

All strains tested are susceptible to chloramphenicol, clindamycin, erythromycin, penicillin G, and tetracycline. They also are susceptible to metronidazole, rifampin, cephalothin, and cefoxitin, but resistant to nalidixic acid and gentamicin.

Culture supernatants are toxic to mice; culture supernatants of strains of type C are toxic to gallinaceous birds; strains are pathogenic for laboratory animals. The toxin of *Clostridium botulinum* type C is inactivated by bacteria in the rumen of cattle and sheep, but the specific organisms involved are not known. Toxin production by *Clostridium botulinum* type C and type D is phage-mediated, and the specific type of toxin produced is determined by the specific phage with which the culture is infected. *Clostridium botulinum* type C can be cured of type C phage and of its toxin, then converted to *Clostridium novyi* type A following infection by a *Clostridium novyi* type A phage.

Other characteristics of these types are given in Table 142.

Source: feces and carcasses of animals and birds, soil, lake mud, rotting vegetation.

DNA G+C content (mol%): is: 26–28 (T_m).

Reference strains: Type C, ATCC 25766 (NCIB 10618); Type D, ATCC 25767 (NCIB 10619).

Further comments: These species can be identified by toxin neutralization testing. They are most easily differentiated from *Clostridium novyi* type A, which they resemble phenotypically, by distinct patterns of cellular proteins produced by polyacrylamide gel electrophoresis. 16S rRNA gene sequence comparisons show these strains to cluster with the type strain of *Clostridium noyi* with greater than 99.0% sequence similarity (Hutson et al., 1993; Figure 128).

Type G. Cells in PYG broth cultures are straight rods, motile and peritrichous, 1.3–1.9 × 1.6–9.4 μm, and occur singly and in pairs. Spores, though rarely seen, are oval, subterminal, and swell the cell. We have not detected spores from any medium. Cell-wall composition has not been reported.

Surface colonies on blood agar plates are β-hemolytic, 1–4 mm in diameter, circular to irregular, lobate to filamentous, raised, translucent, smooth, and shiny. A spreading film may cover the entire plate. Large, rough, fried-egg colonies may be formed.

Cultures in PYG broth are turbid with a smooth white sediment and have a pH of 6.2–6.3 after 5 d incubation.

Optimum temperature for growth is 30–37 °C. Cultures grow almost as well at 25 °C and 45 °C. Growth is inhibited by 6.5% NaCl and by 20% bile. The reference strain is inhibited by three strains of *Clostridium perfringens* isolated from soil. Moderate gas is detected in PYG deep agar cultures.

Products in PY broth are acetic, butyric, isovaleric, isobutyric, and phenylacetic acids and butyl and ethyl alcohols. When products are converted to butyl esters, hydroxyphenylacetic acid is also detected. Abundant H$_2$ is formed. Pyruvate is converted to acetate and ethyl alcohol; lactate and threonine are not utilized. Valine is converted to isobutyric acid and leucine to isovaleric acid. Indoleacetic acid is produced and lysine is utilized.

Gelatin and casein are digested rapidly; milk and meat are digested within 3 weeks. Ammonia and H$_2$S are produced.

The type strain is susceptible to chloramphenicol, clindamycin, erythromycin, penicillin G, and tetracycline. It also is susceptible to metronidazole, rifampin, cephalothin, and cefoxitin, but is resistant to vancomycin, nalidixic acid, and gentamicin.

Culture supernatants are toxic to mice. Monkeys, chickens, guinea pigs, and mice are susceptible to the toxin; sheep and dogs are resistant.

Other characteristics are given in Table 142.

Source: soil, human autopsy specimens.

DNA G+C content (mol%): not reported.

Reference strains: ATCC 27322, NCIB 10714.

GenBank accession number (16S rRNA gene): M59087.

Further comments: Unlike all other types of *Clostridium botulinum*, strains of type G do not produce lipase on egg-yolk agar. They have been distinguished from strains of *Clostridium subterminale*, which they resemble most closely phenotypically, by their toxicity for mice, but they can be differentiated readily by their distinct patterns of soluble cellular protein shown by polyacrylamide gel electrophoresis.

In any other group of organisms, this species would have been divided into four separate species because of the distinct differences in metabolic activity exhibited by strains in the four groups and the lack of DNA homology among groups. However, because of the unique and similar action of the toxins produced by all strains and to facilitate communication between the microbiological and medical professions, they have been retained in one species. 16S rRNA gene sequence comparisons show these strains to cluster with the type strain of *Clostridium argentinense* with greater than 99.0% sequence similarity (Hutson et al., 1993).

29. **Clostridium bowmanii** Spring, Merkhoffer, Weiss, Kroppenstedt, Hippe and Stackebrandt 2003, 1027^{VP}

bow.ma′ni.i. N.L. gen. n. *bowmanii* referring to Bowman, in honor of the microbiologist

John P. Bowman, who has made important contributions to our knowledge of the diversity of psychrophilic bacteria.

Rod-shaped, 1.0–1.2 × 2.0–8.0 μm, occurring singly, in pairs or short chains. Gram-stain-positive. Motile by peritrichous flagella. Endospores are spherical and located in a terminal to subterminal position; sporangium is not swollen. When grown on solid media, cultures frequently produce filamentous cells. Colonies on sheep-blood agar are 1–2 mm in diameter, round with often coarsely granulated margins, smooth, slightly raised, cream-white to grayish, semi-transparent to opaque and nonhemolytic.

The optimum temperature range for growth is 12–16 °C with an upper limit is 20 °C. The pH range for growth is 5.6–7.4 with optimal growth occurring at pH 6.8–7.2. Under optimal conditions, the doubling time is 8.4 hours.

The following carbohydrates are utilized: fructose, galactose, glucose, inulin, maltose, mannose, ribose (weak), salicin, sucrose, trehalose, and xylose. The following carbohydrates are not utilized: amygdalin, arabinose, cellobiose, glycogen, inositol, lactose, mannitol, melezitose, melibiose, raffinose, rhamnose, sorbitol, and starch. Gelatin and starch are not hydrolyzed. The fermentation products formed are butyrate, acetate, formate, ethanol, lactate, 1-butanol, hydrogen, and carbon dioxide. The cell-wall peptidoglycan contains *meso*-diaminopimelic acid. The major cellular fatty acids are $C_{16:1\ \omega 9c}$, $C_{14:0}$, $C_{16:1\ DMA\ \omega 9c}$, $C_{16:0}$, $C_{16:1\ \omega 11c}$, and $C_{16:1\ \omega 7c}$.

Source: a microbial mat sample taken from a moated area around Lake Fryxell, Antarctica.

DNA G+C content (mol%): 32.0 (HPLC).

Type strain: A-1/C-an/C1, ATCC BAA-581, DSM 14206.

GenBank accession number (16S rRNA gene): AJ506119 (clone 11), AJ506120 (clone 16).

Further comments: 16S rRNA gene sequence analysis shows *Clostridium bowmanii* to fall within cluster I of the clostridia as defined by Collins et al. (1994). Two 16S rRNA gene sequence clones were obtained for strain DSM 14206[T] which shared >99.0% sequence similarity (Spring et al., 2003). Sequence similarity values in the range 98.2–98.8% are found between *Clostridium bowmanii* and the species *Clostridium lacusfryxellense*, *Clostridium estertheticum* subsp. *laramiense*, and *Clostridium estertheticum* subsp. *estertheticum*. *Clostridium bowmanii* can be differentiated from these related species based on substrate utilization patterns as well as end products of fermentation. Of these species *Clostridium bowmanii* also has a higher temperature optimum for growth (Spring et al., 2003). DNA–DNA reassociation studies demonstrated the distinct species status of *Clostridium bowmanii* with a values of 28–47% to the related species (Spring et al., 2003).

30. **Clostridium cadaveris**[*] (Klein 1899) McClung and McCoy 1957, 672[AL] (*Bacillus cadaveris* Klein 1899, 280.)

ca.dav′er.is. L. n. *cadaver* dead body; L. gen. n. *cadaveris* of a corpse.

Cells in PYG broth are Gram-stain-positive straight rods, usually motile and peritrichous, 0.5–1.3 × 1.4–9.4 μm, and occur singly or in pairs. Spores are oval, terminal, and swell the cell. Subterminal spores are seen occasionally. Sporulation of most strains occurs most readily on chopped-meat agar slants incubated at 30 °C for 1 week.

Cell walls of most strains contain *meso*-DAP; no DAP was detected in 5 of 12 strains. Surface colonies on blood agar plates are 0.5–3 mm, circular, entire to slightly scalloped, convex, translucent to opaque, smooth, and shiny. Hemolysis is variable.

Cultures in PYG broth are turbid with a smooth sediment and have a pH of 4.9–5.4 after incubation for 5 d.

The optimum temperature for growth is 30–37 °C. Most strains grow nearly as well at 25 °C and 45 °C. Growth is stimulated by fermentable carbohydrate and inhibited by 6.5% NaCl. Most strains are inhibited by 20% bile.

Ammonia and H_2S are produced. Neutral red is reduced; most strains reduce resazurin. Hippurate is hydrolyzed by 5 of 30 strains tested. Abundant gas is detected in PYG deep agar cultures. Deoxyribonuclease is formed.

Fermentation products in PYG broth include large amounts of butyric and acetic acids along with ethyl and butyl alcohols. Traces of other alcohols and small amounts of lactic, formic, propionic, and succinic acids may be detected. Abundant H_2 is formed. In chopped-meat broth or PY broth cultures containing at least 2% peptone, large amounts of acetate and butyrate and moderate amounts of isobutyrate and isovalerate are produced. Lipid fatty acids of the type strain detected in a trypticase-yeast extract-thioglycolate medium without glucose have been identified. Major fatty acids formed are normal saturated C_{16}, C_{14}, C_{18}, and C_{12} acids with smaller amounts of unsaturated C_6 acids with a double bond at the 7 or 9 position. Pyruvate is converted to acetate and butyrate; neither lactate nor threonine is converted to propionate. There is some increase in the amount of isobutyrate produced in valine broth without glucose but no utilization of leucine or isoleucine.

All strains tested are susceptible to penicillin G; of 37 strains, 4 are resistant to erythromycin and clindamycin, 5 are moderately resistant to clindamycin alone, 3 are resistant to tetracycline, and 2 are resistant to chloramphenicol. Three clinical isolates tested are susceptible to cefoxitin (1 μg/ml), moxalactam (4 μg/ml), cefoperazone (4 μg/ml), clindamycin (2 μg/ml), and metronidazole (0.25 μg/ml).

Culture supernatants are nontoxic to mice. Phage-like particles that inhibited growth of a strain of *Clostridium septicum* have been demonstrated in one strain.

Other characteristics of the species are given in Table 142.

Source: soil, marine sediment, animal and human feces, snake venom, human clinical specimens from abscesses, wounds, and blood.

DNA G+C content (mol%): 27 (T_m).

Type strain: ATCC 25783, CCUG 24035, CIP 104314, CNCIB 10676, DSM 1284, JCM 1392.

GenBank accession number (16S rRNA gene): M59086.

Further comments: Fermentation of glucose, production of indole, and lack of lecithinase activity phenotypically differentiate *Clostridium cadaveris* from other clostridial species. The formation of isobutyrate and isovalerate from PY broth but not from PYG broth is most helpful in identification. 16S rRNA gene sequence comparisons show *Clostridium cadaveris* to represent a distinct species within cluster I of the clostridia as defined by Collins et al. (1994). The species to which it shows highest 16S rRNA gene sequence similarity are *Clostridium friridicarnis* (93.6%), *Clostridium diolis* (92.7%), *Clostridium beijerinckii* (92.7%), and *Clostridium subterminale* (92.6%).

31. **Clostridium caminithermale** Brisbarre, Fardeau, Cueff, Cayol, Barbier, Cilia, Ravot, Thomas, Garcia and Ollivier 2003, 1046[VP]

ca.mi.ni.ther.ma′le. L. n. *caminus* chimney; L. pl. n. *thermae* hot springs; N.L. neut.adj. *caminithermale* of a thermal chimney, describing the site of sampling.

Strictly anaerobic rods, 0.4–0.5 × 5–9 μm, and occurring singly or in pairs. Gram-stain-positive. The cell wall is composed of three dense layers with two thick layers and a middle thinner layer separated by two less dense spaces. Motile by a few laterally inserted flagella.

Optimum temperature range for growth is 20–58 °C with an optimum at 45 °C. Grows in the presence of sea salt at 12–55 g/l and optimally at 30 g/l. The optimum pH for growth is 6.6, with growth occurring at pH 5.8–8.2.

Heterotrophic. Yeast extract is required for growth on sugars. Ferments yeast extract, peptone, Bio-trypticase, and Casamino acids into a mixture of volatile fatty acids. Fructose, galactose, glucose, glycerol, maltose, mannose, and ribose are fermented primarily into acetate, butyrate, propionate, and H_2 + CO_2. Succinate, fumarate, and pyruvate are also fermented. The following substrates are not used: arabinose, cellobiose, lactose, mannitol, melibiose, raffinose, rhamnose, starch, sucrose, xylose, dulcitol, sorbitol, lactate, formate, acetate, propionate, and H_2 + CO_2. The following amino acids are used as energy sources in the presence of yeast extract and peptone (0.5 g/l): arginine, cysteine, glycine, praline, and tyrosine (oxidized to acetate); glutamic acid (to propionate); histidine (to propionate and formate); isoleucine (to methyl 2-butyrate); leucine (to isovalerate); lysine (to acetate and butyrate); and methionine (to propionate and acetate). The following amino acids are not used: alanine, asparagine, aspartic acid, glutamine, phenylalanine, serine, threonine, tryptophan, and valine. Performs the Stickland reaction, using isoleucine as electron donor and methionine or betaine as electron acceptors. Does not use elemental sulfur, sulfate, thiosulfate, sulfite, nitrate, or nitrite as an electron acceptors. The following tests were negative: β-galactosidase, arginine dihydrolase, lysine and ornithine decarboxylases, Simmons' citrate, H_2S production, urease, tryptophan deaminase, indole, and acetoin production (Voges–Proskauer reaction). Adverse effects on animals and humans are not known. Because of the ability to degrade amino acids and peptides, the possibility of harmful effects cannot be excluded. Cautious handling and autoclaving of cultures before disposal is recommended.

Source: an Atlantic Ocean hydrothermal chimney.

DNA G+C content (mol%): 33.1 (HPLC).

Type strain: DVird3, CIP 107654, DSM 15212.

GenBank accession number (16S rRNA gene): AF458779.

Further comments: 16S rRNA gene sequence comparisons show *Clostridium caminithermale* to fall within the cluster XI of the clostridia as defined by Collins et al. (1994). Strain DVird3[T] forms a distinct lineage with the closest related *Clostridium* species being *Clostridium halophilum* at on 92.8% sequence similarity. *Clostridium caminithermale* is related at the 91.7% and 91.4% to the species *Caminicella sporogenes* and *Anerovirgula multivorans*, respectively. The addition of this strain to the genus *Clostridium*, although it was not a member of cluster I as defined by Collins et al. (1994), was on the basis of its relatedness to *Clostridium halophilum*. Strain DVird[T] is differentiated from *Clostridium halophilum* on the basis of differences in substrate utilization patterns, temperature, and NaCl range for growth (Brisbarre et al., 2003).

32. **Clostridium carboxidivorans** Liou, Balkwill, Drake and Tanner 2005, 2089[VP]

car.bo.xi.di.vo′rans. N.L. neut. n. *carboxidum* carbon monoxide; L. part. adj. *vorans* devouring; N.L. part. adj. *carboxidivorans* carbon monoxide-devouring.

Rod-shaped cells, 0.5×3.0 μm, occurring singly and in pairs and motile. Gram-stain-positive. Cells rarely sporulate, but spores are subterminal to terminal with slight cell swelling. Obligate anaerobe. Optimum growth temperature is 38 °C and an optimum pH for growth of 6.2.

Grows autotrophically with H_2/CO_2 or CO and chemoorganotrophically with ribose, xylose, fructose, glucose, galactose, L-arabinose, mannose, rhamnose, sucrose, cellobiose, trehalose, melezitose, pectin, starch, cellulose, inositol, mannitol, glycerol, ethanol, propanol, 2-propanol, butanol, citrate, serine, alanine, histidine, glutamate, aspartate, asparagine, Casamino acids, betaine, choline,and syringate. Methanol, D-arabinose, fucose, maltose, lactose, raffinose, melibiose, amygdalin, sorbitol, gluconate, lactate, malate, succinate, and arginine do not support growth. The end products of metabolism are acetic acid, ethanol, butyrate, and butanol. Methyl red positive, but negative for the Voges–Proskauer reaction, esculin hydrolysis, gelatin hydrolysis, nitrate reduction, indole production, catalase, oxidase, and urease.

Source: an agricultural settling lagoon in Oklahoma, USA.

DNA G+C content (mol%): 31–32 (HPLC).

Type strain: P7, ATCC BAA-624, DSM 15243.

GenBank accession number (16S rRNA gene): AY170379.

Further comments: 16S rRNA gene sequence comparison shows *Clostridium carboxidivorans* to be a member of cluster I of the clostridia as defined by Collins et al. (1994). Strain P-7[T] forms a tight cluster with *Clostridium scatologenes* and *Clostridium drakei* and shows highest 16S rRNA gene sequence similarity with these species: *Clostridium scatologenes* (99.7%) and *Clostridium drakei* (99.8%). DNA–DNA reassociation studies demonstrate the species status of strain P-7[T] with reassociation values of 31.8% and 50.2% with *Clostridium drakei* and *Clostridium Scatologenes*, repectively (Liou et al., 2005). *Clostridium carboxidivorans*, *Clostridium Drakei*, and *Clostridium scatologenes* can also be differentiated on the basis of substrate utilization patterns and the fact that *Clostridium scatologenes* grows slowly on H_2/CO_2 (Liou et al., 2005).

33. **Clostridium carnis**[*] (Klein 1904) Spray 1939, 750[AL] (*Bacillus carnis* Klein 1904, 459)

car nis. L. gen. n. *carnis* of flesh.

Cells in PYG broth cultures are Gram-stain-positive, motile and peritrichous, straight to slightly curved rods, 0.5–1.1×1.6–9.9 μm, and occur singly or in pairs. Spores are oval, terminal or subterminal, and swell the cell. Sporulation occurs most readily in chopped-meat broth cultures incubated for 24 h. Cell walls contain LL-DAP and glycine.

Surface colonies on anaerobic blood agar plates are pinpoint–2 mm, circular, entire, convex to slightly peaked, translucent, smooth, grayish white, with a mottled or mosaic internal structure. They may be slightly β-hemolytic. Colonies will grow on blood agar plates incubated aerobically, but spores are not formed under aerobic conditions. Cultures in PYG broth are turbid with a smooth sediment and have a pH of 4.9–5.4 after incubation for 1 week. The optimum temperature for growth is 37 °C. Growth is moderate at 30 °C but little or no growth occurs at 25 °C or 45 °C. Growth is stimulated by fermentable carbohydrate, but inhibited by 6.5% NaCl, by 20% bile, or by a pH of 8.5. Abundant gas is formed in deep agar cultures. Neutral red and resazurin are reduced. DNase activity has been demonstrated.

Products in PYG broth are butyric, acetic, and lactic acids, with small amounts of formic and usually succinic acids. Abundant H_2 is produced. Pyruvate is converted to acetate; neither lactate nor threonine is utilized.

Strains are susceptible to chloramphenicol (5 μg/ml), erythromycin (2 μg/ml), penicillin (2 U/ml), and tetracycline (5 μg/ml). Culture supernatants are toxic to mice.

Other characteristics of the species are given in Table 142.

Source: soil, putrefying meat, soft tissue infections and blood in humans, human feces, clinical specimens in animals.

DNA G+C content (mol%): 28 (T_m).

Type strain: ATCC 25777, CIP 104315, DSM 1293, JCM 1393, NCIB 10670, NCTC 13036.

GenBank accession number (16S rRNA gene): M59091.

Further comments: 16S rRNA gene sequence comparison shows *Clostridium carnis* to be a member of cluster I of the clostridia as defined by Collins et al. (1994). It shows highest 16S rRNA gene sequence similarity with the species *Clostridium tertium* (98.7%), *Clostridium septicum* (98.3%), and *Clostridium chauvoei* (98.2%). This species is most easily differentiated from *Clostridium tertium*, which it resembles most closely phenotypically, by its lack of fermentation of mannitol or melibiose.

34. Clostridium celatum* Hauschild and Holdeman 1974, 479^AL

ce.la′tum. L. neut. part. adj. *celatum* hidden.

This description is based on that by Hauschild and Holdeman (1974) and on study of the type and three other strains received from A.H.W. Hauschild.

Cells in PYG broth are Gram-stain-positive rods, straight or curved, nonmotile, 0.8–3.0 × 6.3 μm, occur singly, in pairs, or in long chains. Long filaments up to 200 μm are common. Spores are oval, terminal, subterminal, or central, and usually do not swell the cell. Sporulation occurs most readily on solid media (blood agar or egg-yolk agar plates or chopped-meat slants). Cell-wall composition has not been reported. Surface colonies on blood agar plates are 2–7 mm in diameter, circular, lobate to erose, flat to low convex, translucent to opaque, white, with a granular or mottled internal structure. Two of four strains tested were α-hemolytic, two were nonhemolytic on rabbit blood.

The optimum temperature for growth is 37 °C. Strains grow nearly as well at 30 °C and poorly, if at all, at 25 °C or 45 °C. Cultures in PYG broth are turbid with a smooth or fluffy sediment and have a pH of 5.2–5.5 after incubation for 1 d. Growth is stimulated by fermentable carbohydrate, but inhibited by 6.5% NaCl or 20% bile. Slight to moderate gas is detected in PYG deep agar cultures. Milk is acidified and slightly curdled. H_2S is produced from the reduction of bisulfite, but not from sulfate, in SIM medium. Nitrate is reduced in Bacto-nitrate broth media supplemented with 0.3% agar, glycerol, and galactose, but two of the four strains did not reduce nitrate in indole-nitrite medium (BBL). Urease is produced. Neutral red is reduced; resazurin is not reduced.

Fermentation products in PYG broth are acetic and formic acids and ethanol, with small amounts of butyric and pyruvic acids; succinic and fumaric acids are sometimes produced. Abundant H_2 is formed. Pyruvate is converted to acetate and formate; neither threonine, lactate, nor gluconate is utilized. Other characteristics of the species are given in Table 142. All strains tested are susceptible to chloramphenicol, clindamycin, erythromycin, penicillin G, and tetracycline. Culture supernatants are nontoxic to mice.

Source: human feces.

DNA G+C content (mol%): not reported.

Type strain: ATCC 27791, CIP 104316, DSM 1785, JCM 1394, NCTC 12746.

GenBank accession number (16S rRNA gene): X77844.

Further comments: 16S rRNA gene sequence comparison shows *Clostridium celatum* to be a member of cluster I of the clostridia as defined by Collins et al. (1994). It shows highest 16S rRNA gene sequence similarity with the species *Clostridium disporicum* (99.4%), *Clostridium quinii* (98.3%), and *Clostridium isatidis* (97.3%). Distinctive characteristics of this species are production of acetic and formic acids and ethanol from glucose, production of urease, and lack of production of lecithinase. *Clostridium celatum* differs from *Clostridium disporicum* in that it produces acetic and formic acids and ethanol while the latter only produces acetate and lactate.

35. Clostridium celerecrescens Palop, Valles, Piñaga and Flors 1989, 70^VP (emend. Chamkha, Garcia and Labat 2001a, 2110.)

cel.er.e.cres′cens. L. adj. *celer -eris -ere* fast; L. v. *crescere* to grow; N.L. part. adj. *celerecrescens* fast-growing.

Straight to slightly curved rods, 0.5–0.8 × 2–4 μm. Anaerobic. Gram-stain-positive. Motile with peritrichous flagella. Form spherical and terminal spores which cause cell swelling. Colonies appear after 10–14 d and are cellulolytic. Deep colonies are smooth, circular, translucent, unpigmented, and 0.5 mm in diameter. The optimal temperature for growth is 30–37 °C at a pH of 7.0.

Ferments cellulose, cellobiose, glucose, fructose, galactose, mannitol, maltose, adonitol, rhamnose, xylose, ribose, mannose, raffinose, trehalose, arabinose, and esculin. Lactose, melezitose, melibiose, sorbose, and sucrose are weakly fermented. Does not ferment dulcitol, erythritol, glycerol, inulin, sorbitol, or starch. The end products of fermentation of cellulose and cellobiose are ethanol, acetate, formate, butyrate, iosbutyrate, isovalerate, caproate, lactate, succinate, H_2, and CO_2. Reduces cinnamic acids. Negative for urease, catalase, lipase, and lecithinase activity. Does not reduce nitrate or hydrolyze casein. Digests milk and forms curd. Gelatin is hydrolyzed. Produces indole. Does not produce acetylmethylcarbinol. Reduces *m*-methoxycinnamic acid to 3-(3-methoxyphenyl) propionic acid, *p*-methoxycinnamic acid to 3-(4-methoxyphenyl) propionic acid, and *p*-methylcinnamic acid to 3-(4-methylphenyl) propionic acid. Addition of glucose markedly increases the yield of conversions. Cinnamic, *o*-, *m*-, *p*-coumaric, *o*-methoxycinnamic, caffeic, ferulic, isoferulic, 3,4,5-trimethoxycinnamic acids, cinnamyl alcohol, and phenylpropionic acids (including 3-phenylpropionic acid and 3,4-dihydroxyphenylpropionic acid) are not metabolized after 2 weeks incubation, with or without addition of glucose.

Source: a methanogenic cellulolytic culture inoculated from cow manure.

DNA G+C content (mol%): 38 (T_m).

Type strain: 18A, ATCC 49205, CECT 954, DSM 5628.

GenBank accession number (16S rRNA gene): X71848.

Further comments: 16S rRNA gene sequence comparisons places *Clostridium celerecresens* in cluster XIVa of the clostridia as defined by Collins et al. (1994). The highest similarities are to the species *Clostridium sphenoides* (99.1%), *Clostridium saccharolyticum* (98.5%), *Clostridium indolis* (98.4%), *Clostridium methoxybenzovorans* (98.1%), *Clostridium aerotolerans* (97.5%), and *Clostridium xylanolyticum* (97.5%).

36. **Clostridium cellobioparum*** Hungate 1944, 503[AL]

cel.lo.bi.o'par.um. N.L. n. *cellobiosum* cellobiose; L. verb. adj. suff. *-parus* producing; N.L. neut. adj. *cellobioparum (sic)* cellobiose-producing.

Cells in PYG broth culture stain Gram-negative and are straight or slightly curved rods, motile and peritrichous, 0.5–0.6 × 1.4–3.3 µm, and occur singly or in pairs. Spores are spherical or oval, terminal, and swell the cell. Sporulation occurs most readily on cellulose agar or in 3-week-old rumen fluid broth cultures. Cell walls contain *meso*-DAP. Surface colonies on blood agar plates are 0.5–1.5 mm in diameter, circular, entire, convex to pulvinate, semiopaque, creamy white to yellowish, shiny, and smooth. On rumen fluid cellobiose agar, they may have a scalloped margin, raised elevation, and translucent appearance.

Cultures in PYG broth supplemented with rumen fluid are turbid with a stringy sediment and have a pH of 5.3–5.5 after incubation for 5 d.

The optimum temperature for growth is 30–37 °C. There is little or no growth at 25 °C and none at 45 °C. Growth is stimulated by rumen fluid, a fermentable carbohydrate, and a gaseous atmosphere of 90% N_2 and 10% CO_2 (rather than 100% CO_2 or 100% N_2). H_2 gas produced by this organism will limit growth unless removed either mechanically or by H_2-utilizing organisms growing concurrently. Neither neutral red nor resazurin is reduced. Cellulose is hydrolyzed to cellobiose; glucose is not formed.

Fermentation products in PYG broth are acetic, lactic, and formic acids, ethanol, CO_2, and large amounts of H_2. Pyruvate is converted to acetate; neither lactate nor threonine is utilized.

The type strain is susceptible to chloramphenicol, clindamycin, erythromycin, penicillin G, and tetracycline. Culture supernatants are nontoxic to mice.

Other characteristics of the species are given in Table 143.

Source: the bovine rumen, human feces.

DNA G+C content (mol%): 28 (T_m).

Type strain: ATCC 15832, NCIB 10669, DSM 1351, LMG 5589.

GenBank accession number (16S rRNA gene): X71856.

Further comments: 16S rRNA gene sequence comparisons show *Clostridium cellobioparum* to fall with the cluster III of the clostridia as defined by Collins et al. (1994). It forms a cluster with the species *Clostridium termitidis*, *Clostridium josui*, *Clostridium hungatei*, *Clostridium papyrosolvens*, and *Clostridium cellulolyticum*. The highest 16S rRNA gene sequence similarity is with *Clostridium termitidis* at 99.1%. *Clostridium cellobioparum* can be differentiated from *Clostridium termitidis* on the basis of arabinose utilization by *Clostridium cellulolyticum* but not by *Clostridium termitidis* as well as *Clostridium cellobioparum* having a negative Gram strain reaction compared to a positive reaction for *Clostridium termitidis*.

37. **Clostridium cellulofermentans** He, Ding and Long 1991, 308[VP]

cell.u.lo.fer.men'tans. N.L. n. *cellulosum* cellulose; L. part. adj. *fermentans* fermenting; N.L. adj. *cellulofermentans* cellulose-fermenting.

Straight or slightly curved rods, 0.4–0.7 × 1.5–7.0 µm; found singly, pairs, or in short or long chains. Motile with peritrichous flagella. Anaerobic. Gram-stain-negative. Form terminal endospores, which cause swelling of the cell. Mature spores are 0.8–1.2 µm in diameter. On cellulose agar clear zones appear within 48–72 hours. Surface colonies are white, opaque, circular, and flat to slightly convex.

Negative for catalase and gelatinase activity. Does not produce acetylmethylcarbinol or indole. Reduces nitrate. Curdles milk. Does not reduce sulfate. Ferments cellulose to hydrogen, carbon dioxide, ethanol, and acetic acid. Slight growth occurs on PY media with no fermentable carbohydrates.

The temperature range for growth is 20–45 °C with an optimum of 37–40 °C. pH Range for growth is 6.0–8.2, with an optimum of 7.0–7.2.

Source: soil at a dairy farm.

DNA G+C content (mol%): 34 (T_m).

Type strain: AS 1.1775 (China Committee for Culture Collection of Microorganisms).

GenBank accession number (16S rRNA gene): not reported.

38. **Clostridium cellulolyticum** Petitdemange, Caillet, Giallo and Gaudin 1984, 157[VP]

cell.u.lo'sum. N.L. n. *cellulosum* cellulose; Gr. adj. *lutikos* loosening, dissolving; N.L. adj. *lyticus -a -um* dissolving; N.L. neut. adj. *cellulolyticum* cellulose-dissolving.

Straight to slightly curved rods, 0.6–1.0 × 3–6 µm. Motile with peritrichous flagella. Anaerobic. Gram-stain-positive. Produce spherical, terminal spores, which cause cell swelling. Spores are 1.5 µm in diameter, and appear on cellulose media that is 3 d or more old. Surface colonies are smooth, circular, translucent, unpigmented, 0.5 mm in diameter, have a butyrous texture, and have undulate margins. Clear zones of cellulose around colonies are 1–4 mm in diameter. The temperature range for growth is 25–45 °C with an optimum at 32–35 °C. Growth occurs in the pH range 5.2–7.0.

Cellulose is fermented to carbon dioxide, hydrogen, acetate, ethanol, lactate, and formate. Utilizes arabinose, cellobiose, cellulose, fructose, glucose, and xylose. Does not grow with adonitol, amygdalin, dulcitol, erythritol, glycerol, glycogen, inositol, inulin, lactose, maltose, mannitol, melezitose, raffinose, rhamnose, salicin, sorbitol, sorbose, sucrose, and trehalose. Tests negative for catalase, urease, lipase, and lecithinase activity. Hydrolyzes esculin. Does not digest milk or casein. Does not hydrolyze starch or gelatin.

Does not produce nitrite, indole, or acetylmethylcarbinol. Does reduce thiosulfate. Lactate and pyruvate are not utilized.

Source: a decayed grass compost that had been packed for 3–4 months.

DNA G+C content (mol%): 41 (T_m).

Type strain: H10, ATCC 35319, DSM 5812.

GenBank accession number (16S rRNA gene): X71847.

Further comments: 16S rRNA gene sequence comparisons show *Clostridium cellulolyticum* to fall within the radiation of cluster III of the clostridia as defined by Collins et al. (1994). It forms a cluster with the species *Clostridium papyrosolvens* (98.1%), *Clostridium josui* (98.1%), *Clostridium termitidis* (97.4%), and *Clostridium hungatei* (97.1%). *Clostridium cellulolyticum* can be differentiated from *Clostridium papyrosolvens* on the basis of mannose and glycerol utilization. In addition, *Clostridium cellulolyticum* shows a Gram-stain-positive straining reaction and *Clostridium papyrosolvens* a Gram-stain-negative staining reaction. There is also a considerable difference in the G+C content of the DNA of these two species; *Clostridium cellulolyticum* has a G+C content of 41 mol% while that of *Clostridium papryrosolvens* is 30 mol%.

39. **Clostridium cellulosi** He, Ding and Long 1991, 309[VP]

cell.u.lo´si. L. gen. n. *cellulosi* of cellulose.

Straight to slightly curved rods, 0.3–0.6 × 2.0–15.0 μm, occur singly, in pairs, or in chains. Motile with lophotrichous flagella. Strict anaerobe. Gram-stain-negative staining reaction. Form terminal, spherical endospores which cause cell swelling. Colonies show clear zones on cellulose agar after 48 hours, and continue expansion afterwards. Tiny colonies appear within centers of the clear zone. Colonies appear white. Surface colonies are watery, irregular, and spread. Deep colonies are round with entire margins.

The temperature range for growth is 40–65 °C with an optimum of 55–60 °C. The pH range for growth is 6.2–8.5 with an optimum of 7.3–7.5.

Ruminal fluid, vitamins, or tryptic peptone not required for growth. Grows on 2 g of yeast per liter. No germination in media with O_2 present. Tests negative for catalase and gelatinase activity. Does not produce indole. Curdles milk. Produces acetylmethylcarbinol. Ferments cellulose into hydrogen, carbon dioxide, ethanol, and acetic acid.

Source: cow manure compost.

DNA G+C content (mol%): 35 (T_m).

Type strain: AS 1.1777 (available from CCCCM but an accession number is not provided).

GenBank accession number (16S rRNA gene): L09177.

Further comments: 16S rRNA gene sequence analysis shows *Clostridium cellulosi* to fall in the cluster IV of the clostridia as defined by Collins et al. (1994). The strain represents a distinct lineage that is not closely related to any species of the genus *Clostridium*. A 16S rRNA gene sequence similarity of <89% is found to *Clostridium Thermocellum*, a thermophilic, cellulolytic species at the root of the cluster III clostridia. The highest 16S rRNA gene sequence similarity is to the species *Ethanoligenens harbinense* at 93.9% similarity.

40. **Clostridium cellulovorans** Sleat, Mah and Robinson 1985, 223[VP] (Effective publication: Sleat, Mah and Robinson 1984, 92.)

cell.u´lo.vor.ans. L. n. *cellula* small cell; L. v. *vorare* to devour; N.L. adj. *cellulovorans* cell-devouring.

Rod-shaped cells, 0.7–0.9 × 2.5–3.5 μm. Anaerobic. Nonmotile. Gram-stain-negative. Forms central to subterminal spores within a sporangium that measures 1.5–2.0 × 4–7 μm. Spores are oblong and measure 1–1.5 × 2–4 μm. Colonies are irregularly shaped with an opaque edge and an empty center. Colonies in cellobiose roll tubes are rhizoid. The temperature range for growth is 20–40 °C with an optimum at 37 °C. The pH range for growth is 6.0–7.8 with an optimum of 7.0.

Ferments cellulose, xylan, pectin, cellobiose, glucose, fructose, galactose, sucrose, lactose, and mannose to H_2, CO_2, acetate, butyrate, formate, and lactate. Gum Arabic, rhamnose, melezitose, sorbitol, threhalose, xylose, arbitol, glycerol, erythritol, lactate, and pyruvate are not fermented. Gelatin is not liquefied. Starch and casein are not hydrolyzed. Yeast extract is required for growth. Vitamins and rumen fluid are not growth requirements.

Source: a batch of methanogenic fermentation of fire divided hybrid poplar wood.

DNA G+C content (mol%): 26–27 (T_m).

Type strain: 743B, ATCC 35296, DSM 3052.

GenBank accession number (16S rRNA gene): X71849, X73438.

Further comments: 16S rRNA gene sequence analysis shows *Clostridium cellulovorans* to fall in the cluster I of the clostridia as defined by Collins et al. (1994). It represents a distinct species not closely related to other cellulolytic species of the genus. The closest relatives within cluster I are the species *Clostridium paraputrificum*, *Clostridium disoricum*, *Clostridium sartagoforme*, and *Clostridium tertium*, with sequence similarities in the range 94.4–94.8%.

41. **Clostridium chartatabidum** Kelly, Asmundson and Hopcroft 1996, 625[VP] (Effective publication: Kelly, Asmundson and Hopcroft 1987, 173.)

char´ta´tab´id´um; L. n. *charta* paper; L. adj. *tabidus* dissolving; N.L. adj. *chartatabidum* paper-dissolving.

Rod-shaped cells 1.0 × 3–6 μm. Obligate anaerobe. Motile with peritrichous flagella. Gram-stain-positive. Central to subterminal rod-shaped spores that measure 0.7 × 2–3.5 μm. Cells lack color. Colonies have orange centers with colorless umbonate to undulated edges.

The temperature range for growth is 28–45 °C with an optimum at 39 °C. Growth occurs in the pH range 5.9–7.0.

Ferments cellulose, cellobiose, sucrose, fructose, glucose, salicin, and xylose to hydrogen, acetate, butyrate, and ethanol. Rumen fluid is required for growth. Cells are cellulolytic. Utilizes glucose, sucrose, cellobiose, cellulose, lactose, fructose, and xylose. Does not utilize amygdalin, galctose, maltose, mannitol, melibiose, rhamnose, starch trehalose, and xylan.

Source: the rumen of sheep and cattle.

DNA G+C content (mol%): 31 (T_m).

Type strain: 163, CIP 104882, DSM 5482.

GenBank accession number (16S rRNA gene): X71850.

Further comments: 16S rRNA gene sequence analysis shows *Clostridium chartatabidum* to fall in the cluster I of the clostridia as defined by Collins et al. (1994). *Clostridium chartatabidum* shares 16S rRNA gene sequence similarities in the range 96.3–96.6% to the species *Clostridium diolis, Clostridium beijerinckii, Clostridium vicentii, Clostridium saccharoperbutylacetonicum,* and *Clostridium puniceum.*

42. **Clostridium chauvoei*** (Arloing, Cornevin and Thomas 1887) Scott 1928, 260[AL] (*Bacterium chauvoei* Arloing, Cornevin and Thomas 1887, 82)

chau'voe.i. N.L. gen. n. *chauvoei* of Chauveau, named after J.A.B. Chauveau, French bacteriologist.

Cells in PYG broth cultures are Gram-stain-positive rods, usually motile and peritrichous. They may be quite pleomorphic with irregular staining, particularly in older cultures. Citron forms are seen often. Cells are 0.5–1.7 × 1.6–9.7 μm and occur singly or in pairs. Spores are oval, central to subterminal, and swell the cell. Sporulation occurs readily both in broth and on solid media. Cell walls contain lysine.

Surface colonies on blood agar plates are β-hemolytic, 0.5–3 mm in diameter, circular, low convex or raised, translucent to opaque, granular, shiny or dull, smooth, and have entire to erose margins.

Cultures in PYG broth are turbid with a smooth sediment and have a pH of 5.0–5.4 after incubation for 4 d.

The optimum temperature for growth is 37 °C. There is poor growth at 25 °C and 30 °C, no growth at 45 °C. Growth is stimulated by fermentable carbohydrate, inhibited by 6.5% NaCl, 20% bile, or a pH of 8.5. Abundant gas is formed in PYG deep agar cultures. Neutral red is reduced; reduction of resazurin is variable.

Products in PYG broth are acetic, butyric, and formic acids, butanol, CO_2, and H_2. Small amounts of lactic, succinic, and pyruvic acids may be detected. Pyruvate is converted to acetate and butyrate; neither lactate nor threonine is utilized.

All strains tested are susceptible to chloramphenicol, clindamycin, erythromycin, penicillin G, and tetracycline. Culture supernatants are nontoxic to mice. Strains produce deoxyribonuclease (β-toxin) and hyaluronidase (γ-toxin). Neuraminidase is also produced. Strains are pathogenic through tissue invasion for mice, guinea pigs, and hamsters. Experimentally, $CaCl_2$ must be injected to provide some tissue destruction before infection can occur. Cattle, sheep, goats, swine, deer, mink, freshwater fish, whales, and frogs are susceptible; humans, birds, cats, dogs, and rabbits are resistant, although the organism has been isolated from wounds in dogs and cats. No well-documented strain of *Clostridium chauvoei* has been isolated from humans. The organism is best known as the cause of blackleg in cattle and sheep. Habitat is probably the soil, and outbreaks of the disease often follow soil excavation in areas where animals graze.

Other characteristics of the species are given in Table 142
Source: infections in cattle, sheep, and other animals; intestinal contents of cattle and dogs.

DNA G+C content (mol%): 27 (T_m).

Type strain: ATCC 10092, CIP 104317, DSM 7528, NCIB 10665, NCTC 13023.

GenBank accession number (16S rRNA gene): U51843.

Further comments: 16S rRNA gene sequence analysis places *Clostridium chauvoei* in the cluster I of the clostridia as defined by Collins et al. (1994). The type strain shares 99.2% 16S rRNA gene sequence similarity with *Clostridium septicum.* Other related species include *Clostridium carnis* (98.2%), and *Clostridium tertium* (98.0%). *Clostridium chauvoei* is most easily differentiated from *Clostridium septicum*, which it resembles most closely phenotypically, by its fermentation of sucrose and lack of fermentation of cellobiose or trehalose. Patterns of soluble cellular proteins are distinctive for each species (Cato et al., 1982) and specific fluorescent antibody is available commercially for identification.

43. **Clostridium citroniae** Warren, Tyrrell, Citron and Goldstein 2007, 893[VP] (Effective publication: Warren, Tyrrell, Citron and Goldstein 2006, 2421.)

ci.tro'ni.i. N.L. gen. n. *citroniae* named after Diane M. Citron for numerous contributions to clinical anaerobic bacteriology as a clinical microbiologist and educator.

Rod-shaped cells, 0.8 to 1.1 × 2 to 5 μm. Gram-stain-negative staining reaction. Colonies are 1–2 mm in diameter, flat, opaque to white, and nonhemolytic on Brucella blood agar plates. Rarely a sporeformer. Optimal temperature for growth is 37 °C.

Produces acid from glucose, maltose, mannose, rhamnose, sucrose, trehalose, and xylose but not from cellobiose, esculin, lactose, mannitol, melezitose, raffinose, salicin, sorbitol, or starch. Does not hydrolyze urea, esculin, starch, or gelatin or reduce nitrate. Fermentations of arabinose are variable. Indole-positive. Alkaline phosphatase is positive, but arginine dihydrolase, α-galactosidase, β-galactosidase, β-galactosidase-6-phosphate, α-glucosidase, β-glucosidase, α-arabinosidase, β-glucuronidase, N-acetyl-β-glucosaminidase, arginine arylamidase, praline arylamidase, leucyl glycine arylamidase, phenylalanine arylamidase, leucine arylamidase, pyroglutamic acid arylamidase, tyrosine arylamidase, alanine arylamidase, and glycine arylamidase are negative.

Source: human clinical infections.

DNA G+C content (mol%): not reported.

Type strain: RMA 16102, ATCC BAA-1317, CCUG 52203.

GenBank accession number (16S rRNA gene): DQ279737.

Further comments: Clostridium citroniae and *Clostridium aldenense* were described after the study of 158 strains previously identified as *Clostridium clostridioforme* strains (Warren et al., 2006). Comparison of the 16S rRNA gene sequence of *Clostridium citroniae* shows it to cluster with the species of the *Clostridium clostridiforme* group. This group was designated cluster XIVa by Collins et al. (1994). 16S rRNA gene sequence similarities are in the range 96.2–97.2% to the species *Clostridium clostridiforme* (97.2%), *Clostridium bolteae* (96.9%), *Clostridium aldenense* (96.7%), and *Clostridium asparagiforme* (96.2%).

Clostridium citroniae strains can be differentiated from the other species in this cluster by their profile numbers obtained by Rapid ID 32A and individual phenotypic tests such as raffinose utilization and indole production (Warren et al., 2006).

44. **Clostridium clostridioforme*** corrig. (Burri and Ankersmit 1906) Kaneuchi, Watanabe, Terada, Benno and Mitsuoka 1976b, 202[AL] (*Bacterium clostridiiforme* Burri and Ankersmit 1906, 115)

clos.tri.di.o.for′me. Gr. n. *kloster* a spindle; Gr. dim. suff. *-idion*; N.L. neut. n. *clostridium* a small spindle; L. suff. *-formis* in the form of, N.L. neut. adj. *clostridioforme* in the form of a small spindle, spindle-shaped.

Cells in PYG broth culture are straight rods with pointed ends, 0.3–0.9 × 1.4–9.0 µm. Gram-stain-negative. Cells usually occur in pairs but they also occur singly or in short chains. Spores are oval, central to subterminal, and swell the cell. They are often difficult to demonstrate, particularly in young cultures or from freshly isolated strains. Sporulation occurs most reliably on chopped-meat agar slants incubated at 30 °C for 2–3 weeks or in 3-week-old chopped-meat broth cultures. All strains resist heating at 70 °C for 10 min, but many do not survive heating at 80 °C. Motility is variable but sluggishly motile peritrichous cells or cells with a subpolar tuft of flagella can be detected in the majority of strains. Cell walls contain *meso*-DAP.

Surface colonies on blood agar plates are nonhemolytic, 0.5–2.0 mm in diameter, entire, slightly scalloped or erose, convex to slightly peaked, translucent to opaque, gray-white, usually with a mottled or mosaic internal structure.

Cultures in PYG broth are turbid with a heavy, sometimes viscous or ropy sediment and have a pH of 5.2–5.5 after incubation for 24 h.

Optimum temperature for growth is 37 °C; many strains grow equally well at 30 °C; growth is moderate at 25 °C, usually poor at 45 °C. Growth is inhibited by 6.5% NaCl or a pH of 8.5; strains vary in their reaction to 20% bile. Abundant gas is detected in PYG deep agar cultures. Ammonia is produced. H_2S is produced by 11 of 21 strains tested. Neutral red is reduced; reduction of resazurin is variable.

Low levels of superoxide dismutase have been reported from two clinical isolates. Deoxyribonuclease is formed by the one strain tested.

Fermentation products in PYG broth cultures are acetic acid, usually formic and lactic acids, and ethanol. Large amounts of H_2 are formed. Pyruvate is converted to acetate, usually with formate and ethanol. Some strains convert threonine to propionate; lactate is not utilized.

Susceptibility to antibiotics is variable. Of 42 strains tested, 28 are susceptible to chloramphenicol, 31 to clindamycin, 19 to erythromycin, 23 to penicillin G, and 27 to tetracycline. Clinical isolates most sensitive to metronidazole have been reported. An active β-lactamase has been demonstrated in one strain. Culture supernatants are nontoxic for mice.

Other characteristics of the species are given in Table 144.

Source: intestinal contents of birds, humans, and other animals; calf rumen contents; turkey liver lesions; abdominal, cervical, scrotal, pleural, and other infections; septicemias, peritonitis, and appendicitis.

DNA G+C content (mol%): 47–49 (T_m).

Type strain: ATCC 25537, BCRC 14545, CCUG 16791, CIP 104318, DSM 933, JCM 1291, NCIMB 11018, NCTC 11224, VIP 0316.

GenBank accession number (16S rRNA gene): M59089.

Further comments: Biochemical characteristics of strains in this species are quite variable, even those of a strain tested at different times. However, because of their morphological similarity and the close agreement in high G+C ratios of DNA found in all strains tested, they have been retained in one species. When spores are not detected, they can be distinguished from other Gram-stain-negative anaerobes by their fermentation products. When spores are detected, the spindle-shaped cells are distinctive. 16S rRNA gene sequence comparisons have shown *Clostridium clostridioforme* to fall within the cluster XIVa of the clostridia as defined by Collins et al. (1994). The closest relatives are *Clostridium bolteae* (98.2%), *Clostridium asparagiforme* (97.3%), and *Clostridium citroniae* (97.2%). *Clostridium clostridioforme* can be differentiated from *Clostridium bolteae* on the basis of acid from lactose, melezitose, salicin, and sorbitol (Song et al., 2003b). In addition, differences are observed in β-glucouronidase production and esculin hydrolysis (Song et al., 2003b).

45. **Clostridium coccoides**[*] Kaneuchi, Benno and Mitsuoka 1976a, 485[AL]

coc.coi′des. Gr. n. *coccos* a berry; Gr. n. *eidos* shape; N.L. adj. *coccoides* berry-shaped.

Cells in PYG broth cultures are Gram-stain-positive, coccoid short rods, 0.6–1.0 × 0.6–1.5 µm and occur singly, in pairs, and in short chains. Nonmotile. Spores are round, central to subterminal, and slightly swell the cell. Sporulation occurs most readily on chopped-meat agar slants or on modified Eggerth–Gagnon agar plates after incubation at 37 °C for 2–5 d. All strains survive heating at 70 °C for 10 min but resistance is variable to 80 °C for 10 min. Cell walls contain *meso*-DAP, lysine, and alanine.

Surface colonies on blood agar plates are punctiform–1 mm in diameter, circular, slightly irregular, slightly undulate, convex, gray-white, shiny, smooth, and nonhemolytic. On modified Eggerth–Gagnon agar plates, colonies are 1.5–2.5 mm in diameter, circular, convex, entire, translucent, yellowish gray, smooth, shiny, and nonhemolytic. Cultures in PYG broth are turbid with a smooth sediment and have a pH of 4.4 after incubation for 24 h. Growth is stimulated by fermentable carbohydrate.

The optimum temperature for growth is 37 °C. Strains grow well at 25 °C and 45 °C but poorly at 15 °C. Abundant gas is detected in PYG deep agar cultures. Resazurin is reduced. Acid is produced from inositol, dulcitol, β-methylglucoside, and α-methylmannoside. Orotic acid and tributyrin are not hydrolyzed. Small amounts of H_2S are produced.

Products in PYG broth culture after 5 d of incubation are acetic and succinic acids and abundant H_2. Neither lactate nor threonine is utilized.

The type strain is susceptible to chloramphenicol, clindamycin, erythromycin, and tetracycline. It is resistant to penicillin G. Culture supernatants are nontoxic to mice.

Other characteristics of the species are given in Table 144.

Source: the cecum of mice fed high-lactose diets.

DNA G+C content (mol%): 43–45 (T_m).

Type strain: ATCC 29236, DSM 935, JCM 1395, NCTC 11035.

GenBank accession number (16S rRNA gene): not reported.

Further comments: This species is readily differentiated from other clostridial species by its morphology, by its production of large amounts of succinic acid, by its fermentation of inositol and by its failure to produce indole. 16S rRNA gene sequence comparisons show *Clostridium coccoides* to fall within the radiation of cluster XIVa of the clostridia as defined by Collins et al. (1994). The highest similarity is to *Ruminococcus productus* at 99.1% and similarities in the range 93.4–95.2% to other *Ruminococcus* species.

46. **Clostridium cochlearium**[*] (Douglas, Fleming and Colebrook 1919) Bergey, Harrison, Breed, Hammer and Huntoon 1923, 333[AL] (*Bacillus cochlearius* Douglas, Fleming and Colebrook in Bulloch et al., 1919, 40; *Clostridium lentoputrescens* Hartsell and Rettger 1934)

coch.le.a′ri.um. L. n. *cochlear* spoon; N.L. neut. adj. *cochlearium* resembling a spoon.

Cells in PYG broth are motile and peritrichous, straight to slightly curved rods with rounded ends, 0.5–1.3 × 1.6–14.1 µm, and occur singly or in pairs. Spores are round to oval, swelling the cell, usually terminal although subterminal spores are seen occasionally. Sporulation occurs readily both in broth media or on solid media (blood agar plates or chopped-meat slants). Three distinct spore coats and tubular appendages attached to one end of the spore have been demonstrated. Cell walls contain *meso*-DAP.

Surface colonies on blood agar plates are 0.5–2.0 mm, circular to irregular, slightly scalloped to lobate, flat to raised, translucent, gray-white, smooth, shiny, with a mottled or mosaic internal structure. Hemolysis is variable.

Cultures in PYG broth are turbid with a smooth sediment, and have a pH of 6.0–6.5 after 1 week of incubation.

Optimum temperature for growth is 37 °C. Most strains grow well at 30 °C and 45 °C but poorly if at all at 25 °C. Growth is inhibited by 6.5% NaCl, by 20% bile, or at a pH of 8.5.

Gelatin may be weakly and slowly digested; neither meat, casein, nor milk is digested. H_2S and ammonia are formed. Hippurate is hydrolyzed by three of eight strains tested. Neutral red is reduced.

One strain isolated from human subgingival plaque degrades fibrinogen, possibly contributing to periodontal disease.

Products in PYG broth are butyrate, acetate, and propionate; small amounts of lactate, formate, succinate, and butanol are sometimes formed; abundant H_2 is produced. Pyruvate is converted to acetate and butyrate; threonine is converted to propionate; lactate is not utilized. Glutamate is fermented to butyrate, acetate, CO_2, and ammonia by the methylaspartate pathway. Serine, aspartate, and histidine are utilized for growth. Phenylalanine is fermented to phenol. Major cellular fatty acids produced in trypticase-yeast extract medium are principally of the straight-chain saturated series including C_{12}, C_{14}, C_{15}, C_{16}, and C_{18} acids; small amounts of straight-chain unsaturated acids, C_{15} and C_{18}, each with one double bond, are also formed.

All strains tested are susceptible to clindamycin, erythromycin, penicillin G, and tetracycline. One of six strains is resistant to chloramphenicol. Culture supernatants are nontoxic to mice.

Other characteristics of the species are given in Table 142.

Source: soil, human oral cavity, human and horse feces, wounds, crabmeat.

DNA G+C content (mol%): 27–28 (T_m).

Type strain: ATCC 17787, CCUG 31665, BCRC 14471, CIP 104319, DSM 1285, NCCB 73036, NCIB 10633, NCTC 13027.

GenBank accession number (16S rRNA gene): M59093.

Further comments: 16S rRNA gene sequence comparisons show *Clostridium cochlearium* to fall within the radiation of cluster I of the clostridia as defined by Collins et al. (1994). The highest 16S rRNA gene sequence similarity is to *Clostridium tetani* (98.1%). This species can be differentiated from *Clostridium tetani*, which it resembles most closely phenotypically, by its lack of toxicity for mice, by its weak to negative digestion of gelatin, and by distinct patterns of soluble cellular proteins of the two species.

47. **Clostridium cocleatum**[*] Kaneuchi, Miyazato, Shinjo and Mitsuoka 1979, 10[AL]

co.cle.a′tum. L. n. *coclea* a snail shell or whirlpool; L. neut. adj. *cocleatum* in the shape of a snail shell or whirlpool.

Cells in PYG broth cultures are Gram-stain-positive rods, nonmotile, 0.3–0.4 × 1.0–5 µm, semicircular, circular or spiral-shaped. They may grow in long chains, and be Gram-stain-negative with filaments or swollen forms. Spores are difficult to demonstrate but can be found most reliably on medium 10 agar or on chopped-meat agar slants incubated at 30 °C. Spores are round, terminal to subterminal, and swell the cell. Cell walls contain *meso*-DAP.

Surface colonies on blood agar plates are pinpoint–2 mm in diameter, circular, entire to erose, low convex to pulvinate, semiopaque to opaque, smooth, shiny, grayish white, and nonhemolytic.

Cultures in PYG broth are turbid with a smooth to mucoid sediment and have a pH of 4.9–5.4 after incubation for 4 d.

The optimum temperature for growth is 37–45 °C. Strains grow nearly as well at 30 °C, poorly at 25 °C and not at all at 15 °C. Growth is stimulated by fermentable carbohydrate and by rumen fluid, but inhibited by 6.5% NaCl or 20% bile. Abundant gas is detected in PYG deep agar cultures. Neutral red and resazurin are reduced. Dextrin, dulcitol, α-methylglucoside, and α-methylmannoside are not fermented.

Products in PYG broth are acetic and formic acids; lactic acid usually is produced. Abundant H_2 is formed. Pyruvate is converted to acetate and formate; neither threonine nor lactate is utilized.

All strains tested are susceptible to chloramphenicol, erythromycin, penicillin G, and tetracycline but resistant to clindamycin. Toxicity and pathogenicity have not been determined.

Other characteristics of the species are given in Table 144.

Source: feces of healthy humans; cecal contents of mice, rats, and chickens.

DNA G+C content (mol%): 28–29 (T_m).

Type strain: I50, ATCC 29902, DSM 1551, JCM 1397, NCTC 11210.

GenBank accession number (16S rRNA gene): Y18188.

Further comments: 16S rRNA gene sequence comparisons show *Clostridium cocleatum* to fall within the radiation of cluster

XVIII. Sequence similarities in the range 96.6–96.9% are found to the species *Clostridium ramosum*, *Clostridium saccharogumia*, and *Clostridium spiroforme*. This species is closely related to *Clostridium spiroforme*, having 46–60% homology with strains of that species and being similar in morphology. *Clostridium cocleatum* ferments galactose, while *Clostridium spiroforme* does not. The patterns of soluble cellular protein as determined by polyacrylamide gel electrophoresis are also helpful in differentiating the two species.

48. **Clostridium colicanis** Greetham, Gibson, Giffard, Hippe, Merkhoffer, Steiner, Falsen and Collins 2003, 261[VP]

co.li.can′is. L. n. *colum* colon, gut; L. gen. n. *canis* of the dog; N.L. gen. n. *colicanis* of the gut of a dog.

Rod-shaped cells, 0.9–1.0 × 3–10 μm. Gram-stain-negative. Oval to oblong shaped spores in PY-starch medium; position varies from subterminal to almost terminal or even central. Nonmotile. Colonies are 3–5 mm in diameter, round, with undulate margin and are slightly convex, opaque, glossy, and grayish-white. Nonhemolytic on Columbia sheep blood agar. Anaerobic.

Growth occurs in the range 30–45 °C, with no growth at 20 or 50 °C and optimum growth at 37–40 °C.

Produces acid from glucose (acidification of PY medium containing 1% glucose, pH 6.8; after 6 d the pH was 4.8). Acid is produced from cellobiose, esculin (weak), fructose, galactose, glucose, lactose, maltose, mannose, ribose, salicin (weak), starch (weak), and sucrose. Acid is not produced from amygdalin, L-arabinose, glycogen, inositol, mannitol, melezitose, melibiose, raffinose, rhamnose, sorbitol, trehalose, or xylose. Esculin and urea, but not gelatin, are hydrolyzed. Lecithinase- and lipase-negative. Indole is not produced. Nitrate is reduced to nitrite. Catalase-negative. Using API Rapid 32AN test system, activity is detected for alkaline phosphatase, arginine arylamidase, arginine dihydrolase, β-galactosidase, glycine arylamidase, histidine arylamidase, leucine arylamidase, leucyl glycine arylamidase, and *N*-acetyl-β-glucosaminidase. No activity is detected for alanine arylamidase, α-arabinosidase, α-fucosidase, α-glucosidase, β-glucosidase, β-glucuronidase, α-galactosidase, β-galactosidase-6-phosphate, glutamic acid decarboxylase, glutamyl glutamic acid arylamidase, phenylalanine arylamidase, proline arylamidase, pyroglutamic acid arylamidase, serine arylamidase, or tyrosine arylamidase.

Source: feces of a male Labrador dog.

DNA G+C content (mol%): 31.7 (HPLC).

Type strain: 3WC2, CCUG 44556, DSM 13634.

GenBank accession number(16S rRNA gene): AJ420008.

Further comments: 16S rRNA gene sequence comparisons show *Clostridium colicanis* to fall within the radiation of cluster I of the clostridia as defined by Collins et al. (1994). The closest relatives based on 16S rRNA gene sequence similarities are the *Eubacterium* species *Eubacterium moniliforme* (96.2%), *Eubacterium multiforme* (95.9%), *Eubacterium budayi* (95.8%), *Eubacterium nitrogenes* (95.5%) as well as *Clostridium baratii* (95.9%) and *Clostridium diolis* (95.5%). The ~3–4% 16S rRNA gene sequence divergence along with some differences at the phenotypic level were used as justification for the description of the species *Clostridium colicanis* (Greetham et al., 2003).

49. **Clostridium colinum**[*] (*ex* Berkhoff, Campbell, Naylor and Smith 1974) Berkhoff 1985, 157[VP] (*Clostridium colinum* Berkhoff, Campbell, Naylor and Smith 1974, 203)

co.li′num. N.L. n. *Colinus* a zoological name; N.L. neut. adj. *colinum* referring to the most susceptible host, the quail (*Colinus virginianus*).

Cells in PYG broth culture are Gram-stain-positive rods that may rapidly become Gram-stain-negative. They are motile, peritrichous, 1 × 3–4 μm, and occur singly or in pairs. Spores are oval and subterminal. They are sparse in most media, but when transferred from chopped-meat glucose medium into PY-1% starch medium will survive heating at 70 °C for 10 min.

Cell-wall composition has not been reported.

Surface colonies on blood agar plates are pinpoint–0.5 mm, circular to slightly irregular, low convex, transparent, grayish white to colorless, shiny, and smooth. Most strains are α-hemolytic.

Cultures in PYG broth have a smooth white sediment without turbidity and a pH of 5.4 after incubation in an atmosphere of 100% CO_2 for 6 d. In brain heart infusion broth with glucose and an atmosphere of 100% N_2, the final pH is 5.2–5.5.

The optimum temperature for growth is 37 °C. Some strains grow equally well at 45 °C. There is little growth at 30 °C and none at 25 °C. Growth is stimulated by fermentable carbohydrate. Tryptose-phosphate-glucose agar with 8% horse plasma has been recommended for isolation. Although isolation may be difficult on other media (e.g., supplemented brain heart infusion agar or rumen fluid-glucose-cellobiose agar), stock cultures usually will grow on these media. Moderate gas is detected in PYG deep agar cultures.

Products in PYG broth are formic, acetic, and small amounts of lactic acids. Propionic acid usually is detected. Abundant H_2 is produced. Neither pyruvate, lactate, nor threonine is utilized.

All strains tested are susceptible to chloramphenicol, clindamycin, erythromycin, penicillin G, and tetracycline.

Strains are pathogenic to quail and chickens, causing ulcerative enteritis or liver necrosis and death in less than 18 h. Guinea pigs are not susceptible.

Other characteristics of the species are given in Table 144.

Source: intestinal tracts of quail, pheasants, grouse, partridge, chickens, and turkeys.

DNA G+C content (mol%): not reported.

Type strain: 72042, ATCC 27770, CCUG 21927, DSM 6011 and JCM 5831.

GenBank accession number (16S rRNA gene): X76748.

Further comments: 16S rRNA gene sequence comparisons show *Clostridium colinum* to fall within the radiation of cluster XIVb of the clostridia as defined by Collins et al. (1994). The highest sequence similarity is found to *Clostridium piliforme* at 95.2%. The next closest relative is *Clostridium lactatifermentans* with a similarity value of 89.9%.

50. **Clostridium collagenovorans** Jain and Zeikus 1988a, 328[VP] (Effective publication: Jain and Zeikus 1988b, 140.)

col.ageno.vor′ans. Fr. n. *collagène* from Gr. n. *kolla* glue; L. v. *vorare* to devour; N.L. adj. *collagenovorans* collagen-devouring.

Rod-shaped cells, 0.8 × 3.0 μm. Obligate proteolytic anaerobe. Motile. Gram-stain-positive. Cells have a monolayer cell wall. Cells form subterminal spores. Colonies are white, circular, irregular in shape, and 1–3 mm in diameter.

The optimum temperature for growth is 30–37 °C with an optimum pH for growth of 6.0–8.0. Utilizes gelatin, collagen, azocoll, peptone, and meat. Slow growth on casein and bovine serum albumin. Does not grow on carbohydrates, and single or mixtures of amino acids. Cells are nonhemolytic. Ferments gelatin to acetate, CO_2, H_2, ethanol, isovalerate, and isobutyrate. No vitamins or growth factors are required for growth. Hydrogen is produced and consumed during growth. Cells produce collagenase. Cysteine inhibits growth. Catalase- and oxidase-negative. Growth is inhibited by penicillin and streptomycin (100 μg/ml)

Source: Nine Springs Sewage digestor in Madison, WI, USA.

DNA G+C content (mol%): 24.22 (T_m).

Type strain: SG, ATCC 49001, DSM 3089.

GenBank accession number (16S rRNA gene): X73439.

Further comments: 16S rRNA gene sequence comparisons show *Clostridium collagenovorans* to fall within the radiation of cluster I of the clostridia as defined by Collins et al. (1994). The closest relatives are *Clostridium sardiniense* (95.1%), *Clostridium acetobutylicum* (94.1%), and *Clostridium lundense* (93.7%).

51. **Clostridium cylindrosporum** (*ex* Barker and Beck 1942) Andreesen, Zindal and Dürre 1985, 207[VP] ("*Clostridium cylindrosporum*" Barker and Beck 1942)

cy.lin.dro.spo′rum. Gr. n. *kylindros* a cylinder; Gr. n. *sporos* a seed; N.L. adj. *cylindrosporum* cylinder-spored.

Cells are straight rods, 0.8 × 3.3 μm. Motile by peritrichous flagella. Colonies are unpigmented, flat, circular and up to 2 mm in diameter. Cylindrical to oval spores in terminal or subterminal position. Only actively growing cultures have a positive Gram-staining reaction.

Strictly anaerobic. The optimum temperature for growth is 40–45 °C, some growth at 19 °C. The pH optimum for growth is 6.7–8.5. This pH optimum varies with substrate utilized. Only uric acid, xanthine, guanine, 6,8-dihydroxypurine, and hypoxanthine are utilized as carbon and energy sources. Glycine is fermented when small amounts of uric acid or purines are present. Marginal growth with 4-aminoimidazole-5-caroxamide. Thiamine and biotin required for growth of some strains. Selenite is required. The end products of fermentation of purines are acetate, formate, CO_2, and ammonia. Glycine in detected in trace amounts under some growth conditions.

Acetoin, butanediol, butyrate, caproate, diacetyl, gas, indole, and sulfide are not produced. Nitrate is not reduced. Esculin, gelatin, and starch are not hydrolyzed. Catalase, lecithinase, and urease activities are not detected. Some strains produce lipase. Nonhemolytic on bovine blood cells. Bile extract (2%) inhibits growth but growth still occurs.

Source: soils and chicken intestines.

DNA G+C content (mol%): 27.2–30.4 (T_m).

Type strain: Barker and Beck HC-1, ATCC 7905, DSM 605, IFO (now NBRC) 13695, NCCB 46096.

GenBank accession number (16S rRNA gene): Y18179.

Further comments: 16S rRNA gene sequence comparisons show *Clostridium cylindrosporum* to fall within the family *Clostridiaceae*. This species clusters with the species of the genus *Caloramator* at the 91–92% 16S rRNA gene sequence similarity level.

52. **Clostridium difficile**[*] (Hall and O'Toole 1935) Prévot 1938, 84[AL] (*Bacillus difficile* Hall and O'Toole 1935, 390)

dif′fi.cile. L. neut. adj. *difficile* difficult (referring to "the unusual difficulty that was encountered in its isolation and study")

Cells are Gram-stain-positive, usually motile in broth cultures, peritrichous, and are 0.5–1.9 × 3.0–16.9 μm. Some strains produce chains consisting of two to six cells aligned end-to-end. Spores are oval, subterminal (rarely terminal), and swell the cell. Sporulation by most strains occurs on Brucella blood agar incubated for 2 d. Sporulation may be enhanced on solid media containing 0.1% sodium taurocholate.

Cell walls contain *meso*-DAP. Surface colonies on blood agar are 2–5 mm, circular, occasionally rhizoid, flat or low convex, opaque, grayish or whitish, and have a matt to glossy surface. All strains produce an evanescent pale green fluorescence under long wavelength ultraviolet light after 48 h incubation on Brucella blood agar supplemented with hemin and vitamin K_1.

Cultures in PYG broth are turbid with a smooth sediment and have a pH of 5.0–5.5 after incubation for 5 d.

The optimum temperature for growth is 30–37 °C; growth also occurs at 25 °C and 45 °C. Proline, aspartic acid, serine, leucine, alanine, threonine, valine, phenylalanine, methionine, and isoleucine are utilized for growth; δ-aminovalerate and α-aminobutyrate are produced. A selective minimal medium described by Hubert et al. (1981), which contains selected amino acids as a source of carbon and energy, only a trace of fructose (0.1%), 2% bile as a growth stimulant, 16 pg/ml cefoxitin, and 500 μg/ml streptomycin for reduction of associated flora, has been found useful for isolation of these organisms from feces.

Abundant gas is produced in PYG deep agar cultures. Abundant H_2 is produced in PYG broth. Ammonia is produced. H_2S is produced by 8 of 17 strains tested.

The type strain metabolizes the bile acids cholic and chenodeoxycholic acids and splits the conjugate taurocholic acid. Hyaluronidase, chondroitin sulfatase, and collagenase are present in the one strain studied. One of two strains tested produces an extracellular β-glucuronidase. Uracil is not utilized.

Products in PYG broth include acetic, isobutyric, butyric, isovaleric, valeric, isocaproic, formic, and lactic acids. Pyruvate is converted to acetate and butyrate; threonine is converted to propionate; lactate is not utilized. Phenylacetic, phenylpropionic, hydroxyphenylacetic, indole acetic acids, and *p*-cresol are produced in trypticase medium supplemented with L-phenylalanine, L-tyrosine, and L-tryptophan. *p*-Hydroxyphenylacetic acid is converted to *p*-cresol. Isocaproate is produced from leucine, isovalerate from leucine and isoleucine, and isobutyrate from valine.

All strains are susceptible to 8 U penicillin G/ml, 4 μg ampicillin/ml, 4 μg vancomycin/ml, <1 μg rifampin/ml, and 2 μg metronidazole/ml; susceptibility to clindamycin, cephalosporins, cephamycins, tetracyclines, chloramphenicol, and erythromycin is variable, whereas all strains are resistant to aminoglycosides. Tetracycline resistance has been demonstrated to be transferable from resistant to sensitive strains; the resistance determinant may be plasmid-mediated or chromosomal.

Pseudomembranous colitis in humans is caused by overgrowth of the organism in the colon, usually after the flora has been disturbed by antimicrobial therapy. A very similar disease can be produced in hamsters and several other rodents by administration of antibiotics, but rats and mice are not affected. Although *Clostridium difficile* has been reported to cause cecitis in rabbits and hares, *Clostridium spiroforme* appears to be a more common cause of the disease in these animals.

Clostridium difficile produces two large protein toxins (toxins A and B), and hamsters must be immunized against both toxins to survive *Clostridium difficile* cecitis. Toxin A is lethal when given orally to hamsters but toxin B is not; either toxin is lethal when injected intraperitoneally into these rodents. Toxins A and B are lethal for mice. 16,16-Dimethyl-prostaglandin E2 may inhibit production or release of cytopathic toxin(s) *in vitro*. Toxin A has been referred to as the enterotoxin because it causes fluid accumulation in the bowel, but the mechanism of action is not through stimulation of adenyl cyclase. Toxin B does not cause fluid accumulation, but is extremely cytopathic for all tissue-cultured cells tested. Exposure to less than a picogram of this toxin causes cells to become round, detach from supports, and slowly die. The mechanism of action is unknown. A motility altering factor also has been described that is different from the two known toxins, but the significance is unknown.

Other characteristics of the species are shown in Table 144.

Source: marine sediment, soil, sand, the hospital environment; camel, horse, and donkey dung; feces of dogs, cats, and domestic birds; the human genital tract; feces of humans without diarrhea; and rarely from blood and pyogenic infections in humans and animals.

DNA G+C content (mol%): 28.

Type strain: AS 1.2184, ATCC 9689, BCRC 10642, CCUG 4938, CIP 104282, DSM 1296, JCM 1296, LMG 15861, NCIB 10666, NCIMB 10666, NTCC 11209.

GenBank accession number (16S rRNA gene): AB075770.

Further comments: 16S rRNA gene sequence comparisons show *Clostridium difficile* to fall within the radiation of cluster XI of the clostridia as defined by Collins et al. (1994). The closest relatives based on 16S rRNA gene sequence similarities are *Eubacterium tenue, Clostridium irregulare, Clostridium sordellii, Clostridium ghonii, Clostridium lituseburense, Clostridium mangenotii,* and *Clostridium Bifermentans,* all having similarity values in the range 94.7–95.4%. This species is most easily distinguished from *Clostridium sporogenes,* which it resembles most closely phenotypically (although phylogenetically they are unrelated), by its ability to ferment mannitol and by its inability to digest meat or milk, or to produce lipase. *Clostridium difficile* is one of the few species that produces isocaproic and valeric acids.

53. **Clostridium diolis** Biebl and Spröer 2003, 627^VP (Effective publication: Biebl and Spröer 2002, 496.)

di.o′lis. N.L. neut. gen. n. *diolis* of a diol (producing a diol).

Cells are 0.7–1.7 × 1.2–8 μm, occur singly or in pairs and occasionally in filaments. Gram-stain-positive becoming Gram-stain-negative in older cultures. Spores are oval, terminal or subterminal, and cause the cell to swell. Motile by peritrichous flagella.

The optimum temperature range for growth is 25–42 °C with an optimum of 37 °C. The pH range for growth is 5.5–8.5 with an optimum at 7.0. Glucose, fructose, cellobiose, lactose, galactose, trehalose, xylose, salicin, amygdalin, melibiose, glycogen, mannitol, and glycerol are fermented. Melizitose, arabinose, ribose, esculin, and sorbitol are weakly used. Raffinose, starch, and *m*-inositol are not used. Gelatin is not liquefied. Esculin is hydrolyzed. Milk is coagulated. Able to grow in a glucose-mineral medium with biotin as the only growth factor. Unable to grow in yeast extract-peptone medium without a fermentable carbon source. Fermentation products from glucose are butyrate and acetate, from glycerol 1,3-propanediol in addition to butyrate and acetate.

Source: decaying straw.

DNA G+C content (mol%): not reported.

Type strain: SH1, DSM 15410, ATCC BAA-557.

GenBank accession number: not reported.

Further comments: Strain SH1 was previously considered a *Clostridium butryicum* strain. Further characterization showed strain SH1 to be more closely related to *Clostridium beijerinckii* than *Clostridium butyricum* (Biebl and Sproer, 2002). Although strain SH1 has >99.0% 16S rRNA gene sequence similarity with that of the type strain of *Clostridium beijerinckii* (X81021) and a corresponding DNA–DNA reassociation value of 67.2%; the fact that strain SH1 and related glycerol fermenting strains did not require complex substances for growth, as well as the fact that these strains did not ferment starch, raffinose, and inositol was used as justification for the description of a new species, *Clostridium diolis,* by Biebl and Spröer (2002). High 16S rRNA gene sequence similarities are also found between *Clostridium diolis* and *Clostridium puniceum* (99.1%) and *Clostridium saccharobutylicum* (99.15%).

54. **Clostridium disporicum** Horn 1987, 400^VP

di.spor′i.cum. Gr. pref. *di* two; Gr. n. *sporos* seed; Gr. suff. *ikos* pertaining to; N.L. neut. adj. *disporicum* pertaining to two spores.

Rod-shaped cells, 1.5 × 2.5–10 μm. Obligate anaerobe. Gram-stain-positive. Nonmotile. Form subterminal oval spores. Some cells form two spores. The temperature range for growth is 22–44 °C with an optimum at 25–26 °C.

Fermentable carbohydrates required for growth. Ferments glucose to acetic and lactic acid with trace amounts of butyrate and succinate. Ethanol is also formed. Utilizes esculin, cellobiose, dulcitol, fructose, galactose, glucose, lactose, maltose, mannitol, mannose, melibiose, raffinose, ribose, salicin, starch, sucrose, trehalose, and xylose. Hydrolyzes esculin and starch. Does not utilize arabinose, inositol, melezitose, rhamnose, and sorbitol. Negative for lecithinase, lipase, and catalase activity. Does not hydrolyze gelatin, digest cooked meat, or casein in milk. Indole is not produced.

Source: a slurry of rat cecal contents.

DNA G+C content (mol%): 40–41 (T_m).

Type strain: DS1, ATCC 43838, DSM 5521, NCIMB 12424.

GenBank accession number (16S rRNA gene): Y18176.

Further comments: 16S rRNA gene sequence comparisons show *Clostridium disporicum* to fall within the radiation of the cluster I of the clostridia as defined by Collins et al. (1994). It shares 99.4% sequence similarity with *Clostridium celatum* and 97.8% with the next closest relative *Clostridium quinii*. *Clostridium celatum* differs from *Clostridium disporicum* in that it produces acetic and formic acids and ethanol while the latter only produces acetate and lactate.

55. **Clostridium drakei** Liou, Balkwill, Drake and Tanner 2005, 2089[VP]

dra′ke.i. N.L. gen. n. *drakei* of Drake, in recognition of Harold L. Drake's contributions to our understanding of the physiology and ecology of acetogens.

Gram-stain-negative, motile rods, 0.6 × 3–4 μm. Terminal spore former. Obligate anaerobe. Optimal growth temperature of 30–37 °C and optimum pH of 5.5–7.5. Grows autotrophically with H_2/CO_2 or CO and chemoorganotrophically with ribose, xylose, fructose, glucose, galactose, fucose, L-arabinose, mannose, rhamnose, sucrose, cellobiose, melezitose, starch, cellulose, inositol, mannitol, gluconate, glycerol, ethanol, propanol, 2-propanol, butanol, citrate, malate, fumarate, lactate, serine, alanine, histidine, glutamate, aspartate, asparagine, arginine, as amino acids, betaine, choline, and syringate. Methanol, D-arabinose, maltose, lactose, trehalose, raffinose, melibiose, amygdalin, sorbitol, and succinate do not support growth. Acetic acid, ethanol, butyrate, and butanol are the end products of metabolism. Cultures are methyl red-positive, but negative for the Voges–Proskauer reaction, esculin hydrolysis, gelatin hydrolysis, nitrate reduction, and indole production.

Source: sediment collected from an acidic coal-mine pond in east-central Germany.

DNA G+C content (mol%): 30–32 (HPLC).

Type strain: SL1, ATCC BAA-623, DSM 12750.

GenBank accession number (16S rRNA gene): Y18813.

Further comments: Strain SL1[T] was originally described as a strain of *Clostridium scatologenes* (Kusel et al., 2000). Further characterization of this strain differentiated it from the type strain of *Clostridium scatologenes* and it was described as *Clostridium drakei* (Liou et al., 2005). 16S rRNA gene sequence comparison shows *Clostridium drakei* to be a member of cluster I of the clostridia as defined by Collins et al. (1994). Strain SL1[T] forms a tight cluster with *Clostridium scatologenes* and *Clostridium carboxidivorans* and shows highest 16S rRNA gene sequence similarity with these species: *Clostridium scatologenes* (99.7%) and *Clostridium carboxidivorans* (99.8%). DNA–DNA reassociation studies demonstrated the species status of strain SL1[T] with reassociation values of 31.8% and 50.2% with *Clostridium carboxidivorans* and *Clostridium Scatologenes*, repectively (Liou et al., 2005). *Clostridium carboxidivorans*, *Clostridium drakei*, and *Clostridium scatologenes* can also be differentiated on the basis of substrate utilization patterns and the fact that *Clostridium scatologenes* grows slowly on H_2/CO_2 (Liou et al., 2005).

56. **Clostridium estertheticum** Collins, Rodrigues, Dainty, Edwards and Roberts 1993, 188[VP] (Effective publication: Collins, Rodrigues, Dainty, Edwards and Roberts 1992, 239.) emend. Spring, Merkhoffer, Weiss, Kroppenstedt, Hippe and Stackebrandt 2003, 1028.

est.er.thet′i.cum. N.L. neut. adj. *estertheticum* ester-producing

Cells are rod-shaped, 1.3–1.5 × 2.4–6.0 μm, occur singly, in pairs or short chains. Motile by peritrichous flagella. Endospores are ellipsoidal and located mainly in a subterminal position, but sometimes also terminal or central; sporangium is slightly swollen. Gram-stain-positive. Colonies on sheep-blood agar are 1–2 mm in diameter, round with often coarsely granulated margins, smooth, slightly raised, cream-white to grayish, and semitransparent to opaque. Psychrophilic. The pH optimum is pH 6.5. Arabinose, cellobiose, fructose, galactose, glucose, inositol, inulin, maltose, mannitol, mannose, melibiose, raffinose, rhamnose, salicin, sorbitol, starch, sucrose, and xylose are utilized.

Amygdalin, lactose, melezitose, ribose, and trehalose are not utilized. Starch is hydrolyzed, but gelatin is not. The end products of fermentation are butyrate, acetate, lactate, formate, 1-butanol, ethanol, hydrogen, and carbon dioxide. The cell-wall peptidoglycan contains *meso*-diaminopimelic acid. The major cellular fatty acids are $C_{14:0}$, $C_{16:1\ DMA\ \omega9c}$, $C_{16:1\ \omega9c}$, $C_{16:0}$, $C_{16:1\ \omega11c}$, and an unknown compound with an equivalent chain-length of 14.777–14.783.

Source: chill-stored vacuum-packed beef.

DNA G+C content (mol%): 33.9 (HPLC).

Type strain: ATCC 51377, CIP 105093, DSM 8809, NCIMB 12511.

GenBank accession number (16S rRNA gene): S46734, X68181.

Further comments: 16S rRNA gene sequence analysis shows the species *Clostridium estertheticum* and both of its subspecies to fall within the radiation of cluster I of the clostridia as defined by Collins et al. (1994). The subspecies of this species share 99.93% 16S rRNA gene sequence similarity and a DNA–DNA reasocciation value of 79%. 16S rRNA gene sequence similarities are in the range of 98.7–99.6% to the species *Clostridium bowmanii*, *Clostridium frigioris*, and *Clostridium lacusfryxellense*. The type strains of the two subspecies can be differemntiated on the basis of *Clostridium estertheticum* subsp. *laramiense* being β-hemolytic on blood agar (Spring et al., 2003) as well as some substrate utilization differences and physiological characteristics. It should be noted that the two subspecies of this species do not cluster together in the phylogenetic tree presented by (Spring et al., 2003) or in the analysis presented in Figure 128.

56a. **Clostridium estertheticum subsp. estertheticum** Collins, Rodrigues, Dainty, Edwards and Roberts 1993, 188[VP] (Effective publication: Collins, Rodrigues, Dainty, Edwards and Roberts 1992, 239.) emend. Spring, Merkhoffer, Weiss, Kroppenstedt, Hippe and Stackebrandt 2003, 1028.

In addition to the characteristics mentioned in the description of *Clostridium estertheticum*, the following distinguishing properties allow identification of this subspecies: colonies on sheep-blood agar are nonhemolytic; temperature optimum for growth is 6–8 °C, the upper limit is 13 °C; pH range for growth is 5.5–7.8 with an optimum of 6.5–7.2;

glycogen cannot be utilized; and, in PYG broth, the most abundant nongaseous fermentation end products are volatile fatty acids including butyrate, acetate, and formate.

Source: chill-stored vacuum-packed beef.

DNA G+C content (mol%): 33.9 (HPLC).

Type strain: ATCC 51377, CIP 105093, DSM 8809, NCIMB 12511.

GenBank accession number (16S rRNA gene): S46734, X68181.

56b. **Clostridium estertheticum subsp. laramiense** Spring, Merkhoffer, Weiss, Kroppenstedt, Hippe and Stackebrandt 2003, 1028^VP

la.ra.mi.en'se. N.L. neut. adj. *laramiense* referring to the city of Laramie, WY, USA.

Basonym: *Clostridium laramiense* Kalchayanand et al. (1993). The original description was given by Kalchayanand et al. (1993).

Medium to thick rod-shaped cells, 0.5 × 10 μm). Gram-stain-positive. Oval, terminal endospores are formed. Straight and tumbling motion by peritricous flagella.

Colonies on blood agar are small, grayish white, smooth, and convex. Good growth with gas formation in broth supplemented with L-cysteine (0.05%), hemin (0.1%), and vitamin K (0.001%).

Growth occurs in the temperature range −3–21 °C with optimum growth at 15 °C. No growth at 25 °C or <−3 °C. The pH range for growth is 4.5–7.5 with optimum growth at pH 6.5.

Glucose, fructose, galactose, sucrose, maltose, melibiose, mannose, rhamnose, raffinose, mannitol, inositol, and gluconate are fermented. Arabinose, cellobiose, and xylose are also utilized. Lactose, ribose, trehalose, and inulin are not fermented. The end products of fermentation are butyric acid, butanol, acetate, formate, propionate, isobutyrate, and ethanol depending on the growth medium and substrate.

Catalase, lecithinase, and indole-negative. Lipase and β-hemolysin positive. Starch is hydrolyzed and meat digested. Gelatin and esculin are not hydrolyzed. Milk reaction negative. Nitrate is reduced. The following antibiotics inhibit growth: ampicillin (4 μg), cephalothin (6 μg), chloramphenicol (12 μg), erythromycin (3 μg), penicillin G (2 U), and tetracycline (6 μg). Nontoxic to mice.

DNA G+C content (mol%): 32.4 (HPLC).

Type strain: NK1, ATCC 51254, DSM 14864.

GenBank accession number (16S rRNA gene): AJ506115.

Further comments: In contrast to the original description (Kalchayanand et al., 1993), Spring et al. (2003) reported that arabinose, cellobiose, and xylose can be utilized as substrates. In addition, Spring et al. (2003) reported the G+C content of the DNA of strain NK^T to be 32.4 mol% (HPLC method) in contrast to the value of 26 mol% (thermal denaturation method) reported by Kalchayanand et al. (1993) in the original description of *Clostridium laramiense*. Strain NK^T can be differentiation from *Clostridium estertheticum* subsp. *estertheticum* on the basis of the colonies on sheep-blood agar being β-hemolytic; having a temperature optimum for growth of 15 °C, with an upper limit of 21 °C; the pH range for growth being 4.5–7.5 with an optimum of 6.5; glycogen

being utilized; and, in PYG broth, the end products of fermentation are butyrate, 1-butanol, and lactate.

57. **Clostridium fallax**[*] (Weinberg and Séguin 1915) Bergey, Harrison, Breed, Hammer and Huntoon 1923, 325^AL (*Bacillus fallax* Weinberg and Séguin 1915, 686.)

fal'lax. L. adj. *fallax* deceptive.

Cells in PYG broth cultures are Gram-stain-positive rods with rounded ends, 0.5–1.4 × 1.6–15.4 μm that occur singly or in pairs. In freshly isolated strains and young cultures, cells are motile and peritrichous, but both motility and flagella may be lost on subsequent transfer. Spores are oval, central to subterminal, and swell the cell. Sporulation of most strains occurs most readily in chopped-meat broth cultures. Cell walls contain LL-DAP and the sugars glucose, galactose, and rhamnose.

Surface colonies on blood agar plates are 1–5 mm in diameter, hemolytic, circular to slightly irregular, with entire to slightly erose margins, raised or convex, translucent, gray, shiny and smooth, often with a mottled or granular internal structure.

Cultures in PYG broth are turbid with a smooth sediment and have a pH of 4.8–5.3 after incubation for 1 week. The optimum temperature for growth is 37 °C. Strains grow nearly as well at 30 °C and 45 °C but poorly if at all at 25 °C. Growth is inhibited by 20% bile, 6.5 NaCl, or a pH of 8.5. Abundant gas is detected in PYG deep agar cultures. Deoxyribonuclease is present. Products in PYG broth are acetic, butyric, and lactic acids; pyruvic and succinic acids are usually detected. Abundant H_2 is produced. Pyruvate is converted to acetate; neither lacetate nor threonine is utilized. Ammonia and H_2S are produced.

All strains tested are susceptible to chloramphenicol, clindamycin, erythromycin, penicillin G, and tetracycline. Culture supernatants are nontoxic to mice. The organism has been reported to be pathogenic for guinea pigs and mice, but pathogenicity is quickly lost.

Other characteristics of the species are given in Table 142.

Source: soil, marine sediments, animal wounds, clinical specimens from soft tissue infections in humans; human feces.

DNA G+C content (mol%): 26 (T_m).

Type strain: ATCC 19400, BCRC 14512, CCUG 4853, DSM 2631, JCM 1398, NCIMB 10634, NCTC 8380.

GenBank accession number (16S rRNA gene): M59088.

Further comments: 16S rRNA gene sequence analysis shows *Clostridium fallax* to fall within the radiation of the cluster I of the clostridia as defined by Collins et al. (1994). The highest sequence similarities are found to the species *Eubacterium tarantellae* (95.6%), *Clostridium polyendosporum* (95.6%), *Clostridium septicum* (95.1%), and *Clostridium chauvoei* (95.0%).

58. **Clostridium felsineum**[*] (Carbone and Tombolato 1917) Spray 1939, 766^AL (*Bacillus felsineus* Carbone and Tombolato 1917, 563)

fel.si'ne.um. L. n. *Felsina* ancient Latin name for Bologna, Italy; N.L. neut. adj. *felsineum* pertaining to Bologna.

Cells in PYG broth cultures are Gram-stain-positive rods; Gram-stain-negative in older cultures, and 0.5–1.3 ×

3.1–25.7 μm. They are motile and peritrichous although motility may be lost in cultures that have been maintained in the laboratory for many years. They are granulose positive in starch medium. Cells occur singly, in pairs, and in short or long chains. Spores are oval, subterminal, and swell the cell. There are no appendages and no exposporium. Spores are difficult to detect, but usually can be found in PYG or PY-starch broth cultures incubated for 3–5 d. Cell walls contain *meso*-DAP, major amounts of galactose and rhamnose, with lesser amounts of glucose and mannose.

These organisms grow very poorly if at all on anaerobic blood agar or egg-yolk agar plates. Surface colonies on brain heart infusion agar roll streak tubes after incubation for 2–4 d are 1–4 mm in diameter, circular, flat to low convex, translucent to opaque, with a mottled or granular surface, a pebbled or mosaic internal structure, and an entire to slightly scalloped or lobate margin. They may be white, yellow, orange, or brownish. When slight growth occurs on anaerobic blood agar plates (3 of 13 strains), colonies are β-hemolytic.

Cultures in PYG broth are turbid with a heavy, ropy, or flocculent, often dark orange sediment, and have a pH of 4.7–5.4 after incubation for 24 h.

The optimum temperature for growth is 37 °C. Most strains grow nearly as well at 30 °C, but poorly if at all at 25 °C or 45 °C. Growth is markedly stimulated by fermen Table carbohydrate but inhibited by 20% bile, 6.5% NaCl, or pH of 8.5. Abundant gas is detected in PYG deep agar cultures. Acetyl methyl carbinol is produced by 7 of 11 strains tested. Neutral red is reduced; resazurin is reduced by 4 of 13 strains tested.

Each of four strains tested, including the type strain, completely digests thin slices of carrot, turnip, and radish in PY-broth cultures within 3 d. PY-pectin is strongly fermented (pH 4.8–5.2), and has pectic lyase, pectic hydrolase, and polygalacturonase but no detectable pectinesterase activity in members of the species.

Spore suspensions of a strain of *Clostridium felsineum* will germinate in and lyse tumor tissue but not healthy tissue in mice. Destruction of the tumor tissue, however, is neither complete nor permanent. Atmospheric N$_2$ is fixed.

Products in PYG broth culture are butyric and acetic acids and butanol; lactic, formic and acetic acids may be detected. Pyruvate is converted to acetate, butyrate, and usually butanol. Neither lactate nor threonine is utilized. Abundant H$_2$ is detected. Acetone, CO$_2$, and ethanol may also be produced.

All strains tested are susceptible to clindamycin, erythromycin, penicillin G, and tetracycline. One of five strains is resistant to chloramphenicol. Culture supernatants are nontoxic to mice.

Other characteristics of the species are given in Table 144.

Source: rotting flax, soil in the USA and Antarctica, human feces.

DNA G+C content (mol%): 26 (T_m).

Type strain: ATCC 17788, DSM 794, JCM 1399.

GenBank accession number (16S rRNA gene): AF270502, X77851.

Further comments: Most of the strains available for study were isolated from enrichment cultures in 5% corn meal mash as described by McClung (1943). Pigmentation of colonies is most pronounced in this medium. Plating medium for purification of cultures was yeast infusion-starch agar and plates were incubated in an oat jar. Colonies of *Clostridium felsineum* under these conditions are yellow and, unlike colonies of *Clostridium roseum*, do not darken on exposure to air. *Clostridium felsineum* may be distinguished from *Clostridium aurantibutyricum* by failure to reduce nitrate and from *Clostridium puniceum* by fermentation of rhamnose, by lack of pectinesterase, and by colony pigmentation on corn meal agar. Although sharing some phenotypic characteristics with these species, 16S rRNA gene sequence comparisons show *Clostridium felsineum* to fall within the radiation of cluster XI of the clostridia as defined by Collins et al. (1994) and it shares highest sequence similarity with the species *Clostridium formicaceticum* (99.3%) and *Clostridium aceticum* (97.7%).

59. **Clostridium fimetarium** Kotsyurbenko, Nozhevnikova, Osipov, Kostrikina and Lysenko 1997, 242^{VP} (Effective publication: Kotsyurbenko, Nozhevnikova, Osipov, Kostrikina and Lysenko 1995, 810.)

fi.me.ta′ri.um. N.L. neut. adj. *fimetarium* inhabiting manure.

Rod-shaped cells 0.5–0.6 × 2.1–5.0 μm long. Gram-stain-positive. Motile by peritrichous flagella. Oligosporus. Oval spores are 1.1 × 1.3 μm and terminal. Growth occurs in the temperature range 1–30 °C with an optimum temperature for growth of 20–25 °C. The pH range for growth is 5.5–8.3 with optimum growth at 6.8. Strict anaerobe.

Saccharolytic. Glucose, fructose, maltose, arabinose, xylose, cellobiose, galactose, and mannose are fermented. Raffinose, sorbose, ribose, lactose, rhamnose, trehalose, fucose, sucrose, melibiose, erythritol, inositol, dulcitol, adonite, mannitol, sorbitol, N-acetyl-D-glucosamine, D-glucosamine HCl, betaine, choline chloride, sarcosine HCl, glycogen, starch, cellulose (MCC), peptone, L-histidine HCl, DL-methionine, DL-serine, L-glutamate, L-glutamine, glycine, N,N-dimethylglycine, pyruvate, succinate, fumarate, malate, lactate, Casamino acids, glycerol, formate, acetate, propionate, butyrate, methanol, ethanol, propanol, butanol, mono-, di-, and trimethylamine, and H$_2$ + CO$_2$. and are not utilized. Polysaccharides are not hydrolyzed. The end products of fermentation are acetate, ethanol, lactate, formate, H$_2$, and CO$_2$. Formate, butyrate, and butanol are not produced. The main fatty acids are C$_{16:0}$ and C$_{18:1 ω9}$.

Source: cattle manure digested at low temperature.

DNA G+C content (mol%): 35.6 (T_m).

Type strain: Z-2189, CIP 105360, DSM 9179.

GenBank accession number (16S rRNA gene): AF126687.

Further comments: When this species was described, the 16S rRNA gene sequence was not determined and the species was assigned to the genus *Clostridium* on the basis of the morphological and physiological characteristics observed. Subsequent determination and analysis of the 16S rRNA gene sequence of *Clostridium fimetarium* placed it outside cluster I of the clostridia as defined by Collins et al. (1994). *Clostridium fimetarium* falls within the radiation of cluster XIVa as defined by Collins et al. (1994).

60. **Clostridium formicaceticum**[*] corrig. Andreesen, Gottschalk and Schlegel 1970, 155[AL]

for.mic.a.ce ti.cum. L. n. *formica* an ant, N.L. adj. *formicus* pertaining to ants, to formic acid; L. n. *acetum* wine-vinegar, N.L. adj. *aceticus* pertaining to vinegar, to acetic acid; N.L. neut. adj. *formicaceticum* pertaining to formic and acetic acids.

Cells in PY-fructose broth cultures are Gram-stain-negative, straight to slightly curved rods, motile and peritrichous, 1.2–2.0 × 5–12 μm, and occur singly or in pairs. Spores are round, terminal or subterminal, and swell the cell. Sporulation occurs readily both in broth media and on chopped-meat slants. Cell walls contain *meso*-DAP.

Surface colonies on brain heart infusion agar streak tubes incubated in an atmosphere of 100% N_2 are 1–3 mm in diameter, circular to slightly irregular, entire, flat to low convex, semiopaque, white, shiny, smooth, with a mosaic internal structure.

Cultures in PY-fructose broth are slightly turbid with a smooth sediment and have a pH of 6.3–6.7 (compared with a pH of 7.2 in PY broth) after incubation for 4 d in an atmosphere of 100% N_2. Ammonia is produced. Resazurin is not reduced.

The optimum temperature for growth is 37 °C. Moderate growth occurs at 28 °C, 32 °C, and 44 °C; there is no growth at 52 °C. Bicarbonate or formate and fermentable carbohydrate are required for growth. Substrates utilized for growth include fructose, ribose, gluconate, glucuronate, galacturonate, 2-keto-3-deoxygluconate, mannonate, galacturonate, glutamate, malate, mannitol, glycerol, lactate, pyruvate, fumarate, and pectin. An α-isopropylmalate synthase has been identified in extracts of *Clostridium formicaceticum* indicating that leucine is synthesized by the isopropylmalate pathway used by many aerobic organisms. Enzymes utilized in the degradation of glutamate and those required for the assimilation of N_2 and NH_4^+ have been identified.

Carbon monoxide is oxidized to CO_2 by cell suspensions from cultures grown in fructose broth and this reaction is coupled with the reduction of CO_2 to acetate. During active growth in an atmosphere of either 100% N_2 or 90% N_2–10% CO_2, acetate is the major product detected in PY-fructose broth cultures; small amounts of succinate also may be formed. In the stationary phase of growth, both acetate and formate are produced. No H_2 is produced. L-malate is converted to acetate and CO_2; fumarate is reduced to succinate, acetate, and CO_2. The electron carriers cytochrome *b* and menaquinone are present in cultures and possibly involved in this reaction.

All strains tested are susceptible to chloramphenicol, clindamycin, erythromycin, penicillin G, and tetracycline. Culture supernatants are nontoxic to mice.

Other characteristics of the species are given in Table 143.

Source: sewage, pond and ditch mud, stagnant river water.

DNA G+C content (mol%): 34 (T_m).

Type strain: ATCC 27076, DSM 92.

GenBank accession number (16S rRNA gene): X77836.

Further comments: This species can be distinguished most readily from *Clostridium aceticum* by its production of formate in later stages of growth and by its inability to utilize H_2 as a reducing agent in the conversion of CO_2 to acetate. *Clostridium formicaceticum* falls within the cluster XI of the clostridia as defined by Collins et al. (1994). It shows highest sequence similarity to *Clostridium felsineum* (99.3%) and *Clostridium aceticum* (97.7%).

61. **Clostridium frigidicarnis** Broda, Lawson, Bell and Musgrave 1999, 1549[VP]

fri.gi.di.car′nis. L. adj. *frigidus* cool; L. n. *caro carnis* meat; N.L. gen. n. *frigidicarnis* of cool meat.

Rod-shaped cells, 1.3–1.6 × 4.5–9.4 μm, occur singly or in pairs. Motile. Gram-stain-positive. Form elliptical spores in early stationary growth phase. Colonies on sheep blood agar are 2.2–7.2 mm in diameter, circular to irregular shape, and creamy-gray to gray in color. Colonies are semi-opaque to translucent with circular or irregular with undulate, lobate, or erose margins. β-Hemolytic. Psychrotolerant growing in the temperature range 3.8–40.5 °C with an optimum of 30.0–38.5 °C. The pH range for growth is 4.7–9.5 with optimum growth at 6.4–7.2.

Ferments fructose, glucose, mannose, maltose, sorbitol, and trehalose. Cellobiose, galactose, lactose, mannitol, and xylose are not fermented. Ferments PYGS broth to produce acetate, ethanol, butyrate, isovalerate, butanol, isobutyrate, oxalacetate, lactate, hydrogen, and carbon dioxide. Indole is not produced. Lecithinase-positive and lipase-negative. Hydrolyzes gelatin. Esculin and starch are not hydrolyzed. Meat and milk are digested.

Source: temperature abused vacuum-packed beef.

DNA G+C content (mol%): 27.3–28.4 (T_m).

Type strain: SPL77A, ATCC BAA-154, DSM 12271.

GenBank accession number (16S rRNA gene): AF069742.

Further comments: 16S rRNA gene sequence analysis places *Clostridium frigidicarnis* in the cluster I of the clostridia as defined by Collins et al. (1994). It represents a distinct lineage within cluster I but shows highest sequence similarities to the species *Clostridium septicum* (95.9%), *Clostridium chauvoei* (95.2%), *Clostridium tertium* (94.7%), and *Clostridium carnis* (94.5%).

62. **Clostridium frigoris** Spring, Merkhoffer, Weiss, Kroppenstedt, Hippe and Stackebrandt 2003, 1026[VP]

fri′go.ris. L. gen. n. *frigoris* of the cold.

Rod-shaped cells, 1.4–1.8 × 2.2–5.0 μm, occurring singly, in pairs or short chains. Filamentous cells are occasionally present, especially in cultures grown on agar plates. Motile by peritrichous flagella. Endospores are spherical and terminal in position; sporangium not swollen. Gram-stain-positive. Cell-wall peptidoglycan contains *meso*-diaminopimelic acid. Colonies on sheep-blood agar are 1–2 mm in diameter, round with often coarsely granulated margins, smooth, slightly raised, cream-white to grayish, semi-transparent to opaque and nonhemolytic. The major cellular fatty acids are $C_{16:1\,\omega9c}$, $C_{14:0}$, $C_{16:1\,DMA\,\omega9c}$, $C_{16:0}$, $C_{16:1\,\omega11c}$, and an unknown compound with an equivalent chain-length of 14.777–14.783. Optimal growth occurs at 5–7 °C with an upper limit of 11 °C. The pH range for growth is 5.5–7.5 with optimal growth in the range 6.8–7.2. Doubling time of 11 h under optimal growth conditions. Amygdalin, arabinose, cellobiose, fructose, galactose, glucose, glycogen, inositol (weak), inulin, lactose, maltose, mannose, melibiose, raffinose, rhamnose, ribose (weak), salicin, starch, sucrose, trehalose, and xylose are utilized. Mannitol, melezitose and sorbitol are not utilized. Gelatin and starch are not hydrolyzed. The end products of fermentation are butyrate, formate, lactate, acetate, ethanol, hydrogen, and carbon dioxide.

Source: a microbial mat from a moated area around Lake Fryxell, Antarctica.

DNA G+C content (mol%): 31.9 (HPLC).

Type strain: D-1/D-an/II, ATCC BAA-579, DSM 14204.

GenBank accession number (16S rRNA gene): AJ506116 (clone 5), AJ506117 (clone 10).

Further comments: 16S rRNA gene sequence comparisons place *Clostridium frigoris* within cluster I of the clostridia as defined by Collins et al. (1994). Two 16S rRNA gene sequence clones of *Clostridium frigoris* were sequenced and found to share 99.4% sequence similarity. *Clostridium frigoris* shares highest sequence similarity with the species *Clostridium estertheticum* subsp. *laramiense* (99.7%), *Clostridium estertheticum* subsp. *estertheticum* (99.6%), *Clostridium lacusfryxellense* (99.2%), and *Clostridium bowmanii* (98.6%). DNA–DNA reassociation values in the range 16–62% between *Clostridium frigoris* and these related species helped define the individual species status of *Clostridium frigoris* (Spring et al., 2003). In addition, substrate utilization patterns and end products of fermentation can be used to differentiate this species.

63. **Clostridium ganghwense** Kim, Jeong, Kim and Chun 2006, 693[VP]

gang.hwen'se. N.L. neut. adj. *ganghwense* named after Ganghwa Island in South Korea, the geographical origin of the type strain.

Rod-shaped cells, 0.7–0.8 × 4–8 μm. Motile with peritrichous flagella. Gram-stain-negative. KOH reaction negative. Colonies are circular and yellowish on RCM. Catalase-negative. Strictly anaerobic. Chemoheterotrophic. Requires 1–9% (w/v) artificial sea salts (optimum 3%). Does not grow on RCM containing 0–5% (w/v) NaCl alone. Growth occurs in the temperature range 15–40 °C with an optimal of 35 °C. Growth occurs in the pH range 5.5–10.0 with optimal growth at pH 7.5.

Glucose, maltose, salicin, cellobiose, and mannose are utilized. Mannitol, lactose, sucrose, xylose, arabinose, glycerol, melezitose, raffinose, sorbitol, rhamnose, and trehalose are not utilized. The end products of fermentation of glucose are glycerol, ethanol, and CO_2. Gelatin and esculin are hydrolyzed. Indole is produced urease is not.

Source: tidal flat sediment of Ganghwa Island, South Korea.

DNA G+C content (mol%): not reported.

Type strain: HY-42-06, IMSNU 40127, JCM 13193, KCTC 5146.

GenBank accession number (16S rRNA gene): AY903294.

Further comments: 16S rRNA gene sequence comparisons place *Clostridium ganghwense* within the radiation of cluster I of the clostridia as defined by Collins et al. (1994). The closest relatives based on 16S rRNA gene sequence similarity values are *Clostridium aestuarii* (96.8%), *Clostridium homopropionicum* (95.0%), and *Clostridium hemolyticum* (94.9%).

Clostridium ganghwense can be differentiated from *Clostridium aestuarii* based on differences in substrate utilization patterns for sucrose, mannose, salicin, gelatin, and cellobiose. Differences also exist in end products formed from fermentation with *Clostridium aestuarii* forming butyrate, propionate, glycerol, and H_2 while *Clostridium ganghwense* forms glycerol, ethanol, and CO_2.

64. **Clostridium gasigenes** Broda, Saul, Lawson, Bell and Musgrave 2000b, 116[VP]

ga.si'ge.nes. N.L. neut. n. *gasum* gas; Gr. v. *gennaio* to produce; N.L. gen. n. *gasigenes* gas-producing.

Rod-shaped cells, 0.4 × 2.0–7.5 μm. Motile. Gram-stain-positive. On blood sheep agar, colonies measure 0.7–3.0 mm in diameter. Colonies are gray-white to gray, with entire margins, circular, convex and shiny. β-Hemolytic. Produce elliptical subterminal spores in late-stationary growth phase. Psychrophilic with a temperature range for growth of 1.5–26 °C and an optimum of 20–22 °C. The pH range for growth is 5.4–8.9 with an optimum at 6.2–8.6.

Requires a fermentable carbohydrate for growth, even in PY broth. Ferments cellobiose, fructose, glucose, inositol, maltose, mannose, salicin, and trehalose. Lactose, xylose, arabinose, mannitol, raffinose, and sorbitol are not fermented. Hydrolyzes gelatin, esculin, and starch. Fermentation end products from PYGS broth are ethanol, acetate, butyrate, lactate, butanol, CO_2, and H_2.

Source: vacuum-packed, chilled lamb that showed "blown-pack" spoilage, which causes gas production and pack distension in vacuum-packed meat.

DNA G+C content (mol%): 28.3–29.4 (T_m).

Type strain: DB1A, ATCC BAA-158, CIP 106517, DSM 12272.

GenBank accession number (16S rRNA gene): AF092548.

Further comments: 16S rRNA gene sequence comparisons shows *Clostridium gasigenes* to fall within cluster I of the clostridia as defined by Collins et al. (1994). The closest relatives based on 16S rRNA gene sequence comparisons are *Clostridium carnis* (96.7%), *Clostridium vincentii* (96.2%), and *Clostridium tertium* (96.1%). *Clostridium gasigenes* is differentiated from these related species on the basis of soluble protein patterns, fatty acid profiles, end products of fermentation, and substrate utilization patterns (Broda et al., 2000b).

65. **Clostridium ghonii*** corrig. Prévot 1938, 83[AL]

gho'ni.i. N.L. gen. *ghonii* pertaining to Anton Ghon, Austrian bacteriologist.

Cells in PYG broth culture are Gram-stain-positive, straight rods, motile and peritrichous, and 0.5–1.4 × 1.6–6.3 μm. They usually occur singly, occasionally in pairs. Spores are central to subterminal, oval, and swell the cell. Sporulation occurs most readily on egg-yolk agar plates or in chopped-meat broth incubated for 2 d. Cell walls contain *meso*-DAP.

Surface colonies on blood agar plates are usually β-hemolytic, 0.5–2.0 mm in diameter, circular to slightly irregular, scalloped to lobate, translucent to semiopaque, flat to raised, white, shiny, and with a granular or mosaic internal structure.

Cultures in PYG broth are turbid with a smooth to ropy sediment and a pH of 6.2–6.4 after incubation for 4 d.

Growth is stimulated by carbohydrate even though no pH depression occurs. Good growth occurs at a pH of 8.5; growth is inhibited by 6.5% NaCl or 20% bile. Abundant gas is detected in PYG deep agar cultures. Gelatin, milk, meat, and casein are rapidly digested. H_2S and ammonia are formed. Neutral red is reduced; two of the five strains reduce resazurin.

The organism contains a wide variety of cellular C_{12}–C_{18} fatty acids with unbranched, saturated fatty acids predominating, but with significant amounts of branched chain acids as well.

Products in PYG broth after 4 d of incubation include acetate, isobutyrate, butyrate, isovalerate, isocaproate, ethanol, butanol, and isobutanol; formate, propionate, and propanol may also be detected.

Moderate amounts of H_2 are formed. *Clostridium ghonii* has been reported to form significant amounts of methane and dithia-2,3 butane, and moderate amounts of propanone and thiocyclopropane in sodium thioglycolate glucose broth cultures, but this finding requires confirmation. Pyruvate is converted to acetate, butyrate, and ethanol; threonine is converted to propionate; lactate is not utilized. Valine is converted to isobutyrate, leucine to isovalerate and isocaproate, isoleucine to isovalerate. Phenylpropionic and phenyllactic acids and indole are formed in trypticase-yeast extract-thioglycollate medium containing phenylalanine, tyrosine, and tryptophan.

All strains, tested are susceptible to chloramphenicol, clindamycin, erythromycin, penicillin G, and tetracycline. Culture supernatants are nontoxic to mice.

Other characteristics of the species are given in Table 145.

Source: soil and marine sediment, soft tissue infections in humans, human feces.

DNA G+C content (mol%): 27 (T_m).

Type strain: ATCC 25757, BCRC 14548, CCUG 9282, DSM 15049, JCM 1400, NCIMB 10636.

GenBank accession number (16S rRNA gene): X73451.

Further comments: 16S rRNA gene sequence comparisons show *Clostridium ghonii* to fall within the cluster XI of the clostridia as defined by Collins et al. (1994). The closest relatives based on sequence similarities are *Eubacterium tenue* (99.0%), *Clostridium sordellii* (98.9%), and *Clostridium bifermentans* (97.2%).

66. **Clostridium glycolicum**[*] Gaston and Stadtman 1963, 356[AL.] emend. Chamkha, Labat, Patel and Garcia 2001b, 2052.

gly.co′li.cum. L. adj. suff. *-icus* related to, belonging to; N.L. neut. adj. *glycolicum* referring to the ability to ferment ethylene glycol.

This description is based on that of Gaston and Stadtman (1963) and the emended description of Chamkha et al. (2001b).

Cells in PYG broth cultures are Gram-stain-positive, straight to slightly curved rods, motile and peritrichous, 0.3–1.3 × 1.8–15.4 μm, and occur singly or in pairs. Spores are oval and usually subterminal, occasionally terminal, often occurring as free spores. Sporulation occurs most readily in chopped-meat broth cultures or on chopped-meat agar slants. Cell walls do not contain DAP. They do contain lysine and iso-D-asparagine.

Surface colonies on blood agar plates are 1–4 mm, circular to irregular, raised to convex, translucent to semiopaque, grayish white, shiny, and smooth with a granular, mottled, or mosaic internal structure, and entire, scalloped, or erose margins.

Cultures in PYG broth are turbid with a smooth to stringy sediment and have a pH of 5.5–5.9 after incubation for 24 h. After 5 d of incubation, the pH is 5.2–5.6.

The optimum temperature for growth is 30–37 °C. Most strains grow moderately well at 25 °C, poorly if at all at 45 °C. Growth is inhibited by 6.5% NaCl, 20% bile, and a pH of 8.5. Abundant gas is detected in PYG deep agar cultures. Deoxyribonuclease is present. Ammonia is produced. H_2S is formed by 9 of 13 strains tested. Four of 13 strains hydrolyze hippurate. This organism can ferment ethylene glycol to acetate and ethanol, and propylene glycol to propionate, propanol, and small amounts of acetate and ethanol. A wide variety of both normal straight-chain saturated and iso-branched chain saturated and unsaturated fatty acids is present in strains of this species. *Clostridium glycolicum* degrades uracil to β-alanine, ammonia, and CO_2. Sugars fermented are glucose, fructose, sorbitol, xylose, maltose, and dulcitol, with acid and gas formation. Succinate is fermented into propionate and CO_2; glycerol is fermented into acetate, ethanol, H_2, and CO_2; yeast extract, Casamino acids, and Biotrypcase are fermented into acetate, isovalerate, ethanol, H_2, and CO_2; and peptone is fermented into acetate, isovalerate, and CO_2. Substrates not fermented include amygdalin, arabinose, cellobiose, galactose, glycogen, inulin, lactose, mannitol, mannose, melezitose, melibiose, raffinose, rhamnose, ribose, salicin, sucrose, and trehalose

Products in PYG broth are acetic, isovaleric, and isobutyric acids, and ethyl, propyl, isobutyl, and isoamyl alcohols; propionic, formic, lactic, and succinic acids also may be detected. Abundant H_2 is produced. Threonine is converted to propionate; lactate is not utilized. Reduces the double bond of the side chain of a wide range of cinnamic acids: cinnamic, *o*-, *m*-, and *p*-coumaric, *o*-, *m*-, and *p*-methoxycinnamic, *p*-methylcinnamic, caffeic, ferulic, and isoferulic acids. The end products of metabolism of cinnamic acids are the corresponding phenylpropionic acids: 3-phenyl, 3-(2-hydroxyphenyl), 3-(3-hydroxyphenyl), 3-(4-hydroxyphenyl), 3-(2-methoxyphenyl), 3-(3-methoxyphenyl), 3-(4-methoxyphenyl), 3-(4-methylphenyl), 3-(3,4-dihydroxyphenyl), 3-(4-hydroxy-3- methoxyphenyl), and 3-(3-hydroxy-4-methoxyphenyl) propionic acids.

All strains are susceptible to penicillin G; of 30 strains tested, five are resistant to erythromycin, two to clindamycin, one to chloramphenicol, and one to tetracycline. Susceptibility to metronidazole is variable. Culture supernatants are nontoxic to mice.

Does not produce lecithinase, lipase, indole, caproic, or butyric acid. Does not hydrolyze starch or gelatin. Sulfate, thiosulfate, sulfite, elemental sulfur, nitrate, and fumarate are not reduced.

Source: soil, mud, snake venom, bovine intestine; human clinical specimens including wounds, abscesses, and peritoneal fluid; human feces.

DNA G+C content (mol%): 29 (T_m).

Type strain: ATCC 14880, DSM 1288, JCM 1401, NCIMB 10632, NCTC 13026.

GenBank accession number (16S rRNA gene): X76750.

Further comments: 16S rRNA gene sequence comparisons show *Clostridium glycolicum* to fall within the radiation of cluster XI of the clostridia as defined by Collins et al. (1994). The closest relatives based on sequence similarity values are *Clostridium mayombei* (98.9%), *Clostridium bartlettii* (96.7%), and *Clostridium irregulare* (96.3%). The production of large amounts of isovaleric acid is most helpful in distinguishing this species from other saccharolytic, nonproteolytic clostridia.

67. **Clostridium glycyrrhizinilyticum** Sakuma, Kitahara, Kibe, Sakamoto and Benno 2006b, 2057[VP] (Effective publication: Sakuma, Kitahara, Kibe, Sakamoto and Benno 2006a, 484.)

gly.cyr.rhi.zi.ni.ly'ti.cum. N.L. neut. n. *glycyrrhizinum* glycyrrhizin (a sugar from the roots of *Glycyrrhizinum* species); N.L. neut. adj. *lyticum* dissolving, able to dissolve; N.L. adj. *glycyrrhizinilyticum* glycyrrhizin-dissolving.

Short rods or rod-shaped cells, 0.5 × 1–2 μm. Colonies are <1 mm in diameter, disc shaped, white-grayish color when grown on GAM blood agar plate for 48 hours. Obligate anaerobe. Nonsporeformer. Gram-stain-positive. Nonmotile. Growth temperature 37 °C.

Acid is produced from L-arabinose, glucose, lactose, maltose, L-rhamnose, and D-xylose. Acid is not produced from glycerol, D-mannitol, sucrose, D-cellobiose, D-mannose, D-melezitose, D-raffinose, D-sorbitol, or D-trehalose. Esculin and gelatin are not hydrolyzed. Indole, catalase, and urease are not produced. Using Rapid ID 32A two strains had positive reactions for β-galactosidase, β-glucosidase, α-arabinosidase, β-glucuronidase and glutamic acid decarboxylase. Two strains had negative reactions on *N*-acetyl-β-glucosaminidase, alkaline phosphatase, leucylglycine arylamidase, alanine arylamidase, glycine arylamidase, glutamylglutamic acid arylamidase, α-galactosidase, β-galactosidasae-6-phosphate, α-glucosidase, α-fucosidase, urease, arginine dihydrorase, praline arylamidase, pyroglutamic acid arylamidase, tyrosine arylamidase, serine arylamidase, arginine arylamidase, phenylalanine arylamidase, leucine arylamidase, and histidine arylamidase.

Source: feces of a healthy human.

DNA G+C content (mol%): 45.7 (HPLC).

Type strain: ZM35, DSM 17593, JCM 13368.

GenBank accession number (16S rRNA gene): AB233029.

Further comments: 16S rRNA gene sequence comparisons show *Clostridium glycyrrhizinilyticum* to fall within the cluster XIVa of the clostridia as defined by Collins et al. (1994). The closest relatives based on sequence similarity values in the range 95.7–96.8% are *Eubacterium contortum*, *Coprococcus comes*, *Clostridium oroticum*, and *Ruminococcus torques*. *Clostridium glycyrrhizinilyticum* can be differentiated from these related taxa on the basis of its negative result for esculin hydrolysis as well as acid production from a number of substrates.

68. **Clostridium grantii** Mountfort, Rainey, Burghardt and Stackebrandt 1996, 625[VP] (Effective publication: Mountfort, Rainey, Burghardt and Stackebrandt 1994, 178.)

grant.i'i. N.L. gen. n. *grantii* pertaining to William Donaldson Grant for his contribution to our understanding of marine polysaccharide degradation.

Slightly curved rod-shaped cells, 0.6 × 3.5–7.0 μm. Obligate anaerobe. Gram-stain-positive. Cell walls are composed of *meso*-diaminopimelic acid. Motile. Forms spores.

The temperature range for growth is 17–45 °C with an optimum at 30 °C. Growth occurs in the pH range 5.5–8.0 with an optimum at 6.5. The salinity range for growth is 1 to 10/10³. Ferments alginate, cellobiose, glucose, maltose, mannose, and fructose to produce acetate, ethanol, formate, and CO_2. Lactate, propionate, butyrate, succinate, and H_2 are not produced in fermentation. Does not utilize lactose, galactose, sucrose, dulcitol, carrageenan, agar, laminarin, cellulose, xylan, fucoidin, isoleucine, valine, esculin, fumarate, lactate, glycerol, malate, yeast extract, and Casamino acids. Nitrate and sulfate are not reduced.

Source: mullet gut contents.

DNA G+C content (mol%): 30.2 (T_m).

Type strain: A-1, CIP 105529, DSM 8605.

GenBank accession number (16S rRNA gene): X75272.

Further comments: 16S rRNA gene sequence comparisons show *Clostridium grantii* to fall within the radiation of cluster I of the clostridia as defined by Collins et al. (1994). The closest relatives based on sequence similarity values are *Clostridium ganghwense* (94.4%), *Clostridium hemolyticum* (94.3%), *Clostridium aestuarii* (94.3%), and *Clostridium homopropionicum* (94.3%).

69. **Clostridium haemolyticum**[*] (Hall 1929) Scott, Turner and Vawter 1935, 1972[AL] (*Bacillus hemolyticus* Hall 1929, 156)

hae.mo.ly'ti.cum. Gr. n. *haema* blood; Gr. adj. *lutikos* dissolving; N.L. neut. adj. *hemolyticum* blood-dissolving, hemolytic.

Cells in PYG broth cultures are motile and peritrichous, 0.6–1.6 × 1.9–17.3 μm, and occur singly or in pairs. In very young cultures, they are Gram-stain-positive but they rapidly become Gram-stain-negative. Spores are oval, subterminal, and swell the cell. Sporulation occurs most readily in 2–3-week-old chopped-meat broth cultures or on chopped-meat slants incubated at 30 °C. Cell walls contain *meso*-DAP, alanine, and glutamic acid.

Surface colonies on blood agar plates are 1–3 mm in diameter, circular, raised to convex, translucent, gray, shiny, with a granular or mosaic surface, and an erose on slightly scalloped margin. Cultures in PYG broth are turbid, usually with a granular or flocculent sediment and have a pH of 5.0–5.5 after incubation for 24 h.

The optimum temperature for growth is 37 °C; growth is slight at 30 °C; there is little or no growth at 25 °C or 45 °C. The organisms are extremely sensitive to oxygen and require both prereduced media and stringent anaerobic conditions. Although they grow moderately well in carbohydrate-free media, growth is stimulated by fermentable carbohydrate. Growth is inhibited by 20% bile, 6.5% NaCl, or a pH of 8.5. Moderate gas is detected in PYG deep agar cultures. Small amounts of acid phosphatase are produced. Fermentation products in PYG broth are propionic, butyric, and acetic acids. Lactate is converted to propionate; threonine is converted to propionate by 9 of 12 strains tested; pyruvate is converted to acetate, propionate, and butyrate. Abundant H_2 is produced. All strains tested are susceptible to chloramphenicol, clindamycin, erythromycin, penicillin G, and tetracycline; the one strain tested is resistant to >100 μg/ml of metronidazole.

Culture supernatants are toxic to mice and cultures are pathogenic for cattle, sheep, and laboratory animals. The major lethal toxin is a phospholipase C (identical with *Clostridium novyi* beta toxin) which hydrolyzes lecithin and sphingomyelin and hemolyzes red blood cells. The organism produces the beta, eta, and theta toxins of *Clostridium novyi* B but not the alpha (necrotizing) toxin, the major lethal toxin of that species. In susceptible animals, *Clostridium haemolyticum* causes fatal bacillary hemoglobinuria. For a thorough review of the properties of *Clostridium haemolyticum*, its toxins, the disease it causes, and its relationship to *Clostridium novyi* B, see Smith (1975).

Other characteristics of the species are given in Table 142.

Source: liver infections, muscle of cattle and sheep, human feces.

DNA G+C content (mol%): 26–27 (T_m).

Type strain: ATCC 9650, DSM 5565, NCTC 13022.

GenBank accession number (16S rRNA gene): AB037910, Y18173.

Further comments: Based on 16S rRNA gene sequence comparisons, *Clostridium haemolyticum* falls within the radiation of cluster I of the clostridia as defined by Collins et al. (1994). The closest relatives based on sequence similarity values are *Clostridium thermobutyricum* (95.6%), *Clostridium thermopalmarium* (95.5%), and *Clostridium homopropionicum* (95.4%).

70. **Clostridium halophilum** corrig. Fendrich, Hippe and Gottschalk 1991, 580^{VP} (Effective publication: Fendrich, Hippe and Gottschalk 1990, 131.)

ha.lo′phi.lum. Gr. n. *hals halos* salt; Gr. adj. *philos* loving; N.L. adj. *halophilum* salt-loving.

Cells are rods, 0.8–1.0 × 2.5–7.0 µm. Strict anaerobe. Chemoorganotroph. Motile. Gram-stain-positive. Halophilic. Form spherical spores that measure 1.2 µm. Temperature range for growth is 18–49 °C with an optimum of 41 °C. Growth occurs in the pH range 6.0–8.0 with an optimum of 7.4.

Ferments D-cellobiose, dulcitol, D-fructose, D-glucose, inulin, D-mannitol, D-mannose, D-ribose, salicin, and D-sorbitol. When grown on pyruvate, cells produce propionate from threonine. Cells undergo the Stickland reaction cleaving betaine into trimethylamine and acetate. Fermentation of complex standard media produces acetate and ethanol. Tween 80 stimulates growth. Cells require 1.5% NaCl for growth, and grow best at 6.0% NaCl. Cells reduce resazurin and neutral red. Esculin is hydrolyzed. Gelatin and starch are not hydrolyzed. Tests negative for catalase, oxidase, lipase, lecithinase, and urease production. Nitrate, sulfate, thiosulfate, or sulfite is not reduced.

Source: anoxic hypersaline sediments of Maledive Islands and Solar Lake Egypt; marine sediments of the North Sea, Germany.

DNA G+C content (mol%): 26.9 (T_m).

Type strain: strain M1, ATCC 49637, DSM 5387.

GenBank accession number (16S rRNA gene): X77837.

Further comments: 16S rRNA gene sequence comparisons show *Clostridium halophilum* to represent a distinct lineage within the cluster XI of the clostridia as defined by Col-

lins et al. (1994). The closest relatives based on sequence similarity values are *Clostridium caminithermale* (92.8%) and *Caminicella sporogenes* (92.7%).

71. **Clostridium hathewayi** Steer, Collins, Gibson, Hippe and Lawson 2002, 685^{VP} (Effective publication: Steer, Collins, Gibson, Hippe and Lawson 2001, 356.)

ha.the.way.i. N.L. gen. n. *hathewayi* of Hatheway, to honor the late American microbiologist Charles L. Hatheway in recognition of his outstanding contributions to the *Clostridium botulinum* group of organisms.

Rod-shaped cells with pointed ends, 0.7–1.0 × 2.0–5.0 µm, occurring in chains up to 30 cells. Strict anaerobe. Gram-stain-negative. May or may not be motile. Cells swell to form oval to round, subterminal spores. Some cells form slim capsules. Colonies are 2–3 mm in diameter, opaque, convex, round, and grayish-white in color. Growth occurs in the temperature range 15–45 °C with an optimum of 37 °C. Growth occurs in the pH range 5.0–8.5 with an optimum of 7.0. Growth occurs in the NaCl range 0–2.5%.

Ferments glucose to acetate, ethanol, CO_2, and H_2. Utilizes lactose, cellobiose, fructose, galactose, mannose, ribose, starch, sucrose, rhamnose, raffinose, melezitose, and xylose. Catalase-negative. Hydrolyzes starch and esculin, but not cellulose, gelatin, or urea. Negative for lipase and lecithinase, indole production, and nitrate reduction.

Source: human feces.

DNA G+C content (mol%): 50.7–50.9 (T_m).

Type strain: 1313, DSM 13479, CCUG 43506.

GenBank accession number (16S rRNA gene): AJ311620.

Further comments: 16S rRNA gene sequence comparisons show *Clostridium hathewayi* to fall within cluster XIVa of the clostridia as defined by Collins et al. (1994). The closest relatives based on sequence similarity are *Clostridium celerecrescens* (95.8%), *Clostridium sphenoides* (95.7%), *Clostridium aerotolerans* (95.3%), *Clostridium indolis* (95.3%), and *Clostridium saccharolyticum* (95.2%). *Clostridium hathewayi* can be differentiated from these related taxa on the basis of differences in end products of fermentation as well as substrate fermentation tests (Steer et al., 2001).

72. **Clostridium herbivorans** Varel, Tanner and Woese 1995, 493^{VP}

her.bi.vo′rans. L. fem. n. *herba* a green plant; L. v. *vorare* to devour; N.L. part. adj. *herbivorans* devouring plants.

Straight rods, 0.7–0.9 × 3.5–4.0 µm. Cells occur singly or in pairs. Motile with 15–20 peritrichous flagella. Gram-stain-positive. Obligate anaerobe. Subterminal to terminal spores are produced rarely. Spores are 1 × 2 µm long and cause the sporagium to swell. Cultures grown on insoluble substrates like cellulose or plant cell walls readily produce spores. The optimum temperature for growth is 39–42 °C and the optimum pH 6.8–7.2.

Fermentable carbohydrates like cellobiose, cellulose, maltose, starch, or glycogen are required for growth. Growth is not supported by amygdalin, arabinose, Casamino acids, erythritol, fructose, glucose, inositol, lactate, lactose, mannitol, mannose, melezitose, melibiose, pectin, pyruvate, raffinose, rhammose, ribose, salicin, sorbitol, sucrose, trehalose, xylose, and xylan. Ruminal fluid (15%)

and yeast extract (1%) stimulate growth. Does not reduce nitrate or sulfate. Negative for catalase, oxidase, and urease activity. Cannot hydrolyze esculin, lecithin, and gelatin. Meat cannot be digested. Indole is not produced. Cellobiose is fermented to produce formate, butyrate, and little ethanol and hydrogen.

Source: pig intestine contents.

DNA G+C content (mol%): 38 (T_m).

Type strain: strain 54408, ATCC 49925, CIP 104610.

GenBank accession number (16S rRNA gene): L34418.

Further comments: 16S rRNA gene sequence analysis shows *Clostridium herivorans* to fall within cluster XIVa of the clostridia as defined by Collins et al. (1994). This species is phylogenetically distinct from the other celulolytic species of the genus *Clostridium* and has as its closest relatives *Clostridium polysaccharolyticum* (96.7%) and *Clostridium populeti* (94.6%). *Clostridium herbivorans* and *Clostridium polysaccharolyticum* both utilize cellulose, cellobiose, maltose, and starch but do not utilize glucose. They can be differentiated based on the fact that *Clostridium polysaccharolyticum* can utilize arbinose, xylan, and xylose but *Clostridium herbivorans* cannot. In addition the DNA G+C content of the two species differs by 4 mol%.

73. **Clostridium hiranonis** Kitahara, Takamine, Imamura and Benno 2001, 43[VP]

hi.ra.no′nis. N.L. masc. gen. n. *hiranonis* of Hirano, after the Japanese microbiologist Seiju Hirano for his contribution to the study of this bacterial isolate.

Straight to slightly curved rods, 0.8 ×1.6–10 µm. Cells occur singly or in pairs. Obligate anaerobe. Gram-stain-positive staining reaction. Colonies on EG blood agar are 1.0–2.0 mm in diameter, disc-shaped, and gray to grayish-white in color. The optimum temperature for growth is 37 °C and the pH 7.5–8.0.

Produces acid from glucose, fructose, mannose, and sucrose. Cannot not produce acid from adonitol, amygdalin, arabinose, cellobiose, dulcitol, erythritol, esculin, galactose, glycerol, glycogen, inositol, inulin, lactose, maltose, mannitol, melezitose, melibiose, raffinose, rhamnose, ribose, salicin, sorbitol, sorbose, starch, trehalose, or xylose. Does not reduce nitrate or sulfate. Cannot produce indole. Does not hydrolyze esculin, gelatin, and starch. Positive for *N*-acetyl-β-D-glucosamidase and proline aminopeptidase using the ANI-DENT test. Peptone-yeast-extract media produce acetic acid and iso-valeric acid, and small amounts of propionic and isobutyric acid.

Source: healthy human feces.

DNA G+C content (mol%): 31.1 (T_m).

Type strain: TO-931, JCM 10541, DSM 13275.

GenBank accession number (16S rRNA gene): AB023970.

Further comments: 16S rRNA gene sequence comparisons show *Clostridium hiranonis* to fall within the radiation of cluster XI of the clostridia as defined by Collins et al. (1994). The closest relative in cluster XI is *Clostridium sordellii* with 94.8% sequence similarity. The lack of relationship between *Clostridium hiranonis* and *Clostridium sordellii* was demonstrated by DNA–DNA reasociation studies. All other species in cluster XI have <94.8% sequence similarity. *Clostridium hiranonis* and *Clostridium sordellii* both have

7α-dehydroxylating activity but that of *Clostridium sordellii* is low by comparison (Kitahara et al., 2001). The two species can be further differentiated based on indole formation, β-acetyl-β-D-glucosamidase, α-glucosidase, and alkaline phosphatase activities.

74. **Clostridium histolyticum**[*] (Weinberg and Séguin 1916) Bergey, Harrison, Breed, Hammer and Huntoon 1923, 328[AL] (*Bacillus histolyticus* Weinberg and Séguin 1916, 449)

his.to.ly′ti.cum. Gr. n. *histos* tissue; Gr. adj. *lutikos* dissolving; N.L. neut. adj. *histolyticum* tissue-dissolving.

Cells in PYG broth culture usually are motile and peritrichous, straight rods, Gram-stain-positive, 0.5–0.9 × 1.3–9.2 µm, and occur singly, in pairs, or in short chains. Spores are oval, central to subterminal, and may slightly swell the cell. There are no exosporium or appendages. Sporulation of most strains occurs most readily in chopped-meat broth cultures. Cell walls contain *meso*-DAP, glutamic acid, and alanine.

Surface colonies on blood agar plates incubated anaerobically are β-hemolytic, 0.5–2 mm in diameter, circular to irregular, flat to low convex, translucent to semiopaque, gray-white, shiny, with a mosaic or granular surface, and an entire to undulate margin. Similar colonies are formed on blood agar plates incubated aerobically, but they are usually smaller and fewer in number. Spores are not formed during aerobic growth.

The optimum temperature for growth is 37 °C; most strains grow nearly as well at 30 °C; there is moderate growth at 25 °C but little or none at 45 °C. Although there is no pH depression, growth is stimulated by carbohydrate. Strains grow well at a pH of 8.5 but are inhibited by 6.5% NaCl or by 20% bile. Moderate amounts of gas are detected in PYG deep agar cultures.

Strains are strongly proteolytic, digesting gelatin, meat, milk, casein, collagen, hemoglobin, fibrin, elastin, egg white, coagulated serum, muscle, liver, brain, and Achilles tendon. Deoxyribonuclease is produced by the one strain tested.

The only major product detected in PYG broth is acetate; traces of formate, lactate or succinate are sometimes present. Small amounts of H_2 are produced; ammonia is produced; H_2S is produced by 4 of 12 strains tested. Pyruvate is converted to acetate; neither lactate nor threonine is converted to propionate. Threonine is utilized, but the product is acetate. Glycine, glutamate, serine, aspartate, and arginine are utilized as sources of energy; valine, leucine, isoleucine, lysine, proline, and alanine are produced. Glycine is fermented to acetate, CO_2, and NH_3.

Only straight-chain fatty acids, principally saturated C_{14}, C_{12}, C_{16}, and C_{18} acids with small amounts of unsaturated acids of the same chain length are present in cells of this species. All strains tested are susceptible to chloramphenicol, clindamycin, erythromycin, penicillin G, and tetracycline. Under anaerobic conditions, strains are susceptible to 0.5–1.0 µg/ml of metronidazole. As the O_2 level increases, they are increasingly resistant.

Supernatant culture fluids are toxic to mice, although the degree of toxigenicity varies among strains and may be lost on subculture. In addition, as the culture ages, the active proteases of the organism may destroy the toxins. *Clostridium histolyticum* is highly pathogenic to laboratory animals. Toxin found in fecal samples from guinea pigs with clindamycin-associated enterocolitis is neutralized by specific antitoxin to *Clostridium histolyticum*. Major lethal toxins are the α-toxin (necrotizing) and at least two collagenases (β-toxin). Collagenase from atoxic *Clostridium histolyticum* strains has been used to digest and remove necrotic tissue from burns. Other characteristics of the species are given in Table 142. Principal habitat is probably the soil.

Source: soil, war wounds, gas gangrene in humans and horses, human intestinal contents; gingival plaque of institutionalized and primitive populations.

DNA G+C content: not reported.

Type strain: ATCC 19401, BCRC 10644, CCUG 4854, CIP 103713, CN 1693, DSM 2158, JCM 1403, NCIMB 503, NCTC 503.

GenBank accession number (16S rRNA gene): M59094.

Further comments: 16S rRNA gene sequence comparisons show that *Clostridium histolyticum* clusters with *Clostridium limosum* (97.2%) and *Clostridium proteolyticum* (96.1%) in cluster II of the clostridia as defined by Collins et al. (1994). *Clostridium histolyticum* can be differentiated readily from *Clostridium limosum*, which it resembles most closely phenotypically, by its production of collagenase and lack of production of lecithinase. Unlike other aerotolerant clostridia, it is extremely proteolytic and does not ferment any carbohydrates tested. Unlike *Clostridium irregulare* it produces only acetic acid in broth cultures.

75. **Clostridium homopropionicum** Dörner and Schink 1991, 580[VP] (Effective publication: Dörner and Schink 1990, 347.)

ho.mo.prop.ion'ic.um. Gr. adj. *homoios* similar; M.L. n. *acidum propionicum* prionic acid; *homopropionicum* referring to metabolic analogy to *Clostridium propionicum*.

Straight rods with rounded ends, 1.2–1.5 × 5.5–10 µm; occur singly or in pairs. Motile with peritrichous flagella. Strict anaerobe. Chemoorganotroph. Gram-stain-negative. KOH test negative. Form spherical terminal to subterminal spores, usually after growth with pyruvate, lactate, or in agar medium. Growth with lactate produces dark inclusions. Grow occurs in the temperature range 4–32 °C with an optimum at 28 °C. pH Range for growth is 6.0–8.5 with an optimum at 7.0.

Fermentation of fructose produces acetate, butyrate, butanol, and H₂. Ferments 2-, 3-, and 4-hydroxybutyrate, 4-chlorobutyrate, crotonate, vinylacetate, and pyruvate to acetate and butyrate. Acetate and propionate are fermentation products of lactate and acrylate. Does not utilize glucose, arabinose, xylose, ethanol, glycerol, 2-aminobutyrate, 4-aminobutyrate, α-alanine, β-alanine, β-hydroxypropionate, serine, aspartate, phenylalanine, malate, fumarate, succinate, mandelate, glycolate, ethylene glycol, trimethoxybenzoate, acetoin, and H_2/CO_2. Does not hydrolyze gelatin. Nitrogen is fixed. Cells do not reduce nitrate, nitrite, sulfite, sulfur, or thiosulfate. Fresh water media produce optimal growth. Cells contain no cytochromes.

Source: anoxic digestor sludge.

DNA G+C content (mol%): 32 (T_m).

Type strain: LuHBu1, ATCC 51426, DSM 5847.

GenBank accession number (16S rRNA gene): X76744.

Further comments: 16S rRNA gene sequence comparisons show *Clostridium homopropionicum* to fall within cluster I as defined by Collins et al. (1994). The closest relatives based on sequence similarity values are *Clostridium thermopalmarium* (95.5%), *Clostridium thermobutyricum* (95.3%), and *Clostridium hemolyticum* (95.3%).

76. **Clostridium hungatei** Monserrate, Leschine and Canale-Parola 2001, 130[VP]

hun.gat'e.i. N.L. gen. n. *hungatei* of Hungate, named after R.E. Hungate, who pioneered the study of the ecology of cellulolytic bacteria.

Slightly curved rods, 0.5 × 2.0–6.0 µm. Motile with subpolar flagella. Obligately anaerobic. Gram-stain-negative. Form round terminal spores, 0.8–1.0 µm. Surface colonies are circular, smooth, slightly raised, unpigmented, and measure 1–2 mm in diameter. Optimum temperature for growth is 30–40 °C. Growth occurs at 45 °C but not at 15 or 50 °C. Optimum pH for growth is 7.2.

Cellulolytic. Fixes nitrogen. Utilizes carbohydrates as energy sources. Ferments cellulose, cellobiose, cellotriose, cellotetraose, cellopentaose, D-glucose, D-fructose, D-mannose, D-xylose, xylan, and gentiobiose. Does not ferment D-ribose, sucrose, maltose, lactose, glycerol, starch, pectin, and polygalacturonic acids. Ferments cellulose producing H₂, CO₂, acetate, ethanol, lactate, and formate. Cells produce an extracellular cellulase system producing a large protein. Vitamins or fatty acids are not required. Grows with rifampin, streptomycin, penicillin, erythromycin, and vancomycin. Growth is inhibited by tetracycline, ampicillin, kanamycin, neomycin, or chloramphenicol.

Source: moist rich soil in decaying plant material.

DNA G+C content (mol%): 40 (T_m).

Type strain: AD, ATCC 700212, DSM 14427.

GenBank accession number (16S rRNA gene): AF020429.

Further comments: 16S rRNA gene sequence comparisons show *Clostridium hungatei* to fall within the radiation of cluster III of the clostridia as defined by Collins et al. (1994). The closest relatives based on sequence similarity are *Clostridium josui* (97.4%), *Clostridium termitidis* (97.1%), *Clostridium cellulolyticum* (97.1%), *Clostridium cellobioparum* (96.9%), and *Clostridium papyrosolvens* (96.4%). *Clostridium hungatei* can be differentiated from these related species based on phenotypic characteristics including Gram staining reaction, flagellar arrangement, and substrate utilization patterns.

77. **Clostridium hylemonae** Kitahara, Takamine, Imamura and Benno 2000, 977[VP]

hai.le.mon'ae. N.L. gen. n. *hylemonae* of Hylemon, after the American microbiologist Phillip B. Hylemon, for his contributions to research on bile acid.

Straight to slightly curved rods, 0.2–0.5 × 1.0–5.3 µm, occur singly or in pairs. Gram-stain-positive. Obligate anaerobe. Nonmotile. Form spores. Colonies are disc-shaped, gray, and 0.5–1.0 mm in diameter. The optimum temperature for growth is 37 °C. Optimum pH for growth is 7.6.

Fermentation of peptone with yeast extract and glucose produces acetic, butyric, propionic, isobutyric, and iso-valeric acids. Two strains ferment galactose to acid. Variable reactions with arabinose, fructose, maltose, ribose, and xylose. None of the strains produce acid from adonitol, amygdalin, cellobiose, dulcitol, erythritol, esculin, glycerol, glycogen, inositol, inulin, lactose, mannitol, mannose, melezitose, melibiose, rhamnose, salicin, sorbitol, sorbose, starch, or trehalose. H_2S and gas are produced. Does not reduce sulfate or nitrate. Does not hydrolyze gelatin, esculin, or starch. Indole not produced. *Source:* healthy adult human feces.

DNA G+C content (mol%): 48.6 (T_m).

Type strain: TN-271, CIP 106689, DSM 15053, JCM 10539.

GenBank accession number (16S rRNA gene): AB023972.

Further comments: 16S rRNA gene sequence comparisons show *Clostridium hylemonae* to fall within the radiation of cluster XIVa as defined by Collins et al. (1994). Previous studies have concentrated on the comparison of *Clostridium hylemonae* and *Clostridium scindens* (Kitahara et al., 2000). However, the 16S rRNA gene sequence similarity between these two species is only 93.2%. Sequence similarities in the range 93.5–94.5% are found with a number of non-clostridial genera in this heterogeneous cluster.

78. **Clostridium indolis**[*] McClung and McCoy 1957, 674[AL]

in.do'lis. N.L. gen. n. *indolis* of indole.

Cells in PYG broth cultures are motile and peritrichous, straight to slightly curved rods, 0.5–0.9 × 1.3–10.2 µm, and occur singly and in pairs. They are Gram-stain-negative in 24 h. Spores are round to oval, swollen, and usually terminal, although subterminal spores are sometimes seen in the same preparation. Spores have no exosporium and no appendages. Sporulation occurs most readily on chopped-meat slants incubated at 30 °C.

Cell walls contain *meso*-DAP. Surface colonies on blood agar plates are nonhemolytic, 0.5–3.0 mm in diameter, circular to slightly irregular, convex, translucent to opaque, white, with a dull, granular surface and an entire to erose margin. Cultures in PYG broth are turbid with a smooth sediment and have a pH of 5.3–5.7 after incubation for 3 d.

Optimum temperature for growth is 37 °C; most strains grow nearly as well at 30 °C; growth is poor at 25 °C and there is little or no growth at 45 °C. Growth is stimulated by fermentable carbohydrate but inhibited by 20% bile or 6.5% NaCl. Abundant gas is detected in PYG deep agar cultures. More than 30% of strains tested can dehydrogenate the steroid nucleus. Polypectate is liquefied; pectin, pectate, and galacturonate are fermented. Citrate is not utilized.

Ammonia and H_2S are produced. Neutral red and resazurin are reduced. Of five strains tested, two produce acetyl methyl carbinol and hydrolyze hippurate.

Products in PYG broth include acetate, formate, and ethanol. Abundant H_2 is detected. Pyruvate is converted to acetate, formate, ethanol, and moderate amounts of butyrate; neither lactate nor threonine is utilized. Galacturonate is converted to acetate, butyrate, CO_2, and H_2. Of five strains tested, three are susceptible to chloramphenicol, four to clindamycin, four to erythromycin, four to tetracycline, and two to penicillin G. Culture supernatant fluids are nontoxic to mice.

Other characteristics of the species are given in Table 142.

Source: soil, human feces, and human clinical specimens from infections associated with the intestinal tract.

DNA G+C content (mol%): 44 (T_m).

Type strain: ATCC 25771, DSM 755, JCM 1380, NCTC 11811.

GenBank accession number (16S rRNA gene): Y18184.

Further comments: Clostridium indolis falls within the radiation of cluster XIVa of the clostridia as defined by Collins et al. (1994). The closest relative based on sequence similarity values is *Clostridium methoxybenzovorans* (99.4%). Other species showing similarity values in the range 98.4–98.5% include *Clostridium celerecrescens*, *Clostridium saccharolyticum*, *Clostridium amygdalium*, and *Clostridium sphenoides*.

Clostridium indolis can be differentiated from *Clostridium sphenoides*, which it resembles most closely phenotypically, by the terminal spores that are produced by all strains; spores of *Clostridium sphenoides* are subterminal. The electrophoretic patterns of soluble cellular proteins of the two species are distinctive. In addition, strains of *Clostridium sphenoides* can ferment citrate with the production of acetate and ethanol, while strains of *Clostridium indolis* do not utilize citrate.

79. **Clostridium innocuum**[*] Smith and King 1962, 939[AL]

in.noc'u.um. L. neut. adj. *innocuum* harmless.

Cells in PYG broth cultures are Gram-stain-positive, nonmotile, straight rods with round or tapered ends, 0.4–1.6 × 1.6–9.4 µm. Sporulating cells tend to be among the larger ones. Cells occur singly or in pairs. Spores are oval, terminal, subterminal, or free and wider than the cell. Sporulation usually occurs readily in chopped-meat broth cultures incubated for 24 h or on egg-yolk agar plates incubated for 48–72 h.

Cell walls contain glucose and galactose but no DAP. They contain a lysine-alanine type of peptidoglycan.

Surface colonies on blood agar plates are 0.5–3 mm in diameter, circular, raised or convex, translucent, gray-white or yellowish, smooth, and shiny, with a mottled or mosaic internal structure and an entire or slightly scalloped margin. Hemolysis is variable between strains. Cultures in PYG broth are turbid with a smooth sediment and a pH of 4.7–5.2 after incubation for 24 h.

Optimum temperature for growth is 37 °C although many strains grow equally well at 25 °C, 30 °C, and 45 °C. Growth is stimulated by fermentable carbohydrate but inhibited by 6.5% NaCl. Abundant gas is detected in PYG deep agar cultures. Neither gelatin, milk, casein, nor meat is digested. Milk may be weakly acidified. Bacteriocin-like particles have been demonstrated in one strain of *Clostridium Innocuum*, but no inhibitory activity was found against any strain tested.

Products in PYG broth are butyric, lactic, and acetic acids; small amounts of formic or succinic acids may be detected. Abundant H_2 is produced. Pyruvate is converted to acetate, butyrate, and sometimes formate. Neither lactate nor threonine is utilized. Ethanol has been detected by high performance liquid chromatography PYG broth cultures. Susceptibility to chloramphenicol, clindamycin, erythromycin, penicillin G, and tetracycline is variable among strains; two strains tested were susceptible to 1.6 μg/ml of metronidazole. Supernatant culture fluids are not toxic to mice, and cultures are not pathogenic in laboratory animals.

Other characteristics of the species are given in Table 144.

Source: human infections, particularly those associated with the intestinal tract; empyema fluids; normal intestinal flora of human infants and adults.

DNA G+C content (mol%): 43–44 (T_m).

Type strain: ATCC 14501, CCUG 36763, BCRC (formerly CCRC) 14517, DSM 1286, JCM 1292, NCIMB 10674.

GenBank accession number (16S rRNA gene): M23732.

Further comments: 16S rRNA gene sequence analysis shows *Clostridium innocum* to fall with cluster XVI of the clostridia as defined by Collins et al. (1994). This species shows sequence similarity value of 94.0 and 93.1% to two *Eubacterium* species, *Eubacterium dolichum* and *Eubacterium tortuosum*. All other taxa show less than 90.0% sequence similarity with *Clostridium innocuum*.

80. **Clostridium intestinale** corrig. Lee, Fujisawa, Kawamura, Itoh and Mitsuoka 1989, 335[VP]

in.test.in.al′e. N.L. neut. adj. *intestinale* pertaining to the intestine.

Straight to slightly curved rods, 0.3–0.4 × 1.4–5.4 μm; occur singly, in pairs, or short chains. Aerotolerant, although cultures grow better anaerobically. Motile with peritrichous flagella. Gram-stain-positive. Form large, terminal, slightly oval spores under anaerobic conditions, measuring 0.1–1.1 × 1.5–2.0 μm. Colonies are circular, convex, rough, translucent, undulate, grayish-white in color, β-hemolytic, and 2–4 mm in diameter under anaerobic conditions. Under aerobic conditions, colonies are circular, convex, smooth, translucent, yellowish white in color, and 1 mm in diameter.

Optimum temperature for growth is 37 °C. Ferments glucose, mannose, fructose, galactose, sucrose, cellobiose, lactose, trehalose, mannitol, sorbitol, salicin, and amygdalin. Does not ferment arabinose, xylose, rhamnose, sorbose, ribose, maltose, melibiose, raffinose, melezitose, starch, glycogen, inulin, glycerol, erythritol, inositol, adonitol, dulcitol, and esculin. Cells produce gas. Fermentation products include acetic acid and butyric acid, with some lactic, formic, and succinic acid. Does not reduce sulfate or nitrate. Hydrolyzes esculin but not starch. Tests negative for catalase, lipase, and lecithinase activity. Does not produce indole.

Source: feces of cattle and pigs.

DNA G+C content (mol%): 26–28 (T_m).

Type strain: Catt39, ATCC 49213, DSM 6191, JCM 7506.

GenBank accession number (16S rRNA gene): X76740.

Further comments: 16S rRNA gene sequence comparisons show *Clostridium intestinale* to fall within cluster I of the clostridia as defined by Collins et al. (1994). Sequence similarity values in the range 94.0–94.9 are found between *Clostridium intestinale* and the species *Anaerobacter polyendosporus*, *Eubacterium tarantellae*, *Clostridium fallax*, and *Clostridium sartagoforme*.

81. **Clostridium irregulare**[*] corrig. (Choukévitch 1911) Prévot 1938, 85[AL] (*Bacillus irregularis* Choukévitch 1911, 348)

ir.reg.u.lar′e. L. neut. adj. *irregulare* irregular, referring to pleomorphic, irregular cells.

Cells in PYG broth culture are Gram-stain-positive, straight to slightly curved rods, motile and peritrichous, and 0.8–1.6 × 3.5–12.6 μm. Cells in week-old cultures may be quite filamentous. They occur singly, in pairs, or in short chains. Spores are oval, central, or subterminal and swell the cell. Sporulation occurs most readily on chopped-meat agar slants incubated at 30 °C.

Surface colonies on blood agar plates are nonhemolytic, pinpoint–0.5 mm, circular to irregular, convex, transparent to translucent, colorless, with a granular or mottled surface and an entire or scalloped margin.

Cultures in PY-glucose broth are turbid with a smooth or ropy sediment and have a pH of 6.0–6.3 after incubation for 4 d. The optimum temperature for growth is between 30 °C and 37 °C. There is little or no growth at 25 °C or 45 °C. Growth is inhibited by 6.5% NaCl and by 20% bile. No gas is detected in PYG deep agar cultures. H_2S is formed in SIM medium. Ammonia is produced. Gelatin is digested but not milk, casein, or meat. The type strain contains a 3-α-hydroxysteroid dehydrogenase. Products detected in PYG broth are acetate and isovalerate; formate, isobutyrate, and propionate also may be detected. H_2 is not formed. Pyruvate is converted to acetate and formate. Neither lactate nor threonine is utilized.

All strains are susceptible to chloramphenicol, clindamycin, erythromycin, penicillin G, and tetracycline.

Supernatant culture fluids are nontoxic for mice, and cultures are not pathogenic for laboratory animals.

Other characteristics of the species are given in Table 145.

Source: normal fecal flora of humans and horses, pond mud, pharmaceutical products prepared in a laboratory adjacent to a stable, human penile lesions, soil.

DNA G+C content (mol%): not reported.

Type strain: ATCC 25756, DSM 2635, JCM 1425.

GenBank accession number (16S rRNA gene): X73447.

Further comments: 16S rRNA gene sequence comparisons show *Clostridium irregulare* to fall within the radiation of cluster XI of the clostridia as defined by Collins et al. (1994). The closest relatives are *Clostridium mayombei* (97.4%), *Clostridium lituseburense* (96.5%), *Clostridium glycolicum* (96.3%), and *Clostridium bartlettii* (96.2%).

82. **Clostridium isatidis** Padden, Dillion, Edmonds, Collins, Alvarez and John 1999, 1030[VP]

i.sa′ti.dis. L. gen. n. *isatidis* of crucifer plant *Isatis*.

Rod-shaped cells, 0.3–0.6 × 1.8–9.1 μm. Cells occur singly, in pairs, or in chains. Gram-stain-positive. Motile. Anaerobic. Produce terminal oval endospores. Colonies are mucoid, white, circular, and convex or raised to a central point. The temperature range for growth is 30–55 °C with an optimum of 49–52 °C. The pH range for growth is 5.6–9.9 with an optimum of 7.2.

Produces acid from amygdalin, cellobiose, fructose, galactose, glucose, maltose, mannose, salicin, sorbitol, sucrose, starch, and trehalose. Arabinose, glycogen, inulin, lactose, mannitol, melezitose, raffinose, rhamnose, ribose, and xylose are not utilized. Nitrate is not reduced.

Growth on PYG produces acetic, lactic, and formic acids with ethanol, carbon dioxide, and hydrogen. Indole is not produced. Hydrolyze esculin. Lectinase-negative. Meat is not digested.

Source: a woad vat.

DNA G+C content (mol%): 27 (T_m).

Type strain: Wv6, CIP 107118, DSM 15098, NCIMB 703071.

GenBank accession number (16S rRNA gene): X98395.

Further comments: 16S rRNA gene sequence comparisons show *Clostridium isatidis* to fall within the cluster I of the clostridia as defined by Collins et al. (1994). The closest relatives based on sequence similarity values are *Clostridium carnis* (97.6%), *Clostridium tertium* (97.4%), *Clostridium celatum* (97.4%), *Clostridium sartagoforme* (97.2%), and *Clostridium septicum* (97.1%). *Clostridium isatidis* can be differentiated from *Clostridium carnis* on the basis of differences in antibiotic resistances and substrate utilization patterns. In addition, *Clostridium carnis* produces butyrate as an end product of fermentation while *Clostridium isatidis* does not.

83. **Clostridium jejuense** Jeong, Yi, Sekiguchi, Muramatsu, Kamagata and Chun 2004, 1467^{VP}

je.ju.en'se. N.L. neut. adj. *jejuense* pertaining to Jeju Island, Korea, geographical origin of the type strain of the species.

Straight to slightly curved rods, 0.5 ×1.8–4.5 µm, with peritrichous flagella. Gram-stain-positive. Strictly anaerobic, mesophilic. Oval to almost spherical terminal endospores. Colonies on RC agar plates are 1.0–2.5 mm in diameter, circular, entire, flat, translucent to opaque, grayish, and smooth after 72 h.

The temperature range for growth is 10–40 °C with an optimum at 30 °C. The pH range for growth is 5.5–9.5 with an optimum at pH 7.0. Growth occurs at NaCl concentrations of 0–0.5% (w/v). Under optimum conditions, the doubling time is 10.5 hours. Utilizes D-cellobiose, D-glucose, lactose, D-maltose, D-mannose, sucrose, D-trehalose, D-fructose, D-galactose, D-ribose, L-arabinose, D-raffinose, D-xylose, and cellulose, but not D-sorbitol, D-mannitol, peptone (0.1%, w/v), or yeast extract (0.2%, w/v). Fermentation end products from glucose include pyruvate, lactate, acetate, formate, and H_2. Catalase- and indole-negative. Esculin is hydrolyzed, but gelatin and urea are not.

Source: soil in Jeju, Korea.

DNA G+C content (mol%): 41 (HPLC).

Type strain: HY-35-12, DSM 15929, IMSNU 40003, KCTC 5026.

GenBank accession number (16S rRNA gene): AY494606.

Further comments: 16S rRNA gene sequence comparison shows *Clostridium jejuense* to fall within the radiation of cluster XIVa as defined by Collins et al. (1994). The closest relatives based on sequence similarities are *Clostridium xylanovorans* (96.6%), *Clostridium aminovalericum* (95.4%), *Clostridium populeti* (93.8%), and *Clostridium amygdalinum* (93.6%). *Clostridium jejuense* can be differentiated from *Clostridium xylanovorans* on the basis of its catalase-negative reaction, utilization of arabinose and xylose but not mannitol. Differentiation of *Clostridium jejuense* from *Clostridium aminovalericum* can be made on the basis of utilization of lactose and D-raffinose but not D-sorbitol, D-mannitol, and yeast extract.

84. **Clostridium josui** Sukhumavasi, Ohmiya, Shimizu and Ueno 1988, 180^{VP}

jo.su'i. N.L. *josui* the first four letters were obtained by combining the initial letters of the names of the authors, and the last letter was added according to the requirements of the *International Code of Nomenclature of Bacteria.*

Rod-shaped cells slightly curved at the end, 0.2–0.3 × 3–5 µm, found singly, or long or short chains. Nonmotile. Anaerobic. Gram-stain-positive, but becomes Gram-stain-negative. Cells swell and form terminal oval spores 0.4–0.5 µm in diameter. Cultures show a clear zone of 10 mm around each colony within 3 d on BMC media. Deep colonies are yellow and spindle shaped. The temperature range for growth is 25–60 °C with an optimum of 45 °C. No growth below 20 °C or above 65 °C. The optimum pH for growth is 7.0. After 3 d of incubation the pH is changed from 7.0 to 5.5–5.1.

Cellulolytic. Does not grow on adonitol, amygdalin, bile, dulcitol, dextrin, erythritol, fructose, galactose, glycerol, glycogen, inositol, inulin, mannitol, mannose, melezitose, pectin, raffinose, rhamnose, salicin, sorbitol, sorbose, starch, sucrose, or trehalose. Esculin is hydrolyzed; starch is not. Tests indole-and nitrate-negative. Casein is not digested, and gelatin is not liquefied. Does not produce lecithinase and lipase. Growth on cellulose produces acetate, ethanol, CO_2, and H_2, with traces of propionate and butyrate.

Source: a compost in Thai.

DNA G+C content (mol%): 40 (T_m).

Type strain: strain III, FERM P-9684.

GenBank accession number (16S rRNA gene): AB011057.

Further comments: 16S rRNA gene sequence comparisons show *Clostridium josui* to fall within the radiation of cluster III of the clostridia as defined by Collins et al. (1994). The closest relatives based on sequence similarity values are *Clostridium papyrosolvens* (98.5%) and *Clostridium cellulolyticum* (98.1%). Similarity vaues in the range 97.4–97.6% are found with the species *Clostridium termitidis*, *Clostridium hungatei*, and *Clostridium cellobioparum*. *Clostridium josui* can be differentiated from *Clostridium papyrosolvens* and *Clostridium cellulolyticum* on the basis of its higher optimum growth temperature as well as Gram-stain-positive staining in young cultures, lack of motility, and utilization of lactose and maltose. The end products of fermentation also differentiate *Clostridium josui* from its closest phylogenetic relatives.

85. **Clostridium kluyveri*** Barker and Taha 1942, 362^{AL}

kluy'ver.i. N.L. gen. n. *kluyveri* of Kluyver, named for A.J. Kluyver, Dutch microbiologist.

Cells in medium containing 20% yeast autolysate, 0.5% ethanol, and inorganic salts are motile and peritrichous, 0.9–1.1 × 3–11 µm, and usually occur singly, occasionally in pairs, or chains. They are weakly Gram-stain-positive, but quickly become Gram-stain-negative. Spores are oval, terminal, or subterminal and swell the cell.

Ultrastructural studies have demonstrated a five-layered cell wall and a three-layered plasma membrane. Cell walls contain *meso*-DAP.

Surface colonies on blood agar plates are nonhemolytic, pinpoint–0.5 mm in diameter, gray, with rhizoid margins. In laked blood streak tubes, colonies are low convex, gray-white, shiny, and smooth, with a scalloped margin and a mosaic internal structure. There is no growth on egg-yolk agar plates. Cultures in PYG-1.5% ethanol broth have a stringy sediment and no turbidity after 5 d incubation in a 90% N_2–10% CO_2 atmosphere, and the pH is 6.8.

The optimum temperature for growth is 35 °C. Growth is slow and occurs between 19 °C and 37 °C. Strains require ethanol, CO_2, or sodium carbonate, and either a high concentration of yeast autolysate or acetate, propionate, or butyrate for growth. A synthetic medium containing inorganic salts, acetate, ethanol, biotin, and p-aminobenzoic acid will support growth. Small amounts of gas are detected in PYG deep agar cultures. Urease is produced; gelatin is not hydrolyzed; carbohydrates are not attacked.

In the presence of CO_2 or carbonate and acetate or propionate, ethanol is converted to butyrate, caproate, and H_2. H_2 gas is formed and small amounts of butyrate may be detected. Neither pyruvate, lactate, nor threonine is utilized. *Clostridium kluyveri* can be adapted to utilize crotonate and produce acetate, butyrate, and caproate.

The type strain is susceptible to chloramphenicol, clindamycin, erythromycin, penicillin G, and tetracycline. Other characteristics of the species are given in Table 142.

Source: fresh water, marine black mud, decaying plants, garden soil.

DNA G+C content (mol%): 30 (T_m).

Type strain: ATCC 8527, DSM 555, NBRC 12016, NCIMB 10680, NCCB 72061, NCCB 46095.

GenBank accession number (16S rRNA gene): M59092.

Further comments: 16S rRNA gene sequence comparisons show *Clostridium kluyveri* to fall within the radiation of cluster I of the clostridia as defined by Collins et al. (1994). The closest relatives based on sequence similarities are *Clostridium ljungdahlii* (93.4%) and *Clostridium tyrobutyricum* (93.1%). This species is distinctive both in its nutritional requirements and in its ability to ferment ethanol to caproate.

86. **Clostridium lactatifermentans** van der Wielen, Rovers, Scheepens and Biesterveld 2002, 925[VP]

lac.ta′ti.fer.men.tans. N.L. n. *lactas -atis* lactate; L. part. adj. *fermentans* fermenting; N.L. adj. *lactatifermnetans* fermenting lactate.

Tapered rods, 2.8–10 × 1.1–1.3 μm wide. Gram-stain-negative, but Gram-positive cell-wall structure. Nonmotile. Nonsporeforming. Obligate anaerobe. Chemoorganoheterotroph. Temperature range for growth is 30–47 °C with an optimum of 41 °C. Growth occurs in the pH range 5.6–8.3 and optimumally at pH 6.4–7.3.

Produces acetate, propionate, butyrate, and isovalerate from glucose and lactate. Utilizes glucose, xylose, pyruvate, DL-lactate, L-alanine, L-serine, L-cysteine, and L-threonine. Slow growth on L-valine, L-leucine, L-isoleucine, and L-aspartate. Does not utilize cellobiose, melibiose, raffinose, sorbitol, lactose, arabinose, starch, succinate, ethanol, acrylate, L-proline, L-lysine, L-arginine, or glycine. Hydrolyzes gelatin, but not esculin or urease. Negative for indole and catalase.

Source: ceca of a 31-d-old broiler chicken.

DNA G+C content (mol%): 44.6 (T_m).

Type strain: G17, DSM 14214, LMG 20954.

GenBank accession number (16S rRNA gene): AY033434.

Further comments: 16S rRNA gene sequence comparisons show *Clostridium lactatifermentans* to fall within the radiation of cluster XIVb of the clostridia as defined by Collins et al. (1994). The closest relatives based on sequence similarities are *Clostridium neopropionicum* (94.8%) and *Clostridium propionicum* (94.7%). *Clostridium lactatifermentans* can be differentiated from its closest relatives on the basis of its ability to utilize glucose and xylose but not ethanol and acrylate. In addition, *Clostridium lactatifermentans* does not form spores, has a higher optimum growth temperature and DNA G+C content than its phylogenetic relatives.

87. **Clostridium lacusfryxellense** Spring, Merkhoffer, Weiss, Kroppenstedt, Hippe and Stackebrandt 2003, 1027[VP]

la′cus.fry.xel.len′se. N.L. neut. adj. *lacusfryxellense* of Lake Fryxell, the lake in Antarctica from which the type strain was isolated.

Rod-shaped, 1.0–1.2 × 2.2–5.0 μm, occur singly, in pairs, or short chains. Gram-stain-positive. Motile by peritrichous flagella. Spherical to slightly ellipsoidal endospores, located at a terminal to subterminal position; sporangium is not swollen. Colonies on sheep-blood agar are 1–2 mm in diameter, round with often coarsely granulated margins, smooth, slightly raised, cream-white to grayish, semi-transparent to opaque, and nonhemolytic. The cell-wall peptidoglycan contains *meso*-diaminopimelic acid. The major cellular fatty acids are $C_{16:1\ \omega9c}$, $C_{16:1\ DMA\ \omega9c}$, $C_{14:0}$, $C_{16:1\ \omega7c}$, $C_{16:0}$ and $C_{16:1\ \omega11c}$.

Temperature optimum for growth is in the range 8–12 °C with an upper limit of 15 °C. The pH range for growth is 6.0–7.3 with optimal grow at pH 6.6–7.1. Under optimal conditions, the doubling time is 10.7 hours.

Carbohydrates utilized include amygdalin, cellobiose, fructose, galactose, glucose, glycogen, inositol, inulin, lactose, mannitol, melezitose, melibiose, raffinose, ribose, salicin, starch, sucrose, trehalose, and xylose. Carbohydrates not utilized include arabinose, maltose, mannose, rhamnose, and sorbitol. Gelatin and starch are not hydrolyzed. End products of fermentation are butyrate, formate, acetate, lactate, ethanol, H_2, and CO_2.

Source: a microbial mat sample taken from a moated area around Lake Fryxell, Antarctica.

DNA G+C content (mol%): 32.1 (HPLC).

Type strain: C/C-an/B1, ATCC BAA-580, DSM 14205.

GenBank accession number (16S rRNA gene): AJ506118.

Further comments: 16S rRNA gene sequence comparisons show *Clostridium lacusfryxellense* to fall within the radiation of cluster I of the clostridia as defined by Collins et al. (1994). The closest relatives of *Clostridium lacusfryxellense* are the two subspecies of *Clostridium estertheticum* (99.7%), *Clostridium frigoris* (98.9%), *Clostridium bowmanii* (98.8%), and *Clostridium psychrophilum* (97.5%). DNA–DNA reassociation studies show *Clostridium lacusfryxellense* to represent a distinct species even though the 16S rRNA gene sequence similarity values are 99.7% with the subspecies of *Clostridium estertheticum*. At the phenotypic level *Clostridium lacusfryxellense* can be differentiated from the two subspecies of *Clostridium estertheticum* on the basis of having spherical spores and utilizing amygdalin, lactose, ribose, and trehalose but not starch.

88. **Clostridium lentocellum** Murray, Hofmann, Campbell and Madden 1987, 179[VP] (Effective publication: Murray, Hofmann, Campbell and Madden 1986, 182.)

lent.to.cel'lum. L. adj. *lentus* slow; N.L. n. *cellulosum* cellulose; N.L. adj. *lentocellum.*

Slightly curved rods, 0.3–0.5 × 2.5–4.0 μm, found singly. Strict anaerobe. Gram-stain-negative. Motile, with fimbriae located at the cell end. Cells form round terminal spores. Colonies are flat, colorless, transparent, have undulate margins, and measure 7–10 mm in diameter. Colonies are cellulolytic.

The temperature range for growth is 15–46.5 °C with an optimum at 40 °C. Growth occurs in the pH range 5.7–9.13 with an optimum at 7.5–7.7.

Ferments cellulose to produce ethanol, acetic acid, hydrogen, and carbon dioxide. Utilizes arabinose, cellobiose, cellulose, esculin, fructose, galactose, glucose, glycogen, lactose, maltose, pyruvate, salicin, sucrose, trehalose, xylose, xylan, milk, starch, bile, and chopped meat for growth. Does not utilize adonitol, amygdalin, arginine, dulcitol, erythritol, glycerol, hippurate, inositol, inulin, lactate, mannitol, mannose, malezitose, melibiose, raffinose, rhamnose, ribose, sorbitol, sorbose, and thronine. Bile inhibits growth. Hemin, vitamin K, and rumen fluid stimulate growth. Cells weakly produce urease, but do not produce indole, catalase, or gelatinase. Sulfate and nitrate are not reduced. Acetylmethylcarbinol is produced.

Source: an estuarine mud bank or a river that was fed with paper-mill and domestic effluent.

DNA G+C content (mol%): 36 (T_m).

Type strain: RHM5, ATCC 49066, DSM 5427, NCIMB 11756.

GenBank accession number (16S rRNA gene): X71851, X76162.

Further comments: When first described, this species was compared at the phenotypic level with other cellulolytic species of the genus *Clostridium* described at that time. The availability of the 16S rRNA gene sequence of this strain showed it to be unrelated to other cellulolytic *Clostridium* species and to represent a distinct lineage within cluster XIVb of the clostridia as defined by Collins et al. (1994). The 16S rRNA gene sequence of *Clostridium lentocellum* shares less than 90% sequence similarity with any other described species.

89. **Clostridium leptum**[*] Moore, Johnson and Holdeman 1976, 250[AL]

lep'tum. Gr. adj. *leptos* thin, delicate; N.L. neut. adj. *leptum* (referring to the morphological appearance of the cells).

Cells in PY-fructose broth cultures are nonmotile, slightly curved, Gram-stain-positive rods, 0.6–0.8 × 1.3–2.8 μm, that occur in pairs or short chains. Spores are oval, nearly terminal, and rarely seen. Heat-resistant cells can be demonstrated from chopped-meat slants incubated at 30 °C for 3 weeks, inoculated into PY-maltose broth, and heated at 80 °C for 10 min. Cell walls contain a lysine-serine-glycine type of peptidoglycan.

Surface colonies on supplemented brain heart infusion roll streak tubes are pinpoint–0.5 mm in diameter, circular, entire, low convex, tan, and translucent. Neither of the 2 of

8 strains tested that grew on the surface of anaerobic blood agar plates was hemolytic. Neither lecithinase, lipase, nor urease was detected from the 5 of 8 strains tested that grew on anaerobic egg-yolk agar plates.

Cultures in PY-maltose broth usually produce a smooth to stringy sediment with little or no turbidity and have a pH of 5.3–5.8 after incubation for 5 d. The optimum temperature for growth is 37 °C; some strains grow at 30 °C and 45 °C. Growth is markedly stimulated by some carbohydrates (glucose, cellobiose, mannitol, ribose) although there is little or no depression of pH of the medium. Growth is inhibited by 6.5% NaCl and by 20% bile. Traces of gas may be detected in PYG deep agar cultures. Glucose may be fermented weakly. A 7 α-dehydroxylase and a 12 α-hydroxysteroid dehydrogenase that may be significant in the breakdown of bile acids in the human colon have been isolated from a strain of this species. Products in PY-maltose broth are acetic acid and abundant H_2; ethanol also may be detected. Neither lactate nor pyruvate is utilized. Other characteristics of the species are given in Table 143.

Source: human feces and colonic contents.

DNA G+C content (mol%): 51–52 (T_m).

Type strain: ATCC 29065, CCUG 48287, DSM 753.

GenBank accession number (16S rRNA gene): AJ305238, M59095.

Further comments: 16S rRNA gene sequence comparisons show *Clostridium leptum* to fall within the radiation of cluster IV of the clostridia as defined by Collins et al. (1994). The closest relative based on sequence similarity values is *Clostridium sporosphaeroides* (93.9%). The low 16S gene sequence similarity value is in correlation with the fact that these species differ extensively including different cell-wall types,.

90. **Clostridium limosum**[*] André in Prévot 1948a, 165[AL]

li.mo'sum. L. neut. adj. *limosum* muddy or slimy.

Cells in PYG broth cultures are Gram-stain-positive straight rods, 0.6–1.6 × 1.7–16 μm, and occur singly, in pairs, or in short chains. Motility is variable; motile cells are peritrichous. Spores are oval, central to subterminal, and usually swell the cell. Sporulation occurs most reliably on chopped-meat slants incubated at 30 °C or on egg-yolk agar plates incubated for 48–72 h.

Cell walls contain *meso*-DAP, alanine, and glutamic acid; cell-wall sugars are glucose, galactose, and small amounts of mannose and rhamnose.

Surface colonies on blood agar plates are β-hemolytic, 1–4 mm in diameter, circular to irregular, raised to convex, translucent, gray, shiny or dull, smooth, with a mosaic or granular internal structure and an entire, scalloped, or undulate edge.

Cultures in PYG broth are turbid with a smooth to ropy sediment and a pH of 6.1–6.5 after incubation for 1 week.

The optimum temperature for growth is 37 °C. Some strains grow as well at 30 °C or 45 °C; growth is poor at 25 °C. Growth is inhibited by 6.5% NaCl, 20% bile, or a pH of 8.5 Trace to moderate gas is produced in PYG deep agar cultures. Ammonia and H_2S are produced. Lecithinase, collagenase, ribonuclease, and deoxyribonuclease are produced. One of nine strains produces an extracellular β-glucuronidase.

Principal cellular lipid fatty acids are the saturated straight-chain C_{16}, C_{14}, and C_{12} acids. The principal product in PYG broth is acetate; small amounts of formate, succinate, or lactate may be detected. Little or no H_2 is produced. Pyruvate is converted to acetate; neither lactate nor threonine is utilized. There is an increase in propionate in a threonine-trypticase-yeast extract-thioglycollate medium over that detected in the basal medium. β-Phenylethylamine, isoamylamine, and usually di-n-butylamine are produced. Glutamate is fermented by the methylaspartate pathway. Indole and phenol are produced from fermentation of tryptophan and tyrosine, respectively, although indole formation is repressed in usual laboratory test media. Glutamic acid and histidine are fermented.

All strains tested are susceptible to chloramphenicol, clindamycin, erythromycin, penicillin G, and tetracycline; the one strain tested was resistant to clindamycin.

Supernatant culture fluids of some strains are weakly toxic to mice; some strains are pathogenic for guinea pigs. Pathogenicity appears to be related to the action of the collagenase and the lecithinase. Toxicity and pathogenicity are readily lost in laboratory cultures. Other characteristics of the species are given in Table 142.

Source: mud, infections in cattle, water buffalo, alligators and chickens; snake venom; home-preserved meat; human feces; human clinical specimens including blood, peritoneal fluid, pleural fluids, and lung biopsy from pulmonary infections.

DNA G+C content (mol%): 24 (T_m).

Type strain: ATCC 25620, BCRC (formerly CCRC) 14513, CCUG 24037, DSM 1400, JCM 1427, NCIMB 10638, VPI 2700.

GenBank accession number (16S rRNA gene): M59096.

Further comments: 16S rRNA gene sequence comparions show *Clostridium limosum* to fall within the cluster II of the clostridia as defined by Collins et al. (1994). The closest relatives based on sequence similarity values are *Clostridium histolyticum* (97.2%) and *Clostridium proteolyticum* (96.9%). *Clostridium limosum* can be differentiated most easily from *Clostridium histolyticum*, which it resembles phenotypically, by its production of lecithinase.

91. **Clostridium litorale** Fendrich, Hippe and Gottschalk 1991, 580[VP] (Effective publication: Fendrich, Hippe and Gottschalk 1990, 131.)

li.to.ra′le. L. adj. n. *litorale* at the beach or coast, referring to the source of samples, the seashore of northern Germany, from which the species was isolated.

Rods with rounded ends, 1.0–1.5 × 2.0–8.0 μm. Motile. Forms subterminal, ovoid spores, 1.5 × 2.0 μm in old cultures. Gram-stain-positive. Strict anaerobe. Chemoorganotroph. Temperature range for growth is 13–38.5 °C with an optimum of 28 °C. The pH range for growth is 6.5–8.4 with an optimum of pH 7.3. Growth occurs at 1.0% NaCl, and is inhibited by 6.0% NaCl.

No complex nutrients required for growth. Complex standard media is fermented to produce butyrate and acetate. Does not ferment carbohydrates. Betaine is reductively cleaved in the Stickland reaction to trimethylamine, acetate, H_2, glycine, L-alanine, L-leucine, L-isoleucine, L-valine, and L-phenylalanine. Sarcosine is fermented with L-alanine and utilizes glycine with L-serine. Does not reduce nitrate, sulfate, thiosulfate, or sulfite. H_2S is not produced. Growth stimulated by Tween 80. Does not digest meat or milk. Starch and gelatin are not hydrolyzed. Does not produce gas, indole, ammonia, or hydrogen. Lipase, lecithinase, and urease are not produced. Tests negative for catalase and oxidase. Cells hydrolyze esculin, and reduce natural red and resazurin.

Source: anoxic marine sediments from the North Sea, Germany.

DNA G+C content (mol%): 26.1 (T_m).

Type strain: W6, ATCC 49638, DSM 5388.

GenBank accession number (16S rRNA gene): X77845.

Further comments: 16S rRNA gene sequence comparisons show *Clostridium litorale* to fall within the radiation of cluster XI of the clostridia as defined by Collins et al. (1994). *Clostridium litorale* represents a distinct lineage within cluster XI and the closest realtive based on sequence similarity values is *Eubacterium acidaminophilum* (94.2%).

92. **Clostridium lituseburense*** (Laplanche and Saissac in Prévot 1948a) McClung and McCoy 1957, 664[AL] (*Inflabilis lituseburense* Laplanche and Saissac in Prévot 1948a, 167.)

li.tus.e.bu.ren′se. L. n. *litus* coast; L. n. *ebur* ivory; N.L. adj. *lituseburense* pertaining to Côte d'Ivoire.

Cells in PYG broth cultures are Gram-stain-positive rods, straight or slightly curved, motile and peritrichous, 1.4–1.7 × 3.1–6.3 μm, and occur singly, in pairs, or in short chains. Spores are oval, central to subterminal, and may swell the cell or occur as free spores. Sporulation occurs readily in chopped-meat broth cultures incubated for 24 h. Cell walls contain L-lysine and aspartic acid.

Surface colonies on blood agar plates are β-hemolytic, 1–3 mm in diameter, circular, low convex, opaque with translucent margins, white, shiny, and smooth, with a coarse granular internal structure and an entire to scalloped margin. Cultures in PYG broth are turbid with a ropy sediment and a pH of 5.3 after incubation for 5 d.

Growth is equally profuse at 35 °C, 30 °C, and 37 °C, slightly less at 45 °C. Growth is inhibited by 6.5% NaCl and 20% bile but unaffected by a pH of 8.5. Traces of gas may be detected in PYG deep agar cultures. Ammonia is produced. Neutral red and resazurin are reduced. Gelatin, chopped meat, and casein are digested; milk is curdled and weakly digested in 3 weeks.

A complex mixture of straight and branched chain saturated and unsaturated cellular fatty acids is present with normal unsaturated C_{16} acid predominating.

Products in PYG broth cultures include large amounts of butyric, acetic, and isovaleric acids and small amounts of formic, propionic, and isobutyric acids. Small amounts of ethyl, propyl, and isobutyl alcohols may be detected. In chopped-meat carbohydrate broth, butyric and isovaleric acids are greatly increased, and moderate isobutyric acid is detected. Little or no H_2 gas is formed. Threonine is converted to propionate; an increased amount of acetate is detected from pyruvate; lactate is not utilized. Phenylalanine, tyrosine, and tryptophan are converted to phenylacetic, hydroxyphenylacetic, and indole acetic acids. Valine is converted to isobutyric acid, leucine and isoleucine to isovaleric acid. *Clostridium lituseburense* utilizes serine, threonine, and arginine, and produces 2-aminobutyric acid.

The type strain is susceptible to chloramphenicol, clindamycin, erythromycin, penicillin G, and tetracycline. Strains are not pathogenic for guinea pigs or mice. Other characteristics of the species are given in Table 144.

Source: soil and humus from Côte d'Ivoire.

DNA G+C content (mol%): 27 (T_m).

Type strain: ATCC 25759, BCRC (formerly CCRC) 14536, CCUG 18920, DSM 797, JCM 1404, NCIMB 10637.

GenBank accession number (16S rRNA gene): M59107.

Further comments: 16S rRNA gene sequence comparisons show *Clostridium lituseburense* to fall within the radiation of cluster XI of the clostridia as defined by Collins et al. (1994). The closest relatives based on sequence similarities are *Clostridium bifermentans* (96.8%), *Clostridium bartlettii* (96.5%), *Clostridium irregulare* (96.5%), *Eubacterium tenue* (96.3%), *Clostridium sordellii* (96.0%), and *Clostridium mayombei* (96.0%).

93. **Clostridium ljungdahlii** Tanner, Miller and Yang 1993, 235[VP]

ljung.dahl'i.i. N.L. gen. n. *ljungdahlii* of Ljungdahl, in recognition of Lars G. Ljungdahl's research contributions in the study of both acetogens and clostridia.

Rod-shaped cells, 0.6×2–$3\,\mu m$. Cells occur singly. Anaerobe. Gram-stain-positive. Motile. Form spores rarely. The temperature range for growth is 30–40 °C with an optimum of 37 °C. The pH range for growth is 4.0–7.0 with an optimum at 6.0.

Utilizes H_2-CO_2 or CO autotrophically. Grows chemoorganotrophically with formate, ethanol, pyruvate, fumarate, erythrose, threose, arabinose, xylose, glucose, and fructose. Does not utilize methanol, ferulate, trimethoxyvenzoate, lactate, glycerol, citrate, succinate, galactose, mannose, sorbitol, sucrose, lactose, maltose, and starch. Major metabolic end product of metabolism is acetic acid with trace amounts of ethanol.

Source: chicken yard waste.

DNA G+C content (mol%): 22–23 (T_m).

Type strain: PETC, ATCC 55383, DSM 13528.

GenBank accession number (16S rRNA gene): M59097.

Further comments: 16S rRNA gene sequence comparisons show *Clostridium ljungdahlii* to fall within the radiation of cluster I of the clostridia as defined by Collins et al. (1994). The closest relatives based on sequence similarity are *Clostridium tyrobutyricum* (95.4%), *Clostridium aciditolerans* (93.7%), *Clostridium scatologenes* (93.6%), *Clostridium kluyveri* (93.4%), *Clostridium drakei* (93.4%), and *Clostridium magnum* (93.4%). It should be noted that the type strain ATCC 49587, cited in the paper by Tanner et al. (1993), has been deaccessioned.

94. **Clostridium lundense** Cirne, Delgado, Marichamy and Mattiasson 2006, 627[VP]

lund.en'se. N.L. neut. adj. *lundense* from Lund, relating to the city where the type strain was isolated.

Rod-shaped, 0.56×2.8–$4.5\,\mu m$ in size. Gram-stain-positive. Nonmotile. Spores are spherical, terminal, and deform the cell shape. Cells form associations of two or more cells in the stationary phase of growth. Colonies are circular, 1 mm in diameter, convex, cream color with entire margins on peptone/yeast extract/glucose medium. Obligately anaerobic. Temperature range for growth is 25–47 °C with optimum of 37 °C. Growth occurs between pH 5.0–8.5 with a broad optimal pH for growth between 5.5 and 7.0. Growth occurs with 0–3% (w/v) NaCl but is optimal in the absence of NaCl.

Catalase-negative, indole-positive, glucosidase positive, sulfide- production positive. Esculin is hydrolyzed. Gelatin is not hydrolyzed. The strain does not utilize glucose, xylose, maltose, sorbitol, salicin, mannitol, lactose, sucrose, arabinose, glycerol, mannose, melezitose, raffinose, rhamnose, or trehalose. The strain shows lipolytic activity and hydrolyzes olive, sesame, and corn oils.

Source: bovine rumen fluid.

DNA G+C content (mol%): 31.2 (HPLC).

Type strain: R1, CCUG 50446, DSM 17049.

GenBank accession number (16S rRNA gene): AY858804.

Further comments: 16S rRNA gene sequence comparisons show *Clostridium lundense* to fall within the radiaton of cluster I of the clostridia as defined by Collins et al. (1994). The closest relatives based on sequence similarities are *Clostridium tetanomorphum* (97.9%), *Clostridium pascui* (95.4%), and *Clostridium peptidivorans* (95.2%). DNA–DNA reassociation studies confirm the species status of strain R1 with values of 61.9% and 54.3% to *Clostridium pascui* and *Clostridium tetanomorphum*, respectively (Cirne et al., 2006). Strain R1 differs from *Clostridium tetanomorphum* in colony morphology, being able to hydrolyze esculin, and in not utilizing glucose, xylose, maltose, sorbitol, and salicin, which the later does.

95. **Clostridium magnum** Schink 1984b, 355[VP] (Effective publication: Schink 1984a, 254.)

mag'num. L. neut. adj. *magnum* big, referring to cell size.

Rods with slightly pointed ends, 1.0–4.0×4–$16\,\mu m$. Gram-stain-negative. No outer membrane in thin sections. Motile with polar and subpolar flagella. Older cells form spindles with dark zones at each end. Cells form central to subterminal elliptical spores, $1.5 \times 2.5\,\mu m$, in ageing cultures, after growth on sugars. Strict anaerobe. Chemoorganotroph. Temperature range for growth is 15–45 °C with an optimum of 30–32 °C. Growth occurs in the pH range 6.0–7.5 with optimal growth at 7.0.

Utilizes fructose, glucose, sucrose, xylose, malate, citrate, 2,3-butanediol, acetoin, and pyruvate. The only fermentation product is acetate. Does not utilize C_1 compounds, ethylene glycol, ethanol, acetate, glyoxylate, glycolate, serine, lactate, oxalate, malonate, fumarate, succinate, oxaloacetate, glutamate, glycerate, glycerol, diacetyl, orotate, maltose, ribose, arabinose, starch, peptone, Casamino acids, and yeast extract. Does not reduce nitrate, sulfate, thiosulfate, or fumarate. Gelatin and urea are not hydrolyzed. Mineral media with a reductant are required for growth. No growth factors or vitamins are required for growth. Cytochromes not detected.

Source: anoxic freshwater sediments, digestor sludge.

DNA G+C content (mol%): 29.1 (T_m).

Type strain: Wo Bd P1, ATCC 49199, DSM 2767.

GenBank accession number (16S rRNA gene): X77835.

Further comments: 16S rRNA gene sequence comparisons show *Clostridium magnum* to fall within the radiation of cluster I of the clostridia as defined by Collins et al. (1994). The closest relatives based on sequence similarities are *Clostridium scatologenes* (96.0%), *Clostridium nitrophenolicum* (95.8%), *Clostridium drakei* (95.8%), *Clostridium aciditolerans* (95.7%), and *Clostridium pascui* (95.2%).

96. **Clostridium malenominatum**[*] (Weinberg, Nativelle and Prévot 1937) Spray 1948, 786[AL] (*Bacillus malenominatus* Weinberg, Nativelle and Prévot 1937, 763.)

ma.le.nom.i.na′tum. L. adv. *male* ill; L. inf. *nominare* to name; N.L. part. adj. *malenominatum* poorly named.

Cells in PYG broth culture are Gram-stain-positive, although usually only Gram-stain-negative cells can be seen in 24-h cultures. Straight rods, 0.3–0.9 × 1.4–10.8 μm, usually motile and peritrichous, and occur singly or in pairs. Spores are oval or round, subterminal or terminal, and swell the cell. Sporulation occurs most readily on egg-yolk agar plates or on chopped-meat slants incubated at 30 °C for 1 week. Cell walls contain *meso*-diaminopimelic acid (Wilde et al., 1997).

Surface colonies on blood agar plates are pinpoint–2 mm, circular or slightly irregular, convex or raised, gray-white, and translucent, with a crystalline, mottled or granular internal structure, and an entire, erose or scalloped margin. The surface may be smooth or lumpy. Hemolysis is variable among strains.

Cultures in PYG broth are turbid with a smooth or ropy sediment and have a pH of 6.1–6.5 after incubation for 1 week. The optimum temperature for growth is 37 °C. Some strains grow equally well at 30 °C. Growth is moderate at 25 °C, poor at 45 °C. Growth is inhibited by 20% bile, 6.5% NaCl, or a pH of 8.5. Moderate amounts of gas are detected in PYG deep agar cultures. Ammonia is produced. Five of 27 strains tested weakly digest gelatin after incubation for 3 weeks. Neither casein, milk, nor meat is digested. H$_2$S is produced in SIM by 22 of the 27 strains. Uric acid, but not urea, is decomposed by one strain, an isolate from the chicken cecum.

Cellular fatty acids detected are straight-chain acids, principally saturated C$_{14}$ and C$_{16}$ and unsaturated C$_{16}$ acids.

Products in PYG broth include butyrate and acetate; propionate, lactate, and formate usually are detected; propyl and butyl alcohols may be present. Abundant H$_2$ is produced. Pyruvate is converted to acetate- and butyrate; 19 strains, including the type strain, convert threonine to propionate, and 5 strains convert lactate to butyrate. Glutamic acid and tyrosine are utilized. Glutamate is converted by the type strain to butyrate and acetate by the methylaspartate pathway. Traces of phenol and indole are formed from phenylalanine and tryptophan.

All strains tested are susceptible to chloramphenicol and erythromycin; susceptibility to clindamycin or tetracycline is variable among strains; one of seven strains tested is resistant to penicillin G. Supernatant fluids are not toxic to mice. Cultures have been reported pathogenic for the guinea pig and rabbit. Other characteristics of the species are given in Table 141.

Source: human feces, chicken cecal contents, soil, human and animal infections.

DNA G+C content (mol%): 28 (*T*$_m$).

Type strain: ATCC 25776, DSM 1127, JCM 1405.

GenBank accession number (16S rRNA gene): M59099.

Further comments: 16S rRNA gene sequence comparisons show *Clostridium malenominatum* to fall within the radiation of the cluster I of the clostridia as defined by Collins et al. (1994). The closest relatives bsased on sequence similarities are *Clostridium lundense* (95.0%), *Clostridium tetanomorphum* (94.6%), *Clostridium pascui* (94.3%), *Clostridium coclearium* (94.3%), and *Clostridium peptidivorans* (94.2%).

97. **Clostridium mangenotii**[*] (Prévot and Zimmés-Chaverou 1947) McClung and McCoy 1957, 664[AL] (*Inflabilis mangenoti* Prévot and Zimmès-Chaverou 1947, 603)

man.ge.no′ti.i. N.L. gen. n. *mangenotii* pertaining to Professor G. Mangenot, Italian bacteriologist.

Cells in PYG broth cultures are Gram-stain-positive, nonmotile, 0.6–0.9 × 3.1–8.2 μm, and occur singly, in pairs, and in short chains. Spores are oval, subterminal, and swell the cell. Sporulation occurs most readily on egg-yolk agar plates incubated for 72 h. Cell walls contain *meso*-DAP.

Surface colonies on blood agar plates are pinpoint–0.5 mm, circular, low convex, translucent, granular, gray-white, and dull, with an entire margin and a grainy surface.

Cultures in PYG broth are turbid and have a stringy sediment and a pH of 6.2 after incubation for 1 week. The optimum temperature for growth is 30–37 °C. Strains grow moderately well at 25 °C but not at all at 45 °C. Growth is good at a pH of 8.5, but there is no growth with 6.5% NaCl or 20% bile. Moderate gas is detected in PYG deep agar cultures. Ammonia is produced in chopped-meat broth and H$_2$S is produced in SIM.

A wide variety of straight and branched chain saturated and unsaturated cellular fatty acids is produced with the straight-chain saturated C$_{16}$ acid predominating.

Products in PYG broth culture include acetate, formate, isocaproate, isovalerate, and isobutyrate; a small amount of propionate is present. A large amount of H$_2$ is formed. Pyruvate is converted to acetate; threonine is converted to propionate; lactate is not utilized. Valine is converted to isobutyrate; leucine to isovalerate and isocaproate; isoleucine to isovalerate. Phenyl propionic and phenyl lactic acids are produced from phenylalanine, tyrosine, and tryptophan. Proline is utilized and 5-aminovaleric acid and 2-aminobutyric acid are produced.

The type strain is susceptible to chloramphenicol, clindamycin, erythromycin, penicillin G, and tetracyline. Supernatant cultures are nontoxic to mice. Other characteristics of the species are given in Table 145.

Source: soil, marine sediments, human feces.

DNA G+C content (mol%): not reported.

Type strain: ATCC 25761, DSM 1289, JCM 1428.

GenBank accession number (16S rRNA gene): M59098.

Further comments: 16S rRNA gene sequence comparisons show *Clostridium mangenotii* to fall within the radiation of the cluster XI of the clostridia as defined by Collins et al. (1994). The closest relatives based on sequence similarities are *Clostridium difficile* (94.7%), *Clostridium irregulare* (94.2%), *Eubacterium tenue* (94.1%), *Clostridium ghonii* (94.0%), and *Clostridium sordellii* (94.0%).

98. Clostridium mayombei Kane, Brauman and Breznak 1992, 191^{VP} (Effective publication: Kane, Brauman and Breznak 1991, 103.)

may.omb'e.i. N.L. gen. n. *mayombei* of the Mayombe tropical rainforest, People's Republic of Congo, which is home to the termite (*Cubitermes speciosus*) from whose gut this bacterium was isolated.

Straight to slightly curved rods, 1 × 2–6 µm, rounded ends, occurring singly or in pairs. Gram-stain-positive. Motile with peritrichous flagella. Form oval, terminal to subterminal heat-resistant spores 1 µm in diameter. Colonies are oval, white to slightly yellow in color, 2 mm in diameter, and have smooth edges when grown on H_2 + CO_2. Strict anaerobe. Temperature range for growth is 15–45 °C with an optimum of 33 °C. The pH range for growth is 5.5–9.3 with an optimum of 7.3. Facultative chemolithotroph.

Ferments glucose, fructose, xylose, maltose, cellobiose, dextrin, starch, sorbitol, dulcitol, glycerol, formate, pyruvate, malate, syringate, alanine, glutamate, serine, salicin, and esculin to acetate as the major end product. Succinate is fermented to propionate and CO_2. Valine is fermented to isobutyrate. No cytochromes detected. Does not reduce nitrate or sulfate. Tests negative for catalase and oxidase activity. Trypticase and yeast extract are required for good growth.

Source: the gut of the soil-feeding termite, *Cubitermes speciosus* found in the Mayombe tropical rainforest in the People's Republic of Congo.

DNA G+C content (mol%): 25.6 (T_m).

Type strain: SFC-5, ATCC 51428, DSM 6539.

GenBank accession number (16S rRNA gene): not reported.

Further comments: 16S rRNA gene sequence comparisons show *Clostridium mayombei* to fall within the radiation of cluster XI of the clostridia as defined by Collins et al. (1994). The closest relatives based on sequence similarities are *Clostridium glycolicum* (98.9%), *Clostridium irregulare* (97.4%), *Clostridium bartlettii* (96.9%), and *Clostridium lituseburense* (96.0%).

99. Clostridium methoxybenzovorans Mechichi, Labat, Patel, Woo, Thomas and Garcia 1999b, 1207^{VP}

me.tho.xy.ben.zo'vo.rans. Fr. n. *methyl* the methyl radical; Gr. n. *oxys* acid, Fr. n. *benzoin* frankincense of Java; Ger. n. *benzoesaure* resin obtained from the tree *Styrax benzoin*; L. v. *vorare* to devour; N.L. neut. adj. *methoxybenzovorans* pertaining to the use of the organic acid methoxybenzoic acid as carbon and energy source, which is characteristic of this organism.

Rod-shaped cells, 3.0–5.0 × 0.4–0.8 µm, occurring singly or in pairs. Gram-stain-positive. Nonmotile. Form terminal, spherical endospores that cause cell swelling. Strict anaerobe. Chemoorganoheterotroph. Optimal growth occurs at 37 °C and pH 7.4, on a glucose medium.

Utilizes glucose, fructose, sorbose, galactose, *myo*-inositol, sucrose, lactose, cellobiose, lactate, betaine, sarcosine, dimethylglycine, methanethiol, dimethylsulfide, alcohols (methanol), and all methoxylated aromatic compounds as sole carbon and energy sources. Ferments carbohydrates to formate, acetate, and ethanol. Ferments lactate

to methanol. Ferments methoxylated aromatics to acetate and butyrate. Betaine, sarcosine, dimethyl-glycine, methanethiol, and dimethylsulfide are fermented to acetate.

Source: anaerobic methanogenic pilot-scale digester fed from olive mill wastewater.

DNA G+C content (mol%): 44 (T_m).

Type strain: SR3, ATCC 700855, DSM 12182.

GenBank accession number (16S rRNA gene): AF067965.

Further comments: 16S rRNA gene sequence comparisons show *Clostridium methoxybenzovorans* to fall within the radiation of cluster XIVa of the clostridia as defined by Collins et al. (1994). The closest relatives based on sequence similarities are *Clostridium indolis* (99.4%), *Desulfotomaculum guttoideum* (98.1%), *Clostridium amygdalinum* (98.1%), *Clostridium sphenoides* (98.0%), and *Clostridium saccharolyticum* (98.0%).

100. Clostridium methylpentosum Himelbloom and Canale-Parola 1989, 495^{VP} (Effective publication: Himelbloom and Canale-Parola 1989, 291.)

me' thyl.pen.to'sum. N.L. neut. adj. *methylpentosum* pertaining to methylpentose.

Cells are rod-shaped, bent in the shape of rings with overlapping ends, 0.3 × 2.0 µm; found singly or in the form of left-handed helical chains. Most chains are 3–4 cells, but can be up to 50 cells. Nonmotile. Terminal, spherical to oval spores form which distend the cells. Spores form on Sweet E agar supplemented with rhamnose. Gram-stain-positive. Surface colonies are 1–2 mm in diameter, white, opaque, smooth, mucoid, and circular. Obligate anaerobe. Chemoorganotroph. Growth occurs optimally at 45 °C with no growth at 25 or 50 °C. Cultures in L-rhamnose broth are at pH range 4.5–5.0 after 16 h at 37 °C.

Ferments only L-rhamnose, L-fucose, L-lyxose, and D-arabinose. Does not ferment, *N*-acetylglucosamine, amino acids, L-arabitol, L-arabinose, L-ascorbate, cellobiose, D-fructose, fucoidan, D-fucose, D-galactose, D-galacturonate, D-gluconate, D-glucosamine, D-glucose, 2-deoxy-D-glucose, D-glucuronate, glycerol, lactate, lactose, D-lyxose, maltose, mannitol, D-mannose, pectin, polygalacturonate, 1,2-propanediol, pyruvate, raffinose, L-sorbose, starch, sucrose, D-xylose, and L-xylose. L-Rhamnose and L-fucose are fermented to propionate, *n*-propanol, acetate, CO_2, and H_2. L-Lyxose and D-arabinose are fermented to acetate, glycolaldehyde, CO_2, and H_2. Exogenous biotin, vitamin B_{12}, and bicarbonate are required for growth. Riboflavin and exogenous folic acid stimulate growth. Grows in 20% bile. Does not reduce sulfate.

Source: human feces.

DNA G+C content (mol%): 46 (T_m).

Type strain: R2, ATCC 43829, DSM 5476.

GenBank accession number (16S rRNA gene): Y18181.

Further comments: 16S rRNA gene sequence comparisons show *Clostridium methylpentosum* to fall within the radiation of cluster IV as defined by Collins et al. (1994). This organism represents a distinct lineage that shares less than 90% with any other species of the genus *Clostridium*.

101. Clostridium neopropionicum Tholozan, Touzel, Samain, Grivet, Prensier and Albagnac 1995, 879^{VP} (Effective publication: Tholozan, Touzel, Samain, Grivet, Prensier and

Albagnac 1992, 256.)

ne.o.pro.pi.o.ni.cum. Gr. adj. *neos* new; N.L. n. *acidum propionicum* propionic acid; N.L. adj. *neopropionicum* a new propionic acid-producing *Clostridium*.

Straight rods with tapered ends, 0.5–0.6 × 1.4–3.0 μm; found singly, in pairs, or short chains. Strict anaerobe. Motile with one side-inserted flagellum. Gram-stain-negative. Large, oval, subterminal spores, measuring 0.7–0.9 μm. Temperature range for growth 15–40 °C with an optimum of 30 °C. Growth occurs in the pH range 6.1–8.2 with an optimum at 7.1–7.6.

Ferments ethanol to propionate, acetate, propanol, and butyrate. Threonine, gluose, and β-alanine are fermented to propionate, acetate, and butyrate. Ferments L-cysteine, L- and D-alanine, and L-serine to propionate and acetate. Growth occurs in freshwater selective media containing ethanol. No cytochromes detected.

Source: an anoxic digestor sludge.

DNA G+C content (mol%): 34.5 (T_m).

Type strain: X4, DSM 3847.

GenBank accession number (16S rRNA gene): X76746.

Further comments: 16S rRNA gene sequence comparisons show *Clostridium neopropionicum* to fall within the radiation of cluster XIVb of the clostridia as defined by Collins et al. (1994). The closest relatives based on sequence similarity values are *Clostridium propionicum* (98.1%) and *Clostridium lactatifermentans* (94.8%). DNA–DNA reassociation demonstrated the species status of *Clostridium neopropionicum* with a value of 18% when compared to *Clostridium propionicum* (Tholozan et al., 1992). Phenotypically *Clostridium neopropionicum* and *Clostridium propionicum* are very similar, only differing in the utilization of acrylate (30 mM), Gram staining reaction, and temperature range for growth. A larger number of characteristics exist for the differentiation of *Clostridium neopropionicum* and *Clostridium lactatifermentans* including utilization of glucose, xylose, ethanol, and acrylate (30 mM) as well as spore formation, optimum temperature for growth, and mol% G+C content of DNA (van der Wielen et al., 2002).

102. **Clostridium nexile*** Holdeman and Moore 1974, 276[AL]

nek′si.le. L. neut. adj. *nexile* tied together (referring to its chain formation).

This description is based on that of Holdeman and Moore (1974) and on their study of the type and 10 other strains.

Cells in PYG broth cultures are Gram-stain-positive ovals or straight rods, nonmotile, 0.8–1.7 × 0.8–6.3 μm, and occur in pairs or chains. Spores are round or oval, subterminal or nearly terminal, but are rarely seen, even in cultures that resist heating at 80 °C for 10 min. Spores have been seen in 6-d-old cultures of PYG broth.

Surface colonies on blood agar plates are 0.5–1.0 mm in diameter, circular, convex to raised, semiopaque or with opaque centers and translucent edges, smooth, shiny, white to yellowish, nonhemolytic, and with entire margins.

Cultures in PYG broth have a smooth or ropy sediment, often with no turbidity, and a pH of 4.9–5.3 after incubation for 5 d. The optimum temperature for growth is 30–37 °C. Most strains grow well at 45 °C but poorly at 25 °C. Growth is stimulated by fermentable carbohydrate, but is inhibited by 20% bile or 6.5% NaCl. Abundant gas is produced in PYG deep agar cultures. H_2S is produced by three of six strains tested. Neutral red and resazurin are reduced.

Products in PYG broth culture are acetic and formic acids and ethanol; moderate amounts of lactate and succinate are usually detected. Pyruvate is converted to acetate and ethanol; increased amounts of formate also are usually formed. Neither threonine nor lactate is utilized. Orotic acid is not hydrolyzed.

Of seven strains tested, all are susceptible to chloramphenicol and clindamycin, five are susceptible to penicillin G, three to erythromycin, and two to tetracycline. Other characteristics of the species are given in Table 144.

Source: human feces as part of the normal flora.

DNA G+C content (mol%): 40–41 (T_m).

Type strain: ATCC 27757, DSM 1787.

GenBank accession number (16S rRNA gene): X73443.

Further comments: 16S rRNA gene sequence comparisons show *Clostridium nexile* to fall within the radiation of cluster XIVa of the clostridia as defined by Collins et al. (1994). Sequence similarities in the range 95.4–96.7% are found to the following taxa: *Coprococcus comes, Eubacterium contortum, Ruminococcus gnavus, Clostridium glycyrrhizinilyticum*, and *Hespellia porcina*. This species can be differentiated from *Clostridium oroticum*, which it resembles most closely phenotypically, by not hydrolyzing orotic acid, and by lack of fermentation of arabinose, maltose, and rhamnose as well as sharing 94.8% similarity in the 16S rRNA gene sequence.

103. **Clostridium nitrophenolicum** Suresh, Prakash, Rastogi and Jain 2007, 1889[VP]

ni.tro.phen.o′li.cum. N.L. n. *nitrophenol* nitrophenol; L. suff. *-icus -a -um* suffix used with the sense of belonging to; N.L. neut. adj. *nitrophenolicum* referring to the substrate nitrophenol that can be utilized by the species.

Rod-shaped cells, 0.6–0.9 × 3.5–5.0 μm. Spores are oval and in a central position. Colonies grown on TSA are circular, smooth and convex with wavy margin. Strictly anaerobic.

Temperature range for growth is 20–45 °C. Does not grow at 15 or 50 °C. pH Range for growth is pH 6.5 and 8.0; no growth is observed below pH 6.5 and above 8.0. Optimum growth is observed at pH 7.2 and 30 °C. Growth occurs at 0–1% NaCl but not at 2.0% NaCl.

Negative for catalase and oxidase; positive for indole and methyl red test. Starch and urea are not hydrolyzed. Does not reduce nitrate to nitrite and is negative for Voges–Proskauer test, citrate utilization, and H_2S production. Using Biolog MicroPlates (AN), the strain showed a positive reaction for the assimilation of D-fructose, L-fucose, D-galactose, D-galacturonic acid, palatinose, and L-rhamnose and a negative reaction for acetic acid, *N*-acetyl-D-galactosamine, *N*-acetyl-D-glucosamine, *N*-acetyl-β-D-mannosamine adonitol, L-alaninamide, L-alanine, L-alanyl L-glutamine, L-alanyl L-histidine, L-alanyl

L-threonine, amygdalin, D-arabitol, arbutin, L-asparagine, D-cellobiose, α-cyclodextrin, β-cyclodextrin, dextrin, dulcitol, *i*-erythritol, formic acid, D-gluconic acid, α-D-glucose 1-phosphate, α-D-glucose 6-phosphate, L-glutamic acid, L-glutamine, glycerol, glyoxylic acid, α-hydroxybutyric acid, β-hydroxybutyric acid, inosine, *myo*-inositol, itaconic acid, α-ketobutyric acid, β-ketobutyric acid, lactic acid, lactulose, malic acid, maltose, maltotriose, D-mannitol, mannose, D-melibiose, L-methionine, methyl β-D-galactoside, methyl α-D-glucoside, methyl β-D-glucoside, 3-methyl-D-glucose, L-phenylalanine, propionic acid, pyruvic acid, D-raffinose, L-rhamnose, D-saccharic acid, salicin, L-serine, D-sorbitol, stachylose, succinic acid, sucrose, *m*-tartaric acid, L-threonine, thymidine, trehalose, turanose, uridine, urocanic acid, and L-valine. It produces acid from glucose, dulcitol, fructose, galactose, maltose, mannose, sucrose, and trehalose, but not from adonitol, arabinose, cellobiose, inositol, mannitol, melibiose, salicin, or sorbitol. Fermentation end products from glucose include acetate, formate, and pyruvate.

Menaquinone present is MK-7 and the cell-wall amino acid is *meso*-diaminopimelic acid. Polar lipids present are DPG, PG, PE, and two unknown phospholipids (UKP1 and UKP2). The major cellular fatty acids are (%); $C_{16:0}$ (28.02), $C_{17:1 iso}$ I/ante B (23.05), $C_{16:1 \omega 7c}$/$C_{15:0 iso 2OH}$ (10.82) and $C_{14:0}$ (10.02).

Source: a subsurface soil sample from a depth of about 3–4 m.

DNA G+C content (mol%): 35.5 (HPLC).

Type strain: 1D, JCM 14030, MTCC 7832.

GenBank accession number (16S rRNA gene): AM261414.

Further comments: 16S rRNA gene sequence comparisons show *Clostridium nitrophenolicum* to fall within the radiation of cluster I of the clostridia as defined by Collins et al. (1994). The closest relatives based on sequence similarities are *Clostridium aciditolerans* (98.7%), *Clostridium madnum* (95.8%), *Clostridium scatologenes* (95.7%), and *Clostridium drakei* (95.7%). DNA–DNA reassociation studies between *Clostridium aciditolerans* and *Clostridium nitrophenolicum* demonstrate their distinct species status with a value of 36.4% (Suresh et al., 2007). Differentiating charcteristics include different fatty acid compositions, different Gram-staining reactions, the utilization of very few substrates by *Clostridium nitrophenolicum*, and different end products of fermentation of glucose (Suresh et al., 2007).

104. **Clostridium novyi**[*] (Migula 1900) Bergey, Harrison, Breed, Hammer and Huntoon 1923, 236[AL] (*Bacillus novyi* Migula 1900, 672)

no'vy.i. N.L. gen. n. *novyi* pertaining to F.G. Novy, American bacteriologist.

This description is based on those of Smith (1975) and of Holdeman et al. (1977a) and on the study of the type and 30 other strains of *Clostridium novyi* type A, 7 strains of type B, and the strain of type C ("*Clostridium bubalorum*") on which the description of this type was based.

Type A. Cells in PYG broth cultures are Gram-stain-positive rods, 0.6–1.4 × 1.6–17 μm, and occur singly or in pairs. They are usually motile and peritrichous. Motility may be difficult to detect in wet mounts, but peritrichous cells can usually be demonstrated. Spores are oval, central, or subterminal, and may swell the cell. There are no appendages or exosporium. Sporulation of most strains occurs most readily on blood agar plates incubated for 3 d. Cell walls contain meso-DAP, glutamic acid, and alanine.

Surface colonies on blood agar plates are β-hemolytic, 1–5 mm in diameter, circular or irregular, flat or raised, translucent or opaque, gray, dull or glistening, with a crystalline or mosaic internal structure, and a scalloped, undulate, lobate, or rhizoid margin. They may appear as a spreading film over the entire plate. Cultures in PYG broth are turbid with a smooth or flocculent sediment and have a pH of 5.1–5.8 after incubation for 1 week.

The optimum temperature for growth is 45 °C. Most strains grow nearly as well at 37 °C and moderately at 30 °C. There is little or no growth at 25 °C. Strains require strictly anaerobic conditions and will not grow in the presence of even traces of O_2. Fermentable carbohydrate greatly stimulates growth. There is no growth in 20% bile, 6.5% NaCl, or at a pH of 8.5. Moderate to abundant gas is produced in PYG deep agar cultures. Both lecithinase and lipase are produced on egg-yolk agar plates. Gelatin is digested by all strains; there is weak and slow digestion of chopped meat by 6 of 31 strains; milk is curdled but neither milk nor casein is digested. Deoxyribonuclease is produced by the type strain and by 16 of 25 other strains tested. One strain, isolated from marine mud, causes the breakdown of chitin to N-acetylglucosamine. Of 25 strains tested, 20 produced protease on casein agar, 4 produced ribonuclease, 3 produced amylase, and 8 produced hyaluronatelyase.

Products in PYG broth culture include major amounts of butyric and propionic, and small amounts of acetic acids; small amounts of valeric acid and propanol may be detected. Abundant H_2 is produced. Pyruvate is converted to acetate, butyrate, and sometimes propionate; lactate is converted to propionate; threonine is not utilized. Large amounts of isoamylamine and phenethylamine are produced in chopped-meat glucose broth cultures.

All strains tested are susceptible to chloramphenicol, clindamycin, erythromycin, and penicillin G. One of 15 strains is resistant to tetracycline. Cultures are pathogenic for guinea pigs, rabbits, mice, rats, and pigeons, and supernatant culture fluids of about one-half of the pathogenic strains are toxic to mice. The principal lethal toxin of both types A and B is the necrotizing alpha toxin. Type A strains also produce gamma (phospholipase C) and epsilon (lipase) toxins; some strains also produce delta (oxygen-labile hemolysin) toxin. The gamma toxin (lecithinase of type A) is active on horse red-blood cells. The epsilon toxin (lipase) produces a pearly layer on colonies on egg-yolk agar plates and is useful in differentiating between type A and type B strains. Also, the type A lecithinase (gamma toxin) is antigenically distinct from the lecithinase (beta toxin) of *Clostridium novyi* type B and *Clostridium hemolyticum*. Nontoxic strains can be converted to toxic strains through the mediation of specific bacteriophages. Plasmids have been found in both toxic and nontoxic variants of the same strains and may be involved in this transfer of toxigenicity. Phage-like particles that inhibit the growth of strains of *Clostridium perfringens* and *Clostridium tertium* have been detected in one strain of *Clostridium novyi* type A; this strain of *Clostridium novyi* type A was inhibited by a strain of *Clostridium bifermentans*.

High levels of rRNA homology exist between *Clostridium botulinum* types C and D, *Clostridium hemolyticum*, and *Clostridium novyi* types A and B (Johnson and Francis, 1975). In addition, some strains of *Clostridium novyi* type A have been shown to share antigens with strains of *Clostridium botulinum* type C. For differentiation of these species, see *Further comments* following their descriptions. Other characteristics of the species are given in Table 142. The phylogenetic relationships are shown in Figure 128.

Source: soil, marine sediments, animal wounds, human wounds including gas gangrene.

DNA G+C content (mol%): 29 (T_m).

Type strain: ATCC 17861, DSM 14992, JCM 1406, NCTC 13029.

GenBank accession number (16S rRNA gene): AB045606, L37594.

Type B. Phenotypic characteristics that help to distinguish strains of type B from those of type A are: cells tend to be larger, 1.1–2.5 × 3.3–22.5 µm; mannose is fermented; milk and chopped meat are digested; lipase is not detected on egg-yolk agar; electrophoretic patterns of soluble cellular proteins are quite distinct. Like type A strains, *Clostridium novyi* type B strains produce the lethal, necrotizing *Clostridium novyi* alpha toxin. However, the lecithinase of type B strains is the beta toxin, which also is produced by strains of *Clostridium hemolyticum*. The beta toxin (lecithinase) can be identified by its hemolytic action on human red blood cells and can be neutralized by either *Clostridium novyi* type B or *Clostridium hemolyticum* antitoxin. Type B strains also produce zeta (hemolysin) and eta (tropomyosinase) toxins. Small amounts of theta toxin (lipase), commonly produced by strains of *Clostridium hemolyticum*, may be produced by some strains of *Clostridium novyi* type B. The lipase of the theta toxin does not produce a reaction on egg-yolk agar.

Type strain: ATCC 25758, NCIB 10626.

GenBank accession number (16S rRNA gene): AB035087.

Further comments: A nontoxigenic, nonpathogenic strain that is otherwise indistinguishable from *Clostridium novyi* type A was isolated from water buffalo with osteomyelitis. It was designated *Clostridium novyi* type C and *Clostridium* "*bubalorum*". It produces only small amounts of lecithinase (gamma toxin) that are not detected on egg-yolk agar. Strains are quite commonly isolated that are phenotypically identical to *Clostridium novyi* type A or to *Clostridium botulinum* types C or D except that they are nontoxigenic, presumably through loss of their infecting phages. It has been shown that the heating process sometimes used in isolation of these strains can select for more actively sporulating strains and that there is an inverse relationship between sporulating potency and toxigenicity. For further discussion of the genetic relationships between these species, see below.

Reference strain: Clostridium novyi type C: ATCC 27323, NCIMB 9747.

GenBank accession number (16S rRNA gene): AB041865.

16S rRNA gene sequence comparisons show *Clostridium novyi* to fall within the radiation of cluster I of the clostridia as defined by Collins et al. (1994). Representatives of each type A, B, and C fall within a cluster that comprises *Clostridium hemolyticum* and *Clostridium botulinum* type C and type D strains (Sasaki et al., 2001). *Clostridium novyi* types B and C share >99.9% 16S rRNA gene sequence similarity with *Clostridium hemolyticum*. The type A *Clostridium novyi*, the actual type strain of the species, shares 98.6% 16S rRNA gene sequence similarity with *Clostridium hemolyticum* and type B and C *Clostridium novyi* strains. It has been shown that strains of *Clostridium novyi* type B have 89–93% DNA–DNA homology to *Clostridium hemolyticum*, 28–37% homology to *Clostridium novyi* type A, and 68–70% homology to *Clostridium botulinum* type C (Nakamura et al., 1983).

105. **Clostridium oceanicum**[*] Smith 1970, 811[AL]

o.ce.an′i.cum. L. neut. adj. *oceanicum* belonging to the sea.

This description is based on that of Smith (1970), and on study of the type and 26 other strains.

Cells in PYG broth culture are Gram-stain-positive rods, 0.3–1.6 × 1.7–25.7 µm, and occur singly or in pairs. They are motile and peritrichous. Some strains are nonmotile when incubated at 37 °C but motile at 25 °C. Long filaments may be seen in older cultures. Spores are oval, terminal, or subterminal, and usually do not swell the cell. There may be two spores, one at each end, in a single cell. Sporulation of most strains occurs most readily in chopped-meat glucose broth incubated at 30 °C. Cell walls contain *meso*-DAP.

Surface colonies on blood agar plates are β-hemolytic, 1–6 mm in diameter, circular to irregular, flat or raised, translucent, gray and shiny, with an undulate or scalloped margin and a crystalline mosaic internal structure.

Cultures in PYG broth are slightly turbid with a smooth or flocculent sediment and a pH of 5.0–5.7 after incubation for 1 week.

The optimum temperature for growth is 30–37 °C. Strains grow poorly at 25 °C and not at all at 45 °C. Growth is stimulated by carbohydrate, by Tween 80, and by a N_2 rather than a CO_2 atmosphere. Strains will grow in media with a pH from 6.5–8.5; there is little or no growth initiation at a pH of 6.0. Growth is inhibited by 20% bile or by 6.5% NaCl; 4% NaCl is not inhibitory. Moderate to abundant gas is produced in PYG deep agar cultures. Ammonia is produced; H_2S is produced in SIM medium by 20 of 26 strains tested. Deoxyribonuclease and ribonuclease are produced.

Products in PYG broth include lactic, butyric, and acetic acids; small amounts of formic, propionic, isobutyric, isovaleric, isocaproic, and succinic acids; and ethanol, propanol, and butanol may be detected. Abundant H_2 is produced. Acid production, both in amount and variety, is enhanced in PY broth without fermentable carbohydrate; production of acetic, propionic, isobutyric, isovaleric, and isocaproic acids is stimulated, and valeric acid usually is detected. Pyruvate is converted to acetate and butyrate. Threonine is converted to propionate by 19 of 26 strains. Utilization of lactate is variable.

All strains tested are susceptible to chloramphenicol, clindamycin, erythromycin, penicillin G, and tetracycline. Culture supernatants are nontoxic to mice. Other characteristics of the species are given in Table 142.

Source: marine sediments, human feces.

DNA G+C content (mol%): 26–28 (T_m).

Type strain: ATCC 25647, DSM 1290, JCM 1407, LMG 3287.

GenBank accession number (16S rRNA gene): M59101.

Further comments: 16S rRNA gene sequence comparisons show *Clostridium oceanicum* to fall within the radiation of cluster I of the clostridia as defined by Collins et al. (1994). The closest realatives based on sequence similarities are *Clostridium sporogenes* (96.2%), *Clostridium botulinum* (96.0%), and *Clostridium novyi* (95.9%). All other species of cluster I share less than 94% 16S rRNA gene sequence similarity.

106. **Clostridium orbiscindens** Winter, Popoff, Grimont and Bokkenheuser 1991, 356[VP]

or.bi.scin′dens. N.L. adj. *orbiscindens* ring-cutting.

Straight rods, 0.9–1.0 × 2–7 μm, occurring singly or in pairs. Motile with peritrichous flagella. Gram-stain-variable. Form round, subterminal, uncapsulated spores, measuring 0.7 × 0.5 μm. Colonies on rabbit blood agar are 1 mm in diameter, circular, convex, slightly irregular, smooth, β-hemolytic, shiny, and gray or white in color. Colonies on sheep-blood agar are not hemolytic and slightly larger in size. Strict anaerobe. Optimal temperature 37 °C.

One of the four strains does not grow on bile. Good growth of all strains when grown anaerobically in broth of brain heart media. Does not ferment amygdalin, arabinose, cellobiose, erythritol, glucose, glycogen, mannitole, melzitose, raffinose, rhamnose, ribose, salicin, sorbitol, stach, sucrose, trehalose, xylan, and xylose. Produces indole (one of four strains) and H_2S (three of four strains). Utilizes fumarate (two of four strains) and pyruvate (three of four strains). Does not digest casein, meat, or gelatin. Does not produce urease, lipase, or lecithinase. Does not reduce nitrate. Produces acetic acid, butyric acid, and some propionic acid when grown on peptone-yeast-glucose broth.

Synthesizes a flavonoid molecule C-ring-cleaving enzyme. Susceptible to cephalothin (30 U), kanamycin (1,000 U), and metronidazole (80 U) but resistant to nalidixic acid (30 U), rifampin (15 U), and vancomycin (5 U). Susceptibility to penicillin is variable (two of four strains are susceptible). Apathogenic in mice.

Source: human feces.

DNA G+C content (mol%): 56–57 (T_m).

Type strain: 265, ATCC 49531, DSM 6740.

GenBank accession number (16S rRNA gene): Y18187.

Further comments: 16S rRNA gene sequence comparisons show *Clostridium orbiscindens* to fall within the radiation of cluster IV of the clostridia as defined by Collins et al. (1994). The closest relatives based on sequence similarities are *Eubacterium plautii* (99.7%) and *Bacteroides capillosus* (97.4%); both of these taxa are clearly misclassified and unrelated to the type species of their respective genera.

107. **Clostridium oroticum**[*] (Wachsman and Barker 1954) Cato, Moore and Holdeman 1968, 9[AL] (*Zymobacterium oroticum* Wachsman and Barker 1954, 400)

o.ro′ti.cum. N.L. n. *acidum oroticum* orotic acid; N.L. neut. adj. *oroticum* pertaining to orotic acid.

Cells in PYG broth cultures are 0.6–1.6 × 1.3–3.9 μm, Gram-stain-positive rods or ovals with tapering ends. They

are nonmotile and occur in long tangled chains. Spores are round to ellipsoidal, central to subterminal, and do not swell the cell. Sporulation occurs most readily in PY broth cultures. Cell walls contain *meso*-DAP.

Surface colonies on blood agar plates are 1–2 mm in diameter, circular, convex, opaque, white or buff-colored, shiny, smooth, and nonhemolytic. No internal structure is visible.

Cultures in PYG broth are not turbid and have a heavy, smooth to mucoid sediment and a pH of 5.1 after incubation for 4 d. The optimum temperature for growth is 30–37 °C. Growth is nearly as good at 45 °C and moderate at 25 °C. Growth is only slightly inhibited by 20% bile or a pH of 8.5 but completely inhibited by 6.5% NaCl.

Abundant gas is detected in PYG deep agar cultures. Orotic acid is utilized with 90% of the substrate degraded in 4 d. The organism contains high levels of the iron-sulfur flavoprotein dihydroorotate dehydrogenase which catalyzes both the synthesis and degradation of pyrimidines. The organism also produces a zinc-containing metalloenzyme, dihydroorotase, which is active in pyrimidine degradation.

Products in PYG broth cultures are acetic and formic acids, ethanol, CO_2, and large amounts of H_2; trace amounts of lactic and succinic acids may be detected. Pyruvate is converted to acetate, formate, and ethanol; neither lactate nor threonine is utilized. Ammonia is produced in orotic acid medium. H_2S is not produced in SIM.

The two strains tested are susceptible to chloramphenicol, erythromycin, penicillin G, and tetracycline, but resistant to clindamycin. One clinical isolate of *Clostridium oroticum* is susceptible to amoxycillin (0.25 μg/ml), carbenicillin (2 μg/ml), cephalexin (0.063 μg/ml), tiberal (0.063 μg/ml), clindamycin (2 μg/ml), metronidazole (0.063 μg/ml), chloramphenicol (2 μg/ml), LY 99638 (0.25 μg/ml), and Searle 28538 (0.125 μg/ml); this strain is moderately resistant to moxalactam, cefoperazone, cefoxitin and doxycycline, and resistant to cefamandole, erythromycin, rosaramicin, and tetracycline. Culture supernatants are nontoxic to mice. Other characteristics of the species are given in Table 144.

Source: black mud from San Francisco Bay, human feces, suprapubic bladder aspirates, a rectal abscess.

DNA G+C content (mol%): 44 (T_m).

Type strain: ATCC 13619, DSM 1287, JCM 1429, LMG 3286, NCCB 73016.

GenBank accession number (16S rRNA gene): M59109.

Further comments: 16S rRNA gene sequence comparions show *Clostridium oroticum* to fall within the radiation of cluster XIVa of the clostridia as defined by Collins et al. (1994). The closest relatives based on sequence similarities are *Eubacterium contortum* (98.1%), *Clostridium glycyrrhizinilyticum* (95.8%), *Ruminococcus torques* (95.5%), and *Ruminococcus gnavus* (94.9%).

108. **Clostridium papyrosolvens**[*] Madden, Bryder and Poole 1982, 90[VP]

pa.py′ro.sol′vens. Gr. n. *papyros* paper; L. v. *soluere* to dissolve; N.L. part. adj. *papyrosolvens* paper-dissolving (intended to reflect the organisms' rapid fermentation of filter paper constituents).

This description is based on that of Madden et al. (1982) and on study of the type strain.

Cells in yeast extract-sea water-mineral solution-cellobiose broth are straight Gram-stain-negative rods with Gram-stain-positive cell-wall structure as shown by electron microscopy. They are motile and peritrichous, 0.5–0.8 × 2.0–5.0 μm, and occur singly and in pairs. Spores are round, terminal, and swell the cell. Spores do not germinate unless they are heat shocked at 70 °C for 15 min. After incubation for 3 weeks, deep colonies in cellulose agar roll tubes are 1–2 mm in diameter, circular, translucent, colorless and granular, surrounded by clear zones of cellulose hydrolysis 1–2 cm in diameter. Surface colonies in rumen fluid agar roll tubes are circular to slightly irregular, entire to erose, convex, translucent, colorless, shiny, and smooth. Cell-wall composition has not been reported.

Optimum temperature for growth is 25–30 °C. The organism grows at 15 °C and 37 °C but only slightly at 45 °C. Best growth occurs in an atmosphere of 90% N_2–10% CO_2.

Cultures in PYG broth are slightly turbid with a granular sediment and have a pH of 5.0 after incubation for 5 d. Fermentable carbohydrate is required for growth. Abundant gas is detected in PYG deep agar cultures. Cellulose, glycerol, and esculin are fermented. Other characteristics of the species are given in Table 144. End products from growth in cellulose broth cultures are ethanol, acetate, lactate, H_2, and CO_2. Products in PYG broth are moderate amounts of lactic, formic, pyruvic, fumaric, and lactic acids, and abundant H_2.

The type strain is susceptible to chloramphenicol, clindamycin, erythromycin, penicillin G, and tetracycline.

Source: estuarine sediments.

DNA G+C content (mol%): 30 (T_m).

Type strain: ATCC 35413, DSM 2782, NCIMB 11394.

GenBank accession number (16S rRNA gene): X71852.

Further comments: 16S rRNA gene sequence comparisons show *Clostridium papyrosolvens* to fall within the radiation of cluster III of the clostridia as defined by Collins et al. (1994). The closest relatives based on sequence similarities are *Clostridium josui* (98.4%), *Clostridium cellulolyticum* (98.1%), *Clostridium termitidis* (97.1%), *Clostridium cellobioparum* (96.7%), and *Clostridium hungatei* (96.4%). *Clostridium papyrosolvens* can be differentiated from *Clostridium josui* and *Clostridium cellulolyticum* on the basis of differences in optimum temperature for growth, Gram staining reaction in young cultures, differences in motility and flagella arrangement, and substrate utilization patterns.

109. **Clostridium paradoxum** Li, Mandelco and Wiegel 1993, 454[VP]

para.dox.um. Gr. adj. *paradoxon* fr. neut. of *paradoxos* contrary to expectation, incredible; N.L. adj. *paradoxum* referring to the unusual property of sporulated cells to be highly motile.

Rod-shaped cells, 0.7–1.1 × 2–4.5 μm, two or more cells in chains in exponential phase. Pleiomorphic in stationary phase. Motile with 2–6 peritrichous flagella. Gram-stain-positive cell wall with Gram-stain-negative staining reaction. Sporulation is pH dependent. Cells form round to oval terminal spores which distend and slightly enlarge cells. Sporulated cells are highly motile. Anaerobe. Growth occurs in the temperature range 30–63 °C with optimum at 56 °C. Alkaliphilic. The pH range for growth 7.3–11.0 with optimal growth at pH 9.8–10.3. Cell wall contains *m*-diaminopimelic acid. Major fatty acid is $C_{15:0\ iso}$.

Growth requires prereduced medium with yeast extract or tryptone. Utilizes glucose, fructose, sucrose, maltose, and pyruvate. Does not utilize galactose, xylose, ribose, L-(+)-arabinose, mannose, mannitol, rhamnose, raffinose, salicin, sorbitol, malonate, glycerol, lactate, formate, pectin, starch, cellulose, ethanol, methanol, H_2-CO_2, and Casamino acids.

Major fermentation end products include acetate, CO_2, and H_2, with trace amounts of isovalerate, succinate, and butyrate. Negative for sulfate reduction. Strong positive results for An-Ident tests (API) incubated at 37 and 50 °C, pH 8.5 and 10.5 include: α-glucosidae, indoxylacetate hydrolysis, arginine aminopeptidase, and arginine utilization. Weak positive results were obtained for indole production and catalase production. Negative results were obtained for *N*-acetylglucosaminidase, L-arabinosidase, β-glucosidase, α-fucosidase, α-galactosidase, β-galactosidase, leucineaminopeptidase, proline aminopeptidase, pyroglutamic acid arylamidase, tyrosine aminopeptidase, alanine aminopeptidase, histidine aminopeptidase, phenylalanine aminopeptidase, and glycine aminopeptidase.

Growth inhibited by chloramphenicol (25 μg/ml), erythromycin (25 μg/ml), tetracycline (25 μg/ml), monesin (25 μg/ml), gramicidin S (25 μg/ml), lasalocid (25 μg/ml), and gentamicin (50 μg/ml).

Source: a sewage plant in Athens, Georgia, USA; aeration pools, anaerobic digestors, digestor effluent.

DNA G+C content (mol%): 30 (T_m).

Type strain: JW-YL-7, ATCC 51510, CIP 105527, DSM 7308.

GenBank accession number (16S rRNA gene): Z69939.

Further comments: 16S rRNA gene sequence comparisons show *Clostridium parodoxum* to fall within the radiation of cluster XI of the clostridia as defined by Collins et al. (1994). The closest relative based on sequence similarities is *Clostridium thermoalcaliphilum* at 98.1% 16S rRNA gene sequence similarity. Other species are related at less than 93.0% similarity.

Mutiple copies of the 16S rRNA gene that differ from each other are found in *Clostridium paradoxum* (Rainey et al., 1996). *Clostridium paradoxum* can be differentiated from *Clostridium thermoalcaliphilum* on the basis of differences in cell-wall type, *Clostridium paradoxum* having *m*-diaminopimelic acid while that of *Clostridium thermoalcaliphilum* contains L-Orn–D-Asp. In addition, viable heat-resistant spores have not been observed in *Clostridium thermoalcaliphilum*. While *Clostridium paradoxum* is a thermophile, *Clostridium thermoalcaliphilum* is considered a thermolerant organism.

110. **Clostridium paraputrificum**[*] (Bienstock 1906) Snyder 1936, 402[AL] (*Bacillus paraputrificus* Bienstock 1906, 413)

pa.ra.pu.tri fi.cum. Gr. pref. *para* beside; N.L. n. *putrificum* a specific epithet; N.L. neut. adj. *paraputrificum* resembling (Clostridium) *putrificum*.

Cells in PYG broth cultures are straight or slightly curved rods, usually motile and peritrichous, 0.5–1.4 × 1.9–17.0 μm, and occur singly or in pairs. They are Gram-stain-positive, but rapidly become Gram-stain-negative. Spores are oval, usually terminal, and swell the cells; subterminal and free spores may be seen in the same preparations. Sporulation occurs readily in chopped-meat or PY broth after incubation for 24 h.

Cell walls do not contain DAP; the peptidoglycan bridge is composed of lysine, serine, and glycine. Glutamic acid and alanine also are present. Traces of galactose are present in the wall of the type strain; other strains have been reported to contain glucose, rhamnose, and mannose as well.

Surface colonies on blood agar plates are nonhemolytic, 1–5 mm in diameter, circular, low convex or flat, translucent or semiopaque, smooth, dull, with a mottled or mosaic internal structure and a scalloped, erose, or undulate margin.

Cultures in PYG broth are turbid with a smooth, ropy, or flocculent sediment and a pH of 4.5–5.0 after incubation for 5 d.

Optimum temperature for growth is 30–37°C. Most strains grow nearly as well at 45°C but poorly at 25°C. Growth is markedly stimulated by fermentable carbohydrate; 20% bile is not inhibitory; there is little or no growth with 6.5% NaCl. Abundant gas is detected in PYG deep agar cultures. Acetyl methyl carbinol is produced by 8 of 16 strains tested.

Strains of this species are active in the metabolism of bile acids and steroids and produce compounds that have been implicated in the incidence of colon and breast cancer. One strain of this species is able to lyse Ehrlich ascites tumor tissue in mice. The lysis, however, is neither complete nor permanent. Other strains have been found to promote the formation of liver tumors in mice. Culture supernatants are not toxic to mice and strains are not pathogenic for guinea pigs or rabbits.

Products in PYG broth cultures include acetic, butyric, and lactic acids; formic, pyruvic, and succinic acids also may be detected. Abundant H_2 is produced. Of 39 strains tested, 1 is resistant to chloramphenicol, 3 are resistant to penicillin G, 3 are resistant to tetracycline, 13 are resistant to erythromycin, and 35 are resistant to clindamycin. Other characteristics of the species are given in Table 142.

Source: soil; marine sediments; avian; human infant feces; human adult feces; porcine and bovine feces; human clinical specimens including blood, peritoneal fluid, wounds, and appendicitis.

DNA G+C content (mol%): 26–27 (T_m).

Type strain: ATCC 25780, CCUG 32755, DSM 2630, JCM 1293, NCIB 10671, NCTC 11833.

GenBank accession number (16S rRNA gene): X73445, X75907.

Further comments: 16S rRNA gene sequence comparions show *Clostridium paraputrificum* to fall within the radiation of cluster I of the clostridia as defined by Collins et al.

(1994). The closest relative based on sequence similarities is *Clostridium vincentii* at 97.1%. *Clostridium paraputrificum* is a mesophile which grows optimally between 30 and 37°C, while *Clostridium vincentii* is a psychrophilic species growing optimally at 12°C.

111. **Clostridium pascui** Wilde, Collins and Hippe 1997, 169[VP]

pas'cu.i. L. gen. n. *pascui* of a pasture, referring to the habitat where the organism was isolated.

Rod-shaped cells, 0.75–1.0 × 3.2–8.0 μm; found singly or in pairs. Gram-stain-negative. Motile. Form subterminal elliptical spores, which cause cell swelling. On blood sheep agar, colonies are nonhemolytic, round, opaque, white-gray in color, and 2 mm in diameter. Cell wall contains *meso*-diaminopimelic acid. The major cellular fatty acids are $C_{14:0}$, $C_{16:0}$, $C_{16:1\ 7c}$, and $C_{16:1\ 9c}$ acids and aldehydes.

Anaerobic. The temperature range for growth is 10–43°C with an optimum of 37–40°C. The pH range for growth is 5.5–9.0 with optimal growth at pH 6.4–7.8. Does not ferment carbohydrates. Ferments glutamate and histidine. Acid and gas produced. Fermentation end products include acetate, butyrate, ethanol, H_2, and CO_2. Positive for indole production. Negative for nitrate and sulfate reduction, lipase and lecithinase production, and the hydrolysis of gelatin.

Source: a donkey pasture in Pakistan; actual habitat unkown and may be the intestinal tract of the donkey.

DNA G+C content (mol%): 27 (T_m).

Type strain: Cm19, CIP 105172, DSM 10365, JCM 11012.

GenBank accession number (16S rRNA gene): X96736.

Further comments: 16S rRNA gene sequence comparions show *Clostridium pascui* to fall within the radiation of cluster I of the clostridia as defined by Collins et al. (1994). The closest relative based on sequence similarities is *Clostridium peptidivorans* at 98.7%. Both of these species share the ability to degrade amino acids but not carbohydrates. *Clostridium pascui* stains Gram-negative while *Clostridium peptidivorans* has Gram-positive staining reaction. They also differ in that *Clostridium peptidivorans* is able to degrade proteinaceous compounds whereas *Clostridium pascui* is not. *Clostridium pascui* produces *n*-butyrate and ethanol in addition to acetate which along with formate are the end products of *Clostridium peptidivorans*.

112. **Clostridium pasteurianum** Winogradsky 1895, 330[AL]

pas.teu.ri.a′num. N.L. neut. adj. *pasteurianum* pertaining to Louis Pasteur, French microbiologist.

Cells in PYG broth culture are Gram-stain-positive, becoming Gram-stain-negative as old cultures, straight to slightly curved rods, 0.5–1.3 × 2.7–13.2 μm, occurring singly or in pairs. Motility is variable and may be lost on subculture; motile cells are peritrichous. Granulose-positive. Spores are oval, subterminal, and swell the cell. Sporulation occurs most readily on chopped-meat slants incubated at 30°C for 1 week. There are no exosporium or appendages.

Cell walls contain *meso*-DAP; wall sugars are glucose, galactose, rhamnose, and mannose. Glutamic acid and alanine also are present. Squalene has been found in cell membranes, with greater amounts being present in sporulating than in vegetative cells. Activity of magnesium-dependent membrane ATP has been described.

Surface colonies on blood agar plates are nonhemolytic, 1–3 mm in diameter, circular to irregular, low convex or flat, translucent to semiopaque, gray, shiny, and smooth with an erose or rhizoid margin and a mosaic internal structure.

Cultures in PYG broth are turbid with a smooth sediment, and have a pH of 4.8–5.0 after incubation for 5 d. The optimum temperature for growth is 37 °C. Strains grow nearly as well at 30 °C, moderately well at 25 °C, and poorly if at all at 45 °C. Growth is stimulated by fermen Table carbohydrate but inhibited by 20% bile, 6.5% NaCl, or a pH of 8.5. Growth occurs in synthetic medium. Abundant gas is detected in PYG deep agar cultures. Deoxyribonuclease is produced. Acetyl methyl carbinol is produced by the type strain.

Atmospheric N_2 is fixed; molybdenum is essential for biosynthesis and activity of the nitrogenase involved. There is an increase in cellular phospholipids with a high proportion of palmitic acid during N_2-fixing growth; during non-N_2-fixing growth, the proportion of palmitic acid decreases, accompanied by marked increases in shorter chain saturated fatty acids. Cellular fatty acids include mainly the C_{16} straight-chain, monounsaturated straight-chain C_{16}, C_{15} cyclopropane, and unsaturated cyclopropane. Leucine is synthesized by the α-iso-propylmalate pathway. The type strain of *Clostridium pasteurianum* can utilize crotonate slowly when the medium is supplemented with peptone and yeast extract. Chlorinated hydrocarbon pesticides can be degraded by this species. Has a nickel-requiring carbon monoxide dehydrogenase.

Products in PYG broth cultures include butyric and acetic acids and small amounts of formic acid; abundant CO_2 and H_2 are detected. Ethanol is produced. Neither ammonia nor H_2S is detected. Pyruvate is converted principally to acetate, CO_2, H_2, and small amounts of butyrate; neither lactate nor threonine is utilized. Gluconate is fermented by way of 2-keto-3-deoxygluconate. The amino acids formed by resting cells in synthetic medium with ammonium salts as sole nitrogen source are alanine, threonine, aspartic acid, arginine, glutamic acid, lysine, and valine, with traces of methionine, isoleucine, and serine.

Strains are susceptible to chloramphenicol, clindamycin, erythromycin, penicillin G, and tetracycline. Culture supernatants are not toxic to mice and strains are not pathogenic for laboratory animals. A bacteriocin produced by the type strain of *Clostridium butyricum* is bactericidal against growing cultures of the type strain of *Clostridium pasteurianum*. Other characteristics of the species are given in Table 142.

Source: soil.

DNA G+C content (mol%): 26–28 (T_m).

Type strain: ATCC 6013, BCRC (formerly CCRC) 10942, CCUG 31328, DSM 525, JCM 1408, KCTC 1674, LMG 3285, NCIB 9486, VKM B-1774.

GenBank accession number (16S rRNA gene): M23930.

Further comments: 16S rRNA gene sequence comparisons show *Clostridium pasteurianum* to fall within the radiation of the cluster I of the clostridia as defined

by Collins et al. (1994). The closest relatives based on sequence similarities are *Clostridium acidisoli* (97.2%) and *Clostridium akagii* (96.4%). *Clostridium pasteurianum* differs from *Clostridium acidisoli* and *Clostridium akagii* in substrate range, in formating ethanol but not lactate, not growing or fixing N_2 at pH 3.7, and having a carbon monoxide dehydrogenase.

113. **Clostridium peptidivorans** Mechichi, Fardeau, Labat, Garcia, Verhé and Patel 2000a, 1263[VP]

pep.ti.di.vo′rans. N.L. n. *peptidum* peptide; L. v. *vorare* to devour; N.L. part. adj. *peptidivorans* peptide-consuming.

Motile rods, 0.6–1.2 × 5–10 μm. Gram-stain-positive. Form subterminal to terminal oval spores, which distend the cell.

Strictly anaerobic. Growth occurs in the temperature range 20–42 °C with optimal growth at 37 °C. The pH range for growth is 6–9 with an optimum at 7.0. Grows at concentrations less than 4% of NaCl. Grows on Biotrypcase, yeast extract, Casamino acids, gelatin, peptone, arginine, lysine, cysteine, methionine, histidine, serine, isoleucine, and crotonate. No growth on the following carbohydrates: glucose, fructose, xylose, ribose, sorbose, sorbitol, sucrose, melibiose, galactose, *myo*-inositol, sucrose, lactose, cellobiose, mannitol, mannose, arabinose, arabitol, raffinose, cellulose, xylan; organic acids: formate, acetate, propionate, *n*-butyrate, valerate, fumarate, malonate, malate, lactate, citrate, or succinate; or other amino acids: alanine, proline, aspartate, glycine, threonine, glutamate, glutamine, leucine, aspartate, asparagine, valine, tyrosine, phenylalanine, or tryptophan.

Reduces thiosulfate. Ferments lysine and cronate to acetate and butyrate; biotrypcase, gelatin, and peptone to acetate, butyrate, H_2 and CO_2; cysteine to acetate, alanine, H_2, and CO_2; serine, cysteine, and yeast extract to H_2 and CO_2; histidine to acetate and formate; methionine to propionate; isoleucine to methyl 2-butyrate, H_2, and CO_2; arginine to acetate and ethanol; and Casamino acids to acetate, propionate, butyrate, methyl 2-butyrate, H_2, and CO_2. Cannot reduce sulfate, elemental sulfur, nitrate, and fumarate

Source: olive mill wastewater treatment digester.

DNA G+C content (mol%): 31 (T_m).

Type strain: TMC4, DSM 12505.

GenBank accession number (16S rRNA gene): AF156796.

Further comments: 16S rRNA gene sequence comparisons show *Clostridium peptidivorans* to fall within the radiation of the cluster I of the clostridia as defined by Collins et al. (1994). The closest relative based on sequence similarities is *Clostridium pascui* (98.7%) with other species of this genus having <95.5% 16S rRNA gene sequence similarity. *Clostridium peptidivorans* and *Clostridium pascui* both ferment histidine but not carbohydrates. However, *Clostridium peptidivorans* utilizes proteinaceous substrates in contrast to *Clostridium pascui* which is not proteolytic and only grows on amino acids. They differ in Gram staining reaction as well as ability to ferment glutamate. *Clostridium peptidivorans* produces just acetate and formate from histidine while *Clostridium pascui* produces acetate, butyrate, ethanol, CO_2, and H_2.

114. **Clostridium perfringens**[*] (Veillon and Zuber 1898) Hauduroy, Ehringer, Urbain, Guillot and Magrou 1937, 119[AL] (*Bacillus perfringens* Veillon and Zuber 1898, 539; "*Bacterium welchii*" Migula 1900, 392)

per.frin'gens. L. part. adj. *perfringens* breaking through.

This species produces a number of soluble substances that cause a variety of toxic effects in *in vitro* or *in vivo* conditions, or both. *Clostridium perfringens* has been divided into five types (A, B, C, D, and E) on the basis of production of major lethal toxins. Type A produces alpha toxin; type B produces alpha, beta, and epsilon toxins; type C produces alpha and beta toxins; type D alpha and epsilon toxins; type E produces alpha and iota toxins. A sixth type (type F) was proposed, but it is now considered type C, and type F designation has been abandoned. The five types of *Clostridium perfringens* cannot be differentiated reliably on the basis of cellular or colonial morphology, biochemical reactions, or gas-liquid-chromatographic analyses of fatty and organic acid end products of metabolism.

This description, unless otherwise indicated, is based on study of strains representing the five types (A, B, C, D, and E) and 285 other strains.

Cells in PYG broth culture are Gram-stain-positive, atrichous, nonmotile, straight rods with blunt ends, that occur singly or in pairs and are $0.6–2.4 \times 1.3–19.0\,\mu m$. Spores are rarely seen *in vivo* or in the usual *in vitro* conditions; when present they are large, oval, central, or subterminal, and distend the cell. There is no exosporium and spores lack appendages. A complex relationship exists between previous heat treatment, the ability to ferment certain sugars, and the ability to sporulate. Spore yield is markedly increased by addition of various methylxanthines to the medium. Spores from heat-resistant strains usually require heat activation to germinate, whereas those from heat-susceptible strains do not; spores from heat-resistant strains are also more resistant to the lethal effects of gamma radiation.

The cell wall contains LL-DAP; cell-wall sugars that may be present are galactose, glucose, and rhamnose; however, cell walls of different strains or types may not possess all three sugars. Approximately three-quarters of strains possess a capsule that is composed largely of polysaccharides; the composition of the capsular polysaccharide may vary among strains.

Colonies on the surface of sheep blood agar are usually 2–5 mm in diameter, circular, entire, dome-shaped, gray to grayish yellow, and translucent with a glossy surface. Several other colonial morphologies (dwarfs, rough colonies with lobate margins, and flat colonies with an irregular surface and filamentous margins) occur occasionally, even in the same culture. The kind and extent of hemolysis present depends on both the species of blood and the type of *Clostridium perfringens* being examined; the three types of hemolysins that may be produced in varying quantities are designated alpha, delta, and theta. On rabbit, sheep, cow, horse, or human blood, most strains produce a narrow zone of complete hemolysis due to the theta toxin and a surrounding zone of incomplete hemolysis due to the alpha toxin. Some type B and C strains may produce a very wide zone of hemolysis on sheep or cow blood due to

delta toxin. Synergistic hemolysis (CAMP phenomenon) between this species and *Streptococcus agalactiae* has been described. Colonies in agar are usually lenticular.

Cultures in PYG broth are turbid with a smooth or occasionally ropy sediment and have a pH of 4.8–5.6 after incubation for 1 week. The temperature for optimum growth of types A, D, and E is 45 °C; types B and C grow equally well at 37 °C and 45 °C. The range of temperatures that will support growth of most strains is 20–50 °C; occasional strains will grow at 6 °C for a limited number of passages, but they are not truly psychrophilic. Growth is stimulated by the presence of a fermentable carbohydrate and is not inhibited by 20% bile. Growth occurs readily from pH 5.5–8.0. Growth is not inhibited by NaCl concentrations up to 2%, but is markedly inhibited by 6.5% NaCl. Growth is not enhanced by addition of CO_2 to the atmosphere of incubation. Hyperbaric oxygen (100% oxygen at 3 atm) is bactericidal for this species; partial protection is conferred by addition of whole blood, presumably because of enzymic destruction of hydrogen peroxide by catalase. Abundant gas is produced in PYG deep agar cultures.

Occasional strains weakly ferment glycerol, inulin, and sorbose; ammonia is produced; acetyl methyl carbinol and H_2S are produced, and hippurate is hydrolyzed by some strains; neutral red is reduced; resazurin is reduced by most strains. Although more than 95% of strains ferment sucrose and produce lecithinase, an occasional isolate will be negative for one of these reactions.

Deoxyribonuclease, acid phosphatase, ribonuclease, elastase, hyaluronidase, amylase, neuraminidase, a hemoagglutinin that is distinct from neuraminidase, exo-β-D-galactosidase, ferredoxin-linked nitrate reductase, and superoxide dismutase are produced by some or all strains.

Bile acid metabolism includes deconjugation of taurocholic and glycocholic acids by types A, B, C, D, and E, degradation of chenodeoxycholic acid and cholic acid, and conversion of a number of 3-α-hydroxy bile acids to 3-β-hydroxy and 3-oxo-bile acids. Occasional strains are reported to possess a Δ^4-steroid dehydrogenase and/or the ability to cause aromatization of the Δ-ring of 4-androsten-3,17-dione; these steps have been postulated to be important in converting steroids to colon carcinogens. Monoacetone glucose is fermented.

Most strains are susceptible to bacteriocins produced by enterococci, particularly *Streptococcus faecium*. Some soil isolates of *Clostridium perfringens* produce inhibitors that are active against some or all of *Clostridium botulinum* types A, B, E, and F. *Clostridium perfringens* also produces bacteriocins that are active against other strains of *Clostridium perfringens*. Bacteriocin produced by one strain has a single-chained polypeptide with a molecular weight of approximately 82,000. The mechanism of action of many bacteriocins involves inhibition of macromolecular synthesis (DNA, RNA, and proteins); an additional (but undefined) mechanism probably also exists. Bacteriocin production and resistance to that bacteriocin by several strains have been shown to be related to the presence of a plasmid. Caseinase activity appears to be related to the presence of a plasmid in the one strain tested.

Bacteriophages of this species have been recognized since 1949. Virulent bacteriophage has been recovered from sewage or river water below sewage discharge points. Some strains of types A, B, and C are lysogenic, whereas lysogeny has not been described for types D and E; in addition, smooth or rough strains usually are phage-susceptible whereas mucoid strains usually are resistant. The rapidity of sporulation and the percentage of spores that were heat-resistant decreased when one strain studied was cured of bacteriophage; these changes were reversed when the cured strain was reinfected with the temperate phage.

Cultures of all five types, when grown in PYG broth, produce large amounts of acetic, butyric, and lactic acids; sometimes smaller amounts of propionic, formic, and succinic acids are detected. Abundant H_2 is produced. Lactate occasionally is converted to butyrate; pyruvate is converted to acetate and butyrate and, occasionally, to formate; threonine is converted to propionate. The production of butyrate by one strain is abolished when nitrate was added to the medium. One strain of type E produces isovaleric acid and larger amounts of propionic acid in a medium containing monoacetone-glucose (rather than glucose). Two of four strains tested produce phenylacetic acid from a trypticase-yeast extract medium.

Most β-lactams, particularly penicillin G, are quite active against *Clostridium perfringens*, but relatively greater resistance of this species to penicillin G has been reported. Chloramphenicol, clindamycin, and metronidazole are active against most isolates, but less so on a weight basis than is penicillin G; erythromycin and tetracycline are generally less active than most of the penicillins, chloramphenicol, clindamycin, and metronidazole. A number of strains isolated from pig feces were reported to be resistant to tetracycline, erythromycin, lincomycin, and clindamycin; this degree of resistance appeared to correlate with the use of antimicrobial-containing animal feed. Aminoglycosides are inactive against *Clostridium perfringens*. Twelve of 25 strains were inhibited by 0.25 mg or less of nalidixic acid/ml whereas the remaining 13 strains required 4–64 µg/ml for inhibition. Plasmid-mediated resistance of *Clostridium perfringens* to clindamycin-erythromycin, to tetracycline-chloramphenicol, and to tetracycline alone has been described.

Clostridium perfringens produces a variety of substances (often referred to as "exotoxins") that have been suggested as possible virulence factors. The four toxins alpha, beta, epsilon, and iota are often referred to as the major lethal toxins. The alpha toxin is a phospholipase C that hydrolyzes lecithin to phosphorylcholine and a diglyceride; it is produced by all five types of *Clostridium perfringens*. Alpha toxin exerts its lethal effects by lysing cell membranes, presumably as a consequence of hydrolysis of membrane lecithin. This toxin alone is the cause of muscle death in myonecrosis in cases of *Clostridium perfringens* gas gangrene in humans and other animals following trauma or abortion. The alpha toxin may also cause intravascular hemolysis. Several other diseases in animals are known or suspected to be caused by type A (presumably by the alpha toxin). These include a fatal enterotoxemia of lambs, newborn alpacas, captive wild goats, reindeer, and possibly chickens, and delayed hypersensitivity to alpha toxin in swine that results in arthritis with eventual joint deformity, parakeratosis, and proliferative glomerulonephritis. In addition to causing myonecrosis and/or intravascular hemolysis in humans via the alpha toxin, type A strains may also produce an enterotoxin that causes a food-poisoning syndrome in humans (see below).

The beta toxin, produced by strains of types B and C, has not been completely characterized, but appears to be a highly trypsin-sensitive, single chain polypeptide with a molecular mass of approximately 30,000. It appears to exert its toxic effect by increasing capillary permeability. Intravenously administered beta toxin appears to induce release of endogenous catecholamines. Beta toxin production is thought possibly to be plasmid-mediated. The diseases produced by beta toxin involve the gastrointestinal tract. In humans a necrotic enteritis (pig-bel) develops soon after ingestion of roast pig that has been accidentally contaminated with feces containing *Clostridium perfringens* type C; it is not known whether toxin is ingested or produced in the gut following ingestion of the organism. A necessary antecedent of pig-bel appears to be ingestion of large quantities of sweet potatoes which contain a trypsin inhibitor. A similar form of necrotic enteritis in humans (Darmbrand) was seen in Europe in the mid- to late 1940s; in this setting the presence of starvation, which tends to produce low gastrointestinal levels of trypsin, was probably an important cofactor. Type C also causes enterotoxemia or necrotic enteritis in lambs, calves, piglets, and sheep, whereas type B causes enterotoxemia or necrotic enteritis in foals, lambs, sheep, and goats. An important cofactor for development of beta toxin-induced necrotic enteritis in newborn or very young animals may be the absence of trypsin in the gut.

Epsilon toxin is elaborated by types B and D in the form of a virtually nontoxic prototoxin that is converted to a potent heat-labile toxin by certain proteolytic enzymes such as trypsin. This toxin appears to increase vascular permeability, possibly by adenyl cyclase activation, that ultimately leads to tissue necrosis. Exposure of the gut to epsilon toxin (to which it is normally impervious) results in greater permeability to proteins, including epsilon toxin; this may be an important factor in pathogenesis of disease because oral administration of a single large dose of epsilon toxin may be without effect, whereas administration of the same total dose in several smaller portions may be lethal. The toxin probably produces an increase in vascular permeability of many organs, the most seriously affected of which is the brain; cerebral edema and necrosis of brain tissue is the most likely cause of death in afflicted animals. The toxin also is cytotoxic to guinea pig and rabbit peritoneal macrophages. Type D causes enterotoxemia of lambs, sheep, goats, cattle, possibly humans (very rarely), and chinchillas.

Iota toxin is produced only by type E. This toxin is elaborated as a prototoxin that is usually activated by proteolytic enzymes produced by the organism. Iota toxin markedly increases vascular permeability, produces necrosis on intradermal injection, and is lethal on intravenous injection. The organism is sometimes carried by normal sheep and cattle, and is purported to be a rare cause of enterotoxemia in calves. Iota toxin has been implicated (by antitoxic neutralization) as a cause of antimicrobial-induced colitis in rabbits, the causative organism probably is *Clostridium spiroforme*; iota-like toxic activity is produced by strains tentatively identified as *Clostridium spiroforme*.

Type A strains and some type C and D strains produce an enterotoxin; *Clostridium perfringens* food poisoning in humans is produced by type A strains. Food poisoning can be caused by strains that produce only heat-sensitive spores, as well as by strains that produce heat-resistant spores. Spores that survive cooking may germinate and proliferate to high counts in food products (usually warm meats or meat products); when ingested, the vegetative cells sporulate in the gut and release enterotoxin. Enterotoxin is a product of the sporulation process and causes fluid accumulation in the small intestine of laboratory animals. Feeding of purified enterotoxin to volunteers reproduces the food poisoning syndrome. Type A enterotoxin as been implicated as the cause of a lethal enteritis in horses.

Clostridium perfringens also produces a variety of other substances that have been referred to as toxins. The role of these factors as regards virulence and pathogenicity either is not significant or is not known. Other possible "virulence factors" include sialidase and a non-alpha-delta-theta hemolysin.

Other characteristics of the species are given in Table 142. Some allegedly lecithinase-negative strains can be shown to produce lecithinase on a modified medium.

This species is more widely spread in nature than any other pathogenic microorganism. Although most investigators do not type their isolates, it has been stated that only type A strains are found as part of the microflora of both soil and intestinal tracts, and that types B, C, D, and E seem to be obligate parasites of animals and occasionally are found in humans. Sources yielding *Clostridium perfringens* include soil and marine sediment samples worldwide, clothing, raw milk, cheese, semipreserved meat products, and venison. *Clostridium perfringens* has been isolated from the intestinal contents of virtually every animal that has been investigated. It has also been isolated from pheasant small intestine and from rattlesnake venom.

In addition to causing specific toxin-induced diseases, *Clostridium perfringens* has also been isolated from a variety of mixed anaerobic/aerobic pyogenic infections of a number of different species of domesticated animals.

Humans frequently carry *Clostridium perfringens* as part of the normal endogenous flora. Although this species can be recovered in a small percentage of patients from the normal oral flora, the normal cervicovaginal flora, from urine (presumably reflecting the flora of the distal urethra), and from the skin of the antecubital fossae of approximately 20% of subjects, the main site of carriage is the distal gastrointestinal tract.

Approximately 80% of cases of gas gangrene (clostridial myonecrosis) involve *Clostridium perfringens*. In addition, this species has been reported to cause bacteremia (with and without intravascular hemolysis).

Clostridium perfringens is the species of *Clostridium* most commonly isolated from infections in humans; such infections are often polymicrobial. Although virtually every type of infection in humans has yielded *Clostridium perfringens* on one or more occasions, it is most commonly recovered from infections derived from the colonic flora (e.g., peritonitis, intra-abdominal abscess, and soft tissue infections below the waist).

Source: infections in humans, especially from colonic flora.

DNA G+C content (mol%): 24–27 (T_m).

Type strain: ATCC 13124, BCRC (formerly CCRC) 10913, CCUG 1795, CIP 103409, DSM 756, JCM 1290, LMG 11264, NCAIM B.01417, NCCB 89165, NCIMB 6125, NCTC 8237.

Reference strains: type B: ATCC 3626, NCIB 10691; type C: ATCC, 3628, NCIB 10662; type D: ATCC 3629, NCIB 10663; type E: ATCC 27324, NCIB 10748.

GenBank accession number (16S rRNA gene): M59103.

Further comments: 16S rRNA gene sequence comparisons show *Clostridium perfringens* to fall within the radiation of the cluster I of the clostridia as defined by Collins et al. (1994). The closest relatives based on sequence similarities include *Clostridium baratii* (95.0%), *Clostridium saccharobutylicum* (94.9%), and the *Eubacterium* species: *Eubacterium budayi*, *Eubacterium tarantellae*, and *Eubacterium moniliforme* (94.4–94.7%). Isolates previously identified as "*Clostridium plagarum*" possess a high degree of DNA homology (Nakamura et al., 1976) and 100% rRNA homology with *Clostridium perfringens*; such isolates should be considered lecithinase-negative, theta "toxin" negative variants of *Clostridium perfringens* (Nakamura et al., 1976).

115. **Clostridium phytofermentans** Warnick, Methé and Leschine 2002, 1158[VP]

phy.to.ferm.men′tans. Gr. n. *phyton* plant; L. part. adj. *fermentans* fermenting; N.L. part. adj. *phytofermentans* plant-fermenting, referring to the wide range of plant polysaccharides that this organism is capable of utilizing as growth substrate.

Straight rods, 0.5–0.8 × 3–15 μm, singly or in pairs. Gram-stain-positive cell-wall type; Gram-stain-negative staining reaction. Motile with one or two flagella per cell. Form round, terminal spores, 0.9–1.5 μm in diameter, and cause sporangium swelling. Colonies are round, glossy, translucent, with slightly raised centers, and 2–5 mm in diameter.

Obligate anaerobe. Temperature range for growth is 15–42 °C with an optimum of 37 °C. The pH range for growth is 6.0–9.0 with optimal growth at pH 8.0. Ferments cellulose, pectin, polygalacturonic acid, starch, xylan, arabinose, cellobiose, fructose, galactose, gentiobiose, glucose, lactose, maltose, mannose, ribose, and xylose. Does not utilize glucose, pyruvate, sucrose, trehalose, and tryptone. Negative for urease, esculin hydrolysis, and nitrate and sulfate reduction. Ferments cellulose to ethanol, acetate, CO_2, H_2, and minor amounts of formate and lactate.

Source: forest soil in Massachusetts, USA.

DNA G+C content (mol%): 35.9 (T_m).

Type strain: ISDg, ATCC 700394, DSM 18823.

GenBank accession number (16S rRNA gene): AF020431.

Further comments: 16 S rRNA gene sequence comparisons show *Clostridium phytofermentans* to fall within the radiation of the cluster XIVa of the clostridia as defined by Collins et al. (1994). *Clostridium phytofermentans* forms a distinct lineage within this cluster and its closest relatives based on sequence similarities include *Anaerosporobacter mobilis* (93.8%) and *Clostridium jejuense* (93.0%); all other *Clostridium* species are related at less than 93% 16S rRNA gene sequence similarity.

116. **Clostridium polyendosporum** comb. nov. (Duda, Lebedinsky, Mushegjan and Mitjushina 1996) Effective publication: Duda, Lebedinsky, Mushegjan and Mitjushina 1987, 126 (*Anaerobacter polyendosporus* Duda, Lebedinsky, Mushegjan and Mitjushin 1996, 625)

po.ly.en.do.spo′rum. Gr. pref. *poly* many; Gr. pref. *endo* within; Gr. n. *spora* spore; N.L. adj. *polyendosporum* (forming) several endospores.

The description is as given previously by Duda et al. (1987) and Siunov et al. (1999).

Cells are oval, 3–4 × 4–8µm or spherical, 4–6µm in diameter on solid media. In liquid media the cells are thick rods with rounded ends, 1.5–3 × 4–8µm. Nonmotile. Surrounded by a polysaccharide capsule. The rod-shaped cells form one or two subterminal spores at the opposite poles. The oval and spherical cells form 1–5 round or oval spores. The endospores possess spore coats, exosporium, inner and outer membranes, cortex, and core. The cell wall consists of one layer and is sensitive to penicillin and lysozyme. Prosthecae-like appendages surrounded by the cytoplasmic membrane and the cell wall and containing cytoplasm are sometimes formed. Extensive lipid leaves form in the cytoplasmic membrane, located between the outer and internal lipid layers of the membrane. Colonies on the surface of solid media are large, up to 5 mm in diameter, round, smooth, opaque, brownish-white, viscous, navel-shaped, with entire margins.

Moderately aerotolerant anaerobe. Growth on solid media containing no reductants is possible at $pO_2 \leq 500\,Pa$. A weak catalase activity can sometimes be detected after growth at pO_2 of 500 Pa. Cytochromes are not present. Neutrophilic and mesophilic. Growth occurs in pH range 5.5–8.5 with an optimum at 6.5–7.5. The temperature range for growth is 15–45 °C with an optimum at 25–35 °C. Growth supported by cellobiose, fructose, galactose, glucose, inulin, maltose, mannitol, mannose, raffinose, salicin, sorbitol, sucrose, trehalose, xylose, and lactate. Starch is hydrolyzed but cannot support growth as the sole source of carbon and energy. Amino acids, alcohols, and organic acids other than lactate do not support growth. Gelatin is not liquefied. Lipase, lecithinase, and urease are not produced. Indol and sulfide are not formed. The end products of glucose fermentation include acetate, lactate, butyrate, ethanol, butanol, H_2, and CO_2. Sulfate and nitrate are not used as electron acceptors, but nitrite is reduced to ammonia in a dissimilatory process.

Source: meadow-gley soil under rice.

DNA G+C content (mol%): 29 (T_m).

Type strain: PS-1, DSM 5272, VKM B-1724.

GenBank accession number (16S rRNA gene): Y18189.

Further comments: 16S rRNA gene sequence comparisons show *Anaerobacter polyendosporus* to fall within the radiation of the cluster I of the clostridia as defined by Collins et al. (1994). The original description of this species as *Anaerobacter polyendosporus* did not include phylogenetic information for analysis (Duda et al., 1987). Subsequent studies demonstrated that based on 16S rRNA gene sequence analyses, *Anaeronacter polyendosporus* did, in fact group, within the cluster I of the clostridia as defined by Collins et al. (1994) (Siunov et al., 1999; Stackebrandt et al., 1999). The original justification for the creation of the genus *Anaerobacter* was based on the morphological characteristics of this organism as well as its ability to form numerous spores within a single cell (Duda et al., 1987). The study of Siunov et al. (1999) suggests that these morphological characteristics as well as the observation of lipid leaves in the cytoplasmic membrane, located between the outer and internal lipid layers of the membrane, justify the existence of the genus *Anaerobacter* even though it fell within the species of the genus *Clostridium sensu stricto*. Stackebrandt et al. (1999a) suggests that *Anaerobacter polyendosporus* should be reclassified as a species of the genus *Clostridium*. *Anerobacter polyendosporus* shows highest 16S rRNA gene sequence similarity to the species *Clostridium fallax* (95.6%), *Eubacterium tarantellae* (95.1%), *Clostridium putrefaciens* (94.9%), *Clostridium intestinale* (94.8%), and *Clostridium chauvoei* (94.7%). In an attempt to further the reorganization of the genus *Clostridium*, the species *Anaerobacter polyendosporus* is transferred to the genus *Clostridium* as *Clostridium polyendosporum* comb. nov.

117. **Clostridium polysaccharolyticum**[*] (van Gylswyk 1980) van Gylswyk, Morris and Els 1983, 438[VP] (Effective publication: van Gylswyk, Morris and Els 1980, 492) (*Fusobacterium polysaccharolyticum* van Gylswyk 1980, 157)

po.ly.sac.ca.ro.ly′ti.cum. Gr. adj. *polys* many; Gr. n. *saccharon* sugar; Gr. adj. *lutikos* dissolving; N.L. neut. adj. *polysaccharolyticum* degrading polysaccharides.

This description is based on that of van Gylswyk (1980) and on study of the type strain.

Cells from cellobiose-rumen fluid agar medium are motile and peritrichous, Gram-stain-negative straight rods with rounded ends, 0.6–1.1 × 2–15µm. In broth cellobiose-rumen fluid cultures, aseptate filaments may be more than 50µm in length. Cells occur singly, in pairs, or in short chains. Spores are usually oval, occasionally spherical and subterminal, although terminal spores may be seen. They swell the cell. Sporulation occurs most readily on cellulose-rumen fluid-agar slants. Although the cells stain Gram-negative, cell walls have a peptidoglycan layer characteristic of Gram-stain-positive cells.

Surface colonies on rumen fluid-glucose-cellobiose agar plates are 1 mm in diameter, circular, low convex, semiopaque, buff colored, shiny, and smooth, with an entire edge and have no visible internal structure.

Cultures in PY-cellobiose broth with 30% rumen fluid are not turbid but have a smooth viscous sediment and a pH of 5.7 after incubation for 1 week. Optimum temperature for growth is 30–38 °C. There is no growth at 22 °C and little or none at 45 °C. Fermentable carbohydrate and CO_2 in the gas phase are required for growth. Acetate stimulates growth. Rumen fluid (30%) is greatly stimulatory. Moderate gas is detected in cellobiose-rumen fluid deep agar cultures after incubation for 1 week. Cellulose, xylan, starch, and cellobiose are fermented consistently. Products in PY-cellobiose broth are formate, butyrate, acetate, propionate, H_2, and small amounts of ethanol. In media containing acetate and propionate, these compounds are utilized to form more butyrate or formate. Antibiotic susceptibility has not been determined.

Other characteristics of the species are given in Table 143.

Source: the sheep rumen.

DNA G+C content (mol%): 42 (T_m).

Type strain: strain B, ATCC 33142, DSM 1801.

GenBank accession number (16S rRNA gene): X71858, X77839.

Further comments: 16S rRNA gene sequence comparisons show *Clostridium polysaccharolyticum* to fall within the radiation of the cluster XIVa of the clostridia as defined by Collins et al. (1994). The closest relatives of *Clostridium polysaccharolyticum* are *Clostridium herbivorans* (96.4%) and *Clostridium populeti* (92.9%). *Clostridium polysaccharolyticum*, like *Clostridium herbivorans*, utilizes cellulose, cellobiose, maltose, and starch but do not utilize glucose. They can be differentiated based on the fact that *Clostridium polysaccharolyticum* can utilize arbinose, xylan, and xylose but *Clostridium herbivorans* cannot. These two species differ in their DNA G+C content by 4 mol%.

118. **Clostridium populeti** Sleat and Mah 1985, 162[VP]

po.pu′le.ti. L. n. *populetum* poplar wood; L. gen. n. *populeti* of poplar wood.

Motile rods, 1–1.5 × 1.7–3.0 μm; found singly or in pairs. Gram-stain-negative staining reaction. Form terminal spores which cause cell swelling and are 1.0–1.2 μm in diameter. Cells are resistant to sodium dodecyl sulfate, but are lysed by lysozyme and EDTA. Deep colonies are irregular, opaque, and yellow. Surface colonies are invisible except for a thin opaque edge. Clear zone of cellulolysis measures 20 mm in diameter.

Anaerobic growth. The temperature range for growth is 20–40 °C with an optimum at 35 °C. No growth at 15 °C. Growth occurs in the pH range 6.4–8.1 with optimum growth at pH 7.0.

Utilizes arabinose, xylose, fuctose, galactose, glucose, cellobiose, maltose, sucrose, cellulose, xylan, and pectin. Does not ferement rhamnose, glycerol, mannose, lactose, trehalose, melezitose, erythritol, arabitol, sorbitol, lactate, and pyruvate.

Ferments glucose and cellulose to produce H_2, CO_2, acetate, butyrate, and lactate. Trace amounts of ethanol and succinate are detected after glucose fermentation but no pyruvate or formate are produced. Produces butyrate and H_2 from cellulose fermentation. Yellow pigment is formed on the pebble-milled cellulose when the organism is growing on it. Hydrolyzes gelatin after extended incubaction. Casein and starch are not hydrolyzed.

Source: a batch methanogenic fermentation of finely divided hybrid poplar wood.

DNA G+C content (mol%): 28 (T_m).

Type strain: 743A, ATCC 35295, DSM 5832.

GenBank accession number (16S rRNA gene): X71853.

Further comments: 16S rRNA gene sequence comparisons show *Clostridium populeti* to fall within the radiation of the cluster XIVa of the clostridia as defined by Collins et al. (1994). The closest relatives of *Clostridium populeti* are *Clostridium herbivorans* (94.6%) and *Clostridium jejuense* (93.8%). *Clostridium populeti* can be differentiated from *Clostridium herbivorans* in that it does not produce formate and has a G+C value 10 mol% lower than *Clostridium herbivorans*.

119. **Clostridium propionicum** Cardon and Barker 1946, 631[AL]

pro.pi.o′ni.cum. N.L. neut. adj. *propionicum* pertaining to propionic acid.

This description is based on that of Cardon and Barker (1946) and on study of the type strain.

Cells in PYG broth cultures are motile and peritrichous, Gram-stain-positive rods that rapidly become Gram-stain-negative, 0.5–0.8 × 1.3–5.0 μm. They are straight or slightly curved with tapered or rounded ends, and usually occur in pairs, occasionally singly or in short chains. Spores are oval, subterminal, and swell the cell. Sporulation occurs most readily on egg yolk agar plates or in PY broth. Cell walls contain *meso*-DAP.

Surface colonies on blood agar plates are pinpoint, circular, convex, translucent, gray, dull with a slightly shiny outer rim, smooth, with an entire to slightly scalloped margin and have no visible internal structure.

Cultures in PYG broth are turbid with a smooth sediment and have a pH of 6.0–6.2 after incubation for 1 week. *Clostridium propionicum* grows well between 25 °C and 30 °C. There is no growth at 45 °C. Alanine, serine, threonine, lactate, pyruvate, or acrylate is required for growth. Moderate amounts of gas are formed in PYG deep agar cultures; resazurin is reduced. Ammonia is produced. H_2S is formed in SIM medium.

A wide variety of lipid fatty acids, both straight and branched chain, saturated and unsaturated, is present in cells of the type strain of this species.

Products in PYG broth are propionate, isovalerate, isobutyrate, butyrate, small amounts of acetate, succinate, and sometimes lactate. Abundant H_2 is produced. Pyruvate is converted to propionate and acetate; threonine is converted to propionate and butyrate, CO_2, and ammonia. Lactate is converted to propionate by the acrylate pathway rather than by the more common fumarate pathway. Acrylate can accumulate in resting cell solutions when alternative electron acceptors are provided. Alanine is converted to propionate, acetate, NH_3, and CO_2. Valine is converted to isobutyrate; leucine and isoleucine are converted to isovalerate. Phenylalanine and tyrosine are oxidized to small amounts of phenylacetic and hydroxyphenylacetic acids.

Antibiotic susceptibility has not been determined. Other characteristics of the species are given in Table 143.

Source: black mud in San Francisco Bay, California, USA.

DNA G+C content (mol%): not reported.

Type strain: ATCC 25522, CCUG 9280, DSM 1682, JCM 1430, NCIB 10656, VPI 5303.

GenBank accession number (16S rRNA gene): X77841.

Further comments: 16S rRNA gene sequence comparisons show *Clostridium propionicum* to fall within the radiation of the cluster XIV of the clostridia as defined by Collins et al. (1994). The closest relatives are *Clostridium neopropionicum* (98.1%) and *Clostridium lactatifermentans* (94.7%).

Clostridium propionicum differs from *Clostridium neopropionicum* in not utilizing glucose, staining Gram-stain-positive, and its temperature range for growth. Tholozan et al. (1992) demonstrated that *Clostridium propionicum* and *Clostridium neopropionicum* are distinct species based on a DNA–DNA reassociation value of 18%.

120. **Clostridium proteoclasticum** Attwood, Reilly and Patel 1996, 755[VP]

pro.te.o.clas′ti.cum. N.L. neut. n. *proteinum* protein; Gr. v. *proteuein* to be first; N.L. adj. *clasticus* breaking; Gr. part. perf. *klastos* broken; N.L. adj. *proteoclasticum* protein-breaking.

Straight to slightly curved rods with tapered ends, 0.4–0.6 × 1.3–3.0 μm; found in short chains. Gram-stain-positive. Nonsporeforming. Cells have a single subterminal flagella, but are not motile in liquid media. Cells aggregate and sediment after overnight incubation, but disperse when shaken. Colonies grown on CC-glucose medium are 0.5 mm in diameter, tan, irregular, convex, viscouse, smooth, and transparent. Colonies grown 2–3 d are 1 mm in diameter and granular. Colonies on blood agar are nonhemolytic.

Strictly anaerobic. Optimum temperature for growth is 39 °C with no growth at 25 or 39 °C. The pH range for growth is 5.8–7.0. Grows on amygdalin, arabinose, cellobiose, fructose, galactose, glucose, inulin, lactose, maltose, melibiose, rhamnose, salicin, starch, sucrose, xylan, and xylose. Weak growth on glycogen and mannose. No growth on adonitol, cellulose, dextrin, dulcitol, erythritol, glacturonic acid, glycerol, inositol, lactate, mannitol, melizitose, sorbitol, sorbose, and trehalose. Fermentation products are formate, butyrate, acetate, propionate, and traces of succinate. Hydrogen produced from growth on CC-glucose medium. Does not reduce nitrate. Does not produce indole. No catalase, lipase, or lecithinase activity. Hydrolyzes casein, gelatin, and esculin. Curd produced from milk.

Growth is inhibited by bile, stimulated by 20% rumen fluid, and unaffected by Tween 80 or hemin. Growth is inhibited on CC-glucose agar plates containing ampicillin, tetracycline, chloramphenicol, gentamicin, and monensin. Strain B316[T] is not inhibited by streptomycin at concentrations up to 100 μg/ml

Source: rumen contents of grazing cattle in New Zealand.

DNA G+C content (mol%): 28 (T_m).

Type strain: B316, ATCC 51982, DSM 14932.

GenBank accession number (16S rRNA gene): U37378.

Further comments: 16S rRNA gene sequence comparisons show *Clostridium proteoclasticum* to fall within the radiation of the cluster XIVa of the clostridia as defined by Collins et al. (1994). The closest relatives are *Butyrivibrio hungatei* (95.9%) and *Butyrivibrio fibrisolvens* (93.1%). The study of Kopečný et al. (2003) demonstrated that *Clostridium proteoclasticum* falls within the radiation of the genus *Butyrivibrio* based on 16S rRNA gene sequence analyzes and shares many similarities in fatty acid composition patterns. Interestingly, the DNA G+C content of 28 mol% given by Attwood et al. (1996) for the type strain is well out of the range of the species of the genus *Butyrivibrio* (40–41 mol%). DNA G+C content values of ~40 mol% were found for other strains of *Clostridium proteoclasticum* (Kopecny et al., 2003). *Clostridium proteoclasticum* differs from *Butyrivibrio* species in showing proteolytic activities.

121. **Clostridium proteolyticum** Jain and Zeikus 1988, 328[VP] (Effective publication: Jain and Zeikus 1988, 140.)

pro.teo.ly′ti.cum Gr. n. *proteo* protein; Gr. adj. *lyticus* dissolving; N.L. neut. adj. *proteolyticum* protein dissolving, proteolytic.

Rod-shaped cells, 0.5 × 2.2 μm. Nonmotile. Obligate anaerobe. Gram-stain-positive. Bilayered cell wall. Sporeforming. Colonies are small, flat, off-white in color, 1–1.5 mm in diameter. Cell walls are difficult to lyse with lysozyme and Pronase. Obligate anaerobe. Optimum growth occurs in the temperature range 30–37 °C and at pH 6.0–8.0.

Proteolytic. Utilizes gelatin, collagen, azocoll, peptone, meat, and poly pep. Grows slowly on casein and serum albumin. Does not grow on carbohydrates, single or mixtures of amino acids. No vitamins or growth factors are required for growth. Acetate is the major end product of fermentation. Ethanol, isobutyrate, isovalerate, and hydrogen are also produced. Significant amounts of butyrate produced from cooked meat medium. Cells are nonhemolytic. Sensitive to streptomycin and penicillin (100 μg/ml). Negative for catalase and oxidase activity.

Source: chicken manure.

DNA G+C content (mol%): 29.54 (T_m).

Type strain: CG, ATCC 49002, DSM 3090.

GenBank accession number (16S rRNA gene): X73448.

Further comments: 16S rRNA gene sequence comparisons show *Clostridium proteolyticum* to fall within the radiation of the cluster II of the clostridia as defined by Collins et al. (1994). The closest relatives are *Clostridium limosum* (97.0%) and *Clostridium histolyticum* (96.1%).

122. **Clostridium psychrophilum** Spring, Merkhoffer, Weiss, Kroppenstedt, Hippe and Stackebrandt 2003, 1027[VP]

psy.chro′phi.lum. Gr. adj. *psychros* cold; Gr. adj. *philos* loving; N.L. neut. adj. *psychrophilum* cold-loving.

Rod-shaped cells, 1.0–1.4 × 2.5–8.0 μm, occur singly, in pairs or short chains. Gram-stain-positive. Motile by peritrichous flagella. Filamentous cells are frequently present in agar grown cultures. Endospores are ellipsoidal, subterminal to terminal position; sporangium is not swollen. Colonies are 1–2 mm in diameter, round with often coarsely granulated margins, smooth, slightly raised, cream-white to grayish, semi-transparent to opaque, and nonhemolytic on sheep-blood agar. The cell-wall peptidoglycan contains *meso*-DAP. The major cellular fatty acids are $C_{16:1\ \omega 9c}$, $C_{14:0}$, $C_{16:1\ DMA\ \omega 9c}$, $C_{16:1\omega 11c}$, and $C_{16:0}$.

Optimum temperature for growth is 4 °C; no growth above 10 °C. Growth occurs in the pH range 5.5–7.5 with optimal growth at 6.5–7.0. Under optimal conditions, the doubling time is 33.9 h.

Carbohydrates utilized include: arabinose, cellobiose, fructose, glucose, inulin, maltose, mannose, sucrose, trehalose, and xylose. The following carbohydrates are not utilized: amygdalin, galactose, glycogen, inositol, lactose, mannitol, melezitose, melibiose, raffinose, rhamnose, ribose, salicin, sorbitol, and starch. Gelatin and starch are not hydrolyzed. The end products of fermentation are lactate, ethanol, 1-butanol, butyrate, hydrogen, and carbon dioxide.

Source: a microbial mat sample taken from a moated area around Lake Fryxell, Antarctica.

DNA G+C content (mol%): 31.8 (HPLC).

Type strain: A-1/C-an/I, ATCC BAA-582, DSM 14207.

GenBank accession number (16S rRNA gene): AJ297443.

Further comments: 16S rRNA gene sequence comparisons show *Clostridium psychrophilum* to fall within the radiation of cluster I of the clostridia as defined by Collins et al. (1994). The closest relatives of *Clostridium psychrophilum* are the two subspecies of *Clostridium estertheticum* (97.8%), *Clostridium frigoris* (97.7%), *Clostridium lacusfryxellense* (97.6%), and *Clostridium bowmanii* (97.0%). DNA–DNA reassociation studies show *Clostridium psychrophilum* to represent a distinct species even though the 16S rRNA gene sequence similarity values are 97.8% with the subspecies of *Clostridium estertheticum*. In addition, *Clostridium psychrophilum* can be differentiated from closely related species on the basis of optimum temperatures for growth, substrate utilization patterns, starch hydrolysis, and end products of fermentation.

123. **Clostridium puniceum** Lund, Brocklehurst and Wyatt 1981b, 216[VP] (Effective publication: Lund, Brocklehurst and Wyatt 1981a, 17.)

pu.ni ce.um. L. neut. adj. *puniceum* purplish, referring to pink color of colonies on potato infusion agar.

This description is based on that of Lund et al. (1981a) and on study of the type and four other strains.

Cells in PYG broth culture are straight or curved rods, motile and peritrichous, 0.6 × 1.8–4.2 μm, granulose positive, and occur singly, in pairs, or in short chains. They are usually Gram-stain-negative, but Gram-stain-positive cells are sometimes present. Spores are oval, subterminal, and do not swell the cell. Free spores are common and have an extensive exosporium. Cell-wall content has not been reported. Surface colonies on blood agar plates are nonhemolytic, pinpoint–2 mm, circular, raised to convex, opaque, white, slightly shiny, and smooth, with an entire margin, and have a granular or mottled internal structure. On potato infusion agar, colonies are similar except that they are pale pink or deep pink and may have undulate or lobate margins.

Cultures in PYG broth are turbid with a viscous or flocculent sediment and have a pH of 5.2–5.5 after incubation

for 6 d. The optimum temperature for growth is 23–33 °C. Growth range is 7–39 °C. Strains grow moderately well in PY broth without fermentable carbohydrate; they do not grow in the defined medium of Lund et al. (1981a) unless fermentable carbohydrate is present. Abundant gas is detected in PYG deep agar cultures.

Slices of potato, carrot, radish, and turnip are digested. Pectin is fermented. Pectate is hydrolyzed; tributyrin is not attacked. Cultures grown in potato tissue form pectate lyase and pectinesterase but no pectic hydrolase. Products in PYG broth include acetic, butyric, and formic acids, and butanol. Abundant H_2 is produced.

All strains tested are susceptible to chloramphenicol, clindamycin, erythromycin, penicillin G, and tetracycline. Culture supernatants are not toxic for mice. Other characteristics of the species are given in Table 142.

Source: rotting potatoes, a cavity spot lesion in a carrot.

DNA G+C content (mol%): 28–29 (T_m).

Type strain: BL 70/20, ATCC 43978, DSM 2619, ICMP 12529, NCIMB 11596.

GenBank accession number (16S rRNA gene): X71857, X73444.

Further comments: 16S rRNA gene sequence comparisons show *Clostridium puniceum* to fall within the radiation of cluster I of the clostridia as defined by Collins et al. (1994). The closest relatives of *Clostridium puniceum* are the two subspecies of *Clostridium saccharoperbutylacetonicum* (99.1%), *Clostridium diolis* (98.8%), *Clostridium beijerinckii* (98.6%), *Clostridium saccharobutylicum* (98.5), and *Clostridium butyricum* (97.5%).

124. **Clostridium purinilyticum** corrig. Dürre, Andersch and Andreesen 1981, 192[VP]

pu.ri.ni.ly'ti.cum. N.L. n. *purum uricum* condensed as "purin", a term proposed by E. Fisher for the basic ring system of uric acid; Gr. adj. *lutikos* dissolving; N.L. neut. adj. *purinilyticum* decomposing the purine ring.

This description is based on that of Dürre et al. (1981) and on study of the type strain.

Cells in PYG broth culture are Gram-stain-positive straight rods, 1.1–1.6 × 2.7–9.6 μm, motile with lateral and subterminal flagella, and occur singly or in pairs. Spores are round, terminal, and swell the cell. Sporulation occurs most readily on blood agar plates incubated anaerobically for 48 h or in chopped-meat broth incubated in an atmosphere of 90% N/10% CO_2. Higher spore yields may be obtained in a medium with added hypoxanthine or guanine. Cell walls contain *meso*-DAP.

Surface colonies on blood-agar plates prepared with rabbit blood are β-hemolytic, 2–3 mm in diameter, irregular, flat, transparent, slightly buff-colored, dull, and slightly pitted with an irregular or occasionally fringed margin, and have no visible internal structure.

Cultures in PYG broth incubated in an atmosphere of 90% N_2/10% CO_2 are turbid with a smooth sediment and have a pH of 7 after incubation for 4 d. Final pH of cultures after growth on purines or glycine is approximately 8.8.

The optimum temperature for growth is 36 °C; growth occurs at 42 °C. Selenite (0.1 μm) bicarbonate (0.2%), and thiamin are required for growth; molybdate, tungsten, and yeast extract are stimulatory. Optimum pH for growth is 7.3–7.8; growth occurs at pH 6.5–9.0. No gas is detected in PYG deep agar cultures.

Utilizes only purines (including adenine, guanine, xanthine, hypoxanthine, uric acid, and others), glycine, and some glycine derivatives for growth. Hippurate (benzoylglycine) is hydrolyzed.

Products from the fermentation of purines are acetate, formate, CO_2, and ammonia. The major product of PYG cultures is acetate; no H_2 is formed. Pyruvate and threonine are converted to acetate; lactate is not utilized.

The type strain is susceptible to chloramphenicol, clindamycin, penicillin G, and tetracycline, but resistant to erythromycin. Other characteristics of the species are given in Table 143.

Source: soils exposed to chicken manure, sewage sludge enriched with adenine.

DNA G+C content (mol%): 29 (T_m).

Type strain: WA-1, ATCC 33906, DSM 1384.

GenBank accession number (16S rRNA gene): M60491.

Further comments: This species is very similar phenotypically to *Clostridium acidurici* and *Clostridium cylindrosporum*. However, Dürre et al. (1981) report that there is very low DNA–DNA homology between the type strain of *Clostridium purinilyticum* (DSM 1384), the type strain of *Clostridium acidurici* (ATCC 7906; DSM 604), and the type strain of *Clostridium cylindrosporum* (ATCC 7905; DSM 605). They may be differentiated by the ability of *Clostridium purinilyticum* to grow readily on adenine and hypoxanthine and to utilize glycine.

16S rRNA gene sequence comparisons show *Clostridium purinilyticum* to fall within the radiation of cluster XII of the clostridia as defined by Collins et al. (1994). The closest relative of *Clostridium puniceum* is *Clostridium acidurici* (93.8%). *Clostridium purinilyticum* is in fact unrelated to *Clostridium cylindrosporum* which is a member of cluster I as defined by Collins et al. (1994).

125. **Clostridium putrefaciens*** (McBryde 1911) Sturges and Drake 1927, 125^AL (*Bacillus putrefaciens* McBryde 1911, 50.)

pu.tre.fa′ci.ens. L. adj. *putrefaciens* putrefying.

Cells in PYG broth cultures are Gram-stain-positive rods, nonmotile, 1.5–1.8 × 7.5->15 µm. They often occur as long curving filaments. Cells occur singly, in pairs, in long chains or in tangled masses. Spores are round or oval, subterminal or terminal, and swell the cell. Sporulation occurs most readily in chopped-meat cultures held at room temperature. Cell walls contain LL-DAP and glycine.

Surface colonies on blood agar plates are β-hemolytic, pinpoint–0.5 mm, circular to irregular, flat to low convex, transparent to translucent, colorless, shiny, and smooth, with a slightly scalloped or rhizoid margin, and a crystalline or mosaic internal structure.

Cultures in PYG broth are only slightly turbid with a smooth sediment and have a pH of 7.0 after incubation under 90% N_2/10% CO_2 gas for 1 week. The optimum temperature for growth is 15–22 °C; growth is good at 25 °C and 30 °C, slow at 5 °C; no growth at 37 °C. There is good growth between pH 6.2 and 7.4, moderate and slow

growth at a pH of 5.8 or 8.5. Growth is inhibited by 6.5% NaCl or 20% bile. No gas is detected in PYG deep agar cultures. Ammonia is formed slowly in chopped meat and gelatin cultures. Traces of H_2S are produced. Neutral red is reduced.

Cells contain both straight and branched chain fatty acids with saturated C_{16}, C_{12}, and unsaturated C_{18} acids predominating.

Products in PYG broth are moderate amounts of acetate, formate, lactate, and succinate; no H_2 is produced. Pyruvate is not utilized; propionate is not formed from lactate or threonine. The amino acids serine, threonine, glycine, and arginine are utilized for growth, and alanine and valine are produced in 3% casein hydrolysate medium. There is no increase in products in threonine medium. Isobutyric acid is produced from valine; isovaleric acid is produced from leucine and isoleucine. Phenylacetic, hydroxyphenylacetic, and indoleacetic acids are produced from phenylalanine, tyrosine, and tryptophan, respectively.

Other characteristics of the species are given in Table 142. The type strain is nontoxic for mice.

Source: spoiled hams, hog muscle, chicken carcasses, human feces, urine specimens from pregnant women with bacteriuria.

DNA G+C content (mol%): 22–25 (T_m).

Type strain: ATCC 25786, BCRC (formerly CCRC) 14480, CCUG 30534, DSM 1291, JCM 1431, NCIMB 11406, NCTC 9836.

GenBank accession number (16S rRNA gene): AF127024, Y18177.

Further comments: 16S rRNA gene sequence comparisons show *Clostridium putrefaciens* to fall within the radiation of cluster I of the clostridia as defined by Collins et al. (1994). The closest relative of *Clostridium putrefaciens* is *Clostridium algidicarnis* (~99.0%); all other species are related at less than 95% 16S rRNA gene sequence similarity. In the paper describing *Clostridium algidicarnis* (Lawson et al., 1994), *Clostridium putrefaciens* was not included in the phylogenetic analysis or any other comparisons. Further phenotypic and genetic comparisons including DNA–DNA reassocciation studies should be made between the type strains of the species *Clostridium algidicarnis* and *Clostridium putrefaciens*.

126. **Clostridium quinii** Svensson, Dubourguier, Prensier and Zehnder 1995, 879^VP (Effective publication: Svensson, Dubourguier, Prensier and Zehnder 1992, 102.)

qui′ni.i. N.L. gen. n. *quinii* of Quin; named after J.I. Quin, a South African microbiologist who first described a large ovoid bacterium.

Straight rods with pointed ends, 1.0 × 2.5–5 µm. Forms large ovoid cells. Strict anaerobe. Gram-stain-positive. Motile with peritrichous flagella but then become immotile. Chemoorganotroph. Form straight to slightly curved spores. Glycogen inclusions 5–10 × 3–6 µm occur in cells. Colonies are translucent and unpigmented.

Growth occurs in the temperature range 15–50 °C with optimal growth at 40–45 °C. The pH range for growth is 6.2–10.5 with optimal growth at 7.4.

Requires carbohydrates for growth. Utilizes glucose, fructose, xylose, galactose, mannose, sucrose, cellobiose, maltose, raffinose, gentiobiose, turanose, mannitol, arbutin, and salicin as carbon and energy sources. Cellulose, starch, ribose, rhamnose, lactose, inositol, xylitol, sorbitol, dulcitol, glycerol, lactate, citrate, tartrate, fumarate, benzoate, pyruvate, hexanol, and acetone are not utilized for growth. Vitamins not required for growth. Biotin stimulates growth. Esculin is hydrolyzed, but gelatin is not. Nitrate and sulfate are not reduced. Indole, lecithinase, and urease are not present. Main fermentation products are formate, acetate, lactate, ethanol, and hydrogen with minor amounts of butyrate.

Source: anaerobic sludge.

DNA G+C content (mol%): 28 (T_m).

Type strain: BS1, DSM 6736.

GenBank accession number (16S rRNA gene): X76745.

Further comments: 16S rRNA gene sequence comparisons shows *Clostridium quinii* to fall within the radiation of cluster I of the clostridia as defined by Collins et al. (1994). The closest relatives of *Clostridium quinii* are *Clostridium celatum* (98.4%) and *Clostridium disporicum* (97.6%). *Clostridium quinii* is differentiated from *Clostridium celatum* on the basis of morphology, endospore position, motility, and urease activity.

127. **Clostridium ramosum**[*] (Veillon and Zuber 1898) Holdeman, Cato and Moore 1971, 39[AL] ("*Bacillus ramosus*" Veillon and Zuber 1898; *Nocardia ramosa* Vuillemin 1931, 32)

ra.mo'sum. L. neut. adj. *ramosum* much-branched.

Cells in PYG broth cultures stain Gram-positive or Gram-stain-negative and are nonmotile, straight rods, 0.5–0.9 × 2–12.8 µm, and occur singly, in pairs or in short chains often in "V" arrangements, with a "rail fence" appearance, or in irregular masses. Cells may have central or terminal swellings up to 1.6 µm in width. Spores are round, thin-walled, usually terminal, and swell the cell, but are very rarely seen and often are difficult to detect by heat tests. They can be demonstrated most readily from 3-week-old chopped-meat agar slants incubated at 30 °C, or in old chopped meat or PYG broth cultures. Cell walls contain *meso*-DAP; glutamic acid and alanine are present as well. Surface colonies on blood agar plates are nonhemolytic, 0.5–2 mm in diameter, circular to slightly irregular, convex or raised, colorless to gray-white, translucent or semiopaque, and smooth, with an entire, scalloped, or erose margin, and a mottled, mosaic, or granular internal structure.

Cultures in PYG broth are turbid with a smooth or ropy sediment and have a pH of 4.4–4.8 after incubation for 5 d. The optimum temperature for growth is 37 °C; most strains grow equally well at 30 °C and 45 °C and grow well at 25 °C. Growth is stimulated by fermentable carbohydrate, inhibited by 6.5% NaCl and reduced in 20% bile. Moderate gas is detected in glucose deep agar cultures. Production of ammonia and acetyl methyl carbinol is variable among strains. Some strains produce an extracellular β-glucuronidase.

Major products in PYG broth are acetic, formic, and lactic acids; small amounts of pyruvic and succinic acids may be detected, and ethanol often is present. Pyruvate is converted to acetate and formate; neither threonine nor lactate is utilized. H_2 production is variable.

All strains tested are susceptible to chloramphenicol; of 61 strains tested, 4 are resistant to penicillin G, 7 to clindamycin, 13 to erythromycin, and 31 to tetracycline. Strains are susceptible to achievable blood levels of carbenicillin and vancomycin but resistant to lincomycin, rifampin, and gentamicin; 37 of 48 strains tested are susceptible to metronidazole; the remaining strains are susceptible to high but achievable levels (25 µg/ml) of metronidazole. Niridazole, chemically similar to metronidazole, is approximately 15 times as effective as metronidazole.

Culture supernatants are not toxic to mice, but strains are pathogenic for guinea pigs; pathogenicity may be lost in laboratory cultures. Other characteristics of the species are given in Table 144.

Source: infant and adult feces; normal human cervix; human infections of the abdominal cavity, genital tract, lung, biliary tract; blood cultures.

DNA G+C content (mol%): 26 (T_m).

Type strain: ATCC 25582, BCRC (formerly CCRC) 14518, CCUG 24038, DSM 1402, JCM 1298, NCIMB 10673, NCTC 11812, VPI 0427.

GenBank accession number (16S rRNA gene): M23731, X73440.

Further comments: 16S rRNA gene sequence comparisons show *Clostridium ramosum* to fall within the radiation of cluster XVIII of the clostridia as defined by Collins et al. (1994). The closest relative of *Clostridium ramosum* is *Clostridium spiroforme* (96.2%); these two species cluster together in a distinct lineage close to the mycoplasmas.

128. **Clostridium rectum**[*] (Heller 1922) Holdeman and Moore 1972, 69[AL] (*Hiblerillus rectus* Heller 1922, 17.)

rec'tum. L. neut. adj. *rectum* straight.

Cells in PYG broth cultures are nonmotile, Gram-stain-positive straight rods, 0.5–1.1 × 1.6–3.1 µm, and occur singly or in pairs. Spores are oval, subterminal, and swell the cell. Sporulation occurs most readily in chopped-meat broth cultures. Cell walls contain *meso*-DAP. Surface colonies on blood agar plates are pinpoint–1 mm in diameter, translucent to semiopaque, convex, grayish white, shiny, and smooth, with entire margins and no visible internal structure. They may be slightly β-hemolytic.

Cultures in PYG broth are turbid with a smooth sediment and have a pH of 5.2–5.7 after incubation for 1 week. The optimum temperature for growth is 37–45 °C; moderate growth occurs at 25 °C and 30 °C. There is good growth at a pH of 8.5 or in 20% bile, no growth in 6.5% NaCl. Abundant gas is detected in PYG deep agar cultures. Ammonia and H_2S are produced. Neutral red and resazurin are reduced. The insecticide lindane can be degraded provided dithiothreitol or a leucine-proline mixture is present; isovaleric acid is formed.

The major product in PYG broth culture is butyric acid with moderate amounts of acetic and propionic, and a trace of valeric acid. Pyruvate is converted to acetate and butyrate; threonine is converted to propionate. Abundant H_2 is produced.

The type strain is susceptible to chloramphenicol, clindamycin, penicillin G, and tetracycline, but resistant to erythromycin. Culture supernatants are nontoxic to mice. Other characteristics of the species are given in Table 144.

Source: horse manure, beet rhizosphere, rice paddy soil.

DNA G+C content (mol%): 26 (T_m).

Type strain: ATCC 25751, DSM 1295, JCM 1412, NCIB 10651.

GenBank accession number (16S rRNA gene): X77850.

Further comments: 16S rRNA gene sequence comparisons show *Clostridium rectum* to fall within the radiation of cluster XIX of the clostridia as defined by Collins et al. (1994). This species is clearly unrelated to the genus *Clostridium* as it shows highest 16S rRNA gene sequence similarity with species of the genus *Fusobacterium*. The closest relative based on 16S rRNA gene sequence similarity is *Fusobacterium necrogenes* (99.8%). This species should be reclassified to the genus *Fusobacterium*.

129. **Clostridium roseum***** (*ex* McCoy and McClung (1935) nom. rev. Cato, George and Finegold 1988, 220^VP (Effective publication: Cato, George and Finegold 1986, 1186.) (*Clostridium roseum* McCoy and McClung 1935, 237)

ro′se.um. L. neut. adj. *roseum* rosy.

Cells in corn mash cultures are 0.7–0.9 × 3.2–4.3 µm, granulose positive in corn mash or glucose-tryptone broth, and occur singly, in pairs, or in short chains. They are Gram-stain-positive, but rapidly become Gram-stain-negative. Vegetative cells are motile and peritrichous; sporulating cells are sluggish or nonmotile. Spores are oval, subterminal, and swell the cell. Sporulation occurs most rapidly in 5% corn meal mash medium. Cell-wall content has not been reported. Surface colonies on blood agar plates are 4 mm in diameter, irregular, flat, grayish-white with raised white centers, a dull rough surface, a rhizoid margin, and are nonhemolytic. On beef PYG agar, surface colonies are raised and smooth, with irregular margins and a pink to orange pigmentation. Colonies become purplish-black after exposure to air. Pigmentation is most pronounced in 5% corn meal mash semisolid medium. Cultures in PYG broth are only slightly turbid with a peach to orange ropy sediment and have a pH of 4.1 after incubation for 6 d.

The optimum temperature for growth is 37 °C; growth occurs between 20 °C and 47 °C. Growth is greatly stimulated by fermentable carbohydrate and an atmosphere of 90% N_2/10% CO_2. Abundant gas is formed in PYG deep agar cultures.

Gelatin is completely digested; there is a stormy fermentation in milk with an acid curd formed that is 50% digested in 3 weeks. H_2S is produced in 0.25% glucose-tryptone-sulfite or 0.25% glucose-tryptone-thiosulfate broth, but not in SIM which contains 0.02% thiosulfate and 0.02% sulfate. Pectin is strongly fermented (final pH 4.8). Gluconate is fermented to pyruvate and glyceraldehyde-3-phosphate by a modified Entner–Doudoroff pathway.

Products in PYG broth are butyric and acetic acids and butanol; a small amount of succinic acid also is detected. Neither pyruvate, lactate, nor threonine is utilized. In 5% corn mash medium, acetone and ethyl alcohol also are detected. Antibiotic susceptibility has not been determined. Strains are not pathogenic for guinea pigs or rabbits. Other characteristics of the species are given in Table 142.

Source: German maize (probable habitat is the soil).

DNA G+C content (mol%): not reported.

Type strain: ATCC 17797, DSM 7320.

GenBank accession number (16S rRNA gene): Y18170.

Further comments: For differentiation of this species from *Clostridium felsineum*, which it most closely resembles, see *Further comments* following the description of that species. In addition, strains of *Clostridium roseum* have been shown to be serologically distinct from strains of *Clostridium felsineum*. 16S rRNA gene sequence comparisons show *Clostridium roseum* to fall within the radiation of the species of cluster I of the clostridia as defined by Collins et al. (1994) and to be unrelated to *Clostridium felsineum* which is a member of cluster XI. The closest relative of *Clostridium roseum* based on 16S rRNA gene sequence comparisons is *Clostridium acetobutylicum* (98.9%); all other species are related at less than 95% sequence similarity.

130. **Clostridium saccharobutylicum** Keis, Shaheen and Jones 2001, 2101^VP

sac.cha.ro.bu.ty′li.cum. Gr. n. *saccharon* sugar juice; N.L. n. *butylum* butanol; N.L. neut. adj. *saccharon* sugar juice; N.L. n. *butylum* butanol; N.L. neut. adj. *saccharobutylicum* denoting the production of butanol from sugar.

Rods with rounded ends, 1.4 µm × 6.3 µm; can be 3.8–10 µm in length; occur singly, in pairs, or short chains. Motile with peritrichous flagella. Gram-positive cell-wall structure, but stain Gram-negative in older cultures. Older cells accumulate granulose and produce an extracellular slime or capsule. Form oval, terminal, to subterminal spores 1.8 × 1.7–3.9 µm. Colonies are smooth, yellow, have irregular margins, and 2–3 mm in diameter. Mesophilic. Optimum growth and solvent production in the temperature range 30–34 °C and pH 6.2–7.0.

Arabinose, xylose, glucose, mannose, cellobiose, lactose, maltose, sucrose, inositol, melibiose, methylglucopyranoside, raffinose, salicin, trehalose, turanose, amygdalin, starch, glycogen, and dextrin are fermented by all strains. Ribose is weakly fermented. Glycerol, dulcitol, sorbitol, melezitose, rhamnose, and pectin are not fermented. Mannitol, inulin, and D- and L-arabitol are fermented by some strains. Acetone, butanol, ethanol, CO_2, H_2, and acetic and butyric acids are fermentation products. Hydrolyzes esculin and gelatin. Tests negative for catalase, urease, and indole production. Curds milk.

DNA G+C content (mol%): 28–32 (genome sequencing).

Type strain: NCP 262, ATCC BAA-117, DSM 13864.

GenBank accession number (16S rRNA gene): U16147.

Further comments: Strains of this species used in industrial fermentations were isolated and patented by Commerical Solvents Corporation under the name "Clostridium saccharo-butyl-acetonicum-liquefaciens" (Jones and Keis, 1995). These strains were later designated strains of the species *Clostridium acetobutylicum* (Keis et al., 2001), but 16S rRNA gene sequence analysis showed them to be unrelated to that species.

16S rRNA gene sequence comparisons show *Clostridium saccharobutylicum* to fall within the radiation of cluster I of the clostridia as defined by Collins et al. (1994). The closest relatives of *Clostridium saccharobutylicum* are

Clostridium saccharoperbutylacetonicum (98.8%), *Clostridium diolis* (98.6%), *Clostridium beijerinckii* (98.6%), *Clostridium puniceum* (98.5%), and *Clostridium butylicum* (97.5%). *Clostridium saccharobutylicum* can be differentiated from *Clostridium saccharoperbutylacetonicum* on the basis of it being susceptibile to rifampin and not utilizing melezitose and pectin.

131. **Clostridium saccharogumia** Clavel, Lippman, Gavini, Doré and Blunt 2007b, 893[VP] (Effective publication: Clavel, Lippman, Gavini, Doré and Blunt 2007a, 23.)

sac.cha.ro.gu′mi.a. Gr. neut. n. *saccharon* sugar, L. fem. n. *gumia* eater, N.L. fem. n. *saccharogumia* sugar eater.

Helically coiled rods, 0.9 × 2.0–3.0 μm, occurring as single cells. Gram-stain-positive in early exponential phase and Gram-stain-variable or Gram-stain-negative in other phases of growth. Strictly anaerobic. Spores not observed. Cells do not survive ethanol or heat treatments. Nonmotile. Colonies on Columbia-agar are circular, entire, up to 3 mm in diameter, convex to umbonate, opaque, grayish-white and nonhemolytic.

The temperature range for growth is 25–45 °C. pH Range for growth is 6–9. Utilizes cellobiose, fructose, galactose, glucose, lactose, mannose, sucrose, and salicin as the sole carbon and energy source. Maltose, melibiose, raffinose, rhamnose, and threhalose are not utilized. Catalase- and indole-negative. The major cellular fatty acids are $C_{16:0}$ (20.2%) and $C_{17:1iso\ \omega5c}$ (18.2%). Tolerant to penicillin and vancomycin at 0.5 μg/ml and to rifampin and metronidazole at concentrations of up to 15 μg/ml.

Source: feces of a healthy human male adult.

DNA G+C content (mol%): 29.4–31.7 (T_m, HPLC).

Type strain: SDG-Mt85-3Db, CCUG 51486, DSM 17460.

GenBank accession number (16S rRNA gene): DQ100445.

Further comments: 16S rRNA gene sequence comparisons show *Clostridium saccharogumia* to fall within the radiation of cluster XVIII of the clostridia as defined by Collins et al. (1994). The closest relatives of *Clostridium saccharogumia* are *Clostridium cocleatum* (96.7%), *Clostridium ramosum* (96.6%), and *Clostridium spiroforme* (95.9%). Clavel et al. (2007a) differentiated *Clostridium saccharogumia* from these related species on the basis of fatty acid profiles and substrate utilization patterns. It is interesting to note that these are recorded as ND (not determined) for the type strains of the related species in Clavel et al. (2007a).

132. **Clostridium saccharolyticum*** Murray, Khan and van den Berg 1982, 135[VP]

sac′cha.ro.ly′ti.cum. Gr. n. *sacchar* sugar; Gr. adj. *lutikos* dissolving; N.L. neut. adj. *saccharolyticum* sugar-dissolving.

This description is based on that by Murray et al. (1982) and on study of the type strain.

Cells are Gram-stain-negative, atrichous, nonmotile, spindle-shaped straight rods, 0.5–0.7 × 3.0 μm. Spores are round, terminal, or subterminal and distend the cell. Sporulation occurs readily in chopped-meat carbohydrate broth. Cell-wall composition has not been reported.

Surface colonies on cellobiose-yeast extract agar after 48 h incubation are 0.5–1.5 mm in diameter, circular with smooth margins, convex and white. On blood agar plates, colonies are pinpoint–1 mm in diameter, circular or slightly irregular, low convex, with entire or slightly scalloped margins and a mottled or crystalline internal structure.

Cultures in PYG broth are turbid with a smooth sediment and have a pH of 5.1 after incubation for 4 d.

The temperature for optimum growth is 37 °C; growth also occurs at 17 °C and 43 °C but not at 14 °C or 45 °C. Optimum pH for growth is 7.4; growth occurs between pH 6.0–8.8. Abundant growth occurs in PY media supplemented with vitamin K and heme, with or without carbohydrate. There is no growth in a defined carbohydrate-mineral salt-vitamin medium even when supplemented with synthetic mixtures of amino acids, purines, and pyrimidines; good growth is obtained in this medium upon addition of yeast extract or in co-culture with *Acetivibrio cellulolyticus*. In minimal medium, 0.1% yeast extract, fermentable carbohydrate, iron, and a reduced form of sulfur are required for growth; growth in this medium is enhanced by addition of B vitamins and phosphate.

Milk is curdled; acetyl methyl carbinol, ammonia, and H_2S are produced; and resazurin is reduced. Starch is hydrolyzed and nitrate reduced.

Products of fermentation in cellobiose-yeast extract broth after 7 d incubation detected by a flame-ionization detector are acetic acid, ethanol, and traces of pyruvic and lactic acids. Products in PYG broth are acetic, formic, and lactic acids. H_2 and CO_2 are produced. In naturally occurring habitats, the acetic acid and H_2 are utilized by methanogens and the ethanol and lactic acid are utilized by sulfate reducers to provide a favorable environment for cellulolytic anaerobes. The presence of calcium carbonate and of higher concentrations of yeast extract, or the presence of added H_2 in the headspace gas, produce a metabolic shift to less acetic acid production and greater ethanol production. Pyruvate is utilized; neither lactate nor threonine is utilized.

The type strain is susceptible to chloramphenicol, clindamycin, erythromycin, penicillin G, and tetracycline. Other characteristics of the species are shown in Table 144.

Source: a methanogenic cellulose-enrichment culture from sewage sludge.

DNA G+C content (mol%): 28 (T_m).

Type strain: WM1, ATCC 35040, DSM 2544, NRC 2533.

GenBank accession number (16S rRNA gene): Y18185.

Further comments: *Clostridium saccharolyticum* falls within the radiation of cluster XIVa (Collins et al., 1994) and shows highest 16S rRNA gene sequence similarity to the *Clostridium* species *Clostridium amygdalinum* (98.9%), *Clostridium celerecrescens* (98.5%), *Clostridium indolis* (98.5%), and *Clostridium aerotolerans* (98.4%). Strain WM1[T] differs from *Clostridium amygdalinum* in its ability to utilize galactose, rhamnose, lactose, and mannose. *Clostridium saccharolyticum* does not degrade starch while *Clostridium amygdalinum* does. *Clostridium amygdalinum* is a moderate thermophile while *Clostridium saccharolyticum* is a mesophile.

133. **Clostridium saccharoperbutylacetonicum** Keis, Shaheen and Jones 2001, 2101[VP]

sa.cha.ro.per.bu.tyl.a.ce.to'ni.cum. Gr. n. *saccharon* sugar juice; Gr. pron. *per* throughout; N.L. n. *butylum* butanol; N.L. adj. *acetonicus* acetonic; N.L. adj. *saccharoperbutylacetonicum* denoting the production of a large amount of butanol and acetone from sugar.

Straight rods, 0.4–0.8 × 3.1–6.2 µm; occurring singly or in pairs. Motile with peritrichous flagella. Gram-stain-positive in young cultures, and Gram-stain-negative in older cultures. Cells accumulate granulose during late exponential growth. Form oval spores, 0.8–1.5 × 1.6–2.2 µm. Colonies on CBM agar are white, smooth, domed, and 2–3 mm in diameter with entire/undulated margins.

Mesophilic. Optimum temperature for growth and solvent production is 25–35 °C; the optimum pH range for growth is 5.6–6.7.

Ferments arabinose, xylose, glucose, mannose, cellobiose, lactose, maltose, sucrose, mannitol, melibiose, raffinose, trehalose, salicin, turanose, amygdalin, starch, glycogen, dextrin, pectin, melezitose, D-arabitol, L-arabitol, and inulin. Ribose and glycerol are not fermented. Sorbitol, dulcitol, and inositol are fermented by the type strain. Rhamnose is fermented weakly. Acetone, butanol, ethanol, CO_2, H_2, and acetic and butyric acid are end products of fermentation.

Does not reduce nitrate. Ammonia produced from nitrite. Hydrogen sulfide is not produced from peptone-containing medium, is not produced or is weakly produced from the reduction of sulfites, but is produced from the reduction of thiosulfates. Does not produce catalase, oxidase, and indole. Hydrolyzes esculin and gelatin. Digestion of coagulated albumin is very weak or negative. Milk is not digested. Resistant to rifampin (100 ng). Used in the industrial setting for the production of the solvents acetone, butanol, and ethanol from sugar- and starch-based substrates.

DNA G+C content (mol%): 31 (T_m).

Type strain: N1-4 (HMT) (derived from strain N1-4), ATCC 27021 (derived from ATCC 13564), DSM 14923, NCIMB 12606.

GenBank accession number (16S rRNA gene): U16122.

Further comments: 16S rRNA gene sequence comparisons show *Clostridium saccharoperbutylacetonicum* to fall within the radiation of cluster I of the clostridia as defined by Collins et al. (1994). The closest relatives based on 16S RNA gene sequence similarity values are *Clostridium diolis* (99.1%), *Clostridium puniceum* (99.1%), *Clostridium beijerinckii* (99.1%), and *Clostridium saccharobutylicum* (98.8%).

134. **Clostridium sardiniense**[*] corrig. Prévot 1938, 81[AL] emend. Wang, Maegawa, Karasawa, Ozaki and Nakamura 2005, 1196.

sar.din.i.en'se. N.L. neut. adj. *sardiniense* from Sardinia.

Cells in PYG broth are Gram-stain-positive, straight or slightly curved rods, motile, and peritrichous. Motility may be lost on subculture. Cells occur singly, in pairs, or in short chains, and are 0.5–1.7 × 1–10.0 µm. Spores are oval and subterminal, or occasionally terminal; very few spores are produced in usual media. Sporulation occurs most readily in 3-week-old chopped-meat broth cultures. Cell walls contain *meso*-DAP.

Surface colonies on blood agar plates are β-hemolytic, 1–3 mm in diameter, circular to irregular, raised or low convex, translucent or semiopaque, gray-white, shiny, and smooth, with a lobate or erose margin and usually with a granular or mottled internal structure. Cultures in PYG broth are turbid with a smooth or stringy sediment, and have a pH of 4.5–5.0 after incubation for 5 d.

The temperature for optimum growth is 25–37 °C; growth is poor at 45 °C. Growth is stimulated by fermentable carbohydrate but inhibited by 6.5% NaCl. Abundant gas is produced in PYG deep agar cultures. Ammonia is produced, neutral red and resazurin are reduced.

Products in PYG broth include large amounts of acetic, butyric, and lactic acids, sometimes with small amounts of formic and propionic acids. Abundant H_2 is produced. Pyruvate is converted to acetate and butyrate. Neither lactate nor threonine is converted to propionate.

All strains tested are susceptible to chloramphenicol, penicillin G, and tetracycline; 4 of 8 strains are resistant to clindamycin; 2 of 8 strains are resistant to erythromycin.

Culture supernatants are nontoxic to mice. Cultures injected subcutaneously or intramuscularly are pathogenic for sheep, guinea pigs, goats, dogs, rabbits, mice, rats, and chickens. Other characteristics of the species are given in Table 142.

Source: lesions of symptomatic anthrax in sheep, soil, water, feces of infants.

DNA G+C content (mol%): not determined.

Type strain: ATCC 33455, DSM 2632, VPI 2971.

GenBank accession number (16S rRNA gene): AB161367, AB161368.

Further comments: Cato et al. (1986) indicated the species is differentiated from *Clostridium absonum* by motility but that some strains designated to the species *Clostridium absonum* have been found to be motile. Subsequent phylogenetic analyzes using the 16S rRNA gene and PLC amino acid sequences show the type strains of *Clostridium sardieniense* and *Clostridium absonum* to be highly similar (Wang et al., 2005). DNA–DNA reassociation values of > 83% between strains of *Clostridium sardieniense* and *Clostridium absonum* indicated that the two species were heterotypic synonyms (Wang et al., 2005). This species is most easily differentiated from *Clostridium perfringens*, which it closely resembles phenotypically, by its motility and its lack of toxicity for mice. However, *Clostridium sardinense* is not closely related to *Clostridium perfringens* (94.8%) based on 16S rRNA gene sequence similarities. The closest relatives of *Clostridium sardinense* are *Clostridium baratii* (98.9%) and the *Eubacterium* species *Eubacterium budayi* (98.6%), *Eubacterium moniliforme* (98.2%), *Eubacterium nitritogenes* (98.2%), and *Eubacterium multiforme* (98.0%).

135. **Clostridium sartagoforme** corrig.[*] Partansky and Henry 1935, 564[AL]

sar.ta.go.for'me. L. n. *sartago* frying pan; L. adj. suffix *-formis* shaped like; N.L. neut. adj. *sartagoforme* intended to mean shaped like a frying pan (in reference to a sporulating cell).

Cells in PYG broth culture are Gram-stain-positive, straight to slightly curved rods that occur singly or in pairs, and are 0.3–0.9 × 2.2–8.0 µm. Motility is variable; cells of motile strains are peritrichous. Spores are oval (occasionally round), terminal, and swell the cell. Sporulation occurs most readily in chopped-meat broth cultures. Cell wall does not contain DAP. The cell wall is susceptible to dissolution by lysozyme.

Surface colonies on blood agar plates are 1–3 mm in diameter, circular, with entire or erose margins, flat or convex, gray, translucent, with a matt surface, and are usually nonhemolytic. Colonies in agar are 1–1.5 mm in diameter and lenticular.

Cultures in PYG broth are turbid with a smooth, or occasionally, flocculent to ropy, sediment and have a pH of 4.8–5.1 after incubation for 5 d. Temperature range for optimum growth is 30–37 °C; growth also occurs at 25 °C and 45 °C. Growth is stimulated by a fermentable carbohydrate and is inhibited by 6.5% NaCl or 20% bile. Abundant gas is produced in PYG deep agar cultures.

Adonitol, dulcitol, erythritol, glycerol, inositol, and sorbose are not fermented. Ammonia is sometimes produced, hippurate is hydrolyzed by 1 of 3 strains tested, neutral red is reduced, and resazurin usually is reduced. Deoxyribonuclease activity is not present.

Products of fermentation in PYG broth are large amounts of acetic, butyric, and formic acids and sometimes large amounts of lactic acid. Abundant H_2 is produced. Neither lactate nor threonine is converted to propionate; pyruvate is converted to acetate and butyrate.

All strains tested are susceptible to chloramphenicol, erythromycin, and tetracycline. Of four strains tested, one is resistant to clindamycin, and a different one is resistant to penicillin G. Some strains are inhibited by 0.1–0.5 µg of cefoxitin/ml. Culture supernatants are nontoxic to mice.

Other characteristics of the species are given in Table 142.

Source: soil, mud, rumen fluid of healthy and bloating calves; the human gingival crevice; feces of neonates and infants; the feces of approximately 5% of adult subjects tested.

DNA G+C content (mol%): 28 (T_m).

Type strain: ATCC 25778, DSM 1292, JCM 1413.

GenBank accession number (16S rRNA gene): Y18175.

Further comments: 16S rRNA gene sequence analysis shows *Clostridium sartagoforme* to fall within the radiation of cluster I of the clostridia as defined by Collins et al. (1994). The closest relatives of *Clostridium sartagoforme* based on 16S rRNA gene sequence similarities are *Clostridium tertium* (98.4%), *Clostridium carnis* (97.5%), *Clostridium isatidis* (97.2%), *Clostridium quinii* (97.0%), and *Clostridium chauvoei* (97.0%).

136. **Clostridium scatologenes**[*] (Weinberg and Ginsbourg 1927) Prévot 1948b, 191[AL] (*Clostridium scatol* Fellers and Clough 1925, 128; *Bacillus scatologenes* Weinberg and Ginsbourg 1927, 54)

sca.to.lo′gen.es. Gr. n. *skatos* dung; Gr. v. *gennaio* to produce; N.L. part. adj. *scatologenes* meaning either an organism that produces a dung-like odor or an organism that produces skatol.

Cells in PYG broth are Gram-stain-positive straight rods that are motile and peritrichous, and are 0.5–1.6 ×

3.1–21.2 µm. Spores are oval, terminal, or occasionally subterminal, and distend the cell slightly. Cell walls contain *meso*-DAP. The cell wall is susceptible to dissolution by lysozyme. Surface colonies on blood agar are α-hemolytic, 0.5–2.0 mm in diameter, circular, convex, translucent, gray, with an entire or scalloped margin, a matt surface, and a granular or mottled internal structure. Cultures in PYG broth are turbid with a smooth sediment and a pH of 5.2–5.4 after incubation for 5 d.

The temperature for optimum growth is 30–37 °C; also grows at 25 °C but not at 45 °C. Grows well in absence of a fermentable carbohydrate. Growth inhibited by 20% bile, 6.5% NaCl, or a pH of 8.5. Abundant gas is produced in PYG deep agar cultures.

Produces ammonia and H_2S; two of three strains hydrolyze hippurate; reduces neutral red and usually reduces resazurin. Uracil is not utilized.

Products of fermentation in PYG broth are large amounts of acetic and butyric acids, and sometimes small amounts of formic, propionic, isobutyric, isovaleric, valeric, caproic, lactic, and succinic acids. Abundant H_2 is produced. Addition of valine to the medium results in increased production of isobutyric acid, whereas addition of either leucine or isoleucine results in production of increased amounts of isovaleric acid. Lactate is converted to acetate, butyrate, and caproate and occasionally to valerate; pyruvate is converted to acetate and butyrate and occasionally caproate; and threonine is converted to propionate. Serine and arginine are utilized and 2-aminobutyric acid and ornithine are produced in an amino acid-trypticase-yeast extract broth culture. Produces skatol.

The major fatty acid is normal, saturated C_{16}; in addition, normal saturated C_{12}, C_{14}, C_{15}, C_{18}, and normal unsaturated C_{16} acids represent appreciable proportions of the total fatty acids detected. Small amounts of branched chain fatty acids are also present.

Two strains tested are susceptible to chloramphenicol, erythromycin, clindamycin; 2 of 3 strains are susceptible to penicillin G. One strain tested is susceptible to the following antimicrobials in the concentrations indicated: penicillin V, 2 U/ml; cephalothin, 64 µg/ml; cephalexin, 32 µg/ml, cephradine, 16 µg/ml; cefoxitin, 128 µg/ml; tetracycline, 64 µg/ml; and doxycycline, 16 µg/ml. Culture supernatants are nontoxic to mice. Reported to be nonpathogenic. Other characteristics of the species are given in Table 142.

Source: soil, contaminated food, feces of infants undersized at birth.

DNA G+C content (mol%): 27 (T_m).

Type strain: ATCC 25775, CCUG 9283, DSM 757, JCM 1414, NCIB 8855 NCTC 12292.

GenBank accession number (16S rRNA gene): M59104.

Further comments: 16S rRNA gene sequence comparisons show *Clostridium scatologenes* to be a member of cluster I of the clostridia as defined by Collins et al. (1994). *Clostridium scatologenes* groups closely with the species *Clostridium carboxidivorans* and *Clostridium drakei* sharing 99.7% 16S rRNA gene sequence similarity with each of these species. The species status of each of three species is supported by DNA–DNA reassociation data, and the species can be differentiated the basis of substrate utilization patterns and the fact that *Clostridium scatologenes* grows slowly on H_2/CO_2 (Liou et al., 2005).

137. **Clostridium schirmacherense** Alam, Dixit, Reddy, Dube, Palit, Shivaji and Singh 2006, 719[VP]

schir.ma.cher.en′se. N.L. neut. adj. *schirmacherense* pertaining to Schirmacher Oasis in Antarctica.

Rods-shaped cells, 0.5–0.7 × 2–4 μm. Motile. Gram-stain-positive. Spores are subterminal and distend the cell. Contains diphosphatidylglycerol as the major phospholipids, and *meso*-diaminopimelic acid is present in the cell wall. The major cellular fatty acids are $C_{15:0}$, $C_{16:0}$, and $C_{17:0}$. Strictly anaerobic and chemoorganotrophic. Temperature range range for growth is 5–35 °C; produces maximum cell mass at 5–10 °C. The pH range for growth is 6–9 with an optimum of pH 8.0. NaCl is not required for growth but is tolerated up to a concentration of 7.5%.

Esculin, gelatin, and casein are hydrolyzed. Does not reduce nitrate to nitrite. Negative for catalase, oxidase, lipase, indole production, and β-hemolysis. Raffinose, glucose, adonitol, rhamnose, and ribose are utilized but not sucrose, arabinose, mannose, mannitol, lactose, inositol, trehalose, maltose, dulcitol, xylan, arabitol, xylose, inulin, amygdalin, glycogen, galactose, or sorbitol. Complex substrates including yeast extract, Casamino acids, peptone, and gelatin are fermented. Arginine, serine, leucine, isoleucine, cysteine, and glutamate, are utilized while asparagine, proline, glutamine, aspartate, glycine, lysine, methionine, threonine, tryptophan, tyrosine, valine, 2-aminobutyric acid, ornithine, and phenylalanine are not utilized. H_2S produced. Sulfite and thiosulfate are reduced but sulfate, elemental sulfur and fumarate are not reduced.

The major fermentation end products are acetate, propionate, H_2, and CO_2 from glutamate; acetate and butyrate from isoleucine; acetate, propionate, H_2, and CO_2 from cysteine; formate, acetate, and propionate from arginine; acetate, propionate, H_2, and CO_2 from leucine; and acetate, isobutyrate, and butyrate from serine. Glucose is fermented to formate, acetate, propionate, butyrate, isovalerate, H_2, and CO_2.

Source: lake sediments of Schirmacher Oasis, Antarctica.

DNA G+C content (mol%): 24 (HPLC).

Type strain: AP15, DSM 17394, JCM 13289.

GenBank accession number (16S rRNA gene): AM114453.

Further comments: 16S rRNA gene sequence analysis shows *Clostridium schirmacherense* to fall within the radiation of cluster I of the clostridia as defined by Collins et al. (1994). The closest relatives based on 16S rRNA gene sequence similarity values are *Clostridium subterminale* (99.6%) and *Clostridium argentinense* (99.4%). *Clostridium schirmacherense* can be differentiated from these close relatives based on its much lower temperature range for growth, esculin hydrolysis, production of formate and propionate, lower G+C content, and lack of β-hemolysis.

138. **Clostridium scindens** Morris, Winter, Cato, Ritchie and Bokkenheuser 1985, 478[VP]

scin′dens L. part. adj. *scindens* splitting, because it produces a desmolase.

Pleomorphic, straight to slightly curved rods, 0.5–0.7 × 1.0–2.5 μm; occur singly or in chains. Nonmotile. Some coccus-shaped cells 0.4–0.6 μm are found. Grows anaerobically. Aerotolerant. Chemorganotrophic. Gram-stain-positive. Forms spores which are terminal and wider than the cells. Cultures survive heat treatment at 80 °C for 10 mins and absolute ethanol exposure for 30 mins. Many cells have fimbriae 3.0–3.5 nm in diameter, in clusters of 2–10. Colonies are circular to slightly irregular, smooth, grayish-white, convex, and 0.3–1 mm diameter. Nonhemolytic to sheep and rabbit erythrocytes.

Temperature optimum for growth is 42 °C with growth at 37 and 45 °C but not at 50 °C. Slow growth occurs at 25 °C. Ferments D-fructose, D-glucose, lactose, D-mannose, D-ribse, and D-xylose. Does not ferment amygdalin, L-arabinose, D-cellobiose, *meso*-erythritol, esculin, glycogen, *myo*-inositol, maltose, D-mannitol, D-melezitose, D-melibiose, D-raffinose, L-rhamnose, salicin, D-sorbitol, starch, sucrose, and D-trehalose. Ferments PYEG broth to acetic acid and ethanol. Converts pyruvate to acetate, and lactate is not used. Has desmolase enzymic activity. Reduces sulfate, but not nitrate. H_2S detected in sulfide-indole-motility medium. Does not digest gelatin, milk, or meat. Tests negative for catalase, indole, lecithinase, or lipase activity. Starch and esculin are not hydrolyzed. Produces at least four enzymes that cleave steroidal substrates. Produces desmolase. Sensitive to penicillin G (2U/ml). Resistant to chloramphenicol 12 μg/ml), clindamycin (1.6 μg/ml), erythromycin (3 μg/ml), and tetracycline (6 μg/ml).

Source: human feces.

DNA G+C content (mol%): 45 (T_m).

Type strain: Bokkenheuser strain 19, ATCC 35704, CIP 106687, DSM 5676, JCM 6567.

GenBank accession number (16S rRNA gene): AB020883, AF262238, Y18186.

Further comments: 16S rRNA gene sequence comparisons show *Clostridium scindens* to fall within the radiation of cluster XIVa as defined by Collins et al. (1994). Previous studies have concentrated on the comparison of *Clostridium scindens* and *Clostridium hylemonae* (Kitahara et al., 2000). However, the 16S rRNA gene sequence similarity between these two species is only 93.0%. The closest relatives of *Clostridium scindens* based on 16S rRNA gene sequences are species of the genus *Dorea* including *Dorea longicatena* and *Dorea formicigenerans* (94.4 and 92.9%), and *Ruminococcus gnavus* (93.6%).

139. **Clostridium septicum**[*] (Macé 1889) Ford 1927, 726[AL] (*Bacillus septicus* Macé 1889, 445)

sep′ti.cum. L. neut. adj. *septicum* putrefactive.

Cells in PYG broth culture are Gram-stain-positive in young cultures but may be Gram-stain-negative in older cultures; staining is often uneven, resulting in intensely Gram-stain-positive bars or spots interspersed with decolorized areas. Cells in broth are straight or curved rods that occur singly or in pairs, are usually motile and peritrichous, and are 0.6–1.9 × 1.9–35.0 μm. Forms long filaments on the peritoneal surface of the liver of infected animals in contrast to other pathogenic species of *Clostridium*. Cells may be extremely pleomorphic under certain conditions.

Spores are oval, subterminal, and distend the cell. There is no exosporium and the spore lacks appendages.

Cell wall contains L-lysine in place of DAP. Glutamic acid and alanine are present; cell-wall sugars are glucose, galactose, rhamnose, and mannose. The cell wall is susceptible to dissolution by lysozyme.

Colonies on the surface of blood agar are 1–5 mm in diameter, circular, with markedly irregular to rhizoid margins, slightly raised, translucent, gray, glossy, and β-hemolytic. Surface growth of most strains on 1.5% agar is an invisible film over the entire agar surface; this swarming often may be prevented by shortening the incubation period or using 4–6% agar in the medium. Subsurface colonies in 1% agar are spherical or lenticular and transparent; in 2% agar, colonies are brownish-yellow and heart shaped.

Cultures in PYG broth are turbid with a smooth sediment and have a pH of 4.7–5.3 after incubation for 5 d.

The temperature range for optimum growth is 37–40 °C; most strains grow well at 44 °C but not at all at 46 °C. CO_2 is not required for growth, but good growth occurs in atmospheres containing up to 100% CO_2. Growth is stimulated by the presence of a fermentable carbohydrate, serum, or peptic digest of blood. In chopped-meat medium, the meat particles commonly turn pink after 48 h incubation. Factors essential for growth include biotin, nicotinic acid, pyridoxine, thiamin, cysteine, tryptophan, iron and, for some strains, pantothenate. Adenine, arginine, aspartic acid, histidine, isoleucine, phenylalanine, serine, threonine, tyrosine, and valine are also required. Exposure of exponential-phase cultures of this species to hyperbaric oxygen (100% O_2 at 3 atm) is relatively nonlethal compared with other species; approximately 50% of cells of *Clostridium septicum* survive this treatment. Abundant gas is produced in PYG deep agar cultures.

Adonitol, dulcitol, erythritol, and glycerol are not fermented; sorbose occasionally is fermented weakly. Neutral red and resazurin are reduced and ammonia is produced by some strains; hippurate is hydrolyzed by 3 of 17 strains tested.

Phage-like particles produced by a strain of *Clostridium septicum* inhibit the growth of a strain of *Clostridium sporogenes*. This strain of *Clostridium septicum* is in turn inhibited by strains of *Clostridium bifermentans*, *Clostridium butyricum*, *Clostridium cadaveris*, *Clostridium perfringens*, *Clostridium tertium*, and *Clostridium tetani*. All of 16 strains of *Clostridium septicum* tested were susceptible to bacteriocins produced by 9 of 21 strains of *Streptococcus faecium* or *Streptococcus faecalis* studied. Bacteriocin produced by *Clostridium septicum* inhibited RNA and protein synthesis, and rapidly kills all 16 strains of *Clostridium septicum* and all 10 strains of *Clostridium chauvoei* tested, but none of a variety of other species of bacteria.

One strain of *Clostridium septicum* has been shown to hydrolyze the conjugated bile acids taurocholate and glycocholate. All of 26 strains tested produce deoxyribonuclease (beta toxin) and hyaluronidase (gamma toxin) in both acidic and alkaline conditions. An oxygen-stable hemolysin with necrotizing properties (alpha toxin), and an oxygen-labile hemolysin (delta toxin) are also produced. A neuraminidase and a hemagglutinin are produced. Chitinase is produced. Uracil is not utilized. Although lipase is not detected on egg-yolk agar plates, traces of lipidolytic activity, detected by a gas-liquid chromatographic method, have been reported in a strain of this species.

Products in PYG broth include large amounts of acetic and butyric acids, and usually large amounts of formic acid. Abundant H_2 is produced. Electron capture gas-liquid chromatography has detected butanol, isoamyl alcohol, and propionic, isobutyric, isovaleric, and isocaproic as well as acetic and butyric acids in chopped-meat glucose cultures of *Clostridium septicum*. Lactate usually is not utilized, but increased butyrate may be formed in PY-lactate broth. Pyruvate is converted to acetate and butyrate and, occasionally, to formate. Threonine is not converted to propionate. Oleic acid is converted to hydroxystearic acid.

Ethylamine was detected in the head-space gas above a chopped-meat glucose culture of one isolate. All strains tested are susceptible to chloramphenicol, clindamycin, erythromycin, penicillin G, and tetracycline.

This species is capable of causing rapidly fatal infections in humans and other animals by production of its lethal hemolytic and necrotizing alpha toxin. A lethal disseminated infection can be produced in mice and guinea pigs by intramuscular injection of fewer than 10 spores plus $CaCl_2$; postmortem examination reveals bloodstained edema fluid, a deep red color of the muscle and soft tissue, gas formation in the inoculated limb, and dissemination of the organism to all parts of the body. In cattle, wound infection produces "malignant edema", a disease similar to the experimental disease in laboratory animals, as described above; and in sheep, braxy, a fatal bacteremic infection following penetration of the abomasal wall. Similar diseases have been noted in other species of domestic and wild animals. *Clostridium septicum*, *Clostridium perfringens*, and *Clostridium novyi* are the three most common causes of clostridial myonecrosis (gas gangrene). Such infection may occur either by direct contamination of an infected traumatic wound or by metastatic infection following bacteremia from the gastrointestinal tract; in the latter case, a breech in the bowel mucosa (such as due to a malignancy) serves as the portal of entry. The mechanism of disease production apparently is similar to that of the animal model described above. Other characteristics of the species are given in Table 142.

Source: soil; probably the gastrointestinal tract contents of many domesticated animals; rattlesnake venom; feces of human infants and adults; animal infections including wound infections; myositis and enterotoxemia in ruminants and bacteremia; human infections including bacteremia, suppurative infections, necrotizing enterocolitis, and myonecrosis or gas gangrene.

DNA G+C content (mol%): 24 (T_m).

Type strain: ATCC 12464, CCM 5743, BCRC 14552, CCUG 4855, CIP 61.10, CN 3790, DSM 7534, JCM 8144, NCCB 47070, NCIB 547, NCTC 547, NRRL B-3197.

GenBank accession number (16S rRNA gene): U59278.

Further comments: 16S rRNA gene sequence analysis shows *Clostridium septicum* to fall within the radiation of cluster I of the clostridia as defined by Collins et al. (1994). The type strain of *Clostridium septicum* shares 99.2% 16S rRNA gene sequence similarity with *Clostridium chauvoei*. Other related species include *Clostridium carnis* (98.2%) and *Clostridium tertium* (97.8%). *Clostridium chauvoei* and *Clostridium septicum* can be differentiated on the basis of substrate utilization patterns.

140. **Clostridium sordellii**[*] (Hall and Scott 1927) Prévot 1938, 83[AL] (*Bacillus sordellii* Hall and Scott 1927, 330)

sor.del′li.i. N.L. gen. n. *sordellii* pertaining to Professor A. Sordelli, an Argentinian bacteriologist.

Cells in PYG broth cultures are Gram-stain-positive straight rods, usually motile and peritrichous, 0.5–1.7 × 1.6–20.6 μm, and occur singly or in pairs. Spores are oval and central or subterminal, and often occur as free spores. They swell the cell slightly. Spores have a thick exosporium; most have tubular appendages. Sporulation occurs readily in chopped-meat broth cultures incubated for 24 h or on blood agar plates incubated 48 h.

Cell walls contain *meso*-DAP, glucose, and a trace of galactose. Glutamic acid and alanine also are present. Urease-negative strains that are homologous by DNA–DNA homology determinations with authentic urease-positive strains of *Clostridium sordellii* do not have glucose in their cell walls. Surface colonies on blood agar plates are 1–4 mm in diameter, circular to irregular, flat or raised, translucent or opaque, gray or chalk-white, with a dull or shiny surface, a granular or mottled internal structure, and a scalloped, lobate, or entire margin. Hemolysis is variable, with most strains being slightly β-hemolytic on rabbit blood agar.

The optimum temperature for growth is 30–37 °C. Strains grow moderately well at 25 °C and 45 °C. Growth is inhibited by 6.5% NaCl, by 20% bile, or by a pH of 8.5. Strains of *Clostridium sordellii* are inhibited by phage-like particles found in strains of *Clostridium bifermentans*; although similar particles have been found in these strains of *Clostridium sordellii*, the inhibition is not reciprocal. Strains of *Clostridium sordellii* share a cross-reactive carbohydrate cell surface antigen with both *Clostridium bifermentans* and *Clostridium difficile*. Abundant gas is detected in PYG deep agar cultures. Ammonia is produced; H$_2$S is produced in SIM by most strains; hippurate is hydrolyzed by most strains. Benzoic acid released from hydrolysis of hippurate can be detected by gas chromatography within 2–4 d.

Gelatin is digested by all strains; milk, casein, and meat are digested by nearly all strains but more slowly. Some strains produce DNase; some strains produce hyaluronatelyase. Supernatant culture fluids hydrolyze bile acid conjugates. Cholic acid is dehydroxylated at the 7α position to deoxycholic acid by the type strain. Oleic acid is converted to hydroxystearic acid *in vitro*, a reaction that may be important in the pathogenesis of diarrhea in humans. Supernatant culture fluids in protease peptone water broth contain substantial amounts of neuraminidase. A wide variety of straight-chain, iso-branched chain, and anteiso-branched chain fatty acids, both saturated and unsaturated, is present in the lipids of this species.

Acetic acid is the major product in PYG broth cultures. Large amounts of formic and moderate amounts of isocaproic acid usually are detected, and trace amounts of propionic, isobutyric, butyric, and isovaleric acids may be present. Ethanol and propanol also are detected. In the absence of glucose, the proportion of acids other than acetic is increased substantially. Abundant H$_2$ is produced. Pyruvate is converted to acetate and formate. Threonine is converted to propionate. Lactate is not utilized. Phenylalanine and tryptophan are converted to phenylacetic, phenylpropionic, phenyllactic acids, and indole. Leucine is converted to isovalerate, isocaproate, and CO$_2$. Isoleucine is converted to isovalerate and valine to isobutyrate. Phenylacetic acid is produced in trypticase-yeast extract medium. Proline, serine, threonine, alanine, aspartic acid, glycine, glutamic acid, and methionine are amino acids utilized for growth in 3% casein hydrolysate medium; γ-aminovalerate, γ-aminobutyrate, and valine are produced.

Of 37 strains tested, all are susceptible to chloramphenicol, erythromycin, and penicillin G; 3, are resistant to tetracycline, 1 is resistant to clindamycin. Culture supernatants of only 9 of 71 strains tested were toxic to mice and this toxicity may be lost on subculture. *Clostridium sordellii* is pathogenic for man, cattle, sheep, guinea pigs, and mice. Pathogenicity also may be ephemeral, and nonpathogenic, urease-positive strains are isolated frequently. Although antitoxin to *Clostridium sordellii* will neutralize the toxins of *Clostridium difficile*, the toxins of the two species have been shown to have different properties. Other characteristics of the species are given in Table 144.

Source: soil; normal human feces; human clinical specimens including wounds, penile lesions, blood cultures, abscesses, and abdominal and vaginal drainage; intestinal tracts of both normal and diseased pheasants; bony tissue from dogs with osteomyelitis; bovine intestinal inflammatory lesions; bovine uterus and muscle; alpaca and sheep infections; chicken skin.

DNA G+C content (mol%): 26 (T_m).

Type strain: ATCC 9714, BCRC 10649, CCUG 9284, CIP 103658, DSM 2141, JCM 3814, LMG 15708, NCIMB 10717.

GenBank accession number (16S rRNA gene): AB075771, M59105.

Further comments: 16S rRNA gene sequence shows *Clostridium sordellii* to fall within the radiation of the cluster XI as defined by Collins et al. (1994). The closest relatives based on 16S rRNA gene sequence similarity values are *Eubacterium tenue* (99.0%), *Clostridium ghonii* (98.6%), and *Clostridium bifermentans* (97.8%).

141. **Clostridium sphenoides**[*] (Douglas, Fleming and Colebrook 1919) Bergey, Harrison, Breed, Hammer and Huntoon 1923, 33[AL] (*Bacillus sphenoides* Douglas, Fleming and Colebrook *in* Bulloch, Bullock, Douglas, Henry, McIntoch, O'Brien, Robertson and Wolf 1919, 43)

sphe.noi′des. Gr. adj. *sphenoides* wedge-shaped.

Cells usually stain Gram-negative, are straight rods with tapered or rounded ends, motile and peritrichous, occur singly, in pairs, or in short chains and are 0.3–1.1 × 1.3–8.6 μm.

Spores are oval and subterminal or occasionally terminal, and swell the cell. Sporulation occurs most readily on blood agar plates incubated 48 h or on chopped-meat slants incubated at 30 °C. Cell wall contains *meso*-DAP. The cell wall is susceptible to dissolution by lysozyme. Colonies on the surface of blood agar are 1–2 mm in diameter, nonhemolytic, circular with an entire or erose margin, low convex, translucent, gray, with a glossy surface, often with a mottled internal structure.

Cultures in PYG broth are turbid with a smooth, or occasionally ropy sediment and a pH of 4.9–5.4 after incubation for 5 d. Optimum temperature for growth is 30–37 °C. There is good growth at 25 °C, little or none at 45 °C. Growth is slightly stimulated by the presence of a fermentable carbohydrate and inhibited by 20% bile. Abundant gas is produced in PYG deep agar cultures.

Adonitol, dulcitol, erythritol, glycerol, and inositol are not fermented; sorbose is weakly fermented. Ammonia and H_2S are produced, acetyl methyl carbinol is produced occasionally, neutral red is reduced, and resazurin is reduced by most strains. Citrate is used as both a carbon and an energy source with production of acetic acid, ethanol, CO_2, and H_2. Citrate utilization by this species may be unique among clostridia. Taurocholic and glycocholic acids are hydrolyzed, deoxyribonuclease is produced. The insecticide hexachlorocyclohexane is converted to tetrachlorocyclohexene. Phage-like particles are produced by one strain tested. Products of metabolism in PYG broth are large amounts of acetic and formic acids, and occasionally, small amounts of lactic acid, succinic acid, and ethanol. Lactate is not utilized, and threonine usually is not converted to propionate. Pyruvate is converted to acetate, occasionally with formate.

Of 20 strains tested, all are susceptible to erythromycin and tetracycline, 15 are susceptible to chloramphenicol, 12 to clindamycin, and 10 to penicillin G. Culture supernatants are nontoxic to mice.

Although *Clostridium sphenoides* occasionally is recovered from polymicrobial infections in humans, pathogenicity has not been reported. Other characteristics of the species are given in Table 144.

Source: soil; marine sediment; dog feces; normal human appendices; feces of from 4–6.4% of adult humans; infections in range animals; blood, bone, and soft tissue infections; intraperitoneal infections; war wounds; visceral gas gangrene; and renal abscess.

DNA G+C content (mol%): 41–42 (T_m).

Type strain: ATCC 19403, CIP 104283, DSM 632, JCM 1415, LMG 10390, NCCB 77027, NCTC 507.

GenBank accession number (16S rRNA gene): AB075772, X73449.

Further comments: Clostridium sphenoides falls within the radiation of cluster XIVa of the clostridia as defined by Collins et al. (1994). The closest relative based on sequence similarity values is the misclassified *Desulfotomaculum guttoideum* (99.5%). Other species showing similarity values in the range 97.5–99.0% include *Clostridium celerecrescens*, *Clostridium indolis*, *Clostridium saccharolyticum*, *Clostridium methoxybenzovorans*, *Clostridium aerotolerans*, and *Clostridium amygdalium*.

142. **Clostridium spiroforme**[*] Kaneuchi, Miyazato, Shinjo and Mitsuoka 1979, 10[AL]

spi.ro.for′me. Gr. n. *spira* a coil; L. n. *forma* shape; N.L. neut. adj. *spiroforme* in the shape of a coil.

Cells on Eggerth–Gagnon agar after 2 d of incubation are Gram-stain-positive, nonmotile, $0.3–0.5 \times 2.0–10.0\,\mu m$, and exhibit various degrees of coiling; long chains of organisms forming tight coils often are seen. After heating at 80 °C for 10 min, cells may be nearly straight and subsequent subcultures do not have coiled cells. Most strains produce round, terminal, or occasionally subterminal spores approximately 0.7 μm in diameter when incubated for 2 weeks at 30 °C on medium 10 (Kaneuchi et al., 1979) or on chopped-meat agar slants. Most strains survive heating to 70 °C for 10 min; survival after heating to 80 °C for 10 min is variable. Spore demonstration is sometimes difficult; fresh isolates from humans may not form spores or survive heating to 70 °C for 10 min.

Colonies on the surface of Eggerth–Gagnon agar after 2 d incubation and on medium 10 agar after 3 d incubation are 0.7–1.5 mm in diameter, have an entire or slightly erose margin, are circular, convex to slightly pulvinate, smooth, shiny, semiopaque to opaque, and are whitish to brownish-gray. Colonies on blood agar plates are nonhemolytic.

PYG broth cultures are slightly turbid with a smooth to flocculent sediment and have a pH of 4.9–5.2 after incubation for 7 d. The temperature for optimum growth is 30–37 °C; most strains grow poorly, if at all, at 25 °C or 45 °C. There is no growth at 15 °C. Growth is enhanced by the presence of a fermentable carbohydrate or of 10% rumen fluid. Growth of most strains is inhibited by 20% bile or 6.5% NaCl. Small to moderate amounts of gas are produced in PYG deep agar cultures. Acetyl methyl carbinol is produced; neutral red and resazurin are reduced. Esculin is fermented occasionally; dextrin, dulcitol, α-methylglucoside, and α-methylmannoside are not fermented. Urease is not produced. Large amounts of acetic and formic acids, and small amounts of lactic acid are produced in PYG broth. Abundant H_2 is produced. Lactate and threonine are not converted to propionate.

The type strain is susceptible to chloramphenicol, clindamycin, erythromycin, penicillin G, and tetracycline. An organism that is identical to, or closely resembles, *Clostridium spiroforme* has been implicated as the cause in rabbits of an apparently toxin-induced diarrhea that occurs spontaneously and in association with antimicrobial therapy. The toxic effect can be neutralized by antitoxin to *Clostridium perfringens* type E iota toxin; this toxic effect is not produced by the type strain of the species. A toxin produced *in vitro* is lethal to mice and causes dermonecrosis in guinea pigs, but toxin is not produced by all strains. Other characteristics of the species are given in Table 144.

Source: the feces of healthy humans; the ceca of healthy chickens and rabbits; the ceca of rabbits with diarrhea.

DNA G+C content (mol%): 27 (T_m).

Type strain: ATCC 29900, CCUG 46510, CIP 106966, DSM 1552, JCM 1432, NCTC 11211, VPI C28-23-1A.

GenBank accession number (16S rRNA gene): X73441, X75908.

Further comments: Clostridium spiroforme can be distinguished from most other species of *Clostridium* by its coiled morphology and its failure to produce butyric acid. It is most readily distinguished from *Clostridium cocleatum*, which it most closely resembles, by its inability to ferment galactose. 16S rRNA gene sequence comparisons show *Clostridium spiroforme* to fall within the radiation of cluster XVIII of the clostridia (Collins et al., 1994). Sequence similarities in the range 95.6–96.4% are found to the species *Clostridium cocleatum*, *Clostridium ramosum*, and *Clostridium saccharogumia*. *Clostridium spiroforme* is most closely related to *Clostridium cocleatum*, but is a distinct species having 46–60% homology with strains of *Clostridium cocleatum* (Kaneuchi et al., 1979).

143. **Clostridium sporogenes*** (Heller 1922) Bergey, Harrison, Breed, Hammer and Huntoon 1923, 329[AL] (*Metchnikovillus sporogenes* Heller 1922, 29)

spo.ro′ge.nes. Gr. n. *sporos* seed; Gr. suff. *-genes* born of; N.L. neut. adj. *sporogenes* spore-producing.

Cells in PYG broth cultures are Gram-stain-positive straight rods, motile and peritrichous, that occur singly and are $0.3–1.4 \times 1.3–16.0\,\mu m$. Spores are oval, subterminal, and distend the cell; sporulation occurs readily on most media. Following sporulation, the vegetative material may disintegrate rapidly to leave only free spores. Outgrowth of spores in meat slurries is inhibited by nitrites and nisin. Cell wall contains *meso*-DAP; galactose is the only cell-wall sugar. Glutamic acid and alanine are present. The cell wall is susceptible to dissolution by lysozyme. Surface colonies on blood agar plates are 2–6 mm in diameter, irregularly circular, possess a coarse rhizoid edge, have a raised yellowish-gray center and a flattened periphery composed of entangled filaments ("Medusa head" colony), are opaque, possess a matt surface, are usually β-hemolytic and are firmly adherent to the agar. Colonies on more moist agar are larger, flatter, gray, and less adherent. Colonies in agar are spherical, with an opaque center, and a woolly semi-translucent periphery; lenticular colonies with fine marginal outgrowths may be produced.

Cultures in PYG broth are turbid, with a smooth or occasionally ropy to flocculent sediment, and have a pH of 5.7–6.4 after incubation for 1 week. In chopped-meat medium with iron filings, there is often marked blackening of the meat particles. The temperature range for optimum growth is 30–40 °C; will grow at 25 °C and 45 °C. Good growth occurs in an atmosphere containing up to 100% CO_2. Growth in 6.5% NaCl, 20% bile, or at a pH of 8.5 is variable; there is no inhibition of some strains, others are completely inhibited. Abundant gas is produced in PYG deep agar cultures.

Adonitol, dulcitol, erythritol, glycerol, and sorbose are not fermented. Ammonia and H_2S are produced, neutral red is reduced and resazurin is reduced by some strains; a few strains produce urease and hydrolyze hippurate. Produces deoxyribonuclease, thiaminase, chitinase, kininase, L-methioninase, hyaluronatelyase, and superoxide dismutase. Uracil is reduced to dihydrouracil. Cells contain ferredoxin. Three flagellar antigens, four somatic antigens, and four spore antigens detected. Some strains of *Clostridium sporogenes* are capable of inhibiting growth and toxin production by *Clostridium botulinum* type A. Only a few strains were lysed by mitomycin C, whereas all strains of *Clostridium botulinum* tested were lysed. Nonlysogenic bacteriophages active against *Clostridium sporogenes* have been isolated from soil, sewage, and chicken feces. Bacteriocin-like substances that are active against other strains of *Clostridium sporogenes* are produced by *Clostridium sporogenes*. The mutagen, 1-nitropyrene, is reduced to 1-aminopyrene with concomitant decrease in mutagenicity by a strain of this species.

Products of metabolism in PYG broth include large amounts of acetic and butyric acids, and small amounts of isobutyric and isovaleric acids; propionic, valeric, isocaproic, lactic, and succinic acids also may be produced. Ethanol and abundant H_2 are produced. Lactate is converted to butyrate; pyruvate is converted to acetate and butyrate; threonine usually is converted to propionate. *Clostridium sporogenes* is capable of carrying out the "Stickland reaction" in which energy is obtained by the coupled oxidation and reduction of various amino acid pairs. Betaine may also serve as an oxidant in this type of reaction. When glycine is used as the oxidant, *Clostridium sporogenes* has an absolute requirement for selenium. Leucine and valine can be interconverted. In a thioglycollate-trypticase-yeast extract medium supplemented with casein hydrolysate, proline is converted to 5-aminovaleric acid, 2-aminobutyric acid is produced, phenylalanine is converted to phenylpropionic acid, tyrosine is converted to *p*-hydroxyphenylpropionic acid, and tryptophan is converted to indole propionic acid. Arginine and serine also can be utilized. Addition of valine to a basal medium results in increased isobutyric acid production; addition of leucine results in increased isovaleric and isocaproic acid production and addition of isoleucine results in increased isovaleric acid production. *Clostridium sporogenes* can metabolize monoacetone glucose.

The major cellular fatty acids produced by this species are normal saturated C_{14}, C_{15}, C_{16}, C_{17}, and C_{18} acids, and normal unsaturated C_{16} acid with a double bond at the 7 or 9 position. Small amounts of other normal and branched chain fatty acids also are present.

Of 105 strains tested by the broth disk method, 1 is resistant to penicillin G, 1 to tetracycline, 4 to chloramphenicol, 5 to erythromycin, and 57 to clindamycin; all others are susceptible. Determinations of MICs of various antibiotics against 18 strains of *Clostridium sporogenes* indicate that all are susceptible to penicillin G (2 U/ml), metronidazole (0.25 μg/ml), tinidazole (1 μg/ml), chloramphenicol (4 μg/ml), tetracycline (0.5 μg/ml), doxycycline (0.25 μg/ml), but resistant to streptomycin, neomycin, kanamycin, tobramycin, and amikacin; susceptibility to cephalothin, erythromycin, clindamycin, lincomycin, and gentamicin is variable between strains. Culture supernatants are not toxic to mice. Although *Clostridium sporogenes* is isolated from infections, these infections are usually polymicrobial and the role, if any, of this species as a pathogen in such infections has not been established.

Untoward effects are not seen when germ-free animals undergo gastrointestinal monoassociation. Self-limited, spontaneously healing abscesses are produced following intramuscular injection of a relatively large inoculum into guinea pigs. A generalized lethal disease, possibly egg-borne, in newly hatched chicks has been attributed to *Clostridium sporogenes*. Cerebrocortical necrosis in ruminant animals is thought to be due to thiamine deficiency; an association between this disease and colonization by thiaminase-producing strains of *Clostridium sporogenes* has been postulated, but not proven. The highly proteolytic nature of *Clostridium sporogenes* is thought possibly to act as an adjuvant and promote invasiveness of other bacteria in various mixed infections of animals and humans. Other characteristics of the species are given in Table 142.

Source: soil throughout the world; marine and fresh water lake sediment; preserved meat and dairy products; snake venom; feces of sheep and dogs; human infant and adult feces; infections in domestic animals; infections in humans including bacteremia, infective endocarditis, central nervous system, and pleuropulmonary infections; penile lesions; abscesses; war wounds; other pyogenic infections.

DNA G+C content (mol%): 26 (T_m).

Type strain: ATCC 3584, BCRC 11259, CCUG 15941, CIP 106155, DSM 795, NBRC 13950, JCM 1416, LMG 8421, NCIMB 10696, NCTC 13020.

GenBank accession number (16S rRNA gene): M59115, X68189.

Further comments: Clostridium sporogenes can be differentiated from proteolytic strains of *Clostridium botulinum* types A, B, and F, which it closely resembles phenotypically, by toxin neutralization in mice, by polyacrylamide gel electrophoretic examination of soluble cellular proteins, or by gas chromatography of trimethylsilyl derivatives of whole-cell hydrolysates (see *Clostridium botulinum*). *Clostridium sporogenes* and *Clostridium difficile* also are morphologically similar and have similar fermentation products; they differ, however, in mannitol fermentation, proteolytic activity in milk and meat, and in lipase production.

16S rRNA gene sequence comparisons show *Clostridium sporogenes* to fall within the radiation of cluster I of the clostridia (Collins et al., 1994). The highest 16S rRNA gene sequence similarities are to the species *Clostridium novyi* (99.8%; accession no. M59100) and *Clostridium botulinum* (99.6%; accession no. L37585). Olsen et al. (1995) proposed that the name *Clostridium sporogenes* be conserved for the nontoxigenic strains of this species cluster.

144. **Clostridium sporosphaeroides**[*] Soriano and Soriano 1948, 39[AL]

spo.ro.sphae.roi'des. Gr. n. *spora* seed; Gr. adj. *sphairoides* globular; N.L. neut. adj. *sporosphaeroides* having spherical spores.

Cells in PYG broth are Gram-stain-positive but easily decolorized, nonmotile, straight rods, 0.5–0.6 × 1.8–5.5 μm. Spores are oval or round, and terminal. Sporulation occurs most readily on chopped-meat slants incubated at 30 °C. Cell wall contains *meso*-DAP. The cell wall is susceptible to dissolution by lysozyme.

Colonies on the surface of blood agar are 1–2 mm in diameter, circular, with a slightly lobate or scalloped margin, slightly raised, with a fried-egg appearance, gray-white, have a glossy surface, a mosaic internal structure, and are nonhemolytic.

Cultures in PYG broth are turbid with a smooth to ropy sediment and have a pH of 6.1 after incubation for 5 d. The temperature range for optimum growth is 37–45 °C. There is slight growth at 30 °C, no growth at 25 °C. Growth is inhibited by 20% bile and by 6.5% NaCl. A moderate amount of gas is produced in PYG deep agar cultures.

Adonitol, dulcitol, erythritol, glycerol, and sorbose are not fermented. Ammonia and H_2S are produced; hippurate is hydrolyzed; neutral red and resazurin are reduced.

An NADP-dependent 7 α-hydroxysteroid dehydrogenase capable of dehydrogenation of bile salts is produced. Deoxyribonuclease activity is not present.

Products of metabolism in PYG broth include large amounts of acetate and butyrate, and small amounts of propionate. Abundant H_2 is produced. Lactate is converted to propionate; pyruvate is converted to acetate; threonine is not utilized. *Clostridium sporosphaeroides* does not metabolize phenylalanine, tyrosine, or tryptophan. Glutamate is metabolized to ammonia, CO_2, acetate, and butyrate via a hydroxyglutarate pathway.

The type strain is susceptible to chloramphenicol, clindamycin, erythromycin, penicillin G, and tetracycline. Culture supernatants of the type strain are nontoxic to mice.

Other characteristics of the species are given in Table 143.

Source: canned food, human feces.

DNA G+C content (mol%): 27 (T_m).

Type strain: ATCC 25781, DSM 1294.

GenBank accession number (16S rRNA gene): M59116, X66002.

Further comments: 16S rRNA gene sequence comparisons show *Clostridium sporosphaeroides* to fall within the radiation of cluster IV of the clostridia as defined by Collins et al. (1994). The closest relative of *Clostridium sporosphaeroides* is *Clostridium leptum* with which it shares 93.9% 16S rRNA gene sequence similarity.

145. **Clostridium stercorarium** Madden 1983, 839[VP] emend. Fardeau, Ollivier, Garcia and Patel 2001, 1130

ster.cor.ar.i'um. N.L. adj. *stercorarium* pertaining to dung, referring to the source of the isolate, a compost heap.

Rod-shaped cells, 0.3–0.4 × 2–4 μm. Gram-stain-negative. Chemoorganotrophic. Motile with peritrichous flagella. Forms oval, terminal spores. Colonies are unpigmented, glossy, flat, and 3–5 mm in diameter.

Obligate anaerobe. Optimum temperature for growth is 65 °C. Optimum pH for growth is 7.3. Ferments cellulose, cellobiose, mellibiose, raffinose, fructose, and sucrose. Does not ferment arabinose, mannitol, sorbitol, gelatin, or casein. The end products of glucose fermentation produces are acetate, lactate, ethanol, H_2, CO_2, and small amounts of L-alanine. No sulfate reduction.

Source: compost.

DNA G+C content (mol%): 39 (T_m).

Type strain: ATCC 35414, DSM 8532, NCIMB 11754.

GenBank accession number (16S rRNA gene): AJ310082, L09174.

Further comments: 16S rRNA gene sequence comparisons show *Clostridium stercorarium* subsp. *stercorarium* to fall within the radiation of group III of the clostridia as defined by Collins et al. (1994). *Clostridium stercorarium* subsp. *stercorarium, Clostridium stercorarium* subsp. *leptospertum,* and *Clostridium stercorarium* subsp. *thermolacticum* share 98.7–99.1% 16S rRNA gene sequence similarity. This subspecies cluster forms a distinct lineage within cluster III sharing less than 92% 16S rRNA gene sequence similarity to other *Clostridium* species of cluster III.

145a. **Clostridium stercorarium subsp. stercorarium** Madden 1983, 840[VP] emend. Fardeau, Ollivier, Garcia and Patel 2001, 1130.)

See description above for *Clostridium stercorarium.*

145b. **Clostridium stercorarium subsp. leptospartum** (Toda, Saiki, Uozumi and Beppu 1988) Fardeau, Ollivier, Garcia, and Patel 2001, 1130[VP] (*Thermobacteroides leptospartum* Toda, Saiki, Uozumi and Beppu 1988, 1343)

lep.to.spar′tum. Gr. adj. *leptos* thin, delicate, small; Gr. n. *sparton* rope; L. n. *spartum* rope; N.L. neut. n. *leptospartum* a thin rope.

Rod-shaped cells, 0.25–0.45 × 4.5–15 μm; found singly or paired. Obligate anaerobe. Gram-stain-negative. Nonmotile. Spores not detected. Colonies are round, flat, colorless, and 2–3 mm in diameter. Temperature range for growth is 45–71 °C with an optimum temperature for growth of 60 °C. pH Range for growth is 6.7–8.9 with optimum growth at pH 7.5.

Utilizes lactose, galactose, cellobiose, xylose, and mannose. Arabinose, ribose, fructose, sucrose, gelatin, and casein are not utilized. Ferments glucose to produce acetate, lactate, and ethanol. Sulfate is not reduced. Cells are cellulolytic. Hydrolyzes esculin. Cells do not utilize fructose or sucrose. Gelatin and casein are not hydrolyzed.

Source: compost.

DNA G+C content (mol%): 43 (T_m).

Type strain: C-17-70, ATCC 51338, DSM 9219, IAM 13499.

GenBank accession number (16S rRNA gene): not reported.

145c. **Clostridium stercorarium subsp. thermolacticum** (Le Ruyet, Dubourguier, Albagnac and Prensier 1985) Fardeau, Ollivier, Garcia and Patel 2001, 1130[VP] (*Clostridium thermolacticum* Le Ruyet, Dubourguier, Albagnac and Prensier 1985, 201)

ther.mo.lac.ti cum. Gr. adj. *thermos* hot; L. n. *lac, lactis* milk; N.L. neut. adj. *thermolacticum* pertaining to the high growth temperature and a growth substrate.

Rod-shaped cells, 0.7–0.8 × 2.7–7.7 μm. Nonmotile or motile by single lateral flagellum. Gram-stain-negative. Forms round, terminal spores. Colonies are translucent, flat, glossy, and 1–1.5 mm in diameter.

Strict anaerobe. Temperature range for growth is 50–70 °C with an optimum of 60–65 °C. pH Range for growth is 6.8–7.4 with an optimum of 7.0. Carbohydrates required for growth. Ferments glucose, starch, cellobiose, fructose, raffinose, ribose, sucrose, mannose,

lactose, xylan, salicin, melibiose, and rhamnose. End products of glucose fermentation are acetate, ethanol, CO_2, H_2, and L(+) lactate. Arabinose, gelatin, and casein are not utilized. Hydrolyzes esculin, but not gelatin. Catalase and urease-negative. Does not reduce nitrate or sulfate.

Source: compost.

DNA G+C content (mol%): 41 (T_m).

Type strain: TX 41, ATCC 43739, DSM 2910.

GenBank accession number (16S rRNA gene): X72870.

146. **Clostridium sticklandii*** Stadtman and McClung 1957, 218[AL]

stick.lan′di.i. N.L. gen. n. *sticklandii* pertaining to L.H. Stickland, British biochemist.

This description, unless stated otherwise, is based on that of Stadtman and McClung (1957) and on study of the type strain.

Cells in PYG broth are straight, slender Gram-stain-positive rods that are motile and peritrichous, 0.3–0.5 × 1.3–3.8 μm, and occur singly, in pairs, and sometimes in short chains. Spores are oval, subterminal, distend the cell slightly, but are rarely seen, and then only in old cultures. Cell-wall composition has not been reported.

Surface colonies on blood agar are 1–2 mm in diameter, circular with an entire or slightly undulated margin, convex, grayish-white, translucent or opaque, possess a glossy surface and a mottled internal structure, and are nonhemolytic. Colonies in agar are 1–2 mm in size and lenticular, becoming lobate.

Cultures in PYG broth are turbid with a smooth sediment and have a pH of 6.0 after incubation for 6 d. The temperature range for optimum growth is 30–37 °C; moderate growth occurs at 25 °C and 45 °C. Moderate gas is produced in PYG deep agar cultures.

Carbohydrates are not appreciably fermented. Ammonia is produced. Energy for growth is obtained by coupled oxidation-reduction reactions between certain amino acid pairs ("Stickland reaction"). Uracil is not utilized. Products in PY broth are large amounts of acetate, butyrate, and isovalerate, small amounts of propionate and isobutyrate, and a small amount of H_2. Lesser amounts of these products are detected when glucose is added to the basal medium. Pyruvate is converted to acetate; lactate is not utilized; threonine is converted to acetate and to propionate. Threonine is reduced to α-aminobutyrate by dried cells and cell-free extracts of *Clostridium sticklandii.* Ornithine is converted to δ-aminovalerate. In amino-acid-supplemented broth cultures, 2-aminobutyric acid is produced; arginine, glycine, lysine, methionine, and serine are utilized; proline is converted to 5-aminovaleric acid, phenylalanine to phenylacetic acid, tyrosine to *p*-hydroxyphenylacetic acid, valine to isobutyric acid, leucine and isoleucine to isovaleric acid, and tryptophan to indole acetic acid. The purines adenine, hypoxanthine, xanthine, and uric acid can serve as hydrogen acceptors. Lysine is converted to acetic and butyric acids and ammonia. Glutamate is fermented slowly by way of the methylaspartate pathway.

The major fatty acids of *Clostridium sticklandii* are normal, saturated C_{14} and C_{16} acids; small amounts of other normal and branched chain fatty acids, both saturated and unsaturated, also are present.

The type strain is susceptible to chloramphenicol, clindamycin, erythromycin, penicillin G, and tetracycline. Culture supernatants of the type strain are not toxic to mice; strains are not pathogenic for guinea pigs. Other characteristics of the species are given in Table 143.

Source: soil, San Francisco Bay black mud, feces of one subject in Uganda.

DNA G+C content (mol%): 31 (T_m).

Type strain: ATCC 12662, BCRC (formerly CCRC) 14485, CCUG 9281, DSM 519, JCM 1433, NCIMB 10654.

GenBank accession number (16S rRNA gene): not reported.

Further comments: 16S rRNA gene sequence comparisons show *Clostridium sticklandii* to fall within the radiation of cluster XI of the clostridia as defined by Collins et al. (1994). *Clostridium sticklandii* forms a distinct lineage and is related at ~90% 16S rRNA gene sequence similarity to the species *Clostridium litorale, Clostridium sordellii, Eubacterium tenue,* and *Clostridium ghonii.*

147. **Clostridium straminisolvens** Kato, Haruta, Cui, Ishii, Yokota and Igarashi 2004, 2046^VP

stra.mi.ni.sol'vens. L. neut. n. *stramen* straw; L. v. *solvere* to dissolve; N.L. part. adj. *straminisolvens* straw-dissolving.

Straight or slightly curved rods, 0.5–1.0 × 3.0–8.0 μm, occurring singly or in pairs. Anaerobic growth. Aerotolerant. Nonmotile. Forms subterminal, oval spores 0.5–1.0 × 1.0–1.5 μm. On cellobiose agar colonies are 1–2 μm in diameter, tan-yellow, round, with entire margins. Growth occurs under a gas phase containing up to 4% O_2. Temperature range for growth is 50–55 °C; no growth occurs at or below 45 °C or at 65 °C and above. The pH range for growth is 6.0–8.5 with an optimum of pH 7.5.

Cellulose and cellobiose are utilized as sole carbon and energy sources. Glucose, fructose, ribose, mannose, mannitol, melibiose, sucrose, xylose, sucrose, lactose, xylan, and starch are not utilized as sole carbon and energy sources. The end products of fermentation are acetate, lactate, ethanol, hydrogen, and carbon dioxide. Esculin is hydrolyzed. Nitrate is not reduced. Casein is not digested. Lipase, lectinase, and indole are not produced.

Source: a cellulose-degrading bacterial community.

DNA G+C content (mol%): 41.3 (HPLC).

Type strain: CSK1, DSM 16021, IAM 15070, JCM 21531, NBRC 103399.

GenBank accession number (16S rRNA gene): AB125279.

Further comments: 16S rRNA gene sequence comparisons show *Clostridium straminisolvens* to fall within the radiation of cluster III of the clostridia as defined by Collins et al. (1994). The closest relatives of *Clostridium straminisolvens* based on 16S rRNA gene sequence similarity values are *Clostridium thermocellum* (97.1%), *Acetivibrio cellulolyticus* (96.3%), *Clostridium alkalicellulosi* (96.2%), and *Clostridium aldrichii* (95.9%). *Clostridium straminisolvens* differs from its closest relatives in its temperature range for growth and its aerotolerance (Kato et al., 2004). A DNA–DNA reassociation value of 23% indicated that *Clostridium thermocellum* and strain CSK1^T are distinct species.

148. **Clostridium subterminale*** (Hall and Whitehead 1927) Spray 1948, 786^AL (*Bacillus subterminalis* Hall and Whitehead 1927, 67)

sub.ter.mi.na'le. L. pref. *sub* under; L. adj. *terminalis* terminal; N.L. neut. adj. *subterminale* near the end, subterminal.

Cells in PYG broth are Gram-stain-positive, straight rods, usually motile and peritrichous, and are 0.5–1.9 × 1.6–11.0 μm. Spores are without appendages and are oval and usually subterminal (occasionally central), and distend the cell. Sporulation occurs readily on blood agar, egg-yolk agar, and chopped-meat agar. Cell walls contain *meso*-DAP. The cell wall is digested by lysozyme.

Surface colonies on blood agar are 1–4 mm in diameter, raised or low convex, translucent, gray, irregularly circular with a lobate or scalloped margin and a matt surface, and usually are β-hemolytic. They often have a crystalline, mottled, or mosaic internal structure.

Cultures in PYG broth are turbid with a smooth or ropy sediment and have a pH of 6.0–6.4 after incubation for 5 d. The temperature for optimum growth is 37 °C; most strains also grow at 25 °C and 45 °C. Growth is inhibited by 20% bile or 6.5% NaCl. Moderate amounts of gas are produced in PYG deep agar cultures. Ammonia is produced, H_2S is produced by most strains, neutral red is reduced, and resazurin occasionally is reduced. An occasional strain may produce a slight amount of lecithinase. Cells possess deoxyribonuclease. Neuraminidase is not produced.

Metabolic products in PYG broth include large amounts of acetate, butyrate, and isovalerate and small amounts of isobutyrate; small amounts of formate, propionate, isocaproate, lactate, and succinate may be produced. Ethanol and traces of other alcohols are usually detected. Phenylacetate is produced by the type strain. Abundant H_2 is produced. Threonine is converted to propionate by 6 of 22 strains tested; pyruvate is converted to acetate and usually to butyrate; lactate is not utilized. Valine is converted to isobutyrate; leucine and isoleucine are converted to isovalerate. In amino acid-containing broth cultures, arginine, glycine, lysine, and serine are utilized, and phenylalanine is converted to phenylacetic acid, tyrosine to *p*-hydroxyphenylacetic acid, and tryptophan to indole acetic acid.

The major fatty acids of *Clostridium subterminale* are normal, saturated C_{12}, C_{14}, and C_{16} acids; lesser amounts of saturated iso-branched C_{15} and normal unsaturated C_{16} acids with a double bond at the 7 or 9 position; small amounts of other normal and branched chain fatty acids also are present.

Of 31 strains tested, all are susceptible to erythromycin, penicillin G, and tetracycline; 25 are susceptible to chloramphenicol, and 22 are susceptible to clindamycin. One strain was inhibited by 0.19 μg of erythromycin/ml. Culture supernatants are nontoxic when injected into mice. Other characteristics of the species are given in Table 142.

Source: marine sediment; soil; bovine feces; the small bowel contents of adult humans; feces of healthy humans, of humans with antimicrobial-associated diarrhea; infection in animals; rarely (and usually as part of a polymicrobial flora) from blood, biliary tract infections, empyema fluid, and soft tissue or bone infections of humans.

DNA G+C content (mol%): 28 (T_m).

Type strain: ATCC 25774, BCRC (formerly CCRC) 14486, CCUG 21841, CDC KA152, DSM 6970, JCM 1417, NCIMB 10746, VPI 20231.

GenBank accession number (16S rRNA gene): AF241844, L37595, M59106, X68451.

Further comments: 16S rRNA gene sequence comparisons show *Clostridium subterminale* to fall within the radiation of cluster I of the clostridia as defined by Collins et al. (1994). The closest relative of *Clostridium subterminale* based on 16S rRNA gene sequence similarity values is *Clostridium thiosulfatireducens* (99.9%). DNA–DNA reassociation data and phenotypic differences including the fact that *Clostridium subterminale* does not use sulfate as an electron acceptor differentiate *Clostridium subterminale* from *Clostridium thiosulfatireducens*.

149. **Clostridium symbiosum*** (Stevens 1956) Kaneuchi, Watanabe, Terada, Benno and Mitsuoka 1976b, 202[AL] (*Bacteroides symbiosus* Stevens 1956, 100)

sym.bi.o′sum. Gr. n. *symbios* a companion; N.L. neut. adj. *symbiosum* living together with, symbiotic (refers to its use as a symbiote for cultivation of *Entamoeba histolytica*).

Cells on Eggerth–Gagnon agar are Gram-stain-negative, are usually motile and peritrichous, straight rods, often with pointed ends, 0.5–0.6 × 1.5–2.0 µm, and occur singly, in pairs, or in chains. Spores are round or oval, and subterminal; spores survive heating to 70 °C for 10 min but may not survive 80 °C for 10 min. Cell walls contain *meso*-DAP. Surface colonies on Eggerth–Gagnon agar after 3 d of incubation are minute–1.0 mm in diameter, circular, entire, low convex, smooth (sometimes with a slightly irregular surface and margin), translucent, grayish to whitish, and have a whitish-gray or reddish mottled center when viewed through a dissecting microscope. On blood agar, strains may be slightly β-hemolytic.

Cultures in PY-Fildes solution-glucose broth are turbid with a smooth sediment, abundant gas, and have a pH of 5.3–6.0 after 10 d incubation. This species grows well at 37 °C. Growth is nearly as good as at 30 °C, but poor at 25 °C or 45 °C. Growth is stimulated by a fermentable carbohydrate and inhibited by 20% bile or 6.5% NaCl.

Dextrin, glycerol, inositol, and sorbose are not fermented. Ammonia and H_2S are produced, acetyl methyl carbinol usually is produced, neutral red is reduced, and resazurin usually is reduced. Phage-like particles have been demonstrated from a strain of *Clostridium symbiosum*.

Large amounts of acetic acid, butyric acid, and lactic acid are produced in PYG broth; ethanol and formic acid may be detected as well. Abundant H_2 is produced. Pyruvate is converted to acetate and butyrate; neither lactate nor threonine is utilized. Glutamate is fermented to ammonia, CO_2, and acetic and butyric acids via a hydroxyglutarate pathway. Other characteristics of the species are given in Table 144.

Source: feces of healthy humans, liver abscesses and blood infections in humans, occasionally from human infections derived from the bowel flora.

DNA G+C content (mol%): 46 (T_m).

Type strain: ATCC 14940, DSM 934, JCM 1297.

GenBank accession number (16S rRNA gene): M59112.

Further comments: 16S rRNA gene sequence comparisons show *Clostridium symbiosum* to fall within the radiation of cluster XIVa of the clostridia as defined by Collins et al. (1994). The closest relative based on 16S rRNA gene sequence similarity values is *Clostridium bolteae* (94.5%).

150. **Clostridium tepidiprofundi** Slobodkina, Kolganova, Tourova, Kostrikina, Jeanthon, Bonch-Osmolovskaya and Slobodkin 2008, 854[VP]

te.pi.di.pro.fun′di. L. adj. *tepidus* moderately warm; L. n. *profundum* the depths of the ocean, N.L. gen. n. *tepidiprofundi* of the warm bottom of the ocean.

Cells are straight to slightly curved rods, 0.4–0.6 × 2.0–3.0 µm. Spore formation is observed only below pH 5.5. Colonies are white, lens shaped, 0.1–0.2 mm in diameter in agar-shake cultures. Anaerobic. The temperature range for growth is 22–60 °C, with optimum growth at 50 °C. The pH range for growth is 4.0–8.5, with optimum growth at pH 6.0–6.8. Growth occurs at NaCl concentrations of 1.0–6.0% (w/v), with optimum growth at 2.5% (w/v). Substrates utilized include casein, peptone, tryptone, yeast extract, beef extract, starch, maltose, and glucose. Pyruvate, L-valine, L-arginine, DL-alanine, L-proline, DL-alanine plus L-proline, glycine, DL-alanine plus glycine, fructose, xylose, cellobiose, sucrose, L-arabinose, glycerol, acetate, butyrate, lactate, formate, methanol, fumarate, betaine, olive oil, xylan, CM-cellulose, filter paper, chitin, and H_2/ CO_2 are not utilized. The end products of glucose fermentation are ethanol, acetate, H_2, formate, and CO_2. Reduces elemental sulfur to hydrogen sulfide. Is not able to utilize nitrate, fumarate, sulfate, sulfite, thiosulfate, amorphous iron (III) oxide, iron(III) citrate, or oxygen (20%,v/v, in the gas phase) as electron acceptors.

Source: a deep-sea hydrothermal vent chimney located at 13° N on the East Pacific Rise.

DNA G+C content (mol%): 30.9 (T_m).

Type strain: SG 508, DSM 19306, VKM B-2459.

GenBank accession number (16S rRNA gene): EF197795.

Further comments: 16S rRNA gene sequence comparisons show *Clostridium tepidiprofundi* to fall within the radiation of cluster I of the clostridia as defined by Collins et al. (1994). The highest sequence similarities were to the species *Clostridium pascui* (93.0%), *Clostridium cochlearium* (92.6%), and *Clostridium peptidivorans* (92.5%). *Clostridium tepidiprofundi* can be differentiated from its closest relatives on the basis of its thermophilic nature, marine origin and subsequent NaCl requirements for optimal growth.

151. **Clostridium termitidis** Hethener, Brauman and Garcia 1992a, 327[VP] (Effective publication: Hethener, Brauman and Garcia 1992b, 57.)

ter.mi′ti.dis. L. n. *tarmes*, *tarmit-* (L.L. var. *termes*, *termit-*) worm that eats wood; N.L. adj. *termitidis* pertaining to the termite.

Straight to slightly curved rods, 0.5 × 4–6 µm. Gram-stain-positive. Motile by peritrichous flagella. Oval terminal spores that cause cell swelling. Deep colonies in roll tubes are small, circular, and slightly yellow. On agar surface colonies are widespread. Obligate anaerobe. Temperature range for growth is 20–48 °C with an optimum of 37 °C. The pH range for growth is 5.0–8.2 with an optimum pH of 7.5.

Ferments cellulose, cellobiose, glucose, fructose, galactose, lactose, mannose, ribose, sorbose, xylose, maltose, melibiose, mannitol, sorbitol, and glycerol. Adonitol, arabinose, dulcitol, melezitose, pectin, raffinose, rhamnose, ribose, salicin, sucrose, and threalose are not utilized. End products of fermentation are acetate, ethanol, H$_2$, and CO$_2$. No growth factors required. Enhanced growth observed with the addition of yeast extract, Biotrypcase, or vitamins.

Source: gut of wood-feeding termite (*Nasutitermes lujae*) from Mayombe tropical rainforest, Congo, Central Africa.

DNA G+C content (mol%): 39.2 (T_m).

Type strain: CT1112, ATCC 51846, DSM 5398.

GenBank accession number (16S rRNA gene): X71854.

Further comments: 16S rRNA gene sequence comparisons show *Clostridium termitidis* to fall within the radiation of cluster III of the clostridia as defined by Collins et al. (1994). The closest relatives based on 16S rRNA gene sequence similarity values are *Clostridium cellobioparum* (99.1%), *Clostridium josui* (97.7%), *Clostridium cellulolyticum* (97.3%), *Clostridium papyrosolvens* (97.1%), and *Clostridium hungatei* (97.1%). *Clostridium termitidis* can be differentiated from *Clostridium cellobioparum* on the basis of Gram-staining reaction, substrate utilization tests, and G+C content of the DNA.

152. **Clostridium tertium**[*] (Henry 1917) Bergey, Harrison, Breed, Hammer and Huntoon 1923, 332[AL] (*Bacillus tertius* Henry 1917, 347)

ter'ti.um. L. neut. adj. *tertium* third, referring to it being the anaerobe third most frequently isolated by Henry (1917) from open war wounds.

Cells in broth cultures are Gram-stain-positive, motile and peritrichous, and are 0.5–1.4 × 1.5–10.2 µm. They occur singly or in pairs. Spores are large, oval, terminal, or occasionally subterminal, and markedly distend the cell. Although optimal growth is achieved by incubation under anaerobic conditions, the species is aerotolerant and will grow on the surface of freshly prepared blood agar incubated in air. Spores are formed readily in most media incubated anaerobically, but not under aerobic conditions. The cell wall contains lysine rather than DAP; glucose alone or glucose and mannose are the cell-wall sugars. Colonies on the surface of blood agar incubated anaerobically are 2–4 mm in diameter, circular, low convex, have slightly irregular margins, are white to gray, and have a matt surface and usually a mottled or granular internal structure. Hemolysis is variable and, when present, colonies may be α- or β-hemolytic. Surface colonies after aerobic incubation are 1 mm, circular with entire edges, are dome shaped, and have an opalescent appearance. Colonies in agar are small and lenticular.

Cultures in PYG broth grow well and are turbid with a smooth, ropy, or flocculent sediment. The optimum

temperature for growth is 37 °C; growth also occurs readily at 50 °C. Growth is stimulated by the presence of a fermentable carbohydrate and inhibited by 20% bile or by 6.5% NaCl. Abundant gas is produced in PYG deep agar cultures. Adonitol, dulcitol, erythritol, glycerol, inositol, and sorbose are not fermented. Ammonia is produced by some strains, and neutral red and resazurin are reduced. Nitrate reduction is dissimilatory rather than assimilatory. Enzymes produced include deoxyribonuclease, neuraminidase, and an inducible chitinase. Bile acid deconjugation and Δ⁴-steroid dehydrogenase activity that may be related to the etiology of colon cancer have been reported.

Products of fermentation of PYG broth include large amounts of acetic, butyric, and lactic acids, and occasionally small amounts of formic and succinic acids and ethanol. Abundant H$_2$ is produced. Neither lactate nor threonine is utilized; pyruvate is converted to acetate and occasionally also to butyrate and formate.

Of 20 strains tested, all are susceptible to chloramphenicol and tetracycline, 19 to erythromycin, 10 to penicillin G, and 7 to clindamycin. *Clostridium tertium* is relatively nonpathogenic. Culture supernatants are not toxic to mice. *Clostridium tertium* increases the severity of a lethal enteritis induced by *Clostridium difficile* in gnotobiotic newborn hares. *Clostridium tertium* was found to be only weakly pathogenic when compared with several other species of *Clostridium* in a study of experimental intestinal strangulation in monoassociated gnotobiotic rats. Pneumatosis cystoides intestinalis can occur in these monoassociated animals. Other characteristics of the species are shown in Table 142.

Source: soil; guano in Antarctica; the nares of a beagle dog; feces of healthy neonates and infants; appendices of healthy adults; the feces or colonic mucosa of healthy adults; osteomyelitis in a dog; infection in a calf; a variety of conditions in humans including brain abscess, the gingival sulcus of two patients with periodontitis, infections related to the intestinal tract, soft tissue infections, war wounds, and blood.

DNA G+C content (mol%): 24–26 (T_m).

Type strain: ATCC 14573, CCUG 4219, DSM 2485, IAM 14196, JCM 6289, NCIMB 10697, NCIMB 541, NCTC 541.

GenBank accession number (16S rRNA gene): AJ245413, Y18174.

Further comments: 16S rRNA gene sequence comparisons show *Clostridium tertium* to fall within the radiation of cluster I of the clostridia as defined by Collins et al. (1994). The closest relatives of *Clostridium tertium* based on 16S rRNA gene sequence similarity values are *Clostridium carnis* (99.8%), *Clostridium sartagoforme* (98.2%), *Clostridium chauvoei* (97.6%), *Clostridium isatidis* (97.4%), and *Clostridium septicum* (97.3%). This species can be differentiated from *Clostridium carnis* on the basis of mannitol and melibiose fermentation.

153. **Clostridium tetani**[*] (Flügge 1886) Bergey, Harrison, Breed, Hammer and Huntoon 1923, 330[AL] (*Bacillus tetani* Flügge 1886, 274)

te'ta.ni. L. n. *tetanus* (from Gr. n. *tetanos*) tension, tetanus; L. gen. n. *tetani* of a tension, of tetanus.

Cells in young cultures are Gram-stain-positive, but become Gram-stain-negative after approximately 24h incubation. Most strains are motile and peritrichous. Cells are 0.5–1.7 × 2.1–18.1 μm and occur singly or in pairs. Spores are usually round and terminal; occasionally they may be oval or subterminal, or both. Spores will germinate in liver broth at a starting O/R potential of up to +580 mV, but vegetative growth does not occur unless the culture is covered with liquid paraffin. When spores of *Clostridium tetani* were introduced orally or intrarectally into germ-free rats, the spores remained viable in the intestinal tract but were unable to germinate. Vegetative cells given by oral swabs did colonize the intestinal tract and stimulated the production of antitoxin, but no toxin was detected. Cell wall contains *meso*-DAP; cell-wall sugars are primarily glucose and rhamnose with traces of galactose and mannose. Cell wall is susceptible to dissolution by lysozyme.

Colonies on the surface of blood agar are 4–6 mm in diameter, flat, translucent, gray, with a matt surface, irregular and rhizoid margins, and usually cause a narrow zone of β-hemolysis. There is a tendency to swarm that is more marked on moist plates. Colonies in agar are transparent and very woolly.

Cultures in PYG broth are turbid with a smooth or ropy sediment and have a pH of 6.1–6.5 after incubation for 5 d. The optimum temperature for growth is 37 °C; there is moderate growth at 30 °C, little or none at 25 °C or 45 °C. Growth is inhibited by 20% bile or 6.5% NaCl. Exposure of exponential phase cultures to hyperbaric oxygen (100% O_2 at 3 atm) resulted in death of more than 99.9% of cells. Moderate to abundant gas is produced by most strains in PYG deep agar cultures. Ammonia and H_2S are produced, neutral red is reduced, and resazurin is usually reduced. Deoxyribonuclease is produced; neuraminidase is not produced.

Products of metabolism in PYG broth are butanol, large amounts of acetate and butyrate, and small amounts of propionate; lactate and succinate may also be produced. Abundant H_2 is produced. Lactate and threonine are converted to propionate, pyruvate is converted to butyrate and acetate. Phenol is produced by all strains and indole by some strains in broth supplemented with L-phenylalanine, L-tryptophan, and L-tyrosine. All strains utilize aspartate, glutamate, histidine, and serine for growth; some strains utilize methionine, threonine, and tyrosine. Glutamate is converted to ammonia, CO_2, acetate and butyrate via a methylaspartate pathway.

All of seven strains tested are susceptible to chloramphenicol, clindamycin, erythromycin, penicillin G, and tetracycline.

Major lipid fatty acids formed in a trypticase-yeast extract-thioglycollate medium without glucose are normal saturated C_{12}, C_{14}, C_{16}, and C_{18} acids and smaller amounts of unsaturated C_{18} and C_{16} acid with a double bond at the 7 or 9 position. Branched chain fatty acids are not present.

Two pathogenic soluble antigens, tetanolysin and tetanospasmin, are produced. Injection of partially purified tetanolysin produces electrocardiographic changes in monkeys and mice and intravascular hemolysis in rabbits and monkeys; there is little evidence, however, that tetanolysin is normally involved in disease production. Tetanospasmin, the cause of clinical tetanus in humans and other animals, is an extremely potent neurotoxin that probably spreads to the central nervous system both by passage up perineural tissue, and by lymphatic and hematogenous routes from a localized site of production; the toxin suppresses central inhibitory influences on motor neurons thereby leading to excessive muscle activity that is manifest by spasticity and tetanic contractions. Localized forms of tetanus may also occur. A "nonspasmogenic toxin" has been detected (in partially purified form) and may interfere with function of motor nerve. A high molecular mass hemolysin that is distinct from tetanolysin has been detected. Although bacteriophages have been recovered from some strains, they do not appear to be involved in toxigenicity. Toxigenicity is probably plasmid-related. Other characteristics of the species are given in Table 142.

This is a ubiquitous organism.

Source: soil throughout the world; human and other animal feces particularly horse feces; the atmosphere in a hospital; occasionally from wounds in animals and wounds in humans including infected gums and teeth, corneal ulcerations, mastoid and middle ear infections, intraperitoneal infection, omphalitis (tetanus neonatorum), postpartum uterine infections, various soft tissue infections related to trauma (including abrasions and lacerations), use of contaminated needles and catgut.

DNA G+C content (mol%): 25–26 (T_m).

Type strain: ATCC 19406, CCUG 4220, NCTC 279.

GenBank accession number (16S rRNA gene): X74770.

*Further comments:*16S rRNA gene sequence comparisons show *Clostridium tetani* to fall within the radiation of cluster I of the clostridia as defined by Collins et al. (1994). The closest relative based on 16S rRNA gene sequence similarity values is *Clostridium cochlearium* (98.1%). Similarity values to other species are below 95%.

154. **Clostridium tetanomorphum** (*ex* Bulloch, Bulloch, Douglas, Henry, McIntosh, O'Brien, Robertson and Wolf 1919) Wilde, Hippe, Tosunoglu, Schallehn, Herwig and Gottschalk 1989, 133[VP] ("*Clostridium tetanomorphum*" Bulloch, Bulloch, Douglas, Henry, McIntosh, O'Brien, Robertson and Wolf 1919)

te.ta.no.mor′phum. L. gen. n. *tetani* a bacterial specific epithet; Gr. n. *morphe* shape; N.L. adj. *tetanomorphum*, [*Clostridium*] *tetani*-shaped.

Slender rods, 0.7–1.0 × 3.0–10.0 μm; found singly or in pairs and sometimes curved. Anaerobic. Motile with peritrichous flagella. Gram-stain-positive. Form terminal to subterminal spherical or elliptical spores, which cause cell swelling. Surface colonies on PY agar are 2–5 mm in diameter, flat, gray, semitranslucent with glossy surfaces, and irregular margins. Colonies on PYG agar are white, opaque, and convex, with a mottled internal structure. Swarm on moist agar surfaces. Cell wall contains *meso*-diaminopimelic acid.

Anaerobic. Growth on media after 5 d reduces pH from 6.9 to 5.0. Optimal temperature for growth is 37 °C. Ferments glucose and maltose to acid and gas. Most strains ferment inositol, ribose, and xylose. Some strains ferment cellobiose, fructose, salicin, and sorbitol. Growth aided with a fermentable carbohydrate. Cannot ferment arabinose, adonitol, galactose, lactose, mannose, rhamnose, sucrose, and sorbose. Fermentable carbohydrates stimulate growth. Fermentation products from glucose are *n*-butanol, ethanol, acetate, and butyrate. Glutamate is fermented via the methylaspartate pathway to acetate, butyrate, CO_2, and NH_3. Histidine is fermented. Threonine is degraded to propionate; tyrosine is degraded to phenol. Ethanolamine is fermented to acetate, ethanol, and NH_3. Betaine, choline, and creatine are not fermented.

Catalase-negative. Gelatin is not hydrolyzed. Indole production and nitrate reduction vary with strains. Lecithinase-negative. Meat, casein, and milk are not digested. Nontoxigenic.

Source: septic wounds in London, UK (in 1920).

DNA G+C content (mol%): 25.4–28.5 (T_m).

Type strain: Robertson 259E, ATCC 49273, DSM 4474, NCTC 543.

GenBank accession number (16S rRNA gene): DQ241819.

Further comments: 16S rRNA gene sequence comparisons show *Clostridium tetanomorphum* to fall within the radiation of cluster I of the clostridia as defined by Collins et al. (1994). The closest relative of *Clostridium tetanomorphum* based on 16S rRNA gene sequence similarities is *Clostridium lundense* (97.9%). *Clostridium tetanomorphum* differs from *Clostridium lundense* in colony morphology and substrate utilization patterns. DNA–DNA reassociation studies have shown the distinct species status of these two species.

155. **Clostridium thermoalcaliphilum** Li, Engle, Weiss, Mandelco and Wiegel 1994, 113[VP]

ther.mo.al.cal.i.phil'um. Gr. adj. *thermos* hot; N.L. n. *alcali* from the arabic *al-qili* the ashes of the saltwort plant; N.L. adj. *philus* from Gr. adj. *philos* dear, loving; N.L. adj. *thermoalcaliphilum* referring to the organism's optimal growth under alkaline growth conditions at elevated temperatures.

Rod-shaped cells, 0.8–1.2 × 3–20 µm in length. Gram-stain-positive cell-wall structure; Gram-stain-negative staining reaction. Nonsporeforming. Motile with 2–12 peritrichous flagella. Motile at pH 7.0–10.6, optimal motility at pH 9.8. Glucose required for motility of almost all cells. Pseudofilaments formed when grown at pH values of 10.9–11.0. At pH values of 10.1–11.0 swollen sporangium-like structures form. Temperature and pH affect motility. Surface colonies are circular and white with a yellow tint. Colonies are lens shaped in agar. The cell-wall type is A4β (L-Orn–D-Asp). The main fatty acids are $C_{15:0\ iso}$, $C_{15:0\ ante}$, $C_{13:0\ iso}$, $C_{16:0}$, $C_{17:0\ iso}$, $C_{14:0}$, and $C_{18:0}$.

Obligate anaerobe. The temperature range for growth at pH 10.1 is 27–57.5 °C with optimal growth at 48–51 °C. The pH range for growth is 7.0–11.0 at 50 °C with optimal growth at 9.5–10.0. Shortest doubling time of 19 min observed at pH 10.1, 50 °C in YTG medium. Yeast extract required for growth. Tryptone but not peptone can be used to substitute for yeat extract. The Na^+ concentration range for growth is 5–1.25 mM. Utilizes glucose, fructose, sucrose, maltose, cellobiose, and Casamino acids.

Galactose, lactose, xylose, ribose, mannose, rhamnose, raffinose, salicin, sorbitol, mannitol, glycerol, lactate, formate, methanol, xylan, cellulose, starch, pectin, and H_2-CO_2 are not utilized. The end products of fermentation from yeast extract and glucose at pH 10.5 and 50 °C are acetate, H_2, isovalerate, lactate, and traces of succinate and malate. No sulfate reduction in presence of glucose, acetate, and lactate.

API AN-IDENT strips gave positive results for indole acetate, α-glucosidase, arginine aminopeptidase, and arginine utilization. Negative results for *N*-acetyl-glucosaminidase, β-glucosidase, α-fucosidase, phosphatase, α-galactosidase, β-galactosidase, leucine aminopeptidase, proline aminopeptidase, pyroglutamic acid aryl-amidase, tyrosine ammopeptidase, alanine aminopeptidase, glycine aminopeptidase, arabinosidase, indole production, and catalase. Cells are susceptible to 50 µg penicillin G/ml, 50 µg ampicillin/ml, 50 µg metroimidazole/ml, 50 µg gentamicin/ml, and 50 µg rifampin/ml.

Source: sewage plant in Atlanta, Georgia, USA.

DNA G+C content (mol%): 32 (HPLC).

Type strain: JW/YL23-2, ATCC 51508, CIP 105528, DSM 7309.

GenBank accession number (16S rRNA gene): L11304.

Further comments: 16S rRNA gene sequence comparisons show *Clostridium thermoalcaliphilum* to fall within the radiation of cluster XI of the clostridia as defined by Collins et al. (1994). The closest relative based on sequence similarities is *Clostridium paradoxum* at 98.1% 16S rRNA gene sequence similarity. *Clostridium thermoalcaliphilum* can be differentiated from *Clostridium paradoxum* on the basis of differences in cell-wall type, the lack of viable heat-resistant spores in *Clostridium thermoalcaliphilum* and the fact that *Clostridium thermoalcaliphilum* is considered a thermolerant organism while *Clostridium paradoxum* is a thermophile.

156. **Clostridium thermobutyricum** Wiegel, Kuk and Kohring 1989a, 200[VP]

ther.mo.bu.ty'ri.cum. Gr. adj. *thermos* hot; Gr. n. *boutyron* butter; N. L. neut. adj. *thermobutyricum* referring to the production of butyrate under thermophilic conditions.

Straight to slightly curved rods, 0.9–1.1 × 2.0–4.5 µm, occur singly, in pairs, or chains (up to 15 cells). Gram-stain-positive cell-wall structure; Gram-negative staining reaction. Some very early exponential-phase cells may stain Gram-positive. Central to subterminal spores form at pH values below 6.2. Cells have 5–12 peritrichous flagella, but pronounced motility is not observed. The organism has an *m*-DPM-direct cell-wall type. Colonies in agar roll tubes are irregular, circular with lobate margins, and white to cream color. Colonies 3 weeks old turn brown, but do not contain a water-soluble pigment.

Anaerobic, showing no growth in medium in which the resazurin is slightly pink. Yeast extract required for growth. Yeast extract cannot be replaced by tryptone, casein, or vitamin mixture. Moderate thermophile. Temperature range for growth is 26–61.5 °C with optimal growth at 55 °C. pH Range for growth is 5.8–9.0, with optimal growth at 6.8–7.1.

Grows with yeast extract (0.3%) alone. The following substrates stimulate growth and result in a pH drop in the culture medium: D-glucose, D-fructose, maltose, D-xylose, D-ribose, cellobiose D-glucuronic acid, D-galacturonic acid, and pyruvate. Growth is not stimulated by D-arabinose, D-galactose, D-mannose, sucrose, xylitol, pectin, polygalacturonic acid, D-raffinose, starch, beech or birch hemicellulose, succinate, and fumarate (with or without formate, and malate). Growth is not stimulated by H_2 plus CO_2, H_2 plus S^0 with or without CO_2, 0.3% (w/v) methanol, gelatin, or ethanol. The addition of 25 mM threonine, 25 mM valine, 25 mM phenylalanine, 25 mM leucine, or 25 mM isoleucine does not support, stimulate, or inhibit growth on medium containing 0.3% yeast extract and 0.25% (w/v) glucose. Fermentation of glucose produces butyrate, H_2, CO_2, and small amounts of acetate and lactate.

API AN-IDENT strips give positive results for L-glucosidase and arginine aminopeptidase. Phosphatase reaction and gelatin hydrolysis give weak positive reactions. The following reactions are negative: ammonia production from arginine; casein hydrolysis; nitrate reduction; denitrification; indole production; N-acetylglucosaminidase; L-arabinosidase; β-glucosidase; L-fucosidase; L-galactosidase; β-galactosidase; pyroglutamic acid arylamidase; indoxyl acetate reaction; leucine, proline, tyrosine, alanine, histidine, phenylalanine, glycine aminopeptidases, and formate-plus fumarate utilization.

Negative reactions for nitrate reduction, indole production, catalase, meat digestion, and denitrification. *Clostridium thermobutyricum* is not susceptible to cycloheximide, amphotericin B, actinomycin D, and novobiocin at concentrations of 100 μg/ml. Susceptible to erythromycin, tetracycline, neomycin sulfate (B + C mixture), polymyxin B sulfate, and metroimidazole at 5 μg/ml.

Source: horse manure compost containing straw and wood shavings.

DNA G+C content (mol%): 37 (T_m).

Type strain: JW171K, ATCC 49875, DSM 4928.

GenBank accession number (16S rRNA gene): X72868.

Further comments: 16S rRNA gene sequence comparisons show *Clostridium thermobutyricum* to fall within the radiation of cluster I of the clostridia as defined by Collins et al. (1994). The closest relative based on 16S rRNA gene sequence similarities is *Clostridium thermopalmarium* (98.5%). These two species differ in that *Clostridium thermopalmarium* produces ethanol from glucose fermentation and ferments sucrose; *Clostridium thermobutyricum* does not produce ethanol or ferment sucrose.

157. **Clostridium thermocellum*** Viljoen, Fred and Peterson 1926, 7[AL]

ther.mo.cel′lum. Gr. adj. *thermos* hot; N.L. n. *cellulosum* cellulose; N.L. neut. adj. *thermocellum* a thermophile that digests cellulose.

This description is based on that by Ng et al. (1977) and on study of the type strain.

Cells in PY-cellobiose broth cultures are nonmotile, 0.5–0.7 × 2.5–5.0 μm and are Gram-stain-negative. They are straight or slightly curved rods, often with tapered ends, and occur singly or in pairs. Spores are oval, terminal, and swell the cell.

Surface colonies are watery, slightly convex, and frequently produce an insoluble yellow pigment. Deep colonies in cellulose-agar roll tubes are tannish yellow, round, and filamentous. Cultures in PY-cellobiose broth are slightly turbid with a smooth sediment and have a pH of 5.6–5.8 after incubation for 4 d. Gelatin is hydrolyzed when cellobiose is included in the medium. Growth on most media requires the presence of cellulose, cellobiose, or one of the hemicelluloses. Medium 122, described in the 1983 DSM catalogue of strains, has been recommended. In a chemically defined medium, growth factors required are biotin, pyridoxamine, vitamin B12, and *p*-aminobenzoic acid.

The optimum temperature for growth is 60–64 °C; there is no growth at 37 °C. Acetyl methyl carbinol is produced and resazurin is reduced. Although xylan does not serve as a carbon source, xylanase is produced when *Clostridium thermocellum* is grown in cellobiose medium.

An extracellular, inducible cellulase that exists as a complex and is stable at 70 °C is produced. One component of this complex, an endo-β-1,4-glucanase, has been isolated. Approximately 5% of the total cellulase is an intracellular β-glucanase. A cell-bound β-glucosidase is present which increases the amount of glucose liberated during cellulose fermentation.

Products of metabolism in PY-cellobiose broth are acetic and lactic acids, ethanol, CO_2, and abundant H_2. *Clostridium thermocellum* is an important organism for its ability to convert cellulosic waste products to ethanol and H_2. Other characteristics are given in Table 145.

Source: sewage digestor sludge, said to be present in "nearly all decaying organic material" (Wiegel, 1980).

DNA G+C content (mol%): 38–39 (T_m).

Type strain: ATCC 27405, DSM 1237, JCM9322, LMG 10435, NCIMB 10682, NBRC 103400, NRRL B-4536.

GenBank accession number (16S rRNA gene): L09173.

Further comments: 16S rRNA gene sequence comparisons show *Clostridium thermocellum* to fall within the radiation of cluster III of the clostridia as defined by Collins et al. (1994). The closest relatives of *Clostridium thermocellum* based on 16S rRNA gene sequence similarity values are *Clostridium straminisolvens* (97.1%), *Clostridium alkalicellulosi* (95.4%), *Acetivibrio cellulolyticus* (94.9%), and *Clostridium aldrichii* (94.4%). A DNA–DNA reassociation value of 23% indicated that *Clostridium thermocellum* and *Clostridium straminisolvens* are distinct species.

158. **Clostridium thermopalmarium** Soh, Ralambotiana, Ollivier, Prensier, Tine and Garcia 1991a, 331[VP] (Effective publication: Soh, Ralambotiana, Ollivier, Prensier, Tine and Garcia 1991b, 138.)

ther.mo.pal.mar′ri.um. Gr. adj. *thermos* hot; L. adj. *palmarius* of palm tree; L. neut. adj. *thermoplamarium* referring to its thermophily and isolation from palm wine.

Straight rods, 0.7–1.0 × 2.0–8.0 μm. Gram-stain-positive cell-wall structure; Gram-stain-negative staining reaction. Motile with peritrichous flagella. Produce subterminal to terminal spores with slightly swollen sporangia.

Strictly anaerobic. Optimum temperature for growth is 50–55 °C, with no growth above 60 °C. The pH range for growth is 6.0–8.2 with optimum growth at 6.6. Ferments glucose, fructose, maltose, pyruvate, xylose, ribose, rahmnose, mannitol, sucrose, and cellobiose. Yeast extract was necessary for growth and could not be replaced by a vitamin solution. Does not utilize H_2-CO_2, acetate, propionate, butyrate, lactate, lactose, sorbose, melibiose, galactose, arabinose, amylose, glycerol, adonitol, dulcitol, methanol, starch, cellulose, and Biotrypcase. The end products of fermentation of carbohydrates are butyrate, H_2, CO_2, and small amounts of acetate, lactate, and ethanol. Yeast extract is required for growth. Sulfate, thiosulfate, and sulfite were not reduced to H_2S.

Source: palm wine in Senegal.

DNA G+C content (mol%): 35.7 (T_m).

Type strain: BVP, ATCC 51427, DSM 5974.

GenBank accession number (16S rRNA gene): X72869.

Further comments: 16S rRNA gene sequence comparisons show *Clostridium thermopalmarium* to fall within the radiation of cluster I of the clostridia as defined by Collins et al. (1994). The closest relative based on 16S rRNA gene sequence similarities is *Clostridium thermobutyricum* (98.5%).

159. **Clostridium thermopapyrolyticum** Méndez, Pettinari, Ivanier, Ramos and Siñeriz 1991, 282[VP]

ther.mo.pa.py.ro.ly'ti.cum. Gr. adj. *thermos* hot; Gr. n. *papyros* paper; Gr. adj. *lutikos* dissolving; N.L. neut. adj. *thermopapyrolyticum* a paper-degrading thermophile.

Straight to slightly curved rods, 0.5–0.6 × 4–7 μm. Gram-stain-positive in early growth stages, and some cells Gram-stain-negative in the late growth stages. Motile. Form subterminal spores on glucose media after 8 d.

Colonies on cellobiose or glucose agar are cream in color and measure 1.5–2.0 mm in diameter. On cellulose agar colonies are 0.5–0.8 mm after 7 d incubation and do not show zones of clearing.

Strictly anaerobic. Temperature range for growth is 45–66 °C with optimal growth at 59 °C. Utilizes filter paper as sole carbohydrate source. Grows on cellobiose, cellulose, arabinose, mannitol, xylose, lactose, maltose, fructose, starch, glucose, and galactose. Acid is produced from all substrates supporting growth. Does not utilize sucrose, melibiose, glycerol, sorbitol, raffinose, or inulin. Hydrolyzes gelatin. Negative for nitrate reduction and indole production. End products of glucose fermentation are ethanol, acetate, butyrate, lactate, butanol, H_2, CO_2, and H_2S.

Source: sediment from a river bank in Buenos Aires, Argentina.

DNA G+C content (mol%): 34 (T_m).

Type strain: SABAMMRCCC UBA 305 (deposited in the South American Biotechnology and Applied Microbiology Microbiological Resource Center Culture Collection in Tucuman, Argentina).

GenBank accession number (16S rRNA gene): not reported.

160. **Clostridium thermosuccinogenes** Drent, Laphor, Wiegant and Gottschal 1995, 879[VP] (Effective publication: Drent, Laphor, Wiegant and Gottschal 1991, 460.)

ther.mo'suc.ci.no'ge.nes. Gr. adj. *thermos* hot; N.L. n. *acidum succinicum* succinic acid; Gr. v. *gennaio* produce; N.L. adj. *thermosuccinogenes* succinic acid producing in heat.

Straight rods, 0.3–0.4 × 2–4 μm. Motile with peritrichous flagella. Gram type positive cell-wall structure; Gram-stain-negative staining reaction. Forms round, terminal spores. Cells contain hexagonally arranged particles. Colonies are 1–2 mm in diameter, white, and round.

Anaerobic. Temperature range for growth is 40–65 °C with an optimum of 58 °C. Optimal pH for growth is 7.0. Utilizes fructose, glucose, galactose, inulin, xylose, ribose, sucrose, lactose, maltose, cellobiose, raffinose, and starch. No growth was observed with methanol (both 10 mM and 30 mM tested), glutamate, glycerol, pyruvate, citrate (30 mM each), arabinose (10 mM), Casamino acids (5 g/l), xylan, pectin, cellulose (1.5 g/l), or H_2-CO_2. Formate, acetate, lactate, succinate, and H_2 are the end products of fermentation of inulin. Nitrate or sulfur are not reduced. Milk is curdled, and esculin is hydrolyzed.

Source: beet pulp at a sugar refinery, soil around a Jerusalem artichoke, fresh cow manure, mud from a tropical pond in a botanical garden.

DNA G+C content (mol%): 35.9 (T_m).

Type strain: IC, DSM 5807.

GenBank accession number (16S rRNA gene): Y18180.

Further comments: 16S rRNA gene sequence comparisons shows *Clostridium thermosuccinogenes* to fall within the radiation of cluster III of the clostridia as defined by Collins et al. (1994). Stackebrandt et al. (1999) indicated that the primary structure of the 16S rRNA was different from other species of this cluster and contained a large insertion of about 200 nucelotides between positions 70 and 100 of the gene.

161. **Clostridium thiosulfatireducens** Hernández-Eugenio, Fardeau, Cayol, Patel, Thomas, Macarie, Garcia and Ollivier 2002, 1466[VP]

thi.o.sul.fa.ti.re.du'cens. N.L. n. *thiosulfas* (-*atis*) thiosulfate; L. v. *reduco* to draw backwards, bring back to a state or condition; N.L. part. adj. *thiosulfatireducens* thiosulfate-reducing.

Rod-shaped cells, 0.5–0.6 × 2–4 μm, occurring singly or in pairs. Gram-stain-positive; Gram-stain-positive cell-wall ultrastructure. Terminal spores with swollen sporangia. Motile with peritrichous flagella. Colonies appearing after 1 d of incubation in peptone-rich medium at 37 °C, are 1 mm in diameter, arborescent, and translucent.

Anaerobic. Temperature range for growth is 18–45 °C with optimal growth at 37 °C. The pH range for growth is 6.0–9.8 with an optimum of 7.4. Tolerates NaCl concentrations up to 60 g/l. Yeast extract is required for growth on amino acids but not for growth on peptides. Gelatin, peptone, bio-Trypcase, and Trypticase soy are used as carbon and energy sources in the absence of thiosulfate. Alanine, histidine, isoleucine, leucine, lysine, methionine, phenylalanine, serine, threonine, and valine are used as carbon and energy sources in the presence of thiosulfate. Pyruvate is converted to acetate. Arginine, asparagine, aspartate, cysteine, glutamine, glycine, tryptophan, tyrosine, ʟ-arabinose, D-fructose, D-galactose, D-glucose, maltose, mannitol, D-ribose, L-xylose, D-xylose, formate, acetate, butyrate, propionate, valerate, ethanol, *n*-butanol, *n*-propanol, fumarate, lactate, malate, and succinate are not utilized., Casamino acids, gelatin, peptone, and Trypticase soy are fermented to acetate, butyrate, isobutyrate, isovalerate or 2-methylbutyrate, CO_2, and sulfide in the presence of thiosulfate. In the absence of thiosulfate, acetate is the major end product of the metabolism of proteinaceous compounds. Acetate is the only fatty acid detected from alanine and threonine oxidation, whereas isoleucine is oxidized to 2-methylbutyrate, leucine to isovalerate, valine to isobutyrate, and phenylalanine to phenylacetate. The isolate ferments serine to acetate, lysine to acetate and butyrate, methionine to propionate, and histidine to an unidentified product. Stickland reaction is carried out using alanine as an electron donor and methionine and serine as electron acceptors. Thiosulfate and sulfur, but not sulfate, sulfite, nitrate, or nitrite, are used as electron acceptors. Adverse effects on animals and humans are not known.

Source: upflow anaerobic sludge blanket digestors in Mexico treating industrial wastewaters.

DNA G+C content (mol%): 31.4 (HPLC).

Type strain: Lup 21, CIP 106908, DSM 13105.

GenBank accession number (16S rRNA gene): AB317650, AF317650, AY024332.

Further comments: 16S rRNA gene sequence comparisons show *Clostridium thiosulfatireducens* to fall within the radiation of cluster I of the clostridia as defined by Collins et al. (1994). The closest relative of *Clostridium thiosulfatireducens* is *Clostridium subterminale* (99.6%). *Clostridium subterminale* (DSM 6970^T) does not reduce thiosulfate, differentiating it from *Clostridium thiosulfatireducens*. DNA–DNA reassociation studies showed *Clostridium thiosulfatireducens* and *Clostridium subterminale* to represent distinct species (Hernandez-Eugenio et al., 2002).

162. **Clostridium tyrobutyricum*** van Beynum and Pette 1935, 205^AL

ty.ro.bu.ty′ri.cum. Gr. n. *tyros* cheese; N.L. n. *acidum butyricum* butyric acid; N.L. neut. adj. *tyrobutyricum* the butyric acid-producing organism from cheese.

Cells are Gram-stain-positive, usually motile and peritrichous, 1.1–1.6 × 1.9–13.3 μm, and occur singly or in pairs. Spores are oval, subterminal, and swell the cell. Acetokinase and phosphotransacetylase are involved in spore germination. Sporulation occurs most readily on chopped-meat agar slants or in old PYG cultures. Cell walls contain *meso*-DAP; glucose is the only cell-wall sugar. Glutamic acid and alanine are present. The cell wall is susceptible to dissolution by lysozyme. Colonies on the surface of blood agar are circular, 0.5 mm in diameter, convex with an entire margin, gray, translucent, have a glossy surface, and are often β-hemolytic.

Cultures in PYG broth are turbid with a smooth sediment and have a pH of 5.0–5.4 after incubation for 1 week. Optimum temperature for growth is 30–37 °C; growth is moderate at 25 °C; poor or no growth at 45 °C. Growth in broth is stimulated by a fermentable carbohydrate but inhibited by 6.5% NaCl or 20% bile. Abundant gas is produced in PYG deep agar cultures. Ammonia occasionally is produced; neutral red is reduced.

Roux and Bergere (1997) noted that of 77 strains studied, 82% fermented mannitol, 56% fermented mannose, and 27% fermented xylose. Kininase-like activity has been detected in one strain. Cells contain rubredoxin which is involved in NAD oxidation-reduction reactions. Immunologically similar enoate reductases have been isolated from the type and three other strains that are 81–89% homologous with the type strain. Deoxyribonuclease is not produced. Products of fermentation in PYG broth are large amounts of butyric and acetic acids, and small amounts of succinic acid; traces of formic acid may be detected. Abundant H_2 is produced. Pyruvate is converted to butyrate and acetate; threonine is not converted to propionate; in the presence of acetate, *Clostridium tyrobutyricum* converts lactate to butyrate, CO_2, and H_2.

One strain tested is susceptible to chloramphenicol, clindamycin, erythromycin, penicillin G, and tetracycline. Culture supernatants are not toxic to mice. Nonpathogenic for humans and other animals. Other characteristics of the species are given in Table 142.

Source: gley soil, dairy products, silage, feces of beagle dogs, bovine feces, human adult and infant feces.

DNA G+C content (mol%): 28 (T_m).

Type strain: ATCC 25755, CCUG 48315, CIP 105092, DSM 2637, JCM 11008, LMG 1285, NCIB 10635).

GenBank accession number (16S rRNA gene): M59113.

Further comments: 16S rRNA gene sequence comparisons show *Clostridium tyrobutyricum* to fall within the radiation of cluster I of the clostridia as defined by Collins et al. (1994). The closest relatives of *Clostridium tyrobutyricum* based on 16S rRNA gene sequence similarity values are *Clostridium carboxidivorans* (94.2%) and *Clostridium aciditolerans* (94.1%).

163. **Clostridium uliginosum** Matthies, Kuhner, Acker and Drake 2001, 1124^VP

u.li.gi.no′sum. L. n. *uligo -inis* moisture of the soil, swamp; L. neut. adj. *uliginosum* swampy.

Rod-shaped cells, 1.0 × 8.0 μm, singly or in chains. Cells in chains are linked by a connecting filament with a core and outer sheath. Terminal spores are formed. Motile with 2–4 peritrichous flagella. Gram-stain-negative. Cells do not contain cytochromes.

Anaerobic. Chemoorganotrophic with fermentative metabolism. Utilizes glucose, cellobiose, sucrose, xylose, mannose, mannitol, and peptone. No growth occurs with lactose, maltose, glycerol, arabinose, raffinose, Casamino acids, H_2/CO_2, CO/CO_2, canillate, or various purines or pyrimidines. Ferments glucose to acetate, butyrate, lactate, H_2, and CO_2. Cannot reduce nitrate or sulfate. Temperature range for growth is 10–30 °C with an optimum of 20–25 °C. Acid tolerant with a pH range for growth of 4.0–9.0 and optimal growth occurring at pH 6.5.

Source: acidic forest bog.

DNA G+C content (mol%): 28.0 (T_m).

Type strain: CK55, ATCC BAA-53, DSM 12992.

GenBank accession number (16S rRNA gene): AJ276992.

Further comments: 16S rRNA gene sequence comparisons show *Clostridium uliginosum* to fall within the radiation of cluster I of the clostridia as defined by Collins et al. (1994). The closest relatives of *Clostridium uliginosum* based on 16S rRNA gene sequence similarity values are *Clostridium saccharobutylicum* (97.3%), *Clostridium punicuem* (96.9%), *Clostridium saccharoperbutylacetonicum* (96.8%), *Clostridium beijerinckii* (96.8%), and *Clostridium diolis* (96.7%). In addition, *Clostridium uliginosum* shares 97% 16S rRNA gene sequence similarity with strains of *Clostridium botulinum* types B, E, and F (a *Clostridium botulinum* strain cluster unrelated to the type strain of *Clostridium botulinum*) (see Figure 128).

164. **Clostridium ultunense** Schnürer, Schink and Svensson 1996, 1151[VP]

ul.tu.nen′se. *Ull* ancient Nordic god associated with winter sports and duels; *tuna* place of worship; L. ending *ense*; N.L. adj. *ultenense* referring to Ultuna, the area where the Swedish University of Agricultural Sciences is located.

Straight to slightly curved rods, 0.5–7.0 × 0.5–7.0 μm; occur singly, in pairs or in chains. Gram-stain-negative. Motile with a polar flagellum in early growth stages; becomes nonmotile. Forms terminal, round spores. Colonies are white, disc shaped, and 0.5–1 mm in diameter.

Strict anaerobe. Temperature range for growth is 15–50 °C with optimal growth at 37 °C. pH Range for growth is 5–10 with optimal growth at 7.0. Utilizes formate, glucose, cysteine, pyruvate, betaine, and ethylene glycol. Does not utilize other sugars, H_2-CO_2, CO, methoxylated aromatic compounds, alcohols, and all amino acids except cysteine. Yeast extract required for growth in mineral medium. Main fermentation product is acetate, but formate can also be formed. Small amounts of alanine are formed from cysteine and betaine. Sulfate, sulfite, thiosulfate, sulfur, nitrate, fumarate, malate, glycine, and ferric ion are not used as electron acceptors. Negative for gelatin hydrolysis, catalase, and oxidase acitivity. Produces indole from tryptophan.

Source: anerobic digester sludge.
DNA G+C content (mol%): 32 (T_m).
Type strain: BS, ATCC 700254, DSM 10521.
GenBank accession number (16SrRNA gene): Z69293.
Further comments: 16S rRNA gene sequence comparisons show *Clostridium ultunense* to fall within the radiation of cluster XII of the clostridia as defined by Collins et al. (1994).

165. **Clostridium vincentii** Mountfort, Rainey, Burghardt, Kaspar and Stackebrandt 1997a, 915[VP] (Effective publication: Mountfort, Rainey, Burghardt, Kaspar and Stackebrandt 1997b, 59.)

vin.cen′ti.i. N.L. masc. gen. n. *vincenti* of Vincent, in honor of Warwick Vincent for his contribution to our understanding of microbial ecosystems in Antarctica.

Slightly curved rods, 0.6 × 5.0 μm. Gram-stain-positive. Motile with peritrichous flagella. Temperature range for growth 2–20 °C with optimum growth at 12 °C. pH range for growth 5.5–8.0 with optimum growth at 6.5. Growth occurs at 0–3% NaCl (w/v) with optimum growth at <0.5%. Obligate anaerobe. Utilizes xylan, fructose, galac-

tose, glucose, *N*-acetylglucosamine, xylose, cellobiose, lactose, maltose, mannose, and sucrose. Esculin supports growth. Starch, cellulose, glycogen, rhamnose, sorbitol, valine, isoleucine, fumarate, lactate, formate, glycerol, malate, yeast extract, and Casamino acids are not utilized as growth substrates. Glucose and lactose are fermented to acetate, butyrate, formate, H_2, and CO_2. Nitrate and sulfate are not reduced.

Source: sediment below the cyanobacterial mat of a low-salinity pond on the McMurdo Ice Shelf in Antarctica.
DNA G+C content (mol%): 33 (T_m).
Type strain: lac-1, DSM 10228.
GenBank accession number (16S rRNA gene): X97432.
Further comments: 16S rRNA gene sequence comparisons show *Clostridium vincentii* to fall within the radiation of cluster I of the clostridia as defined by Collins et al. (1994). The closest relatives based on 16S rRNA gene sequence similaritie values are *Clostridium paraputrificum* (97.1%), *Clostridium aurantibutyricum* (96.7%), and *Clostridium chartatabidum* (96.3%). *Clostridium vincentii* is a psychrophilic species growing optimally at 12 °C while *Clostridium paraputrificum* is a mesophile which grows optimally between 30 and 37 °C.

166. **Clostridium viride** Buckel, Janssen, Schuhmann, Eikmanns, Messner, Sleytr and Liesack 1995, 619[VP] (Effective publication: Buckel, Janssen, Schuhmann, Eikmanns, Messner, Sleytr and Liesack 1994, 393.)

vi′ri.de. L. adj. *viridis, e* green, greenish; in reference to the green-colored cells and green FAD-containing enzymes of this organism.

Cells are pointed-ended to oval rods, 0.8 × 1.2–1.5 μm. Motile by two subpolar flagella. Gram-stain-positive. Cells are yellowish to green in color.

Strict anaerobe. Temperature range for growth is 19–40 °C. No growth was observed at 44 °C. Sample was isolated from anaerobic sewage sludge. Ferments 5-aminovalerate (to acetate, propionate, valerate, and ammonia), 5-hydroxyvalerate (to acetate, propionate, and valerate), crotonate, vinylacetate, and 4-hydroxybutyrate (to acetate and butyrate). No significant H_2 produced from 5-aminovallerate.

Growth is not supported by glucose, fructose, galactose, ribose, xylose, arabinose, melibiose, mannose, cellobiose, maltose, sucrose, lactose, cellulose, starch, pectin, formate, pyruvate, lactate, succinate, malate, fumarate, glycerol, mannitol, DL-3-hydroxybutyrate, tartrate, 1,2-propanediol, acrylate, citrate, betaine, alanine, β-alanine, aspartate, norvaline, 2-aminobutyrate, 4-aminobutyrate, 6-aminocaproate, glutamate, isoleucine, arginine, proline, urea, valine, leucine, tryptophan, trypticase yeast extract, and Casamino acids. 5-Aminovalerate is metabolized via the 2,4-pentadienoyl-CoA and 2-pentenoyl-CoA that utilizes two FAD-containing enzymes that form green crystals. Yeast extract and L-cysteine are required for growth on 5-aminovalerate. Sulfur is reduced. Nitrate, sulfate, amorphous ferric hydroxide, and fumarate are not used as electron acceptors. Does not produce indole. Catalase is not produced. Gelatin, esculin, and urea are not hydrolyzed.

DNA G+C content (mol%): 41.5 (HPLC).

Type strain: T2-7, ATCC 43977, DSM 6836.

GenBank accession number (16S rRNA gene): X81125.

Further comments: 16S rRNA gene sequence comparisons show *Clostridium viride* to fall within the radiation of cluster IV of the clostridia as defined by Collins et al. (1994).

167. **Clostridium xylanolyticum** Rogers and Baecker 1991, 142[VP] (emend. Chamkha, Garcia and Labat 2001a, 2109[VP])

xy.lan.o.ly'ti.cum. Gr. n. *xylanosum* xylan; Gr. adj. *lutikos* dissolving, loosening; N.L. part. adj. *xylanolyticum* xylan-dissolving.

Rod-shaped cells, 1.8–3.0 × 0.5–0.8 μm. Gram-stain-negative. Each cell forms single, terminal, round spores, which distend the cell wall. Motile with peritrichous flagella. Cells have a five layer cell wall.

Obligate anaerobe. Temperature range for growth is 25–55 °C with optimum growth at 35 °C. The pH range for growth is 5.0–8.0 with optimum growth at 7.2. Good growth observed with xylan. Does not ferment amygdalin, arabinose, glycogen, inositol, lactose, mannitol, ribose, sorbitol, and trehalose. Ferments PYG to formic, lactic, and acetic acid. β-hemolytic. Resistant to clindamycin. Reduces the double bond of side chain cinnamic acids: cinnamic, *o*-, *m*-, *p*-coumaric, *o*-, *m*-, *p*-methoxycinnamic, *p*-methylcinnamic, caffeic, ferulic, isoferulic, and 3,4,5- trimethoxycinnamic acids. End products from cinnamic acid metabolism are the corresponding phenylpropionic acids: 3-phenyl, 3-(2-hydroxyphenyl), 3-(3-hydroxyphenyl), 3-(4-hydroxyphenyl), 3-(2-methoxyphenyl), 3-(3-methoxyphenyl), 3-(4-methoxyphenyl), 3-(4-methylphenyl), 3-(3,4-dihydroxyphenyl), 3-(4- hydroxy-3-methoxyphenyl), 3-(3-hydroxy-4-methoxyphenyl), and 3-(3,4,5-trimethoxyphenyl) propionic acids. Addition of glucose markedly increases the rate of conversions. Cinnamyl alcohol and phenylpropionic acids, including 3-phenylpropionic acid and 3,4-dihydroxyphenylpropionic acid, are not transformed. Does not produce lecithinase, urease, indole, catalase, or lipase. Does not hydrolyze gelatin. Nitrate is not reduced.

Source: wood chips from *Pinus patula* that was exposed to the environment for 10 weeks.

DNA G+C content (mol%): 40 (T_m).

Type strain: ATCC 49623, DSM 6555.

GenBank accession number (16S rRNA gene): X71855, X76739.

Further comments: 16S rRNA gene sequence comparisons show *Clostridium xylanolyticum* to fall within the radiation of cluster XIVa of the clostridia as defined by Collins et al. (1994). The closest relative based on 16S rRNA gene sequence similarity values is *Clostridium aerotolerans* (99.4%).

168. **Clostridium xylanovorans** Mechichi, Labat, Garcia, Thomas and Patel 2000b, 3[VP] (Effective publication: Mechichi, Labat, Garcia, Thomas and Patel 1999a, 370.)

xy.la.no.vo'rans. Gr. n. *xylanosum* xylan; L. v. *vorare* devour, to eat; N.L. neut. adj. *xylanovorans* xylan-eating bacterium.

Rod-shaped cells, 0.8–1.0 × 4–10 μm. Gram-stain-positive. Motile with subpolar to laterally inserted flagella. Forms terminal spherical spores.

Strict anaerobe. Chemoorganotrophic. Temperature range for growth is 25–42 °C with an optimum for growth of 37 °C. pH Range for growth is 6–8 with an optimum of 7.0. Grows at NaCl concentrations of 0–1%. Grows on fructose, glucose, lactose, trehalose, maltose, raffinose, sucrose, xylan, mannitol, cellobiose, glactose, mannose, melibiose, and ribose. Does not grow on xylose, sorbose, sorbitol, *myo*-inositol, arabinose, arabitol, cellulose, formate, acetate, propionate, butyrate, valerate, crotonate, malonate, malate, lactate, citrate, succinate, ethanol, propanol, 2-propanol, butanol, isobutanol, biotrypcase, yeast extract, casammo acids, gelatin, and peptone. The major end products from the fermentation of fructose, glucose, lactose, trehalose, maltose, raffinose, sucrose, xylan, mannitol, cellobiose, galactose, mannose, melibiose and ribose are acetate, butyrate, formate, ethanol, H_2, and CO_2. Reduces fumarate to acetate. Does not reduce nitrate, sulfate, thiosulfate, and sulfur.

Source: olive mill wastewater treatment digester.

DNA G+C content (mol%): 40 (T_m).

Type strain: HESP1, DSM 12503.

GenBank accession number (16S rRNA gene): AF116920.

Further comments: 16S rRNA gene sequence comparisons show *Clostridium xylanovorans* to fall within the radiation of cluster XIVa of the clostridia as defined by Collins et al. (1994). The closest relative based on 16S rRNA gene sequence similarity values are *Clostridium jejuense* (96.6%) and *Clostridium aminovalericum* (93.6%). These two species can be differentiated on the basis of substrate utilization patterns.

Genus II. **Alkaliphilus** Takai, Moser, Onstott, Spoelstra, Pfiffner, Dohnalkova and Fredrickson 2001, 1245[VP] emend. Cao, Liu and Dong 2003, 974

KEN TAKAI

Al.ka.li'phil.us. N.L. n. *alkali* alkali; Gr. adj. *philos* loving; N.L. masc. n. *alkaliphilus* bacterium liking alkaline environments.

Straight to slightly curved rods, 0.4–0.7 × 2.0–6.0 μm. Gram-stain-positive and endosporeforming. The cells exhibit motility with flagella. Slightly halophilic and mesophilic with optimum growth at 35–40 °C. **Strictly anaerobic, and neutrophilic to alkaliphilic heterotrophs.** Proteinaceous substrates such as yeast extract, peptone, and tryptone are utilized as energy and carbon sources. **Fermentation and/or respiration of organic substrates in the presence and/** or absence of the electron acceptors such as **S⁰, thiosulfate, fumarate, crotonate, Fe(III), Co(III), or Cr (VI)** supports growth. Predominant cellular fatty acids are variable among the species but $C_{14:0}$, $C_{14:1}$, $C_{15:0\ iso}$, $C_{16:0}$, $C_{16:1\ \omega7c}$ are major components. Cell walls contain *m*-diaminopimelic acid, glycine, aspartate, glutamate, and ribose. Isolated from a deep-subsurface alkaline environment, an aerobic digester of wastewater, and a boron-rich leachate pond.

DNA G+C content (mol%): 30.6–36.4.

Type species: **Alkaliphilus transvaalensis** Takai, Moser, Onstott, Spoelstra, Pfiffner, Dohnalkova and Fredrickson 2001, 1245^VP.

Enrichment and isolation procedures

The type strain was isolated from alkaline water and sedimentary materials (pH 11.6) in a containment dam located at a depth of 3.2 km below surface (kmbls) in a South African gold mine (Takai et al., 2001). The slurried sample was directly inoculated into the standard medium (SM) for anaerobic, alkaliphilic heterotrophs at pH 10.5 (Takai et al., 2001), and the enrichment was performed at 37 °C for 2 d. After 2 d of incubation, growth of motile, straight to slightly curved rods was observed. The isolation of the type strain was performed on SM plates solidified with 1.2% (w/v) gelrite gellan gum. The strains of a second species of this genus (*Alkaliphilus crotonatoxidans*) were isolated from a methanogenic butyrate-degrading consortium (triculture) originally enriched from an aerobic digester for the wastewater treatment of a bean curd farm in Beijing, China (Cao et al., 2003). The strain was isolated with a crotonate-containing, anaerobic heterotrophic medium at 37 °C by the Hungate roll-tube technique (Hungate, 1969). Strains of the third species of this genus ("*Alkaliphilus metalliredigens*") which is not yet validly described were isolated from the partially dried soft sediments around alkaline leachate ponds hosting algal and cyanobacterial blooms. The strains were enriched with an anaerobic heterotrophic medium for soluble iron-reduction using lactate as the energy and carbon source and Fe(III)-citrate as the electron acceptor (Ye et al., 2004). The enrichment was performed at 20 °C, and the growth was observed after 5 d incubation. The isolates were obtained from the same medium for enrichment solidified by 2% (w/v) agar.

Maintenance procedures

All the species produce endospores at the late exponential and stationary growth phases. After growth, the strains are preserved in culture tubes at 4 °C. Lyophilized cultures are also used for a longer maintenance.

Taxonomic comments

16S rRNA gene sequence analysis (Takai et al., 2001) placed this genus within cluster XI of the low G+C Gram-positive group of Collins et al. (1994). According to the revised roadmap for Volume 3 (Figure 5, Ludwig et al., this volume), *Alkaliphilus* is placed in family *Clostridiaceae*, order *Clostridiales*, class *Clostridia*, phylum *Firmicutes*.

Differentiation of the genus *Alkaliphilus* from other genera

Table 146 lists characteristics differentiating *Alkaliphilus* from other genera of the family *Clostridiaceae* that are phylogenetically related to *Alkaliphilus*. The most outstanding physiological property of *Alkaliphilus* distinguishing it from other genera is its versatile utilization of electron acceptors. The potential electron acceptors for *Alkaliphilus* are elemental sulfur, thiosulfate, fumarate, crotonate, Fe(III), Co(III), and Cr(VI). In contrast, none of the *Alkaliphilus* species is able to grow on amino acids by the Stickland reaction. 16S rRNA gene sequences of *Alkaliphilus* species also possess a low level of similarity with members of other *Clostridiaceae* genera (<92%).

Differentiation of the species of the genus *Alkaliphilus*

Table 146 also lists characteristics distinguishing species of the genus *Alkaliphilus*. The pH and NaCl ranges for growth are key to distinguishing the species of the genus *Alkaliphilus*. Within this genus, *Alkaliphilus transvaalensis* is the most alkaliphilic (capable of growth at up to pH 12.4), and it is moderately halophilic. "*Alkaliphilus metalliredigens*" is moderately alkaliphilic and the most halophilic. Lastly, *Alkaliphilus crotonatoxidans* is neutrophilic and the least halophilic. In addition, the *Alkaliphilus* species utilize different electron acceptors. Different cellular fatty acid compositions are also described (Cao et al., 2003; Ye et al., 2004).

List of species of the genus *Alkaliphilus*

1. **Alkaliphilus transvaalensis** Takai, Moser, Onstott, Spoelstra, Pfiffner, Dohnalkova and Fredrickson 2001, 1254^VP

trans.vaa.len′sis. N.L. adj. *transvaalensis* of Transvaal, a region of South Africa.

Gram-stain-positive, straight to slightly curved rods that are 0.4–0.7 × 3–6 μm. Cells occur singly or in pairs. Exhibits motility with flagella. Vegetative cells swell to form terminal spherical spores. Strictly anaerobic. The temperature range for growth is 20–50 °C, with the optimum being 40 °C. The pH range for growth is 8.5–12.5, with the optimum growth occurring at pH 10.0. NaCl is not absolutely required for growth but enhances the growth. Heterotrophic growth with yeast extract, peptone, tryptone, or casein. Elemental sulfur, thiosulfate, fumarate, and crotonate stimulate growth as electron acceptors. The major phospholipid fatty acids are $C_{14:0}$, $C_{15:1\ iso\ \omega7c}$, $C_{15:0\ iso}$, $C_{16:0}$, $C_{17:1\ iso\ \omega7c}$, and $C_{17:0\ iso}$.

DNA G+C content (mol%): 36.4 (HPLC).

Type strain: SAGM1, ATCC 700919, CIP 107133, JCM 10712.

GenBank accession number (16S rRNA gene): AB037677.

2. **Alkaliphilus crotonatoxidans** Cao, Liu and Dong 2003, 973^VP

cro.to.nat.ox′i.dans. N.L. part. adj. *crotonatooxidans* of the one that oxidizes crotonate.

Gram positive, straight to slightly curved rods (0.4–0.6 × 2–3 μm). Cells occur singly or in pairs. Exhibits motility with multiple flagella. Vegetative cells swell to form terminal spherical spores. Strictly anaerobic. The temperature range for growth is 15–45 °C, with the optimum being 37 °C. The pH range for growth is 5.5–9.0, with the optimum growth occurring at pH 7.5. Heterotrophic growth with yeast extract, peptone, fructose, cellobiose, maltose, trehalose, xylose, ribose, citrate, malate, and crotonate, but not with glucose, lactose, sucrose, galactose, lactate, succinate, or butyrate. Crotonate is dismutated to acetate and butyrate. Growth is not stimulated by other electron acceptors. Cell wall contains *m*-diaminopimelic acid, glycine, aspartate, glutamate, and ribose. The major cellular fatty acids are $C_{13:0\ iso}$, $C_{13:1\ iso\ \omega2c}$, $C_{14:0}$, $C_{15:1\ iso\ \omega7c}$, $C_{15:0\ iso}$, $C_{15:0\ ante}$, $C_{16:0}$, $C_{16:1\omega7c}$, and $C_{17:0\ ante}$.

TABLE 146. Differential characteristics of *Alkaliphilus* species and other phylogenetically related genera of the family *Clostridiaceae*[a]

Characteristic	*Alkaliphilus transvaalensis*	*Alkaliphilus crotonatoxidan*	*"Alkaliphilus metalliredigens"*	*Tindallia magadiensis*	*Tindallia californiensis*	*Natronincola histidinovorans*	*Clostridium felsineum*
Motility	+	+	+	+[b]	+	+	+
Spore formation	+	+	+	−	+	d	+
pH optimum for growth	10	7.5	9.5	8.5	8.5	9.4	Neutral pH
Upper limit of pH for growth	12.5	9	10	10.5	10.5	10.5	ND[c]
NaCl (%) range for growth (optimum)	0–2.3 (0.4)	(0.08)	0–6 (2.5)	1–10 (3–6)	1–20 (3)	4–16 (8)	ND
Electron donor/acceptor couple:							
Yeast extract/ elemental sulfur	+	−	ND	ND	ND	ND	ND
Yeast extract/ thiosulfate	+	−	−	ND	ND	ND	ND
Yeast extract/ fumarate	+	−	−	ND	ND	ND	ND
Yeast extract/ crotonate	+	+	ND	ND	ND	ND	ND
Lactate/Fe(III)	ND	ND	+	+	ND	ND	ND
Lactate/Co(III)	ND	ND	+	ND	ND	ND	ND
Lactate/Cr(VI)	ND	ND	+	ND	ND	ND	ND
Respiration by the Stickland reaction	−	−	−	+	+	+	−
DNA G+C content (mol%)	36.4	30.6	ND	37.6	44.4	31.9	26

[a]Symbols: +, >85% positive; d, different strains give different reactions (16–84% positive); −, 0–15% positive; w, weak reaction; ND, not determined.
[b]Described as nonmotile in the original article, but motility has since been observed.

DNA G+C content (mol%): 30.6 (T_m).
Type strain: B11-2, AS 1.2897, JCM 11672.
GenBank accession number (16S rRNA gene): AF467248.

Other organisms

1. "Alkaliphilus metalliredigens"

Described by Ye et al. (2004), this interesting organism has not yet been deposited in culture collections and is not validly published. Nevertheless, it provides insight into the phenotypic diversity of this genus. Cells are Gram positive, straight to slightly curved rods (0.5 × 3–6 μm) and exhibit motility. Vegetative cells swell to form terminal spheri-

cal spores. Strictly anaerobic. The temperature range for growth is 4–45 °C, with the optimum being 35 °C. The pH range for growth is 8.0–10.0, with optimum growth occurring at pH 9.5. NaCl is not absolutely required for growth but enhances the growth optimally at 2% (w/v). Heterotrophic growth with yeast extract, lactate, and acetate. Soluble forms of Fe(III), Co(III), and Cr(VI) serve as electron acceptors. The organism is able to grow in the presence of 10 mM sodium borate.

DNA G+C content (mol%): not determined.
Type strain: QYMF.
GenBank accession number (16S rRNA gene): AY137848.

Genus III. **Anaerobacter** Duda, Lebedinsky, Mushegjan and Mitjushina 1996, 625[VP] (Effective publication: Duda, Lebedinsky, Mushegjan and Mitjushina 1987, 126.)

THE EDITORIAL BOARD

An.ae.ro.bac'ter. Gr. pref. *an-* not; Gr. masc. n. *aer* air; N.L. masc. n. *bacter* the equiv. of Gr. neut. dim. n. *bakterion* small rod, staff; N.L. masc. n. *Anaerobacter* rod not [living] in air.

Cells are **thick rods, ovals or spheres**, depending upon the growth conditions. Oval and spherical cells **form three or more endospores per cell. Gram-stain-positive** cell-wall structure. **Obligate anaerobe that ferments carbohydrates and fixes N₂.** Nonmotile. Sulfate and nitrate are not reduced. Nitrite is reduced to ammonia in a dissimilatory process.

DNA G+C content (mol%): 29.

Type species: **Anaerobacter polyendosporus** Duda, Lebedinsky, Mushegjan and Mitjushina 1996, 625[VP] (Effective publication: Duda, Lebedinsky, Mushegjan and Mitjushina 1987, 127.).

Further descriptive information

This description is as given previously by Duda et al. (1987) and Siunov et al. (1999). Cells are oval (4–8 μm in length and

3–4 μm in diameter) or spherical (4–6 μm in diameter) on solid media. In liquid media, the cells are thick rods with rounded ends (1.5–3 μm in diameter and 4–8 μm in length). Nonmotile. Surrounded by a polysaccharide capsule. The rod-shaped cells form one or two subterminal endospores at opposite poles. The oval and spherical cells form from one up to five round or oval endospores. The endospores possess spore coats, exosporium, inner and outer membranes, cortex and core. The cell wall consists of one layer and is sensitive to penicillin and lysozyme. Prosthecae-like appendages surrounded by cytoplasmic membrane and cell wall and containing cytoplasm are sometimes formed. Extensive lipid leaves form in the cytoplasmic membrane, located between the outer and internal lipid layers.

Colonies on the surface of solid media are large (up to 5 mm in diameter), round, smooth, opaque, brownish-white, viscous, navel-shaped, with entire margins.

Moderately aerotolerant anaerobe. Growth on solid media containing no reductants is possible at $pO_2 \leq 500$ Pa. A weak catalase activity can sometimes be detected after growth at pO_2 of 500 Pa. Cytochromes are not present. Neutrophilic and mesophilic. Growth occurs in the pH range of 5.5–8.5, with an optimum at 6.5–7.5. The temperature range for growth is 15–45 °C, with an optimum at 25–35 °C. Substrates supporting growth include cellobiose, fructose, galactose, glucose, inulin, maltose, mannitol, mannose, raffinose, salicin, sorbitol, sucrose, trehalose, xylose and lactate. Starch is hydrolysed but cannot support growth as the sole source of carbon and energy. Amino acids, alcohols and organic acids other than lactate do not support growth. Gelatin is not liquefied. Lipase, lecithinase and urease are not produced. Indol and sulfide are not formed. The end products of glucose fermentation include acetate, lactate, butyrate, ethanol, butanol, H_2 and CO_2. Sulfate and nitrate are not used as electron acceptors, but nitrite is reduced to ammonia in a dissimilatory process.

Isolated from meadow-gley soil under rice under anaerobic conditions on potato agar containing 0.5% glucose, 0.1% yeast extract, and 0.04% sodium thioglycollate.

Taxonomic comments

Based upon 16S rRNA gene sequence comparisons, *Anaerobacter polyendosporus* falls within the radiation of cluster I of the clostridia as defined by Collins et al. (1994). The original description of *Anaerobacter polyendosporus* did not include phylogenetic analyses, and the justification for the creation of a novel genus was based upon the morphological characteristics as well as the ability to form numerous spores within a single cell (Duda et al., 1987). Subsequent sequencing of its 16S rRNA gene identified a close relationship with cluster I of the clostridia as defined by (Collins et al., 1994; Siunov et al., 1999; Stackebrandt et al., 1999b). Siunov et al. (1999) suggested that the unusual morphological characteristics justified the existence of a novel genus even though the rRNA sequence fell within the species of the genus *Clostridium sensu stricto*. Stackebrandt et al. (1999b) suggested that *Anaerobacter polyendosporus* should be reclassified as a species of the genus *Clostridium*. *Anerobacter polyendosporus* shows highest 16S rRNA gene sequence similarity to the species *Clostridium fallax* (95.6%), *Eubacterium tarantellae* (95.1%), *Clostridium putrefaciens* (94.9%), *Clostridium intestinale* (94.8%) and *Clostridium chauvoei* (94.7%). In an attempt to further the reorganization of the genus *Clostridium*, the species *Anaerobacter polyendosporus* is transferred to the genus *Clostridium* as *Clostridium polyendosporum* comb. nov. (Rainey et al., 2009).

List of species of the genus *Anaerobacter*

1. **Anaerobacter polyendosporus** Duda, Lebedinsky, Mushegjan and Mitjushina 1996, 625 (Effective publication: Duda, Lebedinsky, Mushegjan and Mitjushina 1987, 127.)

po.ly.en.do.spo′rum. Gr. pref. *poly* many; Gr. pref. *endo* within; Gr. n. *spora* spore; N.L. adj. *polyendosporus* (forming) several endospores.

The description is the same as for the genus.
DNA G+C content (mol%): 29 (T_m).
Type strain: PS-1, DSM 5272, VKM B-1724.
GenBank accession number (16S rRNA gene): Y18189.

Genus IV. **Anoxynatronum** Garnova, Zhilina and Tourova 2003a, 1219[VP] (Effective publication: Garnova, Zhilina and Tourova 2003b, 217.)

ELENA S. GARNOVA AND TATJANA N. ZHILINA

An.ox′y.na.tro′num. Gr. pref. *an*-without; Gr. adj. *oxys* acid or sour, and in combined words indicating oxygen; N. Gr. n. *natron* (arbitrarily derived from Ar. n. *natrun* or *natron*) soda; N.L. neut. n. *Anoxynatronum* organism which inhabits anaerobic and soda environment.

Slightly curved rods with pointed ends, 0.5–0.7 × 3.8–5 μm. Gram-stain-positive cell-wall structure. Motile. Nonsporeforming. **Moderately alkaliphilic** with good growth at pH 7.6–9.5. **Growth requires sodium and bicarbonate** but not chloride ions. **Catalase-positive anaerobe** which possesses fermentative metabolism. **Chemoorganoheterotroph**: able to utilize mono- and disaccharides, sugar alcohols, and proteinaceous substrates. The products of glucose fermentation are acetate and ethanol. Yeast extract is required for growth. Habitat is saline-carbonate lakes. The genus is monotypic.

DNA G+C content (mol%): 48.1–48.4.
Type species: **Anoxynatronum sibiricum** Garnova, Zhilina and Tourova 2003a, 1219[VP] (Effective publication: Garnova, Zhilina and Tourova 2003b, 219.).

Further descriptive information

Cells occur singly or in pairs and multiply by binary division (Figure 129). In the stationary phase, very long chains are rarely formed. Actively motile; the flagellation is peritrichous. Sporulation was not observed. Viability is retained upon heating to 70 °C

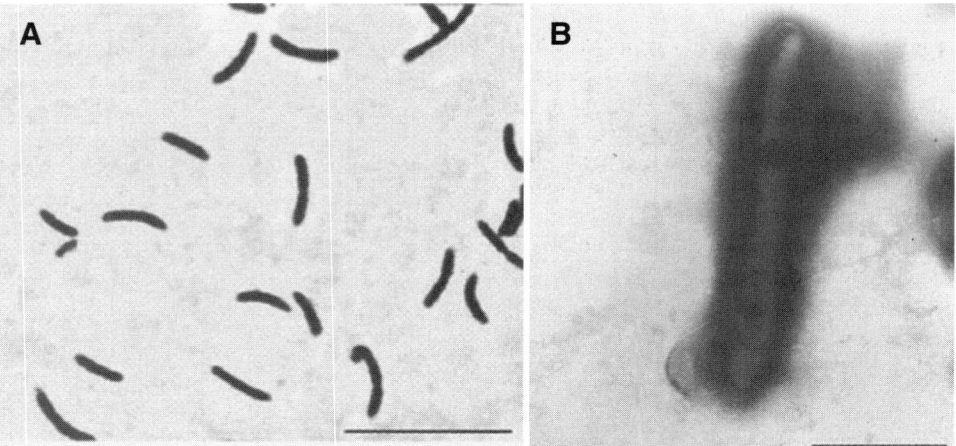

FIGURE 129. Morphology of *Anoxynatronum sibiricum* cells as viewed under phase contrast microscope. Bar = 10 μm.

for 10 or 20 min but not for 50 min. Catalase-positive anaerobe. Tolerates up to 4.5% O_2 in nitrogen gas phase; growth is possible without any reducing agents in the medium.

A typical athalassophile (i.e., adapted to continental waters), it obligately requires sodium carbonate but not chloride ions. The optimum pH for growth is 9.1, with a range of pH 7.1–10.1. Requires Na^+, and no growth is observed after replacing sodium with potassium ions. Grows when the total concentration of Na^+ is in the range of 0.08–1.3 M, with a relatively broad optimum between 0.25–0.86 M Na^+. The broad optimum allows adaption to seasonal changes in salt content. The type species has an optimal growth temperature of 35 °C. Due to the cryoarid climate of the region, the temperature of sediments rarely exceeds 40 °C, which might explain the rapid decline of the growth rate above 41 °C and its upper limit at 46 °C.

Metabolism is fermentative. *Anoxynatronum* is incapable of dissimilatory reduction of nitrite, nitrate, or sulfur compounds (S, SO_4^{2-}, SO_3^{2-}, $S_2O_3^{2-}$, $S_2O_4^{2-}$), but it reduces Fe^{3+} to Fe^{2+} without energy generation. Vitamins are not required, but they improve the growth rate. Does not assimilate N_2. Possible sources of nitrogen are NH_4Cl, $NaNO_3$, and organic nitrogen from yeast extract.

Physiologically, *Anoxynatronum* is a primary anaerobe; in a trophic chain of bacterial community it consumes low-molecular-mass organic compounds produced by hydrolytic microorganisms. Some of *Anoxynatronum* fermentation products are metabolized by second anaerobes, e.g., sulfate-reducing bacteria. *Anoxynatronum* utilizes a wide variety of catabolic substrates including monosaccharides (ribose, glucose, mannose, and fructose), disaccharide (sucrose), sugar alcohols (glycerol and L-inositol), a few amino acids (Table 147), and proteinaceous compounds (peptone, tryptone, meat extract, and yeast extract). It also grows slowly on dried *Spirulina* biomass, which could be a nutritional source in its natural habitat in addition to the microbial products of plant cellulose decomposition. During the fermentation of glucose, it forms 2.5 mol of acetate and 0.48 mol of ethanol per mol of glucose. Because acetate is the major product, *Anoxynatronum* should be considered an acetogen. In the trophic food chain of the anaerobic community in soda lakes, it could be regarded as a copiotrophic component of the saccharolytic pathway of biomass degradation.

Methods of isolation and cultivation of pure cultures

The type strain was isolated from an anaerobic enrichment on glucose originally inoculated with a mixture of mud and surface cyanobacterial mat from a lagoon of Lake Nizhnee Beloe (south-eastern Transbaikal region, Russia). Pure cultures were obtained by serial dilutions in liquid medium under strictly anaerobic conditions. The mineral medium which was used for the enrichment culture mimicked the mineral composition of Lake Nizhnee Beloe water and contained (per liter): NaCl, 3.4 g; Na_2CO_3, 4.45 g; $NaHCO_3$, 5.5 g; and glucose, 5 g. Incubations were performed anaerobically under N_2 gas at pH 9.5 and 36 °C. To obtain a pure culture, colonies were isolated from Hungate roll tubes (Hungate, 1969) containing the same medium plus 3% agar (w/v). The uniform colonial and cellular morphologies confirmed culture purity. After optimization of the growth conditions, the *Anoxynatronum* medium contained (per liter): KH_2PO_4, 0.2 g; $MgCl_2$, 0.1 g; NH_4Cl, 0.5 g; KCl, 0.2 g; Na_2CO_3, 9.3 g; $NaHCO_3$, 44.1 g; $Na_2S·9H_2O$, 0.7 g; yeast extract, 0.2 g; glucose, 5.0 g; trace element solution, 1 ml (Kevbrin and Zavarzin, 1992); vitamin solution, 10 ml (Wolin et al., 1963); and resazurin, 0.001 g. The pH was 9.05, and the gas phase was 100% N_2.

Maintenance procedures

The type strain remains viable when stored in *Anoxynatronum* medium at 4 °C and transferred every six months. Glycerol (15%) stocks are stored in liquid nitrogen.

Differentiation of the genus *Anoxynatronum* from other genera

Based on 16S rRNA analyses, *Anoxynatronum sibiricum* belongs to the *Clostridium felsineum* subgroup of cluster XI of Collins et al. (Collins et al., 1994). In the phylogenetic analysis by Ludwig et al. presented in the roadmap to the current volume, *Anoxynatronum* is placed in the family *Clostridiaceae*, order *Clostridiales*, class *Clostridia*, phylum *Firmicutes* (Figure 5). The closest relative among the validly published alkaliphilic representatives of this subgroup is the anaerobic ammonifier *Tindallia* (Kevbrin et al., 1998), with a 16S rRNA sequence similarity of 94.2%. For the other alkalinophiles, *Alkaliphilus* (Takai et al., 2001) and *Natron-*

TABLE 147. Characteristics differentiating *Anoxynatronum sibiricum* from closely related alkaliphilic anaerobes[a]

Characteristic	Anoxynatronum sibiricum	Alkaliphilus transvaalensis	Natronincola histidinovorans	Tindallia magadiensis
Strain	Z-7981[b]	SAGM1[c]	Z-7940[d]	Z-7934[e]
Morphology: rods	+	+	+	+
Vesicles or minicells	–	+	+	–
Flagellation	Peritrichous	Multiple	Peritrichous	–
Spore shape and location	–	Spherical, terminal	Lacking in type strain	–
Na⁺ range (M)	0.08–1.3	0.77–1.3	0.7–2.7	0.17–1.7
Na⁺ optimum (M)	0.25–0.86	0.84	1.4–1.7	0.7–1.4
pH range	7.1–10.1	8.5–12.5	8.0–10.5	7.5–10.5
pH optimum	9.1	10	9.4	8.5
Temperature range,°C	25–41	20–50	ND	19–47
Temperature optimum,°C	35	40	37–40	37
Electron acceptors:				
Dimethylsulfoxide	ND	–	ND	+
Elemental sulfur	–	+	ND	ND
Fe³⁺	+	ND	ND	+
Fumarate	ND	+	ND	ND
Nitrate	–	–	ND	–
Thiosulfate	–	+	ND	–
Substrates utilized:				
Arginine	–	–	–	+
Carbohydrates	+	–	–	–
Cysteine	+	–	–	–
Glutamine	–	–	–	–
Glutamate	+	–	+	±
Histidine	–	–	+	±
Ornithine	–	–	–	+
Pyruvate	+	–	–	+
Sugar alcohols	+	–	–	–
Spirulina biomass	+	ND	ND	ND
Chitin	+	ND	ND	ND
Proteinaceous substrates	+	+	+	±
Fermentation products	From carbohydrates: A, E	ND	From amino acids: A, F, NH₃	From amino acids: A, P, NH₃, H₂
DNA G+C content (mol%)	48.4	36.4	31.9	37.6

[a]Symbols: +, >85% positive; –, 0–15% positive; w, weak reaction; ND, not determined. A, acetate; E, ethanol; F, formate; P, propionate.
[b]Data from Garnova et al. (2003b).
[c]Data from Takai et al. (2001).
[d]Data from Zhilina et al. (1998).
[e]Data from Kevbrin et al. (1998).

incola (Zhilina et al., 1998), the sequence similarity is 91.7% and 89.5%, respectively (Figure 130). However, the DNA G+C content of *Anoxynatronum sibiricum* differs from that of *Tindallia* by 10 mol%, and they have a very low DNA–DNA similarity value of 11%. Other physiological characteristics which distinguish *Anoxynatronum* from *Tindallia* and related alkaliphiles are its ability to ferment sugars, the spectrum of substrates utilized, and the fermentation products (Table 147).

Taxonomic comments

Phylogenetic analyses of the 16S rRNA gene suggests that *Anoxynatronum sibiricum* forms a separate clade with the alkaliphile "*Clostridium alcalibutyricum*" E2SE1 from Lake Elmenteita, Kenya (Jones et al., 1998; Figure 130). In addition to acetate, "*Clostridium alcalibutyricum*" produces butyrate which was not detected among the fermentation products of *Anoxynatronum sibiricum*. Because very limited information about the taxonomy and physiology of "*Clostridium alcalibutyricum*" is available, it is not possible to make a complete comparison between these two species.

List of species of the genus *Anoxynatronum*

1. **Anoxynatronum sibiricum** Garnova, Zhilina and Tourova 2003a, 1219[VP] (Effective publication: Garnova, Zhilina and Tourova 2003b, 219.)

si.bi′ri.cum. N.L. neut. adj. *sibiricum* pertaining to Siberia, a Russian region.

The description for the type species is as given for the genus and as listed in Table 147. Another strain of *Anoxynatronum sibiricum*, Z-7981, was isolated from Lake Nizhnee Beloe. Aside from small differences in cell diameter, it is similar to the type strain (Tourova et al., 1999). These two strains have 98% DNA–DNA

FIGURE 130. Phylogeny of *Anoxynatronum* and other representatives of the *Clostridium felsineum* subgroup of *Clostridium* cluster XI of Collins et al. (1994). *Clostridium butyricum* was taken as an outgroup. Based on the nucleotide sequence of the 16S rRNA gene, where the bar corresponds to 5 substitutions per 100 positions. Bootstrap values >95 (expressed as percentage of 100 replications) are shown at branch points.

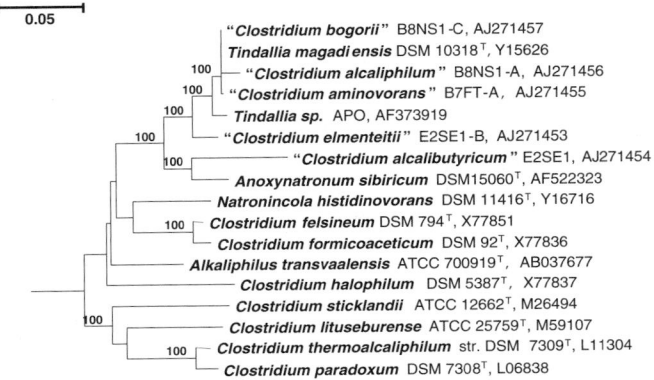

similarity and DNA G+C contents of 48.4 mol% for Z-7981 and 48.1 mol% for Z-7981.

    Habitat: saline-carbonate lake.

DNA G+C content (mol%): 48.4 (T_m).
Type strain: Z-7981, DSM 15060, Uniqem U-262.
GenBank accession number (16S rRNA gene): AF522323.

Genus V. **Caloramator** Collins, Lawson, Willems, Cordoba, Fernández-Garayzábal, Garcia, Cai, Hippe and Farrow 1994, 812[VP] emend. Chrisostomos, Patel, Dwivedi and Denman 1996, 497

SANDRA BAENA AND BHARAT K. C. PATEL

Ca.lora.ma′tor. L. n. *calor* heat; L. masc. n. *amator* lover; N. L. masc. n. *Caloramator* heat lover.

Straight to slightly curved rods or filaments (2–100 × 0.4–0.8 μm), which occur singly or pairs. Cells stain Gram-negative or Gram-positive but possess a Gram-positive cell-wall ultrastructure. May be **motile or nonmotile.** Endospores may or may not be produced and if present, are spherical and located terminally or subterminally and do not distend the sporangia. All species grow at **thermophilic** temperatures (growth temperature optimum between 50 and 68 °C). The metabolism is heterotrophic and **saccharolytic** and the end products from glucose fermentation are acetate, isobutyrate, isovalerate, valerate, lactate and ethanol. Unable to grow autotrophically or produce sulfur from thiosulfate.

    DNA G+C content (mol%): 25–39.

    Type species: **Caloramator fervidus** (Patel, Monk, Littleworth, Morgan and Daniel 1987) Collins, Lawson, Willems, Cordoba, Fernández-Garayzábal, Garcia, Cai, Hippe and Farrow 1994, 812[VP] (*Clostridium fervidus* Patel, Monk, Littleworth, Morgan and Daniel 1987, 125).

Further descriptive information

To date, six species of the genus *Caloramator* have been described. The first member of the genus *Caloramator, Caloramator fervidus* (formerly *Clostridium fervidus*) was isolated from a hot spring in New Zealand and subsequently other members were isolated from volcanic as well as non-volcanically heated environments: *Caloramator indicus* was isolated from an Indian thermal artesian basin (Chrisostomos et al., 1996), "*Caloramator celer*" (formerly *Thermobrachium celere*) from a New Zealand hot spring (Engle et al., 1996), *Caloramator proteoclasticus* from sludge of a mesophilic granular methanogenic digester (Tarlera et al., 1997), *Caloramator coolhaasii* from sludge of a thermophilic granular digeter (Plugge et al., 2000) and *Caloramator viterbiensis* from an Italian hot spring (Seyfried et al., 2002).

Cells of the genus *Caloramator* are straight to slightly curved rods or filaments measuring 2–100 μm × 0.4–0.8 μm, and occur singly, or pairs. Spores have so far only been observed in *Caloramator fervidus* and *Caloramator proteoclasticus*. With the exception of *Caloramator viterbiensis*, all other species stain Gram-negative but ultrathin sections of cells reveal a multilayered, complex, thick cell wall with an external S-layer similar to that of Gram-stain-positive cell-walls. All species are strictly anaerobic chemo-organotrophs, which grow at temperatures ranging from 37 to 80 °C, with an optimum between 50 and 68 °C. No growth is observed below 37 and above 80 °C. The pH range for growth is between pH 5.0 and 9.5 and the optimum pH for growth is between 6.0 and 8.1.

    The presence of thermoresistant endospores has been demonstrated in *Caloramator fervidus* and *Caloramator proteoclasticus*, but not in *Caloramator indicus, Caloramator coolhaasii, Caloramator viterbiensis* and "*Caloramator celer*". The value of spore formation as an important criterion in classification has been questioned (Chrisostomos et al., 1996) as sporulation genes can be identified in non-sporulating bacteria.

    Members of the genus *Caloramator* ferment glucose, starch, amylose, maltose, xylan, mannose, galactose, fructose, ribose, xylose, mannose, sucrose, cellobiose, lactose, galactose, and dextrin (Table 148) but only in the presence of yeast extract and/or peptone suggesting that trace components present in yeast extract and/or peptone are required for growth. Limited growth occurs in yeast extract and/or peptone in the absence of fermentable substrates. *Caloramator* species also utilize a range of amino acids (Table 148). The end products of glucose fermentation are acetate, *i*-butyrate, *i*-valerate, n-valerate, ethanol, lactate, formate, H_2 and CO_2 but the pattern produced depends on the species (Table 148).

TABLE 148. Differential characteristics of the members of the genus *Caloramator*[a]

Characteristic	1. *C. fervidus* (formerly *Clostridium fervidus*)	2. "*C. celer*" (formerly *Thermobrachium celere*)	3. *C. coolhaasii*	4. *C. indicus*	5. *C. proteoclasticus*	6. *C. viterbiensis*
Type strain	Strain Rt4-B1[T]=ATCC 43204[T]=DSM 5463[T]	Strain JW/YL-NZ35[T]=DSM 8682[T]=ATCC 700318[T]	Strain Z[T]=DSM 12679[T]	Strain IndiB4[T]=ACM 3982[T]	Strain U[T]=DSM 10124[T]	Strain JM/MS-VS5[T]=DSM 13723[T] = ATCC PTA 584[T]
Reference	Patel et al., 1987	Engle et al., 1996	Plugge et al., 2000	Chrisostomos et al., 1996	Tarlera et al., 1997	Seyfried et al., 2002
Isolation source	Hot spring, New Zealand	Hot spring, New Zealand	Anaerobic thermophilic granular sludge	Non-volcanically heated waters, India	Mesophilic granular methanogenic sludge	Hot spring, Italy
Morphology and size (μm)	Rods, 2–2.5×0.65–0.75	Rods and branched filaments, 1.5–14×0.5–1.2	Rod-shaped and filaments 2–40×0.5–0.7	Rods and filaments, 10–100×0.6–0.8	Slightly curved rods, 2.4–4.0×0.4	Straight to sligthly curved rods 2.0–3.0×0.4–0.6
Gram stain reaction	Negative	Positive	Negative	Negative	Negative	Positive
DNA G+C content (mol%)	39	31 (HPLC)	31.7	25.6 ± 0.3	31	32
Presence of spores	+	–	–	–	+	–
Motility	+	ND	–	–	+	–
Presence of flagella		ND	–	–	+	–
Temperature growth range (°C)	37–80	43–75	37–65	37–65	30–68	33–64
Temperature optimum (°C)	68	66	50–55	60–65	55	58
pH growth range	5.5–9.0	5.4–9.5	6.0–8.0	6.2–9.2	6.0–9.5	5.0–7.8
pH optimum	7.0–7.5	8.2	7.0–7.5	7.5–8.1	7.0–7.5	6.0–6.5
Colony forming ability in agar medium	+	+	+		+	+
Yeast extract and / or peptone required for growth	+	+	+	+	+	ND
Growth on carbohydrates	+[b]	+[c]	+[d]	+[e]	+[f]	+[g]
Growth on amino acids	Serine used as sole carbon & energy source	ND	Glutamate aspartate, methionine, arginine, alanine	ND	Glutamate, aspartate methionine, arginine, histidine, threonine, leucine, valine, glycine	Serine, glutamate, aspartate, methionine, histidine, threonine, leucine, valine
End-products from glucose fermentation	Acetate (major), i-butyrate, i-valerate, n-valerate, ethanol, lactate, CO_2 + H_2	Acetate, ethanol, formate CO_2 + H_2	Acetate, lactate, CO_2 + H_2	Acetate, ethanol, lactate, CO_2 + H_2	Acetate, ethanol, lactate formate, CO_2 + H_2	Acetate, ethanol, CO_2 + H_2

[a]Symbols: +, positive; –, negative; w, weak reaction; ND, not determined.
[b]Grows well on glucose, maltose, xylose, starch, xylan, mannose and pyruvate but not sucrose, galactose, rhamnose, arabinose and cellulose.
[c]Grows well on glucose, fructose, galactose, maltose and sucrose but not cellobiose, ribose, mannose, arabinose, xylose, glucuronic acid and xylan.
[d]Grows well on glucose, fructose, galactose, sucrose, maltose, ribose, xylose, starch, cellobiose and mannose but not arabinose, rhamnose, lactose and cellulose.
[e]Gows well on glucose, fructose, starch, amylose, dextrin, amylopectin, cellobiose, lactose, mannose and sucrose but not cellulose, dextran and chitin.
[f]Grows well on glucose, fructose, starch, cellobiose and mannose but not lactose, xylose, malate, xylan, and cellulose.
[g]Grows well on glucose, fructose, sucrose, cellobiose, lactose, galactose, starch and mannose but not xylose and arabinose.

Sulfate, thiosulfate, elemental sulfur, sulfite, nitrate and fumarate are not utilized as electron acceptors.

Enrichment and isolation procedures

Various media have been used to enrich, isolate and characterize the members of genus *Caloramator*. All media are prepared anaerobically. *Caloramator fervidus* was enriched at 70 °C on Trypticase peptone-Yeast extract-Xylan (TYEX) medium and isolated by serial dilution in agar shakes (Patel et al., 1987). *Caloramator indicus* was enriched at 70 °C without agitation in a pre-reduced anaerobic Trypticase peptone-Yeast extract-Glucose (TYEG) medium and isolated by end point dilution in TYEG agar (2.5%) shake tubes. A peptone yeast extract medium supplemented with 0.4 g/l glucose and an incubation temperature of 55 °C, was used to enrich *Caloramator proteoclasticus* and isolation of the pure culture was achieved in1.5% agar roll tubes (Holdeman et al., 1977b; Hungate, 1969). A bicarbonate–buffered anaerobic medium supplemented with 0.02% yeast extract was used to enrich *Caloramator coolhaasii* (Stams et al., 1993) and isolation of a pure culture was achieved on Wilkens-Chalgren broth medium (16 g/l) fortified with agar (WC broth, Oxoid). *Caloramator viterbiensis* was enriched at 60 °C in a basal medium containing yeast extract and a dilution series of the enrichment plated on the same medium fortified with 1.5% agar for isolation of a pure culture (Seyfried et al., 2002). "*Caloramator celer*" (formerly *Thermobrachium celere*) was enriched and isolated in a complex medium (Engle et al., 1996). We routinely use pre-reduced anaerobic TYEG medium and an incubation temperature of between 55–65 °C in our laboratory to culture all the six *Caloramator* species and other as yet uncharacterized species (Patel et al., 1985b, 1985a).

Maintenance procedures

In our laboratory, mid-exponential phase grown cultures of *Caloramator* species are resuspended in 10 ml TYEG medium containing 20% glycerol (v/v) and 0.1 mol of amorphous ferrous sulfide and stored at –20 and/or –80 °C. Amorphous ferrous sulfide is prepared according to the method of Brock and Od'ea (1997). Seyfried et al. (2002) have reported that *Caloramator viterbiensis* can be kept at room temperature or preferably at 4 °C for several days in a medium of Kell et al. (1981). For long-term storage, *Caloramator viterbiensis* is grown to mid- to late-exponential growth phase in a modified medium of Kell et al. (1981) containing 2% yeast extract and 3 g/l glycerol; 10% glycerol (v/v, final concentration) and 10% DMSO (v/v, final concentration) added, the culture kept for 15 min at room temperature before being frozen and stored at –75 °C. Cultures were viable under these conditions for more than 28 months.

Differentiation of the genus *Caloramator* from other genera

All six species of the genus *Caloramator* are thermophilic, carbohydrate fermenting, strict anaerobes, and form a cohesive and independent cluster within the radiation of members of the family *Clostridiaceae*, in the order *Clostridiales*, class *Clostridia*, phylum *Firmicutes* (Figure 5). This unique phylogenetic placement differentiates them from the other phenotypically similar members of the genus *Clostridium* which include, for example, *Clostridium thermocellum*, *Clostridium cellulosi*, *Clostridium thermbutyricum*, *Clostridium thermopalmarium*, *Clostridium thermolacticum* and *Clostridium stercorarium*, and also from members of the genera *Thermoanaerobacter*, *Thermoanaerobium*, and *Thermoanaerobacterium*.

Caloramator fervidus, *Caloramator proteoclasticus*, *Caloramator coolhaasi*, and *Caloramator viterbiensis* use a number of amino acids, such as serine, glutamate, threonine, methionime, asparagine, arginine, histidine, leucine, and valine, as a carbon source. In addition, *Caloramator proteoclasticus* and *Caloramator coolhaasii* use glutamate in co-culture with hydrogen-scavenging methanogens. These traits in common with a number of other phylogenetically distant fermentative mesophilic anaerobes which include *Thermanaerovibrio acidaminovorans* (Baena et al., 1999; Cheng et al., 1992) and *Aminobacterium* species (Baena et al., 2000, 1998). Amino acid utilization in *Caloramator indicus* and "*Caloramator celer*" is yet to be investigated and once determined will provide a clearer picture into whether this is a common trait in the genus.

Differentiation of the species of the genus *Caloramator*

Table 148 lists the features that differentiate the six species of the genus. A number of characteristics are useful in distinguishing the species from each other (Table 148). All members of the genus use carbohydrates, but not all use the same carbohydrates and this carbohydrate utilization pattern could be used to differentiate the six species. The use of amino acids by four of six *Caloramator* species could also be considered as characteristic of the species and the range used as a criterion for further differentiation. The data from the remaining two species, namely *Caloramator indicus* and "*Caloramator celer*" should be included once this is available. *Caloramator viterbiensis* but not "*Caloramator celer*" ferments glycerol to 1,3 propanediol, a property and could be used as a distinguishing and differentiating trait provided it has been tested for the remaining four members of the genus. The growth temperature optimum is also a useful criterion for species differentiation: *Caloramator proteoclasticus*, *Caloramator coolhaasii* and *Caloramator viterbiensis*, which grow optimally between 50 and 58 °C, can be considered as moderate thermophiles whereas *Caloramator fervidus*, *Caloramator indicus* and "*Caloramator celer*," which grow optimally at or above 60 °C, can be considered as thermophiles.

Taxonomic comments

The 16S RNA analysis presented in the roadmap to Volume 3 (Figure 5) places the six members of the genus *Caloramator* as a cohesive cluster in the family *Clostridiaceae*, order *Clostridiales*, class *Clostridia*, phylum *Firmicutes*. The closest relative is *Clostridium cylindrosporum* (DSM 605[T], Y18179) with a mean similarity value of 92% (Figure 131).

List of species of the genus *Caloramator*

1. **Caloramator fervidus** (Patel, Monk, Littleworth, Morgan and Daniel 1987) Collins, Lawson, Willems, Cordoba, Fernández-Garayzábal, Garcia, Cai, Hippe and Farrow 1994, 812[VP] (*Clostridium fervidus* Patel, Monk, Littleworth, Morgan and Daniel 1987, 125)
   fervidus. L. adj. *fervidus* hot.

   Sluggishly motile rod-shaped cells (2–2.5 μm by 0.65–0.75 μm) with rounded to slightly tapered ends. Cells occur singly and rarely in pairs on in chains and stain Gram-negative but thin sections reveal a single-layered Gram-positive cell-wall structure. Many cells contain Gram-stain-positive granules which appear as small spherical refractive bodies under a phase-contrast microscope. Spherical subterminal to terminal spores, which do not swell the sporangium are present. Obligate anaerobe. Yeast extract and/or Trypticase peptone can be used as a carbon and energy source and is required for the fermentation of glucose, maltose, mannose, xylose, xylan, starch and pyruvate. Vitamins or Casamino

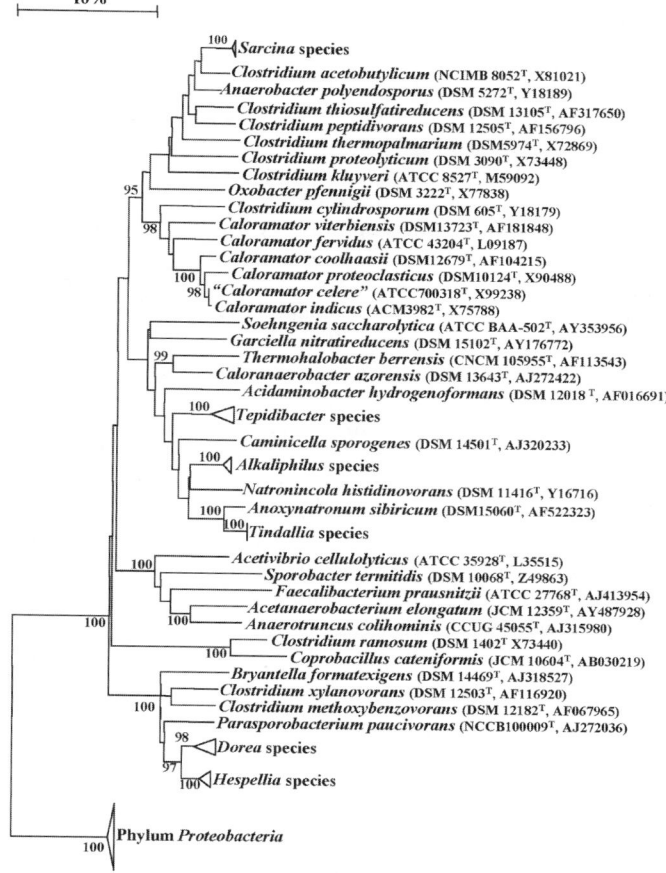

FIGURE 131. Phylogenetic position of *Caloramator* species amongst members of the family *Clostridiaceae*, order *Clostridiales*, class *Clostridia*, phylum *Firmicutes*. Sequences were aligned and manually adjusted according to the 16S rRNA secondary structure using BioEdit (Hall, 1999). Sequence uncertainties were omitted and phylogenetic reconstruction was achieve with 1233 unambiguous nucleotides using TreeCon (Van de Peer and De Wachter, 1994) in which pairwise evolutionary distances were computed from percentage similarities (Jukes and Cantor, 1969) and phylogenetic trees constructed from the evolutionary distances using the neighbor-joining method (Saitou and Nei, 1987). Tree topology was re-examined by the bootstrap method of resampling using 1000 bootstraps (Felsenstein, 1985). The strains used in the analysis, their culture collection numbers and corresponding 16S rRNA gene sequences extracted from GenBank are shown in parentheses. The clusters represented as triangles are for the following species: *Sarcina maxima* (DSM 316^T, X76650), *Sarcina ventriculi* (DSM 286^T, X76649), *Alkaliphilus transvaalensis* (ATCC 700919T^T, AB037677), *Alkaliphilus crotonatoxidans* (JCM 11672^T, AF467248), *Dorea longicatena* (JCM 11232^T, AJ132842), *Dorea formicigenerans* (ATCC 27755^T, L34619), *Hespellia porcina* (ATCC BAA-674^T, AF445239), *Hespellia stercorisuis* (ATCC BAA-677^T, AF445264), *Tepidibacter thalassicus* (DSM 15285^T, AY158079), *Tepidibacter formicigenes* (DSM 15518^T, AY245527), *Tindallia magadiensis* (DSM 10318^T, AY15626) and *Tindallia californiensis* (ATCC BAA-393^T, AF373919). Bootstrap values (≥95) from 100 resampling are shown. Bar indicates 10 nucleotide changes per 100 nucleotides. The following abbreviations have been used: T, Type culture; ATCC, American Type Culture Collection; DSM, Deutsche Sammlung von Mikroorganismen und Zellkulturen; JCM, Japan Collection of Microorganisms; CNCM, Collection Nationale de Cultures Microbiennes; NCCB, Netherlands Culture Collection of Bacteria; ACM, Australain Collection of Microorganisms; NCIMB, National Collections of Industrial Food and Marine Bacteria; and CCUG, Culture Collection of the University of Göteborg, Sweden.

acids cannot replace yeast extract or Trypticase peptone. Serine, but not carbohydrates and other amino acids, can be used as a sole carbon and energy source. The optimum growth temperature and pH are 68 °C and pH 7–7.5, respectively. Glucose fermentation end products in the presence of Trypticase peptone are acetate, ethanol, lactate, *i*-butyrate, *i*-valerate, *n*-valerate, H$_2$ and CO$_2$, Growth is inhibited by tetracycline, penicillium, chloramphenicol, and streptomycin. Isolated from a New Zealand geothermal spring (70 °C, pH 7.2).

DNA G+C content (mol%): 39 (T_m).

Type strain: Rt4-B1, ATCC 43204, DSM 5463.

GenBank accession number (16S rRNA gene): L09187.

2. **Caloramator celer** comb. nov. (basonym *Thermobrachium celere* Engle, Li, Rainey, DeBlois, Mai, Reichert, Mayer, Messner and Wiegel 1996).

ce′ler. L. masc. adj. *celer* fast, referring to the unusual fast growth of the bacterium.

The following description is from Engle et al. (1996) and unpublished results from the laboratory of Juergen Wiegel. Obligate anaerobe which can survive exposure to oxygen for more than 24h but does not grow under microaerophilic or aerobic conditions. Colonies are white and circular. Rod-shaped cells, which measure 1.5–14μm by 0.5–1.2μm. Between 1 to 10% of the cells show true branching but this diminishes to 0.2% with extended periods of subculturing. Cells in the late exponential growth phase are in chains, cells tend to form filaments and L-forms are commonly observed. Cells are peritrichously flagellated and are sluggishly motile. The cells stain Gram-positive and possess a typical Gram-positive cell-wall ultrastructure consisting of an S-layer. Spores have not been observed but sporulation specific genes have been demonstrated. Chemo-organotrophic and proteolytic but not cellulolytic. 0.05% yeast extract is required for visible growth and yeast extract between 0.1–2% can serve as a sole carbon and energy source. Glucose, sucrose, fructose, galactose, and maltose, but not cellobiose, ribose, mannose, arabinose, xylose, glucuronic acid, xylan, Casamino acids, and pyruvate, are utilized. The fermentation products from glucose in the presence of yeast extract are acetate, formate, ethanol, CO$_2$, and H$_2$. The optimum growth temperature is 66 °C (temperature growth range is 43–75 °C) and the optimum pH is 8.2 (pH range growth range is pH 5.4–9.5). A doubling time of 10 min is obtained in a medium (250 ml) containing 0.5% glucose ad 2% yeast extract under pH controlled batch fermentation.

DNA G+C content (mol%): 30–31 (HPLC).

Type strain: JW/YL-NZ35^T, DSM 8683^T, ATCC 700318^T.

GenBank accession number (16S rRNA gene): X99238.

3. **Caloramator coolhaasii** Plugge, Zoetendal and Stams 2000, 1161^{VP}

cool.haas′i.i. N.L. gen. *coolhaasii* named after Caspar Coolhaas, a Dutch microbiologist, who was the first to describe thermophilic protein degradation under methanogenic conditions.

Rods to filamentous cells (2–40μm × 0.5–0.7μm) are nonmotile, do not form spores and stain Gram-negative but have a Gram-positive cell-wall ultrastructure. Moderately thermophilic. Temperature range for growth is 37–65 °C (optimum 50–55 °C). and pH growth range is pH 6.0–8.5

(optimum pH 7.0–7.5). Strictly anaerobic chemoorgano-heterotroph. Utilizes glutamate, aspartate, alanine, arginine, methionine, glucose, fructose, galactose, sucrose, maltose, ribose, xylose, starch, cellobiose and mannose, Casamino acids, yeast extract, gelatin, casitone, peptone, pyruvate, and WC broth for growth. Growth on glutamate, aspartate, alanine, arginine, methionine, mannose, ribose and pyruvate is enhanced in the presence of the methanogen *Methanobacterium thermoautrotrophicum* Z-245^T. Acetate, lactate, CO_2 and H_2 are produced from glucose fermentation. Yeast extract stimulates growth on glutamate, but is not required. No Stickland reaction could be observed. Sulfide-reduced media was required for growth. No reduction of oxygen, nitrate, sulfate, thiosulfate or fumarate.

Habitat: anaerobic thermophilic methanogenic sludge.

DNA G+C content (mol%): 31.7 ± 0.4 (HPLC).

Type strain: Z^T = DSM 12679^T.

GenBank accession number (16S rRNA gene): AF104215.

4. **Caloramator indicus** Chrisostomos, Patel, Dwivedi and Denman 1996, 500^{VP}

in′di.cus. N.L. adj. *indicus* of India, the country from which the organism was isolated.

Nonmotile rod to filamentous cells (10–100 × 0.6–0.8 μm) which stain Gram-negative but possess a Gram-positive cell-wall ultrastructure. Spores have not been found. Chemoorganotrophic and obligately anaerobic. The optimum temperature for growth is between 60–65 °C and no growth occurs at temperatures below 37 °C or above 75 °C. The optimum pH for growth is between 7.5 and 8.1, and no growth occurs at pH values below 6.2 or above 9.2. Utilizes glucose, amylopectin, starch, amylose, dextrin, cellobiose, fructose, lactose, mannose and sucrose but not cellulose, dextran and chitin. Trypticase peptone and/or yeast extract is required for growth on carbohydrates though either of these can be used as a sole carbon and energy source in the absence of carbohydrates. Produces ethanol, acetate, lactate, CO_2 and H_2 from glucose fermentation. Growth is inhibited by tetracycline, penicillium G, chloramphenicol, novobiocin, polymxin B and rifampin.

DNA G+C content (mol%): 25.6 ± 0.3 (T_m).

Type strain: IndiB4, ACM 3982.

GenBank accession number (16S rRNA gene): X75788.

5. **Caloramator proteoclasticus** Tarlera, Muxí, Soubes and Stams 1997, 655^{VP}

pro.te.o.clas′ti.cus. Gr. adj. *protos* first; Gr. adj. *clasticus* breaking, from Gr. part. perf. *klastos* broken; N.L. masc. adj. *proteoclasticus* protein-breaking.

Strictly anaerobic straight or slightly curved rods (2.4–4.0 μm×0.4 μm) with pointed ends which occur singly or in pairs. Surface colonies (4 mm) are golden, circular, smooth, opaque and bluish around the edges. Cells stain Gram-negative and possess an atypical two-layered Gram-positive cell-wall ultrastructure. Tumbling motility by means of peritrichous flagella. Spores are produced in a yeast extract and gelatin medium in the late exponential growth phase.

Cell lysis occurs in carbohydrate grown stationary phase cultures. Moderately thermophilic, with an optimum growth temperature of 55 °C (temperature growth range is between 30 and 68 °C). The pH growth optimum is pH 7.0–7.5 (pH range for growth is 6.0–9.5). Chemo-organotrophic, requiring yeast extract for fermentation of complex proteins (casein, gelatin, peptone, Casamino acids), amino acids (glutamate, arginine, histidine, threonine, methionine, glycine, leucine, valine, aspartate), carbohydrates (glucose, mannose, fructose, cellobiose, starch) and pyruvate. Casein, gelatin, peptone, Casmino Acids are fermented to acetate, propionate, butyrate, branch-chain fatty acids, formate and H_2; glutamate, arginine and histidine to acetate, formate and H_2; threonine and methionine to propionate; glutamate to alanine, formate and branched-chain amino acids; alanine and glycine to branch-chain fatty acids and acetate respectively. A complete Strickland reaction is observed with leucine and glycine. The end products from carbohydrate and pyruvate fermentation include acetate, lactate, ethanol, formate and H_2. Substrates such as ornithine, proline, acetate, H_2-CO_2 (80:20), formate, 2-oxoglutarate, lactose, xylose, malate, xylan, and cellulose are not utilized. Sulfate, thiosulfate, nitrate, and fumarate are not reduced. Isolated from a mesophilic whey-fermenting upflow anerobic sludge blanket reactor.

DNA G+C content (mol%): 31 (HPLC).

Type strain: U^T, DSM 10124^T.

GenBank accession number (16S rRNA gene): X90488.

6. **Caloramator viterbiensis** Seyfried, Lyon, Rainey and Wiegel 2002, 1183^{VP}

vi.ter.ben′sis. L. masc. adj. *viterbiensis* pertaining to Viterbo in Italy (ancient name Viterbium), the region from which the strain was isolated.

Non-sporulating and nonmotile rods occur singly, are straight to slightly curved (0.4–0.6×2.0–3.0 μm) and stain Gram-positive. Ultrathin sections also reveal a Gram-positive cell wall structure. The temperature range for growth is 33–64 °C with an optimum of 58 °C. The pH range for growth is pH 5.0–7.8, with an optimum pH at 6.0–6.5. Growth is observed with glycerol, glucose, fructose, mannose, galactose, sucrose, cellobiose, lactose, starch and yeast extract. Xylose, arabinose, acetate, lactate, formate, methanol, ethanol, n-propanol, iso-propanol, n-butanol, propionate, acetone, succinate, ethylene glycol, 1,2-propanediol, phenol and benzoate are not fermented. Various single amino acids (0.5%, w/v) which include serine, glutamine, threonine, leucine, methionine, aspartate, valine and histidine but not arginine, are utilized. No growth occurs under autotrophic conditions in the presence of H_2/CO_2. Fermentation of glycerol yields acetate, 1,3-propanediol and H_2 whereas glucose fermentation yields ethanol, acetate, CO_2 and H_2. Ampicillin, chloramphenicol, rifampin, kanamycin and erythromycin inhibits growth.

DNA G+C content (mol%): 32 (HPLC).

Type strain: JW/MS-VS5^T= DSM 13723^T, ATCC PTA 584^T.

GenBank accession number (16S rRNA gene): AF181848.

Genus VI. **Caloranaerobacter** Wery, Moricet, Cueff, Jean, Pignet, Lesongeur, Cambon-Bonavita and Barbier 2001, 1795^{VP}

GEORGES BARBIER AND NATHALIE WERY

Ca.lor.an.ae.ro.bac′ter. L. n. *calor* heat; Gr. pref. *an* not; Gr. n. *aer* air; N.L. *bacter* masc. equiv. of Gr. neut. dim. n. *bakterion* small rod or staff; N.L. masc. n. *Caloranaerobacter* a thermophilic, anaerobic rod

Motile rod-shaped cells. Under optimal conditions are short rods, about $0.3–0.5 \times 0.5–2\,\mu m$, found singly, in pairs, or in short chains (less than 5 cells). **Gram-stain-negative. Thermophilic**; temperature range for growth 45–65 °C, optimum growth at 65 °C. **Adapted to the pH and salinity of ocean waters.** pH range 5.5–9.0, optimum pH 7.0. Salinity range 10–100 g sea salts per liter, optimum 30 g sea salts per liter. **Anaerobic, chemo-organotrophic**, able to ferment carbohydrates and proteinaceous substrates. Sulfur not required for growth. On the basis of 16S rRNA gene sequence analysis presented in the roadmap to Volume 3 (Figure 5), *Caloranaerobacter* is located in the family *Clostridiaceae*, order *Clostridiales*, class *Clostridia*, in the phylum *Firmicutes*; formerly, it was classified in cluster XII of the *Clostridium* subphylum (Collins et al., 1994).

Habitat: minerals from a deep-sea hydrothermal diffuser supporting mussels, mid-Atlantic Ridge, site Lucky Strike (32°16′W, 37°17′N, 1650 m depth).

DNA G+C content (mol%): 27 (T_m).

Type species: **Caloranaerobacter azorensis** Wery, Moricet, Cueff, Jean, Pignet, Lesongeur, Cambon-Bonavita and Barbier 2001, 1795^{VP}.

Further descriptive information

Cells are motile round-ended rods, and flagella were observed by negative staining (Figure 132). *Caloranaerobacter* cells stain Gram-negative and ultrathin sections indicate a typical cell envelope with cytoplasmic and outer membranes. Cells more than 20 µm long can sometimes be observed. No spores, but cells able to survive after 20 min at 100 °C. When stored for a few days at 4 °C, inoculated in poor media, or grown at high pH or at high hydrostatic pressure (65 MPa) some terminal, round, external bodies are produced. Under optimal conditions, growth is very fast, generation time is about 15 min and such a culture produces a maximum cell yield of 2×10^9 cells per ml.

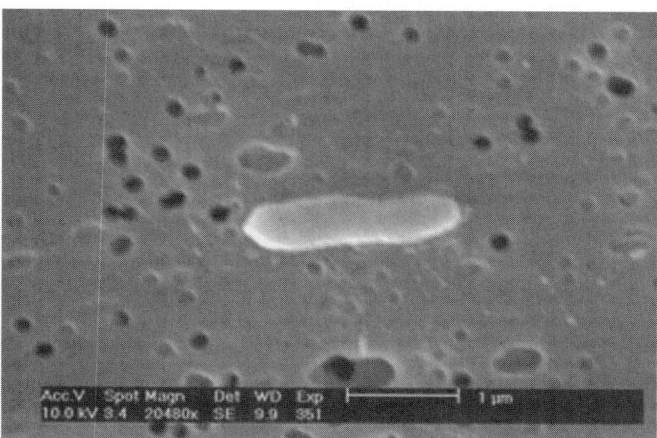

FIGURE 132. Scanning electron micrograph showing a typical cell of *Caloranaerobacter azorensis* with flagella.

Caloranaerobacter is barotolerant; hydrostatic pressure has no significant effect on growth until 40 MPa. At higher pressures growth efficiency decreases; growth stops at 65 MPa.

Anaerobic heterotrophic growth is possible with proteinaceous substrates (gluten, brain heart infusion) and with pyruvate or carbohydrates (starch, glucose, xylan, xylose, arabinose, ribose, galactose, fructose, or sorbose) in the presence of yeast extract and tryptone. Growth was not observed with sucrose, lactose, maltose, cellobiose, cellulose, and other organic acids or alcohols. Sulfur, thiosulfate, nitrate and ferric iron do not enhance growth. No growth in presence of nitrite or sulfite. Significant production of H_2S in presence of sulfur.

Caloranaerobacter was isolated from a deep-sea hydrothermal diffuser supporting mussels that was sampled at mid-Atlantic Ridge, site Lucky Strike. 16S rRNA gene sequences available in databases indicate that the closest studied micro-organism (GenBank accession no. L10086, 98.5% similarity) is an unpublished isolate obtained from the gut content of a polychaete worm sampled at Juan de Fuca Ridge, North East Pacific (J. A. Baross, personal communication). This isolate most likely belongs to the same genus suggesting a wide oceanic distribution.

Enrichment and isolation procedures

Caloranaerobacter was enriched at 65 °C, pH 7.5, at atmospheric pressure, under an anoxic gas phase $N_2/H_2/CO_2$ (90:5:5). YPXS medium (yeast extract, 0.5 g/l; peptone, 1 g/l; oat spelt xylan, 5 g/l; sulfur, 10 g/l; sea salts, 30 g/l; PIPES buffer, 6.05 g/l; resazurin, 1 mg/l) was used for enrichment and isolation and then efficiently replaced by YTG medium (yeast extract, 1 g/l; tryptone, 1 g/l; glucose, 2.5 g/l; sea salts, 30 g/l; PIPES buffer, 6.05 g/l; resazurin, 1 mg/l). Vent minerals were crushed in sterile sea water in anaerobic conditions; this mixture was inoculated. Motile rods were observed after several days of incubation.

Maintenance procedures

Cultures of *Caloranaerobacter* can be maintained for short periods (several months) at 4 °C. Long-term storage of *Caloranaerobacter* is possible in a freezer (−70 °C) and liquid nitrogen after addition of DMSO 5% (v/v) to exponentially growing cultures in anaerobic conditions.

Differentiation of the genus *Caloranaerobacter* from other genera

Comparisons of 16S rRNA gene sequences place *Caloranaerobacter* close to *Thermohalobacter berrensis* (Cayol et al., 2000b), a moderate halophile isolated from a solar saltern, and *Clostridium acidurici* (Barker, 1938; Liebert, 1909), *Clostridium purinilyticum* (Dürre et al., 1981), and *Eubacterium angustum* (Beuscher and Andreesen, 1984), three terrestrial species. 16S rDNA similarity values with these four species are less than 93%. Phenotypic characteristics also differentiate those genera. *Caloranaerobacter* differs from the three terrestrial species because it stains Gram-negative, because of its thermophily and adaption to marine salinity, and because it is metabolically more diverse. *Caloranaerobacter*, which has a DNA G+C content 6 mol% lower, differs from *Thermohalobacter berrensis* because it is less halophilic and it does not use the same carbohydrates.

Acknowledgements

We would like to thank JAMSTEC (Japanese Marine Science and Technology Center), specially Dr C. Kato and Professor K. Horikoshi, for cultures in the Deep Bath System (effect of hydrostatic pressure on growth).

List of species of the genus *Caloranaerobacter*

1. **Caloranaerobacter azorensis** Wery, Moricet, Cueff, Jean, Pignet, Lesongeur, Cambon-Bonavita and Barbier 2001, 1795[VP]

   a.zo.ren'sis. N.L. masc. adj. *azorensis* of the Azores.

   Only one species has been described for this genus; the species description is the same as that of the genus.

DNA G+C content (mol%): 27 (T_m).
Type strain: MV1087, CNCM I-2543, DSM 13643.
GenBank accession number (16S rRNA gene): AJ272422.

Genus VII. **Caminicella** Alain, Pignet, Zbinden, Quillevere, Duchiron, Donval, Lesongeur, Raguenes, Crassous, Querellou and Cambon-Bonavita 2002, 1627[VP]

FRED A. RAINEY

Ca.mi'ni.cel'la. L. gen. n. *camini* of a chimney, relating to the hydrothermal chimney origin; L. fem. n. *cella* cell; N.L. fem. n. *caminicella* cell from a hydrothermal chimney.

Gram-stain-negative. Motile, sporeforming, rod-shaped cells. Thermophilic. Adapted to the pH and salinity of the ocean. Anaerobic and heterotrophic. Ferments proteinaceous substrates and carbohydrates. Member of the family *Clostridiaceae* based on 16S rRNA gene sequence comparisons (Figure 5).

DNA G+C content (mol%): 24.2.

Type species: **Caminicella sporogenes** Alain, Pignet, Zbinden, Quillevere, Duchiron, Donval, Lesongeur, Raguenes, Crassous, Querellou and Cambon-Bonavita 2002, 1627[VP].

Taxonomic comment

The genus *Caminicella* was described on the basis of a single strain that was phenotypically and phylogenetically distinct from related taxa. The 16S rRNA gene sequence of *Caminicella sporogenes* shows highest similarity to the species *Clostridium caminithermale* (93.0%), and *Clostridium halophilum* (92.7%). The phylogenetic relationships are shown in Figure 133. *Caminicella sporogenes* represents a distinct lineage that clusters with *Clostridium caminithermale* and *Clostridium halophilum*.

List of species of the genus *Caminicella*

1. **Caminicella sporogenes** Alain, Pignet, Zbinden, Quillevere, Duchiron, Donval, Lesongeur, Raguenes, Crassous, Querellou and Cambon-Bonavita 2002, 1627[VP]

   spo.rog.en'es. Gr. n. *spora* a spore; Gr. v. *gennaio* produce; N.L. part. adj. *sporogenes* spore-producing.

   Cells are rod-shaped and motile by means of peritrichous flagella. Exhibit a Gram-stain-negative cell-wall ultrastructure. Cells are up to 3.0–10 μm in length and 0.5–0.7 μm in width. Sporulation is observed in the late stationary phase of growth. The temperature range for growth is 45–65 °C; optimum at 55–60 °C. Growth occurs in the pH range 4.5–8.0; optimum at pH 7.5–8.0. Growth occurs at sea salt concentrations in the range 20–60 g/l; optimum at 25–30 g/l. The maximum doubling time is approximately 45 min; the maximum cell yield is

8×10^8 cells/ml. Anaerobic. Ferments yeast extract, brain heart infusion, D(+)-glucose, maltose. Weak growth on peptone, galactose, and a mixture of 20 amino acids. The fermentation products on GYPS medium (Alain et al., 2002) are H_2, CO_2, butyrate, ethanol, acetate, formate, and L-alanine. The 16S rRNA gene sequence similarities to *Clostridium halophilum* and *Clostridium caminithermale* are 92.7% and 93.1%, respectively.

The type strain was from young entire *Alvinella pompejana* tubes attached to small fragments of chimney rocks that were collected from the hydrothermal site Elsa (HOT3) in the East-Pacific Rise (103°56′326 W, 12°48′200 N).

DNA G+C content (mol%) of the type strain: 24.2 (T_m).
Type strain: AM1114, CIP 107141, DSM 14501.
GenBank accession number (16S rRNA gene): AJ320233.

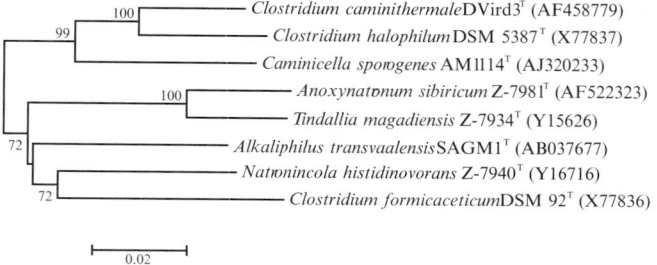

FIGURE 133. *Caminicella* 16S rRNA phylogenetic tree.

Genus VIII. **Natronincola** corrig. Zhilina, Detkova, Rainey, Osipov, Lysenko, Kostrikina and Zavarzin 1999, 1[VP] (Effective publication: Zhilina, Detkova, Rainey, Osipov, Lysenko, Kostrikina and Zavarzin 1998, 183.)

TATJANA N. ZHILINA AND GEORGE A. ZAVARZIN

Nat.ron'in.co.la. N.Gr. n. *natron* (arbitrarily derived from Ar. n. *natrun* or *natron*) soda; L. masc. or fem. n. *incola* inhabitant, dweller; N.L. masc. n. *Natronincola* an organism indigenous to soda deposits.

Cells are rod-shaped, flexible, and motile by peritrichous flagella. The cell wall has a **Gram-stain-positive** structure. **Obligately anaerobic. Chemo-organotrophic** with fermentative metabolism. **Certain amino acids** (and pyruvate by some strains) are **utilized** as a sole source of energy, **using CO-dehydrogenase pathway, with acetate as a main product.** Ammonium is formed in large quantities as the product. **Moderately halophilic and extremely alkaliphilic. Obligately dependent on Na$^+$ and CO$_3^{2-}$ ions. Endospores** may be **produced** by **some strains.**

DNA G+C content (mol%): 31.9–32.3 (T_m).

Type species: **Natronincola histidinovorans** corrig. Zhilina, Detkova, Rainey, Osipov, Lysenko, Kostrikina and Zavarzin 1999, 1[VP] (Effective publication: Zhilina, Detkova, Rainey, Osipov, Lysenko, Kostrikina and Zavarzin 1998, 183.).

Further descriptive information

Phylogenetic position. Currently, a single species is classified within the genus, namely *Natronincola histidinovorans*. Analysis of the 16S rRNA gene sequence of the type species demonstrated that genus *Natronincola* (Zhilina et al., 1998) falls within the radiation of the clostridia *Clostridium felsineum* (Hippe et al., 1992), *Clostridium formicoaceticum* (Andreesen et al., 1970), and *Tindallia magadiensis* (Kevbrin et al., 1998) and belonging to "clostridial group" cluster XI of the low G+C Gram-positive bacteria as defined by Collins et al. (1994). In the phylogenetic analysis presented in the roadmap to this volume (Figure 5), *Natronincola* is placed in family *Clostridiaceae*, order *Clostridiales*, class *Clostridia*, phylum *Firmicutes*.

Cell morphology and fine structure. Cells are flexible motile rods 0.7–1 × 2–6 μm, differing in length, and sometimes forming long filaments. Motile by peritrichous flagella. Multiplication is by fission, often unequal (Figure 134). Formation of terminal round minicells is characteristic; sometimes these minicells retain motility. Side blebs of minicell dimensions are observed (Figure 134). Asporogenous or oligosporic; spores, if present, are round and terminal. Not thermoresistant, but resistant to desiccation. The cell wall is of Gram-stain-positive structure and has no outer membrane.

Growth conditions and nutrition. Truly alkaliphilic and moderately halophilic. Obligately dependent on sodium and carbonate ions. In the high-alkalinity medium 1 (Zhilina and Zavarzin, 1994) with the composition (g/l) NaHCO$_3$:Na$_2$CO$_3$:NaCl = 38.3:68.3:15.7, the pH optimum for growth is pH 9.4 within the range pH 8.0–10.5. Growth occurs at salinities not less than 4% NaCl but not at salinities greater than 16% with a growth optimum around 8–10%. Mesophilic: optimal growth at 37–40 °C, no growth above 45 °C.

Strictly anaerobic. The type strain Z-7940 ferments only histidine or glutamate with acetate as a main end product of fermentation. Ammonium is formed in large quantities. With histidine as the substrate, a minor production of formate was recorded. H$_2$ production was not found. Per 1 mmol of metabolized histidine, 2.34 mmol acetate, 0.27 mmol formate, and 3.15 mmol ammonium are produced. The K_m was 2.13 mM. The optimal concentration of histidine for growth ranges from 4.77–11.93 mM; at higher concentrations growth is inhibited. Moderate growth is observed with media containing peptone, yeast extract, and Casamino acids. In addition, strain Z-7939 is able to utilize pyruvate. Both strains require yeast extract for anabolism. Carbohydrates, other amino acids, alcohols, and organic acids are not utilized.

Metabolic properties. Natronincola histidinovorans uses the noncyclic acetyl-CoA pathway for CO$_2$ fixation with acetate as the main product. The key enzyme is CO-dehydrogenase with optimal pH 9.5. Growth and metabolism are completely inhibited by N,N′-dicyclohexylcarbodiimide (DCCD) indicating a chemiosmotic type of ATP formation by insensitive-to-vanadate F$_1$F$_2$-type ATPase. Monensin (14 μM) completely stops acetate formation but, in contrast to other haloalkliphiles, Natroniella cell suspensions do not undergo lysis at the action of energy uncouplers as 3,5-di-tert-butyl-4-hydroxbenzylidenyl malonitrile (FCCP) or carbonyl cyanide m-chlorphenylhydrazone, 100 μM (CCCP) indicating energy-independent osmotic stability (Pusheva et al., 1999).

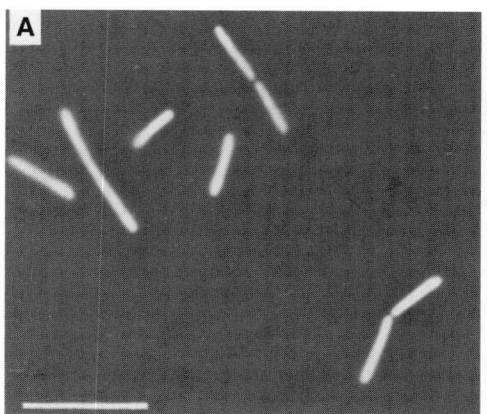

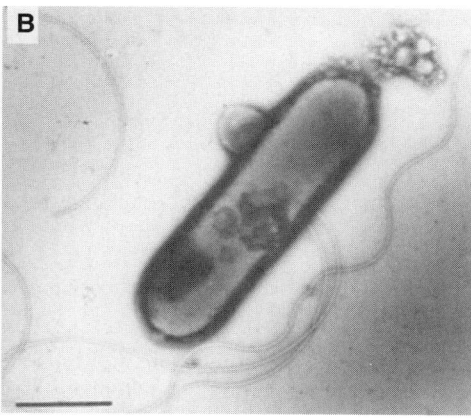

FIGURE 134. Morphology of *Natronincola histidinovorans*. (A) anoptral microscopy (bar = 10 μm); (B) negatively stained 2% phoshotungstic acid cell with flagella (bar 1 μm). Blebs on the left side of the cell.

Ecology

Two strains of *Natronincola histidinovorans*, asporogenous Z-7940[T] and sporogenous Z-7939, were isolated from flowing masses of cyanobacterial bloom in soda-depositing Lake Magadi, Kenya after the rainy season. These bacteria grew rather poorly on peptone but proliferated in media containing certain amino acids and a low concentration of yeast extract for anabolic needs. They are the representatives of the alkaliphilic, aceto-genic, ammonifiers of the proteolytic pathway in the anaerobic community in the soda lakes (Zavarzin and Zhilina, 2000; Zavarzin et al., 1999). They might be involved in decomposition of osmoprotective compounds such as glutamate.

Enrichment and isolation procedures

The type strain Z-7940 was a component of an anaerobic enrichment with H_2 as the substrate in selective carbonate medium 1 with 0.2 g/l yeast extract (Zhilina and Zavarzin, 1994) where acetate was formed as the main end product. However, after isolation of the type strain Z-7940 in pure culture, it was found to be incapable of chemolithotrophic growth with H_2 or CO and was quite restricted in its metabolic capacities to histidine and glutamate and to slow growth with peptone. In addition, strain Z-7939 was isolated by a selective procedure on histidine-containing medium from a desiccated sample that had been stored for 3 years in a refrigerator. This strain showed 94% DNA–DNA homology with the type strain. However, the two strains differed in that strain Z-7939 had round terminal spores in some cells (representing an oligosporous type), utilized pyruvate as an additional substrate, and differed in the fatty acid profile. This demonstrated the use of a selective procedure with histidine as the substrate in alkaline, strictly anaerobic sodium carbonate medium at pH 9–9.5 and using N_2 as gas phase.

Maintenance procedure

A pure culture may be stored in the laboratory with regular transfers at weekly intervals and/or in early exponential phase in a refrigerator for 1–2 months. Old cultures are nonviable, making maintenance a problem. For long-term preservation, storage in liquid nitrogen is recommended.

Differentation of the genus *Natronincola* from other genera

Natronincola histidinivorans can be phylogenetically differentiated from other closely related genera and species of Collins et al. (1994) cluster XI of the low G+C Gram-positive bacteria by 16S rRNA gene sequence comparisons. The closest relatives were found to be *Clostridium felsineum* (92.1% similarity), *Clostridium formicoaceticum* (91.5% similarity), and *Tindallia magadiensis* (90% similarity), but it is phylogenetically unrelated to the type species of the genera *Clostridium* and *Tindallia*. *Natronincola histidinivorans* obligately depends on sodium and carbonate ions. This differentiates it from *Clostridium felsineum* and *Clostridium formicoaceticum*. In contrast to these organisms, *Natronincola histidinivorans* cannot utilize carbohydrates. It can be differentiated from the alkaliphilic *Tindallia magadiensis* on the basis of morphology, lipid profile, and inability to utilize amino acids of the ornithine cycle.

Chemotaxonomically, *Natronincola* Z-7940[T] is differentiated from *Tindallia* based on its fatty acid profile with major membrane fatty acids $C_{16:1\,\omega7}$, $C_{16:0}$, $C_{14:0}$, and $C_{18:0}$ which contribute about 65% of total acids. The major (about 25%) palmitoleic aldehyde was C_{16} with $C_{16:1\,\omega7}$, $C_{18:1\,\omega9}$, and C_{18} as a minor component. Strain Z-7939 contains plasmalogen in the membrane together with palmitine (20%) and palmitoleinic aldehyde (3%) and an unusual combination of aldehydes with branched acids ($C_{15\,iso}$, 11.4%). This combination is not found among previously described bacteria. The lipid profiles alone indicate that strains of *Natronincola* represented a new taxon.

Taxonomic comments

Taxonomically, *Natronincola* (the original spelling, *Natronoincola* (sic), was corrected on validation according to Rule 61, see footnote to Validation List no. 68. Int. J. Syst. Bacteriol. 1999, *49*:1–3) groups within the *Bacteria* as a Gram-positive organism with a low G+C value belonging to the *Firmicutes* and the order *Clostridiales*. It represents a distinct lineage of generic status in the family *Clostridiaceae*.

List of species of the genus *Natronincola*

1. **Natronincola histidinovorans** corrig. Zhilina, Detkova, Rainey, Osipov, Lysenko, Kostrikina and Zavarzin 1999, 1[VP] (Effective publication: Zhilina, Detkova, Rainey, Osipov, Lysenko, Kostrikina and Zavarzin 1998, 183.)

   his.ti.di.no.vo′rans. N.L. n. *histidinum* histidine, L. v. *vorare* to devour; N.L. part. adj. *histidinivorans* utilizing histidine as the main substrate

   The description of this species is the same as that of the genus.

   Habitat: Found in flowing masses of cyanobacterial bloom in soda-deposing Lake Magadi, Kenya.
   DNA G+C content (mol%): 31.9 (T_m).
   Type strain: Z-7940, DSM 11416.
   GenBank accession number (16S rRNA gene): Y16716.
   Additional remarks: The DNA G+C content is 32.3 mol% for strain Z-7939.

Genus IX. **Oxobacter** Collins, Lawson, Willems, Cordoba, Fernandez-Garayzabal, Garcia, Cai, Hippe and Farrow 1994, 822[VP]

Fred A. Rainey

Ok.so.bac′ter. Gr. n. *oxos* vinegar; Gr neut. dim. n. *bakterion* small rod; N.L. masc. n. *bacter* small rod; N.L. masc. n. *Oxobacter* acetogenic rod.

Gram-stain-positive rod-shaped cells. **Spores** are oval, subterminal to terminal. **Obligate anaerobe.** End products from pyruvate catabolism are acetate and CO_2; from catabolism of CO are acetate and butyrate; from catabolism of methoxybenzenoids are butyrate and hydroxybenzenoids. Sugars, amino acids, organic acids and alcohol are not utilized as energy sources.

DNA G+C content (mol%): 38.0 (T_m).

Type species: **Oxobacter pfennigii** (Krumholz and Bryant 1985) Collins, Lawson, Willems, Cordoba, Fernandez-Garayzabal, Garcia, Cai, Hippe and Farrow 1994, 822^{VP} (*Clostridium pfennigii* Krumholz and Bryant 1985, 455).

Taxonomic comments

Oxobacter pfennigii was originally described as *Clostridium pfennigii* on the basis of its ability to produce heat-resistant spores (Krumholz and Bryant, 1985). The determination of the 16S rRNA gene sequence of the type strain of *Clostridium pfennigii* and comparison to a large number of taxa in the low G+C Gram-positive phylum (*Firmicutes*) indicated a novel phylogenetic position of this organism (Collins et al., 1994). *Clostridium pfennigii* fell at the base of, but distinct from, cluster I defined as the group of *Clostridium* species related to the

type species of the genus *Clostridium*, *Clostridium butyricum*, and representing the genus *Clostridium sensu stricto* (Collins et al., 1994). The 16S rRNA gene sequence of *Clostridium pfennigii* shows less than 89% sequence similarity to its nearest neighbors, the *Clostridium* species of cluster I and species of the genus *Caloramator*. This distinct phylogenetic position, unique metabolic characteristics, and lack of relationship to species of the genus *Clostridium sensu stricto* justified the description of *Clostridium pfennigii* as *Oxobacter pfennigii* comb. nov. (Collins et al., 1994). No additional species have been added to this genus and a search of the current GenBank database of 16S rRNA gene sequences shows no cultured strains closely related to this taxon. In the 16S rRNA phylogenetic analysis presented in the roadmap to the current volume, *Oxobacter* is in the family *Clostridiaceae*, order *Clostridiales*, and class *Clostridia* in the phylum *Firmicutes* (Figure 5).

List of species of the genus *Oxobacter*

1. **Oxobacter pfennigii** (Krumholz and Bryant 1985) Collins, Lawson, Willems, Cordoba, Fernandez-Garayzabal, Garcia, Cai, Hippe and Farrow 1994, 822^{VP} (*Clostridium pfennigii* Krumholz and Bryant 1985, 455)

pfen.nig' i.i. N.L. gen. n. *pfennigii* of Pfennig, named after Norbert Pfennig, who first documented the catabolism of methyl groups of benzenoid compounds by an anaerobic bacterium.

Gram-stain-positive, motile, slightly curved rods, with slightly tapered ends, 0.4 μm in diameter and 1.6–3.5 μm in length. Oval, subterminal to terminal endospores formed. Spores are resistant to boiling for 3 mins, slightly swell the sporangium, and are 0.4 μm in diameter and 0.4–0.8 μm in length. Colonies in roll tubes are smooth, convex, entire, opaque and 2–3 mm in diameter after 1 week of incubation in media containing 5 mM vanillate. The optimum temperature for growth is between 36 and 38 °C in the range 29–39 °C. The pH range is 6.3–8.0, with an optimum at pH 7.3. Only methoxybenzenoids

(vanillate, vanillin, ferulate, and syringate), pyruvate, or CO supports growth. Butyrate and hydroxybenzenoids are produced from methoxybenzenoids. CO is catabolized to acetate, butyrate, and CO_2. Pyruvate is fermented to acetate and CO_2. Hydrogen, other organic acids, ethanol, or methanol is not produced. Cell yields from growth are increased when grown on ferulate. Rumen fluid and yeast extract enhance growth. Little growth without rumen fluid. Growth on syringate is not stimulated by the addition of Casitone, Casamino acids, 1,4-naphthoquinone, hemin, or a mixture of volatile fatty acids to media with or without rumen fluid. Growth is not stimulated by sulfate, thiosulfate, nitrate, or fumarate. Nitrate is not reduced. Sulfate does not allow growth on lactate or ethanol. Gelatin is not liquefied. Isolated from rumina of cattle.

DNA G+C content (mol%): 38.0 (T_m).

Type strain: strain V5-2, ATCC 43583, DSM 3222.

GenBank accession number (16S rRNA gene): X77838.

Genus X. **Sarcina** Goodsir 1842, 434^{AL}*

ERCOLE CANALE-PAROLA

Sar.ci′na. L. fem.n. *Sarcina* a package, bundle.

Nearly spherical cells, 1.8–3.0 μm in diameter, **occurring in packets** of eight or more. Some of the cells in cultures may be present singly, or as groups of fewer than eight cells. Generally the cells are flattened in the areas of contact with adjacent cells. **Division occurs in three perpendicular planes.** Spore formation by these organisms has been reported (Knöll, 1965; Knöll and R. Horschak, 1971). **Gram-stain-positive. Nonmotile. Chemo-organotrophic anaerobes, having an exclusively fermentative metabolism.** Relatively aerotolerant. **Carbohydrates are the fermentable substrates.** The main products of glucose fermentation are CO_2, H_2, acetic acid as well as ethanol for *Sarcina ventriculi* and butyric acid for *Sarcina maxima*. Not pigmented. Catalase-negative. The minimal growth requirements include numerous amino acids and few vitamins, in addition to a fermentable substrate and inorganic salts. Grow at pH values near 1 and up to pH 9.8.

DNA G+C content (mol%): 28–31.

Type species: **Sarcina ventriculi** Goodsir 1842, 437^{AL}.

Further descriptive information

Usually, packets of *Sarcina ventriculi* consist of a greater number of cells than those of *Sarcina maxima*. Many of the cells in large packets of *Sarcina ventriculi* (e.g., packets comprising approximately 64 or more cells) have flattened shapes and are irregularly arranged (Canale-Parola et al., 1961; Holt and Canale-Parola, 1967; Smit, 1930). Thus, large packets tend to have a distorted appearance. Holt and Canale-Parola (1967) isolated strains of *Sarcina ventriculi* that consistently form small packets. These strains are especially difficult to distinguish from *Sarcina maxima* by means of light microscopy.

Cells of *Sarcina ventriculi* are surrounded by a thick, fibrous layer of cellulose usually 150–200 nm in thickness (Canale-Parola et al., 1961; Canale-Parola and Wolfe, 1964). The cellulose layer

_____

*This chapter was largely reprinted from the First Edition.

TABLE 149. Differential characteristics of the species of the genus *Sarcina*[a]

Characteristic	1. *Sarcina ventriculi*	2. *Sarcina maxima*
Ethanol production	+	–
Cellulose formation	+	–
Butyrate production	–	+
Fermentation of D-xylose	–	+

[a]Symbols: +, >85% positive; –, 0–15% positive.

around each cell is either continuous with, or attached to, the cellulose layer surrounding adjacent cells in the same packet (Canale-Parola et al., 1961; Holt and Canale-Parola, 1967). Apparently the cellulose layer functions as a matrix or cementing material that holds together cells of *Sarcina ventriculi* into the large packets typical of this bacterium (Canale-Parola et al., 1961). In fact, strains of *Sarcina ventriculi* that produce little or no cellulose form packets consisting of relatively few cells, loosely bound to one another (Canale-Parola et al., 1961; Holt and Canale-Parola, 1967). In contrast, cells of strains that form large packets are surrounded by a thick, fibrous layer of cellulose, with the result that the packets are crisscrossed by an intercellular network of this polymer. It has been suggested that nutrients diffuse from the external environment into the interior regions of the packets via this network of cellulose fibers (Canale-Parola, 1970). In this manner nutrients may reach cells that have no direct contract with the growth medium because they are located in the interior regions of the packets (Canale-Parola, 1970).

The cellulose layer is absent from cells of *Sarcina maxima* (Canale-Parola, 1970).

The peptidoglycans of *Sarcina ventriculi* and *Sarcina maxima* contain LL-diaminopimelic acid. The interpeptide bridge consists of one glycine residue (Kandler et al., 1972). This type of peptidoglycan is not found in aerobic packet-forming cocci (Kandler et al., 1972).

Subsurface colonies of *Sarcina ventriculi* in agar media grow to several millimeters in diameter and are irregularly cubical or star-shaped. *Sarcina maxima* forms smaller subsurface colonies, which may be cuboid with protuberances or shaped as uneven spheres. On the surface of agar media, in an anaerobic atmosphere, both organisms form roundish colonies which frequently have jagged edges.

Knöll (1965) and Knöll and Horschak (1971) reported that both *Sarcina ventriculi* and *Sarcina maxima* form endospores. The procedure they used to induce sporulation involved incubation of growing cells in CO_2 atmosphere, followed by addition of phosphate buffer and alkali to raise the pH of the cultures rapidly to 7.5 for *Sarcina ventriculi* and to 9–10 for *Sarcina maxima.* Spore formation occurred during further incubation of the cultures in N_2. Spherical spores were formed by *Sarcina ventriculi*, oval spores by *Sarcina maxima.* The spores were heat-resistant and stained green with Wirtz's (1908) spore stain (Knöll and Horschak (1971) Knöll, personal communication).

The optimum temperature range for growth is 30–37 °C for *Sarcina ventriculi* and 30–35 °C for *Sarcina maxima.* Growth of both species occurs between pH 1 and 9.8 (Smit, 1930, 1933), an extraordinarily wide pH range. *Sarcina ventriculi* strains freshly isolated from natural environments can be subcultured indefinitely in media of very low pH, but when the strains are subcultured repeatedly at near neutral pH they tend to lose their ability to grow in media of pH 4.5 or lower (Canale-Parola, 1970).

TABLE 150. Characteristics of the species of the genus *Sarcina*[a,b]

Characteristics	1. *Sarcina ventriculi*	2. *Sarcina maxima*
Cell diameter (μm)	1.8–2.4	2–3
Packet information	+	+
Anaerobic	+	+
Pigmentation	–	–
Catalase	–	–
Growth at pH 2	+	+
Cellulose formation	+	–
DNA G+C content (Bd) (mol%)	30.6	28.6
Products from sugars:		
CO_2, H_2 and acetate	+	+
Ethanol	+	–
Butyrate	–	+
Fermentable substrates:[c]		
D-Arabinose	–	d
L-Arabinose	d	+
D-Ribose	–	–
D-Xylose	–	+
D-Fructose	+	+
D-Galactose	+	+
D-Glucose	+	+
D-Mannose	+	+
Cellobiose	d	–
Lactose	+	–[d]
Maltose	+	+
Melibiose	+	–
Sucrose	+	+

[a]Symbols: +, >85% positive; d, different strains give different reactions (16–84% positive); –, 0–15% positive.

[b]The following substrates are not fermented by either species: trehalose, raffinose, dextrin, starch, glycogen, glycerol, dulcitol, mannitol and amino acids. Citrate, gluconate and succinate are not fermented by *Sarcina ventriculi* (fermentation of these compounds by *Sarcina maxima* was not tested).

[c]Data from Smit, 1933; Canale-Parola and Wolfe (1960b); Claus and Wilmanns (1974).

[d]Smit (1933) reported that *Sarcina maxima* ferments lactose.

Sarcina ventriculi and *Sarcina maxima* require for growth a fermentable carbohydrate (Table 150), vitamins, amino acids, as well as inorganic salts. For example, *Sarcina ventriculi* strain EC-1 requires biotin, nicotinic acid and 11 amino acids (serine, histidine, isoleucine, leucine, tyrosine, methionine, tryptophan, phenylalanine, arginine, valine and glutamic acids) (Canale-Parola and Wolfe, 1960b). A chemically defined medium containing glucose, inorganic salts and the above-mentioned vitamins and amino acids supports abundant growth of the organism (Canale-Parola and Wolfe, 1960b). A strain of *Sarcina maxima* studied required thiamine, threonine, alanine and asparatic acid, in addition to the vitamins and amino acids required by *Sarcina ventriculi* EC-1 (Knöll and R. Horschak, 1964).

In complex media (e.g., 2 g each of glucose and yeast extract/100 ml distilled water (DW) *Sarcina ventriculi* EC-1 grew to cell yields of 8–12 g (wet wt)/liter (Canale-Parola et al., 1961). A complex medium for *Sarcina maxima* consists of (g/100 ml DW): glucose and peptone, 1.0 each; yeast extract, 0.5; L-cysteine, 0.05; and $FeSO_4 \cdot 7H_2O$, 0.005 (Kupfer and Canale-Parola, 1967). One liter of this medium yielded approximately 2.4 g (wet wt) of *Sarcina maxima* strain 11 cells. Cultivation procedures for both species of *Sarcina* have been described (Canale-Parola, 1970; Canale-Parola et al., 1961; Kupfer and Canale-Parola, 1967).

TABLE 151. Glucose fermentation by *Sarcina ventriculi* and *Sarcina maxima*[a,b]

Product[c]	S. ventriculi			S. maxima	
	1	2	3	4	5
Carbon dioxide	195	190	190	149	197
Hydrogen	41	170	140	230	223
Formic acid	3	Trace	NR	4	NP[e]
Acetic acid	20	90	60	30	40
Butyric acid	NR	NR	NP	76	77
Lactic acid	NR	NR	10	21	NP
Succinic acid	NR	NR	NR	5	NR
Ethanol	171	80	100	Trace	NP
Acetoin	4	NR	Trace	NR	NP
Neutral volatile (as butanol)	NR	NR	NR	NR	9
Percentage carbon recovered	99.3	88.3	90.0	100.0	103.5
Oxidation-reduction balance	1.0	1.15	1.12	0.80	0.95

[a]Table adapted from Canale-Parola (1970), with permission of the publisher.
[b]Symbols: NR, not reported; and NP, not present in detecTable amounts.
[c]Fermentation products of growing cells, except for the data in column 2 which were obtained with cell suspensions. Data in column 1 are from Kluyver (1931), in column 2 from Milhaud et al. (1956), in column 3 from Canale-Parola and Wolfe (1960a), in column 4 from Smit (1930) and in column 5 from Kupfer and Canale-Parola (1967). Data in columns 1 and 4 were originally reported as percentages of glucose fermented.
[d]Expressed as micromoles of product/100 μM of glucose fermented.
[e]Detected when cells were grown in media containing $CaCO_3$.

Carbohydrate fermentation by *Sarcina ventriculi* yields mainly ethanol, acetate, CO_2 and H_2, whereas *Sarcina maxima* ferments sugars primarily to butyrate, acetate, CO_2, and H_2 (see Table 151). Both species utilize the Embden–Meyerhof pathway for the fermentation of carbohydrates (Canale-Parola, 1970). Two enzymic systems for pyruvate metabolism are present in *Sarcina ventriculi*: a yeast-type decarboxylase that produces acetaldehyde and CO_2 from pyruvate, and a clostridial-type ferredoxin-dependent pyruvate clastic system that yields acetyl phosphate, CO_2 and H_2. Furthermore, CO_2 and H_2 are produced from formate by *Sarcina ventriculi* via a reaction catalyzed by formate hydrogen-lyase (Stephenson and Dawes, 1971). Like *Sarcina ventriculi*, *Sarcina maxima* possesses a clostridial-type pathway for pyruvate cleavage and metabolism. Cleavage of pyruvate by *Sarcina maxima* result in the production of acetyl-CoA, CO_2 and electrons which are transferred to ferredoxin (Kupfer and Canale-Parola, 1967). Phosphotransacetylase (EC 2.3.1.8) catalyzes the conversion of acetyl-CoA to acetyl phosphate, which is metabolized to acetate in a reaction catalyzed by acetate kinase (EC 2.7.2.1). Furthermore, acetyl-CoA is metabolized to butyrate via a pathway similar to that present in saccharolytic clostridia and involving butyryl-CoA dehydrogenase (EC 1.3.99.2), phosphate butyryltransferase (EC 2.3.1.19), and butyrate kinase (EC 2.7.2.7) (Kupfer and Canale-Parola, 1968). *Sarcina maxima* utilizes a hydrogenlyase system to convert formate to CO_2 and H_2 (Kupfer and Canale-Parola, 1967). The occurrence of the latter enzymic activity and the accumulation of small amounts of formate in cultures fermenting glucose suggest that, in addition to the clostridial-type system, *Sarcina maxima* possesses a coliform-type pyruvate clastic system (Kupfer and Canale-Parola, 1967). The latter enzymic system catabolizes pyruvate to acetyl phosphate and formate in coliform bacteria.

Growth of *Sarcina ventriculi* takes place in the human stomach as a result of the development of certain pathological conditions (e.g., pyloric ulceration, stenosis) that retard the flow of food to the intestine. Under these abnormal circumstances, at the acid pH of the stomach and in the presence of carbohydrates and other growth nutrients contained in food, *Sarcina ventriculi* thrives and multiplies rapidly (Smit, 1933). The sarcinae, which occur commonly in soil, are ingested with soil particles present in food.

Sarcina ventriculi has been isolated from soil, mud, contents of diseased human stomach, rabbit and guinea pig stomach contents, elephant dung, human feces and the surface of cereal seeds. *Sarcina maxima* has been isolated from the hull or outer coat of cereal grains, such as wheat, oat, rice and rye. It was also isolated from fresh wheat bran, horse manure and soil.

Enrichment and isolation procedures

Selective isolation procedures for *Sarcina ventriculi* and *Sarcina maxima* are based on the ability of these bacteria to grow anaerobically at very low pH (e.g., 2.0–2.5) in the presence of a fermentable carbohydrates.

The following is a procedure for the isolation of *Sarcina ventriculi* (Canale-Parola, 1970). "The enrichment medium contains (g/100 ml of distilled or tap water): maltose, technical (Pfanstiehl Lab, Inc., Waukegan, IL), 2.0; malt extract broth (powder, BBL or Difco), 5.0; peptone (Difco), 0.5. The pH of the medium is adjusted to 2.2 ± 0.1 with diluted acid (e.g., 1 vol. of H_2SO_4 (specific gravity 1.84) to 9 vols of H_2O). The medium is boiled for 2 or 3 min and, while still hot, it is poured into 60-ml glass-stopper bottles. The bottles, completely filled with medium, are cooled to 40 °C in a cold-water bath. Garden soil (preferably saturated with water or growth medium to decrease the amount of air introduced in the bottles) is added to form a layer (2–4 mm) on the bottom of each bottle. The bottles are stoppered without trapping air bubbles and are incubated at 37 °C.

After 16–48 h, successful enrichments exhibit vigorous gas production. Fine gas bubbles originating from the sarcinae in the sediment rise through the medium and form a layer of foam in the upper part of the bottles. Unless gas production is so active that it resuspends part of the sediment, the supernatant liquid is clear and essentially free of microbial growth since the large sarcina packets remain settled on the bottom. The supernatant liquid of enrichments containing a large number of contaminating organisms (frequently rod-shaped bacteria or yeasts) is turbid; the contamination develops as a result of a rise in pH, generally due to reactions between the acid in the culture and material present in the soil. When very alkaline soil is used, the initial pH of the medium should be lower than 2.2.

Second enrichment cultures, also in bottles, are prepared without delay by using the same procedure, except that 1–2 ml of sediment from the first enrichments is used as the inoculum. After 16–48 h of incubation, cells from the second enrichments may be used to inoculate identical third enrichments for the purpose of accomplishing further dilution of contaminating organisms. Serial dilutions of the growth in these cultures are plated to obtain isolated colonies.

Preparation of third enrichments may not be necessary since the sarcinae in the second enrichments are often so numerically predominant over other organisms that it is advantageous to plate serial dilutions directly from the second enrichments. The following medium (MYA) is used for plating (g/100 ml of distilled water): malt extract broth (powder, BBL or Difco), 2; maltose, technical (Pfanstiehl), 2; yeast extract (Difco), 0.1; agar, 2. The pH of the medium is adjusted to 6.0 ± 0.2 with 5% (w/v) KOH and the medium is sterilized. Serial dilutions are prepared in tubes of melted medium MYA at 45 °C. These are poured in sterile Petri dishes and, after solidification, are incubated anaerobically, or 15 ml of medium MYA is poured on the surface of the medium in each plate (double-layer plates) and the cultures are incubated in air.

Colonies of Sarcina ventriculi appear after 10–32 h. Cells from the colonies are transferred to tubes containing medium MYA from which the agar has been omitted. The medium in the tubes is heated for 5–10 min in a boiling-water bath, then cooled to 40 °C before inoculation. Serial dilutions of the growth in these tubes are plated as described above to obtain pure cultures.

A selective method for the isolation of Sarcina maxima from soil has been described by Claus and Wilmanns (1974). In this method, D-xylose is used as the fermentable substrate in the enrichment medium to exclude growth of Sarcina ventriculi, which does not ferment this pentose. The enrichment medium contains (g/100 ml DW): peptone (Difco), 0.5; yeast extract (Difco), 0.5; D-xylose, 2.0. The medium is adjusted to pH 2 or 3 with HCl. The isolation procedure is similar to that describe above for Sarcina ventriculi.

Maintenance procedures

Cells of Sarcina ventriculi and Sarcina maxima are viable only for 2–4 d in broth cultures in which nutrients are present at concentrations that do not limit growth. Thus, it is prudent to transfer the organisms on alternate days so that viable cells may be maintained in such cultures The physiological bases for the rapid loss of viability have not been elucidated. Viability of Sarcina ventriculi cells can be prolonged by using growth-limiting concentrations of fermentable carbohydrate or of other nutrients in the culture medium (Canale-Parola and Wolfe, 1960a). Furthermore, when K_2HPO_4 or $NaHCO_3$ is added to culture media containing a growth-limiting concentration of fermentable sub-

strate, most strains of Sarcina ventriculi survive for 30 d under anaerobic conditions (Claus et al., 1970).

A convenient method for maintenance of Sarcina ventriculi or Sarcina maxima involves heavy inoculation of the organism in a small well, melted through the surface of a relatively large volume of agar medium contained in an Erlenmeyer flask (Canale-Parola and Wolfe, 1960a). These flask stock cultures must be kept at a temperature (e.g., 30 or 37 °C) that allows continuous growth of the cells. Cells grow only within or near the well and remain viable for approximately 2 months.

Cells of Sarcina ventriculi and Sarcina maxima retain viability for at least several years in liquid N_2 storage (Canale-Parola, 1970).

Procedures for testing special characters

The fact that cells of Sarcina ventriculi are surrounded by a layer of cellulose (see *Further descriptive information*, above), whereas those of Sarcina maxima are not, is a useful characteristic for differentiation. The following staining procedure may be used to demonstrate the presence or absence of the cellulose layer in sarcina packets (E. Canale-Parola, unpublished data).

Approximately 5 g (wet wt) of Sarcina ventriculi or Sarcina maxima cells are refluxed in 100 ml boiling 2% (w/v) KOH for 1 h to remove most of the cytoplasmic material from the cells. Then the cells are harvested by centrifugation and washed four times (centrifugation) with DW. The white, washed pellet consists of almost "empty" cells still arranged in packets. Material from the pellet, from which as much water as possible has been drained, is smeared thickly on a glass microscope slide and the staining solution applied dropwise directly on the smear.

An Iodine-Zinc chloride (I-$ZnCl_2$) solution used for staining is prepared as follows. Twenty grams $ZnCl_2$ are dissolved in 8.5 ml water and the solution is allowed to cool to room temperature. Then, iodine solution (DW, 60 ml; KI, 3 g; I, 1.5 g) is added to the $ZnCl_2$ solution dropwise until iodine begins to precipitate. Only a few ml of iodine solution is required. When the I-$ZnCl_2$ solution is added to the treated Sarcina ventriculi cells smeared on a glass slide, the cellulose present in Sarcina ventriculi packets stains blue and, consequently, the smear on the glass slide appears blue. In contrast, smears of Sarcina maxima cells (refluxed with KOH solution and washed as described above) remain colorless when treated with I-$ZnCl_2$ solution. Cellulose powder (Nutritional Biochemicals), treated with the same procedure, and untreated filter paper stain blue with the I-$ZnCl_2$ solution.

Smit's cellulose-staining solution (Smit, 1930) may be used instead of the I-$ZnCl_2$ solution. Smit's solution consists of: DW, 100 ml; KI, 26 g; I, 0.4 g; and $ZnCl_2$, 85 g. Prior to staining, the cells are refluxed with KOH solution, washed and smeared on a glass slide as described above. Smears of Sarcina ventriculi cells stained with Smit's solution become dark brown with a reddish-purplish tinge. The same color appears when treated cellulose powder or untreated filter paper is stained with Smit's solution. Smears of treated Sarcina maxima cells remain colorless.

Ethanol, which is a major fermentation end product of Sarcina ventriculi but is not formed by Sarcina maxima, may be determined enzymically or by gas-liquid chromatography techniques. Butyrate, produced by Sarcina maxima but not by Sarcina ventriculi, and acetate (produced by both species) are assayed by gas-liquid chromatography.

Taxonomic comments

John Goodsir discovered and named *Sarcina ventriculi* in 1842 (Goodsir, 1842). Subsequently, other investigators described and cultivated various packet-forming cocci, which were assigned to the genus *Sarcina*. Some of these packet formers were aerobes (e.g., *Sarcina lutea*), others anaerobes (e.g., *Sarcina methanica*). Some of the aerobic packet-forming cocci were flagellated and formed spores (e.g., *Sarcina ureae*). In time it became evident that *Sarcina ventriculi*, and the closely related *Sarcina maxima*, are phylogenetically distant from other packet-forming cocci (Canale-Parola, 1970; Canale-Parola et al., 1967). Aerobic, non-sporeforming, nonmotile, packet-forming cocci were assigned to the genus *Micrococcus*; anaerobic, methanogenic, packet-forming cocci to the genus *Methanosarcina*; and aerobic, spore-forming sarcinae to the genus *Sporosarcina*. Only the anaerobic, sugar-fermenting species (*Sarcina ventriculi* and *Sarcina maxima*) were retained in the genus *Sarcina* (Canale-Parola, 1970).

According to 16S rRNA sequence characterization, *Sarcina ventriculi* and *Sarcina maxima* are closely related phylogenetically to *Clostridium butyricum* and other clostridia, but it is distant from packet-forming cocci in the genera *Micrococcus*, *Methanosarcina*, and *Sporosarcina* (Fox et al., 1980). In the 16S rRNA phylogeny presented in the roadmap to this volume, *Sarcina* clusters with *Clostridium* and *Anaerobacter* in the family *Clostridiaceae*, order *Clostridiales*, and class *Clostridia* (Figure 5).

Differentiation and characteristics of the species of the genus *Sarcina*

The differential characteristics of the species of *Sarcina* are listed in Table 149. Other characteristics of the species are indicated in Table 150 and Table 151.

List of species of the genus *Sarcina*

1. **Sarcina ventriculi** Goodsir 1842, 437[AL]

ven.tri′cu.li. L. n. *ventriculus* the stomach; L. gen. n. *ventriculi* of the stomach.

Nearly spherical cells, 1.8–2.4 μm in diameter, occurring in packets of eight to several hundred or more. Large packets (e.g., consisting of approximately 60 or more cells) tend to have an irregular or distorted appearance. Frequently, cells in these packets exhibit flattened shapes and are irregularly arranged. A fibrous layer 150–200 nm thick, composed either totally or in great part of cellulose, is present on the outer surface of the cell wall. This layer is absent in some strains. Reported to form spherical spores (Knöll, 1965). Subsurface colonies in agar media are star-shaped or irregularly cubical, and measure up to several mm in diameter. Surface colonies (anaerobic) are roundish, often with rugged edges.

Carbohydrates are fermented. Amino acids are not fermented. Carbohydrate fermentation patterns and other physiological characteristics are summarized in Table 150. The main products of glucose fermentation are ethanol, acetate, CO_2, and H_2 (Table 151). Two strains studied required for growth two vitamins (biotin, nicotinic acid and 11 amino acids (see *Further descriptive information*, above), in addition to a fermentable carbohydrate and inorganic salt.

Temperature optimum for growth is 30–37 °C.

Isolated from soil, mud, contents of diseased human stomach, rabbit and guinea pig stomach contents, elephant dung, human feces and the surface of cereal seeds.

Phase-contrast photomicrographs and electron micrographs of this species have been published (Canale-Parola, 1970; Canale-Parola et al., 1961).

DNA G+C content (mol%): 30.6 ± 1 (Bd).

Type strain: DSM 286 (ATCC 19633).

GenBank accession number (16S rRNA gene): X77838, X76649.

2. **Sarcina maxima** Lindner 1888, 54[AL]

max′i.ma. L. sup. adj. *maximus* greatest, largest.

Nearly spherical cells, 2–3 μm in diameter, occurring in packets of eight or more. The cells lack a cellulose outer layer. Reported to form oval spores (Knöll, 1965). Subsurface colonies in agar media are cuboid with protuberances or unevenly spherical. Surface colonies (anaerobic) are roundish, often with rugged edges.

Carbohydrates are fermented. Amino acids are not fermented. Carbohydrate fermentation patterns and other physiological characteristics are summarized in Table 150. The main products of glucose fermentation are butyrate, acetate, CO_2, and H_2 (Table 151). A strain studied required for growth three vitamins (thiamine, biotin, nicotinic acid) and 14 amino acids (see *Further descriptive information*, above), in addition to a fermentable carbohydrate and inorganic salts.

Temperature optimum for growth is 30–35 °C.

Isolated from the hull or outer coat of cereal grains, such as wheat, oat, rice and rye. Fresh wheat bran has been used as a source. Also isolated from horse manure, field soil, and garden soil.

Phase-contrast photomicrographs and electron micrographs of this species have been published (Canale-Parola, 1970; Smit, 1930).

DNA G+C content (mol%): 28.6 ± 1 (Bd).

Type strain: DSM 316.

GenBank accession number (16S rRNA gene): X76650.

Genus XI. **Thermobrachium** Engle, Li, Rainey, DeBlois, Mai, Reichert, Mayer, Messner and Wiegel 1996, 1032[VP]

JUERGEN WIEGEL

Ther.mo.bra′chi.um. Gr. adj. *thermos* hot; L. n. *brachium* arm, branch; N.L. neut. n. *Thermobrachium* referring to the branched cells observed frequently with this thermophilic bacterium.

Cells are usually **rod-shaped** and frequently **exhibit true branching. Gram-stain-positive**, although cells of some strains are easily decolorized. An **S-layer with a hexagonal lattice** is present. Cells are flagellated and only sluggishly motile. **Chemoorganoheterotrophic.** Habitats include anthropogenically heated environments (composts) and geothermally heated environments (hot springs and sediments) as well as mesobiotic freshwater sediments.

DNA G+C content (mol%): 30–31.

Type species: **Thermobrachium celere** Engle, Li, Rainey, DeBlois, Mai, Reichert, Mayer, Messner and Wiegel 1996, 1032[VP].

Taxonomic comments

The genus *Thermobrachium* with its type and only species, *Thermobrachium celere* (Engle et al., 1996), was validly published about the same time as the genus *Caloramator* with its type species *Caloramator fervidus*. At that time, the evolutionary distance of 93.3% justified placement of the two taxa in different genera. *Caloramator* was subsequently extended by the description of two other species. The 16S rDNA sequence analyses of all the species, especially the resulting evolutionary distance to *Caloramator indicus*, suggested that *Thermobrachium celere* was within the radiation of the genus *Caloramator*. Thus, it is now assigned to the genus *Caloramator* as *Caloramator celer* comb. nov. despite the relatively large distance of some *Caloramator* species to the type species *Caloramator fervidus*. The description of *Caloramator celer* is included in the genus *Caloramator* chapter, above (Baena and Patel, 2009).

Genus XII. **Thermohalobacter** Cayol, Ducerf, Patel, Garcia, Thomas and Ollivier 2000a, 562[VP]

JEAN-LUC CAYOL, JEAN-LOUIS GARCIA AND BERNARD OLLIVIER

Ther.mo.ha.lo.bac′ter. Gr. adj. *thermos* hot; Gr. n. *hals* salt; N.L. n. *bacter* masc. equiv. of Gr. neut. dim. n. *bakterion* small rod; N.L. masc. n. *Thermohalobacter* a thermophilic fermentative halophile.

Rod-shaped cells, generally occurring singly or in pairs, about 0.5×3–$8\,\mu m$. Motile by laterally inserted flagella. Spores are not observed under different growth conditions and at different growth phases. **Gram-stain-negative.** Obligately anaerobic. Halophilic and thermophilic; optimum growth temperature is $65\,°C$, optimal pH is 7.0, optimum NaCl concentration is 5%. **Chemo-organotrophic growth. Ferments sugars.** Products from glucose fermentation are acetate, ethanol, H_2, and, presumably, CO_2. External electron acceptors are not used.

DNA G+C content (mol%): 33.

Type species: **Thermohalobacter berrensis** Cayol, Ducerf, Patel, Garcia, Thomas and Ollivier 2000a, 562[VP].

Further descriptive information

Thermohalobacter berrensis is the only species described within the genus *Thermohalobacter* so far. It is a rod-shaped bacterium measuring 0.5×3.0–$8.0\,\mu m$ which grows singly or in pairs (Figure 135). The cells are motile with laterally inserted flagella and stain Gram-negative. Electron microscopy of thin sections of *Thermohalobacter berrensis* reveals a cell-wall ultrastructure typical of Gram-stain-negative bacteria. Despite spores not being observed under different growth conditions and at different growth phases, cells survive pasteurization at $100\,°C$ for $20\,min$, thus suggesting the presence of heat-resistant forms.

Thermohalobacter berrensis does not grow in anaerobic medium that contains traces of oxygen and is therefore ascribed as a strict anaerobe. It grows at temperatures ranging from 45–$70\,°C$, with optimum at $65\,°C$ at pH 7.0. The isolate grows in the presence of NaCl concentrations ranging from 2–15%, with an optimum of 5% NaCl at pH 7.0 and $65\,°C$. Growth occurs from pH 5.2–8.8 at $65\,°C$ with an optimum pH of 7.0. The doubling time under optimal growth conditions is $0.46\,h$.

Yeast extract or bio-Trypticase are required for growth on carbohydrates. *Thermohalobacter berrensis* grows on sugars, glycerol, pyruvate, and bio-Trypticase. Acetate, ethanol, H_2, and, presumably, CO_2 are produced during glucose fermentation. *Thermohalobacter berrensis* does not reduce thiosulfate, sulfate, or sulfur into sulfide.

Enrichment and isolation procedures

Enrichment and isolation should be performed using a growth medium which contains (per liter of distilled water): NH_4Cl,

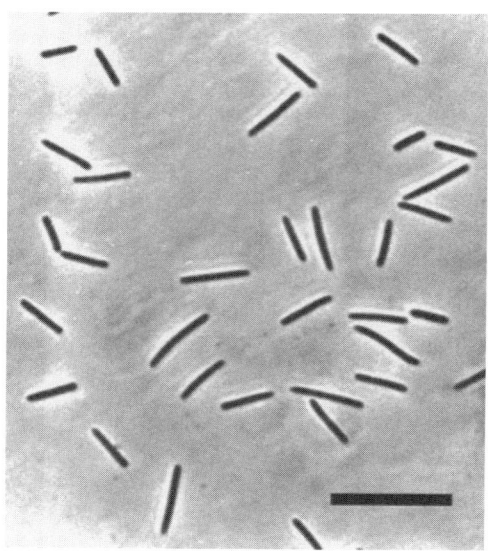

FIGURE 135. Photomicrograph of *Thermohalobacter berrensis*. Bar = $10\,\mu m$.

0.1 g; K_2HPO_4, 0.3 g; KH_2PO_4, 0.3 g; $MgCl_2$, 2 g; $CaCl_2$, 0.2 g; CH_3COONa, 0.5 g; NaCl, 100 g; KCl, 4 g; glucose, 10 g; cysteine-HCl, 0.5 g; yeast extract, 1 g; bio-Trypticase, 1 g; trace mineral element solution (Imhoff-Stuckle and Pfennig, 1983), 1 ml; resazurin, 1 mg. The pH is adjusted to 7.0 with 10 M KOH, after which the medium is boiled under a stream of O_2-free N_2 gas and cooled to room temperature. Five ml or 20 ml aliquots are dispensed into Hungate tubes or serum bottles, respectively, under a stream of N_2/CO_2 (80:20 v/v) gas, and the vessels are autoclaved for 45 min at 110 °C. Prior to inoculation, $Na_2S \cdot 9H_2O$ and $NaHCO_3$ are injected from sterile stock solutions to obtain a final concentration of 0.04% (w/v) and 0.2% (w/v), respectively.

For initiating enrichment cultures, a small portion of the sediment sample is inoculated into growth medium followed by incubation at 62 °C without agitation. The culture is purified by repeated use of the Hungate roll tube method (Hungate, 1969; Macy et al., 1972; Miller and Wolin, 1974) in growth medium amended with 2% (w/v) agar. Enrichment cultures are positive after 3 d of incubation at 62 °C. Colonies 1 mm in diameter develop in roll tubes after 48 h incubation at 62 °C. Single colonies are picked and ten-fold serial dilutions in roll tubes are repeated at least twice before the culture is considered pure.

Maintenance procedures

Stock cultures can be maintained on medium described by Cayol et al. (2000a) by monthly transfers. Liquid cultures retain viability after several weeks storage at 4 °C, at ambient temperature or when lyophilized or after storage at −80 °C in the basal medium containing 20% glycerol (v/v). Viability is best maintained from mid-exponential phase cultures.

Differentiation of the genus *Thermohalobacter* from other genera

The genus *Thermohalobacter*, represented only by one species, *Thermohalobacter berrensis*, is a heterotrophic, moderately halophilic member of domain *Bacteria*. Within this domain, the main characteristic of being moderately halophilic is shared by members of the family *Halanaerobiaceae* with the exception of *Halanaerobium lacusrosei*, an extreme halophilic bacterium isolated from Retba lake in Senegal (Africa) (Cayol et al., 1995). *Thermohalobacter berrensis* is not only halophilic, but is also a strict thermophile growing optimally at 65 °C. The family *Halanaerobiaceae* is comprised exclusively of mesophilic to thermotolerant micro-organisms (Ollivier et al., 1994). *Halothermothrix orenii* is the only true thermophile belonging to this family, growing at temperatures up to 68 °C (Cayol et al., 1994) (Table 152). Due to its typical salt requirement, *Thermohalobacter* is related neither to members of the genus *Thermoanaerobacter*, which are described as halotolerant micro-organisms (Lee et al., 1993), nor to the thermophilic and hyperthermophilic members of the genus *Thermotoga*, order *Thermotogales*, which do not grow at NaCl concentrations above 6–7% (Huber et al., 1986; Windberger et al.,

TABLE 152. Salient features of *Thermohalobacter berrensis* and *Halothermothrix orenii*[a]

Characteristic	*Thermohalobacter berrensis*[b]	*Halothermothrix orenii*[c]
Motility	+	+
NaCl concentration range (%)	2–5	4–16
Optimum NaCl concn (%)	5	6
Temperature range (°C)	45–70	45–64
Optimum temperature (°C)	65	56
pH range	5.2–8.8	5.5–8.4
Optimum pH	7	6.5–7.4
Habitat	Berre Lagoon, France	Chott El Guettar, Tunisia
DNA G+C content (mol%)	33	36
Substrates used:		
N-Acetylglucosamine	Yes	ND
Arabinose	No	Yes
Bio-Trypticase	Yes	No
Cellobiose	Yes	Yes
Fructose	Yes	Yes
Galactose	No	Yes
Glucose	Yes	Yes
Glycerol	Yes	No
Pyruvate	Yes	No
Maltose	Yes	No
Mannitol	Yes	No
Mannose	Yes	Yes
Melibiose	No	Yes
Starch	Yes	Yes
Sucrose	Yes	No
Ribose	No	Yes
Xylose	No	Yes
End products of fermentation	Acetate, ethanol, hydrogen, carbon dioxide	Acetate, ethanol, hydrogen, carbon dioxide

[a]Symbols: +, positive; −, negative; w, weak reaction; ND, not determined.
[b]Data from Cayol et al. (2000a).
[c]Data from Cayol et al. (1994).

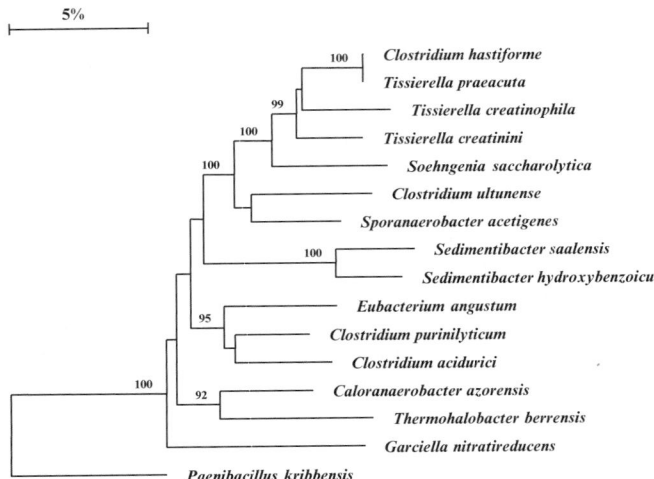

5%

100 — Clostridium hastiforme
99 ┬ Tissierella praeacuta
100 ┬ Tissierella creatinophila
100 ┬ Tissierella creatinini
Soehngenia saccharolytica
Clostridium ultunense
Sporanaerobacter acetigenes
100 ┬ Sedimentibacter saalensis
Sedimentibacter hydroxybenzoicus
95 ┬ Eubacterium angustum
Clostridium purinilyticum
Clostridium acidurici
100 ┬ Caloranaerobacter azorensis
92 ┬ Thermohalobacter berrensis
Garciella nitratireducens
Paenibacillus kribbensis

FIGURE 136. Phylogenetic tree based on 16S rRNA gene sequence comparison and obtained by a neighbor-joining algorithm (PHYLIP package), indicating the position of *Thermohalobacter berrensis* among the members of the cluster XII of the *Firmicutes*. These clusters are defined based on the guidelines described by Collins et al. (1994). Bootstrap values are shown at branching points. Only values above 80% were considered significant and reported. Scale bar indicates 5 nucleotide substitution per 100 nucleotides.

1989). Additionally within this order, the thermophilic *Geotoga*, *Petrotoga*, and *Marinitoga* species do not gow in the presence of 12.5% NaCl (Davey et al., 1993; Postec et al., 2005). Interestingly, the growth temperature optimum for *Thermohalobacter berrensis* (65 °C) is higher than that of any moderately to extremely halophilic microbe reported to date including the extremely halophilic thermophilic archaeon *Methanohalobium evestigatum* (Zhilina and Zavarzin, 1987) and the moderately halophilic thermophilic bacterium, *Halothiobacillus orenii* (Cayol et al., 1994). Taking into account all of these observations, the phenotypically closest relative of *Thermohalobacter berrensis* is *Halothermothrix orenii*. However, the 16S rRNA gene sequence analysis indicates that it is not a member of the *Halanaerobiaceae* but a member of cluster XII of the *Clostridiales* (Collins et al., 1994) (Figure 136). *Thermohalobacter berrensis* also differs phenotypically from *Halothermothrix orenii* (Table 152). Cluster XII of the order *Clostridiales* comprises *Clostridium* species (*Clostridium hastiforme*, *Clostridium acidurici*, *Clostridium purinilyticum*, and *Clostridium ultunense*), *Sedimentibacter* species (*Sedimentibacter saalensis* and *Sedimentibacter hydroxybenzoicus*), *Tissierella* species (*Tissierella creatinophila*, *Tissierella creatini*, and *Tissierella praecuta*), *Soehngenia saccharolytica*, *Eubacterium angustum*, and *Sporanaerobacter acetigenes*. None of these species is reported as thermophilic and/ or halophilic. The only moderately halophilic micro-organism so far described within the genus *Clostridium* is *Clostridium halophilum*, but it does not grow at 50 °C (Fendrich et al., 1990) and is phylogenetically related to cluster XI of the *Clostridiales* (Collins et al., 1994). The only thermophilic micro-organisms belonging to the cluster XII of this latter order are *Garciella nitratireducens* isolated from an oilfield separator (Miranda-Tello et al., 2003), and *Caloranaerobacter azorensis* isolated from a deep-sea hydrothermal vent (Wery et al., 2001) which are thermotolerant and slightly halophilic, respectively. In contrast to *Garciella nitratireducens*, *Thermohalobacter berrensis* does not reduce thiosulfate into sulfide. *Caloranaerobacter azorensis* markedly differs from *Thermohalobacter berrensis* by (i) its optimum NaCl concentration for growth, (ii) the end products of glucose metabolism, and (iii) having a lower G+C content in its DNA (27 mol% for *Caloranaerobacter azorensis* and 33 mol% for *Thermohalobacter berrensis*).

Taxonomic comments

Thermohalobacter is a newly defined genus consisting of one species (*Thermohalobacter berrensis*). Phenotypic and genotypic characteristics clearly place it within the cluster XII of the order *Clostridiales*. Within this cluster, it is the only thermophilic, moderately halophilic microbe. The isolation and characterization of more species representative of this genus should be helpful to confirm the classification of this genus.

List of species of the genus *Thermohalobacter*

1. **Thermohalobacter berrensis** Cayol, Ducerf, Patel, Garcia, Thomas and Ollivier 2000a, 562[VP]

ber.ren′sis. N.L. adj. *berrensis* from Berre, south of France.

Cells are rods (0.5 × 3–8 μm), occurring singly or in pairs, motile by laterally inserted flagella. Spores are not observed under different growth conditions and at different growth phases. The cells stain Gram-negative. Round colonies (diameter, 1 mm) are present after 2 d of incubation at 62 °C. Chemo-organotrophic and obligately anaerobic member of the domain *Bacteria*. The optimum temperature for growth is 65 °C at pH 7.0; temperature range for growth 45–70 °C. The optimum pH is 7.0 at 65 °C; growth occurs between pH 5.2 and pH 8.8. The optimum NaCl concentration for growth is 5% at 65 °C and pH 7.0; growth occurs at NaCl concentration ranging from 2–15%. Ferments cellobiose, fructose, glucose, maltose, mannose, mannitol, sucrose, glycerol, *N*-acetylglucosamine, starch, pyruvate, and bio-Trypticase, but not arabinose, galactose, lactose, melibiose, raffinose, rhamnose, ribose, sorbose, trehalose, xylose, glycine betaine, fatty acids (formate, acetate, fumarate, and lactate), Casamino acids, cellulose, and yeast extract. Acetate, ethanol, H₂, and, presumably, CO_2 are produced during glucose fermentation. Elemental sulfur, sulfate, and thiosulfate cannot be used as electron acceptors. Isolated from sediment of a feeding canal of a solar saltern.

DNA G+C content (mol%): 33 (HPLC).
Type strain: CTT3, CNCM 105955.
GenBank accession number (16S rRNA gene): AF113543.

Genus XIII. **Tindallia** Kevbrin, Zhilina, Rainey and Zavarzin 1999, 2[VP] (Effective publication: Kevbrin, Zhilina, Rainey and Zavarzin 1998, 100.)

ELENA V. PIKUTA

Tin.dal'li.a. N.L. fem. n. *Tindallia* derived from the name of Brian J. Tindall, who pioneered the study of alkaliphilic micro-organisms in Lake Magadi.

Cells are **slightly curved rods,** with **rounded and slightly pointed ends (Figure 137, top)**. Cells occur **singly, in pairs, or in short chains**; size is **0.4–0.7 × 1.2–5.0 μm. Gram-stain-positive. Motile. Asporogenous or oligospore forming. Endospores are round or oval** and **located in cells subterminally, terminally, or centrally. Sporangium is not swollen.** The ultrastructure of cell surfaces has **S-layers. Obligate anaerobes and catalase-negative.** Dissipotrophs and **chemoorganoheterotrophs** with **fermentative metabolism** or **respiration by the Stikland reaction. Halotolerant alkaliphiles, no growth below pH 7.0. Obligately dependent on Na⁺ ions. Mesophilic. Acetogenic ammonifiers.** The homoacetic or acetyl-coenzyme A-CO-dehydrogenase pathway participates in metabolism.

DNA G+C content (mol%): 37.6–44.4.

Type species: **Tindallia magadiensis** corrig. Kevbrin, Zhilina, Rainey and Zavarzin 1999, 2[VP] (Effective publication: *Tindallia Magadii* Kevbrin, Zhilina, Rainey and Zavarzin 1998, 100.).

Further descriptive information of the genus *Tindallia*

The first work that mentioned strain Z-7934 was published in *Current Microbiology* (Zhilina and Zavarzin, 1994), where it was described as a member of an anaerobic haloalkaliphilic microbial community from sediments of soda Lake Magadi in Kenya (Equatorial Africa). This strain was described as a primary anaerobe and dissipotroph, participating in organic decomposition on a level with proteolysis products and amino acids. A complete taxonomic description of this isolate as a new species of a new genus was published in the same journal in 1998 with the name *Tindallia magadii* (Kevbrin et al., 1998); the species name was corrected and validated as *Tindallia magadiensis* (Kevbrin et al., 1999).

The phylogenetic position for strain Z-7934[T] placed it within cluster XI of the low G+C Gram-positive bacteria, as defined on the basis of 16S rDNA analyses of *Clostridium* species and related taxa (Collins et al., 1994). The 16S rRNA phylogenetic analysis presented in the roadmap to this volume classifies *Tindallia* as in the family *Clostridiaceae*, order *Clostridiales*, class *Clostridia* in the phylum *Firmicutes*.

Cell morphology of all species is similar with slight variations in sizes. Species of the genus are asporogenous or oligospore forming and motile: the genus *Tindallia* was originally described as nonsporeforming and nonmotile, but later the motile cells and spores of the type species were detected. Flagella of *Tindallia californiensis* are attached laterally (Figure 137, middle). Cells of *Tindallia texcoconensis* have peretrichous flagella.

The ultrastructure of the cell surface demonstrates the presence of S-layers: *Tindallia californiensis* APO[T] has the outer S-layer with tetragonal texture (Figure 137, bottom). Cells multiply by binary fusion with the formation of two daughter cells. Cells occur singly, in pairs, or in short irregularly curved chains.

Colonies in deep agar were described for *Tindallia californiensis*; they have a typical (for anaerobes) lens shape with smooth edges, 1.0–2.2 mm in diameter, and are matt yellowish. The colonies of *Tindallia texcoconensis* are round, white in color, and 2–4 mm in diameter.

Tindallia species have a truly alkaliphilic nature, not one could grow below pH 7.0, but they are not dependent on car-

bonate ions. Previous diagnosis of *Tindallia magadiensis* tested positive for the dependence on carbonate ions, and since the experiment was performed on tris-buffered medium with different concentrations of carbonates it was concluded that strain Z-7934[T] was obligately dependent upon carbonates. Comparative study of *Tindallia magadiensis* with *Tindallia californiensis* has demonstrated that both species could grow on the L-serine-buffered medium without carbonates at pH 9.0 adjusted by 6N NaOH. All three species of the genus are obligately dependent

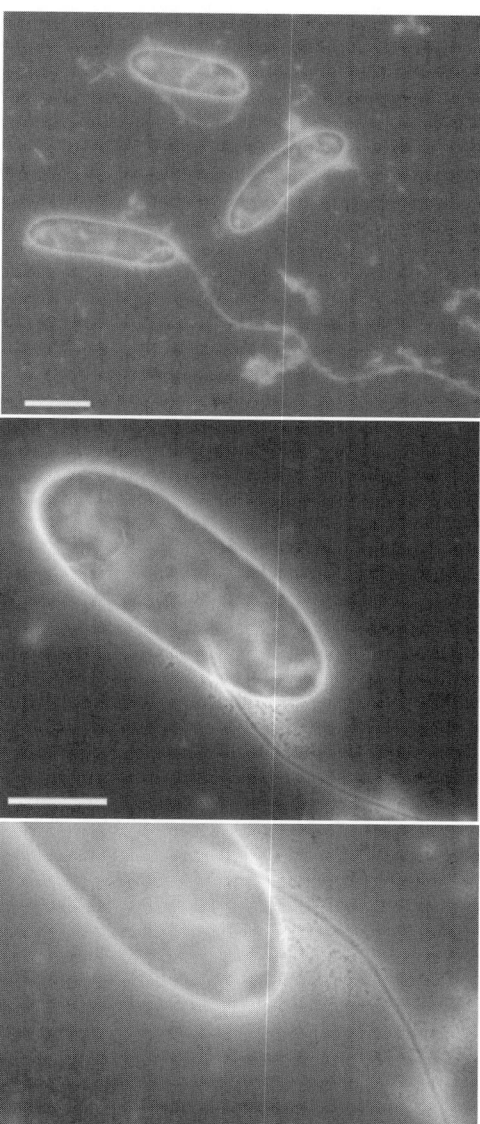

FIGURE 137. Transmission electron microscope image of *Tindallia californiensis* strain APO[T]. (top) Rod-shaped cells. Bar = 1 μm. (middle) Laterally attached flagellum. Bar = 1 μm. (bottom) Cell surface with tetragonal structure of outer S-layer.

upon Na⁺ ions and they are halotolerant with optimum NaCl at about 3–7% (w/v). The halotolerance of *Tindallia californiensis* is twice as high of *Tindallia magadiensis*: it is capable of growing at 20% (w/v) NaCl, but the maximum concentration of NaCl for *Tindallia magadiensis* is only 10% (w/v). *Tindallia texcoconensis* has maximum of NaCl concentration for growth at 25% (w/v). Members of the genus are mesophiles with an optimum at around 35 °C and maximum temperature of 47–48 °C. The temperature characteristic is significantly different at the minimum values: *Tindallia californiensis* could grow at 10 °C, but *Tindallia magadiensis* has a minimum temperature growth at 19 °C and cells of *Tindallia texcoconensis* have minimum growth at 25 °C.

Concerning the oxygen requirement, *Tindallia* species are strictly anaerobic and catalase-negative. Species of the genus are acetogens and have a fermentative metabolism with the functioning of the *homoacetic* or *acetyl-coenzyme A-CO-dehydrogenase* pathway. The key enzymes of this pathway *CO-dehydrogenase* and *hydrogenase* with their high activity were detected in cells of *Tindallia magadiensis* and *Tindallia californiensis*. The species specialize in using proteolysis products as peptone, bacto-tryptone, Casamino acids, yeast extract, some amino acids, and pyruvate. It means that in bacterial communities they could play the role of primary anaerobes and dissipotrophs. Species of the genus cannot use sugars and alcohols. Species of the genus are capable of respiration by the Stickland reaction, but the lists of amino-acid pairs are different (Table 153 and Table 154)

Antibiotic susceptibility was checked for two species and it was shown that *Tindallia californiensis* was sensitive to ampicillin, tetracycline, gentamicin, and chloramphenicol. However, the strain was resistant to kanamycin (typically for anaerobes) and to rifampin. *Tindallia magadiensis* was sensitive to ampicillin, tetracycline, gentamicin, chloramphenicol, and rifampin, but resistant to kanamycin.

Ecology for the *Tindallia* species is similar: they all were isolated from soda lakes but geographically distant continents: Africa and North America. They could be found in anaerobic layers of mud sediments in Lake Magadi (Kenya), soda Mono Lake (North California), and in underground water near athalassic saline alkaline Texcoco lake in Mexico.

The non-validly published species "*Clostridium aminovorans*", "*Clostridium bogorii*", and "*Clostridium alcaliphilum*" by 16S rRNA sequence analysis have about 99.1% similarity to *Tindallia californiensis* (Figure 138); in the future these species probably will be described as *Tindallia* species. From validly published taxa, the most closely related to *Tindallia californiensis* after *Tindallia magadiensis* (with a 98.49% similarity) are: *Natronincola histidinovorans* (90.35% similarity), *Clostridium felsineum* (90.26% similarity), and *Clostridium formicoaceticum* with *Clostridium halophilum* (less then 90% similarity). *Tindallia texcoconensis* has 97.5% similarity with *Tindallia californiensis* and 96.4% with *Tindallia californiensis* (Figure 139).

The DNA–DNA hybridization between *Tindallia magadiensis* and *Tindallia californiensis* showed 55% homology, and between *Tindallia texcoconensis* and *Tindallia californiensis* it was 42.2%. The G+C content of the DNA of *Tindallia magadiensis* is 37.6 mol% (by T_m), for *Tindallia californiensis* it is 44.4 ± 0.2 mol% (by HPLC), and for *Tindallia texcoconensis* it is 40.0 mol% (by HPLC).

Pathogenicity, antigenic structure, mutants, plasmids, phages, and phage typing for all species were not studied.

Enrichment and isolation procedures

Anaerobic techniques should be used for *Tindallia* cultures. Isolation of species was performed with a selective medium at pH 10.0–10.3, and pure cultures were received by the dilution method on peptone-containing media. The medium contains (per l): NaCl, 30.0 g; Na_2CO_3, 2.76 g; $NaHCO_3$, 24.0 g; KCl, 0.2 g; K_2HPO_4, 0.2 g; $MgCl_2 \cdot 6H_2O$, 0.1 g; NH_4Cl, 1.0 g; $Na_2S \cdot 9H_2O$,

TABLE 153. Diagnostic Table for species of the genus *Tindallia*[a]

Characteristic	*T. magadiensis*	*T. californiensis*	*T. texcoconensis*
Motility	+/–	+	+
Growth at 10 °C	–	+	–
Growth at pH 7.5	+	–	+
Growth at 20% NaCl	–	+	+
Growth on:			
ʟ-Aspartate	(+)	–	–
Citrate	+	+[b]	+
ʟ-Glutamine	(+)	–	–
Glycine	(+)	–	–
ʟ-Threonine	(+)	–	–
End products:			
Ethanol	–	+	+[c]
Stickland reaction:			
Glycine + ʟ-isoleucine	+	–	+
Glycine + ʟ-valine	+	–	–
ʟ-Proline + ʟ-isoleucine	+	–	+
ʟ-Proline + ʟ-leucine	+	–	+
ʟ-Proline + ʟ-valine	+	–	–
Growth with rifampin	–	+	ND

[a]Symbols: +, good growth; –, no growth; (+), weak growth; ND, no data.
[b]Growth absent on the citrate (3 g/l) containing medium with low concentrations of yeast extract (50–100 mg/l); but on the medium with citrate (3 g/l) supplemented by 500–1000 mg/l of yeast extract the optic density of growing culture is twice higher then on medium with 500–1000 mg/l of yeast extract without citrate.
[c]Only on peptone, and only as minor product.

TABLE 154. Descriptive table of the *Tindallia* species[a]

	T. magadiensis	*T. californiensis*	*T. texcoconensis*
Morphology:			
Cell sizes (μm)	0.5–0.6 × 1.2–2.5	0.6–0.7 × 2.4–3.0	0.4–0.6×3.0–5.0
Gram-positive cell-wall structure	+	+	+
Motility	−/+[b]	+	+
Spores	+	+	−
Temperature, °C range (optimum)	19–47 (37)	10–48 (37)	25–45 (35)
pH range (optimum)	7.5–10.5 (8.5)	8.0–10.3 (9.5)	7.5–10.5 (9.5)
NaCl, % range (optimum)	1–10 (3–6)	>1–20 (3)	2.5–25 (7.5)
Substrates:			
L-Arginine	+	+	+
L-Aspartate	(+)	−	−
Citrate	+	+[c]	+
Glycine	(+)	−	−
L-Glutamine	(+)	−	−
L-Histidine	(+)	+	+
L-Lysine	ND	+	+
Peptone	+	+	+
Pyruvate	+	+	+
L-Serine	(+)	+	−
L-Threonine	(+)	−	+
End products:			
Acetate	+	+	+
Ethanol	−	+	+
Formate	+	+	−
H₂	+	+	+
Iso-valerate	−	ND	+
Lactate	+	+	−
Propionate	+	+	+
Stickland reaction:			
Gly + L-Ala	+	+	−
Gly + L-Ile	+	−	+
Gly + L-Leu	+	−	−
Gly + L-Val	+	−	−
L-Pro + L-Ala	−	+	−
L-Pro + L-Ile	+	−	+
L-Pro + L-Leu	+	−	+
L-Pro + L-Val	+	−	−
L-Trp + L-Val	+	+	−
Antibiotic reaction:			
Resistant to	Kanamycin	Kanamycin and rifampin	ND
Sensitive to	Tetracycline, gentamycin, ampicillin, rifampin, and chloramphenicol	Tetracycline, gentamycin, ampicillin, and chloram-phenicol	
DNA G+C mol% (*T*ₘ)	37.6	44.4	40.0
Genome size, Da	1.14 × 10⁹	1.02 × 10⁹	ND
Source of isolation (habitat)	Anaerobic sediments of Lake Magadi (Kenya)	Anaerobic sediments of Mono Lake (California)	Underground water sample near Texcoco lake in Mexico
Type strain	Z-7934[T] (DSM 10318)	APO[T] (ATCC BAA-393, DSM 14871, CIP 107910)	IMP-300[T] (DSM 18041, JCM 13990)

[a]+, Good growth; −, no growth; (+), weak growth; ND, not determined.
[b]Motility was observed in the culture from DSM 10318[T], but in original article strain Z-7934[T] was described as nonmotile.
[c]Growth required 0.5–1.0 g/1 of yeast extract.

0.4 g; resazurin, 0.001 g; yeast extract, 0.5; peptone, 5.0 g; vitamin solution (Wolin et al., 1963), 2 ml; and trace mineral solution (Whitman et al., 1982), 1 ml. The final pH is adjusted to 9.0–9.5 by 6 M NaOH. High-purity nitrogen is used for the gas phase. Colonies could be obtained on 3% agar medium by the "roll tubes" method, but carbonate solutions are added separately after sterilization. As a negative control for detection of culture purity (test on contamination), the media (pH 7.0) with a mixture of glucose, peptone, and yeast extract could be applied.

Maintenance procedures

The cultures of both species can be maintained on a liquid or agar-containing anaerobic medium for 4–5 months. The best method for the long-term preservation of *Tindallia* cultures is lyophilization.

Procedures for testing special characters

The detection of a mesophilic nature, the haloalkaliphilic features, and a relationship to oxygen have been performed by traditional techniques. Since the spore formation for species *Tindallia* has

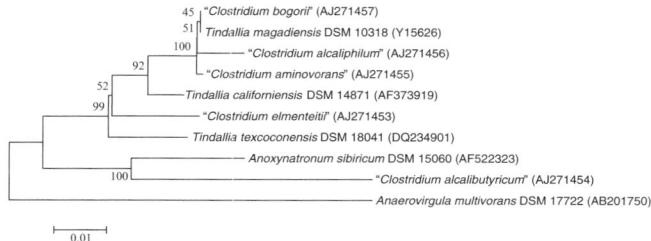

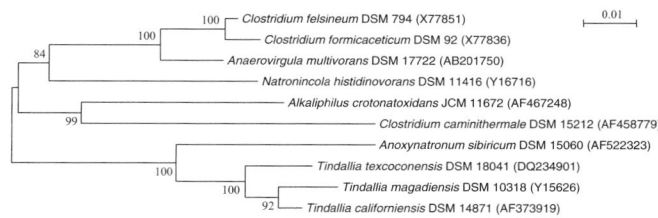

FIGURE 138. Evolutionary relationships in the genus *Tindallia* and closest relatives (including non-validly published names). The bootstrap consensus tree inferred from 2000 replicates is taken to represent the evolutionary history of the taxa analyzed. The percentages of replicate trees in which the associated taxa clustered together in the bootstrap test (2000 replicates) are shown next to the branches.

FIGURE 139. Evolutionary relationships of the genus *Tindallia* and closely related species (validly published names only). The bootstrap consensus tree inferred from 2000 replicates is taken to represent the evolutionary history of the taxa. The percentages of replicate trees in which the associated taxa clustered together in the bootstrap test (2000 replicates) are shown next to the branches.

been detected in exclusively rare cases it requires specific procedures, as an action by acetate buffer (pH 4.0), or by antibiotics, or by transferring cultures from carbonate media to serine-buffered medium without peptone. Isolation of DNA and amplification were performed by the usual methods: phenol/chloroform extraction and PCR (by *Thermus aquaticus* DNA thermostable polymerase).

Differentiation from closely related taxa

Phenotypic and genotypic differences of the genus *Tindallia* and other related taxa shown in Table 155.

Taxonomic comments

The genus *Tindallia* includes three species, *Tindallia magadiensis*, *Tindallia californiensis*, and *Tindallia texcoconensis*. These species have a 98.2% 16S rDNA sequence similarity by analysis but have DNA–DNA hybridization homology of 55%); this and differences in physiological properties allowed them to be separated into different species.

Acknowledgements

Author thanks M. Farmer and J. Shield (Center for Advanced Ultrastructural Research at the University of Georgia in Athens) for transmission electron microscopy and Richard B. Hoover (National Space Sciences and Technologies Center/ NASA) and D. Marsic (University of Alabama in Huntsville) for assistance.

List of species of the genus *Tindallia*

1. **Tindallia magadiensis** corrig. Kevbrin, Zhilina, Rainey and Zavarzin 1999, 2[VP] (Effective publication: *Tindallia magadii* Kevbrin, Zhilina, Rainey and Zavarzin 1998, 100.)

   ma.ga.di.en′sis. N.L. fem. adj. *magadiensis* pertaining to Lake Magadi in Kenya, equatorial Africa.

   Data are from Kevbrin et al. (1998), Pikuta et al. (2003b). Slightly curved rods, 0.5–0.6 (1.2–2.5 μm in size, occurred singly, in pairs, and in short chains. Cells are Gram-stain-positive and motile. Oligospore-forming: forms oval or round endospores. Strictly anaerobic; sulfide is needed only as a reductant.

   Mesophilic. Grows at 19–47 °C, with an optimum at 37 °C. Obligate alkaliphile that cannot grow at pH 7.0; grows in a pH range of 7.5–10.5 with an optimum 8.5. CO_3^{2-} is obligately required. Halotolerant, has optimum growth at 3% NaCl, tolerant to no more then 10% NaCl. Requires Na[+] ions, no growth at 0% NaCl. Chemoorganoheterotrophic. Fermentative, nonhydrolytic, growth on proteolysis products as peptone, bacto-tryptone, yeast extract, Casamino acids, gluconate, pyruvate, citrate, 2-oxoglutarate, 2,3-butanediol, uracil, and amino acids (arginine, ornithine, citrulline, histidine, aspartate, glutamate, glutamine, serine, threonine, glycine). Production of ammonium from arginine and compounds of the ornithine cycle, which are preferred substrates. Ammonium is stimulatory at low concentrations but is inhibitory at concentrations over 9 mM and at high pH. The major fermentation product is acetate. Minor end products are formate, propionate, isovalerate, and hydrogen depending on the substrate. Stickland reaction performs on the following amino acids pairs: L-tryptophan + L-valine; glycine + L-alanine; glycine + L-valine; glycine + L-leucine; glycine + isoleucine; L-proline + L-valine; L-proline + L-leucine; L-proline + L-isoleucine. Reduces DMSO to DMS and Fe(III) to Fe(II). For anabolic needs they utilize amino acids, do not require vitamins. Resistant to kanamycin, but sensitive to ampicillin, rifampin, tetracycline, gentamicin, and chloramphenicol.

   Habitat: anaerobic sediments of soda Lake Magadi, Kenya (equatorial Africa).

   DNA G+C content (mol%): 37.6 (T_m).

   Type strain: Z-7934, DSM 10318.

   GenBank accession number (16S rRNA gene): Y15626.

2. **Tindallia californiensis** Pikuta, Hoover, Bej, Marsic, Detkova, Whitman and Krader 2003b, 1701[VP] (Effective publication: Pikuta, Hoover, Bej, Marsic, Detkova, Whitman and Krader 2003a, 333.)

   ca.li.for.nien′sis. N.L. fem. adj. *californiensis* pertaining to California, the region where Mono Lake is located, from which this organism was isolated.

   Data are from Pikuta et al. (2003b). Slightly curved rods, motile with single laterally attached flagellum; cells sizes are between 0.55 and 0.7 μm wide and 1.7 and 3.0 μm long. Gram-

Table 155. Comparative Table for the genus *Tindallia* and other alkaliphilic, acetogenic, anaerobic, mesophilic genera[a]

Features:	*Tindallia*	*Anaerovirgula*	*Alkaliphilus*	*Anoxynatronum*	*Clostridium*	*Natroniella*	*Natronincola*
	Kevbrin et al., 1998	Pikuta et al., 2006	Takai et al., 2001	Garnova et al., 2003b	Hippe et al., 1992	Zhilina et al., 1996	Zhilina et al., 1998
Motility	+/−	+	+	+	+	+	+
Gram reaction	+	+	+	+	+/−	+	+
Spores	+[b]	+	−	−	+	−	+
Relation to O$_2$[c]	oblig An	oblig An	oblig An	An (atl)[c]	An (atl)	oblig An	oblig An
Reduction of:							
NO$_3^{2-}$ to NO$_2^-$	−	−	−		+/−	ND	ND
SO$_4^{2-}$ to H$_2$S	−	−	−[d]		−	−	−
Activity of:							
Catalase	−	−	ND	+	−	ND	ND
Oxidase	−	−	ND	ND	−	ND	ND
NaCl (3–12%) requirement	+	−	−	+	+/−	+	+
CO$_3^{2-}$ requirement	−	−	−	+	−	+	+
Using as substrates:							
Sugars	−	+	−	+	+/−	−	−
Proteolysis products and amino acids	+	+	+	+	+/−	+[e]	+
Acetogenesis	+	+	ND	+	+/−	+[f]	+

[a]Symbols: +, >85% positive; d, different strains give different reactions (16–84% positive); −, 0–15% positive; w, weak reaction; ND, not determined.
[b]Oligospore-forming.
[c]oblig An, obligate anaerobe; An (atl), aerotolerant anaerobe.
[d]Reduction of sulfur, thiosulfate, and fumarate.
[e]Using only glutamate.
[f]Homoacetogen.

stain-positive structure of cell wall. Oligospore forming (central or subterminal location) without swelling of sporangium.

Haloalkaliphilic: the pH range for growth is 8.0–10.5, with optimum growth at pH 9.5. Growth depends on Na$^+$ but not CO$_3^{2-}$ ions. Range of NaCl is >1–20% (w/v) with optimum growth at 3–5% (w/v). Mesophilic: the temperature range for growth is 10–48 °C, with the optimum being 37 °C.

Strictly anaerobic, catalase-negative, and does not reduce sulfate. Chemorganoheterotroph. Using peptone, bacto-tryptone, Casamino acids, yeast extract, L-serine, L-lysine, L-histidine, L-arginine, pyruvate, and citrate as substrates. Capable of growth by performing the Stickland reaction on the following amino acid pairs: L-proline + L-alanine; glycine + L-alanine; glycine + L-leucine; and L-tryptophan + L-valine. Main end product is acetate; minor end products are propionate, lactate, and ethanol. Requires yeast extract for growth (on citrate no less than 0.5–1.0 g/l). Resistant to kanamycin and rifampin, but sensitive to ampicillin, tetracycline, genta-micin, and chloramphenicol.

Habitat: isolated from the mud sediments of the hypersaline, meromictic, alkaline Mono Lake in California, USA.

DNA G+C content (mol%): 44.4 (HPLC).

Type strain: APO, DSM 14871, CIP 107910, ATCC BAA-393.

GenBank accession number (16S rRNA gene): AF 373919.

3. **Tindallia texcoconensis** Alazard, Badillo, Fardeau, Cayol, Thomas, Roldan, Tholozan and Ollivier 2009, 1555[VP] (Effective publication: Alazard, Badillo, Fardeau, Cayol, Thomas, Roldan, Tholozan and Ollivier 2007, 38.)

tex.co.con.en′sis. N.L. adj. *texcoconensis* pertaining to Texcoco lake in Mexico, from which this micro-organism was isolated.

Data are from Alazard et al. (2007). Thin straight rods, motile by peritrichous flagella; with sizes 0.4–0.6×3.0–5.0 μm. Gram-stain-positive. Asporogenous. Strictly anaerobic.

Haloalkaliphilic: the pH range for growth is 7.5–10.5, with optimum growth at pH 9.5. Obligately depends on Na$^+$ ions. Range of NaCl is 25–250 g/l with optimum growth at 50–100 g/l. Mesophilic: the temperature range for growth is 25–45 °C, with the optimum at 35 °C.

Under optimal growth conditions, the generation time is 28 h. Chemorganoheterotroph. Fermentative; uses only a few organic compounds as substrates: peptone, Casamino acids, arginine, ornithine, alanine, pyruvate, and citrate. Reduces Fe(III) to Fe(II) (with peptone as electron donor), but incapable of dissimilatory reduction of nitrite, nitrate, or thiosulfate and sulfate. Capable of respiration by performing the Stickland reaction on the following amino acid pairs: L-leucine + L-proline, L-leucine + L-tryptophan, L-isoleucine + glycine, L-isoleucine + L-proline, and L-isoleucine + L-tryptophan. Yeast extract stimulate the growth. The main products of amino acids fermentation were acetate, propionate and hydrogen. With peptone as substrate, isovalerate was detected, ethanol was as minor end product.

Source: underground water sample near the athalassic saline alkaline Texcoco lake in Mexico.

DNA G+C content (mol%): 40.0 (HPLC).

Type strain: IMP-300, DSM 18041, JCM 13990.

GenBank accession number (16S rRNA gene): DQ234901.

Other bacteria

Figure 138 shows a phylogenetic tree with three closely related, but not validly named, "*Clostridium*" species. These three species were mentioned in a paper in *Extremophiles* journal (Jones et al., 1998). In the future they could be described as a species of *Tindallia*.

References

Alain, K., P. Pignet, M. Zbinden, M. Quillevere, F. Duchiron, J.P. Donval, F. Lesongeur, G. Raguenes, P. Crassous, J. Querellou and M.A. Cambon-Bonavita. 2002. *Caminicella sporogenes* gen. nov., sp. nov., a novel thermophilic spore-forming bacterium isolated from an East-Pacific Rise hydrothermal vent. Int. J. Syst. Evol. Microbiol. *52*: 1621–1628.

Alam, S.I., A. Dixit, G.S. Reddy, S. Dube, M. Palit, S. Shivaji and L. Singh. 2006. *Clostridium schirmacherense* sp. nov., an obligately anaerobic, proteolytic, psychrophilic bacterium isolated from lake sediment of Schirmacher Oasis, Antarctica. Int. J. Syst. Evol. Microbiol. *56*: 715–720.

Alazard, D., C. Badillo, M.L. Fardeau, J.L. Cayol, P. Thomas, T. Roldan, J.L. Tholozan and B. Ollivier. 2007. *Tindallia texcoconensis* sp. nov., a new haloalkaliphilic bacterium isolated from lake Texcoco, Mexico. Extremophiles *11*: 33–39.

Alazard, D., C. Badillo, M.L. Fardeau, J.L. Cayol, P. Thomas, T. Roldan, J.L. Tholozan and B. Ollivier. 2009. *In* List of new names and new combinations previously effectively, but not validly, published. Validation List no. 128. Int. J. Syst. Evol. Microbiol. *59*: 1555–1556.

Andreesen, J.R., G. Gottschalk and H.G. Schlegel. 1970. *Clostridium formicoaceticum* nov. spec. isolation, description and distinction from *C. aceticum* and *C. thermoaceticum*. Arch. Mikrobiol. *72*: 154–174.

Andreesen, J.R., U. Zindel and P. Dürre. 1985. *Clostridium cylindrosporum (ex* Barker and Beck 1942) nom. rev. Int. J. Syst. Bacteriol. *35*: 206–208.

Arloing, S., Cornevin and Thomas. 1887. Le charbon symptomatique du boeuf. 2nd ed. Asselin and Houzeau, Paris, pp. 1–281.

Attwood, G.T., K. Reilly and B.K.C. Patel. 1996. *Clostridium proteoclasticum* sp. nov., a novel proteolytic bacterium from the bovine rumen. Int. J. Syst. Bacteriol. *46*: 753–758.

Baena, S., M.L. Fardeau, M. Labat, B. Ollivier, P. Thomas, J.L. Garcia and B.K. Patel. 1998. *Aminobacterium colombiense* gen. nov. sp. nov., an amino acid-degrading anaerobe isolated from anaerobic sludge. Anaerobe *4*: 241–250.

Baena, S., M.L. Fardeau, T.H.S. Woo, B. Ollivier, M. Labat and B.K.C. Patel. 1999. Phylogenetic relationships of three amino-acid-utilizing anaerobes, *Selenomonas acidaminovorans*, 'Selenomonas acidaminophila' and *Eubacterium acidaminophilum*, as inferred from partial 16S rDNA nucleotide sequences and proposal of *Thermanaerovibrio acidaminovorans* gen. nov., comb. nov and *Anaeromusa acidaminophila* gen. nov., comb. nov. Int. J. Syst. Bacteriol. *49*: 969–974.

Baena, S., M.L. Fardeau, M. Labat, B. Ollivier, J.L. Garcia and B.K. Patel. 2000. *Aminobacterium mobile* sp. nov., a new anaerobic amino-acid-degrading bacterium. Int. J. Syst. Evol. Microbiol. *50*: 259–264.

Baena, S. and B.K.C. Patel. 2009. Genus V. *Caloramator* Collins, Lawson, Willems, Cordoba, Fernández-Garayzábal, Garcia, Cai, Hippe and Farrow 1994, 812VP emend. Chrisostomos, Patel, Dwivedi and Denman 1996, 497. *In* De Vos, Garrity, Jones, Krieg, Ludwig, Rainey, Schleifer and Whitman (Editors), Bergey's Manual of Systematic Bacteriology, 2nd edn, Vol. 3. Springer, New York, pp. 834–838.

Barker, H.A. 1938. The fermentation of definite nitrogenous compounds by members of the genus *Clostridium*. J. Bacteriol. *36*: 322–323.

Barker, H.A. and J.V. Beck. 1942. *Clostridium acidi-uridi* and *Clostridium cylindrosporum*, organisms fermenting uric acid and some other purines. J. Bacteriol. *43*: 291–304.

Barker, H.A. and S.M. Taha. 1942. *Clostridium kluyveri*, an organism concerned in the formation of caproic acid from ethyl alcohol. J. Bacteriol. *43*: 347–363.

Bergey, D.H., F.C. Harrison, R.S. Breed, B.W. Hammer and F.M. Huntoon. 1923. Bergey's Manual of Determinative Bacteriology, 1st edn. The Williams & Wilkins Co., Baltimore.

Berkhoff, H.A., S.G. Campbell, H.B. Naylor and L.D. Smith. 1974. Etiology and pathogenesis of ulcerative enteritis ("quail disease"). Characterization of the causative anaerobe. Avian Dis. *18*: 195–204.

Berkhoff, H.A. 1985. *Clostridium colinum* sp. nov., nom. rev., the causative agent of ulcerative enteritis (quail disease) in quail, chickens, and pheasants. Int. J. Syst. Bacteriol. *35*: 155–159.

Beuscher, H.U. and J.R. Andreesen. 1984. *Eubacterium angustum* sp. nov., a Gram positive anaerobic, non-sporeforming, obligate purine fermenting organism. Arch. Microbiol. *140*: 2–8.

Biebl and Spröer. 2003. *In* Validation of the publication of new names and new combinations previously effectively published outside the IJSEM. List no. 91. Int. J. Syst. Evol. Microbiol *53*: 627–628.

Biebl, H. and C. Sproer. 2002. Taxonomy of the glycerol fermenting clostridia and description of *Clostridium diolis* sp. nov. Syst. Appl. Microbiol. *25*: 491–497.

Bienstock, B. 1906. *Bacillus putrificus*. Ann. Inst. Pasteur (Paris) *20*: 407–415.

Brill, J.A. and J. Wiegel. 1997. Differentiation between sporeforming and asporogenous bacteria by a PCR and Southern hybridization based method. J. Microbiol. Methods *31*: 29–36.

Brisbarre, N., M.L. Fardeau, V. Cueff, J.L. Cayol, G. Barbier, V. Cilia, G. Ravot, P. Thomas, J.L. Garcia and B. Ollivier. 2003. *Clostridium caminithermale* sp. nov., a slightly halophilic and moderately thermophilic bacterium isolated from an Atlantic deep-sea hydrothermal chimney. Int. J. Syst. Evol. Microbiol. *53*: 1043–1049.

Brock, T.D. and K. Od'ea. 1997. Amorphous ferrous sulfide as a reducing agent for culture of anaerobes. Appl. Environ. Microbiol. *33*: 254–256.

Broda, D.M., P.A. Lawson, R.G. Bell and D.R. Musgrave. 1999. *Clostridium frigidicarnis* sp. nov., a psychrotolerant bacterium associated with 'blown pack' spoilage of vacuum-packed meats. Int. J. Syst. Bacteriol. *49*: 1539–1550.

Broda, D.M., D.J. Saul, R.G. Bell and D.R. Musgrave. 2000a. *Clostridium algidixylanolyticum* sp. nov., a psychrotolerant, xylan-degrading, spore-forming bacterium. Int. J. Syst. Evol. Microbiol. *50*: 623–631.

Broda, D.M., D.J. Saul, P.A. Lawson, R.G. Bell and D.R. Musgrave. 2000b. *Clostridium gasigenes* sp. nov., a psychrophile causing spoilage of vacuumpacked meat. Int. J. Syst. Evol. Microbiol. *50*: 107–118.

Buckel, W., P.H. Janssen, A. Schuhmann, U. Eikmanns, P. Messner, U. Sleytr and W. Liesack. 1994. *Clostridium viride* sp. nov., a strictly anaerobic bacterium using 5-aminovalerate as growth substrate, previously assigned to *Clostridium aminovalericum*. Arch. Microbiol. *162*: 387–394.

Buckel, W., P.H. Janssen, A. Schuhmann, U. Eikmanns, P. Messner, U. Sleytr and W. Liesack. 1995. *In* Validation of the publication of new names and new publications previously effectively published outside the IJSB. List no. 54 Int. J. Syst. Bacteriol *45*: 619–620.

Bulloch, W., W.E. Bulloch, S.R. Douglas, H. Henry, J. McIntosh, R.A. O'Brien, M. Robertson and C.G.L. Wolf. 1919. Report on the anaerobic infections of wounds and the bacteriological and serological problems arising therefrom. Med. Res. Comm. (Gt. Brit.) Spec. Rep. Ser. *39*: 1–182.

Burri, R. and O. Ankersmit. 1906. *Bacterium clostridiiforme*. Untersuchungen über die Bakterien I Verdauungskanal des Rindes. Bakeriol. Parasitenkd. Infektionskr. Hyg. Abt. I Orig. *40*: 100–118.

Canale-Parola, E. and R.S. Wolfe. 1960a. Studies on *Sarcina ventriculi*. I. Stock culture method. J. Bacteriol. *79*: 857–859.

Canale-Parola, E. and R.S. Wolfe. 1960b. Studies on *Sarcina ventriculi*. II. Nutrition. J. Bacteriol. *79*: 860–862.

Canale-Parola, E., R. Borasky and R.S. Wolfe. 1961. Studies on *Sarcina ventriculi*. III. Localization of cellulose. J. Bacteriol. *81*: 311–318.

Canale-Parola, E. and R.S. Wolfe. 1964. Synthesis of Cellulose by *Sarcina ventriculi*. Biochim. Biophys. Acta *82*: 403–405.

Canale-Parola, E., M. Mandel and D.G. Kupper. 1967. The classification of sarcinae. Arch. Mikrobiol. *58*: 30–34.

Canale-Parola, E. 1970. Biology of the sugar-fermenting Sarcinae. Bacteriol. Rev. *34*: 82–97.

Cao, X., X. Liu and X. Dong. 2003. *Alkaliphilus crotonatoxidans* sp. nov., a strictly anaerobic, crotonate-dismutating bacterium isolated from a methanogenic environment. Int. J. Syst. Evol. Microbiol. *53*: 971–975.

Carbone, D. and A. Tombolato. 1917. Sulla macerazione rustica della canapé. Staz. Sper. Agr. Ital. *50:* 563–575.

Cardon, B.P. and H.A. Barker. 1946. Two new amino-acid-fermenting bacteria, *Clostridium propionicum* and *Diplococcus glycinophilus*. J. Bacteriol. *52:* 629–634.

Cato, E.P., W.E.C. Moore and L.V. Holdeman. 1968. *Clostridium oroticum* comb. nov. amended description. Int. J. Syst. Bacteriol. *18:* 9–13.

Cato, E.P., D.E. Hash, L.V. Holdeman and W.E.C. Moore. 1982. Electrophoretic study of *Clostridium* species. J. Clin. Microbiol *15:* 688–702.

Cato, E.P., W.L. George and S.M. Finegold. 1986. Genus *Clostridium* Prazmowski 1880, 23^{AL}. *In* Sneath, Mair, Sharpe and Holt (Editors), Bergey's Manual of Systematic Bacteriology, Vol. 2. The Williams & Wilkins Co., Baltimore, pp. 1141–1200.

Cato, E.P., W.L. George. and S.M.Finegold. 1988. *In* Validation of the publication of new names and new combinations previously effectively published outside the IJSB. List no. 25. Int. J. Syst. Bacteriol. *38:* 220–222.

Cayol, J.L., B. Ollivier, K.C. Patel, G. Prensier, J. Guezennec and J.L. Garcia. 1994. Isolation and characterization of *Halothermothrix orenii* gen. nov., sp. nov., a halophilic, thermophilic, fermentative, strictly anaerobic bacterium. Int. J. Syst. Bacteriol. *44:* 534–540.

Cayol, J.L., B. Ollivier, B.K.C. Patel, E. Ageron, P.A.D. Grimont, G. Prensier and J.L. Garcia. 1995. *Haloanaerobium lacusroseus* sp. nov., an extremely halophilic fermentative bacterium from the sediments of a hypersaline lake. Int. J. Syst. Bacteriol. *45:* 790–797.

Cayol, J.L., S. Ducerf, B.K. Patel, J.L. Garcia, P. Thomas and B. Ollivier. 2000a. *Thermohalobacter berrensis* gen. nov., sp. nov., a thermophilic, strictly halophilic bacterium from a solar saltern. Int. J. Syst. Evol. Microbiol. *50:* 559–564.

Cayol, J.L., S. Ducerf, B.K.C. Patel, J.L. Garcia, P. Thomas and B. Ollivier. 2000b. *Thermohalobacter berrensis* gen. nov., sp. nov., a thermophilic, strictly halophilic bacterium from a solar saltern. Int. J. Syst. Evol. Microbiol. *50:* 559–564.

Chamkha, M., J.L. Garcia and M. Labat. 2001a. Metabolism of cinnamic acids by some *Clostridiales* and emendation of the descriptions of *Clostridium aerotolerans*, *Clostridium celerecrescens* and *Clostridium xylanolyticum*. Int. J. Syst. Evol. Microbiol. *51:* 2105–2111.

Chamkha, M., M. Labat, B.K. Patel and J.L. Garcia. 2001b. Isolation of a cinnamic acid-metabolizing *Clostridium glycolicum* strain from oil mill wastewaters and emendation of the species description. Int. J. Syst. Evol. Microbiol. *51:* 2049–2054.

Chamkha, M., B.K.C. Patel, J.L. Garcia and M. Labat. 2001c. Isolation of *Clostridium bifermentans* from oil mill wastewaters converting cinnamic 3-phenylpropionic acid and emendation of acid to the species. Anaerobe *7:* 189–197.

Cheng, G.S., C.M. Plugge, W. Roelofsen, F.P. Houwen and A.J.M. Stams. 1992. *Selenomonas acidaminovorans* sp. nov., a versatile thermophilic proton reducing, anaerobe able to grow by decarboxylation of succinate to propionate. Arch. Microbiol. *157:* 169–175.

Choukévitch, J. 1911. Étude de la flore bactérienne du gros intestin du cheval. Ann. Inst. Pasteur (Paris) *25:* 245–368.

Chrisostomos, S., B.K.C. Patel, P.P. Dwivedi and S.E. Denman. 1996. *Caloramator indicus* sp. nov., a new thermophilic anaerobic bacterium isolated from the deep-seated non volcanically heated waters of an Indian artesian aquifer. Int. J. Syst. Bacteriol *46:* 497–501.

Cirne, D.G., O.D. Delgado, S. Marichamy and B. Mattiasson. 2006. *Clostridium lundense* sp. nov., a novel anaerobic lipolytic bacterium isolated from bovine rumen. Int. J. Syst. Evol. Microbiol. *56:* 625–628.

Claus, D., A.A. Chowdhury and K.V. Tilak. 1970. On cultivation and preservaation of *Sarcina ventriculi*. Publ. Fac. Sci. Univ. J.E. Purkyne, Brno *47:* 135–142.

Claus, D. and H. Wilmanns. 1974. Enrichment and selective isolation of *Sarcina maxima* Lindner. Arch. Microbiol. *96:* 201–204.

Clavel, T., R. Lippman, F. Gavini, J. Dore and M. Blaut. 2007a. *Clostridium saccharogumia* sp. nov. and *Lactonifactor longoviformis* gen. nov.,

sp. nov., two novel human faecal bacteria involved in the conversion of the dietary phytoestrogen secoisolariciresinol diglucoside. Syst. Appl. Microbiol. *30:* 16–26.

Clavel, T., R. Lippman, F. Gavini, J. Doré and M. Blunt. 2007b. *In* Validation of the publication of new names and new combinations previously effectively published outside the IJSEM. List no. 115. Int. J. Syst. Evol. Microbiol. *30:* 16–26.

Collins, M.D., U.M. Rodrigues, R.H. Dainty, R.A. Edwards and T.A. Roberts. 1992. Taxonomic studies on a psychrophilic *Clostridium* from vacuum-packed beef: description of *Clostridium estertheticum* sp. nov. FEMS Microbiol. Lett. *75:* 235–240.

Collins, M.D., A.M. Williams and S. Wallbanks. 1993. *In* Validation of the publication of new names and new combinations previously effectively published outside the IJSB. List no. 44. Int. J. Syst. Bacteriol. *43:* 188–189.

Collins, M.D., P.A. Lawson, A. Willems, J.J. Cordoba, J. Fernández-Garayzábal, P. Garcia, J. Cai, H. Hippe and J.A. Farrow. 1994. The phylogeny of the genus *Clostridium*: proposal of five new genera and eleven new species combinations. Int. J. Syst. Bacteriol *44:* 812–826.

Davey, M.E., W.A. Wood, R. Key, K. Nakamura and D.A. Stahl. 1993. Isolation of three species of *Geotoga* and *Petrotoga*: two new genera, representing a new lineage in the bacterial line of descent distantly related to the *Thermotogales*. Syst. Appl. Microbiol. *16:* 191–200.

Donker, H.J.L. 1926. Bijdrage tot de Kennis der Boterzuur-, Butylalcoholen acetonigistigen. Diss. Delft. W.E. Meinema, Delft.

Dörner, C. and B. Schink. 1990. *Clostridium homopropionicum* sp. nov., a new strict anaerobe growing with 2-hydroxybutyrate, 3-hydroxybutyrate, or 4-hydroxybutyrate. Arch. Microbiol. *154:* 342–348.

Dörner, C. and B. Schink. 1991. *In* Validation of the publication of new names and new combinations previously effectively published outside the IJSB. List no. 39. Int. J. Syst. Bacteriol *41:* 580–581.

Douglas, S.R., A. Fleming and L. Colebrook. 1919. Studies in wound infections. Med. Res. Comm. (Gt. Brit.) Spec. Rep. Ser. *57:* 1–159.

Drent, W.J., G.A. Lahpor, W.M. Wiegant and J.C. Gottschal. 1991. Fermentation of inulin by *Clostridium thermosuccinogenes* sp. nov., a thermophilic anaerobic bacterium isolated from various habitats. Appl. Environ. Microbiol. *57:* 455–462.

Drent, W.J., G.A. Laphor, W.M. Wiegant and J.C. Gottschal. 1995. *In* Validation of the publication of new names and new combonations previously effectively published outside the IJSB. List no. 55. Int. J. Syst. Bacteriol *45:* 879–880.

Duda, V.I., A.V. Lebedinsky, M.S. Mushegjan and I.L. Mitjushina. 1987. A new anaerobic bacterium, forming up to five endospores per cell - *Anaerobacter polyendosporus* gen. et spec. nov. Arch. Microbiol. *148:* 121–127.

Duda, V.I., A.V. Lebedinsky, M.S. Mushegjan and L.L. Mitjushina. 1996. *In* Validation of the publication of new names and new combinations previously effectively published outside the IJSB. List no. 57. Int. J. Syst. Bacteriol. *46:* 625–626.

Dürre, P., W. Andersch and J.R. Andreesen. 1981. Isolation and characterization of an adenine-utilizing, anaerobic sporeformer, *Clostridium purinolyticum* sp. nov. Int. J. Syst. Bacteriol. *31:* 184–194.

Engle, M., Y.H. Li, F. Rainey, S. DeBlois, V. Mai, A. Reichert, F. Mayer, P. Messner and J. Wiegel. 1996. *Thermobrachium celere* gen. nov., sp. nov., a rapidly growing thermophilic, alkalitolerant, and proteolytic obligate anaerobe. Int. J. Syst. Bacteriol. *46:* 1025–1033.

Euzéby, J.P. 1997. List of bacterial names with standing in nomenclature: a folder available on the internet. Int. J. Syst. Bacteriol.: 590–592.

Euzéby, J.P. 2007. List of Bacterial Names with Standing in Nomenclature (updated after every issue of IJSEM). http://www.bacterio.cict.fr/search.html. [Online.]

Fardeau, M.-L., B. Ollivier, J.-L. Garcia and B.K.C. Patel. 2001. Transfer of *Thermobacterioides leptospartum* and *Clostridium thermolacticum* as *Clostridium stercorarium* subsp. *leptospartum* subsp. nov., comb. nov. and *C. stercorarium* subsp. *thermolacticum* subsp. nov., comb. nov. Int. J. Syst. Evol. Microbiol *51:* 1127–1131.

Fellers, C.R. and R.W. Clough. 1925. Indol and skatol determination of bacterial cultures. J. Bacteriol. *10:* 105–133.

Felsenstein, J. 1985. Confidence limits on phylogenies: an approach using the bootstrap. Evolution *39:* 783–791.

Fendrich, C., H. Hippe and G. Gottschal. 1990. *Clostridium halophilum* sp. nov. and *C. litorale* sp. nov., an obligate halophilic and a marine species degrading betaine in the Strickland reaction. Arch. Microbiol. *154:* 127–132.

Fendrich, C., H. Hippe and G. Gottschalk. 1991. *In* Validation of the publication of new names and new combinatios previously effectively published outside the IJSB. List no. 39. Int. J. Syst. Bacteriol. *41:* 580–581.

Flügge, C. 1886. Die Mikrooganismen. F.C.W. Vogel, Leipzig.

Ford, W.W. 1927. Textbook of bacteriology. Saunders, Philadelphia.

Fox, G.E., E. Stackebrandt, R.B. Hespell, J. Gibson, J. Maniloff, T.A. Dyer, R.S. Wolfe, W.E. Balch, R.S. Tanner, L.J. Magrum, L.B. Zablen, R. Blakemore, R. Gupta, L. Bonen, B.J. Lewis, D.A. Stahl, K.R. Luehrsen, K.N. Chen and C.R. Woese. 1980. The phylogeny of prokaryotes. Science *209:* 457–463.

Garnova, E.S., T.P.T. T.N. Zhilina and A.M. Lysenko. 2003a. *In* Validation of publication of new names and new combinations previously effectively published outside the IJSEM. List no. 93. Int. J. Syst. Evol. Microbiol. *53:* 1219–1220.

Garnova, E.S., T.N. Zhilina, T.P. Tourova and A.M. Lysenko. 2003b. *Anoxynatronum sibiricum* gen.nov., sp.nov. alkaliphilic saccharolytic anaerobe from cellulolytic community of Nizhnee Beloe (Transbaikal region). Extremophiles *7:* 213–220.

Gaston, L.W. and E.R. Stadtman. 1963. Fermentation of ethylene glycol by *Clostridium glycolicum*, sp. n. J. Bacteriol. *85:* 356–362.

Giménez, D.F. and A.S. Ciccarelli. 1970. Another type of *Clostridium botulinum*. Zentralbl. Bakteriol. Parasitenkd. Infektionskr. Hyg. *215:* 221–224.

Goodsir, J. 1842. History of a case in which a fluid periodically ejected from the stomach contained vegetable organisms of an undescribed form. Edinburgh Surg. J. *57:* 430–443.

Gottschalk, G. and M. Braun. 1981. Revival of name *Clostridium aceticum*. Int. J. Syst. Bacteriol. *31:* 476.

Gottschalk, G., J.R. Andreesen and H. Hippe. 1981. The genus *Clostridium* (nonmedical aspects). *In* Starr, Stolp, Trüper, Balows and Schlegel (Editors), The Prokaryotes, a handbook on the habitats, isolation and identification of bacteria. Springer-Verlag, Berlin, Heidelberg, New York, pp. 1767–1803.

Greetham, H.L., G.R. Gibson, C. Giffard, H. Hippe, B. Merkhoffer, U. Steiner, E. Falsen and M.D. Collins. 2003. *Clostridium colicanis* sp. nov., from canine faeces. Int. J. Syst. Evol. Microbiol. *53:* 259–262.

Hall, I.C. and J.P. Scott. 1927. *Bacillus sordellii*, a cause of malignant edema in man. J. Infect. Dis.: *41* 329–335.

Hall, I.C. and R.W. Whitehead. 1927. A pharmaco-bacteriologic study of African poisoned arrows. J. Infect. Dis. *41:* 51–69.

Hall, I.C. 1929. The occurrence of *Bacillus sordelllii* in icterohemoglobinuria of cattle in Nevada. J. Infect. Dis. *45:* 156–162.

Hall, I.C. and E. O'Toole. 1935. Intestinal flora in newborn infants with a description of a new pathogenic anaerobe, *Bacillus difficilis*. Am. J. Dis. Child. *49:* 390–402.

Hall, T.H. 1999. BioEdit: biological sequence alignment editor for Win95/98/NT/2K/XP. Nucleic Acids Symp. Ser. *41:* 95–98.

Hardman, J.K. and T.C. Stadtman. 1960. Metabolism of ω-amino acids. II. Fermentation of Δ-aminovaleric acid by *Clostridium aminobutyricum* n. sp. J. Bacteriol. *79:* 549–552.

Hartsell, S.E. and L.F. Rettger. 1934. A taxonomic study of "*Clostridium putrificum*" and its establishment as a definite entity *Clostridium lentoputrescens* nov. spec. J. Bacteriol. *27:* 497–514.

Hauduroy, P., G. Ehringer, A. Urbain, G. Guillot and J. Magrou. 1937. Dictionnaire des bactéries pathogènes. Masson and Co., Paris, pp. 1–597.

Hauschild, A.H.W. and L.V. Holdeman. 1974. *Clostridium celatum* sp. nov., isolated from normal human feces. Int. J. Syst. Bacteriol. *24:* 478–481.

He, Y.L., Y.F. Ding and Y.Q. Long. 1991. Two cellulolytic *Clostridium* species: *Clostridium cellulosi* sp. nov. and *Clostridium cellulofermentans* sp. nov. Int. J. Syst. Bacteriol. *41:* 306–309.

Heller, H.H. 1922. Certain genera of the *Clostridiaceae*. Studies in pathogenic anaerobes. J. Bacteriol. *7:* 1–38.

Hellinger. 1944. Studies on a pink butyric acid *Clostridium*. Presented at the Commemorative Vol. to Dr. Weizmann's 70th birthday – Private print Nov. 1944, pp. 37–46.

Henry, H. 1917. An investigation of the cultural reactions of certain anaerobes found in war wounds. J. Pathol. Bacteriol. *21:* 344–385.

Hernandez-Eugenio, G., M.L. Fardeau, J.L. Cayol, B.K.C. Patel, P. Thomas, H. Macarie, J.L. Garcia and B. Ollivier. 2002. *Clostridium thiosulfatireducens* sp. nov., a proteolytic, thiosulfate- and sulfur-reducing bacterium isolated from an upflow anaerobic sludge blanket (UASB) reactor. Int. J. Syst. Evol. Microbiol. *52:* 1461–1468.

Hethener, P., A. Brauman and J.-L. Garcia. 1992a. *In* Validation of the publication of new names and new combinations previously published outside the IJSB. List no. 41. Int. J. Syst. Bacteriol *42:* 327–328.

Hethener, P., A. Brauman and J.-L. Garcia. 1992b. *Clostridium termitidis* sp. nov., a cellulolytic bacterium from the gut of the wood-feeding termite, *Nasutitermes lujae*. Syst. Appl. microbiol. *15:* 52–58.

Himelbloom, B. and E. Canale-Parola. 1989. *In* Validation of the publication of new names and new combinations previously effectively published outside th IJSB. List no. 31. Int. J. Syst. Bacteriol *39:* 495–497.

Himelbloom, B.H. and E. Canale-Parola. 1989. *Clostridium methylpentosum* sp. nov., a ring-shaped intestinal bacterium that ferments only methylpentoses and pentoses. Arch. Microbiol. *151:* 287–293.

Hippe, H., J.R. Andreesen and G. Gottschalk. 1992. The genus *Clostridium*: nonmedical. *In* Balows, Trüper, Dworkin, Harder and Schleifer (Editors), The Prokaryotes, 2nd edn, Vol. 2. Springer-Verlag, New York, pp. 1800–1866.

Holdeman, L.V. and J.B. Brooks. 1970. Variation among strains of *Clostridium botulinum* and related clostridia. Presented at the Proc. of 1st U.S.-Japan Conference on Toxic Microorganisms, 1968. US Govt Printing Office, Washington, DC, pp. 278–286.

Holdeman, L.V. and W.E.C. Moore. 1970. Outline of Clinical Methods in Anaerobic Bacteriology, 2nd revision. *In* Cato, Cummins, Holdeman, Johnson, Moore, Smibert and Smith (Editors), Virginia Polytechnic Institute, Anaerobe Laboratory, Blacksburg, Virginia, pp. 57–66.

Holdeman, L.V., E.P. Cato and W.E.C. Moore. 1971. *Clostridium ramosum* (Vuillemin) comb. nov. emended description and proposed neotype strain. Int. J. Syst. Bacteriol *21:* 35–39.

Holdeman, L.V. and W.E.C. Moore. 1972. Anaerobe Laboratory Manual. Anaerobe Laboratory, Virginia Polytechnic Institute and State University, Blacksburg.

Holdeman, L.V. and W.E.C. Moore. 1974. New genus, *Coprococcus*, twelve new species, and emended descriptions of four previously described species of bacteria from human feces. Int. J. Syst. Bacteriol. *24:* 260–277.

Holdeman, L.V., E.P. Cato and W.E.C. Moore (Editors). 1977a. Anaerobe Laboratory Manual, 4th edn. Anaerobe Laboratory, Virginia Polytechnic Institute and State University, Blacksburg, VA.

Holdeman, L.V., E.P. Cato and W.E.C. Moore. 1977b. Culture methods: Use of pre-reduced media. *In* Holdeman, Cato and Moore (Editors), Anaerobic Laboratory Manual, 4th edn. Anaerobe Laboratory, Virginia Polytechnic Institute and State University, Blacksburg, pp. 117–149.

Holt, S.C. and E. Canale-Parola. 1967. Fine structure of *Sarcina maxima* and *Sarcina ventriculi*. J. Bacteriol. *93:* 399–410.

Horn, N. 1987. *Clostridium disporicum* sp. nov., a saccharolytic species able to form two spores per cell, isolated from a rat cecum. Int. J. Syst. Bacteriol. *37:* 398–401.

Huber, R., T.A. Langworthy, H. Konig, M. Thomm, C.R. Woese, U.B. Sleytr and K.O. Stetter. 1986. *Thermotoga maritima* sp. nov. represents a new genus of unique extremely thermophilic eubacteria growing up to 90-degrees C. Arch. Microbiol. *144:* 324–333.

Hubert, J., H. Ionesco and M. Sebald. 1981. Detection de *Clostridium difficile* par isolement sur milieu minimal selectif et par immunofluorescence. Ann. Microbiol. (Inst. Pasteur) *132A:* 149–157.

Hungate, R.E. 1944. Studies on Cellulose Fermentation: I. The culture and physiology of an anaerobic cellulose-digesting bacterium. J. Bacteriol. *48:* 499–513.

Hungate, R.E. 1969. A roll tube method for cultivation of strict anaerobes. *In* Norris and Ribbons (Editors), Methods in Microbiology, Vol. 3B. Academic Press, London and New York, pp. 117–132.

Hutson, R.A., D.E. Thompson, P.A. Lawson, R.P. Schocken-Itturino, E.C. Bottger and M.D. Collins. 1993. Genetic interrelationships of proteolytic *Clostridium botulinum* types A, B, and F and other members of the *Clostridium botulinum* complex as revealed by small-subunit rRNA gene sequences. Antonie van Leeuwenhoek *64:* 273–283.

Imhoff-Stuckle, D. and N. Pfennig. 1983. Isolation and characterization of a nicotinic-acid degrading sulfate-reducing bacterium, *Desulfococcus niacini* sp. nov. Arch. Microbiol. *136:* 194–198.

Jain, M.K. and J.G. Zeikus. 1988a. *In* Validation of the publication of new names and new combinations previously effectively published outside the IJSB. List no. 26. Int. J. Syst. Bacteriol *38:* 328–329.

Jain, M.K. and J.G. Zeikus. 1988b. Taxonomic distinction of two new protein-specific, hydrolytic anaerobes: isolation and characterization of *Clostridium proteolyticum* sp. nov. and *Clostridium collagenovorans* sp. nov. Syst. Appl. Microbiol. *10:* 134–141.

Jeong, H., H. Yi, Y. Sekiguchi, M. Muramatsu, Y. Kamagata and J. Chun. 2004. *Clostridium jejuense* sp. nov., isolated from soil. Int. J. Syst. Evol. Microbiol. *54:* 1465–1468.

Johnson, J.L. and B.S. Francis. 1975. Taxonomy of the clostridia: ribosomal ribonucleic acid homologies among the species. J. Gen. Microbiol. *88:* 229–244.

Jones, B.E., W.D. Grant, A.W. Duckworth and G.G. Owenson. 1998. Microbial diversity of soda lakes. Extremophiles *2:* 191–200.

Jones, D.T. and S. Keis. 1995. Origins and relationships of industrial solvent-producing clostridial strains. FEMS Microbiol. Rev. *17:* 223–232.

Jordan, D.C. and P.J. McNicol. 1979. A new nitrogen-fixing *Clostridium* species from a high Arctic ecosystem. Can. J. Microbiol. *25:* 947–948.

Jukes, T.H. and C. Cantor. 1969. Evolution of protein molecules. *In* Murano (Editor), Mammalian protein metabolism. Academic Press, New York pp. 21–132.

Kalchayanand, N., B. Ray and R.A. Field. 1993. Characteristics of psychrotrophic *Clostridium laramie* causing spoilage of vacuumpackaged refrigerated fresh and roasted beef. J. Food Prot. *56:* 13–17.

Kandler, O., D. Claus and A. Moore. 1972. [Amino acid sequence of the murein of *Sarcina ventriculi* and *Sarcina maxima*]. Arch. Mikrobiol. *82:* 140–146.

Kane, M.D., A. Brauman and J.A. Breznak. 1991. *Clostridium mayombei* sp. nov., an H_2/CO_2 acetogenic bacterium from the gut of the African soil-feeding termite, Cubitermes speciosus. Arch. Microbiol. *156:* 99–104.

Kane, M., A. Brauman and J.A. Breznak. 1992. *In* Validation of the publication of new names and new combinations previously effectively published outside the IJSB. List no. 40. Int. J. Syst. Bacteriol. *42:* 191–192.

Kaneuchi, C., Y. Benno and T. Mitsuoka. 1976a. *Clostridium coccoides*, a new species from feces of mice. Int. J. Syst. Bacteriol. *26:* 482–486.

Kaneuchi, C., K. Watanabe, A. Terada, Y. Benno and T. Mitsuoka. 1976b. Taxonomic study of *Bacteroides clostridiiformis* subsp. *clostridiiformis*

(Burri and Ankersmit) Holdeman and Moore and of related organisms: proposal of *Clostridium clostridiiformis* (Burri and Ankersmit) comb. nov. and *Clostridium symbiosum* (Stevens) comb. nov. Int. J. Syst. Bacteriol. *26:* 195–204.

Kaneuchi, C., T. Miyazato, T. Shinjo and T. Mitsuoka. 1979. Taxonomic study of helically coiled, sporeforming anaerobes isolated from the intestines of humans and other animals: *Clostridium cocleatum* sp. nov. and *Clostridium spiroforme* sp. nov. Int. J. Syst. Bacteriol. *29:* 1–12.

Kato, S., S. Haruta, Z.J. Cui, M. Ishii, A. Yokota and Y. Igarashi. 2004. *Clostridium straminisolvens* sp. nov., a moderately thermophilic, aerotolerant and cellulolytic bacterium isolated from a cellulose-degrading bacterial community. Int. J. Syst. Evol. Microbiol. *54:* 2043–2047.

Keis, S., R. Shaheen and D.T. Jones. 2001. Emended descriptions of *Clostridium acetobutylicum* and *Clostridium beijerinckii*, and descriptions of *Clostridium saccharoperbutylacetonicum* sp. nov. and *Clostridium saccharobutylicum* sp. nov. Int. J. Syst. Evol. Microbiol. *51:* 2095–2103.

Kell, D.B., M.W. Peck, G. Rodger and J.G. Morris. 1981. On the permeability to weak acids and bases of the cytoplasmic membrane of *Clostridium pasteurianum*. Biochem. Biophys. Res. Commun. *99:* 81–88.

Kelly, W.J., R.V. Asmundson and D.H. Hopcroft. 1987. Isolation and characterization of a strictly anaerobic, cellulolytic spore former, *Clostridium chartatabidum* sp. nov. Arch. Microbiol. *147:* 169–173.

Kelly, W.J., R.V. Asmundson and D.H. Hopcroft. 1996. *In* Validation of the publication of new names and new combinations previously effectively published outside the IJSEM. List no. 57. Int. J. Syst. Bacteriol. *46:* 625–626.

Kevbrin, V.V. and G.A. Zavarzin. 1992. Effect of sulfur compounds on the growth of the halophilic homoacetic bacterium *Acetohalobium arabaticum*. Microbiology (En. transl. from Mikrobiologiya) *61:* 563–567.

Kevbrin, V.V., T.N. Zhilina, F.A. Rainey and G.A. Zavarzin. 1998. *Tindallia magadii* gen. nov., sp. nov.: an alkaliphilic anaerobic ammonifier from soda lake deposits. Curr. Microbiol. *37:* 94–100.

Kevbrin, V.V., T.N. Zhilina, F.A. Rainey and G.A. Zavarzin. 1999. *In* Validation of publication of new names and new combinations previously effectively published outside the IJSB. List no. 68. Int. J. Syst. Bacteriol. *49:* 1–3.

Kim, S., H. Jeong and J. Chun. 2006. *Clostridium ganghwense* sp. nov., isolated from tidal flat sediment. Int. J. Syst. Evol. Microbiol. *56:* 691–693.

Kim, S., H. Jeong and J. Chun. 2007. *Clostridium aestuarii* sp. nov., from tidal flat sediment. Int. J. Syst. Evol. Microbiol. *57:* 1315–1317.

Kitahara, M., F. Takamine, T. Imamura and Y. Benno. 2000. Assignment of *Eubacterium* sp. VPI 12708 and related strains with high bile acid 7 alpha-dehydroxylating activity to *Clostridium scindens* and proposal of *Clostridium hylemonae* sp. nov., isolated from human faeces. Int. J. Syst. Evol. Microbiol. *50:* 971–978.

Kitahara, M., F. Takamine, T. Imamura and Y. Benno. 2001. *Clostridium hiranonis* sp. nov., a human intestinal bacterium with bile acid 7 alpha-dehydroxylating activity. Int. J. Syst. Evol. Microbiol. *51:* 39–44.

Klein, E. 1899. Ein Beitrag zur Bakteriologie der Leichenverwesung. Zentralbl. Bakeriol. Parasitenkd. Infektionskr. Hyg. Abt. I Orig. *25:* 278–284.

Klein, E. 1904. Ein neuer tierpathogenen Mikrobe – *Bacillus carnis*. Zentralbl. Bakteriol Parasitenkd. Infektionskr. Hyg., Abt. I. Orig. *35:* 450–461.

Kluyver, A.J. 1931. The chemical activities of micro-organisms. University of London Press, Ltd., London.

Knöll, H. and R. Horschak. 1964. Zur Eranährungs-physiologie der Gärungssarcinen. Montatsber. Dtsch. Akad. Wiss. Berl. *6:* 847–849.

Knöll, H. 1965. Zur Biologie der Gärungssarcinen. Monatsber. Dtsch. Akad. Wiss. Berl. *7:* 475–477.

Knöll, H. and R. Horschak. 1971. Zur Sporulation der Gärungssarcinen. Monatsber. Dtsch. Akad. Wiss. Berl. *13:* 222–224.

Kopečný, J., M. Zorec, J. Mrazek, Y. Kobayashi and R. Marinsek-Logar. 2003. *Butyrivibrio hungatei* sp. nov. and *Pseudobutyrivibrio xylanivorans* sp. nov., butyrate-producing bacteria from the rumen. Int. J. Syst. Evol. Microbiol. *53:* 201–209.

Kotsyurbenko, O.R., A.N. Nozhevnikova, G.A. Osipov, N.A. Kostrikina and A.M. Lysenko. 1995. A new psychoactive bacterium *Clostridium fimetarium*, isolated from cattle manure digested at low temperature. Microbiology (En. transl. from Mikrobiologiya) *64:* 681–686.

Kotsyurbenko, O.R., M.V. Simankova, A.N. Nozhevnikova, T.N. Zhilina, N.P. Bolotina, A.M. Lysenko and G.A. Osipov. 1997. *In* Validation of the publication of new names and new combinations previously effectively published outside the IJSB. List no. 60. Int. J. Syst. Bacteriol. *47:* 242.

Krumholz, L.R. and M.P. Bryant. 1985. *Clostridium pfennigii* sp. nov. uses methoxyl groups of monobenzenoids and produces butyrate. Int. J. Syst. Bacteriol. *35:* 454–456.

Kuhner, C.H., C. Matthies, G. Acker, M. Schmittroth, A.S. Gößner and H.L. Drake. 2000. *Clostridium akagii* sp. nov. and *Clostridium acidisoli* sp. nov.: acid-tolerant, N_2-fixing clostridia isolated from acidic forest soil and litter. Int. J. Syst. Evol. Microbiol. *50:* 873–881.

Kupfer, D.G. and E. Canale-Parola. 1967. Pyruvate metabolism in *Sarcina maxima.* J. Bacteriol. *94:* 984–990.

Kupfer, D.G. and E. Canale-Parola. 1968. Fermentation of glucose by *Sarcina maxima.* J. Bacteriol. *95:* 247–248.

Kusel, K., T. Dorsch, G. Acker, E. Stackebrandt and H.L. Drake. 2000. *Clostridium scatologenes* strain SL1 isolated as an acetogenic bacterium from acidic sediments. Int. J. Syst. Evol. Microbiol. *50:* 537–546.

Lawson, P., R.H. Dainty, N. Kristiansen, J. Berg and M.D. Collins. 1994. Characterization of a psychrotrophic *Clostridium* causing spoilage in vacuum-packed cooked pork: description of *Clostridium algidicarnis* sp. nov. Lett. Appl. Microbiol. *19:* 153–157.

Lawson, P., R.H. Dainty, N. Kristiansen, J. Berg and M.D. Collins. 1995. *In* Validation of the publication of new names and new combinations previously effectively published outside the IJSB. List no. 52. Int. J. Syst. Evol. Bacteriol *45:* 197–198.

Lee, W., T. Fujisawa, S. Kawamura, T. Itoh and T. Mitsuoka. 1989. *Clostridium intestinalis* sp. nov., an aerotolerant species isolated from the feces of cattle and pigs. Int. J. Syst. Bacteriol. *39:* 334–336.

Lee, Y.E., M.K. Jain, C.Y. Lee, S.E. Lowe and J.G. Zeikus. 1993. Taxonomic distinction of saccharolytic thermophilic anaerobes: description of *Thermoanaerobacterium xylanolyticum* gen. nov., sp. nov., and *Thermoanaerobacterium saccharolyticum* gen. nov., sp. nov., reclassification of *Thermoanaerobium brockii, Clostridium thermosulfurogenes,* and *Clostridium thermohydrosulfuricum* E100-69 as *Thermoanaerobacter brockii* comb. nov., *Thermoanaerobacterium thermosulfurigenes* comb. nov., and *Thermoanaerobacter thermohydrosulfuricus* comb. nov., respectively, and transfer of *Clostridium thermohydrosulfuricum* 39e to *Thermoanaerobacter ethanolicus.* Int. J. Syst. Bacteriol. *43:* 41–51.

Lee, Y.J., C.S. Romanek and J. Wiegel. 2007. *Clostridium aciditolerans* sp. nov., an acid-tolerant spore-forming anaerobic bacterium from constructed wetland sediment. Int. J. Syst. Evol. Microbiol. *57:* 311–315.

Leruyet, P., H.C. Dubourguier, G. Albagnac and G. Prensier. 1985. Characterization of *Clostridium thermolacticum* sp. nov., a hydrolytic thermophilic anaerobe producing high amounts of lactate. Syst. Appl. Microbiol. *6:* 196–202.

Li, Y.H., L. Mandelco and J. Wiegel. 1993. Isolation and characterization of a moderately thermophilic anaerobic alkaliphile, *Clostridium paradoxum* sp. nov. Int. J. Syst. Bacteriol. *43:* 450–460.

Li, Y.H., M. Engle, N. Weiss, L. Mandelco and J. Wiegel. 1994. *Clostridium thermoalcaliphilum* sp. nov., an anaerobic and thermotolerant facultative alkaliphile. Int. J. Syst. Bacteriol. *44:* 111–118.

Liebert, F. 1909. Het afbreken van urinezuur door bakterien. Verslagen van de gewone vergadering der wis- en natuurkundige afdeeling. K. Akad. van wetenschappen te Amsterdam *17:* 990–1001.

Linder, P. 1888. Die *Sarcina* Organismen der Gärungsgewerben. Inaugural Dissertation, Friedrich-Wilhems Universitat, 1–58.

Lindner, P. 1888. Die sarcina organismen der gärungsgewerben. Inaugural dissertation. Friedrich-Wilhelms Universitat.

Liou, J.S., D.L. Balkwill, G.R. Drake and R.S. Tanner. 2005. *Clostridium carboxidivorans* sp. nov., a solvent-producing *Clostridium* isolated from an agricultural settling lagoon, and reclassification of the acetogen *Clostridium scatologenes* strain SL1 as *Clostridium drakei* sp. nov. Int. J. Syst. Evol. Microbiol. *55:* 2085–2091.

Ludwig, W., K. H. Schleifer and W. B. Whitman. 2009. Revised road map to the phylum *Firmicutes.* (Editors), Bergey's Manual of Systematic Bacteriology, Vol. 3, 2nd edn, Springer, New York.

Lund, B.M., T.F. Brocklehurst and G.M. Wyatt. 1981a. Characterization of strains of *Clostridium puniceum* sp. nov., a pink-pigmented, pectolytic bacterium. J. Gen. Microbiol. *122:* 17–26.

Lund, B.M., T.F. Brocklehurst and G.M. Wyatt. 1981b. *In* Validation of the publication of new names and new combinations previously effectively published outside the IJSB. List no. 6. Int. J. Syst. Bacteriol *31:* 215–218.

Macé, E. 1889. Traité Pratique de Bactériologie, 1st Ed. edn. Ballière, Paris.

Macy, J.M., J.E. Snellen and R.E. Hungate. 1972. Use of syringe methods for anaerobiosis. Am. J. Clin. Nutr. *25:* 1318–1323.

Madden, R.H., M.J. Bryder and N.J. Poole. 1982. Isolation and characterization of an anaerobic, cellulolytic bacterium, *Clostridium papyrosolvens* sp. nov. Int. J. Syst. Bacteriol. *32:* 87–91.

Madden, R.H. 1983. Isolation and characterization of *Clostridium stercorarium* sp. nov., cellulolytic thermophile. Int. J. Syst. Bacteriol. *33:* 837–840.

Matthies, C., C.H. Kuhner, G. Acker and H.L. Drake. 2001. *Clostridium uliginosum* sp. nov., a novel acid-tolerant anaerobic bacterium with connecting filaments. Int. J. Syst. Evol. Microbiol. *51:* 1119–1125.

McBryde, C.N. 1911. A bacteriological study of ham souring. U.S. Bur. Anim. Ind. *132:* 1–55.

McClung, L.S. 1943. On the enrichment and purification of chromogenic sporeforming anaerobic bacteria. J. Bacteriol. *46:* 507–511.

McClung, L.S. and E. McCoy. 1957. Genus II *Clostridium* Prazmowski 1880. *In* Breed, Murray and Smith (Editors), Bergey's Manual of Determinative Bacteriology, 7th edn. The Williams & Wilkins Co, Baltimore, pp. 634–693.

McCoy, E., E.B. Fred, W.H. Peterson and E.G. Hastings. 1926. A cultural study of the acetone butyl alcohol organisms. J. Infect. Dis. *39:* 457–483.

McCoy, E. and L.S. McClung. 1935. Studies on anaerobic bacteria. VI. The nature and systematic position of a new chromogenic *Clostridium.* Arch. Mikrobiol. *6:* 230–238.

Mechichi, T., M. Labat, J.L. Garcia, P. Thomas and B.K. Patel. 1999a. Characterization of a new xylanolytic bacterium, *Clostridium xylanovorans* sp. nov. Syst. Appl. Microbiol. *22:* 366–371.

Mechichi, T., M. Labat, B.K. Patel, T.H. Woo, P. Thomas and J.L. Garcia. 1999b. *Clostridium methoxybenzovorans* sp. nov., a new aromatic *o*-demethylating homoacetogen from an olive mill wastewater treatment digester. Int. J. Syst. Bacteriol. *49:* 1201–1209.

Mechichi, T., M.L. Fardeau, M. Labat, J.L. Garcia, F. Verhe and B.K. Patel. 2000a. *Clostridium peptidivorans* sp. nov., a peptide-fermenting bacterium from an olive mill wastewater treatment digester. Int. J. Syst. Evol. Microbiol. *50:* 1259–1264.

Mechichi, T., M. Labat, J-L. Garcia, P. Thomas and B.K.C. Patel. 2000b. *In* Validation of the publication of new names and new combinations previously published outside the IJSEM. List no. 72. Int. J. Syst. Evol. Microbiol *50:* 3–4.

Mendez, B.S., M.J. Pettinari, S.E. Ivanier, C.A. Ramos and F. Sineriz. 1991. *Clostridium thermopapyrolyticum* sp. nov., a cellulolytic thermophile. Int. J. Syst. Bacteriol. *41:* 281–283.

Migula, W. 1900. System der Bakterien, Vol. 2. Gustav Fischer, Jena.

Milhaud, G., J.P. Aubert and C.B. Van Niel. 1956. [Study of glycolysis of *Zymosarcina ventriculi*.] Etude de la glycolyse de *Zymosarcina ventriculi*. Ann. Inst. Pasteur (Paris) *91:* 363–368.

Miller, T.L. and M.J. Wolin. 1974. A serum bottle modification of the Hungate technique for cultivating obligate anaerobes. Appl. Microbiol. *27:* 985–987.

Miranda-Tello, E., M.L. Fardeau, J. Sepulveda, L. Fernandez, J.L. Cayol, P. Thomas and B. Ollivier. 2003. *Garciella nitratireducens* gen. nov., sp. nov., an anaerobic, thermophilic, nitrate- and thiosulfate-reducing bacterium isolated from an oilfield separator in the Gulf of Mexico. Int. J. Syst. Evol. Microbiol. *53:* 1509–1514.

Mohan, R., P. Namsolleck, P.A. Lawson, M. Osterhoff, M.D. Collins, C.A. Alpert and M. Blaut. 2006. *Clostridium asparagiforme* sp. nov., isolated from a human faecal sample. Syst. Appl. Microbiol. *29:* 292–299.

Mohan, R., P. Namsolleck, P. A. Lawson, M. Osterhoff, M. D. Collins, C. A. Alpert and M. Blaut. 2007. *In* Validation of the publication of new names and new publications previously effectively published outside the IJSEM. List no. 17. Int. J. Syst. Evol. Microbiol *57:* 1933–1934.

Monserrate, E., S.B. Leschine and E. Canale-Parola. 2001. *Clostridium hungatei* sp. nov., a mesophilic, N_2-fixing cellulolytic bacterium isolated from soil. Int. J. Syst. Evol. Microbiol. *51:* 123–132.

Moore, W.E.C., J.L. Johnson and L.V. Holdeman. 1976. Emendation of *Bacteroidaceae* and *Butyrivibrio* and descriptions of *Desulfomonas* gen. nov. and ten new species in genera *Desulfomonas, Butyrivibrio, Eubacterium, Clostridium,* and *Ruminococcus.* Int. J. Syst. Bacteriol. *26:* 238–252.

Morris, G.N., J. Winter, E.P. Cato, A.E. Ritchie and V.D. Bokkenheuser. 1985. *Clostridium scindens* sp. nov., a human intestinal bacterium with desmolytic activity on corticoids. Int. J. Syst. Bacteriol. *35:* 478–481.

Mountfort, D.O., F.A. Rainey, J. Burghardt and E. Stackebrandt. 1994. *Clostridium grantii* sp. nov., a new obligately anaerobic, alginolytic bacterium isolated from mullet gut. Arch. Microbiol *162:* 173–179.

Mountfort, D.O., F.A. Rainey, J. Burghardt and E. Stackebrandt. 1996. *In* Validation of the publication of new names and new combinations previously effectively published outside the IJSB. List no. 57. Int. J. Syst.Bacteriol. *46:* 625–626.

Mountfort, D.O., F.A. Rainey, J. Burghardt, H.F. Kaspar and E. Stackebrandt. 1997a. *In* Validation of the publication of new names and new combinations previously effectively published outside the IJSB. List no. 62. Int. J. Syst. Bacteriol *47:* 915–916.

Mountfort, D.O., F.A. Rainey, J. Burghardt, H.F. Kaspar and E. Stackebrandt. 1997b. *Clostridium vincentii* sp. nov., a new obligately anaerobic, saccharolytic, psychrophilic bacterium isolated from low-salinity pond sediment of the McMurdo Ice Shelf, Antarctica. Arch. Microbiol. *167:* 54–60.

Murray, W.D., A.W. Khan and L. Vandenberg. 1982. *Clostridium saccharolyticum* sp. nov, a saccharolytic species from sewage sludge. Int. J. Syst. Bacteriol. *32:* 132–135.

Murray, W.D., L. Hofmann, N.L. Campbell and R.H. Madden. 1986. *Clostridium lentocellum* sp. nov., a cellulolytic species from river sediment containing paper mill waste. Syst. Appl. Microbiol. *8:* 181–184.

Murray, W.D., L. Hofmann, N.L. Campbell and R.H. Madden. 1987. *In* Validation pf the publication of new names and new combinatinspreviously effectively published outside the IJSB. List no. 23. Int. J. Syst. Bacteriol *37:* 179–180.

Nakamura, S., K. Tamai and S. Nishida. 1970. Criteria for identification of *Clostridium perfringens*. 6. *Clostridium paraperfringens* sp. nov. Med. Biol. *8:* 137–140.

Nakamura, S., M. Sakurai and S. Nishida. 1976. Lecithinase-negative variants of *Clostridium perfringens*; the identity of *C. plagarum* with *C. perfringens*. Can. J. Microbiol. *22:* 1497–1501.

Nakamura, S., I. Kimura, K. Yamakawa and S. Nishida. 1983. Taxonomic relationships among *Clostridium novyi* types A and B, *Clostridium haemolyticum* and *Clostridium botulinum* type C. J. Gen. Microbiol. *129:* 1473–1479.

Ng, T.K., T.K. Weimer and J.G. Zeikus. 1977. Cellulolytic and physiological properties of *Clostridium thermocellum*. Arch. Microbiol. *114:* 1–7.

Ollivier, B., P. Caumette, J.-L. Garcia and R.A. Mah. 1994. Anaerobic bacteria from hypersaline environments. Microbiol. Rev. *58:* 27–38.

Olsen, I., J.L. Johnston, L.V.H. Moore and W.E.C. Moore. 1995. Rejection of *Clostridium putrificum* and conservation of *Clostridium botulinum* and *Clostridium sporogenes*. Request for an Opinion. Int. J. Syst. Bacteriol. *45:* 414.

Onyenwoke, R.U., J.A. Brill, K. Farahi and J. Wiegel. 2004. Sporulation genes in members of the low G+C Gram-type-positive phylogenetic branch (*Firmicutes*). Arch. Microbiol. *182:* 182–192.

Örlygsson, J., J. Krooneman, M.D. Collins, C. Pascual and J.C. Gottschal. 1996. *Clostridium acetireducens* sp. nov., a novel amino acid-oxidizing acetate-reducing anaerobic bacterium. Int. J. Syst. Bacteriol. *46:* 454–459.

Padden, A.N., V.M. Dillon, J. Edmonds, M.D. Collins, N. Alvarez and P. John. 1999. An indigo-reducing moderate thermophile from a woad vat, *Clostridium isatidis* sp. nov. Int. J. Syst. Bacteriol. *49:* 1025–1031.

Palop, M.L., S. Valles, F. Pinaga and A. Flors. 1989. Isolation and characterization of an anaerobic, celluloytic bacterium, *Clostridium celerecrescens* sp. nov. Int. J. Syst. Bacteriol. *39:* 68–71.

Parshina, S.N., R. Kleerebezem, J.L. Sanz, G. Lettinga, A.N. Nozhevnikova, N.A. Kostrikina, A.M. Lysenko and A.J. Stams. 2003. *Soehngenia saccharolytica* gen. nov., sp. nov. and *Clostridium amygdalinum* sp. nov., two novel anaerobic, benzaldehyde-converting bacteria. Int. J. Syst. Evol. Microbiol. *53:* 1791–1799.

Partansky, A.M. and B.S. Henry. 1935. Anaerobic bateria capable of fermenting sulfite waste liquor. J. Bacteriol. *30:* 559–571.

Paster, B.J., J.B. Russell, C.M. Yang, J.M. Chow, C.R. Woese and R. Tanner. 1993. Phylogeny of the ammonia-producing ruminal bacteria *Peptostreptococcus anaerobius*, *Clostridium sticklandii*, and *Clostridium aminophilum* sp. nov. Int. J. Syst. Bacteriol. *43:* 107–110.

Patel, B.K.C., H.W. Morgan and R.M. Daniel. 1985a. A simple and efficient method for preparing anaerobic media. Biotechnol. Lett. *7:* 227–228.

Patel, B.K.C., H.W. Morgan and R.M. Daniel. 1985b. *Fervidobacterium nodosum* gen. nov. and spec. nov., a new chemoorganotrophic, caldoactive, anaerobic bacterium. Arch. Microbiol. *141:* 63–69.

Patel, B.K.C., C. Monk, H. Littleworth, H.W. Morgan and R.M. Daniel. 1987. *Clostridium fervidus* sp. nov., a new chemoorganotrophic acetogenic thermophile. Int. J. Syst. Bacteriol. *37:* 123–126.

Petitdemange, E., F. Caillet, J. Giallo and C. Gaudin. 1984. *Clostridium cellulolyticum* sp. nov., a cellulolytic, mesophilic species from decayed grass. Int. J. Syst. Bacteriol. *34:* 155–159.

Pikuta, E.V., R.B. Hoover, A.K. Bej, D. Marsic, E.N. Detkova, W.B. Whitman and P. Krader. 2003a. *Tindallia californiensis* sp. nov., a new anaerobic, haloalkaliphilic, spore-forming acetogen isolated from Mono Lake in California. Extremophiles *7:* 327–334.

Pikuta, E.V., R.B. Hoover, A.K. Bej, D. Marsic, E.N. Detkova, W.B. Whitman and P. Krader. 2003b. *In* Validation of publication of new names and new combinations previously effectively published outside the IJSB. List no. 94. Int. J. Syst. Bacteriol. *53:* 1701–1702.

Pikuta, E.V., T. Itoh, P. Krader, J. Tang, W.B. Whitman and R.B. Hoover. 2006. *Anaerovirgula multivorans* gen. nov., sp. nov., a novel spore-forming, alkaliphilic anaerobe isolated from Owens Lake, California, USA. Int. J. Syst. Evol. Microbiol. *56:* 2623–2629.

Plugge, C.M., E.G. Zoetendal and A.J. Stams. 2000. *Caloramator coolhaasii* sp. nov., a glutamate-degrading, moderately thermophilic anaerobe. Int. J. Syst. Evol. Microbiol. *50*: 1155–1162.

Postec, A., C. Le Breton, M.L. Fardeau, F. Lesongeur, P. Pignet, J. Querellou, B. Ollivier and A. Godfroy. 2005. *Marinitoga hydrogenitolerans* sp. nov., a novel member of the order *Thermotogales* isolated from a black smoker chimney on the Mid-Atlantic Ridge. Int. J. Syst. Evol. Microbiol. *55*: 1217–1221.

Prazmowski, A. 1880. Untersuchung über die Entwickelungsgeschichte und Fermentwirkung einiger Bacterien-Arten Hugo Voigt, Leipzig, pg. 1–58.

Prévot, A.R. 1938. Études de systématique bactérienne. IV. Critique de la conception actuelle du genre *Clostridium*. Ann. Inst. Pasteur (Parish) *61*: 72–91.

Prévot, A.R. 1940. Manuel de classification et de determination des bacteries anaerobies. Masson and Co., Paris, 1–223.

Prévot, A.R. and J. Zimmés-Chaverou. 1947. Étude d'une novella espéce anaérobie de Côte d'Ivoire: *Inflablis mangenoti*. Ann. Inst. Pasteur (Paris) *73*: 602–604.

Prévot, A.R. 1948a. Étude des bactéries anaérobies d'Afrique occidentale française (Sénégal, Guinée, Côte d'Ivoire). Ann. Inst. Pasteur (Paris) *74*: 157–170.

Prévot, A.R. 1948b. Manuel de classification et de determination des bactéries anaérobies, 2nd ed. Masson and Co., Paris, 1–290.

Prévot, A.R. 1953. Dictionnaire des Bactéries Pathogènes, 2nd edn. Masson, Paris.

Pribram, E. 1933. Klassification der Schizomyceten. F. Deuticke, Leipzig.

Pusheva, M.A., A.V. Pitryuk, E.N. Detkova and G.A. Zavarzin. 1999. Bioenergetics of acetogenesis in the extremely alkaliphilic homoacetogenic bacteria *Natroniella acetigena* and *Natronoincola histidinovorans*. Mikrobiologiya *68*: 651–656.

Rainey, F.A., N.L. Ward, H.W. Morgan, R. Toalster and E. Stackebrandt. 1993. Phylogenetic analysis of anaerobic thermophilic bacteria: aid for their reclassification. J. Bacteriol *175*: 4772–4779.

Rainey, F.A., N.L. Ward-Rainey, P.H. Janssen, H. Hippe and E. Stackebrandt. 1996. *Clostridium paradoxum* DSM 7308^T contains multiple 16S rRNA genes with heterogeneous intervening sequences. Microbiology *142*: 2087–2095.

Rainey, F.A., B.J. Hollen and A. Small. 2009. Genus I. *Clostridium*. *In* De Vos, Garrity, Jones, Krieg, Ludwig, Rainey, Schleifer and Whitman (Editors), Bergey's Manual of Systematic Bacteriology, 2nd edition, Vol. 3. Springer, New York pp. 736–864.

Rogers, G.M. and A.A.W. Baecker. 1991. *Clostridium xylanolyticum* sp. nov., an anaerobic xylanolytic bacterium from decayed *Pinus patula* wood chips. Int. J. Syst. Bacteriol. *41*: 140–143.

Roux, C. and J.L. Bergere. 1997. Caractéres taxonomiques de *Clostridium tyrobutyricum*. Ann. Microbiol. (Inst. Pasteur) *128A*: 267–276.

Saitou, N. and M. Nei. 1987. The neighbor-joining method: a new method for reconstructing phylogenetic trees. Mol. Biol. Evol. *4*: 406–425.

Sakuma, K, M. Kitahara, R. Kibe, M. Sakamoto and Y. Benno. 2006a. *Clostridium glycyrrhizinilyticum* sp. nov., a glycyrrhizin-hydrolysing bacterium isolated from human faeces. Microbiol. Immunol. *50*: 481–485.

Sakuma, K, M. Kitahara, R. Kibe, M. Sakamoto and Y. Benno. 2006b. *In* Validation of the publication of new names and new combinations previously effectively published outside the IJSEM. List no. 112. Int. J. Syst. Evol. Microbiol *56*: 2507–2508.

Sasaki, Y., N. Takikawa, A. Kojima, M. Norimatsu, S. Suzuki and Y. Tamura. 2001. Phylogenetic positions of *Clostridium novyi* and *Clostridium haemolyticum* based on 16S rDNA sequences. Int. J. Syst. Evol. Microbiol. *51*: 901–904.

Schink, B. 1984a. *Clostridium magnum* sp. nov, a non-autotrophic homoacetogenic bacterium. Arch. Microbiol. *137*: 250–255.

Schink, B. 1984b. *In* Validation of the publication of new names and new combinations previously effectively published outside the IJSB. List no. 15. Int. J. Syst. Bacteriol *34*: 355–357.

Schnurer, A., B. Schink and B.H. Svensson. 1996. *Clostridium ultunense* sp. nov., a mesophilic bacterium oxidizing acetate in syntrophic association with a hydrogenotrophic methanogenic bacterium. Int. J. Syst. Bacteriol. *46*: 1145–1152.

Scott, J., A.W. Turner and L.R. Vawter. 1935. Gas edema diseases. Twelfth Intl. Vet. Congr., pp. 168–182.

Scott, J.P. 1928. The etiology of blackleg and methods of determining *Clostridium chauvoei* from other anaerobic organisms found in cases of blackleg. Cornell Vet. *18*: 249–271.

Seyfried, M., D. Lyon, F.A. Rainey and J. Wiegel. 2002. *Caloramator viterbensis* sp. nov., a novel thermophilic, glycerol-fermenting bacterium isolated from a hot spring in Italy. Int. J. Syst. Evol. Microbiol. *52*: 1177–1184.

Siunov, A.V., D.V. Nikitin, N.E. Suzina, V.V. Dmitriev, N.P. Kuzmin and V.I. Duda. 1999. Phylogenetic status of *Anaerobacter polyendosporus*, an anaerobic, polysporogenic bacterium. Int. J. Syst. Bacteriol. *49*: 1119–1124.

Sleat, R., R.A. Mah and R. Robinson. 1984. Isolation and characterization of an anaerobic, cellulolytic bacterium, *Clostridium cellulovorans* sp. nov. Appl. Environ. Microbiol. *48*: 88–93.

Sleat, R. and R.A. Mah. 1985. *Clostridium populeti* sp. nov., a cellulolytic species from a woody-biomass digester. Int. J. Syst. Bacteriol. *35*: 160–163.

Sleat, R., R.A. Mah and R. Robinson. 1985. *In* Validation of the publication of new names and new combinations previously effectively published outside the IJSB. List no. 17. Int. J. Syst. Bacteriol *35*: 223–225.

Slobodkina, G.B., T.V. Kolganova, T.P. Tourova, N.A. Kostrikina, C. Jeanthon, E.A. Bonch-Osmolovskaya and A.I. Slobodkin. 2008. *Clostridium tepidiprofundi* sp. nov., a moderately thermophilic bacterium from a deep-sea hydrothermal vent. Int. J. Syst. Evol. Microbiol. *58*: 852–855.

Smit, J. 1930. Die Gärungsarcinen. Gustav Fischer, Jena.

Smit, J. 1933. The biology of fermenting sarcinae. J. Pathol. Bacteriol. *36*: 455–468.

Smith, L.D. and E. King. 1962. *Clostridium innocuum*, sp. n., a sporeforming anaerobe isolated from human infections. J. Bacteriol. *83*: 938–939.

Smith, L.D. and G. Hobbs. 1974. Genus III *Clostridium* Prazmowski 1880, 23. *In* Buchanan and Gibbons (Editors), Bergey's Manual of Determinative Bacteriology, 8th edn. The Williams & Wilkins Co, Baltimore, pp. 551–572.

Smith, L.D. 1975. Common mesophilic anaerobes, including *Clostridium botulinum* and *Clostridium tetani*, in 21 soil specimens. Appl. Microbiol. *29*: 590–594.

Smith, L.D.S. 1970. *Clostridium oceanicum*, sp. n., a sporeforming anaerobe isolated from marine sediments. J Bacteriol *103*: 811–813.

Snyder, M.L. 1936. The serologic agglutination of the obligate anaerobes *Clostridium paraputrificum* (Bienstock) and *Clostridium capitovalis* (Synder and Hall). J. Bacteriol. *32*: 401–410.

Soh, A.L., H. Ralambotiana, B. Ollivier, G. Prensier, E. Tine and J.L. Garcia. 1991a. *In* Validation of the publication of new names and new combinations previously effectively published outside the IJSB. List no. 37. Int. J. Syst. Bacteriol *41*: 331.

Soh, A.L., H. Ralambotiana, B. Ollivier, G. Prensier, E. Tine and J.L. Garcia. 1991b. *Clostridium thermopalmarium* sp. nov., a moderately thermophilic butyrate-producing bacterium isolated from palm wine in Senegal. Syst. Appl. Microbiol. *14*: 135–139.

Song, Y., C. Liu, M, D, T. Tomzynski, P. Lawson, M. Collins and S.M. Finegold. 2003a. *In* Validation of the publication of new names and new combinations previously effectively published outside the IJSEM. List no. 92. Int. J. Syst. Evol. Microbiol *53*: 935–937.

Song, Y.L., C.X. Liu, D.R. Molitoris, T.J. Tomzynski, P.A. Lawson, M.D. Collins and S.M. Finegold. 2003b. *Clostridium bolteae* sp. nov., isolated from human sources. Syst. Appl. Microbiol. *26*: 84–89.

Song, Y., C. Liu, M. McTeague, P. Summanen and S.M. Finegold. 2004a. *In* Validation of the publication of new names and new combinations previously effectively published outside the IJSEM. List no. 99. Int. J. Syst. Evol. Microbiol *54:* 1425–1426.

Song, Y.L., C.X. Liu, M. McTeague, P. Summanen and S.M. Finegold. 2004b. *Clostridium bartlettii* sp. nov., isolated from human faeces. Anaerobe *10:* 179–184.

Soriano, S. and A. Soriano. 1948. Nueva bacteria anaerobia productora de una alteracion en sordinas envasadas. Rev. Asoc. Argent. Dietol. *6:* 36–41.

Spray, R.S. 1939. Genus II *Clostridium* Prazmowski. *In* Bergey, Breed, Murray and Hitchens (Editors), Bergey's Manual of Determinative Bacteriology, 5th edn. The Williams & Wilkins Co., Baltimore.

Spray, R.S. and L.S. McClung. 1948. Genus II *Clostridium* Prazmowski. *In* Breed, Murray and Hitchens (Editors), Bergey's Manual of Determinative Bacteriology, 6th edn. The Williams & Wilkins Co., Baltimore, pp. 763–827.

Spring, S., B. Merkhoffer, N. Weiss, R.M. Kroppenstedt, H. Hippe and E. Stackebrandt. 2003. Characterization of novel psychrophilic clostridia from an Antarctic microbial mat: description of *Clostridium frigoris* sp. nov., *Clostridium lacusfryxellense* sp. nov., *Clostridium bowmanii* sp. nov. and *Clostridium psychrophilum* sp. nov. and reclassification of *Clostridium laramiense* as *Clostridium estertheticum* subsp. *laramiense* subsp. nov. Int. J. Syst. Evol. Microbiol. *53:* 1019–1029.

Stackebrandt, E. and F.A. Rainey. 1997. Phylogenetic relationships. *In* Rood, McClane, Songer and Titball (Editors), The Clostridia: Molecular Biology and Pathogenesis. Academic Press, New York.

Stackebrandt, E., I. Kramer, J. Swiderski and H. Hippe. 1999. Phylogenetic basis for a taxonomic dissection of the genus *Clostridium*. FEMS Immunol. Med. Microbiol. *24:* 253–258.

Stadtman, T.C. and L.S. McClung. 1957. *Clostridium sticklandii* nov. spec. J. Bacteriol. *73:* 218–219.

Stams, A.J.M., J.B. Vandijk, C. Dijkema and C.M. Plugge. 1993. Growth of syntrophic propionate-oxidizing bacteria with fumarate in the absence of methanogenic bacteria. Appl. Environ. Microbiol. *59:* 1114–1119.

Steer, T., M.D. Collins, G.R. Gibson, H. Hippe and P.A. Lawson. 2001. *Clostridium hathewayi* sp. nov., from human faeces. Syst. Appl. Microbiol. *24:* 353–357.

Steer, T., M.D. Collins, G.R. Gibson, H. Hippe and P.A. Lawson. 2002. *In* Validation of the publication of new names and new combinations previously effectively published outside the IJSEM. List no. 85. Int. J. Syst. Evol. Microbiol *52:* 685–690.

Stephenson, M.P. and E.A. Dawes. 1971. Pyruvic acid and formic acid metabolism in *Sarcina ventriculi* and the role of ferredoxin. J. Gen. Microbiol. *69:* 331–343.

Stevens, W.C. 1956. Taxonomic studies on the genus *Bacteroides* and similar forms. Vanderbilt University.

Sturges, W.S. and E.T. Drake. 1927. A complete description of *Costridium putrefaciens* (McBryde). J. Bacteriol. *14:* 175–179.

Suen, J.C., C.L. Hatheway, A.G. Steigerwalt and D.J. Brenner. 1988. *Clostridium argentinense* sp. nov.: a genetically homogeneous group composed of all strains of *Clostridium botulinum* toxin type-G and some nontoxigenic strains previously identified as *Clostridium subterminale* or *Clostridium hastiforme*. Int. J. Syst. Bacteriol. *38:* 375–381.

Sukhumavasi, J., K. Ohmiya, S. Shimizu and K. Ueno. 1988. *Clostridium josui* sp. nov., a cellulolytic, moderate thermophilic species from Thai compost. Int. J. Syst. Bacteriol. *38:* 179–182.

Suresh, K., D. Prakash, N. Rastogi and R.K. Jain. 2007. *Clostridium nitrophenolicum* sp. nov., a novel anaerobic *p*-nitrophenol-degrading bacterium, isolated from a subsurface soil sample. Int. J. Syst. Evol. Microbiol. *57:* 1886–1890.

Svensson, B.H., H. Dubourguier, G. Prensier and A.J.B. Zehnder. 1992. *Clostridium quinii* sp. nov., a new saccharolytic anaerobic bacterium isolated from granular sludge. Arch. Microbiol. *157:* 97–103.

Svensson, B.H., H. Dubourguier, G. Prensier and A.J.B. Zehnder. 1995. *In* Validation of the publication of new names and new combinations previously effectively published outside the IJSB. List no. 55. Int. J. Syst. Bacteriol *45:* 879–880.

Swenson, J.M., C. Thornsberry, L.M. McCrosky, C.L. Hatheway and J.V.R. Dowell. 1980. Susceptibility of *Clostridium botulinum* to thieteen antimicrobial agents. Antimicrob. Agents Chemother. *18:* 13–19.

Takai, K., D.P. Moser, T.C. Onstott, N. Spoelstra, S.M. Pfiffner, A. Dohnalkova and J.K. Fredrickson. 2001. *Alkaliphilus transvaalensis* gen. nov., sp. nov., an extremely alkaliphilic bacterium isolated from a deep South African gold mine. Int. J. Syst. Evol. Microbiol. *51:* 1245–1256.

Tanner, R.S., L.M. Miller and D. Yang. 1993. *Clostridium ljungdahlii* sp. nov., an acetogenic species in clostridial ribosomal RNA homology group-I. Int. J. Syst. Bacteriol. *43:* 232–236.

Tarlera, S., L. Muxi, M. Soubes and A.J.M. Stams. 1997. *Caloramator proteoclasticus* sp. nov., a new moderately thermophilic anaerobic proteolytic bacterium. Int. J. Syst. Bacteriol. *47:* 651–656.

Tholozan, J.L., J.P. Touzel, E. Samain, J.P. Grivet, G. Prensier and G. Albagnac. 1992. *Clostridium neopropionicum* sp. nov. a strict anaerobic bacterium fermenting ethanol to propionate through acrylate pathway. Arch. Microbiol. *157:* 249–257.

Tholozan, J.L., J.P. Touzel, E. Samain, J.P. Grivet, G. Prensier and G. Albagnac. 1995. *In* Validation of the publication of new names and new combinations previously effectively published outside the IJSB. List no. 55. Int. J. Syst. Bacteriol. *45:* 879–880.

Toda, Y., T. Saiki, T. Uozumi and T. Beppu. 1988. Isolation and characterization of a protease-producing, thermophilic, anaerobic bacterium, *Thermobacteroides leptospartum* sp. nov. Agric. Biol. Chem. *52:* 1339–1344.

Tourova, T.P., E.S. Garnova and T.N. Zhilina. 1999. Phylogenetic diversity of alkaliphilic anaerobic saccharolytic bacteria isolated from soda lakes. Microbiology (En. transl. from Mikrobiologiya) *68:* 615–622.

van Beynum, J. and J.W. Pette. 1935. Zuckervergärend und Laktat vergärende Buttersäurebakterien. Zentralbl. Bakteriol. Parasitenkd. Infektionskr. Hyg. *93:* 198–212.

Van de Peer, Y. and R. De Wachter. 1994. TREECON for Windows: a software package for the construction and drawing of evolutionary trees for the Microsoft Windows environment. Comput. Appl. Biosci. *10:* 569–570.

van der Wielen, P.W.J.J., G.M. L.L. Rovers, J.M. Scheepens and S. Biesterveld. 2002. *Clostridium lactatifermentans* sp. nov., a lactate-fermenting anaerobe isolated from the caeca of a chicken. Int. J. Syst. Evol. Microbiol. *52:* 921–925.

van Ermengem, E. 1896. Untersuchungen über Fälle von Fleischvergiftung mit Sympotomen von Botulismus. Zentralbl. Bakteriol. Parasitenkd. Infektionskr. Hyg. I. Abt. Orig. *19:* 442–444.

van Gylswyk, N.O. 1980. *Fusobacterium polysaccharolyticum* sp. nov., Gram-negative rod from the rumen that produces butyrate and ferments cellulose and starch. J. Gen. Microbiol. *116:* 157–163.

van Gylswyk, N.O., E.J. Morris and H.J. Els. 1980. Sporulation and cell-wall structure of *Clostridium polysaccharolyticum* comb. nov. (formerly *Fusobacterium polysaccharolyticum*). J. Gen. Microbiol. *121:* 491–493.

van Gylswyk, N.O., E. J. Morris and H.J. Els. 1983. *In* Validation of the publication of new names and new combinations previously effectively published outside the IJSB. List no. 10. Int. J. Syst. Bacteriol *33:* 438–440.

van Gylswyk, N.O. and J.J.K. van der Toorn. 1987. *Clostridium aerotolerans* sp. nov., a xylanolytic bacterium from corn stover and from the rumina of sheep fed corn stover. Int. J. Syst. Bacteriol. *37:* 102–105.

Varel, V.H., R.S. Tanner and C.R. Woese. 1995. *Clostridium herbivorans* sp. nov., a cellulolytic anaerobe from the pig intestine. Int. J. Syst. Bacteriol. *45:* 490–494.

Veillon, A. and A. Zuber. 1898. Recherches sur quelques microbes strictement anaérobies et leur rôle en pathologie. Arch. Med. Exp *10:* 517–545.

Viljoen, J.A., E.B. Fred and W.H. Peterson. 1926. The fermentation of cellulose by thermophilic bacteria. J. Agric. Sci. Camb. *16:* 1–17.

Vuillemin, P. 1931. Les champignons parasites et les mycoses de l'homme. Encylopédie Mycologique II. Paul Le Chevalier and Sons, Paris, pp. 1–290.

Wachsman, J.T. and H.A. Barker. 1954. Characterization of an orotic acid fermenting bacterium, *Zymobacterium oroticum*, nov. gen., nov. spec. J. Bacteriol. *68:* 400–404.

Wang, X., T. Maegawa, T. Karasawa, E. Ozaki and S. Nakamura. 2005. *Clostridium sardiniense* Prevot 1938 and *Clostridium absonum* Nakamura et al. 1973 are heterotypic synonyms: evidence from phylogenetic analyses of phospholipase C and 16S rRNA sequences, and DNA relatedness. Int. J. Syst. Evol. Microbiol. *55:* 1193–1197.

Warnick, T.A., B.A. Methe and S.B. Leschine. 2002. *Clostridium phytofermentans* sp. nov., a cellulolytic mesophile from forest soil. Int. J. Syst. Evol. Microbiol. *52:* 1155–1160.

Warren, Tyrrell, Citron and Goldstein. 2007. *In* Validation of the publication of new names and enw combinations previously effectively published outside the IJSEM. List no. 115. Int. J. Syst. Evol. Microbiol. *57:* 893–897.

Warren, Y.A., K.L. Tyrrell, D.M. Citron and E.J. Goldstein. 2006. *Clostridium aldenense* sp. nov. and *Clostridium citroniae* sp. nov. isolated from human clinical infections. J. Clin. Microbiol. *44:* 2416–2422.

Weinberg, M. and P. Séguin. 1915. Flore microbienne de la gangrene gazeuse. Le *B. fallax.* C.R. Seances Soc. Biol. Filiales *78:* 686–689.

Weinberg, M. and P. Séguin. 1916. Contribution á l'étiologie de la g gangréne gazeuse. C.R. Seances Soc. Biol. Filiales *163:* 449–451.

Weinberg, M. and P. Séguin. 1918. La gangrene gazeuse-bactériologie, reproduction expérimentale, séreothéapie. Masson and Co., Paris.

Weinberg, M. and B. Ginsbourg. 1927. Données récéntes sur les microbes anaérobies et leur role en pathologie. Masson et Cie, Paris.

Weinberg, M., R. Nativelle and A.R. Prévot. 1937. Les microbes anaérobies. Masson and Co., Paris.

Wery, N., J.M. Moricet, V. Cueff, J. Jean, P. Pignet, F. Lesongeur, M.A. Cambon-Bonavita and G. Barbier. 2001. *Caloranaerobacter azorensis* gen. nov., sp. nov., an anaerobic thermophilic bacterium isolated from a deep-sea hydrothermal vent. Int. J. Syst. Evol. Microbiol. *51:* 1789–1796.

Whitman, W.B., E. Ankwanda and R.S. Wolfe. 1982. Nutrition and carbon metabolism of *Methanococcus voltae.* J. Bacteriol. *149:* 852–863.

Wiegel, J. 1980. Formation of ethanol by bacteria. A pledge for the use of extreme thermophilic anaerobic bacteria in industrial ethanol fermentation processes. Experientia *36:* 1434–1446.

Wiegel, J. 1981. Distinction between the Gram reaction and the Gram type of bacteria. Int. J. Syst. Bacteriol. *31:* 88.

Wiegel, J., S. Kuk and G.W. Kohring. 1989a. *Clostridium thermobutyricum* sp.nov., a moderate thermophile isolated from a cellulolytic culture, that produces butyrate as the major product. Int. J. Syst. Bacteriol. *39:* 199–204.

Wiegel, J., S.U. Kuk and G.W. Kohring. 1989b. *Clostridium thermobutyricum* sp. nov., a moderate thermophile isolated from a cellulolytic culture, that produces butyrate as the major product. Int. J. Syst. Bacteriol. *39:* 199–204.

Wiegel, J., R. Tanner and F.A. Rainey. 2006. An introduction to the family *Clostridiaceae. In* Dworkin, Falkow, Rosenberg, Schleifer and Stackebrandt (Editors), The Prokaryotes, A Handbook on the Biology of Bacteria, Bacteria: *Firmicutes, Cyanobacteria*, Vol. 4. Springer, New York, pp. 654–678.

Wieringa, K.T. 1940. The formation of acetic acid from carbon dioxide and hydrogen by anaerobic spore-forming bacteria. Antonie van Leeuwenhoek J. Microbiol. Serol. *6:* 251–262.

Wilcke, B.W.J., T.F. Midura and S.S. Arnon. 1980. Quantitative evidence of intestinal colonization by *Clostridium botulinum* in four cases of infant botulism. J. Infect. Dis. *141:* 419–423.

Wilde, E., H. Hippe, N. Tosunoglu, G. Schallehn, K. Herwig and G. Gottschalk. 1989. *Clostridium tetanomorphum* sp. nov., nom. rev. Int. J. Syst. Bacteriol. *39:* 127–134.

Wilde, E., M.D. Collins and H. Hippe. 1997. *Clostridium pascui* sp. nov., a new glutamate-fermenting sporeformer from a pasture in Pakistan. Int. J. Syst. Bacteriol. *47:* 164–170.

Windberger, E., R. Huber, A. Trincone, H. Fricke and K.O. Stetter. 1989. *Thermotoga thermarum* sp. nov. and *Thermotoga neapolitana* occurring in African continental solfataric springs. Arch. Microbiol. *151:* 506–512.

Winogradsky, S. 1895. Recherches sur l'assimilation de l'azote libre de l'atmosphère par les microbes. Arch. Sci. Biol. St. Pétersb. *3:* 297–352.

Winter, J., M.R. Popoff, P. Grimont and V.D. Bokkenheuser. 1991. *Clostridium orbiscindens* sp. nov., a human intestinal bacterium capable of cleaving the flavonoid C-ring. Int. J. Syst. Bacteriol. *41:* 355–357.

Wirtz, R. 1908. Eine einfache Art der Sporenfärbung. Zentralbl. Bakteriol. Parasitenkd. Infektionskr. Hyg. Abt. I Orig. *46:* 727–728.

Wolin, E.A., M.G. Wolin and R.S. Wolfe. 1963. Formation of methane by bacterial extracts. J. Biol. Chem. *238:* 2882–2886.

Yang, J.C., D.P. Chynoweth, D.S. Williams and A. Li. 1990. *Clostridium aldrichii* sp. nov., a cellulolytic mesophile inhabiting a wood-fermenting anaerobic digester. Int. J. Syst. Bacteriol. *40:* 268–272.

Ye, Q., Y. Roh, S.L. Carroll, B. Blair, J.Z. Zhou, C.L. Zhang and M.W. Fields. 2004. Alkaline anaerobic respiration: isolation and characterization of a novel alkaliphilic and metal-reducing bacterium. Appl. Environ. Microbiol. *70:* 5595–5602.

Zavarzin, G.A., T.N. Zhilina and V.V. Kevbrin. 1999. The alkaliphilic microbial community and its functional diversity. Microbiology (En. transl. from Mikrobiologiya) *68:* 503–521.

Zavarzin, G.A. and T.N. Zhilina. 2000. Anaerobic chemotrophic alkaliphiles. *In* Seckbach (Editor), Journey to Diverse Microbial Worlds. Kluwer Academic Publishing, The Netherlands, pp. pp.191–208.

Zhilina, T.N. and G.A. Zavarzin. 1987. *Methanohalobium evestigatum* gen. nov., sp.nov., extremely halophilic methane-producing archaebacteria. Dokl. Akad. Nauk. SSSR. *293:* 464–468 (Rus.).

Zhilina, T.N. and G.A. Zavarzin. 1994. Alkaliphilic anaerobic community at pH 10. Curr. Microbiol. *29:* 109–112.

Zhilina, T.N., G.A. Zavarzin, E.N. Detkova and F.A. Rainey. 1996. *Natroniella acetigena* gen. nov. sp. nov., an extremely haloalkaliphilic, homoacetic bacterium: a new member of *Haloanaerobiales.* Curr. Microbiol. *32:* 320–326.

Zhilina, T.N., E.N. Detkova, F.A. Rainey, G.A. Osipov, A.M. Lysenko, N.A. Kostrikina and G.A. Zavarzin. 1998. *Natronoincola histidinovorans* gen. nov., sp. nov., a new alkaliphilic acetogenic anaerobe. Curr. Microbiol. *37:* 177–185.

Zhilina, T.N., E.N. Detkova, F.A. Rainey, G.A. Osipov, A.M. Lysenko, N.A. Kostrikina and G.A. Zavarin. 1999. *In* Validation of publication of new names and new combinations previously effectively published outside the IJSEM. List no. 68. Int. J. Syst. Evol. Microbiol. *49:* 1–3.

Zhilina, T.N., V.V. Kevbrin, T.P. Tourova, A.M. Lysenko, N.A. Kostrikina and G.A. Zavarzin. 2005. *Clostridium alkalicellum* sp. nov., an obligately alkaliphilic cellulolytic bacterium from a soda lake in the Baikal region. Microbiology, Mikrobiologiya *74:* 557–566.

Zhilina, T.N., V.V. Kevbrin, T.P. Tourova, A.M. Lysenko, N.A. Kostrikina and G.A. Zavarzin. 2006. *In* Validation of the publication of new names and new combinations previously effectively published outside the IJSEM. List no. 109. Int. J. Syst. Evol. Microbiol *56:* 925–927.

Family II. **Eubacteriaceae** fam. nov.

Wolfgang Ludwig, Karl-Heinz Schleifer and William B. Whitman

Eu.bac.te.ri.a′ce.ae. N.L. neut. n. *Eubacterium* type genus of the family; suff. *-aceae* ending denoting family; N.L. fem. pl. n. *Eubacteriaceae* the *Eubacterium* family.

The family *Eubacteriaceae* is circumscribed for this volume on the basis of phylogenetic analyses of the 16S rRNA gene sequences and includes the genus *Eubacterium* and its close relatives. However, large numbers of validly published species have historically been misclassified within this genus. They continue to be described in the *Eubacterium* chapter although, based on their phylogenetic assignment, they belong to other families (Wade, this volume). More recent investigations suggest that the genus *Eubacterium* should be restricted to the type species *Eubacterium limosum* and it close relatives, such as *Eubacterium aggregans*, *Eubacterium barkeri*, and *Eubacterium callanderi*.

The family is then restricted to genera related to these species, such as *Acetobacterium*, *Alkalibacter*, *Anaerofustis*, *Pseudoramibacter*, and possibly *Garciella*. The family *Eubacteriaceae* is then composed of Gram-stain-positive rods, which are frequently thin (~0.5 μm). Formation of endospores is rare. May be motile by a single flagellum or nonmotile. Obligately anaerobic, catalase-negative. Chemo-organotrophs, many of which ferment sugars and utilize proteinaceous nitrogen sources. Mesophilic to moderately thermophilic. Neutrophilic to alkaliphilic. Some species are halotolerant.

Type genus: **Eubacterium** Prévot 1938, 294[AL].

Genus I. **Eubacterium** Prévot 1938, 294[AL]

William G. Wade

Eu.bac.te′ri.um. Gr. pref. *eu-* good-, well-, beneficial (not as opposed to pseudo-); Gr. neut. dim. n. *bakterion* a small rod; N.L. neut. n. *Eubacterium* beneficial bacterium.

Uniform or pleomorphic **nonspore-forming**, Gram-stain-positive **rods**. Nonmotile or motile. **Obligately anaerobic. Chemoorganotrophs**, saccharoclastic or nonsaccharoclastic. Usually **produce mixtures of organic acids from carbohydrates** or peptone, often including large amounts of **butyric, acetic, or formic acids**. Do not produce:

1. Propionic as a major acid product (see *Propionibacterium*).
2. Lactic as the sole major acid product (see *Lactobacillus*).
3. Succinic (in the presence of CO_2) and lactic acids with small amounts of acetic or formic acids (see *Actinomyces*).
4. Acetic and lactic (acetic > lactic) acids, with or without formic acid, as the sole major acid products (see *Bifidobacterium*).

   DNA G+C content (mol%): 47 (T_m).
   DNA G+C content (mol%) of the other species examined: 30–57 (determined by a variety of methods).

   Type species: **Eubacterium limosum** (Eggerth 1935) Prévot 1938, 295[AL] (*Bacteroides limosus* Eggerth 1935; *Butyribacterium rerrgeri* Barker and Haas 1944, 303).

Further descriptive information

The species and strains within species vary in sensitivity to oxygen; some can be cultured only in prereduced media. Catalase usually is not produced (trace amounts are detected in some strains). Hippurate usually is not hydrolyzed. Growth usually is most rapid at 37°C and pH near 7.

Found in cavities of man and other animals, animal and plant products, infections of soft tissue, and soil. Some species may be pathogenic.

Enrichment and isolation procedures

Strains of *Eubacterium* species found in clinical infections can be isolated on usual complex media appropriate for culture of anaerobes. Such media include blood agar plates or prere-

duced media in roll tubes made with a peptone base including or supplemented with 0.5% (w/v) yeast extract, 0.2% glucose, 0.001 μl/ml vitamin K_1, and 0.005 μg/ml (w/v) hemin. Species from the rumen or feces usually are isolated in rumen fluid-glucose-cellobiose agar (RGCA), RGCA supplemented with 1% peptone, or similar medium containing hemin and volatile fatty acids. The addition of 0.5% arginine to culture media stimulates the growth of a number of species. Some rumen bacteria grow better in media containing 1,4-naphthoquinone than in media with menadione, vitamin K_1, or vitamin K_5 (Gomez-Alarcon et al., 1982). Although this effect was observed with a Gram-stain-negative anaerobe (*Succinivibrio dextrinosolvens*), it also might apply to the Gram-stain-positive anaerobes.

Maintenance procedures

Lyophilization of cultures in the early stationary phase of growth in a medium containing no more than 0.1–0.2% fermentable carbohydrate is recommended for long-term storage of most strains. For short-term storage, cultures should be grown in a nutritionally adequate medium containing no, or minimal, fermentable carbohydrate and stored at room temperature (~22°C). Saccharoclastic strains should be transferred every 7–10 d, nonsaccharoclastic strains every 2–3 weeks.

Procedures and methods for characterization tests

The description given here is taken from the previous edition of the *Manual* (Moore and Holdeman Moore, 1986). These remain the most robust methods for use in the characterization of this group of organisms. Some *Eubacterium* species require specific nutrients and conditions for growth; these are given in the species descriptions, where appropriate.

Use of heavy inoculum (~5% v/v of broth culture) of a young culture (late log or early stationary phase of growth) gives the most reliable results for biochemical reactions based upon growth of the organism in the substrate. Results of tests for constitutive enzymes with heavy inocula on dehydrated substrates may differ from results based upon growth of the organism in the substrate where adaptive enzymes are also detected. Unless otherwise cited, the characteristics listed here for the species were determined with prereduced anaerobically sterilized media by the methods described in Holdeman et al. (1977), and susceptibility to antimicrobial agents was tested by the broth-disk method of Wilkins and Thiel (1973). The basal prereduced peptone-yeast extract (PY) medium contained (per 100 ml): 0.5 g trypticase, 0.5 g peptone, 1 g yeast extract, 100 μg vitamin K_1, 500 μg hemin, 0.05 g cysteine hydrochloride, and 4 ml salts solution (0.02% anhydrous $CaCl_2$, 0.02% anhydrous $MgSO_4$, 0.1% K_2HPO_4, 0.1% KH_2PO_4, 1% $NaHCO_3$, and 0.2% NaCl).

Morphology and Gram reaction. For determination of Gram reaction, a buffered Gram's stain (such as Kopeloff's modification) of cells from young cultures (exponential phase of growth) is recommended. Cells of many Gram-stain-positive species stain Gram-negative in older cultures or where acid has been produced in the culture medium.

Hydrogen production. H_2 production was determined by gas chromatography of the headspace gas of cultures grown in tubes closed with rubber stoppers (Holdeman et al., 1977). Production of H_2 may not disrupt the agar in PY-glucose (PYG) agar deep cultures.

Gas production. Gas production was determined by disruption of agar in loosely covered PYG agar deep cultures. Accumulation of small (lenticular) bubbles near the colonies in the agar deep tube was recorded as "1" gas, a small split of the agar column across the tube as "2" gas, agar separated and displaced in one or more places as "3" gas, and agar forced to the top of the tube as "4" gas.

Esculin hydrolysis and esculin fermentation. These are independent reactions. Hydrolytic products are detected by the appearance of a black color upon addition of ferric ammonium citrate reagent to esculin-PY broth cultures. Acid production is determined by pH measurement.

Indole. Cultures in media containing sufficient tryptophan (chopped meat medium, tryptone-yeast extract medium, etc.) and no fermentable carbohydrate are required. The culture medium should be extracted with xylene and the reagent (either Ehrlich's or Kovac's) poured down the side of the tube to layer next to the xylene. Do not shake the tube after the reagent has been added. Alternatively, a loopful of growth from a pure culture can be smeared on filter paper saturated with 1% *p*-dimethylaminocinnamaldehyde in 10% (v/v) HCl (Sutter et al., 1980). Development of a blue color indicates indole. This method may not detect weak indole production.

Gelatin digestion. Incubated cultures of PY medium containing 10% gelatin and 1% glucose and uninoculated tubes of the same medium are chilled until the originally liquid controls are solid (about 15 min at 4°C). Failure of the gelatin culture to solidify at 4°C indicates complete digestion (+). Liquefaction of cultures at room temperature within 30 min, or in less than half the time of the control tube, indicates partial liquefaction (w).

Milk proteolysis. Acid production in milk often causes protein coagulation. The curd may or may not show evidence of streaks caused by evolution of gas. Shrinking of the curd may leave a clear whey. This liquid is sometimes mistaken for digestion (proteolysis) of the milk. Digestion of curd is evidenced by dissolution of the curd (first) and increasing turbidity of the whey and often takes many days. Digestion may occur without previous curd formation and results (usually slowly) in clear liquid after extended incubation (up to 3 weeks). (*Clostridium sporogenes* is a good positive control culture for milk proteolysis.)

Acid production. Positive reactions listed for carbohydrate fermentation represent a pH below 5.5 and a decrease in pH of at least 0.5 pH units below the control PY basal medium culture. Weak (w) acid production represents a pH of 5.5–5.9 and at least 0.3 pH units below the PY culture control. A pH of 5.9 or above, or within 0.3 pH unit of the PY control culture pH, is considered negative. When in doubt, examination of the amount of growth in the sugar-containing medium vs that in medium without sugar can help in interpretation of weak reactions, i.e., with slight pH decrease, if growth is much better in the sugar-containing medium, the carbohydrate probably was fermented. For rapid-growing saccharoclastic species, the final pH is reached after incubation for 18–24 h in nutritionally adequate media. Uninoculated xylose and arabinose, under CO_2, are often pH 5.9. For cultures in these media, a final pH of 5.4 or below is considered strong acid production.

Differentiation of the genus *Eubacterium* from other genera

The genera of anaerobic, nonspore-forming, Gram-stain-positive, rod-shaped bacteria are differentiated according to the major metabolic pathways and products of these organisms as described above. The current classification, based on assignment of species to the genus *Eubacterium* by a process of elimination from other anaerobic genera, has led to a taxon that includes somewhat diverse but well-recognized species in three subgroups: those that produce butyric acid, usually in combination with other volatile fatty acids and sometimes with alcohols; those that produce various combinations of lactate, acetate, and formate together with H_2 gas; and those that produce little if any detectable fermentation acid. Hydrogen gas is not produced by anaerobic species or strains of *Lactobacillus*, *Bifidobacterium*, or *Actinomyces* that may also produce combinations of lactate, acetate, and formate. The genus *Eubacterium* includes a wide range of species, many of which are phylogenetically distant from each other. Although previously the genus has served a useful purpose for effective scientific communication, clarification of the taxonomy is now required, particularly the relationships between members of this and other genera. The establishment of a robust taxonomy will allow the determination of the relationship between phylogeny and function and further understanding of the role of these organisms in bacterial communities in the environment and associated with mammalian hosts.

All descriptions are based on strains in which no spores have been detected. Heat-resistant spores, often difficult to detect, have been found in organisms similar to some described species, especially motile or filamentous species; these strains are members of the genus *Clostridium*.

Taxonomic comments

The genus *Eubacterium* is highly heterogeneous, largely because, as mentioned above, it is defined by default. Phylogenetic studies have shown that species within the genus are widely distributed among the phylum *Firmicutes*. Indeed, some taxa have even been found within the *Actinobacteria*, such as the former

Eubacterium lentum, now renamed *Eggerthella lenta* (Wade et al., 1999a). Other novel genera within the *Actinobacteria*, formerly classified as "*Eubacterium*", include *Slackia* (Wade et al., 1999a) and *Cryptobacterium* (Nakazawa et al., 1999, 2000). However, 16S rRNA gene sequence analysis indicates that 42 of the 44 current *Eubacterium* species fall within the phylum *Firmicutes*. The phylogenetic position of *Eubacterium coprostanoligenes* remains to be determined while that of *Eubacterium combesii* is uncertain. A 16S rRNA gene sequence for this species has been deposited in GenBank under the accession number L34614. Comparative analysis of this sequence indicates that *Eubacterium combesii* is closely related to members of the genus *Propionibacterium*. However, since *Eubacterium combesii* does not produce propionic acid as a major end product of metabolism, this identification should be regarded with caution.

It has been suggested that *Eubacterium sensu stricto* should be restricted to *Eubacterium limosum*, the type species, *Eubacterium callanderi*, and *Eubacterium barkeri* (Willems and Collins, 1996). Phylogenetic analysis suggests that these species should be joined by *Eubacterium aggregans* (Figure 140), although the G+C content of its DNA at 55 mol% is significantly different from those of *Eubacterium limosum*, *Eubacterium barkeri*, and *Eubacterium callanderi* at 47–48 mol% (Mechichi et al., 1998).

The remaining *Eubacterium* species should therefore be incorporated into either existing or novel genera. Many species have, as their closest phylogenetic relatives, members of the genus *Clostridium*, another highly heterogeneous genus. One of the principal definitions for *Clostridium* is the formation of spores. However, many lineages within the *Firmicutes* contain both spore-forming and nonspore-forming taxa which suggests that the formation of spores is not a reliable evolutionary marker. It could be that the ability to form spores has been acquired by horizontal gene transfer or that some nonspore-forming species have nonfunctioning sporulation genes. Whatever the reason, if phylogeny is used as a basis for taxonomy, the importance of spore formation as a taxonomic character should be downplayed.

Eubacterium angustum is most closely related to *Clostridium purinilyticum* and both species are obligate purine fermenters that produce acetate, formate, CO_2, and ammonia as metabolic end products. However, despite their phenotypic similarities, they share only 92% 16S rRNA gene sequence similarity, and the G+C content of their DNA differs by 11 mol%: 40 mol% for *Eubacterium angustum* and 29 mol% for *Clostridium purinilyticum*. Further work is required to determine if these two species should be combined in a single novel genus.

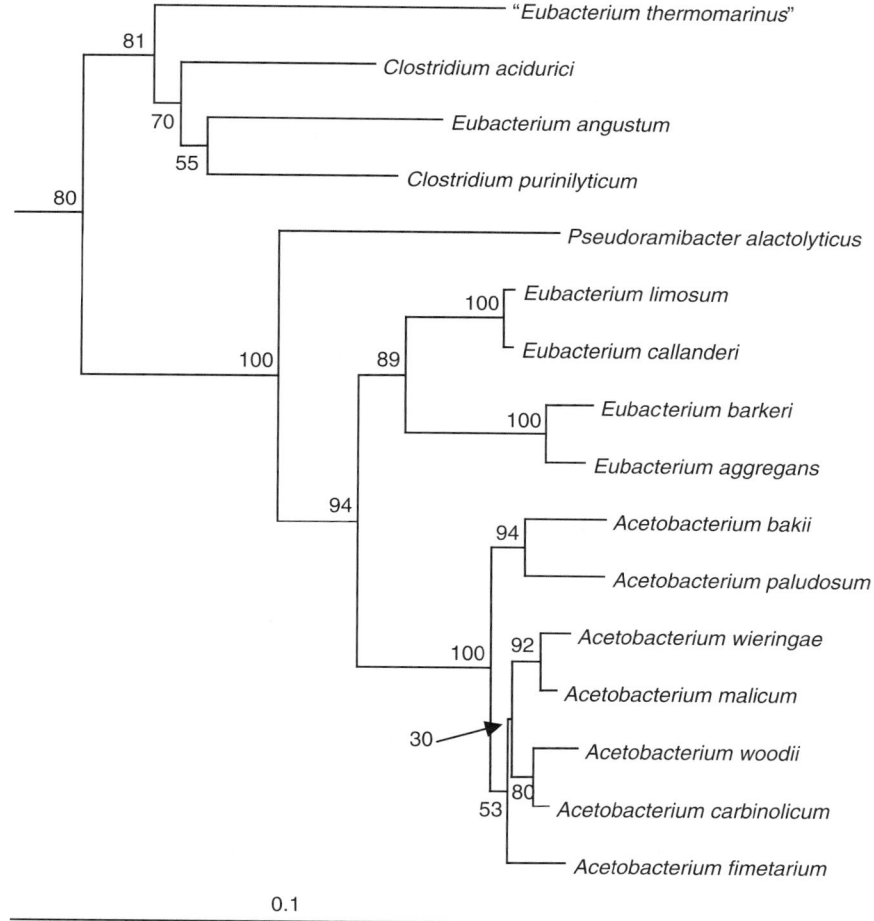

FIGURE 140. Phylogenetic tree based on 16S rRNA gene sequence comparisons over 1354 aligned bases showing relationship between *Eubacterium limosum* and related species. Tree was constructed using the neighbor-joining method following distance analysis of aligned sequences. Numbers represent bootstrap values for each branch based on data for 100 trees.

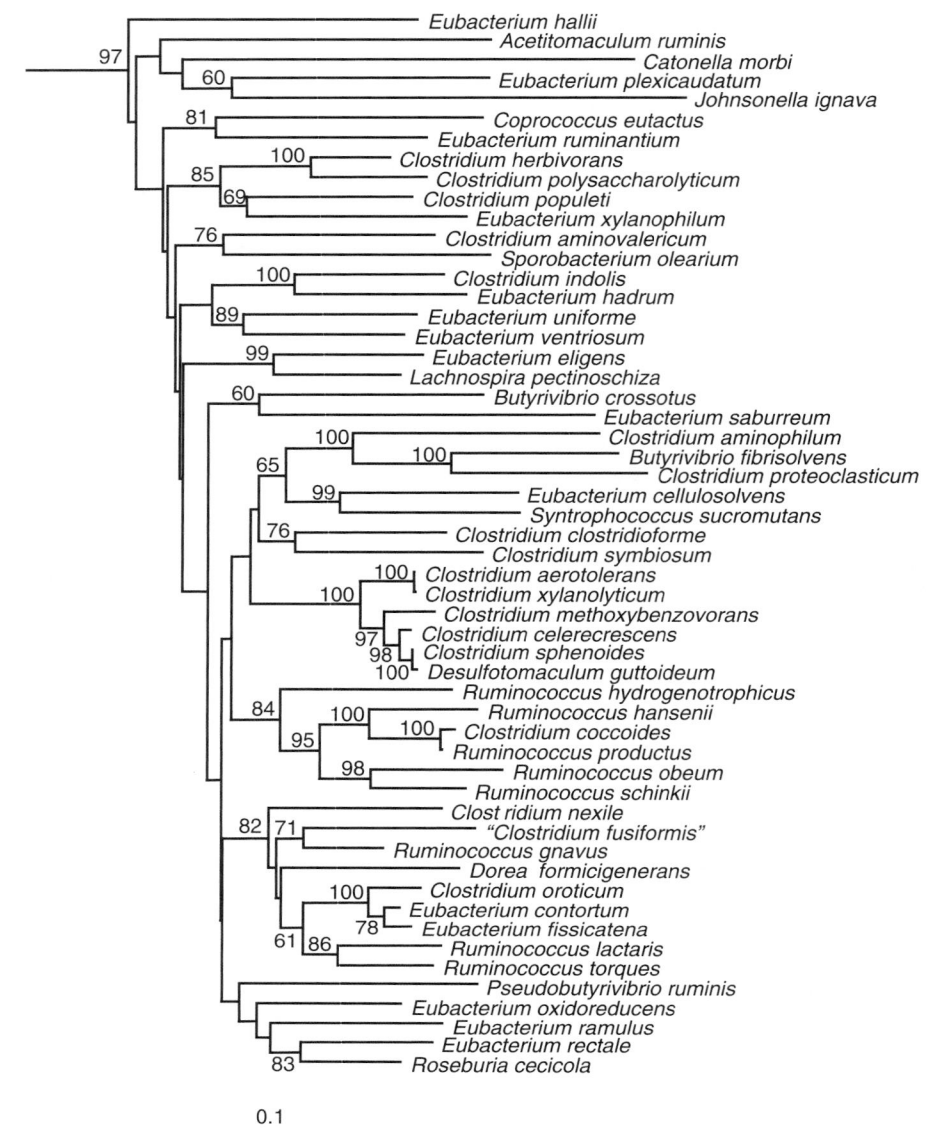

FIGURE 141. Phylogenetic tree based on 16S rRNA gene sequence comparisons over 1230 aligned bases showing relationship between *Eubacterium cellulosolvens* and related species. See legend to Figure 140 for methodology and bootstrap labels.

Figure 141 shows the phylogeny of 15 *Eubacterium* species, all of which can be provisionally assigned to the family *Lachnospiraceae*. *Eubacterium contortum* and *Eubacterium fissicatena* form a tight cluster with *Clostridium oroticum*. These three species are phenotypically similar, all producing acetate, formate, and ethanol as principal metabolic end products and the G+C content of their DNA is in the range of 44–45 mol%. A new genus should be created to include these species.

A looser phylogenetic group is formed by *Eubacterium oxidoreducens*, *Eubacterium ramulus*, *Eubacterium rectale*, *Roseburia cecicola*, and *Roseburia intestinalis* (not shown). All these species produce butyrate and with the exception of *Eubacterium oxidoreducens* are saccharolytic. The taxonomic relationships among members of this group are discussed by Duncan et al. (2002); it would appear that some, at least, should be brought together in a new genus.

Eubacterium eligens and *Lachnospira pectinoschiza* cluster together in the phylogenetic tree and are both motile rods with peritrichous flagella that require fermentable carbohydrate for good growth in broth cultures. It would appear then that *Eubacterium eligens* should be transferred to the genus *Lachnospira*.

The remaining species in this group, *Eubacterium cellulosolvens*, *Eubacterium hallii*, *Eubacterium indolis*, *Eubacterium plexicaudatum*, *Eubacterium ruminantium*, *Eubacterium saburreum*, *Eubacterium uniforme*, *Eubacterium ventriosum*, and *Eubacterium xylanophilum* appear to be sufficiently distinct (on the basis of phenotypic and phylogenetic evidence) from other named taxa to warrant the creation of a novel genus for each one. Polyphasic taxonomic studies are required to confirm this.

The phylogeny of the mainly oral human species related to *Eubacterium minutum*, within the family *Eubacteriaceae*, is shown in Figure 142. On the basis of their phylogeny and phenotypes,

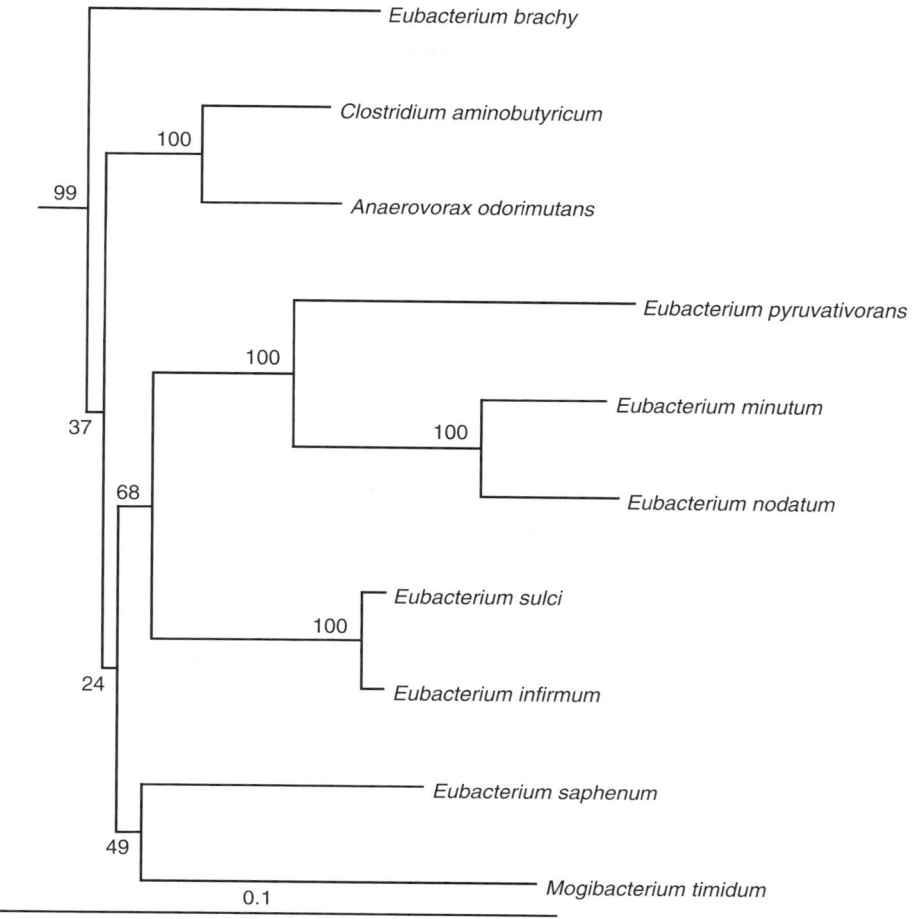

FIGURE 142. Phylogenetic tree based on 16S rRNA gene sequence comparisons over 1238 aligned bases showing relationship between *Eubacterium minutum* and related species. See legend to Figure 140 for methodology and bootstrap labels.

novel genera could be created for the majority of the *Eubacterium* species here with the exception of the pairs of species *Eubacterium minutum* and *Eubacterium nodatum*, and *Eubacterium sulci* and *Eubacterium infirmum*, which are closely related. The former *Eubacterium* species, *Mogibacterium timidum* (Nakazawa et al., 2000), belongs to this group.

Eubacterium acidaminophilum, *Eubacterium tenue*, and *Eubacterium yurii* fall into the clostridial cluster XI described by Collins et al. (1994) (Figure 143). *Eubacterium acidaminophilum* is most closely related to *Clostridium litorale* while *Eubacterium yurii* clusters with, but is distinct from, the genus *Filifactor*. *Eubacterium tenue* is closely related to *Clostridium ghonii* and *Clostridium sordellii*; they produce similar metabolic end products and the DNA of *Clostridium ghonii* and *Clostridium sordellii* have similar G+C contents (26–27 mol%). The G+C content of the DNA of *Eubacterium tenue* has not been determined. Comparative genetic and phenotypic studies are required to determine whether these species constitute a novel genus.

Four species including *Eubacterium biforme*, *Eubacterium cylindroides*, *Eubacterium dolichum*, and *Eubacterium tortuosum* fall within the family *Erysipelotrichaceae* (Figure 144). *Eubacterium biforme* and *Eubacterium cylindroides* form a cluster with *Streptococcus pleomorphus*, while *Eubacterium dolichum* and *Eubacterium*

tortuosum are grouped with *Clostridium innocuum*. As is the case with many *Eubacterium* species, additional phenotypic and genetic characterization is required to confirm the validity of the phylogenetic grouping.

Eubacterium budayi, *Eubacterium moniliforme*, *Eubacterium multiforme*, and *Eubacterium nitritogenes* are found in clostridial Cluster 1 (Collins et al., 1994) and are related to *Clostridium absonum* and *Clostridium barati* (Figure 145). The relationship between *Eubacterium budayii* and *Eubacterium nitritogenes* is not clear as, phenotypically, they are virtually indistinguishable. *Eubacterium tarantellae* is a member of the same cluster and most closely related phylogenetically to the genus *Sarcina* but is phenotypically quite distinct.

Eubacterium desmolans, *Eubacterium plautii*, and *Eubacterium siraeum* cluster in two branches of a group related to *Clostridium leptum* and which includes a number of *Ruminococcus* species (Figure 146). *Eubacterium plautii* and *Clostridium orbiscindens* share 99% 16S rRNA gene sequence identity; further studies are required to determine if they are the same species. *Eubacterium desmolans* and *Eubacterium siraeum* have no close relatives and each would appear to represent a novel genus.

It can be seen, therefore, that substantial attention to the taxonomy of the genus *Eubacterium* is required. The benefit of

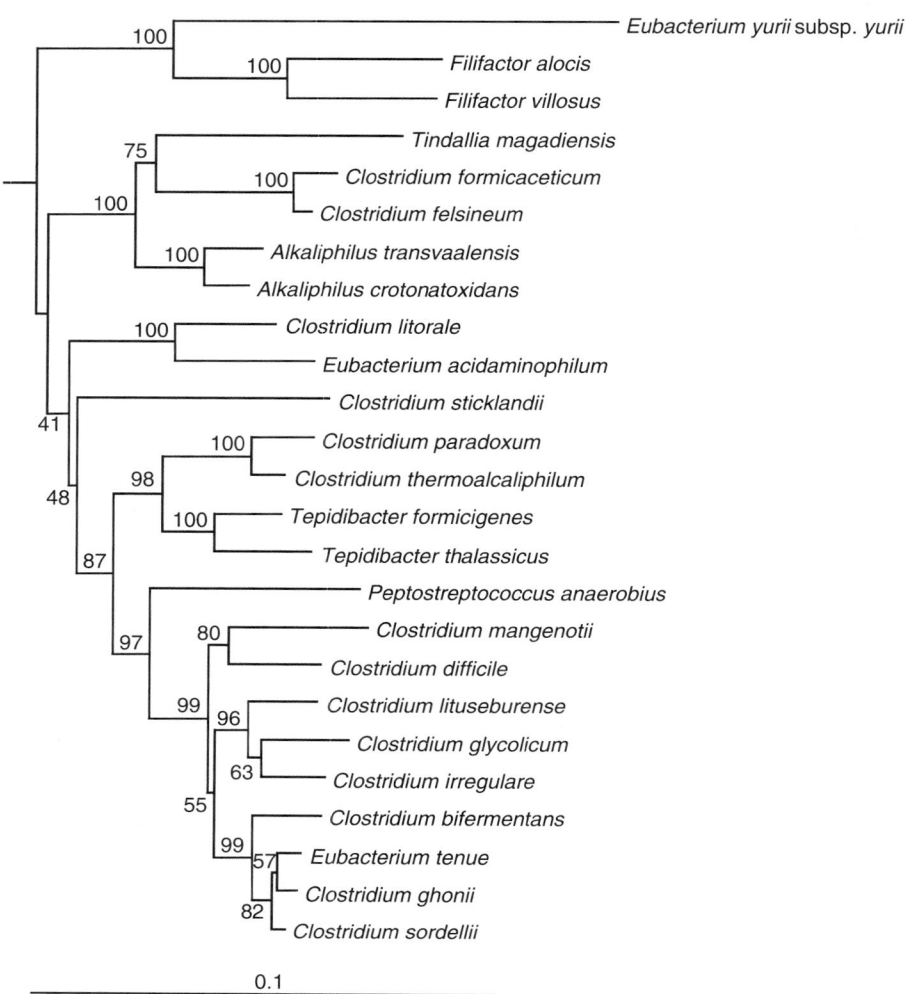

FIGURE 143. Phylogenetic tree based on 16S rRNA gene sequence comparisons over 1251 aligned bases showing relationship between *Eubacterium acidaminophilum* and related species. See legend to Figure 140 for methodology and bootstrap labels.

the phylogenetic approach has been to identify closely related species hitherto classified in disparate genera. The clarification of the taxonomy of this group will be extremely beneficial to the understanding of the physiology of its members and their role in the environment and in human and animal disease.

Acknowledgements

Much of the data given below in the species descriptions were generated by Drs W.E.C. Moore and L.V. Holdeman Moore for the last edition of the *Manual*. Their contribution to this chapter, and anaerobic microbiology in general, is gratefully acknowledged.

I thank my colleague Julia Downes for numerous helpful discussions and for her contribution to the recent work described in this chapter. I should also like to thank the following individuals who have contributed to work on "*Eubacterium*" species in my laboratory over many years: Sarah Cheeseman, Sarah Hiom, Susan Milsom, David Dymock, David Spratt, and Mark Munson. I should also like to thank the following individuals who have collaborated with me in this work: Andrew Weightman, Anne Tanner, Floyd Dewhirst, and Bruce Paster.

Differentiation of the species of the genus *Eubacterium*

Characteristics by which species in the genus can be differentiated are given in Table 156. Acid production in the key refers to both strong and weak fermentations, i.e., a species is positive if the terminal pH is below 5.9 and 0.3 pH units below the control basal medium culture. Additional characteristics concerning each species are given in the text. A number of recently proposed species have unusual and specific growth requirements. These often prevent the performance of "standard" biochemical and physiological tests; the results of such tests that have been reported are included for completeness, but, inevitability, the usefulness of the table for the identification of these species is limited. In my own laboratory, 16S rRNA gene sequence analysis (augmented by relevant phenotypic tests) is the primary method used for preliminary identification. This method is now becoming widely available and allows the easy recognition of currently un-named taxa, of which there are a large number within this genus.

FIGURE 144. Phylogenetic tree based on 16S rRNA gene sequence comparisons over 1355 aligned bases showing relationship between *Eubacterium dolichum* and related species. See legend to Figure 140 for methodology and bootstrap labels.

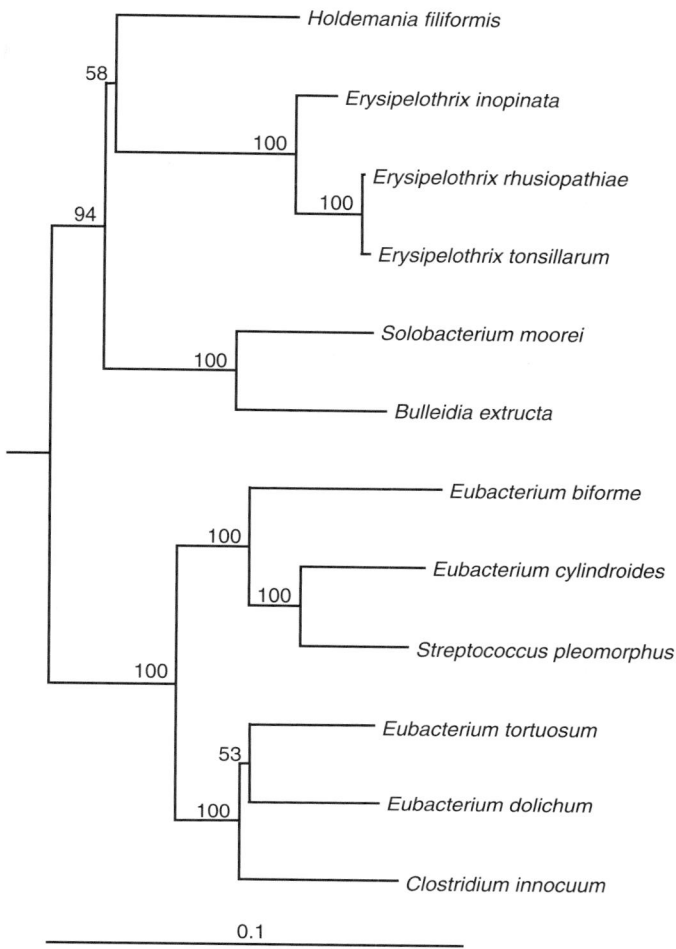

Colony descriptions are from blood agar plates or streak tube cultures incubated anaerobically for 2 d, unless otherwise indicated. Reported G+C contents of the DNAs are expressed as the nearest whole number. Unless specified otherwise in the text, cells are nonmotile.

List of species of the genus *Eubacterium*

1. **Eubacterium limosum** (Eggerth 1935) Prévot 1938, 295[AL] (*Bacteroides limosus* Eggerth 1935, 290; *Butyribacterium rettgeri* Barker and Haas 1944, 303)

li.mo'sum. L. neut. adj. *limosum* full of slime, slimy.

Type species of the genus *Eubacterium* (Cato et al., 1981; Wayne, 1982). This description is based on examination of the type and 50 similar strains (Moore and Holdeman Moore, 1986). Cells of the type strain grown in PYG broth are nonmotile, 0.6–0.9 × 1.6–4.8 μm, often with swollen ends and bifurcations, and occur singly, and in pairs and small clumps. Cell walls contain peptidoglycan of the group B type (Guinand et al., 1969; Schleifer and Kandler, 1972). Surface colonies on blood agar are punctiform to 2 mm in diameter, circular, entire, convex, translucent to slightly opaque, and sometimes with mottled appearance when viewed by obliquely transmitted light. Glucose broth cultures are turbid with viscous, stringy, or smooth sediment and a terminal pH of 4.5–5.2. Growth often is better with a fermentable carbohydrate. The optimum temperature for growth is 37°C. Most strains grow well at temperatures between 30 and 45°C, but growth often is inhibited at 25°C.

Abundant gas is produced in PYG agar deep cultures. Adonitol and pectin are fermented; some strains ferment galactose. Dextrin, dulcitol, glycerol, inulin, and sorbose are not fermented. Hippurate is not hydrolyzed. Some strains produce ammonia from arginine. Catalase and urease are not produced. H_2S may be produced. Synthesizes vitamin B_{12}. Products (milliequivalents per 100 ml of culture) in PYG cultures are acetic (1–3), butyric (0.5–2.5), and lactic (1–3) acids, occasionally with small amounts of succinic, iso-butyric and iso-valeric acids. Carbon dioxide and copious amounts of H_2 are produced in PYG broth cultures. Products in PY broth cultures are acetate, iso-butyrate, butyrate (small amounts), and iso-valerate, sometimes with traces of propionate. Glucose spares amino acid (peptone) metabolism, and the branched chain acids (iso-butyric and iso-valeric) usually are not detected in PYG cultures. Lactate and pyruvate are converted to acetate and butyrate.

Carbon monoxide can be used as the sole energy source; acetate and CO_2 are the major products (Genthner and Bryant, 1982). Other energy sources include methanol, H_2 (in the presence of CO_2), adonitol, arabitol, erythritol,

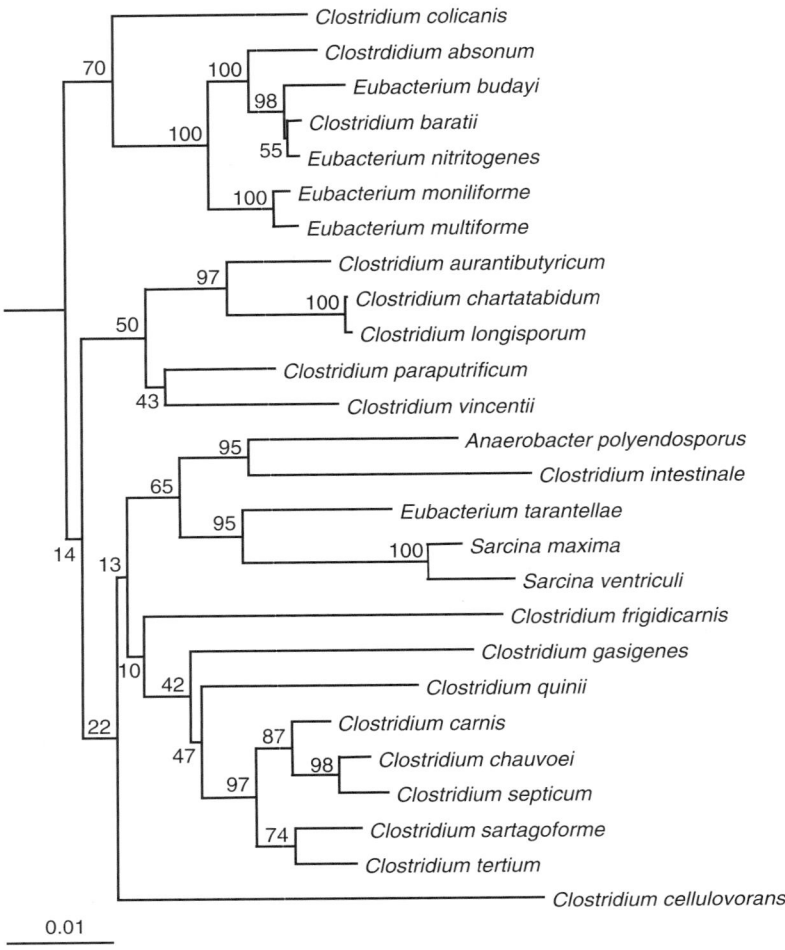

FIGURE 145. Phylogenetic tree based on 16S rRNA gene sequence comparisons over 1322 aligned bases showing relationship between *Eubacterium tarantellae* and related species. See legend to Figure 140 for methodology and bootstrap labels.

fructose, glucose, isoleucine, lactate, mannitol, ribose, and valine. Ammonia, or each of several amino acids, serves as the main nitrogen source. Acetate, cysteine, CO_2, calcium-D-pantothenate, and lipoic acid are required for growth on a chemically defined methanol medium. Acetate, butyrate, and caproate are produced from methanol. Ammonia and branched chain fatty acids are produced from amino acids. Arabinose, galactose, galacturonic acid, gluconate, mannose, rhamnose, cellobiose, lactose, maltose, melibiose, sucrose, trehalose, melezitose, raffinose, dextrin, pectin, starch, dulcitol, salicin, xylitol, ethanol, propanol, butanol, inositol, acetate, fumarate, malate, succinate, adenine, thymine, cytosine, uracil, allantoin, urea, methylamine, and esculin are not utilized as energy sources although esculin is hydrolyzed (Genthner et al., 1981). Although *Eubacterium limosum* may grow slightly when pectin is autoclaved in the medium, the organism probably is not using pectin but is growing on methanol that is hydrolyzed from the pectin during autoclaving (Rode et al., 1981).

Most strains are susceptible to chloramphenicol (12 μg/ml) and clindamycin (1.6 μg/ml). Some strains are resistant to erythromycin (3 μg/ml), penicillin G (2 units/ml), or tetracycline (6 μg/ml).

Isolated from feces of man; rumen; intestinal contents of man, rats, poultry, and fish; various human and animal infections (rectal and vaginal abscesses, blood, wounds); sewage sludge; and mud. Other characteristics of the species are given in Table 156.

DNA G+C content (mol%): 47 (T_m).

Type strain: ATCC 8486, CCUG 16793, CIP 104169, DSM 20543, JCM 6421, JCM 9978, NCIB 9763.

GenBank accession number (16S rRNA gene): M59120.

2. **Eubacterium acidaminophilum** Zindel, Freudenberg, Rieth, Andreesen, Schnell and Widdel 1989, 93^VP (Effective publication: Zindel, Freudenberg, Rieth, Andreesen, Schnell and Widdel 1988, 264.)

a.cid.a.mi.no′phi.lum. N.L. n. *acidum* acid; N.L. pref. *amino* amino group containing; N.L. neut. adj. *philum* from Gr. adj. *philos* loving; N.L. neut. adj. *acidaminophilum* loving amino acids.

Rod-shaped cells, 0.7–1.0 × 1.3–6 μm, mostly with pointed ends, usually single, sometimes in chains; slow tumbling motility by means of a subpolar to polar flagellum. Gram-stain-positive.

Fast growth (doubling time about 1 h) occurs by fermentation of glycine to acetate, ammonia, and CO_2. Glycylglycine,

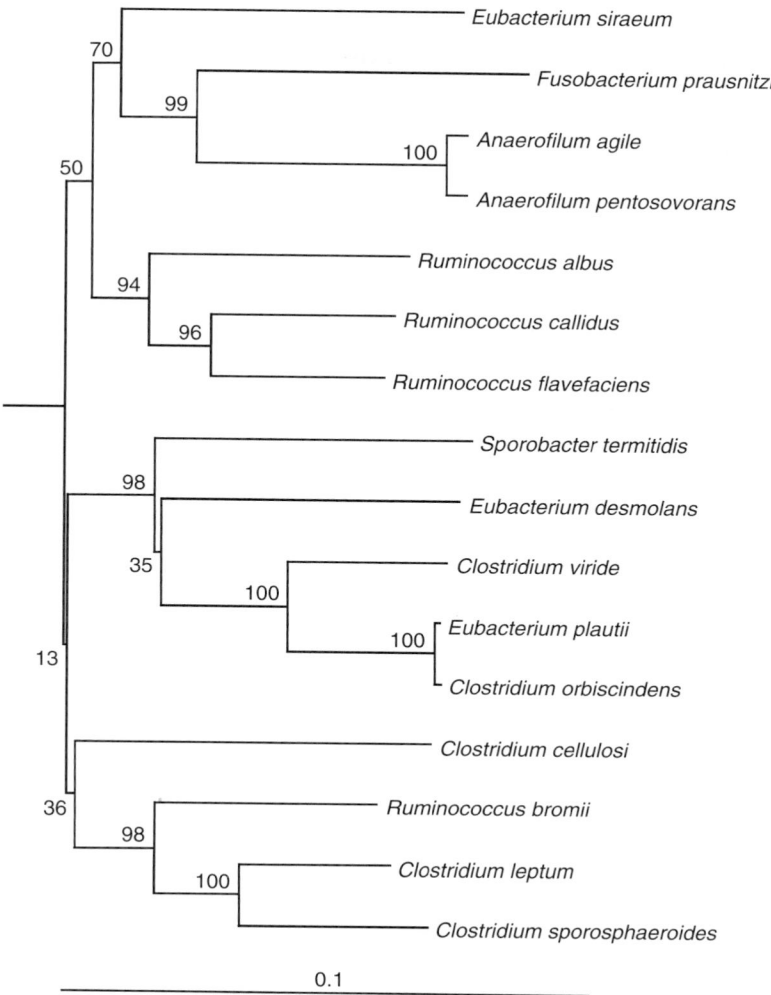

FIGURE 146. Phylogenetic tree based on 16S rRNA gene sequence comparisons over 1300 aligned bases showing relationship between *Eubacterium desmolans* and related species. See legend to Figure 140 for methodology and bootstrap labels.

glycylglycylglycine, *N*-carbamoylglycine, hydantoin, the glycine moiety of hippurate and glutathione, serine, and pyruvate are also fermented. Glycine derivatives such as betaine, sarcosine, and creatine serve as substrates only in the presence of an electron donor such as formate. Pure cultures grow very poorly by fermentation of alanine or aspartate to acetate, ammonia, CO_2, and H_2. Good growth on alanine or aspartate occurs in co-cultures with H_2-utilizing, sulfate-reducing, methanogenic or homoacetogenic bacteria. Valine, leucine, and malate are also oxidized by the co-cultures or in the presence of an electron-accepting substrate. Glutamate, glucose, and fructose are not fermented. Sulfate, thiosulfate, sulfur, and nitrate are not used as electron acceptors.

Grows in defined sulfide-reduced, bicarbonate-buffered media with biotin as a vitamin. NaCl concentrations higher than approximately 30 g/l are inhibitory. Enrichment is possible with alanine at low H_2 partial pressure, e.g., in the presence of sulfate and *Desulfovibrio vulgaris*. Selenite (≥0.1 µM) is required for growth on glycine, betaine, sarcosine, or creatine, but not on other substrates. Temperature range 10–40°C, optimum at 32–36°C; pH range 6.5–8.5, optimum 7.1–7.4.

Isolated from black, anaerobic mud.

DNA G+C content (mol%): 44 (T_m).

Type strain: a1-2, ATCC 49065, DSM 3953.

GenBank accession number (16S rRNA gene): AF071416.

3. **Eubacterium aggregans** Mechichi, Labat, Woo, Thomas, Garcia and Patel 2000, 1699[VP] (Effective publication: Mechichi, Labat, Woo, Thomas, Garcia and Patel 1998, 290.)

ag′gre.gans. L. v. *aggregare* to flock or band together; L. pres. part. *aggregans* assembling (aggregating).

Cells are nonspore-forming, 2–3 µm × 0.4–0.8 µm, nonmotile and stain Gram-positive. The temperature range for growth is 20–42°C and the optimal temperature is 35°C. The pH range for growth is 6–9 and the optimum is 7.2. There is growth at NaCl concentrations from 0–3.5%. Utilizes glucose, fructose, sucrose, lactate, formate, methanol, $H_2 + CO_2$, betaine, crotonate, and methoxyl groups from the following compounds: 2-methoxyphenol, 3,4- and 3,5-dimethoxybenzoates, 3,4-dimethoxybenzaldehyde, 3,4,5-trimethoxycinnamate, 3,4,5-trimethoxybenzoate, 4-hydroxy-3,5-dimethoxybenzoate, 4-hydroxy-3,5-dimethoxycinnamate, 3-hydroxy-4-methoxybenzoate,

TABLE 156. Biochemical reactions of species in the genus *Eubacterium*[a]

Species columns (left to right):
E. linosum, *E. acidaminophilum*, *E. aggregans*, *E. angustum*, *E. barkeri*, *E. biforme*, *E. brachy*, *E. budayi*, *E. callanderi*, *E. cellulosolvens*, *E. combesii*, *E. contortum*, *E. coprostanoligenes*, *E. cylindroides*, *E. desmolans*, *E. dolichum*, *E. eligens*, *E. fissicatena*, *E. hadrum*, *E. hallii*, *E. infirmum*, *E. minutum*, *E. moniliforme*, *E. multiforme*, *E. nitritogenes*, *E. nodatum*, *E. oxidoreducens*, *E. plautii*, *E. plexicaudatum*, *E. pyruvativorans*, *E. ramulus*, *E. rectale*, *E. ruminantium*, *E. saburreum*, *E. saphenum*, *E. siraeum*, *E. sulci*, *E. tarantellae*, *E. tenue*, *E. tortuosum*, *E. uniforme*, *E. ventriosum*, *E. xylanophilum*, *E. yurii* subsp. *yurii*, *E. yurii* subsp. *margaretiae*, *E. yurii* subsp. *schtitka*

Characteristic (rows):

Cells motile

Acid produced from:
Amygdalin, Arabinose, Cellobiose, Esculin, Fructose, Glucose, Glycogen, Lactose, Maltose, Mannitol, Mannose, Melezitose, Melibiose, Raffinose, Rhamnose, Ribose, Salicin, Sorbitol, Starch, Sucrose, Trehalose, Xylose

Esculin hydrolyzed
Starch hydrolyzed
Gelatin digested
Milk reaction
Meat digested
Indole produced
Nitrate reduced
H₂ produced
Butyrate produced
Other (see footnote)

[a]Symbols: −, negative reaction for 90–100% of strains; +, positive reaction for 90–100% of strains; ±, see footnote b; A, acid (pH below 5.5); w, weak reaction (pH 5.5–5.9); numbers (hydrogen) represent abundant (4) to negative on a "−" to "4+" scale; nr, not reported. When two reactions are given, the usual reaction is given first. Unless otherwise noted (in the footnotes), erythritol and inositol are not fermented and lecithinase is not produced.

[b]±, pH 6.0–7.0; − (for this species only, pH above 7.0).

[c]Erythritol fermented.

[d]Lecithinase produced.

[e]Cellulose digested.

[f]Inositol fermented.

[g]Lecithinase may be produced.

4-hydroxy-3-methoxybenzoate, 4-hydroxy-3-methoxycinnamate, 3-hydroxy-3-methoxycinnamate, 3-hydroxy-4-methoxybenzaldehyde, 4-hydroxy-3-methoxybenzaldehyde, and 4-hydroxy-3-methooxycinnamylalcohol. No growth occurs with xylose, sorbose, galactose, *myo*-inositol, lactose, cellobiose, xylan, and cellulose.

Isolated from a digestor loaded with olive mill wastewater.

DNA G+C content (mol%): 55 (HPLC).

Type strain: SR12, DSM 12183.

GenBank accession number (16S rRNA gene): AF073898.

4. **Eubacterium angustum** Beuscher and Andreesen 1985, 535^{VP} (Effective publication: Beuscher and Andreesen 1984, 7.)

an.gus′tum. L. adj. *angustum* restricted, scanty, narrow (substrate spectrum).

Straight, angular rods, 1.1–1.5 μm wide and 3–6.5 μm long, older cultures form chains, often clumps. Nonmotile but has flagella, mostly laterally inserted. Colonies are nonpigmented, flat, circular, and 0.5–1.5 mm in diameter. Gram-stain-positive.

Strictly anaerobic; optimal temperature for growth is 37°C although growth occurs at 18 and 45°C; pH range for growth is 6.5–10; optimum range, pH 8.0–8.2. Chemoorganotrophic. For growth, utilizes only uric acid, xanthine, and guanine, hypoxanthine in connection with uric acid, requires thiamine. Acetate, formate, CO_2, and NH_3 are metabolic end products. Hydrogen, succinate, butyrate, acetoin, diacetyl, 2,3-butanediol, indole, and sulfide are not produced. Nitrate and sulfate are not reduced. Hydrolysis of gelatin, starch, and esculin does not occur. No hemolysis. Catalase, lipase, lecithinase, and urease are not produced. No cytochromes are detected. Growth not influenced by the presence of 2% bile extract. Peptidoglycan is directly cross-linked via *meso*-diaminopimelic acid.

Isolated from sludge from sewage plants.

DNA G+C content (mol%): 40 (T_m).

Type strain: MK-1, ATCC 43737, DSM 1989.

GenBank accession number (16S rRNA gene): L34612.

5. **Eubacterium barkeri** (Stadtman, Stadtman, Pastan and Smith 1972) Collins, Lawson, Willems, Cordoba, Fernandez-Garayzabal, Garcia, Cai, Hippe and Farrow 1994, 824^{VP} (*Clostridium barkeri* Stadtman, Stadtman, Pastan and Smith 1972, 760)

bar′ker.i. N.L. gen. n. *barkeri* of Barker named for H.A. Barker, American biochemist.

Based on the description by Stadtman et al. (1972), Holdeman et al. (1977), and study of the type strain (Moore and Holdeman Moore, 1986). Cells in PYG broth are Gram-stain-positive, nonmotile, and 0.3–0.5 by 1.6–9.7 μm, and occur singly or in pairs. Spores are oval, terminal, and swell the cell. Sporulation occurs most reliably on chopped meat slants incubated at 30°C. Cell walls do not contain diaminopimelic acid (DAP). They do contain the B type of peptidoglycan as described by Schleifer and Kandler (1972) and the cross-linking amino acids D-lysine and small amounts of D-ornithine. Surface colonies on blood agar are 0.5–1.0 mm, circular, entire, convex, translucent to slightly opaque, mosaic, white, shiny, smooth, and nonhemolytic. Broth cultures are turbid with smooth sediment. After 5 d incubation, the pH of PYG cultures is 4.5. Optimum temperature

for growth is 37°C. Growth is good at 30°C, fair at 25°C and poor at 45°C. Good growth at a pH of 8.5. No growth in 6.5% NaCl or 20% bile. Ammonia is not produced. Neutral red, but not resazurin, is reduced.

Fermentation products from glucose include butyric and lactic acids and large amounts of H_2; moderate amounts of acetic acid may be formed. Pyruvate is converted to acetate and butyrate. Neither threonine nor lactate is converted to propionate. Nicotinic acid is fermented to propionic and acetic acids, CO_2, and ammonia (Stadtman et al., 1972). Vitamin B_{12} is formed by a pathway similar to that of *Eubacterium limosum*, with glycine and methionine as precursors (Hollriegl et al., 1982).

The type strain is sensitive to chloramphenicol, clindamycin, erythromycin, and tetracycline, but resistant to penicillin G. Culture supernatants are nontoxic to mice.

Isolated from Potomac River mud and human feces (Finegold et al., 1983).

DNA G+C content (mol%): 45 (T_m).

Type strain: ATCC 25849, CIP 104307, DSM 1223, JCM 1389, NCIB 10623, VKM B-1775.

GenBank accession number (16S rRNA gene): M23927.

6. **Eubacterium biforme** (Eggerth 1935) Prévot 1938, 295^{AL} (*Bacteroides biformis* Eggerth 1935, 283)

bi.for′me. L. adj. *biformis* two-formed (pertaining to cellular morphology).

Cells from PYG broth cultures are nonmotile, 0.6–2.7 μm in diameter × 1.6–15 μm long in pairs and chains. Definite rods and coccoid forms often occur in the same chain. Central swellings ("pelton de jardinier") are present in some cultures, particularly among cells from growth on solid media. After incubation for 5 d in RGCA, subsurface colonies are 0.5–1.0 mm in diameter, white to tan, and usually lenticular and translucent, opaque or translucent with opaque centers, sometimes with diffuse edges. Occasional colonies look like "woolly balls". Surface colonies on blood agar plates incubated for 2 d are 0.5–2.0 mm in diameter, circular, entire to slightly erose, flat to low convex, white, shiny, smooth, translucent with dense centers, and nonhemolytic. The temperature for optimal growth is 37°C; some strains grow at 45°C. Most strains do not grow at 30°C.

PYG-0.2% Tween 80 broth cultures have abundant growth with smooth sediment, little or no turbidity, and a pH of 4.5–5.0 in 1–2 d. Growth of most strains is stimulated by 0.02% Tween 80 but not by rumen fluid and may be inhibited by 20% bile. Growth is markedly stimulated by fermentable carbohydrate. Variable amounts of gas are detected in glucose agar deep cultures. Most strains produce acid from galactose. Neutral red is reduced. A small amount of ammonia may be produced in chopped meat cultures.

Products (milliequivalents per 100 ml of culture) in PYG-0.02% Tween 80 cultures are butyric (0.4–1.2), caproic (0.01–0.07), and lactic (2.0–5.0) acids. Moderate to abundant H_2 is produced from PYG-Tween 80 cultures. Caproic acid is produced irregularly and probably is dependent upon the amount of growth and the age of the culture analyzed. Pyruvate is converted to butyrate, often with lactate and acetate, sometimes with formate, and a small amount of caproate. Lactate and gluconate are not utilized.

The type strain is susceptible to 12 μg/ml of chloramphenicol, 1.6 μg/ml of clindamycin, 3 μg/ml of erythromycin, 2 units/ml of penicillin G, and 6 μg/ml of tetracycline. Other characteristics of the species are given in Table 156.

Isolated from human feces at 1–3% of the flora (Holdeman et al., 1976; Moore and Holdeman, 1974).

DNA G+C content (mol%): 32 (T_m).

Type strain: ATCC 27806, CCUG 28091, DSM 3989.

GenBank accession number (16S rRNA gene): M59230.

Further comments: Moore and Holdeman Moore (1986) reported that definite rod-shaped cells did not appear in all cultures of each strain. The production of caproic acid in small amounts was a distinguishing characteristic, but was sometimes detected only on repeated culture. The morphology of the cells and the production of H_2 and of small amounts of caproic acid by *Eubacterium biforme* were useful characters in distinguishing this species from *Eubacterium cylindroides*.

7. **Eubacterium brachy** Holdeman, Cato, Burmeister and Moore 1980, 167[VP]

bra′chy. N.L. neut. adj. *brachy* (from Gr. neut. adj. *brachu*) short; referring to the length of the cells.

Cells from PY broth cultures are 0.4–0.8 × 1.0–3.0 μm and occur predominantly in long chains. Although some cells in a chain might be coccoid, definite rods can be seen upon close examination of the smears. After 5 d of incubation, subsurface colonies in supplemented brain heart infusion agar (BHIA-S, Holdeman et al., 1977) enriched with 5% v/v rabbit serum and 0.02 g each of sodium formate and ammonium fumarate and 400 μg thiamine pyrophosphate/100 ml are 0.5–2.0 mm in diameter, lenticular to trifoliate, and transparent to opaque. After incubation for 2 d, surface colonies on BHIA-S streak tubes are less than 0.5 mm in diameter, circular, entire, low convex, white, shiny, and smooth. Surface colonies on chopped meat broth agar are similar to those on BHIA-S but are larger (1.0 mm in diameter). Surface growth is obtained more reliably on chopped meat broth agar than on BHIA-S tubes or anaerobic blood agar plates.

Broth cultures have a small amount of fine, granular to smooth sediment, often without turbidity. Broth cultures often require incubation for 2–3 d to attain maximum growth, which at best is 2+, as graded on a scale of negative to 4+. Carbohydrates are not fermented.

The temperature for optimal growth is 37°C; most strains do not grow at 30 or 45°C. Growth is inhibited by 6.5% NaCl and slightly inhibited by 2% oxgall (20% bile). Growth is neither inhibited nor stimulated by the addition of 0.02% Tween 80, 5% (v/v) rumen fluid, or 10% (v/v) serum with thiamine pyrophosphate.

A small amount of gas is produced in glucose agar deep cultures. Ammonia is produced in PY broth cultures.

Products (milliequivalents per 100 ml of culture) in PYG broth are acetate (0.1–0.4), iso-butyrate (0.1–0.4), iso-valerate (0.2–0.4), and iso-caproate (0.2–0.5), often with trace amounts of formate, propionate, butyrate, or valerate. Headspace gas of 5-d-old PY broth cultures contains 2–3% H_2.

Strains tested are susceptible to chloramphenicol (12 μg/ml), clindamycin (1.6 μg/ml), erythromycin (3 μg/ml), penicillin G (2 units/ml), and tetracycline (6 μg/ml). Other characteristics of the species are given in Table 156.

Isolated from a variety of oral infections including periodontitis, endodontic infections, and dentinal caries. One isolate from a human lung abscess has been reported (Rochford, 1980).

DNA G+C content (mol%): not determined because of the sparse amount of growth.

Type strain: ATCC 33089, CIP 104210, DSM 3990, VPI D6B-23.

GenBank accession number (16S rRNA gene): U13038.

Further comments: Because of the fermentation products and the occurrence of short cells in chains, *Eubacterium brachy* resembles *Peptostreptococcus anaerobius*. It does not grow as well in standard media as *Peptostreptococcus anaerobius*, and definite rod-shaped cells can be seen in Gram stains. Additionally, growth of *Eubacterium brachy* is not inhibited by sodium polyanetholsulfonate.

8. **Eubacterium budayi** (Le Blaye and Guggenheim 1914) Holdeman and Moore 1970, 23[AL] (*Bacterium budayi* Le Blaye and Guggenheim 1914, 402)

bu.day′i. N.L. gen. n. *budayi* of Buday; named for K. Buday, the bacteriologist who first isolated the organism.

The description is from Prévot et al. (1967) and study of the type strain (Moore and Holdeman Moore, 1986). Cells in PYG broth cultures are 0.8–1.7 μm in diameter × 3.0 to more than 78 μm in length and are straight or slightly curved. Cells occur singly and in pairs with angular and parallel arrangements. Short forms may occur in media without a fermentable carbohydrate. Surface colonies on McClung-Toabe egg yolk agar (EYA) are 4–5 mm, irregular, low convex, scalloped, yellowish, and are surrounded by a small opaque zone in the agar. Colonies in deep agar look like snowflakes, often with rhizoids. Glucose broth cultures are turbid with smooth sediment and pH of 5.0.

The optimal temperature for growth is 37°C. Moderately good growth occurs at 25–30°C but not at 45°C. Growth is stimulated by 0.02% Tween 80. Moderate growth occurs in peptone-yeast extract broth; growth usually is stimulated by a fermentable carbohydrate. Growth is inhibited by 6.5% NaCl. Growth occurs in PYG with 20% bile, but fermentation may be suppressed.

Abundant gas is produced in PYG agar deep cultures. Galactose is fermented. Adonitol, dulcitol, inulin, pectin, and sorbose are not fermented; little or no ammonia is produced from peptone, arginine, threonine, or chopped meat medium; H_2S is not produced in SIM medium. Slight lecithinase activity may be observed on EYA.

Products (milliequivalents per 100 ml of culture) in PYG cultures are acetic (0.4), butyric (0.8), L-(+)-lactic (1.4), and pyruvic (0.1) acids. Copious H_2 is produced. Other characteristics of the species are given in Table 156.

Isolated from a cadaver by Buday and from poorly sterilized catgut, mud, and soil by Prévot et al.

DNA G+C content (mol%): not determined.

Type strain: ATCC 25541, CIPP ECI, DSM 3981, JCM 9989.

GenBank accession number (16S rRNA gene): AB018183.

9. **Eubacterium callanderi** Mountfort, Grant, Clarke and Asher 1988, 257[VP]

call.and′er.i. N.L. gen. n. *callanderi* of Callander, named in honor of John Callander for his pioneering work on anaerobic digestion of wood-based materials.

Cells are obligately anaerobic, nonmotile, nonspore-forming, Gram-stain-postive rods, 0.9 × 1.5–3.0 μm. Cell walls contain the peptidoglycan components muramic acid, glucosamine, lysine, ornithine, serine, glutamic acid, and alanine, but diaminopimelic acid is not present.

Cleaves the phenylether bonds of the methoxylated aromatic substrates ferulate, sinapate, syringate, vanillate, 3,4,5-trimethoxycinnamate, and vanillin to give the corresponding hydroxyaromatic derivatives and acetate, butyrate, and formate. In addition, lactate and H_2 are produced from the fermentation of glucose. No reduction of SO_4^{2-} or NO_3^- occurs during fermentation of glucose or methoxylated aromatic compounds. Methanol, H_2-CO_2, formate, carbon monoxide, and isoleucine are not utilized for growth.

The best conditions for growth are in basal medium supplemented with 5% rumen fluid. No growth occurs on glucose in defined medium (rumen fluid absent) unless acetate is present. The pH range for growth is 6.0 to at least 8.5, and the pH optimum is 7.0. The temperature range for growth is 25–43°C and the optimum temperature is 37°C.

Isolated from an anaerobic digestor fed with the contents of a wood to fiber to alcohol fermentation plant.

DNA G+C content (mol%): 47 (T_m).

Type strain: FD, ATCC 49165, DSM 3662, JCM 10284.

GenBank accession number (16S rRNA gene): X96961.

10. **Eubacterium cellulosolvens** (Bryant, Small, Bouma and Robinson 1958) Holdeman and Moore 1972, 39[AL] (*Cillobacterium cellulosolvens* Bryant, Small, Bouma and Robinson 1958, 529)

cel.lu.lo.sol′vens. N.L. n. *cellulosum* cellulose; L. part. adj. *solvens* dissolving; N.L. part. adj. *cellulosolvens* cellulose-dissolving.

The description is from Bryant et al. (1958), van Gylswyk and Hoffman (1970), and from study of one strain from van Gylswyk and Hoffman (1970) and one isolate from the intestinal tract of a hog (Moore and Holdeman Moore, 1986). Cells in PYG broth usually are motile with peritrichous flagella and are 0.3–0.8 × 0.8–3.1 μm. They occur singly, in pairs, and in clumps.

Surface colonies on anaerobic streak tubes of BHIA-S are 3–5 mm, circular, entire, flat to slightly convex, translucent, and light tan to colorless. Subsurface colonies are lenticular. No growth is produced on the surface of blood agar plates incubated in anaerobic jars. PYG-Tween 80 broth cultures are turbid with smooth sediment and have a terminal pH of 4.9–5.4.

The temperature for optimal growth is 37°C. Growth is slightly inhibited at 45°C and markedly inhibited at 30°C. Growth is stimulated by 0.02% Tween 80. Growth is inhibited by 20% bile or 6.5% NaCl and at pH 9.6. Poor to moderate growth occurs without fermentable carbohydrate.

Variable amounts of gas are detected in glucose agar deep cultures of the strains available for study. Cellulose is digested. Pectin is fermented by some strains; xylan usually is not fermented or only weakly fermented; dextrin and gum arabic are not fermented. Adonitol, dextrin, dulcitol, galactose, glycerol, inulin, and sorbose are not fermented. Resazurin is not reduced. Hippurate is not hydrolyzed and acetylmethylcarbinol is not produced. Ammonia is not produced from arginine or threonine and only small amounts are produced from peptone. H_2S is not detected in SIM medium.

Products (milliequivalents per 100 ml of culture) in PYG broth are acetic (0.1–0.3), butyric (0.1–0.3), and lactic (3.0–5.0) acids; trace amounts of formic or succinic acids also may be detected. Little or no H_2 is detected in headspace gas of cultures of strains available for study. Pyruvate and lactate are not utilized; threonine is not converted to propionate. Other characteristics of the species are given in Table 156.

Isolated from rumen contents of sheep and cows.

DNA G+C content (mol%): not determined.

Type strain: van Gylswyk and Hoffman strain 6, ATCC 43171, JCM 9499.

GenBank accession number (16S rRNA gene): L34613, X71860.

Further comments: The type strain Bryant B348 has been lost; the type strains listed above are neotype strains. The strains studied by Bryant et al. (1958) produce primarily lactic acid with small amounts of acetic and formic but no butyric, propionic, and succinic acids or H_2. Strains isolated by van Gylswyk and Hoffman (1970) also produce small amounts of butyric and valeric acids. Moore and Holdeman Moore (1986) detected butyric acid (0.25–0.40 meq/100 ml) in one strain from van Gylswyk and Hoffman. Van Gylswyk and Hoffman suggested that the species description of Bryant et al. (1958) be emended to include organisms that produce a small amount of butyrate and valerate.

11. **Eubacterium combesii** (Prévot and Laplanche 1947) Holdeman and Moore 1970, 23[AL] (*Cillobacterium combesi* Prévot and Laplanche 1947, 688)

com.be′si.i. N.L. gen. n. *combesii* of Combes; named for Combes.

This description is mainly from Prévot et al. (1967) and the study of the type strain and 12 other strains. Cells are motile with peritrichous flagella. Cells in PYG broth cultures are 0.6–0.8 × 3.0–10.0 μm and occur as single cells and in pairs, short chains, and palisade arrangements. Surface colonies of the type strain on horse blood agar are 0.5–2.0 mm, circular, entire to irregular, convex, semi-opaque, whitish yellow, shiny, and smooth with fine granular or mosaic appearance when viewed by obliquely transmitted light. The type and most other strains are β-hemolytic.

Glucose broth cultures are turbid with flocculent sediment and pH of 6.4. The optimum temperature for growth is 35–37°C. The range of temperatures at which growth occurs is variable. Moderate to abundant gas is produced in glucose agar deep cultures. Ammonia is produced from peptone; neutral red and resazurin are reduced. H_2S usually is produced in SIM medium.

Products (milliequivalents per 100 ml of culture) in PYG broth cultures are acetate (1.5–3.8), iso-butyrate (0.05–0.37), butyrate (trace to 1.8), iso-valerate (0.1–1.3), and lactate (0.1–0.7), usually with ethanol, propanol, iso-butanol, butanol, and iso-amyl alcohols. Variable amounts of formate and small amounts of propionate may be detected. Abundant H_2 is produced. Pyruvate is converted to acetate.

Lactate is not utilized and threonine is not converted to propionate.

Four clinical isolates tested are susceptible to chloramphenicol (12 μg/ml), clindamycin (1.6 μg/ml), penicillin G (2 units/ml), and tetracycline (6 μg/ml). One of these strains was resistant to erythromycin (3 μg/ml). Culture supernatants of 1-d-old chopped meat broth cultures of two strains tested are not toxic for mice (0.5 ml intraperitoneally). Other characteristics of the species are given in Table 156.

Isolated from human infections and African soil.

DNA G+C content (mol%): not reported.

Type strain: ATCC 25545, CIPP A13D, DSM 20696, JCM 9988.

GenBank accession number (16S rRNA gene): AY533380.

Further comments: Although no spores have been detected in the strains on which this description is based, some strains that are otherwise like *Eubacterium combesii* produce spores and were identified by Moore and Holdeman Moore (1986) as *Clostridium subterminale*. Analysis of the original 16S rRNA gene sequence (L34614) indicated that *Eubacterium combesii* belonged to the genus *Propionibacterium*. However, *Eubacterium combesii* does not resemble *Propionibacterium* phenotypically. The strain was re-sequenced for this study and found to be a close relative of *Clostridium botulinum*.

12. **Eubacterium contortum** (Prévot 1947) Holdeman, Cato and Moore 1971, 306^AL (*Catenabacterium contortum* Prévot 1947, 414)

con.tor'tum. L. neut. adj. *contortum* twisted.

This description is from study of the type and ten other strains (Moore and Holdeman Moore, 1986). Cells in PYG broth cultures are ovoid, 0.4–0.9 × 1.4–2.3 μm and occur in pairs and short or long twisted chains. Surface colonies on horse blood agar are 0.5 mm, circular, entire to erose, low convex, and translucent. There may be slight clearing of the blood under the area of heavy growth. Glucose broth cultures usually are turbid with heavy sediment (gelatinous or granular) and a terminal pH of 4.8–5.0.

The optimum temperature for growth is 35–37°C. Strains grow well at 30 and 45°C but poorly at 25°C. Growth is not affected by 0.02% Tween 80; growth of some strains is inhibited by 20% bile.

Abundant gas is detected in PYG agar deep cultures. Acid is produced from galactose. No acid is produced from adonitol, dextrin, dulcitol, glycerol, inulin, or sorbose. Little or no ammonia is produced from peptone and no ammonia is produced from arginine or threonine. Urease and acetylmethylcarbinol are not produced. H_2S is not produced in SIM medium. Hippurate is not hydrolyzed.

Products (milliequivalents per 100 ml of culture) in PYG broth are acetic (0.5–4.6) and formic (0.1–5.3) acids, with large amounts of ethanol. Trace amounts of lactic and succinic acids may be detected.

The type and seven other strains tested are susceptible to chloramphenicol (12 μg/ml). Two of these strains are resistant to clindamycin (1.6 μg/ml), four are resistant to erythromycin (3 μg/ml), one is resistant to penicillin G (2 units/ml), and three are resistant to tetracycline (6 μg/ml). Other characteristics of the species are given in Table 156.

Isolated from human feces, vaginal swab, and human clinical specimens, including blood (post-kidney transplant), abdominal aortic aneurism, and wounds. Sera from 36–50% of patients with Crohn's disease are reported to contain agglutinating antibodies to strains of *Eubacterium contortum* (van de Merwe, 1981).

DNA G+C content (mol%): 45 (method unknown).

Type strain: ATCC 25540, DSM 3982.

GenBank accession number (16S rRNA gene): L34615.

13. **Eubacterium coprostanoligenes** Freier, Beitz, Li and Hartman 1994, 141^VP

co.pro.stan.ol.i'gen.es. N.L. n. *coprostanol* coprostanol; Gr. v. *gennaô* produce; N.L. part. adj. *coprostanoligenes* producing coprostanol.

Cells are Gram-stain-positive, nonmotile, nonspore-forming coccobacilli, 0.5–0.7 μm × 0.7–1.0 μm, occurring singly and in pairs. Surface colonies on anaerobically incubated MLA plates are small, white, and circular, with a powdery texture. Growth and coprostanol production are optimal at pH 7.0–7.5 and at 35°C. Growth and coprostanol production do not occur at pH 5.5 or 8.0 or at incubation temperatures of 25 or 45°C.

Aerotolerant, anaerobic, and chemo-organotroph. Phosphatidyl choline is metabolized and is required for growth. Cholesterol is reduced to coprostanol, but is not required for growth. Nitrate is not reduced and indole is not produced. Starch and gelatin are not hydrolyzed and cells produce β-glucosidase. Acid is produced from amygdalin, lactose, and salicin. L-arabinose, cellobiose, fructose, glucose, mannose, and melibiose are weakly fermented. Moderate amounts of H_2 and small amounts of CO_2 are produced. Acetic, formic, and succinic acids are produced. No alcohols are produced.

DNA G+C content (mol%): 41 (T_m).

Type strain: HL, ATCC 51222.

GenBank accession number (16S rRNA gene): not determined.

14. **Eubacterium cylindroides** (Rocchi 1908) Holdeman and Moore 1970, 23^AL (*Bacterium cylindroides* Rocchi 1908, 479)

cy.lin.dro.i'des. L. n. *cylindrus* a cylinder; L. suff. *-oides* resembling, similar; N.L. neut. adj. *cylindroides* cylinder-shaped.

The description is from Cato et al. (1974) and study of the type and 20 other strains (Moore and Holdeman Moore, 1986). Cells in PYG broth cultures are 0.7–1.0 μm in diameter by 1.5–18.0 μm long and occur singly, in pairs, and in long chains. Gram-stain-positive cells usually are seen only in young cultures (exponential phase or early stationary phase of growth); cells of older cultures destain very easily. Surface colonies on horse blood agar incubated for 2 d are punctate to 2 mm, circular to slightly irregular, entire to diffuse, flat to low convex, and translucent. They sometimes have a mottled appearance when viewed by obliquely transmitted light. Glucose broth cultures are turbid with smooth or ropy sediment and terminal pH of 4.8–5.5.

Growth is most rapid at 37–45°C; most strains grow at 25°C. Glucose or fructose enhances growth. Cell density is not markedly enhanced by 0.02% Tween 80. Rumen fluid

(5% final concentration) often increases the amount of acid produced. Comparatively less growth or acid production occurs in glucose media containing 20% bile.

Little or no gas is detected in glucose agar deep cultures. The type strain ferments inulin and pectin (weakly) and produces ammonia from peptone. Neutral red is reduced. The type strain does not ferment adonitol, dextrin, dulcitol, galactose, glycerol, or sorbose and does not hydrolyze hippurate or produce acetylmethylcarbinol. Acetylmethylcarbinol is produced by 8 of 21 strains tested (Cato et al., 1974).

Products (milliequivalents per 100 ml of culture) in PYG broth cultures are butyric (0.3–1.4) and DL-lactic (1.0–3.0) acids, often with smaller amounts of acetic, formic, and succinic acids. Little or no H_2 is detected in headspace gas. Pyruvate is converted to acetate, formate, and butyrate. Lactate is not utilized; threonine is not converted to propionate.

The type strain is susceptible to chloramphenicol (12 mg/ml), clindamycin (1.6 μg/ml), erythromycin (3 μg/ml), and penicillin G (2 units/ml). Susceptibility to tetracycline (6 μg/ml) is variable. Other characteristics of the species are given in Table 156.

Isolated from human feces.

DNA G+C content (mol%): 31 (T_m).

Type strain: ATCC 27803, DSM 3983, JCM 10261.

GenBank accession number (16S rRNA gene): L34617.

15. **Eubacterium desmolans** Morris, Winter, Cato, Ritchie and Bokkenheuser 1986, 184[VP]

des'mo.lans. Gr. n. *desmos* a bond, mod. chem. term desmolase, an enzyme that splits a carbon–carbon bond; N.L. part. adj. *desmolans* making desmolase.

Cells are Gram-stain-positive plump rods that are 0.8–1.1 μm × 1.7–2.3 μm and occur singly or in short chains. Some cells are capsulated. Tumbling motility, due to four to six long peritrichous flagellae. Internal mesosomes are present, especially in dividing cells. Spores are not formed, and the organism does not survive exposure to 80°C for 10 min or treatment with absolute alcohol for 30 min. Primary colonies on Columbia agar plates supplemented with 5% sheep blood (BAP) incubated at 37°C are invisible to the naked eye. Subcultures on the same medium yield circular to slightly irregular, convex, shiny, entire, semi-opaque, white to colorless colonies with a diameter of 0.6–0.8 mm. Neither sheep nor rabbit erythrocytes are hemolyzed. Growth in supplemented peptone broth (SPB) is light and slightly turbid, with a fine precipitate after 3–4 d. Addition of 1% inositol enhances growth; 20% bile (2% oxgall) is tolerated but has no effect on multiplication.

Acid is formed from inositol. Indole is produced. H_2S is not produced on sulfide indole motility medium. Gelatin, milk, and meat are not digested. Acetate and butyrate and trace amounts of succinate and lactate are formed as end products of the fermentation of inositol. Hydrogen is not produced. Acid is not formed from amygdalin, arabinose, cellobiose, erythritol, esculin, glycogen, fructose, glucose, lactose, altose, mannitol, mannose, melezitose, melibiose, pectin, raffinose, rhamnose, ribose, salicin, sorbitol, sucrose, trehalose, and xylose. Starch is not hydrolyzed.

Lipase, lecithinase, oxidase, and catalase are not produced. Nitrate and resazurin are not reduced. At least two enzymes with activity on corticoids are produced: a desmolase which cleaves the side chain of 17α-hydroxysteroids between C_{17} and C_{20} and a 20-β-hydroxysteroid dehydrogenase which reduces the 20-keto group of both 17α-hydroxy- and 17-deoxysteroids. Many C_{20} steroids are also reduced.

Eubacterium desmolans is susceptible to chloramphenicol (12 μg/ml), erythromycin (3 μg/ml) and clindamycin (1.6 μg/ml) but resistant to penicillin G (2 U/ml) and tetracycline (6 μg/ml).

Isolated from cat feces.

DNA G+C content (mol%): 35 (T_m).

Type strain: ATCC 43058, CCUG 27818, JCM 6566.

GenBank accession number (16S rRNA gene): L34618.

16. **Eubacterium dolichum** Moore, Johnson and Holdeman 1976, 246[AL]

do'li.chum. N.L. neut. adj. *dolichum* (Gr. neut. adj. *dolichon*) long, referring to the long chains formed in broth cultures.

Cells from PYG broth cultures are nonmotile, thin rods in long chains. The ends of the cells usually are slightly tapered. Cells are 0.4–0.6 × 1.6–6.0 μm. Subsurface colonies in RGCA incubated for 5 d are 0.2–0.5 mm in diameter, lenticular, and transparent to translucent. Surface colonies on BHIA-S streak tubes are 0.5–1.5 mm in diameter, circular, entire, convex, opaque to translucent, granular, dull, and smooth. None of the four strains tested grew on anaerobically incubated blood agar plates. Three of four strains tested grew on EYA plates incubated anaerobically. PY broth cultures have a stringy to ropy sediment without marked turbidity. In general, growth is poor in most broth media. Growth often is enhanced in a broth medium containing fructose, glucose, maltose, starch, or sucrose, although the pH is not necessarily lowered.

Best growth is obtained at 37°C, although most strains grow at 45°C and, to a limited extent, at 30°C. Growth usually is stimulated by 0.02% Tween 80 and sometimes is inhibited by 20% bile.

No gas is produced in glucose agar deep cultures. The type strain does not ferment adonitol, dextrin, dulcitol, galactose, glycerol, inulin, or sorbose. Ammonia is not produced from peptone, arginine, or chopped meat. Hippurate is not hydrolyzed. The type strain reduces neutral red and resazurin.

The product (milliequivalents per 100 ml of culture) in PYG broth is butyrate (0.2–1.7), often with small amounts of acetate and lactate. Little or no H_2 is detected in the headspace gas. Lactate and pyruvate are not utilized. Threonine is not converted to propionate. Other characteristics of the species are given in Table 156.

Isolated from human feces.

DNA G+C content (mol%): not determined because of the sparse growth (Moore et al., 1976).

Type strain: ATCC 29143, DSM 3991.

GenBank accession number (16S rRNA gene): L34682.

17. **Eubacterium eligens** Holdeman and Moore 1974, 273[AL]

el'i.gens. L. adj. *eligens* choosy, referring to its generally poor growth without fermentable carbohydrate.

Cells from PYG broth cultures are straight to slightly curved rods, occurring singly or in pairs, occasionally in short chains of three to six cells, and are 0.3–0.8 × 1.9–4.9 μm. Central or eccentric swellings predominate in some strains. Twenty-four of 33 strains tested were motile and flagella were peritrichous. Cells decolorize easily in older cultures, but some Gram-stain-positive cells usually are seen. After incubation for 5 d in RGCA, subsurface colonies are 0.5–1.0 mm in diameter, white to tan, and lenticular. Surface colonies on BHIA-S are punctiform, circular, entire, transparent to translucent, white to tan, and smooth. Surface colonies on 2-d-old anaerobic BHIA-S blood agar plates are 0.5–1.0 mm in diameter, circular, entire, convex, smooth, translucent to semi-opaque, shiny, white, and not hemolytic (sheep blood). PY-fructose broth cultures are turbid with a smooth sediment and pH of 4.6–5.8 in 5 d.

The temperature for optimal growth is 37°C. Most strains grow well at 45°C; little or no growth occurs at 30°C. Tween 80 (0.02%) sometimes enhances growth. Growth usually is not effected by addition of 10–30% (v/v) rumen fluid. Bile (20%) inhibits acid production and may inhibit growth. There is limited, if any, growth in PY broth without a fermentable carbohydrate. Little or no gas is produced in PYG agar deep cultures.

Neutral red is reduced. Adonitol, dextrin, dulcitol, galactose, glycerol, inulin, and sorbose are not fermented. Ammonia is not produced from PY broth or arginine. Acetyl methyl carbinol is not produced. H₂S is not produced in SIM medium.

Products (milliequivalents per 100 ml of culture) in PYG broth cultures are acetic (0.7–2.3), formic (0.8–3.7), and lactic (0.1–0.9) acids and ethanol, occasionally with a trace of succinic acid. No H₂ is detected in headspace gas. Pyruvate and gluconate are converted to acetate, formate, lactate, and ethanol. Lactate is not utilized. Other characteristics of the species are given in Table 156.

Isolated from human feces.

DNA G+C content (mol%): 36 (T_m).

Type strain: ATCC 27750, DSM 3376, VPI C15-48.

GenBank accession number (16S rRNA gene): L34420.

18. **Eubacterium fissicatena**   Taylor 1972, 462[AL]

fiss.i.ca.te′na. L. n. *fissum* a cleft; L. n. *catena* a chain; N.L. n. *fissicatena* a broken chain.

The description is from Taylor (1972) and study of the type and seven other strains (Moore and Holdeman Moore, 1986). Cells in glucose broth cultures are motile, 0.3–0.8 × 1.4–8.0 μm, and occur singly and in pairs, often with parallel arrangement. Surface colonies on blood agar are 0.5 mm, circular, entire to slightly scalloped, low convex, opaque, grayish white, shiny, and smooth. Glucose broth cultures are turbid with smooth sediment and terminal pH of 5.0–5.5.

The optimum temperature for growth is 37°C. All strains grow at 25°C but rarely grow at 13°C. There usually is no growth at 45°C. Growth occurs at pH 9.6 and is unaffected by Tween 80. Although growth is not inhibited in PY broth containing 20% bile, acid production is decreased. Growth is inhibited by 6.5% NaCl.

Abundant gas is detected in PYG agar deep cultures. Galactose is fermented. Neutral red is reduced. Small amounts of ammonia are produced from peptone. All strains produce H₂S from yeast extract broth (Taylor, 1972). Some strains produce H₂S in SIM medium. Adonitol, dextrin, dulcitol, glycerol, inulin, and sorbose are not fermented. Hippurate is not hydrolyzed. Acetylmethylcarbinol is not produced. Hydroxyethylflavine is produced from riboflavin.

Products (milliequivalents per 100 ml of culture) in PYG broth are acetic (0.7–2.5) and formic (0.4–2.1) acids and large amounts of ethanol. Trace amounts of lactic and succinic acids may be detected. Carbon dioxide and abundant H₂ are produced. Pyruvate is converted to acetate, formate, and ethanol. Some cultures produce methane in media without glucose. Lactate is not utilized. Threonine is not converted to propionate. Other characteristics of the species are given in Table 156.

Isolated from the alimentary tract of goats.

DNA G+C content (mol%): 46 (reported by Taylor, 1972, but method not given).

Type strain: ATCC 33661, DSM 3598, JCM 9983, NCIMB 10446.

The 16S rRNA gene sequence of the type strain is available from the Ribosomal Database Project.

19. **Eubacterium hadrum**   Moore, Johnson and Holdeman 1976, 247[AL]

had′rum. N.L. neut. adj. *hadrum* (from Gr. n. adj. *hadron*) thick, bulky (referring to the relatively large size of the cell).

The description is from Moore et al. (1976), who studied the type and 17 other similar strains (Moore and Holdeman Moore, 1986). Cells from PYG broth cultures are nonmotile rods of uniform width with rounded ends, occurring in pairs and short chains, and are 0.7–1.0 μm in diameter and 3.0–10.0 μm long. Subsurface colonies in RGCA incubated for 5 d are woolly balls 1–2 mm in diameter. Surface colonies on BHIA-S streak tubes or BHIA-S blood agar plates incubated anaerobically are 2–3 mm in diameter, circular, entire to erose, convex, opaque to translucent, and smooth. Some strains produce slight greening hemolysis on rabbit blood. The type and one other strain tested did not grow on EYA. Glucose broth cultures have abundant growth and are turbid with smooth, sometimes ropy, sediment. After incubation for 1 d, the pH in glucose broth cultures is 4.9–5.4.

The optimum temperature for growth is 37°C; good growth also occurs at 45°C but lesser growth is obtained at 30°C. Growth is stimulated by 0.02% Tween 80. No growth occurs in medium containing 6.5% NaCl. Growth occurs in PYG with 20% bile, but no acid is produced. Abundant gas is observed in glucose agar deep cultures.

Products (milliequivalents per 100 ml of culture) from PYG broth cultures are butyric (1–6) and lactic (0.1–2) acids with little or no acetic acid; trace amounts of pyruvic and succinic acids sometimes may be detected. Abundant H₂ is produced from fermentation of carbohydrates. Pyruvate is converted to butyrate and acetate; little or no lactate is utilized.

The type strain ferments galactose and sorbose but does not ferment adonitol or glycerol. Neutral red and resazurin are reduced. Hippurate is not hydrolyzed. Ammonia is not produced from peptone or arginine. Other characteristics of the species are given in Table 156.

Isolated from human feces.

DNA G+C content (mol%): 32–33 (T_m).

Type strain: ATCC 29173, DSM 3319, JCM 9980, VPI B2-52.

The 16S rRNA gene sequence of the type strain is available from the Ribosomal Database Project.

Further comments: Moore and Holdeman Moore (1986) noted that although *Eubacterium hadrum* has large cells typical of those found in many species of *Clostridium*, and even though slight swellings in the cells are very occasionally observed, no typical spores are seen, and cultures with swellings do not survive heating at 80°C for 10 min.

20. **Eubacterium hallii** Holdeman and Moore 1974, 275[AL]

hall'i.i. N.L. gen. n. *hallii* of Hall, named for Ivan C. Hall, an American bacteriologist.

Cells from PYG broth cultures are nonmotile rods occurring singly and in pairs, occasionally in short chains. Cells are 0.8–2.4 µm in diameter × 4.7 µm to more than 25.0 µm in length. Although subterminal and terminal swellings sometimes are observed, cultures do not survive heating at 80°C for 10 min. Subsurface colonies in RGCA incubated for 5 d are 0.5–1.0 mm in diameter and lenticular or woolly balls. Surface colonies in BHIA-S streak tubes or BHIA-S blood agar plates are 1–2 mm in diameter, circular to slightly irregular, entire to slightly erose, low convex, semi-opaque, white to yellowish, smooth, shiny, and not hemolytic (rabbit blood). Glucose broth cultures have abundant growth and smooth or ropy sediment with little or no turbidity. The pH of 5 d-old cultures in PYG is 4.7–5.5. Moderate growth occurs in PY broth without a fermentable carbohydrate.

The temperature for optimum growth is 37°C. Most strains grow at 30°C, but less well than at 37°C. Strains usually do not grow at 25 and 45°C. Growth is inhibited by 6.5% NaCl, may be inhibited by 20% bile, but is unaffected by 0.02% Tween 80. Growth is poor in the absence of fermentable carbohydrate.

Abundant gas is produced in glucose agar deep cultures. The type strain and five other strains produce acid from galactose and reduce neutral red. Strains do not produce acetylmethylcarbinol, ammonia, urease, or H_2S in SIM medium and do not hydrolyze hippurate.

Products (milliequivalents per 100 ml of culture) from PYG broth are acetic (0.1–0.8), formic (0.1–0.8), and butyric (2.0–4.0) acids and large amounts of butanol. Trace amounts of lactic and succinic acids sometimes are detected. Abundant H_2 is produced from fermentation of carbohydrates. The type strain converts pyruvate to butyrate and acetate. Lactate and gluconate are not utilized; threonine is not converted to propionate.

The type and one other strain tested are susceptible to chloramphenicol (12 µg/ml), clindamycin (1.6 µg/ml), erythromycin (3 µg/ml), penicillin G (2 units/ml), and tetracycline (6 µg/ml). Other characteristics of the species are given in Table 156.

Isolated from human feces.

DNA G+C content (mol%): 38 (T_m).

Type strain: ATCC 27751, DSM 3353, VPI B4-27.

GenBank accession number (16S rRNA gene): L34621.

21. **Eubacterium infirmum** Cheeseman, Hiom, Weightman and Wade 1996, 958[VP]

in.fir'mum. L. neut. adj. *infirmum* delicate, referring to the delicate growth of the organism.

Cells are obligately anaerobic, nonspore-forming, nonmotile, Gram-stain-positive, short rods (0.5 µm by 1–2 µm). Cells occur singly. After incubation for 7 d on fastidious anaerobe agar plates, colonies are approximately 1 mm in diameter, circular, convex, and translucent. No hemolysis occurs on blood-containing media.

No acid is produced from arabinose, cellobiose, galactose, glucose, lactose, mannitol, raffinose, ribose, salicin, sucrose, or xylose. Moderate amounts of acetic and butyric acids are produced in PYG. Catalase, urease, and indole are not produced, and ammonia is not produced from arginine.

Isolated from human periodontal pockets.

DNA G+C content (mol%): 38 (T_m).

Type strain: W 1471, NCTC 12940.

GenBank accession number (16S rRNA gene): U13039 (from strain W 1471).

Further comments: *Eubacterium sulci* is a closely related species. There are no phenotypic characters that distinguish *Eubacterium sulci* from *Eubacterium infirmum*. The two species may represent subspecies of the same taxon.

22. **Eubacterium minutum** Poco, Nakazawa, Sato and Hoshino 1996, 33[VP]

mi.nu'tum. L. neut. adj. *minutum* minute, small, referring to the minute colonies formed by the organism.

Cells are obligately anaerobic, nonspore-forming, nonmotile, Gram-stain-positive, short rods. Cells are 0.5 µm by 1.0–1.5 µm and occur singly, in pairs, and in clumps. Cells in actively growing cultures are Gram-stain-positive, but often cells in older cultures are Gram-stain-negative. Growth in broth media is poor, and growth is moderately enhanced by the addition of 5% bovine serum and 0.3% $MgSO_4$. After 1 week of anaerobic incubation, colonies on BHI-blood agar plates are 0.3–0.5 mm in diameter, circular, convex, entire, and translucent. No hemolysis occurs on blood-containing agar plates.

Starch and esculin are not hydrolyzed, nitrate is not reduced, no liquefaction of gelatin occurs, ammonia is not produced from peptone or arginine and indole, urease and catalase are not produced. No acid is produced from adonitol, amygdalin, arabinose, cellobiose, erythritol, esculin, fructose, galactose, glucose, glycogen, inositol, lactose, maltose, mannitol, mannose, melezitose, melibiose, rhamnose, ribose, salicin, sorbitol, starch, sucrose, trehalose, or xylose. Moderate amounts of butyrate are produced in peptone-yeast extract medium supplemented with glucose or peptone-yeast extract-glucose broth.

Isolated from human periodontal pockets.

DNA G+C content (mol%): 38–40 (HPLC).

Type strain: M-6, ATCC 700079, CIP 104795.

GenBank accession number (16S rRNA gene): AJ005636.

Further comments: *Eubacterium minutum* is a senior heterotypic synonym of "*Eubacterium tardum*" (Wade et al., 1999b).

23. **Eubacterium moniliforme** (Repaci 1910) Holdeman and Moore 1970, 23[AL] (*Bacillus moniliformis* Repaci 1910, 412)

mo.ni.li.for′me. L. n. *monile* a necklace; L. n. *forma* shape; N.L. neut. adj. *moniliforme* necklace-shaped.

The description is from study of the type and 20 other strains (Moore and Holdeman Moore, 1986). Cells of young cultures are motile with peritrichous flagella. Cells of the type strain are 0.6–0.9 μm in diameter × 1.7–9.4 μm long and occur singly and in short chains, often in palisade arrangement. Surface colonies on horse blood agar (2 d) are 2–8 mm, circular with irregular edges, pulvinate or umbonate, and opaque. Glucose broth cultures are turbid with a smooth sediment and a terminal pH of 5.0–5.6.

All strains grow at 45°C, most grow at 30°C, and some grow at 25°C. Moderate to good growth occurs in PY broth cultures. Growth is slightly greater in media with a fermentable carbohydrate. Growth is not affected by 0.02% Tween 80 or 20% bile.

Abundant gas is produced in glucose agar deep cultures. The type strain does not produce lecithinase, but lecithinase activity is detected in some other strains. Ammonia is produced from peptone and arginine. The type strain ferments galactose, produces H_2S in SIM medium, and reduces neutral red and resazurin. The type strain does not produce acetylmethylcarbinol, does not hydrolyze hippurate, and does not produce acid from adonitol, dextrin, dulcitol, glycerol, inulin, and sorbose.

Products (milliequivalents per 100 ml of culture) in PYG are acetic (0.5–2.5), butyric (1.3–2.3), and L-(+) (or DL-)lactic (1.0–4.9) acids and butanol, sometimes with trace amounts of formic, propionic, and succinic acids. Abundant H_2 is produced. The type strain converts pyruvate principally to acetate. Lactate is not utilized, and threonine is not converted to propionate.

Of nine strains tested, four were resistant to erythromycin (3 μg/ml) and one was resistant to tetracycline (6 μg/ml). All were susceptible to chloramphenicol (12 μg/ml), clindamycin (1.6 μg/ml), and penicillin G (2 units/ml). Other characteristics of the species are given in Table 156.

Isolated from blood, various kinds of human clinical infections, intestinal tract, and soil.

The mol% G+C of the DNA: has not been determined.

Type strain: ATCC 25546, CCUG 28088, CIPP 2055, DSM 3984, JCM 9990.

GenBank accession number (16S rRNA gene): L34622.

24. **Eubacterium multiforme** (Distaso 1911) Holdeman and Moore 1970, 23[AL] (*Bacillus multiformis* Distaso 1911, 101)

mul.ti.for′me. L. adj. *multus* much, many; L. n. *forma* shape; N.L. neut. adj. *multiforme* many-shaped.

The description is based on Distaso (1911), Prévot et al. (1967), and study of the type and four other strains (Moore and Holdeman Moore, 1986). Cells are 0.6–0.8 × 0.8–8.0 μm and are motile with single or multiple subpolar flagella. One strain is not motile. Cells occur singly and in pairs, often in a palisade arrangement. *meso*-DAP is present in the cell walls of the type and one other strain. Surface colonies on blood agar incubated for 2 d are 1-2 mm, circular, erose, convex, translucent, gray-white, smooth, and slightly shiny

with mosaic appearance when viewed by obliquely transmitted light. Glucose broth cultures are turbid with a smooth, granular, or flocculent sediment and pH of 5.3–5.6.

The optimal temperature for growth is 37°C. Strains grow well at temperatures between 25 and 45°C. Growth may be slightly stimulated by Tween 80 and is inhibited by 6.5% NaCl. Growth and acid production may be inhibited by 20% bile.

Abundant gas is produced in PYG agar deep cultures. Galactose may be fermented. Adonitol, dextrin, dulcitol, inulin, pectin, and sorbose are not fermented. Ammonia is produced from peptone. Hippurate may be hydrolyzed. Acetylmethylcarbinol generally is not produced. Neutral red and resazurin are reduced.

Products (milliequivalents per 100 ml of culture) in PYG cultures are acetic (1.0–3.0), butyric (1.5–4.3) and lactic (0.6–3.6) acids, sometimes with trace amounts of formic, propionic, or succinic acids. Copious H_2 is produced. Pyruvate is converted to acetate, formate, and butyrate. Little or no lactate is used.

Two of three strains tested were resistant to erythromycin (3 μg/ml). All three strains were susceptible to chloramphenicol (12 μg/ml), clindamycin (1.6 μg/ml), penicillin G (2 units/ml), and tetracycline (6 μg/ml). Other characteristics of the species are given in Table 156.

Isolated from soil of the Ivory Coast (Africa) and a gunshot wound.

DNA G+C content (mol%): not determined.

Type strain: ATCC 25552, CCUG 27817, DSM 20694, JCM 6484, Prévot collection 06A.

GenBank accession number (16S rRNA gene): AB018184.

25. **Eubacterium nitritogenes** Prévot 1940, 355[AL]

ni.tri.to′ge.nes. N.L. n. *nitritum* nitrite; N.L. verbal suff. *-genes* from Gr. v. *gennaio* beget, produce; N.L. adj. *nitritogenes* nitrite-producing.

The description is from Prévot et al. (1967) and study of the type strain and three similar strains (Moore and Holdeman Moore, 1986). Cells of the type strain are straight rods with blunt ends and are 0.8–1.6 × 1.6–16.0 μm. Cells occur singly and in short chains and may occur in palisade arrangement. Occasionally have central swellings. In many cultures, approximately one-half of the cells stain uniformly Gram-positive, a few stain Gram-positive only at one end, and the remainder stain Gram-negative. Surface colonies on horse blood agar (2 d) are 0.5–2.0 mm, circular to slightly irregular with scalloped edge, low convex, translucent to opaque, sometimes with mottled appearance when viewed by obliquely transmitted light. Glucose broth cultures are turbid with sediment and terminal pH of 5.1–5.3.

Growth is most rapid at a pH of 6.5–7.8. Optimum temperature is 37°C but good growth generally occurs at temperatures between 30 and 45°C; three of four strains tested grew at 25°C. Growth is not affected by 0.02% Tween 80.

Abundant gas is produced in PYG agar deep cultures. Galactose is fermented by the type and one other strain. Adonitol, dulcitol, dextrin, inulin, and sorbose are not fermented. Growth is inhibited in media containing 6.5% NaCl or 20% bile. Small amounts of ammonia are produced from PY broth. Hippurate hydrolysis is variable for the species. Lecithinase is not produced by the type and one other strain but is produced by two of the strains. Neutral red and resazurin are reduced.

Products (milliequivalents per 100 ml of culture) in PYG broth cultures are acetic (0.7–2.4), butyric (1.0–3.0), and DL-lactic (1.0–3.0) acids; small amounts of formic or succinic acids may be detected. Abundant H_2 is produced. Other characteristics of the species are given in Table 156.

Isolated from soil and human infections.

DNA G+C content (mol%): not determined.

Type strain: ATCC 25547, DSM 3985, JCM 6485.

GenBank accession number (16S rRNA gene): AB018185.

Further comments: The relationship between *Eubacterium nitritogenes* and *Eubacterium budayi* is not clear. They are exceedingly difficult, if not impossible, to differentiate by the usual phenotypic tests.

26. **Eubacterium nodatum** Holdeman, Cato, Burmeister and Moore 1980, 167^{VP}

no.da′tum. L. neut. adj. *nodatum* entangled, referring to the tangled arrangement of the cells.

The phenotypic description is from Holdeman et al. (1980), who studied the type and 49 other isolates. Cells from PY broth are nonmotile and occur in clumps. Individual cells are 0.5–0.9 × 2.0–12.0 μm and appear branched, somewhat filamentous, or club shaped. Subsurface colonies on BHIA-S enriched with 5% (v/v) rabbit serum, 0.02 g each of formate and fumarate, and 400 μg thiamine pyrophosphate/100 ml, are 0.5–2.0 mm in diameter, translucent to opaque, and raspberry shaped. After incubation for 2–4 d, surface colonies on BHIA-S streak tubes or anaerobically incubated blood agar plates are less than 0.5–1.0 mm in diameter, generally circular, entire to lobate, and heaped or berry-like in appearance. There is no hemolytic action on rabbit blood cells. No growth occurs on EYA plates incubated anaerobically. PY broth cultures are not turbid and have a small to moderate amount of flocculent, granular or bread crumb-like sediment.

The temperature for optimum growth is 37°C. Most strains grow at 30 and 45°C, but only an occasional strain grows at 25°C. Growth is inhibited by 20% bile and 6.5% NaCl, may be stimulated by 0.02% Tween 80, and is neither inhibited nor enhanced by the addition of 5% rumen fluid, 10% serum with thiamine pyrophosphate, or 0.05% sodium polyanetholsulfonate.

Little or no gas is detected in agar deep cultures. Neutral red is reduced. The type strain does not produce H_2S in SIM medium. Ammonia is produced from peptone and may be produced from arginine. Carbohydrates are not fermented.

Products (milliequivalents per 100 ml of culture) in PYG broth are acetate (0.04–0.36) and butyrate (0.3–1.6), often with trace amounts of formate, lactate, and succinate. No H_2 is detected in headspace gas of broth cultures. Neither lactate nor pyruvate is utilized.

The 32 strains tested were susceptible to chloramphenicol (12 μg/ml), clindamycin (1.6 μg/ml), erythromycin (3 μg/ml), penicillin G (2 units/ml), and tetracycline (6 μg/ml). Other characteristics of the species are given in Table 156.

Isolated from subgingival samples and from supragingival tooth scrapings from persons with periodontal disease.

DNA G+C content (mol%): 36–38 (T_m).

Type strain: ATCC 33099, CCUG 15996, CIP 104213, DSM 3993, JCM 9977, VPI D6A-5.

GenBank accession number (16S rRNA gene): U13041.

27. **Eubacterium oxidoreducens** corrig. Krumholz and Bryant 1986a, 489^{VP} (Effective publication: Krumholz and Bryant 1986b, 13.)

ox.i.do.re.du′cens. *oxido* combining form of modern chemical term, oxide, L. part. adj. *reducens* reducing; N.L. part. adj. *oxidoreducens* reducing compounds (containing) oxygen.

Rod-shaped curved cells; 0.45 × 1.5–22 μm in size, with rounded ends; singles, or in pairs or in small clumps. Nonmotile. No spore formation (pasteurized cultures are not viable). Gram-stain-positive.

Strictly anaerobic chemo-organotroph. Requires formate or hydrogen as electron donor to catabolize approximately equimolar gallate, pyrogallol, phloroglucinol, or quercetin to acetate, butyrate, and sometimes CO_2. No exogenous electron donor is required for catabolism (fermentation) of crotonate (growth rate much faster with 60 mM than with 30 mM or less) to acetate and butyrate. No other compounds are used as energy sources with or without formate present. These include rutin, hesperidin, monobenzenoids with or without methoxyl groups, fatty acids, citrate, acrylate, lactate, pyruvate, dicarboxylic acids, alcohols, sugars, amino acids, and peptides. It also does not grow in co-culture with *Desulfovibrio* species (with sulfate and with or without formate) with butyrate, protocatechuate, 3,5-dihydroxybenzoate, benzoate, phenol, 4-hydroxybenzoate, hydroquinone, or caffeate as substrate. Sulfate does not serve as a dissimilatory electron acceptor with formate in the medium. Nitrate is not reduced. It grows well in defined medium containing utilizable energy sources, minerals, including NH_4Cl, CO_2-bicarbonate (required), B-vitamins, sulfide, and cysteine. Gelatin is not hydrolyzed, ammonia is not produced from arginine or Casitone, but ammonia is essential as the main nitrogen source. The pH range for growth is 6.9–7.8 with the optimum at 7.4. Temperature range: 30–43°C, optimum 39–41°C.

Isolated from the bovine rumen

DNA G+C content (mol%): 36 (T_m).

Type strain: G41, ATCC 43585, DSM 3217.

GenBank accession number (16S rRNA gene): AF202258 (from strain DAS110, not the type strain).

28. **Eubacterium plautii** (Séguin 1928) Hofstad and Aasjord 1982, 347^{VP} (*Fusobacterium plauti* Séguin 1928, 439)

plau′ti.i. N.L. gen. n. *plautii* of Plaut; named for Hugo Carl Plaut, the bacteriologist who first described this organism.

This description is from study of the type strain (Moore and Holdeman Moore, 1986) and from Hofstad and Aasjord (1982). Cells of the type strain are straight rods with rounded ends and are 0.4–0.8 μm in diameter × 2.0–10.0 μm in length. Cells are motile with peritrichous flagella and occur singly, and in pairs, or short chains. Cells stain Gram-negative with very occasional weak Gram-stain-positive areas. Surface colonies on horse blood agar (2 d) are 0.5 mm, circular with diffuse edges, gray-white, dull, smooth, and translucent with mottled appearance when viewed by obliquely transmitted light. Glucose broth cultures are moderately turbid with a smooth (occasionally flocculent) sediment and final pH of 5.2. Small amounts of gas may be present in PYG agar deep cultures.

Products (milliequivalents per 100 ml of culture) in PYG broth cultures are lactic (2), butyric (0.7), and acetic (0.2) acids, often with a trace amount of succinic acid. Pyruvate is converted to acetate, butyrate, and lactate. Lactate is not utilized; threonine is not converted to propionate. No H_2 is detected in headspace gas of PYG cultures. Other characteristics of the species are given in Table 156.

Isolated from cultures of *Entamoeba histolytica*.

DNA G+C content (mol%): not determined.

Type strain: Prévot S1, ATCC 29863, CCUG 28093, DSM 4000, VPI 0310.

GenBank accession number (16S rRNA gene): AY724678.

Further comments: Moore and Holdeman Moore (1986) reported that motility was difficult to demonstrate and found that best results were obtained from microscopic examination of cells in the water of syneresis of a PYG agar slant.

29. **Eubacterium plexicaudatum** Wilkins, Fulghum and Wilkins 1974, 408[AL]

plex.i.cau.da′tum. Gr. adj. *plectos* twisted or braided; N.L. adj. *caudatus* with a tail; N.L. neut. adj. *plexicaudatum* with a braided tail, referring to the tuft of subpolar flagella that are twisted together to form the "large flagellum" often visible by darkfield or phase-contrast microscopy.

This description is from Wilkins et al. (1974) and the study of the type and 16 other strains (Moore and Holdeman Moore, 1986). Cells in PYG broth cultures are motile, have bipolar tufts of flagella, and are 0.8–1.6 × 4.0–10.0 μm; often have tapered ends. By phase-contrast or darkfield microscopy, the tufts of bipolar flagella may appear to be single polar flagella. Upon initial isolation, the cells are slightly curved and may have a double curvature. After several transfers in culture media, the cells often are thinner and usually straight. Cells may decolorize easily. They may have refractile areas or swellings but do not survive heating at 70°C for 10 min or treatment with ethanol. Surface colonies on RGCA or BHIA-S in anaerobic streak tubes incubated for 5 d are 0.5–1 mm, circular to slightly irregular, convex, translucent, dull, smooth, and white to light gray. In the lower portion of the streak tubes, the organisms often grow as a translucent film. No growth occurs on agar plates incubated in anaerobic jars. Cultures in PYG broth incubated in O_2-free N_2 are turbid, usually with no sediment, and have a terminal pH of 6.0–6.8. The pH of PY cultures is 7.0–7.6. Optimum growth occurs at a pH near neutrality. Growth often is stimulated by 20% bile or 15% rumen fluid.

Moderate to abundant gas is produced in glucose agar deep cultures. Growth in media containing galactose and various carbohydrates decreases the pH to between 6.0 and 7.0 (but not below 6.0) in any sugar medium tested. Neutral red is reduced; resazurin is not reduced. The type strain does not produce acid from adonitol, dextrin, glycerol, inulin, and sorbose. Ammonia is not produced from peptone or arginine. H_2S is not produced in SIM medium. Hippurate is not hydrolyzed.

Products (milliequivalents per 100 ml of culture) in PYG broth are butyrate (1.0–2.5), usually with acetate (0.02–0.12), and butanol. Trace amounts of pyruvate and

succinate may be detected. Hydrogen is produced. Pyruvate and lactate are not utilized. Threonine is not converted to propionate. Other characteristics of the species are given in Table 156.

Isolated from the ceca of mice or rats.

DNA G+C content (mol%): 44 (T_m).

Type strain: VPI 7582.

GenBank accession number (16S rRNA gene): AF157058.

Further comments: Moore and Holdeman Moore (1986) found that many strains of *Eubacterium plexicaudatum* were difficult to grow, which might explain the variation seen in carbohydrate fermentation reactions. They further found that the organisms were quite susceptible to oxidation and therefore difficult to preserve in a lyophilized or frozen state. Cultures streaked on BHIA-S streak tubes survived if stored at 37°C and transferred monthly.

30. **Eubacterium pyruvativorans** Wallace, McKain, McEwan, Miyagawa, Chaudhary, King, Walker, Apajalahti and Newbold 2003, 969[VP]

pyr.uv.at′i.vor.ans. N.L. n. *pyruvatum* pyruvate; L. part. *vorans* devouring, eating greedily; N.L. part. adj. *pyruvativorans* pyruvate-devouring.

Cells are straight to slightly curved rods, 0.3–0.5 μm wide × 1.0–1.5 μm long, occurring in short chains. Colonies on M2 agar are light tan in color and 2 mm in diameter with slightly irregular edges after 72 h anaerobic incubation. Growth is supported by pyruvate and, to a lesser extent, lactate. Amino acids can be used as the sole source of carbon and energy, with yields much lower than those obtained with pyruvate. Sugars are not fermented. Caproate is the principal metabolic end product from a complex rumen fluid containing medium.

Isolated from the sheep rumen.

DNA G+C content (mol%): 57 (differential dye-binding).

Type strain: I-6, ATCC BAA-574, NCIMB 13911.

GenBank accession number (16S rRNA gene): AJ310135 (from isolate 6).

31. **Eubacterium ramulus** Moore, Johnson and Holdeman 1976, 249[AL]

ra′mu.lus. L. n. *ramulus* twig, referring to the shape of the cell.

Cells are regular rods with rounded ends, 0.5–0.9 μm in diameter × 1.0–5.0 μm long with filaments exceeding 25.0 μm in length. Cells usually stain boldly in young cultures and are arranged in pairs or short chains. The filaments present in some cultures appear to be either undivided cells or long chains of distinct cells. Cells in the long chains often are of unequal length and occasionally have marked swellings. Cultures having cells with swellings do not survive heating at 80°C for 10 min.

Subsurface colonies in RGCA incubated for 5 d are 1–4 mm in diameter and have the appearance of woolly balls or balls of fuzz or are cauliflower-like, or sometimes are of such indefinite form that there is doubt that the area picked (under × 10 magnification) truly contained a colony. Surface colonies on BHIA-S roll streak tubes are 1–4 mm in diameter, circular to slightly irregular, entire or slightly lobate, raised to low convex or umbonate, translucent, and white to beige. Three of eight strains tested did not grow on the surface of anaerobically incubated blood agar or EYA plates. When there is growth, there is no reaction on EYA and no hemolytic activity on rabbit blood. Glucose broth cultures have stringy or flocculent sediment, usually without turbidity. The pH of 1-d-old cultures in PYG is 4.8–5.3.

Best growth is obtained reliably at 37°C, although most strains grow equally well at 30 and 45°C. Strains grow not at all or poorly at 25°C. Growth is not stimulated by 0.02% Tween 90; 20% bile may inhibit growth or fermentation. Growth usually is best in media containing fermentable carbohydrate. Abundant gas is observed in glucose agar deep cultures.

Products (milliequivalents per 100 ml of culture) in PYG cultures are acetic (0.1–1.0), formic (0.5–2.3), butyric (1.6–3.0), and lactic (0.1–0.7) acids; trace amounts of succinic acid sometimes are detected. Abundant H_2 is produced from fermentation of carbohydrates.

The type strain is susceptible to chloramphenicol (12 μg/ml), clindamycin (1.6 μg/ml), penicillin G (2 units/ml), and tetracycline (6 μg/ml), and is resistant to erythromycin (3 μg/ml). Other characteristics of the species are given in Table 156.

Isolated from human feces.

DNA G+C content (mol%): 39 (T_m).

Type strain: ATCC 29099, DSM 3995.

GenBank accession number (16S rRNA gene): L34623.

32. **Eubacterium rectale** (Hauduroy, Ehringer, Urbain, Guillot and Magrou 1937) Prévot 1938, 294[AL] (*Bacteroides rectalis* Hauduroy, Ehringer, Urbain, Guillot and Magrou 1937, 72)

rec.ta′le. N.L. n. *rectum* the straight bowel; N.L. neut. adj. *rectale* rectal.

The phenotypic description is from Prévot et al. (1967), Holdeman and Moore (1974), and study of the type and 22 other strains (Moore and Holdeman Moore, 1986). Cells in PYG broth cultures are 0.5–0.6 × 1.7–4.7 μm and occur singly and in short chains and small clumps. Cells may be slightly curved and may have central or terminal swellings. Some strains are motile with peritrichous flagella. Surface colonies are 0.5–2.0 mm, circular to irregular, entire to scalloped, convex, translucent, smooth, and shiny. They may be mottled when viewed by obliquely transmitted light. The type strain is nonhemolytic on horse blood. Some strains will not grow on the surface of blood agar plates incubated in an anaerobe jar. Glucose broth cultures are turbid with a smooth or flocculent sediment and have a terminal pH of 4.7–5.5, usually around 5.0. Growth in prereduced PY broth is questionable or very sparse.

The optimal temperature for growth is 37°C. Most strains grow at 25–45°C. Growth is stimulated markedly by a fermentable carbohydrate. Growth usually is not stimulated by 0.02% Tween 80 and is inhibited by 20% bile.

Moderate to large amounts of gas are produced in PYG agar deep cultures. The type strain ferments dextrin, galactose, inulin, and pectin and reduces neutral red. The type strain does not ferment adonitol, dulcitol, glycerol, or sorbose; does not produce acetylmethylcarbinol or catalase; does not produce ammonia from peptone, arginine, or threonine; does not produce H_2S in SIM medium, does not grow in medium containing 6.5% NaCl, and does not hydrolyze hippurate.

Products (milliequivalents per 100 ml of culture) in PYG cultures are butyric (0.5–1.5), lactic (1.5–5.5), and acetic (0–0.4) acids, occasionally with a trace of propionate or succinate. Copious H_2 is produced. Pyruvate is converted to acetate, butyrate, and lactate. Lactate is not utilized. Other characteristics of the species are given in Table 156.

Isolated from human colon and feces.

DNA G+C content (mol%): 30 (T_m).

Type strain: VPI 0989, ATCC 33656.

GenBank accession number (16S rRNA gene): L34627.

Further comments: VPI 0989, the type strain in the Approved Lists (Skerman et al., 1980) was deposited in ATCC (25578) but was lost in both collections. A different isolate, VPI 0990, from the same fecal sample was deposited to represent the type strain as ATCC 33656. Moore and Holdeman Moore (1986) believed that, because these isolates came from the same sample, they are the same strain.

Strains of fecal bacteria with the general characteristics of *Eubacterium rectale* comprise at least five distinct groups (Moore and Holdeman, 1974). All strains ferment cellobiose, fructose, glucose, maltose, and starch. They decolorize readily, and Gram-stain-positive cells cannot be demonstrated in some strains. The rods are usually curved and motile with flagella singly, in pairs, or in tufts at one or both ends of the cells.

Strains of *Eubacterium rectale*, referred to as "*Eubacterium rectale*-I" in Moore and Holdeman (1974), are slender curved rods, generally longer than the other phenotypes, frequently with large central or terminal swellings. They uniformly ferment arabinose, melezitose, melibiose, raffinose, sucrose, and xylose, and produce large quantities of H_2. Strains of *Fusobacterium mortiferum* may be similar to *Eubacterium rectale* except that they have no flagella and are thicker rods showing more pleomorphism, especially when stained from growth on blood agar. Strains designated "*Eubacterium rectale*-II" differ from "*Eubacterium rectale*-I" strains in that they fail to ferment melezitose and may or may not ferment melibiose and sucrose. Cells are generally shorter, more uniform, curved rods. Strains of "*Eubacterium rectale*-III-H" differ from "*Eubacterium rectale*-II" strains in that they fail to ferment raffinose. Strains designated "*Eubacterium rectale*-III-F" have the same reactions as do strains of "III-H" except that they produce major amounts of formic acid and no H_2. Strains designated "*Eubacterium rectale*-IV" are similar to strains of "III-HV" except that they fail to ferment xylose, may or may not ferment raffinose, and never reduce the pH in arabinose to below 5.5. Heat resistant spores have been detected in some strains otherwise similar to *Eubacterium rectale* subgroups I, II, and III-H. These strains were called unnamed "*Clostridium* species A" (Moore and Holdeman, 1974). However, not all strains with swellings resist heating at 80°C for 10 min or treatment with absolute ethanol for 30 min.

The relationship between *Eubacterium rectale* and *Butyrivibrio fibrisolvens* is in question and has been discussed by

Moore and Holdeman (1974). They report that strains of *Butyrivibrio fibrisolvens* stain only Gram-negative and are monotrichously flagellated and generally are more fastidious than strains of *Eubacterium rectale*.

33. **Eubacterium ruminantium** Bryant 1959, 140[AL]

ru.mi.nan'ti.um. N.L. pl. n. *ruminantia* ruminants; N.L. gen. pl. n. *ruminantium* of ruminants.

The phenotypic description is from Bryant (1959) who studied 20 strains and from study of the type and one other strain (Moore and Holdeman Moore, 1986). Cells in PYG broth cultures are 0.2–0.3 × 0.8–2.5 μm and occur singly and in pairs. Cells decolorize readily. Surface colonies on RGCA are entire, low convex, smooth, translucent to opaque, and light buff-colored. Subsurface colonies are lenticular and do not produce gas. No growth occurs on the surface of BHIA-S with 5% blood, even when rumen fluid is added to the medium. EYA does not support growth of the type strain. PYG broth cultures are turbid in 18 h with smooth or ropy sediment; the terminal pH (5 d) is 5.0–5.5.

The optimal temperature for growth is 37°C, little or no growth occurs at 30 or 45°C, no growth occurs at 22 or 50°C. Less growth is produced when rumen fluid in the medium is replaced by 0.5% yeast extract and 1.5% trypticase. Growth is unaffected by heme (0.5 mg/100 ml), 20% bile, or 0.02% Tween 80. Growth of a test strain was inhibited by 20 μg/ml of either $HgCl_2$ or $CuCl_2$ but not by 100 μg/ml of $CdCl_2$ (Forsberg, 1978). No growth occurs at Na^+ concentrations of 3.1 mM or less; best growth occurs with at least 91 mm Na^+ (Caldwell and Hudson, 1974).

No gas is detected in glucose agar deep cultures. The type and one other strain tested do not ferment adonitol, dextrin, dulcitol, galactose, glycerol, inulin, or sorbose. Pectin is weakly fermented. Hydrogen sulfide is not produced in SIM medium. Ammonia is not produced from arginine or threonine, but small quantities may be detected in peptone cultures. The type strain does not produce acetylmethylcarbinol or hydrolyze hippurate. Neutral red and resazurin are not reduced. Nine of 20 strains tested ferment xylan; none ferment gum arabic.

Products (milliequivalents per 100 ml of culture) in PYG-10% rumen fluid broth cultures are formic (0.5–2.0), acetic (0.01–2.5), butyric (0.3–0.6), DL-lactic (0.4–1.5), and succinic (0.05–0.15) acids. Small amounts of CO_2 are produced. Little or no H_2 is detected in headspace gas. Galacturonic acid is utilized (Tomerska and Wojciechowicz, 1973). Pyruvate and lactate are not utilized and threonine is not converted to propionate.

Exogenous ammonia is required for growth and is the preferred nitrogen source, even in very complex media (Bryant and Robinson, 1962) and amino acids (mainly alanine, valine, and isoleucine) are excreted into the medium during the exponential phase of growth (Stevenson, 1978). One or more of *n*-valerate, iso-valerate, 2-methyl-*n*-butyrate, or iso-butyrate, but not amino acids, are required as carbon sources for growth (Bryant and Robinson, 1962, 1963). Other characteristics of the species are given in Table 156.

Isolated from bovine rumen contents where it represents up to 7.3% of the total isolates.

DNA G+C content (mol%): not determined.

Type strain: ATCC 17233, Bryant GA 195, DSM 20704. *GenBank accession number (16S rRNA gene):* AB008552.

Further comments: Bryant described two biotypes, with eight strains in biotype 1 and five in biotype 2. The other seven strains seemed to be intermediates between the two biotypes.

In differentiating human fecal isolates of *Eubacterium ruminantium* from *Eubacterium ventriosum*, Moore and Holdeman (1974) designated *Eubacterium ruminantium* those strains that ferment cellobiose and produce no H_2; these strains did not produce acid from mannose but usually produced acid from salicin. Strains designated *Eubacterium ventriosum* either did not ferment cellobiose or they produced H_2; they often fermented mannose and did not ferment salicin.

Moore and Holdeman Moore (1986) noted that although the reactions of *Eubacterium ruminantium* are similar to those of *Gemmiger formicilis* (Gossling and Moore, 1975), the cells and attached "buds" of *Gemmiger formicilis* are more nearly spherical than are the cells of *Eubacterium ruminantium* (Moore and Holdeman, 1974).

34. **Eubacterium saburreum** (Prévot 1966) Holdeman and Moore 1970, 23[AL] (*Catenabacterium saburreum* Prévot 1966, 171)

sa.bur're.um. L. n. *saburra* sand; N.L. neut. adj. *saburreum* sandy.

Nineteen strains were isolated and described but not named by Theilade and Gilmour. This description is from Theilade and Gilmour (1961), Hofstad (1967), and study of the type and 37 other strains (Moore and Holdeman Moore, 1986).

Cells of the type strain in PYG broth cultures are 0.7–1.1 × 6–18 μm and occur in pairs and short chains of cells, often in parallel arrangement. Curving filaments, sometimes with swellings, often occur. Cells often stain very weakly Gram-positive or Gram-negative with a few Gram-stain-positive spots or areas within some cells (Kopeloff's modification of the Gram stain). Cells from broth cultures containing fermentable carbohydrate usually stain Gram-negative. Cell walls contain glucose and rhamnose but no galactose. Surface colonies on blood agar are 1–4 mm and flat with interlaced filamentous rhizoid edges and small slightly raised granular centers which penetrate the agar. Cultures in broth produce slight turbidity with flocculent or granular sediment that may adhere to the tube. There is moderate growth in prereduced PY broth and heavy growth in prereduced broth containing a fermentable carbohydrate. The terminal pH in culture medium with a fermentable carbohydrate is 4.7–5.5.

Growth of most strains is stimulated by 0.02% Tween 80. The optimal temperature for growth is 37°C. The type strain grows moderately well at 30°C and poorly at 25 and 45°C.

Abundant gas is produced in PYG agar deep cultures. Urease is not produced by the type strain. H_2S is not produced in SIM medium. α-Methyl glucoside is fermented, sometimes only weakly.

Products (milliequivalents per 100 ml of culture) in PYG cultures are acetic (0.3–1.4), butyric (0.2–1.0), and lactic

(0.1–3.5) acids, occasionally with trace amounts of formic or succinic acid. Copious H_2 and moderate amounts of CO_2 are produced.

Tested strains are susceptible to chloramphenicol (12 µg/ml), clindamycin (1.6 µg/ml), erythromycin (3 µg/ml), and penicillin G (2 units/ml). Some strains are resistant to tetracycline (6 µg/ml).

Based on double diffusion in agar gel, strains of *Eubacterium saburreum* have been classified into serotypes 1, 2, and 3 (Kondo et al., 1979). The type-specific polysaccharide antigens representative of these and other groups apparently are located on the surface of the cell and contain heptose and *O*-acetyl as major constituents (Hoffman et al., 1976, 1980, 1974; Hofstad, 1972, 1975, 1977, 1978; Kondo et al., 1979; Skaug and Hofstad, 1979). Other characteristics of the species are given in Table 156.

Isolated from human dental plaque and gingival crevice.

DNA G+C content (mol%): not determined.

Type strain: ATCC 33271, CCUG 28089, CIP 105341, DSM 3986, JCM 11021, VPI 11763.

GenBank accession number (16S rRNA gene): not available.

The 16S rRNA gene sequence of the type strain is available from the Ribosomal Database Project.

35. **Eubacterium saphenum**   corrig. Uematsu, Nakazawa, Ikeda and Hoshino 1993, 303[VP]

sa.phen'um. L. adj. *saphenus* (probably from Gr. n. *saphenes* the plain truth) hidden; referring to the fact the organisms had been hidden in a bacterial flora.

Cells are obligately anaerobic, nonspore-forming, non-motile, Gram-stain-positive short rods (0.5 × 1.0–1.2 µm). Cells occur singly, in pairs, and sometimes in chains. Cells from older cultures often stain Gram-negative. The growth is poor in broth media and moderately enhanced in the presence of 5% bovine serum, 0.2% lysine, or 0.2% arginine. Strictly anaerobic conditions are required for growth. After incubation of 7 d in an atmosphere of 80% N_2, 10% H_2, and 10% CO_2, colonies on BHI-blood agar plates are 0.3–0.5 mm in diameter, circular, convex, and translucent. After 14 d, they are approximately 1 mm in diameter. No hemolysis is produced around colonies on BHI-blood agar plates.

Strains do not produce acid from adonitol, amygdalin, arabinose, cellobiose, erythritol, esculin, fructose, galactose, glucose, glycogen, inositol, lactose, maltose, mannose, mannitol, melezitose, melibiose, raffinose, rhamnose, ribose, salicin, sorbitol, starch, sucrose, trehalose, or xylose. Moderate amounts (approx. 5 mM each) of acetate and butyrate are produced in peptone-yeast extract or PYG. Esculin and starch are not hydrolyzed, nitrate is not reduced, no liquefaction of gelatin occurs, ammonia is not produced from arginine and indole, and catalase and urease are not produced.

Isolated from human periodontal pockets.

DNA G+C content (mol%): 44–48 (HPLC).

Type strain: U 164-47, ATCC 49989.

GenBank accession number (16S rRNA gene): U65987.

36. **Eubacterium siraeum** Moore, Johnson and Holdeman 1976, 250[AL]

si.rae'um. Gr. adj. *siraeum* (probably from Gr. n. *siraion* new wine boiled down) sluggish, referring to the relative inactivity of this organism in most substrates tested.

Cells are 0.5–0.6 × 1.3–3.0 µm and occur singly, in pairs, or short chains, sometimes in "V" or "flying gull" arrangements. Some strains are motile and have one or two subpolar flagella. After incubation for 5 d in RGCA, subsurface colonies are 0.5–1.0 mm in diameter, lenticular, translucent, and tan or white. The larger colonies often have a dense center. Surface colonies on BHIA-S in roll streak tubes or on blood agar plates are 0.5 mm in diameter, circular, entire, low convex, smooth, shiny, and transparent to translucent. Three of eight strains tested that grew on anaerobically incubated blood agar plates did not lyse the rabbit red blood cells. There is little growth in PY broth without fermentable carbohydrate. Cultures in PY broth with a fermentable carbohydrate are slightly turbid with abundant smooth sediment and pH of 5.3–5.8.

Most strains grow equally well at 37 and 45°C; little growth occurs at 30°C. Addition of 10–30% (v/v) rumen fluid stimulates growth of most strains. Growth is not affected by 0.02% Tween 80 and is inhibited by 20% bile. Gas production in glucose agar deep cultures is variable.

Products (milliequivalents per 100 ml of culture) in PY-cellobiose broth are acetic acid (1.3–4.1) and large amounts of ethanol. Trace amounts of lactic, butyric, or succinic acids also may be detected. Abundant H_2 (greater than 3% of the headspace gas) is produced from cellobiose, fructose, or starch. Other characteristics of the species are given in Table 156.

Isolated from human feces.

DNA G+C content (mol%): 45 (T_m).

Type strain: ATCC 29066, DSM 3996.

GenBank accession number (16S rRNA gene): L34625.

37. **Eubacterium sulci** (Cato, Moore and Moore 1985) Jalava and Eerola 1999, 1378[VP] (*Fusobacterium sulci* Cato, Moore and Moore 1985, 476)

sul'ci. L. gen. n. *sulci* of a furrow, referring to its habitat, the human gingival sulcus.

Cells are Gram-stain-positive, nonmotile, nonspore-forming, obligately anaerobic, straight rods, 0.4–0.7 × 1.0–7.0 µm, with rounded ends. They occur singly, in pairs, or in short chains, often with cells of different sizes in the same chain. Surface colonies on blood agar plates are pinpoint-size to 1.00 mm, circular, entire, flat to low convex, translucent, colorless, shiny, smooth, and nonhemolytic. No internal colony structure is visible. Cultures in PYG broth have stringy or granular sediment and may be slightly turbid. After 5–7 d of incubation under CO_2/N_2, the pH of PYG and basal medium broth cultures is 6.9–7.1; under 100% CO_2, the pH of each is 6.0–6.2.

The optimum temperature for growth is 37°C, at which growth is only moderate at best. No growth occurs at 25 or 45°C. Growth can be stimulated by the addition of rabbit serum (10–15%) and cocarboxylase (5 µg/ml) and in an atmosphere of 10% CO_2/90% N_2.

Gas is not detected in PYG deep agar cultures; trace to moderate amounts of H$_2$ may be produced. H$_2$S is not detected in sulfide-indole-motility medium. No acid is produced from adonitol, amygdalin, L-arabinose, cellobiose, dextrin, dulcitol, *meso*-erythritol, esculin, D-fructose, D-galactose, D-glucose, glycerol, glycogen, inositol, inulin, lactose, maltose, D-mannitol, D-mannose, melezitose, melibiose, pectin, raffinose, rhamnose, D-ribose, salicin, D-sorbitol, L-sorbose, starch, sucrose, trehalose, or D-xylose. Indole, acetylmethylcarbinol, catalase, lecithinase, lipase, urease, oxidase, and deoxyribonuclease are not produced. Esculin, starch, and hippurate were not hydrolyzed. Nitrate is not reduced; neither resazurin nor neutral red is reduced. There is no reaction in milk and no digestion of gelatin or chopped meat. The products of fermentation in PYG broth are moderate to large amounts of butyrate and trace to small amounts of acetate; traces of succinate are occasionally detected. Pyruvate, DL-lactate, D-gluconate, and L-threonine are utilized.

Strains are susceptible to chloramphenicol (12 μg/ml), clindamycin (1.6 μg/ml), erythromycin (3 μg/ml), penicillin (2U/ml), and tetracycline (6 μg/ml).

Isolated from the human gingival sulcus.

DNA G+C content (mol%): 39(T_m).

Type strain: ATCC 35585, CCUG 20560, VPI D45A-29A.

GenBank accession number (16S rRNA gene): AJ006963.

Further comments: Eubacterium sulci is closely related to *Eubacterium infirmum* and the two species may represent subspecies of the same taxon.

38. **Eubacterium tarantellae** corrig. Udey, Young and Sallman 1977, 407[AL]

tar.an.tel'l. ae. N.L. n. *tarantellus*; It. n. *tarantella* a fast, whirling dance; referring to the disease symptoms of the fish from which the species was isolated.

This description is from Udey et al. (1977) and studies of the type strain (Moore and Holdeman Moore, 1986). In tissue and when initially isolated, cells are long unbranched filaments. After several transfers in the laboratory, the cells are 1.3–1.6 × 10.0–17.0 μm. No motility has been detected. Surface colonies on BHIA are 2–5 mm, flat, translucent, colorless, rhizoid, soft, and slightly mucoid, with a distinct "pinwheel" appearance. A large zone of β-hemolysis is present around colonies on sheep blood agar plates. Lecithinase is produced on EYA medium. In Brewer's thioglycollate medium, the species grows as discrete clusters which are fluffy and filamentous. The clusters of growth often are surrounded by a slimy layer. After 48 h incubation, PY broth cultures have smooth to cottony sediment, no turbidity, and terminal pH of 5.3–5.8.

Strains grow well at temperatures between 25 and 37°C. Growth at 15 or 45°C is marginal. Cells survive for 2 weeks at 4°C, but there is no evidence of growth at this temperature. Cultures grow within 72 h in brain heart infusion broth (BHIB) adjusted to an initial pH from 5.6–8.0 but do not grow at pH extremes outside this range. The species does, however, survive for 24 h at pH 3.4 and 8.8. Growth occurs in Brewer's thioglycollate medium with 2% (w/v) NaCl. Cells can be recovered after 24 h incubation in concentrations of NaCl up to 10% (w/v).

Deoxyribonuclease is produced. Moderate to abundant gas is produced in PYG agar deep cultures.

Products (milliequivalents per 100 ml of culture) in PYG broth are acetic (1.5–5.0) and formic (1.0–5.0) acids often with trace amounts of lactic or succinic acids. Abundant H$_2$ is detected in headspace gas.

Strains are sensitive (zones of inhibition surrounding the disk) to erythromycin, chloramphenicol, penicillin, tetracycline, and novobiocin (concentrations not given). Strains grow well in BHIB containing 100 μg/ml gentamicin, 100 μg/ml neomycin, or 300 units/ml polymyxin B. Vancomycin is inhibitory at 7.5 μg/ml. Blood agar with 100 μg/ml gentamicin is a highly selective isolation medium. The type strain is susceptible to chloramphenicol (12 μg/ml), clindamycin (1.6) μg/ml), erythromycin (3 μg/ml), penicillin G (2 units/ml), and tetracycline (6 μg/ml).

Not toxigenic for mice (intraperitoneal inoculation) or pathogenic for guinea pigs (intramuscular inoculation with CaCl$_2$). Pathogenic for channel catfish (*Ictalurus punctatus*) by intraperitoneal inoculation. Other characteristics of the species are given in Table 156.

Isolated from brains of dead or moribund striped mullet (*Mugil cephalus*) in Biscayne Bay, Florida, that have evidence of a neurological disease.

DNA G+C content (mol%): not determined.

Type strain: UM-87, ATCC 29255, DSM 3997.

GenBank accession number (16S rRNA gene): L34624.

Further comments: Henley and Lewis (1976) reported isolation of two strains with characteristics similar, but not identical to those of *Eubacterium tarantellae* from moribund fish from the Texas coast. These strains share antigens in common with *Eubacterium tarantellae* (Udey et al., 1977).

39. **Eubacterium tenue** (Bergey, Harrison, Breed, Hammer and Huntoon 1923) Holdeman and Moore 1970, 23[AL] (*Bacteroides tenuis* Bergey, Harrison, Breed, Hammer and Huntoon 1923, 263)

te'nu.e. L. neut. adj. *tenue* slender (originally used with *spatuliformis* to indicate forms like slender spatulas).

The description is from study of the type and three other strains (Moore and Holdeman Moore, 1986). Cells in PYG broth cultures are 0.5–0.8 × 4.9–20.0 μm and occur singly and in pairs and short chains. Individual cells often have slightly widened and blunt ends. Cells of young cultures (exponential phase or early stationary phase of growth) of the type strain are motile by microscopic examination, but no flagella have been seen with Leifson's flagella stain. One other strain has peritrichous flagella. Motile cells have not been seen in two strains. Surface colonies on horse blood agar (2 d) are 4–6 mm, slightly irregular with lobate to diffuse edges, flat, translucent, gray-white, smooth, and dull, with granular or mottled appearance when viewed by obliquely transmitted light. Strains are either not hemolytic or produce slight clearing of the blood beneath heavy growth. Glucose broth cultures are turbid with sediment and terminal pH of 5.5–5.9. They have a putrid odor. Growth is most rapid at 24–37°C. Growth is stimulated by 0.02% Tween 80 and inhibited by 6.5% NaCl.

Moderate to abundant gas is detected in PYG agar deep cultures. Lecithinase is produced on McClung-Toabe EYA. H₂S is produced in SIM medium. Ammonia is produced from peptone. Neutral red is reduced. Hippurate is not hydrolyzed and acetylmethylcarbinol is not produced.

Products (milliequivalents per 100 ml of culture) in PYG broth are acetate (1.0–4.7) and formate (0.4–0.8), often with small amounts of propionate, iso-butyrate, iso-valerate, iso-caproate, ethanol, propanol, iso-butanol, and butanol.

One human clinical isolate from blood is susceptible to chloramphenicol (12 g/ml), clindamycin (1.6 μg/ml), erythromycin (3 μg/ml), penicillin G (2 units/ml), and tetracycline (6 μg/ml). The type strain is nontoxic for mice (0.5 ml of 24 h chopped meat culture supernatant fluid injected intraperitoneally). Other characteristics of the species are given in Table 156.

Isolated from abscess following abortion, from knee synovial fluid, and from blood.

DNA G+C content (mol%): not determined.

Type strain: ATCC 25553, DSM 20695, JCM 6486.

GenBank accession number (16S rRNA gene): M59118.

40. **Eubacterium tortuosum** (Debonono 1912) Prévot 1938, 295[AL] (*Bacillus tortuosus* Debonono 1912, 233)

tor.tu.o′sum. L. neut. adj. *tortuosum* full of windings.

This description is from the study of the type and 15 other strains (Moore and Holdeman Moore, 1986). Cells in PYG broth cultures are 0.5–0.6 × 2.4–5.0 μm in long chains of 50 or more elements. Surface colonies on horse blood agar (2 d) are 0.5–4.0 mm, circular, entire to erose to diffuse, convex to umbonate, translucent, gray to white, and smooth to slightly rough. Glucose broth cultures have flocculent to gelatinous (occasionally granular) sediment, no turbidity and terminal pH of 5.3–5.6.

The temperature for optimal growth is 37–41°C. Most strains grow at 30 and 45°C, some grow at 25°C. Growth is inhibited by 20% bile but not affected by 0.02% Tween 80.

Variable amounts of gas are produced in glucose agar deep cultures. Neutral red is reduced; resazurin is not reduced. Adonitol, dextrin, dulcitol, galactose, glycerol, inulin, and sorbose are not fermented; ammonia is not produced from peptone, arginine, or threonine. Hippurate is not hydrolyzed. Little or no acetylmethylcarbinol is produced. Hydrogen sulfide is not produced in SIM medium.

Products (milliequivalents per 100 ml of culture) in PYG broth are acetic (0.05–0.45), butyric (0.3–0.8), DL-lactic (2.1–4.6), and succinic (0.06–0.5) acids, with trace to moderate amounts of formic acid. Moderate to abundant amounts of H₂ are detected in headspace gas. Pyruvate is converted primarily to acetate and butyrate. Lactate is not utilized. Threonine is not converted to propionate. Other characteristics of the species are given in Table 156.

Isolated from turkey liver granulomas, turkey enteritis, human feces, soil, and fresh water.

DNA G+C content (mol%): not determined.

Type strain: ATCC 25548, DSM 3987.

GenBank accession number (16S rRNA gene): L34683.

Further comments: Eubacterium tortuosum (referred to as "*Catenabacterium*," Moore and Gross, 1968) is believed to be a causative agent of turkey liver granulomas.

41. **Eubacterium uniforme** van Gylswyk and van der Toorn 1985, 324[VP]

u.ni.for′me. L. neut. adj. *uniforme* uniform, denoting unusual uniformity among strains.

Cells are obligately anaerobic and nonmotile. The majority of cells from 16-h cultures on xylan agar stain Gram-positive. Most cells are short rods (coccoid forms are present) and have rounded or somewhat blunt ends (Figure 140). They occur singly, in pairs, and sometimes in short chains. The cells are usually about 0.4 μm wide, but some are wider (up to 0.6 μm), and the length varies from 0.6–1.5 μm. No differences in cell size occur on xylan or cellobiose agar medium. No spores are produced. After incubation for 3 d on films of xylan (3%) agar medium in roll bottles, surface colonies appear mucoid and round with smooth to wavy or irregular edges. Colony diameters vary from 2 to 8 mm. The colonies are white and show strong bluish-green iridescence when they are viewed obliquely in transmitted light. Clearings surrounding the colonies and resulting from xylan solubilization in the opaque medium vary from 15 to 25 mm in diameter. The degradation of xylan within the clearings is complete in the vicinity of the colonies. Submerged colonies are lens shaped.

Neither rumen fluid nor carbon dioxide-bicarbonate is required for growth. All strains grow well in medium which lacks added reducing agent (sulfide and cysteine) but which is nevertheless reduced with respect to resazurin. Injection of 5 ml of sterile air into 30-ml bottles containing either 4 or 10 ml of medium results in at least partial oxidation of the medium, as indicated by the color of the resazurin indicator, and causes complete inhibition of growth. The majority of strains grow at 45°C but there is no growth 22°C.

Gelatin is liquefied. Indole, α-methylindole, hydrogen sulfide, and catalase are not produced. Nitrate is not reduced. The major products of xylan fermentation are formate, acetate, lactate, and ethanol.

Isolated from the sheep rumen.

DNA G+C content (mol%): 35 (determined spectrophotometrically).

Type strain: X3C39, ATCC 35992.

GenBank accession number (16S rRNA gene): L34626.

42. **Eubacterium ventriosum** (Tissier 1908) Prévot 1938, 295[AL] (*Bacillus ventriosus* Tissier 1908, 204)

ven.tri.o′sum. L. neut. adj. *ventriosum* pot-bellied.

The description is from Eggerth (1935), Weinberg et al. (1937), Prévot et al. (1967) and study of the type and 20 similar strains (Moore and Holdeman Moore, 1986). Cells of the type strain are 0.8–1.3 × 1.9–5.0 μm and occur singly and in pairs and short chains of pairs. Surface colonies on blood agar are 0.5–3.0 mm, circular, entire-diffuse, convex, translucent, smooth, and shiny, with slightly mottled or granular appearance when viewed by obliquely transmitted light. Colonies in agar are 1.0–2.0 mm in diameter, lenticular, and translucent. Glucose broth cultures are turbid with smooth, ropy, or granular sediment and terminal pH of 4.6–5.4.

The optimum temperature for growth is 37°C. Strains grow well at 45°C and moderately well at 30°C. Some strains grow at 25°C. Growth is stimulated by a fermentable carbohydrate and by Tween 80. Growth is inhibited by 6.5% NaCl. Growth and fermentation are partially inhibited by 20% bile.

Little or no gas is produced by PYG agar deep cultures. The type strain ferments pectin and galactose and weakly ferments dextrin. Adonitol, glycerol, inulin, and sorbose are not fermented. Acetyl methyl carbinol is produced. Neutral red and resazurin are reduced. Ammonia is not produced from arginine, threonine, or peptone. H_2S is not produced in SIM medium.

Products (milliequivalents per 100 ml of culture) in PYG cultures are acetic (0.1–0.5), formic (0.8–3.0), butyric (0.5–1.5), and lactic (0.8–2.7) acids; trace amounts of succinic acid may be detected. Little or no H_2 is detected. Pyruvate is partially converted to acetate and formate. Lactate, threonine, and glucose are not utilized.

The type strain is susceptible to chloramphenicol (12 µg/ml), clindamycin (1.6 µg/ml), erythromycin (3 µg/ml), penicillin G (2 units/ml), and tetracycline (6 µg/ml). Other characteristics of the species are given in Table 156. Available strains are isolated from human feces. Isolations from dog feces, mouth abscess, neck infection, purulent pleurisy, pulmonary abscesses, and material from a bronchiectasis have been reported.

DNA G+C content (mol%): not determined.

Type strain: ATCC 27560, DSM 3988.

GenBank accession number (16S rRNA gene): L34421.

Further comments: Holdeman et at. (1976) differentiated *Eubacterium ventriosum* from *Eubacterium ruminatium* on the basis of H_2 production in at least small amounts by strains of *Eubacterium ventriosum*. These authors recognized two biogroups (I and II) that differed from strains of *Eubacterium ventriosum*.

43. **Eubacterium xylanophilum** van Gylswyk and van der Toorn 1985, 325[VP]

xy.lan.o.phil′um. N.L. n. *xylanum* xylan; Gr. part. *philos* loving; N.L. neut. adj. *xylanophilum* xylan-loving.

Cells are obligately anaerobic, nonspore-forming and motile, exhibiting a rapid corkscrew-like motion. Flagella are observed by transmission electron microscopy, but their points of insertion and their number (one or two per cell) have not been determined with certainty. After growing for 16 h on xylan agar medium, fewer than one half of the cells stain Gram-positive. The cells are straight rods or coccoid with rounded ends. They occur singly, in pairs, and sometimes in short chains. The cell width varies from 0.4–0.6 µm, and the cell length varies from 0.5–2.0 µm (sometimes up to about 3 µm). Surface colonies in roll bottles containing xylan (3%) agar medium incubated for 3 d are 2–4 mm wide, circular, entire, smooth, white, and iridescent in obliquely transmitted light. Submerged colonies are lenticular and 0.5–1 mm in diameter. Clearings due to solubilization of xylan surrounding surface colonies vary in diameter from 6 to 20 mm. Xylan is extensively degraded within the clearings.

Rumen fluid is not required for growth, but growth is slightly enhanced in the presence of carbon dioxide and bicarbonate. There is no growth at either 22 or 45°C.

Xylan and cellobiose are fermented when the inoculum for the tests is grown on xylan agar slopes. However, when the inoculum is grown on cellobiose agar slopes, esculin is also (but poorly) fermented (pH drop of 0.2–0.3 U) by all three strains. The final pH in poorly buffered medium containing xylan or cellobiose does not drop below 5.6. These bacteria utilize cellobiose but not glucose or xylose, which suggests that they may utilize xylobiose derived from xylan. Indole and α-methylindole are not produced. Gelatin is not liquefied, nitrate is not reduced, catalase is not produced, and starch is not hydrolyzed. The fermentation products are formate, acetate, and butyric acids.

Isolated from the sheep rumen.

DNA G+C content (mol%): 39 (determined spectrophotometrically).

Type strain: X6C58, ATCC 35991.

GenBank accession number (16S rRNA gene): L34628.

44. **Eubacterium yurii** Margaret and Krywolap 1986, 147[VP]

yur′i.i. N.L. gen. n. *yurii* of Yuri (author).

Obligately anaerobic, nonspore-forming, straight Gram-stain-positive rods with slightly rounded ends. Individual cells are motile by means of a single subpolar flagellum. The cells form three dimensional brush-like aggregates held together by an amorphous extracellular substance. Cells from 48-h Schaedler broth cultures measure approximately 0.5 µm × 4 µm. After 48 h of incubation at 37°C on Schaedler blood agar, colonies measure 1 mm in diameter, with delicate spreading margins. Some strains exhibit a pale yellow pigment. Growth on MM10 blood agar is similar, although the colonies are smaller. Colonies do not spread on media without blood. In Schaedler broth, growth is typically as a granular sediment, with colonies adhering to the glass; the supernatant fluid remains clear; gas may be formed. Growth is poor in chopped-meat medium. Vitamin K is not required for growth.

The cultures grow well at 34–37°C but not at 25 or 40°C. All strains produce H_2S, indole, and RNase and do not hydrolyze esculin, gelatin, hippurate, or starch. Catalase and acetylmethylcarbinol are not produced. Most strains are asaccharolytic. Acid production from glucose, maltose, and sucrose and production of DNase and phosphatase is variable. In peptone-yeast extract-glucose broth, the major product is butyrate, with minor production of acetate and propionate. All strains are susceptible to penicillin, erythromycin, tetracycline, clindamycin, and chloramphenicol.

Isolated from subgingival dental plaque.

DNA G+C content (mol%): 32 (T_m).

Based upon production of phosphatase, saccharolytic activity, and DNA renaturation studies, three subspecies of *Eubacterium yurii* have been validly published.

44a. **Eubacterium yurii subsp. yurii** Margaret and Krywolap 1986, 148[VP]

Phosphatase-positive; most strains are asaccharolytic; an occasional strain may produce weak acid in sucrose and may be strongly DNase-positive.

Type strain: SM14, ATCC 43714.

GenBank accession number (16S rRNA gene): L34629.

44b. Eubacterium yurii subsp. margaretiae Margaret and Krywolap 1986, 148[VP]

mar.ga.ret′i.ae. N.L. gen. n. *margaretiae* of Margaret (author).

All strains are phosphatase-negative; most are asaccharolytic; some may produce weak acid in sucrose, maltose or glucose.

Type strain: SM65, ATCC 43715.
GenBank accession number (16S rRNA gene): AY533381.

44c. Eubacterium yurii subsp. schtitka Margaret and Krywolap 1988, 207[VP]

schtit′ka. *schtitka* from Ukrainian n. *schtitka* brush.

Strains are phosphatase-negative, asaccharolytic, and do not stimulate bone resorption *in vitro*.

Type strain: III, ATCC 43716, SMN.
GenBank accession number (16S rRNA gene): AY533382.

Unnamed species of the genus *Eubacterium*. Characteristics of 19 groups of unnamed species of eubacteria isolated from human feces are given in Moore and Holdeman (1974) and Holdeman et al. (1976). In addition, molecular analysis of bacterial communities in the environment and associated with animals have revealed an extremely large number of cloned 16S rRNA genes that do not correspond to existing species. Some of these may be unculturable using standard culture techniques. However, experience in the oral cavity (Munson et al., 2002) suggests that the identification of isolates by 16S rRNA gene sequence analysis will reveal numerous unnamed species.

Genus II. **Acetobacterium** Balch, Schoberth, Tanner and Wolfe 1977, 360[AL]

MARIA V. SIMANKOVA AND OLEG R. KOTSYURBENKO

A.ce.to.bac.te′ri.um. L. n. *acetum* vinegar; Gr. neut. n. *bakterion* a small rod; N.L. neut. n. *Acetobacterium* vinegar rod.

Oval-shaped, short rods. Gram-stain-positive. Motile. Endospores not formed. **Strictly anaerobic.** Optimal temperature 27–30°C for mesophilic species, 20–30°C for psychrotolerant species. Optimal pH 7.0–8.0. Colonies are convex, white, slightly yellow, or brownish, 0.6–1.0 mm in diameter. **Autotrophic** growth occurs by anaerobic oxidation of H_2 and reduction of CO_2 to acetic acid. **Chemo-organotrophic,** carrying out **homoacetogenic** fermentation of reduced substrates, such as fructose and some other monomeric sugars, as well as pyruvate, lactate, glycerol, and methanol; methyl groups of phenyl methyl ethers and betaine are converted to acetate. The acetyl-CoM pathway serves as an energy-conserving process and as a mechanism for autotrophic assimilation of carbon. Cytochromes have not been detected.

DNA G + C content (mol%): 39–45.8.

Type species: **Acetobacterium woodii** Balch, Schoberth, Tanner and Wolfe 1977, 360.

Further descriptive information

Oval-shaped, short rods 0.7–1.0 × 2.0–4.0 μm, single or in pairs (Figure 147). Motile by means of one or two subterminal flagella or peritrichous flagella. Swollen and elongated cells can appear under nonoptimal growth conditions. Psychrotolerant species of the genus form swollen cells at temperatures higher than optimal ones, and usually the number of the swollen cells reaches 5% of the cell number. However, at 27–30°C swollen cells of the psychrotolerant species *Acetobacterium tundrae* represent 90–95% of the cell number; the size of these cells reaches 2–3 μm in width and 10–15 μm in length (Simankova et al., 2000). At temperatures higher than 25°C, the cell size increases, but cell division does not occur. The breakage of the cell wall of swollen cells is sometimes observed, occurring when the cell content reaches a critical amount. If swollen cells grown at 25–30°C are transferred to a new fresh medium and are cultivated at 20°C, normal-sized cells develop. Analysis of the lipid complex of the cell membrane was carried out with *Acetobacterium bakii, Aceto-*

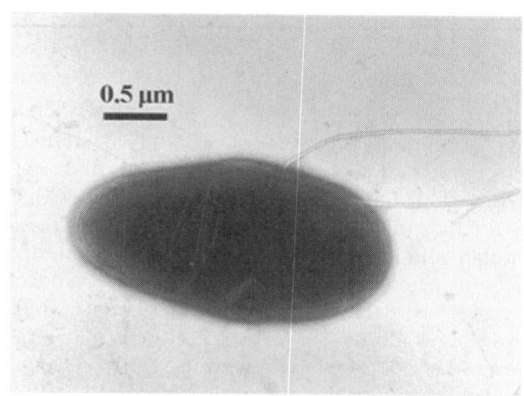

FIGURE 147. Micrograph of *Acetobacterium woodii*. Bar = 0.5 μm.

bacterium fimetarium, Acetobacterium paludosum, and *Acetobacterium tundrae.* A peculiarity of the lipid complex structure is the presence of a large amount of plasmalogens, which are detected as aldehydes released after sample processing (Kotsyurbenko et al., 1995; Simankova et al., 2000). The main components of the lipid complex of the species studied are the saturated fatty acid $C_{16:0}$, the unsaturated fatty acids $C_{16:1\Delta9}$ and $C_{16:1\Delta11}$, as well as the saturated aldehyde $C_{16:0}$; however, the latter is absent from the lipid complex of *Acetobacterium fimetarium.*

The cell wall has a Gram-stain-positive structure and contains the more rare peptidoglycan of the cross-linkage type B (Braun and Gottschalk, 1982; Eichler and Schink, 1984), in which a L-seryl residue replaces the L-alanyl residue in position 1 of the peptide subunit, and the ornithinyl residues function as interpeptide bridges (Kandler and Schoberth, 1979).

Colonies are convex, white, 0.6–1.0 mm in diameter. Some species produce slightly yellow (*Acetobacterium woodii, Acetobacterium carbinolicum*) or brownish (*Acetobacterium wieringae*) colonies.

The members of the genus *Acetobacterium* are metabolically versatile organisms. They are able to grow both autotrophically and chemo-organotrophically, catalyzing the formation of acetate from C_1 units as the sole or the major end product of their energy metabolism, proceeding through the acetyl-CoA (the Wood–Ljungdahl) pathway.

$H_2 + CO_2$. All members of the genus use H_2 as the electron donor for CO_2 reduction to acetate. Some species can also similarly utilize formate and CO, which are intermediates in acetate formation from CO_2. In the acetyl-CoA pathway, the methyl group of acetate is derived from CO_2 via formate and tetrahydrofolate-bound C_1 intermediates and is then transferred to corrinoid enzyme prior to incorporation into the methyl group of acetyl-CoA. The first step of carboxyl group synthesis is CO_2 reduction to a carbonyl group by means of a carbon monoxide dehydrogenase, the key enzyme of this metabolic pathway. It also catalyzes the acetyl-CoA formation from the carbonyl group, the methyl group of methyl-tetrahydrofolate, and coenzyme A. In the experiments with *Acetobacterium woodii*, it was found that the acetyl-CoA pathway is coupled with the generation of a primary sodium ion potential, which in turn drives ATP synthesis via Na^+-translocating ATP synthase (Aufurch et al., 2000; Heise et al., 1992; Müller et al., 2001).

Chemo-organotrophic growth is provided by homoacetogenic fermentation of different methylated compounds, alcohols, monomeric sugars, hydroxyacids, and some other substrates.

Methyl compounds. Methanol, as well as methyl groups of phenyl methyl ethers, methyl chlorides, and betaine is converted to acetate. Methanol is converted to acetate with CO_2 as a co-substrate. The methyl group of methanol is transferred to tetrahydrofolate and is further transformed into the methyl group of the acetate, whereas the carboxyl group is derived from CO_2. Methoxylated aromatic compounds are demethylated to the corresponding phenols; the methyl residue is fermented analogously to methanol (Bache and Pfennig, 1981). o-Demethylase catalyzes the cleavage of the ether bond of phenyl methyl ethers and the transfer of the methyl group to tetrahydrofolate (Kaufmann et al., 1997; Messmer et al., 1996). *Acetobacterium woodii* dehalogenates chloromethanes (CH_2Cl_2 and CH_3Cl), converting the methyl groups to acetate in an energy-yielding reaction. It can also catalyze the reductive dechlorination of CCl_4 and its substitutive transformation to CO_2 (Egli et al., 1988; Stromeyer et al., 1991, 1992). CH_3Cl and CO_2 as a co-substrate can also be converted to acetate by "*Acetobacterium dehalogenans*" (Traunecker et al., 1991). The methyl transfer reaction is mediated by a methyl chloride dehalogenase (Messmer et al., 1996; Messmer et al., 1993; Wohlfarth and Diekert, 1997). Betaine is only demethylated to acetate and dimethylglycine (Eichler and Schink, 1984).

Alcohols. *Acetobacterium carbinolicum* is able to utilize aliphatic alcohols (C_2–C_5), which are fermented to the corresponding fatty acids and acetate (Eichler and Schink, 1984). *Acetobacterium woodii* can also convert these substrates if the bicarbonate buffer concentration is at least 100 mM; only weak growth is supported by propanol, butanol, and pentanol (Buschhorn et al., 1989).

Some species consume diols and acetoin, a component of the butanediol cycle in microorganisms. 1,2-Propanediol is fermented to propionate and acetate by *Acetobacterium carbinolicum*, *Acetobacterium wieringae*, and *Acetobacterium woodii* (Eichler and Schink, 1984; Schink and Bomar, 1992) or to propanol, propionate, and acetate by *Acetobacterium malicum* (Tanaka and Pfennig, 1988). 2,3-Butane, as well as acetoin, is converted to acetate by *Acetobacterium carbinolicum* and *Acetobacterium woodii*. A diol dehydratase is considered to be an enzyme participating in diols cleavage. *Acetobacterium malicum*, *Acetobacterium bakii*, *Acetobacterium fimetarium*, and *Acetobacterium paludosum* can also utilize 2-methoxyethanol, converting this substrate to methanol and acetaldehyde via a diol dehydratase-analogous reaction; the latter product is then oxidized to acetate (Tanaka and Pfennig, 1988).

Sugars and hydroxyacids. One mol of fructose or any other monomeric sugar is fermented via the Embden–Meyerhof–Parnas pathway to 2 mol of pyruvate, which is further oxidized to 2 mol of acetate and 2 mol of CO_2. The latter is also reduced to acetate. Similarly, pyruvate and lactate are converted to acetate. Lactate is first oxidized to pyruvate by a lactic dehydrogenase. Malate is fermented to acetate by *Acetobacterium malicum*, *Acetobacterium bakii*, *Acetobacterium fimetarium*, and *Acetobacterium paludosum* presumably through an NAD-dependent malic enzyme (Strohhäcker and Schink, 1991).

Other substrates. *Acetobacterium woodii* saturates the carbon-carbon double bond of the acrylate side chain of caffeate derivatives and converts these compounds to the corresponding hydrocaffeates (Bache and Pfennig, 1981). The process of caffeate reduction is coupled to energy conservation (Hansen et al., 1988; Tschech and Pfennig, 1984). The latter investigation of hydrogen-dependent caffeate reduction by *Acetobacterium woodii* established chemiosmotic mechanism of ATP synthesis by means of a transmembrane Na^+ gradient generation (Imkamp and Müller, 2002).

Mandelate (phenylglycolate) is oxidized by *Acetobacterium* strain LuPhe1 via benzoyl-CoA to benzoate; carboxy carbon of mandelate serves as one-carbon substrate for acetate formation (Dorner and Schink, 1991). *Acetobacterium* strain LuPhe1 also converts 2-phenoxyethanol to phenol and acetate through acetaldehyde formation (Speranza et al., 2003)

The cytochromes have not been found in *Acetobacterium woodii* (Tschech and Pfennig, 1984), *Acetobacterium carbinolicum* (Eichler and Schink, 1984), and *Acetobacterium malicum* (Tanaka and Pfennig, 1988). Neither oxidase activity has ever been reported. Catalase and superoxide dismutase (SOD) activities were investigated in *Acetobacterium woodii*, *Acetobacterium paludosum*, and *Acetobacterium wieringae* (Brioukhanov et al., 2002). All studied species were defined as catalase-positive, but only *Acetobacterium wieringae* exhibited a high catalase activity. High specific activities of SOD were detected in *Acetobacterium woodii* and *Acetobacterium wieringae*, whereas *Acetobacterium paludosum* demonstrated a low SOD activity.

All *Acetobacterium* species are able to fix molecular nitrogen, but this process has not been studied in detail (Schink and Bomar, 1992).

The 16S rRNA gene sequence analysis shows that *Acetobacterium* species form a tight phylogenetic group exhibiting a high level (7% and higher) of sequence divergence with most closely-related species of the genus *Eubacterium* (Willems and Collins, 1996). The level of sequence similarity between species of the genus *Acetobacterium* ranges between 96.2 and 99.4% (Simankova et al., 2000; Willems and Collins, 1996). Psychrotolerant species do not exhibit a closer phylogenetic relationship with each other than with mesophilic species. DNA–DNA hybridization experiments between *Acetobacterium* species show that hybridization levels are less than 37% for all species combinations (Simankova et al., 2000). The only exception is *Acetobacterium woodii* and *Acetobacterium carbinolicum* demonstrating 69% of DNA–DNA homology. However, 16S rRNA gene sequence analysis shows a sequence similarity of 98.4% between these species.

Streptomycin, benzylpenicillin, vancomycin, rifampin, and bacitracin completely inhibit growth of *Acetobacterium bakii*, *Acetobacterium fimetarium*, and *Acetobacterium paludosum*.

All strains of the genus *Acetobacterium* were isolated from strictly anoxic environments. Most of *Acetobacterium* species inhabit fresh water ecosystems: anoxic pond sediments, ditches, wetlands, and anoxic sewage sludge. However, *Acetobacterium woodii* strain WB1 was isolated from a marine estuary, and laboratory experiments with *Acetobacterium* strains demonstrated their equally good growth in fresh water, brackish water, and salt-water medium (Schink, 1994). In recent years, it was discovered that the acetyl-CoM pathway is coupled to a chemiosmotic mechanism of ATP synthesis, namely, to the presence of sodium-proton antiporters in the cytoplasmic membrane. This finding explains the capacity of *Acetobacterium* species to well adapt to environments with periodical changes in salinity or pH, such as estuaries, soils, or sewage sludge digestors. However, *Acetobacterium* species seem to be more important for fresh water terrestrial ecosystems, where they can produce acetate from a variety of different substrates. Representatives of genus *Acetobacterium* are among the most typical microorganisms in many cold fresh water ecosystems (Kotsyurbenko et al., 1993a; Kotsyurbenko et al., 1993b; Nozhevnikova et al., 1994). All four new psychrotolerant species of homoacetogenic bacteria isolated over the last decade belong to this genus. They play an important ecological role in different cold anoxic environments as strong competitors of methanogenic archaea for hydrogen (Conrad et al., 1989; Kotsyurbenko et al., 1996), the key intermediate in the anaerobic microbial community. Under conditions of H_2 partial pressures higher than 10 Pa, psychrotolerant homoacetogens are able to outcompete methanogens owing to higher growth rates at low temperature (Kotsyurbenko et al., 2001). It can result in essential changes in the trophic structure of the microbial community and in production of acetate as one of the end products.

Enrichment and isolation procedures

Enrichment is carried out using a liquid medium and H_2:CO_2 (80:20) as gas phase. Medium contains minerals, vitamins, yeast extract, sodium sulfide as reducing agent, sodium bicarbonate (2.5 g/l) as buffer, and resazurin (Kotsyurbenko et al., 1995). The vitamin and microelement solutions described by Wolin et al. (1963) and Pfennig and Lippert (1966) are used, respectively. Yeast extract is not an obligatory requirement of any species but enhances growth yield. Methanogens can also be enriched under anoxic conditions and an atmosphere of H_2/CO_2. However, they are outgrown by faster-growing homoacetogenic bacteria when transferred repeatedly on liquid medium using tenfold serial dilutions. Moreover, methanogens can be inhibited by the addition of 50 mg/l of sodium dithionite (Balch et al., 1977) or 10–50 mM

bromoethanesulfonate (Smith and Mah, 1981). Growth of *Acetobacterium* is judged from acetate production and by observation of characteristic cell morphology under the microscope. The last positive dilution is used to inoculate roll-tubes containing H_2/CO_2 as gas phase. Colonies that developed in the roll-tubes are transferred to liquid medium and cultivated under an atmosphere of H_2/CO_2. Some species of the genus *Acetobacterium* can be enriched using selective substrates, such as methoxylated aromatic compounds (syringate, vanillate, and trimethoxycinnamate ferulate) for *Acetobacterium woodii* isolation (Bache and Pfennig, 1981), ethanol, propanol, or butanol for *Acetobacterium carbinolicum* enrichment (Eichler and Schink, 1984), and 2-methoxyethanol for *Acetobacterium malicum* isolation (Tanaka and Pfennig, 1988).

Maintenance procedures

Storage is possible anaerobically in liquid medium at 4°C, under these conditions, cells remain viable for 6 months. *Acetobacterium* may be stored after lyophilization according to the procedure for strict anaerobes (Hippe, 1991).

Differentiation of the genus *Acetobacterium* from other genera

The genus *Acetobacterium* is phylogenetically differentiated from other genera of homoacetogenic anaerobic members of the *Firmicutes* (Gram-positive bacteria with a low DNA G + C content) on the basis of oligonucleotide composition of the 16S rRNA gene. The phylogenetic relationship with the most closely related species of the genus *Eubacterium* is at the level of not more than 93% sequence similarity (Willems and Collins, 1996). *Acetobacterium* is also distinguished from other homoacetogenic anaerobic Gram-stain-positive bacteria on the basis of morphological and physiological properties. In contrast to homoacetogenic bacteria of the genus *Clostridium*, the members of the genus *Acetobacterium* do not form endospores. *Acetobacterium* differ in their cell morphology and peptidoglycan composition from the genera *Peptostreptococcus*, *Syntrophococcus*, and *Acetitomaculum*. The members of the genus *Acetobacterium* are distinguished from the only known species of the genus *Acetogenium* on the basis of their substrate spectrum, peptidoglycan composition, and inability to grow at a high temperature. The members of the genus *Acetobacterium* differ from homoacetogenic bacteria of the genus *Eubacterium* by their ability to carry out only homoacetogenic fermentation of reduced substrates.

Taxonomic comments

In the original description of the genus *Acetobacterium* it was tentatively placed in the family *Propionibacteriaceae* (Balch et al., 1977). Later, according to 16S rRNA gene sequence data, the genus *Acetobacterium* was placed in the family *Eubacteriaceae* belonging to cluster XV of the *Clostridium* subphylum of the Gram-positive bacteria (Willems and Collins, 1996). According to the 16S rRNA phylogenetic analysis presented in the roadmap to this volume (Figure 5), the genus *Acetobacterium* is a member of the family *Eubacteriaceae*, order *Clostridiales*, class *Clostridia* in the phylum *Firmicutes*.

Acknowledgements

We are grateful to N.A. Kostrikina, who took the electron micrograph presented in this description.

Further reading

Dierert, G. and G. Wohlfarth. 1994. Metabolism of homoaceto gens. Antonie van Leeuwenhoek. *66*: 209–221.

Drake, H.L. 1994. Acetogenesis, acetogenic bacteria, and the Acethyl-CoA "Wood/Ljungdahl" pathway: past and current perspectives. *In* Drake (Editor), Acetogenesis, Chapman & Hall, New York, London, pp. 3–60.

Schuppert, B. and B. Schink. 1990. Fermentation of methoxyac etate to glycolate and acetate by newly isolated strains of *Ace tobacterium* sp. Arch. Microbiol. *153*: 200–204.

List of the species of the genus *Acetobacterium*

1. **Acetobacterium woodii** Balch, Schoberth, Tanner and Wolfe 1977, 360[AL.]

wood'i.i. N.L. gen. n. *woodii* of Wood; named for H.G. Wood for his pioneering work on the total synthesis of acetate from CO_2 by bacteria.

Cells are oval-shaped, short rods $1.0 \times 2.0\,\mu m$. Motile by means of one or two subterminal flagella. Colonies on plates are circular, convex, and white; 1 mm in diameter. Older colonies may show a slight yellow pigmentation. Optimal growth occurs at 30°C. Autotrophic growth on H_2/CO_2 and formate; acetate is formed. Methanol, 2,3-butanediol, acetoin, glycerol, ethylene glycol, lactate, pyruvate, fructose, as well as methyl groups of methoxylated aromatic compounds and betaine are fermented to acetate. Ethanol, propanol, butanol, and pentanol are utilized if medium contains at least 100 mM bicarbonate buffer. Ethanol is converted to acetate; propanol, butanol, pentanol, and 1,2-propanediol are fermented to corresponding fatty acids and acetate. Caffeate derivates are reduced to the corresponding hydrocaffeate. Chloromethanes (CH_2Cl_2, CH_3Cl, and CCl_4) are dehalogenated; CH_2Cl_2 and CH_3Cl are then fermented to acetate, whereas CCl_4 is either reduced to CH_3Cl or transformed to CO_2.

Habitat: marine and fresh water sediments, and sewage sludge.

DNA G + C content (mol %): 39 (Bd).

Type strain: WB1, DSM 1030, ATCC 29683.

GenBank accession number (16S rRNA gene): X96954.

2. **Acetobacterium bakii** Kotsyurbenko, Simankova, Nozhevnikova, Zhilina, Bolotina, Lysenko and Osipov 1997, 242[VP] (Effective publication: Kotsyurbenko, Simankova, Nozhevnikova, Zhilina, Bolotina, Lysenko and Osipov 1995, 33.)

ba'ki.i. N.L. gen. n. *bakii* of Bak, named for F. Bak who isolated the first psychrotolerant homoacetogenic bacterium.

Short rods, $0.9–1.5 \times 1.5–2.7\,\mu m$. Motile by means of two subterminal flagella. Colonies on agar are white and 0.6–1.0 mm in diameter. Psychrotolerant. Temperature range for growth is 1–30°C, with an optimum at 20°C. Growth at pH values 5.5–8.5, with an optimum at 6.5. Autotrophic growth on H_2/CO_2, CO, and formate; acetate is formed. Fructose, lactate, malate, as well as methanol and methyl groups of vanillate and betaine are fermented to acetate. Weak growth on maltose, glucose, xylose, and 2-methoxyethanol. The main components of the lipid complex of the cell membrane are the saturated fatty acid $C_{16:0}$, the saturated aldehyde $C_{16:0}$, and the unsaturated fatty acids $C_{16:1\Delta9}$.

Isolated from anoxic sediments of a pond polluted by paper-mill wastewater.

DNA G + C content (mol %): 42.1 (T_m).

Type strain: Z-4391, DSM 8239, ATCC 51794.

GenBank accession number (16S rRNA gene): X96960.

3. **Acetobacterium carbinolicum** Eichler and Schink 1985, 375[VP] (Effective publication: Eichler and Schink 1984, 152.)

car.bi.no'li.cum. N.L. adj. *carbinolicum* metabolizing alcohols.

Rod-shaped cells, $0.8–1.0 \times 1.5–2.5\,\mu m$, with slightly pointed ends, single or in pairs. Motile. Colonies on agar are white or slightly yellow. Temperature range for growth is 15–40°C, with an optimum at 20°C. Growth at pH range 6.0–8.0, with an optimum at 7.0. Autotrophic growth on H_2/CO_2 and formate; acetate is formed. Methanol, ethanol, 2,3-butanediol, acetoin, glycerol, ethylene glycol, lactate, pyruvate, glucose, fructose, methyl groups of methoxylated aromatic compounds, and betaine are fermented to acetate. Propanol, butanol, pentanol, and 1,2-propanediol are fermented to the corresponding fatty acids and acetate. Does not require any vitamins. Sulfate, thiosulfate, elemental sulfur, and nitrate are not reduced. No cytochromes.

Isolated from anoxic freshwater mud.

DNA G + C content (mol %): 38.5.

Type strain: WoProp1, DSM 2925.

GenBank accession number (16S rRNA gene): X96956.

4. **Acetobacterium fimetarium** Kotsyurbenko, Simankova, Nozhevnikova, Zhilina, Bolotina, Lysenko and Osipov 1997, 242[VP] (Effective publication: Kotsyurbenko, Simankova, Nozhevnikova, Zhilina, Bolotina, Lysenko and Osipov 1995, 34.)

fi.me.ta'ri.um. L. n. *fimetum* a dung-hill; N.L. neut. adj. *fimetarium* inhabiting manure.

Short rods, $0.8–1.1 \times 1.5–2.6\,\mu m$. Motile by means of peritrichous flagella. Colonies on agar are white and 0.6–1.0 mm in diameter. Temperature range for growth is 1–35°C, with an optimum at 30°C. Growth at pH values 6.0–8.5, with an optimum at 7.5. Autotrophic growth on H_2/CO_2, CO, and formate with acetate production. Fructose, lactate, malate and methyl groups of vanillate and betaine are utilized for growth and fermented to acetate. 2,3-Butanediol and 2-methoxyethanol are weakly used. The main components of the lipid complex of the cell membrane are the saturated aldehyde $C_{16:0}$ and the unsaturated fatty acids $C_{16:1\Delta9}$ and $C_{16:1\Delta11}$.

Isolated from digested manure.

DNA G + C content (mol %): 45.8 (T_m).

Type strain: Z-4290, DSM 8238, ATCC 51795.

GenBank accession number (16S rRNA gene): X96959.

5. **Acetobacterium malicum** Tanaka and Pfennig 1990, 470[VP] (Effective publication: Tanaka and Pfennig 1988, 186.)

ma'li.cim. N.L. neut. adj. *malicum* pertaining to malic acid.

Rod-shaped cells, $1.0–1.3 \times 1.8–4.0\,\mu m$, with slightly pointed ends, single or in pairs. Motile. Colonies on agar are white. Optimal temperature is 30°C, no growth at 16°C and 40°C. Optimal pH is 7.5–8.0, no growth at initial pH 6.0 and 9.5. Autotrophic growth on H_2/CO_2 and formate; acetate is formed. Fructose, lactate, pyruvate, malate, glycerol, acetoin, ethylene glycol, 2-methoxyethanol, 2-ethoxyethanol, as well as methyl groups of methoxylated aromatic compounds and betaine are fermented to acetate. 1,2-Propanediol is fermented to propanol, propionate, and acetate. Under optimal conditions, doubling time on 2-methoxyethanol and malate is 22 h and 7.5 h, respectively. Sulfate, sulfite, thiosulfate, elemental sulfur, and nitrate are not reduced. No cytochromes.

Isolated from anoxic freshwater sediment of a ditch.

DNA G + C content (mol%): 44.1 (Bd).

Type strain: MuME1, DSM 4132, ATCC 51201.

GenBank accession number (16S rRNA gene): X96957.

6. **Acetobacterium paludosum** Kotsyurbenko, Simankova, Nozhevnikova, Zhilina, Bolotina, Lysenko and Osipov 1997, 242[VP] (Effective publication: Kotsyurbenko, Simankova, Nozhevnikova, Zhilina, Bolotina, Lysenko and Osipov 1995, 34.)

pa.lu.do'sum. L. neut. adj. *paludosum* inhabiting a fen.

Short rods, 0.8–1.1 × 1.3–2.9 μm. Motile by means of peritrichous flagella. Colonies in roll tube agar are white and 0.6–1.0 mm in diameter. Psychrotolerant. Temperature range for growth is 1–30°C, with an optimum at 20°C. Growth at pH values 5.0–8.0, with an optimum at 7.0. Autotrophic growth on H_2/CO_2, CO, and formate; acetate is formed. Maltose, fructose, glucose, lactate, malate, methanol, as well as methyl groups of betaine are fermented to acetate. 2-Methoxyethanol, xylose, and cellobiose are weakly used. The main components of the lipid complex of the cell membrane are the saturated fatty acid $C_{16:0}$, the saturated aldehyde $C_{16:0}$, and the unsaturated fatty acids $C_{16:1\Delta9}$ and $C_{16:1\Delta11}$. Habitat: anoxic freshwater sediments.

Isolated from anoxic sediments of a fen.

DNA G + C content (mol%): 41.7 (T_m).

Type strain: Z-4092, DSM 8237, ATCC 51793.

GenBank accession number (16S rRNA gene): X96958.

7. **Acetobacterium tundrae** Simankova, Kotsyurbenko, Stackebrandt, Kostrikina, Lysenko, Osipov and Nozhevnikova 2001, 793[VP] (Effective publication: Simankova, Kotsyurbenko, Stackebrandt, Kostrikina, Lysenko, Osipov and Nozhevnikova 2000, 446.)

tun'drae. N.L. fem. gen. n. *tundrae* from the tundra (a cold treeless zone located in the north of Eurasia and North America).

Cells are short rods, 0.7–1.1 × 1.1–4.0 μm, with slightly pointed ends. Motile by means of peritrichous flagella. The irregular swollen cells 2–3 μm × 10–15 μm appear at temperatures 25–30°C as a result of a defect in cell division. Colonies in agar roll tubes are white and 1.0 mm in diameter. Psychrotolerant. Temperature range for growth is 1–30°C, with an optimum 20°C. Growth at pH values 6.0–8.0, with an optimum at 7.0. Autotrophic growth on H_2/CO_2, CO, and formate with acetate production. Maltose, mannose, fructose, glucose, xylose, pyruvate, lactate, methanol, and methyl groups of betaine are utilized for growth and fermented

to acetate. The main components of the lipid complex of the cell membrane are the saturated aldehyde $C_{16:0}$ and the unsaturated fatty acid $C_{16:1\Delta9}$.

Isolated from tundra soil.

DNA G + C content (mol%): 39.2 (T_m).

Type strain: Z-4493, DSM 9173.

GenBank accession number (16S rRNA gene): AJ297449.

8. **Acetobacterium wieringae** Braun and Gottschalk 1983, 438[VP] (Effective publication: Braun and Gottschalk 1982, 374.)

wie.rin'gae. N.L. gen. n. *wieringae* of Wieringa; named for K. T. Wieringa who isolated and described *Clostridium aceticum*, the first known acetogenic bacterium.

Cells are oval-shaped, short rods 1.0 × 2.0 μm. Motile by means of one or two subterminal flagella. Colonies are circular, smooth, and brownish; 3 mm in diameter. Optimal growth occurs at temperature 30°C and pH 7.2–7.4. Autotrophic growth on H_2/CO_2 and formate; acetate is formed. Fructose, lactate, ethylene glycol, acetoin, glycerol, and ethanol are fermented to acetate. 1,2-Propanediol is fermented to acetate and propionate. Under optimal conditions, doubling time on H_2/CO_2 mixture corresponds to 10 h.

Isolated from sewage sludge.

DNA G + C content (mol%): 43 (T_m).

Type strain: C, DSM 1911, JCM 2380, ATCC 43740.

GenBank accession number (16S rDNA gene): X96955.

Other organisms

1. **"Acetobacterium dehalogenans"** Traunecker, Preuß and Diekert 1991, 417

de.ha.lo'ge.nans. N.L. adj. part. *dehalogenans* dehalogenating.

The description of this species is based on the phenotypic features and G + C content of the DNA. The data of phylogenetic analysis of the 16S rRNA gene is lacking. Cells are elongated cocci 1.1 × 1.5 μm arranged in chains. Nonmotile. Gram-stain-positive. Endospores not formed. Strictly anaerobic. Colonies are smooth, convex, and yellow, about 1 mm in diameter. Optimal growth occurs at temperature 25°C and pH 7.3–7.7. Autotrophic growth on H_2/CO_2 and CO with acetate production. Fructose, glucose, ribose, pyruvate, lactate, glycerol, as well as methyl groups of phenyl methyl ethers and methyl chloride are converted to acetate.

Isolated from a sewage digestor.

DNA G + C content (mol%): 47.5 (HPLC).

Type strain: MC.

2. **"Acetobacterium psammolithicum"** Krumholz, Harris, Tay and Suflita 1999, 2302

psam.mo.li'thi.cum. Gr. fem. n. *psammos* sand; Gr. masc. n. *lithos* stone. N.L. neut. adj. *lithicum* of stone; N.L. neut. adj. *psammolithicum* of sandstone.

The taxonomic status of this organism is uncertain because its detailed description is incomplete. Cells are elongated rods 1.1 × 1.7–3.3 μm. Flagella are not observed. Gram-stain-negative. Endospores not formed. Strictly anaerobic. The substrates, which support growth include H_2, formate, methanol, glucose, syringate, pyruvate, lactate, betaine, ethanol, propanol, glycerol, and acetoin. According to 16S rDNA sequence data and substrate specificity, the organism could be affiliated with the genus *Acetobacterium*. However, its morphology is not typical for the genus *Acetobacterium*: the organism was described as Gram-stain-negative, and flagella were not observed. Temperature and pH

characteristics of growth, as well as the G+C content of the DNA are lacking.

Isolated from a subsurface sandstone.

Type strain: CN-E.

Genus III. **Alkalibacter** Garnova, Zhilina, Tourova, Kostrikina and Zavarzin 2005, 983[VP] (Effective publication: Garnova, Zhilina, Tourova, Kostrikina and Zavarzin 2004, 314)

Elena S. Garnova and Tatjana N. Zhilina

Al.ka.li.bac'ter. Ar. def. art. *al* the; Ar. n. *qaliy* ashes of saltwort; N.L. masc. n. *bacter* (from Gr. n. *baktron*) a rod; N.L. masc. n. *Alkalibacter* alkaliphilic rod.

Rod-shaped cells, 0.5 × 1.5–2.5 μm, with **Gram-stain-positive** cell-wall structure. Nonmotile. Asporogenous. **Strictly anaerobic,** catalase negative. **Obligately alkaliphilic and halotolerant**; growth obligately depends on sodium carbonates or chlorides. **Chemoorganoheterotroph**; consumes mono- and disaccharides, sugar alcohols, and proteinaceous substrates as energy sources. Glucose is fermented to acetate, ethanol, formate, H_2, and CO_2. Requires yeast extract as a possible source of nitrogen and sulfur. Habitats are saline-carbonate lakes. The genus is monotypic.

DNA G + C content (mol%): 40.8–42.1.

Type species: **Alkalibacter saccharofermentans** Garnova, Zhilina, Tourova, Kostrikina and Zavarzin 2005, 983[VP] (Effective publication: Garnova, Zhilina, Tourova, Kostrikina and Zavarzin 2004, 315.).

Further descriptive information

Cells are short rods with pointed ends (Figure 148). The cellular morphology varies somewhat with the age of the culture. Cells taken from the exponential growth phase usually occur singly or in pairs (Figure 148); sometimes they are at an angle to each other, or, rarely, in short chains of 3–5 cells. In the late stationary phase, most cells are swollen, irregular in shape, and arranged in chains. Cells divide by septum formation, and the cell wall has a Gram-stain-positive structure (Figure 148). Although sporulation has never been observed, cells are ther-

moresistant to pasteurization at 70°C for 50 min but not at 80°C for 10 min.

Alkalibacter possesses a remarkable adaptation to wide ranges of environmental conditions. Even though the growth optimum is narrowly defined at pH 9.0, 35°C, and 0–4% NaCl; the growth rate remains significant over broad ranges in pH (7.2–10.2), temperature (6–50°C), and salt concentration (0–10% NaCl) (Garnova et al., 2004). Thus, the organism appears to be adapted to a highly unstable environment in a cryoarid climate where seasonal rains and evaporation cause wide fluctuations in salinity and pH. Under optimal growth conditions, the generation time is 5 h.

A catalase-negative, obligate anaerobe, *Alkalibacter* does not grow under aerobic or microaerobic conditions in the presence of >0.9% O_2 in a N_2 gas phase. Growth occurs without reducing agents in the presence of glucose.

The organism is chemoorganoheterotrophic and nonhydrolytic. It utilizes the monosaccharides ribose, xylose, glucose, mannose, and fructose; the disaccharide sucrose; the sugar alcohol mannitol; and peptone and tryptone. Yeast extract is required for growth and cannot be replaced by Casamino acids. A vitamin solution is not required for growth, but it significantly improves the growth rate. Glucose is fermented to acetate, ethanol, formate, H_2, and CO_2. *Alkalibacter* is incapable of the dissimilatory reduction of sulfate, sulfite, thiosulfate, dithionite,

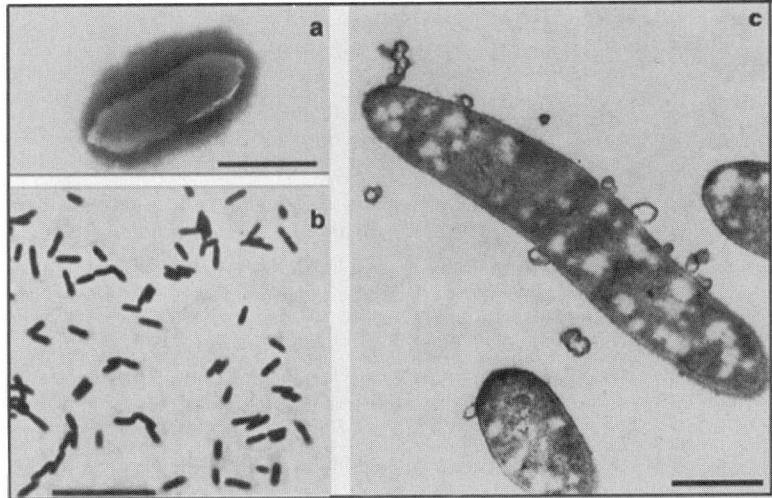

FIGURE 148. Morphology of *Alkalibacter saccharofermentans*. (a) Negatively stained cell. Flagella are not observed in these nonmotile bacteria. Bar = 1 μm. (b) Cells from the exponential growth phase as viewed by phase-contrast microscopy. Bar = 10 μm. (c) Thin-section demonstrates the Gram-positive type of cell wall as well as the septum formed during cell division. Bar = 0.5 μm. (Used with permission from Garnova et al., 2004, *Extremophiles* 8: 309–316.)

NO_3^-, or NO_2^-. NO_2^- completely inhibits the growth while sulfur significantly improves it. Other electron acceptors do not influence the growth rate.

Methods of isolation and cultivation of pure cultures

The type strain was isolated from an anaerobic enrichment on glucose originally inoculated with a mixture of mud and surface cyanobacterial mat from a lagoon of Lake Nizhnee Beloe (southeastern Transbaikal region, Russia). The mineral medium which was used for the enrichment culture contained (per liter): NaCl, 3.4 g; Na_2CO_3, 4.45 g; $NaHCO_3$, 5.5 g; and glucose, 5 g. Incubations were performed anaerobically under N_2 gas at pH 9.5 and 36°C. To obtain a pure culture, the enrichment culture was serially diluted in mineral medium, and colonies were isolated from the highest dilution yielding growth in the same medium containing 3% agar (w/v). The uniform colonial and cellular morphologies confirmed culture purity. After optimization of the growth conditions, the *Alkalibacter* medium contained (per liter): KH_2PO_4, 0.2g; $MgCl_2$, 0.1g; NH_4Cl, 0.5g; KCl, 0.2g; NaCl, 6.0g; Na_2CO_3, 15.5g; $NaHCO_3$, 3.5g; yeast extract, 0.2g; glucose, 5.0g; trace element solution, 1 ml (Kevbrin and Zavarzin, 1992), vitamin solution, 10 ml (Wolin et al., 1963); and resazurin, 0.001g. The pH was 9.0, and the gas phase was N_2.

Maintenance procedure

The type strain remains viable when stored in *Alkalibacter* medium at +4°C and transferred every 6 months. Glycerol (15%) stocks are stored in liquid nitrogen.

Differentiation of the genus *Alkalibacter* from other genera

The genus *Alkalibacter* is differentiated from other members of the *Eubacteriaceae* by its obligate alkaliphily, dependence on Na^+ and ability to grow at high salinities (Table 157). It differs from the homoacetogenic species of the genera *Acetobacterium* and *Eubacterium* in its inability to grow autotrophically and ferment sugars with acetate as the sole product. In addition, they have different carbohydrate specificity. *Eubacterium aggregans* (Mechichi et al., 1998) and *Eubacterium limosum* (Moore and Holdeman Moore, 1986) also produce different products from that of *Alkalibacter* when they ferment carbohydrates during chemoorganotrophic growth. *Alkalibacter* differs from the other chemoorganoheterotrophs, *Anaerofustis stercorihominis* (Finegold et al., 2004), *Eubacterium barkeri* (Moore and Holdeman Moore, 1986), *Eubacterium callanderi* (Mountfort et al., 1988), *Garciella nitratireducens* (Miranda-Tello et al., 2003), and *Pseudoramibacter alactolyticus* (Willems and Collins, 1996) in its substrate specificity and fermentation products (Table 157).

Ecophysiological features of *Alkalibacter* are consistent with its isolation from soda lakes, and it possesses some characteristics that are unusual for the other members of this family, which were isolated from sources at neutral pH. It is the first true alkaliphile in this group.

Taxonomic comments

On the basis of phylogenetic analyses of the 16S rRNA gene, *Alkalibacter* is related to cluster XV of the low G + C content Gram-positive group (Collins et al., 1994; Garnova et al., 2004). According to the 16S rRNA phylogenetic analysis presented in the roadmap to this volume (Figure 5), the genus *Alkalibacter* is a member of the family *Eubacteriaceae*, order *Clostridiales*, class *Clostridia* in the phylum *Firmicutes* (Ludwig et al., 2009). Among the neutrophilic species of the genera *Acetobacterium*, *Anaerofustis*, *Eubacterium*, *Garciella*, and *Pseudoramibacter*, *Alkalibacter* forms (however with low bootstrap value) a separate clade together with the thermophilic thiosulfate and nitrate reductant *Garciella nitratireducens* (Figure 149) and exhibits at least 10% sequence divergence in its 16S rRNA gene from other members of this family. Together with the physiological and genotypic differences, this level of divergence supports its status as a separate genus.

List of species of the genus *Alkalibacter*

1. **Alkalibacter saccharofermentans** Garnova, Zhilina, Tourova, Kostrikina and Zavarzin 2005, 983^VP (Effective publication: Garnova, Zhilina, Tourova, Kostrikina and Zavarzin 2004, 315.) sac.cha.ro.fer.men'tans. Gr. n. *sakkharon* sugar; L. v. *fermentare* to ferment; N.L. part. adj. *saccharofermentans* sugar-fermenting.

The description for the type species is the same as given for the genus and listed in Table 157. Three strains of *Anoxynatronum saccharofermentans*, Z-7980, Z-79820, and Z-7983, have been isolated from an anaerobic microbial community decomposing cellulose (Garnova et al., 2004). While these strains have a uniform cellular morphology in liquid medium, they form colonies that differed in shape, diameter, consistency, and edge on solid medium. All three strains possess a similar mol% G + C content (40.8–42.1) and high levels of DNA–DNA similarity (96–100%) (Tourova et al., 1999). Thus, they belong to the same species, and the morphology of the colonies is not a differentiating characteristic. Strain Z-79820 was described as the type strain.

DNA G + C content (mol%): 42.1 (T_m).

Type strain: Z-79820, DSM 14828, UNIQEM U-218, VKM B-2308.

GenBank accession number (16S rRNA gene): AY312403.

Genus IV. **Anaerofustis** Finegold, Lawson, Vaisanen, Molitoris, Song, Liu and Collins 2004, 1005^VP (Effective publication: Finegold, Lawson, Vaisanen, Molitoris, Song, Liu and Collins 2004, 44.)

PAUL A. LAWSON

An.a.e.ro.fus'tis. Gr. pref. *an* without; Gr. masc. n. *aer* air; L. masc. n. *fustis* stick; N.L. masc. n. *Anaerofustis* stick living without air.

Cells are rod-shaped, stain Gram-positive, and do not form spores. **Strictly anaerobic and catalase-negative.** Resistant to 20% bile. Some carbohydrates are fermented. **End products of** metabolism from glucose peptone-yeast extract broth are **acetic and butyric acids.** Esculin is not hydrolyzed. **Lipase-, lecithinase-, and urease-negative.** Nitrate is not reduced to nitrite.

TABLE 157. Characteristics differentiating *Alkalibacter* and other representatives of the *Eubacteriaceae*[a]

Characteristic	*Alkalibacter saccharofermentans*[b]	*Acetobacterium bakii*[c]	*Acetobacterium carbinolicum*[d]	*Acetobacterium fimetarium*[e]	*Acetobacterium malicum*[f]	*Acetobacterium paludosum*[g]	*Acetobacterium psammolithicum*[h]	*Acetobacterium tundrae*[i]	*Acetobacterium wieringae*[d]	*Acetobacterium woodii*	*Anaerofustis stercorihominis*[j]	*Eubacterium aggregans*[k]	*Eubacterium barkeri*[l]	*Eubacterium callanderi*[m]	*Eubacterium limosum*[f]	*Garciella nitratireducens*[n]	*Pseudoramibacter alactolyticus*[o]
Growth substrate:																	
H$_2$+CO$_2$[p]	–	+	+	+	+	+	+	+	+	+	ND	+	ND	–	+	ND	–
Other C$_1$ compounds	–	+	+	+	+	+	+	+	+	+	ND	+	ND	+	+	–	ND
Alcohols	–	+	+	+	+	+	+	+	+	+	ND	+	+	–	+	–	+
Carboxylic acids	–	+	+	+	+	+	+	+	+	+	ND	–	+	ND	+	–	+
Ribose	+	–	ND	–	ND	–	ND	–	–	–	+	+	+	ND	ND	+	+
Xylose	+	+	ND	–	ND	+	ND	+	–	–	+	+	ND	–	ND	+	+
Glucose	+	+	+	–	–	+	+	+	+	+	+	+	+	+	+	+	+
Fructose	+	+	+	+	+/–	+	–	+	+	+	–/+	+	ND	+	+	+	+
Maltose	–	+	ND	+	ND	+	ND	+	–	–	ND	ND	–	–	–	+	+
Cellobiose	–	–	ND	–	ND	+	ND	+	ND	–	–/+	–	–	–	–	+	+
Mannose	+	ND	ND	ND	ND	ND	ND	+	ND	–	+	–	–	ND	–	+	+
Sucrose	+	–	ND	–	ND	–	ND	–	–	–	–	+	–	–	–	–	+
Products of C$_1$ fermentation	–	A	A	A	A	A	ND	A	A	A	A, B	A, B, F	–	ND	A, B	–	–
Products of carbohydrates fermentation	A, E, F, H$_2$, CO$_2$	A	A, Fa	A	A, Pa, Po	A	ND	A	A	A, S	ND	A, B, F, H$_2$, CO$_2$	B, L, H$_2$, CO$_2$	A, B, F, L, H$_2$	L, H$_2$	A, B, L, H$_2$, CO$_2$	A, B, F, C, H$_2$
pH range, optimum	7.2–10.2, 9.0	5.5–8.5, 6.5	6.0–8.0, 7.0–7.2	6.0–8.5, 7.5	6.0–9.5, 7.5–8.0	5.0–8.0, 7.0	ND	6.0–8.0, 7.0	ND, 7.2–7.8	ND	ND	6.0–9.0, 7.2	ND	6.0–8.5, 7.0	4.5–5.2	5.5–9.0, 7.5	5.0–6.0
NaCl range, %	0–10	ND	ND	ND	ND	ND	ND	ND	ND	ND	ND	1–4	0–6.5	ND	ND	0–10	ND

[a]Symbols: +, >85% positive; d, different strains give different reactions (16–84% positive); –, 0–15% positive; w, weak reaction; ND, not determined. A, acetate; B, butyrate; C, caproate; E, ethanol; F, formate; Fa, fatty acids; L, lactate; Pa, propionate; Po, propanol; S, succinate.

[b]Data from Garnova et al. (2004).

[c]Data from Kotsyurbenko et al. (1995).

[d]Data from Eichler and Schink (1984).

[e]Data from Tanaka and Pfennig (1988).

[f]Data from Krumholz et al. (1999).

[g]Data from Simankova et al. (2000).

[h]Data from Braun and Gottschalk (1982).

[i]Data from Balch et al. (1977).

[j]Data from Finegold et al. (2004).

[k]Data from Mechichi et al. (1998).

[l]Data from Moore and Holdeman Moore (1986).

[m]Data from Mountfort et al. (1988).

[n]Data from Miranda-Tello et al. (2003).

[o]Data from Willems and Collins (1996).

[p]+, Homoacetogens.

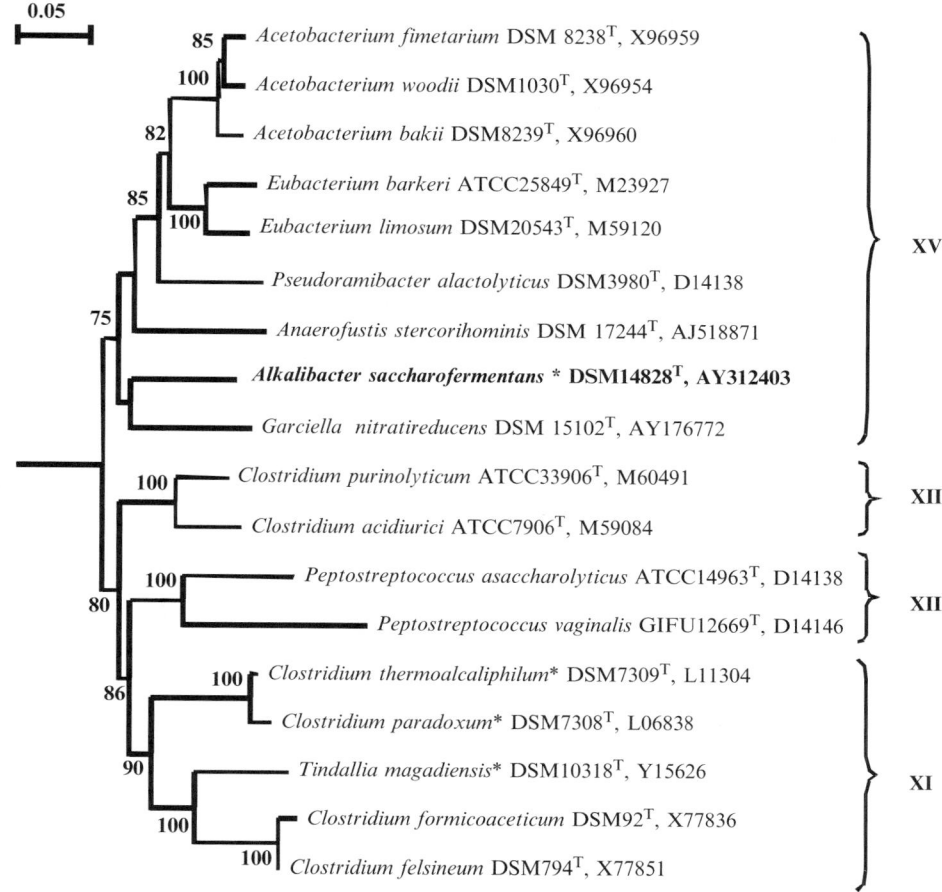

0.05

85 ┌ *Acetobacterium fimetarium* DSM 8238^T, X96959

100 ├ *Acetobacterium woodii* DSM1030^T, X96954

82 └ *Acetobacterium bakii* DSM8239^T, X96960

85 ┌ *Eubacterium barkeri* ATCC25849^T, M23927

100 └ *Eubacterium limosum* DSM20543^T, M59120

Pseudoramibacter alactolyticus DSM3980^T, D14138

75 *Anaerofustis stercorihominis* DSM 17244^T, AJ518871

Alkalibacter saccharofermentans * **DSM14828^T, AY312403**

Garciella nitratireducens DSM 15102^T, AY176772

XV

100 ┌ *Clostridium purinolyticum* ATCC33906^T, M60491

└ *Clostridium acidiurici* ATCC7906^T, M59084

XII

100 ┌ *Peptostreptococcus asaccharolyticus* ATCC14963^T, D14138

80 └ *Peptostreptococcus vaginalis* GIFU12669^T, D14146

XIII

100 ┌ *Clostridium thermoalcaliphilum** DSM7309^T, L11304

86 └ *Clostridium paradoxum** DSM7308^T, L06838

90 *Tindallia magadiensis** DSM10318^T, Y15626

100 ┌ *Clostridium formicoaceticum* DSM92^T, X77836

100 └ *Clostridium felsineum* DSM794^T, X77851

XI

FIGURE 149. Phylogeny of *Alkalibacter saccharofermentans* and representatives of the *Eubacteriaceae* (rRNA cluster XV) and *Clostridiales* (rRNA clusters XI, XII, and XIII). Based upon the nucleotide sequence of the 16S rRNA gene, where the bar corresponds to five substitutions per 100 positions. *Bacillus subtilis* was the outgroup. Bootstrap values >75 (expressed as percentage of 100 replications) are shown at branch points. Asterisks indicate alkaliphiles.

Indole-negative. The predominant long-chain cellular fatty acids consist of a complex mixture of straight-chain saturated and monounsaturated fatty acids and dimethyl acetals (DMAs), with $C_{14:0}$, $C_{16:0}$, $C_{18:1cis9}$, and $C_{18:1cis9}$ DMA predominating.

Isolated from human feces.

DNA G+C content (mol%): 70.

Type species: **Anaerofustis stercorihominis** Finegold, Lawson, Vaisanen, Molitoris, Song, Liu and Collins 2004, 1005^{VP} (Effective publication: Finegold, Lawson, Vaisanen, Molitoris, Song, Liu and Collins 2004, 44.)

Further descriptive information

The genus contains only one species, *Anaerofustis stercorihominis*, and therefore the characteristics provided below refer to this species. The organism was recovered from human feces. The colonies are 1.5 mm in diameter, gray, circular, raised, smooth, with entire edges after 48 h of anaerobic incubation at 37°C. The colonies display a light yellow fluorescence under long-wave UV light. Resistant to 20% bile. Strictly anaerobic and catalase-negative. Acetic acid is produced in peptone–yeast extract broth, but in peptone–yeast extract glucose broth, acetic and butyric acids are the end products of metabolism. Grows in peptone–yeast extract broth supplemented with 1% of fructose, glucose, mannitol,

ribose, salicin, sucrose, and xylose but grows poorly with lactose, maltose, and mannose. Fructose is fermented with production of acid and gas; weak reactions occur with glucose and xylose. Lecithinase-, lipase-, and urease-negative. Cells are indole-negative, and nitrate is not reduced to nitrite. Using the commercially available API ZYM kit, weak positive reactions are obtained for esterase C4, esterase lipase C8, acid phosphatase, and naphthol-AS-BI-phosphohydrolase. Alkaline phosphatase, lipase C14, leucine arylamidase, valine arylamidase, cystine arylamidase, trypsin, α-chymotrypsin, α- and β-galactosidase, β-glucuronidase, α- and β-glucosidase, *N*-acetyl-β-glucosaminidase, α-mannosidase or α-fucosidase are not detected. The predominant long-chain cellular fatty acids are $C_{14:0}$, $C_{16:0}$, $C_{18:1\ cis9}$, and $C_{18:1\ cis9\ DMA}$, together with other minor straight-chain saturated and monounsaturated fatty acids and DMAs. Resistant to colistin sulfate (10 mg) but sensitive to vancomycin (5 mg) and kanamycin (1000 mg) on the basis of disk diffusion testing.

Isolation procedures

Anaerofustis stercorihominis was recovered from a stool specimen of a 5-year 11-month-old boy with late onset autism before treatment with vancomycin. The entire stool specimen was homogenized using a sterile stainless steel blender with one to three volumes of peptone (0.05%) added as diluent. An aliquot of the specimen (approximately 1 g) was used to make serial ten-fold dilutions in Pre Reduced Anaerobically Sterilized (PRAS) dilution blanks (Anaerobe Systems, Morgan Hill, CA, USA). Various dilutions (100 μl) were plated onto Brucella blood agar (BAP, Anaerobe Systems) and incubated anaerobically at 37°C under a gas phase comprising N_2, CO_2, and H_2 (86:7:7% by vol.).

Maintenance procedures

Strains grow well on Brucella agar base supplemented with 5% horse blood at 37°C under anaerobic conditions. For long-term preservation, strains can be maintained in the same medium containing 15–20% glycerol at –70°C or lyophilized.

Taxonomic comments

The genus *Anaerofustis* was proposed in 2004 to accommodate a phylogenetically distinct Gram-stain-positive, strictly anaerobic, catalase-negative, rod-shaped organism originating from human feces (Finegold et al., 2004). According to the 16S rRNA phylogenetic analysis presented in the roadmap to this volume (Figure 5), the genus *Anaerofustis* is a member of the family *Eubacteriaceae*, order *Clostridiales*, class *Clostridia* in the phylum *Firmicutes*.

Differentiation of the genus *Anaerofustis* from other genera

In addition to differentiation by 16S rRNA gene sequence analysis, *Anaerofustis stercorihominis* can be readily distinguished on the basis of biochemical, chemotaxonomic, and morphological criteria. For example, it may be differentiated from the genera *Acetobacterium* and *Eubacterium sensu stricto* by its end products of glucose fermentation. It produces acetic and butyric acids whereas *Acetobacterium* produces only acetic acid, and *Eubacterium sensu stricto* forms acetic, butyric, and lactic acids (some species may, in addition, produce formic acid) (Collins et al., 1994). Similarly the unidentified bacterium can be distinguished from *Pseudoramibacter alactolyticus* by its cellular morphology and resistance to 20% bile. *Pseudoramibacter alactolyticus* also produces rod-shaped cells that occur in pairs, resembling flying birds or Chinese characters. In addition, *Pseudoramibacter alactolyticus* produces formic, acetic, butyric, and caproic acids as end products of glucose metabolism (Moore and Holdeman Moore, 1986; Willems and Collins, 1996).

List of species of the genus *Anaerofustis*

1. **Anaerofustis stercorihominis** Finegold, Lawson, Vaisanen, Molitoris, Song, Liu and Collins 2004, 1005^{VP} (Effective publication: Finegold, Lawson, Vaisanen, Molitoris, Song, Liu and Collins 2004, 44.)

   ster.co.ri.ho′mi.nis. L. neut. n. *stercus -oris* dung, excrements, ordure; L. gen. n. *hominis* of human; N.L. gen. n. *stercorihominis* of human feces.

   Cells stain Gram-positive and consist of thin rods (0.5 × 1–3 μm). Colonies after 48 hours of anaerobic incubation at 37°C are 1.5 mm in diameter, gray, circular, raised, smooth, with entire edges. Fluoresces a light yellow under UV light. Acetic acid is produced in peptone-yeast extract broth, but in peptone-yeast extract glucose broth, acetic and butyric acids are the end products of metabolism. Resistant to 20% bile. Grows in peptone-yeast extract broth supplemented with 1% of fructose, glucose, mannitol, ribose, salicin, sucrose, and xylose but grows poorly with lactose, maltose, and mannose.

   Isolated from human feces. Habitat is not known, but this organism probably is a member of the human gut microflora. Other characteristics are as given for the genus.

   DNA G+C content (mol%): 70.0 (HPLC).
   Type strain: ATCC BAA-858, CCUG 47767, WAL 14563.
   GenBank accession number (16S rRNA gene): AJ518871.

Genus V. **Garciella** Miranda-Tello, Fardeau, Sepulveda, Fernandez, Cayol, Thomas and Ollivier 2003, 1512^{VP}

Elizabeth Miranda-Tello, Marie-Laure Fardeau, Bernard Ollivier and Didier Alazard

Gar.ci.el′la. L. dim. ending -*ella*; N.L. fem. n. *Garciella* named in honor of the French microbiologist Jean-Louis Garcia for his important contribution to the taxonomy of anaerobes.

Cells are **straight rods**, 0.5–0.7 μm × 1.4–2.8 μm, occurring singly or in pairs. Motile by means of one subpolar flagellum. **Terminal spores** appear in old cultures. **Gram-stain-positive. Obligately anaerobic. Moderately thermophilic and halotolerant.** The optimum growth temperature is 55°C, optimum NaCl concentration is 1%, and optimum pH is 7.5. **Chemo-organotrophic growth.** Ferments sugars. Lactate, acetate, butyrate, H_2, and CO_2 are products of glucose metabolism. **Thiosulfate is reduced to sulfide, and nitrate is reduced to ammonium.** Elemental sulfur, sulfite, sulfate, fumarate, and nitrite are not used as electron acceptors.

DNA G+C content (mol%): 30.9.

Type species: **Garciella nitratireducens** Miranda-Tello, Fardeau, Sepulveda, Fernandez, Cayol, Thomas and Ollivier 2003, 1512^{VP}.

Further descriptive information

Round, gray colonies (2 mm in diameter) develop in roll tubes under anaerobic conditions after 1 week of incubation at 40°C. Cells are straight rods, 0.5–0.7 μm × 1.4–2.8 μm (Figure 150) and motile by a subpolar flagellum. When grown under optimal conditions in reduced medium in the presence of air, cells form irregular gas vacuoles (Figure 150). Spores are formed in stationary phase cultures or after thermal stress (Figure 150).

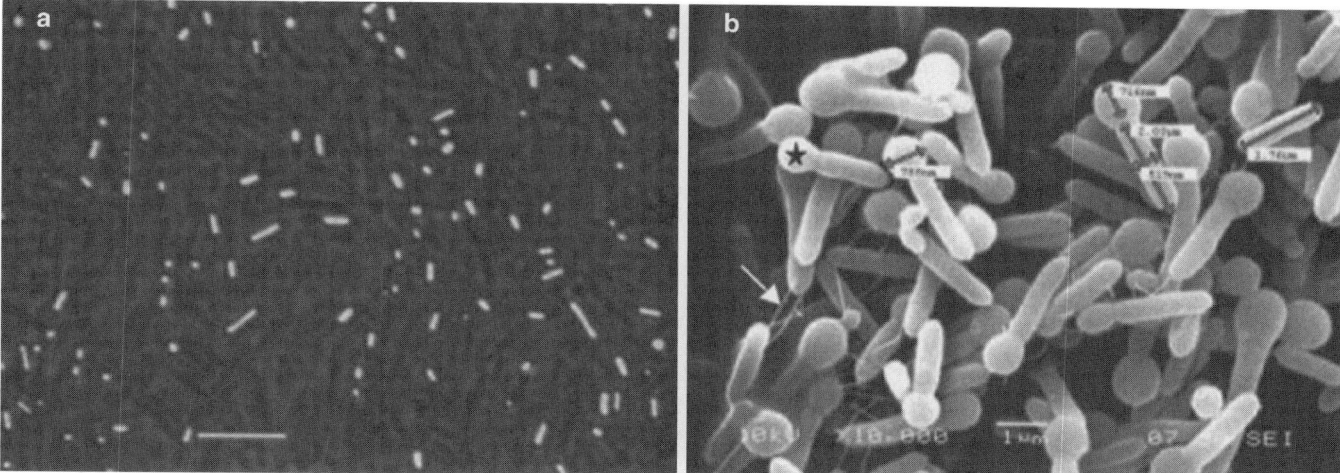

FIGURE 150. Cellular morphology of *Garciella nitratireducens*. (a) Phase-contrast micrograph of cells grown in optimal conditions in reduced medium in the presence of air. Intracellular gas vacuoles are the bright bodies within some cells. Bar = 2 μm. (b) Scanning electron microscopy of cells after thermal stress, showing spherical, terminal spores, swollen sporangia (black star) and a subpolar flagellum (arrow). Bar = 1 μm. (Reprinted with permission from E. Miranda-Tello et al., 2003. *International Journal of Systematic and Evolutionary Microbiology, 53*: 1509; © International Union of Microbiological Societies.)

Garciella nitratireducens ferments cellobiose, fructose, galactose, glucose, lactose, maltose, mannose, ribose, sucrose, D-xylose, glycerol, mannitol, fumarate, lactate, malate, pyruvate, and Casamino acids. Compounds not used as energy sources include arabinose, raffinose, rhamnose, starch, cellulose, xylan, acetate, butyrate, propionate, methanol, ethanol, butanol, propanol, formate, methylamine, *p*-coumaric acid, ferulic acid, peptone, bio-Trypticase, and H_2. Thiosulfate or nitrate enhances growth. Yeast extract is required for growth and cannot be replaced by vitamins. *Garciella nitratireducens* can grow in the presence of 25 μg/ml kanamycin or chloramphenicol, 100 μg/ml ampicillin, but not in the presence of 10 μg/ml vancomycin or rifampicin.

Enrichment and isolation procedures

Garciella nitratireducens was originally enriched under anaerobic conditions from production fluids treated in an oil–water separation tank in the Gulf of Mexico using a basal medium as described by Miranda-Tello et al. (2003). Enrichments were serially diluted to extinction three times before isolation. Strains were isolated by the Hungate roll tube technique with medium solidified with 1.6% agar (Hungate, 1969). Serial dilution in roll tubes was performed twice in order to purify the culture.

Maintenance procedures

For long-term preservation, cells can be stored at −80°C in the basal medium plus 20% glycerol.

Differentiation of the genus *Garciella* from other genera

Based on 16S rRNA gene sequence analyses, *Garciella nitratireducens* is closely related to *Caloranaerobacter azorensis* (Wery et al., 2001), *Alkalibacter saccharofermentans* (Garnova et al., 2004), and *Thermohalobacter berrensis* (Cayol et al., 2000). Characteristics differentiating *Garciella* from these genera belonging to cluster XII (*Caloranaerobacter* and *Thermohalobacter*) and cluster XV (*Alkalibacter*) within the order *Clostridiales* are presented in Table 158. *Garciella* differs in the site of isolation, cell-wall structure, and spore formation. *Caloranaerobacter* was isolated from a deep-sea hydrothermal vent in the Pacific Ocean, *Alkalibacter* was isolated from an alkaline lake with low salt concentration in Russia, whereas *Thermohalobacter* was isolated from sediments of a solar saltern in France. *Garciella* also possesses a thick, homogeneous S-layer not seen in the other genera. In contrast to *Garciella*, neither *Caloranaerobacter*, *Alkalibacter*, nor *Thermohalobacter* form endospores. *Garciella* may be also distinguished by its ability to reduce thiosulfate into sulfide, which appears to be a relatively common metabolic feature for heterotrophic anaerobes isolated from oilfield ecosystems (Ollivier and Cayol, 2005).

Taxonomic comments

The closest relatives of *Garciella* are *Caloranaerobacter*, *Alkalibacter*, and *Thermohalobacter*, with 16S rRNA gene sequence similarities of 88.7%, 88.3%, and 86.1%, respectively. They all belong to the *Clostridium* subphylum of the order *Clostridiales* (Figure 151). However, the genus *Garciella* does not possess high similarity to any of the type genera of the named families within this order. Therefore, its classification is uncertain. Detailed phylogenetic analyses presented in the roadmap to this volume of the *Clostridiales* suggest that *Garciella* may represent a deep branch associated with the *Eubacteriaceae*, while *Caloranaerobacter* and *Thermohalobacter* represent an independent lineage (Figure 5; Ludwig et al., 2009).

List of species of the genus *Garciella*

1. **Garciella nitratireducens** Miranda-Tello, Fardeau, Sepulveda, Fernandez, Cayol, Thomas and Ollivier 2003, 1509[VP].

ni.tra.ti.re.du'cens. N.L. n. *nitratum* nitrate; L. v. *reduco* to draw backwards, bring back to a state or condition; N.L. part. adj. *nitratireducens* nitrate-reducing.

The characteristics are as described for the genus. The type strain was isolated from a water separator collecting fluids produced from different oil wells located in SAMIII oilfield, Gulf of Mexico.

DNA G+C content (mol%): 30.9 (HPLC).
Type strain: MET79, DSM 15102, CIP 107615.
GenBank accession number (16S rRNA gene): AY176772.

Genus VI. **Pseudoramibacter** Willems and Collins 1996, 1086[VP]

ANNE WILLEMS AND MATTHEW D. COLLINS

Pseu.do.ra.mi.bac'ter. Gr. adj. *pseudes* false; L. masc. n. *ramus* a branch; N.L. masc. n. *bacter* the equivalent of Gr. neut. n. *baktron* rod, staff; N.L. masc. n. *Pseudoramibacter* false branching rod.

Strictly anaerobic, catalase-negative, Gram-stain-positive, nonsporeforming rods. Nonmotile. Cells occur in pairs resembling flying birds, short-chains, clumps, or Chinese characters. Growth is stimulated by fermentable carbohydrates. **Fermentation end products are formate, acetate, butyrate, caproate and hydrogen gas. The cell-wall murein is of type A**, containing *meso*-diaminopimelic acid as the dibasic acid. *Pseudoramibacter* belongs to cluster XV of the *Clostridium* subphylum of the Gram-positive bacteria (Collins et al., 1994) together with *Acetobacterium* and *Eubacterium*.

Type species: **Pseudoramibacter alactolyticus** (Prévot and Taffanel 1942) Willems and Collins 1996, 1086[VP].

Further descriptive information

Because the genus contains only one species, all the characteristics given for the genus also describe the species *Pseudoramibacter alactolyticus*.

Phylogenetic treatment. On the basis of 16S rDNA phylogeny, *Pseudoramibacter* is closely related to the genus *Eubacterium sensu stricto* (approx. 91% sequence similarity) and the genus *Acetobacterium* (approx. 90% sequence similarity) in the *Clostridium* subphylum of the phylum *Firmicutes* (Willems and Collins, 1996).

Cell morphology. Cells grown in peptone-yeast extract-glucose (PYG) broth* (Holdeman et al., 1977) are rods 0.3–0.6 by 1.6–7.5 μm. They occur in pairs resembling flying birds, short-chains, clumps or Chinese characters (Moore and Holdeman Moore, 1986); Y-shaped forms give appearance of false branching.

Cell-wall composition. *Pseudoramibacter* has Gram-stain-positive type cell walls. Cell-wall murein is directly cross-linked with *meso*-diaminopimelic acid as the dibasic acid (Andreesen, 1992; Severin et al., 1989) (type A1γ, nomenclature of Schleifer and Kandler, 1972). This is a significant difference with its closest phylogenetic relatives *Eubacterium sensu stricto* and *Acetobacterium* which both have mureins of type B (Andreesen, 1992; Schink and Bomar, 1992; Tanner et al., 1981).

Colonial characteristics. After 2–3 d of incubation on horse blood agar, colonies are punctate to 0.5 mm in diameter, round with an entire margin, convex to pulvinate, smooth and shiny. Smaller colonies can be translucent and larger colonies are often opaque (Moore and Holdeman Moore, 1986). Deep agar colonies are lenticular.

Nutrition and growth conditions. Optimal growth temperature is 35–37°C. Most strains grow at 30°C and some grow at 25 and 45°C. Optimal pH is between 6 and 8. Growth is stimulated by fermentable carbohydrates, but not by 0.02% Tween 80 or 5% rumen fluid and is inhibited by 20% bile. Glucose broth cultures are usually turbid with formation of granular or smooth (sometimes ropy or flocculent) sediment; occasionally sediment formation occurs without turbidity in broth cultures. The terminal pH in glucose cultures is 5.0–5.6 or, occasionally, 5.8–6.0 (Moore and Holdeman Moore, 1986).

Metabolism and metabolic pathways. *Pseudoramibacter* has a strictly anaerobic, fermentative metabolism. End products produced from PYG broth are (in milliequivalents/100 ml of culture) formate, 0.01–1.0; acetate, 0.01–0.5; butyrate, 0.1–0.5; and caproate, 0.2–2.0. Abundant hydrogen gas is produced and small amounts of caprylate may also be produced. Pyruvate is converted to acetate and formate. Lactate is not used. Threonine is not converted to propionate. In the presence of glucose and propionic acid, valeric and heptanoic acids are produced (Moore and Holdeman Moore, 1986).

Acid (pH <5.5) is produced from fructose and glucose (but weak by some strains; (Holdeman et al., 1977), but not from amygdalin, arabinose, cellobiose, dulcitol, galactose, glycerol, glycogen, inositol, inulin, lactose, maltose, mannose, melezitose (weak by some strains; Holdeman al., 1977), melibiose, raffinose, rhamnose, ribose, salicin, sorbose, starch, sucrose, trehalose and xylose. Some strains produce acid from amygdalin, mannitol or sorbitol (pH <5.5) or from erythritol or esculin (weak reaction). Acid production from adonitol is variable (Holdeman et al., 1977; Moore and Holdeman Moore, 1986).

Resazurin is usually not reduced. The following characteristics are negative: catalase, lecithinase, lipase, hemolysis,

---

*Composition of PYG: 0.5 g peptone, 0.5 g trypticase, 1 g yeast extract, 0.4 ml resazurin solution (25 mg in 100 ml of distilled water), 4 ml salts solution (0.2 g CaCl₂, 0.2 g MgSO₄, 1 g K₂HPO₄, 1 g K₂HPO₄, 10 g NaHCO₃, 2 g NaCl; dissolve CaCl₂, and MgSO₄ in 300 ml distilled water; add 500 ml water and while swirling, slowly add remaining salts, when dissolved, add 200 ml distilled water and mix; store at 4°C), 100 ml distilled water, 1 ml hemin solution (dissolve 50 mg hemin in 1 ml 1 N NaOH; make to 100 ml with distilled water, autoclave for 15 min at 121°C), 0.02 ml vitamin K1 solution (Dissolve 0.15 ml of vitamin K1 in 30 ml of 95% ethanol; keep for max. 1 month, at 4°C in a brown bottle), 0.05 g cysteine-HCl H₂O, 1 g glucose; vitamin K1, hemin and cysteine are added after the medium is boiled, but before dispensing and autoclaving (Holdeman et al., 1977).

TABLE 158. Major characteristics that discriminate *Garciella nitratireducens* from *Caloranaerobacter azorensis*, *Alkalibacter saccharofermentans*, and *Thermohalobacter berrensis*[a]

Characteristic	*Garciella nitratireducens*[b]	*Caloranaerobacter azorensis*[c]	*Alkalibacter saccharofermentans*[d]	*Thermohalobacter berrensis*[e]
Motility	Polar flagella	ND	–	Lateral flagella
Sporulation	+	–	–	
Gram reaction	+	–	–	+
Temp. range (°C)	25–60	45–65	6–50	45–70
Optimum temp. (°C)	55	65	35	65
NaCl range (%)	0–10	0.6–6.5	0–10	0–15
Optimum NaCl (%)	1	2	0–4	5
G + C content (mol %)	30.9	26–28	42.1	33
Reduction of S^0	–	+	ND	–
Reduction of $S_2O_3^{2-}$	+	–	–	–
Reduction of NO_3	+	–	–	ND

[a]Symbols: +, positive; –, negative; ND, not determined.
[b]Data from Miranda-Tello et al. (2003).
[c]Data from Wery et al. (2001).
[d]Data from Garnova et al. (2004).
[e]Data from Cayol et al. (2000).

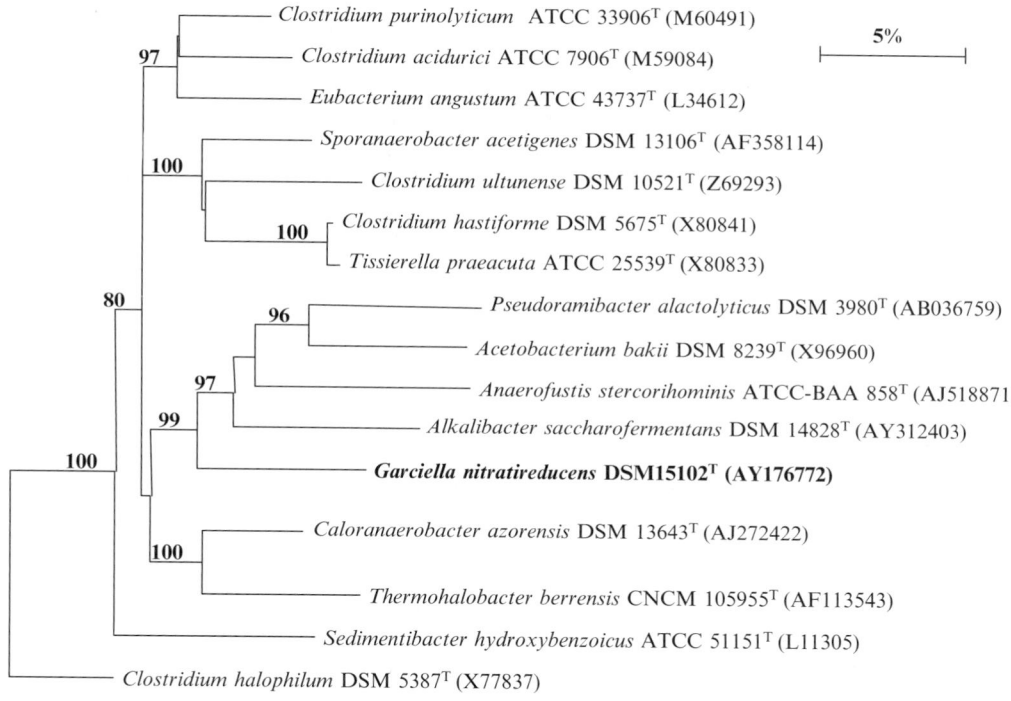

FIGURE 151. Phylogeny of *Garciella nitratireducens* and closely related members of *Clostridiales* based upon the 16S rRNA gene. A neighbor-joining dendrogram is shown with *Clostridium halophilum* as the outgroup. Bootstrap values from 100 replicates are shown at the nodes. Only values greater than 95 are reported. The bar represents 5 inferred substitutions per 100 positions.

hydrolysis of hippurate, esculin and starch, H_2S production in sulfide-indole motility medium, production of indole and acetylmethylcarbinol, digestion of meat and gelatin, and nitrate reduction. Milk is not coagulated. Some strains can grow in bile medium and some strains produce ammonia (Holdeman et al., 1977; Moore and Holdeman Moore, 1986).

Antibiotic sensitivity. Twenty-five strains of *Pseudoramibacter alactolyticus*, including the type strain, were susceptible to chloramphenicol (12 μg/ml), clindamycin (1.6 μg/ml), erythromycin (3 μg/ml), penicillin G (2 U/ml) and tetracycline (6 μg/ml) (Moore and Holdeman Moore, 1986).

Ecology. *Pseudoramibacter alactolyticus* is a member of the normal mouth flora (Holdeman et al., 1977) and has also been identified among isolates associated with periodontitis in young adults (Moore et al., 1985; Moore et al., 1982) where *Pseudoramibacter alactolyticus* was identified in 57% of samples taken from diseased sites (Moore et al., 1985). It occurs in gingival crevices and periodontal pockets of humans with periodontal disease

(Moore and Holdeman Moore, 1986). It has been identified from infected dental root canals (Piovano, 1999; Siqueira and Rocas, 2004; Sundqvist, 1992), cases of advanced caries (Chhour et al., 2005)and in microflora of nifedipine-induced gingival overgrowth in patients with cardiovascular disorders (Nakou et al., 1998). Like several *Eubacterium* species, *Pseudoramibacter alactolyticus* is able to coaggregate with *Fusobacterium nucleatum*, another member of the human oral microflora often associated with infections. The ability to coaggregate may be important in the infection process (George and Falkler, 1992). *Pseudoramibacter alactolyticus* has also been identified in the subgingival microflora of healthy children with primary dentition (4–5-year-olds) and mixed dentition (7–8-year-olds) where it is associated with non-bleeding sites (Kamma et al., 2000b, a). It was also detected among isolates from odontogenic infections, infections associated with dental implants or saliva from healthy subjects (Downes et al., 2001).

Pseudoramibacter also occurs in various infected materials and abscesses: purulent pleurisy, jugal cellulitis, postoperative wounds, abscesses of the brain, lung, intestinal tract and mouth (Moore and Holdeman Moore, 1986).

Enrichment and isolation procedures

No specific enrichment and isolation procedures selecting for *Pseudoramibacter* have been described. These bacteria can be isolated from the gingival crevices and periodontal pockets of humans suffering from periodontitis by using standard procedures for the isolation of anaerobic bacteria (Holdeman et al., 1977).

Maintenance procedures

Cultures can be maintained using prereduced media and standard procedures for culturing anaerobic bacteria (Holdeman et al., 1977). Strains can be lyophilized for long-term preservation.

Differentiation from other Gram-stain-positive, anaerobic, nonsporeforming bacilli

Pseudoramibacter can be differentiated from its nearest phylogenetic neighbors, *Acetobacterium* and *Eubacterium sensu stricto* (comprising *Eubacterium barkeri*, *Eubacterium callanderi* and

Eubacterium limosum) by its high G+C content, its murein type, fermentation end products and its inability for autotrophic growth with H_2 and CO_2 (Table 159). Its presence in cases of advanced caries (Chhour et al., 2005) and periradicular lesions of root-filled teeth (Siqueira and Rocas, 2004) has been demonstrated with PCR and sequence analysis of 16S rRNA genes. A nested PCR test specifically targeting *Pseudoramibacter ramibacter* 16S rDNA has been used to detect the species in samples from endodontic infections (Siqueira and Rocas, 2003).

Taxonomic comments

Ramibacterium alactolyticum was described in 1942 for anaerobic, Gram-stain-positive, nonmotile, pleiomorphic, nonsporeforming rods (Prévot and Taffanel, 1942). Later two further species, *Ramibacterium pleuriticum* and *Ramibacterium dentium* were described for similar organisms (Prévot et al., 1947; Vinzent and Reynes, 1947). None of the strains used for the original descriptions of these species were preserved, but in a study of the other named strains of these species available from the Institute Pasteur, Holdeman et al. (1967) showed *Ramibacterium alactolyticum*, *Ramibacterium pleuriticum* and *Ramibacterium dentium* to form a single group. They were therefore regarded as objective synonyms with the oldest name, *Ramibacterium alactolyticum*, having priority. Strain DO-4 (= VPI 1416) was designated neotype strain (Holdeman et al., 1967).

Ramibacterium alactolyticus was later transferred to the genus *Eubacterium* as *Eubacterium alactolyticum* (Holdeman and Moore, 1970). On the basis of 16S rRNA gene sequences it was confirmed that *Eubacterium alatcolyticum* was indeed phylogenetically close to the type species *Eubacterium limosum* and also to *Clostridium barkeri* and *Acetobacterium woodii* (Weizenegger et al., 1992) and along with these organisms belonged to cluster XV of the *Clostridium* subphylum of the Gram-positive bacteria (Collins et al., 1994). Analysis of additional *Acetobacterium* species and *Eubacterium callanderi*, showed that *Eubacterium alactolyticum* in fact occupies a separate position in cluster XV, and therefore it was removed from the genus *Eubacterium* and transferred to a new genus *Pseudoramibacter* as *Pseudoramibacter alactolyticus* (Willems and Collins, 1996). According to the 16S rRNA

TABLE 159. Differentiation of *Pseudoramibacter* from its nearest phylogenetic neighbors and some other genera Gram-stain-negative, nonsporeforming rods

Genus	*Acetobacterium*[a]	*Eubacterium sensu stricto*[b]	*Pseudoramibacter*[c]	*Lactobacillus*[d]	*Carnobacterium*[d]
H_2-CO_2 autotrophic growth	+	v[e]	−		
Murein[f]	Type B	Type B2α	Type A1γ	Type A (Lys or Dmp or Orn)	Type A1γ
Fermentation end products from glucose	Acetate	Acetate, butyrate, lactate, formate[g], hydrogen	Formate, acetate, butyrate, caproate, hydrogen	Lactate (homofermenters) or lactate, acetate, ethanol and CO_2 (heterofermenters)	Lactate, acetate, ethanol and CO_2
DNA G+C content (mol%)	39–46	45–50	61	32–55	33–37

[a]Data from Kotsyurbenko et al. (1995), Schink and Bomar (1992) and Tanner (1986).
[b]Data from Andreesen (1992), Cato et al. (1986), Moore and Holdeman Moore (1986), Mountfort et al. (1988) and Tanner et al. (1981).
[c]Data from Andreesen (1992), Moore and Holdeman Moore (1986) and Nakazawa and Hoshino (1994).
[d]Data from Hammes et al. (1992).
[e]v, variable: *Eubacterium limosum* is positive, *Eubacterium barkeri* and *Eubacterium callanderi* are negative.
[f]Types as defined by Schleifer and Kandler (1972).
[g]Only produced by *Eubacterium callanderi*.

phylogenetic analysis presented in the roadmap to this volume (Figure 5), the genus *Pseudoramibacter* is a member of the family *Eubacteriaceae*, order *Clostridiales*, class *Clostridia* in the phylum *Firmicutes*.

Acknowledgements

A.W. is indebted to the Fund for Scientific Research – Flanders for a position as post-doctoral research fellow.

List of species of the genus *Pseudoramibacter*

1. **Pseudoramibacter alactolyticus** (Prévot and Taffanel 1942) Willems and Collins 1996, 1086[VP] (*Eubacterium alactolyticum* Prévot and Taffanel 1942) Holdeman and Moore 1970, 23[AL]; *Ramibacterium alactolyticum* Prévot and Taffanel 1942, 261)

a.lac.to.ly′ti.cus. Gr. pref. *a* not; L. n. *lac, lactis* milk; N.L. adj. *lyticus* dissolving; N.L. masc. adj. *alactolyticus* not milk-digesting.

Characteristics of the species are the same as those given for the genus.

Isolated from dental callus and the gingival crevices of patients with periodontal disease, from dental root canals, and from patients with various infections, including purulent pleurisy, jugal cellulitis, postoperative wounds, and abscesses of the brain, lung, intestinal tract, and mouth.

The type strain is ATCC 23263 and originated from purulent pleurisy. Its characteristics are the same as those of the species. This strain does not ferment adonitol, dextrin, dulcitol, galactose, glycerol, inulin, or sorbose. It does not produce ammonia from peptone, arginine, or threonine.

DNA G+C content (mol%): 61 (HPLC; Nakazawa and Hoshino, 1994).

EMBL/GenBank accession number (16S rRNA gene) for the type strain: AB036759.

References

Andreesen, J.R. 1992. The genus *Eubacterium*. *In* Balows, Trüper, Dworkin, Harder and Schleifer (Editors), The Prokaryotes. A Handbook on the Biology of Bacteria: Ecophysiology, Isolation, Identification, Applications, 2nd edn. Springer-Verlag, New York, pp. 1914–1924.

Aufurch, S., H. Schagger and V. Müller. 2000. Identification of subunits a, b, and c_1 from *Acetobacterium woodii* Na$^+$ -F_1 F_0-ATPase subunits c_1, c_2, and c_3 constitute a mixed c-oligomer. J. Biol. Chem. *275*: 33397–33301.

Bache, R. and N. Pfennig. 1981. Selective isolation of *Acetobacterium woodii* on methoxylated aromatic acids and determination of growth yields. Arch. Microbiol. *130*: 255–261.

Balch, W.E., S. Schoberth, R.S. Tanner and R.S. Wolfe. 1977. *Acetobacterium*, a new genus of hydrogen-oxidizing, carbon dioxide-reducing, anaerobic bacteria. Int. J. Syst. Bacteriol. *27*: 355–361.

Barker, H.A. and V. Haas. 1944. Butyribacterium, a new genus of gram-positive, non-sporulating bacteria of intestinal origin. J. Bacteriol. *47*: 301–305.

Bergey, D.H., F.C. Harrison, R.S. Breed, B.W. Hammer and F.M. Huntoon. 1923. Bergey's Manual of Determinative Bacteriology, 1st edn. The Williams & Wilkins Co., Baltimore.

Beuscher, H.U. and J.R. Andreesen. 1984. *Eubacterium angustum* sp. nov., a Gram positive anaerobic, non-sporeforming, obligate purine fermenting organism. Arch. Microbiol. *140*: 2–8.

Beuscher, H.U. and J.R. Andreesen. 1985. *In* Validation of the publication of new names and new combinations previously effectively published outside the IJSB. List no. 19. Int. J. Syst. Bacteriol. *35*: 535.

Braun, M. and G. Gottschalk. 1982. *Acetobacterium wieringae* sp. nov., a new species producing acetic acid from molecular hydrogen and carbon dioxide. Zentbl. Bakteriol. Mikrobiol. Hyg. I Abt. Orig. C *3*: 368–376.

Braun, M. and G. Gottschalk. 1983. *In* Validation of the publication of new names and new combinations previously effectively published outside the IJSB. List no. 10. Int. J. Syst. Bacteriol. *33*: 438–440.

Brioukhanov, A.L., R.K. Thauer and A. Netrusov. 2002. Catalase and superoxide dismutase in the cells of strictly anaerobic microorga-nisms. Microbiology (En. transl. from Mikrobiologiya) *71*: 281–285.

Bryant, M.P., N. Small, C. Bouma and I.M. Robinson. 1958. Characteristics of ruminal anaerobic celluloytic cocci and Cillobacterium cellulosolvens n. sp. J. Bacteriol. *76*: 529–537.

Bryant, M.P. 1959. Bacterial species of the rumen. Bacteriol. Rev. *23*: 125–153.

Bryant, M.P. and I.M. Robinson. 1962. Some nutritional characteristics of predominant culturable ruminal bacteria. J. Bacteriol. *84*: 605–614.

Bryant, M.P. and I.M. Robinson. 1963. Apparent incorporation of ammonia and amino acid carbon during growth of selected species of ruminal bacteria. J. Dairy Sci. *46*: 150–154.

Buschhorn, H., P. Durre and G. Gottschalk. 1989. Production and utilization of ethanol by the homoacetogen *Acetobacterium woodii*. Appl. Environ. Microbiol. *55*: 1835–1840.

Caldwell, D.R. and R.F. Hudson. 1974. Sodium, an obligate growth requirement for predominant rumen bacteria. Appl. Microbiol. *27*: 549–552.

Cato, E.P., C.W. Salmon and L.V. Holdeman. 1974. *Eubacterium cylindroides* (Rocchi) Holdeman and Moore: emended description and designation of neotype strain. Int. J. Syst. Bacteriol. *24*: 256–259.

Cato, E.P., L.V. Holdeman and W.E.C. Moore. 1981. Designation of *Eubacterium limosum* (Eggerth) Prévot as the type species of *Eubacterium*. Request for an Opinion. Int. J. Syst. Bacteriol. *31*: 209–210.

Cato, E.P., L.V.H. Moore and W.E.C. Moore. 1985. *Fusobacterium alocis* sp. nov. and *Fusobacterium sulci* sp. nov. from the human gingival sulcus. Int. J. Syst. Bacteriol. *35*: 475–477.

Cato, E.P., W.L. George and S.M. Finegold. 1986. Genus *Clostridium* Prazmowski 1880, 23[AL]. *In* Sneath, Mair, Sharpe and Holt (Editors), Bergey's Manual of Systematic Bacteriology, Vol. 2. The Williams & Wilkins Co., Baltimore, pp. 1141–1200.

Cayol, J.L., S. Ducerf, B.K. Patel, J.L. Garcia, P. Thomas and B. Ollivier. 2000. *Thermohalobacter berrensis* gen. nov., sp. nov., a thermophilic, strictly halophilic bacterium from a solar saltern. Int. J. Syst. Evol. Microbiol. *50*: 559–564.

Cheeseman, S.L., S.J. Hiom, A.J. Weightman and W.G. Wade. 1996. Phylogeny of oral asaccharolytic *Eubacterium* species determined by 16S ribosomal DNA sequence comparison and proposal of *Eubacterium infirmum* sp. nov. and *Eubacterium tardum* sp. nov. Int. J. Syst. Bacteriol. *46*: 957–959.

Chhour, K.L., M.A. Nadkarni, R. Byun, F.E. Martin, N.A. Jacques and N. Hunter. 2005. Molecular analysis of microbial diversity in advanced caries. J. Clin. Microbiol. *43*: 843–849.

Collins, M.D., P.A. Lawson, A. Willems, J.J. Cordoba, J. Fernandez-Garayzabal, P. Garcia, J. Cai, H. Hippe and J.A. Farrow. 1994. The phylogeny of the genus *Clostridium*: proposal of five new genera and eleven new species combinations. Int. J. Syst. Bacteriol *44*: 812–826.

Conrad, R., F. Bak, H.J. Seitz, B. Thebrath, H.P. Mayer and H. Schutz. 1989. Hydrogen turnover by psychrotrophic homoacetogenic and mesophilic methanogenic bacteria in anoxic paddy soil and lake sediment. FEMS Microbiol. Ecol. *62*: 285–294.

Debonono, M. 1912. On some anaerobical bacteria of the normal human intestine. Zentralbl. Bakteriol. Parasitenkd. Infektionskr. Hyg. Abt. I Orig. *62*: 229–234.

Distaso, A. 1911. Sur les microbes proteolytiques de la flore intestinale de l'homme et des animaux. Zentralbl. Bakteriol. Parasitenkd. Infektionskr. Hyg. Abt. I Orig *59*: 97–103.

Dorner, C. and B. Schink. 1991. Fermentation of mandelate to benzoate and acetate by a homoacetogenic bacterium. Arch. Microbiol. *156*: 302–306.

Downes, J., M.A. Munson, D.A. Spratt, E. Kononen, E. Tarkka, H. Jousimies-Somer and W.G. Wade. 2001. Characterisation of *Eubacterium*-like strains isolated from oral infections. J. Med. Microbiol. *50*: 947–951.

Duncan, S.H., G.L. Hold, A. Barcenilla, C.S. Stewart and H.J. Flint. 2002. *Roseburia intestinalis* sp. nov., a novel saccharolytic, butyrate-producing bacterium from human faeces. Int. J. Syst. Evol. Microbiol. *52*: 1615–1620.

Eggerth, A.H. 1935. The gram-positive non-spore-bearing anaerobic bacilli of human feces. J. Bacteriol. *30*: 277–290.

Egli, C., T. Tschan, R. Scholtz, A.M. Cook and T. Leisinger. 1988. Transformation of tetrachloromethane to dichloromethane and carbon dioxide by *Acetobacterium woodii*. Appl. Environ. Microbiol. *54*: 2819–2824.

Eichler, B. and B. Schink. 1984. Oxidation of primary aliphatic alcohols by *Acetobacterium carbinolicum* sp. nov., a homoacetogenic anaerobe. Arch. Microbiol. *140*: 147–152.

Eichler, B. and B. Schink. 1985. *In* Validation of the publication of new names and new combinations previously effectively published outside the IJSB. List no. 18. Int. J. Syst. Bacteriol. *35*: 375–376.

Finegold S. M, D. Molitoris, Y. Song, M. D. Collins and P.A. Lawson. 2004. *In* Validation of publication of New Names and New Combinations Previously Effectively Published Outside the IJSEM Validation List no 98 Int. J. Syst. Evol. Microbiol. *54*: 1005–1006.

Finegold, S.M., V.L. Sutter and G.E. Mathisen. 1983. Normal indigenous intestinal flora. *In* Hentges (Editors), Human Intestinal Microflora in Health and Disease. Academic Press, New York, pp. pp. 3–31.

Finegold, S.M., P.A. Lawson, M.L. Vaisanen, D.R. Molitoris, Y. Song, C. Liu and M.D. Collins. 2004. *Anaerofustis stercorihominis* gen. nov., sp. nov., from human feces. Anaerobe *10*: 41–45.

Forsberg, C.W. 1978. Effects of heavy metals and other trace elements on the fermentative activity of the rumen microflora and growth of functionally important rumen bacteria. Can. J. Microbiol. *24*: 298–306.

Freier, T.A., D.C. Beitz, L. Li and P.A. Hartman. 1994. Characterization of *Eubacterium coprostanoligenes* sp. nov. a cholesterol-reducing anaerobe. Int. J. Syst. Bacteriol. *44*: 137–142.

Garnova, E.S., T.N. Zhilina, T.P. Tourova, N.A. Kostrikina and G.A. Zavarzin. 2004. Anaerobic, alkaliphilic, saccharolytic bacterium *Alkalibacter saccharofermentans* gen. nov., sp. nov. from a soda lake in the Transbaikal region of Russia. Extremophiles *8*: 309–316.

Garnova, E.S., T.N. Zhilina, T.P. Tourova, N.A. Kostrikina and G.A. Zavarzin. 2005. *In* Validation of publication of new names and new combinations previously effectively published outside the IJSEM. List no. 103 Int. J. Syst. Evol. Microbiol. *55*: 983–985.

Genthner, B.R., C.L. Davis and M.P. Bryant. 1981. Features of rumen and sewage sludge strains of *Eubacterium limosum*, a methanol- and H_2-CO_2-utilizing species. Appl. Environ. Microbiol. *42*: 12–19.

Genthner, B.R.S. and M.P. Bryant. 1982. Growth of *Eubacterium limosum* with carbon monoxide as the energy source. Appl. Environ. Microbiol. *43*: 70–74.

George, K.S. and W.A. Falkler, Jr. 1992. Coaggregation studies of the *Eubacterium* species. Oral Microbiol. Immunol. *7*: 285–290.

Gomez-Alarcon, R.A., C. O'Dowd, A.Z. Leedle and M.P. Bryant. 1982. 1,4-Napthoquinone and other nutrient requirements of *Succinivibrio dextrinosolvens*. Appl. Environ. Microbiol. *44*: 346–350.

Gossling, J. and W.E.C. Moore. 1975. *Gemmiger formicilis*, n. gen., n. sp., an anaerobic budding bacterium from intestines. Int. J. Syst. Bacteriol. *25*: 202–207.

Guinand, M., J.M. Ghuysen, K.H. Schleifer and O. Kandler. 1969. The peptidoglycan in walls of *Butyribacterium rettgeri*. Biochemistry *8*: 200–207.

Hammes, W.P., N. Weiss and W. Holzapfel. 1992. The genera *Lactobacillus* and *Carnobacterium*. *In* Balows, Trüper, Dworkin, Harder and Schleifer (Editors), The Prokaryotes. A Handbook on the Biology of Bacteria: Ecophysiology, Isolation, Identification, Applications, 2nd edn. Springer-Verlag, New York, pp. 1535–1594.

Hansen, B., M. Bokranz, P. Schonheit and A. Kroger. 1988. ATP formation coupled to caffeate reduction by H_2 in *Acetobacterium woodii* NZVA16. Arch. Microbiol. *150*: 447–451.

Hauduroy, P., G. Ehringer, A. Urbain, G. Guillot and J. Magrou. 1937. Dictionnaire des bactéries pathogènes. Masson and Co., Paris, pp. 1–597.

Heise, R., V. Müller and G. Gottschalk. 1992. Presence of a sodium-translocating ATPase in membrane vesicles of the homoacetogenic bacterium *Acetobacterium woodii*. Eur. J. Biochem. *206*: 553–557.

Henley, M.W. and D.H. Lewis. 1976. Anaerobic bacteria associated with epizootics in grey mullet (*Mugil cephalus*) and red fish (*Sciaenops ocellata*) along the Texas Gulf coast. J. Wildl. Dis. *12*: 448–453.

Hippe, H. 1991. Maintenance of methanogenic bacteria. *In* Kirsop and Doyle (Editors), Maintenance of microorganisms and cultured cells, 2nd edn. Academic Press, London, pp. 101–113.

Hoffman, J., B. Lindberg, S. Svensson and T. Hofstad. 1974. Structure of the polysaccharide antigen of *Eubacterium saburreum*, strain L44. Carbohydr. Res. *35*: 49–53.

Hoffman, J., B. Lindberg and J. Lonngren. 1976. Structural studies of the polysaccharide antigen of *Eubacterium saburreum*, strain 49. Carbohydr. Res. *47*: 261–267.

Hoffman, J., B. Lindberg, N. Skaug and T. Hofstad. 1980. Structural studies of the *Eubacterium saburreum* strain O2 antigen. Carbohydr. Res. *84*: 181–183.

Hofstad, T. 1967. An anaerobic oral filamentous organism possibly related to *Leptotrichia buccalis*. 1. Morphology, some physiological and serological properties. Acta Pathol. Microbiol. Scand. *69*: 543–548.

Hofstad, T. 1972. A polysaccharide antigen of an anaerobic oral filamentous microorganism (*Eubacterium saburreum*) containing heptose and O-acetyl as main constituents. 3. Some serological properties. Acta Pathol. Microbiol. Scand. [B] Microbiol. Immunol. *80*: 609–614.

Hofstad, T. 1975. Immunochemistry of a cell wall polysaccharide isolated from *Eubacterium saburreum*, strain L49. Acta Pathol. Microbiol. Scand. [B] *83*: 471–476.

Hofstad, T. and H. Lygre. 1977. Composition and antigenic properties of a surface polysaccharide isolated from *Eubacterium saburreum*, strain L452. Acta Pathol. Microbiol. Scand. [B] *85B*: 14–17.

Hofstad, T. and N. Skaug. 1978. A polysaccharide antigen from the Gram-positive organism *Eubacterium saburreum* containing dideoxyhexose as the immunodominant sugar. J. Gen. Microbiol. *106*: 227–232.

Hofstad, T. and P. Aasjord. 1982. *Eubacterium plautii* (Seguin 1928) comb. nov. Int. J. Syst. Bacteriol. *32*: 346–349.

Holdeman, L.V., E.P. Cato and W.E.C. Moore. 1967. Amended description of *Ramibacterium alactolyticum* Prévot and Taffanel with proposal of a neotype strain. Int. J. Syst. Bacteriol. *17*: 323–341.

Holdeman, L.V. and W.E.C. Moore. 1970. *Eubacterium. In* Cato, Cummins, Holdeman, Johnson, Moore, Smibert and Smith (Editors), Outline of clinical methods in anaerobic bacteriology, 2nd revision. Virginia Polytechnic Institute-Anerobe Laboratory, Blacksburg, pp. 23–30.

Holdeman, L.V., E.P. Cato and W.E.C. Moore. 1971. *Eubacterium contortum* (Prévot) comb. nov.: emendation of description and designation of the type strain. Int. J. Syst. Bacteriol. *21*: 304–306.

Holdeman, L.V. and W.E.C. Moore. 1972. Anaerobe Laboratory Manual. Anaerobe Laboratory, Virginia Polytechnic Institute and State University, Blacksburg.

Holdeman, L.V. and W.E.C. Moore. 1974. New genus, *Coprococcus*, twelve new species, and emended descriptions of four previously described species of bacteria from human feces. Int. J. Syst. Bacteriol. *24*: 260–277.

Holdeman, L.V., I.J. Good and W.E. Moore. 1976. Human fecal flora: variation in bacterial composition within individuals and a possible effect of emotional stress. Appl. Environ. Microbiol. *31*: 359–375.

Holdeman, L.V., E.P. Cato and W.E.C. Moore (Editors). 1977. Anaerobe Laboratory Manual, 4th edn. Anaerobe Laboratory, Virginia Polytechnic Institute and State University, Blacksburg, VA.

Holdeman, L.V., E.P. Cato, J.A. Burmeister and W.E.C. Moore. 1980. Descriptions of *Eubacterium timidum* sp. nov. *Eubacterium brachy* sp. nov. and *Eubacterium nodatum* sp. nov. isolated from human periodontitis. Int. J. Syst. Bacteriol. *30*: 163–169.

Hollriegl, V., L. Lamm, J. Rowold, J. Horig and P. Renz. 1982. Biosynthesis of vitamin B_{12}: different pathways in some aerobic and anaerobic microorganisms. Arch. Microbiol. *132*: 155–158.

Hungate, R.E. 1969. A roll tube method for cultivation of strict anaerobes. *In* Norris and Ribbons (Editors), Methods in Microbiology, Vol. 3B. Academic Press, London and New York, pp. 117–132.

Imkamp, F. and V. Müller. 2002. Chemiosmotic energy conservation with Na(+) as the coupling ion during hydrogen-dependent caffeate reduction by *Acetobacterium woodii*. J. Bacteriol. *184*: 1947–1951.

Jalava, J. and E. Eerola. 1999. Phylogenetic analysis of *Fusobacterium alocis* and *Fusobacterium sulci* based on 16S rRNA gene sequences: proposal of *Filifactor alocis* (Cato, Moore and Moore) comb. nov and *Eubacterium sulci* (Cato, Moore and Moore) comb. nov. Int. J. Syst. Bacteriol. *49*: 1375–1379.

Kamma, J.J., A. Diamanti-Kipioti, M. Nakou and F.J. Mitsis. 2000a. Profile of subgingival microbiota in children with primary dentition. J. Periodontal Res. *35*: 33–41.

Kamma, J.J., A. Diamanti-Kipioti, M. Nakou and F.J. Mitsis. 2000b. Profile of subgingival microbiota in children with mixed dentition. Oral Microbiol. Immunol. *15*: 103–111.

Kandler, O. and S. Schoberth. 1979. Murein structure of *Acetobacterium woodii*. Arch. Microbiol. *120*: 181–183.

Kaufmann, F., G. Wohlfarth and G. Diekert. 1997. Isolation of *O*-demethylase, an ether-cleaving enzyme system of the homoacetogenic strain MC. Arch. Microbiol. *168*: 136–142.

Kevbrin, V.V. and G.A. Zavarzin. 1992. Effect of sulfur compounds on the growth of the halophilic homoacetic bacterium *Acetohalobium arabaticum*. Microbiology (En. transl. from Mikrobiologiya) *61*: 563–567.

Kondo, W., N. Sato and T. Ito. 1979. Chemical structure of the polysaccharide antigen of *Eubacterium saburreum*, strain O2. Carbohydr. Res. *70*: 117–123.

Kotsyurbenko, O.R., A.N. Nozhevnikova, S.V. Kalyuzhny and G.A. Zavarzin. 1993a. Methanogenic digestion of cattle manure under psychrophilic conditions. Microbiology (En. transl. from Mikrobiologiya) *62*: 462–467.

Kotsyurbenko, O.R., A.N. Nozhevnikova and G.A. Zavarzin. 1993b. Methanogenic degradation of organic matter by anaerobic bacteria at low temperature. Chemosphere *27*: 1745–1761.

Kotsyurbenko, O.R., M.V. Simankova, A.N. Nozhevnikova, T.N. Zhilina, N.P. Bolotina, A.M. Lysenko and G.A. Osipov. 1995. New species of psychrophilic acetogens *Acetobacterium bakii* sp. nov., *A. paludosum* sp.nov., *A. fimetarium* sp. nov. Arch. Microbiol. *163*: 29–34.

Kotsyurbenko, O.R., A.N. Nozhevnikova, T.I. Soloviova and G.A. Zavarzin. 1996. Methanogenesis at low temperatures by microflora of tundra wetland soil. Antonie van Leeuwenhoek *69*: 75–86.

Kotsyurbenko, O.R., M.V. Simankova, A.N. Nozhevnikova, T.N. Zhilina, N.P. Bolotina, A.M. Lysenko and G.A. Osipov. 1997. *In* Validation of the publication of new names and new combinations previously effectively published outside the IJSB. List no. 60. Int. J. Syst. Bacteriol. *47*: 242.

Kotsyurbenko, O.R., M.V. Glagolev, A.N. Nozhevnikova and R. Conrad. 2001. Competition between homoacetogenic bacteria and methanogenic archaea for hydrogen at low temperature. FEMS Microbiol. Ecol. *38*: 153–159.

Krumholz, L.R. and M.P. Bryant. 1986a. *In* Validation of the publication of new names and new combinations previously effectively published outside the IJSB. List no. 21. Int. J. Syst. Bacteriol. *36*: 489.

Krumholz, L.R. and M.P. Bryant. 1986b. *Eubacterium oxidoreducens* sp. nov. requiring H_2 or formate to degrade gallate, pyrogallol, phloroglucinol and quercetin. Arch. Microbiol. *144*: 8–14.

Krumholz, L.R., S.H. Harris, S.T. Tay and J.M. Suflita. 1999. Characterization of two subsurface H_2-utilizing bacteria, *Desulfomicrobium hypogeium* sp. nov. and *Acetobacterium psammolithicum* sp. nov., and their ecological roles. Appl. Environ. Microbiol. *65*: 2300–2306.

Le Blaye, R. and H. Guggenheim. 1914. Manuel pratique de diagnostic bacteriologique et de technique applique a la determination des bacteries. Vigot Freres Edition, Paris.

Ludwig, W., K.-H. Schleifer and W.B. Whitman. 2009. Revised road map to the phylum *Firmicutes. In* De Vos, Garrity, Jones, Krieg, Ludwig, Rainey, Schleifer and Whitman (Editors). Bergey's Manual of Systematic Bacteriology. 2nd edn, Vol. 3. The *Firmicutes*, Springer, New York. pp. 1–14.

Margaret, B.S. and G.N. Krywolap. 1986. *Eubacterium yurii* subsp yurii sp. nov. and *Eubacterium yurii* subsp. *margaretiae* subsp. nov., test tube brush bacteria from subgingival dental plaque. Int. J. Syst. Bacteriol. *36*: 145–149.

Margaret, B.S. and G.N. Krywolap. 1988. *Eubacterium yurii* subsp. *schtitka* subsp. nov., test tube brush bacteria from subgingival dental plaque. Int. J. Syst. Bacteriol. *38*: 207–208.

Mechichi, T., M. Labat, T.H. S. Woo, P. Thomas, J.L. Garcia and B.K. C. Patel. 1998. *Eubacterium aggregans* sp. nov., a new homoacetogenic bacterium from olive mill wastewater treatment digestor. Anaerobe *4*: 283–291.

Mechichi, T., M. Labat, T.H.S. Woo, P. Thomas, J.L. Garcia and B.K.C. Patel. 2000. *In* Validation of publication of new names and new combinations previously effectively published outside the IJSEM. List no. 76. Int. J. Syst. Evol. Microbiol. *50*: 1699–1700.

Messmer, M., G. Wohlfarth and G. Diekert. 1993. Methyl chloride metabolism of the strictly anaerobic, methyl chloride-utilizing homoacetogen strain MC. Arch. Microbiol. *160*: 383–387.

Messmer, M., S. Reinhardt, G. Wohlfarth and G. Diekert. 1996. Studies on methyl chloride dehalogenase and O-demethylase in cell extracts of the homoacetogen strain MC based on a newly developed coupled enzyme assay. Arch. Microbiol. *165*: 18–25.

Miranda-Tello, E., M.L. Fardeau, J. Sepulveda, L. Fernandez, J.L. Cayol, P. Thomas and B. Ollivier. 2003. *Garciella nitratireducens* gen. nov., sp. nov., an anaerobic, thermophilic, nitrate- and thiosulfate-reducing bacterium isolated from an oilfield separator in the Gulf of Mexico. Int. J. Syst. Evol. Microbiol. *53*: 1509–1514.

Moore, W.E. and W.B. Gross. 1968. Liver granulomas of turkeys-causative agents and mechanism of infection. Avian Dis. *12*: 417–422.

Moore, W.E. and L.V. Holdeman. 1974. Human fecal flora: the normal flora of 20 Japanese-Hawaiians. Appl. Microbiol. *27*: 961–979.

Moore, W.E., L.V. Holdeman, R.M. Smibert, D.E. Hash, J.A. Burmeister and R.R. Ranney. 1982. Bacteriology of severe periodontitis in young adult humans. Infect. Immun. *38*: 1137–1148.

Moore, W.E., L.V. Holdeman, E.P. Cato, R.M. Smibert, J.A. Burmeister, K.G. Palcanis and R.R. Ranney. 1985. Comparative bacteriology of juvenile periodontitis. Infect. Immun. *48*: 507–519.

Moore, W.E.C., J.L. Johnson and L.V. Holdeman. 1976. Emendation of *Bacteroidaceae* and *Butyrivibrio* and descriptions of *Desulfomonas* gen. nov. and ten new species in genera *Desulfomonas*, *Butyrivibrio*, *Eubacterium*, *Clostridium*, and *Ruminococcus*. Int. J. Syst. Bacteriol. *26*: 238–252.

Moore, W.E.C. and L.V. Holdeman Moore. 1986. Genus *Eubacterium* Prévot 1938. *In* Sneath, Mair, Sharpe and Holt (Editors), Bergey's Manual of Systematic Bacteriology, Vol. 2. The Williams & Wilkins Co., Baltimore, pp. 1353–1373.

Morris, G.N., J. Winter, E.P. Cato, A.E. Ritchie and V.D. Bokkenheuser. 1986. *Eubacterium desmolans* sp. nov., a steroid desmolase-producing species from cat fecal flora. Int. J. Syst. Bacteriol. *36*: 183–186.

Mountfort, D.O., W.D. Grant, R. Clarke and R.A. Asher. 1988. *Eubacterium callanderi* sp. nov. that demethoxylates o-methoxylated aromatic acids to volatile fatty acids. Int. J. Syst. Bacteriol. *38*: 254–258.

Müller, V., S. Aufurth and S. Rahlfs. 2001. The Na$^+$ cycle in *Acetobacterium woodii*: identification and characterization of a Na$^+$ translocating F$_1$F$_0$-ATPase with a mixed oligomer of 8 and 16 kDa proteolipids. Biochim. Biophys. Acta *1505*: 108–120.

Munson, M.A., Pitt-Ford, T., B. Chong, A. Weightman and W.G. Wade. 2002. Molecular and cultural analysis of the microflora associated with endodontic infections. J. Dent. Res. *81*: 761–766.

Nakazawa, F. and E. Hoshino. 1994. Genetic relationships among *Eubacterium* species. Int. J. Syst. Bacteriol. *44*: 787–790.

Nakazawa, F., S.E. Poco, T. Ikeda, M. Sato, S. Kalfas, G. Sundqvist and E. Hoshino. 1999. *Cryptobacterium curtum* gen. nov., sp. nov., a new genus of Gram-positive anaerobic rod isolated from human oral cavities. Int. J. Syst. Bacteriol. *49*: 1193–1200.

Nakazawa, F., M. Sato, S.E. Poco, T. Hashimura, T. Ikeda, S. Kalfas, G. Sundqvist and E. Hoshino. 2000. Description of *Mogibacterium pumilum* gen. nov., sp. nov. and *Mogibacterium vescum* gen. nov., sp. nov., and reclassification of *Eubacterium timidum* (Holdeman et al. 1980) as *Mogibacterium timidum* gen. nov., comb. nov. Int. J. Syst. Evol. Microbiol. *50*: 679–688.

Nakou, M., J.J. Kamma, A. Andronikaki and F. Mitsis. 1998. Subgingival microflora associated with nifedipine-induced gingival overgrowth. J. Periodontol *69*: 664–669.

Nozhevnikova, A.N., O.R. Kotsyurbenko and M.V. Simankova. 1994. Acetogenesis at low temperature. *In* Drake (Editor), Acetogenesis. Chapman & Hall, New York and London, pp. 416–431.

Ollivier, B. and J.-L. Cayol. 2005. Fermentative, iron-reducing, and nitrate-reducing micro-organisms. *In* Ollivier and Magot (Editors), Petroleum Microbiology. ASM Press, Washington, pp. 71–88.

Pfennig, N. and K.D. Lippert. 1966. Über das Vitamin B$_{12}$-Bedurfnis phototropher Schwefelbakterien. Arch. Mikrobiol. *55*: 245–246.

Piovano, S. 1999. Bacteriology of most frequent oral anaerobic infections. Anaerobe *5*: 221–227.

Poco, S.E., F. Nakazawa, M. Sato and E. Hoshino. 1996. *Eubacterium minutum* sp. nov., isolated from human periodontal pockets. Int. J. Syst. Bacteriol. *46*: 31–34.

Prévot, A.R. 1938. Études de systématique bactérienne. III. Invalidité du genre *Bacteroides* Castellani et Chalmers démembrement et reclassification. Ann. Inst. Pasteur *20*: 285–307.

Prévot, A.R. 1940. Un anaérobie strict réduisant les nitrates en nitrites *Eubacterium nitritogenes* n. sp. C. R. Séances Soc. Biol. (Paris) *134*: 353–355.

Prévot, A.R. and J. Taffanel. 1942. Recherches sur une nouvelle espèce anaérobie *Ramibacter alactolyticum* (nov. spec.). Ann. Inst. Pasteur *68*: 259–262.

Prévot, A.R. 1947. Étude de quelques bacteries anaérobies nouvelles ou mal connues. Ann. Inst. Pasteur (Paris) *73*: 409–418.

Prévot, A.R. and J. Laplanche. 1947. Étude d'une bactérie anaérobie nouvelle de Guinée Française *Cillobacterium combesi* n. sp. Ann. Inst. Pasteur (Paris) *73*: 687–688.

Prévot, A.R., M. Raynaud and M. Digeon. 1947. Sur une espèce anaérobie: *Ramibacter pleuriticum*. Ann. Inst. Pasteur *73*: 481–483.

Prévot, A.R. 1966. Manual for the classification and determination of the anaerobic bacteria, 1st Amer. Ed. edn. Lea and Febiger, Philadelphia.

Prévot, A.R., A. Turpin and P. Kaiser. 1967. Les bactéries anaérobies. Dunod, Paris.

Repaci, G. 1910. Contribution à l'étude de la flore bactérienne anaérobic des gangrènes pulmonaires. Un streptobacille anaérobie. C. R. Soc. Biol. Paris *68*: 410–412.

Rocchi, G. 1908. Lo stato attuale delle nostre cognizioni sui germi anaerobi. Bull. Sci. Med. *8*: 457–528.

Rochford, J.C. 1980. Pleuropulmonary infection associated with *Eubacterium brachy*, a new species of *Eubacterium*. J. Clin. Microbiol. *12*: 722–723.

Rode, L.M., B.R.S. Genthner and M.P. Bryant. 1981. Synthrophic association by cocultures of the methanol- and CO$_2$-H$_2$-utilizing species, *Eubacterium limosum*, and pectin-fermenting *Lachnospira multiparus* during growth in a pectin medium. Appl. Environ. Microbiol. *42*: 20–22.

Schink, B. and M. Bomar. 1992. The genera *Acetobacterium*, *Acetogenium*, *Acetoanaerobium*, and *Acetitomaculum*. *In* Trüper, Dworkin, Harder and Schleifer (Editors), The Prokaryotes, 2nd edn, Vol. 2. Springer-Verlag, New York, pp. 1925–1936.

Schink, B. 1994. Diversity, ecology, and isolation of acetogenic bacteria. *In* Drake (Editor), Acetogenesis. Chapman & Hall, New York, pp. 197–235.

Schleifer, K.H. and O. Kandler. 1972. Peptidoglycan types of bacterial cell walls and their taxonomic implications. Bacteriol Rev *36*: 407–477.

Séguin, P. 1928. Culture du *Fusobacterium plautii* forme mobile dul bacille fusiforme. C. R. Soc. Biol. (Paris) *99*: 439–442.

Severin, A., S. Kokeguchi and K. Kato. 1989. Chemical composition of *Eubacterium alactolyticum* cell wall peptidoglycan. Arch. Microbiol. *151*: 348–352.

Simankova, M.V., O.R. Kotsyurbenko, E. Stackebrandt, N.A. Kostrikina, A.M. Lysenko, G.A. Osipov and A.N. Nozhevnikova. 2000. *Acetobacterium tundrae* sp. nov., a new psychrophilic acetogenic bacterium from tundra soil. Arch. Microbiol. *174*: 440–447.

Simankova, M.V., O.R. Kotsyurbenko, E. Stackebrandt and N.A. Kostrikina. 2001. *In* Validation of publication of new names and new combinations previously effectively published outside the IJSEM. List no. 80. Int. J. Syst. Evol. Microbiol. *51*: 793–794.

Siqueira, J.F., Jr. and I.N. Rocas. 2003. *Pseudoramibacter alactolyticus* in primary endodontic infections. J. Endodont. *29*: 735–738.

Siqueira, J.F., Jr. and I.N. Rocas. 2004. Polymerase chain reaction-based analysis of microorganisms associated with failed endodontic treatment. Oral Surg Oral Med Oral Pathol Oral Radiol Endod *97*: 85–94.

Skaug, N. and T. Hofstad. 1979. Immunochemical characterization of a polysaccharide antigen isolated from the oral microorganism *Eubacterium saburreum*, strain O2/725. Curr. Microbiol. *2*: 369–373.

Skerman, V.B.D., V. McGowan and P.H.A. Sneath. 1980. Approved lists of bacterial names. Int. J. Syst. Bacteriol *30*: 225–420.

Smith, M.R. and A.R. Mah. 1981. 2-Bromoethanesulfonate: a selective agent for isolating resistant *Methanosarcina* mutant. Curr. Microbiol. *6*: 321–326.

Speranza, G., B. Mueller, M. Orlandi, C.F. Morelli, P. Manitto and B. Schink. 2003. Stereochemistry of the conversion of 2-phenoxyethanol into phenol and acetaldehyde by *Acetobacterium* sp. Helvetica Chimica Acta *86*: 2629–2636.

Stadtman, E.R., T.C. Stadtman, I. Pastan and L.D.S. Smith. 1972. *Clostridium barkeri* sp. n. J. Bacteriol. *110*: 758–760.

Stevenson, I.L. 1978. The production of extracellular amino acids by rumen bacteria. Can. J. Microbiol. *24*: 1236–1241.

Strohhäcker, S.A. and B. Schink. 1991. Energetic aspects of malate and lactate fermentation by *Acetobacterium malicum*. FEMS Microbiol. Lett. *90*: 83–80.

Stromeyer, S.A., W. Winkelbauer, H. Kohler, A.M. Cook and T. Leisinger. 1991. Dichloromethane utilized by an anaerobic mixed culture: acetogenesis and methanogenesis. Biodegradation *2*: 129–137.

Stromeyer, S.A., W. Winkelbauer, H. Kohler, A.M. Cook and T. Leisinger. 1992. Anaerobic degradation of tetrachloromethane by *Acetobacterium woodii*. Biodegradation *3*: 113–123.

Sundqvist, G. 1992. Associations between microbial species in dental root canal infections. Oral Microbiol. Immunol. *7*: 257–262.

Sutter, V.L., D.M. Citron and S.M. Finegold. 1980. Wadsworth anaerobic bacteriology manual, 3rd edn. C.V. Mosby Co., St. Louis, Mo.

Tanaka, K. and N. Pfennig. 1988. Fermentation of 2-methoxyethanol by *Acetobacterium malicum* sp. nov. and *Pelobacter venetianus*. Arch. Microbiol. *149*: 181–187.

Tanaka, K. and N. Pfennig. 1990. *In* Validation of publication of new names and new combinations previously effectively published outside the IJSEM. List no. 35. Int. J. Syst. Bacteriol. *40*: 470–471.

Tanner, R.S., E. Stackebrandt, G.E. Fox and C.R. Woese. 1981. A phylogenetic analysis of *Acetobacterium woodi*, *Clostridium barkeri*, *Clostridium butyricum*, *Clostridium lituseburense*, *Eubacterium limosum*, and *Eubacterium tenue*. Curr. Microbiol. *5*: 35–38.

Tanner, R.S. 1986. Genus *Acetobacterium* Balch, Schoberth, Tanner and Wolfe 1977. *In* Sneath, Mair, Sharpe and Holt (Editors), Bergey's Manual of Systematic Bacteriology, Vol. 2. The Williams & Wilkins Co., Baltimore, pp. 1373–1375.

Taylor, M.M. 1972. *Eubacterium fissicatena* sp. nov. isolated from the alimentary tract of the goat. J. Gen. Microbiol. *71*: 457–463.

Theilade, E. and M.N. Gilmour. 1961. An anaerobic oral filamentous microorganism. J. Bacteriol. *81*: 661–666.

Tissier, H. 1908. Recherches sur la flore intestinale normale des enfants agés d'un an à cinq ans. Ann. Inst. Pasteur (Paris) *22*: 189–208.

Tomerska, H. and M. Wojciechowicz. 1973. Utilization of the intermediate products of the decomposition of pectin and of galacturonic acid by pure strains of rumen bacteria. Acta Microbiol. Pol. B. *5*: 63–69.

Tourova, T.P., E.S. Garnova and T.N. Zhilina. 1999. Phylogenetic diversity of alkaliphilic anaerobic saccharolytic bacteria isolated from soda lakes. Microbiology (En. transl. from Mikrobiologiya) *68*: 615–622.

Traunecker, J., A. Preuß, G. Diekert and A. Preuß. 1991. Isolation and characterization of a methyl chloride utilizing, strictly anaerobic bacterium. Arch. Microbiol. *156*: 416–421.

Tschech, A. and N. Pfennig. 1984. Growthyield increase linked to caffeate reduction in *Acetobacterium woodii*. Arch. Microbiol. *137*: 163–167.

Udey, L.R., E. Young and B. Sallman. 1977. Isolation and characterization of an anaerobic bacterium, *Eubacterium tarantellus* sp. nov., associated with striped mullet (*Mugil cephalus*) mortality in Biscayne Bay, Florida. J. Fish. Res. Board Can. *34*: 402–409.

Uematsu, H., F. Nakazawa, T. Ikeda and E. Hoshino. 1993. *Eubacterium saphenum* sp. nov., isolated from human periodontal pockets. Int. J. Syst. Bacteriol. *43*: 302–304.

van de Merwe, J.P. 1981. Agglutination of *Eubacterium* and *Peptostreptococcus* species as a diagnostic test for Crohn's disease. Hepatogastroenterology *28*: 155–156.

van Gylswyk, N.O. and J.P. Hoffman. 1970. Characteristics of cellulolytic cillobacteria from the rumens of sheep fed teff (*Eragrostis tef*) hay diets. J. Gen. Microbiol. *60*: 381–386.

van Gylswyk, N.O. and J.J.T.K. van der Toorn. 1985. *Eubacterium uniforme* sp. nov. and *Eubacterium xylanophilum* sp. nov., fiber-digesting bacteria from the rumina of sheep fed corn stover. Int. J. Syst. Bacteriol. *35*: 323–326.

Vinzent, R. and V. Reynes. 1947. Etude d'un nouvel anaérobie: *Ramibacterium dentium*. Ann. Inst. Pasteur *73*: 594–595.

Wade, W.G., J. Downes, D. Dymock, S.J. Hiom, A.J. Weightman, F.E. Dewhirst, B.J. Paster, N. Tzellas and B. Coleman. 1999a. The family *Coriobacteriaceae*: reclassification of *Eubacterium exiguum* (Poco *et al.* 1996) and *Peptostreptococcus heliotrinreducens* (Lanigan 1976) as *Slackia exigua* gen. nov., comb. nov. and *Slackia heliotrinireducens* gen. nov., comb. nov., and *Eubacterium lentum* (Prevot 1938) as *Eggerthella lenta* gen. nov., comb. nov. Int. J. Syst. Bacteriol. *49*: 595–600.

Wade, W.G., J. Downes, M.A. Munson and A.J. Weightman. 1999b. *Eubacterium minutum* is an earlier synonym of *Eubacterium tardum* and has priority. Int. J. Syst. Bacteriol. *49*: 1939–1941.

Wallace, R.J., N. McKain, N.R. McEwan, E. Miyagawa, L.C. Chaudhary, T.P. King, N.D. Walker, J.H. Apajalahti and C.J. Newbold. 2003. *Eubacterium pyruvativorans* sp. nov., a novel non-saccharolytic anaerobe from the rumen that ferments pyruvate and amino acids, forms caproate and utilizes acetate and propionate. Int. J. Syst. Evol. Microbiol. *53*: 965–970.

Wayne, L.G. 1982. Actions of the Judicial Commission of the International Committee on Systematic Bacteriology on requests for opinions published between July 1979 and April 1981. Int. J. Syst. Bacteriol. *32*: 464–465.

Weinberg, M., R. Nativelle and A.R. Prévot. 1937. Les microbes anaérobies. Pp 1–1186. Masson and Co., Paris.

Weizenegger, M., M. Neumann, E. Stackebrandt, N. Weiss and W. Ludwig. 1992. *Eubacterium alactolyticum* phylogenetically groups with *Eubacterium limosum*, *Acetobacterium woodii* and *Clostridium barkeri*. Syst. Appl. Microbiol. *15*: 32–36.

Wery, N., J.M. Moricet, V. Cueff, J. Jean, P. Pignet, F. Lesongeur, M.A. Cambon-Bonavita and G. Barbier. 2001. *Caloranaerobacter azorensis* gen. nov., sp. nov., an anaerobic thermophilic bacterium isolated from a deep-sea hydrothermal vent. Int. J. Syst. Evol. Microbiol. *51*: 1789–1796.

Wilkins, T.D. and T. Thiel. 1973. A modified broth-disk method for testing the antibiotic susceptibility of anaerobic bacteria. Antimicrob. Agents Chemother. *3*: 350–356.

Wilkins, T.D., R.S. Fulghum and J.H. Wilkins. 1974. *Eubacterium plexicaudatum* sp. nov., an anaerobic bacterium with a subpolar tuft of flagella, isolated from a mouse cecum. Int. J. Syst. Bacteriol. *24*: 408–411.

Willems, A. and M.D. Collins. 1996. Phylogenetic relationships of the genera *Acetobacterium* and *Eubacterium* sensu stricto and reclassification of *Eubacterium alactolyticum* as *Pseudoramibacter alactolyticus* gen. nov., comb. nov. Int. J. Syst. Bacteriol. *46*: 1083–1087.

Wohlfarth, G. and G. Diekert. 1997. Anaerobic dehalogenases. Curr. Opin. Biotechnol. *8*: 290–295.

Wolin, E.A., M.G. Wolin and R.S. Wolfe. 1963. Formation of methane by bacterial extracts. J. Biol. Chem. *238*: 2882–2886.

Zindel, U., W. Freudenberg, M. Rieth, J.R. Andreesen, J. Schnell and F. Widdel. 1988. *Eubacterium acidaminophilum* sp. nov., a versatile amino acid-degrading anaerobe producing or utilizing H_2 or formate: description and enzymatic studies. Arch. Microbiol. *150*: 254–266.

Zindel, U., W. Freundenberg, M. Rieth, J.R. Andressen, J. Schnell and F. Widdel. 1989. *In* Validation of the publication of new names and new combinations previously effectively published outside the IJSB. List no. 28. Int. J. Syst. Bacteriol. *39*: 93–94.

Family III. **Gracilibacteraceae** fam. nov.

Yong-Jin Lee, Rob U. Onyenwoke and Juergen Wiegel

Gra.ci.li.bac.ter.a′ce.ae. N.L. masc. n. *Gracilibacter* type genus of the family; suff. *-aceae* ending to denote a family; N.L. fem. pl. n. *Gracilibacteraceae* the *Gracilibacter* family.

Straight to curved rods that are 0.2–0.4 µm in diameter × 2.0–7.0 µm in length. Elongate infrequently up to 45 µm. Cells occur singly or form chains. Gram-positive wall structure, but stain Gram-negative. **Nonsporeforming.** Cells in old cultures become like **autoplasts** (L-shaped cells). Obligately anaerobic. **Thermotolerant.** Chemoorganotrophs.

DNA G+C content (mol%): 43.

Type genus: **Gracilibacter** Lee, Romanek, Mills, Davis, Whitman and Wiegel 2006, 2092[VP].

Further descriptive information

The family *Gracilibacteraceae* belongs to the order *Clostridiales*. To date, it is represented by only one genus containing one species. Its closest neighbor in an inferred phylogenetic tree is *Clostridium thermosuccinogenes*, a member of the new family "*Ruminococcaceae*" (Ludwig et al., 2009). The delineation of the novel family "*Gracilibacteriaceae*" was primarily based on phylogenetic criteria and a few metabolic and physiological characteristics observed, with *Gracilibacter* as the type for the family. From the phylogenetic analyses (Figure 6), *Gracilibacter* is distantly related to members of the *Clostridiaceae* (formerly Collins' *Clostridium* cluster I/II) and the new family "*Ruminococcaceae*" (formerly Collins' *Clostridium* cluster III). In addition to the phylogenetic evidence, *Gracilibacter* does not form endospores as tested under various conditions. The present member of the genus *Gracilibacter* is not cellulolytic, which distinguishes it from the cellulolytic species in the "*Ruminococcaceae*". The genus *Mahella* is presently not included in the family "*Gracilibacteraceae*" nor in the family "*Thermoananerobacterceae*" as originally suggested due to variation in its position in different phylogenetic trees (Ludwig et al., 2009).

Key to the genera of the family *Gracilibacteraceae*

1. The DNA G+C content is around 43 mol%. Growth only under anaerobic conditions. Optimal growth is at 42–47 °C. No indication of chemolithoautotrophic growth with H_2/CO_2. Indole is produced. No production of H_2S. Major branched-chain fatty acids of the lipids are $C_{15:0\ iso}$, $C_{15:0\ ante}$, $C_{16:0\ iso}$, and $C_{17:0\ iso}$.

Genus I. **Gracilibacter** Lee, Romanek, Mills, Davis, Whitman and Wiegel 2006, 2092[VP]

Yong-Jin Lee and Juergen Wiegel

Gra.ci.li.bac′ter. L. adj. *gracilis* slender; N.L. masc. n. *bacter* equivalent of Gr. neut. n. *baktron* rod or staff; N.L. masc. n. *Gracilibacter* slender rod, referring to its cell shape.

Gram-positive cell-wall structure, but stains Gram-negative. Slightly polymorphic rods. Obligately anaerobic, chemoorganotrophic, and thermotolerant. No spores have been observed under various conditions tested. Not resistant to heating at 80 °C for 10 min.

DNA G+C content (mol%): 43.

Type species: **Gracilibacter thermotolerans** Lee, Romanek, Mills, Davis, Whitman and Wiegel 2006, 2092[VP].

Further descriptive information

The genus *Gracilibacter* is to date represented by only one species, and this description reflects the properties of that species. Based on an inferred phylogenetic tree, the closest cultured relative is *Clostridium thermosuccinogenes*, a member of the new family "*Ruminococcus*" (formerly *Clostridium* cluster III; Collins et al., 1994; Rainey et al., 2006; Ludwig et al., 2009). BLAST analysis yields *Clostridium pascui* (*Clostridiaceae sensu stricto*) with about 92% similarity as the taxon with the highest similarity in the 16S rRNA gene (Figure 152). Specific habitats are unknown, however, 16S rRNA genes with similar sequences have been found in various environments including hydrocarbon-degrading methanogenic consortia, rice paddy soils, oil reservoirs, and urban aerosols.

Cells are straight to curved rods, 0.2–0.4 µm × 2.0–7.0 µm, with infrequently elongated cells of up to 45 µm. Cells are either single or form chains. During the late-stationary growth phase, autoplasts (L-shaped-like cells) can occur. Endospores have not been observed. Cells are sluggishly motile or nonmotile depending upon the growth conditions. Cells have 1–5 peritrichous flagella (Figure 153).

Growth characteristics

Growth conditions. The temperature range for growth is 25–54 °C, with an optimum at 42.5–46.5 °C. Thus, the type strain is a thermotolerant anaerobe. The pH^{25C} range for growth is 6.0–8.25 with an optimum at 6.8–7.75. The salinity range for growth is from 0–1.5% (w/v) with an optimum at 0.5%.

Saccharolytic and peptidolytic growth. *Gracilibacter thermotolerans* requires yeast extract for growth. In the presence of 0.02% yeast extract, Casamino acids, tryptone, peptone, maltose, sucrose, arabinose, fructose, galactose, glucose, mannose, xylose, mannitol, and sorbitol serve as carbon and energy sources. No growth has been observed with cellobiose, lactose, raffinose, ribose, trehalose, inositol, mannitol, xylitol, acetate, lactate, pyruvate, methanol, and carboxymethyl cellulose (1.0% w/v, CMC 7LT or 7M; Hercules) as carbon and energy

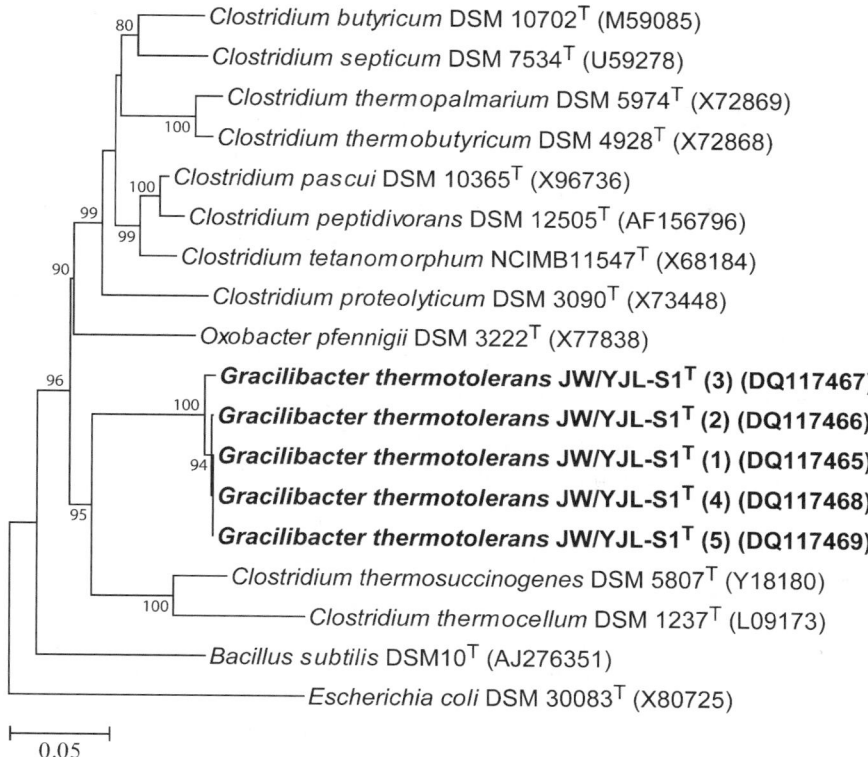

FIGURE 152. Phylogenetic dendrogram based on 16S rRNA gene sequences showing the relationship of five clones of *Gracilibacter thermotolerans* strain JW/YJL-S1ᵀ to representatives of the *Clostridiales*. Neighbor-joining method with Jukes and Cantor distance corrections. Numbers at the nodes represent the bootstrap values (1000 replicates). Bar = 5 nucleotides substitutions per 100 nucleotides. (From Lee et al., 2006; Int. J. Syst. Evol. Microbiol. *56*: 2089–2093. Used with permission of the publisher.)

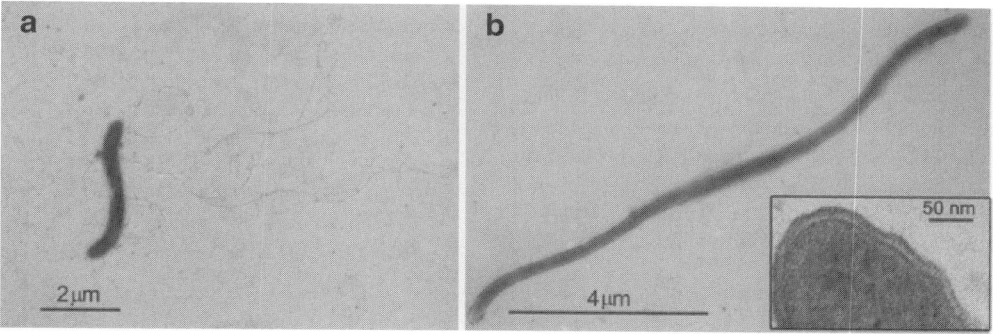

FIGURE 153. Electron micrographs of *Gracilibacter thermotolerans* showing cells with 1–5 peritrichously inserted flagella (a) and a representative image of the cell (b). The inset in (b) shows a thin section demonstrating the Gram-positive cell-wall structure and S-layer. (From Lee et al., 2006; Int. J. Syst. Evol. Microbiol. *56*: 2089–2093. Used with permission of the publisher.)

sources. The main fermentation end products from glucose are acetate, lactate, and ethanol. No chemolithoautotrophic growth has been detected using H_2/CO_2 (80:20, v/v). No indication of growth has been observed using Fe(III), MnO_2, nitrate, thiosulfate, elemental sulfur, sulfate, sulfite, or fumarate as electron acceptors. The doubling time for *Gracilibacter*

thermotolerans is 3.1 h at 42 °C and pH²⁵ᶜ 6.5 with 0.3% yeast extract as the substrate.

Antibiotic resistance. Growth of *Gracilibacter thermotolerans* at pH²⁵ᶜ 6.8 and 42 °C is resistant to streptomycin at 10 μM but inhibited by 10 μM of ampicillin, chloramphenicol, erythromycin, rifampin, and tetracycline and by 100 μM streptomycin.

Enrichment and habitat

Gracilibacter thermotolerans was isolated from an enrichment in a carbonate-buffered basal medium (Widdel and Bak, 1992) supplemented with 20 mM acetate and 0.1 mM ferric citrate at pH^{25C} 6.8 and 37 °C under anaerobic conditions (N$_2$, 100%). The strain also grows in a phosphate-buffered medium and classical peptone-sugar media including peptone-yeast extract, peptone-yeast extract-glucose, reinforced clostridial medium (Difco), and thioglycolate broth (Difco).

Gracilibacter thermotolerans was isolated from the uppermost sediment in a constructed wetland system at the Savannah River Site (Aiken, SC, USA). The organic substrate used for the constructed treatment wetland was composed primarily of composted stable wastes and spent brewing grains therefore, no defined habitat can be given. However, 16S rRNA genes with similar sequences have been found in various environments including hydrocarbon-degrading methanogenic consortia, rice paddy soils, oil reservoirs, and urban aerosols.

Maintenance procedures

The best method of preservation is storage in liquid nitrogen. Long-term storage is presently done as a freeze-dried culture (DSMZ). Short-term storage (up to 3 years) is performed in prereduced medium containing 12–15% (v/v) glycerol at −20 °C (for frequent sampling since the mixture stays liquid) and −80 °C for longer and uninterrupted storage. Storage in liquid medium at room temperature is possible for several months.

Differentiation from closely related taxa

Phylogenetic analyses of its 16S rRNA genes failed to place *Gracilibacter thermotolerans* within any of the currently described families. It is outside the radiation of the *Clostridiaceae sensu stricto* as well as the "*Ruminococcaceae*" (formerly Collins' *Clostridium* cluster III). The more complete phylogenetic analyses of the *Firmicutes* places it as a unique lineage within the *Clostridiales* (Ludwig et al., 2009) as the separate family "*Gracilibacteraceae*". Besides the phylogenetic evidence, *Gracilibacter thermotolerans* has different physiological properties from the members of the genus *Clostridium sensu stricto* and *Oxobacter pfennigii* (genus IX in the *Clostridiaceae*). Whereas most *Clostridium* species and *Oxobacter pfennigii* form endospores, endospores have not been detected in *Gracilibacter thermotolerans*. Unlike the members of the family "*Ruminococcaceae*", *Gracilibacter thermotolerans* is not cellulolytic, does not produce succinic acid, and does not grow above 60 °C.

Miscellaneous comments

The 16S rRNA genes of strain JW/YJL-S1^T are polymorphic. Among five clones of the 16S rRNA gene sequenced, four possessed >99% sequence similarity, while the fifth clone possessed about 2% divergence from the rest.

List of species of the genus *Gracilibacter*

1. **Gracilibacter thermotolerans**   Lee, Romanek, Mills, Davis, Whitman and Wiegel 2006, 2092^{VP}

   ther.mo.to'le.rans. Gr. n. *thermê* heat; L. pres. part. *tolerans* tolerating; N.L. part. adj. *thermotolerans* heat-tolerating.

   Obligatory anaerobe, thermotolerant, and asporogenic (Onyenwoke et al., 2004). Endospores have not been observed by microscopy or by heat treatment (10 min. at 80 °C) under all tested growth conditions. The phospholipid fatty acid composition is dominated by C$_{16:0}$ fatty acid (29.0%) and the branched-chained fatty acids (C$_{15:0\ iso}$, C$_{15:0}$ ante, and C$_{17:0\ iso}$) which account for 39.2% of the total phospholipid fatty acids. In addition, small amounts of the unsaturated C$_{16:1}$, C$_{17:1}$, and C$_{18:1}$ fatty acids (5.4%, 5.4%, and 2.5%, respectively) are also present. Positive for esterase, leucine arylamidase, acid phosphatase, naphthol-AS-BI-phosphohydrolase, β-galactosidase, α-glucosidase, and β-glucosidase. Produces indole, but not H$_2$S.

   DNA G+C content (mol%): 42.8 (HPLC).
   Type strain: JW/YJL-S1, ATCC BAA-1219, DSM 17427.
   GenBank accession number (16S rRNA gene): DQ117465, DQ117466, DQ117467, DQ117468, DQ117469.

References

Collins, M.D., P.A. Lawson, A. Willems, J.J. Cordoba, J. Fernández-Garayzábal, P. Garcia, J. Cai, H. Hippe and J.A.E. Farrow. 1994. The phylogeny of the genus *Clostridium*: proposal of five new genera and eleven new species combinations. Int. J. Syst. Bacteriol. *44*: 812–826.

Lee, Y.J., C.S. Romanek, G.L. Mills, R.C. Davis, W.B. Whitman and J. Wiegel. 2006. *Gracilibacter thermotolerans* gen. nov., sp. nov., an anaerobic, thermotolerant bacterium from a constructed wetland receiving acid sulfate water. Int. J. Syst. Evol. Microbiol. *56*: 2089–2093.

Ludwig, W., K.-H. Schleifer and W.B. Whitman. 2009. Revised road map to the phylum *Firmicutes. In* De Vos, Garrity, Jones, Krieg, Ludwig, Rainey, Schleifer and Whitman (Editors). Bergey's Manual of Systematic Bacteriology, 2nd edn, Vol. 3, The *Firmicutes*. Springer, New York, pp. 1–14.

Onyenwoke, R.U., J.A. Brill, K. Farahi and J. Wiegel. 2004. Sporulation genes in members of the low G+C Gram-type-positive phylogenetic branch (*Firmicutes*). Arch. Microbiol. *182*: 182–192.

Rainey, F.A., R. Tanner and J. Wiegel. 2006. Family *Clostridiaceae In* The Prokaryotes: A Handbook on the Biology of Bacteria: *Bacteria: Firmicutes, Cyanobacteria*, 3rd edn release 3.20, Vol. 4. Springer-Verlag, New York, pp. 654–678.

Widdel, F. and F. Bak. 1992. Gram-negative mesophilic sulfate-reducing bacteria. *In* Balows, Trüper, Dworkin, Harder and Schleifer (Editors), The Prokaryotes, Vol. 4. Springer, New York, pp. 3352–3378.

Family IV. **Heliobacteriaceae** Madigan 2001, 625

Marie Asao and Michael T. Madigan

He.li.o.bac.te.ri.a′ce.ae. N.L. neut. n. *Heliobacterium* type genus of the family; -*aceae* ending to denote a family; N.L. fem. pl. n. *Heliobacteriaceae* the *Heliobacterium* family.

Species of the family *Heliobacteriaceae* (the heliobacteria) are **anoxygenic phototrophs** that have **rod-shaped or spirillum-shaped cells** and multiply by **binary fission**. Rod-shaped cells are sometimes curved or twisted. In the genus *Heliophilum*, cells are straight and tapered and group together to form bundles that are motile as a unit. Currently, four genera with a total of ten species have been formally described.

Cells of heliobacteria **stain Gram-negative, although an outer membrane is lacking**. The heliobacteria form a distinct clade within the phylum *Firmicutes* (Madigan and Ormerod, 1995; Ormerod et al., 1996a; Woese et al., 1985). The closest relatives of the heliobacteria are species of *Desulfitobacterium*, chemotrophic bacteria that carry out anaerobic respiration using various electron acceptors, including sulfite, nitrate, Fe^{3+}, or halogenated organic compounds (Figure 154). **Motility in the heliobacteria is by gliding or flagellar means**; if the latter, then by polar, subpolar, or peritrichous flagella.

Heliobacteria **contain bacteriochlorophyll *g*** as sole bacteriochlorophyll (Brockmann and Lipinski, 1983; Michalski et al., 1987) and diaponeurosporene (Takaichi et al., 1997) as the major carotenoid; small amounts of 8′-OH-chlorophyll *a* is also present (Amesz, 1995). Alkaliphilic species of heliobacteria contain a variety of glycosylated carotenoids (Takaichi et al., 2003). Bacteriochlorophyll *g* and 8′-OH-chlorophyll *a* are esterified with the C-15 alcohol farnesol. The **color of phototrophically grown (anoxic/light) cultures of heliobacteria is brownish-green** although cultures turn emerald green in stationary phase or if exposed to air. **Intracytoplasmic membranes of the kind observed in phototrophic purple bacteria or chlorosomes characteristic of the green sulfur or green nonsulfur bacteria are not observed** in heliobacteria (Figure 155 and Figure 156). In addition, **gas vesicles have never been found**.

Metabolism of the heliobacteria is strictly anaerobic; cells grow as **photoheterotrophs** on a limited number of organic carbon sources or **chemo-organotrophically** (in darkness) by fermentation of pyruvate or other simple organic compounds (Kimble et al., 1994). Pyruvate is used by all heliobacteria. Aerobic or microaerobic dark growth has not been observed. In addition, **photoautotrophic growth on CO_2 + H_2 or on CO_2 + H_2S has not been observed** in any species of heliobacteria, and the genome of *Heliobacterium modesticaldum* (which has been sequenced, Sattley et al., 2008) lacks genes encoding proteins of any known autotrophic pathway. Sulfide tolerance varies among heliobacteria; some species are quite sulfide sensitive whereas others are quite tolerant. However, in media containing sulfide, the **sulfide is oxidized and elemental sulfur is formed and deposited outside the cells**. Optimal pH for growth of most heliobacteria is 7, with a **pH growth range of 5.5–8**, depending on the species. Species of the genus *Heliorestis* are **alkaliphilic and show pH optima near 9** with growth occurring from pH 8–10. **Optimal temperature is 37–42 °C for most heliobacteria (52 °C for the single known thermophilic species) and 30–32 °C for alkaliphilic species**. Ammonium salts and glutamine are used as nitrogen sources by all heliobacteria, and certain species can use other amino acids as well. All heliobacteria that have been tested **fix nitrogen gas** (Kimble and Madigan, 1992; Madigan, 1995). **Biotin is required** for growth by most heliobacteria, and some species also require a reduced sulfur compound (sulfide, thiosulfate, or cysteine) for biosynthetic purposes. Low levels (0.025–0.05%) of yeast extract typically stimulate growth. **Neither poly-β-hydroxybutyrate nor glycogen, common carbon storage polymers in other anoxygenic phototrophs, have been observed in heliobacteria**.

Endospores containing dipicolinic acid and elevated levels of Ca^{2+} are produced by some heliobacteria (Ormerod et al., 1996a) and this property is likely to be universal (Kimble-Long and Madigan, 2001). However, loss of the ability to form endospores in laboratory cultures is commonly observed. Heliobacteria are found **primarily in soil** (Stevenson et al., 1997) and in this regard differ significantly in their ecology from purple and green anoxygenic phototrophs. *Heliobacterium modesticaldum* is thermophilic and inhabits neutral to alkaline hot springs worldwide at temperatures below 60 °C.

Recently, two new strains of heliobacteria were isolated from alkaline soda lakes, Lake El Hamra (Wadi El Natroun, Egypt) and Soap Lake (Washington, USA) (unpublished results). Preliminary characterization of the strain isolated from soil of the shoreline of Lake El Hamra showed that this bacterium was an alkaliphilic, motile rod and clearly a member of the genus *Heliorestis* based on 16S rRNA gene analysis. However, unlike other heliobacteria, all of which utilize only a restricted set of carbon sources, the El Hamra strain photoassimilated a wide variety of carbon substrates, including several amino acids.

The heliobacterium isolated from Soap Lake was remarkable from the standpoint of its habitat and phylogeny. This organism was isolated from lake sediment at a depth of 23 m, which indicates that heliobacteria are not totally restricted to terrestrial habitats (Stevenson et al., 1997). The Soap Lake heliobacterium is an alkaliphilic, motile, curved rod. Although all of the alkaliphilic species of heliobacteria previously isolated from soda lakes are phylogenetically closely related, forming a monophyletic clade of the genus *Heliorestis* (Figure 154), the Soap Lake isolate is outside this clade, possibly representing a novel genus basal to the nonalkaliphilic heliobacteria (data not shown). These results suggest that the phylogeny of heliobacteria may be more extensive than previously acknowledged and that other aquatic heliobacteria exist.

DNA G+C content (mol%): 45–57.

Type genus: **Heliobacterium** Gest and Favinger 1985, 223[VP] (Effective publication: Gest and Favinger 1983, 15.).

Further descriptive information

The family *Heliobacteriaceae* contains all of the anoxygenic phototrophic bacteria that synthesize bacteriochlorophyll *g*. This structurally and spectrally unique bacteriochlorophyll distinguishes these organisms from the purple bacteria, which

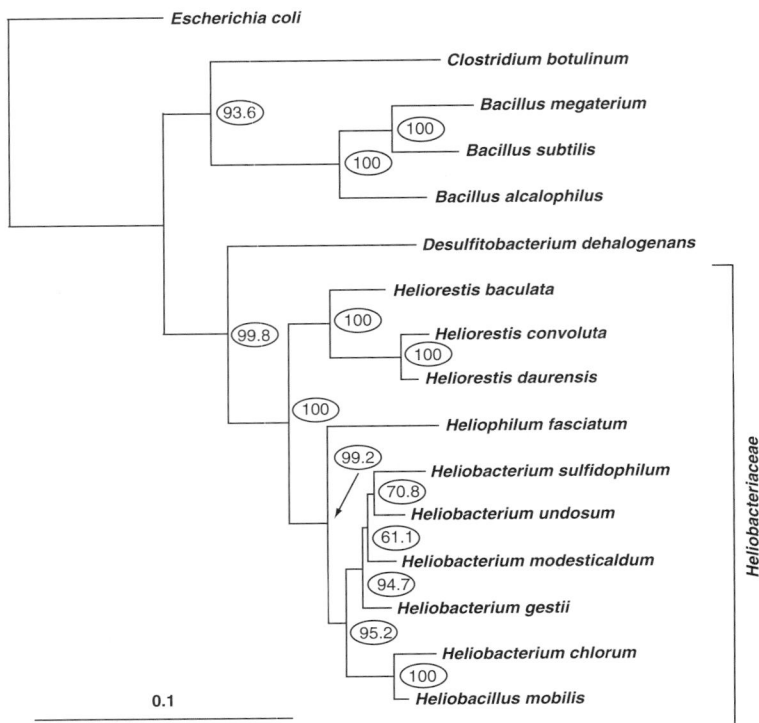

FIGURE 154. Phylogenetic tree of the family *Heliobacteriaceae* based on 16S rRNA gene sequences. 1301 nucleotide positions were compared. The tree was computed using PHYLIP version 3.66 (Felsenstein, 1989). Alignment was converted to a distance matrix using F84 algorithm (transition/transversion ratio = 2.0, empirical base frequencies) in the program DNADIST. The tree was based on the neighbor-joining method using the program NEIGHBOR. Bootstrap analysis was conducted on 1000 replications; the confidence values >50% are indicated at the nodes. The tree was rooted using *Escherichia coli* (Gamma proteobacteria) as the outgroup. All the sequences have been deposited in GenBank as follows: *Escherichia coli* (J01859), *Clostridium botulinum* (X68187), *Bacillus megaterium* (X60629), *Bacillus subtilis* (AJ276351), *Bacillus alcalophilus* (X76436), *Desulfitobacterium dehalogenans* (L28946), *Heliorestis baculata* (AF249680), *Heliorestis convoluta* (DQ266255), *Heliorestis daurensis* (AF079102), *Heliophilum fasciatum* (L36197), *Heliobacterium sulfidophilum* (AF249678), *Heliobacterium undosum* (AF249679), *Heliobacterium modesticaldum* (U14559), *Heliobacterium gestii* (L36198), *Heliobacterium chlorum* (M11212); *Heliobacillus mobilis* (U14560).

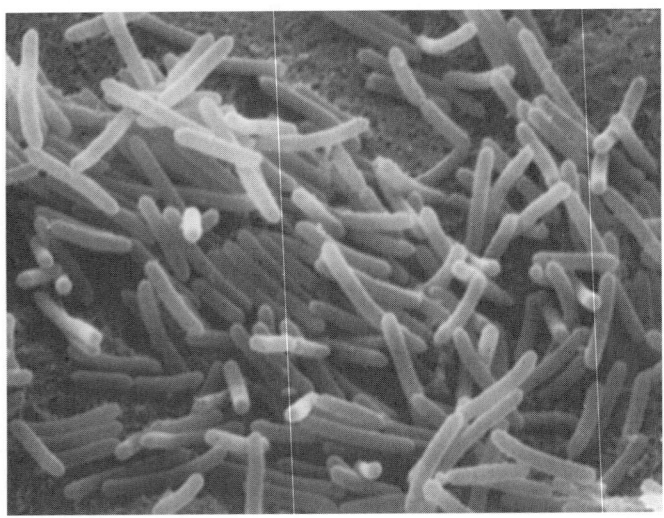

FIGURE 155. Scanning electron micrograph of cells of *Heliobacterium chlorum* ATCC 35205^T. Magnification, ×3800. (Courtesy of F. Rudy Turner, Indiana University.)

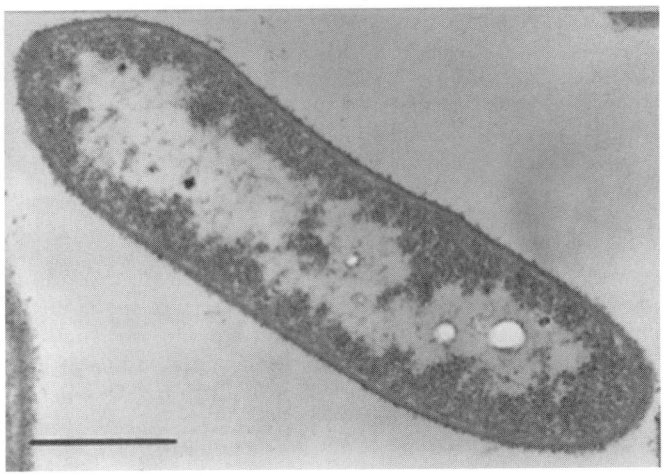

FIGURE 156. Thin section transmission electron micrograph of the thermophile *Heliobacterium modesticaldum* strain Ice1^T. Bar = 1 μm. (Reprinted with permission from Kimble et al., 1995. Archives of Microbiology *163*: 259–267.)

TABLE 160. Characteristics differentiating heliobacteria[a,b]

Characteristic	Heliobacterium chlorum	Heliobacterium gestii	Heliobacterium modesticaldum	Heliobacterium sulfidophilum	Heliobacterium undosum	Heliobacillus mobilis	Heliophilum fasciatum	Heliorestis daurensis	Heliorestis baculata	Heliorestis convoluta
Habitat	Temperate soil	Tropical paddy soil	Neutral/alkaline hot springs and volcanic soils	Alkaline sulfidic hot springs	Alkaline sulfidic hot springs	Tropical paddy soil	Tropical paddy soil	Soda lake shore line soil	Soda lake shoreline soil	Soda lake shore line soil
Morphology	Rod	Spirillum	Rod/curved rod	Rod	Rod/slightly twisted spirillum	Rod	Straight rods with tapered ends grouped in bundles of two to many cells	Coil/bent filament	Rod/curved rod	Coil
Dimensions (µm)	1×7–9	1×7–10	0.8–1×2.5–9	0.6–1×4–7	0.8–1.2×7–20	1×7–10	0.8–1×5–8	0.8–$1.2 \times {<}20$	0.6–1×6–10	$0.6 \times$ variable
Motility	Gliding	Multiple subpolar flagella	Flagella or none	Peritrichous flagella	Peritrichous flagella	Peritrichous flagella	Polar to subpolar flagella; cell bundles move as a unit	Peritrichous flagella	Peritrichous flagella	Unknown
Major carotenoids	4,4'-diaponeurosporene	4,4'-diaponeurosporene	4,4'-diaponeurosporene	Neurosporene[c]	Neurosporene[c]	4,4'-diaponeurosporene	4,4'-diaponeurosporene	OH-diaponeurosporene gluco side esters	OH-diaponeurosporene gluco side esters	OH-diaponeurosporene gluco side esters
Vitamin requirement	Biotin	Biotin	Biotin	Biotin	Biotin	Biotin	Biotin	Biotin	Biotin	None
Carbon sources photometabolized	P, L, YE	P, L, fructose, glucose, ribose; A, B, ethanol ($+CO_2$); YE	P, L, A, YE	P, L, A; B ($+CO_2$); C, malate, YE	P, L, A, C, Pr, YE	P, L; A, B ($+CO_2$); YE	P, L; A, B ($+CO_2$)	P, A, Pr ($+HCO_3^-/CO_3^{2-}$)	P, L, A ($+CO_3^{2-}$)	P, A, B, Pr ($+HCO_3^-/CO_3^{2-}$)
Pyruvate fermentation	+	+	+	+	+	+	+	–	–	–
Nitrogen source[d]	NH_3, Gln, N_2	NH_3, Gln, N_2	NH_3, Gln, N_2	NH_3, C, Gln, N_2	ND	NH_3, Gln, N_2	NH_3, Gln, N_2	NH_3	NH_3	NH_3, Gln, Asn, YE, N_2
Biosynthetic sulfur source[e]	SO_4^{2-}, $S_2O_3^{2-}$, Met, Cys	$S_2O_3^{2-}$, Met, Cys	$S_2O_3^{2-}$, H_2S, Met, Cys	H_2S, S^0	H_2S	SO_4^{2-}, $S_2O_3^{2-}$, Met, Cys	SO_4^{2-}, $S_2O_3^{2-}$	H_2S	H_2S	SO_4^{2-}, H_2S
Endospores produced	N/O	+	+	+	N/O	N/O	+	+[f]	+	N/O
Optimum temperature (°C)	37–42	37–42	50–52	32	31–36	38–42	37–40	25–30	30	30–35
Optimum pH	6.2–7	6.2–7	6–7	7–8	7–8	6.2–7	7	9	8.5–9	8.5
DNA G + C content (mol%)	52	54.8	54.6–55	51.3	57.2–57.7	50.3	51.8	44.9	45	ND

[a]Symbols: +, >85% positive; d, different strains give different reactions (16–84% positive); –, 0–15% positive; w, weak reaction; ND, not determined; N/O, none observed; A, acetate; B, butyrate; C, casein hydrolysate; L, lactate; P, pyruvate; Pr, propionate; YE, yeast extract.

[b]Data obtained from Gest and Favinger (1983), Beer-Romero and Gest (1987), Beer-Romero et al. (1988), Kimble and Madigan (1992), Kimble et al. (1995), Ormerod et al. (1996a), Stevenson et al. (1997), Takaichi et al. (1997), Bryantseva et al. (1999), Bryantseva et al. (2000a, 2001a), Takaichi et al. (2003), and Asao et al. (2006).

[c]HPLC analysis was not conducted in determining the type of carotenoid. Bryantseva et al. (2000b) indicated the presence of neurosporene based on the in vivo absorption spectrum peak at 412 nm. However, it is likely that this is 4,4'-diaponeurosporene, as for other heliobacteria that grow at neutral pH.

[d]The standard three-letter amino acid abbreviations were used where appropriate. Nitrogen fixation was not tested for Heliorestis baculata and Heliorestis daurensis.

[e]The standard three-letter amino acid abbreviations were used where appropriate. H_2S may exist in its anionic form (HS^-) at high pH, in which the species of Heliorestis are normally cultured.

[f]Phase-bright structures in the coiled cells were observed in the enrichment cultures; whether these structures were indeed endospores is unknown (Bryantseva et al., 1999). Kimble-Long and Madigan (2001) present molecular evidence that endospore formation is a universal property of species of Heliobacteriaceae.

contain bacteriochlorophyll *a* or *b*, and the green sulfur (*Chlorobium* and relatives) and green nonsulfur (*Chloroflexus* and relatives) bacteria, which contain bacteriochlorophylls *c*, *d*, or *e*, or bacteriochlorophyll c_s, respectively. *In vivo* absorption spectra (performed anoxically) of heliobacteria show a major peak at 784–792 nm (depending on the species) due to absorption by bacteriochlorophyll *g* and comparatively small peaks at about 575 and 670 nm; the latter is due to a small amount of 8′-OH-chlorophyll *a* present in the reaction center. Exposure of cultures of heliobacteria to air causes the oxidation of bacteriochlorophyll *g* to chlorophyll *a*, a major increase in absorbance at 670 nm coupled to a corresponding loss of absorbance at 788 nm, and loss of cell viability. Fermentatively (anoxic/dark) grown cells of heliobacteria produce bacteriochlorophyll and carotenoids (Kimble et al., 1994), showing that pigment synthesis does not require light.

The cell wall of heliobacteria is fragile, and stationary phase cells of many species form sphaeroplasts. The peptidoglycan of the heliobacteria, which is present in cells in small amounts, contains L,L-diaminopimelic acid instead of *meso*-diaminopimelic acid as the diamino acid. Heliobacterial cell walls also contain a considerable amount of lipid, although this lipid is not lipopolysaccharide (Beck et al., 1990) and is not arranged in an outer membrane. Heliobacteria are unusually sensitive to penicillin; for example, growth of *Heliobacterium chlorum* is inhibited by 2 ng/ml of penicillin G, a level 1000-fold lower than the growth-inhibitory level for *Escherichia coli* (Beer-Romero et al., 1988).

The heliobacteria are differentiated from all other anoxygenic phototrophs by virtue of their production of bacteriochlorophyll *g*.

A summary of the properties of the four genera and ten species described within the family *Heliobacteriaceae* is given in Table 160.

Enrichment, isolation, and ecology

Enrichment of heliobacteria takes advantage of their ability to form endospores and their presence in soils; rice soils are particular good sources of these phototrophs (Stevenson et al., 1997). A soil sample is pasteurized (80 °C for 10 min) in a tube of anoxically prepared growth medium (for enrichment, a medium composed of 0.25% (w/v) yeast extract at pH 7 works well), sealed under an atmosphere of N_2:CO_2 (95:5 v/v), and incubated in incandescent light (> 25 μE m^{-2} s^{-1}) at 38–40 °C. It should be emphasized that all heliobacteria isolated to date are strict anaerobes. Thus, extreme care should be taken to ensure that media and growth conditions for the isolation of heliobacteria are strictly anoxic.

Successful enrichments for heliobacteria typically show a green film of cells atop the soil in the tube. This material can be removed with a sterile Pasteur pipette for streaking plates of prereduced enrichment medium (or defined media, see Kimble and Madigan, 1992; Kimble et al., 1995; Stevenson, 1997) inside an anoxic glove bag. Streaked plates are incubated inside illuminated anoxic jars (e.g., Gas-Pak jars), and green colonies with irregular edges form within 2–3 d. Higher light intensities are tolerated by heliobacteria than by purple and green bacteria (Kimble-Long and Madigan., 2002), and high light stimulates rapid enrichment. Pure cultures of heliobacteria can be obtained by repeated picking and restreaking of well-isolated colonies. Alternatively, isolated colonies can be obtained in anoxic agar dilution tubes and picked for transfer to liquid media.

In a survey of soils for the presence of heliobacteria, it was found that these phototrophs are common in rice paddy soils in both traditional rice-growing regions (such as Southeast Asia) and in modern cultivated rice fields (such as those in the southern United States). Other agricultural or garden soils only infrequently yielded heliobacteria (Stevenson et al., 1997). Aquatic habitats, with the exception of neutral to alkaline hot springs (below 70 °C) and soda lake sediments (see above), have not yielded heliobacteria. The connection between heliobacteria and rice plants suggest that a plant-bacterium association may exist, perhaps with the photoheterotrophic heliobacteria assimilating organic compounds excreted by rice plant roots and the plants benefiting from fixed nitrogen excreted by the nitrogen-fixing heliobacteria. All species of heliobacteria strongly fix molecular nitrogen, and some species contain alternative (non-molybdenum) nitrogenases (Kimble and Madigan, 1992; Madigan, 1995).

Genus I. **Heliobacterium** Gest and Favinger 1985, 223[VP] (Effective publication: Gest and Favinger 1983, 15.)

Marie Asao and Michael T. Madigan

He.li.o.bac.te′ri.um. Gr. n. *helios* sun; Gr. neut. n. *bakterion* a small rod; N.L. neut. n. *Heliobacterium* sun bacterium.

Rod-shaped cells, slightly curved, 0.8–1 × 4–10 μm (Figure 155 and Figure 156) or **spirilla**, 1–1.2 × 4–10 μm (Figure 157). Cells lack intracytoplasmic membranes or chlorosomes, and thus electron micrographs of thin sections give no indication that the cells are phototrophic (Figure 156). Growth occurs both **photoheterotrophically (anoxic/light) or fermentatively (anoxic/dark). Pyruvate metabolized both phototrophically and chemotrophically.** Colonies on a yeast extract/mineral salts medium (medium PYE, Kimble et al., 1995) frequently have irregular cell margins because of wavy protrusions of cell masses in palisade formations similar to what is seen in gliding bacteria. Motility variable but usually by **polar or subpolar flagella**; *Heliobacterium chlorum* may show slow gliding movement. Some species, particularly *Heliobacterium chlorum*, lyse readily at stationary phase in rich or defined media. **One species is thermophilic.**

The major differentiating properties for species of *Heliobacterium* are listed in Table 160.

DNA G+C content (mol%): 51–57.

Type species: **Heliobacterium chlorum** Gest and Favinger 1985, 223[VP] (Effective publication: Gest and Favinger 1983, 15.).

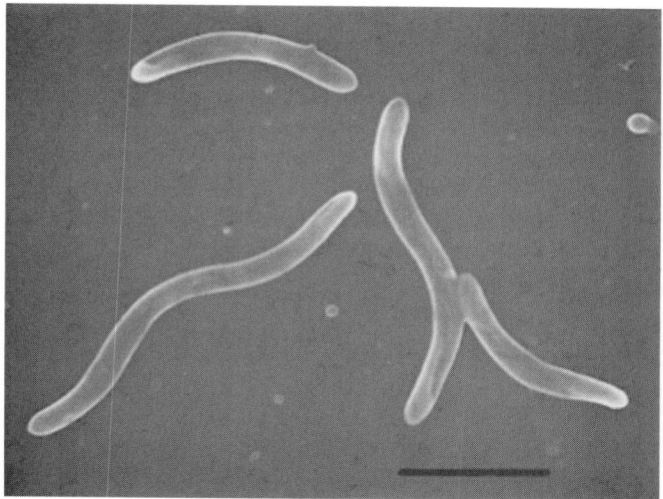

FIGURE 157. Scanning electron micrograph of cells of *Heliobacterium gestii*. Bar = 5 μm.

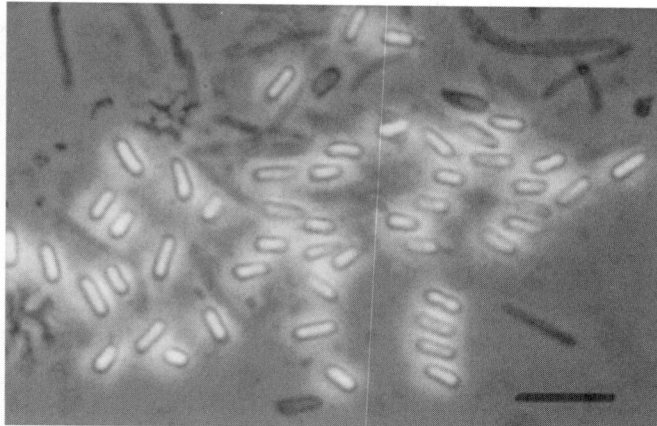

FIGURE 158. Phase-contrast photomicrograph of endospores produced in a culture of *Heliobacterium gestii*. Bar = 5 μm. (Reprinted with permission from Ormerod et al., 1996a. Archives of Microbiology *165*: 226–234.)

Differentiation of the genus *Heliobacterium* from other genera

The genus *Heliobacterium* is differentiated from other genera by its distinct 16S rRNA gene phylogeny. In addition, species of *Heliobacterium* are motile by gliding or subpolar flagella while species of *Heliobacillus* possess peritrichous flagella, and species of *Heliophilum* are motile as cell bundles that move as a unit. Its pH optimum of 7 distinguishes it from *Heliorestis*, which has a pH optimum of 9.

List of species of the genus *Heliobacterium*

1. **Heliobacterium chlorum** Gest and Favinger 1985, 223[VP] (Effective publication: Gest and Favinger 1983, 15.)

    chlo′rum. N.L. neut. adj. *chlorum* from the Gr. adj. *chloros* green.

    Cells are rod shaped; other characteristics are the same as those described for the genus. Isolated from garden soil on the campus of Indiana University, Bloomington, USA.
    DNA G+C content (mol%): 52 (T_m).
    Type strain: ATCC 35205, DSM 3682.
    GenBank accession number (16S rRNA gene): M11212.

2. **Heliobacterium gestii** Ormerod, Kimble, Nesbakken, Torgersen, Woese and Madigan 1996b, 1189[VP] (Effective publication: Ormerod, Kimble, Nesbakken, Torgersen, Woese and Madigan 1996a, 233.)

    gest.i.i. N.L. gen.n. *gestii* of Gest, named for Howard Gest, an American microbiologist.

    Spirillum (Figure 157) that is motile by polar or subpolar flagella. Subterminal, cylindrical endospores (Figure 158) formed in freshly isolated strains. Optimal pH is 7. Incapable of assimilatory sulfate reduction. NaCl not required and growth inhibitory above 1% (w/v). This is the only species of heliobacteria known to use sugars as carbon sources. Isolated from soil of a Thailand rice paddy.
    DNA G+C content (mol%): 55.1 (T_m).
    Type strain: Chainat, ATCC 43375, DSM 11169.
    GenBank accession number (16S rRNA gene): AB100837, L36198.

3. **Heliobacterium modesticaldum** Kimble, Mandelco, Woese and Madigan 1996, 1189[VP] (Effective publication: Kimble, Mandelco, Woese and Madigan 1995, 266.)

    mo.des.ti.cal′dum. L. neut. adj. *modesticum*; L. neut. adj. *caldum*; N.L. neut. adj. *modesticaldum* moderately hot; named for its thermophilic character.

    Cells are rod shaped or slightly curved, 0.8–1 × 2.5–9 μm (Figure 157). Motile by flagella or nonmotile. Forms cylindrical, subterminal endospores. Thermophilic; optimum growth temperature is 52 °C (temperature range 30–56 °C). NaCl not required for growth, and growth inhibited by 1% (w/v) NaCl. Optimal pH is 6.5. Fixes molecular nitrogen up to its maximum growth temperature. Isolated from neutral to alkaline hot spring microbial mats and volcanic soils. Type strain isolated from soil obtained near alkaline hot springs at Reykjanes, Iceland.
    DNA G+C content (mol%): 54.6–55 (T_m).
    Type strain: Ice1, ATCC 51547, DSM 9504.
    GenBank accession number (16S rRNA gene): AB100836, U14559.

4. **Heliobacterium sulfidophilum** Bryantseva, Gorlenko, Tourova, Kuznetsov, Lysenko, Bykova, Gal'chenko, Mityushina and Osipov 2001b, 1[VP] (Effective publication: Bryantseva, Gorlenko, Tourova, Kuznetsov, Lysenko, Bykova, Gal'chenko, Mityushina and Osipov 2000b, 333.)

    sul.fi.do.phil.um. N.L. adj, *sulfidophilum* sulfide-loving.

    Cells are rod shaped and motile by peritrichous flagella. Produces endospores. Optimal pH is 7.5, and optimal growth temperature is 35 °C. Sulfide tolerant (up to 2 mM) and oxidizes sulfide to elemental sulfur. Photoheterotrophic growth on acetate, pyruvate, or lactate. Isolated from a hot spring near the Bol'shaya River, Russia.
    DNA G+C content (mol%): 51.3 (T_m).
    Type strain: BR4, UNIQEM 113.
    GenBank accession number (16S rRNA gene): AF249678.

5. **Heliobacterium undosum** Bryantseva, Gorlenko, Tourova, Kuznetsov, Lysenko, Bykova, Gal'chenko, Mityushina and Osipov 2001b, 1[VP] (Effective publication: Bryantseva, Gorlenko, Tourova, Kuznetsov, Lysenko, Bykova, Gal'chenko, Mityushina and Osipov 2000b, 333.)

un.do′sum. L. neut. adj. *undosum* curving.

Cells are short rods to twisted spirilla; motile by peritrichous flagella. No evidence of endospores. Optimal pH is 7.8, and optimal growth temperature is 35 °C. Highly sulfide tolerant and oxidizes sulfide to elemental sulfur. Photoheterotrophic growth on pyruvate, lactate, or propionate. Isolated from Garginskii sulfidic hot spring, Russia.

DNA G+C content (mol%): 57.2 (T_m).
Type strain: BG29, DSM 13378, UNIQEM 114.
GenBank accession number (16S rRNA gene): AF249679.

Genus II. **Heliobacillus** Beer-Romero and Gest 1998, 627[VP] (Effective publication: Beer-Romero and Gest 1987, 113.)

MARIE ASAO AND MICHAEL T. MADIGAN

He.li.o.ba.cil′lus. Gr. n. *helios* sun; L. dim. n. *bacillum* a small rod; N.L. masc. n. *Heliobacillus* sun rod.

Highly motile, straight rod-shaped cells containing peritrichous flagella. For other properties, see Table 160. *Heliobacillus* may be differentiated from other genera of heliobacteria by its morphology and motility. *Heliobacillus* cells are straight rods and not curved like most species of *Heliobacterium*. Cells in young cultures are highly motile and swim much faster than cells of *Heliobacterium*.

DNA G+C content (mol%): 50.3.
Type species: **Heliobacillus mobilis** Beer-Romero and Gest 1998, 627[VP] (Effective publication: Beer-Romero and Gest 1987, 113.).

List of species of the genus *Heliobacillus*

1. **Heliobacillus mobilis** Beer-Romero and Gest 1998, 627[VP] (Effective publication: Beer-Romero and Gest 1987, 113.)

mo′bil.is. L. masc. adj. *mobilis* movement; named for its rapid motility.

Cells are rod-shaped, 1 × 7–10 μm, and highly motile in young cultures. Produce peritrichous flagella. Optimal pH is 7. NaCl not required. Capable of assimilatory sulfate reduction. Isolated from dry soil from Thailand.

DNA G+C content (mol%): 50.3 (T_m).
Type strain: 6, ATCC 43427, DSM 6151.
GenBank accession number (16S rRNA gene): U14560.

Genus III. **Heliophilum** Ormerod, Kimble, Nesbakken, Torgersen, Woese and Madigan 1996b, 1189[VP] (Effective publication: Ormerod, Kimble, Nesbakken, Torgersen, Woese and Madigan 1996a, 233.)

MARIE ASAO AND MICHAEL T. MADIGAN

He.li.o′phi.lum. Gr. n. *helios* sun; *philos* loving; N.L. neut. n. *Heliophilum* sun lover.

Cells are straight rods with tapered ends and subpolar flagella. They do not exist singly but form into bundles, typically of 8–16 cells, that swim slowly as a collective unit. Entire bundles show scotophotophobic response. Heat-resistant endospores are produced in newly isolated cultures. Capable of assimilatory sulfate reduction; growth is inhibited by sulfide at levels above 0.1 mM. Biotin required. NaCl is not required for growth and is inhibitory above 1% (w/v). Other properties are described in Table 160 or are the same as those described for the family. The genus *Heliophilum* is distinguished from other heliobacteria by its collective motility and unusual sensitivity to sulfide. Most heliobacteria are quite sulfide tolerant. However, it should be cautioned that sulfide sensitivity may be strain specific, so this trait may be of limited diagnostic value.

DNA G+C content (mol%): 51.8 (T_m).
Type species: **Heliophilum fasciatum** Ormerod, Kimble, Nesbakken, Torgersen, Woese and Madigan 1996b, 1189[VP] (Effective publication: Ormerod, Kimble, Nesbakken, Torgersen, Woese and Madigan 1996a, 233.).

List of species of the genus *Heliophilum*

1. **Heliophilum fasciatum** Ormerod, Kimble, Nesbakken, Torgersen, Woese and Madigan 1996b, 1189[VP] (Effective publication: Ormerod, Kimble, Nesbakken, Torgersen, Woese and Madigan 1996a, 233.)

fas.ci.a′tum. L. neut. part. adj. *fasciatum* bundled; named for the fact that cells form into bundles that move as a unit.

The characteristics are the same as those described for the genus. Isolated from rice soil from Tanzania, continent of Africa.

DNA G+C content (mol%): 51.8 (T_m).
Type strain: Tanzania, ATCC 51790, DSM 11170.
GenBank accession number (16S rRNA gene): L36197.

Genus IV. **Heliorestis** Bryantseva, Gorlenko, Kompantseva, Achenbach and Madigan 2000c, 949[VP] emend. Bryantseva, Gorlenko, Kompantseva, Tourova, Kuznetsov and Osipov 2000a, 289 (Effective publication: Bryantseva, Gorlenko, Kompantseva, Achenbach and Madigan 1999, 173.)

MARIE ASAO AND MICHAEL T. MADIGAN

He.li.o.res'tis. Gr. n. *helios* sun; L. fem. n. *restis* a rope; N.L. fem. n. *Heliorestis* sun rope.

Cells are coils, bent filaments, or rods. Alkaliphilic (optimum pH 8.5–9) and some species can grow at a pH as high as 10.5. No growth occurs at pH 7. Not halophilic. Sulfide tolerant, growing in media containing up to 3–10 mM sulfide. Optimal growth temperature is 30 °C; no growth above 40 °C. Obligately photoheterotrophic; chemoorganotrophy in darkness with pyruvate fermentation does not occur. Major carotenoids are OH-diaponeurosporene glucoside esters, although 4,4'-diaponeurosporene is also present as a minor component (Takaichi et al., 2003). Other properties are described in Table 160 or are the same as those described for the family. Species of *Heliorestis* are distinguished by their alkaliphily and the presence of glycosylated carotenoids.

DNA G+C content (mol%): 45.

Type species: **Heliorestis daurensis** Bryantseva, Gorlenko, Kompantseva, Achenbach and Madigan 2000c, 949[VP] (Effective publication: Bryantseva, Gorlenko, Kompantseva, Achenbach and Madigan 1999, 173.)

List of species of the genus *Heliorestis*

1. **Heliorestis daurensis** Bryantseva, Gorlenko, Kompantseva, Achenbach and Madigan 2000c, 949[VP] (Effective publication: Bryantseva, Gorlenko, Kompantseva, Achenbach and Madigan, 1999, 173.)

dau.ren'sis. N.L. adj. *daurensis* the name of the geographic region Dauria (Daur Steppe, Russia), from which the type strain was isolated.

Cells are 0.8–1.2 μm wide, forming a twisted or bent filament with a length up to 20 μm. Slowly motile by peritrichous flagella. Optimal growth pH is 9; pH range from 7.5–10.5. NaCl not required and growth inhibitory above 1% (w/v). Biotin required; yeast extract at 0.025–0.05% (w/v) is highly stimulatory for growth. Acetate, pyruvate, or propionate used as carbon sources in the presence of sulfide/HCO_3^-/CO_3^{2-}. Very sulfide tolerant; growth occurs in media containing up to 10 mM sulfide; sulfide is oxidized to elemental sulfur and polysulfide. Other characteristics are as described for the genus. Isolated from microbial mats on the shoreline and soils near Lake Barun Torey which is located on the Daur Steppe (South Chita region, Southeast Siberia, Russia).

DNA G+C content (mol%): 44.9 (T_m).

Type strain: BT-H1, ATCC 700798.

GenBank accession number (16S rRNA gene): AF079102.

2. **Heliorestis baculata** Bryantseva, Gorlenko, Kompantseva, Tourova, Kuznetsov and Osipov 2001a, 264[VP] (Effective publication: Bryantseva, Gorlenko, Kompantseva, Tourova, Kuznetsov and Osipov 2000a, 290.)

ba.cu.la'ta. L. adj. *baculatus* rod-shaped; named for the morphological characteristic of the cells.

Cells are straight or slightly curved rods, 0.6–1 × 6–10 μm. Motile by peritrichous flagella. Optimum growth pH is 8.5–9; pH range from 8–10.2. NaCl is not required. Growth observed in media containing up to 3% (w/v) NaCl. Biotin required; low levels of yeast extract are highly stimulatory. Sodium carbonate required (optimum, 0.5–1% (w/v); range, 0.2–2%). Acetate, pyruvate, or lactate is used as carbon source in the presence of sulfide and carbonate. Growth in media containing up to 3 mM sulfide; sulfide is oxidized to

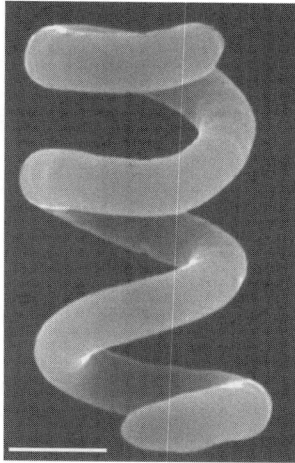

FIGURE 159. Scanning electron micrograph of *Heliorestis convoluta.* Bar = 1 μm. (Reprinted with permission from Asao et al., 2006. Extremophiles *10*: 403–410.)

sulfur and polysulfide. Other characteristics are as described for the genus. Isolated from shoreline soil of Lake Ostozhe, located in the steppe of South Chita region (Southeast Siberia, Russia).

DNA G+C content (mol%): 45 (T_m).

Type strain: OS-H1, DSM 13446.

GenBank accession number (16S rRNA gene): AB100838, AF249680.

Other organisms

1. **"Heliorestis convoluta"** Asao, Jung, Achenbach and Madigan 2006, 409.

con.vo.lu'ta. L. fem. adj. *convoluta* coiled; named for the unique morphology of the cells.

Cells are 0.6 μm wide, forming a coil of variable length (Figure 159). The coil begins as a ring-shaped cell that elongates, forms a septum, but fails to divide. Motile. Optimum growth pH is 8.5–9; pH range is 8–10. NaCl is not required. Growth observed in media containing up to 3% (w/v) NaCl.

No growth factors are required. Acetate, pyruvate, propionate, or butyrate is used as carbon sources in the presence of sulfide/HCO_3^-/CO_3^{2-}. Ammonia, glutamine, asparagine, yeast extract, or N_2 is used as nitrogen source. Very sulfide tolerant; growth occurs in media containing up to 10 mM sulfide; sulfide is oxidized to sulfur. Other characteristics are as described for the genus. Isolated from shoreline soil and water of Lake El Hamra, Wadi El Natroun, Egypt.

DNA G+C content (mol%): not reported.

Type strain: HH, ATCC BAA-1281.

GenBank accession number (16S rRNA gene): DQ266255.

References

Amesz, J. 1995. The antenna-reaction center complex of heliobacteria. *In* Blankenship, Madigan and Bauer (Editors), Anoxygenic Photosynthetic Bacteria. Kluwer Academic Publishers, Dordrecht, The Netherlands, pp. 687–697.

Asao, M., D.O. Jung, L.A. Achenbach and M.T. Madigan. 2006. *Heliorestis convoluta* sp. nov., a coiled, alkaliphilic *Heliobacterium* from the Wadi El Natroun, Egypt. Extremophiles *10*: 403–410.

Beck, H., G.D. Hegeman and D. White. 1990. Fatty acid and lipopolysaccharide analyses of three *Heliobacterium* spp. FEMS Microbiol. Lett. *57*: 229–232.

Beer-Romero, P. and H. Gest. 1987. *Heliobacillus mobilis*, a peritrichously flagellated anoxyphototroph containing bacteriochlorophyll *g*. FEMS Microbiol. Lett. *41*: 109–114.

Beer-Romero, P., J.L. Favinger and H. Gest. 1988. Distinctive properties of bacilliform photosynthetic heliobacteria. FEMS Microbiol. Lett. *49*: 451–454.

Beer-Romero, P. and H. Gest. 1998. *In* Validation of publication of new names and new combinations previously effectively published outside the IJSB. List no. 65. Int. J. Syst. Evol. Microbiol. *48*: 627.

Brockmann, H. and A. Lipinski. 1983. Bacteriochlorophyll *g*. A new bacteriochlorophyll from *Heliobacterium chlorum*. Arch. Microbiol. *136*: 17–19.

Bryantseva, I.A., V.M. Gorlenko, E.I. Kompantseva, L.A. Achenbach and M.T. Madigan. 1999. *Heliorestis daurensis*, gen. nov. sp. nov., an alkaliphilic rod-to-coiled-shaped phototrophic *Heliobacterium* from a Siberian soda lake. Arch. Microbiol. *172*: 167–174.

Bryantseva, I.A., V.M. Gorlenko, E.I. Kompantseva, T.P. Tourova, B.B. Kuznetsov and G.A. Osipov. 2000a. Alkaliphilic heliobacterium *Heliorestis baculata* sp. nov. and emended description of the genus *Heliorestis*. Arch. Microbiol. *174*: 283–291.

Bryantseva, I.A., V.M. Gorlenko, T.P. Tourova, B.B. Kuznetsov, A.M. Lysenko, S.A. Bykova, V.F. Gal'chenko, L.L. Mityushina and G.A. Osipov. 2000b. *Heliobacterium sulfidophilum* sp. nov. and *Heliobacterium undosum* sp. nov.: sulfide-oxidizing heliobacteria from thermal sulfidic springs. Microbiology (En. transl. from Mikrobiologiya) *69*: 325–334.

Bryantseva, I.A., V.M. Gorlenko, E.I. Kompantseva, L.A. Achenbach and M.T. Madigan. 2000c. *In* Validation of publication of new names and new combinations previously effectively published outside the IJSEM. List no. 74. Int. J. Syst. Evol. Microbiol *50*: 949–950.

Bryantseva, I.A., V M. Gorlenko, E.I. Kompantseva, T.P. Tourova, B.B. Kuznetsov and G.A. Osipov. 2001a. *In* Validation of publication of new names and new combinations previously effectively published outside the IJSEM. List no. 79. Int. J. Syst. Evol. Microbiol. *51*: 263–265.

Bryantseva, I.A., V M. Gorlenko, E.I. Kompantseva, B.B.K. T.P. Tourova, A.M. Lysenko, A.M. Bykova, V.F. Galchenko, L.L. Mityushina and G.A. Osipov. 2001b. *In* Validation of publication of new names and new combinations previously effectively published outside the IJSEM. List no. 78. Int. J. Syst. Evol. Microbiol. *51*: 1–2.

Felsenstein, J. 1989. PHYLIP-Phylogeny inference package (Version 3.2). Cladistics *5*: 164–166.

Gest, H. and J.L. Favinger. 1983. *Heliobacterium chlorum*, an anoxygenic brownish-green photosynthetic bacterium containing a "new" form of bacteriochlrophyll. Arch. Microbiol. *136*: 11–16.

Gest, H. and J.L. Favinger. 1985. *In* Validation of publication of new names and new combinations previously effectively published outside the IJSB. List no. 17. Int. J. Syst. Bacteriol *35*: 223–225.

Kimble-Long, L.K. and M.T. Madigan. 2001. Molecular evidence that the capacity for endosporulation is universal among phototrophic heliobacteria. FEMS Microbiol. Lett. *199*: 191–195.

Kimble-Long, L.K. and M.T. Madigan. 2002. Irradiance effects on growth and bacteriochlorophyll content of phototrophic heliobacteria, purple and green photosynthetic bacteria. Photosynthetica *40*: 629–632.

Kimble, L.K. and M.T. Madigan. 1992. Nitrogen fixation and nitrogen metabolism in heliobacteria. Arch. Microbiol. *158*: 155–161.

Kimble, L.K., A.K. Stevenson and M.T. Madigan. 1994. Chemotrophic growth of heliobacteria in darkness. FEMS Microbiol. Lett. *115*: 51–55.

Kimble, L.K., L. Mandelco, C.R. Woese and M.T. Madigan. 1995. *Heliobacterium modesticaldum*, sp. nov., a thermophilic *Heliobacterium* of hot springs and volcanic soils. Arch. Microbiol. *163*: 259–267.

Kimble, L.K., L. Mandelco, C.R. Woese and M.T. Madigan. 1996. *In* Validation of publication of new names and new combinations previously effectively published outside the IJSB. List no. 59. Int. J. Syst. Bacteriol. *46*: 1189–1190.

Madigan, M.T. 1995. Microbiology of nitrogen fixation by photosynthetic bacteria. *In* Blankenship, Madigan and Bauer (Editors), Anoxygenic Photosynthetic Bacteria. Kluwer Academic Publishers, Dordrecht, The Netherlands, pp. 915–924.

Madigan, M.T. and J.G. Ormerod. 1995. Taxonomy, physiology and ecology of heliobacteria. *In* Blankenship, Madigan and Bauer (Editors), Anoxygenic Photosynthetic Bacteria. Kluwer Academic Publishers, Dordrecht, The Netherlands, pp. 17–30.

Madigan, M.T. 2001. Family VI. "*Heliobacteriaceae*" Beer-Romero and Gest 1987, 113. *In* Boone, Castenholz and Garrity (Editors), Bergey's Manual of Systematic Bacteriology, Vol. 1, The *Archaea* and the Deeply Branching and Phototrophic *Bacteria*. Springer, New York, pp. 625–626.

Michalski, T.J., J.E. Hunt, M.K. Bowman, U. Smith, K. Bardeen, H. Gest, J.R. Norris and J.J. Katz. 1987. Bacteriopheophytin *g*: properties and some peculations on a possible primary role for bacteriochlorophylls *b* and *g* in the biosynthesis of chlorophylls. Proc. Natl. Acad. Sci. U.S.A. *84*: 2570–2574.

Ormerod, J.G., L.K. Kimble, T. Nesbakken, Y.A. Torgersen, C.R. Woese and M.T. Madigan. 1996a. *Heliophilum fasciatum* gen. nov. sp. nov. and *Heliobacterium gestii* sp. nov.: endospore-forming heliobacteria from rice field soils. Arch. Microbiol. *165*: 226–234.

Ormerod, J.G., L.K. Kimble, T Nesbakken, Y.A. Torgersen, C.R. Woese and M.T. Madigan. 1996b. *In* Validation of the publication of new names and new combinations previously effectively published outside the IJSB. List no. 59. Int. J. Syst. Bacteriol *46*: 1189–1190.

Sattley, W.M., M.T. Madigan, W.D. Swingley, P.C. Cheung, K.M. Clocksin, A.L. Conrad, L.C. Dejesa, B.M. Honchak, D.O. Jung, L.E. Karbach, A. Kurdoglu, S. Lahiri, S.D. Mastrian, L.E. Page, H.L. Taylor, Z.T. Wang, J. Raymond, M. Chen, R.E. Blankenship and J.W. Touchman. 2008. The genome of *Heliobacterium modesticaldum*, a phototrophic representative of the *Firmicutes* containing the simplest photosynthetic apparatus. J. Bacteriol. *190*: 4687–4696.

Stevenson, A.K., L.K. Kimble, C.R. Woese and M.T. Madigan. 1997. Characterization of new phototrophic heliobacteria and their habitats. Photosynth. Res. *53*: 1–12.

Takaichi, S., K. Inoue, M. Akaike, M. Kobayashi, H. Ohoka and M.T. Madigan. 1997. The major carotenoid in all known species of heliobacteria is the C-30 carotenoid 4,4'-diaponeurosporene, not neurosporene. Arch. Microbiol. *168*: 277–281.

Takaichi, S., H. Oh-Oka, T. Maoka, D.O. Jung and M.T. Madigan. 2003. Novel carotenoid glucoside esters from alkaliphilic heliobacteria. Arch. Microbiol. *179*: 95–100.

Woese, C.R., B.A. Debrunner-Vossbrinck, H. Oyaizu, E. Stackebrandt and W. Ludwig. 1985. Gram-positive bacteria: possible photosynthetic ancestry. Science *229*: 762–765.

Family V. **Lachnospiraceae** fam. nov.

Lach.no.spi.ra′ce.ae. N.L. masc. n. *Lachnospira* type genus of the family; *-aceae* ending to denote family; N.L. fem. pl. n. *Lachnospiraceae* the *Lachnospira* family.

The family *Lachnospiraceae* is described on the basis of phylogenetic analyses of 16S rRNA gene sequences (Figure 160); the family contains the genera *Lachnospira* (type genus), *Acetitomaculum*, *Anaerostipes*, *Bryantella*, *Butyrivibrio*, *Catonella*, *Coprococcus*, *Dorea*, *Hespellia*, *Johnsonella*, *Lachnobacterium*, *Moryella*, *Oribacterium*, *Parasporobacterium*, *Pseudobutyrivibrio*, *Roseburia*, *Shuttleworthia*, *Sporobacterium*, and *Syntrophococcus*.

The family is morphologically diverse and includes rods, vibrios, and cocci. All species are anaerobes.

Type genus: **Lachnospira** Bryant and Small 1956a, 24[AL].

Genus I. **Lachnospira** Bryant and Small 1956a, 24[AL]

NANCY A. CORNICK AND THADDEUS B. STANTON

Lach. no. spi′ra. Gr. n. *lachnos* woolly hair, down; L. n. *spira* a coil; N.L.fem. n. *Lachnospira* named for the filamentous or "woolly" colonies formed in agar by curved or helical cells of *Lachnospira multipara*.

Straight or slightly curved rod-shaped cells. May appear helical. Single cells measure 0.35–0.6×2.0–4.0 μm. Cells occur in pairs and occasionally in long chains. Gram-stain-variable or -positive. Cells possess Gram-positive ultrastructure. **Actively motile** by monotrichous or peritrichous flagella, depending on the species. **Strictly anaerobic.** Grows at 30–45 °C. Grows in anaerobically prepared media containing rumen fluid or yeast extract, inorganic salts, and pectin or polygalacturonic acid. May also grow in a chemically defined medium. Chemo-organotrophic metabolism. **Pectinolytic.** Ferments pectin, polygalacturonic acid, fructose, and cellobiose. Acetate, formate, ethanol, and CO_2 are the major end products from polygalacturonic acid and pectin fermentation. Small amounts of H_2 may also be produced. Methanol is also produced when pectin is fermented. Succinate, butyrate, and propionate are not produced. Galacturonic acid (the monomer of polygalacturonic acid) is not fermented. Does not produce indole, catalase, or H_2S. Does not reduce nitrate. Does not hydrolyze gelatin or starch. **Indigenous to mammalian intestinal tract.** Isolated from bovine rumen contents and from swine feces and cecal contents.

Two species of *Lachnospira* have been characterized. They can be differentiated by characteristics given in Table 161.

DNA G+C content (mol%): 38–45.

Type species: **Lachnospira multipara** Bryant and Small 1956a, 24[AL].

Further descriptive information

Cell walls have a Gram-positive structure but *Lachnospira multipara* may appear to stain as Gram-negative (Cheng et al., 1979). Formerly, analysis of the 16S rRNA grouped *Lachnospira* within the XIVa cluster of the *Clostridium* subphylum, which includes several diverse Gram-positive genera (Figure 160) (Collins et al., 1994). According to the 16S rRNA phylogenetic analysis presented in the roadmap to this volume (Figure 5), the genus *Lachnospira* is a member of the family *Lachnospiraceae*, order *Clostridiales*, class *Clostridia* in the phylum *Firmicutes*. *Lachnospira multipara* and *Lachnospira pectinoschiza* form a monophyletic subset within this group and *Lachnospira pectinoschiza* shows 94% similarity to *Lachnospira multipara*. The major constituents of the peptidoglycan of *Lachnospira multipara* are muramic acid, glucosamine, alanine, glutamic acid, and *meso*-diaminopimelic acid (Hespell et al., 1993).

Lachnospira multipara is considered one of the major pectin degraders in the rumen even though it accounts for only a small fraction of the total microflora. When cattle are fed diets rich in pectin, such as clover or fruit pulp, the population of *Lachnospira multipara* increases significantly (Bryant et al., 1960). *In vitro* incubation of *Lachnospira multipara* with clover leaflets results in extensive maceration of the clover (Cheng et al., 1979). The ecological role of *Lachnospira pectinoschiza* in the pig gastrointestinal tract is not known but presumably is involved in pectin degradation. Both species degrade pectin by using pectin methyl esterase and extracellular pectate lyase (Preston et al., 1991; Silley, 1985; Wojciechowicz et al., 1980). Polygalacturonic acid is degraded primarily to trigalacturonic acid and some digalacturonic acid by *Lachnospira multipara* and to digalacturonic acid only by *Lachnospira pectinoschiza*.

Glucose is fermented by *Lachnospira multipara* but not by *Lachnospira pectinoschiza*. Arabinose, galactose, glycerol, glycogen, maltose, mannitol, mannose, raffinose, rhamnose, ribose, salicin, sorbitol, sucrose, trehalose, amino acids, arabinoglactan, carboxymethyl cellulose, cellulose, dextrin, gum arabic, inositol, inulin, and xylan do not support growth. The major long-chain fatty acids extracted from whole cells are palmitic acid (16:0) and myristic acid (14:0) (Cornick et al., 1994).

Enrichment and isolation procedures

The isolation of *Lachnospira* requires strict anaerobic conditions and the presence of a fermentable carbohydrate in the medium. Both medium RGM (Hespell et al., 1987) and medium PF (Jensen and Canale-Parola, 1986) support heavy growth. *Lachnospira multipara* may also be grown in a defined medium (Bryant and Robinson, 1962). A semiselective medium containing energy-depleted rumen fluid and polygalacturonic acid as the sole carbon source aids in isolation, but other anaerobic bacteria that utilize pectin, such as *Butrivibrio fibrosolvens* and *Prevotella ruminicola*, will also grow on this medium (Cornick et al., 1994). The distinctive filamentous colony morphology of *Lachnospira multipara* in agar-roll tubes aids in isolate selection.

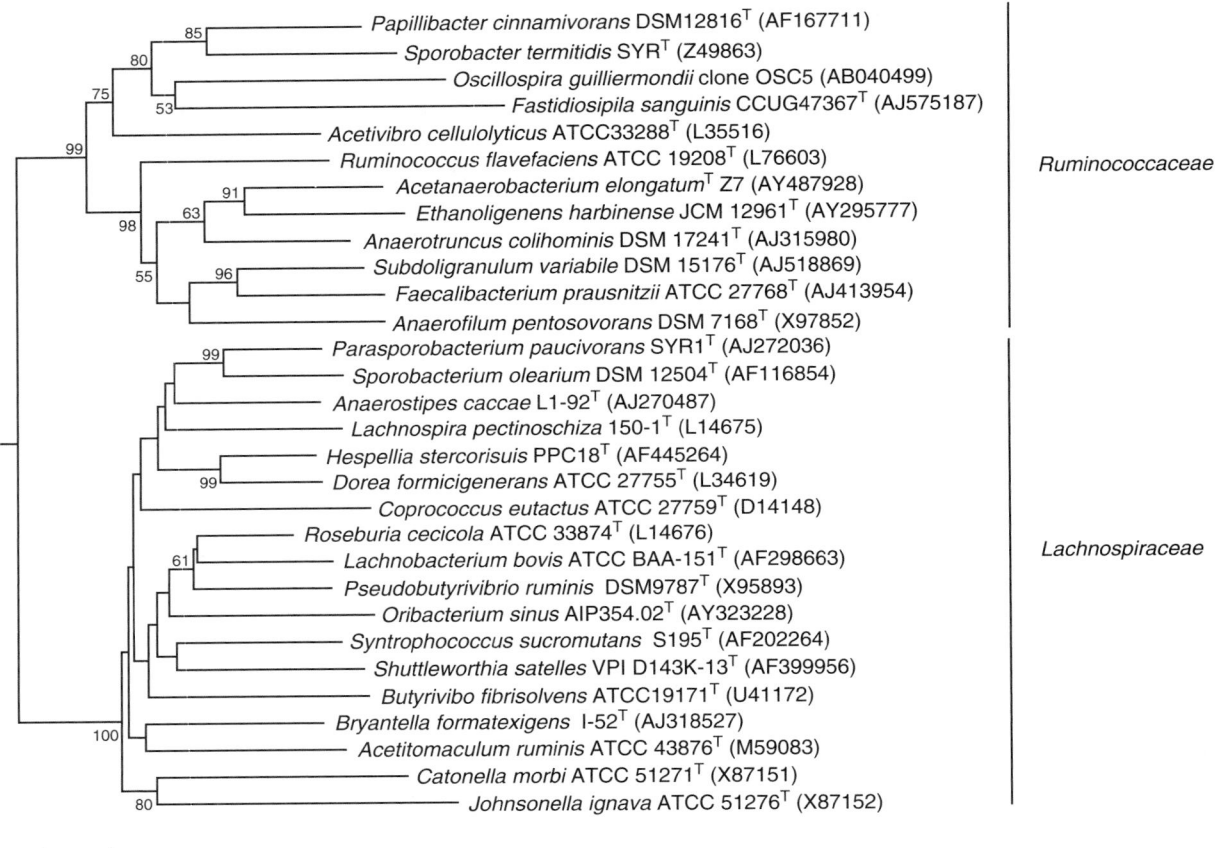

FIGURE 160. 16S rRNA gene sequence based phylogenetic analysis showing the relationships between the genera of the families *Lachnospiraceae* and *Ruminococcaceae*. Each genus is represented by the type species. The tree was reconstructed from distance matrices using the neighbor-joining method. The exception to this is the genus *Lachnospira* for which only a short and low quality sequence is available for the type species *Lachnospira multipara*. The species *Lachnospira pectinoschiza* has been used to represent this genus. The genus *Oscillospira* for which no type material exists is represented by a clone sequence from an organism with the morphological characteristics of the organism originally described as *Oscillospira guilliermondii*. The scale bar represents 2 inferred nucleotide changes per 100 nucleotides.

TABLE 161. Differentation of *Lachnospira* species[a]

Characteristic	*L. multipara*	*L. pectinoschiza*
Flagella	Monotrichous	Peritrichous
Fermentation of:		
Gluconic acid	–	+
Glucose	+	–
Lactose	–	+
Salicin	+	–
Sucrose	+	–
Esculin hydrolysis	+	–

[a]Symbols: +, >85% positive; –, 0–15% positive.

Isolates that grow on polygalacturonic acid but not on glucose are suggestive of *Lachnospira pectinoschiza*.

Maintenance procedures

Both *Lachnospira multipara* and *Lachnospira pectinoschiza* can be frozen at −70 °C in culture broth containing 10% glycerol. *Lachnospira multipara* can also be lyophilized.

Differentiation of the genus *Lachnospira* from other genera

Lachnospira are distinguished from other genera of anaerobic, Gram-stain-positive, non-spore-forming rods primarily by their 16S rRNA sequence. DNA–DNA relative reassociation has also been used to differentiate *Lachnospira* from *Eubacterium*, *Butyrivibrio*, and *Roseburia* (Mannarelli et al., 1990b).

Further reading

Cotta, M. and R. Forster. 2000. The family *Lachnospiraceae*, including the genera *Butyrivibrio*, *Lachnospira* and *Roseburia*. *In* Dworkin (Editor), The Prokaryotes, Springer-Verlag, New York.

List of species of the genus *Lachnospira*

1. **Lachnospira multipara** corrig. Bryant and Small 1956a, 24[AL] (*Lachnospira multiparus* Bryant and Small (1956a)

mul.ti.par′a. L. adj. *multus* much, many; L. suff. *para* from L. v. *pario* to produce; N.L. adj. *multipara* many products produced.

Curved rods 0.4–0.6 × 2.0–4.0 μm in single, pairs, and occasional long chains, weakly Gram-stain-positive, monotrichous subterminal flagella. Colonies in agarose are white and filamentous in appearance. Colonies embedded in agar are wooly in appearance. Good growth at 30–45 °C, no growth at 20 or 50 °C. Glucose, fructose, esculin, cellobiose, sucrose, and salicin are fermented. Xylose fermentation is variable, acetylmethylcarbinol may be produced. Major fermentation products from glucose are acetate, formate, ethanol, and CO_2. Minor amounts of lactate and H_2. Methanol is produced when pectin is fermented. Isolated from the bovine rumen.

DNA G+C content (mol%): 38–39 (Bd).

Type strain: D32, ATCC 19207.

GenBank accession number (16S RNA gene): L14674.

2. **Lachnospira pectinoschiza** Cornick, Jensen, Stahl, Hartman and Allison 1994, 92[VP]

pec′ti.no.schiz.a. N.L. neut. n. *pectinum* pectin; Gr. n. *schizein* to split; N.L. fem. adj. *pectinoschiza* pectin-splitting.

Gram-stain-positive rod 0.36–0.56 × 2.4–3.1 μm in single, pairs, and long chains. Strictly anaerobic. Colonies on agar are circular, umbonate, 3–5 mm in diameter. Grows at 30–45 °C, but not at 25 or 50 °C. Ferments pectin, polygalacturonic acid, D-gluconate, D-fructose, lactose, and cellobiose. Growth on pectin, polygalacturonic acid and D-gluconate is rapid with a doubling time of approximately 55 min. Growth is much slower on cellobiose, lactose, and D-fructose (doubling time, 4–10 h). Other carbohydrates and amino acids are not utilized. Major end products from polygalacturonic acid fermentation are formate, acetate, ethanol, and CO_2. Methanol is produced when pectin is fermented. Ethanol is the major end product when cellobiose, lactose, or fructose are utilized. Bile does not inhibit growth. Spore-like structures may be observed by electron microscopy from cells grown on egg yolk agar supplemented with polygalacturonic acid. However, cells do not survive heating at 60 °C for 5 min or exposure to 50% ethanol for 45 min. Isolated from the cecum and colon of pigs. The 16S rRNA sequence signature of *Lachnospira pectinoschiza* has also been detected in human colonic samples (Wilson and Blitchington, 1996).

DNA G+C content (mol%): 42–45 (T_m).

Type species: strain 150-1.

GenBank accession number (16S RNA gene): L14675.

Genus II. **Acetitomaculum** Greening and Leedle 1995, 879[VP] (Effective publication: Greening and Leedle 1989, 405.)

FRED A. RAINEY

A.ceti.to.ma′cu.lum. L. n. *acetum* vinegar; L. n. *tomaculum* sausage; N.L. neut. n. *Acetitomaculum* vinegar sausage.

Curved rods, 0.8–1.0 μm in diameter and 2.0–4.0 μm in length. Occurring in singles, pairs, or small clumps. **Gram-stain-positive**. Older cultures stain Gram-stain-negative. Cells may be flagellated. **Endospores not observed. Obligate anaerobe.** Strains of this organism are catalase-negative. Mesophilic with growth in the temperature range 34–43 °C. Growth occurs in the pH range 6.4–7.3. Cell wall contains L-serine, D-glutamate, D-alanine, *m*-diaminopimelic acid, D-ornithine, and D-lysine. **Acetogenic, fermentative metabolism.** Ferments formate, glucose, cellobiose, fructose, and esculin to acetic acid. **Oxidizes hydrogen** and **reduces carbon dioxide and carbon monoxide.** Based on 16S rRNA gene sequence analysis, *Acetitomaculum* belongs to the family *Lachnospiraceae*.

DNA G+C content (mol%): 32–36 (T_m).

Type species: **Acetitomaculum ruminis** Greening and Leedle 1995, 879[VP] (Effective publication: Greening and Leedle 1989, 405.)

Further descriptive information

The genus *Acetitomaculum* contains a single species *Acetitomaculum ruminis*, which is represented by five strains including the type strain 139B. These strains are strict anaerobes and grow only in reduced medium. All strains are curved rods and two of them show motility (strains 190A4 and 139B[T]). On AC-11.1 agar, colonies of this species are 2–3 mm in diameter, opaque to translucent, circular, buff or tan in color, and have a butyrous consistency with entire margins. With the exception of strain 40C, which has a granular surface, colonies of all strains have smooth surfaces.

In the paper describing these strains, the temperature range for growth is 37–42 °C (Greening and Leedle, 1995). However, in the description of the genus and species in the same paper, the temperature range for growth is 34–43 °C with an optimum for growth of 38 °C. All of the strains grow on formate, CO, CO_2:H_2, and glucose. In addition to these substrates, the type strain 139B grows on cellobiose, esculin, glucose, fructose, ferulic acid, and syringic acid. Strain 190A4 differs from the other four strains in its ability to grow on vanillic acid. The following substrates were not utilized by any of the strains of this species: adonitol, erythritol, glycerol, maltose, mannitol, pectin, ribose, starch, DL-alanine, betaine, methylamine, lactate, pyruvate, succinate, anisic acid, 3,4-dimethoxybenzoate, gallic acid, *p*-hydroxybenzoate, 3, 4, 5-trimethoxybenzoate, methanol, and phytic acid. Under a headspace of H_2:CO_2 in AC-11.1 medium containing 5% clarified rumen fluid (CRF), the fermentation acids include acetic, propionic, isobutyric, butyric, isovaleric, and valeric acids. The oxidation of hydrogen and reduction of carbon dioxide is according to the following equation: $4 H_2 + 2 CO_2 \rightarrow CH_3COOH + 2 H_2O$. The doubling times of the stains are in the range 1.6–2.8 h. Since the concentration of sulfate does not change during growth, it was considered that strains of this species do not reduce sulfate (Greening and Leedle, 1989). The ratio of serine:glutamic acid: alanine:*m*-diaminopimelic acid:ornithine:lysine was 1.0:2.6:2.3:0.4:0:1.8. The presence of both lysine and *m*-diaminopimelic acid in the cell wall differentiated this strain from other acetogenic bacteria.

Enrichment, isolation, and growth conditions

Acetitomaculum ruminis was isolated from rumen fluid recovered from a mature, rumen-fistulated Hereford crossbred steer. The isolation medium designated AC-11.1 contained (in g/l) KH_2PO_4, 0.28; K_2HPO_4, 0.94; NaCl, 0.14; KCl, 0.16; $MgSO_4$ ·$7H_2O$, 0.2; NH_4Cl, 0.5; $CaCl_2$·$2H_2O$, 0.001; 10 ml of trace mineral solution per liter as described by Balch et al. (1979), modified by the addition of 0.1 g of $NiCl_2$·$6H_2O$ and 0.01 g of Na_2SeO_3; 10 ml of vitamin solution (in mg/l: sodium ascorbate, 2; biotin, 2; folic acid, 2; pyridoxine hydrochloride, 10; thiamine hydrochloride, 5; riboflavin, 5; nicotinic acid, 5; DL-calcium pantothenate, 5; vitamin B$_{12}$, 0.1; *p*-aminobenzoic acid, 5; lipoic (thioctic) acid, 5; choline chloride, 5; *myo*-inositol, 5; niacinamide, 5; pyridoxal hydrochloride, 5); Na$_2$WO$_4$·2H$_2$O, 3 × 10^{-4} mM; resazurin, 0.001 g/l; yeast extract (Difco), 0.5 g/l; NaHCO$_3$, 6 g/l; reducing agent, 10 ml; distilled water, 660 ml; and incubated, clarified rumen fluid (Leedle and Hespell, 1980), 100 ml. The medium was boiled under a headspace of 80% N$_2$ and 20% CO$_2$ before dispensing. For enrichments, filter-sterilized 2-bromoethanesulfonic acid (sodium salt) was added to a final concentration of 50 mM. The final pH of the medium was adjusted to 7.15. For solid medium the NaHCO$_3$ and reducing agent were added as sterile solutions after autoclaving. Plates were poured in an anaerobic chamber with an atmosphere of 85% N$_2$, 10% H$_2$, and 5% CO$_2$. The strains of *Acetitomaculum* were isolated from serial dilutions of enrichment cultures using rumen fluid as the inoculum. The H$_2$:CO$_2$ enrichments incubated for 7 d were streaked on plates of AC-11.1 isolation medium and incubated at 37 °C. Colonies were picked after incubation of streak plates for 7 d. Isolates were inoculated into liquid medium under H$_2$:CO$_2$ or N$_2$:CO$_2$. Cultures showing a twofold increase in acetate production under H$_2$:CO$_2$ as compared to N$_2$:CO$_2$ were selected for further study. The organisms were routinely cultured in AC-11.1 liquid medium containing 5% CRF and an organic substrate.

Maintenance procedures

Acetitomaculum ruminis can be stored long-term lyophilized in AC-11.1 medium containing glycerol (Teather, 1982a).

Taxonomic comments

In the original description of *Acetitomaculum ruminis* the justification for the new genus was based on a comparison of the new strains to other acetogenic anaerobes (Greening and Leedle, 1989). The differences from previously described organisms included lack of endospores, Gram-stain-positive staining,

mol% G+C values, end products of fermentation and the ratio of amino acids in the cell wall. It was indicated that analysis of the 16S rRNA of the type strain 139B demonstrated a relationship to the clostridia, but no specifics were provided. Analysis of the 16S rRNA gene sequence of *Acetitomaculum ruminis* strain 139B^T (M59083) shows it to have highest similarity with the type strains of the species *Clostridium polysaccharolyticum* (91.0%), *Ruminococcus luti* (90.5%), and *Bryantella formatexigens* (90.2%). 16S rRNA gene sequence similarities to all other validly named species are below 90.0%. The species *Clostridium polysaccharolyticum* and *Ruminococcus luti* are misclassified species that are unrelated to the type strains of the genera to which they have been assigned. With 16S rRNA gene sequence similarities of 90–91%, these species would not be considered species of the genus *Acetitomaculum*. The analysis places the genus *Acetitomaculum* in the family *Lachnospiraceae*, with the most closely related genus being *Bryantella* (Figure 160.). The 16S rRNA gene sequence is only available for the type strain 139B and so the species status of the other four strains (20A, 40, 40C, and 190A4) is presently not determined. The thought that these strains are members of the same species is based on their metabolic characteristics as well as similarities in their mol% G+C values and cell-wall amino acids (Greening and Leedle, 1989).

Differentiation of the genus *Acetitomaculum* from other genera

As indicated by 16S rRNA gene sequence analysis, the closest relative of *Acetitomaculum ruminis* that is not a misclassified taxon is *Bryantella formatexigens*. *Acetitomaculum ruminis* differs from *Bryantella formatexigens* in a number of characteristics including cell shape, lack of growth on H$_2$:CO$_2$, mol% G+C content of the DNA, and the source of isolates. *Bryantella formatexigens* does not have the curved rod morphology of *Acetitomaculum ruminis* and has a DNA G+C content of 50.3 mol%, which is very different from that of the *Acetitomaculum ruminis* strains, which are in the range 32–36 mol%.

List of species of the genus *Acetitomaculum*

1. **Acetitomaculum ruminis** Greening and Leedle 1995, 879^{VP} (Effective publication: Greening and Leedle 1989, 405.)

   ru.min.is. L. gen. n. *ruminis* of the rumen, for its source.

   Colonies are convex, circular, 2–3 mm in diameter, entire, translucent, buff or tan in color, with a smooth or granular surface. Cells are curved rods. Optimum temperature for growth is 38 °C. Optimal pH for growth is 6.8. Acetogenic. Ferments formate, glucose, cellobiose, fructose, and esculin to acetic acid. Oxidizes hydrogen and reduces carbon

   dioxide according to the following equation: 4 H$_2$ + 2 CO$_2$ → CH$_3$COOH + 2 H$_2$O. Grows in a mineral media with H$_2$ and CO$_2$ or organic substrate. Rumen fluid is stimulatory and may be required. The type strain was isolated from the rumen of a mature Hereford × Angus crossbred steer fed a high-forage diet.

   DNA G+C content (mol%): 34.0 (T_m).
   Type strain: 139B, ATCC 43876, DSM 5222, UC 12185.
   GenBank accession number (16S rRNA gene): M59083.

Genus III. **Anaerostipes** Schwiertz, Hold, Duncan, Gruhl, Collins, Lawson, Flint and Blaut 2002a, 1437^{VP} (Effective publication: Schwiertz, Hold, Duncan, Gruhl, Collins, Lawson, Flint and Blaut 2002b, 50.)

THE EDITORIAL BOARD

An.ae.ro.sti′pes. Gr. pref. *an* not; Gr. n. *aer* air; L. masc. n. *stipes* club of stick; *Anaerostipes* a stick not living in air.

Nonsporeforming, nonmotile rods, sometimes in chains of 2–4 cells. **Gram positive in exponential phase** but not stationary-phase cultures. Nonmotile. **Strictly anaerobic** and both catalase- and oxidase-negative. **Nonhemolytic.** Cells **produce mainly acetic**, lactic, and butyric acids from glucose. **Arginine dihydrolase,** phosphoam-

idase, and α-galactosidase are produced. **Gelatin and urea are not hydrolyzed; indole is not produced.** Positive for **nitrate reduction.**
DNA G + C content (mol%): 45.5–46.
Type species: **Anaerostipes caccae** Schwiertz, Hold, Duncan, Gruhl, Collins, Lawson, Flint and Blaut 2002a, 1437 (Effective

publication: Schwiertz, Hold, Duncan, Gruhl, Collins, Lawson, Flint and Blaut 2002b, 50.).

Further descriptive information

Although phenotypically similar, this new genus cannot be identified with *Eubacterium sensu stricto* with which it has 20% 16S rRNA sequence divergence. Formerly a member of the *Clostridium coccoides* group (*Clostridium* 16S rRNA cluster XIVa), it possesses no less than 10% sequence divergence with previously described species. According to the 16S rRNA phylogenetic analysis presented in the roadmap to this volume (Figure 5), the genus *Anaerostipes* is a member of the family *Lachnospiraceae*, order *Clostridiales*, class *Clostridia* in the phylum *Firmicutes*.

Cells form ropy sediment with little or no turbidity in ST medium (Schwiertz et al., 2000), and form colonies (1–3 mm in diameter) that are circular, smooth, convex, shiny, white, opaque, and nonhemolytic on Columbia blood agar and Wilkins–Chalgren agar.

Enrichment and isolation procedures

Strains L1-92^T and P2 were from human fecal samples: the former isolated as described by Barcellina et al. (2000) and the latter as described below. In an anaerobic chamber, human feces was serially diluted (tenfold) to 10^{10} in Sorensen buffer (25 mM KH_2PO_4, 33 mM $Na_2HPO_4 \cdot 12 H_2O_4$, 0.04% (v/v) thioglycolic acid, 0.06% (w/v) cysteine, pH 6.8) and 1 ml aliquots were plated onto starch agar (4 g $NaHCO_3$, 0.348 g K_2HPO_4, 0.227 g KH_2PO_4, 0.5 g NH_4Cl, 2.25 g NaCl, 0.5 g $MgCO_3 \cdot 7 H_2O$, 0.15 mg $NaSeO_3$, 2 g tryptically digested peptone from casein, 2 g yeast extract, 2 g soluble starch, 15 g agar, 1 mg resazurin, 3 ml trace element solution SL 10 (Widdel et al., 1983), and 0.5 mg cysteine HCl, at pH 7.0. Following autoclaving, 1 l of medium was supplemented with 20 ml of a filter-sterilized vitamin solution (Wolin et al., 1964). Single colonies were restreaked until pure cultures were obtained.

Maintenance procedures

Anaerostipes caccae can be grown under strictly anaerobic conditions at 37 °C either on Columbia blood agar (bioMérieux), in ST broth (Schwiertz et al., 2000), or in a medium used for culturing acetogenic bacteria (Kamlage et al., 1997). For long term storage, cultures have to be transferred every second day. Best results for long term storage can be obtained by maintaining the culture in broth medium as mentioned above.

List of species of the genus *Anaerostipes*

1. **Anaerostipes caccae** Schwiertz, Hold, Duncan, Gruhl, Collins, Lawson, Flint and Blaut 2002a, 1437^VP (Effective publication: Schwiertz, Hold, Duncan, Gruhl, Collins, Lawson, Flint and Blaut 2002b, 50.)

cac'cae. Gr. n. *kakkê* feces: N.L. gen. n. *caccae* of feces, pronounced kak'ka.

Colonies (1–3 mm in diameter) are as described for the genus. Cells are rod-shaped (0.5–0.6 μm × 2.0–4.0 μm), nonmotile, Gram positive (Gram negative after 16 h), and occur in chains up to 4 cells. Strict anaerobes, they are also oxidase- and catalase-negative. Butyrate, acetate, and lactate are produced from glucose, and acetate is utilized.

Traditional tests detect acid production from D-fructose, fructooligosaccharides (FOS), D-glucose, D-galactose, inositol, maltose, D-mannose, ribose (weak), soluble starch, sucrose, L-sorbose, and sorbitol, but not from L-arabinose, cellobiose, glycerol, inulin, lactose, lactulose, melibiose, melezitose, L-rhamnose, D-trehalose, or D-xylose. Detection of acid production from salicin is variable. The API 50 system detects acid production from adonitol, D-arabitol, L-arabitol, D-arabinose, dulcitol, erythritol, galactose, D-glucose, D-fructose, inositol, D-lyxose, maltose, D-mannose, mannitol, melibiose, α-methyl-D-glucoside, N-acetylglucosamine, D-raffinose, (weak), ribose (weak), L-sorbose, sorbitol, sucrose, D-tagatose, D-turanose, and xylitol but not from amygdalin, L-arabinose, cellobiose, D-fucose, β-gentiobiose, gluconate, glycerol, glycogen, inulin, lactose, 5-ketogluconate, melezitose, α-methyl-D-mannoside, β-methyl-D-xyloside, rhamnose, trehalose, D-xylose, or L-xylose. Acid production from arbutin, salicin, and 2-keto-gluconate as well as esculin hydrolysis is variable. The API ZYM system detects acid phosphatase and phosphoamidase activity, while the API rapid ID32A system detects arginine dihydrolase, α-galactosidase, and nitrate reductase (weak reaction) activity. Neither system detects activity for alanine arylamidase, alkaline phosphatase, arginine arylamidase, α-arabinosidase, chymotrypsin, cystine arylamidase, esterase C4, ester lipase C8, α-fucosidase, β-galactosidase, β-galactosidase-6-phosphate, α-glucosidase, β-glucosidase, β-glucuronidase, glycine arylamidase, glutamylglutamic acid arylamidase, histidine arylamidase, leucine arylamidase, leucylglycine arylamidase, lipase C14, α-mannosidase, N-acetyl-α-glucosaminidase, phenylalanine arylamidase, pyroglutamic acid arylamidase, proline arylamidase, serine arylamidase, trypsin, tyrosine arylamidase, valine arylamidase, or urease. Negative for gelatin hydrolysis and indole production.

For all other characteristics, refer to the genus description.
Source: human feces.
DNA G+C content (mol%): 45.5–46.0 (HPLC).
Type strain: L1-92, DSM 14662, JCM 13470, NCIMB 13811.
GenBank accession number (16S rRNA gene): AJ270487.

Genus IV. **Bryantella** Wolin, Miller, Collins and Lawson 2004, 1^VP (Effective publication: Wolin, Miller, Collins and Lawson 2003, 6325.)

PAUL A. LAWSON

Bry.an.tel'la. N.L. fem. n. *Bryantella* named after the American microbiologist Marvin P. Bryant in recognition of his outstanding contributions to the microbial ecology of anaerobic ecosystems.

Short, rod-shaped cells that occur in pairs and short chains. Gram-stain-positive. Nonmotile. Does not form spores. **Strictly anaerobic chemo-organotroph. Catalase- and oxidase-negative. Acid but not gas is produced from glucose.** Does not require rumen fluid for growth. Indole-negative. Nitrate is not reduced. Acetate is the sole product of glucose fermentation when grown in the presence of high concentrations of formate. Isolated from human feces.

DNA G+C content (mol%): 50.3.

Type species: **Bryantella formatexigens** Wolin, Miller, Collins and Lawson 2004, 1[VP] (Effective publication: Wolin, Miller, Collins and Lawson 2003, 6325.).

Further descriptive information

The genus contains only one species, *Bryantella formatexigens*, and therefore the characteristics provided below refer to this species. Cells are nonmotile, short rods that occur mainly in pairs and short chains. Strictly anaerobic chemo-organotroph. Acetate is the sole product of glucose fermentation when grown in the presence of high concentrations of formate. Glucose fermentation with low concentrations of formate yield succinate, lactate, and acetate. In the presence of 54 mM formate, it grows with added glucose, purified cabbage cellulose, stachyose, sucrose, lactose, maltose, galactose, mannose, or xylose. In the presence of formate, it does not grow with Avicel (FMC Corp.), lactate, starch, pectin, vanillate, syringate, methanol, or ethanol. It does not grow with 0.25% formate in the absence of carbohydrates. Indole-negative. Nitrate is not reduced.

Enrichment and isolation procedures

The organism was isolated from human feces using the serum bottle modification of the Hungate technique (Miller and Wolin, 1974). B1C medium was used and contained (per liter): $NaHCO_3$, 5.0 g; K_2HPO_4, 0.3 g; KH_2PO_4, 0.3 g; $(NH_4)_2SO_4$, 0.3 g; NH_4Cl, 1 g; NaCl, 0.61 g; $MgSO_4 \cdot 7H_2O$, 0.15 g; $CaCl_2 \cdot 2H_2O$, 80 mg; $MnSO_4 \cdot H_2O$, 4.5 mg; $FeSO_4 \cdot 7H_2O$, 3.0 mg; $CoSO_4 \cdot H_2O$, 1.8 mg; $ZnSO_4 \cdot 7H_2O$, 1.8 mg; $CuSO_4 \cdot 5H_2O$, 100 µg; $AlK(SO_4)_2 \cdot 12H_2O$, 180 µg; $Na_2MoO_4 \cdot 2H_2O$, 100 µg; H_3BO_3, 100 µg; Na_2SeO_4, 1.9 mg; $NiCl_2 \cdot 6H_2O$, 92 µg; nitrilotriacetic acid, 15 mg; thiamine HCl, 2.0 mg; D-pantothenic acid, 2.0 mg; nicotinamide, 2.0 mg; riboflavin, 2.0 mg; pyridoxine HCl, 2.0 mg; biotin, 10.0 mg; cyanocobalamin, 20 µg; *p*-aminobenzoic acid, 100 µg; folic acid, 50 µg; cysteine HCl, 0.5 g; rumen fluid, 100 ml; sodium acetate, 2.5 g; sodium formate, 2.5 g; trypticase, 2.0 g. WM medium was also used and contained (per liter): $NaHCO_3$, 3.5 g; KH_2PO_4, 0.5 g; NH_4Cl, 1 g; NaCl, 0.4 g; $MgCl_2 \cdot 6H_2O$, 0.33 g; $CaCl_2 \cdot 2H_2O$, 50 mg; $FeCl_2 \cdot 4H_2O$, 1.5 mg; $CoCl_2 \cdot 6H_2O$, 0.2 mg; $ZnSO_4 \cdot 7H_2O$, 0.1 mg; $MnCl_2 \cdot 4H_2O$, 0.03 mg; $CuCl_2 \cdot 2H_2O$, 0.01 mg; $Na_2MoO_4 \cdot 2H_2O$, 0.03 mg; H_3BO_3, 0.3 mg; Na_2SeO_4, 1.9 mg; $NiCl_2 \cdot 6H_2O$, 0.02 mg; the same vitamins as B1C medium; cysteine HCl, 0.5 g; sodium acetate, 1.0 g; isobutyric acid, 0.54 ml; 2-methylbutyric, valeric, and isovaleric acids (0.6 ml each); casein hydrolysate, 2.0 g (Type I, No. C-9386, Sigma Chemical Co., St. Louis, MO). Resazurin (1 mg/l) was added to both media as an oxidation-reduction potential indicator. Both media were adjusted to pH 7 with NaOH prior to gassing with 100% CO_2 and the addition of $NaHCO_3$. After autoclaving under a CO_2 atmosphere, a sterile solution containing 0.125 g/l each of cysteine and sodium sulfide (30 µl per ml of medium) was added prior to inoculation. All incubations were performed at 37 °C.

Serial dilutions of feces in B1C medium plus 0.8% VCP (cellulose-enriched fiber from cabbage) yielded the original enrichments. After transfer to the same medium, cultures were plated on VM medium plus 2% agar and 0.6% carboxymethylcellulose (CMC) in roll tubes. An isolated colony was transferred to B1C broth plus 0.6% CMC and replated on VM medium. Once isolated in pure culture, the organism could be grown anaerobically on blood-based agar.

Maintenance procedures

Grows well anaerobically on blood-based agar at 37 °C. For long-term preservation, strains are maintained in media containing 15–20% glycerol at −70 °C or lyophilized.

Taxonomic comments

The genus *Bryantella* was created to accommodate a phylogenetically distinct nonmotile, Gram-stain-positive, rod-shaped organism originating from human feces (Wolin et al., 2003). The genus is monospecific, and phylogenetic analyses of the 16S rRNA gene demonstrates a relationship between *Bryantella* and the *Clostridium coccoides* group (clostridial rRNA cluster XIVa; Collins et al., 1994). This diverse assortment of organisms is classified in this volume of Bergey's in the novel family *Lachnospiraceae* and includes the genera *Acetitomaculum*, *Anaerostipes*, *Bryantella*, *Butyrivibrio*, *Catonella*, *Coprococcus*, *Dorea*, *Hespellia*, *Johnsonella*, *Lachnospira*, *Lachnobacterium*, *Moryella*, *Oribacterium*, *Parasporobacterium*, *Pseudobutyrivibrio*, *Roseburia*, *Shuttleworthia*, *Sporobacterium*, and *Syntrophococcus* in addition to many misclassified clostridial species (see treatment of the family *Lachnospiraceae* in this volume). However, *Bryantella* forms a distinct lineage within this large grouping and does not display a particularly close affinity with any of the aforementioned taxa. It is clear from phylogenetic molecular profiling studies of this large family that human feces contains additional species of *Bryantella* that remain to be isolated and described (Barcenilla et al., 2000; Hold et al., 2002; Namsolleck et al., 2004; Suau et al., 1999).

Differentiation of the genus *Bryantella* from other genera

Bryantella can be readily distinguished from its closest phylogenetic relatives by a combination of morphological, biochemical, and chemotaxonomic criteria. Its unusual homoacetogenic fermentation distinguishes it from the other members of the family *Lachnospiraceae*. Further distinctive features include the following. While the morphology of *Bryantella* somewhat resembles that of *Syntrophococcus sucromutans*, its Gram reaction is different. *Syntrophococcus sucromutans* also requires large amounts of rumen fluid for growth (Krumholz and Bryant, 1986a). *Bryantella* can be distinguished from *Clostridium* spp. and *Sporobacterium* because it does not produce endospores. Its morphology distinguishes it from *Lachnospira*. The absence of motility and butyrate as a product of the glucose fermentation distinguishes it from *Roseburia* and *Butyrivibrio*. Differences in cellular morphology and products of glucose metabolism distinguish it from *Coprococcus* and *Ruminococcus*.

Importantly, in this diverse group of organisms, an accurate identification relies to a great extent upon molecular genetic techniques such as 16S rRNA gene sequence comparisons. Once the sole domain of specialized research facilities, these high-throughput methodologies are becoming increasingly automated, reducing costs, and making them routinely accessible. The availability of these methods facilitates more rapid and accurate identification of hitherto unknown taxa.

List of species of the genus *Bryantella*

1. **Bryantella formatexigens** Wolin, Miller, Collins and Lawson 2004, 1[VP] (Effective publication: Wolin, Miller, Collins and Lawson 2003, 6325.)

for.mat.ex′i.gens. N.L. n. *formas -atis* formate; L. part. adj. *exigens* demanding; N.L. fem. part. adj. *formatexigens* formate-demanding.

Characteristics are as given for the genus with the following information. Using the commercially available API Rapid ID32A system, activity is detected for α-arabinosidase, α-galactosidase, β-galactosidase, β-galactosidase-6-phosphate, α-glucosidase, β-glucosidase, β-glucuronidase, and *N*-acetyl-β-glucosaminidase. No activity is detected for alkaline phosphatase, arginine arylamidase, arginine dihydrolase, alanine arylamidase, α-fucosidase, glutamic acid decarboxylase, glutamyl glutamic acid arylamidase, glycine arylamidase, histidine arylamidase, leucine arylamidase, leucyl glycine arylamidase, phosphoamidase, phenylalanine arylamidase, proline arylamidase, pyroglutamic acid arylamidase, serine arylamidase, tyrosine arylamidase, or urease. Isolated from human feces.

DNA G+C content (mol%): 50.3 (HPLC).

Type species: I-52, CCUG 46960, DSM 14469.

GenBank accession number (16S rRNA gene): AJ318527.

Genus V. **Butyrivibrio** Bryant and Small 1956b, 18, emend. Moore, Johnson and Holdeman 1976, 241[AL]

ANNE WILLEMS AND MATTHEW D. COLLINS

Bu.ty.ri.vib′ri.o. N.L. adj. *butyricus* butyric; L. v. *vibro* to vibrate; N.L. n. *Vibrio* that which vibrates, a bacterial genus name; N.L. masc. n. *Butyrivibrio* a butyric vibrio.

Gram-stain-negative, but structurally Gram-positive, non-spore-forming, strictly anaerobic straight to curved rods, 0.3–0.8 × 1.0–5.0 μm, occurring singly or in chains or filaments that may be helical. **Motile by monotrichous or lophotrichous, polar or subpolar flagella or nonmotile. Chemo-organotrophic, fermentative metabolism** with carbohydrates as the main substrates. **Glucose or maltose are fermented with butyrate as a major end product.** Under some conditions some strains may produce large amounts of lactate and little butyrate. On the basis of 16S rRNA gene sequences, *Butyrivibrio* belongs to cluster XIVa of the *Clostridium* subphylum of the Gram-positive bacteria. It is not monophyletic: *Butyrivibrio fibrisolvens* strains form two distinct subgroups which correspond to different genera and *Butyrivibrio crossotus* occupies a separate position (Willems et al., 1996). *Butyrivibrio hungatei* belongs to the *Butyrivibrio fibrisolvens* subgroup that also comprises the type strain of *Butyrivibrio fibrisolvens* (Kopečný et al., 2003). According to the 16S rRNA phylogenetic analysis presented in the roadmap to this volume (Figure 5), the genus *Butyrivibrio* is a member of the family *Lachnospiraceae*, order *Clostridiales*, class *Clostridia* in the phylum *Firmicutes*. Isolated from the rumen of ruminants and in human, rabbit, and horse feces.

DNA G+C content (mol%): 36–45 mol% (T_m and HPLC).

Type species: **Butyrivibrio fibrisolvens** Bryant and Small 1956b, 19[AL].

Further descriptive information

Cell morphology. Cells are straight to curved rods with tapered or round ends. They occur singly or in short or long chains that may show helical arrangement. Pairs of cells may occur in an "S" arrangement. Cells sometimes appear as filaments. Cells of *Butyrivibrio crossotus* tend to be larger in diameter than those of *Butyrivibrio fibrisolvens* (Bryant, 1986b). Motile by means of polar or subpolar monotrichous (*Butyrivibrio fibrisolvens* and *Butyrivibrio hungatei*) or lophotrichous flagellation (*Butyrivibrio crossotus*). Often only a few cells in a culture are motile. Nonmotile strains have been reported. Flagellar filaments of *Butyrivibrio fibrisolvens* strain OR77 have been characterized biochemically and some of the genes encoding flagellins (*flaA* and *flaB*) characterized (Kalmokoff et al., 2000).

Cell-wall composition. *Butyrivibrio* strains stain Gram-negative but lack the trilamellar outer membrane structure commonly associated with Gram-negative bacteria. Some strains (including the type strain) possess an atypical, very thin Gram-positive cell-wall ultrastructure (Cheng and Costerton, 1977; Dibbayawan et al., 1985). The thin walls (approx. 12–20 nm) compared with those of normal Gram-positive walls (approx. 30–50 nm) probably account for the Gram-negative staining reactions of strains. The peptidoglycans of *Butyrivibrio* strains contain muramic acid, glucosamine, alanine, glutamic acid, and *meso*-diaminopimelic acid (Hespell et al., 1993). These components are consistent with a *meso*-diaminopimelic acid direct cross-linked structure of the A type (variation A1γ; Schleifer and Kandler, 1972). However, although *meso*-diaminopimelic acid is the major component involved in cross-linking, it appears that cross-linking is not extensive in many strains (Hespell et al., 1993). In addition, small amounts of glycine and aspartic acid are present in *Butyrivibrio* mureins, indicating some cross-links may involve these amino acids (Hespell et al., 1993). Further studies are needed to determine the precise structures and nature of linkages in *Butyrivibrio* peptidoglycans. Lipoteichoic acids, comprising 1,3-phosphodiester-linked polyglycerophosphate chain linked to a glycolipid, and deacylated lipoteichoic acids have been isolated from *Butyrivibrio fibrisolvens* strains (Hewitt et al., 1976).

Extracellular polysaccharides. Many *Butyrivibrio fibrisolvens* strains produce extracellular polysaccharides with rheological properties similar to those of xanthan gum (Ha et al., 1991; Wachenheim and Patterson, 1992). The structure of the exopolysaccharides of several strains has been extensively studied. In strain 49, the structure of the capsular polysaccharide consists of pentasaccharide repeating units with one of the *O*-acetyl groups substituted to O-3 of the 4-substituted α-D-galactopyranoside (Ferreira et al., 1995). In addition, the acid sugars 4-*O*-(1-carboxyethyl)-D-galactose and 4-*O*-(1-carboxyethyl)-L-rhamnose were reported for the first time in this strain (Stack et al., 1988b; Stack and Weisleder, 1990). In strain X6C61, the structure of the capsular polysaccharide consists of hexasaccharide repeating units with approximately 70% of *O*-acetyl groups substituted to O-3 of β-D-glucopyranoside

(Andersson et al., 1993). L-Iduronic acid and 6-deoxy-D-talose were identified as components of the polysaccharides of this strain (Stack, 1989; Stack et al., 1988a). In strain CF3, L-altrose was identified in the extracellular polysaccharide (Stack, 1987) that consists of pentasaccharide repeating units (Ferreira et al., 1997).

The neutral sugar composition and other compositional features of extracellular polysaccharides have been used to sort *Butyrivibrio fibrisolvens* isolates into groups (Stack, 1988). Some of these groups may be taxonomically significant, but more comparative phylogenetic data are needed to establish this clearly.

Colonial characteristics. In rumen fluid-glucose-cellobiose-agar (RGCA) roll-tubes (Bryant and Small, 1956b), surface colonies of *Butyrivibrio fibrisolvens* are usually smooth, entire, slightly convex, translucent, light brown with a diameter of 2–4mm. Some strains have rough, more flat colonies that are paler and have filamentous margins. Deep colonies are usually lenticular or Y-shaped but some form compound lenticular colonies (Bryant, 1986b). In rumen fluid-cellulose agar, cellulolytic strains have a diverse colony morphology, varying from lens shaped over triangular to compound lenticular or rhizoidal. Zones of cellulose digestion around colonies vary from very narrow for strains with slow digestion to broad for strains with rapid and complete digestion (Bryant, 1986b).

On RGCA roll tubes, *Butyrivibrio crossotus* subsurface colonies are 0.5–1 mm in diameter, lenticular, and translucent to transparent (Bryant, 1986b). On supplemented brain heart infusion agar roll streaks of these organisms are 0.2–1 mm in diameter, circular, entire, convex, translucent to semiopaque, and smooth (Moore et al., 1976).

Nutrition and growth conditions. *Butyrivibrio* strains are strict anaerobes, occurring in the gastrointestinal system of various domestic animals and humans. They have a fermentative metabolism and therefore growth is poor in media lacking a carbohydrate source. As fiber degraders, they are able to utilize various hemicelluloses and xylans. Most strains can degrade starch and pectines, but few strains can grow on cellulose. Most strains can ferment a variety of soluble sugars, disaccharides, or oligosaccharides to form butyrate, as well as lactate and acetate (Hespell, 1991). *Butyrivibrio* strains can use various components of plant cell walls but are also inhibited by some of these compounds. For example, growth of *Butyrivibrio fibrisolvens* strain 49 is limited by ester-linked feruloyl and *p*-coumaroyl groups. Growth limitation by phenolic acid-carbohydrate complexes varies with the ability to hydrolyze carbohydrate linkages (Akin et al., 1993).

Most *Butyrivibrio* strains can use ammonia as sole nitrogen source; many strains can also use urea. Peptides are not required but often stimulate growth on ammonia-containing media (Hespell, 1991). Strain E14 is reported to require methionine for growth (Nili and Brooker, 1995, 1997) and strain S2 requires fatty acids for growth (Hazlewood and Dawson, 1979). Growth of *Butyrivibrio fibrisolvens* strain TC33 in clarified rumen fluid with glucose is stimulated by the addition of a vitamin/casein hydrolysate mixture or yeast extract. Most rapid growth is obtained by the addition of a combination of folic acid, pyridoxamine.2HCl, and an enzymic hydrolysate of casein (Wejdemar, 1996).

Best growth temperature is 37 °C, and little or no growth occurs at 30 or 45 °C (Hespell, 1991).

Metabolism of carbohydrates. *Butyrivibrio* strains are able to metabolize a variety of carbohydrates and contribute significantly to rumen processes. Many studies have focussed on the carbohydrate metabolism of particular model strains.

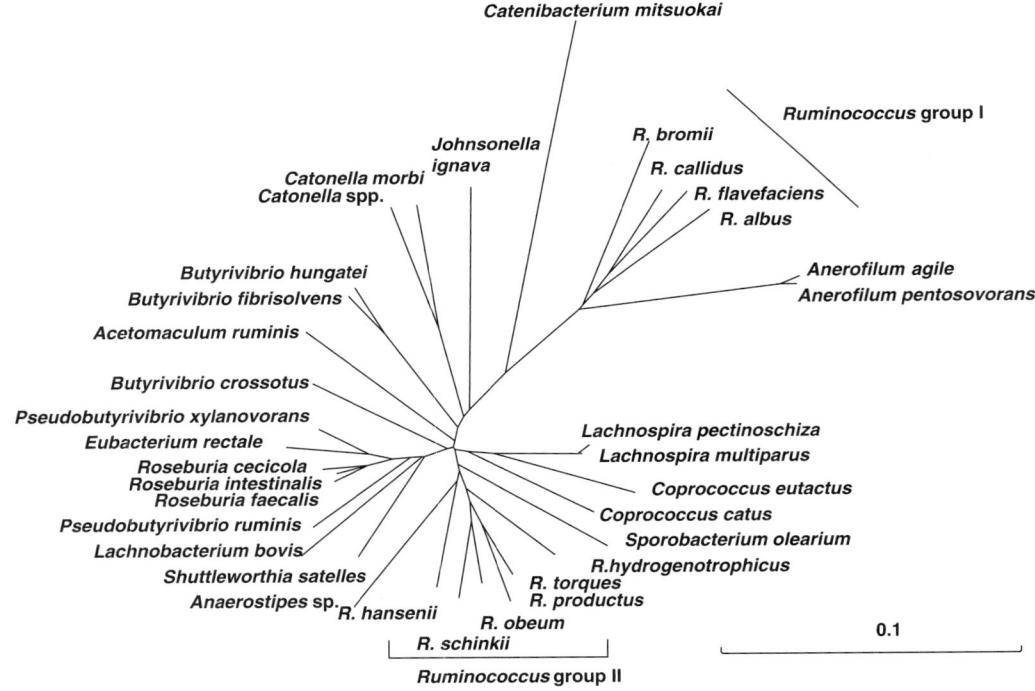

FIGURE 161. Phylogenetic position of ruminococci and coprococci among the family *Lachnospiraceae*.

In *Butyrivibrio fibrisolvens* strains ATCC 19171 and CE 51, production of cell dry matter and growth rate are higher in cultures with glucose than xylose. In the latter cultures, more carbon is converted to metabolites and less to cell material (Marounek and Petr, 1995). When provided with both glucose and xylose, *Butyrivibrio fibrisolvens* strains ATCC 19171 and 86 used both sugars simultaneously, strains X1 and CE 51 showed classical diauxic growth using glucose first, whereas strain X2D62 slowly used xylose first until depletion and then rapidly used glucose. ATP-dependent phosphorylation of glucose occurred in all these strains and, in addition, strain CE 51 also had phosphoenolpyruvate-dependent phosphorylation of glucose (Marounek and Kopečný, 1994). Strain D1 couses glucose and xylose but has a clear preference for glucose or xylose over arabinose. It coutilizes xylose or arabinose and cellobiose, but prefers either pentose over maltose. Strain A38 showed a strong preference for disaccharides over xylose or arabinose (Strobel and Dawson, 1993). Pentose transport in strain D1 depends on a high-affinity, ATP-dependent mechanism and hexose sugars affect the use of xylose and arabinose (Strobel, 1994).

Butyrivibrio fibrisolvens strain 49 rapidly hydrolyzes starch to produce glucose, maltose, maltotriose, maltotetrose, and maltopentaose (Cotta, 1992). The α-amylase gene from strain H17c has been cloned and sequenced and the enzyme shown to be a calcium metalloenzyme (Rumbak et al., 1991).

Butyrivibrio fibrisolvens fermentation end products from glucose include formate, butyrate, acetate, and varying amounts of lactate and succinate (Holdeman et al., 1977b). However, considerable variation among strains has been reported: strains D1 and A38 were shown to produce little lactate, whereas strains 49 and nor 37 produced a lot of lactate, but this could be inhibited and butyrate would be produced instead by the presence of acetate. In the latter strains a butyryl-CoA/acetate-CoA transferase system converts butyryl-CoA into butyrate, using acetate as an acceptor. The former strains do not have this system but possess butyrate kinase instead (Diez-Gonzalez et al., 1999).

Metabolism and degradation of plant material. The role of *Butyrivibrio* in the degradation of plant material from various sources has been extensively studied because of the importance to rumen metabolism. Many studies into the solubilization of plant cell walls and the digestion cell-wall monosaccharides by co-cultures of various rumen bacteria have been published. In co-cultures, *Butyrivibrio fibrisolvens* strain D1 solubilized cell-wall carbohydrates less well than *Fibrobacter succinogenes* but did interact complementarily with this species to utilize solubilized carbohydrates and improve co-culture growth. Alfalfa hays were used to a lesser extent than ryegrass (Miron and Benghedalia, 1993a). co-cultures of the same *Butyrivibrio fibrisolvens* strain with other ruminal bacteria are capable of degrading barley straw (Miron et al., 1994), sorghum straw (Benghedalia et al., 1993), panicum hay and vetch hay (Miron and Benghedalia, 1993c), cotton stalks (Miron and Benghedalia, 1993b), wheat straw (Miron and Benghedalia, 1992), and lucerne (Miron, 1991). In of all these reports, the role of *Butyrivibrio fibrisolvens* was mainly to use solubilized carbohydrates. Cellulose-binding proteins were eluted from a cell lysate of a *Butyrivibrio fibrisolvens* strain using SDS, but not carboxymethylcellulose (Mitsumori and Minato, 1995). In a comparative study of the use of pectin from different sources, pure cultures of *Butyrivibrio fibrisolvens* used citrus pectin best and sugar beet pectin was used better than pectin from apple

or lucerne; with mixed cultures of rumen micro-organisms, *Butyrivibrio fibrisolvens* used pectin less than *Prevotella ruminicola* and *Lachnospira multiparus* (Kasperowicz, 1994). *Butyrivibrio fibrisolvens* strain 787 grown on pectin produced significantly more acetate and less butyrate, lactate, succinate, and hydrogen than cultures grown on L-arabinose and D-glucose. Pectin-grown cells have 2-keto-3-deoxy-6-phosphogluconate aldolase and fructose-diphosphate aldolase activity but not phosphoketolase activity (Marounek and Duskova, 1999).

In a comparison of the degradation of fructan from timothy grass by rumen bacteria, *Butyrivibrio fibrisolvens* strain 3 was more effective than *Streptococcus bovis*, *Bacteroides ruminicola*, *Selenomonas ruminantium*, and two *Treponema* species (Ziolecki et al., 1992).

Metabolism and use of xylan. Xylans are hemicelluloses that constitute a major component of the ruminants diet. The xylanolytic activity of *Butyrivibrio fibrisolvens* contributes significantly to the degradation of these substrates and results in a pool of degradation intermediates including low-molecular-weight xylooligosaccharides that can be used by *Butyrivibrio fibrisolvens* and other species in the rumen microflora (Cotta, 1993; Hespell et al., 1987). For example, xylanolytic *Butyrivibrio fibrisolvens* strain H17c rapidly degrades oat spelt xylan and *Selenomonas ruminantium* strain GA 192 grown in co-culture will use the resulting xylooligosaccharides, although residual carbohydrates do remain after growth (Cotta and Zeltwanger, 1995). The ability of strain H17c to degrade various chemically and physically different xylans was studied and it was shown that the type and distribution of xylan side chains are more important for xylan utilization than the extent of water insolubility (Hespell, 1992, 1995). In strain NCFB 2249 xylanolytic enzymes were induced by xylan, xylooligosaccharides, and xylase under anaerobiosis. Highest activity was observed at pH 6.5–7 and with substrate concentrations of 2 mg/ml or higher. Enzymes involved are subject to catabolite regulation and include α-L-arabinofuranosidase, β-D-xylosidase, and xylanase. The β-xylosidase of *Butyrivibrio fibrisolvens* belongs to family 43 of a general classification of glycosyl hydrolases based on amino acid sequences and its molecular mechanism of action has been elucidated (Braun et al., 1993). In the presence of readily metabolized sugars such as glucose and arabinose, diauxic growth was observed with delayed xylanolysis (Williams and Withers, 1992a, 1992b). Xylanase genes of *Butyrivibrio fibrisolvens* were cloned, sequenced, and expressed in *Escherichia coli* (Lin and Thomson, 1991; Mannarelli et al., 1990a; Sewell et al., 1989; Utt et al., 1991). A *xynA* gene was described from strain Bu49, whereas *xynB* was described form strain H17c. Strain Bu49 harbors two enzymes with xylanase activity, XynA and a second, smaller enzyme (Hespell and Whitehead, 1990; Mannarelli et al., 1990a). The extracellular xylanase component of strain H17c is thought to be a multisubunit protein aggregate of xylanases, some with cellulase activity (Lin and Thomson, 1991). A more extensive study of the diversity and distribution of xylanase genes in *Butyrivibrio* strains revealed two gene families, *xynA* (with two subfamilies) and *xynB* and evidence that at least one additional xylanase gene family remains to be identified (Dalrymple et al., 1999).

Butyrivibrio fibrisolvens strains are able to hydrolyze ester-linked ferulic and *p*-coumaric acid, present in the arabinoxylans of dry season tropical grasses through extracellular cinnamoyl esterase activity (McSweeney et al., 1998).

About 33% of *Butyrivibrio hungatei* strains had xylanase extivity (Kopečný et al., 2003).

Chemotaxonomic data. The cellular fatty acids of *Butyrivibrio fibrisolvens* have been characterized (Ifkovits and Ragheb, 1968; Miyagawa, 1982; Moore et al., 1994). *Butyrivibrio fibrisolvens* contain fatty acids of the straight-chain saturated, monounsaturated, and branched types. Moore et al. (1994) reported the major components of the type strain of *Butyrivibrio fibrisolvens* to be $C_{16:0}$ fatty acid (45%), $C_{18:1 \text{-}cis11}$ dimethyl acetal (15%), and $C_{16:0}$ dimethyl acetal (11%). Smaller amounts of the following were also detected: $C_{12:0}$ fatty acid (1%), $C_{14:0}$ fatty acid (4%), $C_{14:0}$ dimethyl acetal (2%), $C_{16:0}$ iso fatty acid (1%), $C_{16:1 \text{-}cis9}$ fatty acid (1%), $C_{18:1 \text{-}cis9}$ fatty acid (2%), $C_{18:0}$ fatty acid (3%), $C_{17:1 \text{-}cis9}$ (3%), and $C_{18:1 \text{-}cis11/9/6}$ fatty acid or an unknown component with an equivalent chain-length of 17.834 (4%). These data are essentially consistent with the earlier findings of Ifkovits and Ragheb (1968) except the latter workers did not report the presence of large amounts of fatty aldehyde components nor did Kopečný et al. (2003). Miyagawa (1982) also detected fatty aldehydes in *Butyrivibrio* strains. In the study of Miyagawa (1982), *Butyrivibrio* strains formed two broad groups on the basis of their fatty acid profiles; group 1 contained branched-chain fatty acids and fatty aldehydes as major components, whereas group 2 had predominantly straight and monounsaturated chains. Group 1 consisted of nine strains and contained $C_{15:0 \text{ anteiso}}$ fatty acid and $C_{15:0 \text{ anteiso}}$ aldehyde in relatively large amounts; some strains had iso-$C_{14:0}$ aldehyde, and some had $C_{16:0}$ iso fatty acid and $C_{17:0 \text{ anteiso}}$ fatty acid in large amounts. Group 2 consisted of 12 strains that displayed considerable variation in their fatty acids and aldehydes; for example, some strains contained $C_{16:0}$ fatty acid, $C_{18:1}$ aldehyde, and $C_{16:0}$ aldehyde, one strain had $C_{16:0}$ fatty acid, $C_{16:0}$ aldehyde, and $C_{17:0 \text{ iso}}$ fatty acid, whereas some others had $C_{15:0}$ fatty acid, $C_{16:0}$ fatty acid, $C_{15:0 \text{ iso}}$ fatty acid, $C_{15:0 \text{ iso}}$ aldehyde, and/or $C_{14:0}$ fatty acid. The fatty acid compositional studies of Miyagawa (1982) revealed considerable taxonomic diversity within the *Butyrivibrio* strains examined. Marinšek-Logar et al. (2001) examined the fatty acid profiles of 45 *Butyrivibrio* strains. In the study of Marinšek-Logar et al. (2001) the *Butyrivibrio* strains fell into four major and two minor groups; the major fatty acids for the bulk of the strains were reported to be $C_{14:0}$, $C_{15:1}$, and $C_{16:1}$. These findings are at variance with earlier investigations (Miyagawa, 1982; Moore et al., 1994). *Butyrivibrio hungatei* contained relatively large amounts of $C_{16:0 \text{ iso}}$ fatty acid (15%), $C_{16:1}$ fatty acid (13%), $C_{17:0 \text{ anteiso}}$ fatty acid (17%), $C_{16:0}$ fatty acid (9%), $C_{16:0 \text{ 2-OH}}$ fatty acid (7%), $C_{14:0 \text{ 2-OH}}$ fatty acid (4%) and $C_{17:1}$ iso fatty acid (3%) as well as significant amounts (1–3%) of the following fatty acids: $C_{12:1}$, $C_{14:0}$ iso, $C_{14:0}$, -$C_{13:0 \text{ iso 3-OH}}$, $C_{15:1 \text{ iso}}$ or $C_{13:0 \text{ 3-OH}}$, $C_{15:1 \text{ anteiso}}$, $C_{15:0 \text{ anteiso}}$, $C_{15:1} {}^{c}_{10}$, $C_{15:0}$, $C_{14:0 \text{ 3-OH}}$ or -$C_{16:1 \text{ iso}}$, $C_{17:0}$, $C_{18:1 \text{ iso}}$, $C_{18:0}$ and $C_{19:1 \text{ iso}}$ (Kopečný et al., 2003). No fatty acid data for *Butyrivibrio crossotus* have been reported.

Plasmids. With the aim of optimizing the activity of the rumen microflora, many studies have explored the occurrence of plasmids in *Butyrivibrio* strains and their use in developing vector systems for the transformation of rumen strains. Screening of rumen bacteria for the presence of plasmids, revealed only 0.2% of *Butyrivibrio* strains bear plasmids (Ogata et al., 1996). An earlier study, screening specifically *Butyrivibrio fibrisolvens* strains from the bovine rumen, found plasmids to be common features of these organisms, with large plasmids (250 kDa) present in all strains tested (Teather, 1982b).

A small cryptic plasmid pOM1 was characterized from strain Bu49 (Mann et al., 1986). It was later sequenced and has been used to construct an *Escherichia coli–Butyrivibrio fibrisolvens* shuttle vector pSMerm1 (Hefford et al., 1997). Derived from this system, plasmid pBHE can be used to characterize promoters in *Butyrivibrio* (Beard et al., 2000). Several other small cryptic plasmids have been characterized from different *Butyrivibrio* strains (Hefford et al., 1993; Kobayashi et al., 1995; Ware et al., 1992). Plasmid pRJF1 from strain OB156 was used in combination with pUC118 and pAMβ1 to construct a shuttle vector (Beard et al., 1995).

An electroporation protocol was optimized to introduce cloned genes into *Butyrivibrio fibrisolvens* (Whitehead, 1992). A conjugative transfer system has been developed using tetracycline-resistant transposon Tn916, pUB110 from *Staphylococcus aureus* and pUBLRS, a pUB110-based shuttle vector (Clark et al., 1994).

Stable transfer of a plasmid (pAM-β-1) and transposons (Tn916, Tn916-δ-E) has been shown from the nonruminal species *Enterococcus faecalis*, as well as from the ruminal species *Streptococcus bovis* to various *Butyrivibrio fibrisolvens* strains (Hespell and Whitehead, 1991b, 1991a).

Antibiotic sensitivity. *Butyrivibrio fibrisolvens* strains are resistant to high levels of nalidixic acid, a DNA gyrase inhibitor (30–500 µg/ml) (Hespell, 1991, 1993). They are generally sensitive to many other antibiotics, especially those affecting cell-wall synthesis such as β-lactams, cephalothin, and bacitracin (Fulghum et al., 1968; Hespell et al., 1993). Five ruminal *Butyrivibrio fibrisolvens* strains were found to be very sensitive to ionophores and inhibitors of protein synthesis, except for auromycin. They were relatively insensitive to inhibitors of carbohydrate metabolism and uncouplers and sensitivity varied to salinomycin, aureomycin, and bacitracin (Marounek and Savka, 1994).

Horizontal transfer of tetracycline resistance on a chromosomal element was reported among *Butyrivibrio fibrisolvens* strains (Scott et al., 1997). Evidence supporting intergeneric exchange of tetracycline resistance genes has also been reported (Melville et al., 2001). Two types of tetracycline resistance genes are present in *Butyrivibrio fibrisolvens*: one nontransmissible with a sequence identical with that of *Streptococcus pneumoniae tet(O)* and a transmissible resistance gene, *tet(W)*, representing a new class of ribosomal protection determinants. This gene has a higher DNA GC content than other *Butyrivibrio fibrisolvens* genes and similar sequences are also present in other rumen anaerobes, suggesting recent intergeneric transfer among ruminal bacteria (Barbosa et al., 1999; Billington et al., 2002).

Tannins, as present in the leafs of sainfoin (*Onobrychis viciifolia*), inhibit growth and protease activity in *Butyrivibrio fibrisolvens* strain A38 (Jones et al., 1994).

Butyrivibrio fibrisolvens strain 49 is sensitive to the ionophore antibiotic monensin, which is used as a feed additive (Callaway et al., 1999).

Bacteriocide activity. Many *Butyrivibrio fibrisolvens* strains (over 50% of strains isolated from sheep, deer, and cattle) have bacteriocide activities, probably due to different inhibitory compounds they produce (Kalmokoff et al., 1996). The bacteriocin-like activity from strain AR10 is due to a single peptide showing no homology with previously reported bacteriocins (Kalmokoff and Teather, 1997). Strain OR79 produces a bacteriocin-like activity with a broad spectrum activity. Two very similar peptide inhibitors were isolated from spent culture liquid and partially

characterized; their genes were located. Amino acid sequence comparisons indicate these molecules represent a new type of antibiotic (Kalmokoff et al., 1999). *Butyrivibrio fibrisolvens* strain JL5 from bovine rumen produces a bacteriocin which inhibits some *Butyrivibrio fibrisolvens* but not others (Rychlik and Russell, 2002). The bacteriocin from JL5 inhibits a variety of Gram-stain-positive organisms, such as obligate amino acid-fermenting species *Clostridium sticklandii* and *Clostridium aminophilum*, and may play a role in regulating ammonia production *in vivo* (Rychlik and Russell, 2002).

Ecology. *Butyrivibrio* is a component of the normal rumen microflora in ruminants all over the world. This bacterial community is responsible for the primary degradation of raw plant material, converting it to simpler metabolites and bacterial protein that feed the animal. In the rumen ecosystem, cross-feeding between species is an important factor in the efficient degradation of plant material, as illustrated in various studies using mixed cultures (see section on degradation of plant material). Experiments using various plants have shown that *Butyrivibrio fibrisolvens* alone is unable to degrade straw extensively. In co-cultures with other ruminal bacteria it increases solublization of cell-wall carbohydrates and significantly increases the use of arabinose, xylose, and glucose (Miron et al., 1994). In co-cultures on starch, *Selenomonas ruminantium*, a nonamylolytic species, was able to cross-feed on maltooligosaccharides produced by *Butyrivibrio fibrisolvens* (Cotta, 1992).

Genetic modification of rumen organisms opens the prospect of improving rumen fermentation and reducing economic losses of livestock industries (for reviews see Kobayashi and Onodera, 1999; Forsberg et al., 1999). A xylanase gene of *Eubacterium ruminantium* was successfully transferred to and expressed in *Butyrivibrio fibrisolvens* (Kobayashi et al., 1998). Similarly, a fungal xylanase gene has been transferred to *Butyrivibrio fibrisolvens* and its product is secreted by the bacterium, even after prolonged cultivation (Xue et al., 1997). However, *in vivo* experiments suggest that persistence in the rumen system is limited (Krause et al., 2001). Strains of *Butyrivibrio fibrisolvens* genetically modified to carry genes that detoxify fluoroacetate, a poisonous component of some trees and shrubs, have been produced and fed to sheep that subsequently suffered less toxicological symptoms when challenged with fluoroacetate (Gregg, 1995, 1998). To follow recombinant strains in the rumen, a competitive PCR assay has been tested (Kobayashi et al., 2000).

Butyrivibrio fibrisolvens may provide selenium, an important trace element for grazing ruminants, in a form utilizable by the animals by converting selenite into seleno-amino acids (Hudman and Glenn, 1985).

In a survey of 50 strains of *Butyrivibrio* of various origins, half of the strains were found to produce bacteriocins with a broad spectrum of activity. These components are thought to contribute significantly to the composition and structure of the ruminal ecosystem (Kalmokoff et al., 1996).

Butyrivibrio strains isolated from human feces are similar to *Butyrivibrio fibrisolvens* but have been placed in a separate species, *Butyrivibrio crossotus*, because they are lophotrichous, whereas *Butyrivibrio fibrisolvens* is monotrichous or occasionally has two polar flagella (Moore et al., 1976).

Enrichment and isolation procedures

A wide variety of media and growth conditions have been used for the isolation of *Butyrivibrio* strains, but no specific selective isolation or enrichment procedures have been published.

Butyrivibrio strains can be isolated by nonselective procedures in rumen fluid-glucose-cellobiose-agar (RGCA) roll-tubes among the predominant bacteria in rumen fluid (Bryant and Small, 1956b) or from human feces (Moore et al., 1976). *Butyivibrio fibrisolvens* can be isolated from high dilutions of rumen fluid by using rumen fluid agar roll-tube media with finely ground cellulose. Small zones of cellulose digestion may be seen around colonies (Hungate, 1966; Shane et al., 1969) compared with larger zones of clearing caused by other cellulolytic bacteria (Shane et al., 1969). It can also be isolated on RGCA roll-tube medium containing 0.2% rutin as the sole energy source. Colonies from clear zones in the rutin and a yellow quercetin precipitate may form around some (Cheng et al., 1969; Leedle and Hespell, 1980). A medium containing xylan as sole carbohydrate source and nalidixic acid is suggested as selective for *Butyrivibrio fibrisolvens* because of the high resistance of some strains to this antibiotic (Hespell, 1991).

Maintenance procedures

Strains can be maintained by storage of cultures in liquid nitrogen or ultracold freezers (Hespell and Canale-Parola, 1970). Strains can also be preserved by lyophilization and storage at 4 °C. For short-term storage (6–15 months) glycerol-containing cultures can be kept at −20 °C (Teather, 1982a).

Differentiation of the genus *Butyrivibrio* from other genus

Butyrivibrio can be distinguished from most other genera of anaerobic bacteria on the basis of its curved rod-shaped cells, absence of spore formation, Gram-negative cell walls, its polar or subpolar flagella, and its fermentative metabolism with butyric acid as a major end product (Holdeman et al., 1977b, 1986). *Butyrivibrio fibrisolvens* can be distinguished from *Fusobacterium* species which also produce butyric acid, by its inability to produce indole, its motility, and its ability to hydrolyze esculin and ferment mannose (Holdeman et al., 1977b). A PCR assay to distinguish *Butyrivibrio* from *Pseudobutyrivibrio* has been developed (Kopečný et al., 2003; Mrázek and J. Kopečný, 2001).

Taxonomic comments

Butyrivibrio was proposed as a genus for butyric acid producing anaerobic, Gram-stain-negative, monotrichous, curved rods by Bryant and Small (1956b). Only one species, *Butyrivibrio fibrisolvens*, was proposed at the time, but the authors recognized the diversity within the genus: "Because of the variability in characteristics found between strains of this group of organisms, it will be difficult to determine what characteristics logically can be used to define natural species specific patterns". A second species, "*Butyrivibrio alacticidigens*" was proposed later for those strains that do not produce lactate (Hungate, 1966), but this was not included in *Bergey's Manual of Determinative Bacteriology* (Bryant, 1974), nor on the Approved Lists (Skerman et al., 1980). Shane et al. (1969), in a study of cellulolytic bacteria from the rumen of sheep, described two distinct groups of curved rods within the genus *Butyrivibrio*. Their group 1 produced appreciable amounts of lactate and removed acetate during cellobiose fermentation, whereas group 2 produced acetate but little or no lactate and was nutritionally more fastidious than group 1. The former group was different from "*Butyrivibrio alacticidigens*". In 1976, Moore et al. described a new species, *Butyrivibrio crossotus*, for isolates from human feces that resemble *Butyrivibrio fibrisolvens* but are lophotrichous.

Using DNA–DNA hybridizations, Mannarelli (1988) demonstrated a large diversity among *Butyrivibrio fibrisolvens* strains: five groups of related strains formed at least five distinct genospecies and 19 other strains (including the type strain) were not closely related to any other strain. This work was subsequently extended (Mannarelli et al., 1990b) by including more strains, also from related taxa, and three additional new *Butyrivibrio* genospecies were demonstrated. One of these groups contained the type strain of *Butyrivibrio crossotus* and strains recovered from the rumen of bison. The same strains were also studied in extracellular polysaccharide analyses and this produced groupings similar to those of DNA–DNA hybridization (Mannarelli et al., 1990b; Stack, 1988). Monoclonal antibodies have been successfully used to identify some *Butyrivibrio* strains from the rumen, but insufficient strains were included to establish the taxonomic potential of this technique inside this group (Hazlewood et al., 1986).

A phylogenetic study of the 16S rDNA genes of *Butyrivibrio* strains (Willems et al., 1996) revealed 12 rRNA types which formed three distinct lineages in cluster XIVa of the *Clostridium* subphylum as defined by Collins et al. (1994). An overview of their position is given in Figure 162 and a more detailed view of these groups is presented in Figure 163. Ribosomal RNA types 1–7 formed a first group in which rRNA type 1 (which includes the type strain of *Butyrivibrio fibrisolvens*) was the most peripheral. It showed 93–94% 16S rDNA sequence similarity with rRNA types 2–7, which showed 96–98.5% sequence similarity with each other. A second group was composed of rRNA types 8–11 which showed 98–99% 16S rDNA sequence similarity with each other and 87–88% with members of the first group. Finally, rRNA type 12, containing the type strain of *Butyrivibrio crossotus*, formed a completely separate line with 88–91% 16S rDNA sequence similarity with the first two clusters. Other studies, using 16S rDNA sequencing (Forster et al., 1996) and 16S rDNA-based hybridization probes (Forster et al., 1997), defined three further *Butyrivibrio* groups, only one of which was equivalent to one of the preceding rRNA types (type 8 = probe 49 group). Strains representing rRNA types 1, 2, 8, and 9 were previously included in DNA–DNA hybridizations (Mannarelli, 1988; Mannarelli et al., 1990b) where they formed separate genospecies. Apart from *Butyrivibrio crossotus*, in total at least 13 separate *Butyrivibrio* genotypic groups were thus reported in two distinct phylogenetic lineages. An overview of the strains that belong to each of these groups is given in Table 162. According to the 16S rRNA phylogenetic analysis presented in the roadmap to this volume (Figure 5), the genus *Butyrivibrio* is a member of the family *Lachnospiraceae*, order *Clostridiales*, class *Clostridia* in the phylum *Firmicutes*.

At about the same time, a new genus and species, *Pseudobutyrivibrio ruminis*, was described for a strain from the rumen of a cow (van Gylswyk et al., 1996). The strain was phenotypically very similar to *Butyrivibrio fibrisolvens*, but its 16S rDNA differed from that of the type strains of both *Butyrivibrio* species by at

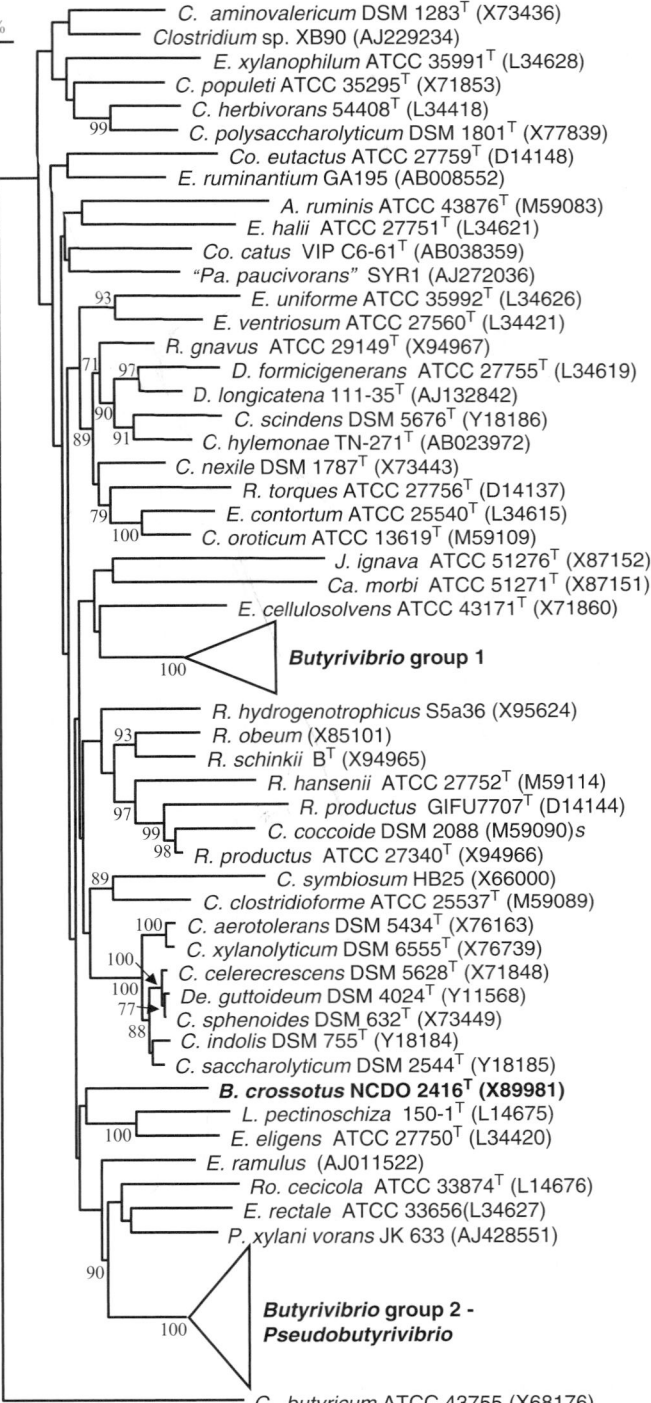

FIGURE 162. Phylogenetic neighbor-joining tree showing the position of *Butyrivibrio* groups in *Clostridium* cluster XIVa based on 16S rDNA sequences. The bar represents 1% estimated substitutions. Bootstrap values of 70% or higher (based on 500 repetitions) are given at branching points. *Abbreviations*: *A., Acetitomaculum; B., Butyrivibrio; C., Clostridium; Ca., Catonella; Co., Coprococcus, D., Dorea; De., Desulfotomaculum; E., Eubacterium; J., Johnsonella; L., Lachnospira; P., Pseudobutyrivibrio; Pa., Parasporobacterium; R., Ruminococcus; Ro., Roseburia.*

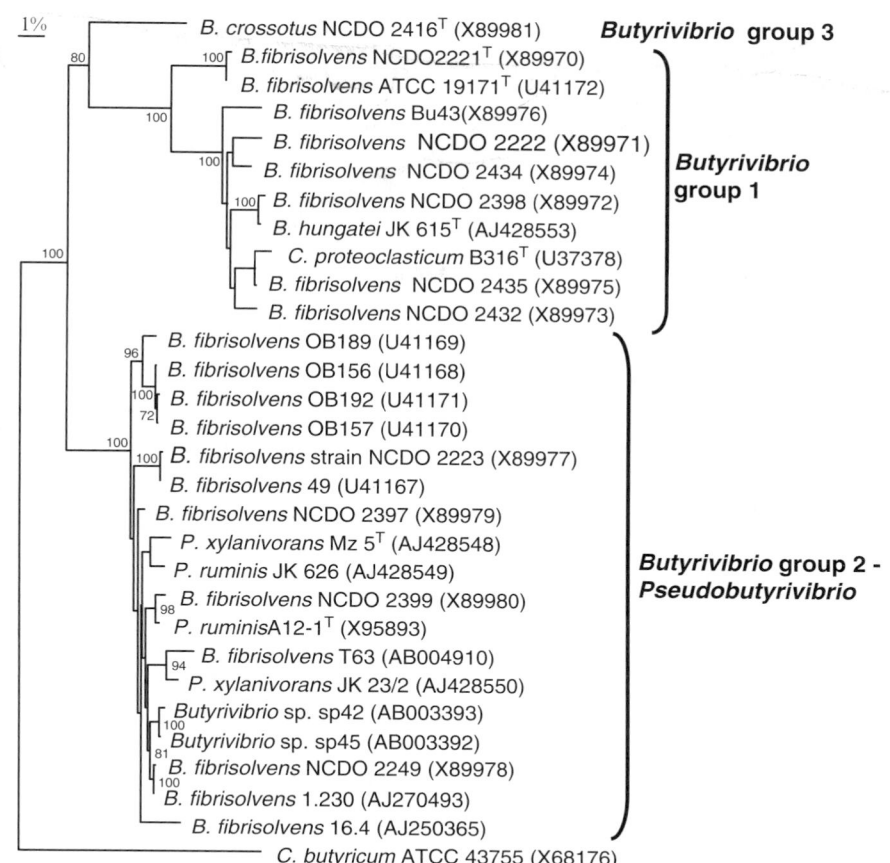

FIGURE 163. Phylogenetic neighbor-joining tree showing the detailed composition of the different *Butyrivibrio* groups. The bar represents 1% estimated substitutions. Bootstrap values of 70% or higher (based on 500 repetitions) are given at branching points. Abbreviations: *B., Butyrivibrio; C., Clostridium; P., Pseudobutyrivibrio.*

TABLE 162. Strains assigned to different *Butyrivibrio* groups on the basis of genetic information[a]

Group	Strains
Mannarelli group 1 = Willems rRNA type 8 = Forster probe 49 group	49=NDCO 2223=Bu49, H17c, CE 51=Bu37, CE 52=Bu38, 12=Bu12=K29CF=12, UC12254, UC12487, UC12494, UC12492, UC12493, UC12491, NCDO 2418=Bu27=19, NCDO 2419=Bu31=CXS 18, NCDO 2421=Bu41=CE 58, Bu15=K29HF=15, Bu17=K13KF=17, Bu26=18, Bu28=20, Bu29=21, Bu30=24, Bu32=CXS13, Bu33=22, Bu34=CE 36, Bu35=CE 46, Bu39=CE 53, Bu40=CE56
Mannarelli group 2	CF2d, CF3c, CF3a, CF3, CF1b, CF4c
Manarelli group 3 = Willems rRNA type 9	E21c, E9a, B-835, NOR-37=NCDO 2249, 1L6-31=NCDO 2400, NCDO 2438=CE 74
Mannarelli group 4	X6C61, D30g
Mannarelli group 5	E46a, H4a
Mannarelli group 6 = Willems rRNA type 12 = *B. crossotus*	VV1[b], VV2, VV3, VV4, VV5, T9-40A[T]=NCDO 2416=Bu46, NCDO 2415=Bu45=C3824
Mannarelli group 7	ARD-27b, ARD-31a
Mannarelli group 8	GS114, AcTF2
Willems rRNA type 1 = *B. fibrisolvens* s.s.	NCDO 2221=ATCC 19171=D1[T]
Willems rRNA type 2	NCDO 2222=A38=ATCC 27208, Bu25=K202-21-07-6D
Willems rRNA type 3	NCDO 2398=B835, NCDO 2417=B834, Bu22=K27BF1=1
Willems rRNA type 4	NCDO 2432=CE 65
Willems rRNA type 5	NCDO 2434=Bu42=K13-26-037-A
Willems rRNA type 6	NCDO 2435=Bu24=K10-02-04-6B, NCDO 2436=Bu44=CE 78, Bu36=CE 47, H17c(SA)[c]
Willems rRNA type 7	Bu43=CE 64
Willems rRNA type10	NCDO 2397=JL
Willems rRNA type 11	NCDO 2399=Bu21=RO3-21-08-7A, CF3
Forster probe 156 group	OB143, OB148, OB149, OB150, OB153, OB155, OB156, OB157, OB192, OB194, OB204, OB205, OB206, OB209, OB224
Forster probe 189 group	OB188, OB189, OB190, OB196, OB202, OB203, OB218

[a]Data from Mannarelli (1988), Mannarelli et al. (1990b), Willems et al. (1996), Forster et al. (1997), Dalrymple et al. (1999).
[b]According to Forster et al. (1997) this strain belongs with Mannarelli group 1.
[c]Dalrymple et al. (1999) demonstrated that two different strains labeled H17c appear to have been used by different research groups: the original H17c belongs to rRNA type 8, whereas another isolate, labeled H17c(SA), belongs to rRNA type 6.

least 10% and it was therefore described as a separate genus. Its closest neighbors were *Roseburia cecicola*, *Eubacterium rectale*, and *Lachnospira pectinoschiza* (van Gylswyk et al., 1996). When comparing the 16S rDNA sequence of *Pseudobutyrivibrio ruminis* with the *Butyrivibrio* rRNA types (Figure 163), it is clear that this species belongs to the second *Butyrivibrio fibrisolvens* rRNA cluster as defined by Willems et al. (1996). The type strain of *Pseudobutyrivibrio ruminis* shows 97.5–99% 16S rDNA sequence similarity with rRNA types 8–11 which make up this cluster. From this comparison and from the limited DNA–DNA hybridizations between rRNA type 8 and rRNA type 9 strains (Mannarelli, 1988), it seems justified to conclude that the second *Butyrivibrio* rRNA group could be considered as a separate genus, *Pseudobutyrivibrio*, and that rRNA types 8–11 probably represent several genospecies within *Pseudobutyrivibrio*. It is possible that one of these rRNA types is identical with *Pseudobutyrivibrio ruminis* (e.g., rRNA type 11 shows 99% 16S rDNA sequence similarity with the type strain), but this needs to be confirmed by DNA–DNA hybridizations. The genus *Butyrivibrio* should be restricted to the first *Butyrivibrio* rRNA group with *Butyrivibrio fibrisolvens* rRNA types 1–7 (with type 1 including the type strain) probably representing separate species. A new *Clostridium* species, *Clostridium proteoclasticum*, which was described for an isolate from the rumen of New Zealand cattle (Attwood et al., 1996), groups within *Butyrivibrio* rRNA group 1 (Figure 163). It is phenotypically similar to *Butyrivibrio fibrisolvens*, although it can be clearly distinguished by its low GC content of 28% (Attwood et al., 1996). On the basis of its 16S rDNA phylogeny, *Butyrivibrio crossotus* could represent a separate new genus provided sufficient additional criteria supporting its separateness becomes available.

In a study of isolates from rumen fluid of cow and sheep, together with reference strains, Kopečný et al. (2003) defined five taxa among the strains: *Butyrivibrio fibrisolvens*, *Clostridium proteoclasticum*, *Pseudobutyrivibrio ruminis* and two new species, *Butyrivibrio hungatei* and *Pseudobutyrivibrio xylanivorans*. The assignment of strains to these five taxa is based mostly on the similarity of RFLP patterns of 16S rRNA genes and fatty acid compositions. This may be insufficiently precise for definite species allocation. Indeed, using RFLP criteria several of the *Butyrivibrio* rRNA types described previously based on nearly full

gene sequence data (Willems et al., 1996) would be comprised in existing species with distinct rRNA types: rRNA type 7 (represented by Bu43) would be included in *Clostridium proteoclasticum*, rRNA types 3 (NCDO 2398) and 5 (NCDO 2434) would be included in *Butyrivibrio hungatei*, rRNA types 10 (NCDO 2397) and 11 (NCDO 2399) would be included in *Pseudobutyrivibrio xylanivorans* and rRNA type 9 (NCDO 2249) would be included in *Pseudobutyrivibrio ruminis* (Kopečný et al., 2003). From the position of these strains in the 16S rDNA phylogeny (Figure 163), it would seem that finer methods such as DNA–DNA hybridizations or phylogeny of housekeeping genes may help resolve this issue in future.

At present data allowing the phenotypic differentiation between *Pseudobutyrivibrio* and *Butyrivibrio sensu stricto* are incomplete. However, the division between both groups that has emerged from phylogenetic data appeared to be consistent with the division of *Butyrivibrio* isolates into two groups of curved rods by Shane et al. (1969). Their group 1 contains mostly strains belonging to the second *Butyrivibrio* rRNA group, equivalent to *Pseudobutyrivibrio*, whereas their group 2 comprises mostly strains of the first *Butyrivibrio* rRNA group. As mentioned previously, group 1 produces lactate and utilizes acetate during fermentation, whereas group 2 produces acetate but little or no lactate and is nutritionally more fastidious than group 1. These may be useful phenotypic features to differentiate *Butyrivibrio* from *Pseudobutyrivibrio*. An overview of phenotypic data reported for these groups is presented in Table 163. In addition, it has been suggested that the method of butyrate production, using either butyryl-CoA/acetate-CoA transferase, as in strains 49 and 37, or butyrate kinase, as in strains D1 and A38, is a phylogenetically significant characteristic (Diez-Gonzalez et al., 1999). Since on the basis of 16S rDNA sequences, the former strains may represent *Pseudobutyrivibrio* and the latter strains represent *Butyrivibrio*, this characteristic could be a potential differentiating feature between these genera, if confirmed in additional strains.

Acknowledgements

A.W. is indebted to the Fund for Scientific Research–Flanders for a position as a Postdoctoral Research Fellow.

List of species of the genus *Butyrivibrio*

There are at present three species recognized within *Butyrivibrio*, *Butyrivibrio fibrisolvens*, *Butyrivibrio crossotus* and *Butyrivibrio hungatei*. However, it is now clear (see previous taxonomic comments) that the originally described species *Butyrivibrio fibrisolvens* in fact represents several species and even several genera. The phenotypic description of this species as given in the previous version of this manual (Bryant, 1986b), in fact, comprises all of these groups. However, because of lack of a recent comprehensive phenotypic study it is not possible at present to provide reliable phenotypic descriptions for more restricted genotypically defined groups, with the exception of the recently described *Butyrivibrio hungatei* (Kopečný et al., 2003). The species description of *Butyrivibrio fibrisolvens* that follow is therefore based on that in the previous version of the manual (Bryant, 1986b). It is probably based on strains of *Butyrivibrio* rRNA group 1 and on strains that would now be classified as *Pseudobutyrivibrio*.

1. **Butyrivibrio fibrisolvens**   Bryant and Small 1956b, 19[AL]

fi.bri.sol'vens. L. n. *fibra* fiber; L. part. adj. *solvens* dissolving; N.L. part. adj. *fibrisolvens* fiber-dissolving.

Cell morphology is the same as that presented in the genus description. Cells stain Gram-negative but have a very thin cell wall with a Gram-positive ultrastructure. Colonial characteristics are as present previously. Growth in liquid glucose medium varies from a uniform turbidity to a flocculent or granular sediment. Good growth at 37 °C and generally at 45 °C, but slower at 30 °C. No growth occurs at 22 or 50 °C.

In poorly buffered glucose medium the final pH mostly ranges from 5.0–5.6.

Most strains will grow in chemically defined media containing glucose or cellobiose as the energy source, amino acid mixtures and ammonium salts as nitrogen source, minerals, B vitamins, and cysteine. Many strains will grow with

TABLE 163. Phenotypic features of *Butyrivibrio* groups[a]

	Butyrivibrio crossotus[b]	*Butyrivibrio fibrisolvens*[c]	*Butyrivibrio hungatei*[a]	Curved rods group 2[e]	Curved rods group1[e]	*Pseudobutyrivibrio ruminis*[f]	*Pseudobutyrivibrio xylanivorans*[d]
Saccharolytic metabolism	+		+			+	+
Poor growth without fermentable carbohydrate	+		+				+
Fermentation end products:							
Formate	+	+	+	+	+	+	+
Butyrate	+	+	+	+	+	+	+
Lactate	+	d	−	Tr	+	+	d
Acetate	+	d	+	+	Used	Tr	−
Propionate	Tr						
Pyruvate	Tr						
Succinate	Tr		−	−	−		+
Ethanol		−[c]	d	+	Tr		d
Hydrogen		+		+	+		
Flagella	Lophotrichous, polar or sub-polar	Monotrichous polar or subpolar	Monotrichous polar or sub-polar	Monotrichous polar or subpolar	Monotrichouspolar or subpolar	Monotrichous polar or subpolar	
Motility	+		+	+	+	+	
Optimal growth temperature (°C)	37			37	37	37–39	
Growth at 30 °C	None or poor			d	+		
Growth in PY with lactate, pyruvate or threonine	−						
Ammonia from PY or CM medium	−						
pH in PY-maltose cultures (5 d)	4.7–5.2						
DNA G+C content (mol%)	36–37	41–42	40–45			40–41	41–44
Fermentation of:							
Adonitol	−						
Amygdalin	−						
Arabinose	−		+			+	
Aspartate						−	
Cellobiose	−	+	+	+	+	+	+
Cellulose		d		d	d		
Dextrin	+						
Esculin	−	d	+	d	d		+
Erythritol	−						
Fructose	−/w	+		d	+	+	
Fumarate						−	
Galactose	−	+				+	
Glucose	−/w	+	+	d	+	+	+
Glutamate						−	
Glycerol	−	−	−	−	−	w	−
Glycogen	+						
Hemicellulose		d					
Inositol	−	−					
Inulin	−	+		d	+		
Lactose	v	+	+	d	+	+	+
Lactate	−					−	
Malate						−	
Malonate						−	
Maltose	+	+	+	d	+	+	+
Mannitol	−	−	−	−	−	−	
Mannose	−		+	d	+	+	+
Melezitose	−		−				+

(continued)

TABLE 163. (continued)

	Butyrivibrio crossotus[b]	Butyrivibrio fibrisolvens[c]	Butyrivibrio hungatei[a]	Curved rods group 2[e]	Curved rods group1[e]	Pseudobutyrivibrio ruminis[f]	Pseudobutyrivibrio xylanivorans[d]
Melibiose	–						
Pectin		d					
Oxalate	–					–	
Pyruvate						–	
Raffinose	–		+	d	+		–
Rhamnose	–	+[c]	d	–	–	–[c]	d
Ribose	–					–	
Salicin	–	+	+	d	+		+
Sorbitol	–		–			–	–
Sorbose	–						
Starch	+	d				–	
Succinate						–	
Sucrose	–	+	+			+	+
Trehalose	–	d	–	d	–	+	
Xylan		d				–	
Xylose	–	+	+	d	+	+	+
Starch hydrolysis	+	–[c]	–			–	d
Hydrolysis of esculin	v	d					
Milk	Curd						
Ammonia from peptone	–						
H2S production	–	–	–	–	–		–
Gas from glucose agar	–						
Growth on PY	1						
Growth on glucose-bile	–/v						
Gelatin hydrolysis	–	v	–	d	d	–	
Digestion of meat	–						
Digestion of milk	–						
Nitrate reduction	–	–	–	–	–		–
Catalase	–	–	–				–
Indole	–	–					
Hydrolysis of hippurate	–						
Growth with 6.5% NaCl	–						
Aerobic growth	–		–			–	–

[a]Symbols: +, >85% positive; d, different strains give different reactions (16–84% positive); –, 0–15% positive; w, weak reaction; Tr, trace.
[b]Data from Moore et al. (1976).
[c]Data from Bryant (1986b).
[d]Data from Kopečný et al. (2003).
[e]Data from Shane et al. (1969). Curved rods group 1 contains many strains that are known to belong to *Butyrivibrio* group 2 that also contains *Pseudobutyrivibrio ruminis*. Curved rods group 2 contains many strains now known to belong to *Butyrivibrio* group 1 that contains the type strain of the type species *Butyrivibrio fibrisolvens*.
[f]Data taken from van Gylswyck et al. (1996).

only ammonium salts as nitrogen source, but amino acid mixtures are often growth promoting. Acetate often stimulates growth, as do propionate or branched-chain volatile acids, but for some strains only. Additional properties are given in Table 163 and Table 164.

Butyrivibrio fibrisolvens is a nonpathogenic member of the normal rumen microflora.

DNA G+C content (mol%): 42 (Bd) (Mannarelli, 1988).

Type strain: ATCC 19171 (=D1).

EMBL/GenBank accession number (16S rDNA): X89970 and U41172.

2. **Butyrivibrio crossotus** Moore, Johnson and Holdeman 1976, 241[AL]

cros.so′tus. Gr. adj. *crossotus* tasseled.

Cell morphology is the same as that presented in the genus description and colonial characteristics are as presented previously. Only a few strains grow as surface colonies on anaerobically incubated blood agar plates and no hemolysis is observed. Few strains grow on egg yolk agar and no lecithinase of lipase activity is observed. Growth is poor in peptone-yeast extract broth unless a fermentable carbohydrate is present. In maltose broth, growth is abundant with

TABLE 164. Differentiating features of the three *Butyrivibrio* species[a]

Characteristic	*B. fibrisolvens*[b]	*B. crossotus*[b]	*B. hungatei*[c]
Flagella	Monotrichous	Lophotrichous	Monotrichous
H$_2$ produced from glucose or maltose	+	–	
Fermentation of glucose	+	w or –	+
Fermentation of fructose	+	w or –	
Fermentation of sucrose, cellobiose, and xylose	+	–	+
DNA G+C content (mol%)	41–42[d]	36–37[e]	40–45

[a]Symbols: +, >85% positive; –, 0–15% positive; w, weak reaction.
[b]Data from Bryant (1986b). *Butyrivibrio fibrisolvens* probably includes strains that would now be classified as *Pseudobutyrivibrio ruminis*.
[c]Data from Kopečný et al. (2003).
[d]Calculated from buoyant density (Mannarelli, 1988).
[e]Calculated from thermal melting point (Moore et al., 1976).

a smooth, flocculent, or ropy sediment and usually some uniform turbidity. Optimum growth temperature is 37 °C. Growth is slow at 30 °C and some strains may grow at 45 °C.

Metabolic properties and features differentiating *Butyrivibrio crossotus* from *Butyrivibrio fibrisolvens* are given in Table 163 and Table 164.

Butyrivibrio crossotus strains are isolated from human feces or rectal contents and are not known to be pathogenic.

DNA G+C content (mol%): 37 (T_m) (Moore et al., 1976).

Type stain: ATCC 29175 (=T9-40A).

EMBL/GenBank accession number (16S rDNA): X89981.

3. **Butyrivibrio hungatei** Kopečný, Zorec, Mrázek, Kobayashi and Marinšek-Logar 1976, 207[VP]

hun.ga'te.i. N.L. gen. n. *hungatei* named after Robert E. Hungate, an American microbiologist who isolated similar strains in the 1960s.

Cell morphology is the same as that presented in the genus description. Motile by a single polar or subpolar flagellum. Anaerobic growth at 39 °C, but not at 25 °C. Limited growth at 45 °C. Metabolic properties and features of

Butyrivibrio hungatei are given in Table 163 and Table 164. Tryptophan, urea and gelatin are not utilized. Does not show significant fibrolytic or proteolytic activity, utilizes mainly oligo- and monosaccharides as growth substrates. Produces α-galactosidase, α-arabinosidase, phenylalanine arylamidase and leucine arylamidase. No amylase, xylanase, β-endoglucanase, laminarinase, pectin hydrolase, proteinase or Dnase activity detected. Branched chain fatty acids represent the prevailing proportion of fatty acids. In DSM medium 330 without rumen fluid, the type strain produces as major cellular fatty acids anteiso-C$_{17:0}$ (17.4%), iso-C$_{16:0}$ (14.5%) and C$_{16:1}$ (13.2%) and also contains C$_{16:0}$ 2-OH (6.9%), C$_{14:0}$ 2-OH (4.3%) and iso-C$_{17:1}$ (3.4%).

Butyrivibrio hungatei strains are isolated from the rumen fluid of cow and sheep where they participate in the utilization of intermediates of fibre degradation.

DNA G+C content (mol%): 44.8 (HPLC) (Moore et al., 1976).

Type stain: JK 615 (= DSM 14810 = ATCC BAA-456).

EMBL/GenBank accession number (16S rDNA): AJ428553.

Genus VI. **Catonella** Moore and Moore 1994, 189[VP]

ANNE WILLEMS AND MATTHEW D. COLLINS

Ca.to.nel'la. N.L. fem. n. *Catonella* in honor of Elizabeth P. Cato, an American microbiologist.

Obligately anaerobic, catalase-negative, Gram-stain-negative, non-spore-forming rods. Nonmotile. End products of carbohydrate fermentation include major amounts of acetate and smaller amounts of formate and lactate. Major cellular fatty acids when grown in peptone-yeast extract-glucose broth include **C$_{14:0}$ fatty acid, C$_{14:0}$ dimethyl acetal, and C$_{16:0}$ fatty acid.** *Catonella* is associated with periodontitis in human gingival crevices and periodontal pockets. The mol% G+C content of DNA is 34. Based on 16S rDNA sequence of its type strain, *Catonella* belongs to cluster XIVa (Collins et al., 1994) of the *Clostridium* subphylum of the Gram-positive bacteria.

Type species: **Catonella morbi** Moore and Moore 1994, 189[VP].

Further descriptive information

The genus contains only one species and therefore all of the characteristics provided below describe the species *Catonella morbi*.

Phylogenetic treatment. According to the 16S rRNA phylogenetic analysis presented in the roadmap to this volume (Figure 5), the genus *Catonella* is a member of the family *Lachnospiraceae*, order *Clostridiales*, class *Clostridia* in the phylum *Firmicutes*; in particular, it belongs to a subgroup with *Butyrivibrio*, *Coprococcus*, *Ruminococcus*, *Lachnospira*, and several *Eubacterium* and *Clostridium* species. Closest relatives are *Johnsonella ignava* and *Eubacterium saburreum* (approx. 85% sequence similarity) (Willems and Collins, 1995b).

Cell morphology. Cells of the type strain grown in peptone-yeast extract-glucose (PYG) broth* are 0.7–1.2 μm wide × 1.6–4.0 μm long. Central swelling may occur in media containing a fermentable carbohydrate. Cells occur in pairs and short chains (Moore and Moore, 1994).

Colonial characteristics. On blood agar, incubated for 2 d at 37 °C, colonies are 1 mm in diameter, round with an entire margin, convex, and opaque. No hemolysis on rabbit blood agar is observed and no dark pigments are produced (Moore and Moore, 1994).

Nutrition and growth conditions. Because of the human origin of the strains, cultures are grown at 37 °C. No data on minimal, maximal, or optimal growth temperatures were reported. Serum (10%, v/v) improves growth and is usually required to detect carbohydrate fermentation. Moderate turbidity with limited sediment is produced in peptone-yeast extract-serum broth and moderately heavy turbidity with a smooth sediment is produced in PYG-serum broth. The final pH of PYG-serum broth cultures is 5.0–5.6 (Moore and Moore, 1994).

Metabolism and metabolic pathways. *Catonella* has a strictly anaerobic, saccharolytic metabolism. The end products of glucose fermentation in PYG-serum broth were reported as (in milliequivalents per 100 ml of culture; mean ± standard error of mean) acetic acid, 2.2 ± 0.2; formic acid, 0.6 ± 0.1; and lactic acid, 0.4 ± 0.1. Abundant hydrogen gas is produced (Moore and Moore, 1994). Additional metabolic characteristics are presented in Table 165.

Chemotaxonomic characteristics. The cellular fatty acid composition of *Catonella morbi*, based on data for eight strains grown in 10 ml of PYG broth cultures at 37 °C, was as follows: 3% $C_{12:0}$, 42% $C_{14:0}$, 14% $C_{14:0}$ dimethyl acetal, 2% $C_{16:1-cis9}$, 12% $C_{16:0}$, 1% $C_{16:0}$ dimethyl acetal, 5% $C_{18:1-cis9}$, 4% $C_{18:0}$ (Moore and Moore, 1994).

Antibiotic susceptibility. *Catonella morbi* is susceptible (nine strains tested) to chloramphenicol (12 μg/ml), clindamycin (1.6 μg/ml), erythromycin (3 μg/ml), penicillin G (2 U/ml), and tetracycline (6 μg/ml) (Moore and Moore, 1994).

Ecology. *Catonella morbi* occurs in human gingival crevice and periodontal pockets and is associated with adult periodontitis.

Enrichment and isolation procedures

No specific enrichment procedures for *Catonella* have been described. *Catonella* have been isolated and identified in studies of the anaerobic bacterial flora associated with human gingivitis and periodontitis (Moore et al., 1985, 1982b). In these studies, samples from supragingival plaque and periodontal pockets were taken with sterile paper tips, toothpicks, or scalers and immediately placed in prereduced, anaerobically sterilized broth under an atmosphere of CO_2 (Moore et al., 1982a). The samples were then shaken with 100-μm-diameter glass beads and dilutions were cultured in roll tubes of D4 medium with rabbit serum (Moore et

TABLE 165. Metabolic characteristics of *Catonella morbi* strains[a]

Characteristic	ATCC 51271[T]	Eight other strains[b]
Acid from:		
Amygdalin	−	25
Arabinose	−	12
Cellobiose	+	62
Dextrin	+	60
Glucose	+	86
Glycogen	−	62[c]
Lactose	+	71
Maltose	+	71[c]
Mannose	−	14
Melezitose	−	14
Melibiose	−	28
Rhamnose	+	88
Salicin	−	37
Starch	−[c]	12[c]
Trehalose	−	25[c]
Xylose	−	12
Starch hydrolysis	−[c]	80[c]
Digestion of gelatin	−	38

[a]Data taken from Moore and Moore (1994). All strains produce acid from raffinose and sucrose, hydrolyze esculin, and curd milk. No acid is produced from following substrates: erythritol, esculin, fructose, gum arabic, inulin, larch arabino-galactan, mannitol, pectin, ribose, and sorbitol. Indole and catalase are not produced; nitrate is not reduced; no lecithinase or lipase is detected on egg yolk agar; no digestion of milk or meat; no H_2S production in sulfide-indole motility medium; no growth in PYG-20% bile.

[b]Percentages of strains positive.

[c]A positive (acid) reaction occurs in the presence of serum (10% v/v) in the medium; usually reaction is negative in the absence of serum.

al., 1982a) and on anaerobic blood agar plates (D4 medium with 5% [v/v] rabbit blood) (Moore et al., 1985). The D4 role tube isolation medium (Moore et al., 1982a) contained 3.7 g brain heart infusion broth base (BBL), 0.5 g yeast extract (Difco), 0.5 ml of 6% (wt/v) ammonium formate solution, 0.4 ml of resazurin (25 mg in 100 ml of distilled water), and 100 ml of distilled water. After this medium is boiled and cooled under CO_2, 0.05 g cysteine-HCl H_2O, 0.02 ml of vitamin K_1 stock solution (0.15 ml of vitamin K_1 in 30 ml of 95% ethanol; store in the dark, for maximum of 1 month under refrigeration) and 1 ml of hemin solution (50 mg hemin in 1 ml of 1 N NaOH, make to 100 ml with distilled water, autoclave 15 min, 121 °C) were added and the pH was adjusted to 7. The broth was dispensed (12 ml per tube) under an atmosphere of oxygen-free nitrogen and the stoppered tubes were autoclaved. For sample collection, agar medium in roll tubes was melted and cooled to 56 °C, 1.2 ml of serum-yeast autolysate-TPP mixture (containing 50 ml of filter-sterilized, inactivated [56 °C for 30 min], nonhemolytic rabbit serum, 50 ml of yeast autolysate [prepared by incubating 1 ounce of Fleishman's yeast powder in 100 ml of water at 56 °C for 72 h; the mixture is filtered through cheese-cloth and filters of decreasing pore size

---

*Composition of PYG: 0.5 g peptone, 0.5 g trypticase, 1 g yeast extract, 0.4 ml of resazurin solution (25 mg in 100 ml of distilled water), 4 ml of salts solution (0.2 g $CaCl_2$, 0.2 g $MgSO_4$, 1g K_2HPO_4, 1 g KH_2PO_4, 10 g $NaHCO_3$, 2 g NaCl; dissolve $CaCl_2$ and $MgSO_4$ in 300 ml of distilled water; add 500 ml of water and, while swirling, slowly add remaining salts; when dissolved, add 200 ml of distilled water and mix; store at 4 °C), 100 ml of distilled water, 1 ml of hemin solution (dissolve

50 mg hemin in 1 ml of 1 N NaOH; make to 100 ml with distilled water; autoclave for 15 min at 121 °C), 0.02 ml of vitamin K1 solution (dissolve 0.15 ml of vitamin K1 in 30 ml of 95% ethanol; keep for maximum of 1 month, at 4 °C in a brown bottle), 0.05 g cysteine-HCl·H_2O, 1 g glucose; vitamin K1, hemin, and cysteine are added after the medium is boiled, but before dispensing and autoclaving (Holdeman et al., 1977b).

until it is sterilized by passing through a 0.2 μm pore size filter] and 0.5 ml of a thiamine pyrophosphate (TPP) stock [0.5 g of TPP in 100 ml of distilled water, filter-sterilized]) was added to each tube under N_2-CO_2-H_2 (85:12:3) atmosphere. Tubes were inoculated under anaerobic gas, restoppered, spun and chilled until the agar solidified and incubated. The blood agar plates contained the same medium except that 5.2 g of brain heart infusion agar (BBL) replaced the brain heart infusion broth base and agar, 4 ml of defibrinated rabbit blood replaced the serum and no resazurin was added. The basal medium was autoclaved, cooled to 56 °C and blood, yeast autolysate, TPP, vitamin K_1, and hemin were added and plates were poured in an anaerobic atmosphere. Plates can be stored before use at room temperature under an atmosphere of 10% CO_2 and 90% H_2. After 5 d of incubation at 37 °C, colonies were randomly picked and purified. This procedure resulted in the isolation of *Catonella* as well as numerous other anaerobic bacteria of the periodontal microflora (e.g., *Bacteroides, Eubacterium, Fusobacterium, Selenomonas, Streptococcus, Lactobacillus* etc., which were identified using a combination of morphological, biochemical, and fatty acid characteristics (Moore et al., 1985, 1982b).

Maintenance procedures

Strains can be maintained in prereduced, anaerobically sterilized D5 broth (Moore et al., 1982a), which contains 3.7 g of brain heart infusion broth base (BBL), 0.5 g yeast extract (Difco), 0.05 ml pyruvic acid, 0.5 ml ammonium formate (6% [wt/v] in water), 0.2 g pectin, 2 ml IsoVitaleX (BBL), 1 g agar, 0.1 g KNO_3, 0.4 ml resazurin (25 mg in 100 ml of distilled water), and 100 ml of distilled water. This basal medium is boiled and cooled under CO_2 and 0.05 g cysteine-HCl·H_2O, 0.02 ml vitamin K_1 (0.15 ml vitamin K_1 in 30 ml of 95% ethanol; store in the dark, for maximum of 1 month under refrigeration) and 1 ml hemin (50 mg hemin in 1 ml of 1 N NaOH, make to 100 ml with distilled water, autoclave 15 min, 121 °C) are added. The pH is adjusted to 7.0 with 1 N NaOH and the medium (3 ml per tube) is dispensed under oxygen-free nitrogen gas and sterilized. Before use, 0.3 ml of sterile serum-yeast autolysate-TPP solution (see above) is added aseptically to each tube (Moore et al., 1982a).

For long-term storage, cultures can be preserved by lyophilization.

Differentiation of the genus saccharolytic bacilli from other genera

Catonella can readily be distinguished from most genera of Gram-stain-negative, anaerobic, saccharolytic rods (including *Bacteroides, Prevotella, Oribaculum, Butyrivibrio, Ruminobacter, Hallela, Fibrobacter, Selenomonas, Pectinatus,* and *Mitsuokella*) by the G+C content of its DNA (34 mol%). It can be differentiated from other genera of Gram-stain-negative, anaerobic, saccharolytic rods with a similar G+C content by end products of glucose fermentation and by its cellular fatty acid composition (Table 166) (Moore et al., 1994; Moore and Moore, 1994).

Taxonomic comments

Catonella morbi was originally known as "*Bacteroides* D42", one of a group of unnamed bacterial species that occur in the human gingival crevice and periodontal pockets that were first recognized and isolated at the Anaerobe Laboratory of the Virginia Polytechnic Institute and State University in Blacksburg (Moore et al., 1985, 1982b). It was formally described as *Catonella morbi*, a new genus and species of anaerobic, Gram-stain-negative, nonsporing bacilli and therefore placed in the family *Bacteroidaceae* (Moore and Moore, 1994). Later, by 16S rDNA sequence analysis, it was shown to belong to the *Clostridium* subphylum of the Gram-positive bacteria. It is a member of *Clostridium* cluster XIVa as defined by Collins et al. (1994). This cluster consists of species from several Gram-stain-positive (*Clostridium, Coprococcus, Eubacterium,* and *Ruminococcus*) and Gram-stain-negative genera (*Butyrivibrio, Johnsonella,* and *Catonella*). A similar occurrence of Gram-stain-negative and Gram-stain-positive bacteria in one cluster is only observed in *Clostridium* cluster XII, which contains the Gram-stain-negative *Tissierella* (Farrow et al., 1995) and in the *Sporomusa* branch (= *Clostridium* cluster IX; Collins et al., 1994). Some of these groups, e.g., *Butyrivibrio*, are Gram-negative in traditional staining procedures, but by ultrastructural electron microscopy they were shown to possess a very thin but Gram-positive type of cell wall (Cheng and Costerton, 1977; Hespell et al., 1993). Similar observations for *Catonella* have not been reported.

Acknowledgements

A.W. is indebted to the Fund for Scientific Research–Flanders for a position as postdoctoral research fellow.

Further reading

A comparison of biochemical and fatty acid characteristics of 32 genera of anaerobic Gram-stain-negative rods is presented by Moore et al. (1994).

TABLE 166. Differentiation of *Catonella* from other Gram-stain negative, anaerobic, saccharolytic genera with similar G+C content[a]

Characteristic	*Catonella*	*Anaerorhabdus*	*Fusobacterium*	*Megamonas*
DNA G+C content (mol%)	34	34	26–34	35
Major fermentation end product	Acetic acid	Lactic acid	Butyric acid	Propionic acid
Major cellular fatty acids[b]	$C_{14:0}$ FA, $C_{14:0}$ DMA, $C_{16:0}$ FA	$C_{16:0}$ FA, $C_{18:1\ cis9}$ FA, $C_{18:0}$ FA, $C_{18:1\ cis11/t9/t6}$ FA or unknown (ECL 17.8)	$C_{14:0}$ FA, $C_{16:1\ cis9}$ FA, $C_{16:0}$ FA	$C_{11:0}$ FA, $C_{14:0}$ DMA, $C_{15:0}$ FA, $C_{17:1\ cis9}$ FA or $C_{17:2}$ FA (ECL 16.8)

[a]Data taken from Moore and Moore (1994).

[b]Components making up more than 10% of fatty acids; FA, fatty acid; DMA, dimethyl acetal; ECL, equivalent chain-length.

List of species of the genus *Catonella*

1. **Catonella morbi** Moore and Moore 1994, 189[VP]

mor'bi. L. gen. n. *morbi* of disease, because originally the organism was isolated from diseased periodontal pockets.

The morphological and cellular characteristics are as described for the genus. Additional descriptive informa-

tion is presented in Table 165 which is based on data from Moore and Moore (1994) and Moore et al. (1994).

Isolated from gingival crevices of humans with adult periodontitis.

DNA G+C content (mol%): 34.

Type strain: ATCC 51271 (= VPI D154F-12).

GenBank accession number (16S rRNA gene): X87151.

Genus VII. **Coprococcus** Holdeman and Moore 1974, 260[AL]

TAKAYUKI EZAKI

Co'pro.coc'cus. Gr. n. *kopros* feces; Gr. n. *kokkos* berry; N.L. masc. n. *Coprococcus* fecal coccus.

Cocci that are **Gram-stain-positive,** nonmotile, and **obligately anaerobic chemoorganotrophs.** All species in the genus are isolated from human feces (Moore and Holdeman, 1974) but rarely isolated from human clinical specimens. Strains were originally isolated on an anaerobically sterilized rumen-fluid-glucose-cellobiose agar (Holdeman et al., 1977a) roll tube. However, pure cultured strains grow on anaerobically incubated blood agar plates (supplemented with brain heart infusion agar with 5% sheep blood). Cells may occur as pairs or chains of pairs. Cells of the some species are slightly elongated, in particular, when grown in medium containing fermentable carbohydrates. **Fermentable carbohydrates** are either **required** or are **highly stimulatory for growth. Major fermentation products include butyric and acetic acids, with formic or propionic acid.** Characteristics to differentiate species of the genus *Coprococcus* and biochemically closely related species of the *Ruminococcus* are given in Table 167. The genus phylogenetically belongs to family *Lachnospiraceae* and the phylogenetic position within the family is shown in Figure 161 (Ezaki et al., 1994; Rainey and Janssen, 1995; Willems and Collins, 1995a). In the family *Lachnospiraceae*, members of genus *Coprococcus* are phylogenetically closely related to anaerobic curved bacteria, genus *Lachnospira* as in Figure 161.

DNA G+C content (mol%): 39–42.

Type species: **Coprococcus eutactus** Holdeman and Moore 1974, 261[AL].

TABLE 167. Biochemical characteristics of species of the genus *Coprococcus*[a]

Organism	*C. eutactus*	*C. catus*	*C. comes*
DNA G+C content (mol%)	41	39–41	40–42
Major PYG fermentation product[b]	F, B, l, a	B, P, a	L, B, a
Fermentation of:			
Arabinose	–	–	+
Cellobiose	+	–	–
Glucose	+	w/–	+
Lactose	+	–	+
Mannose	+	–	w/–
Maltose	+	–	+
Mannitol	–	+	d
Raffinose	+	–	+
Sucrose	+	–	+
Xylose	–	–	+

[a]Symbols: +, >85% positive; d, different among different strains (16–84% positive); –, 0–15% positive; w, weak reaction.

[d]A/a, acetate; F, formate; L/l, lactate; P, propionate. Upper- and lower-case letters indicate ≥1 and <1 meq/100 ml of culture, respectively.

List of species of the genus *Coprococcus*

1. **Coprococcus eutactus** Holdeman and Moore 1974, 261[AL]

eu'tac'tus. Gr. adj. *eutactos* orderly, well-disciplined (referring to the uniform reactions of the different strains).

Cells are usually round and 0.7–1.3 μm in diameter. They have abundant growth and are often elongated in a medium with fermentable carbohydrates. However, their growth is very poor or absent in peptone-yeast medium without fermentable carbohydrates.

Growth occurs equally well at 37 °C and 45 °C but growth is poor to moderate at 25 °C and 35 °C.

This species is easily recognized by its production of formic, lactic, and butyric acids from glucose (Holdeman et al., 1977a). This species is isolated from human feces. Needs anaerobically sterilized rumen-fluid-glucose-cellobiose agar for the primary isolate. Differential characteristics of the

species from other carbohydrate-requiring anaerobic cocci are given in Table 167.

DNA G+C content (mol%): 41–42.

Type strain: ATCC 27759, VPI C33-22.

GenBank accession number (16S rRNA gene): D14148.

2. **Coprococcus catus** Holdeman and Moore 1974, 263[AL]

ca'tus. L. adj. *catus* clever, referring to unusual property of producing large quantities of both propionate and butyrate

Cells usually occur in pairs that form long chains. Coccoid and oval cells range in size from 0.8 to 1.4 μm in diameter by 1.6–1.9 μm in length Holdeman and Moore (1974). Slight and moderate growth in peptone-yeast (PY) medium. PY-fructose cultures have abundant growth. Their growth is very poor or absent in peptone-yeast medium without fermentable carbohydrates. Growth occurs well at 37 °C and

45 °C but growth is poor to moderate at 25 °C and 35 °C. This species is isolated from human feces. Need anaerobically sterilized rumen-fluid-glucose-cellobiose agar for the primary isolate.

Differential characteristics of the species from other carbohydrate-requiring anaerobic cocci are given in Table 167.

DNA G+C content (mol%): 41–42.

Type strain: ATCC 27761, VPI C6-61, NCTC 11835.

GenBank accession number (16S rRNA gene): AB038359.

3. **Coprococcus comes** Holdeman and Moore 1974, 263[AL]

co'mes. L. n. comes comparison, fellow traveller (referring to the presence of the species in human feces).

Cells usually occur in pairs that form long chains. Coccoid and oval cells, 0.8–1.4 µm in diameter by 1.6–1.9 µm in length (Holdeman and Moore (1974). Slight and moderate growth in peptone-yeast (PY) medium. PY-fructose cultures have abundant growth. Their growth is very poor or absent in peptone-yeast medium without fermentable carbohydrates. Growth occurs well at 37 °C and 45 °C but growth is poor to moderate at 25 °C and 35 °C. This species is isolated from human feces. Differential characteristics of the species from other carbohydrate-requiring anaerobic cocci are given in Table 167.

DNA G+C content (mol%): 40–41.

Type strain: ATCC 27758, VPI C1-38.

GenBank accession number (16S rRNA gene): EF031542.

Genus VIII. **Dorea** Taras, Simmering, Collins, Lawson and Blaut 2002, 426[VP]

Michael Blaut, Matthew D. Collins and David Taras

Do.ré.a. N.L. fem. n. *Dorea* named in honor of the French microbiologist Joel Doré, in recognition of his many contributions to gut microbiology.

Nonsporeforming, Gram-stain-positive rods. Nonmotile. **Obligately anaerobic. Chemo-organotrophic**. Major end products of glucose metabolism are **ethanol, formate, acetate, H_2, and CO_2**. Butyrate is not produced.

DNA G+C content (mol%): 40–45.6 (T_m).

Type species: **Dorea formicigenerans** (Holdeman and Moore 1974) Taras, Simmering, Collins, Lawson and Blaut 2002, 426[VP] (*Eubacterium formicigenerans* Holdeman and Moore 1974, 274).

Further descriptive information

Dorea species generally stain Gram-positive, but some cells from old cultures may decolorize easily and give Gram-variable or negative reactions. Cells are rod-shaped; size of cells may vary considerably but are generally 0.5–0.8 by 1.0–4.5 µm and occur in pairs or chains. Strictly anaerobic, catalase- and oxidase-negative. Glucose, fructose, galactose, lactose, maltose, and some other sugars are fermented. Starch, cellulose, and gelatin are not hydrolyzed. Nitrate is not reduced. Growth is most rapid at 37 °C and pH near 7. *Dorea* species have been isolated from human feces and are inhabitants of the human gastrointestinal tract.

Enrichment and isolation procedures

Species of the genus *Dorea* found in fecal or intestinal samples can be isolated on usual complex media appropriate for culture of anaerobes using strict anoxic techniques (Hungate, 1969). Such media include blood agar plates, Wilkins–Chalgren anaerobic medium or other peptone and yeast extract containing complex media (e.g., ST medium; Schwiertz et al., 2000). Growth of *Dorea longicatena* in ST medium is supported by the addition of 0.15% agar.

Maintenance procedures

Short-term preservation is achieved by growing cultures in the above-mentioned media, storage at 4 °C up to room temperature (22 °C), and approximately weekly transfers. For long-term preservation, strains can be stored on cryogenic beads at –70 °C or lyophilized.

Procedures for testing special characters

The characteristics of *Dorea formicigenerans* were determined by Moore and Holdeman Moore (1986) using methods described in Holdeman et al. (1977b) in a basal prereduced peptone-yeast extract (PY) medium. Reactions for *Dorea longicatena* were assessed by Taras et al. (2002) with ST medium (Schwiertz et al., 2000) and in a medium used for culturing acetogenic bacteria (Kamlage et al., 1997) but modified by adding 0.5 g Proteose-Peptone No. 2 (Difco) per liter (HA medium). Growth and fermentation of substrates was monitored by changes in optical density measured at 600 nm and the pH of the medium. A pH below 5.5 and a decrease in pH of at least 0.5 pH units below the basal control ST or HA medium were scored as a positive reaction. Results were confirmed by comparison of the growth amount in carbohydrate supplemented medium and unsupplemented control medium. In addition, biochemical features were determined with the API 50 CHL system (bioMérieux). Hydrogen production was determined by gas chromatography of the headspace gas of cultures grown in tubes closed with rubber stoppers (Hartmann et al., 2000). For identification and quantification of *Dorea longicatena* in fecal samples by whole-cell *in situ* hybridization, a species-specific oligonucleotide probe (5′-CTCAGCAGTTCCAAATGC-3′) targeting a hypervariable region of the 16S rRNA has been designed and validated (D. Taras and M. Blaut, unpublished).

Differentiation of the genus *Dorea* from other genera

The genus *Dorea* can be distinguished from most clostridial species and other spore-forming taxa (e.g., *Sporobacterium*) in not producing endospores, from *Lachnospira* by the absence of curved cellular shapes, from *Acetobacterium* in being nonmotile, from *Coprococcus* and *Ruminococcus* by cellular morphology and end products of glucose fermentation, and from *Roseburia* and *Butyrivibrio* in end products of glucose metabolism (i.e., not producing butyric acid) and in being nonmotile. More demanding is the differentiation from members of the genus *Eubacterium* (see also Taxonomic comments) which share, to some extent, cell mor-

TABLE 168. Characteristics differentiating *Dorea formicigenerans* and *Dorea longicatena* from other nonspore-forming Gram-stain-positive rods that do not produce butyrate[a,b]

Characteristic	*Dorea formicigenerans*	*Dorea longicatena*	*Collinsella aerofaciens*	*Eggerthella lenta*	*Eubacterium contortum*	*Eubacterium eligens*	*Eubacterium fissicatena*	*Eubacterium hadrum*
Motility	–	–	–	–	–	+	–	–
Utilization of:								
Amygdalin	–	+	–	–	–	–	–	–
Arabinose	d	+	–	–	+	–	–	–
Esculin	–	+	–	–	–	–	–	–
Glucose	+	+	+	–	+	v	+	+
Inositol	–	+	–	–	NR	–	+	NR
Inulin	–	w	–	–	–	NR	–	NR
Mannose	–	–	+	–	d	–	w	+
Maltose	+	+	+	–	+	–	+	v
Melibiose	–	–	–	–	+	–	–	–
Raffinose	–	w	–	–	+	–	–	–
Rhamnose	–	–	–	–	v	–	+	–
Sorbitol	–	+	–	–	–	–	–	v
Trehalose	–	–	–	–	–	–	w	–

[a]Symbols: +, >85% positive; d, different strains give different reactions (16–84% positive); –, 0–15% positive; w, weak reaction; v, variable; NR, not reported.
[b]Data for species not members of the genus *Dorea* were taken from Moore and Holdeman Moore, (1986).

phology and many biochemical characteristics with species of the genus *Dorea*. Tests which are useful in differentiating *Dorea* species from some other nonspore-forming Gram-positive rods that also fail to produce butyrate and indole are given in Table 168.

Differentiation of the species of the genus *Dorea*

Biochemical tests which may be helpful in differentiating *Dorea formicigenerans* and *Dorea longicatena* from each other are given in Table 169. Currently the most reliable means of differentiating these species is by 16S rRNA gene sequence analysis.

Taxonomic comments

The genus *Dorea* was created to accommodate the species formerly designated *Eubacterium formicigenerans* and some Gram-stain-positive, asporogenous, rod-shaped organisms isolated from human feces (Taras et al., 2002). The two currently defined *Dorea* species form a distinct subline within a supra-generic rRNA cluster referred to as the *Clostridium coccoides* group (rRNA cluster XIVa) (Collins et al., 1994). Phylogenetically, the genus *Dorea* does not display a particularly close affinity to any other taxon within this group. The two *Dorea* species display a 16S rRNA sequence divergence value of approximately 6%, which clearly shows that they are closely related, but nevertheless, different species. A tree showing the position of *Dorea* species within the *Clostridium coccoides* rRNA group is shown in Figure 164. According to the 16S rRNA phylogenetic analysis presented in the roadmap to this volume (Figure 5), the genus *Dorea* is a member of the family *Lachnospiraceae*, order *Clostridiales*, class *Clostridia* in the phylum *Firmicutes*.

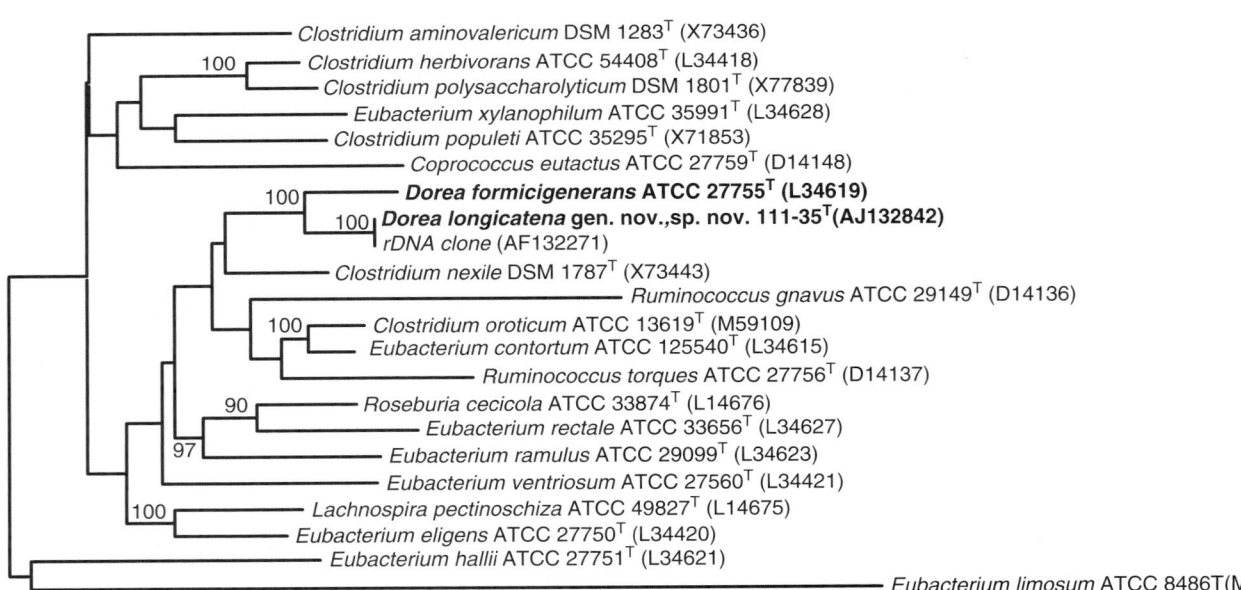

FIGURE 164. Unrooted tree showing the phylogenetic relationship of *Dorea formicigenerans* and *Dorea longicatena* with some other members of the *Clostridium coccoides* group. The tree was constructed using the neighbor-joining method and is based on a comparison of 1330 nucleotides. Bootstrap values, each expressed as a percentage of 500 replications, are given at branching points. (Reprinted with permission from Taras, D., R. Simmering, M.D. Collins, P.A. Lawson, M. Blaut. 2002. Int. J. Syst. Evol. Microbiol. *52:* 423-428).

Before the advent of 16S rRNA gene sequencing, taxonomically the two species of the genus *Dorea* would be considered to conform to the genus *Eubacterium*, which acted as a repository for strictly anaerobic, asporogenous rod-shaped organisms. During the past decade, there have been great advances in unraveling the taxonomic complexities within the *Clostridium* subphylum of the Gram-positive bacteria, and it is universally acknowledged that the eubacteria represent a phylogenetically very heterogeneous group of organisms. There is now a growing consensus that the genus *Eubacterium sensu stricto* should be restricted to the type species *Eubacterium limosum*, and its close phylogenetic relatives (*Eubacterium barkeri*, *Eubacterium callanderi*, and *Eubacterium aggregans*) (Kageyama et al., 1999; Willems and Collins, 1996). Members of the genus *Dorea* are, however, phylogenetically far removed from *Eubacterium limosum* and related species (approx. 20% 16S rRNA sequence divergence). Furthermore, the members of the genus *Dorea* are incompatible with the definition of *Eubacterium sensu stricto* in not producing butyrate as a major fermentation product.

It is pertinent to note that several rDNA clones derived directly from human fecal and colonic tissue samples have been shown to correspond phylogenetically to *Dorea longicatena* and *Dorea formicigenerans* (Hold et al., 2002; Suau et al., 1999). *Dorea formicigenerans*-like rDNA clones derived from the gastrointestinal tract of pigs have also been reported (Lester et al., 2002). As living organisms giving rise to these rDNA clones have not been isolated, it is not known if they correspond to *Dorea formicigenerans* or to genomically closely related species.

List of species of the genus *Dorea*

1. **Dorea formicigenerans** (Holdeman and Moore 1974) Taras, Simmering, Collins, Lawson and Blaut 2002, 426[VP] (*Eubacterium formicigenerans* Holdeman and Moore 1974, 274)

for.mi.ci.ge′ne.rans. N.L. adj. *formicigenerans* formic acid producing; referring to its production of large amounts of formic acid form carbohydrate fermentation.

The description is based on that of Moore and Holdeman Moore (1986).

Cells are 0.6–1.4 wide × 0.8–4.7 μm long and occur in pairs or chains. After 5 d incubation on rumen fluid-glucose-cellobiose agar (RGCA), colonies are 0.5–1.0 mm in diameter, white to tan, circular to lenticular, and often have fuzzy edges or a woolly ball appearance. Colonies on blood agar or BHIA plates are 0.5–3.0 mm in diameter, circular to slightly irregular, entire to slightly erose, convex to umbonate, opaque, white to tan, shiny, and smooth. Some strains may show slight greening on sheep blood agar. PYG broth cultures have little or no turbidity, stringy or flocculent (occasionally smooth) sediment, and pH of 4.7–5.0 in 5 d. Growth in PY broth is increased by the presence of fermentable carbohydrate. Growth in PYG is generally not enhanced by the addition of 0.02% Tween 80 or 10% (v/v) rumen fluid. Growth is inhibited by 6.5% NaCl. Optimum growth temperature is 37 °C; most strains grow moderately well at 30 and 45 °C, but usually not at 25 °C. Moderate to abundant gas is produced in glucose agar deep cultures. Acetic, formic, lactic acids, and ethanol are the major products of glucose metabolism. Pyruvate is converted to acetate, formate, and ethanol, usually with a trace of lactate and sometimes succinate. Lactate and gluconate are not utilized. Threonine is not converted to propionate. Acid is produced from fructose, galactose, lactose, and maltose. Acid may or may not be formed from arabinose, ribose, and xylose. Acid is not produced from adonitol, amygdalin, cellobiose, cellulose, dextrin, dulcitol, erythritol, glycerol, glycogen, inositol, inulin, mannitol, mannose, melezitose, melibiose, raffinose, rhamnose, salicin, sorbitol, L-sorbose, starch, or trehalose. Hippurate hydrolysis is variable. Indole is not produced. Meat is not digested. The type strain does not produce ammonia from peptone or arginine.

DNA G + C content (mol%): 40–44 mol% (T_m).

Type strain: ATCC 27755, DSM 3992.

GenBank accession number (16S rRNA gene): L34619.

2. **Dorea longicatena** Taras, Simmering, Collins, Lawson and Blaut 2002, 427[VP]

lon.gi.ca.te′na. L. adj. *longus* long; L. fem. n. *catena* chain; N.L. fem. n. *longicatena* long chain, referring to the long chains, this organism develops in culture medium.

Cells are 0.5–0.6 × 2.0–4.3 μm and occur in chains of 4–200 cells. Cells are Gram-stain-positive but sometimes in old cultures stain Gram-negative. The cell-wall murein is directly cross-linked, based upon *meso*-diaminopimelic acid (type A1γ) (Schleifer and Kandler, 1972). On Columbia blood and WCA agar, cells form opaque white colonies that are 1–3 mm in diameter, circular, convex, smooth, shiny, and sticky. Nonhemolytic. Cultures in ST medium supplemented with 0.15% agar exhibit turbidity with dense areas of fluffy, "woolly" appearance. Ethanol, formate, and acetate are the major products of glucose metabolism; H_2 is produced. Acid is produced from amygdalin, L-arabinose, L-arabitol (weak), arbutin, fructose, fructo-oligosaccharides, galactose, β-gentiobiose, D-glucosamine, inositol, inulin, lactose, lactulose, D-lyxose (weak), maltose, raffinose (weak), salicin, sorbitol, sucrose, xylitol (weak), and xylose. Acid is not produced from adonitol, DL-arginine, D-arabitol, D-arabinose, cellobiose, dulcitol, erythritol, glycerol, glycogen, methyl β-xyloside, D-fucose, L-fucose, gluconate, 2-ketogluconate, 5-ketogluconate, mannitol, mannose, melezitose, melibiose, methyl β-D-mannoside, methyl β-D-glucoside, rhamnose, ribose, sorbose, starch, trehalose, D-turanose, *N*-acetylglucosamine, pyruvate, or D-tagatose. Esculin and gelatin are hydrolyzed, but hippurate is not hydrolyzed. Isolated from human feces.

DNA G+C content (mol%): 43.8–45.6 (HPLC) (Mesbah and Whitman, 1989).

Type strain: 111-35, CCUG 45247, DSM 13814, JCM 11232.

GenBank accession number (16S rRNA gene): AJ132842.

Genus IX. **Hespellia** Whitehead, Cotta, Collins and Lawson 2004, 244[VP]

TERENCE R. WHITEHEAD, PAUL A. LAWSON AND MICHAEL A. COTTA

Hes.pel'li.a. N.L. fem. n. *Hespellia* named to honor the late American microbiologist Robert B. Hespell, in recognition of his many contributions to anaerobic microbiology.

Short, rod-shaped cells that occur singly and in pairs or short chains. Gram-stain-positive. Nonmotile. Does not form spores. **Strictly anaerobic and negative for catalase- and oxidase. Acid is produced from D-glucose and some other carbohydrates.** Major end products of glucose metabolism are formic, acetic, lactic, and propionic acids. **Butyric acid is not formed.** Esculin and starch are hydrolyzed, but gelatin is not. Negative for indole and does not reduce nitrate. Isolated from swine manure slurry.

DNA G + C content (mol%): 43.7–43.8.

Type species: **Hespellia stercorisuis** Whitehead, Cotta, Collins and Lawson 2004, 244[VP].

Further descriptive information

The genus contains two species, *Hespellia stercorisuis* and *Hespellia porcina.* Cells are nonmotile rods. After 48 h anaerobic incubation at 37 °C under a gas phase of N_2/CO_2 (80:20, v/v) on RGM-glucose-agar (Hespell et al., 1987), colonies are gray, convex, smooth, shiny, and translucent. Optimum growth temperature is 37 °C. Acid is produced from glucose; H_2 is produced. The long-chain cellular fatty acids (FAs) consist of complex mixtures of FAs and dimethylacetals (DMAs), together with small amounts of aldehydes.

Enrichment and isolation procedures

Hespellia species were recovered from a manure storage pit at a swine facility near Peoria, IL, USA. Isolations and enumerations were performed by plating serial dilutions in anaerobic buffer onto habitat-simulating media containing either 40% (v/v) substrate-depleted rumen fluid (RF medium; Dehority and Grubb, 1976; Leedle and Hespell (1980) or 80% (v/v) clarified swine manure slurry (Slurry medium, 8000 g, 20 min, 4 °C; (Cotta et al., 2003). The media used in these experiments were prepared anaerobically using the method of Hungate, as modified by Bryant (1972). The basic media contained macro-minerals, microminerals, buffers, reducing agents, and other components as in the RGM medium described by Hespell et al. (Hespell et al., 1987) or anaerobic brain-heart infusion (BHI) medium described by Whitehead and Flint (1995). No additional volatile FAs were added to slurry-containing media. Glucose, xylose, cellobiose, maltose, starch (0.05%, w/v each), and peptone (0.3%, w/v) were provided as complex carbon, nitrogen, and energy sources. Plates were incubated initially at room temperature for manure slurry samples to simulate the pit environment and anaerobically in an atmosphere of 96% CO_2 and 4% H_2 (Cotta et al., 2003). The *Hespellia* isolates grew equally well at 37 °C. Single colonies were picked and repeatedly streaked out until pure cultures were obtained.

Maintenance procedures

Good growth is obtained anaerobically on RGM (Hespell et al., 1987) or BHI at 37 °C. For long-term preservation, strains are maintained in media containing 15–20% glycerol at −70 °C or lyophilized.

Taxonomic comments

The genus *Hespellia* was described in 2004 to accommodate a phylogenetically distinct catalase-negative, Gram-positive, rod-shaped bacterium originating from stored swine manure (Cotta et al., 2003; Whitehead et al., 2004). Phylogenetic analysis of the 16S rRNA gene demonstrates a relationship between *Hespellia* and the *Clostridium coccoides* group (clostridial rRNA cluster XIVa; Collins et al., 1994). This diverse assortment of organisms is classified in the novel family *Lachnospiraceae* and includes the genera *Acetitomaculum, Anaerostipes, Bryantella, Butyrivibrio, Catonella, Coprococcus, Dorea, Hespellia, Johnsonella, Lachnospira, Lachnobacterium, Oribacterium, Parasporobacterium, Pseudobutyrivibrio, Roseburia, Shuttleworthia, Sporobacterium,* and *Syntrophococcus,* in addition to many misclassified clostridial species (see the family *Lachnospiraceae,* above). However, *Hespellia* species form a distinct lineage within this large grouping and do not display a particularly close affinity with any of the aforementioned taxa. According to the 16S rRNA phylogenetic analysis presented in the roadmap to this volume (Figure 5), the genus *Hespellia* is a member of the family *Lachnospiraceae,* order *Clostridiales,* class *Clostridia* in the phylum *Firmicutes.*

In common with many organisms in this large family, phylogenetic molecular profiling studies of both animal and human sources demonstrate that additional species of *Hespellia* remain to be isolated and described (Barcenilla et al., 2000; Hold et al., 2002; Leser et al., 2002; Namsolleck et al., 2004; Suau et al., 1999).

Differentiation of the genus *Hespellia* from other genera

Hespellia can be readily distinguished from its closest phylogenetic relatives by a combination of morphological, biochemical, and chemotaxonomic criteria. It differs phenotypically from *Eubacterium* and related species because it does not produce butyrate as a major fermentation product. *Hespellia* also differs phenotypically from all other genera within this supra-generic grouping. It can be distinguished from clostridial species and other sporeforming taxa (e.g., *Sporobacterium*) in not producing endospores, from *Lachnospira* by the absence of curved cellular shapes, from *Dorea* by hydrolyzing starch and its fermentation products, from *Coprococcus* and *Ruminococcus* by cellular morphology and its fermentation products, and from *Roseburia* and *Butyrivibrio* by its fermentation products (i.e., no butyric acid production) and lack of motility. Based on the taxonomic and phylogenetic findings, it is evident that the isolates form a distinct group and do not display a close affinity with any recognized genus within this rRNA cluster. In this diverse group of organisms, it is clear that no one organism processes a completely unique biochemical profile; thus, an accurate identification, especially at the laboratory bench is becoming ever more reliant on molecular genetic techniques, in particular 16S rRNA gene sequence comparisons.

List of species of the genus *Hespellia*

1. **Hespellia stercorisuis** Whitehead, Cotta, Collins and Lawson 2004, 244[VP]

ster.co.ri.su′is. L. masc. n. *stercus, -oris* feces, manure; L. gen. n. *suis* of a pig; N.L. gen. n. *stercorisuis* from pig feces/manure, referring to the isolation of the type strain from swine manure.

Characteristics are as given for the genus with the following information. Cells are 0.5–1.0 × 1.5–5.0 μm and occur singly and in pairs or short chains. H_2 is formed after growth on glucose. The fermentation products following growth on BHI broth are formic, acetic, lactic, and propionic acids. Glucose, lactose, cellobiose, trehalose, amygdalin, sorbitol, maltose, mannose, sucrose, fructose, and xylose are utilized as energy sources, but not arabinose, inositol, raffinose, rhamnose, or inulin. Esculin and starch are hydrolyzed, but gelatin is not. The long-chain cellular FAs consist of complex mixtures of FAs and DMAs, together with small amounts of aldehydes; the predominant components are $C_{14:0}$ FA, $C_{16:0}$ FA, $C_{16:1\ cis9}$ DMA, and $C_{18:1\ cis11}$ DMA. Isolated from pig manure.

DNA G + C content (mol%): 43.7 (T_m).

Type strain: PC18, NRRL B-23456, CCUG 46279, ATCC BAA-677.

GenBank accession number (16S rRNA gene): AF445264.

2. **Hespellia porcina** Whitehead, Cotta, Collins and Lawson 2004, 244[VP]

por.ci′na. L. fem. adj. *porcina* of pigs, pertaining to pigs, referring to the isolation of the type strain from swine manure.

Characteristics are as given for the genus with the following information. Cells are 0.5–1.0 × 1.5–4.0 μm and occur singly and in pairs or short chains. H_2 is formed after growth on glucose. The fermentation products following growth on BHI broth are formic, acetic, lactic, and propionic acids. Glucose, arabinose, inositol, maltose, mannose, sucrose, fructose, and xylose are utilized as energy sources, but not amygdalin, cellobiose, lactose, raffinose, rhamnose, sorbitol, trehalose, or inulin. Esculin and starch are hydrolyzed, but gelatin is not. The long-chain cellular FAs consist of complex mixtures of FA and DMA, together with small amounts of aldehydes; the predominant components are $C_{14:0}$ FA, $C_{14:0}$ DMA, $C_{16:0}$ FA, and $C_{16:1\ cis9}$ DMA. Isolated from pig manure.

DNA G + C content (mol%): 43.8 (T_m).

Type strain: PC80, NRRL B-23458, ATCC BAA-674.

GenBank accession number (16S rRNA gene): AF445239.

Genus X. **Johnsonella** Moore and Moore 1994, 190[VP]

ANNE WILLEMS AND MATTHEW D. COLLINS

John.son.el′la. N.L. fem. n. *Johnsonella* in honor of John L. Johnson, a microbiologist from the United States.

Obligately anaerobic, catalase-negative, Gram-stain-negative, non-spore-forming rods. Nonmotile. Nonfermentative. In peptone-yeast extract-glucose broth a moderate amount of acetate and isovalerate, and traces of lactate, succinate, isobutyrate and butyrate are produced. **Major cellular fatty acids** when grown in peptone-yeast extract-glucose broth include **an unidentified compound with an equivalent chain-length of 9.740, $C_{14:0}$ and $C_{16:0}$ fatty acids.** *Johnsonella* is associated with gingivitis and periodontitis in human gingival crevice and periodontal pockets.

DNA G+C content (mol%): 32.

Type species: **Johnsonella ignava** Moore and Moore 1994, 190[VP].

Further descriptive information

The genus contains only one species and therefore all of the characteristics provided below describe the species *Johnsonella ignava*.

Phylogeny. According to the 16S rRNA phylogenetic analysis presented in the roadmap to this volume (Figure 5), the genus *Johnsonella* is a member of the family *Lachnospiraceae*, order *Clostridiales*, class *Clostridia* in the phylum *Firmicutes*. Closest relatives are *Catonella morbi* (approx. 85% sequence similarity) and *Eubacterium saburreum* (approx. 87%) (Willems and Collins, 1995b).

Colony and cell morphology. On blood agar colonies are minute to 2 mm in diameter, round with an entire margin, low convex, white to tan, shiny, smooth and sometimes mottled. No hemolysis on rabbit blood agar is observed and no dark pigments are produced. Cells of the type strain grown in peptone-yeast extract-glucose (PYG) broth* are 0.8 μm wide by 3.7–6.4 μm long. They occur singly and in pairs; some cells display central swellings (Moore and Moore, 1994).

Growth conditions. Because of the human origin of the strains, cultures are grown at 37 °C. No data on minimal, maximal or optimal growth temperatures have been reported. In broth cultures, a small amount of granular, smooth or stringy sediment without turbidity is produced (Moore and Moore, 1994).

---

*Composition of PYG: 0.5 g peptone, 0.5 g trypticase, 1g yeast extract, 0.4 ml resazurin solution (25 mg in 100 ml of distilled water), 4 ml salts solution (0.2 g $CaCl_2$, 0.2 g $MgSO_4$, 1g K_2HPO_4, 1 g KH_2PO_4, 10 g $NaHCO_3$, 2 g NaCl; dissolve $CaCl_2$ and $MgSO_4$ in 300 ml distilled water; add 500 ml water and while swirling, slowly add remaining salts; when dissolved, add 200 ml distilled water and mix; store at 4°C), 100 ml distilled water, 1 ml hemin solution (dissolve 50 mg hemin in 1 ml 1 M NaOH; make to 100 ml with distilled water; autoclave for 15 min at 121°C), 0.02 ml vitamin K1 solution (dissolve 0.15 ml of vitamin K1 in 30 ml of 95% ethanol; keep for max. 1 month, at 4°C in a brown bottle), 0.05 g cysteineHCl·H_2O, 1 g glucose ; vitamin K1, hemin and cysteine are added after the medium is boiled, but before dispensing and autoclaving.

Nutrition and metabolism. *Johnsonella* has a nonfermentative, nonproteolytic metabolism. Fermentation end products of PYG broth cultures are (in milliequivalents per 100 ml of culture; mean ± standard error of mean) acetate, 0.7 ± 0.1 and isovalerate 0.2 ± 0.04. Traces of lactate, succinate, butyrate and isobutyrate are also detected. No hydrogen gas is detectable (Moore and Moore, 1994). The following substrates are not fermented: amygdalin, arabinose, cellobiose, erythritol, esculin, fructose, glucose, glycerol, glycogen, gum arabic, inositol, lactose, larch arabinogalactan, maltose, mannitol, mannose, melezitose, melibiose, raffinose, rhamnose, ribose, salicin, sorbitol, starch, sucrose, trehalose, xylan and xylose. No H_2S production in sulfide-indole motility medium. No growth in PYG with 20% bile or in deoxycholate broth. Indole, oxidase, DNase and catalase are not produced. Lipase or lecithinase not detected on egg yolk agar. No reduction of resazurin.

Chemotaxonomic characteristics. The cellular fatty acid composition of *Johnsonella ignava*, based on data for 11 strains grown in 10 ml PYG both cultures at 37 °C included 45% of an unidentified compound with equivalent chain-length of 9.74, 8% $C_{14:0}$, 2% $C_{16:1\text{-}cis9}$, 15% $C_{16:0}$, 5% $C_{16:0}$ dimethyl acetal, 5% $C_{18:1\text{-}cis9}$, 4% $C_{18:0}$, 2% $C_{18:1\text{-}cis11}$ dimethyl acetal, 3% $C_{18:1\text{-}cis11/t9/t6}$ or an unidentified compound with ECL 17.834 and 1% iso-3-OH-$C_{17:0}$ or $C_{18:2}$ dimethyl acetal (Moore et al., 1994).

Enrichment and isolation procedures

No specific isolation procedures specific for *Johnsonella* have been described. These organisms have been isolated in studies of the anaerobic bacterial flora of gingivitis and periodontitis in children and young adults in which randomly selected strains were characterized and identified (Moore et al., 1985, 1984, 1982a). The isolation procedure and media used in these studies have been described in detail in the chapter on the genus *Catonella*.

Maintenance procedures

Strains can be maintained in prereduced, anaerobically sterilized D5-broth (Moore et al., 1982a), the composition of which is described in the chapter on *Catonella*. Cultures can also be lyophilized for long term preservation.

Differentiation from other Gram-stain-negative anaerobic nonfermentative bacilli

Johnsonella can readily be distinguished from most genera of Gram-stain-negative, anaerobic, nonfermentative rods (including *Porphyromonas*, *Desulfomonas*, *Dialister*, and *Dichelobacter*) by the G+C content of its DNA (except for *Dialister*, the G+C content of which has not been reported), its fermentation end products in PYG broth and its fatty acid composition (Moore et al., 1994; Moore and Moore, 1994). The organism with phenotypic characteristics most similar to those of *Johnsonella* is *Tissierella praeacuta*, but its fatty acid composition is clearly different since it does not contain the unidentified compound with ECL 9.74 that makes up 45% of fatty acids in *Johnsonella ignava* (Moore et al., 1994).

Taxonomic comments

Johnsonella ignava was originally known as "*Bacteroides* D19", one of a group of unnamed bacterial species that occur in the human gingival crevice and periodontal pockets that were first recognized and isolated at the Anaerobe Laboratory of the Virginia Polytechnic Institute and State University in Blacksburg (Moore et al., 1985, 1984, 1982a). It was formally described as *Johnsonella ignava*, a new genus and species of anaerobic, Gram-stain-negative, nonsporing bacilli and therefore placed in the family *Bacteroidaceae* (Moore and Moore, 1994). Later, by 16S rDNA sequence analysis, it was shown to belong to the *Clostridium* subphylum of the *Firmicutes* (Willems and Collins, 1995b). It is a member of *Clostridium* cluster XIVa as defined by Collins et al. (1994), and has now been classified as a member of the family *Lachnospiraceae*, order *Clostridiales*, class *Clostridia* in the phylum *Firmicutes* according to the phylogenetic outlinefor the this volume (Figure 5).

Acknowledgements

A.W. is indebted to the Fund for Scientific Research - Flanders for a position as post-doctoral research fellow.

List of species of the genus *Johnsonella*

1. **Johnsonella ignava** Moore and Moore 1994, 190[VP]

   ig.na′va. L. fem. adj. *ignava* sluggish, because of the inactivity of the organism *in vitro*.

   The morphological and cellular characteristics are as described for the genus. Isolated from gingival crevices of humans with gingivitis and periodontitis.

   DNA G+C content (mol%): 32.
   Type strain: ATCC 51276 (= VPI D94B-12).
   EMBL accession number (16S rRNA gene): X87152.

Genus XI. **Lachnobacterium** Whitford, Yanke, Forster and Teather 2001, 1980[VP]

Fred A. Rainey

Lach.no.bac.te′ri.um. Gr. n. *lachnos* woolly hair, down; L. dim. n. *bakterion* a small rod; N. L. neut. n. *Lachnobacterium* woolly rod, after its colonial morphology on agar.

Long, straight rods occurring as chains or filaments. **Gram-stain-positive. Anaerobic. Spores not formed.** Products from glucose fermentation are lactic acid with very minor amounts of acetic and butyric acids.

DNA G + C content (mol%): 33.9 (HPLC).
Type species: **Lachnobacterium bovis** Whitford, Yanke, Forster and Teather 2001, 1980[VP].

Further descriptive information

The genus *Lachnobacterium* contains a single species *Lachnobacterium bovis* represented by four strains (Whitford et al., 2001). These organisms are anaerobes but grow weakly under aerobic conditions in liquid cultures and in agar stabs. No growth is observed on plates incubated aerobically.

The strains stain weakly Gram-positive and transmission electron microscopy shows a typical Gram-positive cell wall composed of a single layer surrounding the cytoplasmic membrane. Growth is flocculent in liquid culture, while colonies growing on agar media penetrate the surface and have a woolly morphology. The temperature range for growth is 27–42 °C with no growth occurring at 22 or 50 °C. The four strains produce acid from glucose, lactose, cellobiose, sucrose, maltose, fructose, and arabinose, but not pectin, mannitol, glycerol, galactose, starch, xylan, barley straw, or cellulose. Xylose was fermented by strain YZ 63 only. All strains produced lactic acid (71–112 mM) as the major end product when grown on glucose. Minor amounts of acetic and butyric acids are also produced (1.5–3.0 mM). The major fatty acids found in all four strains of this species are $C_{16:0}$, $C_{16:0\ DMA}$, $C_{18:0}$, $C_{18:1\ cis\ 11\ DMA}$, and an unknown fatty acid with a retention time of 17.834 min. All four strains of *Lachnobacterium bovis* produce a protease-sensitive bacteriocin-like inhibitory substance (BLIS). This compound was shown to have inhibitory effects on a number of Gram-positive rumen bacteria. Compounds from the different strains of *Lachnobacterium bovis* were shown to have different target specificities. The BLIS produced by strain YZ 87[T] inhibited strains YZ 39 and YZ 63 (Whitford et al., 2001).

Enrichment, isolation, and growth conditions

Lachnobacterium bovis was isolated from rumen fluid and feces of cattle. Two of the four strains (YZ 87T and YZ 63) were isolated from rumen fluid while strains YZ 140 and YZ 39 were isolated from feces. The isolation was as described by El-Meadaway et al. (1998). For the study of the bacteriocin-like inhibitory substance the strains were grown in L10 broth (Caldwell and Bryant, 1966). Cultures were incubated at 39 °C with a H_2:CO_2 (10:90) headspace.

Taxonomic comments

16S rRNA gene sequences are available for all four strains (YZ 140, YZ 87[T], YZ 39, and YZ 63; AF298662, AF298663, AF298664,

TABLE 169. Tests differentiating *Dorea formicigenerans* and *Dorea longicatena*[a]

Characteristic	D. formicigenerans	D. longicatena
Acid produced from:		
Amygdalin	–	+
L-Arabinose	d	+
Esculin	–	+
Inositol	–	+
Inulin	–	+
Raffinose	–	w
Ribose	d	–
Salicin	–	+
Sorbitol	–	+
Sucrose	–	+

[a]*Symbols*: +, >85% positive; d, different strains give different reactions (16–84% positive); –, 0–15% positive; w, weak reaction.

and AF298665, respectively) and they share 100% similarity. Phylogenetic analyses show them to have highest sequence similarity to species of the genus *Roseburia* (92.5–93.4%) and the misclassified *Eubacterium rectale* (92.1%). The genus *Lachnobacterium* falls within the radiation of the genera comprising the family *Lachnospiraceae* (see Figure 160.). The majority of genera in the family *Lachnospiraceae* represent distinct lineages and their clustering with the lineages of other genera is generally not supported by bootstrap analysis. This is the case for the clustering of the genera *Lachnobacterium*, *Roseburia*, and *Pseudobutyrivibrio* (see Figure 160).

Differentiation of the genus *Lachnobacterium* from other genera

The genus *Lachnobacterium* can be differentiated from the genus *Roseburia* on the basis of differences in mol% G + C content of the DNA and the end products of glucose fermentation. *Lachnobacterium bovis* has a mol% G + C of 33.9, which is much lower than the values reported for *Roseburia* species (41–47 mol%). The major end product of glucose fermentation by *Lachnobacterium bovis* is lactic acid; only small amounts of lactate are produced by *Roseburia* species, which produce butyrate as the main end product of glucose fermentation.

List of the species of the genus *Lachnobacterium*

1. **Lachnobacterium bovis** Whitford, Yanke, Forster and Teather 2001, 1980[VP]

bo'vis. L. n. *bos* cow; L. gen. n. *bovis* of a cow.

Straight rod-shaped cells, 2.0–3.0 μm in length and 0.6–0.75 μm in diameter. Cells occur in chains or filaments. Generally nonmotile, but rare motile cells are observed in old cultures. Colonies penetrate the agar surface and have a woolly morphology. Strains grow well overnight at 39 and 42 °C, with limited growth at 27 °C and no growth at 22 or 50 °C. Strains ferment, with acid production (final pH in parentheses), glucose (5.3), lactose (5.9), cellobiose (6.1), sucrose (5.6), maltose (6.0), fructose (6.1), and arabinose (6.1). No growth on pectin, mannitol, glycerol, galactose, starch, xylan, barley straw, or cellulose. The end products from glucose fermentation (55 mM) are lactic acid (71–112 mM) and very minor

amounts of acetic and butyric acids (1.5–3.0 mM). Positive for α-glucosidase, β-glucosidase, and α-galactosidase, but not N-acetyl-glucosaminidase, α-arabinosidase, α-fucosidase, β-galactosidase, phosphatase, arginine utilization, indole production, leucine aminopeptidase, proline aminopeptidase, pyroglutamic acid arylase, tyrosine aminopeptidase, arginine aminopeptidase, alanine aminopeptidase, histidine aminopeptidase, phenylalanine aminopeptidase, or glycine aminopeptidase. Produces a temperature-sensitive bacteriocin-like inhibitory substance. The most abundant fatty acids are $C_{16:0}$ and isomers of $C_{18:1}$. The type strain was isolated from rumen fluid of a rumen-cannulated steer.

DNA G + C content (mol%): 33.9 (HPLC).

Type strain: YZ 87, ATCC BAA-151, DSM 14045, LRC 5382.

GenBank accession number (16S rRNA gene): AF298663.

Genus XII. **Moryella**.Carlier, K'ouas and Han 2007, 726[VP]

JEAN-PHILIPPE CARLIER

Mo.ry.el'la. L. fem. dim. ending -*ella*; N.L. fem. n. *Moryella* named in honor of the French microbiologist Francine Mory.

Elongated rod with **pointed ends**, about 0.8–1.7 μm × 0.5–0.6 μm, usually occurring singly or in pairs and occasionally in short chains. **Nonmotile. Gram-stain-positive,** but cells may decolorize easily. **Strictly anaerobic.** Does not form spores. **Catalase- and urease-negative.** Nitrate is not reduced. **Indole-positive** and weakly saccharolytic. Traces of gas may be produced in trypticase-glucose-yeast extract (TGY, see below) deep agar cultures. The major metabolic end products in TGY broth are **acetic, butyric, and lactic acids.** Its habitat is not known, but it is probably a resident of the human digestive tract.

DNA G + C content (mol%): 50.2.

Type species: **Moryella indoligenes** Carlier, K'ouas and Han 2007, 728[VP].

Further descriptive information

Morphology and ultrastructure. Cells are elongated, sometimes warped rods with pointed ends. Although they stain Gram-variable, ultrathin sections viewed by electron microscopy reveal a typical Gram-positive type of cell wall (Figure 165).

Colonial and cultural characteristics. Colonies appear on Wilkins–Chalgren blood agar (Wilkins and Chalgren, 1976) after 24–48 h of incubation. They are circular, convex, and about 0.5–1.0 mm in diameter, nonpigmented, and nonhemolytic. TGY broth cultures form a flocculent sediment with or without turbidity. Traces of gas may be detected in TGY deep agar cultures. The strains are susceptible to bile and to 1 mg kanamycin, 10 μg colistin, 5 μg vancomycin special-potency antibiotic disks (Engelkirk et al., 1992).

Nutrition and metabolism. Based on a study of three strains, glucose, galactose, maltose, and ribose fermentations are variable. Acid is not produced from raffinose, sucrose, esculin, arabinose, cellobiose, fructose, glycerol, inositol, lactose, mannitol, mannose, melezitose, melibiose, rhamnose, salicin, sorbitol, starch, trehalose, and xylose. Esculin is not hydrolyzed. Catalase activity and nitrate and nitrite reduction are not detected. Gelatin is not liquefied, and milk is not modified. The major metabolic end products from TGY broth are acetic, butyric, and lactic acids.

Fatty acid composition. Major fatty acids from TGY broth cultures are $C_{13:0 \text{ iso}}$ 2OH (14%), $C_{14:0}$ (47%), and $C_{16:0}$ (9%). Percentages are mean values. The following minor fatty acids are also detected: $C_{12:0}$, $C_{14:0iso}$, $C_{15:0iso}$, $C_{15:0 \text{ anteiso}}$, $C_{16:0iso}$, $C_{16:1 \text{ ω9c}}$, and $C_{18:0}$.

Antibiotic susceptibility. The type strain is susceptible to penicillin G, ampicillin, amoxycillin, imipenem, cefalotin, cefotaxime, cefoxitin, latamoxef, and metronidazole, moderately resistant to tetracycline, and resistant to trimethoprim–sulfamethoxazole, erythromycin, and rifampin. Strains AIP241.03 and MDA2477 are resistant only to trimethoprim–sulfamethoxazole and erythromycin.

Pathogenicity. Two strains were recovered from a buttock abscess and from an intra-abdominal abscess. A third strain was from a polymicrobial thigh abscess in a man with a history of adenocarcinoma of the prostate, local radiotherapy, and complications of chronic osteomyelitis of the symphysis pubis. Bacteria similar to *Moryella indoligenes* strain MDA 2477 have been detected by terminal restriction fragment length polymorphism analyzes (T-RFLP) of symptomatic endodontic infections clinically diagnosed as acute abscesses (Sakamoto et al., 2006).

Enrichment and isolation procedures

Moryella can be isolated on Columbia or Wilkins–Chalgren sheep-blood agar after 1–2 d of incubation at 37 °C. Strictly anaerobic conditions are required for growth. Good growth can be obtained in an anaerobic jar containing a mixture of H_2:CO_2:N_2 (5:5:90 by vol.).

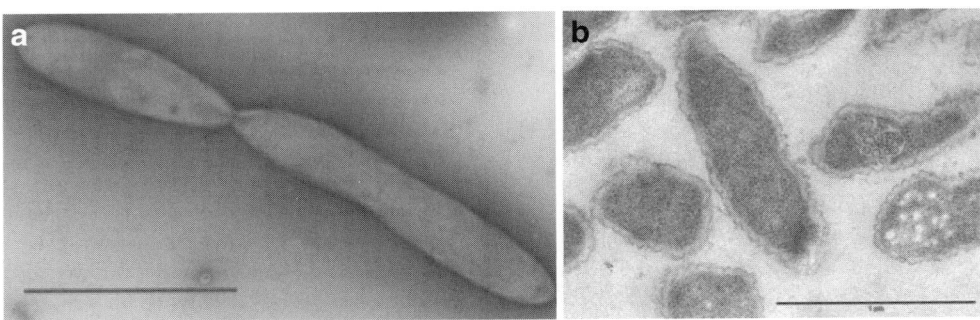

FIGURE 165. Morphology of *Moryella indoligenes* AIP 220.04[T]. Transmission electron micrographs after negative staining (a) and ultrathin sectioning (b) shows the Gram-positive type of cell wall (CW) and the cytoplasmic membrane (CM). Bars = 1 μm.

TABLE 170. Characteristics differentiating *Moryella indoligenes* from some phylogenetically closely related bacteria[a,b]

Characteristics	*Moryella indoligenes*	*Butyrivibrio crossotus*	*Butyrivibrio fibrisolvens*	*Clostridium clostridioforme*	*Clostridium bolteae*	*Clostridium asparagiforme*	*Lachnospira multipara*	*Lachnospira pectinoschiza*	*Oribacterium sinus*	*Shuttleworthia satelles*
Cell shape	Elongated, with pointed ends	Curved	Curved	Straight	Straight	Straight with tapered ends	Curved	Straight	Ovoid	Slightly curved
Motility	–	+	+	+	ND	ND	+	+	+	–
Spore formation	–	–	–	+	+	+	–	+	–	–
Indole production	+	ND	–	v	–	ND	–	ND	+	+
Acid from:										
Lactose	–	d	+	v	–	ND	–	+	–	v
Maltose	–	–	+	+	+	ND	–	–	–	+
Metabolic end products[d]	A, B, L	F, A, B, L	F, B, (a, l)	A	A, L	A, L, E	F, A, L, E	F, A, E	A, L	A, B, l
DNA G+C content (mol%)	50.2	36–37	41	47–49	50.5	53	ND	51.4	42.4	51
Source	Human buttock, pubic and intra-abdominal abscesses	Human feces	Rumen; human, rabbit, horse feces	Humans and animals	Human feces, blood, intra-abdominal abscess	Human feces	Rumen	Pig intestine	Human oral cavity	Human oral cavity
Reference	Carlier et al. (2007)	Bryant (1986b)	Bryant (1986b)	Cato et al. (1986)	Song et al. (2003)	Mohan et al. (2006)	Bryant (1986a)	Cornick et al. (1994)	Carlier et al. (2004)	(Downes et al. 2002)

[a]Symbols: +, >85% positive; –, 0–15% positive; d, different reactions in different strains; v, variable; ND, not determined.
[b]Adapted with permission from Carlier et al. (2007).
[c]Formation of spore-like structures.
[d]F, formate; A, acetate; B, butyrate; E, ethanol; L, lactate. Parentheses indicate an inconstant production. Capital letters indicate major products.

Maintenance procedures

The strain may be maintained in trypticase-glucose-yeast extract (TGY) medium consisting of: 3% (w/v) biotrypcase, 0.5% glucose, 2% yeast extract, 0.05% L-cysteine hydrochloride, and 5 mg hemin/min (Carlier et al., 2002) or in Wilkins–Chalgren Anaerobe Broth (Oxoid) under anaerobic conditions as described above. Cultures of *Moryella* will usually not survive longer than 1–2 weeks without transferring. However, it can be preserved for several years by freeze-drying or in liquid nitrogen.

Differentiation of the genus *Moryella* from other genera

Table 170 lists characteristics differentiating *Moryella* from other closely related species. The pointed ends of these rod-shaped cells, the absence of motility, and the production of indole and butyrate are the main features that distinguish the genus *Moryella* from related genera.

Taxonomic comments

Phylogenetically, the genus belongs to the *Clostridium coccoides* rRNA cluster (rRNA cluster XIVa, Collins et al., 1994) and is classified within the family *Lachnospiraceae* (Figure 166). However, it forms a distinct lineage within this phenotypically heterogeneous collection of organisms that includes sporulating and nonsporulating species. Several butyrate-producing organisms are also widely distributed within this group (Barcenilla et al., 2000). The genus *Moryella* has 16S rRNA gene sequence similarity values of approximately 91% with neighboring groups including *Clostridium clostridioforme*, *Clostridium bolteae*, and *Clostridium asparagiforme*. However, it differs from these organisms because it is not a sporeformer and it produces butyrate. On the other hand, it is phylogenetically too distant from the butyrate-producing members of this family to belong to those genera.

List of species of the genus *Moryella*

1. **Moryella indoligenes.** Carlier, K'ouas and Han 2007, 728[VP]

   in.dol.i'ge.nes. N.L. n. *indolum* indole; N.L. suff. *-genes* producing from; Gr. v. *gennaio* to produce; N.L. fem. adj. *indoligenes* indole-producing.

   The biochemical characteristics are as described for the genus.

   Habitat: probable resident of human intestinal tract.
   DNA G + C content (mol%): 50.2 (T_m).
   Type strain: AIP 220.04, CCUG 52648, CIP 109174.
   GenBank accession number (16S rRNA gene): DQ377947.

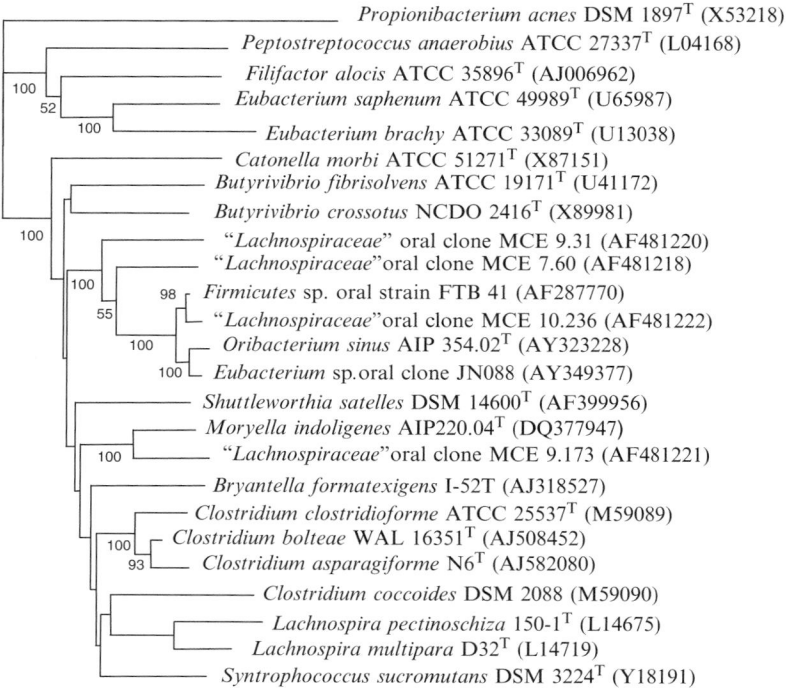

FIGURE 166. Phylogenetic relationships of *Moryella indoligenes, Oribabacterium sinus,* and related genera. The neighbor-joining method was used to construct the tree from 16S rRNA genes. *Propionibacterium acnes* was the outgroup. Bootstrap confidence levels (shown as percentages above the nodes) were determined from 100 replicates. Scale bar indicates 10% difference in 16S rDNA nucleotide sequences.

Genus XIII. **Oribacterium** Carlier, K'ouas, Bonne, Lozniewski and Mory 2004, 1614[VP]

JEAN-PHILIPPE CARLIER

O.ri.bac.te′ri.um. L. gen. n. *oris* of the mouth; N.L. neut. n. *bacterium* from Gr. dim. n. *bakterion* a small rod; N.L. neut. n. *Oribacterium* small rod from the mouth.

Elongated ovoid rods, about $1.7–2.2\,\mu m \times 0.8–1.0\,\mu m$, usually occurring singly or in pairs and occasionally in short chains. **Motile** with laterally inserted flagella. **Gram-stain-positive** but may stain Gram-negative. **Strictly anaerobic**. Does not form spores. **Catalase** is not produced. Weakly fermentative. **Abundant gas is produced** in glucose broth cultures; **nitrate reduction is negative; indole production is positive**. Major metabolic end products in trypticase-glucose-yeast extract (TGY, see below) broth are **acetic and lactic acids**. Habitat: mouth, upper respiratory tract.

DNA G + C content (mol%): 42.4.

Type species: **Oribacterium sinus** Carlier, K'ouas, Bonne, Lozniewski and Mory 2004, 1614[VP].

Further descriptive information

As seen by electron microscopy, cells possess a Gram-positive type of cell wall and 2–4 lateral flagella (Figure 167). Colonies on Wilkins–Chalgren sheep-blood agar are circular, convex, 1–1.5 mm in diameter, nonpigmented, and nonhemolytic. Strictly anaerobic. Abundant gas is produced in TGY deep agar cultures. By presumptive identification tests (Engelkirk et al.,

1992), *Oribacterium* is resistant to 1 mg kanamycin and 10 μg colistin disks, but susceptible to 5 μg vancomycin, 4 μg metronidazole, and bile disks. Catalase and nitrate reduction are negative. Indole is produced. Gelatin is not liquefied, and milk is not modified. Acid is produced from glucose, galactose, raffinose, and sucrose. Acid is not produced from esculin, arabinose, cellobiose, fructose, glycerol, inositol, lactose, maltose, mannitol, mannose, melezitose, melibiose, rhamnose, ribose, salicin, sorbitol, starch, trehalose, or xylose. Esculin is not hydrolyzed. The metabolic end products are acetic and lactic acids. Major cellular fatty acids that are present in substantial amounts are $C_{14:0}$, $C_{15:0\,anteiso}$, $C_{15:0}$, $C_{16:1\,\omega 9c}$, and $C_{16:0}$. Smaller amounts of $C_{17:0\,ante}$ and $C_{18:1\,\omega 9t}$ are also present.

Enrichment and isolation procedures

Oribacterium can be isolated on Columbia or Wilkins–Chalgren sheep-blood agar after incubating at 37 °C for 2 d. Strictly anaerobic conditions are required for growth. Good growth can be obtained in an anaerobic jar containing a mixture of H_2:CO_2:N_2 (5:5:90 by vol.).

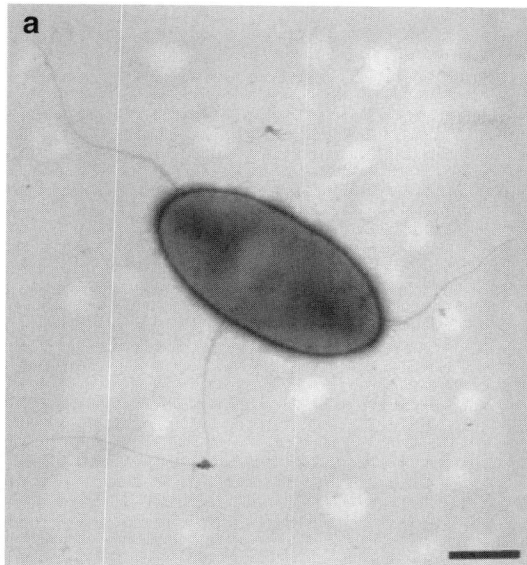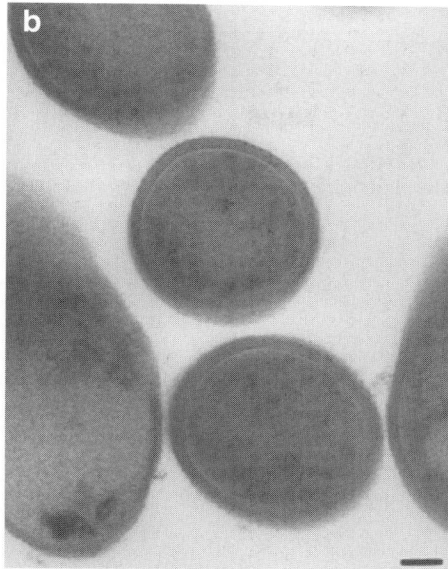

FIGURE 167. Morphology of *Oribacterium sinus* AIP354.02[T]. Transmission electron micrographs showing the (a) general morphology after negative staining and (b) ultrathin sections demonstrating the Gram-positive type of cell wall and the cytoplasmic membrane. Bars, (a) 500 nm; (b) 100 nm. (Reproduced with permission from Carlier et al., 2004. Int. J. Syst. Evol. Microbiol. *54:* 1611–1615.)

Maintenance procedures

The strain may be maintained in trypticase-glucose-yeast extract (TGY) medium consisting of: 3% (w/v) biotrypcase, 0.5% glucose, 2% yeast extract, 0.05% L-cysteine hydrochloride, and 5 mg hemin/ml or in Wilkins–Chalgren Anaerobe Broth (Oxoid) under anaerobic conditions as described above. Cultures of *Oribacterium* will usually survive for no longer than 1–2 weeks without transferring. The organism may not survive freeze-drying procedures, but it can be preserved in liquid nitrogen for several years.

Differentiation of the genus *Oribacterium* from other genera

Oribacterium can be distinguished by its ovoid morphology, rapid motility, and ability to produce gas and indole. Other characteristics that distinguish *Oribacterium* from related organisms are listed Table 170.

Taxonomic comments

Based on 16S rRNA gene sequence analysis, the genus *Oribacterium* belongs to the family "*Lachnospiraceae*" located within the *Clostridium coccoides* rRNA cluster (cluster XIVa, (Collins et al., 1994). This cluster is a large group of disparate organisms including a large number of genera such as *Butyrivibrio, Clostridium, Lachnospira, Eubacterium, Filifactor, Moryella, Shuttleworthia*, and many other species. Among these genera, *Oribacterium* forms a distinct subcluster with several oral clones described by Munson (2002) and Paster (2001). (Figure 166). The most closely related valid species are *Shuttleworthia satelles* and *Butyrivibrio crossotus* each with less than 90% 16S rRNA gene sequence similarity.

Acknowledgements

I would like to express my gratitude to Francine Mory for her critical comments on this chapter.

List of species of the genus *Oribacterium*

1. **Oribacterium sinus** Carlier, K'ouas, Bonne, Lozniewski and Mory 2004, 1614[VP]

sin'us. L. gen. n. *sinus* of the sinus, referring to the anatomical site from which the type strain was isolated.

Colonies appear on Wilkins–Chalgren blood agar after 2 d incubation. They are circular, convex, about 1–1.5 mm in diameter, nonpigmented, and nonhemolytic. The biochemical characteristics are as described for the genus. The type strain is susceptible to penicillin G, ampicillin, amoxycillin, ticarcillin, mezlocillin, imipenem, cefalotin, cefoxitin, cefotaxime, latamoxef, chloramphenicol, clindamycin, and rifampin, moderately resistant to tetracycline, and resistant to erythromycin and trimethoprim sulfamethoxazole.

Habitat: isolated from a human maxillary sinus.

DNA G + C content (mol%): 42.4 (T_m).

Type strain: AIP 354.02, CCUG 48084, CIP 107991, DSM 17245.

GenBank accession number (16S rRNA gene): AY323228.

Genus XIV. **Parasporobacterium** Lomans, Leijdekkers, Wesselink, Bakkes, Pol, van der Drift and Op den Camp 2004, 307[VP] (Effective publication: Lomans, Leijdekkers, Wesselink, Bakkes, Pol, van der Drift and Op den Camp 2001, 4022.)

THE EDITORIAL BOARD

Pa.ra.spo.ro.bac.te′ri.um. Gr. prep. *para* besides, next to; N.L. neut. n. *Sporobacterium* name of a bacterial genus; N.L. neut. n. *Parasporobacterium* a genus similar to *Sporobacterium*.

Strictly anaerobic rods. Stains Gram-negative. Nonsporeforming. Unusual metabolism that **produces dimethyl sulfide and methanethiol** from methoxy-containing aromatic compounds and sulfide.

DNA G + C content (mol%): not reported.

Type species: **Parasporobacterium paucivorans** Lomans, Leijdekkers, Wesselink, Bakkes, Pol, van der Drift and Op den Camp 2004, 307[VP] (Effective publication: Lomans, Leijdekkers, Wesselink, Bakkes, Pol, van der Drift and Op den Camp 2001, 4022.).

Further descriptive information

Syringate, 3,4,5-trimethoxybenzoate, and gallate serve as sole carbon and energy sources (Lomans et al., 2001). Growth on the methoxylated aromatic compounds, syringate and 3,4,5-trimethoxybenzoate, requires the presence of sulfide, which is methylated to methanethiol and dimethyl sulfide (Lomans et al., 2001). Sulfide is not required for growth on gallate. Acetate and butyrate are also formed. No growth was observed on 3,4-dimethyloxybenzoate, 3,5-dimethyloxybenzoate, vanillate, pyrogallol, phloroglucinol, benzoate, pyruvate, H_2-CO_2, methanol, malate, fructose, glucose, pectin, or glycinebetaine. Optimal growth occurs at 34–37 °C, pH 6.5–7.0, and less than 1.8 g/l of NaCl.

Enrichment and isolation procedures

Parasporobacterium can be enriched in a chemostat using syringate as the sole carbon and energy source and sodium sulfide as reducing agent and methyl-group acceptor. After inoculation with freshwater lake sediment, the chemostat is allowed to run until one morphotype predominates. A pure culture is then obtained following dilution in agar deeps.

Maintenance procedures

Stock cultures must be transferred into fresh medium once a month. Cultures can also be stored in glass ampules at –80 °C under N_2/CO_2 (80/20%, v/v) after addition of glycerol (final concentration 5%).

Differentiation of the genus *Parasporobacterium* from other genera

Unlike *Parasporobacterium*, the most closely related genus, *Sporobacterium olearium*, utilizes a wider range of substrates, produces spores, and stains Gram-positive. It also only produces methanethiol from syringate plus sulfide. Unlike many related anaerobes in the *Clostridiales*, both *Parasporobacterium* and *Sporobacterium* are unable to grow with either glucose or fructose as carbon and energy sources.

Taxonomic comments

Based upon rRNA gene sequence analyses, the nearest relative of *Parasporobacterium paucivorans* is *Sporobacterium olearium* (similarity 91.8%). Both genera are members of the *Lachnospiraceae* within the *Clostridiales* (Ludwig et al., 2009).

List of species of the genus *Parasporobacterium*

1. **Parasporobacterium paucivorans** Lomans, Leijdekkers, Wesselink, Bakkes, Pol, van der Drift and Op den Camp 2004, 307[VP] (Effective publication: Lomans, Leijdekkers, Wesselink, Bakkes, Pol, van der Drift and Op den Camp 2001, 4022.)

   pau.ci.vo′rans. L. subst. *pauci* a few; L. adj. part. *vorans* devouring; N.L. neut. part. adj. *paucivorans* degrading a limited number of substrates.

   Rod-shaped cells, 1.5–2 µm × 0.3–0.5 µm, form chains of two cells. They are insensitive to lysis by 0.1 g/l of SDS. In the presence of syringate and sulfide, they form into white, deep-agar, circular colonies (1–2 mm in diameter) within 7 d.

   The type strain was originally isolated from slurry of an eutrophic lake sediment (campus of the Dekkerswald Institute, Nijmegen, The Netherlands).

   For all other characteristics, refer to the genus description.

   DNA G + C content (mol%): not reported.

   Type strain: SYR1, DSM 15970, NCCB100052.

   GenBank accession number (16S rRNA gene): AJ272036.

Genus XV. **Pseudobutyrivibrio** van Gylswyk, Hippe and Rainey 1996, 561[VP]

FRED A. RAINEY

Pseu.do.bu.ty.ri.vib.ri.o. Gr. adj. *pseudes* false; *butyrivibrio* from the genus *Butyrivibrio*; N.L. masc. n. *Pseudobutyrivibrio* not a true butyribvibrio.

Non-spore forming, Gram-stain-negative, anaerobic rods. Ferment a variety of carbohydrates. Major end products of fermentation are formate, butyrate, and lactate. The major cellular fatty acid is $C_{16:0}$.

DNA G + C content (mol%): 40.5–42.7 (HPLC).

Type species: **Pseudobutyrivibrio ruminis** van Gylswyk, Hippe and Rainey 1996, 561[VP].

Further descriptive information

The genus *Pseudobutyrivibrio* contains two species, the type species *Pseudobutyrivibrio ruminis* (van Gylswyk et al., 1996) and an additional species *Pseudobutyrivibrio xylanivorans* (Kopečný et al., 2003). The genus was initially described on the basis of a single strain, which showed many phenotypic characters similar to those described for *Butyrivibrio fibrisolvens* but was phylogenetically distinct. The lack of xylanase activity and C_{18} fatty acids in *Pseudobutyrivibrio ruminis* were considered differentiating characteristics when compared to *Butyrivibrio fibrisolvens* (van Gylswyk et al., 1996). However, the study of a large number of butyrate-producing rumen bacteria resulted in the description of the second species of the genus *Pseudobutyrivibrio xylanivorans* which was xylanolytic and contained C_{18} fatty acids (Kopečný et al., 2003). Currently the two genera can be differentiated on the basis of 16S rRNA gene sequence comparisons as well as PCR assays based on the 16S rRNA gene sequences (Mrázek and Kopečný, 2001). The type strains of *Pseudobutyrivibrio ruminis* and *Pseudobutyrivibrio xylanivorans* share 97.8% 16S rRNA gene sequence similarity. Similarity values in the range 91.2–93.1% are found to the related species of the genera *Roseburia* and *Lachnobacterium* and the misclassified *Eubacterium rectale*. The phenotypically similar species of the genus *Butyrivibrio* share <90.0% 16S rRNA gene sequence similarity with *Pseudobutyrivibrio* species. On the basis of 16S rRNA gene sequence comparisons the genus *Pseudobutyrivibrio* falls within the radiation of the family *Lachnospiraceae* (see Figure 160). The species of the genus *Pseudobutyrivibrio* can be separated on the basis of differences in their 16S rRNA gene sequences as well as enzymic activities, end products of glucose fermentation, and fatty acid profiles.

List of the species of the genus *Pseudobutyrivibrio*

1. **Pseudobutyrivibrio ruminis** van Gylswyk, Hippe and Rainey 1996, 561[VP]

ru′mi.nis. L. neut. gen. n. *ruminis* of the rumen.

Cells are Gram-stain-negative, curved rods, often with tapering ends. Cells are 1–3 μm long and 0.3–0.5 μm in diameter. Very long cells often occur when grown on solid medium. Spores are not produced. Motile by single polar or subpolar flagellum. Colonies are 1.5–2 mm in diameter, convex, round, smooth edged, and opaque with radiating striations. Submerged colonies are lens-shaped discs with diameters of 0.5 mm. Growth occurs anaerobically at 39 °C, with little growth 22 or 45 °C. Substrates fermented include arabinose, cellobiose, fructose, galactose, glucose, lactose, maltose, mannose, sucrose, trehalose, and xylose. Glycerol weakly fermented. Mannitol, ribose, sorbitol, soluble starch, xylan, lactate, succinate, fumarate, oxalate, pyruvate, malonate, aspartate, and glutamate do not support growth. Xylanase, amylase, proteinase, pectinase, and laminarinase activities are not detected. The end products of glucose (50 mM) fermentation are formate (29 mM), butyrate (19 mM), lactate (40 mM), and acetate (<1 mM). The major cellular fatty acid components of cells grown in DSM medium no. 330 without rumen fluid are $C_{16:0}$ (27.6%), $C_{14:0}$ (14.2%), $C_{15:1iso}/C_{13:0\ 3-OH}$ (12.7%), and $C_{17:1\ anteiso}$ (5.2%). Type strain was isolated from whole rumen contents of a pasture-grazing cow.

DNA G + C content (mol%): 40.5 (HPLC).

Type strain: A12-1, DSM 9787.

GenBank accession number (16S rRNA gene): X95893.

2. **Pseudobutyrivibrio xylanivorans** Kopenčný, Zorec, Mrázek, Kobayashi and Marinšek-Logar 2003, 207[VP]

xy.la.ni.vo′rans. N.L. n. *xylanum* xylan; L. part. adj. *vorans* devouring, digesting; N.L. part. adj. *xylanivorans* xylan-digesting.

Cells are Gram-stain-negative, straight to slightly curved rods 1.5–3.0 μm long and 0.4–0.6 μm in diameter. Endospores are not formed. Cells are motile by means of a single polar or subpolar flagellum. Growth occurs anaerobically at 39 °C, but not at 25 or 45 °C. No growth is observed in the presence of oxygen. Exposure of liquid cultures to air for several hours does not kill cells. Substrates that support growth of the type strain are cellobiose, esculin, glucose, lactose, maltose, mannose, melezitose, salicin, sucrose, and xylose. Tryptophan, urea, and gelatin are not utilized. The following substrates do not support growth of the type strain: glycerol, mannitol, farinose, sorbitol, rhamnose, and trehalose. Highly xylanolytic. Most strains produce amylase, proteinase, and DNase. Laminarinase is detected in some strains. Medium M10 with glucose (22 mM) is fermented to formate (15.1 mM), butyrate (17.6 mM), lactate (10.9 mM), succinate (0.3 mM), and ethanol (1.1 mM). Acetate is utilized. No growth occurs in the absence of fermentable carbohydrates. The type strain does not liquefy gelatin or produce hydrogen sulfide. Nitrate is not reduced; catalase and hydrolysis of urea are negative. The major cellular fatty acid components of the type strain grown in DSM medium no. 330 without rumen fluid are $C_{16:0}$ (39.2%) and $C_{17:1\ anteiso}$ (6.4%), followed by $C_{17:0\ iso\ 3-OH}$ (3.9%) and $C_{14:0}$ (4.3%). This species is characterized by a higher content of straight-chain cellular fatty acids and a lower content of branched-chain fatty acids. This species represents the most commonly isolated butyrate-producing anaerobic bacterium from the rumen of sheep and cow. The type strain was isolated from the rumen of a cow.

DNA G + C content (mol%): 42.1 (HPLC).

Type strain: Mz 5, DSM 14809, ATCC BAA-455.

GenBank accession number (16S rRNA gene): AJ428548.

Genus XVI. **Roseburia** Stanton and Savage 1983a, 626[VP]

THADDEUS B. STANTON, SYLVIA H. DUNCAN AND HARRY J. FLINT

Rose.bur'i.a. N.L. fem. n. *Roseburia* named in honor of Theodor Rosebury, an American microbiologist who studied and described micro-organisms indigenous to humans.

Slightly curved, rod-shaped cells ($0.5 \times 1.5–5\,\mu m$) occur singly and in (dividing) pairs. Nonsporulating. Gram-negative to Gram-variable staining reaction. Taxonomically grouped within cluster XIVa of the *Clostridium* subphylum on the basis of 16S rDNA sequence (see below). **Actively motile at 37 °C by means of multiple flagella inserted along the concave side and occasionally along one end of the cell. The flagella appear as a subterminal bundle** when cells are examined by scanning electron microscopy or phase-contrast light microscopy (Figure 168). **Strictly anaerobic.** Chemo-organotrophic. Uses the carbohydrates D-glucose, cellobiose, D-maltose, D-raffinose, sucrose, and D-xylose as carbon and energy sources. **Hydrolyzes and ferments starch.** Grows in anaerobically prepared media containing volatile fatty acids, yeast extract, trypticase peptone, inorganic salts, hemin, glucose, and vitamins, beneath 95% $N_2/5\%$ CO_2 or 100% CO_2 atmosphere. **Produces H_2, CO_2, and large amounts of butyrate from fermentation of glucose and acetate.** May produce lactate, formate, and trace amounts of ethanol. **Consumes acetate, which may be stimulatory during growth on carbohydrates. Indigenous to mammalian intestinal tract.** Catalase-negative. Isolated from mouse cecal mucosa and from human feces. Uncultured bacteria with partial 16S rDNA sequences 95–97% similar to those of *Roseburia* (GenBank nos AJ312385 and L14676) have been detected in 16S rDNA analyses of swine gastrointestinal tract samples (Leser et al., 2002) and adult human fecal samples (Hold et al., 2002). Two species of *Roseburia* have been characterized. They can be differentiated by characteristics given in Table 171.

DNA G+C content (mol%): 29–42% (T_m).

Type species: **Roseburia cecicola** Stanton and Savage 1983a, 626[VP].

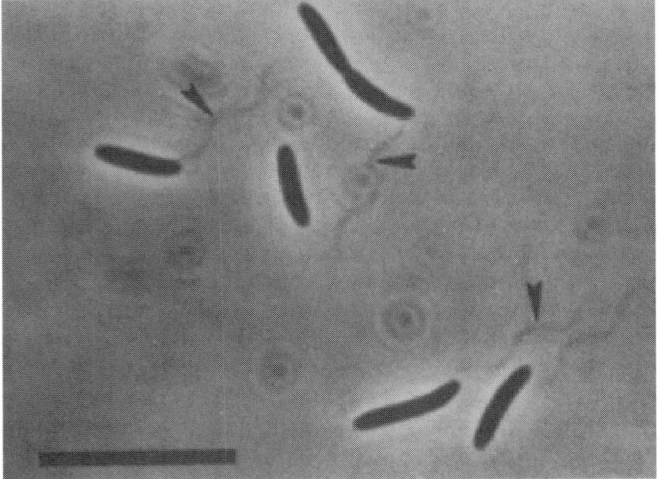

FIGURE 168. Phase-contrast micrograph of *Roseburia cecicola* GM[T] cells. Each cell has a bundle of flagella (arrow). Bar = $5\,\mu m$. (Reprinted with permission from Stanton, T.B. and D.C. Savage. 1983. Appl. Environ. Microbiol., 45: 1677–1684).

TABLE 171. Characteristics useful for differentiating *Roseburia* species[a]

Characteristic	*R. cecicola*	*R. intestinalis*
Host	Mouse	Human (infant)
16S rDNA sequence similarity to *Roseburia cecicola* GM[T]	100%	94.9–97.1 %[b]
Esculin hydrolysis	–	+
Fructose fermentation	–	+
Formate production	–	+
Lactate production	–	+
Ethanol production	+ (trace amts)	–
DNA G+C content (mol%) (T_m)	42	29–31

[a]Symbols: +, >85% positive; –, 0–15% positive.

[b]The identifiable nucleotides of *Roseburia cecicola* GM[T] (L14676) and *Roseburia intestinalis* LC1-82T (AJ312385) 16S rRNAs are 94.9% identical over 1357 base positions. The overall sequences would be 97.1% identical, if ambiguous nucleotides were identical.

Further descriptive information

By the Gram-staining reaction, *Roseburia* cells are Gram-stain-negative to Gram-stain-variable. *Roseburia cecicola* cells have an apparent trilaminar outer envelope, resembling that of Gram-stain-negative bacteria (*Proteobacteria*) (Stanton and Savage, 1983a). Nevertheless, analyses of *Roseburia* species 16S rDNA indicate the 16S rRNA gene has signature nucleotide bases typical of Gram-stain-positive bacteria (Martin and Savage, 1988). Comparisons of flagellin amino acid sequences indicate the *Roseburia cecicola* flagellar protein shares greatest sequence similarity with flagellins of *Bacillus*, *Clostridium*, and spirochete species (Martin and Savage, 1988; Stanton, unpublished observations). According to the 16S rRNA phylogenetic analysis presented in the roadmap to this volume (Figure 5), the genus *Roseburia* is a member of the family *Lachnospiraceae*, order *Clostridiales*, class *Clostridia* in the phylum *Firmicutes*. Phylogenetically close relatives of *Roseburia* species are strains of *Eubacterium rectale* and *Eubacterium oxidoreducens* (Figure 169), both heterogeneous taxa (Duncan et al., 2002). DNA–DNA reassociation studies would undoubtedly be useful for establishing definitive taxonomic relationships among these diverse taxa.

Enrichment and isolation procedures

Roseburia can be isolated by selecting for motile bacteria or for butyrate-forming bacteria from intestinal samples. *Roseburia cecicola* was isolated by inoculating mouse cecal tissue scrapings into one end of a capillary tube filled with sterile culture broth. After 24h incubation at 37 °C, bacteria that had migrated through the broth were recovered from the opposite end of the capillary tube and cultured as isolated colonies in semisolid agar-containing medium. *Roseburia cecicola* strain GM was obtained from one of the colonies (Stanton and Savage, 1983b). *Roseburia intestinalis* strains were obtained from bacteria

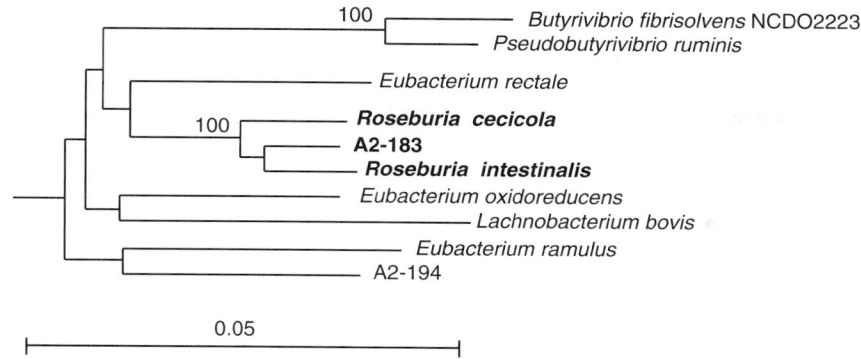

FIGURE 169. Phylogenetic tree based on 16S rDNA sequences for *Roseburia* species and their closest relatives. Sequence accession numbers are as follows: *Roseburia intestinalis* L1-82^T, AJ312385; *Roseburia cecicola* GM^T ATCC 33874, L14676; *rectale* ATCC 33656, L34627; *Eubacterium oxidoreducens* G2–2, AF202259; *Lachnobacterium bovis* LRC 5436, AF298665; *Eubacterium ramulus* ATCC 29099, L34623; *Pseudobutyrivibrio ruminis* pC-XS7, AF202262, *Butyrivibrio fibrisolvens* NCDO 2223, X89977; butyrate-producing human fecal bacterium *Roseburia* sp. A2–183, AJ270482; butyrate-producing human fecal bacterium *Eubacterium* sp. A2–194, AJ270473. The scale bar represents genetic distance (substitutions per nucleotide position). Bootstrap values exceeding 97% are also indicated.

forming colonies in roll tubes containing a rumen-fluid-based agar medium and inoculated with dilutions of human infant fecal homogenates. *Roseburia intestinalis* was selected after screening for strains producing large amounts of butyrate (Barcenilla et al., 2000).

Maintenance procedures

Roseburia cecicola cultures to which the cryoprotectant DMSO (final concentration, 10% v/v) has been added, can be stored frozen at −80 °C for many years. *Roseburia intestinalis* cultures can be stored frozen at −20 °C on rumen fluid medium containing agar (0.75% final concentration).

Differentiation of the genus *Roseburia* from other genera

Roseburia species can be differentiated from related taxa by 16S rDNA sequence analysis. *Roseburia cecicola* can be differentiated from various *Butyrivibrio*, *Lachnospira*, and *Eubacterium* strains by DNA–DNA relative reassociation (Mannarelli et al., 1991).

Further reading

Cotta, M.A. and R.J. Forster. 2000. The family *Lachnospiraceae* including the genera *Butyrivibrio*, *Lachnospira*, and *Roseburia*. *In* Dworkin (Editor), The Prokaryotes (Web version), Springer-Verlag, New York.

List of species of the genus *Roseburia*

1. **Roseburia cecicola** Stanton and Savage 1983a, 627^{VP}

ce.ci.co′la. N.L. n. *cecum* blind pouch; L.v. suff. *cola* from L. v. *colo* to dwell; N.L. n. *cecicola* cecum-dweller.

The type strain GM is the only characterized strain. Slightly curved, rod-shaped cells, 0.5 μm in diameter and 2.5–5.0 μm in length (Figure 168). Actively motile by means of 20–35 flagella inserted into the concave side and occasionally into the end of each cell, forming a flagellar bundle visible by phase-contrast light microscopy (Figure 168). Transmission electron micrographs of a negative stained cell and of ultra-thin sections of cells have been published (Stanton and Savage, 1983a). Gram-stain-negative in staining reaction and by electron microscopy appearance. Based on phylogenetic analyses of 16S rRNA and flagellin genes, however, the species is phylogenically associated with Gram-stain-positive genera (Collins et al., 1994; Martin and Savage, 1988; Willems et al., 1996).

Obligately anaerobic, having a fermentative type of metabolism. Nutritional requirements are undefined. Grows readily in VTY broth (Table 172). Various carbohydrates are growth substrates (Table 173) (Stanton and Savage, 1983a). GM cells constitutively produce extracellular amylase activity. Products

of glucose plus acetate metabolism are H$_2$, CO$_2$, *n*-butyrate, and trace amounts of ethanol (Table 174) (Stanton and Savage, 1983a). Catalase-negative.

In VTY agar medium supplemented with 0.7% Noble agar, after 48 h of incubation at 37 °C, subsurface colonies are 1 mm in diameter, brownish white, lens shaped, and occasionally surrounded by gas pockets that split the agar medium. Surface colonies are white, mucoid, granular in appearance, 1.5–3 mm in diameter, and circular with smooth edges.

In assays using antibiotic disks on VTY agar medium, *Roseburia cecicola* cells are strongly inhibited by chloramphenicol, erythromycin, novobiocin, and vancomycin. Slightly to moderately sensitive to penicillin, polymyxin B, tetracycline, and neomycin. Insensitive to nalidixic acid. An uncharacterized *Roseburia* strain carries a tetracycline resistance gene, *tetW* (GenBank AJ421625).

Roseburia cecicola nonmotile strains produced by UV mutagenesis are impaired in their ability to colonize the mouse cecum in competition with other intestinal micro-organisms (Stanton and Savage, 1984).

Optimal growth (5×10^8 cells/ml; 2 h population-doubling time) at 37°C in VTY broth (Table 172) containing glucose. Grows slowly at 30°C but not at 22°C or 45°C.

Isolated from mouse cecal mucosa.

DNA G+C content (mol%): 42 (T_m).

Type strain: ATCC 33874 (strain GM).

2. **Roseburia intestinalis** Duncan, Hold, Barcenilla, Stewart and Flint 2002, 1619[VP]

in.tes.ti.na'lis. L. adj. *intestinalis* of the intestine, the presumed habitat of the isolates.

The type strain L1-82 is one of five strains characterized. Strains initially categorized as one of 18 ribogroups of butyrate-producing bacteria based on RFLP profiles following cleavage with restriction enzyme *Alu*I (Barcenilla et al., 2000). Slightly curved, rod-shaped cells, 0.4–0.5 μm in width and 1.0–5.0 μm in length. Actively motile by means of flagella inserted on the concave side of the cells. In M2GSC roll tubes surface colonies are white, mucoid, and circular with smooth edges.

TABLE 172. Growth media for *Roseburia* species[a]

Medium component	VTY Medium (*R. cecicola*)	YCFA Medium (*R. intestinalis*)
Distilled water	80 ml	100 ml
Glucose	0.2 g	0.5 g
Yeast extract	0.5 g	0.25 g
Trypticase peptone	1.0 g	
Casitone		1.0 g
Hemin	0.5 mg	1.0 mg
Resazurin	0.1 mg	0.1 mg
L-Cysteine-HCl	0.1 g	0.1 g
NaHCO₃	0.05 g	0.4 g
CaCl₂	9 mg	9 mg
MgSO₄	9 mg	4.5 mg
Potassium phosphate buffer, 0.5 M, pH 7.4	20 ml	
K₂HPO₄; KH₂PO₄		0.045 g each
NaCl	0.09 g	0.09 g
(NH₄)₂SO₄	0.09 g	
Acetic acid	0.16 ml	0.19 ml
Propionic acid	0.06 ml	0.07 ml
n-Butyric acid	0.04 ml	
iso-Butyric acid	9 μl	9 μl
n-Valeric acid	9 μl	10 μl
iso-Valeric acid	9 μl	10 μl
DL-α-Methylbutyric acid	9 μl	
Biotin		1 μg
Cobalamin		1 μg
p-Aminobenzoic acid		3 μg
Folic acid		5 μg
Pyridoxamine		15 μg
Culture atmosphere	95%N₂:5%CO₂	100% CO₂

[a]Heat-labile medium components and glucose are filter-sterilized and added separately. Preparation of media under anaerobic conditions as described (Duncan et al., 2002; Stanton and Savage, 1983a).

Obligately anaerobic, surviving less than 2 min on exposure to air on the surface of agar plates; have a fermentative type of metabolism. Grows readily on a rumen-based medium-supplemented medium (M2GSC) and on a semidefined medium (YCFA) supplemented with glucose (Table 172). Growth is stimulated when acetate is present in the medium.

Hydrolyzes esculin. Arylamidase activity not detected. Various carbohydrates are growth substrates (Table 173). Products of glucose and acetate metabolism are H_2, CO_2, formate, butyrate, and lactate (Table 174). Catalase- and urease-negative. Strain L1-82 is tetracycline resistant.

Optimal growth occurs at 37°C with little to no growth at 22°C, 30°C and slower growth at 45°C. Isolated from a healthy infant (11 mo) fecal sample.

DNA G+C content (mol%): 29–31% (T_m).

Type strain: NCIMB 13810 (strain L1-82).

TABLE 173. Utilization of carbon compounds for growth by *Roseburia* species[a,b]

Compound	*R. cecicola* GM[T]	*R. intestinalis*
Cellobiose, D-glucose D-maltose, raffinose, starch, sucrose, D-xylose	+	+
D-Galactose D-glucuronic acid, glycerol, glycogen, sorbitol	+	
L-Arabinose, melibiose, xylan (oat spelt)		+
Esculin, D-fructose	–	+
Inulin, D-ribose, trehalose	–	–
Cellulose, dulcitol, D-galacturonic acid, gelatin, lactate, mucin (pig), N-acetyl-D-galactosamine, N-acetyl-D-glucosamine, pectin, salicin	–	–
Mannitol, melezitose, rhamnose		–

[a]Symbols: +, >85% positive; –, 0–15% positive.
[b]Data are for *Roseburia cecicola* strain GM[T] (Stanton and Savage, 1983a) and for *Roseburia intestinalis* strains L1-82[T], L1-952, L1-8151, L1-81, L1-93 (Duncan et al., 2002).

TABLE 174. Products of glucose and acetate metabolism by growing cells of *Roseburia* species[a,b]

Substrate/product	*Roseburia cecicola* Amount consumed	*Roseburia cecicola* Amount produced	*Roseburia intestinalis* Amount consumed	*Roseburia intestinalis* Amount produced
Glucose	6.4		NR	
Acetate	6.5		9.1	
Formate				4.7
n-Butyrate		9.9		18.5
Lactate		ND		10.2
Ethanol		0.8		ND
H₂		+		+
CO₂		15.6		+

[a]Values in table expressed as μmol/ml medium. ND, not detected; NR, not reported; +, product detected but not quantified.
[b]Data are based on *Roseburia cecicola* strain GM[T] (Stanton and Savage, 1983a) and *Roseburia intestinalis* L1-82[T] (Duncan et al., 2002)

Genus XVII. **Shuttleworthia** Downes, Munson, Radford, Spratt and Wade 2002, 1473[VP]

WILLIAM G. WADE AND JULIA DOWNES

Shutt.le.worth′ia. N.L. fem. n. *Shuttleworthia* named to honor Cyril Shuttleworth, the distinguished British microbiologist.

Slightly curved short rods which occur singly, in pairs, short chains, or diphtheroidal arrangements. Rods 0.4–0.6 μm in diameter and 1.0–2.5 μm in length. **Gram-stain-positive, non-spore-forming, and nonmotile.** Obligately **anaerobic.** Optimal growth temperature 30–37 °C. Several different colony morphologies exist.

Saccharolytic. Principal end products of glucose fermentation are acetic, butyric, and lactic acids. Growth in broth media is good and is stimulated by the presence of fermentable carbohydrates. **Esculin is hydrolyzed and indole is produced.** Nitrate is not reduced; arginine and urea are not hydrolyzed. Gelatin is not liquefied and H$_2$S and catalase are not produced. No growth in 20% bile.

Occurs in the oral cavity of man and has been isolated from periodontal pockets and subgingival plaque.

DNA G+C content (mol%): 50–51 (HPLC).

Type species: **Shuttleworthia satelles** Downes, Munson, Radford, Spratt and Wade 2002, 1474[VP].

Further descriptive information

The colony morphology of the type species, *Shuttleworthia satelles*, is variable with three distinct phenotypes observable. The majority of colonies are 0.7–0.9 mm in diameter, circular, entire, with a gray, low convex center surrounded by a flat, translucent margin after 5 d incubation on Fastidious Anaerobe Agar (FAA). On further incubation, smaller satellite colonies arise on the periphery or within the original colonies (Figure 170). These satellite colo-

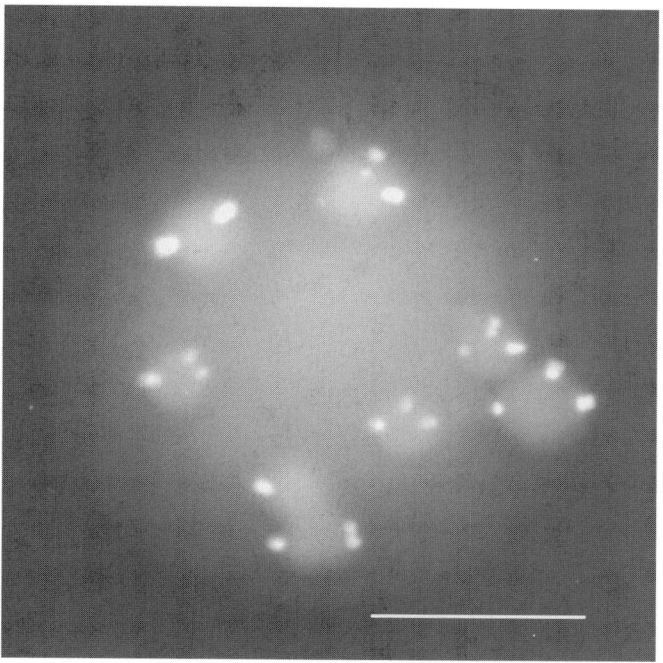

FIGURE 170. Colony morphology of 10-d-old culture of *Shuttleworthia satelles* DSM 14600 showing satellite colonies arising within the primary colonies. Bar = 0.5 mm.

TABLE 175. Characteristics differentiating *Shuttleworthia satelles* and other closely related taxa[a]

Characteristic	*Shuttleworthia satelles*	*Butyrivibrio fibrisolvens*	*Eubacterium rectale*	*Eubacterium ramulus*	*Pseudobutyrivibrio ruminis*	*Roseburia cecicola*
Cell shape	Slightly curved	Curved	Slightly curved	Straight	Curved	Slightly curved
Gram reaction	+	–	d	+	–	–
Motility	–	+	d	–	+	+
Indole production	+	–	–	–	ND	ND
Esculin hydrolysis	+	d	+	+	ND	–
Fructose fermentation	+	+	+	+	+	–
Sorbitol fermentation	–	–	d	d	–	+
Trehalose fermentation	+	d	d	–	+	–
Fermentation products[b]	a, b, l	F, B, l, (a)	a, B, L, (s)	a, f, B, l, (s)	F, B, L, (a)	B
DNA G+C content (mol%)	50–51	41	38	39	40–41	42
Source	Human oral	Rumen of ruminants; human, rabbit, horse feces	Human feces and colon	Human feces	Cow rumen	Mouse cecum
Reference	Downes et al. (2002)	Bryant (1986b)	Moore and Holdeman (1986)	Moore et al. (1976)	van Gylswyk et al. (1996)	Stanton and Savage (1983a)

[a]Symbols: +, 90% or more of strains are positive; –, 90% or more of strains are negative; d, different strains give different reactions (11–89% positive); ND, data not available or incomplete.

[b]a, acetic acid; f, formic acid; b, butyric acid; l, lactic acid; s, succinic acid; (), strain variation or trace amounts detected. Capital letters indicate major products.

nies are 0.2–0.3 mm in diameter, convex, circular, entire, off-white, semi-opaque, and shiny. A third colony type appears as irregular, cream-colored colonies after 6 d of incubation. On subculture, the third colony type is 0.6 mm in diameter, circular, entire, high convex, cream-colored, and opaque after 5 d incubation on FAA.

The fermentation of certain sugars was the only biochemical characteristic that differed in the different colony types. Cells forming the cream-colored, high convex colonies ferment arabinose, cellobiose, melibiose, and salicin while those cells forming the gray, low convex colonies usually do not ferment these sugars.

All nine strains of *Shuttleworthia satelles* tested in the Rapid ID32A anaerobe identification kit (bioMérieux) were positive for α-galactosidase, β-galactosidase, α-glucosidase, β-glucosidase, and indole in the panel, while reactions to β-glucuronidase were weak and variable.

Enrichment and isolation procedures

Strains of *Shuttleworthia* species can be isolated on complex media appropriate for the culture of anaerobes including blood agar, enriched brain heart infusion agar, and Fastidious Anaerobe Agar (FAA) (Lab M).

Maintenance procedures

Lyophilization of cultures in the early stationary phase of growth is recommended for long term storage of most strains.

TABLE 176. Characteristic features of *Shuttleworthia satelles*[a]

Characteristic	
Cell shape	Slightly curved
Gram reaction	+
Motility	–
Esculin hydrolysis	+
Arginine hydrolysis	–
Growth in 20% bile	–
Catalase production	–
Gelatin hydrolysis	–
H₂S production	–
Indole production	+
Nitrate reduction	–
Urea hydrolysis	–
Fermentation of:	
Arabinose	v
Cellobiose	v
Fructose	+
Glucose	+
Lactose	d
Maltose	+
Mannitol	d
Melezitose	+
Melibiose	v
Rhamnose	+
Salicin	v
Sorbitol	–
Sucrose	+
Trehalose	+
Fermentation end products[b]	a, b, l
Enzyme profile[c]	45 1/5 0 2000 00

[a]Symbols: +, >85% positive; d, different strains give different reactions (16–84% positive); –, 0–15% positive; v, variation among different colony morphotypes.
[b]a, Acetate; b, butyrate; l, lactate.
[c]Enzyme profile obtained with Rapid ID32A anaerobe identification kit (bioMérieux).

Strains can also be stored at –70 °C in brain heart infusion broth supplemented with 10% glycerol.

Procedures for testing special characters

The general methods described for the characterization of anaerobes in the VPI Anaerobe Laboratory Manual (Holdeman et al., 1977b) and the Wadsworth-KTL Anaerobic Laboratory Manual (Jousimies-Somer et al., 2002) are suitable for the study of members of this genus.

Differentiation of the genus *Shuttleworthia* from other genera

Characteristics that differentiate *Shuttleworthia satelles* from related taxa are shown in Table 176.

Taxonomic comments

Phylogenetic analysis of the 16S rRNA gene sequence reveals the single species of the genus, *Shuttleworthia satelles*, to be a deep-branching member of the family *Lachnospiraceae* (Figure 171), most closely related to *Eubacterium ramulus* and *Eubacterium rectale* but sharing less than 90% sequence

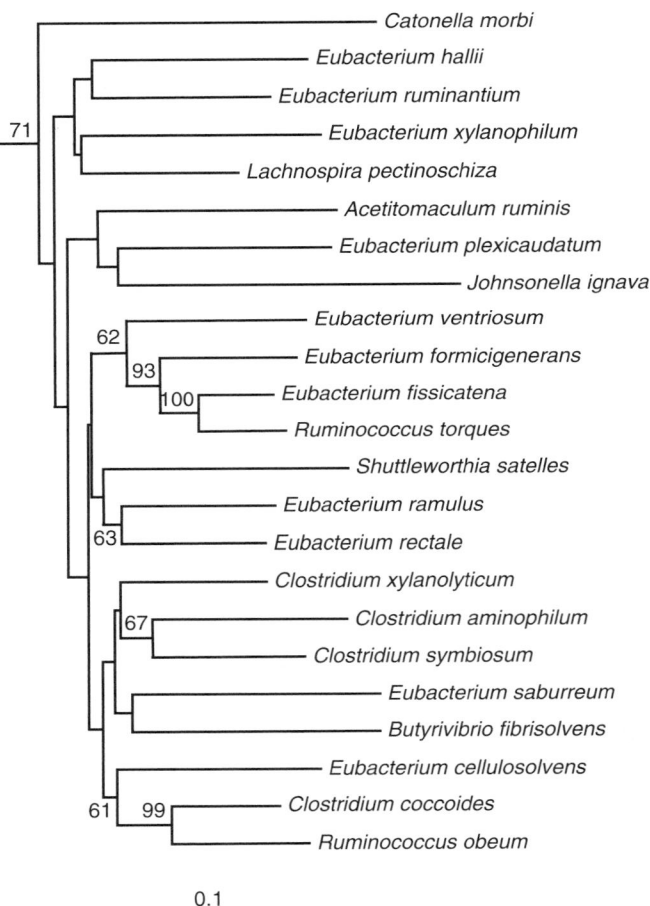

FIGURE 171. Phylogenetic tree based on 16S rRNA gene sequence comparisons over 1362 aligned bases showing relationships between *Shuttleworthia satelles* and related species. Tree was constructed using the neighbor-joining method following distance analysis of aligned sequences. Numbers represent bootstrap values for each branch based on data for 100 trees; only values greater than 50% are shown.

identity with these species. Bootstrap values for the tree are generally low indicating that the topology shown should be interpreted with caution.

Acknowledgements

The work reported in this chapter was funded, in part, by Wellcome Trust grant ref. 058950.

List of species of the genus *Shuttleworthia*

1. **Shuttleworthia satelles** Downes, Munson, Radford, Spratt and Wade 2002, 1474[VP]

   sat.ell'es. L. n. *satelles* a satellite or attendant upon a distinguished person, referring to the satelliting appearance of older cultures.

   This species is the only member of the genus, thus the species description is the same as that for the genus and as listed in Table 176. The description is based on nine strains isolated from human subgingival plaque and periodontal pockets in patients with periodontitis.

   DNA G+C content (mol%): 51 (HPLC).
   Type strain: CCUG 45864, DSM 14600, VPI D143K-13.
   GenBank accession number (16S rRNA gene): AF399956.

Genus XVIII. **Sporobacterium** Mechichi, Labat, Garcia, Thomas and Patel 1999, 1747[VP]

TAHAR MECHICHI AND BHARAT K. C. PATEL

Spo.ro.bac.ter'ium. Gr. n. *spora* spore; Gr. neut. dim. n. *bakterion* rod; N.L. neut. n. *Sporobacterium* spore-forming rod.

Slightly curved rods (5–10 μm × 0.4–0.8 μm). **Gram-stain-positive. Terminal spores** distend the cells. Motile by **peritrichous flagella. Strictly anaerobic. Chemo-organotrophic.** Utilizes a wide range of aromatic compounds, crotonate, and methanol. Growth occurs between 25–45 °C and at pH 6.5–8.5. Isolated from an anaerobic methanogenic waste treatment digester fed with olive mill wastewater.

DNA G+C content (mol%): 38.

Type species: **Sporobacterium olearium** Mechichi, Labat, Garcia, Thomas and Patel 1999, 1747[VP].

Further descriptive information

Aromatic compounds are structurally complex and represent one of the most diverse groups of organic substrates for microorganisms. These compounds are widespread in the environment and are natural products of plants (Harborne, 1980), micro-organisms, and animals and are also released from industrial processes. Tarvin and Buswell (1934) were the first to show that monocyclic aromatic compounds were degraded to methane and CO_2 under anaerobic conditions. Since their findings, numerous investigations have demonstrated that aromatic compounds can be degraded to CO_2 or volatile fatty acids under anoxic conditions by sulfate-reducing, nitrate-reducing, ferric iron-reducing, photosynthetic, and fermenting bacteria. The trait of aromatic compound degradation by fermenting anaerobes can occur via ring cleavage or by the transformation of the side chains. The latter trait appears to be widely distributed in the domain *Bacteria*, but that of aromatic ring cleavage is more restricted and is only known to occur in eight genera and eleven species: *Pelobacter acidigallici, Pelobacter massiliensis, Syntropus buswellii, Syntrophus gentianae,* and *Syntrophus aciditrophicus* which are members of *Deltaproteobacteria; Holophaga foetida,* a member of phylum *Acidobacteria;* and *Parasporobacterium paucivorans, Sporobacter termitidis, Eubacterium oxidoreducens, Sporotomaculum hydroxybenzoicum,* and *Sporobacterium olearum* which are members of the phylum *Firmicutes.* A few other *Eubacterium* species, *Ruminococcus schinkii* (Rieu-Lesme et al., 1996), and *Coprococcus* species, which are also members of phylum *Firmicutes,* can ferment aromatic compounds but can be differentiated from the latter group, as they are also able to use carbohydrates. The five aromatic-ring cleaving members of the phylum *Firmicutes* can be differentiated on a number of traits as indicated in Table 177.

Under anaerobic conditions, aromatic compounds are degraded via three different pathways, namely, the phloroglucinol pathway, the resorcinol pathway, and the benzoyl-CoA pathway. The phloroglucinol and resorcinol pathways are utilized by pure cultures of fermenting bacteria whereas the benzoyl-CoA pathway is utilized by pure cultures of fermenting bacteria or by syntrophic co-cultures. At least two different metabolic groups of bacteria that degrade aromatic compounds via the phloroglucinol pathway can be distinguished. The first group includes *Eubacterium oxidoreducens* (Krumholz and Bryant, 1986b), *Pelobacter acidigallici* (Schnell et al., 1991), and *Pelobacter massiliensis* (Schink and Pfennig, 1982), all of which degrade aromatic compounds containing two or three hydroxyl groups in the *meta*-position. Such compounds include trihydroxylated benzenes (gallate, pyrogallol, phloroglucinol, hydroxyquinone, and 2,4,6-trihydroxybenzoate) and dihydroxybenzenes (resorcinol), which are degraded to acetate or acetate and butyrate, respectively. *Eubacterium oxidoreducens* but not *Pelobacter acidigallici* or *Pelobacter massiliensis* requires hydrogen or formate for growth and utilizes quercitin. The second group, which includes *Holophaga foetida* (Liesack et al., 1994), *Sporobacter termitidis* (Grech-Mora et al., 1996), *Sporobacterium olearium* (Mechichi et al., 1999), and *Parasporobacterium paucivorans* (Lomans et al., 2001), transforms the methoxylated aromatic compounds, 3,4,5-trimethoxybenzoate, 3,4,5-trimethoxycinnamate, syringate, sinapate, and 5-hydroxyvanillate, to their corresponding hydroxylated derivatives, namely 3,4,5-trihydroxybenzoate (gallate) and trihydroxycinnamate and then to acetate and acetate and butyrate, respectively, via the phloroglucinol pathway.

Enrichment and isolation procedures

Samples from a methanogenic digester that had been fed with olive mill wastes were collected anaerobically using N_2-flushed

TABLE 177. Differential characteristics of the genus *Sporobacterium* from its phylogenetic and aromatic ring cleaving physiological relatives[a]

Characteristic	*Sporobacterium*	*Parasporobacterium*	*Sporobacter*	*Eubacterium*	*Sporotomaculum*
Representative species	*Sporobacterium olearium*	*Parasporobacterium paucivorans*	*Sporobacter termitidis*	*Eubacterium oxidoreducens*	*Sporotomaculum hydroxybenzoicum*
Cell size (μm)	5–10 × 0.4–0.8	1.5–2 × 0.3–0.5	1.2 × 0.3–0.4	NR	2–3 × 0.6–0.8
Cell shape	Curved rods	Rods in pairs	Slightly curved rods	Rods	Thick rods
Spores	+	–	+	–	+
Motility	+	–	+	–	+
Fermentation end products	Acetate, butyrate, methanethiol	Acetate, butyrate, methanethiol, dimethylsulfide	Acetate, methanethiol, dimethylsulfide	Acetate, butyrate	Acetate, butyrate
DNA G+C content (mol%)	38	NR	57	35.7	48
Habitat	Anaerobic digester fed with olive mill wastewater	Eutrophic lake sediment	Termite (*Nasutitermes lujae*) gut	Rumen of cattle	Termite (*Cubitermes speciosus*) gut
Temperature optimum (°C)	37–40	34–37	32–35	39–41	30
pH optimum	7.2	6.5–7	6.7–7.2	7.4	7.3–7.6
Carbon sources	Aromatic compounds, methanol and crotonate[b]	Aromatic compounds[c]	Aromatic compounds[d]	Aromatic compounds and quercitin[e]	Aromatic compounds[f]
Family classification	*Lachnospiraceae*	*Lachnospiraceae*	*Clostridiaceae*	*Eubacteriaceae*	*Peptococcaceae*
References	Mechichi et al. (1999)	Lomans et al. (2001)	Grech-Mora et al. (1996)	Krumholz and Bryant (1986b)	Brauman et al. (1998)

[a]All are anaerobes with fermentation as the mode of respiration and cleave the aromatic ring. In addition, *Ruminococcus schinkii* (Rieu-Lesme et al., 1996) and *Coprococcus* (Tsai et al., 1976), which are members of family *Lachospiraceae*, also cleave the armoatic ring but also ferment carbohydrates. NR, Not reported.

[b]3,4,5-Trimethoxybenzoate, 3,4,5-trimethoxycinnamte, 3,4,5-trimethoxyphenylacetate, 3,4,5-trimethoxyphenylpropionate, syringate, ferulate, sinapate, vanillate, 3,4-dimethoxybenzoate, 2,3-dimethoxybenzoate, gallate, 2,4,6-trihydroxybenzoate, pyrogallol, phloroglicinol and quercetin.

[c]3,4,5-Trimethoxybenzoate, syringate, 5-hydroxyvanillate, sinapate and ferulate.

[d]3,4,5-Trimethoxybenzoate, 3,4,5-trimethoxycinnamate, syringate, 3,4-dimethoxycinnamate, vanillate, sinapate and ferulate/

[e]Gallate, pyrogallol, phloroglucinol.

[f]3-Hydroxybenzoate

syringes and injected into N_2-flushed 100-ml sterile serum vials. For enrichments, a 0.5-ml sample is inoculated into 5-ml basal medium containing 5 mM syringate. The basal medium is prepared using the anaerobic techniques described by Hungate (1969) as modified for use with syringes (Macy et al., 1972; Miller and Wolin, 1974) and contains (per liter of deionized water): 1 g NH_4Cl, 0.3 g K_2HPO_4, 0.3 g KH_2PO_4, 0.6 g NaCl, 0.1 g $CaCl_2\cdot2H_2O$, 0.2 g $MgCl_2\cdot6H_2O$, 0.1 g KCl, 0.5 g cysteine-HCl, 1 g yeast extract (Difco), 1.5 ml trace-element solution (Widdel and Pfennig, 1981), and 1 mg resazurin. The pH is adjusted to 7.0 with 10 M KOH solution and the medium boiled under a stream of O_2-free N_2 gas and cooled to room temperature; 5 ml aliquots are dispensed into Hungate tubes under N_2:CO_2 (80:20%, v/v) and subsequently sterilized by autoclaving at 110 °C for 45 min. Prior to use, 0.2 ml 5% (w/v) $NaHCO_3$ and 0.05 ml 2.5% (w/v) $Na_2S\cdot9H_2O$ are injected from sterile stock solutions into presterilized basal medium. A positive enrichment culture obtained after two weeks incubation at 37 °C is subcultured several times under the same conditions and should consist of a stable morphologically dominant microbial consortium that degrades syringate with concomitant production of acetate and butyrate. For isolation, a serial 10-fold dilution of the enrichment culture is inoculated into roll tubes containing basal medium supplemented with 5 mM syringate and 1.6% (w/v) purified agar. Several single well-isolated colonies that developed are picked and resuspended and the purification procedure repeated several times.

Maintenance procedures

Cultures can be maintained at room temperature for at least a month. Stock cultures are best stored for long term at −20 °C in anaerobic medium containing glycerol (50:50 v/v).

Differentiation of the genus *Sporobacterium* from other genera

The five aromatic-ring cleaving members of the phylum *Firmicutes* can be differentiated on a number of traits as indicated in Table 177.

The common features of *Sporobacterium olearium*, *Sporobacter termitidis*, and *Parasporobacterium paucivorans* (as well as *Holophaga foetida*) are the ring cleavage of polyphenolic compounds and the production of methanethiol or dimethylsufide. However, several metabolic differences distinguish the three strains from *Sporobacterium olearium*. *Holophaga foetida* and *Sporobacterium olearium*, but not *Sporobacter termitidis*, utilize gallate, 2,4,6-trihydroxybenzoate, pyrogallol, and phloroglucinol as carbon and energy sources. *Holophaga foetida* and *Sporobacter termitidis* produce both methanethiol and dimethylsulfide from the methoxyl group, and sulfide from cysteine in the presence of Na_2S which is used as a reducing agent in the culture media. However, *Sporobacterium olearium* produces only methanethiol and is also able to produce methanethiol from methanol. *O*-demethylation in *Parasporobacterium paucivorans* appears to be strictly sulfide or methanethiol (MT) dependent, whereas this is not the case in *Holophaga foetida*, *Sporobacter termitidis*, and *Sporobacterium olearium*.

Taxonomic comments

Sequence analysis of 16S rDNA indicates that *Sporobacterium olearium* and *Parasporobacterium paucivorans* are phylogenetically close and are members of the family *Lachnospiraceae*, order *Clostridiales* in phylum *Firmicutes*. The vast majority of the members of family *Lachnospiraceae* have been isolated from digestive tracts of herbivores where abundant aromatic compounds are present, but only a limited number of strains have so far been tested for their ability to ferment aromatic compounds.

Differentiation of the species of the genus Sporobacterium

Only one species, *Sporobacterium olearium*, is described.

List of species of the genus *Sporobacterium*

1. **Sporobacterium olearium** Mechichi, Labat, Garcia, Thomas and Patel 1999, 1747[VP]

o.le.a'ri.um. L. adj. *olearius* of oil. N.L. neut. n. *olearium* related to olive oil.

A slightly curved rod with terminal spores. Stains Gram-stain-positive. Motile, with peritrichous flagella. Strictly anaerobic, chemo-organotrophic. Grows on crotonate, methanol, **3,4,5-trimethoxybenzoate (TMB) 3,4,5-trimethoxycinnamate** (TMC), syringate, **3,4,5-trimethoxyphenylacetate** (TMPA), **3,4,5-trimethoxyphenylpropionate** (TMPP), ferulate, sinapate, vanillate, 3,4-dimethoxybenzoate, 2,3-dimethoxybenzoate, gallate, 2,4,6-trihydroxybenzoate, pyrogallol, phloroglycinol, and quercetin. Does not ferment benzoate, 2-, 3-,4-methoxybenzoates, 2,4-, 2,5-, 2,6-, 3,5-dimethoxybenzoates, glucose, fructose, sucrose, xylose, sorbose, galactose, *myo*-inositol, lactose, ribose, mannitol, cellobiose, formate, fumarate, pyruvate, malonate, succinate, ethanol, propanol, butanol, or grows on $H_2 + CO_2$. Yeast extract stimulates growth, but is not required. TMB, TMC, TMPA, TMPP, syringate, sinapate, 2,3-dimethoxybenzoate, 3,4-dimethoxybenzoate, vanillate, ferulate, and methanol are fermented to acetate, butyrate, and MT. Gallate, 2,4,6-trihydroxybenzoate, pyrogallol, phloroglucinol, and crotonate are fermented to acetate and butyrate. H_2 is not required for growth but is required for complete degradation of gallate, 2,4,6-trihydroxybenzoate, pyrogallol, and phloroglucinol. Sulfate, thiosulfate, sulfite, nitrate, elemental sulfur, and fumarate are not reduced. Optimum growth temperature is between 37 and 40 °C (temperature growth range is 25–45 °C). Optimum pH for growth is 7.2 (pH growth range 6.5–8.5). Grows with NaCl concentrations 0–30 g/l but not with 35 g/l.

Habitat: anaerobic methanogenic digester fed with olive mill wastewater (Tunisia).

DNA G+C content (mol%): 38 (T_m).

Type strain: SR1, DSM 12504.

GenBank accession number: AF116854.

Genus XIX. **Syntrophococcus** Krumholz and Bryant 1986c, 489[VP] (Effective publication: Krumholz and Bryant 1986a, 317.)

Yuji Sekiguchi

Syn.tro.pho.coc'cus. Gr. adj. *syn* together with; Gr. n. *trophos* one who feeds; Gr. n. *kokkos* a grain, berry; N.L. masc. n. *Syntrophococcus* coccus which feeds together with (another species).

Coccus-shaped cells. Gram-stain-negative. Nonmotile. Non-spore-forming. **Strictly anaerobic. Sugars are oxidized to acetate when an electron-accepting system is present, such as formate, methoxyl-containing benzoids, or a hydrogenotrophic methanogen.** Habitats are mesophilic, anaerobic, microbial ecosystems where complex organic matter is degraded.

DNA G+C content (mol%): 52.

Type species: **Syntrophococcus sucromutans** Krumholz and Bryant 1986c, 489 (Effective publication: Krumholz and Bryant 1986a, 317.)

Further descriptive information

Cells are nonmotile cocci with diameters of about 1 μm (Krumholz and Bryant, 1986a). Gram staining is variable. Electron micrographs show a distinctly bilayered envelope, the outer membrane of which reacts with polymyxin B indicating the presence of lipids. The molar ratio of the components of the cell wall are: Glu:*m*-DAP:Ala:GluN:Mur = 1:0.95:1.65:0.9:0.85. The major cellular fatty acids are $C_{14:0}$, $C_{16:0}$, and $C_{18:0}$ (Dore and Bryant, 1989).

Comparative 16S rRNA gene sequence analysis of *Syntrophococcus* places the genus within the phylum *Firmicutes* and the family *Lachnospiraceae*.

Syntrophococcus is a strictly anaerobic chemo-organotroph that grows at mesophilic temperatures (30–44 °C) and nearly neutral pH (6.0–7.6). It grows well in a basal medium containing 30% rumen fluid, 0.5% yeast extract, and 0.5% Casitone when methoxylated benzenoids are added (Krumholz and Bryant, 1986a). This organism possesses an unusual one-carbon metabolism using sugars as electron donors and forming acetate. Formate, the methoxyl groups of benzenoids, or hydrogenotrophic methanogens serve as electron-accepting systems. Formate is reduced to acetate. The methoxyl-containing benzenoids, such as syringate, vanillin, vanillate, caffeate, ferulate, and 3,4,5-trimethoxybenzoate, are demethoxylated to the hydroxyl compounds while the methoxyl groups are utilized for concomitant acetate production. The biochemistry and metabolism of these acetogenic transformations have been studied in detail (Dore and Bryant, 1990). Lastly, hydrogenotrophic organisms such as the methanogen *Methanobrevibacter smithii*

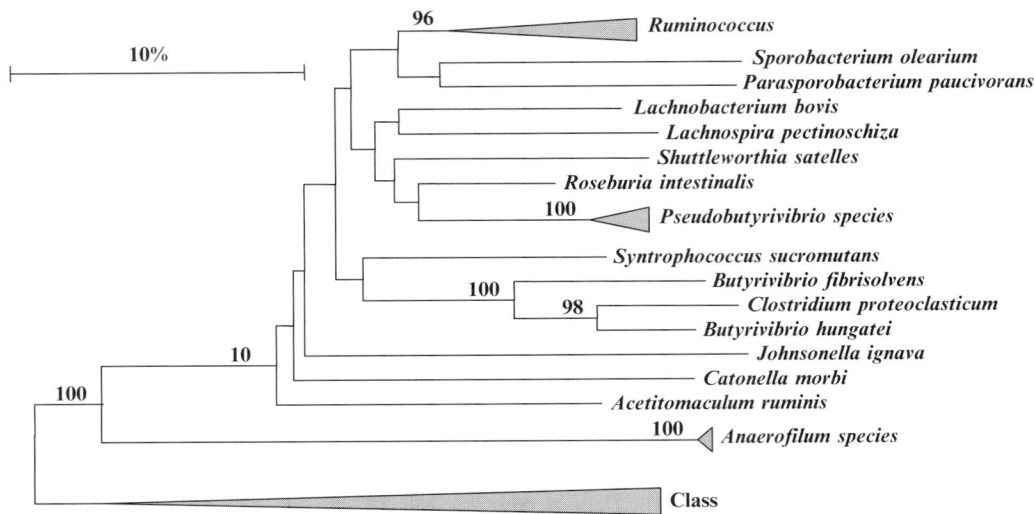

FIGURE 172. The phylogenetic position of genus *Sporobacterium* as a member of the family *Lachnospiraceae*, order *Clostridiales* is shown. The strains used in the analysis, their culture collection numbers and corresponding 16S rRNA gene sequences extracted from GenBank are listed below: *Lachnospira pectinoschiza* (ATCC 49827[T], L14675), A*cetitomaculum ruminis* (ATCC 43876[T], M59083), *Parasporobacterium paucivorans* (DSM 15970[T], AJ272036), *Butyrivibrio hungatei* (ATCC BAA-456[T], AJ428553), "*Clostridium proteoclasticum*" (ATCC 51982[T], U37378), *Butyrivibrio fibrisolvens* (ATCC 19171[T], U41172), *Catonella morbi* (ATCC 51271[T], X87151), *Johnsonella ignava* (ATCC 51276[T], X87152), *Lachnobacterium bovis* (ATCC BAA-151[T], AF298662), *Roseburia intestinalis* (DSM 14610[T], AJ312385), *Syntrophococcus sucromutans* (DSM 3224[T], AF202264), *Shuttleworthia satelles* (DSM 14600[T], AF399956), *Sporobacterium olearium* (DSM 12504[T], AF116854), and the cluster represented as triangle for *Pseudobutyrivibrio* species includes *Pseudobutyrivibrio ruminis* (DSM 9787[T], X95893)and *Pseudobutyrivibrio xylanivorans* (DSM 14809[T], AJ428548), for *Ruminococcus* species includes *Ruminococcus gnavus* (ATCC 29149[T], L76597), *Ruminococcus productus* (ATCC 27340[T], D14144), *Ruminococcus hydrogenotrophicus* (DSM 10507[T], X95624) and *Ruminococcus schinkii* (DSM 10518[T], X94965) and for *Anaerofilum* species includes *Anaerofilum pentosovorans* (DSM 7168[T], X97852) and *Anaerofilum agile* (DSM 4272[T], X98011). The outgroup triangle is represented by members of the class *Bacilli* and includes *Mycoplasma hominis* (ATCC 23114[T], AJ002265), *Mycoplasma canis* (ATCC 19525[T], AF412972), *Bacillus halmapalus* (DSM 8723[T], X76447), *Bacillus horikoshii* (DSM 8719[T], X76443), *Geobacillus uzenensis* (DSM 13551[T], AF276304), *Paenibacillus macerans* (ATCC 8244[T], X57306), *Thermobacillus xylanilyticus*(CNCM I-1017[T], AJ005795), *Marinococcus halophilus* (DSM 20408[T], X90835), *Jeotgalicoccus psychrophilus* (JCM 11199[T], AY028926), *Lactobacillus hilgardii* (ATCC 8290[T], M58821), *Leuconostoc gelidum* (DSM 5578[T], AF175402) and *Turicibacter sanguinis* (DSM 14220[T], AF349724). Phylogenetic analysis was performed on 1212 unambiguous nucleotides using DNADIST and neighbor-joining programs which form part of the PHYLIP suite of software. Bar indicates 10 nucleotide changes per 100 nucleotides. The following abbreviations have been used: ATCC, American Type Culture Collection; DSM - DSMZ, Deutsche Sammlungvon Mikroorganismen und Zellkulturen GmbH; JCM, Japan Collection of Microorganisms; and CNCM, Collection Nationale de Cultures Microbiennes.

DSM 861 can also serve as the electron-accepting system using H_2 and/or formate as electron carriers (Krumholz and Bryant, 1986a). This reaction is similar to syntrophic fatty acid oxidation (Schink, 1997). Electron donors include various carbohydrates such as fructose, galactose, and maltose (Krumholz and Bryant, 1986a). Acetate is the sole end product of the oxidation. These sugars support only a small amount of growth in the absence of the electron-accepting systems. Addition of rumen fluid in medium is required for good growth. The type species of *Syntrophococcus* was isolated from the rumen of a hay-fed steer.

Enrichment and isolation procedures

Syntrophococcus is selectively enriched with the basal medium containing 5 mM syringate, 5% rumen fluid with a bicarbonate buffer, and N_2:CO_2 (4:1) gas phase at 37 °C (Krumholz and Bryant, 1986a). The rumen contents of a steer fed a diet of 30% grain and 70% forage is serially diluted in the medium and incubated anaerobically. After growth has been observed, typically after 3–4 weeks, cultures from the highest positive dilutions are further enriched by successive transfers. The medium for the serial dilution and enrichments is initially reduced with 1 mM Na_2S, but, after several transfers, cysteine-sulfide is added as the reducing agent. The roll-tube method is used for the final isolation of axenic cultures. Under these conditions, *Syntrophococcus* forms round colonies that are convex with entire edges.

Taxonomic comments

Upon its initial isolation, *Syntrophococcus* was placed in the family *Veillonellaceae* because it stained Gram-negative and was an anaerobic coccus (Krumholz and Bryant, 1986a). However, comparative analyses of 16S rRNA genes indicated that the genus should be placed in the phylum *Firmicutes* (Stackebrandt et al., 1999) within the family *Lachnospiraceae* (Figure 5; Ludwig et al., 2009). At present, the genus *Syntrophococcus* encompasses only one species, *Syntrophococcus sucromutans*.

Differentiation of the genus *Syntrophococcus* from other genera

The genus *Syntrophococcus* is unique in its metabolic capability of fermenting carbohydrates using formate, methoxy-monobenzenoids, or a methanogenic bacterium as electron-accepting systems. Good growth occurs only in the presence of these electron acceptors, a property which distinguishes it from phylogenetically related homoacetogens within the *Firmicutes*.

List of species of the genus *Syntrophococcus*

1. **Syntrophococcus sucromutans** Krumholz and Bryant 1986c, 489[VP] (Effective publication: Krumholz and Bryant 1986a, 317.)

su.cro.mu′tans. N.L. n. *sucro* modern chemical combined form referring to all sugars; L. part. adj. *mutans* changing or converting; N.L. part. adj. *sucromutans* converting sugar.

Coccus-shaped cells. 1.0–1.3 μm in diameter, with rounded to flattened plane of division, forming short chains. Nonmotile. Does not form spores. Stains Gram-negative. Strictly anaerobic chemo-organotroph. Electron donors for growth include pyruvate, glucose, fructose, galactose, maltose, cellobiose, lactose, arabinose, maltose, ribose, xylose, salicin, and esculin. The major organic product of these carbohydrates is acetate. Compounds not used as electron donors include hydrocaffeate, gallate, protocatechuate, citrate, mannitol, fumarate, succinate, malate, aspartate, glutamate, glycine, 3-hydroxybutyrate, valine, serine, isoleucine, butyrate, pectin, and starch. Electron acceptors with the major organic products in parentheses are: formate (acetate), caffeate (hydrocaffeate), ferulate (caffeate, hydrocaffeate, acetate), syringate and 3,4,5-trimethoxybenzoate (gallate and acetate), vanillin (protocatechuic aldehyde, protocatechuate, acetate), and vanillate (protocatechuate and acetate). The following compounds were not used as electron acceptors or as sole energy sources: methanol, formaldehyde, CO, ethanol, fumarate, malate, crotonate, lactate, aspartate, glutamate, glycine, valine, serine, isoleucine, 1-methoxybenzene, 4-methoxybenzoate, cinnamate, 4-hydroxycinnamate, H_2:CO_2 (4:1 v/v), and H_2:CO_2 (4:1 v/v) plus fumarate, crotonate, or caffeate. Growth also occurs with electron donors such as fructose when an H_2-CO_2-utilizing methanogen is the electron-accepting system.

Good growth occurs in an anaerobic medium containing N_2:CO_2 (4:1 v/v) gas phase, minerals, 40 mM NaHCO$_3$, 2 mM cysteine, 6 mM sodium formate, 6 mM fructose or xylose, 0.5% Casitone, and 30 (v/v) rumen fluid. The temperature range for growth is 30–44 °C, with an optimum at 35–42 °C. The pH range is 6.0–7.6, with an optimum of 6.4. Habitats are the rumens of hay-fed cattle.

DNA G+C content (mol%): 52 (T_m).

Type strain: S195, ATCC 43584, DSM 3224.

GenBank accession number (16S rRNA gene): AF202264, Y18191.

References

Akin, D.E., W.S. Borneman, L.L. Rigsby and S.A. Martin. 1993. *p*-Coumaroyl and feruloyl arabinoxylans from plant cell walls as substrates for ruminal bacteria. Appl. Environ. Microbiol. *59*: 644–647.

Andersson, M., S. Ratnayake, L. Kenne, L. Ericsson and R.J. Stack. 1993. Structural studies of the extracellular polysaccharide from *Butyrivibrio fibrisolvens* strain X6C61. Carbohydr. Res. *246*: 291–301.

Attwood, G.T., K. Reilly and B.K.C. Patel. 1996. *Clostridium proteoclasticum* sp. nov., a novel proteolytic bacterium from the bovine rumen. Int. J. Syst. Bacteriol. *46*: 753–758.

Balch, W.E., G.E. Fox, L.J. Magrum, C.R. Woese and R.S. Wolfe. 1979. Methanogens: reevaluation of a unique biological group. Microbiol. Rev. *43*: 260–296.

Barbosa, T.M., K.P. Scott and H.J. Flint. 1999. Evidence for recent intergeneric transfer of a new tetracycline resistance gene, *tet(W)*, isolated from *Butyrivibrio fibrisolvens*, and the occurence of *tet(O)* in ruminal bacterial. Environ. Microbiol. *1*: 53–64.

Barcenilla, A., S.E. Pryde, J.C. Martin, S.H. Duncan, C.S. Stewart, C. Henderson and H.J. Flint. 2000. Phylogenetic relationships of butyrate-producing bacteria from the human gut. Appl. Environ. Microbiol. *66*: 1654–1661.

Beard, C.E., M.A. Hefford, R.J. Forster, S. Sontakke, R.M. Teather and K. Gregg. 1995. A stable and efficient transformation system for *Butyrivibrio fibrisolvens* OB156. Curr. Microbiol. *30*: 105–109.

Beard, C.E., K. Gregg, M. Kalmokoff and R.M. Teather. 2000. Construction of a promoter-rescue plasmid for *Butyrivibrio fibrisolvens* and its use in characterization of a flagellin promoter. Curr. Microbiol. *40*: 164–168.

Benghedalia, D., J. Miron and R. Solomon. 1993. The degradation and utilization of structural polysaccharides of sorghum straw by defined ruminal bacteria. Animal Feed Sci. Technol. *42*: 283–295.

Billington, S.J., J.G. Songer and B.H. Jost. 2002. Widespread distribution of a *tetW* determinant among tetracycline-resistant isolates of the animal pathogen *Arcanobacterium pyogenes*. Antimicrob. Agents Chemother. *46*: 1281–1287.

Brauman, A., J.A. Muller, J.L. Garcia, A. Brune and B. Schink. 1998. Fermentative degradation of 3-hydroxybenzoate in pure culture by a novel strictly anaerobic bacterium, *Sporotomaculum hydroxybenzoicum* gen. nov., sp. nov. Int. J. Syst. Bacteriol. *48*: 215–221.

Braun, C., A. Meinke, L. Ziser and S.G. Withers. 1993. Simultaneous high-performance liquid chromatographic determination of both the cleavage pattern and the stereochemical outcome of the hydrolysis reactions catalyzed by various glycosidases. Anal. Biochem. *212*: 259–262.

Bryant, M.P. and N. Small. 1956a. Characteristics of two new genera of anaerobic curved rods isolated from the rumen of cattle. J. Bacteriol. *72*: 22–26.

Bryant, M.P. and N. Small. 1956b. The anaerobic monotrichous butyric acid-producing curved rod-shaped bacteria of the rumen. J. Bacteriol. *72*: 16–21.

Bryant, M.P., B.F. Barrentine, J.F. Sykes, I.M. Robinson, V. Shawver and L.W. Williams. 1960. Predominant bacteria in the rumen of cattle on bloat-provoking ladino clover pasture. J. Dairy Sci. *43*: 1435–1444.

Bryant, M.P. and I.M. Robinson. 1962. Some nutritional characteristics of predominant culturable ruminal bacteria. J. Bacteriol. *84*: 605–614.

Bryant, M.P. 1972. Commentary on the Hungate technique for culture of anaerobic bacteria. Am. J. Clin. Nutr. *25*: 1324–1328.

Bryant, M.P. 1974. Genus *Butyrivibrio*. *In* Buchanan and Gibbons (Editors), Bergey's Manual of Determinative Bacteriology. The Williams & Wilkins Co., Baltimore, pp. 420.

Bryant, M.P. 1986a. Genus XIII. *Lachnospira* Bryant and Small 1956, 24[AL]. *In* Sneath, Sharpe and Holt (Editors), Bergey's Manual of Systematic Bacteriology, Vol. 2. The Williams & Wilkins Co., Baltimore, pp. 1375–1376.

Bryant, M.P. 1986b. Genus *Butyrivibrio*. *In* Sneath, Sharpe and Holt (Editors), Bergey's Manual of Systematic Bacteriology, Vol. 2. The Williams & Wilkins Co., Baltimore, pp. 1376–1379.

Caldwell, D.R. and M.P. Bryant. 1966. Medium without rumen fluid for nonselective enumeration and isolation of rumen bacteria. Appl. Microbiol. *14*: 794–801.

Callaway, T.R., K.A. Adams and J.B. Russell. 1999. The ability of "low G + C gram-positive" ruminal bacteria to resist monensin and counteract potassium depletion. Curr. Microbiol. *39*: 226–230.

Carlier, J.P., H. Marchandin, E. Jumas-Bilak, V. Lorin, C. Henry, C. Carriere and H. Jean-Pierre. 2002. *Anaeroglobus geminatus* gen. nov., sp. nov., a novel member of the family *Veillonellaceae*. Int. J. Syst. Evol. Microbiol. *52*: 983–986.

Carlier, J.P., G. K'Ouas, I. Bonne, A. Lozniewski and F. Mory. 2004. *Oribacterium sinus* gen. nov., sp. nov., within the family 'Lachnospiraceae' (phylum *Firmicutes*). Int. J. Syst. Evol. Microbiol. *54*: 1611–1615.

Carlier, J.P., G. K'Ouas and X.Y. Han. 2007. *Moryella indoligenes* gen. nov., sp. nov., an anaerobic bacterium isolated from clinical specimens. Int. J. Syst. Evol. Microbiol. *57*: 725–729.

Cato, E.P., W.L. George and S.M. Finegold. 1986. Genus *Clostridium* Prazmowski 1880, 23^AL^. *In* Sneath, Mair, Sharpe and Holt (Editors), Bergey's Manual of Systematic Bacteriology, Vol. 2. The Williams & Wilkins Co., Baltimore, pp. 1141–1200.

Cheng, K.-J. and J.W. Costerton. 1977. Ultrastructure of *Butyrivibrio fibrisolvens*: a gram-positive bacterium? J. Bacteriol. *129*: 1506–1512.

Cheng, K.J., G.A. Jones, F.J. Simpson and M.P. Bryant. 1969. Isolation and identification of rumen bacteria capable of anaerobic rutin degradation. Can. J. Microbiol. *15*: 1365–1371.

Cheng, K.J., D. Dinsdale and C.S. Stewart. 1979. Maceration of clover and grass leaves by *Lachnospira multiparus*. Appl. Environ. Microbiol. *38*: 723–729.

Clark, R.G., K.J. Cheng, L.B. Selinger and M.F. Hynes. 1994. A conjugative transfer system for the rumen bacterium, *Butyrivibrio fibrisolvens*, based on Tn*916*-mediated transfer of the *Staphylococcus aureus* plasmid pUB110. Plasmid *32*: 295–305.

Collins, M.D., P.A. Lawson, A. Willems, J.J. Cordoba, J. Fernández-Garayzábal, P. Garcia, J. Cai, H. Hippe and J.A.E. Farrow. 1994. The phylogeny of the genus *Clostridium*: proposal of five new genera and eleven new species combinations. Int. J. Syst. Bacteriol. *44*: 812–826.

Cornick, N.A., N.S. Jensen, D.A. Stahl, P.A. Hartman and M.J. Allison. 1994. *Lachnospira pectinoschiza* sp. nov. an anaerobic pectinophile from the pig intestine. Int. J. Syst. Bacteriol. *44*: 87–93.

Cotta, M.A. 1992. Interaction of ruminal bacteria in the production and utilization of maltooligosaccharides from starch. Appl. Environ. Microbiol. *58*: 48–54.

Cotta, M.A. 1993. Utilization of xylooligosaccharides by selected ruminal bacteria. Appl. Environ. Microbiol. *59*: 3557–3563.

Cotta, M.A. and R.L. Zeltwanger. 1995. Degradation and utilization of xylan by the ruminal bacteria *Butyrivibrio fibrisolvens* and *Selenomonas ruminantium*. Appl. Environ. Microbiol. *61*: 4396–4402.

Cotta, M.A., T.R. Whitehead and R.L. Zeltwanger. 2003. Isolation, characterization and comparison of bacteria from swine faeces and manure storage pits. Environ. Microbiol. *5*: 737–745.

Dalrymple, B.P., Y. Swadling, I. Layton, K.S. Gobius and G.P. Xue. 1999. Distribution and evolution of the xylanase genes *xynA* and *xynB* and their homologues in strains of *Butyrivibrio fibrisolvens*. Appl. Environ. Microbiol. *65*: 3660–3667.

Dehority, B.A. and J.A. Grubb. 1976. Basal medium for the selective enumeration of rumen bacteria utilizing specific energy sources. Appl. Environ. Microbiol. *32*: 703–710.

Dibbayawan, T., G. Cox, K.Y. Cho and D.M. Dwarte. 1985. Cell wall and plasma membrane architecture of *Butyrivibrio* spp. J. Ultrastruct. Res. *90*: 286–293.

Diez-Gonzalez, F., D.R. Bond, E. Jennings and J.B. Russell. 1999. Alternative schemes of butyrate production in *Butyrivibrio fibrisolvens* and their relationship to acetate utilization, lactate production, and phylogeny. Arch. Microbiol. *171*: 324–330.

Dore, J. and M.P. Bryant. 1989. Lipid growth requirement and influence of lipid supplement on fatty acid and aldehyde composition of *Syntrophococcus sucromutans*. Appl. Environ. Microbiol. *55*: 927–933.

Dore, J. and M.P. Bryant. 1990. Metabolism of one-carbon compounds by the ruminal acetogen *Syntrophococcus sucromutans*. Appl. Environ. Microbiol. *56*: 984–989.

Downes, J., M.A. Munson, D.R. Radford, D.A. Spratt and W.G. Wade. 2002. *Shuttleworthia satelles* gen. nov., sp. nov., isolated from the human oral cavity. Int. J. Syst. Evol. Microbiol. *52*: 1469–1475.

Duncan, S.H., G.L. Hold, A. Barcenilla, C.S. Stewart and H.J. Flint. 2002. *Roseburia intestinalis* sp. nov., a novel saccharolytic, butyrate-producing bacterium from human faeces. Int. J. Syst. Evol. Microbiol. *52*: 1615–1620.

El-Meadaway, A., Z. Mir, P. S. Mir, M. S. Zaman and L. J. Yanke. 1998. Relative efficacy of inocula for rumen fluid or faecal solution for determining in vitro digestibility and gas production. Can. J. Anim. Sci. *78*: 673–679.

Engelkirk, P.G., J. Duben-Engelkirk and V.R. Dowell. 1992. Principles and Practice of Clinical Anaerobic Bacteriology. Star Publishing Company, Belmont, Ca.

Ezaki, T., N. Li, Y. Hashimoto, H. Miura and H. Yamamoto. 1994. 16s Ribosomal DNA sequences of anaerobic cocci and proposal of *Ruminococcus hansenii* comb. nov. and *Ruminococcus productus* comb. nov. Int. J. Syst. Bacteriol. *44*: 130–136.

Farrow, J.A.E., P.A. Lawson, H. Hippe, U. Gauglitz and M.D. Collins. 1995. Phylogenetic evidence that the Gram-negative nonsporulating bacterium *Tissierella (Bacteroides) praeacuta* is a member of the *Clostridium* subphylum of the gram-positive bacteria and description of *Tissierella creatinini* sp. nov. Int. J. Syst. Bacteriol. *45*: 436–440.

Ferreira, F., M. Andersson, L. Kenne, M.A. Cotta and R.J. Stack. 1995. Structural studies of the extracellular polysaccharide from *Butyrivibrio fibrisolvens* strain 49. Carbohydr. Res. *278*: 143–153.

Ferreira, F., L. Kenne, M.A. Cotta and R.J. Stack. 1997. Structural studies of the extracellular polysaccharide from *Butyrivibrio fibrisolvens* strain CF3. Carbohydr. Res. *301*: 193–203.

Forsberg, C.W., E.E. Egbosimba and S. MacLellan. 1999. Recent advances in biotechnology of rumen bacteria - Review. Asian-Australas. J. Anim. Sci. *12*: 93–103.

Forster, R.J., R.M. Teather and J. Gong. 1996. 16S rDNA analysis of *Butyrivibrio fibrisolvens*: phylogenetic position and relation to butyrate-producing anaerobic bacteria from the rumen of white-tailed deer. Lett. Appl. Microbiol. *23*: 218–222.

Forster, R.J., J.H. Gong and R.M. Teather. 1997. Group-specific 16S rRNA hybridization probes for determinative and community structure studies of *Butyrivibrio fibrisolvens* in the rumen. Appl. Environ. Microbiol. *63*: 1256–1260.

Fulghum, R.S., B.B. Baldwin and P.P. Williams. 1968. Antibiotic susceptibility of anaerobic ruminal bacteria. Appl. Microbiol. *16*: 301–307.

Grech-Mora, I., M.L. Fardeau, B.K.C. Patel, B. Ollivier, A. Rimbault, G. Prensier, J.L. Garcia and E. Garnier-Sillam. 1996. Isolation and characterization of *Sporobacter termitidis* gen nov sp. nov., from the digestive tract of the wood-feeding termite *Nasutitermes lujae*. Int. J. Syst. Bacteriol. *46*: 512–518.

Greening, R.C. and J.A.Z. Leedle. 1989. Enrichment and isolation of *Acetitomaculum ruminis*, gen. nov., sp. nov., acetogenic bacteria from the bovine rumen. Arch. Microbiol. *151*: 399–406.

Greening, R.C. and J.A.Z. Leedle. 1995. *In* Validation of the publication of new names and new combinations previously effectively published outside the IJSB. List no. 55. Int. J. Syst. Bacteriol. *45*: 879–880.

Gregg, K. 1995. Engineering gut flora of ruminant livestock to reduce forage toxicity: progress and problems. Trends Biotechnol. *13*: 418–421.

Gregg, K., B. Hamdorf, K. Henderson, J. Kopečný and C. Wong. 1998. Genetically modified ruminal bacteria protect sheep from fluoroacetate poisoning. Appl. Environ. Microbiol. *64*: 3496–3498.

Harborne, J.B. 1980. Plant phenolics. *In* Bell and Charlwood (Editors), Secondary Plant Products. Springer-Verlag, New York, pp. 329–402.

Hartmann, L., D. Taras, B. Kamlage and M. Blaut. 2000. A new technique to determine hydrogen excreted by gnotobiotic rats. Lab. Anim. *34*: 162–170.

Hazlewood, G. and R.M. Dawson. 1979. Characteristics of a lipolytic and fatty acid-requiring *Butyrivibrio* sp. isolated from the ovine rumen. J. Gen. Microbiol. *112*: 15–27.

Hazlewood, G., M.K. Theodorou, A. Hutchings, D.J. Jordan and G. Galfre. 1986. Preparation and characterization of monoclonal antibodies to a *Butyrivibrio* sp. and their potential use in the identification of rumen butyrivibrios, using an enzyme-linked immunosorbent assay. J. Gen. Microbiol. *132*: 43–52.

Hefford, M.A., Y. Kobayashi, S.E. Allard, R.J. Forster and R.M. Teather. 1997. Sequence analysis and characterization of pOM1, a small cryptic plasmid from *Butyrivibrio fibrisolvens*, and its use in construction of a new family of cloning vectors for butyrivibrios. Appl. Environ. Microbiol. *63*: 1701–1711.

Hespell, R.B. and E. Canale-Parola. 1970. *Spirochaeta litoralis* sp.n., a strictly anaerobic marine spirochete. Arch. Mikrobiol. *74*: 1–18.

Hespell, R.B., R. Wolf and R.J. Bothast. 1987. Fermentation of xylans by *Butyrivibrio fibrisolvens* and other ruminal bacteria. Appl. Environ. Microbiol. *53*: 2849–2853.

Hespell, R.B. and T.R. Whitehead. 1990. Physiology and genetics of xylan degradation by gastrointestinal tract bacteria. J. Dairy Sci. *73*: 3013–3022.

Hespell, R.B. 1991. The genera *Butyrivibrio, Lachnospira,* and *Roseburia. In* Balows, Truper, Dworkin, Harder and Schleifer (Editors), The Prokaryotes: A Handbook on the Biology of Bacteria: Ecophysiology, Isolation, Identification and Applications. Springer-Verlag, New York, pp. 2022–2033.

Hespell, R.B. 1992. Fermentation of xylans by *Butyrivibrio fibrisolvens* and *Thermoanaerobacter* strain B6A: utilization of uronic acids and xylanolytic activities. Curr. Microbiol. *25*: 189–195.

Hespell, R.B., K. Kato and J.W. Costerton. 1993. Characterization of the cell wall of *Butyrivibrio* species. Can. J. Microbiol. *39*: 912–921.

Hespell, R.B. and M.A. Cotta. 1995. Degradation and utilization by *Butyrivibrio fibrisolvens* H17c of xylans with different chemical and physical properties. Appl. Environ. Microbiol. *61*: 3042–3050.

Hewitt, M.J., A.J. Wicken, K.W. Knox and M.E. Sharpe. 1976. Isolation of lipoteichoic acids from *Butyrivibrio fibrisolvens*. J. Gen. Microbiol. *94*: 126–130.

Hold, G.L., S.E. Pryde, V.J. Russell, E. Furnie and H.J. Flint. 2002. Assessment of microbial diversity in human colonic samples by 16S rDNA sequence analysis. FEMS Microbiol. Ecol. *39*: 33–39.

Holdeman, L.V. and W.E.C. Moore. 1974. New genus, *Coprococcus*, twelve new species, and emended descriptions of four previously described species of bacteria from human feces. Int. J. Syst. Bacteriol. *24*: 260–277.

Holdeman, L.V., E.P. Cato and W.E.C. Moore. 1977a. Culture methods: Use of pre-reduced media. *In* Holdeman, Cato and Moore (Editors), Anaerobic Laboratory Manual, 4th edn. Anaerobe Laboratory, Virginia Polytechnic Institute and State University, Blacksburg, pp. 117–149.

Holdeman, L.V., E.P. Cato and W.E.C. Moore (Editors). 1977b. Anaerobe Laboratory Manual, 4th edn. Anaerobe Laboratory, Virginia Polytechnic Institute and State University, Blacksburg, VA.

Holdeman, L.V., R.W. Kelley and W.E.C. Moore. 1986. Family I. *Bacteroidaceae* Pribram 1933, 10^AL^. *In* Sneath, Mair, Sharpe and Holt (Editors), Bergey's Manual of Systematic Bacteriology, Vol. 2. The Williams & WIlkins Co., Baltimore, pp. 602–603.

Hudman, J.F. and A.R. Glenn. 1985. Selenium Uptake by *Butyrivibrio fibrisolvens* and *Bacteroides ruminicola*. FEMS Microbiol. Lett. *27*: 215–220.

Hungate, R.E. 1966. The Rumen and Its Microbes. Academic Press, New York.

Hungate, R.E. 1969. A roll tube method for cultivation of strict anaerobes. *In* Norris and Ribbons (Editors), Methods in Microbiology, Vol. 3B. Academic Press, London and New York, pp. 117–132.

Ifkovits, R.W. and H.S. Ragheb. 1968. Cellular fatty acid composition and identification of rumen bacteria. Appl. Microbiol. *16*: 1406–1413.

Jensen, N.S. and E. Canale-Parola. 1986. *Bacteroides pectinophilus* sp. nov. and *Bacteroides galacturonicus* sp. nov., two pectinolytic bacteria from the human intestinal tract. Appl. Environ. Microbiol. *52*: 880–887.

Jones, G.A., T.A. Mcallister, A.D. Muir and K.J. Cheng. 1994. Effects of sainfoin (*Onobrychis viciifolia* Scop.) condensed tannins on growth and proteolysis by four strains of ruminal bacteria. Appl. Environ. Microbiol. *60*: 1374–1378.

Jousimies-Somer, H.R., P. Summanen, D.M. Citron, E.J. Baron, H.M. Wexler and S.M. Finegold. 2002. Wadsworth-KTL anaerobic bacteriology manual. Star Publishing Company, Belmont, CA.

Kageyama, A., Y. Benno and T. Nakase. 1999. Phylogenetic and phenotypic evidence for the transfer of *Eubacterium aerofaciens* to the genus *Collinsella* as *Collinsella aerofaciens* gen. nov., comb. nov. Int. J. Syst. Bacteriol. *49*: 557–565.

Kalmokoff, M., F. Bartlett and R.M. Teather. 1996. Are ruminal bacteria armed with bacteriocins? J. Dairy Sci. *79*: 2297–2306.

Kalmokoff, M.L. and R.M. Teather. 1997. Isolation and characterization of a bacteriocin (Butyrivibriocin AR10) from the ruminal anaerobe *Butyrivibrio fibrisolvens* AR10: evidence in support of the widespread occurrence of bacteriocin-like activity among ruminal isolates of *B. fibrisolvens*. Appl. Environ. Microbiol. *63*: 394–402.

Kalmokoff, M.L., D. Lu, M.F. Whitford and R.M. Teather. 1999. Evidence for production of a new lantibiotic (butyrivibriocin OR79A) by the ruminal anaerobe *Butyrivibrio fibrisolvens* OR79: characterization of the structural gene encoding butyrivibriocin OR79A. Appl. Environ. Microbiol. *65*: 2128–2135.

Kalmokoff, M.L., S. Allard, J.W. Austin, M.F. Whitford, M.A. Hefford and R.M. Teather. 2000. Biochemical and genetic characterization of the flagellar filaments from the rumen anaerobe *Butyrivibrio fibrisolvens* OR77. Anaerobe *6*: 93–109.

Kamlage, B., B. Gruhl and M. Blaut. 1997. Isolation and characterization of two new homoacetogenic hydrogen-utilizing bacteria from the human intestinal tract that are closely related to *Clostridium coccoides*. Appl. Environ. Microbiol. *63*: 1732–1738.

Kasperowicz, A. 1994. Comparison of utilization of pectins from various sources by pure cultures of pectinolytic rumen bacteria and mixed cultures of rumen microorganisms. Acta Microbiol. Pol. *43*: 47–56.

Kobayashi, Y., N. Okuda, M. Matsumoto, K. Inoue, M. Wakita and S. Hoshino. 1998. Constitutive expression of a heterologous *Eubacterium ruminantium* xylanase gene (*xynA*) in *Butyrivibrio fibrisolvens*. FEMS Microbiol. Lett. *163*: 11–17.

Kobayashi, Y. and R. Onodera. 1999. Application of molecular biology to rumen microbes - Review. Asian-Australas. J. Anim. Sci. *12*: 77–83.

Kobayashi, Y., R.J. Forster and R.M. Teather. 2000. Development of a competitive polymerase chain reaction assay for the ruminal bacterium *Butyrivibrio fibrisolvens* OB156 and its use for tracking an OB156-derived recombinant. FEMS Microbiol. Lett. *188*: 185–190.

Kopečný, J., M. Zorec, J. Mrázek, Y. Kobayashi and R. Marinšek-Logar. 2003. *Butyrivibrio hungatei* sp. nov. and *Pseudobutyrivibrio xylanivorans* sp. nov., butyrate-producing bacteria from the rumen. Int. J. Syst. Evol. Microbiol. *53*: 201–209.

Krause, D.O., R.J. Bunch, B.D. Dalrymple, K.S. Gobius, W.J. Smith, G.P. Xue and C.S. McSweeney. 2001. Expression of a modified *Neocallimastix patriciarum* xylanase in *Butyrivibrio fibrisolvens* digests more fibre but cannot effectively compete with highly fibrolytic bacteria in the rumen. J. Appl. Microbiol. *90*: 388–396.

Krumholz, L.R. and M.P. Bryant. 1986a. *Syntrophococcus sucromutans* sp. nov. gen. nov. uses carbohydrates as electron donors and formate, methoxymonobenzenoids or *Methanobrevibacter* as electron acceptor systems. Arch. Microbiol. *143*: 313–318.

Krumholz, L.R. and M.P. Bryant. 1986b. *Eubacterium oxidoreducens* sp. nov. requiring H₂ or formate to degrade gallate, pyrogallol, phloroglucinol and quercetin. Arch. Microbiol. *144*: 8–14.

Krumholz, L.R. and M.P. Bryant. 1986c. *In* Validation of the publication of new names and new combinations previously effectively published outside the IJSB. List no. 21. Int. J. Syst. Bacteriol. *36*: 489.

Leedle, J.A. and R.B. Hespell. 1980. Differential carbohydrate media and anaerobic replica plating techniques in delineating carbohydrate-utilizing subgroups in rumen bacterial populations. Appl. Environ. Microbiol. *39*: 709–719.

Leser, T.D., J.Z. Amenuvor, T.K. Jensen, R.H. Lindecrona, M. Boye and K. Moller. 2002. Culture-independent analysis of gut bacteria: the pig gastrointestinal tract microbiota revisited. Appl. Environ. Microbiol. *68*: 673–690.

Lester, T.D., J. Z. Amenuvor, T. K. Jensen, R. H. Lindecrona, M. Boye and K. Moller. 2002. Culture-independent analysis of gut bacteria: the pig gastrointestinal tract microbiota revisted. Appl. Environ. Microbiol. *68*: 673–690.

Liesack, W., F. Bak, J.U. Kreft and E. Stackebrandt. 1994. *Holophaga foetida* gen. nov., sp. nov., a new, homoacetogenic bacterium degrading methoxylated aromatic compounds. Arch. Microbiol. *162*: 85–90.

Lin, L.L. and J.A. Thomson. 1991. An analysis of the extracellular xylanases and cellulases of *Butyrivibrio fibrisolvens* H17c. FEMS Microbiol. Lett. *68*: 197–203.

Lomans, B.P., P. Leijdekkers, J.J. Wesselink, P. Bakkes, A. Pol, C. van der Drift and H.J.M. op den Camp. 2001. Obligate sulfide-dependent degradation of methoxylated aromatic compounds and formation of methanethiol and dimethyl sulfide by a freshwater sediment isolate, *Parasporobacterium paucivorans* gen. nov., sp. nov. Appl. Environ. Microbiol. *67*: 4017–4023.

Lomans, B.P., Leijdekkers P., Wesselink J.J., Bakkes P., Pol A., van der Drift C. and Op den Camp H.J.M. 2004. *In* Validation of publication of new names and new combinations previously effectively published outside the IJSEM. List no. 96. Int. J. Syst. Evol. Microbiol. *54*: 307–308.

Ludwig, W., K.-H. Schleifer and W. B. Whitman. 2009. Revised road map to the phylum *Firmicutes*. *In* De Vos, Garrity, Jones, Krieg, Ludwig, Rainey, Schleifer and Whitman (Editors), Bergey's Manual of Systematic Bacteriology, 2nd edn, Vol. 3, The *Firmicutes*. Spinger, New York, pp. 1–14.

Macy, J.M., J.E. Snellen and R.E. Hungate. 1972. Use of syringe methods for anaerobiosis. Am. J. Clin. Nutr. *25*: 1318–1323.

Mann, S.P., G.P. Hazlewood and C.G. Orpin. 1986. Characterization of a cryptic plasmid (pOM1) in *Butyrivibrio fibrisolvens* by restriction endonuclease analysis and its cloning in *Escherichia coli*. Curr. Microbiol. *13*: 17–22.

Mannarelli, B.M. 1988. Deoxyribonucleic acid relatedness among strains of the species *Butyrivibrio fibrisolvens*. Int. J. Syst. Bacteriol. *38*: 340–347.

Mannarelli, B.M., S. Evans and D. Lee. 1990a. Cloning, sequencing, and expression of a xylanase gene from the anaerobic ruminal bacterium *Butyrivibrio fibrisolvens*. J. Bacteriol. *172*: 4247–4254.

Mannarelli, B.M., R.J. Stack, D. Lee and L. Ericsson. 1990b. Taxonomic relatedness of *Butyrivibrio*, *Lachnospira*, *Roseburia*, and *Eubacterium* species as determined by DNA hybridization and extracellular-polysaccharide analysis. Int. J. Syst. Bacteriol. *40*: 370–378.

Mannarelli, B.M., L.D. Ericsson, D. Lee and R.J. Stack. 1991. Taxonomic relationships among strains of the anaerobic bacterium *Bacteroides ruminicola* determined by DNA and extracellular polysaccharide analysis. Appl. Environ. Microbiol. *57*: 2975–2980.

Marinšek-Logar, R., M. Zorec and J. Kopečný. 2001. Reliable identification of *Prevotella* and *Butyrivibrio* spp. from rumen by fatty acid methyl ester profiles. Folia Microbiol. *46*: 57–59.

Marounek, M. and J. Kopečný. 1994. Utilization of glucose and xylose in ruminal strains of *Butyrivibrio fibrisolvens*. Appl. Environ. Microbiol. *60*: 738–739.

Marounek, M. and O.G. Savka. 1994. Antimicrobial susceptibility of ruminal strains of *Butyrivibrio fibrisolvens*. Acta Vet. Brno *63*: 129–132.

Marounek, M. and O. Petr. 1995. Fermentation of glucose and xylose in ruminal strains of *Butyrivibrio fibrisolvens*. Lett. Appl. Microbiol. *21*: 272–276.

Marounek, M. and D. Duskova. 1999. Metabolism of pectin in rumen bacteria *Butyrivibrio fibrisolvens* and *Prevotella ruminicola*. Lett. Appl. Microbiol. *29*: 429–433.

Martin, J.H. and D.C. Savage. 1988. Cloning, nucleotide sequence, and taxonomic implications of the flagellin gene of *Roseburia cecicola*. J. Bacteriol. *170*: 2612–2617.

McSweeney, C.S., A. Dulieu and R. Bunch. 1998. *Butyrivibrio* spp. and other xylanolytic microorganisms from the rumen have cinnamoyl esterase activity. Anaerobe *4*: 57–65.

Mechichi, T., M. Labat, J.L. Garcia, P. Thomas and B.K.C. Patel. 1999. *Sporobacterium olearium* gen. nov., sp. nov., a new methanethiol-producing bacterium that degrades aromatic compounds, isolated from an olive mill wastewater treatment digester. Int. J. Syst. Bacteriol. *49*: 1741–1748.

Melville, C.M., K.P. Scott, D.K. Mercer and H.J. Flint. 2001. Novel tetracycline resistance gene, *tet(32)*, in the *Clostridium*-related human colonic anaerobe K10 and its transmission in vitro to the rumen anaerobe *Butyrivibrio fibrisolvens*. Antimicrob. Agents Chemother. *45*: 3246–3249.

Mesbah, M. and W.B. Whitman. 1989. Measurement of deoxyguanosine/thymidine ratios in complex mixtures by high-performance liquid chromatography for determination of the mole percentage guanine + cytosine of DNA. J. Chromatogr. *479*: 297–306.

Miller, T.L. and M.J. Wolin. 1974. A serum bottle modification of the Hungate technique for cultivating obligate anaerobes. Appl. Microbiol. *27*: 985–987.

Miron, J. 1991. The hydrolysis of lucerne cell-wall monosaccharide components by monocultures or pair combinations of defined ruminal bacteria. J. Appl. Bacteriol. *70*: 245–252.

Miron, J. and D. Benghedalia. 1992. The degradation and utilization of wheat straw cell wall monosaccharide components by defined ruminal cellulolytic bacteria. Appl. Microbiol. Biotechnol. *38*: 432–437.

Miron, J. and D. Benghedalia. 1993a. Digestion of cell-wall monosaccharides of ryegrass and alfalfa hays by the ruminal bacteria *Fibrobacter succinogenes* and *Butyrivibrio fibrisolvens*. Can. J. Microbiol. *39*: 780–786.

Miron, J. and D. Benghedalia. 1993b. Untreated and delignified cotton stalks as model substrates for degradation and utilization of cell-wall monosaccharide components by defined ruminal cellulolytic bacteria. Bioresource Technol. *43*: 241–247.

Miron, J. and D. Benghedalia. 1993c. Digestion of structural polysaccharides of panicum and vetch hays by the rumen bacterial strains *Fibrobacter succinogenes* BL2 and *Butyrivibrio fibrisolvens* D1. Appl. Microbiol. Biotechnol. *39*: 756–759.

Miron, J., S.H. Duncan and C.S. Stewart. 1994. Interactions between rumen bacterial strains during the degradation and utilization of the monosaccharides of barley straw cell-walls. J. Appl. Bacteriol. *76*: 282–287.

Mitsumori, M. and H. Minato. 1995. Distribution of cellulose-binding proteins among the representative strains of rumen bacteria. J. Gen. Appl. Microbiol. *41*: 297–306.

Miyagawa, E. 1982. Cellular fatty acid and fatty aldehyde composition of rumen bacteria. J. Gen. Appl. Microbiol. *28*: 389–408.

Mohan, R., P. Namsolleck, P.A. Lawson, M. Osterhoff, M.D. Collins, C.-A. Alpert and M. Blaut. 2006. *Clostridium asparagiforme* sp. nov., isolated from a human faecal sample. Syst. Appl. Microbiol *29*: 292–299.

Moore, L.V.H., D.M. Bourne and W.E.C. Moore. 1994. Comparative distribution and taxonomic value of cellular fatty acids in 33 genera of anaerobic gram-negative bacilli. Int. J. Syst. Bacteriol. *44*: 338–347.

Moore, L.V.H. and W.E.C. Moore. 1994. *Oribaculum catoniae* gen. nov., sp. nov., *Catonella morbi* gen. nov., sp. nov., *Hallella seregens* gen. nov., sp. nov., *Johnsonella ignava* gen. nov., sp. nov., and *Dialister pneumosintes* gen. nov., comb. nov., nom. rev., anaerobic Gram-negative bacilli from the human gingival crevice. Int. J. Syst. Bacteriol. *44*: 187–192.

Moore, W.E. and L.V. Holdeman. 1974. Human fecal flora: the normal flora of 20 Japanese-Hawaiians. Appl. Microbiol. *27*: 961–979.

Moore, W.E., L.V. Holdeman, R.M. Smibert, I.J. Good, J.A. Burmeister, K.G. Palcanis and R.R. Ranney. 1982a. Bacteriology of experimental gingivitis in young adult humans. Infect. Immun. *38*: 651–667.

Moore, W.E., L.V. Holdeman, R.M. Smibert, D.E. Hash, J.A. Burmeister and R.R. Ranney. 1982b. Bacteriology of severe periodontitis in young adult humans. Infect. Immun. *38*: 1137–1148.

Moore, W.E., L.V. Holdeman, R.M. Smibert, E.P. Cato, J.A. Burmeister, K.G. Palcanis and R.R. Ranney. 1984. Bacteriology of experimental gingivitis in children. Infect. Immun. *46*: 1–6.

Moore, W.E., L.V. Holdeman, E.P. Cato, R.M. Smibert, J.A. Burmeister, K.G. Palcanis and R.R. Ranney. 1985. Comparative bacteriology of juvenile periodontitis. Infect. Immun. *48*: 507–519.

Moore, W.E.C., J.L. Johnson and L.V. Holdeman. 1976. Emendation of *Bacteroidaceae* and *Butyrivibrio* and descriptions of *Desulfomonas* gen. nov. and ten new species in genera *Desulfomonas, Butyrivibrio, Eubacterium, Clostridium*, and *Ruminococcus*. Int. J. Syst. Bacteriol. *26*: 238–252.

Moore, W.E.C. and L.V. Holdeman Moore. 1986. Genus *Eubacterium* Prévot 1938. *In* Sneath, Mair, Sharpe and Holt (Editors), Bergey's Manual of Systematic Bacteriology, Vol. 2. The Williams & Wilkins Co., Baltimore, pp. 1353–1373.

Mrázek, J. and J. Kopečný. 2001. Development of competitive PCR for detection of *Butyrivibrio fibrisolvens* in the rumen. Folia Microbiol. (Praha) *46*: 63–65.

Munson, M.A., Pitt-Ford, T., B. Chong, A. Weightman and W.G. Wade. 2002. Molecular and cultural analysis of the microflora associated with endodontic infections. J. Dent. Res. *81*: 761–766.

Namsolleck, P., R. Thiel, P. Lawson, K. Holmstrøm, M. Rajilic, E. E. Vaughan, L. Rigottier-Gois, M. D. Collins, W. De Vos and M. Blaut. 2004. Molecular methods for the analysis of gut microbiota. Microb. Ecol. Health Dis. *19*: 71–85.

Nili, N. and J.D. Brooker. 1995. A defined medium for rumen bacteria and identification of strains impaired in de novo biosynthesis of certain amino acids. Lett. Appl. Microbiol. *21*: 69–74.

Nili, N. and J.D. Brooker. 1997. Lack of methionine biosynthesis de novo in *Butyrivibrio fibrisolvens* strain E14. Lett. Appl. Microbiol. *25*: 85–90.

Ogata, K., R.I. Aminov, T. Nagamine, Y. Benno, T. Sekizaki, M. Mitsumori, H. Minato and H. Itabashi. 1996. Structural organization of pRAM4, a cryptic plasmid from *Prevotella ruminicola*. Plasmid *35*: 91–97.

Paster, B.J., S.K. Boches, J.L. Galvin, R.E. Ericson, C.N. Lau, V.A. Levanos, A. Sahasrabudhe and F.E. Dewhirst. 2001. Bacterial diversity in human subgingival plaque. J. Bacteriol. *183*: 3770–3783.

Preston, J.F., J.D. Rice, M.C. Chow and B.J. Brown. 1991. Kinetic comparisons of trimer-generating pectate and alginate lyases by reversed-phase ion-pair liquid chromotography. Carbohydr. Res. *215*: 147–157.

Rainey, F.A. and P.H. Janssen. 1995. Phylogenetic analysis by 16S ribosomal DNA sequence comparison reveals two unrelated groups of species within the genus *Ruminococcus*. FEMS Microbiol. Lett. *129*: 69–73.

Rieu-Lesme, F., B. Morvan, M.D. Collins, G. Fonty and A. Willems. 1996. A new H_2/CO_2-using acetogenic bacterium from the rumen: description of *Ruminococcus schinkii* sp. nov. FEMS Microbiol. Lett. *140*: 281–286.

Rumbak, E., D.E. Rawlings, G.G. Lindsey and D.R. Woods. 1991. Cloning, nucleotide-sequence, and enzymatic characterization of an alpha-amylase from the ruminal bacterium *Butyrivibrio fibrisolvens* H17c. J. Bacteriol. *173*: 4203–4211.

Rychlik, J.L. and J.B. Russell. 2002. Bacteriocin-like activity of *Butyrivibrio fibrisolvens* JL5 and its effect on other ruminal bacteria and ammonia production. Appl. Environ. Microbiol. *68*: 1040–1046.

Sakamoto, M., I.N. Rôças, J.F. Siqueira Jr and Y. Benno. 2006. Molecular analysis of bacteria in asymptomatic and symptomatic endodontic infections. Oral Microbiol. Immunol. *21*: 112–122.

Schink, B. and N. Pfennig. 1982. Fermentation of trihydroxybenzenes by *Pelobacter acidigallici* gen. nov. sp. nov. a new strictly anaerobic, non-sporeforming Bacterium. Arch. Microbiol. *133*: 195–201.

Schink, B. 1997. Energetics of syntrophic cooperation in methanogenic degradation. Microbiol. Mol. Biol. Rev. *61*: 262–280.

Schleifer, K.H. and O. Kandler. 1972. Peptidoglycan types of bacterial cell walls and their taxonomic implications. Bacteriol. Rev. *36*: 407–477.

Schnell, S., A. Brune and B. Schink. 1991. Degradation of hydroxyhydroquinone by the strictly anaerobic fermenting bacterium *Pelobacter massiliensis* sp. nov. Arch. Microbiol. *155*: 511–516.

Schwiertz, A., G. Le Blay and M. Blaut. 2000. Quantification of different *Eubacterium* spp. in human fecal samples with species-specific 16S rRNA-targeted oligonucleotide probes. Appl. Environ. Microbiol. *66*: 375–382.

Schwiertz, A., G.L.Hold, S.H. Duncan, B. Gruhl, M.D. Collins, P.A. Lawson, H.J. Flint and M. Blaut. 2002a. *In* Validation of publication of new names and new combinations previously effectively published outside the IJSEM. List no. 87 Int. J. Syst. Evol. Microbiol. *52*: 1437–1438.

Schwiertz, A., G.L. Hold, S.H. Duncan, B. Gruhl, M.D. Collins, P.A. Lawson, H.J. Flint and M. Blaut. 2002b. *Anaerostipes caccae* gen. nov., sp. nov., a new saccharolytic, acetate-utilising, butyrate-producing bacterium from human faeces. Syst. Appl. Microbiol. *25*: 46–51.

Scott, K.P., T.M. Barbosa, K.J. Forbes and H.J. Flint. 1997. High-frequency transfer of a naturally occurring chromosomal tetracycline resistance element in the ruminal anaerobe *Butyrivibrio fibrisolvens*. Appl. Environ. Microbiol. *63*: 3405–3411.

Sewell, G.W., E.A. Utt, R.B. Hespell, K.F. Mackenzie and L.O. Ingram. 1989. Identification of the *Butyrivibrio fibrisolvens* xylosidase gene (*xylB*) coding region and its expression in *Escherichia coli*. Appl. Environ. Microbiol. *55*: 306–311.

Shane, B.S., L. Gouws and A. Kitner. 1969. Cellulolytic bacteria occurring in the rumen of sheep conditioned to low-protein teff hay. J. Gen. Microbiol. *55*: 445–457.

Silley, P. 1985. A note on the pectinolytic enzymes of *Lachnospira multiparus*. J. Appl. Bacteriol. *58*: 145–149.

Skerman, V.B.D., V. McGowan and P.H.A. Sneath. 1980. Approved lists of bacterial names. Int. J. Syst. Bacteriol. *30*: 225–420.

Song, Y., C. Liu, D.R. Molitoris, T.J. Tomzynski, P.A. Lawson, M.D. Collins and S.M. Finegold. 2003. *Clostridium bolteae* sp. nov., isolated from human sources. Syst. Appl. Microbiol. *26*: 84–89.

Stack, R.J. 1987. Identification of L-altrose in the extracellular polysaccharide from *Butyrivibrio fibrisolvens* strain Cf3. FEMS Microbiol. Lett. *48*: 83–87.

Stack, R.J. 1988. Neutral sugar composition of extracellular polysaccharides produced by strains of *Butyrivibrio fibrisolvens*. Appl. Environ. Microbiol. *54*: 878–883.

Stack, R.J., R.D. Plattner and G.L. Cote. 1988a. Identification of L-iduronic acid as a constituent of the major extracellular polysaccharide produced by *Butyrivibrio fibrisolvens* strain-X6c61. FEMS Microbiol. Lett. *51*: 1–5.

Stack, R.J., T.M. Stein and R.D. Plattner. 1988b. 4-*O*-(1-Carboxyethyl)-D-galactose - a new acidic sugar from the extracellular polysaccharide produced by *Butyrivibrio fibrisolvens* strain 49. Biochem. J. *256*: 769–773.

Stack, R.J. 1989. Identification of 6-deoxy-D-talose in an extracellular polysaccharide produced by *Butyrivibrio fibrisolvens* strain-X6C61. Carbohydr. Res. *189*: 281–288.

Stack, R.J. and D. Weisleder. 1990. 4-*O*-(1-Carboxyethyl)-L-rhamnose, a second unique acidic sugar found in an extracellular polysaccharide from *Butyrivibrio fibrisolvens* strain 49. Biochem. J. *268*: 281–285.

Stackebrandt, E., I. Kramer, J. Swiderski and H. Hippe. 1999. Phylogenetic basis for a taxonomic dissection of the genus *Clostridium*. FEMS Immunol. Med. Microbiol. *24*: 253–258.

Stanton, T.B. and D.C. Savage. 1983a. *Roseburia cecicola* gen. nov., sp. nov., a motile, obligately anaerobic bacterium from a mouse cecum. Int. J. Syst. Bacteriol. *33*: 618–627.

Stanton, T.B. and D.C. Savage. 1983b. Colonization of gnotobiotic mice by *Roseburia cecicola*, a motile, obligately anaerobic bacterium from murine ceca. Appl. Environ. Microbiol. *45*: 1677–1684.

Stanton, T.B. and D.C. Savage. 1984. Motility as a factor in bowel colonization by *Roseburia cecicola*, an obligately anaerobic bacterium from the mouse caecum. J. Gen. Microbiol. *130*: 173–183.

Strobel, H.J. and K.A. Dawson. 1993. Xylose and arabinose utilization by the rumen bacterium *Butyrivibrio fibrisolvens*. FEMS Microbiol. Lett. *113*: 291–296.

Strobel, H.J. 1994. Pentose transport by the ruminal bacterium *Butyrivibrio fibrisolvens*. FEMS Microbiol. Lett. *122*: 217–222.

Suau, A., R. Bonnet, M. Sutren, J.J. Godon, G.R. Gibson, M.D. Collins and J. Doré. 1999. Direct analysis of genes encoding 16S rRNA from complex communities reveals many novel molecular species within the human gut. Appl. Environ. Microbiol. *65*: 4799–4807.

Taras, D., R. Simmering, M.D. Collins, P.A. Lawson and M. Blaut. 2002. Reclassification of *Eubacterium formicigenerans* Holdeman and Moore 1974 as *Dorea formicigenerans* gen. nov., comb. nov., and description of *Dorea longicatena* sp. nov., isolated from human faeces. Int. J. Syst. Evol. Microbiol. *52*: 423–428.

Tarvin, D. and A.M. Buswell. 1934. The methane fermentation of organic acids and carbohydrates. J. Am. Chem. Soc. *56*: 1751–1755.

Teather, R.M. 1982a. Maintenance of laboratory strains of obligately anaerobic rumen bacteria. Appl. Environ. Microbiol. *44*: 499–501.

Teather, R.M. 1982b. Isolation of plasmid DNA from *Butyrivibrio fibrisolvens*. Appl. Environ. Microbiol. *43*: 298–302.

Tsai, C.G., D.M. Gates, W.M. Ingledew and G.A. Jones. 1976. Products of anaerobic phloroglucinol degradation by *Coprococcus* sp. Pe15. Can. J. Microbiol. *22*: 159–164.

Utt, E.A., C.K. Eddy, K.F. Keshav and L.O. Ingram. 1991. Sequencing and expression of the *Butyrivibrio fibrisolvens xylB* gene encoding a novel bifunctional protein with beta-D-xylosidase and alpha-L-arabinofuranosidase activities. Appl. Environ. Microbiol. *57*: 1227–1234.

van Gylswyk, N.O., H. Hippe and F.A. Rainey. 1996. *Pseudobutyrivibrio ruminis* gen. nov., sp. nov., a butyrate-producing bacterium from the rumen that closely resembles *Butyrivibrio fibrisolvens* in phenotype. Int. J. Syst. Bacteriol. *46*: 559–563.

Wejdemar, K. 1996. Some factors stimulating the growth of *Butyrivibrio fibrisolvens* TC33 in clarified rumen fluid. Swed. J. Agric. Res. *26*: 11–18.

Whitehead, T.R. 1992. Genetic transformation of the ruminal bacteria *Butyrivibrio fibrisolvens* and *Streptococcus bovis* by electroporation. Lett. Appl. Microbiol. *15*: 186–189.

Whitehead, T.R. and H.J. Flint. 1995. Heterologous expression of an endoglucanase gene (*endA*) from the ruminal anaerobe *Ruminococcus flavefaciens* 17 in *Streptococcus bovis* and *Streptococcus sanguis*. FEMS Microbiol. Lett. *126*: 165–169.

Whitehead, T.R., M.A. Cotta, M.D. Collins and P.A. Lawson. 2004. *Hespellia stercorisuis* gen. nov., sp. nov. and *Hespellia porcina* sp. nov., isolated from swine manure storage pits. Int. J. Syst. Evol. Microbiol. *54*: 241–245.

Whitford, M.F., L.J. Yanke, R.J. Forster and R.M. Teather. 2001. *Lachnobacterium bovis* gen. nov., sp. nov., a novel bacterium isolated from the rumen and faeces of cattle. Int. J. Sys. Evol. Microbiol. *51*: 1977–1981.

Widdel, F. and N. Pfennig. 1981. Studies on dissimilatory sulfate-reducing bacteria that decompose fatty acids. 1. Isolation of new sulfate-reducing bacteria enriched with acetate from saline environments: description of *Desulfobacter postgatei* gen. nov., sp. nov. Arch. Microbiol. *129*: 395–400.

Widdel, F., G.W. Kohring and F. Mayer. 1983. Studies on dissimilatory sulfate-reducing bacteria that decompose fatty acids. 3. Characterization of the filamentous gliding *Desulfonema limicola* gen. nov. sp. nov., and *Desulfonema magnum* sp. nov. Arch. Microbiol. *134*: 286–294.

Wilkins, T.D. and S. Chalgren. 1976. Medium for use in antibiotic susceptibility testing of anaerobic bacteria. Antimicrob. Agents Chemother. *10*: 926–928.

Willems, A. and M.D. Collins. 1995a. Phylogenetic analysis of *Ruminococcus flavefaciens*, the type species of the genus *Ruminococcus*, does not support the reclassification of *Streptococcus hansenii* and *Peptostreptococcus productus* as ruminococci. Int. J. Syst. Bacteriol. *45*: 572–575.

Willems, A. and M.D. Collins. 1995b. Evidence for the placement of the Gram-negative *Catonella morbi* (Moore and Moore) and *Johnsonella ignava* (Moore and Moore) within the *Clostridium* subphylum of the Gram-positive bacteria on the basis of 16S rRNA sequences. Int. J. Syst. Bacteriol. *45*: 855–857.

Willems, A., M. Amat-Marco and M.D. Collins. 1996. Phylogenetic analysis of *Butyrivibrio* strains reveals three distinct groups of species within the *Clostridium* subphylum of the gram-positive bacteria. Int. J. Syst. Bacteriol. *46*: 195–199.

Willems, A. and M.D. Collins. 1996. Phylogenetic relationships of the genera *Acetobacterium* and *Eubacterium sensu stricto* and reclassification of *Eubacterium alactolyticum* as *Pseudoramibacter alactolyticus* gen. nov., comb. nov. Int. J. Syst. Bacteriol. *46*: 1083–1087.

Williams, A.G. and S.E. Withers. 1992a. The regulation of xylanolytic enzyme formation by *Butyrivibrio fibrisolvens* NCFB 2249. Lett. Appl. Microbiol. *14*: 194–198.

Williams, A.G. and S.E. Withers. 1992b. Induction of xylan-degrading enzymes in *Butyrivibrio fibrisolvens*. Curr. Microbiol. *25*: 297–303.

Wilson, K.H. and R.B. Blitchington. 1996. Human colonic biota studied by ribosomal DNA sequence analysis. Appl. Environ. Microbiol. *62*: 2273–2278.

Wojciechowicz, M., K. Heinrichova and A. Ziolecki. 1980. A polygalacturonate lyase produced by *Lachnospira multiparus* isolated from the bovine rumen. J. Gen. Microbiol. *117*: 193–199.

Wolin, E.A., R.S. Wolfe and M.J. Wolin. 1964. Viologen dye inhibition of methane formation by *Methanobacillus omelianskii*. J. Bacteriol. *87*: 993–998.

Wolin, M.J., T.L. Miller, M.D. Collins and P.A. Lawson. 2003. Formate-dependent growth and homoacetogenic fermentation by a bacterium from human feces: description of *Bryantella formatexigens* gen. nov., sp. nov. Appl. Environ. Microbiol. *69*: 6321–6326.

Wolin, M.J., T.L. Miller, M.D. Collins and P.A. Lawson. 2004. *In* Validation of publication of new names and new combinations previously effectively published outside the IJSEM Validation list no 95. Int. J. Syst. Evol. Microbiol. *54*: 1–2.

Xue, G.P., J.S. Johnson, K.L. Bransgrove, K. Gregg, C.E. Beard, B.P. Dalrymple, K.S. Gobius and J.H. Aylward. 1997. Improvement of expression and secretion of a fungal xylanase in the rumen bacterium *Butyrivibrio fibrisolvens* OB156 by manipulation of promoter and signal sequences. J. Biotechnol. *54*: 139–148.

Ziolecki, A., W. Guczynska and M. Wojciechowicz. 1992. Some rumen bacteria degrading fructan. Lett. Appl. Microbiol. *15*: 244–247.

Family VI. **Peptococcaceae** Rogosa 1971, 235[AL]

TAKAYUKI EZAKI

Pepto.coc.ca.ce′ae. *Peptococcus* type genus of the family; *-aceae* ending to denote a family; N.L. fem. pl. n. *Peptococcaceae* the family of *Peptococcus*.

Coccoid, curved rods, or straight rods. Motile or nonmotile; motility may be lost during cultivation. All members of the family are obligate anaerobes. Gram-positive cell-wall structure but some genera stain negative. Members of this family are chemoorganoheterotrophs, chemolithoheterotrophs or chemolithoautotrophs. Fermentative or respiratory type metabolism; some members in syntrophy with hydrogenotrophs.

Species of type genus is isolated from human intestine, vagina, and umbilicus. Most previous members of the type genus *Peptococcus* found in human and animal intestine and oral and vaginal cavities were reclassified as members of the Family XI *Incertae Sedis* as *Anaerococcus*, *Peptoniphilus*, *Finegoldia*, and *Parvimonas* (Ludwig et al., 1990; Ezaki et al., 2001). Other members of the family *Peptococcaceae* are found in various habitats including soil, terrestrial and marine sediments, sewage, or extremely thermophilic environments. Differential characteristics are given in Table 178.

Type genus: **Peptococcus** Kluyver and van Niel 1936, 400[AL].

Taxonomic comments

Currently, this family can be divided into at least three separate phylogenetic groups (Figure 6). The first group is comprised of *Peptococcus*, *Dehalobacter*, *Desulfitobacterium*, *Desulfonispora*, *Desulfosporosinus*, and *Syntrophobotulus*. This monophyletic group probably represents the *Peptococcaceae stricto sensu*. The second group includes *Cryptanaerobacter*, *Desulfotomaculum*, *Pelotomaculum* and *Sporotomaculum*. These genera are deep branches of the *Clostridiales* and probably represent a novel family. The last group includes only the genus *Thermincola*. Phylogenetic analyses of its relationship to the family *Peptococcaceae strict sensu* position are ambiguous. In the absence of additional information, it was retained within this family.

Key to the genera of the family *Peptococcaceae*

Gram-stain-positive coccus, 1–2 μm in diameter. Spore formation absent. Fermentative metabolism. Optimal growth at 30–40°C in strictly anaerobic conditions. Produce caproic acid as a terminal metabolite of glucose. Found in human vagina and naval cavity. DNA G+C content is 50–51 mol% (T_m).

Genus I. *Peptococcus*

Gram-stain-positive, short rods, 1 × 2 μm. Spores not observed. Obligately anaerobic, not motile. Not fermentative, requires complex media. Phenol or 4-hydroxybenzoate required for growth. Optimal growth at 30–37°C. Isolated from a methanogenic coculture enriched from a mixture of swamp water, sewage sludge, swine waste and soil. DNA G+C content is 51 mol% (HPLC).

Genus II. *Cryptanaerobacter*

Gram-stain-negative rods, 0.5 × 2–3 μm. Spore formation absent. Obligately anaerobic flagellated bacterium. Chemolithoheterotroph, acetate is required as carbon source. Dechlorinates tetra- and trichloroethene with hydrogen as electron donor. Grows at 25–30°C. Found in soil and sediment. DNA G+C content is 45.0–45.6 mol% (HPLC).

Genus III. *Dehalobacter*

Gram-stain-variable, curved, nonmotile rods, 0.5–0.7 × 2–5 μm. Cell wall is of Gram positive structure. Spore formation variable. Obligate anaerobe, some strains can grow under microaerophilic conditions. Fermentative and respiratory type of metabolism. Reduces halogenated alkenes and aromatic compounds, sulfite, thiosulfate, sulfur, fumarate, nitrate, various metals, and humic acids. Dechlorinates chlorophenolic compounds. Optimal growth temperature is 20–37°C. Found in soil and aquifer sediment. DNA G+C content is 45.8–48.8 mol% (T_m, HPLC).

Genus IV. *Desulfitobacterium*

Gram-stain-positive, motile rods, 0.7–1.0 × 2–5 μm. Subterminal spore formation observed. Obligate anaerobe. Fermentative metabolism. Fermentation of taurine to acetate, ammonia and thiosulfate. Optimal growth temperature is 20–37°C. Found in sewage sludge and lake sediments. DNA G+C content is 52 mol% (HPLC).

Genus V. *Desulfonispora*

Slightly curved, motile or nonmotile rods, 0.4–1.0 × 2.3–5 μm. Subterminal to terminal endospore. Obligate anaerobe. Fermentative and respiratory type of metabolism. Reduction of sulfate, thiosulfate to sulfide. Growth temperature range is 4–42°C. Found in soil and freshwater sediments. DNA G+C content is 41.6–46.9 mol% (T_m, HPLC).

Genus VI. *Desulfosporosinus*

Gram-positive cell wall structure but stains negative. Straight or curved rods, 0.3–2.5 × 2–15 μm. Central, subterminal, or terminal spores. Obligate anaerobe. Fermentative and respiratory type of metabolism. Reduction of sulfate, thiosulfate to sulfide. Growth temperature range is 30–75°C. Found in soil, freshwater and marine sediments, and hot sulfur springs. DNA G+C content is 37.5–56.3 mol% (T_m, HPLC).

Genus VII. *Desulfotomaculum*

Gram-positive cell wall structure but stains negative. Curved rods, 0.7–1.0 × 1.7–4. Central spherical endospores. Obligate anaerobe. Growth by fermentation or in syntrophy with hydrogenotrophs. Found in mesophilic or thermophilic anaerobic sludge. DNA G+C content is 52.8–53.6 mol% (T_m, HPLC).

Genus VIII. *Pelotomaculum*

Gram-positive rods, 0.6–1.0 × 1–3 μm. Central spherical endospore. Obligate anaerobe. Fermentative type of metabolism. Some organic compounds are metabolized only in syntrophy with hydrogenotrophs. Found in termite guts and anaerobic sludge. Optimal growth temperature is 30–40°C. DNA G+C content is 46.8–48.0 mol% (HPLC).

Genus IX. *Sporotomaculum*

Gram-positive cell-wall structure but stains negative. Slightly curved rods, 0.5 × 2.5–3.5 μm. Terminal spores observed. Strict anaerobe. Fermentative type of metabolism. Glycolate is oxidized

TABLE 178. Characteristics of the genera of the family *Peptococcaceae*

Characteristic	*Peptococcus* niger (X55797)	*Cryptanaerobacter* phenolicus (AY327251)	*Dehalobacter* restrictus (U84497)	*Desulfitobacterium* dehalogenans (L28946)	*Desulfitospora* thiosulfatigenes (Y18214)	*Desulfosporosinus* orientis (M34417)	*Desulfotomaculum* nigrificans (X62176)	*Pelotomaculum* thermopropionicum (AB035723)	*Sporotomaculum* Hydroxybenzoicum (Y14845)	*Syntrophobotulus* glycolicus (of strain SiGlym X9970)	*Thermincola* carboxydiphila (AY603000)
Type species (GenBank accesion no.)											
G+C content (mol%)	50–51	51	45.0–45.6	45.8–48.8	52	41.6–46.9	37.5–56.3	52.8–53.6	46.8–48.0	46.7	45–48
Spore formation	–	–	–	+/–	+	+	+	+	+	+	+
Habitats (source of isolate)	Human intestine, vagina, umbilicus	Sediment and soil, sewage sludge (mixture)	Soil, freshwater sediment	Aquifer sediment	Sewage sludge, lake sediment	Soil, freshwater sediments	Freshwater and marine sediment, soil, terrestrial hot springs	Anaerobic sludge	Termite gut, anaerobic sludge	Freshwater sediment	Terrestrial hot springs
Metabolism	Fermentative	Phenol or 4-hydroxy benzoate required for growth	Respiratory (dehalo respiration). Dechlorinates tetra- and trichloroethene	Respiratory (dehalo respiration, sulfite reduction), some species also fermentative. Dehalogenation of various organic compounds	Fermentative (fermentation of taurine)	Respiratory (sulfate reduction) and fermentative; some species homoacetogenic	Respiratory (sulfate reduction) and fermentative; some species homoacetogenic	Fermentative, syntrophy with hydrogenotrophs	Fermentative; inorganic electron acceptors not used	Fermentative, syntrophy with hydrogenotrophs	Respiratory (CO metabolism or Fe(III) reduction)
Temperature range for growth	Mesophilic	Mesophilic	Mesophilic	Mesophilic	Mesophilic	Mesophilic	Mesophilic or thermophilic	Mesophilic or thermophilic	Mesophilic	Mesophilic	Thermophilic

in syntrophy with hydrogenotrophs. Found in freshwater sediment. Optimal growth temperature is 15–37°C. DNA G+C content is 46.7 mol% (HPLC).

Genus X. *Syntrophobotulus*

Gram-positive rods, 0.4–0.5 × 0.6–3. Spore formation. Strict anaerobe. Chemolithotrophic growth on CO and acetate as carbon source of the type strain. Other can grow chemolithoautotrophically with hydrogen and carbon dioxide, or chemoorganoheterotrophically linked to the reduction of metals [Fe(III), Mn(IV)] and humic acids. Optimal growth temperature is 55–60°C. Found in terrestrial hot springs. DNA G+C content is 45–48 mol% (T_m).

Genus XI. *Thermincola*

Genus I. **Peptococcus** Kluyver and van Niel 1936, 400[AL]

TAKAYUKI EZAKI

Pep.to.coc′cus. Gr. adj. *peptos* cooked, digested; Gr. n. *kokkos* a grain berry; N.L. masc. n. *Peptococcus* the digesting coccus.

Cells are Gram-stain-positive, non-spore-forming anaerobic cocci that occur singly, in pairs, and irregular mass. Chemo-organotrophs. Found in human intestine, vagina, or human umblicus. Metabolize peptone and amino acids to acetic acid, butyric acid, isocaproic acid. Nonsaccharolytic. Cell-wall diamino acid is lysine.

DNA G+C content (mol%): 50–51 (T_m) (Ezaki et al., 1983).

Type species: **Peptococcus niger** (Hall 1930) Kluyver and van Niel 1936, 400[AL].

Further descriptive information

Old members of the genus *Peptococcus* isolated from human intestine, vagina, or oral cavity were transferred to family *Peptostreptococcaceae* fam. nov. or to family *Bacillaceae* (Ezaki et al., 2001; Ezaki et al., 1994; Ezaki et al., 1983). *Peptococcus asaccharoliticus* and *Peptococcus indolicus* reclassified as members of a new genus *Peptinophilus*, *Peptococcus magnus* as *Finegoldia magna*, *Peptococcus prevotii* as a member of genus *Anaerococcus*, and *Peptococcus saccharolyticus* as a member of genus *Staphylococcus*.

Differentiation of the species of the genus *Peptococcus* from other genera

Differential characteristics of *Peptococcus niger* from other anaerobic cocci isolated from humans (Murdoch and Magee, 1995; Murdoch and Mitchelmore, 1991) are given in Table 179.

List of species of the genus *Peptococcus*

1. **Peptococcus niger** (Hall 1930) Kluyver and van Niel 1936, 400[AL] ("*Micrococcus niger*" Hall 1930, 409)

   ni′ger. L. adj. *niger* black.

   Cells are about 0.3–1.3 μm in diameter and occur singly and in pairs, tetrads, and irregular masses. Stain Gram-stain-positive. Catalase-negative. Colonies on blood agar plate are black, minute to 0.5 mm, circular with entire margins. After several transfers, the black colonies become light gray and may not produce black color. The optimum temperature for growth is 37°C, growth may occur at 25°C and 45°C. Carbohydrates are not fermented. Hydrogen sulfide and ammonia are produced. Metabolize peptone and amino acids to acetic, butyric, isobutyric, isovaleric, and caproic acid.

   DNA G+C content (mol%): 50–51 (T_m) (Ezaki et al., 1983).

   Type strain: ATCC 27731, DSM 20475, NCTC 11805, VPI 7953.

   GenBank accession number (16S rDNA gene): AB036759 (VPI 7953).

Genus II. **Cryptanaerobacter** Juteau, Côté, Duckett, Beaudet, Lépine, Villemur and Bisaillon 2005, 248[VP]

PIERRE JUTEAU

Crypt.an.ae′ro.bac.ter. Gr. adj. *kryptos* hidden; Gr. pref. *an* not; Gr. n. *aer* air; *anaero* not (living) in air; N.L. masc. n. *bacter* rod; N.L. masc. n. *Cryptanaerobacter* referring to the fact that this anaerobic rod was hidden within a culture of another organism taken for pure.

Short, rod-shaped cells, 1 μm × 2 μm. **Gram-stain-positive**. Not flagellated. **Strictly anaerobic**. Sulfate, thiosulfate, nitrate, nitrite, FeCl₃, fumarate, and arsenate are not used as electron acceptors. Sulfite is not normally used even though growth stimulation has been noted in certain culture conditions. Not fermentative. **Phenol or 4-hydroxybenzoate (4-OHB) is required for growth**. 4-OHB is believed to be an electron acceptor for anaerobic respiration. Isolated from a methanogenic consortium derived from a mixture of swamp water, sewage sludge, swine waste, and soil.

TABLE 179. Characteristics differentiating the genus *Peptococcus* from other anaerobic cocci found in human specimens[a]

Characteristic	*Peptococcus*	*Anaerococcus*	*Finegoldia*	*Gallicola*	*Helcococcus*	*Parvimonas*	*Peptoniphilus*	*Peptostreptococcus*
DNA G+C content (mol%)	50–51	30–35	32–34	32–34	29–29.5	28–30	30–34	34–36
Growth conditions	Obligate anaerobe	Obligate anaerobe	Obligate anaerobe	Obligate anaerobe	Facultative anaerobe	Obligate anaerobe	Obligate anaerobe	Obligate anaerobe
Peptidoglycan (Pos3, bridge)	Lys, D-Asp	Lys, D-Glu	Lys, Gly	Orn, D-Asp	No data	Lys	Orn, D-Glu	Lys, D-Asp
Sugar fermented	–	w	–	–	D	–	–	w
Major fermentation product	Butyric, caproic acids	Butyric acid	Acetic acid	Butyric, acetic acids	No data	Acetic acid	Butyric acid	iso-Caproic, isovalelic acids

[a]Symbols: +, >85% positive; –, 0–15% positive; D, different among species; w, weak acid produced.

DNA G+C content (mol%): 51.

Type species: **Cryptanaerobacter phenolicus** Juteau, Côté, Duckett, Beaudet, Lépine, Villemur and Bisaillon 2005, 249[VP].

Further descriptive information

Cells stain Gram-positive, and electron microscopy of thin sections reveals a typical Gram-positive cell wall. When observed by electron microscopy, cells appear particularly dark (electron-dense) (Letowski et al., 2001). The major membrane fatty acid is $C_{15:0\ ante}$.

Colonies are obtained only in semi-solid medium (0.3% w/v agar). Colonies are 1 mm in diameter with diffuse margins and are brownish after 10 d of incubation. Growth in pure culture requires a complex medium, SBM4, composed in part of a conditioned medium. While medium conditioned by a culture of *Clostridium sporogenes* M55 (ATCC 13732, DSM 754) works best (see below), conditioned media from cultures of strain 6, the *Clostridium*-like organism from which *Cryptanaerobacter phenolicus* LR7.2 was separated, and *Enterococcus faecalis* can also be used. Optimum growth occurs under an atmosphere consisting of $H_2:CO_2:N_2$ (10:10:80 by vol.) at 30–37°C and pH 7.5–8.0. Growth is also possible under 100% nitrogen atmosphere or with H_2/N_2 gas mixtures (10:90, 20:80, 80:20 v/v). Even in the best culture conditions, growth is weak, and the final cell yield is ~4×10^6 cells/ml.

Spores have not been observed, and both pure and co-cultures are killed when heated at 70°C for 10 minutes. However, because pasteurization was used during the enrichment of *Cryptanaerobacter phenolicus*, it is likely that cells can sporulate under certain environmental conditions that are not reproduced in pure cultures or in defined co-cultures.

An important characteristic of *Cryptanaerobacter phenolicus* is its growth dependency on phenol and 4-OHB. When phenol is present, it is first carboxylated into 4-OHB, which is dehydroxylated into benzoate (phenol → 4-OHB → benzoate). Phenol carboxylation is a reversible reaction in which the decarboxylation is thermodynamically favored at equilibrium. The overall reaction is pulled in the direction of benzoate formation because the dehydroxylation reaction maintains a very low concentration of 4-OHB. As a consequence, when 4-OHB is given to *Cryptanaerobacter phenolicus* instead of phenol, most of it is first transformed into phenol until the 4-OHB concentration is low enough to enable the transformation of phenol (4-OHB → phenol → 4-OHB → benzoate). Phenol and 4-OHB are not a source of carbon since they are stochiometrically transformed into benzoate, which is not further metabolized. Nevertheless, no growth is observed in the absence of phenol or 4-OHB, which means that the strain cannot grow fermentatively. Moreover, the cell yield is proportional to the concentration of these compounds up to the optimal values of 6.5 mM for 4-OHB and 3.5 mM for phenol. These observations indicate that *Cryptanaerobacter phenolicus* uses these transformations as an energy source for growth. Presumably, these transformations represent an unusual anaerobic respiration in which 4-OHB is an electron acceptor. However, neither the electron donor nor the carbon source has been identified.

Based on analysis of its 16S rRNA gene sequence, *Cryptanaerobacter* is related to the genus *Pelotomaculum* (Figure 173).

Enrichment and isolation procedures

The isolation of the type strain is quite a long story, and nearly 20 years separate the first report of the enrichment and the defining publication. The original consortium was made by mixing swamp water, sewage sludge, swine waste, and soil (Beaudet et al., 1986). Major steps toward the isolation included pasteurization of the consortium (Létourneau et al.,

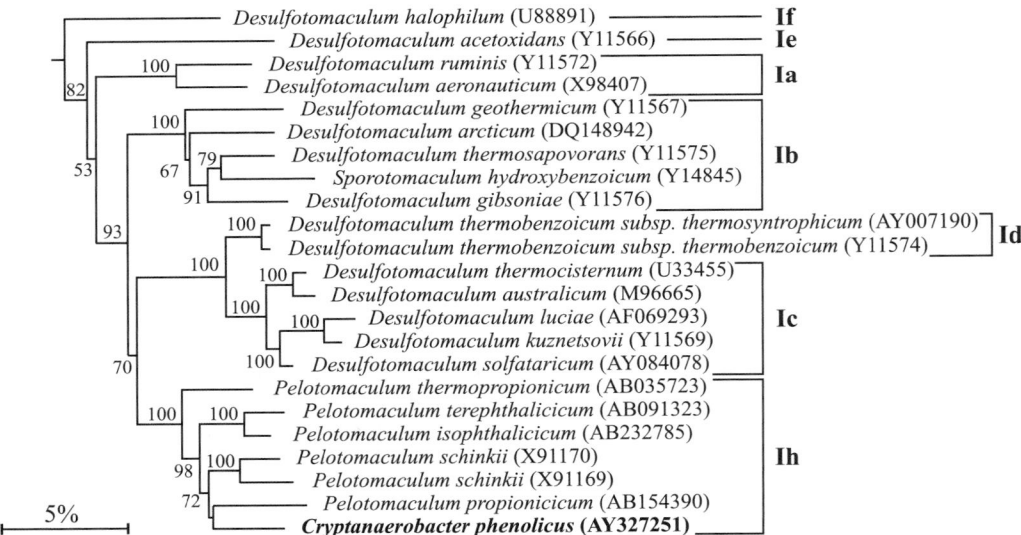

FIGURE 173. Phylogeny of *Cryptanaerobacter*. The tree shows the 16S rRNA gene sequence analysis of *Desulfotomaculum* cluster I, as first defined by Stackebrandt et al. (1997) and amended by Kuever et al. (1999)and Imachi et al. (2006). It includes all *Pelotomaculum* species, which form with *Cryptanaerobacter phenolicus* the subcluster Ih, and representatives of the other subclusters, except subcluster Ig that does not contain any validly named species. The tree was constructed by the Fitch–Margoliash method using a DNA distance matrix, itself based on the Kimura two-parameter model, with PHYLIP 3.66 package programs. One thousand bootstrap trees were generated, and bootstrap confidence levels (shown as percentages above nodes) were determined. *Bacillus subtilis* (Z99104, position 171497–173046) was used as the outgroup organism. In order to include organisms for which the available 16S rDNA sequences are shorter (particularly *Pelotomaculum schinkii*), only 1357 nucleotides (*Escherichia coli* position 124–1488) were used for *Cryptanaerobacter phenolicus* (of 1529 available). Bar = 5 nucleotide substitutions per 100 nucleotides.

1995), use of antibiotics (Li et al., 1996), separation of the two bacteria (strain 6 and strain 7) of the resulting co-culture by Percoll density-gradient centrifugation (Letowski et al., 2001), and finally the use of a mixture of fresh medium and conditioned medium from strain 6 or *Clostridium sporogenes* to obtain stable liquid cultures and colonies in soft (0.3%) agar (Juteau et al., 2005). For some time during this process, the co-culture of strains 6 and 7 was thought to be one axenic culture (Li et al., 1996).

Best growth of *Cryptanaerobacter* is obtained with a mixture of fresh and conditioned medium. The growth medium contains (per liter): yeast extract, 5 g; $NaHCO_3$, 1.5 g; KH_2PO_4, 0.27 g; K_2HPO_4, 0.35 g; NH_4Cl, 0.53 g; resazurin, 0.6 mg; $MgCl_2·6H_2O$, 0.1 g; $CaCl_2·2H_2O$, 77 mg; $FeCl_2·4H_2O$, 2 mg; pyridoxine, 0.15 mg; biotin, 30 μg; folic acid, 30 μg; riboflavin, 76 μg; thiamine, 76 μg; nicotinic acid, 76 μg; pantothenic acid, 76 μg; *p*-aminobenzoic acid, 76 μg; thioctic acid, 76 μg; vitamin B_{12}, 2 μg. The medium is made anaerobic by boiling, sparging with a $H_2:CO_2:N_2$ 10:10:80 (by vol.) gas mixture and adding $Na_2S·9H_2O$ (53 mg/L). The pH is then adjusted to 7.5–7.8. The medium is then conditioned by growing *Clostridium sporogenes* M55 for 5–8 d. After growth, the culture is autoclaved, suspended solids are removed by centrifugation and filtration on cheesecloth, and the broth is sterilized by autoclaving. The final growth medium (SBM4) contains 62% of the fresh medium (prepared without Na_2S), 38% of conditioned medium, and 3 mM of 4-hydroxybenzoate.

Maintenance procedures

The type strain can be maintained at –80°C in SBM4 medium with 10% glycerol or as lyophilized culture. However, revival of preserved stock cultures can be lengthy, so the strain is also maintained by subculturing in SBM4 medium.

Differentiation of the genus *Cryptanaerobacter* from other genera

Table 180 lists physiological characteristics differentiating *Cryptanaerobacter* from its closest relatives, which are species of the genus *Pelotomaculum*. The ability to transform phenol

and 4-OHB to benzoate and the requirement of these compounds for growth are the most important distinguishing characters of *Cryptanaerobacter*. Unlike the *Pelotomaculum* species, sporulation has not been observed for *Cryptanaerobacter*. However, this criterion should not be the sole basis for differentiation because *Cryptanaerobacter* was isolated following pasteurization, suggesting that spores can be formed under certain environmental conditions. The highest 16S rRNA gene sequence similarity is with *Pelotomaculum schinkii rrnB* (94.5% over 1360 nt) and *Pelotomaculum isophthalicicum* (94.3% over 1460 nt).

Taxonomic comments

At the time *Cryptanaerobacter* was described, just one species was known in the genus *Pelotomaculum*, *Pelotomaculum thermopropionicum*. It possessed only 90% 16S rRNA sequence similarity with *Cryptanaerobacter phenolicus*. Since then, other species of *Pelotomaculum* have been described which appear much more similar to *Cryptanaerobacter*. The most closely related species possesses 94.5% 16S rRNA gene sequence similarity (Figure 173). As a group, *Cryptanaerobacter* and *Pelotomaculum* form the *Desulfotomaculum* subcluster Ih, as proposed by Imachi et al. (2006). Although this group is genetically related, its unusual energy metabolism still distinguishes *Cryptanaerobacter*. The sole energy source is the transformation of 4-OHB into benzoate. In contrast, all *Pelotomaculum* species are capable of syntrophic growth by using propionate, primary alcohols, low molecular weight aromatic compounds, lactate, or other compounds in association with hydrogenotrophic methanogens. Some *Pelotomaculum* species are also capable of fermentative growth. However, none of them can use phenol or 4-OHB. While the ability of *Cryptanaerobacter* to grow syntrophically has not been tested, indirect evidence suggests that it does not. It grows under an atmosphere of 80% H_2. Within the original consortium, the transformation of phenol was insensitive to the methanogen inhibitor 2-bromoethanosulfonic acid (Béchard et al., 1990). Given the importance of this property to classification within the genus *Pelotomaculum*, additional tests should be performed to confirm this conclusion.

TABLE 180. Characteristics differentiating *Cryptanaerobacter* from some closely related bacteria[a]

	Cryptanaerobacter phenolicus	*Pelotomaculum isophthalicicum*	*Pelotomaculum propionicicum*	*Pelotomaculum schinkii*	*Pelotomaculum terephthalicicum*	*Pelotomaculum thermopropionicum*
Gram reaction	+	−	+	ND	−	−
Growth temperature (°C):						
Range	20–42	25–45	25–45	ND	30–40	45–65
Optimal	30–37	37	37	37	37	55
Spore formation	−[b]	+	+	+	+	+
Fermentative growth	−	−	−	−	+	+
Substrate utilization in pure culture:						
Phenol	+[c]	−	−	−	−	−
4-hydroxybenzoate (4-OHB)	+[c]	−	−	−	−	−
Sulfite utilization	−[d]	−	−	−	−	−

[a]Symbols: +, >85% positive; −, 0–15% positive; ND, not determined.

[b]No spore has been observed and cultures are not heat resistant. However, pasteurization has been used during enrichment, which suggests that *Cryptanaerobacter phenolicus* could sporulate in nature.

[c]Phenol and 4-OHB are stochiometrically transformed into benzoate, which is not further metabolized.

[d]Sulfite not used in SBM4 medium. Stimulation by sulfite has been reported by Letowski et al. (2001) with another medium, but they did not confirm that the bacterium used it as an electron acceptor.

List of species of the genus *Cryptanaerobacter*

1. **Cryptanaerobacter phenolicus** Juteau, Côté, Duckett, Beaudet, Lépine, Villemur and Bisaillon 2005, 249[VP]

phe.no′li.cus. N.L. n. *phenol -olis* phenol; N.L. masc. adj. *phenolicus* relating to phenol.

Since it is the only species of *Cryptanaerobacter*, all the characteristics described for the genus apply to *Cryptanaerobacter phenolicus*. None of the following compounds structurally related to 4-OHB or phenol are transformed or used for growth: 4-hydroxybenzamide, 4-hydroxysulfonic acid, 4-hydroxyacetophenone, 4-hydroxybenzoic alcohol, hydroquinone, 4-chlorophenol, 4-hydroxycinnamic acid, 4-hydroxybenzoic hydrazide, 4-hydroxybenzaldehyde, 4-hydroxyphenyl pyruvic acid, 3-(4-hydroxyphenyl) propionic acid, *p*-cresol, 3-OHB, 4-hydroxypyridine, catechol, 2-bromophenol, 2-chlorophenol, 2-fluorophenol, 2-aminophenol.

DNA G+C content (mol%): 51 (HPLC).

Type strain: LR7.2, ATCC BAA-820; DSM 15808.
GenBank accession number (16S rRNA gene): AY327251.

Acknowledgements

Most of the research that led to the isolation and description of *Cryptanaerobacter phenolicus* was directed by Dr Jean-Guy Bisaillon (now retired) at the INRS – Institut Armand-Frappier (Laval, Québec, Canada) over a 15-year period, with the technical assistance of Louis Racine. Dr Pierre Juteau, author of this text, joined them at the end. The following graduate students were also involved: Geneviève Béchard, Valérie Côté, Marie-France Duckett, Lynda Létourneau, Jaroslaw Letowski, and Tong Li. This work was supported by grants from the Natural Sciences and Engineering Research Council of Canada (NSERC) and the Fonds pour la Formation des Chercheurs et l'Aide à la Recherche (FCAR).

Genus III. **Dehalobacter** Holliger, Hahn, Harmsen, Ludwig, Schumacher, Tindall, Vazquez, Weiss and Zehnder 1998a, 631[VP] (Effective publication: Holliger, Hahn, Harmsen, Ludwig, Schumacher, Tindall, Vazquez, Weiss and Zehnder 1998b, 319.)

CHRISTOF HOLLIGER

De.ha.lo.bac′ter. L. pref. *de* from; Gr. n. *halo*, *halos* the sea, salt; N.L. masc. n. *bacter* equivalent of Gr. neut. n. *baktron* a rod; N.L. masc. n. *Dehalobacter* a halogen-removing rod-shaped bacterium.

Cells with a diameter of 0.3–0.5 μm and 2–3 μm long are **rod-shaped with tapered ends**. Appear singly or in pairs; nonsporeforming. Cells stain Gram-negative. **Motile**, having one lateral flagellum per cell. **Strictly anaerobic**. Electron micrographs indicate that cells are surrounded by an S-layer. Cell wall contains the **peptidoglycan type A3γ**. Menaquinones and *b*-type cytochromes are present. **Mesophilic** with optimal growth 25–35°C; no growth at 37°C. **Chemolithoheterotrophic**, using dehalorespiration as method of metabolism and acetate as carbon source.

DNA G+C content (mol%): 45.

Type species: **Dehalobacter restrictus** Holliger, Hahn, Harmsen, Ludwig, Schumacher, Tindall, Vazquez, Weiss and Zehnder 1998a, 631[VP] (Effective publication: Holliger, Hahn, Harmsen, Ludwig, Schumacher, Tindall, Vazquez, Weiss and Zehnder 1998b, 319.).

Further descriptive information

Phylogenetic analysis based on 16S rRNA gene sequences (Figure 6) shows that the genus *Dehalobacter* is a member of the family *Peptococcaceae*, order *Clostridiales*, class *Clostridia* in the phylum *Firmicutes*. The closest relative of the genus *Dehalobacter* is *Syntrophobotulus glycolicus* with an overall 16S rRNA sequence identity of 92.9%. Interestingly, the genus *Desulfitobacterium*, containing several species that reductively dechlorinate chlorinated compounds such as chloroethenes and chlorophenols, also affiliates with the family *Peptococcaceae* and has 16S rRNA sequence identities of 90.8% and lower compared with the genus *Dehalobacter*.

Although *Dehalobacter* cells stain Gram-negative, electron microscopy of sections did not indicate the presence of an outer membrane, but rather an S-layer. Freeze-etch preparation showed a hexagonal structure on the cell surface presumably caused by S-layer proteins. The peptidoglycan contains L-alanine, D-glutamic acid, LL-diaminopimelic acid, and glycine, and are cross-linked between positions 3 and 4 with one glycine as interpeptide bridge. Only some strains of the genus *Propionibacterium* have this type of peptidoglycan.

Enrichment and isolation procedures

The type strain *Dehalobacter restrictus* strain PER-K23 was isolated from a tetrachloroethene-dechlorinating packed-bed column which was filled with river Rhine sediment and ground anaerobic granular sludge from a sugar refinery. Tests with different substrates showed that hydrogen resulted in the highest dechlorination rates in batch enrichment cultures. Hydrogen-consuming dehalorespiring bacteria have to compete with methanogens and homoacetogens for the electron donor. To create advantageous conditions for tetrachloroethene-dehalorespiring bacteria and at the same time avoid too high and toxic concentrations of tetrachloroethene, a two-liquid phase system (water/hexadecane) was used. This allowed the addition of large amounts of tetrachloroethene to the batch culture with tetrachloroethene concentrations in the medium below saturation, and having a continuous supply of the aqueous phase tetrachloroethene from the "reservoir" present in the organic phase. Dechlorination was followed by monitoring chloride concentration in the specially designed medium low in chloride ions. Enrichment was obtained with serial dilutions. Colonies for pure culture isolation were only obtained if special care was taken to use soft agar at a temperature just above solidification (approx. 42°C) for agar shake production and rapid cooling of the inoculated tubes.

Maintenance procedure

No specific information on suitable procedures for the maintenance of *Dehalobacter* species is available. It may be assumed that procedures satisfactory for other strictly anaerobic bacteria will also be suitable for the maintenance of *Dehalobacter* isolates.

List of species of the genus *Dehalobacter*

1. **Dehalobacter restrictus** Holliger, Hahn, Harmsen, Ludwig, Schumacher, Tindall, Vazquez, Weiss and Zehnder 1998a, 631[VP] (Effective publication: Holliger, Hahn, Harmsen, Ludwig, Schumacher, Tindall, Vazquez, Weiss and Zehnder 1998b, 319.)

re.stric'tus. L. adj. *restrictus* limited, confined, referring to the limited substrate range utilized.

Cells are rod-shaped and have tapered ends, are 0.3–0.5 µm in diameter and 2–3 µm long. They appear singly or in pairs, are motile, and nonspore-forming. Cells stain Gram-negative and are surrounded by an S-layer. Cell wall consists of type A3γ peptidoglycans (L-alanine, D-glutamic acid, LL-diaminopimelic acid, and glycine; cross-linked between positions 3 and 4 with a glycine interpeptide bridge). Menaquinones (MK-7 and MK-8 mainly, minor amounts of MK-6 and MK-9) and *b*-type cytochromes are present. Strictly anaerobic with no catalase and no superoxide dismutase.

Chemolithoheterotrophic using only H_2 as electron donor and tetra- or trichloroethene as terminal electron acceptors. *cis*-1,2-Dichloroethene as dechlorination product. Substrates such as lactate, pyruvate, propionate, butyrate, acetate, formate, succinate, fumarate, glycine, alanine, aspartate, glutamate, methanol, ethanol, propanol, glucose, fructose, xylose, glycerol, acetoin, and CO are not utilized in the presence or in the absence of tetrachloroethene. Nitrate, nitrite, fumarate, dimethyl sulfoxide, trimethylamine *N*-oxide, Fe(III), Mn(IV), sulfate, sulfite, thiosulfate, and sulfur are not reduced. Acetate serves as carbon source. Growth factors required in a defined medium are ferrous iron, thiamine, cyanocobalamin, arginine, histidine, and threonine. Optimal growth at 25–35°C and pH 6.8–7.6.

DNA G+C content (mol%): 45.3 ± 0.3 (HPLC).

Type strain: PER-K23, DSM 9455.

GenBank accession number (16S rRNA gene): U84497.

Genus IV. **Desulfitobacterium** Utkin, Woese and Wiegel 1994, 615[VP]

Bogusław Lupa and Juergen Wiegel

De.sul.fi.to.bac.te'ri.um. L. pref. *de* from, off, away; N.L. n. *sulfis* sulfite; N.L. masc. n. *bacter* rod; N.L. neut. n. *Desulfitobacterium* rod-shaped bacterium that reduces sulfite.

Cell wall is of **Gram-positive** structure but may stain **Gram-negative or -positive** depending on the strain. Exponential-growth phase cells are **straight or curved rods**, 2–5 µm in length, depending on the species and strain. Although the physiology is obligately anaerobic; certain species tolerate microaerophilic culture conditions (< 5% air in N_2 head gas phase); **mesophilic** and **heterotrophic**. *Desulfitobacterium* species use a variety of chlorinated phenols and/or alkenes as electron acceptors during dehalorespiration (also called halorespiration or chloridogenesis). Furthermore, they reduce sulfite, thiosulfate, sulfur, fumarate (forming succinate), and nitrate, but not sulfate, in the presence of various electron donors. Yeast extract as growth supplement is required; some strains grow only using pyruvate (forming lactate + acetate + CO_2), others can utilize various sugars. Four species are validly published, *Desulfitobacterium dehalogenans*, *Desulfitobacterium chlororespirans*, *Desulfitobacterium hafniense*, and *Desulfitobacterium metallireducens*. For differentiation of the species, see Table 181.

DNA G+C content (mol%): 45–48.8.

Type species: **Desulfitobacterium dehalogenans** Utkin, Woese and Wiegel 1994, 615[VP].

Further descriptive information

Most species are motile, usually equipped with one to several terminal, rarely lateral, flagella. Most species and strains are spore-forming. Spores are located terminally or subterminally and usually appear in the late-exponential or early stationary growth phase (Bouchard et al., 1996; Christiansen et al., 1998; Finneran et al., 2002; Sanford et al., 1996; Utkin et al., 1994). The genus is recognized as one of the most important groups of anaerobic dehalogenating bacteria (Villemur et al., 2006). Members of the genus dechlorinate aromatic and

alkyl compounds, including some of the most problematic pollutants, e.g., chlorinated phenols, chlorinated ethenes, and hydroxy polychlorinated biphenyls (HOPCBs) (Christiansen and Ahring, 1996; Finneran et al., 2002; Sanford et al., 1996; Utkin et al., 1994; Wiegel et al., 1991).

Molecular investigations. The organization of the genes encoding the reductive dehalogenases has been characterized in many of species of *Desulfitobacterium*. The majority of the described chloroaromatic dehalogenases belong to the CprA family and are encoded by the *cprA* gene. These genes are usually linked to the *cprB* gene, which encodes an intrinsic membrane protein. These genes form an apparent operon *cprBA*. Usually, the *cprB* gene is upstream of *cprA*, except for *cprA₅B₅* from *Desulfitobacterium hafniense* PCP-1 (Thibodeau et al., 2004). The chloroethyl reductive dehalogenases, which constitute the PceA enzyme family, are encoded by the *pceA* genes and are located upstream to another intrinsic membrane protein encoded by the *pceB* gene. Genes of both dehalogenase families, *cprBA* and *pceAB*, are believed to be co-expressed (for a review, see Villemur et al., 2006). In addition, *Desulfitobacterium* possesses reductive dehalogenases that are not members of either family. Genomic sequencing has also identified genes with high similarity to either *cprA* and *pceA*. However, the function of these genes has not been confirmed experimentally. In all these cases, the genes are denoted as *rdhA*.

In *Desulfitobacterium dehalogenans*, *cprBA* appears to be part of a larger operon containing five additional ORFs: *cprKZEBACD*. A sixth gene, *cprT*, is located immediately upstream and is divergently transcribed (Smidt et al., 2000). CprC possesses six predicted transmembrane domains. It and CprK are involved in the regulation of *cprA* expression (Gabor et al., 2006; Pop et al., 2004, 2006). CprD, CprE, and CprT are predicted to be

TABLE 181. Comparative characteristics of the validly published *Desulfitobacterium* species

Characteristic	*Desulfitobacterium dehalogenans*	*Desulfitobacterium chlororespirans*	*Desulfitobacterium hafniense*	*Desulfitobacterium metallireducens*
Shape	Straight or slightly curved rod	Curved rod	Curved rod	Curved rod
Size (μm)	2.5–4 × 0.5–0.7	3–5 × 0.5–0.7	3.3–6 × 0.6–0.7	2–5 × 0.5
Motility	+	+	+	–
Flagella (no./position)	1–4/terminal	1–2/terminal	1–2/terminal	–
Gram stain	+	–		–
Spores observed	–	Terminal	Terminal	–
Colony morphology	Spherical or slightly irregular, translucent, white 2–3 mm in diameter	White, round, smooth, 1–2 mm	No growth on aerobic blood agar plates	Round, smooth, domed, white, entire
O_2 relationship	Obligately anaerobic, slightly aerotolerant	Obligately anaerobic	Obligately anaerobic, some strains aerotolerant	Obligately anaerobic
Isolated from	Methanogenic freshwater sediment	Residential compost soil	Municipal sludge, feces	Uranium-contaminated shallow aquifer sediment
DNA G+C content (mol%)	45	48.8	47	ND
Growth optima:				
pH	7.5 (6.0–9.0)	6.8–7.5	7.0	7.0
Temp. (°C)	38 (13–45)	37 (15–37)	37	30 (20–37)
Electron donors[a]	H_2, for, lac, pyr, YE	H_2, for, lac, pyr, but, mal, cro, YE	H_2, for, cro, eth, lac, pyr, but, YE, van, syr, ver, ser	lac, for, but, eth, bOH, pyr, mal
Electron acceptors[b]	SO_3^{2-}, $S_2O_3^{2-}$, S, NO_3^-, fum, HA, AQDS, IT, cys, Fe(III) PP, Se(VI), MnO_2,	SO_3^{2-}, $S_2O_3^{2-}$, S, Fe(III) PP, Se(VI), MnO_2	SO_3^{2-}, $S_2O_3^{2-}$, NO_3^-, fum, As(V), Fe(III), IT	Fe(III), Mn(IV), Cr(VI), AQDS, HA, S, $S_2O_3^{2-}$, fum
Dechlorinated compounds[c]	CP, DCP, TCP, TeCP, PCP, BP, DBP, 2B4CP at *ortho* positions, 3Cl4HOPA, HOP-ClBs, TCMP, TCHQ, PCE, 2,6DC4RPs, 2C4RPs	DCP; TCP at *ortho* positions, TBP, 3Cl4HOB, 3Cl4HOPA, TCMP, TCHQ, BXN, IXN, DBHB	PCP, TeCP, TCP at *ortho*; DCP at *ortho* and *meta*; 3Cl4HOPA, PCE, TCMP, TCHQ, ATIA	3Cl4HOPA, TCE, PCE
Carbon source(s)[a]	pyr, lac	pyr	pyr, trp, van, lac	lac

[a]Abbreviations are: bOH, butanol; but, butyrate; cro, crotonate; eth, ethanol; for, formate; lac, lactate; mal, malate; pyr, pyruvate; ser, serine; syr, syringate; trp, tryptophan; van, vanillate; ver, verartol; YE, yeast extract.

[b]Abbreviations are: AQDS, anthraquinone-2,6-disulfonate; cys, cysteate (alanine-3-sulfonate); fum, fumarate; HA, humic acid; IT, isethionate (2-hydroxyethanesulfonate); PP, pyrophosphate.

[c]Abbreviations are: ATIA, 5-amino-2,4,6-triiodoisophtalic acid; BP, 2-bromophenol; BXN, bromoxynil (3,5-dibromo-4-hydroxybenzonitrile); CP; 2-chlorophenol, DBHB, 3,5-dibromo-4-hydroxybenzoate; DBP, 2,6-dibromophenol; DCP, dichlorophenol, including: 2,3-DCP; 2,4-DCP; 2,6-DCP; 3,5-DCP; HOPClB, hydroxy polychlorinated biphenyl, including: 3,5-Cl-4-HOBP; 3,4′,5-Cl-4-HOBP; 3,3′,5,5′-Cl-4,4′-diHOBP; IXN, ioxynil (3,5-diiodo-4-hydroxybenzonitrile); PCE, perchloroethene (tetrachloroethene); PCP, pentachlorophenol; TBP, 2,4,6-tribromophenol; TCE, trichloroethene; TCHQ, tetrachlorohydroquinone; TCMP, 2,3,5,6-tetrachloro-4-methoxyphenol; TCP, trichlorophenol, including: 2,3,4-TCP; 2,3,6-TCP; 2,4,6-TCP; TeCP, tetrachlorophenol, including: 2,3,4,6-TCP; 2,3,4,5-TeCP; 2,3,5,6-TeCP; 2,6DCl4RPs, 2,6-dichloro-4-R-phenols, (where R is -H, -F, -Cl, -NO_2, -CO_2, or -$COOCH_3$); 2B4CP, 2-bromo-4-chlorophenol; 2C4RPs, 2-chloro-4-R-phenols (where R is -H, -F, -Cl, -Br, -NO_2, -CO_2, -CH_2CO_2, or -$COOCH_3$); 3Cl4HOPA, 3-chloro-4-hydroxyphenylacetate.

involved in proper folding of the dehalogenase and insertion of the cofactors (for a review, see Villemur et al., 2006). The reductive dehalogenase of *Desulfitobacterium chlororespirans* is encoded by an operon with the organization of *cprKBABAC*, where the individual genes are homologous to those in *Desulfitobacterium dehalogenans* (Villemur et al., 2006). *Desulfitobacterium hafniense* DCB-2 contains multiple copies of the *cpr* genes. One cluster is similar in organization to that found in *Desulfitobacterium dehalogenans*. The other gene clusters are unlinked and include *cprBA*, *cprKBA₄rdhB*, *cprKBA₂BA₃KZDCD*, *rdhACABCTK*, and *rdhKABC*. *Desulfitobacterium hafniense* (basonym *Desulfitobacterium frapieri*) PCP-1 harbors two dehalogenase gene clusters, *cprABA* and *cprA₅B₅*, and is thus able to dehalogenate chlorophenols substituted in both the *ortho*- and *meta*- positions. Strains, *Desulfitobacterium hafniense* TCE-1 and Y51, contain *pceABCT* gene clusters, and strain *Desulfitobacterium hafniense* PCE-S only contains the *pceAB* genes (for a review, see Villemur et al., 2006).

Enzymology. The reductive dehalogenases are indispensable for the ability of the genus *Desulfitobacterium* to respire halogenated compounds. On the basis of their substrate specificity, two groups of dehalogenases are distinguished: chlorophenol reductive dehalogenases (the CprA family) and chloroalkyl reductive dehalogenases (the PceA family). Strains which dehalogenate both chloroaromatic and chloroaliphatic compounds possess two separate enzyme systems, one for each class of substrate. The majority of the reported dehalogenases are membrane-associated. The B subunits of both types of dehalogenases contain predicted hydrophobic domains and are relatively small, with less than 100 amino acids. The A subunits do not contain transmembrane motifs, but their N-terminal amino acid sequences possess signal peptides with a TAT motif. Thus, it is likely that the dehalogenases are anchored in the membrane through the B subunits, with the A subunits on the outside of the cytoplasmic membrane facing the cell wall. All the dehalogenases tested also contain corrinoid prosthetic groups and Fe-S clusters involved in electron transport (for a review, see Villemur et al., 2006).

The *ortho*-chlorophenol reductive dehalogenase, which catalyzes the dehalogenation of 3-chloro-4-hydroxyphenylacetate (3Cl4HOPA), was purified from membrane fractions of

Desulfitobacterium dehalogenans under strictly anaerobic conditions (van de Pas et al., 1999). The enzyme has a molecular mass of 45,300, which is consistent with the removal of the 42 amino acid leader sequence from the translated polypeptide. The primary sequence and the redox properties are similar to those of the haloalkene reductive dehalogenase purified from *Dehalobacter restrictus*. *Desulfitobacterium hafniense* strains, such as strain PCP-1, which carry out the dechlorination of *ortho-*, *meta-*, and *para-*chlorosubstituted aromatic compounds, harbor separate, inducible enzyme systems for the *ortho* as well as for *meta-* and *para-*dechlorination reactions (Bouchard et al., 1996; Gauthier et al., 2006). Additional dehalogenases have been purified and characterized. A 3Cl4HOPA dehalogenase with a subunit molecular mass of 47,000 has been purified from the membrane fraction of *Desulfitobacterium hafniense* strain DCB-2 (Christiansen et al., 1998). This enzyme is inhibited by sulfite, azide, and nitrate but not by sulfate (Boyer et al., 2003). A CprA-type enzyme with a subunit molecular mass of 50,000 has also been purified from *Desulfitobacterium hafniense* strain Co23 (Loffler et al., 1996). This enzyme dechlorinates various *ortho-*chlorophenolic compounds and has a substrate specificity similar to that of the 3Cl4HOPA dehalogenase from *Desulfitobacterium dehalogenans*. *Desulfitobacterium hafniense* strain PCP-1 contains a membrane-associated CprA-type 3,5-dichlorophenol (DCP) reductive dehalogenase, with a molecular mass of 57,000 Da, which is sensitive to inhibition by 2.5 mM sulfite and 10 mM cyanide but not by sulfate and nitrate (Thibodeau et al., 2004). The phenolic hydroxyl group is required for binding of the substrate (Krasotkina et al., 2001). This strain also contains CrdA, which is a representative of a different family of reductive dehalogenases. This enzyme is an oxygen-sensitive 2,4,6-trichlorophenol (TCP) dehalogenase with a subunit molecular mass of 37,000 Da (Boyer et al., 2003).

Enzymes that dechlorinate aliphatic compounds belong to the PceA family of reductive dehalogenases and have been purified from different *Desulfitobacterium hafniense* strains. The substrate-inducible PceA derived from strain Y51 is found in the periplasm. The translation product has a molecular mass of 61,000 Da, but it is 58,000 Da after removal of the signal peptide (Suyama et al., 2002). A similar enzyme with a molecular mass of 59,000 Da was found in strain TCE-1 (van de Pas et al., 2001a). The enzyme from strain PCE-S has a molecular mass of 65,000 Da and is inhibited by sulfite (Miller et al., 1997).

The dissimilatory sulfite reductases (DsrAB) are conserved siroheme-containing enzymes (Klein et al., 2001) and catalyze the reduction of sulfite to sulfide. This enzyme is essential for sulfur respiration. In *Desulfitobacterium dehalogenans* and *Desulfitobacterium hafniense* DCB-2, it is a heterotetramer ($\alpha_2\beta_2$) encoded by *dsrAB* (Klein et al., 2001). The α-subunit contains the (Cys-X_5-Cys)-X_n-(Cys-X_3-Cys) sequence motif required for binding of the [Fe$_4$S$_4$]-siroheme cofactor (Crane et al., 1995).

Energy metabolism. *Desulfitobacterium* species are capable of dehalorespiration or halorespiration, i.e., the process coupling reductive dechlorination (dehalogenation) with energy metabolism and growth. The term "chloridogenesis" is also proposed for this process. This term may be more accurate, because the redox state of chloride anion released during dehalorespiration does not change. Energy conservation during halorespiration was demonstrated by growth yield studies and stoichiometric analyses (Mackiewicz and Wiegel, 1998). The reductive dechlorination

of 3Cl4HOPA is coupled with ATP synthesis and pyruvate, formate, or H$_2$ oxidation. During pyruvate oxidation to acetate, 1 ATP is obtained per mol of acetate through substrate level phosphorylation, and 1 ATP is obtained per chloride ion formed via chloridogenesis (dehalorespiration) and electron transport phosphorylation (Mackiewicz and Wiegel, 1998). Dehalogenation with formate as the electron donor yields only 1 ATP via electron transport phosphorylation. Chlorophenol respiration in *Desulfitobacterium dehalogenans* is not an efficient respiratory pathway since cells recover only a small fraction of the energy theoretically available (van de Pas et al., 2001c). Free energy calculations suggest that 2 ATPs could be formed per chlorine removed with H$_2$ as electron donor (El Fantroussi et al., 1998). However, the cell yields are consistent with only one-half of that value. Even though the efficiency is low, the haloaromatic compounds are the preferred electron acceptors. Thus, when *Desulfitobacterium dehalogenans* is provided with equimolar concentrations of various electron acceptors, 3Cl4HOPA is reduced first. As the concentration drops, it is utilized concomitantly with the other electron acceptors (Mackiewicz and Wiegel, 1998).

Genetics. Tools for gene cloning and inactivation for *Desulfitobacterium dehalogenans* have been developed, and halorespiration-deficient mutants have been isolated (Smidt et al., 1999, 2001, 2000).

Industrial applications. Members of the genus are of industrial importance for degradation of toxic environmental pollutants. All the reported species, with the exception of *Desulfitobacterium hafniense* strain DP7 (van de Pas et al., 2001b), carry out reductive dechlorinations in bioremediation reactors (Holliger et al., 1999; Tartakovsky et al., 1999). Members of the genus also have features desired for *in situ* bioremediation, such as relatively rapid growth, broad substrate specificities for halogenated compounds (Utkin et al., 1995), and long-term survival in the environment due to spore formation (Sanford et al., 1996). In addition, some of the species respire toxic metals and metalloids, and they can use MnO$_2$, As(V), Fe(III) gel, Fe(III) pyrophosphate, Fe(III) citrate, Cr(VI), Se(VI), and As(V) as terminal electron acceptors (Niggemyer et al., 2001).

Desulfitobacterium dehalogenans forms a biofilm on rotating pads in a continuous flow fermenter (Knoblich, 1996; Wiegel et al., 1991). The rotating pads give this fermenter an exceptionally high surface area. Upon formation of the biofilm, the transformation rates of the chlorinated compounds are 10–20 times higher than in the batch culture. This biofilm is stable at dilution rates greater than the generation time of the bacterium and even when diluted sewage sludge contaminated with chlorophenol is used as the feedstock. This system could be used as a biofilter for bioremediation of other compounds (Knoblich and Wiegel, unpublished data; Knoblich, 1996). *Desulfitobacterium hafniense* PCP-1 was the major constituent (20%) of the microbial community in a biofilm formed in a pentachlorophenol (PCP)-degrading, methanogenic reactor and was indispensible for the efficient operation of the reactor (Lanthier et al., 2005).

Degradation of 2,4-DCP in anaerobic environments, such as methanogenic freshwater sediment, is accomplished by a consortium of at least five sequentially interacting species of bacteria and archaea. The initial step is a reductive dehalogenation catalyzed by bacteria such as *Desulfitobacterium dehalogenans*, which forms 4-chlorophenol (4-CP). The *para*-dehalogenation of 4-CP is carried out by *Desulfitobacterium hafniense*-like strains (Christiansen and Ahring, 1996). In an enrichment from a freshwater sediment, 2,4-DCP and 4-CP dechlorination occurs in the temperature range of 18–40°C (Kohring et al., 1989b). Nitrate inhibits this process. Adding sufficient lactate to completely reduce the nitrate to N_2 and ammonia alleviates inhibition by preventing the accumulation of lethal concentrations of nitrite. Addition of sulfate to enrichments reduces the dehalogenation rate only to a small extent but strongly inhibits methanogenesis, presumably because the sulfate reducers outcompete the methanogens for H_2 (Kohring et al., 1989b).

Natural microbial communities may be adapted to utilize particular chlorophenols. Following adaption, these communities utilize these specific chlorophenols at a high rate without a lag phase (Hale et al., 1990, 1991; Kohring et al., 1989a, 1989b; Zhang and Wiegel, 1992). However, the enrichments for specific chlorophenols are usually not adapted for utilizing other isomers of chlorophenol, presumably because a different bacterium is required to use each isomer, except when strains similar to *Desulfitobacterium hafniense* PCP-1 are present.

Desulfitobacterium hafniense strain DCB-2 reduces humic acids or their analog anthraquinone-2,6-disulfonate (AQDS). This process may be utilized to generate electricity for more than 24 h in a microbial fuel cell. H_2, formate, lactate, pyruvate, and ethanol sustain electricity generation, but not acetate, propionate, and butyrate (Milliken and May, 2007).

Enrichment and isolation procedures

The majority of strains were isolated either from methanogenic freshwater sediments, soils, compost soils, sewage sludge, aquifers, or groundwater polluted with chlorinated organic compounds or from bioreactors (El Fantroussi et al., 1998; Fetzner, 1998; Gerritse et al., 1999; Holliger et al., 1998b; Utkin et al., 1994). An exception is *Desulfitobacterium hafniense* strain DP7, which was isolated from a human fecal sample that was not contaminated with chlorinated phenolic or aliphatic compounds (van de Pas et al., 2001b). *Desulfitobacterium dehalogenans* was enriched from an apparently pristine methanogenic freshwater sediment in Sandy Creek Nature Park in Athens, GA, USA. Isolation was performed using anaerobic mineral medium supplemented with pyruvate, formate, and 3Cl4HOPA, at concentrations of 2, 2, and 1 mM, respectively. Single colonies were collected from 0.6% agar shake culture tubes and transferred to the enrichment medium supplemented with 0.1% yeast extract, 20 mM pyruvate, and 10 mM 3Cl4HOPA (Utkin et al., 1994). In the original enrichments, agar concentrations above 1% were inhibitory; however, after several years of cultivation in the laboratory, the strain adapted to form colonies on 1.5% agar plates. *Desulfitobacterium chlororespirans* was isolated from a Michigan residential compost soil based on its ability to dechlorinate 2,3-DCP. Anaerobic agar (1.5%) shake cultures that were inoculated with the secondary enrichments yielded isolated colonies. The product of 2,3-DCP dechlorination, 3-chlorophenol (3-CP), inhibited growth, and 3-chloro-4-hydroxybenzoate (3Cl4HOB) was a better substrate since its dehalogenation product 4-hydroxybenzoate (4HOB) was not inhibitory at concentrations below 20 mM (Sanford et al., 1996). *Desulfitobacterium metallireducens* was enriched from anaerobic sediment collected from the floodplain of the San Juan River at the uranium mill tailings site of the Department of Energy in Shiprock, New Mexico, USA. The enrichment medium was a defined freshwater medium that contained 10 mM lactate as the electron donor and 5 mM AQDS as the electron acceptor. Colonies were picked from agar-solidified (1.5%) medium and restreaked on solid agar slants before they were resuspended in liquid medium. The isolate was grown in medium that was supplemented with 1% yeast extract, 10 mM lactate, and 5 mM AQDS (Finneran et al., 2002).

Maintenance procedures

The anaerobic culture tubes can be stored at room temperature or at 4°C for up to 6 weeks. For longer-term storage anaerobic, 30–50% glycerol stocks are kept at–20°C (for more frequent sampling) and at–70°C (for the main stock culture). The strains in culture collections are freeze-dried.

Differentiation of the genus *Desulfitobacterium* from closely related genera

The phylogenetic analysis based on 16S rRNA gene sequences reveal that the *Desulfitobacterium* species are closely related to the *Dehalobacter–Syntrophobotulus–Desulfosporosinus* cluster (Figure 174). *Desulfosporosinus* is a novel sulfate-reducing genus (Stackebrandt et al., 1997) encompassing species recently excluded from the genus *Desulfotomaculum* (Robertson et al., 2001; Stackebrandt et al., 2003; Stackebrandt et al., 1997). Despite common morphological features of micro-organisms within the *Desulfitobacterium* and *Dehalobacter–Syntrophobotulus–Desulfosporosinus* group, the genera are readily distinguished by physiological criteria. All members of *Desulfosporosinus* reduce sulfate in the presence of lactate or pyruvate, which differentiates this group from *Desulfitobacterium* (Robertson et al., 2001; Stackebrandt et al., 2003; Stackebrandt et al., 1997; Utkin et al., 1994). *Dehalobacter* and *Syntrophobotulus*, unlike *Desulfitobacterium* and *Desulfosporosinus*, are unable to reduce sulfite, sulfate, thiosulfate, and sulfur (Friedrich et al., 1996; Holliger et al., 1998b). Species that belong to the genus *Desulfitobacterium* differ in Gram staining, motility, and/or spore formation. However, phylogenetic analyses of the 16S rRNA genes strongly support placing all *Desulfitobacterium* strains within one genus. Common physiological characteristics shared by the genus include the utilization of fumarate as an electron acceptor and ability to reduce sulfite and thiosulfate. Moreover, strains are positive for utilization of pyruvate as an electron donor, except "*Desulfitobacterium dichloroeliminans*" DCA1 (De Wildeman et al., 2004), and negative for utilization of sulfate as an electron acceptor (Bouchard et al., 1996; Breitenstein et al., 2001; Christiansen and Ahring, 1996; Finneran et al., 2002; Niggemyer et al., 2001; Sanford et al., 1996, 1995; Utkin et al., 1994; van de Pas et al., 2001b).

Taxonomic comments

The genus *Desulfitobacterium* belongs to the family *Peptococcaceae*. In contrast to the type genus giving the family its name, cells of this genus (as well as some other genera in this family) are not cocci. Instead, they are rod-shaped throughout all growth phases. Similar to many other *Firmicutes*, some of the *Desulfitobacterium* species are Gram-staining negative.

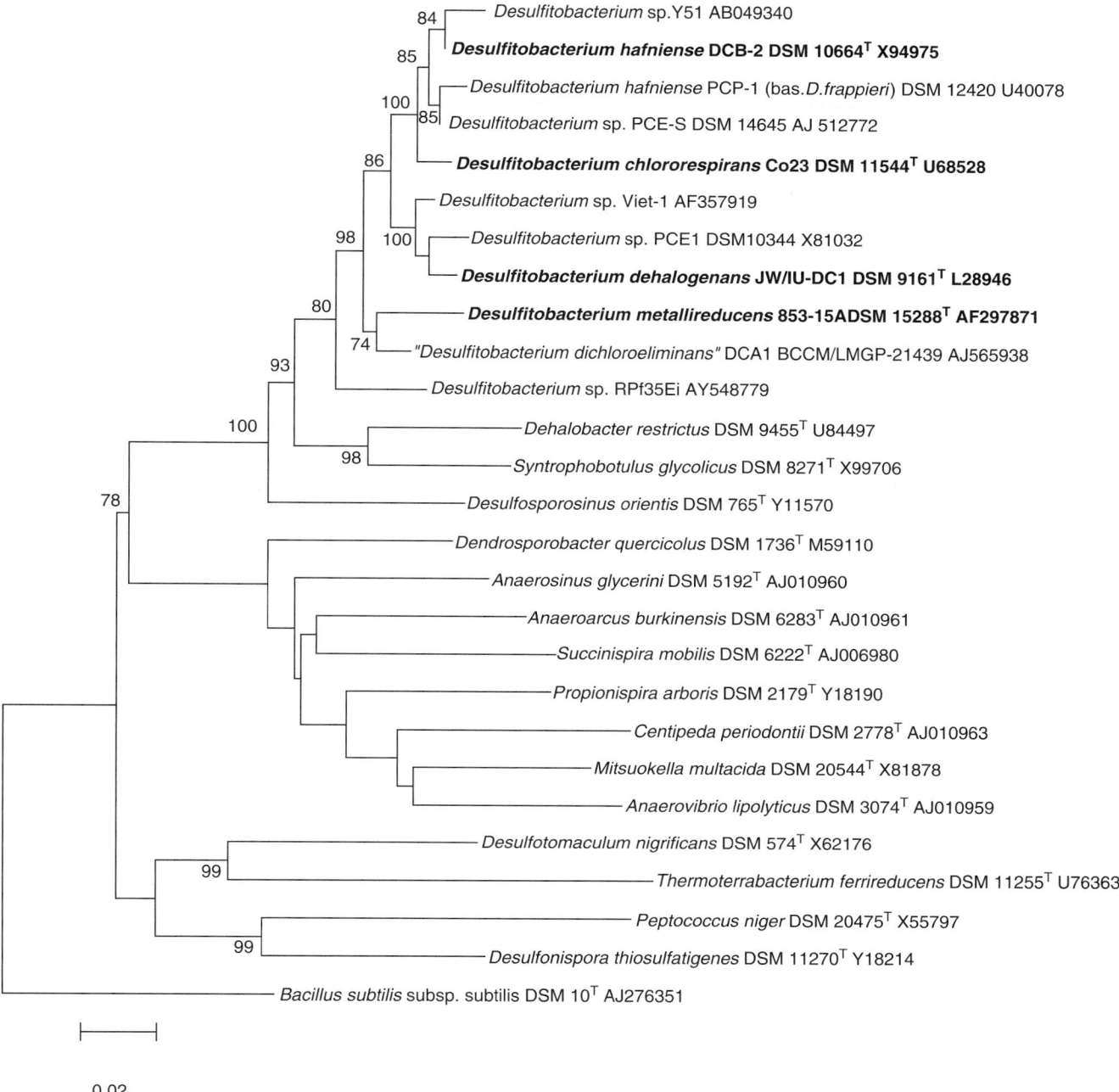

FIGURE 174. Phylogeny of 16S rRNA genes of *Desulfitobacterium* strains within the family *Peptococcaceae*. Validly published *Desulfitobacterium* species are in bold. Includes the type species of the validly published genera within the *Peptococcaceae*, with *Bacillus subtilis* as an outgroup. The tree was calculated according to the neighbor-joining method using the MEGA 3.1 software package. The alignment performed with the NAST Aligner (http://greengenes.lbl.gov/cgi-bin/nph-NAST_align.cgi). The 16S rRNA gene sequence of *Carboxydothermus hydrogenoformans* Z-2901 (GeneID: 3728363) was extracted from the complete genome sequence. The bar indicates 2% inferred substitutions. The accession numbers of the sequences are given next to the species name, strain designation, and DSM number (where applicable). The strains of *Desulfitobacterium hafniense* ATIA-3 (AY223537), ATIA-6 (AY223534), and ATIA-12 (AY223535) were omitted owing to the short lengths of their sequences. (Courtesy of Rob Onyenewoke.)

Desulfitobacterium hafniense strain PCP-1 (Bouchard, Beaudet, Villemur, McSween, Lepine and Bisaillon, 1996), formerly the type strain of the basonym *Desulfitobacterium frappieri*, is classified as *Desulfitobacterium hafniense*. On the basis of DNA–DNA hybridization and comparative physiological studies, *Desulfitobacterium hafniense* and *Desulfitobacterium frappieri* belong to the same species, with *Desulfitobacterium hafniense* having seniority (Niggemyer et al., 2001).

Further reading

Hale, D.D, W. Reineke, and J. Wiegel. 1994. Chlorophenol degradation, Chapter 4. *In* Chaudry (Editor). Biological Degradation and Bioremediation Technologies of Toxic Chemicals, Timber Press, Portland, OR. pp. 74–91.

Häggblom, M.M. and I.D. Bossert (Editors). 2003. Dehalogenation. Microbial Processes and Environmental Applications. Kluwer Academic Publishers, Boston.

List of species of the genus *Desulfitobacterium*

1. **Desulfitobacterium dehalogenans** Utkin, Woese and Wiegel 1994, 615[VP]

de.ha.lo′ge.nans. L. pref. *de* off, away; Gr. n. *hals* salt, sea; F. n. *halogen* referring to the group VII elements; L. pres. part. *dehalogenans* dehalogenating, split off halogens, referring to the characteristic property of the micro-organism to dehalogenate various chlorophenolic compounds.

The description of the species is based on the type strain JW/IU-DC1[T] (DSM 9161[T]). Four other strains of *Desulfitobacterium dehalogenans* or *Desulfitobacterium dehalogenans*-like strains have been described, including strain PCE-1 (DSM 10344; (Gerritse et al., 1996; Villemur et al., 2006), strain KBC1 (Tsukagoshi et al., 2006; Villemur et al., 2006), strain Viet-1 (Villemur et al., 2006), and strain XZ-1 (ATCC 700041; Wiegel et al., 1999). Except for strain PCE-1, which stains Gram-negative, other strains are Gram-stain positive regardless of the growth phase. Moreover, lipopolysaccharide is absent based on absence of formation of a complex with polymyxin B (Wiegel and Quandt, 1982). Cell-wall ultrastructure is typical for Gram-positive bacteria. Exponential-phase cells are slightly curved rods, 0.5–0.7 × 2.5–6 μm. In the early stationary phase, cells are more pleiomorphic, thicker, and shorter. Physiology is obligately anaerobic, heterotrophic, and mesophilic; the growth temperature range is 13–45°C, with an optimum of 37–38°C for JW/IU-DC1[T] and 10–40°C with an optimum at 34°C for strain KBC1. The pH growth range is 6.0–9.2, with an optimum at 7.5. Spore formation was not observed (Utkin et al., 1994). However, major sporulation genes are present (Onyenwoke et al., 2004). *Desulfitobacterium dehalogenans* tolerates microaerophilic conditions (<0.5% v/v, air in N₂). Growth and dechlorination of 3Cl4HOPA occurs between 13–45°C and pH of 6–9, with optima at 37°C and a pH of 7.5. NaCl, 1 M but not 0.5 M, completely inhibits dechlorination of 3Cl4HOPA. In the absence of electron acceptors, pyruvate is fermented to lactate, acetate, and CO₂. In the presence of pyruvate or formate as electron donors, 3Cl4HOPA is dehalogenated to 4-hydroxyphenylacetate (4HOPA) during an anaerobic respiration that yields 1 ATP equivalent per 1 mol of 4HOPA (Mackiewicz and Wiegel, 1998). For the spectrum of dehalogenated haloaromatic compounds, see Table 181. In addition to pyruvate and formate, lactate and H₂ are electron donors for 3Cl4HOPA reduction. Arabinose, cellobiose, mannitol, raffinose, rhamnose, ribose, sucrose, xylose, ethanol, methanol, butyrate, isobutyrate, propionate, isovalerate, acetate, and formate do not support growth in anaerobic conditions in the presence of 0.1% yeast extract. The highest rate of dehalogenation is observed with 2,3-DCP, but concentrations above 2 mM are inhibitory, In contrast, 3Cl4HOB is metabolized at concentrations of 20 mM (Utkin et al., 1995; Zhang and Wiegel, 1992). Sulfur compounds are reduced to sulfide. Dissimilatory reduction of sulfate does not occur (Mackiewicz and Wiegel, 1998, 1995; Utkin et al., 1994). Reduction of nitrate with low concentrations of electron donors (e.g., <20 mM lactate) leads to accumulation of lethal concentrations of nitrite, whereas in the presence of >20 mM of lactate, nitrite is reduced to ammonia without accumulation of nitrite. Approximately 2.5 mM lactate was transformed to acetate during the reduction of 1 mM nitrate to ammonia (Knoblich, 1996; Knoblich and Wiegel, unpublished data). *Desulfitobacterium dehalogenans* reduces humic acids when either lactate or H₂ is present as an electron donor. Moreover it reduces the AQDS with lactate as an electron donor. Reducing of both chemicals is coupled to bacterial growth (Cervantes et al., 2002). Isethionate (2-hydroxyethanesulfonate) also serves as terminal electron acceptor for strain PCE-1 (Lie et al., 1999). Strain PCE-1 dehalogenates PCE and several *ortho*-chlorinated phenolic compounds (Gerritse et al., 1996).

DNA G+C content (mol%): 45.8–46 (HPLC).
Type strain: JW/IU-DC1, ATCC 51507, DSM 9161.
GenBank accession number (16S rRNA gene): L28946.

2. **Desulfitobacterium chlororespirans** Sanford, Cole, Löffler and Tiedje 2001, 793[VP] (Effective publication: Sanford, Cole, Löffler and Tiedje 1996, 3806.)

chlor.o.resp.i′rans. N.L. part. adj. *chloro* referring to the group VII element chlorine; fr. L. *respirare* to blow, breathe; N.L. part. adj. *chlororespirans* breathing chlorine, referring to the characteristic of coupling oxidation of electron donors to reductive removal of chlorines from various chlorophenolic compounds via a respiratory process used for obtaining energy for growth.

Description is based on the type strain Co23[T] (DSM 11544[T]; Sanford et al., 1996). Exponential growth phase cells are slightly curved rods, 3–5 × 0.5–0.7 μm. Cells stain Gram-negative. Terminal spores appear in late-exponential growth phase causing swelling. Metabolism is obligately anaerobic. Colonies are round with a diameter of 1–2 mm, smooth and white on R2A agar medium (Sanford et al., 1996). The optimal temperature for growth is 37°C, and the pH range is 6.8–7.5. The strain utilizes lactate, pyruvate, H₂ + acetate, formate, butyrate, and yeast extract as electron donors, and sulfite, thiosulfate, sulfur, and *ortho*-substituted haloaromatic compounds as terminal electron acceptors (Table 181). 3Cl4HOB is the superlative substrate for dehalorespiration. The cell density increases in lactate-fed cultures in the presence of 3Cl4HOB. However, no increase in turbidity occurs with lactate alone or with lactate + sulfate. Pyruvate supports growth in the absence of an electron acceptor and is fermented stoichiometrically to acetate, CO₂, and H₂.

Compounds not utilized include: 2,4-DCP, 2,5-DCP, and PCP (Sanford et al., 1996). In addition to *ortho*-substituted haloaromatic compounds, *Desulfitobacterium chlororespirans* debrominates and deiodinates the polysubstituted herbicides, such as bromoxynil (3,5-dibromo-4-hydroxybenzonitrile), ioxynil (3,5-diiodo-4-hydroxybenzonitrile), and the bromoxynil metabolite 3,5-dibromo-4-hydroxybenzoate in the presence of lactate as an electron donor (Cupples et al., 2005). This strain was isolated from a Michigan residential compost soil.

DNA G + C content (mol%): 48.8 (T_m).
Type strain: Co23, ATCC 700175, DSM 11544.
GenBank accession number (16S rRNA gene): U68528.

3. **Desulfitobacterium hafniense** Christiansen and Ahring 1996, 446[VP] emend. Niggemyer, Spring, Stackebrandt and Rosenzweig 2001, 5578.

haf.ni.en'se. L. n. *Hafnia* Copenhagen; L. suff. *-ensis* native of; L. neut. adj. *hafniense* native of Copenhagen, referring to the place of isolation.

Description of the species is based on the type strain DCB-2[T] (DSM 10664[T]); however, some features of other strains are indicated. Exponential growth phase cells are curved rods 2–6 × 0.5–0.8 µm and occur singly, in pairs, or in small chains. Cells are Gram staining negative (Christiansen and Ahring, 1996; Niggemyer et al., 2001), some strains stain positive (TCP-A) (Breitenstein et al., 2001) or react positively with a fluorescent probe against Gram-type-positive bacteria. The electron microscopic analysis reveals a Gram-positive type of cell wall (Bouchard et al., 1996; Gerritse et al., 1999). The cells are motile and contain one or rarely two terminal and up to six lateral flagella (strains GBFH and TCE-1). Strain PCP-1 (basonym *Desulfitobacterium frapieri* PCP-1) is nonmotile. Spores are formed and located terminally, causing the swelling of the cells; some strains (TCE-1, DP7) do not sporulate (Gerritse et al., 1999; van de Pas et al., 2001b). Growth is obligately anaerobic (Bouchard et al., 1996; Breitenstein et al., 2001; Christiansen and Ahring, 1996; Gerritse et al., 1999; Niggemyer et al., 2001). Strain GBFH is resistant to 2% air in the head space of the liquid cultures (Niggemyer et al., 2001). Optima for growth are pH 7.0 and 37°C, but growth occurs in pH range 6.5–7.5 and temperature range 23–40°C. Pyruvate, lactate, formate, fumarate, and tryptophan support growth of all strains. Butyrate, succinate, malate, and ethanol support growth of the type strain DCB-2 and strain PCP-1 (Niggemyer et al., 2001). Growth does not occur with L-arabinose, D-glucose, D-fructose, D-galactose, L-rhamnose, D-ribose, cellobiose, xylose, lactose, maltose, mannose, sucrose, raffinose, trehalose, D-mannitol, benzoate, acetate, propionate, sorbitol, salicilin, inulin, starch, glycerol, esculin, L-histidine, glutamine, L-threonine, and phenylalanine. Thiosulfate, sulfite, and Fe(III) are reduced in the presence of pyruvate as an electron donor. In addition, isethionate (2-hydroxyethanesulfonate) serves as a terminal electron acceptor (Lie et al., 1999). Sulfate is not reduced in the presence of the variety of carbon sources (Christiansen and Ahring, 1996). Strain DP7 does not carry out dechlorination of either polychlorinated ethenes or phenolic compounds (van de Pas et al., 2001b). Strain PCE-S can oxidize phenyl methyl ethers such as vanillate, syringate, and veratrol for fumarate reduction and growth. *O*-Demethylase activity for utilization of phenyl methyl ethers is produced only when cells are growing on syringate, indicating that this enzyme might be substrate induced. PCE strongly inhibits the enzyme activity (Neumann et al., 2004).

Desulfitobacterium hafniense type strain DCB-2[T] (JGI) and strain Y51 (Nonaka et al., 2006) are the only representatives of the genus with nearly completed genome sequence. Additional strains of the species include: PCP-1 (basonym *Desulfitobacterium frapieri*; DSM 12420; Bouchard et al., 1996), PCE-S (DSM 14645; Miller et al., 1997), TCE-1 (DSM 12704; Gerritse et al., 1999), TCP-A (DSM 13557; Breitenstein et al., 2001), GBFH (Niggemyer et al., 2001), DP7 (van de Pas et al., 2001b), Y51 (Suyama et al., 2001), ATIA-3 (Lecouturier et al., 2003), ATIA-6 (Lecouturier et al., 2003), and G2 (Genbank accession number for the 16S rRNA gene sequence: AF320982).

DNA G+C content (mol%): 47 (HPLC)
Type strain: DCB-2, DSM 10664.
GenBank accession number (16S rRNA gene): X94975.

4. **Desulfitobacterium metallireducens** Finneran, Forbush, Van-Praagh and Lovley 2002, 1934[VP]

me.tal.li.re.du'cens. L. n. *metallum* metal; L. part. adj. *reducens* converting to a different state; N.L. part. adj. *metallireducens* reducing metal, referring to the ability to couple growth to respiration of several metals.

The description of the species is based on the type strain 853-15A[T] (DSM 15288[T]). Cells during the exponential growth-phase are nonmotile, slightly curved rods, 2–5 × 0.5 µm. In stationary phase, they become shorter and eventually "C" shaped. Stains Gram-positive in all growth phases. In micrographs of ultrathin sections, the presence of a thick peptidoglycan layer characteristic for Gram-type positive bacteria is observed. Does not form spores (Finneran et al., 2002). *Desulfitobacterium metallireducens* is not motile and does not possess flagella. It reduces metal cations and dehalogenates polychlorinated ethenes. The 3Cl4HOPA is the only haloaromatic compound utilized (Finneran et al., 2002). Optimal growth is at 30°C and pH 7.5. Electron donors for respiration include lactate, formate, ethanol, butanol, butyrate, pyruvate, and malate. Fe(III), Mn(IV), sulfur, thiosulfate, 3Cl4HOPA, trichloroethylene, PCE, AQDS, humic acids, Cr(VI), and selenite are electron acceptors (Table 181). AQDS reduction results in an increase in cell number and accumulation of acetate from organic substrates. Good growth is also obtained with lactate and Fe(III) citrate as the electron acceptor. The strain does not utilize yeast extract, acetate, H_2, methanol, 2-propanol, benzoate, peptone, Casamino acids, isobutyrate, valerate, benzyl alcohol, salicylic acid, phenol, benzene, glucose, fructose, fumarate, glycerol, nicotinate, or caproate. Glucose, lactate, and citrate are not fermented (Finneran et al., 2002). The strain was enriched from anaerobic sediment collected from the floodplain of the San Juan River at the uranium mill tailings site of the Department of Energy in Shiprock, NM, USA.

DNA G+C content (mol%): not determined.
Type strain: 853–15A, ATCC BAA-636, DSM 15288.
GenBank accession number (16S rRNA gene): AF297871.

Other available strains

1. **"Desulfitobacterium dichloroeliminans"** Effective publication: De Wildeman, Diekert, Van Langenhove and Verstraete 2003, 5645.

    The strain DCA1 is available as BCCM/LMG P-21439. Cells are motile, curved rods, 2–5 × 0.5–0.7 μm. Gram-stain-positive, spore formation does not occur. Optimal temperature and pH are 25–30°C and 7.2–7.8, respectively. H$_2$, formate, and lactate serve as electron donors during 1,2-dichloroethane (1,2-DCA) dechlorination. Sulfite, thiosulfate, and nitrate are electron acceptors. Aerotolerant and able to survive the presence of O$_2$ for at least 24 h. Strain completely dechlorinates 1,2-DCA, and dichlorinated derivatives of propane and butane, but not highly chlorinated alkanes such as hexa-, penta-, and tetrachloroethanes. The mechanism of dechlorination does not rely on the reductive hydrogenolysis, rather it exclusively employs dichloroelimination of various substrates.

    The process of dichloroelimination can be utilized in bioremediation, since addition of the cells to 1,2-DCA-contaminated groundwater yielded complete detoxification with high efficiency (De Wildeman et al., 2003; Maes et al., 2006). The real-time PCR quantification method in groundwater for the strain DCA1 was developed and may be useful to study the kinetics and abundance of this halorespiring strain in the field (Van Raemdonck et al., 2006).

2. **Desulfitobacterium sp. strain RPf35Ei**
    Cells are rod-shaped, 2.5–4 × 0.3–0.5 μm. The strain forms spores, oxidizes ethanol and lactate, and reduces sulfite but not sulfate. It was isolated from a sulfate-reducing fluidized-bed reactor (Kaksonen et al., 2004). Phylogenetic analyses of the 16S rRNA gene indicate that this strain may represent a novel species of this genus (Figure 174; Kaksonen et al., 2004; Villemur et al., 2006).

Genus V. **Desulfonispora** Denger, Stackebrandt and Cook 1999, 1602^{VP}

THE EDITORIAL BOARD

De.sul.fo.ni.spo′ra. N.L. pref. *desulfono* desulfonating; Gr. fem. n. *spora* spore; N.L. fem. n. *Desulfonispora* desulfonating spore (-former).

Rods with subterminal spores. Gram-stain-positive. Motile. **Strictly anaerobic. Ferments taurine to acetate, ammonia, and thiosulfate**. Oxidase negative.

DNA G+C content (mol%): 52.

Type species: **Desulfonispora thiosulfatigenes** Denger, Stackebrandt and Cook 1999, 1602^{VP}.

Further descriptive information

Phylogeny. Phylogenetic analysis of the 16S rRNA gene sequence places this genus within the family *Peptococcaceae*, of the order *Clostridiales* (Figure 6). Its closest relatives (with 86–88% sequence similarity) are *Desulfotomaculum, Desulfitobacterium, Desulfosporosinus,* and *Peptococcus*. The level of relatedness even to its closest relative (*Peptococcus niger*) is low.

Desulfonispora specializes in taurine fermentation and synthesizes thiosulfate from sulfite. This synthetic transformation is proposed to be mediated by sulfite reductase and a putative electron transport chain containing membrane-bound cytochrome *b*. None of the other naturally occurring organosulfonates (cysteate, isethionate, and coenzyme M) are utilized for growth. Bile salts, such as taurocholate, do not support growth. Similarly, other common carbon sources, tryptone, and yeast extract do not support growth (Denger et al., 1997).

Enrichment and isolation procedures

Enrichment was performed under strictly anoxic conditions in bicarbonate-buffered, titanium(III) nitrilotriacetate-reduced mineral salts medium at pH 7.0 containing 10 mM taurine as the sole source of carbon and energy, as described by Denger et al. (1997). Incubation was at 30°C in the dark, and cultures were shaken occasionally. Subcultures of putative enrichments were isolated by the agar shake method (Pfennig, 1978).

List of species of the genus *Desulfonispora*

1. **Desulfonispora thiosulfatigenes** Denger, Stackebrandt and Cook 1999, 1602^{VP}

    thi.o.sul.fa.ti′ge.nes. N.L. n. *thiosulfas* thiosulfate; N.L. suff. *genes* -producing; N.L. part. adj. *thiosulfatigenes* thiosulfate-producing.

    Cells are rods (0.7–1.0 × 2–5 μm) with subterminal spores. Motile. All are oxidase negative and most are catalase negative.

    Membranes contain a high level of cytochrome *b*. Growth occurs at mesophilic temperatures with taurine consumption and concomitant production of acetate, ammonia, and thiosulfate (1:1:0.5). Sulfate, sulfite, and nitrate are not reduced.

    DNA G+C content (mol%): 52 (HPLC).

    Type strain: GKNTAU, ATCC 700533, DSM 11270.

    GenBank accession number (16S rRNA gene): Y18214.

Genus VI. **Desulfosporosinus** Stackebrandt, Sproer, Rainey, Burghardt, Päuker and Hippe 1997, 1138^VP emend. Robertson, Bowman, Franzmann and Mee 2001, 139, emend. Stackebrandt, Schumann, Schüler and Hippe 2003, 1441.

HANS HIPPE AND ERKO STACKEBRANDT

De.sul.fo.spo.ro.si'nus. L. pref. *de* from; L.n. *sulfur* sulfur; Gr. n. *spora* spore; L. n. *sinus* bend; N.L. masc. n. Desulfosporosinus, a spore-forming curved (organism) that reduces sulfur compounds.

Gram-negative rods that have a multilayered cell-wall structure. Endospores present, oval and subterminal to (almost) terminal, causing the cells to swell slightly. Motile with lateral or peritrichous flagella or nonmotile. Strictly anaerobic. If determined, desulfoviridin and cytochrome c_3 absent and bisulfite reductase P582 present. Sulfate and thiosulfate are reduced to sulfide in the presence of lactate but not in the presence of acetate or fructose. Incomplete oxidation of organic compounds to acetate. Acetate is the fermentation end product. Autotrophic growth with hydrogen plus sulfate. LL-Diaminopimelic acid is the diagnostic diamino acid of peptidoglycan. Contains menaquinone with a side chain with seven isoprene units (MK-7 type) as major component. Predominant fatty acids are even-numbered, saturated and unsaturated fatty acids; significant amounts of aldehydes, detected as 1,1-dimethylacetals. Phylogenetically, a member of the *Clostridium/Bacillus* subphylum of Gram-positive bacteria.

DNA G+C content (mol%): 41.6–45.9 (T_m).

Type species: **Desulfosporosinus orientis** (Campbell and Postgate 1965) Stackebrandt, Sproer, Rainey, Burghardt, Päuker and Hippe 1997, 1138^VP (*Desulfotomaculum orientis* Campbell and Postgate 1965, 361; "*Desulfovibrio orientis*" Adams and Postgate 1959, 256).

Taxonomic comments

The type species of *Desulfosporosinus*, *Desulfosporosinus orientis*, was originally described as *Desulfovibrio orientis* (Adams and Postgate, 1959). The authors indicated in the original publication, describing strain NCIB 8323^T from soil in Singapore as a new species, that the nomenclature was not in total agreement with the taxonomic judgment of A.R. Prévot, who had also examined this strain. As Prévot found cells to contain spores, the affiliation of strain NCIB 8323^T to *Desulfovibrio* did not match the genus description. According to the classification used by Prévot (1957), it would have been classified as "*Sporovibrio orientis*". As already mentioned by Adams and Postgate (1959), *Desulfovibrio orientis* differs in morphology and physiology from other *Desulfovibrio* species, and the authors indicated that this species may phylogenetically not be related to other members of the genus.

Desulfovibrio orientis was reclassified as *Desulfotomaculum orientis* (Campbell and Postgate, 1965) and was included in the Approved Lists of Bacterial Names (Skerman et al., 1980) as one of four species of this genus. Vainshtein et al. (1995) isolated a mesophilic, spore-forming sulfate-reducer from ancient permafrost in Russia, which was classified as a strain of the species *Desulfotomaculum orientis* (DSM 8344) on the basis of phenotypic characteristics.

The reclassification of *Desulfotomaculum orientis* NCIB 8323^T (DSM 765^T) as *Desulfosporosinus orientis* was proposed in 1997 (Stackebrandt et al., 1997). The placement of strains DSM 765^T and DSM 8344 was based on phylogenetic analysis in which they formed a separate cluster from other *Desulfotomaculum*, being more closely related to members of *Desulfitobacterium* than to the other *Desulfotomaculum* species. The second species, *Desulfosporosinus meridiei*, was described for eight strains of spore-forming, sulfate-reducing bacteria isolated from a gasoline contaminated shallow aquifer in the sandy soils at Eden Hill, on the Swan Coastal Plain near Perth in the south-west of Australia (Robertson et al., 2001).

A third species has recently been transferred to the genus by reclassifying *Desulfotomaculum auripigmentum* as *Desulfosporosinus auripigmenti* (Stackebrandt et al., 2003). *Desulfotomaculum auripigmentum* ATCC 700205^T (DSM 13351^T) was originally affiliated to *Desulfotomaculum* mainly on the basis of 16S rDNA analysis (Newman et al., 1997). This nonmotile, sausage shaped, arsenate and sulfate reducing, Gram-positively staining bacterium, for which spore formation has not been reported, was placed phylogenetically adjacent to *Desulfotomaculum orientis* in the 16S rDNA dendrogram of relationship (96.2% similarity). The description by Newman et al. (1997) overlapped with the reclassification of *Desulfotomaculum orientis* as *Desulfosporosinus orientis* (Stackebrandt et al., 1997) and therefore neither of the two research groups was aware of the other group's work. In the publication of Robertson et al. (2001), *Desulfosporosinus meridiei* DSM 13257^T branched adjacent to *Desulfotomaculum auripigmentum* DSM 13351^T (97.6% 16S rRNA gene sequence similarity), but, despite the grouping of a *Desulfotomaculum* species between *Desulfosporosinus meridiei* and *Desulfosporosinus orientis*, the generic affiliation of *Desulfosporosinus auripigmentum* remained unchallenged. Morphological and physiological traits, phylogenetic position, results of DNA–DNA reassociation studies, and chemotaxonomic properties of *Desulfotomaculum auripigmentum*, including properties supplementary to the original description, led at the end to the transfer of this species to *Desulfosporosinus*.

Further descriptive information

Cell morphology. In the original description of *Desulfovibrio orientis* (Adams and Postgate, 1959) the type strain NCIB 8382^T was described as a fat rod, often slightly curved, sometimes paired or in short chains; the mean dimensions were 1.4 × 4.8 μm. Cells stained Gram-negatively and no evidence for a capsule was given. Culturing for 5 or more days (the description of Adams and Postgate, (1959), did not indicate the growth medium) contained a significant number of cells showing a granulated cytoplasm; forms with a single central or near-central granule was sometimes seen. Figure 175 displays cells of *Desulfosporosinus orientis* DSM 765[T] grown on pyruvate medium, containing these central granules. Occasionally, long filament-like cell chains are observed (Figure 176). Nonprogressive "twisting and turning" motions were sometimes seen, particularly in cells taken from colonies on agar media. Cells contained one or two polar or near-polar flagella. In contrast to the observation of Adams and Postgate (1959), Prévot reported infrequent sporulation of this strain with spores occurring central or paracentral (cited Adams and Postgate, 1959), later also detected by the latter authors. Later reports described the location of spores to be subterminal or terminal, causing the cells to swell slightly (Figure 177), and motility by peritrichous flagellation (Campbell and Postgate, 1965; Campbell and Singleton, 1986).

Strains of *Desulfosporosinus meridiei* investigated by Robertson et al. (2001) differed significantly in size; while cells of most strains are 0.7–1.1 μm in width and 2.3–4.2 μm in length, some cells measure up to 13.0 μm in length. Cells stained Gram-negatively or were Gram-variable and possessed a multilayered cell wall. Cells possessed a single lateral flagellum (only demonstrated for one strain) and were motile only in the early exponential phase of growth.

The only known strain of *Desulfosporosinus auripigmenti*, ATCC 700205[T], is a Gram-positive and nonmotile rod, 0.4 μm in width × 2.5 μm in length, with an S-layer (hexagonal surface array) attached to the thin multilayered peptidoglycan (Newman et al., 1997). Spores, originally not observed, were detected in cells (Stackebrandt et al., 2003) grown on mineral-vitamin medium no 641 (DSM catalog of strains, 2001) (Figure 177).

Morphological and cultural characteristics of species of the genus *Desulfosporosinus* and of strain DSM 8344 are listed in Table 182.

Phylogenetic position. The rationale for describing *Desulfotomaculum orientis* as the type of a new genus was based on the phylogenetic position derived from analysis of 16S rRNA gene sequences from members of *Desulfotomaculum* (Stackebrandt et al., 1997). While the majority of species formed several phylogenetic lineages, *Desulfotomaculum orientis* represented yet another lineage branching closely to the genus *Desulfitobacterium* (93.1-94.4% sequence similarity). This pattern was recovered consistently with different treeing algorithms contained in the PHYLIP package (Felsenstein, 1993), including neighbor-joining (Saitou and Nei, 1987) and maximum-parsimony analyses and the distance matrix algorithm of De Soete (1983). Most of these clusters were recovered in a high proportion of the trees generated, as demonstrated by the high bootstrap values for these groups (Figure 178).

While *Desulfosporosinus orientis* DSM 765[T] (Singapore I) (GenBank acesssion no. Y11570) and strain DSM 7493 (Singapore II) (GenBank accession no. AJ493052) are highly related (99.5% similarity), strain DSM 8344 (GenBank accession no. Y11571) is only moderately related to the type strain (96.2% similarity). This was confirmed by DNA–DNA binding similarity values of 39% (Stackebrandt et al., 2003). Strain DSM 8344 forms a

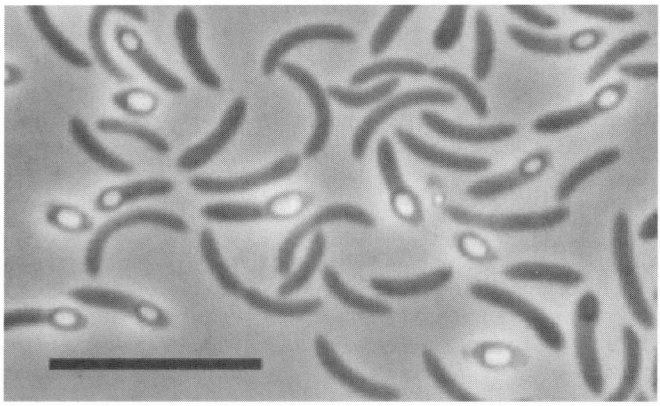

FIGURE 175. Sporulating cells of *Desulfosporosinus orientis* DSM 765[T] grown in sulfate-limited chemostat culture with H₂ as electron donor. Bar = 10 μm. (Courtesy of H. Cypionka.)

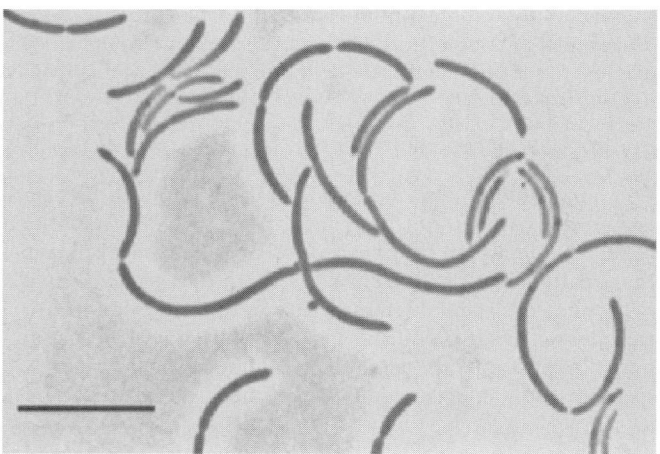

FIGURE 176. Vegetative cells of *Desulfosporosinus orientis* DSM 765[T] grown in DSMZ medium 641 with pyruvate, showing near-central granule. Bar = 10 μm.

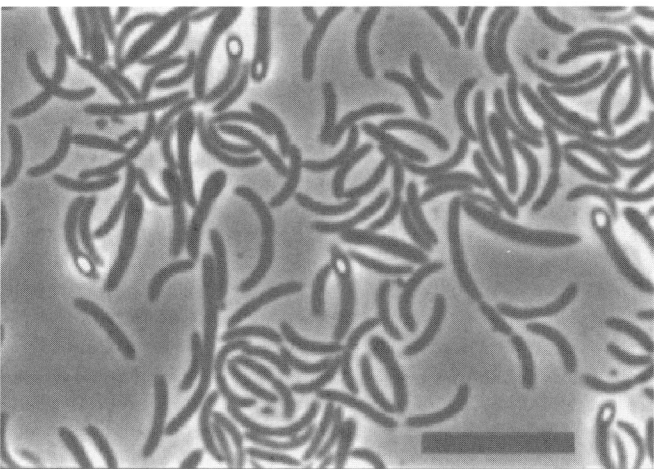

FIGURE 177. Sporulating cells of *Desulfosporosinus auripigmenti* DSM 13351[T], grown in DSMZ medium 641. Bar = 10 μm. (Reprinted with permission from Stackebrandt et al., 2003. International Journal of Systematic and Evolutionary Microbiology *53*: 1439–1443.)

TABLE 182. Morphological and cultural characteristics of species of the genus *Desulfosporosinus* and related strains[a]

Property	*D. orientis* DSM 765[T] (Singapore I[T])	*D. auripigmenti* DSM 13351[T] (OREX-4[T])	*D. meridiei* DSM 13257[T] (S10[T])	*Desulfosporosinus* sp. DSM 8344
Cell shape	Curved rod	Curved rod	Curved rod	Curved rod
Cell size (μm)	0.7–1 × 3–5 (filaments 10–15 μm may occur)	0.4 × 2.5	0.7–1.1 × 2.3–4.2 (3.5–13[b])	1–1.2 × 4.5–5.5 (filaments 10–15 μm may occur)
Endospore morphology	Oval	Oval	Oval	Spherical
Endospore location[c]	ST to T	ST to T	ST	T
Motility	+	–	+	+
Flagella type	Peritrichous	–	Single lateral	NR
pH optimum	~7	6.4–7.0	~7	~7
NaCl optimum (%)	NR	NR	NR	0–0.1
NaCl range (%)	0–<5	NR	0–<4	0–<2.5
Temperature optimum (°C)	37	25–30	28	28
Temperature range (°C)	30–42	NR	10–37	NR

[a]Symbols: +, >85% positive; d, different strains give different reactions (16–84% positive); –, 0–15% positive; w, weak reaction; NR, not recorded.
[b]Reached by some strains.
[c]ST, subterminal; T, terminal.

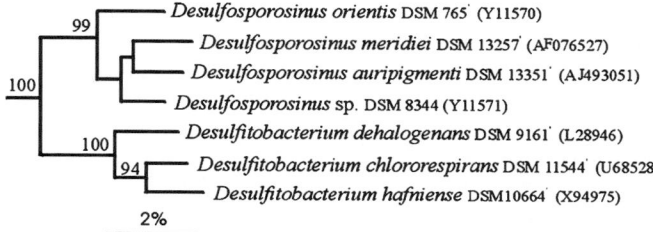

FIGURE 178. Dendrogram of 16S rRNA gene sequence relationship (De Soete, 1983), displaying the phylogenetic position of species of *Desulfosporosinus* and of strain DSM 8344. Numbers at branching points refer to bootstrap values (1000 resamplings). The tree was rooted with the 16S rRNA gene sequences of *Desulfotomaculum* species. Scale bar = 2 nucleotide substitutions per 100 sequence positions. Modified from Stackebrandt et al. (2003).

cluster with *Desulfosporosinus meridiei* DSM 13257[T] (GenBank accession no. AF076527) and *Desulfosporosinus auripigmenti* DSM 13351[T] (GenBank accession no. AJ493051) (which share about 97.5% sequence similarity). The phylogenetic distance between the type strains of *Desulfosporosinus orientis* and *Desulfosporosinus meridiei* was confirmed by a DNA–DNA binding value of 37.7% (Robertson et al., 2001) and 54% (Stackebrandt et al., 2003), respectively, while strain DSM 765[T] and *Desulfosporosinus auripigmenti* DSM 13351[T] showed 30% DNA–DNA reassociation, thus confirming the status of all three species. The distinctness of the species has also been confirmed by riboprint analyses (Stackebrandt et al., 2003) and, for *Desulfosporosinus orientis* DSM 765[T] and strains of *Desulfosporosinus meridiei*, by RAPD-PCR analyses (Robertson et al., 2001). As a result of the latter study the grouping of *Desulfosporosinus meridiei* strains into subclusters I-III was changed to group A (for subcluster I plus strain S4) and group B (for subcluster III plus strain S7; the latter subcluster contains the type strain).

16S rDNA sequence comparison (Robertson et al., 2000) indicated that subcluster I, containing 3 strains, was most closely related to DSM 8344 (~98.8% similarity). One representative each of subclusters I and III had a DNA–DNA similarity of 81%.

Riboprint analyses (Stackebrandt et al., 2003) showed similar *Eco*RI restriction patterns for the type strain of *Desulfosporosinus orientis* DSM 765[T] and strain DSM 7439, while the type strains of *Desulfosporosinus meridiei* and *Desulfosporosinus auripigmenti* and strain DSM 8344 were characterized by individual patterns.

Chemotaxonomy. Cellular fatty acids profiles of *Desulfosporosinus orientis* are dominated by even carbon, straight chain saturated ($C_{14:0}$, $C_{16:0}$) and mono-unsaturated ($C_{16:1\ \omega7c}$; $C_{18:1\ \omega9c}$; $C_{18:1\ \omega8c}$) fatty acids. In addition, significant amounts of $C_{16:0}$ dimethylacetals (8.7%) and traces of aldehydes (2.8%) were found (Table 183). The profiles of strains NCIMB 8445 (strain Singapore II) and *Desulfosporosinus* species DSM 8344 were almost identical to that of *Desulfosporosinus orientis* DSM 765[T] (Hippe et al., 1995; Vainshtein et al., 1995). Neither cyclopropane fatty acids, nor iso- and anteiso fatty acids were detected. This is in contrast to the studies of Robertson et al. (2000) and Ueki and Suto (1979), who reported on the presence of these fatty acids in *Desulfosporosinus orientis*. These differences may be caused by the use of different growth media and analytical equipment.

The fatty acid profiles of strains of *Desulfosporosinus meridiei* are also dominated by even-carbon, straight-chain, saturated ($C_{14:0}$; $C_{16:0}$) and monounsaturated ($C_{16:1\ \omega7c}$; $C_{18:1\ \omega7c}$) fatty acids. In contrast to the finding of a significant percentage (29% in the type strain) accounting for iso- and anteiso branched fatty acids (Robertson et al., 2000), the emendation of the genus *Desulfosporosinus* (Robertson et al., 2001) indicates the presence of minor amounts of branched-chain and cyclopropane fatty acids. Major fatty acids of *Desulfosporosinus auripigmenti* DSM 13351[T] are $C_{16:0}$ and the monounsaturated $C_{16:1\ \omega7c}$ and $C_{18:1\ \omega9c}$. Dimethylacetals (DMA), such as $C_{DMA-18:1\ \omega9c}$, $C_{DMA-16:1\ \omega7c}$, and $C_{DMA-18:0}$, as well as traces of $C_{17:0\ cyclo}$ occur. 10-Methyl branched fatty acids, considered chemotaxonomic markers in members of the *Desulfobacteriaceae*, and $C_{17:1\ iso\ \omega10c}$ fatty acids, occurring in many species of the genus *Desulfovibrio* but also in *Desulfomicrobium* and in several species of *Desulfotomaculum*, are absent in strains of *Desulfosporosinus* (Hagenauer et al., 1997a; Hippe et al., 1995; Vainshtein et al., 1992).

The peptidoglycan contains LL-diaminopimelic acid (A_2pm) as the diamino acid (Table 183) This diagnostic feature has also

TABLE 183. Biochemical and chemotaxonomic characteristics of species of the genus *Desulfosporosinus* and strain DSM 8344[a]

Property	*D. orientis* DSM 765[T] (Singapore I[T])	*D. meridiei* DSM 13257[T] (S10[T])	*D. auripigmenti* DSM 13351[T] (OREX-4[T])	*Desulfosporosinus* sp. DSM 8344
Gram-stain	Negative	Negative/variable	Negative	Negative
Type of oxidation	Incomplete	Incomplete	Incomplete	Incomplete
Chemoautotrophic growth on H_2+CO_2	+	+	+	NR
Homoacetogenic growth on	Formate, methanol, ethanol trimethoxybenzoate	Methanol, ethanol	Pyruvate	NR
DNA G+C content (mol%)	45 (Bd), 45.9 (T_m)	46.9 (T_m)	41.6 (HPLC)	42.1 (HPLC)
Menaquinone	MK-7	MK-7	MK-7	NR
Cytochrome type	*b*	NR	NR	*b*
Cellular fatty acids[b]	u, s, dma, ald	u, s, cp	s, u, dma, cp, ald	u, s, dma, ald

[a]Symbols: +, >85% positive; NR, not recorded.

[b]u, Unsaturated; s, saturated; dma, dimethylacetals; cp, cyclopropane (traces); ald, aldehydes (traces).

been found in members of the phylogenetically neighboring genus *Desulfitobacterium*, e.g., *Desulfitobacterium* species PCE1 (Gerritse et al., 1996), *Desulfitobacterium dehalogenans* DSM 9161[T], and *Desulfitobacterium hafniense* DSM 10664[T] (Stackebrandt et al., 2003). Also, some species of the genus *Desulfotomaculum*, e.g., *Desulfotomaculum thermoacetoxidans* DSM 5813[T] and *Desulfotomaculum thermobenzoicum* subspecies *thermobenzoicum* DSM 6193[T], contain LL-A$_2$pm.

Cells of *Desulfosporosinus orientis* contain spermidine and spermine as the major polyamines (Hamana, 1999). The same polyamines were detected in the phylogenetically related *Desulfitobacterium dehalogenans* but also in *Desulfotomaculum acetoxidans*. Other species of the genus *Desulfotomaculum* (*Desulfotomaculum nigrificans*, *Desulfotomaculum ruminis*, and *Desulfotomaculum thermobenzoicum*) differed from this composition (Hamana, 1999). The taxonomic significance of polyamines in Gram-positive sulfate- and sulfite-reducers needs to be placed on a broader basis by being included in a detailed study of all extant species.

The major isoprenoid quinone is a menaquinone of the MK-7 type (Collins and Widdel, 1986; Stackebrandt et al., 2003). *Desulfotomaculum auripigmenti* also contains significant amounts of MK-5, while MK-6 was found in traces in *Desulfosporosinus orientis* and *Desulfotomaculum auripigmenti* (Table 183).

Cytochromes have only been determined for *Desulfosporosinus orientis* DSM 765[T] and strain DSM 8344 (Table 183). In both strains, cytochrome *b* has been found as the principal component, with traces of cytochrome *c* in strain DSM 765[T].

Physiological and metabolic properties. Oxygen is inhibitory to *Desulfosporosinus orientis* as a strictly anaerobic bacterium. Even in the absence of sulfide, survival of *Desulfosporosinus orientis* after oxygen stress induced by 3 h exposure was found to be very low (survival fraction 0.01%) compared with other sulfate-reducing bacteria (e.g., species of *Desulfovibrio*, *Desulfococcus*, and *Desulfotomaculum*) (Cypionka et al., 1985). Reducing agents even increased the oxygen toxicity. Enhanced sulfide concentrations (exceeding 2 mmol Na$_2$S/l) in growth media are toxic to cells of *Desulfosporosinus orientis* (Klemps et al., 1985). Growth yields of *Desulfosporosinus orientis* obtained in a sulfide-controlled continuous culture under sulfate-, hydrogen-, and nonlimited conditions were considerably higher (max. 12.4 g dry mass per mol sulfate reduced) than determined in batch culture (7.5 g per mol) (Cypionka, 1986).

Reinvestigation of physiological properties of the type strain of *Desulfosporosinus orientis* (Klemps et al., 1985) led to the unexpected finding that this species, in contrast to *Desulfotomaculum* strains, was able to grow chemoautotrophically with hydrogen, carbon dioxide, and sulfate. It was also able to grow in the absence of sulfate with formate, methanol, ethanol, lactate, pyruvate, or trimethoxybenzoate. Homoacetogenic growth on methanol and ethanol has also been reported for *Desulfosporosinus meridiei* (Robertson et al., 2001). Newman et al. (1997) report in the original description of *Desulfotomaculum auripigmentum*, that the type strain grows autotrophically on H$_2$ and produces acetate from organic substrates. It can thus be concluded that all species of the genus can be considered homoacetogenic bacteria.

Diazotrophic growth has been demonstrated in *Desulfosporosinus orientis* (Lespinat et al., 1985). During studies on the removal of SO$_2$ from flue gas by sulfate-reducing bacteria, growth of *Desulfosporosinus orientis* on H$_2$, CO$_2$, and SO$_2$ with the production of H$_2$S has been observed (Deshmane et al., 1993).

In contrast to the species *Desulfosporosinus orientis* and *Desulfosporosinus auripigmenti*, intraspecific diversity has been determined for several species. Considering the range of genomic diversity within a species, it is not surprising to find a substantial degree of intraspecific physiological diversity, as laid down in the description of the species *Desulfosporosinus meridiei* (Robertson et al., 2001). For example, nitrate reduction is variable (3 out of 7 strains, type strain negative). In the presence of sulfate and thiosulfate, the following compounds are used as carbon sources and electron donors: H$_2$/CO$_2$ + 1 mM acetate, pyruvate, lactate, ethanol, formate (3 out of 7 strains, type strain positive), caproate (1 out of 7 strains, type strain negative), caprylate (4 out of 7 strains, type strain negative), methanol (2 out of 7 strains, type strain positive), syringic acid (3 out of 7 strains, type strain positive), H$_2$/CO (only type strain tested), butyrate (2 out of 2 strains, including type strain), caprate (2 out of 2 strains, including type strain), and laurate (2 out of 2 strains, including type strain). The following compounds are not used as carbon sources and electron donors in the presence of sulfate and thiosulfate: acetate, propionate, fumarate, malate, benzoate, valerate (2 strains tested, including type strain), fructose (2 strains tested, including type strain) and 3,4,5-trimethoxybenzoic acid (2 strains tested, including type strain).

The following compounds are used as carbon and energy sources: in the absence of an electron acceptor (2 strains tested, including type strain): lactate, pyruvate, ethanol, methanol (1 out of 2 strains, type strain positive); perchloroethene is reduced in the presence of sulfate and pyruvate (1 strain tested).

Enrichment and isolation procedures

Procedures and media for the enrichment and isolation of sulfate-reducing, spore-forming, anaerobic bacteria, including *Desulfosporosinus orientis*, were described by Postgate (1979) and Widdel (1992). Toluene or benzoate has been used as electron donor and dithionite as reducing agent in sulfate-containing enrichment and isolation media (in agar roll tubes) for the isolation of eight strains of *Desulfosporosinus meridiei* (Robertson et al., 2000). *Desulfosporosinus auripigmenti* was enriched in a freshwater mineral medium with lactate, sulfate, and cysteine. After several transfers into medium with dibasic sodium arsenate replacing the sulfate, isolation was done by colony transfers in agar shake tubes. Reduction of arsenate was visible by formation of a yellow precipitate, As_2S_3 (Newman et al., 1997). All members of the genus grow well and sporulate in medium 641 (DSMZ Catalog of Strains, 2001) at 30°C, in which, for growth of *Desulfosporosinus meridiei*, Na-malate is replaced by 1.0 g/l Na-pyruvate.

Occasionally, *Desulfosporosinus orientis* shows very poor growth in media which have been reduced with commercial sodium sulfide. It is likely that certain impurities in the commercial Na_2S are inhibitory (Widdel and Hansen, 1992). Media which are prereduced with ascorbate and thioglycollate or reduced with sodium dithionite immediately before use may be better suited for cultivation (Widdel and Hansen, 1992).

Maintenance procedures

Vegetative cells of *Desulfosporosinus* species usually die within few days after growth has ceased, especially if grown under optimal conditions and kept at optimal temperature. The best method for maintenance is in the sporulated state. Spore formation is oftern induced in colonies in agar deeps. *Desulfosporosinus orientis* sporulates well if grown under sulfate limitation with hydrogen as electron donor (Cypionka and Pfennig, 1986) (Figure 175). Agar tubes with spore containing colonies may survive for several months if refrigerated. Vegetative cells of *Desulfosporosinus* can be preserved long-term by deep freezing in liquid nitrogen, using methods described for methanogens and other anaerobes (Hippe, 1991). Sporulated cultures survive freeze-drying and long-term storage after applying described methods (Hippe, 1991).

Differentiation of the genus *Desulfosporosinus* from other genera

Members of *Desulfosporosinus* can be differentiated from members of the neighboring genera *Desulfotomaculum* and *Desulfitobacterium* by the properties listed in Table 185. Analysis of 16S rRNA gene sequences is probably the most direct way to affiliate a strain to the genus. A set of signature nucleotides of 16S rRNA genes demonstrate the clear demarcation of members of *Desulfosporosinus* and *Desulfitobacterium*. (Stackebrandt et al., 2003). The phylogenetic heterogeneity of the genus *Desulfotomaculum* excluded provision of a clear-cut set of genus-specific signatures.

Further reading

Widdel, F. 1988. Microbiology and ecology of sulfate- and sulfur-reducing bacteria. *In* Zehnder (Editor), Biology of Anaerobic Micro-organisms, John Wiley & Sons, New York. pp. 469–586.

List of species of the genus *Desulfosporosinus*

1. **Desulfosporosinus orientis** (Campbell and Postgate 1965) Stackebrandt, Sproer, Rainey, Burghardt, Päuker, and Hippe 1997, 1138[VP] (*Desulfotomaculum orientis* Campbell and Postgate 1965, 361; "*Desulfovibrio orientis*" Adams and Postgate 1959, 256)

   or.i.en'tis. L. part. adj. *oriens* rising (sun), hence the orient; L. gen. n. *orientis* of the orient.

   The description of *Desulfosporosinus orientis* is identical to that given by Campbell and Postgate (1965), supplemented with data from Campbell and Singleton, Jr. (1986), Klemps et al. (1985), Robertson et al. (2001), and Stackebrandt et al. (2003). Table 182, Table 183, and Table 184 list additional cultural, physiological, and chemotaxonomic properties.

   Gram-negative, fat, curved rods (1.5 × 5 μm), sometimes in short chains or filamentous. Tumbling and twisting motility by means of peritrichous flagella. Spores are oval and subterminal (on rare occasions terminal), slightly swelling the cells. Sulfate, thiosulfate, sulfite, sulfur, DMSO, and Fe (III) are electron acceptors in the presence of lactate.

   Mesophilic; temperature range 30–42°C. Optimal temperature between 30 and 37°C. Growth between 0 and nearly 5% NaCl.

   Growth is inhibited by <0.1μg/ml hibitane.

   Major fatty acids are unsaturated ($C_{16:1\,\omega9c}$, $C_{16:1\,\omega7c}$, $C_{18:1\,\omega9c}$, and $C_{18:1\,\omega7c}$), and saturated ($C_{14:0}$ and $C_{16:0}$) fatty acids; dimethylacetals ($C_{16:0}$), and traces of aldehydes ($C_{16:0}$) are also present. Isolated from soil of a rising main at Sungei Whampoa, Rangoon Road, Singapore.

   DNA G+C content (mol%): 45 (Bd)–45.9 (T_m).

   GenBank accession number (16S rRNA gene): Y11570.

   Type strain: Singapore I, ATCC 19365, DSM 765, NCIMB 8382, VKM B-1628.

2. **Desulfosporosinus auripigmenti** corrig. (Newman, Kennedy, Coates, Ahmann, Ellis, Lovley and Morel 1997) Stackebrandt, Schumann, Schüler and Hippe 2003, 1441[VP] (*Desulfotomaculum auripigmentum* Newman, Kennedy, Coates, Ahmann, Ellis, Lovley and Morel 1997, 387)

   au.ri.pig.men'ti. L. neut. n. *aurum* gold; L. neut. n. *pigmentum* pigment; N.L. gen. n. *auripigmenti* of golden pigment, due to As_2S_3 precipitation.

   The emended description of Stackebrandt et al. (2003) is based on the description of Newman et al. (1997). Table 182, Table 183, and Table 184 list additional cultural, physiological, and chemotaxonomic properties.

   Gram-negative staining, nonmotile. Sausage-shaped cells 2.5 × 0.4μm. The murein sacculus is thinner than usual for Gram-positive bacteria; a hexagonal S-layer is attached to its

TABLE 184. Summary of physiological properties of species of the genus *Desulfosporosinus*[a,b]

Property	*D. orientis* DSM 765[T] (Singapore I[T])	*D. auripigmenti* DSM 13351[T] (OREX-4[T])	*D. meridiei* DSM 13257[T] (S10[T])
Electron donor:			
Butyrate	+[c]	+w	+
Caprate	−	NR	+
Caproate	+	NR	−
Caprylate	+	NR	−
Fumarate	+[d]	−	−
Malate	−	+	−
Syringate	+	NR	+
3,4,5-Trimethoxybenzoate	+	NR	−
Methanol	+	−	+
Propanol	+[c]	NR	NR
Glycerol	NR	+	NR
Electron acceptor:			
Sulfur	+[d]	NR	+
As(V)	−	+	−
Fumarate	−	+	NR
DMSO	+[d]	−	+
Fe(III)	+[d]	−	+
Fermentative growth on:			
Pyruvate	+[d]	+	+
Lactate	+	NR	+
Ethanol	+	NR	+
Methanol	+	NR	+
Formate	+	NR	NR
3,4,5-Trimethoxybenzoate	+	NR	NR

[a]Symbols: +, >85% positive; −, 0–15% positive; w, weak reaction; NR, not recorded.

[b]All strains use as electron donor: H_2, H_2 with acetate, formate, pyruvate, lactate, and ethanol, but not acetate or propionate. Electron acceptors are sulfate, sulfite, and thiosulfate, but not nitrate.

[c]Production of sulfide without turbidity increase.

[d]Positive reported by Robertson et al. (2001).

TABLE 185. Distinguishing characteristics of *Desulfosporosinus* and phylogenetically related genera[a]

Characteristic	*Desulfosporosinus*	*Desulfitobacterium*	*Desulfotomaculum*
Cell shape	Curved rods; occasionally waves, spirals, snake-like cell chains, and filaments formed	Curved rods	Generally straight rods
Sulfate reduction	+	−	+
Autotrophic growth on H_2+CO_2	+	−	+[b]
Cell-wall peptidoglycan[c]	LL-A$_2$pm	LL-A$_2$pm	meso-A$_2$pm, some LL-A$_2$pm
Cytochrome type	$b(c)$	c	b, $b(c)$, or c
Substrate oxidation type	Incomplete	Incomplete	Incomplete or complete
Acetate as electron donor	−	−	− or +
Sulfur as electron acceptor	+[d]	+[e]	−
Reductive dechlorination of chlorophenols	−	+	+

[a]Symbols: +, >85% positive; −, 0–15% positive.

[b]Most species positive; some require acetate in addition; *Desulfotomaculum acetoxidans* does not utilize H_2.

[c]LL-A$_2$pm, LL-diaminopimelic acid.

[d]As reported for *Desulfosporosinus orientis* and *Desulfosporosinus meridiei* by Robertson et al. (2001).

[e]Two out of four species tested.

cell wall. Cells occasionally form oval, subterminal to terminal spores.

Growth by oxidation of H_2, lactate, pyruvate, butyrate, ethanol, glycerol, and malate with concomitant reduction of sulfate. Incomplete oxidation, forming acetate from organic substrates. Sulfate, thiosulfate, sulfite, arsenate, and fumarate are used as electron acceptors. Temperature optimum of 25–30°C and a pH optimum of 6.4–7.0 for growth. Arsenate [As(V)] is reduced to arsenite [As(III)]. In the presence of arsenate and sulfur, arsenate is preferred as electron acceptor; golden-colored As$_2$S$_3$ precipitate is produced

The diagnostic amino acid of peptidoglycan is LL-diaminopimelic acid. MK-7 and MK-5 are the predominant iso-

prenoid quinones; MK-6 is a minor component. Major fatty acids (>5%) are unsaturated (C$_{16:1\ \omega7c}$ and C$_{18:1\ \omega9c}$), and saturated (C$_{16:0}$) fatty acids as well as C$_{DMA-16:0}$ (13.4%) and C$_{DMA-18:1\ \omega9c}$. Isolated from freshwater sediment samples taken from Upper Mystic Lake in Woburn, MA, USA.

DNA G+C content (mol%): 41.6% (HPLC).

Type strain: OREX-4, ATCC 700205, DSM 13351.

GenBank accession number (16S rRNA gene): AJ493051.

3. **Desulfosporosinus meridiei** Robertson, Bowman, Franzmann and Mee 2001, 139[VP]

me.ri.di.e'i. L. n. *meridies* south; L. fem. gen. n. *meridiei* of the south, referring to its isolation in the Southern Hemisphere.

The description has been taken from Robertson et al. (2001). Table 182, Table 183, and Table 184 list additional cultural, physiological, and chemotaxonomic properties.

Gram-negative and Gram-variable curved rods with multilayered cell wall. Cells are 0.7–1.1 μm in width and generally 2.3–4.2 μm in length, cells of some strains reaching up to 13 μm in length. Cells singly or in chains of two or more cells, motile by means of a single lateral flagellum. Oval and subterminal endospores, sometimes causing the cells to swell.

Sulfate, sulfite, thiosulfate, elemental sulfur, DMSO, Fe(III), and nitrate (some strains) serve as electron acceptors in the presence of lactate. Manganese (IV) and arsenic (V) are not used as electron acceptors. Mesophilic, temperature range for growth between 10 and 37°C, optimum at 28°C. The upper limit of salt tolerance is 4% NaCl. The predominant whole-cell fatty acids are unsaturated ($C_{16:1\ \omega7c}$ and $C_{18:1\ \omega9c}$) and saturated fatty acids ($C_{14:0}$ and $C_{16:0}$); minor amounts of iso- and anteiso-branched-chain fatty acids, as well as trace amounts of cyclic fatty acids ($C_{17:0\ cyclo}$ and $C_{19:0\ cyclo}$) may occur. Isolated from a gasoline contaminated shallow aquifer in the sandy soils at Eden Hill, on the Swan Coastal Plain near Perth.

DNA G+C content (mol%): 46.8–46.9 (T_m).

Type strain: S10, ATCC BAA-275, DSM 13257, NCIMB 13706.

GenBank accession number (16S rRNA gene): AF076527.

Genus VII. **Desulfotomaculum** Campbell and Postgate 1965, 361[AL]

JAN KUEVER AND FRED A. RAINEY

De.sul.fo.to.ma′cu.lum. L. pref. *de* from; L. n. *sulfo* sulfur; L. n. *tomaculum* a kind of sausage; N.L. neut. n. *Desulfotomaculum* a sausage-shaped sulfate reducer.

Straight or curved rod-shaped cells, 0.3–2.5 × 2.5–15 μm with rounded or pointed ends. Occur singly or in pairs. Spores oval or round, terminal to central, causing swelling of the cells. Stain Gram-positive, although often only detected by electron microscopy. Cells are motile by single polar or peritrichous flagella, but motility can be lost during cultivation. **Strict anaerobe** with a respiratory type of metabolism. **Chemo-organotrophs** or **chemoautotrophs**. Simple organic compounds are used as electron donor and carbon sources, and are either completely oxidized to CO_2 or incompletely to acetate. Some species can grow on H_2 autotrophically with CO_2 as the sole carbon source; one species can grow on CO alone. Sulfate, and usually sulfite and thiosulfate, serve as terminal electron acceptors and are reduced to H_2S. Sulfur and nitrate are not used as electron acceptors. Fermentative growth has been observed for some species. Desulfoviridin is absent. The major **menaquinone is MK-7.** Growth occurs in simple defined media containing sulfide as a reductant. Some species require vitamins or yeast extract. Some species can fix N_2. The optimum temperature range for growth is 30–37°C for mesophilic and 50–65°C for thermophilic species. The optimum pH range for growth is 6.5–7.5. Carbon monoxide dehydrogenase activity has been demonstrated for some species. *Desulfotomaculum* species are common in anoxic freshwater, brackish or marine sediments.

DNA G+C content (mol%): 37.1–56.3.

Type species: **Desulfotomaculum nigrificans** (Werkman and Weaver 1927) Campbell and Postgate 1965, 361[AL] (*Clostridium nigrificans* Werkman and Weaver 1927, 63).

Further descriptive information

Many *Desulfotomaculum* strains have typical cell morphology which allows them to be distinguished from other sulfate-reducing bacteria (SRB); this is especially true for sporulating cells. Some strains form longer rods or curved cells; even clumps of cells have been observed (A. Galushko, personal communication). In the first enrichment passages, cells are often motile. After several transfers, motility may be lost.

The optimum growth temperature for mesophilic strains is usually 30–37°C and 50–65°C for thermophilic species. The minimum temperature for mesophilic species might be lower than 10°C and the maximum temperature for thermophilic species can exceed 85°C. The pH range for growth is 5.5–8.9, with an optimum of 6.5–7.5.

Growth occurs in the presence of sulfate, sulfite, or thiosulfate as an electron acceptor. Several *Desulfotomaculum* strains can use a large variety of organic compounds as electron donor and carbon sources (see Table 186). Some strains are restricted and oxidize organic compounds incompletely to acetate; others are capable of complete oxidation even though substrates may be partially converted to acetate, depending on substrate concentration (Kuever et al., 1999). Some *Desulfotomaculum* strains can grow autotrophically with H_2 as an electron donor and CO_2 as carbon source.

In the absence of sulfate, some species can grow by fermentation of fructose, glucose, pyruvate, or lactate. Certain *Desulfotomaculum* strains can grow by homoacetogenesis and convert methanol, methoxyl groups of aromatic compounds, formate, and H_2 and CO_2 to acetate. Although growth by homoacetogenesis is weak, it allows survival in a sulfate-free habitat. Some strains reduce metal ions. In this case, growth would be linked to fermentation using certain substrates; electron transport phosphorylation is unlikely. For some strains, carbon monoxide dehydrogenase activity has been demonstrated. It can be considered that strains which show a complete substrate oxidation or autotrophic growth must have this enzyme.

Desulfotomaculum strains contain membrane-bound and soluble cytochromes of the *b* and *c* type. Desulfoviridin and desulforubidin have never been found in the genus *Desulfotomaculum*, however, the sulfite reductase P582 is present. All members of the genus investigated so far contain MK-7 as the major menaquinone (Collins and Widdel, 1986). Major cellular fatty acids of members of the genus vary between and within the phylogenetic subclusters.

Desulfotomaculum strains occur in black anoxic sediment from freshwater, brackish water, and marine habitats. Due to spore formation, they might be dominating SRB in habitats of varying redox conditions, such as groundwater, soil, or paddy fields.

Enrichment and isolation procedures

Selective enrichment of *Desulfotomaculum* species is possible from various sources using different electron donors and sulfate as the electron acceptor in simple mineral media – see *Desulfobacter* chapter in Volume 2 of the *Systematics* (Kuever et al., 2005). The formation of heat- and drought-resistant spores in *Desulfotomaculum* species allows them to survive well in oxic and varying redox regimes. Growth of nonspore-forming, sulfate-reducing bacteria can be eliminated by pasteurization at 80°C for 10–20 min. The only other spore-forming, sulfate-reducing bacteria known are *Desulfosporosinus* species and *Thermacetogenium phaeum*, but their physiological properties are much more restricted. Enrichments at elevated temperature (50–65°C) generally result in the isolation of *Desulfotomaculum* species because only a few other sulfate-reducing bacteria can grow at this higher temperature range. Isolation of pure cultures from such enrichments is achieved via the agar tube serial dilution technique as described in Volume 2 of the *Systematics* ((Kuever et al., 2005). In contrast to spores, vegetative cells are very oxygen sensitive. Therefore, the addition of sodium dithionite as an additional reducing agent is helpful. After inoculation from a spore suspension, long lag phases sometimes occur. Germination times for spores can also be shortened by the addition of dithionite (10–30 mg/l medium).

Maintenance procedures

Cultures can be stored at low temperature for several months. After sporulation occurs, cultures can be stored for years. Sporulated liquid cultures can be freeze-dried and used for long term conservation. For some species, the formation of spores depends on the growth substrate. In these cases, incubation at suboptimal temperatures and the use of a substrate which allows only slow growth (e.g., acetate) can be helpful.

Differentiation of the genus *Desulfotomaculum* from other genera

Together with *Desulfosporosinus* species and *Thermacetogenium phaeum*, members of the genus *Desulfotomaculum* can be distinguished from other sulfate-reducing bacteria by their ability to form spores. A unique physiological feature of some *Desulfotomaculum* and *Desulfosporosinus* is their use of methoxylated aromatic compounds, an ability shared only by *Desulfomonile* species. A clear differentiation from other spore-forming genera is only possible by comparative 16S rRNA gene analysis.

Taxonomic comments

On the basis of the analysis of 16S rRNA gene sequence data, many of the species of the *Desulfotomaculum* have been shown to have less 16S rRNA gene sequence similarity to each other than they do to other genera within the *Clostridiales*. Stackebrandt et al. (1997) addressed this by designating clusters Ia through Ie for the various species clusters into which the true species of the genus *Desulfotomaculum* fell. Two species were shown to fall outside the radiation of the majority of the species of the genus *Desulfotomaculum* as defined by the position of the type species *Desulfotomaculum nigrificans* and were assigned to clusters II and III (Stackebrandt et al., 1997). This resulted in the reclassification of *Desulfotomaculum orientis* as *Desulfosporosinus orientis* (Stackebrandt et al., 1997). The species *Desulfotomaculum guttoideum* was shown to be unrelated to the other species of the genus *Desulfotomaculum* and grouped with the *Clostridium* species *Clostridium sphenoides*, *Clostridium celerecresens*, *Clostridium aerotolerans*, and *Clostridium xylanolyticum* in a cluster designated III (Stackebrandt et al., 1997). This *Desulfotomaculum* cluster III corresponds to the cluster XIVa of Collins et al. (1994). As new species have been added to the genus *Desulfotomaculum*, the clusters Ia through Ie have expanded and an additional cluster designated If is formed by the addition of the species *Desulfotomaculum alkaliphilum* and *Desulfotomaculum halophilum* (Figure 179). The bootstrap values shown in Figure 179 indicate good support for the branching patterns of the 6 *Desulfotomaculum* clusters, all being supported by values greater than 98%. As pointed out by Stackebrandt et al. (1997), each of these clusters have individual genus status on the basis of 16S rRNA gene sequence analysis and the corresponding similarity values between the sequences of the *Desulfotomaculum* species as well as between the related genera. These findings are also supported by signature sequences that have been defined for a number of the *Desulfotomaculum* species and the clusters to which they have been shown to group (Hristova et al., 2000; Stackebrandt et al., 1997). It has been reported that several *Desulfotomaculum* strains contain multiple copies of the 16S rRNA gene that show heterogeneity within the multiple copies of the 16S rRNA gene sequences (Stackebrandt et al., 1997; Tourova et al., 2001). Stackebrandt et al. (1997) also reported the presence of large inserts in the helical region 73–82/87–97 of the 16S rRNA gene. From the alignment of the 16S rRNA gene sequences of all 23 validly named species of the genus used for the construction of the phylogenetic dendrogram (Figure 179), such large inserts are found in the all nine species of clusters Ic and Id as well as *Desulfotomaculum geothermicum*. Further detailed studies of the chemotaxonomic and physiological properties of all species of the genus *Desulfotomaculum* are required before new genera can be proposed for the clusters other than cluster Ia which contains the type species of the genus *Desulfotomaculum nigrificans*.

List of species of the genus *Desulfotomaculum*

1. **Desulfotomaculum nigrificans** (Werkman and Weaver 1927) Campbell and Postgate 1965, 361[AL] (*Clostridium nigrificans* Werkman and Weaver 1927, 63)

ni.gri'fi.cans. L. part. adj. *nigrificans* blackening.

Characteristics are summarized in Table 186. The type strain grows at 45–70°C. Can be adapted to grow slowly at 30–37°C. The original strain was isolated from spoiled canned sweet corn (Werkman and H.J.Weaver, 1927). The deposited type strain was isolated from freshwater in Delft, Netherlands.

DNA G+C content (mol%): 48.5–49.9 (Bd, HPLC, T_m).

Type strain: strain Delft 74, ATCC 19858, ATCC 19998, DSM 574, NBRC 13698, NCIMB 8395.

GenBank accession number (16S rRNA gene): X62176.

2. **Desulfotomaculum acetoxidans** Widdel and Pfennig 1977a, 306[VP] (Effective publication: Widdel and Pfennig 1977b, 121.)

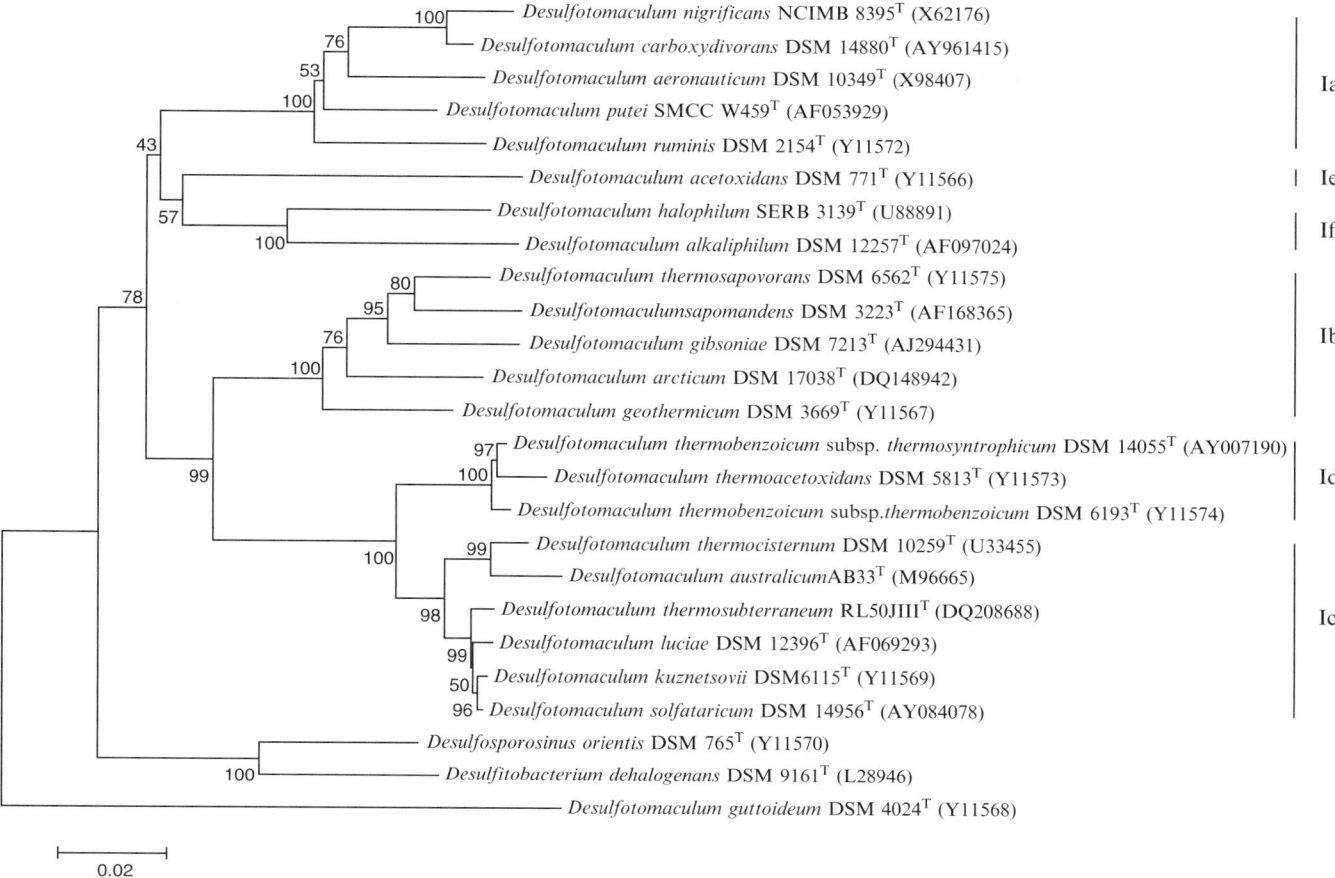

FIGURE 179. Neighbor-joining 16S rRNA gene sequence-based phylogeny of the species of the genus *Desulfotomaculum* and related genera. Numbers at branching points indicate bootstrap values from 1000 data samplings. The scale represents 2 inferred nucleotide changes per 100 nucleotides.

a.cet.o′xi.dans. L. n. *acetum* vinegar; N.L. n. *acidum aceticum* acetic acid; N.L. v. *oxido* oxidize from Gr. adj. *oxys* sour, acid; L. neut. adj. *acetoxidans* oxidizing acetic acid.

Characteristics are summarized in Table 186. The type strain grows at 20–40°C, and over a pH range of 6.6–7.6 with an optimum at 7.1. Isolated from piggery waste.

DNA G+C content (mol%): 37.5 (T_m).

Type strain: "Göttingen" 5575, ATCC 49208, DSM 771, VKM B-1644.

GenBank accession number (16S rRNA gene): Y11566.

3. **Desulfotomaculum aeronauticum** Hagenauer, Hippe and Rainey 1997b, 915[VP] (Effective publication: Hagenauer, Hippe and Rainey 1997a, 70.)

ae.ro.nau′ti.cum. Gr. n. *aer* air; Gr. adj. *nautikos* nautical, concerning ship/navigation; N.L. neut. adj. *aeronauticum* navigation in air.

Characteristics are summarized in Table 186. The type strain grows at 20–42°C, and over a pH range of 6.0–9.0 with an optimum for growth at 7.0. Isolated from a corroded stringer made of aluminum alloy 2024 in the luggage area of an aircraft.

DNA G+C content (mol%): 43.8 (HPLC).

Type strain: 9, DSM 10349.

GenBank accession number (16S rRNA gene): X98407.

4. **Desulfotomaculum alkaliphilum** Pikuta, Lysenko, Suzina, Osipov, Kuznetsov, Tourova, Akimenko and Laurinavichius 2000, 32[VP]

al.ka.li.phi′lum N.L. fem. n. *alkali* alkali, Gr. adj. *philos* loving; N.L. neut. adj. *alkaliphilum* alkali-loving.

Characteristics are summarized in Table 186. The type strain grows at 30–60°C, and over a pH range of 8.0–9.15 with an optimum for growth at 8.7. Requires carbonate anion. Isolated from a cow/pig manure mixture at neutral pH.

DNA G+C content (mol%): 40.9 (T_m).

Type strain: S1, ATCC 700784, DSM 12257, VKM B-2192.

GenBank accession number (16S rRNA gene): AF097024.

5. **Desulfotomaculum antarcticum** (*ex* Iizuka, Okazaki and Seto 1969) Campbell and Singleton 1988, 220[VP] (Effective publication: Campbell and Singleton 1986, 1202.)

ant.arc′ti.cum. N.L. adj. *antarcticum* pertaining to Antarctica.

Only a few characteristics are available and they summarized in Table 186.

TABLE 186. Characteristics differentiating the species of the genus *Desulfotomaculum*[a]

Species	*Desulfotomaculum nigrificans*	*Desulfotomaculum acetoxidans*	*Desulfotomaculum aeronauticum*	*Desulfotomaculum alkaliphilum*	*Desulfotomaculum antarcticum*	*Desulfotomaculum arcticum*	*Desulfotomaculum australicum*	*Desulfotomaculum carboxydivorans*	*Desulfotomaculum geothermicum*	*Desulfotomaculum gibsoniae*	*Desulfotomaculum halophilum*	*Desulfotomaculum kuznetsovii*
Cluster	1a	1e	1a	1f	Unknown	1b	1c	1a	1b	1b	1f	1c
Width × length (µm)	0.5–0.7 × 2–4[b]	1.0–1.5 × 3.5–9	0.5–0.8 × 2.2–4.5	0.6–0.7 × 3.0–3.5	1.0–1.2 ×4–6	2.0–3.0 × 1.0	0.8–1 × 3–6	0.5–1.5 × 5–15	0.5–0.8 × 2.3–2.5	1.5–2.5 × 4–7	0.5 × 3–6	1–1.4 × 3.5–5
Spore shape	Oval	Oval	Oval	Oval	Oval	Sphere	Sphere	Oval	Sphere	Sphere	Oval	Sphere
Spore orientation	ST	T to PC	ST or C	T	C	C	C	ST	ST	C	T	C
Gas vesicles	−	+	−	NR	NR	−	−	−	+	−	+	−
Motility (flagella)	+ (pe)	+ (sp)	+	−	+ (pe)	−	+	+	+	+[c]	+	+ (pe)
Major menaquinone	MK–7	MK–7	MK–7	NR	NR	NR	NR	NR	NR	NR	NR	NR
Optimal growth temperature (°C)	55	37	37	50–55	20–30	44	68	55	54	35–37	35	60–65
Oxidation of substrate	I	C	I	I	I	C	C	I	C[d]	C	I	C
Electron donors used:												
Acetate	−	+	−	−	−		(+)	−	−	−	−	+
Alanine	+	NR	(+)	NR	NR	+	NR	+	NR	NR	−	−
Benzoate	−	−	−	NR	NR	NR	(+)	NR	−	+	−	−
n-Butanol	+	+	(+)[c]	NR	NR	+	NR		NR	+	(+)	+
Ethanol	+	+	(+)	+	NR	+	+	+	+	+	+	+
Fatty acids (carbon atoms)	−	4, 5	−	−	NR	NR	−	−	3, 4, 6, 8, 9, 16, 18	3–8	−	3, 4, 5, 6, 8, 16
Formate	+[e]	(+)[e]	+[e]	+[e]	−	+	−		+[e]	+	+[e]	+
Fructose	+	−	−	NR	NR	−	NR	+	+	−	−	−
Fumarate	−	−	−	NR	NR	+	−	−	NR	+	−	+
Glycerol	NR	NR	+	−	NR	−	NR	(+)	NR	+	−	NR
H₂	+[e]	−	+[c]	+[c]	NR	+	(+)	+[e]	+	+	+[e]	+
Lactate	+	−	+	+	+	+	+	+	+	−	+	+
Malate	−	−	−	NR	NR	+	−	−	NR	+	(+)	+
Methanol	−	NR	NR	−	NR	+	−	−	−	−	−	+
Methoxylated aromatic compounds	−	NR	−	NR	NR	NR	NR	NR	NR	+	−	NR
n-Propanol	+[e]	−	(+)[e]	NR	NR	+	NR		NR	+	−	+
Pyruvate	+	−	+	+	+	+	+	+	NR	+	+	+
Succinate	−	−	NR	NR	NR	+	−	−	NR	+	−	+
Fermentative growth on:												
Fructose	+	NR	NR	NR	NR	−	NR	+	+	−	NR	NR
Glucose	+[g]	NR	NR	NR	NR	−	NR	+	NR	NR	NR	NR
Lactate	−	−	−	−	NR	−	+[h]	+	NR	−	−	−
Pyruvate	+	−	+	(+)	NR	+	+	+	NR	+	(+)	+
Electron acceptors used:												
Fumarate	−	−	−	−	NR	−	NR	NR	−	−		NR
Nitrate	−	−	−	−	−	−	NR	−	−	−	−	−
Sulfate	+	+	+	+	+	+	+	+	+	+	+	+
Sulfite	+	−	+	+	NR	+	NR	+	+	+	+	+
Sulfur	+	−	(+)	−	NR	−	NR	−	NR	−	−	−
Thiosulfate	+	−	+	+	NR	+	NR	+	NR	+	+	+
Homoacetogenic growth	NR	NR	NR	NR	NR	NR	NR	NR	−	+	−	NR
Growth factor requirement	Unknown, yeast extract	Biotin	Menadione, vitamins	Yeast extract or vitamins	NR	−	NR	Yeast extract	−	−	Yeast extract	None
NaCl requirement (g/l)	−		NR	1	NR	−	+ (not given)	−	2	−	10	−
G+C content (mol%)	48.5–49.9	37.5	43.8	40.9	NR	48.9	48.1	45.6	50.4	54.8	56.3	49
Literature	Campbell and Singleton (1986); Klemps et al. (1985); Liu et al. (1997)	Widdel and Pfennig (1977b, 1981)	Hagenauer et al. (1997a); Hippe et al.	Pikuta et al. (2000)	Iizuka et al. (1969)	Vandieken et al. (2006)	Love et al. (1993b)	Parshina et al. (2005)	Daumas et al. (1988)	Kuever et al. (1993, 1999)	Tardy-Jacquenod et al. (1998)	Nazina et al. (1989)

[a]Symbols: Spore orientation: ST, subterminal; T, terminal; PC, paracentral; C, central; Motility: pe, peritrichous; sp, single or polar; Oxidation of substrate: I, incomplete; C, complete; +, positive; − negative; (+), poorly utilized; NR, not reported.

[b]Liu et al. (1997) determined a cell width of 1–1.2 µm and a cell length of 3–6 µm for this species.

[c]Motility was lost during cultivation.

[d]Substrate oxidation should be regarded as complete. The high substrate concentration used may result in misinterpretation.

[e]Only in the presence of acetate or yeast extract as carbon sources.

[f]In the original publication, fructose utilization without sulfate reduction is described in the text, but not listed in the table.

[g]Klemps et al. (1985) observed only growth with autoclaved glucose, but not with filter-sterilized glucose.

[h]Description given in the text and table of the original publication is varying on this point.

[i]Data retrieved from whole-genome sequencing.

Species	*Desulfotomaculum luciae*	*Desulfotomaculum putei*	*Desulfotomaculum ruminis*	*Desulfotomaculum sapomandens*	*Desulfotomaculum solfataricum*	*Desulfotomaculum subterraneum*	*Desulfotomaculum thermoacetoxidans*	*Desulfotomaculum thermobenzoicum* subsp. *thermobenzoicum*	*Desulfotomaculum thermobenzoicum* subsp. *syntrophicum*	*Desulfotomaculum thermocisternum*	*Desulfotomaculum thermosapovorans*	*Desulfotomaculum "reducens"*
Cluster	1c	1a	1a	1b	1c	1c	1d	1d	1d	1c	1b	1a
Width × length (μm)	1 × 3	1–1.1 × 2–5	0.5–0.7 × 2–4	1.2–2.0 × 5–7	1.5–2.5 × 3.5–5	0.8–1.0 × 3–10	0.7 × 2–5	1.5–2 × 5–8	1 × 3–11	0.7–1 × 2–5.2	1.5–2.0 × 5–8	0.8–1.0 × 5–10
Spore shape	Sphere	Oval	Oval	Sphere	Sphere	Sphere	Oval	Oval	Oval	Sphere	Sphere	NR
Spore orientation	C	PC	ST	C	C	C to ST	C	C	C	C	PC	NR
Gas vesicles	–	–	–	+	–	NR	–	–	–	–	–	NR
Motility (flagella)	+	+	+ (pe)	+	–	+	+	+	+	+	+	+
Major menaquinone	NR	NR	MK–7	NR	NR	MK–7	NR	NR	NR	NR	NR	NR
Optimal growth temperature (°C)	60–65	30–36	37	38	60	61–66	55–60	62	55	62	50	37
Oxidation of substrate	NR	I	I	C	C	C	C	C	I	I	C[d]	I
Electron donors used:												
Acetate	–	–	–	(+)	+	–	+	–	–	–	–	–
Alanine	–	NR	+	NR	NR	+	+	NR	(+)	NR	NR	NR
Benzoate	NR	NR	–	+	–	–	–	+	+	–	–	NR
n-Butanol	NR	+	+[e]	+	+	+	+	+	–	+	+	+
Ethanol	+	+	–	+	+	+	–	+	–	+	+	+
Fatty acids (carbon atoms)	NR	–	–	4, 6–8, 10, 12, 16, 18	3, 4	4–10, 16	3–5	3–6	3	3–10, 12	4–9, 12, 16, 18, 20, 22	3–5
Formate	+	+	+[e]	(+)[e]	+	+	+	(+)	NR	–	+	NR
Fructose	–	+[f]	–	–	+	–	NR	–	–	–	–	NR
Fumarate	NR	–	–	(+)	+	+	NR	+	+	NR	+	–
Glycerol	NR	NR	NR	–	NR	NR	NR	NR	NR	NR	–	NR
H₂	+	+	+[e]	–	+	+	+	+	+	+	+	–
Lactate	+	+	+	–	+	+	+	+	+	+	+	+
Malate	–	–	–	(+)	NR	+	+	+	(+)	NR	+	NR
Methanol	–	(+)	–	–	+	–	NR	–	–	–	+	+
Methoxylated aromatic compounds	NR	NR	–	–	NR	NR	NR	+	NR	NR	NR	NR
n-Propanol	NR	NR	+[e]	(+)	+	+	(+)	+	–	+	+	+
Pyruvate	+	+	+	+	NR	+	+	+	–	+	+	+
Succinate	–	NR	–	(+)	+	+	+	–	–	NR	–	NR
Fermentative growth on:												
Fructose	NR	+[f]	–	–	NR	NR	NR	NR	NR	NR	–	NR
Glucose	NR	–	–	NR	NR	NR	NR	NR	NR	NR	NR	NR
Lactate	+	–	+	–	+	+	–	+	+	–	+	NR
Pyruvate	+	+	+	+	NR	+	+	+	+	+	+	NR
Electron acceptors used:												
Fumarate	NR	NR	–	–	NR	NR	–	NR	NR	NR	NR	NR
Nitrate	NR	–	–	–	–	–	–	(+)	–	–	NR	NR
Sulfate	+	+	+	+	+	+	+	+	+	+	+	+
Sulfite	–	+	+	+	+	+	+	+	–	+	+	+
Sulfur	NR	–	–	+	NR	+	+	–	NR	–	–	+
Thiosulfate	+	+	+	+	+	+	NR	(+)	–	+	+	NR
Homoacetogenic growth	NR	NR	NR	–	NR	NR	H₂ + CO₂	Methoxylated aromatic compounds	H₂ + CO₂	NR	NR	NR
Growth factor requirement	None	–	pa, bi	Vitamins	Vitamins	NR	Vitamins	Yeast extract	–	Vitamins	Vitamins	NR
NaCl requirement (g/l)	+	–	–	–	–	–	–	–	–	–	–	NR
G+C content (mol%)	51.4	47.1	48.5–49.9	48	48.3	54.4	49.7	52.8	53.7	56	51.2	42[i]
Literature	Karnauchow et al. (1992); Liu et al. (1997)	Liu et al. (1997)	Campbell and Singleton (1986); Klemps (1985); Daumas et al. (1985); Liu et al. (1997)	Cord-Ruwisch and Garcia (1985); et al. (1988)	Goorisen et al. (2003)	Kaksonen et al. (2006b)	Min and Zinder (1990)	Tasaki et al. (1991a, 1992, 1993)	Plugge et al. (2002)	Rosnes et al. (1991); Nilsen et al. (1996)	Fardeau et al. (1995)	Tebo and Obraztsova (1998)

DNA G+C content (mol%): unknown.
Type strain: IAM 64.
GenBank accession number (16S rRNA gene): not available.

6. **Desulfotomaculum arcticum** Vandieken, Knoblauch and Jørgensen 2006, 689[VP]

arc′ti.cum. L. neut. adj. *arcticum* from the Arctic, referring to the place from which the strain was isolated.

Characteristics are summarized in Table 186. The type strain grows at 26–46.5°C, and over a pH range of 6.8–7.5 with an optimum for growth at 7.1–7.5. Isolated from a permanently cold fjord sediment in Svalbard.
DNA G+C content (mol%): 48.9 (T_m).
Type strain: 15, DSM 17038, JCM 12923.
GenBank accession number (16S rRNA gene): DQ148942.

7. **Desulfotomaculum australicum** Love, Patel, Nichols and Stackebrandt 1993a, 864[VP] (Effective publication: Love, Patel, Nichols and Stackebrandt 1993b, 250.)

au.stra′li.cum. L. n. *australicum* south, pertaining to Australia.

Characteristics are summarized in Table 186. The type strain grows at 40–74°C, and over a pH range of 5.5–8.5 with an optimum for growth at 7.0–7.4. Isolated from 60°C water obtained from a depth of 914 meters in the Great Artesian Basin Australia (borehole 94).
DNA G+C content (mol%): 48.1.
Type strain: AB33, ACM 3917, DSM 11792.
GenBank accession number (16S rRNA gene): M96665.

8. **Desulfotomaculum carboxydivorans** Parshina, Sipma, Nakashimada, Henstra, Smidt, Lysenko, Lens, Lettinga and Stams 2005, 2164[VP]

car.bo.xy.di.vor′ans. N.L. n. *carboxydum* carbon monoxide; L. part. adj. *vorans* devouring; N.L. part. adj. *carboxydivorans* carbon monoxide-digesting.

Characteristics are summarized in Table 186. The type strain grows at 30–68°C, and over a pH range of 6.0–8.0 with the optimum pH for growth at 6.8–7.2. Can grow at high concentrations of carbon monoxide (100%). Isolated from sludge from an anaerobic bioreactor treating paper mill waster water.
DNA G+C content (mol%): 45.6 (HPLC).
Type strain: CO-1-SRB, DSM 14480, VKM B-2319.
GenBank accession number (16S rRNA gene): AY961415.

9. **Desulfotomaculum geothermicum** Daumas, Cord-Ruwisch and Garcia 1990, 105[VP] (Effective publication: Daumas, Cord-Ruwisch and Garcia 1988, 177.)

geo.ther′mi.cum. Gr. n. *gê* earth, Gr. fem. n. *therme* heat; N.L. neut. adj. *geothermicum* geothermal.

Characteristics are summarized in Table 186. The type strain grows at 37–56°C, and over a pH range of 6.0–8.0 with an optimum for growth at 7.2–7.4. Isolated from anoxic geothermal ground water (from a depth of 2500 m) used for a geothermal heating plant.
DNA G+C content (mol%): 50.4 (Bd).
Type strain: BSD, ATCC 49053, DSM 3669.
GenBank accession number (16S rRNA gene): AJ294428, X80789, Y11567.

10. **Desulfotomaculum gibsoniae** Kuever, Rainey and Hippe 1999, 1807[VP]

gib.so′ni.ae. N.L. gen. n. *gibsoniae* of Gibson; named after Jane Gibson, a British-American microbiologist and bio-

chemist who made important contributions to the field of anaerobic degradation of aromatic compounds.

Characteristics are summarized in Table 186. Grows on a large variety of aromatic compounds including lignite monomers. The type strain grows at 20–42°C, and over a pH range of 6.0–8.0 with an optimum for growth at 6.9–7.2. Isolated from anoxic mud of a freshwater ditch.
DNA G+C content (mol%): 54.8 (T_m).
Type strain: Groll, DSM 7213.
GenBank accession number (16S rRNA gene): Y11576.

11. **Desulfotomaculum halophilum** Tardy-Jacquenod, Magot, Patel, Matheron and Caumette 1998, 337[VP]

ha.lo′phi.lum. Gr. n. *hals, halos* salt; Gr. adj. *philos* loving; N.L. neut. adj. *halophilum* salt-loving.

Characteristics are summarized in Table 186. The type strain grows at 30–40°C, and over a pH range of 6.9–8.0 with an optimum for growth at 7.3. Isolated from production fluid of an oil-producing well in France.
DNA G+C content (mol%): 56.3 (HPLC).
Type strain: SEBR 3139, ATCC 700650, DSM 11559.
GenBank accession number (16S rRNA gene): U88891.

12. **Desulfotomaculum kuznetsovii** Nazina, Ivanova, Kanchaveli and Rozanova 1990, 470[VP] (Effective publication: Nazina, Ivanova, Kanchaveli and Rozanova 1989, 662.)

kuz. net.so′vi. N.L. gen. n. *kuznetovii* of Kuznetsov, named in honor of the leading Soviet microbiologist S.I. Kuznetsov who has made a significant contribution to the study of geochemical activity of micro-organisms.

Characteristics are summarized in Table 186. The type strain grows at 50–85°C. Isolated from a water sample obtained from spontaneous effusion of a rift in the Sukhums deposit containing subsurface thermal mineral waters.
DNA G+C content (mol%): 49 (T_m).
Type strain: 17, DSM 6115, VKM B-1805.
GenBank accession number (16S rRNA gene): AJ294427, Y11569.

13. **Desulfotomaculum luciae** Liu, Karnauchow, Jarrell, Balkwill, Drake, Ringelberg, Clarno and Boone 1997, 620[VP]

lu′ci.ae. N.L. gen. n. *luciae* of Lucia, referring to the source of the type strain, a hot spring in St Lucia.

Characteristics are summarized in Table 186. The type strain grows at 50–70°C, and over a pH range of 6.2–8.3 with an optimum for growth at 6.3–7.8. Isolated from a hot spring located on the island St. Lucia.
DNA G+C content (mol%): 51.4 (HPLC).
Type strain: SLT, ATCC 700428, DSM 12396, SMCC W644.
GenBank accession number (16S rRNA gene): AF069293.

14. **Desulfotomaculum putei** Liu, Karnauchow, Jarrell, Balkwill, Drake, Ringelberg, Clarno and Boone 1997, 619[VP]

pu′te.i. L. gen. n. *putei* of a pit or well, referring to the source of the type strain, an exploratory gas well.

Characteristics are summarized in Table 186. The type strain grows at 40–65°C, and over a pH range of 6.0–8.4 with an optimum for growth at 7.0–7.9. Isolated from an exploratory gas well.
DNA G+C content (mol%): 47.1 (HPLC).
Type strain: TH-11, ATCC 700427, DSM 12395, SMCC W459.
GenBank accession number (16S rRNA gene): AF053929.

15. **Desulfotomaculum ruminis** (Adams and Postgate 1959) Campbell and Postgate 1965, 361[AL] (*Desulfovibrio orientis* Adams and Postgate 1959, 256)

ru'mi.nis. L. n. *rumen* throat, adopted for first stomach (rumen) of a ruminant; L. gen. n. *ruminis* of a rumen.

Characteristics are summarized in Table 186.
DNA G+C content (mol%): 48.5–49.9 (Bd, HPLC, T_m).
Type strain: strain DL, ATCC 23193, DSM 2154.
GenBank accession number (16S rRNA gene): AB294140, M34418, Y11572.

16. **Desulfotomaculum sapomandens** Cord-Ruwisch and Garcia 1990, 105[VP] (Effective publication: Cord-Ruwisch and Garcia 1985, 329.)

sa.po.man'dens. L. n. *sapo* soap; L. v. *mando* to eat, to consume; N.L. part. adj. *sapomandens* eating soap.

Characteristics are summarized in Table 186. Grows on phenyl-substituted organic acids. The type strain grows at 20–43°C, and over a pH range of 6.3–8.5 with an optimum at 7.0. Isolated from oxic gasoline-contaminated soil from a gasoline station.
DNA G+C content (mol%): 48 (Bd).
Type strain: "Pato", DSM 3223.
GenBank accession number (16S rRNA gene): AF168365.

17. **Desulfotomaculum solfataricum** Goorissen, Boschker, Stams and Hansen 2003, 1227[VP]

sol.fa.ta'ri.cum. N.L. neut. adj. *solfataricum* pertaining to solfatares, derived from solfatara (field of hot sulfur springs and fumaroles), referring to the original habitat of the organism.

Characteristics are summarized in Table 186. The type strain grows at 48–65°C, and over a pH range of 6.4–7.9 with an optimum for growth at 7.3. Isolated from hot solfataric fields in northeast Iceland.
DNA G+C content (mol%): 48.3 (method unknown).
Type strain: V21, CIP 107984, DSM 14956.
GenBank accession number (16S rRNA gene): AY084078.

18. **Desulfotomaculum thermoacetoxidans** Min and Zinder 1995, 879[VP] (Effective publication: Min and Zinder 1990, 403.)

ther.mo.a.cet.o'xi.dans. Gr. adj. *thermos* hot; L. n. *acetum* vinegar; N.L. n. *acidum aceticum* acetic acid; N.L. v. *oxido* make oxide, oxidize; N.L. part. adj. *thermoacetoxidans* oxidizing acetate under hot conditions.

Characteristics are summarized in Table 186. The type strain grows between 45 and 65°C, and over a pH range of 6.0 to 7.5 with an optimum for growth at 6.5. Isolated from a thermophilic anaerobic digestor converting cellulosic wastes to methane.
DNA G+C content (mol%): 49.7 (T_m).
Type strain: CAMZ, DSM 5813.
GenBank accession number (16S rRNA gene): Y11573.

19. **Desulfotomaculum thermobenzoicum** Tasaki, Kamagata, Nakamura and Mikami 1991b, 580[VP] emend. Plugge, Balk and Stams 2002, 397 (Effective publication: Tasaki, Kamagata, Nakamura and Mikami 1991a, 351.)

ther.mo.ben.zo'i.cum. Gr. adj. *thermos* hot; N.L. *benzoicum* pertaining to benzoate; N.L. neut. adj. *thermobenzoicum* oxidizes benzoate under thermophilic conditions.

See following description of *Desulfotomaculum thermobenzoicum* subsp. *thermobenzoicum*.

DNA G+C content (mol%): 52.8 (HPLC).
Type strain: TSB, ATCC 49756, DSM 6193.
GenBank accession number (16S rRNA gene): Y11574.

19a. **Desulfotomaculum thermobenzoicum subsp. thermobenzoicum** Tasaki, Kamagata, Nakamura and Mikami 1991b, 581[VP] emend. Plugge, Balk and Stams 2002, 397 (Effective publication: Tasaki, Kamagata, Nakamura and Mikami 1991a, 351.)

Characteristics are summarized in Table 186. The type strain grows at 40–70°C, and a pH range of 6.0–8.0 with an optimum for growth at 7.2. Isolated from sludge of a thermophilic, methanogenic reactor treating wastewater from a kraft pulp production process.
DNA G+C content (mol%): 52.8 (HPLC).
Type strain: TSB, ATCC 49756, DSM 6193.
GenBank accession number (16S rRNA gene): Y11574.

19b. **Desulfotomaculum thermobenzoicum subsp. thermosyntrophicum** Plugge, Balk and Stam 2002, 398[VP]

ther.mo.syn.tro'phi.cum Gr. adj. *thermos* hot; Gr. pref. *syn* together, Gr. v. *trophein* to eat; *syntrophos* nourished together; N.L. neut. adj. *thermosynthrophicum* referring to the capacity of the organism to grow at elevated temperatures on propionate in the presence of a partner organism.

Characteristics are summarized in Table 186. Strain TPO could grow fermentatively on benzoate by an unknown pathway. Grows syntrophically on propionate in co-culture with H_2-scavenging methanogens. The type strain grows at 4–62°C, and over a pH range of 6.0–8.0 with an optimum for growth at 7.0. Isolated from granular methanogenic sludge.
DNA G+C content (mol%): 53.7 (HPLC).
Type strain: strain TPO, ATCC BAA-281, DSM 14055.
GenBank accession number (16S rRNA gene): AY007190.

20. **Desulfotomaculum thermocisternum** Nilsen, Torsvik and Lien 1996, 401[VP]

ther.mo.cis.ter'num. Gr. adj. *thermos* hot; L. fem. n. *cisterna* reservoir; N.L. adj. *thermocisternum* hot reservoir, referring to the original habitat of the organism.

Characteristics are summarized in Table 186. The type strain grows at 41–75°C, and over a pH range of 6.2–8.9 with an optimum for growth at 6.7. Isolated from pure formation water that originated from the subterranean Brent Group oil formation 2.6 km below the sea floor of the Norwegian sector of the North Sea.
DNA G+C content (mol%): 56 (HPLC), 57 (T_m).
Type strain: ST90, DSM 10259.
GenBank accession number (16S rRNA gene): U33455.

21. **Desulfotomaculum thermosapovorans** Fardeau Ollivier, Patel, Dwivedi, Ragot and Garcia 1995, 221[VP]

ther.mo.sa.po.vo'rans. Gr. adj. *thermos* hot; L. masc. n. *sapo* soap; L. v. *voro* to devour; N.L. part. adj. *thermosapovorans* thermophilic and soap-devouring.

Characteristics are summarized in Table 186. The type strain grows at 35–60°C. The optimum pH for growth is 7.2–7.5. Isolated from a thermophilic, anaerobic enrichment culture growing on rice hulls. The initial inoculum was a mixed compost containing rice hulls and peanut shells.
DNA G+C content (mol%): 51.2 (Bd).
Type strain: MLF, DSM 6562.

GenBank accession number (16S rRNA gene): Y11575.

22. **Desulfotomaculum thermosubterraneum** Kaksonen Spring, Schumann, Kroppenstedt and Puhakka 2006b, 2606^VP ther.mo.sub.ter.ra'ne.um. Gr. adj. *thermos* hot; L. neut. adj. *subterraneum* subterranean, underground, below the Earth's surface; *thermosubterraneum* thermophilic inhabitant of the Earth's subsurface.

Characteristics are summarized in Table 186. The type strain grows at 50–72°C and over a pH range of 6.4–7.8 with an optimum for growth at 7.2–7.4. Isolated from a geothermally active underground mine in Japan.

DNA G+C content (mol%): 54.4 (HPLC).
Type strain: RL50JIII, DSM 16057, JCM 13837.
GenBank accession number (16S rRNA gene): DQ208688.

Species *incertae sedis*

1. **"Desulfotomaculum reducens"** (not formally described and not validly published)

re.du'cens. L. part. adj. *reducens* converting to a different state.

Characteristics are summarized in Table 186. Incomplete description is found in Tebo and Obraztsova, (1998). Strain MI-1 grows with Cr(VI), Mn(IV) Fe(III),

and U(VI) as electron acceptors. Isolated from sediment of a shipyard located in the San Francisco Bay estuary, CA, USA.

DNA G+C content (mol%): 42 (whole-genome sequence).
Reference strain: strain MI-1 (not deposited in any service collection)
GenBank accession number (16S rRNA gene): U95951.

Genus VIII. **Pelotomaculum** Imachi, Sekiguchi, Kamagata, Hanada, Ohashi and Harada 2002, 1734^VP emend. deBok, Harmsen, Plugge, de Vries, Akkermans, de Vos and Stams 2005, 1702 emend. Qiu, Sekiguchi, Hanada, Imachi, Tseng, Cheng, Ohashi, Harada and Kamagata 2006a, 180

FRED A. RAINEY

Pe.lo.to.ma'cu.lum. Gr. adj. *pelos* dark-colored, hence anaerobic mud; L. neut. n. *tomaculum* sausage; N.L. neut. n. *Pelotomaculum* sausage-shaped bacteria living in anaerobic environments.

Nonmotile, sausage-shaped cells with spherical endospores. Gram reaction negative but has Gram-positive cell-wall structure. Strictly anaerobic. Mesophilic and thermophilic. Growth by syntrophy with hydrogenotrophs or by fermentation. Can use a limited number of compounds including propionate, primary alcohols, low molecular mass aromatics, or lactate in association with hydrogenotrophic methanogens. Organic compounds such as fumarate are used by some species as alternative electron acceptors. Sulfate, sulfite, thiosulfite, elemental sulfur, nitrate, and ferric ion are not reduced. The major cellular fatty acid is $C_{15:0\ iso}$. Major quinones are MK-7, MK-7(H_4), or MK-9(H_4). Member of the family *Peptococcaceae* on the basis of 16S rRNA gene sequence comparisons (Figure 6).

DNA G+C content (mol%): 52.8–53.6.

Type species: **Pelotomaculum thermopropionicum** Imachi, Sekiguchi, Kamagata, Hanada, Ohashi and Harada 2002, 1734^VP.

Taxonomic comment

The genus *Pelotomaculum* comprises five species that form a phylogenetically coherent cluster within the family *Peptococcaceae*. The genus falls in a large cluster that includes the species of the genera *Desulfotomaculum* and *Sporotomaculum* (see Figure 180). The genus *Cryptanaerobacter* (Juteau et al., 2005) falls within the *Pelotomaculum* cluster and was described after the genus *Pelotomaculum* on the basis of differences in temperature optima for growth, fatty acid composition, and substrate utilization patterns. Since the description of *Cryptanaerobacter*, four additional species of the genus *Pelotomaculum* have been described and fall

on either side of the branch representing *Cryptanaerobacter phenolicus* (see Figure 180). The 16S rRNA gene sequence similarities between the species of the genus *Pelotomaculum* are in the range 92.7–97.5% and between *Pelotomaculum* species and *Cryptanaerobacter phenolicus* are 93.4–94.4%. All species of the genus are unable to reduce sulfate as is also the case with *Cryptanaerobacter phenolicus* (Juteau et al., 2005). However, the *dsrAB* genes have been found in one of the species, *Pelotomaculum propionicicum* (Imachi et al., 2007). The fact that the species of the genus *Pelotomaculum* are unable to reduce sulfate differentiates them from the related *Desulfotomaculum* species in the family *Peptococcaceae*. The species of the genus are differentiated on the basis of phylogenetic position, all sharing less than 95% 16S rRNA gene sequence similarity with the exception of *Pelotomaculum isophthalicicum* and *Pelotomaculum terephthalicicum* which share 97.5% sequence similarity. The species *Pelotomaculum schinkii* has been shown to contain at least two copies of the 16S rRNA gene that differ in sequence 3% (de Bok et al., 2005). The species *Pelotomaculum isophthalicicum* and *Pelotomaculum terephthalicicum* can be differentiated on the basis that *Pelotomaculum terephthalicicum* is able to grow in pure culture on crotonate, while *Pelotomaculum isophthalicicum* only grows syntrophically. The species *Pelotomaculum isophthalicicum*, *Pelotomaculum propionicicum*, and *Pelotomaculum schinkii*, can only grow syntrophically in co-culture with hydrogenotrophic methanogens. *Pelotomaculum thermopropionicum* and *Pelotomaculum terephthalicicum* can grow in pure culture on a limited number of substrates. *Cryptanaerobacter phenolicus* grows in pure culture, but syntrophic growth has not been tested (Juteau et al., 2005).

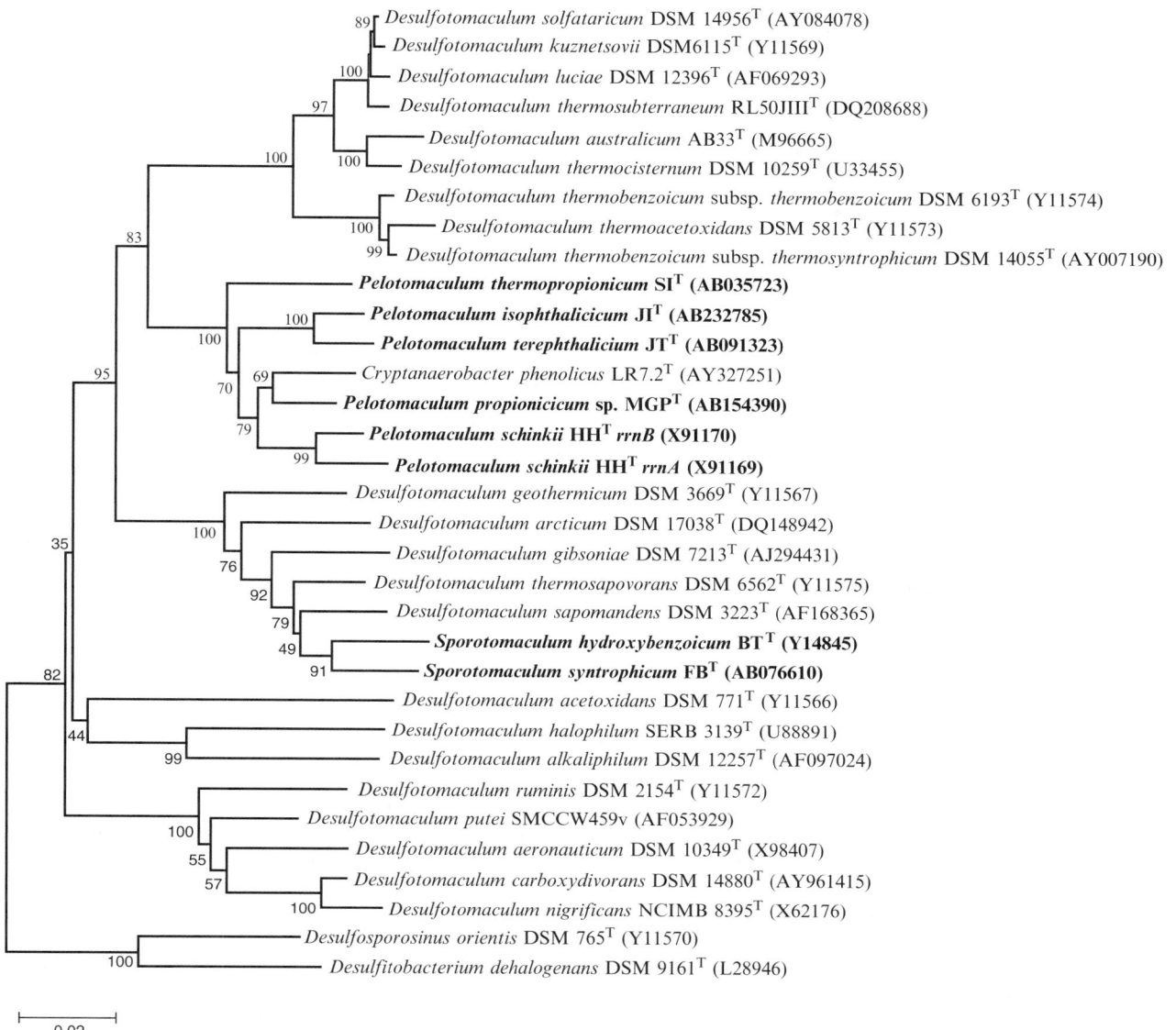

FIGURE 180. 16S rRNA gene sequence based phylogeny indicating the position of the genera *Pelotomaculum* and *Sporotomaculum* within the radiation of the species of the genus *Desulfotomaculum*. The scale bar represents 2 inferred nucleotide substitutions per 100 nucleotides.

List of species of the genus *Pelotomaculum*

1. **Pelotomaculum thermopropionicum** Imachi, Sekiguchi, Kamagata, Hanada, Ohashi and Harada 2002, 1734[VP]

   ther.mo.pro.pi.o′ni.cum. Gr. adj. *thermos* hot; N.L. n. *propionatum* propionate; L. suff. *-icus* pertaining to; N.L. neut. adj. *thermopropionicum* thermophilic and pertaining to propionate.

   Cells are 1.7–2.8 μm × 0.7–0.8 μm. Spores are spherical and central. Utilizes propionate, ethanol, lactate, ethylene glycol, 1-butanol, 1-propanol, 1-pentanol, and 1,3-propanediol in syntrophic association with hydrogenotrophic methanogens. Ferments pyruvate and fumarate in pure culture. Fumarate used as an electron acceptor in the presence of propionate, ethanol, or lactate as an electron donor. Growth occurs

   0–0.4% NaCl but not at 0.5% NaCl. Growth occurs between 45–65°C, optimum 55°C. pH range for growth is 6.7–7.5, optimum 7.0. The type strain was isolated from granular sludge in a thermophilic upflow anaerobic sludge blanket reactor.

   DNA G+C content (mol%): 52.8 (T_m).

   Type strain: SI, DSM 13744, JCM 10971.

   GenBank accession number (16S rRNA gene): AB035723.

2. **Pelotomaculum isophthalicicum** corrig. Qiu, Sekiguchi, Hanada, Imachi, Tseng, Cheng, Ohashi, Harada and Kamagata 2006b, 2026[VP] (Effective publication: Qiu, Sekiguchi, Hanada, Imachi, Tseng, Cheng, Ohashi, Harada and Kamagata 2006a, 181.)

i.so.phtha.li'ci.cum. N.L. neut. n. *acidum isophthalicum* isophthalic acid; L. suff. *-icus -a -um*, suffix used in adjectives with the sense of belonging to; N.L. neut. adj. *isophthalicicum* referring to the substrate isophthalic acid, which can be utilized by the species.

Cells are 0.8–1.0 μm × 2.0–3.0 μm, occurring singly or in pairs. Spores are spherical and central. Mesophilic. Strictly anaerobic. The organism only shows syntrophic growth with a methanogen under anaerobic (methanogenic) conditions. In syntrophic association with hydrogenotrophic methanogens, the strain can utilize ortho-phthalate, isophthalate, terephthalate, benzoate, and 3-hydroxybenzoate. No substrates tested support growth in pure culture. The temperature range for growth is 25–45°C, optimum 37°C. Growth occurs in the pH range 6.8–7.2, optimum 7.0. Growth occurs in the presence of 0–0.5% NaCl. Growth is inhibited at 0.75% NaCl. Sulfate, sulfite, thiosulfate, nitrate, elemental sulfur, fumarate, ferric iron, and 4-hydroxybenzoate cannot be utilized as electron acceptors. The type strain was isolated from granular sludge from an upflow anaerobic sludge bed reactor treating wastewater from the manufacturing of terephthalic and isophthalic acids. This strain has been deposited in the culture collections as a co-culture with *Methanospirillum hungatei*.

DNA G+C content (mol%): not determined.

Type strain: JI, ATCC BAA-1053, JCM 12282.

GenBank accession number (16S rRNA gene): AB232785.

3. **Pelotomaculum propionicicum** Imachi, Sakai, Ohashi, Harada, Hanada, Kamagata and Sekiguchi 2007, 1491[VP]

pro.pi.o.ni'ci.cum. N.L. neut. adj. *propionicicum* referring to the substrate propionic acid, which can be metabolized by the species.

Cells are Gram-positive, 1.0 μm × 2.0–4.0 μm, occurring singly or in pairs. Spores are spherical and central. Mesophilic. Strictly anaerobic. The organism only grows syntrophically on propionate with hydrogenotrophic methanogens. Does not demonstrate sulfate reduction but has *dsrAB* genes that are essential for sulfate respiration. Temperature range for growth is 25–45°C, optimum 37°C. Growth occurs in the pH range 6.5–7.5, optimum 6.5–7.2.

Growth occurs in the presence of 0–0.5% NaCl. Growth is inhibited at >1.0% NaCl. The type strain was isolated from anaerobic, mesophilic, granular sludge. This strain has been deposited in the culture collections as a co-culture with *Methanospirillum hungatei*.

DNA G+C content (mol%): not determined.

Type strain: MGP, DSM 15578, JCM 11929.

GenBank accession number (16S rRNA gene): AB154390.

4. **Pelotomaculum schinkii** de Bok, Harmsen, Plugge, de Vries, Akkermans, de Vos and Stams 2005, 1702[VP]

schin'ki.i. N.L. gen. n. *schinkii* named after Bernhard Schink, who studied several syntrophic conversions. The first spore-forming syntrophic bacterium, *Syntrophospora bryantii*, was isolated by his group.

Cells are Gram-stain-positive, 1.0 μm × 2.0–2.5 μm. Spores are spherical in late exponential phase. Nonmotile. Strictly anaerobic. Grows syntrophically on propionate with methanogens that utilize both hydrogen and formate, such as *Methanospirillum hungatei* JF-1[T] and *Methanobacterium formicicum* MF[NT]. Organic supplements not required for growth. Yeast extract (0.1%) and fumarate (10–20 mM) enhance growth. The type strain was isolated from freeze-dried granular sludge from a UASB reactor treating sugar beet waste in co-culture with *Methanospirillum hungatei* JF-1[T].

DNA G+C content (mol%): Not determined.

Type strain: HH, ATCC BAA-615, DSM 15200.

GenBank accession number (16S rRNA gene): X91169, X91170.

5. **Pelotomaculum terephthalicicum** corrig. Qiu, Sekiguchi, Hanada, Imachi, Tseng, Cheng, Ohashi, Harada and Kamagata 2006b, 2026[VP] (Effective publication: Qiu, Sekiguchi, Hanada, Imachi, Tseng, Cheng, Ohashi, Harada and Kamagata 2006a, 181.)

te.re.phtha.li'ci.cum. N.L. neut. n. *acidum terephthalicum* terephthalic acid; L. suff. *-icus -a -um*, suffix used in adjectives with the sense of belonging to; N.L. neut. adj. *terephthalicicum* referring to the substrate, terephthalic acid, which can be utilized by the species.

Cells are 0.8–1.0 μm × 2.0–3.0 μm, occurring singly or in pairs. Spores are spherical and central. Mesophilic. Strictly anaerobic. Grows syntrophically and in pure culture. In syntrophic association with hydrogenotrophic methanogens, the strain can utilize terephthalate, isophthalate, benzoate, hydroquinone, 2-hydroxybenzoate, 3-hydroxybenzoate, 2,5-dihydroxybenzoate, 3-phenylpropionate, and crotonate. In pure culture ferments crotonate, hydroquinone, and 2,5-dihydroxybenzoate. The temperature range for growth is 30–40°C, optimum 37°C. Growth occurs in the pH range 6.5–7.5, optimum 7.0. Growth occurs in the presence of 0–0.5% NaCl. Growth is inhibited at >2.0% NaCl. Sulfate, sulfite, thiosulfate, nitrate, elemental sulfur, fumarate, ferric iron, and 4-hydroxybenzoate cannot be utilized as electron acceptors. The major cellular fatty acids are $C_{15:0\,iso}$ and $C_{14:0}$. Major quinone is MK-9 (H_4). The type strain was isolated from granular sludge from an upflow anaerobic sludge bed reactor treating wastewater from the manufacture of terephthalic and isophthalic acids.

DNA G+C content (mol%): 53.6 (HPLC).

Type strain: JT, DSM 16121, JCM 11824, NBRC 100523.

GenBank accession number (16S rRNA gene): AB091323.

Genus IX. **Sporotomaculum** Brauman, Müller, Garcia, Brune and Schink 1998, 219[VP]

FRED A. RAINEY

Spo.ro.to.ma′cu.lum. Gr. n. *spora* spore; L. n. *tomaculum* sausage. N.L. neut. n. *Sporotomaculum* a spore-forming sausage-shaped organism.

Gram positive, strictly anaerobic. Metabolism fermentative, inorganic electron acceptors not used. Member of the family *Peptococcaceae* on the basis of 16S rRNA gene sequence comparisons (Figure 6).

DNA G+C content (mol%): 46.8–48.0.

Type species: **Sporotomaculum hydroxybenzoicum** Brauman, Müller, Garcia, Brune and Schink 1998, 219[VP].

Taxonomic comments

The genus *Sporotomaculum* comprises two species that fall within the radiation of a number of species of the genus *Desulfotomaculum* (see Figure 180) in the family *Peptococcaceae*. Both species of the genus *Sporotomaculum* fall within the phylogenetic cluster of the genus *Desulfotomaculum* that contains *Desulfotomaculum geothermicum*, *Desulfotomaculum arcticum*, *Desulfotomaculum gibsoniae*, *Desulfotomaculum thermosapovorans*, and *Desulfotomaculum sapomandens* (see Figure 180). The 16S rRNA gene sequence similarity values between the *Sporotomaculum* species and these related *Desulfotomaculum* species are in the range 92.3–95.2% and it is interesting to note that the two species of the genus *Sporotomaculum* share the same 16S rRNA gene sequence similarity (95.2%) as *Sporotomaculum syntrophicum* and *Desulfotomaculum sapomandens*.

The genus was described based on its differentiation from *Desulfotomaculum* species due to its inability to reduce sulfate, sulfite, or thiosulfate even in the presence of added hemin or 1,4-naphthoquinone. In addition, *Sporotomaculum hydroxybenzoicum* was limited in its substrate range, using only 3-hydroxybenzoate as a sole carbon and energy source (Brauman et al., 1998). *Sporotomaculum syntrophicum* was added to the genus *Sporotomaculum* on the basis of its close phylogenetic relationship to *Sporotomaculum hydroxybenzoicum*; they share 95.2% 16S rRNA gene sequence similarity and lack the ability to reduce sulfate. The two species can be differentiated on the basis of a number of characteristics including motility (*Sporotomaculum hydroxybenzoicum* is motile). *Sporotomaculum syntrophicum* shows syntrophic growth on benzoate, crotonate, and butyrate in co-culture with *Methanospirillum hungatei*, and grows in pure culture on crotonate and benzoate. *Sporotomaculum hydroxybenzoicum* grows in pure culture and only on 3-hydroxybenzoate.

List of species of the genus *Sporotomaculum*

1. **Sporotomaculum hydroxybenzoicum** Brauman, Müller, Garcia, Brune and Schink 1998, 219[VP]

hy.dro.xy.ben.zo′i.cum. N.L. neut. adj. *hydroxybenzoicum* referring to hydroxybenzoic acid which is used as sole carbon and energy source.

Rod-shaped cells, 2.0–3.0 μm × 0.6–0.8 μm, motile, with pointed ends, occurring singly or in pairs. Gram positive and slightly motile in the early exponential phase. Endospores are central and spherical. Strictly anaerobic. Chemoorganoheterotroph. Grows on 3-hydroxybenzoate as sole source of carbon and energy. No growth with pyruvate, DL-lactate, succinate, fumarate, malate, citrate, valerate, pimelate, crotonate, crotonate with H_2/CO_2 atmosphere (80:20), caproate, adipate, hexanoate, heptanoate, cyclohexane carboxylate, ethanol, methanol, D-fructose, glucose, benzoate, 2-hydroxybenzoate, 4-hydroxybenzoate, 3-aminobenzoate, 4-aminobenzoate, 2,3- and 2,5-dihydroxybenzoate, α-, β-, and γ-resorcylate, protocatechuate, gallate, syringate, ferulate, caffeate, hydroxycinnamate, terephthalate, phenylacetate, 2-, 3-, and 4-chlorobenzoate, phenol, catechol, resorcinol, hydroquinone, phloroglucinol, pyrogallol, 2-, 3- and 4-cresol. 3-Hydroxybenzoate is fermented to butyrate, acetate, and CO_2. No reduction of nitrate, sulfate, thiosulfate, sulfite, ferric iron, oxygen, or fumarate. Growth requires sulfide-reduced mineral media. Addition of small amounts of dithionite shortens the lag phase. Yeast extract and tryptone stimulate growth but are not required for growth. Cytochromes not detected. The temperature range for growth is 24–37°C, optimum 30°C. pH range for growth is 6.8–8.1, optimum 7.3–7.6.

The type strain was isolated from a selective enrichment in sulfide-reduced freshwater mineral medium with 3-hydroxybenzoate as substrate using gut homogenates of soil-feeding termites (*Cubitermes speciosus*).

DNA G+C content (mol%): 48 (HPLC).

Type strain: BT, ATCC 700645, DSM 5475.

GenBank accession number (16S rRNA): Y14845.

2. **Sporotomaculum syntrophicum** Qiu, Sekiguchi, Imachi, Kamagata, Tseng, Cheng, Ohashi and Harada 2003b, 937[VP] (Effective publication: Qiu, Sekiguchi, Imachi, Kamagata, Tseng, Cheng, Ohashi and Harada 2003a, 248.)

syn.tro′phi.cum. Gr. pref. *syn* together with; Gr. v. *trophein* nourish; N.L. suff. *icus* pertaining to; N.L. neut. adj. *syntrophicum* pertaining to syntrophic substrate utilization.

Rod-shaped cells, 1.0–2.0 μm × 0.8–1.0 μm, occurring singly or in pairs. Spores are spherical and central. Gram-stain-positive. Strictly anaerobic. Utilizes benzoate, butyrate, and crotonate in syntrophic association with hydrogenotrophic methanogens. Ferments crotonate in pure culture. Growth occurs in the presence of 0–1% NaCl but does not occur in the presence of more than 2% NaCl. The temperature range for growth is 28–45°C, optimum 35–40°C. The pH range for growth is 6.0–7.5, optimum 7.0–7.2. Sulfate, sulfite, thiosulfate, nitrate, elemental sulfur, fumarate, or ferric iron cannot be used as electron acceptors. The type strain was isolated from methanogenic sludge from a two-phase anaerobic treatment system.

DNA G+C content (mol%): 46.8 (HPLC).

Type strain: FB, DSM 14795, JCM 11495.

GenBank accession number (16S rRNA): AB076610.

Genus X. **Syntrophobotulus** Friedrich, Springer, Ludwig and Schink 1996, 1068[VP]

BERNHARD SCHINK AND MICHAEL FRIEDRICH

Syn.tro.pho.bo'tu.lus. Gr. pref. *syn* together; Gr. v. *trophein* to nourish; L. masc. n. *botulus* sausage; N.L. masc. n. *Syntrophobotulus* a syntrophic, sausage-like bacterium.

Straight to slightly curved rods. Spores formed at the cell ends. Stains Gram-stain-negative; ultrastructural analysis shows Gram-positive cell-wall architecture. **Strictly anaerobic; fermentative metabolism**. Chemo-organotrophic. Cytochromes and catalase activity absent. Isolated from sewage sludge.

DNA G+C content (mol%): 46.7 ± 0.15 % (HPLC).

Type species: **Syntrophobotulus glycolicus** Friedrich, Springer, Ludwig and Schink 1996, 1068[VP].

Further descriptive information

According to the 16S rRNA phylogenetic analysis presented in the roadmap to this volume (Figure 6), the genus *Syntrophobotulus* is a member of the family *Peptococcaceae*, order *Clostridiales*, class *Clostridia* in the phylum *Firmicutes*. *Desulfitobacterium dehalogenans* and *Desulfotomaculum orientis* share 91.2% and 88.6% 16S rRNA sequence similarity, respectively. More distant relationships were found to other *Desulfotomaculum* species, to *Heliobacterium* sp., and to *Selenomonas* sp. (Friedrich et al., 1996).

Syntrophobotulus glycolicus was originally isolated in a defined co-culture with the methanogenic partner organism *Methanospirillum hungatei* using glycolate as the sole organic substrate (Friedrich et al., 1991). Cells are thin, slightly curved rods, 0.5 μm in diameter and 2.5–3.5 μm in length. Cells stain Gram-negative but do not form slime in the KOH-test that is typical of Gram-stain-negative bacteria. Electron microscopic examination of ultrathin sections reveals a cell-wall structure typical of Gram-positive bacteria. Oval spores are formed in ageing cultures at the cell ends. The bacterium depends on cooperation with a methanogenic or a homoacetogenic partner organism if grown on glycolate. *Syntrophobotulus glycolicus* strain FlGlyR was isolated later in pure culture with glyoxylate as the sole substrate (Friedrich and Schink, 1995).

Syntrophobotulus glycolicus is unique in its energy metabolism because it uses only glyoxylate in pure culture or glycolate in syntrophic co-culture with hydrogen-scavenging partners such as *Methanospirillum hungatei* or *Acetobacterium woodii*. Glyoxylate is condensed with acetyl-CoA to form malyl-CoA which is converted to malate in an ATP-yielding reaction. The malic enzyme converts malate to pyruvate with concomitant decarboxylation and NADP reduction, and pyruvate is oxidized to acetyl-CoA with ferredoxin as electron acceptor, thus closing the reaction cycle. Glycolate oxidation in syntrophic cultures is catalyzed by a membrane-bound glycolate dehydrogenase enzyme which forms glyoxylate and cooperates with a membrane bound hydrogenase system. Hydrogen release from glycolate oxidation requires ATP investment via a proton-motive force (Friedrich and Schink, 1993) and glyoxylate reduction with molecular hydrogen can be coupled in membrane preparations with ATP synthesis via electron transport phosphorylation (Friedrich and Schink, 1995). Thus, during growth with glycolate in syntrophic co-culture, ATP is synthesized only by substrate level phosphorylation, and part of this energy is reinvested into a reversed electron transport driven by ATP hydrolysis, whereas during growth in pure culture on glyoxylate ATP is produced by glyoxylate dismutation to glycolate and CO_2, employing substrate-level phosphorylation and electron-transport phosphorylation (Friedrich and Schink, 1995). As a result, *Syntrophobotulus glycolicus* has become a model organism for the study of syntrophic energy metabolism because all biochemical components of this unusual energy metabolism system can be identified in cell-free extracts.

Enrichment and isolation procedures

Syntrophobotulus glycolicus was enriched with glycolate as sole carbon source under strictly anoxic conditions in bicarbonate-buffered mineral medium at pH 7.2 and 30°C. A primary co-culture contained a homoacetogenic partner bacterium which could be replaced by *Methanospirillum hungatei* to form a defined co-culture (Friedrich et al., 1991). From this co-culture, *Syntrophobotulus glycolicus* was isolated with glyoxylate as sole organic substrate.

Maintenance procedures

Pure cultures may be stored anoxically in liquid medium at 4°C in the dark. Stock cultures should be transferred every 1–2 months. For long-term preservation, a dense cell suspension of *Syntrophobotulus glycolicus* may be made in anoxic medium containing glycerol or dimethyl sulfoxide, sealed in sterile glas capillaries, and stored in liquid nitrogen.

Differentiation of the genus S*yntrophobotulus* from other genera

Syntrophobotulus glycolicus differs from all other described bacteria by its unusual energy metabolism which is based exclusively on utilization of glycolate and glyoxylate. Glyoxylate is dismutated in pure culture to glycolate and CO_2; glycolate is oxidized to CO_2 only in syntrophic co-culture with hydrogen-scavenging partner organisms.

List of species of the genus *Syntrophobotulus*

1. **Syntrophobotulus glycolicus** Friedrich, Springer, Ludwig and Schink 1996, 1068[VP]

gly.co'li.cus. N.L. n. *acidum glycolicum* glycolic acid; N.L. masc. adj. *glycolicus* referring to glycolic acid, the key substrate of this species.

Rod-shaped bacterium, slightly curved, 0.5 μm by 2.5–3.5 μm in size, occurring typically as single cells or in short chains or small aggregates. Nonmotile. Oval spores formed in ageing cultures at the cell ends. Cells stain Gram-stain-negative; ultrastructural analysis shows Gram-positive cell-wall architecture.

Strictly anaerobic, grow chemotrophically in pure culture by fermentative oxidation of glyoxylate. Glycolate is oxidized in syntrophic co-culture with, e.g., *Methanospirillum hungatei* or *Acetobacterium woodii* as a partner. No other organic or inorganic substrates used. Glycolic acid is converted to carbon dioxide and hydrogen in syntrophic culture; glyoxylic acid is fermented in pure culture to carbon dioxide, hydrogen, and glycolic acid. Does not reduce sulfate, sulfite, thiosulfate, elemental sulfur, or nitrate. Cells contain menaquinones 7–10, with MK-9 as major fraction. No cytochromes present.

pH range 6.7–8.3; optimum pH 7.3. Temperature range, 15–37°C; optimum temperature 28°C. Growth optimal in freshwater medium. Growth possible also in brackish-water medium with 110 mM NaCl and 5 mM $MgCl_2$; no growth in marine medium. Habitats are sewage sludge and anoxic freshwater sediments.

DNA G+C content (mol%): 46.7 ± 0.15% (HPLC).

Type strain: FlGlyR, DSM 8271.

GenBank accession number (16S rRNA gene): not available for the type strain; X99706 for strain SlGlym.

Genus XI. **Thermincola** Sokolova, Kostrikina, Chernyh, Kolganova, Tourova and Bonch-Osmolovskaya 2005, 2072[VP]

TATYANA G. SOKOLOVA AND DARIA G. ZAVARZINA

Therm.in′co.la. Gr. adj. *thermos* hot; L. fem. n. *incola* inhabitant; N.L. fem. n. *Thermincola* inhabitant of a hot spring.

Cells are rods with **Gram-positive** type cell walls. Anaerobic thermophile **capable of chemolithotrophic growth by anaerobic CO oxidation coupled to molecular hydrogen and CO_2 production**. Does not ferment organic substrates. Found in terrestrial hot springs.

DNA G+C content (mol%): 45–48.

Type species: **Thermincola carboxydiphila** Sokolova, Kostrikina, Chernyh, Kolganova, Tourova and Bonch-Osmolovskaya 2005, 2072[VP].

Further descriptive information

Members of the genus *Thermincola* are geographically widespread. The type species *Thermincola carboxydiphila* was isolated from a slightly alkaline hot spring (pH 8.5, 55°C) on the bank of Bolshaya River, Baikal Lake region (Sokolova et al., 2005). The second characterized species, *Thermincola ferriacetica*, was isolated from ferric deposits at a terrestrial hot spring (pH 6.8–7.0, 65°C) of the Stolbovskie group, Kunashir Island, Kurils (Zavarzina et al., 2007). Additional isolates have been isolated from a hot spring in Iceland (personal communication from N. K. Birkland). There is also evidence that the habitat of *Thermincola* is not limited to terrestrial hot springs. Genes of 16S rRNA with high similarity to *Thermincola* have been detected in oil reservoirs by Liew and Jong (GenBank accession no. EF095439), in waste waters (personal communication from A. I. Slobodkin), and in the enrichments from a geothermally active mine in Japan (Kaksonen et al., 2006a).

Enrichment and isolation procedures

Thermincola species were enriched and isolated with anaerobically prepared media. *Thermincola carboxydiphila* was isolated in mineral carbonate-bicarbonate buffered medium (pH 9.0) under 100% CO in the gas phase. The medium was supplemented with vitamins and sodium acetate (0.2 g/l) and incubated at 55°C (Sokolova et al., 2005). For the enrichment and isolation of *Thermincola ferriacetica*, bicarbonate-CO_2 buffered medium (pH 7.0) under N_2/CO_2 (80:20 v/v) was used. The medium contained sodium acetate (20 mM) and amorphous Fe(III) oxide (90 mM) and was incubated at 60°C (Zavarzina et al., 2007). Enrichments were serially diluted in the growth medium prior to isolation of single colonies in roll tubes.

Maintenance procedures

Cultures may be stored at room temperature for 1–2 weeks, at 4°C for 3–5 months, or in liquid nitrogen.

Taxonomic comments

On the basis of 16S rRNA sequence analyses, the genus *Thermincola* is a deep lineage within the family *Peptococcaceae* in the order *Clostridiales* (Zavarzina et al., 2007) (see Figure 5 and Figure 6). *Thermincola* is currently represented by two described species, *Thermincola carboxydiphila* and *Thermincola ferriacetica*, which share 98% sequence similarity of the 16S rRNA gene and 27% DNA–DNA hybridization (Zavarzina et al., 2007). Outside the genus, the mostly closely related organism is *Pelotomaculum thermopropionicum*, with 95% similarity of its 16S rRNA gene (Imachi et al., 2002).

List of species of the genus *Thermincola*

1. **Thermincola carboxydiphila** Sokolova, Kostrikina, Chernyh, Kolganova, Tourova and Bonch-Osmolovskaya 2005, 2072[VP]

car.bo.xy.di′phi.la. N.L. neut. n. *carboxydum* carbon monoxide, Gr. adj. *philos* loving, N.L. fem. adj. *carboxydiphila* loving carbon monoxide.

Has the characteristics of the genus. Cells are straight, thick rods, with rounded ends, about 0.5 μm by 0.6–3.0 μm. Motile due to one or two lateral flagella. The temperature range for growth is 37–68°C, and the optimum is 55°C. Alkalitolerant, with a pH range for growth of 6.7–9.5 and an optimum at 8.0. The presence of 0.2 g/l of yeast extract or acetate is required for growth with CO. Elemental sulfur, thiosulfate, sulfate, or nitrate do not stimulate the growth and are not reduced during the growth on CO. Does not grow organotrophically on peptone, yeast extract, starch, cellulose, cellobiose, sucrose, maltose, ribose, xylose, lactose, glucose, galactose, fructose, mannitol, sorbitol, pyruvate, acetate, formate, lactate, succinate, methanol, ethanol, and glycerol. $H_2 + CO_2$ (80/20 v/v) and H_2 or CO with ferric iron does not support growth. Does not grow on peptone, yeast extract, sucrose, pyruvate, acetate, formate, lactate, succinate, methanol, ethanol, and glycerol in the presence of elemental sulfur, sulfate, thiosulfate, or ferric iron. Growth is inhibited by penicillin, erythromycin, streptomycin, rifampin, vancomycin, and tetracycline. Isolated from a mixture of water, mud, and cyanobacterial mat from a hot spring of the Baikal Lake region.

DNA G+C content (mol%): 45.4±1.0 (T_m).

Type strain: 2204, DSM 17129, JCM 13258, VKM B-2283.

GenBank accession number (16S rRNA gene): AY603000.

2. **Thermincola ferriacetica** Zavarzina, Sokolova, Tourova, Chernyh, Kostrikina and Bonch-Osmolovskaya 2007, 894[VP] (Effective publication: Zavarzina, Sokolova, Tourova, Chernyh, Kostrikina and Bonch-Osmolovskaya 2007, 5.)

fer.ri.ace.ti.ca. L. n. *ferrum* iron; N.L. n. *acetas -atis* acetate; N.L. fem. n. *ferriacetica* iron oxide- and acetate-utilizing.

Thermincola ferriacetica has the characteristics of the genus. Cells are straight to slightly curved rods, 0.4–0.5 × 1.0–3.0 μm, occurring singly or in large chains (Figure 182). They exhibit a slight tumbling motility by means of 1–4 peritrichous flagella (Figure 181). Some of the cells in chains form round, clostridial-type spores that are considerably larger in diameter than the vegetative cells and located in the middle of the chain (arrows in Figure 182). Spores are exceptionally thermoresistant and survive heating at 121°C for 30 min. In roll tubes, colonies are single lens-shaped, white-cream in color, and 0.2–0.4 mm in diameter. Multiplication occurs by binary fission. Growth occurs from 45–70°C, with an optimum at 57–60°C, and in a pH range of 5.9–8.0, with an optimum at 7.0–7.1. Reduces amorphous Fe(III)-oxide by oxidation of molecular hydrogen, acetate, peptone, yeast and beef extracts, glycogen, glycolate, pyruvate, betaine, choline, *N*-acetyl-D-glucosamine and Casamino acids. No growth occurs on adonite, arginine, butyrate, citrate, formate, glutamate,

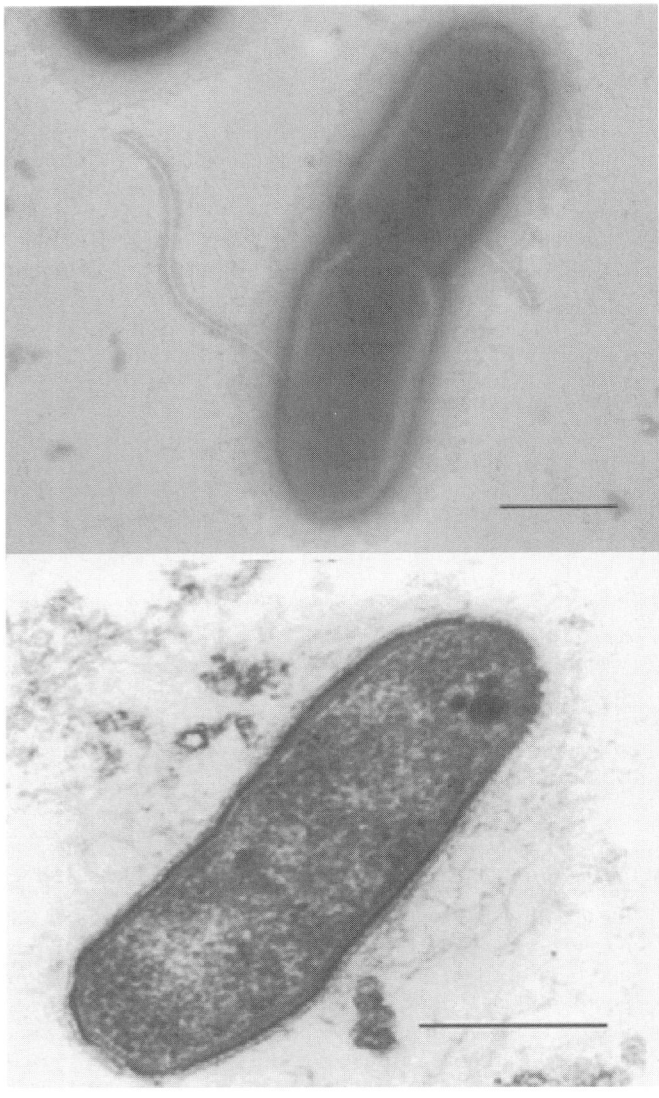

FIGURE 181. Morphology of *Thermincola carboxydiphila* strain 2204[T]. Negative stained (a) and thin section (b) electron micrographs. Bars = 0.5 μm. (Reprinted with permission from Sokolova et al., 2005. International Journal of Systematic and Evolutionary Microbiology 55: 2069–2073.)

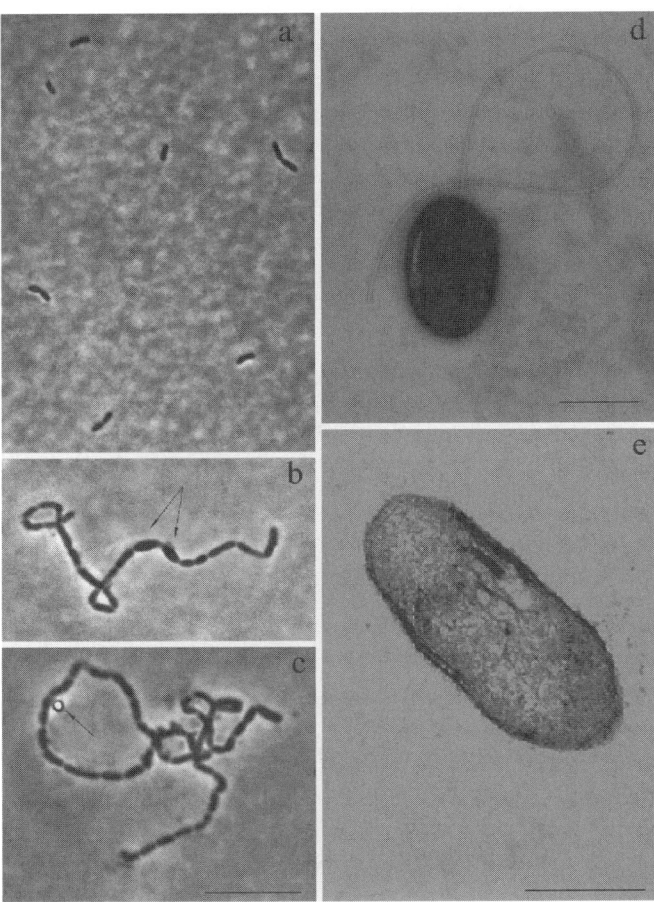

FIGURE 182. Morphology of *Thermincola ferriacetica*. Light micrographs showing the single cell morphology (a) and spore-forming chains (b, c). The arrows indicate the presence of forespores (b) and free spores (c). Bar = 5 μm. Negatively stained (d) and thin-section (e) electron micrographs. The single flagellum is apparent in (d), and the Gram-positive type cell wall is apparent in (e). Bars = 5 μm.

lactate, malate, propionate, serine, succinate, tartrate, ethanol, mannitol, methanol, propanol, L-sorbitol, L-histidine, glycerol, DL-lysine, sarcosine, tryptone, cellulose, chitin, and starch. Does not grow by fermentation of sugars, peptone, and yeast and beef extracts. With acetate, it uses amorphous Fe(III) oxide, magnetite, 2,6-anthraquinone disulfonate (AQDS), and MnO_2 as the electron acceptors, but not citrate, fumarate, sulfate, sulfite, thiosulfate, dithionite, elemental sulfur, and nitrate. With molecular hydrogen and yeast extract (0.2 g/l), it reduces Fe(III) oxide, AQDS, MnO_2, and thiosulfate. Grows chemolithoautotrophicaly with hydrogen as the energy source, Fe(III) as the electron acceptor, and CO_2 as carbon source. Yeast extract stimulates growth. Requires 0.2 g/l both of yeast extract and acetate for growth on CO. Chloramphenicol, neomycin, penicillin, kanamycin, but not polymyxin B and streptomycin, inhibit growth. The type strain was isolated from ochre deposits in a hot spring in Stolbovskiye, Kuril Islands, Russia.

DNA G+C content (mol%): 47.8±1.0 (T_m).

Type strain: Z-0001, DSM 14005, VKM B-2307.

GenBank accession number (16S rRNA gene): AY631277.

References

Adams, M.E. and J.R. Postgate. 1959. A new sulfate-reducing vibrio. J. Gen. Microbiol. *20*: 252–257.

Beaudet, R., J.G. Bisaillon, M. Ishaque and M. Sylvestre. 1986. Isolation of an anaerobic bacterial consortium degrading phenolic compounds - Assay in swine waste. Agric. Waste *17*: 131–140.

Béchard, G., J.G. Bisaillon, R. Beaudet and M. Sylvestre. 1990. Degradation of phenol by a bacterial consortium under methanogenic conditions. Can. J. Microbiol. *36*: 573–578.

Bouchard, B., R. Beaudet, R. Villemur, G. McSween, F. Lepine and J.G. Bisaillon. 1996. Isolation and characterization of *Desulfitobacterium frappieri* sp. nov., an anaerobic bacterium which reductively dechlorinates pentachlorophenol to 3-chlorophenol. Int. J. Syst. Bacteriol. *46*: 1010–1015.

Boyer, A., R. Page-BeLanger, M. Saucier, R. Villemur, F. Lepine, P. Juteau and R. Beaudet. 2003. Purification, cloning and sequencing of an enzyme mediating the reductive dechlorination of 2,4,6-trichlorophenol from *Desulfitobacterium frappieri* PCP-1. Biochem. J. *373*: 297–303.

Brauman, A., J.A. Muller, J.L. Garcia, A. Brune and B. Schink. 1998. Fermentative degradation of 3-hydroxybenzoate in pure culture by a novel strictly anaerobic bacterium, *Sporotomaculum* hydroxybenoicum gen. nov., sp. nov. Int. J. Syst. Bacteriol. *48*: 215–221.

Breitenstein, A., A. Saano, M. Salkinoja-Salonen, J.R. Andreesen and U. Lechner. 2001. Analysis of a 2,4,6-trichlorophenol-dehalogenating enrichment culture and isolation of the dehalogenating member *Desulfitobacterium frappieri* strain TCP-A. Arch. Microbiol. *175*: 133–142.

Campbell, L.L. and J.R. Postgate. 1965. Classification of the spore-forming sulfate-reducing bacteria. Bacteriol. Rev. *29*: 359–363.

Campbell, L.L. and R. Singleton, Jr. 1986. Genus *Desulfotomaculum* Campbell and Postgate 1965, 361^AL. *In* Sneath, Mair, Sharpe and Holt (Editors), Bergey's Manual of Systematic Bacteriology, Vol. 2. The Williams & Wilkins Co., Baltimore, pp. 1200–1202.

Campbell, L.L. and R.S. Jr. 1988. *In* Validation of the publication of new names and new combinations previously effectively published outside the IJSB. List no. 25. Int. J. Syst. Bacteriol. *38*: 220–222.

Cervantes, F.J., F.A. de Bok, T. Duong-Dac, A.J. Stams, G. Lettinga and J.A. Field. 2002. Reduction of humic substances by haloreviring, sulphate-reducing and methanogenic microorganisms. Environ. Microbiol. *4*: 51–57.

Christiansen, N. and B.K. Ahring. 1996. *Desulfitobacterium hafniense* sp. nov., an anaerobic, reductively dechlorinating bacterium. Int. J. Syst. Bacteriol. *46*: 442–448.

Christiansen, N., B.K. Ahring, G. Wohlfarth and G. Diekert. 1998. Purification and characterization of the 3-chloro-4-hydroxy-phenylacetate reductive dehalogenase of *Desulfitobacterium hafniense*. FEBS Lett. *436*: 159–162.

Collins, M.D. and F. Widdel. 1986. Respiratory quinones of sulphate-reducing and sulphur-reducing bacteria: a systematic investigation. Syst. Appl. Microbiol. *8*: 8–18.

Collins, M.D., P.A. Lawson, A.Willems, J.J. Cordoba, J. Fernandez-Garayzabal, P. Garcia, J. Cia, H. Hippe and J.A.E. Farrow. 1994. The phylogeny of the genus *Clostridium*: proposal of five new genera and eleven new species combination. Int. J. Syst. Bacteriol. *44*: 812–826.

Cord-Ruwisch, R. and J.L. Garcia. 1985. Isolation and characterization of an anaerobic benzoate-degrading spore-forming sulfate-reducing bacterium, *Desulfotomaculum sapomandens* sp. nov. FEMS Microbiol. Lett. *29*: 325–330.

Cord-Ruwisch, R. and J.L. Garcia. 1990. *In* Validation of the publication of new names and new combinations previously effectively published outside the IJSB. List no. 32. Int. J. Syst. Bacteriol. *40*: 105–106.

Crane, B.R., L.M. Siegel and E.D. Getzoff. 1995. Sulfite reductase structure at 1.6 Å: evolution and catalysis for reduction of inorganic anions. Science *270*: 59–67.

Cupples, A.M., R.A. Sanford and G.K. Sims. 2005. Dehalogenation of the herbicides bromoxynil (3,5-dibromo-4-hydroxybenzonitrile) and ioxynil (3,5-diiodino-4-hydroxybenzonitrile) by *Desulfitobacterium chlororespirans*. Appl. Environ. Microbiol. *71*: 3741–3746.

Cypionka, H., F. Widdel and N. Pfennig. 1985. Survival of sulfate-reducing bacteria after oxygen stress, and growth in sulfate-free oxygen-sulfide gradients. FEMS Microb. Ecol. *31*: 39–45.

Cypionka, H. 1986. Sulfide-controlled continuous culture of sulfate-reducing bacteria. J. Microbiol. Lett. *5*: 1–9.

Cypionka, H. and N. Pfennig. 1986. Growth yield of *Desulfotomaculum orientis* with hydrogen in chemostat culture. Arch. Microbiol. *143*: 396–399.

Daumas, S., R. Cordruwisch and J.L. Garcia. 1988. *Desulfotomaculum geothermicum* sp. nov., a thermophilic, fatty acid-degrading, sulfate-reducing bacterium isolated with H_2 from geothermal ground-water. Antonie van Leeuwenhoek *54*: 165–178.

Daumas, S., R. Cord-Ruwisch and J.L. Garcia. 1990. *In* Validation of the publication of new names and new combinations previously effectively published outside the IJSB. List no. 32. Int. J. Syst. Bacteriol. *40*: 105–106.

de Bok, F.A., H.J. Harmsen, C.M. Plugge, M.C. de Vries, A.D. Akkermans, W.M. de Vos and A.J. Stams. 2005. The first true obligately syntrophic propionate-oxidizing bacterium, *Pelotomaculum schinkii* sp. nov., co-cultured with *Methanospirillum hungatei*, and emended description of the genus *Pelotomaculum*. Int. J. Syst. Evol. Microbiol. *55*: 1697–1703.

De Soete, G. 1983. A least squares alogorithm for fitting additive trees to proximity data. Psychometrika *48*: 621–626.

De Wildeman, S., G. Diekert, H. Van Langenhove and W. Verstraete. 2003. Stereoselective microbial dehalorespiration with vicinal dichlorinated alkanes. Appl. Environ. Microbiol. *69*: 5643–5647.

De Wildeman, S., G. Linthout, H. Van Langenhove and W. Verstraete. 2004. Complete lab-scale detoxification of groundwater containing 1,2-dichloroethane. Appl. Microbiol. Biotechnol. *63*: 609–612.

Denger, K., H. Laue and A.M. Cook. 1997. Thiosulfate as a metabolic product: the bacterial fermentation of taurine. Arch. Microbiol. *168*: 297–301.

Denger, K., E. Stackebrandt and A.M. Cook. 1999. *Desulfonispora thiosulfatigenes* gen. nov., sp. nov., a taurine-fermenting, thiosulfate-producing anaerobic bacterium. Int. J. Syst. Bacteriol. *49*: 1599–1603.

Deshmane, V., C.M. Lee and K.L. Sublette. 1993. Microbial reduction of sulfur dioxide with pretreated sewage sludge and elemental hydrogen as electron donors. Appl. Biochem. Biotechnol. *39–40*: 739–752.

DSMZ. 2001. Catalog of Strains, Seventh edn, DSMZ, Mascheroder Weg 1b, D-38124 Braunschweig, Germany.

El Fantroussi, S., H. Naveau and S.N. Agathos. 1998. Anaerobic dechlorinating bacteria. Biotechnol. Prog. *14*: 167–188.

Ezaki, T., N. Yamamoto, K. Ninomiya, S. Suzuki and E. Yabuuchi. 1983. Transfer of *Peptococcus indolicus, Peptococcus asaccharolyticus, Peptococcus prevotii*, and *Peptococcus magnus* to the genus *Peptostreptococcus* and proposal of *Peptostreptococcus tetradius* sp. nov. Int. J. Syst. Bacteriol. *33*: 683–698.

Ezaki, T., N. Li, Y. Hashimoto, H. Miura and H. Yamamoto. 1994. 16s Ribosomal DNA sequences of anaerobic cocci and proposal of *Ruminococcus hansenii* comb. nov. and *Ruminococcus productus* comb. nov. Int. J. Syst. Bacteriol. *44*: 130–136.

Ezaki, T., Y. Kawamura, N. Li, Z.Y. Li, L. Zhao and S. Shu. 2001. Proposal of the genera *Anaerococcus* gen. nov., *Peptoniphilus* gen. nov. and *Gallicola* gen. nov. for members of the genus *Peptostreptococcus*. Int. J. Syst. Evol. Microbiol. *51*: 1521–1528.

Fardeau, M.L., B. Ollivier, B.K.C. Patel, P. Dwivedi, M. Ragot and J.L. Garcia. 1995. Isolation and characterization of a thermophilic sulfate-reducing bacterium, *Desulfotomaculum thermosapovorans* sp. nov. Int. J. Syst. Bacteriol. *45*: 218–221.

Felsenstein, D. 1993. PHYLIP (Phylogeny Inference Package) 3.57 edn. Department of Genetics, University of Washington, Seattle.

Fetzner, S. 1998. Bacterial dehalogenation. Appl. Microbiol. Biotechnol. *50*: 633–657.

Finneran, K.T., H.M. Forbush, C.V. VanPraagh and D.R. Lovley. 2002. *Desulfitobacterium metallireducens* sp. nov., an anaerobic bacterium that couples growth to the reduction of metals and humic acids as well as chlorinated compounds. Int. J. Syst. Evol. Microbiol. *52*: 1929–1935.

Friedrich, M., U. Laderer and B. Schink. 1991. Fermentative degradation of glycolic acid by defined syntrophic cocultures. Arch. Microbiol. *156*: 398–404.

Friedrich, M. and B. Schink. 1993. Hydrogen formation from glycolate driven by reversed electron transport in membrane vesicles of a syntrophic glycolate-oxidizing bacterium. Eur. J. Biochem. *217*: 233–240.

Friedrich, M. and B. Schink. 1995. Electron transport phosphorylation driven by glyoxylate respiration with hydrogen as electron donor in membrane vesicles of a glyoxylate-fermenting bacterium. Arch. Microbiol. *163*: 268–275.

Friedrich, M., N. Springer, W. Ludwig and B. Schink. 1996. Phylogenetic positions of *Desulfofustis glycolicus* gen. nov., sp. nov., and *Syntrophobotulus glycolicus* gen. nov., sp. nov., two new strict anaerobes growing with glycolic acid. Int. J. Syst. Bacteriol. *46*: 1065–1069.

Gabor, K., C.S. Verissimo, B.C. Cyran, P. Ter Horst, N.P. Meijer, H. Smidt, W.M. de Vos and J. van der Oost. 2006. Characterization of CprK1, a CRP/FNR-type transcriptional regulator of halorespiration from *Desulfitobacterium hafniense*. J. Bacteriol. *188*: 2604–2613.

Gauthier, A., R. Beaudet, F. Lepine, P. Juteau and R. Villemur. 2006. Occurrence and expression of *crdA* and *cprA5* encoding chloroaromatic reductive dehalogenases in *Desulfitobacterium* strains. Can. J. Microbiol. *52*: 47–55.

Gerritse, J., V. Renard, T.M.P. Gomes, P.A. Lawson, M.D. Collins and J.C. Gottschal. 1996. *Desulfitobacterium* sp. strain PCE1, an anaerobic bacterium that can grow by reductive dechlorination of tetrachloroethene or ortho-chlorinated phenols. Arch. Microbiol. *165*: 132–140.

Gerritse, J., O. Drzyzga, G. Kloetstra, M. Keijmel, L.P. Wiersum, R. Hutson, M.D. Collins and J.C. Gottschal. 1999. Influence of different electron donors and accepters on dehalorespiration of tetrachlo-

roethene by *Desulfitobacterium frappieri* TCE1. Appl. Environ. Microbiol. *65*: 5212–5221.

Goorissen, H.P., H.T. Boschker, A.J. Stams and T.A. Hansen. 2003. Isolation of thermophilic *Desulfotomaculum* strains with methanol and sulfite from solfataric mud pools, and characterization of *Desulfotomaculum solfataricum* sp. nov. Int. J. Syst. Evol. Microbiol. *53*: 1223–1229.

Hagenauer, A., H. Hippe and F.A. Rainey. 1997a. *Desulfotomaculum aeronauticum* sp. nov., a sporeforming, thiosulfate-reducing bacterium from corroded aluminium alloy in an aircraft. Syst. Appl. Microbiol. *20*: 65–71.

Hagenauer, A., H. Hippe and F.A. Rainey. 1997b. *In* Validation of the publication of new names and new combinations previously effectively published outside the IJSB. List no. 62. Int. J. Syst. Bacteriol. *47*: 915–916.

Hale, D.D., J.E. Rogers and J. Wiegel. 1990. Reductive dechlorination of dichlorophenols by nonadapted and adapted microbial communities in pond sediments. Microb. Ecol. *20*: 185–196.

Hale, D.D., J.E. Rogers and J. Wiegel. 1991. Environmental factors correlated to dichlorophenol dechlorination in anoxic freshwater sediments. Environ. Toxicol. Chem. *10*: 1255–1265.

Hall, I.C. 1930. *Micrococcus* niger, a new pigment-forming, anaerobic coccus recovered from urine in a case of general arteriosclerosis. J. Bacteriol. *20*: 407–415.

Hamana, K. 1999. Polyamine distribution catalogues of clostridia, acetogenic anaerobes, *Actinobacteria*, bacilli, heliobacteria and haloanaerobes within Gram-positive eubacteria - distribution of spermine and agmatine in thermophiles and halophiles. Microbiol. Cult. Coll. *15*: 9–28.

Hippe, H. 1991. Maintenance of methanogenic bacteria. *In* Kirsop and Doyle (Editors), Maintenance of Micoororganisms and Cultured Cells, 2nd edn. Academic Press, London, pp. 101–113.

Hippe, H., A. Hagenauer and R.M. Kroppenstedt. 1997. Menadione requirement for sulfate-reduction in *Desulfotomaculum aeronauticum*, and emended species description. Syst. Appl. Microbiol. *20*: 554–558.

Hippe, N., M. Vainshtein and R.M. Kroppenstedt. 1995. Fatty acid composition and taxonomic significance in the genus *Desulfotomaculum*. Spring meeting of the Vereinigung für Allgemeine und Angewandte Mikrobiologie (VAAM), Stuttgart, Germany.

Holliger, C., D. Hahn, H. Harmsen, W. Ludwig, W. Schumacher, B. Tindall, F. Vazquez, N. Weiss and A.J.B. Zehnder. 1998a. *In* Validation of publication of new names and new combinations previously effectively published outside the IJSB. List no. 66. Int. J. Syst. Bacteriol. *48*: 631–632.

Holliger, C., D. Hahn, H. Harmsen, W. Ludwig, W. Schumacher, B. Tindall, F. Vazquez, N. Weiss and A.J.B. Zehnder. 1998b. *Dehalobacter restrictus* gen. nov. and sp. nov., a strictly anaerobic bacterium that reductively dechlorinates tetra- and trichloroethene in an anaerobic respiration. Arch. Microbiol. *169*: 313–321.

Holliger, C., G. Wohlfarth and G. Diekert. 1999. Reductive dechlorination in the energy metabolism of anaerobic bacteria. FEMS Microbiol. *22*: 383–398.

Hristova, K.R., M. Mau, D. Zheng, R.I. Aminov, R.I. Mackie, H.R. Gaskins and L. Raskin. 2000. *Desulfotomaculum* genus- and subgenus-specific 16S rRNA hybridization probes for environmental studies. Environ. Microbiol. *2*: 143–159.

Iizuka, H., H. Okazaki and N. Seto. 1969. A new sulfate-reducing bacterium from Antarctica. J. Gen. Microbiol. *15*: 11–18.

Imachi, H., Y. Sekiguchi, Y. Kamagata, S. Hanada, A. Ohashi and H. Harada. 2002. *Pelotomaculum thermopropionicum* gen. nov., sp. nov., an anaerobic, thermophilic, syntrophic propionate-oxidizing bacterium. Int. J. Syst. Evol. Microbiol. *52*: 1729–1735.

Imachi, H., Y. Sekiguchi, Y. Kamagata, A. Loy, Y.L. Qiu, P. Hugenholtz, N. Kimura, M. Wagner, A. Ohashi and H. Harada. 2006. Non-sulfate-reducing, syntrophic bacteria affiliated with *Desulfotomaculum* cluster I are widely distributed in methanogenic environments. Appl. Environ. Microbiol. *72*: 2080–2091.

Imachi, H., S. Sakai, A. Ohashi, H. Harada, S. Hanada, Y. Kamagata and Y. Sekiguchi. 2007. *Pelotomaculum propionicicum* sp. nov., an anaerobic, mesophilic, obligately syntrophic, propionate-oxidizing bacterium. Int. J. Syst. Evol. Microbiol. *57*: 1487–1492.

Juteau, P., V. Cote, M.F. Duckett, R. Beaudet, F. Lepine, R. Villemur and J.G. Bisaillon. 2005. *Cryptanaerobacter phenolicus* gen. nov., sp. nov., an anaerobe that transforms phenol into benzoate via 4-hydroxybenzoate. Int. J. Syst. Evol. Microbiol. *55*: 245–250.

Kaksonen, A.H., J.J. Plumb, W.J. Robertson, P.D. Franzmann, A.E. Gibson and J.A. Puhakka. 2004. Culturable diversity and community fatty acid profiling of sulfate-reducing fluidized-bed reactors treating acidic, metal-containing wastewater. Geomicrobiol. J. *21*: 469–480.

Kaksonen, A.H., J.J. Plumb, W.J. Robertson, S. Spring, P. Schumann, P.D. Franzmann and J.A. Puhakka. 2006a. Novel thermophilic sulfate-reducing bacteria from a geothermally active underground mine in Japan. Appl. Environ. Microbiol. *72*: 3759–3762.

Kaksonen, A.H., S. Spring, P. Schumann, R.M. Kroppenstedt and J.A. Puhakka. 2006b. *Desulfotomaculum thermosubterraneum* sp. nov., a thermophilic sulfate-reducer isolated from an underground mine located in a geothermally active area. Int. J. Syst. Evol. Microbiol. *56*: 2603–2608.

Karnauchow, T.M., S.F. Koval and K.F. Jarrell. 1992. Isolation and characterization of three thermophilic anaerobes from a St. Lucia hot spring. Syst. Appl. Microbiol. *15*: 296–310.

Klein, M., M. Friedrich, A.J. Roger, P. Hugenholtz, S. Fishbain, H. Abicht, L.L. Blackall, D.A. Stahl and M. Wagner. 2001. Multiple lateral transfers of dissimilatory sulfite reductase genes between major lineages of sulfate-reducing prokaryotes. J. Bacteriol. *183*: 6028–6035.

Klemps, R., H. Cypionka, F. Widdel and N. Pfennig. 1985. Growth with hydrogen, and further physiological characteristics of *Desulfotomaculum* species. Arch. Microbiol. *143*: 203–208.

Kluyver, A.J. and C.B. van Neil. 1936. Prospects for a natural classification of bacteria. Zentbl. Bacteriol. Parasitenkd. Infektionskr. Hyg. Abt. II *94*: 369–403.

Knoblich, P. 1996. Growth of *Desulfitobacterium dehalogenans* an anaerobic bacterium capable of reductive dehalogenation, in continuous culture with increased surface area. Universität Karlsruhe.

Kohring, G.W., J.E. Rogers and J. Wiegel. 1989a. Anaerobic biodegradation of 2,4-dichlorophenol in freshwater lake sediments at different temperatures. Appl. Environ. Microbiol. *55*: 348–353.

Kohring, G.W., X.M. Zhang and J. Wiegel. 1989b. Anaerobic dechlorination of 2,4-dichlorophenol in freshwater sediments in the presence of sulfate. Appl. Environ. Microbiol. *55*: 2735–2737.

Krasotkina, J., T. Walters, K.A. Maruya and S.W. Ragsdale. 2001. Characterization of the B12- and iron-sulfur-containing reductive dehalogenase from *Desulfitobacterium chlororespirans*. J. Biol. Chem. *276*: 40991–40997.

Kuever, J., J. Kulmer, S. Jannsen, U. Fischer and K.H. Blotevogel. 1993. Isolation and characterization of a new spore-forming sulfate-reducing bacterium growing by complete oxidation of catechol. Arch. Microbiol. *159*: 282–288.

Kuever, J., F. Rainey and H. Hippe. 1999. Description of *Desulfotomaculum* sp. Groll as *Desulfotomaculum gibsoniae* sp. nov. Int. J. Syst. Evol. Microbiol. *49*: 1801–1808.

Kuever, J., F.A. Rainey and F.W. Widdel. 2005. Genus I. *Desulfobacter* Widdel 1981, 382[VP] (Effective publication: Widdel 1980, 376). *In* Brenner, Krieg, Staley and Garrity (Editors), Bergey's Manual of Systematic Bacteriology, 2nd edn, Vol. 2, The *Proteobacteria*, Part C, *Alpha-*, *Beta-*, *Delta-*, and *Epsilonproteobacteria*. Springer, New York, pp. 961–964.

Lanthier, M., P. Juteau, F. Lepine, R. Beaudet and R. Villemur. 2005. *Desulfitobacterium hafniense* is present in a high proportion within the biofilms of a high-performance pentachlorophenol-degrading, methanogenic fixed-film reactor. Appl. Environ. Microbiol. *71*: 1058–1065.

Lecouturier, D., J.J. Godon and J.M. Lebeault. 2003. Phylogenetic analysis of an anaerobic microbial consortium deiodinating 5-amino-2,4,6-triiodoisophthalic acid. Appl. Microbiol. Biotechnol. *62*: 400–406.

Lespinat, P.A., G. Denariaz, G. Fauque, R. Toci, Y. Berlier and J. LeGall. 1985. Fixation de l'azote atmosphérique et métabolism de l'hydrogène chez une bactérie sulfato-réductrice sporulante, *Desulfotomaculum orientis*. C.R. Acad. Sci. Paris *301*: 707–710.

Létourneau, L., J.-G. Bisaillon, F. Lépine and R. Beaudet. 1995. Spore-forming bacteria that carboxylate phenol to benzoic acid under anaerobic conditions. Can. J. Microbiol. *41*: 266–272.

Letowski, J., P. Juteau, R. Villemur, R. Beaudet, F. Lépine and J.-G. Bisaillon. 2001. Separation of a phenol carboxylating organism from a two-member, strict anaerobic coculture. Can. J. Microbiol. *47*: 373–381.

Li, T., J.G. Bisaillon, R. Villemur, L. Létourneau, K. Bernard, F. Lépine and R. Beaudet. 1996. Isolation and characterization of a new bacterium carboxylating phenol to benzoic acid under anaerobic conditions. J. Bacteriol. *178*: 2551–2558.

Lie, T.J., W. Godchaux and E.R. Leadbetter. 1999. Sulfonates as terminal electron acceptors for growth of sulfite-reducing bacteria (*Desulfitobacterium* spp.) and sulfate-reducing bacteria: effects of inhibitors of sulfidogenesis. Appl. Environ. Microbiol. *65*: 4611–4617.

Liu, Y.T., T.M. Karnauchow, K.F. Jarrell, D.L. Balkwill, G.R. Drake, D. Ringelberg, R. Clarno and D.R. Boone. 1997. Description of two new thermophilic *Desulfotomaculum* spp., *Desulfotomaculum putei* sp. nov., from a deep terrestrial subsurface, and *Desulfotomaculum luciae* sp. nov., from a hot spring. Int. J. Syst. Bacteriol. *47*: 615–621.

Loffler, F.E., R.A. Sanford and J.M. Tiedje. 1996. Initial characterization of a reductive dehalogenase from *Desulfitobacterium chlororespirans* CO23. Appl. Environ. Microbiol. *62*: 3809–3813.

Love, C.A., B.K.C. Patel, P.D. Nichols and E. Stackebrandt. 1993a. *In* Validation of the publication of new names and new combinations previously effectively published outside the IJSB. List no. 47. Int. J. Syst. Bacteriol. *43*: 864–865.

Love, C.A., B.K.C. Patel, P.D. Nichols and E. Stackebrandt. 1993b. *Desulfotomaculum australicum*, sp. nov., a thermophilic sulfate-reducing bacterium isolated from the Great Artesian Basin of Australia. Syst. Appl. Microbiol. *16*: 244–251.

Ludwig, W., M. Weizenegger, S. Dorn, J. Andreesen and K.H. Schleifer. 1990. The phylogenetic position of *Peptococcus niger* based on 16S rRNA sequence studies. FEMS Microbiol. Lett. *59*: 139–143.

Mackiewicz, N. and J. Wiegel. 1998. Comparison of energy and growth yields for *Desulfitobacterium dehalogenans* during utilization of chlorophenol and various traditional electron aeceptors. Appl. Environ. Microbiol. *64*: 352–355.

Maes, A., H. Van Raemdonck, W.O. K. Smith, L. Lebbe and W. Verstraete. 2006. Transport and activity of *Desulfitobacterium* dichloroeliminans strain DCA1 during bioaugmentation of 1,2-DCA-contaminated groundwater. Environ. Sci. Technol. *40*: 5544–5552.

Miller, E., G. Wohlfarth and G. Diekert. 1997. Comparative studies on tetrachloroethene reductive dechlorination mediated by *Desulfitobacterium* sp. strain PCE-S. Arch. Microbiol. *168*: 513–519.

Milliken, C.E. and H.D. May. 2007. Sustained generation of electricity by the spore-forming, Gram-positive, *Desulfitobacterium hafniense* strain DCB2. Appl. Microbiol. Biotechnol. *73*: 1180–1189.

Min, H. and S.H. Zinder. 1990. Isolation and characterization of a thermophilic sulfate-reducing bacterium *Desulfotomaculum thermoacetoxidans* sp. nov. Arch. Microbiol. *153*: 399–404.

Min, H. and S.H. Zinder. 1995. *In* Validation of the publication of new names and new combinations previously effectively published outside the IJSB. List no. 55. Int. J. Syst. Bacteriol. *45*: 879–880.

Murdoch, D.A. and I.J. Mitchelmore. 1991. The laboratory identification of gram-positive anaerobic cocci. J. Med. Microbiol. *34*: 295–308.

Murdoch, D.A. and J.T. Magee. 1995. A numerical taxonomic study of the gram-positive anaerobic cocci. J. Med. Microbiol. *43*: 148–155.

Nazina, T.N., A.E. Ivanova, L.P. Kanchaveli and E.F. Rozanova. 1989. A new spore-forming thermophilic methylotrophic sulfate-reducing bacterium, *Desulfotomaculum kuznetsovii* sp. nov. Microbiology (En. transl. from Mikrobiologiya) *57*: 659–663.

Nazina, T.N., A.E. Ivanova, L.P. Kanchaveli and E.F. Rozanova. 1990. *In* Validation of the publication of new names and new combinations previously effectively published outside the IJSB. List no. 35. Int. J. Syst. Bacteriol. *40*: 470–471.

Neumann, A., T. Engelmann, R. Schmitz, Y. Greiser, A. Orthaus and G. Diekert. 2004. Phenyl methyl ethers: novel electron donors for respiratory growth of *Desulfitobacterium hafniense* and *Desulfitobacterium* sp. strain PCE-S. Arch. Microbiol. *181*: 245–249.

Newman, D.K., E.K. Kennedy, J.D. Coates, D. Ahmann, D.J. Ellis, D.R. Lovley and F.M.M. Morel. 1997. Dissimilatory arsenate and sulfate reduction in *Desulfotomaculum auripigmentum* sp. nov. Arch. Microbiol. *168*: 380–388.

Niggemyer, A., S. Spring, E. Stackebrandt and R.F. Rosenzweig. 2001. Isolation and characterization of a novel As(V)-reducing bacterium: Implications for arsenic mobilization and the genus *Desulfitobacterium*. Appl. Environ. Microbiol. *67*: 5568–5580.

Nilsen, R.K., T. Torsvik and T. Lien. 1996. *Desulfotomaculum thermocisternum* sp. nov., a sulfate reducer isolated from a hot North Sea oil reservoir. Int. J. Syst. Bacteriol. *46*: 397–402.

Nonaka, H., G. Keresztes, Y. Shinoda, Y. Ikenaga, M. Abe, K. Naito, K. Inatomi, K. Furukawa, M. Inui and H. Yukawa. 2006. Complete genome sequence of the dehalorespiring bacterium *Desulfitobacterium hafniense* Y51 and comparison with *Dehalococcoides ethenogenes* 195. J. Bacteriol. *188*: 2262–2274.

Onyenwoke, R.U., J.A. Brill, K. Farahi and J. Wiegel. 2004. Sporulation genes in members of the low G+C Gram-type-positive phylogenetic branch (*Firmicutes*). Arch. Microbiol. *182*: 182–192.

Parshina, S.N., J. Sipma, Y. Nakashimada, A.M. Henstra, H. Smidt, A.M. Lysenko, P.N. Lens, G. Lettinga and A.J. Stams. 2005. *Desulfotomaculum carboxydivorans* sp. nov., a novel sulfate-reducing bacterium capable of growth at 100% CO. Int. J. Syst. Evol. Microbiol. *55*: 2159–2165.

Pfennig, N. 1978. *Rhodocyclus purpureus* gen. nov. and sp. nov. a ring-shaped, vitamin-B12-requiring member of family *Rhodospirillaceae*. Int. J. Syst. Bacteriol. *28*: 283–288.

Pikuta, E., A. Lysenko, N. Suzina, G. Osipov, B. Kuznetsov, T. Tourova, V. Akimenko and K. Laurinavichius. 2000. *Desulfotomaculum alkaliphilum* sp. nov., a new alkaliphilic, moderately thermophilic, sulfate-reducing bacterium. Int. J. Syst. Evol. Microbiol. *50*: 25–33.

Plugge, C.M., M. Balk and A.J. Stams. 2002. *Desulfotomaculum thermobenzoicum* subsp. *thermosyntrophicum* subsp. nov., a thermophilic, syntrophic, propionate-oxidizing, spore-forming bacterium. Int. J. Syst. Evol. Microbiol. *52*: 391–399.

Pop, S.M., R.J. Kolarik and S.W. Ragsdale. 2004. Regulation of anaerobic dehalorespiration by the transcriptional activator CprK. J. Biol. Chem. *279*: 49910–49918.

Pop, S.M., N. Gupta, A.S. Raza and S.W. Ragsdale. 2006. Transcriptional activation of dehalorespiration. Identification of redox-active cysteines regulating dimerization and DNA binding. J. Biol. Chem *281*: 26382–26390.

Postgate, J.R. 1979. The sulphate-reducing bacteria. Cambridge University Press, Cambridge.

Prévot, A.R. 1957. Manual de classification et de détermination des bactéries anaérobies. Masson, Paris.

Qiu, L.Y., Y. Sekiguchi, H. Imachi, Y. Kamagata, I.C. Tseng, S.S. Cheng, A. Ohashi and H. Harada. 2003a. *Sporotomaculum syntrophicum* sp. nov., a novel anaerobic, syntrophic benzoate-degrading bacterium isolated from methanogenic sludge treating wastewater from terephthalate manufacturing. Arch. Microbiol. *179*: 242–249.

Qiu, Y.-L., Y. Sekiguchi, H. Imachi, Y. Kamagata, I.-C. Tseng, S.S. Cheng, A. Ohashi and H. Harada. 2003b. *In* Validation of publication of new names and new combinations previously effectively published outside the IJSB. List no. 92. Int. J. Syst. Evol. Microbiol. *53*: 935–937.

Qiu, Y.L., Y. Sekiguchi, S. Hanada, H. Imachi, I.C. Tseng, S.S. Cheng, A. Ohashi, H. Harada and Y. Kamagata. 2006a. *Pelotomaculum terephthalicum* sp. nov. and *Pelotomaculum isophthalicum* sp. nov.: two anaerobic bacteria that degrade phthalate isomers in syntrophic association with hydrogenotrophic methanogens. Arch Microbiol *185*: 172–182.

Qiu, Y.L., Y. Sekiguchi, S. Hanada, H. Imachi, I. C. Tseng, S. S. Cheng, A. Ohashi, H. Harada and K. Kamagata. 2006b. *In* Validation of publication of new names and new combinations previously effectively published outside the IJSB. List no. 111. Int. J. Syst. Evol. Microbiol. *56*: 2025–2027.

Robertson, W.J., P.D. Franzmann and B.J. Mee. 2000. Spore-forming, *Desulfosporosinus*-like sulphate-reducing bacteria from a shallow aquifer contaminated with gasoline. J. Appl. Microbiol. *88*: 248–259.

Robertson, W.J., J.P. Bowman, P.D. Franzmann and B.J. Mee. 2001. *Desulfosporosinus meridiei* sp. nov., a spore-forming sulfate-reducing bacterium isolated from gasolene-contaminated groundwater. Int. J. Syst. Evol. Microbiol. *51*: 133–140.

Rogosa, M. 1971. *Peptococcaceae*, a new family to include the gram positive, anaerobic cocci of the genera *Peptococcus*, *Peptostreptococcus*, and *Ruminococcus*. Int. J. Syst. Bacteriol. *21*: 234–237.

Rosnes, J.T., T. Torsvik and T. Lien. 1991. Spore-forming thermophilic sulfate-reducing bacteria isolated from North Sea oil field waters. Appl. Environ. Microbiol. *57*: 2302–2307.

Saitou, N. and M. Nei. 1987. The neighbor-joining method: a new method for reconstructing phylogenetic trees. Mol. Biol. Evol. *4*: 406–425.

Sanford, R.A., J.R. Cole, F.E. Loffler and J.N. Tiedje. 1996. Characterization of *Desulfitobacterium chlororespirans* sp. nov., which grows by coupling the oxidation of lactate to the reductive dechlorination of 3-chloro-4-hydroxybenzoate. Appl. Environ. Microbiol. *62*: 3800–3808.

Sanford, R.A., J.R. Cole, F.E. Loffler and J.M. Tiedje. 2001. *In* Validation of publication of new names and new combinations previously effectively published outside the IJSEM. List no. 80. Int. J. Syst. Evol. Microbiol. *51*: 793–794.

Skerman, V.B.D., V. McGowan and P.H.A. Sneath. 1980. Approved lists of bacterial names. Int. J. Syst. Bacteriol *30*: 225–420.

Smidt, H., D. Song, J. van Der Oost and W.M. de Vos. 1999. Random transposition by Tn*916* in *Desulfitobacterium dehalogenans* allows for isolation and characterization of halorespiration-deficient mutants. J. Bacteriol. *181*: 6882–6888.

Smidt, H., M. van Leest, J. van der Oost and W.M. de Vos. 2000. Transcriptional regulation of the cpr gene cluster in ortho-chlorophenol-respiring *Desulfitobacterium dehalogenans*. J. Bacteriol. *182*: 5683–5691.

Smidt, H., J. van der Oost and W.M. de Vos. 2001. Development of a gene cloning and inactivation system for halorespiring *Desulfitobacterium dehalogenans*. Appl. Environ. Microbiol. *67*: 591–597.

Sokolova, T.G., N.A. Kostrikina, N.A. Chernyh, T.V. Kolganova, T.P. Tourova and E.A. Bonch-Osmolovskaya. 2005. *Thermincola carboxydiphila* gen. nov., sp. nov., a novel anaerobic, carboxydotrophic, hydrogenogenic bacterium from a hot spring of the Lake Baikal area. Int. J. Syst. Evol. Microbiol. *55*: 2069–2073.

Stackebrandt, E., C. Sproer, F.A. Rainey, J. Burghardt, O. Pauker and H. Hippe. 1997. Phylogenetic analysis of the genus *Desulfotomaculum*: evidence for the misclassification of *Desulfotomaculum guttoideum* and description of *Desulfotomaculum orientis* as *Desulfosporosinus orientis* gen. nov., comb. nov. Int. J. Syst. Bacteriol. *47*: 1134–1139.

Stackebrandt, E., P. Schumann, E. Schuler and H. Hippe. 2003. Reclassification of *Desulfotomaculum auripigmentum* as *Desulfosporosinus auripigmenti* corrig., comb. nov. Int. J. Syst. Evol. Microbiol. *53*: 1439–1443.

Suyama, A., R. Iwakiri, K. Kai, T. Tokunaga, N. Sera and K. Furukawa. 2001. Isolation and characterization of *Desulfitobacterium* sp. strain Y51 capable of efficient dehalogenation of tetrachloroethene and polychloroethanes. Biosci. Biotechnol. Biochem. *65*: 1474–1481.

Suyama, A., M. Yamashita, S. Yoshino and K. Furukawa. 2002. Molecular characterization of the PceA reductive dehalogenase of *Desulfitobacterium* sp. strain Y51. J. Bacteriol. *184*: 3419–3425.

Tardy-Jacquenod, C., M. Magot, B.K.C. Patel, R. Matheron and P. Caumette. 1998. *Desulfotomaculum halophilum* sp. nov., a halophilic sulfate-reducing bacterium isolated from oil production facilities. Int. J. Syst. Bacteriol. *48*: 333–338.

Tartakovsky, B., M. Levesque, R. Dumortier, R. Beaudet and S.R. Guiot. 1999. Biodegradation of pentachlorophenol in a continuous anaerobic reactor augmented with *Desulfitobacterium frappieri* PCP-1. Appl. Environ. Microbiol. *65*: 4357–4362.

Tasaki, M., Y. Kamagata, K. Nakamura and E. Mikami. 1991a. Isolation and characterization of a thermophilic benzoate-degrading, sulfate-reducing bacterium, *Desulfotomaculum thermobenzoicum* sp. nov. Arch. Microbiol. *155*: 348–352.

Tasaki, M., Y. Kamagata, K. Nakamura and E. Mikami. 1991b. *In* Validation of the publication of new names and new combinations previously effectively published outside the IJSB. List no. 39. Int. J. Syst. Bacteriol. *41*: 580–581.

Tasaki, M., Y. Kamagata, K. Nakamura and E. Mikami. 1992. Utilization of methoxylated benzoates and formation of intermediates by *Desulfotomaculum thermobenzoicum* in the presence or absence of sulfate. Arch. Microbiol. *157*: 209–212.

Tasaki, M., Y. Kamagata, K. Nakamura, K. Okamura and E. Mikami. 1993. Acetogenesis from pyruvate by *Desulfotomaculum thermobenzoicum* and differences in pyruvate metabolism among three sulfate-reducing bacteria in the absence of sulfate. FEMS Microbiol. Lett. *106*: 259–264.

Tebo, B.M. and A.Y. Obraztsova. 1998. Sulfate-reducing bacterium grows with Cr(VI), U(VI), Mn(IV), and Fe(III) as electron acceptors. FEMS Microbiol. Lett. *162*: 193–198.

Thibodeau, J., A. Gauthier, M. Duguay, R. Villemur, F. Lepine, P. Juteau and R. Beaudet. 2004. Purification, cloning, and sequencing of a 3,5-dichlorophenol reductive dehalogenase from *Desulfitobacterium frappieri* PCP-1. Appl. Environ. Microbiol. *70*: 4532–4537.

Tourova, T.P., B.B. Kuznetzov, E.V. Novikova, A.B. Poltaraus and T.N. Nazina. 2001. Heterogeneity of the nucleotide sequences of the 16S rRNA genes of the type strain of *Desulfotomaculum kuznetsovii*. Microbiology (En. transl. from Mikrobiologiya) *70*: 678–684.

Tsukagoshi, N., S. Ezaki, T. Uenaka, N. Suzuki and R. Kurane. 2006. Isolation and transcriptional analysis of novel tetrachloroethene reductive dehalogenase gene from *Desulfitobacterium* sp. strain KBC1. Appl. Microbiol. Biotechnol. *69*: 543–553.

Ueki, A. and T. Suto. 1979. Cellular fatty acid composition of sulfate reducing bacteria. J. Gen. Appl. Microbiol. *25*: 185–196.

Utkin, I., C. Woese and J. Wiegel. 1994. Isolation and characterization of *Desulfitobacterium dehalogenans* gen. nov., sp. nov., an anaerobic bacterium which reductively dechlorinates chlorophenolic compounds. Int. J. Syst. Bacteriol. *44*: 612–619.

Utkin, I., D.D. Dalton and J. Wiegel. 1995. Specificity of reductive dehalogenation of substituted ortho-chlorophenols by *Desulfitobacterium dehalogenans* JW/IU-DC1. Appl. Environ. Microbiol. *61*: 346–351; erratum p. 1677.

Vainshtein, M., H. Hippe and R.M. Kroppenstedt. 1992. Cellular fatty acid composition of *Desulfovibrio* species and its use in classification of sulfate-reducing bacteria. Syst. Appl. Microbiol. *15*: 554–566.

Vainshtein, M.B., G. Gogotova and H. Hippe. 1995. A sulfate reducing bacterium from permafrost. Microbiology (En. transl. from Mikrobiologiya) *64*: 514–518.

van de Pas, B.A., H. Smidt, W.R. Hagen, J. van der Oost, G. Schraa, A.J. Stams and W.M. de Vos. 1999. Purification and molecular characterization of ortho-chlorophenol reductive dehalogenase, a key enzyme of halorespiration in *Desulfitobacterium dehalogenans*. J. Biol. Chem. *274*: 20287–20292.

van de Pas, B.A., J. Gerritse, W.M. de Vos, G. Schraa and A.J. Stams. 2001a. Two distinct enzyme systems are responsible for tetrachloroethene and chlorophenol reductive dehalogenation in *Desulfitobacterium* strain PCE1. Arch. Microbiol. *176*: 165–169.

van de Pas, B.A., H.J.M. Harmsen, G.C. Raangs, W.M. de Vos, G. Schraa and A.J.M. Stams. 2001b. A *Desulfitobacterium* strain isolated from human feces that does not dechlorinate chloroethenes or chlorophenols. Arch. Microbiol. *175*: 389–394.

van de Pas, B.A., S. Jansen, C. Dijkema, G. Schraa, W.M. de Vos and A.J.M. Stams. 2001c. Energy yield of respiration on chloroaromatic compounds in *Desulfitobacterium dehalogenans*. Appl. Environ. Microbiol. *67*: 3958–3963.

Van Raemdonck, A.M. H., W. Ossieur, K. Verthe, T. Vercauteren, W. Verstraete and N. Boon. 2006. Real-time PCR quantification in groundwater of the dehalorespiring *Desulfitobacterium* dichloroeliminans strain DCA1. J. Microbiol. Methods. *67*: 294–303.

Vandieken, V., C. Knoblauch and B.B. Jorgensen. 2006. *Desulfotomaculum arcticum* sp. nov., a novel spore-forming, moderately thermophilic, sulfate-reducing bacterium isolated from a permanently cold fjord sediment of Svalbard. Int. J. Syst. Evol. Microbiol. *56*: 687–690.

Villemur, R., M. Lanthier, R. Beaudet and F. Lépine. 2006. The *Desulfitobacterium* genus. FEMS Microbiol. Rev. *30*: 706–733.

Werkman, C.H. and H.J.Weaver. 1927. Studies in the bacteriology of sulfur stinker spoilage of canned sweet corn. Iowa State J. Sci. *2*: 57–67.

Widdel, F. and N. Pfennig. 1977a. *In* Announcement of new names and new combinations previously effectively published outside the IJSB. List no. 1. Int. J. Syst. Bacteriol. *27*: 306.

Widdel, F. and N. Pfennig. 1977b. New anaerobic, sporing, acetate-oxidizing, sulfate-reducing bacterium, *Desulfotomaculum* (emend.) *acetoxidans*. Arch. Microbiol. *112*: 119–122.

Widdel, F. and N. Pfennig. 1981. Sporulation and further nutritional characteristics of *Desulfotomaculum acetoxidans*. Arch. Microbiol. *129*: 401–402.

Widdel, F. and T.A. Hansen. 1992. The dissimilatory sulfate- and sulfur bacteria. *In* Balows, Trüper, Dworkin, Harder and Schleifer (Editors), The Prokaryotes. Springer-Verlag, New York, pp. 583–624.

Wiegel, J. and L. Quandt. 1982. Determination of the gram type using the reaction between polymyxin B and lipopolysaccharides of the outer cell wall of whole bacteria. J. Gen. Microbiol. *128*: 2261–2270.

Wiegel, J., L.H. Carreira, R. Garrison, N. Rabek and L.G. Ljungdahl. 1991. Calcium magnesium acetate (CMA) manufacture from glucose by fermentation with thermophilic homoacetogenic bacteria. *In* Wise, Levendis and Metghalchi (Editors), Calcium Magnesium Acetate. Elsevier Science, Amsterdam, pp. 359–418.

Wiegel, J., X. Zhang and Q. Wu. 1999. Anaerobic dehalogenation of hydroxylated polychlorinated biphenyls by *Desulfitobacterium dehalogenans*. Appl. Environ. Microbiol. *65*: 2217–2221.

Zavarzina, D.G., T.G. Sokolova, T.P. Tourova, N.A. Chernyh, N.A. Kostrikina and E.A. Bonch-Osmolovskaya. 2007. *Thermincola ferriacetica* sp. nov., a new anaerobic, thermophilic, facultatively chemolithoautotrophic bacterium capable of dissimilatory Fe(III) reduction. Extremophiles *11*: 1–7.

Zhang, X. and J. Wiegel. 1992. The anaerobic degradation of 3-chloro-4-hydroxybenzoate in freshwater sediment proceeds via either chlorophenol or hydroxybenzoate to phenol and subsequently to benzoate. Appl. Environ. Microbiol. *58*: 3580–3585.

Family VII. Peptostreptococcaceae fam. nov.

TAKAYUKI EZAKI

Pep.to.strep.to.coc.ca′ce.ae. N.L. n. *Peptostreptococcus* a bacterial genus, the type genus of the family; *-aceae* ending to denote a family; N.L. fem. pl. n. *Peptostreptococcaceae* the family of *Peptostreptococcus*.

Obligately or facultatively anaerobic, Gram-stain-positive cocci or filamentous rods. Catalase-negative or occasionally weakly positive. Species of the genus *Filifactor* are endosporeforming. Family *Peptostreptococcaceae* contains the type genus *Peptostreptococcus*, *Filifactor*, and *Tepidibacter*. According to the 16S rRNA phylogenetic analysis presented in the roadmap to this volume (Figure 5), the genus the family *Peptostreptococcaceae* is in order *Clostridiales*, class *Clostridia* in the phylum *Firmicutes*. The phylogenetic relationships among the members of the family *Peptostreptococcaceae* and closely related genera are given in Figure 183. The differential characteristics and features of the family are summarized in Table 187.

DNA G+C content (mol%): 28–43.

Type genus: **Peptostreptococcus** Kluyver and van Niel 1936.

Genus I. Peptostreptococcus Kluyver and van Niel 1936, 401[AL] emend. Ezaki, Kawamura, Li, Li, Zhao and Shu 2001, 1527

TAKAYUKI EZAKI

Pep.to.strep.to.cocc′cus. Gr. adj. *peptos* cooked, digested; N.L. masc. n. *Streptococcus* a bacterial genus name; N.L. masc. n. *Peptostreptococcus* the digesting streptococcus.

Non-spore-forming obligately anaerobic Gram-stain-positive cocci. Cells may occur in pairs, irregular masses, or chains. Chemo-organotrophs. The optimum temperature for growth is 37°C. Metabolize peptone and amino acids to acetic, butyric, isobutyric, caproic, and isocaproic acid (Holdeman et al., 1986). Carbohydrates are weakly fermented. Found in human intestine, vagina, and various abscesses. Diamino acid of peptidoglycan is lysine.

DNA G+C content (mol%): 34–36.

Type species: **Peptostreptococcus anaerobius** (Natvig 1905) Kluyver and van Niel 1936, 401[AL] (*Streptococcus anaerobius* Natvig 1905, 724).

Taxonomic comments

After 16S rRNA gene sequence analysis, old members of the genus *Peptostreptococcus* were reclassified as *Anaerococcus*, *Peptoniphilus*, *Finegoldia*, *Micromonas*, *Gallicola*, *Slackia*, *Ruminococcus*, and *Atopobium*.

Nonsaccharolytic members of the genus *Peptostreptococcus* (*Peptostreptococcus asaccharolyticus*, *Peptostreptococcus harei*, *Peptostreptococcus indolicus*, *Peptostreptococcus ivorii*, and *Peptostreptococcus lacrimalis*) were reclassified as members of the genus *Peptoniphilus* (Ezaki et al., 2001). Saccharolytic members of the genus *Peptostreptococcus* (*Peptostreptococcus prevotii*, *Peptostreptococcus tetradius*, *Peptostreptococcus hydrogenalis*, *Peptostreptococcus lactolyticus*, and *Peptostreptococcus octavius*) were reclassified as members of the genus *Anaerococcus* (Ezaki et al., 2001). *Peptostreptococcus magnus* was transferred to the genus *Finegoldia*, *Peptostreptococcus micros* to the genus *Micromonas micros* (Murdoch and Shah, 1999), since renamed *Parvimonas micra* by Tindall and Euzéby (2006) because the genus name was illegitimate. *Peptostreptococcus heliotrinreducens* was transferred to the genus *Slackia* (Wade et al., 1999). *Peptostreptococcus barnesae* was transferred to the genus *Gallicola* (Ezaki et al., 2001), *Peptostreptococcus productus* to the genus *Ruminococcus* (Ezaki et al., 1994), and *Peptostreptococcus parvulus* to the genus *Atopobium* (Collins and Wallbanks, 1992).

List of species of the genus *Peptostreptococcus*

1. **Peptostreptococcus anaerobius** (Natvig 1905) Kluyver and van Niel 1936, 401[AL] (*Streptococcus anaerobius* Natvig, 1905, 724)

an.a.e.ro′bi.us. Gr. pref. *an* not; Gr. n. *aer* air; Gr. n. *bios* life; N.L. adj. *anaerobius* not living in air, anaerobic.

Nonsporeforming obligately anaerobic Gram-stain-positive cocci. Cells may occur in pairs, irregular masses, or chains. Chemo-organotrophs. The optimum temperature for growth is 37°C. Metabolizes peptone and amino acids to acetic, butyric, isobutyric, caproic, and isocaproic acid. Weak acid is produced from glucose and mannose. No acid is produced from raffinose, lactose, sucrose, mannitol, sorbitol, or arabi-nose. Found in human intestine, vagina, and various abscesses. Differential characteristics of the *Peptostreptococcus anaerobius* from other anaerobic cocci from human sources are given in Table 188.

Peptidoglycan position 1, position 3, and interpeptide bridge are alanine, lysine, and D-aspartate. (Ezaki et al., 1983). Major cellular fatty acids are $C_{18:1}$, C_{16}, and C_{18} (Ezaki et al., 1983; O'Leary and Wilkinson, 1988).

DNA G+C content (mol%): 34–36.

Type strain: ATCC 27337, CCUG 7835, CIP 104411, CIPP 4372, DSM 2949, LMG 15865, NCTC 11460, VPI 4330.

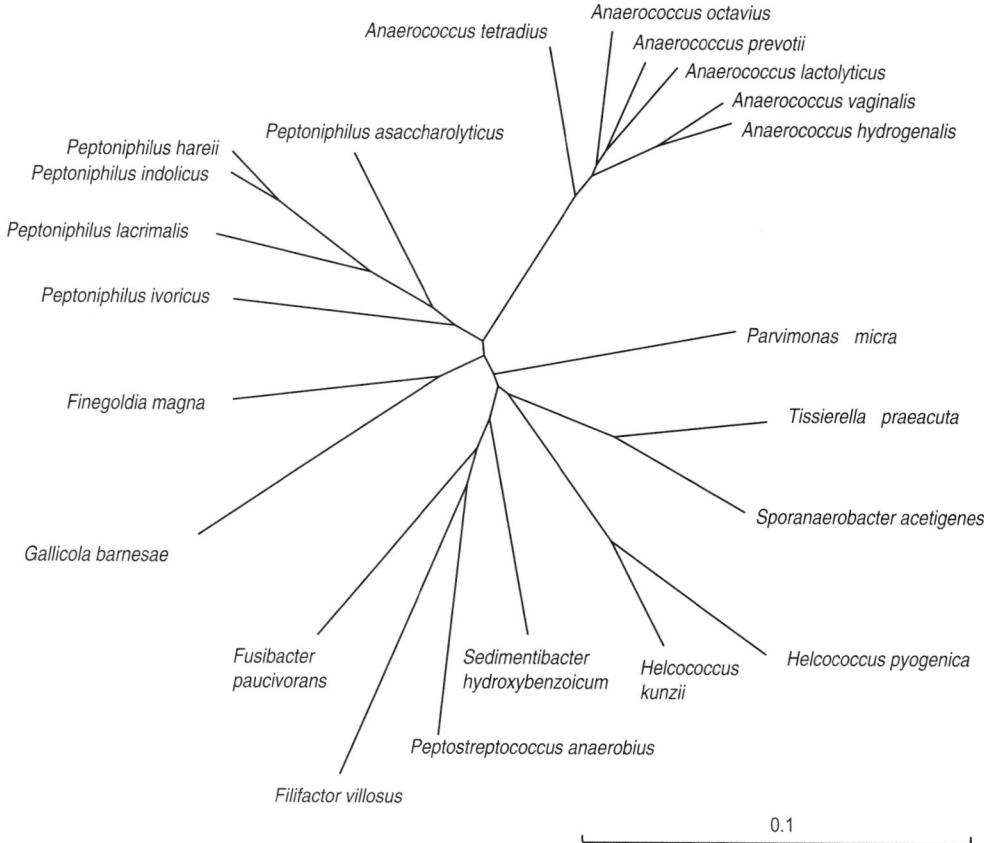

FIGURE 183. Phylogenetic relationships among type species of the family *Peptostreptococaceae* and closely related genera from *Clostridiales* families *incertae sedis* XI and XII.

GenBank accession number (16S rRNA gene): AY326462, D14150, L04168.

2. **Peptostreptococcus stomatis** Downes and Wade 2006, 753[VP]

sto.ma′tis. N.L. gen. n. *stomatis* of the mouth from Gr. n. *stoma* mouth.

Obligately anaerobic, Gram-stain-positive cocci, 0.8×0.8–0.9 μm, occurring in pairs and short chains. After 5 d incubation on fastidious anaerobe agar (LabM) plates, colonies are 0.8–1.8 mm in diameter, circular, entire, high convex to pyramidal, opaque, shiny and cream to off-white in colour with a narrow, grey, peripheral outer ring. Moderate growth is obtained in broth media and growth is further enhanced by the addition of fermentable carbohydrates. Weakly saccharolytic and ferments fructose, glucose and maltose weakly; arabinose, cellobiose, lactose, mannitol, mannose, melezitose, melibiose, raffinose, rhamnose, ribose, salicin, sorbitol, sucrose and trehalose are not fermented. Major amounts of acetic and isocaproic acids, minor amounts of isobutyric and isovaleric acids and trace to minor amounts of butyric acid are produced as end products of metabolism in PYG. Esculin, arginine, gelatin and urea are not hydrolyzed. Indole and catalase are not produced and nitrate is not reduced.

Source: human oral cavity infections.

DNA G+C content (mol%): 36 (HPLC).

Type strain: W2278, DSM 17678, CCUG 51858.

GenBank accession number (16S rRNA gene): DQ160208.

Genus II. **Filifactor** Collins, Lawson, Willems, Cordoba, Fernández-Garayzábal, Garcia, Cai, Hippe and Farrow 1994, 822[VP]

JARI JALAVA AND ERKKI EEROLA

Fi.li.fac′tor. L. n. *filum* thread; L. masc. n. *factor* maker; N.L. masc. n. *Filifactor* thread-maker.

Rods, which are obligately **anaerobic** chemo-organotrophs. Cells do not have flagella. Produce **acetate** and **butyrate**. *DNA G+C content (mol%):* 28–34.

Type species: **Filifactor villosus** Collins, Lawson, Willems, Cordoba, Fernández-Garayzábal, Garcia, Cai, Hippe and Farrow 1994, 822[VP] (*Clostridium villosum* Love, Jones and Bailey 1979, 242).

TABLE 187. Characteristics differentiating members of the family *Peptostreptococcus* fam. nov. and related genera[a]

Characteristic	Peptostreptococcus	Filifactor	Tepidibacter	Anaerococcus	Finegoldia	Fusibacter	Gallicola	Helcococcus	Micromonas	Peptoniphilus	Sedimentibacter	Sporanaerobacter	Tissierella
GenBank accession number (16S rRNA)[b]	D14150	X73452	AY158079	D14139	D14149	AF050099	AB038361	X69837	AF542231	D14138	L11305	AF358114	X80833
DNA G+C content (mol%)	34–36	34	24–29	30–35	32–34	43	32–34	29–29.5	28–30	30–34	34	32.2	28–32
Growth temperature (optimum) (°C)	30–39 (37)	(37)	50	30–39 (37)	30–39 (37)	20–40 (37)	30–39 (37)	30–39 (37)	30–39 (37)	30–39 (37)	(37)	25–50 (40)	20–39 (37)
Habitats	Human intestine, vagina, abscess	Cat skin abscess, human gingival sulcus	Hydrothermal vent	Human intestine, vagina, abscess	Human intestine, vagina, abscess	Oil-producing well	Chicken intestine, human abscess	Human abscess, otorrhea	Human oral cavity, human abscess	Human intestine, vagina, abscess	Freshwater sediment	Anaerobic sludge	Human intestine, vagina, abscess
Morphology	Gram-stain-positive cocci (pairs, chains)	Gram-stain-negative, filamentous, spore variable	Gram-stain-positive rods (single, in pairs or short chains)	Gram-stain-positive cocci (pairs, tetrads)	Large Gram-stain-positive cocci (pairs, mass)	Spindle-shaped rod, motile, rod	Gram-stain-positive cocci (pairs, mass)	Gram-stain-positive cocci (pairs, mass)	Small Gram-stain-positive cocci (pairs, mass)	Gram-stain-positive cocci (pairs, chains)	Gram-stain-positive, curved rods, motile, sporeforming	Gram-stain-positive, rod, motile, sporeforming	Short rods, pairs, often stained negative
Oxygen requirement	Obligate anaerobe	Obligate anaerobe	Anaerobic	Obligate anaerobe	Obligate anaerobe	Obligate anaerobe	Obligate anaerobe	Facultative anaerobe	Obligate anaerobe	Obligate anaerobe	Obligate anaerobe	Obligate anaerobe	Obligate anaerobe
Peptidoglycan (position 3, bridge)	Lys, D-Asp	Orn, D-Asp	ND	Lys, D-Glu	Lys, Gly	No data	Orn, D-Asp	ND	Lys, D-Asp	Orn, D-Glu	No data	No data	m-DAP, Orn, D-Glu
Sugar fermented	w	–	+	w	–	w	–	D	–	+	–	+	–
Major metabolic product from PYG	Butyric acid, caproic acid	ND	ND	Butyric acid	Acetic acid	No data	Acetic acid, butyric acid	Acetic acid, lactic acid	Acetic acid	Butyric acid, acetic acid	No data	No data	Acetic acid

[a]Symbols: +, >85% positive; d, different strains give different reactions (16–84% positive); –, 0–15% positive; D, different among species; w, weak acid produced; ND, not determined.
[b]GenBank accession number given for type species of the genus.

TABLE 188. Differentiation of butyrate-producing anaerobic Gram-stain-positive cocci found in human specimens[a]

	Genus *Peptostreptococcus*	*Peptostreptococcus anaerobius*	Genus *Anaerococcus*	*Anaerococcus prevotii*	*Anaerococcus tetradius*	*Anaerococcus lactolyticus*	*Anaerococcus hydrogenalis*	*Anaerococcus octavius*	*Anaerococcus vaginalis*	Genus *Peptoniphilus*	*Peptoniphilus asaccharolyticus*	*Peptoniphilus lacrimalis*	*Peptoniphilus harei*	*Peptoniphilus ivorii*	*Peptoniphilus indolicus*
16S rRNA GenBank accession no.		L04168		D14139	D14142	D14154	D14140	Y07841	D14146		D14138	D14141	Y07839	Y07840	D14147
DNA G+C content (mol%)	33–34	33–34	26–34	29–33	30–32	30–34	30–34	26–31	28–30	25–34	31–32	30–34	25	29	32–34
Product from PYG		C, a, b, ib, iv		B, l, a, p	B, L, a	B, L, a, p	B, l, a, p	B, l, a, c	B, l, a		B, a	B, a	B, a	B, a	B, a
Peptidoglycan type:															
Position 1		Ala		Gly	Gly	Gly	Ala	Ala	Ala		Ala	Ala	Ala	Ala	Ala
Position 2		Lys		Lys	Lys	Lys	Lys	Lys	Lys		Orn	Orn	Orn	Orn	Orn
Bridge		D-Asp		D-Glu	D-Glu	D-Glu	D-Glu	D-Asp	D-Glu		D-Glu	D-Glu	D-Glu	D-Glu	D-Glu
Production of:															
ALP		–		–	–	–	d	–	d		–	–	–	–	+
Coagulase		–		–	–	–	–	–	–		–	–	–	–	+
Indole			–	–	–	–	–	–	d		d	–	d	–	+
Urease		–		d	+	+	d	–	–		–	–	–	–	–
Sugar fermentation:															
Glucose		+		d	d	+	+	+	+		–	–	–	–	–
Lactose		–		–	–	+	+	+	+		–	–	–	–	–
Mannose		w		+	+	+	+	+	d		–	–	–	–	–
Raffinose		–		+	–	–	+	–	–		–	–	–	–	–
Production of saccharolytic and proteolytic enzymes:															
ArgA		–		+	+	+	–	–	+		+	+	+	–	+
α-GAL		–		+	–	–	–	–	–		–	–	–	–	+
β-GAL		–		–	–	+	–	–	–		–	–	–	–	+
α-GLU		+		+	+	–	d	–	–		–	–	–	–	–
β-GUR		–		+	+	–	–	–	–		–	–	–	–	–
HisA		–		+	w	–	–	–	+		w	+	+	–	+
LeuA		–		–	+	–	–	–	+		d	+	d	–	+
PheA		–		–	w	–	–	–	–		–	+	–	+	+
ProA		+		–	–	–	–	+	–		–	–	–	+	–
PyrA		–		+	w	–	–	w	–		–	–	–	–	–

[a]Symbols: +, >85% positive; –, 0–15% positive; d, different among strains; w, weak reaction; ND, not determined. FA, fatty acid: upper-case letter = major product, lower-case letter = minor product; B, butyrate; C, capronate; iC, iso-capronate; iv, iso-valerate; A, acetate; L, lactate; P, propionate; ALP, alkaline phosphatase; α-GAL, α-galactosidase; β-GAL, β-galactosidase; α-GLU, α-glucosidase; β-GUR, β-glucronidase; ArgA, arginine arylamidase(AMD); ProA, proline AMD; PheA, phenylalanine AMD; LeuA, leucine AMD; PryA, pyloglutamyl AMD; HisA, histidine AMD.

Further descriptive information

Cells occur singly, in pairs, and occasionally in short chains or filaments. Cells have variable Gram-staining properties. Cells have rounded to tapered ends (Cato et al., 1985; Love et al., 1979). Spores may be formed. They are single, oval, and subterminal (Love et al., 1979). Cells are nonmotile, but some of the strains have been shown to have twitching or end-over-end type of motility. Colonies are small (0.5–1.0 mm) and nonhemolytic (Cato et al., 1985; Love et al., 1979).

No acid is produced from esculin, fructose, glucose, maltose, mannitol, mannose, melibiose, ribose, sucrose, or xylose. Threonine and lactate are not utilized (Cato et al., 1985; Love et al., 1979).

The 16S rRNA sequence homology between the two *Filifactor* species described thus far is 92.6 % (Jalava and Eerola, 1999).

All tested strains are susceptible to chloramphenicol (12 μg/ml) and erythromycin (3 μg/ml) and most of the strains are susceptible to penicillin (2 U/ml) (Cato et al., 1985; Love et al., 1979).

Filifactor species have been isolated from human gingival sulcus of patients with gingivitis or periodontitis, from oral cavities of cats, and from subcutaneous wound abscesses of cats. They might have a pathogenic role in mixed anaerobic infections (Cato et al., 1985; Downes et al., 2001; Love et al., 1987, 1979). Diagnostic characteristics of the two species are given in the Table 189.

Enrichment and isolation procedures

The species can be isolated and grown on sheep blood agar plates and brain heart infusion agar plates incubated anaerobically (Love et al., 1979).

Maintenance procedures

Cooked meat plus peptic digest of meat broth and cooked meat plus peptic digest of meat broth supplemented with 0.4 % glucose, 0.1 % cellobiose, 0.1 % maltose and 0.1 % starch can be used for cultivation of pure cultures (Love et al., 1979). Fastidious anaerobic agar plates with and without 7% (w/v) bovine blood have been used for cultivation of pure cultures in anaerobic conditions (Jalava and Eerola, 1999).

Taxonomic comments

The species belonging to the genus *Filifactor* are biochemically relatively inert. Also Gram-staining might give variable reactions. Because of these difficulties, proper classification of the species in the genus *Filifactor* based only on phenotypic properties is difficult. According to the 16S rRNA phylogenetic analysis presented in the roadmap to this volume (Figure 5), the genus *Filifactor* is a member of the family *Peptostreptococcaceae*, order *Clostridiales*, class *Clostridia* in the phylum *Firmicutes*.

List of species of the genus *Filifactor*

1. **Filifactor villosus** Collins, Lawson, Willems, Cordoba, Fernández-Garayzábal, Garcia, Cai, Hippe and Farrow 1994, 822[VP] (*Clostridium villosum* Love, Jones and Bailey 1979, 242)

vil.los′us. L. adj. hairy, shaggy, rough-haired.

This description is based on studies of Love et al. (1979), comparative 16S rRNA sequence analyses by Collins et al. (1994), and the description of the *Clostridium villosum* by Cato et al. (1986) in the first edition of the Manual.

Cells are rods with parallel sides and rounded ends. Cell size is 0.6 × 4.0–6.0 μm, but up to 24- to 30-μm-long filaments are common in old cultures (24–48 h). Cells are Gram-stain-positive if cultured on agar or in broth not more than 18–24 h but after this Gram-staining gives variable or negative

reaction. Gram reaction of the old cultures (7 d) is variable and most cells are Gram-stain-negative with unstained gaps existing along cells. However, thin-section electron microscopy studies have shown that the structure of the cell wall

TABLE 190. Descriptive characteristics of the species of the genus *Filifactor*[a]

Characteristic	F. alocis	F. villosus
Production of:		
Acetate	+[b]	+
Butyrate	+[b]	+
Isobutyrate	–	+
Formate	–	+
Isovalerate	–	+
Lactate	–	+[c]
Succinate	–	+[c]
Methylmalonate	–	+[c]
Spore formation	–	+
Gram reaction	–	+
Hemolysis on blood agar plates	–	–
Utilization of pyruvate	–	+
DNA G+C content (mol%)	28–34	NT
Growth at 37°C	+	+

[a]NT, Not tested.
[a]Small amounts.
[b]Small to moderate amounts.

TABLE 189. Diagnostic characteristics of the species of the genus *Filifactor*

Characteristic	F. alocis	F. villosus
Production of:		
Isobutyrate	–	+
Formate	–	+
Isovalerate	–	+
Spore formation	–	+
Utilization of pyruvate	–	+
Gram reaction	–	+

and the mode of the division are consistent with the Gram-positive bacteria. Cells are nonmotile and flagella have not been detected. Spores are single, oval, and subterminal, and they slightly swell the cell. Spores are not always detected but chopped meat-carbohydrate slant cultures are heat resistant (Love et al., 1979).

Young (24 h) colonies on sheep blood agar and brain heart infusion agar are 0.5 mm in diameter. The diameter of the colony increases to approximately 5 mm by 3 d. Colonies are irregular, rhizoid, convex, whitish to yellow, rough, have a matt surface, adhere strongly to the medium. Colonies do not hemolyze blood (Love et al., 1979).

Cultures in chopped meat-carbohydrate (CMC) broth grow in a delicate membrane near the surface of the broth, with bacteria clumped as discrete colonies throughout the membrane; later the membrane collapses and settles to the bottom of the tube. Cultures in peptone-yeast-extracted glucose (PYG) broth have a ropy or flaky sediment and no turbidity. Growth at 37°C is stimulated by 5% horse serum. No gas is detected in PYG deep agar cultures. Ammonia is produced in chopped meat and in peptone-yeast extract broth containing 5% horse serum (SPY) cultures (Cato et al., 1986).

Products of fermentation in CMC and PY broth supplemented with 5% horse serum are acetate, butyrate, isobutyrate, formate, and isovalerate. Trace amounts of lactate, succinate, and methylmalonate may be detected. Small amounts of H_2 are produced. Pyruvate is converted to butyrate; lactate is not utilized; increased amounts of acetate and butyrate are produced in threonine cultures but propionate is not produced. Indole, lecithinase, or lipase are not produced. Esculin is not hydrolyzed. Nitrate is not reduced. There is no milk reaction and meat is not digested. Gelatin is liquefied weakly (Cato et al., 1986).

Other characteristics are given in the Table 190.

Culture supernatants are nontoxic to mice (Cato et al., 1986).

The strains are susceptible to chloramphenicol (12 μg/ml), erythromycin (3 μg/ml), penicillin (2 U/ml), amoxicillin (2.5 μg/ml), carbenicillin (100 μg/ml) and doxycycline (6 μg/ml) (Love et al., 1979).

Isolated from subcutaneous wound abscesses of cats (Love et al., 1979).

DNA G+C content (mol%): not reported.

Type strain: NCTC 11220.

EMBL/GenBank accession number (16S rRNA gene): X73452.

2. **Filifactor alocis** Jalava and Eerola 1999, 1378[VP] (*Fusobacterium alocis* Cato, Moore and Moore 1985, 475)

a′lo.cis. Gr. n. *alox* a furrow; N.L. gen. n. *alocis* of a furrow, referring to its isolation from a crevice of the gums.

This description is based on study of Cato et al. (1985) and comparative 16S rRNA sequence analyses (Jalava and Eerola, 1999).

Cells are Gram-stain-negative, obligately anaerobic rods. The size of the cells is 0.4–0.7 × 1.5–7.0 μm. Normally they occur singly, in pairs, or sometimes in short chains and have rounded to tapered ends. Cells are nonmotile and do not have any flagella but some of the strains show slow twitching or end-over-end type of motility. Spores are not formed.

Optimal growth temperature is 37°C. Colonies are small (up to 1.0 mm), circular, entire, flat to low convex, translucent to transparent, shiny and smooth. Colonies do not produce hemolysis on blood agar plates.

Little or no gas is detected in PYG deep agar cultures. H_2 is detected in broth cultures if the growth is sufficient. Acid is not produced from adonitol, amygdalin, L-arabinose, cellobiose, dextrin, dulcitol, *meso*-erythritol, esculin, D-fructose, D-galactose, D-glucose, glycerol, glycogen, inositol, inulin, lactose, maltose, D-mannitol, D-mannose, melezitose, melibiose, pectin, raffinose, rhamnose, D-(−)-ribose, salicin, D-sorbitol, L-sorbose, starch, sucrose, trehalose, or D-xylose. Esculin, starch, and hippurate are not hydrolyzed. Nitrate, resazurin, or neutral red are not reduced. Indole, acetylmethylcarbinol, catalase, lecithinase, lipase, urease, oxidase, or deoxyribonuclease are not produced. Pyruvate, DL-lactate, D-gluconate, or DL-threonine are not utilized. Milk, gelatin, or chopped meat are not digested. Small to moderate amounts of butyrate and acetate are produced in PYG broth.

Other characteristics are given in the Table 190.

The strains are susceptible to chloramphenicol (12 μg/ml), clindamycin (1.6 μg/ml), erythromycin (3 μg/ml), and tetracycline (6 μg/ml). Most of the strains are susceptible to penicillin (2 U/ml).

Isolated from human gingival sulcus of patients with gingivitis or periodontitis, from oral cavities of cats, and soft tissue infections of cats caused by contamination from oral cavities.

DNA G+C content (mol%): 34 (T_m).

Type strain: ATCC 35896.

EMBL/GenBank accession number (16S rRNA gene): AJ006962.

Genus III. **Tepidibacter** Slobodkin, Tourova, Kostrikina, Chernyh, Bonch-Osmolovskaya, Jeanthon and Jones 2003, 1133[VP]

ALEXANDER SLOBODKIN

Te.pi.di.bac′ter. L. adj. *tepidus* warm; N.L. *bacter* masc. equivalent of Gr. neut. dim. n. *bakterion* rod; N.L. masc. n. *Tepidibacter* a warm rod.

Straight to slightly curved **rods** 0.7–0.9 μm in diameter and 3.5–6.0 μm in length. **Gram-positive type cell wall.** Cells occur singly, in pairs or in short chains and exhibit tumbling motility due to peritrichous flagellation. **Forms** round or ovoid refractile **endospores** in terminally or subterminally swollen sporangia. **Anaerobic. Moderately thermophilic,** the temperature range for growth is 30–60°C. Neutrophilic. Optimal growth at marine salinity. **Grows organotrophically** on a number of proteinaceous substrates and carbohydrates. Elemental sulfur may be reduced, but sulfur reduction does not stimulate growth.

DNA G+C content (mol%): 24–29 (T_m).

Type species: **Tepidibacter thalassicus** Slobodkin, Tourova, Kostrikina, Chernyh, Bonch-Osmolovskaya, Jeanthon and Jones 2003, 1133[VP].

Further descriptive information

16S rRNA gene sequence analysis places the genus *Tepidibacter* within cluster XI of the clostridia (nomenclature of Collins et al., 1994). Members of the genus *Tepidibacter* are currently represented by two species, *Tepidibacter thalassicus* and *Tepidibacter formicigenes*, which share 95% of 16S rRNA gene sequence similarity.

Both species of the genus *Tepidibacter* form terminal or subterminal endospores. In the late-exponential phase of growth, up to 30% of the cells contain spores.

The best growth of *Tepidibacter* can be obtained on complex proteinaceous substrates such as tryptone, casein, and peptone. *Tepidibacter thalassicus* is able to perform the Stickland reaction with alanine and proline. Carbohydrates, with an exception of starch, slightly stimulate growth of *Tepidibacter thalassicus*. Growth of *Tepidibacter formicigenes* on sugars is more efficient. Both species produce acetate and ethanol from glucose, and either H_2/CO_2, (*Tepidibacter thalassicus*) or formate (*Tepidibacter formicigenes*). External electron acceptors do not enhance growth, however, elemental sulfur can be reduced to hydrogen sulfide.

Both members of the genus *Tepidibacter* were isolated from deep-sea hydrothermal vents. *Tepidibacter thalassicus* was isolated from the outer wall of a 'black smoker' covered with the polychetous annelid *Alvinella* species (13° N hydrothermal field on the East-Pacific Rise, depth 2650 m); *Tepidibacter formicigenes* was isolated from hydrothermal fluid (the Menez-Gwen hydrothermal site, Mid-Atlantic Ridge, depth 800–1000 m). Location of these vents, one of which is in Pacific and other in Atlantic Ocean, suggests wide geographical distribution of *Tepidibacter* in marine hydrothermal environments where they probably function as decomposers of organic matter produced by deep-sea biota.

Table 187 shows differential characteristics for *Tepidibacter* from other members of *Peptostreptococcaceae* and related genera.

Enrichment and isolation procedures

Members of the genus *Tepidibacter* have been isolated from deep-sea hydrothermal vents (Slobodkin et al., 2003; Urios et al., 2004). There are no reports that members of this genus occur in other environments. Employment of the marine anaerobic media rich in proteinaceous substrates and incubation in the temperature range of 45–55°C may favor the enrichment of *Tepidibacter* species. *Tepidibacter thalassicus* rapidly hydrolyzes casein (Hammerstein grade) that results in visual disappearance of the casein flocks and may help in the detection of growth. *Tepidibacter* forms colonies in 1.5% (w/v) agar.

Maintenance procedures

Members of the genus *Tepidibacter* may be maintained on the medium of Slobodkin et al. (2003) with peptone or casein as a substrate or on the glucose/yeast extract/peptone medium of Urios et al. (2004). Good and reproducible growth can also be obtained in liquid medium lacking sulfide as a reducing agent and prepared anaerobically (Slobodkin et al., 2003). Freeze-drying of the cultures results in good recovery. Liquid cultures may be stored at +4°C for 10–12 months without loss of viability.

List of species of the genus *Tepidibacter*

1. **Tepidibacter thalassicus** Slobodkin, Tourova, Kostrikina, Chernyh, Bonch-Osmolovskaya, Jeanthon and Jones 2003, 1133[VP]

tha.las′si.cus. Gr. fem. n. *thalassa* the sea; N.L. masc. adj. *thalassicus* of the sea.

Cells are straight to slightly curved rods, 0.7–0.9 μm in diameter and 3.5–6.0 μm in length, which form round, refractile endospores in terminally swollen sporangia. Cells occur singly or in short chains and exhibit tumbling motility due to peritrichous flagellation. The temperature range for growth is 33–60°C, with an optimum at 50°C. The pH range for growth is 4.8–8.5, with an optimum at 6.5–6.8. Growth occurs at NaCl concentrations in the range 1.5–6% (w/v). Anaerobic. Substrates utilized include casein, peptone, albumin, yeast extract, beef extract, alanine plus praline, and starch. Glucose, maltose, pyruvate, valine, and arginine slightly stimulate growth in the presence of yeast extract. Fructose, sucrose, xylose, cellobiose, L-arabinose, glycerol, sorbitol, acetate, butyrate, lactate, formate, methanol, fumarate, glycine, alanine, proline, alanine plus glycine, betaine, olive oil, xylan, carboxymethylcellulose, filter paper, chitin, keratin, and H_2/CO_2 are not utilized. The products of glucose fermentation are ethanol, acetate, and molecular hydrogen. Reduces elemental sulfur to hydrogen sulfide. Does not use nitrate, fumarate, sulfate, sulfite, thiosulfate, amorphous Fe(III) oxide, Fe(III) citrate, or oxygen as electron acceptors.

DNA G+C content (mol%): 24 (T_m).
Type strain: SC 562, DSM 15285, UNIQEM 215.
GenBank accession number (16S rRNA gene): AY158079.

2. **Tepidibacter formicigenes** Urios, Cueff, Pignet and Barbier 2004, 442[VP]

for.mi.ci′ge.nes. N.L. adj. *formicicum* from L. n. *formica* ant; Gr. v. *gennaio* produce; N.L. adj. *formicigenes* producing formic acid.

Rod-shaped cells, 0.8 μm in diameter and 4.0 μm in length; motile by means of peritrichous flagella. Cells stain Gram-stain-positive. Forms ovoid refractile subterminal endospore. Growth occurs between 35 and 55°C (optimum, 45°C), between pH 5.0 and 8.0 (optimum, 6.0) and at 2–6% (w/v) sea salts (optimum, 3%). Anaerobic, able to ferment mainly complex proteinaceous substrates and carbohydrates. Grows on glucose/yeast extract/peptone medium. Tryptone, glucose, sucrose, fructose, maltose, and pyruvate support growth in the presence of yeast extract. Poor growth on ethanol, mannose, and peptone. Cellobiose, xylose, starch, cellulose, dextran, xylan, succinate, lactate, trehalose, lactose, arabinose, galactose, ribose, rhamnose, mannitol, sorbitol, glycerol, urea, and olive oil are not utilized. The products of glucose fermentation (in decreasing order) are formate, acetate, and ethanol. Elemental sulfur, polysulfides, thiosulfate, sulfite, sulfate, nitrite, nitrate, and $FeCl_3$ do not enhance growth.

DNA G+C content (mol%): 29 (T_m).
Type strain: DV1184, CIP 107893, DSM 15518.
GenBank accession number (16S rRNA gene): AY245527.

References

Cato, E.P., L.V.H. Moore and W.E.C. Moore. 1985. *Fusobacterium alocis* sp. nov. and *Fusobacterium sulci* sp. nov. from the human gingival sulcus. Int. J. Syst. Bacteriol. *35*: 475–477.

Cato, E.P., W.L. George and S.M. Finegold. 1986. Genus *Clostridium* Prazmowski 1880, 23[AL]. *In* Sneath, Mair, Sharpe and Holt (Editors), Bergey's Manual of Systematic Bacteriology, Vol. 2. The Williams & Wilkins Co., Baltimore, pp. 1141–1200.

Collins, M.D. and S. Wallbanks. 1992. Comparative sequence analyses of the 16S rRNA genes of *Lactobacillus minutus*, *Lactobacillus rimae* and *Streptococcus parvulus*: proposal for the creation of a new genus *Atopobium*. FEMS Microbiol. Lett. *74*: 235–240.

Collins, M.D., P.A. Lawson, A. Willems, J.J. Cordoba, J. Fernández-Garayzábal, P. Garcia, J. Cai, H. Hippe and J.A.E. Farrow. 1994. The phylogeny of the genus *Clostridium*: proposal of five new genera and eleven new species combinations. Int. J. Syst. Bacteriol *44*: 812–826.

Downes, J., M.A. Munson, D.A. Spratt, E. Kononen, E. Tarkka, H. Jousimies-Somer and W.G. Wade. 2001. Characterisation of *Eubacterium*-like strains isolated from oral infections. J. Med. Microbiol. *50*: 947–951.

Downes, J. and W.G. Wade. 2006. *Peptostreptococcus stomatis* sp. nov., isolated from the human oral cavity. Int. J. Syst. Evol. Microbiol. *56*: 751–754.

Ezaki, T., N. Yamamoto, K. Ninomiya, S. Suzuki and E. Yabuuchi. 1983. Transfer of *Peptococcus indolicus*, *Peptococcus asaccharolyticus*, *Peptococcus prevotii*, and *Peptococcus magnus* to the genus *Peptostreptococcus* and proposal of *Peptostreptococcus tetradius* sp. nov. Int. J. Syst. Bacteriol. *33*: 683–698.

Ezaki, T., N. Li, Y. Hashimoto, H. Miura and H. Yamamoto. 1994. 16s Ribosomal DNA sequences of anaerobic cocci and proposal of *Ruminococcus hansenii* comb. nov. and *Ruminococcus productus* comb. nov. Int. J. Syst. Bacteriol. *44*: 130–136.

Ezaki, T., Y. Kawamura, N. Li, Z.Y. Li, L. Zhao and S. Shu. 2001. Proposal of the genera *Anaerococcus* gen. nov., *Peptoniphilus* gen. nov. and *Gallicola* gen. nov. for members of the genus *Peptostreptococcus*. Int. J. Syst. Evol. Microbiol. *51*: 1521–1528.

Holdeman, L.V., J.L. Johnson and W.E.C. Moore. 1986. Genus *Peptostreptococcus* Kluyver and van Neil 1936, 401[AL]. *In* Sneath, Mair and Holt (Editors), Bergey's Manual of Systematic Bacteriology, Vol. 2. The Williams & Wilkins Co., Baltimore, pp. 1083–1092.

Jalava, J. and E. Eerola. 1999. Phylogenetic analysis of *Fusobacterium alocis* and *Fusobacterium sulci* based on 16S rRNA gene sequences: proposal of *Filifactor alocis* (Cato, Moore and Moore) comb. nov and *Eubacterium sulci* (Cato, Moore and Moore) comb. nov. Int. J. Syst. Bacteriol. *49*: 1375–1379.

Kluyver, A.J. and C.B. van Neil. 1936. Prospects for a natural classification of bacteria. Zentbl. Bacteriol. Parasitenkd. Infektionskr. Hyg. Abt. II *94*: 369–403.

Love, D.N., R.F. Jones and M. Bailey. 1979. *Clostridium villosum* sp. nov. from subcutaneous abscesses in Cats. Int. J. Syst. Bacteriol. *29*: 241–244.

Love, D.N., E.P. Cato, J.L. Johnson, R.F. Jones and M. Bailey. 1987. Deoxyribonucleic acid hybridization among strains of fusobacteria isolated from soft tissue infections of cats: comparison with human and animal type strains from oral and other sites. Int. J. Syst. Bacteriol. *37*: 23–26.

Murdoch, D.A. and H.N. Shah. 1999. Reclassification of *Peptostreptococcus magnus* (Prevot 1933) Holdeman and Moore 1972 as *Finegoldia magna* comb. nov. and *Peptostreptococcus micros* (Prevot 1933) Smith 1957 as *Micromonas micros* comb. nov. Anaerobe *5*: 555–559.

Natvig, H. 1905. Bakteriologische Verhältnesse in weiblichen Genitalsekreten. Erste Mittheilung. Studien über Streptokokken der weiblichen Genitalien in Partus und Puerperium. Arch. Gynekol. (Berlin) *76*: 701–858.

O'Leary, W.M. and S.G. Wilkinson. 1988. Gram-positive bacteria. *In* Ratledge and Wilkinson (Editors), Microbial Lipids. Academic Press, London, pp. 117–201.

Slobodkin, A.I., T.P. Tourova, N.A. Kostrikina, N.A. Chernyh, E.A. Bonch-Osmolovskaya, C. Jeanthon and B.E. Jones. 2003. *Tepidibacter thalassicus* gen. nov., sp. nov., a novel moderately thermophilic, anaerobic, fermentative bacterium from a deepsea hydrothermal vent. Int. J. Syst. Evol. Microbiol. *53*: 1131–1134.

Tindall, B.J. and J.P. Euzéby. 2006. Proposal of *Parvimonas* gen. nov. and *Quatrionicoccus* gen. nov. as replacements for the illegitimate, prokaryotic, generic names *Micromonas* Murdoch and Shah 2000 and *Quadricoccus* Maszenan *et al.* 2002, respectively. Int. J. Syst. Evol. Microbiol. *56*: 2711–2713.

Urios, L., V. Cueff, P. Pignet and G. Barbier. 2004. *Tepidibacter formicigenes* sp. nov., a novel spore-forming bacterium isolated from a Mid-Atlantic Ridge hydrothermal vent. Int. J. Syst. Evol. Microbiol. *54*: 439–443.

Wade, W.G., J. Downes, D. Dymock, S.J. Hiom, A.J. Weightman, F.E. Dewhirst, B.J. Paster, N. Tzellas and B. Coleman. 1999. The family *Coriobacteriaceae*: reclassification of *Eubacterium exiguum* (Poco et al. 1996) and *Peptostreptococcus heliotrinreducens* (Lanigan 1976) as *Slackia exigua* gen. nov., comb. nov. and *Slackia heliotrinireducens* gen. nov., comb. nov., and *Eubacterium lentum* (Prevot 1938) as *Eggerthella lenta* gen. nov., comb. nov. Int. J. Syst. Bacteriol. *49*: 595–600.

Family VIII. **Ruminococcaceae** fam. nov.

FRED A. RAINEY

Ru.mi.no.coc.ca′ce.ae. N.L. masc. n. *Ruminococcus* type genus of the family; *-aceae* ending to denote family; N.L. fem. pl. n. *Ruminococcaceae* the *Ruminococcus* family.

The family *Ruminococcaceae* is described on the basis of phylogenetic analyses of 16S rRNA gene sequences (Figure 160); the family contains the genera *Ruminococcus* (type genus), *Acetanaerobacterium*, *Acetivibrio*, *Anaerofilum*, *Anaerotruncus*, *Ethanoligenens*, *Faecalibacterium*, *Fastidiosipila*, *Oscillospira*, *Papillibacter*, *Sporobacter*, and *Subdoligranulum*. The genus *Ethanoligenens* was proposed after finalization of the content of this volume and is not described here (see Xing et al., 2006).

Family is morphologically diverse and includes long thin rods, rods, cocci and pleomorphic forms. All species are obligate anaerobes.

Type genus: **Ruminococcus** Sijpesteijn 1948, 152[AL].

Genus I. **Ruminococcus** Sijpesteijn 1948, 152[AL]

TAKAYUKI EZAKI

Ru.min.o.coc′cus. L. adj. *ruminalis* of the rumen; Gr. n. *kokkos* a grain, berry; N.L. masc. n. *Ruminococcus* coccus of the rumen.

Cells are coccoid, usually 0.3–1.5 × 0.7–1.8 μm. **Cells are in pairs and chains. A few are motile with 1–3 flagella. Gram-stain-positive cell-wall structure but many stain Gram-negative. Optimal temperature, 37–42°C. Chemo-organotrophic. Strictly anaerobic and require fermentable carbohydrates to grow.** Fermentation of carbohydrate yields various proportions of **acetate, formate, succinate, lactate, and ethanol. Amino acid and peptides are not fermented. Indole is not produced (Bryant, 1986). Isolated from rumen, large bowel, or cecum of many animals and humans.** Some are isolated from human clinical specimens. Rumen fluid agar is effective to isolate and characterize strains (Holdeman et al., 1977). Some strains use cellulose on rumen fluid cellobiose agar plate (Holdeman et al., 1977) and require ammonia as a nitrogen source.

DNA G+C content (mol%): 39–47.

Type species: **Ruminococcus flavefaciens** Sijpesteijn 1948, 152[AL].

Further taxonomic comments

Species in the genus *Ruminococcus* were divided into two phylogenetically different groups within a family *Lachnospiraceae* (Figure 161). The type species of the genus *Ruminococcus*, *Ruminococcus flavefaciens*, belongs to group I *Streptococcus hansenii* Holdeman and Moore (1974) and *Peptostreptococcus productus* Prévot (1941) are reclassified to group II ruminococci (Ezaki et al., 1994). Group II ruminococci are phylogenetically close to the sugar-requiring anaerobic cocci, genus *Coprococcus*. *Ruminococcus pasteurii* and *Ruminococcus palustris* were transferred to genus *Trichococcus* by Liu et al. (2002).

List of species of the genus *Ruminococcus*

1. **Ruminococcus flavefaciens** Sijpesteijn 1948, 152[AL]

   fla.ve.fac′i.ens. L. adj. *flavus* yellow. L. pres. part. *faciens* producing: N.L. part. adj. *flavefaciens* yellow-producing.

   Strictly anaerobic, Gram-stain-positive, nonsporulating, chain-forming cocci. Cellulose and hemicellulose fermenter and pectin hydrolyzer. Catalase-negative. Nonmotile. Isolated from the rumen of many different species (Hungate, 1957). Features are listed in Table 191. Belongs to ruminococcus group I (Figure 161).

   DNA G+C content (mol%): 39–44.

   Type strain: Bryant C94, ATCC 19208, NCDO 2213.

   GenBank accession number (16S rRNA gene): L76603.

2. **Ruminococcus albus** Hungate 1957, 307[AL]

   al′bus. L. adj. *albus* white.

   Strictly anaerobic, Gram-stain-positive, nonsporulating, chain-forming cocci. Found in human feces and rumen of cattle and other animals. Cellobiose-fermenting rumen bacteria. Catalase-negative. Nonmotile. Produces more ethanol, H_2, and CO_2 than *Ruminococcus flavefaciens*. Other features are listed in Table 191. Belongs to ruminococcus group I (Figure 161). Isolated from human feces.

   DNA G+C content (mol%): 42.6–45.8.

   Type strain: Hungate 1957-7, ATCC 27210, DSM 20455, NCDO 2250.

   GenBank accession number (16S rRNA gene): L76598.

3. **Ruminococcus bromii** Moore, Cato and Holdeman 1972, 80[AL]

   brom′i.i. L. gen. n. *bromii* of Bromius, god of alcohol.

   Strictly anaerobic, Gram-stain-positive, nonsporulating, chain-forming cocci. Isolated from human feces. Cellobiose is not fermented. Catalase-negative. Nonmotile. Features are listed in Table 191. Belongs to ruminococcus group I (Figure 161).

   DNA G+C content (mol%): 39–40.

   Type strain: ATCC 27255, VPI 6883.

   GenBank accession number (16S rRNA gene): L76600.

4. **Ruminococcus callidus** Holdeman and Moore 1974, 264[AL]

   cal′li.dus. L. adj. *callidus* clever, expert.

   Strictly anaerobic, Gram-stain-positive, nonsporulating cocci. Catalase-negative. Nonmotile. Ferment cellobiose. Features are listed in Table 191. Belongs to ruminococcus group I (Figure 161). Isolated from human feces.

TABLE 191. Characteristics of species of the genus *Ruminococcus*[a]

Characteristic	*R. flavefaciens*	*R. albus*	*R. bromii*	*R. callidus*	*R. gnavus*	*R. hansenii*	*R. hydrogenotrophicus*	*R. lactaris*	*R. luti*	*R. obeum*	*R. productus*	*R. schinkii*	*R. torques*
16S rRNA gene accession no.	X83430	X85098	X85099	X85100	D14136	D14155	X95624	L76602	AJ133124	X85101	D14144	X94965	L76604
DNA G+C content (mol%)	39–44	43–46	39–40	42	41	37–38	45	43	43.3	45	44–45	46–47	43
Major PYG product	A, F, S	A, F	A	S, a	A, F	L, a	A	A, F	A	A	L, a	A	L, a
Fermentation of:													
Arabinose	–	–	–	–	+	–	–	–	+	+	+	+	–
Cellobiose	+	+	–	+	–	–	+	–	+	+	+	+	–
Glucose	–	+	+	+	+	+	ND	+	+	+	+	+	+
Lactose	+	+	–	–	–	+	d	w/–	+	+	+	ND	+
Mannose	–	+	w/–	+	–	–	ND	d	+	+	+	+	w/–
Maltose	–	–	+	–	+	+	–	+	+	+	+	ND	w
Mannitol	–	–	–	–	–	–	ND	–	–	–	+	+	–
Raffinose	–	–	–	+	+	+	–	–	+	+	+	+	–
Sucrose	–	+	–	+	+	–	–	–	+	–	+	+	–
Xylose	–	–	–	w/–	+	–	ND	–	+	+	+	+	–

[a]Symbols: +, >85% positive; d, different strains give different reactions; –, 0–15% positive; w, weak reaction; ND, not determined; A, acetate; F, formate; S, succinate; L, lactate.

DNA G+C content (mol%): 43.

Type strain: ATCC 27760, VPI 57-31.

GenBank accession number (16S rRNA gene): L76596.

5. **Ruminococcus gnavus** Moore, Johnson and Holdeman 1976, 243[AL]

gna'vus. L. adj. *gnavus* busy, active.

Strictly anaerobic, Gram-stain-positive, nonsporulating cocci. Cellobiose is not fermented. Catalase-negative. Nonmotile. Features are listed in Table 191. Belongs to ruminococcus group I. Isolated from human feces.

DNA G+C content (mol%): 43.

Type strain: ATCC 29149, VPI C7-9.

GenBank accession number (16S rRNA gene): D14136, L76597, X94967.

6. **Ruminococcus hansenii** (Holdeman and Moore 1974) Ezaki, Li, Hashimoto, Miura and Yamamoto 1994, 134[VP] (*Streptococcus hansenii* Holdeman and Moore 1974, 266)

han.sen'i.i. N.L. gen. n. *hansenii* of Hansen named after P. Arne Hansen, a Danish-American bacteriologist.

Strictly anaerobic, Gram-stain-positive, nonsporulating cocci. Cells occur in pairs and chains. Catalase-negative. Nonmotile. Features are listed in Table 191. Belongs to ruminococcus group II (Figure 161). Isolated from human feces.

DNA G+C content (mol%): 44–45.

Type strain: ATCC 27752, CIP 104219, DSM 20583, VPI C7-24.

GenBank accession number (16S rRNA gene): D14155, M59114.

7. **Ruminococcus hydrogenotrophicus** Bernalier Willems, Leclerc, Rochet and Collins 1997, 601[VP] (Effective publication: Bernalier Willems, Leclerc, Rochet and Collins 1996, 182.)

hy.dro.ge.no.tro'phi.cus. Gr. n. *hydor* water; Gr. n. *genus* race, offspring; N.L. n. *hydrogenum* hydrogen, that which produces water; Gr. n. *trophos* one who feeds. N.L. masc. adj. *hydrogenotrophicus* one who feeds on hydrogen, referring to the ability of the micro-organism to grow with H_2/CO_2 as energy source.

Strictly anaerobic, Gram-stain-positive, nonsporulating coccobacilli. Isolated from human feces. Features are listed in Table 191. Belongs to ruminococcus group II (Figure 161).

DNA G+C content (mol%): 45.2.

Type strain: S5a33, DSM 10507.

GenBank accession number (16S rRNA gene): X95624.

8. **Ruminococcus lactaris** Moore, Johnson and Holdeman 1976, 244[AL]

lac.ta'ris. L. adj. *lactaris* milk-drinking.

Strictly anaerobic, Gram-stain-positive, nonsporulating cocci. Features are listed in Table 191. Belongs to ruminococcus group II (Figure 161). Isolated from human fecal samples.

DNA G+C content (mol%): 45.

Type strain: ATCC 29176, VPI X6-29.

GenBank accession number (16S rRNA gene): L76602.

9. **Ruminococcus luti** Simmering, Taras, Schwiertz, Le Blay, Gruhl, Lawson, Collins and Blaut 2002b, 1915[VP] (Effective publication: Simmering, Taras, Schwiertz, Le Blay, Gruhl, Lawson, Collins and Blaut 2002a, 192.)

lu'ti. L. gen. neut. n. *luti* of mud (feces).

Strictly anaerobic, Gram-stain-positive, nonsporulating coccobacilli. Cells often appear in long chains, nonmotile. Catalase-negative. Features are listed in Table 191. Belongs to ruminococcus group II (Figure 161). Found in human fecal samples.

DNA G+C content (mol%): 43.3.

Type strain: BInIX, CCUG 45635, DSM 14534.

GenBank accession number (16S rRNA gene): AJ133124.

10. **Ruminococcus obeum** Moore, Johnson and Holdman 1976, 245[AL]

o'be.um. Gr. n. *obeum* egg.

Strictly anaerobic, Gram-stain-positive, nonsporulating coccobacilli. Found in human fecal samples. Features are listed in Table 191. Belongs to ruminococcus group II (Figure 161).

DNA G+C content (mol%): 42 (T_m).

Type strain: ATCC 29174, VPI B3-21.

GenBank accession number (16S rRNA gene): L76601.

11. **Ruminococcus productus** (Prévot 1941) Ezaki, Li, Hashimoto, Miura and Yamamoto 1994, 135[VP] (*Streptococcus productus* Prévot 1941, 105; *Peptostreptococcus productus* Smith 1957, 536)

pro.duc'tus L. adj *productus* produced.

Strictly anaerobic, Gram-stain-positive, nonsporulating coccobacilli. Catalase-negative. Nonmotile. Isolated from human feces. Features are listed in Table 191. Belongs to ruminococcus group II (Figure 161).

DNA G+C content (mol%): 44–45.

Type strain: ATCC 27340, CCUG 9990, CCUG 10976, DSM 2950, JCM 1471, VPI 4299.

GenBank accession number (16S rRNA gene): D14144, L76595, X94966.

12. **Ruminococcus schinkii** Rieu-Lesme, Morvan, Collins, Fonty and Willems 1997, 242[VP] (Effective publication: Rieu-Lesme, Morvan, Collins, Fonty and Willems 1996, 286.)

schink.i.i. N.L. gen. n. *schinkii* of Schink named after Bernhard Schink, a German bacteriologist.

Strictly anaerobic, Gram-stain-positive, nonsporulating coccobacilli. Cells occur in pairs or short chains. Mean size is 0.8–1.1 × 1.4–2.8 µm. Temperature range from 20 to 45°C. Optimum growth at 39°C. Isolated from human feces. Indole is not formed. Oxidase- and catalase-negative. Nitrate is not reduced. Nonmotile. Features are listed in Table 191. Belongs to ruminococcus group II (Figure 161).

DNA G+C content (mol%): 46–47.

Type strain: B, CIP 105464, DSM 10518.

GenBank accession number (16S rRNA gene): X94965.

13. **Ruminococcus torques** Holdeman and Moore 1974, 265[AL]

tor'ques. from L. n. *torquis* a twisted necklace.

Strictly anaerobic, Gram-stain-positive, nonsporulating coccobacilli. Catalase-negative. Nonmotile. Features are listed in Table 191. Belongs to ruminococcus group II (Figure 161). Isolated from human feces.

DNA G+C content (mol%): 40–42.

Type strain: ATCC 27756, VPI B2-51.

GenBank accession number (16S rRNA gene): D14137, L76604.

Genus II. **Acetanaerobacterium** Chen and Dong 2004, 2261[VP]

Xiuzhu Dong

A.cet.an.ae.ro.bac.te′ri.um. L. neut. n. *acetum* vinegar; Gr. pref. *an* not; Gr. n. *aer* air; *anaero* not (living) in air; Gr. neut. dim. n. *bakterion* small rod; N.L. neut. n. *Acetanaerobacterium* vinegar-producing anaerobic small rod.

Straight, thin rod-shaped cells. $0.2–0.4 \times 4–8\,\mu m$. Gram-stain-positive. Motile. Nonspore-forming. **Obligately anaerobic. No microaerophilic or aerobic growth occurs.** Mesophilic; grows at 20–42°C. Grows at neutral pH. Chemo-organotrophic. **Oxidase and catalase are not produced.** Amino acids and peptides may serve as nitrogen sources. A variety of mono-, di-, and oligo-saccharides are fermented. Gelatin and esculin are hydrolyzed. **The major fermentation products from glucose include acetate, ethanol, hydrogen, and carbon dioxide**; lactic acid, propionate, and succinate are not produced. Sulfate is not reduced. 16S rRNA sequence analysis indicates this genus belonging to the *Clostridium leptum* rRNA subgroup (Collins et al., 1994) within the family *Clostridiaceae*. Isolated from the anaerobic sludge of paper mill waste water.

DNA G+C content (mol%): 48.6–50.4.

Type species: **Acetanaerobacterium elongatum** Chen and Dong 2004, 2261[VP].

Further descriptive information

The bacterium possesses one polar flagellum (Figure 184). Colonies on PYG agar are white, translucent, smooth, circular, entire, and slightly convex. They reach 1.5–3 mm in diameter after cultivation at 37°C for 72 h. Cell-wall peptidoglycan of *Acetanaerobacterium* contains ʟʟ-diaminopimelic acid. Cellular fatty acids consist mainly of iso-branched fatty acids, predominantly $C_{15:0\,iso}$ and $C_{14:0\,iso}$.

The only species, *Acetanaerobacterium elongatum*, was isolated from anaerobic sludge of paper mill waste water. *Acetanaerobacterium elongatum* requires amino acids and peptides as the sole nitrogen sources, but not the inorganic nitrogen sources such as NH_4Cl, $(NH_4)_2SO_4$, $(NH_4)_2HPO_4$, and KNO_3. Although the genus includes only one species so far, closely related organisms have been detected by molecular techniques in a biological hydrogen production reactor (Ren et al., 2007).

Enrichment and isolation procedures

The type strain was isolated by subculturing the sludge in pre-reduced peptone-yeast extract-glucose (PYG) broth (Holdeman et al., 1977). After enrichment, the strains were isolated on PYG using the Hungate roll-tube technique.

Maintenance procedures

The type strain is maintained in PYG broth plus 50% glycerol at –80°C. Lyophilized cultures are also used.

Differentiation of the genus *Acetanaerobacterium* from other genera

Table 192 lists the characteristics that differentiate strains of *Acetanaerobacterium* from the phylogenetically related genera. They differ from *Anaerotruncus colihominis* by the latter's production of butyric acid during glucose fermentation, production of indole, and the wide pH range for growth (pH 5.5–11.0). Unlike related *Clostridium* species, *Acetanaerobacterium* strains are nonspore-forming, hydrolyze gelatin, and have different sugar

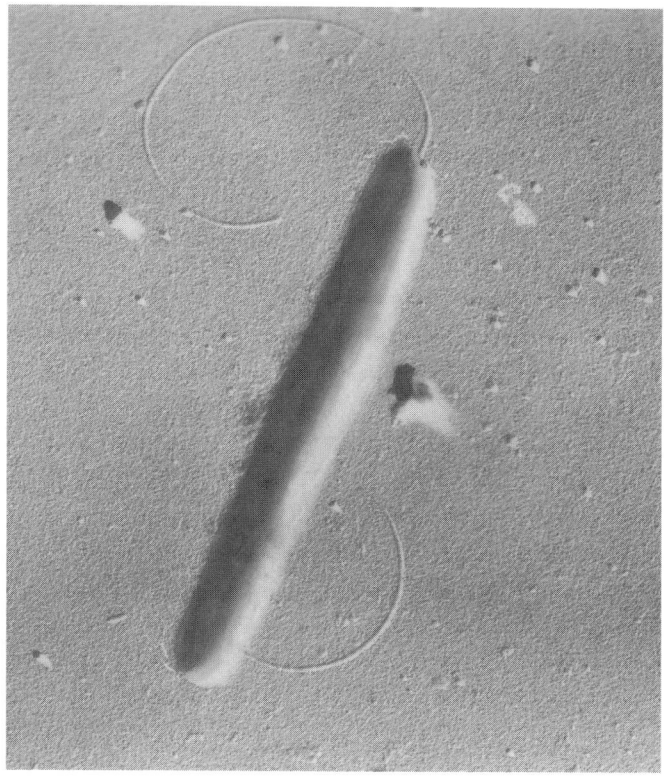

FIGURE 184. Electron micrograph of *Acetanaerobacterium elongatum* strain JCM 12359[T] showing the thin-rod cell with a single polar flagellum. An unattached flagellum is also visible next to the cell shown. Transmission electron micrograph of PYG-grown cells coated with palladium/iridium alloy. Bar = 0.5 μm.

fermentation profiles. They differ from *Ruminococcus* species in their thin rod shape, different biochemical traits, and 6–11 higher mol% G+C content of their DNA. They produce a large amount of hydrogen, but no lactic acid during glucose fermentation, enabling them to be distinguished from *Anaerofilum* species. They differ from *Eubacterium siraeum* by the latter's inability to ferment glucose. *Fusobacterium prausnitzii* has a Gram-stain-negative cell wall and produces butyrate, ᴅ-lactate, and formate, but no hydrogen during glucose fermentation, enabling it to be differentiated from the *Acetanaerobacterium* strains. The 16S rRNA gene sequence of *Acetanaerobacterium elongatum* exhibits <91% similarity to other currently described species.

Taxonomic comments

Based on 16S rRNA gene sequence analysis, the genus *Acetanaerobacterium* belongs to the *Clostridium leptum* rRNA subgroup within the family *Clostridiaceae*; the closest related genera are *Clostridium*, *Ruminococcus*, *Anaerofilum*, and *Eubacterium*.

TABLE 192. Characteristics differentiating *Acetanaerobacterium elongatum* from its phylogenetic relatives[a]

Characteristic	*Acetanaerobacterium elongatum*	*Anaerofilum agile*	*Anaerofilum pentosovorans*	*Anaerotruncus colihominis*	*Clostridium cellulosi*	*Clostridium leptum*	*Clostridium methylpentosum*	*Clostridium sporosphaeroides*	*Eubacterium siraeum*	*Fusobacterium prausnitzii*	*Ruminococcus flavefaciens*
DNA G+C content (mol%)	50.5	54.5	55	54	35	51–52	46	27	45	52–57	39–44
Cell morphology	Rod	Rod	Rod	Rod	Rod	Rod	Ring	Rod	Rod	Rod	Coccus
Spore formation	–	–	–	–	+	+	+	+	–	–	–
Products from PYG[b]	A2	LA2F	LA2F	AB	A2	A(2)	–	ABp	–	FBAL	S2(L)
H_2 produced[c]	4	–	–	+	4	4	–	4	–	–	4
Motility	+	+	–	–	+	–	–	–	–	–	–
Optimum growth temperature (°C)	37	37	25–40	36–40	55–60	37	45	37–45	37–45	37	37–42

[a]Symbols: +, >90% of strains positive; –, <10% of strains positive.

[b]Products from PYG (peptone-yeast extract-glucose; given in decreasing order of amounts usually detected): A, acetate; B, butyrate; F, formate; L, lactate; P, propionate; S, succinate; 2, ethanol. Upper-case letters and arabic numerals indicate at least 1 mg/ml of culture, and lower-case letters and number within parentheses indicate less than 1 mg/ml of culture. Products in parentheses are not detected uniformly.

[c]On a scale of – (negative) to 4 (abundant); +, H_2 produced.

List of species of the genus *Acetanaerobacterium*

1. **Acetanaerobacterium elongatum** Chen and Dong 2004, 2261[VP]

e.lon.ga′tum. L. neut. adj. *elongatum* elongated, referring to the cell shape.

Morphology and general characters are as described for the genus. Colonies on PYG agar are circular, slightly convex, white, and translucent, reaching 1.5–3 mm in diameter after 3 d incubation at 37°C. Optimal growth occurs at 37°C. The pH range for growth is 5.0–7.5 with an optimum at 6.5–7.0. Acid is produced from a few sugars, such as D-glucose, D-fructose, D-galactose, L-arabinose, D-xylose, cellobiose, D-maltose, sucrose, raffinose, inulin, and salicin. Acid is not produced from sorbose, ribose, D-lactose, mannose, melibiose, rhamnose, trehalose, starch, glycogen, amygdalin, adonitol, dulcitol, erythritol, inositol, mannitol, sorbitol, and ribitol.

No acid is produced from the following compounds: methanol, ethanol, 1-propanol, pyruvate, citrate, fumarate, malate, succinate, malonic acid, hippurate, sodium gluconate, butanedioic acid, β-hydroxybutyric acid, phenylacetic acid, cellulose, and xylan. Milk is curdled. Starch is hydrolyzed and arginine dihydrolase is produced. Urease, lecithinase, and lipase are not produced. Methyl red and Voges–Proskauer tests are negative. Nitrate is not reduced. No H_2S is produced from peptone and thiosulfate. The major cellular fatty acids are $C_{15:0\ iso}$ (43%), $C_{14:0\ iso}$ (32%), $C_{15:0\ ante}$ (6%,) and $C_{14:0\ 2OH}$ (8%).

DNA G+C content (mol%): 50.5 (T_m).

Type strain: Z7, AS 1.5012, JCM 12359.

GenBank accession number (16S rRNA gene): AY487928.

Genus III. **Acetivibrio** Patel, Khan, Agnew and Colvin 1980, 184[VP] (emend. Robinson and Ritchie 1981, 335; emend. Khan, Meek, Sowden and Colvin 1984, 420; emend. Murray 1986, 316)

FRED A. RAINEY

A.cet.i.vib.ri′o. L. n. *acetum* vinegar; N.L. masc. n. *Vibrio* genus of bacteria. N.L. masc. n. *Acetivibrio* vinegar (acetic) vibrio.

Gram-stain-negative rods or slightly curved rods. Spores are not formed. Obligate anaerobe. Chemoorganotroph. End products of fermentation include acetic acid, ethanol, H_2, and CO_2. Member of the family *Ruminococcaceae* on the basis of 16S rRNA gene sequence comparisons.

DNA G+C content (mol%): 38–44 (T_m/HPLC).

Type species: **Acetivibrio cellulolyticus** Patel, Khan, Agnew and Colvin 1980, 184[VP].

List of species of the genus *Acetivibrio*

1. **Acetivibrio cellulolyticus** Patel, Khan, Agnew and Colvin 1980, 184[VP]

cell.u.lo.ly′ti.cus. N.L. neut. n. *cellulosum* cellulose; N.L. adj *lyticus* able to dissolve; N.L. adj. *cellulolyticus* cellulose-dissolving.

Cells are straight to slightly curved rods, 4.0–10.0 μm in length and 0.5–0.8 μm wide. Cells occur in chains up to 40 μm in length. Gram-stain-negative. Motile by means of a single flagellum attached to the cell about one-third of the distance from pole to pole. Exhibits tumbling motility. Colonies are 1–2 mm in diameter, cream colored, round, and raised with an undulate margin. A clear zone indicating cellulose digestion extends from the periphery of the colony on cellulose agar. Obligately anaerobic. Cellulose, cellobiose, and salicin are utilized. Temperature range for growth is 20–40°C; optimum is 35°C. pH range for growth is 6.5–7.7, optimum is pH 7.0. The type strain was isolated from a methanogenic culture from sewage sludge.

DNA G+C content (mol%): 38 (T_m).

Type strain: CD2, NRC 2248, ATCC 33288.

GenBank accession number (16S rRNA gene): L35516.

2. **Acetivibrio cellulosolvens** Khan, Meek, Sowden and Colvin 1984, 420[VP]

cell.u.lo.sol′vens. L. v. *solvere* to dissolve; N.L. adj. *cellulosolvens* cellulose dissolving, indicating the ability of the organism to ferment cellulosic substrates.

Cells are rods, 2.0–3.0 μm in length and 0.3–0.7 μm wide, Gram-stain-negative, and occur singly, in pairs, or in chains of three or more cells. No spores are detected. Cultures grown on cellobiose (72 h) or cellulose medium (2 weeks) fail to survive heating at 80°C for 10 min or exposure to 50% ethanol. Cells are nonmotile. Electron micrographs show that one end of each cell is round and uniform and the other end is rough and uneven. Cells have an outer membrane. Colonies on cellulose agar are about 2–3 mm in diameter, entire, raised, and cream colored, showing no clear zones surrounding the colonies. Obligately anaerobic. Growth in cellulose-containing broth under an N_2/CO_2 gas phase has little or no color. The optimum growth temperature is 35–37°C; no growth occurs at 20 or 45°C. The optimum pH is 6.5–7.5; no growth occurs at pH 6.0 or 8.0. Growth is supported only on cellulose, cellobiose, esculin, and salicin. Yeast extract and rumen fluid are not required for growth on these carbon sources. Bile (2% Oxgall; Difco) inhibits growth. No growth is observed on adonitol, amygdalin, arabinose, Casamino acids, casein, chopped meat broth, dulcitol, erythritol, fructose, galactose, glucose, glycerol, glycogen, inositol, inulin, lactate, lactose, maltose, mannitol, mannose, melezitose, melibiose, milk, peptone, pyruvate, raffinose, rhamnose, ribose, sorbitol, sorbose, starch, sucrose, trehalose, urea, xylose, or yeast extract. Esculin is hydrolyzed to give a positive reaction with ferric ammonium citrate. Sulfate is not reduced to sulfite, gelatin is not hydrolyzed, and catalase is not produced. Indole and nitrate reduction tests are negative. The metabolic products of cultures in basal medium containing cellobiose incubated for 4–8 d under 100% N_2 or 80% N_2/20% CO_2 are acetic acid (18 mM), ethanol (3 mM), CO_2, and H_2. Trace amounts of lactic acid also are produced. In cellulose-containing medium, glucose (8 mM) is also produced. The type strain was isolated from domestic sewage sludge.

DNA G+C content (mol%): 41 (T_m).

Type strain: BAS, NRC 2936, ATCC 35928.

GenBank accession number (16S rRNA gene): L35515.

3. **Acetivibrio ethanolgignens** Robinson and Ritchie 1981, 335[VP]

e.tha.nol.gig′nens. N.L. n. *ethanol* ethanol; L. part. adj. *gignens* giving birth to, producing; N.L. part. adj. *ethanolgignens* producing ethanol.

Cells are rod-shaped, 1.5–2.5 μm in length and 0.5–0.9 μm wide, Gram-stain-negative, non-spore-forming, motile, slightly curved rods, occurring singly, in pairs, and often in short chains. Flagella (10–15 per cell) are arranged linearly on the concave surface of each cell. Colonies are 0.5–1.5 mm in diameter, circular, convex, smooth, and translucent. Obligately anaerobic. Optimal temperature for growth is 37°C; no growth occurs at 15 or 45°C. Growth is enhanced by the presence of fermentable carbohydrates. Fructose, galactose, glucose, lactose, maltose, mannitol, mannose, and pyruvate are fermented. Salicin and starch are weakly fermented. Arabinose, cellobiose, esculin, glycerol, inositol, inulin, lactate, raffinose, rhamnose, ribose, sorbitol, sucrose, trehalose, and xylose are not fermented. Acetic acid and ethanol are major products of fermentation. Gelatin is hydrolyzed. Nitrate is reduced to nitrite. Cellulose is not degraded. Bile inhibits growth. The type strain was isolated from the colon of a pig with dysentery.

DNA G+C content (mol%): 40 (T_m).

Type strain: 77-6, ATCC 33324.

GenBank accession number (16S rRNA gene): not determined.

4. **Acetivibrio multivorans** Tanaka, Nakamura and Mikami 1992, 191[VP] (Effective publication: Tanaka, Nakamura and Mikami 1991, 123.)

mul.ti.vo′rans. L. adj. *multus* many, numerous; L. v. *voro* to devour, eat greedily; N.L. part. adj. *multivorans* devouring many kinds of substrates.

Cells are straight to curved rods, 0.8–3.5 μm in length and 0.5–0.9 μm wide, occurring singly or in pairs. Motile. Gram-stain-negative. Surface colonies are colorless, translucent, up to 2 mm in diameter, and circular in form with convex elevation and entire margins. Strictly anaerobic. Temperature range for growth is 20–35°C and optimal growth occurs at 30°C; no growth is observed at 15 or 40°C. pH range for growth is 7.2–8.6 and optimal growth occurs around pH 7.8; no growth is observed at pH 6.8. Cinnamate, crotonate, pyruvate, sucrose, trehalose, and other carbohydrates are utilized for growth. Cellulose is not utilized. Cinnamate is fermented to 3-phenylpropionate, benzoate, and acetate. Crotonate is fermented to butyrate and acetate. Pyruvate is fermented to lactate and acetate. Carbohydrates are fermented to ethanol, formate, and acetate. A trace amount of hydrogen is produced during growth on cinnamate. Nitrate, sulfate, and other sulfur compounds are not utilized as electron acceptors. The type strain was isolated from sludge from an oil refinery wastewater treatment facility.

DNA G+C content (mol%): 44 (HPLC).

Type strain: PeC1, DSM 6139.

GenBank accession number (16S rRNA gene): Not determined.

Genus IV. **Anaerofilum** Zellner, Stackebrandt, Nagel, Messner, Weiss and Winter 1996, 874[VP]

FRED A. RAINEY

An.ae.ro′fi.lum. Gr. pref. *an* not; Gr. n. *aer* air; Gr. adj. *anaero* absence of air, referring to the anaerobic mode of living; L. neut. n. *filum* thread, referring to the very thin rod-shaped cells; N.L. neut. n. *Anaerofilum* anaerobic thin rods.

Thin, straight rods, 0.2–0.6 μm in diameter and 3–6 μm in length, occurring singly or in pairs. Filaments up to 30 μm observed occasionally. Spores are not formed. Cell wall contains L-glycine, D-glutamate, L-lysine, and D-alanine. Serine forms linkage in interpeptide bridge. Sphaeroplasts are formed in the stationary phase. Some strains are motile. Obligate anaerobe. Strains of this organism are catalase negative. Mesophilic with temperature range for growth between 18 and 44°C. Chemoorganotrophic. Ferments mono- and disaccharides. Major end products of fermentation are lactate, acetate, ethanol, formate, 2,3-butanediol, and carbon dioxide. Sulfate is not reduced. Hydrogen and hydrogen sulfide are not produced. Found in anaerobic sewage sludge of municipal and industrial wastewater treatment plants.

DNA G+C content (mol%): 54–55 (T_m/HPLC).

Type species: **Anaerofilum pentosovorans** Zellner, Stackebrandt, Nagel, Messner, Weiss and Winter 1996, 874[VP].

Further descriptive information

The genus *Anaerofilum* contains two species, *Anaerofilum pentosovorans* and *Anaerofilum agile*, which are represented by the single strains Fae[T] and F[T], respectively (Zellner et al., 1996). These organisms are strict anaerobes and only grow in reduced medium. Both species require 0.2% yeast extract for growth and this cannot be replaced by trypticase peptone. The species differ in their response to temperature with *Anaerofilum pentosovorans* having a broad optimal temperature for growth of 25–40°C; *Anaerofilum agile* grows optimally at 37°C. No growth is observed for either species below 15°C or above 45°C. Growth of both strains occurs in the pH range 6.5–8.0 with optimal growth at pH 7.0–7.4. Cellobiose, fructose, galactose, glucose, maltose, ribose, trehalose, and xylose act as growth substrates for both strains. Additionally, *Anaerofilum pentosovorans* can utilize arabinose, mannose, and sorbose, whereas *Anaerofilum agile* utilizes D-salicin and D-sorbitol. Lactose, melibiose, raffinose, rhamnose, sucrose, methanol, ethanol, 1-propanol, citrate, fumarate, malate, succinate, L-lactate, cellulose, starch, xylan, pectin A, L-glutamate, L-tryptophan, L-phenylalanine, L-alanine, and L-glycine are not utilized by either species. Growth of *Anaerofilum agile* on all substrates is much less than that of *Anaerofilum pentosovorans*. *Anaerofilum agile* grows very weakly on L-valine and L-isoleucine. The main products of glucose fermentation by *Anaerofilum agile* are L-(+)-lactate, acetate, formate, ethanol, and carbon dioxide. *Anaerofilum pentosovorans* ferments glucose to lactic acid [both isomers, although the L-(+)-lactate isomer is dominant], ethanol, acetate, formate, carbon dioxide, and a small amount of 2,3-butanediol. Hydrogen and hydrogen sulfide are not produced by either species. Growth of *Anaerofilum agile* is not inhibited when hydrogen is added to the gas phase of the culture vessel.

Enrichment, isolation and growth conditions

Anaerofilum pentosovorans was isolated from sludge of a Biohochreaktor at the Hoechst Chemical Company (Kelsterbach, Frankfurt, Germany), containing wastewater. The sludge was serially diluted in 120 ml serum vials containing 20 ml pre-reduced medium with a headspace of 80% N_2 and 20% CO_2. In the only paper describing the isolation, characterization, and growth of these strains the composition of the pre-reduced medium was not given (Zellner et al., 1996). *Anaerofilum agile* was isolated from an anaerobic biofilm in a whey treatment reactor to which anaerobic sewage sludge from a municipal sewage treatment plant had been added.

The organism can be grown on DSMZ medium 119 (*Methanobacterium* medium; http://www.dsmz.de/microorganisms/html/media/medium000119.html).

To obtain single colonies, cultures can be streaked on agar plates of the pre-reduced medium containing 0.5% glucose and 2% agar (Oxoid). Colonies take 14 d to develop and are white and 1–2 mm in diameter with lobate margins and raised centers.

Taxonomic comments

In the initial study, this genus was shown to represent a distinct lineage close to *Fusobacterium prausnitzii* [later reclassified as *Faecalibacterium prausnitzii* (Duncan et al., 2002)] and *Ruminococcus flavefaciens* (Zellner et al., 1996). With the description in recent years of a number of new genera in this area of the *Firmicutes*, the two species of the genus *Anaerofilum* can now be considered to cluster within the family *Ruminococcaceae* with the genera *Ruminococcus*, *Acetivibrio*, *Acetanaerobacterium*, *Anaerotruncus*, *Ethanoligenens*, *Faecalibacterium*, *Fastidiosipila*, *Oscillospira*, *Papillibacter*, *Sporobacter*, and *Subdoligranulum* (see Figure 185). The two species share 98.7% 16S rRNA gene sequence similarity with each other and sequence similarities in the range 88.0–91.4% to type strains of the type species of the genera *Ruminococcus*, *Acetanaerobacterium*, *Anaerotruncus*, *Ethanoligenens*, *Faecalibacterium*, and *Subdoligranulum*, as well as the misclassified species *Clostridium methylopentosum* and *Eubacterium siraeum*. Although the highest 16S rRNA gene sequence similarities are to *Subdoligranulum variabile* and *Faecalibacterium prausnitzii*, these relationships are not supported by the bootstrap values (see Figure 185). Comparison of the *Anaerofilum pentosovorans* and *Anaerofilum agile* 16S rRNA gene sequences with the current GenBank database of 16S rRNA gene sequences shows that no further sequences of isolates of this genus are currently available. However, there are a number of environmental clone sequences (DQ011251, DQ011252, and AF150697) that cluster with these species at the 97–98% similarity level. These clone sequences originate from stratum sewage and "river snow" aggregates.

Although strains Fae[T] and F[T] share 98.7% 16S rRNA gene sequence similarity, they have been shown by DNA–DNA hybridization studies to share only 48% relatedness and, thus, represent distinct species (Zellner et al., 1996).

Differentiation of the genus *Anaerofilum* from other genera

Strains of the genus *Anaerofilum* can be differentiated from the most closely related genera (*Ruminococcus*, *Acetanaerobacterium*, *Anaerotruncus*, *Ethanoligenens*, *Faecalibacterium*, and *Subdoligranulum*)

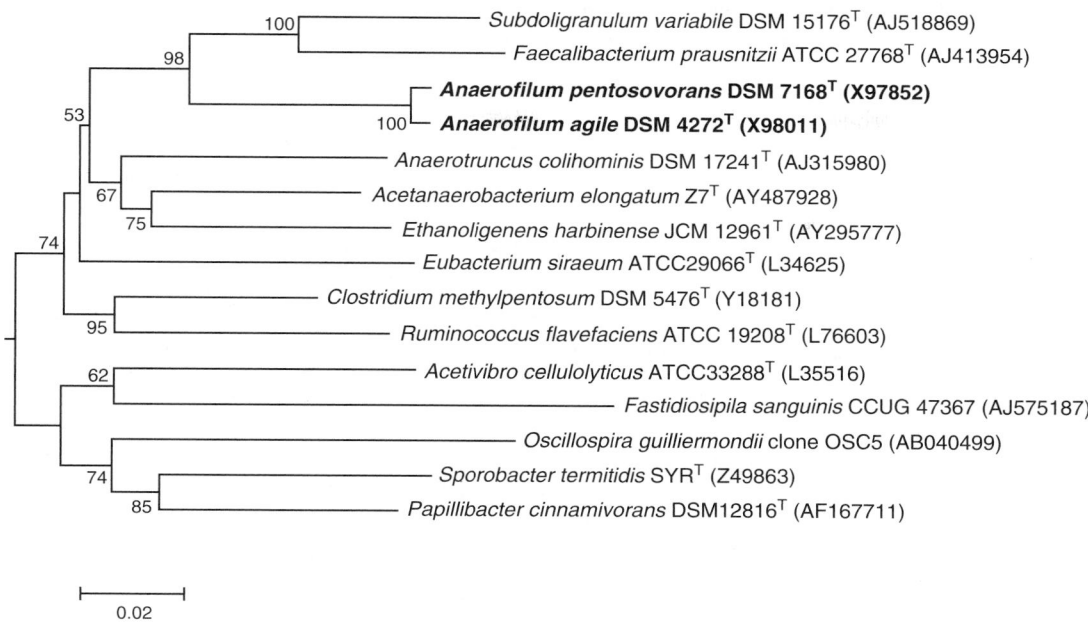

FIGURE 185. 16S rRNA gene sequence based-phylogeny showing the position of the two species of the genus *Anaerofilum* within the radiation of the family *Ruminococcaceae*. The tree was reconstructed from distance matrices using the neighbor-joining method. The genus *Oscillospira*, for which no type material exists, is represented by a clone sequence from an organism with the morphological characteristics of the organism originally described as *Oscillospira guilliermondii*. Bar = 2 inferred nucleotide changes per 100 nucleotides.

on the basis of cell morphology. The *Anaerofilum* strains are thin long rods, whereas other species of related genera are represented by rods, cocci, and pleomorphic forms. In addition, the wide range of end products of fermentation produced by *Anaerofilum* strains differentiates this genus from related taxa. The end products of fermentation are lactate, acetate, formate, ethanol, and 2,3-butanediol. Related genera produce different and a less diverse range of end products of fermentation.

List of the species of the genus *Anaerofilum*

1. **Anaerofilum pentosovorans** Zellner, Stackebrandt, Nagel, Messner, Weiss and Winter 1996, 874[VP]

pen.to.so′vo.rans. N.L. neut. n. *pentosum* sugar with five carbon atoms; L. v. *vorare* to eat; L. adj. *pentosovorans* fermenting pentose.

Forms thin, long, rod-shaped cells, 3–6 μm in length and 0.2–0.6 μm in diameter. Occasionally motile. Sphaeroplasts form in stationary phase. Colonies are white and translucent with lobate margins and raised centers on glucose-containing media. The type strain was isolated from an industrial wastewater bioreactor.

DNA G+C content (mol%) of the type strain: 55.0 (HPLC).

Type strain: Fae, DSM 7168.

GenBank accession number (16S rRNA gene): X97852.

2. **Anaerofilum agile** Zellner, Stackebrandt, Nagel, Messner, Weiss and Winter 1996, 874[VP]

a′gi.le. L. neut. adj. *agile* rapidly moving.

Forms thin, long, rod-shaped cells 3–6 μm in length and 0.2–0.6 μm in diameter. Motile by means of peritrichous flagella. Sphaeroplasts form in stationary phase. Colonies are white, circular, slightly convex, and translucent on glucose-containing media. The type strain was isolated from an industrial wastewater bioreactor.

DNA G+C content (mol%): 54.5 (T_m).

Type strain: F, DSM 4272.

GenBank accession number (16S rRNA gene): X98011.

Genus V. **Anaerotruncus** Lawson, Song, Liu, Molitoris, Vaisanen, Collins and Finegold 2004, 415[VP]

Paul A. Lawson

An.ae.ro.trun′cus. Gr. pref. *an* without; Gr. masc. n. *aer* air; L. masc. n. *truncus* stick; N.L. masc. n. *Anaerotruncus* a stick that lives without air.

Thin rod-shaped cells. Gram positive and sporeforming. **Strictly anaerobic and catalase negative.** Glucose is fermented. **End products of metabolism from peptone/yeast extract broth are** acetic and butyric acids. Gelatin is not hydrolyzed. Urease negative. Nitrate is not reduced to nitrite. Indole positive. Based on 16S rRNA sequence and phylogenetic analysis, this genus belongs to the *Clostridium leptum* rRNA super cluster (Cluster IV). Isolated from human feces and blood.

DNA G+Ccontent (mol%): 53–54.

Type species: **Anaerotruncus colihominis** Lawson, Song, Liu, Molitoris, Vaisanen, Collins and Finegold 2004, 415[VP].

Further descriptive information

The genus contains only one species, *Anaerotruncus colihominis* and, therefore, the characteristics provided below refer to this species. Cells are Gram positive, thin rods of approximately 0.5 × 2–5 μm. After 48 h of anaerobic incubation at 37°C under an N_2/CO_2 (80:20, v/v) gas phase, colonies are 2–3 mm in diameter, gray, entire-edged, irregularly shaped, low pyramidal in profile, and translucent. The authors of the original description (Lawson et al., 2004) did not observe spores, but Lau et al. (2006) observed that oval, terminal spores are occasionally formed. The pH range for growth is 5.5–11, and the growth temperature range is 36–40°C. Esculin, gelatin, and urea are not hydrolyzed. It is able to grow in peptone–yeast broth supplemented with 1% glucose, fructose, mannose, or cellobiose, but does not grow in peptone–yeast broth alone or peptone–yeast broth supplemented with arabinose, inositol, lactose, maltose, mannitol, melezitose, melibiose, raffinose, rhamnose, ribose, salicin, sorbitol, starch, or xylose. With the Biolog system, *N*-acetyl-D-glucosamine, *N*-acetyl-β-D-mannosamine, arbutin, D-cellobiose, dextrin, D-fructose, D-galactose, D-galacturonic acid, α-D-glucose, maltose, maltotriose, D-mannose, methyl-3-D-glucose, methyl-β-D-glucose, methyl-β-D-galactoside, methyl-β-D-glucoside, palatinose, D-trehalose, and turanose are also used. α-Ketobutyric acid, α-ketovaleric acid, L-malic acid, pyruvic acid, and pyruvic acid methyl ester are also used. Serine, L-valine, 2′-deoxyadenosine, inosine, thymidine, and uridine are used for growth. The organism is isolated from human clinical specimens (Lau et al., 2006; Lawson et al., 2004).

Isolation procedures

Anaerotruncus colihominis was isolated from both human fecal samples and a human blood sample. Fecal samples were homogenized by using a sterile stainless steel blender with one to three volumes of peptone (0.05 %) added as diluents. An aliquot of the specimen was used to make serial tenfold dilutions in pre-reduced, anaerobically sterilized (PRAS) dilution blanks. A 100 μl aliquot of each dilution was plated onto Brucella blood agar (Anaerobe Systems) and incubated anaerobically at 37°C under an N_2/CO_2 (80:20, v/v) gas phase (Lawson et al., 2004). The strain from blood was recovered using the BACTEC 9240 blood culture system (Becton Dickinson, MD, USA) and grown on Brucella blood agar (Lau et al., 2006).

Maintenance procedures

Strains grow well on *Brucella* blood agar base supplemented with 5% horse blood at 37°C under anaerobic conditions. For long term preservation, strains can be maintained in medium containing 15–20% glycerol at –70°C or lyophilized

Taxonomic comments

The genus *Anaerotruncus* was created to accommodate a phylogenetically distinct Gram positive, anaerobic, rod-shaped organism originating from human feces (Lawson et al., 2004). The genus is monospecific and belongs to the *Firmicutes* and, in particular, is a member within a suprageneric grouping (rRNA cluster IV, (Collins et al., 1994) that includes a diverse assortment of organisms such as *Acetanaerobacterium elongatum*, *Anaerofilum* species, *Clostridium* species, *Ethanoligenens harbinense*, *Eubacterium siraeum*, *Faecalibacterium prausnitzii*, *Ruminococcus* species, and *Subdoligranulum variabile*. In this volume, it is classified with the "*Ruminococcaceae*" (Ludwig et al., 2009). *Anaerotruncus colihominis* forms a distinct lineage within this family and does not display a particularly close affinity with any of the aforementioned taxa.

Differentiation of the genus *Anaerotruncus* from other genera

In addition to differentiation by 16S rRNA gene sequence analysis, *Anaerotruncus colihominis* can be readily distinguished from its closest phylogenetic relatives *Acetanaerobacterium elongatum*, *Anaerofilum* species, *Clostridium* species, *Ethanoligenens harbinense*, *Eubacterium siraeum*, *Faecalibacterium prausnitzii*, *Ruminococcus* species, and *Subdoligranulum variabile* using a combination of morphology, biochemical, and chemotaxonomic criteria (see Table 193).

List of species of the genus *Anaerotruncus*

1. **Anaerotruncus colihominis** Lawson, Song, Liu, Molitoris, Vaisanen, Collins and Finegold 2004, 415[VP]

co.li.ho′mi.nis. L. n. *colum* colon; L. gen. n. *hominis* of humans; N.L. gen. n. *colihominis* of the gut of humans.

Cells stain Gram positive and are thin rods approximately 0.5 × 2–5 μm in size, that form oval, terminal spores. The cells are catalase negative; nitrate is not reduced to nitrite, but indole is produced. The end products of metabolism from peptone–yeast broth are acetic and butyric acids. Other characteristics are as given for the genus with the following information. With the API ZYM and Rapid ID 32A systems, acid phosphatase and indole activities are detected, but *N*-acetyl-β-glucosaminidase, alanine arylamidase, alkaline phosphatase, arginine arylamidase, arginine dehydrolase, α-arabinosidase, ester lipase C8, α-galactosidase, β-galactosidase, α-glucosidase, β-glucuronidase, glutamyl glutamic acid arylamidase, α-mannosidase, α-fucosidase, chymotrypsin, alanine arylamidase, leucyl glycine arylamidase, phosphoamidase, trypsin, cystine arylamidase, esterase C4, β-galactosidase-6-phosphate, glutamic acid decarboxylase, glycine arylamidase, histidine arylamidase, lipase C14, leucine arylamidase, phenylalanine arylamidase, proline arylamidase, pyroglutamic acid arylamidase, serine arylamidase, tyrosine arylamidase, and valine arylamidase activities are not detected. Cells are sensitive to vancomycin (5 mg) and kanamycin (1000 mg), but resistant to colistin sulfate (10 mg) identification discs. Isolated from human feces and blood.

DNA G+C content (mol%): 53.0–54.0 (T_m and HPLC).

Type strain: WAL 14565, CCUG 45055, CIP 107754, DSM 17241.

GenBank accession number (16S rRNA gene): AJ315980.

TABLE 193. Characteristics differentiating the genus *Anaerotruncus* from phylogenetically closely related taxa[a,b]

Characteristic	*Anaerotruncus*	*Acetanaerobacterium*	*Anaerofilum*	*Clostridium cellulosi*	*Clostridium leptum*	*Clostridium sporosphaeroides*	*Ethanoligenens*	*Eubacterium siraeum*	*Faecalibacterium*	*Ruminococcus*	*Subdoligranulum*
Cellular morphology	Thin rods	Thin rods	Thin long rods	Rods	Rods	Rods	Rods	Rods	Rods	Coccoid	Coccoid-droplet, highly pleomorphic
Spores	+	−	−	+	+	+	−	−	−	−	−
Motility	−	+	d	+	−	−	+	d	−	−	−
End products of metabolism	A, B	A, E	L, A, E, F, 2,3-butanediol	A, E[c]	A, e	A, B, p	A, E	A, E, l, b, s[d]	B, L, F,s,(p)	A,E,L,S,f, various (depending on species)[e]	B, L, a, s
Hydrolysis of esculin	−	+	d	+	d	−	+	d	+	d	+
Hydrolysis of starch	−	−	−	+	−	−	w	d	d	d	−
Cellulose degradation	−	−	−	+	−	−	ND	−	−	d	−
Growth at 60 °C	−	−	−	+	−	−	−	−	−	−	−
DNA G + C content (mol%)	53–54	49–50	54–55	35	51–52	27	48–49	45	47–57	39–46	52

[a]Symbols: +, >85% positive; d, different strains give different reactions −, 0–15% positive; w, weak reaction; ND, not determined; A, acetate; B, butyrate; E, ethanol; F, formate; L, lactate; P, propionate; S, succinate; minor products are indicated by lower-case letters; products in parentheses may or may not be formed.

[b]Data from *Acetanaerobacterium* (Chen and Dong, 2004), *Anaerophilum* (Zellner et al., 1996), *Anaerotruncus* (Lau et al., 2006; Lawson et al., 2004), *Clostridium cellulosi* (Yanling et al., 1991), *Clostridium leptum* (Cato et al., 1986), *Clostridium sporosphaeroides* (Cato et al., 1986), *Ethanoligenens* (Moore and Moore, 1985), *Eubacterium siraeum* (Moore and Moore, 1985), *Faecalibacterium* (Duncan et al., 2002), *Ruminococcus* (Bryant, 1985), and *Subdoligranulum* (Holmstrom et al., 2004).

[c]From cellulose fermentation.

[d]PY-cellobiose broth.

[e]Genus *Ruminococcus* is phylogenetically heterogenous; data refer to *Ruminococcus albus*, *Ruminococcus bromii*, *Ruminococcus callidus*, and *Ruminococcus flavescens*.

Genus VI. **Faecalibacterium** Duncan, Hold, Harmsen, Stewart and Flint 2002, 2145[VP]

SYLVIA H. DUNCAN AND HARRY J. FLINT

Fae.ca.li.bac.te′ri.um. L. adj. *faecalis* pertaining to feces; Gr. dim. n. *bakterion* a small rod; N.L. neut. n. *Faecalibacterium* rod from feces, as this bacterium is abundant in feces, with the colon its presumed habitat.

Variable length, straight, rod-shaped cells (0.5–0.8 × 2.0–14.0μm) mainly occur singly and sometimes seen as hourglass-shaped. Nonsporulating. Usually gives a Gram-negative staining reaction. **Cells are nonmotile.**

Strictly anaerobic. Chemo-organotrophic. Uses the carbohydrates fructose, glucose, and starch as carbon and energy sources. Strains differ in their ability to ferment cellobiose, maltose, and melezitose and fail to ferment raffinose, rhamnose, and ribose. Acetate is strongly stimulatory during growth on carbohydrates, when **net acetate utilization** is detected. Hydrogen production not detected. Grows in an anaerobically prepared medium containing volatile fatty acids, yeast extract, casein hydrolysate, inorganic salts, hemin, glucose, and vitamins at 37°C under a gas atmosphere of 100% CO_2. **Produces butyrate, formate, and D-lactate from fermentation of glucose and acetate.** Indigenous to mammalian intestinal tract. In M2GSC roll tubes colonies are 1–2 mm in diameter, circular, and entire, translucent to transparent, colorless to pale white. Catalase-negative. Isolated from human or animal feces.

DNA G+C content (mol%): 47–57 (T_m).

Type species: **Faecalibacterium prausnitzii** (Hauduroy, Ehringer, Urbain, Guillot and Magrou 1937) Duncan, Hold, Harmsen, Stewart and Flint 2002, 2145[VP] (*Bacteroides praussnitzii* (*sic*) Hauduroy, Ehringer, Urbain, Guillot and Magrou 1937, 68; *Fusobacterium prausnitzii* Cato, Salmon and Moore 1974, 45).

Further descriptive information

Faecalibacterium prausnitzii cells usually give Gram-negative staining; they were formerly classified within the genus *Fusobacterium*. *Faecalibacterium prausnitzii* strains are not phylogenetically related to true *Fusobacterium* species, however, and form a distinct group within the clostridial cluster IV (also referred to as the *Clostridium leptum* cluster) (Collins et al., 1994) based on 16S rRNA sequences (Figure 186). According to the 16S rRNA phylogenetic analysis presented in the roadmap to this volume (Figure 6), the genus *Faecalibacterium* is a member of the family *Ruminococcaceae*, order *Clostridiales*, class *Clostridia* in the phylum *Firmicutes*. *Faecalibacterium prausnitzii*

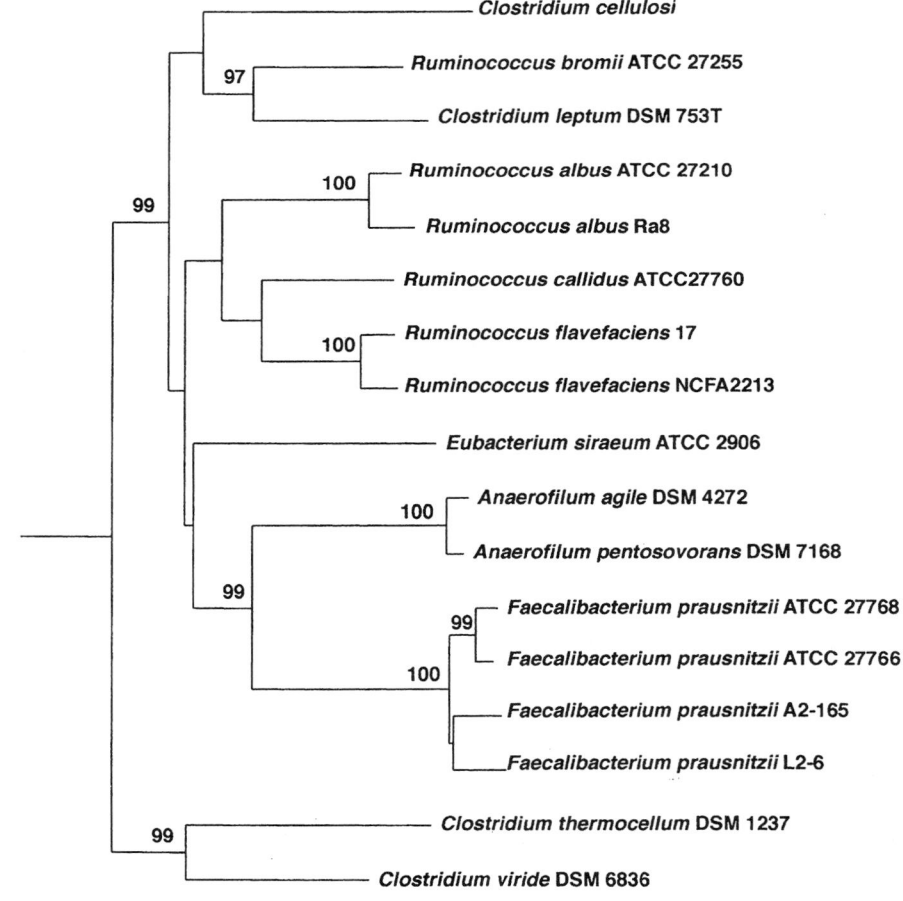

FIGURE 186. Phylogenetic relatedness of *Faecalibacterium prausnitzii* to members of the closely related cluster IV (*Clostridium leptum* group) based on a comparison of approximately 1340 nt.

comprises approximately 13% of the fecal microbiota of healthy adults when analyzed by fluorescent *in situ* hybridization (Lay et al., 2005).

Enrichment and isolation procedures

Faecalibacterium can be isolated by screening for butyrate-forming bacteria from intestinal samples. *Faecalibacterium prausnitzii* strains (A2-165 and L2-6) were obtained from colonies formed in roll tubes containing a rumen fluid-based M2GSC agar medium that were inoculated with dilutions of infant or adult human fecal homogenates maintained anaerobically. Butyrate-producing isolates belonging to ribotype 7 (i.e., those producing fragments of 797, 464, 168, and 75 bp upon *Alu*I cleavage of 16S

rRNA sequences amplified with FD1 and RP2 primers) were found to correspond to *Faecalibacterium prausnitzii* (Barcenilla et al., 2000).

Maintenance procedures

Faecalibacterium prausnitzii cultures to which the cryoprotectant glycerol (final concentration, 20 %) has been added can be stored at −20°C for several years.

Differentiation of the genus *Faecalibacterium* from other closely related genera

Faecalibacterium species can be differentiated from related taxa by 16S rRNA sequence analysis (Duncan et al., 2002).

List of species of the genus *Faecalibacterium*

1. **Faecalibacterium prausnitzii** (Hauduroy, Ehringer, Urbain, Guillot and Magrou 1937) Duncan, Hold, Harmsen, Stewart and Flint 2002, 2145[VP] (*Bacteroides praussnitzii* (*sic*) Hauduroy, Ehringer, Urbain, Guillot and Magrou 1937, 68; *Fusobacterium prausnitzii* Cato, Salmon and Moore 1974, 45)

praus.nit′zi.i. N.L. gen. n. *prausnitzii* of Prausnitz; named for C. Prausnitz, the bacteriologist who first isolated this organism.

Four strains have been characterized including the type strain ATCC 27768[T]. All strains are nonmotile straight rods 0.5–0.8 μm in diameter and 2.0–14.0 μm in length (Figure 187). Cells give a Gram-negative staining reaction, but, based on phylogenetic analyses, the species clusters among *Firmicutes* in clostridial cluster IV (Figure 186).

Obligately anaerobic, having a fermentative type of metabolism. Various carbohydrates are growth substrates

(Table 194). Products of glucose plus acetate metabolism are butyrate, formate, and D-lactate (Table 195). Catalase-negative.

In M2GSC roll tubes colonies are 1–2 mm in diameter, circular, and entire, translucent to transparent, colorless to pale white. Isolated from animal and human fecal samples.

DNA G+C content (mol%): 47–57 (T_m).

Type strain: ATCC 27768[T].

GenBank accession numbers (16S rRNA gene): L2-6, A2-165, ATCC 27766, and ATCC 27768[T] are AJ270470, AJ270469, M58682, and AJ413954, respectively.

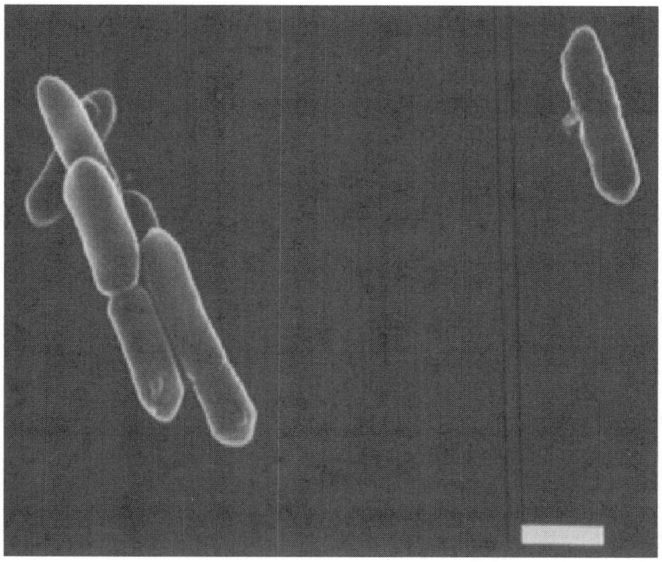

FIGURE 187. Scanning electron micrograph of *Faecalibacterium prausnitzii* strain A2-165. Bar = 1 μm.

TABLE 194. Phenotypic properties of *Faecalibacterium prausznitzii* strains[a]

Carbohydrate substrate	A2-165	L2-6	27766	27768[T]
Arabinose	−	−	−	−
Cellobiose	+	−	−	w
Fructose	+	+	+	+
Maltose	+	+	−	w
Mannitol	−	−	−	ND
Melezitose	−	−	+	−
Melibiose	−	−	−	−
Raffinose	−	−	−	−
Rhamnose	−	−	−	−
Ribose	−	−	−	−
Sucrose	−	w	−	w
Trehalose	−	−	−	−/w
Xylose	−	−	−	−
Starch, soluble	+	+	+	w

[a]Symbols: +, >85% positive; −, 0–15% positive; w, weak reaction; ND, not determined.

TABLE 195. Short-chain fatty acid formed and utilized (mM) by *Faecalibacterium prausnitzii* strains on M2G medium (Duncan et al., 2002)

Strain	A2-165	L2-6	27766	27768[T]
Formate	17.3	7.6	4.7	10.3
Acetate	−5.3	−12.3	−10.2	−9.7
Butyrate	13.8	18.6	13.2	18.6
D-Lactate	1.7	5.3	2.6	5.5

Genus VII. **Fastidiosipila** Falsen, Collins, Welinder-Olsson, Song, Finegold and Lawson 2005, 856[VP]

Paul A. Lawson

Fas.ti.di.o.si.pi'la. L. adj. *fastidiosus* fastidious; L. fem n. *pila* ball; N.L. fem. n. *fastidiosipila* a fastidious ball, because the organisms are difficult to grow.

Cells are cocci (approx. 0.5 μm diameter) that stain Gram-stain-positive and do not form spores. **Anaerobic but can grow in the presence of 2% and 6% O₂**. **Catalase-negative. Carbohydrates are not fermented.** Small amounts of acetic and butyric acids are detected in Fastidious Anaerobe Broth with meat granules. **Lecithinase-, lipase-, and urease-negative. Non-cellulolytic.** Cells are indole-negative and do not reduce nitrate to nitrite. Based on 16S rRNA gene sequence and phylogenetic analyses, this genus is loosely associated with *Clostridium* rRNA cluster III of the *Firmicutes*. Isolated from human blood.

DNA G+C content (mol%): 32.9.

Type species: **Fastidiosipila sanguinis** Falsen, Collins, Welinder-Olsson, Song, Finegold and Lawson 2005, 856[VP].

Further descriptive information

The genus contains only one species, *Fastidiosipila sanguinis,* and therefore the characteristics provided below refer to this species. After 48 h of anaerobic incubation at 37°C, colonies on chocolate agar and anaerobic blood agar are small, pinpoint, gray in color and non-hemolytic. Grows poorly on Fastidious Anaerobe Agar (Oxoid) and Fastidious Anaerobe Broth with meat granules. In the latter, small amounts of acetic and butyric acids are detected. Lipase, lecithinase, and urease are not produced. Gelatin is not hydrolyzed. In the API Rapid ID 32S system, activity is detected for α- and β-galactosidase; β-glucosidase; alanine, phenylalanine, and proline arylamidase; and β-mannosidase. No activity is detected for alkaline phosphatase, arginine dihydrolase, glycyl tryptophan arylamidase, pyroglutamic acid arylamidase, or urease; β-glucuronidase activity may or may not be detected. Hippurate is hydrolyzed and acetoin is not produced. In the commercially available API Rapid ID 32A kit, positive reactions are obtained for α- and β-galactosidase, α- and β-glucosidase, N-acetyl-β-glucosaminidase, alanine arylamidase (weak), arginine arylamidase, proline arylamidase, leucine arylamidase (weak), or α-fucosidase. Reactions for serine arylamidase, histi-dine arylamidase, glycine arylamidase, and tyrosine aryla-midase are either weakly positive or negative. Activity is not detected for alkaline phosphatase, α-arabinosidase, arginine dihydrolase, β-galactosidase-6-phosphate, glutamic acid decar-boxylase, glutamyl glutamic acid arylamidase, β-glucuronidase, leucyl glycine arylamidase, phenylalanine arylamidase, pyro-glutamic acid arylamidase, or urease. Sensitive to kanamycin (1000 mg) and vancomycin (5 mg) but resistant to colistin sulfate (10 mg) and metronidazole (5 mg) on the basis of disk diffusion testing.

Isolation procedures

Fastidiosipila sanguinis was recovered from the blood of two elderly men in Sweden. The strains were cultivated on chocolate agar (Difco) and anaerobic blood agar (Oxoid) and incubated anaerobically at 37°C under a gas phase comprising N₂/H₂/CO₂ (86:7:7% v/v). There is no selection or enrichment medium for this species. Aerotolerance was determined by incubating the organism on Brucella agar plates (Difco) with and without 5% (v/v) laked sheep blood in an atmosphere containing either 2% or 6% O₂, as described by Wexler et al. (1996).

Maintenance procedures

Grows anaerobically on chocolate agar (Difco), blood-based agars and fastidious anaerobe broth (Oxoid) with meat gran-ules at 37°C. For long-term preservation, strains are maintained in media containing 15–20% glycerol at −70°C or lyophilized.

Taxonomic comments

The genus *Fastidiosipila* was proposed to accommodate a phylo-genetically distinct nonmotile, Gram-stain-positive, coccus orig-inating from human blood (Falsen et al., 2005). The genus is monospecific, and phylogenetic analyses of the 16S rRNA gene demonstrate it is located at the base of the clostridial rRNA cluster III of the *Firmicutes* (Collins et al., 1994). However, *Fasti-diosipila* forms a distinct lineage within this grouping and does

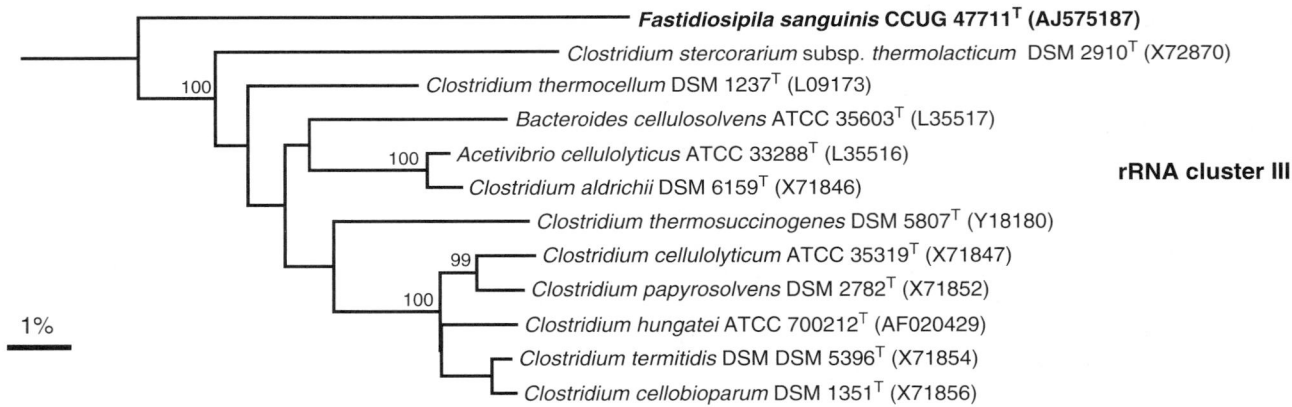

FIGURE 188. Unrooted phylogenetic tree based on 16S rRNA gene sequences showing the association of *Fastidiosipila sanguinis* with *Clostridium* rRNA cluster III. The tree was constructed using the neighbor-joining method and was based on a comparison of approximately 1330 nucleotides. Bootstrap values, expressed as a percentage of 1000 replications, are given at branching points. Scale bar = 1% sequence divergence.

not display a particularly close affinity with any of the organisms present within this cluster (see Figure 188). In the current treatment, rRNA clusters III and IV of Collins et al. (1994) are combined in the family *Ruminococcaceae* (Ludwig et al., 2009).

The anaerobic, Gram-stain-positive cocci represent a taxonomically heterogeneous group of organisms. These organisms were at one time classified in a single family, "*Peptostreptococcaceae*" (Rogosa, 1974), which included the genera *Ruminococcus, Coprococcus, Sarcina, Peptococcus,* and *Peptostreptococcus.* During the past decade, knowledge of the phylogenetic inter-relationships of these organisms has been greatly clarified through the use of 16S rRNA gene sequencing, and it is now known that the various genera of the anaerobic Gram-positive cocci are scattered throughout the order *Clostridiales* (Collins et al., 1994; Li et al., 1994; Murdoch et al., 1997). Phylogenetically, *Peptostreptococcus anaerobius,* the type species of the genus *Peptostreptococcus,* is a member of *Clostridium* rRNA group XI (see Collins et al., (1994), for group designations) or the *Peptostreptococcaceae,* whereas other peptostreptococci are found in *Clostridium* rRNA group XIII and now have been assigned to distinct genera in Family XI *Incertae Sedis* (*Anaerococcus, Finegoldia, Gallicola, Peptoniphilus;* Murdoch and Shah, 1999; Ezaki et al, 2001). *Sarcina ventriculi* is a member of *Clostridium* rRNA group I of the *Clostridiaceae,* coprococci are related to the members of *Clostridium* rRNA group XVI or the *Lachnospiraceae,* whereas ruminococci

are located in two different rRNA lines (Collins et al., 1994; Rainey and Janssen, 1995), now the *Ruminococcaceae* and the *Lachnospiraceae* (Ludwig et al., 2009). In addition to these changes, a plethora of novel species of strictly anaerobic, Gram-stain-positive cocci has been described (Ezaki et al., 1990; Li et al., 1994; Murdoch et al., 1997).

In addition to using classical methods in combination with 16S rRNA gene sequence analysis to aid taxonomy in the reclassification of the aforementioned taxa, culture-independent studies have shown that much new diversity remains to be discovered, especially from human sources (Namsolleck et al., 2004). The use of high-throughput methods are becoming increasingly automated, reducing costs, and making them routinely accessible; the availability of these methods facilitates more rapid and accurate identification of hitherto unknown taxa that are important components of the particular ecosystem being studied.

Differentiation of the genus *Fastidiosipila* from other genera

Fastidiosipila can be readily distinguished from its closest phylogenetic relatives by a combination of chemotaxonomic, biochemical, and morphological criteria (see Table 196). However in light of its coccal morphology it is pertinent to compare *Fastidiosipila* with other anaerobic Gram-stain-positive cocci (see Table 197).

TABLE 196. Characteristics which are useful in differentiating *Fastidiosipila sanguinis* from other taxa in rRNA cluster III and some other related organisms[a,b]

Characteristic	*Fastidiosipila sanguinis*	*Acetivibrio cellulolyticus*	*Anaerofilum*	*Bacteroides cellulosolvens*	*Clostridium aldrichii*	*Clostridium cellobioparum*	*Clostridium cellulolyticum*	*Clostridium hungatei*	*Clostridium papyrosolvens*	*Clostridium stercorarium*	*Clostridium thermocellum*	*Clostridium thermosuccinogenes*	*Faecalibacterium prausnitzii*	*Papillibacter cinnaminivorans*
Cellular morphology	Cocci	Rods	Rods	Rods	Rods	Rods	Rods	Rods	Rods	Rods	Rods	Rods	Rods	Rods
Spores	−	−	−	−	+	+	+	+	+	d	+	+	−	−
Gram-stain	+	−	+	−	+	−	+	−	−	−	−	−	−	−
Motility	−	+	d	−	+	+	+	+	+	d	−	+	−	−
Flagella	−	M	P	−	PB	P	P	SP	P	P/SL	−	P	−	−
End products of metabolism	A, B	A, B, P[c]	A, E, F, L, 2,3-buta-nediol	A, E, L[c]	A, B, P, iB, iV, L, S	A, E, L, F[d]	A, E, F[d]	A, E, F, L	A, E, L, F[d]	A, E, L[d]	A, L, E	A, E, L[c]	A, E, F, L, S	B, L, F, S	ND
Cellulose degradation	−	+	−	+	+	+	+	+	+	+	+	−	−	−
Growth at 60°C	−	−	−	−	−	−	−	−	−	+	+	+	−	−
Mol% G+C	33	38	54–55	43	40	28	41	40	30	39–43	38–39	36	47–57	56

[a]Characteristics are scored as: +, positive; −, negative; d, strain-dependent; ND, not determined; M, monotrichous; P, peritrichous, PB, polar bundle; SL, single lateral; SP, subpolar; End products, A, acetate; B, butyrate; E, ethanol; F, formate; iB, isobutyrate; iV, isovalerate; L, lactate; S, succinate.
[b]Data from Cato et al. (1986); Defnoun et al. (2000); Duncan et al. (2002); Falsen et al. (2005); Madden et al. (1982); Monserrate et al. (2001); Murray et al. (1984); Patel et al. (1980); Petitdemange et al. (1984); Yang et al. (1990); Zellner et al. (1996).
[c]Cellobiose fermentation.
[d]Cellulose fermentation.

TABLE 197. Characteristics which are useful in differentiating *Fastidiosipila sanguinis* from some other Gram-positive coccus-shaped taxa[a,b]

Characteristic	*Fastidiosipila sanguinis*	*Anaerococcus prevotii*	*Anaerococcus tetradius*	*Anaerococcus lactolyticus*	*Anaerococcus hydrogenalis*	*Anaerococcus octavius*	*Anaerococcus vaginalis*	*Peptoniphilus asaccharolyticus*	*Peptoniphilus lacrimalis*	*Peptoniphilus harei*	*Peptoniphilus ivorii*	*Peptoniphilus indolicus*	*Gallicola barnesae*	*Finegoldia magna*	*Micromonas micros*	*Peptostreptococcus anaerobius*	*Peptococcus niger*
Major terminal VFA[c]	A, B	B	B	B	B	B, C	B	B	B	B	B	B	A, B	A	A	IC, IV	C
Production of:																	
Indole	–	–	–	–	+	–	d	d	–	d	–	+	W	–	–	–	–
Urease	–	d	+	+	d	–	–	–	–	–	–	+	–	–	+	–	–
Alkaline phosphatase	–	–	–	–	d	–	d	–	–	–	–	+	–	d	–	–	–
Fermentation of:																	
Glucose	–	d	d	+	+	+	+	–	–	–	–	–	–	–	–	+	–
Lactose	–	–	–	+	+	–	–	–	–	–	–	–	–	–	–	–	–
Raffinose	–	+	–	–	+	–	–	–	–	–	–	–	–	–	–	–	–
Mannose	–	+	+	+	+	+	d	–	–	–	–	–	–	–	–	–	–
Activity for:																	
α-Galactosidase	+	+	–	–	–	–	–	–	–	–	–	–	–	–	–	–	–
β-Galactosidase	+	–	+	+	–	–	–	–	–	–	–	–	–	–	–	–	–
α-Glucosidase	+	+	+	–	d	–	–	–	–	–	–	–	–	–	–	–	–
β-Glucosidase	+	+	+	–	–	–	–	–	–	–	–	–	–	–	–	–	–
Arginine arylamidase	d	+	+	+	–	–	+	+	+	+	–	+	–	+	+	–	–
Proline arylamidase	+	–	W	–	–	+	–	–	+	–	+	+	–	–	+	+	–
Phenylalanine arylamidase	–	–	+	–	–	–	+	d	+	d	+	+	–	+	+	+	–
Leucine arylamidase	d	–	W	–	–	–	+	d	+	d	–	+	–	+	+	+	–
Pyroglutamic acid arylamidase	–	+	W	–	–	W	+	W	+	+	–	+	–	+	+	–	–
Histidine arylamidase	d	+	W	–	–	–	+	W	+	+	–	+	–	d	+	–	–
G+C content (mol%)	33	29–33	30–32	30–34	30–34	26–31	30–34	31–32	30–34	25	29	32–35	32–34	32–34	28–30	34–36	50–51

[a] Characteristics are scored as: +, positive; –, negative; d, strain-dependent; W, weak.
[b] Data from Ezaki et al. (2001) and Falsen et al. (2005).
[c] VFA, volatile fatty acids; B, butyrate; C, caproate; IC, isocaproate; IV, isovalerate; A, acetate.

List of species of the genus *Fastidiosipila*

1. **Fastidiosipila sanguinis** Falsen, Collins, Welinder-Olsson, Song, Finegold and Lawson 2005, 856[VP]

san′gui.nis. L. gen. n. *sanguinis* of blood, referring to the source of the organism.

Characteristics are as given for the genus. Habitat is not known.

DNA G+C content (mol%): 32.9 (HPLC).
Type strain: CCUG 47711, CIP 108292, DSM 6.
GenBank accession number (16S rRNA gene): AJ575187.

Genus VIII. **Oscillospira** Chatton and Pérard 1913, 1159[AL]

YOICHI KAMAGATA

Os.cil.lo.spi′ra. L. n. *oscillum* a swing; L. n. *spira* a spiral; N.L. fem. n. *Oscillospira* the oscillating spiral.

Large rods or filaments 3–6 μm in diameter and 10–40 μm in length, divided by closely spaced cross-walls into numerous disk-shaped cells. Reproduction by transverse fission. Sometimes motile by means of numerous lateral flagella. Endospores occasionally formed. **Growth in pure culture has not been reported. Found only in herbivorous animals.** *Oscillospira guilliermondii* is the only species of this genus.

Type species: **Oscillospira guilliermondii** corrig. Chatton and Pérard 1913, 1159[AL].

Further descriptive information

Oscillospira is a large bacterium (3–6 × 10–40 μm) often observed in the rumen contents of sheep and cattle as well as the alimentary tract of other herbivorous animals (Figure 189). *Oscillospira* is characterized as a Gram-stain-positive bacterium, with closely spaced transverse septa and endospores (Gibson, 1974; Grain and J. Senaud., 1976; Sneath, 1986; Stewart et al., 1997). Although this bacterium was first described almost a century ago (Chatton and Pérard, 1913), growth in pure culture has not been reported (Gibson, 1974; Sneath, 1986; Stewart et al., 1997), hence little is known of its ecological role and physiological properties in the intestinal tract. Clarke (1979) found that *Oscillospira* and other large bacteria attach rapidly to the cuticular surface of clover and grass leaves in the rumen suggesting that the cuticle of green leaves constitutes a specific niche for

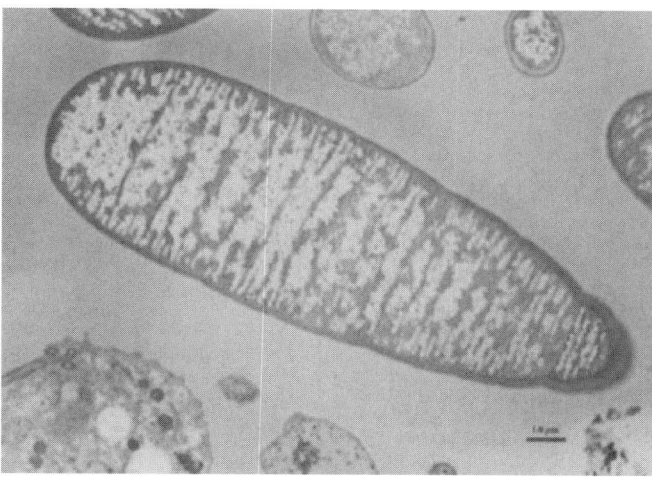

FIGURE 189. Transmission electron micrograph of an *Oscillospira* cell. Bar = 1 μm.

these bacteria. Warner (1966) investigated the relation between grazing behavior and changes in rumen microbial populations in sheep and found that the population of *Oscillospira* in the rumen fluctuates and the length of the trichome also changes depending upon the amount of feed consumed. Kurihara et al. (1968) reported that the total counts of *Oscillospira* in the sheep rumen tend to decrease in the presence of ciliates. Although the number of *Oscillospira* cells is relatively low compared to other bacterial cells, they may make a significant contribution to rumen fermentation because of their large biomass, roughly equivalent to that of ruminal ciliate protozoa (Clarke, 1979; Williams and Coleman, 1997).

Recently, *Oscillospira* cells were sorted by flow cytometry based on cell size, and the 16S rRNA gene was amplified, cloned, and sequenced (Yanagita et al., 2003). The phylogenetic analysis indicates that *Oscillospira* is affiliated with the low G+C subclass of the Gram-positive bacteria (*Firmicutes*) (Figure 190). The *Oscillospira* sequences are closely affiliated with uncultured bacterial clones within the clostridial cluster IV (Collins' nomenclature) obtained from the ceca of broiler chickens and rumen contents of cows, forming a coherent group with sequence similarities of 90.2–91.1% (Yanagita et al., 2003). Interestingly, a clonal sequence from a human fecal sample also fell into this cluster, suggesting that micro-organisms within this cluster are widespread in the alimentary tracts not only of herbivores but also of omnivores. It is also distantly related to *Sporobacter termitidis* and *Papillibacter cinnaminovorans* with 86.3–88.1% sequence similarities. *Sporobacter termitidis* is, as its name indicates, an anaerobic bacterium isolated from the paunch of the wood-feeding termite *Nasutitermes lujae*. *Sporobacter termitidis* is a chemo-organotroph that grows exclusively on a limited range of methoxylated aromatic compounds including 3,4,5-trimethoxybenzoate and 3,4,5-trimethoxycinnamate, with ring cleavage and production of acetate as a major end product, suggesting that this bacterium may contribute to the degradation of lignocellulosic matter in the digestive tract of the termite (Grech-Mora et al., 1996). *Papillibacter cinnaminovorans* is a strictly anaerobic bacterium which has recently been isolated from anaerobic digester sludge and is capable of metabolizing several methoxycinnamates (Defnoun et al., 2000).

Whether or not *Oscillospira* is capable of utilizing these substrates in the rumen ecosystem is unknown. To date, all attempts to grow the flow cytometry-sorted *Oscillospira* cells anaerobically on various substrates have been unsuccessful. However, very recently emergence of *Oscillospira* was found to be dependent on the diets of its host animals (Mackie et al., 2003). For instance, distinct *Oscillospira* morphotypes were consistently observed in samples from cattle fed

FIGURE 190. Phylogenetic position of *Papillibacter cinnamivorans* as a member of the family *Clostridiaceae*, order *Clostridiales*, class *Clostridia* in the phylum *Firmicutes*. The closest phylogenetic neighbor of *Papillibacter cinnamate* is *Sporobacter termitidis* (similarity value of 91%) and *Clostridium orbiscindens* (similarity value of 89%). The tree was constructed using 35 representative 16S rRNA sequences of the family *Clostridiaceae*, extracted from GenBank, release 151 (http://www.ncbi.nlm.nih.gov/) and Ribosomal Database II project, release 9.35 (http://rdp.cme.msu.edu/). The analysis was performed with 1150 aligned, unambiguous nucleotides using DNADIST and neighbor-joining programs which form part of the PHYLIP suit of software. Bar indicates 10 nucleotide changes per 100 nucleotides. The numeric indicates bootstrap values out of 100 resampling.

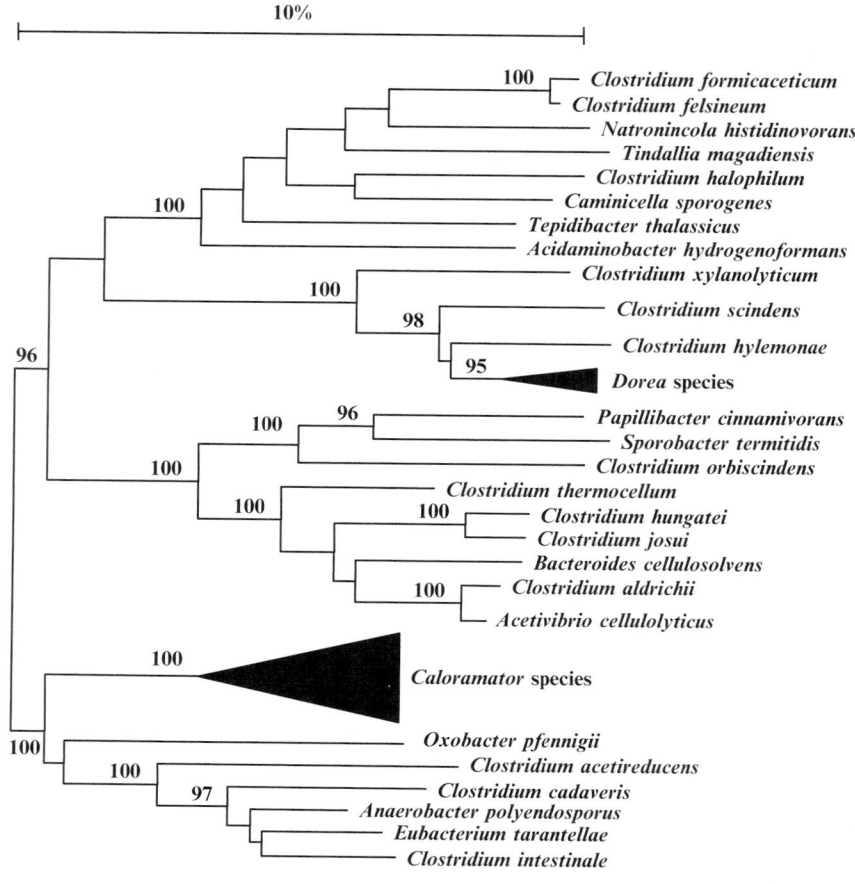

on green pasture. However, two weeks after the transition of the cattle from being pasture-fed to being indoor-fed, no typical morphotypes could be detected microscopically. These morphotypes reappeared 18 d after the cattle began pasture-feeding again. It took more than two weeks for the reappearance of *Oscillospira*, but it quickly reached the level characteristic of pasture-fed animals (1.2×10^5 cells/g rumen content) 1 week after the first detection. Similar results were obtained in reindeer and sheep.

Differentiation of the genus *Oscillospira* from other genera

Oscillospira is a large morphologically conspicuous micro-organism, which can be easily differentiated from other microbes.

Taxonomic comments

Oscillospira is an uncultured bacterium, but its phylogenetic position was determined using collected cells by flow cytometry sorting. It is affiliated with uncultured bacterial clones within the clostridial cluster IV. Recent molecular analysis of rumen samples from reindeer, cattle, and sheep strongly suggests that the genus *Oscillospira* contains at least three different groups at species level (Mackie et al., 2003). According to the 16S rRNA phylogenetic analysis presented in the roadmap to this volume (Figure 6), the genus *Oscillospira* is a member of the family *Ruminococcaceae*, order *Clostridiales*, class *Clostridia* in the phylum *Firmicutes*.

List of species of the genus *Oscillospira*

1. **Oscillospira guilliermondii** corrig. Chatton and Pérard 1913, 1159[AL]

guil.lier.mon'di.i. N.L. gen. n. *guilliermondii* of Guilliermond; named for A. Guilliermond, a French biologist.

Large rods or filaments, $3–6 \times 10–40\,\mu$m in size. Larger or smaller forms may be present. Rods have rounded ends and may taper to one pole. Closely spaced cross-walls formed by diaphragm-like ingrowth from the outer wall divide the rods into disk-shaped cells not more than 2 μm long.

Endospores (approx. $2.5 \times 4\,\mu$m in size) may be formed and found lying longitudinally in rods, occupying as much space as several disk-shaped cells. Rarely, there may be two spores in a single rod. Spores are refractile. The occurrence of sporulation is variable; the host's diet is thought to be a controlling factor. The cells frequently contain much polysaccharide that gives a reddish to mauve color in the presence of iodine.

Originally described in cecal contents of the guinea pig. Organisms that have the internal structure of *Oscillospira guilliermondii* have been found in the alimentary tract (mainly in the rumen or cecum) of several species of herbivorous animals. Some differ from the original description of the species in certain morphological features. An organism in the rumen of sheep that had a spherical, not an elliptical, spore was identified as *Oscillospira guilliermondii* by Moir and Masson (1952).

DNA G+C content (mol%): not determined.

Type strain: the organism has not been cultivated.

GenBank accession number (16S rRNA gene): not available for type strain.

Genus IX. **Papillibacter** Defnoun, Labat, Ambrosio, Garcia and Patel 2000, 1227[VP]

BHARAT K. C. PATEL

Pa.pil.li.bac′ter. L. fem. n. *papilla* teat; N.L. n. *bacter* masc. equivalent of Gr. neut. dim. n. *bakterion* rod or staff; N.L. masc. n. *Papillibacter* a rod with ends looking like a teat.

Cells are **rod-shaped** and **stain Gram-positive. Nonmotile. Nonsporulating Strict anaerobic.** Grows on a limited number of aromatic compounds and crotonate but not on carbohydrates, organic acids and alcohols.

Type species: **Papillibacter cinnamivorans** Defnoun, Labat, Ambrosio, Garcia and Patel 2000, 1227[VP].

Further descriptive information

Papillibacter utilizes a very limited range of aromatic compounds (cinnamate, 3-methoxycinnamate, and 4-methoxycinnamate) and crotonate by transforming a double bond in the C_3-aliphatic side chain without ring cleavage. Acetate and benzoate are produced from the transformation of cinnamate, acetate, and 3-methoxybenzoate from 3-methoxycinnamate transformation, acetate and 4-methoxybenzoate from 4-methoxycinnamate, and acetate and butyrate from the degradation of crotonate. Such a characteristic has not been reported for any other isolates and hence is a new physiological trait. The transformation of 3-methoxycinnamate and 4-methoxycinnamate but not 2-methoxycinnamate is also an interesting property of the isolate and suggests stereospecific selection by the species.

Enrichment and isolation procedures

The natural habitat for the genus *Papillibacter* is yet to be defined, but anaerobic digesters, which are routinely supplemented with the tannin- and aromatic-rich shea cakes, and inoculated with sludge from the pit of slaughter-houses, are considered to be ideal habitats. A 0.5-ml aliquot of the sample is inoculated into an anaerobic basal medium emended with 5 mM cinnamate to initiate enrichment. The basal medium is prepared using the anaerobic technique described by Hungate (1969) and modified for use with syringes (Macy et al., 1972; Miller and Wolin, 1974) and contains (g/l deionized water): 0.2 g K_2HPO_4; 1 g NaCl; 0.15 g $CaCl_2 \cdot 2H_2O$; 0.4 g $MgCl_2 \cdot 6H_2O$; 0.5 g KCl; 0.5 g cysteine-HCl; 0.5 g yeast extract; 1.5 ml trace element solution (Widdel and Pfennig, 1981), and 1 mg resazurin. The pH is adjusted to 7.0 with a 10 M NaOH solution, the medium boiled under a stream of O_2-free N_2 gas, cooled to room temperature, 5-ml aliquots dispensed into Hungate tubes under a $N_2:CO_2$ (80:20) gas mixture and sterilized by autoclaving at 121°C for 15 min. Prior to use, 0.2 ml 5% (w/v) $NaHCO_3$ and 0.05 ml 2% (w/v) $Na_2S \cdot 9H_2O$ were injected from anaerobic sterile stock solutions into the presterilized basal medium.

A stable microbial consortium that degrades cinnamate with concomitant production of acetate and benzoate is established after 3 weeks incubation at 37°C. The consortium is then subcultured several times under the same conditions and subsequently; serial tenfold dilutions are prepared and inoculated into roll tubes containing the basal medium supplemented with 5 mM cinnamate and

TABLE 198. Characteristics that differentiate *Papillibacter cinnamivorans* from *Sporobacter termitidis*[a]

Characteristic	*Papillibacter cinnamivorans*	*Sporobacter termitidis*
Habitat	Shea cake fed anaerobic digester	Digestive tract of the wood-feeding termite, *Nasutitermes lujae*
Morphology	Rod-shaped with pointy ends, occur singly, in pairs, or in chains	Slightly curved rods
Size (μm)	0.5–0.6 × 1.3–3.0	0.3–0.4 × 1.2
Spore	−	+
Motility	−	+
Temperature optimum (°C)	37	32–35
pH optimum	7.5	6.7–7.2
NaCl optimum (g/l)	5–10	0–5
Growth on aromatic compounds:[b]		
Cinnamate	+	−
2-Methoxycinnamate	−	ND
3-Methoxycinnamate	+	ND
4-Methoxycinnamate	+	−
3,4,5-Trimethoxycinnamate	ND	+
Sinapate	−	+
3,4-Dimethoxycinnamate	ND	+
3,4,5-Trimethoxybenzoate	−	+
Ferulate	−	+
Syringate	−	+
Vanillate	−	+
DNA G+C content (mol%)	56	57
Reference	Defnoun et al. (2000)	Grech-Mora et al. (1996)

[a]Symbols: +, positive; −, negative; ND, not determined.

[b]Substrates tested but not used by both strains include the following: aromatic compounds (benzoate, phenylacetate, caffeate, gallate, phloroglucinol, pyrogallol, catechol, p-coumarate, phenol, 4-hydroxybenzoate), organic acids (pyruvate), carbohydrates (glucose, fructose, ribose, xylose, maltose, galactose, lactose), and alcohols (methanol, ethanol).

1.6% (w/v) agar. Single, well-isolated colonies are picked and serially diluted in the anaerobic medium; the procedure was repeated at least three times before an isolate was deemed to be pure. The isolates fail to grow in anaerobic basal medium amended with 0.5% glucose and 0.5% Biotrypcase, which is used to check routinely for fermentative heterotrophic contaminants.

Maintenance procedures

Papillibacter cinnamivorans is maintained by subculturing once every four weeks in the anaerobic growth medium. The culture can also be stored anaerobically at −20°C in a glycerol and growth medium (50:50 v/v).

Differentiation of the genus *Papillibacter* from other genera

This trait of cinnamate transformation is common to *Syntrophus buswellii* and *Rhodopseudomonas palustris*. *Syntrophus buswellii* requires a syntrophic partner such as *Desulfovibrio vulgaris* to degrade cinnamate and produces acetate and H_2S without the accumulation of benzoate (Auburger and Winter, 1995). However, *Papillibacter* transforms cinnamate to produce acetate and benzoate, which accumulates without further degradation. *Rhodopseudomonas palustris* requires light for cinnamate degradation (Harwood and Gibson, 1988) whereas *Papillibacter* is a nonphotosynthetic bacterium.

Taxonomic comments

Analysis of 16S RNA sequence data indicates that *Papillibacter* is a member of the family *Clostridiaceae*, order *Clostridiales*, class *Clostridia*, in the phylum *Firmicutes*. The genus *Papillibacter* includes a single species, *Papillibacter cinnamivorans*. The closest phylogenetic neighbor of *Papillibacter cinnamivorans* is *Sporobacter termitidis* (Grech-Mora et al., 1996) (similarity value of 91%), and, though both transform aromatic compounds, significant differences in their traits are notable (Table 198). *Papillibacter cinnaminivorans* is also phylogenetically related to the flavonoid ring-cleaving anaerobe, *Clostridium orbiscindens* (similarity value of 89%), and distantly to *Clostridium scindens* (similarity value of 85%), and *Eubacterium ramulus*, a member of the family *Eubacteriaceae*, all of which have been isolated from human gastrointestinal tracts (Schneider et al., 1999; Winter et al., 1989; Winter et al., 1991). Flavonoids are polyphenolic compounds, widely believed to have beneficial effects on human health based on their anti-inflammatory, antioxidant, vasodilatory, anticarcinogenic, and antibacterial properties. They are widely distributed in plants and over 5,000 different natural flavonoids have been described so far.

Differentiation of the species of the genus *Papillibacter*

Only one species of *Papillibacter* has been described so far.

List of species of the genus *Papillibacter*

1. **Papillibacter cinnamivorans** Defnoun, Labat, Ambrosio, Garcia and Patel 2000, 1227^VP

cin.na.mi.vo′rans. N.L. n. *cinnamum* cinnamic acid; L. part. adj. *vorans* devouring, digesting; N.L. part. adj. *cinnamivorans* referring to the ability to digest cinnamic acid.

Rod-shaped cells (1.3–3.0 × 0.5–0.6 μm), usually arranged singly, in pairs, or in chains. Stain Gram-stain-positive. Nonmotile. Nonsporulating. Strictly anaerobic. Growth occurs at 15–40°C (optimum 37°C) and at pH 6.9–8.5 (optimum pH 7.5). Chemo-organotroph. Requires yeast extract for growth and biotrypcase and casein stimulates growth. Grows on a limited range of aromatic compounds and crotonate but not on carbohydrates, organic acids, and alcohols. Sulfate, thiosulfate, sulfite, nitrate, elemental sulfur, or fumarate are not utilized as electron acceptors. Isolated from an anaerobic digester in Burkina Faso. The digester had been inoculated with sludge from a pit of a slaughter house and had been routinely supplemented with tannin and aromatic compound containing shea cake.

DNA G+C content (mol%): 56 (HPLC).
Type strain: Strain CIN1, DSM 12816, ATCC 700879.
GenBank accession number (16S rRNA gene): AF167711.

Genus X. **Sporobacter** Grech-Mora, Fardeau, Patel, Ollivier, Rimbault, Prensier, Garcia and Garnier-Sillam 1996, 517^VP

Isabelle Grech-Mora, Marie-Laure Fardeau, Jean-Louis Garcia and Bernard Ollivier

Spo.ro.bac′ter. Gr. n. *spora* a spore; N.L. n. *bacter* masc. equivalent of Gr. neut. dim. n. *bakterion* rod; N.L. masc. n. *Sporobacter* a spore-forming rod.

Slightly curved rods that are 0.2–0.4 × 1–2 μm and occur singly or in pairs. **Motile by peritrichous flagella.** Gram-stain-positive. Terminal spore swelling the sporange. Chemo-organotrophic and **obligately anaerobic** member of the domain *Bacteria*. Yeast extract (0.01%) is required for growth. Produces methylated sulfides from methylated aromatic compounds (e.g., 3,4,5-trimethoxycinnamate) if sulfide or cysteine is present in the medium with acetate being the only volatile fatty acid produced from this metabolism. External electron acceptors are not used. **Mesophilic and neutrophilic.**

DNA G+C content (mol%): 57 (HPLC).

Type species: **Sporobacter termitidis** Grech-Mora, Fardeau, Patel, Ollivier, Rimbault, Prensier, Garcia and Garnier-Sillam 1996, 517^VP.

Further descriptive information

Sporobacter termitidis possesses peritrichous flagella (Figure 191a) and a Gram-positive type of cell wall (Figure 191b). Acetate is not only the major product of metabolism from 3,4,5-trimethoxycinnamate (TMC) but also from sinapate, syringate, and 3,4,5-trimethoxybenzoate. The amount of TMC degraded depends on the initial substrate concentration in the growth medium. For concentrations of TMC <5.45 mM, the substrate is

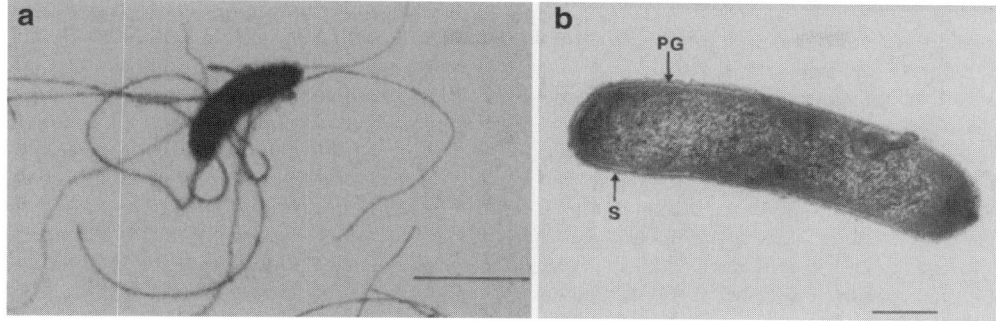

FIGURE 191. (a) electron micrographs of *Sporobacter termitidis* negatively stained showing peritrichous flagellation. Bar = 1 μm. (b) Ultrathin section of *Sporobacter termitidis* showing the S Layer (S), and peptidoglycan (PG). Bar = 0.2 μm.

degraded into acetate (approx. 4 mol/mol of TMC). At concentrations between 5.45 and 10.82 mM, approximately 1/4 of the substrate remains unused but the amount of acetate produced per mol of TMC degraded is still approximately 4. With initial TMC concentrations higher than 10.82 mM, a greater amount of TMC remains unused and the ratio of acetate/TMC diminishes to 2.57. Initial concentrations of TMC greater than 40 mM inhibit growth of *Sporobacter termitidis* completely. Optical density increases until concentration of TMC reaches 10.82 mM (μ=0.027/h). The maximum growth rate is obtained in the presence of 8.39 mM of TMC (μ=0.035/h). When *Sporobacter termitidis* grows on 3,4-dimethoxycinnamate, or vanillate, acetate and catechol accumulate in the medium after 3 months of incubation. Vanillate is detected as an intermediary product from 3,4-dimethoxycinnamate metabolism. After 7 weeks of incubation, acetate and vanillate are produced from ferulate degradation, the latter being poorly used. Utilization of sulfide or cysteine as a methyl acceptor during growth on TMC results in production of dimethylsulfide (DMS), the major volatile sulfur compound, with methanethiol being an intermediate product of metabolism. However, growth on TMC also occurs in the absence of sulfide, suggesting that CO derived from the CO_2 of the medium probably acts as the methyl acceptor (Diekert and Wohlfarth, 1994). During fermentation of methoxylated aromatic substrates by *Sporobacter termitidis*, no H_2 is evolved and therefore *Sporobacter termitidis* does not possess any H_2-evolving hydrogenase. Accordingly, no H_2 interspecies transfer can result from its association with an hydrogenotrophic methanogen (*Methanospirillum hungateii*). Mineralization of the aromatic nucleus is achieved with tri- or disubstituted compounds (TMC, sinapate, syringate, 3,4,5 trimethoxybenzoate) and not with monosubstituted methoxy aromatic compounds (ferulate, vanillate). In addition, the ring of 3,4-dimethoxycinnamate is not cleaved by *Sporobacter termitidis*, indicating that the position of methoxy substitutes on the aromatic compound is of importance for cleavage.

Enrichment and isolation procedures

Nests of the wood feeding termite, *Nasutitermes lujae* (*Termitidae*) were collected from M'Balmayo forest (Cameroon, Central Africa) and maintained at 27°C. After dipping termites in ethanol, guts were extracted and served as inoculum. Guts of 10 workers were removed with a pair of forceps and placed into a drop of 0.9 % NaCl solution. The contents were added to a tube containing 5 ml of culture medium and transferred into an anaerobic glove box where they were homogenized. Anoxic media were prepared by using the techniques of Hungate (1969). The growth medium for enrichment, isolation, and most routine culture work contained (per liter) NH_4Cl, 1.0 g; K_2HPO_4, 0.3 g; KH_2PO_4, 0.3 g; $MgCl_2 \cdot 6H_2O$, 0.2 g; $CaCl_2 \cdot 2H_2O$; 0.1 g; KCl, 0.1 g; $CH_3COONa \cdot 3H_2O$, 0.5 g; NaCl, 0.6 g; yeast extract (Difco), 0.5 g; resazurin, 0.001 g; cysteine·HCl, 0.5 g; and 1.5 ml of the trace element solution of Imhoff-Stuckle and Pfennig (1983) and 3,4,5-trimethoxycinnamate (TMC). The pH was adjusted to 7 with 10 M KOH. The medium was boiled under a stream of O_2-free N_2 gas and cooled to room temperature. Five-milliliter aliquots were dispensed into Hungate tubes under a stream of N_2-CO_2 (4:1, v/v), and the tubes were autoclaved for 45 min at 110°C. Prior to inoculation, 0.05 ml of 2% $Na_2S \cdot 9H_2O$ and 0.2 ml of 10% $NaHCO_3$ were injected to all the anoxic media from sterile anaerobic stock solutions.

Hindgut samples were inoculated into TMC-containing (20 mM) growth medium to initiate enrichments. After several transfers, the enrichment cultures were serially diluted and pure cultures were isolated using the roll tube technique. For preparing roll tubes, the growth medium was solidified with 1.6 % (w/v) Noble agar (Difco). After incubation, colonies that developed in the roll tube containing the most diluted enrichment inoculum were picked and the process of serial dilution in roll tubes repeated at least twice to purify the cultures. Purity of the isolates was routinely checked by microscopical examination and by culturing in the growth medium lacking TMC, but containing glucose.

Differentiation of the genus *Sporobacter* from other genera

A great variety of aerobic and anaerobic bacteria are able to produce DMS from methionine degradation. DMS can also be formed from S-methylmethionine present in various terrestrial plants. Bak et al. (1992) were the first to report on the production of DMS from aromatic compound degradation by a new anaerobic strain isolated from anoxic sediments, *Holophaga foetida* (Liesack et al., 1994). This trait is similar to that observed for *Sporobacter termitidis*. However dimethylsulfide and methanethiol cannot be formed only from inorganic sulfide, but also from organic sulfide such as cysteine in *Sporobacter termitidis*. The biochemical mechanism by which the methylated sulfur compounds are formed is still unknown. Other similarities between *Holophaga foetida* and *Sporobacter termitidis* include the ability to ferment methoxylated aromatic compounds such as 3,4,5-trimethoxybenzoate, syringate, sinapate, ferulate, but not H_2 plus CO_2, methanol or other alcohols. However, unlike *Holophaga foetida*, *Sporobacter termitidis* is a Gram-stain-positive spore-former and does not use phenolic compounds, in particular, gallate and pyrogallol, nor pyruvate, and has a G+C content of 57mol%, which is lower than that of *Holophaga foetida* (63mol%). Phylogenetically *Holophaga foetida* and *Sporobacter termitidis* are also different; *Holophaga foetida* is placed in family "Acidobacteriaceae" of the phylum *Acidobacteria*, whereas *Sporobacter termitidis* is clearly a member of the sub-branch of the low G+C Gram-stain-positive clostridial group in the family *Clostridiaceae*, phylum *Firmicutes*. Brauman et al. (1998) reported on *Sporotomaculum hydroxybenzoicum*, a new sporulating anaerobe isolated from the soil-feeding termite, *Cubitermes speciosus*, which ferments 3-hydroxybenzoate to acetate and butyrate. In contrast, *Sporobacter termitidis* which has been isolated from the hindgut of a wood-feeding termite, produces only acetate by cleaving the TMC aromatic ring in the absence of oxygen. *Sporobacter termitidis* clearly differs from the former as it does not ferment 3-hydroxybenzoate.

Phylogenetically, *Sporobacter termitidis* together with *Papillibacter cinnaminovorans* (Defnoun et al., 2000), its closest phylogenetic relative, are members of cluster IV of the phylum *Firmicutes* branch (Collins et al., 1994) (Figure 190). However, *Sporobacter termitidis* appears to be affiliated with, but not specifically related to, *Papillibacter cinnaminovorans* as the evolutionary distance separating them is 8%. Physiologically, *Papillibacter cinnaminovorans* is very different from *Sporobacter termitidis*. It grows on aromatic compounds (e.g., cinnamate) without the ability to cleave the aromatic ring or to demethoxylate (Defnoun et al., 2000). Acetate and benzoate are the end products of cinnamate metabolism. Phenotypically, *Sporobacter termitidis* is also distinct from *Papillibacter cinnaminovorans* (Table 198).

Taxonomic and ecological comments

Alignment of the sequence (positions 8–1542, *Escherichia coli* numbering according to Winker and Woese (1991) consisting of 1507 bases of the 16S rRNA gene of *Sporobacter termitidis* with representatives of the various phyla of domain *Bacteria*, showed that *Sporobacter termitidis*, isolated from the digestive tract of the wood-feeding termite *Nasitutermes lujae*, was consistently placed as a member of cluster IV, in the sub-branch of the low G+C *Firmicutes* clostridial group as defined by Collins et al. (1994).

The hindgut microbial consortium of wood-feeding termites in contrast to soil-feeding termites, is inhabited by cellulose-degrading or homoacetogenic hydrogen-oxidizing bacteria such as *Clostridium termitidis* (Hethener et al., 1992), *Acetonema longum* (Kane and Breznak, 1991) and *Sporomusa termitida* (Breznak et al., 1988). The microflora therefore contributes significantly in the overall anaerobic digestion of the ingested organic plant lignocellulosic matter. However, relatively little is known about the degradation of lignin monomers and lignin-related compounds. Nevertheless, it has been suggested that microbial symbionts could play a role in the metabolism of some lignin compounds (Grasse and Noirot, 1959). Of a diverse range of available lignin-derived compounds, the fermentation of aromatic compounds in termite gut has been the most widely studied (Brauman et al., 1992; Breznak and Brune, 1994). In almost all cases studied to date, bacteria (e.g., *Sporomusa termitida*) are involved in the demethylation of methoxylated aromatic compounds (Breznak and Switzer, 1986). By using ¹⁴C-(lignin)-lignocelluloses, Butler and Buckerfield (1979) and Cookson (1988) demonstrated that the ring (nucleus) was cleaved. Cookson (1988) suggested that this degradation process could have been the result of anaerobic bacteria. A contrasting report indicated that oxic conditions are required to completely oxidize lignin-derived phenylpropanoids and other monoaromatic compounds (Brune et al., 1995). In *Nasutitermes lujae*, *Sporobacter termitidis* is a strictly anaerobic bacterium that oxidizes methoxyaromatic compounds mainly into acetate and DMS with methanethiol produced as an intermediary product of metabolism, in the presence of sulfide or cysteine. In the absence of MPN (Most Probable Number) counts, it cannot be predicted with certainty whether *Sporobacter termitidis* and other physiologically related bacteria play a major ecologically important role in the hindgut of *Nasutitermes lujae*.

Acknowledgements

Many thanks to P. Thomas for photographs and C. Lesaulnier for improving the manuscript.

List of species of the genus *Sporobacter*

1. **Sporobacter termitidis** Grech-Mora, Fardeau, Patel, Ollivier, Rimbault, Prensier, Garcia and Garnier-Sillam 1996, 517ᵛᴾ

   ter.mi″ti.dis. L. n. masc. *termes*, *termitis* wood-eating worm; N.L. masc. adj. *termitidis* pertaining to the termite.

   cells are slightly curved rods (0.3–0.4 × 1–2μm), occur singly or in pairs. They are not motile when viewed under a phase-contrast microscope but electron microscopy of negatively stained cells revealed the presence of peritrichous flagella (Figure 191a). Ultrathin sections reveal an atypical Gram-stain-positive cell wall composed of a thin (presumably peptidoglycan layer) and an S-layer (Figure 191b). Spores are sparse especially in liquid cultures; they are spherical to ellipsoid and slightly swell the sporange.

   Chemo-organotrophic and obligately anaerobic member of the domain *Bacteria*. Oxidizes methoxylated aromatic compounds, including 3,4,5-trimethoxycinnamate (TMC), sinapate (3,5-dimethoxy-4-hydroxycinnamate), 3,4-dimethoxycinnamate, 3,4,5-trimethoxybenzoate, ferulate, syringate (3,5-dimethoxy-4-hydroxybenzoate), and vanillate (4-hydroxy-3-methoxy benzoate) only in the presence of yeast extract (0.01%). No growth observed with the following substrates: veratrate, isovanillate, benzoate, β-resorcylate (2,4-dihydroxybenzoate), gentisate (2,5-dihydroxybenzoate), vanillin (4-hydroxy-3-methoxybenzaldehyde), 4-hydroxybenzoate, 2,4,6-trihydroxybenzoate, protocatechuate, cinnamate, *p*-coumarate, 4-methoxycinnamate, pyrogallol, gallate, caffeate, phenol, *o*-cresol, *p*-cresol, catechol, salicin, arbutin, phloroglucinol, cyclohexanecarboxylate, hydroxycinnamate, phenylacetate (5mM), fructose, galactose, glucose, ribose, xylose, lactose, maltose, sucrose, trehalose, mannitol, rhamnose, sorbose, melibiose, adonitol, dulcitol, sorbitol, pyruvate (20mM), lactate, formate (10mM), ethanol, methanol, glycerol (40mM), and H_2/CO_2 (4:1, v/v; 2 bars). Acetate is the only end product from TMC or syringate oxidation and H_2 or CO_2 are not produced. co-culture of *Sporobacter termitidis* and the hydrogenotrophic methanogen *Methanospirillum hungatei* on methoxyaromatic compounds does not lead to methane formation, thus indicating the absence of interspecies hydrogen transfer.

Growth requires the presence of a reducing agent in the medium such as dithionite, sulfide, or cysteine. Sulfide and cysteine act as a methyl acceptor which results in the production of dimethylsulfide (DMS) and methanethiol as an intermediary product of metabolism. Oxidation of 5.45 mM TMC led to the production of 20.25 mM acetate and 0.93 mM DMS. Nitrate, sulfur, sulfate, thiosulfate, or sulfite cannot be used as alternate electron acceptors.

Growth between 20 and 40°C with the optimum at about 35°C. The pH growth range is 5.9–8.8, and the optimum pH for growth is between 6.7 and 7.2. Growth occurs at the same rate with NaCl concentrations between 0 and 5 g/l but growth inhibition occurs at concentrations > 12.5 g/l. Isolated from the hindgut of the wood-feeding termite, *Nasutitermes lujae*.

DNA G+C content (mol%): 57 (HPLC).

Type strain: SYR[T], DSM 10068[T].

EMBL accession number (16S rRNA gene): Z49863.

Genus XI. **Subdoligranulum** Holmstrøm, Collins, Møller, Falsen and Lawson 2004, 1909[VP] (Effective publication: Holmstrøm, Collins, Møller, Falsen and Lawson 2004, 201.)*

KIM HOLMSTRØM AND PAUL A. LAWSON

Sub.do.li.gra′nu.lum. L. adj. *subdolus* deceptive, alludes to the somewhat deceptive and unusual coccoid form; L. neut. n. *granulum* a small grain; N.L. neut. n. *Subdoligranulum* a deceptive grain.

Cells are coccoid in shape but may show some pleomorphism. Gram-stain-negative. Nonmotile and nonspore-forming. **Strictly anaerobic; no growth occurs in 2% oxygen.** Cells do not survive heating at 80°C for 10 min. **Catalase-negative.** Glucose and some other carbohydrates are fermented. **The major acid products formed in PYG broth are butyric and lactic acids** together with minor amounts of acetic and succinic acids. Cells are indole-negative and nitrate is not reduced to nitrite. **Esculin is hydrolyzed, but starch is not. Lecithinase-, urease-, and lipase-negative.** Based on 16S rRNA gene sequence and phylogenetic analysis, this genus belongs to the *Clostridium leptum* rRNA super cluster (cluster IV). Isolated from human feces. A 16S rRNA-derived probe, Svariab_645: 5′-TGCACTACTCAAGGCCAG-3′, has been designed that specifically detects this genus by *in situ* hybridization to human fecal samples.

DNA G+C content (mol%): 52.2.

Type species: **Subdoligranulum variabile** Holmstrøm, Collins, Møller, Falsen and Lawson 2004, 1909[VP] (Effective publication: Holmstrøm, Collins, Møller, Falsen and Lawson 2004, 202.).

Further descriptive information

The genus contains only one species, *Subdoligranulum variabile*, and therefore the characteristics provided below refer to this species. The organism recovered from human feces is strictly anaerobic and fails to grow in 2% or 6% (v/v) oxygen. The colonies of the isolate on Biolog Universal Anaerobe agar supplemented with 5% (v/v) defibrinated calf's blood (BUA+B) (Biolog Inc., Hayward, CA, USA) grow to approximately 1 mm in diameter after 4–5 d of strictly anaerobic incubation at 37°C. The colonies have a grayish-white appearance and are circular and concave. It is not possible to grow the strain on TPY or BHI agar. The cells are Gram-stain-negative, nonmotile, and nonspore-forming. Figure 192 shows a phase-contrast micrograph of the cells obtained from BUA+B agar. The cells have a tendency to associate in pairs or even, in some cases, triplets. The cells range in size between 0.6–2.5 μm and can be variable in shape (from circular, ellipsoid to drop-like). Growth is obtained in M2GSC broth (Barcenilla et al., 2000) and in fastidious anaerobe broth with cooked meat granules (Oxoid). TPY and BHI broths do not sustain growth. Growth on solid or liquid medium produces *Subdoligranulum variable* cells of similar morphology as observed when inspected by phase-contrast microscopy. The organism requires rich media for growth, and it is not possible to grow the *Subdoligranulum variabile* on defined media. The optimum growth temperature of the organism is 37°C. It fails to grow at 30°C but grows weakly at 45°C.

The major end products of glucose metabolism are butyric and lactic acids together with minor amounts of acetic and succinic acids. The organism hydrolyzes esculin but fails to hydrolyze starch. It is lecithinase, lipase, and urease-negative. Using the AN-MicroLog system, the isolate utilizes a wide range of carbohydrate substrates (*N*-acetyl-D-glucosamine, *N*-acetyl-D-mannosamine, amygdalin, arbutin, D-cellobiose, dextrin, D-fructose, L-fucose, D-galactose, D-galacturonic acid, gentiobiose, α-D-glucose, glucose-1-phosphate, glucose-6-phosphate, α-D-lactose, lactulose, maltose, maltotriose, D-mannose, D-melibiose, 3-methyl-D-glucose, α-methyl-D-galactoside, β-methyl-D-galactoside, palatinose, D-raffinose, L-rhamnose, salicin, sucrose, D-trehalose, and turanose). Using the API Rapid ID 32A kit, activity is detected for α-galactosidase, β-galactosidase, β-glucosidase, β-glucuronidase, arginine arylamidase, leucine arylamidase, and histidine arylamidase; activity for α-glucosidase is weak or negative. All other tests give negative results using this test system. Employing the API ZYM test system, the organism yields positive reactions for α-galactosidase, β-galactosidase, β-glucosidase, β-glucuronidase, with α-glucosidase displaying a weak reaction. All other tests in this kit give negative reactions. The long-chain cellular fatty acids consist primarily of the straight-chain saturated and monounsaturated types, with $C_{16:0}$ (29.4%), $C_{18:0}$ (10.3%), and $C_{18:1\,\omega 9c}$ (34.3%) predominating. Other minor components corresponded to $C_{10:0\,iso}$ (0.2%), $C_{10:0}$ (0.4%), $C_{12:0}$ (0.5%), $C_{14:1\,\omega 5c}$ (0.6%), $C_{14:0}$ (5.5%), $C_{14:0\,DMA}$ (0.5%), $C_{15:0\,ante}$ (0.2%), $C_{16:0\,ALD}$ (2.8%), $C_{15:0}$ (0.4%), $C_{16:0\,iso}$ (0.2%), $C_{16:1\,\omega 7c}$ (4.7%), $C_{16:0\,DMA}$ (4.6%), $C_{17:0\,iso}$ (0.3%), $C_{17:0\,ante}$ (0.6%), $C_{17:0}$ (1.1%), $C_{18:1\,\omega 9c\,DMA}$ (1.7%), and $C_{18:0\,DMA}$ (0.8%).

---

* Parts of this chapter were published in Holmstrøm, K., M.D. Collins, T. Møller, E. Falsen and P.A. Lawson. *Subdoligranulum variabile* gen. nov., sp. nov. from human feces, Anaerobe *10*: 197–203, © Elsevier (2004).

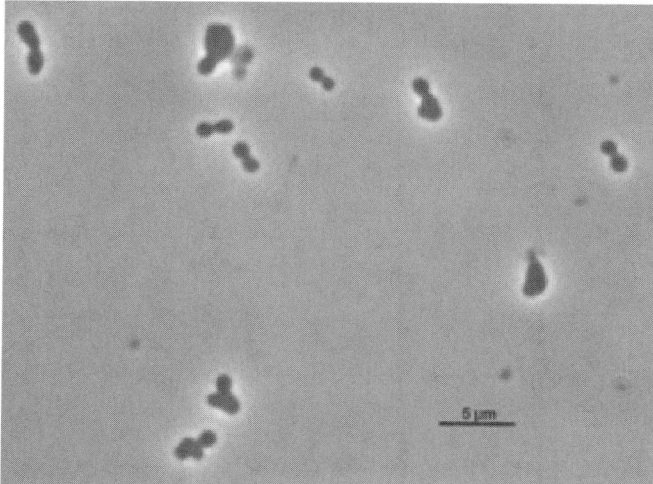

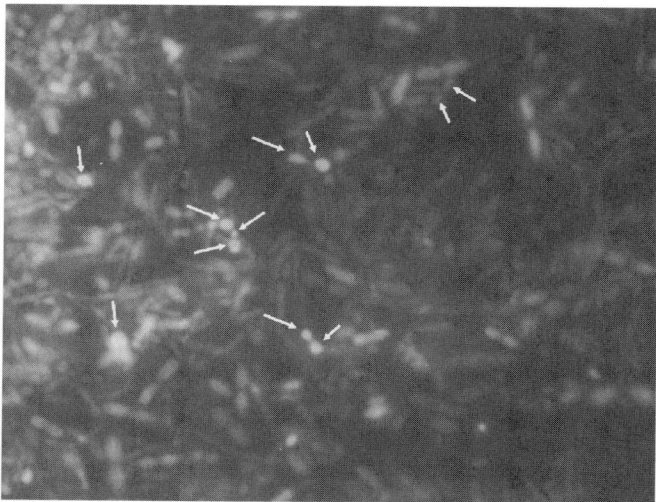

FIGURE 192. Phase-contrast micrograph showing the pleiomorphic morphology of *Subdoligranulum variabile*. Reprinted with permission from Holmstrøm, K., Collins, M.D., Møller, T., Falsen, E. and Lawson, P.A. Anaerobe, 2004, *10*: 197–203.

FIGURE 193. *In situ* visualization of *Subdoligranulum variabile* cells (white arrows) by FISH in a human fecal sample. Notice the similar cellular morphology of putative *Subdoligranulum variabile* positive cells in these samples compared to the pleiotrophic morphology of *Subdoligranulum variabile* following *in vitro* cultivation (Figure 192).

Unpublished analyses on the prevalence of *Subdoligranulum variabile* in human fecal samples using FISH (fluorescent *in situ* hybridization) with Svariab_645 have shown a relatively high proportion of positively hybridizing cells in relation to the total number of cells in fecal samples. In order to quantify the proportion of the gut microflora corresponding to *Subdoligranulum variabile*, a screening of fecal samples from 11 different individuals was conducted using a Cy3-labeled Svariab_645 probe, a Cy5-labeled Bif-164 probe (i.e., detecting the genus *Bifidobacterium*), and a FITC-labeled Eub-338 probe (i.e., a general bacterial probe). By processing the images from the FISH-analyses, the areas covered by the Cy5-, the Cy3-, and the FITC-labeled cells, respectively, were determined. The relative proportion of bifidobacteria and *Subdoligranulum variabile* cells was calculated by dividing the area (assumed to be proportional to cell volume) of FITC-labeled cells with the area of Cy5- and Cy3-labeled cells, respectively. Svariab_645 positive cells were shown to be present in all samples with a mean proportion of 1.7% and a range of 0.3% to 4.4%. Bif-164 positive cells were present with a mean proportion of 3.0% and a range of 0–11.4%. The proportion of bifidobacteria also varied considerably from one sample to another, while the distribution of *Subdoligranulum variabile* cells was more uniform. In five samples, the numbers of *Subdoligranulum variabile* cells actually exceeded the numbers of Bif-164-positive cells. In conclusion, the new subgroup of gut bacteria represented by the Svariab_645 probe constitutes a major proportion of the bacteria in feces. In Figure 193, a black and white representation of a micrograph of a fecal sample after FISH analyses is depicted, and cells that gave strong fluorescent signal from the Cy3-labeled Svariab_645 probe are identified by white arrows.

Based on sequence comparisons between the 16S rDNA sequence of *Subdoligranulum variabile* and previously published *Clostridium leptum* subgroup associated probes (Clep866, Cvir1414, Edes635, Rbro730, Rfla729, Rcal733, and Fprau645; Lay et al., 2005), Clep866 is the only probe likely to hybridize efficiently with *Subdoligranulum variabile* cells. Therefore, the Svariab_645

positive cells are likely to represent a major fraction of the previously unidentified Clep866-positive cells (Lay et al., 2005).

The wide distribution of *Subdoligranulum variabile* has recently been confirmed by molecular profiling of the *Clostridium leptum* subgroup in human fecal samples. Using a combination of group specific PCR and denaturing gradient gel electrophoresis (DGGE), Shen et al. (2006) demonstrated that, in the majority of subjects sampled, approximately 6% of clones isolated within the *Clostridium leptum* group displayed high 16S rRNA sequence similarity with *Subdoligranulum variablile*. Of the remaining clones, 64% were related to *Faecalibacterium prausnitzii*, 2% to butyrate-producing bacteria, and 28% were not identified to species level. Similarly, Lay et al. (2007) employed both fluorescence-activated cell sorting (FACS) and group-specific PCR to investigate the *Clostridium leptum* subgroup. In addition to clones related to *Faecalibacterium prausnitzii* and butyrate-producing bacteria, both methodologies yielded clones that were highly related to *Subdoligranulum variabile*. These results provide additional evidence that *Subdoligranulum variabile* is an important member of the human gut community which is cultivable, well characterized, and named.

Isolation procedures

Subdoligranulum variabile was recovered from a fecal specimen of a 47 year-old female from Hørsholm, Denmark. The organism was isolated on the commercially available Biolog Universal Anaerobe agar supplemented with 5% (v/v) defibrinated calF's blood (BUA+B) (Biolog Inc., Hayward, CA, USA) and incubated anaerobically at 37°C under N_2, CO_2, and H_2 (80:10:10% [v/v]) gas phase. There is no enrichment or selective medium for this species.

Maintenance procedures

Subdoligranulum variabile can be grown in liquid medium M2GSC (including rumen fluid) prepared according to Barcenilla

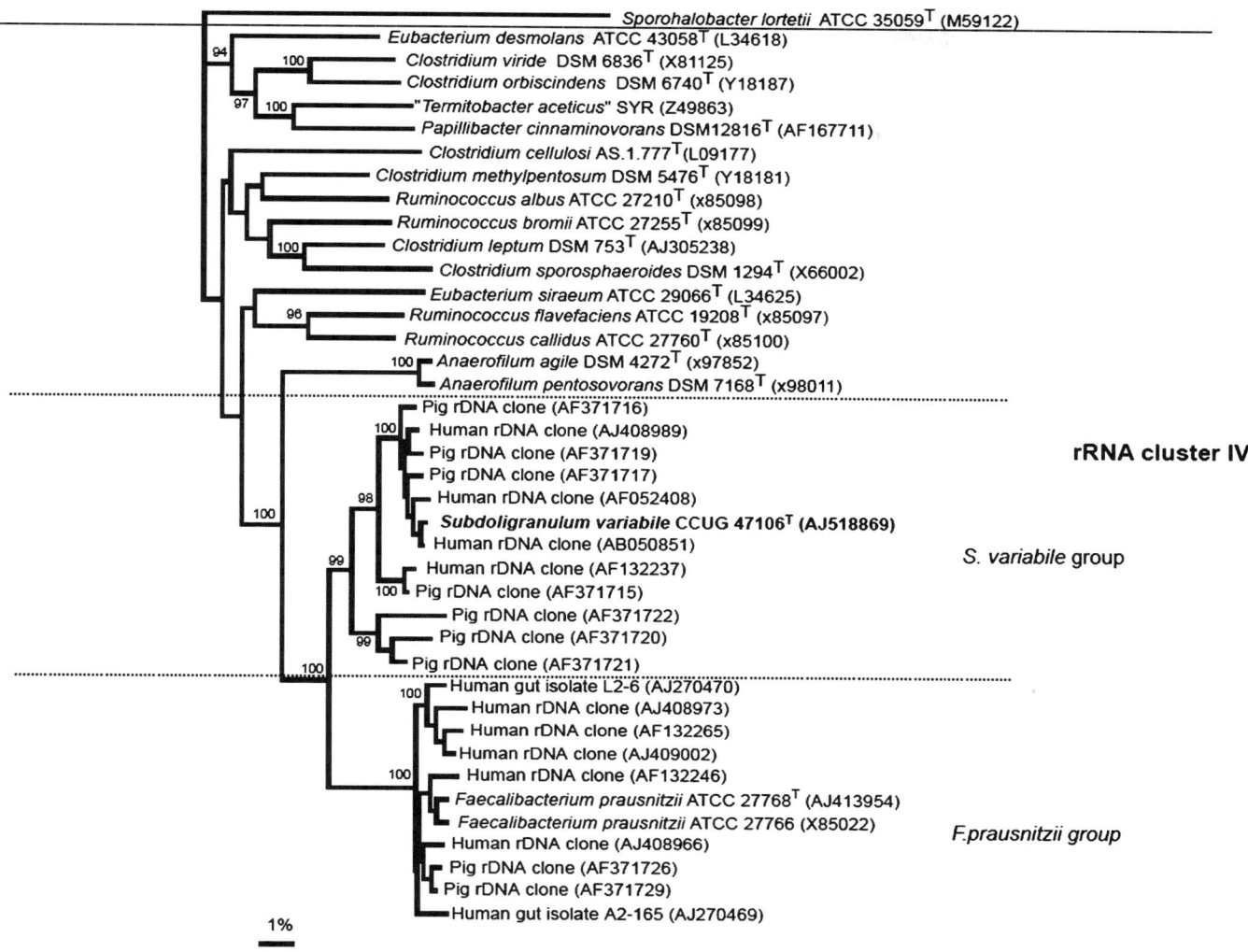

FIGURE 194. Unrooted phylogenetic tree of the 16S rRNA genes of the *Clostridium leptum* supra-generic rRNA cluster indicating the position of *Subdoligranulum variabile*. The phylogenetic tree was calculated by the neighbor-joining method. Bootstrap values greater than 90 are indicated. Bar = 1% sequence divergence.

et al. (2000) and in fastidious anaerobe broth with cooked meat granules (Oxoid). For long term preservation, strains can be maintained in media containing 15–20% glycerol at –70°C or lyophilized.

Taxonomic comments

The genus *Subdoligranulum* was created in 2004 to accommodate a phylogenetically distinct Gram-stain-negative, asporogenous, strictly anaerobic organism recovered from human feces (Holmstrøm et al., 2004). Phylogenetic analysis using 16S rDNA sequences shows that *Subdoligranulum variabile* clusters within a supra-generic grouping (rRNA cluster IV; Collins et al., 1994) that includes a diverse assortment of organisms such as *Anaerofilum* species, *Clostridium* species, *Eubacterium siraeum*, *Faecalibacterium prausnitzii*, and *Ruminococcus* species. *Subdoligranulum variabile* forms a distinct lineage within this grouping and does not display

a particularly close affinity with any of the aforementioned taxa (Figure 194). It is clear from phylogenetic molecular profiling studies of the *Clostridium leptum* group in human feces that, in addition to the diverse group of named organisms, some isolates and many cloned sequences corresponding to as yet uncultivated organisms are present. Furthermore, many of the sequences cloned from human feces possess high similarity to those cloned from other animals such as swine (Leser et al., 2002). These sequences may represent novel taxa including additional species of *Subdoligranulum*, that have not yet been described (Holmstrøm et al., 2004; Lay et al., 2007; Shen et al., 2006).

Differentiation of the genus *Subdoligranulum* from other closely related genera

In addition to differentiation by 16S rRNA gene sequence analysis, *Subdoligranulum variabile* can be readily distinguished from

TABLE 199. Characteristics differentiating the genus *Subdoligranulum* from phylogenetically closely related taxa[a,b]

Characteristic	*Subdoligranulum*	*Anaerofilum*	*Clostridium cellulosi*	*Clostridium leptum*	*Clostridium sporosphaeroides*	*Eubacterium siraeum*	*Faecalibacterium*	*Ruminococcus*[c]
Cellular morphology	Coccoid-droplet, highly pleomorphic	Thin long rods	Rods	Rods	Rods	Rods	Rods	Coccoid
Spores	–	–	+	+	+	–	–	–
Motility	Nonmotile	Motile	Motile	Nonmotile	Nonmotile	Motile	Nonmotile	Nonmotile
End products of metabolism	B, L, a, s	L, A, E, F, 2,3-butanediol	A, E[d]	A, e	A, B, p	A, E, l, b, s[e]	B, L, F	Various (depending on species)[c]
Hydrolysis of esculin	+	ND	+	D	–	D	+	D
Hydrolysis of starch	–	–	+	–	–	D	D	D
Cellulose degradation	–	–	+	–	–	–	–	V
Growth at 60°C	–	–	+			–	–	–
DNA G+C content (mol%)	52	54–55	35	51–52	27	45	47–57	39–46

[a]Symbols: +, >85% positive; d, different strains give different reactions (16–84% positive); –, 0–15% positive; w, weak reaction; ND, not determined. A (a), Acetic acid; B (b), Butyric acid; E (e), Ethanol; F, Formic acid; L, Lactic acid; P (p), Propionic acid; S (s), Succinic acid. Upper-case letter indicate major end products; lower-case letter indicate minor end products.

[b]Data from *Anaerophilum* (Zellner et al., 1996), *Clostridium cellulosi* (Yanling et al., 1991), *Clostridium leptum* (Cato et al., 1986), *Clostridium sporosphaeroides* (Cato et al., 1986), *Eubacterium siraeum* (Moore and Moore, 1985), *Faecalibacterium* (Duncan et al., 2002) and *Ruminococcus* (Bryant, 1985).

[c]Genus *Ruminococcus* is phylogenetically heterogeneous; data refers to *Ruminococcus albus*, *Ruminococcus bromii*, *Ruminococcus callidus*, and *Ruminococcus flavescens*.

[d]From cellulose fermentation.

[e]PY-cellobiose broth.

its closest phylogenetic relatives *Anaerofilum* species, *Clostridium* species, *Eubacterium siraeum*, *Faecalibacterium prausnitzii*, and *Ruminococcus* species using a combination of morphology and biochemical and chemotaxonomic criteria (see Table 199).

List of species of the genus *Subdoligranulum*

1. **Subdoligranulum variabile** Holmstrøm, Collins, Møller, Falsen and Lawson 2004, 1909[VP] (Effective publication: Holmstrøm, Collins, Møller, Falsen and Lawson 2004, 202.)

va.ri.a′bi.le. L. neut. adj. *variabile* because the cells are varied in shape.

Cells stain Gram-negative, are nonmotile and nonspore-forming. Colonies on BUA+B agar grow to a size of approximately 1 mm in diameter after 4–5 d incubation at 37°C and are grayish-white in appearance, circular, and concave. Cells range in size between 0.6–2.5 μm and exhibit variable forms (from circular, ellipsoid to drop-like shapes). Strictly anaerobic; no growth in 2% or 6% oxygen. Optimum growth temperature is 37°C; grows weakly at 45°C. Catalase-negative. The major products of glucose fermentation are butyric and lactic acids; small amounts of acetic and succinic acids are also formed. Using traditional methods, esculin is hydrolyzed but starch is not. Lecithinase, lipase, and urease are not produced. Using the AN-MicroLog system, *N*-acetyl-D-glucosamine, *N*-acetyl-D-mannosamine, amygdalin, arbutin, D-cellobiose, dextrin, D-fructose, L-fucose, D-galactose, D-galacturonic acid, gentiobiose, α-D-glucose, glucose-1-phosphate, glucose-6-phosphate, α-D-lactose, lactulose, maltose, maltotriose, D-mannose, D-melibiose, 3-methyl-D-glucose, α-methyl-D-galactoside, β-methyl-D-galactoside, palatinose, D-raffinose, L-rhamnose, salicin, sucrose, D-trehalose, and turanose are utilized. *N*-acetyl-galactosamine, adonitol, D-arabitol, α-cyclodextrin, β-cyclodextrin, dulcitol, erythritol, D-gluconic acid, D-glucosaminic acid, glycerol, DL-α–glycerol phosphate, inositol, D-mannitol, melezitose, α-methyl-D-glucoside, β-methyl-D-glucoside, D-sorbitol, and stachyose are not utilized. Using the API Rapid ID 32A system, acid is not formed from mannose or raffinose. Activity is observed for α-galactosidase, β-galactosidase, β-glucosidase, β-glucuronidase, arginine arylamidase, leucine arylamidase, and histidine arylamidase but not for alkaline phosphatase, α -arabinosidase, arginine dihydrolase, α-fucosidase, β-galactosidase-6-phosphate, N-acetyl-β-glucosaminidase, glutamic acid decarboxylase, glutamyl glutamic acid arylamidase, glycine arylamidase, leucyl glycine arylamidase, proline arylamidase, phenyl alanine arylamidase, pyroglutamic acid arylamidase, serine arylamidase, tyrosine arylamidase, or urease. Activity for α-glucosidase is either absent or weak. Indole is not produced and nitrate is not reduced. Using the API ZYM test system, positive reactions are observed for α-galactosidase, β-galactosidase, β-glucosidase, β-glucuronidase, and α-glucosidase (weak reaction). Alka-

line phosphatase, esterase C4, esterase lipase C8, lipase C14, valine arylamidase, cystine arylamidase, trypsin, chymotrypsin, acid phosphatase, α-fucosidase, α-mannosidase, and N-acetyl-β-glucosaminidase are not produced. The long-chain cellular fatty acids are mainly of the straight-chain saturated and monounsaturated types, with $C_{16:0}$ and $C_{18:1\ \omega9c}$ predominating; minor amounts of aldehydes and dimethyl acetals are also detected. Habitat is not known but probably resides in the human gut.

DNA G+C content (mol%): 52.2 (T_m).

Type strain: BI 114, CCUG 47106, DSM 15176.

GenBank accession number (16S rRNA gene): AJ518869.

References

Auburger, G. and J. Winter. 1995. Isolation and physiological characterization of *Syntrophus buswellii* strain GA from a syntrophic benzoate-degrading, strictly anaerobic coculture. Appl. Environ. Microbiol. *44*: 241–248.

Bak, F., K. Finster and F. Rothfuss. 1992. Formation of dimethylsulfide and methanethiol from methoxylated aromatic compounds and inorganic sulfide by newly isolated anaerobic bacteria. Arch. Microbiol. *157*: 529–534.

Barcenilla, A., S.E. Pryde, J.C. Martin, S.H. Duncan, C.S. Stewart, C. Henderson and H.J. Flint. 2000. Phylogenetic relationships of butyrate-producing bacteria from the human gut. Appl. Environ. Microbiol. *66*: 1654–1661.

Bernalier, A., A. Willems, M. Leclerc, V. Rochet and M.D. Collins. 1996. *Ruminococcus hydrogenotrophicus* sp. nov., a new H_2/CO_2-utilizing acetogenic bacterium isolated from human feces. Arch. Microbiol. *166*: 176–183.

Bernalier, A., A. Willems, M. Leclerc, V. Rochet and M.D. Collins. 1997. *In* Validation of the publication of new names and new combinations previously effectively published outside the IJSB. List no. 61. Int. J. Syst. Bacteriol. *47*: 601–602.

Brauman, A., M.D. Kane, M. Labat and J.A. Breznak. 1992. Genesis of acetate and methane by gut bacteria of nutritionally diverse termites. Science *257*: 1384–1387.

Brauman, A., J.A. Muller, J.L. Garcia, A. Brune and B. Schink. 1998. Fermentative degradation of 3-hydroxybenzoate in pure culture by a novel strictly anaerobic bacterium, *Sporotomaculum hydroxybenzicum* gen. nov., sp. nov. Int. J. Syst. Bacteriol. *48*: 215–221.

Breznak, J.A. and J.M. Switzer. 1986. Acetate synthesis from H_2 plus CO_2 by termite gut microbes. Appl. Environ. Microbiol. *52*: 623–630.

Breznak, J.A., J.M. Switzer and H.J. Seitz. 1988. *Sporomusa termitida* sp. nov., an H_2/CO_2-utilizing acetogen Iisolated from termites. Arch. Microbiol. *150*: 282–288.

Breznak, J.A. and A. Brune. 1994. Role of microorganisms in the digestion of lignocellulose by termites. Annu. Rev. Entomol. *39*: 453–487.

Brune, A., E. Miambi and J.A. Breznak. 1995. Roles of oxygen and the intestinal microflora in the metabolism of lignin-derived phenylpropanoids and other monoaromatic compounds by termites. Appl. Environ. Microbiol. *61*: 2688–2695.

Bryant, M.P. 1985. Genus *Ruminococcus*. *In* Sneath, Mair, Sharpe and Holt (Editors), Bergey's Manual of Systematic Bacteriology, Vol. 2. Williams & Wilkins Co., Baltimore, pp. 1093–1097.

Bryant, M.P. 1986. Genus *Ruminococcus* Sijpesteijn 1948, 152[AL]. *In* Sneath, Mair, Sharpe and Holt (Editors), Bergey's Manual of Systematic Bacteriology, Vol. 2. The Williams & WIlkins Co., Baltimore, pp. 1093–1099.

Butler, J.H. and J.C. Buckerfield. 1979. Digestion of lignin by termite. Soil Biol. Biochem. *11*: 507–513.

Cato, E.P., C.W. Salmon and L.V. Holdeman. 1974. *Eubacterium cylindroides* (Rocchi) Holdeman and Moore: emended description and designation of neotype strain. Int. J. Syst. Bacteriol. *24*: 256–259.

Cato, E.P., W.L. George and S.M. Finegold. 1986. Genus *Clostridium* Prazmowski 1880, 23[AL]. *In* Sneath, Mair, Sharpe and Holt (Editors), Bergey's Manual of Systematic Bacteriology, Vol. 2. The Williams & Wilkins Co., Baltimore, pp. 1141–1200.

Chatton, E. and C. Pérard. 1913. Schizophytes du caecum du cobaye. I. *Oscillospira guilliermondii* n. g., n. sp. CR Soc. Biol. Paris *74*.

Chen, S. and X. Dong. 2004. *Acetanaerobacterium elongatum* gen. nov., sp. nov., from paper mill waste water. Int. J. Syst. Evol. Microbiol. *54*: 2257–2262.

Clarke, R.T.J. 1979. Niche in pasture-fed ruminants for the large rumen bacteria *Oscillospira, Lampropedia,* and Quin's and Eadie's Ovals. Appl. Environ. Microbiol. *37*: 654–657.

Collins, M.D., P.A. Lawson, A.Willems, J.J. Cordoba, J. Fernandez-Garayzabal, P. Garcia, J. Cia, H. Hippe and J.A.E. Farrow. 1994. The phylogeny of the genus *Clostridium*: proposal of five new genera and eleven new species combination. Int. J. Syst. Bacteriol. *44*: 812–826.

Cookson, L.J. 1988. The site of mechanism 14C lignin degradation by *Nasutitermes exitiosus*. J. Insect. Physiol. *34*: 409–414.

Defnoun, S., M. Labat, M. Ambrosio, J.L. Garcia and B.K. Patel. 2000. *Papillibacter cinnamivorans* gen. nov., sp. nov., a cinnamate-transforming bacterium from a shea cake digester. Int. J. Syst. Evol. Microbiol. *50*: 1221–1228.

Diekert, G. and G. Wohlfarth. 1994. Metabolism of homoacetogens. Antonie Van Leeuwenhoek *66*: 209–221.

Duncan, S.H., G.L. Hold, H.J. Harmsen, C.S. Stewart and H.J. Flint. 2002. Growth requirements and fermentation products of *Fusobacterium prausnitzii*, and a proposal to reclassify it as *Faecalibacterium prausnitzii* gen. nov., comb. nov. Int. J. Syst. Evol. Microbiol. *52*: 2141–2146.

Ezaki, T., S.L. Liu, Y. Hashimoto and E. Yabuuchi. 1990. *Peptostreptococcus hydrogenalis* sp.nov from human fecal and vaginal flora. Int. J. Syst. Bacteriol. *40*: 305–306.

Ezaki, T., N. Li, Y. Hashimoto, H. Miura and H. Yamamoto. 1994. 16S Ribosomal DNA sequences of anaerobic cocci and proposal of *Ruminococcus hansenii* comb. nov. and *Ruminococcus productus* comb. nov. Int. J. Syst. Bacteriol. *44*: 130–136.

Ezaki, T., Y. Kawamura, N. Li, Z.Y. Li, L. Zhao and S. Shu. 2001. Proposal of the genera *Anaerococcus* gen. nov., *Peptoniphilus* gen. nov. and *Gallicola* gen. nov. for members of the genus *Peptostreptococcus*. Int. J. Syst. Evol. Microbiol. *51*: 1521–1528.

Falsen, E., M.D. Collins, C. Welinder-Olsson, Y. Song, S.M. Finegold and P.A. Lawson. 2005. *Fastidiosipila sanguinis* gen. nov., sp. nov., a new Gram-positive, coccus-shaped organism from human blood. Int. J. Syst. Evol. Microbiol. *55*: 853–858.

Gibson, T. 1974. Genus *Sporosarcina* Kluyver and van Neil. *In* Buchanan and Gibbons (Editors), Bergey's Manual of Determinative Bacteriology. The WIlliams & Wilkins Co., Baltimore, pp. 573–574.

Grain, J. and J. Senaud. 1976. *Oscillospira guilliermondii*, a rumen bacteria: ultrastructural study of the trichome and of sporulation. J. Ultrastruct. Res. *55*: 228–244.

Grasse, P.P. and C. Noirot. 1959. L'évolution de la symbiose chez les Isotères. Experientia *15*: 365–372.

Grech-Mora, I., M.L. Fardeau, B.K.C. Patel, B. Ollivier, A. Rimbault, G. Prensier, J.L. Garcia and E. Garnier-Sillam. 1996. Isolation and characterization of *Sporobacter termitidis* gen nov sp. nov., from the digestive tract of the wood-feeding termite *Nasutitermes lujae*. Int. J. Syst. Bacteriol. *46*: 512–518.

Harwood, C.S. and J. Gibson. 1988. Anaerobic and aerobic metabolism of diverse aromatic compounds by the photosynthetic bacterium *Rhodopseudomonas palustris*. Appl. Environ. Microbiol. *54*: 712–717.

Hauduroy, P., G. Ehringer, A. Urbain, G. Guillot and J. Magrou. 1937. Dictionnaire des bactéries pathogènes. Masson and Co., Paris, pp. 1–597.

Hethener, P., A. Brauman and J.L. Garcia. 1992. *Clostridium termitidis* sp. nov., a cellulolytic bacterium from the gut of the wood-feeding termite, *Nasutitermes lujae*. Syst. Appl. Microbiol. *15*: 52–58.

Holdeman, L.V. and W.E.C. Moore. 1974. New genus, *Coprococcus*, twelve new species, and emended descriptions of four previously described species of bacteria from human feces. Int. J. Syst. Bacteriol. *24*: 260–277.

Holdeman, L.V., E.P. Cato and W.E.C. Moore (Editors). 1977. Anaerobe Laboratory Manual, 4th edn. Anaerobe Laboratory, Virginia Polytechnic Institute and State University, Blacksburg, VA.

Holmstrøm, K., M.D. Collins, T. Moller, E. Falsen and P.A. Lawson. 2004. *Subdoligranulum variabile* gen. nov., sp. nov. from human feces. Anaerobe *10*: 197–203.

Holmstrøm, K., M. D. Collins, T. Møller, E. Falsen and P.A. Lawson. 2004. *In* Validation of publication of new names and new combinations previously effectively published outside the IJSEM Validation List no100 Int. J. Syst. Evol. Microbiol. *54*: 1909–1910.

Hungate, R.E. 1957. Microorganisms in the rumen of cattle fed a constant ration. Can. J. Microbiol. *3*: 289–311.

Hungate, R.E. 1969. A roll tube method for cultivation of strict anaerobes. *In* Norris and Ribbons (Editors), Methods in Microbiology, Vol. 3B. Academic Press, London and New York, pp. 117–132.

Imhoff-Stuckle, D. and N. Pfennig. 1983. Isolation and characterization of a nicotinic-acid degrading sulfate-reducing bacterium, *Desulfococcus niacini* sp. nov. Arch. Microbiol. *136*: 194–198.

Kane, M.D. and J.A. Breznak. 1991. *Acetonema longum* gen. nov. sp. nov., an H_2/CO_2 acetogenic bacterium from the termite, *Pterotermes occidentis*. Arch. Microbiol. *156*: 91–98.

Khan, A.W., E. Meek, L.C. Sowden and J.R. Colvin. 1984. Emendation of the genus *Acetivibrio* and description of *Acetivibrio cellulosolvens* sp. nov., a nonmotile cellulolytic mesophile. Int. J. Syst. Bacteriol. *34*: 419–422.

Kurihara, Y., J.M. Eadie, P.N. Hobson and S.O. Mann. 1968. Relationship between bacteria and ciliate protozoa in the sheep rumen. J. Gen. Microbiol. *51*: 267–288.

Lau, S.K., P.C. Woo, G.K. Woo, A.M. Fung, A.H. Ngan, Y. Song, C. Liu, P. Summanen, S.M. Finegold and K. Yuen. 2006. Bacteraemia caused by *Anaerotruncus colihominis* and emended description of the species. J. Clin. Pathol. *59*: 748–752.

Lawson, P.A., Y. Song, C. Liu, D.R. Molitoris, M.L. Vaisanen, M.D. Collins and S.M. Finegold. 2004. *Anaerotruncus colihominis* gen. nov., sp. nov., from human faeces. Int. J. Syst. Evol. Microbiol. *54*: 413–417.

Lay, C., M. Sutren, V. Rochet, K. Saunier, J. Dore and L. Rigottier-Gois. 2005. Design and validation of 16S rRNA probes to enumerate members of the *Clostridium leptum* subgroup in human faecal microbiota. Environ. Microbiol. *7*: 933–946.

Lay, C., J. Dore and L. Rigottier-Gois. 2007. Separation of bacteria of the *Clostridium leptum* subgroup from the human colonic microbiota by fluorescence-activated cell sorting or group-specific PCR using 16S rRNA gene oligonucleotides. FEMS Microbiol. Ecol. *60*: 513–520.

Leser, T.D., J.Z. Amenuvor, T.K. Jensen, R.H. Lindecrona, M. Boye and K. Moller. 2002. Culture-independent analysis of gut bacteria: the pig gastrointestinal tract microbiota revisited. Appl. Environ. Microbiol. *68*: 673–690.

Li, N., Y. Hashimoto and T. Ezaki. 1994. Determination of 16S ribosomal RNA sequences of all members of the genus *Peptostreptococcus* and their phylogenetic position. FEMS Microbiol. Lett. *116*: 1–5.

Liesack, W., F. Bak, J.U. Kreft and E. Stackebrandt. 1994. *Holophaga foetida* gen. nov., sp. nov., a new, homoacetogenic bacterium degrading methoxylated aromatic compounds. Arch. Microbiol. *162*: 85–90.

Liu, J.R., R.S. Tanner, P. Schumann, N. Weiss, C.A. McKenzie, P.H. Janssen, E.M. Seviour, P.A. Lawson, T.D. Allen and R.J. Seviour. 2002. Emended description of the genus *Trichococcus*, description of *Trichococcus collinsii* sp. nov., and reclassification of *Lactosphaera pasteurii* as *Trichococcus pasteurii* comb. nov. and of *Ruminococcus palustris* as *Trichococcus palustris* comb. nov. in the low-G+C Gram-positive bacteria. Int. J. Syst. Evol. Microbiol. *52*: 1113–1126.

Ludwig, W., K.-H. Schleifer and W.B. Whitman. 2009. Revised road map to the phylum *Firmicutes*. *In* De Vos, Garrity, Jones, Krieg, Ludwig, Rainey, Schleifer and Whitman (Editors), Bergey's Manual of Systematic Bacteriology, 2nd edn, Vol. 3, The *Firmicutes*, Springer, New York, pp. 1–14.

Mackie, R.I., R.I. Aminov, W.P. Hu, A.V. Klieve, D. Ouwerkerk, M.A. Sundset and Y. Kamagata. 2003. Ecology of uncultivated *Oscillospira* species in the rumen of cattle, sheep, and reindeer as assessed by microscopy and molecular approaches. Appl. Environ. Microbiol. *69*: 6808–6815.

Macy, J.M., J.E. Snellen and R.E. Hungate. 1972. Use of syringe methods for anaerobiosis. Am. J. Clin. Nutr. *25*: 1318–1323.

Madden, R.H., M.J. Bryder and N.J. Poole. 1982. Isolation and characterization of an anaerobic, cellulolytic bacterium, *Clostridium papyrosolvens* sp. nov. Int. J. Syst. Bacteriol. *32*: 87–91.

Miller, T.L. and M.J. Wolin. 1974. A serum bottle modification of the Hungate technique for cultivating obligate anaerobes. Appl. Microbiol. *27*: 985–987.

Moir, R.J. and M.J. Masson. 1952. An illustrated scheme for the microscopic identification of the rumen microorganisms of sheep. J. Pathol. Bacteriol. *64*: 343–350.

Monserrate, E., S.B. Leschine and E. Canale-Parola. 2001. *Clostridium hungatei* sp. nov., a mesophilic, N_2-fixing cellulolytic bacterium isolated from soil. Int. J. Syst. Evol. Microbiol. *51*: 123–132.

Moore, W.E.C., E.P. Cato and L.V. Holdeman. 1972. *Ruminococcus bromii* sp. n. and emendation of the description of *Ruminococcus* Sijpestein. Int. J. Syst. Bacteriol. *22*: 78–80.

Moore, W.E.C., J.L. Johnson and L.V. Holdeman. 1976. Emendation of *Bacteroidaceae* and *Butyrivibrio* and descriptions of *Desulfomonas* gen. nov. and ten new species in genera *Desulfomonas*, *Butyrivibrio*, *Eubacterium*, *Clostridium*, and *Ruminococcus*. Int. J. Syst. Bacteriol. *26*: 238–252.

Moore, W.E.C. and L.V.H. Moore. 1985. Genus *Eubacterium*. *In* Sneath, Mair, Sharpe and Holt (Editors), Bergey's Manual of Systematic Bacteriology, Vol. 2. The Williams & Wilkins Co., Baltimore, pp. 1353–1373.

Murdoch, D.A., M.D. Collins, A. Willems, J.M. Hardie, K.A. Young and J.T. Magee. 1997. Description of three new species of the genus *Peptostreptococcus* from human clinical specimens: *Peptostreptococcus harei* sp. nov., *Peptostreptococcus ivorii* sp. nov., and *Peptostreptococcus octavius* sp. nov. Int. J. Syst. Bacteriol. *47*: 781–787.

Murdoch, D.A. and H.N. Shah. 1999. Reclassification of *Peptostreptococcus magnus* (Prevot 1933) Holdeman and Moore 1972 as *Finegoldia magna* comb. nov. and *Peptostreptococcus micros* (Prevot 1933) Smith 1957 as *Micromonas micros* comb. nov. Anaerobe *5*: 555–559.

Murray, W.D., L.C. Sowden and J.R. Colvin. 1984. *Bacteroides cellulosolvens* sp. nov., a cellulolytic species from sewage sludge. Int. J. Syst. Bacteriol. *34*: 185–187.

Murray, W.D. 1986. *Acetivibrio cellulosolvens* is a synonym for *Acetivibrio cellulolyticus*: emendation of the genus *Acetivibrio*. Int. J. Syst. Bacteriol. *36*: 314–316.

Namsolleck, P., R. Thiel, P. Lawson, K. Holmstrøm, M. Rajilic, E. E. Vaughan, L. Rigottier-Gois, M. D. Collins, W. De Vos and M. Blaut. 2004. Molecular methods for the analysis of gut microbiota. Microb. Ecol. Health Dis. *19*: 71–85.

Patel, G.B., A.W. Khan, B.J. Agnew and J.R. Colvin. 1980. Isolation and characterization of an anaerobic, cellulolytic microorganism, *Acetivibrio cellulolyticus* gen. nov. sp. nov. Int. J. Syst. Bacteriol. *30*: 179–185.

Petitdemange, E., F. Caillet, J. Giallo and C. Gaudin. 1984. *Clostridium cellulolyticum* sp. nov., a cellulolytic, mesophilic species from decayed grass. Int. J. Syst. Bacteriol. *34*: 155–159.

Prévot, A.R. 1941. Sur une nouvelle espéces de streptococque anaérobie gazogéne: *Streptococcus productus* nov. spec. C.R. Séances Soc. Biol. Fil. (France) *135*: 105–107.

Rainey, F.A. and P.H. Janssen. 1995. Phylogenetic analysis by 16S ribosomal DNA sequence comparison reveals two unrelated groups of species within the genus *Ruminococcus*. FEMS Microbiol. Lett. *129*: 69–73.

Ren, N., D. Xing, B.E. Rittmann, L. Zhao, T. Xie and X. Zhao. 2007. Microbial community structure of ethanol type fermentation in biohydrogen production. Environ Microbiol *9*: 1112–1125.

Rieu-Lesme, F., B. Morvan, M.D. Collins, G. Fonty and A. Willems. 1996. A new H$_2$/CO$_2$-using acetogenic bacterium from the rumen: description of *Ruminococcus schinkii* sp. nov. FEMS Microbiol. Lett. *140*: 281–286.

Rieu-Lesme, F., B. Morvan, M.D. Collins, G. Fonty and A. Willems. 1997. *In* Validation of the publication of new names and new combinations previously effectively published outside the IJSB. List no. 60. Int. J. Syst. Bacteriol. *47*: 242.

Robinson, I.M. and A.E. Ritchie. 1981. Emendation of *Acetivibrio* and description of *Acetivibrio ethanolgignens*, a new species from the colons of pigs with dysentery. Int. J. Syst. Bacteriol. *31*: 333–338.

Rogosa, M. 1974. Family III. *Peptostreptococcaceae* Rogosa 1971. *In* Buchanan and Gibbons (Editors), Bergey's Manual of Determinative Bacteriology, 8th edn. The Williams & Wilkins Co., Baltimore, pp. 517–528.

Schneider, H., A. Schwiertz, M.D. Collins and M. Blaut. 1999. Anaerobic transformation of quercetin-3-glucoside by bacteria from the human intestinal tract. Arch. Microbiol. *171*: 81–91.

Shen, J., B. Zhang, G. Wei, X. Pang, H. Wei, M. Li, Y. Zhang, W. Jia and L. Zhao. 2006. Molecular profiling of the *Clostridium leptum* subgroup in human fecal microflora by PCR-denaturing gradient gel electrophoresis and clone library analysis. Appl. Environ. Microbiol. *72*: 5232–5238.

Sijpesteijn, A.K. 1948. Cellulose-decomposnig bacteria from the rumen of cattle. Leiden University, Eduard Ijdo N.V., Leiden.

Simmering, R., D. Taras, A. Schwiertz, G. Le Blay, B. Gruhl, P.A. Lawson, M.D. Collins and M. Blaut. 2002a. *Ruminococcus luti* sp. nov., isolated from a human faecal sample. Syst. Appl. Microbiol. *25*: 189–193.

Simmering, R., D. Taras, A. Schwiertz, G. Le Blay, B. Gruhl, P.A. Lawson, M.D. Collins and M. Blaut. 2002b. *In* Validation of publication of new names and new combinations previously effectively published ouside the IJSEM. List no. 88. Int. J. Syst. Evol. Microbiol. *52*: 1915–1916.

Smith, L.D.S. 1957. *Peptostreptococcus* Kluyver and Van Neil, 1936. *In* Breed, Murray and Smith (Editors), Bergey's Manual of Determinative Bacteriology, 7th edn. The Williams & Wilkins Co., Baltimore, pp. 533–541.

Sneath, P.H.A. 1986. Endospore-forming gram positive rods and cocci. *In* Sneath, Mair, Sharpe and Holt (Editors), Bergey's Manual of Systematic Bacteriology, Vol. 2. The Williams & Wilkins Co., Baltimore, pp. 1104–1207.

Stewart, C.S., H.J. Flint and M.P. Bryant. 1997. The rumen bacteria. *In* Hobson and Stewart (Editors), The Rumen Microbial Ecology, 2nd edn. Blackie Academic & Professional, London & New York, pp. 10–72.

Tanaka, K., K. Nakamura and E. Mikami. 1991. Fermentation of cinnamate by a mesophilic strict anaerobe, *Acetivibrio multivorans* sp. nov. Arch. Microbiol. *155*: 120–124.

Tanaka, K., K. Nakamura and E. Mikami. 1992. *In* Validation of the publication of new names and new combinations previously effectively published outside the IJSB. List no. 40. Int. J. Syst. Bacteriol. *42*: 191–192.

Warner, A.C. 1966. Diurnal changes in the concentrations of micro-organisms in the rumens of sheep fed to appetite in pens or at pasture. J. Gen. Microbiol. *45*: 243–251.

Wexler, H.M., D. Reeves, P.H. Summanen, E. Molitoris, M. McTeague, J. Duncan, K.H. Wilson and S.M. Finegold. 1996. *Sutterella wadsworthensis* gen. nov, sp. nov., bile-resistant microaerophilic *Campylobacter gracilis*-like clinical isolates. Int. J. Syst. Bacteriol. *46*: 252–258.

Widdel, F. and N. Pfennig. 1981. Studies on dissimilatory sulfate-reducing bacteria that decompose fatty acids. 1. Isolation of new sulfate-reducing bacteria enriched with acetate from saline environments: description of *Desulfobacter postgatei* gen. nov., sp. nov. Arch. Microbiol. *129*: 395–400.

Williams, A.G. and G.S. Coleman. 1997. The rumen protozoa. *In* Hobson and Stewart (Editors), The Rumen Microbial Ecology, 2nd edn. Blackie Academic & Professional, London, pp. 73–139.

Winker, S. and C.R. Woese. 1991. A definition of the domains *Archaea*, *Bacteria* and *Eucarya* in terms of small subunit ribosomal RNA characteristics. Syst. Appl. Microbiol. *13*: 161–165.

Winter, J., L.H. Moore, V.R.J. Dowell and V.D. Bokkenheuser. 1989. C-ring cleavage of flavonoids by human intestinal bacteria. Appl. Environ. Microbiol. *55*: 1203–1208.

Winter, J., M.R. Popoff, P. Grimont and V.D. Bokkenheuser. 1991. *Clostridium orbiscindens* sp. nov., a human intestinal bacterium capable of cleaving the flavonoid C-ring. Int. J. Syst. Bacteriol. *41*: 355–357.

Xing, D., N. Ren, Q. Li, M. Lin, A. Wang and L. Zhao. 2006. *Ethanoligenens harbinense* gen. nov., sp. nov., isolated from molasses wastewater. Int. J. Syst. Evol. Microbiol. *56*: 755–760.

Yanagita, K., A. Manome, X.Y. Meng, S. Hanada, T. Kanagawa, T. Tsuchida, R.I. Mackie and Y. Kamagata. 2003. Flow cytometric sorting, phylogenetic analysis and in situ detection of *Oscillospira guilliermondii*, a large, morphologically conspicuous but uncultured ruminal bacterium. Int. J. Syst. Evol. Microbiol. *53*: 1609–1614.

Yang, J.C., D.P. Chynoweth, D.S. Williams and A. Li. 1990. *Clostridium aldrichii* sp. nov., a cellulolytic mesophile inhabiting a wood-fermenting anaerobic digester. Int. J. Syst. Bacteriol. *40*: 268–272.

Yanling, H., D. Youfang and L. Yanquan. 1991. Two cellulolytic *Clostridium* species: *Clostridium cellulosi* sp. nov. and *Clostridium cellulofermentans* sp. nov. Int. J. Syst. Bacteriol. *41*: 306–309.

Zellner, G., E. Stackebrandt, D. Nagel, P. Messner, N. Weiss and J. Winter. 1996. *Anaerofilum pentosovorans* gen. nov., sp. nov., and *Anaerofilum agile* sp. nov., two new, strictly anaerobic, mesophilic, acidogenic bacteria from anaerobic bioreactors. Int. J. Syst. Bacteriol. *46*: 871–875.

Family IX. **Syntrophomonadaceae** Zhao, Yang, Woese and Bryant 1993a, 284[VP]

YUJI SEKIGUCHI

Syn.tro.pho.mon.a.da′ce.ae. N.L. fem. n. *Syntrophomonas* type genus of the family; suff. *-aceae* ending to denote family; N.L. fem. pl. n. *Syntrophomonadaceae* the *Syntrophomonas* family.

Cells have various rod shapes, often curved, with or without flagella, and with or without endospores. Multiplication is by binary fission. Cells may have distinctive outer membranes or no outer membranes and stain either Gram-positive or Gram-negative. Cell walls contain muramic and *meso*-diaminopimelic acids. Poly-β-hydroxyalkanoates are the reserve material. The *Syntrophomonadaceae* are very **strict anaerobes** and heterotrophs; most of them (except for *Pelospora*) anaerobically **β-oxidize saturated fatty acids containing 4–18 straight- or branched-chained carbon atoms**, depending on the strain, in **syntrophic association with hydrogenotrophic micro-organisms** such as methanogenic archaea of the genera *Methanospirillum* and *Methanobacterium*. Organic end products from co-cultures include acetate, propionate, and other fatty acids, depending on the fatty acid energy source. Except for members of the genus *Pelospora*, strains can **grow in axenic culture without a hydrogenotrophic micro-organism with a few types of substrates, such as crotonate, *trans*-2-pentenoate, *trans*-2-hexenoate, *trans*-3-hexenoate, and *trans*,trans-2,4-hexadienoate**. These compounds are reduced to saturated fatty acids such as C_2 to C_6 fatty acids and little hydrogen (or formate) is produced. **Carbohydrates, proteinaceous materials, alcohols, or other organic compounds do not support growth of most members of the *Syntrophomonadaceae*.** Members of the genus *Pelospora* do not require syntrophic association and ferment a few compounds such as glutarate. Common electron acceptors such as **fumarate, malate, nitrate, nitrite, oxygen, sulfate, sulfite, sulfur, and thiosulfate are not used.** The *Syntrophomonadaceae* may be either marine, requiring at least 1% NaCl, or freshwater, requiring considerably lower salt concentrations. Some require type B vitamins. These bacteria are present in anaerobic microbial ecosystems, such as marine and freshwater sediments, sewage digester sludge, and the rumen, where organic matter is degraded to CO_2 and CH_4. Phylogenetically, as determined by 16S rRNA gene sequencing, this family can be distinguished from other members of the phylum *Firmicutes*. The following genera belong to this family: *Syntrophomonas, Syntrophospora, Pelospora, Thermosyntropha,* and *Syntrophothermus*. Differentiating features for species of these taxa are presented in Table 200.

DNA G+C content (mol%): 37–48.

Type genus: **Syntrophomonas** McInerney, Bryant, Hespell and Costerton 1982, 267[VP] (Effective publication: McInerney, Bryant, Hespell and Costerton 1981, 1037.) emend. Lorowitz, Zhao and Bryant 1989, 126, emend. Wu, Liu and Dong 2006, 2334.

Taxonomic comments

The family *Syntrophomonadaceae* is a phylogenetically distinct group of anaerobes that generally grows in syntrophic association with hydrogenotrophic organisms (such as methanogens) (Zhao et al., 1993a). It is one of the most deeply branching phylogenetic groups currently classified within the bacterial phylum *Firmicutes* (Figure 195 and Figure 7), and future classifications may move it to another phylum (see the road map chapter, Ludwig et al., 2009). The family encompasses physiologically

related organisms: most of the members (all known species of the genera *Syntrophomonas, Syntrophospora, Thermosyntropha,* and *Syntrophothermus*) syntrophically oxidize monocarboxylic acids with 4 (butyrate) to 18 carbons. The one exception is the genus *Pelospora* (Table 200).

While 16S rRNA gene-based phylogenetic analyses support the existence of the family *Syntrophomonadaceae sensu stricto,* many of the genera previously assigned to this family are widely dispersed and not closely related (Garrity et al., 2005). Because these organisms are only distantly related to the remainder of the *Firmicutes*, future reclassifications may necessitate the creation of more than one higher taxon, possibly even at the phylum level. Given this uncertain state of affairs, these other genera are described in subsequent chapters as "Family *incertae sedis*". Only two genera (*Anaerobranca* and *Carboxydocella*) clearly represent lineages of the *Firmicutes* and even these genera are not closely related to each other. The remaining three groups are only distantly related. Not only can they not be assigned definitively to any of the established phyla, but they are not closely related to each other. They include the following genera. One deep lineage is represented by *Aminobacterium, Aminomonas, Anaerobaculum, Dethiosulfovibrio, Thermanaerovibrio,* and *Thermovirga*. The second deep lineage includes the genera *Thermaerobacter, Symbiobacterium,* and *Sulfobacillus* (formerly assigned to the family "*Alicyclobacillaceae*" in the class "*Bacilli*"). Lastly, *Caldicellulosiruptor* represents the final deep lineage. This genus is related to some other unclassified genera, i.e., *Tepidanaerobacter, Thermosediminibacter,* and *Thermovenabulum*. The genus *Thermohydrogenium* is not described in the current edition because a type strain has not been located (F. Rainey, personal communication).

Further descriptive information

The physiological properties of the *Syntrophomonadaceae sensu stricto* are quite distinctive and deserve special explanation. Under methanogenic conditions, butyrate and longer chain fatty acids are degraded by syntrophic association with two or three trophic groups of anaerobes, namely proton-reducing, syntrophic monocarboxylic-acid-oxidizing bacteria, hydrogenotrophic methanogens, and aceticlastic methanogens (Schink, 1997; Stams, 1994). Monocarboxylic acids are converted by syntrophic acid-oxidizers to acetate and H_2 (and/or formate) and these products are subsequently utilized by the latter two trophic groups to form methane. Without this food chain, acid oxidation cannot proceed because the first step of the reaction is endergonic without coupling to H_2- and acetate-consuming reactions. Of the two intermediates, H_2 is the most important intermediate and H_2 consumption makes the whole process energetically feasible (Schink and Friedrich, 1994). Hence, this mutual relationship between the substrate oxidizers and hydrogenotrophic anaerobes is called "syntrophy" and the anaerobes of the former group are often called "syntrophs". This terminology is justified because growth of the syntrophs in their natural environments is obligately dependent on the activity of hydrogenotrophic organisms.

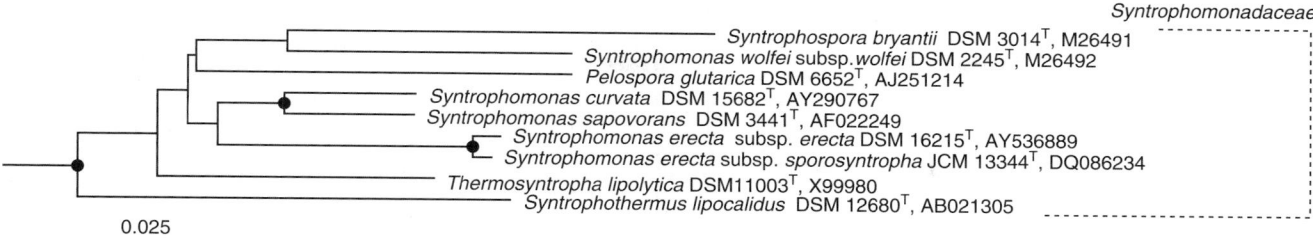

Syntrophomonadaceae

Syntrophospora bryantii DSM 3014ᵀ, M26491
Syntrophomonas wolfei subsp.wolfei DSM 2245ᵀ, M26492
Pelospora glutarica DSM 6652ᵀ, AJ251214
Syntrophomonas curvata DSM 15682ᵀ, AY290767
Syntrophomonas sapovorans DSM 3441ᵀ, AF022249
Syntrophomonas erecta subsp. erecta DSM 16215ᵀ, AY536889
Syntrophomonas erecta subsp. sporosyntropha JCM 13344ᵀ, DQ086234
Thermosyntropha lipolytica DSM11003ᵀ, X99980
Syntrophothermus lipocalidus DSM 12680ᵀ, AB021305

0.025

FIGURE 195. Phylogenetic neighbor-joining tree showing the relationship among the members of the family *Syntrophomonadaceae sensu stricto*. Bar = 25 nt substitutions per 1000 nt. Nodes supported by bootstrap values higher than 95 % are shown by solid circles. 16S rRNA gene sequences of some *Desulfotomaculum* species were used to root the tree.

Most of the organisms described that catalyze the syntrophic degradation of saturated fatty acids are members of this family, although a few exceptions exist, such as *Syntrophus aciditrophicus* (Jackson et al., 1999) and *Smithella propionica* (both in the *Deltaproteobacteria*) (Liu et al., 1999). The first member of the family, *Syntrophomonas wolfei*, was isolated from anaerobic digester sludge as "pure" co-cultures with either the hydrogenotrophic methanogen *Methanospirillum hungatei* or with the hydrogenotrophic sulfate-reducer *Desulfovibrio* sp. (McInerney et al., 1981, 1979). For that reason, the syntrophic fatty-acid-oxidizing anaerobes were originally thought to be obligately syntrophic anaerobes that could grow only in co-culture with hydrogenotrophic organisms. However, *Syntrophomonas wolfei* has been grown subsequently in pure culture with crotonate (Beaty and McInerney, 1987) and most of the *Syntrophomonadaceae* are now known to grow in axenic culture without a hydrogenotroph with a few substrates such as crotonate or pentenoate plus butyrate.

Genus I. **Syntrophomonas** McInerney, Bryant, Hespell and Costerton 1982, 267ᵛᴾ (Effective publication: McInerney, Bryant, Hespell and Costerton 1981, 1037.) emend. Lorowitz, Zhao and Bryant 1989, 126, emend. Wu, Liu and Dong 2006, 2334

Yᴜᴊɪ Sᴇᴋɪɢᴜᴄʜɪ

Syn.tro.pho.mon'as. Gr. adj. *syn* together with; Gr. n. *trophos* one who feeds; Gr. n. *monas* a unit, monad; N.L. fem. n. *Syntrophomonas* monad which feeds together with (another species).

Slightly helical rods, 0.5–1.0 by 2–7 μm, with slightly tapered rounded ends. Most cells occur singly or in pairs with helical chains of three or more. Multiplication is by binary fission. Cells possess 2–8 flagella (about 20 nm in diameter) that are inserted laterally in a linear fashion on the concave side of the cell, about 130 nm or more apart. Under most conditions, cells usually exhibit only sluggish twitching motility. Cells have an **unusual multilayered cell wall that stains Gram-negative.** Muramic and *meso*-diaminopimelic acids are present and the organism is susceptible to penicillin. Poly-β-hydroxybutyrate is present. Cells **β-oxidize saturated fatty acids anaerobically with protons serving as the electron acceptors in co-cultures with a hydrogenotroph.** Growth is much faster with *Methanospirillum hungatei*, which utilizes both H_2 and formate, than with methanogens such as *Methanobacterium bryantii* that utilize only H_2. In co-culture, cells produce acetate and H_2 from butyrate and longer straight-chain, even-numbered-carbon fatty acids and acetate, propionate, and H_2 from n-valerate and longer straight-chain, odd-numbered-carbon fatty acids. At least some strains produce acetate, isovalerate, and H_2 from iso acids such as isoheptanoate and some strains produce acetate, propionate, and H_2 from 2-methylbutyrate. It is possible, but has not yet been determined, that anteiso acids such as anteisooctanoate are degraded to acetate, anteisohexanoate, and H_2. The strains studied so far can, with considerable difficulty, be adapted to **grow slowly in axenic culture on crotonate, with acetate and butyrate as the products.**

Carbohydrates, proteinaceous materials, alcohols, and other organic compounds do not support growth. Common electron acceptors such as **fumarate, malate, nitrate, oxygen, sulfate, sulfite, sulfur, and thiosulfate are not utilized with butyrate as the electron donor.** Growth may be stimulated in co-cultures by factors in rumen fluid or mixtures of vitamins or both. Isolated as co-cultures from anaerobic ecosystems, such as aquatic sediments, sewage digester sludge, the rumen, and rice field mud, where organic matter is degraded, with CO_2 and CH_4 as major products. The temperature range for growth of co-cultures with *Methanospirillum hungatei* is in the broad mesophilic range, with an optimum of 30–37°C. Phylogenetically, as determined by 16S rRNA gene sequencing, the genus *Syntrophomonas* forms a deeply branched lineage in the bacteria related to the *Firmicutes*.

DNA G+C content (mol%): 40–47.

Type species: **Syntrophomonas wolfei** McInerney, Bryant, Hespell and Costerton 1982, 267ᵛᴾ (Effective publication: McInerney, Bryant, Hespell and Costerton 1981, 1037.) emend. Lorowitz, Zhao and Bryant 1989, 126.

Further descriptive information

Cell morphology. Cells are slightly curved rods with round ends, occurring singly or in pairs (Figure 196). Tumbling motility is observed in most strains. *Syntrophomonas wolfei* strains possess 2–8 flagella that are inserted laterally in a linear fashion about 130 nm or more apart on the concave side of the cell

FIGURE 196. Phase-contrast (a) and fluorescence (b) micrograph of *Syntrophomonas wolfei* subsp. *wolfei* DSM 2245B[T] (co-culture with *Methanospirillum hungatei* strain DSM 864[T]) grown with butyrate (bar = 10 μm). Both micrographs show an identical field, indicating the F_{420}-autofluorescence of the methanogens.

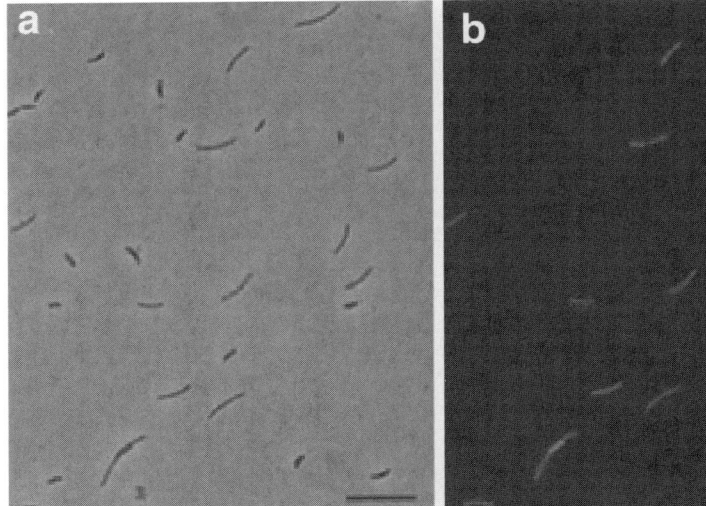

(McInerney et al., 1981). *Syntrophomonas sapovorans* possesses 2–4 flagella (Roy et al., 1986). *Syntrophomonas erecta* possesses 2–5 flagella (Wu et al., 2006). Among *Syntrophomonas* strains, only "*Syntrophomonas erecta* subsp. *sporosyntropha*" is characterized as nonmotile without flagella. Spore formation is not observed in most strains of the genus *Syntrophomonas*, with the exception of the spore-forming subspecies of "*Syntrophomonas erecta* subsp. *sporosyntropha*" (Wu et al., 2006).

Cell-wall composition and chemotaxonomic data. *Syntrophomonas* strains generally have an unusual, multilayered cell wall that stains Gram-negative. A similar envelope is also seen in phylogenetically related taxa such as those in the family "*Acidaminococcaceae*". The major membrane phospholipid fatty acids of *Syntrophomonas wolfei* are $C_{16:1}$, $C_{16:0}$, and $C_{15:0}$ (Henson et al., 1988). Similarly, $C_{14:0}$, OH-$C_{14:0}$, $C_{15:0}$, $C_{16:0}$, $C_{16:1}$, and iso-$C_{17:1}$ are the major cellular fatty acids in *Syntrophomonas curvata* and "*Syntrophomonas erecta* subsp. *erecta*" (see Table 200). Muramic acid and diaminopimelic acid have been demonstrated in *Syntrophomonas wolfei* (McInerney et al., 1981) and *Syntrophomonas erecta* (Zhang et al., 2005b).

Poly-β-hydroxyalkanoate is observed in *Syntrophomonas wolfei* (Amos and McInerney, 1989, 1991; McInerney et al., 1981) and "*Syntrophomonas erecta* subsp. *sporosyntropha*" (Wu et al., 2006).

Nutrition and growth conditions. *Syntrophomonas* strains grow on saturated fatty acids containing 4–18 carbon atoms in syntrophic association with hydrogenotrophic organisms. The range of degradable fatty acids varies with the species (see Table 200). In addition to saturated fatty acids, *Syntrophomonas wolfei* subsp. *saponavida* and *Syntrophomonas sapovorans* degrade some unsaturated fatty acids, such as oleate (*cis*-9-octadecenoate) and elaidate (*trans*-9-octadecenoate), when grown in co-culture with hydrogenotrophs (Lorowitz *et al.*, 1989; Roy *et al.*, 1986). Most *Syntrophomonas* strains grow in axenic culture without a hydrogenotroph with crotonate (Beaty and McInerney, 1987). *Syntrophomonas wolfei* also grows in axenic culture with some unsaturated short-chain fatty acids, such as *trans*-2-pentenoate, *trans*-2-hexenoate, *trans*-3-hexenoate, and *trans*,*trans*-2,4-hexadienoate (Amos and McInerney, 1990). Media supplemented with butyrate plus pentenoate or with butyrate plus dimethyl sulfoxide

(DMSO) also support growth of axenic cultures of some *Syntrophomonas* strains (Wu et al., 2006; Zhang et al., 2005b). Rumen fluid and B vitamins (thiamine, lipoic acid, biotin, and cyanocobalamin) stimulate the growth of *Syntrophomonas wolfei* in co-culture with *Methanospirillum hungatei* (McInerney et al., 1981). Axenic cultures of *Syntrophomonas wolfei* grow poorly in a defined crotonate medium in the absence of rumen fluid (Beaty and McInerney, 1990). However, addition of rumen fluid or B vitamins stimulates growth of the axenic culture as well as the co-culture (Beaty and McInerney, 1990). Iron and cobalt are required for the growth of *Syntrophomonas wolfei* in chemically defined medium (Beaty and McInerney, 1990). Amino acids and the B vitamin mixture are required for growth of *Syntrophomonas sapovorans* (Roy et al., 1986). Optimum growth temperature is 35–40°C.

Metabolism. *Syntrophomonas* strains have a unique anaerobic metabolism catalyzing β-oxidation of fatty acids (C_4–C_{18}). This reaction normally couples to the production of H_2 (and/or formate), which is subsequently consumed by partner hydrogenotrophic organisms such as methanogens. For energetic reasons, the H_2 partial pressure in cultures must be 10 to 100 Pa for β-oxidation to proceed. The partner hydrogenotrophs maintain this low partial pressure of H_2. The pathway of the β-oxidation by *Syntrophomonas wolfei* has been studied in detail (Amos and McInerney, 1993; Wofford et al., 1986). The pathway of crotonate metabolism has also been elucidated in *Syntrophomonas wolfei* (McInerney and Wofford, 1992; Sobieraj and Boone, 2002). A *c*-type cytochrome was detected in *Syntrophomonas wolfei* in pure culture with crotonate or in co-culture with methanogens, indicating the presence of an electron transport system (McInerney and Wofford, 1992).

Long-chain fatty acids (longer than C_8) are degraded by *Syntrophomonas wolfei* subsp. *saponavida*, *Syntrophomonas sapovorans*, and *Syntrophomonas curvata*. An equimolar concentration of calcium is required for the degradation of long-chain fatty acids. The degradation may involve several rounds of β-oxidation with the concomitant release of electrons as H_2, analogous to the pathway of syntrophic butyrate degradation (Sobieraj and Boone, 2002).

TABLE 200. Characteristics that differentiate species in the family *Syntrophomonadaceae*[a]

Characteristic	Syntrophomonas wolfei subsp. wolfei	Syntrophomonas wolfei subsp. saponavida	Syntrophomonas saporovorans	Syntrophomonas curvata	"Syntrophomonas erecta subsp. erecta"	"Syntrophomonas erecta subsp. sporosyntropha"	Pelospora glutarica	Syntrophospora bryantii	Syntrophothermus lipocalidus	Thermosyntropha lipolytica
Cell width (μm)	0.5–1.0	0.4–0.6	0.4–0.6	0.5–0.7	0.6–0.8	0.5–0.7	0.8	0.4	0.4–0.5	0.3–0.4
Cell length (μm)	2.0–7.0	2.0–4.0	2.0–4.0	2.3–4.0	2.0–8.0	4–14	4.5–6.5	4.5–6	2.0–4.0	2.0–3.5
Temperature range (°C)	25–45	ND	25–45	20–42	25–47	20–48	20–37	28–34	45–60	52–70
Optimum growth temp. (°C)	35–37	ND	35–37	35–37	37–40	37–40	37	ND	55	60–66
pH range	6.2–8.0	ND	6.3–8.1	6.3–8.4	6.0–8.8	5.5–8.4	>6.0	6.5–7.5	5.8–7.5	7.0–9.5
Optimum pH	7.0–7.5	ND	7.3	7.5	7.8	7.0	7.1–8.2	ND	6.5–7.0	8.1–8.9
Motility	+	+	+	+	+	+	+	–	+	–
Spore formation	–	–	–	–	–	ND	+	+	–	ND
Major cellular fatty acids	$C_{16:1}$, $C_{16:0}$, $C_{15:0}$	ND	ND	$C_{14:0}$, $C_{15:0}$, OH-$C_{14:0}$	$C_{14:0}$, $C_{16:0}$, $C_{16:1}$, iso-$C_{17:1}$	ND	ND	ND	ND	ND
DNA G+C content (mol%)	45.1	ND	ND	46.6	43.2	40.6–40.9	49.0	37.6	51	43.6
Substrate utilization in pure culture:										
Yeast extract	–	ND	ND	–	–	–	–	–	–	+
Tryptone	–	–	ND	–	–	–	ND	ND	–	+
Casamino acids	–	ND	ND	ND	ND	ND	–	ND	+	+
Crotonate	+	+	–	+	+/–	+	–	+	–	+
Betaine	ND	ND	ND	ND	ND	ND	ND	ND	–	+
Pyruvate	–	ND	–	–	–	–	–	–	–	±
Ribose	–	ND	–	–	–	–	ND	–	–	±
Xylose	–	ND	–	–	–	–	–	ND	–	±
Glutarate	ND	ND	–	+	ND	ND	+	ND	ND	ND
Butyrate plus pentenoate	–	ND	–	–	+/–	–	ND	+	ND	ND
Butyrate plus dimethyl sulfoxide	ND	ND	ND	ND	+	+	ND	ND	ND	ND

(continued)

TABLE 200. (continued)

Characteristic	*Syntrophomonas wolfei* subsp. *wolfei*	*Syntrophomonas wolfei* subsp. *saponavida*	*Syntrophomonas sapovorans*	*Syntrophomonas curvata*	"*Syntrophomonas erecta* subsp. *erecta*"	"*Syntrophomonas erecta* subsp. *sporosyntropha*"	*Pelospora glutarica*	*Syntrophospora bryantii*	*Syntrophothermus lipocalidus*	*Thermosyntropha lipolytica*
Substrate utilization in co-culture with methanogens:										
Acetate $C_{2:0}$	−	ND	−	−	−	−	ND	ND	−	−
Propionate $C_{3:0}$	−	−	−	−	−	+	ND	−	−	−
Butyrate to caprylate $C_{4:0-C8:0}$ [b]	+	+	+	+	+	ND	−	+	+	+
Pelargonate $C_{9:0}$	ND	+	ND	ND	ND	−	ND	+	+	+
Caprate $C_{10:0}$	−	+	+	+	−	−	ND	+	+	+
Laurate $C_{12:0}$	−	+	+	+	−	−	ND	−	+	+
Myristate $C_{14:0}$	−	+	+	+	−	−	ND	−	−	+
Palmitate $C_{16:0}$	−	+	+	+	−	−	ND	−	−	+
Stearate $C_{18:0}$	−	+	+	+	−	−	ND	−	−	+
Oleate $C_{18:1}$	−	−	+	−	−	−	ND	ND	−	+
Linoleate $C_{18:2}$	−	−	+	−	−	−	ND	ND	−	+
Isobutyrate	−	−	−	−	−	−	ND	−	+	−
Isovalerate	−	−	−	−	−	−	ND	−	−	−
Triacylglycerides	ND	−	−	ND	ND	ND	ND	ND	−	+
Habitat	Anaerobic digester sludge	Anaerobic digester sludge	Anaerobic digester sludge	Wastewater treatment sludge	Wastewater treatment sludge	Wastewater treatment sludge	Anoxic freshwater sediment	Aquatic and marine anaerobic sediments	Wastewater treatment sludge	Alkaline hot springs

[a]Only differences found among the shown species are listed in this table. *Syntrophomonas wolfei* subsp. *wolfei*: type strain is DSM 2245; data are from McInerney et al. (1981, 1979), Beaty & McInerney (1987), Henson et al. (1988), Lorowitz et al. (1989), Zhang et al. (2005a). *Syntrophomonas wolfei* subsp. *saponavida*: type strain is DSM 4212; data are from Lorowitz et al. (1989). *Syntrophomonas sapovorans*: type strain is DSM 3441; data are from Roy et al. (1986). *Syntrophomonas curvata*: type strain is DSM 15682; data are from Zhang et al. (2004, 2005a). "*Syntrophomonas erecta* subsp. *erecta*" and "*Syntrophomonas erecta* subsp. *sporosyntropha*": data are from Zhang et al. (2005a) and Wu et al. (2006). *Syntrophospora bryantii*: type strain is DSM 3014; data are from Stieb & Schink (1985), Zhao et al. (1990), Dong et al. (1994). *Pelospora glutarica*: type strain is DSM 6652; data are from Matthies & Schink (1992a), Matthies et al. (2000). *Thermosyntropha lipolytica*: type strain is DSM 11003; data are from Svetlitshnyi et al. (1996). *Syntrophothermus lipocalidus*: type strain is DSM 12680; data are from Sekiguchi et al. (2000). −, Negative; ±, weak; +/−, variable among strains; +, positive; ND, not determined.

[b]Straight-chain fatty acids with C4 (butyrate) to C8 (caprylate).

Ecology. *Syntrophomonas* strains have been isolated from anaerobic digester and anaerobic wastewater treatment sludges, where methanogenic degradation of organic compounds is the dominant metabolism. Anaerobic digestion of lipids and other organic compounds, such as carbohydrates, often results in the production of large amounts of fatty acids as intermediates. *Syntrophomonas* species, like other members of the *Syntrophomonadaceae*, play an important role by scavenging these fatty acids under methanogenic conditions. Small-subunit rRNA-targeted oligonucleotide probes for members of the family *Syntrophomonadaceae* have been used for the quantification of these organisms in anaerobic digester sludge decomposing organic suspended solids (swine and cattle manure with industrial organic wastes). In these sludges, 0.2–1% of the total rRNA was attributed to members of the *Syntrophomonadaceae*, of which the majority was accounted for by *Syntrophomonas* species (Hansen et al., 1999).

Enrichment and isolation procedures

Syntrophomonas is selectively enriched using a defined anaerobic basal medium with butyrate as the sole energy source. Enrichment of *Syntrophomonas* strains with the ability to degrade long-chain fatty acids is performed with basal medium supplemented with a long-chain fatty acid and equimolar concentrations of $CaCl_2$ (Roy et al., 1986). Bottles or tubes (100, 50, or 10 ml vials) suitable for culturing strict anaerobes can be used. The medium is inoculated with methanogenic sludge. Enrichments are incubated in the dark at 35–40°C. After growth and methane production have been observed (1–2 weeks, or much longer in case of long-chain fatty acids), the cultures are further enriched by successive transfers. The enrichment should be stable and convert the fatty acid into acetate and methane. It should contain at least two major cell morphotypes. One is a hydrogenotrophic methanogen that can be recognized by its autofluorescence due to the abundance of co-enzyme F_{420}. The other morphotype may be *Syntrophomonas*.

For isolation of *Syntrophomonas*, roll tubes are inoculated with a mesophilic hydrogenotrophic methanogen, e.g., *Methanospirillum hungatei* cells, prior to inoculation with the enrichment culture. Pinpoint colonies are picked from the highest dilution, transferred into the liquid medium containing the substrate used for the enrichment, and incubated again.

This isolation step is repeated several times and cultures are then further purified by using the liquid basal medium containing crotonate and bromoethanesulfonate. Growth of *Syntrophomonas* occurs after a week of incubation. Three successive transfers

into the crotonate-bromoethanesulfonate liquid medium result in the disappearance of F_{420}-autofluorescent methanogens and cells in the culture are homogeneous. Then, roll tube isolation using crotonate as the sole substrate is performed. Pinpoint colonies are again picked from the highest dilution, transferred into the liquid medium containing crotonate, and incubated again. This step is repeated several times to obtain axenic cultures of *Syntrophomonas*.

Maintenance procedures

Stocks of axenic cultures may be maintained in the liquid basal medium with crotonate as the sole energy source. In the case of co-culture with methanogens, cultures may be maintained in the liquid basal medium with butyrate or longer chain fatty acids. After growth, cultures may be stored in the dark at room temperature for at least a month.

Taxonomic comments

In an earlier study (Roy et al., 1986), the genus *Syntrophomonas* was placed in the family *Bacteroidaceae*. However, comparative analyses of 16S rRNA gene sequences indicate that the genus *Syntrophomonas* encompasses physiologically and phylogenetically similar anaerobes in the deeply branching phylogenetic group (*Syntrophomonadaceae*) within the phylum *Firmicutes* (Zhao et al., 1993b). According to the 16S rRNA phylogenetic analysis presented in the roadmap to this volume (Figure 7), the genus *Syntrophomonas* is the type genus of the family *Syntrophomonadaceae*, order *Clostridiales*, class *Clostridia* in the phylum *Firmicutes*.

Differentiation of the genus *Syntrophomonas* from other genera

Characteristics that differentiate genera of the family *Syntrophomonadaceae* are listed in Table 200. The most notable features that distinguish the genus *Syntrophomonas* are: (1) they are mesophilic (optimum growth temperature of 35–40°C) anaerobes catalyzing syntrophic β-oxidation of fatty acids (C_4–C_{18}); (2) they possess an unusual multilayered cell wall that stains Gram-negative; and (3) they do not form spores (with the exception of "*Syntrophomonas erecta* subsp. *sporosyntropha*").

Differentiation of the species of the genus *Syntrophomonas*

Characteristics that differentiate species of the genus Syntrophomonas are also listed in Table 200. The most striking differences are found in substrate ranges. In addition, no two species of the genus possess 16S rRNA gene sequence similarities greater than 96%.

List of species of the genus *Syntrophomonas*

1. **Syntrophomonas wolfei** McInerney, Bryant, Hespell and Costerton 1982, 267[VP] (Effective publication: McInerney, Bryant, Hespell and Costerton 1981, 1037.) emend. Lorowitz, Zhao and Bryant 1989, 126.

    wolf′e.i. N.L. gen. n. *wolfei* of Wolfe, to honor Ralph S. Wolfe for his devotion towards the understanding of the biology of anaerobic bacteria.

    Surface colonies in roll tubes of *Syntrophomonas wolfei* co-cultures with methanogens are smooth, convex, and circular

with entire edges. Colonies may be dark to black when the organisms are co-cultured with H_2-utilizing *Desulfovibrio* species with sulfate. Grows in co-cultures with hydrogenotrophs using straight-chain fatty acids containing 4 (butyrate) to 8 (octanoate) carbon atoms as energy sources. Strains utilize isoheptanoate and some strains also use straight-chain fatty acids containing 9 (nonanoate) to 18 carbon atoms (stearate) or 2-methylbutyrate. It is not known whether some strains use anteiso or other iso fatty acids.

DNA G+C content (mol%): 45.1 (T_m).

Type strain: Göttingen G311, DSM 2245A (in co-culture with *Desulfovibrio* sp. strain G-11), DSM 2245B (in co-culture with *Methanospirillum hungatei* strain DSM 864[T]).

GenBank accession number (16S rRNA gene): M26492.

1a. Syntrophomonas wolfei subsp. wolfei Lorowitz, Zhao and Bryant 1989, 126[VP]

The size of the cells is in the higher range for the genus. In co-culture with hydrogenotrophs, utilizes straight-chain fatty acids with 4 (butyrate) to 8 (octanoate) carbon atoms and isoheptanoate as energy sources; in axenic culture, also uses crotonate.

DNA G+C content (mol%): 45.1 (T_m).

Type strain: Göttingen G311, DSM 2245A (in co-culture with *Desulfovibrio* sp. strain G-11), DSM 2245B (in co-culture with *Methanospirillum hungatei* strain DSM 864[T]).

GenBank accession number (16S rRNA gene): M26492.

1b. Syntrophomonas wolfei subsp. saponavida Lorowitz, Zhao and Bryant 1989, 126[VP]

sa.po.na.vi′da. L. n. *sapo* soap; L. adj. *avida* greedy; L. fem. adj. *saponavida* greedy for soap.

The size of the cells is in the lower range for the genus. In co-culture with hydrogenotrophs, utilizes straight-chain fatty acids with 4 (butyrate) to 18 (stearate) carbon atom as energy sources; also probably uses acids such as isoheptanoate and longer-chain iso acids. Grows on crotonate without a hydrogenotroph.

Type strain: SD2, DSM 4212 (in co-culture with *Desulfovibrio* sp. strain G-11).

2. Syntrophomonas curvata Zhang, Liu and Dong 2004, 972[VP]
cur.va′ta. L. fem. adj. *curvata* curved.

Cells stain Gram-negative and are curved rods, 0.5–0.7 by 2.3–4.0 μm, non-spore-forming, with 1–3 polar or subpolar flagella. Straight-chain C_4–C_{18} fatty acids serve as substrates in co-culture with *Methanobacterium formicicum* DSM 1535[T]. Even-numbered fatty acids are degraded into acetate and presumably H_2, whereas odd-numbered ones are degraded into propionate, acetate, and H_2. Acetate, propionate, isobutyrate, isovalerate, and benzoate do not support growth of the co-culture. Fumarate, sulfate, thiosulfate, sulfur, and nitrate cannot act as electron acceptors for butyrate oxidation. Crotonate is the only substrate tested that enables growth in axenic culture; yeast extract, tryptone, glucose, ribose, xylose, pyruvate, and fumarate do not. Cellular fatty acids comprise mainly $C_{14:0}$ (28%), $C_{15:0}$ (19%), $C_{14:0}$ 3-OH (11%), and an unknown component of ECL 14.503 (21%). Isolated from the granular sludge of an up-flow anaerobic sludge blanket (UASB) reactor treating beer wastewater in Beijing, China.

DNA G+C content (mol%): 46.6 (T_m).

Type strain: GB8-1, CGMCC 1.5010, DSM 15682.

GenBank accession number (16S rRNA gene): AY290767.

3. Syntrophomonas erecta Zhang, Liu and Dong 2005b, 802[VP]
e.rec′ta. L. fem. adj. *erecta* erect.

Cells are straight or slightly curved rods, 0.5–1.0 by 2.0–14.0 μm, with Gram-negative cell walls, tapered rounded ends, and 2–5 flagella, usually occurring in the subpole of

cells. Some strains form spores when co-cultured with methanogens on butyrate, but not in monoculture on crotonate. Straight-chain fatty acids with 4–8 carbon atoms serve as substrates in co-culture with *Methanospirillum hungatei* DSM 864[T]. Even-numbered fatty acids are degraded into acetate and presumably H_2, whereas odd-numbered ones are degraded into propionate, acetate, and H_2. Straight-chain fatty acids shorter than C_4 (acetate and propionate), longer than C_8 (caprate, laurate, myristate, palmitate, stearate, oleate, linoleate, and arachidate), branched-chain fatty acids (isobutyrate and isovalerate), and benzoate do not support the co-culture. Fumarate, sulfate, thiosulfate, sulfur, and nitrate do not act as electron acceptors for butyrate oxidation. Crotonate supports growth in axenic culture, whereas butyrate plus pentenoate or butyrate plus DMSO supports growth of some strains. Yeast extract, tryptone, peptone, maltose, glucose, fructose, ribose, xylose, pentenoate, fumarate, and pyruvate alone do not support growth. The type strain can grow at pH 6.0–8.8, 25–47°C and in 0–500 mM NaCl. The cellular fatty acids of the type strain contain mainly $C_{14:0}$ (30%), $C_{16:0}$ (17%), $C_{16:1}$ω5*c* (17%), and iso-$C_{17:1}$ I (15%). LL-Diaminopimelic acid exists in the cellular peptidoglycan. Some strains contain poly-β-hydroxyalkanoate inside cells. Isolated from the granular sludge of a UASB reactor for treating bean-curd farm wastewater in Beijing, China.

DNA G+C content (mol%): 40.6–43.9 (T_m).

Type strain: GB4-38, CGMCC 1.5013, DSM 16215.

GenBank accession number (16S rRNA gene): AY536889.

3a. "Syntrophomonas erecta subsp. erecta" Wu, Liu and Dong 2006, 460

Cells are Gram-negative, straight rods, motile by means of 2–5 flagella. Spores are never observed. Utilizes C_4–C_8 fatty acids as carbon and energy sources when co-cultured with methanogens. Crotonate can serve as sole carbon and energy source for the axenic culture. Utilizes pentenoate and DMSO as electron acceptors.

DNA G+C content (mol%): 43.2–43.9 (T_m).

Type strain: GB4-38, CGMCC 1.5013, DSM 16215.

GenBank accession number (16S rRNA gene): AY536889.

3b. "Syntrophomonas erecta subsp. sporosyntropha" Wu, Liu and Dong 2006, 460

spo.ro.syn′tro.ph.a. Gr. n. *spora* seed; Gr. adj. *syn* together with; Gr. n. *trophos* one who feeds; N.L. fem. n. *sporosyntropha* forms spores when feeds together with others.

Spores are observed in co-culture with methanogens when growing on saturated short-chain fatty acids. Spores are not formed in axenic culture during growth on crotonate. Poly-β-hydroxyalkanoate is produced when grown on crotonate, but not on butyrate or other saturated fatty acids. Utilizes DMSO as an electron acceptor. Isolated from methanogenic environments, such as river sediment, rice field mud, and UASB sludge.

DNA G+C content (mol%): 40.6–40.9 (T_m).

Type strain: 5-3-Z, CGMCC 1.5032, JCM 13344.

GenBank accession number (16S rRNA gene): DQ086234.

4. Syntrophomonas sapovorans Roy, Samain, Dubourguier and Albagnac 1987, 179[VP] (Effective publication: Roy, Samain, Dubourguier and Albagnac 1986, 146.)

sa.po'vo.rans. L. n. *sapo* soap; L. v. *voro* to devour; N.L. part. adj. *sapovorans* devouring soap (i.e., long-chain fatty acids).

Cells possess 2–4 flagella. Linoleate, oleate, elaidate, and saturated linear fatty acids up to 18 carbon atoms are β-oxidized. Fatty acids are degraded to propionate, acetate, and H_2. Linolenate and branched-chain and substituted fatty acids are not used. Both amino acids and B vitamins are required for growth. For degradation of fatty acids longer than 8 carbon atoms, calcium is required. Grows at pH 6.3–8.1 and 25–45°C, with optimum growth at pH 7.3 and 35°C.

Type strain: OM, DSM 3441 (in co-culture with *Methanospirillum hungatei* strain JF-1^T).

GenBank accession number (16S rRNA gene): AF022249.

Genus II. **Pelospora** Matthies, Springer, Ludwig and Schink 2000, 647^{VP}

BERNHARD SCHINK

Pe.lo.spo'ra. Gr. masc. n. *pelos* mud; Gr. fem. n. *spora* a seed, spore; N.L. fem. n. *Pelospora* a spore-forming bacterium originating from mud.

Strictly anaerobic bacteria forming spores. Organic substrates fermented. No reduction of external electron acceptors such as sulfate, nitrate, or Fe(III)-oxides.

Chemo-organotrophic, fermentative metabolism using few simple organic compounds as substrate. Carbohydrates not utilized. Media containing a reductant, e.g., sulfide, are necessary for growth. Catalase-negative. Isolated from sediments of limnic or marine origin.

On the basis of 16S rDNA gene sequence analysis, grouped with the family *Syntrophomonadaceae*, which is comprised of *Pelospora* and the genera *Syntrophomonas*, *Syntrophospora*, and *Thermosyntropha* within the phylum *Firmicutes* (Figure 7).

DNA G+C content (mol%): 49.0 (HPLC).

Type species: **Pelospora glutarica** Matthies, Springer, Ludwig and Schink 2000, 647^{VP}.

Further descriptive information

Pelospora glutarica, the only species described so far, was isolated as a glutarate-fermenting, strictly anaerobic bacterium. Its metabolism is unique because it grows only by decarboxylation of dicarboxylic acids such as glutarate, methylsuccinate, or succinate, which are converted to the corresponding monocarboxylic acids (Matthies and Schink, 1992b). Glutarate is first activated to glutaryl-CoA, oxidized to glutaconyl-CoA, and decarboxylated to crotonyl-CoA by a membrane-bound glutaconyl-CoA decarboxylase which conserves the decarboxylation energy by sodium ion transfer across the cytoplasmic membrane. The sodium ion gradient established this way drives ATP synthesis through a membrane-bound ATPase system. The resulting crotonyl-CoA is reduced to butyryl-CoA and partly isomerized to isobutyryl-CoA, and the bacterium produces a mixture of butyrate and isobutyrate as fermentation products (Matthies and Schink, 1992b). The reasons for isomerization of butyrate to isobutyrate by this bacterium are not clear. Nonetheless, this capacity allows the bacterium to convert isobutyrate to butyrate, and thus to help convert isobutyrate finally to methane and CO_2 in a methanogenic environment deprived of external electron acceptors (Matthies and Schink, 1992a). Succinate and methyl succinate are fermented to propionate or butyrate plus isobutyrate, probably through the methylmalonyl-CoA pathway described for the succinate-decarboxylating bacterium *Propionigenium modestum* (Dimroth and Schink, 1998; Hilpert et al., 1984).

No other substrates have so far been found to be utilized by *Pelospora glutarica*.

Enrichment and isolation procedures

Pelospora glutarica strains have been enriched and isolated from sediments of limnic or marine origin in simple mineral medium with glutarate as the sole source of carbon and energy. The mineral medium for enrichment must be low in phosphate (1–2 mM KPO_4) and rich in CO_2/HCO_3^- (>20 mM). Sulfide is usually used as a reductant. Since *Pelospora glutarica* grows with substrates that yield almost only C2 degradation intermediates, it must form pyruvate and sugars via reductive carboxylation of acetyl CoA and therefore needs carbon dioxide for this reaction. The enrichment medium used successfully for these organisms is derived from that used for the cultivation of sulfate-reducing bacteria (Widdel and Pfennig, 1981) but without sulfate, and is described in detail in the original publication (Matthies and Schink, 1992b). Two versions have been used, one for freshwater and one for marine enrichments. The respective substrates are applied in a concentration range of 10–20 mM, and inocula of at least 2–5 ml are used to secure sufficient attachment surfaces, structure-bound sulfides, and trace cosubstrates in the initial growth steps.

Maintenance procedures

Cultures are maintained either by repeated transfer at intervals of 2–3 months or by freezing in liquid nitrogen using techniques common for strictly anaerobic bacteria. No information exists about survival upon lyophilization.

Differentiation of the genus *Pelospora* from other genera

The genus *Pelospora* differs from all other strictly anaerobic bacteria in its capacity for fermentative growth with glutarate as sole source of carbon and energy, and its lack of utilization of other, more complex substrates. The type strain has been isolated in defined mineral medium with glutarate as the sole substrate and does not require complex, undefined medium additions.

Taxonomic comments

The genus *Pelospora* has been proposed as a taxonomic entity consisting of strictly anaerobic bacteria able to grow by fermentative degradation of glutarate. In this, it differs from all other described genera. Although the cells stain Gram-negative, they form spores and belong to the predominantly Gram-stain-positive phylum *Firmicutes* (Figure 7). *Pelospora* groups with the genera *Syntrophomonas*, *Syntrophospora*, and *Thermosyntropha* which all have been described as syntrophically fatty-acid-oxidizing strict anaerobes. When more strains of *Pelospora* are described, the taxonomic position of this genus and its relationship to other genera may need to be re-examined.

Further reading

Matthies, C. and B. Schink. 1992. Energy conservation in fermentative glutarate degradation by the bacterial strain WoGl3. FEMS Microbiol. Lett. *100*: 221–226.

List of species of the genus *Pelospora*

1. **Pelospora glutarica** Matthies, Springer, Ludwig and Schink 2000, 647[VP]

glu.ta′ri.ca. N.L. n. *acidum glutaricum* glutaric acid; *glutarica* referring to glutarate as the key substrate of this species.

Long, rod-shaped cells, 4.5–6.5 × 0.8 μm in size, motile by one subpolar flagellum, Gram-negative staining, formation of terminal oval spores.

Chemo-organotrophic, fermentative metabolism. Contains no cytochromes. Glutarate, methylsuccinate, and succinate are the only substrates. No growth with more than 30 different substrates such as sugars, organic acids, alcohols, amino acids, primary amines, or other dicarboxylic acids

tested. Products of glutarate and methylsuccinate fermentation are butyrate, isobutyrate, and CO_2; succinate is decarboxylated to propionate.

Growth rate with glutarate at 37°C is 0.062 μ/h. Optimum pH for growth is pH 7.1–8.2; no growth below pH 6.0. Temperature optimum 37°C; no growth below 20°C or above 37°C. Growth in medium containing salt concentrations of 0.1% NaCl and 0.04% $MgCl_2$·$6H_2O$ (w/v); no growth in salt-water medium at 2% NaCl and 0.3% $MgCl_2$·$6H_2O$ (w/v).

Isolated from anoxic freshwater sediment.

DNA G+C content (mol%): 49.0 (HPLC).

Type strain: WoGl3, ATCC BAA-162, DSM 6652.

GenBank accession number (16S rRNA gene): AJ251214.

Genus III. **Syntrophospora** Zhao, Yang, Woese and Bryant 1990, 43[VP]

Yuji Sekiguchi

Syn.tro.pho.spo′ra. Gr. adj. *syn* together with; Gr. fem. n. *trophos* one who feeds; Gr. fem. n. *spora* a spore, seed; N.L. fem. n. *Syntrophospora* a spore former which feeds together with (another species).

Rod-shaped cells. Gram-stain reaction is variable. **Gram-positive cell-wall ultrastructure. Oval, terminal endospores that swell the sporangium are usually formed. Strictly anaerobic chemoorganotroph. Saturated fatty acids, butyrate, and longer chain fatty acids are utilized for growth and are β-oxidized** to acetate and H_2 or, along with odd-numbered straight-chain fatty acids, to acetate, propionate, and H_2 in syntrophic association with H_2-scavenging anaerobes. **Some strains can be adapted to grow in pure culture on crotonate** with acetate and butyrate as major products. No other organic acids, sugars, or alcohols are utilized. Sulfate, sulfur, thiosulfate, nitrate, and fumarate are not reduced. Habitats are aquatic and marine anaerobic sediments.

DNA G+C content (mol%): 37.6 (T_m).

Type species: **Syntrophospora bryantii** (Stieb and Schink 1985) Zhao, Yang, Woese and Bryant 1990, 43[VP] (*Clostridium bryantii* Stieb and Schink 1985, 390).

Further descriptive information

Cells are slender, slightly curved, nonmotile rods with round ends, 0.4 by 3–6 μm, occurring singly or in clumps (Stieb and Schink, 1985). Gram staining is variable and electron micrographs show no indication of an outer cell membrane in sections of sporulating cells. Oval spores, 0.75 by 1.5 μm in size, are formed. Spores are heat-resistant and survive pasteurization (15 min at 80°C).

Comparative 16S rRNA gene sequence analysis of members of the genus *Syntrophospora* shows that the genus falls into the clade representing previously characterized syntrophic butyrate-degrading bacteria (i.e., the family *Syntrophomonadaceae*) (Zhao et al., 1990).

Members of the genus *Syntrophospora* are obligately anaerobic chemoorganotrophs that grow at mesophilic temperatures (28–34°C) and nearly neutral pH (6.5–7.5). In pure culture, growth is observed only with crotonate as energy source. In addition, axenic growth is also found with butyrate oxidation coupled with the reduction of pentenoate to valerate (Dong et al., 1994). In syntrophic association with hydrogen-utilizing microbes (i.e., hydrogenotrophic methanogens), only fatty acids with 4–11 carbon atoms, including 2-methylbutyrate, are utilized. The doubling time of the co-culture (with *Methanospirillum hungatei*) estimated based on acetate formation is about 72–96 h. *Syntrophospora* does not utilize oxygen, sulfate, nitrate, sulfur, thiosulfate, or fumarate as electron acceptors. The type and only species, *Syntrophospora bryantii*, can grow on butyrate with H_2- and formate-utilizing methanogens, whereas no growth is observed with methanogens that metabolize only H_2, suggesting that formate may be an important electron carrier in this organism (Dong et al., 1994; Dong and Stams, 1995; Stams and Dong, 1995).

Syntrophospora can be cultured in a defined anaerobic basal medium (Stieb and Schink, 1985) with caproate. The type strain of *Syntrophospora* was isolated from marine anoxic mud.

Enrichment and isolation procedures

Syntrophospora is selectively enriched using the defined anaerobic basal medium with caproate as the sole energy source at 28°C (Stieb and Schink, 1985). Bottles or tubes (100, 50, or 10 ml vials) may be used as culture vessels as with other strict anaerobes. The medium is inoculated with anoxic mud. After growth and methane production have been observed (2–3 weeks), the cultures are further enriched by successive transfers. A series of transfers should be made. After several transfers, the subcultures may contain two major cell morphotypes: F_{420}-autofluorescent cells resembling hydrogenotrophic methanogens; and nonmotile spore-forming rods (*Syntrophospora*). For isolation of *Syntrophospora*, the co-culture isolation method using agar shake cultures in mineral medium containing 10 mM caproate, 20 mM sulfate, and 5 mM acetate, in which all tubes have been previously inoculated with cells of *Desulfovibrio* sp. is used. After 3 weeks of incubation, yellow, disk-shaped colonies are formed. After two subsequent dilution series in agar shake culture, the defined co-culture can be isolated.

For further isolation in pure culture, *Syntrophospora* co-culture [with methanogens (Stieb & Schink, 1985)] is grown in medium supplemented with 20 mM crotonate and 200 µM bromoethanesulfonate (Zhao et al., 1990). After 3 weeks of incubation, growth of *Syntrophospora* is observed. Three successive transfers into the crotonate-bromoethanesulfonate liquid medium are performed to eliminate methanogens. Then, grown cells are streaked onto bottle plates containing crotonate medium. Colonies that are grayish yellow, round, and 0.5–2 mm in diameter, are formed after 7 weeks of incubation. Pinpoint colonies are again picked, transferred into the liquid medium containing crotonate, and incubated again. Consequently, pure *Syntrophospora* cells are obtained.

Maintenance procedures

Stock cultures (pure culture) may be maintained in the liquid basal medium with crotonate as the sole energy source. In case of co-cultures with *Methanospirillum hungatei*, cultures may be maintained in the liquid basal medium with butyrate. Grown cultures may be stored in the dark at room temperature. Stock cultures are transferred monthly.

Taxonomic comments

In an earlier study (Stieb and Schink, 1985), the type species of the genus *Syntrophospora* was placed in the genus *Clostridium* as "*Clostridium bryantii*". However, comparative analyses of 16S rRNA gene sequences indicated that the species should be placed in the family *Syntrophomonadaceae* as a new genus (Zhao et al., 1990, 1993b). At present, the genus *Syntrophospora* encompasses only one species, *Syntrophospora bryantii*. According to the 16S rRNA phylogenetic analysis presented in the roadmap to this volume (Figure 7), the genus *Syntrophospora* is a member of the family *Syntrophomonadaceae*, order *Clostridiales*, class *Clostridia* in the phylum *Firmicutes*.

Differentiation of the genus *Syntrophospora* from other genera

Table 200 lists characteristics that differentiate the genus *Syntrophospora* from other related genera. Phenotypically, the most closely related genus to *Syntrophospora* is *Syntrophomonas*, since they share common physiological traits such as syntrophic growth, substrate range, and growth temperature. However, the most notable feature that distinguishes the genus *Syntrophospora* from species of the genus *Syntrophomonas* is its Gram-positive cell-wall structure with distinct spore formation. They are also closely related to each other phylogenetically: the 16S rRNA gene sequence of *Syntrophospora bryantii* is most closely related to that of *Syntrophomonas wolfei*. However, comparative 16S rRNA gene analysis indicates that no sequence similarities greater than 94 % were identified among near-full length 16S rRNA genes of other members of the family *Syntrophomonadaceae*, including members of the genus *Syntrophomonas*.

List of species of the genus *Syntrophospora*

1. **Syntrophospora bryantii** (Stieb and Schink 1985) Zhao, Yang, Woese and Bryant 1990, 43^{VP} (*Clostridium bryantii* Stieb and Schink 1985, 390)

bry.an′ti.i. N.L. gen. n. *bryantii* named after Marvin P. Bryant, who pioneered studies on syntrophic methanogenic associations.

Rod-shaped cells, 0.4 by 4.5–6 µm in size with rounded end. Nonmotile, slightly curved. Gram-reaction negative to weakly positive; no outer cell membrane. Spores are terminal, oval, and 0.75 by 1.5 µm in size. Strictly anaerobic chemoorganotroph. Fatty acids with 4–11 carbon atoms and 2-methylbutyrate are utilized for growth and fermented to acetate and H_2 or to acetate, propionate, and H_2 in syntrophic association with hydrogen-scavenging anaerobes. No other organic acids, sugars, or alcohols are metabolized. Sulfate, sulfur, thiosulfate, nitrate, and fumarate are not reduced. Grows at pH 6.5–7.5 and 28–34°C.

DNA G+C content (mol%): 37.6 (T_m).

Type strain: CuCal, DSM 3014A (in co-culture with *Desulfovibrio* sp. strain E70), DSM 3014B (in co-culture with *Methanospirillum hungatei* strain M1h).

GenBank accession number (16S rRNA gene): M26491.

Genus IV. **Syntrophothermus** Sekiguchi, Kamagata, Nakamura, Ohashi and Harada 2000, 778^{VP}

Yuji Sekiguchi

Syn.tro.pho.ther′mus. Gr. adj. *syn* together with; Gr. fem. n. *trophos* one who feeds; Gr. adj. *thermos* hot; N.L. masc. n. *Syntrophothermus* thermophilic syntroph, referring to growth in syntrophic association with hydrogenotrophic organisms at high temperature of around 55°C.

Slightly **curved rods**, 0.4–0.5 × 2–4 µm that often show binary fission. **Gram-stain-negative.** Weakly motile by means of flagella. Spore formation is not observed. **Obligately anaerobic.** Fastest growth at around 55°C. Energy metabolism by reduction of protons with butyrate and higher homologs as electron donors in the presence of hydrogenotrophic microbes. Capable of growth in pure culture with crotonate.

DNA G+C content (mol%): 51 (HPLC).

Type species: **Syntrophothermus lipocalidus** Sekiguchi, Kamagata, Nakamura, Ohashi and Harada 2000, 778[VP].

Further descriptive information

Cells are slightly curved rods with round ends, occurring singly or in pairs. Tumbling motility is observed. Some flagella are seen by electron microscopy (Figure 197). Gram staining is negative, although electron micrographs show that cells possess a typical Gram-positive type cell-wall structure.

Comparative 16S rRNA gene sequence analysis of the genus *Syntrophothermus* shows that the genus falls into the clade representing previously characterized syntrophic butyrate-degrading bacteria (i.e., the family *Syntrophomonadaceae*) and that the genus is, however, phylogenetically distinct, with only a moderate relationship to the genus *Thermosyntropha*.

Syntrophothermus is an obligately anaerobic chemo-organotroph that grows at high temperatures (45–60°C) and nearly neutral pH (6.0–7.5). In pure culture, the growth is observed only with crotonate as energy source. Fermentation products from crotonate are equimolar amounts of acetate and butyrate. In syntrophic association with hydrogen-utilizing microbes (i.e., hydrogenotrophic methanogens), only fatty acids including butyrate and higher homologs are utilized. In the co-culture, fatty acids with an even number of carbon atoms are metabolized to form acetate and methane, while fatty acids having an odd number of carbon atoms are degraded to acetate, propionate, and methane. Similarly, 2-methylpropionate (isobutyrate) is degraded to form acetate and methane. In the conversion of 2-methylpropionate, butyrate can be detected as an intermediate indicating that 2-methylpropionate is isomerized to butyrate in the first step of the metabolism. Other carboxylic acids, sugars, alcohols, amino acids, and aromatic hydrocarbons are not utilized either in pure culture or in co-culture with methanogens. *Syntrophothermus* does not utilize oxygen, sulfate, nitrate, sulfite, thiosulfate, fumarate, or Fe(III) as an electron acceptor.

Syntrophothermus can be cultured in a defined anaerobic basal medium (Sekiguchi et al., 2000) with crotonate. The optimum temperature for growth is 55°C. Optimal growth occurs in freshwater medium, although the genus grows in up to 1% NaCl. The growth of *Syntrophothermus* is inhibited by ampicillin, chloramphenicol, kanamycin, neomycin, rifampin, and vancomycin.

The type strain of *Syntrophothermus* was isolated from an anaerobic wastewater treatment sludge from a thermophilic (55°C) upflow anaerobic sludge blanket (UASB) reactor.

Enrichment and isolation procedures

Syntrophothermus is selectively enriched using the defined anaerobic basal medium with butyrate as the sole energy source at a high temperature (55°C). Bottles or tubes (100, 50, or 10 ml vials) can be used as culture vessels, similarly to other strict anaerobes. The medium is inoculated with anaerobic sludge. Enrichments are incubated in the dark at 55°C. After growth and methane production have been observed (1–2 weeks), the cultures are further enriched by successive transfers (1% inoculum). A series of at least five transfers should be made. The fifth enrichment may still convert butyrate into acetate and methane stably and contain at least two major cell morphotypes; one is F_{420}-autofluorescent cells resembling hydrogenotrophic methanogens, and the other is *Syntrophothermus*. For isolation of *Syntrophothermus*, the co-culture isolation method using roll tubes, in which all tubes were inoculated beforehand with *Methanothermobacter thermautotrophicus* cells, is used. After two weeks of incubation, white to brownish colonies 0.5–1 mm in diameter are formed. Pinpoint colonies are picked from the highest dilution, transferred into the liquid medium containing butyrate, and incubated again. This isolation step is repeated several times, and then the cultures are further purified by using the liquid basal medium containing crotonate and bromoethanesulfonate (BES). After a week of incubation, the growth of *Syntrophothermus* occurs. Three successive transfers into the crotonate-BES liquid medium results in disappearance of F_{420}-autofluorescent *Methanothermobacter* cells, and the cells in the culture seem to be homogeneous. When the roll tube isolation technique using crotonate as the sole substrate is used, very small colonies that are white, lens-shaped, and 0.1–0.2 mm in diameter are formed after 2 weeks of incubation. Pinpoint colonies are again picked from the highest dilution, transferred into the liquid medium containing crotonate, and incubated again. This step is repeated several times, and pure *Syntrophothermus* cells are obtained.

Maintenance procedures

Stock cultures (pure culture) may be maintained in the liquid basal medium with crotonate as the sole energy source. In case of co-culture with *Methanothermobacter thermautotrophicus*, cultures may be maintained in the liquid basal medium

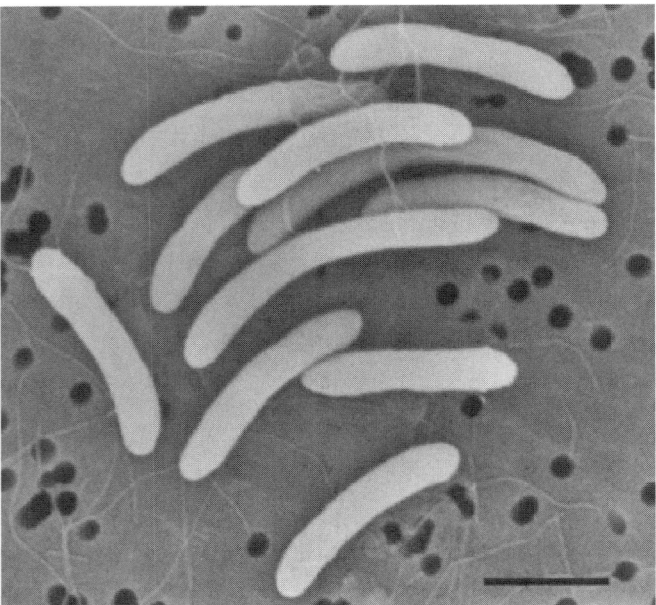

FIGURE 197. Scanning electron micrograph of *Syntrophothermus* cells grown with crotonate in pure culture (bar = 1 μm).

with butyrate. Grown cultures are stored in the dark at room temperature for at least a month. Stock cultures are transferred monthly.

Differentiation of the genus *Syntrophothermus* from other genera

The most convincing evidence for uniqueness of the genus is based on comparative 16S rRNA gene analysis. No sequence similarities greater than 89% were identified among near-full-length 16S rRNA genes of other members of the family *Syntrophomonadaceae*. Physiologically, the genus *Syntrophothermus* represents syntrophic butyrate-degrading bacteria growing at high temperatures and nearly neutral pH; no genera

are known to show the same physiological traits. The closest genus, in terms of both phylogeny and physiology, is *Thermosyntropha*, which is also capable of degrading butyrate in syntrophy at high temperatures (Svetlitshnyi et al., 1996). However, the most striking differences between the two genera are the followings: (1) *Thermosyntropha* exhibits heterotrophic growth on substrates such as yeast extract and tryptone in addition to lipolytic growth, while *Syntrophothermus* grows only on limited fatty acids; (2) *Thermosyntropha* can grow at higher pH (growth range: 7.15–9.5) while *Syntrophothermus* is nearly neutrophilic; (3) the DNA base composition of *Syntrophothermus* is significantly higher than that of *Thermosyntropha*.

List of species of the genus *Syntrophothermus*

1. **Syntrophothermus lipocalidus** Sekiguchi, Kamagata, Nakamura, Ohashi and Harada 2000, 778[VP]

lip.o.cal'id.us Gr. neut. *lipos* fat; L. adj. *calidus* expert; N.L. adj. *lipocalidus* fatty acid-specific, i.e., specifically utilizing fatty acids.

Cells have slightly curved rods with flagella. Gram stain negative. The dimensions of single cells are 2–4 × 0.4–0.5 μm. Weakly motile. Spores are never observed.

Strictly anaerobic chemo-organotroph. Growth occurs only in crotonate in pure culture. Saturated fatty acids with 4–10 carbon atoms are utilized in syntrophic association with hydrogenotrophic methanogens. 2-Methylpropionate is also utilized through isomerization to butyrate. Optimal growth

occurs in freshwater medium; 0.5% NaCl is slightly inhibitory for growth, and 1.5–2.0% NaCl completely inhibits the growth. Sensitive to ampicillin, chloramphenicol, kanamycin, neomycin, rifampin, and vancomycin. Cells grow between 45 and 60°C (optimum: 55°C), and pH 5.8–7.5 (optimum: 6.5–7.0). Occur in thermophilic (55°C) anaerobic wastewater treatment sludge.

DNA G+C content (mol%): 51.0 (HPLC).
Type strain: TGB-C1, DSM 12680.
GenBank accession number (16S rRNA gene): AB021305.
Note: DSM 12681 is *Syntrophothermus lipocalidus* in co-culture with *Methanothermobacter thermautotrophicus* strain ΔH.

Genus V. **Thermosyntropha** Svetlitshnyi, Rainey and Wiegel 1996, 1135[VP]

JUERGEN WIEGEL

Ther.mo.syn.tro'pha. Gr. adj. *thermos* hot; Gr. prefix *syn* with, together; Gr. v. *trophein* to eat; Gr. masc. n. *syntrophos* foster brother or sister; N.L. fem. n. *thermosyntropha* "foster sisters liking it hot", referring to the fact that the bacterium grows at elevated temperatures on fatty acids only in syntrophic cultures with H₂-utilizing micro-organisms.

Gram-positive cell-wall type, but stains Gram-negative, slightly polymorphic rods. Obligately **anaerobic, thermophilic, and heterotrophic**. Members are able to utilize otherwise thermodynamically unfavorable substrates such as fatty acids in **syntrophic** cultures with H₂-utilizing sulfate-reducing bacteria or methane-producing archaea. Axenic cultures grow with crotonate or peptones as a sole carbon and energy sources.

DNA G+C content (mol): 43–44 (HPLC).

Type species: **Thermosyntropha lipolytica** Svetlitshnyi, Rainey and Wiegel 1996, 1136[VP].

Further descriptive information

The genus *Thermosyntropha*, belonging to the *Syntrophomonadaceae*, is to date only represented by one species. Other similar but slightly different strains have been isolated, but have been lost before publication. The most closely related bacteria are the thermophilic neutrophile

Syntrophothermus lipocalidus, the 16S rRNA gene sequence of which possesses an evolutionary distance of 0.11–0.12 (Sekiguchi et al., 2000), and mesophilic syntrophs such as members of the genera *Syntrophospora* and *Syntrophomonas*, which, like *Thermosyntropha*, also use crotonate as the sole carbon and energy source for axenic cultures (Figure 198). So far, the only known habitats are alkaline hot springs from Lake Bogoria (Kenya). Hatamoto et al. (2007) also described thermophilic syntrophs that are able to degrade long-chain fatty acids in methanogenic sludge. Their work, using enrichments, 16S rRNA sequencing, and stable isotope analyses suggests that this phenotype may be shared by diverse bacteria from outside the *Syntrophomonadaceae* and even *Firmicutes*, taxa which remain to be isolated.

Maintenance procedures

Cultures have been successfully preserved in liquid nitrogen and upon freeze-drying. Storage for short times in 30–50%

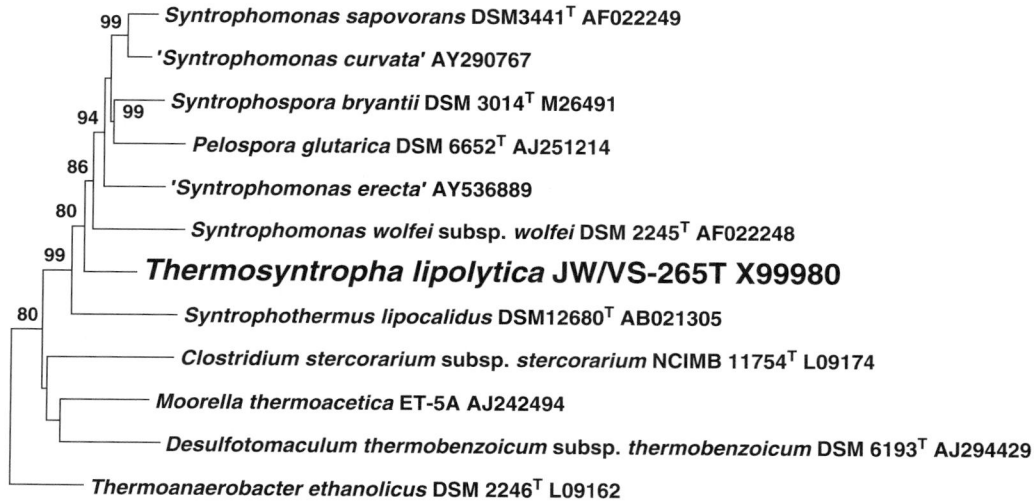

FIGURE 198. Phylogenetic dendrogram based on 16S rRNA sequence analysis demonstrating the position of *Thermosyntropha lipolytica* within the radiation of the *Syntrophomonadaceae* using *Thermoanaerobacter* (of the order "*Thermoanaerobacterales*") as an outgroup. The 16S rRNA data used represents *Escherichia coli* DSM 30083[T] nucleotide positions 1–1499. The tree was constructed using the neighbor-joining method with Jukes and Cantor distance corrections. Numbers at the nodes represent the bootstrap values (%) of 1000 replicates.

pre-reduced glycerol at –20 and –80°C has been successful. However, these cultures were lost within 4 years. Cultures growing on lipids quickly lose viability in the absence of equimolar concentrations of Ca^{2+} ions. Cultures grown on peptones remain viable at room temperature for 8–12 months if the pH at 25°C ($pH^{25°C}$) is kept between 7.0 and 8.5.

Differentiation of the genus *Thermosyntropha* from other genera

Beside its unique position in a 16S rRNA-based phylogenetic tree, *Thermosyntropha lipolytica*, the type species of this genus, is currently the only species with a validly published name. It is an alkaliphilic thermophile that hydrolyzes long-chain (C_8–C_{20}) fatty-acid-containing lipids in syntrophic co-culture with H_2-utilizing methane-producing archaea or sulfate-reducing bacteria. The only other thermophilic fatty acid β-oxidizer known is *Syntrophothermus lipocalidus*, which is a neutrophile, has a much higher G+C content (51 mol% vs 43–44 mol%), cannot hydrolyze lipids such as tributyrin or olive oil, and only utilizes C_4–C_{10} fatty acids. *Thermosyntropha* can be further distinguished from known mesophilic syntrophs by its growth temperature, pH range, rapid growth on crotonate, and substrate utilization spectrum.

Thermosyntropha also differs from many other anaerobic thermophiles in that it cannot utilize carbohydrates. Its lipolytic capability also differentiates it from other peptidolytic alkalithermophiles such as *Clostridium paradoxum*, which forms spores, and species of *Anaerobranca* and *Thermobrachium*, which form branched cells.

Further reading

Kevbrin, V.V., C.S. Romanek and J. Wiegel. 2003. Alkalithermophiles: a double challenge from extreme environments. *In* Seckbach (Editor), Cellular Origins, Life in Extreme Habitats and Astrobiology (COLE), Section VI: Extremophiles and Biodiversity, Vol. 6 "Origins: Genesis, Evolution and the Biodiversity of Life", Kluwer Academic Publishers, Dordrecht, NL, pp. 395–412.

Salameh, M. and J. Wiegel. 2007. Lipases from extremophiles and potential for industrial applications. Adv. Appl. Microbiol. *61*: 253–283.

Wiegel, J. 1998. Anaerobic alkalithermophiles, a novel group of extremophiles. Extremophiles *2*: 257–267.

Wiegel, J. and V.V. Kevbrin. 2004. Alkalithermophiles. Biochem. Soc. Trans. *32*: 193–198.

List of species of the genus *Thermosyntropha*

1. **Thermosyntropha lipolytica** Svetlitshnyi, Rainey and Wiegel 1996, 1136[VP]

li.po.ly'ti.ca. Gr. n. *lipos* fat; Gr. adj. *lutikos* able to loosen; N.L. fem. adj. *lipolytica* referring to the ability to hydrolyze lipids to glycerol and fatty acids.

An alkalithermophilic, lipolytic anaerobe, cells are straight, slightly curved, or polymorphic rods, 2.0–3.5 × 0.3–0.4 μm, that stain Gram-negative. Strains grow in agar-shake-roll tubes as small (<0.5 mm diameter), lens-shaped, white colonies. During exponential growth, pairs and chains of cells are observed. Sporulation and motility has never been observed under any conditions tested. Only the type strain is available; other strains, which were also lipolytic and peptidolytic, have been lost due to freezer malfunction.

All strains in co-culture with an H_2-utilizing organism utilize fatty acids ranging in size from butyrate (very slowly) to long-chain fatty acids such as stearate as well as the unsaturated fatty acids oleate and linoleate. Longer fatty acids such as arachidate ($C_{20:0}$) and behenate ($C_{22:0}$) do not support growth. The liberated glycerol is not metabolized in either pure or syntrophic cultures, which is unusual for lipolytic bacteria. Fatty acids are metabolized only in the presence of equimolar concentrations of Ca^{2+} ions and H_2-utilizing micro-organisms such as *Methanothermobacter thermautotrophicus* (or its heterotypic synonym *Methanobacterium thermalcaliphilum*) and *Desulfotomaculum nigrificans* strains that are able to grow above pH 8.0. These strains must also lower the H_2 levels produced during fatty acid oxidation to concentrations rendering the fatty acid β-oxidation exothermic. Substrates such as tributyrin, trilaurin, tripalmitin, tristearin, and triolein support, in the presence of Ca^{2+} ions, growth of the methanogenic or sulfidogenic co-cultures.

Strain JW/VS-265^T contains two constitutively expressed lipases termed LipA and LipB. They possess differences in their N-terminal amino acid sequences as well as temperature optima (between 90 and 98°C) and pH optima (pH$^{25°C}$ 9.4 and 9.6, respectively). Both enzymes hydrolyze the ester bond in positions 1 and 2, leaving the monosubstituted glycerol. In nonaqueous solutions, synthesis of various glycerides and other esters are catalyzed by these enzymes (Salameh, 2006).

Axenic growth also occurs with yeast extract. Cell yields are proportional to the concentrations of yeast extract up to 1% (w/v) and 1.4×10^8 cells/ml. Yeast extract can be replaced with tryptone, Casamino acids, beef extract, heart infusion, or nutrient broth. In the presence of 0.1% yeast extract, pyruvate, ribose, and xylose support weak growth beyond that of yeast extract alone. None of the strains are able to utilize hexoses, grow chemolithoautotrophically with H_2/CO_2, or grow on fatty acids in the absence of H_2-utilizing micro-organisms.

Crotonate supports growth within 3 d, producing about 2 mol acetate and 1 mol butyrate per 2 mol crotonate. Crotonate was first observed to support axenic growth of syntrophic bacteria by Beaty and McInerney (1987). All strains are thermophilic with a growth temperature range of 50–70°C (optimum around 60–66°C). Cultures do not grow at 43 or 73°C. All isolates are moderate alkaliphiles, growing at pH$^{25°C}$ 7.15–9.5, with optimum growth at pH$^{25°C}$ 8.1–8.9.

NaCl concentrations above 1% (w/v) are inhibitory.

Growth of strain JW/VS265^T at pH$^{25°C}$ 8.4 and 60°C is inhibited by 50 µg/ml ampicillin, chloramphenicol, kanamycin, neomycin, rifampin, and vancomycin.

All strains were isolated on commercial olive oil from enrichments inoculated with mixed sediment/water samples from unnamed hot springs of Lake Bogoria (Kenya) in the presence of 0.05% (w/v) yeast extract.

DNA G+C content (mol%): 43–44 (HPLC).

Type strain: JW/VS-265, ATCC 700317, DSM 11003.

GenBank accession number (16S rRNA gene): X99980.

References

Amos, D.A. and M.J. McInerney. 1989. Poly-β-hydroxyalkanoate in *Syntrophomonas wolfei*. Arch. Microbiol. *152*: 172–177.

Amos, D.A. and M.J. McInerney. 1990. Growth of *Syntrophomonas wolfei* on unsaturated short chain fatty acids. Arch. Microbiol. *154*: 31–36.

Amos, D.A. and M.J. McInerney. 1991. Composition of poly-β-hydroxyalkanoate from *Syntrophomonas wolfei* grown on unsaturated fatty-acid substrates. Arch. Microbiol. *155*: 103–106.

Amos, D.A. and M.J. McInerney. 1993. Formation of D-3-hydroxybutyryl coenzyme A by an acetoacetyl coenzyme A reductase in *Syntrophomonas wolfei* subsp *wolfei*. Arch. Microbiol. *159*: 16–20.

Beaty, P.S. and M.J. McInerney. 1987. Growth of *Syntrophomonas wolfei* in pure culture on crotonate. Arch. Microbiol. *147*: 389–393.

Beaty, P.S. and M.J. McInerney. 1990. Nutritional features of *Syntrophomonas wolfei*. Appl. Environ. Microbiol. *56*: 3223–3224.

Dimroth, P. and B. Schink. 1998. Energy conservation in the decarboxylation of dicarboxylic acids by fermenting bacteria. Arch. Microbiol. *170*: 69–77.

Dong, X., G. Cheng and A.J.M. Stams. 1994. Butyrate oxidation by *Syntrophospora bryantii* in coculture with different methanogens and in pure culture with pentenoate as electron acceptor. Appl. Microbiol. Biotechnol. *42*: 647–652.

Dong, X.Z. and A.J.M. Stams. 1995. Evidence for H_2 and formate formation during syntrophic butyrate and propionate degradation. Anaerobe *1*: 35–39.

Garrity, G.M., J.A. Bell and T. Lilburn. 2005. The Revised Road Map to the Manual. *In* Brenner, Krieg, Staley and Garrity (Editors), Bergey's Manual of Systematic Bacteriology, 2nd edn, Vol. 2, The *Proteobacteria*, Part A, Introductory Essays. Springer, New York, pp. 159–220.

Hansen, K.H., B.K. Ahring and L. Raskin. 1999. Quantification of syntrophic fatty acid-beta-oxidizing bacteria in a mesophilic biogas reactor by oligonucleotide probe hybridization. Appl. Environ. Microbiol. *65*: 4767–4774.

Hatamoto, M., H. Imachi, A. Ohashi and H. Harada. 2007. Identification and cultivation of anaerobic, syntrophic long-chain fatty acid-degrading microbes from mesophilic and thermophilic methanogenic sludges. Appl Environ Microbiol *73*: 1332–1340.

Henson, J.M., M.J. McInerney, P.S. Beaty, J. Nickels and D.C. White. 1988. Phospholipid fatty acid composition of the syntrophic anaerobic bacterium *Syntrophomonas wolfei*. Appl. Environ. Microbiol. *54*: 1570–1574.

Hilpert, W., B. Schink and P. Dimroth. 1984. Life by a new decarboxylation-dependent energy conservation mechanism with Na⁺ as coupling ion. EMBO J. *3*: 1665–1670.

Jackson, B.E., V.K. Bhupathiraju, R.S. Tanner, C.R. Woese and M.J. McInerney. 1999. *Syntrophus aciditrophicus* sp. nov., a new anaerobic bacterium that degrades fatty acids and benzoate in syntrophic association with hydrogen-using microorganisms. Arch. Microbiol. *171*: 107–114.

Liu, Y.T., D.L. Balkwill, H.C. Aldrich, G.R. Drake and D.R. Boone. 1999. Characterization of the anaerobic propioate-degrading syntrophs *Smithella propionica* gen. nov., sp. nov. and *Syntrophobacter wolinii*. Int. J. Syst. Bacteriol. *49*: 545–556.

Lorowitz, W.H., H. Zhao and M.P. Bryant. 1989. *Syntrophomonas wolfei* subsp. *saponavida* subsp. nov., a long-chain fatty-acid-degrading, anaerobic, syntrophic bacterium; *Syntrophomonas wolfei* subsp. *wolfei* subsp. nov.; and emended descriptions of the genus and species. Int. J. Syst. Bacteriol. *39*: 122–126.

Ludwig, W., K.-H. Schleifer and W.B. Whitman. 2009. Revised road map to the phylum *Firmicutes*. *In* De Vos, Garrity, Jones, Krieg, Ludwig, Rainey, Schleifer and Whitman (Editors), Bergey's Manual of Systematic Bacteriology, 2nd edn, Vol. 3, The *Firmicutes*. Springer, New York, pp. 1–14.

Matthies, C. and B. Schink. 1992a. Reciprocal isomerization of butyrate and isobutyrate by the strictly anaerobic bacterium strain-WOG13 and methanogenic isobutyrate degradation by a defined triculture. Appl. Environ. Microbiol. *58*: 1435–1439.

Matthies, C. and B. Schink. 1992b. Fermentative degradation of glutarate via decarboxylation by newly isolated strictly anaerobic bacteria. Arch. Microbiol. *157*: 290–296.

Matthies, C., N. Springer, W. Ludwig and B. Schink. 2000. *Pelospora glutarica* gen. nov., sp. nov., a glutarate-fermenting, strictly anaerobic, spore-forming bacterium. Int. J. Syst. Evol. Microbiol. *50*: 645–648.

McInerney, M.J., M.P. Bryant and N. Pfennig. 1979. Anaerobic bacterium that degrades fatty acids in syntrophic association with methanogens. Arch. Microbiol. *122*: 129–135.

McInerney, M.J., M.P. Bryant, R.B. Hespell and J.W. Costerton. 1981. *Syntrophomonas wolfei* gen. nov. sp. nov., an anaerobic, syntrophic, fatty acid oxidizing bacterium. Appl. Environ. Microbiol. *41*: 1029–1039.

McInerney, M.J., M.P. Bryant, R.B. Hespell and J.W. Costerton. 1982. *In* Validation of the publication of new names and new combinations previously effectively published outside the IJSB. List no. 8. Int. J. Syst. Bacteriol. *32*: 266–268.

McInerney, M.J. and N.Q. Wofford. 1992. Enzymes involved in crotonate metabolism in *Syntrophomonas wolfei*. Arch. Microbiol. *158*: 344–349.

Roy, F., E. Samain, H. Dubourguier and G. Albagnac. 1986. *Syntrophomonas sapovorans* sp. nov., a new obligately proton reducing anaerobe oxidizing saturated and unsaturated long chain fatty acids. Arch. Microbiol. *145*: 142–147.

Roy, F., E. Samain, H. Dubourguier and G. Albagnac. 1987. *In* Validation of the publication of new names and new combinations previously effectively published outside the IJSB. List no. 23. Int. J. Syst. Bacteriol. *37*: 179–180.

Salameh, M.A. 2006. Alkalithermophilic lipases from *Thermosyntropha lipolytica*. Dissertation. University of Georgia, Athens, GA, USA.

Schink, B. and M. Friedrich. 1994. Energetics of syntrophic fatty acid oxidation. FEMS Microbiol. Rev. *15*: 85–94.

Schink, B. 1997. Energetics of syntrophic cooperation in methanogenic degradation. Microbiol. Mol. Biol. Rev. *61*: 262–280.

Sekiguchi, Y., Y. Kamagata, K. Nakamura, A. Ohashi and H. Harada. 2000. *Syntrophothermus lipocalidus* gen. nov., sp. nov., a novel thermophilic, syntrophic, fatty-acid-oxidizing anaerobe which utilizes isobutyrate. Int. J. Syst. Evol. Microbiol. *50*: 771–779.

Sobieraj, M. and D.R. Boone. 2002. The family *Syntrophomonadaceae*. *In* Dworkin, Falkow, Rosenberg, Schleifer and Stackebrandt (Editors), The Prokaryotes: An Evolving Electronic Resource for the Microbiological Community, release 3.11, November 22, 2002 edn. Springer-Verlag, New York.

Stams, A.J.M. 1994. Metabolic interactions between anaerobic bacteria in methanogenic environments. Antonie van Leeuwenhoek *66*: 271–294.

Stams, A.J.M. and X.Z. Dong. 1995. Role of formate and hydrogen in the degradation of propionate and butyrate by defined suspended cocultures of acetogenic and methanogenic bacteria. Antonie van Leeuwenhoek Int. J. Gen. Mol. Microbiol. *68*: 281–284.

Stieb, M. and B. Schink. 1985. Anaerobic oxidation of fatty acids by *Clostridium bryantii* sp. nov., a sporeforming, obligately syntrophic bacterium. Arch. Microbiol. *140*: 387–390.

Svetlitshnyi, V., F. Rainey and J. Wiegel. 1996. *Thermosyntropha lipolytica* gen. nov. sp. nov., a lipolytic, anaerobic, alkalitolerant, thermophilic bacterium utilizing short- and long-chain fatty acids in syntrophic coculture with a methanogenic archaeum. Int. J. Syst. Bacteriol. *46*: 1131–1137.

Widdel, F. and N. Pfennig. 1981. Studies on dissimilatory sulfate-reducing bacteria that decompose fatty acids. 1. Isolation of new sulfate-reducing bacteria enriched with acetate from saline environments: description of *Desulfobacter postgatei* gen. nov., sp. nov. Arch. Microbiol. *129*: 395–400.

Wofford, N.Q., P.S. Beaty and M.J. McInerney. 1986. Preparation of cell-free extracts and the enzymes involved in fatty acid metabolism in *Syntrophomonas wolfei*. J. Bacteriol. *167*: 179–185.

Wu, C., X. Liu and X. Dong. 2006. *Syntrophomonas erecta* subsp. *sporosyntropha* subsp. nov., a spore-forming bacterium that degrades short chain fatty acids in co-culture with methanogens. Syst. Appl. Microbiol. *29*: 457–462.

Zhang, C., X. Liu and X. Dong. 2004. *Syntrophomonas curvata* sp. nov., an anaerobe that degrades fatty acids in co-culture with methanogens. Int. J. Syst. Evol. Microbiol. *54*: 969–973.

Zhang, C., X. Liu and X. Dong. 2005a. *Syntrophomonas erecta* sp. nov., a novel anaerobe that syntrophically degrades short-chain fatty acids. Int. J. Syst. Evol. Microbiol. *55*: 799–803.

Zhang, C.Y., X.L. Liu and X.X. Dong. 2005b. *Syntrophomonas erecta* sp. nov., a novel anaerobe that syntrophically degrades short-chain fatty acids. Int. J. Syst. Evol. Microbiol. *55*: 799–803.

Zhao, H., D. Yang, C.R. Woese and M.P. Bryant. 1993a. Assignment of fatty acid-beta-oxidizing syntrophic bacteria to *Syntrophomonadaceae* fam. nov. on the basis of 16S rRNA sequence analyses. Int. J. Syst. Bacteriol. *43*: 278–286.

Zhao, H.X., D.C. Yang, C.R. Woese and M.P. Bryant. 1990. Assignment of *Clostridium bryantii* to *Syntrophospora bryantii* gen. nov., comb. nov. on the basis of a 16S ribosomal RNA sequence: analysis of its crotonate-grown pure culture. Int. J. Syst. Bacteriol. *40*: 40–44.

Zhao, H.X., D.C. Yang, C.R. Woese and M.P. Bryant. 1993b. Assignment of fatty acid-beta-oxidizing syntrophic bacteria to *Syntrophomonadaceae* fam. nov. on the basis of 16S ribosomal RNA sequence analyses. Int. J. Syst. Bacteriol. *43*: 278–286.

Family X. **Veillonellaceae** Rogosa 1971b, 232[AL]

FRED A. RAINEY

Veil.lo.nel.la'ce.ae. N.L. fem. n. *Veillonella* type genus of the family; *-aceae* ending to denote family; N.L. fem. pl. n. *Veillonellaceae* the *Veillonella* family.

The family *Veillonellaceae* can be described on the basis of phylogenetic analyses of 16S rRNA gene sequences (Figure 199); the family contains the genera *Veillonella* (type genus), *Acetonema*, *Acidaminococcus*, *Allisonella*, *Anaeroarcus*, *Anaeroglobus*, *Anaeromusa*, *Anaerosinus*, *Anaerovibrio*, *Centipeda*, *Dendrosporobacter*, *Dialister*, *Megasphaera*, *Mitsuokella*, *Pectinatus*, *Pelosinus*, *Phascolarctobacterium*, *Propionispira*, *Propionispora*, *Quinella*, *Schwartzia*, *Selenomonas*, *Sporomusa*, *Sporotalea*, *Succiniclasticum*, *Succinispira*, *Thermosinus* and *Zymophilus*. The genera *Pelosinus* and *Sporotalea* were proposed after finalization of the content of this volume and are not described here (see Boga et al. (2007); Shelobolina et al. (2007).

Members of the family are Gram-stain-negative, morphologically diverse and include short rods, curved rods, pleomorphic rods, cocci and vibrioid forms. All species are obligate anaerobes.

Type genus: **Veillonella** Prévot 1933, 118[AL].

Taxonomic comments

The family *Veillonellaceae* was described by Rogosa (1971b) to accommodate the genera *Veillonella*, *Acidominococcus* and *Megasphaera*. The genera *Veillonella* and *Acidominococcus* had previously been placed in the family *Neisseriaceae* (Prévot, 1933; Rogosa, 1969). The family description was very specific and based on the characteristics of the three genera assigned to it (Rogosa, 1971b). These properties included coccoid morphology, lack of endospores, nonmotile, chemoorganotrophy and parasites of homothermic animals (Rogosa, 1971b). As new genera have been described and shown by phylogenetic analysis to fall within the radiation of the three genera that had been assigned to the family *Veillonellaceae* the characteristics and thus the description of the family broadened so that many of the family characteristics defined by Rogosa (1971b) were no longer applicable. The emended family description given above is broad to include the diverse characteristics of the organisms of the family *Veillonellaceae*. More specific details of the characteristics of the members of the family *Veillonellaceae* are found in the chapters describing these taxa in this volume.

Genus I. **Veillonella** Prévot 1933, 118[AL] emend. Mays, Holdeman, Moore, Rogosa and Johnson 1982, 35

JEAN-PHILIPPE CARLIER

Veil.lo.nel'la. N.L. dim. suff. *-ella*; N.L. fem. n. *Veillonella* named after Adrien Veillon, the French microbiologist who isolated the type species.

Cocci, 0.3–0.5 μm in diameter; cells are usually arranged in pairs, masses, or short chains. Stains **Gram negative.** Nonmotile, nonsporeforming. **Anaerobic.** Optimum temperature, 30–37°C. Optimum pH, 6.5–8.0. Oxidase negative. **Nitrate is reduced.** Some species produce an atypical catalase lacking porphyrin. **Gas is produced by most strains. Pyruvate, lactate,** malate, fumarate, and oxaloacetate are fermented. **Carbohydrates and polyols are not fermented by most strains.** Major metabolic end products in trypticase-glucose-yeast extract (TGY, see below) broth are **acetic and propionic acids.** In addition, **CO_2 and H_2** are produced from lactate (Rogosa, 1964). Resident of oral cavity, genito-urinary, respiratory, and intestinal tracts of humans and animals, but can also cause severe human infections such as bacteremia, endocarditis, osteomyelitis, and prosthetic joint infection.

DNA G+C content (mol%): 36–43 (T_m) or 40–44 (Bd).

Type species: **Veillonella parvula** (Veillon and Zuber 1898) Prévot 1933, 119[AL] (*Staphyloccocus parvulus* Veillon and Zuber 1898, 542).

Further descriptive information

The genus *Veillonella* is represented by 10 species.

Morphology and ultrastructure. They are small, nonmotile, nonsporeforming organisms. Ultrastructure of thin sections viewed by electron microscopy reveal typical Gram-negative surface layers consisting of an outer membrane composed of two dense layers 3 nm wide separated by a less dense layer of 2 nm, a thin peptidoglycan layer, and a cytoplasmic membrane (Figure 200). By negative staining, the outer membrane appears extremely convoluted, whereas the peptidoglycan layer tightly follows the periphery of the protoplasm (Bladen and Mergenhagen, 1964).

Colonial characteristics. Colonies on blood agar are 1–3 mm in diameter and appear smooth, opaque, and grayish-white. Nonhemolytic. *Veillonella*, unlike *Acidaminococcus*, *Anaeroglobus*, and *Megasphaera*, exhibits a pink to red fluorescence on brain heart infusion agar (BHI) containing either sheep or horse blood (Brazier and Riley, 1988). Fluorescence appears under UV light at 366 nm and fades rapidly after 5–10 min of exposure to air. However, the reaction depends upon the medium and varies among the species. The fluorescent pigment is a porphyrin.

Nutrition and metabolism. The species of *Veillonella* are not clearly distinguished by their phenotypic characteristics. Most species, except *Veillonella criceti* and an unidentified phylogenetically related strain (Marchandin et al., 2005), are unable to ferment carbohydrates or amino acids (Rogosa, 1964). Some strains require putrescine or cadaverine (Rogosa and Bishop, 1964a). H_2S is produced from cysteine, cystine, glutathione, thiosulfate, thiocyanate, and thioglycolate (Rogosa and Bishop, 1964b). Indole is not produced; gelatin is not liquefied. Lactate enhances the growth with accompanying production of acetic and propionic acids, CO_2, and H_2. Rogosa and Bishop (1964b)

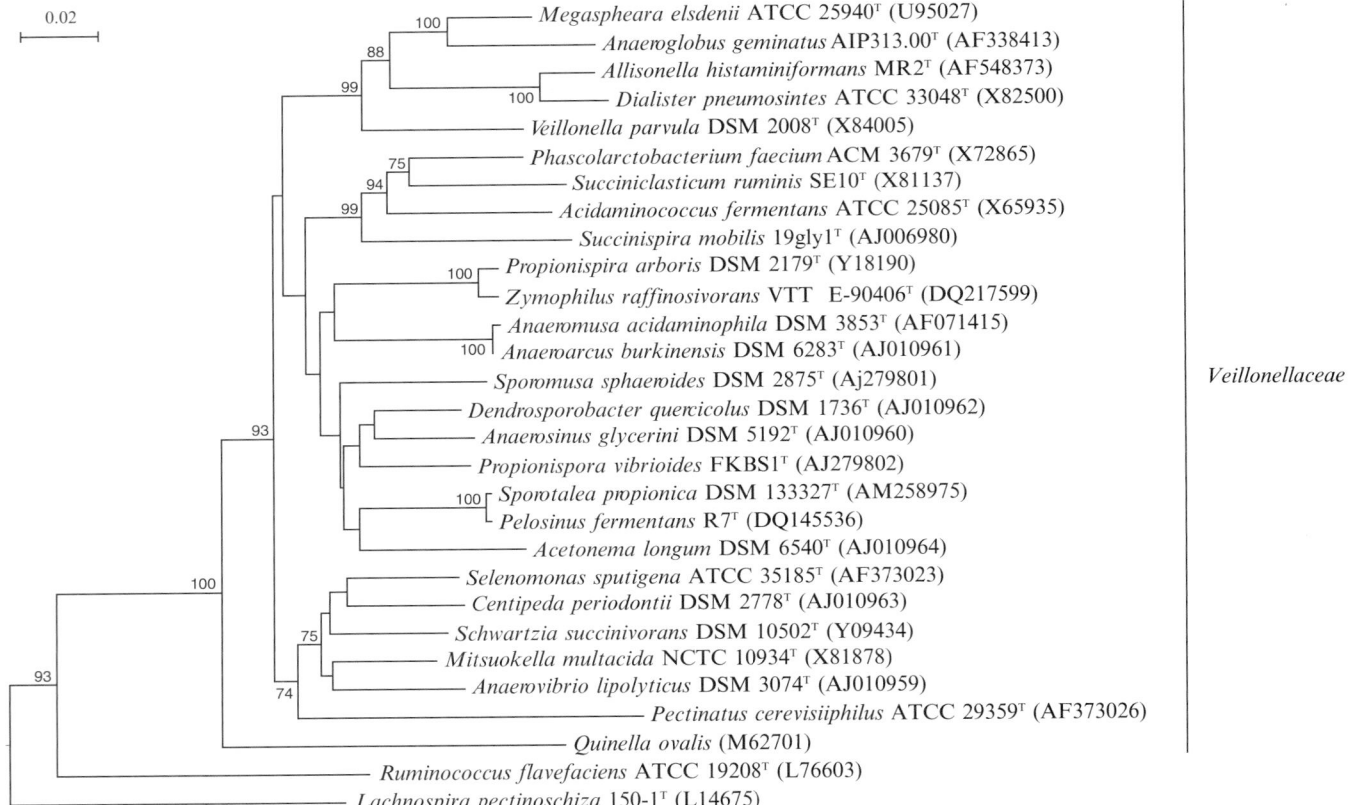

0.02

Megaspheara elsdenii ATCC 25940^T (U95027)
Anaeroglobus geminatus AIP313.00^T (AF338413)
Allisonella histaminiformans MR2^T (AF548373)
Dialister pneumosintes ATCC 33048^T (X82500)
Veillonella parvula DSM 2008^T (X84005)
Phascolarctobacterium faecium ACM 3679^T (X72865)
Succiniclasticum ruminis SE10^T (X81137)
Acidaminococcus fermentans ATCC 25085^T (X65935)
Succinispira mobilis 19gly1^T (AJ006980)
Propionispira arboris DSM 2179^T (Y18190)
Zymophilus raffinosivorans VTT E-90406^T (DQ217599)
Anaeromusa acidaminophila DSM 3853^T (AF071415)
Anaeroarcus burkinensis DSM 6283^T (AJ010961)
Sporomusa sphaeroides DSM 2875^T (Aj279801)
Dendrosporobacter quercicolus DSM 1736^T (AJ010962)
Anaerosinus glycerini DSM 5192^T (AJ010960)
Propionispora vibrioides FKBS1^T (AJ279802)
Sporotalea propionica DSM 133327^T (AM258975)
Pelosinus fermentans R7^T (DQ145536)
Acetonema longum DSM 6540^T (AJ010964)
Selenomonas sputigena ATCC 35185^T (AF373023)
Centipeda periodontii DSM 2778^T (AJ010963)
Schwartzia succinivorans DSM 10502^T (Y09434)
Mitsuokella multacida NCTC 10934^T (X81878)
Anaerovibrio lipolyticus DSM 3074^T (AJ010959)
Pectinatus cerevisiiphilus ATCC 29359^T (AF373026)
Quinella ovalis (M62701)
Ruminococcus flavefaciens ATCC 19208^T (L76603)
Lachnospira pectinoschiza 150-1^T (L14675)

Veillonellaceae

FIGURE 199. 16S rRNA gene sequence based phylogenetic analysis showing the relationships between the genera of the family *Veillonellaceae*. Each genus is represented by its type species. The outgroups are the type species of the type genera of the families *Ruminococcaceae* and *Lachnospiraceae*. The tree was reconstructed from distance matrices using the neighbor-joining method. The scale bar represents 2 inferred nucleotide changes per 100 nucleotides.

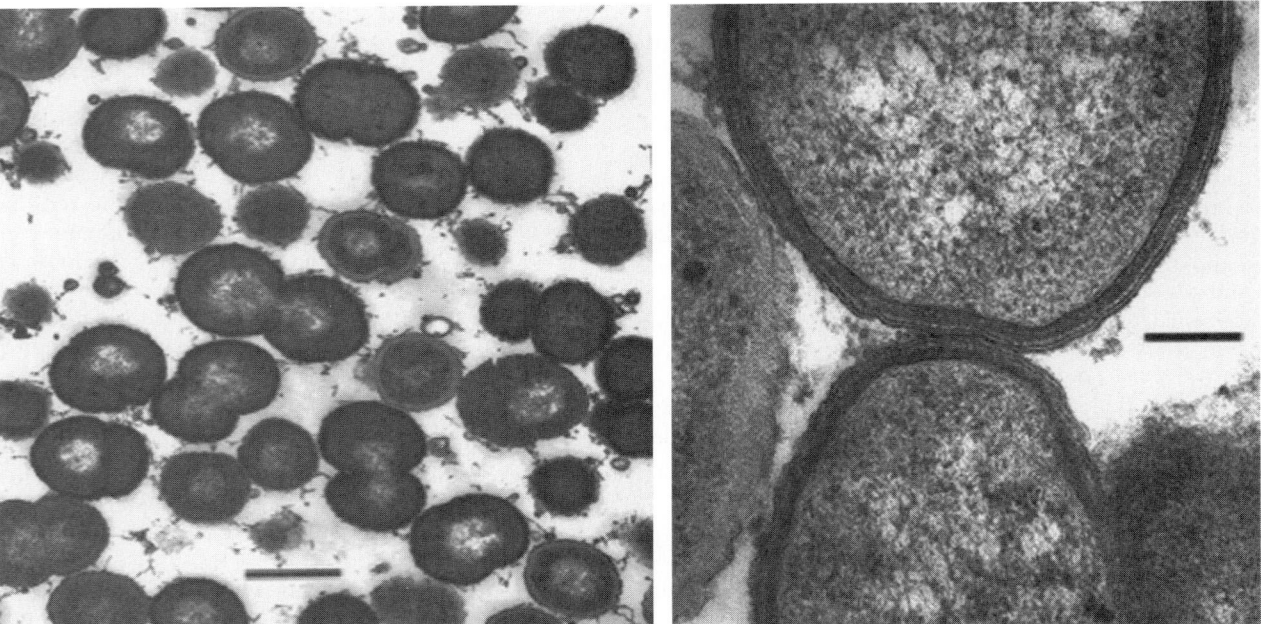

FIGURE 200. Morphology and cell-wall structure of *Veillonella dispar* ATCC 17748. Transmission electron micrographs showing the general morphology after negative staining (left) and ultrathin section showing cell wall and membranes (right). Bars: left = 666 nm, right = 100 nm. (Courtesy of H. Marchandin.)

reported that succinate is metabolized by resting cells, whereas the organisms cannot grow with succinate as an energy source in basal medium. However, most strains of *Veillonella* species ferment succinate naturally present within TGY broth. In addition, when succinate is added to trypticase-yeast extract broth, it is metabolized to propionate. In fact, succinate is co-metabolized with lactate, and the decarboxylation energy is mainly used for nongrowth purposes (Samuelov et al., 1990).

About 34% of strains are catalase positive, and 99% of strains are susceptible to kanamycin and colistin special-potency antibiotic disks (Jousimies-Somer et al., 1999). Resistance to colistin has been observed for *Veillonella montpellierensis*, the type strain of *Veillonella ratti* (Jumas-Bilak et al., 2004), and an unidentified phylogenetically related strain (Marchandin et al., 2005).

Major metabolic end products from TGY broth are acetic and propionic acids, which are produced in approximately equal quantities.

Fatty acid composition. Major cellular fatty acids are $C^{13:0}$ and $C^{17:1}$ ω^8. Minor cellular fatty acids include $C^{11:0}$, $C^{12:0}$, $C^{14:0}$, $C^{15:0}$, $C^{16:1}$ $\omega9c$, $C^{16:0}$, $C^{17:0}$, $C^{18:1}$ $\omega9c$, $C^{18:0}$. Other compounds are present in trace amounts (Table 201).

Genetics. *Veillonella* species can be differentiated by DNA–DNA hybridization (Mays et al., 1982), 16S rRNA gene sequence comparison (Figure 201), and restricted fragment-length polymorphism (RFLP) analysis of 16S rRNA genes (Sato et al., 1997). However, the utility of the last two methods is limited by the low level of sequence variation among some *Veillonella*

TABLE 201. Relative content (%) of cellular fatty acids of *Veillonella* species[a]

Fatty acid	*V. parvula* ATCC 17745[T]	*V. atypica* ATCC 17744[T]	*V. caviae* DSM 20738[T]	*V. criceti* DSM 20734[T]	*V. denticariosi* RBV 106[T]	*V. dispar* ATCC 17748[T]	*V. montpellierensis* ADV 281.99[T]	*V. ratti* ATCC 17746[T]	*V. rodentium* ATCC 17743[T]
11:00	1.8	1.1	1.4	2	0.5	0.8	2.3	2.1	1.6
12:00	4.9	3.8	4	1.4	3	3.8	7	4.5	5
13:00	24	20.4	7.3	16.3	12.9	23.5	6.3	20	25.3
14:0 iso	1.9	0.6	1.2	–	5.1	1.5	–	0	–
14:00	6.5	5.7	3.1	1.2	3.9	4.9	5	2	9.2
15:00	12	17.7	7	12.7	4.4	7	6.3	10.3	11.6
14:2 2OH	–	0.3	–	0.3	–	–	–	1	–
16:1 ω9c	5.2	5.2	10.6	4.2	4.5	7.7	8.4	4.8	5.5
16:00	7	7.5	10.3	3.7	8.6	8.4	12.9	4.6	6.7
17:1 ω8	22.3	25.7	29.9	48.9	10.1	25.6	20.7	36.9	17
17:00	4.1	2.4	2.5	2.6	3.2	2.5	2	2.4	4.2
18:1 ω9*c*	6.2	4.3	14.3	2.8	5.7	8	17.4	6.1	6.7
18:1 ω9*t*	–	–	1.9	1.2	3.5	–	2	–	–
18:00	3.9	2.7	6.2	2.1	6.6	5.7	11.1	5.8	6.7

[a]Fatty acid nomenclature: unsaturated fatty acids, the position of the double bond can be located by counting from the methyl (ω) end of the carbon chain; OH, hydroxylated fatty acids; *cis*, *trans* isomers are indicated by the suffix *c* and *t*, respectively.

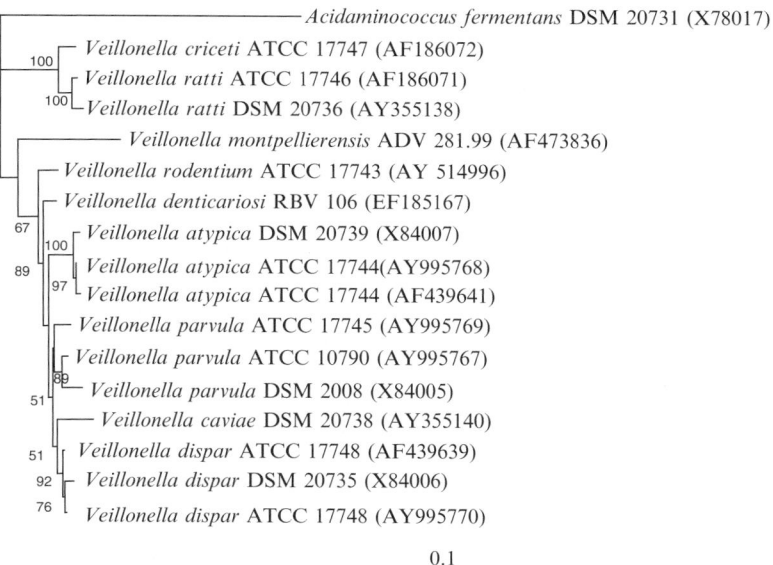

FIGURE 201. Phylogeny of *Veillonella* species. The phylogenetic relationships of all sequences of type strains of the genus *Veillonella* available from public databases are shown above. The sequence of *Acidaminococcus fermentans* DSM 20731 was used as an outgroup. Bar = 10% difference in 16S rRNA gene sequences. The neighbor-joining method was used for tree construction. One hundred bootstrap trees were generated, and bootstrap confidence levels are shown as percentages on the nodes.

species and the frequent intrachromosomal heterogeneity between copies of the 16S rRNA genes in human isolates (Marchandin et al., 2003c). Four 16S ribosomal operons are present in reference strains of *Veillonella* parvula, *Veillonella dispar*, *Veillonella atypica*, and 27 clinical isolates. The variability observed between these operons is higher than that between the 16S rRNA gene sequences of Veillonella parvula and *Veillonella dispar*. Thus, 16S rDNA-based methods may be not suitable for the identification of these species. A similar situation occurs within the type strains of *Veillonella ratti* and *Veillonella criceti* which form a very tight phylogenetic group. Thus, analyses of 16S rRNA gene sequences can not identify species within this group (Marchandin et al., 2005). The limitations of 16S rRNA gene sequence analyses within *Veillonella* and related genera caused by the variability of the 16S rRNA operons have been reviewed in detail (Marchandin and Jumas-Bilak, 2006). The 70kDa heat-shock protein gene *(dnaK)* is suggested to be an effective alternative for discriminating between species because of its higher interspecies variability.

The G + C values for the DNA are 36–43 mol%; the majority clustering around 39 mol% when tested by the thermal denaturation method (Mays et al., 1982).

Genomic structure. The genome of *Veillonella* species consists of one circular chromosome of 2.15 Mb (Marchandin et al., 2003a).

Plasmids and bacteriophages. Plasmids have been found in about 50% of the human oral strains. Their molecular sizes range from 1.1 to 28 MDa (Arai et al., 1984). Mays et al. (1982) have suggested that the fructose fermentation by some strains of *Veillonella criceti* may be associated with the presence of a plasmid.

A temperate bacteriophage specific for genus *Veillonella* has been reported (Shimizu, 1968). In addition, 25 virulent phages were isolated from washings of the oral cavity of 200 individuals (Hiroki et al., 1976; Totsuka, 1976). These phages were classified into two major groups on the basis of their morphology and serological characteristics. Seven phages infected strain ATCC 17743, now classified as *Veillonella rodentium*. The lipopolysaccharide of *Veillonella rodentium* ATCC 17743 acts as receptor for one of the bacteriophages (Totsuka, 1988; Totsuka and Ono, 1989).

Antibiotic susceptibility. *Veillonella* species are considered susceptible to ampicillin (0.12–4 µg/ml), amoxycillin/clavulanic acid (0.06–2 µg/ml), piperacillin/tazobactam (0.12–32 µg/ml), cefoxitin (0.12–8 µg/ml), cefotetan (0.12–2 µg/ml), ceftriaxone (0.12–8 µg/ml), imipenem (0.06–1 µg/ml), meropenem (0.06–0.12 µg/ml), clindamycin (0.06–0.12 µg/ml), bacitracin (1–4 units/ml), and metronidazole (0.5–4 µg/ml) and are resistant to ramoplanin (512 µg/ml), trimethoprim/sulfamethoxazole (8–256 µg/ml), and vancomycin (32–256 µg/ml) (Finegold et al., 2004; Roberts et al., 2006).

In the first edition of *Bergey's Manual of Systematic Bacteriology* (Rogosa, 1984b), veillonellae are considered susceptible to penicillin G. However, based on the National Committee for Clinical Laboratory Standards (NCCLS, now called Clinical and Laboratory Standards Institute) recommendations (National Committee for Clinical Laboratory Standards, 2003), most strains should be considered resistant. Several studies have shown that a large population of *Veillonella* species is penicillin resistant with MIC over 2 µg/ml (the breakpoint for susceptibility according to NCCLS criteria). Resistance to amoxicillin and penicillin has also been reported for *Veillonella dispar* and *Veillonella parvula* (Ready et al., 2004). In another study on 40 clinical strains tested by agar dilution method, 80% were found resistant to penicillin (Reig et al., 1997). The strains for which penicillin MICs were >8 µg/ml were also less susceptible to ampicillin and cefoxitin (MIC >0.25 µg/ml and >1 µg/ml, respectively). The mechanism involved in high levels of penicillin resistance in *Veillonella* remains to be elucidated. Although occasional strains of *Veillonella parvula* produce β-lactamase (Valdes et al., 1982), the main resistance mechanism of veillonellae may be due to alterations in a penicillin-binding protein (Theron et al., 2003).

Antibiotic resistance genes. Tetracycline resistance (Tcr) genes are present within about 10–12.5% of *Veillonella* species isolated from the oral cavity (Lancaster et al., 2005, 2003; Ready et al., 2006). Five different Tc^r genes have been detected: *tetA*, *tetL*, *tetM*, *tetO*, and *tetS*. The most common is *tetM*, followed by *tetS*. Some strains harbor two Tc^r genes (Ready et al., 2006). The *tetM* gene is located on a Tn*916*-like element.

A nitroimidazole resistance gene *(nimE)* is present in a metronidazole-susceptible *Veillonella* species (Marchandin et al., 2004). This gene is probably located on a low copy number plasmid which can explain the susceptibility observed.

Pathogenicity. Veillonellae are commensals in the oropharynx, gastrointestinal tract, and uro-genital tract of humans and animals. Of the ten recognized species, only *Veillonella parvula*, *Veillonella dispar*, *Veillonella atypica*, *Veillonella denticariosi*, and *Veillonella montpellierensis* have been isolated from humans. However, a strain genetically closely related to the *Veillonella criceti*–*Veillonella ratti* group has been isolated from a human clinical specimen (Marchandin et al., 2005). Veillonellae are implicated in several systemic diseases such as meningitis (Bhatti and Frank, 2000), bone infections (Isner-Horobeti et al., 2006; Singh and Yu, 1992), prosthetic joint infections (Marchandin et al., 2001; Zaninetti-Schaerer et al., 2004), endocarditis (Boo et al., 2005; Houston et al., 1997; Rovery et al., 2005), bacteremia (Liu et al., 1998), and in various infections in children (Brook, 1996).

Pathogenicity in animals is unknown.

Virulence factors. Mergenhagen et al. (1961) established that a phenol-water extract of *Veillonella* cells exhibited biological and immunological activities characteristic of endotoxins and that this water-soluble material was composed mainly of lipid and polysaccharide. The lipid moiety accounting for approximately two-thirds of the complex contains glucosamine, tridecanoic acid, 3-hydroxytridecanoic acid and 3-hydroxypentadecanoic acid. (Hewett et al., 1971; Tortorello and Delwiche, 1984). Tridecanoic acid and 3-hydroxytridecanoic acid represent about 50% of the total lipopolysaccharide fatty acids (Bishop et al., 1971). The polysaccharide moiety contains 2-keto-3-deoxyoctonic acid (KDO), glyceromannoheptose, galactose, rhamnose, glucose, and glucosamine. (Hewett et al., 1971; Hofstad and Kristoffersen, 1970). However, other hexoamines may be present (Hewett et al., 1971). The *Veillonella* LPS is highly endotoxic. Sveen showed that *Veillonella* LPS induces skin inflammation and the Schwartzman phenomenon in rabbits (Sveen, 1977). Moreover, phenol-water extracted lipopolysaccharide is lethal at nanogram levels for mice, chick embryos, pyrogenic in rabbits, and gelated *Limulus* amoebocyte lysate. The endotoxin activity of *Veillonella* LPS is comparable to that from *Fusobacterium* and *Salmonella enteritidis* (Sveen et al., 1977). LPS from both *Veillonella parvula* and *Veillonella atypica* reduce microbial activity of human neutrophils but do not affect monocyte function (Focà et al., 1990).

Ecology. Veillonellae colonize the human mouth within the three first months of life (Kononen et al., 1992). The colonization is not dependent on the presence of erupted teeth (Kononen et al., 1999). The organisms persist in the oral flora, and *Veillonella* species remain the dominant anaerobic bacteria in the saliva of elderly edentulous persons (Sato et al., 1993). Dental plaque, buccal mucosa, and the tongue are the main ecological niches of *Veillonella* species.

During the plaque development, the dominant strains of *Veillonella* change in their phenotypic and genotypic characteristics (Palmer et al., 2006). Within the oral cavity, *Veillonella* species form coaggregates with many other bacteria (Hughes et al., 1988). Coaggregation promotes formation of mixed-species bacterial colonies between veillonellae and other oral bacteria. Both the rapid succession within the *Veillonella* population and coaggregation are important processes in establishment of a stable oral ecosystem (Hughes et al., 1988; Palmer et al., 2006).

Enrichment and isolation procedures

Veillonellae can be isolated by streaking samples onto Columbia or Wilkins-Chalgren sheep-blood agar and incubating at 37°C for 2 d. Lactate agar medium containing vancomycin (7.5µg/ml) favors isolation. Strict anaerobic conditions are required for growth.

Maintenance procedures

Strains may be maintained in trypticase-glucose-yeast extract (TGY) medium (Carlier et al., 2002) under anaerobic conditions as described above. Cultures of veillonellae will usually survive for no longer than 1 week without subcultivation. Strains in the exponential growth stage in BHI or TGY broth can be stored at −80°C or in liquid nitrogen. Cells suspended in skim milk can be successfully lyophilized.

Differentiation of the genus *Veillonella* from closely related taxa

The differentiation of the genus *Veillonella* from other genera of anaerobic Gram-negative cocci can be done primarily on the basis of poor growth with gas formation on the usual media (except for *Veillonella denticariosi* which does not produce gas), nitrate reduction, and lactate fermentation. Other characteristics are given in Table 201 of the chapter describing the genus *Anaeroglobus*.

Upon routine laboratory examination, *Veillonella* is frequently misidentified as *Dialister*, a tiny Gram-negative rod. Nevertheless, a positive nitrate reaction and production of large amounts of propionic acid are the key features differentiating *Veillonella* species from most *Dialister* species which are negative for these tests. Although *Dialister propionicifaciens* produces small amounts of propionic acid, it does not reduce nitrate to nitrite. Upon misinterpretation of the Gram reaction, *Parvimonas micra*, which is a very small anaerobic Gram-positive coccus, can also be misidentified as *Veillonella*. *Parvimonas micra* does not produce propionic acid and does not reduce nitrate.

Taxonomic comments

The type species of the genus *Veillonella* was first isolated by Veillon and Zuber in 1898 and designated *Staphylococcus parvulus*. Prévot (1933) further described these bacteria and proposed the genus *Veillonella*, with two species, *Veillonella parvula* and *Veillonella alcalescens*. He placed the new genus in the family *Neisseriaceae*. Based on serological reactions, Rogosa (1965) divided veillonellae into seven distinct subspecies, which were subsequently shown to be species by DNA–DNA hybridization (Mays et al., 1982; Rogosa, 1984b). In 1971, the genus *Veillonella* was excluded from the family *Nesseriaceae* and transferred into the new family *Veillonellaceae* (Rogosa, 1971b). Subsequently, it was proposed to include the genus *Veillonella* in a new family "*Acidaminococcaceae*" (Garrity and Holt, 2001). The family "*Acidaminococcaceae*" was created to include members of the *Firmicutes* or low G + C Gram-positive bacteria that possessed a Gram-negative type of cell wall. *Acidaminococcus* was selected as the type genus. However, the placement of the genus *Veillonella* (Prévot, 1933) in the family "*Acidaminococcaceae*" is not in accordance with the Rules 23a and 56a of the International Code of Nomenclature of Bacteria (1990 Revision) (Lapage et al., 1992). Moreover, the name *Veillonellaceae* has priority over "*Acidaminococcaceae*". Therefore, in the current volume, the family "*Acidaminococcaceae*" is not used, and the family *Veillonellaceae* is retained.

Today, the identification of *Veillonella* at the species level remains difficult. Because of the lack of phenotypic criteria, the commonly used biochemical tests are useless. Serological tests (Rogosa, 1965) are no longer available. DNA–DNA hybridization (Mays et al., 1982), which seems to be a good method, is laborious, difficult to interpret and therefore not well-suited for routine identification. Molecular genetic methods, such as RFLP analysis and 16S rRNA sequencing, are valuable means to distinguish some species (Kolenbrander, 2006). However, these methods, though helpful, are not absolute. There is doubt as to whether *Veillonella parvula* and *Veillonella dispar* (Marchandin et al., 2003c; Palmer et al., 2006)as well as *Veillonella criceti* and *Veillonella ratti* (Marchandin et al., 2005) are truly separate species. Alternative molecular markers to 16S rRNA, such as other housekeeping genes, could clarify these interspecies relationships. For instance, three highly conserved genes, *rpoB*, *dnaK*, and *gyrB*, had greater discriminatory power than the 16S rRNA gene in differentiating between *Veillonella* strains. Of these, the *rpoB* gene showed the highest level of interspecies but the lowest intraspecies diversity (Byun et al., 2006). Thus, multilocus Sequencing Typing (MLST) of highly conserved genes could be an attractive method to differentiate *Veillonella* species (Stackebrandt et al., 2002).

Acknowledgements

I would like to thank Hélène Marchandin for her critical reading of this chapter and providing the transmission electron micrographs and Jean Euzéby for his advice on etymology.

List of species of the genus *Veillonella*

1. **Veillonella parvula** (Veillon and Zuber 1898) Prévot 1933, 119[AL] (*Staphyloccocus parvulus* Veillon and Zuber 1898, 542)

par'vu.la. L. fem. dim. adj. *parvula* very small.

The characteristics are as described for the genus. Some strains of *Veillonella parvula* can grow aerobically in static culture on lactate (Kolenbrander, 2006). The type strain of *Veillonella parvula* subsp. *parvula* ATCC 10790 has been found to have 74% homology with type strain of *Veillonella alcalescens* subsp. *alcalescens* ATCC 17745 and, therefore, they were considered as synonyms (Mays et al., 1982). *Veillonella parvula* has priority.

Antigenic properties: serogroup VI (*Veillonella parvula* subsp. *parvula*); serogroup IV (*Veillonella alcalescens* subsp. *alcalescens*) (Rogosa, 1965).

Pathogenicity: *Veillonella parvula* has been associated with osteomyelitis, prosthetic valve endocarditis, and bacteremia. Two fatal cases of *Veillonella parvula* bacteremia have been reported (Liu et al., 1998). Isolated from oral cavity and intestinal tract of humans and rodents.

DNA G+C content (mol%): 37–40, mean 38 (T_m) and 41 (Bd).

Type strain: ATCC 10790, ATCC 17742, CCUG 5123, DSM 2008, JCM 12972, KCTC 5019, NCTC 11810, VPI 11221, Prévot Te 3, ATCC 17745, VPI 11224.

GenBank accession number (16S rRNA gene): AY995767, X84005 (*Veillonella parvula* subsp. *parvula*); AY995769 (*Veillonella parvula* subsp. *alcalescens*).

2. **Veillonella atypica** (Rogosa 1965) Mays, Holdeman, Moore, Rogosa and Johnson 1982, 35[VP] (*Veillonella parvula* subsp. *atypica* Rogosa 1965, 707)

a.ty'pi.ca. Gr. pref. *a* not; L. adj. *typicus -a -um* typical; N.L. fem. adj. *atypica* not typical.

The characteristics are as described for the genus.
Antigenic properties: Serogroups V and VI (Rogosa, 1965).
Pathogenicity: not reported; isolated from human saliva.
DNA G+C content (mol%): 36–40, mean 39 (T_m).
Type strain: ATCC 17744, DSM 20739, NCTC 11830, VPI 11220 (strain KON).
GenBank accession number (16S rRNA gene): AF439641, AY995768, X84007.

3. **Veillonella caviae** Mays, Holdeman, Moore, Rogosa and Johnson 1982, 35[VP]

ca.vi'a.e. N.L. gen. n. *caviae* of *Cavia*, the zoological genus name of the guinea pig.

The characteristics are as described for the genus.
Antigenic properties: Serogroup VIII (Rogosa, 1984b).
Pathogenicity: not reported; isolated from guinea pig mouth.
DNA G+C content (mol%): 37–39, mean 39 (T_m).
Type strain: ATCC 33540, DSM 20738, NCTC 12021, VPI 12140, Rogosa strain PV1.
GenBank accession number (16S rRNA gene): AY355140.

4. **Veillonella criceti** (Rogosa 1965) Mays, Holdeman, Moore, Rogosa and Johnson 1982, 35[VP] (*Veillonella alcalescens* subsp. *criceti* Rogosa 1965, 708)

cri.ce'ti. N.L. gen. n. *criceti* of *Cricetus*, the zoological genus name of the hamster.

The characteristics are as described for the genus. Nine out of ten strains fermented fructose (Rogosa, 1984b). A human clinical isolate phylogenetically closely related to *Veillonella criceti* and fermenting fructose has been reported (Marchandin et al., 2005).

Antigenic properties: Serogroup I (Rogosa, 1965).
Pathogenicity: not reported; isolated from hamster mouth.
DNA G+C content (mol%): 38–40, mean 39 (T_m).
Type strain: ATCC 17747, DSM 20734, VPI 11226, NCTC 12020, Rogosa strain HV1.
GenBank accession number (16S rRNA gene): AF186072.

5. **Veillonella denticariosi** Byun, Carlier, Jacques, Marchandin and Hunter 2007, 2847[VP]

den.ti.car.i.o'si. L. n. *dens dentis* tooth; L. adj. *cariosus* rotten, decayed; N.L. gen. n. *denticariosi* of a decayed tooth.

Does not produce gas from TGY broth. Other characteristics are as described for the genus. Can be differentiated from other species of the genus *Veillonella* by 16S DNA and *dnaK* sequencing.

Pathogenicity: Isolated from human carious dentin.
DNA G+C content (mol%): unknown.
Type strain: RBV 106, CIP 109448, CCUG 54362, DSM 19009.
GenBank accession number (16S rRNA gene): EF185167.

6. **Veillonella dispar** (Rogosa 1965) Mays, Holdeman, Moore, Rogosa and Johnson 1982, 35[VP] (*Veillonella alcalescens* subsp. *dispar* Rogosa 1965, 708)

dis'par. L. fem. adj. *dispar* dissimilar, different.

The characteristics are as described for the genus.
Antigenic properties: Serogroup VII (Rogosa, 1965).
Pathogenicity: *Veillonella dispar* has been associated with prosthetic valve endocarditis and prosthetic joint infections. Isolated from mouth and respiratory tract of humans.
DNA G+C content (mol%): 38–40, mean 39 (T_m) and 42 (Bd).
Type strain: ATCC 17748, DSM 20735, NCTC 11831, VPI 11223, Rogosa strain ERN.
GenBank accession number (16S rRNA gene): AF439639, AY995770, X84006.

7. **Veillonella montpellierensis** Jumas-Bilak, Carlier, Jean-Pierre, Teyssier, Gay, Campos and Marchandin 2004, 1315[VP]

mont.pel.li.er.en'sis. N.L. fem. adj. *montpellierensis* pertaining to Montpellier, where the type strain and two other strains were isolated.

The characteristics are as described for the genus. Cells are coccoid (0.3–0.5 mm in diameter) and occur singly, in pairs, or in short chains. Colonies on Columbia blood agar are 1–3 mm in diameter and appear smooth, opaque, and grayish-white. Can be differentiated from other species of the genus *Veillonella* by 16S DNA and *dnaK* gene sequencing.

Pathogenicity: *Veillonella montpellierensis* has been isolated from the gastric fluid of a newborn and from the amniotic fluid of 2 women. Association with endocarditis has been reported (Rovery et al., 2005).

DNA G+C content (mol%): unknown.

Type strain: ADV 281.99, CCUG 48299, CIP 107992, DSM 17217.

GenBank accession number (16S rRNA gene): AF473836.

8. **Veillonella ratti** (Rogosa 1965) Mays, Holdeman, Moore, Rogosa and Johnson 1982, 35^{VP} (*Veillonella alcalescens* subsp. *ratti* Rogosa 1965, 708)

rat′ti. N.L. gen. n. *ratti* of *Rattus*, the zoological genus name of the rat.

> The characteristics are as described for the genus. Antigenic properties: Serogroup III (Rogosa, 1965). Pathogenicity: not reported; isolated from rat mouth.
>
> *DNA G+C content (mol%):* 41–43, mean 42 (T_m) and 44 (Bd).
>
> *Type strain:* ATCC 17746, DSM 20736, VPI 11225, NCTC 12019, Rogosa strain RV-12X.
>
> *GenBank accession number (16S rRNA gene):* AF186071, AY355138.

9. **Veillonella rogosae** Arif, Do, Byun, Sheehy, Clarke, Gilbert and Beighton 2007, 584^{VP}

rog.o.sae. N.L. gen. n. *rogosae* of Rogosa, named in honor of the American microbiologist Morrison Rogosa.

> The characteristics are as described for the genus. Using the rapid ID 32A system, alkaline phosphatase and pyroglutamic acid arylamidase activities are detected. Can be differentiated from other species of the genus *Veillonella* by 16S rRNA and *rpoB* sequencing.
>
> Pathogenicity: not reported. Found in supra-gingival plaque of caries-free children.
>
> *DNA G+C content (mol%):* unknown.
>
> *Type strain:* 100CF, CCUG 54233.
>
> *GenBank accession number (16S rRNA gene):* EF108443.

10. **Veillonella rodentium** (Rogosa 1965) Mays, Holdeman, Moore, Rogosa and Johnson 1982, 35^{VP} (*Veillonella parvula* subsp. *rodentium* Rogosa 1965, 707)

ro.den′ti.um. N.L. pl. gen. n. *rodentium* (from L. part. adj. *rodens -entis* gnawing), of rodents.

> The characteristics are as described for the genus. Antigenic properties: Serogroup II (Rogosa, 1965). Pathogenicity: not reported; isolated from hamster mouth.
>
> *DNA G+C content (mol%):* 42–43, mean 43 (T_m) and 44.4 (Bd).
>
> *Type strain:* ATCC 17743, DSM 20737, VPI 11222, NCTC 12018, Rogosa strain HV 19.
>
> *GenBank accession number (16S rRNA gene):* AY514996.

Genus II. **Acetonema** Kane and Breznak 1992, 191^{VP} (Effective publication: Kane and Breznak 1991, 97.)

FRED A. RAINEY

A.ce.to.ne′ ma. L. n. *acetum* vinegar; Gr. neut n. *nema* thread; N. L. neut. n. *Acetonema* vinegar-forming thread.

Thin, straight rods, 0.3–0.4 μm in diameter and 6–60 μm in length. Motile by means of peritrichous flagella. **Gram-negative cell wall**. Gram-negative staining. Lipopolysaccharide present. Heat resistant, dipicolinic acid containing **endospores are formed**. Catalase-positive, oxidase-negative. **Strict anaerobe. Facultative chemolithotroph.** Mesophiles. **Ferments H$_2$ + CO$_2$ to acetate.** Sugars and organic acids are fermented to acetate, propionate, butyrate, succinate, and H$_2$ and CO$_2$ depending on the substrate. Do not respire anaerobically with nitrate or sulfate. Cytochromes not detected. Isolated from gut contents of the termite *Pterotermes occidentis*.

DNA G+C content (mol%): 55.1 (Bd).

Type species: **Acetonema longum** Kane and Breznak 1992, 191^{VP} (Effective publication: Kane and Breznak 1991, 97.).

Further descriptive information

The genus *Acetonema* contains the single species *Acetonema longum* the description of which is based on the characterization of a single strain APO-1^T (Kane and Breznak, 1991). Cells of this species stain Gram-negative, and transmission electron microscopy of thin sections shows the presence of a distinct inner cytoplasm and outer membrane as found in Gram-negative bacteria. The Gram-negative nature of the cell wall was confirmed using the polymyxin B test as described by Wiegel and Quandt (1982). In addition, a significant amount of lipopolysaccharide is determined in the cell lipid fraction. The endospores formed by this organism are spherical and terminal and swell the sporangium. Viable cells can be recovered from cultures that have been heat treated (80°C, 10 min). The length of the cells varies with phase of growth as well as substrate. Cells grown on H$_2$:CO$_2$ are 20 μm long during the exponential phase of growth and decrease in length by at least 50% at the end of the exponential phase. At the end of the exponential phase, 75% of the cells have formed endospores; these sporulating cells are all ≤12 μm long. Cells growing on organic substrates are generally longer (up to 60 μm) and form fewer endospores than cells grown on H$_2$:CO$_2$.

Growth of strain APO-1^T is stimulated by the addition of trypticase and yeast extract, but rumen fluid had little effect on growth. Cells grow in the temperature range 19–40°C and the pH range 6.4–8.6. With H$_2$:CO$_2$ as the growth substrate, maximum growth rates were obtained at 30–33°C and pH 7.8. The pH of the culture drops to 6.7 during growth. Doubling times vary with growth substrate–H$_2$:CO$_2$ (36 h), glucose (8 h), rhamnose (15 h), and fumarate (40 h). When grown on H$_2$:CO$_2$, only acetate is produced, but on organic substrates a number of major end products are detected. On glucose, the major products of fermentation are acetate, butyrate, CO$_2$, and H$_2$. Fermentation of fructose, mannose, mannitol, oxaloacetate, pyruvate, and ribose yields butyrate and acetate as the main acidic end products. The major product of rhamnose fermentation is 1, 2-propanediol while that of fumarate dissimilation is propionate. Substrates used as energy sources by strain APO-1^T include: H$_2$:CO$_2$, glucose, fructose, mannose, rhamnose, ribose, mannitol, pyruvate, oxaloacetate, and fumarate. Poor growth is observed on citrate, propanol, ethylene glycol, and 3,4,5-trimethylbenzoate. No growth is observed on the following substrates: melibiose, raffinose, maltose, cellobiose, arabinose, galactose, lactose, xylose, sucrose, trehalose, starch, L-fucose, formate, lactate, malate, D-gluconate, acetate, oxalate, succinate, gallate, syringate, caffeate, 3-hydroxybenzoate, benzoate, pyrogallol, methanol, ethanol, glycerol, adonitol, sorbitol, erythritol, butanol, isobutanol, dulcitol, pectin, xanthine, dextrin, betaine, salicin, esculin, *N,N*-dimethylglycine, and casamino acids.

Enrichment, isolation, and growth conditions

Acetonema longum was isolated from homogenates of the hind gut contents of the termite *Pterotermes occidentis*. The homogenates were prepared under anaerobic conditions using buffered salts solution (Breznak and Switzer, 1986). Serially diluted homogenates were inoculated into anaerobic AC-K1 or AC-K2 medium with a H_2:CO_2 headspace (Kane and Breznak, 1991). Isolates were obtained from tubes of AC-K2 showing bacterial growth, gas consumption, and acetate production using the roll tube technique in which AC-K2 medium was solidified with 2% agar. Strains were routinely cultured in AC-K1 medium. From the initial enrichment studies using AC-K2 and a H_2:CO_2 headspace, three strains were obtained and strain APO-1[T] was further characterized. AC-K3 was used to culture strain APO-1[T] for growth and nutrition studies (Kane and Breznak, 1991). For the determination of growth on non-gaseous substrates, tubes were filled completely with liquid medium without a headspace. A 100% N_2 atmosphere was used in the headspace for studies of the fermentation of organic substrates using AC-K3 buffered with 3-(*N*-morpholino)propanesulfonic acid rather than the $NaHCO_3$ buffering system.

Taxonomic comments

When this genus was described in 1991 (Kane and Breznak, 1991) there were a limited number of 16S rRNA gene sequences available for comparison. However, the comparison to available sequences showed it to represent a distinct lineage that fell within the radiation of the species of the genera *Sporomusa*, *Selenomonas*, and *Megasphaera* and had highest similarity to *Sporomusa termitida* and *Sporomusa paucivorans*. Since then, a number of new genera have been described in this group and *Acetonema longum* now clusters with species of the genera *Pelosinus*, *Sporotalea*, *Propionispora*, *Anaerosinus*, *Dendrosporobacter*, and *Sporomusa* (see Figure 199) The 16S rRNA gene sequence similarities between *Acetonema longum* and these related taxa are in the range 90–92.5%. The genera *Pelosinus* and *Sporotalea* were proposed after finalization of the content of this volume and are not described here (see Shelobolina et al., 2007; Boga et al., 2007). The genus *Acetonema* is a member of the family *Veillonellaceae*.

Differentiation of the genus *Acetonema*

Acetonema longum can be differentiated from related taxa on the basis of a number of characteristics (Table 202). These include cell morphology, spore formation, flagellation, temperature optimum for growth, substrate utilization, end products of fermentation, and mol% G+C content of DNA. A number of characteristics are shared with only some related genera. Morphologically *Acetonema* is similar to *Sporotalea* and *Dendrosporobacter* in forming straight rods although the length of the rods of the latter two genera is much shorter. *Acetonema* can be differentiated from *Anaerosinus*, which does not have flagella or form endospores. The two characteristics that distinguish *Acetonema* from all of the related genera are its abilities to grow on H_2:CO_2 and produce butyrate as the major end product of glucose fermentation.

TABLE 202. Differentiation of *Acetonema* from related genera

	Acetonema[a]	*Pelosinus*[b]	*Sporotalea*[c]	*Propionispora*[d]	*Anaerosinus*[e]	*Dendrosporobacter*[f]
Cell shape	Straight rods	Slightly curved rods	Straight rods	Curved rods	Curved rods	Straight rods
Cell size	0.4 × 6–60 μm	0.6 × 2–6 μm	0.6 × 2–12 μm	0.6 × 2–6 μm	0.5 × 2–10 μm	0.5 × 3 μm
Endospores	+	+	+	+	−	+
Flagella	1–3 peritrichous	1–6 peritrichous	Peritrichous	7–10 concave side of cell	Not found	1–3 peritrichous
Temp opt (°C)	30–33	22–30	19–35	35–37	37	25–30
Growth on:						
H_2:CO_2	+	−	−	−	−	−
Fructose	+	+	+	+	−	+
Fumarate	+	+	+	ND[g]	−	ND
Glucose	+	+	+	+/−	−	ND
Glycerol	−	−	+	+	+	+
Lactate	−	+	+	−	−	ND
Malate	−	+	+	ND	−	ND
Pyruvate	+	+	ND	−	−	ND
Succinate	−	+	−	ND	−	ND
End products of fermentation[h]	B, H_2, A, P (from glucose)	A (from citrate)	P, A (from glucose)	P, A (from fructose)	P (from glycerol)	A, P, propanol, H_2 (from fructose)
DNA G+C content (mol%)	51.5	41.0	ND	42–48	35.0	52.0–54.0

[a]Kane and Breznak (1991).
[b]Shelobolina et al. (2007).
[c]Boga et al. (2007).
[d]Biebl et al. (2000) and Abou-Zeid et al. (2004).
[e]Schauder and Schink (1989) and Strömpl et al. (1999).
[f]Strömpl et al. (2000).
[g]ND = Not determined.
[h]Major products: A, Acetate; B, Butyrate; P, Propionate.

List of species of the genus *Acetonema*

1. **Acetonema longum** Kane and Breznak 1992, 191[VP] (Effective publication: Kane and Breznak 1991, 97.)

lon'gum. L. neut. adj. *longum* long in shape.

Forms thin straight cells with rounded ends. Cells are 6–60 μm in length and 0.3–0.4 μm in diameter. Motile by means of peritrichous flagella. Endospores are spherical, terminal, and swell the sporangium. Subsurface colonies are 1–2 mm in diameter, circular with uneven edges and opaque, slightly brown in color when grown in 2% agar on H_2:CO_2. Strict anaerobe. Catalase-positive, oxidase-negative. Obtains energy from H_2:CO_2 by acetogenesis. Growth occurs on glucose, fructose, mannose, ribose, rhamnose, mannitol, pyruvate, and oxaloacetate, which are fermented to butyrate and acetate. H_2, CO_2, and propionate are formed from glucose fermentation: H_2, CO_2, succinate and 1,2-propanediol are formed from rhamnose fermentation. Acetate is the sole acid end product from the fermentation of ethylene glycol and 3,4,5-trimethoxybenzoate. The pH range for growth is 6.4–8.6 with an optimum at 7.8. The temperature range for growth is 19–40°C with an optimum in the range 30–33°C. Trypticase or yeast extract is required for good growth. Isolated from the gut contents of the termite *Pterotermes occidentis* (Walker) (Kalotermitidae).

DNA G+C content (mol%): 55.1 (Bd).

Type strain: APO-1, DSM 6540, ATCC 51454.

GenBank accession number (16S rRNA gene): AJ010964.

Genus III. **Acidaminococcus** Rogosa 1969, 765[AL] emend. Cook, Rainey, Chen, Stackebrandt and Russell 1994a, 577

JAMES B. RUSSELL

A.cid.a.min.o.coc'cus. L. n. *acidum* acid; N.L. neut. n. *amino* amino; Gr. n. *kokkos* a grain, berry; N.L. masc. n. *Acidaminococcus* the amino acid coccus.

Cells coccoid, 0.6–1.0 μm in diameter, often occurring as oval or kidney-shaped diplococci. Nonspore-forming. **Gram-stain-negative**. Nonmotile; flagella are not present. The cellwall contains ***meso*-diaminopimelic acid**; whole cells contain galactose, glucose, and ribose. Menaquinones and ubiquinones are absent. **Anaerobic**; no growth on the surface of agar media incubated in the air. Optimal growth at 30–37°C and pH 7.0. Oxidase and catalase-negative. **Chemo-organotrophic**: D- and L-Glutamate, *trans*-aconitate, and citrate are the known energy sources. Other amino acids, pyruvate, lactate, fumarate, malate, and succinate are not used as energy sources. Approximately 40% of the strains catabolize glucose, from which little acid is produced. On complex media, ammonia, CO_2, acetate, butyrate, and hydrogen are produced, but propionate and valerate have not been detected. Glutamate is fermented to about 1.0 CO_2, 1.0 ammonia, 1.2 acetate, 0.4 butyrate, and up to 10 kPa H_2; citrate and *trans*-aconitate are fermented to about 2.0 CO_2, 1.8 acetate, 0.1 butyrate, and 90 kPa H_2. Nutritional requirements are complex. Isolated from the intestinal tract of the pig and humans. Has also been isolated from cattle rumen, but not usually a predominant ruminal bacterium.

DNA G+C content (mol%): 54.7–57.4.

Type species: **Acidaminococcus fermentans** Rogosa 1969, 765[AL] emend. Cook, Rainey, Chen, Stackebrandt and Russell 1994a, 577.

Further descriptive information

An outer cell-wall membrane is demonstrable in thin sections by electron microscopy. Lipopolysaccharide (endotoxin) is present, and a Shwartzman reaction occurs in rabbits.

Optimal growth is obtained on media containing 0.1 M sodium glutamate, 0.5% yeast extract, 20 mM potassium phosphate (pH 7.4), VRB salts (Rogosa, 1969), and 20 mM thioglycollate as a reducing agent under N_2 at 37°C. On liquid media, cell densities up to OD_{600} = 3.0 are reached. In freshly prepared semisolid media (0.2% agar) in 16 × 160 mm sealed tubes under air, growth starts at the site of inoculation with a syringe and spreads over the entire medium except about 5 mm below the surface. H_2 up to 50 kPa does not inhibit growth. On agar plates under an anoxic atmosphere consisting of 95 kPa N_2 and 5 kPa H_2, after 48 h at 37°C colonies, are obtained which are 1–2 mm in diameter, round, entire, slightly raised, and whitish gray or nearly transparent. The plating efficiency is very low.

Growth on citrate is less efficient (OD_{600} ≤ 1.5) than on *trans*-aconitate and glutamate. After transfer from glutamate to citrate medium, growth may not commence for as long as one week. Growth on glutamate requires ≥ 5 mM Na^+, whereas for the tricarboxylates > 0.1 mM Na^+ is necessary (Cook et al., 1994a; Härtel and Buckel, 1996; Wohlfahrt and Buckel, 1985).

Growth is poor or absent at 25 and 45°C. Cells do not survive 60°C for 30 min. Growth occurs at initial pH values between 6.2 and 7.5, although best growth occurs at a neutral reaction. Final pH values in media initially at pH 7.5 range ~ 6.1–6.7.

Nutritional requirements are multiple. Tryptophan, glutamate, valine, and arginine are required by all strains. Cysteine and histidine are required by 93% of strains, tyrosine by 75%, phenylalanine and serine by 50%. Glycine is sometimes stimulatory. Alanine, leucine, isoleucine, proline, threonine, methionine, lysine, and aspartate are not required for growth. In amino acid containing media, vitamin B_{12}, pyridoxal, pantothenate, and biotin are indispensable for growth; *p*-aminobenzoic acid is essential or highly stimulatory. Exogenous putrescine, folic acid, folinic acid, thiamine, niacin, and riboflavin are not required. On glutamate/yeast extract medium, growth is completely inhibited by avidin (100 μg/ml). Pyruvate is utilized by some strains.

Polyols including adonitol, dulcitol, erythritol, glycerol, inositol, mannitol, and sorbitol are not attacked. Amygdalin, arabinose, fructose, galactose, insulin, maltose, mannose, melezitose, α-methyl-D-glucoside, α-methyl-D-mannoside, raffinose, salicin, sorbose, sucrose, trehalose, xylose, erythrose, and esculin are not attacked. Ambiguous, extremely weak, or negative reactions occur with cellobiose, fucose, lactose, melibiose, rhamnose, and ribose. The type strain does not ferment glucose. Reference strains fermenting glucose weakly include ATCC 25086 (strain VR7) and ATCC 25087 (strain VR11).

Ammonia is produced. Gelatin is generally not liquefied, although slow and partial liquefaction may sometimes occur. H_2S is produced by some strains. Indole is generally not produced. Oxidase and catalase-negative. The benzidine test for porphyrin is negative. Nitrate is not reduced. Sulfophthalein indicators are not reduced.

Resistant to vancomycin (7.5 µg/ml), kanamycin (100 µg/ml), and streptomycin (100 µg/ml). Sensitive to ampicillin (20 µg/ml), chloramphenicol (25 µg/ml), colistin (10 µg/ml), erythromycin (2 µg/ml), neomycin (20 µg/ml), spectinomycin (20 µg/ml), and tetracycline (6 µg/ml). Cultures grown on and trans-aconitate were resistant to the ionophore, monensin, but those grown on citrate were sensitive (Cook and Russell, 1994).

Enrichment and isolation procedures

Acidaminococcus strains have been isolated from intestinal tracts of pig and humans, in relatively large numbers from 25% of normal human feces, from a closed abdominal abscess, and from a putrid lung abscess as part of mixed anaerobic and facultative flora (Sugihara et al., 1974). They may be widespread in the intestinal tracts of various homothermic animals. In a study of the isolation of Acidaminococcus from human feces, Sugihara et al. (1974) found that the highest recovery was obtained from dilutions of the fecal samples when kanamycin-vancomycin blood agar, neomycin blood agar, or nonantibiotic-containing blood agar were used. More recently, Acidaminococcus fermentans strain AO, which is "virtually identical to the type strain" was isolated from the rumen, but it is not normally a predominant ruminal bacterium (Cook et al., 1994a). The ability of Acid-aminococcus fermentans to convert trans-aconitate to acetate may prevent tricarballylate accumulation and grass tetany (magnesium deficiency) in ruminants (Cook et al., 1994b).

Maintenance procedures

Cells suspended in skim milk may be lyophilized successfully.

Biochemical comments

Organisms of the order Clostridales use two different pathways of glutamate fermentation which lead to identical products (Braune et al., 1999; Buckel, 1980): The methylaspartate pathway, which involves coenzyme B_{12}, is found in the genus Clostridium and in enterobacteria. Acidaminococcus fermentans, Peptostreptococcus asaccharolyticus, Fusobacterium nucleatum, and Clostridium symbiosum ferment glutamate via the (R)-2-hydroxglutarate pathway, in which following unusual enzymes are involved: 2-hydroxyglutarate dehydrogenase, glutaconate CoA-transferase, 2-hydroxyglutaryl-CoA dehydratase, and glutaconyl-CoA decarboxylase, a biotin containing Na^+ pump. All genes encoding these enzymes from Acidaminococcus fermentans have been cloned and sequenced. Deduced proteins with the highest sequence identities have been found in other Clostridiales (e.g., Veillonella) and in the archaeon Archaeoglobus fulgidus (Braune et al., 1999; Hans et al., 1999). There is no serological cross-reaction between strains of Acidaminococcus and either Veillonella serovars or Peptostreptococcus asaccharolyticus (formerly called Peptococcus aerogenes). The newly proposed family "Acidaminococcaceae" also contains organisms which ferment glutamate to propionate instead of butyrate, e.g., Selenomonas acidaminophila.

List of species of the genus Acidaminococcus

1. **Acidaminococcus fermentans** Rogosa 1969, 765[AL] emend. Cook, Rainey, Chen, Stackebrandt and Russell 1994a, 577

   fer.men′tans. L. part. adj. fermentans fermenting.
   Description is the same as for the genus.

   DNA G+C content (mol%): 54.7–57.4 (T_m).
   Type strain: VR4, ATCC 25085, CCUG 9996, CIP 106432, DSM 20731.
   GenBank accession number (16S rRNA gene): X65935, X78017.

Genus IV. **Allisonella** Garner, Flint and Russell 2003, 373[VP] (Effective publication: Garner, Flint and Russell 2002, 504.)

FRED A. RAINEY

Al.li.son.ella. N.L. dim. suff. -ella; N.L. fem. n. Allisonella named after the American microbiologist Milton J. Allison, a prominent rumen microbiologist who isolated Oxalobacter formigenes, a ruminal bacterium that decarboxylates oxalate.

Cells are **ovoid shaped**, **nonmotile**, and form diploids or chains. **Spores not observed**. Cells stain **Gram-negative**. Resistant to ionophore and monensin. **Catalase-, oxidase-, and indole-negative**. Nonfermentative, facultative anaerobic, and chemoheterotrophic. Utilizes histidine as sole energy source. Produces histamine and CO_2 as end products from histidine. Growth occurs in medium containing yeast extract, butyrate, and histidine. On the basis of 16S rRNA gene sequence comparisons, the genus Allisonella is a member of the family Veillonellaceae.

DNA G+C content (mol%): 45–48 (HPLC).

Type species: **Allisonella histaminiformans** Garner, Flint and Russell 2003, 373[VP] (Effective publication: Garner, Flint and Russell 2002, 504.).

Enrichment and isolation procedures

Allisonella histaminiformans was isolated in a study looking at the component of a mixed rumen bacterial population that could convert histidine to histamine (Garner et al., 2002). Rumen contents from cows on three different diets were studied. Rumen bacteria capable of converting histidine to histamine were detected in cows fed 3.4 kg of grain and 9.8 kg of grain. The highest numbers of bacteria that utilized histidine as a sole energy source were found in the cow fed 9.8 kg of grain. When enriched in MRS medium, histamine-producing bacteria can be detected but are not selectively enriched, and a number of colony morphologies are observed on plating. MRS enrichments showing histamine production are transferred to carbonate medium with added histidine (50 mM). A variety of colonies are also produced when these carbonate medium cultures are streaked on agar containing histidine. Not all of these colonies are histamine producers and only the small translucent colonies are capable of histamine production when transferred to carbonate broth plus histidine. *Allisonella histaminiformans* can be easily isolated from the rumen of cattle fed grain, but it cannot be detected in the grain itself or in cattle fed hay. Also, *Allisonella histaminiformans* can be isolated at lower numbers from bovine feces and the cecum of a horse.

Taxonomic comments

The genus *Allisonella* was described on the basis of its phenotypic, chemotaxonomic, and phylogenetic characteristics. *Allisonella histaminiformans* is considered to be the first histamine-producing bacterium that can only utilize histidine as a sole energy source. The fatty acid profile is dominated by unsaturated $C_{16:1}$ and $C_{18:1}$ species as well as $C_{14:0}$ and was shown to differ from the profiles of the classical rumen bacteria *Megasphaera elsdenii* and *Selenomonas ruminantium*. Comparison of the 16S rRNA gene sequence of *Allisonella histaminiformans* with those of other members of the *Firmicutes* lineage shows the genus *Allisonella* falls within the radiation of the species of the genera *Dialister, Anaeroglobus, Megasphaera,* and *Veillonella* (see Figure 199). The clustering of these genera is supported by a bootstrap value of 99%. The highest 16S rRNA gene sequence similarities are found between *Allisonella histaminiformans* and species of the genus *Dialister* (*Dialister pneumosintes*, 94.7%; *Dialister invisus*, 93.5%; *Dialister propionicifaciens*, 93.1%; *Dialister micraerophilus*, 91.4%). The 16S rRNA gene sequence similarity to other genera of this cluster is lower and in the range 88.0–90.3%.

List of species of the genus *Allisonella*

1. **Allisonella histaminiformans** Garner, Flint and Russell 2003, 373[VP] (Effective publication: Garner, Flint and Russell 2002, 504.)

   his.ta.min.i.for.mans. N.L. neut n. *histaminum* histamine; L. part. adj. *formans* forming; N.L. part. adj. *histaminiformans* histamine-forming.

   Cells are ovoid shaped, nonmotile and occur in diploids or chains. Cells are 1.0–8.0 μm in length and 0.4–0.8 μm in diameter. Spores are not observed. Stains Gram-negative. Highly resistant to ionophore and monensin. Facultative anaerobe, chemoheterotroph, utilizes histidine producing histamine and CO_2 as end products. Cellular yield is 1.5 mg of protein per mmol of histidine. Growth occurs in medium containing yeast extract, butyrate, and histidine. Volatile fatty acids are not produced and butyric acid is required for growth. Lysine is utilized, but lysine alone does not support growth. Does not utilize carbohydrates or organic acids. Optimum temperature for growth is 39°C with a doubling time of 110 min. Cellular fatty acids profiles are dominated by $C_{14:0}$, $C_{16:1}$, and $C_{18:1}$. $C_{16:0}$ saturated fatty acids account for less than 5% of the total. The type strain was isolated from dairy cow rumen content.

   DNA G+C content (mol%): 46.8 (HPLC).
   Type strain: MR2, ATCC BAA-610, DSM 15230.
   GenBank accession number (16S rRNA gene): AF548373.

Genus V. **Anaeroarcus** Strömpl, Tindall, Jarvis, Lünsdorf, Moore and Hippe 1999, 1870[VP]

CARSTEN STRÖMPL AND HEINRICH LÜNSDORF

An.ae.ro.ar'cus. Gr. pref. *an* not; Gr. n. *aer* air, anaero not (living) in air; L. masc. n. *arcus* a bow, arc; N.L. masc. n. *Anaeroarcus* a bow not living in air.

Cells are **curved** or spiral-shaped **rods**, motile and **Gram-stain-negative**. **Endospores are not formed**. Obligately **anaerobic chemo-organotroph**. A limited range of organic acids, amino acids, carbohydrates, and alcohols are fermented mainly to acetate, propionate, succinate, and propanol. Yeast extract, Casamino acids, peptone, and Biotrypticase support growth. The modified **naphthoquinone** "Lipid F", with octa- and nona-prenologues (ratio 4:1) as the predominating isoprenologues, and *b*-type **cytochromes** are present. Ferric iron, but not sulfate or nitrate, is reduced. **Mesophilic.**

DNA G+C content (mol%): ~48.

Type species: **Anaeroarcus burkinensis** (Outtara, Traore and Garcia 1992) Strömpl, Tindall, Jarvis, Lünsdorf, Moore and Hippe 1999, 1870[VP] (*Anaerovibrio burkinabensis* Outtara, Traore and Garcia 1992, 395).

Further descriptive information

Phosphatidyl ethanolamine and phosphatidyl serine are the major polar lipids, together with an unidentified, probably atypical glycolipid. Major fatty acids are $C_{11:0\ iso}$, $C_{15:1}$, $C_{17:1}$, and $C_{13:0\ iso\ 3OH}$. Minor variations in sequence primary structure and length were observed in particular in helix 6 of the 16S rRNA gene, indicating multiple gene copies.

Enrichment and isolation procedures

The sole strain characterized has been isolated from water-logged, iron-rich rice field soils by use of a basal medium (Outtara et al., 1992) and standard anaerobic techniques. Pure cultures have been obtained by repeated application of the roll tube method (Hungate, 1969).

Maintenance and storage

Active cultures can be maintained in the laboratory by frequent transfer in PY broth supplemented with lactate (0.25%) or DSM medium 575. Lyophilized cultures (Hippe, 1991) have been viable after 12 years of storage.

Taxonomic comments

When Outtara et al. (1992) characterized an isolate from anoxic rice field soils, they compared their strain among others with the type species of *Anaerovibrio*, *Anaerovibrio lipolytica*, and found sufficient similarities to describe their isolate as a new species of *Anaerovibrio*, *Anaerovibrio burkinabensis*. Strömpl et al. (1999) re-evaluated the species of *Anaerovibrio* and concluded that the differences in 16S rRNA gene sequence, chemical composition, and metabolic traits between the two species *Anaerovibrio lipolyticus* and *Anaerovibrio burkinabensis* justified the assignment of *Anaerovibrio burkinabensis* to a new genus and combination, *Anaeroarcus burkinensis* (corrig.).

The genera *Anaeroarcus* (with the single species, *Anaerovibrio burkinensis*) and *Anaeromusa* (with *Anaerovibrio acidaminophila* Baena et al., 1999) are very closely related. 16S rRNA gene sequence similarity values indicate a relationship at genus level. The two organisms share, among other features, very similar pH ranges and optima, growth temperatures, DNA base ratios, and fermentation profiles. One major difference observed so far lies in the inability of *Anaeromusa* to utilize carbohydrates (Table 203). This bacterium has also not been tested for its ability to reduce ferric iron, nor has its lipid composition or its lipoquinone compo-sition been determined. *Anaeromusa* possesses several (up to 16), lateral flagella mainly emerging from the concave side of the cell. Re-examination of the type strain of *Anaeroarcus burkinensis* (Figure 202) confirmed the observation reported by Outtara et al. (1992), that this species normally possesses a single, subpolar flagellum. If future analyses support the view that the two genera should be merged in one genus, the name *Anaeromusa* would have priority.

Further comments

Gu et al. (2004) reported on 16S rRNA gene clone sequences (cluster TCE33, GenBank accession number AF349762) from trichloroethene and *cis*-1,2-dichloroethene degrading enrichments which may be related both to *Anaeroarcus* and *Anaeromusa* at generic level. The enrichments had initially been supplied with lactate as an energy source, but the function of these uncultivated organisms in the final, dechlorinating culture remained unclear. Similar 16S rDNA sequences were obtained from a chlorobenzene degrading reactor (GenBank accession nos AJ488080 and AJ488092).

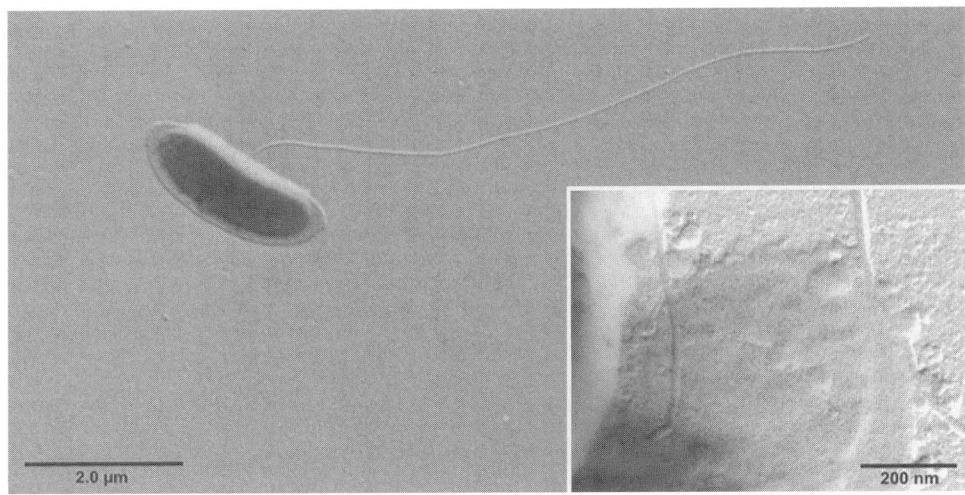

FIGURE 202. Transmission electron micrograph of an individual cell of *Anaeroarcus burkinensis* $B_4B_0^T$, unfixed and shadow-casted with Pt/C-coating (see Golyshina et al., 2000). The slightly vibrioid-shaped cell shows the outer membrane as a flattened contour of the darker cell body, caused by cytoplasmic shrinkage during preparation. A single flagellum of about 8 μm in length is inserted subpolarly as it can be recognized from the flagellum insertion (see inset, lower left). Cell width is 0.70–0.98 μm and cell length ranges from 2.3–2.9 μm.

TABLE 203. Characteristics differentiating the genera *Anaeroarcus* and *Anaeromusa*

Characteristic	*Anaeroarcus*[a]	*Anaeromusa*[b]
Flagella	Subpolar, single	Up to 16, predominantly on the concave side of the cell
Substrates utilized	Citrate, fumarate, malate, lactate, pyruvate, dihydroxyacetone, fructose, 1,2-propanediol, aspartate, glutamate. Slight degradation of glycerol; glycerol 3-phosphate degraded in the presence of yeast extract	Lactate, pyruvate, aspartate, glutamate
Fermentation products	Acetate and propionate from all substrates; succinate from fumarate, malate and aspartate; H_2 (traces), 1,3-propanediol from glycerol; propanol from 1,2-propanediol; traces of lactate from sugars	Acetate and propionate from all substrates; succinate from aspartate
Ferric iron reduction	+	ND[c]
Cytochromes	*b*	*b, c*
16S rDNA sequence similarity (%)	99.5	

[a]Data from Outtara et al. (1992).
[b]Data from Nanninga et al. (1987).
[c]ND, Not determined.

List of species of the genus *Anaeroarcus*

1. **Anaeroarcus burkinensis** (Outtara, Traore and Garcia 1992) Strömpl, Tindall, Jarvis, Lünsdorf, Moore and Hippe 1999, 1870[VP] (*Anaerovibrio burkinabensis* Outtara, Traore and Garcia 1992, 395)

bur.kin.en'sis. N.L. adj. *burkinensis* pertaining to Burkina Faso, the place from which the organism was isolated.

Strictly anaerobic chemo-organotroph. Substrates used as energy sources are citrate, fumarate, malate, lactate, pyruvate, dihydroxyacetone, fructose, 1,2-propanediol, aspartate, glutamate. Slight degradation of glycerol; glycerol 3-phosphate degraded in the presence of yeast extract. Acetate and propionate are the main fermentation products; succinate formed from malate, fumarate, and aspartate, propanol from 1,2-propanediol, and 1,3-propanediol from glycerol. Vitamins required for growth, NaCl not required. The pH range for growth is 5.3–8.4 with the optimum around 6.8. The temperature range is 13–43°C with the optimum around 35°C.

DNA G+C content (mol%): 48.5 ±0.4 (HPLC).
Type strain: B_4B_0, ATCC 51455, CIP 105409, DSM 6283.
GenBank accession number (16S rRNA gene): AJ010961.

Genus VI. **Anaeroglobus** Carlier, Marchandin, Jumas-Bilak, Lorin, Henry, Carriere and Jean-Pierre 2002, 986[VP]

JEAN-PHILIPPE CARLIER

An.ae.ro. glo′bus. Gr. pref. *an* not; Gr. n. *aer* air; N.L. *anaero* not (living) in air; L. masc. n. *globus* sphere; N.L. masc. n. *Anaeroglobus* a sphere not living in air.

Anaerobic, Gram-stain-negative cocci, 0.5–1.1 µm in diameter, nonmotile, endospores not formed. Cells are coccoid to ellipsoidal and found in pairs or short chains. Most of the usual carbohydrates are not fermented. **No gas is produced** in glucose agar deeps. **Catalase and nitrate reduction tests are negative**. The metabolic end products are acetic, propionic, isobutyric, butyric, and isovaleric acids. Occurs in gastrointestinal tract and presumably in oral cavity of man.

DNA G+C content (mol%): 51.8 (T_m).

Type species: **Anaeroglobus geminatus** Carlier, Marchandin, Jumas-Bilak, Lorin, Henry, Carriere and Jean-Pierre 2002, 986[VP].

Further descriptive information

The cells are coccoid to ellipsoidal. Individual cells are about 0.5–1.1 µm in diameter (Figure 203). Ultrastructural examination reveals distinct inner and outer membranes of 3.6 and 4.9 nm in width, respectively. The membranes are separated by approximately 24 nm (Marchandin, 2001). The nonpigmented and nonhemolytic colonies formed on blood agar after 2 d incubation are small, circular, convex, and translucent with a smooth surface. *Anaeroglobus geminatus* is unreactive in most conventional biochemical tests. Indole is not produced. Gelatin and milk are not modified and lactate is not fermented. Acid is produced from galactose and mannose. Acid is not produced from esculin, arabinose, cellobiose, fructose, glucose, glycerol, inositol, lactose, maltose, mannitol, melezitose, melibiose, raffinose, rhamnose, ribose, salicin, sorbitol, starch, sucrose, trehalose, and xylose. Esculin and starch are not hydrolyzed. The type strain has been tested for enzymic activity using the Rapid ID32A kit (API bioMérieux), and a positive reaction was obtained for the β-glucosidase test (API code 0010 0000 00). The strain is resistant to 5 µg vancomycin discs, but is susceptible to 1 mg kanamycin, 10 µg colistin, 4 µg metronidazole, and bile discs. The major metabolic end products are acetic, propionic, isobutyric, butyric, and isovaleric acids. Lactic and succinic acids are not produced. The 16S rRNA gene sequence analysis reveals that *Anaeroglobus* belongs to the family *Veillonellaceae* (Carlier et al., 2002) which includes all members of the *Sporomusa* sub-branch of the class *Clostridia* of the phylum *Firmicutes* (Garrity and Holt, 2001). The closest phylogenetic neighbors of *Anaeroglobus* are *Megasphaera cerevisiae*, *Megasphaera elsdenii*, *Veillonella parvula*, and *Acidaminococcus fermentans* with sequence identities of 91, 87, 87, and 84%, respectively. In addition, several sequences (AF287783, AF287784, AF287785 (Paster et al., 2001) and AF481223 (Munson et al., 2002) which have been deposited in the public databases as oral clones of

TABLE 204. Characteristics useful in differentiation of the genus *Anaeroglobus* from other closely related genera[a,b]

Characteristics	*Anaeroglobus*	*Megasphaera*	*Veillonella*	*Acidaminococcus*
Cell diameter (µm)	0.5–1.1	0.4–2.6[c]	0.3–0.5	0.6–1.0
Gas production	−	+	+	+
Acid production from:				
Galactose	+	−	−	−
Mannose	+	−	−	−
Fermentation of lactate	−	+	+	−
Decarboxylation of succinate	−	−	+	−
Metabolic end products[d]	A, P, iB, B, iV	A, P, B, V, C	A, P	A, B
DNA G+C content (mol%)	51.8 (T_m)	53.1–54.1 (Bd), 42.4.–46.4 (T_m)	40.3–44.4 (T_m)	56.6 (T_m)

[a]Symbols: +, >85% positive; −, 0–15% positive.
[b]Adapted from Carlier et al. (2002) and Marchandin et al. (2003).
[c]Cell size varies according to species.
[d]A, acetic acid; P, propionic acid; iB, isobutyric acid; B, butyric acid; iV, isovaleric acid; V, valeric acid; C, caproic acid.

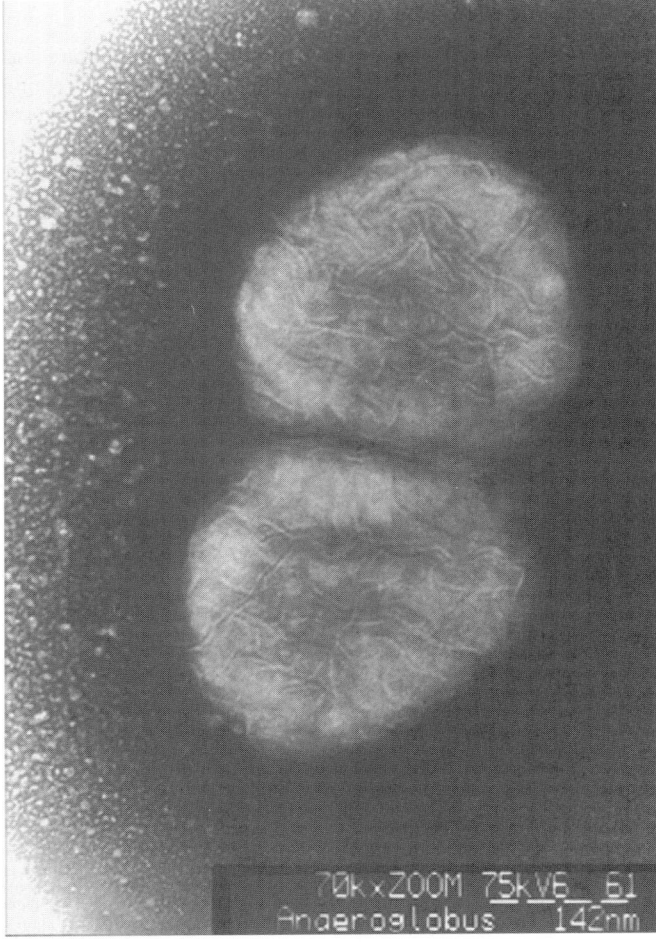

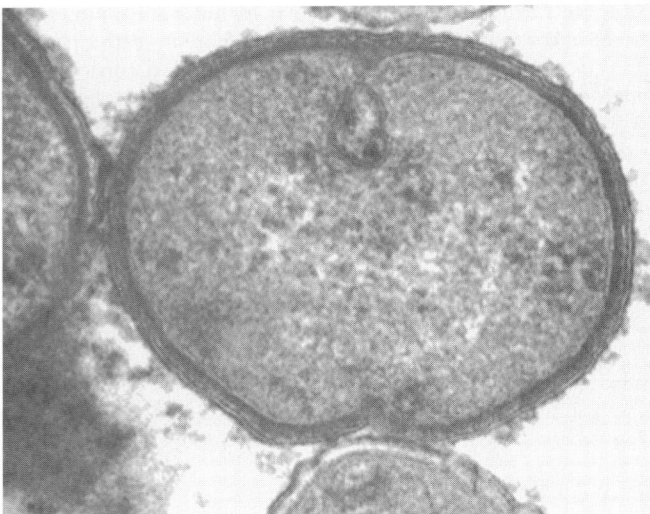

FIGURE 203. Transmission electron micrographs of *Anaeroglobus geminatus* AIP 313.00[T]. (top) General morphology after negative staining and (bottom) cell wall and membranes after ultrathin sectioning.

Megasphaera species exhibit high similarity values (93–99%) with the sequence of *Anaeroglobus geminatus*. However, these sequences were obtained from cloning and sequencing of 16S rRNA genes from oral flora without cultures, and thus the phenotypic characteristics of these micro-organisms are unknown.

Genomic mapping with the endonuclease I-CeuI revealed that *Anaeroglobus geminatus* possesses a genome of about 1800 kb with four *rrn* operons. This genomic structure is similar to those of *Veillonella* species and *Dialister pneumosintes*, which are 2191 kb and 1445 kb carrying 4 and 5 *rrn* copies, respectively. In contrast, the chromosome of *Megasphaera elsdenii* is 2878 kb in size and contains seven *rrn* copies (Marchandin, 2001).

Enrichment and isolation procedures

Anaeroglobus can be isolated by streaking samples onto Columbia or Wilkins-Chalgren sheep-blood agar and incubating at 37°C for 2 d. Strict anaerobic conditions are required for growth of *Anaeroglobus*, and good growth can be obtained in an anaerobic jar containing either an Anaerogen System (Oxoid Unipath) or a mixture of 5% H_2, 5% CO_2 and 90% N_2 (by vol.).

Maintenance procedures

The strain may be maintained in trypticase-glucose-yeast extract (TGY) medium (Carlier et al., 2002) under anaerobic conditions as described above. Under these conditions, cultures of *A. geminatus* will usually survive for no longer than 1 week without subcultivation. The organism may not survive freeze-drying procedures, but can be preserved in liquid nitrogen for several years.

Differentiation from closely related taxa

The differentiation of *Anaeroglobus* from other closely related anaerobic Gram-stain-negative cocci can be done primarily on the basis of poor growth on the usual media and absence of gas. Other characteristics are given in Table 204.

Taxonomic comments

The anaerobic Gram-stain-negative cocci were originally classified in a single family, the *Veillonellaceae* (Rogosa, 1984a), belonging to the *Sporomusa* sub-branch of the class *Clostridia* of the phylum *Firmicutes*. Three genera were included into this family: *Veillonella*, the type genus of the family, *Acidaminococcus*, and *Megasphaera*. In the first volume of this edition of *Bergey's Manual of Systematic Bacteriology* (Garrity and Holt, 2001), all members of the *Sporomusa* sub-branch were grouped in the family "*Acidaminococcaceae*", and the family *Veillonellaceae* was not recognized. However, *Veillonellaceae* has priority over "*Acidaminococcaceae*", so this error was corrected in the revised road map of this volume (Ludwig et al., this volume). Phylogenetically, *Anaeroglobus* and the closely related oral clones identified by Paster et al. (2001) fall within the family *Veillonellaceae* and branch together with the *Dialister* and *Megasphaera* genera (Figure 204). The *Anaeroglobus–Dialister–Megasphaera* group forms a cluster that is distinct from *Veillonella* and *Acidaminococcus*. In this subgroup, *Megasphaera cerevisiae* and the new species *Megasphaera micronuciformis* group together and form a separate branch from *Anaeroglobus* and related clones. On the other hand, *Dialister pneumosintes* and *Megasphaera elsdenii* individually are more deeply branched. However, in a larger tree comprising 41 representative sequences of the *Sporomusa* sub-branch, *Megasphaera elsdenii* clustered with *Megasphaera micronuciformis*, whereas *Anaeroglobus geminatus* and *Megasphaera cerevisiae* represented two distinct lineages (Marchandin et al., 2003b). Thus, the relationship between each species of *Megasphaera* and *Anaeroglobus geminatus* remains questionable.

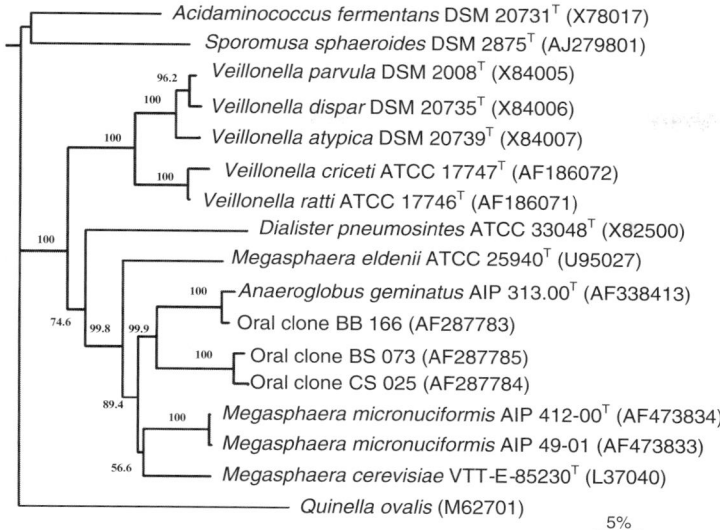

FIGURE 204. Phylogenetic relationships of *Anaeroglobus geminatus* and other closely related genera, inferred from 16S rRNA gene sequence analysis. The sequence of *Quinella ovalis* was used as an outgroup. Numbers at the nodes represent bootstrap percentage values from 1000 resampled datasets. Accession numbers are given in parentheses. The bar represents a 5% sequence difference.

List of species of the genus *Anaeroglobus*

1. **Anaeroglobus geminatus**   Carlier, Marchandin, Jumas-Bilak, Lorin, Henry, Carriere and Jean-Pierre 2002, 986[VP]

ge.min'at.us. L. adj. masc. *geminatus* paired, double, referring to twin cells of this organism.

The characteristics are as described for the genus. Isolated from human intestinal tract.

DNA G+C content (mol%): 51.8 % (T_m).
Type strain: AIP 313.00, CIP 106856, CCUG 44773.
GenBank accession number (16S rRNA gene): AF338413.

Genus VII. **Anaeromusa** Baena, Fardeau, Woo, Ollivier, Labat and Patel, 1999, 973[VP]

CARSTEN STRÖMPL

An.ae.ro'mu.sa. L. v. Gr. pref. *an* not; Gr. n. *aer* air; N.L. n. *musa* a banana; N.L. fem. n. *Anaeromusa* an anaerobic banana.

Cells are **curved Gram-stain-negative rods**, usually single or in pairs. Motile by means of lateral flagella. **Nonspore-former.** Obligately **anaerobic chemo-organotroph.** A limited range of amino acids are fermented mainly to acetate and propionate. Sulfate, thiosulfate, and nitrate are not reduced. Mesophilic.

DNA G+C content (mol%): ~48.

Type species: **Anaeromusa acidaminophila** (Nanninga, Drent and Gottschal 1987) Baena, Fardeau, Woo, Ollivier, Labat and Patel, 1999, 973[VP] (*Selenomonas acidaminophila* Nanninga, Drent and Gottschal 1987, 156).

Further descriptive information

Nanninga et al. (1987) studied the amino acid metabolism of the type strain, DKglu16. A distinguishing feature of this strain is the formation of propionate from the fermentation of glutamate. The arrangement of the flagella is somewhat intermediate between *Pectinatus* and *Selenomonas*; it is neither comb-like nor tufted.

Enrichment and isolation procedures

The type strain was isolated from a glutamate plus aspartate-limited chemostat culture which was inoculated with a sample from an anaerobic purification plant of a potato starch producing factory. Similar organisms have been enriched, but not obtained in pure culture, in several types of dehalogenating reactors (Gu et al., (2004), accession numbers AF349762, AJ488080 and AJ488092).

Maintenance and storage

DSM medium 339 or the medium of Nanninga et al. (1985) may be used for routine culturing of the strain. Lyophilization is recommended for long term storage (Hippe, 1991).

Taxonomic comments

Nanninga et al. (1987) effectively described "*Selenomonas acidaminophila*" based on similarities in cell shape, temperature range and optimum, and fermentation products as a new member of the genus *Selenomonas*, but never validated their description. Baena et al. (1999), studying the 16S rRNA gene sequence-derived phylogenetic relationships of some amino acid-utilizing anaerobes, then renamed and validly described this strain as *Anaeromusa acidaminophila*. This genus is closely related to the genus *Anaeroarcus*.

List of species of the genus *Anaeromusa*

1. **Anaeromusa acidaminophila** (Nanninga, Drent and Gottschal 1987) Baena, Fardeau, Woo, Ollivier, Labat and Patel, 1999, 973[VP] (*Selenomonas acidaminophila* Nanninga, Drent and Gottschal 1987, 156)

a.ci.dam.in.o′phi.la. N.L. neut. n. *acidum* acid; N.L. neut. n. *aminum* amine; Gr. adj. *philos* loving; N.L. part. adj. *acidaminophila* amino acid-loving.

Cell size is approximately 0.3–0.5 × 0.7–1.8 μm, with rounded ends. Cells are motile by means of several lateral flagella emerging mainly from the concave side of the cell. Glutamate, aspartate, lactate, and pyruvate are utilized as energy substrates. Acetate and propionate are the main fermentation products; succinate is formed from aspartate. Growth occurs in mineral medium with a reducing agent. Indole is not formed; gelatin is hydrolyzed, but urea and esculin are not. Catalase-negative. Cells contain cytochromes *b* and *c*. No requirement for sodium. The temperature range is from 25–46°C, with the optimum around 38°C. The pH range is from 5.0–8.5, optimum 6.8–7.2.

DNA G+C content (mol%): 48.0 ± 1 (T_m).
Type strain: DKglu16, ATCC 43704, DSM 3853.
GenBank accession number (16S rRNA gene): AF071415.

Genus VIII. **Anaerosinus** Strömpl, Tindall, Jarvis, Lünsdorf, Moore and Hippe 1999, 1870[VP]

CARSTEN STRÖMPL AND HANS HIPPE

An.ae.ro.si′nus. Gr. pref. *an* not; Gr. n. *aer* air; *anaero* not (living) in air; L. masc. n. *sinus* bend; N. L. masc. n. *Anaerosinus* a curved organism not living in air.

Cells are curved rods or spirals exhibiting a **Gram-stain-negative cell wall**. Motile, but motility may be lost in culture. **Endospores are not formed. Obligately anaerobic. Chemo-organotrophic.** Catalase-negative. **Cytochrome** *b* and the modified **naphthoquinone**, "lipid F", with octa-and nonaprenologues (ratio 1:2) as the predominating isoprenologues, are present. **Mesophilic.** Phosphatidyl serine and phosphatidyl ethanolamine are the **major polar lipids**. Major fatty acids are $C_{15:1}$, $C_{15:0}$, $C_{17:1}$, and $C_{17:0}$, and, of the hydroxy fatty acids, $C_{11:0\ 3OH}$, $C_{12:0\ 3OH}$, and $C_{13:0\ 3OH}$ dominate.

DNA G+C content (mol%): ~35.

Type species: **Anaerosinus glycerini** (Schauder and Schink 1996) Strömpl, Tindall, Jarvis, Lünsdorf, Moore and Hippe 1999, 1870[VP] (*Anaerovibrio glycerini* Schauder and Schink 1996, 625).

Further descriptive information

Anaerosinus glycerini was originally described and thereafter validly published as *Anaerovibrio glycerini* (Schauder and Schink, 1989, 1996). The allocation of strain LG4 to the genus *Anaerovibrio* was based on common properties, including production of propionate as fermentation product, a similar DNA G+C content, a slight lipolytic activity (degradation of diolein), and the lack of spore formation.

Phylogenetic data derived from 16S rRNA gene sequence analyses confirmed that strain LG4 clustered with the *Sporomusa-Pectinatus-Selenomonas* phyletic group. The sequence similarity between strain LG4 and *Anaerovibrio lipolyticus* was, however, remote and therefore justified the creation of a new genus, *Anaerosinus*, for strain LG4 (Strömpl et al., 1999).

Anaerosinus glycerini proved to be the predominant glycerol degrader in anoxic sediments; 10^8 cells per ml were found in various freshwater sediment samples (Schauder and Schink, 1989). In contrast to batch culture, enrichment in continuous culture at low dilution rates, a method which favors bacteria with high substrate affinities, allowed the development and isolation of this highly specialized glycerol degrader. Besides glycerol, the glycerol residue of diolein and perhaps other glycerolesters is used as

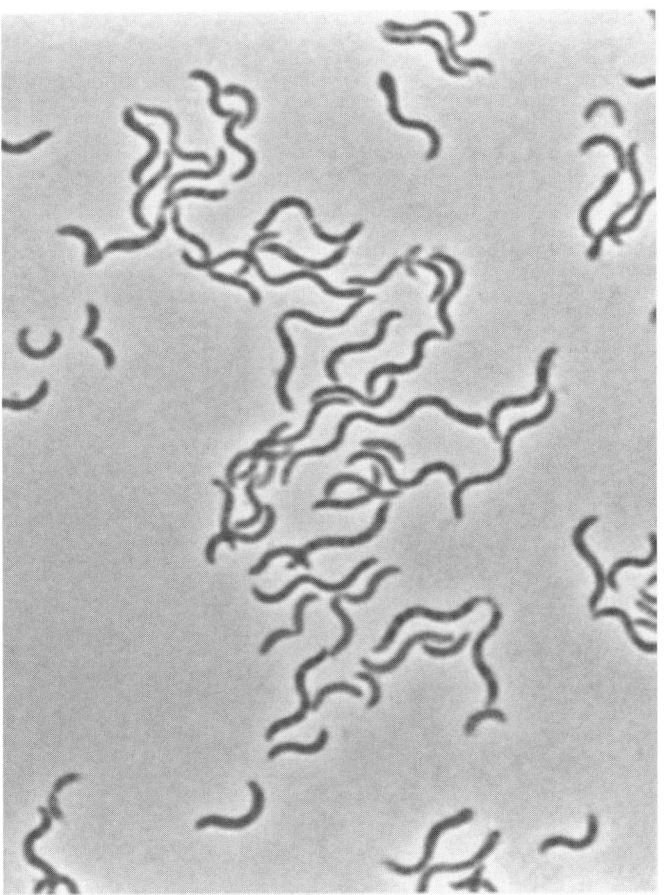

FIGURE 205. Phase-contrast micrograph of cells of *Anaerosinus glycerini* strain LGS4[T] in the late exponential phase of growth (× 2000). Note the dark granular inclusion bodies in the cells and the tapering cell tips (courtesy of B. Schink, University of Konstanz, Germany).

sole source of carbon and energy. *Anaerosinus glycerini* grows in mineral medium without vitamins. Yeast extract increases the cell yield by about 40% but does not support growth itself. The metabolic pathway of glycerol fermentation and energy metabolism in *Anaerosinus glycerini* appear to be basically understood. It has been assumed that glycerol is activated to glycerophosphate (consumes 1 ATP) which is converted to pyruvate via glyceraldehyde phosphate, phosphoglycerate, and phosphoenolpyruvate (yields 2 ATP). Pyruvate then is converted to propionate via the methylmalonyl-CoA pathway, yielding further ATP by electron transport phosphorylation in fumarate reduction with electrons derived from glycerophosphate dehydrogenation. However, enzyme studies show no or only poor activities of some enzymes involved. Cell-free extracts convert succinate to propionate. Cytochrome *b* in the cell fraction is reduced by glycerophosphate or glycerol in the presence of ATP and reoxidized by fumarate (Schauder and Schink, 1989).

Enrichment and isolation procedures

Anaerosinus glycerini was isolated from a continuous culture inoculated with black freshwater sediments and anoxic sewage sludge with glycerol as the sole source of carbon and energy (Schauder and Schink, 1989).

Maintenance and storage

Anaerosinus may be maintained in PY-broth with 0.5% glycerol (Holdeman et al., 1977). Freezing in liquid nitrogen or lyophilization is recommended for long-term storage (Hippe, 1991).

Further comments

Strain SB90 (Chin et al., 1999) exhibits almost 95% 16S rDNA sequence similarity to strain LG4, but it does not belong to the genus *Anaerosinus* due to its ability to ferment sugars to acetate and propionate. Furthermore, it was reported to stain Gram-positive.

List of species of *Anaerosinus*

1. **Anaerosinus glycerini** (Schauder and Schink 1996) Strömpl, Tindall, Jarvis, Lünsdorf, Moore and Hippe 1999, 1870[VP] (*Anaerovibrio glycerini* Schauder and Schink 1989, 477)

gly.ce.ri′ni. N.L. neut. n. *glycerinum* glycerol, *glycerini* of glycerol, referring to utilization of glycerol as sole substrate.

Cell size is approximately 0.5 × 2–10 μm. Cells exhibit slightly pointed ends (Figure 205). Colonies on agar are small, cone-shaped, and yellow to orange with a slightly rough surface. Growth occurs between 10 and 42°C but optimally at 30°C to 37°C. Optimal pH is 6.5–7.5, with limits at pH 5.0

and 8.5. The type strain grows equally well in the absence and presence of 2% NaCl. Sulfate, sulfite, thiosulfate, sulfur, and nitrate are not reduced. Molecular nitrogen is not fixed. No vitamins required for growth. Glycerol and diolein used as sole source of carbon and energy. Propionate is the only organic fermentation product. Hydrogen is formed from glycerol. Indole is not formed; gelatin and urea are not hydrolyzed.

DNA G+C content (mol%): 35.0 (HPLC).
Type strain: LGS4, ATCC 51177, CIP 105408, DSM 5192.
GenBank accession number (16S rRNA gene): AJ010960.

Genus IX: **Anaerovibrio** Hungate 1966, 80[AL]

CARSTEN STRÖMPL AND GRAEME N. JARVIS

An.aer.o.vib′rio. Gr. pref. *an* not; Gr. n. *aer* air; *anaero* not (living) in air; L. v. *vibrio* to move rapidly to and fro, vibrate; N.L. masc. n. *vibrio* that which vibrates; N.L. masc. n. *Anaerovibrio* vibrio not living in air.

Slightly curved rods usually 0.5 μm × 1.2–3.6 μm. Capsules not formed. **Nonspore-forming.** No resting stage known. **Gram-stain-negative. Motile with a single polar flagellum.** Obligately **anaerobic. Chemo-organotrophic**; no scavenging system for oxygen is present. Growth occurs at 38°C, but not at 20°C, 30°C, or 50°C. Optimum pH for growth, 6.3; no growth occurs below pH 5.9 or above 7.0. **Lipolytic.** A limited range of sugars can be utilized as a carbon source. Glycerol is fermented mainly to propionate. Fermentation products from growth on DL-lactate, ribose, and fructose are acetate, propionate, CO_2, and traces of H_2 and succinate. Peptides, triacylglycerol, and phospholipids also support growth. An organic nitrogen source and vitamins (folic acid, pantothenate and pyridoxal-HCl) are required.

DNA G+C content (mol%): 44.0 ± 0.3 (HPLC).

Type species: **Anaerovibrio lipolyticus** corrig. Hungate 1966, 80[AL].

Further descriptive information

Large inserts occurring at the 5′ end in the helix 6 region (Neefs et al., 1991) of the 16S rRNA gene lead to multiple 16S rRNA genes of different length and/or sequence. 16S rRNA data indicate that the genus falls within the radius of the *Veillonellaceae*, but specifically within a lineage including species of the genera *Selenomonas, Schwartzia, Centipeda, Mitsuokella, Pecti-*

natus, and *Zymophilus* (Strömpl et al., 1999). Closest 16S rRNA gene sequence similarity (92–93%) is with the species *Selenomonas ruminantium* subsp. *ruminantium, Mitsuokella multacida,* and *Schwarzia succinovorans* (Strömpl et al., 1999). Phylogenetic and chemotaxonomic analyses have indicated that this genus is restricted to only one species, *Anaerovibrio lipolyticus* (corrig.) (Strömpl et al., 1999). There are no other described species in this genus.

The morphology of the slightly curved, Gram-stain-negative cells alters with age. Young cells are generally 0.5 μm × 1.2–1.8 μm in liquid media, but with age become more difficult to stain, develop a central granule, and are generally longer (3.0–3.6 μm). With further ageing, the cells seem to disintegrate leaving only a granular mass (Hobson and Mann, 1961). Although noncapsulated, a large amount of slime formation is noted during active growth (Hobson and Mann, 1961).

The cell-wall peptidoglycan contains no lipoproteins, but its major polyamines are spermidine and cadaverine; the latter is a common peptidoglycan constituent for members of the *Sporomusa–Pectinatus–Selenomonas* phyletic group (Hamana et al., 2002; Hirao et al., 2000). These compounds are covalently linked to the peptidoglycan and are essential for both cell surface integrity and normal cell growth (Hirao et al., 2000).

The respiratory lipoquinones of *Anaerovibrio* are primarily represented by the compound "lipid F" with the predominant form being the "lipid F" with eight isoprenologues. Interestingly, "lipid F" is now thought to be a modified naphthoquinone with an isoprenoid side chain (Strömpl et al., 1999). No evidence exists for the presence of ubiquinones or menaquinones.

The polar lipids phosphatidyl serine and phosphotidyl ethanolamine predominate, with the principal fatty acids being $C_{15:0}$, $C_{15:1}$, $C_{17:0}$, and $C_{17:1}$ (Strömpl et al., 1999; van Golde et al., 1975; Verkley et al., 1975). A third constituent, an atypical glycolipid, has also been reported (Strömpl et al., 1999). The majority of fatty acids present are straight chained, even and odd numbered, and saturated and unsaturated. No iso-branched fatty acids occur. The main 3-OH fatty acids present are $C_{13:0\ 3OH}$ and $C_{15:0\ 3OH}$ (Strömpl et al., 1999).

Initial isolations of *Anaerovibrio* employed linseed oil as the carbon source with 30–40% (v/v) rumen fluid (Hobson and Mann, 1961). Although it was later established that good growth could be achieved in a glycerol-based medium supplemented with yeast extract and trypticase, data also indicted that growth (albeit poorer) was also possible if glycerol was omitted and yeast extract and trypticase alone were employed (Prins et al., 1975).

Substrates utilized for growth include glycerol, ribose, DL-lactate, trypticase, triglycerides, and phospholipids (Prins et al., 1975). Strains do not utilize the long chain fatty acids resulting from lipid hydrolysis for growth, and their growth is not adversely affected by their presence (Henderson, 1973a; Hobson and Mann, 1961). Hydrogen sulfide is produced, a weak urease activity occurs, and both esterase and lipase activity are present. Neither gelatin nor esculin is hydrolyzed, nitrate is not reduced, and the indole test is negative (Counotte, 1981; Hobson and Mann, 1961).

Fermentation products from glycerol are propionate, succinate, and traces of H_2 and L-lactate, whereas fermentation of ribose, fructose, and DL-lactate produce acetate, propionate, CO_2, traces of succinate and H_2, and traces of L-lactate (when sugars are fermented) (Hobson and Summers, 1967; Prins et al., 1975). The formation of propionate from DL-lactate occurs via succinate using the dicarboxylic (succinate-propionate) pathway (Gottschalk, 1986; Prins et al., 1975). Glycerol is fermented via glycerol kinase in a manner similar to that of the propionic acid bacteria (de Vries et al., 1974). *Anaerovibrio* is also capable of reducing fumarate to succinate via the presence of a membrane-bound fumarate reductase (Henderson, 1980). This enzyme is considered to be markedly different in properties from that of other bacterial fumarate reducers in the rumen (Asanuma and Hino, 2000).

Lipolytic activity is due to a constitutive enzyme which is associated with extracellular membranous structures known as "blebs" (Henderson and Hodgkiss, 1973). Much information has been published on the properties and purification of the lipase (Henderson, 1970, 1971, 1968; Hobson and Summers, 1966, 1967; Prins et al., 1975).

Ecology

Anaerovibrio has only been found in the rumens of sheep and cattle. Its ability to hydrolyze lipids and utilize lactate probably indicates its key role in this ecosystem. It has been isolated from sheep fed a linseed cake meal (Hobson and Mann, 1961) and from cattle during transition from forage to concentrate feeding (Slyter et al., 1976) at levels of at least 10^8/ml of rumen contents. Lower levels ($0.5–1.6 \times 10^7$/ml) have been found in sheep and cows fed hay at maintenance levels (Prins et al., 1975). The relative sensitivity of *Anaerovibrio* to low pH values is probably indicative of only a modest role in ruminal lactate fermentation, and there is some contention over its importance in ruminal lipolysis (Harfoot and Hazlewood, 1997; Prins et al., 1975).

Enrichment and isolation procedures

Rumen contents are serially diluted in a specialized linseed oil-agar medium such as Medium 13 of Hobson and Mann (1961)* in roll tubes using the Hungate technique (1966). Clearing zones will develop around colonies positive for lipolysis. Final purification occurs in an agar medium in which the linseed oil is replaced by glycerol (0.2%) as the substrate. Henderson (1973b) suggested the use of trilaurin agar medium (TAM). This non-rumen fluid based agar contained trilaurin as the lipid source as well as yeast extract and casein hydrolysate. The benefits were noted as being a more rapid colony formation and more defined zones of clearing around bacterial colonies.

Maintenance procedures

Stock cultures of *Anaerovibrio* can be maintained in the medium of Henderson (1971) or in PY-broth (Holdeman et al., 1977) supplemented with 0.5% (v/v) glycerol with biweekly transfers. Long-term storage of cultures of this genus is best achieved by lyophilization using common procedures employed for obligate anaerobes (Hippe, 1991).

Differentiation of the genus *Anaerovibrio* from other closely related genera

Anaerovibrio can be differentiated from other genera within the *Veillonellaceae* based on chemotaxonomic, 16S rRNA gene sequence analysis, cell morphology, narrow range of growth substrates, and lipolytic ability. Species of the genus *Selenomonas* are generally nonlipolytic, have different substrate ranges, and/or have low 16S rRNA gene sequence similarity values compared to *Anaerovibrio*. The single polar flagellum and narrow substrate range differentiate *Anaerovibrio* from members of the genera *Zymophilus* and *Pectinatus*. A recently described rumen organism *Schwartzia succinovorans* (van Gylswyk et al., 1997), while having similar mol% G+C content, has a different substrate range and 16S rRNA gene sequence from that of *Anaerovibrio*.

Taxonomic comments

The taxonomic relationship of *Anaerovibrio* to other genera of anaerobic Gram-stain-negative bacteria has recently been

---

*Medium 13 of Hobson and Mann (1961) contains (per liter): mineral solution A (KH_2PO_4, 3.0g; $(NH_4)_2SO_4$, 6.0g; NaCl, 6.0g; $MgSO_4$, 0.6g; $CaCl_2$, 0.6g; distilled water, 1000 ml), 150 ml; mineral solution B (K_2HPO_4, 3.0g; distilled water, 1000 ml), 150 ml; rumen fluid (clarified through cheesecloth and centrifuged at $62,000 \times g$ for 10 minutes), 400 ml; distilled water, 290 ml; resazurin (0.1% (w/v) aqueous solution), 1 ml; 1.0 ml $NaHCO_3$, 4.0g; 1.0 ml cysteine-HCl, 0.5g; linseed oil (50% (v/v) emulsion in sterile rumen fluid), 20.0 ml; and 20 g agar. The minerals, rumen fluid, water, and resazurin are autoclaved together at 120°C for 15 min. The $NaHCO_3$ and cysteine-HCl are added aseptically from filter-sterilized solutions, and the linseed oil from a sterile emulsion. The medium is shaken immediately after the linseed oil addition to facilitate oil-agar emulsion forming on setting, and it is kept at 50°C until just prior to inoculation.

resolved (Strömpl et al., 1999). There is only one member of this genus, *Anaerovibrio lipolyticus* (Hungate, 1966; Strömpl et al., 1999). According to rule 65 of the Bacteriological Code (1990 Revision, Lapage et al., 1992), the name *Anaerovibrio lipolytica* Hungate (1966) was corrected to *Anaerovibrio lipolyticus* (corrig.) (Strömpl et al., 1999).

List of species of the genus *Anaerovibrio*

1. **Anaerovibrio lipolyticus** corrig. Hungate 1966, 80[AL]

li.po.ly'ti.cus. Gr. n. *lipos* fat; Gr. adj. *lutikos* dissolving; N.L. adj. *lipolyticus* fat-dissolving.

The description is as given for the genus. Minute brownish colonies develop in 5–7 d in linseed oil agar and produce zones of clearing in agar where the triacylglycerols have been hydrolyzed. In glycerol agar, the colonies are lens-shaped, mucoid, whitish discs. Glycerol, fructose, ribose, and DL-lactate can be utilized as carbon sources. Resistant to ionophoric antibiotics such as monensin or lasalocid, and fungal mycotoxins of the trichothecene group (Dennis et al., 1981; Westlake et al., 1987). Found in the rumen of sheep and cattle.

DNA G+C content (mol%): 44.0 ± 0.3 (HPLC).

Type strain: ATCC 33276, CIP 105407, DSM 3074, 5S, VPI 7553.

GenBank accession number (16S rRNA gene): AB034191, AJ010959.

Genus X. **Centipeda** Lai, Males, Dougherty, Berthold and Listgarten 1983, 631[VP]

ALANNA M. SMALL AND FRED A. RAINEY

Cen.ti' pe.da. L. fem. n. *centipede* a centipede.

Gram-stain-negative, nonspore-forming, **serpentine**, and **rod-shaped** with three or more curves. The cells are **motile** by means of flagella, which are inserted in a spiral path along the cell body. The movement of the cells occurs by flexion of the entire cell and rotation around its long axis. **Anaerobic**, chemoorganotrophic. Saccharoclastic. Propionic acid is a major end product of growth; acetic acid, succinic acid, and lactic acid are also produced from carbohydrate fermentation.

DNA G+C content (mol%): 51.4–53.6 (T_m).

Type species: **Centipeda periodontii** Lai, Males, Dougherty, Berthold and Listgarten 1983, 631[VP].

Further descriptive information

Cells of *Centipeda periodontii* are slightly serpentine, long, and rod-shaped with three or more curves with bluntly tapered ends. The degree of cell curvature is variable among strains and is lessened after repeated passage in culture. The cell diameter is 0.65 μm, and the cell length ranges from 4 to 17 μm or longer in older cultures. Gram-stain-negative. The cells occur singly or occasionally in chains. The occasional budding of the cells gives rise to branched forms. Multiple sets of flagellar bundles are present on each side of the cell with the number of bundles increasing with cell length. The bundles begin alternately from the concave areas along the curved cell body. Higher magnifications of negatively stained cells show a spiral path of insertion of the individual flagella from which the bundles are formed. This flagellar pattern is distinct from the one-sided patterns of *Selenomonas* and *Pectinatus* species.

Cells are actively motile in broth cultures 24h after incubation. The cell's movement is described as flexion of the entire cell in a snakelike motion with rotation around the long axis of the cell body and occasional tumbling is observed. Before the cells are able to change direction they exhibit a twitching motion in which the cells are nearly stationary, then the flagellar bundles on the cells fully extend and rotate erratically. After 5-d growth in liquid media few cells are motile, and cytoplasmic granules are visible when cells are examined by dark-field microscopy.

Cells stain Gram negative. Typical brain-like surface contour, similar to descriptions of other Gram-stain-negative bacteria, is found in negatively stained preparations of cell surfaces. A typical Gram-negative cell-wall structure is seen on cellular cross-sections. The outer and inner membranes consist of two-electron-dense layers separated by an electron-lucent layer. The outer membrane has an irregular, undulating profile, whereas the cytoplasmic membrane is relatively smooth and in direct contact with the moderately electron-dense peptidoglycan layer. Nucleoid areas are dispersed throughout the cytoplasm, which contains rod-shaped, electron-dense inclusions. Specialized regions of the peripheral cytoplasm approximately 30 × 130–220 nm occur along the inner membrane.

During initial isolation on blood agar, strains form flat, transparent colonies approximately 2 mm in diameter with irregular borders. After subculturing, the colonies are gray, transparent, and flat or thinly raised with finely granular surfaces. The colonies grow quickly and, after only 3 d of incubation, the entire surface of a plate is covered. In liquid culture, growth in tubes is turbid forming dense, white, sediment, which may be 2–3 cm deep after 24h.

Cultures grown on complex media (e.g., blood agar and Todd–Hewitt broth) are strict anaerobes and produce no growth aerobically or in atmospheres containing increased concentrations of CO_2. The temperature range for growth is 32°C and 37°C; the optimum growth temperature is 35°C. Growth is not enhanced by bile, but fermentable carbohydrates enhance growth in liquid cultures. Acid without gas is produced from the fermentation of carbohydrates. All strains ferment fructose, galactose, glucose, lactose, maltose, mannitol, melibiose, farinose, sorbitol, and sucrose. Fermentation of adonitol, cellobiose, mannose, melezitose, rhamnose, salicin, trehalose, and xylose is variable among strains. The final pH after growth in peptone-yeast extract broth containing 1% glucose is 4.2–5.0.

Centipeda periodontii strains produce 1.2 and 2.2 meq of propionic acid per 100 ml of peptone-yeast extract broth and peptone-yeast extract-glucose broth, respectively; smaller amounts

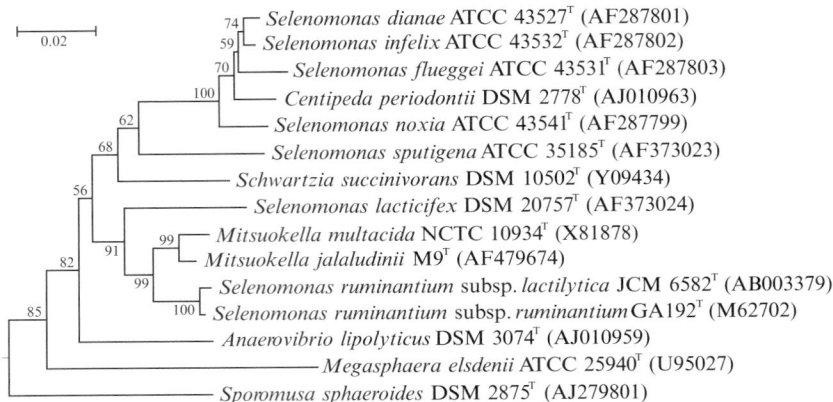

FIGURE 206. 16S rRNA gene sequence based phylogeny showing the position of the genus *Centipeda* within the radiation of the species of the genus *Selenomonas* and related genera. The tree was reconstructed from distance matrices using the neighbor-joining method. The species *Selenomonas lipolytica* was not included owing to availability of only a short (~250-nucleotide) 16S rRNA gene sequence. The scale bar represents 2 inferred nucleotide changes per 100 nucleotides.

of succinic acid and lactic acid are also produced. Lactate may be converted to succinate. Propionate is not produced from the threonine. The pathogenicity of *Centipeda periodontii* has not been determined.

The lipopolysaccharide of *Centipeda periodontii* is composed of hexose (21.7%), hexosamine (9.1%), fatty acid (10.3%), protein (0.5%), phosphorous (2.5%), 2-keto-3-deoxyoctonate (4.3%), and heptose (2.3%) (Kokeguchi et al., 1990).

Enrichment and isolation procedures

Strains of *Centipeda* species can be isolated from subgingival debris dispersed and diluted in prereduced dispersion media, plated on trypticase-soy agar containing 5% sheep blood (BBL) and incubated at 35°C under anaerobic conditions for 5–7 d. The atmosphere of the anaerobic chamber in which the plates are incubated is N_2:CO_2:H_2 (80:10:10). Strains can subsequently be cultivated on blood agar plates.

Maintenance procedures

Active cultures can be maintained on blood agar and subcultured weekly. Strains can be grown on blood agar and pieces of the agar can be placed in Socransky complex broth supplemented with 5% DMSO and stored in liquid nitrogen. Strains can be lyophilized in brain heart infusion supplemented medium containing 12% sucrose. Strains can be grown in Todd–Hewitt broth (BBL) for inocula or large volume cultures.

Procedures and methods for characterization tests

Members of this genus can be characterized according to the general methods outlined in the Virginia Polytechnic Institute Anaerobic Laboratory Manual (Holdeman and Moore, 1972).

Taxonomic comments

The characteristics of the genus *Centipeda* and the single species *Centipeda periodontii* are based on the study of nine strains (Lai et al., 1983). The cells have a centipede-like appearance due to the numerous sets of flagellar bundles found on both sides of the cells, and this is considered the main differentiating characteristic of this genus from related genera, namely, *Selenomonas*. The first phylogenetic study of *Centipeda periodontii* placed it close to the genus *Selenomonas* (Sawada et al., 1999). Phylogenetic analysis based on the current 16S rRNA gene sequence database shows the genus *Centipeda* falls within the radiation of the genera *Selenomonas*, *Schwartzia*, *Anaerovibrio*, and *Mitsuokella* (Figure 206). It shares 96.1–97.8% 16S rRNA gene sequence similarity with a cluster of *Selenomonas* species, namely, *Selenomonas dianae*, *Selenomonas flueggei*, *Selenomonas infelix*, and *Selenomonas noxia* that is supported by a bootstrap value of 100%. Interestingly *Centipeda periodontii* shares only 91% 16S rRNA gene sequence similarity with *Selenomonas sputigena*—the type species of the genus *Selenomonas*—which does not cluster with the majority of the *Selenomonas* species. Considering the lack of relationship between *Selenomonas sputigena* and the *Selenomonas* species closely related to *Centipeda periodontii*, it could be considered that the *Selenomonas* species *Selenomonas dianae*, *Selenomonas flueggei*, *Selenomonas infelix*, and *Selenomonas noxia* are in fact species of the genus *Centipeda*. A 16S rRNA gene sequence based PCR assay has been developed for the detection of *Centipeda periodontii* and for differentiating it from *Selenomonas sputigena* (Sawada et al., 2000).

List of species of the genus *Centipeda*

1. **Centipeda periodontii** Lai, Males, Dougherty, Berthold and Listgarten 1983, 631[VP]

    per.i.o.don′ti.i. Gr. prop. *peri* around; Gr. masc. n. *odon* tooth; N.L. neut. n. *periodontium* periodontium; N.L. gen. n. *periodontii* of the periodontium.

    Cells are Gram-stain-negative, slightly serpentine, long, and rod-shaped with bluntly tapered ends. The cell diameter is 0.65 μm, and the cell length ranges from 4 to >17 μm. Multiple sets of flagellar bundles are present on each side of the cell. Actively motile with occasional tumbling. Older cultures lack motility, and cytoplasmic granules are visible. Nucleoid areas. Cytoplasm contains rod-shaped, electron-dense inclusions. Colonies on isolation on blood agar are flat, transparent, and approximately 2 mm in diameter with irregular borders. Subcultured colonies are gray, transparent, and flat or thinly raised with finely granular surfaces. Growth on complex media. Strict anaerobes. Growth occurs between 32°C and 37°C; the optimum temperature is 35°C. Fructose, galactose, glucose, lactose, maltose, mannitol, melibiose, farinose, sorbitol, and sucrose are fermented by all strains. Fermentation of adonitol, cellobiose, mannose, melezitose, rhamnose, salicin, trehalose, and xylose is variable among strains. Amygdalin, arabinose, dulcitol, esculin, glycogen, inositol, inulin, sorbose, and starch are not fermented. Starch is not hydrolyzed; esculin hydrolysis varies among strains. Arginine is not utilized. The major end product of growth on peptone-yeast extract broth and peptone-yeast extract-glucose broth is propionic acid; succinic acid and lactic acid are also produced. Catalase, oxidase, urease, H_2S, and indole are not produced. Nitrate is reduced, and o-nitrophenyl-β-(D)-galactopyranoside is utilized. Propionate is not produced from threonine, and gelatin is not liquefied. No hemolysis occurs on sheep blood agar. Isolated from human periodontal lesions.

DNA G+C content (mol%): 51.4–53.6 (T_m).

Type strain: LL2383, ATCC 35019, CCUG 44586, CIP 105322, DSM 2778.

GenBank accession number (16S rRNA gene): AJ010963.

Genus XI. **Dendrosporobacter** Strömpl, Tindall, Lünsdorf, Wong, Moore and Hippe 2000, 105[VP]

CARSTEN STRÖMPL

Den.dro.spo.ro.bac'ter. Gr. n. *dendro* tree; Gr. n. *spora* seed, spore; N.L. masc. n. *bacter* equivalent of Gr. neut. dim. n. *bakterion* rod, staff; N.L. masc. n. *Dendrosporobacter* a spore-bearing rod from a tree.

Cells are straight rods with **Gram-stain-negative** cell walls. Cells occur singly or in pairs and often show unequal division. Motile by one to three lateral flagella. Cell size is $0.4–0.6 \times 1.2–2.7\,\mu m$. **Spores** are round and terminal or oval and central, and swell the cell. **Obligately anaerobic. Mesophilic** with optimal growth occurring at 25–30°C. Catalase and oxidase-negative. **Chemoorganotrophic.** Major products in Peptone/Yeast extract (PY) broth with 0.5% (w/v) fructose. are acetate, propionate, propanol, and hydrogen. The modified **naphthoquinone** "lipid F", with octa- and nonaprenologues (ratio approx. 1:1) as the predominant isoprenologues, is present, but **no cytochromes.** Predominant polar lipids are phosphatidyl ethanolamine, phosphatidyl serine, and an unidentified glycolipid, together with a number of minor phospholipids and unidentified components. 3- Hydroxy fatty acids

($C_{11:0\ 3OH}$, $C_{12:0\ 3OH}$, and $C_{13:03OH}$) are present, approximately half of which are not esterified. The predominant non- hydroxylated fatty acids are $C_{11:0}$, $C_{15:1}$, $C_{15:0}$ and $C_{17:1}$.

DNA G+C content (mol%): 48.5 (HPLC) or 52–54 (T_m).

Type species: **Dendrosporobacter quercicolus** (Stankewich, Cosenza and Shigo 1971) Strömpl, Tindall, Lünsdorf, Wong, Moore and Hippe 2000, 105[VP] (*Clostridium quercicolum* Stankewich, Cosenza and Shigo 1971, 302).

Further descriptive information

Lipoquinones are present at relatively low concentrations. Multiple heterogeneous copies of the 16S rRNA gene may be present. Cadaverine is the main polyamid in cell-wall peptidoglycan; minor polyamids are spermidine and putrescine (Hamana et al., 2002).

Enrichment and isolation procedures

Dendrosporobacter has been isolated from the ooze of discolored oak tree tissue. This organism has been isolated only once and has not been detected from other habitats in culture-independent studies suggesting a high specialization of this fastidious anaerobe.

Maintenance procedures

Dendrosporobacter may be maintained in PY-fructose broth (Holdeman et al., 1977). Long-term storage by lyophilization (Hippe, 1991) is successfully applied at the DSMZ.

Taxonomic comments

Stankewich et al. (1971) assigned their isolate to the genus *Clostridium*, although it possesses some characteristics which are not typical for the genus (Cato et al., 1986). Besides its restricted physiological activities, it exhibits a Gram-stain-negative cell wall and an unusually high G+C content of the DNA. The 16S rRNA sequence (M59110) published by Woese grouped this species remote from the type species, *Clostridium butyricum*. Consequently, Collins et al. (1986) created their "clostridial cluster IX" exclusively for *Clostridium quercicolum* during their phylogenetic analysis of *Clostridium sensu lato*. It became apparent that *Clostridium quercicolum* is related to the genera in the *Veillonellaceae* and not to the typical *Clostridium*. The additional chemotaxonomic data obtained by Strömpl et al. (2000) led them to recognize a new genus and combination for *Clostridium quercicolum*, as *Dendrosporobacter quercicolus*. The next closest relatives by 16S rRNA gene sequence similarity are the genera *Anaerosinus* and *Propionispora*. *Propionispira arboris*, isolated from wetwoods of living trees, shows very different metabolic properties, a significantly lower G+C content of the DNA, and possesses cytochromes.

List of species of the genus *Dendrosporobacter*

1. **Dendrosporobacter quercicolus** (Stankewich, Cosenza and Shigo 1971) Strömpl, Tindall, Lünsdorf, Wong, Moore and Hippe 2000, 105[VP] (*Clostridium quercicolum* Stankewich, Cosenza and Shigo 1971, 302[AL])

   quer.ci'co.lus. L. n. *quercus* oak, L. masc. adj. *quercicolus* associated with oak.

The species description is identical with that of the genus with the following additions.

Surface colonies on blood agar plates are irregular, raised, gray, and dull, with a lobate or rhizoid margin. Gelatin, esculin, and starch are not hydrolyzed. Lipase and lecithinase are not produced. Indole is not produced. Nitrate is not reduced. Acid is formed from fructose, but not from glucose, amygdalin, cellobiose, glycogen, lactose, maltose, ribose, starch, sucrose, trehalose, and xylose. Weak fermentation of glycerol, inositol, and ribose may occur in trypticase-yeast extract broth. The type strain is susceptible to chloramphenicol, clindamycin, erythromycin, and tetracycline but resistant to penicillin G. Culture supernatants are nontoxic to mice.

DNA G+C content (mol%): 48.5 (HPLC) or 52–54 (T_m).

Type strain: ATCC 25974, DSM 1736.

GenBank accession number (16S rRNA gene): M59110, AJ010962.

Genus XII. **Dialister** (*ex* Bergey, Harrison, Breed, Hammer and Huntoon 1923) Moore and Moore 1994, 191[VP] emend. Downes, Munson and Wade 2003, 1939 emend. Jumas-Bilak, Jean-Pierre, Carlier, Teyssier, Bernard, Gay, Campos, Morio and Marchandin 2005, 2478

WILLIAM G. WADE

Di.a.lis'ter. The etymology, and therefore the gender, of the genus name is unknown. Downes et al. (2003) proposed that the gender of the genus name be assigned as masculine, as allowed under Rule 65(3) of the Bacteriological Code.

Cells are **obligately anaerobic** or **microaerophilic**, nonmotile, nonspore-forming, **nonfermentative**, small Gram-stain-negative **coccobacilli** (0.2–0.4 × 0.3–0.7 μm). Growth in broth media is only slightly turbid at best. Esculin and urea are not hydrolyzed; indole and catalase are not produced. There is no growth in 20% bile. **Metabolic end products are variable amounts of acetate, lactate, and propionate.** The major cellular constituents of the type species *Dialister pneumosintes, Dialister micraerophilus*, and *Dialister propionicifaciens* include the fatty acids $C_{18:1\ \omega9c}$, $C_{16:0}$, $C_{18:0}$, and $C_{16:1\ \omega7c}$, but have not been determined for *Dialister invisus*.

DNA G+C content (mol%): 35–46; type species, 35.

Type species: **Dialister pneumosintes** (Olitsky and Gates 1921) (*ex* Bergey, Harrison, Breed, Hammer and Huntoon 1923, Moore and Moore 1994, 191[VP] emend. Downes, Munson and Wade 2003, 1940 emend. Jumas-Bilak, Jean-Pierre, Carlier, Teyssier, Bernard, Gay, Campos, Morio and Marchandin 2005, 2477 (*Bacterium pneumosintes* Olitsky and Gates 1921, 727; *Bacteroides pneumonsintes* Holdeman and Moore 1970, 33).

Further descriptive information

Dialister species are found in the oral cavity of humans. They have been isolated from healthy subjects and from a variety of oral infections. They have also been isolated from clinical specimens including blood cultures and abscesses at various body sites including the brain, infected cysts, and wound infections.

All species are predominantly unreactive in conventional biochemical physiological tests, but some useful differential characteristics are given in Table 205.

Enrichment and isolation procedures

Dialister species grow slowly and form small colonies (<0.5 mm in diameter) on solid media even after extended incubation, but are able to grow on standard blood agar base media supplemented with 5% horse or sheep blood. Growth in peptone/yeast extract broth is poor, showing faint turbidity at best, although this may be partly due to the small size and consequent lack of light refraction of the organisms. Supplementation of broth media with carbohydrates does not improve growth.

Maintenance procedures

Strains can be maintained on blood agar incubated anaerobically at 37°C and subcultured weekly. Lyophilization of cultures in the early stationary phase of growth is recommended for long term storage of most strains. Strains can also be stored at –70°C in Brain heart Infusion broth supplemented with 10% glycerol.

Procedures for testing special characters

The general methods described for the characterization of anaerobes in the VPI Anaerobe Laboratory Manual (Holdeman et al., 1977) and the Wadsworth-KTL Anaerobic Laboratory

TABLE 205. Descriptive and differential characteristics of *Dialister* species and related taxa[a]

Characteristic	D. pneumosintes	D. invisus	D. micraerophilus	D. propionicifaciens
Growth in microaerophilic conditions	–	–	+	d
Colistin (10 μg)	R	S	R	R
Growth stimulation by sodium succinate	–	–	–	+
Metabolic end products[b]	a (trace)	a, p (trace)	Not detected	a, l, p
Rapid ID32A profile[c]	0000 0124 01	0000 0000 00	2000 0133 05	0000 0000 00
DNA G+C content (mol%)	35	45–46	36	nd

[a]Symbols: +, 90% or more of strains are positive; –, 90% or more of strains are negative; d, 11–89% of strains are positive; S, sensitive; R, resistant.

[b]a, Acetate; l, lactate; p, propionate.

[c]Enzyme profile generated by RapidID 32A anaerobe identification kit (bioMerieux).

Manual (Jousimies-Somer et al., 2002) are suitable for the study of members of this genus. The RapidID 32A anaerobe identification kit (bioMerieux) is of limited value for the identification of members of this genus because two species, *Dialister invisus* and *Dialister propionicifaciens*, give uniformly negative results in each of the constituent tests (Jumas-Bilak et al., 2005).

Differentiation of the genus *Dialister* from closely related taxa

The genera most closely related to *Dialister* are *Veillonella*, *Anaeroglobus*, and *Megasphaera*. *Veillonella* species resemble *Dialister* in being Gram-stain-negative anaerobic cocci, It is difficult to reliably distinguish members of the two genera on the basis of cellular or colonial morphology, although *Veillonella* species exhibit faster and relatively more luxuriant growth on solid media. *Anaeroglobus geminatus* produces acid from galactose and mannose while all species of *Dialister*, *Megasphaera*, and *Veillonella* do not.

The most reliable way to distinguish *Dialister* from related genera is 16S rRNA gene sequence analysis (Downes et al., 2003; Jumas-Bilak et al., 2005; Willems and Collins, 1995a). 16S rRNA comparisons also allow the differentiation of *Dialister* species, as does comparative analysis of *dnaK* sequences (Jumas-Bilak et al., 2005).

Taxonomic comments

Dialister belongs to a group of genera, sometimes referred to as the *Sporomusa* sub-branch, within the predominantly Gram-stain-positive phylum *Firmicutes*, that have true Gram-negative cell walls (Figure 207). Designated Cluster IX in the clostridial classification described by Collins et al. (1994), this group contains a number of genera in addition to those mentioned above including *Sporomusa*, *Selenomonas*, and *Acidaminococcus*.

Dialister species vary considerably in the mol% G+C content of their DNAs, from 35% for the type species *Dialister pneumonsintes* to 46% for *Dialister invisus*, an unusually wide range for members of the same genus. However, Jumas-Bilak et al. (2005) have shown that the species also vary markedly in their genome size with *Dialister invisus* having a genome of 1.9 Mb, almost 50% larger than that of *Dialister pneumosintes* at 1.34 Mb. It would appear then, that a large scale horizontal transfer event or deletion has occurred during the evolution of this genus, resulting in chromosomes of differing G+C content.

Acknowledgements

Julia Downes is thanked for her contribution to the data reported in this chapter and for numerous helpful discussions.

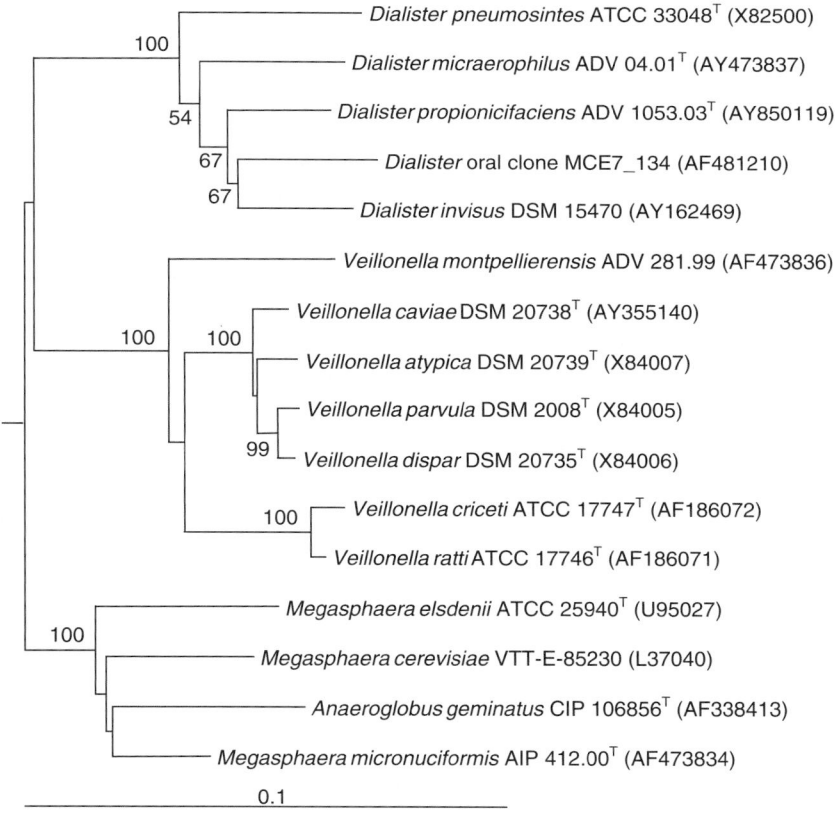

FIGURE 207. Phylogenetic tree based on 16S rRNA gene sequence comparisons over 1296 aligned bases showing relationship between *Dialister* species and related taxa. Tree was constructed using the neighbor-joining method following distance analysis of aligned sequences. Numbers represent bootstrap values for each branch based on data for 100 trees.

List of species of the genus *Dialister*

1. **Dialister pneumosintes** (Olitsky and Gates 1921) (*ex* Bergey, Harrison, Breed, Hammer, Huntoon 1923) Moore and Moore 1994, 191^VP emend. Downes, Munson and Wade 2003, 1940 emend. Jumas-Bilak, Jean-Pierre, Carlier, Teyssier, Bernard, Gay, Campos, Morio and Marchandin 2005, 2477 (*Bacterium pneumosintes* Olitsky and Gates 1921, 727; *Bacteroides pneumonsintes* Holdeman and Moore 1970, 33)

pneu.mo.sint'es. Gr. n. *pneuma* air; Gr. n. *sintes* a spoiler, thief; N.L. adj. *pneumosintes* breath destroying.

Cells of the type strain are 0.2–0.4 × 0.3–0.6 μm, arranged singly, in pairs, or in very short chains. Deep agar colonies are punctiform, granular, and white with no evidence of gas production. Surface colonies on blood agar are punctiform, circular, entire, convex, clear, transparent, shiny, and smooth. Growth in broth exhibits faint turbidity at best. Equivalent growth at 30 and 37°C; no growth at 25 or 45°C. Isolated from the nasopharynx and mouth in healthy individuals and from clinical specimens from periodontitis, endodontic infections, blood, respiratory tract, and brain abscesses. No DNase or phosphatase detected. Pathogenic for rabbits when injected intratracheally.

DNA G+C content (mol%): 35 (HPLC).

Type strain: ATCC 33048, CCUG 21025, CIP 107041, DSM 11619, JCM 10004.

GenBank accession number (16S rRNA gene): X82500.

2. **Dialister invisus** Downes, Munson and Wade 2003, 1939^VP

in.vis'us. L. adj. *invisus* unseen, referring to the lack of turbidity of broth cultures of this organism.

Description is based on six strains isolated from the human oral cavity. Cells are obligately anaerobic, nonmotile, Gram-stain-negative, small or ovoid cocci (0.3–0.4 × 0.3–0.6 μm) occurring singly, in pairs, short chains, and small clumps. After 7 d incubation on FAA plates (Fastidious Anaerobe Agar, LabM, Bury, UK), colonies are 0.5–0.7 mm in diameter, circular, entire, and either translucent and umbonate or transparent and low convex with a narrow marginal, translucent fringe. Growth in broth media produces only a slight turbidity and is not visibly stimulated by the addition of 1% carbohydrates. Cells are asaccharolytic and only trace amounts of acetate and propionate are detected as end products of metabolism in PYG (peptone yeast extract glucose broth). Esculin, arginine, and urea are not hydrolyzed. Indole and catalase are not produced. There is no growth in 20% bile. Isolated from the human oral cavity in patients with endodontic and periodontal infections.

DNA G+C content (mol%): 45 (HPLC).

Type strain: E7.25, CCUG 47026, DSM 15470.

GenBank accession number (16S rRNA gene): AY162469.

3. **Dialister micraerophilus** Jumas-Bilak, Jean-Pierre, Carlier, Teyssier, Bernard, Gay, Campos, Morio and Marchandin 2005, 2476^VP

micr.aer'o.phi.lus. Gr. adj. *mikros* small; Gr. n. aer, air; N.L. adj. *philus* from Gr. adj. *philos* loving; N.L. masc. adj. *micraerophilus* slightly air-loving, referring to the ability of the species to grow in microaerophilic conditions.

Cells are Gram-stain-negative, nonmotile, nonspore-forming, coccoid to coccobacillary (0.2–0.4 × 0.3–0.6 μm) occurring singly, in pairs or in clumps. After prolonged incubation, on Columbia blood agar, colonies are less than 0.5 mm in diameter, circular, convex, and translucent. Growth occurs under anaerobic and microaerophilic conditions. Members of this species are unreactive in most conventional tests, and metabolic end products are not produced in any significant quantities. Isolated from human clinical samples.

DNA G+C content (mol%) of the type strain: 36 (HPLC).

Type strain: ADV04.01, AIP 25.04, CIP 108278, CCUG 48837.

GenBank accession number (16S rRNA gene): AF473837.

4. **Dialister propionicifaciens** Jumas-Bilak, Jean-Pierre, Carlier, Teyssier, Bernard, Gay, Campos, Morio and Marchandin 2005, 2477^VP

pro.pi.o'ni.ci.fa.ci'ens. N.L. n. *acidum propionicum* propionic acid; L. v. *facio ere* to produce; N.L. part. adj. *propionicifaciens* propionic acid-producing.

Cells are Gram-stain-negative, nonmotile, nonspore-forming, coccoid to coccobacillary (0.2–0.4 × 0.3–0.6 μm) occurring singly, in pairs, or in clumps. After prolonged incubation, on Columbia blood agar, colonies are less than 0.5 mm in diameter, circular, convex, and translucent. Growth occurs under anaerobic and microaerophilic conditions. Members of this species are unreactive in most conventional tests. Small amounts of acetic, propionic, and lactic acids are produced as end products of metabolism. Propionate is produced when growth medium is supplemented with sodium succinate. Isolated from human clinical samples.

DNA G+C content (mol%): not determined because of the poor growth of the organism.

Type strain: ADV 1053.03, AIP 26.04, CIP 108336, CCUG 49291.

GenBank accession number (16S rRNA gene): AY850119.

Genus XIII. **Megasphaera** Rogosa 1971a, 187^AL emend. Engelmann and Weiss 1985; emend. Marchandin, Jumas-Bilak, Gay, Teyssier, Jean-Pierre, Siméon de Buochberg, Carrière and Carlier 2003b, 552

HÉLÈNE MARCHANDIN, RIIKKA JUVONEN AND AULI HAIKARA

Me.ga.sphae'ra. Gr. adj. *megas* big; Gr. n. *sphaera* a sphere; N.L. fem. n. *Megasphaera* big sphere.

Anaerobic. Gram-stain-negative cocci, 0.4–2.0 μm or more in diameter. Nonmotile. Endospores not formed. Glucose, fructose, and lactate may or may not be fermented. Gas may or may not be produced. Gelatin or milk is not hydrolysed and nitrate is not reduced. Pyruvate but not succinate is utilized. Common metabolic end products of all five species described are acetic, propionic, butyric, and valeric acid. Found in the rumen of cattle and sheep, in the feces and intestine of man, in human clinical samples, and in spoiled bottled beer.

DNA G+C content (mol%): 42.4–46.4 (T_m), 53.6 (Bd).

Type species: **Megasphaera elsdenii** (Gutierrez, Davis, Lindahl and Warwick 1959) Rogosa 1971a, 187[AL] (*Peptostreptococcus elsdenii* Gutierrez, Davis, Lindahl and Warwick 1959, 20; organism LC Elsden and Lewis 1953, 183; rumen organism LC Elsden, Volcani, Gilchrist and Lewis 1956, 686).

Further descriptive information

All strictly anaerobic, Gram-stain-negative cocci were originally classified in a single family, the *Veillonellaceae*, including the genera *Veillonella*, *Acidaminococcus*, and *Megasphaera* (Rogosa, 1971b, 1984a). An additional genus, i.e., the genus *Anaeroglobus*, was subsequently characterized within this family (Carlier et al., 2002). According to phylogenetic taxonomy, these four genera, despite the presence of a cell envelope typical for Gram-stain-negative bacteria, belong to the *Sporomusa* sub-branch of the *Clostridium* subphylum (Schleifer et al., 1990; Stackebrandt et al., 1985) of the *Firmicutes* (Carlier et al., 2002). Further phylogenetic analyses have supported the relationships among the species of genera belonging to the *Sporomusa* sub-branch (Both et al., 1992; Willems and Collins, 1995a; Marchandin et al., 2003b). More recently, in *Bergey's Manual of Systematic Bacteriology*, all members of the *Sporomusa* sub-branch have been grouped in the family *Veillonellaceae* (Figure 6; Ludwig et al., 2009). The genus *Megasphaera* created by Rogosa (1971a) presently includes five species: *Megasphaera elsdenii* (the type species), *Megasphaera cerevisiae*, *Megasphaera micronuciformis*, *Megasphaera paucivorans*, and *Megasphaera sueciensis*. Based on the 16S rRNA gene sequence analysis, the similarity between *Megasphaera elsdenii* and *Megasphaera cerevisiae* was 92%. *Megasphaera micronuciformis* displayed 94.5% and 93.8% 16S rRNA gene sequence identity with *Megasphaera cerevisiae* and *Megasphaera elsdenii*, respectively (Marchandin et al., 2003b). *Megasphaera sueciensis* and *Megasphaera paucivorans* shared almost identical sequences and showed 93.9%, 93.2% and 89.8% sequence identity with *Megasphaera cerevisiae*, *Megasphaera micronuciformis* and *Megasphaera elsdenii*, respectively (Juvonen and Suihko, 2006). The nearest other species (having similarities with the type strain *Megasphaera elsdenii* ATCC 25940[T]) are *Anaeroglobus geminatus* (92.4%), *Allisonella histaminiformans* (88.9%), and two *Dialister* species,

Dialister pneumosintes and *Dialister invisus* (88–89%). Doyle et al. (1995) have first demonstrated the phylogenetic position of *Megasphaera cerevisiae* as a sister taxon of *Megasphaera elsdenii* by determining the nucleotide sequence of the small subunit 16S rRNA of the bacterium. The phylogenetic tree (obtained by applying the neighbor-joining method and parsimony analyses) also showed the relationship between *Megasphaera cerevisiae* and organisms of the *Sporomusa* group (Doyle et al., 1995). Marchandin et al. (2003b) conducted a general phylogenetic analysis of the *Sporomusa* sub-branch based on 16S rRNA gene sequences and showed that members of the genus *Megasphaera* grouped together with *Anaeroglobus geminatus* and then with members of the genus *Dialister*. However, an absolute branching order in the *Megasphaera–Anaeroglobus* group still remains uncertain owing to the low bootstrap values obtained for the corresponding nodes (Juvonen and Suihko, 2006; Marchandin et al., 2003b).

The species of the genus *Megasphaera* can be differentiated from each other based on cell size with the exception of *Megasphaera sueciensis* and *Megasphaera paucivorans*. *Megasphaera elsdenii* cells are the largest, up to 2.4–2.6 μm in diameter, occurring in pairs or occasionally in chains of 8–20 cells (Figure 208) (Elsden et al., 1956; Rogosa, 1971b). Extremely long chains are formed in the rumen (Gutierrez et al., 1959). In stained preparations, the adjacent sides of pairs of cells tend to be flattened and *Megasphaera elsdenii* cells were 1.2–1.8 × 1.7 μm (Elsden et al., 1956). *Megasphaera cerevisiae* cells are spherical or slightly oval, 1.3–2.1 μm in diameter, occur singly, in pairs, or occasionally in short chains (Figure 209 and Figure 210) (Engelmann and Weiss, 1985; Juvonen and Suihko, 2006). In fixed or stained preparations, the cell diameter is 1.0–1.2 μm (Haikara and Lounatmaa, 1987). The ultrastructure of the cell surface of the *Megasphaera cerevisiae* brewery isolates and *Megasphaera elsdenii* is very uniform (Haikara and Lounatmaa, 1987). The cells of *Megasphaera sueciensis* and *Megasphaera paucivorans* are on mean smaller than those of *Megasphaera cerevisiae* and occur mostly in pairs. Stationary phase cells of *Megasphaera paucivorans* may form clumps and chains of 20–25 diplococci (Juvonen and Suihko, 2006). *Megasphaera micronuciformis* is the smallest representative of the genus *Megasphaera* and consists of coccoid cells showing a convoluted surface after negative staining. *Megasphaera micronuciformis* cells are usually single and their diameter varies from 0.4 to 0.6 μm (Figure 211) (Marchandin

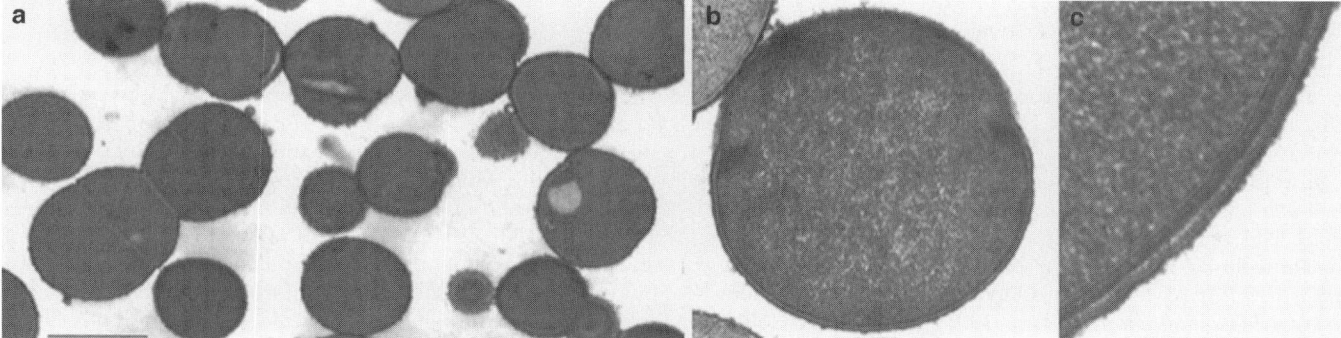

FIGURE 208. Electron micrograph of *Megasphaera elsdenii* strain AIP 10100[T] (=NCIB 8927[T]) (Printed with permission from Marchandin, H.). a) general morphology after negative staining; b) general morphology after electron microscopy (EM) of ultrathin sections; c) cell wall and membrane after EM of ultrathin sections. Bars, (a) 1.6 μm, (b) 250 nm; (c) 100 nm.

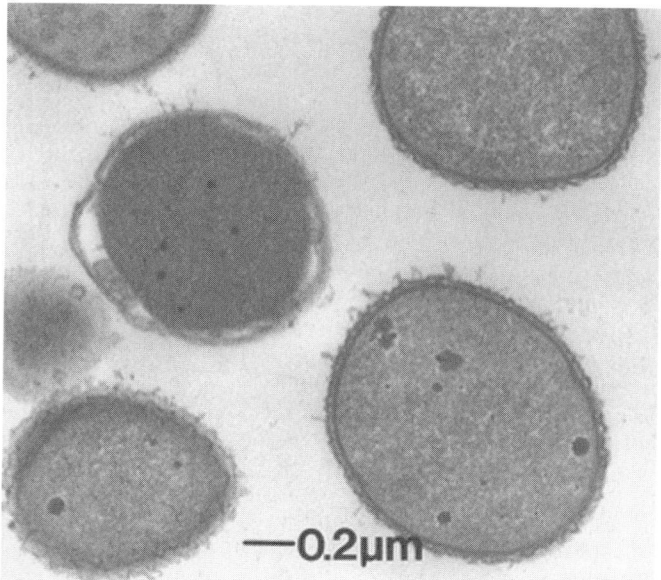

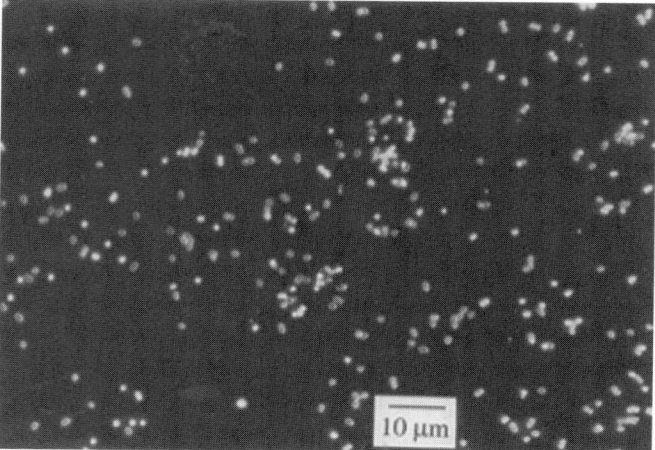

FIGURE 210. Darkfield micrograph of *Megasphaera cerevisiae*. (Reprinted with permission from Haikara, A. 1991. *In* Balows, Trüper, Dworkin, Harder, and Schleifer (Editors). The Prokaryotes, 2nd edn Vol. II, Springer, New York, pp. 1993–2004.)

FIGURE 209. Electron micrograph of *Megasphaera cerevisiae*. (Reprinted with permission from Haikara, A. 1991. *In* Balows, Trüper, Dworkin, Harder, and Schleifer (Editors). The Prokaryotes, 2nd edn, Vol. II, Springer, New York, pp. 1993–2004.)

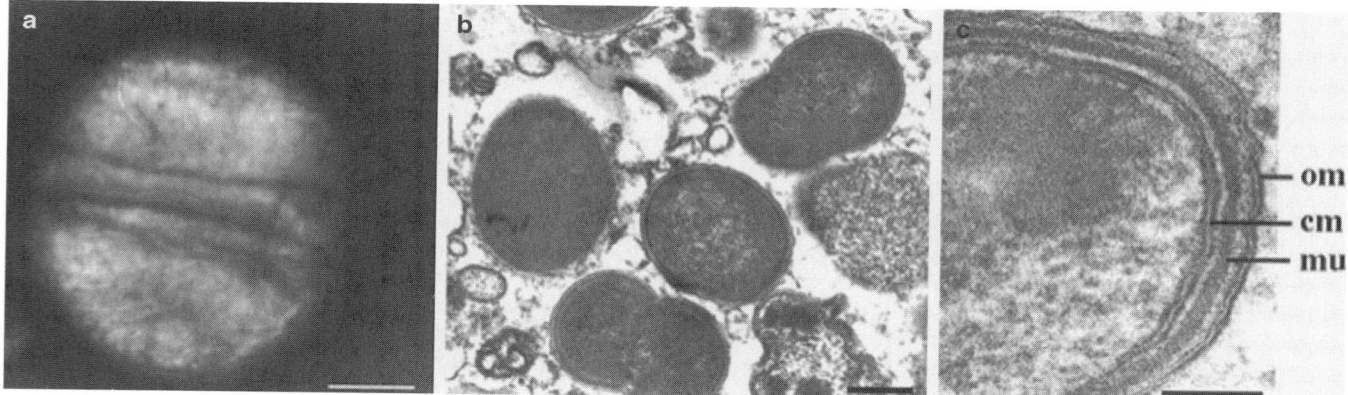

FIGURE 211. Ultrastructure of *Megasphaera micronuciformis* strain AIP 412.00[T] (CIP 107280[T], CCUG 45952[T]). (Reprinted with permission from Marchandin, H., E. Jumas-Bilak, B. Gay, C. Teyssier, H. Jean-Pierre, M. Siméon de Buochberg, C. Carrière, and J.-P. Carlier. Int. J. Syst. Evol. Microbiol., 2003b, 53: 547–553.) (a) general morphology after negative staining; (b) general morphology after electron microscopy (EM) of ultrathin sections; (c) cell wall and membrane after EM of ultrathin sections. OM, outer membrane; MU, peptidoglycan layer; CM, cytoplasmic membrane. Bars, (a) 178 nm; (b) 333 nm; (c) 66 nm.

et al., 2003b). Electron microscopy of ultrathin sections reveals that the *Megasphaera* species, despite their phylogenetic affiliation to the Gram-stain-positive bacteria, possess a triple-layered cell wall typical of Gram-stain-negative bacteria with an outer cell-wall membrane (Engelmann and Weiss, 1985; Marchandin et al., 2003b; Rogosa, 1971a).

The cellular fatty acids of *Megasphaera cerevisiae* and *Megasphaera elsdenii* have been investigated, and shown to be almost identical (Helander and Haikara, 1995; Johnston and Goldfine, 1982). The main fatty acid components are $C_{12:0}$, $C_{16:0}$, $C_{16:1}$, $C_{18:1}$, $C_{17\ cyclo}$, $C_{19\ cyclo}$, $C_{12:0\ 3OH}$, and $C_{14:0\ 3OH}$. Alk-1-enyl chains instead of acyl chains are detected to a considerable extent (14 % of total fatty acids); alk-1-enyl chains are indicative of plasmalo-

gen lipids (1-alkenyl-2-acyl glycerolipids), which are confined to strict anaerobes. *Megasphaera elsdenii* is particularly rich in the plasmalogen forms of phosphatidylethanolamine and phosphatidylserine, which together represent approximately 50% of the total phospholipid (Kaufman et al., 1990; Kaufman et al., 1988; van Golde et al., 1973). This plasmalogen content is not affected by growth temperature (Johnston and Goldfine, 1982) but can dramatically decrease when the interval between serial subcultures is prolonged to 3–6 weeks (Kaufman et al., 1988). The cellular fatty acids of the three other *Megasphaera* species are unknown.

The cell-wall peptidoglycan of *Megasphaera cerevisiae* as well as that of *Megasphaera elsdenii* is of the *meso*-diaminopimelic acid direct type and contains putrescine residues (Engelmann and Weiss, 1985). No information is available concerning the cell surface protein pattern and the lipopolysaccharide (LPS) composition of the genus *Megasphaera*.

Several growth media have been used for culturing *Megasphaera elsdenii* including the semi-defined medium described by Scott and Dehority (1965) (Cheng et al., 1988; Forsberg, 1978) and the medium described by Van Golde et al. (1973) (Johnston and Goldfine, 1982). However, the most commonly used media for cultivation and isolation of *Megasphaera elsdenii* strains are peptone-yeast extract (PY) media (Holdeman et al., 1977) supplemented with lactate (PYL) or glucose (PYG) (1% w/v, final concentration), lactate-based growth media (ATCC medium 566), and the Reinforced Clostridial Medium (ATCC medium 1053) (Kaufman et al., 1988). Moreover, Sugihara et al. showed that *Megasphaera elsdenii* can be recovered on *Bifidobacterium* selective agar or Eugonagar with maltose and on media containing neomycin (*Veillonella* agar with neomycin, egg yolk agar with neomycin, or blood agar with neomycin) (Sugihara et al., 1974). Owing to the inability of *Megasphaera cerevisiae* to utilize glucose, replacement of glucose with fructose or lactate in peptone-yeast extract medium (PYF*, PYL) is necessary for maximal growth (Engelmann and Weiss, 1985). *Megasphaera micronuciformis* can grow on Columbia sheep blood agar (Marchandin et al., 2003b) as well as in trypticase/glucose/yeast extract (TGY) medium (Carlier et al., 2002). Excellent growth of *Megasphaera paucivorans* and *Megasphaera sueciensis* (4+ on a scale from 0 to 4+) is obtained in PY supplemented with pyruvate or gluconate (1 %, v/v, final concentration). PYG and PYF support moderate to good (1+ or 2+) growth (Juvonen and Suihko, 2006). Because the *Megasphaera* species are strict anaerobes, anaerobic cultivation conditions are required (see *Pectinatus*). *Megasphaera elsdenii* colonies on yeast extract peptone agar supplemented with sodium lactate are round, smooth, slightly raised, and have a glistening, mucoid appearance. At 48 h, the colonies were 0.2–1.0 mm in diameter and older colonies were as large as 3–4 mm. Deep colonies are lenticular, 1–4 mm in diameter, and tan with soft butyrous texture (Gutierrez et al., 1959; Rogosa, 1971a). Colonies of *Megasphaera cerevisiae* on PYL or PYF are whitish, smooth, opaque, flat, shiny, and 2–5 mm in diameter (Engelmann and Weiss, 1985; Weiss et al., 1979). Colonies of *Megasphaera micronuciformis* appeared on Columbia sheep blood agar after 2–3 d incubation at 37°C and are circular, convex, shiny, and translucent with a smooth surface. The colonies are approximately 0.5–1.0 mm in diameter, nonpigmented and nonhemolytic (Marchandin et al., 2003b). Colonies of *Megasphaera paucivorans* and *Megasphaera sueciensis* after 7 d incubation have a diameter of 1–1.5 mm and 0.5–0.8 mm, respectively, and are circular, convex, glossy and opaque with entire margins and yellowish color. The optimum growth temperature for the brewery-related *Megasphaera* species is around 30°C (28–37°C) (Engelmann and Weiss, 1985; Haikara and Lounatmaa, 1987; Juvonen and Suihko, 2006), whereas for *Megasphaera elsdenii* it is 37–40°C (25–40°C) (Rogosa, 1971a). A few *Megasphaera elsdenii* strains can grow at 45°C, but no culture can be obtained at 50°C. Moreover, *Megasphaera elsdenii* does not grow at room temperature, whereas the brewery isolates can

grow at 15°C. A selective medium is designed for the detection of *Megasphaera cerevisiae* (see *Enrichment and isolation procedures*, below) that also supports the growth of the two other brewery-related *Megasphaera* species (Juvonen and Suihko, 2006).

Effect of pH on the growth rates of *Megasphaera elsdenii* showed that it grew over a pH range of 4.6–7.8 with an optimum pH for growth about 6.05 (Therion et al., 1982). *Megasphaera elsdenii* is therefore considered as relatively acid-tolerant. In comparison to acid-intolerant species, this acid tolerance could be attributed to a larger amount of H+-ATPase at neutral pH and to a higher capacity to enhance the amount of H+-ATPase at low pH (Miwa et al., 1997). *Megasphaera cerevisiae* is sensitive to the normal low pH of beer. Increase of pH from 4.1 to 4.7 accelerates growth, and no growth occurs in beer of pH 4.0–4.1 (Haikara, 1984; Haikara and Lounatmaa, 1987; Seidel et al., 1979). Acid tolerance of *Megasphaera paucivorans* and *Megasphaera sueciensis* is unknown but the bottled beers spoiled by these organisms had a pH of 4.3–4.9 (Juvonen and Suihko, 2006).

In addition to low pH, alcohol is the important factor inhibiting the growth of *Megasphaera cerevisiae* in beer. Beer with low alcohol content (<2.25% w/v) is more prone to spoilage by *Megasphaera cerevisiae* than beers with higher alcohol content (Haikara, 1984; Haikara and Lounatmaa, 1987). The growth of *Megasphaera cerevisiae* is restricted in commercial beer with an alcohol content of 3.5% (w/v) and is totally prevented by an alcohol content above 4.3% (w/v) (Back, 1981; Haikara, 1991; Seidel et al., 1979). *Megasphaera sueciensis* has hitherto only been found in low-alcohol beer (2.2 %, w/v) whereas *Megasphaera paucivorans* has spoilt beer with an alcohol content of 3.9 % (w/v) (Juvonen and Suihko, 2006). *Megasphaera cerevisiae* is rather tolerant to hop bitter substances of beer (Back, 1981; Kirchner et al., 1980; Seidel et al., 1979), and due to its sensitivity to alcohol and low pH, the respective contaminations are much more uncommon than those caused by *Pectinatus* species.

Utilization of different carbon sources by the five *Megasphaera* species is presented in Table 206. *Megasphaera elsdenii* possesses a fermentative type of metabolism and can use both carbohydrates and organic acids. Good growth and gas production are observed with lactate, glucose, and fructose. Growth and fermentation are variable with maltose, sorbitol, and mannitol. No growth and no fermentation occur with arabinose, cellobiose, dextrin, galactose, glycerol, inulin, lactose, mannose, raffinose, rhamnose, salicin, starch, sucrose, trehalose, or xylose. Due to the different metabolic pathways involved, the composition of the fermentation end products is variable depending on the energy source present in the medium. Products from lactate fermentation are acetate, propionate, C4 straight- and branched-chain fatty acids, valerate, little or no caproate and formate according to the strain tested, a large quantity of CO_2, and small amounts of H_2 (Marounek et al., 1989). C4 straight- and branched-chain volatile fatty acids produced by *Megasphaera elsdenii* are butyric, isobutyric, iso-valeric, and 2-methylbutyric acids (Allison, 1978). Glucose is fermented producing different end products. First observations reported by Elsden et al. (1956) and then by Rogosa (1971a, 1984a) indicated that propionate was produced from glucose and that caproate was the most copious product from glucose fermentation (about 60% or more of the total). Several further studies showed that butyrate was the main fermentation product

---

*PYF consists of (per liter): fructose, 5 g; peptone, 5 g; tryptone, 5 g; yeast extract, 10 g; Na$_2$HPO$_4$, 2 g; Tween 80, 1 ml; cysteine-HCl, 0.5 g; pH 7.0.

TABLE 206. Utilization of different carbon sources by *Megasphaera* species[a,b]

Carbon source	*M. elsdenii* ATCC 25940[T]	*M. cerevisiae* DSM 20462[T]	*M. cerevisiae* DSM 20461	*M. cerevisiae* VTT E-84195, VTT E-85230	*M. micronuciformis* AIP 412.00[T]	*M. paucivorans* VTT E-032341[T], VTT E-042576	*M. sueciensis* VTT E-97791[T]
Arabinose	–	+	–	+	–	–	–
Fructose	+	+	+	+	–	–	–
Glucose	+	–	–	–	–	–	–
Mannitol	+	–	–	–	–	–	–
Maltose	+	–	–	–	–	–	–
Sucrose	+	–	–	–	–	ND	ND
Lactate	+	+	+	+	–	–	–
Gluconate	+	+	ND	ND	–	+	+
None	(+)	(+)	(+)	(+)	ND	(+)	(+)

[a]+, pH ≤5.5 or enhanced growth in comparison to medium w/o the added carbon source; (+), pH 5.5–6.0 or slight growth; –, no additional growth or pH decrease caused by the carbon source; ND, not determined.

[b]Test results (except for the strains in parentheses): none of the strains used adonitol, esculin, D-cellobiose, dulcitol, erythritol, glycerol (AIP 412.00[T]), i-inositol, inulin, lactose, α-D-melibiose, melezitose (AIP 412.00[T]), N-acetylglucosamine, raffinose, rhamnose, D-ribose, D-salicin, sorbitol (VTT E-97791[T], VTT E-032341[T]), succinate, trehalose (VTT E-97791[T], VTT E-032341[T]), D-xylose or xylitol. All strains used pyruvate (data compiled from Haikara, 1991; Marchandin et al., 2003b; Juvonen and Suihko, 2006).

and that no production of propionate from glucose is observed (Forsberg, 1978; Marounek et al., 1989). This latter observation was further attributed to the lack of lactate racemase synthesis by strains fermenting glucose (Hino and Kuroda, 1993). Indeed, lactate racemase synthesis is induced by lactate and the metabolism from D-lactate to propionate is suppressed when *Megasphaera elsdenii* is fermenting glucose (Hino and Kuroda, 1993). Finally, products from glucose differed from those from lactate by formate and caproate production, a lesser production of valerate, little or no production of acetate (according to the strain tested), no production of propionate, and the major metabolic end product being butyric acid (Marounek et al., 1989). Succinate, fumarate, and malate are not attacked. Many studies on the different transport mechanisms and metabolic pathways in *Megasphaera elsdenii* showed that (i) lactate is fermented and propionate is produced via the acrylate pathway, and lactate transport across the membrane appears to occur by active transport that is not dependent on Na+ or K+ (Martin, 1994), (ii) glucose and fructose uptake involve an inducible phosphoenolpyruvate phosphotransferase (PEP-PTS) system (Dills et al., 1981), and (iii) maltose metabolism involves inducible maltose phosphorylase and maltase (Martin and Wani, 2000). Based on its metabolic characteristics, particularly its capacity to ferment 74–97% of ruminal lactate, *Megasphaera elsdenii* is considered as a probiotic micro-organism useful in reducing acidosis due to lactic acid accumulation in cattle introduced to a high grain diet (Counotte et al., 1981; Ouwerkerk et al., 2002). The carbohydrate spectrum of *Megasphaera cerevisiae* is very narrow compared with that of *Megasphaera elsdenii*. The isolates do not utilize glucose or maltose, but like *Megasphaera elsdenii*, they can grow in sugar-free medium, although growth is very poor (Engelmann and Weiss, 1985; Weiss et al., 1979). With respect to the fermentation of carbohydrates, *Megasphaera cerevisiae* strains form a highly uniform group. Like *Pectinatus*, they can utilize lactate. Metabolic end products are acetic, propionic, iso- and n-butyric, iso- and n-valeric, and caproic acids (Engelmann and Weiss, 1985; Haikara, 1985; Weiss et al., 1979). Depending on the carbon source used, the predominant end product is butyrate, valerate or caproate. Of the carbon sources tested, *Megasphaera paucivorans* and *Megasphaera sueciensis* only

utilize the organic acids gluconate and pyruvate. Gluconate is mainly fermented to acetate and n-butyrate whereas degradation of pyruvate additionally leads to the synthesis of propionate (Juvonen, unpublished data). The major metabolite produced by *Megasphaera paucivorans* and *Megasphaera sueciensis* in autoclaved PYF is iso-valerate but acetate, propionate, iso- and n-butyrate, iso- and n-valerate and caproate are also formed. Propionate production by *Megasphaera paucivorans* is, however, minor or negligible. In beer, the main metabolic product of both *Megasphaera cerevisiae* and *Megasphaera paucivorans* is n-butyric acid (Haikara and Lounatmaa, 1987; Juvonen and Suihko, 2006). The organic acids produced by *Megasphaera sueciensis* in beer have not been determined. All the *Megasphaera* species with the exception of *Megasphaera micronuciformis* produce H₂S (Engelmann and Weiss, 1985; Juvonen and Suihko, 2006; Weiss et al., 1979). Owing to the mixture of various fatty acids and H₂S, the flavor of contaminated beer is particularly unpleasant. *Megasphaera micronuciformis* does not produce gas or ferment any of the carbohydrates tested, and acid is not produced from arabinose, cellobiose, fructose, galactose, glucose, glycerol, inositol, lactose, maltose, mannitol, mannose, melezitose, melibiose, raffinose, rhamnose, ribose, salicin, sorbitol, starch, sucrose, trehalose, or xylose. Its major metabolic end products are acetic, propionic, butyric, iso-valeric, and 2-phenylacetic acids (Marchandin et al., 2003b).

The five *Megasphaera* species are catalase and indole-negative and do not hydrolyze urea, arginine, esculin, milk or gelatin. They do not reduce nitrate (Gutierrez et al., 1959; Juvonen and Suihko, 2006; Marchandin et al., 2003b; Weiss et al., 1979), but *Megasphaera elsdenii* can metabolize nitrite (Cheng et al., 1988). Voges–Proskauer reaction (production of acetoin) is negative or weakly positive (Juvonen and Suihko, 2006). *Megasphaera elsdenii* and *Megasphaera cerevisiae* are known to be also benzidine negative. Resistance pattern of the type strains of the five *Megasphaera* species to vancomycin (5 µg) and colistin (10 µg) differ. *Megasphaera paucivorans* and *Megasphaera sueciensis* are resistant to vancomycin and colistin, *Megasphaera micronuciformis* is sensitive to both of these antibiotics whereas *Megasphaera cerevisiae* and *Megasphaera elsdenii* are resistant to vancomycin but sensitive to colistin (Juvonen and Suihko, 2006).

The *Megasphaera* species can be differentiated from each other based on the G+C content of their DNA with the exception of *Megasphaera cerevisiae* and *Megasphaera sueciensis*. The G+C content of the DNA of *Megasphaera elsdenii* is 53.6 mol% (Bd) and that of *Megasphaera cerevisiae* is 42.4–44.8 mol% (T_m) (Engelmann and Weiss, 1985), indicating that there is no close genomic relationship between them. The G+C content of the DNA of *Megasphaera micronuciformis*, *Megasphaera paucivorans* and *Megasphaera sueciensis* is 46.4, 40.5 and 43.1 mol% (T_m), respectively (Juvonen and Suihko, 2006; Marchandin et al., 2003b). According to DNA–DNA hybridization results, 12 *Megasphaera cerevisiae* strains belong to a single genospecies, the similarity being 72–100%, whereas a low degree of similarity, i.e., 22–38% exists between *Megasphaera cerevisiae* strains and *Megasphaera elsdenii* (Engelmann and Weiss, 1985). The DNA–DNA hybridization has confirmed that *Megasphaera sueciensis* and *Megasphaera paucivorans* (41.0 %) are not related to each other or to (respectively) *Megasphaera micronuciformis*[T] (28.9 % and 17.1 %), *Megasphaera elsdenii*[T] (7.2 % and 23.6 %) or *Megasphaera cerevisiae*[T] (22.0 % and 3.1 %) at species level (Stackebrandt and Goebel, 1994; Wayne et al., 1987). The characteristics useful for differentiating the species belonging to the genus *Megasphaera* (except the different carbon sources utilized which are summarized in Table 206) are given in Table 207.

Based on the 16S rRNA gene sequences, real-time *Taq* nuclease PCR assays have been developed to enumerate *Megasphaera elsdenii* in the complex rumen microbial system (Ouwerkerk et al., 2002). The combined primer and probe set used are not specific for *Megasphaera elsdenii* since they also recognize *Megasphaera cerevisiae*. However, since *Megasphaera cerevisiae* was never isolated from the rumen, the assay was considered to be specific for *Megasphaera elsdenii* in the rumen fluid (Ouwerkerk et al., 2002). Specific primers based on the 16S rRNA gene sequence have been designed for early detection of *Megasphaera cerevisiae*

in beer by PCR technique (Doyle et al., 1995; Juvonen et al., 2003; Sakamoto et al., 1997; Satokari et al., 1998), and commercial ready-to-use kits for the analysis of this species in brewery samples are also available (Braune and Eidtmann, 2003; Homann et al., 2002).

Automated ribotyping with *Eco*RI, *Pst*I, and *Pvu*II has been used for the characterization of 13 *Megasphaera cerevisiae* strains (Suihko and Haikara, 2001). Using a combination of these enzymes, the strains could be divided into six ribotypes. Ribotyping with *Eco*RI also distinguished between *Megasphaera sueciensis* and *Megasphaera paucivorans* strains although their 16S rRNA gene sequences are nearly identical (Juvonen and Suihko, 2006; Suihko and Haikara, 2001). Serological characteristics of *Megasphaera cerevisiae* have been studied by producing peptidoglycan-specific and bacterial surface-reactive Mabs and by producing antibodies against the 40–50 kDa surface protein (Hakalehto, 2000; Ziola et al., 1999, 2000).

Megasphaera elsdenii was shown to be resistant to monensin (growth in the presence of 10 µM), an ionophore antibiotic used as antimicrobial feed additive in cattle diets (Callaway et al., 1999). This observation is in agreement with the negative Gram staining and the presence of an outer membrane in *Megasphaera elsdenii* and contrasts with the phylogenetic placement of *Megasphaera elsdenii* in the low G+C Gram-positive bacteria (*Firmicutes*). Indeed, Gram-stain-negative ruminal bacteria with an outer membrane were shown to be more resistant to monensin than were Gram-stain-positive species (Callaway et al., 1999). Further studies on antimicrobial susceptibility patterns of *Megasphaera elsdenii* revealed that a great strain-to-strain variability can be observed (Marounek et al., 1989; Piriz et al., 1992) and that some strains displayed resistance to several antimicrobial cattle feed additives (Nagaraja and Taylor, 1987). In particular, tetracycline-resistant *Megasphaera elsdenii* strains harboring *tet* genes were isolated from cecal contents and tissues

TABLE 207. Characteristics differentiating the species of the genus *Megasphaera*[a,b]

Characteristic	M. elsdenii	M. cerevisiae	M. micronuciformis	M. paucivorans	M. sueciensis
Cell size (µm)	1.6–2.6	1.3–2.1	0.4–0.6	1.2–1.9 × 1.0–1.4	1.0–1.4 × 0.8–1.2
Acid production from:					
L-Arabinose	–	±	–	–	–
D-Fructose	+	+	–	–	–
D-Glucose	+	–	–	–	–
D-Maltose	±	–	–	–	–
Mannitol	±	–	–	–	–
Susceptibility to:					
5 µg Vancomycin	R	R	S	R	R
10 µg Colistin	S	S	S	R	R
Lactate fermentation	+	+	–	–	–
Gluconate fermentation	+	+	–	+	+
Gas production	+	+	–	+	+
Volatile fatty acids produced[c]	A, (P), (iB), <u>B</u>, iV, V, <u>C</u>	A, P, (iB), B, <u>iV</u>, V, <u>C</u>	A, P, (iB), B, iV, (V), PhA	A, (P), iB, B, <u>iV</u>, V, (iC), <u>C</u>	(A), P, iB, <u>B</u>, <u>iV</u>, <u>V</u>, C
G+C content of DNA (mol%)	53.1–54.1 (Bd)	42.4–44.8 (T_m)	46.4 (T_m)	40.5 (T_m)	43.1 (T_m)

[a]±, Variable; R; resistant, S; susceptible. A, acetic acid; P, propionic acid; iB, iso-butyric acid; B, butyric acid; iV, iso-valeric acid; V, valeric acid; C, caproic acid; PhA, 2-phenylacetic acid. For other symbols, see footnote of Table 206. Adapted from Marchandin et al. (2003b) and Juvonen and Suihko (2006).

[b]Taxa (reference): *Megasphaera elsdenii* (Juvonen and Suihko, 2006; Rogosa, 1984a); *Megasphaera cerevisiae* (Engelmann and Weiss, 1985; Juvonen and Suihko, 2006); *Megasphaera micronuciformis* (Juvonen and Suihko, 2006; Marchandin et al., 2003b), *Megasphaera paucivorans* and *Megasphaera sueciensis* (Juvonen and Suihko, 2006).

[c]Parentheses indicate that production is not constant. Major products are underlined.

of pigs (Stanton and Humphrey, 2003). *Megasphaera elsdenii* was therefore considered to play a role not only in the preservation and dissemination of antibiotic resistance in the intestinal tract but also in the evolution of resistance. *Megasphaera elsdenii* has been detected in intestinal contents and feces of cattle, sheep (Elsden et al., 1956; Gutierrez et al., 1959), and pigs (Giesecke et al., 1970). *Megasphaera elsdenii* is considered one of the most important micro-organisms in the rumen (Marounek et al., 1989) where it can play a major role in preventing or controlling acidosis by removing lactic acid through catabolic action (Stewart and Bryant, 1988). Human isolates were from normal feces (Sugihara et al., 1974; Werner, 1973), from fecal samples of adults and children suffering from gastrointestinal disorders (Haralambie, 1983), from conjunctiva (Thiel and Schumacher, 1994), from gastric and amniotic fluid samples (Marchandin, unpublished data) and from a putrid lung abscess as part of mixed flora (Sugihara et al., 1974). More rarely, *Megasphaera elsdenii* is associated with disease and was implicated in a case of human endocarditis (Brancaccio and Legendre, 1979) and in foot rot lesions in goats (Duran et al., 1990).The natural habitats of the *Megasphaera* species remain unknown. Most of the *Megasphaera cerevisiae* strains and both of the *Megasphaera paucivorans* and the *Megasphaera sueciensis* strain described and characterized have been isolated from spoiled, unpasteurized beer. The sporadic occurrence of *Megasphaera cerevisiae* in pitching yeast and in brewery environment has also been reported (Haikara, 1989, 1991; Seidel, 1985). The *Megasphaera micronuciformis* currently reported are of human origin and were isolated from a liver abscess, pus samples (Marchandin et al., 2003b), sinus and throat (Marchandin, unpublished data). "*Candidatus* Megasphaera micronuciformis" oral clone BU057 (GenBank accession no. AF385566) was described from human purified crevicular cells (Paster et al., unpublished data; Marchandin et al., 2003b). Moreover, *Megasphaera* genomospecies C1, which displayed 99.4% of 16S rRNA gene sequence identity with the type strain of *Megasphaera micronuciformis* and could therefore be considered as "*Candidatus* Megasphaera micronuciformis" oral clone C3MLM013 (GenBank accession no. AY278622), was recovered from the human mouth in microflora associated with dental caries (Wade and Munson, unpublished data).

Enrichment and isolation procedures

Strictly anaerobic conditions are required for growth of the *Megasphaera* species. For example, *Megasphaera elsdenii* and *Megasphaera micronuciformis* grew well in anaerobic conditions obtained in an anaerobic jar with Anaerogen System (Oxoid). Many agar media and broths (see description above) containing fermentable sugars or lactate can be used for cultivation and isolation of *Megasphaera elsdenii*. No enrichment method and no specific selective medium exist for growth of *Megasphaera elsdenii*, however, Sugihara et al. obtained high recovery of *Megasphaera elsdenii* on several media containing neomycin (Sugihara et al., 1974). In any cultivation medium, fructose or lactate (instead of glucose) is required for *Megasphaera cerevisiae* whereas pyruvate or gluconate (1 %, v/v, final concentration) is required for good growth of *Megasphaera paucivorans* and *Megasphaera sueciensis* (Juvonen and Suihko, 2006). In the quality control of beer, an enrichment method is used (Anonymous, 2001). In this method concentrated culture medium is added to the headspace of the bottle immediately after filling

and capping. Development of turbidity is monitored for three to four weeks and the presence of *Megasphaera* is confirmed microscopically and by smell (H_2S, fatty acids).

A selective medium (SMMP) for enrichment of *Megasphaera* and *Pectinatus* in beer has been developed (the recipe is given in the chapter on *Pectinatus*). The change in medium color from purple to yellow indicates the presence of *Megasphaera* bacteria (Anonymous, 1998; Juvonen and Suihko, 2006).

Maintenance procedures

Megasphaera elsdenii cultures can be maintained in lactate-based media at 4°C in anaerobic conditions if they are routinely transferred each week or at least fortnightly (Furtado et al., 1994; Rogosa, 1971a). They can also be frozen at –70°C with either dimethylsulfoxide (10 % v/v, final concentration) (Stanton and Humphrey, 2003) or 30% glycerol. *Megasphaera elsdenii* may not tolerate a number of freeze-drying procedures but survives in liquid nitrogen. *Megasphaera micronuciformis* and *Megasphaera elsdenii* can also be frozen at –80°C on glass beads. See the chapter on *Pectinatus* for maintenance of *Megasphaera cerevisiae*.

Differentation of the genus Megasphaera from other closely related taxa

The genus *Megasphaera* can be differentiated from the three other genera of anaerobic Gram-stain-negative cocci of the family *Veillonellaceae*, i.e., the genera *Veillonella*, *Acidaminococcus*, and *Anaeroglobus*, on the basis of major metabolic end products, physiological tests, DNA G+C content and 16S rRNA gene sequence. The characteristics useful in differentiating the genus *Megasphaera* from these other genera are given in Table 208.

Taxonomic comments

The genus *Megasphaera* was created by Rogosa in (1971a) to reclassify *Peptostreptococcus elsdenii* (Gutierrez et al., 1959), formerly known as Organism LC (Elsden et al., 1956). This organism showed many features uncharacteristic of peptostreptococci, particularly its Gram-stain-negative stain and the outer membrane observed in electron microscopy. Rogosa included the *Megasphaera* genus and its type species *Megasphaera elsdenii*, together with the genera *Veillonella* and *Acidaminococcus*, in a novel family, the *Veillonellaceae* (Rogosa, 1971b). Phylogenetic analysis revealed that this family belongs to the *Sporomusa* subbranch of the Gram-stain-positive bacteria, corresponding to clostridial cluster IX of Collins et al. (1994), and the family *Veillonellaceae* was enlarged to include most of these and related genera (Ludwig et al., this volume). In 1979, Weiss et al. isolated anaerobic Gram-stain-negative cocci from beer (Weiss et al., 1979). On the basis of differential characteristics of the genera *Veillonella*, *Acidaminococcus*, and *Megasphaera*, the beer isolates could be assigned to the genus *Megasphaera* (Weiss et al., 1979). Based on G+C content of the DNA and low similarity in DNA–DNA hybridization with *Megasphaera elsdenii* and *Veillonella parvula*, the beer isolates were assigned to a new species, *Megasphaera cerevisiae* (Engelmann and Weiss, 1985). The third species of the genus *Megasphaera*, *Megasphaera micronuciformis*, has been characterized from human clinical specimens on the basis of cell size, biochemical tests (particularly its metabolic end products), G+C content of the DNA, and 16S rRNA gene sequence (Marchandin et al., 2003b). Two novel *Megasphaera*

TABLE 208. Characteristics useful in differentiating the genus *Megasphaera* from other genera of anaerobic Gram-stain-negative cocci belonging to the family *Veillonellaceae*[a,b]

Characteristic	*Megasphaera*	*Acidaminococcus*	*Anaeroglobus*	*Veillonella*
Cell size (μm)	0.4–0.6 × 1.3–2.6	0.6–1	0.5–1.1	0.3–0.5
Acid production from:				
Galactose	–	–	+	–
Mannose	–	–	+	–
Decarboxylation of succinate	–	–	–	+
Reduction of nitrate	–	–	–	+
Volatile fatty acids produced in sugar-containing or sugar-free medium	A, (P), (iB), B, iV, (V)	A, B	A, P, iB, B, iV	A, P
DNA G+C content (mol%)	53.1–54.1(Bd), 40.5–46.4 (T_m)	56.6	51.8	40.3–44.4

[a]For symbols, see standard footnote of Table 207.

[b]Adapted from Carlier et al., (2002). Data for *Megasphaera elsdenii, Acidaminococcus,* and *Veillonella* were taken from Rogosa, (1984a). Data for *Megasphaera cerevisiae* are given by Engelmann and Weiss, (1985). Data for *Megasphaera micronuciformis* were from Marchandin et al., (2003b). Data for *Megasphaera paucivorans* and *Megasphaera sueciensis* are from Juvonen and Suihko (2006).

species have been isolated from spoiled bottled beers. Based on cell size, physiological tests, volatile fatty acid profiles, DNA–DNA hybridization, 16S rRNA gene sequence and G+C content of the DNA, *Megasphaera paucivorans* and *Megasphaera sueciensis* were described. In contrast to *Megasphaera elsdenii* and *Megasphaera cerevisiae, Megasphaera micronuciformis, Megasphaera sueciensis* and *Megasphaera paucivorans* are negative in most of the conventional biochemical and physiological tests. The characteristics useful in differentiating between the five *Megasphaera* species are shown in Table 208. Besides automated ribotyping with *Eco*RI is a useful technique to identify and differentiate the five *Megasphaera* species (Juvonen and Suihko, 2006). Moreover, *Megasphaera micronuciformis* differs from *Megasphaera elsdenii* at the genomic level (Marchandin and Jumas-Bilak, 2006). A genomic organization study of *Megasphaera micronuciformis* reveals a genome of about 1800 kb showing four *rrn* operons whereas the genome of *Megasphaera elsdenii* is about 2600 kb and comprises seven *rrn* operons (Marchandin et al., 2003b). The ability to produce volatile

fatty acids containing 4–6 carbon atoms is a distinctive characteristic of the five species of *Megasphaera*. Biodiversity in the genus *Megasphaera* still seems to be underestimated. Culture-independent DNA-based methods used to study vaginal microbial flora in five adult healthy women by Zhou et al. (2004) revealed numerous identical uncultured *Megasphaera* species clones (GenBank accession nos from AY271931 to AY271953) that could represent a potential novel species. On another hand, Fredricks et al. (2005) demonstrated an association between potential novel *Megasphaera* species and bacterial vaginosis.

Further reading

Haikara, A. and I. Helander. 2002. *Pectinatus, Megasphaera,* and *Zymophilus. In* Dworkin (Editor), The Prokaryotes: An Evolving Electronic Database for the Microbiological Community, 3rd edn (release 3.5), Springer Verlag, New York. ISBN 0-387-14254-1.

List of species of the genus *Megasphaera*

1. **Megasphaera elsdenii** (Gutierrez, Davis, Lindahl and Warwick 1959) Rogosa 1971a, 187[AL] (*Peptostreptococcus elsdenii* Gutierrez, Davis, Lindahl and Warwick 1959, 20; organism LC Elsden and Lewis 1953, 183; rumen organism LC Elsden, Volcani, Gilchrist and Lewis 1956, 686)

els.de'ni.i. N.L. gen. n. *elsdenii* of Elsden; named after S.R. Elsden who first isolated the organism.

Cocci 2.0 μm or more in diameter, in pairs or occasionally in chains. Nonsporulating. Gram-stain-negative. Nonmotile. Anaerobic. Colonies in yeast extract peptone agar supplemented with sodium lactate are round, smooth, slightly raised, and have a glistening, mucoid appearance. At 48 h, the colonies are 0.2–1.0 mm in diameter with older colonies 3–4 mm. Growth occurs from 25 to 40°C but generally not at 45°C. Catalase and indole-negative. Gelatin is not liquefied. Nitrate is not reduced. H_2S is produced. Chemoorganotrophic. Gas is produced. Lactate is fermented with the production of acetic, propionic, butyric, iso-butyric, isovaleric, 2-methylbutyric, and valeric acids, little or no caproic and formic acids, a large quantity of CO_2, and small amounts

of H_2. Products from glucose fermentation are different from those from lactate: butyrate is the most copious product, some formate and caproate are produced, less valerate is formed, little or no acetate is produced, and propionate is not produced. Pyruvate is utilized, but succinate, fumarate, and malate are not attacked. Nutritional requirements are complex. Found in the rumen of cattle and sheep and in the feces and intestine of man and pigs.

DNA G+C content (mol%): 53.1–54.1 (Bd).

Type strain: ATCC 25940, CCUG 6199, CIP 106852, DSM 20460, JCM 1772, LC1, NCBI 8927.

GenBank accession number (16S rRNA gene): U95027.

2. **Megasphaera cerevisiae** Engelmann and Weiss 1986, 355[VP] (Effective publication: Engelmann and Weiss 1985, 290.)

ce.re. vi' si. ae. L. n. *cerevisia* beer; L. gen. n. *cerevisiae* of beer.

Slightly elongated cocci, 1.3–2.1 μm in diameter, occurring singly, in pairs, and occasionally in short chains. Gram-stain-negative. Nonsporulating. Nonmotile. Strictly anaerobic. Colonies on PYL or PYF agar are whitish, smooth and flat, 2–3 mm in diameter. Growth occurs between 15 and 37°C, but not at 40°C. Catalase and benzidine test negative. H₂S is produced. Nitrate is not reduced to nitrite. Indole is not produced. Gas, acetic, propionic, iso-and *n*-butyric, iso- and *n*-valeric, and caproic acid are produced. Utilization of carbon sources is shown in Table 206. Cell-wall peptidoglycan is of the *m*-Dpm-direct type. The peptide subunit contains one putrescine residue covalently bound to the α-carbonyl group of the glutamic acid. Habitat unknown. Strains of this species mainly isolated from spoiled beer.

> *DNA G+C content (mol%):* 42.4–44.8 (T_m).
> *Type strain:* PAT 1, ATCC 43254, DSM 20462, JCM 6130.
> *GenBank accession number (16S rRNA gene):* L37040.

3. **Megasphaera micronuciformis** Marchandin, Jumas-Bilak, Gay, Teyssier, Jean-Pierre, Siméon de Buochberg, Carrière and Carlier 2003b, 552[VP]

mic.ro.nu.ci.for'mis. Gr. adj. *micros* small; L. fem. gen. n. *nucis* a nut; L. adj. suff. *-formis* is shaped like; N.L. fem. adj. *micronuciformis* small walnut-shaped (referring to the morphology of bacterial cells and cell surface).

Cells are Gram-stain-negative, coccoid, usually single, 0.4–0.6 mm in diameter, nonmotile. Endospores are not formed. Strictly anaerobic. Colonies appear on Columbia sheep blood agar after 2–3 d incubation at 37°C and are circular, convex, shiny, and translucent with a smooth surface. Colonies are approximately 0.5–1.0 mm in diameter, non-pigmented and non-hemolytic. Non-fermentative and non-proteolytic (gelatin and milk negative). Gas is not produced. Nitrate is not reduced, and indole and catalase are not produced. Desulfoviridin may be produced. Metabolic end products are acetic, propionic, butyric, iso-valeric, and 2-phenylacetic acids. Valeric acid and trace amounts of isobutyric acid may be produced. Utilization of carbon sources is shown in Table 206. Differentiated from other *Megasphaera* species by size, DNA G+C content, metabolic end products, and 16S rRNA gene sequence. Habitat is unknown.

> *DNA G+C content (mol%):* 46.4 (T_m).
> *Type strain:* AIP 412.00, CCUG 45952, CIP 107280.
> *GenBank accession number (16S rRNA gene):* AF473834.

4. **Megasphaera paucivorans** Juvonen and Suihko 2006, 700[VP]

pau.ci.vo'rans. L. adj. *paucus* few, little, L. part. adj. *vorans* devouring; N.L. part. adj. *paucivorans* devouring a few substrates.

Gram-stain-negative, non-spore-forming and nonmotile cocci with a mean size of 1.5 μm × 1.2 μm, and mainly arranged in pairs. Stationary phase cells may form cell clumps and chains of 20–25 diplococci. Strictly anaerobic. Moderate to good growth (2+ or 3+ on a scale of 0 to 4+) is obtained in autoclaved PYF, PYG, MRS and SMMP media. Growth in SMMP is accompanied by color change from violet to yellow. Poor growth in PY medium (1+), but the addition of 1 % (v/v) pyruvate or gluconate results in heavily turbid suspension (4+). Colonies on PYF plates appear after 3 d at 30°C and after 7 d are yellowish, circular, convex, glossy and opaque with entire margins and a diameter of 1–1.5 mm. Grows at 15–37°C, with an optimum at around 30°C, but not at 10°C or 45°C. Major volatile fatty acids produced in beer are butyric and iso-valeric acid. H₂S and minor amounts of propionic, isobutyric, valeric and caproic acid are also formed. Other phenotypic properties and the characteristics differentiating the species from the other *Megasphaera* species are presented in Table 206 and Table 207. The type strain was isolated from spoiled Italian beer.

> *DNA G+C content (mol%):* 46.4 (T_m).
> *Type strain:* VTT E-032341[T], DSM 16981[T]).
> *GenBank accession number (16S rRNA gene):* DQ223730.

5. **Megasphaera sueciensis** Juvonen and Suihko 2006, 700[VP]

sue.ci.en'sis. N.L. fem. adj. *sueciensis* pertaining to Sweden.

Strictly anaerobic, Gram-stain-negative, non-spore-forming and nonmotile cocci, mainly arranged in pairs and occasionally in short chains. Mean cell size is 1.2 μm × 1.0 μm. Moderate growth (2+ on a scale of 0 to 4+) is obtained in autoclaved PYF, PYG, MRS and SMMP media at 30°C. The addition of 1 % (v/v) pyruvate or gluconate to the PY medium markedly stimulates the growth. Colonies on PYF and PYG plates appear after 4 d at 30°C and after 7 d are slightly yellowish, glossy, convex, opaque, smooth and circular with entire edges and a diameter of 0.5–0.8 mm. Grows at 15–37°C, with an optimum at around 30°C, but not at 10°C or 45°C. Other physiological properties and characteristics differentiating this species from the other *Megasphaera* species are shown in Table 206 and Table 207. The type strain was isolated from spoiled Swedish beer.

> *DNA G+C content (mol%):* 43.1 (T_m).
> *Type strain:* VTT E-97791, DSM 17042.
> *GenBank accession number (16S rRNA gene):* DQ223729.

Genus XIV. **Mitsuokella** Shah and Collins 1983, 439[VP]

ANNE WILLEMS AND MATTHEW D. COLLINS

Mit.su.o.kel'la. N.L. fem. n. *Mitsuokella* named after Tomotari Mitsuoka, the Japanese bacteriologist who first described the organism.

Gram-stain-negative, obligately anaerobic, non-spore-forming, nonmotile rods. Stout rods of regular shape with rounded ends. **Fermentative metabolism.** Growth is enhanced by glucose or other fermentable carbohydrates. **Major end products from glucose fermentation are acetate, lactate and succinate.** No copious gas formation from glucose. Gelatinase and catalase absent; amylase present. Indole-negative. Nitrate is reduced to nitrite.

DNA G+C composition is 55.9–58.2 mol% (T_m). On the basis of 16S rRNA gene sequences, *Mitsuokella* belongs to the *Sporomusa* branch of the *Clostrium* subphylum of the *Firmicutes* where it is closely related to *Selenemonas ruminantium* (Lan et al., 2002b; Willems and Collins, 1995b).

> *Type species:* **Mitsuokella multacida** corrig. (Mitsuoka, Terada, Watanabe and Uchida 1974) Shah and Collins 1983, 439[VP] (Effective publication: Shah and Collins 1982, 493.).

Further descriptive information

Phylogenetic treatment. The 16S rRNA gene sequences of *Mitsuokella multacida* and *Mitsuokella jalaludinii* share 98.7% similar-

TABLE 209. Differentiation of *Mitsuokella* and other genera of Gram-stain-negative, anaerobic, non-spore-forming rods[a]

Genus	Mitsuokella	Bacteroides	Prevotella	Porphyromonas	Selenomonas	Eikenella	Hallella
DNA G+C content (mol%)	56–58	40–48	40–60	46–51	54–61	56–58	58
Saccharolytic	+	+	+	−	+	−	+
Major end products[b]	L, A/a, S/s	S, A	S, A	A, B, iV	P, A	a	S, a, l
Predominant fatty acid	C16:1[c]	anteiso-C15:0	anteiso-C15:0	iso-C15:0	C14:0 DMA	C16:0	C16:0
Peptidoglycan type	mDpm[c]	Dpm	mDpm	No Dpm	ND	ND	ND
Menaquinones	−[c]	+	+	+	−	ND	ND
Motility	−	−	−	−	+	−	−

[a]Data from Shah and Collins (1982), Moore et al. (1994), Willems and Collins (1995b).

[b]A and a, acetic acid; B, butyric acid; L and l, lactic acid; P, propionic acid; S and s, succinic acid; iV, isovalerate. Upper-case letters indicate large amounts produced; lower-case letters indicate smaller amounts produced.

[c]No information available for *Mitsuokella jalaludinii*.

tity. According to the 16S rRNA phylogenetic analysis presented in the roadmap to this volume (Figure 6), the genus *Mitsuokella* is a member of the family *Veillonellaceae*, order *Clostridiales*, class *Clostridia* in the phylum *Firmicutes*. This family contains several other Gram-stain-negative genera (e.g., *Sporomusa, Selenomonas, Veillonella*). Closest phylogentic relative is *Selenomonas ruminantium* with approximately 96% 16S rDNA sequence similarity (Willems and Collins, 1995b). Despite this relatively close relatedness, *Selenomonas* differs from *Mitsuokella* in a number of phenotypic and chemotaxonomic features (Table 209).

Cell morphology. Cells of *Mitsuokella multacida* are nonmotile, large, regular rods, 0.8–1.5 μm wide by 3.0–20 μm long, with rounded ends (Shah and Collins, 1982). Cells of *Mitsuokella jalaludinii* are smaller: 0.6–0.8 μm by 1.2–2.4 μm. Cells occur singly, in short chains or irregular groups (Holdeman et al., 1986; Lan et al., 2002b).

Cell-wall composition. Cell-wall murein of *Mitsuokella multacida* is of the A1γ type (Schleifer and Kandler, 1972) containing meso-diaminopimelic acid as the dibasic amino acid (Shah et al., 1983).

Colonial characteristics. After 2–3 d of incubation at 37°C, colonies of *Mitsuokella multacida* on blood agar plates are circular, 3–8 mm in diameter, convex with an irregular surface and edge and are grayish-white in color. Some strains are slightly hemolytic (Shah and Collins, 1982). Colonies of *Mitsuokella jalaludinii* were circular with regular margins, grayish-white, convex with smooth surface and 1–4 mm in diameter after 2 d at 39°C on PYG agar (Lan et al., 2002b).

Nutrition and growth conditions. A strictly anaerobic atmosphere is required, e.g., of nitrogen, CO$_2$ (10%) and hydrogen (20%). Optimum growth temperature is 37°C and 42°C for *Mitsuokella multacida* and *Mitsuokella jalaludinii*, respectively. There is usually no growth of *Mitsuokella multacida* at 45°C, however, *Mitsuokella jalaludinii* is capable of growth at 45°C and at 47°C. For both species growth is markedly enhanced by the presence of glucose or another fermentable carbohydrate (Shah and Collins, 1982) but not by bile (Lan et al., 2002b). The final pH in glucose broth is 4.1–4.3 for *Mitsuokella multacida* (Shah and Collins, 1982) and 3.8–4.0 for *Mitsuokella jalaludinii* (Lan et al., 2002b).

Metabolism and metabolic pathways. *Mitsuokella* has a fermentative metabolism, using a variety of carbohydrates. Major end products from glucose fermentation are acetic and lactic acid with small amounts of succinic acid for *Mitsuokella*

multacida (Mitsuoka et al., 1974; Shah et al., 1983) and lactic and succinic acid with some acetic acid for *Mitsuokella jalaludinii* (Lan et al., 2002b). Additional biochemical characteristics are listed in Table 210.

Chemotaxonomic data. No chemotaxonomic data are available for *Mitsuokella jalaludinii*. The data in this section only characterize *Mituokella multacida*. Fatty acids consist of non-hydroxylated (ca. 80%) and 3-hydroxy (ca. 20%) types. The non-hydroxylated fatty acids are predominantly straight-chain and monounsaturated types with C$_{16:1}$ constituting the major compound, C$_{12:0}$ and C$_{16:0}$ also present in substantial amounts and C$_{14:0}$ present in smaller amounts. The major 3-hydroxy fatty acid is 3-OH-C$_{14:0}$. Small amounts of branched-chain fatty acids are also present (Shah et al., 1983).

Menaquinones and ubiquinones are not present (Shah et al., 1983). Sphingolipids are absent (Miyagawa et al., 1978). Polar lipids comprise major amounts of phosphatidylethanolamine and an unidentified amino-containing phospholipid that does not contain sugar residues. Small amounts of four other unidentified components are also present. Some strains posses diphosphatidylglycerol (Shah et al., 1983). *Mitsuokella multacida* does not contain polyamines (Hamana et al., 1995).

Plasmids. Some strains of *Mitsuokella multacida* contain plasmids (size 11.7 kb) (Flint and Stewart, 1987).

Antibiotic sensitivity. Antibiotic resistance data for both species are listed in Table 210. *Mitsuokella multacida* strains vary in their resistance to tetracycline (Flint and Stewart, 1987). Several *Mitsuokella multacida* strains contain tetracycline resistance genes: TetQ was identified in one strain (Leng et al., 1997) and TetW, representing a new class of tetracycline resistance determinants, has been identified in both porcine and bovine strains (Barbosa et al., 1999).

Ecology. *Mitsuokella* species are strict anaerobes that inhabit the intestinal tract of humans and livestock animals. The acids they produce may be used by acid utilizing bacteria to form butyrate. *Mitsuokella multacida* belongs to the normal anaerobic rumen and colon microflora and has been isolated from feces of humans and domestic animals (Al Jassim, 2003; Mitsuoka et al., 1974; Sirotek et al., 2003; Tsukahara et al., 2002). *Mitsuokella jalaudinii* was originally reported from the rumen of cattle in Malaysia (Lan et al., 2002b) and has also been recovered from the equine gastro-intestinal tract (Al Jassim et al., 2005). Both species have a strong phytase activity; this enzyme catalyzes the release of phosphate from phytate, the predominant

TABLE 210. Phenotypic characteristics of *Mitsuokella* species[a, b]

Characteristic	*M. jalaludinii*	*M. multacida*
Methyl red test, Voges–Proskauer test	+	+[c]
Growth in 4.5% NaCl	–	–[c]
Growth at 45°C	+	V
Growth at 47°C	+	–[c]
Urease, hydrolysis of Tween 80, arginine decarboxylase	–	–[c]
Malate dehydrogenase	ND	+
Glutamate, glucose-6-phosphate and 6-phosphogluconate dehydrogenases	ND	–
Meat digestion, lecithinase	ND	–
Coagulation of milk, hemolysis	ND	V
Growth in 20% bile	ND	+
Acid from:		
Cellobiose, trehalose	+	+[d]
Starch	+	+[e]
Inositol	+	v[f]
Melezitose	–	+[g]
Sorbitol, ribose, esculin	+	v
Glycerol	+	–
Glycogen	–	v[h]
Mannitol, rhamnose	–	v
Amygdalin	–	–[d]
Dextrin	ND	+
Erythritol	ND	v, w
Sorbose	ND	–
Arbutin, D-turanose, D-arabitol, 5-ketogluconate	+	+[c]
D-Tagatose	–	+[c]
D-Arabinose, L-xylose, methyl-β-xyloside, L-sorbose, *N*-acetylglucosamine, xylitol, β-gentiobiose, D-lyxose, D-fucose, L-fucose, L-arabitol, gluconate, 2-ketogluconate	–	–[c]
Resistance to:		
Kanamycin (100 μg/ml)	+	–
Penicillin (10 μg/ml), erythromycin (5 μg/ml), bacitracin (3 μg/ml)	+	v
Brilliant green (0.001%)	–	+
Rifampin (10 μg/ml), sodium propionate (1.5%)	ND	+
Polymyxin B (10 μg/ml), colistin (10 μg/ml)	ND	–

[a]Data from Mitsuoka et al. (1974), Shah and Collins (1982), Shah et al. (1983) and Lan et al. (2002b).

[b]Both species hydrolyse starch and esculin and reduce nitrate to nitrite. All strains produce acid from glucose, mannose, fructose, galactose, L-arabinose, D-xylose, sucrose, maltose, lactose, salicin, melibiose and raffinose. No acid is produced from adonitol, dulcitol, inulin, meso-erythritol, α-methyl mannoside or α-methyl glucoside. Neither species produces gelatinase, indole, H$_2$S or catalase and grows at 20°C. Both species are resistant to neomycin and sensitive to 0.005% crystal violet.

[c]Result only reported for the type strain (Lan et al., 2002b).

[d]Variable according to Holdeman et al. (1977).

[e]The type strain is negative according to Lan et al. (2002b).

[f]Positive according to Holdeman et al. (1986).

[g]Weak or negative response according to Holdeman et al. (1986).

[h]Negative according to Holdeman et al. (1986).

form of phosphorous in seeds, cereals and legumes (Lan et al., 2002a; Yanke et al., 1998). The phytase of *Mitsuokella multacida* is located in the outer membrane (D'Silva et al., 2000). *Mitsuokella multacida* strains from the rabbit caecum were mucinolytic (Sirotek et al., 2003).

The role of the microflora in the development of cancers of the colon remains unclear. In a study of the effects of intestinal microflora on the development of colonic neoplasm induced by 1,2-dimethylhydrazine in mice, the incidence of colonic adenoma in mice mono-associated with *Mitsuokella multacida* was less (68%) then in germ-free mice (74%), but higher than in conventionalized mice (58%). The adenoma in germ-free mice was smaller than in conventionalized mice, suggesting that micro-organisms may have two different effects: on the one hand they may inhibit incidence of

adenoma, but on the other hand, they may promote tumour growth (Horie et al., 1999).

Genetic characteristics. In *Mitsuokella multacida* two GATC-specific DNA methyltransferases and *Sau*3AI isoschizomeric restriction endonucleases were partially characterized (Piknova et al., 2005).

Isolation procedures

The strains used for the original description of *Bacteroides multiacidus* (Mitsuoka et al., 1974) were isolated by plating out fecal material, emulsified in an anaerobic buffer solution, on modified Eggerth-Gagnon (EG) agar, glucose blood liver agar and neomycin brilliant green taurocholate blood agar (Mitsuoka et al., 1964, 1965). Plates were incubated for 3 d at 37°C under an atmosphere of 100% CO$_2$ and strains were checked for purity and routinely maintained on EG agar (Mitsuoka et al., 1974). The more recently isolated *Mitsuokella jalaludinii* was obtained from rumen fluid by serial dilution and inoculation into roll-tubes of modified phytase-screening (MPS) agar using Hungate anaerobic techniques (Lan et al., 2002b).

Maintenance procedures

Cultures can be maintained on blood agar plates by twice-weekly subculture in an atmosphere of nitrogen plus 10% CO_2 and 20% H_2 at 37°C (Shah et al., 1983). According to the original description of *Bacteroides multiacidus* (Mitsuoka et al., 1974), strains can be maintained on slants of EG agar H_2CO_3–CO_2 buffer and stored at 4°C for up to 3 months. For long term preservation strains can be lyophilized using standard procedures.

Differentiation of the genus *Mitsuokella* from other Gram-stain-negative, anaerobic, non-spore-forming, saccharolytic bacilli

On phenotypic grounds, *Mitsuokella* was previously regarded as a member of the genus *Bacteroides* (Mitsuoka et al., 1974). It can be differentiated from *Bacteroides* and *Prevotella*, the two saccharolytic genera of the *Bacteroidaceae*, by its higher G+C content, the production of lactic acid in addition to acetate and succinate as end products from glucose fermentation. The absence of respiratory menaquinones, and a distinctive fatty acid composition have only been reported for *Mitsuokella multacida* (Table 209). Features for the differentiation of *Mitsuokella* from *Selenomonas*, its closest phylogenetic neighbor, and other Gram-stain-negative, anaerobic, non-spore-forming bacilli with a similar G+C content are given in Table 209.

In a system using oligonucleotide primers targeting the 16S rRNA gene to monitor species in the rumen ecosystem by real-time PCR, one primer-set was used to amplify both *Mitsuokella multacida* and *Selenomonas ruminantium*. It was not possible to design primers with a sufficiently high melting temperature to distinguish both species in the experimental set-up used (Tajima et al., 2001), although on the basis of the level of sequence similarity between both species (approx. 96%) it should be possible to select species-specific primers.

Taxonomic comments

Bacteroides multiacidus was originally proposed to accommodate a group of Gram-stain-negative, strictly anaerobic, saccharolytic, large rods isolated from the feces of humans and pigs (Mitsuoka et al., 1974). Subsequently several chemotaxonomic and biochemical studies provided strong evidence to exclude *Bacteroides multiacidus* from the genus *Bacteroides* because of differences in cell-wall peptidoglycan structure, fatty acid and lipid composition, DNA base composition, fermentation end products and dehydrogenase activities (Miyagawa et al., 1978, 1979; Shah et al., 1983). Because these characteristics indicated that *Bacteroides multiacidus* was unlike any other existing Gram-stain-negative anaerobic bacillus, a new genus, *Mitsuokella*, was proposed with one species, *Mitsuokella multiacidus* (Shah and Collins, 1982), later corrected to *Mitsuokella multacida* (Euzéby, 1998). Sequencing of the 16S rRNA gene of the type strain of *Mitsuokella multacida* confirmed its separateness from the genus *Bacteroides* and placed it in the *Sporomusa* cluster of the *Clostridium* subphylum, close to *Selenomonas ruminantium* (Willems and Collins, 1995b).

A second species, *Mitsuokella dentalis*, which was described for isolates from human dental root canals (Haapasalo et al., 1986a), differs from *Mitsuokella multacida* in its fermentation end products, carbohydrate fermentation patterns, starch hydrolysis, resistance to bile and fatty acid composition (Haapasalo et al., 1986b; Shah and Collins, 1982). On the basis of its 16S rRNA gene sequence, *Mitsuokella dentalis* belongs to the genus *Prevotella* in the *Bacteroides–Flavobacterium* phylum of the Gram-stain-negative bacteria. Since it is distinct from the other *Prevotella* species, it was transferred to this genus as *Prevotella dentalis* comb. nov. Its is highly related (99.8% 16S rDNA sequence similarity (Willems and Collins, 1995b) to *Hallella seregens*, a group of Gram-stain-negative, anaerobic, rod-shaped bacteria from the gingival crevices of humans with gingivitis or periodontitis (Moore and Moore, 1994). Phenotypically, both species are very similar, with the only reported difference being the fermentation of trehalose and the hydrolysis of starch by *Hallella seregens* and not by *Prevotella dentalis*. DNA–DNA hybridizations should establish whether *Hallella seregens* represents a separate *Prevotella* species or is a later subjective synonym of *Prevotella dentalis* (Willems and Collins, 1995b).

Mitsuokella jalaludinii, created for strains from cattle in Malaysia, is genetically highly related to *Mitsuokella multacida*: a DNA–DNA hybridization value of 63.8% was reported between both type strains (Lan et al., 2002b). However, the species can be distinguished by the ability of *Mitsuokella jalaludinii* to grow at 47°C, to produce acid from glycerol but not from melezitose or D-tagatose and to resist kanamycin (100 µg/ml) but not brilliant green (0.001%).

Acknowledgements

A.W. is indebted to the Fund for Scientific Research - Flanders for a position as post-doctoral research fellow.

List of species of the genus *Mitsuokella*

1. **Mitsuokella multacida** corrig. (Mitsuoka, Terada, Watanabe and Uchida 1974) Shah and Collins 1983, 439[VP] (Effective publication: *Mitsuokella multiacidus* Shah and Collins 1982, 493.) (*Bacteroides multiacidus* Mitsuoka, Terada, Watanabe and Uchida 1974, 40)

   mul.ta'ci.da. L. adj. *multus -a -um* many; L. adj. *acidus -a -um* sour; N.L. fem. adj. *multacida* producing much acid.

   Characteristics of the species are the same as those given for the genus. Additional characteristics are given in Table 210. The G+C content of the DNA is 55.9–58.2 mol% (T_m) (Shah et al., 1983). DNA is modified by Dam-methylation (Pristas et al., 2001).

   Isolated from the feces of humans and pigs and from rabbit caecum and sheep rumen.

   DNA G+C content (mol%): 57.9 (Shah et al., 1983).
   Type strain: ATCC 27723, NCTC 10934, A405-1.
   EMBL accession number (16S rRNA gene): X81878.

2. **Mitsuokella jalaludinii** Lan, Ho and Abdullah 2002b, 717[VP]

   jal.al.u.di'ni.i. N.L. gen. n. *jalaludinii* of Jalaludin, in honor of S. Jalaludin, an animal nutritionist and Vice-Chancelor of

University Putra Malaysia, who has contributed significantly to rumen microbiology.

Characteristics of the species are the same as those given for the genus. Additional characteristics are given in Table 210. The G+C content of the DNA is 56.8 mol% (HPLC) (Lan et al., 2002b).

Isolated from the rumen fluid of cattle in Malaysia (Lan et al., 2002b) and the equine gastro-intestinal tract (Al Jassim et al., 2005).

Type strain: DSM 13811 (= ATCC BAA-307 = M9).
DNA G+C content (mol%): 56.8 (HPLC).
EMBL accession number (16S rRNA gene): AF479674.

Genus XV. **Pectinatus** Lee, Mabee and Jangaard 1978, 582[AL] emend. Juvonen and Suihko 2006, 700

AULI HAIKARA AND RIIKKA JUVONEN

Pec.ti.na'tus. L. part. adj. *pectinatus* combed; N.L. masc. n. *Pectinatus* combed (bacteria).

Slightly curved rods, 0.4–0.9 μm in diameter and 2–50 μm or more in length, with rounded ends. They occur singly, in pairs, and only rarely in short chains. In old cultures, very elongated cells with a helical shape can be found and round cell forms are also observed. Gram-stain-negative. According to the cell envelope structure, intermediates between Gram-stain-positive and Gram-stain-negative bacteria. Cadaverine or putrescine is found in the cell-wall peptidoglycan. **Motile,** young cells form an "X" shape during movement whereas old cells have characteristically slow snakelike movements. **Flagella emanate from only one side of the cell body.** The organisms are **obligately anaerobic, non-spore-forming mesophiles.** They are cytochrome oxidase-negative and do not produce indole, liquefy gelatin, reduce nitrate, or hydrolyze arginine. Urease and catalase may or may not be produced. The main metabolic products from glucose are **propionic and acetic acid. Hydrogen sulfide and acetoin are also produced and succinic and lactic acids may be produced.** Originally isolated from **spoiled, packaged beer.** The habitat outside the brewery is not defined.

DNA G+C content (mol%): 38–41.

Type species: **Pectinatus cerevisiiphilus** Lee, Mabee and Jangaard 1978, 582[AL] emend. Schleifer, Leuteritz, Weiss, Ludwig, Kirchhof and Seidel-Rüfer 1990, 25.

Further descriptive information

The genus *Pectinatus* was originally suggested to belong to the family *Bacteroidaceae* (Lee et al., 1978). The phylogenetic position of *Pectinatus* has established during recent years when the 16S rRNA, 23S rRNA and the ATPase β-subunit gene sequences, the 16S–23S rRNA intergenic spacer regions, and/or DNA–DNA hybridization have been used as markers (Both et al., 1992; Juvonen and Suihko, 2006; Klugbauer et al., 1992; Ludwig et al., 1992; Motoyama and Ogata, 2000a; Sawada et al., 1999; Schleifer et al., 1990; Willems and Collins, 1995a). The genus *Pectinatus*, along with *Selenomonas, Megasphaera, Sporomusa, Veillonella,* and *Zymophilus,* has been placed to the *Sporomusa* sub-branch of the *Clostridium* subphylum of Gram-stain-positive bacteria (Collins et al., 1994; Schleifer et al., 1990; Willems and Collins, 1995a). The cell envelope composition of *Pectinatus* has further supported its new taxonomic position (Helander et al., 1983; Helander et al., 1992; Schleifer et al., 1990; Strömpl et al., 1999). More recently, all members of the *Sporomusa* sub-branch have been grouped in the family *Veillonellaceae* of the class *Clostridia* of the phylum *Firmicutes* (Garrity and Holt, 2001). The genus *Pectinatus* currently comprises three validly named species, *Pectinatus cerevisiiphilus* (Lee et al., 1978; Schleifer et al.,

1990), *Pectinatus frisingensis* (Schleifer et al., 1990) and *Pectinatus haikarae* (Juvonen and Suihko, 2006). Based on the 16S rRNA gene sequence analysis, the similarity between *Pectinatus frisingensis*[T] and *Pectinatus haikarae*[T] is 93.6%. *Pectinatus cerevisiiphilus*[T] has 95.6% and 94.3% sequence identity to *Pectinatus haikarae*[T] and *Pectinatus frisingensis*[T], respectively. Based on the sequence similarity calculations of Willems and Collins (1995a) on fourteen species of the *Sporomusa* sub-branch the nearest species with *Pectinatus frisingensis* and *Pectinatus cerevisiiphilus* were three *Selenomonas* species (88–89%), *Zymophilus paucivorans* (88%), *Clostridium quercicolum* (87–88%), *Sporomusa paucivorans* (86–88%), *Veillonella parvula* (86–87%) and *Megasphaera elsdenii* (86%). The phylogenetic study of Motoyama and Ogata (2000a) on anaerobic beer spoilage bacteria showed that the 16S–23S rDNA intergenic spacer regions of *Pectinatus frisingensis* and *Pectinatus cerevisiiphilus, Zymophilus raffinosivorans* and *Zymophilus paucivorans* and *Selenomonas lacticifex* were of two molecular sizes (long and short). The dendrogram of the short spacer region corresponded with that of the 16S rRNA gene sequence (Schleifer et al., 1990). On the contrary, the order of the tRNA genes in the long spacer region and the DNA sequences of the long spacer region indicated that *Pectinatus* was more closely related to the *Selenomonas lacticifex* than to the *Zymophilus* species, which according to the authors better reflects the phylogenetic positions of the species. Notably, the order of the alanine tRNA/isoleucine tRNA genes in the long spacer regions of *Pectinatus* and *Selenomonas lacticifex* was the reverse of what has previously been reported for other bacteria (Gürtler and Stanisich, 1996).

Pectinatus cells typically appear as curved rods (Figure 212). Very elongated cells up to 50 μm or more with helical shape can be found in old cultures (Haikara, 1991; Juvonen and Suihko, 2006). Ultrastructure of several *Pectinatus cerevisiiphilus* and *Pectinatus frisingensis* strains has been characterized. The unique comb-like flagellar arrangement of *Pectinatus cerevisiiphilus* observed by Lee et al. (1978) using scanning electron microscopy has been confirmed by other workers in negatively stained preparations (Back et al., 1979; Haikara et al., 1981b; Kirchner et al., 1980). The cell envelope structure seen in such sections has revealed a multilayered cell wall typical of Gram-stain-negative bacteria (Lee et al., 1978). The peptidoglycan layer is very thick (30 nm), almost filling the periplasmic space of the cell envelope (Haikara et al., 1981a, 1981b). The freeze-fracture technique has demonstrated that the cell envelope structure in different strains is similar (Haikara et al., 1981a, 1981b). The thick peptidoglycan layer and the invaginations of

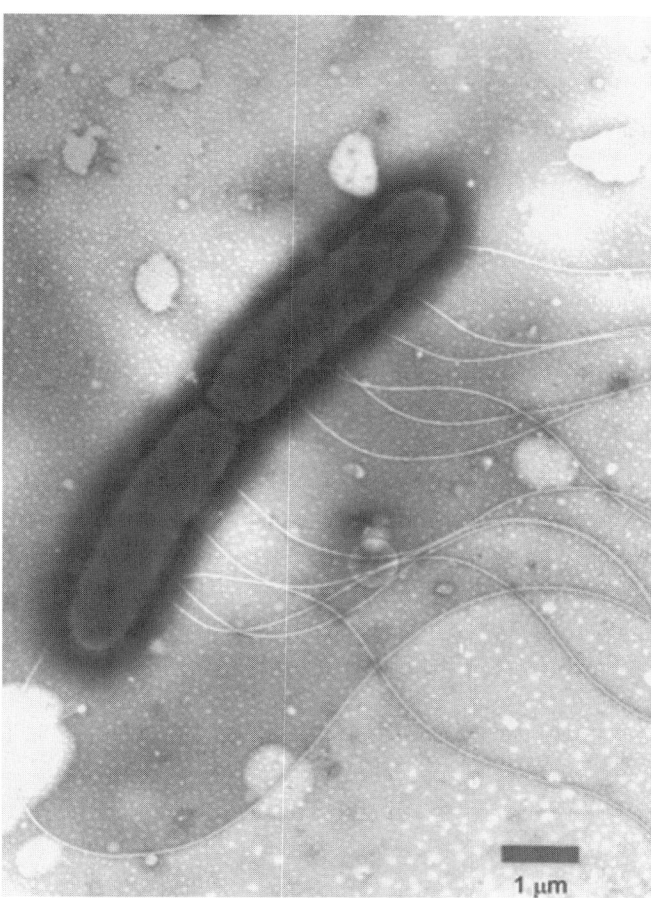

FIGURE 212. Electron micrograph of *Pectinatus frisingensis* illustrates flagella on one side of the cell body. Bar = 1 μm. (Reprinted with permission from Haikara, A. 1991. The genera *Pectinatus* and *Megasphaera*. In Balows, Trüper, Dworkin, Harder and Schleifer (Editors). The Prokaryotes, 2nd edn, Vol. II, Springer-Verlag, New York, pp. 1993–2004.)

the cytoplasmic membrane are typical features of Gram-stain-positive bacteria whereas the presence of an outer membrane is typical of Gram-stain-negative bacteria (Figure 213). The outer membrane of *Pectinatus*, however, is exceptional since it lacks the permeability barrier function normally assigned to the outer membrane of Gram-stain-negative (e.g., the enteric) bacteria (Helander et al., 1994). The cellular fatty acid composition is similar in type strains of *Pectinatus cerevisiiphilus* and *Pectinatus frisingensis*. The fatty acids are heavily dominated by odd-numbered fatty acids, $C_{11:0}$, $C_{15:0}$, $C_{17:1}$, $C_{18:0\ cyclo}$, and $C_{13:0\ 3OH}$, which are the main fatty acids detected in both species (Helander and Haikara, 1995). Alk-1-enyl chains with chain lengths of 15–18 are also found, indicating the presence of plasmalogens common in strict anaerobic bacteria. In addition to the major hydroxy acid component $(R)13:0(3OH)$, both *Pectinatus* species are shown to contain, in minor amounts, five 3-hydroxy acids with chain lengths of 11–15 carbons. These include the unsaturated acid $C_{13:1\ 3OH}$, the occurrence of which is limited to *Pectinatus*. As discussed by Helander and Haikara (1995) and Strömpl et al. (1999), because fatty acid $C_{13:0\ 3OH}$ is not in the

MIDI database utilized for fatty acid-based identification of anaerobic bacteria, this fatty acid has been erroneously identified as $C_{14:0}$ dimethylacetal (DMA) in several bacterial genera, including *Pectinatus* (Moore et al., 1994).

Pectinatus possesses directly cross-linked *meso*-diaminopimelic acid-containing peptidoglycan, with covalently linked diamine (cadaverine or, rarely, putrescine) in the peptide subunit of the peptidoglycan (Schleifer et al., 1990). With the aid of monoclonal antibodies isolated on the basis of their binding to *Pectinatus cerevisiiphilus* peptidoglycan, Ziola et al. (1999) found a peptidoglycan structure that they suggested was common to several species of anaerobic beer spoilage bacteria, including *Pectinatus frisingensis*, *Selenomonas lacticifex*, *Zymophilus paucivorans* and *Zymophilus raffinosivorans*.

The cell surface lipopolysaccharides (LPS) of the type strains *Pectinatus cerevisiiphilus* ATCC 29359 and *Pectinatus frisingensis* ATCC 33332 have been characterized in detail. Several exceptional properties are assigned to *Pectinatus* LPS, including the production of at least two distinct types of LPS by one strain (Helander et al., 1992), the presence of the phosphorylated disaccharide α-D-Gl$_{cp}$N-(1'-4)-Kdo (3-deoxy-D-*manno*-oct-2-ulopyranosonic acid) in the LPS core (Helander et al., 1993), the resistance of the lipid A-polysaccharide linkage to acid (Helander et al., 1994) and the predominance of furanosidic 6-deoxyhexose in the O-specific chains (Senchenkova et al., 1995). The lipid A backbone of *Pectinatus cerevisiiphilus* and *Pectinatus frisingensis* is composed of the common bisphosphorylated $\beta1'$-6-linked glucosamine (GlcN) disaccharide, with almost quantitative substitution of the ester-linked phosphate by 4-amino-4-deoxyarabinose (Helander et al., 1994). The glycosidically linked phosphate of *Pectinatus* also carries minor amounts of this aminopentose, analogous to the structure found in *Klebsiella pneumoniae* O3 lipid A (Helander et al., 1996). The fatty acids in *Pectinatus* lipid A comprise two amide-linked and two ester-linked $C_{13:0\ 3OH}$. The hydroxyl groups of the fatty acids linked to the nonreducing GlcN carry the fatty acid $C_{11:0}$ or, to a small extent, $C_{13:0}$ in acyloxyacyl linkage. Whereas $C_{13:0\ 3OH}$ is the main hydroxy acid in *Pectinatus*, the minor 3-hydroxy acids are also most probably constituents of the lipid A, although their position has not been determined. Complete structures of the core oligosaccharides in *Pectinatus* LPS are not known, but each strain presumably elaborates two structurally distinct cores, only one of which carries polymeric O-specific chains; these two LPS populations can be separated from each other during processing of initial phenol/chloroform/petroleum ether extracts (Helander et al., 1992). The O-specific chain of *Pectinatus cerevisiiphilus* is composed of repeating disaccharides of 1→2 linked α-D-fucofuranose and α-D-glucopyranose, whereas the repeating unit of *Pectinatus frisingensis* O-specific chain is a branched tetrasaccharide consisting of a single sugar type, the rare 6-deoxy-L-altrofuranose (Senchenkova et al., 1995). It is worthy of note that the O-specific chains of *Pectinatus frisingensis* and *Pectinatus cerevisiiphilus* are highly labile towards acid, whereas the normally acid-labile lipid A-core linkage is exceptionally stable. The reason for the latter property is unknown at present, but the presence of acid-stable linking sugars such a 2-octulosonic acid has been excluded in their LPS (Helander et al., 1994). The LPS of both species exhibit the biological potency of classical endotoxins (Helander et al., 1984).

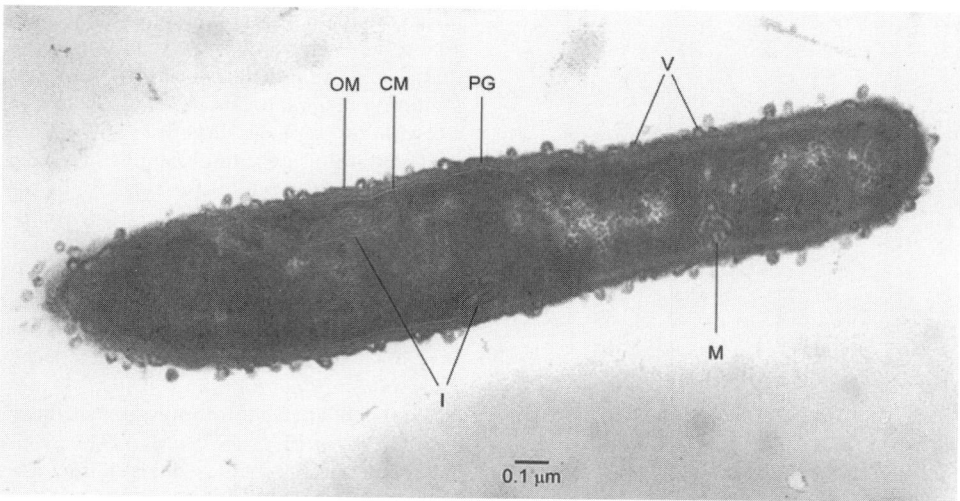

FIGURE 213. Electron micrograph of thin section of *Pectinatus frisingensis*. The cell envelope structure, the cytoplasmic membrane (CM), the thick peptidoglycan layer (PC), the outer membrane (OM) with numerous vesicles (V) are around the cytoplasm (C). Inside the cytoplasm invaginations (I) of the CM and the mesosomes (M), most likely artefacts of preparation for electron microscopy are seen. Bar = 0.1 μm. (Reprinted with permission from Haikara, A., Enari,T.-M. and Lounatmaa, K. (1981a). Proceedings of the 18th Congress, European Brewery Convention, Copenhagen, pp. 229–240.)

The most commonly used media for the cultivation and isolation of *Pectinatus* strains are de Man–Rogosa–Sharpe (MRS) lactobacilli broth or agar, modified MRS medium (Schleifer et al., 1990), thioglycollate medium with 1% glucose and 1.5% agar (Lee, 1984), and peptone-yeast extract-fructose (PYF) medium (the recipe given in *Megasphaera* chapter, above) (Engelmann and Weiss, 1985). Colonies of *Pectinatus* species on these agars are circular, entire, low convex to pulvinate or pyramidal, beige to white or grayish in color, glistening, and opaque (Haikara et al., 1981b; Juvonen and Suihko, 2006; Schleifer et al., 1990). Back et al. (1979) have reported a rough colony produced by one strain. A selective medium for the isolation and differentiation of *Pectinatus* and *Megasphaera* from other brewery organisms has been developed by Lee (1994) (see *Enrichment and isolation procedures*).

The acid tolerance of *Pectinatus cerevisiiphilus* and *Pectinatus frisingensis* in beer has been studied in growth media and beer. The strains of these species grow well in the pH range 4.5–8.5 (Back et al., 1979; Haikara et al., 1981b) the optimum being 6.0–7.0 (Kirchner et al., 1980; Takahashi, 1983). In beer some retardation of growth has been measured at pH 4.0–4.1 (Haikara, 1984; Kirchner et al., 1980; Seidel et al., 1979). According to Tholozan et al. (1997), *Pectinatus frisingensis* copes better with low pH and is able to grow over a wider pH range than *Pectinatus cerevisiiphilus*; for the latter, pH 6.2 is optimal for growth and biomass production, whereas for the former, pH of approximately 4.9 is optimal at glucose concentrations below 20 mM.

In the utilization of carbon sources, *Pectinatus cerevisiiphilus* and *Pectinatus haikarae* more closely resemble each other than *Pectinatus frisingensis* but clear differences between the species still exist (Haikara, 1991; Juvonen and Suihko, 2006; Schleifer et al., 1990) (Table 211). *Pectinatus cerevisiiphilus* can be distinguished from the other two species by its inability to ferment inositol. *Pectinatus frisingensis* utilizes *N*-acetylglucosamine and cellobiose but not melibiose or xylose whereas for the *Pecti-*

natus cerevisiiphilus and *Pectinatus haikarae* the situation is the reverse. *Pectinatus haikarae*, on the other hand, can be easily differentiated from the other two species by its positive catalase reaction. Furthermore, it is able to ferment lactose and does not grow at +37°C. In contrast to other brewery contaminants, most *Pectinatus* strains are incapable of utilizing maltose, the main carbohydrate of wort. On the other hand, they can utilize lactate. The major metabolic end products of the *Pectinatus* species from glucose or fructose are propionic and acetic acids and acetoin. *Pectinatus cerevisiiphilus* and *Pectinatus frisingensis* also produce succinic and lactic acid from glucose (Back et al., 1979; Haikara et al., 1981b; Juvonen and Suihko, 2006; Lee et al., 1978; Tholozan et al., 1994). The most abundant acid produced is propionic acid; more than 1000 mg/l has been detected in contaminated beer. Similar to the propionibacteria, *Pectinatus frisingensis* and *Pectinatus cerevisiiphilus* have been shown to use the succinate pathway for the production of propionic acid (Haikara et al., 1981a, 1981b; Tholozan et al., 1994). This observation further supports the assignment of *Pectinatus* to the family *Veillonellaceae* instead of *Bacteroidaceae*, as the *Bacteroides* species are known to use the acrylate pathway for propionate synthesis. A very specific feature of the metabolism of *Pectinatus* bacteria is the production of organic sulfur compounds (Haikara et al., 1981a). In addition to H_2S detected by Lee et al. (1980), production of methyl mercaptan and dimethyltrisulfide has been observed.

According to DNA–DNA hybridization, the 16S rRNA and 23S rRNA gene sequences, and the 16S rRNA–23S rDNA intergenic spacer regions, *Pectinatus cerevisiiphilus* and *Pectinatus frisingensis* can be genotypically distinguished (Haikara, 1989; Motoyama and Ogata, 2000a; Schleifer et al., 1990; Willems and Collins, 1995a). Nearly complete 16S rRNA gene sequence of *Pectinatus haikarae* has been determined which can be used to differentiate it from the other *Pectinatus* species (Juvonen and Suihko, 2006). PCR has been applied to the identification and

TABLE 211. Differential characteristics of closely related anaerobic Gram-stain-negative rod-shaped bacteria isolated from breweries[a,b]

Characteristic	Pectinatus cerevisiiphilus[c]	Pectinatus frisingensis[c]	Pectinatus haikarae[c]	Zymophilus paucivorans	Zymophilus raffinosivorans	Selenomonas lacticifex
Acid produced from:						
N-Acetylglucosamine	–	+	–	–	+	ND
Cellobiose	–	+	–	+	+	+
Esculin	–	–	d	ND	ND	ND
Inositol	–	+	+	–	+	–
Lactose	–	–	+	–	+	+
Maltose	–	–	–	+	+	d
Mannitol	+	+	+	+	+	–
Melibiose	+	–	+	–	+	+
Raffinose	–	–	–	–	+	+
Rhamnose	+	+	+	–	+	–
Sorbitol	+	+	ND	+	+	–
Sucrose	–	–	ND	+	+	+
Xylitol	–	d	+	–	+	–
Xylose	+	–	+	–	+	+
Acetoin production	+	+	+	–	–	ND
Catalase activity	–	–	+	ND	ND	ND
Urease activity	d	–	–	ND	ND	ND
Growth at 37°C	+	+	–	ND	ND	ND
Lactic acid as a main fermentation product	–	–	–	–	–	+
DNA G+C content (mol%)	38–41	38–41	39	39–41	38–41	51–52

[a]Symbols: +, >85% positive; d, different strains give different reactions (16–84% positive); –, 0–15% positive; ND, not determined.

[b]Adapted from Schleifer et al. (1990); Haikara (1991); Juvonen and Suihko (2006).

[c]Combined data for eleven, fourteen and four *Pectinatus cerevisiiphilus*, *Pectinatus frisingensis* and *Pectinatus haikarae*, respectively. All strains are positive for acid production from adonitol, L-arabinose, DL-erythritol, D-fructose, D-galactose, D-glucose, glycerol, mannose, D-ribose and utilization of DL-lactate. All strains are negative for acid production from glycogen, inulin and melezitose, for the utilization of succinate and for the production of oxidase, desulfoviridin and indole, and for the hydrolysis of arginine and gelatin and for the reduction of nitrate.

detection of *Pectinatus* by using species-specific, genus-specific and group-specific primers (Juvonen et al., 2003, 2008; Motoyama and Ogata, 2000b; Sakamoto et al., 1997; Satokari et al., 1997; Satokari et al., 1998). The ribotyping fingerprinting technique has been used for the characterization and identification of *Pectinatus* species (Juvonen and Suihko, 2006; Motoyama et al., 1998; Suihko and Haikara, 2001). Motoyama et al. (1998) employed three different restriction enzymes (i.e., *Eco*RI, *Hin*dIII, and *Bam*HI) for ribotyping *Pectinatus cerevisiiphilus* and *Pectinatus frisingensis* strains. By using a combination of all these enzymes, all 34 *Pectinatus frisingensis* strains fell into one of 17 ribotypes. Five *Pectinatus cerevisiiphilus* strains were defined as one of three types with single enzymes, the composite ribotypes not showing greater definition. Suihko and Haikara (2001) used the *Eco*RI enzyme for the characterization of 24 *Pectinatus frisingensis* and eight *Pectinatus cerevisiiphilus* strains. *Pectinatus frisingensis* strains were divided into nine different ribotypes and *Pectinatus cerevisiiphilus* strains into five, allowing identification below the species level. Ribotyping with *Eco*RI divided the Finnish and German isolates of *Pectinatus haikarae* into separate ribotypes that could also be clearly distinguished from the ribotypes of *Pectinatus frisingensis* and *Pectinatus cerevisiiphilus* strains (Juvonen and Suihko, 2006; Suihko and Haikara, 2001).

The immunological characteristics of *Pectinatus frisingensis* and *Pectinatus cerevisiiphilus* have been studied using polyclonal and monoclonal antibodies for purposes of detection, subgrouping and identification (Gares et al., 1993; Haikara, 1983; Hakalehto, 2000; Hakalehto and Finne, 1990; Hakalehto et al., 1984; Ziola et al., 1999).

Unlike most bacterial species with a Gram-negative outer membrane structure, *Pectinatus frisingensis* and *Pectinatus cerevisiiphilus* have been found to be sensitive to cationic agents such as polymyxin B and nisin and to antibacterial agents with fairly large molecular mass such as bacitracin (Chihib et al., 1999; Helander et al., 1994). The sensitivity of *Pectinatus* can be related to an abnormal outer membrane which does not form an effective barrier and might have cracks permitting the entry of various compounds (Haikara et al., 1981b; Lee et al., 1978). In disc tests, *Pectinatus haikarae* and *Pectinatus frisingensis* are resistant to 5 µg vancomycin, whereas *Pectinatus cerevisiiphilus* is sensitive (Juvonen and Suihko, 2006).

The original source of *Pectinatus*, i.e., its habitat outside the brewery is not known. It has been speculated that *Pectinatus* bacteria have been carried to breweries with the various plant materials utilized (cereals, rice, corn, hops) because the lipopolysaccharide of *Pectinatus cerevisiiphilus* contains O-chains with D-6-deoxyhexose that is often found in plant-associated bacteria (Helander, 2003; Helander et al., 1992; Senchenkova et al., 1995). Most of the strains described and characterized have been isolated exclusively from spoiled, unpasteurized packaged beer. *Pectinatus* has sporadically been found in lubrication oil mixed with beer, in drainage and water pipe systems, in the air of the filling hall, in the filling machine, on the floor of the filling hall, in condensed water on the ceiling, in chain lubricants and in the steeping water before milling (Back et al., 1988; Dürr, 1983; Haikara, 1991; Haukeli, 1980; Juvonen and Suihko, 2006; Lee et al., 1980; Soberka and Warzecha, 1986; Suihko and Haikara, 2001). It is obvious that *Pectinatus*, despite its anaerobic nature, can survive in aerosols in the filling hall and be transferred via air into beer. Catalase activity of *Pectinatus haikarae* may help the survival of this species in aerobic conditions. In fact, two of the four characterized strains were isolated from the air of a brewery bottling hall (Juvonen and Suihko, 2006).

The growth of *Pectinatus frisingensis* and *Pectinatus cerevisiiphilus* in beer is dependent on the combined effects of several factors such as pH, alcohol and oxygen content, and hop bitter substances. In addition to the tolerance of *Pectinatus* bacteria towards acid and hops, they are also rather alcohol-tolerant, growing in beer with an alcohol content of 3.7–4.4% w/v although the growth is slower than in low-alcohol beer (< 2.25% w/v) (Haikara, 1984). Alcohol content exceeding 5.2% w/v prevents growth (Haikara et al., 1981b; Haukeli, 1980; Kirchner et al., 1980; Seidel et al., 1979). In culture medium, ethanol concentrations of 6.0–7.8% w/v (1.3–1.7 M) totally inhibit the growth of *Pectinatus frisingensis* and *Pectinatus cerevisiiphilus*, the former appearing less sensitive (Tholozan et al., 1997; Tholozan et al., 1996; Watier et al., 1996). The oxygen content of beer is the most decisive factor regarding its susceptibility to spoilage by *Pectinatus frisingensis* and *Pectinatus cerevisiiphilus*. The dissolved oxygen content as well as the volume of air in the headspace has decreased considerably in recent years due to advances in filling technology resulting in improved chemical stability of beer, but, on the other hand, facilitating the growth of strictly anaerobic bacteria such as *Pectinatus* and *Megasphaera* in unpasteurized beer. Factors affecting the growth of *Pectinatus haikarae* in beer have not been studied but all the spoilage incidents from which detailed information is available concern low-alcohol beer (Juvonen and Suihko, 2006; Juvonen unpublished data).

Enrichment and isolation procedures

As mentioned above, many agar media and broths containing fermentable sugars can be used for the cultivation and isolation of *Pectinatus* strains. Lee (1994) developed a "Selective Medium Megasphaera, Pectinatus" (SMMP)* for the isolation and differentiation of *Pectinatus* and *Megasphaera* species from other brewery organisms. The SMMP medium is beer-based, contains 1% lactate (v/v, final concentration) as a sole carbon source and is supplemented with reducing agents, various nutrients, 20 ppm actidione (cycloheximide) to inhibit yeasts, and 5 ppm crystal violet and 25 ppm sodium fusidate to inhibit Gram-stain-

positive bacteria. Enterobacteria are suppressed by alcohol and hop compounds present in beer, and hence appropriate levels of ethanol must be incorporated into SMMP when nonalcoholic beer is assayed (Lee, 1994).

The maintenance of strictly anaerobic conditions is important for the cultivation of *Pectinatus* bacteria. The anaerobic glove box and the GasPak system provide optimal growth conditions (Back et al., 1979; Haikara et al., 1981a, 1981b; Kirchner et al., 1980; Lee et al., 1978). Using reduced media and special reducing agents can enhance growth. In quality control of beer, forcing tests and enrichment are practical methods (Anonymous, 2001). In these methods, the development of turbidity of beer without or with added liquid medium, such as concentrated MRS broth, is followed at 27–30°C for 2–6 weeks and the presence of *Pectinatus* is confirmed microscopically (motile) and by smell (H_2S).

Maintenance procedures

Working cultures of *Pectinatus* can be maintained by subculturing (30°C, 2–3 d) at least every 2 weeks in PYF broth (0.5% peptone, 0.5% tryptone, 1% yeast extract, 0.5% fructose, 0.2% Na_2HPO_4, 0.1% Tween 80, and 0.05% cysteine hydrochloride added as a reducing agent). After incubation, the cultures are stored at 4°C in anaerobic jars. The working cultures can also be maintained in 5% dimethylsulfoxide in plastic screw-cap ampoules frozen at −70°C. For long-term preservation, conventional freeze-drying methods using 20% skim milk as protective agent as well as freezing in drinking straws in a −150°C deep freezer or in liquid nitrogen at −196°C using 5% dimethylsulfoxide as protective agent have been used. The polypropylene straws are packed after filling (0.1 ml) into 2-ml plastic screw-cap ampoules (e.g., Nunc) and placed directly into liquid nitrogen (Suihko and Haikara, 1990).

Differentation of the genus *Pectinatus* from other closely related taxa

The anaerobic, Gram-stain-negative rod-shaped bacteria originating from beer or the brewery environment belong to the genera *Pectinatus*, *Zymophilus*, *Zymomonas* and *Selenomonas*. *Pectinatus* can be differentiated from these other genera on the basis of its major metabolic end products. *Pectinatus* species produce propionic and acetic acid, H_2S, and acetoin; *Zymomonas* species produce acetaldehyde in addition to H_2S and the brewery *Selenomonas* species produce lactic acid. The differential characteristics of the brewery-related species belonging to the *Veillonellaceae* are given in Table 211.

Taxonomic comments

Pectinatus does not belong to the family *Bacteroidaceae*, as initially suggested by Lee et al. (1978), since the genus *Bacteroides* and related genera represent a distinct phylum within the phylogenetic tree of bacteria (Woese, 1987). According to *Bergey's Manual of Systematic Bacteriology* (Garrity and Holt, 2001; Ludwig et al., 2009), *Pectinatus* species, despite their negative Gram stain, belong to the class *Clostridia* of the phylum *Firmicutes* in the family *Veillonellaceae*. Cellular ultrastructure as well as several chemotaxonomic markers further support the inclusion of *Pectinatus* species into this family. At present three validly named *Pectinatus* species have been described. Very soon after the description of *Pectinatus cerevisiiphilus*

---

*Yeast extract, 75 g; Bacto-peptone, 75 g; DL-lactic acid, sodium salt (60% syrup), 75 ml; sodium thioglycollate, 0.75 g; L-cysteine·HCl, 0.75 g; K_2HPO_4·$3H_2O$, 7.5 g; KH_2PO_4, 7.5 g; NaCl, 7.5 g; $(NH_4)_2 HPO_4$, 7.5 g; $NaC_2H_3O_2$·$3H_2O$, 7.5 g; H_2O, 728.5 ml. B1 selective stock solution: sodium fusidate (Sigma or equivalent), 0.75 g; cycloheximide (Upjohn or equivalent), 0.60 g; crystal violet (Sigma or equivalent), 0.075–0.15 g; absolute ethanol, 100 ml.

(Lee et al., 1978), another strain (VTT E-79100, later *Pectinatus frisingensis* ATCC 33332[T]) was isolated and identified as the same species on the basis of its physiological and biochemical characteristics (Haikara et al., 1981b). However, this strain as well as many other new isolates, when compared to *Pectinatus cerevisiiphilus*, showed clear-cut differences in serological characteristics, cell surface-protein patterns and DNA–DNA hybridization (Haikara, 1983; Haikara, 1989; Hakalehto and Finne, 1990; Hakalehto et al., 1984). Finally, Schleifer et al. (1990) described the new species, *Pectinatus frisingensis*, using the 16S rRNA gene sequence analysis and confirmatory physiological tests. They also emended the description of *Pectinatus cerevisiiphilus*. In 1987, a *Pectinatus* strain DSM 20764 (= VTT E-97914) which was phenotypically similar to *Pectinatus cerevisiiphilus* but exhibited less than 20% DNA similarity with *Pectinatus frisingensis* and *Pectinatus cerevisiiphilus* was deposited from spoilt German beer to the DSMZ. Further genetic analyses performed on this and similar strains from spoiled beer or from the air of a brewery bottling hall (Sakamoto et al., 1997; Suihko and Haikara, 2001) supported their inclusion into a new species. In 2006, a novel species, *Pectinatus haikarae*, was described based on the sequence of the 16S rRNA gene sequence and unique biochemical and physiological properties. In addition, the description of the genus *Pectinatus* was emended (Juvonen and Suihko, 2006).

Further reading

Haikara, A. and I. Helander. 2002. *Pectinatus, Megasphaera,* and *Zymophilus*. *In* Dworkin (Editor), The Prokaryotes: An Evolving Electronic Database for the Microbiological Community, 3rd Ed. (release 3.5), Springer Verlag, New York. ISBN 0-387-14254-1.

List of species of the genus *Pectinatus*

1. **Pectinatus cerevisiiphilus**   Lee, Mabee and Jangaard 1978, 582[AL] emend. Schleifer, Leuteritz, Weiss, Ludwig, Kirchhof and Seidel-Rüfer 1990, 25

ce.re.vi.si.i′phi.lus. L. n. *cerevisia* beer; Gr. adj. *philos* loving N.L. adj. *cerevisiiphilus* beer-loving (bacteria).

The strains of this species form slightly curved rods that are 0.4–0.9 μm in diameter and 2–30 μm in length. They occur singly, in pairs, and only rarely in short chains. The occurrence of longer, helical filaments is characteristic for older cells. The cells are usually motile. Flagella emanate on only one side of the cell (comb-like). Cells stain Gram-stain-negative and do not form endospores. Colonies are circular, entire, beige to white, glistening, and opaque. Obligately anaerobic mesophiles with the optimum temperature for growth of 30–32°C. Glucose is fermented to acetic and propionic acid. Organic sulfur compounds produced. Catalase and cytochrome-oxidase-negative. Cadaverine and putrescine are found as characteristic components of the cell-wall peptidoglycan type, directly cross-linked *meso*-diaminopimelic acid. Lipid F is also found as a characteristic cellular compound. Strains of this species have been isolated from spoiled beer. The physiological characteristics useful for differentiating *Pectinatus cerevisiiphilus* from other species of the genus *Pectinatus* are given in Table 211.

DNA G+C content (mol%): 38–41 (T_m).

Type strain: ATCC 29359, DSM 20467, HAMBI 1348.

GenBank accession number (16S rRNA gene): AF373026.

2. **Pectinatus frisingensis**   Schleifer, Leuteritz, Weiss, Ludwig, Kirchhof and Seidel-Rüfer 1990, 25[VP]

fri.sin.gen′sis. L. adj. *frisingensis,* of Frisinga, the Latin name of Freising, the German town where the organism was isolated.

The phenotypic characteristics of *Pectinatus frisingensis* are similar to those described for *Pectinatus cerevisiiphilus* except that it ferments cellobiose, inositol, and *N*-acetylglucosamine but not xylose and melibiose. *Pectinatus frisingensis* can also be distinguished from *Pectinatus cerevisiiphilus* serologically, by cell-surface-protein patterns, lipopolysaccharide O-chain composition, DNA–DNA hybridization (similarity value 16%), and 16S rRNA gene sequencing. The physiological characteristics useful for differentiating *Pectinatus frisingensis* from other species of the genus *Pectinatus* are given in Table 211. Isolated from beer and brewery environment, the type strain is isolated in 1978 from Finnish beer.

DNA G+C content (mol%): 38–41 (T_m).

Type strain: VTT E-79100, ATCC 33332, DSM 6306, HAMBI 1347.

GenBank accession number (16S rRNA gene): AF373027.

3. **Pectinatus haikarae** Juvonen and Suihko 2006, 701[VP]

Pectinatus haikarae (hai.ka′ rae. N.L. gen. fem. n. *haikarae* of Haikara, named after Auli Haikara for her many contributions to the characterization and detection of *Pectinatus* species).

Cells are 0.6–0.8 μm by 3–50 μm or more in size but otherwise resemble morphologically *Pectinatus cerevisiiphilus*. Growth occurs at 15–30°C, but not at 10°C or 37°C. The optimum temperature lies between 20 and 30°C. Good growth (3+ or 4+ on a scale of 0 to 4+) in PYF, PYG, SMMP and MRS media after 1–2 d at 30°C. Weak growth also observed on supplemented Brucella Blood agar and PY medium. Colonies on PYF and PYG plates after 3 d at 30°C are from convex to pyramidal, glossy, opaque, cream to gray in color, and circular with entire margins and a diameter of 0.5–2.5 mm. Major products of fructose fermentation are propionic and acetic acids. Acetoin and H_2S are also produced. Differential characteristics compared to closely related anaerobic Gram-stain-negative rod-shaped bacteria isolated from breweries are shown in Table 211.

Isolated from spoiled German and Finnish beer and from the air of a brewery bottling hall in Finland.

DNA G+C content (mol%): 38.8 (T_m).

Type strain: VTT E-88329[T], DSM 16980[T].

GenBank accession number (16S rRNA gene): DQ223731.

Genus XVI. **Phascolarctobacterium** Del Dot, Osawa and Stackebrandt 1994, 182[VP]
(Effective publication: Del Dot, Osawa and Stackebrandt 1993, 383.)

ERKO STACKEBRANDT AND RO OSAWA

Phas.co.larc′to.bac.te′ri.um. Genus name *Phascolarctos* Koala; Gr. neut. dim. n. *bakterion* rod; N.L. neut. n. *Phascolarctobacterium* bacterium of Koalas.

Pleomorphic rods, $0.5 \times 2.0\,\mu m$ to $0.5 \times 5-20\,\mu m$. Pleomorphism is medium dependent; in the presence of succinate small rods are transformed into elongated and fragmented rods with multiple branches. Gram-stain-negative, nonmotile, nonspore-forming chemo-organotroph. **Propionic acid is the main end product of succinate fermentation**. Growth inhibited by fumarate. Growth of satellite colonies around colonies of succinate-producing *Escherichia coli* cells. **Obligately anaerobic.** Growth most rapid at 30–37°C. Phylogenetically a member of the family *Veillonellaceae*, phylum *Firmicutes*. So far *Succiniclasticum ruminis* is the closest phylogenetic neighbor.

DNA G+C content (mol%): 41.4–42.3 (T_m).

Type species: **Phascolarctobacterium faecium** Del Dot, Osawa and Stackebrandt 1994, 182[VP] (Effective publication: Del Dot, Osawa and Stackebrandt 1993, 383.).

Further descriptive information

On Wilkins–Chalgren Anaerobe agar (WCA), *Phascolarctobacterium* strains showed satellite growth around colonies of strains of *Escherichia coli* but not around colonies of strains of *Streptococcus bovis* (Figure 214). Enhanced growth of highly pleomorphic cells was due to the presence of succinic acid produced by *Escherichia coli* (Osawa et al., 1992) (Figure 215). Although a maximum growth response of *Phascolarctobacterium faecium* strains to succinate was reached at 50–100 mmol/l, growth was completely inhibited at 200 mmol or above. Fumarate has a strong inhibitory effect on growth. Active production of propionic acid from succinate-supplemented broth also suggests that AK strains are capable of using the succinate-propionate pathway, which is employed by most propionate-producing organisms including *Propionibacterium* spp., *Selenomonas ruminantium*, and *Veillonella alcalescens* (de Vries et al., 1973; Hilpert and Dimroth, 1984).

Phylogenetic studies by 16S rDNA (Del Dot et al., 1993) revealed that the organism is related to members of cluster IX of clostridia (Collins et al., 1994). While the original description indicated *Acidaminococcus fermentans* as the nearest neighbor (92.8% similarity), recent expansion of the 16S rDNA database also shows *Succiniclasticum ruminis* (van Gylswyk, 1995) and *Succinispira mobilis* (Janssen and O'Farrell, 1999) to be similarly closely related (93.3% similarity) (Figure 216).

Enrichment and isolation procedures

Enrichment of *Phascolarctobacterium* can be accomplished by emulsifying freshly voided feces in 0.25-strength Ringer solution. Serial 10-fold dilutions are plated onto Wilkins-Chalgren Anaerobe (WCA) agar (Oxoid) supplemented with sodium succinate (20 mmol/l). Dilutions are incubated anaerobically at 37°C for 5 d in an atmosphere of $CO_2/H_2/N_2$ (10:10:80, by volume), using a Bio-Bag (Becton Dickinson Co., Cockeysville, MD, USA) onto which cultures of *Escherichia coli* are stamped separately with the tip of a sterile cotton swab. Small transparent colonies grow predominantly on the plates inoculated with higher dilutions (10^{-5} and 10^{-6}) of koala feces adjacent to colonies of *Escherichia coli*.

Maintenance procedures

Phascolarctobacterium faecium is maintained anaerobically on WCA plate medium supplemented with sodium succinate (20 mmol/l).

Differentiation of *Phascolarctobacterium* from other taxa

The distant 16S rDNA sequence similarity determined for *Phascolarctobacterium faecium* and its closest phylogenetic neighbors (<94%) makes determination of this sequence the most straightforward approach to identification (see Figure 216 for accession numbers). Table 212 lists characteristics features that can be used to differentiate cells from those of phylogenetically related genera. Cluster IX of clostridia contains a broad spectrum of Gram-negatively and Gram-positively staining genera, none of which closely matches the description of *Phascolarctobacterium faecium*.

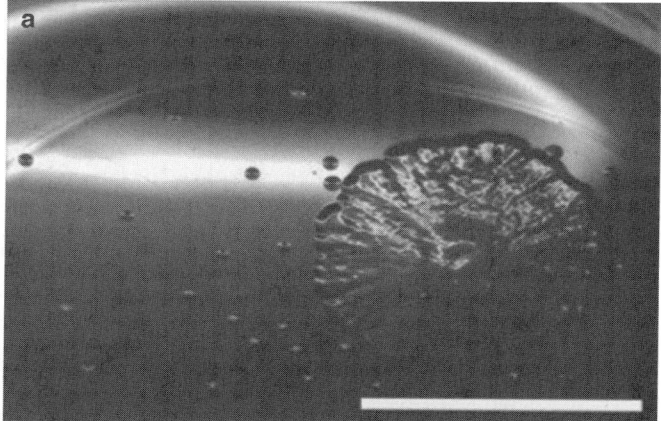
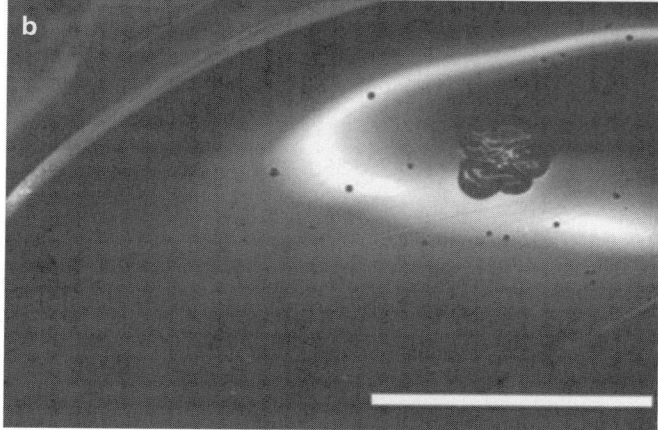

FIGURE 214. Stimulated growth of *Phascolarctobacterium* AK strain near a colony of *Escherichia coli* on WCA with 5% defibrinated horse blood. Colonies of *Phascolarctobacterium faecium* strain ACM 3679[T] around a colony of *Escherichia coli* (a) and a colony of *Streptococcus bovis* (b).

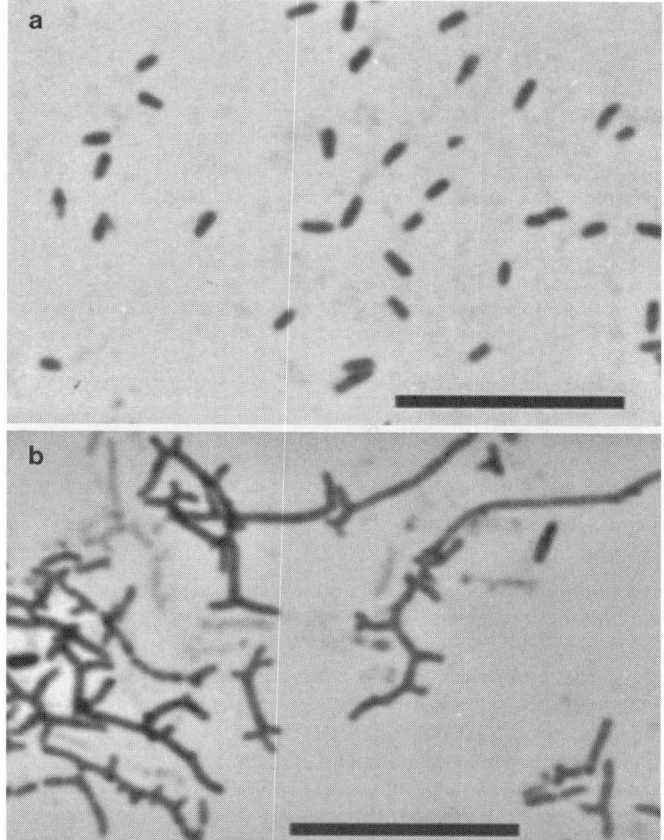

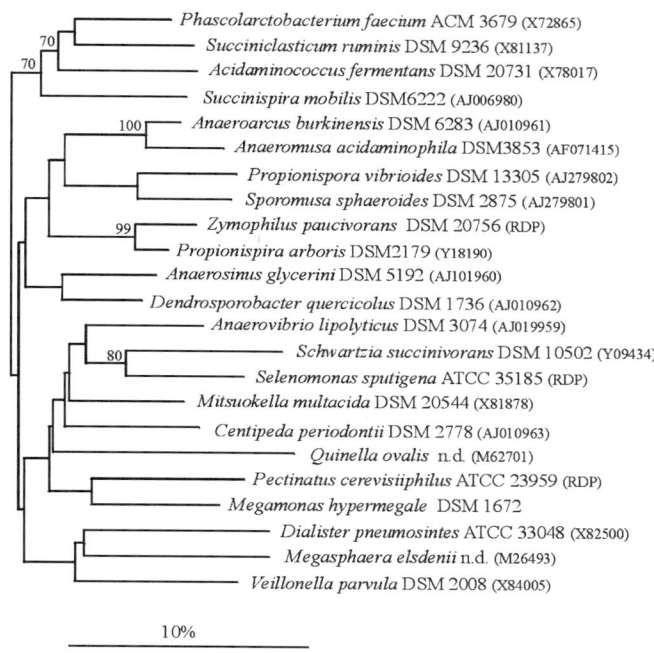

FIGURE 216. Phylogenetic position of *Phascolarctobacterium faecium* to some related bacteria of the *Sporomusa* subbranch as defined by the taxonomic browser of NCIB (http://www.ncib.nlm.nih.gov). The dendrogram was constructed by the neighbor-joining analyisis of 16S rDNA sequences. Bar represents 10% sequence divergence.

FIGURE 215. Photomicrographs of cells of *Phascolarctobacterium faecium* strain ACM 3679[T] from 3-d-old surface colonies on WCA with 5% defibrinated horse blood (a) and WCA supplemented with succinate (20mmol/l) (b). Bars = 10 μm.

TABLE 212. Differentiation of *Phascolarctobacterium* from phylogenetic neighboring genera

Characteristics	*Phascolarctobacterium*	*Succiniclasticum*	*Acidaminococcus*	*Succinispira*
Morphology	Rods, pleomorphic, branching	Short rods	Coccus, diplococci	Curved rods, spiral
Motility	−	−	−	+
DNA G+C content (mol%)	41–42	52	56	36
Major fermentation end products	Propionate[a]	Propionate, CO_2	Acetate, butyrate, CO_2	Formate, acetate, propioniate, malate, CO_2, H_2
Ability to ferment carbohydrates	−	−	−	−
Peptides, amino acids as main energy source	−	−	+	−

[a] Production of CO_2 has not been reported by Del Dot et al. (1993).

Further reading

Osawa, R. and T. Mitsuoka.1990. Fecal microflora of captive koalas, *Phascolarctos cinereus* (MARSUPIALIA; PHASCOLARCTIDAE). Aust. Mammal. *13*: 141–147.

Kanegasaki, S. and H. Takahashi. 1978. Function of growth factors for rumen micro-organisms. II. Metabolic fate of incorporated fatty acids in *Selenomonas ruminantium*. Biochim. Biophys. Acta *152*: 40–49.

Stackebrandt, E. and H. Hippe. 2001. Taxonomy and Systematics. *In* Bahl and Dürre (Editors), Clostridia, Biotechnology and Medical Applications, Wiley-VHC, Weinheim, pp. 19–48.

List of species of the genus *Phascolarctobacterium*

1. **Phascolarctobacterium faecium** Del Dot, Osawa and Stackebrandt 1994, 182[VP] (Effective publication: Del Dot, Osawa and Stackebrandt 1993, 383.)

fae′ci.um N.L. gen. pl. n. *faecium* of feces, isolated from the feces of koala.

The characteristics are the same as those described for the genus. Colonies grown on succinate are transparent, convex, smooth, 0.5 mm in diameter. Nonhemolytic on blood agar with 5% defibrinated horse blood. Cells grown in Wilkins-Chalgren medium and peptone/yeast-extract medium have uniform small rods ($0.5 \times 2.0 \mu m$), while pleomorphic nocardoid forms (0.5×5–$20 \mu m$) develop in the presence of succinate. Grows in 20% bile. Catalase-negative, nitrate reduction negative. No production of indole, urease, alkaline phosphatase, or pyroglutamic acid arylamidase. No acid from fructose, glucose, or mannose. Kanamycin and colistin sensitive, vancomycin-resistant. Isolated from feces of Koala.

DNA G+C content (mol%): 41.4–42.3 (T_m).

Type strain: ACM 3679, DSM 14760.

GenBank accession number (16S rRNA gene): X72865.

Genus XVII. **Propionispira** Schink, Thompson and Zeikus 1983, 673[VP] (Effective publication: Schink, Thompson and Zeikus 1982, 2778.)

BERNHARD SCHINK

Pro.pi.o.ni.spi′ra. N.L. n. *acidum propionicum* propionic acid; L. fem. n. *spira* a coil; N.L. fem. n. *Propionispira* a propionic acid forming coil.

Curved to helical rods, $1 \mu m \times 5$–$9 \mu m$. Spore formation not observed. Stains Gram-negative. **Motile by means of one terminally inserted flagellum. Strictly anaerobic, fermentative metabolism.** Optimum temperature 30–35°C. Chemo-organotrophic, using a wide variety of compounds as energy sources including lactate, glucose, lactose, sorbitol, xylose, and amygdalin. Fermentation end products are propionic acid, acetic acid, and carbon dioxide. Ethanol and succinate are formed under special conditions. Fumarate is reduced to succinate or propionate; sulfate and nitrate not reduced. Cytochrome *b* and nitrogenase are present; catalase is absent. Occurs in anoxic wetwoods in softwood trees such as cottonwood, poplar, or willows. The type strain was isolated from wetwood of a mature cottonwood tree located in Columbus, WI, USA.

DNA G+C content (mol%): 36.7 ± 1 (T_m).

Type species: **Propionispira arboris** Schink, Thompson and Zeikus 1983, 673[VP] (Effective publication: Schink, Thompson and Zeikus 1982, 2778.).

Further descriptive information

Propionispira arboris clusters with Gram-positive bacteria of the low G + C group within the family *Veillonellaceae*, in the order *Clostridiales* (Stackebrandt et al., 1999). According to the 16S rRNA phylogenetic analysis presented in the roadmap to this volume (Figure 6), the genus *Propionispira* is a member of the family *Veillonellaceae*, order *Clostridiales*, class *Clostridia* in the phylum *Firmicutes*.

Propionispira arboris stains Gram-negative, but belongs to the Gram-positive bacteria on the basis of 16S rRNA sequence analysis (Stackebrandt et al., 1999). Cells are coiled to curved rods and are motile by means of terminally inserted flagella. The temperature range for growth is 4–46°C with an optimum temperature at 30–35°C for the type strain. The pH range for growth is 4.5–8.0 with an optimum of 6.0–6.5.

Propionispira arboris ferments arabinose, cellobiose, galactose, glucose, lactose, mannose, mannitol, raffinose, galacturonate, sorbitol, sucrose, glycerol, lactate, and amygdalin. Inositol, maltose, melibiose, melizitose, rhamnose, trehalose, arabinogalactan, cellulose, pectin, polygalacturonic acid, starch, xylan, citrate, pyruvate, tartrate, esculin, salicin, Casamino acids, gelatin, and tryptone are not fermented.

Growth with glucose is substantially faster in complex media than with ammonium chloride or N_2 as nitrogen source. The cell yield on any carbon substrate tested is highest in complex medium and lowest if N_2 is the sole source of nitrogen.

Propionispira arboris is so far the only representative among the strictly anaerobic propionic-acid-forming bacteria that is able to fix molecular nitrogen. Addition of ammonium chloride to cultures inhibits nitrogenase activity. After growth in the presence of bound nitrogen, no nitrogenase activity is detectable.

The main products of carbohydrate fermentation are propionate, acetate, and CO_2. Ethanol and succinate are formed in trace quantities only when glucose is fermented. Ethanol is a more significant product of xylose metabolism, and succinate is a major end product when fumarate is added to glucose medium. H_2, lactate, and other soluble and gaseous compounds were not detected as fermentation products. Sugars are degraded through the Embden–Meyerhof pathway; propionic acid is formed via the methylmalonyl-CoA pathway (Thompson and Zeikus, 1988).

Enrichment and isolation procedures

Propionispira arboris can be isolated selectively with glucose or other carbohydrates as the carbon and energy source and N_2 as the sole source of nitrogen in reduced media under strictly anoxic conditions. Wetwoods of softwood trees appear to be a natural source of this bacterium. Enrichments are incubated at 30°C. After 1–2 transfers, typical coil-shaped cells will dominate and can be selectively purified under anoxic conditions on agar plates or in agar shake dilution series.

Maintenance procedures

Pure cultures may be stored anoxically in liquid medium at 4°C in the dark. Stock cultures should be transferred every 1–2 months. For long-term preservation, a dense cell suspension of *Propionispira arboris* may be suspended in anoxic medium containing glycerol or dimethyl sulfoxide, sealed in sterile glass capillaries, and stored in liquid nitrogen.

Differentiation of the genus *Propionispira* from other genera

The genus *Propionispira* is distinguished from other fermentative bacteria by its phylogenetic position, its unusual cell shape, and its capacity to ferment carbohydrates exclusively to propionate, acetate, and CO_2, and to fix molecular nitrogen in purely synthetic mineral media. With these properties, *Propionispira* differs substantially from organisms of similar morphology and metabolic capacities such as *Pectinatus* (which cannot fix N_2), from the classical club-shaped propionic acid bacteria, and from most members of the *Bacteroidetes*.

List of species of the genus *Propionispira*

1. **Propionispira arboris** Schink, Thompson and Zeikus 1983, 673[VP] (Effective publication: Schink, Thompson and Zeikus 1982, 2778.)

   ar′ bo.ris. L. masc. n. *arbor* tree; gen. *arboris* of a tree, referring to the occurrence of this bacterium in wetwood.

   The description is as for the genus. Isolated with glucose as only electron donor and N_2 as only nitrogen source. Occurs in anoxic wetwoods of cottonwood trees.

   DNA G+C content (mol%): 36.7 ± 1 (T_m).
   Type strain: 12B4, ATCC 33732, DSM 2179.
   GenBank accession number (16S rRNA gene): Y18190.

Genus XVIII. **Propionispora** Biebl, Schwab-Hanisch, Spröer and Lünsdorf 2001, 793[VP]
(Effective publication: Biebl, Schwab-Hanisch, Spröer and Lünsdorf 2000, 246.)

HANNO BIEBL

Pro.pi.on.i.spo′ra. N.L. n. *acidum propionicum* propionic acid; Gr. n. *spora* a seed; N.L. fem. n. *Propionispora* a propionic-acid-forming, spore-forming organism.

Curved or spiral-shaped rods. Motile by a **tuft of flagellae inserted at the concave side** of the cell (Figure 217). Forms round, terminally located **endospores**. **Gram-stain-negative.** Chemo-organotrophic and anaerobic. **Ferments sugars and sugar alcohols to propionic acid, acetic acid, and CO_2.** Requires yeast extract.

DNA G+C content (mol%): 42.3–48.7.

Type species: **Propionispora vibrioides** Biebl, Schwab-Hanisch, Spröer and Lünsdorf 2001, 793[VP] (Effective publication: Biebl, Schwab-Hanisch, Spröer and Lünsdorf 2000, 246.).

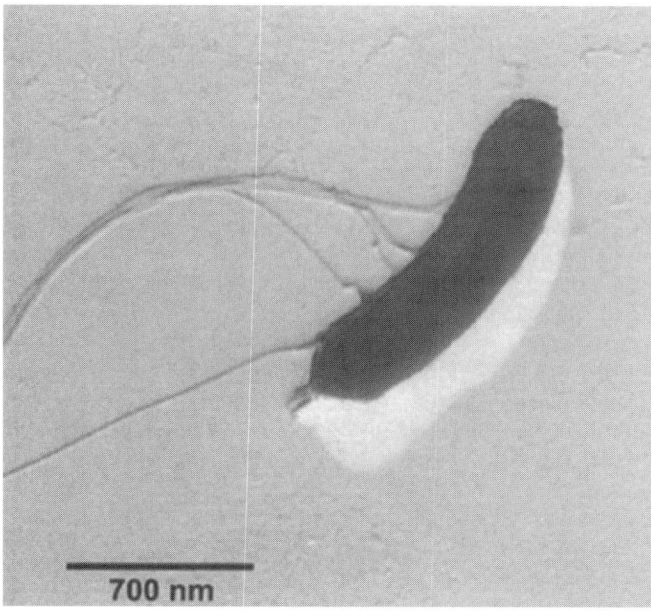

FIGURE 217. *Propionispora vibrioides.* Platinum-carbon shadow-cast preparation of a non-sporing cell showing six flagella inserted at the concave

Further descriptive information

Colonies and centrifuge pellets appear slightly pink from unknown pigments. The number of fermentable sugars and sugar alcohols is limited (see list of species); pyruvate and lactic acid are not used. In addition to propionic and acetic acid, some hydrogen is also produced to balance redox differences between substrate and cell mass. Cultures tolerate exposure to air.

Enrichment and isolation procedures

A common enrichment procedure cannot be given for the genus because the two existing strains were obtained in very different manners.

Maintenance procedures

Both species can be maintained in the anaerobically prepared medium of Biebl et al. (2000) containing erythritol or fructose as substrate. Long-term storage can take place in 10% glycerol at–70°C or by freeze-drying.

Differentiation from other related genera

Propionispora is distinguished from most of the other Gram-stain-negative anaerobic genera by its unique morphology (see generic definition) which is shared with *Selenomonas* and *Sporomusa* (Biebl et al., 2000; Kingsley and Hoeniger, 1973; Möller et al., 1984). It differs from *Selenomonas* by forming spores and from *Sporomusa* by not being acetogenic. *Dendrosporobacter* and *Acetonema*, which are also anaerobic Gram-stain-negative spore-formers, are straight rods and exhibit peritrichous flagellation; *Acetonema* performs an acetogenic type of fermentation.

Taxonomic comments

16S rRNA gene sequences indicate some relationship to *Dendrosporobacter quercicolus* (93% similarity) and to *Sporomusa* species (91%). Other members of the *Sporomusa–Pectinatus–Selenomonas* group are relatively close to these species (88–90%).

List of species of the genus *Propionispora*

1. **Propionispora vibrioides** Biebl, Schwab-Hanisch, Spröer and Lünsdorf 2001, 793[VP] (Effective publication: Biebl, Schwab-Hanisch, Spröer and Lünsdorf 2000, 246.)

vi.brio.i′des. N.L. masc. n. *vibrio* that which vibrates, a generic name; Gr. n. *eidos* shape; N.L. adj. *vibrioides* vibrio-shaped.

Cell shape, flagella insertion, and spore characteristics as in the genus description. Cell size 0.6×2.2–$6\,\mu m$. Cells only slightly swollen during sporulation. Growth and fermentation substrates are polyols such as erythritol, xylitol, mannitol, sorbitol; fructose is also fermented, but not glucose, xylose, or glycerol. Biotin allows weak growth in synthetic medium, but does not entirely replace yeast extract. Good growth between 30 and 40°C, optimum at 37°C; pH range 5.0–8.5, optimum at 7.5.

The existing strain was isolated from an enrichment culture for anaerobic degradation of erythritol using composted fatty material as an inoculum.

DNA G+C content (mol%): 48.7 (HPLC).

Type strain: FKBS1, DSM 13305.

GenBank accession number (16S rRNA gene): AJ279802.

2. **Propionispora hippei** Abou-Zeid, Biebl, Spröer and Müller 2004, 953[VP]

hip′pe.i. N.L. n. *hippei* in honor of Hans-H. Hippe, a German microbiologist and anaerobe specialist.

Cell shape, structure, and size as for *Propionispora vibrioides*. Spore-forming cells distinctly swollen. Fermented substrates as listed for *Propionispora vibrioides*, but glucose and glycerol are used in addition. The temperature range for growth is between 20 and 50°C with an optimum at 40°C; the pH ranges between 5.0 and 8.2 with an optimum at 7.0.

The existing strain was isolated from an enrichment culture for anaerobic degradation of polypropylene adipate using sewage from a German treatment plant as inoculum. The strain is able to depolymerize the polyester to 1,3-propanediol and adipic acid, but cannot attack these products. DNA–DNA hybridization with *Propionispora vibrioides* FKBS1 is 47%.

DNA G+C content (mol%): 42.3 (HPLC).

Type strain: KS, ATCC BAA-665, DSM 15287.

GenBank accession number (16S rRNA gene): AJ508927, AJ508928.

Genus XIX. **Quinella** Krumholz, Bryant, Brulla, Vicini, Clark and Bryant 1993, 295[VP]

Lee R. Krumholz, Wolfgang Buckel, Gregory M. Cook and James B. Russell

Quin.el′la. N.L. fem. n. Quinella named for the pioneering ruminologist, J.I. Quin, who described the organism in some detail (Quin, 1943).

Cells are **Gram-stain-negative** and **oval shaped**. Size is 3–4×5–$8\,\mu m$ in size but can be considerably smaller in the rumen if crowding occurs. **Tumbling motility** with linear tufts of flagella on one side of the cells. **Endospores are not formed. Heterotrophic** and **mesophilic**. Reproduces by binary fission, and it usually occurs as singles and pairs. A small number of **carbohydrates are fermented** mainly to lactate, acetate, propionate, and CO_2. Contains a glycogen-like reserve material.

DNA G+C content (mol%): unknown.

Type species: **Quinella ovalis** Krumholz, Bryant, Brulla, Vicini, Clark and Bryant 1993, 295[VP].

Further descriptive information

This genus has not been axenically cultivated and therefore this description is based on studies with highly purified cell suspensions obtained through differential centrifugation (Wicken and Howard, 1967). Small-subunit rRNA sequence information has shown that the closest cultivated relatives of Quin's oval are members of the genus *Selenomonas*, including rumen strains of *Selenomonas ruminantium* and oral strains of *Selenomonas gingivalis*. However, the evolutionary distance separating these organisms is too large to justify placing them in the same genus.

In addition to being observed in rumens of sheep fed molasses (Vicini et al., 1987) and in rumens of llamas (Orpin, 1972), cloned rRNA genes from members of the genus *Quinella* have been retrieved from rumens of a yak as well as a gaur (*Bos frontalis*). The cell density observed in the rumen is strongly influenced by the amount of fermentable sugars (e.g., glucose or sucrose) in the animal's diet (Brough et al., 1970; Orpin, 1972; Quin, 1943; Vicini et al., 1987), with numbers ranging from 10^5–10^8 cells/ml in hay-fed sheep to 10^{11} cells/ml in sheep fed mainly molasses (Vicini et al., 1987).

Photomicrographs that represent the morphology have been published (Quin, 1943; Vicini et al., 1987; Wicken and Howard, 1967). The cell wall contains a distinct outer membrane as well as muramic and *m*-diaminopimelic acids (Wicken and Howard, 1967).

When sugars do not limit growth, lactic acid is produced as the major fermentation product. However when present in the rumen of molasses-fed sheep, when sugars are limiting and growth is slow, acetate, propionate, and CO_2 are the major fermentation products (Vicini et al., 1987). Under these conditions, no lactate is detectable at any time during the 24-h feeding cycle.

List of species of the genus *Quinella*

1. **Quinella ovalis** Krumholz, Bryant, Brulla, Vicini, Clark and Bryant 1993, 295[VP]

o.val′is. L. fem. adj. *ovalis* egg-shaped.

Cells are oval and are about 3–$4\,\mu m \times 5$–$8\,\mu m$ in size but can be considerably smaller in the rumen if crowding occurs (Vicini et al., 1987). Highly enriched and partially purified cell suspensions ferment glucose, fructose, sucrose, and mannitol and slowly metabolize maltose. Polysaccharides, amino acids, proteins, lactate, melibiose, mannose, glucosamine, galactose, rhamnose, cellobiose, lactose, glucuronic acid, xylose, arabinose, and soluble starch are not fermented or are fermented at extremely low rates (Brough et al., 1970; Orpin, 1972).

Rapid growth occurs at 37–39°C, and slow growth occurs at 44°C. No growth occurs at 25 or 50°C (Orpin, 1972). It grows in the rumen at pH values at least as low as 6.0. The higher pH limit is not known.

Species have been obtained from the rumen of molasses-fed sheep.

DNA G+C content (mol%): unknown.

Type strain: the species has not been axenically cultivated, and therefore this description is the type of the species (in accordance with Rule 18f of the *International Code of Nomenclature of Bacteria*).

GenBank accession number (16S rRNA gene): M62701.

Acknowledgements

Marvin Bryant's love of anaerobic bacteria, his insistence on taxonomic classification, and his hard work are responsible for this description.

Genus XX. **Schwartzia** van Gylswyk, Hippe and Rainey 1997, 158[VP]

FRED A. RAINEY

Schwart'zi.a. L. fem. n. *Schwartzia* named in memory of Helen M. Schwartz, a South African rumen physiologist who had a keen interest in rumen microbiology.

Gram-stain-negative, curved rods, motile by lateral flagella. **Non-spore-forming. Anaerobic.**

Asaccharolytic and do not ferment amino acids or peptides. Ferments succinic acid with the production of propionic acid. Non-proteolytic, catalase-negative, urease-negative. Nitrate is not reduced. **Mesophilic.**

DNA G+C content (mol%): 46 (UV).

Type species: **Schwartzia succinivorans** van Gylswyk, Hippe and Rainey 1997, 158[VP].

Further descriptive information

The description of the genus *Schwartzia* and the single species *Schwartzia succinivorans* is based on the properties of four strains S1-1[T] (DSM 10502[T]), strain S2-3 (DSM 10503), strain S3-2 (DSM 10504), and strain S4-2 (DSM 10505). The cells stain Gram-negative, are slightly curved to curved rods that have tapered ends. The strains are motile by means of flagella, which can be single or in tufts and are found in the middle of the concave side of the cells. The cells vary in width (0.35–0.6 μm) and length (1.6–3.3 μm). S-shaped cells can be up to 5 μm long. Long filaments greater than 30 μm might be undivided cells. Colonies on the surface of agar slopes (40% rumen fluid, 0.2% yeast extract, 1% succinate, and 1.5% agar) are 0.25 mm in diameter, colorless, round, raised, and smooth edged after 24 h of incubation. Subsurface colonies are 0.1 mm in diameter and round discs. After 48 h of incubation, surface colonies are 5 mm in diameter while subsurface colonies are 0.3 mm wide.

Substrate utilization is limited to succinate for all of the four strains tested. Succinate is fermented to propionic acid. These strains failed to ferment galactose, glucose, fructose, mannose, arabinose, xylose, cellobiose, maltose, sucrose, glycerol, mannitol, the sodium salts of DL-lactic and pyruvic acids, the disodium salts of oxalic, malic, malonic, methylmalonic, fumaric, oxaloacetic, and glutaric acids, and the trisodium salts of citric, *trans*-aconitic, and tricarballylic acids. Growth is not supported by amino acids and peptides even when digests of casein are added to the medium. Rumen fluid and yeast extract are required to obtain good growth on succinate. A reducing agent is not required to obtain good growth if the medium is prepared under anaerobic conditions. If air is added to the headspace, growth is not good if the medium has not been reduced using

a reducing agent. No growth is obtained for the four strains at 22°C. Different amounts of growth are observed for each of the four strains when incubated 45°C.

Enrichment and isolation procedures

The basal medium used for the isolation of *Schwartzia* contains (per liter) 0.225 g of K_2HPO_4, 0.225 g of KH_2PO_4, 0.45 g of NaCl, 0.45 g of $(NH_4)_2SO_4$, 0.045 g of $CaCl_2$ (anhydrous), 0.09 g of $MgSO_4 \cdot 7H_2O$, 6.36 g of $NaHCO_3$, 0.25 g of cysteine hydrochloride $\cdot H_2O$, 0.25 g of Na_2S (hydrated), 0.005 g of indigo carmine, and 400 ml of clarified rumen fluid. This basal medium with 2% agar added is preincubated at 39°C for one week after which 0.5% succinate, $NaHCO_3$ and reducing agent are added before sterilization. Samples of the rumen ingesta are diluted to 10^{-7}, and 0.5-ml aliquots are injected into sterile CO_2-purged roll bottles to which 2.5 ml of the molten preincubated and sterilized succinate agar is added. After 5 d incubation, large colonies are selected and their growth on succinate further tested on basal medium containing 40% clarified rumen fluid and 0.5% yeast extract with or without 1% succinate. Microscopic examination of the cultures allows for selection of cells with curved rod morphologies.

Taxonomic comments

The genus *Schwartzia* was described on the basis of its phenotypic, chemotaxonomic, and phylogenetic properties. Based on 16S rRNA gene sequence comparisons, strain S1-1[T] was shown to fall within the radiation of the Gram-stain-negative bacteria of the genera *Selenomonas*, *Pectinatus*, and *Zymophilus* and represents a distinct lineage with no strong affiliation to any previously described genus (van Gylswyk et al., 1997). Comparison of the 16S rRNA gene sequence of strain S1-1[T] with the currently available database confirms this original finding (see Figure 206). 16S rRNA gene sequence similarities are in the range 90.6–92.4% to species of the genera *Selenomonas*, *Mitsuokella*, and *Centipeda*. *Schwartzia succinivorans* falls outside the cluster comprising most species of the genus *Selenomonas*, and its relationship to any of the species/genera clusters is not supported by bootstrap analysis. The genus *Schwartzia* can be differentiated from related genera by its utilization of a limited range of substrates; the majority of species of related genera can utilize a wide range of substrates. The only other Gram-stain-negative anaerobe that is limited to succinate as a fermentable carbon source is *Succiniclasticum ruminis*, but this taxon is phylogenetically unrelated to strain S1-1^T (see *Succiniclasticum* chapter, below). In the original description of the genus *Schwartzia* the case was made that it had a different fatty acid pattern than species of related genera. These data need to be redetermined with all strains grown under the same conditions to further evaluate this claim. Differences in mol% G+C content of the DNA have also been considered to be a differentiating characteristic.

List of species of the genus *Schwartzia*

1. **Schwartzia succinivorans** van Gylswyk, Hippe and Rainey 1997, 158^{VP}

suc.cin.i.vo′rans. L. neut. n. *succinicum* amber; N.L. neut. n. *acidum succinum* succinic acid; L. part. *vorans* devouring; L. part. adj. *succinivorans* succinic acid-devouring.

Cells are curved motile rods with one or more flagella emanating from the concave side in the middle of the cell. Flagella occur in tufts if more than one is present. Cells are 1.6–3.3 μm in length and 0.35–0.6 μm in diameter. S-shaped cells and helical filaments up to 30 μm long occur. Colonies on succinate agar are colorless, slightly raised, and have multiple lobes and smooth edges. Colonies are up to 5 mm in diameter after 48 h growth. The optimum temperature for growth is 39°C, and no growth occurs at 22°C and limited growth at 45°C. Succinate is fermented to propionate. Rumen fluid and yeast extract are required for good growth. Glycerol, lactate, pyruvate, oxalate, malate, malonate, methylmalonate, fumarate, oxalacetate, glutarate, citrate, aconitate, tricarballylate, and amino acids and peptides present in acid or enzymic hydrolysates of casein are not utilized. Catalase, urease, sulfide production, nitrate reduction, and gelatin liquefaction negative. The major cellular fatty acids are: $C_{15:0}$ aldehyde, $C_{16:1}$ *cis*7, $C_{12:0}$, $C_{14:0}$, $C_{15:0}$, and $C_{16:0}$. Branched-chain fatty acids are absent. Type strain was isolated from rumen ingesta of cows on pasture.

DNA G+C content (mol%): 46.0 (UV).

Type strain: S1-1, DSM 10502.

GenBank accession number (16S rRNA gene): Y09434.

Genus XXI. **Selenomonas** Von Prowazek 1913, 36^{AL}

YOGESH S. SHOUCHE, ABHIJIT S. DIGHE, DHIRAJ P. DHOTRE, MILIND S. PATOLE AND DILIP R. RANADE

Se.le.no.mo′nas. Gr. n. *selene* the moon; Gr. n. *monas* a unit, monad; N.L. fem. n. *Selenomonas* moon (-shaped) monad.

Curved to helical rods, usually 0.9–1.1 × 3.0–6.0 μm. The ends are usually tapered and rounded to give short kidney- to crescent-shaped or **vibrioid cells.** Long cells and chains of cells are often helical. Capsules are not formed. Resting stages are not known. Gram-stain-negative. **Motile with active tumbling; flagella (up to 16) are arranged linearly as a tuft near the center of the concave side in the area of cell fission. Strictly anaerobic.** Optimum temperature 35–40°C; maximum 45°C; minimum 20–30°C. Chemoorganotrophic, **having a fermentative type of metabolism. Carbohydrates and lactate can serve as fermentable substrates. Fermentation of glucose chiefly yields acetate, propionate, CO_2, and/or lactate.** Small amounts of H_2 and succinate may be produced. Catalase negative.

DNA G+C content (mol%): 40–61.

Type species: **Selenomonas sputigena** (Flügge 1886) Boskamp 1922, 70^{AL} (*Spirillum sputigenum* Flügge 1886, 387).

Further descriptive information

This description is based in part upon the chapter in the first edition of the *Systematics* by Bryant (1984). Cells show varying morphology; typically they are vibrioid to helical to crescent shape. They occur singly, in pairs, and short chains and may occur in clumps, especially in *Selenomonas sputigena* (Kingsley and Hoeniger, 1973). The cells are usually 1 μm × 2.0–4.0 μm long, however, some could be larger, 2.0–3.0 × 5–10 μm (Prins, 1971). The ultrastructure and flagella have been studied extensively for *Selenomonas sputigena*, *Selenomonas Palpitans*, and *Selenomonas ruminantium*. The flagella are present as a tuft arranged linearly near the center of the concave side of the organism.

The number can vary from 4–5 (*Selenomonas lipolytica*) to 22 (*Selenomonas palpitans*). A typical flagellum is 17 nm thick and 7 μm long, with a center to center distance between adjacent flagella of 68 nm.

Much information has been published on the cell envelope, membrane structure, and lipid chemistry of *Selenomonas ruminantium* (see Takatsuka and Kamio, 2004 for further references). Under the electron microscope, the *Selenomonas ruminantium* outer membrane is wrinkled and has several irregular blebs. These are formed at several sites where the membrane is pinched off, forming vesicles of various sizes. The cell wall is not as complex as that seen in the *Proteobacteria*, and the addition of very low concentrations of lysozyme leads to rapid cell lysis. *Selenomonas ruminantium* does not have Braun lipoprotein which has an important role in the maintenance of the structural integrity of proteobacterial outer membranes (Kamio and Takahashi, 1980). Instead, *Selenomonas ruminantium* has cadaverine, which is covalently linked to the peptidoglycan and is an essential component for cell division (Kamio et al., 1981b). The major phospholipids of both outer and inner membranes from *Selenomonas* are phosphatidyl ethanolamine, ethanolamine plasmalogens, and glyceryl ether type ethanolamines, while phosphatidyl glycerol or cardiolipins are absent in both membranes. Types of phosphatidyl ethanolamine plasmalogen present in outer and inner membranes are quite different. One of the important lipid fractions present in *Selenomonas* membranes is 2-keto-3-deoxyoctulosonic acid-lipid A (Kamio and Takahashi, 1980). This lipid has phosphorus, 2-keto-3-deoxyoctulosonic acid and 3-OH fatty acid but no detectable levels of glycerol.

The outer membrane from both glucose- and lactate-grown cells contain two major peptidoglycan-associated proteins with apparent molecular masses of 42,000 and 40,000. The amount of each protein varies considerably, depending upon the culture conditions. These are novel lysine decarboxylases that are responsible for synthesis of cadaverin (Takatsuka and Kamio, 2004). A regulatory protein "P22" is found to be associated with lysine decarboxylase, and it is a stimulatory factor for the degradation of the enzyme. Lipopolysaccharides from cell envelopes of *Selenomonas* also play a role as the virulence factors in human periodontal diseases (Kumada et al., 1997).

Selenomonas species show wide variation in colony morphology. Colonies typically are visible after 48–60 h of growth. *Selenomonas ruminantium* var. *bryanti* produces glassy, round, lenticular, transluscent water-droplet-like colonies, which become opaque and larger as they grow, becoming 6 mm in diameter after 5–6 d. *Selenomonas ruminantium* strains are strongly iodophillic and accumulate reserve carbohydrate granules in the cytoplasm on addition of fermentable sugars to growth media. *Selenomonas noxia* colonies on rabbit blood agar after 48 h are minute to 0.5 mm in diameter, circular, entire, low convex to flat, transparent, smooth, shiny, and colorless to yellowish. *Selenomonas fluggei* colonies are 0.5–1.0 mm in diameter, circular, entire or spreading, opaque, white, shiny, and smooth. They sometimes form a spreading growth over the entire plate. *Selenomonas infelix* colonies are 0.5–1.0 mm, circular, entire, pulvinate, transluscent, shiny, and smooth. It may also produce a spreading growth. *Selenomonas dianae* colonies are minute, entire, circular, shiny, smooth, and gray to white. *Selenomonas artemidis* colonies are minute to 0.5 mm in diameter, circular, entire, low convex, transluscent, smooth, shiny, and colorless. *Selenomonas sputigena* colonies on blood-agar media are generally 0.5–1.2 mm in diameter, smooth, convex, and gray to gray-yellow in color. *Selenomonas lipolytica* colonies on PYG agar are light brown, spherical, 2 mm in diameter, opaque, and convex with defined borders. Fully grown colonies possess elevated darker pointed centers.

Selenomonas species except *Selenomonas lipolytica* require volatile fatty acids, such as *n*-valerate, for growth when glucose is the energy source (Kanegasaki and Takahashi, 1967). CO_2 is required for growth on lactate or glycerol but not on glucose in complex media containing trypticase and yeast extract. Strains of *Selenomonas ruminantium* require L-aspartate, CO_2, *p*-aminobenzoic acid, and biotin for growth in a lactate-salts medium, and aspartate can be replaced by malate or fumarate (Evans and Martin, 1997). Medium containing small amounts of dithiothreitol is reported to be better for the growth of *Selenomonas*. *Selenomonas ruminantium* cannot grow on aspartate, fumarate, or malate in the absence of lactate, but it can grow on all three organic acids in the presence of extracellular hydrogen gas (Martin, 1998). Sulfide can be the sole source of sulfur (Linehan et al., 1978).

Much information has been published on *Selenomonas* metabolism and metabolic pathways – see Ricke et al. (1996) for further references. *Selenomonas* cannot degrade complex polysaccharides but can metabolize simpler carbohydrates released from the initial hydrolysis of these polymers. Selenomonads have multiple routes for carbon flow during carbohydrate catabolism and ATP generation. Most *Selenomonas* species, like *Selenomonas ruminantium*, *Selenomonas sputigena*, and *Selenomonas lipolytica* utilize glucose by the Embden–Meyerhof–Parnas glycolytic pathway (EMP) and produce acids, such as acetate, lactate, propionate, and succinate (Melville et al., 1988). *Selenomonas ruminantium* ferments glucose to lactate, acetate, propionate, and CO_2. The amounts and proportions of these end products depend upon the growth and nutritional conditions. In batch culture, with glucose as the carbon source, the cells first produce DL-lactate. Some strains of *Selenomonas* can utilize lactate and produce acetate and propionate (Nisbet and Martin, 1994).

The pathways for these transformations are complex. Glucose is first catabolized to phosphoenolpyruvate, which is converted to either pyruvate or oxaloacetate. Pyruvate is either reduced to L-lactate or oxidatively decarboxylated to acetyl coenzyme A and CO_2. Acetyl-CoA is converted to acetate and CoA by acetate thiokinase with formation of ATP (Melville et al., 1988). Oxaloacetate is converted to succinate that is either excreted or decarboxylated to propionate. Enzymes such as pyruvate kinase, acetate thiokinase, PEP carboxykinase, fumarate reductase, and methylmalonyl-CoA decarboxylase are involved in energy conservation in *Selenomonas ruminantium* (Melville et al., 1988). Strains of *Selenomonas ruminantium* can ferment xylose and xylo-oligosaccharides, and they possess xylosidase and arabinosidase that are induced by these sugars (Cotta and Whitehead, 1998). Different *Selenomonas* strains have the ability to produce trace amounts of hydrogen gas. *Selenomonas ruminantium* accumulates intracellular polysaccharide as an energy reserve during starvation (Russell, 1998).

Selenomonas can utilize ammonia, protein, and amino acids as a nitrogen source for protein synthesis. Selenomonads can use urea as a source of nitrogen, and most *Selenomonas* species produce urease (Smith et al., 1981). *Selenomonas* strains are known to utilize glutamate dehydrogenase, glutamine synthetase–glutamate synthase pathways for ammonia assimilation (Smith et al., 1981).

The phylogenetic structure of the genus has been studied using 16S rRNA gene sequences, and the genus was found to be phylogenetically coherent with interspecies similarity levels of 90–99% (Figure 218). *Selenomonas dianae*, *Selenomonas infelix*, *Selenomonas flueggei*, *Selenomonas noxia*, and *Selenomonas artemedis* form a tight cluster with sequence similarities in the range of 96–99%. *Selenomonas sputigena* and *Selenomonas ruminantium* have a mean sequence similarity of 94% with members of this cluster. DNA hybridization has been used to delineate oral species of *Selenomonas* which have a narrow range of G + C content of their DNA (53–58 mol%).

Selenomonas ruminantium contains several plasmids, ranging in size from 1.4 kb to >30 kb. At least seven of these have been sequenced (Al-Khaldi et al., 1999; Sprincova et al., 2005). Two phages, one temperate and one lysogenic, have been reported from this organism (Cheong and Brooker, 1998; Lockington et al., 1988).

The antigenicity of *Selenomonas* has not been studied in detail for either identification or classification. Monoclonal antibodies against *Selenomonas ruminantium* have been used for ELISA detection (Broker and Stokes, 1990). Antisera against the 42 kDa heat modifiable protein from cell envelopes of *Selenomonas ruminantium* cross-react with other *Selenomonas* species (Kalmokoff et al., 2000). This antigen forms a regularly ordered array in the outer membrane and is very similar to outer-membrane proteins in certain *Proteobacteria*.

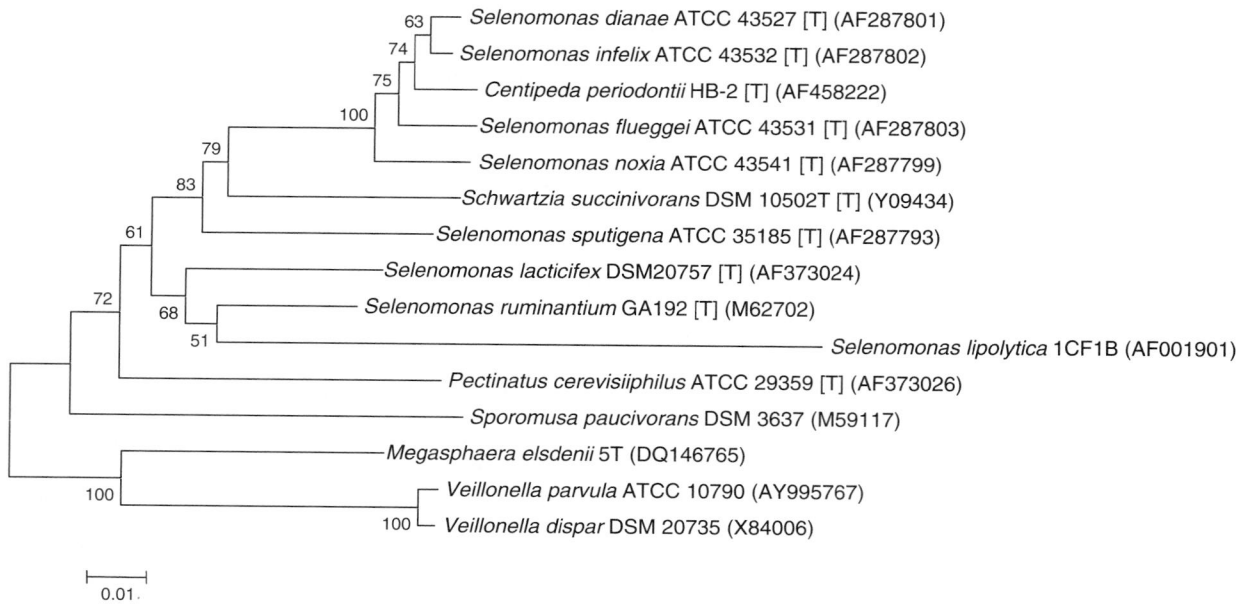

FIGURE 218. Rooted phylogenetic relationships of members of the genus *Selenomonas* and related species based on 16S rRNA gene sequences. The topology was inferred by MEGA3.1 using neighbor-joining method with Kimura 2-parameter as a model of nucleotide substitution with 1000 bootstrap replicates. Tree was rooted using *Veillonella* as an outgroup. Bootstrap values are shown at each internal node. Accession numbers are shown after each species name in parentheses. Scale bar represents substitutions per site.

Most *Selenomonas* species are sensitive to various antibiotics that act against protein and cell-wall biosynthesis. *Selenomonas sputigena* is resistant to penicillin G (Walker et al., 1985), while *Selenomonas lipolytica* is resistant to erythromycin (Dighe et al., 1998).

McCarthy et al. (1981) reported the first isolation of *Selenomonas sputigena* from blood culture of a patient with septicemia. Clones representing 16S rRNA gene sequences from *Selenomonas* have since been found associated with endodontic infections, dental caries, and in subgingival plaques (Munson et al., 2004; Paster et al., 2001; Rolph et al., 2001).

Selenomonads are believed to be one of the first bacteria observed by Antonie van Leeuwenhoek in gingival scrapings from the human mouth (Dobell, 1960, as quoted by Bryant, 1984), and selenomonads are considered as part of the normal indigenous microflora of human gingival crevices. The type species *Selenomonas sputigena* was isolated from the human oral cavity by Flügee ((1886) as quoted by Bryant, 1984). Originally named *Spirillum sputigenum*, the generic name *Selenomonas* was suggested by Von Prowazek ((1913), as quoted by Bryant, 1984), and *Spirillum sputigenum* was renamed as *Selenomonas sputigena* by Boskamp (1922). Although its function in the oral cavity is not readily understood, its saccharolytic activity presumably plays some role. The selenomonads are often more abundant in individuals with clinically detectable gingivitis or periodontal disease. The pure lipopolysaccharide from selenomonads possesses endotoxin properties in mice (Kurimoto et al., 1986). Some occurrences of *Selenomonas sputigena* and other selenomonads in human septicemia also have been reported (McCarthy and Carlson, 1981; Pomeroy et al., 1987).

Selenomonas ruminantium is a very common species in the rumens of cows and sheep, and many strains of the species are described (Bryant, 1956; Hobson and Mann, 1961; Prins, 1971).

Animal feed containing rapidly fermentable carbohydrates such as grains leads to more rapid growth of selenomonads in the rumen than feed low in rapidly fermentable carbohydrates, such as straw or silage (Caldwell and Bryant, 1966). These bacteria are also responsible for glycerol fermentation in the sheep and cattle rumen (Bryant, 1956; Hobson and Mann, 1961). Only a few species of ruminal bacteria ferment lactate. The lactate-fermenting strain of *Selenomonas ruminantium* is designated *Selenomonas ruminantium* subsp. *lactilytica* (Bryant, 1956). Animal bloat resulting from feed high in grain is associated with lower ruminal pHs, increased lactate formation, and substantial increases in the number of selenomonads.

Selenomonads are also present in swine intestine. Approximately 21% of the total bacteria isolated from the cecal contents of healthy swine are strains of *Selenomonas ruminantium* (Robinson et al., 1981). These organisms are also found in swine feces (Salanitro et al., 1977). Presumably, like in the rumen, a functional role for selenomonads in the swine intestine involves fermentation of soluble sugars. The dominance of *Selenomonas ruminantium* strains in the swine intestine may be related to the availability of lactate.

Selenomonads are also found in the cecum of a number of small rodents. About 5% of the total bacterial isolates from rat cecal contents are selenomonads (Ogimoto, 1972). Morphologically typical selenomonads have been observed in cecal contents of squirrels. As many as 10^9–10^{10} cells per gram of cecal contents were isolated from cecal contents of the ground squirrel, *Citellus tridecemlineatus* (Barnes and Burton, 1970). Selenomonads are also reported in the cecum of guinea pigs (Kingsley and Hoeniger, 1973; Robinow, 1954). Simons reported the presence of "*Selenomonas palpitans*" from the cecum and feces of guinea pigs. The electron micrographs of this isolate reveal that it differs in some respects in cell morphology from *Selenomonas ruminantium* (Kingsley and Hoeniger, 1973). However, in the absence of a pure culture, the existence of this species is questionable.

Other environments from which selenomonads have been isolated include river water (Leifson, 1960) and ditch water from a bog habitat (Harborth and H. H. Hanert, 1982). A chemostat culture developed with anaerobic sewage sludge possessed selenomonads (Nanninga et al., 1987). *Selenomonas lipolytica*, which possesses lipolytic activity, was isolated from a ditch in an oil mill (Dighe et al., 1998). All these studies suggest that selenomonads are present in many naturally anaerobic habitats.

Isolation

Selenomonads are obligately anaerobic bacteria. The exclusion of oxygen at every stage of isolation, transfer, and preservation is an essential prerequisite. The principles of anaerobic techniques described by Hungate (1969) and subsequently modified by Miller and Wolin (1974) are employed.

The nutritional requirements of selenomonads are simple. These bacteria grow well in medium containing glucose, ammonia, minerals, vitamins, and sulfide. Addition of *n*-valeric acid and trypticase enhances growth. An O_2-free atmosphere of N_2 or CO_2 or mixture of both of these gases is essential for isolation and growth of these bacteria. Media like PYG (Holdeman et al., 1977), MPB (Kingsley and Hoeniger, 1973), and OA-1 (Dighe et al., 1998) are useful in isolation of selenomonads. Hespell et al. (1992) suggest a modified selective medium (SS) for isolation of *Selenomonas ruminantium*. Tiwari et al. (1969) used SS medium for isolation of ruminal selenomonads. The medium is selective because of the acidic pH (6.0) and presence of mannitol as the sole carbohydrate. This medium allows growth of selenomonads and inhibits growth of many other ruminal bacteria that are unable to use mannitol. Isolation with glycerol or lactate allows preferential growth of *Selenomonas ruminantium* subsp. *lactilytica*. MacDonald and Madlener (1957) recommend the use of complex medium containing sodium lauryl sulfate (0.01%) or sodium oleate (0.15%) and sheep serum (10%). These additions suppress the growth of many micro-organisms present in gingival scrapings without affecting selenomonad growth. The complex media such as brain heart infusion agar are used for such isolations.

The OA-1 medium described by Dighe et al. (1998) seems to be a good choice for enrichment and isolation of lipid-degrading selenomonads. The medium has less protein and more edible oil. Anaerobic lipolytic bacteria therefore get better opportunity to grow in this medium.

Maintenance procedures

Cultures of *Selenomonas* can be preserved for long periods in liquid nitrogen and for 18 months in medium with 6% (w/v) glycerol at −20°C. For some strains, storage by lyophilization under anaerobic conditions is possible.

Differentiation of the genus *Selenomonas* from closely related taxa

Two major characteristics of the genus *Selenomonas*, flagellar arrangement and production of propionate from glucose fermentation are useful in differentiating the genus from many other genera of anaerobic, curved, motile, Gram-stain-negative, nonsporulating rods. *Selenomonas* possess a tuft of flagella

arranged in a closely spaced linear fashion near the middle of the concave side of the cell. Propionate is the major end product of glucose fermentation. It differs from the genus *Pectinatus* in that the latter has linearly arranged flagella that may be placed all along one side of the cell. *Selenomonas* species do not grow at 15°C, whereas *Pectinatus* grows between 15 and 40°C. *Selenomonas* species usually ferment esculin, lactose, raffinose and sucrose, whereas *Pectinatus* strains do not. Also, the mol% G+C of the DNA of *Selenomonas* is 54–61, in contrast to a value of ~40 for *Pectinatus*. The only species that has mol% G+C of the DNA similar to *Pectinatus* is *Selenomonas lipolytica*. The two genera also have markedly different cellular fatty acid profiles; C_{12} fatty acids are absent in *Pectinatus*, and the ratios of other fatty acids are also very different in the two genera.

Miscellaneous comments

In the 16S rRNA gene sequence based phylogenetic analyses, two single species genera, *Centipeda* (Lai et al., 1983) and *Schwartzia* (van Gylswyk et al., 1997) also fall within the radiation that includes species of *Selenomonas* (Figure 218). However, these species are defined as separate genera based on low levels of DNA hybridization with members of *Selenomonas* and on other morphological and chemotaxonomic characters.

Similarly, *Selenomonas acidaminovorans* was found to be distantly placed from other members of the genus with low level of sequence similarity. In the phylogenetic analysis it formed an independent line of descent. This, together with its thermophilic nature and other metabolic properties, led to the creation of the new genus *Thermanaerovibrio*, and it is now called *Thermanaerovibrio acidaminovorans* (Baena et al., 1999).

Taxonomic comments

Current phylogenetic analyses of 16S rRNA genes of *Selenomonas* suggest that it is a member of the family *Veillonellaceae* (Ludwig et al., 2009). The detailed comparative studies of *Selenomonas sputigena* and *Selenomonas ruminantium* by Kingsley and Hoeniger (1973) indicate that these bacteria should not be placed in *Spirillum* or *Aquaspirillum* (MacDonald et al., 1959).

Differentiation of the species of the genus *Selenomonas*

Some differential features of the species of *Selenomonas* are indicated in Table 213.

Acknowledgements

This chapter is based in part on the genus *Selenomonas* chapter of the first edition of *Bergey's Manual of Systematic Bacteriology*, authored by the late Dr M.P. Bryant. The authors recognize his contributions to this manuscript. The authors also wish to acknowledge Dr M.V. Deshpande, National Chemical Laboratory, Pune, India for careful reading of the manuscript and constructive criticism.

Further reading

Johnson, J.L., Holdeman, L.V. and Moore, W.E.C. 1985. Replacement of the type strain of *Selenomonas sputigena* under Rule 18g. Request for an Opinion. Int. J. Syst. Bacteriol. *35*: 371–374

TABLE 213. Characteristics differentiating the species of the genus *Selenomonas*[a]

Characteristic	S. sputigena	S. artemidis	S. dianae	S. flueggei	S. infelix	S. lacticifex	S. noxia	S. ruminantium
Requirement for valerate	+	+	+	+	+	−	+	+
DNA G + C content (mol%)	57	58	53	56	58	51–52	57	48–53
Acid production from:								
Cellobiose	−	−	−	−	+	−	−	+
Lactose	+	−	+	+	+	−	−	v
Mannitol	−	+	+	+	+	−	−	+
Sucrose	+	+	+	+	+	+	+	+
Lactic acid as major fermentation product	−	−	−	−	−	+	−	v

[a]Symbols: +, >85% positive; −, 0–15% positive; v, variable.

Wayne, L.G. 1994 Actions of the Judicial Commission of the International Committee on Systematic Bacteriology on Requests for Opinions published between January 1985 and July 1993. Int. J. Syst. Bacteriol. *44*: 177–178

List of species of the genus *Selenomonas*

1. **Selenomonas sputigena** (Flügge 1886) Boskamp 1922, 70[AL] (*Spirillum sputigenum* Flügge 1886, 387)

spu.ti´ge.na. L. n. *sputum* spit, sputum; L. v. *gigno* to produce; N.L. fem. adj. sputigena sputum-produced.

Colonies on anaerobic blood agar are generally smooth, convex, grayish yellow in color, sometimes mottled, and sometimes with a more opaque center. Usually the diameter is less than 0.5 mm. Sometimes the colonies are smooth, granular, flat, grayish, and translucent with an irregular border, and up to 2 mm in diameter.

In Brewer's thioglycolate broth (Difco), growth occurs as heavy floccules and coarse granulates (MacDonald et al., 1953). In MPB broth (Kingsley and Hoeniger, 1973) growth is turbid, often with a granular sediment.

Selenomonas sputigena is a strong fermenter of certain carbohydrates. Its ability to utilize amino acids or many other potential energy sources has not been determined. It does not usually produce H_2S from cysteine, suggesting that it does not degrade cysteine.

Glucose is fermented with production of propionate and acetate (Loesche et al., 1965), and may produce some lactate and small amounts of succinate. No clear cut evidence on the ability of the strain to produce H_2 is reported.

Selenomonas sputigena appears not to be pathogenic for mice or guinea pigs. Intravenous injection into rabbits has caused some symptoms and even death in some cases but not in others (MacDonald et al., 1959). Although *Selenomonas sputigena* is found in the human gingival crevice, its relationship to the etiology of periodontal disease has not been established.

DNA G+C content (mol%): 61 (Bd).

Type strain: VPI 10068, ATCC 35185, CCUG 44933, DSM 20758, VPI D19B-28.

GenBank accession number (16S rRNA gene): AF287793, AF373026.

2. **Selenomonas artemidis** Moore, Johnson and Moore 1987, 279[VP]

ar.te´mi.dis Gr. gen. n. *aretemidos* of Artemis, goddess of the moon. N.L. gen n. *artemidis* of Artemis, referring to crescent shape of the cells.

Colonies on blood agar plate after 48 h are minute to 0.5 mm, circular, entire, low convex, translucent, smooth, shiny, and colorless. Abundant growth with slight turbidity and granular to nonflaky sediment in PYG cultures, whereas slight growth is seen in PY broth. Acetate and propionate are produced in PY as well as PYG media. No ammonia is detected in either medium, but a small amount is detected in chopped meat cultures. Threonine is not converted to propionate and lactate is not utilized.

DNA G+C content (mol%): 56–59 (T_m).

Type strain: ATCC 43528, JCM 8543, VPI D22B-14.

GenBank accession number (16S rRNA gene): not reported.

3. **Selenomonas dianae** Moore, Johnson and Moore 1987, 279[VP]

di.a´nae. L gen. n. *dianae* of Diana, goddess of the moon, referring to crescent shape of the cells.

Colonies on rabbit blood agar plate are minute, circular, entire, convex to flat and slightly spreading, transparent, shiny, smooth, and gray-white. Strains produce excellent growth in PYG broth and granular to flaky sediments and good growth in PY broth. Acetate, propionate, pyruvate, and lactate are produced in PYG medium; acetate and propionate are produced in PY medium. No ammonia is detected in either medium, but moderate amounts are detected in chopped meat cultures. Threonine is not converted to propionate and neither lactate nor pyruvate is utilized.

DNA G+C content (mol%): 53–59 (T_m).

Type strain: ATCC 43527, JCM 8542.

GenBank accession number (16S rRNA gene): AF287801.

4. **Selenomonas flueggei** Moore, Johnson and Moore 1987, 274[VP]

fluegge´i. N.L. gen. n. *flueggei* of Flügge, an early German bacteriologist.

Colonies on rabbit blood agar plate after 48 h are 0.5–1 mm in diameter, circular, entire to spreading, opaque, white, smooth, and shiny. Sometimes growth spreads over entire plate. Moderate turbidity is produced in PY broth and abundant growth with smooth sediment and clear broth is seen in PYG cultures. Acetate, lactate, and propionate are produced in both PY and PYG media, although molar ratios in each medium could be different. No H2 is produced. In PYG medium, fermentation products are acetate and propioniate, whereas in PY medium the products are acetate, propionate, and lactate. Moderate ammonia is produced in chopped meat cultures but none in PY or PYG cultures. Threonine is not converted to propionate, and lactate is not utilized.

DNA G+C content (mol%): 54–60 (T_m).
Type strain: ATCC 43531, JCM 8544.
GenBank accession number (16S rRNA gene): AF287803.

5. **Selenomonas infelix** Moore, Johnson and Moore 1987, 277[VP]

in.fe'lix. L adj. *infelix* causing harm, unlucky.

Colonies on rabbit blood agar plate after 48 h are 0.5–1 mm in diameter, circular, entire, pulvinate, transluscent, shiny, smooth, and white. Sometimes growth spreads as a thin film over entire plate. Moderate turbidity is produced in PY broth and abundant growth with turbidity and smooth sediment in PYG broth. Acetate and propionate are produced in PY and PYG broths with different proportions. Threonine is not converted to propionate, and lactate is not utilized. Slight ammonia production is detected in these as well as in chopped meat cultures.

DNA G+C content (mol%): 56–58 (T_m).
Type strain: ATCC 43532, JCM 8545.
GenBank accession number (16S rRNA gene): AF287802.

6. **Selenomonas lacticifex** Schleifer, Leuteritz, Weiss, Ludwig, Kirchhof and Seidel-Rüfer 1990, 26[VP]

lac.ti'ci.fex. N.L. n. *acidum lacticum* lactic acid; L. suff. *fex* a maker; N.L. n. *lacticifex* a maker of lactic acid.

Colonies on modified MRS medium are smooth, opaque, circular, yellowish, and 2–3 mm in diameter after 72 h at 30°C. Glucose is fermented to acetic, lactic, and propionic acids. It is distinguished from other lactic acid-producing species by lower G + C content and its inability to utilize mannitol and dulcitol.

DNA G+C content (mol%): 51–52 (T_m).
Type strain: VB4b, ATCC 49690, DSM 20757.
GenBank accession number (16S rRNA gene): AF373024.

7. **Selenomonas lipolytica** Dighe, Shouche and Ranade 1998, 790[VP]

li.po.ly'ti.ca. Gr. n. *lipos* fat; Gr. adj. *lutikos* dissolving; N.L. fem. adj. *lipolytica* fat-dissolving.

Acetate and propionate are the only volatile fatty acids produced after fermentation of glucose, with propionate as the major end product. The strain has strong and true lipolytic activity, with degradation zones on tributyrin, triolein, and groundnut oil in qualitative plate clearance assays. The strain utilizes glycerol but not lactate. Does not require n-valerate for growth in glucose.

DNA G+C content (mol%): 40 (T_m).
Type strain: CF1B, MCMB 505.
GenBank accession number (16S rRNA gene): AF001901.

8. **Selenomonas noxia** Moore, Johnson and Moore 1987, 273[VP]

no'xia L. fem. adj. *noxia* harmful.

Colonies on rabbit blood agar plates appear after 48 h and are minute to 0.5 mm in diameter, circular, entire, low convex to flat, transparent, smooth, shiny, and colorless to yellowish. Strains produce slight turbidity in Peptone Yeast extract broth but abundant growth in PYG with smooth sediment. Growth in PYG medium produces acetate and propionate and in PY medium acetate, propionate, and lactate. Pyruvate and lactate are not utilized. Trace to abundant H2 production is observed.

DNA G+C content (mol%): 56–58 (T_m).
Type strain: ATCC 43541, JCM 8546, VPI D9B-5.
GenBank accession number (16S rRNA gene): AF287799.

9. **Selenomonas ruminantium** (Certes 1889) Wenyon 1926, 311[AL] (*Ancyromonas ruminantium* Certes 1889, 70)

ru.mi.nan'ti.um. N.L. pl. gen. n. *ruminantium* of ruminants.

On RGCA-medium roll tubes, colonies are entire, slightly convex, translucent, lightly tan, and 2–4 mm in diameter after incubation for 3 d at 37°C; deep colonies are thin and lenticular. Growth in liquid medium is heavily turbid, often with some lightly flocculent sediments (Bryant, 1956).

All strains of the species so far studied can utilize glucose as the energy source, ammonia as the nitrogen source, and sulfide as the sulfur source. They grow very well in a chemically defined media containing these three components along with B-vitamins, n-valerate, and minerals (Linehan et al., 1978). Some strains require only biotin among the B-vitamins, and some strains do not require *n*-valerate. Some strains can utilize amino acids such as aspartate, histidine, serine, or cysteine as sole nitrogen sources. Cysteine can also serve as a sole sulfur source. A few strains contain urease and can utilize urea as a sole nitrogen source (John et al., 1974; Robinson et al., 1981). *Selenomonas ruminantium* subsp. *lactilytica* strains, which require *n*-valerate when grown on glucose, do not require it for growth on lactate but have added requirements for aspartate and *p*-aminobenzoate (Kanegasaki and Takahashi, 1967; Linehan et al., 1978). Nitrate is reduced to ammonia in a few strains and thus serves as a sole nitrogen source (John et al., 1974), but other strains reduce nitrate only to nitrite or do not reduce nitrate at all.

Propionate, acetate, CO_2, and lactate (L- and sometimes a mixture of D- and L-; Scheifinger et al., 1975) are major end products of glucose fermentation while small amounts of H_2 are also produced. The amount of lactate produced as compared to propionate and acetate depends on the strain, the amount of glucose in the medium and, in chemostat cultures, the growth rate (Scheifinger et al., 1975). Strains may also produce some succinate. The amount of H_2 produced is more in presence of H_2-utilizing methane-producing bacteria (Scheifinger et al., 1975).

DNA G+C content (mol%): 54 (Bd).
Type strain: ATCC 12561, DSM 2150, NADL GA-192, JCM 6583.
GenBank accession number (16S rRNA gene): M62702.

9a. **Selenomonas ruminantium subsp. ruminantium** Bryant 1974, 425[AL]

The description is as for the species. Differs from the subspecies *lactilytica* by being unable to ferment lactate and glycerol.

DNA G+C content (mol%): 54 (Bd).
Type strain: ATCC 12561, DSM 2150.

GenBank accession number (16S rRNA gene): M62702.

9b. **Selenomonas ruminantium subsp. lactilytica** (Hungate 1966) Bryant 1956, 165[AL] (*Selenomonas lactilytica* Hungate 1966, 68)

lac.ti.ly′ti.ca. L. n. *lac, lactis* milk; Gr. adj. *lutikos* dissolving; N.L. adj. *lactilytica* milk (lactate) dissolving.

This subspecies possesses the same characteristics as the species except that lactate and glycerol are fermented. Lactate is fermented with the production of propionate, acetate, and CO_2 (Bryant, 1956). Glycerol is fermented mainly to propionate with small amounts of lactate, succinate, and acetate (Hobson and Mann, 1961).

"O" antisera from sheep strains have been found to react with a bovine strain of the same subspecies but not with a bovine strain of the subspecies *ruminantium*, suggesting that the two subspecies may differ in "O" antigens (Hobson et al., 1962).

Type strain: ATCC 19205, DSM 2872, JCM 6582.

Other organisms

The following organisms either have no available strains or have not been obtained in pure culture.

1. **"Selenomonas ruminantium subsp. bryanti"** Prins 1971, 825

bry.an′ti. N.L. gen. n. *bryanti* of Bryant.

This subspecies possesses characteristics similar to subspecies *ruminantium* except that the cells in pure culture are larger, measuring 2.0–3.0–5.0–10.0 μm. H_2S is not produced from cysteine, and arabinose, xylose, galactose, lactose, and dulcitol are not fermented.

This subspecies was isolated using anaerobic rumen-fluid agar medium containing minerals similar to those of RGCA medium and deep agar tubes rather than roll tubes. The rumen fluid source material was differentially centrifuged to enrich the large selenomonads before culture (Prins, 1971).

No type strain has been designated and none of the originally described strains are now available.

2. **"Selenomonas palpitans"** Simons 1920, 50

pal′pi.tans. L. part. adj. *palpitans* trembling.

Simons (1920) first observed an organism similar to Selenomonas in the cecum of guinea pigs and it was later named Selenomonas palpitans (Boskamp, 1922; Simons, 1920). It has not been obtained in pure culture, however, and in a study of the cytology of the organism from guinea pig cecal contents, Kinglsey and Hoeniger (1973) found that the flagella of the tuft were bunched together in a single area rather than forming a line of insertions as is the case in *Selenomonas*. The further taxonomic status of this organism must be delayed until it is isolated in pure culture.

Genus XXII. **Sporomusa** Möller, Ossmer, Howard, Gottschalk and Hippe 1985, 224[VP] (Effective publication: Möller, Ossmer, Howard, Gottschalk and Hippe 1984, 394.)

HAROLD L. DRAKE AND ANITA S. GÖSSNER

Spo.ro.mu′sa. Gr. fem. n. *spora* a spore; N.L. fem. n. *musa* a banana; N.L. fem. n. *Sporomusa* spore-bearing banana.

Mesophilic, **acetogenic, anaerobic**. Cells are **banana shaped**, flagellated on the concave side of the cell, and form **endospores** that tend to be subterminal (Figure 219). Gram-stain-negative or weakly Gram-stain-positive. Growth can occur under both **chemo-organotrophic** and **chemolithoautotrophic** conditions.

DNA G+C content (mol%): 42.0–48.6.

Type species: **Sporomusa sphaeroides** Möller, Ossmer, Howard, Gottschalk and Hippe 1985, 224[VP] (Effective publication: Möller, Ossmer, Howard, Gottschalk and Hippe 1984, 395.).

Further descriptive information

An overview of the general characteristics of the nine different species of the genus *Sporomusa* are presented in Table 214. Species of *Sporomusa* have been obtained from a variety of habitats, including silage (Möller et al., 1984), the termite gut (Boga et al., 2003; Breznak et al., 1988), freshwater sediments (Dehning et al., 1990; Möller et al., 1984), aerated soil (Kuhner et al., 1997), and roots of the estuarine plant black needlerush (*Juncus roemerianus*) (Gössner et al., 2006b). This diverse habitat range illustrates the versatility of this spore-forming genus.

The main physiological feature that unites all species of the genus *Sporomusa* is their utilization of the acetyl-CoA "Wood-Ljungdahl" pathway (Wood and Ljungdahl, 1991) as an energy-conserving, terminal electron-accepting process (Figure 220). Reductant derived from the oxidation of a large variety of substrates is funneled toward the reductive synthesis of acetate from CO_2. The pathway also assimilates CO_2 and thus provides sporomusal species with a mechanism for autotrophic fixation of carbon. Lithotrophy occurs via the utilization of H_2. Membranous *b*-type cytochrome(s) is typical of the genus *Sporomusa* (note: some sporomusal species have not been analyzed for cytochromes), and it is likely that energy conservation occurs via the electron transport mediated translocation of protons and the synthesis of ATP by membranous ATPases (Müller et al., 2004).

There are 21 bacterial genera in which species of acetogens have been isolated (Drake et al., 2004). *Sporomusa* is functionally a monophyletic genus in regard to the ability of all nine sporomusal species to engage the acetyl-CoA pathway. *Acetobacterium* and *Moorella* are other examples of monophyletic acetogenic genera. However, certain genera that contain acetogens also contain species that lack the acetyl-CoA pathway (e.g., *Clostridium* and *Ruminococcus*).

H_2-CO_2, formate, methanol, lactate, pyruvate, and betaine are commonly utilized substrates of sporomusal species (Table 214). Several sporomusal species have the capacity to utilize methoxyl groups of aromatic compounds and to tolerate and consume (i.e., reduce) small amounts of O_2; these properties are shared with other acetogens in other genera (Drake et al., 2004). Certain sporomusal species (e.g., *Sporomusa*

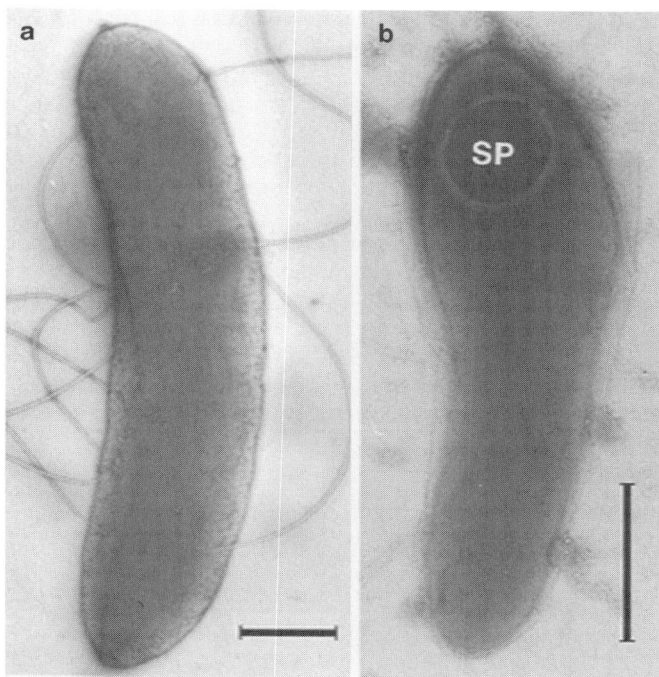

FIGURE 219. Electron micrographs of (a) *Sporomusa silvacetica* DSM 10669. Bar = 0.5 μM. (From Kuhner et al., 1997; used with permission.) and (b) *Sporomusa rhizae* DSM 16652. Bar = 1.0 μM. (From Gössner et al., 2006b; used with permission.) Abbreviation: SP, spore.

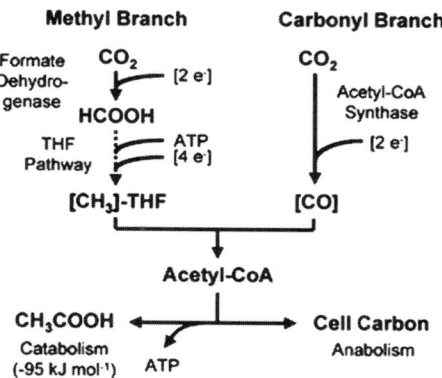

FIGURE 220. Acetyl-CoA "Wood–Ljungdahl" pathway. Abbreviations: THF, tetrahydrofolate; CoA, coenzyme A; ATP, adenosine triphosphate; e–, reducing equivalent. A corrinoid enzyme mediates a transfer of the methyl unit of methyl-THF prior to the formation of acetyl-CoA. See Drake et al. (2004), for details.

acidovorans and *Sporomusa malonica*) conserve energy via the decarboxylation of dicarboxylic acids (Dehning et al., 1989). For example, the change in Gibb's free energy from the decarboxylation of succinate to propionate is approximately −20 kJ per mol reaction, and this small amount of available energy is adequate for the growth of certain sporomusal species. The subterminal endospores typical of the genus can also be observed as free spores. One type strain (*Sporomusa paucivorans*) does not form spores. Certain species require yeast extract or a source of amino acids for growth (together with a supplemental substrate).

TABLE 214. Utilization of substrates by species of the genus *Sporomusa*[a]

Utilization of	1. *S. sphaeroides*	2. *S. acidovorans*	3. *S. aerivorans*	4. *S. malonica*	5. *S. ovata*	6. *S. paucivorans*	7. *S. rhizae*	8. *S. silvacetica*	9. *S. termitida*
N-Acetylglucosamine	NR	NR	NR	NR	NR	NR	+	NR	NR
Betaine	+	NR	NR	+	+	+	+	+	+
2,3-Butanediol	+	NR	NR	+	+	+	NR	+	NR
Citrate	–	–	+	+	–	–	+	–	+
Ethanol	+	NR	+	+	+	+	–	+	+
Ferulate	NR	NR	NR	NR	NR	NR	–	+	NR
Formate	+	+	+	+	+	+	+	+	+
Fructose	–	+	–	+	+	–	–	+	–
Fumarate	–	+	+	+	–	–	NR	+	–
Glyercol	+	+	–	–	–	+	NR	+	–
H₂–CO₂	+	+	+	+	+	+	+	+	+
Lactate	+	NR	+	+	+	+	+	+	+
L-Malate	–	+	+	+	–	–	NR	NR	–
Mannitol	–	NR	+	NR	–	–	NR	–	+
Methanol	+	+	+	+	+	+	NR	+	+
Pyruvate	+	+	+	+	+	+	NR	+	+
Succinate	–	+[b]	+	+	–	–	NR	–	+
Syringate	NR	NR	+	NR	NR	NR	+	NR	NR
Vanillate	NR	NR	+	NR	NR	NR	+	+	NR

[a]Abbreviations: +, positive;–, negative; NR, not reported.

[b]Produces acetate from succinate; other species form propionate as the major succinate-derived product.

Enrichment and isolation procedures

Nine species of *Sporomusa* have been isolated to date, but a selective procedure that is specific for the enrichment of *Sporomusa* has not been reported. H_2-CO_2, methanol, betaine, dimethylglycine, and glutarate have been used in primary enrichments. Mesophilic, anoxic, near pH neutral conditions, and a source of CO_2, can be regarded as essential. For cultivation medium, see following section.

Maintenance procedures

Species of *Sporomusa* are anaerobes and must therefore be cultivated under anoxic conditions. Near pH neutral media used for the cultivation of *Sporomusa* contain minerals, trace elements, vitamins, reducing agent (e.g., sodium sulfide, cysteine, dithi-onite, or dithiothreitol), and a source of CO_2 (e.g., bicarbonate and/or CO_2 in the gas phase); yeast extract or Casamino acids can enhance growth and are required for certain strains of *Sporomusa*. ATCC Medium 1425 [http://www.lgcpromochem-atcc.com] and DSMZ Medium 311 [http://www.dsmz.de/]) are examples of *Sporomusa* media. A more detailed statement on media and cultivation methods for acetogens can be found in Drake et al., (2004).

Species of *Sporomusa* are stable to long term storage conditions under both lyophilized and liquid conditions. This stability is likely due to the durability of sporomusal endospores. For example, endospores of *Sporomusa acidovorans* survive 20 min at 80°C (Ollivier et al., 1985), and endospores of *Sporomusa rhizae* can be revived after 1 year of storage under oxic, aqueous conditions (Gössner et al. 2006).

List of species of the genus *Sporomusa*

1. **Sporomusa sphaeroides** Möller, Ossmer, Howard, Gottschalk and Hippe 1985, 224[VP] (Effective publication: Möller, Ossmer, Howard, Gottschalk and Hippe 1984, 395.)

   sphae.roi′des. Gr. fem. adj. *sphaeroides* spherical, referring to the shape of the endospore.

   Cells are Gram-stain-negative, spore-forming, motile, curved rods, 2–4×0.5–$0.8 \mu m$. Growth-supportive substrates include H_2-CO_2, pyruvate, lactate, alcohols, glycerol, serine, ethyleneglycol, betaine, and other *N*-methyl compounds. Yeast extract is not required for growth. Acetate is the main reduced end product; traces of ethanol are also produced. Methylamines are formed from *N*-methyl compounds. Cultures demethylate the osmolytes dimethylsulfoniopropionate and glycine-betaine to methylthiopropionate and dimethylglycine, respectively, however, only the demethylation of glycine-betaine supports growth (Jansen and Hansen, 2001). Catalase-positive. Nitrate, sulfate, thiosulfate, and sulfite are not utilized as terminal electron acceptors. Growth is optimum at pH 6.5 (range 5.7–8.7) and 34°C (range 15–45°C). Contains membranous *b*-type cytochrome(s); might also contain a *c*-type cytochrome. The type strain (which is also the type species of the genus) was isolated from river mud; primary enrichment was on betaine (Möller et al., 1984).

   DNA G+C content (mol%): 46.7.

   Type strain: E, ATCC 35900, DSM 2875.

   GenBank accession number (16S rRNA gene): AJ279801.

2. **Sporomusa acidovorans** Ollivier, Cord-Ruwisch, Lombardo and Garcia 1990, 105[VP] (Effective publication: Ollivier, Cord-Ruwisch, Lombardo and Garcia 1985, 310.)

   a.ci.do′vo.rans. L. neut. n. *acidum* acid; L. v. *voro* to devour; N.L. part. adj. *acidovorans* acid devouring.

   Cells are Gram-stain-negative, spore-forming, motile, curved rods, 2–8×0.7–$1 \mu m$. Growth-supportive substrates include H_2-CO_2, organic acids (e.g., fumarate and oxaloacetate), methanol, glycerol, fructose, and ribose. Yeast extract is required for growth. Acetate is the sole reduced end product. Decarboxylates succinate to propionate (Dehning et al., 1989). Nitrate, sulfate, and sulfite are not utilized as terminal electron acceptors. Growth is optimum at pH 6.5 (range 5.4–7.5) and 35°C (range 20–40°C). Type strain was isolated from a distillation wastewater bioreactor; primary enrichment was on methanol (Ollivier et al., 1985).

   DNA G+C content (mol%): 42.

   Type strain: Mol, ATCC 49682, DSM 3132.

   GenBank accession number (16S rRNA gene): AJ279798.

3. **Sporomusa aerivorans** Boga, Ludwig and Brune 2003, 1403[VP]

   ae.ri.vo′rans. L. n. *aer* air; L. pres. part. *vorans* digesting, devouring; N.L. pres. part. *aerivorans* devouring air (O_2), referring to the organism's ability to reduce O_2.

   Cells are Gram-stain-variable, spore-forming, motile, curved rods, 1.3–7.0×0.6–$0.7 \mu m$. Growth supportive substrates include H_2-CO_2, formate, methanol, ethanol, lactate, pyruvate, mannitol, citrate, and methoxylated aromatic compounds. Yeast extract or Casamino acids are required for growth. Acetate is the main reduced end product. Tolerates and reduces small amounts of O_2; catalase-positive. Nitrate and sulfate are not utilized as terminal electron acceptors. Growth is optimum at pH 7 (range 6.2–8.2) and 30°C (range 19–35°C). Contains membranous *b*-type cytochrome(s). Type strain was isolated from the soil-feeding termite *Thoracotermes macrothorax*; primary enrichment was on H_2-CO_2 and lactate (Boga et al., 2003).

   DNA G+C content (mol%): not reported.

   Type strain: TmAO3, ATCC BAA-625, DSM 13326.

   GenBank accession number (16S rRNA gene): AJ506191.

4. **Sporomusa malonica** Dehning, Stieb and Schink 1990, 321[VP] (Effective publication: Dehning, Stieb and Schink 1989, 425.)

   ma.lo′ni.ca. N.L. n. *acidum malonicum* malonic acid; N.L. fem. adj. *malonica* referring to organism's ability to metabolize malonic acid.

   Cells are Gram-stain-negative, spore-forming, motile, curved rods, 2.6–$4.8 \times 0.7 \mu m$. Growth-supportive substrates include H_2-CO_2 and numerous organic compounds, including formate, pyruvate, alcohols, dicarboxylic acids, fructose, and trimethoxycinnamate. Yeast extract is required for growth. Acetate is the reduced end product when H_2-CO_2, formate, methanol, fructose, pyruvate, or the *O*-methyl groups of trimethoxycinnamate are metabolized. Alcohols yield acetate and the respective fatty acids, and crotonate and 3-hydroxybutyrate yield acetate and butyrate. Dicarboxylic acids (e.g., malonate and succinate) are decarboxylated to the respective fatty acids. Growth is optimal at pH 7.3 (range 6.0–8.5) and 30°C (range 15–38°C). Contains membranous *b*-type cytochrome(s). Type strain was isolated from freshwater sediment; primary enrichment was on glutarate (Dehning et al., 1989).

DNA G+C content (mol%): 44.1.

Type strain: WoG12, ATCC 49684, DSM 5090.

GenBank accession number (16S rRNA gene): AJ279799.

5. **Sporomusa ovata** Möller, Ossmer, Howard, Gottschalk and Hippe 1985, 224[VP] (Effective publication: Möller, Ossmer, Howard, Gottschalk and Hippe 1984, 395.)

o.va′ta. L. fem. adj. *ovata* egg shaped, referring to the shape of the endospore.

Cells are Gram-stain-negative, spore-forming, motile, curved rods, 1–5 × 0.7–1.0 μm. Growth-supportive substrates include H_2–CO_2, pyruvate, lactate, alcohols, fructose, betaine, dimethylglycine, and sarcosine. Yeast extract is not required for growth. Acetate is the main reduced end product; traces of ethanol are also produced. Methylamines are formed from *N*-methyl compounds. Reductively dechlorinates tetrachloroethylene to trichloroethylene (Terzenbach and Blaut, 1994). Cultures demethylate the osmolytes dimethylsulfoniopropionate and glycine-betaine to methylthiopropionate and dimethylglycine, respectively, however, only the demethylation of glycine-betaine supports growth (Jansen and Hansen, 2001). Catalase-positive. Nitrate and sulfate are not utilized as terminal electron acceptors. Growth is optimum at pH 6.3 (range 5.0–8.1) and 34°C (range 15–43°C). Contains membranous *b*-type cytochrome(s); might also contain a *c*-type cytochrome. Type strain isolated from sugar beet leaf silage; primary enrichment was on dimethylglycine (Möller et al., 1984).

DNA G+C content (mol%): 42.2.

Type strain: H1, ATCC 35899, DSM 2662, HAMBI 2226.

GenBank accession number (16S rRNA gene): AJ279800.

6. **Sporomusa paucivorans** Hermann, Popoff and Sebald 1987, 97[VP]

pau.ci.vo′rans. L. n. *pauca* a few; L. v. *voro* to devour; N.L. part. adj. *paucivorans* referring to the relatively small substrate range of the organism.

Cells are Gram-stain-negative, motile, slightly curved rods, 2–3 × 0.4–0.7 μm. This is the only species of *Sporomusa* that has not been observed to form spores. H_2–CO_2, formate, methanol, pyruvate, serine, betaine, alcohols, and ethylene glycol support growth. Yeast extract is required for growth. Acetate is the sole reduced end product. Oxidation of alcohols yields the corresponding fatty acids. Nitrate, sulfate, thiosulfate, and sulfite are not utilized as terminal electron acceptors. Grows syntrophically with *Methanobacterium formicicum* on alcohols. The type strain was isolated from lake sediment; primary enrichment was on methanol (Hermann et al., 1987).

DNA G+C content (mol%): 47.1.

Type strain: X, DSM 3697.

GenBank accession number (16S rRNA gene): not available for type strain. (M59117 is the number given by author for strain DSM 3637.)

7. **Sporomusa rhizae** Gössner, Küsel, Schulz, Trenz, Acker, Lovell and Drake 2006a, 1460[VP] (Effective publication: Gössner, Küsel, Schulz, Trenz, Acker, Lovell and Drake 2006, 1217.)

rhi′zae. N.L. gen. fem. n. *rhizae* (from Gr. fem. n. *rhiza* root) of the root; to indicate a spore-forming banana (per epithet of *Sporomusa*) from roots (the origin of the type strain).

Cells are weakly Gram-stain-positive, spore-forming, motile, curved rods. Cells mean 3.5 × 0.8 μm. Growth supportive substrates include H_2–CO_2, formate, lactate, vanillate, syringate, citrate, *N*-acetylglucosamine, and betaine. Acetate is the primary reduced end product; traces of butyrate, iso-butyrate, and propionate are also formed. Tolerates and reduces small amounts of O_2; spores withstand oxic aqueous conditions for 1 year. Peroxidase and NADH oxidase-positive. Growth is optimum at pH 7.5 (range 5.5–9.0) and 35°C (range 15–40°C). Contains membranous *b*-type cytochrome(s). Grows by trophic interaction with the aerotolerant fermentative anaerobe *Clostridium intestinale* RC (fortuitously co-isolated with the type strain). The type strain was isolated from a root homogenate prepared from the black needlerush *Juncus roemerianus* obtained from a pristine salt marsh; primary enrichment was on H_2–CO_2 (Gössner et al., 2006b).

DNA G+C content (mol%): not reported.

Type strain: RS, ATCC BAA-1028, DSM 16652.

GenBank accession number (16S rRNA gene): AM158322.

8. **Sporomusa silvacetica** Kuhner, Frank, Griesshammer, Schmittroth, Acker, Gössner and Drake 1997, 357[VP]

sil.va.ce′ti.ca. L. n. *silva* forest; L. fem. adj. *acetica* pertaining to vinegar, acetic acid; N.L. fem. adj. *silvacetica* a forest organism producing acetic acid.

Cells are weakly Gram-stain-positive, spore-forming, motile, slightly curved rods. Mean cell dimensions are 3.5 × 0.7 μm. Growth-supportive substrates include H_2–CO_2, formate, methanol, pyruvate, vanillate, ferulate, fructose, betaine, fumarate, 2,3-butanediol, ethanol, lactate, and glycerol. Yeast extract and sodium are not required for growth. With most substrates, acetate is the main reduced end product. Traces of methane are produced under certain conditions. Nitrate and sulfate are not utilized as terminal electron acceptors. Tolerates and reduces small amounts of O_2 (Drake et al., 2002; Karnholz et al., 2002). Fumarate is dismutated to acetate and succinate. The acrylate side chain of ferulate is reduced. Growth is optimum at pH 7.2 (range 5.9–7.7) and 30°C (range 10–35°C). Contains membranous *b*-type cytochrome(s); might also contain a cytoplasmic *c*-type cytochrome. The type strain was isolated from forest soil; primary enrichment was on H_2–CO_2 (Kuhner et al., 1997). Strain DR5, phylogenetically closely related to the type strain, was isolated from flooded rice paddy soil and enriched on trimethoxybenzoate (Rosencrantz et al., 1999).

DNA G+C content (mol%): 42.7.

Type strain: DG-1, ATCC 700346, DSM 10669.

GenBank accession number (16S rRNA gene): Y09976.

9. **Sporomusa termitida** Breznak, Switzer and Seitz 1990, 212[VP] (Effective publication: Breznak, Switzer and Seitz 1988, 287.)

ter.mi′ti.da. L. n. *tarmes, tarmit-* (L.L. var. *termes, termit-*) worm that eats wood; N.L. adj. *termitida* referring to termites, from whose intestinal tract the type strain was isolated.

Cells of *Sporomusa termitida* are Gram-stain-negative, spore-forming, motile, straight to slightly curved rods, 2–8 × 0.5–0.8 μm. Substrates include H_2–CO_2, CO, formate, methanol, ethanol, betaine, sarcosine, lactate, pyruvate, oxaloacetate, citrate, malonate, succinate, mannitol, and trimethoxybenzoate. Yeast extract or Casamino acids is required for growth. Acetate is the main reduced end product. Catalase-positive. Decarboxylates succinate to propionate. Nitrate and sulfate are not utilized as terminal electron acceptors. Growth is optimal at pH 7.2 (no growth at 6.2 or 8.1) and 30°C (range 19–37°C). Contains membranous *b*-type cytochrome(s). The type strain was isolated from the gut of the wood-feeding termite *Nasutitermes nigriceps*; primary enrichment was on H2–CO2 (Breznak et al., 1988).

DNA G+C content (mol%): 48.6.

Type strain: JSN-2, ATCC 49683, DSM 4440.

GenBank accession number (16S rRNA gene): M61920.

Genus XXIII. **Succiniclasticum** van Gylswyk 1995, 298^{VP}

FRED A. RAINEY

Suc.cin.i.clas′ti.cum. N.L. n. *acidum succinicum* succinic acid; Gr. adj. *klastos* broken; N.L. adj. *clasticum* breaking; N.L. neut. n. *Succiniclasticum* breaking or splitting succinic acid.

Gram-stain-negative. Anaerobic. Nonmotile. **Non-spore-forming**, rod-shaped, branching cells. Mesophilic. Ferments succinate to propionate. Other carbohydrates, amino acids, or mono-, di-, and other tricarboxylic acids are not fermented. Non-proteolytic, catalase- and urease-negative. Nitrate not reduced.

DNA G+C content (mol%): 52.0 (UV).

Type species: **Succiniclasticum ruminis** van Gylswyk 1995, 299^{VP}.

Further descriptive information

The description of the genus *Succiniclasticum* and the single species *Succiniclasticum ruminis* is based on the properties of a single strain SE10^T (DSM 9236^T). The cells are short, Gram-stain-negative, nonmotile rods and do not produce spores. Cells are up to 1.8 μm in length and 0.3–0.5 μm in diameter. Thicker rods tend to have rounded ends, while the narrower ones often have tapered ends and can be slightly curved. Cells form large clumps in liquid culture or in the liquid of agar slants.

On succinate agar, the colonies grow up to 5 mm in diameter, are practically colorless, convex, irregularly round with smooth edges and slightly opaque with a glistening appearance. Colonies are lens-shaped discs when growing below the agar surface.

Of the substrates tested, succinate is the only one fermented by strain SE10^T. Growth is slower with 40 mM succinate and 10 mM fumarate than with succinate alone although the same maximum optical density can be obtained. Increasing the amount of fumarate to 40 mM greatly reduces the amount of growth. The omission of either rumen fluid or yeast extract, results in very little growth. Increasing the amount of succinate in the culture medium reduces the growth rate but results in an increase in final biomass. Succinate is decarboxylated to produce propionate with no other end products detected although it is assumed that CO_2 is released. Growth in medium without CO_2 and $NaHCO_3$ is less than when these are available and the medium is highly buffered. A reducing agent is not required to obtain good growth if the medium is prepared under anaerobic conditions. If air is added to the headspace, growth is not good if the medium has not been reduced using a reducing agent.

Enrichment and isolation procedures

All media are prepared under strictly anoxic conditions and the head space gas has O_2-free CO_2. The basal medium used for the isolation of *Succiniclasticum* contains (per liter) 0.225 g of K_2HPO_4, 0.225 g of KH_2PO_4, 0.45 g of NaCl, 0.45 g of $(NH_4)_2SO_4$, 0.045 g of $CaCl_2$ (anhydrous), 0.09 g of $MgSO_4 \cdot 7H_2O$, 6.36 g of $NaHCO_3$, 0.25 g of cysteine hydrochloride·H_2O, 0.25 g of Na_2S (hydrated), 0.005 g of indigo carmine, and 400 ml of clarified rumen fluid. Diluent contains the same ingredients except that rumen fluid and sulfide are omitted and the concentration of cysteine is increased to 0.5 g/l. For enrichment cultures, basal media with 20 mM succinate is used. Samples of whole rumen ingesta (obtained from lactating, rumen-fistulated cows) are serially diluted and an aliquot of the 10^8 dilution added to 5 ml of enrichment medium and incubated at 39°C. After 1–4 d of incubation, aliquots of the enrichment culture are diluted and added to roll bottles containing succinate agar. Colonies that have grown after 10 d incubation are restreaked and their growth in liquid succinate medium tested.

Taxonomic comments

The description of the genus *Succiniclasticum* is based on its phylogenetic position as well as its inability to utilize substrates other than succinate. Comparisons based on 16S rRNA gene sequences show strain SE10^T to group in a cluster with the species *Phascolarctobacterium*, *Acidaminococcus*, and *Succinispira* (see family Veillonellaceae tree Figure 199). The similarities of the 16S rRNA gene sequences of the species of these genera to that of strain SE10^T are in the range 89.4–91.9%. The highest similarity is between the 16S rRNA gene sequences of strain SE10^T and *Phascolarctobacterium faecium*. The cluster comprising *Succiniclasticum ruminis*, *Phascolarctobacterium faecium*, *Acidaminococcus fermentans* and *Succinispira mobilis* is supported by a 99% bootstrap value. In addition to the low 16S rRNA gene sequence similarities between strain SE10^T and these related taxa differences in cell shape, mol% G+C content of the DNA and substrate utilization patterns differentiate *Succiniclasticum ruminis* from related taxa. *Succiniclasticum ruminis* forms short rods with limited branching while the cells of *Phascolarctobacterium faecium* are pleomorphic, *Acidaminococcus fermentans* are cocci, and *Succinispira mobilis* are curved rods. The mol% G+C of the DNA of *Succiniclasticum ruminis* is 52 compared to 41–42 for *Phascolarctobacterium faecium*, 56 for *Acidaminococcus fermentans*, and 36 for *Succinispira mobilis*. *Succiniclasticum ruminis* and *Succinispira mobilis* but not *Acidaminococcus fermentans* utilize succinate, and in *Phascolarctobacterium faecium* succinate only stimulates growth. In addition, amino acids support growth of *Acidaminococcus fermentans* and *Succinispira mobilis* but not *Succiniclasticum ruminis*; this characteristic was not determined for *Phascolarctobacterium faecium*.

List of species of the genus *Succiniclasticum*

1. **Succiniclasticum ruminis** van Gylswyk 1995, 299[VP]

   ru′mi.nis. L. neut. gen. n. *ruminis* of the rumen.

   Cells are short, Gram-stain-negative, nonmotile rods. Spores are not produced. Cells are up to 1.8 μm in length and 0.3–0.5 μm in diameter. Colonies are up to 5 mm in diameter on succinate agar slopes; irregularly round with smooth edges, convex, practically colorless, and slightly opaque with a glistening appearance. Subsurface colonies are lens-shaped discs. The maximum optical density is obtained at 39°C, with half as much growth at 45°C and no growth at 22°C. Succinate is the only substrate tested that is fermented. Rumen fluid and yeast extract are required for good growth. The end product of growth on succinate is propionate. Glucose, galactose, mannose, rhamnose, fructose, xylose, arabinose, maltose, cellobiose, lactose, sucrose, trehalose, mannitol, glycerol, aspartic acid, the sodium salts of DL-lactic and pyruvic acids, the disodium salts of oxalic, malonic, oxaloacetic, malic, fumaric, glutaric, and methylmalonic acids, the trisodium salts of citric and *trans*-aconitic acids, and yeast extract (which contains amino acids) do not support growth. Prolonged incubation (9 d) with fumarate, malate, aspartate, citrate, or *trans*-aconitate does not induce utilization. Tween 80 has no effect on growth. Non-proteolytic, does not liquefy gelatin or hydrolyze casein, H_2S is not produced. Catalase and urease are negative. Nitrate is not reduced. Type strain was isolated from rumen ingesta of a dairy cow.

   DNA G+C content (mol%): 52.0 (UV).

   Type strain: SE10, DSM 9236.

   GenBank accession number (16S rRNA gene): X81137.

Genus XXIV. **Succinispira** Janssen and O'Farrell 1999, 1012[VP]

PETER H. JANSSEN

Suc.ci.ni.spi′ra. L. n. *succinum* amber; N.L. n. *acidum succinicum* succinic acid (derived from amber); Gr. n. *spira* coil, spiral; N.L. fem. n. *Succinispira* succinate (utilizing) spiral (shaped bacterium).

Curved rods, 0.5 μm in diameter and 2–10 μm long. Longer cells can be spirals of up to 3 or 4 turns with an amplitude of 1.5–2.0 μm, and a wavelength of 3–4 μm. Gram-stain-negative. Aminopeptidase-negative. Cell-wall structure appears to be of a Gram-negative type with an outer membrane. **Highly motile** with laterally inserted flagella. Cytochromes are not formed. **No endospores**. Strict anaerobe; media containing a suitable reductant are required for growth. Chemo-organotrophic metabolism. **Organic and amino acids are fermented**, while carbohydrates and alcohols are not. **Formate, acetate, propionate, malate, CO_2, and H_2 are the fermentation end products**, depending on the substrate. Sulfate, sulfur, thiosulfate, nitrate, and fumarate are not used as terminal electron acceptors.

DNA G+C content (mol%): 36.

Type species: **Succinispira mobilis** Janssen and O'Farrell 1999, 1012[VP].

Enrichment and isolation procedures

Janssen (1990) enriched a mixed culture using an anaerobic freshwater medium with sulfide and cysteine as reductants and glycolate and yeast extract as growth substrates. *Succinispira mobilis* was isolated from this mixed culture after transferring it to a similar medium with citrate and yeast extract (Janssen, 1991). A pure culture of *Succinispira mobilis* was obtained from the mixed culture using agar deep culture under anaerobic conditions to yield well-separated colonies, which were then removed from the agar using sterile, drawn out glass Pasteur pipettes. Pure cultures grew in the citrate-yeast extract medium within 24 h of inoculation after incubation at 34°C. Media containing a suitable reductant are required for growth. At this time, there is no known selective medium for *Succinispira* strains. There are a number of physiologically similar anaerobes, some within the family *Veillonellaceae*, and others with as-yet unknown phylogenetic affiliations, making selective enrichment difficult.

Maintenance procedures

Cultures of *Succinispira* can be stored for at least one year at 4°C in anaerobic media in sealed bottles or tubes either completely filled or under a nitrogen plus carbon dioxide headspace. Lyophilized cultures have been stored for a number of years without loss of viability.

Taxonomic comments

The genus is represented by only one isolate from a pond receiving effluent from an anaerobic solids digester. This means that the assignment of characteristics to the generic and specific descriptions is somewhat arbitrary. To date, no closely related 16S rRNA genes have been reported in the published literature, and so no information on the distribution of members of this genus is available.

List of species of the genus *Succinispira*

1. **Succinispira mobilis** Janssen and O'Farrell 1999, 1012[VP]

   mo′bi.lis. L. adj. *mobilis* movable, motile.

   The cells of this species are curved rods, 0.5 μm in diameter and 2–10 μm long. Longer cells can be spirals of up to 3 or 4 turns with an amplitude of 1.5–2.0 μm, and a wavelength of 3–4 μm. The cells are highly motile with 4 laterally inserted flagella. Cytochromes are not formed. Colonies in agar-deep cultures are white and lens shaped. Liquid cultures grow with a uniform turbidity. The species displays a chemo-organotrophic metabolism and probably uses a decarboxylase-driven transmembrane sodium gradient for part of its ATP synthesis. Pyruvate, citrate, oxaloacetate, maleate, succinate, fumarate, glutamate, and aspartate are fermented, while carbohydrates and alcohols are not. Formate, acetate, propionate, malate, CO_2, and H_2 are the fermentation end products, depending on the substrate. Hydrogenase activity is present. Sulfate, sulfur, thiosulfate, nitrate, and fumarate are not used as terminal electron acceptors. No esculin hydrolysis, urea hydrolysis, indole formation from L-tryptophan, or sulfide formation from L-cysteine. Growth factors are required and can be supplied by the addition of yeast extract. The range of pH for growth is 6.7–8.5, with an optimum of 7.4–7.7. The optimum growth temperature is 37°C. No growth is possible at 40°C. NaCl at 0.2 g/l is sufficient for growth. Growth is possible at 19 g NaCl per liter, but not at 24 g/l.

DNA G+C content (mol%): 36 (T_m).

Type strain: 19gly1 (= DSM 6222).

GenBank accession number (16S rRNA gene): AJ006980.

Genus XXV. **Thermosinus** Sokolova, González, Kostrikina, Chernyh, Slepova, Bonch-Osmolovskaya and Robb 2004, 2357[VP]

TATYANA G. SOKOLOVA

Ther.mo.sin'us. Gr. adj. *thermos* hot; L. masc. n. *sinus* bend; N.L. masc. n. *Thermosinus* thermophilic curved rod.

Cells are **motile, curved rods** that do not form spores. Cell wall is **Gram-negative type.** Divides by binary transverse fission. **Obligate anaerobe. Thermophilic** neutrophile that ferments carbohydrates. **Grows chemolithotrophically on CO with the formation of equimolar amounts of H$_2$ and CO$_2$.** Found in terrestrial hot springs.

DNA G+C content (mol%): 51.7.

Type species: **Thermosinus carboxydivorans** Sokolova, González, Kostrikina, Chernyh, Slepova, Bonch-Osmolovskaya and Robb 2004, 2358[VP].

Further descriptive information

Cells are curved rods of about 0.5 μm by 2.6–3 μm. Motile by means of lateral flagella. Thermophilic; it grows in the temperature range 40–68°C, with optimum growth at 60°C. Neutrophilic; it grows in the pH range 6.5–7.6, with optimum growth at pH 6.8–7.0. Grows on glucose, sucrose, lactose, arabinose, maltose, fructose, xylose, and pyruvate, but not on cellobiose, galactose, peptone, yeast extract, lactate, acetate, formate, ethanol, methanol, or sodium citrate. During glucose fermentation, acetate, H$_2$, and CO$_2$ are produced. Grows chemolithotrophically on CO with the formation of equimolar amounts of H$_2$ and CO$_2$. During growth on CO, sucrose or lactose, ferric iron is reduced. Reduces selenite to elemental selenium during growth on CO. Neither ferric iron nor selenite cause a significant shift in the ratio of H$_2$ and CO$_2$ produced. Elemental sulfur, thiosulfate, sulfate, and nitrate do not stimulate growth during growth on CO and are not reduced. However, thiosulfate enhances the growth rate and cell yield during growth on glucose, sucrose, or lactose; in this case, the fermentation products are acetate, H$_2$S, and CO$_2$. Lactate, acetate, formate, and H$_2$ are not utilized, either in the absence or presence of ferric iron, thiosulfate, sulfate, sulfite, elemental sulfur, or nitrate. Growth is completely inhibited by penicillin, ampicillin, streptomycin, kanamycin, and neomycin.

Enrichment and isolation procedures

The only representative of the genus *Thermosinus, Thermosinus carboxydivorans* Nor1[T], was isolated from a sample of mud and water from a small hot pool in a wooded part of Norris Basin at Yellowstone National Park. The sample temperature and pH were 50°C and 7.5, respectively. The enrichment medium was prepared anaerobically and contained yeast extract (0.5 g/l) and ferric citrate or amorphous ferric oxide under a 100% CO atmosphere (Sokolova et al., 2004). After incubation at 60°C, growth of curved rod-shaped cells was observed accompanied by ferrous iron formation and CO oxidation to equimolar amounts of CO$_2$ and H$_2$. After several passages by serial 10-fold dilutions, the culture was transferred to a similar medium without ferric citrate or amorphous ferric oxide. After several passages by serial 10-fold dilutions in this new medium, colonies were obtained on the same medium solidified with 5% agar in roll-tubes under 100% CO (Sokolova et al., 2004). A pure culture was isolated from a single colony.

Maintenance procedures

The type strain is maintained on medium supplemented with ferric citrate under 100% CO (Sokolova et al., 2004). Liquid cultures may be stored at 4°C for 8–10 months without loss of viability. Good recovery is also obtained following vacuum drying by the DSMZ, Braunschweig, Germany.

Taxonomic comments

On the basis of 16S rRNA sequence analyses and the Gram-negative nature of the cell wall, the genus *Thermosinus* was originally placed within the so-called *Sporomusa–Selenomonas* group (Sokolova et al., 2004). Currently, members of this group have been reclassified in the family *Veillonellaceae*, which was created within the class *Clostridia* to contain species possessing Gram-negative cell-wall structures but phylogenetically related to *Firmicutes* (see Ludwig et al., 2009). Within this family, the closest relatives of *Thermosinus* are *Propionispora hippei* and *Propionispora vibrioides* (91% 16S rRNA gene sequence similarity). *Thermosinus* is the only genus within this family that has been shown to oxidize CO.

List of species of the genus *Thermosinus*

1. **Thermosinus carboxydivorans** Sokolova, González, Kostrikina, Chernyh, Slepova, Bonch-Osmolovskaya and Robb 2004, 2358[VP]

car.bo.xy.di.vo'rans. N.L. n. *carboxydum* carbon monoxide; L. part. adj. *vorans* devouring, digesting; N.L. part. adj. *carboxydivorans* digesting carbon monoxide.

Has the characteristics of the genus. Grows chemolithotrophically on CO with the formation of equimolar amounts of H$_2$ and CO$_2$. The type strain was isolated from a hot pool with neutral pH at Norris Basin, Yellowstone National Park, USA.

DNA G+C content (mol%): 51.7±1 (T_m).

Type strain: Nor1, DSM 14886, VKM B-2281.

GenBank accession number (16S rRNA gene): AY519200.

Genus XXVI. **Zymophilus** Schleifer, Leuteritz, Weiss, Ludwig, Kirchhof and Seidel-Rüfer 1990, 26[VP]

KARL-HEINZ SCHLEIFER

Zy.mo'phi.lus. Gr. n. *zyme* leaven, yeast; Gr. adj. *philos* lover; N.L. masc. n. *Zymophilus* yeast-lover.

Slightly curved, helical, or crescent-shaped **rods** (0.7–1.0 × 3–30 μm) with rounded ends, occurring singly, in pairs, or in short chains. Gram-stain-negative. Motile. Mobility can be lost after several subcultivations. Nonspore-forming. **Strictly anaerobic.** Chemo-organotrophic, fermentative metabolism. **Glucose is fermented to acetic and propionic acids. Acetoin is not produced.** Occur in pitching yeast or brewery waste. The peptidoglycan type is directly cross-linked *meso*-diaminopimelic acid (variation A1γ; Schleifer and Kandler, 1972). Diamines cadaverine and rarely putrescine are found as characteristic components of peptidoglycan. Lipid F is found as characteristic cellular compound. Phylogenetically a member of the family *Veillonellaceae* (low G+C Gram-stain-positives with a Gram-negative cell-wall type). The closest phylogenetic neighbors are *Pectinatus*, *Megasphaera*, and *Selenomonas*.

DNA G+C content (mol%): 38–41 (T_m).

Type species: **Zymomonas raffinosivorans** Schleifer, Leuteritz, Weiss, Ludwig, Kirchhof and Seidel-Rüfer 1990, 26[VP].

Further taxonomic information

Phylogenetic analysis of 16S rDNA indicates that *Zymophilus paucivorans* is equally related to *Pectinatus cerevisiiphilus*, *Pectinatus frisingensis*, and *Selenomonas lacticifex* with 88% similarity (Willems and Collins, 1995a). An unrooted phylogenetic tree of members of the family *Veillonellaceae* based on 16S rRNA gene sequences is shown in Figure 199. There is a very close phylogenetic relationship between *Zymophilus paucivorans* and *Propionispira arboris*. The latter organism is an obligately anaerobic, Gram-stain-negative bacterium with a G+C content (36–37 mol%) similar to that of members of the genus *Zymophilus*. It forms long spiral filaments and can fix nitrogen. It ferments lactate and a variety of saccharides to propionate, acetate, and CO_2 as major end products. It has been isolated from alkaline wetwoods of poplar trees as described by Schink et al. (1982). Based on their close phylogenetic relatedness and their similar fermentation products, the two genera may be combined in one genus.

Comparative DNA sequence analysis of long 16S–23S intergenic spacer regions showed that two *Zymophilus* species are more closely related to each other (77% similarity) than to *Pectinatus* spp. and *Selenomonas lacticifex* (40.3%; Motoyama and Ogata, 2000a). A comparison of the shorter 16S–23S intergenic spacer regions indicates an even higher relationship among the two *Zymophilus* species (96.7% similarity) and a slightly closer relationship to *Selenomonas lacticifex* (45%) than to the *Pectinatus* spp. (39.3%; Motoyama and Ogata, 2000a). *Zymophilus raffinosivorans* and *Selenomonas lacticifex* as well as the *Pectinatus* spp. contained two tRNA genes (alanine, isoleucine) within the longer spacer, whereas *Zymophilus paucivorans* contained only the alanine tRNA gene (Motoyama and Ogata, 2000a).

Strictly anaerobic Gram-stain-negative rods were isolated from pitching yeast and brewery waste and described as a new genus comprising the two species *Zymophilus raffinosivorans* and *Zymophilus paucivorans* (Schleifer et al., 1990).

Isolation, cultivation, and maintenance procedures

Strictly anaerobic conditions are required for the isolation and cultivation of *Zymophilus* species. Modified MRS (de Man–Rogosa–Sharpe) and MRS medium at 30°C have been used (Schleifer et al., 1990; Ziola et al., 1999). Working cultures can be maintained by subculturing (30°C, 2–3 d) in peptone yeast extract fructose broth (0.5% peptone, 0.5% tryptone, 1% yeast extract, 0.5% fructose, 0.2% Na_2HPO_4, 0.15 Tween 80, and 0.05% cysteine hydrochloride). After incubation, the cultures should be stored at 4°C in anaerobic jars. However, the cultures should be stored in the refrigerator no longer than 14 d because the cultures are no longer viable after four weeks of storage (Haikara, 1991; Seidel-Rüefer, 1990).

For long-term preservation, conventional freeze-drying methods using 20% skim milk as a protective agent and freezing in drinking straws in a –150°C deep freezer or in liquid nitrogen at –196°C using 5% dimethyl sulfoxide as a protective agent have been used (Suihko and Haikara, 1990).

Differentiation of the genus *Zymophilus* from other anaerobic, Gram-stain-negative, rod-shaped bacteria

Zymophilus morphologically resembles other anaerobic, Gram-stain-negative, rod-shaped bacteria isolated from breweries. Most of them were initially classified as *Pectinatus*. However, detailed chemotaxonomic and molecular taxonomic studies have indicated that these bacteria can be divided into three genera: *Pectinatus*, *Selenomonas*, and *Zymophilus* (Schleifer et al., 1990). They also differ in terms of their beer spoilage ability. Strains of *Pectinatus* are typical beer spoilage organisms, whereas *Selenomonas* and *Zymophilus* grow poorly in beer (Haikara, 1989, 1991). The G+C contents of the DNAs of *Pectinatus* and *Zymophilus* species are rather similar (38–41 mol%), whereas that of *Selenomonas lacticifex* is significantly higher (51–52 mol%; Schleifer et al., 1990). All of them contain the directly cross-linked *meso*-diaminopimelic acid peptidoglycan type as well as cadaverine (or, rarely, putrescine) as typical substituents of the α-carboxyl group of D-glutamic acid in the peptidoglycan (Schleifer et al., 1990). Such a peptidoglycan type containing diamine-substituted glutamic acid was also found in strains of *Megasphaera*, other *Selenomonas* species, and *Sporomusa* (Engelmann and Weiss, 1985; Kamio et al., 1981a). Neither menaquinones nor ubiquinones were detected. However, as in *Megasphaera*, *Pectinatus*, *Selenomonas*, and *Sporomusa*, lipid F (Stackebrandt et al., 1985) was also detected in *Zymophilus*.

Zymophilus and *Pectinatus* species produce acetic and propionic acids as major fermentation products from glucose, but, in contrast to *Pectinatus* species, *Zymophilus* species are not able to produce acetoin. The major fermentation product of *Selenomonas lacticifex* is lactic acid. *Zymophilus* species can also be distinguished from *Pectinatus* species by their ability to ferment maltose and sucrose. Major differences between *Zymophilus* and other similar genera are summarized in Table 215.

TABLE 215. Characteristics differentiating *Zymophilus* from other related taxa[a]

Characteristic	*Zymophilus*	*Pectinatus*	*Propionispira*	*Selenomonas*
Acetoin production	–	+	ND	ND
Acid produced from:				
Maltose	+	–	ND	+
Sucrose	+	–	ND	+
Lactic acid as major fermentation product	–	–	–	– or +
DNA G+C content (mol%)	38–41	38–41	36–37	48–58
Nitrogen fixation	ND	ND	+	ND

[a]Symbols: +, >85% positive; –, 0–15% positive; ND, not determined.

TABLE 216. Characteristics differentiating *Zymphilus raffinosivorans* and *Zymophilus paucivorans*[a]

Characteristic	*Z. raffinosivorans*	*Z. paucivorans*
Acid produced from:		
Dulcitol	+	–
Inositol	+	–
Lactose	+	–
Melibiose	+	–
Raffinose	+	–
Rhamnose	+	–
Xylitol	+	–
Xylose	+	–

[a]Symbols: +, >85% positive; –, 0–15% positive.

Differentiation of the species of the genus *Zymophilus*

The two species are morphologically quite different. Cells of *Zymophilus paucivorans* are curved, helical, or crescent-shaped, motile, Gram-stain-negative rods with rounded ends. Motility can disappear in some strains after several subcultivations. They are 0.8–1.0 × 5–30 μm and occur singly, in pairs, or in short chains. Colonies on modified MRS agar medium are circular, smooth, opaque, slightly yellow, and 1–2 mm in diameter after 3 d at 30°C.

Zymophilus raffinosivorans cells are straight to slightly curved, motile Gram-stain-negative rods (0.7–0.9 × 3–15 μm in size) with rounded ends. They occur predominantly singly and sometimes in pairs or short chains. Colonies on modified MRS agar resemble those of *Zymophilus paucivorans*.

The two species also differ in the utilization of carbohydrates (Table 216). In contrast to *Zymophilus paucivorans*, *Zymophilus raffinosivorans* can utilize the following carbohydrates: dulcitol, inositol, lactose, melibiose, *N*-acetylglucosamine, raffinose, and xylose. All strains of the two species produced acid from cellobiose, fructose, glucose, maltose, mannitol, mannose, ribose, and sucrose. None of the strains utilized glycogen, inulin, melizitose, or trehalose.

List of species of the genus *Zymophilus*

1. **Zymophilus raffinosivorans** Schleifer, Leuteritz, Weiss, Ludwig, Kirchhof and Seidel-Rüfer 1990, 26[VP]

   raf.fi.no.si′vo.rans. N.L. n. *raffinosum* raffinose; L. v. vorare to eat; N.L. part. adj. *raffinosivorans* raffinose- devouring.

   The description is based on ten strains isolated from pitching yeast and four strains isolated from brewery waste.

   Straight to slightly curved rods, 0.7–0.9 × 3–15 μm, with rounded ends. Occur predominantly singly or sometimes in pairs or short chains. Gram-stain-negative. Endospores not formed. Motile. Colonies on modified MRS agar medium are circular, smooth, opaque, slightly yellow, and 1–2 mm in diameter after 3 d at 30°C. Optimum temperature for growth is 30°C. Strictly anaerobic. Fermentative metabolism. Glucose is fermented to acetic and propionic acids. Acid is produced from arabinose, cellobiose, dulcitol, fructose, galactose, glucose, lactose, maltose, mannitol, mannose, sorbitol, sucrose, xylitol, and xylose. Most of the strains produce acid from adonitol, inositol, melibiose, raffinose, rhamnose, and salicin. No acid is produced from inulin, glycogen, melezitose, starch (with one exception), and trehalose.

   The peptidoglycan type is directly cross-linked *meso*-diaminopimelic acid. Cadaverine and, rarely, putrescine are linked to the α-carboxyl group of D-glutamic acid in the peptidoglycan. Lipid F is also found as a characteristic compound.

   DNA G+C content (mol%): 38–41 (T_m).
   Type strain: SH2, ATCC 49691, DSM 20765.

   GenBank accession number (16S rRNA gene): DQ217599.

2. **Zymophilus paucivorans** Schleifer, Leuteritz, Weiss, Ludwig, Kirchhof and Seidel-Rüfer 1990, 27[VP]

   pau.ci′vo.rans. L. adj. *paucus* little; L. v. *vorare* to eat; N.L. adj. *paucivorans* utilizing relatively few carbohydrates.

   The description is based on four strains that were isolated from pitching yeast.

   Curved or helically shaped rods, 0.8–1.0 × 5–30 μm, with rounded ends. Occur singly, in pairs, or in short chains. Gram-stain-negative. Endospores not formed. Motile. In two of the strains, motility disappeared after several subcultivations. Colonies on modified MRS agar medium are circular, smooth, slightly yellow, and 1–2 mm in diameter after 3 d at 30°C. Strictly anaerobic. Fermentative metabolism. Glucose is fermented to acetic and propionic acids and trace amounts of lactic acid. Acid is produced from cellobiose, fructose, glucose, maltose, mannitol, mannose, ribose, and sucrose. Most strains produce acid from arabinose. One strain produced acid from adonitol, glycerol (weak), rhamnose, and xylitol. No acid was produced from amygdalin, dulcitol, erythritol, glycogen, inositol, inulin, melezitose, melibiose, raffinose, starch, trehalose, and xylose. Same peptidoglycan type and substitution as *Zymophilus raffinosivorans*. Lipid F is present.

   DNA G+C content (mol%): 39–41 (T_m).
   Type strain: AA1, ATCC 49689, DSM 20756.
   GenBank accession number (16S rRNA gene): AF373025.

References

Abou-Zeid, D.M., H. Biebl, C. Sprör and R.J. Muller. 2004. *Propionispora hippei* sp. nov., a novel Gram-negative, spore-forming anaerobe that produces propionic acid. Int. J. Syst. Evol. Microbiol. *54*: 951–954.

Al-Khaldi, S.F., J.D. Evans and S.A. Martin. 1999. Complete nucleotide sequence of a cryptic plasmid from the ruminal bacterium *Selenomonas ruminantium* HD4 and identification of two predicted open reading frames. Plasmid *42*: 45–52.

Al Jassim, R.A., P.T. Scott, A.L. Trebbin, D. Trott and C.C. Pollitt. 2005. The genetic diversity of lactic acid producing bacteria in the equine gastrointestinal tract. FEMS Microbiol. Lett. *248*: 75–81.

Al Jassim, R.A.M. 2003. *Lactobacillus ruminis* is a predominant lactic acid producing bacterium in the caecum and rectum of the pig. Lett. Appl. Microbiol. *37*: 213–217.

Allison, M.J. 1978. Production of branched-chain volatile fatty acids by certain anaerobic bacteria. Appl. Environ. Microbiol. *35*: 872–877.

Anonymous. 1998. Report of subcommittee on SMMP medium for selective isolation of *Megasphaera* and *Pectinatus*. J. Am. Soc. Brew. Chem. *56*: 212–214.

Anonymous. 2001. Analytica Microbiologica - EBC. Section 4. Detection of contaminants. CD-ROM. Fachverlag Hans Carl, Germany.

Arai, T., A. Kusakabe and S. Komatsu. 1984. A survey of plasmids in *Veillonella* strains isolated from human oral cavity. Kitasato Arch. Exp. Med. *57*: 233–237.

Arif, N., T. Do, R.Byun, E. Sheehy, D. Clarke, S.C. Gilbert and D. Beighton. 2007. *Veillonella rogosae* sp. nov., an anaerobic, Gram-negative coccus isolated from the dental plaque of caries-free children. *In* press. Int. J. Syst. Evol. Microbiol.

Asanuma, N. and T. Hino. 2000. Activity and properties of fumarate reductase in ruminal bacteria. J. Gen. Appl. Microbiol. *46*: 119–125.

Back, W., N. Weiss and H. Seidel. 1979. Isolation and systematic classification of Gram negative bacteria which are harmful to beer. II. Gram negative anaerobic rods (En. transl.). Brauwissenschaft *32*: 233–238.

Back, W. 1981. Nachweis und Identifizierung gramnegativer bierschädlicher Bakterien. Brauwissenschaft. *34*: 197–204.

Back, W., S. Breus and C. Weigand. 1988. Infektionsursachen im Jahre 1987. Brauwelt *128*: 1358–1362.

Baena, S., M.L. Fardeau, T.H.S. Woo, B. Ollivier, M. Labat and B.K.C. Patel. 1999. Phylogenetic relationships of three amino-acid-utilizing anaerobes, *Selenomonas acidaminovorans*, 'Selenomonas acidaminophila' and *Eubacterium acidaminophilum*, as inferred from partial 16S rDNA nucleotide sequences and proposal of *Thermanaerovibrio acidaminovorans* gen. nov., comb. nov and *Anaeromusa acidaminophila* gen. nov., comb. nov. Int. J. Syst. Bacteriol. *49*: 969–974.

Barbosa, T.M., K.P. Scott and H.J. Flint. 1999. Evidence for recent intergeneric transfer of a new tetracycline resistance gene, *tet(W)*, isolated from *Butyrivibrio fibrisolvens*, and the occurence of *tet(O)* in ruminal bacterial. Environ. Microbiol. *1*: 53–64.

Barnes, E.M. and G.C. Burton. 1970. The effect of hibernation on the caecal flora of the thirteen-lined ground squirrel (*Citellus tridecemlineatus*). J. Appl. Bacteriol. *33*: 505–514.

Bergey, D.H., F.C. Harrison, R.S. Breed, B.W. Hammer and F.M. Huntoon. 1923. Bergey's Manual of Determinative Bacteriology. The Williams & Wilkins Co., Baltimore.

Bhatti, M.A. and M.O. Frank. 2000. *Veillonella parvula* meningitis: case report and review of *Veillonella* infections. Clin. Infect. Dis. *31*: 839–840.

Biebl, H., H. Schwab-Hanisch, C. Sproer and H. Lunsdorf. 2000. *Propionispora vibrioides*, nov. gen., nov. sp., a new gram-negative, spore-forming anaerobe that ferments sugar alcohols. Arch. Microbiol. *174*: 239–247.

Biebl, H., H. Schwab-Hanisch, C. Sproer and H. Lunsdorf. 2001. *In* Validation of publication of new names and new combinations previously effectively published outside the IJSEM. List no. 80. Int. J. Syst. Evol. Microbiol. *51*: 793–794.

Bishop, D.G., M.J. Hewett and K.W. Knox. 1971. Occurrence of 3-hydroxytridecanoic and 3-hydroxypentadecanoic acids in the lipopolysaccharides of *Veillonella*. Biochim. Biophys. Acta *231*: 274–276.

Bladen, H.A. and S.E. Mergenhagen. 1964. Ultrastructure of *Veillonella* and morphological correlation of an outer membrane with particles associated with endotoxic activity. J. Bacteriol. *88*: 1482–1492.

Boga, H.I., W. Ludwig and A. Brune. 2003. *Sporomusa aerivorans* sp. nov., an oxygen-reducing homoacetogenic bacterium from the gut of a soil-feeding termite. Int. J. Syst. Evol. Microbiol. *53*: 1397–1404.

Boga, H.I., R. Ji, W. Ludwig and A. Brune. 2007. *Sporotalea propionica* gen. nov. sp. nov., a hydrogen-oxidizing, oxygen-reducing, propionigenic firmicute from the intestinal tract of a soil-feeding termite. Arch. Microbiol. *187*: 15–27.

Boo, T.W., B. Cryan, A. O'Donnell and G. Fahy. 2005. Prosthetic valve endocarditis caused by *Veillonella parvula*. J. Infect. *50*: 81–83.

Boskamp, E. 1922. Ueber Bau und Lebensweise und systematische Stellung von *Selenomonas palpitans* (Simons). Zentralbl. Bakteriol. Parasitenk. Infektionskr. Abt. I. Orig. *88*: 58–73.

Both, B., W. Buckel, R. Kroppenstedt and E. Stackebrandt. 1992. Phylogenetic and chemotaxonomic characterization of *Acidaminococcus fermentans*. FEMS Microbiol. Lett. *76*: 7–11.

Brancaccio, M. and G.G. Legendre. 1979. *Megasphaera elsdenii* endocarditis. J. Clin. Microbiol. *10*: 72–74.

Braune, A., K. Bendrat, S. Rospert and W. Buckel. 1999. The sodium ion translocating glutaconyl-CoA decarboxylase from *Acidaminococcus fermentans*: cloning and function of the genes forming a second operon. Mol. Microbiol. *31*: 473–487.

Braune, A. and A. Eidtmann. 2003. First experiences using realtime-PCR as a rapid detection method for brewery process control at Beck & Co. Proc. 29th Eur. Brew. Con. Congr. Dublin, pp. 1128–1135 [CD-ROM].

Brazier, J.S. and T.V. Riley. 1988. UV red fluorescence of *Veillonella* spp. J. Clin. Microbiol. *26*: 383–384.

Breznak, J.A. and J.M. Switzer. 1986. Acetate synthesis from H_2 plus CO_2 by termite gut microbes. Appl. Environ. Microbiol. *52*: 623–630.

Breznak, J.A., J.M. Switzer and H.J. Seitz. 1988. *Sporomusa termitida* sp. nov., an H_2/CO_2-utilizing acetogen Iisolated from termites. Arch. Microbiol. *150*: 282–288.

Breznak, J.A., J. M. Switzer and H.-J. Seitz. 1990. *In* Validation of the publication of new names and new combinations previously effectively published outside the IJSB. List no. 33. Int. J. Syst. Bacteriol. *40*: 212.

Broker, J.D. and B. Stokes. 1990. Monoclonal antibodies against the ruminal bacterium *Selenomonas ruminantium*. Appl. Environ. Microbiol. *56*: 2193–2199.

Brook, I. 1996. *Veillonella* infections in children. J. Clin. Microbiol. *34*: 1283–1285.

Brough, B.E., T.C. Reid and B.H. Howard. 1970. The biochemistry of the rumen bacterium "Quin's oval". Part 1. Fermentation of carbohydrates. N. Z. J. Sci. *13*: 570–575.

Bryant, M.P. 1956. The characteristics of strains of *Selenomonas* isolated from bovine rumen contents. J. Bacteriol. *72*: 162–167.

Bryant, M.P. 1974. Genus *Selenomonas* von Prowazek 1913, 36 Nom. cons. Opin. 21. Jud. Comm. 1958, 163. *In* Buchanan and Gibbons (Editors), Bergey's Manual of Determinative Bacteriology, 8th edn. The Williams & Wilkins Co, Baltimore, pp. 424–426.

Bryant, M.P. 1984. Genus IX. *Selenomonas* Von Prowazek 1913, 36^AL^. *In* Krieg and Holt (Editors), Bergey's Manual of Sytematic Bacteriology, Vol. 1. Springer, New York, pp. 650–653.

Buckel, W. 1980. Analysis of the fermentation pathways of clostridia using double labelled glutamade. Arch. Microbiol. *127*: 167–169.

Byun, R., M. Nadkarni, N.A. Jacques and N. Hunter. 2006. Alternatives to 16S rRNA for the identification of *Veillonella* species. J. Dent. Res. *85*: Special issue B, abstract 1844.

Byun, R., J.P. Carlier, N.A. Jacques, H. Marchandin and N. Hunter. 2007. *Veillonella denticariosi* sp. nov., isolated from human carious dentine. Int. J. Syst. Evol. Microbiol. *57*: 2844–2848.

Caldwell, D.R. and M.P. Bryant. 1966. Medium without rumen fluid for nonselective enumeration and isolation of rumen bacteria. Appl. Microbiol. *14*: 794–801.

Callaway, T.R., K.A. Adams and J.B. Russell. 1999. The ability of "low G + C Gram-positive" ruminal bacteria to resist monensin and counteract potassium depletion. Curr. Microbiol. *39*: 226–230.

Carlier, J.P., H. Marchandin, E. Jumas-Bilak, V. Lorin, C. Henry, C. Carriere and H. Jean-Pierre. 2002. *Anaeroglobus geminatus* gen. nov., sp. nov., a novel member of the family *Veillonellaceae*. Int. J. Syst. Evol. Microbiol. *52*: 983–986.

Cato, E.P., W.L. George and S.M. Finegold. 1986. Genus *Clostridium* Prazmowski 1880, 23^AL^. *In* Sneath, Mair, Sharpe and Holt (Editors), Bergey's Manual of Systematic Bacteriology, Vol. 2. The Williams & Wilkins Co., Baltimore, pp. 1141–1200.

Certes, A. 1889. Note sur les micro-organismes de la panse des ruminants. Bull. Soc. Zool. France *14*: 70–73.

Cheng, K.J., R.C. Phillippe and W. Majak. 1988. Identification of rumen bacteria that anaerobically degrade nitrite. Can. J. Microbiol. *34*: 1099–1102.

Cheong, J.P. and J.D. Brooker. 1998. Lysogenic bacteriophage M1 from *Selenomonas ruminantium*: isolation, characterization and DNA sequence analysis of the integration site. Microbiology *144*: 2195–2202.

Chihib, N., L. Monnerat, J.M. Membre and J. Tholozan. 1999. Nisin, temperature and pH effects on growth and viability of *Pectinatus frisingensis*, a gram-negative, strictly anaerobic beer-spoilage bacterium. J. Appl. Microbiol. *87*: 438–446.

Chin, K.J., D. Hahn, U. Hengstmann, W. Liesack and P.H. Janssen. 1999. Characterization and identification of numerically abundant culturable bacteria from the anoxic bulk soil of rice paddy microcosms. Appl. Environ. Microbiol. *65*: 5042–5049.

Collins, M.D., P.A. Lawson, A. Willems, J.J. Cordoba, J. Fernandez-Garayzabal, P. Garcia, J. Cia, H. Hippe and J.A.E. Farrow. 1994. The phylogeny of the genus *Clostridium*: proposal of five new genera and eleven new species combination. Int. J. Syst. Bacteriol. *44*: 812–826.

Cook, G.M., F.A. Rainey, G.J. Chen, E. Stackebrandt and J.B. Russell. 1994a. Emendation of the description of *Acidaminococcus fermentans*, a trans-aconitate-oxidizing and citrate-oxidizing bacterium. Int. J. Syst. Bacteriol. *44*: 576–578.

Cook, G.M. and J.B. Russell. 1994. Dual mechanisms of tricarboxylate transport and catabolism by *Acidaminococcus fermentans*. Appl. Environ. Microbiol. *60*: 2538–2544.

Cook, G.M., J.E. Wells and J.B. Russell. 1994b. Ability of *Acidaminococcus fermentans* to oxidize trans-aconitate and decrease the accumulation of tricarballylate, a toxic end product of ruminal fermentation. Appl. Environ. Microbiol. *60*: 2533–2537.

Cotta, M.A. and T.R. Whitehead. 1998. Xylooligosaccharide utilization by the ruminal anaerobic bacterium *Selenomonas ruminantium*. Curr. Microbiol. *36*: 183–189.

Counotte, G.H.M. 1981. Regulation of lactate metabolism in the rumen. PhD thesis, University of Utrecht.

Counotte, G.H.M., R.A. Prins, R.H.A.M. Janssen and M.J.A. Debie. 1981. Role of *Megasphaera elsdenii* in the fermentation of DL-[2-¹³C] lactate in the rumen of dairy cattle. Appl. Environ. Microbiol. *42*: 649–655.

D'Silva, C.G., H.D. Bae, L.J. Yanke, K.J. Cheng and L.B. Selinger. 2000. Localization of phytase in *Selenomonas ruminantium* and *Mitsuokella multiacidus* by transmission electron microscopy. Can. J. Microbiol. *46*: 391–395.

de Vries, W., W.M. van Wyck-Kapteyn and A.H. Stouthamer. 1973. Generation of ATP during cytochrome-linked anaerobic electron transport in propionic acid bacteria. J. Gen. Microbiol. *76*: 31–41.

de Vries, W., W.M. van Wijck-Kapteyn and S.K. Oosterhuis. 1974. The presence and function of cytochromes in *Selenomonas ruminantium, Anaerovibrio* lipolytica and *Veillonella alcalescens*. J. Gen. Microbiol. *81*: 69–78.

Dehning, I., M. Stieb and B. Schink. 1989. *Sporomusa malonica* sp. nov., a homoacetogenic bacterium growing by decarboxylation of malonate or succinate. Arch. Microbiol. *151*: 421–426.

Dehning, I., M. Stieb and B. Schink. 1990. *In* Validation of the publication of new names and new combinations previously effectively published outside the IJSB. List no. 34. Int. J. Syst. Bacteriol. *40*: 320–321.

Del Dot, T., R. Osawa and E. Stackebrandt. 1993. *Phascolarctobacterium faecium* gen. nov., spec. nov., a novel taxon of the *Sporomusa* group of bacteria. Syst. Appl. Microbiol. *16*: 380–384.

Del Dot, T., R. Osawa and E. Stackebrandt. 1994. *In* Validation of the publication of new names and new combinations previously effectively published outside the IJSB. List no. 48. Int. J. Syst. Bacteriol. *44*: 182–183.

Dennis, S.M., T.G. Nagaraja and E.E. Bartley. 1981. Effects of lasalocid or monensin on lactate-producing or -using rumen bacteria. J. Anim. Sci. *52*: 418–426.

Dighe, A.S., Y.S. Shouche and D.R. Ranade. 1998. *Selenomonas lipolytica* sp. nov., an obligately anaerobic bacterium possessing lipolytic activity. Int. J. Syst. Bacteriol. *48*: 783–791.

Dills, S.S., C.A. Lee and M.H. Saier, Jr. 1981. Phosphoenolpyruvate-dependent sugar phosphotransferase activity in *Megasphaera elsdenii*. Can. J. Microbiol. *27*: 949–952.

Downes, J., M. Munson and W.G. Wade. 2003. *Dialister invisus* sp. nov., isolated from the human oral cavity. Int. J. Syst. Evol. Microbiol. *53*: 1937–1940.

Doyle, L.M., J.O. McInerney, J. Mooney, R. Powell, A. Haikara and A.P. Moran. 1995. Sequence of the gene encoding the 16S rRNA of the beer spoilage organism *Megasphaera cerevisiae*. J. Ind. Microbiol. *15*: 67–70.

Drake, H.L., K. Küsel and C. Matthies. 2002. Ecological consequences of the phylogenetic and physiological diversities of acetogens. Antonie van Leeuwenhoek *81*: 203–213.

Drake, H.L., K. Küsel and C. Matthies. 2004. Acetogenic prokaryotes. The Prokaryotes, 3rd edn: An Evolving Electronic Resource for the Microbiological Community, release 3.17, August 2004. Springer (http://springeronline.com), New York.

Duran, S.P., J.V. Manzano, R.C. Valera and S.V. Machota. 1990. Obligately anaerobic bacterial species isolated from foot-rot lesions in goats. Br. Vet. J. *146*: 551–558.

Dürr, P. 1983. Luftkeimindikation bierschädlicher Bakterien, neue Methode mittels Luftkeimsammelgerät und Luftkeimindikator. Brauwelt *123*: 1652–1655.

Elsden, S.R. and D. Lewis. 1953. The production of fatty acids by a Gram-negative coccus. Biochem. J. *55*: 183–189.

Elsden, S.R., B.E. Volcani, F.M.C. Gilchrist and D. Lewis. 1956. Properties of a fatty acid forming organism isolated from the rumen of sheep. J. Bacteriol. *72*: 681–689.

Engelmann, U. and N. Weiss. 1985. *Megasphaera cerevisiae* sp. nov., a new Gram-negative obligately anaerobic coccus isolated from spoiled beer. Syst. Appl. Microbiol. *6*: 287–290.

Engelmann, U. and N. Weiss. 1986. *In* Validation of the publication of new names and new combinations previously effectively published outside the IJSB. List no. 20. Int. J. Syst. Bacteriol. *36*: 354–356.

Euzéby, J.P. 1998. Taxonomic note: necessary correction of specific and subspecific epithets according to Rules 12c and 13b of the International Code of Nomenclature of Bacteria (1990 revision). Int. J. Syst. Bacteriol. *48*: 1073–1075.

Evans, J.D. and S.A. Martin. 1997. Factors affecting lactate and malate utilization by *Selenomonas ruminantium*. Appl. Environ. Microbiol. *63*: 4853–4858.

Finegold, S.M., S.S. John, A.W. Vu, C.M. Li, D. Molitoris, Y. Song, C. Liu and H.M. Wexler. 2004. *In vitro* activity of ramoplanin and comparator drugs against anaerobic intestinal bacteria from the perspective of potential utility in pathology involving bowel flora. Anaerobe *10*: 205–211.

Flint, H.J. and C.S. Stewart. 1987. Antibiotic resistance patterns and plasmids of ruminal strains of *Bacteroides ruminicola* and *Bacteroides multiacidus*. Appl. Microbiol. Biotech. *26*: 450–455.

Flügge, C. 1886. Die Mikrooganismen. F.C.W. Vogel, Leipzig.

Focà, A., G. Matera, M.C. Liberto, M.C. Berlinghieri and S.K. Ng. 1990. *Veillonella* lipopolysaccharides reduce microbial activity of human neutrophils, but do not affect monocyte function. J. Immunol. Immunopharmacol. *10*: 151–152.

Forsberg, C.W. 1978. Nutritional characteristics of *Megasphaera elsdenii*. Can. J. Microbiol. *24*: 981–985.

Fredricks, D.N., T.L. Fiedler and J.M. Marrazzo. 2005. Molecular identification of bacteria associated with bacterial vaginosis. N. Engl. J. Med. *353*: 1899–1911.

Furtado, A.F., T.A. McAllister, K.J. Cheng and L.P. Milligan. 1994. Production of 2-aminobutyrate by *Megasphaera elsdenii*. Can. J. Microbiol. *40*: 393–396.

Gares, S.L., M.S. Whiting, W.M. Ingledew and B. Ziola. 1993. Detection and identification of *Pectinatus cerevisiiphilus* using surface-reactive monoclonal antibodies in a membrane filter-based fluoroimmunoassay. J. Am. Soc. Brew. Chem. *51*: 158–163.

Garner, M.R., J.F. Flint and J.B. Russell. 2002. *Allisonella histaminiformans* gen. nov., sp. nov. A novel bacterium that produces histamine, utilizes histicline as its sole energy source, and could play a role in bovine and equine laminitis. Syst. Appl. Microbiol. *25*: 498–506.

Garner, M.R., J. F. Flint and J.R. Russell. 2003. *In* Validation of publication of new names and new combinations previously effectively published outside the IJSB. List no. 90 Int. J. Syst. Bacteriol. *53*: 373.

Garrity, G.M. and J.G. Holt. 2001. Taxonomic outline of the Archaea and Bacteria. *In* Boone and Castenholz (Editors), Bargey's Manual of Systematic Bacteriology, 2nd edn, Vol. 1. Springer, New York, pp. 161–162.

Giesecke, D., S. Wiesmayr and M. Ledinek. 1970. *Peptostreptococcus elsdenii* from the caecum of pigs. J. Gen. Microbiol. *64*: 123–126.

Golyshina, O.V., T.A. Pivovarova, G.I. Karavaiko, T.F. Kondrat'eva, E.R.B. Moore, W.R. Abraham, H. Lunsdorf, K.N. Timmis, M.M. Yakimov and P.N. Golyshin. 2000. *Ferroplasma acidiphilum* gen. nov., sp. nov., an acidophilic, autotrophic, ferrous-iron-oxidizing, cell-wall-lacking, mesophilic member of the *Ferroplasmaceae* fam. nov., comprising a distinct lineage of the *Archaea*. Int. J. Syst. Evol. Microbiol. *50*: 997–1006.

Gössner, A.S., K. Küsel, Schulz, S. Trenz, G. Acker, C. R. Lovell and H.L. Drake. 2006a. *In* Validation of publication of new names and new combinations previously effectively published outside the IJSEM. List no. 110. Int. J. Syst. Evol. Microbiol. *56*: 1459–1460.

Gössner, A.S., K. Kusel, D. Schulz, S. Trenz, G. Acker, C.R. Lovell and H.L. Drake. 2006b. Trophic interaction of the aerotolerant anaerobe *Clostridium intestinale* and the acetogen *Sporomusa rhizae* sp. nov. isolated from roots of the black needlerush *Juncus roemerianus*. Microbiology *152*: 1209–1219.

Gottschalk, G. 1986. Bacterial Metabolism, 2nd edn. Springer-Verlag, New York.

Gu, A.Z., B.P. Hedlund, J.T. Staley, S.E. Strand and H.D. Stensel. 2004. Analysis and comparison of the microbial community structures of two enrichment cultures capable of reductively dechlorinating TCE and *cis*-DCE. Environ. Microbiol. *6*: 45–54.

Gürtler, V. and V.A. Stanisich. 1996. New approaches to typing and identification of bacteria using the 16S–23S rDNA spacer region. Microbiology *142*: 3–16.

Gutierrez, J., R.E. Davis, I.H. Lindahl and E.J. Warwick. 1959. Bacterial changes in the rumen during the onset of feed-lot bloat of cattle and characteristics of *Peptostreptococcus elsdenii* n. sp. Appl. Microbiol. *7*: 16–22.

Haapasalo, M., H. Ranta, H. Shah, K. Ranta, K. Lounatmaa and R.M. Kroppenstedt. 1986a. *Mitsuokella dentalis* sp. nov. from dental root canals. Int. J. Syst. Bacteriol. *36*: 566–568.

Haapasalo, M., H. Ranta, H. Shah, K. Ranta, K. Lounatmaa and R.M. Kroppenstedt. 1986b. Biochemical and structural characterization of an unusual group of Gram-negative, anaerobic rods from human periapical osteitis. J. Gen. Microbiol. *132*: 417–426.

Haikara, A., T.M. Enari and K. Lounatmaa. 1981a. The genus *Pectinatus*, a new group of anaerobic beer spoilage bacteria. Proc. 18th Congr. Eur. Brew. Conv., pp. 229–240, Copenhagen.

Haikara, A., L. Penttilä, T.-M. Enari and K. Lounatmaa. 1981b. Microbiological, biochemical and electron microscopic characterization of a *Pectinatus* strain. Appl. Environ. Microbiol. *41*: 511–517.

Haikara, A. 1983. Immunological characterization of *Pectinatus cerevisiophilus* strains. Appl. Environ. Microbiol. *46*: 1054–1058.

Haikara, A. 1984. Beer spoilage organisms. Occurrence and detection with particular reference to a new genus *Pectinatus*. PhD thesis, Technical Research Centre of Finland, Publication 14.

Haikara, A. 1985. Detection of anaerobic, Gram-negative bacteria in beer. Monatsschr. Brauwiss. *38*: 239–243.

Haikara, A. and K. Lounatmaa. 1987. Characterization of *Megasphaera* sp, a new anaerobic beer spoilage coccus. Proc. 21st Congr. Eur. Brew. Conv., pp. 473–480, Madrid.

Haikara, A. 1989. Invasion of anaerobic bacteria into pitching yeast. Proc. 22nd Congr. Eur. Brew. Conv., pp. 537–544, Zürich.

Haikara, A. 1991. The genera *Pectinatus* and *Megasphaera*. *In* Balows, Trüper, Dworkin, Harder and Schleifer (Editors), The Prokaryotes, 2nd edn, Vol. 2. Springer-Verlag, New York, pp. 1993–2004.

Hakalehto, E., A. Haikara, T.M. Enari and K. Lounatmaa. 1984. Hydrochloric acid extractable protein patterns of *Pectinatus cerevisiophilus* strains. Food Microbiol. *1*: 209–216.

Hakalehto, E. and J. Finne. 1990. Identification by immunoblot analysis of major antigenic determinants of the anaerobic beer spoilage bacterium genus *Pectinatus*. FEMS Microbiol. Lett. 67: 307–312.

Hakalehto, E. 2000. Characterization of *Pectinatus cerevisiiphilus* and *P. frisingiensis* surface components. Use of synthetic peptides in the detection of some gram-negative bacteria. PhD thesis, Kuopio University C. Natural and Environmental Sciences.

Hamana, K., Y. Nakagawa and K. Yamasato. 1995. Chemotaxonomic significance of polyamine distribution patterns in the *Flavobacterium-Cytophaga* complex and related genera. Microbios *81*: 135–145.

Hamana, K., T. Saito, M. Okada, A. Sakamoto and R. Hosoya. 2002. Covalently linked polyamines in the cell wall peptidoglycan of *Selenomonas, Anaeromusa, Dendrosporobacter, Acidaminococcus* and *Anaerovibrio* belonging to the *Sporomusa* subbranch. J. Gen. Appl. Microbiol. *48*: 177–180.

Hans, M., J. Sievers, U. Muller, E. Bill, J.A. Vorholt, D. Linder and W. Buckel. 1999. 2-Hydroxyglutaryl-CoA dehydratase from *Clostridium symbiosum*. Eur. J. Biochem. *265*: 404–414.

Haralambie, E. 1983. *Megasphaera elsdenii*, occurrence in 2255 fecal samples from men, chimpanzees and mice. Zentbl. Bakteriol. Mikrobiol. Hyg. Ser. A *253*: 489–494.

Harborth, P.B. and H. H. Hanert. 1982. Isolation of *Selenomonas ruminantium* from an aquatic ecosystem. Arch. Microbiol. *132*: 135–145.

Harfoot, C.G. and G.P. Hazlewood. 1997. Lipid metabolism in the rumen. *In* Hobson and Bryant (Editors), The Rumen Microbial Ecosystem, 2nd edn. Elsevier Applied Science Publishers, London, pp. 382–426.

Härtel, U. and W. Buckel. 1996. Fermentation of *trans*-aconitate via citrate, oxaloacetate, and pyruvate by *Acidaminococcus fermentans*. Arch. Microbiol. *166*: 342–349.

Haukeli, A.D. 1980. En ny ølskadelig bakterie i tappet øl. Referat från det 18. Skandinaviska Bryggeritekniska Mötet, Stockholm, pp. 112–122.

Helander, I., E. Hakalehto, J. Ahvenainen and A. Haikara. 1983. Characterization of lipopolysaccharides of *Pectinatus cerevisiophilus*. FEMS Microbiol. Lett. *18*: 223–226.

Helander, I., K. Saukkonen, E. Hakalehto and M. Vaara. 1984. Biological activities of lipopolysaccharides from *Pectinatus cerevisiophilus*. FEMS Microbiol. Lett. *24*: 39–42.

Helander, I.M., R. Hurme, A. Haikara and A.P. Moran. 1992. Separation and characterization of two chemically distinct lipopolysaccharides in two *Pectinatus* species. J. Bacteriol. *174*: 3348–3354.

Helander, I.M., H. Moll and U. Zähringer. 1993. 4-O-(2-amino-2-deoxy-α-D-glucopyranosyl)-3-deoxy-D-manno-2-octulosonic acid, a constituent of lipopolysaccharides of the genus *Pectinatus*. Eur. J. Biochem. *213*: 377–381.

Helander, I.M., I. Kilpeläinen, M. Vaara, A.P. Moran, B. Lindner and U. Seydel. 1994. Chemical structure of the lipid A component of lipopolysaccharides of the genus *Pectinatus*. Eur. J. Biochem. *224*: 63–70.

Helander, I.M. and A. Haikara. 1995. Cellular fatty acyl and alkenyl residues in *Megasphaera* and *Pectinatus* species: contrasting profiles and detection of beer spoilage. Microbiology *141*: 1131–1137.

Helander, I.M., Y. Kato, I. Kilpeläinen, R. Kostiainen, B. Lindner, K. Nummila, T. Sugiyama and T. Yokochi. 1996. Characterization of lipopolysaccharides of polymyxin-resistant and polymyxin-sensitive *Klebsiella pneumoniae* O3. Eur. J. Biochem. *237*: 272–278.

Helander, I.M. 2003. Lipopolysaccharides of anaerobic beer spoilage bacteria of the genus *Pectinatus*. PhD thesis, University of Helsinki.

Henderson, C. and P.N. Hobson. 1968. The lipase of a rumen bacterium. J. Gen. Microbiol. *53*: Suppl:8.

Henderson, C. 1970. The lipases produced by *Anaerovibrio lipolytica* in continuous culture. Biochem. J. *119*: 5P–6P.

Henderson, C. 1971. A study of the lipase produced by *Anaerovibrio lipolytica*, a rumen bacterium. J. Gen. Microbiol. *65*: 81–89.

Henderson, C. 1973a. The effects of fatty acids on pure cultures of rumen bacteria. J. Agric. Sci. Camb. *81*: 107–112.

Henderson, C. 1973b. An improved method for enumerating and isolating lipolytic rumen bacteria. J. Appl. Bacteriol. *36*: 187–188.

Henderson, C. and W. Hodgkiss. 1973. An electron microscopic study of *Anaerovibrio lipolytica* (strain 5S) and its lipolytic enzyme. J. Gen. Microbiol. *76*: 389–393.

Henderson, C. 1980. The influence of extracellular hydrogen on the metabolism of *Bacteroides ruminicola*, *Anaerovibrio lipolytica* and *Selenomonas ruminantium*. J. Gen. Microbiol. *119*: 485–491.

Hermann, M., M.R. Popoff and M. Sebald. 1987. *Sporomusa paucivorans* sp. nov, a methylotrophic bacterium that forms acetic acid from hydrogen and carbon dioxide. Int. J. Syst. Bacteriol. *37*: 93–101.

Hespell, R.B., B.J. Paster and F.E. Dewhirst. 1992. The genus *Selenomonas*. *In* Balows, Truper, Dworkin, Harder and Schleifer (Editors), The Prokaryotes, Vol. II. Springer, New York, pp. 2005–2013.

Hewett, M.J., K.W. Knox and D.G. Bishop. 1971. Biochemical studies on lipopolysaccharides of *Veillonella*. Eur. J. Biochem. *19*: 169–175.

Hilpert, W. and P. Dimroth. 1984. Conversiojn of the chemical energy of methylmalonyl-CoA decarboxylation into a Na-gradient. Nature (Lond) *296*: 584–585.

Hino, T. and S. Kuroda. 1993. Presence of lactate Dehydrogenase and lactate racemase in *Megasphaera elsdenii* grown on glucose or lactate. Appl. Environ. Microbiol. *59*: 255–259.

Hippe, H. 1991. Maintenance of methanogenic bacteria. *In* Kirsop and Doyle (Editors), Maintenance of micoorganisms and cultured cells, nd edn. Academic Press, London, pp. 101–113.

Hirao, T., M. Sato, A. Shirahata and Y. Kamio. 2000. Covalent linkage of polyamines to peptidoglycan in *Anaerovibrio lipolytica*. J. Bacteriol. *182*: 1154–1157.

Hiroki, H., J. Shiiki, A. Handa, M. Totsuka and O. Nakamura. 1976. Isolation of bacteriophages specific for the genus *Veillonella*. Arch. Oral. Biol. *21*: 215–217.

Hobson, P.N. and S.O. Mann. 1961. The isolation of glycerol-fermenting and lipolytic bacteria from the rumen of the sheep. J. Gen. Microbiol. *25*: 227–240.

Hobson, P.N., S.O. Mann and W. Smith. 1962. Serological tests of a relationship between rumen selenomonads *in vitro* and *in vivo*. J. Gen. Microbiol. *29*: 265–270.

Hobson, P.N. and R. Summers. 1966. Effect of growth rate on the lipase activity of a rumen bacterium. Nature *209*: 736–737.

Hobson, P.N. and R. Summers. 1967. The continuous culture of anaerobic bacteria. J. Gen. Microbiol. *47*: 53–65.

Hofstad, T. and T. Kristoffersen. 1970. Chemical composition of endotoxin from oral *Veillonella*. Acta Path. Microbiol. Scand. Section B 78: 760–764.

Holdeman, L.V. and W.E.C. Moore (Editors). 1970. *Bacteroides*. Virginia Polytechnic Institute, Anaerobe Laboratory, Blacksburg, VA.

Holdeman, L.V. and W.E.C. Moore (Editors). 1972. Anaerobe Laboratory Manual. Anaerobe Laboratory, Virginia Polytechnic Institute and State University, Blacksburg, VA.

Holdeman, L.V., E.P. Cato and W.E.C. Moore (Editors). 1977. Anaerobe Laboratory Manual, 4th edn. Anaerobe Laboratory, Virginia Polytechnic Institute and State University, Blacksburg, VA.

Holdeman, L.V., R. W. Kelley and W. E. C. Moore. 1986. Genus *Bacteroides* Castellani and Chalmers 1919. *In* Sneath, Mair, Sharpe and Holt (Editors), Bergey's Manual of Systematic Bacteriology, Vol. 2. The Williams & Wilkins Co., Baltimore, pp. 604–631.

Homann, F., Z. Bremer and G. Möller-Hergt. 2002. LightCycler™ PCR establishing itself as a rapid detection method. Beer-spoilage bacteria in microbiological routine analysis. Brauwelt Int. V: 292–296.

Horie, H., K. Kanazawa, M. Okada, S. Narushima, K. Itoh and A. Terada. 1999. Effects of intestinal bacteria on the development of colonic neoplasm: an experimental study. Eur. J. Cancer Prev. *8*: 237–245.

Houston, S., D. Taylor and R. Rennie. 1997. Prosthetic valve endocarditis due to *Veillonella dispar*: successful medical treatment following penicillin desensitization. Clin. Infect. Dis. *24*: 1013–1014.

Hughes, C.V., P.E. Kolenbrander, R.N. Andersen and L.V. Moore. 1988. Coaggregation properties of human oral *Veillonella* spp.: relationship to colonization site and oral ecology. Appl. Environ. Microbiol. *54*: 1957–1963.

Hungate, R.E. 1966. The Rumen and Its Microbes. Academic Press, New York.

Hungate, R.E. 1969. A roll tube method for cultivation of strict anaerobes. *In* Norris and Ribbons (Editors), Methods in Microbiology, Vol. 3B. Academic Press, London and New York, pp. 117–132.

Isner-Horobeti, M.E., J. Lecocq, A. Dupeyron, S.J. De Martino, P. Froehlig and P. Vautravers. 2006. *Veillonella discitis*. A case report. Joint Bone Spine *73*: 113–115.

Jansen, M. and T.A. Hansen. 2001. Non-growth-associated demethylation of dimethylsulfoniopropionate by (homo)acetogenic bacteria. Appl. Environ. Microbiol. *67*: 300–306.

Janssen, P.H. 1990. Fermentation of glycollate by a mixed culture of anaerobic bacteria. Syst. Appl. Microbiol. *13*: 327–332.

Janssen, P.H. 1991. Characterization of a succinate-fermenting anaerobic bacterium isolated from a glycolate-degrading mixed culture. Arch. Microbiol. *155*: 288–293.

Janssen, P.H. and K.A. O'Farrell. 1999. *Succinispira mobilis* gen. nov., sp. nov., a succinate-decarboxylating anaerobic bacterium. Int. J. Syst. Bacteriol. *49*: 1009–1013.

John, A., H.R. Isaacson and M.P. Bryant. 1974. Isolation and characteristics of a ureolytic strain of *Selenomonas ruminantium*. J. Dairy Sci. *57*: 1003–1014.

Johnston, N.C. and H. Goldfine. 1982. Effects of growth temperature on fatty acid and alk-1-enyl group compositions of *Veillonella parvula* and *Megasphaera elsdenii* phospholipids. J. Bacteriol. *149*: 567–575.

Jousimies-Somer, H.R., A. Bryk, S. Asikainen, A. Kanervo, A. Takala and E. Könönen. 1999. Oral colonization of infants with *Veillonella* species. Anaerobe *5*: 251–253.

Jousimies-Somer, H.R., P. Summanen, D.M. Citron, E.J. Baron, H.M. Wexler and S.M. Finegold. 2002. Wadsworth-KTL anaerobic bacteriology manual. Star Publishing Company, Belmont, CA.

Jumas-Bilak, E., J.P. Carlier, H. Jean-Pierre, C. Teyssier, B. Gay, J. Campos and H. Marchandin. 2004. *Veillonella montpellierensis* sp. nov., a novel, anaerobic, Gram-negative coccus isolated from human clinical samples. Int. J. Syst. Evol. Microbiol. *54*: 1311–1316.

Jumas-Bilak, E., H. Jean-Pierre, J.P. Carlier, C. Teyssier, K. Bernard, B. Gay, J. Campos, F. Morio and H. Marchandin. 2005. *Dialister micraerophilus* sp. nov. and *Dialister propionicifaciens* sp. nov., isolated from human clinical samples. Int. J. Syst. Evol. Microbiol. *55*: 2471–2478.

Juvonen, R., T. Koivula and A. Haikara. 2003. PCR detection of Gram-negative brewery contaminants - present state. Proc. 29th Congr. Eur. Brew. Conv., pp. 1047–1056, Dublin.

Juvonen, R. and M.L. Suihko. 2006. *Megasphaera paucivorans* sp. nov., *Megasphaera sueciensis* sp. nov. and *Pectinatus haikarae* sp. nov., isolated from brewery samples, and emended description of the genus *Pectinatus*. Int. J. Syst. Evol. Microbiol. *56*: 695–702.

Juvonen, R., T. Koivula and A. Haikara. 2008. Group-specific PCR-RFLP and real-time PCR methods for detection and tentative discrimination of strictly anaerobic beer-spoilage bacteria of the class *Clostridia*. Int. J. Food Microbiol. *125*: 162–169.

Kalmokoff, M.L., J.W. Austin, M.F. Whitford and R.M. Teather. 2000. Characterization of a major envelope protein from the rumen anaerobe *Selenomonas ruminantium* OB268. Can. J. Microbiol. *46*: 295–303.

Kamio, Y. and H. Takahashi. 1980. Outer membrane proteins and cell surface structure of *Selenomonas ruminantium.* J. Bacteriol. *141*: 899–907.

Kamio, Y., Y. Itoh and Y. Terawaki. 1981a. Chemical structure of peptidoglycan in *Selenomonas ruminantium*: cadaverine links covalently to the D-glutamic acid residue of peptidoglycan. J. Bacteriol. *146*: 49–53.

Kamio, Y., Y. Itoh, Y. Terawaki and T. Kusano. 1981b. Cadaverine is covalently linked to peptidoglycan in *Selenomonas ruminantium.* J. Bacteriol. *145*: 122–128.

Kane, M. and J.A. Breznak. 1992. *In* Validation of the publication of new names and new combinations previously effectively published outside the IJSB. List no. 40. Int. J. Syst. Bacteriol. *42*: 191–192.

Kane, M.D. and J.A. Breznak. 1991. *Acetonema longum* gen. nov. sp. nov., an H$_2$/CO$_2$ acetogenic bacterium from the termite, *Pterotermes occidentis.* Arch. Microbiol. *156*: 91–98.

Kanegasaki, S. and H. Takahashi. 1967. Function of growth factors for rumen microorganisms. I. Nutritional characteristics of *Selenomonas ruminantium.* J. Bacteriol. *93*: 456–463.

Karnholz, A., K. Kusel, A. Gössner, A. Schramm and H.L. Drake. 2002. Tolerance and metabolic response of acetogenic bacteria toward oxygen. Appl. Environ. Microbiol. *68*: 1005–1009.

Kaufman, A.E., J.N. Verma and H. Goldfine. 1988. Disappearance of plasmalogen-containing phospholipids in *Megasphaera elsdenii.* J. Bacteriol. *170*: 2770–2774.

Kaufman, A.E., H. Goldfine, O. Narayan and S.M. Gruner. 1990. Physical studies on the membranes and lipids of plasmalogen-deficient *Megasphaera elsdenii.* Chem. Phys. Lipids *55*: 41–48.

Kingsley, V.V. and J.F. Hoeniger. 1973. Growth, structure, and classification of *Selenomonas.* Bacteriol. Rev. *37*: 479–521.

Kirchner, G., R. Lurz and K. Matsuzawa. 1980. Biertrübungen durch Bakterien der Gattung *Bacteroides.* Monatsschr. Brauwiss. *33*: 461–467.

Klugbauer, N., W. Ludwig, E. Bäuerlein and K.H. Schleifer. 1992. Subunit b of adenosine triphosphate synthase of *Pectinatus frisingensis* and *Lactobacillus casei.* Syst. Appl. Microbiol. *15*: 323–330.

Kokeguchi, S., O. Tsutsui, K. Kato and T. Matsumura. 1990. Isolation and characterization of lipopolysaccharide from *Centipeda periodontii* ATCC 35019. Oral Microbiol. Immunol. *5*: 108–112.

Kolenbrander, P. 2006. The genus *Veillonella. In* Dworkin, Rosenberg and Stackebrandt (Editors), The Prokaryotes, a handbook on the biology of bacteria, *Bacteria: Firmicutes, Cyanobacteria*, 3rd edn, Vol. 4. Springer, New York, pp. 1022–1040.

Kononen, E., S. Asikainen and H. Jousimies-Somer. 1992. The early colonization of gram-negative anaerobic bacteria in edentulous infants. Oral Microbiol. Immunol. *7*: 28–31.

Kononen, E., A. Kanervo, A. Takala, S. Asikainen and H. Jousimies-Somer. 1999. Establishment of oral anaerobes during the first year of life. J. Dent. Res. *78*: 1634–1639.

Krumholz, L.R., M.P. Bryant, W.J. Brulla, J.L. Vicini, J.H. Clark and D.A. Stahl. 1993. Proposal of *Quinella ovalis* gen. nov., sp.nov., based on phylogenetic analysis. Int. J. Syst. Bacteriol. *43*: 293–296.

Kuhner, C.H., C. Frank, A. Griesshammer, M. Schmittroth, G. Acker, A. Gössner and H.L. Drake. 1997. *Sporomusa silvacetica* sp. nov., an acetogenic bacterium isolated from aggregated forest soil. Int. J. Syst. Bacteriol. *47*: 352–358.

Kumada, H., K. Watanabe, A. Nakamu, Y. Haishima, S. Kondo, K. Hisatsune and T. Umemoto. 1997. Chemical and biological properties of lipopolysaccharide from *Selenomonas sputigena* ATCC 33150. Oral Microbiol. Immunol. *12*: 162–167.

Kurimoto, T., C. Tachibana, M. Suzuki and T. Watanabe. 1986. Biological and chemical characterization of lipopolysaccharide from *Selenomonas* spp. in human periodontal pockets. Infect. Immun. *51*: 969–971.

Lai, C.H., B.M. Males, P.A. Dougherty, P. Berthold and M.A. Listgarten. 1983. *Centipeda periodontii* gen. nov., sp. nov. from human periodontal lesions. Int. J. Syst. Bacteriol. *33*: 628–635.

Lan, G.Q., N. Abdullah, S. Jalaludin and Y.W. Ho. 2002a. Culture conditions influencing phytase production of *Mitsuokella jalaludinii*, a new

bacterial species from the rumen of cattle. J. Appl. Microbiol. *93*: 668–674.

Lan, G.Q., Y.W. Ho and N. Abdullah. 2002b. *Mitsuokella jalaludinii* sp. nov., from the rumens of cattle in Malaysia. Int. J. Syst. Evol. Microbiol. *52*: 713–718.

Lapage, S.P., P.H.A. Sneath, E.F. Lessel, V.B.D. Skerman, H.P.R. Seeliger and W.A. Clark. 1992. International Code of Nomenclature of Bacteria (1990 Revision). Bacteriological Code. Published for IUMS by the American Society for Microbiology, Washington, DC.

Lee, S.Y., M.S. Mabee and N.O. Jangaard. 1978. *Pectinatus*, a new genus of family *Bacteroidaceae.* Int. J. Syst. Bacteriol. *28*: 582–594.

Lee, S.Y., M.S. Mabee, N.O. Jangaard and E.K. Horiuchi. 1980. *Pectinatus*, a new genus of bacteria capable of growth in hopped beer. J. Inst. Brew. *86*: 28–30.

Lee, S.Y. 1984. Genus XI. *Pectinatus. In* Krieg and Holt (Editors), Bergey's Manual of Systematic Bacteriology, Vol. 1. The Williams & Wilkins Co., Baltimore, pp. 655–658.

Lee, S.Y. 1994. SMMP: A medium for selective isolation of *Megasphaera* and *Pectinatus* from the brewery. J. Am. Soc. Brew. Chem *52*: 115–119.

Leifson, E. 1960. Atlas of bacterial flagellation. Academic Press, New York.

Leng, Z., D.E. Riley, R.E. Berger, J.N. Krieger and M.C. Roberts. 1997. Distribution and mobility of the tetracycline resistance determinant *tetQ.* J. Antimicrob. Chemother. *40*: 551–559.

Linehan, B., C.C. Scheifinger and M.J. Wolin. 1978. Nutritional requirements of *Selenomonas ruminantium* for growth on lactate, glycerol, or glucose. Appl. Environ. Microbiol. *35*: 317–322.

Liu, J.W., J.J. Wu, L.R. Wang, L.J. Teng and T.C. Huang. 1998. Two fatal cases of *Veillonella* bacteremia. Eur. J. Clin. Microbiol. Infect. Dis. *17*: 62–64.

Lockington, R.A., G.T. Attwood and J.D. Brooker. 1988. Isolation and characterization of a temperate bacteriophage from the ruminal anaerobe *Selenomonas ruminantium.* Appl. Environ. Microbiol. *54*: 1575–1580.

Loesche, W.J., R.J. Gibbons and S.S. Socransky. 1965. Biochemical characteristics of *Vibrio sputorum* and relationship to *Vibrio bubulus* and *Vibrio fetus.* J. Bacteriol. *89*: 1109–1116.

Ludwig, W., G. Kirchhof, N. Klugbauer, M. Weizenegger, D. Betzl, M. Ehrmann, C. Hertel, S. Jilg, R. Tatzel, H. Zitzelberger, S. Liebl, M. Hochberger, J. Shah, D. Lane, P.R. Wallnofer and K.H. Schleifer. 1992. Complete 23S ribosomal RNA sequences of Gram-positive bacteria with a low DNA G+C content. Syst. Appl. Microbiol. *15*: 487–501.

Ludwig, W., K.H. Schleifer and W.B. Whitman. 2009. Revised road map to the phylum *Firmicutes. In* De Vos, Garrity, Jones, Krieg, Ludwig, Rainey, Schleifer and Whitman (Editors), Bergey's Manual of Systematic Bacteriology, 2nd edn, Vol. 3, The *Firmicutes*, Springer, New York, pp. 1–14.

MacDonald, J.B., E.B. Madlener and S.S. Socransky. 1953. Observations on *Spirillum sputigenum* and its relationship with *Selenomonas* species with special reference to flagellation. J. Bacteriol. *77*: 559.

MacDonald, J.B. and E. M. Madlener. 1957. Studies on the isolation of *Spirillum sputigenum.* J. Dent. Res. *34*: 709–714.

MacDonald, J.B., S. Socransky and S. Sawyer. 1959. A survey of the bacterial flora of the periodontium in the rice rat. Arch. Oral Biol. *1*: 1–7.

Marchandin, H. 2001. Organisation génomique, phylogéenie et taxonomie polyphasique des bactéries du genre *Veillonella* et des genres apparentés du sous-groupe *Sporomusa.* Université Montpelier I, Montpelier, France.

Marchandin, H., H. Jean-Pierre, C. Carriere, F. Canovas, H. Darbas and E. Jumas-Bilak. 2001. Prosthetic joint infection due to *Veillonella dispar.* Eur. J. Clin. Microbiol. Infect. Dis. *20*: 340–342.

Marchandin, H., C. Teyssier, M. Siméon de Buochberg, H. Jean-Pierre and E. Jumas-Bilak. 2003a. Genome sizing and rrn copy numbering in some representative members of the *Sporomusa* sub-branch. FEMS Microbiol. Lett. *222*: 173 abstract P3-16.

Marchandin, H., E. Jumas-Bilak, B. Gay, C. Teyssier, H. Jean-Pierre, M.S. de Buochberg, C. Carriere and J.P. Carlier. 2003b. Phylogenetic analysis of some *Sporomusa* sub-branch members isolated from human clinical specimens: description of *Megasphaera micronuciformis* sp. nov. Int. J. Syst. Evol. Microbiol. *53*: 547–553.

Marchandin, H., C. Teyssier, M. Simeon De Buochberg, H. Jean-Pierre, C. Carriere and E. Jumas-Bilak. 2003c. Intra-chromosomal heterogeneity between the four 16S rRNA gene copies in the genus *Veillonella*: implications for phylogeny and taxonomy. Microbiology *149*: 1493–1501.

Marchandin, H., H. Jean-Pierre, J. Campos, L. Dubreuil, C. Teyssier and E. Jumas-Bilak. 2004. nimE gene in a metronidazole-susceptible *Veillonella* sp. strain. Antimicrob. Agents Chemother. *48*: 3207–3208.

Marchandin, H., C. Teyssier, E. Jumas-Bilak, M. Robert, A.C. Artigues and H. Jean-Pierre. 2005. Molecular identification of the first human isolate belonging to the *Veillonella ratti–Veillonella criceti* group based on 16S rDNA and *dnaK* gene sequencing. Res. Microbiol. *156*: 603–607.

Marchandin, H. and E. Jumas-Bilak. 2006. 16S rRNA gene sequencing: Interest and limits for identification and characterization of novel taxa within the family *Acidaminococcaceae*. *In* McNamara (Editors), Trends in RNA Research. Nova Science, New York,, pp. 225–251.

Marounek, M., K. Fliegrova and S. Bartos. 1989. Metabolism and some characteristics of ruminal strains of *Megasphaera elsdenii*. Appl. Environ. Microbiol. *55*: 1570–1573.

Martin, S.A. 1994. Nutrient transport by ruminal bacteria: a review. J. Anim. Sci. *72*: 3019–3031.

Martin, S.A. 1998. Manipulation of ruminal fermentation with organic acids: a review. J. Anim. Sci. *76*: 3123–3132.

Martin, S.A. and E.L. Wani. 2000. Factors affecting glucose and maltose phosphorylation by the ruminal bacterium *Megasphaera elsdenii*. Curr. Microbiol. *40*: 387–391.

Mays, T.D., L.V. Holdeman, W.E.C. Moore, M. Rogosa and J.L. Johnson. 1982. Taxonomy of the genus *Veillonella* Prévot. Int. J. Syst. Bacteriol. *32*: 28–36.

McCarthy, L.R. and J.R. Carlson. 1981. *Selenomonas sputigena* septicemia. J. Clin. Microbiol. *14*: 684–685.

Melville, S.B., T.A. Michel and J.M. Macy. 1988. Pathway and sites for energy conservation in the metabolism of glucose by *Selenomonas ruminantium*. J. Bacteriol. *170*: 5298–5304.

Mergenhagen, S.E., E.G. Hampp and H.W. Scherp. 1961. Preparation and biological activities of endotoxins from oral bacteria. J. Infect. Dis. *108*: 304–310.

Miller, T.L. and M.J. Wolin. 1974. A serum bottle modification of the Hungate technique for cultivating obligate anaerobes. Appl. Microbiol. *27*: 985–987.

Mitsuoka, T., T. Sega and S. Yamamoto. 1964. A new selective medium for *Bacteroides*. Zentralbl. Bakteriol. [Orig.] *195*: 69–79.

Mitsuoka, T., T. Sega and S. Yamamoto. 1965. Improved methodology of qualitative and quantitative analysis of the intestinal flora of man and animals. Zentralbl. Bakteriol. [Orig.] *195*: 455–469.

Mitsuoka, T., A. Terada, K. Watanabe and K. Uchida. 1974. *Bacteroides multiacidus*, a new species from feces of humans and pigs. Int. J. Syst. Bacteriol. *24*: 35–41.

Miwa, T., H. Esaki, J. Umemori and T. Hino. 1997. Activity of H⁺-ATPase in ruminal bacteria with special reference to acid tolerance. Appl. Environ. Microbiol. *63*: 2155–2158.

Miyagawa, E., R. Azuma and T. Suto. 1978. Distribution of sphingolipids in *Bacteroides* species. J. Gen. Appl. Microbiol. *24*: 341–348.

Miyagawa, E., R. Azuma and T. Suto. 1979. Cellular fatty acid composition in Gram-negative obligately anaerobic rods. J. Gen. Microbiol. *25*: 41–51.

Möller, B., R. Ossmer, B.H. Howard, G. Gottschalk and H. Hippe. 1984. *Sporomusa*, a new genus of Gram-negative anaerobic bacteria including *Sporomusa sphaeroides* spec. nov. and *Sporomusa ovata* spec. nov. Arch. Microbiol. *139*: 388–396.

Möller, B., R. Ossmer, B. H. Howard, G. Gottschalk and H. Hippe. 1985. *In* Validation of the publication of new names and new combinations previously effectively published outside the IJSB. List no. 17. Int. J. Syst. Bacteriol. *35*: 223–225.

Moore, L.V.H., J.L. Johnson and W.E.C. Moore. 1987. *Selenomonas noxia* sp. nov, *Selenomonas flueggei* sp. nov., *Selenomonas infelix* sp. nov, *Selenomonas dianae* sp. nov, and *Selenomonas artemidis* sp. nov, from the human gingival crevice. Int. J. Syst. Bacteriol. *37*: 271–280.

Moore, L.V.H., D.M. Bourne and W.E.C. Moore. 1994. Comparative distribution and taxonomic value of cellular fatty acids in 33 genera of anaerobic gram-negative bacilli. Int. J. Syst. Bacteriol. *44*: 338–347.

Moore, L.V.H. and W.E.C. Moore. 1994. *Oribaculum catoniae* gen. nov., sp. nov., *Catonella morbi* gen. nov., sp. nov., *Hallella seregens* gen. nov., sp. nov., *Johnsonella ignava* gen. nov., sp. nov., and *Dialister pneumosintes* gen. nov., comb. nov., nom. rev., anaerobic Gram-negative bacilli from the human gingival crevice. Int. J. Syst. Bacteriol. *44*: 187–192.

Motoyama, Y., T. Ogata and K. Sakai. 1998. Characterization of *Pectinatus cerevisiiphilus* and *P. frisingensis* by ribotyping. J. Am. Soc. Brew. Chem. *56*: 19–23.

Motoyama, Y. and T. Ogata. 2000a. 16S-23S rDNA spacer of *Pectinatus*, *Selenomonas* and *Zymophilus* reveal new phylogenetic relationships between these genera. Int. J. Sys. Evol. Microbiol. *50*: 883–886.

Motoyama, Y. and T. Ogata. 2000b. Detection of *Pectinatus* spp. by PCR using 16S-23S rDNA spacer regions. J. Am. Soc. Brew. Chem. *58*: 4–7.

Müller, V., F. Inkamp, A. Rauwolf, K. Küsel and H.L. Drake. 2004. Molecular and cellular biology of acetogenic bacteria. *In* Nakano and Zuber (Editors), Strict and Facultative Anaerobes: Medical and Environmental Aspects, Horizon Bioscience, Norfolk, United Kingdom, pp. 251–281.

Munson, M.A., T. Pitt-Ford, B. Chong, A. Weightman and W.G. Wade. 2002. Molecular and cultural analysis of the microflora associated with endodontic infections. J. Dent. Res. *81*: 761–766.

Munson, M.A., A. Banerjee, T.F. Watson and W.G. Wade. 2004. Molecular analysis of the microflora associated with dental caries. J. Clin. Microbiol. *42*: 3023–3029.

Nagaraja, T.G. and M.B. Taylor. 1987. Susceptibility and resistance of ruminal bacteria to antimicrobial feed additives. Appl. Environ. Microbiol. *53*: 1620–1625.

Nanninga, H.J. and J.C. Gottschal. 1985. Amino acid fermentation and hydrogen transfer in mixed cultures. FEMS Microbiol. Ecol. *31*: 261–269.

Nanninga, H.J., W.J. Drent and J.C. Gottschal. 1987. Fermentation of glutamate by *Selenomonas acidaminophila* sp. nov. Arch. Microbiol. *147*: 152–157.

National Committee for Clinical Laboratory Standards. 2003. Methods for antimicrobial susceptibility testing of anaerobic bacteria, 6th edn, Approved standards M11-A6. National Committee for Clinical Laboratory Standards, Wayne, PA.

Neefs, J.M., Y. Van de Peer, P. De Rijk, A. Goris and R. De Wachter. 1991. Compilation of small ribosomal subunit RNA sequences. Nucleic Acids Res. *19 Suppl.*: 1987–2015.

Nisbet, D.J. and S.A. Martin. 1994. Factors affecting L-lactate utilization by *Selenomonas ruminantium*. J. Anim. Sci. *72*: 1355–1361.

Ogimoto, K. 1972. Uber *Selenomonas* aus dem Caecum von Ratten. Zentralbl. Bakteriol. Parasitenk. Infektionskr. Abt. I. Orig. *221*: 467–473.

Olitsky, P.K. and F.L. Gates. 1921. Experimental studies of the nasopharyngeal secretions from influenza patients. J. Exp. Med. *33*: 713–729.

Ollivier, B., R. Cord-Ruwisch, A. Lombardo and J.L. Garcia. 1985. Isolation and characterization of *Sporomusa acidovorans* sp. nov., a methylotrophic homoacetogenic bacterium. Arch. Microbiol. *142*: 307–310.

Ollivier, B., R. Cord-Ruwisch, A. Lombardo and J. -L. Garcia. 1990. *In* Validation of the publication of new names and new combinations previously effectively published outside the IJSB. List no. 32. Int. J. Syst. Bacteriol. *40*: 105–106.

Orpin, C.G. 1972. The culture *in vitro* of the rumen bacterium Quin's oval. J. Gen. Microbiol. *73*: 523–530.

Osawa, R., T. Fujisawa and T. Mitsuoka. 1992. Characterization of Gramnegative anaerobic strains, isolated from koala feces, which exhibit satellite growth and pleomorphism. Syst. Appl. Microbiol. *15*: 628–635.

Outtara, A.S., A.S. Traore and J.L. Garcia. 1992. Characterization of *Anaerovibrio burkinabensis* sp. nov., a lactate-fermenting bacterium isolated from rice field soils. Int. J. Syst. Bacteriol. *42*: 390–397.

Ouwerkerk, D., A.V. Klieve and R.J. Forster. 2002. Enumeration of *Megasphaera elsdenii* in rumen contents by real-time *Taq* nuclease assay. J. Appl. Microbiol. *92*: 753–758.

Palmer, R.J., Jr., P.I. Diaz and P.E. Kolenbrander. 2006. Rapid succession within the *Veillonella* population of a developing human oral biofilm in situ. J. Bacteriol. *188*: 4117–4124.

Paster, B.J., S.K. Boches, J.L. Galvin, R.E. Ericson, C.N. Lau, V.A. Levanos, A. Sahasrabudhe and F.E. Dewhirst. 2001. Bacterial diversity in human subgingival plaque. J. Bacteriol. *183*: 3770–3783.

Piknova, M., M. Filova, P. Javorsky and P. Pristas. 2005. A unique pair of GATC specific DNA methyltransferases in *Mitsuokella multiacida*. Mol. Biol. Rep. *32*: 281–284.

Piriz, S., R. Cuenca, J. Valle and S. Vadillo. 1992. Susceptibilities of anaerobic bacteria isolated from animals with ovine foot rot to 28 antimicrobial agents. Antimicrob. Agents Chemother. *36*: 198–201.

Pomeroy, C., C.J. Shanholtzer and L.R. Peterson. 1987. *Selenomonas* bacteraemia–case report and review of the literature. J. Infect. *15*: 237–242.

Prévot, A.R. 1933. Etudes de systematique bactérienne. I. Lois générales. II. Cocci anaérobies. Ann. Sci. Nat. Bot. *15*: 23–261.

Prins, R.A. 1971. Isolation, culture, and fermentation characteristics of *Selenomonas ruminantium* var. bryantivar. n. from the rumen of sheep. J. Bacteriol. *105*: 820–825.

Prins, R.A., A. Lankhorst, P. van der Meer and C.J. Van Nevel. 1975. Some characteristics of *Anaerovibrio lipolytica* a rumen lipolytic organism. Antonie van Leeuwenhoek *41*: 1–11.

Pristas, P., V. Molnarova and P. Javorsky. 2001. Restriction and modification systems of ruminal bacteria. Folia. Microbiol. *46*: 71–72.

Quin, J.I. 1943. Studies on the alimentary tracts of Merino sheep in South Africa. VII Fermentation in forestomachs of sheep. Onderstepoort J. Vet. Sci. Anim. Ind. *18*: 91–112.

Ready, D., H. Lancaster, F. Qureshi, R. Bedi, P. Mullany and M. Wilson. 2004. Effect of amoxicillin use on oral microbiota in young children. Antimicrob. Agents Chemother. *48*: 2883–2887.

Reig, M., N. Mir and F. Baquero. 1997. Penicillin resistance in *Veillonella*. Antimicrob. Agents Chemother. *41*: 1210.

Ricke, S.C., S.A. Martin and D.J. Nisbet. 1996. Ecology, metabolism, and genetics of ruminal selenomonads. Crit. Rev. Microbiol. *22*: 27–56.

Roberts, S.A., K.P. Shore, S.D. Paviour, D. Holland and A.J. Morris. 2006. Antimicrobial susceptibility of anaerobic bacteria in New Zealand: 1999–2003. J. Antimicrob. Chemother. *57*: 992–998.

Robinow, C.F. 1954. Addendum to *Selenomonas* Boskamp, 1922 – a genus that includes species showing an unusual type of flagellation, by E. F. Lessel and R. S. Breed. Bacteriol. Rev. *18*: 169–170.

Robinson, I.M., M.J. Allison and J.A. Bucklin. 1981. Characterization of the cecal bacteria of normal pigs. Appl. Environ. Microbiol. *41*: 950–955.

Rogosa, M. 1964. The genus *Veillonella*. I. General cultural, ecological, and biochemical considerations. J. Bacteriol. *87*: 162–170.

Rogosa, M. and F.S. Bishop. 1964a. The genus *Veillonella*. II. Nutritional Studies. J. Bacteriol. *87*: 574–580.

Rogosa, M. and F.S. Bishop. 1964b. The genus *Veillonella*. III. Hydrogen sulfide production by growing cultures. J. Bacteriol. *88*: 37–41.

Rogosa, M. 1965. The genus *Veillonella*. IV. Serological groupings, and genus and species emendations. J. Bacteriol. *90*: 704–709.

Rogosa, M. 1969. *Acidaminococcus* gen. n., *Acidaminococcus fermentans* sp. n., anaerobic gram-negative diplococci using amino acids as the sole energy source for growth. J. Bacteriol. *98*: 756–766.

Rogosa, M. 1971a. Transfer of *Peptostreptococcus elsdenii* Gutierrez et al. to a new genus, *Megasphaera* [*M. elsdenii* (Gutierrez et al.) comb. nov.]. Int. J. Syst. Bacteriol. *21*: 187–189.

Rogosa, M. 1971b. Transfer of *Veillonella* Prévot and *Acidaminococcus* Rogosa from *Neisseriaceae* to *Veillonellaceae* fam. nov. and the inclusion of *Megasphaera* Rogosa in *Veillonellaceae*. Int. J. Syst. Bacteriol. *21*: 231–233.

Rogosa, M. 1984a. Anaerobic Gram-negative cocci. Genus III *Megasphaera*. *In* Krieg and Holt (Editors), Bergey's Manual of Sytematic Bacteriology, Vol. 1. The Williams & Wilkins Co., Baltimore, pp. 685.

Rogosa, M. 1984b. Family I. *Veillonellaceae* Rogosa 1971, 232^AL. *In* Krieg and Holt (Editors), Bergey's Manual of Systematic Bacteriology, 2nd edn, Vol. 1. The Williams & Wilkins Co., Baltimore, pp. 680–685.

Rolph, H.J., A. Lennon, M.P. Riggio, W.P. Saunders, D. MacKenzie, L. Coldero and J. Bagg. 2001. Molecular identification of microorganisms from endodontic infections. J. Clin. Microbiol. *39*: 3282–3289.

Rosencrantz, D., F.A. Rainey and P.H. Janssen. 1999. Culturable populations of *Sporomusa* spp. and *Desulfovibrio* spp. in the anoxic bulk soil of flooded rice microcosms. Appl. Environ. Microbiol. *65*: 3526–3533.

Rovery, C., A. Etienne, C. Foucault, P. Berger and P. Brouqui. 2005. *Veillonella montpellierensis* endocarditis. Emerg. Infect. Dis. *11*: 1112–1114.

Russell, J.B. 1998. Strategies that ruminal bacteria use to handle excess carbohydrate. J. Anim. Sci. *76*: 1955–1963.

Sakamoto, K., W. Funahashi, H. Yamashita and E. Masakazu. 1997. A reliable method for detection and identification of beer-spoilage bacteria with internal positive control PCR (IPC-PCR). Proc. 26th Congr. Eur. Brew. Conv., pp. 631–638, Maastricht.

Salanitro, J.P., I.G. Blake and P.A. Muirhead. 1977. Isolation and identification of fecal bacteria from adult swine. Appl. Environ. Microbiol. *33*: 79–84.

Samuelov, N.S., R. Datta, M.K. Jain and J.G. Zeikus. 1990. Microbial decarboxylation of succinate to propionate. Ann. N. Y. Acad. Sci. *589*: 697–704.

Sato, M., E. Hoshino, S. Nomura and K. Ishioka. 1993. Salivary microflora of geriatric edentulous persons wearing dentures. Microb. Ecol. Health Dis. *6*: 293–299.

Sato, T., J. Matsuyama, M. Sato and E. Hoshino. 1997. Differentiation of *Veillonella atypica*, *Veillonella dispar* and *Veillonella parvula* using restricted fragment-length polymorphism analysis of 16S rDNA amplified by polymerase chain reaction. Oral Microbiol. Immunol. *12*: 350–353.

Satokari, R., R. Juvonen, A. von Wright and A. Haikara. 1997. Detection of *Pectinatus* beer spoilage bacteria by using the polymerase chain reaction. J. Food Prot. *60*: 1571–1573.

Satokari, R., R. Juvonen, K. Mallison, A. von Wright and A. Haikara. 1998. Detection of beer spoilage bacteria *Megasphaera* and *Pectinatus* by polymerase chain reaction and colorimetric microplate hybridization. Int. J. Food Microbiol. *45*: 119–127.

Sawada, S., S. Kokeguchi, F. Nishimura, S. Takashiba and Y. Murayama. 1999. Phylogenetic characterization of *Centipeda periodontii*, *Selenomonas sputigena* and *Selenomonas* species by 16S rRNA gene sequence analysis. Microbios *98*: 133–140.

Sawada, S., S. Kokeguchi, S. Takashiba and Y. Murayama. 2000. Development of 16S rDNA-based PCR assay for detecting *Centipeda periodontii* and *Selenomonas sputigena*. Lett Appl. Microbiol. *30*: 423–426.

Schauder, R. and B. Schink. 1989. *Anaerovibrio glycerini* sp. nov., an anaerobic bacterium fermenting glycerol to propionate, cell matter, and hydrogen. Arch. Microbiol. *152*: 473–478.

Schauder, R. and B. Schink. 1996. *Anaerovibrio glycerini* sp. nov. Validation of the publication of new names and new combinations previously effectively published outside the IJSB. List no. 57. Int. J. Syst. Bacteriol. *46*: 625–626.

Scheifinger C.C, Linehan B. and Wolin M.J. 1975. H$_2$ production by *Selenomonas ruminantium* in the absence and presence of methanogenic bacteria. Appl. Microbiol. *29*: 480–483.

Schink, B., T.E. Thompson and J.G. Zeikus. 1982. Characterization of *Propionispira arboris* gen. nov. sp. nov., a nitrogen-fixing anaerobe common to wetwoods of living trees. J. Gen. Microbiol. *128*: 2771–2779.

Schink, B., T.E. Thompson and J.G. Zeikus. 1983. *In* Validation of the publication of new names and new combinations previously effectively published outside the IJSB. List no. 11. Int. J. Syst. Bacteriol. *33*: 672–674.

Schleifer, K.H. and O. Kandler. 1972. Peptidoglycan types of bacterial cell walls and their taxonomic implications. Bacteriol. Rev. *36*: 407–477.

Schleifer, K.H., M. Leuteritz, N. Weiss, W. Ludwig, G. Kirchhof and H. Seidel-Rüfer. 1990. Taxonomic study of anaerobic, Gram-negative, rod-shaped bacteria from breweries: emended description of *Pectinatus cerevisiiphilus* and description of *Pectinatus frisingensis* sp. nov., *Selenomonas lacticifex* sp. nov., *Zymophilus raffinosivorans* gen. nov., sp. nov., and *Zymophilus-paucivorans* sp. nov. Int. J. Syst. Bacteriol. *40*: 19–27.

Scott, H.W. and B.A. Dehority. 1965. Vitamin requirements of several cellulolytic rumen bacteria. J. Bacteriol. *89*: 1169–1175.

Seidel-Rüefer, H. 1990. *Pectinatus* und andere morphologisch ähnliche Gram-negative, anaerobe Stäbchen aus dem Brauereibereich. Monatsschrift. Brau. *3*: 101–105.

Seidel, H., W. Back and N. Weiss. 1979. Isolierung und systematische Zuordnung bierschädlicher gramnegativer Bakterien III: Welche Gefahr stellen die in den beiden vorausgegangenen Mitteilungen vorgestellten gramnegativen Kokken und Stäbchen für das Bier dar? Brauwissenschaft. *32*: 262–270.

Seidel, H. 1985. 100 jahre biologische Brauerei-Betriebskontrolle, alte und neue problem. Brauwelt *125*: 1954–1958.

Senchenkova, S.N., A.S. Shashkov, A.P. Moran, I.M. Helander and Y.A. Knirel. 1995. Structures of the O-specific polysaccharide chains of *Pectinatus cerevisiiphilus* and *Pectinatus frisingensis* lipopolysaccharides. Eur. J. Biochem. *232*: 552–557.

Shah, H.N. and M.D. Collins. 1982. Reclassification of *Bacteroides multiacidus* (Mitsuoka, Terada, Watanabe and Uchida) in a new genus *Mitsuokella*, as *Mitsuokella multiacidus* comb. nov. Zentralbl. Bakteriol. Parasitenkd. Infektionskr. Hyg. I Abt. Orig. C *3*: 491–494.

Shah, H.N. and M.D. Collins. 1983. *In* Validation of the publication of new names and new combinations previously effectively published outside the IJSB. List no. 10. Int. J. Syst. Bacteriol. *33*: 438–440.

Shah, H.N., M.D. Collins and R.M. Kroppenstedt. 1983. Biochemical and chemical studies on *Bacteroides multiacidus* and *Bacteroides hypermegas*. J. Appl. Bacteriol. *55*: 151–158.

Shelobolina, E.S., K.P. Nevin, J.D. Blakeney-Hayward, C.V. Johnsen, T.W. Plaia, P. Krader, T. Woodard, D.E. Holmes, C.G. Vanpraagh and D.R. Lovley. 2007. *Geobacter pickeringii* sp. nov., *Geobacter argillaceus* sp. nov. and *Pelosinus fermentans* gen. nov., sp. nov., isolated from subsurface kaolin lenses. Int. J. Syst. Evol. Microbiol. *57*: 126–135.

Shimizu, Y. 1968. [Experimental studies on the bacterio phages of the *Veillonella* strains isolated from the oral cavity]. Shigaku *55*: 533–541.

Simons, H. 1920. Eine saprophytische Oscillarie in Darm des Meerschweinchens. Zentralbl. Bakteriol. Parasitenk. Infektionskr. Abt. I. Orig. *5*: 356–364.

Singh, N. and V.L. Yu. 1992. Osteomyelitis due to *Veillonella parvula*: case report and review. Clin. Infect. Dis. *14*: 361–363.

Sirotek, K., E. Santos, V. Benda and M. Marounek. 2003. Isolation, identification and characterization of rabbit caecal mucinolytic bacteria. Acta Veterin. Brno *72*: 365–370.

Slyter, L.L., D.L. Kern and J.M. Weaver. 1976. Effect of pH on ruminal lactic acid utilization and accumulation in vitro. J. Anim. Sci. *43*: 333–334.

Smith, C.J., R.B. Hespell and M.P. Bryant. 1981. Regulation of urease and ammonia assimilatory enzymes in *Selenomonas ruminantium*. Appl. Environ. Microbiol. *42*: 89–96.

Soberka, R. and A. Warzecha. 1986. Influence de certains facteurs sur le taux d'oxygène dissous au cours de la fabrication de la bière. Bios *17*: 31–40.

Sokolova, T.G., J.M. Gonzalez, N.A. Kostrikina, N.A. Chernyh, T.V. Slepova, E.A. Bonch-Osmolovskaya and F.T. Robb. 2004. *Thermosinus carboxydivorans* gen. nov., sp. nov., a new anaerobic, thermophilic, carbon-monoxide-oxidizing, hydrogenogenic bacterium from a hot pool of Yellowstone National Park. Int. J. Syst. Evol. Microbiol. *54*: 2353–2359.

Sprincova, A., P. Javorsky and P. Pristas. 2005. pSRD191, a new member of RepL replicating plasmid family from *Selenomonas ruminantium*. Plasmid *54*: 39–47.

Stackebrandt, E., H. Pohla, R. Kroppenstedt, H. Hippe and C.R. Woese. 1985. 16S rRNA analysis of *Sporomusa*, *Selenomonas*, and *Megasphaera*: on the phylogenetic origin of Gram-positive eubacteria. Arch. Microbiol. *143*: 270–276.

Stackebrandt, E. and B.M. Goebel. 1994. Taxonomic note: a place for DNA–DNA reassociation and 16S rRNA sequence analysis in the present species definition in bacteriology. Int. J. Syst. Bacteriol. *44*: 846–849.

Stackebrandt, E., I. Kramer, J. Swiderski and H. Hippe. 1999. Phylogenetic basis for a taxonomic dissection of the genus *Clostridium*. FEMS Immunol. Med. Microbiol. *24*: 253–258.

Stackebrandt, E., W. Frederiksen, G.M. Garrity, P.A. Grimont, P. Kämpfer, M.C. Maiden, X. Nesme, R. Rosselló-Mora, J. Swings, H.G. Truper, L. Vauterin, A.C. Ward and W.B. Whitman. 2002. Report of the ad hoc committee for the re-evaluation of the species definition in bacteriology. Int. J. Syst. Evol. Microbiol. *52*: 1043–1047.

Stankewich, J.P., B.J. Cosenza and A.L. Shigo. 1971. *Clostridium quercicolum* sp.n., isolated from discolored tissues in living oak trees. Antonie van Leeuwenhoek *37*: 299–302.

Stanton, T.B. and S.B. Humphrey. 2003. Isolation of tetracycline-resistant *Megasphaera elsdenii* strains with novel mosaic gene combinations of *tet(O)* and *tet(W)* from swine. Appl. Environ. Microbiol. *69*: 3874–3882.

Stewart, S.C. and M.P. Bryant. 1988. The rumen bacteria. *In* Hobson (Editor), The Rumen Microbial Ecosystem. Elsevier Applied Science, London, pp. 21–71.

Strömpl, C., B.J. Tindall, G.N. Jarvis, H. Lunsdorf, E.R.B. Moore and H. Hippe. 1999. A re-evaluation of the taxonomy of the genus *Anaerovibrio*, with the reclassification of *Anaerovibrio glycerini* as *Anaerosinus glycerini* gen. nov., comb. nov., and *Anaerovibrio burkinabensis* as *Anaeroarcus burkinensis* corrig. gen. nov., comb. nov. Int. J. Syst. Bacteriol. *49*: 1861–1872.

Strömpl, C., B.J. Tindall, H. Lunsdorf, T.Y. Wong, E.R. Moore and H. Hippe. 2000. Reclassification of *Clostridium quercicolum* as *Dendrosporobacter quercicolus* gen. nov., comb. nov. Int. J. Syst. Evol. Microbiol. *50*: 101–106.

Sugihara, P.T., V.L. Sutter, H.R. Attebery, K.S. Bricknell and S.M. Finegold. 1974. Isolation of *Acidaminococcus fermentans* and *Megasphaera elsdenii* from normal human feces. Appl. Microbiol. *27*: 274–275.

Suihko, M.-L. and A. Haikara. 1990. Maintenance of the anaerobic beer spoilage bacteria *Pectinatus* and *Megasphaera*. Food Microbiol. *7*: 33–42.

Suihko, M.-L. and A. Haikara. 2001. Characterization of *Pectinatus* and *Megasphaera* strains by automated ribotyping. J. Inst. Brew. *107*: 175–184.

Sveen, K. 1977. The capacity of lipopolysaccharides from *Bacteroides*, *Fusobacterium* and *Veillonella* to produce skin inflammation and the local and generalized Shwartzman reaction in rabbits. J. Periodont. Res. *12*: 340–350.

Sveen, K., T. Hofstad and K.C. Milner. 1977. Lethality for mice and chick embryos, pyrogenicity in rabbits and ability to gelate lysate from amoebocytes of *Limulus polyphemus* by lipopolysaccharides from *Bacteroides*, *Fusobacterium* and *Veillonella*. Acta. Pathol. Microbiol. Scand. [B] *85B*: 388–396.

Tajima, K., R.I. Aminov, T. Nagamine, H. Matsui, M. Nakamura and Y. Benno. 2001. Diet-dependent shifts in the bacterial population of the rumen revealed with real-time PCR. Appl. Environ. Microbiol. *67*: 2766–2774.

Takahashi, N. 1983. Presumed *Pectinatus* strain isolated from Japanese beer. Bull. Brew. Sci. *28*: 11–14.

Takatsuka, Y. and Y. Kamio. 2004. Molecular dissection of the *Selenomonas ruminantium* cell envelope and lysine decarboxylase involved in the biosynthesis of a polyamine covalently linked to the cell wall peptidoglycan layer. Biosci. Biotechnol. Biochem. *68*: 1–19.

Terzenbach, D.P. and M. Blaut. 1994. Transformation of tetrachloroethylene to trichloroethylene by homoacetogenic bacteria. FEMS Microbiol. Lett. *123*: 213–218.

Therion, J.J., A. Kistner and J.H. Kornelius. 1982. Effect of pH on growth rates of rumen amylolytic and lactilytic bacteria. Appl. Environ. Microbiol. *44*: 428–434.

Theron, M.M., M.N. van Rensburg and L.J. Chalkley. 2003. Penicillin-binding proteins involved in high-level piperacillin resistance in *Veillonella* spp. J. Antimicrob. Chemother. *52*: 120–122.

Thiel, H.J. and U. Schumacher. 1994. Normal flora of the human conjunctiva: a study of 135 persons of various ages. Klin. Monatsbl. Augenheilkd. *205*: 348–357.

Tholozan, J.L., J.P. Grivet and C. Vallet. 1994. Metabolic pathway to propionate of *Pectinatus frisingensis*, a strictly anaerobic beer-spoilage bacterium. Arch. Microbiol. *162*: 401–408.

Tholozan, J.L., J.M. Membre and M. Kubaczka. 1996. Effects of culture conditions on *Pectinatus cerevisiiphilus* and *Pectinatus frisingensis* metabolism: a physiological and statistical approach. J. Appl. Bacteriol. *80*: 418–424.

Tholozan, J.L., J.M. Membré and J.P. Grivet. 1997. Physiology and development of *Pectinatus cerevisiiphilus* and *Pectinatus frisingensis*, two strict anaerobic beer spoilage bacteria. Int. J. Food. Microbiol. *18*: 29–39.

Thompson, T.E. and J.G. Zeikus. 1988. Regulation of carbon and electron flow in *Propionispira arboris*: relationship of catabolic enzyme levels to carbon substrates fermented during propionate formation via the methylmalonyl coenzyme A pathway. J. Bacteriol. *170*: 3996–4000.

Tiwari, A.D., M. P. Bryant and R. S. Wolf. 1969. Simple method for isolation of *Selenomonas ruminantium* and some nutritional characteristics of the species. J. Dairy Sci. *52*: 2054–2056.

Tortorello, M.L. and A. Delwiche. 1984. Characterization of the lipopolysaccharide of *Veillonella parvula* ATN and growth conditions affecting its yield. Current Microbiol. *11*: 107–112.

Totsuka, M. 1976. Studies on veillonellophages isolated from washings of human oral cavity. Bull. Tokyo Med. Dent. Univ. *23*: 261–273.

Totsuka, M. 1988. Phage-receptor on the cell wall of *Veillonella rodentium*. Antonie van Leeuwenhoek *54*: 229–233.

Totsuka, M. and T. Ono. 1989. Purification and characterization of bacteriophage receptor on *Veillonella rodentium* cells. Phage-receptor on *Veillonella*. Antonie van Leeuwenhoek *56*: 263–271.

Tsukahara, T., H. Koyama, M. Okada and K. Ushida. 2002. Stimulation of butyrate production by gluconic acid in batch culture of pig cecal digesta and identification of butyrate-producing bacteria. J. Nutr. *132*: 2229–2234.

Valdes, M.V., P.M. Lobbins and J. Slots. 1982. Beta-lactamase producing bacteria in the human oral cavity. J. Oral Pathol. *11*: 58–63.

van Golde, L.M., R.A. Prins, W. Franklin-Klein and J. Akkermans-Kruyswijk. 1973. Phosphatidylserine and its plasmalogen analogue as major lipid constituents in *Megasphaera elsdenii*. Biochim. Biophys. Acta *326*: 314–324.

van Golde, L.M.G., J. Akkermans-Kruyswijk, W. Franklin-Klein, A. Lankhorst and R.A. Prins. 1975. Accumulation of phosphatidylserine in strictly anaerobic lactate fermenting bacteria. FEBS Lett. *53*: 57–60.

van Gylswyk, N.O. 1995. *Succiniclasticum ruminis* gen. nov., sp. nov., a ruminal bacterium converting succinate to propionate as the sole energy-yielding mechanism. Int. J. Syst. Bacteriol. *45*: 297–300.

van Gylswyk, N.O., H. Hippe and F.A. Rainey. 1997. *Schwartzia succinivorans* gen. nov., sp. nov., another ruminal bacterium utilizing succinate as the sole energy source. Int. J. Syst. Bacteriol. *47*: 155–159.

Veillon, A. and A. Zuber. 1898. Recherches sur quelques microbes strictement anaérobies et leur rôle en pathologie. Arch. Med. Exp. *10*: 517–545.

Verkley, A.J., P.H. Ververgaert, R.A. Prins and L.M. van Golde. 1975. Lipid-phase transitions of the strictly anaerobic bacteria *Veillonella parvula* and *Anaerovibrio lipolytica*. J. Bacteriol. *124*: 1522–1528.

Vicini, J.L., W.J. Brulla, C.L. Davis and M.P. Bryant. 1987. Quin's oval and other microbiota in the rumens of mollasses-fed sheep. Appl. Environ. Microbiol. *53*: 1273–1276.

Von Prowazek, S. 1913. Zur Parasitologie von Westafrika. Zentralbl. Bakteriol. Parasitenk. Infektionskr. Abt. I. Orig. *70*: 32–36.

Walker, C.B., J.D. Pappas, K.Z. Tyler, S. Cohen and J.M. Gordon. 1985. Antibiotic susceptibilities of periodontal bacteria. *In vitro* susceptibilities to eight antimicrobial agents. J. Periodontol. *56*: 67–74.

Watier, D., H.C. Dubourguier, I. Lequerinel and J.P. Hornez. 1996. Response surface models to describe the effects of temperature, pH and ethanol concentration on growth kinetics and fermentation end products of a *Pectinatus* sp. Appl. Environ. Microbiol. *62*: 1233–1237.

Wayne, L.G., D.J. Brenner, R.R. Colwell, P.A.D. Grimont, O. Kandler, M.I. Krichevsky, L.H. Moore, W.E.C. Moore, R.G.E. Murray, E. Stackebrandt, M.P. Starr and H.G. Trüper. 1987. Report of the ad hoc committee on the reconciliation of approaches to bacterial systematics. Int. J. Syst. Evol. Microbiol. *37*: 463–464.

Weiss, N., H. Seidel and W. Back. 1979. Isolation and systematic identification of Gram-negative bacteria which are harmful to beer. I: Gram-negative cocci which are strictly anaerobic. Brauwissenschaft *32*: 189–194.

Wenyon, C.M. 1926. Protozoology, Vol. 1. Bailliere, Tindall and Cox, London.

Werner, H. 1973. *Megasphaera elsdenii* – a normal inhabitant of the human intestines. Zentralbl. Bakteriol. Parasitenkd. Infektionskr. Hyg. Abt. I Orig. A *223*: 343–347.

Westlake, K., R.I. Mackie and M.F. Dutton. 1987. T-2 toxin metabolism by ruminal bacteria and its effect on their growth. Appl. Environ. Microbiol. *53*: 587–592.

Wicken, A.J. and B.H. Howard. 1967. On the taxonomic status of "Quin's oval" organisms. J. Gen. Microbiol. *47*: 207–211.

Wiegel, J. and L. Quandt. 1982. Determination of the gram type using the reaction between polymyxin B and lipopolysaccharides of the outer cell wall of whole bacteria. J. Gen. Microbiol. *128*: 2261–2270.

Willems, A. and M.D. Collins. 1995a. Phylogenetic placement of *Dialister pneumosintes* (formerly *Bacteroides pneumosintes*) within the *Sporomusa* subbranch of the *Clostridium* subbphylum of the gram-positive bacteria. Int. J. Syst. Bacteriol. *45*: 403–405.

Willems, A. and M.D. Collins. 1995b. 16S ribosomal RNA gene similarities indicate that *Hallella seregens* (Moore and Moore) and *Mitsuokella dentalis* (Haapasalo et al.) are genealogically highly related and are members of the genus *Prevotella*: emended description of the genus *Prevotella* (Shah and Collins) and description of *Prevotella dentalis* comb. nov. Int. J. Syst. Bacteriol. *45*: 832–836.

Woese, C.R. 1987. Bacterial evolution. Microbiol. Rev. *51*: 221–271.

Wohlfahrt, G. and W. Buckel. 1985. A sodium ion gradient as energy source for *Peptostreptococcus asaccharolyticus*. Arch. Microbiol. *142*: 128–135.

Wood, H.G. and L.G. Ljungdahl. 1991. Autotrophic character of the acetogenic bacteria. *In* Shively and Barton (Editors), Variations in Autotrophic Life. Academic Press, San Diego, pp. 201–250.

Yanke, L.J., H.D. Bae, L.B. Selinger and K.J. Cheng. 1998. Phytase activity of anaerobic ruminal bacteria. Microbiology *144*: 1565–1573.

Zaninetti-Schaerer, A., C. Van Delden, S. Genevay and C. Gabay. 2004. Total hip prosthetic joint infection due to *Veillonella* species. Joint Bone Spine *71*: 161–163.

Zhou, X., S.J. Bent, M.G. Schneider, C.C. Davis, M.R. Islam and L.J. Forney. 2004. Characterization of vaginal microbial communities in adult healthy women using cultivation-independent methods. Microbiology *150*: 2565–2573.

Ziola, B., S.L. Gares, B. Lorrain, L. Gee, W.M. Ingledew and S.Y. Lee. 1999. Epitope mapping of monoclonal antibodies specific for the directly cross-linked mesodiaminopimelic acid peptidoglycan found in the anaerobic beer spoilage bacterium *Pectinatus cerevisiiphilus*. Can. J. Microbiol. *45*: 779–785.

Ziola, B., L. Gee, N.N. Berg and S.Y. Lee. 2000. Serogroups of the beer spoilage bacterium *Megasphaera cerevisiae* correlate with the molecular weight of the major EDTA-extractable surface protein. Can. J. Microbiol. *46*: 95–100.

Family XI. **Incertae Sedis**

Previously assigned to the "*Peptostreptococcaceae*" by Garrity et al. (2005), subsequent analyses suggest that these genera, while closely related to each other, are not closely related to *Peptostrep-* *tococcus* or a member of any other previously described family. Until their taxonomic status has been clarified, they are classified within a family *incertae sedis*.

Genus I. **Anaerococcus** Ezaki, Kawamura, Li, Li, Zhao and Shu 2001b, 1526^{VP}

TAKAYUKI EZAKI AND KIYOFUMI OHKUSU

An.ae.ro.coc'cus. Gr. prep. *an* without; Gr. n. *aer* air; Gr. masc. n. *kokkos* berry, coccus; N.L. masc. n. *Anaerococcus* anaerobic coccus.

Cocci, occurring in pairs, tetrads, irregular masses or chains. Gram-stain-positive. Nonmotile. Nonsporeforming. Strictly anaerobic. Metabolize peptones and amino acids; the major metabolic end products are butyric acid, lactic acid and small amounts of propionic and succinic acids. Most species are able to ferment several carbohydrates, but are weakly fermentative. Glucose, fructose, sucrose and lactose are major sugars fermented. Most species do not produce indole. The position 1, position 3, and interpeptide bride of peptidoglycan are gly-cine or alanine, L-lysine, and D-glutamic acid or D-aspartic acid. Major cellular fatty acid are $C_{18:1}$, $C_{16:1}$, C_{18}, and C_{16} (Ezaki et al., 1983; O'Leary and Wilkinson, 1988). Members of the genus are typically isolated from the human vagina and various purulent secretions.

DNA G+C content (mol%): 30–35.

Type species: **Anaerococcus prevotii** (Foubert and Douglas 1948) Ezaki, Kawamura, Li, Li, Zhao and Shu 2001b, 1526^{VP}.

List of species of the genus *Anaerococcus*

1. **Anaerococcus prevotii** (Foubert and Douglas 1948) Ezaki, Kawamura, Li, Li, Zhao and Shu 2001b, 1526^{VP} (*Peptostrepto-cocus prevotii* Foubert and Douglas 1948, Ezaki, Yamamoto, Ninomiya, Suzuki and Yabuuchi 1983, 693; Peptococcus prevotii Foubert and Douglas 1948, Douglas, 1957, 477^{AL}; *Micrococcus prevotii* Foubert and Douglas 1948, 31)

pre.vo'ti.i. N.L. gen. n. *prevotii* of Prévot, named after A.R. Prévot, a French microbiologist.

The characteristics are as described for the genus and as listed in Table 188, with the following additional information. Endospores not formed Cells are 0.6–0.9 µm in diameter and found in pairs and irregular masses. Negative for indole, alkaline phosphatase and coagulase. Some strains produce urease. Most strains ferment glucose and mannose. Acid is also produced from raffinose, ribose and mannose but not from lactose, cellobiose, maltose, sucrose or xylose. The major metabolic end product from peptone-yeast- glucose medium is butyric acid. Position 1, position 3, and the interpeptide bridge of the peptidoglycan are glycine, L-lysine, and D-glutamic acid, respectively. Descriptions of the major saccharolytic and proteolytic enzymes are given by Murdoch (1988). Differential characteristics from other members of the genus and closely related species of the genus *Peptoniphilus* are given in Table 188. Often isolated from human clinical specimens such as vaginal discharges and ovarian, peritoneal and sacral abscesses.

DNA G+C content (mol%): 29–33 (T_m).

Type strain: ATCC 9321, CCUG 41932, CIP 105881, DSM 20548, JCM 6508, NCTC 11806.

GenBank accession number (16S rRNA gene): D14139.

2. **Anaerococcus hydrogenalis** (Ezaki, Liu, Hashimoto and Yabuuchi 1990) Ezaki, Kawamura, Li, Li, Zhao and Shu 2001b, 1526^{VP} (*Peptostreptococcus hydrogenalis* Ezaki, Liu, Hashimoto and Yabuuchi 1990, 306)

hyd.ro.ge.na'lis. Chem. term *hydrogen*; L. suff. *-alis* pertaining to; N.L. masc. adj. *hydrogenalis* pertaining to hydrogen, because this organism produces hydrogen.

The characteristics are as described for the genus and as listed in Table 188, with the following additional information. Cells are 0.6–0.9 µm in diameter and occur in pairs, chains, and irregular masses. Endospores are not formed. Produces abundant H_2 from peptone-yeast-glucose medium. Indole-positive. Some strains produce urease and alkaline phosphatase. Acid is produced from glucose, lactose, raffinose and mannose. Negative for arginine dihydrolase, and coagulase. The major metabolic end product from PYG is butyric acid. Position 1, position 3, and the interpeptide bridge of peptidoglycan are ala-nine, L-lysine, and D-glutamic acid, respectively. Descriptions of major saccharolytic and proteolytic enzymes are given by Murdoch (1988). Differential characteristics from other members of the genus are given in Table 188. The organism is isolated from vaginal discharges and ovarian abscesses.

DNA G+C content (mol%): 30–34 (method not known).

Type strain: ATCC 49630, DSM 7454, JCM 7635, GIFU 7662.

GenBank accession number (16S rRNA gene): D14140.

3. **Anaerococcus lactolyticus** (Li, Hashimoto, Adnan, Miura, Yamamoto and Ezaki 1992) Ezaki, Kawamura, Li, Li, Zhao and Shu 2001b, 1527^{VP} (*Peptostreptococcus lactolyticus* Li, Hashimoto, Adnan, Miura, Yamamoto and Ezaki 1992, 604)

lac.to.ly'ti.cus. L. n. *lac, lactis* milk; Gr. adj. *lutikos* dissolving; N.L. adj. *lactolyticus* milk-dissolving.

The characteristics are as described for the genus and as listed in Table 188, with the following additional information. Cells occur in short chains or irregular masses. Endospores are not formed. Strains produce urease, β-galactosidase, and alkaline phosphatase but not indole or coagulase. Strong acid is produced from glucose, lactose and mannose. Raffnose and ribose are not fermented. The major metabolic end products from PYG is butyric, lactic and acetic acid. Position 1, position 3, and interpeptide bride of peptidoglycan are glycine, L-lysine, and D-glutamic acid, respectively. Ppeptones and oligopeptides are used as major energy sources. Leucine arylamidase (AMD), alanine AMD, histidine AMD, glycine AMD, pyrrolidone AMD are present. Differential characteristics from other butyrate producing members of the genus are given in Table 188. Isolated from vaginal discharges and ovarian abscesses.

DNA G+C content (mol%): 30–34 (method not known).

Type strain: ATCC 51172, CIP 103725, DSM 7456, JCM 8140, GIFU 8586.

GenBank accession number (16S rRNA gene): D14154.

4. **Anaerococcus octavius** (Murdoch, Collins, Willems, Hardie, Young and Magee 1997) Ezaki, Kawamura, Li, Li, Zhao and Shu 2001b, 1527[VP] (*Peptostreptococcus octavius* Murdoch, Collins, Willems, Hardie, Young and Magee 1997, 785)

oc.ta′vi.us. L. adj. *octavius* eight, referring to the fact that the organism was previously assigned to Hare group VIII.

The characteristics are as described for the genus and as listed in Table 188, with the following additional information. Cells occur in short chains or irregular masses. Endospores are not formed. Indole, urease, alkaline phosphatase, arginine dihydrolase and coagulase are negative. Glucose, ribose and mannose are fermented. Lactose and raffnose are not fermented. The major metabolic end products from PYG are butyric acid and caproic acid. Position 1, position 3, and interpeptide bridge of the peptidoglycan are alanine, L-lysine, and D-aspartic acid. Descriptions of major saccharolytic and proteolytic enzymes are given by Murdoch (1988). Differential characteristics from other members of the genus are given in Table 188. Isolated from vaginal discharges and ovarian abscesses. The type strain was isolated from a human nasal cavity. Often isolated from skin, vagina and nasal cavity.

DNA G+C content (mol%): 26–31 (method not known).

Type strain: Vavey 1, DSM 11663, NCTC 9810.

GenBank accession number (16S rRNA gene): Y07841.

5. **Anaerococcus tetradius** (Ezaki, Yamamoto, Ninomiya, Suzuki and Yabuuchi 1983) Ezaki, Kawamura, Li, Li, Zhao and Shu 2001b, 1526[VP] (*Peptostreptococcus tetradius* Ezaki, Yamamoto, Ninomiya, Suzuki and Yabuuchi 1983, 696)

te.tra′di.us. Gr. n. *tetradion* group of four; N.L. adj. *tetradius* occurring in groups of four.

The characteristics are as described for the genus and as listed in Table 188, with the following additional information. Cells are 0.6–1.0 μm in diameter and found in tetrads, pairs, and irregular masses. Endospores not formed. Urease-positive. Negative for indole, alkaline phosphatase, arginine dihydrolase and coagulase. Glucose and mannose are fermented. Acid is not produced from lactose, raffnose, ribose, cellulose, maltose, sucrose, or xylose. The major metabolic end product from peptone-yeast- glucose medium is butyric acid and lactic acid. Position 1, position 3, and the interpeptide bride of peptidoglycan are glycine, L-lysine, and D-glutamic acid, respectively. Descriptions of major saccharolytic and proteolytic enzymes are given by Murdoch (1988). Differential characteristics from other members of the genus and closely related species of the genus Peptoniphilus are given in Table 188. Isolated from vaginal discharges and ovarian abscesses.

DNA G+C content (mol%): 30–32 (Tm).

Type strain: ATCC 35098, CCM 3634, CCUG 17637, CIP 103927, DSM 2951, GIFU 7672, JCM 1964, LMG 14264.

GenBank accession number (16S rRNA gene): D14142.

6. **Anaerococcus vaginalis** (Li et al., 1992). Ezaki, Kawamura, Li, Li, Zhao and Shu 2001b, 1527[VP]. (*Peptostreptococcus vaginalis* Li, Hashimoto, Adnan, Miura, Yamamoto and Ezaki 1992, 604)

va.gi.na′lis. L. adj. *vaginalis* of the vagina.

The characteristics are as described for the genus and as listed in Table 188, with the following additional information. Cells are 0.7–0.9 μm in diameter and occur in short chains or irregular masses. Produces arginine dihydrolase. Some strains are positive for indole and alkaline phosphatase. Coagulase and urease are negative. Major metabolic end products from PYG are butyric and acetic acid. Position 1, position 3, and the interpeptide bride of peptidoglycan are alanine, L-lysine, and D-glutamic acid, respectively. Weak acid is produced from glucose and maltose but not from lactose cellobiose, sucrose, arabinose, ribose, mannitol, trehalose, sorbitol, and raffnose. Peptones and oligopeptides are used as major energy sources. Leucine arylamidase (AMD), alanine AMD, histidine AMD, and glycine AMD are present. Differential characteristics from other members of the genus are given in Table 188 Isolated from vaginal discharges and ovarian abscesses.

DNA G+C content (mol%): 28–30 (method not known).

Type strain: ATCC 51170, CIP 103621, DSM 7457, GIFU 12669, JCM 8138.

GenBank accession number (16S rRNA gene): D14146.

Genus II. **Finegoldia** Murdoch and Shah 2000, 1415[VP] (Effective publication: Murdoch and Shah 1999, 556.)

TAKAYUKI EZAKI

Fine.gol′dia. N.L. fem. n. *Finegoldia* named after Sydney M. Finegold, an American microbiologist.

Non-spore-forming, obligately anaerobic, Gram-stain-positive cocci. Cells may occur in pairs, tetrads, and irregular masses and are 0.7–1.5 μm in diameter. Carbohydrates are not fermented. Indole is negative. The optimum temperature for growth is 37°C. Metabolize peptone and amino acids to acetic acid. Alkaline phosphatase test is variable among strains. Found in vagina, oral cavity, and various abscess-forming human specimens. Differential characteristics from closely related genera are given in Table 187 and Table 217.

Peptidoglycan diamino acid is lysine and interpeptide bridge is D-aspartic acid.

Major cellular fatty acids are $C_{18:1}$, $C_{16:1}$, C_{18}, and C_{16} (Ezaki et al., 1983; O'Leary and Wilkinson, 1988), 1988).

Type species: **Finegoldia magna** (Prévot 1933) Murdoch and Shah 2000, 1415[VP] (Effective publication: Murdoch and Shah 1999, 557.) (*Diplococcus magnus* Prévot 1933, 140; *Peptococcus magnus* Holdeman and Moore 1972; *Peptostreptococcus magnus* Ezaki, Yamamoto, Ninomiya, Suzuki and Yobucchi 1983, 696).

List of species of the genus *Finegoldia*

1. **Finegoldia magna** (Prévot 1933) Murdoch and Shah 2000, 1415[VP] (Effective publication: Murdoch and Shah 1999, 557.) (*Diplococcus magnus* Prévot 1933, 140; *Peptococcus magnus* Holdeman and Moore 1972; *Peptostreptococcus magnus* Ezaki, Yamamoto, Ninomiya, Suzuki and Yobucchi 1983, 696)

Mag'na. L. adj. *magna* large.

The species description is same to the genus description. Differential characteristics from closely related genera are given in Table 187 and Table 217.

DNA G+C content (mol%): 32–34 (T_m).

Type strain: ATCC 15794, CCUG 17636, DSM 20470, GIFU 7629, NCTC 11804.

GenBank accession number (16S rRNA gene): AF542227, D14149.

Genus III. **Gallicola** Ezaki, Kawamura, Li, Li, Zhao and Shu 2001a, 1527[VP]

TAKAYUKI EZAKI

Ga.li'co.la. L. n. *gallus* rooster/chicken; L. masc. suff. *-cola* inhabitant; N.L.masc. n. *Gallicola* inhabitant of chickens, referring to the isolation of the type species from chicken feces.

Non-spore-forming, obligately anaerobic, Gram-stain-positive cocci. Cells may occur in pairs and as irregular masses. Carbohydrates are not fermented. Indole is negative. The optimum temperature for growth is 37°C. Metabolize peptone and amino acids to acetic and butyric acid. Found in chicken feces.

Peptidoglycan diamino acid is ornithine and the interpeptide bridge is D-aspartic acid.

Major cellular fatty acids are $C_{18:1}$, $C_{16:1}$, C_{18}, and C_{16} (Ezaki et al., 1983; O'Leary and Wilkinson, 1988).

DNA G+C content (mol%): 32–34.

Type species: **Gallicola barnesae** (Scheifer-Ullrich and Andreesen 1985) Ezaki, Kawamura, Li, Li, Zhao and Shu 2001a, 1527[VP] (*Peptostreptococcus barnesae* Scheifer-Ullrich and Andreesen 1985, 30).

List of species of the genus *Gallicola*

1. **Gallicola barnesae** (Scheifer-Ullrich and Andreesen 1985) Ezaki, Kawamura, Li, Li, Zhao and Shu 2001a, 1527[VP] (*Peptostreptococcus barnesae* Scheifer-Ullrich and Andreesen 1985, 30)

barne'sae. N.L. gen. n. *barnesae* of Barnes named after E.M. Barnes, a microbiologist.

Description of the type species is same to the genus description.

Isolated from chicken feces. Negative for the production of indole, urease, alkaline phosphatase, arginine dihydrolase, and coagulase. A full description is given by Schleifer-Ullrich and Andreesen (1985). Glucose, lactose, raffinose,

ribose, and mannose are not fermented. Occasional strains produce weak indole. The metabolic end products from PYG medium are acetic and butyric acids. Descriptions of major saccharolytic and proteolytic enzymes are given by Murdoch (1988). Differential characteristics from closely related genera are given in Table 187 and Table 217.

Major cellular fatty acids are $C_{18:1}$, $C_{16:1}$, C_{18}, and C_{16} (Ezaki et al., 1983; O'Leary and Wilkinson, 1988).

DNA G+C content (mol%): 32–34.

Type strain: HKB-5, ATCC 49795, DSM 3244.

GenBank accession number (16S rRNA gene): AB038361.

Genus IV. **Helcococcus** Collins, Facklam, Rodrigues and Ruoff 1993, 427[VP]

KATHRYN L. RUOFF

Hel.co.coc'cus. Gr. masc. n. *helkos* wound; Gr. masc. n. *kokkos* berry, sphere; N.L. masc. n. *Helcococcus* a sphere found in wounds.

Gram-stain-positive cocci that are **catalase-negative** and **nonmotile.** Cells are arranged in **irregular groups, pairs or short chains.** Helcococci form **small pinpoint non-hemolytic colonies** on blood agar after 24 hours of incubation at 35–37°C. **Facultative anaerobes that produce acid, but not gas from glucose. No**

hydrolysis of urea, hippurate or gelatin. Arginine dihydrolase is not produced. Nitrate is not reduced.

DNA G+C content (mol%): 29–30 mol% (T_m).

Type species: **Helcococcus kunzii** Collins, Facklam, Rodrigues and Ruoff 1993, 427[VP].

TABLE 217. Characteristics differentiating *Finegoldia*, *Gallicola*, and *Parvimonas* species[a]

Characteristic	Finegoldia magna	Gallicola barnesae	Parvimonas micra
16S rRNA GenBank accession number	D14149	AB038361	D14143
DNA G+C content (mol%)	32–34	32–34	27–28
Major product from PYG	Acetate	Acetate, butyrate	Acetate
Cell size (μm)	0.7–1.5	0.7–1.2	0.3–0.6
Peptidoglycan (position 3, bridge)	Lys, Gly	Orn, D-Asp	Lys, D-Asp
Production of:			
Alkaline phosphatase	d	–	+
Indole	–	w	–
Production of saccharolytic and proteolytic enzymes:			
ArgA	+	–	+
HisA	d	–	+
LeuA	+	–	+
ProA	–	–	+
PyrA	+	–	+

[a]Symbols: +, >85% positive; d d, different among strains; –, 0–15% positive; w, weak reaction; ND, not determined. ArgA, arginine arylamidase(AMD); ProA, proline AMD; PheA, phenylalanine AMD; LeuA, leucine AMD; PryA, pyloglutamyl AMD; HisA, histidine AMD.

Further descriptive information

Helcococcus strains form Gram-stain-positive coccoid cells that are typically arranged in groups, pairs or short chains. Stavri and coworkers (2002) described a Pronase-, heat- and mutanolysin-sensitive hemagglutinin-lectin in *Helcococcus kunzii* cells with specificity for *N*-acetylglucosamine and lactose. The absence of fimbriae and fibrillae on the surface of *Helcococcus kunzii* cells suggested that the lectin is associated with peptidoglycan or the cell membrane.

Helcococci produce tiny colonies on blood agar and may be difficult to culture on non-blood-containing media. Growth of *Helcococcus kunzii* on brain heart infusion media is stimulated by the addition of 1% horse serum or 0.1% Tween 80, suggesting that this species is lipophilic (Collins et al., 1993). A strain of *Helcococcus ovis* exhibited satelliting growth around *Staphylococcus aureus* colonies on initial isolation, but adapted to independent growth after repeated subculture onto Columbia agar containing 7% sheep blood incubated in a candle jar (Collins et al., 1999a). Incubation of cultures in the ambient atmosphere appears to be sufficient for routine culture of helcococci after initial isolation. Strains of *Helcococcus kunzii* grow equally well on brucella agar with 5% horse blood when incubated in an atmosphere containing 5% CO_2, an anaerobic atmosphere, or the ambient atmosphere (Collins et al., 1993).

Helcococci are fairly inactive metabolically when tested on conventional media or with commercially available identification kits. Glucose and a small number of other carbohydrates are acidified and few enzymic activities are detected. Helcococci are Voges–Proskauer-negative. Growth is variable in serum-supplemented brain heart infusion broth containing 6.5% NaCl.

Analysis of the 16S rRNA gene sequence of *Helcococcus kunzii* (GenBank accession no. X69837) revealed only a remote relationship between *Helcococcus* and other low-G+C content Gram-stain-positive bacteria with similar phenotypic characteristics (Collins et al., 1993). *Helcococcus ovis* and *Helcococcus kunzii* 16S rRNA gene sequences display 96% similarity (Collins et al., 1999a).

Although different methods have been employed, all *Helcococcus* strains tested have exhibited susceptibility to vancomycin. *Helcococcus kunzii* strains are susceptible to penicillin (Caliendo et al., 1995; Chagla et al., 1998; Peel et al., 1997). Chagla and associates (1998) reported that 8 of the 9 *Helcococcus kunzii* strains in their collection were resistant to erythromycin and all were susceptible to clindamycin.

Helcococcus kunzii is isolated from human clinical material, often in mixed cultures of wounds (Caliendo et al., 1995), but the bacterium has been recovered as the sole isolate from an infected sebaceous cyst (Peel et al., 1997) and a breast abscess (Chagla et al., 1998). Lomholt and Killian (2000) did not detect the virulence factor immunoglobulin A1 protease in the single strain of *Helcococcus kunzii* included in their survey of this factor in various Gram-stain-positive bacteria. *Helcococcus ovis* has been recovered from postmortem specimens of an adult male sheep and from the milk of a sheep with subclinical mastitis. As with most reports of *Helcococcus kunzii* isolation, other bacteria were also present in cultures yielding *Helcococcus ovis* (Collins et al., 1999a). Although more information is needed concerning the clinical significance of *Helcococcus ovis*, *Helcococcus kunzii* appears to function as an opportunistic pathogen.

In a study by Haas and colleagues (1997) *Helcococcus kunzii* was recovered from swabs of the intact skin of the feet of both diabetic and non-diabetic podiatry patients, and from both the foot and hand skin of a healthy volunteer. These observations suggest that *Helcococcus kunzii* is part of normal human skin flora. The *Helcococcus kuzii* lectin described by Stavri and coworkers (2002) is hypothesized to play a role in the adhesion and colonization activities of this bacterium.

Species of the genus *Helcococcus* may not be included in the databases of commercially available identification products. The profile produced by most strains of *Helcococcus kunzii* examined with the API 20S Strep system (bioMérieux Vitek) was 4100413, corresponding to a doubtful identification of *Aerococcus viridans* (Chagla et al., 1998; Collins et al., 1993). The *Helcococcus kunzii* profile observed with this system corresponds to positive reactions for pyrrolidonyl arylamidase, esculin hydrolysis, and acidification of lactose, trehalose, starch and glycogen.

Enrichment and isolation procedures

Helcococci can be recovered from clinical specimens on blood agar media incubated at 35–37°C in a CO_2-enriched (5%) atmosphere. In specimens yielding mixed bacterial growth, the tiny colonies of *Helcococcus* strains may be overgrown by other more vigorously growing bacteria (e.g., staphylococci, *Enterobacteriaceae* strains). The selective agar plate medium described by Haas and associates (1997) may be employed to facilitate isolation of *Helcococcus kunzii*. The medium consists of brain heart infusion broth (37 g), agar (15 g), and Tween 80 (1 ml) added to 990 ml of distilled water. After autoclaving and tempering the medium to 50°C, 10 ml of a filter-sterilized aqueous solution containing novobiocin (1 mg/ml) and colistin (750 µg/ml) are added. The selective medium supports good growth of *Helcococcus kunzii*, but inhibits the growth of a number of Gram-stain-positive and Gram-stain-negative bacteria that might be encountered in clinical specimens containing mixed flora.

Maintenance procedures

After initial isolation from clinical specimens, *Helcococcus* strains can be routinely cultured on blood agar media at 35–37°C in the ambient atmosphere or an aerobic atmosphere with 5% CO_2. Long term storage of isolates can be accomplished by suspending growth from blood agar plates in a few ml of horse or other animal blood and freezing suspensions at –70°C. For retrieval of isolates, portions of the frozen suspensions are thawed, streaked onto blood agar and incubated at 35–37°C either in the ambient atmosphere or in an atmosphere containing 5% CO_2.

Procedures for testing special characteristics

The lipophilic nature of *Helcococcus kunzii* can be demonstrated by enhancement of growth in brain heart infusion or Todd–Hewitt broth supplemented with 0.1% Tween 80 (Chagla et al., 1998; Collins et al., 1993). Chagla and coworkers (1998) advocated use of the heart infusion broth-based media described by Facklam and Elliott (1995) for studying acidification of carbohydrates by *Helcococcus kunzii*. These authors noted that although isolates grew well in these media, acidification as indicated by a color change in the bromcresol purple indicator occurred only when the media were supplemented with 0.1% Tween 80.

Differentiation of the genus *Helcococcus* from other genera

The genera *Helcococcus* and *Aerococcus* share some phenotypic characteristics (catalase-negative Gram-stain-positive cocci arranged in clusters or groups) even though they are not closely related (see Table 187). *Helcococcus kunzii* and *Aerococcus viridans* are both pyrrolidonyl arylamidase-positive and leucine aminopeptidase-negative. These organisms can be differentiated on the basis of colony size and hemolysis (*Aerococcus viridans* forms alpha-hemolytic colonies that are larger than the tiny non-hemolytic colonies of *Helcococcus kunzii*) and preferred incubation conditions (*Aerococcus viridans* prefers aerobic conditions but *Helcococcus kunzii* grows equally well aerobically or anaerobically; Collins et al., 1993).

The pyrrolidonyl arylamidase-positive species *Aerococcus sanguicola* is leucine aminopeptidase and hippurate hydrolysis-positive in contrast to *Helcococcus kunzii* (Lawson et al., 2001b).

Helcococcus kunzii can be differentiated from the pyrrolidonyl arylamidase-positive, leucine aminopeptidase-negative genera *Globicatella* and *Dolosicoccus* on the basis of chain formation by cells of these organisms (Ruoff, 2002).

Both *Helcococcus ovis* and *Aerococcus urinae* are pyrrolidonyl arylamidase-negative and leucine aminopeptidase-positive, but *Aerococcus urinae* and other pyrrolidonylaryl amidase-negative *Aerococcus* species (*Aerococcus urinaehominis*, *Aerococcus christensenii*) hydrolyze hippurate, while both *Helcococcus* species are negative for this activity (Aguirre and Collins, 1992; Christensen et al., 1991; Collins et al., 1999b; Lawson et al., 2001a). *Helcococcus ovis* cells are arranged singly and in pairs or short chains, in contrast to the pairs, tetrads and groups of cells exhibited by *Aerococcus* species.

The tiny non-hemolytic colonies formed by *Helcococcus ovis* may help to distinguish this species from strains of streptococci that are pyrrolidonyl arylamidase-negative and leucine aminiopeptidase-positive.

Positive results for both pyrrolidonyl arylamidase and leucine aminopeptidase or resistance to vancomycin will help to differentiate *Helcococcus* strains from many other currently described genera of catalase-negative Gram-stain-positive cocci (Ruoff, 2002).

Taxonomic comments

Although the genus *Helcococcus* bears phenotypic similarities with members of the lactic acid bacteria, its proposed phylogenetic relationship with these organisms diverges at the taxonomic class level. The genus *Helcococcus* is currently included within the clostridial branch of Gram-stain-positive bacteria with low G+C content and is classified within a family *incertae sedis* with a number of other genera formerly classified in the family *Peptostreptococcaceae*.

List of species of the genus *Helcococcus*

1. **Helcococcus kunzii** Collins, Facklam, Rodrigues and Ruoff 1993, 427[VP]

kunzi.i. N.L. gen. n. *kunzii* of Kunz named after Lawrence J. Kunz, an American bacteriologist.

Gram-stain-positive cocci arranged in pairs and groups. Cells are nonmotile and non-pigmented. Non-hemolytic colonies formed on blood agar. No growth on bile esculin agar; growth may or may not occur in serum-supplemented 6.5% NaCl. Catalase-negative and facultatively anaerobic.

Lipohilic; growth stimulated by supplementation of non-blood containing media with 1% horse serum or 0.1% Tween 80. Acid but not gas produced from glucose and some other carbohydrates. The major end products of glucose metabolism are lactate and acetate. Acid is produced from lactose and trehalose, and by some strains from glycogen. No acid production from arabinose, inulin, raffinose, ribose, or sorbitol. Hippurate, urea and gelatin are not hydrolyzed. Esculin is hydrolyzed in the API 20Strep system, but not in conventional test media. Arginine is not deaminated. Pyrrolidonyl arylamidase is produced. Pyrazinamidase is produced by some strains. No production of alkaline phosphatase, α-galactosidase, β-galactosidase, β-glucuronidase or leucine aminopeptidase. Voges–Proskauer-negative. Nitrate not reduced. Vancomycin-susceptible.

Isolated from human sources (wounds, abscesses, intact skin). Habitat thought to be human skin.

DNA G+C content (mol%): 30 (T_m, Collins et al., 1993).

Type strain: NCFB 2900 (IFO 15552; ATCC 51366).

GenBank accession number (16S rRNA gene): X69837.

2. **Helcococcus ovis** Collins, Falsen, Foster, Monasterio, Dominguez and Fernandez-Garazabal 1999a, 1432[VP]

o'vis. L. gen. n. *ovis* of a sheep.

Gram-stain-positive cocci arranged singly, in pairs and short chains. Cells are nonmotile and non-pigmented. Nonhemolytic colonies formed on blood agar. Catalase-negative and facultatively anaerobic. Acid but not gas produced from glucose. Acid may or may not be produced from cyclodextrin, glycogen and maltose. No acid production from D-arabitol, L-arabinose, inulin, lactose, mannitol, melezitose, melibiose, methyl β-D-glucopyranoside, pullulan, raffinose, ribose, sorbitol starch, sucrose, tagatose or D-xylose. Esculin, hippurate, urea, and gelatin are not hydrolyzed. Arginine is not deaminated. Alkaline phosphatase, acid phosphatase, β–glucuronidase and leucine aminopeptidase are produced. No production of pyrrolidonyl arylamidase, alanyl-phenylalanine-proline arylamidase, glycyl-tryptophan arylamidase, chymotrypsin, esterase C4, ester lipase C8, α-fucosidase, α-galactosidase, β-glucosidase, valine arylamidase, lipase C14, α-mannosidase, β-mannosidase, trypsin and N-acetyl-β-glucosaminidase. Variable production of cystine arylamidase, β-galactosidase and phosphoamidase. Voges–Proskauer-negative. Nitrate not reduced. Vancomycin-susceptible.

Isolated from sheep. Habitat unknown.

DNA G+C content (mol%): 29 (T_m; Collins et al., 1999a).

Type strain: CCUG 37441 (CIP 106312; ATCC BAA-59).

GenBank accession number (16S rRNA gene): Y16279.

Genus V. **Parvimonas** Tindall and Euzéby 2006, 2712[VP]

THE EDITORIAL BOARD

Par.vi.mo′nas. L. adj. *parvus* little, small; Gr. fem n. *monas* a unit, monad; N.L. fem. n. *Parvimonas* a small monad.

Nonsporeforming, obligately **anaerobic** Gram-stain-positive **cocci**. Cells may occur in pairs, chains, and masses and are 0.3–0.7 μm in diameter. **Carbohydrates are not fermented.** Indole is negative. Alkaline phosphatase is positive. The optimum temperature for growth is 37°C. Metabolize peptone and amino acids to acetic acid. Found in oral cavity and various human abscesses.

Peptidoglycan diamino acid is **lysine** and the interpeptide bridge is **glycine**. Major cellular fatty acids are $C_{18:1}$, C_{16}, C_{18}, and $C_{16:1}$.

DNA G+C content (mol%): 27–28.

Type species: **Parvimonas micra** (Prévot 1933) Tindall and Euzéby 2006, 2712[VP] (*Streptococcus micros* Prévot 1933, 195; *Peptostreptococcus micros* Smith 1957, 537; *Micromonas micros* Murdoch and Shah 2000, 1415).

Further descriptive information

This description of *Parvimonas* is based upon Ezaki et al. (1983) and Murdoch and Shah (1999). *Parvimonas* does not ferment carbohydrates, and fructose, glucose, lactose, mannose, raffinose, ribose, and sucrose are not metabolized. Likewise, the enzymes β-galactosidase, β-glucuronidase, and α-gluconsidase are not produced. Instead, peptones and amino acids are fermented, with acetate being the major acid formed. Arylamidases for arginine, histidine, leucine, phenylalanine, proline, pyroglutamate, serine and tyrosine are found in most if not all strains. Catalase, coagulase, urease and arginine dihyrdrolase are not produced. Nitrate is not reduced to nitrite, and gelatin is not liquified. See Table 187 and Table 217 for differential characteristics with related genera.

Colonies form on blood agar plates vary in size from 1–2 mm in diameter after 5 d. They are circular, entire, white or gray, glistening and domed. They are frequently surrounded by a yellow-brown halo of discolored agar, up to 2 mm wide. Colonies are β-hemolytic.

Major cellular fatty acids are $C_{18:1}$, $C_{16:1}$, C_{18}, and C_{16} (Ezaki et al., 1983; Wells and Field, 1976).

This organism is part of the normal flora of the human mouth and probably the gastrointestinal and female genitourinary tracts. Sometimes it is isolated from pathological specimens.

Taxonomic comments

Originally described as *Streptococcus micros* (Prévot, 1933), this species was transferred to *Peptostreptococcus* by Smith (1957). Upon sequencing of the 16S rRNA gene, it was discovered to be a representative of new genus (Li et al., 1994; Song et al., 2003). Originally, it was reclassified into a novel genus, *Micromonas*, by Murdoch and Shah (1999). However, this name was illegitimate because it was also in use within botany (*Micromonas* Manton and Parke 1960). Hence, *Parvimonas* was proposed to replace *Micromonas* (Tindall and Euzéby, 2006).

List of species of the genus *Parvimonas*

1. **Parvimonas micra** (Prévot 1933) Tindall and Euzéby 2006, 2712[VP] (*Streptococcus micros* Prévot 1933, 195; *Peptostreptococcus micros* Smith 1957, 537; *Micromonas micros* Murdoch and Shah 2000, 1415)

mi′cra. Gr. adj. *micros -ê -on* small, little; N.L. fem. adj. *micra* small, little.

The species description is the same as the genus description.

DNA G+C content (mol%): 27–28 (T_m).

Type strain: ATCC 33270, CCUG 17638 A, CIP 105294, DSM 20468, GIFU 7824, NCTC 11808, VPI 5464.

GenBank accession number (16S rRNA gene): AF542231, AY323523.

Note: The name *Micromonas* Murdoch and Shah 2000 is illegitimate because of precedence of a microalga *Micromonas* I. Manton and M. Parke (Eukaryota; Viridiplantae; Chlorophyta; Prasinophyceae; Mamiellales; Mamiellaceae; *Micromonas*; *Micromonas pusilla*). See: Principle 2 and Rule 51b(4) of the Bacteriological Code (Lapage et al., 1992).

Genus VI. **Peptoniphilus** Ezaki, Kawamura, Li, Li, Zhao and Shu 2001a, 1524[VP]

TAKAYUKI EZAKI AND YOSHIAKI KAWAMURA

Pep.to.ni.phil′us. N.L. neut. n. *peptonum* peptone; Gr. adj. *philos* liking, friendly to; N.L. masc. n. *Peptoniphilus* friend of peptone, referring to the use of peptone as a major energy source).

Non-spore-forming, obligately anaerobic, Gram-stain-positive, cocci. Cells may occur in pairs, short chains, tetrads, or small masses. Nonmotile. Optimal growth temperature is 37°C. The major metabolic end product from peptone yeast extract glucose (PYG) medium is butyric acid. Carbohydrates are not fermented. Use peptones and oligopeptide as major energy source. The position 1, position 3, and interpeptide bridge of peptidoglycan are alanine, L-ornithine, and D-glutamic acid (Li et al., 1992; Schleifer and Nimmermann, 1973; Weiss, 1981).

Major cellular fatty acid: $C_{18:1}$, C_{16}, C_{18}, and $C_{16:1}$ (Ezaki et al., 1983; O'Leary and Wilkinson, 1988).

DNA G+C content (mol%): 25–34 (Ezaki, 1982).

Type species: **Peptoniphilus asaccharolyticus** (Distaso 1912) Ezaki, Kawamura, Li, Li, Zhao and Shu 2001a, 1525 (*Peptococcus asaccharolyticus* Distaso 1912, Douglas 1957; *Peptostreptococcus asaccharolyticus* Ezaki, Yamamoto, Ninomiya, Suzuki and Yabuuchi 1983, 690).

List of species of the genus *Peptoniphilus*

1. **Peptoniphilus asaccharolyticus** (Distaso 1912) Ezaki, Kawamura, Li, Li, Zhao and Shu 2001a, 1525[VP] (*Peptococcus asaccharolyticus* Distaso 1912, Douglas 1957; *Peptostreptococcus asaccharolyticus* Ezaki, Yamamoto, Ninomiya, Suzuki and Yabuuchi 1983, 690)

a.sac.cha.ro.ly′ti.cus. Gr. pref. *a* not; Gr. n. *sackhar* sugar; Gr. adj. *lutikos* able to loosen; N.L. adj. *asaccharolyticus* not digesting sugar.

Obligately anaerobic, Gram-stain-positive cocci that appear in pairs, chains, and irregular masses.

Cells are 0.5–1.6μm in diameter. Often isolated from various human clinical specimens such as vaginal discharges and ovarian and peritoneal abscesses. Most strains produce indole. Negative for urease, alkaline phosphatase, arginine dihydrolase, and coagulase. Does not produce acid from carbohydrates.

Major metabolic end product from PYG is butyric acid. Carbohydrates are not fermented. Use peptones and oligopeptide as major energy source. The position 1, position 3, and interpeptide bridge of peptidoglycan are alanine, L-ornithine, and D-glutamic acid. Production of major saccharolytic and proteolytic enzymes (Murdoch et al., 1988) and differential characteristics are summarized in Table 188.

The cell-wall peptidoglycan (position 1, position 3, and interpeptide bridge): alanine, ornithine, D-glutamate.

DNA G+C content (mol%): 31–32.

Type strain: ATCC 14963, CCUG 9988, CIP 74.17, DSM 20463, GIFU 7656, NCTC 11461.

GenBank accession number (16S rRNA gene): AF542228, D14138.

2. **Peptoniphilus harei** (Murdoch, Collins, Willems, Hardie, Young and Magee 1997) Ezaki, Kawamura, Li, Li, Zhao and Shu (2001a) 1526[VP] (*Peptostreptococcus harei* Murdoch, Collins, Willems, Hardie, Young and Magee 1997, 783)

Ha′re.i. N.L. gen. n. *harei* of R. Hare, a British microbiologist.

Obligately anaerobic, Gram-stain-positive cocci that appear in pairs and irregular masses. Cell size varies considerably (diameter, 0.5–1.5μm).

The type strain was isolated from pus of a human sacral ulcer. Indole and catalase production are variable among strains. Negative for urease, arginine dihydrolase, alkaline phosphatase, and coagulase. Carbohydrates are not fermented. Use peptones and oligopeptide as major energy source. Major metabolic end product from PYG is butyric acid. Production of major saccharolytic and proteolytic enzymes (Murdoch et al., 1988) and differential characteristics are summarized in Table 188.

The position 1, position 3, and interpeptide bridge of peptidoglycan are alanine, L-ornithine, and D-glutamic acid.

DNA G+C content (mol%): 25.

Type strain: CIP 105325, DSM 10020, NCTC 13076, SBH 093.

GenBank accession number (16S rRNA gene): Y07839.

3. **Peptoniphilus indolicus** (Christiansen 1934) Ezaki, Kawamura, Li, Li, Zhao and Shu 2001a, 1525[VP] ("*Micrococcus indolicus*" Christiansen 1934, 366; *Peptococcus indolicus* Høi Sorensen 1975, 221; *Peptostreptococcus indolicus* Ezaki, Yamamoto, Ninomiya, Suzuki and Yabuuchi 1983, 692)

in.do′li.cus. Chem. term *indole*, L. suff. *-icus* related to; N.L. n. *indolicus* related to indole, referring to the ability of the organism to produce indole.

Obligately anaerobic, Gram-stain-positive cocci that appear in pairs, chains, and irregular masses.

Isolated from summer mastitis of cattle. Occasionally isolated from human clinical specimen. Most strains are coagulase, alkaline phosphatase, and indole-positive. Negative for urease and arginine dihydrolase. No sugars are fermented. Use peptones and oligopeptide as major energy source. Major metabolic end product from PYG is butyric acid. Production of major saccharolytic and proteolytic enzymes (Murdoch et al., 1988) and differential characteristics are summarized in Table 188.

The position 1, position 3, and interpeptide bridge of peptidoglycan are alanine, L-ornithine, and D-glutamic acid.

DNA G+C content (mol%): 32–34.

Type strain: ATCC 29427, CCM 5987, CCUG 17639, DSM 20464, NCTC 11088.

GenBank accession number (16S rRNA gene): AY153430, D14147.

4. **Peptoniphilus ivorii** (Murdoch, Collins, Willems, Hardie, Young and Magee 1997) Ezaki, Kawamura, Li, Li, Zhao and Shu 2001a, 1526^{VP} (*Peptostreptococcus ivorii* Murdoch, Collins, Willems, Hardie, Young and Magee 1997, 785)

i.vo'ri.i. N.L. gen. n. *ivorii* of Ivor, a British microbiologist who first isolated the organism.

Obligately anaerobic, Gram-stain-positive cocci that appear in pairs and irregular masses.

The type strain was isolated from a human leg ulcer. Negative for urease, arginine dihydrolase, alkaline phosphatase, and coagulase. Carbohydrates are not fermented. Use peptones and oligopeptide as major energy source. Major metabolic end product from PYG is butyric acid. Production of major saccharolytic and proteolytic enzymes (Murdoch et al., 1988) and differential characteristics are summarized in Table 188.

The position 1, position 3, and interpeptide bride of peptidoglycan are alanine, L-ornithine, and D-glutamic acid

DNA G+C content (mol%): 29%.

Type strain: CIP 105325, DSM 10022, NCTC 13078, SBH 093.

GenBank accession number (16S rRNA gene): Y07840.

5. **Peptoniphilus lacrimalis** (Li, Hashimoto, Adnan, Miura, Yamamoto and Ezaki 1992) Ezaki, Kawamura, Li, Li, Zhao and Shu 2001a, 1526^{VP} (*Peptostreptococcus lacrimalis* Li, Hashimoto, Adnan, Miura, Yamamoto and Ezaki 1992, 604)

la.cri.ma'lis. L. fem. n. *lacrima* a tear; N.L. adj. *lacrimalis* referring to the lacrimal gland, where the organism was isolated.

Obligately anaerobic, Gram-stain-positive cocci that appear in short chains and irregular masses.

Isolated from abscess of human lachrymal gland. Occasionally from ottorhea and abscesses from other areas of the human body. Negative for indole, urease, arginine dihydrolase, alkaline phosphatase, and coagulase. Carbohydrates are not fermented. Use peptones and oligopeptide as major energy source. Major metabolic end product from PYG is butyric acid. The following arylamidases (AMD) are strong: leucine AMD, alanine AMD, histidine AMD, glycine AMD, and phenylalanine AMD. The differential characteristics are summarized in Table 188.

The position 1, position 3, and interpeptide bridge of peptidoglycan are alanine, L-ornithine, and D-glutamic acid.

DNA G+C content (mol%): 30–34%.

Type strain: ATCC 51171, CCUG 31350, CIP 103724, DSM 7455, JCM 8139, GIFU 7667, NCTC 13149.

GenBank accession number (16S rDNA): AF542230, D14141.

Genus VII. **Sedimentibacter** Breitenstein, Wiegel, Haertig, Weiss, Andreesen and Lechner 2002, 806^{VP}

UTE LECHNER

Se.di.men.ti.bac'ter. N.L. masc. n. *sedimentum* sediment; N.L. n. *bacter* masc. equivalent of Gr. neut. dim. n. *bakterion* rod or staff; N.L. masc. n. *Sedimentibacter* rod from sediment, referring to its origin.

Slightly curved rod-shaped cells found singly or in chains, 0.35 × 7 μm, oval spores formed from terminal swollen sporangia. Gram-stain-positive or -negative. Motile by means of **peritrichous flagella.** Strict anaerobe. Growth between 12 and 41°C with optimal growth at 33–37°C. pH optimum 7–8.2. Growth requires yeast extract and is supported by the **fermentation of pyruvate or of amino acids in a Stickland-type reaction. H₂ not produced. Carbohydrates not fermented. Purines including uric acid, adenine, hypoxanthine, guanine, and xanthine are not utilized.** Catalase and urease are absent. The **cell-wall type is A1α** (L-lysine, direct). Menaquinones are not present. The genus represents a separate line of descent within the *Peptostreptococcaceae* according to 16S rRNA gene sequence analysis.

DNA G+C content (mol%): 34–35.5.

Type species: **Sedimentibacter hydroxybenzoicus** (Zhang, Mandelco and Wiegel 1994) Breitenstein, Wiegel, Haertig, Weiss, Andreesen and Lechner 2002, 806^{VP} (*Clostridium hydroxybenzoicum* Zhang, Mandelco and Wiegel 1994, 218).

Further descriptive information

Based on 16S rRNA gene sequences, the genus *Sedimentibacter* forms a distinct line of descent within a cluster of genera previously assigned to the family *Peptostreptococcaceae* and additional

bacteria formerly belonging to the cluster XII-clostridia including *Clostridium purinolyticum, Clostridium acidiurici, Clostridium hastiforme,* and *Tissierella praeacuta* (Collins et al., 1994; Garrity et al., 2005) and the recently described genus *Soehngenia* (Parshina et al., 2003) (Figure 221). In phylogenetic trees prepared for this volume, it appears as a neighboring lineage to Family XI *Incertae Sedis,* and it was retained within that group until the taxonomic ambiguities are resolved. Sequence identities are highest (87–89%) with rod-shaped bacteria in this phylogenetic branch (*Clostridium purinolyticum, Clostridium acidiurici, Sporanaerobacter acetigens, Soehngenia saccharolytica, Tissierella praeacuta, Clostridium hastiforme,* and *Fusibacter paucivorans*). 16S rRNA genes of coccoid members of the Family XI *Incertae Sedis* (*Helcococcus kunzii, Anaerococcus prevotii, Peptoniphilus asaccharolyticus, Finegoldia magna, Gallicola barnesae, Parvimonas micra* and *Peptostreptococcus anaerobius*) are less similar (82–86% identity) and form a distinctly different phylogenetic branch (Figure 221). Morphologically the cells of *Sedimentibacter* are rod-shaped, peritrichously flagellated, and form spores. These features are common in the above-mentioned cluster XII-associated group, with the exception of *Tissierella,* where spore formation was not detected (Farrow et al., 1995). However, evidence was recently provided that the type strains of *Tissierella praeacuta* and of the

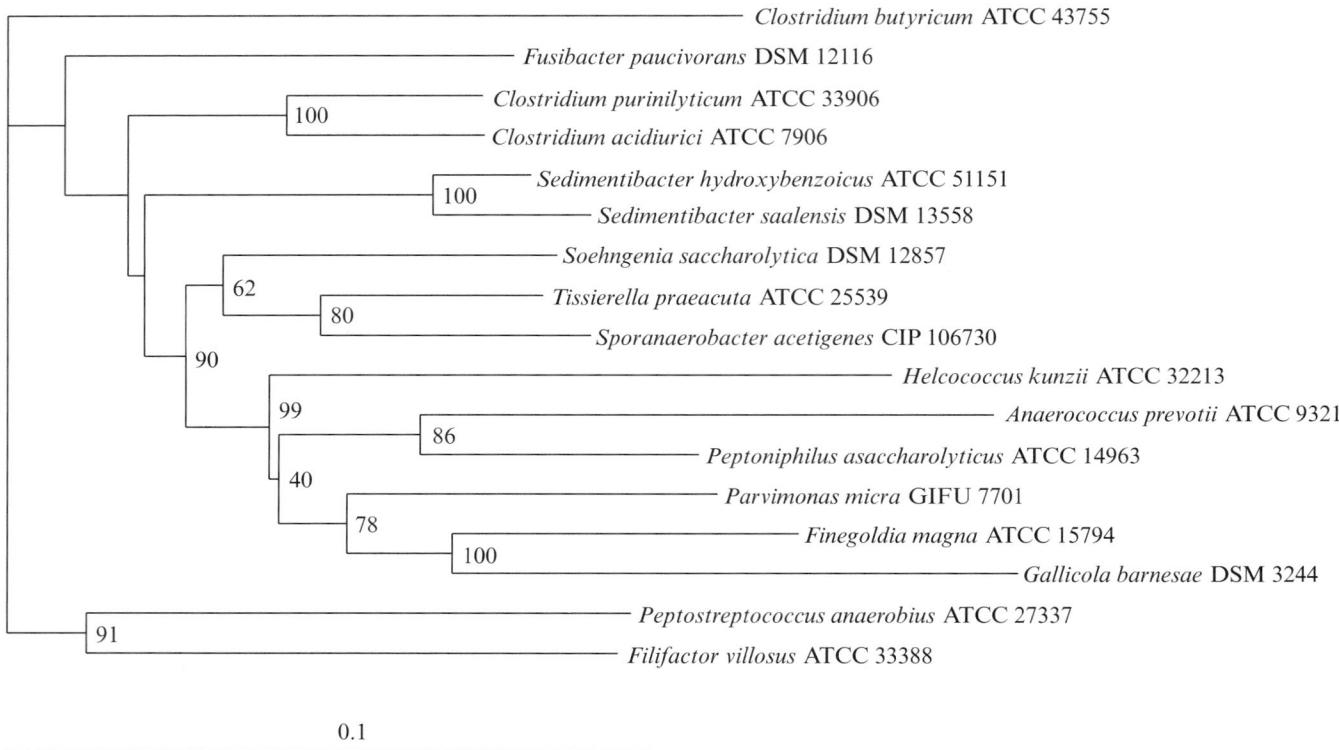

FIGURE 221. Phylogenetic neighbor-joining dendrogram showing the relationships of *Sedimentibacter* species with representatives of the family *Peptostreptococcaceae* and clostridia formerly assigned to cluster XII (Collins et al., 1994). Percentage bootstrap values are based on 500 replicates and are indicated at the branching points. Bar = 0.1 nucleotide substitutions per site. *Clostridium butyricum* was used as the outgroup.

spore-forming *Clostridium hastiforme* are members of the same species (Bae et al., 2004) and that *Tissierella praeacuta* contains sporulation genes (Onyenwoke et al., 2004; Jürgen Wiegel, personal communication). Cells of *Sedimentibacter* have pointed ends. Cells stain Gram-stain-negative or Gram-stain-positive, but exhibit the typical cell-wall structure of Gram-stain-positive organisms. The cell-wall type is A1α (L-lysine, direct) in contrast to most genera within the *Peptostreptococcaceae* which often contain interpeptide bridges of dicarboxylic amino acids or glycine. *Tissierella praeacuta* and *Clostridium purinolyticum* contain *meso*-diaminopimelic acid (*meso*-DAP) or D-ornithine instead of L-lysine (Table 218). The dominant lipid fatty acids are $C_{14:0}$, $C_{16:0}$, and dimethyl acetals (for more details see footnote of Table 219). Methyl branched fatty acids as observed for *Peptostreptococcus anaerobius* and *Tissierella praeacuta* are not present.

White, disc-shaped colonies (0.5–2 mm in diameter) are formed in minimal medium with 10 mM pyruvate and 0.05% yeast extract solidified with 1% agar. The bacteria can be grown in an anaerobic minimal medium containing (per liter): $(NH_4)HCO_3$, 0.45 g; $CaCl_2 \cdot 2H_2O$, 0.12 g; $MgSO_4 \cdot 6H_2O$, 0.13 g; 1 ml trace element solution SL-10 (25% HCl, 10 ml; $FeCl_2 \cdot 4H_2O$, 1.5 g; $ZnCl_2$, 70 mg; $MnCl_2 \cdot 4H_2O$, 100 mg; H_3BO_3, 6 mg; $CoCl_2 \cdot 6H_2O$, 190 mg; $CuCl_2 \cdot 2H_2O$, 2 mg; $NiCl_2 \cdot 6H_2O$, 24 mg; $Na_2MoO_4 \cdot 2H_2O$, 36 mg; bidistilled water to 1000 ml); 1 ml selenite and tungstate solution (0.5 g NaOH, 3 mg $Na_2SeO_3 \cdot 5H_2O$; 4 mg $Na_2WO_4 \cdot 2H_2O$, bidistilled water to 1000 ml); resazurin, 1 mg; yeast extract, 3 g. After boiling and gassing with N_2/CO_2 (80:20), the pH is adjusted to 7.3 by the addition of $NaHCO_3$

(ca. 3.8 g) under continuous stirring. After autoclaving, the following compounds are added from anaerobic stock solutions: phosphate buffer (pH 7.3), 10 mM; glycine, 10 mM; arginine, 10 mM; Na_2S, 1 mM; Ti (III) citrate, 0.15 mM; vitamin solution, 1 ml/l, containing the following vitamins (in mg/l): L-lipoic acid, 5; D-(+)biotin, 2; folic acid, 2; riboflavin, 5; thiamine, 5; nicotinamide, 5; cyanocobalamin, 5; *p*-aminobenzoic acid, 5; pyridoxine, 10; pantothenic acid, 5; 1,4-naphthoquinone, 10.

Growth of the strictly anaerobic bacterium in minimal medium depends on yeast extract (at least 0.01%, optimum growth at 1.5% w/v for the type species). Carbohydrates are not fermented. Pyruvate is fermented to acetate and butyrate. Electrons produced during oxidative metabolism were probably transferred to acetate to form butyrate as described for *Clostridium acetireducens* (Girbal et al., 1997). Hydrogen is not formed either from pyruvate or yeast extract, in contrast to *Soehngenia* or *Sporanaerobacter* species. *Sedimentibacter* ferments the amino acids glycine and L-lysine to acetate and acetate plus butyrate, respectively. In the presence of both amino acids, an increased amount of acetate is formed at the expense of butyrate indicating that glycine functions as the electron acceptor in a Stickland-type reaction (Andreesen et al., 1989). Arginine alone does not support growth in contrast to many clostridia, which can use the energy-providing arginine deiminase pathway (Andreesen et al., 1989). The addition of arginine to glycine leads to an increase in growth and acetate formation. Arginine is converted to ornithine (Zhang et al., 1994) which probably serves as electron donor for the reduction of glycine to acetate (Andreesen et al.,

TABLE 218. Comparison of *Sedimentibacter* with some Gram-stain-positive, rod-shaped bacteria from the family *Peptostreptococcaceae*, former clostridial cluster XII (Collins et al., 1994), and *Clostridiales* Family *Incerta Sedis* XI (Ludwig et al., 2009)[a]

Characteristic	*Sedimentibacter hydroxybenzoicus*	*Clostridium acidurici*	*Clostridium purinolyticum*	*Filifactor villosus*	*Fusibacter paucivorans*	*Soehngenia saccharolytica*	*Sporanaerobacter acetigens*	*Tissierella praeacuta*[b]
Shape of rods	Slightly curved, pointed ends	Straight	Straight	Rounded ends	Spindle shaped	Slightly thickened in the middle	Straight	Rounded ends, swelling toward the end
Cell dimensions (μm)	0.3–0.8 × 2.5–5.1	0.5–0.7 × 2.5–4	1.1–1.6 × 2.7–9.6	0.6 × 4–6	1 × 2–5 μm	0.5–0.7 × 2–11	0.4–0.5 × 3–5	0.6–0.9 × 2–8
Flagella	+	+	+	–	+	+	+	+
Endospores	+	+	+	+	–	+	+	+
DNA G+C content (mol%)	35.5	28	29	NR	43	42	32	28
Gram stain reaction	–	Variable	+	+	+	NR	NR	–
Cell wall contains	L-Lysine, direct	*meso*-DAP	*meso*-DAP	Ornithine-iso-D-asparagine	NR	NR	NR	*meso*-DAP
Temperature optimum (°C)	33–34	NR	36	NR	37	30–37	40	37
pH optimum	7.2–8.2	7.6–8.1	7.3–7.8	NR	7.3	7	7.5	NR
Requirement of yeast extract for growth	+	–	Stimulatory	NR	+	+	+	NR
Decarboxylation of 4-OHB and 3,4-OHB	+	–[c]	–[c]	–[c]	NR	NR	NR	–[c]
Fermentation of:								
Glycine	+	+[d]	+	NR	NR	–	+	NR
Purines	+	+	+	NR	NR	NR	NR	NR
Pyruvate[e]	+ (A, B)	–	–	+ (B, A)	NR	+	+ (A, H₂, CO₂)	–
Carbohydrates	–	–	–	–	+	+	+	–
H₂ production	–	–	–	–	+	+	+	–
Formation of H₂S	+ (from $S_2O_3^{2-}$)	NR	NR	NR	+ (from $S_2O_3^{2-}$, sulfur)	+ (from $S_2O_3^{2-}$, SO_3^{2-})	+ (from sulfur)	+ (from $S_2O_3^{2-}$)
Indole formation	+	–	–	NR	NR	+	NR	–
Relation to oxygen	Strictly anaerobic	Strictly anaerobic	Strictly anaerobic	Strictly anaerobic	Strictly anaerobic	Anaerobic, but aerotolerant	Strictly anaerobic	Strictly anaerobic

[a]Symbols and abbreviations: +/–, character reported as positive or negative, respectively, for the type strain; ; NR, not reported; *meso*-DAP, *meso*-diaminopimelic acid; 4-OHB, 4-hydroxybenzoate; 3,4-OHB, 3,4-dihydroxybenzoate.

[b]Spore formation and pyruvate fermentation as reported for *Clostridium hastiforme* (later synonym of *Tissierella praeacuta*) (Bae et al., 2004).

[c]Data from Zhang (1993).

[d]Utilization in the presence of a cosubstrate.

[e]Major end products in parentheses (A, acetate; B, butyrate).

TABLE 219. Characteristics for differentiation of *Sedimentibacter hydroxybenzoicus* and *Sedimentibacter saalensis*[a]

Characteristic[b]	S. hydroxybenzoicus[c]	S. saalensis[d]
Cell size (μm)[e]	$0.35–0.67 \times 2.5–5.1$	$0.5–0.7 \times 3.7–6$
Formation of filaments	Rare	Frequent
Spore shape	Round	Oval
Gram-stain reaction	–	+
G + C content (mol%)	35.5	34.0
Optimum temperature (°C)	33–34	37
Optimum pH	7.2–8.2	6.8–7.3
Indole production	+	–
Decarboxylation of:		
4-Hydroxybenzoate	+	–
3,4-Dihydroxybenzoate	+	–

[a]Symbols: +, >85% positive; –, 0–15% positive.

[b]The following characteristics tested positive for both strains: fermentation of pyruvate, glycine, lysine, and arginine plus glycine and the formation of H_2S in SIM medium. Both contained the peptidoglycan type A1α, L-Lys direct and the cellular fatty acids $C_{14:0}$, $C_{16:0}$, $C_{16:1\ \omega7c}$, C_{9c}, $C_{18:1\ \omega7c}$ and the dimethylacetals $C_{16:0}$ and $C_{18:0}$. The following characteristics were negative for both strains: catalase, gelatin and esculin hydrolysis, nitrate reduction, fermentation of carbohydrates (cellobiose, maltose, arabinose, glucose, mannose, xylose, trehalose, and sorbitol) and purines (uric acid, hypoxanthine, xanthine, guanine, and adenine), and production of H_2. The following characteristics tested negative for one or both strains: lipase reaction, hydrolysis of starch, lecithinase activity, acid formation from fructose, ribose, and glycogen (JW/Z-1[T]), urease activity, and acid formation from mannitol, lactose, sucrose, salicin, glycerol, melizitose, raffinose, and rhamnose (ZF2[T]).

[c]Data from strain JW/Z-1[T] (Zhang et al., 1994).

[d]Data from strain ZF2[T] (Breitenstein et al., 2002).

[e]Cells of both strains were slightly curved rods motile by peritrichous flagella.

1989). *Clostridium purinolyticum*, *Clostridium acidiurici*, *Gallicola barnesae*, *Peptoniphilus asaccharilyticus*, and *Anaerococcus prevotii* are able to use purines as growth substrates (Andreesen, 1994), whereas *Sedimentibacter* does not ferment any purines tested (uric acid, hypoxanthine, xanthine, guanine, and adenine). Sodium ions stimulate growth in sodium-deficient medium and monensin inhibits growth and amino acid utilization suggesting that *Sedimentibacter* utilizes a sodium-proton antiport and a sodium-amino acid symport (Zhang et al., 1994). The reversible decarboxylation of 4-hydroxy- and 3,4-dihydroxybenzoate to phenol and catechol, respectively, is a characteristic feature of *Sedimentibacter hydroxybenzoicus* (Zhang and Wiegel, 1990; Zhang et al., 1994). The products are not further metabolized. The respective enzymes, 4-hydroxy- and 3,4-dihydroxybenzoate decarboxylases, were purified and characterized (He and Wiegel, 1995, 1996). They differ from known decarboxylases in that they are neither biotin, pyridoxal-5′-phosphate, nor thiamine diphosphate dependent. The gene of 4-hydroxybenzoate decarboxylase is the first characterized representative of a novel gene family which is distributed within *Bacteria* and *Archaea* (Huang et al., 1999). The physiological and ecological meaning of these reactions is not yet clear. Originally it was assumed that *Sedimentibacter hydroxybenzoicus* catalyzed the carboxylation of phenol in an enrichment culture which degraded 2,4-dichlorophenol via phenol (Zhang and Wiegel, 1990). However, the equilibrium of the reversible decarboxylation reaction is on the side of phenol (Zhang and Wiegel, 1994). In contrast to *Sedimentibacter hydroxybenzoicus*, *Sedimentibacter saalensis* neither decarboxylates 4-hydroxy- and 3,4-dihydroxybenzoate nor carboxylates phenol and catechol.

Sedimentibacter occurs in methanogenic freshwater sediments from which the strains *Sedimentibacter hydroxybenzoicus* JW/Z-1[T] and *Sedimentibacter saalensis* ZF2[T] were both isolated during attempts to enrich chlorophenol-dechlorinating communities. Another bacterium, strain BRS2, is related to *Sedimentibacter* on the basis of its 16S rDNA sequence (94% identity; GenBank accession number AY221992). It was isolated from a vinyl chloride-dechlorinating enrichment culture (Frank Löffler, personal communication) further indicating that the genus grows well under conditions used to enrich for reductively dechlorinating bacteria, but in no case was reductive dechlorination observed.

Isolation procedure

Sedimentibacter hydroxybenzoicus was obtained from a sediment which exhibited sequential degradation of 2,4-dichlorophenol via 4-chlorophenol and phenol to CH_4 and CO_2 (Zhang and Wiegel, 1990). Enrichment and isolation was achieved by feeding a phenol-degrading enrichment culture with 4-hydroxybenzoate which led to an increase of its decarboxylation from less then 1 mM per 24 h to over 100 mM of hydroxybenzoate per 24 h with a subsequent degradation to carbon dioxide, hydrogen, and acetate. Although *Sedimentibacter hydroxybenzoicus* does not metabolize phenol or hydroxybenzoate itself, the enrichment of this bacterium probably occurred due to the subsequent production of acetate by other members in the mixed culture. Acetate stimulates growth of *Sedimentibacter hydroxybenzoicus*. Final isolation occurred on yeast extract containing mineral media using dilution rows in anaerobic agar-shake-roll tubes inoculated with pasteurized enrichment samples (Zhang and Wiegel, 1990; J. Wiegel, personal communication).

Sedimentibacter saalensis was obtained from an anaerobic, 2,4,6-trichlorophenol-dehalogenating, mixed culture enriched from freshwater sediment by isolation of a single colony from agar-shake dilution series, which contained 10 mM pyruvate, 0.01% yeast extract, and 100 μM 2,4,6-trichlorophenol (Breitenstein et al., 2002). Further purification was achieved by consecutive serial dilutions in agar shakes containing 25 mM pyruvate and 0.05% yeast extract.

Maintenance procedures

Cultures can be maintained on either liquid medium (see above) or solidified by 1% agar. Cultures grow in liquid medium without shaking. Liquid cultures or colonies in agar shake tubes were observed to survive at 4°C for several months. Liquid cultures grown to the late exponential phase can be dispensed into 2 ml aliquots of N_2/CO_2 (80:20)-gassed Hungate tubes and mixed with 0.6 ml of anaerobic glycerol (87% v/v gassed with N_2, autoclaved and reduced with 1 mM Na_2S). Following incubation at 30°C for 2 h, cultures are frozen in liquid nitrogen and stored at −80°C. *Sedimentibacter* can also be lyophilized and is provided by DSMZ and ATCC in this form.

Characteristics for *Sedimentibacter* are listed in Table 219. All methods for analyzes of these properties are presented in Zhang and Wiegel (1990), Zhang et al. (1994), or Breitenstein et al. (2002).

Differentiation of the genus *Sedimentibacter* from other genera

Sedimentibacter can be distinguished from other rod-shaped members of the Family XI *Incertae Sedis* and the purinolytic clostridia *Clostridium acidiurici* and *Clostridium purinolyticum* by a combination of properties such as the inability to ferment carbohydrates and purines and by the formation of acetate plus butyrate, but not hydrogen, from amino acids, pyruvate, or yeast extract (Table 218). So far, the observed type of cell-wall cross-linking A1α (L-lysine, direct) of *Sedimentibacter* has not been described for the other genera mentioned. In addition, its phylogeny based on 16S rRNA sequence places it in a relatively deep branching lineage within the Family XI *Incertae Sedis*.

Taxonomic comments

Sedimentibacter hydroxybenzoicus JW/Z-1^T was originally assigned to the clostridia (Zhang et al., 1994), however, the polyphyletic character of the genus *Clostridium* is now well established (Collins et al., 1994). Based on the 16S rRNA sequences of *Sedimentibacter hydroxybenzoicus* JW/Z-1^T and *Sedimentibacter saalensis* ZF2^T, which are 94% identical and show an affiliation to clostridial cluster XII and the *Peptostreptococcaceae*, the new genus *Sedimentibacter* was proposed and characterized by morphological, physiological, and chemotaxonomic studies.

Differentiation of the species of the genus *Sedimentibacter*

Differential characteristics of the species of *Sedimentibacter* are indicated in Table 219.

List of species of the genus *Sedimentibacter*

1. **Sedimentibacter hydroxybenzoicus** (Zhang, Mandelco and Wiegel 1994) Breitenstein, Wiegel, Haertig, Weiss, Andreesen and Lechner 2002, 806^{VP} (*Clostridium hydroxybenzoicum* Zhang, Mandelco and Wiegel 1994, 218)

hy.drox.y.ben.zo'i.cus. Gr. n. *hydor* water; Gr. n. *oxys* acid; Ger. and Fr. n. *benzoin* frankincense of Java; Ger. n. *Benzoesäure* resin obtained from the tree *Styrax benzoin*; N.L. adj. *hydroxybenzoicus* pertaining to the organic acid hydroxybenzoic acid, referring to the characteristic feature of this organism, the reversible decarboxylation of 4-hydroxybenzoate and 3,4-dihydroxybenzoate.

Morphological, cultural and biochemical characteristics are the same as described for the genus (Table 218). Other characteristics are described in Table 219.

Originally isolated from a methanogenic sediment from a freshwater pond of the Cherokee Trailer Park, Athens, GA (USA)

DNA G+C content (mol%): 35.5 (HPLC).
Type strain: JW/Z-1, ATCC 51151, DSM 7310.
GenBank accession number (16S rRNA gene): L11305.

2. **Sedimentibacter saalensis** Breitenstein, Wiegel, Haertig, Weiss, Andreesen and Lechner 2002, 806^{VP}

saa.len'sis. N.L. adj. *saalensis* referring to the German River Saale, from which the organism was isolated.

Morphological, cultural and biochemical characteristics are essentially the same as described for the genus (Table 218) with some deviations and further properties given in Table 219. Striking differences from *Sedimentibacter hydroxybenzoicus* are the inability of *Sedimentibacter saalensis* to decarboxylate hydroxybenzoates and to form indole and the frequent occurrence of long filaments (up to 130 μm with only a few or no septations in *Sedimentibacter saalensis* cultures). The filaments contain nucleoids every 4–6 μm as indicated by staining with 4 , 6-diamidino-2-phenylindole (DAPI).

Originally isolated from freshwater sediment of the River Saale, Germany.

DNA G+C content (mol%): 34 (HPLC).
Type strain: ZF2, ATCC BAA-283, DSM 13558.
GenBank accession number (16S rRNA gene): AJ404680.

Genus VIII. **Soehngenia** Parshina, Kleerebezem, Sanz, Lettinga, Nozhevnikova, Kostrikina, Lysenko and Stams 2003, 1796^{VP}

SOFIYA N. PARSHINA AND ALFONS J. M. STAMS

Soehn.ge'ni.a. N.L. fem. n. *Soehngenia* named in honor of Nicolas L. Soehngen, founder and first head (1911–1937) of the Laboratory of Microbiology of Wageningen University, The Netherlands, where this strain was isolated and described.

Rod-shaped cells, 0.5–0.7 × 2–11 μm (Figure 222). **Gram-stain-positive**. In the early exponential phase of growth, cells are slightly motile by means of peritrichous flagella (Figure 223); older cells lose their motility. Colonies are rhizoid, resemble a snowflake, dark, and creamy. Rare terminal and subterminal spore formation. **Mesophilic**. Anaerobic, but **aerotolerant**; it can grow with 50% air in the gas phase. Fixes molecular nitrogen. **Chemo-organotrophic**. Saccharolytic and weakly proteolytic. Growth substrates are a wide range of carbohydrates and some other carbon sources, including yeast extract, cysteine, and serine. Isolated from a laboratory anaerobic digester sludge with benzaldehyde as the substrate.

DNA G+C content (mol%): 43.0 (T_m).
Type species: **Soehngenia saccharolytica** Parshina, Kleerebezem, Sanz, Lettinga, Nozhevnikova, Kostrikina, Lysenko and Stams 2003, 1797^{VP}.

Further descriptive information

Because only one species in the genus has been described, the description of the genus is the same as that for the species.

Enrichment and isolation procedures

A stable enrichment culture was initiated from methanogenic granular sludge from a laboratory-scale UASB reactor treating

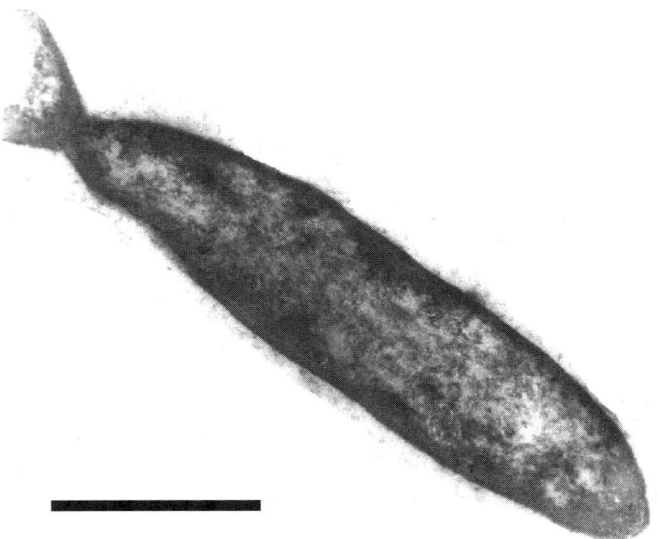

FIGURE 222. Electron micrograph of a thin section of a cell of *Soehngenia saccharolytica* strain BOR-Y^T. Bar = 1 μm.

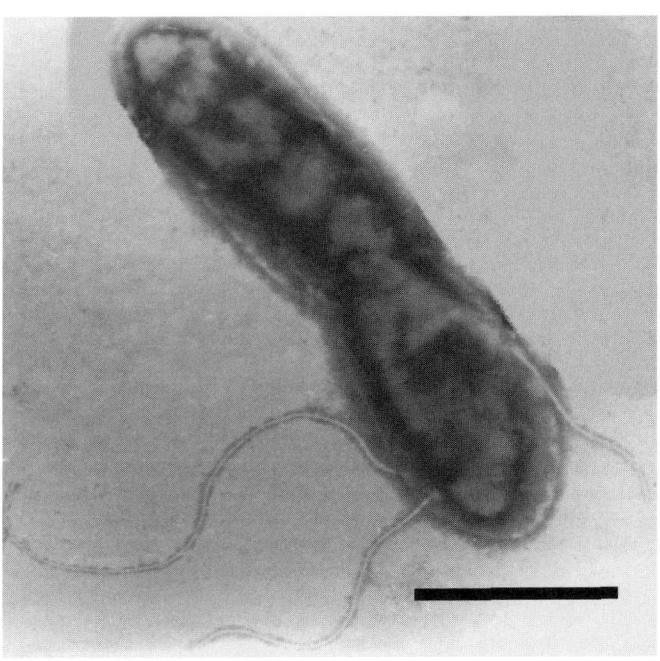

FIGURE 223. Electron micrograph of a negatively stained whole cell of *Soehngenia saccharolytica* strain BOR-Y^T showing the flagella. Bar = 1 μm.

benzaldehyde as the sole substrate. The reactor treating benzaldehyde was itself inoculated with granular sludge from an industrial UASB reactor treating potato starch wastewater. The presence of yeast extract (routinely 0.2 g/l) was required to obtain a stable enrichment. A pure culture was isolated upon repeated serial dilutions in liquid medium with 5–15 mM benzaldehyde and yeast extract.

Maintenance procedures

The type strain is maintained in the bicarbonate-phosphate buffered medium supplemented with 1.0 g/l yeast extract and 10 mmol/l benzaldehyde (Parshina et al., 2003; Parshina et al., 2000). Cultures can be lyophilized. In medium supplemented

with carbohydrates, cells lyse rapidly in the stationary phase of growth.

Differentiation of the genus *Soehngenia* from other genera

Table 220 lists characteristics differentiating *Soehngenia* from phylogenetically related genera. *Soehngenia* does not grow on creatine and creatinine like *Tissierella creatinini* and *Tissierella praeacuta*.

Tissierella creatinophila and *Clostridium hastiforme* do not grow on carbohydrates. *Soehngenia saccharolytica* does not grow on for-

TABLE 220. Characteristics differentiating *Soehngenia* from phylogenetically related bacteria[a]

Characteristic	*Soehngenia saccharolytica*	*Clostridium hastiforme*	*Clostridium ultunense*	*Tissierella creatinini*	*Tissierella creatinophila*	*Tissierella praeacuta*
Growth temperature (°C)	30–37	37	37	37	30–34	37
pH optimum	7	7.5	7	8.3	7.4	7.5
Utilization of:						
Arginine	−	+		−	−	
Betaine	−		+	−	−	
Creatine	−	−		−	+	
Creatinine	−	−		+	+	
Cysteine	+		+			
Ethylene glycol	−		+			
Glucose	+	−	+	−	−	−
Formate	−		+	−	+ (with creatinine)	−
Fructose	+	−	−	−	−	−
Pyruvate	+	−	+	−	−	
Serine	+	+	−	−	−	
Starch	+	−				−
DNA G+C content (mol%)	43	28	32	32	30	28

[a]Symbols: +, >85% positive; d, different strains give different reactions (16–84% positive); −, 0–15% positive; w, weak reaction; ND, not determined.

mate, but it grows on fructose; *Clostridium ultunense* does not grow on fructose, but utilizes formate.

Strain *Soehngenia saccharolytica* has a high G+C content (43 mol%), which distinguishes it from species of the other related genera (28–32 mol%). DNA–DNA hybridization of *Soehngenia saccharolytica* BOR-Y with *Clostridium ultunense* BS and *Tissierella creatinini* DSM 9508, performed by method of De Ley (1970), revealed reassociation values of 9 and 13%, respectively.

Taxonomic comments

Based on detailed phylogenetic analyses of 16S rRNA gene sequences, *Soehngenia* is a deep branch of a clade currently assigned to Family XI *Incertae Sedis* that includes many genera formerly classified within the family *Peptostreptococcaceae* (Ludwig et al., 2009). This clade includes the genus *Tissierella*, with which it possesses the highest sequence similarity, as well as *Parvimonas*, *Anaerococcus*, *Helcococcus*, *Finegoldia*, *Gallicola*, *Peptoniphilus*, *Sporanaerobacter*, *Tepidimicrobium*, and *Sedimentibacter*.

List of species of the genus Soehngenia

1. **Soehngenia saccharolytica** Parshina, Kleerebezem, Sanz, Lettinga, Nozhevnikova, Kostrikina, Lysenko and Stams 2003, 1797^{VP}

sac.cha.ro.ly'ti.ca. Gr. n. *sakkharos* sugar; Gr. adj. *lutikos* loosening, dissolving; N.L. fem. adj. *saccharolytica* sugar-diss-olving.

Colonies on agar are rhizoid, resemble a snow-flake, and are dark and creamy.

Rod-shaped cells are straight or slightly thickened at the middle and 0.5–0.7 μm × 2–11 μm; they occur singly, in pairs, or in chains. Chains of cells are formed in the stationary growth phase. Terminal or subterminal spores are formed occasionally on pyruvate or in nitrogen-free medium. Anaerobic with fermentative metabolism. Aerotolerant. Able to grow with 50% air. Fixes molecular nitrogen. Sulfite and thiosulfate are weakly used as electron acceptors. Sulfate, dithionite, disulfite, sulfur, and nitrate are not reduced. Benzaldehyde is dismutated to benzoate and benzylalcohol. Catalase-negative. Produces indole and does not liquefy gelatin. Substrates used as carbon and energy sources include yeast extract, glucose, fructose, sucrose, xylose, arabinose, rhamnose, mannose, ribose, maltose, cellobiose, galactose, melibiose, lactose, cellulose, xylan, mannitol, pyruvate, malate, starch, cysteine, and serine. Vitamins are required for optimal growth. Moderate growth occurs in mineral medium supplemented with 0.2 g/l yeast extract. Abundant growth occurs in the mineral medium with 2 g/l yeast extract plus 10 mmol/l glucose or some other carbohydrates. In the medium supplemented with carbohydrates, cells lyse rapidly in the stationary phase of growth.

Growth temperature range is 15–40°C. The optimum temperature is 30–37°C. The growth pH range 6.0–7.5, and the optimum is around 7.0.

Habitat: anaerobic digester sludge.

DNA G+C content (mol%): 43.0 (T_m).

Type strain: BOR-Y, ATCC BAA-502, DSM 12858.

GenBank accession number (16S rRNA gene): AY353956.

Genus IX. **Sporanaerobacter** Hernandez-Eugenio, Fardeau, Cayol, Patel, Thomas, Macarie, Garcia and Ollivier 2002, 1221^{VP}

GUADALUPE HERNANDEZ-EUGENIO, MARIE-LAURE FARDEAU, JEAN-LOUIS GARCIA AND BERNARD OLLIVIER

Spor.an.a.e.ro.bac'ter. Gr. fem. n. *spora* spore; Gr. pref. *an* not; Gr. n. *aer* air; *anaero* not (living) in air; N.L. n. *bacter* masc. equivalent of Gr. neut. dim. n. *bakterion* rod, staff; N.L. masc. n. *Sporanaerobacter* a spore-forming anaerobic rod.

Cells are strictly anaerobic **rods**, occurring singly or in pairs and motile by a few laterally inserted flagella. Spherical, terminal, oval **spores** which swell the sporangium appear in the cells. **Gram-positive type** cell wall. Mesophilic and moderately thermophilic, growing at up to 50°C. Neutrophilic. **Heterotrophic.** Yeast extract is required for growth on sugars. **Ferments peptides and amino acids** in the presence of yeast extract. Glucose is fermented to acetate, H$_2$, and CO$_2$ as the major end products of metabolism. Performs the Stickland reaction using isoleucine as electron donor and glycine or serine as electron acceptors. **Uses elemental sulfur** but not sulfate, thiosufate, sulfite, nitrate, or nitrite as an electron acceptor.

DNA G+C content (mol%): 32.2 (HPLC).

Type species: **Sporanaerobacter acetigenes** Hernandez-Eugenio, Fardeau, Cayol, Patel, Thomas, Macarie, Garcia and Ollivier 2002, 1221^{VP}.

Further descriptive information

Sporanaerobacter cells are 0.4–0.5 μm in width and 3–5 μm in length and occur singly or in pairs (Figure 224). Cells are motile by a few laterally inserted flagella. In media appropriate for spore induction, spherical, terminal, oval spores which swell the sporangium appear in the cells. Electron microscopy of sections of cells exhibits a 33 nm thick stratified Gram-stain-positive type cell wall, composed of three dense layers (two thick layers and a thinner middle layer) separated by two light spaces. *Sporanaerobacter* cells grow at temperatures ranging from 25 to 50°C, with an optimum at 40°C, and in the presence of NaCl concentrations ranging from 0 to 4% NaCl, with optimum growth in the absence of NaCl. The optimum pH for growth is around 7.5 and growth occurs between pH 5.5 and 8.5.

Yeast extract is required for growth on sugars. Gelatin and casein are not used as energy sources, but peptone, bio-Trypcase and Trypticase soy are fermented. Small amounts of the following amino acids (around 1 mM) are used as energy sources in the presence of yeast extract (2 g/l): DL-histidine, L-isoleucine, DL-leucine, L-methionine, L-phenylalanine, DL-tryptophan, and DL-valine. In the presence of elemental sulfur as a terminal electron acceptor, the oxidation of the latter amino acids increases significantly (mean 2–3 mM). Lysine (1.7 mM) is only oxidized

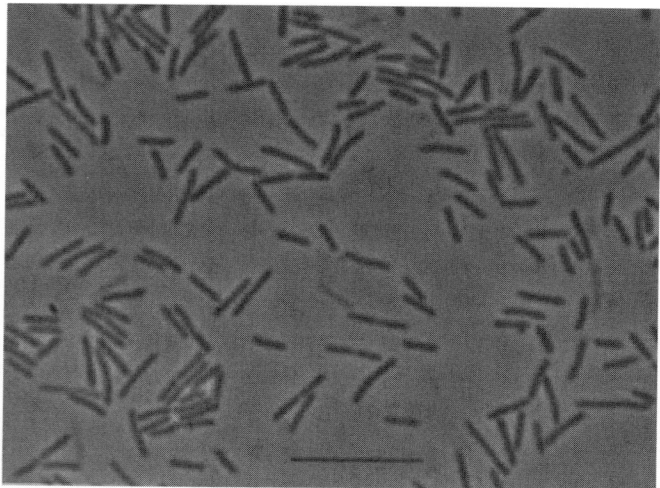

FIGURE 224. Photomicrograph of *Sporanaerobacter acetigenes*. Bar = 10 μm.

in the presence of elemental sulfur. Arginine (4.6 mM) is fermented, and the use of arginine is not improved in the presence of elemental sulfur. Valine is oxidized to isobutyrate, leucine to isovalerate, isoleucine to methyl-2-butyrate, phenylalanine to phenyl acetate, lysine to acetate, and histidine to acetate together with an unidentified product. Glycine and serine may serve as electron acceptors for oxidizing isoleucine. Peptone is fermented into acetate, hydrogen, and carbon dioxide as the major end products of metabolism. Traces of isobutyrate and isovalerate are also detected. Pyruvate is converted to acetate, H_2, and CO_2. *Sporanaerobacter* cells can utilize D-glucose, D-ribose, and fumarate, but not D-arabinose, D-fructose, D-galactose, maltose, D- and L-xylose, acetate, propionate, butyrate, valerate, ethanol, *n*-butanol, *n*-propanol, malate, and succinate. Acetate is the only fatty acid produced from glucose metabolism. No H_2 is detected from glucose metabolism in the presence of elemental sulfur as the electron acceptor.

Enrichment and isolation procedures

The Hungate technique (Hungate, 1969) is used to isolate *Sporanaerobacter* relatives. *Sporanaerobacter acetigenes* was isolated from an upflow anaerobic sludge blanket (UASB) reactor in Mexico. Enrichments were performed in 120 ml serum bottles using the following basal medium (BM) containing (per liter of distilled water): 1 g NH_4Cl, 3 g K_2HPO_4, 3 g KH_2PO_4, 0.2 g $MgCl_2 \cdot 6H_2O$, 0.1 g $CaCl_2 \cdot 2H_2O$, 0.1 g KCl, 0.6 g NaCl, 0.5 g cysteine-HCl, 1 mg resazurin and 10 ml trace mineral element solution (Balch et al., 1979). The pH was adjusted to 7.4 with 10 M KOH. The medium was boiled under a stream of O_2-free N_2 gas and cooled to room temperature. Five ml aliquots were dispensed into Hungate tubes and 20 ml in serum bottles, under a stream of N_2-CO_2 (80:20, v/v) gas and the sealed vessels were autoclaved for 45 min at 110 °C. Prior to inoculation, $Na_2S \cdot 9H_2O$ and $NaHCO_3$ were injected from sterile stock solutions to final concentrations of 0.04 and 0.2% (w/v), respectively. Peptone (5 g/l) and elemental sulfur (20 mM) were added to basal medium as electron donor and acceptor, respectively. The serum bottle was inoculated with 4 ml of

sludge, corresponding to 10% (v/v) of the final liquid volume (40 ml). After inoculation, the serum bottle atmosphere was changed to H_2 at a final pressure of 2.026×10^5 Pa. The bottles were incubated at 35 °C in a controlled temperature room for 2–3 weeks. Isolation was performed in the same medium with N_2 instead of H_2 in the gas phase. Four enrichment series were performed before isolation.

Enrichment cultures were incubated at 35 °C for 2–3 weeks. Growth was regarded as positive on the basis of H_2S production. Colonies, 2–3 mm in diameter, appeared after 2 d of incubation at 37 °C in roll tubes containing peptone-rich agar medium. Single colonies were picked using the techniques developed by Hungate (1969) and the process of serial dilution in roll tubes was repeated at least twice in order to purify the cultures.

Maintenance procedures

Stock cultures can be maintained on medium described by Hernandez-Eugenio et al. (2002) by monthly transfers. Liquid cultures retained viability after several weeks storage at 4 °C, or when lyophilized, or after storage at –80 °C in the basal medium containing 20% glycerol (v/v). Viability is best maintained from mid-exponential phase cultures.

Differentiation of the genus *Sporanaerobacter* from other genera

The genus *Sporanaerobacter* represented by only one species, *Sporanaerobacter acetigenes*, is an anaerobic spore-forming microorganism of the domain *Bacteria* that grows heterotrophically on carbohydrates, peptones, and amino acids. A mixture of volatile fatty acids, including acetate, isovalerate, and isobutyrate, together with H_2 and CO_2, are produced from peptone fermentation, but acetate is the only fatty acid produced from glucose metabolism. In this respect, this bacterium is an acetogen. Its inability to use H_2 as the electron donor and CO_2 as the electron acceptor to produce acetate indicates that it is not a homoacetogenic bacterium (Ljungdahl et al., 1989). Analysis of the 16S rRNA gene sequence of this isolate indicates that it is a member of the order *Clostridiales*, cluster XII (Figure 225), as defined by Collins et al. (1994). This cluster comprises *Clostridium* species (e.g., *Clostridium hastiforme*, *Clostridium acidurici*, *Clostridium purinilyticum*, and *Clostridium ultunense*), *Sedimentibacter* species (e.g., *Sedimentibacter saalensis* and *Sedimentibacter hydroxybenzoicus*), *Tissierella* species, *Soehngenia saccharolytica*, *Eubacterium angustum*, and the recently described thermophilic anaerobes isolated from extreme environments which include *Thermohalobacter berrensis*, *Caloranaerobacter azorensis*, and *Garciella nitratireducens* (Cayol et al., 2000; Miranda-Tello et al., 2003; Wery et al., 2001). In the current volume, many members of this cluster, including *Sporanaerobacter*, are classified within Family XI *Incertae Sedis*. Interestingly, this cluster includes acetogenic bacteria (e.g., *Clostridium acidurici*, *Clostridium purinilyticum*, and *Eubacterium angustum*) that produce acetate, ammonium, and CO_2 from purines but, in contrast to *Sporanaerobacter*, these micro-organisms are described as asaccharolytic (Beuscher and Andreesen, 1984; Ljungdahl et al., 1989). Similarly, close relatives of *Sporanaerobacter acetigenes*, *Clostridium hastiforme*, *Tissierella praeacuta*, and *Sedimentibacter* species are also unable to use sugars (Breitenstein et al., 2002; Cato et al., 1986; Farrow et al., 1995; Hippe et al., 1992). *Sporanaerobacter acetigenes* is a moderately thermophilic

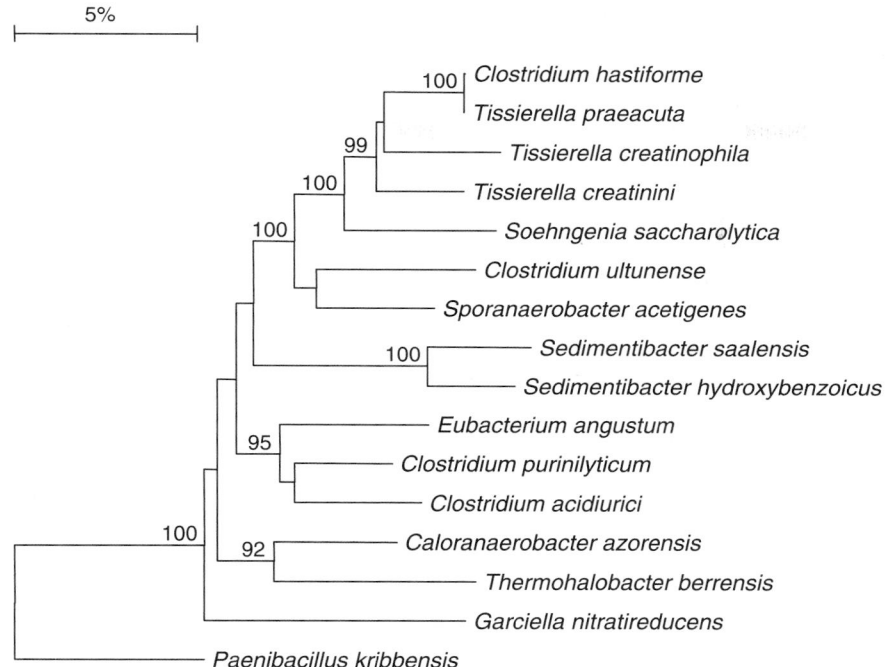

5%

100 — *Clostridium hastiforme*
100 — *Tissierella praeacuta*
99 — *Tissierella creatinophila*
100 — *Tissierella creatinini*
100 — *Soehngenia saccharolytica*
— *Clostridium ultunense*
— *Sporanaerobacter acetigenes*
100 — *Sedimentibacter saalensis*
— *Sedimentibacter hydroxybenzoicus*
95 — *Eubacterium angustum*
— *Clostridium purinilyticum*
— *Clostridium acidiurici*
100 — 92 — *Caloranaerobacter azorensis*
— *Thermohalobacter berrensis*
— *Garciella nitratireducens*
— *Paenibacillus kribbensis*

FIGURE 225. Phylogenetic tree based on 16S rRNA gene sequence comparison and obtained by a neighbor-joining algorithm (PHYLIP package), indicating the position of *Sporanaerobacter acetigenes* among members of the former cluster XII of the *Clostridiales*. *Paenibacillus kribbensis* was used as outgroup. Scale bar represents 5 substitutions per hundred nucleotides.

micro-organism but, unlike *Thermohalobacter berrensis* which was isolated from a solar saltern, *Garciella nitratireducens* which was isolated from an oilfield separator, and *Caloranaerobacter azorensis* which was isolated from a deep-sea hydrothermal, it does not grow at temperatures higher than 50°C (Cayol et al., 2000; Miranda-Tello et al., 2003; Wery et al., 2001). *Sporanaerobacter acetigenes* differs from *Clostridium ultunense* (Schnurer et al., 1996) in its ability to reduce elemental sulfur and in the range of sugars and amino acids used. In contrast to *Sporanaerobacter acetigenes*, *Clostridium ultunense* produces formate as well as acetate from glucose fermentation. Finally the DNA G+C content of *Soehngenia saccharolytica* (42 mol%) (Parshina et al., 2003) is higher than that of *Sporanaerobacter acetigenes* (32.2 mol%).

Sporanaerobacter acetigenes facultatively uses elemental sulfur as a terminal electron acceptor, producing sulfide. The ability to use inorganic sulfur-containing compounds (e.g., thiosulfate and/or elemental sulfur) has been reported for several members of the *Clostridiales*. They include species of *Thermoanaerobacter* and *Thermoanaerobacterium* described as using thiosulfate as an electron acceptor (Fardeau et al., 1994; Fardeau et al., 1993; Fardeau et al., 1997; Faudon et al., 1995; Lee et al., 1993; Schink and Zeikus, 1983). *Thermoanaerobacterium* species reduce thiosulfate to elemental sulfur, whereas *Thermoanaerobacter* spe-

cies reduce both thiosulfate and elemental sulfur to sulfide. Other fermentative bacteria also belonging to the *Clostridiales* that have the ability to reduce thiosulfate and/or elemental sulfur to sulfide have been isolated successfully, particularly from oilfield environments (Fardeau et al., 1993; Magot et al., 1997; Ravot et al., 1999; Ravot et al., 1997) but also from freshwater sediments (Hermann et al., 1987). None of these micro-organisms are phylogenetically closely related to *Sporanaerobacter acetigenes*. These results also suggest that thiosulfate and/or sulfur reduction might be a quite widespread metabolic trait within members of the order *Clostridiales*, which gives them a metabolic advantage through their ability to oxidize amino acids, as reported already for other micro-organisms belonging to this order (Fardeau et al., 1997; Faudon et al., 1995; Magot et al., 1997).

Taxonomic comments

Sporanaerobacter is a recently defined genus consisting of one species (*Sporanaerobacter acetigenes*). Phenotypic, genotypic, and phylogenetic characteristics clearly place it within Family XI *Incertae Sedis* of the *Clostridiales*. However, further analysis of this strain, as well as the isolation and characterization of more species which are representative of this genus, are required to confirm the taxonomic position of this genus.

List of species of the genus *Sporanaerobacter*

1. **Sporanaerobacter acetigenes** Hernandez-Eugenio, Fardeau, Cayol, Patel, Thomas, Macarie, Garcia and Ollivier 2002, 1221[VP]

a.ce.ti.ge′nes. L. n. *acetum* vinegar, acetic acid; Gr. v. *gennaio* produce; N.L. adj. *acetigenes* producing acetate.

In addition to the properties listed in the genus description (Hernandez-Eugenio et al., 2002), the following properties are reported. Cells are 0.4–0.5 × 3–5 μm. Electron microscopy of sections of cells reveals a 33-nm-thick, layered Gram-positive type cell wall, composed of three dense layers (two thick layers and a middle thinner layer) separated by two light spaces. Grows at temperatures ranging from 25 to 50°C, with optimum growth at 40°C. Grows in the presence of NaCl at concentrations ranging from 0 to 4% NaCl, with optimum growth in the absence of NaCl. The optimum pH for growth is around 7.5 and growth occurs between pH 5.5 and 8.5.

Heterotrophic. Gelatin and casein are not used as energy sources, but peptone, bio-Trypcase and Trypticase soy are fermented. The following amino acids are used as energy sources in the presence of yeast extract (2 g/l): arginine, histidine, isoleucine, leucine, methionine, phenylalanine, tryptophan, and valine. In the presence of elemental sulfur as a terminal electron acceptor, the use of the amino acids cited above, with the exception of arginine, is improved. Lysine is only oxidized in the presence of elemental sulfur. Valine is oxidized to isobutyrate, leucine to isovalerate, isoleucine to methyl-2-butyrate, phenylalanine to phenyl acetate, lysine to acetate, and histidine to acetate and an unidentified compound. Peptone is fermented to acetate as the major end product of metabolism. Traces of isobutyrate and isovalerate are also detected. Pyruvate is converted to acetate, H_2, and CO_2. Utilizes D-glucose, D-ribose, and fumarate, but not D-arabinose, D-fructose, D-galactose, maltose, D- or L-xylose, acetate, propionate, butyrate, valerate, ethanol, n-butanol, n-propanol, malate or succinate. Acetate is the only fatty acid produced from glucose metabolism. Adverse effects on animals and humans are not known. Because of its ability to degrade amino acids and peptides, the possibility of harmful effects cannot be excluded. Cautious handling and autoclaving of cultures before disposal is recommended. Isolated from a UASB reactor in Mexico.

DNA G+C content (mol%): 32.2 (HPLC).
Type strain: Lup 33, CIP 106730, DSM 13106.
GenBank accession number (16S rRNA gene): AF358114.

Genus X. **Tissierella** Collins and Shah 1986, 463[VP], emend. Farrow, Lawson, Hippe, Gauglitz and Collins 1995, 439, emend. Bae, Park, Chang, Rhee and Park 2004, 948

HAROUN N. SHAH AND JOHN V. HOOKEY

Tis.si.er.el.la. N.L. dim. ending -ella; N.L. fem. dim. n. *Tissierella* named after Henry Tissier, who first described the organism that was originally isolated from the feces of an infant.

Rod-shaped nonsporulating cells, 2–8 × 0.6–1.1 μm. Gram-stain-positive or -negative. Cells are either nonmotile, or motile via peritrichous flagella that occur singly, in pairs, or in short chains. **Obligately anaerobic.** Colonies are distinct, flat or low convex, white to gray and grow optimally at 37°C. **Weakly or non-fermentative. Creatine or creatinine are required for growth** and typically produce acetate, ammonia, and CO_2. Predominant metabolic end products in peptone-yeast extract-glucose broth are acetic, butyric, and isovaleric acids. Nonhydroxylated and 3-hydroxylated long-chain fatty acids that are straight-chained saturated and *iso*-methyl branched chain types. *meso*-Diaminopimelic acid or D-ornithine are present in the cell-wall peptidoglycan. **Respiratory menaquinones are absent.** Isolated from various infected sites in humans, human feces, and environmental sources. The 16S rRNA gene sequence similarity between the species varies from ~95 to 89%.

DNA G+C content (mol%): 28–32.
Type species: **Tissierella praeacuta** (Tissier 1908) Collins and Shah 1986, 463[VP].

List of species of the genus *Tissierella*

1. **Tissierella praeacuta** (Tissier 1908) Collins and Shah 1986, 463[VP] (*Bacteroides praeacutus* Tissier 1908, 193)

prae.cu.ta. L. fem. adj. *praeacuta* sharpened to a point, sharpened.

Although well characterized, it has variously been assigned to diverse genera as *Bacteroides* (Moore and Holdeman, 1973), *Coccobacillus* (Tissier, 1908), *Fusibacterium* (Hoffman, 1957), and *Zuberella* (Prévot, 1933).

Cells (2–8 × 0.6–0.9 μm) are motile via peritrichous flagella that occur singly or in pairs. Gram-stain-positive. On blood agar, colonies are small, circular, low convex, smooth, and gray in color. Optimum temperature for growth is 37°C. Nitrate reduction, hippurate hydrolysis, and gelatin liquefaction are variable. Urease and indole are not produced. Esculin is not hydrolyzed and H_2S is produced in SIM (Difco) broth. 6-Phosphogluconate dehydrogenase is present, whereas the dehydrogenases (glucose-6-phosphate, glutamate, and malate) are not. Nonhydroxylated and 3-hydroxylated long-chain fatty acids (which are straight-chained, saturated, and iso-methyl branched chain types) and 3-methyltetradecanoic (iso-$C_{15:0}$) acid predominate. *meso*-Diaminopimelic acid is present (Hammann and Werner, 1981), and 2-keto-3-deoxyoctulosonic acid or heptose are absent (Hofstad, 1974) in the cell-wall peptidoglycan. Menaquinones are absent.

Isolated from feces of infants and adults, gangrenous lesions, lung abscesses, and blood.

The previously proposed species *Clostridium hastiforme* and *Tissierella praeacuta* appear to be similar (Bae et al., 2004). The 16S rDNA sequence similarity to "*Clostridium hastiforme*" is 95.2%. There has been considerable progress made in recent years in establishing the phylogenetic relationships within the genus *Bacteroides* and related taxa (Paster et al., 1994). A comparative analysis demonstrated that *Tissierella praeacuta* was a member of the *Clostridium* subphylum of the Gram-positive bacteria (Farrow et al., 1995), and a polyphasic taxonomic study later concluded that the type strains were similar (Bae et al., 2004). The description of the type strain is that of the species, except gelatin is hydrolyzed and nitrate reduced.

DNA G+C content (mol%): 28.0 (T_m).

Type strain: ATCC 25539, CCUG 27825, LMG 8203, NCIMB 703038, NCTC 11158.

GenBank accession number (16S rRNA gene): X80833.

2. **Tissierella creatinini** Farrow, Lawson, Hippe, Gauglitz and Collins 1995, 439[VP]

cre.a.ti.ni′ni. N.L. gen. n. *creatinini* of creatinine.

Cells are rod-shaped, 3.5 × 1.0 μm in size, and occur mostly singly or in pairs. Nonmotile and non-sporeforming. Gram-stain-positive and surrounded by a slime capsule. Growth occurs at 20–39°C, and the optimum pH is 8.3. Sodium chloride and sodium bicarbonate are required for good growth. Creatinine and related nitrogen-containing compounds, with the notable exception of creatine, are metabolized. Biochemical tests for gelatin hydrolysis, production of lipase, lecithinase, indole and urease, reduction of nitrate, and hemolysis are negative. Unsaturated fatty acids ($C_{18:1\ c11/12}$) predominate, with minor amounts of straight-chained saturated, methyl-branched, and 3-hydroxylated long-chain fatty acids. Unlike the cell wall of *Tissierella praeacuta*, which is based on *meso*-diaminopimelic acid, that of strain BN11 contains D-ornithine as the dibasic amino acid. The cell-wall murein is type A4β (with interpeptide bridge composed of L-ornithine with D-glutamic acid). Respiratory menaquinones are not produced. Shares 90.6% and 90.1% 16S rDNA sequence similarity to *Tissierella praeacuta* and "*Clostridium hastiforme*,"

respectively. Isolated from a sugar-refinery wastewater pond.

DNA G+C content (mol%): 32.0 (T_m)

Type strain: BN11, CIP 104584, DSM 9508.

GenBank accession number (16S rRNA gene): X75909.

3. **Tissierella creatinophila** Harms, Schleicher, Collins and Andreesen 1998, 991[VP]

cre.at.in.o′phil.la. Gr. adj. *kreatinos* creatine, referring to meat; Gr. adj. *philos* loving: N.L. fem. adj. *creatinophila* creatine-loving.

Rod-shaped, 2–6 × 0.7–1.1 μm, motile via peritrichous flagella. Gram-stain-negative, either single cells or in chains. Non-sporeforming and non-fermentative. Colonies on PYG (peptone yeast glucose) agar are white to gray, small, circular, flat with a rough surface.

Unlike *Tissierella creatinini*, there is no growth on nitrogenous compounds other than creatinine and creatine. Degrades creatinine via creatine, sarcosine, and glycine to produce acetate, monomethylamine, ammonia, and CO_2. Creatine aminohydrolase is highly significant in medical diagnostics since the rate of creatinine clearance is an indicator of renal function (Siedel et al., 1988). Nitrate, sulfur and sulfate are not reduced. Cell-wall chemistry is unknown. No respiratory menaquinones. This species shares 95% 16S rDNA sequence similarity to other members of the genus *Tissierella*. Isolated from anaerobic human sewage sludge plant. Pathogenicity is unknown.

DNA G+C content (mol%): 30.0 (HPLC).

Type strain: KRE 4, DSM 6911.

GenBank accession number (16S rRNA gene): X80227.

References

Aguirre, M. and M.D. Collins. 1992. Phylogenetic analysis of some *Aerococcus*-like organisms from urinary tract infections: description of *Aerococcus urinae* sp. nov. J. Gen. Microbiol. *138*: 401–405.

Andreesen, J.R., H. Bahl and Gottschalk. 1989. Introduction to the physiology and biochemistry of the genus *Clostridium*. *In* Minton and Clarke (Editors), *Clostridia*, Biotechnology Handbooks 3. Plenum, New York, pp. 27–62.

Andreesen, J.R. 1994. Acetate via glycine: a different form of acetogenesis. *In* Drake (Editor), Acetogenesis. Chapman & Hall, New York, pp. 568–629.

Bae, J.W., J.R. Park, Y.H. Chang, S.K. Rhee, B.C. Kim and Y.H. Park. 2004. *Clostridium hastiforme* is a later synonym of *Tissierella praeacuta*. Int. J. Syst. Evol. Microbiol. *54*: 947–949.

Balch, W.E., G.E. Fox, L.J. Magrum, C.R. Woese and R.S. Wolfe. 1979. Methanogens: reevaluation of a unique biological group. Microbiol. Rev. *43*: 260–296.

Beuscher, H.U. and J.R. Andreesen. 1984. *Eubacterium angustum* sp. nov., a Gram positive anaerobic, non-sporeforming, obligate purine fermenting organism. Arch. Microbiol. *140*: 2–8.

Breitenstein, A., J. Wiegel, C. Haertig, N. Weiss, J.R. Andreesen and U. Lechner. 2002. Reclassification of *Clostridium hydroxybenzoicum* as *Sedimentibacter hydroxybenzoicus* gen. nov., comb. nov., and description of *Sedimentibacter saalensis* sp. nov. Int. J. Syst. Evol. Microbiol. *52*: 801–807.

Caliendo, A.M., C.D. Jordan and K.L. Ruoff. 1995. *Helcococcus*, a new genus of catalase-negative, Gram-positive cocci isolated from clinical specimens. J. Clin. Microbiol. *33*: 1638–1639.

Cato, E.P., W.L. George and S.M. Finegold. 1986. Genus *Clostridium* Prazmowski 1880, 23[AL]. *In* Sneath, Mair, Sharpe and Holt (Editors),

Bergey's Manual of Systematic Bacteriology, Vol. 2. The Williams & Wilkins Co., Baltimore, pp. 1141–1200.

Cayol, J.L., S. Ducerf, B.K. Patel, J.L. Garcia, P. Thomas and B. Ollivier. 2000. *Thermohalobacter berrensis* gen. nov., sp. nov., a thermophilic, strictly halophilic bacterium from a solar saltern. Int. J. Syst. Evol. Microbiol. *50*: 559–564.

Chagla, A.H., A.A. Borczyk, R.R. Facklam and M. Lovgren. 1998. Breast abscess associated with *Helcococcus kunzii*. J. Clin. Microbiol. *36*: 2377–2379.

Christensen, J.J., H. Vibits, J. Ursing and B. Korner. 1991. *Aerococcus*-like organism, a newly recognized potential urinary tract pathogen. J. Clin. Microbiol. *29*: 1049–1053.

Christiansen, M. 1934. Position de *M. indolicus* dans la systématique bactériénne. Acta Pathol. Microbiol. Scand. *11*: 363–366.

Collins, M.D. and H.N. Shah. 1986. Reclassification of *Bacteroides praeacutus* Tissier (Holdeman and Moore) in a new genus, *Tissierella*, as *Tissierella praeacuta* comb. nov. Int. J. Syst. Bacteriol. *36*: 461–463.

Collins, M.D., R.R. Facklam, U.M. Rodrigues and K.L. Ruoff. 1993. Phylogenetic analysis of some *Aerococcus*-like organisms from clinical sources: description of *Helcococcus kunzii* gen. nov., sp. nov. Int. J. Syst. Bacteriol. *43*: 425–429.

Collins, M.D., P.A. Lawson, A.Willems, J.J. Cordoba, J. Fernández-Garayzábal, P. Garcia, J. Cia, H. Hippe and J.A.E. Farrow. 1994. The phylogeny of the genus *Clostridium*: proposal of five new genera and eleven new species combination. Int. J. Syst. Bacteriol. *44*: 812–826.

Collins, M.D., E. Falsen, G. Foster, L.R. Monasterio, L. Dominguez and J.F. Fernández-Garazábal. 1999a. *Helcococcus ovis* sp. nov., a Gram positive organism from sheep. Int. J. Syst. Bacteriol. *49*: 1429–1432.

Collins, M.D., M.R. Jovita, R.A. Hutson, M. Ohlén and E. Falsen. 1999b. *Aerococcus christensenii* sp. nov., from the human vagina. Int. J. Syst. Bacteriol. *49*: 1125–1128.

De Ley, J., H. Cattoir and A. Reynaerts. 1970. The quantitative measurement of DNA hybridization from renaturation rates. Eur. J. Biochem. 12: 133–142.

Distaso, A. 1912. Contribution à l'étude sur l'intoxication intestinale. Zentbl. Bacteriol. Parasitenkd. Infektionskr. Hyg. Abt. I Orig. 62: 433–468.

Douglas, H.C. 1957. Genus VI. Peptococcus Kluyver and van Niel 1936. In Breed, Murray and Smith (Editors), Bergey's Manual of Determinative Bacteriology. The WIlliams & Wilkins Co., Baltimore, pp. 474–480.

Ezaki, T. 1982. Mole % guanine plus cytosine of butyrate-producing anaerobic cocci and DNA-DNA relationship among them. Nippon Saikingaku Zasshi 37: 607–613.

Ezaki, T., N. Yamamoto, K. Ninomiya, S. Suzuki and E. Yabuuchi. 1983. Transfer of Peptococcus indolicus, Peptococcus asaccharolyticus, Peptococcus prevotii, and Peptococcus magnus to the genus Peptostreptococcus and proposal of Peptostreptococcus tetradius sp. nov. Int. J. Syst. Bacteriol. 33: 683–698.

Ezaki, T., S.L. Liu, Y. Hashimoto and E. Yabuuchi. 1990. Peptostreptococcus hydrogenalis sp.nov from human fecal and vaginal flora. Int. J. Syst. Bacteriol. 40: 305–306.

Ezaki, T., Y. Kawamura, N. Li, Z.Y. Li, L. Zhao and S. Shu. 2001a. Proposal of the genera Anaerococcus gen. nov., Peptoniphilus gen. nov. and Gallicola gen. nov. for members of the genus Peptostreptococcus. Int. J. Syst. Evol. Microbiol. 51: 1521–1528.

Ezaki, T., Y. Kawamura, N. Li, Z.Y. Li, L.C. Zhao and S.E. Shu. 2001b. Proposal of the genera Anaerococcus gen. nov., Peptoniphilus gen. nov and Gallicola gen. nov. for members of the genus Peptostreptococcus. Int. J. Syst. Evol. Microbiol. 51: 1521–1528.

Facklam, R. and J.A. Elliott. 1995. Identification, classification, and clinical relevance of catalase-negative, gram-positive cocci, excluding the streptococci and enterococci. Clin. Microbiol. Rev. 8: 479–495.

Fardeau, M.L., J. L. Cayol, M. Magot and B. Ollivier. 1993. H₂ oxidation in the presence of thiosulfate, by a Thermoanaerobacter strain isolated from an oil-producing well. FEMS Microbiol. Lett. 113: 327–332.

Fardeau, M.L., J.L. Cayol, M. Magot and B. Ollivier. 1994. Hydrogen oxidation abilities in the presence of thiosulfate as electron acceptor within the genus Thermoanaerobacter. Curr. Microbiol. 29: 269–272.

Fardeau, M.L., B.K. Patel, M. Magot and B. Ollivier. 1997. Utilization of serine, leucine, isoleucine, and valine by Thermoanaerobacter brockii in the presence of thiosulfate or Methanobacterium sp. as electron acceptors. Anaerobe 3: 405–410.

Farrow, J.A.E., P.A. Lawson, H. Hippe, U. Gauglitz and M.D. Collins. 1995. Phylogenetic evidence that the Gram-negative nonsporulating bacterium Tissierella (Bacteroides) praeacuta is a member of the Clostridium subphylum of the Gram-positive bacteria and description of Tissierella creatinini sp. nov. Int. J. Syst. Bacteriol. 45: 436–440.

Faudon, C., M.L. Fardeau, J. Heim, B.K.C. Patel, M. Magot and B. Ollivier. 1995. Peptide and amino acid oxidation in the presence of thiosulfate by members of the genus Thermoanaerobacter. Curr. Microbiol. 31: 152–157.

Foubert, E.L., Jr and H.C. Douglas. 1948. Studies on the anaerobic micrococci. I. Taxonomic considerations. J. Bacteriol. 56: 25–34.

Garrity, G.M., J.A. Bell and T. Lilburn. 2005. The Revised Road Map to the Manual. In Brenner, Krieg, Staley and Garrity (Editors), Bergey's Manual of Systematic Bacteriology, 2nd edn, Vol. 2, The Proteobacteria, Part A, Introductory Essays. Springer, New York, pp. 159–220.

Girbal, L., J. Orlygsson, B.J. Reinders and J.C. Gottschal. 1997. Why does Clostridium acetireducens not use interspecies hydrogen transfer for growth on leucine? Curr. Microbiol. 35: 155–160.

Haas, J., S.L. Jernick, R.J. Scardina, J. Teruya, A.M. Caliendo and K.L. Ruoff. 1997. Colonization of skin by Helcococcus kunzii. J. Clin. Microbiol. 35: 2759–2761.

Hammann, R. and H. Werner. 1981. Presence of diaminopimelic acid in propionate-negative Bacteroides species and in some butyric acid-producing strains. J. Med. Microbiol. 14: 205–212.

Harms, C., A. Schleicher, M.D. Collins and J.R. Andreesen. 1998. Tissierella creatinophila sp. nov., a Gram-positive, anaerobic, non-spore-forming, creatinine-fermenting organism. Int. J. Syst. Bacteriol. 48: 983–993.

He, Z. and J. Wiegel. 1995. Purification and characterization of an oxygen-sensitive reversible 4-hydroxybenzoate decarboxylase from Clostridium hydroxybenzoicum. Eur. J. Biochem. 229: 77–82.

He, Z. and J. Wiegel. 1996. Purification and characterization of an oxygen-sensitive, reversible, 3,4-dihydroxybenzoate decarboxylase from Clostridium hydroxybenzoicum. J. Bacteriol. 178: 3539–3543.

Hermann, M., M.R. Popoff and M. Sebald. 1987. Sporomusa paucivorans sp. nov, a methylotrophic bacterium that forms acetic acid from hydrogen and carbon dioxide. Int. J. Syst. Bacteriol. 37: 93–101.

Hernandez-Eugenio, G., M.L. Fardeau, J.C. Cayol, B.K.C. Patel, P. Thomas, H. Macarie, J.L. Garcia and B. Ollivier. 2002. Sporanaerobacter acetigenes gen. nov., sp. nov., a novel acetogenic, facultatively sulfur-reducing bacterium. Int. J. Syst. Evol. Microbiol. 52: 1217–1223.

Hippe, H., J.R. Andreesen and G. Gottschalk. 1992. The genus Clostridium: nonmedical. In Balows, Trüper, Dworkin, Harder and Schleifer (Editors), The Prokaryotes, 2nd edn, Vol. 2. Springer-Verlag, New York, pp. 1800–1866.

Hoffman, H. 1957. Genus II. Fusobacterium Knorr. In Breed, Murray and Smith (Editors), Bergey's Manual of Determinative Bacteriology, 7th edn. The Williams & Wilkins Co., Baltimore, pp. 436–440.

Hofstad, T. 1974. The distribution of heptose and 2-keto-3-deoxy-octonate in Bacteroidaceae. J. Gen Microbiol. 85: 314–320.

Høi Sorensen, G. 1975. Peptococcus (s. Micrococcus) indolicus. The demonstration of two varieties of hemolysis forming strains. Acta Vet. Scand. 16: 218–225.

Holdeman, L.V. and W.E.C. Moore. 1972. Anaerobe Laboratory Manual. Anaerobe Laboratory, Virginia Polytechnic Institute and State University, Blacksburg.

Huang, J., Z. He and J. Wiegel. 1999. Cloning, characterization, and expression of a novel gene encoding a reversible 4-hydroxybenzoate decarboxylase from Clostridium hydroxybenzoicum. J. Bacteriol. 181: 5119–5122.

Hungate, R.E. 1969. A roll tube method for cultivation of strict anaerobes. In Norris and Ribbons (Editors), Methods in Microbiology, Vol. 3B. Academic Press, London and New York, pp. 117–132.

Lapage, S.P., P.H.A. Sneath, E.F. Lessel, V.B.D. Skerman, H.P.R. Seeliger and W.A. Clark. 1992. International Code of Nomenclature of Bacteria (1990 Revision). Bacteriological Code. American Society for Microbiology, Washington, DC.

Lawson, P.A., E. Falsen, M. Ohlén and M.D. Collins. 2001a. Aerococcus urinaehominis sp. nov., isolated from human urine. Int. J. Syst. Evol. Microbiol. 51: 683–686.

Lawson, P.A., E. Falsen, K. Truberg-Jensen and M.D. Collins. 2001b. Aerococcus sanguicola sp. nov., isolated from a human clinical source. Int. J. Syst. Evol. Microbiol. 51: 475–479.

Lee, Y.E., M.K. Jain, C.Y. Lee, S.E. Lowe and J.G. Zeikus. 1993. Taxonomic distinction of saccharolytic thermophilic anaerobes: description of Thermoanaerobacterium xylanolyticum gen. nov., sp. nov., and Thermoanaerobacterium saccharolyticum gen. nov., sp. nov., reclassification of Thermoanaerobium brockii, Clostridium thermosulfurogenes, and Clostridium thermohydrosulfuricum E100-69 as Thermoanaerobacter brockii comb. nov., Thermoanaerobacterium thermosulfurigenes comb. nov., and Thermoanaerobacter thermohydrosulfuricum comb. nov., respectively, and transfer of Clostridium thermohydrosulfuricum 39e to Thermoanaerobacter ethanolicus. Int. J. Syst. Bacteriol. 43: 41–51.

Li, N., Y. Hashimoto, S. Adnan, H. Miura, H. Yamamoto and T. Ezaki. 1992. Three new species of the genus Peptostreptococcus isolated from humans: Peptostreptococcus vaginalis sp. nov., Peptostreptococcus lacrimalis sp. nov., and Peptostreptococcus lactolyticus sp. nov. Int. J. Syst. Bacteriol. 42: 602–605.

Li, N., Y. Hashimoto and T. Ezaki. 1994. Determination of 16S ribosomal RNA sequences of all members of the genus Peptostreptococcus and their phylogenetic position. FEMS Microbiol. Lett. 116: 1–5.

Ljungdahl, L.G., J. Hugenholz and J. Wiegel. 1989. Acetogenic and acid-producing clostridia. *In* Minton and Clarke (Editors), Clostridia. Plenum Press, New York, pp. 145–191.

Lomholt, J.A. and M. Kilian. 2000. Immunoglobulin A1 protease activity in *Gemella haemolysans*. J. Clin. Microbiol. *38*: 2760–2762.

Ludwig, W., K.H. Schleifer and W.B. Whitman. 2009. Revised road map to the phylum *Firmicutes*. *In* De Vos, Garrity, Jones, Krieg, Ludwig, Rainey, Schleifer and Whitman (Editors), Bergey's Manual of Systematic Bacteriology, 2nd edn, Vol. 3, The *Firmicutes*, Springer, New York, pp. 1–14.

Magot, M., G. Ravot, X. Campaignolle, B. Ollivier, B.K.C. Patel, M.L. Fardeau, P. Thomas, J.L. Crolet and J.L. Garcia. 1997. *Dethiosulfovibrio peptidovorans* gen. nov., sp. nov., a new anaerobic, slightly halophilic, thiosulfate-reducing bacterium from corroding offshore oil wells. Int. J. Syst. Bacteriol. *47*: 818–824.

Miranda-Tello, E., M.L. Fardeau, J. Sepulveda, L. Fernandez, J.L. Cayol, P. Thomas and B. Ollivier. 2003. *Garciella nitratireducens* gen. nov., sp. nov., an anaerobic, thermophilic, nitrate- and thiosulfate-reducing bacterium isolated from an oilfield separator in the Gulf of Mexico. Int. J. Syst. Evol. Microbiol. *53*: 1509–1514.

Moore, W.E.C. and L.V. Holdeman. 1973. New names and combinations in genera *Bacteroides* Castellani and Chalmers, *Fusobacterium* Knorr, *Eubacterium* Prevot, *Propionibacterium* Delwich, and *Lactobacillus* Orla-Jensen. Int. J. Syst. Bacteriol. *23*: 69–74.

Murdoch, D.A., I.J. Mitchelmore and S. Tabaqchali. 1988. Identification of gram-positive anaerobic cocci by use of systems for detecting pre-formed enzymes. J. Med. Microbiol. *25*: 289–293.

Murdoch, D.A., M.D. Collins, A. Willems, J.M. Hardie, K.A. Young and J.T. Magee. 1997. Description of three new species of the genus *Peptostreptococcus* from human clinical specimens: *Peptostreptococcus harei* sp. nov., *Peptostreptococcus ivorii* sp. nov., and *Peptostreptococcus octavius* sp. nov. Int. J. Syst. Bacteriol!. *47*: 781–787.

Murdoch, D.A. and H.N. Shah. 1999. Reclassification of *Peptostreptococcus magnus* (Prevot 1933) Holdeman and Moore 1972 as *Finegoldia magna* comb. nov. and *Peptostreptococcus micros* (Prevot 1933) Smith 1957 as *Micromonas micros* comb. nov. Anaerobe *5*: 555–559.

Murdoch, D.A. and H.N. Shah. 2000. *In* Validation of publication of new names and new combinations previously effectively published outside the IJSEM. List no. 75. Int. J. Syst. Evol. Microbiol. *50*: 1415–1417.

O'Leary, W.M. and S.G. Wilkinson. 1988. Gram-positive bacteria. *In* Ratledge and Wilkinson (Editors), Microbial Lipids. Academic Press, London, pp. 117–201.

Onyenwoke, R.U., J.A. Brill, K. Farahi and J. Wiegel. 2004. Sporulation genes in members of the low G+C Gram-type-positive phylogenetic branch (*Firmicutes*). Arch. Microbiol. *182*: 182–192.

Parshina, S.N., R. Kleerebezem, E. van Kempen, A.N. Nozhevnikova, G. Lettinga and A.J.M. Stams. 2000. Benzaldehyde conversion by two anaerobic bacteria isolated from an upflow anaerobic sludge bed reactor. Process Biochem. *36*: 423–429.

Parshina, S.N., R. Kleerebezem, J.L.S. Sanz, G. Lettinga, A.N. Nozhevnikova, N.A. Kostrikina, A.M. Lysenko and A.J.M. Stams. 2003. *Soehngenia saccharolytica* gen. nov., sp. nov and *Clostridium amygdalinum* sp. nov., two novel anaerobic, benzaldehyde-converting bacteria. Int. J. Syst. Evol. Microbiol. *53*: 1791–1799.

Paster, B.J., F.E. Dewhirst, I. Olsen and G.J. Fraser. 1994. Phylogeny of *Bacteroides*, *Prevotella*, and *Porphyromonas* spp. and related bacteria. J. Bacteriol. *176*: 725–732.

Peel, M.M., J.M. Davis, K.J. Griffin and D.L. Freedman. 1997. *Helcococcus kunzii* as sole isolate from an infected sebaceous cyst. J. Clin. Microbiol. *35*: 328–329.

Prévot, A.R. 1933. Etudes de systematique bacterienne. I. Lois générales. II. Cocci anaérobies. Ann. Sci. Nat. Bot. *15*: 23–261.

Ravot, G., M. Magot, B. Ollivier, B.K. Patel, E. Ageron, P.A. Grimont, P. Thomas and J.L. Garcia. 1997. Haloanaerobium congolense sp. nov., an anaerobic, moderately halophilic, thio-

sulfate- and sulfur-reducing bacterium from an African oil field. FEMS Microbiol. Lett. *147*: 81–88.

Ravot, G., M. Magot, M.L. Fardeau, B.K.C. Patel, P. Thomas, J.L. Garcia and B. Ollivier. 1999. *Fusibacter paucivorans* gen. nov., sp. nov., an anaerobic, thiosulfate-reducing bacterium from an oil-producing well. Int. J. Syst. Bacteriol. *49*: 1141–1147.

Ruoff, K.L. 2002. Miscellaneous catalase-negative, gram-positive cocci: Emerging opportunists. J. Clin. Microbiol. *40*: 1129–1133.

Schiefer-Ullrich, H. and J.R. Andreesen. 1985. *Peptostreptococcus barnesae* sp. nov., a Gram-positive, anaerobic, obligately purine utilizing coccus from chicken feces. Arch. Microbiol. *143*: 26–31.

Schink, B. and J.G. Zeikus. 1983. *Clostridium thermosulfurogenes* sp. nov., a new thermophile that produces elemental sulfur from thiosulfate. J. Gen. Microbiol. *129*: 1149–1158.

Schleifer, K.H. and E. Nimmermann. 1973. Peptidoglycan types of strains of the genus *Peptococcus*. Arch. Mikrobiol. *93*: 245–258.

Schnurer, A., B. Schink and B.H. Svensson. 1996. *Clostridium ultunense* sp. nov., a mesophilic bacterium oxidizing acetate in syntrophic association with a hydrogenotrophic methanogenic bacterium. Int. J. Syst. Bacteriol. *46*: 1145–1152.

Siedel, J., R. Deeg., H. Seidel., H. Mollering., J. Staepels., H. Gauhl and J. Ziegenhorn. 1988. Fully enzymatic colorimetric assay of serum and urine creatinine which obviates the need for sample blank measurements. Anal. Lett. *21*: 1009–1017.

Smith, L.D.S. 1957. *Peptostreptococcus* Kluyver and Van Neil, 1936. *In* Breed, Murray and Smith (Editors), Bergey's Manual of Determinative Bacteriology, 7th edn. The Williams & Wilkins Co., Baltimore, pp. 533–541.

Song, Y.L., C.X. Liu, M. McTeague and S.M. Finegold. 2003. 16S ribosomal DNA sequence-based analysis of clinically significant gram-positive anaerobic cocci. J. Clin. Microbiol. *41*: 1363–1369.

Stavri, H., T.J. Beveridge, D. Moyles, A. Athamna and R.J. Doyle. 2002. Hemagglutinin of unusual specificity from *Helcococcus kunzii*. Arch. Microbiol. *177*: 197–199.

Tindall, B.J. and J.P. Euzéby. 2006. Proposal of *Parvimonas* gen. nov. and *Quatrionicoccus* gen. nov. as replacements for the illegitimate, prokaryotic, generic names *Micromonas* Murdoch and Shah 2000 and *Quadricoccus* Maszenan *et al.* 2002, respectively. Int. J. Syst. Evol. Microbiol. *56*: 2711–2713.

Tissier, H. 1908. Recherches sur la flore intestinale normale des enfants agés d'un an à cinq ans. Ann. Inst. Pasteur (Paris) *22*: 189–208.

Weiss, N. 1981. Cell wall structure of anaerobic cocci. Rev. Inst. Pasteur Lyon *14*: 53–59.

Wells, C.L. and C.R. Field. 1976. Long-chain fatty acids of peptococci and peptostreptococci. J. Clin. Microbiol. *4*: 515–521.

Wery, N., J.M. Moricet, V. Cueff, J. Jean, P. Pignet, F. Lesongeur, M.A. Cambon-Bonavita and G. Barbier. 2001. *Caloranaerobacter azorensis* gen. nov., sp. nov., an anaerobic thermophilic bacterium isolated from a deep-sea hydrothermal vent. Int. J. Syst. Evol. Microbiol. *51*: 1789–1796.

Zhang, X. and J. Wiegel. 1990. Isolation and partial characterization of a *Clostridium* species transforming para-hydroxybenzoate and 3,4-dihydroxybenzoate and producing phenols as the final transformation products. Microb. Ecol. *20*: 103–121.

Zhang, X. 1993. Sequential degradation of chlorophenols in anaerobic freshwater sediments. PhD thesis, University of Georgia, Athens, GA.

Zhang, X. and J. Wiegel. 1994. Reversible conversion of 4-hydroxybenzoate and phenol by *Clostidium hydroxybenzoicum*. Appl. Environ. Microbiol. *60*: 4182–4185.

Zhang, X.M., L. Mandelco and J. Wiegel. 1994. *Clostridium hydroxybenzoicum* sp. nov., an amino acid-utilizing, hydroxybenzoate-decarboxylating bacterium isolated from methanogenic freshwater pond sediment. Int. J. Syst. Bacteriol. *44*: 214–222.

Family XII. **Incertae Sedis**

Members of this group were previously assigned to the "*Peptostreptococcaceae*" or *Clostridiaceae* by Garrity et al. (2005). Subsequent analyses suggest that these genera are not closely related to either group or any other previously described family. For that reason, they are assigned to their own family *incertae sedis*.

Genus I. **Acidaminobacter** Stams and Hansen 1985, 223[VP] (Effective publication: Stams and Hansen 1984, 335.)

THEO A. HANSEN

A.cid.a.min.o.bac'ter. L. n. *acidum* acid; N.L. neut. n. *aminum* amine; N.L. n. *bacter* the masc. equivalent of Gr. neut. n. *baktron* a staff or rod; N.L. masc. n. *Acidaminobacter* the amino acid rod.

Rod-shaped bacteria with pointed ends; Gram-stain-negative. Endospores not formed. **Chemo-organotrophic fermentative metabolism. Strictly anaerobic.** Amino acids are major energy sources. Acetate is the major fermentation product. Lactate, succinate, butyrate, and ethanol not formed. Growth on many substrates stimulated by or dependent on the presence of hydrogen-utilizing bacteria or archaea. Mesophilic.

DNA G+C content (mol%): 48.1 (T_m).

Type species: **Acidaminobacter hydrogenoformans** Stams and Hansen 1985, 223[VP] (Effective publication: Stams and Hansen 1984, 336.).

Further descriptive information

When it was first described, *Acidaminobacter* was unique because its growth on several amino acids was shown to be stimulated by or dependent on the presence of hydrogen-utilizing bacteria or archaea (such as sulfate reducers or methanogens). Growth on alanine, for instance, was only possible if the reducing equivalents (hydrogen and possibly formate as well) were consumed by a syntrophic partner (Stams and Hansen, 1984). The thermodynamically difficult step is the conversion of the amino acid to the corresponding keto acid with the formation of hydrogen; even with the assumption of a high ratio of the concentration of the amino acid over the keto acid, only at hydrogen concentrations in the order of 1–10 Pa does the free energy change become negative.

α-Amino acid → α-Keto acid + H_2 + NH_4^+ standard free energy change approximately +55–60 kJ

In pure culture, *Acidaminobacter hydrogenoformans* fermented 1 mol of glutamate to approximately 2 mol of acetate, 1 mol of bicarbonate, 1 mol of NH_4^+, and 1 mol of H_2 via a route with 3-methylaspartate as an intermediate. In mixed cultures with a hydrogenotrophic partner, the fermentation yielded less acetate and up to 0.47 mol of propionate per mol of glutamate; most likely, propionate formation occurred via 2-oxoglutarate and succinyl-CoA with the concomitant production of 2 H_2.

Positive effects of hydrogenotrophic anaerobes on fermentations of amino acids have since been described for a number of mesophilic bacteria including *Clostridium sporogenes* (Wildenauer and Winter, 1986; Winter et al., 1987), *Eubacterium acidaminophilum* (Zindel et al., 1988), *Aminobacterium* spp. (Baena et al., 2000, 1998), *Aminomonas paucivorans* (Baena et al., 1999a), and some thermophiles, e.g., *Gelria glutamica* (Plugge et al., 2002).

Enrichment and isolation procedures

Acidaminobacter has been obtained by direct anaerobic isolation from black surface mud of the Ems-Dollard estuary at the border between the Netherlands and Germany. Roll tubes with glutamate medium were used (Stams and Hansen, 1984). Strain glu 65 was the most numerous glutamate-fermenting bacterium.

Maintenance procedures

DSMZ medium 292 containing glycine or another suitable energy source can be used for routine cultivation. For long-term preservation, the organism can be stored in ampoules in medium supplemented with 8% glycerol under liquid nitrogen or at −80°C.

Differentiation of the genus *Acidaminobacter* from other genera

Strictly anaerobic, mesophilic, nonsporeforming, rod-shaped bacteria that mainly ferment amino acids (often dependent on or stimulated by hydrogenotrophic anaerobes) are found in the genera *Acidaminobacter* (Stams and Hansen, 1984), *Eubacterium* (Zindel et al., 1988), *Aminobacterium* (Baena et al., 2000, 1998), and *Aminomonas* (Baena et al., 1999a); see Table 221. Among these taxa, on the basis of 16S rRNA gene sequence comparisons, only *Acidaminobacter* (Meijer et al., 1999) and *Eubacterium acidaminophilum* (Baena et al., 1999b) are members of cluster XI of the *Clostridium* subphylum of the *Firmicutes* (low G+C Gram-positives) (Collins et al., 1994).

Taxonomic comments

The phylogenetic position of *Acidaminobacter* in cluster XI of the *Clostridium* subphylum is such that it is clearly a separate genus (Meijer et al., 1999). Both phenotypically and phylogenetically, the differences with other members of cluster XI are rather large. *Acidaminobacter* shares several characteristics with *Eubacterium acidaminophilum*, but it is not its closest relative (Baena et al., 1999b). *Eubacterium acidaminophilum* does not group with the type species of *Eubacterium*, *Eubacterium limosum*, which is a member of cluster XV (Collins et al., 1994). Other species, *Clostridium halophilum* and *Clostridium litorale*, which group with *Acidaminobacter* have considerable phenotypic differences with *Acidaminobacter* (Fendrich et al., 1990; Table 221). For these reasons it has been placed in a family *incertae sedis* in the current volume along with *Fusibacter* and *Guggenheimella*.

TABLE 221. Differential characteristics of *Acidaminobacter* and phylogenetically or physiologically related bacteria[a]

Characteristic	Acidaminobacter	Eubacterium acidaminophilum	Clostridium litorale	Clostridium halophilum	Aminobacterium	Aminomonas	Gelria
Rod morphology	Straight	Straight	Straight	Straight	Straight	Curved	Straight
Spores	–	–	+	+	–	–	+
Gram stain	–	+	–	+	–	–	–
DNA G+C content (mol%)	48	44	26	27	44–46	43	34
NaCl required for optimal growth	No	No	Yes (1 %)	Yes (6 %)	No	No	No
Temperature optimum (°C)	30	32–36	28	41	37	35	50
Amino acids main energy sources	+	+	+	?	+	+	?
Glutamate fermented pure cultures (products)[b]	+ (A, H$_2$, p)	–	?	?	–	+ (A, F, H$_2$, p)	–
Glutamate fermented cocultures (products)[c]	+ (A, P)	–	?	?	+ (P, a)	+ (P, A)	+ (P, s)
Alanine fermented	Only in co-cultures with H$_2$-utilizer	Only in co-cultures with H$_2$-utilizer	Stickland donor	Stickland donor	Only in co-cultures with H$_2$-utilizer	–	–
Sugars fermented	–	–	–	+	–	–	+
Isolated from	Black estuarine mud	Black mud from waste water ditch	Marine mud	Hypersaline mud	Anaerobic waste water treatment	Anaerobic waste water treatment	Methano-genic sludge

[a]For references, see text.

[b]Product abbreviations: A, a high and low concentrations of acetate; P, p high and low concentrations of propionate; F formate; s low concentration of succinate.

[c]Cocultures with hydrogen-utilizing methanogens or sulfate reducers.

List of species of the genus *Acidaminobacter*

1. **Acidaminobacter hydrogenoformans** Stams and Hansen 1985, 223[VP] (Effective publication: Stams and Hansen 1984, 336.)

hy.dro.ge.no.form′ans. N.L. n. *hydrogenium* hydrogen; L. part. adj. *formans* forming; N.L. adj. *hydrogenoformans* hydrogen-forming.

Characteristics are as described for the genus, with the following additional features. Rod-shaped cells, 0.5–0.6 × 1.5–3.7 μm, with pointed ends, singly or in pairs. Glutamate, histidine, cysteine, serine, glycine, adenine, pyruvate, and citrate fermented in pure culture. Acetate is the major fermentation product. In mixed cultures with hydrogen-utilizing sulfate-reducing bacteria (e.g., *Desulfovibrio*) or hydrogen-utilizing archaea (e.g., *Methanospirillum*), several other amino acids and malate utilized as well. Branched-chain fatty acids formed from branched-chain amino acids. Propionate, in addition to acetate, formed from glutamate and histidine; lower pH limit for growth on glutamate is 6.7. Lactate, ethanol, succinate, glucose, fructose, and aspartate are not utilized. Sulfate and nitrate not reduced. Sulfide required. Indole formed. No catalase; no cytochromes. Temperature range for growth is 15–42°C, optimum 30°C.

DNA G+C content (mol%): 48.1 (T_m).

Type strain: glu 65, CIP 106102, DSM 2784.

GenBank accession number (16S rRNA gene): AF016691.

Genus II. **Fusibacter** Ravot, Magot, Fardeau, Patel, Thomas, Garcia and Ollivier 1999, 1145[VP]

Gilles Ravot, Jean-Louis Garcia, Michel Magot and Bernard Ollivier

Fu.si.bac′ter. L. n. *fusus* a spindle; N.L. n. *bacter* masc. equivalent of Gr. neut. dim. n. *bakterion* rod; N.L. masc. n. *Fusibacter* a small spindle-shaped rod.

Spindle-shaped rods with cells that are 1 × 2–5 μm, and occurring singly or in pairs. **Motile by one to four peritrichous flagella.** Spores are not formed. Gram-stain-positive. Chemoorganotrophic and **obligately anaerobic.** Requires yeast extract for growth. Oxidizes a few carbohydrates to butyrate, acetate, CO$_2$, and H$_2$. Uses thiosulfate and sulfur as electron acceptors during glucose fermentation with production of H$_2$S. **Mesophilic, neutrophilic, and halotolerant.**

DNA G+C content (mol%): 43 (HPLC).

Type species: **Fusibacter paucivorans** Ravot, Magot, Fardeau, Patel, Thomas, Garcia and Ollivier 1999, 1145[VP]

Further descriptive information

Fusibacter paucivorans, isolated from an oil reservoir water, reduces thiosulfate to sulfide. The use of thiosulfate as an electron acceptor causes a shift in the flow of electrons, favoring H$_2$S production as already reported for *Thermoanaerobacter* species (Fardeau et al., 1996). In *Fusibacter paucivorans*, this results in channeling of the electrons away from butyrate to acetate, thereby increasing its concentration with concomitant increases in H$_2$S production (Table 222). Thiosulfate reduction is a common physiological trait exhibited by many fermentative mesophilic, thermophilic, and hyperthermo-

TABLE 222. Fermentation of glucose in the presence and absence of thiosulfate by *Fusibacter paucivorans*

Culture conditions[a]	Amount of substrate used (mM)	Amount of end products formed (mM)					Ratio of acetate produced/sugar consumed
		Acetate	Butyrate	CO_2[b]	H_2[b]	H_2S	
Glucose	5.0	1.6	4.0	10.2	15.2	0	0.32
Glucose + thiosulfate	9.9	12.2	3.0	20.4	7.1	11.5	1.23

[a]Sodium thiosulfate was added at a final concentration of 20 mM.
[b]Millimolar equivalents.

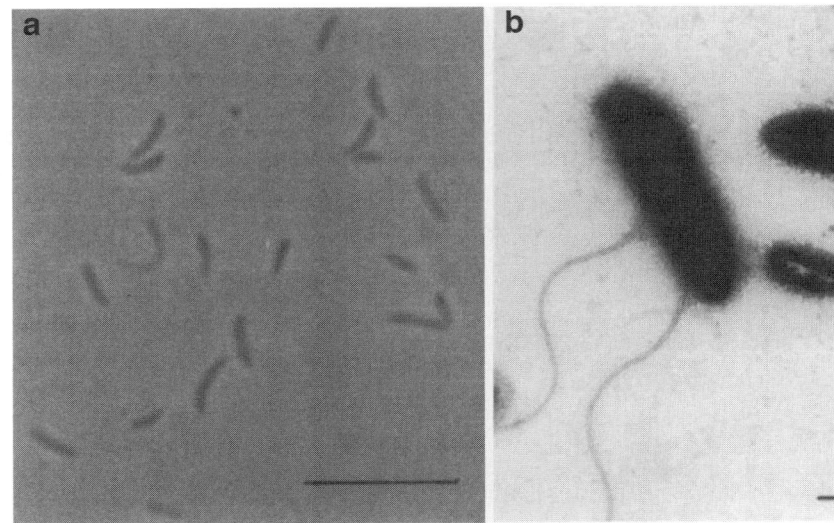

FIGURE 226. (a) Phase-contrast micrograph of *Fusibacter paucivorans*. Bar = 10 μm. (b) Electron micrograph of a negatively stained culture of *Fusibacter paucivorans* showing one or two pertrichous flagella. Bar = 2 μm. (Reprinted with permission from Ravot et al., 1999. Int. J. Syst. Bacteriol. *49*: 1141–1147.)

philic micro-organisms of the domain *Bacteria* also originating from oilfield ecosystems (Fardeau et al., 1993, 1997; Jeanthon et al., 1995; Magot et al., 1994, 1997a, 1997b; Ravot et al., 1995a, 1997, 1995b). Significant amounts of thiosulfate have been detected in oilfield facilities where it is thought to increase biocorrosion (Crolet and Magot, 1996; Magot et al., 1994), but the ecological significance of thiosulfate reduction in oil reservoirs is so far unknown.

Enrichment and isolation procedures

Fusibacter paucivorans was isolated from a reservoir water sample from an offshore oil-producing well (Emeraude oilfield) in Congo, central Africa. The *in situ* temperature was 35–40°C and the concentration of sodium chloride was 40 g/l (total salinity, 53 g/l). A one-liter sample was collected at the well head as described by Bernard et al. (1992), and culture broths were immediately inoculated. Enrichments were at 30°C without agitation by inoculating directly a 2-ml sample from the reservoir water into a basal medium containing (per liter of distilled water) 1 g NH_4Cl, 0.3 g K_2HPO_4, 0.3 g KH_2PO_4, 3 g $MgCl_2 \cdot 6H_2O$, 0.1 g $CaCl_2 \cdot 2H_2O$, 40 g NaCl, 1 g KCl, 0.5 g cysteine-HCl, 0.5 g CH_3COONa, 1 g yeast extract (Difco Laboratories), 1 g bio-Trypticase (bioMérieux), 10 ml of the trace element solution of Balch et al. (1979), and 0.1 mg resazurin.

The pH was adjusted to 7.3 with 10 M KOH. The medium was boiled under a stream of O_2-free N_2 gas and cooled to room temperature. Five- or twenty-milliliter aliquots were dispensed into Hungate tubes or serum bottles, respectively, under a stream of N_2/CO_2 (4:1, v/v), and the vessels were sealed and autoclaved for 45 min at 110°C. Prior to inoculation, $Na_2S \cdot 9H_2O$, $NaHCO_3$, and sodium thiosulfate were injected from sterile anaerobic stock solutions, to final concentrations of 0.04% (w/v), 0.2%, and 20 mM, respectively. The final pH was 7.2. The strain was purified by repeated isolation on Petri dishes containing the same medium solidified with 1.6% (w/v) agar as already reported (Magot et al., 1997a, 1997b; Ravot et al., 1997). Incubation was performed in an anaerobic glove box at 30°C.

Differentiation of the genus *Fusibacter* from other genera

Fusibacter paucivorans is an anaerobic, nonsporulating, spindle-shaped mesophilic flagellated rod (Figure 226) which possesses a Gram-positive cell-wall ultrastructure and reduces thiosulfate and sulfur to sulfide. Phylogenetic analysis of 16S rRNA indicates that it is a member of the *Clostridium* phylum and belongs to cluster XI, which includes *Clostridium* spp. (Collins et al., 1994) together with members of genera *Acidaminobacter*, *Tindallia*, *Alka-*

TABLE 223. Characteristics differentiating *Fusibacter paucivorans* from some close relatives[a]

Characteristics	*Fusibacter paucivorans*	*Acidaminobacter hydrogenoformans*[b]	*Alkaliphilus transvaalensis*[c]	*Clostridium halophilum*[d]
Morphology	Spindle-shaped rod	Rod-shaped with pointed ends	Slightly curved rod	Rod-shaped
Size (μm)	1×2–5	0.5–0.6 × 1.5–3.7	0.4–0.7 × 3–6	0.8–1 × 2.5–7
Spores	–	–	+	+
Motility	+	–	+	+
Gram reaction	+	–	+	+
Temperature optimum (°C)	37	30	40	41
pH optimum	7.3	ND	10	7.4
NaCl optimum (g/l)	0–30	0	5	60
Reduction of thiosulfate	+	–	+	–
Fermentation of glucose	+	–	–	+
G+C content (mol%)	43	48.1	36.4	26.9

[a]Symbols: +, positive; –, negative; ND, not determined.
[b]Data from Stams and Hansen (1984).
[c]Data from Takai et al. (2001).
[d]Data from Fendrich et al. (1990).

liphilus, and *Caminicella* (Alain et al., 2002; Kevbrin et al., 1998; Stams and Hansen, 1984; Takai et al., 2001). This cluster is taxonomically heterogeneous, containing non-spore-forming species such as *Acidaminobacter hydrogenoformans* and *Tindallia magadiensis*. The closest phylogenetic relatives of *Fusibacter paucivorans* are *Acidaminobacter hydrogenoformans* (92% similarity; Stams and Hansen, 1984) followed by *Clostridium halophilum* (90% similarity; Fendrich et al., 1990) and *Alkaliphilus transvaalensis* (mean similarity of 89%; Kevbrin et al., 1998). In the analyses performed for this volume, *Fusibacter* is classified in a family *incertae sedis* with *Acidaminobacter* and *Guggenheimella*. In addition to the significant phylogenetic difference (mean of 10%), the following genotypic and phenotypic traits distinguish *Fusibacter paucivorans* from its three nearest taxonomically validated phylogenetic neighbors (Table 223): (i) In contrast to *Clostridium halophilum* and *Acidaminobacter tranvaalensis*, *Fusibacter paucivorans* failed to produce spores; (ii) the G+C content of DNA from *Fusibacter paucivorans* is 43 mol% whereas that of *Clostridium halophilum* and *Acidaminobacter transvaalensis* DNA is 27 and 36.4 mol%, respectively; (iii) the substrate range of *Fusibacter paucivorans* differed makedly from that of *Acidaminobacter hydrogenoformans* and *Acidaminobacter transvaalensis*, both micro-organisms being unable to ferment sugars; (iv) the end products of carbohydrate fermentation are different from that of *Clostridium halophilum*; in contrast to *Fusibacter paucivorans*, ethanol and lactate are produced by *Clostridium halophilum*.

Acknowledgements

Many thanks to P. Thomas for photographs and C. Lesaulnier for improving the manuscript.

List of species of the genus *Fusibacter*

1. **Fusibacter paucivorans** Ravot, Magot, Fardeau, Patel, Thomas, Garcia and Ollivier 1999, 1145[VP]

   pau.ci.vo′rans. L. adj. *paucus* few; L. v. *voro* to devour; N.L. part. adj. *paucivorans* intended to mean a bacterium that utilizes few substrates.

   Cells are spindle-shaped rods (1 × 2–5 μm), occur singly or in pairs, and possess one to four peritrichous flagella. Cells stain Gram-positive.

   Round colonies (diameter, 1 mm) are present after 8 d incubation at 37°C.

   Chemo-organotrophic and obligately anaerobic member of the domain *Bacteria*. Utilizes D-cellobiose, fructose, glucose, D-mannitol, and D-ribose but not arabinose, dulcitol, galactose, lactose, maltose, mannose, melibiose, raffinose, rhamnose, sorbitol, sorbose, sucrose, trehalose, xylose, cellulose, starch, formate, acetate, butyrate, lactate, propionate, methanol, bio-Trypticase, Casamino acids, or gelatin. Requires yeast extract for growth. The end products of glucose fermentation are butyrate, acetate, CO_2, and H_2. Uses thiosulfate and sulfur as electron acceptors during glucose fermentation. H_2S is produced from thiosulfate and sulfur reduction.

   The optimum temperature for growth is 37°C; temperature range between 20 and 45°C. The optimum pH is 7.3; growth occurs between pH 5.7 and pH 8.0. Halotolerant; the optimum sodium chloride concentration for growth is between 0 and 3 % (w/v); growth occurs at NaCl concentrations ranging between 0 and 10%.

   Isolated from an oil-producing well.

   DNA G+C content (mol%): 43 (HPLC).

   Type strain: SEBR 4211, DSM 12116.

   GenBank accession number (16S rRNA gene): AF050099.

Genus III. **Guggenheimella** Wyss, Dewhirst, Paster, Thurnheer and Luginbühl 2005, 669[VP]

Gug.gen.heim.el'la. L. dim. suff. *-ella*; N.L. fem. n. *Guggenheimella* after the Swiss microbiologist Bernhard Guggenheim.

Short to coccoid, non-spore-forming rods. Gram-stain-positive. Nonmotile. **Obligately anaerobic.** Assacharolytic. Cell wall contains the diamino acid ornithine.

DNA G+C content (mol%): 44.4.

Type species: **Guggenheimella bovis** Wyss, Dewhirst, Paster, Thurnheer and Luginbühl 2005, 669[VP].

Further descriptive information

Phylogenetic analysis of the 16S rRNA gene sequence places the genus within the class *Clostridia* and phylum *Firmicutes*. Its closest relative is *Tindallia magadiensis*. The major fatty acids are $C_{16:0}$ (33%), $C_{18:1cis9}$ (24%), $C_{18:2cis9'12}$ (22%), and $C_{18:0}$ (16%). The cell-wall murein is type A4β containing L-Orn–D-Asp.

Complex components (yeast extract, peptone, and serum) are required for growth.

The two reported strains (OMZ 913[T] and OMZ 915) were isolated from diseased tissue surgically removed from the hooves of heifers with dermatitis digitalis in Switzerland—the only known habitat. Isolation was accomplished by dilution in OMIZ-Pat medium containing (per liter) 10 ml heat-inactivated human serum, 1 mg rifampin, 100 mg fosfomycin, 5 mg polymyxin, and 30 mg nalidixic acid. Stocks cultures are maintained by the addition of glycerol (15% v/v) to well grown liquid cultures and stored frozen at −80° C or in liquid nitrogen.

Differentiation of the genus *Guggenheimella* from other closely related genera

Unlike its relatives (*Eubacterium brachy, Eubacterium limosum, Eubacterium nodatum, Eubacterium saphenum, Eubacterium sulci, Mogibacterium timidum,* and *Tindallia magadiensis*), *Guggenheimella* has higher mol% G+C and displays chymotrypsin-like activity. It is more fastidious than *Tindallia magadiensis* (which does not require yeast extract, peptone, and serum) and less fastidious than *Eubacterium saphenum* (which does require L-lysine, instead of L-arginine, and fetal bovine serum). There is also a difference in fatty acid production between *Guggenheimella* (butyrate and minor amounts of isovalerate, isobutyrate, propionate and acetate) and *Tindallia magadiensis* (mainly acetate, propionate, and isovalerate).

List of species of the genus *Guggenheimella*

1. **Guggenheimella bovis** Wyss, Dewhirst, Paster, Thurnheer and Luginbühl 2005, 669[VP]

bo'vis. L. gen. n. *bovis* of the cow, referring to the source of isolation.

Cells are short rods (0.4 × 0.5–1.5 µm) that grow into shiny, white opaque colonies (0.5 mm in diameter) on OMIZ-Pat agar or Columbia blood agar in Gas-Pak jars after 7–10 d at 37°C, or as a suspension in OMIZ-Pat liquid, with turbidity visible within 3 d and marked change in phenol red indicator to violet.

Growth occurs at mesophilic temperatures (but not at 25 or 45°C) and pH 6.5–9.0, with the optimum being pH 7.5–8.0. API ZYM detects acid and alkaline phosphatases, C4 and C8 esterases, leucine arylamidase, chymotrypsin, and naphthol phosphohydrolase, but not trypsin, α- and β-galactosidases, α- and β-glucosidases, lipase C14, cystyl arylamidase, valine arylamidase, β-glucuronidase, *N*-acetyl-β-glucosaminidase, α-mannosidase, and α-fucosidase. Growth is not affected by D-arabinose, L-arabinose, D-cellobiose, D-fructose, D-fucose, L-fucose, D-galactose, D-galacturonic acid, D-glucose, D-glucuronic acid, glycogen, D-lactose, D-maltose, D-mannitol, D-mannose, D-melibiose, L-rhamnose, D-ribose, L-sorbose, starch, D-sucrose, D-trehalose, D-xylose, and L-xylose. Growth in OMIZ-Pat is resistant to fosfomycin (100 mg/l), rifampin (1 mg/l), polymyxin (5 mg/l), and nalidixic acid (30 mg/) alone or in combination.

For all other characteristics, refer to the genus description.

DNA G+C content (mol%): 44.4 (HPLC).

Type strain: OMZ 913, CIP 108087, DSM 15657.

GenBank accession number (16S rRNA gene): AY272039.

References

Alain, K., P. Pignet, M. Zbinden, M. Quillevere, F. Duchiron, J.P. Donval, F. Lesongeur, G. Raguenes, P. Crassous, J. Querellou and M.A. Cambon-Bonavita. 2002. *Caminicella sporogenes* gen. nov., sp. nov., a novel thermophilic spore-forming bacterium isolated from an East-Pacific Rise hydrothermal vent. Int. J. Syst. Evol. Microbiol. *52*: 1621–1628.

Baena, S., M.L. Fardeau, M. Labat, B. Ollivier, P. Thomas, J.L. Garcia and B.K. Patel. 1998. *Aminobacterium colombiense* gen. nov. sp. nov., an amino acid-degrading anaerobe isolated from anaerobic sludge. Anaerobe *4*: 241–250.

Baena, S., M.L. Fardeau, B. Ollivier, M. Labat, P. Thomas, J.L. Garcia and B.K. Patel. 1999a. *Aminomonas paucivorans* gen. nov., sp. nov., a mesophilic, anaerobic, amino-acid-utilizing bacterium. Int. J. Syst. Bacteriol. *49Pt 3*: 975–982.

Baena, S., M.L. Fardeau, T.H.S. Woo, B. Ollivier, M. Labat and B.K.C. Patel. 1999b. Phylogenetic relationships of three amino-acid-utilizing anaerobes, *Selenomonas acidaminovorans*, '*Selenomonas acidaminophila*' and *Eubacterium acidaminophilum*, as inferred from partial 16S rDNA nucleotide sequences and proposal of *Thermanaerovibrio acidaminovorans* gen. nov., comb. nov and *Anaeromusa acidaminophila* gen. nov., comb. nov. Int. J. Syst. Bacteriol. *49*: 969–974.

Baena, S., M.L. Fardeau, M. Labat, B. Ollivier, J.L. Garcia and B.K.C. Patel. 2000. *Aminobacterium mobile* sp. nov., a new anaerobic amino-acid-degrading bacterium. Int. J. Syst. Evol. Microbiol. *50*: 259–264.

Balch, W.E., G.E. Fox, L.J. Magrum, C.R. Woese and R.S. Wolfe. 1979. Methanogens: reevaluation of a unique biological group. Microbiol. Rev. *43*: 260–296.

Bernard, F.P., J. Connan and M. Magot. 1992. Indigenous microorganisms in connate water of many oil fields: a new tool in exploration and production techniques. Proc. 67th Annual Technical Conference and Exhibition, Richardson, Texas, pp. 1–10.

Collins, M.D., P.A. Lawson, A. Willems, J.J. Cordoba, J. Fernández-Garayzábal, P. Garcia, J. Cai, H. Hippe and J.A. Farrow. 1994. The phylogeny of the genus *Clostridium*: proposal of five new genera and eleven new species combinations. Int. J. Syst. Bacteriol. *44*: 812–826.

Crolet, J.-L. and M. Magot. 1996. Non SRB sulfidogenic bacteria from oil field production facilities. Mater. Perform. *March*: 60–64.

Fardeau, M.L., J. L. Cayol, M. Magot and B. Ollivier. 1993. H$_2$ oxidation in the presence of thiosulfate, by a *Thermoanaerobacter* strain isolated from an oil-producing well. FEMS Microbiol. Lett. *113*: 327–332.

Fardeau, M.L., C. Faudon, J.L. Cayol, M. Magot, B.K. Patel and B. Ollivier. 1996. Effect of thiosulphate as electron acceptor on glucose and xylose oxidation by *Thermoanaerobacter finnii* and a *Thermoanaerobacter* sp. isolated from oil field water. Res. Microbiol. *147*: 159–165.

Fardeau, M.L., B. Ollivier, B.K.C. Patel, M. Magot, P. Thomas, A. Rimbault, F. Rocchiccioli and J.L. Garcia. 1997. *Thermotoga hypogea* sp. nov., a xylanolytic, thermophilic bacterium from an oil-producing well. Int. J. Syst. Bacteriol. *47*: 1013–1019.

Fendrich, C., H. Hippe and G. Gottschal. 1990. *Clostridium halophilum* sp. nov. and *C. litorale* sp. nov., an obligate halophilic and a marine species degrading betaine in the Strickland reaction. Arch. Microbiol. *154*: 127–132.

Garrity, G.M., J.A. Bell and T. Lilburn. 2005. The Revised Road Map to the Manual. *In* Brenner, Krieg, Staley and Garrity (Editors), Bergey's Manual of Systematic Bacteriology, 2nd edn, Vol. 2, The *Proteobacteria*, Part A, Introductory Essays. Springer, New York, pp. 159–220.

Jeanthon, C., A.L. Reysenbach, S. L'Haridon, A. Gambacorta, N.R. Pace, P. Glénat and D. Prieur. 1995. *Thermotoga subterranea* sp. nov., a new thermophilic bacterium isolated from a continental oil reservoir. Arch. Microbiol. *164*: 91–97.

Kevbrin, V.V., T.N. Zhilina, F.A. Rainey and G.A. Zavarzin. 1998. *Tindallia magadii* gen. nov., sp. nov.: an alkaliphilic anaerobic ammonifier from soda lake deposits. Curr. Microbiol. *37*: 94–100.

Magot, M., L. Carreau, J.-L. Cayol, B. Ollivier and J.-L. Crolet. 1994. Sulphide-producing, not sulphate-reducing anaerobic bacteria presumptively involved in bacterial corrosion. Proc. 3rd European Federation of Corrosion Workshop on Microbial Corrosion, London.

Magot, M., M.L. Fardeau, O. Arnauld, C. Lanau, B. Ollivier, P. Thomas and B.K. Patel. 1997a. *Spirochaeta smaragdinae* sp. nov., a new mesophilic strictly anaerobic spirochete from an oil field. FEMS Microbiol. Lett. *155*: 185–191.

Magot, M., G. Ravot, X. Campaignolle, B. Ollivier, B.K.C. Patel, M.L. Fardeau, P. Thomas, J.L. Crolet and J.L. Garcia. 1997b. *Dethiosulfovibrio peptidovorans* gen. nov., sp. nov., a new anaerobic, slightly halophilic, thiosulfate-reducing bacterium from corroding offshore oil wells. Int. J. Syst. Bacteriol. *47*: 818–824.

Meijer, W.G., M.E. Nienhuis-Kuiper and T.A. Hansen. 1999. Fermentative bacteria from estuarine mud: phylogenetic position of *Acidaminobacter hydrogenoformans* and description of a new type of Gram-negative, propionigenic bacterium as *Propionibacter pelophilus* gen. nov., sp. nov. Int. J. Syst. Bacteriol. *49*: 1039–1044.

Plugge, C.M., M. Balk, E.G. Zoetendal and A.J. Stams. 2002. *Gelria glutamica* gen. nov., sp. nov., a thermophilic, obligately syntrophic, glutamate-degrading anaerobe. Int. J. Syst. Evol. Microbiol. *52*: 401–407.

Ravot, G., M. Magot, M.L. Fardeau, B.K.C. Patel, G. Prensier, A. Egan, J.L. Garcia and B. Ollivier. 1995a. *Thermotoga elfii* sp. nov., a novel thermophilic bacterium from an African oil producing well. Int. J. Syst. Bacteriol. *45*: 308–314.

Ravot, G., B. Ollivier, M. Magot, B.K.C. Patel, J.L. Crolet, M.L. Fardeau and J.L. Garcia. 1995b. Thiosulfate reduction, an Important physiological feature shared by members of the order *Thermotogales*. Appl. Environ. Microbiol. *61*: 2053–2055.

Ravot, G., M. Magot, B. Ollivier, B.K. Patel, E. Ageron, P.A. Grimont, P. Thomas and J.L. Garcia. 1997. *Haloanaerobium congolense* sp. nov., an anaerobic, moderately halophilic, thiosulfate- and sulfur-reducing bacterium from an African oil field. FEMS Microbiol. Lett. *147*: 81–88.

Ravot, G., M. Magot, M.L. Fardeau, B.K.C. Patel, P. Thomas, J.L. Garcia and B. Ollivier. 1999. *Fusibacter paucivorans* gen. nov., sp. nov., an anaerobic, thiosulfate-reducing bacterium from an oil-producing well. Int. J. Syst. Bacteriol. *49*: 1141–1147.

Stams, A.J.M. and T.A. Hansen. 1984. Fermentation of glutamate and other compounds by *Acidaminobacter hydrogenoformans* gen. nov. sp. nov., an obligate anaerobe isolated from black mud: studies with pure cultures and mixed cultures with sulfate-reducing and methanogenic bacteria. Arch. Microbiol. *137*: 329–337.

Stams, A.J.M. and T.A. Hansen. 1985. *In* Validation of the publication of new names and new combinations previously effectively published outside the IJSB. List no. 17. Int. J. Syst. Bacteriol. *35*: 223.

Takai, K., D.P. Moser, T.C. Onstott, N. Spoelstra, S.M. Pfiffner, A. Dohnalkova and J.K. Fredrickson. 2001. *Alkaliphilus transvaalensis* gen. nov., sp. nov., an extremely alkaliphilic bacterium isolated from a deep South African gold mine. Int. J. Syst. Evol. Microbiol. *51*: 1245–1256.

Wildenauer, F.X. and J. Winter. 1986. Fermentation of isoleucine and arginine by pure and syntrophic cultures of *Clostridium sporogenes*. FEMS Microbiol. Ecol. *38*: 373–379.

Winter, J., F. Schindler and F.X. Wildenauer. 1987. Fermentation of alanine and glycine by pure and syntrophic cultures of *Clostridium sporogenes*. FEMS Microb. Ecol. *45*: 153–161.

Wyss, C., F.E. Dewhirst, B.J. Paster, T. Thurnheer and A. Luginbühl. 2005. *Guggenheimella bovis* gen. nov., sp. nov., isolated from lesions of bovine dermatitis digitalis. Int. J. Syst. Evol. Microbiol. *55*: 667–671.

Zindel, U., W. Freudenberg, M. Rieth, J.R. Andreesen, J. Schnell and F. Widdel. 1988. *Eubacterium acidaminophilum* sp. nov., a versatile amino acid-degrading anaerobe producing or utilizing H$_2$ or formate: description and enzymatic studies. Arch. Microbiol. *150*: 254–266.

Family XIII. **Incertae Sedis**

Previously assigned to the "*Eubacteriaceae*" by Garrity et al. (2005), subsequent analyses suggest that these genera, while closely related to each other, are not closely related to *Eubacterium* or a member of any other previously described family. For that reason, they are assigned to their own family *incertae sedis*.

Genus I. **Anaerovorax** Matthies, Evers, Ludwig and Schink 2000, 1593[VP]

BERNHARD SCHINK

An.a.e.ro.vo′rax. Gr. suff. *an* non; Gr. n. *aer* air; L. adj. *vorax* voracious; N.L. masc. n. *Anaerovorax* an anaerobic voracious bacterium.

Strictly anaerobic chemoorganotrophic bacteria of fermentative metabolism, nonphotosynthetic; inorganic electron acceptors not used. Typical cell-wall structure of Gram-positive bacteria without an outer membrane but Gram-negative staining. Non-sporeforming.

Chemoorganotrophic, fermentative type of metabolism, using preferentially amino acid derivatives as substrate. Media containing a reductant are necessary for growth. Catalase negative. Isolated from anoxic freshwater or marine sediments.

DNA G+C content (mol%): 29.6 ± 1.0.

Type species: **Anaerovorax odorimutans** Matthies, Evers, Ludwig and Schink 2000, 1593[VP].

Further descriptive information

The only described species so far, *Anaerovorax odorimutans*, was isolated as a strictly anaerobic bacterium growing with the biogenic amine putrescine (1,4-diaminobutane) as the only source of carbon and energy. Such primary amines are formed during oxygen-limited decomposition of organic matter rich in protein. Clostridia, pseudomonads, lactic acid bacteria, and some enterobacteria produce biogenic amines, e.g., putrescine or cadaverine (1,5-diaminopentane), by decarboxylation of ornithine or lysine (Andreesen et al., 1989; Geornaras et al., 1995; Madigan et al., 1997). These putrid-smelling and often toxic compounds ("ptomaines") are also released in food, e.g., in ripening cheese (Stratton et al., 1991; Ten Brink et al., 1990).

In pure culture, the isolated strains of *Anaerovorax odorimutans* ferment putrescine to acetate, butyrate, and hydrogen (Matthies et al., 2000, 1989) according to the following equation: $10 \text{ putrescine}^{2+} + 26H_2O \rightarrow 6 \text{ acetate}^- + 7 \text{ butyrate}^- + 20NH_4^+ + 16H_2 + 13H^+$.

In co-culture with a hydrogen-scavenging methanogenic partner such as *Methanospirillum hungatei*, the fermentation balance was shifted to more acetate production, according to: $2 \text{ putrescine}^{2+} + HCO_3^- + 3H_2O \rightarrow 2 \text{ acetate}^- + \text{butyrate}^- + CH_4 + 4NH_4^+ + 2H^+$. A similar shift was obtained with *Acetobacterium woodii* as partner. Excess accumulation of hydrogen gas inhibited growth.

In addition to putrescine, only 4-aminobutyrate and 4-hydroxybutyrate were used as substrates and fermented to analogous product mixtures. No other substrates were utilized. The pathway of putrescine degradation as analyzed by enzyme measurements in cell-free extracts includes deamination and oxidation to 4-aminobutyrate and 4-hydroxybutyrate, activation and dehydration to vinylacetyl CoA, isomerization to crotonyl CoA, and subsequent dismutation to acetate and butyrate.

The cell-wall architecture is typical of Gram-positive bacteria. Spore formation was not observed, either in defined medium or in specific sporulation media.

Enrichment and isolation procedures

The only described species so far, *Anaerovorax odorimutans*, was isolated from anaerobic enrichment cultures with putrescine (1,4-diaminobutane) as the only source of carbon and energy. The type strain, strain NorPut1[T], was isolated from an anoxic brackish sediment (Norsminde fjord, Aarhus, Denmark); other similar isolates (strains FrPut1, MaPut1) were obtained from anoxic freshwater or marine sediments. These strains did not differ substantially in their metabolism and showed high tolerance toward the salt content of the medium. The comparably high growth optimum of 37°C indicates that this bacterium has its real habitat in intestinal tracts of warm-blooded animals where primary amines formed from incomplete degradation of amino acids may be common substrates.

Anaerovorax odorimutans strain NorPut1[T] was cultivated with 10 mM putrescine in a sulfide-reduced, bicarbonate-buffered mineral medium which contained trace element solution SL10, selenite tungstate solution (Widdel et al., 1983), and 7-vitamin solution (Widdel and Pfennig, 1981) under a N_2/CO_2 (90%/10%) atmosphere. Details of cultivation and physiological characterization are given in the original description (Matthies et al., 1989). The salt content of the medium was adapted to the salinity of the source of isolation, i.e., freshwater, brackish, or saltwater medium. Since the produced hydrogen gas can be inhibitory to growth, sufficient headspace (at least equal to the medium volume) has to be provided in the culture bottles.

Maintenance procedures

Cultures are maintained either by repeated transfer at intervals of 2–3 months or by freezing in liquid nitrogen using techniques common for strictly anaerobic bacteria. No information exists about survival upon lyophilization.

Differentiation of the genus *Anaerovorax* from other genera

As indicated by morphological and physiological characteristics and corroborated by phylogenetic analysis of 16S rDNA sequences, strain NorPut1[T] is a member of the phylum of the Gram-positive bacteria with a low DNA G+C content. Its closest relative among the bacteria is the not yet validly published species "*Clostridium aminobutyricum*" (Collins et al., 1994; Hardman and Stadtman, 1960), with 94.6% overall 16S rRNA sequence similarity. A moderate relationship of this pair and several *Eubacterium* species is indicated by similarity values of 90.5% and lower. Based upon analyses of the currently available 16S rRNA sequence data set, these organisms represent a monophyletic group (Family XIII *Incertae Sedis* in this volume). The separation into two subclusters was supported by the majority of the various phylogenetic analyses. Strain NorPut1[T] clusters with "*Clostridium aminobutyricum*", *Eubacterium brachy, Eubacterium infirmum, Eubacterium saphenum,* and *Eubacterium timidum* (Cheeseman et al., 1996). The second subcluster comprises *Eubacterium nodatum, Eubacterium tardum, Eubacterium minutum* and an unnamed isolate, C2 (Attwood et al., 1998; Cheeseman et al., 1996). Strain Nor-Put1[T] was separated and described as a new genus *Anaerovorax* because the relationship to the *Eubacterium* species is only moderate and the current genus *Eubacterium* combines a phylogenetically diverse collection of species. "*Clostridium aminobutyricum*" could perhaps be added later to the genus *Anaerovorax* as a renamed species, *Anaerovorax aminobutyricus.* Its sequence similarity to *Anaerovorax odorimutans* is just at the borderline to allow assignment to the same genus (around 95% sequence similarity as suggested by Ludwig et al. (1998). Other *Eubacterium* species are more distinct and have to be grouped as a separate genus.

List of species of the genus *Anaerovorax*

1. **Anaerovorax odorimutans** Matthies, Evers, Ludwig and Schink 2000, 1593[VP]

o.do.ri.mu′tans. L. masc. n. *odor* smell, odor; L. v. *mutare* to change, *mutans* changing; L. adj. *odorimutans* odor-changing, referring to the degradation of the odorous compound putrescine to form another odorous one, butyric acid.

Slightly curved rods, 0.7–0.8 × 1.9–2.7 μm, motile by 3–5 flagella inserted on the concave side of the cell; typical cell wall of Gram-positive bacteria without an outer membrane but it stains Gram-negative, and it is nonsporeforming.

Chemoorganotrophic, fermentative metabolism; external electron acceptors are not used. Contains no cytochromes. Putrescine, 4-aminobutyrate, and 4-hydroxybutyrate are the only substrates utilized. No growth with more than 30 different substrates such as sugars, organic acids, alcohols, amino acids, or other amines tested. Products of putrescine fermentation were acetate, butyrate, NH_4^+, and H_2; 4-aminobutyrate was fermented to acetate, butyrate, and NH_4^+; 4-hydroxybutyrate to acetate and butyrate.

Growth in freshwater and saltwater media up to 2% NaCl and 0.3% $MgCl_2 \cdot 6H_2O$ (w/v), pH 7.2–7.6. Reducing agents such as sulfide required. Growth possible at 15–45°C with an optimum at 37°C; no growth at 12 or 50°C. Habitat: anoxic freshwater, marine, or brackish water sediment. Isolated from anoxic brackish water sediment.

DNA G+C content (mol%): 29.6 ± 1.0 (HPLC).
Type strain: NorPut1, ATCC BAA-160, DSM 5092.
GenBank accession number (16S rRNA gene): AJ251215.

Genus II. **Mogibacterium** Nakazawa, Sato, Poco, Hashimura, Ideka, Kalfas, Sundqvist and Hoshino 2000, 686[VP]

THE EDITORIAL BOARD

Mogi.bac.te′ri.um. Gr. n. *mogos* effort; Gr. dim. n. *bakterion* a small rod; N.L. neut. n. *Mogibacterium* a difficult-to-culture, rod-shaped bacterium.

Gram-positive rods, cells are 0.2–0.8 × 1.0–3.1 μm, occurring as single cells, in short chains or in clumps. No spores are formed. Non-motile. **Strictly anaerobic and asaccharolytic.** Very poor growth in broth medium. Colonies on brain heart infusion agar (BHI)-blood agar plates are minute (<1 mm in diameter), circular, convex and translucent even after prolonged incubation in an anaerobic glove box (7–10 d). No haemolysis occurs around colonies on BHI-blood agar plates. Catalase-negative. **Metabolic end product in peptone/yeast extract/glucose medium is phenylacetate.** Members of the genus *Mogibacterium* can be distinguished from each other by 16S rRNA sequences and DNA–DNA homology values.

DNA G+C content (mol%): 41–50.

Type species: **Mogibacterium pumilum** Nakazawa, Sato, Poco, Hashimura, Ideka, Kalfas, Sundqvist and Hoshino 2000, 686[VP].

Further descriptive information

Mogibacterium currently contains five species. Four species, *Mogibacterium diversum, Mogibacterium neglectum, Mogibacterium pumilum* and *Mogibacterium vescum* were originally placed in the genus *Mogibacterium* (Nakazawa et al., 2002, 2000). *Mogibacterium timidum* (originally described as *Eubacterium timidum* by Holdeman et al., 1980) was transferred from *Eubacterium* to *Mogibacterium* by Nakazawa et al. (2000).

Sources and clinical significance. Strains of *Mogibacterium* were from human oral cavities. *Mogibacterium diversum* (three strains studied) was isolated from human tongue plaque. *Mogibacterium neglectum* (two strains) was isolated from necrotic dental pulp. *Mogibacterium pumilum* (two strains) was isolated from an infected root canal and from a periodontal pocket. *Mogibacterium vescum* (one strain) was isolated from a periodontal pocket. No information is given on the distribution or incidence of these four species in human mouths (Nakazawa et al., 2002). Isolates of *Mogibacterium timidum* (67 strains) were from areas associated with periodontitis (Holdeman et al., 1980). The only available antimicrobial susceptibility data were given by Holdeman et al. (1980) for 30 strains of *Mogibacterium timidum.*

Colony and cell morphology. Strains are rod-shaped organisms occurring as single cells, in short chains or in clumps. Electron micrographs of ultra-thin sections of the type strains of *Mogibacterium diversum, Mogibacterium neglectum, Mogibacterium pumilum,* and *Mogibacterium vescum* reveal a cell wall typical of Gram-positive bacteria, containing a plasma membrane layer, a thin peptidoglycan layer in the middle and a thicker outer layer. No pili- or flagella-like structures were found in the type strains of *Mogibacterium diversum* and *Mogibacterium neglectum*. All of the organisms form tiny, non-haemolytic colonies on BHI-blood agar plates. According to Holdeman et al. (1977) blood agar plates were prepared by adding 5 ml of defibrinated rabbit's blood to 100 ml of supplemented BHI agar; however, sheep and horse blood have also been used. The colonies are 0.3–0.6 mm in diameter, and even after prolonged incubation the colonies are less than 1 mm in diameter. The temperature of incubation is 37°C.

Phenotypic analysis. *Mogibacterium* strains do not ferment glucose or other carbohydrates and are inert in most biochemical tests (using the methods described by Holdeman et al., 1977). The following tests were negative: aesculin, arginine and starch hydrolysis, nitrate reduction, gelatin liquefaction and ammonia, catalase, indole and urease production. They produce phenylacetate (<10 mM) as a metabolic end product in peptone/yeast extract (PY) and PYG broth (as assayed by gas chromatography). Enzyme profiles of *Mogibacterium* type strains of the species have been determined by Rapid ID 32A Kit (API bioMérieux), according to the manufacturer's instructions, except for incubation under strictly anaerobic conditions. Results revealed an API Code 0000020000 for *Mogibacterium pumilum*, *Mogibacterium timidum* and *Mogibacterium vescum*, which includes a positive reaction for the proline arylamidase test, and API Code 0000000000 for *Mogibacterium diversum* and *Mogibacterium neglectum* (Nakazawa et al., 2002).

Strains of *Mogibacterium diversum*, *Mogibacterium neglectum*, *Mogibacterium pumilum*, and *Mogibacterium vescum* do not utilize adonitol, amygadalin, arabinose, cellobiose, erythritol, aesculin, fructose, galactose, glucose, glycogen, inositol, lactose, maltose, mannitol, mannose, melezitose, melibiose, rhamnose, ribose, salicin, sorbitol, starch, sucrose, trehalose or xylose.

In whole-cell protein profiles examined using SDS-PAGE, each type strain showed typical protein profiles. There were no major bands in common, indicating great heterogeneity in the whole-cell protein components. Western immunoblotting reactions with rabbit antisera showed that antigens from the type strains of *Mogibacterium* species were recognized by their respective antiserum.

Genotypic analysis. The G+C content of the type strains of each species ranges from 41–50 (HPLC). DNA–DNA relatedness between the type strains of the species of *Mogibacterium* is 16–39 %. Relatedness of *Mogibacterium* to three species of *Eubacterium* (also asaccharolytic, anaerobic, Gram-positive rods) namely *Eubacterium brachy*, *Eubacterium nodatum*, and *Eubacterium saphenum* is negligible (1–3%).

Phylogenetic analysis. The 16S rRNA similarity between type strains of species of *Mogibacterium* is 99–100% with the exception of *Mogibacterium diversum*, which joins the *Mogibacterium* cluster at 90% similarity (Nakazawa et al., 2002). The closest relatives tested by Nakazawa et al. (2002) were *Eubacterium* species (79–87% similarity).

Enrichment, isolation and maintenance procedures

Holdeman et al. (1980) describe their isolation procedures using supplemented brain heart infusion agar and supplemented chopped meat broth agar. Cultures may be stored at −70°C suspended in 10% skim milk or lyophilized.

Differentiation of the genus *Mogibacterium* from other genera

Table 224 gives the characteristics for differentiation of *Mogibacterium* from phenotypically related species (asaccha-

TABLE 224. Differential characteristics for *Mogibacterium* and phenotypically related genera[a]

Characteristic	*Mogibacterium pumilum* ATCC 700696[T]	*Mogibacterium diversum* ATCC 700923[T]	*Mogibacterium neglectum* ATCC 700924[T]	*Mogibacterium timidum* ATCC 33093[T]	*Mogibacterium vescum* ATCC 700697[T]	*Slackia exigua* ATCC 700122[T]	*Cryptobacterium curtum* ATCC 700683[T]	*Eggerthella lenta* ATCC 25559[T]	*Eubacterium minutum* ATCC 70079[T]	*Eubacterium nodatum* ATCC 33099[T]	*Eubacterium saphenum* ATCC 49989[T]	*Eubacterium brachy* ATCC 33089
End products from peptone/yeast extract/glucose broth[b]	phe-a	phe-a	phe-a	phe-a	phe-a	−	−	−	B or b	a,B	a,b	ib, ic, iv, phe-p
Arginine hydrolysis	−	−	−	−	−	+	+	+	−	+	Not given	−
Nitrate reduction	−	−	−	−	−	−	−	+	−	−	−	−

[a]Taken from Nakazawa et al. (2002).
[b]a, Acetate; b, butyrate; ib, isobutyrate; ic, isocaproate; iv, isovalerate; phe-a, phenylacetate; phe-p, phenylpropionate. Upper-case letters indicate > or = 10 mM of product and lower-case indicates <10 mM of product.

rolytic, anaerobic, Gram-positive rods). The closest relative by 16S rRNA gene sequencing is *Anaerovorax* (Ludwig et al., 2009), which can be differentiated from *Mogibacterium* because it is fermentative and has a DNA G+C content of about 30 mol% (Matthies et al., 2000). *Mogibacterium* species could only be differentiated from each other by DNA–DNA hybridization and 16S rRNA gene sequences (Nakazawa et al., 2002).

Further reading

Hori, R., M. Sato, S. Kohno and E. Hoshino. 1999. Tongue microflora in edentulous geriatric denture-wearers. Microb. Ecol. Health Dis. *11*: 89–95.

Katayama-Fujimura, Y., Y. Komatsu, H. Kuraishi and T. Kanedo. 1984. Estimation of DNA base composition by high performance liquid chromatography of its nuclease P1 hydrolysate. Agric. Biol. Chem. *48*: 3169–3172.

List of species of the genus *Mogibacterium*

1. **Mogibacterium pumilum** Nakazawa, Sato, Poco, Hashimura, Ideka, Kalfas, Sundqvist and Hoshino 2000, 686[VP]

   pu.mi′lum. L. adj. *pumilum* small or tiny, referring to the tiny colonies formed by this organism.

   Cell and colony morphology, phenotypic characteristics and genotypic and phylogenetic analyses are given in the genus description. Individual cells are 0.2–0.3 × 1.0 μm and occur singly, in short chains or in clumps. Strains produce approximately 3 mM phenylacetate as the sole metabolic end product in PY and PYG broth.

   DNA G+C content (mol%): 45–46 (HPLC).
   Type strain: D2-18, ATCC 700696.
   GenBank accession number (16S rRNA gene): AB021701.

2. **Mogibacterium diversum** Nakazawa, Poco, Sato, Ikeda, Kalfas, Sundqvist and Hoshino 2002, 121[VP]

   di.ver′sum. L. adj. d*iversum* diverse, referring to its low level of rRNA gene sequence similarity to the other *Mogibacterium* species.

   Cell and colony morphology, phenotypic characteristics and genotypic and phylogenetic analyses are given in the genus description. Individual cells are 0.2–0.3 × 1.0 μm and occur singly, in short chains or in clumps. Strains produce approximately 2 mM phenylacetate as the sole metabolic end product in PY and PYG broth.

   DNA G+C content (mol%): 42 (HPLC).
   Type strain: HM-7, ATCC 700923, JCM 11205.
   GenBank accession number (16S rRNA gene): AB037874.

3. **Mogibacterium neglectum** Nakazawa, Poco, Sato, Ikeda, Kalfas, Sundqvist and Hoshino 2002, 121[VP]

   neg.lect′um. L. adj. *neglectum* neglected, referring to the poor growth and tiny colonies that caused this organism to be neglected for a long time.

   Cell and colony morphology, phenotypic characteristics and genotypic and phylogenetic analyses are given in the genus description. Individual cells are 0.2–0.3 × 1.5 μm and occur singly or in clumps. Strains produce approximately 2 mM phenylacetate as the sole metabolic end product in PY and PYG broth.

   DNA G+C content (mol%): 41–42 (HPLC).
   Type strain: P9a-h, ATCC 700924, JCM 11204.
   GenBank accession number (16S rRNA gene): AB037875.

4. **Mogibacterium timidum** Nakazawa, Sato, Poco, Hashimura, Ideka, Kalfas, Sundqvist and Hoshino 2000, 686[VP] (*Eubacterium timidum* Holdeman, Cato, Burmeister and Moore 1980, 164)

ti′mi.dum. L. neut. adj. *timidum* fearful, timid, referring to the slight or slow growth in clumps.

Cells from PYG broth are 0.8–1.6 × 1.6–3.1 μm. Cells are arranged singly and often in clumps. Only 1 of 6 strains of this species tested by Holdeman et al. (1980) grew on BHI-blood agar. Colonies were <1mm in diameter. Optimum temperature for growth is 37°C.

Sixty-seven *Mogibacterium timidum* strains isolated by Holdeman, et al. (1980) were from subgingival and supragingival samples. It was detected in 42% of the subgingival samples and made up 3–57% of the cultivable flora in specimens from which it was isolated. It was isolated from 19% of the supragingival samples examined and represented 3–13% of the flora. Samples were taken from patients with periodontitis. The type strain was isolated from a subgingival area. Downes et al. (2001) isolated 13 strains of *Mogibacterium timidum*, 11 of which were isolated from odontogenic infections and two from peri-implantitis.

Partial description of the phenotypic characteristics is given in the genus description. In addition Holdeman et al. (1980) reported that 67 strains tested produced no acid from amygdalin, arabinose, cellobiose, erythritol, esculin, fructose, glucose, glycogen, inositol, lactose, maltose, mannitol, mannose, melezitose, melibiose, raffinose, rhamnose, ribose, salicin, sorbitol, starch, sucrose, trehalose, or xylose. In addition, the type strain and five other strains tested did not ferment adonitol, dextrin, dulcitol, galactose, glycerol, inulin, pectin or sorbose. All strains tested by Holdeman et al. (1980) produced trace amounts of acetate, formate or sorbate as the metabolic end product in PY and PYG broth. Thirteen strains isolated by Downes et al. (2001) produced phenylacetate and trace amounts of acetate.

Thirty strains of *Mogibacterium timidum* tested by Holdeman et al. (1980) were susceptible to chloramphenical, clindamycin, and erythromycin. Two were resistant to 2 U/ml of penicillin G and 8 were resistant to 6 μg/ml of tetracycline.

Genotypic and phylogenetic analyses are given in the genus description. The strains isolated by Downes et al. (2001) were identified by 16S rRNA gene sequencing.

DNA G+C content (mol%): 50 (HPLC).
Type strain: VPI D1B-22, ATCC 33093.
GenBank accession number (16S rRNA gene): Z36296.

5. **Mogibacterium vescum** Nakazawa, Sato, Poco, Hashimura, Ideka, Kalfas, Sundqvist and Hoshino 2000, 686[VP]

ves'cum. L. adj. *vescum* weak, referring to the poor growth of this organism.

Cell and colony morphology, phenotypic characteristics and genotypic and phylogenetic analyses are given in the genus description. Individual cells are 0.2–0.3 × 1.5 µm and occur singly, in short chains or in clumps. Strains produce approximately 4 mM phenylacetate as the sole metabolic end product in PY and PYG broth.

Downes et al. (2001) isolated six strains from oral infections that were identified as *Mogibacterium vescum* by 16S rDNA.

DNA G+C content (mol%): 46 (HPLC).
Type strain: D5-2, ATCC 700697.
GenBank accession number (16S rRNA gene): AB021702.

Other organisms

Saito et al. (2006) described a possible novel species of *Mogibacterium* designated "uncultured clone AF_H06" (GenBank accession no. AY821870). This clone clustered with the five species of *Mogibacterium*. It was identified in a study to investigate the bacterial diversity of seven infected root canals by the analysis of 16S rDNA clone libraries. This broad-based cultivation-free approach has enabled detection of bacteria when culture generates negative results.

References

Andreesen, J.R., H. Bahl and Gottschalk. 1989. Introduction to the physiology and biochemistry of the genus *Clostridium*. *In* Munson and Clarke (Editors), *Clostridia*, Vol. Biotechnology Handbooks 3. Plenum, New York, pp. 27–62.

Attwood, G.T., A.V. Klieve, D. Ouwerkerk and B.K. Patel. 1998. Ammonia-hyperproducing bacteria from New Zealand ruminants. Appl. Environ. Microbiol. *64*: 1796–1804.

Cheeseman, S.L., S.J. Hiom, A.J. Weightman and W.G. Wade. 1996. Phylogeny of oral asaccharolytic *Eubacterium* species determined by 16S ribosomal DNA sequence comparison and proposal of *Eubacterium infirmum* sp. nov. and *Eubacterium tardum* sp. nov. Int. J. Syst. Bacteriol. *46*: 957–959.

Collins, M.D., P.A. Lawson, A. Willems, J.J. Cordoba, J. Fernández-Garayzábal, P. Garcia, J. Cai, H. Hippe and J.A. Farrow. 1994. The phylogeny of the genus *Clostridium*: proposal of five new genera and eleven new species combinations. Int. J. Syst. Bacteriol. *44*: 812–826.

Downes, J., M.A. Munson, D.A. Spratt, E. Kononen, E. Tarkka, H. Jousimies-Somer and W.G. Wade. 2001. Characterisation of *Eubacterium*-like strains isolated from oral infections. J. Med. Microbiol. *50*: 947–951.

Garrity, G.M., J.A. Bell and T. Lilburn. 2005. The Revised Road Map to the Manual. *In* Brenner, Krieg, Staley and Garrity (Editors), Bergey's Manual of Systematic Bacteriology, 2nd edn, Vol. 2, The *Proteobacteria*, Part A, Introductory Essays. Springer, New York, pp. 159–220.

Geornaras, I., G.A. Dykes and A. von Holy. 1995. Biogenic amine formation by poultry-associated spoilage and pathogenic bacteria. Lett. Appl. Microbiol. *21*: 164–166.

Hardman, J.K. and T.C. Stadtman. 1960. Metabolism of omega-amino acids. I. Fermentation of gamma-aminobutyric acid by *Clostridium aminobutyricum* n. sp. J. Bacteriol. *79*: 544–548.

Holdeman, L.V., E.P. Cato and W.E.C. Moore (Editors). 1977. Anaerobe Laboratory Manual, 4th edn. Anaerobe Laboratory, Virginia Polytechnic Institute and State University, Blacksburg, VA.

Holdeman, L.V., E.P. Cato, J.A. Burmeister and W.E.C. Moore. 1980. Descriptions of *Eubacterium timidum* sp. nov., *Eubacterium brachy* sp. nov. and *Eubacterium nodatum* sp. nov., isolated from human periodontitis. Int. J. Syst. Bacteriol. *30*: 163–169.

Ludwig, W., O. Strunk, S. Klugbauer, N. Klugbauer, M. Weizenegger, J. Neumaier, M. Bachleitner and K.H. Schleifer. 1998. Bacterial phylogeny based on comparative sequence analysis. Electrophoresis *19*: 554–568.

Ludwig, W., K.-H. Schleifer and W.B. Whitman. 2009. Revised road map to the phylum *Firmicutes*. *In* De Vos, Garrity, Jones, Krieg, Ludwig, Rainey, Schleifer and Whitman (Editors), Bergey's Manual of Systematic Bacteriology, 2nd edn, Vol. 3, The *Firmicutes*, Springer, New York, pp. 1–14.

Madigan, M.T., J. M. Martinko and J. Parker. 1997. Brock Biology of Microorganisms, 8th edn. Prentice Hall, Upper Saddle River, NJ.

Matthies, C., F. Mayer and B. Schink 1989. Fermentative degradation of putrescine by new strictly anaerobic bacteria. Arch. Microbiol. *151*: 498–505.

Matthies, C., S. Evers, W. Ludwig and B. Schink. 2000. *Anaerovorax odorimutans* gen. nov., sp. nov., a putrescine-fermenting, strictly anaerobic bacterium. Int. J. Syst. Evol. Microbiol. *50*: 1591–1594.

Nakazawa, F., M. Sato, S.E. Poco, T. Hashimura, T. Ikeda, S. Kalfas, G. Sundqvist and E. Hoshino. 2000. Description of *Mogibacterium pumilum* gen. nov., sp. nov. and *Mogibacterium vescum* gen. nov., sp. nov., and reclassification of *Eubacterium timidum* (Holdeman et al. 1980) as *Mogibacterium timidum* gen. nov., comb. nov. Int. J. Syst. Evol. Microbiol. *50*: 679–688.

Nakazawa, F., S.E. Poco, M. Sato, T. Ikeda, S. Kalfas, G. Sundqvist and E. Hoshino. 2002. Taxonomic characterization of *Mogibacterium diversum* sp. nov. and *Mogibacterium neglectum* sp. nov., isolated from human oral cavities. Int. J. Syst. Evol. Microbiol. *52*: 115–122.

Saito, D., R. de Toledo Leonardo, J.L.M. Rodrigues, S.M. Tsai, J.F. Höfling and R.B. Gonçalves. 2006. Identification of bacteria in endodontic infections by sequence analysis of 16S rDNA clone libraries. J. Med. Microbiol. *55*: 101–107.

Stratton, J.E., R.W. Hutkins and S. L. Taylor. 1991. Biogenic amines in cheese and other fermented foods: a review. J. Food Protect. *54*: 460–470.

Ten Brink, B., C.Damik, H.M.L.J. Joosten and J.H.J.H. in 't Veld. 1990. Occurrence and formation of biologically active amines in foods. . Int. J. Food Microbiol. *11*: 73–84.

Widdel, F. and N. Pfennig. 1981. Studies on dissimilatory sulfate-reducing bacteria that decompose fatty acids. 1. Isolation of new sulfate-reducing bacteria enriched with acetate from saline environments: description of *Desulfobacter postgatei* gen. nov., sp. nov. Arch. Microbiol. *129*: 395–400.

Widdel, F., G.W. Kohring and F. Mayer. 1983. Studies on dissimilatory sulfate-reducing bacteria that decompose fatty acids. 3. Characterization of the filamentous gliding *Desulfonema limicola* gen. nov. sp. nov., and *Desulfonema magnum* sp. nov. Arch. Microbiol. *134*: 286–294.

Family XIV. Incertae Sedis

Previously assigned to the "*Syntrophomonadaceae*" by Garrity et al. (2005), subsequent analyses suggest that this genus is not closely related to *Syntrophomonas* or a member of any other pre-viously described family. For that reason, it is assigned to its own family *incertae sedis*.

Genus I. **Anaerobranca** Engle, Li, Woese and Wiegel 1995, 459[VP]

JUERGEN WIEGEL

An.ae.ro.bran'ca. Gr. pref. *an* not; Gr. n. *aer* air; N.L. fem. n. *branca* claw, paw, the root of the En. word branch, an arm-like part diverging from a main axis; N.L. n. *Anaerobranca* referring to the branched cell shape of the obligately anaerobic bacterium.

Gram-positive type cell-wall ultrastructure but, depending on the species, cells stain **Gram-positive or -negative**. The cell wall is either thick (*Anaerobranca horikoshii*) or thin. To date, the genus contains only **thermophilic** species growing in the range 30–67°C (no growth at 70°C) and pH 10.3 at 60°C. As indicated by the genus name, all three species exhibit the occurrence of truly branched cells with no septa; up to 10 % of the cells in a culture are branched, depending on strain, growth stage, and conditions. General physiology is **organo-heterotrophic** (proteolytic and glucolytic), but all species also can reduce Fe(III) to Fe(II) with magnetite formation, selenite to elemental selenium (red precipitate) and selenide (Se^{2-}), elemental sulfur to sulfide, and thiosulfate to sulfide. Some of the species can reduce (dissimilatory) fumarate to succinate. Proteinaceous compounds are the preferred sub-strates. The presence of thiosulfate stimulates the slow utiliza-tion of selected carbohydrates by functioning as an efficient electron acceptor. Spore formation not observed, but sporula-tion genes are present.

The genus *Anaerobranca* belongs to the phylum *Firmicutes* (low G+C Gram-positive bacteria) in the order *Clostridiales* within the class *Clostridia*. *Anaerobranca* was placed in Family VIII *Syntrophomonadaceae* as genus VI by Garrity et al. (2005). Despite the indication by the family name, but similar to situ-ations of other genera within this family, none of the *Anaero-branca* strains isolated so far are known to be syntrophic. The most closely related taxa are relatively distant (Figure 227), with *Alkalibacter saccharofermentans* being the closest relative. BLAST searches identify many *Paenibacillus* and some *Brevibacil-lus* species as close relatives, all belonging to a different class of the *Firmicutes*. However, this is misleading due to short branch lengths of these taxa as depicted in the phylogenetic tree; only two species of *Paenibacillus* and *Brevibacillus* were included as representatives of the cluster. In addition, the tree includes *Syn-trophomonas wolfei* subsp. *wolfei* (the type species of the family *Syntrophomonadaceae*), some thermophilic species of this fam-ily, and a few other taxa indicated by the BLAST search as close relatives e.g., *Clostridium tyrobutyricum*. *Clostridium butyricum* is included as a reference because it is the type species of the genus *Clostridium* and the type genus of the class *Clostridiales*. If more taxa in the close vicinity of *Anaerobranca* are isolated in the future, it is possible that they and *Anaerobranca*, as well as other nonsyntrophic taxa, will be placed in novel families. In light of these ambiguities, this genus is assigned to Family XIV *incertae sedis* in the current volume.

DNA G+C content (mol%): 30–34.

Type species: **Anaerobranca horikoshii** Engle, Li, Woese and Wiegel 1995, 459[VP].

Further descriptive information

The biogeography of the species appears to be restricted since species were only isolated from specific geothermally heated environments on the North American and African continents. For example, *Anaerobranca horikoshii* could only be isolated from a very restricted area of Yellowstone National Park but not from other areas of the park.

Three species have been validly published. Differentiating properties for the species are given in Table 225. All species require a growth factor from yeast extract (unidentified) which could not be substituted by vitamin supplementation. Late exponential growth phase cells tend to be pleomorphic includ-ing the appearance of round autoplast-like cells and irregular chains with cells of various lengths. This phenotype is frequently observed among alkalithermophilic aerobic and anaerobic ther-mophiles and was first described among anaerobic thermophiles for the neutrophilic thermophile, *Thermoanaerobacter ethanolicus* (Wiegel and Ljungdahl, 1981). This property is probably due to the inability to maintain an intact rigid cell-wall structure, pos-sibly due to expression of lysogenic enzyme(s), a remnant from the sporulation process, although sporulation has not been observed in *Thermoanaerobacter ethanolicus* JW200 or in the strains of *Anaerobranca* described here (unpublished results).

All strains of *Anaerobranca* belong to the extremophile group of alkalithermophiles, thus they are thermophiles (i.e., optimum growth temperature at or above 50°C) and alkaliphiles (i.e., opti-mum at around pH[20C] of 9.0–9.5 or at pH[60C] around 8.5; see Wie-gel, (1998), for comments on pH measurements and proposed form of reporting pH data for alkaline or acidic media incubated at elevated temperatures). Thus, these strains have to cope with the energetic problems caused by two extremes, i.e., problems caused by elevated temperatures and problems caused by alkaline pH values in the growth medium. Elevated temperatures cause problems that include membranes that become more permeable for protons, making it more difficult for the cells to maintain the necessary membrane energization. Problems arising from alka-line pH values in the growth medium include that, in contrast to the Mitchel theory, the internal pH is kept more acidic than the outside media pH, so that cells have a reversed intracellular/extracellular pH gradient. (Krulwich et al., (1990) and literature cited therein; Speelmans et al., 1995; Prowe et al., 1996).

Further reading

Grant, W.D. and K. Horikoshi. 1992. Alkaliphiles: ecology and biotechnological applications. *In* Herbert and Sharp (Edi-tors), Molecular Biology and Biotechnology of Extremo-philes. London, Blackie, pp. 143–162.

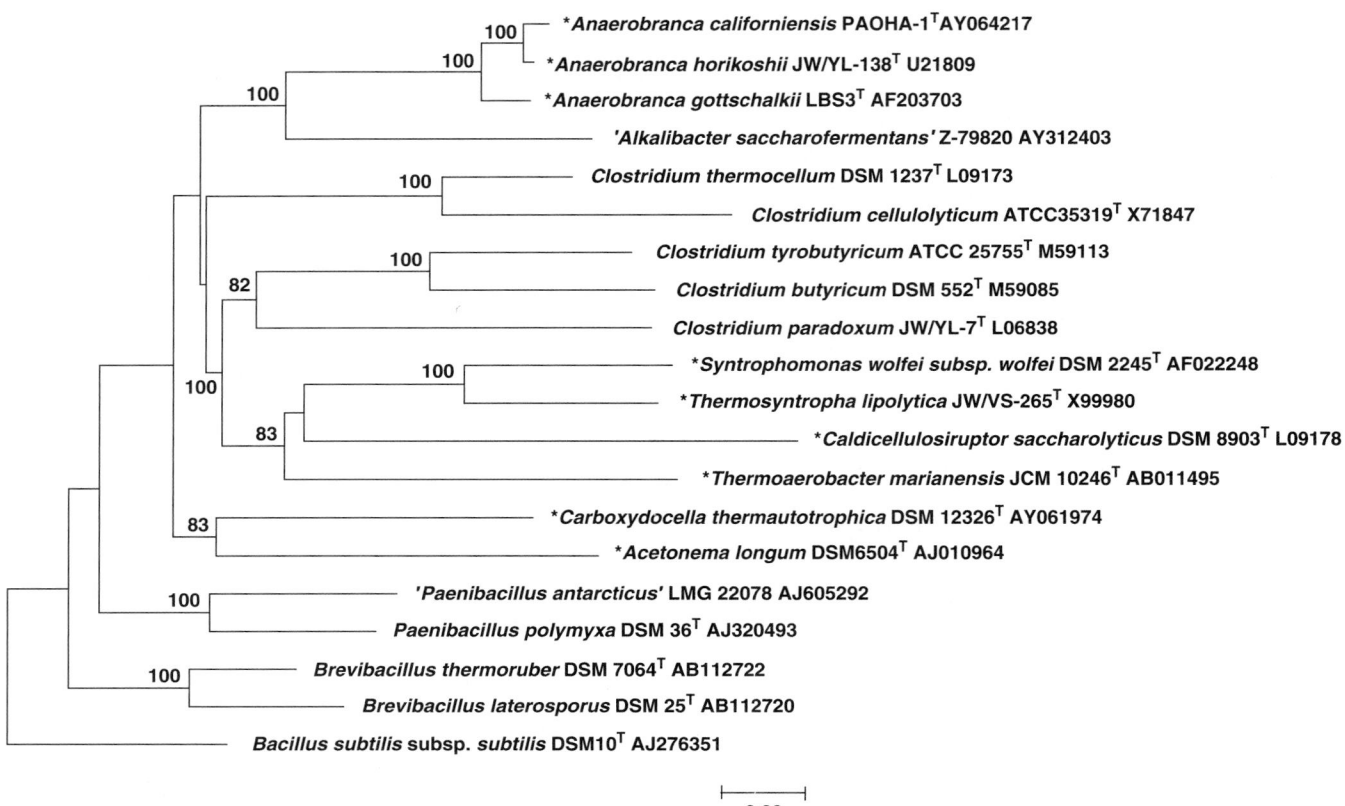

FIGURE 227. Phylogenetic tree (neighbor-joining) based on 16S rRNA sequences depicting selected members of the family *Syntrophomonadaceae* indicated by a BLAST search as closest related taxa. *Bacillus subtilis* functions as the outgroup. After sequence alignment was performed, only nucleotides in positions 87–1450 (*Escherichia coli* numbering) were used to build the tree. *Denotes members of the family *Syntrophomonadaceae*. (Courtesy of Rob Onyenwoke.)

TABLE 225. General characteristics to differentiate the three known species of the genus *Anaerobranca*[a,b]

Characteristic	A. californiensis	A. gottschalkii	A. horikoshii
Habitat	Hot springs of Paoha Island in Mono Lake (CA, USA)	Hot inlet of Lake Bogoria (Kenya)	Hot springs at Old Faithful Hotel (Yellowstone National Park, USA)
Cell size (μm)	0.26–0.3 × 2.4–5	0.3–0.5 × 3–5	0.5–0.65 × 8–22
Gram reaction	Negative at all phases	Negative at all phases	Positive at all phases
Cell wall	Gram-positive type, but thin	Gram-positive type, but thin	Gram-positive type, typical thickness
Growth temperature (optimum) (°C)	45–67 (58)	30–65 (50–55)	30–66 (55–58)
pH range (optimum)	pH[20C] 8.6–10.4 (9.0–9.5)	pH[20C] 6.0–10.5 (9.5)	pH[60C] 6.5–10.3 (8.5)
NaCl range (optimum) (%, w/v)	0–6 (1–2.5)	0–4 (1), at pH 9.0, 50°C	ND
Main metabolism	Proteolytic and glycolytic	Glycolytic and proteolytic	Proteolytic (no carbohydrate utilization)
$Na_2S_2O_4$ as electron acceptor (=> S^{2-})	Yes	Yes	No
Fumarate as electron acceptor (=> succinate)	No	ND	Yes
Cellulose hydrolysis	No	Yes	No
G+C (mol%) content of DNA	30	30	32–34 (various strains)

[a]Data were obtained from Gorlenko et al., (2004) (*Anaerobranca californiensis*), Prowe et al. (2001) (*Anaerobranca gottschalkii*) and Engle et al. (1995) (*Anaerobranca horikoshii*). All three species have rod-shaped, motile, peritrichously flagellated cells, sometimes showing branching. All are obligate anaerobes and can reduce selenite to selenide.
[b]ND, Not determined.

Horikoshi, K. 1991a. Micro-organisms in Alkaline Environments. Weinheim, VCH Verlagsgesellschaft.

Horikoshi, K. 1991b. General view of alkaliphiles and thermophiles. *In* Horikoshi and Grant (Editors), Superbugs: Microorganisms in Extreme Environments. Springer Verlag, Berlin, pp. 3–14.

Kevbrin, V.V., C.S. Romanek and J. Wiegel. 2004. Alkalithermophiles: A double challenge from extreme environments. *In* Seckbach (Editor), Origins: Genesis, Evolution and Diversity of Life. Kluwer Academic Publishers, Dordrecht, pp. 1–16.

Wiegel, J. and V. Kevbrin. 2004. Alkalithermophiles. Biochem. Soc. Trans. *32*: 193–198.

List of species of the genus *Anaerobranca*

1. **Anaerobranca horikoshii** Engle, Li, Woese and Wiegel 1995, 459[VP]

hor.i.kosh'i.i. N.L. gen. n. *horikoshii* of Horikoshi, in honor of Koki Horikoshi, a pioneer in the study of the microbiology of alkaliphilic bacteria.

Description is mainly based on characterization of strain JW/YL-138[T] (from a pH 6.7 and 92°C pool) and JW/YL-268 (from a pH 5.8 and 50°C pool) in Engle et al. (1995).

Cells are usually rod-shaped, 0.5–0.65 μm in diameter and 8–22 μm in length; peritrichously flagellated, but usually showing only tumbling motility. Between about 0.5–10% of the cells in a culture (pending on the strain and culture conditions) form one to three branches without septation. With continued subculturing over the years, the occurrence of branched cells in a culture has decreased. Cells stain Gram-positive during exponential growth phase. Spores have not been observed, although major sporulation specific genes have been detected in the type strain (Onyenwoke et al., 2004). Colonies in agar-shake-roll tubes are lens shaped and of whitish color; surface colonies are more or less circular (entire) and slightly convex.

Habitats: Using casein as growth substrate, *Anaerobranca horikoshii* has only been isolated from the geothermally heated area behind the Old Faithful Hotel in Yellowstone National park, but could not be isolated from other hot springs within Yellowstone National Park or other hot springs in New Zealand, Italy, and Japan. However, strains were isolated from mixed water-sediment samples of hot springs with either slightly alkaline pH (8.7) or slightly acidic pH (5.8) and with water temperature 50–92°C. The nine isolated strains differ slightly in size, marginal growth data, substrate utilization, and the protein band pattern (slight variations) on SDS gels. Minimum pH[25C] for growth for most strains is between 6.7 (no growth) and 7.0 (growth) and the maximum is between pH 9.4 (growth) and 10.3 (no growth), with the pH[25C] optimum for growth around 8.5–8.7. The temperature optimum is around 55–58°C. The minimal growth temperature was between 34°C (for one strain 30°C) (no growth) and 37°C (34°C) (growth) and the upper temperature range between 64°C (growth) and 66°C (no growth). These marginal data classify the bacterium as a facultative alkaliphile and a moderate thermophile. At optimal conditions (57°C, pH[20C] 9.0) the shortest observed doubling times were 36–70 min depending on the strain.

Anaerobranca horikoshii isolates are obligate anaerobes, however, non-growing cells are insensitive to exposure to oxygen (up to 25 h exposure to air tested) at 25 and 60°C. Strain JW/YL-138 and JW/YL-268 (others not tested) cannot utilize carbohydrates (glucose, sucrose, fructose, galactose, maltose, cellobiose, lactose, xylose, ribose, rhamnose, raffinose, arabinose, starch, or pectin), the alcohols mannitol, xylitol, methanol, or ethanol, or the acids acetate, lactate, and formate. Filter-sterilized pyruvate is a weak substrate (in contrast to the original description). Fermentation products from yeast extract are acetate, CO_2, and H_2 in an approximate ratio of 1:1:1. Selenite, thiosulfate, and sulfur can serve as electron acceptors. While growing on yeast extract, fumarate (12 mM) is stoichiometrically reduced to succinate by all except one strain. Interestingly, in the absence of fumarate, 3 mM succinate inhibits growth completely. Dissimilatory sulfate reduction has not been observed.

All strains are peptidolytic (since isolated on casein) and contain metalloproteases (EDTA sensitive) as their main proteases. Peptone and tryptone can substitute for the requirement of yeast extract supplement (0.2% w/v is sufficient). The variation in enzyme activities tested with API-test strips are given in the original species description (Engle et al., 1995). Differences were noted depending on whether or not the cells were incubated aerobically or anaerobically. A pullulanase type I was isolated based on analogies to the gene sequence from *Anaerobranca gottschalkii* genome sequence (Bertoldo et al., 2004). The profile of fatty acid from lipids is shown in Table 226.

DNA G+C content (mol%): 34 for the type strain; 33 for all other strains (HPLC).

Type strain: JW/YL-138, ATCC 700319, DSM9786.

GenBank accession number (16S rRNA gene): U21809.

2. **Anaerobranca californiensis** Gorlenko, Tsapin, Namsaraev, Teal, Tourova, Engler, Mielke and Nealson 2004, 742[VP]

ca.li.for.ni.en'sis. N.L. fem. adj. *californiensis* referring to California, the location of the hot spring from which the microorganism was isolated.

The description is based on Gorlenko et al. (2004).

Cells are rod-shaped, 0.26–0.3 μm in diameter and 2.4–5 μm in length. Branching cells and longer cells are observed within the culture as well as retarded peritrichous flagellation leading to motility in the early to mid exponential growth phase. Division occurs via binary fission. Spores have not been observed. Colony morphology on agar plates incubated in anaerobic jars is circular, entire, convex, and of a whitish color.

The growth ranges for pH[25C] were between 8.6 and 10.4 with an optimum between 9.0 and 9.5. (for a comparison with data from *Anaerobranca horikoshii:* if the pH would be determined at 60°C (pH[60C]), the maximum and optimum pH will be approximately 0.8–1.0 pH unit lower (Wiegel, unpublished)). The temperature range is 45–70°C with an optimum around 58°C. *Anaerobranca californiensis* requires between 0.5% and 6% NaCl for growth with a broad optimum at 47–58°C, while a sharp optimum (2.5% w/v) was observed at 70°C. The shortest observed doubling time is 40 min at 58°C and pH[20C] 9.5.

The metabolism is obligately anaerobic, however, exposure to oxygen does not kill the cells. Exposure to air at 25°C is tolerated for several months without loss of viability. Yeast extract addition to the medium is required for growth and cannot be substituted with vitamin additions. The best growth promoting substrates were yeast extract, peptone, tryptone, malt extract, Casamino acids, and soy tone peptone. In the presence of yeast extract *Anaerobranca californiensis* can use

TABLE 226. Comparison of percentage lipid fatty acid profiles of two *Anaerobranca* species

Fatty acid	A. horikoshii	A. gottschalkii
$C_{13:0}$		2.6.
$C_{13:0\ iso}$		4.3
$C_{14:0}$	26	
$C_{15:0\ iso}$	7.5	42
$C_{15:0\ ante}$		12
$C_{15:0}$	11	
$C_{16:0}$	28	11
$C_{17:0\ ante}$		3.4

fructose, sucrose, maltose, starch, glycogen, cellobiose, pyruvate (filter-sterilized), and glycerol as carbon sources.

During heterotrophic growth, thiosulfate selenite, sulfur, polysulfide (formed at the high pH and elevated temperatures) amorphous Fe(III) hydroxide, and Fe(III) citrate can serve as electron acceptors. Addition of thiosulfate, which is reduced to sulfide, lead to elevated final cell densities. The type strain can tolerate up to 40 mM sulfide. Additional isolates include strain PAOHA-2 (Gorlenko et al., 2004).

DNA G+C content (mol%): 30 (T_m).
Type strain: PAOHA-1, DSM 14826, UNIQEM 227.
GenBank accession number (16S rRNA gene): AY064218.

3. **Anaerobranca gottschalkii** Prowe and Antranikian 2001, 464[VP]

gott.schalk'i.i. N.L. gen. n. *gottschalkii* of Gottschalk, after Gerhard Gottschalk, in recognition of his pioneering contributions to our knowledge of the physiology and metabolism of anaerobes.

Description of the species is based mainly on Prowe and Antranikian (2001).

Cells are rod-shaped, 0.3–0.5 μm in diameter and 3–5 μm long. Early exponential cells are motile. Cells stain Gram-negative, probably because the cell wall is uncharacteristically thin for a *Firmicutes* bacterium (Prowe and Antranikian, 2001; Gorlenko et al., 2004; unpublished results) and thus easily destained, especially when using ethanol in the destaining procedure. Surface colonies on agar plates (0.5% starch) are circular, entire (even), convex, and of a pale whitish color. Truly branched cells make up 2–10% of the total number of cells. Formation of spores was not detected using examination by light microscopy and a heat resistance assay (90°C for up to 10 min).

So far only one isolated strain has been characterized. It was isolated from enrichments inoculated with a sample from a hot lake inlet of Lake Bogoria (Kenya) using liquid enrichment media of pH[20C] 9.0 containing starch and glucose and an incubation temperature of 50°C. Growth range is 30–65°C with an optimum between 50 and 55°C. At growth temperatures of 50°C the pH[20C] range is 6.0–10.5 with optimum at pH[20C] 9.5.

Growth is obligately anaerobic. Utilized substrates include carbohydrates (glucose, fructose, mannose, galactose, ribose, xylose, cellobiose, lactose, maltose, sucrose, starch, and pullulan), proteins and peptides (tryptone, peptone, and yeast extract), and glycerol. The main fermentation product from glucose is acetate with small amounts of ethanol (no fermentation balance available). The addition of thiosulfate stimulates growth on glucose. The addition of thiosulfate or sulfur leads to the formation of sulfide which is inhibited by adding H_2 into the gas phase, indicating that both function as electron acceptor. No indication for dissimilatory sulfate reduction is observed (Gorlenko et al., 2004; Prowe and Antranikian, 2001). Also Fe(III), amorphous Fe(III) hydroxide, Fe(III) citrate, and selenite (Na_2SeO_3) can serve as electron donors forming Fe(II) ion (→ magnetite formation) and elemental selenium (red precipitation), and polyselenite as well as colorless Na_2Se, respectively. Bertoldo et al. (2004) described the characterization of a type I pullulanase (predicted molecular mass 98 kDa) identified from the genome sequence. Pullulan hydrolysis exhibits a pH optimum at pH 8.0 and highest activity at 70°C. Under these conditions the enzyme has a half life of 22 h.

Growth requires presence of Na ion which cannot be replaced by K^+ ions, suggesting that Na^+ ions are required for energy transduction. The uptake of several amino acids exhibits a strict requirement for Na^+, and for L-leucine it was shown that the uptake occurs in symport with Na^+ ions (Prowe et al., 1996). The pH and temperature optima for the leucine transport closely matches the optimal growth conditions for strain LBS3. ATPase activity using inside-out membrane vesicles was stimulated by both Na^+ and Li^+ ions. Consequently, Prowe et al. (1996) concluded that the primary mechanism of energy transduction in strain LBS3 is dependent on sodium cycling.

Antibiotic resistance: 25 μg/ml of chloramphenicol and monensin, 50 μg/ml nalidix acid, and 250 μg/ml streptomycin are inhibitory, but 50 μg/ml of rifampin and 100 μg/ml penicillin are not (see Peteranderl et al., 1990) for heat stability of antibiotics in solution). The profile of fatty acid from lipids is depicted in Table 226.

The complete genome sequence of the type strain is under analysis (Antranikian, private communication)

DNA G+C content (mol%): 30 (HPLC).
Type strain: LBS3, ATCC BAA-51, DSM 13577.
GenBank accession number (16S rRNA gene): AF203703

References

Bertoldo, C., M. Armbrecht, F. Becker, T. Schafer, G. Antranikian and W. Liebl. 2004. Cloning, sequencing, and characterization of thermoalkalistable type I pullulanase from *Anaerobranca gottschalkii*. Appl. Environ. Microbiol. *70*: 3407–3416.

Engle, M., Y.H. Li, C. Woese and J. Wiegel. 1995. Isolation and characterization of a novel alkalitolerant thermophile, *Anaerobranca horikoshii* gen. nov., sp. nov. Int. J. Syst. Bacteriol. *45*: 454–461.

Garrity, G.M., J.A. Bell and T. Lilburn. 2005. The Revised Road Map to the Manual. *In* Brenner, Krieg, Staley and Garrity (Editors), Bergey's Manual of Systematic Bacteriology, 2nd edn, Vol. 2, The *Proteobacteria*, Part A, Introductory Essays. Springer, New York, pp. 159–220.

Gorlenko, V., A. Tsapin, Z. Namsaraev, T. Teal, T. Tourova, D. Engler, R. Mielke and K. Nealson. 2004. *Anaerobranca californiensis* sp. nov., an anaerobic, alkalithermophilic, fermentative bacterium isolated from a hot spring on Mono Lake. Int. J. Syst. Evol. Microbiol. *54*: 739–743.

Krulwich, T.A., A.A. Guffanti and D. Seto-Young. 1990. pH Homeostasis and bioenergetic work in alkalophiles. FEMS Microbiol. Rev. *75*: 271–278.

Onyenwoke, R.U., J.A. Brill, K. Farahi and J. Wiegel. 2004. Sporulation genes in members of the low G+C Gram-type-positive phylogenetic branch (*Firmicutes*). Arch. Microbiol. *182*: 182–192.

Peteranderl, R., E.B. Shotts, Jr. and J. Wiegel. 1990. Stability of antibiotics under growth conditions for thermophilic anaerobes. Appl. Environ. Microbiol. *56*: 1981–1983.

Prowe, S.G., J.L. van de Vossenberg, A.J. Driessen, G. Antranikian and W.N. Konings. 1996. Sodium-coupled energy transduction in the newly isolated thermoalkaliphilic strain LBS3. J. Bacteriol. *178*: 4099–4104.

Prowe, S.G. and G. Antranikian. 2001. *Anaerobranca gottschalkii* sp. nov., a novel thermoalkaliphilic bacterium that grows anaerobically at high pH and temperature. Int. J. Syst. Evol. Microbiol. *51*: 457–465.

Speelmans, G., B. Poolman and W.N. Konings. 1995. Na+ as coupling Ion in energy transduction in extremophilic bacteria and archaea. World J. Microbiol. Biotechnol. *11*: 58–70.

Wiegel, J. and L.G. Ljungdahl. 1981. *Thermoanaerobacter ethanolicus* gen. nov., spec. nov, a new, extreme thermophilic, anaerobic bacterium. Arch. Microbiol. *128*: 343–348.

Wiegel, J. 1998. Anaerobic alkalithermophiles, a novel group of extremophiles. Extremophiles *2*: 257–267.

Family XV. **Incertae Sedis**

Previously assigned to the *Syntrophomonadaceae* by Garrity et al. (2005), subsequent analyses suggest that these genera, while closely related to each other, are not closely related to *Syntroph-* *omonas* or a member of any other previously described family. For that reason, they are assigned to their own family *incertae sedis* (Figure 6).

Genus I. **Aminobacterium** Baena, Fardeau, Labat, Ollivier, Thomas, Garcia and Patel 1999c, 1325[VP] (Effective publication: Baena, Fardeau, Labat, Ollivier, Thomas, Garcia and Patel 1998, 249.)

Sandra Baena, Jean-Louis Garcia, Jean-Luc Cayol and Bernard Ollivier

A.min.o.bac′te.ri.um. N.L. n. *aminum* amine; Gr. dim. n. *bakterion* a small rod; N.L. neut. n. *Aminobacterium* the amino acid rod.

Cells are strictly anaerobic, nonsporeforming, nonmotile or motile by means of one or two lateral flagella, slightly curved **rods**, occurring singly, in pairs, or, rarely, as chains. **Gram-stain-negative**. Mesophilic, neutrophilic. **Heterotrophic**, asaccharolytic. Growth by fermentation of a limited range of amino acids only in the presence of yeast extract.

DNA G+C content (mol%): 44–46.

Type species: **Aminobacterium colombiense** Baena, Fardeau, Labat, Ollivier, Thomas, Garcia and Patel 1999c, 1325[VP] (Effective publication: Baena, Fardeau, Labat, Ollivier, Thomas, Garcia and Patel 1998, 249.).

Further descriptive information

Aminobacterium cells are 0.2–0.3 μm in width × 3–5 μm in length and occur singly, in pairs or, rarely, as chains (Figure 228) when grown on a medium containing serine and yeast extract. Spores are not observed from cells grown under various conditions. Ultrathin sections of cells reveal a multilayered, complex, thick cell wall with an external S-layer similar to that of Gram-positive type cell walls. They are strictly anaerobic, chemo-organotrophic bacteria. Cells grow at temperatures ranging from 20–42°C, with an optimum at 37°C. No growth is observed at 18 and 45°C. Cells do not require NaCl for growth, but tolerate less than 1.5% NaCl, with optimal growth occurring in the presence of 0.05–0.50% NaCl. The optimum pH for growth is around 7.3 and growth occurs between pH 6.6–8.5.

Aminobacterium cells do not grow in basal medium without yeast extract, and 0.2% yeast extract was routinely used for substrate utilization tests. Cells ferment serine, glycine, threonine, and pyruvate. Serine is fermented to acetate or to acetate and alanine. Threonine and glycine are degraded to acetate as the only fatty acid produced. Acetate and H_2 are the end products of metabolism from Casamino acids, peptone, biotrypcase, and cysteine, but these compounds are poorly used. Propionate is also produced in the case of α-ketoglutarate and 2-oxoglutarate. *Aminobacterium* cells do not perform the Stickland reaction when alanine is provided as an electron donnor and glycine, serine, arginine, or proline are provided as electron acceptors. Carbohydrates, gelatin, casein, glycerol, ethanol, acetate, propionate, butyrate, lactate, citrate, fumarate, malate, and succinate are not utilized. Sulfate, thiosulfate, elemental sulfur, sulfite, nitrate, and fumarate are not utilized as electron acceptors.

Alanine, glutamate, valine, isoleucine, leucine, and aspartate are oxidized only by a mixed culture of *Aminobacterium* species and *Methanobacterium formicicum*. The end products resulting from the oxidation of alanine, leucine, isoleucine, and valine were acetate, isobutyrate, 2-methyl-butyrate, and isovalerate, respectively.

Enrichment and isolation procedure

The Hungate technique (Hungate, 1969) should be used to isolate *Aminobacterium* species. The basal medium (BM) contains (per liter of distilled water): 0.3 g NH_4Cl, 0.2 g KH_2PO_4, 0.4 g $MgCl_2·6H_2O$, 0.15 g $CaCl_2·2H_2O$, 0.5 g KCl, 1.0 g NaCl, 1 ml 0.1% resazurin, and 1 ml trace mineral element solution (Imhoff-Stuckle and Pfennig, 1983). The pH is adjusted to 7.2 with 10 M KOH. The medium is boiled under a stream of O_2-free N_2 gas and cooled to room temperature. Five ml aliquots are dispensed into Hungate tubes and 20 ml in serum bottles under a stream of O_2-free N_2 gas. The gas phase is subsequently replaced with N_2/CO_2 (80:20, v/v) gas and the sealed vessels are autoclaved for 45 min at 110°C. Prior to inoculation, 0.15 ml 2% $Na_2S·9H_2O$, 0.25 ml 10% $NaHCO_3$, and 0.05 ml Balch vitamin solution (Balch et al., 1979) are injected from sterile stock solutions. Amino acids are added from sterile (heat or filter-sterilized) anaerobic stock solutions.

For isolation, a tenfold serial dilution is prepared and inoculated into the Hungate tubes containing 5 ml basal medium with serine and yeast extract at final concentration of 10 mM and 0.2%, respectively. The highest dilution showing growth after 3–4 weeks incubation at 37°C is used for further isolation. Positive cultures are serially diluted and inoculated into roll tubes containing enrichment medium with 2% agar to get axenic cultures. Colonies (small, white, lens shaped) develop after 2 weeks incubation at 37°C. The process of serial dilution in roll tubes is repeated at least twice in order to purify the cultures.

Maintenance procedure

Stock cultures can be maintained on medium described by Baena et al. (1998) by monthly transfers. Liquid cultures retained viability after several weeks storage at 4°C, or when lyophilized, or after storage at −80°C in the basal medium containing 20% glycerol (v/v). Viability is best maintained from mid-exponential phase cultures.

Differentiation of the genus *Aminobacterium* from other genera

The genus *Aminobacterium* is represented by two species, *Aminobacterium colombiense* and *Aminobacterium mobile*. They are mesophilic and strictly anaerobic bacteria which ferment serine, threonine, glycine, and pyruvate but not carbohydrates. These characteristics are in common with the nonsaccharolytic

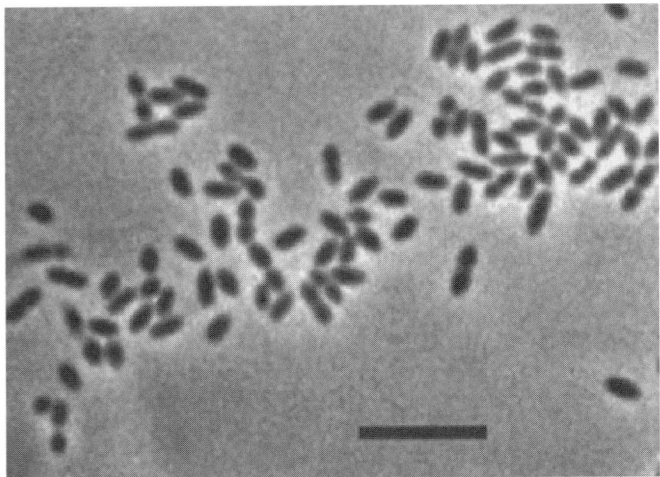

FIGURE 228. Photomicrograph of *Aminobacterium colombiense*. Bar = 5 μm.

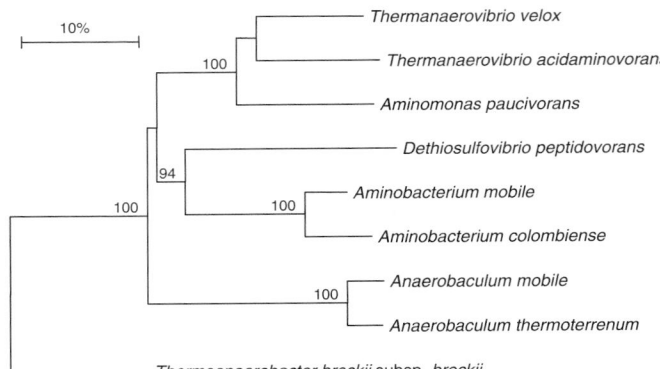

FIGURE 229. Phylogenetic tree based on 16S rRNA gene sequence comparison, and obtained by a neighbor-joining algorithm (PHYLIP package), indicating the position of *Aminobacterium* species and *Aminomonas paucivorans* among members of the low G+C Gram-positive bacteria (*Firmicutes*). Bootstrap values are shown at branching points. Only values above 80% were considered significant and reported. *Thermoanaerobacter brockii* subsp. *brockii* was used as outgroup. Bar = 10 substitutions per 100 nucleotides.

aminolytic species *Acidaminobacter hydrogenoformans* (Stams and Hansen, 1984), *Eubacterium acidaminophilum* (Zindel et al., 1988), *Thermanaerovibrio acidaminovorans* (Baena et al., 1999b) (formerly *Selenomonas acidaminovorans*; Guangsheng et al. (1992), *Peptostreptococcus anaerobius* (Paster et al., 1993), and several species of the genus *Clostridium* including *Clostridium sticklandii*, *Clostridium aminophilum*, *Clostridium morale*, *Clostridium pascui*, *Clostridium hydrobenzoicum*, and *Clostridium acetireducens*. Phylogenetically, the nonsaccharolytic aminolytic species are members of the low G+C containing Gram-positive clostridial branch which is currently comprised of 19 defined clusters and several distinct independent lineages (Collins et al., 1994). *Aminobacterium* species are also members of the low G+C containing Gram-positive clostridial branch and are therefore related to them, but only distantly. Further phylogenetic analysis revealed that *Aminobacterium* species form an independent line of descent together with *Aminomonas paucivorans* (Baena et al., 1999a), *Dethiosulfovibrio peptidovorans* (Magot et al., 1997), *Dictyoglomus thermophilum* (Saiki et al., 1985), *Anaerobaculum* species (Menes and Muxí, 2002; Rees et al., 1997), and *Thermanaerovibrio* species (Baena et al., 1999b; Zavarzina et al., 2000) (Figure 229) in the vicinity of cluster V (Collins et al., 1994) which currently consists of members of the genus *Thermoanaerobacter*.

Like *Aminobacterium* species, *Aminomonas paucivorans*, *Thermanaerovibrio* species, *Anaerobaculum* species, and *Dethiosulfovibrio peptidovorans* use amino acids. However, *Thermanaerovibrio*, and *Anaerobaculum* species are thermophiles which also utilize carbohydrates. Furthermore, *Aminobacterium* species differ from *Anaerobaculum thermoterrenum* and *Anaerobaculum mobile* as both microorganisms ferment several organic acids (Menes and Muxí, 2002; Rees et al., 1997). In contrast to *Dethiosulfovibrio peptidovorans*, *Aminomonas paucivorans* uses thiosulfate or elemental sulfur as terminal electron acceptors. Finally, *Amin-*

omonas paucivorans does not use pyruvate and the same range of amino acids as reported for *Aminobacterium* species.

It is noteworthy that one species of the genus *Aminobacterium*, *Anaerobaculum mobile* ferments serine into acetate and alanine (Baena et al., 2000). Alanine production from amino acid fermentation has rarely been reported. *Caloramator proteoclasticus* (Tarlera et al., 1997) was shown to produce this amino acid during glutamate fermentation, whereas *Clostridium ultunense* (Schnurer et al., 1996) produced it during cysteine degradation.

Threonine is fermented to acetate similarly to that described for some *Clostridium* species which include *Clostridium histolyticum*, *Clostridium sporospheroides*, *Clostridium sticklandii*, and *Clostridium subterminale* (Barker, 1981). However, most members of this genus produce propionate and n-butyrate and a few also produce 2-aminobutyrate. The conversion of glycine to acetate by *Aminobacterium* species suggests a pattern similar to that reported for *Eubacterium acidaminophilum* (Zindel et al., 1988), *Peptococcus anaerobius*, *Peptococcus magnus*, *Clostridium histolyticum*, and *Clostridium purinolyticum* (McInerney, 1988). *Aminobacterium* cells ferment amino acids such as serine, threonine, and glycine in pure cultures and is also able to oxidize additional amino acids such as alanine, valine, leucine, and isoleucine only in the presence of *Methanobacterium formicicum*, a well-known H$_2$ scavenger.

Alanine is oxidized to acetate, valine to isobutyrate, leucine to isovalerate, isoleucine to 2-methylbutyrate. Metabolism of these amino acids in mixed culture resembles that reported for the mesophiles *Acidaminobacter hydrogenoformans* (Stams and Hansen, 1984), *Thermanaerovibrio acidaminovorans* (Guangsheng et al., 1992), and *Eubacterium acidaminophilum* (Zindel et al., 1988), and also the thermophile *Thermoanaerobacter brockii* (Fardeau et al., 1997).

Differentiation between *Aminobacterium* species

The genus *Aminobacterium* comprises two species, *Aminobacterium mobile* and *Aminobacterium colombiense*. The former, unlike *Aminobacterium colombiense*, has a slightly lower DNA G + C content (44 mol% vs 46 mol%), is motile, and ferments serine to acetate and alanine.

Taxonomic comments

Aminobacterium is a genus consisting of two species (*Aminobacterium colombiense* and *Aminobacterium mobile*). Phenotypic and genotypic characteristics clearly place it within a cluster of low G+C Gram-positive bacteria (designated *Clostridiales* Family XV *Incertae Sedis* in this volume) in the phylum *Firmicutes* containing *Thermanaerovibrio* species, *Aminobacterium* species, *Anaerobaculum* species, *Aminomonas paucivorans*, and *Dethiosulfovibrio peptidovorans*. This cluster is adjacent to *Thermoanaerobacter* (cluster V) based on 16S rRNA sequence analysis. For phylogenetic position see Figure 229.

List of species of the genus *Aminobacterium*

1. **Aminobacterium colombiense** Baena, Fardeau, Labat, Ollivier, Thomas, Garcia and Patel 1999c, 1325^{VP} (Effective publication: Baena, Fardeau, Labat, Ollivier, Thomas, Garcia and Patel 1998, 249.)

co.lom.bi.en'se. N.L. neut. adj. *colombiense* pertaining to Columbia, the origin of the isolate.

In addition to the properties listed in the genus description, the following properties are reported. Cells are 0.2–0.3 × 3–4 μm. Nonmotile. Colonies (up to 1.0 mm) are round, smooth, lens shaped and white. Grows at temperatures ranging from 20–42°C, with optimum growth at 37°C. Grows in the presence of NaCl at concentrations ranging from 0–1.5% NaCl, with optimum growth at 0.0–0.50% NaCl. The optimum pH for growth is 7.3 and growth occurs between pH 6.6–8.5. Heterotrophic, asaccharolytic. Yeast extract is required for growth. It ferments serine, threonine, glycine, and pyruvate and uses several other amino acids in mixed culture with *Methanobacterium formicicum* (alanine, glutamate, valine, leucine, isoleucine, aspartate, cysteine, and methionine). No growth is observed on succinate, malate, fumarate, citrate, lactate, glucose, sucrose, ribose, xylose, cellobiose, melobiose, maltose, galactose, mannose, arabinose, rhamnose, lactose, sorbose, mannitol, acetate, propionate, butyrate, glycerol, ethanol, gelatin, and casein. Sulfate, thiosulfate, elemental sulfur, sulfite, nitrate, and fumarate are not utilized as electron acceptors.

Isolated from anaerobic sludge of a dairy wastewater treatment plant in SantaFe de Bogota, Colombia. Adverse effects on animals and humans are not known. Because of the ability of *Aminobacterium colombiense* to degrade amino acids and peptides, the possibility of harmful effects cannot be excluded. Cautious handling and autoclaving of cultures before disposal is recommended.

DNA G+C content (mol%): 46 (HPLC).

Type strain: ALA-1, DSM 12261.
Genbank accession number (16S rRNA gene): AF069287.

2. **Aminobacterium mobile** Baena, Fardeau, Labat, Ollivier, Garcia and Patel 2000, 263^{VP}.

mo'bi.le. L. neut. adj. *mobile* motile.

In addition to the properties listed in the genus description, the following properties are reported. Cells are 0.2–0.3 × 4–5 μm. Cells are motile by means of one or two lateral flagella. Colonies (up to 1.0 mm) are round, smooth, lens shaped and white. Grows at temperatures ranging from 35–42°C, with optimum growth at 37°C. Grows in the presence of NaCl at concentrations ranging from 0–1.5% NaCl, with optimum growth between 0.05–0.50% NaCl. The optimum pH for growth is 7.4 and growth occurs between pH 6.7–8.3. Heterotrophic, asaccharolytic. Yeast extract is required for growth. It ferments serine, threonine, glycine, and pyruvate and uses alanine, glutamate, valine, leucine, isoleucine, and aspartate only in co-culture with *Methanobacterium formicicum*. No growth is observed on succinate, malate, fumarate, citrate, lactate, glucose, saccharose, ribose, xylose, cellobiose, melobiose, maltose, galactose, mannose, arabinose, rhamnose, lactose, sorbose, mannitol, acetate, propionate, butyrate, glycerol, ethanol, gelatin, and casein. Sulfate, thiosulfate, elemental sulfur, sulfite, nitrate, and fumarate are not utilized as electron acceptors. Isolated from anaerobic sludge of a dairy wastewater treatment plant in SantaFe de Bogota, Colombia. Adverse effects on animals and humans are not known. Because of the ability of *Aminobacterium mobile* to degrade amino acids and peptides, the possibility of harmful effects cannot be excluded. Cautious handling and autoclaving of cultures before disposal is recommended.

DNA G+C content (mol%): 44 (HPLC).
Type strain: ILE-3, ATCC BAA-7, DSM 12262.
GenBank accession number (16S rRNA gene): AF073521.

Genus II. **Aminomonas** Baena, Fardeau, Ollivier, Labat, Thomas, Garcia and Patel 1999a, 981^{VP}

Sandra Baena, Jean-Louis Garcia, Jean-Luc Cayol and Bernard Ollivier

A.mi.no.mo'nas. N.L. neut. n. *aminum* amine; Gr. n. *monas* a unit, monad; N.L. fem. n. *Aminomonas* amine-degrading monads.

Cells are strictly anaerobic, nonsporeforming, nonmotile, slightly curved rods, occurring singly or in pairs. **Gram-stain-negative**. Mesophilic, neutrophilic. Optimum growth at 35°C and pH 7.5 on arginine with a generation time of 16 h. **Heterotrophic**. Good growth on arginine, histidine, threonine, and glycine. Acetate is the end product formed from all these substrates; in addition, trace amount of formate is detected from arginine and histidine. Ornithine is produced from arginine. Slow growth on glutamate with production of acetate, carbon dioxide, formate, hydrogen, and traces of propionate as the end products.

In syntrophic association with *Methanobacterium formicicum*, methane is produced from arginine, histidine, and glutamate which are oxidized into propionate as the major product together with acetate and carbon dioxide. Alanine and the branched-chain amino acids valine, leucine, and isoleucine are not degraded either in pure culture or in association with *Methanobacterium formicicum*.

DNA G+C content (mol%): 43.

Type species: **Aminomonas paucivorans** Baena, Fardeau, Ollivier, Labat, Thomas, Garcia and Patel 1999a, 981[VP].

Further descriptive information

Aminomonas paucivorans is the only species described so far within the genus *Aminomonas*. Cells are 0.3–0.4 μm in width and 4–6 μm in length and occur singly or in pairs when grown on a medium containing arginine and yeast extract. Motility is not observed and electron microscopy of negatively stained cells reveals the absence of flagella. Spores are not observed from cells grown under various conditions and cells are not heat resistant. Ultrathin sections of cells reveal a cytoplasmic membrane and a complex cell wall layer characteristic of a Gram-negative cell. *Aminomonas paucivorans* is a strictly anaerobic, chemoorganotrophic bacterium. Cells grow at temperatures ranging from 20–40°C, with an optimum at 35°C. No growth is observed at 18 and 42°C. Cells do not require NaCl for growth, but tolerate up to 2.0% NaCl, with optimal growth occurring in the presence of 0.05–0.50% NaCl. The optimum pH for growth is around 7.5, and growth occurs between pH 6.7–8.3.

Aminomonas paucivorans does not grow in basal medium without yeast extract, and 0.2% yeast extract was routinely used for substrate utilization tests. Cells ferment arginine, histidine, glutamate, threonine, and glycine (Table 227). Arginine is fermented to ornithine, acetate, and formate; histidine is fermented to acetate and formate; glutamate is fermented to acetate and, to a minor extent, propionate and formate. Threonine and glycine are degraded to acetate. Acetate is also the major end product of metabolism from Casamino acids, peptone, and cysteine, but they are poorly used. As 0.2% yeast extract has been used to test for the utilization of amino acids, the possibility that degradation occurs via the Stickland reaction cannot be ruled out. Sugars are not used. Sulfate, thiosulfate, elemental sulfur, sulfite, nitrate, and fumarate are not utilized as electron acceptors.

A mixed culture of *Aminomonas paucivorans* and *Methanobacterium formicicum* is unable to extend the range of utilizable substrates. However, compared to pure cultures, association of *Aminomonas* with *Methanobacterium formicicum* results in a shift in end product formation during degradation of arginine, histidine, and glutamate with methane and propionate as the major fermentation products (Table 227). In contrast to the pure culture, ornithine does not accumulate during arginine degradation by the mixed culture. Methane is not detectable in mixed culture with glycine and threonine, and the same end product profile as for a pure culture is observed (data not shown).

Cultivation of *Aminomonas paucivorans* under a H$_2$/CO$_2$, (80:20) atmosphere does not affect the fermentation of arginine, histidine, glutamate, threonine, or glycine; the end products are similar to those found without a hydrogen atmosphere.

Enrichment and isolation procedure

The Hungate technique (Hungate, 1969) should be used to isolate *Aminomonas paucivorans* and relatives. The basal medium (BM) contains (per liter of distilled water): 0.3 g NH$_4$Cl, 0.2 g KH$_2$PO$_4$, 0.4 g MgCl$_2$·6H$_2$O, 0.15 g CaCl$_2$·2H$_2$O, 0.5 g KCl, 1.0 g NaCl, 1 ml 0.1% resazurin, and 1 ml trace mineral element solution (Imhoff-Stuckle and Pfennig, 1983). The pH is adjusted to 7.2 with 10 M KOH. The medium is boiled under a stream of O$_2$-free N$_2$ gas and cooled to room temperature. Five ml aliquots are dispensed into Hungate tubes and 20 ml in serum bottles, under a stream of O$_2$-free N$_2$ gas. The gas phase is subsequently replaced with N$_2$-CO$_2$ (80:20, v/v) gas and the sealed vessels are autoclaved for 45 min at 110°C. Prior to inoculation, 0.15 ml 2% Na$_2$S·9H$_2$O, 0.25 ml 10% NaHCO$_3$, and 0.05 ml Balch vitamin solution (Balch et al., 1979) are injected from sterile stock solutions. Amino acids are added from sterile (heat or filter-sterilized) anaerobic stock solutions.

For isolation, a tenfold serial dilution is prepared and inoculated into the Hungate tubes containing 5 ml basal medium with arginine and yeast extract at final concentrations of 10 mM and 0.2%, respectively. The highest dilution showing growth after 3–4 weeks incubation at 37°C is used for further isolation. This positive culture is serially diluted and inoculated into roll tubes containing basal medium, yeast extract (0.2%), arginine (10 mM), and agar (2%) to get axenic cultures. Only one type of colony (small, round, whitish with smooth edges) develops after 4 weeks incubation at 37°C. The process of serial dilution in roll tubes is repeated at least twice in order to purify the cultures.

TABLE 227. Fermentation of substrates that supported growth of *Aminomonas* in pure culture[a]

| Substrate[c] | Amino acid degraded (mM) | Products formed (mM)[b] | | | | | |
		Acetate	Propionate	Formate	Ornithine	H$_2$	ΔOD$_{580}$
Arginine	6.1	1.0	0.0	2.5	4.7	+	0.240
Histidine	ND	9.2	0.0	4.8	0.0	ND	0.220
Glutamate	7.7	7.5	2.0	4.0	0.0	+	0.080
Threonine	ND	9.4	0.0	0.0	0.0	ND	0.110
Glycine	3.0	4.5	0.0	0.0	0.0	ND	0.082

[a]Symbols: ND, not determined. Results were recorded after 3 weeks incubation at 37°C. The basal medium contained 0.2% yeast extract. (Reprinted with publisher's permission from Baena et al. 1999a. Int. J. Syst. Bacteriol. *49*: 975–982.)

[b]Tubes containing basal medium with 0.2% yeast extract but lacking substrates were used as control. All values were corrected for a small amount of acetate (2 mM) formed in the control tubes.

[c]Poor growth with Casamino acids, peptone, and cysteine was observed, and acetate levels above those of the control were present.

Maintenance procedure

Stock cultures can be maintained on medium described by Baena et al. (1999a) by monthly transfers. Liquid cultures retain viability after several weeks storage at 4°C, or when lyophilized, or after storage at −80°C in the basal medium containing 20% glycerol (v/v). Viability is best maintained from mid-exponential phase cultures.

Differentiation of the genus *Aminomonas* from other genera

Aminomonas represented by only one species, *Aminomonas paucivorans* is a Gram-stain-negative, slightly curved, nonsporeforming, obligate amino acid-degrading anaerobe, and therefore cannot be assigned to the aminolytic members of the genus *Clostridium* which include *Clostridium sticklandii* and *Clostridium aminophilum* (Paster et al., 1993), *Clostridium litorale*, *Clostridium pascui*, and *Clostridium hydroxybenzoicum* (Fendrich et al., 1990; Wilde et al., 1997; Zhang et al., 1994), and *Clostridium acetireducens* (Örlygsson et al., 1996). These species are Gram-stain-positive, sporeforming, straight rods. This conclusion is confirmed by phylogenetic analysis which places *Aminomonas paucivorans* in the *Firmicutes* (low G+C Gram-positives) in the vicinity of *Thermanaerovibrio acidaminovorans* (Baena et al., 1999b) (formerly *Selenomonas acidaminovorans*; Guangsheng et al. (1992), *Thermanaerovibrio velox* (Zavarzina et al., 2000), *Aminobacterium* species (Baena et al., 2000, 1998), *Anaerobaculum species* (Menes and Muxí, 2002; Rees et al., 1997), and *Dethiosulfovibrio peptidovorans* (Magot et al., 1997), with *Thermanaerovibrio acidaminovorans* and *Thermanaerovibrio velox* being its closest relatives (Figure 229). Like *Aminomonas paucivorans*, *Thermanaerovibrio species*, *Anaerobaculum* species, and *Dethiosulfovibrio peptidovorans* use amino acids. However, *Thermanaerovibrio* and *Anaerobaculum* species are thermophiles which also utilize carbohydrates. Furthermore, *Aminomonas paucivorans* differs from *Anaerobaculum thermoterrenum* and *Aminobacterium mobile* as both microorganisms ferment several organic acids (Menes and Muxí, 2002; Rees et al., 1997). In contrast to *Dethiosulfovibrio peptidovorans*, *Aminomonas paucivorans* uses thiosulfate or elemental sulfur as terminal electron acceptors.

Aminomonas paucivorans ferments histidine to acetate and formate. This fermentative pathway is similar to that used by anaerobic bacteria as suggested by McSweeney et al. (1993) and may also be operating in *Aminomonas paucivorans*. The latter degrades arginine to ornithine and this characteristic is similar to that reported for *Selenomonas acidaminovorans* and *Synergistes jonesii* (McSweeney et al., 1993). This suggests that the arginine deaminase pathway may be operational in *Aminomonas paucivorans*.

In pure culture, *Aminomonas paucivorans* ferments glutamate to acetate, formate, and trace amounts of propionate, a property that closely resembles *Acidaminobacter hydrogenoformans* in pure culture (Stams and Hansen, 1984). This trait differentiates it from other amino acid-degrading members of the genera *Acidaminococcus*, *Peptostreptococcus*, *Fusobacterium*, and *Clostridium* (Barker, 1981; Rogosa, 1969; Wilde et al., 1997), which produce acetate, butyrate, carbon dioxide, ammonium, and hydrogen. In contrast to *Aminomonas paucivorans*, *Thermanaerovibrio acidaminovorans* (Guangsheng et al., 1992) and *Anaeromusa*

acidaminophila (Baena et al., 2000) (formerly *Selenomonas acidaminophila*; Nanninga et al., 1987) produce propionate as a major end product from glutamate fermentation. Acetate is the only fatty acid produced from threonine and glycine fermentation by *Aminomonas paucivorans*. In the case of threonine degradation, this end product is unusual as most species that degrade threonine, with the exception of a few examples such as *Clostridium sticklandii* and *Clostridium subterminale* (Barker, 1981), are known to produce acetate and propionate.

An increase in the range of amino acids utilized is not observed in mixed culture of *Aminomonas paucivorans* with *Methanobacterium formicicum*, in contrast to the situation observed for *Acidaminobacter hydrogenoformans* (Stams and Hansen, 1984), *Thermanaerovibrio acidaminovorans*, and *Eubacterium acidaminophilum* (Zindel et al., 1988). However, the metabolism of all the amino acids used by *Aminomonas paucivorans*, except threonine and glycine, is clearly influenced by the presence of the hydrogenotrophic methanogen. No increase in acetate production, no change in the end product profile, and the lack of methane production in the presence of the hydrogen scavenger suggest that threonine and glycine degradation occurs via a reductive rather than an oxidative process. In addition, a hydrogen gas phase did not alter the growth of *Aminomonas paucivorans* on these two substrates, further strengthening this hypothesis.

Co-culture of *Aminomonas paucivorans* with *Methanobacterium formicicum* shifted the metabolic end products of arginine, histidine, and glutamate degradation from acetate to propionate as the major end product. This trait is similar to that of *Acidaminobacter hydrogenoformans* when it is co-cultured on glutamate and histidine with a hydrogen scavenger (Stams and Hansen, 1984). This may occur because the partial pressure of H_2 is reduced due to consumption by the hydrogen scavenger, with propionate production becoming thermodynamically more favorable (McInerney, 1988; Plugge et al., 2001).

In spite of observed similarities between *Aminomonas paucivorans* and *Acidaminobacter hydrogenoformans* with respect to glutamate metabolism, the latter is inhibited by a hydrogen atmosphere when grown on this amino acid. In addition, there are marked differences in the amino acids utilized. For example, *Acidaminobacter hydrogenoformans* utilized arginine and threonine only in mixed culture with hydrogenotrophic bacteria (Stams and Hansen, 1984), whereas *Aminomonas paucivorans* could ferment both of these amino acids in pure culture. In addition, unlike *Acidaminobacter hydrogenoformans*, *Aminomonas paucivorans* does not utilize alanine, valine, leucine, or isoleucine in the presence of hydrogen scavenging bacteria.

Taxonomic comments

The genus *Aminomonas* consists of only one species (*Aminomonas paucivorans*). Phenotypic and genotypic characteristics clearly place it within a cluster (designated *Clostridiales* Family XV *Incertae Sedis* in this volume) in the phylum *Firmicutes* containing *Thermanaerovibrio* species, *Aminobacterium* species, *Anaerobaculum* species, and *Dethiosulfovibrio peptidovorans*. For phylogenetic position see Figure 229. Further isolation and characterization of more species which are representative of this genus is required to confirm the classification of this genus.

List of species of the genus *Aminomonas*

1. **Aminomonas paucivorans** Baena, Fardeau, Ollivier, Labat, Thomas, Garcia and Patel 1999a, 981[VP].

pau′ci.vor′ans. L. adj. *paucus* few, little; L. pres. part. *vorans* devouring, digesting; N.L. part. adj. *paucivorans* digesting little.

In addition to the properties listed in the genus description, the following properties are reported. Cells are 0.3 × 4–6 μm. Colonies (up to 1.0 mm) are round, smooth, and white. Grows at temperatures ranging from 20–40°C, with optimum growth at 35°C. Grows in the presence of NaCl at concentrations ranging from 0–2% NaCl, with optimum growth at 0.05–0.50% NaCl. The optimum pH for growth is 7.5 and growth occurs at pH 6.7–8.3. Heterotrophic, asaccharolytic. Yeast extract is required for growth. No growth is observed on carbohydrates, gelatin, casein, pyruvate, succinate, malate, fumarate, α-ketoglutarate, mesaconate, β-methylaspartate, oxaloacetate, glycerol, ethanol, acetate, propionate, butyrate, lactate, citrate, leucine, lysine, alanine, valine, proline, serine, methionine, asparagine, phenylalanine, and aspartate. Sulfate, thiosulfate, elemental sulfur, sulfite, nitrate, and fumarate are not utilized as electron acceptors. Isolated from anaerobic sludge of a dairy wastewater treatment plant in SantaFe de Bogota, Colombia. Adverse effects on animals and humans are not known. Because of the ability of *Aminomonas paucivorans* to degrade amino acids and peptides, the possibility of harmful effects cannot be excluded. Cautious handling and autoclaving of cultures before disposal is recommended.

DNA G+C content (mol%): 43 (HPLC).

Type strain: GLU-3, ATCC BAA-6, DSM 12260.

GenBank accession number (16S rRNA gene): AF072581.

Genus III. **Anaerobaculum** Rees, Patel, Grassia and Sheehy 1997, 153[VP] emend. Menes and Muxí 2002, 163

BHARAT K. C. PATEL AND PHILIP HUGENHOLTZ

An.ae.ro.ba′cu.lum. Gr. pref. *an* not; Gr. n. *aer* air; L. neut. n. *baculum* small stick; N.L. neut. n. *Anaerobaculum* rod which grows in the absence of air.

Straight to slightly curved rods. Occur singly or in pairs. Moderately thermophilic, chemorganotrophic anaerobes. May be motile by a single polar flagellum. In complex media, cells may or may not grow with sheathlike material that extends past the cell poles. Endospores have not been observed. Cells stain Gram-negative. Ferment organic acids, protein extracts, and a limited range of carbohydrates. Sulfur, thiosulfate, and cystine are reduced to hydrogen sulfide. Sulfate, sulfite, and nitrate are not reduced. Isolated from production waters of a petroleum reservoir and an anaerobic lagoon treating wool-scouring wastewater.

DNA G+C content (mol%): 44–51.5.

Type species: **Anaerobaculum thermoterrenum** Rees, Patel, Grassia and Sheehy 1997, 153[VP].

Further descriptive information

The two validly published species, *Anaerobaculum thermoterrenum* and *Aminobacterium mobile*, ferment tryptone, Casamino acids, yeast extract, starch, malate, tartrate, pyruvate, glycerol, glucose, and fructose but not lactose, xylose, cellulose, CM-cellulose, gelatin, maltose, sucrose, galactose, rhamnose, raffinose, gum arabic, malonate, lactate, or succinate. Both species reduce thiosulfate, sulfur, and cystine. Both species utilize tryptone and yeast extract as electron donors with thiosulfate, sulfur, and cystine as electron acceptors, which are reduced to sulfide. Butyrate is produced from crotonate reduction by both species, but *Aminobacterium mobile* is more efficient at crotonate reduction than is *Anaerobaculum thermoterrenum*.

Enrichment and isolation procedures

Anaerobaculum thermoterrenum can be enriched from production waters of petroleum reservoirs using a citrate-based, sulfate-free, brackish, bicarbonate-buffered, sulfide-reduced medium (Widdel and Bak, 1992) that is fortified with yeast extract, vitamins, and trace elements (Rees et al., 1997). Isolation is achieved by inoculating enrichment medium, solidified with agar (2%), in an anaerobic chamber followed by the transfer of the plates into an anaerobic jar and incubation at 50°C until colonies develop, usually 3 weeks later. Single colonies that develop are picked and grown on a complex MMB medium (g/l) containing 0.1 g of Na_2SO_4, 0.5 g of NH_4Cl, 0.3 g of K_2HPO_4, 0.3 g of KH_2PO_4, 0.2 g of $MgCl_2·6H_2O$, 0.2 g of $CaCl_2·2H_2O$, 1 g of NaCl, 0.1 g of KCl, 0.8 g of sodium acetate, 0.5 g of cysteine-HCl, 5 g of yeast extract, 5 g of tryptone, 0.001 g of resazurin, and 1 ml of SL-10 trace element solution (Imhoff-Stuckle and Pfennig, 1983). The pH of the medium is adjusted to 7.0, the medium rendered anaerobic, and it is dispensed under oxygen-free nitrogen gas by using the method of Patel et al. (1985a, 1985b).

The isolation source for *Aminobacterium mobile* was a mixed microbial consortium that had been previously enriched in an oleic-acid containing medium inoculated with a wool-scouring wastewater treatment lagoon sludge sample. This mixed microbial consortium was then used to inoculate BCYT-crotonate medium, and incubated at 55°C to enrich for a crotonate-degrading consortium (Menes et al., 2001; Touzel and Albagnac, 1983). BCYT contains (g/l) 1 g of NH_4Cl, 0.6 g of K_2HPO_4, 0.3 g of KH_2PO_4, 0.1 g of $MgCl_2·6H_2O$, 0.08 g of $CaCl_2·2H_2O$, 0.08 g of NaCl, 20 ml of cysteine-HCl (1.25% w/v), 10 g of yeast extract, 5 g of tryptone, 0.002 g of resazurin, 3.6 g of $KHCO_3$, 10 ml of trace element solution, 11.92 g of HEPES, and crotonate (15 mM). The medium is prepared anaerobically and dispensed into 10-ml anaerobic tubes under oxygen-free nitrogen gas. The gas phase is then replaced with N_2:CO_2 (70:30, v:v), the pH adjusted to pH 7.0, and it is autoclaved for 15 min at 121°C. Prior to inoculation, 0.01 ml vitamin solution and 0.2 ml filter-sterilized sulfide-cysteine solution [1.25% (w/v) $Na_2S·9H_2O$, 1.25% (w/v) cysteine-HCl] are injected into the tube. A stable enrichment culture that developed after several subcultures at 55°C was serially diluted in BCYT-crotonate tubes containing 2% agar to isolate *Aminobacterium mobile* (Menes and Muxí, 2002).

Differentiation of the genus *Anaerobaculum* from other genera

Members of the genus *Anaerobaculum* are straight to slightly curved rods that grow under anaerobic conditions and ferment amino acids. These are physiological traits that are shared by members of the phylogenetically related genera *Aminomonas*, *Aminobacterium*, *Dethiosulfovibrio*, *Thermoanaerovibrio*, and *Synergistes*. A number of criteria that can be used to differentiate these genera are listed in Table 228.

FIGURE 230. Phylogenetic position of *Anaerobaculum thermoterrenum* and *Aminobacterium mobile* relative to some members of the family *Syntrophomonadaceae* and Family XV *Incertae Sedis*, order *Clostridiales*, class *Clostridia*, phylum *Firmicutes*.

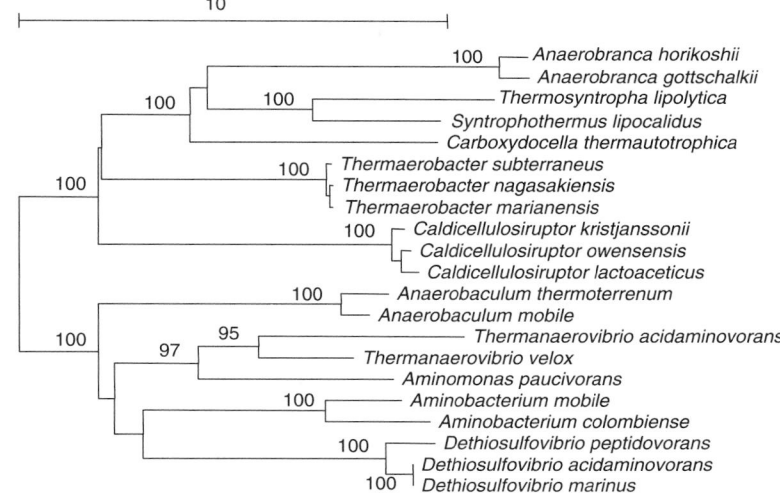

TABLE 228. Differential characteristics of the genus *Anaerobaculum* from other physiologically similar and phylogenetically related genera[a]

Characteristics	Anaerobaculum	Aminobacterium	Aminomonas	Dethiosulfovibrio	Synergistes	Thermoanaerovibrio
Species represented in the genera (references)	A. thermoterrenum (Rees et al., 1997), A. mobile (Menes and Muxí, 2002)	A. colombiense (Baena et al., 1998), A. mobile (Baena et al., 2000)	A. paucivorans (Baena et al., 1999a)	D. peptidovorans (Magot et al., 1997), and D. russensis, D. acidaminovorans D. marinus (Surkov et al., 2001).	S. jonesii (Allison et al., 1992).	T. acidaminovorans (Baena et al., 1999b), T. velox (Zavarzina et al., 2000)
Habitat	Petroleum reservoir fluids and anaerobic wool-scouring wastewater lagoon	Anaerobic lagoon of dairy wastewater	Anaerobic lagoon of dairy wastewater	Saline sulfur mats and oil-producing well.	Rumen of goat	Methanogenic digester and thermophilic cyanobacterial mats
Morphology	Straight to slightly curved with or without sheath	Straight to slightly curved to rod shaped	Straight to slightly curved to rod shaped	Straight to slightly curved to rod shaped and spirals	Oval rods	Curved
Motility	Nonmotile or motile with single polar flagellum	Nonmotile or motile by 1 to 2 lateral flagella	Nonmotile	Motile with a tuft of laterally inserted flagella	Nonmotile	Motile by tuft of lateral flagella
DNA G+C content (mol%)	44–51.5	44–46	43	51–56	57–59	54.6–56.5
Temperature range for growth (optimum) (°C)	28–65 (55–60)	20–42 (37)	20–40 (35)	4–45 (25–42)	37–39	40–70 (50–65)
NaCl growth range (optimum) (% w/v)	0–2 (0.008–1%)	0–1.5 (0.05–0.5)	0–2 (0.05–0.5)	0.5–7 (1–2)	Not required	Not required
pH range for growth (optimum pH)	5.4–8.7 (6.6–7.6)	6.6–8.5 (7.3)	6.7–8.3 (7.5)	5.5–8.8 (6.5–7.0)	Not reported	4.5–8.0 (7.3)
Utilization of carbohydrates	+	–	–	–	–	+
Utilization of peptides (P) and/or amino acids (AA)	AA, P	AA, P	AA, P	AA, P	AA, P	AA
Sulfur compounds used as electron acceptors	Thiosulfate, S⁰, cystine	–	–	Thiosulfate, S⁰	Not reported	–
Oxidation of amino acids in coculture with hydrogen-scavenging methanogens	–	+	+	–	+	+

[a]Symbols: +, >85% positive; d, different strains give different reactions (16–84% positive); –, 0–15% positive; w, weak reaction; ND, not determined.

Taxonomic comments

A number of early phylogenetic analyses showed that *Anaerobaculum thermoterrenum* and *Aminobacterium mobile*, together with members of the genera *Dethiosulfovibrio*, *Aminobacterium*, *Thermanaerovibrio*, and *Aminomonas* formed a deep and equidistant branch in the vicinity of the members of the genus *Thermoanaerobacterium* (phylum *Firmicutes*) and members of the genus *Dictyoglomus* with a mean similarity of 82% (Menes et al., 2001; Rees et al., 1997; Zavarzina et al., 2000) (Figure 230). Consequently, all the members of these genera together with *Anaerobaculum thermoterrenum* and *Aminobacterium mobile* were placed in the family *Syntrophomonadaceae*, order *Clostridiales*, class "*Clostridia*", phylum *Firmicutes*. A subsequent phylogenetic analysis has suggested that the genus *Dictyoglomus* forms a new line of descent and it therefore should be accorded a phylum status, phylum *Dictyoglomi* (Garrity and Holt, 2001a). However, a more recent phylogenetic evaluation of the genus *Anaerobaculum* suggests that it forms a reproducible, monophyletic line of descent with members of the genera *Synergistes*, *Dethiosulfovibrio*, *Aminobacterium*, "*Aminiphilus*", *Thermanaerovibrio*, *Aminomonas*, and *Thermovirga*. Therefore, we propose that a new phylum be created to represent this line of descent: phylum "*Aminanaerobia*" (Figure 231). This phylum has a 15% 16S rRNA divergence, which is on a par with a number of recognized bacterial phyla, e.g., *Cyanobacteria*, *Firmicutes*, and *Dictyoglomi*. Currently, the genus *Synergistes* has been placed as a genus *incertae sedis* in the phylum *Deferribacteres* (Garrity and Holt, 2001b). In light of the analysis, it should be transferred to phylum "*Aminanaerobia*". All characterized members of the proposed phylum reported to date are obligate anaerobes and have the ability to utilize amino acids and/or peptides (Table 228), and environmental 16S rRNA gene surveys have only identified representatives of this phylum in anaerobic habitats (termite hindgut, rumen, anaerobic digesters, subgingival crevice). The metabolic conformity among the members of the proposed phylum both supports and lends credence to the phylogenetic data.

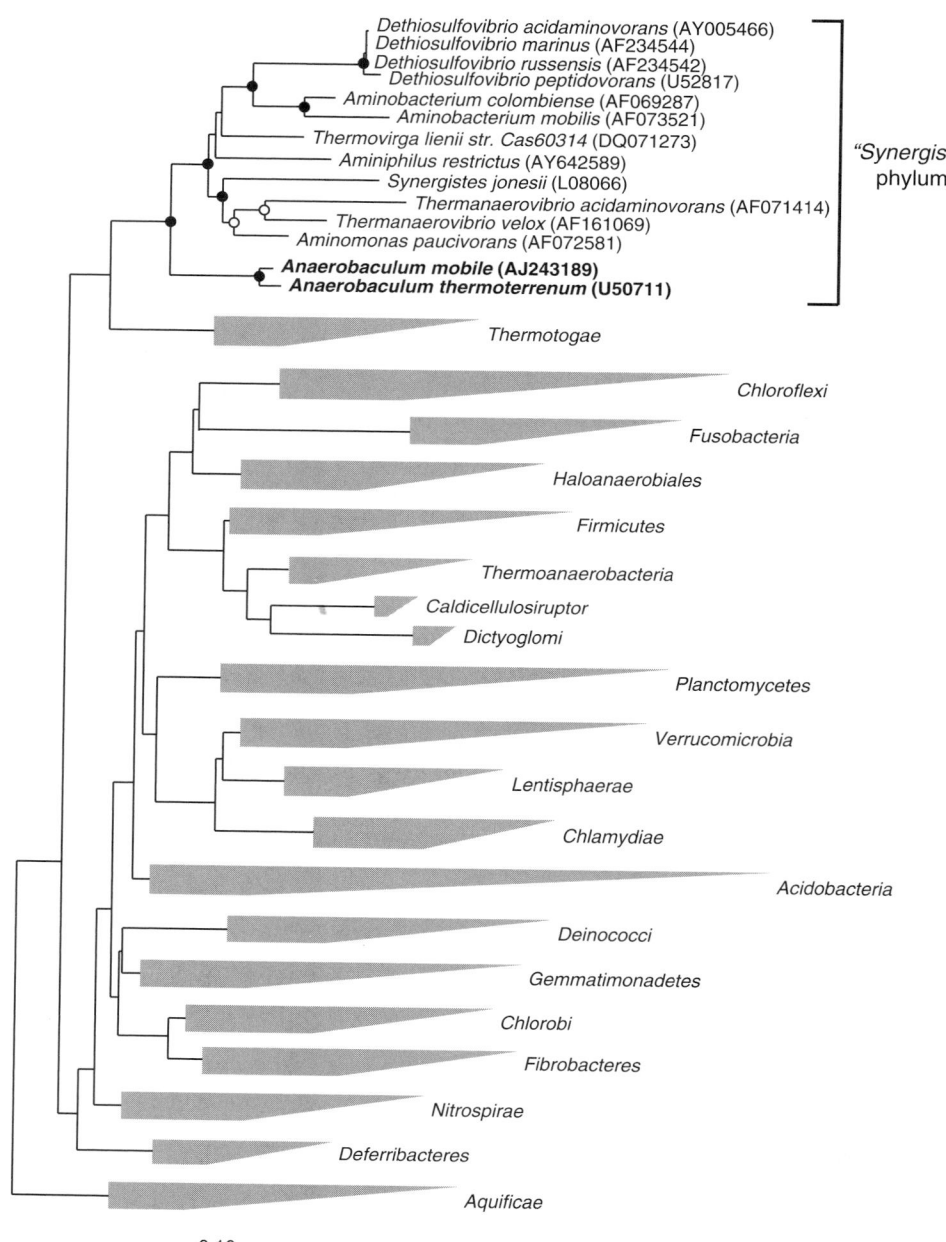

FIGURE 231. Maximum-likelihood tree of 16S rRNA gene sequences obtained from members of the "*Synergistes*" phylum and representatives of other bacterial phyla. *Anaerobaculum* species are bolded. Support for interior nodes is indicated by circles (filled, >90% bootstrap resampling support; open, >75% bootstrap resampling support). Data were obtained from the greengenes database (greengenes.lbl.gov).

Differentiation of the species of the genus *Anaerobaculum*

The characteristics that differentiate species of *Anaerobaculum* are listed in Table 229.

TABLE 229. Differential characteristics of the species of the genus *Anaerobaculum*[a]

Characteristic	A. thermoterrenum	A. mobile
Type strain	Isolate RWcit2^T=ACM 5076^T	Isolate NGA^T=ATCC BAA-54^T=DSM 13181^T
Isolation source	Production waters of a petroleum reservoir	Sludge of anaerobic lagoon treating wool-scouring wastewater
Morphology	Single or paired straight to slightly curved rods. Complex medium grown cells possess a sheath-like material which is absent in citrate grown cells.	Single or paired straight rods
Size (μm)	0.75 × 2 μm	0.5–1 × 2–4 μm
DNA G+C content (mol%)	44 (T_m)	51.5 (HPLC)
Motility	–	+
Presence of flagella	–	+
Temperature growth range (°C)	28–60	35–65
Temperature optimum (°C)	55	55–60
pH Growth range	5.5–8.6	5.4–8.7
pH Optimum	7.0–7.6	7.3
NaCl requirements	+	–
NaCl range for growth (%)	0–2 (optimum 1)	0–1.5 (optimum = 0.008)
Utilization of:		
Adonitol	NR	–
Arabinose	NR	–
Butyrate	NR	–
Carboxymethyl cellulose	–	NR
Cellobiose	NR	–
Citrate	+	–
Dextrin	NR	–
Fumarate	+	–
Gelatin	NR	–
Gluconate	NR	+
Glutamate	+	–
Gum arabic	–	NR
Inositol	+	NR
Inulin	NR	–
Mannose	+	–
Melibiose	NR	–
Oleate	NR	–
2-Oxoglutarate	+	–
Pectin	+	–
Polygalacturonate	NR	–
Xylan	NR	–

[a]Both strains stain Gram-negative and utilize malate, pyruvate, tartrate, starch, glucose, fructose, inositol, glycerol, and Casamino acids but not xylose, galactose, lactose, sucrose, maltose, rhamnose, raffinose, cellulose, malonate, succinate, and lactate. NR, Not reported.

Key to the species of the genus *Anaerobaculum*

1. Single or paired straight to slightly curved rods that are surrounded by a sheathlike material when grown in complex medium. Ferments citrate, fumarate, glutamate, 2-oxoglutarate, pectin, and mannose. Requires 1% NaCl for optimal growth. Nonmotile and flagella are absent.

    Anaerobaculum thermoterrenum

2. Single or paired straight rods. Does not ferment citrate, fumarate, glutamate, 2-oxoglutarate, pectin, and mannose. There is no requirement of NaCl for growth. Motile by means of a single flagellum.

    Aminobacterium mobile

List of species of the genus *Anaerobaculum*

1. **Anaerobaculum thermoterrenum** Rees, Patel, Grassia and Sheehy 1997, 153^{VP}

ther.mo.ter.re′num. Gr. adj. *thermos* warm, hot; L. neut. adj. *terrenum* earthen, belonging to the earth; N.L. neut. adj. *thermoterrenum* from hot earth, describing the site of isolation.

Cells are 0.75 × 2 μm and octcur singly or in pairs. A sheath-like material is observed in cells grown in complex medium but is absent in cells grown on citrate. Obligately anaerobic. The temperature range for growth is between 28 and 60°C, with an optimum of 55°C. Growth occurs in a medium containing 0–20 g/l of NaCl with an optimum of 10 g/l. The pH range is from 5.5–8.6, with an optimum between 7 and 7.6. Ferments citrate, fumarate, malate, pyruvate, glutamate, 2-oxoglutarate, tartrate, starch, pectin, glucose, fructose, mannose, inositol, glycerol, protein extracts, and Casamino acids. Compounds that are not used are xylose, galactose, lactose, sucrose, maltose, rhamnose, raffinose, cellulose, carboxymethyl cellulose, gum arabic, malonate, succinate, glutarate, and lactate. Sulfide is produced from elemental sulfur, thiosulfate, and cystine. Sulfate, sulfite, and nitrate are not reduced. Growth is inhibited by H_2. Cytochromes are not present. Isolated from production fluid of a petroleum reservoir.

DNA G+C content (mol%): 44 (T_m).

Type strain: RWcit2, ACM 5076, DSM 13490.

GenBank accession number (16S rRNA gene): U50711.

2. **Anaerobaculum mobile** Menes and Muxí 2002, 163[VP]

mo′bi.le. L. neut. adj. *mobile* motile.

Motile, straight rods, 0.5–1 × 2–4 µm, occurring singly or in pairs. Possess a single laterally inserted flagellum. Stains Gram-negative. Strictly anaerobic and chemo-organotrophic.

Spores have not been observed. Moderately thermophilic with a growth range between 35 and 65°C, and an optimum between 55 and 60°C. The pH growth range is pH 5.4–8.7, with an optimum between 6.6 and 7.3. The optimum NaCl concentration is 0.08 g/l but growth occurs with up to 15 g/l Uses malate, tartrate, pyruvate, glycerol, starch, glucose, fructose, gluconate, Casamino acids, tryptone, and yeast extract. Carbohydrates and organic acids are converted to acetate, hydrogen, and CO_2. Oxidizes leucine to isovalerate with crotonate as electron acceptor, but not in co-culture with a methanogenic partner. Thiosulfate, sulfur, and cystine are reduced to sulfide. Crotonate is reduced to butyrate. Sulfate, fumarate, acetate, and nitrate are not reduced. No growth occurs on citrate, 2-oxoglutarate, glutamate, mannose, pectin, lactose, xylose, galactose, maltose, sucrose, rhamnose, raffinose, malonate, lactate, succinate, xylan, dextrin, inulin, melibiose, adonitol, cellobiose, arabinose, polygalacturonate, cellulose, gelatin, butyrate, or oleate.

DNA G+C content (mol%): 51.5 (HPLC).

Type strain: NGA, DSM 13181, ATCC BAA-54.

GenBank accession number (16S rRNA gene): AJ243189.

Genus IV. **Dethiosulfovibrio** Magot, Ravot, Campaignolle, Ollivier, Patel, Fardeau, Thomas, Crolet and Garcia 1997, 822[VP]

WILLIAM B. WHITMAN

De.thi.o.sul.fo.vi′bri.o. L. pref. *de* from; Gr. n. *thion* sulfur, L. n. *sulfur* sulfur, *thiosulfo* thiosulfate; L. v. *vibrio* to vibrate, N.L. masc. n. *vibrio* that vibrates, a generic name; N.L. masc. n. *Dethiosulfovibrio* a vibrio that reduces thiosulfate.

Vibrios or curved rods. **Strictly anaerobic** heterotrophs that ferment peptides and amino acids. Sugars are not utilized. Reduces thiosulfate and elemental sulfur but not sulfate to hydrogen sulfide. Cells stain **Gram-negative** and possess multilayered cell walls. **Invaginations of the wall** are observed in some species. Endospores are not formed. **Motile** by means of **lateral flagella.** Isolated from the production water of oilfields and "*Thiodendron*" mats in consortia with spirochetes.

DNA G+C content (mol%): 51–56.

Type species: **Dethiosulfovibrio peptidovorans** Magot, Ravot, Campaignolle, Ollivier, Patel, Fardeau, Thomas, Crolet and Garcia 1997, 823[VP].

Further descriptive information

The description of *Dethiosulfovibrio* is based upon Magot et al. (1997) and Surkov et al. (2001). Phylogenetic analyses of 16S rRNA genes place *Dethiosulfovibrio* within Family XV *Incertae Sedis* of the order *Clostridiales* (Ludwig et al., 2009). This well delineated, monophyletic group includes the other strict anaerobes *Anaerobaculum, Aminobacterium, Aminomonas,* and *Thermoanaerovibrio.* These genera all include straight to slightly curved rods that ferment amino acids. Some genera also utilize peptides or carbohydrates. Thiosulfate and/or elemental sulfur is also an electron acceptor for growth of some genera. Endospores are not formed, and these cells stain Gram-negative and possess multilayered cell walls.

The genus *Dethiosulfovibrio* is similar to other members of this family. Cells are curved rods or vibrioid in shape. They often occur in pairs and may form long, helical filaments during some phases of growth. Cells stain Gram-negative and possess

multilayered cell walls. In species where it has been examined, cells from the stationary growth phase form invaginations or incomplete cross-septa composed of in-growths of the cytoplasmic membrane and the peptidoglycan layer but not the outer wall (Surkov et al., 2001). Cells are motile by a tumbling movement and possess 1–6 flagella arranged laterally along the concave side of the cell. See Table 230 for further descriptive information.

The species of *Dethiosulfovibrio* were isolated for their ability to utilize either thiosulfate or elemental sulfur as an electron acceptor. Subsequently, it has been found that all species reduce both of these electron acceptors to sulfide, but they do not utilize sulfate. Moreover, thiosulfate and elemental sulfur are not disproportionated. When tested, sulfate, sulfite, fumarate, nitrate, Fe_2O_3, MnO_2, DMSO and Se^0 were not reduced. Cytochromes are also absent. As carbon sources and electron donors, all species utilize peptides such as those found in Trypticase, peptone, and meat extract. Some individual amino acids, such as serine, are fermented. However, utilization of most amino acids requires an electron acceptor, either thiosulfate or elemental sulfur, and the specific amino acids used depends greatly upon the strain. Amino acids utilized by most if not all species include alanine, cysteine, glutamate, histidine, serine, threonine, and valine. Other amino acids utilized by at least one species are arginine, asparagine, isoleucine, leucine, lysine, and methionine. The branched-chain amino acids are oxidized to the branched-chain fatty acids. Methionine is oxidized to propionate. Alanine and asparagine are oxidized to acetate. Fatty acids and alcohols are not utilized, including: formate, acetate, propionate, butyrate, lactate, fumarate, succinate, ethanol, propanol, glycerol, and

TABLE 230. Descriptive characteristics of the species of the genus *Dethiosulfovibrio*

Characteristic	1. *D. peptidovorans*	2. "*D. acidamionovorans*"	3. "*D. marinus*"	4. "*D. russensis*"
Morphology	Vibrioid	Curved rod or spirals	Curved rod or spirals	Curved rod or spirals
Dimensions (µm)	$1 \times 3–5$	$0.9 \times 3–5$	$0.9 \times 4–8$	$0.9 \times 3–5$
Carbon sources:				
Gelatin	–	+	+	+
Amino acids	+	+	+	+
Carbohydrates	–	–	–	–
Acetate and other organic acids	–	–	–	–
Electron acceptors:				
Thiosulfate	+	+	+	+
S^0	+	+	+	+
Sulfate	–	–	–	–
Requires yeast extract	–	+	+	+
Temperature range (optimum) (°C)	20–45 (42)	4–40 (25–30)	4–37 (28)	4–37 (28)
pH range (optimum)	5.5–8.8 (7.0)	5.5–8.0 (6.5–7.0)	5.5–8.0 (6.5–7.0)	5.5–8.0 (6.5–7.0)
NaCl range (optimum) (% w/v)	1–10 (3)	0.5–7 (2)	0.5–5 (2)	0.5–5 (2)
DNA G+C content (mol%)	56	51	52	51

mannitol. However, citrate, malate, pyruvate and 2-oxoglutarate are used by some species. Sugars are not utilized, including: arabinose, fructose, galactose, glucose, lactose, maltose, mannose, ribose, sucrose, trehalose, and xylose.

Dethiosulfovibrio peptidovorans may have a significant role in pipeline corrosion in oilfields where thiosulfate is present (Magot et al., 1997). Isolated from an oilfield with corroded pipes, it accelerates the rate of corrosion by at least an order of magnitude. In this regard, its affect is comparable to that of the sulfate-reducing bacteria. The remaining species were all isolated from "*Thiodendron*" mats (Surkov et al., 2001). These mats are common in shallow waters where H_2S is abundant (Dubinina et al., 1994b). Typical habitats also possess a salinity of 2 %, depths of 0.2–2 m, and mesophilic temperatures. The mats are composed of a co-culture of a spirochete and one or more sulfidogenic bacteria such as *Dethiosulfovibrio*. The spirochete oxidizes H_2S to elemental sulfur, which often appears as globules along its filaments. The sulfidogenic bacteria reduce the elemental sulfur back to H_2S using organic electron donors (Dubinina et al., 1994a).

Enrichment and isolation procedures

Dethiosulfovibrio peptidovorans was isolated from the production water of the Emeraude oilfield in the Congo. The temperature of the oilfield was 38°C, and its waters possessed a total salinity of 52 g/l and 0.5 mM thiosulfate (Magot et al., 1997). The type strain was isolated following serial dilution and plating on half-strength tryptone-yeast extract-glucose medium under conditions for the cultivation of fermentative organisms.

"*Desulfonispora acidamionovorans*" and "*Dethiosulfovibrio russensis*" were isolated from sulfur "*Thiodendron*" mats in mineral springs at the Staraja, Russia, health resort. "*Dethiosulfovibrio marinus*" was isolated from similar mats from the littoral zone at Kandalaksha Bay in the White Sea near Murmansk, Russia. Enrichments were performed anaerobically in medium containing yeast extract, peptone, citrate, and elemental sulfur.

Differentiation from closely related taxa

Dethiosulfovibrio may be differentiated by other members of Family XV *Incertae Sedis* of the order *Clostridiales* by its 16S rRNA gene sequence and a variety of phenotypic properties. Within this group, it and *Anaerobaculum* are the only genera that contain thiosulfate-reducing species. However, *Anaerobaculum* metabolizes sugars and is moderately thermophilic. *Dethiosulfovibrio* does not metabolize sugars and is mesophilic. While other members of this family tolerate low concentrations of NaCl, *Dethiosulfovibrio* is the only genus that requires 0.5 % (w/v) or greater NaCl for growth.

Taxonomic comments

Although their descriptions were published in the *International Journal of Systematic and Evolutionary Microbiology*, the strains of "*Desulfonispora acidamionovorans*", "*Dethiosulfovibrio marinus*" and "*Dethiosulfovibrio russensis*" have only been deposited in one culture collection, the DSMZ. The revised Rule 27 of the Bacteriological Code requires that all type strains described after December 14, 2000, be deposited in two publicly accessible service collections in different countries (Euzéby and Tindall, 2004). Therefore, these names are not validly published. In contrast, *Dethiosulfovibrio peptidovorans*, which is also only deposited in the DSMZ, was published in 1997 and remains validly published.

List of species of the genus *Dethiosulfovibrio*

1. **Dethiosulfovibrio peptidovorans** Magot, Ravot, Campaignolle, Ollivier, Patel, Fardeau, Thomas, Crolet and Garcia 1997, 823[VP]

pep.ti.do.vo′rans. Gr. adj. *peptos* cooked, L. v. *voro* to devour, N.L. part. adj. *peptidovorans* devouring peptides.

Properties are described in the genus description and Table 230.

DNA G+C content (mol%): 56 (HPLC).
Type strain: SEBR 4207, DSM 11002.
GenBank accession number (16S rRNA gene): U52817.

2. **"Dethiosulfovibrio acidaminovorans"** Surkov, Dubinina, Lysenko, Glöchner and Kuever 2001, 335

a.cid.a.mi.no.vo′rans. N.L. n. *acidum aminum* amino acid; L. v. *vorare* to devour or swallow; N.L. part. adj. *acidaminovorans* devouring amino acids.

Properties are described in the genus description and Table 230.

DNA G+C content (mol%): 51 (T_m).
Type strain: SR15, DSM 12590.
GenBank accession number (16S rRNA gene): AY005466.

3. **"Dethiosulfovibrio marinus"** Surkov, Dubinina, Lysenko, Glöchner and Kuever 2001, 335

ma′ri.nus. L. adj. *marinus* marine.

Properties are described in the genus description and Table 230.

DNA G+C content (mol%): 52 (T_m).
Type strain: WS100, DSM 12537.
GenBank accession number (16S rRNA gene): AF234544.

4. **"Dethiosulfovibrio russensis"** Surkov, Dubinina, Lysenko, Glöchner and Kuever 2001, 335

rus.sen′sis. N.L. adj. *russensis* pertaining to Staraja Russia.

Properties are described in the genus description and Table 230.

DNA G+C content (mol%): 51 (T_m).
Type strain: SR12, DSM 12538.
GenBank accession number (16S rRNA gene): AF234542.

Genus V. **Thermanaerovibrio** Baena, Fardeau, Woo, Ollivier, Labat and Patel 1999b, 973[VP] emend. Zavarzina, Zhilina, Tourova, Kuznetsov, Kostrikina and Bonch-Osmolovskaya 2000, 1293

DARIA G. ZAVARZINA

Therm.an.ae.ro.vib′ri.o. Gr. adj. *thermos* hot; Gr. pref. *an* not; Gr. n. *aer* air; N.L. masc. n. *vibrio* that vibrates; N.L. masc. n. *Thermanaerovibrio* a thermophilic vibrating anaerobe.

Cells are curved rods. Cell wall has **Gram-negative** structure. **Motile** by means of **lateral flagella located on the concave side** of the cells. Nonsporeforming. Obligate anaerobe. Neutrophilic. **Thermophilic. Chemo-organotrophic;** grows fermentatively with some amino acids, organic acids, and carbohydrates as substrates, or **lithoheterotrophically** with molecular **hydrogen and elemental sulfur.**

DNA G+C content (mol%): 54.5–56.5 (T_m).

Type species: **Thermanaerovibrio acidaminovorans** (Guangsheng, Plugge, Roelofsen, Houwen and Stams 1992) Baena, Fardeau, Woo, Ollivier, Labat and Patel 1999b, 973[VP] (*Selenomonas acidaminovorans* Guangsheng, Plugge, Roelofsen, Houwen and Stams 1992, 174).

Further descriptive information

Based upon recent 16S rRNA analyses, *Thermanaerovibrio* has been classified in Family XV *Incertae Sedis* in this volume within the order *Clostridiales*. Previously, it had been classified within the family *Syntrophomonadaceae* and related genera of cluster V of the *Clostridium* group (Baena et al., 1999b; Collins et al., 1994). The genus includes two species, *Thermanaerovibrio acidaminovorans* and *Thermanaerovibrio velox*.

Phenotypically, members of the genus are moderately thermophilic, neutrophilic, obligate anaerobes. The most striking feature is the cell morphology – curved rods with a tuft of flagella located on the concave side of the cell (see Figure 232).

The glutamate catabolism of *Thermanaerovibrio acidaminovorans* was studied by the combined use of [^{13}C]glutamate NMR measurements and enzyme activity determinations (Plugge et al., 2001). *Thermanaerovibrio acidaminovorans* converts glutamate to acetate, propionate, CO_2, NH_4^+, and H_2. NMR spectra do not show any labeled acetate. The presence of key enzymes from the β-methylaspartate pathway indicates that, most likely, acetate forms through this pathway. *Thermanaerovibrio acidaminovorans* has a highly active glutamate dehydrogenase, both in the NAD-dependent and NADH-dependent directions, even when it is grown on glucose. Propionate formation occurs through the direct oxidation of glutamate via succinyl-CoA and methylmalonyl-CoA.

Members of the genus are organotrophic fermentative bacteria with different substrates utilized – many for *Thermanaerovibrio acidaminovorans* and more restricted for *Thermanaerovibrio velox*. In addition to fermentative capacity, both species are able to use an oxidative pathway with hydrogen as an electron donor and elemental sulfur as acceptor. Other acceptors including sulfate, sulfite, thiosulfate, nitrate, fumarate, and Fe(III) are not reduced by both species. Members of the genus are heterotrophic; the ability to grow autotrophically is not recorded.

Habitats are granular methanogenic sludge or neutral hot springs.

Enrichment and isolation procedures

Anaerobically prepared liquid bicarbonate-buffered medium and incubation at 55°C were used for the enrichments of *Thermanaerovibrio acidaminovorans* and *Thermanaerovibrio velox* (Guangsheng et al., 1992; Zavarzina et al., 2000). Pure cultures were obtained by isolation of single colonies in roll-tubes.

Maintenance procedures

Maintenance is not difficult. Cultures can be stored without transfer at room temperature for 1–2 weeks or in a refrigerator for 3–5 months. DSMZ maintains cultures in liquid nitrogen.

Differentiation of the genus *Thermanaerovibrio* from other genera

The differentiating characteristics of *Thermanaerovibrio* that distinguish it from the closely phylogenetically related genera *Dethiosulfovibrio* and *Anaerobaculum* are summarized in Table 231.

Taxonomic comments

The phylogenetic status of *Thermanaerovibrio* is based on 16S rRNA gene sequence analysis. The type species, *Thermanaerovibrio acidaminovorans*, was first assigned to the genus *Selenomonas* as a new species, *Selenomonas acidaminovorans*, based on its morphological characteristics, i.e., Gram-stain-negative, nonsporeforming, curved rods, motile by means of flagella located on the concave side of the cells (Guangsheng et al., 1992). The phylo-

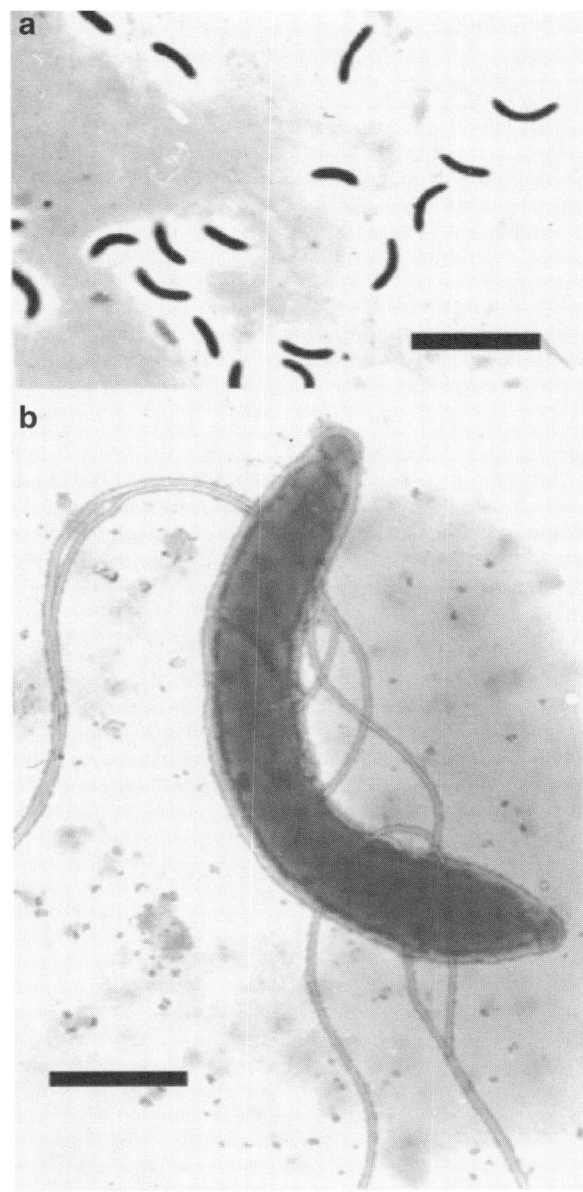

FIGURE 232. Morphology of *Thermanaerovibrio velox.* (a) Phase-contrast light micrograph. Bar = 10 μm. (b) Flagella localization. Negatively stained cell. Bar = 1 μm.

genetic analysis of *Selenomonas acidaminovorans* was completed by Baena et al. (1999b). It revealed that *Selenomonas acidaminovorans* was placed distantly from the members of the genus *Selenomonas* with a similarity of only 80%. This organism was placed equidistantly between *Dethiosulfovibrio peptidovorans* and *Anaerobaculum thermoterrenum* with similarity of 85%. All three species share the common properties of requiring strict anaerobiosis for growth and curved cell morphology; each member forms an independent line of descent in vicinity of cluster V of *Clostridia* group (Collins et al., 1994; Magot et al., 1997; Rees et al., 1997). Based on phylogenetic separation alone, it was proposed to transfer *Selenomonas acidaminovorans* to a new genus, *Thermanaerovibrio* gen. nov., as *Thermanaerovibrio acidaminovorans* comb.

nov. In addition, some phenotypic characteristics (see Table 231) were the argument for the creation of a new genus (Baena et al., 1999b). In the current volume, *Thermanaerovibrio* is classified in Family XV *Incertae Sedis* along with the related genera *Aminobacterium, Aminomonas, Anaerobaculum,* and *Dethiosulfovibrio* (Figure 6).

The second species of the genus, *Thermanaerovibrio velox,* was placed in the genus *Thermanaerovibrio* according to its significant phenotypic similarity to *Thermanaerovibrio acidaminovorans.* The two species have similar morphology and physiology, and close DNA G+C content. However the level of DNA–DNA hybridization is rather low (15 ± 1%). The degree of similarity of 16S rRNA sequences of two species is at the level of generic differentiation 92.2% (Zavarzina et al., 2000). Table 232 shows some differential characterstics for the two *Thermanaerovibrio* species.

An inhibitory effect of elemental sulfur on growth of *Thermanaerovibrio acidaminovorans* growing on glucose was found. At the same time, both species of the genus *Thermanaerovibrio* are able to grow lithotrophically with molecular hydrogen and elemental sulfur. This feature was emended to description of genus *Thermanaerovibrio* (Zavarzina et al., 2000).

TABLE 231. Characteristics differentiating the genus *Thermanaerovibro* from its closest phylogenetic relatives[a]

Characteristic	*Thermanaerovibrio*	*Anaerobaculum*	*Dethiosulfovibrio*
Morphology:			
Curved rod	+	+	+
Flagella	+	−	+
Motile	+	−	+
Substrates:			
Carbohydrates	+	+	−
Growth requirements:			
NaCl (%)	−	1	3
Sulfur sources used as electron acceptor:			
Elemental sulfur	+	+	+
Thiosulfate	−	+	+
DNA G+C content (mol%)	54.5–56.5	44	56

[a]Symbols: +, >85% positive; −, 0–15% positive.

TABLE 232. Characteristics differentiating *Thermanaerovibrio* species[a]

Characteristic	*T. acidaminovorans*	*T. velox*
Growth temperature (°C):		
Optimum	55	60–65
Maximum	58	70
Minimum	40	45
Optimum pH	6.5–8.1	7.3
Growth with succinate	+	−
Organotrophic growth with S[0]	−	+
Glucose fermentation products:		
Acetate	+	+
CO_2	+	+
Ethanol	−	+
H_2	+	+
Lactate	−	+
DNA G+C content (mol%)	56.5	54.6

[a]Symbols: +, >85% positive; −, 0–15% positive; w, weak reaction; ND, not determined.

List of species of the genus *Thermanaerovibrio*

1. **Thermanaerovibrio acidaminovorans** (Guangsheng, Plugge, Roelofsen, Houwen and Stams 1992) Baena, Fardeau, Woo, Ollivier, Labat and Patel 1999b, 973[VP] (*Selenomonas acidaminovorans* Guangsheng, Plugge, Roelofsen, Houwen and Stams 1992, 174)

a.cid.a.mi.no.vo′rans. L. neut. n. *acidum* acid; N.L. neut. n. *aminum* amine; L. part. adj. *vorans* devouring; N.L. part. adj. *acidaminovorans* amino acid-digesting.

Curved cells 0.5–0.6 × 2.5–3.0 μm (when grown on succinate media) with rounded ends, single or in pairs, or in long chains when grown in complex media. On the concave side of the cell, 6–8 flagella are present.

Substrates used are listed in Table 233. Succinate is decarboxylated to propionate. Amino acids (see Table 233) are fermented to propionate, acetate, and hydrogen. Branched amino acids are degraded to branched-chain fatty acids. No growth occurs on glycine, aspartate, acetoin, malonate, and oxalate. Growth and substrate conversion are enhanced by co-cultivation with methanogens (*Methanobacterium thermoautotrophicum* ΔH) (Guangsheng et al., 1992). Elemental sulfur (1%) inhibits growth on medium containing glucose (Zavarzina et al., 2000). Contains *b*-type cytochromes. Growth occurs in a pH range 6.5–8.1. The habitat is methanogenic granular sludge.

DNA G+C content (mol%): 56.5 ± 0.3 (T_m).
Type strain: Su883, ATCC 49978, DSM 6589.
GenBank accession number (16S rRNA gene): AF071414.

2. **Thermanaerovibrio velox** Zavarzina, Zhilina, Tourova, Kuznetsov, Kostrikina and Bonch-Osmolovskaya 2000, 1293[VP]

ve′lox. L. adj. *velox* quick, fast, motile.

Curved rods with tapering ends, 0.5–0.7 × 2.5–5.0 μm, with fast wave-like movement (Figure 232). Cells occur singly or in pairs. Ultrathin sections showed a typical Gram-stain-negative cell envelope profile with a multilayered cell wall (Figure 233). Colonies are small, white, irregular or round, 0.2 mm in diameter, with an even edge. Multiplication by binary fission. Growth occurs in a pH range 4.5–8.0.

Substrates used are listed in Table 233. No growth observed on sorbose, cellobiose, maltose, melibiose, raffinose, trehalose, acetate, ascorbate, butyrate, formate, glycolate, propionate, tartrate, L-dulcitol, L-inositol, manitol, propanol, L-sorbitol, glycogen, DL-lysine, sarcosine, tryptone, choline, cellulose, chitin, starch, or H$_2$ (in the absence of elemental sulfur). Yeast extract (0.25 g/l) and peptone (0.25 g/l) stimulate organotrophic growth on glucose. Yeast extract (0.1 g/l) is required for lithotrophic growth with H$_2$ and S^0. Elemental sulfur stimulates organotrophic growth with glucose, peptone, yeast extract, trypticase, or Casamino acids and was reduced to H$_2$S. Isolated from thermophilic cyanobacterial mat from caldera Uzon, Kamchatka, Russia.

DNA G+C content (mol%): 54.6 ±0.3 (T_m).
Type strain: Z-9701, DSM 12556.
GenBank accession number (16S rRNA gene): AF161069.

TABLE 233. Descriptive features of *Thermanaerovibrio* species[a,b]

Characteristic	T. acidaminovorans	T. velox
Morphology:		
Curved cells	+	+
Flagella located on concave side	+	+
Gram stain	–	–
Sporeforming	–	–
Utilization compounds as carbon and energy sources:		
N-Acetyl-D-glucosamine	ND	+
Adonite	–	+
Alanine	+	–
Arginine	+	+
Casamino acids	+	+
Citrate	+	–
Citrulline	+	ND
Fructose	+	+
Glucose	+	+
Glutamate	+	–
Histidine	+	–
2-Oxoglutarate	+	ND
Malate	+	–
Mannose	–	+
Ornithine	+	ND
Peptone	ND	+
Pyruvate	+	–
Serine	+	+
Threonine	+	ND
Xylose	+	–
Yeast extract	+	+
NaCl required for growth	–	–
Lithotrophic growth with H$_2$ and S^0	+	+

[a]Symbols: +, >85% positive; –, 0–15% positive; ND, not determined.
[b]For both species, no growth occurs on: xylose, ribose, glactose, lactose, sucrose, lactate, ethanol, mathanol, and betaine.

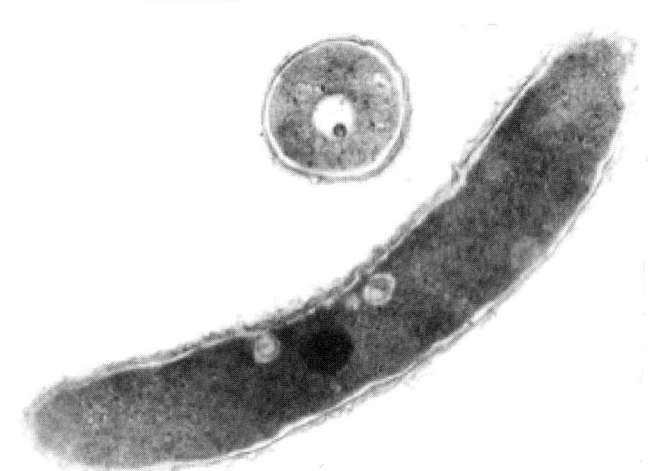

FIGURE 233. Longitudinal and cross sections demonstrating Gram-negative structure of cell wall of *Thermanaerovibrio velox* (bar = 1 μm).

References

Allison, M.J., W.R. Mayberry, C.S. McSweeney and D.A. Stahl. 1992. *Synergistes jonesii*, gen. nov., sp. nov., a rumen bacterium that degrades toxic pyridinediols. Syst. Appl. Microbiol. *15*: 522–529.

Baena, S., M.L. Fardeau, M. Labat, B. Ollivier, P. Thomas, J.L. Garcia and B.K. Patel. 1998. *Aminobacterium colombiense* gen. nov. sp. nov., an amino acid-degrading anaerobe isolated from anaerobic sludge. Anaerobe *4*: 241–250.

Baena, S., M.L. Fardeau, B. Ollivier, M. Labat, P. Thomas, J.L. Garcia and B.K. Patel. 1999a. *Aminomonas paucivorans* gen. nov., sp. nov., a mesophilic, anaerobic, amino-acid-utilizing bacterium. Int. J. Syst. Bacteriol. *49*: 975–982.

Baena, S., M.L. Fardeau, T.H.S. Woo, B. Ollivier, M. Labat and B.K.C. Patel. 1999b. Phylogenetic relationships of three amino-acid-utilizing anaerobes, *Selenomonas acidaminovorans*, '*Selenomonas acidaminophila*' and *Eubacterium acidaminophilum*, as inferred from partial 16S rDNA nucleotide sequences and proposal of *Thermanaerovibrio acidaminovorans* gen. nov., comb. nov and *Anaeromusa acidaminophila* gen. nov., comb. nov. Int. J. Syst. Bacteriol. *49*: 969–974.

Baena, S., M.L. Fardeau, M. Labat, B. Ollivier, P. Thomas, J. L. Garcia and B.K.C. Patel. 1999c. *In* Validation of publication of new names and new combinations previously effectively published outside the IJSB. List no. 71. Int. J. Syst. Bacteriol. *49*: 1325–1326.

Baena, S., M.L. Fardeau, M. Labat, B. Ollivier, J.L. Garcia and B.K.C. Patel. 2000. *Aminobacterium mobile* sp. nov., a new anaerobic amino-acid-degrading bacterium. Int. J. Syst. Evol. Microbiol. *50*: 259–264.

Balch, W.E., G.E. Fox, L.J. Magrum, C.R. Woese and R.S. Wolfe. 1979. Methanogens: reevaluation of a unique biological group. Microbiol. Rev. *43*: 260–296.

Barker, H.A. 1981. Amino acid degradation by anaerobic bacteria. Annu. Rev. Biochem. *50*: 23–40.

Collins, M.D., P.A. Lawson, A.Willems, J.J. Cordoba, J. Fernandez-Garayzabal, P. Garcia, J. Cia, H. Hippe and J.A.E. Farrow. 1994b. The phylogeny of the genus *Clostridium*: proposal of five new genera and eleven new species combination. Int. J. Syst. Bacteriol. *44*: 812–826.

Dubinina, G.A., M.Yu. Grabovich and N.V. Leshcheva. 1994a. Occurrence, structure, and metabolic activity of "*Thiodendron*" sulfur mats in various salt-water environments. Microbiology (En. transl. from Mikrobiologiya) *62*: 450–456.

Dubinina, G.A., N.V. Leshcheva and M.Yu. Grabovich. 1994b. The colorless sulfur bacterium *Thiodendron* is actually a symbiotic association of spirochetes and sulfidogens. Microbiology (En. transl. from Mikrobiologiya) *62*: 432–444.

Euzéby, J.P. and B.J. Tindall. 2004. Status of strains that contravene Rules 27(3) and 30 of the bacteriological code. Request for an Opinion. Int. J. Syst. Evol. Microbiol. *54*: 293–301.

Fardeau, M.L., B.K. Patel, M. Magot and B. Ollivier. 1997. Utilization of serine, leucine, isoleucine, and valine by *Thermoanaerobacter brockii* in the presence of thiosulfate or *Methanobacterium* sp. as electron acceptors. Anaerobe *3*: 405–410.

Garrity, G.M. and J.G. Holt. 2001a. The Road Map to the *Manual*. *In* Boone, Castenholz and Garrity (Editors), Bergey's Manual of Systematic Bacteriology, Vol. 1, The *Archaea* and the Deeply Branching and Phototrophic *Bacteria*. Springer, New York, pp. 119–166.

Garrity, G.M. and J.G. Holt. 2001b. Phylum BIX. *Deferribacteres* phy. nov. *In* Boone, Castenholz and Garrity (Editors), Bergey's Manual of Systematic Bacteriology, Vol. 1, The *Archaea* and the Deeply Branching and Phototrophic *Bacteria*. Springer, New York, pp. 465.

Garrity, G.M., J.A. Bell and T. Lilburn. 2005. The Revised Road Map to the Manual. *In* Brenner, Krieg, Staley and Garrity (Editors), Bergey's Manual of Systematic Bacteriology, 2nd edn, Vol. 2, The *Proteobacteria*, Part A, Introductory Essays. Springer, New York, pp. 159–220.

Guangsheng, C., C.M. Plugge, W. Roelofsen, F.P. Houwen and A.J.M. Stams. 1992. *Selenomonas acidaminovorans* sp. nov., a versatile thermophilic proton-reducing anaerobe able to grow by decarboxylation of succinate to propionate. Arch. Microbiol. *157*: 169–175.

Hungate, R.E. 1969. A roll tube method for cultivation of strict anaerobes. *In* Norris and Ribbons (Editors), Methods in Microbiology, Vol. 3B. Academic Press, London and New York, pp. 117–132.

Imhoff-Stuckle, D. and N. Pfennig. 1983. Isolation and characterization of a nicotinic-acid degrading sulfate-reducing bacterium, *Desulfococcus niacini* sp. nov. Arch. Microbiol. *136*: 194–198.

Ludwig, W., K.-H. Schleifer and W.B. Whitman. 2009. Revised road map to the phylum *Firmicutes*. *In* De Vos, Garrity, Jones, Krieg, Ludwig, Rainey, Schleifer and Whitman (Editors), Bergey's Manual of Systematic Bacteriology, 2nd edn, Vol. 3, The *Firmicutes*. Springer, New York, pp. 1–14.

Magot, M., G. Ravot, X. Campaignolle, B. Ollivier, B.K.C. Patel, M.-L. Fardeau, P. Thomas, J.-L. Crolet and J.-L. Garcia. 1997. *Dethiosulfovibrio peptidovorans* gen. nov., sp. nov., a new anaerobic, slightly halophilic, thiosulfate-reducing bacterium from corroding offshore oil wells. Int. J. Syst. Bacteriol. *47*: 818–824.

McInerney, M.J. 1988. Anaerobic hydrolysis and fermentation of fats and proteins. *In* Zehnder (Editor), Biology of Anaerobic Organisms. John Wiley, New York, pp. 373–409.

McSweeney, C.S., M.J. Allison and R.I. Mackie. 1993. Amino acid utilization by the ruminal bacterium *Synergistes jonesii* strain 78-1. Arch. Microbiol. *159*: 131–135.

Menes, R.J., A. Fernández and L. Muxí. 2001. Physiological and molecular characterization of an anaerobic thermophilic oleate degrading enrichment culture. Anaerobe *7*: 17–24.

Menes, R.J. and L. Muxí. 2002. *Anaerobaculum mobile* sp. nov., a novel anaerobic, moderately thermophilic, peptide-fermenting bacterium that uses crotonate as an electron acceptor, and emended description of the genus *Anaerobaculum*. Int. J. Syst. Evol. Microbiol. *52*: 157–164.

Nanninga, H.J., W.J. Drent and J.C. Gottschal. 1987. Fermentation of glutamate by *Selenomonas acidaminophila* sp. nov. Arch. Microbiol. *147*: 152–157.

Örlygsson, J., J. Krooneman, M.D. Collins, C. Pascual and J.C. Gottschal. 1996. *Clostridium acetireducens* sp. nov., a novel amino acid-oxidizing, acetate-reducing anaerobic bacterium. Int. J. Syst. Bacteriol. *46*: 454–459.

Paster, B.J., J.B. Russell, C.M.J. Yang, J.M. Chow, C.R. Woese and R. Tanner. 1993. Phylogeny of the ammonia-producing ruminal bacteria *Peptostreptococcus anaerobius*, *Clostridium sticklandii*, and *Clostridium aminophilum* sp. nov. Int. J. Syst. Bacteriol. *43*: 107–110.

Patel, B.K.C., H.W. Morgan and R.M. Daniel. 1985a. *Fervidobacterium nodosum* gen. nov. and spec. nov., a new chemoorganotrophic, caldoactive, anaerobic bacterium. Arch. Microbiol. *141*: 63–69.

Patel, B.K.C., H.W. Morgan and R.M. Daniel. 1985b. A simple and efficient method for preparing anaerobic media. Biotechnol. Lett. *7*: 227–228.

Plugge, C.M., J.M. van Leeuwen, T. Hummelen, M. Balk and A.J.M. Stams. 2001. Elucidation of the pathways of catabolic glutamate conversion in three thermophilic anaerobic bacteria. Arch. Microbiol. *176*: 29–36.

Rees, G.N., B.K.C. Patel, G.S. Grassia and A.J. Sheehy. 1997. *Anaerobaculum thermoterrenum* gen. nov., sp. nov., a novel, thermophilic bacterium which ferments citrate. Int. J. Syst. Bacteriol. *47*: 150–154.

Rogosa, M. 1969. *Acidaminococcus* gen. n., *Acidaminococcus fermentans* sp. n., anaerobic gram-negative diplococci using amino acids as the sole energy source for growth. J. Bacteriol. *98*: 756–766.

Saiki, T., Y. Kobayashi, K. Kawagoe and T. Beppu. 1985. *Dictyoglomus thermophilum* gen. nov., sp. nov., a chemoorganotrophic, anaerobic, thermophilic bacterium. Int. J. Syst. Bacteriol. *35*: 253–259.

Schnurer, A., B. Schink and B.H. Svensson. 1996. *Clostridium ultunense* sp. nov., a mesophilic bacterium oxidizing acetate in syntrophic association with a hydrogenotrophic methanogenic bacterium. Int. J. Syst. Bacteriol. *46*: 1145–1152.

Stams, A.J.M. and T.A. Hansen. 1984. Fermentation of glutamate and other compounds by *Acidaminobacter hydrogenoformans* gen. nov. sp. nov., an obligate anaerobe Isolated from black mud: studies with pure cultures and mixed cultures with sulfate-reducing and methanogenic bacteria. Arch. Microbiol. *137*: 329–337.

Surkov, A.V., G.A. Dubinina, A.M. Lysenko, F.O. Glöckner and J. Kuever. 2001. *Dethiosulfovibrio russensis* sp. nov., *Dethiosulfovibrio marinus* sp. nov. and *Dethiosulfovibrio acidaminovorans* sp. nov., novel anaerobic, thiosulfate- and sulfur-reducing bacteria isolated from '*Thiodendron*' sulfur mats in different saline environments. Int. J. Syst. Evol. Microbiol. *51*: 327–337.

Tarlera, S., L. Muxí, M. Soubes and A.J.M. Stams. 1997. *Caloramator proteoclasticus* sp. nov., a new moderately thermophilic anaerobic proteolytic bacterium. Int. J. Syst. Bacteriol. *47*: 651–656.

Touzel, J.P. and G. Albagnac. 1983. Isolation and characterization of *Methanococcus mazei* strain Mc3. FEMS Microbiol. Lett. *16*: 241–245.

Widdel, F. and F. Bak. 1992. Gram-negative mesophilic sulfate-reducing bacteria. *In* Balows, Trüper, Dworkin, Harder and Schleifer. (Editors), The Prokaryotes, Vol. 4. Springer, New York, pp. 3352–3378.

Wilde, E., M.D. Collins and H. Hippe. 1997. *Clostridium pascui* sp. nov., a new glutamate-fermenting sporeformer from a pasture in Pakistan. Int. J. Syst. Bacteriol. *47*: 164–170.

Zavarzina, D.G., T.N. Zhilina, T.P. Tourova, B.B. Kuznetsov, N.A. Kostrikina and E.A. Bonch-Osmolovskaya. 2000. *Thermanaerovibrio velox* sp. nov., a new anaerobic, thermophilic, organotrophic bacterium that reduces elemental sulfur, and emended description of the genus *Thermanaerovibrio*. Int. J. Syst. Evol. Microbiol. *50*: 1287–1295.

Zindel, U., W. Freudenberg, M. Rieth, J.R. Andreesen, J. Schnell and F. Widdel. 1988. *Eubacterium acidaminophilum* sp. nov., a versatile amino acid-degrading anaerobe producing or utilizing H_2 or formate: description and enzymatic studies. Arch. Microbiol. *150*: 254–266.

Family XVI. **Incertae Sedis**

Previously assigned to the *Syntrophomonadaceae* by Garrity et al. (2005), subsequent analyses suggest that this genus is not closely related to *Syntrophomonas* or a member of any other previously described family. For that reason, it is assigned to its own family *incertae sedis*.

Genus I. **Carboxydocella** Sokolova, Kostrikina, Chern yh, Tourova, Kolganova and Bonch-Osmolovskaya 2002, 1965[VP]

TATYANA G. SOKOLOVA

Car.bo.xy.do.cel′la. N.L. neut. n. *carboxydum* carbon monoxide; L. fem. n. *cella* cell; N.L. fem. n. *Carboxydocella* carbon monoxide-utilizing bacterium.

Cells are short **straight rods**. Cell wall of Gram-positive type. Cells divide by binary transverse fission. Obligate anaerobe. Colonies under 100% CO gas phase on solidified agar medium are round, white, 0.5–1 mm in diameter. **Hydrogenic CO-oxidizing bacteria that grow chemolithotrophically on CO. Utilize CO as the sole energy source with equimolar formation of H_2 and CO_2 according to the equation: $CO + H_2O \cdot CO_2 + H_2 \cdot$** Neutrophilic, pH range for growth is 6.2–8.0, with an optimum of 6.8–7.0. Thermophilic, temperature range for growth 40–70°C.

DNA G+C content (mol%): 46–49.

Type species: **Carboxydocella thermautotrophica** Sokolova, Kostrikina, Chernyh, Tourova, Kolganova and Bonch-Osmolovskaya 2002, 1965[VP].

Further descriptive information

Species of the genus *Carboxydocella* are the members of the *Firmicutes*, but they do not possess strong similarities to any of the previously described families. The highest 16S rRNA gene sequence similarity is less than 89% with members of *Thermoanaerobacter–Syntrophomonas* group, but phylogenetic trees do not support a strong affiliation with either of these organisms. For this reason, it is classified with Family XVI *Incertae Sedis* in this volume (Figure 7).

Members of the genus *Carboxydocella* are currently represented by two species, *Carboxydocella thermautotrophica* and *Carboxydocella sporoproducens*, which share 99.5% 16S rRNA gene sequence similarity. The level of DNA–DNA hybridization between type strains of *Carboxydocella thermautotrophica* and *Carboxydocella sporoproducens* is 45% (determined spectrophotometrically). Both species were isolated from hot springs of Kamchatka peninsula (Slepova et al., 2006; Sokolova et al., 2002). *Carboxydocella* species are widely spread in the Kamchatkan Geyzer Valley and the Uzon Caldera hot springs. These hot springs have temperatures and pH ranges of 50–70°C and 5.5–8.6, respectively.

Enrichment and isolation procedures

Members of the genus *Carboxydocella* have been isolated from terrestrial hot springs. There are no reports of members of this genus occurring in other environments. The methods of isolation are serial dilution transfers and single-colony isolations employing mineral medium supplemented with vitamins under 100% CO in the gas phase (Sokolova et al., 2002).

Maintenance procedures

Members of the genus *Carboxydocella* may be maintained on the medium of Sokolova et al. (2002). Liquid cultures may be stored at 4°C for 10–12 months without loss of viability. Lyophilization of the cultures performed in DSMZ resulted in good long-term preservation and recovery.

List of species of the genus *Carboxydocella*

1. **Carboxydocella thermautotrophica** Sokolova, Kostrikina, Chernyh, Tourova, Kolganova and Bonch-Osmolovskaya 2002, 1965[VP]

therm.au.to.tro′phi.ca. Gr. adj. *thermos* hot; Gr. adv. *autos* self; Gr. n. *trophos* food; N.L. fem. adj. *thermautotrophica* indicating that the organism grows at elevated temperatures and uses carbon monoxide as sole source for carbon and energy.

Cells are short straight rods of about 0.4–0.5 μm width and varying length from 1 to 3 μm, arranged singly or in short chains of 3–5 bacteria. Nonsporeforming. Cell wall of Gram-stain-positive type. Motile due to lateral flagella. Cells divide by binary transverse fission. Obligate anaerobe. On solid medium, produces round, white, translucent colonies. Grows chemolithoautotrophically on CO. Utilizes CO as the sole energy source with equimolar formation of H_2 and CO_2 according to the equation: $CO + H_2O \rightarrow CO_2 + H_2$. Does not grow on peptone, yeast extract, starch, cellobiose, sucrose, lactose, glucose, maltose, galactose, arabinose, fructose, acetate, formate, pyruvate, ethanol, methanol, H_2/CO_2 gas mixture (4:1, v/v), or on H_2, acetate, ethanol, or lactate in the presence or absence of elemental sulfur or sulfate. Does not reduce elemental sulfur or sulfate during growth with CO. Thermophile. Grows within the temperature range 40–68°C with an optimum at 58°C. pH for growth ranges from 6.5 to 7.6, with an optimum of 7.0. Growth and CO consumption are inhibited by penicillin, ampicillin, streptomycin, kanamycin, and neomycin. The habitat is terrestrial hot spring.

DNA G+C content (mol%): 46 ± 1 (T_m).

Type strain: 41, DSM 12326, VKM B-2282.

GenBank accession number (16S rRNA gene): AY061974.

2. **Carboxydocella sporoproducens** Slepova, Sokolova, Lysenko, Tourova, Kolganova, Kamzolkina, Karpov and Bonch-Osmolovskaya 2006, 799[VP]

References

Garrity, G.M., J.A. Bell and T. Lilburn. 2005. The Revised Road Map to the Manual. *In* Brenner, Krieg, Staley and Garrity (Editors), Bergey's Manual of Systematic Bacteriology, 2nd edn, Vol. 2, The *Proteobacteria*, Part A, Introductory Essays. Springer, New York, pp. 159–220.

Slepova, T.V., T.G. Sokolova, A.M. Lysenko, T.P. Tourova, T.V. Kolganova, O.V. Kamzolkina, G.A. Karpov and E.A. Bonch-Osmolovskaya. 2006.

spo.ro.pro.du′cens. Gr. n. *spora* a seed and, in biology, a spore; L. part. adj. *producens* producing; N.L. part. adj. *sporoproducens* spore-producing.

Cells are short straight rods of about 0.5 μm in width and varying length from 1 to 6 μm. Forms round or ovoid refractile endospores in terminally swollen sporangia. Nonmotile. Cell wall is of the Gram-stain-positive type. Cells divide by binary transverse fission. Grows within the temperature range of 50–70°C with an optimum at 60°C. Growth pH ranges from 6.2 to 8.0, with an optimum of 6.8. Grows chemolithoautotrophically on CO. Grows organoheterotrophically with yeast extract, sucrose, or pyruvate under N_2 in the gas phase. Does not grow on peptone, starch, cellobiose, glucose, arabinose, fructose, xylose, galactose, lactose, maltose, glycerol, acetate, citrate, succinate, formate, ethanol, methanol, $H_2:CO_2$ gas mixture (4:1, v/v). Does not reduce elemental sulfur, sulfate, or thiosulfate in the presence of H_2, acetate, lactate, glycerol, or xylose or during growth with CO. Elemental sulfur, sulfide, and nitrate inhibit growth. Growth is inhibited by penicillin, novobiocin, streptomycin, kanamycin, and neomycin. The habitat is a terrestrial hot spring.

DNA G+C content (mol%): 49.5 ± 1 (T_m).

Type strain: Kar, DSM 16521, VKM B-2358.

GenBank accession number (16S rRNA gene): AY673988.

Carboxydocella sporoproducens sp. nov., a novel anaerobic CO-utilizing/H_2-producing thermophilic bacterium from a Kamchatka hot spring. Int. J. Syst. Evol. Microbiol. *56*: 797–800.

Sokolova, T.G., N.A. Kostrikina, N.A. Chernyh, T.P. Tourova, T.V. Kolganova and E. Bonch-Osmolovskaya. 2002. *Carboxydocella thermautotrophica* gen. nov., sp. nov., a novel anaerobic, CO-utilizing thermophile from a Kamchatkan hot spring. Int. J. Syst. Evol. Microbiol. *52*: 1961–1967.

Family XVII. **Incertae Sedis**

Previously assigned to the "*Alicyclobacillaceae*" or *Syntrophomonadaceae* by Garrity et al. (2005), subsequent analyses suggest that these genera represent a very deep group within the *Firmicutes* or, possibly, a novel phylum. Until their taxonomic status has been clarified, they are assigned to a family *incertae sedis*.

Genus I. **Sulfobacillus** Golovacheva and Karavaiko 1991, 179[VP] (Effective publication: Golovacheva and Karavaiko 1978, 815.)

Milton S. da Costa, Fred A. Rainey and Luciana Albuquerque

Sul.fo.ba.cil′lus. L. neut. n. *sulfur* sulfur; L. n. *bacillus* small rod; N.L. masc. n. *Sulfobacillus* small sulfur-oxidizing rod.

Straight rods, 0.5–0.8 μm in diameter by 3.0–5.0 μm in length. Terminal or subterminal ovoid or spherical **endospores**. In some strains, the sporangium is swollen. Gram-stain-positive. Strains are generally **nonmotile; a few strains may be motile. Colonies are not pigmented. Mesophilic or slightly thermophilic** with optimum growth temperatures between about 40 and 55°C; the temperature range for growth is between 20 and 60°C. **Acidophilic**, the pH range for growth is from about 1.5 to 5.5; the optimum is pH 1.9–2.4. **Menaquinone-7** is the predominant respiratory quinone. Fatty acids are primarily **iso-** and **anteiso-branched**; straight-chain fatty acids are also present. ω-Cyclohexane or ω-cycloheptane fatty acids are the major components of some strains. **Aerobic** with a strictly respiratory type of metabolism; some strains grow under **anaerobic** conditions with ferric iron as electron acceptor. Species are **mixotrophic, with limited autotrophic and chemoorganotrophic growth.** Mixotrophic growth occurs on Fe^{2+}, S^0, $S_2O_3^{2-}$, $S_4O_6^{2-}$, and sulfide minerals in the presence of yeast extract or single organic compounds. Yeast extract and a few sugars and organic acids are used as sole carbon and energy sources. Strains require yeast extract for growth. Found in acidic **soils and water of geothermal areas, mineral deposits, and ores.**

DNA G+C content (mol%): 47–57 (T_m).

Type species: **Sulfobacillus thermosulfidooxidans** Golovacheva and Karavaiko 1991, 179[VP].

Further descriptive information

Four species are currently classified in the genus *Sulfobacillus*: *Sulfobacillus thermosulfidooxidans*, *Sulfobacillus acidophilus*, *Sulfobacillus sibiricus*, and *Sulfobacillus thermotolerans* (Bogdanova et al., 2006; Golovacheva and Karavaiko, 1978; Melamud et al., 2003; Norris et al., 1996a). The vast majority of the sulfobacilli produce endospores, which tend to be terminal or subterminal and have an ovoid morphology. Strains of the genus *Sulfobacillus* stain Gram-positive and the vast majority are not motile, although motility has been observed in strain L15, but flagella were not observed (Yahya et al., 1999).

The strains currently assigned to this genus are mesophilic or slightly thermophilic; the growth temperature range of these organisms varies between about 20 and 60°C, with optimum growth temperatures ranging between about 40 and 55°C (Bogdanova et al., 2006; Yahya et al., 1999). Organisms of the genus *Sulfobacillus* are extremely acidophilic with optimum pH of about 2.0. Some strains grow in media with pH as high as 5.0 and as low as 1.2 (Bogdanova et al., 2006).

Species of the genus *Sulfobacillus* are generally viewed as strict aerobes; however, several strains identified as *Sulfobacillus acidophilus* and *Sulfobacillus thermosulfidooxidans* are capable of growth under oxygen limitation or anaerobic conditions with ferric iron as terminal electron acceptor and glycerol as electron donor (Bridge and Johnson, 1998). Two of the four species of *Sulfobacillus* examined also appear to couple tetrathionate oxidation to ferric iron reduction and to be autotrophic under anaerobic conditions.

All strains of the genus *Sulfobacillus* have a mixotrophic metabolism on Fe^{2+}, S^0, $S_4O_6^{2-}$, $S_2O_3^{2-}$, and sulfide minerals such as pyrite, arsenopyrite, and pyrotite in the presence of low levels of yeast extract (0.01–0.2%). Other organic compounds such as some sugars, amino acids, or reduced glutathione are also used as carbon sources under mixotrophic growth. Autotrophic growth on Fe^{2+} is generally poor, ceasing after a few transfers of the culture, and it is possible that autotrophic growth in some strains is due to carry over of organic compounds from mixotrophically grown inocula. Organotrophic growth is poor in all strains examined and generally limited to a few culture transfers. However, *Sulfobacillus thermosulfidooxidans* VKM B-1269[T] grows indefinitely in medium containing 0.05% yeast extract as long as the culture is transferred daily, but *Sulfobacillus sibiricus* N1[T] does not grow heterotrophically and appears not to grow under autotrophic conditions with elemental sulfur as the energy source (Melamud et al., 2003; Norris et al., 1996a). The energy metabolism of species of the genus *Sulfobacillus* has not been examined in any detail. It is known that *Sulfobacillus thermosulfidooxidans* VKM B-1269[T] and *Sulfobacillus* strain 41 metabolize glucose primarily via the Entner–Doudoroff pathway,

but after enrichment of the medium with 5% CO_2, glucose is metabolized via the Embden–Meyerhof pathway (Krasil'nikova et al., 2001). *Sulfobacillus sibiricus* N1[T], on the other hand, is reported to assimilate glucose via the fructose bisphosphate pathway and the pentose phosphate pathway (Zakharchuk et al., 2003). These authors also report the presence of ribulose-bisphosphate carboxylase indicating carbon dioxide fixation and autotrophic growth. Enzymes of sulfur metabolism, namely sulfite oxidase, sulfur oxygenase, adenylyl sulfate reductase, rhodonase, sulfur:Fe(III) oxidoreductase, and sufite:Fe(III) oxidoreductase have been found in *Sulfobacillus sibiricus* N1[T] and *Sulfobacillus* strain SSO (Krasil'nikova et al., 2004).

The sulfobacilli possess relatively high proportions of iso- and anteiso-branched fatty acids under mixotrophic growth conditions; ω-cyclohexane and ω-cycloheptane fatty acids are also present in some strains examined grown under organotrophic conditions. These ω-cyclic fatty acids have not been detected in *Sulfobacillus sibiricus* strain N1[T], probably because this organism does not grow heterotrophically (Melamud et al., 2003; Tsaplina et al., 1994). The type strain of *Sulfobacillus acidophilus* possesses relatively large proportions of cycloheptane and cyclohexane fatty acids, namely 18:0 ω-cycloheptane and 18:0 ω-cyclohexane (L. Albuquerque and M. S. da Costa, unpublished results), when grown on heterotrophic *Acidomicrobium* medium (Norris et al., 1996a). Terminal cyclic fatty acids are an important hallmark characteristic of 14 of the 17 described species of the genus *Alicyclobacillus*, but it should be noted that the presence of ω-cyclic fatty acids in the sulfobacilli does not necessarily imply a close phylogenetic relationship with the alicyclobacilli, since other unrelated bacteria, such as *Curtobacterium pusillum* (Suzuki et al., 1981) and *Propionibacterium cyclohexanicum* (Kusano et al., 1997) also possess these fatty acids. The major respiratory quinone is menaquinone-7 (Bogdanova et al., 2006).

Strains of the genus *Sulfobacillus* have been isolated from acidic environments associated with mineral deposits and ores, mineral processing mills, and acidic geothermal areas. The type strain of *Sulfobacillus acidophilus* and other closely related strains were isolated from coal spoil heaps in the UK, thermal springs in Iceland, and Yellowstone National Park (Norris et al., 1996a). The type strain of *Sulfobacillus thermosulfidooxidans* was isolated from a copper–zinc deposit in Kasakhastan (Golovacheva and Karavaiko, 1978), whereas the type strain of *Sulfobacillus thermotolerans* and other closely related strains (L15, RIV14, and Y0017) have been isolated during laboratory oxidation of a gold-containing concentrate, as well as from the acidic geothermal environments on the Island of Montserrat and Yellowstone National Park (Atkinson et al., 2000; Yahya et al., 1999). Three strains, designated M-13, M-16, and M-17, were isolated from uranium mines and uranium enrichment plants in Pakistan (Ghauri et al., 2003). Strain N1[T], the type strain of *Sulfobacillus sibiricus*, was isolated from a gold–arsenic concentrate obtained from an ore deposit in East Siberia (Melamud et al., 2003). Another strain, designated Fras1 (AF213055), was isolated from an unexpected environment for members of the genus *Sulfobacillus*, namely an acidic cave wall biofilm in Italy, but no further information regarding this strain has been published yet.

Enrichment, isolation and growth conditions

Enrichments of sulfobacilli and growth of the isolates are carried out in media containing low levels of yeast extract supplemented with reduced inorganic compounds, such as Fe^{2+}. One medium used extensively for the enrichment and cultivation of these organisms is modified 9K medium, also known as *Sulfobacillus* medium (Melamud and Pivovarova (1998); DSMZ medium 665). This medium is made up of three solutions. Solution A contains (g, all in 700 ml water adjusted to pH 1.8–2.0 with 5 M H_2SO_4): $(NH_4)_2SO_4$, 3.0; KCl, 0.1; K_2HPO_4, 0.5; $MgSO_4 \cdot 7H_2O$, 0.5; and $Ca(NO_3)_2$, 0.01. Solution B contains 2.0 g ferrous iron in 300 ml water acidified with 1 ml 5 M H_2SO_4. Solution C is 20 ml yeast extract solution (1%, w/v, in water). The three solutions are mixed after autoclaving. This medium can be supplemented with reduced sulfur compounds or sulfide ores (Bogdanova et al., 2006; Melamud et al., 2003; Norris et al., 1996a).

Maintenance procedures

Long-term cultures can be maintained frozen at –70 °C in cryotubes containing liquid media, mentioned above, supplemented with a final concentration of 15% (v/v) glycerol. Freeze-dried and liquid nitrogen storage cultures have been maintained for several years without loss of viability.

Taxonomic comments

The taxonomy of the genus *Sulfobacillus* has suffered from the description of species whose names have not been validated and which, in some cases, have been poorly characterized (Johnson et al., 2003, 2005; Yahya et al., 1999). Phylogenetic analysis of 16S rRNA gene sequences has shown that some of these organisms, namely strains L15 (AY007663), RIV14 (AY007664), and YTE1 (AY007665), probably represent true genomic species of the genus *Sulfobacillus* (Figure 22). However, few taxonomic characteristics are available for these three strains.

Two strains formerly classified in the genus *Sulfobacillus* are now assigned as species of *Alicyclobacillus*. Strain SD-11[T]

was classified as *Sulfobacillus disulfidooxidans* on the basis of its ability to utilize elemental sulfur and pyrite as sole source of energy under mixotrophic growth conditions (Dufresne et al., 1996). Phylogenetic analysis of the 16S rRNA gene sequence showed that strain SD-11[T] fell within the radiation of species of the genus *Alicyclobacillus* (Goto et al., 2002, 2003; Matsubara et al., 2002) and it was classified as *Alicyclobacillus disulfidooxidans* (Figure 22) (Karavaiko et al., 2005). Another strain, designated K1[T], isolated from lead–zinc ores and initially named "*Sulfobacillus thermosulfidooxidans* subsp. *thermotolerans*" because it also had the ability to oxidize iron, elemental sulfur, and sulfides (Kovalenko and Malakhova, 1983), was reclassified as *Alicyclobacillus tolerans*, because of its close phylogenetic relationship to other species of *Alicyclobacillus* (Figure 22) (Karavaiko et al., 2005). It should be noted that *Sulfobacillus thermosulfidooxidans* was formerly divided into two subspecies informally named "*Sulfobacillus thermosulfidooxidans* subsp. *asporogenes*" for strain 41 and "*Sulfobacillus thermosulfidooxidans* subsp. *thermotolerans*" for strain K1[T] (Kovalenko and Malakhova, 1983). As mentioned above, strain K1[T] was reclassified as the type strain of the species *Alicyclobacillus tolerans* (Karavaiko et al., 2005), while the taxonomy of strain 41 has not been examined further.

The species of the genus *Sulfobacillus* form a monophyletic group supported by bootstrap analyses (Figure 22). *Sulfobacillus acidophilus* is the deepest branch of the genus cluster and shows 16S rRNA gene sequence similarity in the range 91.4–92.7% to other strains of this genus. The 16S rRNA gene sequence similarity range between members of the *Sulfobacillus* with validly published names is 91.4–98.8%, with the highest similarity found between *Sulfobacillus thermosulfidooxidans* and *Sulfobacillus sibiricus*. Strains L15 and RIV14, isolated from acidic geothermal environments on the island of Montserrat (Atkinson et al., 2000), are most closely related to *Sulfobacillus thermotolerans*, sharing 16S rRNA gene sequence similarities of 99.0 and 98.3%, respectively.

List of species of the genus *Sulfobacillus*

1. **Sulfobacillus thermosulfidooxidans** Golovacheva and Karavaiko 1991, 179[VP] (Effective publication: Golovacheva and Karavaiko 1978, 815.)

ther.mo.sul.fi.do.ox.i'dans. Gr. adj. *thermos* hot; L. neut. n. *sulfur* sulfur; N.L. v. *oxidans* oxidize; N.L. part. adj. *thermosulfidooxidans* thermophilic sulfide oxidizing.

Strains of this species stain Gram-positive and form nonmotile, rod-shaped cells, 1–6 μm in length and 0.6–0.8 μm in diameter with rounded or tapered ends. Cells occur singly, in pairs, or in short chains. Sporangia are slightly or markedly swollen; spherical or slightly oval endospores are located subterminally, terminally, or paracentrally. Colonies on agar media with Fe^{2+} are rounded, shining, initially yellowish and then reddish-brown in color. The pH range for growth is 1.5–5.5, optimum pH is around 1.9–2.4; the temperature range for growth is 20–60 °C, optimum temperature is around 50–55 °C. Strictly aerobic; anaerobic growth does not occur in medium contain-

ing nitrate. Mixotrophic and facultatively organotrophic; Fe^{2+}, S^0, $S_2O_3^{2-}$, $S_4O_6^{2-}$, and sulfide minerals are oxidized in the presence of organic substrates. Strains grow poorly organotrophically on yeast extract, casein, glucose, sucrose, fructose, trehalose, mannose, raffinose, glutamate, and reduced glutathione. Autotrophic growth is very poor and limited to a small number of transfers. Strains of this species have been isolated from zones of spontaneous heating of ores in the Nikolaev copper–zinc–pyrite deposit.

DNA G+C content (mol%): 47.2–47.5 (T_m; type strain).

Type strain: AT-1, DSM 9293, VKM B-1269.

GenBank accession number (16S rRNA gene): AB089844.

2. **Sulfobacillus acidophilus** Norris, Clark, Owen and Waterhouse 1996b, 1189[VP] (Effective publication: Norris, Clark, Owen and Waterhouse 1996a, 781.)

a.ci.do'phi.lus. L. n. *acidum* acid; Gr. adj. *philos* loving; N.L. adj. *acidophilus* acid-loving.

Strains of this species stain Gram-positive and form rod-shaped cells, 3–5 µm in length and 0.5–0.8 µm in diameter. Spherical endospores are located terminally. The optimum pH is around 2.0; the optimum temperature is around 45–50 °C. Aerobic, mixotrophic, autotrophic, and organotrophic. Some strains are facultatively anaerobic. Autotrophic growth occurs with ferrous iron and elemental sulfur as substrates. Mixotrophic; Fe^{2+}, S^0, $S_4O_6^{2-}$, and sulfide minerals are oxidized in the presence of organic substrates. Strains utilize yeast extract for organotrophic growth; some strains use glucose, sucrose, fructose, and ribose. Strains of this species have been isolated from various acidic environments rich in iron, sulfur, or mineral sulfides.

DNA G+C content (mol%): 55–57 (T_m; type strain).
Type strain: NAL, DSM 10332.
GenBank accession number (16S rRNA gene): AF050169.

3. **Sulfobacillus sibiricus** Melamud, Pivovarova, Tourova, Kolgonova, Osipov, Lysenko, Kondrat'eva and Karavaiko 2006, 499[VP] (Effective publication: Melamud, Pivovarova, Tourova, Kolgonova, Osipov, Lysenko, Kondrat'eva and Karavaiko 2003, 611.)

si.bi'ri.cus. N.L. masc. adj. *sibiricus* pertaining to Siberia.

Strains of this species stain Gram-positive and form nonmotile, rod-shaped cells, 1.0–3.0 µm in length and 0.7–1.1 µm in diameter. Spherical endospores are located subterminally. The pH range for growth is 1.1–2.6; optimum pH is around 2.0. The temperature range for growth is 17–60 °C; optimum temperature is around 55 °C. Mixotrophic. Branched-chain iso- and anteiso-fatty acids are produced under mixotrophic conditions. Stable growth is achieved with the simultaneous utilization of ferrous iron, elemental sulfur, or sulfide minerals and reduced glutathione, yeast extract, glucose, fructose, sucrose, sorbitol, glutamate, and alanine.

DNA G+C content (mol%): 48.2 (T_m; type strain).
Type strain: N1, DSM 17363, VKM B-2280.
GenBank accession number (16S rRNA gene): AY079150.

4. **Sulfobacillus thermotolerans** Bogdanova, Tsaplina, Kondrat'eva, Duda, Suzina, Melamud, Tourova and Karavaiko 2006, 1041[VP]

ther.mo.to'le.rans. Gr. adj. *thermos* hot; L. part. adj. *tolerans* tolerating; N.L. part. adj. *thermotolerans* hot-tolerating.

Strains of this species stain Gram-positive and form nonmotile, rod-shaped cells, 1.5–4.5 µm in length and 0.8–1.2 µm in diameter with rounded ends. Spherical endospores are located subterminally. The pH range for growth is 1.2–2.4; optimum pH is around 2.0. The temperature range for growth is 18–60 °C; optimum temperature is around 40–42 °C. Major respiratory quinone is menaquinone-7. Aerobic. Autotrophic and organotrophic growth is possible for a limited number of passages onto fresh medium. Mixotrophic; Fe^{2+}, S^0, $S_2O_3^{2-}$, $S_4O_6^{2-}$, and sulfide minerals are oxidized in the presence of organic substrates. Strains utilize yeast extract, malate, glucose, sucrose, fructose, and reduced glutathione for organotrophic growth. Strains of this species have been isolated during the course of pilot tests when a gold-containing sulfide concentrate was oxidized under mesophilic conditions.

DNA G+C content (mol%): 48.1 (T_m; type strain).
Type strain: Kr1, DSM 17362, VKM B-2339.
GenBank accession number (16S rRNA gene): DQ124681.

Genus II. **Thermaerobacter** Takai, Inoue and Horikoshi 1999, 625[VP] emend. Spanevello, Yamamoto and Patel 2002, 799[VP]

MARK D. SPANEVELLO AND BHARAT K. C. PATEL

Therm.ae.ro.bac'ter. Gr. adj. *thermos* hot; Gr. n. *aer* air; L. *bacter* masc. equivalent of Gr. neut. n. *bakterion* rod or staff; N.L. masc. n. *Thermaerobacter* rod, which grows at high temperatures in the presence of air.

Rods, 2–10 × 0.2–0.6 µm usually **arranged singly or in pairs. Stain Gram-negative or Gram-variable. Motile with monopolar or bipolar flagella.** Nonmotile species do not possess flagella. **May or may not form spores. Oxidase-positive. Strict aerobes**, having a strictly respiratory type of metabolism with oxygen as the terminal electron acceptor. **Thermophilic**, with growth occurring between 50 and 80 °C with optimum between 70 and 75 °C. **Neutrophiles or alkalophiles** with growth occurring between pH 5 and 10 with an optimum pH between 7 and 8.5. **Chemoheterotrophic.** May utilize organic substrates such as yeast extract, peptone, cellulose, starch, chitin, casein, Casamino acids, and a variety of sugars, carboxylic acids, and amino acids. Isolated from nonthermal sediments of the Mariana Trench, bore runoff channels of the thermal bore wells of Great Artesian Basin, and shallow marine hydrothermal vents.

DNA G+C content (mol%): 69–72.

Type species: **Thermaerobacter marianesis** Takai, Inoue and Horikoshi 1999, 625[VP].

Further descriptive information

The first member of the genus *Thermaerobacter, Thermaerobacter marianesis*, was isolated from the nonthermal sediments of Mariana Trench Challenger Deep, the world's deepest seafloor (10,897 m) (Takai et al., 1999) and the second member, *Thermaerobacter nagasakiensis*, was isolated from shallow marine hydrothermal vents (Nunoura et al., 2002). In addition, as of December 2005, "*Thermaerobacter riparius*" strain KW1 and "*Thermaerobacter*" strain C4-1 were isolated from coastal hydrothermal fields in Asia. 16S rRNA gene sequences of these isolates were deposited in GenBank under accession numbers AY936496 and AY094621, respectively, but phenotypic characteristics have not yet been reported. This could lead one to speculate that the most likely primary habitat of *Thermaerobacter* species is the hydrothermal vents and that mantle subduction events led to the deposition of *Thermaerobacter marianesis* in the Mariana Trench Challenger Deep sediments. However, the isolation of the third species, *Thermaerobacter subterraneus*, from sediments of nonvolcanically heated subsurface freshwaters of an Australian bore well runoff channel suggests that *Thermaerobacter* may have a more diverse habitat than previously thought (Spanevello et al., 2002).

Enrichment and isolation procedures

The marine isolate *Thermaerobacter marianensis* was enriched at 75 °C on MJP medium and isolated by dilution to extinction as it did not form colonies on agar-fortified MJP medium (Sako et al., 1996a, 1996b) and *Thermaerobacter nagasakiensis* was enriched on JX medium at 70 °C and isolated by streaking on agar-amended JX medium (Sako et al., 1996b). Both these marine isolates can be grown on Medium 514, Bacto Marine Broth (BD, Inc., Franklin Lakes, NJ, USA) as recommended by the DSMZ (Deutsche Sammlung von Mikroorganismen und Zellkulturen), Braunschweig, Germany. The freshwater isolate, *Thermaerobacter subterraneus*, was enriched in Castenholz TYE media (ATCC medium 461) at 70 °C and isolated by streaking on agar-amended medium

Maintenance procedures

Thermaerobacter species can be stored at room temperature for up to 3 months. Cells can be suspended in a preservation broth, which consists of a 50:50 (v/v) mixture of growth medium and glycerol, and stored at –80 °C for long-term preservation.

Differentiation of *Thermaerobacter* from other genera

In this volume, the genus *Thermaerobacter* is classified with *Sulfobacillus* in Family XVII *Incertae Sedis* in recognition of the low sequence similarity of its 16S rRNA genes with those of other *Firmicutes* (Ludwig et al., 2009). While previous phylogenetic analyses placed the genus within the family *Syntrophomonadaceae* and order *Clostridiales* (Garrity et al., 2005), this conclusion is not consistent with the large phenotypic differences between these groups and examination of additional rRNA gene sequences (Figure 7). For instance, members of the order *Clostridiales* possess an anaerobic metabolism (Collins et al., 1994) with the exception of the three *Thermaerobacter* species and *Sulfobacillus*, which are all aerobes.

Taxonomic comments

The three validly published, named strains of *Thermaerobacter* species exhibit a high 16S rDNA similarity (98.3–99.7%) to one another, but phenotypic characteristics and DNA–DNA hybridization data justify recognition of *Thermaerobacter nagasakiensis* and *Thermaerobacter marianensis* as separate species. However, *Thermaerobacter nagasakienesis* and *Thermaerobacter subterraneus*, which have similar pH and temperature optima, can be distinguished on the basis of the presence or absence of flagella, motility, and spore-forming abilities. DNA–DNA hybridization studies between *Thermaerobacter marienesis* with *Thermaerobacter subterraneus* show only 5% homology and those between *Thermaerobacter nagasakieneis* and *Thermaerobacter marienesis*, 40%. DNA hybridization studies have not yet been performed between *Thermaerobacter subterraneus* and *Thermaerobacter nagasakiensis*. "*Thermaerobacter riparius*" and "*Thermaerobacter strain C4-1*" have a lower 16S rDNA similarity to the three described species (98% and 97%, respectively).

Spores have not been observed in *Thermaerobacter marianesis* and *Thermaerobacter nagasakiensis* though the sporulation gene *spo0A*, but not gene *ssp*, was detected by PCR. This is in direct contrast to *Thermaerobacter subterraneus*, which forms terminal ellipsoidal spores. This lack of congruence of phylogeny with physiology is notable, but not unusual. For example, all members of the family *Bacillaceae* are strict aerobes or facultative anaerobes with the exception of *Bacillus infernus* which is an obligate anaerobe, which necessitated amending the genus description (Boone et al., 1995).

Further reading

Spanevello, M.D. (2001). The phylogeny of prokaryotes associated with Australia's Great Artesian Basin. PhD thesis, Griffith University, published online at http://www4.gu.edu.au:8080/adt-root/uploads/approved/adt-QGU20030303.094942/public/02Whole.pdf.

Differentiation of the species of the genus *Thermaerobacter*

Three species of *Thermaerobacter* are currently described. Table 234 lists the features that differentiate these species.

List of species of the genus *Thermaerobacter**

1. **Thermaerobacter marianensis** Takai, Inoue and Horikoshi 1999, 625[VP]

   ma.ri.a.nen′sis. N.L. masc. adj. *marianensis* pertaining to the Mariana Trench, the source of the type strain.

   Rod-shaped cells, 2–7 × 0.3–0.6 μm. Gram-stain-negative or variable. Colonies are not formed on agar. Does not form spores and is nonmotile. Strictly aerobic and chemoheterotrophic. Growth occurs between 50 and 80 °C (optimum is 70 °C), pH 5.4–9.5 (optimum is 7–7.5), and 0.5–5% NaCl (optimum, 2%).

   DNA G+C content (mol%): 72.5 (HPLC).
   Type strain: 7p75a, JCM 10246, DSM 12885, ATCC 700841.
   GenBank accession number (16S rRNA gene): AB011495.

2. **Thermaerobacter nagasakiensis** Nunoura, Akihara, Takai and Sako 2002, 344[VP]

   na.ga.sa.ki.en′sis. N.L.adj. *nagasakiensis* pertaining to Nagasaki Prefecture, the source of the type strain.

   Rod-shaped cells, 1.5–4 × 0.3–0.5 μm. Gram-stain-negative. Colonies are ivory-white on JX medium amended with agar. Does not form spores and is motile by monopolar or dipolar

---

*Since acceptance of this chapter, a fourth species, *Thermaerobacter litoralis*, has been validly published by Tanaka et al. (2006).

TABLE 234. Characteristics differentiating the species of the genus *Thermaerobacter*[a]

Characteristic	*T. marianesis*	*T. nagasakiensis*	*T. subterraneus*
Isolation source	Mud sample from the bottom of the 10,897 m. deep Mariana Trench	Shallow marine hydrothermal vent, Tachibana Bay, Nagasaki Prefecture, Japan	Water-sediment slurries of the runoff channel of New Lorne Bore (number 17263), Great Artesian Basin of Australia
Morphology and size (μm)	Curved rods, single or in pairs; 2–7 × 0.3–0.6 μm in exponential phase and 10 μm in stationary phase	Rods 1–4 × 0.2–0.5 μm	Rods occur singly or in pairs 2–10 × 0.3 μm
Gram stain	Positive in exponential phase and negative in stationary phase	Negative	Negative
DNA G+C content (mol%)	72.5 (HPLC)	68 (HPLC)	70 (T_m)
Spore-forming ability	–	–	+
Motility	–	+	–
Presence of flagella	–	+ (Monopolar or bipolar)	–
Temperature growth range (°C)	50–80	52–78	55–80
Temperature optimum (°C)	75	70	70
pH growth range	5.4–9.5	5–8	6–10
pH optimum	7.0–7.5	7	8.5
NaCl requirements	+	+	–
NaCl range for growth (optimum) (%)	0.5–5 (2)	0–4.5 (1)	0–1 (0)
Colony forming ability on agar	–	+	+
Requirement for yeast extract and peptone for growth	–	+	+
Growth on carbohydrates	+[b]	–	–
Growth on amino acids	+[c]	–	–
Growth on carboxylic acids	+[d]	–	–

[a]Symbols: +, >85% positive; –, 0–15% positive.

[b]Grows well on starch, xylan, chitin, maltose, maltotriose, cellobiose, lactose, trehalose, sucrose, glucose, galactose, xylose, mannitol, inositol, and mannitol, with weak growth on cellulose.

[c]Grows well on casein, Casamino acids, valine, isoleucine, cysteine, proline, serine, threonine, asparagine, glutamine, aspartate, glutamate, lysine, arginine, histidine but weakly on glycine, alanine, leucine, methionine, and phenylalanine.

[d]Grows well on propionate, 2-aminobutyric acid, malate, pyruvate, tartarate, succinate, lactate, acetate, and glycerol.

flagella. Strictly aerobic and chemoheterotrophic. Growth between 52 and 78 °C (optimum is 70 °C), pH 5–8 (optimum is 7) and 0–4.5% NaCl (optimum is 1%).

DNA G+C content (mol%): 69 (HPLC).

Type strain: Ts1a, JCM 11223, DSM 14512.

GenBank accession number (16S rRNA gene): AB011495.

3. **Thermaerobacter subterraneus** Spanevello, Yamamoto and Patel 2002, 799[VP]

sub.ter.ran'e.us. L. masc. adj. *subterraneus* referring to under the earth, the source of the type strain.

Rod-shaped cells, 2–10 × 0.3 μm. Gram-stain-negative. Colonies are ivory-white on Castenholz TYE media (ATCC medium 461) amended with agar. Forms spores and is nonmotile. Strictly aerobic and chemoheterotrophic. Growth between 55 and 80 °C (optimum is 70 °C), pH 6.5–10.5 (optimum is 8.5), and 0–1% NaCl (optimum, 0%).

DNA G+C content (mol%): 71 (T_m).

Type strain: C21, ATCC BAA-137, DSM 13965.

GenBank accession number (16S rRNA gene): AF343566.

References

Atkinson, T., S. Gairns, D.A. Cowan, M.J. Danson, D.W. Hough, D.B. Johnson, P.R. Norris, N. Raven, C. Robinson, R. Robson and R.J. Sharp. 2000. A microbiological survey of Montserrat Island hydrothermal biotopes. Extremophiles *4*: 305–313.

Bogdanova, T.I., I.A. Tsaplina, T.F. Kondrat'eva, V.I. Duda, N.E. Suzina, V.S. Melamud, T.P. Tourova and G.I. Karavaiko. 2006. *Sulfobacillus thermotolerans* sp. nov., a thermotolerant, chemolithotrophic bacterium. Int. J. Syst. Evol. Microbiol. *56*: 1039–1042.

Boone, D.R., Y.T. Liu, Z.J. Zhao, D.L. Balkwill, G.R. Drake, T.O. Stevens and H.C. Aldrich. 1995. *Bacillus infernus* sp. nov., an Fe(III)-reducing and Mn(IV)-reducing anaerobe from the deep terrestrial subsurface. Int. J. Syst. Bacteriol. *45*: 441–448.

Bridge, T.A.M. and D.B. Johnson. 1998. Reduction of soluble iron and reductive dissolution of ferric iron-containing minerals by moderately thermophilic iron-oxidizing bacteria. Appl. Environ. Microbiol. *64*: 2181–2186.

Collins, M.D., P.A. Lawson, A. Willems, J.J. Cordoba, J. Fernández-Garayzábal, P. Garcia, J. Cai, H. Hippe and J.A.E. Farrow. 1994. The phylogeny of the genus *Clostridium*: proposal of five new genera and eleven new species combinations. Int. J. Syst. Bacteriol. *44*: 812–826.

Dufresne, S., J. Bousquet, M. Boissinot and R. Guay. 1996. *Sulfobacillus disulfidooxidans* sp. nov., a new acidophilic, disulfide-oxidizing, gram-positive, spore-forming bacterium. Int. J. Syst. Bacteriol. *46*: 1056–1064.

Garrity, G.M., J.A. Bell and T. Lilburn. 2005. The Revised Road Map to the Manual. *In* Brenner, Krieg, Staley and Garrity (Editors), Bergey's Manual of Systematic Bacteriology, 2nd edn, Vol. 2, The *Proteobacteria*, Part A, Introductory Essays. Springer, New York, pp. 159–220.

Ghauri, M.A., A.M. Khalid, S. Grant, S. Heaphy and W.D. Grant. 2003. Phylogenetic analysis of different isolates of *Sulfobacillus* spp. isolated from uranium-rich environments and recovery of genes using integron-specific primers. Extremophiles *7*: 341–345.

Golovacheva, R.S. and G.I. Karavaiko. 1978. [*Sulfobacillus*, a new genus of thermophilic sporulating bacteria]. Mikrobiologiia *47*: 815–822.

Golovacheva, R.S. and G.I. Karavaiko. 1991. *In* Validation of the publication of new names and new combinations previously effectively published outside the IJSB. List no. 36. Int. J. Syst. Bacteriol. *41*: 178–179.

Goto, K., H. Matsubara, K. Mochida, T. Matsumura, Y. Hara, M. Niwa and K. Yamasato. 2002. *Alicyclobacillus herbarius* sp. nov., a novel bacterium containing ω-cycloheptane fatty acids, isolated from herbal tea. Int. J. Syst. Evol. Microbiol. *52*: 109–113.

Goto, K., K. Mochida, M. Asahara, M. Suzuki, H. Kasai and A. Yokota. 2003. *Alicyclobacillus pomorum* sp. nov., a novel thermo-acidophilic, endospore-forming bacterium that does not possess omega-alicyclic fatty acids, and emended description of the genus *Alicyclobacillus*. Int. J. Syst. Evol. Microbiol. *53*: 1537–1544.

Johnson, D.B., N. Okibe and F.F. Roberto. 2003. Novel thermo-acidophilic bacteria isolated from geothermal sites in Yellowstone National Park: physiological and phylogenetic characteristics. Arch. Microbiol. *180*: 60–68.

Johnson, D.B., N. Okibe and K.B. Hallberg. 2005. Differentiation and identification of iron-oxidizing acidophilic bacteria using cultivation techniques and amplified ribosomal DNA restriction enzyme analysis. J. Microbiol. Methods *60*: 299–313.

Karavaiko, G.I., T.I. Bogdanova, T.P. Tourova, T.F. Kondrat'eva, I.A. Tsaplina, M.A. Egorova, E.N. Krasil'nikova and L.M. Zakharchuk. 2005. Reclassification of 'Sulfobacillus thermosulfidooxidans subsp. thermotolerans' strain K1 as *Alicyclobacillus tolerans* sp. nov. and *Sulfobacillus disulfidooxidans* Dufresne *et al.* 1996 as *Alicyclobacillus disulfidooxidans* comb. nov., and emended description of the genus *Alicyclobacillus*. Int. J. Syst. Evol. Microbiol. *55*: 941–947.

Kovalenko, E.V. and P.T. Malakhova. 1983. The spore-forming iron-oxidizing bacterium *Sulfobacillus thermosulfidooxidans*. Mikrobiologiya *52*: 962–966.

Krasil'nikova, E.N., I.A. Tsaplina, L.M. Zakharchuk and T.I. Bogdanova. 2001. Effects of exogenous factors on the activity of enzymes involved in carbon metabolism in thermoacidophilic bacteria of the genus *Sulfobacillus* (in Russian). Prikl. Biokhim. Mikrobiol *37*: 418–423.

Krasil'nikova, E.N., T.I. Bogdanova, L.M. Zakharchuk and I.A. Tsaplina. 2004. Sulfur metabolism enzymes in the thermoacidophilus bacteria *Sulfobacillus sibiricus*. (in Russian). Prikl. Biokhim. Mikrobiol. *40*: 62–65.

Kusano, K., H. Yamada, M. Niwa and K. Yamasato. 1997. *Propionibacterium cyclohexanicum* sp. nov, a new acid-tolerant omega-cyclohexyl fatty acid-containing *Propionibacterium* isolated from spoiled orange juice. Int. J. Syst. Bacteriol. *47*: 825–831.

Ludwig, W., K.-H. Schleifer and W.B. Whitman. 2009. Revised road map to the phylum *Firmicutes*. *In* De Vos, Garrity, Jones, Krieg, Ludwig, Rainey, Schleifer and Whitman (Editors), Bergey's Manual of Systematic Bacteriology, 2nd edn, Vol. 3, The *Firmicutes*. Springer, New York, pp. 1–14.

Matsubara, H., K. Goto, T. Matsumura, K. Mochida, M. Iwaki, M. Niwa and K. Yamasato. 2002. *Alicyclobacillus acidiphilus* sp. nov., a novel thermo-acidophilic, omega-alicyclic fatty acid-containing bacterium isolated from acidic beverages. Int. J. Syst. Evol. Microbiol. *52*: 1681–1685.

Melamud, V.S. and T.A. Pivovarova. 1998. Specific features of the growth of the type strain of *Sulfobacillus thermosulfidooxidans* in the 9K medium. (in Russian). Prikl. Biokhim. Mikrobiol *34*: 309–315.

Melamud, V.S., T.A. Pivovarova, T.P. Tourova, T.V. Kolganova, G.A. Osipov, A.M. Lysenko, T.F. Kondrat'eva and G.I. Karavaiko. 2003. *Sulfobacillus sibiricus* sp. nov., a new moderately thermophilic bacterium. Microbiology (En. transl. from Mikrobiologiya) *72*: 605–612.

Melamud, V.S., T.A. Pivovarova, T.P. Tourova, T.V. Kolganova, G.A. Osipov, A.M. Lysenko, T.F. Kondrat'eva and G.I. Karavaiko. 2006. *In* Validation of the publication of new names and new combinations previously effectively published outside the IJSEM. List no. 108. Int. J. Syst. Evol. Microbiol. *56*: 499–500.

Norris, P.R., D.A. Clark, J.P. Owen and S. Waterhouse. 1996a. Characteristics of *Sulfobacillus acidophilus* sp. nov. and other moderately thermophilic mineral-sulphide-oxidizing bacteria. Microbiology *142*: 775–783.

Norris, P.R., D.A. Clark, J.P. Owen and S. Waterhouse. 1996b. *In* Validation of the publication of new names and new combinations previously effectively published outside the IJSB. List no. 59. Int. J. Syst. Bacteriol *46*: 1189–1190.

Nunoura, T., S. Akihara, K. Takai and Y. Sako. 2002. *Thermaerobacter nagasakiensis* sp. nov., a novel aerobic and extremely thermophilic marine bacterium. Arch. Microbiol. *177*: 339–344.

Sako, Y., N. Nomura, A. Uchida, Y. Ishida, H. Morii, Y. Koga, T. Hoaki and T. Maruyama. 1996a. *Aeropyrum pernix* gen. nov., sp. nov., a novel aerobic hyperthermophilic archaeon growing at temperatures up to 100 °C. Int. J. Syst. Bacteriol. *46*: 1070–1077.

Sako, Y., K. Takai, Y. Ishida, A. Uchida and Y. Katayama. 1996b. *Rhodothermus obamensis* sp. nov., a modern lineage of extremely thermophilic marine bacteria. Int. J. Syst. Bacteriol. *46*: 1099–1104.

Spanevello, M.D., H. Yamamoto and B.K.C. Patel. 2002. *Thermaerobacter subterraneus* sp. nov., a novel aerobic bacterium from the Great Artesian Basin of Australia, and emendation of the genus *Thermaerobacter*. Int. J. Syst. Evol. Microbiol. *52*: 795–800.

Suzuki, K.I., K. Saito, A. Kawaguchi, S. Okuda and K. Komagata. 1981. Occurrence of ω-cyclohexyl fatty acids in *Curtobacterium pusillum* strains. J. Gen. Appl. Microbiol. *27*: 261–266.

Takai, K., A. Inoue and K. Horikoshi. 1999. *Thermaerobacter marianensis* gen. nov., sp. nov., an aerobic extremely thermophilic marine bacterium from the 11000 m deep Mariana Trench. Int. J. Syst. Bacteriol. *49*: 619–628.

Tanaka, R., S. Kawaichi, H. Nishimura and Y. Sako. 2006. *Thermaerobacter litoralis* sp. nov., a strictly aerobic and thermophilic bacterium isolated from a coastal hydrothermal field. Int. J. Syst. Evol. Microbiol. *56*: 1531–1534.

Tsaplina, I.A., G.A. Osipov, T.I. Bogdanova, T.P. Nedorezova and G.I. Karavaiko. 1994. Fatty acid composition of lipids in thermoacidophilic bacteria of the genus *Sulfobacillus* (in Russian). Mikrobiologiya *63*: 821–830.

Yahya, A., F.F. Roberto and D.B. Johnson. 1999. Novel mineral-oxidizing bacteria from Montserrat (W.I.): physiological and phylogenetic characteristics. *In* Amils and Ballester (Editors), Biohydrometallurgy and the environment toward the mining of the 21st century: Process Metallurgy 9A. Elsevier, Amsterdam, pp. 729–740.

Zakharchuk, L.M., M.A. Egorova, I.A. Tsaplina, T.I. Bogdanova, E.N. Krasil'nikova, V.S. Melamud and G.I. Karavaiko. 2003. Activity of the enzymes of carbon metabolism in *Sulfobacillus sibiricus* under various conditions of cultivation (in Russian). Mikrobiologiya *72*: 621–626.

Family XVIII. **Incertae Sedis**

Symbiobacterium represents a very deep group within the *Firmicutes* or, possibly, a novel phylum. Until its taxonomic status has been clarified, it is assigned to its own family *incertae sedis*.

Genus I. **Symbiobacterium** Ohno, Shiratori, Park, Saitoh, Kumon, Yamashita, Hirata, Nishida, Ueda and Beppu 2000, 1832[VP]

TERUHIKO BEPPU AND KENJI UEDA

Sym.bi.o.bac.te′ri.um. Gr. adj. *symbiotikos* symbiotic; Gr. dim. n. *bakterion* a small rod; N.L. neut. n. *Symbiobacterium* symbiotic small rod, referring to the growth dependence upon co-culture with other bacteria.

Straight, rod-shaped cells, $0.25–0.35 \times 1.7–7\,\mu m$, with a multilayered cell wall that stains Gram-negative. Exhibits low optical density. **Endospore-like structures are formed.** Peritrichously flagellated. Facultatively anaerobic. **Optimum growth occurs in broth co-culture with a thermophilic *Geobacillus* sp., but limited axenic growth is possible under a CO_2 atmosphere or with bicarbonate.** Small, translucent colonies are formed on gellan gum plates. **Analyses of the 16S rRNA gene indicate that this genus represents a deep bacterial lineage that is distinct from the *Firmicutes*, as well as from the *Actinobacteria*.**

DNA G+C content (mol%): 65–69.

Type species: **Symbiobacterium thermophilum** Ohno, Shiratori, Park, Saitoh, Kumon, Yamashita, Hirata, Nishida, Ueda and Beppu 2000, 1832[VP].

Further descriptive information

Cells are long, thin rods (Figure 234). The cell envelope is complex and composed of a multilayered structure outside the cytoplasmic membrane. The structure consists of an innermost electron-dense layer, an electron-transparent layer, and an outermost electron-dense layer (Ohno et al., 2000). The unit structure resembles the S-layer of *Bacillaceae*.

iso-Branched $C_{15:0}$ and $C_{17:0}$ acids are the major components of the cellular fatty acids. Menaquinone-6 is the major respiratory quinone. Indole is produced, as well as tryptophanase and tyrosine-phenyl lyase.

Co-culture with a *Geobacillus* sp. stimulates growth in a number of ways (see *Enrichment and Isolation Procedures*, below). *Geobacillus* produces bicarbonate, ammonia, and some amino acids that stimulate growth of *Symbiobacterium*. The stimulation by bicarbonate may be due to the absence of carbonic anhydrase in *Symbiobacterium* (Watsuji et al., 2006). Additionally, *Symbiobacterium* produces several indole derivatives that inhibit its own growth; co-cultivation with *Geobacillus* alleviates their toxicity (Watsuji et al., 2007).

The genome of the type species of the genus has been sequenced (Ueda et al., 2004). It contains a complete set of genes involved in endospore formation and an endospore-like structure has been observed (Ueda et al., 2004). However, the frequency of sporulation is extremely low and optimal conditions for endospore formation are not known. The genome also encodes many membrane transporters involved in peptide and amino acid uptake, most of the enzymes involved in primary metabolism, and a variety of respiratory systems including Nap nitrate reductase. However, genes for some biosynthetic enzymes and carbonic anhydrase are not present.

Symbiobacterium is widely distributed in compost, soil, animal intestines, and animal feeds (Ueda et al., 2001). It has also been isolated from seashells and other marine samples. Oyster shell is a frequent isolation source. Cells are probably attached to the shell surface; they cannot be isolated from the digestive organs.

Enrichment and isolation procedures

The type species of the genus, *Symbiobacterium thermophilum*, was isolated from compost collected at Fukuyama (Hiroshima, Japan) as a result of its thermostable tryptophanase activity. The original thermophilic, tryptophanase-positive culture contained *Symbiobacterium thermophilum* in co-culture with a bacillus, strain S. The bacillus was subsequently identified by 16S rRNA gene sequence similarity analysis as a strain of *Geobacillus stearothermophilus* (GenBank accession no. AB051200). The co-culture grew well when incubated without shaking (Suzuki et al., 1988).

It was very difficult to isolate *Symbiobacterium thermophilum* from the co-culture. Subsequent studies obtained pure cultures by growing *Symbiobacterium* separated from strain S by only a dialysis membrane or in medium supplemented with culture broth of *Geobacillus stearothermophilus* strain S (Ohno et al., 1999). Colonies are obtained at an extremely low efficiency on agar medium, but at a higher frequency on gellan gum medium. Optimum growth is obtained by co-culture of the *Symbiobacterium thermophilum* and *Geobacillus stearothermophilus* strains without shaking. Axenic cultures can also be grown in dialysis culture flasks (Ohno et al., 1999). Limited growth also occurs in a liquid culture using LB medium with an anaerobic atmosphere containing CO_2 ($N_2/CO_2/H_2$; 8:1:1, by vol.) (Watsuji et al., 2006). Enrichment cultures containing *Symbiobacterium* cells can be obtained by inoculating a compost sample into LB liquid medium and incubating at $60\,°C$ without shaking for 2–6 d (Ueda et al., 2001).

Maintenance procedures

The type strain is stored in LB medium without glycerol at $-80\,°C$. Cultures may also be lyophilized. LB medium is composed of (g/l distilled water): yeast extract, 5.0; tryptone (Difco), 10.0; and NaCl, 5.0. Solid medium is prepared by the addition of $10\,g/l$ gellan gum. LB liquid medium conditioned by *Geobacillus stearothermophilus* strain S (called "conditioned medium") is prepared by adding filtrate of a fully grown culture of *Geobacillus stearothermophilus* strain S to fresh LB medium in a 1:1 ratio. The culture of *Geobacillus stearothermophilus* strain S is grown at $60\,°C$ for 17 h in LB medium without shaking.

Taxonomic comments

This genus was originally described as a Gram-negative bacterium due to several features observed in traditional physiological characterization studies (Hirahara et al., 1993, 1992; Suzuki

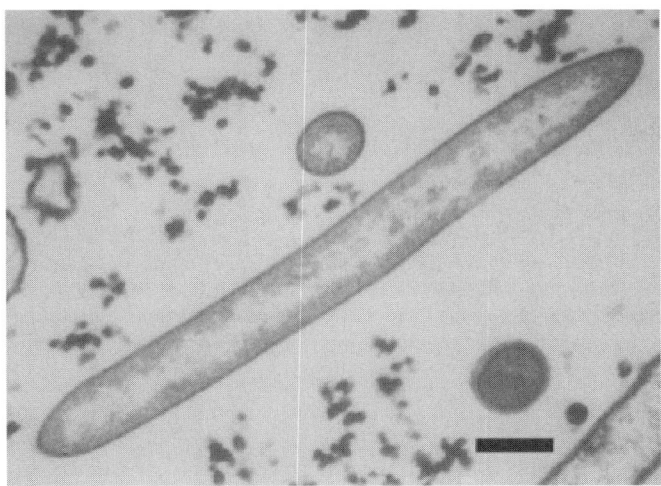

FIGURE 234. Morphology of *Symbiobacterium thermophilum* in a thin-section from an electron micrograph. Bar = 0.4 μm. (Courtesy of A. Hirata.)

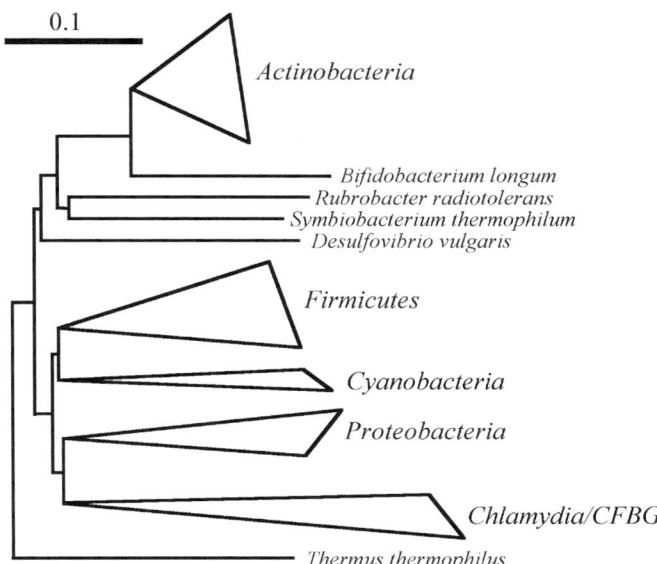

FIGURE 235. Phylogeny of *Symbiobacterium* based on 16S rRNA gene sequences reveals its distinct position outside the major bacterial lineages. The tree is a modification of the neighbor-joining distance tree by Gao and Gupta (2005). Bar = 10% sequence divergence.

et al., 1988), However, in 16S rRNA gene trees, the genus *Symbiobacterium* is more closely related to the *Actinobacteria* and some *Firmicutes* and it forms a distinct branch associated with *Thermaerobacter* and *Sulfobacillus* (Figure 235; Ohno et al., 2000; see Ludwig et al., 2009). The complete genome sequence also reveals a number of features that indicate a stronger affiliation with the *Firmicutes*, in spite of the high G + C content of its DNA (Ueda et al., 2004). The absence of signature sequences that are characteristic of *Actinobacteria* also indicates that the

genus should not be placed in the phylum *Actinobacteria* (Gao and Gupta, 2005). Phylogenetic analysis based on ribosomal protein sequences shows the affiliation of the genus with the class *Clostridia* (Nishida et al., 2009).

List of species of the genus *Symbiobacterium*

1. **Symbiobacterium thermophilum** Ohno, Shiratori, Park, Saitoh, Kumon, Yamashita, Hirata, Nishida, Ueda and Beppu 2000, 1832[VP]

   ther.mo′phil.um. Gr. n. *therme* heat: Gr. adj. *philos* friendly, loving; N.L. neut. n. *thermophilum* heat-loving, referring to the optimum growth at a high temperature.

   Characteristics are as described for the genus with the following additional information. Colonies on LB medium solidified with 1% gellan gum are small (1–2 mm in diameter) and translucent. The efficiency of colony formation on agar medium is extremely low (<1%). Broth cultures have a very low optical density of only 10% of that *Escherichia coli* (Ohno et al., 1999). Resistant to kanamycin (100 μg/ml), but sensitive to neomycin. The maximum cellular yield of axenic cultures is about 1×10^7 cells/ml in LB medium under an anaerobic atmosphere of $N_2/CO_2/H_2$ (8:1:1, by vol.). A 3- to 5-fold increase in cellular yield is achieved by using LB medium supplemented with culture broth of *Geobacillus stearothermophilus* strain S (conditioned medium). Cells occur singly or in pairs. Growth occurs between pH 7.0 and 9.0, with optimum growth at pH 7.5. The temperature range for growth is 45–65 °C, with optimum growth at 60 °C. Minor fatty acids are $C_{17:0 \text{ anteiso}}$, $C_{15:0}$

   anteiso, $C_{16:0 \text{ iso}}$, and $C_{16:0}$. Catalase-positive. Produces thermostable tryptophanase and tyrosine-phenyl lyase.

   The complete genome sequence is known. It is 3.57 Mb in size, contains more than 3300 protein coding sequences, and harbors many class C group II introns (Ueda et al., 2004).

   DNA G+C content (mol%): 68.7 (genome sequencing method).

   Type strain: T, JCM 14929.

   GenBank accession number (16S rRNA gene): AB004913.

   GenBank accession number (genome): AP006840.

2. **"Symbiobacterium toebii"** Sung, Bae, Kim, Kim, Song, Rhee, Jeon, Choi, Hong, Lee, Ha and Kang 2003, 1016

   to.e′bi.i. N.L. neut. n. *toebii* referring to toebi, the farmland compost in Korea, the original source of the organism.

   Characteristics are as described for the genus with the following additional information (Sung et al., 2003). Cells occur singly or in pairs. Growth occurs between pH 6.0 and 9.0, with optimum growth at pH 7.5. The temperature range for growth is 45–70 °C, with optimum growth at 60 °C. Minor fatty acids are $C_{16:0 \text{ iso}}$, $C_{16:0}$, $C_{17:0 \text{ anteiso}}$, $C_{15:0 \text{ anteiso}}$, and $C_{18:0}$. Resistant to kanamycin (200 μg/ml). Contains menaquinone-7 as a minor respiratory quinone. The 16S rRNA gene sequence similarity and DNA–DNA hybridization with *Symbiobacterium thermophilum* are 98.5 and 30%, respectively. Enzymes produced include alkaline phosphatase, esterase (C4), leucine arylamidase, α-chymotrypsin, acid phosphatase, naphthol-phosphohydrolase, α-galactosidase, and α- and β-glucosidase.

Nitrate is reduced to nitrite. Isolated from farmland compost in Gongju, Korea.

DNA G+C content (mol%): 65 (HPLC).

Type strain: SC-1, KCTC 0307BP, DSM 15906.

GenBank accession number (16S rRNA gene): AF190460.

References

Gao, B. and R.S. Gupta. 2005. Conserved indels in protein sequences that are characteristic of the phylum *Actinobacteria*. Int. J. Syst. Evol. Microbiol. *55*: 2401–2412.

Hirahara, T., S. Horinouchi and T. Beppu. 1993. Cloning, nucleotide sequence, and overexpression in *Escherichia coli* of the β-tyrosinase gene from an obligately symbiotic thermophile, *Symbiobacterium thermophilum*. Appl. Microbiol. Biotechnol. *39*: 341–346.

Hirahara, T., S. Suzuki, S. Horinouchi and T. Beppu. 1992. Cloning, nucleotide sequences, and overexpression in *Escherichia coli* of tandem copies of a tryptophanase gene in an obligately symbiotic thermophile, *Symbiobacterium thermophilum*. Appl. Environ. Microbiol. *58*: 2633–2642.

Ludwig, W., K.-H. Schleifer and W.B. Whitman. 2009. Revised road map to the phylum *Firmicutes*. *In* De Vos, Garrity, Jones, Krieg, Ludwig, Rainey, Schleifer and Whitman (Editors), Bergey's Manual of Systematic Bacteriology, 2nd edn, Vol. 3, The *Firmicutes*. Springer, New York, pp. 1–14.

Nishida, H., T. Beppu and K. Ueda. 2009. *Symbiobacterium* lost carbonic anhydrase in the course of evolution. J. Mol. Evol. *68*: 90–96.

Ohno, M., I. Okano, T. Watsuji, T. Kakinuma, K. Ueda and T. Beppu. 1999. Establishing the independent culture of a strictly symbiotic bacterium *Symbiobacterium thermophilum* from its supporting *Bacillus* strain. Biosci. Biotechnol. Biochem. *63*: 1083–1090.

Ohno, M., H. Shiratori, M.J. Park, Y. Saitoh, Y. Kumon, N. Yamashita, A. Hirata, H. Nishida, K. Ueda and T. Beppu. 2000. *Symbiobacterium thermophilum* gen. nov., sp. nov., a symbiotic thermophile that depends on co-culture with a *Bacillus* strain for growth. Int. J. Syst. Evol. Microbiol. *50*: 1829–1832.

Sung, M.-H., J.-W. Bae, J.-J. Kim, K. Kim, J.-J. Song, S.-K. Rhee, C.-O. Jeon, Y.-H. Choi, S.-P. Hong, S.-G. Lee, J.-S. Ha and G.-T. Kang. 2003. *Symbiobacterium toebii* sp. nov., commensal thermophile isolated from Korean compost. J. Microbiol. Biotechnol. *13*: 1013–1017.

Suzuki, S., S. Horinouchi and T. Beppu. 1988. Growth of a tryptophanase-producing thermophile, *Symbiobacterium thermophilum* gen. nov., sp. nov., is dependent on co-culture with a *Bacillus* sp. J. Gen. Microbiol. *134*: 2353–2362.

Ueda, K., M. Ohno, K. Yamamoto, H. Nara, Y. Mori, M. Shimada, M. Hayashi, H. Oida, Y. Terashima, M. Nagata and T. Beppu. 2001. Distribution and diversity of symbiotic thermophiles, *Symbiobacterium thermophilum* and related bacteria, in natural environments. Appl. Environ. Microbiol. *67*: 3779–3784.

Ueda, K., A. Yamashita, J. Ishikawa, M. Shimada, T.O. Watsuji, K. Morimura, H. Ikeda, M. Hattori and T. Beppu. 2004. Genome sequence of *Symbiobacterium thermophilum*, an uncultivable bacterium that depends on microbial commensalism. Nucleic Acids Res. *32*: 4937–4944.

Watsuji, T.O., T. Kato, K. Ueda and T. Beppu. 2006. CO₂ supply induces the growth of *Symbiobacterium thermophilum*, a syntrophic bacterium. Biosci. Biotechnol. Biochem. *70*: 753–756.

Watsuji, T.O., S. Yamada, T. Yamabe, Y. Watanabe, T. Kato, T. Saito, K. Ueda and T. Beppu. 2007. Identification of indole derivatives as self-growth inhibitors of *Symbiobacterium thermophilum*, a unique bacterium whose growth depends on coculture with a *Bacillus* sp. Appl. Environ. Microbiol. *73*: 6159–6165.

Family XIX. **Incertae Sedis**

The 16S rRNA gene sequence of *Acetoanaerobium* is not available. While phenotypic characteristics suggest an affiliation to this order, it is not possible to assign it to a family. For that reason, it is classified within its own family *incertae sedis*.

Genus I. **Acetoanaerobium** Sleat, Mah and Robinson 1985, 13^VP

FRED A. RAINEY

A.ce.to.an.ae.ro'bi.um. L. n. *acetum* vinegar; Gr. pref. *an* not; Gr. n. *aer* air; Gr. n. *bios* life; N.L. neut. n. *Acetoanaerobium* vinegar anaerobe.

Cells are nonsporeforming rods. Cells stain Gram-negative but have an atypical Gram-negative wall structure. Obligate anaerobe. Chemo-organotrophic. Ferment carbohydrates, producing acetate and sometimes other volatile acids. Ferment yeast extract, producing acetate and several volatile acids. At slower growth rates produce acetate from H₂ and CO₂. May require yeast extract for growth.

DNA G+C content (mol%): 36.8 (Bd).

Type species: **Acetoanaerobium noterae** Sleat, Mah and Robinson 1985, 13^VP.

Taxonomic comments

The genus *Acetoanaerobium* was described for a strain that forms acetate from H₂ and CO₂ but differed from previously described acetogenic species in the genera *Clostridium*, *Acetobacterium*, and *Acetogenium*. These differences were for the most part phenotypic in nature including lack of spore formation, Gram staining and cell-wall structure. To date a 16S rRNA gene sequence has not been determined for *Acetoanaerobium noterae* and so the phylogenetic position of this genus remains unknown as does its true taxonomic status.

List of species of the genus *Acetoanaerobium*

1. **Acetoanaerobium noterae** Sleat, Mah and Robinson 1985, 13^VP

   no'ter.ae. L. gen. n. *noterae* of Notera; named for its source, the Notera oil exploration site in Israel.

   Cells are straight rods, 1.0–5.0 μm long and 0.8 μm wide. Motile by three or four peritrichous flagella. Cells stain Gram-negative. Gram-positive cell-wall structure as determined by transmission electron microscopy, is composed of two distinct layers, a darker inner layer and lighter outer layer. Colonies are rhizoid, opaque, and granular. Young colonies are white, older colonies are brownish and up to 2 cm in diameter after 1 month of incubation. Yeast extract, maltose, and glucose are used for heterotrophic growth. Compounds not supporting growth include arabinose, rhamnose, ribose, xylose, fructose, galactose, cellobiose, lactose, mannose, sucrose, melezitose, trehalose, erythritol, adonitol, dulcitol, inositol, mannitol, sorbitol, formate, acetate, pyruvate, lactate, malate, fumarate, succinate, citrate, glutamate, methylamine, trimethylamine, and methanol. Yeast extract is required for growth and H_2 utilization. Growth on yeast extract and H_2-CO_2 is biphasic, with an initial rapid growth phase independent of the presence of H_2, followed by H_2-dependent acetate production during the second slower growth phase. Produces acetate, propionate, isobutyrate, butyrate, and isovalerate (and little or no H_2) during growth on yeast extract alone. Acetate is the only fermentation product from glucose or maltose. Optimum temperature for growth is 37 °C. Growth occurs in the pH range 6.6–8.4, optimum 7.6. Doubling times are 2.8 h for heterotrophic growth and 27 h for H_2-dependent growth. Vitamins are not required. Growth is inhibited by erythromycin, chloramphenicol, penicillin, cephalosporin, and cycloserine at concentrations of 100 ng/liter. The type strain was isolated from sediment collected near the Notera oil drilling site in the Hula swamp area of Galilee, Israel.

DNA G+C content (mol%) of the type strain: 36.8 (Bd).

Type strain: NOT-3, ATCC 35199.

GenBank accession number (16S rRNA gene): not determined.

References

Sleat, R., R.A. Mah, and R. Robinson. 1985. *Acetoanaerobium noterae* gen. nov., sp. nov. an anaerobic bacterium that forms acetate from H_2 and CO_2. Int. J. Syst. Bacteriol. *35*: 10–15.

Order II. **Halanaerobiales** corrig. Rainey and Zhilina 1995a, 879[VP] (Effective publication: Rainey, Zhilina, Boulygina, Stackebrandt, Tourova and Zavarzin 1995b, 193.)

Aharon Oren

Hal.an.ae.ro.bi.a′les. N.L. fem. pl. n. *Halanaerobiaceae* type family of the order; *-ales* ending to denote an order; N.L. fem. pl. n. *Halanaerobiales* the *Halanaerobiaceae* order.

Cells are rod-shaped and **Gram-stain-negative**. Endospores are produced by some species. **Strictly anaerobic.** Oxidase- and catalase-negative. Most species ferment carbohydrates to products including acetate, ethanol, hydrogen, and carbon dioxide. Some species may grow fermentatively on amino acids, others have a homoacetogenic metabolism or may grow by anaerobic respiration on nitrate, trimethylamine *N*-oxide or selenate.

Moderately halophilic. NaCl concentrations between 0.5–3.4 M are required for optimal growth and no growth is observed below 0.3–1.7 M NaCl, depending on the species.

DNA G+C content (mol%): 27–40.

Type genus: **Halanaerobium** corrig. Zeikus, Hegge, Thompson, Phelps and Langworthy 1984, 503[VP] (Effective publication: Zeikus, Hegge, Thompson, Phelps and Langworthy 1983, 232.).

Further descriptive information

All known members of the *Halanaerobiales* are strictly anaerobic and moderately halophilic. One species (*Halothermothrix orenii*, isolated from Chott El Guettar, a warm saline lake in Tunisia) is thermophilic, growing optimally at 60 °C and tolerating up to 68 °C (Cayol et al., 1994a). Some species are moderately alkaliphilic or alkalitolerant. *Natroniella acetigena*, isolated from the alkaline hypersaline Lake Magadi, Kenya, is an obligate alkaliphile and grows between pH 8.1–10.7 with an optimum at 9.7–10.0 (Zhilina et al., 1996). The alkaline (pH 10.2) Lake Magadi was shown to harbor a varied anaerobic community including cellulolytic, proteolytic, saccharolytic, and homoacetogenic bacteria (Zhilina and Zavarzin, 1994) and it may therefore be expected that additional alkaliphilic members of the *Halanaerobiales* may be isolated from this interesting ecosystem in the future.

All species show a negative Gram-stain reaction. Heat-resistant endospores are produced by a number of species belonging to the family *Halobacteroidaceae*, one of the two families classified within the order. Endospore formation has not been reported in any representative of the family *Halanaerobiaceae*. A phenotypic test that may correlate between phylogenetic position in the *Halanaerobiales* within the *Firmicutes* and the ability to form endospores is hydrolysis of the D-isomer of *N*′-benzoyl-arginine-*p*-nitroanilide (BAPA). Four representatives of the *Halanaerobiales* were tested for D-BAPA and L-BAPA hydrolysis and three of them (*Halobacteroides halobius*, *Orenia marismortui*, and *Halanaerobium praevalens*, the last belonging to the family *Halanaerobiaceae*, in which endospore formation has never been observed as yet) hydrolyzed D-BAPA, whereas L-BAPA was not hydrolyzed. *Sporohalobacter lortetii* degrades neither of the BAPA stereoisomers (Oren et al., 1989).

Most species belonging to the *Halanaerobiales* obtain their energy by fermenting simple sugars. This type of metabolism is the only one detected thus far in the family *Halanaerobiaceae*. Within the family *Halobacteroidaceae*, the metabolic diversity appears to be much greater than that observed within the *Halanaerobiaceae*. In addition to sugar fermenters, there are species within the family *Halobacteroidaceae* that can ferment amino acids, either alone or using the Stickland reaction. The neutrophilic *Acetohalobium arabaticum* and the alkaliphilic *Natroniella acetigena* are homoacetogens. Finally, *Selenihalanaerobacter shriftii* oxidizes glycerol or glucose by anaerobic respiration with nitrate, trimethylamine *N*-oxide, or selenate as electron acceptor (Switzer Blum et al., 2001). It is the only representative of the order in which the presence of cytochromes has been documented thus far; no cytochromes have been detected in *Halanaerobium acetethylicum* (Rengpipat et al., 1988a), *Halobacteroides halobius* (Oren et al., 1984), *Sporohalobacter lortetii* (Oren, 1983), *Orenia marismortui* (Oren et al., 1987), or *Acetohalobium arabaticum* (Pusheva and Detkova, 1996).

Most halophilic and halotolerant representatives of the domain *Bacteria* (aerobic heterotrophs, aerobic and anaerobic phototrophs) use organic osmotic solutes such as glycine betaine, ectoine, and others to provide osmotic balance of the cytoplasm with the surrounding medium, while excluding salts from the cytoplasm to a large extent. No such organic osmotic solutes have been detected in most members of the *Halanaerobiales* tested. However, glycine betaine was found to be accumulated to 1.9–2.2 µmol/mg protein when these bacteria were grown in medium containing yeast extract (Mouné et al., 2000). On the other hand, high concentrations of Na$^+$, K$^+$, and Cl$^-$, i.e., high enough to be at least isotonic with the medium, have been measured inside the cells of *Halanaerobium praevalens, Halanaerobium acetethylicum,* and *Halobacteroides halobius* (Oren, 1986; Oren et al., 1997; Rengpipat et al., 1988b). Using X-ray microanalysis with the transmission electron microscope, it was shown that in exponentially growing cells of *Halanaerobium praevalens,* K$^+$ was the major cation (70% of the cation sum). Stationary phase cells showed a high variability among individual cells, with NaCl often replacing KCl (Oren et al., 1997).

The intracellular enzymic machinery was found to be well adapted to function in the presence of the high salt concentrations found in the cytoplasm. The enzymes tested (including glyceraldehyde-3-phosphate dehydrogenase, NAD-linked alcohol dehydrogenase, pyruvate dehydrogenase, methyl viologen-linked hydrogenase from *Halanaerobium acetethylicum,* the fatty acid synthetase complex of *Halanaerobium praevalens,* hydrogenase and CO dehydrogenase activities of *Acetohalobium arabaticum*) functioned better in the presence of molar concentrations of salts than in salt-free medium (Oren and Gurevich, 1993; Rengpipat et al., 1988b; Zavarzin et al., 1994). A high content of acidic amino acids was found in the bulk cellular protein of *Halanaerobium praevalens, Halobacteroides halobius,* and *Sporohalobacter lortetii* (Oren, 1986). However, the ribosomal A-protein of *Halanaerobium praevalens* did not show an especially high content of acidic amino acids (Matheson et al., 1987) and proteins of *Halothermothrix orenii* also are not especially acidic, as proven by partial genomic sequence analysis (Mijts and Patel, 2001). The presence of high intracellular K$^+$ and Cl$^-$ concentrations and highly acidic proteins adapted to function in the presence of these high ionic concentrations in the *Halanaerobiales* brings to mind the similar strategy of adaptation to salt in the aerobic *Archaea* of the order *Halobacteriales*.

Enrichment and isolation procedures

Species belonging to the order *Halanaerobiales* can probably be found in any hypersaline anaerobic environment in which simple organic compounds such as sugars and amino acids are available. They have been isolated from hypersaline lakes, both with thalassohaline and athalassohaline ionic composition, including Great Salt Lake, Utah (Tsai et al., 1995; Zeikus et al., 1983), Salton Sea, California (Shiba, 1991; Shiba and Horikoshi, 1988; Shiba et al., 1989), the Dead Sea (Oren, 1983, 1987; Oren et al., 1984), hypersaline lakes and lagoons in the Crimea (Simankova et al., 1991, 1993, 1997; Zhilina and Zavarzin, 1990) and Senegal (Cayol et al., 1995, 1994b), saltern evaporation ponds in California (Liaw and Mah, 1992) and France (Mouné et al., 2000, 1999), a hot hypersaline lake in Tunisia (Cayol et al., 1994a), and the alkaline hypersaline lakes Magadi, Kenya (Shiba and Horikoshi, 1988; Zhilina et al., 1996) and Big Soda Lake, Nevada (Shiba and Horikoshi, 1988; Shiba et al., 1989). Brines associated with oil wells and petroleum reservoirs have also yielded a number of interesting species (Bhupathiraju et al., 1991, 1993, 1994, 1999; Ravot et al., 1997, 1988a). Finally, they may be present in salted fermented foods (Kobayashi et al., 2000a, 2000b).

Any anoxic reducing medium containing high salt concentrations (1–4 M) and a suitable carbon source is a potential enrichment and growth medium for members of the *Halanaerobiales*. Simple sugars are used by most species. Strict anaerobic techniques should be used, including boiling the media under nitrogen or nitrogen/CO$_2$ (80:20) and adding reducing agents such as sulfide, cysteine, dithionite, or ascorbate to the boiled media. Protocols for the preparation of media have been compiled by Oren (2001); details can be found in the original species descriptions.

Taxonomic comments

The order *Halanaerobiales* was created in 1995 during the course of a comprehensive in-depth study of the known anaerobic halophilic *Bacteria*. Based on 16S rRNA gene sequence comparisons, extensive taxonomic rearrangements were proposed. These included a reclassification of the species of the former family *Halanaerobiaceae* over two families: the *Halanaerobiaceae* and the newly created family *Halobacteroidaceae* (Rainey et al., 1995b). Differential characteristics of the genera of the order *Halanaerobiales* are given in Table 235. At the same time, many species were reclassified in new or extant genera on the basis of their 16S rRNA gene sequences. Physiologically, the group is coherent, to the extent that, as yet, no aerobes or non-halophiles are known to cluster phylogenetically within the order. There are, however, a few fermentative, anaerobic, moderately halophilic *Bacteria* that cluster outside the order. Examples are *Clostridium halophilum* (Fendrich et al., 1990) and *Thermohalobacter berrensis* (Cayol et al., 2000).

On the basis of their 16S rRNA gene sequences, the halophilic anaerobic bacteria should be classified in the domain *Bacteria* within the phylum *Firmicutes* (Figure 7). They form a coherent cluster close to the bifurcation point that separates the actinomycete subphylum and the *Bacillus/Clostridium* subphylum. Sequences of the halophilic anaerobes contain all of the few signature nucleotides that have been defined as characteristic of members of the *Bacillus/Clostridium* subphylum, whereas they lack any of the actinomycete-specific nucleotides. It was therefore suggested that the halophilic anaerobes be classified in the *Bacillus/Clostridium* subphylum (Patel et al., 1995; Rainey et al., 1995b; Tourova et al., 1995). Table 236 presents the signature nucleotides defining the *Halanaerobiales* within the *Bacillus/Clostridium* subphylum of the *Firmicutes*. The phy-

TABLE 235. Differential characteristics of the genera of the order *Halanaerobiales* (families *Halanaerobiaceae* and *Halobacteroidaceae*)[a]

| | Halanaerobiaceae | | | Halobacteroidaceae | | | | | | | |
Characteristic	*Halanaerobium*	*Halocella*	*Halothermothrix*	*Halobacteroides*	*Acetohalobium*	*Halanaerobacter*	*Halonatronum*	*Natroniella*	*Orenia*	*Selenihalanaerobacter*	*Sporohalobacter*
Endospores	–	–	–	D	v	–	+	+	+	–	+
Sphaeroplast formation	–	+	NR	+	NR	+	+	+	–	NR	+
Sugars fermented	+	+	+	+	–	+	+	–	+	–	w
Cellulose degraded	–	+	–	–	–	–	–	–	–	–	–
Presence of cytochromes	–	NR	NR	–	–	NR	NR	NR	–	+	–
Homoacetogenic metabolism	–	–	–	–	+	–	NR	+	–	–	–
Stickland reaction	NR	NR	NR	–	NR	D	–	–	NR	–	NR
Reduction of selenate and nitrate	NR	NR	NR	NR	NR	NR	NR	NR	NR	+	NR
Thermophily	–	–	+	–	–	–	Moderate	–	–	–	–
DNA G+C content (mol%)	27–34	29	39.6	30–31	33.6	31.6–34.8	34.4	31.9	28.6–33.7	31.2	31.5

[a]Symbols: +, >85% positive; –, 0–15% positive; D, different taxa give different reactions; w, weak reaction; NR, not reported.

TABLE 236. 16S rRNA signature nucleotides defining the *Halanaerobiales* within the *Bacillus/Clostridium* subphylum of the *Firmicutes*[a,b]

Position (*E. coli* nomenclature)	*Halanaerobiales* (>90% of species)	Majority of the members of the *Bacillus/Clostridium* subphylum (>90%)
94	One-base insertion	No insertion
771–808	U–A	G–C
772–807	R–Y	U–A
784–798	G–C	A–U
890	Mainly U	G
1059–1198	Mainly C–G	U–A
1115	Mainly C	U
1415–1485	Y–R	G–U

[a]Y = Pyrimidine; R = purine.

[b]Table taken from Rainey et al. 1995b. Anaerobe *1*: 185–199. Reproduced with permission.

logenetic affiliation of *Halanaerobium praevalens* (the type species of the type genus of the type family) with the *Bacillus/Clostridium* group was confirmed by the amino acid sequence of its ribosomal A-protein (Matheson et al., 1987). The location of the branching point of the halophilic anaerobes close to the root of this subphylum is further evidence that certain descendants of the ancestors of the "Gram-positive bacteria" still maintain their Gram-negative wall type, as is the case with *Sporomusa* and its relatives. The deep branching justifies classification in a separate order (Rainey et al., 1995b). The order *Halanaerobiales* has been used as a paradigm to demonstrate the application of 16S rRNA gene sequencing and DNA–DNA hybridization in bacterial taxonomy (Tourova, 2000).

According to Rule 61 of the Bacteriological Code, the original spelling of the name of the order *Haloanaerobiales* (Rainey et al., 1995b) has been changed to *Halanaerobiales* (Oren, 2000).

References

Bhupathiraju, V.K., M.J. McInerney and R.M. Knapp. 1993. Pretest studies for a microbially enhanced oil recovery field pilot in a hypersaline oil reservoir. Geomicrobiol. J. *11*: 19–34.

Bhupathiraju, V.K., M.J. McInerney, C.R. Woese and R.S. Tanner. 1999. *Haloanaerobium kushneri* sp. nov., an obligately halophilic, anaerobic bacterium from an oil brine. Int. J. Syst. Bacteriol. *49*: 953–960.

Bhupathiraju, V.K., A. Oren, P.K. Sharma, R.S. Tanner, C.R. Woese and M.J. McInerney. 1994. *Haloanaerobium salsugo* sp. nov., a moderately halophilic, anaerobic bacterium from a subterranean brine. Int. J. Syst. Bacteriol. *44*: 565–572.

Bhupathiraju, V.K., P.K. Sharma, M.J. McInerney, R.M. Knapp, K. Fowler and W. Jenkins. 1991. Isolation and characterization of novel halophilic anaerobic bacteria from oil field brines. Dev. Petrol. Sci. *31*: 132–143.

Cayol, J.-L., S. Ducerf, B.K. Patel, J.-L. Garcia, P. Thomas and B. Ollivier. 2000. *Thermohalobacter berrensis* gen. nov., sp. nov., a thermophilic, strictly halophilic bacterium from a solar saltern. Int. J. Syst. Evol. Microbiol. *50*: 559–564.

Cayol, J.-L., B. Ollivier, B.K.C. Patel, E. Ageron, P.A.D. Grimont, G. Prensier and J.-L. Garcia. 1995. *Haloanaerobium lacusroseus* sp. nov., an extremely halophilic fermentative bacterium from the sediments of a hypersaline lake. Int. J. Syst. Bacteriol. *45*: 790–797.

Cayol, J.-L., B. Ollivier, B.K.C. Patel, G. Prensier, J. Guezennec and J.-L. Garcia. 1994a. Isolation and characterization of *Halothermothrix orenii* gen. nov., sp. nov., a halophilic, thermophilic, fermentative, strictly anaerobic bacterium. Int. J. Syst. Bacteriol. *44*: 534–540.

Cayol, J.-L., B. Ollivier, A.L.A. Soh, M.L. Fardeau, E. Ageron, P.A.D. Grimont, G. Prensier, J. Guezennec, M. Magot and J.-L. Garcia. 1994b. *Haloincola saccharolytica* subsp. *senegalensis* subsp. nov., isolated from the sediments of a hypersaline lake, and emended description of *Haloincola saccharolytica*. Int. J. Syst. Bacteriol. *44*: 805–811.

Fendrich, C., H. Hippe and G. Gottschalk. 1990. *Clostridium halophilum* sp. nov. and *C. litorale* sp. nov., an obligate halophilic and a marine species degrading betaine in the Stickland reaction. Arch. Microbiol. *154*: 127–132.

Kobayashi, T., B. Kimura and T. Fujii. 2000a. *Haloanaerobium fermentans* sp. nov., a strictly anaerobic, fermentative halophile isolated from fermented puffer fish ovaries. Int. J. Syst. Evol. Microbiol. *50*: 1621–1627.

Kobayashi, T., B. Kimura and T. Fujii. 2000b. Strictly anaerobic halophiles isolated from canned Swedish fermented herrings (Surströmming). Int. J. Food Microbiol. *54*: 81–89.

Liaw, H.J. and R.A. Mah. 1992. Isolation and characterization of *Haloanaerobacter chitinovorans* gen. nov., sp. nov., a halophilic, anaerobic, chitinolytic bacterium from a solar saltern. Appl. Environ. Microbiol. *58*: 260–266.

Matheson, A.T., K.A. Louie, B.D. Tak and M. Zuker. 1987. The primary structure of the ribosomal A-protein (L12) from the halophilic eubacterium *Haloanaerobium praevalens*. Biochimie *69*: 1013–1020.

Mijts, B.N. and B.K. Patel. 2001. Random sequence analysis of genomic DNA of an anaerobic, thermophilic, halophilic bacterium, *Halothermothrix orenii*. Extremophiles *5*: 61–69.

Mouné, S., C. Eatock, R. Matheron, J.C. Willison, A. Hirschler, R. Herbert and P. Caumette. 2000. *Orenia salinaria* sp. nov., a fermentative bacterium isolated from anaerobic sediments of Mediterranean salterns. Int. J. Syst. Evol. Microbiol. *50*: 721–729.

Mouné, S., M. Manac'h, A. Hirschler, P. Caumette, J.C. Willison and R. Matheron. 1999. *Haloanaerobacter salinarius* sp. nov., a novel halophilic fermentative bacterium that reduces glycine-betaine to trimethylamine with hydrogen or serine as electron donors; emendation of the genus *Haloanaerobacter*. Int. J. Syst. Bacteriol. *49*: 103–112.

Oren, A. 1983. *Clostridium lortetii* sp. nov., a halophilic obligatory anaerobic bacterium producing endospores with attached gas vacuoles. Arch. Microbiol. *136*: 42–48.

Oren, A. 1986. Intracellular salt concentrations of the anaerobic halophilic eubacteria *Haloanaerobium praevalens* and *Halobacteroides halobius*. Can. J. Microbiol. *32*: 4–9.

Oren, A. 1987. A procedure for the selective enrichment of *Halobacteroides halobius* and related bacteria from anaerobic hypersaline sediments. FEMS Microbiol. Lett. *42*: 201–204.

Oren, A. 2000. Change of the names *Haloanaerobiales*, *Haloanaerobiaceae* and *Haloanaerobium* to *Halanaerobiales*, *Halanaerobiaceae* and *Halanaerobium*, respectively, and further nomenclatural changes within the order *Halanaerobiales*. Int. J. Syst. Evol. Microbiol. *50*: 2229–2230.

Oren, A. 2001. The order *Haloanaerobiales*. *In* Dworkin, Falkow, Rosenberg, Schleifer and Stackebrandt (Editors), The Prokaryotes: A Handbook on the Biology of Bacteria: Ecophysiology, Isolation, Identification, Applications, 3rd edn. Springer-Verlag, New York, chapter 238.

Oren, A., L.V. Gofshtein-Gandman and A. Keynan. 1989. Hydrolysis of *N*-benzoyl-D-arginine-*p*-nitroanilide by members of the *Haloanaerobiaceae*: additional evidence that *Haloanaerobium praevalens* is related to endospore-forming bacteria. FEMS Microbiol. Lett. *58*: 5–10.

Oren, A. and P. Gurevich. 1993. The fatty acid synthetase complex of *Haloanaerobium praevalens* is not inhibited by salt. FEMS Microbiol. Lett. *108*: 287–290.

Oren, A., M. Heldal and S. Norland. 1997. X-ray microanalysis of intracellular ions in the anaerobic halophilic eubacterium *Haloanaerobium praevalens*. Can. J. Microbiol. *43*: 588–592.

Oren, A., H. Pohla and E. Stackebrandt. 1987. Transfer of *Clostridium lortetii* to a new genus *Sporohalobacter* gen. nov. as *Sporohalobacter lortetii* comb. nov., and description of *Sporohalobacter marismortui* sp. nov. Syst. Appl. Microbiol. *9*: 239–246.

Oren, A., W.G. Weisburg, M. Kessel and C.R. Woese. 1984. *Halobacteroides halobius* gen. nov., sp. nov., a moderately halophilic anaerobic bacterium from the bottom sediments of the Dead Sea. Syst. Appl. Microbiol. *5*: 58–70.

Patel, B.K.C., K.T. Andrews, B. Ollivier, R.A. Mah and J.-L. Garcia. 1995. Reevaluating the classification of *Halobacteroides* and *Haloanaerobacter* species based on sequence comparisons of the 16S ribosomal RNA gene. FEMS Microbiol. Lett. *134*: 115–119.

Pusheva, M.A. and E.N. Detkova. 1996. Bioenergetic aspects of acetogenesis on various substrates by the extremely halophilic acetogenic bacterium *Acetohalobium arabaticum*. Mikrobiologiya *65*: 516–520.

Rainey, F.A., T.N. Zhilina, E.S. Boulygina, E. Stackebrandt, T.P. Tourova and G.A. Zavarzin. 1995a. In Validation of the publication of new names and new combinations previously effectively published outside the IJSB. List no. 55. Int. J. Syst. Bacteriol. *45*: 879–880.

Rainey, F.A., T.N. Zhilina, E.S. Boulygina, E. Stackebrandt, T.P. Tourova and G.A. Zavarzin. 1995b. The taxonomic status of the fermentative halophilic anaerobic bacteria: description of *Haloanaerobiales* ord. nov., *Halobacteroidaceae* fam. nov., *Orenia* gen. nov. and further taxonomic rearrangements at the genus and species level. Anaerobe *1*: 185–199.

Ravot, G., M. Magot, B. Ollivier, B.K. Patel, E. Ageron, P.A. Grimont, P. Thomas and J.L. Garcia. 1997. *Haloanaerobium congolense* sp. nov., an anaerobic, moderately halophilic, thiosulfate- and sulfur-reducing bacterium from an African oil field. FEMS Microbiol. Lett. *147*: 81–88.

Rengpipat, S., T.A. Langworthy and J.G. Zeikus. 1988a. *Halobacteroides acetoethylicus* sp. nov., a new obligately anaerobic halophile isolated from deep subsurface hypersaline environments. Syst. Appl. Microbiol. *11*: 28–35.

Rengpipat, S., S.E. Lowe and J.G. Zeikus. 1988b. Effect of extreme salt concentrations on the physiology and biochemistry of *Halobacteroides acetoethylicus*. J. Bacteriol. *170*: 3065–3071.

Shiba, H. 1991. Anaerobic halophiles. *In* Horikoshi and Grant (Editors), Superbugs: Microorganisms in Extreme Environments. Tokyo/Springer-Verlag, Berlin, pp. 191–211.

Shiba, H. and K. Horikoshi. 1988. Isolation and characterization of novel anaerobic, halophilic eubacteria from hypersaline environments of western America and Kenya, pp. 371–373, Proceedings of the FEMS Symposium on the Microbiology of Extreme Environments and its Biotechnological Potential, Estoril.

Shiba, H., H. Yamamoto and K. Horikoshi. 1989. Isolation of strictly anaerobic halophiles from the aerobic surface sediments of hypersaline environments in California and Nevada. FEMS Microbiol. Lett. *57*: 191–195.

Simankova, M.V., N.A. Chernych, G.A. Osipov and G.A. Zavarzin. 1993. *Halocella cellulolytica* gen nov, sp nov, a new obligately anaerobic, halophilic, cellulolytic bacterium. Syst. Appl. Microbiol. *16*: 385–389.

Switzer Blum, J., J.F. Stolz, A. Oren and R.S. Oremland. 2001. *Selenihalanaerobacter shriftii* gen. nov., sp. nov., a halophilic anaerobe from Dead Sea sediments that respires selenate. Arch. Microbiol. *175*: 208–219.

Tourova, T.P. 2000. The role of DNA-DNA hybridization and 16S rRNA gene sequencing in solving taxonomic problems by the example of the order *Haloanaerobiales*. Microbiology (En. transl. from Mikrobiologiya) *69*: 623–634.

Tourova, T.P., E.S. Boulygina, T.N. Zhilina, R.S. Hanson and G.A. Zavarzin. 1995. Phylogenetic study of haloanaerobic bacteria by 16S ribosomal RNA sequences analysis. Syst. Appl. Microbiol. *18*: 189–195.

Tsai, C.R., J.L. Garcia, B.K.C. Patel, J.L. Cayol, L. Baresi and R.A. Mah. 1995. *Haloanaerobium alcaliphilum* sp. nov., an anaerobic moderate halophile from the sediments of Great Salt Lake, Utah. Int. J. Syst. Bacteriol. *45*: 301–307.

Zavarzin, G.A., T.N. Zhilina and M.A. Pusheva. 1994. Halophilic acetogenic bacteria. *In* Drake (Editor), Acetogenesis. Chapman and Hall, New York, pp. 432–444.

Zeikus, J.G., P.W. Hegge, T.E. Thompson, T.J. Phelps and T.A. Langworthy. 1983. Isolation and description of *Haloanaerobium praevalens* gen. nov. and sp. nov., an obligately anaerobic halophile common to Great Salt Lake sediments. Curr. Microbiol. *9*: 225–234.

Zeikus, J.G., P.W. Hegge, T.E. Thompson, T.J. Phelps and T.A. Langworthy. 1984. In Validation of the publication of new names and new combinations previously effectively published outside the IJSB. List no. 16. Int. J. Syst. Bacteriol. *34*: 503–504.

Zhilina, T.N., V.V. Kevbrin, A.M. Lysenko and G.A. Zavarzin. 1991. Isolation of saccharolytic anaerobes from a halophilic cyanobacterial mat. Microbiology (En. transl. from Mikrobiologiya) *60*: 101–107.

Zhilina, T.N., T.P. Turova, A.M. Lysenko and V.V. Kevbrin. 1997. Reclassification of *Halobacteroides halobius* Z-7287 on the basis of phylogenetic analysis as a new species *Halobacteroides elegans* sp. nov. Microbiology (En. transl. from Mikrobiologiya) *66*: 97–103.

Zhilina, T.N. and G.A. Zavarzin. 1990. A new extremely halophilic homoacetic bacterium *Acetohalobium arabaticum*, gen. nov., sp. nov. Dokl. Akad. Nauk. SSSR. *311*: 745–747.

Zhilina, T.N. and G.A. Zavarzin. 1994. Alkaliphilic anaerobic community at pH 10. Curr. Microbiol. *29*: 109–112.

Zhilina, T.N., G.A. Zavarzin, E.N. Detkova and F.A. Rainey. 1996. *Natroniella acetigena* gen. nov. sp. nov., an extremely haloalkaliphilic, homoacetic bacterium: a new member of *Haloanaerobiales*. Curr. Microbiol. *32*: 320–326.

Family I. **Halanaerobiaceae** corrig. Oren, Paster and Woese 1984b, 503[VP] (Effective publication: Oren, Paster and Woese, 1984a, 79.)

AHARON OREN

Hal.an.ae.ro.bi.a′ce.ae. N.L. neut. n. *Halanaerobium* type genus of the family; -*aceae* ending to denote a family; N.L. fem. pl. n. *Halanaerobiaceae* the *Halanaerobium* family.

Cells are rod-shaped and **Gram-stain-negative**. Endospore formation is never observed. **Strictly anaerobic**. Oxidase- and catalase-negative. Carbohydrates are fermented to products including acetate, ethanol, hydrogen, and carbon dioxide. **Moderately halophilic**. NaCl concentrations in the range 1.7–2.6 M are required for optimal growth and no growth is observed below 0.3–1.7 M NaCl, depending on the species.

DNA G+C content (mol%): 27–40.

Type genus: **Halanaerobium** corrig. Zeikus, Hegge, Thompson, Phelps and Langworthy 1984, 503[VP] (Effective publication: Zeikus, Hegge, Thompson, Phelps and Langworthy 1983, 232.).

Further descriptive information

All known members of the *Halanaerobiaceae* are moderately halophilic, but the most halophilic species known is *Halanaerobium lacusrosei*, which grows between 1.0 and 5.8 M NaCl and has its optimum at 3.4 M (Cayol et al., 1995). Some species are moderately alkaliphilic or alkalitolerant; *Halanaerobium alcaliphilum* grows optimally at neutral pH, but can grow at up to pH 10.0.

All species show a negative Gram-stain reaction. A typical Gram-negative type cell wall with an outer membrane and periplasmic space is often seen in electron micrographs of thin sections. *meso*-Diaminopimelic acid was detected in the peptidoglycan of *Halanaerobium saccharolyticum* subsp. *saccharolyticum* (Zhilina et al., 1992).

All known species belonging to the *Halanaerobiaceae* obtain their energy by fermenting simple sugars. Fermentation products typically include acetate, hydrogen, and carbon dioxide. In addition, some strains produce butyrate, lactate, propionate, and formate. Ethanol is produced by many, but not all, fermentative strains. Thus, *Halanaerobium congolense* does not form ethanol (Ravot et al., 1997). Cellulose is degraded anaerobically by *Halocella cellulosilytica* (Simankova et al., 1993). *Halanaerobium saccharolyticum* ferments glycerol (Cayol et al., 1994b; Zhilina et al., 1992) and so does *Halanaerobium lacusrosei* (Cayol et al., 1995). No details have been published on the fermentation pathway and the products made during growth on glycerol. Glycerol oxidation by anaerobic halophiles was reported to be markedly improved through interspecies hydrogen transfer when glycerol fermenters were grown in co-culture with H_2-consuming, sulfate-reducing bacteria (Cayol et al., 1995).

Several interesting features of the assimilatory and dissimilatory metabolism of sulfur compounds in different representatives of the *Halanaerobiaceae* have been reported. Methanethiol can be used as an assimilatory sulfur source by *Halanaerobium saccharolyticum* (Kevbrin and Zavarzin, 1992; Zhilina et al., 1991b, 1997, 1992). *Halanaerobium congolense* uses thiosulfate and elemental sulfur as electron acceptors; the addition of thiosulfate or sulfur increases the growth yield by 6- and 3-fold, respectively, and growth rates were enhanced (Ravot et al., 1997). *Halanaerobium praevalens* can use nitro-substituted aromatic compounds such as nitrobenzene, nitrophenols, 2,4-dinitrophenol and 2,4-dinitroaniline as electron sinks (Oren et al., 1991).

Few biotechnological uses have been found thus far for the *Halanaerobiaceae*. The use of anaerobic halophilic bacteria in the industrial fermentation of complex organic matter and the production of organic solvents has been proposed (Lowe et al., 1993; Wise, 1987). It has also been suggested that certain halophilic anaerobes may be used in microbially enhanced oil recovery by plugging of porous reservoirs and by anaerobically metabolizing nutrients with the production of gases, biosurfactants, and polymers (Bhupathiraju et al., 1991). The recent isolation of *Halanaerobium fermentans* from traditionally fermented puffer fish ovaries in Japan (Kobayashi et al., 2000b) suggests that members of the *Halanaerobiaceae* have been involved in applied microbiological processes for centuries and that only now has their role started to become recognized.

Enrichment and isolation procedures

Species belonging to the family *Halanaerobiaceae* can probably be found in any hypersaline anaerobic environment in which simple sugars are available. They have been isolated from Great Salt Lake, Utah (Tsai et al., 1995; Zeikus et al., 1983), hypersaline lakes and lagoons in the Crimea (Simankova et al., 1993; Zhilina et al., 1992) and Senegal (Cayol et al., 1994b), a hot hypersaline lake in Tunisia (Cayol et al., 1994a), and from brines associated with oil wells and petroleum reservoirs (Bhupathiraju et al., 1993, 1999, 1994, 1991; Ravot et al., 1997; Rengpipat et al., 1988a). Finally, they may be present in salted fermented foods (Kobayashi et al., 2000a, 2000b).

Any anoxic reducing medium containing high salt concentrations (1–4 M) and simple sugars as carbon and energy sources is a potential enrichment and growth medium for members of the *Halanaerobiales*. Strict anaerobic techniques should be used, including boiling the media under nitrogen or nitrogen/CO_2 (80:20) and adding reducing agents such as cysteine, dithionite, or ascorbate plus thioglycollate to the boiled media. Protocols for the preparation of media have been compiled by Oren (2001); details can be found in the original species descriptions. For the enrichment of thermophiles such as *Halothermothrix*, the incubation temperature should be adjusted to that of the natural environment. No specific enrichment procedures that exclude development of members of the *Halobacteroidaceae* have been devised as yet.

Differentiation of the family *Halanaerobiaceae* from other families

The only sure way to distinguish members of the family *Halanaerobiaceae* from the sugar-fermenting members of the *Halobacteroidaceae* is by 16S rRNA relatedness. There are no known morphological, physiological, or biochemical tests that enable an unequivocal classification in either of the two families. Table 237 presents information on the 16S rRNA gene signature nucleotides defining the two families.

TABLE 237. 16S rRNA gene signature nucleotides defining the two families within the order *Halanaerobiales*[a,b]

Position (*Escherichia coli* nomenclature)	*Halanaerobiaceae*	*Halobacteroidaceae*
70	C	A
73	Y	A
75	C	G
Variable region I (73–97)	Long stem	Short stem
90	U	A
98	G	U
100	Y	R
135	U	C
Variable region II (184–193)	Long stem	Short stem
233	U	C
241–285	A–U	C–G
242–284	G–C	C–G
274	G	A
284	C	G
291–309	U–A	C–G
294–303	U–A	C–G
293–304	G–C	A–U
353	A	U
453	C	A
459–473	U–A/A–U	G–C
467	A	U
457–475	G–Y	Y–R
479	C	U
589–650	R–U	U–A
590–649	U–A	C–G
591–648	A–U	U–A
657–749	U–A	R–Y
896–903	U–A	C–G
943–1340	U–A	C–G
986–1219	G–C	A–U
987–1218	A–U	G–C
1168	–	A/C
1210	U	C
1245–1292	R–Y	U–A

[a]Y = Pyrimidine; R = purine.
[b]Table taken from Rainey et al. 1995b. Anaerobe *1*: 185–199. Reproduced with permission.

Taxonomic comments

The family *Halanaerobiaceae* (basonym: *Haloanaerobiaceae*) was created in 1984, when the first halophilic anaerobes became known. At that time, two genera were classified within the family: *Halanaerobium* (basonym *Haloanaerobium*) and *Halobacteroides* (Oren et al., 1984a, 1984b). With the rearrangement of the taxonomic status of many of the halophilic anaerobes and the creation of the order *Halanaerobiales* (basonym *Haloanaerobiales*) (Rainey et al., 1995a, 1995b), some genera previously classified within the family *Halanaerobiaceae* were transferred to the new family *Halobacteroidaceae*. These included the genera *Halobacteroides*, *Acetohalobium*, *Halanaerobacter* (basonym *Haloanaerobacter*), and *Sporohalobacter*. Presently, three genera are recognized within the *Halanaerobiaceae*: *Halanaerobium*, *Halothermothrix*, and *Halocella*.

According to Rule 61 of the Bacteriological Code, the original spelling of the name of the family *Haloanaerobiaceae* has been changed to *Halanaerobiaceae* (Oren, 2000).

Further reading

Ollivier, B., P. Caumette, J.-L. Garcia and R.A. Mah. 1994. Anaerobic bacteria from hypersaline environments. Microbiol. Rev. *58*: 27–38.

Oren, A. 1986. The ecology and taxonomy of anaerobic halophilic eubacteria. FEMS Microbiol. Rev. *39*: 23–29.

Oren, A. 1992. The genera *Haloanaerobium*, *Halobacteroides* and *Sporohalobacter*. *In* Balows, Trüper, Dworkin, Harder and Schleifer (Editors), The Prokaryotes. A Handbook on the Biology of Bacteria: Ecophysiology, Isolation, Identification, Applications, 2nd edn, vol. 2. Springer-Verlag, New York.

Genus I. **Halanaerobium** corrig. Zeikus, Hegge, Thompson, Phelps and Langworthy 1984, 503[VP] (Effective publication: Zeikus, Hegge, Thompson, Phelps and Langworthy 1983, 232.)

Aharon Oren

Hal.an.ae.ro′bi.um. Gr. n. *hals* salt; Gr. pref. *an* not; Gr. n. *aer* air; Gr. n. *bios* life; N.L. neut. n. *Halanaerobium* salt organism which grows in the absence of air.

Cells rod-shaped, non-motile or motile by peritrichous flagella, and Gram-stain-negative. **Strictly anaerobic,** chemo-organotrophic with fermentative metabolism. Carbohydrates are fermented with the production of acetate, H_2, and CO_2. In some species, ethanol, formate, propionate, butyrate, and lactate are also found. Thiosulfate and elemental sulfur may be used as electron acceptors in certain species. **Halophilic,** growing optimally at NaCl concentrations around 1.7–2.5 M and requiring a minimum of 0.3–1.7 M NaCl for growth, depending on the species. Neutral or slightly alkaline pH values are preferred. Endospore formation is never observed.

DNA G+C content (mol%): 27–37 (T_m, HPLC).

Type species: **Halanaerobium praevalens** corrig. Zeikus, Hegge, Thompson, Phelps and Langworthy 1984, 503[VP] (Effective publication: Zeikus, Hegge, Thompson, Phelps and Langworthy 1983, 232.).

Further descriptive information

All *Halanaerobium* species are moderately halophilic. NaCl concentrations of up to saturation support growth of certain isolates. The most halophilic representative described thus far is *Halanaerobium lacusrosei*, which grows between 1.0 and 5.8 M NaCl with an optimum at 3.4 M (Cayol et al., 1995). It is also the most halophilic species within the order *Halanaerobiales*.

Most species of the genus *Halanaerobium* grow fermentatively on sugars, producing acetate, ethanol, hydrogen, and carbon dioxide. Some strains make additional products such as formate, propionate, butyrate, and lactate. *Halanaerobium congolense* does not form ethanol (Ravot et al., 1997). *Halanaerobium alcaliphilum* can also ferment glycine betaine with the formation of acetate and trimethylamine (Tsai et al., 1995).

The ability to use substrates such as glycerol, glucosylglycerol, and trehalose by different representatives of the genus *Halanaerobium* may be of particular ecological importance. These compounds are accumulated at high concentrations as organic osmotic solutes by aerobic, photosynthetic, halophilic microorganisms inhabiting salt lakes (glycerol by the green unicellular alga *Dunaliella*, glucosylglycerol and trehalose by a variety of cyanobacteria). Such compounds can thus be expected to be available to the anaerobic bacterial community in the bottom sediments of these lakes. *Halanaerobium saccharolyticum* (both subsp. *saccharolyticum* and subsp. *senegalense*) ferments glycerol (Cayol et al., 1994b; Zhilina et al., 1992) and so does *Halanaerobium lacusrosei* (Cayol et al., 1995). No details have been published on the fermentation products made when glycerol serves as the energy source. Glycerol oxidation by anaerobic halophiles was reported to be markedly improved through interspecies hydrogen transfer when glycerol fermenters were grown in co-culture with H_2-consuming sulfate-reducing bacteria (Cayol et al., 1995). *Halanaerobium saccharolyticum* was isolated from a cyanobacterial mat dominated by *Microcoleus chthonoplastes*, covering the bottom of a hypersaline lagoon in the Crimea. Its ability to use glucosylglycerol, the osmotic solute produced by *Microcoleus* and many other cyanobacteria, may be of great ecological importance. *Halanaerobium saccharolyticum* also degrades trehalose, produced by other cyanobacteria for similar purposes (Zhilina et al., 1992).

Several *Halanaerobium* species can use oxidized sulfur compounds as electron acceptors or electron sinks. Thus, *Halanaerobium congolense* uses thiosulfate and elemental sulfur as electron acceptors. Thiosulfate improved carbohydrate utilization and enhanced growth rates. Addition of thiosulfate or sulfur increased the growth yield by 6- or 3-fold, respectively. In addition, the presence of thiosulfate alleviated growth inhibition by accumulating hydrogen (Ravot et al., 1997). Reduction of elemental sulfur to sulfide was also reported in *Halanaerobium saccharolyticum* subsp. *saccharolyticum* (Zhilina et al., 1992). Nitrosubstituted aromatic compounds such as nitrobenzene, nitrophenols, 2,4-dinitrophenol, and 2,4-dinitroaniline are reduced to the amino derivatives by *Halanaerobium praevalens* (Oren et al., 1991).

Some *Halanaerobium* species may have biotechnological applications. *Halanaerobium fermentans* was isolated from "fugunoko nukaduke", a traditional Japanese food prepared from fermented salted puffer fish ovaries (Kobayashi et al., 2000b). Puffer fish ovaries are salted for at least 6 months and the ovaries are then fermented naturally with rice-bran, fish sauce, and koji for several years. *Halanaerobium fermentans* may be one of the main bacteria involved in the fermentation process (Kobayashi et al., 2000b). Other *Halanaerobium* species may find uses in microbially enhanced recovery of oil from oil reservoirs by plugging of porous reservoirs and by anaerobically metabolizing nutrients with the production of useful products such as gases, biosurfactants, polymers, etc., under the environmental conditions that exist in such reservoirs (Bhupathiraju et al., 1991).

Enrichment and isolation procedures

Species of the genus *Halanaerobium* can probably be found in any hypersaline anaerobic environment in which simple sugars are available. They have been isolated from Great Salt Lake, Utah (Tsai et al., 1995; Zeikus et al., 1983), Lake Sivash in the Crimea (Zhilina et al., 1992), Lake Retba in Senegal (Cayol et al., 1994b), from brines associated with oil wells and petroleum reservoirs (Bhupathiraju et al., 1993, 1999, 1994, 1991; Ravot et al., 1997; Rengpipat et al., 1988a), and from salted fermented foods (Kobayashi et al., 2000b, 2000a).

Any anoxic reducing medium containing high salt concentrations (0.8–4.2 M NaCl) and containing glucose or other simple sugars as a carbon source is a potential enrichment and growth medium for anaerobic halophilic bacteria of the genus *Halanaerobium*. Details on the composition of suitable growth media can be found in Oren (2001) and in the original publications on the isolation of species. Strict anaerobic techniques should be used, including boiling the media under nitrogen or nitrogen/CO_2 (80:20) and adding reducing agents such as cysteine, thioglycollate, or sulfide to the boiled media.

No specific enrichment procedures have been designed as yet that enable growth of members of the genus *Halanaerobium* with the exclusion of other members of other genera of the families *Halanaerobiaceae* and *Halobacteroidaceae*.

Maintenance procedures

Lyophilization has proved satisfactory for the preservation of *Halanaerobium* species. *Halanaerobium acetethylicum* was successfully preserved by freezing anaerobic suspensions in 20% glycerol at –80 °C (Rengpipat et al., 1988a).

Differentiation of the species of the genus *Halanaerobium*

Identification of isolates and their assignment to the genus *Halanaerobium* should be based preferentially on the determination of their 16S rRNA gene sequences. In addition, phenotypic characterization should be performed and the properties of the strains should be compared with those of described species (see Table 238).

Taxonomic comments

The genus *Halanaerobium* (originally named *Haloanaerobium* and renamed *Halanaerobium* in accordance with Rule 61) (Oren, 2000) was the first effectively published member of the order *Halanaerobiales* (Zeikus et al., 1983). With its nine species with validly published names, it is now the largest genus within the order. Based on 16S rRNA gene sequence comparisons (Rainey et al., 1995b), a number of species formerly classified in other genera have been transferred to the genus *Halanaerobium*: the former *Halobacteroides acetoethylicus* (Rengpipat et al., 1988a) was reclassified as *Halanaerobium acetethylicum* (Oren, 2000; Patel et al., 1995; Rainey et al., 1995b); and the former *Haloincola saccharolyticus* (originally described under the name *Haloincola saccharolytica*) (Euzeby, 1998; Zhilina et al., 1991a) has been renamed as *Halanaerobium saccharolyticum*, with two subspecies, *saccharolyticum* and *senegalense* (Cayol et al., 1994b; Oren, 2000; Rainey et al., 1995b).

TABLE 238. Differential characteristics of the type strains of species of the genus *Halanaerobium*[a,b]

Characteristic	*H. praevalens* DSM 2228[T]	*H. acetethylicum* DSM 3532[T]	*H. saccharolyticum* subsp. *saccharolyticum* DSM 6643[T]	*H. saccharolyticum* subsp. *senegalense* DSM 7379[T]	*H. salsuginis* ATCC 51327[T]	*H. alcaliphilum* DSM 8275[T]	*H. lacusrosei* DSM 10165[T]	*H. congolense* DSM 11287[T]	*H. kushneri* ATCC 700103[T]	*H. fermentans* JCM 10494[T]
Cell size (μm)	0.9–11 × 2–2.6	0.2–0.7 × 1–1.6	0.5–0.7 × 1–1.5	0.4–0.6 × 2–5	0.3–0.4 × 2.6–4	0.8 × 3.3–5	0.5 × 2–3	0.5–1 × 2–4	0.7 × 2–3.3	1–1.2 × 2.7–3.3
Motility	–	+	+	+	+	+	+	–	+	+
DNA G+C content (mol%)	27	32	31	32	34	31	32	34	34	33
NaCl concentration for growth (M):										
Range	0.3–5.1	1.0–3.4	0.5–5.1	0.9–4.3	1.0–4.1	0.4–4.3	1.3–5.8	0.7–4.1	1.5–3.1	1.2–4.3
Optimum	2.1	1.7	1.7	1.3–2.1	1.5	1.7	3.1–3.4	1.7	2.1	1.7
Growth temperature (°C):										
Range	5–50	15–45	15–47	20–47	22–51	25–50	20–50	20–45	20–45	15–45
Optimum	37	34	37–40	40	40	37–40	40	42	40	35
Growth pH:										
Range	6.0–9.0	5.4–8.0	6.0–8.0	6.3–8.7	5.6–8.0	5.8–10	ND	6.3–8.5	6.0–8.0	6.0–9.0
Optimum	7.0–7.4	6.3–7.4	7.5	7.0	6.1	6.7–7.0	7.0	7.0	6.5–7.5	7.5
Utilization of:										
L-Arabinose	NR	NR	+	–	+	–	–	–	–	–
Cellobiose	–	+	+	+	–	–	+	NR	+	+
Galactose	–	–	+	–	+	–	+	+	+	+
Glycerol	–	+	+	+	–	–	+	NR	–	–
Lactose	–	+	+	+	+	–	–	–	+	+
D-Mannose	+	+	+	NR	+	+	+	+	+	–
Pyruvate	–	+	+	+	+	+	NR	NR	+	+
Raffinose	NR	NR	–	+	+	Slight	NR	NR	NR	+
D-Ribose	NR	NR	+	–	+	–	+	+	NR	–
Rhamnose	NR	NR	–	–	+	–	–	–	NR	–
Starch	–	–	NR	NR	–	–	+	NR	–	+
Sucrose	–	+	+	+	+	+	+	+	+	–
D-Xylose	–	+	+	–	+	–	+	NR	–	+
Pectin	+	–	–	–	NR	NR	+	NR	–	NR
L-Sorbose	–	–	–	–	–	NR	–	NR	–	NR
Fermentation products	Acetate, butyrate, propionate, H₂, CO₂	Acetate, ethanol, H₂, CO₂	Acetate, H₂, CO₂	Acetate, H₂, CO₂	Acetate, ethanol, H₂, CO₂	Acetate, lactate, butyrate, H₂, CO₂	Acetate, ethanol, H₂, CO₂	Acetate, H₂, CO₂	Acetate, formate, ethanol, H₂, CO₂	Acetate, formate, lactate, ethanol, H₂, CO₂

[a]Symbols: +, >85% positive; –, 0–15% positive; w, weak reaction; NR, not reported.

[b]Table based in part on tables provided by Kobayashi et al. (2000b) and Bhupathiraju et al. (1999)

List of species of the genus *Halanaerobium*

1. **Halanaerobium praevalens** corrig. Zeikus, Hegge, Thompson, Phelps and Langworthy 1984, 503[VP] (Effective publication: Zeikus, Hegge, Thompson, Phelps and Langworthy 1983, 232.)

prae.va'lens L. adj. *praevalens* prevalent.

Cells are non-motile rods, 0.9–1.1 × 2.0–2.6 µm in size. Halophilic, requiring between 0.3–5.1 M NaCl. Optimal growth occurs in 2.2 M NaCl and at pH 7.0–7.4. Temperature range for growth is 5–50 °C. Optimum temperature is 37 °C. Strictly anaerobic. Ferments carbohydrates with production of acetate, propionate, butyrate, H_2, and CO_2. Grows on a range of carbohydrates, including *N*-acetylglucosamine, fructose, D-glucose, D-mannose, and pectin. No growth has been found on cellobiose, chitin, galactose, glycerol, lactose, pyruvate, starch, sucrose, or xylose. The major fatty acids are $C_{14:0}$, $C_{16:0}$, and $C_{16:1}$. Methylmercaptan is produced during degradation of methionine.

Source: sediments of Great Salt Lake, UT, USA.
DNA G+C content (mol%): 27±1 (T_m).
Type strain: GSL, ATCC 33744, DSM 2228.
GenBank accession number (16S rRNA gene): AB022034, M59123.

2. **Halanaerobium acetethylicum** corrig. (Rengpipat, Langworthy and Zeikus 1988a) Rainey, Zhilina, Boulygina, Stackebrandt, Tourova and Zavarzin 1995a, 879[VP] (Effective publication: Rainey, Zhilina, Boulygina, Stackebrandt, Tourova and Zavarzin 1995b, 197.) (*Halobacteroides acetoethylicus* Rengpipat, Langworthy and Zeikus 1988a, 33)

a.cet.e.thy'li.cum. L. n. *acetum* vinegar; N.L. n. *ethyl* the ethyl radical; N.L. adj. *acetethylicum* producing vinegar and ethanol.

Cells are rods, motile by peritrichous flagella, 0.4–0.7 × 1.0–1.6 µm in size. Halophilic, requiring between 0.8–3.8 M NaCl. Optimal growth occurs in 1.7 M NaCl and at pH 6.3–7.4. Temperature range for growth: 15–45 °C. Optimum temperature: 34 °C. Strictly anaerobic. Ferments carbohydrates with production of ethanol, acetate, H_2, and CO_2. Grows on a range of carbohydrates and other compounds, including *N*-acetylglucosamine, cellobiose, fructose, D-glucose, lactose, D-mannose, pyruvate, sucrose, and D-xylose. No growth on galactose, glycerol, pectin, or starch. The major fatty acids are $C_{14:0}$, $C_{16:0}$, and $C_{16:1}$.

Source: brines associated with offshore oil rigs in the Gulf of Mexico.
DNA G+C content (mol%): 32.0 (T_m).
Type strain: EIGI, ATCC 43120, DSM 3532.
GenBank accession number (16S rRNA gene): X89071.

3. **Halanaerobium alcaliphilum** corrig. Tsai, Garcia, Patel, Cayol, Baresi and Mah 1995, 305[VP]

al.ca.li'phi.lum. N.L. n. *alcali* from Ar. *al-qaliy* the soda ash; Gr. adj. *philos* loving; N.L. neut. adj. *alcaliphilum* liking alkaline media.

Cells are rods, motile by peritrichous flagella, 0.8 × 3.3–5 µm in size. Halophilic, requiring 0.4–4.3 M NaCl. Can be grown over a wide range of pH values, from 5.8 to up to 10.0. Optimal growth occurs in 1.7 M NaCl and at pH 6.7–7.0. Temperature range for growth: 25–50 °C.

Optimum temperature: 37–40 °C. Strictly anaerobic. Ferments carbohydrates with production of acetate, butyrate, lactate, H_2, and CO_2. Grows on a range of carbohydrates, including *N*-acetylglucosamine, fructose, D-glucose, maltose, D-mannose, pyruvate, and sucrose. Slight growth on raffinose. No growth on adonitol, L-arabinose, cellobiose, dulcitol, erythritol, galactose, glycerol, glycogen, inositol, lactose, mannitol, pectin, D-ribose, rhamnose, sorbitol, starch, trehalose, or D-xylose. Glycine betaine is fermented to acetate and trimethylamine.

Source: sediments of Great Salt Lake, UT, USA.
DNA G+C content (mol%): 31.0 (HPLC).
Type strain: GSLS, DSM 8275.
GenBank accession number (16S rRNA gene): X81850.

4. **Halanaerobium congolense** corrig. Ravot, Magot, Ollivier, Patel, Ageron, Grimont, Thomas and Garcia 1998, 1083[VP] (Effective publication: Ravot, Magot, Ollivier, Patel, Ageron, Grimont, Thomas and Garcia 1997, 87.)

con.go.len'se. N.L. neut. adj. *congolense* pertaining to Congo, Central Africa.

Cells are short, non-motile rods, 0.5–1 × 2–4 µm in size. Halophilic, requiring 0.3–4.1 M NaCl. Optimal growth occurs in 1.7 M NaCl and at pH 7.0. Temperature range for growth: 20–45 °C. Optimum temperature: 42 °C. Strictly anaerobic. Ferments carbohydrates with production of acetate, H_2, and CO_2. Grows on a range of carbohydrates including fructose, galactose, D-glucose, maltose, D-mannose, D-ribose, sucrose, and trehalose. No growth on D-arabinose, dulcitol, lactate, lactose, propionate, rhamnose, or D-xylose. Thiosulfate and elemental sulfur are used as electron acceptors, enabling improved use of carbohydrates and enhanced biomass yields. The doubling time under optimal conditions in the presence of glucose and thiosulfate is about 2.5 h.

Source: oil injection water from an offshore oilfield, Congo, Central Africa.
DNA G+C content (mol%): 34 (HPLC).
Type strain: SEBR 4224, DSM 11287.
GenBank accession number (16S rRNA gene): U76632.

5. **Halanaerobium fermentans** corrig. Kobayashi, Kimura and Fujii 2000b, 1626[VP]

fer.men'tans. L. part. adj. *fermentans* fermenting.

Cells are rods, motile by peritrichous flagella, 1.0–1.2 × 2.7–3.3 µm in size. Halophilic, requiring 1.2–4.3 M NaCl. Optimal growth occurs in 1.7 M NaCl and at pH 7.5. Temperature range for growth: 15–45 °C. Optimum temperature: 35 °C. Strictly anaerobic. Ferments carbohydrates with production of acetate, ethanol, formate, lactate, H_2, and CO_2. Grows on a range of carbohydrates, including *N*-acetylglucosamine, cellobiose, fructose, galactose, D-glucose, lactose, maltose, D-mannose, raffinose, D-ribose, sucrose, and D-xylose. No growth on L-arabinose, glycerol, pyruvate, rhamnose, or starch.

Source: fermenting salted puffer fish ovaries.
DNA G+C content (mol%): 33.3 (HPLC).
Type strain: R-9, JCM 10494.
GenBank accession number (16S rRNA gene): AB023308.

6. **Halanaerobium kushneri** corrig. Bhupathiraju, McInerney, Woese and Tanner 1999, 958[VP]

kush'ner.i. N.L. gen. n. *kushneri* named after Donn J. Kushner, a Canadian microbiologist and author of children's books.

Cells are rods, motile by peritrichous flagella or nonmotile, 0.5–0.8 × 0.7–3.3 μm in size. Halophilic, requiring 1.5–3.1 M NaCl. Optimal growth occurs in 2.1 M NaCl and at pH 6.5–7.5. Temperature range for growth: 20–45 °C. Optimum temperature: 35–40 °C. Strictly anaerobic. Ferments carbohydrates with production of acetate, ethanol, formate, H_2, and CO_2. Grows on a range of carbohydrates and proteinaceous compounds including L-arabinose, cellobiose, fructose, galactose, D-glucose, inulin, maltose, D-mannose, peptone, sucrose, and trypticase. Growth on fucose, glucosamine, glycerol, lactose, mannitol, pyruvate, sorbose, and xylose is variable. No growth on L-alanine, L-ascorbate, adonitol, butyrate, cellulose, chitin, dulcitol, fumarate, glutamate, glycine, glycine betaine, glycogen, inositol, L-lysine, malate, pectin, starch, or sorbitol. Mixed amino acid fermentation (the Stickland reaction) is not used. The major fatty acids are $C_{16:0}$, $C_{16:1\ \omega 7c}$, $C_{16:1\ \omega 9c}$, and $C_{14:0}$.

Source: brines associated with petroleum reservoirs, Oklahoma, U.S.A.

DNA G+C content (mol%): 32–37 (HPLC); 34.1 (HPLC) for type strain.

Type strain: VS-751, ATCC 700103.

GenBank accession number (16S rRNA gene): U86446.

7. **Halanaerobium lacusrosei** corrig. Cayol, Ollivier, Patel, Ageron, Grimont, Prensier and Garcia 1995, 796[VP]

la.cus.ro'se.i. L. n. *lacus* lake; L. adj. *roseus* rose-colored; N.L. gen. n. *lacusrosei* of Rose Colored Lake, another name for Retba Lake in Senegal.

Cells are rods, motile by peritrichous flagella, 0.4–0.6 × 2–3 μm in size. Halophilic, requiring 1.3–5.8 M NaCl. Optimal growth occurs in 3.1–3.4 M NaCl and at pH 7.0. Temperature range for growth: 20–50 °C. Optimum temperature: 40 °C. Strictly anaerobic. Ferments carbohydrates with production of acetate, ethanol, H_2, and CO_2. Grows on a range of carbohydrates, including cellobiose, D-fructose, galactose, D-glucose, glycerol, lactose, maltose, D-mannitol, D-mannose, D-ribose, starch, sucrose, and D-xylose. No growth on L-arabinose, melibiose, rhamnose, or L-sorbose. The major fatty acids are $C_{15:1}$, $C_{16:0}$, and $C_{16:1}$.

Source: sediments of Lake Retba, Senegal.

DNA G+C content (mol%): 32 (HPLC).

Type strain: H200, ATCC 700560, CIP 105492, DSM 10165.

GenBank accession number (16S rRNA gene): L39787.

8. **Halanaerobium saccharolyticum** corrig. (Zhilina, Zavarzin, Bulygina, Kevbrin, Osipov and Chumakov 1992) Rainey, Zhilina, Boulygina, Stackebrandt, Tourova and Zavarzin 1995a, 879[VP] (Effective publication: Rainey, Zhilina, Boulygina, Stackebrandt, Tourova and Zavarzin 1995b, 197.) (*Haloincola saccharolytica* Zhilina, Zavarzin, Bulygina, Kevbrin, Osipov and Chumakov 1991b, 283)

sac.cha.ro.ly'ti.cum. L. n. *saccharum* sugar; Gr. adj. *lytikos* dissolving; N.L. neut. adj. *saccharolyticum* sugar-dissolving.

Cells are rods, motile by peritrichous flagella, 0.4–0.7 × 1–5 μm in size. Halophilic, requiring 0.5–5.1 M NaCl. Optimal growth occurs in 1.3–2.1 M NaCl and at pH 7.0–7.5. Temperature range for growth: 15–47 °C. Optimum temperature: 37–40 °C. Strictly anaerobic. Ferments carbohydrates with production of acetate, H_2, and CO_2.

Source: sediments of hypersaline lakes.

DNA G+C content (mol%): 31.3–32 (HPLC or T_m).

Two subspecies have been described: *Halanaerobium saccharolyticum* subsp. *saccharolyticum* and *Halanaerobium saccharolyticum* subsp. *senegalense*. The main differences between the subspecies are: the use of raffinose by the first and the use of galactose, gluconate, and L-xylose by the second; a difference in NaCl concentration range; and a different distribution of fatty acids in the membrane phospholipids.

8a. **Halanaerobium saccharolyticum subsp. saccharolyticum** corrig. (Zhilina, Zavarzin, Bulygina, Kevbrin, Osipov and Chumakov 1992) Rainey, Zhilina, Boulygina, Stackebrandt, Tourova and Zavarzin 1995a, 879[VP] (Effective publication: Rainey, Zhilina, Boulygina, Stackebrandt, Tourova and Zavarzin 1995b, 197.) (*Haloincola saccharolytica* Zhilina, Zavarzin, Bulygina, Kevbrin, Osipov and Chumakov 1992, 283) (*Haloincola saccharolytica* subsp. *saccharolytica* Cayol, Ollivier, Soh, Fardeau, Ageron, Grimont, Prensier, Guezennec, Magot and Garcia 1994b, 810)

Cells are rods, motile by peritrichous flagella, 0.5–0.7 ×μ 1–5 μm in size. Halophilic, requiring 0.5–5.1 M NaCl. Optimal growth occurs in 1.7 M NaCl and at pH 7.5. Temperature range for growth: 15–47 °C. Optimum temperature: 37–40 °C. Strictly anaerobic. Ferments carbohydrates with production of acetate, H_2, and CO_2. Grows on a range of carbohydrates and other compounds including L-arabinose, N-acetylglucosamine, cellobiose, erythritol, fructose, galactose, D-glucose, glucosylglycerol, glycerol, lactose, maltose, mannitol, D-mannose, melibiose, pyruvate, D-ribose, sucrose, trehalose, and D-xylose. No growth on adonitol, chitin, dulcitol, fucose, inositol, pectin, raffinose, rhamnose, sorbitol, L-sorbose, or starch. The major fatty acids are $C_{12:1\ 3-OH}$, $C_{14:0}$, $C_{15:1}$, and $C_{16:1}$.

Source: sediments of Lake Sivash, Crimea.

DNA G+C content (mol%): 31.3 (T_m).

Type strain: Z-7787, DSM 6643.

GenBank accession number (16S rRNA gene): L37424, X89069, Z49115.

8b. **Halanaerobium saccharolyticum subsp. senegalense** corrig. (Zhilina, Zavarzin, Bulygina, Kevbrin, Osipov and Chumakov 1992) Rainey, Zhilina, Boulygina, Stackebrandt, Tourova and Zavarzin 1995a, 880[VP] (Effective publication: Rainey, Zhilina, Boulygina, Stackebrandt, Tourova and Zavarzin 1995b, 197.) (*Haloincola saccharolytica* Zhilina, Zavarzin, Bulygina, Kevbrin, Osipov and Chumakov (Zhilina et al., 1992), 283) (*Haloincola saccharolytica* subsp. *senegalensis* Cayol, Ollivier, Soh, Fardeau, Ageron, Grimont, Prensier, Guezennec, Magot and Garcia 1994b, 810)

se.ne.ga.len'se. N.L. neut. adj. *senegalense* pertaining to Senegal.

Cells are rods, motile by peritrichous flagella, 0.4–0.6 × 2–5 μm in size. Halophilic, requiring 0.8–4.3 M NaCl. Optimal growth occurs in 1.3–2.1 M NaCl and at pH 7.0. Temperature range for growth: 20–47 °C. Optimum temperature: 40 °C. Strictly anaerobic. Ferments carbohydrates with production of acetate, H_2, and CO_2. Grows on a range of carbohydrates including cellobiose, fructose, D-glucose, glycerol, lactose, maltose, mannitol, D-mannose, raffinose, D-ribose, and sucrose. No growth on adonitol, L-arabinose, galactose, gluconate, pectin, rhamnose, L-sorbose, or D-xylose. The major fatty acids are $C_{12:0\ 3\text{-OH}}$, $C_{14:0}$, $C_{15:1}$, and $C_{16:1}$.

Source: sediments of Lake Retba, Senegal.

DNA G+C content (mol%): 31.7 (HPLC).

Type strain: H150, CIP 104784, DSM 7379.

GenBank accession number (16S rRNA gene): X89070, Z49116.

9. **Halanaerobium salsuginis** corrig. Bhupathiraju, Oren, Sharma, Tanner, Woese and McInerney 1994, 570[VP]

sal.su′gi.nis. L. n. *salsugo* water impregnated with salt; L. n. gen. *salsuginis* referring to isolation of this strain from an oilfield brine.

Cells are non-motile rods, 0.3–0.4 × 2.6–4 μm in size. Halophilic, requiring 1.0–4.1 M NaCl. Optimal growth occurs in 1.5 M NaCl and at pH 6.1. Temperature range for growth: 22–51 °C. Optimum temperature: 40 °C. Strictly anaerobic. Ferments carbohydrates with production of ethanol, acetate, H_2, and CO_2. Grows on a range of carbohydrates including L-arabinose, N-acetylglucosamine, fructose, galactose, glucosamine, D-glucose, lactose, maltotriose, melezitose, maltose, D-mannose, melibiose, pyruvate, raffinose, D-ribose, rhamnose, L-sorbose, sucrose, D-xylose, and trehalose. Yeast extract also supports growth. No growth on L-alanine, L-ascorbate, adonitol, betaine, butyrate, chitin, crotonate, Casamino acids, dextran, dulcitol, ethylene glycol, formate, fucose, fumarate, glutamate, glycine, glycogen, D-gluconate, inositol, inulin, L-lysine, lactate, malate, methanol, mannitol, methionine, glycerol, pectin, proline, propionate, peptone, starch, succinate, sarcosine, sorbitol, salicin, trypticase, or xylan. Amino acid mixtures such as alanine-glycine, leucine-proline, and other combinations were not utilized. The major fatty acids are $C_{14:0}$, $C_{16:0}$, $C_{16:1}$, and $C_{17:0\ cyclo}$.

Source: brines associated with petroleum reservoirs, OK, USA.

DNA G+C content (mol%): 34.0 (HPLC).

Type strain: VS-752, ATCC 51327.

GenBank accession number (16S rRNA gene): L22890.

Genus II. **Halocella** Simankova, Chernych, Osipov and Zavarzin 1994, 182[VP] (Effective publication: Simankova, Chernych, Osipov and Zavarzin 1993, 389.)

GEORGE A. ZAVARZIN

Ha.lo.cel′la. Gr. n. *hals* salt; L. n. fem. *cella* cell; N.L. fem. n. *Halocella* salt cell.

Cells are straight or slightly curved rods, non-sporulating, **motile** by means of peritrichous flagella. **Cell wall of Gram-negative** structure. **Obligately anaerobic.** Moderately **halophilic. Ferment carbohydrates, including cellulose**, producing acetate, ethanol, lactate, hydrogen, and carbon dioxide. Peptides and amino acids are not utilized.

DNA G+C content (mol%): 29 (T_m).

Type species: **Halocella cellulosilytica** corrig. Simankova, Chernych, Osipov and Zavarzin 1994, 182[VP] (Effective publication: Simankova, Chernych, Osipov and Zavarzin 1993, 389.).

Further descriptive information

On the basis of 16S rRNA gene sequence analysis, the genus *Halocella* belongs to the nonsporeforming branch of haloanaerobic eubacteria of the order *Haloanaerobiales* and the family *Haloanaerobiaceae* (Rainey et al., 1995b) emended by Oren (2000) to *Halanaerobiales* and *Halanaerobiaceae*, respectively. The genus includes the single species *Halocella cellulosilytica*. Specific information on the species is given below.

Enrichment and isolation procedures

Samples of decomposing alga *Cladophora sivashensis* were the source of enrichment in the following mineral solution (g/l, except where stated otherwise): NaCl, 150; NaHCO₃, 1.5; MgCl₂·2H₂O, 33; KH₂PO₄, 0.33; CaCl₂·6H₂O, 0.33; NH₄Cl, 0.33; KCl, 0.33; Na₂S·9H₂O, 0.5; yeast extract (Difco), 0.2; trace element solution (Pfennig and Lippert, 1966), 2 ml/l; vitamins (Wolin et al., 1963), 10 ml/l; filter paper, 10 (w/v). Pure culture was obtained by serial dilutions and single colony isolation on 2% agar in roll tubes with microcrystalline cellulose (Simankova et al., 1993).

Maintenance procedures

Cultures are maintained by regular transfer on cellulose. Lysis of cells at the end of active growth results in loss of viability. Cultures may be stored in liquid nitrogen.

Differentiation of the genus *Halocella* from other genera

The genus is differentiated from other genera of the family *Halanaerobiaceae* by its ability to decompose cellulose. *Halocella* differs from members of the genus *Halanaerobium* by its inability to utilize peptides and amino acids and from *Halothermothrix* species by its inability to grow at high temperatures.

Taxonomic comments

Analysis of 5S rRNA showed considerable differences between *Halocella* and other haloanaerobes (Zhilina et al., 1992). The generic status of *Halocella* is based on 16S rRNA gene sequences. The genus represents a distinct lineage in the family *Halanaerobiaceae* with sequence similarity in the range 89.6–89.9% to the other taxa. Its closest neighbor is *Halothermothrix orenii* (89.9% sequence similarity) (Rainey et al., 1995b). The lipid composition is made up of fatty acids, hydroxy fatty acids, and aldehydes. Both straight and branched lipid components are present. The main fatty acid components are $C_{14:0}$ and $C_{16:0}$. The major hydroxy fatty acid is $C_{12:0\ 3\text{-OH}}$ and the main aldehyde is $C_{16:0}$ (Simankova et al., 1993).

List of species of the genus *Halocella*

1. **Halocella cellulosilytica** corrig. Simankova, Chernych, Osipov and Zavarzin 1994, 182^VP (Effective publication: Simankova, Chernych, Osipov and Zavarzin 1993, 389.)

cel.lu.lo.si.ly'ti.ca. N.L. masc. adj. *lyticus* dissolving; N.L. n. *cellulosum* cellulose; N.L. fem. adj. *cellulosilytica* organism which dissolves cellulose.

Cells are rods, straight or slightly curved in old cultures. Under optimal conditions cells are 0.4–0.6 × 3.8–12 µm, but they are shorter under suboptimal conditions. Motile by means of peritrichous flagella. Cell wall of Gram-negative structure. Cells occur singly or in pairs or short chains. Multiply by constriction. Palisade arrangement due to copious slime production. In stationary phase, cells lyse with the formation of non-viable spheres.

Strictly anaerobic. Chemo-organotrophic. Utilizes cellulose in various forms including filter paper, microcrystalline cellulose, and carboxymethylcellulose. Good growth on dead mass of *Cladophora*. The cellulase complex in *Halocella cellulosilytica* contains only one endogluconase with optimal activity at pH 7.5 and 0.5 M NaCl of about 9.4×10^{-3} units of enzyme activity/ml, which is of the same order as that in non-halophilic bacteria, e.g., *Clostridium thermocellum* (Bolobova et al., 1992). Cellulose, cellobiose, and mannose are favored substrates and galactose, glucose, sucrose, sorbitol, and starch are utilized. Pentoses (arabinose, ribose, xylose), xylan and fructose are not utilized. During the initial phase of growth on cellulose, sugars are released into the medium. Products of fermentation are acetate, ethanol, lactate, hydrogen, and carbon dioxide. The organism is highly resistant to H_2S up to 8 mM. Growth is completely inhibited by streptomycin, benzylpenicillin, vancomycin, rifampin, and bacitracin. Moderately halophilic with optimum salinity at 2.5 M NaCl, growth occurs in range 0.8–3.4 M NaCl. Neutrophilic with pH 7.0 being optimum for growth; no growth observed below pH 5.5 or above pH 8.5.

Halocella develops in saline lagoons where decomposition of the decaying dead mass of filamentous green alga *Cladophora* after algal blooms leads to sulfide production in Lake Sivash ("Rotten Sea") by the halophilic microbial community (Zhilina and Zavarzin, 1991). *Halocella* participates in the initial stages of decomposition. The dead alga might be brought by storms, diluted by seepage from irrigating channel water of Sivash of approximately 5% salinity, and deposited in lagoons with higher salinity where cyanobacterial mats develop. Several forms of anaerobic cellulolytic bacteria are found in the hypersaline lagoons of Lake Sivash including the extremely halophilic strain Z-41 at 4.2 M NaCl (Siman'kova and Zavarzin, 1992).

Source: anaerobic sediments from the lagoons of Lake Sivash that had variable salinity and dense cyanobacterial mats of *Microcoleus chthonoplastes*.

DNA G+C content (mol%): 29 (T_m).

Type strain: Z-10151, ATCC 700086, DSM 7362.

GenBank accession number (16S rRNA gene): X89072.

Genus III. **Halothermothrix** Cayol, Ollivier, Patel, Prensier, Guezennec and Garcia 1994a, 538^VP

JEAN-LUC CAYOL, BERNARD OLLIVIER AND JEAN-LOUIS GARCIA

Ha.lo.ther'mo'thrix. Gr. n. *hals, halos* salt; Gr. adj. *thermos* hot; Gr. n. *thrix* hair; N.L. fem. n. *Halothermothrix* a thermophilic fermentative halophile.

Long rod-shaped bacteria with cells that are 0.4–0.6 × 10–20 µm, occurring mainly singly (Figure 236). **Motile by peritrichous flagella.** Non-sporulating. Gram-stain-negative. **Strictly anaerobic.** Chemo-organotrophic; oxidize carbohydrates to acetate, ethanol, H_2, and CO_2. NaCl and yeast extract are required for growth. **Grow in thermophilic conditions** (optimum temperature for growth is 60 °C).

DNA G+C content (mol%): 39.6 (HPLC).

Type species: **Halothermothrix orenii** Cayol, Ollivier, Patel, Prensier, Guezennec and Garcia 1994a, 538^VP.

Further descriptive information

Halothermothrix orenii belongs to the order *Halanaerobiales*. Members of this order comprise the families *Halanaerobiaceae* and *Halobacteroidaceae*, which promote their survival and proliferation in high-salt-containing environments by maintaining comparable high salt levels within their cytoplasm (Oren, 1986, 1999; Rengpipat et al., 1988b). Micro-organisms from these families differ from most other halophilic and halotolerant members of the domain *Bacteria*, which maintain low intracellular salt levels and synthesize or take up small organic molecules such as polyols or quaternary amines when available from the medium (Oren, 1999). This approach to maintaining cell integrity requires both extracellular and intracellular enzymes and cell structures to be active or stable under salt conditions (Oren, 1999). In order to gain information about the genetic structure of *Halothermothrix orenii*, a recent analysis of the genome of this bacterium was performed by Mijts and Patel (2001). A pBluescriptSK+ vector library consisting of 3360 clones with a

FIGURE 236. Phase-contrast photomicrograph of *Halothermothrix orenii*. Bar = 10 µm.

mean insert size of 3.5 kb was constructed. Seventy-seven clones were sequenced from both ends using T3 and T7 vector primers generating 154 sequence tags, representing approximately 85 kb of the genome. Comparison of sequence tags against the Gen-Bank database using BLASTX identified 66 known proteins and 15 conserved hypothetical proteins. The putative proteins included a V-ATPase, hydrogenases and enzymes with potential in industrial applications such as peptidases and esterases. These enzymes would be of particular use in applications where activity or stability at a range of salt levels and high temperatures is required. High levels of excess acidic amino acids were not detected in the putative proteins of *Halothermothrix orenii* as compared to the mesophilic halanaerobes. This may be the result of reduced activity of acidic, halophilic enzymes at high temperatures and intermediate salt concentrations (Mijts and Patel, 2001).

Enrichment and isolation procedures

Halothermothrix orenii has been isolated from the sediment of a Tunisian hypersaline lake (Chott El-Guettar) in samples collected at 20 cm intervals down to 1.20 m. It was only present in the 40–60 cm layer. It was isolated at 60 °C in a medium containing (per liter): NH$_4$Cl, 1.0 g; KH$_2$PO$_4$, 0.3 g; K$_2$HPO$_4$, 0.3 g; MgCl$_2$·6H$_2$O, 2.0 g; CaCl$_2$·2H$_2$O, 0.2 g; KCl, 4.0 g; CH$_3$COONa·3H$_2$O, 1.0 g; glucose, 10.0 g; NaCl, 100 g; Biotrypticase (bioMérieux), 3.0 g; yeast extract (Difco), 3.0 g; resazurin, 0.1% (w/v); trace element solution (Imhoff-Stuckle and Pfennig, 1983), 1 ml. The medium was adjusted to pH 7.0 with 10 M KOH, boiled under a stream of O$_2$-free N$_2$ and cooled to room temperature. Aliquots (20 ml) of medium were distributed into 60 ml serum bottles that were closed with butyl rubber stoppers according to the Hungate anaerobic technique (Hungate, 1969). The serum bottles were gassed with N$_2$/CO$_2$ (80:20) and sterilized for 45 min at 110 °C. After sterilization, 0.2 ml of 2% Na$_2$S·9H$_2$O, 1 ml of 10% NaHCO$_3$ (sterile anaerobic stock solutions), and 0.1 ml of a 0.2% sodium dithionite solution (filter-sterilized solution) were injected into the bottles. For preparing roll-tubes, 2% agar (Difco) was added to the medium and 5 ml portions of medium were distributed into Hungate tubes as described above. Pure cultures were obtained by repeated application of the agar shake dilution method in anaerobic Hungate tubes as described previously (Ollivier et al., 1991). Purity was checked by lack of growth in a complex rich NaCl-free medium at mesophilic and thermophilic temperatures.

Differentiation of the genus *Halothermothrix* from other genera

Research on hypersaline ecosystems has led to the description of two families: *Halanaerobiaceae*, which includes three genera and eleven species; and *Halobacteroidaceae*, which includes eight genera and thirteen species. These families comprise exclusively mesophilic and strictly anaerobic bacteria with the exception of *Halobacteroides lacunaris*, a moderately thermophilic bacterium with an upper temperature limit of 52 °C (Zhilina et al., 1991b). The characterization of *Halothermothrix orenii* extends the temperature limit for growth of halophilic anaerobic members of the order *Halanaerobiales* to 68 °C. To date, two moderately thermophilic, phylogenetically distinct halophiles have been described (Table 239), both of which are members of the *Firmicutes*. *Halothermothrix orenii*, a member of the order *Halanaerobiales*, grows optimally at 60 °C in the presence of 10 % NaCl, whereas *Thermohalobacter berrensis*, a member of the order *Clostridiales*, grows optimally at 70 °C

TABLE 239. Main characteristics that differentiate *Halothermothrix orenii* from *Thermohalobacter berrensis*

Characteristic	*H. orenii*[a]	*T. berrensis*[b]
NaCl range (%)	4–20	2–15
NaCl optimum (%)	10	5
Temperature range (°C)	45–68	45–70
Temperature optimum (°C)	60	65
pH range	5.5–8.2	5.2–8.8
pH optimum	6.5–7.0	7.0
DNA G+C content (mol%)	40	33
Substrates used:		
Arabinose	+	–
Galactose	+	–
Melibiose	+	–
Maltose	–	+
Mannitol	–	+
Sucrose	–	+
Ribose	+	–
Xylose	+	–
Glycerol	–	+
Pyruvate	–	+
Bio-Trypticase	–	+

[a]Data from Cayol et al. (1994a).
[b]Data from Cayol et al. (2000).

TABLE 240. Fatty acid profile of *Halothermothrix orenii*

Fatty acid	Amount (%)
C$_{13:0}$	2.29
C$_{14:0\ iso}$	1.54
C$_{14:0}$	7.96
C$_{15:0\ iso}$	54.3
C$_{15:0\ anteiso}$	9.79
C$_{15:0}$	1.57
C$_{16:0\ iso}$	2.68
C$_{16:0}$	9.94
C$_{17:0\ iso}$	4.99
C$_{17:0\ anteiso}$	2.34
C$_{17:0}$	0.35
C$_{18:0\ iso}$	0.20
C$_{18:0}$	1.22
Total	99.07
Branched	75.64

in the presence of 15 % NaCl (Cayol et al., 2000). The phospholipid fatty acid (PLFA) composition of *Halothermothrix orenii* contains a preponderance of branched saturated fatty acids (75.6%) of which C$_{15:0\ iso}$, C$_{15:0\ anteiso}$, and C$_{16:0}$ dominate (Table 240). This is a trait commonly found in genuine thermophilic bacteria due to growth at high temperatures. *Halothermothrix orenii* is a thermophile, but its obligate salt requirement indicates that it is unrelated to other thermophilic bacteria, e.g., *Thermanaerobacter* (Lee et al., 1993). The absence of lactate production from glucose by *Halothermothrix orenii* also rules out any similarity with members of this genus for which lactate production is a characteristic.

Taxonomic comments

Halothermothrix orenii is a rod-shaped, anaerobic, and halophilic bacterium and therefore resembles members of the family *Hala-*

TABLE 241. Main characteristics that differentiate the genera *Halothermothrix*, *Halocella*, and *Halanaerobium* of the family *Halanaerobiaceae*[a]

Characteristic	*Halothermothrix*	*Halocella*	*Halanaerobium*
Cell dimensions (μm)	0.4–0.6 × 10–20	0.4–0.6 × 3.8–12	0.3–1.2 × 0.5–5
Flagella	Peritrichous	Peritrichous	Peritrichous or no flagella
Temperature optimum (°C)	60	39	34–42
Temperature range (°C)	45–68	20–50	>5–<60
pH optimum	6.5–7.0	7.0	6.1–7.5
pH range	5.5–8.2	5.5–8.5	5.4–10
NaCl optimum (%)	5–10	15	7.5–20
NaCl range (%)	4–20	5–20	2.5–34
DNA G+C content (mol%)	39.6	29	26–37
Substrates used:			
L-Arabinose	+	–	+/–
Fructose	+	–	+
D-Ribose	+	–	+
Sucrose	–	+	+/–
D-Xylose	+	–	+/–
Peptides	–	–	+
Production of lactate from glucose	–	+	+/–

[a]Symbols: +, >85% positive; –, 0–15% positive; +/–, possibly used by some species.

naerobiaceae. Considerable biochemical and physiological differences have been observed between *Halothermothrix orenii* and other members of the family *Halanaerobiaceae*, with *Halocella cellulosilytica* and species of *Halanaerobium* being its closest phylogenetic relatives (similarities of 89% and 85–87%, respectively) (Table 241). The relationship between *Halothermothrix orenii* and members of this family is further strengthened by the presence of a common stretch of sequence at positions 821–842 (*Escherichia coli* numbering, according to Winkler and Woese, 1991) and the presence of a common shortened secondary structural variant in part of helix 47 at positions 1435–1466 (Figure 237). In contrast to some *Halanaerobium* species, *Halothermothrix orenii* oxidizes L-arabinose, sucrose, and D-xylose, it does not use peptides or produce lactate from glucose fermentation, and it grows at high temperatures (Table 241). *Halothermothrix orenii* differs from *Halocella cellulosilytica* as it oxidizes L-arabinose, fructose, D-ribose, and D-xylose, but not sucrose, and grows at high temperatures, whereas *Halocella cellulosilytica* produces lactate from glucose fermentation.

Acknowledgements

Many thanks to J. Guezennec for PLFA analysis, B.K.C. Patel for 16S rRNA structure, F. Verhé for storage and cultivation of the bacterium, and C. Lesaulnier for improving the manuscript.

Further reading

Oren, A. 2001. The order *Haloanaerobiales*. *In* Dworkin, Falkow, Rosenberg, Schleifer and Stackebrandt (Editors), The Prokaryotes: An Evolving Electronic Database for the Microbiological Community, 3rd ed. (release 3.6). Springer-Verlag, New York.

List of species of the genus *Halothermothrix*

1. **Halothermothrix orenii** Cayol, Ollivier, Patel, Prensier, Guezennec and Garcia 1994a, 538[VP]

o.re'ni.i. N.L. gen. n. *orenii* of Oren, named after Aharon Oren who has made important contributions to the knowledge of halophilic anaerobic bacteria.

Cells are long, flexible rods (Figure 236), occurring mainly singly (10–20 by 0.4–0.6 μm) and cytoplasmic density was higher at the periphery of the cells. The cell wall is typical of Gram-stain-negative bacteria. Irregularities in the diameters of cells are observed. Motile with peritrichous flagella. Colonies are yellow, flat, and circular with diameters ranging from 0.5 to 1.0 mm, depending on incubation time.

Obligate anaerobe that oxidizes several carbohydrates including arabinose, cellobiose, fructose, galactose, glucose, mannose, melibiose, ribose, starch, and xylose. Does not use lactose, maltose, rhamnose, sucrose, sorbose, cellulose, formate, acetate, butyrate, propionate, fumarate, lactate, malate, succinate, adonitol, dulcitol, glycerol, mannitol, Casamino acids, trimethylamine, or Biotrypticase. Yeast extract is required for growth. Products of glucose fermentation are ethanol, acetate, H_2, and CO_2.

Moderate halophile, thermophile, and neutrophile. The optimum NaCl concentration is between 5 and 10%, with the lower and upper limits of growth around 4 and 20%, respectively. The optimum temperature for growth is 60°C; growth occurs between 45 and 68°C. The pH range for growth is 5.5–8.2; the optimum pH is 6.5–7.0. Does not require $MgCl_2$, but tolerates it up to 0.7M when 10% NaCl is present in the medium.

Does not possess unsaturated fatty acids in the lipid cell wall composition. Branched-chain saturates account for 75% of the total fatty acids, with $C_{15:0\ iso}$ fatty acid predominating.

Source: sediments of a Tunisian salted lake (Chott El-Guettar).

DNA G+C content (mol%): 39.6 (HPLC).

Type strain: H168, OCM 544, DSM 9562.

GenBank accession number (16S rRNA gene): L22016.

References

Bhupathiraju, V.K., M.J. McInerney and R.M. Knapp. 1993. Pretest studies for a microbially enhanced oil recovery field pilot in a hypersaline oil reservoir. Geomicrobiol. J. *11*: 19–34.

Bhupathiraju, V.K., M.J. McInerney, C.R. Woese and R.S. Tanner. 1999. *Haloanaerobium kushneri* sp. nov., an obligately halophilic, anaerobic bacterium from an oil brine. Int. J. Syst. Bacteriol. *49*: 953–960.

Bhupathiraju, V.K., A. Oren, P.K. Sharma, R.S. Tanner, C.R. Woese and M.J. McInerney. 1994. *Haloanaerobium salsugo* sp. nov., a moderately halophilic, anaerobic bacterium from a subterranean brine. Int. J. Syst. Bacteriol. *44*: 565–572.

Bhupathiraju, V.K., P.K. Sharma, M.J. McInerney, R.M. Knapp, K. Fowler and W. Jenkins. 1991. Isolation and characterization of novel halophilic anaerobic bacteria from oil field brines. Dev. Petrol. Sci. *31*: 132–143.

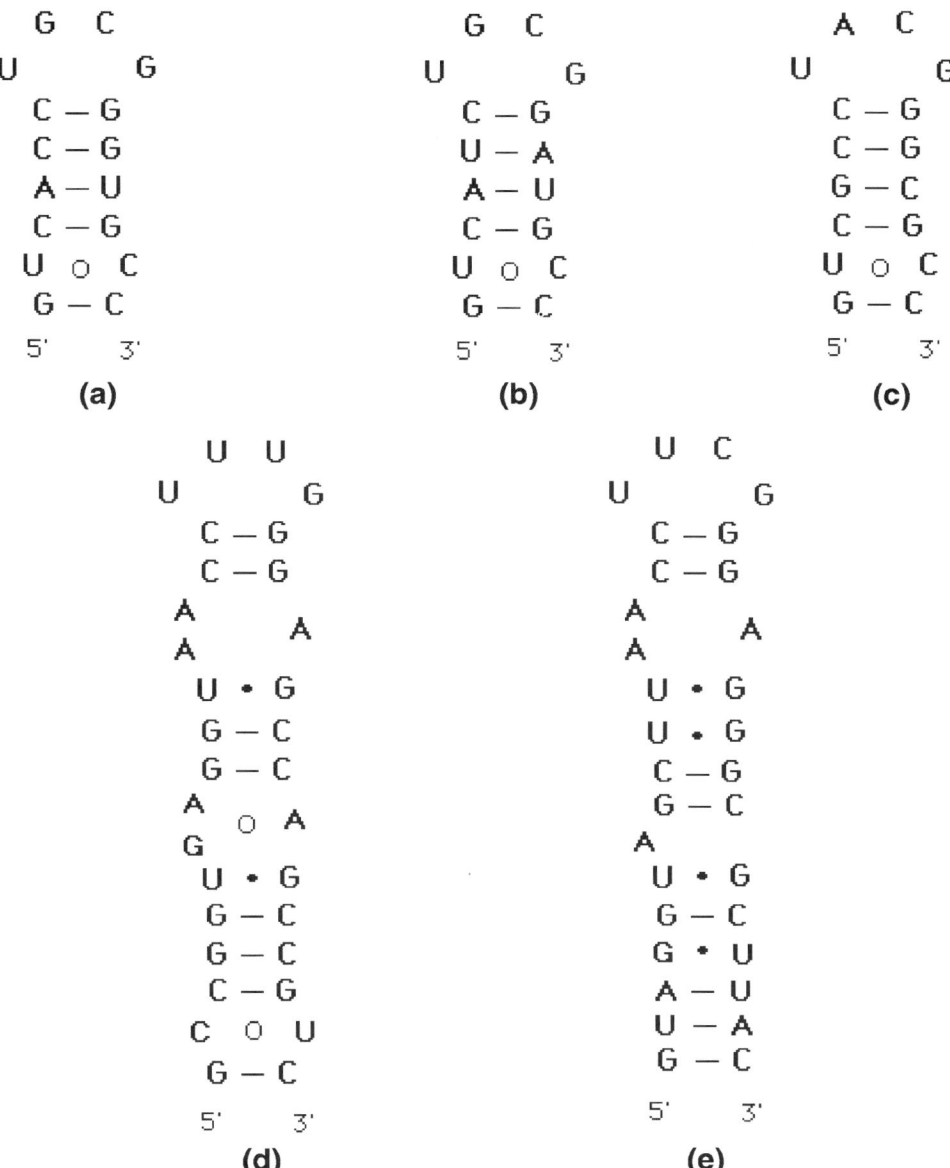

FIGURE 237. Secondary structure of 16S rRNA of halophilic anaerobes, showing the common shortened structure of part of helix 47 consisting of 16 nt in *Halothermothrix orenii* (a), *Halobacteroides praevalens* (b), and *Sporohalobacter lortetii* (c) compared with those of other Gram-positive members which consist of more than 32 nt; the structure of *Megasphaera elsdenii* (d) is shown as an example. The corresponding region of helix 47 in *Escherichia coli* (positions 1435–1466) (e) is shown for comparison.

Bolobova, A.V., M.V. Siman'kova and N.A. Markovich. 1992. Cellulolytic complex of new halophilic bacterium *Halocella cellulolytica*. Mikrobiologiya *61*: 804–811.

Cayol, J.-L., S. Ducerf, B.K.C. Patel, J.-L. Garcia, P. Thomas and B. Ollivier. 2000. *Thermohalobacter berrensis* gen. nov., sp. nov., a thermophilic, strictly halophilic bacterium from a solar saltern. Int. J. Syst. Evol. Microbiol. *50*: 559–564.

Cayol, J.-L., B. Ollivier, B.K.C. Patel, E. Ageron, P.A. Grimont, G. Prensier and J.-L. Garcia. 1995. *Haloanaerobium lacusroseus* sp. nov., an extremely halophilic fermentative bacterium from the sediments of a hypersaline lake. Int. J. Syst. Bacteriol. *45*: 790–797.

Cayol, J.-L., B. Ollivier, B.K.C. Patel, G. Prensier, J. Guezennec and J.-L. Garcia. 1994a. Isolation and characterization of *Halothermothrix orenii* gen. nov., sp. nov., a halophilic, thermophilic, fermentative, strictly anaerobic bacterium. Int. J. Syst. Bacteriol. *44*: 534–540.

Cayol, J.-L., B. Ollivier, A.L.A. Soh, M.-L. Fardeau, E. Ageron, P.A.D. Grimont, G. Prensier, J. Guezennec, M. Magot and J.-L. Garcia. 1994b. *Haloincola saccharolytica* subsp. *senegalensis* subsp. nov., isolated from the sediments of a hypersaline lake, and emended description of *Haloincola saccharolytica*. Int. J. Syst. Bacteriol. *44*: 805–811.

Euzéby, J.P. 1998. Taxonomic note: necessary correction of specific and subspecific epithets according to Rules 12c and 13b of the International Code of Nomenclature of Bacteria (1990 Revision). Int. J. Syst. Bacteriol. *48*: 1073–1075.

Hungate, R.E. 1969. A roll tube method for cultivation of strict anaerobes. *In* Norris and Ribbons (Editors), Methods in Microbiology, vol. 3B. Academic Press, London and New York, pp. 117–132.

Imhoff-Stuckle, D. and N. Pfennig. 1983. Isolation and characterization of a nicotinic-acid degrading sulfate-reducing bacterium, *Desulfococcus niacini* sp. nov. Arch. Microbiol. *136*: 194–198.

Kevbrin, V.V. and G.A. Zavarzin. 1992. Methanethiol utilization and sulfur reduction by anaerobic halophilic saccharolytic bacteria. Curr. Microbiol. *24*: 247–250.

Kobayashi, T., B. Kimura and T. Fujii. 2000a. Strictly anaerobic halophiles isolated from canned Swedish fermented herrings (Surströmming). Int. J. Food Microbiol. *54*: 81–89.

Kobayashi, T., B. Kimura and T. Fujii. 2000b. *Haloanaerobium fermentans* sp. nov., a strictly anaerobic, fermentative halophile isolated from fermented puffer fish ovaries. Int. J. Syst. Evol. Microbiol. *50*: 1621–1627.

Lee, Y.E., M.K. Jain, C. Lee, S.E. Lowe and J.G. Zeikus. 1993. Taxonomic distinction of saccharolytic anaerobes: description of *Thermoanaerobacterium xylanolyticum* gen. nov., sp. nov., and *Thermoanaerobacterium saccharolyticum* gen. nov., sp. nov.; reclassification of *Thermoanaerobium brockii*, *Clostridium thermosulfurogenes*, and *Clostridium thermohydrosulfuricum* E100-69 as *Thermoanaerobacter brockii* comb. nov., *Thermoanaerobacterium thermosulfurogenes* comb. nov., and *Thermoanaerobacter thermohydrosulfuricus* comb. nov., respectively; and transfer of *Clostridium thermohydrosulfuricum* 39E to *Thermoanaerobacter ethanolicus*. Int. J. Syst. Bacteriol. *43*: 41–51.

Lowe, S.E., M.K. Jain and J.G. Zeikus. 1993. Biology, ecology, and biotechnological applications of anaerobic bacteria adapted to environmental stresses in temperature, pH, salinity, or substrates. Microbiol. Rev. *57*: 451–509.

Mijts, B.N. and B.K.C. Patel. 2001. Random sequence analysis of genomic DNA of an anaerobic, thermophilic, halophilic bacterium, *Halothermothrix orenii*. Extremophiles *5*: 61–69.

Ollivier, B., C.E. Hatchikian, G. Prensier, J. Guezennec and J.-L. Garcia. 1991. *Desulfohalobium retbaense* gen. nov., sp. nov., a halophilic sulfate-reducing bacterium from sediments of a hypersaline lake in Senegal. Int. J. Syst. Bacteriol. *41*: 74–81.

Oren, A. 1986. Intracellular salt concentrations of the anaerobic halophilic eubacteria *Haloanaerobium praevalens* and *Halobacteroides halobius*. Can. J. Microbiol. *32*: 4–9.

Oren, A. 1999. Bioenergetic aspects of halophilism. Microbiol. Mol. Biol. Rev. *63*: 334–348.

Oren, A. 2000. Change of the names *Haloanaerobiales, Haloanaerobiaceae* and *Haloanaerobium* to *Halanaerobiales, Halanaerobiaceae* and *Halanaerobium*, respectively, and further nomenclatural changes within the order *Halanaerobiales*. Int. J. Syst. Evol. Microbiol. *50*: 2229–2230.

Oren, A. 2001. The order *Halanaerobiales*. *In* Dworkin, Falkow, Rosenberg, Schleifer and Stackebrandt (Editors), The Prokaryotes: A Handbook on the Biology, of Bacteria: Ecophysiology, Isolation, Identification, Applications, 3rd edn. Springer-Verlag, New York, pp. chapter 238.

Oren, A., P. Gurevich and Y. Henis. 1991. Reduction of nitrosubstituted aromatic compounds by the halophilic anaerobic eubacteria *Haloanaerobium praevalens* and *Sporohalobacter marismortui*. Appl. Environ. Microbiol. *57*: 3367–3370.

Oren, A., B.J. Paster and C.R. Woese. 1984a. *Haloanaerobiaceae*: a new family of moderately halophilic, obligatory anaerobic bacteria. Syst. Appl. Microbiol. *5*: 71–80.

Oren, A., B.J. Paster and C.R. Woese. 1984b. *In* Validation of the publication of new names and new combinations previously effectively published outside the IJSB. List no. 16. Int. J. Syst. Bacteriol. *34*: 503–504.

Patel, B.K.C., K.T. Andrews, B. Ollivier, R.A. Mah and J.L. Garcia. 1995. Reevaluating the classification of *Halobacteroides* and *Haloanaerobacter* species based on sequence comparisons of the 16S ribosomal RNA gene. FEMS Microbiol. Lett. *134*: 115–119.

Pfennig, N. and K.D. Lippert. 1966. Über das Vitamin B$_{12}$-Bedürfnis phototropher Schwefelbakterien. Arch. Mikrobiol. *55*: 245–246.

Rainey, F.A., T.N. Zhilina, E.S. Boulygina, E. Stackebrandt, T.P. Tourova and G.A. Zavarzin. 1995a. *In* Validation of the publication of new names and new combinations previously effectively published outside the IJSB. List no. 55. Int. J. Syst. Bacteriol. *45*: 879–880.

Rainey, F.A., T.N. Zhilina, E.S. Boulygina, E. Stackebrandt, T.P. Tourova and G.A. Zavarzin. 1995b. The taxonomic status of the fermentative halophilic anaerobic bacteria: description of *Haloanaerobiales* ord. nov., *Halobacteroidaceae* fam. nov., *Orenia* gen. nov. and further taxonomic rearrangements at the genus and species Level. Anaerobe *1*: 185–199.

Ravot, G., M. Magot, B. Ollivier, B.K.C. Patel, E. Ageron, P.A. Grimont, P. Thomas and J.-L. Garcia. 1997. *Haloanaerobium congolense* sp. nov., an anaerobic, moderately halophilic, thiosulfate- and sulfur-reducing bacterium from an African oil field. FEMS Microbiol. Lett. *147*: 81–88.

Ravot, G., M. Magot, B. Ollivier, B.K.C. Patel, E. Ageron, P.A.D. Grimont, P. Thomas and J.-L. Garcia. 1998. *In* Validation of publication of new names and new combinations previously effectively published outside the IJSB. List no. 67. Int. J. Syst. Bacteriol. *48*: 1083–1084.

Rengpipat, S., T.A. Langworthy and J.G. Zeikus. 1988a. *Halobacteroides acetoethylicus* sp. nov., a new obligately anaerobic halophile isolated from deep subsurface hypersaline environments. Syst. Appl. Microbiol. *11*: 28–35.

Rengpipat, S., S.E. Lowe and J.G. Zeikus. 1988b. Effect of extreme salt concentrations on the physiology and biochemistry of *Halobacteroides acetoethylicus*. J. Bacteriol. *170*: 3065–3071.

Siman'kova, M.V. and G.A. Zavarzin. 1992. The anaerobic cellulose degradation in the Sivash Lake and hypersaline lagoons of the Arabatskaya spit. Mikrobiologiya *61*: 288–293.

Simankova, M.V., N.A. Chernych, G.A. Osipov and G.A. Zavarzin. 1993. *Halocella cellulolytica* gen-nov, sp-nov, a new obligately anaerobic, halophilic, cellulolytic bacterium. Syst. Appl. Microbiol. *16*: 385–389.

Simankova, M.V., N.A. Chernych, G.A. Osipov and G.A. Zavarzin. 1994. *In* Validation of the publication of new names and new combinations previously effectively published outside the IJSB. List no. 48. Int. J. Syst. Bacteriol. *44*: 182–183.

Tsai, C.R., J.-L. Garcia, B.K. Patel, J.-L. Cayol, L. Baresi and R.A. Mah. 1995. *Haloanaerobium alcaliphilum* sp. nov., an anaerobic moderate halophile from the sediments of Great Salt Lake, Utah. Int. J. Syst. Bacteriol. *45*: 301–307.

Winkler, S. and C.R. Woese. 1991. A definition of the domain *Archaea, Bacteria* and *Eucarya* in terms of small subunit ribosomal RNA characteristics. Syst. Appl. Microbiol. *14*: 305–310.

Wise, D.L. 1987. Meeting report - first international workshop on biogasification and biorefining of Texas lignite. Res. Conserv. *15*: 229–247.

Wolin, E.A., M.G. Wolin and R.S. Wolfe. 1963. Formation of methane by bacterial extracts. J. Biol. Chem. *238*: 2882–2886.

Zeikus, J.G., P.W. Hegge, T.E. Thompson, T.J. Phelps and T.A. Langworthy. 1983. Isolation and description of *Haloanaerobium praevalens* gen. nov. and sp. nov., an obligately anaerobic halophile common to Great Salt Lake sediments. Curr. Microbiol. *9*: 225–234.

Zeikus, J.G., P.W. Hegge, T.E. Thompson, T.J. Phelps and T.A. Langworthy. 1984. *In* Validation of the publication of new names and new combinations previously effectively published outside the IJSB List no. 16. Int. J. Syst. Bacteriol. *34*: 503–504.

Zhilina, T.N., V.V. Kevbrin, A.M. Lysenko and G.A. Zavarzin. 1991a. Isolation of saccharolytic anaerobes from a halophilic cyanobacterial mat. Microbiology (En. transl. from Mikrobiologiya) *60*: 101–107.

Zhilina, T.N., L.V. Miroshnikova, G.A. Osipov and G.A. Zavarzin. 1991b. *Halobacteroides lacunaris* sp. nov., new saccharolytic, anaerobic, extremely halophilic organism from the lagoon-like hypersaline Lake Chokrak. Microbiology (En. transl. from Mikrobiologiya) *60*: 495–503.

Zhilina, T.N., T.P. Tourova, A.M. Lysenko and V.V. Kevbrin. 1997. Reclassification of *Halobacteroides halobius* Z-7287 on the basis of phylogenetic analysis as a new species *Halobacteroides elegans* sp. nov. Microbiology *66*: 97–103.

Zhilina, T.N. and G.A. Zavarzin. 1991. Anaerobic bacteria participating in organic matter destruction in halophilic cyanobacterial community. Zhurn. Obzhei Biol. *52*: 302–318.

Zhilina, T.N., G.A. Zavarzin, E.S. Bulygina, V.V. Kevbrin, G.A. Osipov and K.M. Chumakov. 1992. Ecology, physiology and taxonomy studies on a new taxon of *Haloanaerobiaceae, Haloincola saccharolytica* gen. nov., sp. nov. Syst. Appl. Microbiol. *15*: 275–284.

Family II. **Halobacteroidaceae** Zhilina and Rainey 1995, 879[VP] (Effective publication: Rainey, Zhilina, Boulygina, Stackebrandt, Tourova and Zavarzin 1995a, 193.)

AHARON OREN

Ha.lo.bac.te.ro.i.da′ce.ae. N.L. masc. n. *Halobacteroides* type genus of the family; *-aceae* ending to denote a family; N.L. fem. pl. n. *Halobacteroidaceae* the *Halobacteroides* family.

Cells are rod-shaped and **Gram-stain-negative**. Endospores are produced by some species. **Strictly anaerobic.** Oxidase- and catalase-negative. Most species ferment carbohydrates to products including acetate, ethanol, hydrogen, and carbon dioxide. Some species may grow fermentatively on amino acids; others have a homoacetogenic metabolism or grow by anaerobic respiration while reducing nitrate, trimethylamine N-oxide, or selenate. **Moderately halophilic.** NaCl concentrations in the range 1.7–2.5 M are required for optimal growth and no growth is observed below 0.3–1.7 M NaCl, depending on the species.

DNA G+C content (mol%): 30–50.

Type genus: **Halobacteroides** Oren, Weisburg, Kessel and Woese 1984a, 355[VP] (Effective publication: Oren, Weisburg, Kessel and Woese 1984b, 68.).

Further descriptive information

All known members of the *Halobacteroidaceae* are strictly anaerobic and moderately halophilic. Most grow optimally at NaCl concentrations around 1.7–2.5 M. A minimal NaCl concentration of 0.3–1.7 M is required, depending on the species. *Natroniella acetigena*, isolated from the alkaline hypersaline Lake Magadi, Kenya, is an obligate alkaliphile, growing between pH 8.1–10.7 with optimum growth at pH 9.7–10.0 (Zhilina et al., 1996b).

All species show a Gram-stain-negative reaction. A typical Gram-stain-negative type of cell wall with an outer membrane and periplasmic space is often seen in electron micrographs of thin sections.

Heat-resistant endospores are produced by a number of species belonging to the family. These include *Sporohalobacter lortetii* (Oren, 1983) and *Orenia marismortui* (Oren et al., 1987). *Natroniella acetigena* produces spores infrequently (Zhilina et al., 1996b). When initially isolated, *Acetohalobium arabaticum* produced spores, but sporulation was not observed during subsequent transfers (Zavarzin et al., 1994). Special conditions may be required for induction of endospore formation. Growth on solid media or in nutrient-poor liquid media may enhance sporulation in certain species (Oren, 1983, 1987).

Most species belonging to the *Halobacteroidaceae* obtain their energy by fermenting simple sugars. Complex polysaccharides such as chitin may be used by certain species such as *Halanaerobacter chitinivorans*. Fermentation products typically include acetate, hydrogen, and carbon dioxide. Additionally, some strains produce butyrate, lactate, propionate, and formate. In addition to sugar fermenters, some species can ferment amino acids, either alone or by using the Stickland reaction. For example, *Halanaerobacter salinarius* and *Halanaerobacter chitinivorans* can use serine as an electron donor while reducing glycine betaine, with the formation of acetate, trimethylamine, CO_2, and NH_3 (Mouné et al., 1999). The neutrophilic *Acetohalobium arabaticum* and the alkaliphilic *Natroniella acetigena* are homoacetogens that do not use carbohydrates but grow chemoheterotrophically on substrates such as lactate, ethanol, pyruvate, glutamate, propanol, and glycine betaine. *Acetohalobium arabaticum* can also grow

chemoautotrophically on hydrogen plus carbon dioxide with the formation of acetate or as a methylotroph on trimethylamine, forming acetate and NH_3 (Zavarzin et al., 1994; Zhilina and Zavarzin, 1990c; Zhilina et al., 1996b).

Methanethiol can be used as an assimilatory sulfur source by several species, including *Halobacteroides halobius*, *Halobacteroides elegans*, and *Halanaerobacter lacunarum* (Kevbrin and Zavarzin, 1992b; Zhilina et al., 1991b, 1997b). Some species can use oxidized sulfur compounds as electron acceptors or electron sinks. *Acetohalobium arabaticum* slowly reduces sulfur to sulfide, but no growth enhancement has been observed (Kevbrin and Zavarzin, 1992c; Zavarzin et al., 1994). *Orenia marismortui* produces small amounts of sulfide from thiosulfate (Oren et al., 1987). Other compounds that may serve as electron sinks are nitrosubstituted aromatic compounds such as nitrobenzene, nitrophenols, 2,4-dinitrophenol, and 2,4-dinitroaniline, which are found to be reduced by *Orenia marismortui* (Oren et al., 1991).

Enrichment and isolation procedures

Species belonging to the family *Halobacteroidaceae* can probably be found in any hypersaline anaerobic environment in which simple organic compounds such as sugars and amino acids are available. They have been isolated from hypersaline lakes, both with thalassohaline and athalassohaline ionic composition, including the Dead Sea (Oren, 1983, 1987; Oren et al., 1984b), Salton Sea, California (Shiba, 1991; Shiba and Horikoshi, 1988, 1989), hypersaline lakes and lagoons in the Crimea (Zhilina et al., 1991a, 1997b, 1990c), saltern evaporation ponds in California (Liaw and Mah, 1992) and France (Mouné et al., 2000, 1999), and the alkaline hypersaline lakes Magadi, Kenya (Shiba and Horikoshi, 1988; Zhilina et al., 1996b) and Big Soda Lake, Nevada (Shiba and Horikoshi, 1988, 1989).

Any anoxic reducing medium containing high salt concentrations (0.8–4.2 M) and a suitable carbon source is a potential enrichment and growth medium for members of the *Halobacteroidaceae*. Simple sugars are used by most species. For the isolation of amino-acid-fermenting or homoacetogenic species, the medium composition should be adapted accordingly. For the isolation of *Selenihalanaerobacter*, selenate is the preferred electron acceptor as nitrate and trimethylamine N-oxide enable anaerobic growth of a variety of facultative anaerobes. The formation of heat-resistant endospores has been exploited in a selective enrichment procedure for *Halobacteroides halobius*-like bacteria, based on negative selection by pasteurization of the inoculum for 10–20 min at 80–100 °C (Oren, 1987). In view of the number of endospore-forming genera within the family (*Halobacteroides*, *Orenia*, *Sporohalobacter*, *Acetohalobium*, and *Natroniella*), such an enrichment strategy could be useful for the isolation of other members of the family.

Strictly anaerobic techniques should be used, including boiling the media under nitrogen or nitrogen/CO_2 (80:20) and adding reducing agents such as cysteine, thioglycollate, or sulfide to the boiled media. High pH media should be used for the enrichment and isolation of alkaliphiles. Protocols for the preparation of media have been compiled by Oren (2001); details can be found in the original species descriptions. No specific enrichment procedures have been designed as yet that enable growth of the *Halobacteroidaceae* with the exclusion of the *Halanaerobiaceae*.

Properties that distinguish the family *Halobacteroidaceae* from other families

The only sure way to distinguish members of the family *Halobacteroidaceae* from the members of the *Halanaerobiaceae* is by 16S rRNA gene sequence relatedness. There are no known morphological, physiological, or biochemical tests that enable an unequivocal classification in either of the two families. Table 236 presents information on the 16S rRNA signature nucleotides defining the two families.

Taxonomic comments

The family *Halobacteroidaceae* was created in 1995 in the course of a comprehensive in-depth study of the anaerobic halophilic *Bacteria*. Based on 16S rRNA gene sequence comparisons, extensive taxonomic rearrangements were proposed. These included a reclassification of the species of the former family *Haloanaerobiaceae* over two families: the *Haloanaerobiaceae* (renamed *Halanaerobiaceae*) and the newly created family *Halobacteroidaceae* (Rainey et al., 1995a). On that occasion, the genera *Halobacteroides*, *Acetohalobium*, *Halanaerobacter*, and *Sporohalobacter*, which were previously classified within the family *Halanaerobiaceae*, were transferred to the *Halobacteroidaceae*. Patel et al. (1995) then proposed the reclassification of *Haloanaerobacter chitinovorans* (*Halanaerobacter chitinivorans*) within the genus *Halobacteroides* on the basis of 16S rRNA gene sequence analyses. However, the dendrogram presented by Mouné et al. (1999) supports the recognition of *Halanaerobacter* as a separate genus.

Further reading

Oren, A. 1986. The ecology and taxonomy of anaerobic halophilic eubacteria. FEMS Microbiol. Rev. *39*: 23–29.

Oren, A. 1992. The genera *Haloanaerobium*, *Halobacteroides*, and *Sporohalobacter*. *In* Balows, Trüper, Dworkin, Harder and Schleifer (Editors), The Prokaryotes. A Handbook on the Biology of Bacteria: Ecophysiology, Isolation, Identification, Applications, 2nd edn, vol. II. Springer-Verlag, New York, pp. 1893–1900.

Genus I. **Halobacteroides** Oren, Weisburg, Kessel and Woese 1984a, 355[VP] (Effective publication: Oren, Weisburg, Kessel and Woese 1984b, 68.)

AHARON OREN

Ha.lo.bac.te.ro′i.des. Gr. n. *hals, halos* salt; N.L. masc. n. *bacter* the masc. equivalent of Gr. neut. n. *baktron* a staff or rod; Gr. n. *eidos* form, shape; N.L. masc. n. *Halobacteroides* rod-like salt organism.

Cells are long, thin, often flexible rods, motile by peritrichous flagella, Gram-stain-negative. Endospores may be formed. **Strictly anaerobic**, chemo-organotrophic with fermentative metabolism. Carbohydrates are fermented with production of acetate, ethanol, H_2, and CO_2. **Halophilic**, growing optimally at NaCl concentrations around 1.7–2.6 M and requiring a minimum of 1.2–1.7 M NaCl for growth.

DNA G+C content (mol%): 30.5–30.7 (Bd, HPLC).

Type species: **Halobacteroides halobius** Oren, Weisburg, Kessel and Woese 1984a, 355[VP] (Effective publication: Oren, Weisburg, Kessel and Woese 1984b, 68.).

Further descriptive information

The two presently recognized species within the genus *Halobacteroides* ferment a variety of carbohydrates to acetate, ethanol, H_2, and CO_2. Both *Halobacteroides halobius* and *Halobacteroides elegans* can use methanethiol as the sole source of assimilatory sulfur for growth (Kevbrin and Zavarzin, 1992b; Zhilina et al., 1997b). Reduction of elemental sulfur to sulfide was reported in *Halobacteroides elegans* (Kevbrin and Zavarzin, 1992b; Zhilina et al., 1997b).

Electron microscopic examination of thin sections of cells shows a typical Gram-stain-negative type of cell wall. At the end of the exponential growth phase, cells rapidly degenerate to sphaeroplast-like structures and die.

Measurements of intracellular ionic concentrations in *Halobacteroides halobius* have shown the presence of molar concentrations of K[+] and Cl[−] (Oren, 1986).

Endospores may be formed. *Halobacteroides elegans* produces round, terminal endospores in culture. No endospores have been observed in cultures of *Halobacteroides halobius*, but bacteria resembling this species could be grown from sediment samples obtained from the Dead Sea and other hypersaline anaerobic sediments following pasteurization for 10–20 min at 80–100 °C (Oren, 1987), thus suggesting that heat-resistant structures may be produced.

Enrichment and isolation procedures

Any anoxic reducing medium containing a high salt concentration (0.8–4.3 M NaCl) and containing glucose or other simple sugars as a carbon source is a potential enrichment and growth medium for anaerobic halophilic bacteria of the genus *Halobacteroides*. Details on the composition of suitable growth media can be found in Oren (2001) and in the original publications on the isolation of species. Strictly anaerobic techniques should be used, including boiling the media under nitrogen or nitrogen/ CO_2 (80:20) and adding cysteine as reducing agent to the boiled media. Enrichment cultures should be checked frequently for growth of *Halobacteroides*, as growth may be very rapid, followed by sphaeroplast formation and death of the cells.

The existence of heat-resistant endospores has been exploited in a selective enrichment procedure for *Halobacteroides halobius*-like bacteria, based on negative selection by pasteurization of the inoculum for 10–20 min at 80–100 °C (Oren, 1987). It remains to be determined to what extent this procedure is specific for *Halobacteroides*, excluding other endospore-forming, sugar-fermenting halophiles such as *Orenia* and *Sporohalobacter*. No other specific enrichment procedures for representatives of the genus *Halobacteroides* have been devised thus far.

Maintenance procedures

Lyophilization has proved satisfactory for the preservation of *Halobacteroides* species. Both *Halobacteroides halobius* and *Halobacteroides elegans* easily undergo autolysis, during which spherical degeneration forms can be observed (Oren et al., 1984b; Zhilina et al., 1997b). Lysis starts at the end of the exponential growth phase, especially when relatively high growth temperatures are employed. Death of cultures may be delayed by employment of media with a reduced nutrient content and the use of lower growth temperatures (15–25 °C). Weekly transfers may then suffice to maintain viable cultures.

Taxonomic comments

Presently, two species are recognized within the genus *Halobacteroides: Halobacteroides halobius* and *Halobacteroides elegans*. *Halobacteroides elegans* was originally described as a strain of *Halobacteroides halobius* (Zhilina et al., 1991a), but was subsequently described

as a new species on the basis of 16S rRNA gene sequence analysis (Zhilina et al., 1997b).

The former *Halobacteroides acetoethylicus* (Rengpipat et al., 1989) has been transferred to the new genus *Halanaerobium* as *Halanaerobium acetethylicum* comb. nov., nom. corrig. (Oren, 2000; Rainey et al., 1995b), and the former *Halobacteroides lacunaris* (Zhilina et al., 1992a) has been transferred to the genus *Halanaerobacter* as *Halanaerobacter lacunarum* comb. nov., nom. corrig. (Oren, 2000; Rainey et al., 1995a).

Differentiation of the species of the genus *Halobacteroides*

Identification of isolates and their assignment to the genus *Halobacteroides* should be based preferentially on the determination of their 16S rRNA gene sequences. In addition, phenotypic characterization should be performed and the properties of the strains should be compared with those of the described species.

List of species of the genus *Halobacteroides*

1. **Halobacteroides halobius** Oren, Weisburg, Kessel and Woese 1984a, 355[VP] (Effective publication: Oren, Weisburg, Kessel and Woese 1984b, 68.)

ha.lo′bi.us. Gr. n. *hals* salt; Gr. n. *bios* life; N.L. adj. *halobius* living on salt.

Cells are thin, flexible rods, $0.5–0.6 \times 10–20\,\mu m$ in size. Motile by peritrichous flagella. Sphaeroplasts are formed in ageing cultures. Halophilic, requiring 1.2–3.2 M NaCl for growth, and growing optimally at 1.5–2.6 M NaCl. Temperature range for growth: 30–47 °C. Optimum temperature: 37–42 °C. Strictly anaerobic. Ferments carbohydrates with production of acetate, ethanol, H_2, and CO_2. Grows on a range of carbohydrates and other compounds including fructose, galactose, D-glucose, maltose, pyruvate, raffinose, and starch. No growth on L-arabinose, cellobiose, L-fucose, D-gluconate, glycerol, lactose, D-mannitol, D-melezitose, D-melibiose, D-ribose, L-rhamnose, D-sorbitol, L-sorbose, or D-xylose. The major fatty acids in the lipids are $C_{14:0}$, $C_{16:0}$, and $C_{16:1}$.

Source: sediments of the Dead Sea, Israel.

DNA G+C content (mol%): 30.7 (T_m).

Type strain: MD-1, ATCC 35273, DSM 5150.

GenBank accession number (16S rRNA gene): U32595, X89074.

2. **Halobacteroides elegans** Zhilina, Turova, Lysenko and Kevbrin 1997a, 1274[VP] (Effective publication: Zhilina, Turova, Lysenko and Kevbrin 1997b, 97.)

e′le.gans. L. adj. *elegans* choice, elegant.

Cells are curved rods, $0.3–0.5 \times 2–10\,\mu m$ in size. Motile by peritrichous flagella. Produces terminal round endospores, $0.12–0.15\,\mu m$ in diameter. Sphaeroplasts are formed in ageing cultures. Halophilic, requiring 1.7–5.1 M NaCl for growth, and growing optimally at 1.7–2.6 M NaCl. Temperature range for growth: 28–47 °C. Optimum temperature: 40 °C. Strictly anaerobic. Ferments carbohydrates with production of acetate, ethanol, H_2, and CO_2. Grows on a range of carbohydrates including fructose, D-glucose, mannose, starch, sucrose, and trehalose. Weak growth on cellobiose, galactose, maltose, mannitol, and pyruvate. No growth on *N*-acetylglucosamine, chitin, glycogen, raffinose, D-ribose, or sorbitol. The major fatty acids in the lipids are $C_{14:0}$, $C_{16:0}$, and $C_{16:1}$.

Source: cyanobacterial mats, Lake Sivash, Crimea.

DNA G+C content (mol%): 30.5 (T_m).

Type strain: Z-7287, DSM 6639.

GenBank accession number (16S rRNA gene): AJ238119, L37423.

Genus II. **Acetohalobium** Zhilina and Zavarzin 1990b, 470[VP] (Effective publication: Zhilina and Zavarzin 1990c, 747.)

GEORGE A. ZAVARZIN AND TATJANA N. ZHILINA

A.ce.to.ha.lo′bi.um. N.L. neut. n. *acetum* vinegar; Gr. n. *hals, halos* salt; Gr. n. *bios* life; N.L. neut. n. *Acetohalobium* acetate-producing organism living in salt.

Rod-shaped cells. Motile with 1–2 subterminal flagella. Multiplication by binary fission is by constriction rather than septation. **Gram-stain-negative** cell wall structure. Thermoresistant endospores formed by some strains. **Strictly anaerobic.** Possess a respiratory type of **homoacetogenic metabolism. Extremely halophilic,** growing at 1.7–4 M NaCl. Obligately dependent on

sodium chloride. Neutrophilic. Mesophilic. **Metabolism variable; lithoheterotrophic,** utilizing hydrogen, formate, and carbon monoxide, **methylotrophic,** utilizing methylamines and betaine, **or chemo-organotrophic,** fermenting some amino acids and organic acids. **Acetate is the end product** with all substrates utilized.

DNA G+C content (mol%): 33–35 (T_m).

Type species: **Acetohalobium arabaticum** Zhilina and Zavarzin 1990b, 470^{VP} (Effective publication: Zhilina and Zavarzin 1990c, 747.).

Further descriptive information

Phylogenetically, the genus *Acetohalobium* belongs to the spore-forming branch of the haloanaerobic eubacterial family *Halobacteroidaceae* (Rainey et al., 1995b; Tourova et al., 1995), order *Haloanaerobiales*, now spelled *Halanaerobiales* (Oren, 2000). The genus includes a single species, *Acetohalobium arabaticum*. Information regarding the species and, therefore, the genus is given in the species description.

Enrichment and isolation procedures

Strains of *Acetohalobium* can be enriched and isolated from extremely saline lagoons or salinas of thalassic origin with variable salinity under strictly anaerobic conditions under an H_2 + CO_2 atmosphere which can be substituted by N_2 + CO_2 in a selective mineral medium supplemented with trimethylamine (TMA) chloride or betaine and with inhibition of methylotrophic methanogens by 10 mM bromethanesulfonate (Zhilina and Zavarzin, 1990c). The medium has the following composition (per liter of distilled water): NaCl, 150 g; $MgCl_2 \cdot 6H_2O$, 4 g; NH_4Cl, 0.33 g; KCl, 0.33 g; KH_2PO_4, 0.33 g; $Na_2S \cdot 9H_2O$, 0.5 g; yeast extract 0.1 g; 1 ml trace element solution (Kevbrin and Zavarzin, 1992a); and 10 ml vitamin solution (Wolin et al., 1963). Pure cultures are obtained by serial dilution in liquid medium supplemented with TMA chloride (3 g/l) as substrate followed by isolation of single colonies in Hungate roll tubes. Colonies are whitish yellow, lens shaped, and 0.5–1 mm in diameter.

Maintenance procedures

Acetohalobium is maintained by regular subculturing in liquid medium and stored in a refrigerator at 6 °C. In broth cultures, rapid lysis occurs at the end of active growth. Consequently, preservation and maintenance of the culture is tedious. Cultures should be transferred every 2–3 weeks, but lysis of cells may occur. Strains of *Acetohalobium* may be stored in liquid nitrogen.

Differentiation of the genus *Acetohalobium* from other genera

Acetohalobium can be differentiated from other genera of haloanaerobic, saccharolytic bacteria of the family *Halobacteroidaceae* [*Halobacteroides, Halanaerobacter, Orenia, Sporohalobacter* (Oren, 2000; Rainey et al., 1995a), and *Halonatronum* (Zhilina et al., 2001b)] by its homoacetogenic type of metabolism and its inability to utilize sugars. In common with some of the above genera, *Acetohalobium* is extremely halophilic, with obligate dependence on NaCl at high concentrations. It differs from the haloalkaliphilic, homoacetogenic genus *Natroniella* (Zhilina et al., 1996b) by its inability to grow at high pH and the substrates utilized.

Taxonomic comments

The phylogenetic position of the genus was established first by E.S. Bulygina who showed by 5S rRNA gene sequence analysis that *Acetohalobium* clusters with the group of haloanaerobes (Zhilina et al., 1992b). Its generic status is based on 16S rRNA gene phylogeny. It represents a distinct lineage that clusters with halophilic eubacteria of the genus *Sporohalobacter* with levels of 16S rRNA gene sequence similarity of 96.1% (Rainey et al., 1995a; Tourova et al., 1995).

Cell lipids contain aliphatic fatty acids and β-hydroxy acids characteristic of Gram-stain-negative eubacteria. About 94% of the fatty acids are unsaturated straight-chain $C_{16:1 \omega 7}$ and $C_{16:0 \omega 7}$; β-hydroxy fatty acids are represented by $C_{12:1}$ and $C_{12:0}$, and sugars by pentoses and hexoses (Zhilina et al., 1992b).

List of species of the genus *Acetohalobium*

1. **Acetohalobium arabaticum** Zhilina and Zavarzin 1990b, 470^{VP} (Effective publication: Zhilina and Zavarzin 1990c, 747.)

a.ra.ba′ti.cum. N.L. neut. adj. *arabaticum* from Arabat, a peninsula between the Sea of Azov and Sivash.

Cells are rods, sometimes slightly bent, usually occurring singly or in pairs (Figure 238) and short chains. Motile with 1–2 subterminal flagella, 0.7–1 × 2–5 µm in size. Cell wall has typical Gram-stain-negative structure (Figure 239).

Multiplication by binary fission is often unequal, with formation of mini-cells on the end. Spores are rare, absent for the type strain in pure culture. Strain Z-7984 retains spore formation upon laboratory cultivation; spores are terminal and round (Figure 239).

Extremely halophilic; grows in 1.7–4.2 M NaCl with optimum for type strain at 2.5–3 M. Neutrophilic; grows at pH 5.6–8.4, pH optimum for the type strain is 7.8. Mesophilic; grows at 30–60 °C, optimum for the type strain is 38–40 °C, maximum 47 °C.

Strictly anaerobic. Fermentative with most substrates. O_2, NO_3^-, SO_3^{2-}, $S_2O_3^{2-}$, SO_3^{2-}, and S^0 do not serve as electron acceptors for catabolism. Homoacetogenic. Does not ferment carbohydrates or alcohols. *Acetohalobium arabaticum* uses the following substrates: H_2 + CO_2, CO, betaine, TMA, formate, lactate, pyruvate, histidine, aspartate, asparagine, and glutamate. Lactate, aspartate, and histidine are preferred substrates. The ability to utilize TMA chloride may be lost after cultivation on other substrates. Acetate is the product of fermentation. With betaine, the only products are acetate and TMA (Kevbrin et al., 1995, 1990a; Zhilina and Zavarzin, 1990c). Trace amounts of hydrogen are formed during the lag phase. Growth on all substrates is significantly stimulated by additives of protein origin. Yeast extract is needed for growth. Growth is initiated in the presence of bicarbonate (Kevbrin et al., 1995). *Acetohalobium* slowly reduces sulfur to sulfide and dimethylsulfoxide to dimethylsulfide, but without coupling of energy generation. Sulfite or dithionite at 1 mM concentration completely inhibits growth of the type strain. Sulfide is tolerated up to 12 mM (Kevbrin and Zavarzin, 1992a). Growth is inhibited by streptomycin, kanamycin, erythromycin, benzylpenicillin, gentamicin, vancomycin, and tetracycline.

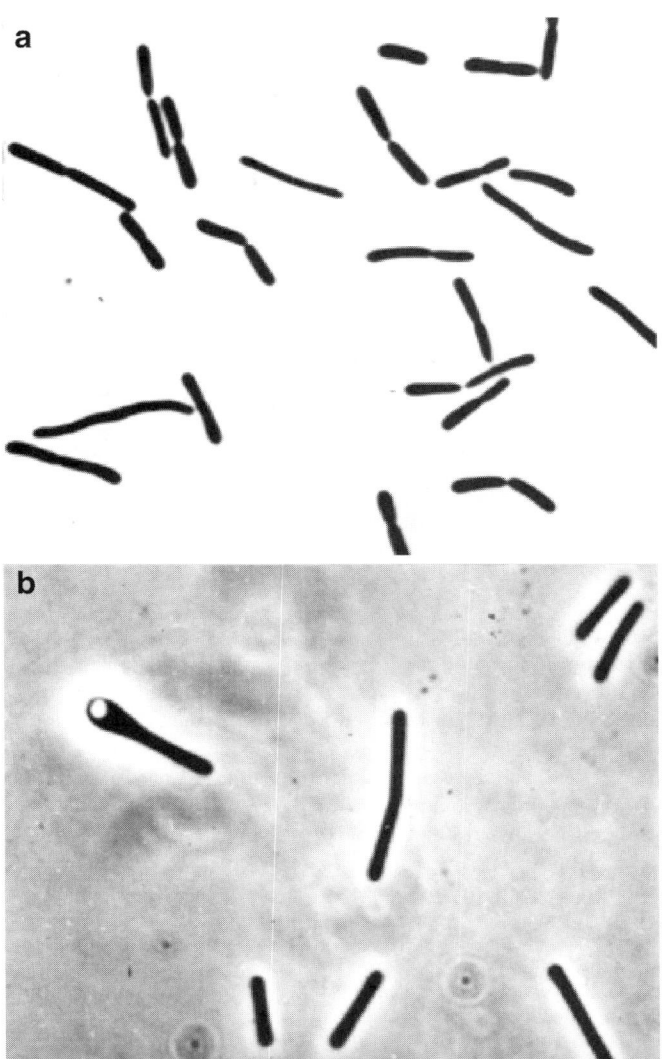

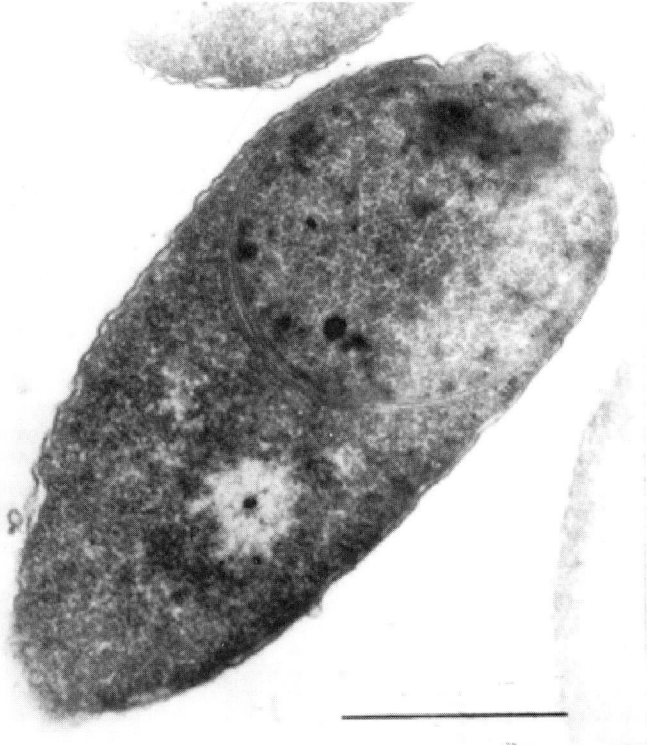

FIGURE 239. Electron micrograph of an ultrathin section of spore-forming cells of *Acetohalobium* strain Z-7492. Bar = 1 μm.

FIGURE 238. Morphology of *Acetohalobium* strains as seen by light microscopy (bar = 10 μm). (a) Strain Z-7288ᵀ on medium containing formate and 15% NaCl; (b) strain Z-7492 on medium containing TMA and 25% NaCl.

The metabolism of *Acetohalobium*, like that of other homoacetogens, is based on CO dehydrogenase, which is found in both periplasmic and soluble cell fractions. Salt concentrations up to 5 M do not inhibit the activity of the enzyme. Hydrogenase activity is found in the periplasmic and membrane fractions, but is mostly in cytoplasmic fractions and is very high. It is stimulated by KCl and NaCl at concentrations of 3.0–4.5 M (Pusheva, 1992). Neither cytochromes nor quinones are found in membranes of the type strain. Flavoproteins and various folates and corrinoids have been detected in the cells, causing their autofluorescence (Bykhovsky et al., 1994; Zavarzin et al., 1994). Growth and acetogenesis on all substrates involves an Na⁺/H⁺ antiporter capable of Na⁺ extrusion from the cell against an NaCl concentration gradient as high as 2.56 M (Pusheva and Detkova, 1996).

Acetohalobium was isolated from cyanobacterial mats in hypersaline lagoons where its trophic position was established during a search for micro-organisms responsible for the so-called non-competitive pathway of methylotrophic methanogenesis in halophilic microbial communities (Zhilina and Zavarzin, 1990a, 1990c). This pathway is driven by the decomposition of osmoprotective compounds and it is initiated by saccharolytic haloanaerobes decomposing sugars like trehalose and heterosides. The other step is caused by decomposition of betaine as osmolyte of proteobacteria with release of TMA and acetate. This step is performed by *Acetohalobium*, which is capable of utilizing TMA, a product of betaine decomposition. However, at this stage it is outcompeted by methylotrophic methanogens such as *Methanohalobium* species. In binary culture of *Acetohalobium* and *Methanohalobium evestigatum*, methane is produced from betaine (Zhilina and Zavarzin, 1990c). Hydrogen might serve as the substrate for *Acetohalobium*, but its consumption is outcompeted by hydrogenotrophic sulfate reducers. Trophic relationships in the anaerobic, halophilic, cyanobacterial community with emphasis on anaerobic decomposition of osmolytes have been discussed by Zhilina and Zavarzin (1991) and Zavarzin et al. (1994).

Source: cyanobacterial mats in hypersaline lagoons.

DNA G+C content (mol%): 33.6 (T_m).

Type strain: Z-7288, ATCC 49924, DSM 5501.

GenBank accession number (16S rRNA gene): L37422, X89077.

Additional remarks: Acetohalobium arabaticum strain Z-7492 differs in a number of traits from the type strain Z-7288ᵀ. It has a DNA G+C content of 34.6 mol%, retains sporulation in pure culture, has temperature optimum of 55 °C and range of 30–60 °C, and optimal salinity at 3.4–4 M NaCl in a range 2.5–4.2 M NaCl. Strain Z-7492 uses the same set of substrates as the type strain. DNA–DNA hybridization of strain Z-7492 with strain Z-7288^T is 78% (Kevbrin et al., 1995; Zavarzin et al., 1994).

Genus III. **Halanaerobacter** Liaw and Mah 1996, 362^{VP} (Effective publication: Liaw and Mah 1992, 265.) (*Haloanaerobacter* [sic] Liaw and Mah 1992, 265)

NOHA M. MESBAH

Hal.an.ae.ro.bac'ter. Gr. n. *hals* salt; Gr. pref. *an* not; Gr. n. *aer* air; N.L. masc. n. *bacter* rod; N.L. masc n. *Halanaerobacter* salt rod which grows in the absence of air.

Cells are rod-shaped or slightly curved, motile by means of peritrichous flagella. Gram-stain-negative. **Strictly anaerobic. Chemo-organotrophic** with fermentative metabolism; some strains can **utilize amino acids** in Stickland reactions or with H$_2$ as electron donor. Carbohydrates are fermented with production of acetate, H$_2$, and CO$_2$. In some species, ethanol, propionate, formate, and isobutyrate are also formed. Elemental sulfur can be used as an electron acceptor in a certain species. **Halophilic;** optimal growth occurs at NaCl concentrations around 2.0–3.0 M. Cells require a minimum of 0.5–1.6 M NaCl for growth. Neutral to slightly alkaline pH values required for optimal growth (6.5–7.8). Mesophilic to slightly thermotolerant, and optimal growth occurs between 35 and 45 °C. Endospores have not been observed; short degenerate cells and sphaeroplasts occur in stationary phase. **Catalase- and oxidase-negative. Found in anoxic sediments in hypersaline environments.**

DNA G+C content (mol%): 32–35.

Type species: **Halanaerobacter chitinivorans** corrig. Liaw and Mah 1996, 362^{VP} (Effective publication: Liaw and Mah 1992, 265.).

Further descriptive information

Phylogenetically, the genus *Halanaerobacter* belongs to the moderately halophilic family *Halobacteroidaceae*, order *Halanaerobiales* (Rainey et al., 1995a), and is closely related to the genus *Halanaerobium* (Figure 240). The genus includes three species.

Cells of all three species are long thin rods (0.5–8.0 µm × 0.3–0.5 µm). Cells are flexible; rods of *Halanaerobacter lacunarum* curl up into circular shapes during the late exponential growth phase (Zhilina et al., 1991b). After the end of exponential growth, cells of all species rapidly form sphaeroplasts and then lyse. All three species are motile by means of peritrichous flagella.

All *Halanaerobacter* species are halophilic. They grow optimally at 2.0–3.0 M NaCl. Depending on the species, a minimum of 0.5–1.6 M NaCl is required for growth. All species tolerate 5.0 M NaCl, which is close to saturation, and are neutrophilic, growing optimally at a pH around 7.0. *Halanaerobacter salinarius* is alkalitolerant and can grow at pH values up to 8.5 (Mouné et al., 1999). *Halanaerobacter salinarius* is also the most thermotolerant, growing optimally at 45 °C. *Halanaerobacter* species are

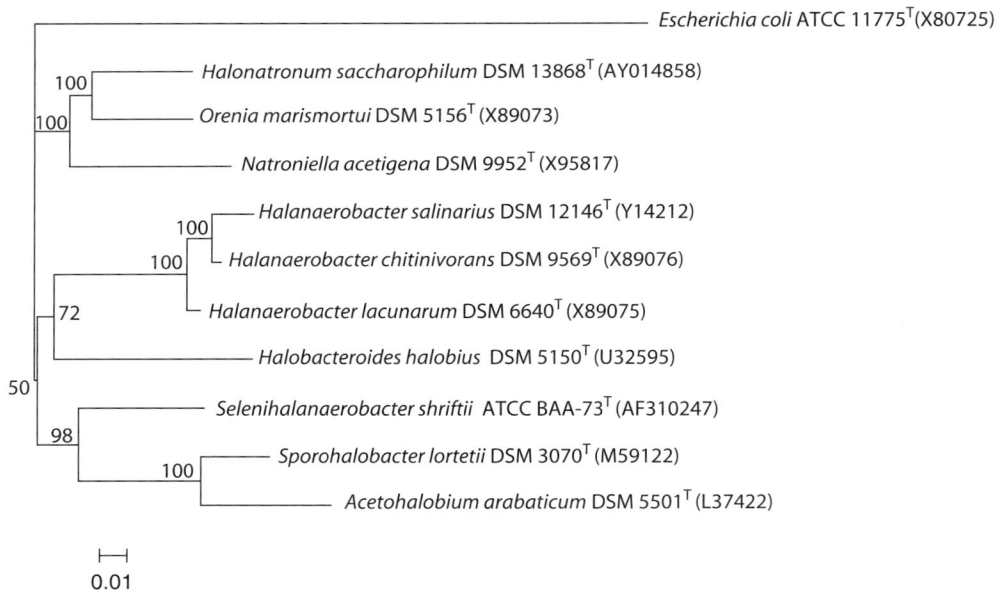

FIGURE 240. Neighbor-joining tree based on 16S rRNA gene sequences showing the position of species of the genus *Halanaerobacter* in relation to type species of genera within the family *Halobacteroidaceae*. Sequences were aligned with the CLUSTAL X program. The tree was constructed using the PHYLIP software package. Distances were calculated using the Jukes–Cantor algorithm of DNADIST, and branching order was determined via the neighbor-joining algorithm of NEIGHBOR. GenBank accession numbers are shown in parentheses. The tree is rooted with the 16S rRNA of *Escherichia coli* ATCC 11775^T as outgroup. Numbers at nodes denote bootstrap values based on 100 replicates; only values 50 or greater are shown. Bar, one nucleotide substitution per 100 nt.

distinguished from other genera among the *Halobacteroidaceae* by their ability to grow at 50 °C (Rainey et al., 1995a).

All *Halanaerobacter* species grow fermentatively on sugars, producing acetate, H_2, and CO_2. Additional products made by some strains include ethanol, propionate, formate, and isobutyrate (Table 242). *Halanaerobacter chitinivorans* strain W5C8 displays chitinolytic activity and is capable of using chitin as a carbon and energy source (Liaw and Mah, 1992). *Halanaerobacter lacunarum* shows enhanced growth on sucrose and trehalose (Zhilina et al., 1991b), two frequently used osmotic solutes.

Halanaerobacter lacunarum is also able to use sodium sulfide both as an electron donor and a sulfur source for growth. When supplemented with 5–7 mM of elemental sulfur, *Halanaerobacter lacunarum* produces hydrogen sulfide. *Halanaerobacter lacunarum* has the highest tolerance to sulfide, and the presence of 50 mM of sulfide in the medium does not reduce the growth rate (Zhilina et al., 1991b). *Halanaerobacter salinarius* can tolerate 14 mM sulfide (Mouné et al., 1999). Tolerance to high sulfide concentrations is most likely an adaptation to the high sulfide concentrations commonly encountered in anoxic sediments present in hypersaline environments.

Neither *Halanaerobacter lacunarum* nor *Halanaerobacter chitinivorans* utilize amino acids as sole substrates. *Halanaerobacter salinarius* is distinguished from these two species by the ability to reduce glycine betaine to trimethylamine using either serine or hydrogen as electron donors (Mouné et al., 1999). This occurs via a Stickland reaction. In contrast, *Halanaerobacter lacunarum* is not capable of glycine betaine reduction and *Halanaerobacter chitinivorans* can reduce glycine betaine only in the presence of hydrogen. None of the *Halanaerobacter* species are capable of dissimilatory reduction of inorganic nitrogen (NO_3^-).

The ability to use substrates such as glycine betaine, sucrose, and trehalose is of ecological importance. These compounds are accumulated at high concentrations as organic osmotic solutes by aerobic halophilic micro-organisms inhabiting saline and hypersaline lakes (Oren, 2008). It follows that these compounds are widely distributed in hypersaline ecosystems and are available to the anaerobic bacterial community inhabiting the bottom sediments of these lakes.

Enrichment and isolation procedures

Species of the genus *Halanaerobacter* are found in anaerobic hypersaline environments where simple sugars are available. The three species available have been isolated from sediments of solar salterns in France, southern California, and Ukraine (Liaw and Mah, 1992; Mouné et al., 1999; Zhilina et al., 1991b). An anoxic reducing medium containing high NaCl concentrations (2.0–3.0 M) and simple sugars as a carbon source is a potential enrichment and growth medium for the anaerobic halophilic bacteria of the genus *Halanaerobacter*. Media suitable for growth of *Halanaerobacter* species are described by Oren (2006) and in the original publications on the species isolations. The *Halanaerobacter* species are obligate anaerobes; strict anaerobic techniques should be used, including boiling the media under nitrogen or nitrogen-CO_2, and adding reducing agents such as cysteine, sulfide, or thioglycollate to boiled media.

Isolation from enrichment cultures can be achieved by agar dilution series using 1.0–1.3% (w/v) agar. Higher concentrations of agar should be avoided, as the high salinity of the medium could result in a decrease in water activity. Use of anaerobic jars and/or agar-roll tubes sealed under nitrogen is necessary to maintain anaerobicity.

No specific enrichment or culture media have been designed that can select for members of the *Halanaerobacter* species over other members of the families *Halanaerobiaceae* and *Halobacteroidaceae*. Members of the *Halanaerobacter* genus can grow optimally at temperatures between 43 and 45 °C, so incubation at higher temperatures can provide a selective advantage. However, other species within the families *Halanaerobiaceae* and *Halobacteroidaceae* can survive temperatures as high as 52 °C, so multiple dilutions are necessary to ensure pure isolates.

TABLE 242. Differential characteristics of *Halanaerobacter* species

Characteristic	1. *H. chitinivorans* OGC 229	2. *H. lacunarum* Z-7888	3. *H. salinarius* SG 3903
Cell size (μm)	0.5 × 1.4–8.0	0.7–1.0 × 0.5–6.0	0.3–0.4 × 5.0–8.0
Cell morphology	Long, flexible rods	Short, flexible rods	Long, flexible rods
NaCl range (M)	0.5–5.0	1.6–5.0	0.8–5.0
NaCl optimum (M)	2.0–3.0	2.5–3.0	2.3–2.5
pH range	ND[a]	6.0–8.0	5.5–8.5
pH optimum	7.0	6.5–7.0	7.4–7.8
Temperature range (°C)	23–50	5–52	10–50
Temperature optimum (°C)	30–45	35–40	45
Doubling time (h)	2.5	2.9–4.5	12
DNA G + C content (mol%)	34.8 (Bd)	32.4 (T_m)	31.6 (HPLC)
Chitin degradation	+	−	−
Sulfur reduction	−	+	−
Tolerance to sulfide (mM)	ND	50	14
Amino acids utilized	−	−	−
Stickland reaction:			
Glycine betaine + serine	−	−	+
Glycine betaine + H_2	+	−	+
Products from glucose fermentation	CO_2, H_2, acetate, isobutyric acid	Acetate, ethanol, H_2, CO_2	Ethanol, propionate, acetate, formate, CO_2, H_2

[a]ND, Not determined.

Maintenance procedures

Members of the genus *Halanaerobacter* undergo spherical degeneration and autolysis at the end of exponential growth phase, particularly when grown at higher temperatures. For short-term preservation, cultures may be inoculated, incubated for a few hours till slight turbidity appears (OD$_{600}$ 0.3–0.4), and then stored at 6 °C. These cultures will remain viable for a few weeks. Long-term preservation should be under liquid nitrogen (T. Zhilina, personal communication).

Differentiation of the species of the genus *Halanaerobacter*

Characteristic properties for differentiation of *Halanaerobacter* species are summarized in Table 242. The genetic relationship

of *Halanaerobacter* species based on 16S rRNA sequences is shown in Figure 240.

Taxonomic comments

The genus *Halanaerobacter* was originally named *Haloanaerobacter* and renamed *Halanaerobacter* in accordance with Rule 61, Appendix 9 of the Bacteriological Code (Oren, 2000). Based on 16S rRNA sequence comparisons, the former *Halobacteroides lacunaris* was reclassified as *Halanaerobacter lacunaris* (Rainey et al., 1995a). Patel et al. (1995) proposed the reclassification of *Halanaerobacter chitinivorans* within the genus *Halobacteroides* on the basis of 16S rRNA sequence analysis, but phylogenetic analysis presented by Mouné et al. (1999) supports classification of *Halanaerobacter* as a separate genus.

List of species of the genus *Halanaerobacter*

1. **Halanaerobacter chitinivorans** Liaw and Mah 1996, 362^{VP} (Effective publication: Liaw and Mah, 1992, 265.)

   chi.tin.i.vo′rans. N.L. n. *chitinum* chitin; L. v. *vorare* devour; N.L. part. adj. *chitinivorans* chitin-devouring.

   Colonies are opaque and glossy with entire edges, colony diameters range from 0.5 to 1.0 mm. Cells are long thin rods, 1.4–8.0 μm by 0.5 μm in exponential growth phase, and short degenerate cells in stationary phase. Motile by means of peritrichous flagella. Gram-stain-negative. Endospores not observed. Halophilic. Growth occurs at 0.5–5.0 M NaCl and optimal growth at 2.0–3.0 M NaCl. Temperature range for growth and optimum temperature range are 23–50 °C and 30–45 °C, respectively. Optimal pH for growth is 7.0. Strictly anaerobic. Oxidase- and catalase-negative. Ferments carbohydrates. Major fermentation products from glucose are acetate, H$_2$, CO$_2$, and isobutyrate. The type strain, W5C8, uses chitin as a carbon source. Grows on a range of carbohydrates including glucose, fructose, mannose, acetylglucosamine, sucrose, maltose, and cellobiose. No growth occurs on acetate, glycerol, pyruvate, D-raffinose, cellulose, or pectin. Cells are resistant to penicillin, carbenicillin, D-cycloserine, streptomycin, and tetracycline, and are susceptible to chloramphenicol.

   Isolated from organic sediment of a solar saltern in Southern California.

   DNA G+C content (mol%): 34.8 ± 1 (Bd).

   Type strain: W5C8, OGC 229; CIP 105156; DSM 9569.

   GenBank accession number (16S rRNA gene): U32596, X89076.

2. **Halanaerobacter lacunarum** (Zhilina, Miroshnikova, Osipov and Zavarzin 1991b) Rainey, Zhilina, Boulygina, Stackebrandt, Tourova and Zavarzin 1995b, 879^{VP} (Effective publication: Rainey, Zhilina, Boulygina, Stackebrandt, Tourova and Zavarzin 1995a, 197.) (*Halobacteroides lacunaris* Zhilina, Miroshnikova, Osipov and Zavarzin 1991b, 503)

   la.cu.na′rum. L. n. *lacuna* hole; L. gen. pl. n. *lacunarum* of holes; referring to the source of isolation.

   Colonies are lens-shaped with a straight edge, light yellow, 0.5–1.0 mm in diameter. Cells (0.5–6.0 μm by 0.7–1.0 μm) are slightly bent flexible rods, single, paired or in short chains. Motile by means of peritrichous flagella. Gram-stain-negative.

Endospores not observed. Halophilic. Optimal growth occurs at 2.5–3.0 M NaCl, and NaCl range for growth is 1.6–5.0 M. Temperature range for growth is 5–52 °C, and optimal growth occurs at 35–40 °C. Optimal pH for growth is 6.5–7.0, and the pH range is 6.0–8.0. Strictly anaerobic. Oxidase- and catalase-negative. Ferments carbohydrates. Fermentation products from glucose are acetate, ethanol, H$_2$, and CO$_2$. Grows on wide range of carbohydrates including sucrose, trehalose, starch, mannitol, glycogen, glucitol, glucose, fructose, maltose, mannitol, and starch. Enhanced growth occurs with sucrose and trehalose. No growth was obtained with arabinose, xylose, rhamnose, ribitol, lactose, mellibiose, fucose, raffinose, cellulose, erythritol, dulcitol, inositol, organic acids, formate, acetate, propionate, butyrate, glycolate, lactate, gluconate, oxalate, succinate, adipic acid, fumaric acid, tartrate, citrate, ascorbic acid, amino acids, methylglycine, dimethylglycine, betaine, mono-, di- and trimethylamine, choline, N-acetylglucosamine, yeast extract, peptone, trypticase, pectin, chitin, methanol, and glycerol. Forms hydrogen sulfide in the presence of elemental sulfur. Can tolerate up to 50 mM of sulfide with no reduction in growth rate. The major fatty acids are Δ9-C$_{16:1}$ and C$_{16:0}$.

   Isolated from silt of saline Lake Chokrak, Kerch peninsula, Crimea, Ukraine.

   DNA G+C content (mol%): 32.4 (T$_m$).

   Type strain: Z-7888, ATCC 49944; DSM 6640.

   GenBank accession number (16S rRNA gene): U32593, L37421, X89075.

3. **Halanaerobacter salinarius** Mouné, Manac'h, Hirschler, Caumette, Willison and Matheron 1999, 111^{VP}

   sa.li.na′ri.us. L. adj. *salinarius* pertaining to *salinae* salterns; referring to the source of isolation.

   Surface colonies are circular, glossy, translucent, and slightly yellow with entire edges. Colony diameters are 1–2 mm. Agar-embedded colonies are opaque. Cells are long flexible rods (5–8 μm by 0.3–0.4 μm). Motile by means of peritrichous flagella. Gram-stain-negative. Endospores not observed. Halophilic. Growth occurs at 0.8–5 M NaCl, and optimal growth at 2.3–2.5 M. Temperature range for growth is 10–50 °C with an optimum at 45 °C. The pH range for growth is 5.5–8.5, and optimal growth occurs at pH 7.4–7.8.

Strictly anaerobic. Oxidase- and catalase-negative. Ferments carbohydrates. Fermentation products from glucose are ethanol, propionate, acetate, formate, CO_2, and H_2. Grows on a range of carbohydrates including glucose, fructose, galactose, mannose, trehalose, sucrose, maltose, raffinose, cellobiose, glucosamine, N-acetylglucosamine, and mannitol. No growth with starch, chitin, pyruvate, malate, lactate, glutamate, serine, and trimethylamine. Glycine betaine is reductively cleaved via a Stickland reaction into trimethylamine and acetate; hydrogen and serine serve as electron donors. Incapable of dissimilatory reduction of nitrate, or the sulfur compounds SO_4^{2-}, SO_3^{2-}, and $S_2O_3^{2-}$. Sulfate and nitrate cannot serve as sulfur or nitrogen sources. Cysteine can be used as a nitrogen and sulfur source. Cells are resistant to anisomycin, kanamycin, tetracycline, and Na-taurocholate and susceptible to chloramphenicol and erythromycin.

Isolated from black sediment below a gypsum crust in hypersaline ponds of salterns of Salin-de-Giraud, France.

DNA G + C content (mol%): 31.6 (HPLC).

Type strain: SG 3903, DSM 12146, ATCC 700559, CIP 107181.

GenBank accession number (16S rRNA gene): Y14212.

Genus IV. **Halonatronum** Zhilina, Garnova, Tourova, Kostrikina and Zavarzin 2001a, 263[VP]
(Effective publication: Zhilina, Garnova, Tourova, Kostrikina and Zavarzin 2001b, 70.)

TATJANA N. ZHILINA, GEORGE A. ZAVARZIN AND AHARON OREN

Ha.lo.na.tro′num. Gr. n. *hals, halos* salt; Ar. n. *natron* soda; N.L. neut. n. *Halonatronum* an organism growing with salt and soda.

Cells are rod-shaped, flexible, and motile by peritrichous flagella. The cell wall has a Gram-stain-negative structure. Strictly anaerobic. Chemo-organotrophic with fermentative metabolism. Carbohydrates, including soluble polysaccharides, are fermented to acetate, ethanol, formate, H_2, and CO_2. **Halophilic and alkaliphilic. Endospores produced.**

DNA G+C content (mol%): 34.4 (T_m).

Type species: **Halonatronum saccharophilum** Zhilina, Garnova, Tourova, Kostrikina and Zavarzin 2001a, 263[VP] (Effective publication: Zhilina, Garnova, Tourova, Kostrikina and Zavarzin 2001b, 71.).

Further descriptive information

Currently, the genus contains a single species, namely *Halonatronum saccharophilum*. This bacterium was isolated from sediments collected from a coastal lagoon of Lake Magadi, Kenya (Zhilina et al., 2001b). The species is halophilic, alkaliphilic, moderately thermophilic, and produces endospores. Optimal growth is achieved in medium containing about 1 M Na_2CO_3 + $NaHCO_3$ in addition to 0.85 M NaCl. High concentrations of Na_2CO_3 + $NaHCO_3$ are specifically required; high NaCl media buffered with serine at the optimal pH (8.0–8.4) without carbonates do not support growth.

Enrichment and isolation procedures

The only isolate of the single recognized species within the genus was obtained from a mud sample collected from a coastal lagoon of Lake Magadi, Kenya. It was isolated from an anaerobic enrichment culture in hypersaline alkaline medium containing sucrose and yeast extract as organic nutrients. Selective isolation procedures for *Halonatronum* may possibly be designed on the basis of its requirement for high pH and high salt concentrations, its ability to grow at relatively high temperatures, and its formation of endospores, but such procedures have not yet been tested.

Maintenance procedures

Cells in the early exponential phase may be stored at 4 °C for up to one month at least. For long-term preservation, storage in liquid nitrogen is recommended.

Taxonomic comments

Analysis of the 16S rRNA gene of the single species classified within the genus showed that *Halonatronum* belongs to the order *Halanaerobiales*, family *Halobacteroidaceae*. Phylogenetically, it is most closely related to members of the genus *Orenia* (91.4–92.8% similarity) and *Natroniella acetigena* (90.3% similarity).

List of species of the genus *Halonatronum*

1. **Halonatronum saccharophilum** Zhilina, Garnova, Tourova, Kostrikina and Zavarzin 2001a, 263[VP] (Effective publication: Zhilina, Garnova, Tourova, Kostrikina and Zavarzin 2001b, 71.)

sac.cha.ro′phi.lum. Gr. n. *sakchâr* sugar; Gr. adj. *philos* loving; N.L. neut. adj. *saccharophilum* sugar-loving.

Cells are rod-shaped, 0.4–0.6 × 3.5–10 µm in size in young cultures. In older cultures, long thickened cells and sphaeroplasts occur. Motile by means of peritrichous flagella. Terminal endospores are round, 1.25 µm in diameter.

Halophilic and alkaliphilic, requiring 0.5–2.9 M NaCl for growth, and growing optimally at 1.2–2.0 M NaCl. Cells are lysed at NaCl concentrations below 0.5 M. The pH range for growth is 7.7–10.3, with an optimum at pH 8.0–8.4.

High levels of Na_2CO_3 + $NaHCO_3$ are required. Moderately thermophilic with optimum growth at 36–55 °C; capable of growing up to 60 °C. The doubling time under optimal conditions is 2.5 h. Strictly anaerobic. Ferments carbohydrates with production of acetate, ethanol, formate, H_2, and CO_2. Grows on a range of carbohydrates including glucose, fructose, sucrose, maltose, starch, glycogen, and N-acetyl-D-glucosamine. No growth on other hexoses, pentoses, disaccharides, sugar alcohols, or cellulose. Elemental sulfur is used as electron acceptor without energy generation.

Source: sediments of Lake Magadi, Kenya.

DNA G+C content (mol%): 34.4 (T_m).

Type strain: Z-7986, DSM 13868, UNIQEM 211.

GenBank accession number (16S rRNA gene): AY014858.

Genus V. **Natroniella** Zhilina, Zavarzin, Detkova and Rainey 1996a, 1189[VP] (Effective publication: Zhilina, Zavarzin, Detkova and Rainey 1996b, 324.)

GEORGE A. ZAVARZIN AND TATJANA N. ZHILINA

Na.tro.ni'el.la. Gr. n. *natron,* derived from Ar. *natrun* soda (sodium carbonate); N.L. fem. n. *Natroniella* organism growing in soda deposits.

Cells are rods, flexible, motile by peritrichous flagella. Spores may be formed. **Cell wall** has **Gram-stain-negative** structure. **Strictly anaerobic. Possess a respiratory type of homoacetogenic metabolism. Extremely alkaliphilic,** developing in soda brines at pH 9–10. **Halophilic,** growing at 1.7–4.4 M NaCl. **Obligately dependent on Na$^+$, Cl$^-$, and CO$_3$$^{2-}$ ions.** Mesophilic. **Chemo-organotrophic;** some organic acids, amino acids, and alcohols are fermented. **Acetate is the product of fermentation.**

DNA G+C content (mol%): 32 (T_m).

Type species: **Natroniella acetigena** Zhilina, Zavarzin, Detkova and Rainey 1996a, 1189[VP] (Effective publication: Zhilina, Zavarzin, Detkova and Rainey 1996b, 324.).

Further descriptive information

On the basis of 16S rRNA gene sequence analysis, *Natroniella* belongs to the main spore-forming branch of the order *Haloanaerobiales* Rainey and Zhilina (Rainey et al., 1995a), emended to *Halanaerobiales* by Oren (2000), family *Halobacteroidaceae* Zhilina and Rainey (Rainey et al., 1995a; Zhilina et al., 1996b). The genus is monospecific, with *Natroniella acetigena* as the type species. Information regarding the species is given in the species description.

Enrichment and isolation procedures

Natroniella acetigena was isolated from the sediment of a lagoon of Lake Magadi, Kenya, during the dry period from brine with a heavy bacterial population under the trona (Na$_2$CO$_3$-NaHCO$_3$ 2H$_2$O) cover. A pure culture was obtained by anaerobic enrichment culture and by serial dilution in selective liquid medium 1 (Zhilina and Zavarzin, 1994) containing 2 M Na$^+$ (1 M sodium carbonate and 1 M sodium chloride) with lactate (0.5%) as the substrate, incubated at 37 °C under N$_2$. The composition of the medium after its optimization is given by Zhilina et al. (1996b).

Maintenance procedures

Maintenance presents difficulties due to the rapid and complete lysis of non-actively metabolizing cells. Cultures in early exponential phase may be stored in a refrigerator, temperature 2–6 °C, for 3–4 months or, alternatively, strains can be subcultured at weekly intervals. Cultures can be preserved in liquid nitrogen.

Differentiation of the genus of *Natroniella* from other genera

The genus *Natroniella* can be differentiated from other genera of the *Halobacteroidaceae* including *Halobacteroides, Halanaerobacter, Orenia, Sporohalobacter* (Oren, 2000; Rainey et al., 1995a), and *Halonatronum* (Zhilina et al., 2001b) by its type of homoacetogenic metabolism and its inability to utilize sugars. In common with some of these genera, it exhibits extreme alkaliphily (pH 9–10) and obligate dependence on Na$_2$CO$_3$ + NaHCO$_3$ at high concentrations. It differs from the halophilic and homoacetogenic genus *Acetohalobium* by its inability to grow in the absence of high alkalinity and by the substrates utilized.

Taxonomic comments

The generic status *Natroniella* is based on 16S rRNA gene sequence data. It represents a distinct lineage within the family *Halobacteroidaceae* with sequence similarities in the range 86.8–90.5% to the other taxa. Its closest neighbor is *Orenia marismortui* (90.5% sequence similarity) (Zhilina et al., 1996b).

Cellular lipids contain aliphatic fatty acids and β-hydroxy acids characteristic of Gram-stain-negative eubacteria. However, the unusual combination of hydroxyacids and palmitoleic aldehydes found in *Natroniella* has, until now, been found only among Gram-stain-positive bacteria. The major fatty acids present in the type strain are C$_{14:0}$, C$_{16:1\,\omega7}$, and C$_{16:0}$, which contribute about 67% of total fatty acids. The major (about 18%) aldehyde is C$_{16:1\,\omega7}$, with C$_{16}$ and C$_{18:1\,\omega9}$ as minor components (Zhilina et al., 1998). These lipid profiles suggest that *Natroniella* represents a separate taxon.

List of species of the genus *Natroniella*

1. **Natroniella acetigena** Zhilina, Zavarzin, Detkova and Rainey 1996a, 1189[VP] (Effective publication Zhilina, Zavarzin, Detkova and Rainey 1996b, 324.)

   a.ce.ti'ge.na. L. neut. n. *acetum* acetic acid; N.L. fem. *gena* from N.L. fem. suff. *gena* producing; N.L. fem. adj. *acetigena* producing acetic acid.

   Cells are large rods with rounded ends 1–1.2 by 6–15 μm. Motile by means of peritrichous flagella. Unequal fission leads to variation in length of cells with formation of filaments. Cell wall has Gram-stain-negative structure (Figure 241). Cell wall flexibility is characteristic of haloanaerobes. Spore formation is rare; when formed, spores are round and located at the end of non-swollen cells. In pure cultures, rapid and complete lysis is characteristic of the type strain.

   Halophilic; growth occurs in highly alkaline NaCl/soda brines at salinities of 1.7–4.4 M with the optimum at 2–2.5 M

total salinity. Obligate dependence on Na$^+$, Cl$^-$, and CO$_3$$^{2-}$ ions. Alkaliphilic; grows at pH 8.1–10.7, optimum at pH 9.7–10.0. Mesophilic. Growth temperature optimum at 37 °C and maximum growth temperature of 42 °C.

   Chemo-organotrophic, obligate anaerobe that ferments a very limited number of substrates to acetic acid. Only lactate, pyruvate, glutamate, ethanol, and propanol are utilized. When growing on propanol, propionate is produced in addition to acetate. Yeast extract and vitamins stimulate growth. Non-cyclic homoacetogenic acetyl-CoA pathway is used for CO$_2$ fixation and high CO-dehydrogenase activity is present in cell extracts. Enzyme activity increases fourfold upon an increase in pH from 7.0 to 9.0 and is independent of the presence of sodium chloride and sodium bicarbonate ions. Rhodamine (3 μM) inhibits ATP synthesis and the addition of 100 μM monensin results in rapid lysis of the

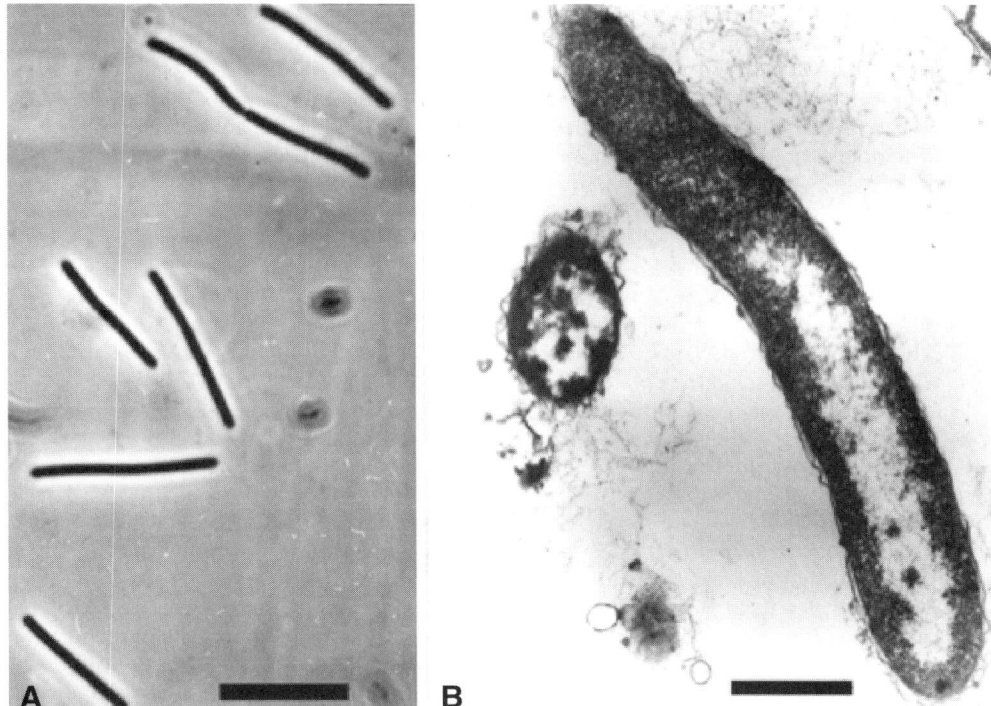

FIGURE 241. Morphology of *Natroniella acetigena*. (A) Light microscopy; bar = 10 μm. (B) Electron microscopy; longitudinal and cross sections demonstrating Gram-negative type of cell-wall structure; bar = 1 μm.

cells, indicating inhibition of Na⁺/H⁺ antiport; Na⁺ ions enter the cells causing death. ATP synthesis in *Natroniella* occurs in response to an artificial proton gradient upon addition of H⁺ ions (Pusheva et al., 1999a, 1999b, 2000). *Natroniella* may be considered as a halophile that has adapted to an alkaline environment. *Natroniella acetigena* is a peculiar organism, not only as an extremophile growing at high salinity and pH, but also as a member of alkaliphilic communities because of its

very restricted trophic needs that are oriented toward lactate and ethanol, the products of primary anaerobes, e.g., alkaliphilic spirochetes from the same Lake Magadi (Zavarzin and Zhilina, 2000; Zavarzin et al., 1999; Zhilina et al., 1996c).

Source: sediment of Lake Magadi, Kenya.

DNA G+C content (mol%): 32 (T_m).

Type strain: Z-7937, CIP 105131, DSM 9952.

GenBank accession number (16S rRNA gene): X95817.

Genus VI. **Orenia** Rainey, Zhilina, Boulygina, Stackebrandt, Tourova and Zavarzin 1995b, 880^VP (Effective publication: Rainey, Zhilina, Boulygina, Stackebrandt, Tourova and Zavarzin 1995a, 197.)

Fred A. Rainey

O.re′ni.a. N.L. fem. n. *Orenia* named after Aharon Oren, an Israeli microbiologist.

Gram negative. Rods, 2.5–13 μm in length with round ends. Motile by peritrichous flagella. **Spores** are round, terminal or subterminal. Gas vacuoles detected in some species. Forms sphaeroplasts. **Strictly anaerobic**. Catalase- and oxidase-negative. **Halophilic**; optimum NaCl concentration for growth ranges 3–12%; no growth below 2% or above 25%. Mesophilic to slightly thermophilic. Optimum temperature range for growth 36–45 °C, maximum below 52 °C, minimum 25 °C. **Chemoorganotrophic**. End products of glucose fermentation include H_2, CO_2, formate, lactate, acetate, butyrate, and ethanol.

DNA G + C content (mol%): 28.6–33.7.

Type species: **Orenia marismortui** (Oren, Pohla and Stackebrandt 1987) Rainey, Zhilina, Boulygina, Stackebrandt, Tourova and Zavarzin 1995b, 880^VP (Effective publication: Rainey, Zhilina, Boulygina, Stackebrandt, Tourova and Zavarzin 1995a, 197.) (*Sporohalobacter marismortui* Oren, Pohla and Stackebrandt 1987, 245).

The description of the genus *Orenia* presented here is based on data published by Oren et al. (1987), Bhupathiraju et al. (1994), Zhilina et al., (1999), and Mouné et al. (2000).

Further descriptive information

The genus *Orenia* contains three species: *Orenia marismortui* (Oren et al., 1987; Rainey et al., 1995a), *Orenia sivashensis* (Zhilina et al., 1999), and *Orenia salinaria* (Mouné et al., 2000). Cells of the strains of this genus stain Gram negative and electron microscopy has revealed a typical Gram-negative cell wall structure. The cells are rod shaped, have rounded ends, and range in diameter from 0.5–0.75 μm and in length from 2.5–13 μm. Zhilina et al. (1999) indicated that cells of *Orenia sivashensis* increase in diameter to 0.7–2.5 μm during spore formation. Endospores are formed by all species of this genus. In the case of *Orenia marismortui*, endospore formation is not observed when grown on standard growth medium, but spores are seen occasionally when grown on solidified medium. Cultures transferred to standard medium in which the yeast extract has been replaced with NH_4Cl (1 g/l) and L-leucine (1 g/l) sporulate within 2–3 d (Oren et al., 1987). *Orenia sivashensis* forms endospores when grown on solidified medium or when transferred from glucose to a glycogen containing medium (Zhilina et al., 1999). Gas vacuoles have been observed in *Orenia sivashensis* but not in *Orenia marismortui* or *Orenia salinaria*. Sphaeroplasts have been observed in all species of this genus forming both during the active growth phase in the case of *Orenia sivashensis* (Zhilina et al., 1999) and at the end of the exponential phase of growth for all three species (Mouné et al., 2000; Zhilina et al., 1999). Colony descriptions are available for *Orenia sivashensis* and *Orenia salinaria* but not for *Orenia marismortui*. Agar surface colonies of *Orenia sivashensis* are slightly elevated, white, even margined, slightly granular, and 5 mm in diameter after 4 d incubation; colonies in agar are lenticular (Zhilina et al., 1999). Colonies of *Orenia salinaria* are opaque in agar deeps and glossy on the surface of agar plates; they are 1–2 mm in diameter with entire margins and white to slightly yellow coloration (Mouné et al., 2000).

The species of the genus *Orenia* are moderate but obligate halophiles. *Orenia marimortui* grows at NaCl concentration of 0.5–3.0 M (~3–18%) and optimally at 0.5–2.0 M (~3–12%); no growth is observed below 0.5 M or above 3.0 M; at concentrations below 0.25 M (~1.5%) cell lysis occurs (Oren et al., 1987). Media containing 5–25% NaCl supports growth of *Orenia sivashensis* with optimal growth observed at 7–10%; cells lyse at NaCl concentrations of less than 5% (Zhilina et al., 1999). NaCl concentrations between 2 and 30% support the growth of *Orenia salinaria* as indicated in the main text of Mouné et al., (2000) or 2–25% as indicated in the actually species description of *Orenia salinaria* (Mouné et al., 2000). Cells of SG 3902^T at NaCl concentrations below 5% and above 10% are distorted and irregu-

lar, becoming elongated and rapidly forming sphaeroplasts at a concentration 2% NaCl (Mouné et al., 2000). On the basis of the temperature ranges for growth and optimal temperature ranges for growth, strains of the *Orenia* species fall at the upper end of the mesophilic range and could be considered moderately thermophilic (Zhilina et al., 1999). *Orenia marismortui* and *Orenia sivashensis* grow in the range 25–50 °C while *Orenia salinaria* grows between 10 and 50 °C. Optimal temperature ranges for growth are 36–45 °C for *Orenia marismortui* and 40–45 °C for both *Orenia sivashensis* and *Orenia salinaria*. The pH range and optima for growth have been determined for *Orenia sivashensis* and *Orenia salinaria* but not for *Orenia marismortui*. The former species have pH ranges for growth of 5.5–7.8 and 5.5–8.5, respectively. The optimal pH range for growth of *Orenia sivashensis* is 6.3–6.6 (Zhilina et al., 1999) and that of *Orenia salinaria* is in the range 7.2–7.4 (Mouné et al., 2000). Although the optimal pH for growth of *Orenia marismortui* was not recorded, it was routinely grown in medium at a pH 6.5 (Oren et al., 1987).

All species of the genus are obligate anaerobes; their relationship to oxygen has been tested in different ways for each species. *Orenia marismortui* is unable to grow in aerated medium; its sensitivity to oxygen is low with some growth in test tube cultures open to the air but containing a reducing agent. When cysteine is not included in the medium or cultures are shaken in cotton plugged flasks, no growth is observed (Oren et al., 1987). Reduced medium with a cotton plug or in a Hungate tube to which air has been added do not support growth of *Orenia sivashensis* (Zhilina et al., 1999). Mouné et al. (2000) indicate that *Orenia salinaria* does not grow in the presence of oxygen when the reducing agent is absent and tubes are open to the air or plugged with cotton-wool stoppers. The species of the genus *Orenia* are chemoorganotrophic with fermentative metabolism. Differences are observed in the substrate utilization patterns of the three species of the genus *Orenia*. These are given in Table 243, and the complete list of substrates utilized or not utilized is provided in the species descriptions below. Unfortunately, a fully comparable dataset for carbon and energy sources is not available due to incomplete datasets presented in the papers describing the species of the genus (Mouné et al., 2000; Oren et al., 1987; Zhilina et al., 1999).

TABLE 243. Differential characteristics of *Orenia* species[a,b]

Characteristic	1. *O. marismortui*	2. *O. salinaria*	3. *O. sivashensis*
Cell size (μm)	0.6 × 3–13	1.0 × 6–10	0.5–0.75 × 2.5–10
Gas vesicles	–	–	+
NaCl range (optimum) (% w/v)	3–18 (3–12)	2–25 (5–10)	5–25 (7–12)
Temperature range (optimum) (°C)	25–50 (36–45)	10–50 (40–45)	25–50 (40–45)
pH, range (optimum)	ND	5.5–8.5 (7.2–7.4)	5.5–7.8 (6.3–6.6)
Energy sources:			
N-Acetylglucosamine	–	–	+
Casamino acids	–	–	+/–
D-Cellobiose	–	+	+
D-Fructose	+	+	–
L-Glutamate	–	–	+
Glycogen	+	–	+
D-Mannose	+	–	+
Pyruvate	–	–	+
Starch	+	–	+

[a]+, Used as energy source; –, not used as energy source; +/–, poorly used for growth, ND, not determined.
[b]Data obtained from Oren et al. (1987), Zhilina et al. (1999), and Mouné et al. (2000).

The end products from the fermentation of glucose for all species of the genus include acetate, ethanol, formate, CO_2, and H_2. In addition, *Orenia marismortui* and *Orenia sivashensis* produce butyrate and *Orenia salinaria* produces lactate. *Orenia sivashensis* cannot reduce thiosulfate, but it does reduce elemental sulfur with the formation of H_2S, although growth was not stimulated by the addition of S^0 (Zhilina et al., 1999). The species *Orenia marismortui* produces small amounts of sulfide, the production of which was stimulated by the addition of thiosulfate but not by the addition of sulfate (Oren et al., 1987). Sulfide concentrations up to 20 mM are tolerated by *Orenia salinaria*, but sulfur containing compounds are not reduced (Mouné et al., 2000). Interestingly, *Orenia salinaria* does not use nitrate or cysteine as a nitrogen source but can utilize and sustain growth on dinitrogen and ammonia in media lacking other sources of nitrogen (Mouné et al., 2000).

DNA G + C content (mol%): 28.6–33.7.

Enrichment, isolation and growth conditions

The species of the genus *Orenia* have been isolated from saline environments. In the case of *Orenia marismortui*, it was isolated from anaerobic black sediment collected from a salt flat on the Western shore of the Dead Sea near Massada, Israel (Oren et al., 1987). The Dead Sea mud samples were inoculated into medium that had been designed for the isolation of sulfate reducing bacteria. The medium contained (g/l): NaCl, 125; $MgCl_2 \cdot 6H_2O$, 50; $MgSO_4 \cdot 7H_2O$, 1.0; $CaSO_4 \cdot 2H_2O$, 0.5; K_2HPO_4, 1.0; NH_4Cl, 1.0; $FeSO_4 \cdot 7H_2O$, 0.1; Na-lactate, 2.0; Na-thioglycolate, 1.0; yeast extract, 0.5, pH 7–7.2 (Oren et al., 1987). The enrichment cultures were set up in completely filled 150 ml stoppered bottles. The resulting enriched cultures were transferred several times before being streaked on plates of the same medium containing 2% agar and 0.5% $CaCO_3$ and incubated at 37 °C (Oren et al., 1987). *Orenia sivashensis* was isolated from a cyanobacterial mat that had anaerobic mud under it collected from the saline lagoons of Lake Sivash, the Crimea (Zhilina et al., 1999). The enrichment culture was established by inoculating 2 g of cyanobacterial mat into 50 ml of bicarbonate medium containing 150 g/l NaCl and 2.5 g/l peptone; after 4 d of incubation, the enrichment culture was pasteurized at 75 °C for 30 min and used to inoculate a fresh culture in the same medium with glucose (5 g/l) and yeast extract (0.5 g/l) added. A pure culture was isolated after a second pasteurization step at 80 °C for 2 h and subsequent dilution and colony selection on solidified medium (Zhilina et al., 1999). Black anoxic sediment from below the gypsum and halite crust of hypersaline (20–34% total salinity) ponds in the Salin-de-Giraud, Camargue, France was used as an inoculum for the isolation of *Orenia*

salinaria (Mouné et al., 2000). The enrichment medium used for the isolation of *Orenia salinaria* contained (g/l): NaCl, 150; $MgCl_2 \cdot 6H_2O$, 15; KCl, 3; NH_4Cl, 0.5; KH_2PO_4, 0.33; $CaCl_2 \cdot H_2O$, 0.05; yeast extract, 0.1; 0.01% resazurin solution, 1 ml; trace-element solution SL12 (Overmann et al., 1992), 1 ml; selenite/tungstate solution ($Na_2SeO_3 \cdot H_2O$, 6 mg/l; $NaWO_4 \cdot 2H_2O$, 8 mg/l; NaOH, 0.4 g/l), 1 ml; $NaHCO_3$, 2 g; $NaS \cdot 9H2O$, 0.5 g; vitamin solution V7 (Pfennig and Trüper, 1981), 1 ml; pH 7.2–7.4. The medium was prepared under a headspace of N_2/CO_2 (90:10). Strain SG 3902^T was obtained in pure culture using agar deeps with glucose as the substrate and incubation at 30 °C in the dark (Mouné et al., 2000).

Taxonomic comments

Orenia marismortui was originally described as *Sporohalobacter marismortui* due to its spore formation and some indication of a relationship to *Sporohalobacter lortetii* based on oligonucleotide catalogues (Oren et al., 1987). Subsequent phylogenetic ana-lysis of the halophilic anaerobes demonstrated that *Sporohalobacter marismortui* and *Sporohalobacter lortetii* were not closely related, sharing only 88% 16S rRNA gene sequence similarity. On the basis of the phylogenetic position and lack of gas vesicle formation in *Sporohalobacter marismortui* as compared to *Sporohalobacter lortetii*, the genus *Orenia* was established with *Orenia marismortui* as the type and only species of the new genus (Rainey et al., 1995b, 1995a). Subsequently, two species *Orenia sivashensis* and *Orenia salinaria* were added to the genus (Mouné et al., 2000; Zhilina et al., 1999). The genus comprises a phylogenetic coherent cluster within the family *Halobacteroidaceae* (Rainey et al., 1995b, 1995a; Zhilina and Rainey, 1995) supported by a 100% bootstrap value (Figure 242). 16S rRNA gene sequence similarities between the species of the genus *Orenia* are in the range 93.6–94.5 with similarities of <92% to the related species *Halobacteroides halobius* (91–92%) and *Natroniella acetigena* (89.6–90.2). In addition to the <97% 16S rRNA gene sequence similarity between *Orenia marismortui* and *Orenia sivashensis*, the distinct species status of these two species was demonstrated by DNA–DNA hybridization studies in which a reassociation value of 44% was determined (Zhilina et al., 1999). In addition, phenotypic differences in substrate utilization and NaCl requirements further differentiate these species. *Orenia salinaria* was described after the publication of *Orenia sivashensis* and no direct comparison of these two species has been made in the literature. *Orenia salinaria* and *Orenia sivashensis* share only 93.6% 16S rRNA gene sequence similarity and differ in a number of phenotypic characteristics including gas vesicle formation, lower NaCl requirements for growth, substrate utilization patterns, and the mol% G + C content of their DNA (Table 243).

List of species of the genus *Orenia*

1. **Orenia marismortui** (Oren, Pohla and Stackebrandt 1987) Rainey, Zhilina, Boulygina, Stackebrandt, Tourova and Zavarzin, 1995b, 880^{VP} (Effective publication: Rainey, Zhilina, Boulygina, Stackebrandt, Tourova and Zavarzin 1995a, 197.) (*Sporohalobacter marismortui* Oren, Pohla and Stackebrandt 1987, 245)

ma.ris.mor'tu.i. L. gen. n. *maris* of the sea; L. adj. *mortuus* dead: N.L. gen. n. *marismortui* of the Dead Sea.

Gram negative. Rods with rounded ends, 0.6 μm in diameter and 3–13 μm in length. Motile by means of peritrichous flagella. Endospores formed. Gas vacuoles not detected. Sphaeroplasts are formed. Moderately halophilic; optimum NaCl concentration for growth ranges between 3 and 12%. No growth occurs below 3% or above 18% NaCl. The optimum temperature for growth is between 36 and 45 °C in the range

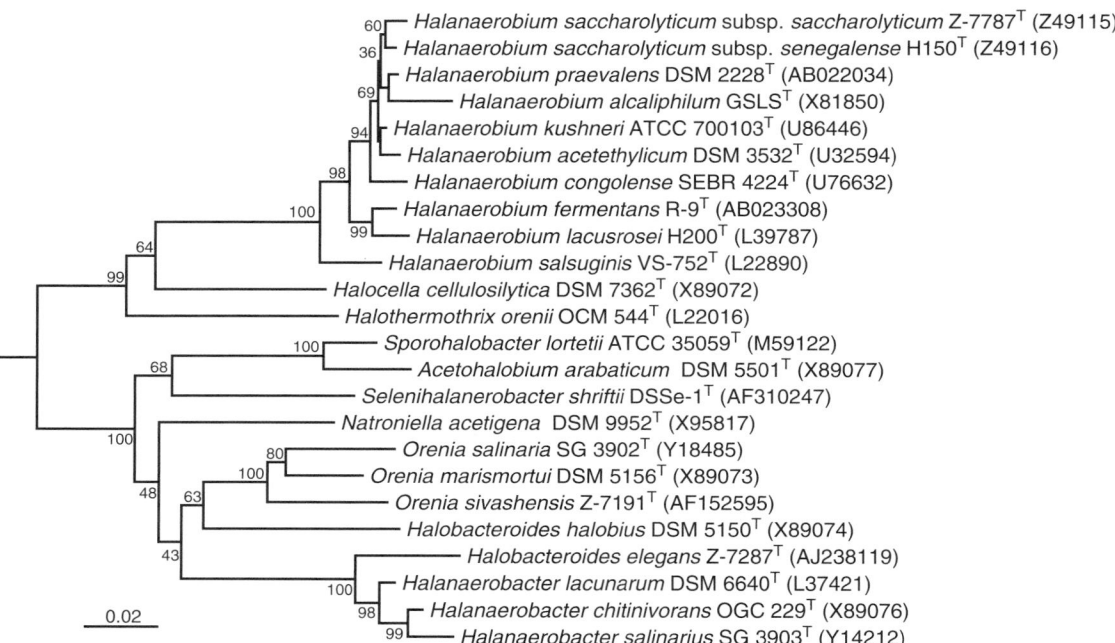

FIGURE 242. 16S rRNA based phylogeny of the order *Halanaerobiales*. The tree was constructed from distance matrices using the neighbor-joining method. The scale bar represents 2 nucleotide substitutions per 100 nucleotides.

25–50 °C. Major fatty acids include $C_{14:0}$, $C_{16:0}$ $C_{16:1}$, and $C_{18:0}$; minor compounds are C_{12}, $C_{14:0}$, and $C_{18:1}$. Carbon sources utilized include: D-glucose, D-mannose, D-maltose, D-ribose, sucrose, D-fructose, starch, and glycogen. Substrates not utilized include D-cellobiose, *N*-acetylglucosamine, pyruvate, Casamino acids, L-arginine, and L-glutamate. End products of glucose fermentation include H_2, CO_2, formate, acetate, butyrate, and ethanol. Isolated from Dead Sea sediment.

DNA G+C content (mol%): 29.6 (T_m).

Type strain: DY-1, ATCC 35420, DSM 5156.

GenBank accession number (16S rRNA gene): X89073.

2. **Orenia salinaria** Mouné, Eatock, Matheron, Willison, Hirschler, Herbert and Caumette 2000, 728[VP]

sa.li.na'ria. L. adj. *salinaria* pertaining to *salinae* salterns, saltworks.

Gram negative. Rods with rounded ends, 1.0 μm in diameter and 6–10 μm in length. Motile by means of peritrichous flagella. Endospores formed and are spherical and subterminal in old cultures. Large degenerate cells and sphaeroplasts are observed in old cultures. Surface colonies are 1–2 mm in diameter, circular, translucid, glossy, and are white to slightly yellow with entire edges. Obligately halophilic, optimum NaCl concentration for growth ranges between 5 and 10%. No growth occurs below 2% or above 25% NaCl. The cells accumulate glycine-betaine. The optimum temperature for growth is between 40 and 45 °C in the range 10–50 °C. The pH range is 5.5–8.5, with an optimum at pH 7.2–7.4. Cells are susceptible to chloramphenicol, erythromycin, and tetracycline and resistant to anisomycin, kanamycin, and Na-taurocholate. Fermentative metabolism. Glucose, fructose, trehalose, sucrose, maltose, cellobiose, and mannitol are fermented. D-Mannose, starch, glycogen, pyruvate, Casamino acids, and L-glutamate are not utilized. The major products of glucose fermentation are ethanol,

formate, acetate, lactate, CO_2, and H_2. Growth occurs with N_2 as the sole nitrogen source. Isolated from sediment of salt ponds in the salterns of Salin-de-Giraud (Camargue, France).

DNA G+C content (mol%): 33.7 (HPLC).

Type strain: SG 3902, ATCC 700911.

GenBank accession number (16S rRNA gene): Y18485.

3. **Orenia sivashensis** Zhilina, Tourova, Kuznetsov, Kostrikina and Lysenko 2000, 3[VP] (Effective publication: Zhilina, Tourova, Kuznetsov, Kostrikina and Lysenko 1999, 456.)

si.va.shen'sis. N.L. gen. n. *sivashensis* inhabiting Lake Sivash.

Gram negative. Rods with rounded ends, 0.5–0.75 μm in diameter and 2.5–10 μm in length; single, paired, or in short chains and filaments. Motile by means of peritrichous flagellation. Sporeforming. The spores are thermostable, round, terminal. Gas vacuoles are present in the cells and on mature spores. The colonies are white, 0.5 mm in diameter, lenticular with an even margin. Obligately anaerobic. Chemoorganotrophic. The substrates utilized include D-glucose, D-mannose, mannitol, D-maltose, sucrose, L-sorbitol, trehalose, D-cellobiose, D-ribose, starch, glycogen, *N*-acetylglucosamine, pyruvate, citrate, ascorbate, yeast extract, Casamino acids, L-arginine, L-glutamate, and DL-lysine. Yeast extract is necessary for growth on carbohydrates. Obligately halophilic; optimum NaCl concentration for growth ranges between 7 and 10% in the range 5–25%. The optimal temperature for growth is 40–45 °C, in the growth temperature range 25–50 °C. The range of growth pH is 5.5–7.8, with an optimum pH at 6.3–6.6. The habitat is sea lagoons with variable salinity, up to halite deposition. Isolated from a cyanobacterial mat of Lake Sivash lagoons (the Crimea).

DNA G + C content (mol%): 28.6 (T_m).

Type strain: Z-7191, DSM 12596.

GenBank accession number (16S rRNA gene): AF152595.

Genus VII. **Selenihalanaerobacter** Switzer Blum, Stolz, Oren and Oremland 2001b, 1229^VP (Effective publication: Switzer Blum, Stolz, Oren and Oremland 2001a, 217.)

AHARON OREN, RONALD S. OREMLAND AND JOHN F. STOLZ

Se.le.ni.hal.an.ae.ro.bac′ter. Gr. n. *selene* moon; N.L. n. *selenium* element 34; Gr. n. *hals, halos* salt; Gr. pref. *an* not; Gr. n. *aer* air; N.L. masc. n. *bacter* the masc. equivalent of Gr. neut. n. *baktron* a staff or rod; N.L. masc. n. *Selenihalanaerobacter* the salty anaerobic selenium rod.

Gram-stain-negative, rod-shaped cells, non-motile. Halophilic, growing optimally at 3.6 M NaCl and requiring a minimum of 1.7 M NaCl for growth. Temperature optimum about 38 °C. **Strictly anaerobic.** Grow by anaerobic respiration on organic electron donors, using selenate and other electron acceptors. Fermentative growth not observed. **Endospores not produced.**

DNA G+C content (mol%): 31.2 (HPLC).

Type species: **Selenihalanaerobacter shriftii** Switzer Blum, Stolz, Oren and Oremland 2001b, 1229^VP (Effective publication: Switzer Blum, Stolz, Oren and Oremland 2001a, 217.).

Further descriptive information

The genus contains a single species, *Selenihalanaerobacter shriftii*. This bacterium was isolated from sediments of the Dead Sea from an anaerobic enrichment culture that contained diluted Dead Sea water, glycerol, and selenate (Switzer Blum et al., 2001a).

Phylogenetically, *Selenihalanaerobacter* can unequivocally be classified within the order *Halanaerobiales*, family *Halobacteroidaceae*. However, its physiology is very different from that of the other known representatives of the order. No fermentative or homoacetogenic metabolism has been demonstrated in the genus and its only mode of energy generation appears to be anaerobic respiration. Selenate, nitrate, and trimethylamine *N*-oxide are the only suitable electron acceptors and the range of electron donors supporting growth of the single described species is limited. Of a wide range of compounds tested, glycerol and glucose were the only compounds capable of serving as an energy source. Both are partially oxidized to acetate + CO_2, while selenate is reduced to a mixture of selenite and elemental selenium. In accordance with its respiratory mode of metabolism, *Selenihalanaerobacter shriftii* contains cytochromes and *Selenihalanaerobacter* is thereby the only genus within the *Halanaerobiales* with a representative in which the presence of cytochromes has been detected thus far.

Enrichment and isolation procedures

The single isolate of the only recognized species within the genus was isolated from a sediment sample collected from a depth of 18 m from the Dead Sea, Israel. It was obtained from an anaerobic enrichment culture containing Dead Sea water diluted to 75% of its original salinity and amended with glycerol, selenate, and cysteine as a reducing agent. Formation of an orange-red precipitate of elemental selenium can be used as evidence for growth. The organism can then be isolated by streaking on agar plates containing NaCl (3 M), other salts, glycerol or glucose, selenate, yeast extract, and cysteine, followed by incubation under nitrogen. Strictly anaerobic techniques should be used, including boiling the media under nitrogen and adding cysteine as reducing agent to the boiled media. The composition of the medium for growth of *Selenihalanaerobacter shriftii* is given by Switzer Blum et al. (2001a).

Although nitrate and trimethylamine *N*-oxide are suitable electron acceptors for the growth of *Selenihalanaerobacter shriftii*, use of these oxidants in the enrichment medium instead of selenate is not to be recommended as facultatively anaerobic halophilic *Bacteria* or *Archaea* that can use these electron acceptors may outcompete the relatively slowly growing *Selenihalanaerobacter*.

Maintenance procedures

Selenihalanaerobacter cultures can be maintained by freezing in liquid nitrogen or by lyophilization.

Taxonomic comments

Analysis of the 16S rRNA gene sequence of the single species classified within the genus showed it to be most closely related to *Acetohalobium arabaticum* (90% sequence similarity) and *Sporohalobacter lortetii* (Switzer Blum et al., 2001a).

List of species of the genus *Selenihalanaerobacter*

1. **Selenihalanaerobacter shriftii** Switzer Blum, Stolz, Oren and Oremland 2001b, 1229^VP (Effective publication: Switzer Blum, Stolz, Oren and Oremland 2001a, 217.)

shrif′ti.i. N.L. gen. n. *shriftii* of Shrift, named after Alex Shrift, an American microbiologist.

Cells are rod-shaped, 0.6 × 2–6 µm in size. Non-motile. Halophilic, requiring 1.7–4.1 M NaCl for growth, and growing optimally at 3.6 M NaCl. Temperature range for growth: 16–42 °C. Optimum temperature: 38 °C. Strictly anaerobic. Oxidizes glycerol and glucose to acetate + CO_2 while reducing selenate to selenite and elemental selenium.

Also uses nitrate and trimethylamine *N*-oxide as electron acceptors. Nitrite, arsenate, fumarate, dimethylsulfoxide, thiosulfate, elemental sulfur, sulfite, and sulfate do not serve as electron acceptors. No growth is observed on short-chain fatty acids, alcohols, amino acids, formate, lactate, pyruvate, or other organic acids. Fermentative growth is not observed.

Source: sediments of the Dead Sea, Israel.
DNA G+C content (mol%): 31.2 (HPLC).
Type strain: DSSe-1, ATCC BAA-73.
GenBank accession number (16S rRNA gene): AF310247.

Genus VIII. **Sporohalobacter** Oren, Pohla and Stackebrandt 1988, 136^{VP} (Effective publication: Oren, Pohla and Stackebrandt 1987, 245.)

AHARON OREN

Spo.ro.ha.lo.bac′ter. Gr. n. *spora* seed; Gr. n. *hals, halos* salt; N.L. masc. n. *bacter* the masc. equivalent of Gr. neut. n. *baktron* a staff or rod; N.L. masc. n. *Sporohalobacter* spore-producing salt rod.

Gram-stain-negative, rod-shaped cells, motile by peritrichous flagella. Halophilic, growing optimally at 1.4–1.5 M NaCl and requiring a minimum of 0.7 M NaCl for growth. Temperature optimum about 40 °C. **Strictly anaerobic.** Ferments amino acids with production of acetate, propionate and other acids, H$_2$, and CO$_2$. Sugars poorly used. **Endospores produced.** Gas vesicles are attached to the endospores in the single species described.

DNA G+C content (mol%): 31.5 (Bd).

Type species: **Sporohalobacter lortetii** (Oren 1983) Oren, Pohla and Stackebrandt 1988, 136^{VP} (Effective publication: Oren, Pohla and Stackebrandt 1987, 245.) (*Clostridium lortetii* Oren 1983, 47).

Further descriptive information

The genus contains one species, namely *Sporohalobacter lortetii.* This bacterium was isolated from sediments of the Dead Sea from an anaerobic enrichment culture that contained lactate and yeast extract as carbon sources. On solid media, endospores are produced. Sporulating cells are terminally swollen. Adjacent to the endospores, gas vesicles are produced and these remain attached to the mature endospores after degeneration of the cells, possibly to enable dispersion of the endospores.

The metabolism of *Sporohalobacter* has not been characterized in depth. Amino acids appear to be the preferred substrate for fermentation. Glucose added to the medium was utilized only after other, more easily fermentable substrates, were exhausted (Oren, 1983).

Enrichment and isolation procedures

The single isolate of the only recognized species within the genus was isolated from a sediment sample collected from a depth of 60 m from the Dead Sea, Israel. It was obtained from an anaerobic enrichment culture containing Dead Sea water diluted to 80% of its original salinity, supplemented with lactate and yeast extract, and with thioglycollate and ascorbate as reducing agents, followed by serial dilution in agar tubes containing similar medium with 70% Dead Sea water and FeSO$_4$ (Oren, 1983).

The production of heat-resistant endospores may enable the development of a more selective enrichment procedure for *Sporohalobacter*, based on negative selection by pasteurization of the inoculum. However, such a strategy may also yield other endospore-forming halophilic anaerobic genera such as *Halobacteroides* (Oren et al., 1987), *Orenia*, and possibly others. Successful isolation of *Sporohalobacter* using such a negative selection procedure has not yet been reported.

Strictly anaerobic techniques should be used, including boiling the media under nitrogen or nitrogen/CO$_2$ (80:20) and adding cysteine as a reducing agent to the boiled media. The composition of the medium for growth of *Sporohalobacter lortetii* can be found in Oren (1983, 2006).

Maintenance procedures

The most reliable way to maintain cultures is by lyophilization; they are maintained thus by culture collections. No data on the long-term viability of the mature endospores are yet available.

Taxonomic comments

The single species classified within the genus was originally described as *Clostridium lortetii* (Oren, 1983, 1984). After it became obvious that the isolate was not closely related phylogenetically to members of the genus *Clostridium*, the isolate was transferred to the genus *Sporohalobacter*, which was newly created to accommodate endospore-forming members of the family *Halobacteroidaceae* (Oren et al., 1987). *Sporohalobacter marismortui*, the second species at that time classified within the genus *Sporohalobacter* (Oren et al., 1987, 1988), has been transferred to the new genus *Orenia* as *Orenia marismortui* comb. nov. on the basis of 16S rRNA gene sequence data (Rainey et al., 1995a).

List of species of the genus *Sporohalobacter*

1. **Sporohalobacter lortetii** (Oren 1983) Oren, Pohla and Stackebrandt 1988, 136^{VP} (Effective publication: Oren, Pohla and Stackebrandt 1987, 245.) (*Clostridium lortetii* Oren 1983, 47)

lor.tet′i.i. N.L. gen. n. *lortetii* of Lortet, named after M.L. Lortet, a French microbiologist.

Cells are rod-shaped, 0.5–0.6 × 2.5–10 μm in size. Motile by peritrichous flagella. Halophilic, requiring 0.7–2.6 M NaCl for growth and growing optimally at 1.4–1.5 M NaCl. Temperature range for growth: 25–52 °C. Optimum temperature: 37–45 °C. Strictly anaerobic. Ferments amino acids with production of acetate, propionate, isobutyrate, isovalerate, H$_2$, and CO$_2$ in medium containing L-glutamic acid, Casamino acids, yeast extract, and nutrient broth. Certain sugars (fructose, D-glucose, maltose, starch, and sucrose) stimulate growth. No growth stimulation by cellobiose, galactose, lactose, or mannose. Endospores produced in terminally swollen cells. Gas vesicles are attached to the endospores. The major fatty acids in the lipids are C$_{16:0}$ and C$_{16:1}$.

Source: sediments of the Dead Sea, Israel.

DNA G+C content (mol%): 31.5 (Bd).

Type strain: MD-2, ATCC 35059, DSM 3070.

GenBank accession number (16S rRNA gene): M59122.

References

Bhupathiraju, V.K., A. Oren, P.K. Sharma, R.S. Tanner, C.R. Woese and M.J. McInerney. 1994. *Haloanaerobium salsugo* sp. nov., a moderately halophilic, anaerobic bacterium from a subterranean brine. Int. J. Syst. Bacteriol. *44*: 565–572.

Bykhovsky, V.Y., M.A. Pusheva, N.I. Zaitseva, T.N. Zhilina, D.B. Pankowskii and E.N. Detkova. 1994. Biosynthesis of corrinoids and its possible precursors in extremely halophilic homoacetogenic bacterium *Acetohalobium arabaticum*. Mikrobiologiya *30*: 97–105.

Kevbrin, V.V. and G.A. Zavarzin. 1992a. The effect of sulfur compounds on growth of halophilic homoacetic bacterium *Acetohalobium arabaticum*. Microbiologiya *62*: 812–817.

Kevbrin, V.V. and G.A. Zavarzin. 1992b. Methanethiol utilization and sulfur reduction by anaerobic halophilic saccharolytic bacteria. Curr. Microbiol. *24*: 247–250.

Kevbrin, V.V. and G.A. Zavarzin. 1992c. Effect of sulfur compounds on the growth of the halophilic homoacetic bacterium *Acetohalobium arabaticum*. Microbiology (En. transl. from Mikrobiologiya) *61*: 563–567.

Kevbrin, V.V., T.N. Zhilina and G.A. Zavarzin. 1995. Physiology of the halophilic homoacetic bacterium *Acetohalobium arabaticum* (in Russian). Mikrobiologiya *64*: 134–138

Liaw, H.J. and R.A. Mah. 1992. Isolation and characterization of *Haloanaerobacter chitinovorans* gen. nov., sp. nov., a halophilic, anaerobic, chitinolytic bacterium from a solar saltern. Appl. Environ. Microbiol. *58*: 260–266.

Liaw, H.J. and R.A. Mah. 1996. *In* Validation of the publication of new names and new combinations previously effectively published outside the IJSB. List no. 56. Int. J. Syst. Bacteriol. *46*: 362–363.

Mouné, S., C. Eatock, R. Matheron, J.C. Willison, A. Hirschler, R. Herbert and P. Caumette. 2000. *Orenia salinaria* sp. nov., a fermentative bacterium isolated from anaerobic sediments of Mediterranean salterns. Int. J. Syst. Evol. Microbiol. *50*: 721–729.

Mouné, S., M. Manac'h, A. Hirschler, P. Caumette, J.C. Willison and R. Matheron. 1999. *Haloanaerobacter salinarius* sp. nov., a novel halophilic fermentative bacterium that reduces glycine-betaine to trimethylamine with hydrogen or serine as electron donors; emendation of the genus *Haloanaerobacter*. Int. J. Syst. Bacteriol. *49*: 103–112.

Oren, A. 1983. *Clostridium lortetii* sp. nov., a halophilic obligatory anaerobic bacterium producing endospores with attached gas vacuoles. Arch. Microbiol. *136*: 42–48.

Oren, A. 1984. *In* Validation of the publication of new names and new combinations previously effectively published outside the IJSB. List no. 14. Int. J. Syst. Bacteriol. *34*: 270–271.

Oren, A., W.G. Weisburg, M. Kessel and C.R. Woese. 1984a. *In* Validation of the publication of new names and new combinations previously effectively published outside the IJSB. List no. 15. Int. J. Syst. Bacteriol. *34*: 355–357.

Oren, A., W.G. Weisburg, M. Kessel and C.R. Woese. 1984b. *Halobacteroides halobius* gen. nov., sp. nov., a moderately halophilic anaerobic bacterium from the bottom sediments of the Dead Sea. Syst. Appl. Microbiol. *5*: 58–70.

Oren, A. 1986. Intracellular salt concentrations of the anaerobic halophilic eubacteria *Haloanaerobium pravalens* and *Halobacteroides halobius*. Can. J. Microbiol. *32*: 4–9.

Oren, A. 1987. a procedure for the selective enrichment of *Halobacteroides halobius* and related bacteria from anaerobic hypersaline sediments. FEMS Microbiol. Lett. *42*: 201–204.

Oren, A., H. Pohla and E. Stackebrandt. 1987. Transfer of *Clostridium lortetii* to a new genus *Sporohalobacter* gen. nov. as *Sporohalobacter lortetii* comb. nov., and description of *Sporohalobacter marismortui* sp. nov. Syst. Appl. Microbiol. *9*: 239–246.

Oren, A., H. Pohla and E. Stackebrandt. 1988. *In* Validation of the publication of new names and new combinations previously published outside the IJSB. List no. 24. Int. J. Syst. Bacteriol. *38*: 136–137.

Oren, A., P. Gurevich and Y. Henis. 1991. Reduction of nitrosubstituted aromatic compounds by the halophilic anaerobic eubacteria *Haloanaerobium praevalens* and *Sporohalobacter marismortui*. Appl. Environ. Microbiol. *57*: 3367–3370.

Oren, A. 2000. Change of the names *Haloanaerobiales*, *Haloanaerobiaceae* and *Haloanaerobium* to *Halanaerobiales*, *Halanaerobiaceae* and *Halanaerobium*, respectively, and further nomenclatural changes within the order *Halanaerobiales*. Int. J. Syst. Evol. Microbiol. *50*: 2229–2230.

Oren, A. 2006. The order *Halanaerobiales*. *In* Dworkin, Falkow, Rosenberg, Schleifer and Stackebrandt (Editors), The Prokaryotes: A Handbook on the Biology of Bacteria, 3rd edn, Vol. 4. Springer, New York, pp. 809–822.

Oren, A. 2008. Microbial life at high salt concentrations: phylogenetic and metabolic diversity. Saline Systems *4*: 2.

Overmann, J., U. Fischer and N. Pfennig. 1992. A new purple sulfur bacterium from saline littoral sediments, *Thiorhodovibrio winogradskyi* gen. nov. and sp. nov. Arch. Microbiol. *157*: 329–335.

Patel, B.K.C., K.T. Andrews, B. Ollivier, R.A. Mah and J.-L. Garcia. 1995. Reevaluating the classification of *Halobacteroides* and *Haloanaerobacter* species based on sequence comparisons of the 16S ribosomal RNA gene. FEMS Microbiol. Lett. *134*: 115–119.

Pfennig, N. and H.G. Trüper. 1981. Isolation of members of the families *Chromatiaceae* and *Chlorobiaceae*. *In* Starr, Stolp, Trüper, Balows and Schlegel (Editors), The Prokaryotes, vol. 1. Springer, Berlin, pp. 279–289.

Pusheva, M.A. 1992. The properties of periplasmic hydrogenase from extremely halophilic homoacetogenic bacterium *Acetohalobium arabaticum*. Mikrobiologiya *61*: 5–10.

Pusheva, M.A. and E.N. Detkova. 1996. Bioenergetic aspects of acetogenesis on various substrates by the extremely halophilic acetogenic bacterium *Acetohalobium arabaticum*. Mikrobiologiya *65*: 516–520.

Pusheva, M.A., A.V. Pitryuk, E.N. Detkova and G.A. Zavarzin. 1999a. Bioenergetics of acetogenesis in the extremely alkaliphilic homoacetogenic bacteria *Natroniella acetigena* and *Natronoincola histidinovorans*. Mikrobiologiya *68*: 651–656.

Pusheva, M.A., A.V. Pitryuk and A. Netrusov. 1999b. Inhibitory analysis of the energy metabolism of the extremely haloalkaliphilic homoacetic bacterium *Natroniella acetigena*. Mikrobiologiya *68*: 647–650.

Pusheva, M.A., A.V. Pitryuk and G.A. Zavarzin. 2000. Na$^+$- and H$^+$-dependent ATP synthesis in extremely alkaliphilic anaerobes. Dokl. Biol. Sci. *374*: 546–548.

Rainey, F.A., T.N. Zhilina, E.S. Boulygina, E. Stackebrandt, T.P. Tourova and G.A. Zavarzin. 1995a. The taxonomic status of the fermentative halophilic anaerobic bacteria: description of *Haloanaerobiales* ord. nov., *Halobacteroidaceae* fam. nov., *Orenia* gen. nov. and further taxonomic rearrangements at the genus and species level. Anaerobe *1*: 185–199.

Rainey, F.A., T.N. Zhilina, E.S. Boulygina, E. Stackebrandt, T.P. Tourova and G.A. Zavarzin. 1995b. *In* Validation of the publication of new names and new combinations previously effectively published outside the IJSB. List no. 55. Int. J. Syst. Bacteriol. *45*: 879–880.

Rengpipat, S., T.A. Langworthy and J.G. Zeikus. 1989. *In* Validation of the publication of new names and new combinations previously effectively published outside the IJSB. List no. 29. Int. J. Syst. Bacteriol. *39*: 205–206.

Shiba, H. and K. Horikoshi. 1988. Isolation and characterization of novel anaerobic, halophilic eubacteria from hypersaline environments of western America and Kenya. Presented at the Proceedings of the FEMS Symposium on the Microbiology of Extreme Environments and its Biotechnological Potential, Estoril.

Shiba, H., H. Yamamoto and K. Horikoshi. 1989. Isolation of strictly anaerobic halophiles from the aerobic surface sediments of hypersaline environments in California and Nevada. FEMS Microbiol. Lett. *57*: 191–195.

Shiba, H. 1991. Anaerobic halophiles. *In* Horikoshi and Grant (Editors), Superbugs: Microorganisms in Extreme Environments. Springer-Verlag, Berlin, pp. 191–211.

Switzer Blum, J., J.F. Stolz, A. Oren and R.S. Oremland. 2001a. *Selenihalanaerobacter shriftii* gen. nov., sp. nov., a halophilic anaerobe from Dead Sea sediments that respires selenate. Arch. Microbiol. *175*: 208–219.

Switzer Blum, J., J.F. Stolz, A. Oren and R.S. Oremland. 2001b. *In* Validation of publication of new names and new combinations previously effectively published outside the IJSEM. List no. 81. Int. J. Syst. Bacteriol. *51*: 1229.

Tourova, T.P., E.S. Boulygina, T.N. Zhilina, R.S. Hanson and G.A. Zavarzin. 1995. Phylogenetic study of haloanaerobic bacteria by 16S ribosomal RNA sequences analysis. Syst. Appl. Microbiol. *18*: 189–195.

Wolin, E.A., M.G. Wolin and R.S. Wolfe. 1963. Formation of methane by bacterial extracts. J. Biol. Chem. *238*: 2882–2886.

Zavarzin, G.A., T.N. Zhilina and M.A. Pusheva. 1994. Halophilic acetogenic bacteria. *In* Drake (Editor), Acetogenesis. Chapman and Hall, New York, pp. 432–444.

Zavarzin, G.A., T.N. Zhilina and V.V. Kevbrin. 1999. The alkaliphilic microbial community and its functional diversity. Microbiology (En. transl. from Mikrobiologiya) *68*: 503–521.

Zavarzin, G.A. and T.N. Zhilina. 2000. Anaerobic chemotrophic alkaliphiles. *In* Seckbach (Editor), Journey to Diverse Microbial Worlds. Kluwer Academic Publishers, The Netherlands, pp.191–208.

Zhilina, T.N. and G.A. Zavarzin. 1990a. Extremely halophilic, methylotrophic, anaerobic bacteria. FEMS Microbiol. Rev. *87*: 315–322.

Zhilina, T.N. and G.A. Zavarzin. 1990b. *In* Validation of the publication of new names and new combinations previously effectively published outside the IJSB. List. no. 35. Int. J. Syst. Bacteriol. *40*: 470–471.

Zhilina, T.N. and G.A. Zavarzin. 1990c. A new extremely halophilic homoacetic bacterium *Acetohalobium arabaticum*, gen. nov., sp. nov. Dokl. Akad. Nauk. SSSR. *311*: 745–747.

Zhilina, T.N., V.V. Kevbrin, A.M. Lysenko and G.A. Zavarzin. 1991a. Isolation of saccharolytic anaerobes from a halophilic cyanobacterial mat. Microbiology (En. transl. from Mikrobiologiya) *60*: 101–107.

Zhilina, T.N., L.V. Miroshnikova, G.A. Osipov and G.A. Zavarzin. 1991b. *Halobacteroides lacunaris* sp. nov., new saccharolytic, anaerobic, extremely halophilic organism from the lagoon-like hypersaline Lake Chokrak. Microbiology (En. transl. from Mikrobiologiya) *60*: 495–503.

Zhilina, T.N. and G.A. Zavarzin. 1991. Anaerobic bacteria participating in organic matter destruction in halophilic cyanobacterial community. Zhurn. Obzhei Biol. *52*: 302–318.

Zhilina, T.N., L.V. Miroshnikova, G.A. Osipov and G.A. Zavarzin. 1992a. *In* Validation of the publication of new names and new combinations previously effectively published outside the IJSB. List. no. 41. Int. J. Syst. Bacteriol. *42*: 327–328.

Zhilina, T.N., G.A. Zavarzin, E.S. Bulygina, V.V. Kevbrin, G.A. Osipov and K.M. Chumakov. 1992b. Ecology, physiology and taxonomy studies on a new taxon of *Haloanaerobiaceae, Haloincola saccharolytica* gen. nov., sp. nov. Syst. Appl. Microbiol. *15*: 275–284.

Zhilina, T.N. and G.A. Zavarzin. 1994. Alkaliphilic anaerobic community at pH 10. Curr. Microbiol. *29*: 109–112.

Zhilina, T.N. and F.A. Rainey. 1995. *In* Validation of the publication of new names and new combinations previously effectively published outside the IJSB. List. no. 55. Int. J. Syst. Bacteriol. *45*: 879–880.

Zhilina, T.N., G.A. Zarvarzin, E.N. Detkova and F.A. Rainey. 1996a. *In* Validation of the publication of new names and new combinations previously effectively published outside the IJSB. List no. 59. Int. J. Syst. Bacteriol. *46*: 1189–1190.

Zhilina, T.N., G.A. Zavarzin, E.N. Detkova and F.A. Rainey. 1996b. *Natroniella acetigena* gen. nov. sp. nov., an extremely haloalkaliphilic, homoacetic bacterium: a new member of *Haloanaerobiales*. Curr. Microbiol. *32*: 320–326.

Zhilina, T.N., G.A. Zavarzin, F. Rainey, V.V. Kevbrin, N.A. Kostrikina and A.M. Lysenko. 1996c. *Spirochaeta alkalica* sp. nov., *Spirochaeta africana* sp. nov., and *Spirochaeta asiatica* sp. nov., alkaliphilic anaerobes from the Continental soda lakes in Central Asia and the East African Rift. Int. J. Syst. Bacteriol. *46*: 305–312.

Zhilina, T.N., T.P. Tourova, A.M. Lysenko and V.V. Kevbrin. 1997a. *In* Validation of the publication of new names and new combinations previously effectively published outside the IJSB. List no. 63. Int. J. Syst. Bacteriol. *47*: 1274.

Zhilina, T.N., T.P. Tourova, A.M. Lysenko and V.V. Kevbrin. 1997b. Reclassification of *Halobacteroides halobius* Z-7287 on the basis of phylogenetic analysis as a new species *Halobacteroides elegans* sp. nov. Microbiology *66*: 97–103.

Zhilina, T.N., E.N. Detkova, F.A. Rainey, G.A. Osipov, A.M. Lysenko, N.A. Kostrikina and G.A. Zavarzin. 1998. *Natronoincola histidinovorans* gen. nov., sp. nov., a new alkaliphilic acetogenic anaerobe. Curr. Microbiol. *37*: 177–185.

Zhilina, T.N., T.P. Tourova, B.B. Kuznetsov, N.A. Kostrikina and A.M. Lysenko. 1999. *Orenia sivashensis* sp. nov., a new moderately halophilic anaerobic bacterium from Lake Sivash lagoons. Microbiology (En. transl. from Mikrobiologiya) *68*: 452–459.

Zhilina, T.N., T.P. Tourova, B.B. Kuznetsov, N.A. Kostrikina and A.M. Lysenko. 2000. *In* Validation of the publication of new names and new combinations previously effectively published outside the IJSEM. List. no. 72. Int. J. Syst. Evol. Microbiol. *50*: 3–4.

Zhilina, T.N., E.S. Garnova, T.P. Tourova, N.A. Kostrikina and G.A. Zavarzin. 2001a. *In* Validation of publication of new names and new combinations previously effectively published outside the IJSEM. List no. 79. Int. J. Syst. Evol. Microbiol. *51*: 263–265.

Zhilina, T.N., E.S. Garnova, T.P. Tourova, N.A. Kostrikina and G.A. Zavarzin. 2001b. *Halonatronum saccharophilum* gen. nov. sp. nov.: a new haloalkaliphilic bacterium of the order *Haloanaerobiales* from Lake Magadi. Microbiology (En. transl. from Mikrobiologiya) *70*: 77–85.

Order III. **Thermoanaerobacterales** ord. nov.

JUERGEN WIEGEL

Ther.mo.an.ae.ro.bac.ter.a′les. N.L. masc. n. *Thermoanaerobacter* type genus of the order; -*ales* ending to denote an order; N.L. fem. n. *Thermoanaerobacterales* the order *Thermoanaerobacter*.

The order presently consists of the families *Thermoanaerobacteraceae* and *Thermodesulfobiaceae*, plus two families *incertae sedis*.

Type genus: **Thermoanaerobacter** Wiegel and Ljungdahl 1982, 384[VP] emend. Lee, Dashti, Prange, Rainey, Rohde, Whitman and Wiegel 2007, 1433 (Effective publication: Wiegel and Ljungdahl 1981, 348.).

References

Lee, Y.J., M. Dashti, A. Prange, F.A. Rainey, M. Rohde, W.B. Whitman and J. Wiegel. 2007. *Thermoanaerobacter sulfurigignens* sp. nov., an anaerobic thermophilic bacterium that reduces 1 M thiosulfate to elemental sulfur and tolerates 90 mM sulfite. Int. J. Syst. Evol. Microbiol. *57*: 1429–1434.

Taxonomic comment

Presently, there is some uncertainty whether the members of this order belong to the phylum *Firmicutes* and class *Clostridia* or whether they should constitute a separate novel phylum (Ludwig and Schleifer, personal communication).

Wiegel, J. and L.G. Ljungdahl. 1981. *Thermoanaerobacter ethanolicus* gen. nov., spec. nov, a new, extreme thermophilic, anaerobic bacterium. Arch. Microbiol. *128*: 343–348.

Wiegel, J. and L.G. Ljungdahl. 1982. *In* Validation of the publication of new names and new combinations previously effectively published outside the IJSB. List no. 9. Int. J. Syst. Bacteriol. *32*: 384–385.

Family I. Thermoanaerobacteraceae fam. nov.

JUERGEN WIEGEL

Ther.mo.an.ae.ro.bac.ter.a′ce.ae. N.L. masc. n. *Thermoanaerobacter* type genus of the family; -*aceae* ending to denote family; N.L. fem. n. *Thermoanaerobacteraceae* the *Thermoanaerobacter* family.

The family presently contains eight genera of thermophilic anaerobic species, mainly heterotrophic; however, some are able to grow chemolithoautotrophically (e.g., members of the genus *Moorella*). As of August 2007, the recognized genera of this family are *Thermoanaerobacter, Ammonifex, Caldanaerobacter, Carboxydothermus, Gelria, Moorella, Thermacetogenium,* and *Thermanaeromonas*.

The type species of the genera *Thermoanaerobium* (*sic*) and *Thermobacteroides* (*sic*) have been transferred to the genus *Thermoanaerobacter*.

Type genus: **Thermoanaerobacter** Wiegel and Ljungdahl 1982, 384^{VP} emend. Lee, Dashti, Prange, Rainey, Rohde, Whitman and Wiegel 2007a, 1433 (Effective publication: Wiegel and Ljungdahl 1981, 348.).

Genus I. Thermoanaerobacter Wiegel and Ljungdahl 1982, 384^{VP} emend. Lee, Dashti, Prange, Rainey, Rohde, Whitman and Wiegel 2007a, 1433 (Effective publication: Wiegel and Ljungdahl 1981, 348.)

ROB U. ONYENWOKE AND JUERGEN WIEGEL

Ther.mo.an.ae.ro.bac′ter. Gr. adj. *thermos* hot; Gr. pref. *an* not; Gr. n. *aer* air; M.L. *bacter* masc. equivalent of Gr. neut. dim. n. *bakterion* rod staff; N.L. masc. n. *Thermoanaerobacter* rod which grows in the absence of air at elevated temperatures.

All species are **obligately anaerobic thermophiles** and have a **Gram-positive cell wall**, i.e., although the **Gram-stain reaction varies** among the individual species and even within strains from the same species, they have a Gram-positive cell-wall structure. Most species exhibit a **sluggish motility** and a peritrichous arrangement of 2–6 flagella (retarded peritrichous flagellation). Cells are **rod shaped** and occur in various arrangements that vary by species. Many of the *Thermoanaerobacter* species form **pleomorphic cells** during the late exponential and stationary growth phases assembling through nonsymmetric cell division chains of alternating rod-shaped and coccoid cells. Other types of pleomorphic rods are formed, presumably through weakening of the cell wall by lysozyme-like enzymes that are the remnants of the sporulation process (**Figure 243**). Growth temperature optima range from 55–75 °C, with **growth ranges of 35–78 °C** and a **temperature span over 35 °C**. The pH for growth ranges from **pH^{25C} 4.0–9.9**, with a pH optimum of 5.8–8.5. Several species also exhibit a wide pH optimum of about 3 units without a sharp peak. All species are able to grow **organoheterotrophically** using various fermentation pathways including the homoacetogenic or Wood–Ljungdahl pathway, or using inorganic electron acceptors (e.g., *Thermoanaerobacter sulfuriphilus* using S^0 as an electron acceptor and lactate as an electron donor and carbon source). Some *Thermoanaerobacter* strains are also able to grow **chemolithoheterotrophically by coupling H$_2$ oxidation** to growth (Fardeau et al., 1994). Furthermore, others are **facultative chemolithoautotrophs**, growing with H$_2$ + CO$_2$ (such as *Thermoanaerobacter kivui*) or Fe(III) + H$_2$ + CO$_2$ (such as *Thermoanaerobacter siderophilus*). Generally, typical fermentation products from hexoses are ethanol, acetate, lactate, H$_2$, and CO$_2$.

DNA G+C content (mol%): 30–39.

Type species: **Thermoanaerobacter ethanolicus** Wiegel and Ljungdahl 1982, 384^{VP} (Effective publication: Wiegel and Ljungdahl 1981, 348.).

Further descriptive information

Biochemical analysis. Isoprenoid quinones, lipid fatty acids (PFLAs) and the cell-wall diaminopimelic acid isomers have been analyzed for a few species (Yamamoto et al., 1998). The menaquinone MK-7 has been found in the type strains of *Thermoanaerobacter brockii, Thermoanaerobacter thermohydrosulfuricus,* and *Thermoanaerobacter thermocopriae*. However, MK-7 is absent in the type strain of the type species, *Thermoanaerobacter ethanolicus* JW200. A similar variation in the presence of MK-7 is observed in the physiologically similar species of *Thermoanaerobacterium*. The main lipid fatty acids (PLFAs) observed in the above strains of *Thermoanaerobacter* (as well as in some strains of *Thermoananerobacterium*) are C$_{15:0 \text{ iso}}$ (45–55%) and C$_{17:0 \text{ iso}}$ (20–25%). The cell walls from *Thermoanaerobacter thermohydrosulfuricus, Thermoanaerobacter brockii* subsp. *finnii* and *Thermoanaerobacter thermocopriae* contain *meso*-diaminopimelic acid (Yamamoto et al., 1998).

Endospores. Endospore formation has been observed for all species except *Thermoananerobacter acetoethylicus, Thermoanaerobacter ethanolicus, Thermoanaerobacter kivui, Thermoanaerobacter mathranii* subsp. *alimentarius,* and *Thermoanaerobacter sulfurophilus*. However, the presence of the characteristic, spore-specific genes has been demonstrated for several species in which no spores have been observed. These species are regarded as asporogenic, meaning that most of the characteristic sporulation genes can be identified in the genome even though spores have not been observed. This distinguishes them from non-sporogenic species that lack all sporulation genes (Onyenwoke et al., 2004). Spore formation is highly variable. In some species, it is rarely observed, e.g., *Thermoanaerobacter brockii* subsp. *brockii* (Cook et al., 1991). In contrast, strains of *Thermoanaerobacter thermohydrosulfuricus* readily form spores, with greater than 50% of the cells sporulating upon reaching the stationary growth phase (Wiegel et al., 1979).

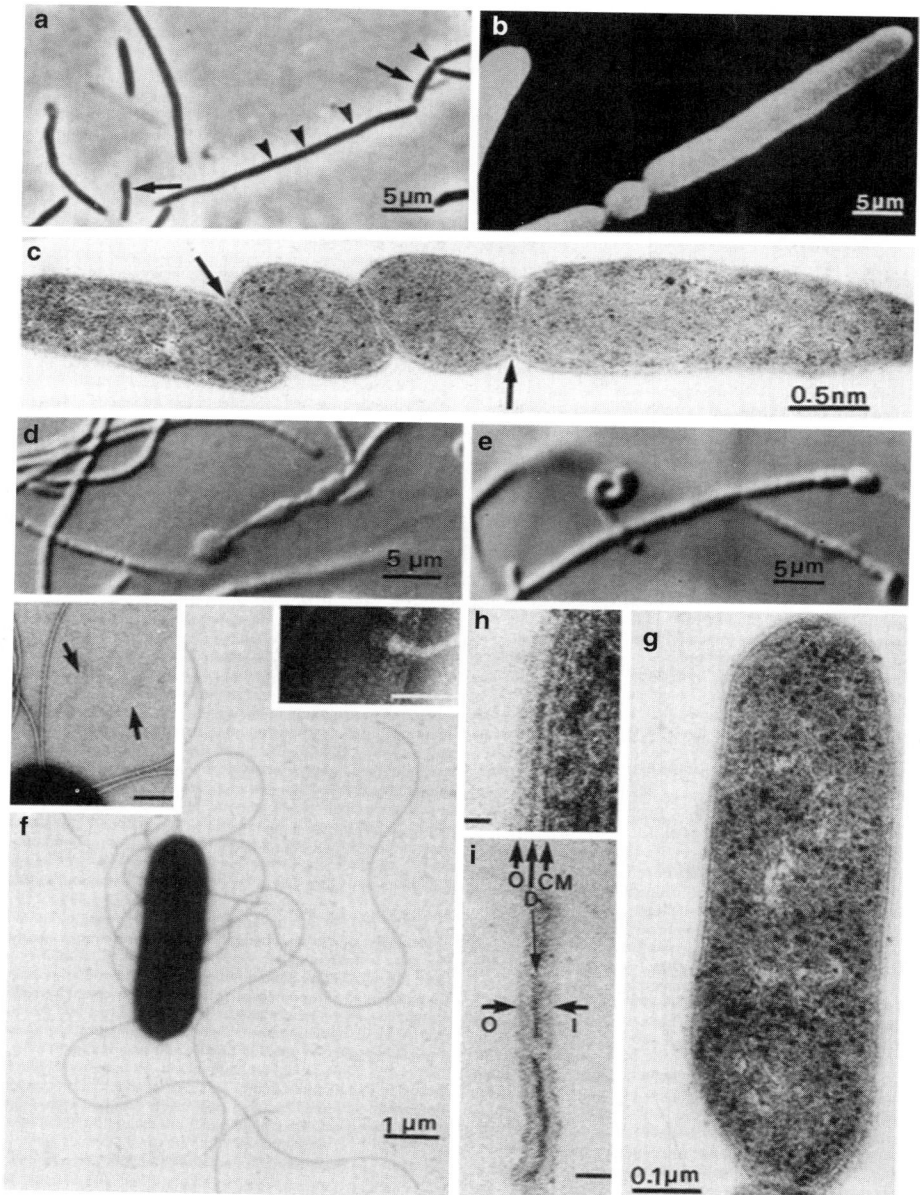

FIGURE 243. Light microscopic and electron microscopic (EM) observations of *Thermoanaerobacter ethanolicus* strain JW200. (a) Normal size (*arrow*) and longer, filamentous cells which may divide to give long chains of bacteria (*arrow heads*) (light microscopy). (b) Coccoid cell between rods (scanning EM). (c) Uneven length of dividing cells; notice incomplete crosswall formation (*arrows*) (EM). (d and e) Protuberances or sphaeroplast type cells (light microscopy, Nomarski technique).(f) Single cell with flagella; right inset: hexagonal pattern of the outer cell-wall layer and the insertion point of a flagellum; left inset: pili (*arrows*) and flagella; bars = 0.1 μm. (g) Ultrathin section showing cell envelope layers (EM). (h) Higher magnification of the cell envelope layers (*arrows*); outer cell-wall layer (*O*); dense layer (*D*); cytoplasmic membrane (*CM*); bar = 2.5 nm (EM). (i) Cell wall from a lysed cell showing three layers (*arrows*): outer cell-wall layer (*O*); dense layer (*D*), and an inner cell-wall layer (*I*); bar = 25 nm (EM). (Reproduced with permission from Wiegel and Ljungdahl. 1981. Arch. Microbiol. *128*: 343–348.)

Conversion of thiosulfate to H₂S or S⁰. One of the characteristics previously thought to be diagnostic of the genus was the reduction of thiosulfate to H$_2$S (Lee et al., 1993). However, recently characterized species were found to produce from thiosulfate either H$_2$S + S^0 (*Thermoanaerobacter italicus*; Kozianowski et al., (1997) or exclusively S^0 (*Thermoanaerobacter sulfurigignens*; Lee et al., 2007a; Table 244). Sulfur accumulates in the medium and in cells of *Thermoanaerobacter italicus* and *Thermoanaerobacter sulfurigignens* (Figure 244 and Figure 245). The latter organism

tolerates high concentrations of thiosulfate and converts up to 1 M thiosulfate to sulfur. It also grows in the presence of up to 90 mM sulfite (Lee et al., 2007b). Several other species tolerate thiosulfate concentrations around 250 mM and sulfite concentrations about 30 mM (*Thermoanaerobacter uzonii*; Wagner et al., 2008). The different effects of thiosulfate on H$_2$ oxidation and glucose and xylose metabolism in *Thermoanaerobacter brockii* subsp. *lactiethylicus* and *Thermoanaerobacter brockii* subsp. *finnii* have been studied to understand the breakdown

TABLE 244. Characteristics differentiating species of the genus *Thermoanaerobacter*[a]

Characteristic	*T. ethanolicus*	*T. acetoethylicus*	*T. brockii* subsp. *brockii*	*T. brockii* subsp. *finnii*	*T. brockii* subsp. *lactiethylicus*	*T. italicus*	*T. kivui*	*T. mathranii* subsp. *mathranii*	*T. mathranii* subsp. *alimentarius*	*T. pseudethanolicus*	*T. siderophilus*	*T. sulfurigignens*	*T. sulfurophilus*	*T. thermocopriae*	*T. thermohydrosulfuricus*	*T. wiegelii*	*T. 'keratinophilus'*
Spores	-	-	+	+	+	+	-	+	-	+	+	+	-	+	+	+	-
Motility	+	+	-	+	+	-	-	+	+	+	+	+	+	+	+	+	NR
Inhibition by H_2	-	+	ND	ND	ND	ND	ND	-	ND	ND	-	ND	ND	ND	+	ND	ND
Inhibition by 2% NaCl	ND	+	+	ND	-	ND	ND	v	ND	ND	-	ND	ND	ND	ND	ND	ND
Gram stain	v	-	+	V	+	+	-	+	+	v	+	v	ND	+	v	+	-
Temperature range (°C)	37–78	40–80	35–85	40–75	40–75	45–78	50–72	47–78	50–75	30–80	39–78	34–74	44–75	47–74	37–76	38–78	50–80
Optimum temp. (°C)	69	65	65–70	65	55–60	70	66	70–75	55–60	65	69–71	63–65	55–60	60	67–69	65–68	70
pH range	4.4–9.9	5.5–8.5	5.5–9.5	NR	5.6–8.8	NR	5.3–7.3	4.7–8.8	NR	NR	4.8–8.2	4.0–8.0	4.5–8.0	6.0–8.0	5.5–9.2	5.0–7.25	5.0–9.0
pH optimum	5.8–8.5	NR	7.5	6.5–6.8	7.3	7	6.4	6.8–7.8	NR	NR	6.3–6.5	5.8–6.5	6.8–7.2	6.5–7.3	6.9–7.5	6.8	7
DNA G+C content (mol%)	32	31	30–31	32	35	34.4	38	37	31.5	34	32	34.5	30.3	37.2	35–37	35.6	37.6
Products of fermentation:[b]																	
Acetic acid	+	+	+	+	+	+	+	+	+	+	-	+	+	+	+	+	ND
Butyric acid	-	+	-	-	-	-	-	-	-	-	-	-	-	-	-	-	ND
CO_2	+	+	+	+	+	+	-	+	+	+	+	+	+	+	+	+	ND
Ethanol	+	+	+	+	+	+	-	+	+	+	+	+	+	+	+	+	ND
H_2	+	+	+	+	+	+	-	+	+	+	+	+	+	+	+	+	ND
Isobutyric acid	-	+	-	-	-	-	-	-	-	-	-	-	-	-	-	-	ND
Lactic acid	+	-	+	+	+	+	-	+	+	+	-	+	-	-	-	+	ND
Succinate	-	-	-	-	-	+	-	-	-	-	-	-	-	-	-	-	ND
Mannose fermented	+	+	+	+	+	+	+	+	ND	ND	ND	+	+	+	-	+	+
Pectin fermented	+	-	-	ND	ND	+	-	-	+	+	ND	ND	+	-	+	-	+
Pentoses fermented	+	-	ND	+	+	+	ND	+	ND	+	+	ND	+	ND	+	+	+[c]
Xylose fermented	+	-	-	+	+	+	ND	+	ND	ND	+	+	+	ND	+	+	-
Thiosulfate reduced to:																	
H_2S	+	+	+	+	+	+	ND	+	ND	ND	-	-	+	+	+	+	ND
S^0	-	ND	-	-	-	+	ND	-	ND	ND	+	+	-	-	-	-	ND

[a]Symbols: +, >85% positive; -, 0–15% positive; w, weak reaction; v, variable; ND, not determined; NR, not reported.

[b]Fermentation of glucose.

[c]Thiosulfate significantly stimulated growth

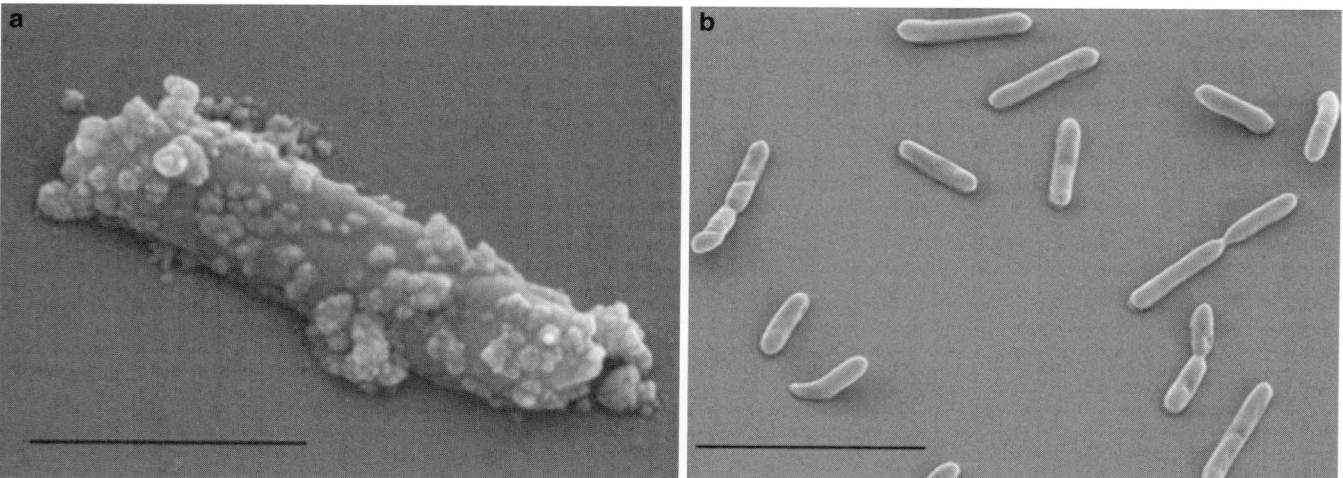

FIGURE 244. Structure of *Thermoanaerobacter sulfurigignens* strain JW/SL-NZ826[T]. (a) Scanning electron micrographs of cells grown in the presence of 50 mM thiosulfate producing sulfur globules outside the cells. Bar = 1 µm. (b) Electron micrograph of cells grown in the absence of thiosulfate. Bar = 5 µm. (Courtesy of Manfred Rohde.)

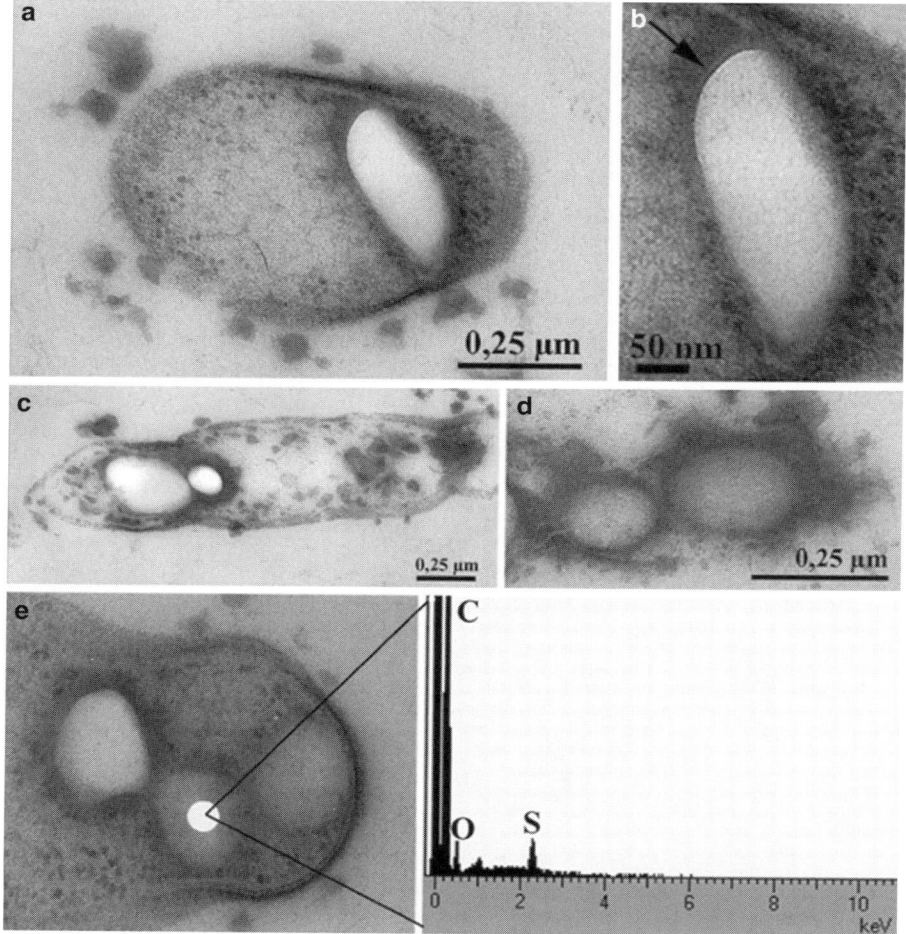

FIGURE 245. Electron micrographs of thin sections of *Thermoanaerobacter sulfurigignens* strain JW/SL-NZ826[T] showing (a) a sulfur globule inside the cell, (b) the membrane around the sulfur globule, (c) a cell containing sulfur globules in the process of lysis, (d) sulfur globules in the culture from lysed cells and (e) Energy Dispersive X-ray (EDX) analysis indicating that the globules inside the cell contain sulfur. (Reproduced with permission from Lee et al., 2007. *J. Bacteriol. 189*: 7525–7529.)

of carbohydrates found in sulfide-, elemental sulfur-, or sulfate-rich thermal hot springs and oilfields (Fardeau et al., 1996) and literature cited therein). Variations in peptide and amino acid oxidation in the presence of thiosulfate by members of the genus have also been reported by Faudon et al. (1995).

Temperature and pH ranges. Growth temperature optima vary from 55–75 °C, with growth ranges of 35–78 °C. Many species also exhibit a wide temperature span for growth of over 35 °C (Wiegel, 1990, 1998b). The temperature versus growth rate plots for these species exhibit biphasic curves with two maxima (the first more or less pronounced) and broken Arrhenius plots, indicating changes in the rate-limiting steps. Wiegel (1990, 1998b) suggested an evolutionary relevance and that these bacteria contain, for some critical metabolic steps, two different enzymes: one for growth at the lower temperature and a second one for growth at the higher temperature. Several species also exhibit a wide pH optimum of about 3 pH units without a specific peak, e.g., *Thermoanaerobacter ethanolicus* JW200. However, for comparison of pH ranges, one must take into consideration that different authors measured the pH differently. It is frequently not known whether or not they corrected for the temperature effect on the pK_a values of various medium components and correspondingly calibrated the pH meter at the elevated growth temperature (i.e., calibration and analysis temperature indicated as superscript; Wiegel, 1998a). *Thermoanaerobacter wiegelii* was used to study the effect of the medium pH on ATP synthesis, the proton motive force, and membrane potential (Cook, 2000).

Ethanol fermentation. *Thermoanaerobacter ethanolicus* was the first wild-type bacterium to receive protection by a U.S. patent. This patent was based upon the nearly quantitative conversion of hexose to ethanol, a property common to both *Thermoanaerobacter ethanolicus* and *Thermoanaerobacter pseudethanolicus* (Larsen et al., 1997; Lovitt et al., 1984; Wiegel et al., 1983; Wiegel and Ljungdahl, 1981). The generalized fermentation balance for these organisms is (in moles): 1.0 glucose + 0.1 H_2O → 1.8 ethanol + 0.1 acetic acid + 0.1 lactic acid + 1.9 CO_2 + 0.2 H_2. Variations in end products are observed. The amount of ethanol formed depends upon the pH. Higher ethanol yields are obtained when the culture is started above pH^{25C} 7.2, and then the pH is allowed to decrease to 6.8 or below during the fermentation. Cultures kept at a constant pH^{25C} produce only 0.8–1.1 mol of ethanol/mol of glucose consumed with a corresponding increase in formation of acetate and lactate. A decrease in the ratio of yeast extract to fermented carbohydrate also leads to a decrease in ethanol production and an increase in acetate and lactate production (Hild et al., 2003). The ethanol yield also depends on the growth conditions and the substrate combinations, such as starch and glucose (Parkkinen, 1986) or glucose and xylose (Carreira et al., 1983; Wiegel et al., 1983; Patel et al., 1988; Lacis and Lawford, 1992, and literature cited therein; Cook and Morgan, 1994; Heitmann et al., 1996). In species such as Thermoanaerobacter thermohydrosulfuricus and Thermoanaerobacter ethanolicus, yeast extract is not only required for growth but also for ethanol production by resting cells (Wiegel et al., 1979).

Several *Thermoanaerobacter* species are suitable for ethanol production from pretreated hemicellulosic hydrolysates of wheat straw (Ahring et al., 1996), steam exploded wood extracts (Wiegel et al., 1983, 1986) and other hemicellulosic waste material

(Sommer et al., 2004). Although the strains produce lower concentrations, a relatively high tolerance to ethanol of 8% v/v and 5% have been obtained for *Thermoanaerobacter ethanolicus* JW 200 and *Thermoanaerobacter mathranii* subsp. *mathranii*, respectively, by stepwise adaption (Wiegel et al., unpublished results; Ahring et al., 1999).

Co-cultures of *Thermoanaerobacter ethanolicus* and *Thermoanaerobacter pseudethanolicus* with cellulolytic bacteria such as *Clostridium thermocellum* have been used to obtain saccharification of cellulosic material and conversion to ethanol in a single vessel (Ljungdahl and Wiegel, 1981; Ng et al., 1981). *Thermoanaerobacter* strain YM3 requires yeast extract for growth unless it is grown in co-culture with *Clostridium thermocellum* strain YM4 or the cell-free broth of YM4 (Mori, 1995).

Instead of ethanol, L-lactate was the main fermentation product in *Thermoanaerobacter* mutants defective in either acetate kinase or phosphotransacetylase (Mayer et al., 1995; Ljungdahl et al., unpublished results). Several groups are working presently to develop bioengineered strains lacking the enzymic activities leading to the formation of the minor products acetate and lactate. Under various conditions and after extended culturing, these side products (especially acetate) can become major products. The formation of acetate also yields one more ATP per acetate formed through substrate level phosporylation. These studies are aided by the genome sequences for strains of *Thermoanaerobacter pseudethanolicus* and for *Thermoanaerobacter ethanolicus* JW200, which have recently become available (see below).

Enzymes. Because of their thermophily (Vieille and Zeikus, 2001; Zamost et al., 1991), *Thermoanaerobacter* species have been exploited as a source of a wide variety of thermostable enzymes for carbohydrate breakdown and alcohol production (Alister et al., 1990; Canganella and Wiegel, 1993; Fardeau et al., 1996; Kuriki and Imanaka, 1999; Svensson, 1994; Zeikus et al., 1991).

Carbohydrate-hydrolyzing enzymes. Enzymes involved in the transformation of starchy and hemicellulosic biomass that have been studied include saccharolytic, cellulolytic, hemicellulolytic, and other sugar-metabolizing enzymes.

Some of the saccharolytic enzymes studied are the following. The thermostable amylolytic and pullulolytic enzymes have been studied from *Thermoanaerobacter italicus* (Ahring et al., 1996, 1999), *Thermoanaerobacter brockii* subsp. *finnii* (Antranikian, 1989), and *Thermoanaerobacter thermohydrosulfuricus* (Antranikian et al., 1987; Melasniemi, 1987). The recombinant *Thermoanaerobacter thermohydrosulfuricus* enzyme has also been heterologously expressed in *Escherichia coli* (Melasniemi and Paloheimo, 1989). Some of these enzymes possess temperature optima of 80–85 °C. The biochemistry of the active site and the regulation of expression have been studied for glucoamylase and pullulanase from *Thermoanaerobacter pseudethanolicus* (Hyun and Zeikus, 1985; Lin and Leu, 2002; Mathupala et al., 1993, 1994). The α-glucosidase from *Thermoanaerobacter thermohydrosulfuricus* exhibits maltase, glucohydrolase, and "maltodextrinohydrolase" activity (Wimmer et al., 1997). The β-1,4-glucosidase from an unidentified species was characterized (Mitchell et al., 1982). Two different β-glucosidases were studied from *Thermoanaerobacter brockii* subsp. *brockii* (Breves et al., 1997). A recombinant cyclodextrin glycosyltransferase was studied from an unidentified species (Jorgensen et al., 1997). The catalytic center of a cyclomaltodextrinase was studied from *Thermoanaerobacter pseudethanolicus* (Podkovyrov et al., 1993). A novel thermostable dextranase was studied from a *Thermoanaerobacter* species (Wynter et al., 1997).

Some biosynthetic capabilities of *Thermoanaerobacter* enzymes have been studied including the production of cyclodextrin from raw corn starch by cyclodextrin glycosyltransferase (Kim et al., 1997), the synthesis of glycosyl glycerol by cyclodextrin glucanotransferases (Nakano et al., 2003), and the synthesis of oligosaccharides by immobilized cyclodextrin glycosyltransferases (Martin et al., 2001, 2004).

The cellulolytic and hemicellulolytic enzymes that have been studied include an extracellular cellulase from *Thermoanaerobacter thermocopriae* (Chaen et al., 2001), the thermostable xylanolytic enzymes from *Thermoanaerobacter italicus* (Antranikian, 1989), a xylulose kinase from *Thermoanaerobacter pseudethanolicus* 39E (Erbeznik et al., 1998a; Jones et al., 2002b), and a β-xylosidase and a bifunctional xylosidase-arabinosidase (*xarB*) from *Thermoanaerobacter ethanolicus* JW200 (Mai et al., 2000; Shao and Wiegel, 1992). The genes for a xylose ABC transport operon (Erbeznik et al., 2004), xylose-binding protein XylF, and the maltose ABC transport system (Jones et al., 2002a) have been characterized in *Thermoanaerobacter pseudethanolicus* 39E (Erbeznik et al., 1998b).

Other sugar metabolizing enzymes that have been studied include a β-galactosidase from an unidentified species (Lind et al., 1989) and *Thermoanaerobacter pseudethanolicus* (Fokina and Velikodvorskaia, 1997), xylose(glucose) isomerases from *Thermoanaerobacter pseudethanolicus* 39E (Erbeznik et al., 1998a, 1998b) and *Thermoanaerobacter brockii* subsp. *brockii* (Okada et al., 2003), the arabinose isomerase from *Thermoanaerobacter mathrani* (Jorgensen et al., 2004), and two thermoactive pectate lyases from *Thermoanaerobacter italicus* (Kozianowski et al., 1997). The kojibiose phosphorylase of *Thermoanaerobacter brockii* subsp. *brockii* has been used for the synthesis of oligosaccharides (Okada et al., 2003). The kojibiose phosphorylase activities from *Thermoanaerobacter brockii* subsp. *brockii* were studied further (Maruta et al., 2006; Yamamoto et al., 2006, and literature cited therein), and chimeras of the kojibiose and trehalose phosphorylase were constructed (Yamamoto et al., 2006). The conversion of D-galactose into D-tagatose has been studied in *Thermoanaerobacter mathranii* subsp. *mathranii* (Jorgensen et al., 2004).

Alcohol dehydrogenases and other enzymes. The ethanol formation in *Thermoanerobacter* species from hexoses includes two reversible redox reactions: acetyl-CoA to acetaldehyde and acetaldehyde to ethanol. Several studies focused on the acetaldehyde dehydrogenase (ALDH, EC 1.2.1.10) and alcohol dehydrogenases (ADH). Two alcohol dehydrogenases, a primary alcohol dehydrogenase (EC 1.1.1.1) and a secondary alcohol dehydrogenase (EC 1.1.1.2.) were purified and characterized from several of the species. However, their functions in the alcohol formation were not unequivocally elucidated. Because of the early formation of the secondary alcohol dehydrogenase during growth, it was suggested that it is mainly involved in alcohol formation. Bryant et al. (1988) showed that this enzyme from *Thermoanaerobacter ethanolicus* JW200 also had a high affinity for acetaldehyde, and Burdette and Zeikus (1994) demonstrated that the enzyme from *Thermoanaerobacter pseudethanolicus* 39E has reductive thioesterase activity suggesting that it is a bifunctional enzyme that forms ethanol from acetyl-CoA and acetaldehyde released by an aldehyde dehydrogenase. The unequivocally role and interactions of these three enzymes

in vivo needs to be studied further using molecular approaches. The three terms: primary and secondary alcohol dehydrogenase and aldehyde dehydrogenase are not always used consistently in the literature.

The primary alcohol dehydrogenase (encoded by *adh*A) is a zinc-containing, NADP(H)-dependent, tetrameric enzyme and has been purified from *Thermoanaerobacter ethanolicus* JW200 and its mutants (Bryant et al., 1988, 1992) and from *Thermoanaerobacter pseudethanolicus* (Burdette and Zeikus, 1994). The enzyme from JW200 was heterologously expressed in *Escherichia coli* (Holt et al., 2000). It oxidizes primary alcohols up to heptanol, but it is inactive with secondary alcohols.

The secondary alcohol dehydrogenase, encoded by *adh*B, has been studied from *Thermoanaerobacter ethanolicus* JW200 (Pham and Phillips, 1990), *Thermoanaerobacter brockii*, (Korkhin et al., 1998) and *Thermoanaerobacter pseudethanolicus* 39E. These studies of the enzymes from *Thermoanaerobacter ethanolicus* JW200 and *Thermoanaerobacter pseudethanolicus* 39E include various biophysical analyses and comparisons of the affects of site specific mutations (Burdette et al., 1997; Heiss et al., 2001; Tripp et al., 1998). Arni et al. (1996) determined the crystal structure. Phillips (2002b, and literature cited therein) used amino acid substitutions in the active site to change the substrate specificity. Burdette et al. (2000) explored the physiological role of the enzyme in the alcohol tolerance of the bacterium. The enzyme can be encapsulated for reactions in organic solvents and also used for the asymmetric reduction of aldehydes and ketones (Musa et al., 2007a, 2007b; Phillips, 1996). A modification of the enzyme has also been developed for the reduction of the solvents phenylacetone and benzylacetone (Ziegelmann-Fjeld et al., 2007). Interestingly, the stereochemistries of the primary and secondary alcohol dehydrogenases from *Thermoanaerobacter ethanolicus* JW200 and *Thermoanaerobacter pseudethanolicus* 39E are temperature dependent and switch around 35 °C which is close to the minimal growth temperature for both species (for reviews, see Phillips, 1996; Yang et al., 1997). The stereospecificity is also dependent on the pH, the active site water content (Phillips, 2002a) and the choice of substrates and cofactor analogs (Zheng and Phillips, 1992). The secondary alcohol dehydrogenase from *Thermoanaerobacter brockii* subsp *brockii*.is a NADPH-dependent enzyme, contains zinc, and possesses a tetrameric quaternary structure. It was isolated and fully characterized (including amino acid sequence and crystal structure) by Peretz and Burstein (1989) and Li et al. (1999) and isolated by affinity chromatography (McMahon and Mulcahy, 2002). The obtained information allowed the characterization of the zinc-containing active site (Bogin et al., 1997), the effect of substitution of cobalt which leads to a lower energy barrier for catalysis (Kleifeld et al., 2004), modification of specific amino acids such as cysteine and proline (Peretz et al., 1997a, 1997b), and a detailed correlation of the structure and activity with temperature stability (Korkhin et al., 1999). This information enabled the contruction of chimeras (Korkhin et al., 1998) and mutations of critical amino acids (Bogin et al., 1998a, 1998b) in the alcohol dehydrogenase from the mesophilic *Clostridium beijerinkii* to enhance its temperature stability. Miroliaei and Nemat-Gorgani (2002) and Olofsson et al. (2005) studied the reactivity in water-permissible solvents. Furthermore, the alcohol dehydrogenase of *Thermoanaerobacter brockii* subsp. *brockii* has been included in studies of disulfide modification and antioxidant properties of *S*-allylmercaptoglutathione (Rabinkov et al., 2000).

The acetaldehyde dehydrogenase has been studied in *Thermoanaerobacter pseudethanolicus* 39E (Burdette and Zeikus, 1994). The lactate dehydrogenase has been studied from *Thermoanaerobacter pseudethanolicus* (Germain et al., 1986) and *Thermoanaerobacter thermohydrosulfuricus* (Turunen et al., 1987). A number of other types of enzymes have been examined. The intracellular protease and extracellular keratinolytic enzymes from *Thermoanaerobacter keratinophilus* 2KXI effectively degrade feathers (Riessen and Antranikian, 2001). The type and only strain of *Thermoanaerobacter keratinophilus* is a patent strain, similar to the keratinolytic *Thermoanaerobacter* strain S290 (French Patent Application FR-04 50513; Tsiroulnikov et al., 2004) and therefore could not be validly published. Thus *Thermoanaerobacter keratinophilus* is described below in the section *Other named isolates*. Other purified and characterized enzymes include the trehalose phosphorylase (Maruta et al., 2006) and the chaperonin 60 and 10 from *Thermoanaerobacter brockii* subsp. *brockii* (Todd et al., 1995; Truscott et al., 1994) and GroE from strain Rt6.G4 isolated from Rotorrua, New Zealand (Truscott and Scopes, 1998). The thermostable DNA polymerase (Sha et al., 1997) and its regulation by protein phosphorylation (Londesborough, 1986) have been studied in *Thermoanaerobacter thermohydrosulfuricus*. The kinetics of the temperature dependence of the 3-phosphoglycerate kinase from a *Thermoanaerobacter* species has been compared to that of the mesophilic enzyme (Thomas and Scopes, 1998). The histidine biosynthetic pathway of *Thermoanaerobacter pseudethanolicus* has also been studied (Erbeznik et al., 2000),

Industrial applications. The hexagonal S-layer lattice of *Thermoanaerobacter thermohydrosulfuricus* L111-69 has been tested as a surface for binding of various enzymes (e.g., Kupcu et al., 1995) and several immunological reactions such as immunoassays requiring immobilization of human IgG (Kupcu et al., 1996; Weber et al., 2001). Other possible uses include immobilization of cell-wall fragments (Sleytr et al., 1999, and literature cited therein), macromolecules such as ferritin and invertase (Sára and Sleytr, 1989), and recombinant major birch pollen allergen Bet v 1 (Jahn-Schmid et al., 1996).

Thermoanaerobacter species have also been screened for their potential in solid substrate cultivation (an alternative culture method using wetted solid substrates) for biomass-supported bioremediation and animal feed enrichments (Chinn et al., 2006).

Thermoanaerobacter mathranii subsp. *mathranii* has been utilized in an upflow anaerobic sludge blanket (UASB) purification reactor for the detoxification of water derived from an industrial ethanol fermentation (Rabinkov et al., 2000; Torry-Smith et al., 2003).

Biotechnological processes have been developed that rely upon *Thermoanaerobacter brockii* subsp. *brockii* to reduce β-oxoesters (Seebach et al., 1984) or *Thermoanaerobacter kivui* to produce Ca-Mg-acetate as an alternate road deicer (Kevbrina et al., 1996; Ljungdahl et al., 1985; Wiegel et al., 1991). *Thermoanaerobacter brockii* subsp. *finnii* has been used to develop a bioluminescence assay for pyridine nucleotide that detects levels as low as 1 pmol in cell extracts (Schmid et al., 1989). The alcohol dehydrogenases from *Thermoanaerobacter ethanolicus* and *Thermoanaerobacter pseudethanolicus* have been used for asymmetric reduction of aliphatic and cyclic ketones (Zheng et al., 1994).

Molecular and genetic studies. Genus-specific oligonucleotide probes for the detection of *Thermoanaerobacter* species are available (Subbotina et al., 2003).

Peteranderl et al. (1993) developed a system for regeneration of wall-less autoplasts of *Thermoanaerobacter, Thermoanaerobacterium*, and related strains to cell-wall-containing rods. This methodology led to the development of efficient electroporation and genetic systems for these anaerobic thermophiles (Mai et al., 2000). These methods were later extended to *Clostridium thermocellum* (Tyurin et al., 2006, and literature cited therein). The first, successful gene transfer to *Thermoanaerobacter ethanolicus* JW200[T] has recently been documented by Peng et al. (2006).

The genome sequences of *Thermoanaerobacter ethanolicus* JW200 and *Thermoanaerobacter pseudethanolicus* strains 39E and X514 have been completed by the Joint Genome Institute.

Enrichment and isolation procedures

Many *Thermoanaerobacter* species have been isolated from a great variety of geothermally heated sources, including slightly alkaline and slightly acidic hot springs (Wiegel and Ljungdahl, 1981), the Australian Great Artesian Basin (Wynter et al., 1996), and high temperature petroleum reservoirs (Grassia et al., 1996; Slobodkin et al., 1999b). They can also be isolated from many mesobiotic environments (Wiegel et al., 1979). For example, the sporulating species *Thermoanaerobacter thermohydrosulfuricus* and *Thermoanaerobacter thermosaccharolyticus* (basonym *Clostridium thermosaccharolyticum*) can be isolated nearly from all types of soil and sediments including permanently frozen soils and snow of Antarctica. Culture-independent methods, such as cloning and sequencing of 16S rRNA genes, have identified *Thermoanaerobacter* as the dominant microorganism in many environments (Leu et al., 1998; Orphan et al., 2000). Szewzyk et al. (1994) have isolated glucose- and starch-degrading strains related to the *Thermoanaerobacter* from a deep borehole in granite. Slobodkin et al. (1999a, 1999b), Zhou et al. (2001) and Roh et al. (2002) have reported on the isolation and characterization of metal-reducing *Thermoanaerobacter pseudethanolicus*-like strains from deep subsurface environments (Jizhong Zhou, personal communication). On the other hand, the type species (type strain JW200) was isolated only once from a slightly alkaline hot spring from Yellowstone National Park (Wiegel and Ljungdahl, 1981), and all other attempts to reisolate this species have yielded only strains of *Thermoanaerobacter pseudethanolicus* (Onyenwoke et al., 2007) or different, but novel, species. This includes strain JW 201 which was originally described prior to 16S rRNA sequencing as a second strain of *Thermoanaerobacter ethanolicus* (Wagner and Wiegel, unpublished results).

Media for enrichments and isolations are typically mineral-based, phosphate-and/or carbonate-buffered, supplemented with 0.05–0.2% yeast extract, and incubated at 60–70 °C. Species have also been enriched from petroleum reservoirs by the direct supplementation of production waters with glucose and either yeast extract, peptone, tryptone, or Casamino acids (Grassia et al., 1996; Slobodkin et al., 1999a), or xylose plus thiosulfate (*Thermoanaerobacter sulfurigignens*; Lee et al., 2007a), or on crotonate (*Thermoanaerobacter uzonensis*; Wagner and Wiegel unpublished results). After enrichment in broth cultures, axenic cultures can be obtained by plating in anaerobic jars or agar-shake-roll tubes prepared with 1.5–2.2% (w/v) agar. Specifics on anaerobic techniques are further described in Ljungdahl and Wiegel (1986).

Maintenance procedures

Freeze-dried cultures have been stored for many years. In addition, some species, i.e., *Thermoanaerobacter ethanolicus* JW200 and *Thermoanaerobacter thermohydrosulfuricus* JW102, have been stored in pre-reduced medium containing 40–45% (v/v) glycerol as cryoprotectant at –20 °C for more than 28 years. This method is convenient for frequent sampling because the mixture remains liquid and small portions can be removed with a syringe without changing the temperature of the stock. For longer storage, the culture is kept at –80 °C. Working cultures can be stored at room temperature and transferred at regular intervals of about 2 weeks, and several species need only be transferred every 3–6 months (e.g., *Thermoanaerobacter ethanolicus*). Sporulating strains can be maintained on agar slants after growth on low concentrations, i.e., less then 0.15% (w/v) of slowly utilizable substrates. Alternatively, the spores may be harvested by centrifugation, resuspended in phosphate buffer or medium, and stored at either 4 °C or room temperature. Spores of *Thermoanaerobacter thermohydrosulfuricus* JW 102 retain viability under aerobic conditions at 4–25 °C for more than 30 years (Wiegel, unpublished results).

Taxonomic comments

The genus *Thermoanaerobacter* is a member of the *Thermoanaerobacterales* and the family *Thermoanaerobacteraceae* (Figure 246). However, in contrast to all previous analyses, the detailed phylogenetic 16S rRNA tree for the *Firmicutes* suggests that this group (which is not monophylogenetic) may constitute its own class (Ludwig et al., 2009). Thus, further phylogenetic analyses are warranted.

The validly published species *Thermoanaerobacter subterraneus* (Fardeau et al., 2000), *Thermoanaerobacter tengcongensis* (Xue et al., 2001), and *Thermoanaerobacter yonseiensis* (Kim et al., 2001) have been reassigned to the genus *Caldanaerobacter* as *Caldanaerobacter subterraneus*, *Caldanaerobacter subterraneus* subsp. *tengcongensis*, and *Caldanaerobacter subterraneus* subsp. *yonseiensis*, respectively (Fardeau et al., 2004). These reassignments were based on phylogenetic and metabolic differences between these strains and *Thermoanaerobacter* species. For instance, these microorganisms produce L-alanine as a major fermentation product from glucose and grow above 80 °C, while *Thermoanaerobacter* species do not.

Further reading

Mai, V. and J. Wiegel. 1999. Recombinant DNA applications in thermophiles. *In* Demain and Davis (Editors) and Hershberger (Section Editor), ASM Manual of Industrial Microbiology and Biotechnology, 2nd edition. ASM Press Washington, DC, pp. 511–519.

Radianingtyas, H. and P.C. Wright. 2003. Alcohol dehydrogenases from thermophilic and hyperthermophilic archaea and bacteria. FEMS Microbiol. Rev. *27*: 593–616.

Scopes, R.K. and K. Truscott. 1998. Chaperonins from *Thermoanaerobacter* species. Methods Enzymol. *290*: 161–169.

List of species of the genus *Thermoanaerobacter*

1. **Thermoanaerobacter ethanolicus** Wiegel and Ljungdahl 1982, 384^{VP} (Effective publication: Wiegel and Ljungdahl 1981, 348.)

   e.tha.no′li.cus. N.L. masc. n. *ethanol* corresponding alcohol of ethane (ethane + ol); N.L. masc. adj. *ethanolicus* indicating the production of ethanol.

   Description is based on the type strain JW200. To date, the species is represented by only one strain, and this descrip-

tion reflects the properties of that strain. Cells in the early exponential growth phase are rods, 0.3–0.8 × 4–8 μm, sometimes with pointed ends. In older cultures, cells can be long and filamentous, up to 100 μm in length without detectable septation, or form chains of rod-shaped cells frequently interspersed with coccoid cells, 0.8–1.5 μm in diameter. In addition, cells with protuberances are frequently observed in older cultures (Figure 243). Cells have 1–12 peritrichous

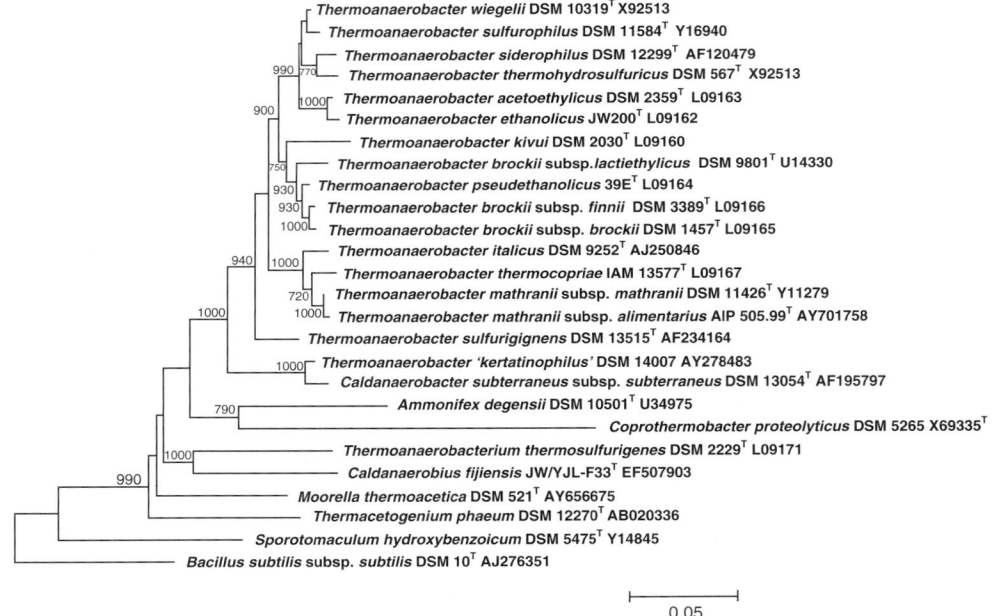

FIGURE 246. Phylogenetic tree for the genus *Thermoanaerobacter*. Neighbor-joining tree showing the estimated phylogenetic relationships for the type strains of each species within the genus *Thermoanaerobacter* based on 16S rRNA gene sequence data with Jukes–Cantor correction for synonymous changes. Numbers at nodes indicate bootstrap values for 1000 replicates. Bar = 0.05 nucleotide substitutions per site. The superscript "T" denotes the sequence from the type strain for the species. GenBank accession numbers are indicated after the strain designations.

flagella that are up to 50 μm in length and 4–8 pili. Spores are not observed, although sporulation specific genes are present. Cells are Gram-stain-variable. However, the cell-wall structure is Gram-positive, as observed by electron microscopy, and contains *meso*-diaminopimelic acid (Aγl-cell-wall type). Strain JW200 is able to form stable, wall-free cells following induction of autolysis. These cells can regenerate back to a walled form (Peteranderl et al., 1993). Colonies that form in deep agar (2% w/v) at 60 °C are 0.5–1.0 mm in diameter, lenticular and white. After 40 h at 60 °C, colonies become 2–4 mm in diameter, white, smooth, round to irregular, and flat. In older cultures, the colonies turn brown, but no pigment has ever been identified.

Cells are obligate anaerobes that grow from pH 4.4–9.9 with a broad pH_{opt} of 5.8–8.5. These values possess a small systematic error because, although the pH was measured at 70 °C with a pH meter equipped with a temperature probe, the pH electrode was calibrated at room temperature. (For a more thorough discussion on the accurate determination of pH at elevated temperatures, see Wiegel, 1998a). The growth temperature range is from 37–78 °C (optimum at 69 °C). Fermented carbohydrates include glucose, fructose, galactose, mannose, ribose, xylose, cellobiose, lactose, maltose, sucrose, starch, and pyruvate but not cellulose, raffinose, rhamnose, fucose, *m*-erythritol, *m*-inositol, xylitol, glycerol, mannitol, sorbitol, trehalose, melezitose, melibiose, niacinamide, or amygdalin (Lacis and Lawford, 1985; Wiegel and Ljungdahl, 1981). Esculin and gelatin are hydrolyzed, but cells are negative for catalase, indole production, and lipase and cellulase activity. The major end products of glucose fermentation are ethanol and CO_2. The generalized fermentation balance (in moles) is: 1.0 glucose + 0.1 H_2O → 1.8 ethanol + 0.1 acetic acid + 0.1 lactic acid + 1.9 CO_2 + 0.2 H_2. Methanol is a major metabolic end product of pectin fermentation (Schink and Zeikus, 1980).

Cells in the late (but not early) exponential growth phase lyse on exposure to 100 μg/ml of polymyxin B for 30 min at 37 °C or 50 °C (pH 7.2). Cells are inhibited by chloramphenicol but are resistant to erythromycin, tetracycline, and penicillin G. H_2 is inhibitory at 10% (v/v), and growth does not occur at 75% or higher concentrations. Lactic acid, acetic acid, and ethanol are inhibitory at 65 °C above 100 mM, 200 mM, and 500 mM, respectively. However, the type strain can easily be adapted to grow in the presence of 1.7 M ethanol (Wiegel, unpublished results). Strain JW200 was isolated from water and mud samples with a pH of about 8.8 and temperature of 45–50 °C that were collected from the slightly alkaline hot spring in the Octopus pool area of White Creek at Fire Hole Lake Drive in the Yellowstone National Park, Wyoming, USA.

In contrast to other species, strain JW200 does not contain menaquinone MK-7 (Yamamoto et al., 1998).

Taxonomic note: No other strains of *Thermoanaerobacter ethanolicus* are currently available. Strain 39E (Lee et al., 1993) has been reassigned to *Thermoanaerobacter pseudethanolicus* (Onyenwoke et al., 2007). Recent analysis of the 16S rRNA gene of strain JW201, which was isolated from Yellowstone National Park at the same time as JW200 (Wiegel and Ljungdahl, 1981), suggests that it is not closely related despite the similarity in physiology and that it may represent a novel species (Wiegel et al., unpublished results).

DNA G+C content (mol%): 32 (T_m); 39 (Bd); 38 (chemical).

Type strain: JW200, ATCC 31550, DSM 2246.
GenBank accession number (16S rRNA gene): L09162.

2. **Thermoanaerobacter acetoethylicus** (Ben-Bassat and Zeikus 1981) Rainey and Stackebrandt 1993, 857^{VP} (*Thermobacteroides acetoethylicus* Ben-Bassat and Zeikus 1981, 367)

a.ce.to.e.thy.licus. L. n. *acetum* vinegar. N.L. adj. *ethylicus* of ethyl alcohol; N.L. masc. n. *acetoethylicus* producing acetate and ethanol.

This description is based on the type strain HTB2 and isolate HTD1 (Zeikus et al., 1980; Zeikus et al., 1979). Cells are motile rods, 0.6 × 1.5–2.5 μm, that occur singly or in pairs. Flagella are arranged peritrichously. Spores have never been observed. The cells have a multi-layered, Gram-positive cell-wall structure; even though it stains Gram-stain-negative, it lacks an outer membrane. Colonies are uniformly round, flat, mucoid, and nonpigmented, with a diameter of 3 mm after 48 h. Cytochrome pigment and catalase are both absent.

The optimum temperature for growth is 65 °C. The maximum and minimum growth temperatures are <80 °C and >40 °C, respectively. The pH range for growth is 5.5–8.5 at 65 °C. Cells are obligate anaerobes that use the glycolytic Embden–Meyerhof–Parnas pathway. The major end products of glucose fermentation are equal amounts of acetic acid and ethanol, lesser amounts of isobutyric and butyric acid, but no lactic acid. Fermentation analysis as reported (unbalanced, given in μmol): 130 glucose → 102 ethanol + 126 acetic acid + 157 CO_2 + 165 H_2 + biomass. The lower yield for ethanol is one of the reasons to distinguish *Thermoanaerobacter acetoethylicus* from *Thermoanaerobacter ethanolicus* despite the high (> 99%) sequence similarity of their 16S rRNAs. DNA–DNA hybridization studies are needed for further clarification.

Other carbohydrates fermented by *Thermoanaerobacter acetoethylicus* include glucose, mannose, cellobiose, maltose, sucrose, lactose, and starch. Fermentation products include ethanol, acetate, lactate, and CO_2. Exogenous H_2 does not inhibit growth, but the addition of 0.4–1.0 atm H_2 to a cellobiose fermentation increases the ratio of ethanol to acetate produced (Lamed and Zeikus, 1980).

Sulfate is not reduced. Growth is inhibited by air; NaCl (2% w/v); and tetracycline, streptomycin, penicillin G, vancomycin, and neomycin (all at 10 μg/ml). Strain HTB2 was isolated from an algal-bacterial mat at Octopus Spring in Yellowstone National Park, Wyoming, USA on trypticase peptone-yeast extract glucose (TYEG) medium (pH 7.2).

DNA G+C content (mol%): 31 (T_m).
Type strain: HTB2, HTB2/W, ATCC 33265, DSM 2359.
GenBank accession number (16S rRNA gene): L09163, X69336.

3. **Thermoanaerobacter brockii** (Zeikus, Hegge and Anderson 1979) Lee, Jain, Lee, Lowe and Zeikus 1993, 49^{VP} emend. Cayol, Ollivier, Patel, Ravot, Magot, Ageron, Grimont and Garcia 1995, 787 (*Thermoanaerobium brockii* Zeikus, Hegge and Anderson 1979, 47)

brock'i.i. N.L. gen. n. *brockii* of Brock, named for Thomas Dale Brock who pioneered studies on the physiology and ecology of thermophiles.

The species is composed of three subspecies, *Thermoanaerobacter brockii* subsp. *brockii*, *Thermoanaerobacter brockii* subsp.

finnii, and *Thermoanaerobacter brockii* subsp. *lactiethylicus*. The description of the species is the same as for the subspecies *Thermoanaerobacter brockii* subsp. *brockii*.

3a. **Thermoanaerobacter brockii subsp. brockii** (Zeikus, Hegge and Anderson 1979) Lee, Jain, Lee, Lowe and Zeikus 1993, 49[VP] emend. Cayol, Ollivier, Patel, Ravot, Magot, Ageron, Grimont and Garcia 1995, 787 (*Thermoanaerobium brockii* Zeikus, Hegge and Anderson 1979, 47)

Cells are Gram-stain-positive short rods, 1.0×2–$20 \mu m$ and exhibit a Gram-positive cell-wall structure. Cell length frequently varies, and minicells are common. Cells occur in pairs, chains, and filaments. Round, heat-resistant terminal spores are observed, but only rarely (Cook et al., 1991). Cytochromes and catalase are absent. Colonies on agar plates are circular, 0.2–0.3 mm in diameter, flat, mucoid, and nonpigmented.

Cells grow chemoorganotrophically with yeast extract, and a fermentable carbohydrate is required for growth. The optimum temperature for growth is 65–70 °C, with a temperature range of >35 °C–77 °C (J. Wiegel unpublished results). The pH[25C] range is 5.5–9.5, with an optimum of 7.5. (J. Wiegel, unpublished results). The doubling time at optimal conditions is about 1 h. Cells are obligate anaerobes and ferment a wide variety of substrates, including glucose, maltose, sucrose, lactose, cellobiose, starch, and pyruvate but not xylose, cellulose, arabinose, mannose, lactate, tartrate, ethanol, tryptone, Casamino acids, or pectin. Serine, leucine, isoleucine, and valine are utilized (Fardeau et al., 1997). Oxygen, sulfate, nitrate, and fumarate are not reduced. Protein is not hydrolyzed. Cell growth is inhibited by penicillin, cycloserine, streptomycin, tetracycline, and chloramphenicol (all at $100 \mu g/ml$); exposure to air (21% O_2); and 2% NaCl.

Strain HTD4 was isolated from Washburn pool B spring sediment at Yellowstone National Park, Wyoming, USA using TYEG medium (pH 7.2–7.4). In addition to the type strain, other strains include HTA1, HTD4, HTD6 and HTR1 = DSM 2599, ATCC 35047 (Zeikus et al., 1979).

DNA G+C content (mol%): 30–31 (T_m).

Type strain: HTD4, ATCC 33075, ATCC 53556, DSM 1457.

GenBank accession number (16S rRNA gene): L09165.

3b. **Thermoanaerobacter brockii subsp. finnii** Schmid, Giesel, Schoberth and Sahm 1986) Cayol, Ollivier, Patel, Ravot, Magot, Ageron, Grimont and Garcia 1995, 788[VP] (*Thermoanaerobacter finnii* Schmid, Giesel, Schoberth and Sahm 1986, 84)

fin'ni.i. N.L. gen. n. *finnii* of Finn, named for Robert K. Finn who made important contributions to the development of the ethanol vacuum fermentation process.

The description is based on the type and only available strain. Short rods, 0.4–0.6×1–$4 \mu m$, occurring singly, in pairs, and short chains. Occasionally forms coccoid cells. Motile. Cells have Gram-positive cell-wall structure, but are Gram-stain-variable. Forms heat-resistant terminal spores. Colonies on agar plates are circular, 1–3 mm in diameter, smooth, white, and round. The cell wall contains peptidoglycan of *meso*-diaminopimelic acid and is of the A1γ type. Temperature optimum for growth is 65 °C, with a temperature

range from 40–75 °C. The optimum pH is between 6.5–6.8. An obligate anaerobe and chemoorganotroph. Fermented sugars include glucose, fructose, galactose, mannose, cellobiose, maltose, melibiose, sucrose, lactose, xylose, ribose, mannitol, and pyruvate. Fermentation occurs according to the following equation (in moles): 1 glucose → 1.45 ethanol + 0.25 acetic acid + 0.3 lactic acid + 1.7 CO_2 + 0.5 H_2. Penicillin G and tetracycline (both at $6 \mu g/ml$) inhibit growth. Strain AKO-1 was isolated from sediment sludge from Lake Kivu in East Africa using M-1 media (which included lactic acid, glucose, and yeast extract) at 60 °C.

DNA G+C content (mol%): 32 (HPLC).

Type strain: AKO-1, ATCC 43586, DSM 3389.

GenBank accession number (16S rRNA gene): L09166.

3c. **Thermoanaerobacter brockii subsp. lactiethylicus** Cayol, Ollivier, Patel, Ravot, Magot, Ageron, Grimont and Garcia 1995, 788[VP]

lac.ti.e.thy'li.cus. N.L. n. *acidum lacticum* lactic acid; N.L. n. *ethylicus* ethyl alcohol; N.L. masc. n. *lactiethylicus* referring to the production of both lactic acid and ethanol.

Description is based on the type strain and strains SEBR 7311 and SEBR 7312 (Cayol et al., 1995). Cells are straight rods, 0.5×2–$3 \mu m$, motile by means of peritrichous flagella and occur singly or in pairs in young cultures. Pleomorphic filamentous cells, $15 \mu m$ in length, appear in older cultures. Cells are Gram-stain-positive. They form spores in a medium containing D-xylose as an electron donor and thiosulfate as an electron acceptor. Spores are not formed in medium with D-glucose. Colonies in roll tubes are 4 mm in diameter after 2 d of incubation at 60 °C, smooth, uniformly round, mucoid, nonpigmented, and flat. Temperature range for growth is 40–75 °C, with a temperature optimum of 55–60 °C. The pH range is 5.6–8.8, and the optimum is around 7.3. Fermentable substrates are glucose, fructose, galactose, mannose, D-ribose, D-xylose, cellobiose, lactose, maltose, sucrose, mannitol, starch, yeast extract, and pyruvate but not L-arabinose, cellulose, L-rhamnose, glycerol, ribitol, galactitol, sorbose, or melibiose. Under optimal conditions, the doubling time with glucose is 2 h. Sulfite and elemental sulfur are reduced to hydrogen sulfide. Sulfate, nitrate and fumarate are not reduced. The optimum NaCl concentration for growth is 1% (w/v). Concentrations up to 4.5% are tolerated. Strain SEBR 5268[T] was isolated from French oil samples that had been geothermally heated to 92 °C. Enrichments were performed in anaerobic, glucose-based medium at 60 °C.

DNA G+C content (mol%): 35 (HPLC).

Type strain: SEBR 5268, DSM 9801.

GenBank accession number (16S rRNA gene): U14330.

4. **Thermoanaerobacter italicus** Kozianowski, Canganella, Rainey, Hippe and Antranikian 1998, 1083[VP] (Effective publication: Kozianowski, Canganella, Rainey, Hippe and Antranikian 1997, 179.)

i.ta.li.cus. L. masc. adj. *italicus* pertaining to Italy, where the organism was isolated.

Description is based primarily on the properties of the type strain, although other strains have been isolated. Cells are nonmotile rods, 0.4–0.75×2–$6 \mu m$, that form chains up to $50 \mu m$ in length during growth on glucose. Cells stain

Gram-stain-negative but contain Gram-positive cell-wall structure. The peptidoglycan contains *meso*-diaminopimelic acid, and the cell wall is of the A1γ type. Spherical, terminal spores are produced with xylose as substrate. Colonies on agar plates with glucose are 2–3 mm in diameter, round with entire margins, grayish-white and opaque with a glassy surface. Growth occurs in the temperature range of 45–78 °C, with a temperature optimum of 70 °C. The pH optimum is around 7.0. Fermentable carbohydrates include amygdalin, arabinose, cellobiose, esculin, fructose, glucose, galactose, lactose, maltose, mannose, melezitose, melibiose, mannitol, raffinose, sucrose, trehalose, starch, xylan, glycogen, D-glucosamine, sucrose, inulin, pectin, and xylose but not cellulose. At 70 °C, the doubling times during growth on glucose and pectin are 2.1 h and 3 h, respectively. End products of the fermentation of 28 mM glucose are ethanol (26 mM), lactic acid (12.4 mM), acetic acid (2.2 mM), succinate (0.3 mM), CO_2, and H_2 (not quantified). Similar products are obtained from pectin. NaCl, 1% (w/v), did not inhibit growth. Cephalosporin, erythromycin, kanamycin, and rifampin (all at 10 μg/ml) totally inhibit growth. Strains were isolated from water and mud samples (40–70 °C) from thermal spas at Abano Terme, Calzignano Terme, Montegrotto Terme, Battaglia Terme, Sirmione, and Agano Terme in northern Italy. Pectin and pectate (polygalacturonic acid) were the substrates used for the enrichments. Strain Ab9 was isolated from medicinal mud (fango) from Abano Terme, Italy.

DNA G+C content (mol%): 34.4 (HPLC).

Type strain: Ab9, DSM 9252.

GenBank accession number (16S rRNA gene): AJ250846.

5. **Thermoanaerobacter kivui** Leigh, Mayer and Wolfe 1981) Collins, Lawson, Willems, Cordoba, Fernández-Garayzábal, Garcia, Cai, Hippe and Farrow 1994, 824^VP (*Acetogenium kivui* Leigh, Mayer and Wolfe 1981, 279)

ki'vui. N.L. gen. n. *kivuus* pertaining to Kivu, named for its source, Lake Kivu.

The only strain available is the type, upon which this description is based. Cells are rods, 0.7–2–3.5 μm during growth on H_2 and CO_2 and 0.7–0.8 × 5.5–7.5 μm during growth on glucose. Cells occur in pairs or chains and are not flagellated. Gram-stain-negative, but the cell wall is of the Gram-positive structure and covered by a hexagonal S-layer composed of 80 kDa subunits (Lupas et al., 1994). Endospores not observed. Colonies on agar plates are convex, circular, entire, translucent, tan in color, and have a smooth, shiny surface. After 1 week, colonies are 2 mm in diameter. Cells are catalase negative.

The temperature optimum is 66 °C, with growth occurring from 50–72 °C. Growth is obligately anaerobic and requires a reducing agent such as cysteine-sulfide in the medium. Small amounts of O_2 in semisolid and liquid media caused a lag phase but did not alter the ability of the bacterium to synthesize acetate via the Wood–Ljungdahl pathway of autotrophic acetyl-coenzyme A biosynthesis (Karnholz et al., 2002). The pH optimum is around 6.4, and the range is 5.3–7.3. Fermentable carbohydrates include glucose, mannose, fructose, and pyruvate. Acetic acid is the sole fermentation product, according to the reaction (Ryabokon et al., 1995): 1 glucose → (2 acetate + 2 CO_2 + 4 H_2) → 3 acetate. Galactose, malt-

ose, raffinose, ribose, sucrose, lactose, trehalose, cellobiose, cellulose, pectin, starch, mannitol, and inositol cannot serve as substrates. Chemolithoautotrophic growth is also possible with H_2 and CO_2, forming acetate as the product of CO_2 reduction. Under optimal conditions, the doubling time for chemolithoautotrophic growth is 1.75 h at 60 °C. Poor cell yields are obtained when H_2 is replaced with formate. Yeast extract and trypticase increase growth. Currently, *Thermoanaerobacter kivui* has only been isolated from the sediments of Lake Kivu, Africa, and was enriched with H_2 and CO_2 (67%: 33%, v/v) at 60 °C.

DNA G+C content (mol%): 38 (Bd).

Type strain: LKT-1, ATCC 33488, DSM 2030.

GenBank accession number (16S rRNA gene): L09160.

6. **Thermoanaerobacter mathranii** Larsen, Nielsen and Ahring 1998, 328^VP emend. Carlier, Bonne and Bedora-Faure 2006, 157 (Effective publication: Larsen, Nielsen and Ahring 1997, 118.)

ma.thra'ni.i. N.L. gen. n. *mathranii* of Mathrani, in honor of Indra M. Mathrani, who contributed to the understanding of thermophilic anaerobes from hot springs during his short career.

The species is composed of two subspecies, *Thermoanaerobacter mathranii* subsp. *mathranii* and *Thermoanaerobacter mathranii* subsp. *alimentarius*. The description of the species is the same as the subspecies.

6a. **Thermoanaerobacter mathranii subsp. mathranii** Larsen, Nielsen and Ahring 1998, 328^VP emend. Carlier, Bonne and Bedora-Faure 2006, 157 (Effective publication: Larsen, Nielsen and Ahring 1997, 118.)

Cells are straight rods, occurring singly and, under suboptimal conditions, in long chains, and are Gram-stain-variable although cell-wall structure is Gram positive. In the exponential growth phase, cells are 0.7 × 1.8–3.9 μm and motile. They form terminal, spherical spores that swell the sporangium. Cells are typically longer when either sporulating (6.4–8.2 μm) or growing under suboptimal conditions, e.g., at a temperature of 75 °C. Colonies on agar plates are white and irregular, with a diameter of 1 mm after growth on beech wood xylan for 7 d. The surface of the colonies is granulated with "top formation" (Sonne-Hansen et al., 1993). The optimum growth temperature is around 70 °C, and no growth is observed at 47 or 78 °C. The pH optimum is 6.8–7.8. The lowest pH for which growth has been observed is 4.7, the highest pH is 8.8. Cells are catalase negative. The doubling time on xylose is 74 min at 69 °C and pH 7.0. Sulfide is produced from casein-peptone and sulfate. Carbohydrates that support growth include amygdalin, L-arabinose, cellobiose, D-fructose, D-glucose, glycogen, lactose, maltose, D-mannitol, mannose, melezitose, melibiose, raffinose, D-ribose, sucrose, trehalose, xylan, and D-xylose but not avicel, casein-peptone, cellulose, D-galactose, glycerol, inulin, pectin, L-rhamnose, salicin, sorbitol, or yeast extract. Its carbohydrate fermentation is represented by the simplified equation (Ahring et al., 1999): 1 xylose + H_2O → 1.1 ethanol + 0.4 acetic acid + 0.06 lactic acid + 1.81 CO_2 + 0.9 H_2. Growth is inhibited by tetracycline, chloramphenicol, penicillin G, neomycin, or vancomycin (all at 100 mg/l). Growth occurs in the presence of 10 μg/l of chloramphenicol or neomycin. *Thermoanaerobacter mathranii* subsp. *mathranii* is not inhibited by 51 kPa overpressure of H$_2$, 2% NaCl, or addition of 5% (w/v) ethanol. When *Thermoanaerobacter mathranii* subsp. *mathranii* is grown with a wet-oxidized wheat straw hydrolysate, growth is inhibited, presumably due to aromatic monomers and breakdown products such as phenolic aldehydes and, to a lesser extent, phenol ketones (Klinke et al., 2001). Strain A3 was isolated from a mixed biomat and sediment sample taken from a slightly alkaline (pH 8.5) hot spring (70 °C) in Hverðagerdi-Hengil, Iceland (Sonne-Hansen et al., 1993) using beech wood xylan supplemented mineral medium at 68 °C and pH 8.4.

DNA G+C content (mol%): 37 (HPLC).

Type strain: A3, AIP 130-05, CIP 108742, DSM 11426.

GenBank accession number (16S rRNA gene): Y11279.

6b. Thermoanaerobacter mathranii subsp. alimentarius Carlier, Bonne and Bedora-Faure 2007, 1^{VP} (Effective publication: Carlier, Bonne and Bedora-Faure 2006, 157.)

a.li.men.ta′ri.us. L. adj. *alimentarius* related to food.

Cells stain Gram-stain-negative at all growth phases and are straight rods, 1.4–2.7 × 0.3–0.6 µm. Cells are motile by peritrichous flagella and occur singly or in long chains. Spores are never observed. Strict anaerobe. The optimum growth temperature is 55–60 °C, and the temperature range is 50–75 °C. Cells are catalase and nitrate reduction negative. Milk is coagulated. Gas is produced. Acid is produced from lactose and salicin. Substrates include arabinose, cellobiose, esculin, fructose, galactose, glucose, lactose, maltose, ribose, salicin, sucrose, and starch but not cellulose (Avicel), glycerol, inositol, mannitol, melezitose, melibiose, raffinose, or trehalose. Metabolic end products of carbohydrate fermentation in 3% trypticase-peptone/2% yeast extract medium are (in mM); ethanol (16), acetic acid (8), and lactic acid (55). Growth is observed at 2% (w/v) ethanol but not at 4%. The type strain AIP 505.99 was isolated at 55 °C from spoiled meat. In addition to the type, strain AIP 431.03 (= CIP 108191 = CCUG 49567) is also available (Carlier et al., 2006). GenBank accession numbers for *recA*, *hsp70* and *hsp60* are DQ422958, DQ431192 and DQ439967, respectively.

DNA G+C content (mol%): 31.5 (HPLC).

Type strain: AIP 505.99, CCUG 49566, CIP108280.

GenBank accession numbers (16S rRNA gene): AY701758.

7. Thermoanaerobacter pseudethanolicus Onyenwoke, Kevbrin, Lysenko and Wiegel 2007, 2192^{VP}

pseud′e.tha.no′li.cus. Gr. adj. *pseudes* false; N.L. adj. *ethanolicus* a bacteria-specific epithet; N.L. masc. adj. *pseudethanolicus* a false (*Thermoanaerobacter*) *ethanolicus*.

The description is based upon the properties of the type strain 39E as described by Lee et al. (1993). Cells are rod shaped and sporulate when grown with xylose. Cells stain Gram-stain-variable, but the wall is Gram positive in structure. Cells are motile. Strain 39E is usually grown in Tryptone-Yeast Extract-Glucose (TYEG) medium, pH 7.0. Details of the pH and temperature ranges not reported. The temperature optimum is around 65 °C. Substrates allowing heterotrophic fermentation include glucose, maltose, xylose, cellobiose, sucrose, and starch. H$_2$/CO$_2$ does not support growth. Thiosulfate is reduced to H$_2$S. The generation time is 75 min.

The Embden–Meyerhof–Parnas and the pentose-phosphate pathways are used during the catabolism of hexoses and pentoses to pyruvate, respectively (Zeikus et al., 1981). Fermented carbohydrates include xylose, cellobiose, starch, glucose, maltose, and sucrose. Glucose is consumed in preference to cellobiose (Ng and Zeikus, 1982). Zeikus et al. (1980) reported that the generalized fermentation products (µmol/ml) from 0.5% glucose growing in 10 ml TYEG medium are: 54.9 ethanol + 3.1 acetic acid + 5.0 lactic acid + 3.1 H$_2$. CO$_2$ was not reported, and the equation was not balanced.

Alpha, omega-dicarboxylic acids constitute 40% of the fatty acyl components of the membrane lipids (Jung et al., 1994). Yamamoto et al. (1998) reported the presence of MK-7 as major (>90%) menaquinone.

Ethanol tolerance is not temperature-dependent in the parental strain 39E, but it is in mutants (Burdette et al., 2002). Glucose is not fermented at an ethanol concentration of 2.0% (w/v) by strain 39E, but mutants ferment glucose at 45 °C and 8.0% ethanol and at 68 °C and 3.3% ethanol (Lovitt et al., 1984). The type strain 39E was isolated from the Octopus Spring algal-bacterial mat in Yellowstone National Park, Wyoming, USA at the same pool as *Thermoanaerobacter ethanolicus* JW200. The enrichment was performed at 65 °C in TYEG medium and modified TYEG medium, respectively, that contained 5% xylose instead of glucose.

In addition, strains X514, X531 and 561, which were previously designated *Thermoanaerobacter ethanolicus*, belong to this species (unpublished results; J. Zhou, personal communication; Roh et al., 2002). The genome sequences of 39E and X514 are available from the Joint Genome Institute.

DNA G+C content (mol%): 34 (from the genome sequence).

Type strain: 39E, ATCC 33223, DSM 2355.

GenBank accession number (16S rRNA gene): L09164.

8. Thermoanaerobacter siderophilus Slobodkin, Tourova, Kuznetsov, Kostrikina, Chernyh and Bonch-Osmolovskaya 1999b, 1477^{VP}

si.de.ro′phil.us. Gr. n. *sideros* iron; Gr. adj. *philos* loving; N.L. adj. *siderophilus* iron-loving.

Description is based on the type strain, which is the only available strain. Cells are straight to curved rods, 0.4–0.6 × 3.5–9.0 µm, that occur singly or in short chains. They stain Gram-positive and have a Gram-positive cell-wall structure. Round, refractile, heat-resistant spores are formed in terminally swollen sporangia. Maximum sporulation occurs during growth with 9,10-anthraquione 2,6-disulfonic acid (AQDS) as an electron acceptor. Colonies in agar-shake-roll tubes are uniformly round, 0.5–1.0 mm in diameter, and white. Cells exhibit slight tumbling motility due to peritrichous flagellation. Obligate anaerobe.

Growth temperature range and optimum are 39–78 °C and 69–71 °C, respectively. The pH range and optimum for growth are 4.8–8.2 and 6.3–6.5, respectively. Substrates utilized in both the presence and absence of the electron acceptor Fe(III) include peptone, yeast extract, beef extract, casein, starch, glycerol, pyruvate, glucose, sucrose, fructose, maltose, xylose, cellobiose, and sorbitol but not formate, acetate, lactate, methanol, ethanol, propanol, 2-propanol, butanol, propionate, *n*-butyrate, succinate, malate, maleate, glycine, alanine, arginine, L-arabinose, olive oil, xylan, and cellulose. Electron acceptors include amorphous iron (III) oxide, AQDS, sulfite, thiosulfate, elemental sulfur, and MnO_2. The products of amorphous iron (III) oxide reduction are magnetite and siderite. Molecular hydrogen and CO_2 can be utilized for growth in the presence of Fe (III). Acetic acid is not produced during glucose fermentation. Sulfite and elemental sulfur are reduced to hydrogen sulfide. Nitrate, sulfate, and O_2 are not electron acceptors for growth. Growth is inhibited by chloramphenicol, neomycin, polymyxin B, and kanamycin (all at 100 μg/ml) but not by penicillin, ampicillin, streptomycin, or novobiocin (all at 100 μg/ml). Growth occurs with NaCl concentrations from 0–3.5% (w/v).

The Fe (III) reduction in *Thermoanaerobacter siderophilus* prevents the production of inhibitory concentrations of H_2 (Gavrilov et al., 2003). The type strain SR4 was isolated from a hydrothermal vent with a temperature of 70–94 °C in the area of the Karymsky volcano on the Kamchatka peninsula, Far East Russia. The anaerobic medium contained amorphous iron (III) oxide and peptone at a pH of 6.8–6.9 and 70 °C.

DNA G+C content (mol%): 32 (T_m).

Type strain: SR4, DSM 12299.

GenBank accession number (16S rRNA gene): AF120479.

9. **Thermoanaerobacter sulfurigignens** Lee, Dashti, Prange, Rainey, Rohde, Whitman and Wiegel 2007a, 1433[VP]

sul.fur.i′gig.nens. L. n. *sulfur* sulfur, brimstone; L. part. adj. *gignere* producing; N.L. part. adj. *sulfurigignens* relating to the formation of sulfur droplets from thiosulfate.

This description is based on the type strain JW/SL-NZ826 and a second isolate, JW/SL-NZ824, from the same vicinity. Cells are rods, 0.3–0.8 × 1.2–4.0 μm during exponential growth, motile by peritrichous flagella, and nonpigmented. They form round, terminal spores, 0.4–0.85 μm in diameter, during the late exponential or early stationary growth phase. Cells tend to be longer during the stationary phase, with lengths up to 35 μm being observed. Cells stain Gram-negative but have a Gram-positive cell-wall structure. Colonies in agar-shake-roll tubes are creamy white, circular, and 1–2 mm in diameter. The temperature range and optimum for growth at pH^{25C} are 34–74 °C and 63–65 °C, respectively. The pH^{25C} range and optimum for growth at 60 °C is 4.0–8.0 and 5.8–6.5, respectively. It grows at 90% of the optimal growth rate within the pH range of 4.8–6.5. Thus, it is one of the most acidophilic, thermophilic, anaerobic bacteria, comparable to *Thermoanaerobacterium acidotolerans* and *Thermoanaerobacterium aotearoense* (Wagner and Wiegel, 2008). Xylose, glucose, starch, lactose, galactose, maltose, fructose, sucrose, mannose, cellobiose, raffinose, pyruvate, methanol, and mannitol are fermented in the presence of 0.3% yeast extract. A generalized fermentation balance (in the presence of 0.3% yeast extract) is: 1 glucose → 0.7 ethanol + 0.4 acetic acid + 0.9 lactic acid + 1.1 CO_2 + 0.8 H_2. Ribose, arabinose, dextran, xylan, cellulose, glycerol, xylitol, formate, and gluconate are not fermented. The addition of 0.5% peptone, tryptone, casein hydrolysate, or Casamino acids in the presence of 0.3% yeast extract does not increase growth. Weak growth is observed with fumarate and succinate. Cells are catalase and indole negative. Sulfate and sulfite are not reduced.

Growth is inhibited by neomycin and chloramphenicol (100 μg/ml) and gramicidin (10 μg/ml). Cells are resistant to vancomycin, bacitracin, tetracycline, and ampicillin (all at 10 μg/ml), and kanamycin, streptomycin, cycloheximide, and cycloserine (all at 100 μg/ml). The strains tolerate up to 1.1 M thiosulfate and 90 mM sulfite, presumably a response to the conditions of the sulfurous environment from which they were isolated. The type strain JW/SL-NZ826 was isolated from an acidic volcanic stream outlet on White Island, New Zealand, at 60 °C and pH^{25C} of 5.0–5.5 using a prereduced mineral medium supplemented with 20 mM thiosulfate, 0.5% (w/v) yeast extract and 0.5% (w/v) xylose.

DNA G+C content (mol%): 34.5 (HPLC).

Type strain: JW/SL-NZ826, ATCC 700320, DSM 17917.

GenBank accession number (16S rRNA gene): AF234164.

10. **Thermoanaerobacter sulfurophilus** Bonch-Osmolovskaya, Miroshnichenko, Chernykh, Kostrikina, Pikuta and Rainey 1998, 631[VP] (Effective publication: Bonch-Osmolovskaya, Miroshnichenko, Chernykh, Kostrikina, Pikuta and Rainey 1997, 488.)

sul.fu.ro.phi′lus. L. n. *sulfur* sulfur; Gr. masc. adj. *philos*-liking; N.L. masc. adj. *sulfurophilus* liking elemental sulfur.

The description is based on the type strain L-64 and other isolates from the same origin, including P-82, G-1, and L-64. Cells stain Gram-stain-positive and have Gram-positive wall structure. They are curved rods, 0.5 × 3–7 μm. In older or nutrient deficient cultures, cells occur in long chains with interspersed coccoid cells. Spores are not observed. Sluggish motility through peritrichous flagella. The temperature range and optimum for the growth are 44–75 °C and 55–60 °C, respectively. The pH range and optimum for growth are 4.5–8.0 and 6.8–7.2, respectively. Obligate anaerobe. Fermentable carbohydrates include glucose, fructose, lactose, rhamnose, arabinose, xylose, sorbitol, inositol, mannitol, sucrose, cellobiose, maltose, starch, pectin, pyruvate, and lactate. Cellulose, succinate, citrate, formate, acetate, propionate, butyrate, methanol, ethanol, and H_2 (in the presence or absence of elemental sulfur) do not support growth. Elemental sulfur stimulates growth and is reduced to sulfide. Fermentation products while growing on 0.5% (w/v) lactate (utilized amount not determined) using S^0 as the electron acceptor were (in μmol/ml of culture): 3.0 H_2S + 0.14 acetic acid + 0.2 H_2 + CO_2 (not quantified).

Nitrate, sulfate, and sulfite are not reduced. The type strain L-64 was isolated from a sulfur-containing cyanobacterial mat occurring along the rim of a hot (53–58 °C) pond at the Uzon caldera, Kamchatka. The enrichment medium contained peptone, glucose, and lactate and was incubated at 55 °C and pH 7.0 (Bonch-Osmolovskaya et al., 1997).

DNA G+C content (mol%): 30–31 (T_m).

Type strain: L-64, DSM 11584.

GenBank accession number (16S rRNA gene): Y16940.

11. **Thermoanaerobacter thermocopriae** (Jin, Yamasato and Toda 1988) Collins, Lawson, Willems, Cordoba, Fernández-Garayzábal, Garcia, Cai, Hippe and Farrow 1994, 824[VP] (*Clostridium thermocopriae* Jin, Yamasato and Toda 1988, 280)

ther.mo.co′pri.ae. Gr. adj. *thermos* hot; Gr. n. *kopria* compost; N.L. gen. n. *thermocopriae* relating to the isolation from compost at elevated temperature.

Description is based upon the type strain (Jin et al., 1988). Cells are sluggishly motile, straight rods, 0.5–0.7 × 2.2–6.0 μm, and produce terminal, spherical spores with a diameter of 1.2–1.6 μm that swell the sporangium. Cells stain Gram-negative but have Gram-positive wall structure. The temperature range and optimum for growth is 47–74 °C and 60 °C, respectively. The pH range and optimum for growth is 6.0–8.0 and 6.5–7.3, respectively. The bacterium is an obligate anaerobe and has a broad substrate specificity for numerous carbohydrates, including cellulose, hemicellulose, cellobiose, glucose, fructose, maltose, arabinose, lactose, trehalose, glycogen, starch, amygaldin, xylan, and mannose but not melibiose, pectin, inulin, or mannitol. The ratio of fermentation products from cellobiose is 0.8–4.0 butyric acid + 0.3–1.7 acetic acid + <0.8 lactic acid + 2–5 ethanol + H_2S (not quantified).

Esculin is hydrolyzed. Indole is not produced, gelatin is not digested, and nitrate is not reduced. The type strain JT3-3 was isolated from a compost of camel feces. Other strains have been isolated from soil and hot springs in Japan (The University of Tokyo and Kinugawa) following anaerobic enrichment with cellobiose- or cellulose-containing medium at 60 °C and pH 7.1. In addition to the type, other strains that have been described include strains JT1, JT2, JT3-1, JT3-2, JT5 (Jin and Toda, 1987, 1988).

DNA G+C content (mol%): 37.2 (T_m).

Type strain: JT3-3, ATCC 51646, IAM 13577, JCM 7501.

GenBank accession number (16S rRNA gene): L09167.

12. **Thermoanaerobacter thermohydrosulfuricus** (Klaushofer and Parkkinen, 1965) Lee, Jain, Lee, Lowe and Zeikus 1993, 49[VP] (*Clostridium thermohydrosulfuricum* Klaushofer and Parkkinen 1965, 448)

ther.mo.hy.dro.sul.fur′i.cus. Gr. masc. adj. *thermos* hot; N.L. masc. adj. *hydrosulfuricum* pertaining to hydrogen sulfide formation; N.L. masc. adj. *thermohydrosulfuricus* indicating that the organism grows at high temperatures and reduces sulfite to H_2S.

This description is based on the type strain E100-69 (Hollaus and Sleytr, 1972) and strain JW102 (Wiegel et al., 1979). Cells are 0.3–0.6 × 2.0–13.0 μm and motile by peritrichous flagella. Cells occur singly, in short chains, or, in some strains, as long, filaments with cells up to 40 μm in length. As shown by freeze fracture planes, the core of flagella are empty, possibly for the transport of flagellin molecules during assembly (Sleytr and Glauert, 1973). Spores are spherical and terminal. The mother cells are terminally swollen and elongated, resembling a drum stick. Cells stain Gram-variable but have a Gram-positive cell wall composed of two layers and *meso*-diaminopimelic acid. The cell wall is covered by a hexagonal S-layer consisting of 14.2 nm center-to-center spaced gly-

coprotein units (Bock et al., 1994; Crowther and Sleytr, 1977; Messner et al., 1995; Sára and U.B. Sletyr, 1996). The molecular mass of the glycoprotein is about 20,000 (Christian et al., 1988; Sára et al., 1989). The cells are catalase negative.

Obligate anaerobe. Growth occurs at a pH range and optimum of 5.5–9.2 and 6.9–7.5, respectively. The optimum growth temperature is 67–69 °C, with no growth occurring at 76–78 °C. Growth at 37 °C is poor, and no growth is observed at or below 35 °C. H_2 at 10% (v/v) in the head space and lactate at >30 mM inhibit growth (Wiegel et al., 1979). Fructose, galactose, glucose, mannose, xylose, cellobiose, maltose, sucrose, trehalose, pectin, esculin, and salicin are fermented. Yeast extract is required for growth and for metabolic activity (Wiegel et al., 1979). Growth with PYG medium yields H_2, CO_2, acetic acid, lactic acid, and ethanol as well as trace amounts of formate, butyrate, isovalerate, isocaproate, propanol, and 2-propanol. Glucose fermentation in the presence of 0.2% yeast extract is described by the simplified equation: 1 glucose + 0.5 H_2O → 1 ethanol + 0.5 acetic acid + 0.5 lactic acid + 1.5 CO_2 + H_2. Fermentation of dextrin, potato starch, mannitol, dulcitol, and sorbitol and coagulation of litmus milk are variable. Inositol, erythritol, glycerol, lactate, tartrate, and cellulose are not fermented. Sulfite is reduced to hydrogen sulfide, but sulfate is not reduced. H_2S is produced from tryptone, peptone, and yeast extract. Nitrate, but not nitrite, is reduced. Acetyl methyl carbinol and indole are not produced. Strains of *Thermoanaerobacter thermohydrosulfuricus* have been isolated from many sources, including extraction juices of beet sugar factories (E100-69[T]; L 110-69 = DSM 568; L 77-66 = DSM 569; S 100–69 = DSM 570; Klaushofer and Parkkinen, 1965; Hollaus and Klaushofer, 1973), mesobiotic mud and soil (Klingeberg et al., 1990; Wiegel et al., 1979), hot springs in Utah and Wyoming, USA, a sewage plant in Georgia, USA (JW102 = DSM 2247; Wiegel et al., 1979), a sugar refinery in Germany (DSM 7021 and DSM 7022; Klingeberg et al., 1990), household dust (Wiegel, unpublished results) and in co-culture with *Clostridium thermocellum* (YM3; Mori, (1990, 1995).

Major lipid fatty acids found in the type strain are 42% $C_{15:0\ iso}$ and 23% $C_{17:0\ iso}$. MK-7 is the major menaquinone.

The ubiquitous distribution of *Thermoanaerobacter thermohydrosulfuricus* in mesobiotic and in geothermally and anthropongenically heated or thermobiotic environments may be partly due to its ability to form spores and its wide substrate spectrum.

DNA G+C content (mol%): 35–37 (T_m, Bd, HPLC).

Type strain: E100-69, ATCC 35045, DSM 567, LMG 6659, NCIMB 10956.

GenBank accession number (16S rRNA gene): L09161.

13. **Thermoanaerobacter wiegelii** Cook, Rainey, Patel and Morgan 1996, 126[VP]

wie.gel′i.i. N.L. gen. n. *wiegelii* of Juergen Wiegel, in recognition of his contributions to the study of thermophilic anaerobes.

Cells grown on solid trypticase peptone-yeast extract-glucose (TYEG) medium produce nonpigmented colonies, 0.5–2.0 mm in diameter, that are smooth and uniformly round. Cells stain Gram-stain-negative but have Gram-positive cell-wall structure. Rod-shaped cells, 0.4–0.6 μm × 4–10 μm, that occur singly, in pairs, or less frequently in chains depending upon growth conditions. Electron micrographs of the cell wall reveal a two-layer structure. The inner layer, which is adjacent to the cytoplasmic membrane, stains intensely, whereas the outer layer is less dense. Cells are usually sluggishly motile by less than 8 peritrichous flagella. Cells growing on TYEG medium do not sporulate. However, in a minimal medium at 65 °C, round spores are produced in elongated and filamentous, terminally distended cells. Spore suspensions remain viable after 80 min at 115 °C.

Obligate anaerobe. Growth occurs at pH 5.5–7.2 but not at pH 5.0 or 7.25. The pH optimum is 6.8. The growth temperature range and optimum are 38–78 °C and 65–68 °C, respectively. Cultures do not grow at 34 or 80 °C. Yeast extract and trypticase are not required for growth and can be replaced by vitamin-free Casamino acids and vitamins. Trypticase, peptone, and yeast extract do not serve as sole carbon sources. Carbohydrates that are utilized include glucose, xylose, maltose, lactose, cellobiose, raffinose, glucosamine, galactose, fructose, mannose, sucrose, glycerol, soluble starch, pectin, and chitin. Sorbitol, mannitol, and trehalose are also fermented, but ethanol, DL-lactate, sodium citrate, sodium succinate, transaconitate, malonate, glutamate, glutamine, sodium pyruvate, 2-deoxyglucose, α-methyl-glucoside, L-arabinose, α-L-rhamnose, dulcitol, m-inositol, ribose, α-L-fucose, and L-sorbose are not. The doubling time on glucose is 72 min. Similar to *Thermoanaerobacter ethanolicus* and *Thermoanaerobacter thermohydrosulfuricus*, glucose and xylose are used simultaneously when supplied together (Carreira et al., 1983; Cook et al., 1996). Propionate, an unusual fermentation product for a member of this genus, is formed during growth on xylose and cellobiose.

Sulfite is reduced to hydrogen sulfide. Nitrate, oxygen, sulfate, and sulfur are not reduced. Indole is not produced, and esculin and gelatin are not hydrolyzed. Cells do not accumulate anthrone-reactive material when they are grown with glucose.

Cephalosporin C, erythromycin, bacitracin, tetracycline, and polymyxin B completely inhibit growth (all at 10 μg/ml). Trimethoprim, rifampin, amphotericin B, D-cycloserine, penicillin G, streptomycin sulfate, chloramphenicol, and ampicillin do not inhibit growth at concentrations up to 100 μg/ml. Monensin (100 mM), 2,4-dinitrophenol (500 mM), tetrachlorosalicylanilide (10 mM), N,N-dicyclohexylcarboiimide (500 mM) and iodoacetate (500 mM), sodium azide, sodium fluoride, potassium cyanide, and sodium arsenate (all at final concentrations of 5 mM) inhibit growth of exponentially growing cultures with glucose as carbon and energy source. The presence of oxygen completely inhibits growth.

The type strain Rt8.B1 was isolated from neutral to alkaline, geothermally heated water (56–69 °C) in Government Gardens, Rotorua, New Zealand. The primary enrichment cultures were prepared by adding 0.2 ml of pool water to 10 ml of prereduced TYEG medium under an N₂ atmosphere and were incubated at 70 °C. Purification involved the use of TYEG agar deeps.

DNA G+C content (mol%): 35.6 (HPLC).
Type strain: Rt8.B1, DSM 10319.
GenBank accession number (16S rRNA gene): X92513.

Other named isolates

1. "**Thermoanaerobacter keratinophilus**" Riessen and Antranikian 2001, 406

   ke.ra.ti.no′phil.us. Gr. n. *keras* keratin; Gr. adj. *philos* loving; N.L. adj. *keratinophilus* keratin-loving.

   Taxonomic note: The only strain of this species is covered by a patent with a restricted distribution and thus cannot be a type strain. Hence, the name is not validly published.

   Exponential growth phase cells are slightly curved rods, 0.2–0.3 × 1–3.0 μm, and occur singly, in pairs or in short chains. Cells stain Gram-negative regardless of the growth phase but have Gram-positive cell-wall structure. Stationary growth phase cells are pleomorphic, and some form filaments 10–15 μm in length. Asymmetric cell division leads to formation of coccoid cells. Does not form spores. Temperature range and optimum for growth are 50 to just below 80 °C and 70 °C, respectively. The pH range and optimum are pH 5.0–9.0 and 7.0, respectively. The range and optimal NaCl concentration for growth are 0–30 g/l and 5–10 g/l, respectively. Growth is fermentative and obligately anaerobic. Yeast extract plus tryptone, both at 0.05% (w/v), are required for growth on carbohydrates. Under optimal conditions in the presence of 0.5% each of yeast extract and tryptone, the doubling time is 67 min. Utilized substrates include casein, bactopeptone, yeast extract, tryptone, collagen, gelatin, starch, pectin, glucose, fructose, galactose, mannose, pyruvate, maltose, and cellobiose but not xylan, cellulose, pullulan, xylose, arabinose, lactose, or olive oil. Growth can be achieved solely with merino wool or chicken feathers. Thiosulfate and sulfate both serve as electron acceptors. Thiosulfate stimulates growth to approximately threefold that of sulfate but is not required for growth. Ampicillin, kanamycin, and streptomycin inhibit growth at 10 μg/ml. Strain 2KXI was isolated from a hydrothermal vent (74 °C, pH 6.0) in the area of Furnas on the Azorean island São Miguel using merino wool and chicken feathers as substrates. The incubation temperature was 70 °C.

   DNA G+C content (mol%): 37.6 (HPLC).
   Proposed type strain: 2KXI, DSM 14007.
   GenBank accession number (16S rRNA gene): AY278483.

Genus II. **Ammonifex** Huber and Stetter 1996, 836[VP] (Effective publication: Huber, Rossnagel, Woese, Rachel, Langworthy and Stetter 1996, 47.)

ROBERT HUBER

Am.mo′ni.fex. L. neut. n. *sal ammoniacum* salt of Ammon (NH₄Cl); L. v. *facere* to make; N.L. masc. n. *Ammonifex* the ammonium-maker.

Rod-shaped, straight or slightly curved cells with rounded ends, usually about 2–3× 0.6 μm. The cells occur singly, in pairs, and in short chains of up to six cells. Complex cell envelope. **Muramic acid present; diaminopimelic acid absent.** Nonsporulating. **Gram-stain-negative.** Motile by terminal and lateral flagellation. Core lipids consist mainly of (non-phytanyl) ether lipids and fatty acid methyl esters. Sensitive to lysozyme and antibiotics. Extremely thermophilic. Growth between 57 °C and 77 °C (optimum 70 °C), pH 5.0 and 8.0 (optimum 7.5), and 0–1.5% NaCl (optimum 0.1%). Growth occurs on solid surfaces. **Strictly anaerobic. Autotrophic growth by oxidation of hydrogen and reduction of nitrate, sulfate, or sulfur.** Nitrate ammonification. Facultatively chemolithoautotrophic. Based on 16S rDNA analysis, *Ammonifex* belongs to the low G+C subgroup of the Gram-stain-positive bacteria.

DNA G+C content (mol%): 54.

Type species: **Ammonifex degensii** Huber and Stetter 1996, 836[VP] (Effective publication: Huber, Rossnagel, Woese, Rachel, Langworthy and Stetter 1996, 47.).

Further descriptive information

Phylogenetic analysis by 16S rRNA gene sequence comparison shows that the genus *Ammonifex* is a member of the Gram-stain-positive bacteria where it represents a separate lineage (Huber et al., 1996).

Cells of *Ammonifex degensii* were Gram-stain-negative, straight or slightly curved rods with rounded ends, and exhibited terminal and lateral flagellation (Figure 247). After treatment with lysozyme, the rods appear granular within one minute, however, the cell morphology remains stable. Ultrathin sections revealed a cytoplasmic membrane (width: 5.0 nm) and a rather complex cell wall with a total width of 40 nm. The cell wall consists of a peptidoglycan layer (width: 10 nm) and a surface layer with a zig-zag profile (width: 15 nm; periodicity 9.0 nm) covered by a 15 nm thick strongly contrasted surface coat. The isolated surface layer was composed of protein complexes arranged on a p4 lattice (center to center distance: 11.0 nm).

Three polar lipid compounds have been identified in *Ammonifex degensii:* glycerol diethers (85%), glycerol monoethers, and fatty acid methyl esters. In total, a series of nine glycerol diethers with C_{16}:C_{16} (34%), C_{16}:C_{17} (18%), and C_{17}:C_{17} (20%) side chains as major components have been detected. In addition, a series of monoethers with C_{15} (21%) and C_{16} (62%) alkyl chain monoethers as the major constituents have been identified. The identity and distribution of the *O*-alkyl side chains of the diethers and monoethers is shown in Table 245. The fatty acids have a composition similar to the *O*-alkyl chains of the diethers and monoethers consisting of $C_{14:0}$ (0.6%), $C_{15:0\ iso}$ (13.9%), $C_{15\ ante}$ (5.4%), $C_{15:0}$ (0.6%), $C_{16:0\ iso}$ (12.5%), $C_{16:0}$ (25.5%), $C_{17:0\ iso}$ (13.5%), $C_{17:0\ ante}$ (20.0%), $C_{17:0}$ (1.4%), $C_{18:0\ ante}$ (1.2%), and $C_{18:0}$ (4.2%) fatty acids. The structure of the glycerol mono- and diethers of *Ammonifex degensii* were very similar

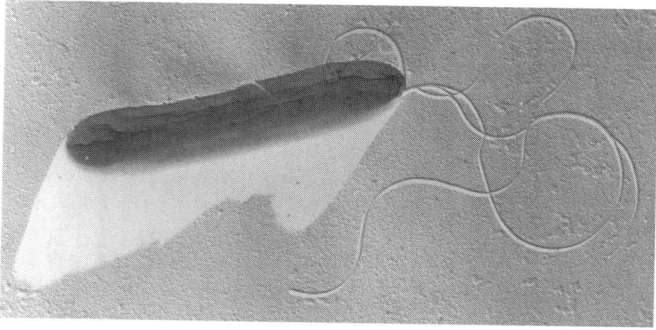

FIGURE 247. Transmission electron micrograph of a single, flagellated cell of *Ammonifex degensii*, platinum-shadowed. Bar = 1 μm.

TABLE 245. Distribution of the *O*-alkyl hydrocarbon chains of the glycerol diethers and glycerol monoethers of *Ammonifex degensii*

Carbon number	Compound	Diether (relative %)	Monoether (relative %)
C14	iso-Tetradecane	0.2	0.6
C14	Tetradecane	1.4	4.4
C15	iso-Pentadecane	7.7	14.2
C15	anteiso-Pentadecane	4.1	6.4
C15	Pentadecane	1.7	2.9
C16	iso-Hexadecane	27.5	32.9
C16	Hexadecane	32	29.1
C17	iso-Heptadecane	6.1	2.4
C17	anteiso-Heptadecane	14.9	4.8
C17	Heptadecane	1.2	0.7
C18	anteiso-Octadecane	0.7	0.4
C18	Octadecane	2.5	1.2

to the lipids known from the extremely thermophilic sulfate-reducer *Thermodesulfobacterium commune* (Langworthy et al., 1983). No isoprenoid glycerol ethers or isoprenoid side chains characteristic of archaeal glycerol ether lipids were detected in *Ammonifex degensii*.

Ammonifex degensii grows autotrophically by oxidation of hydrogen and reduction of nitrate with maximal cell densities up to 5×10^8 cells/ml (10 mmol/l KNO_3, final concentration). Stoichiometric amounts of nitrate and hydrogen consumption versus ammonium production have been determined (Thauer et al., 1977) according to the equation: $NO_3^- + 2\ H^+ + 4\ H_2 \rightarrow NH_4^+ + 3\ H_2O$. NO_2^-, NO, N_2O, and N_2 are not detected. Instead of nitrate, sulfate or sulfur is used as the electron acceptor and H_2S is formed as the end product (3–5 μmol $H_2S/10^8$ cells in the presence of sulfate). Instead of hydrogen, *Ammonifex degensii* is able to use formate as the electron donor in the presence of nitrate or sulfate. Depending on the electron acceptor, ammonium or H_2S is formed.

In the absence of a suitable external electron acceptor, pyruvate is fermented to hydrogen and to stoichiometric amounts of acetate and CO_2. Addition of 0.1% potassium phosphate stimulates growth and CO_2 in the medium is essential. The same products are formed when nitrate or sulfate is added to the medium indicating a fermentation of pyruvate also in the presence of an external electron acceptor.

In combination with formate, thiosulfate, pyruvate (each 0.05%, final concentration), or hydrogen (300 kPa H_2/CO_2; 80:20; v/v), no growth is observed under microaerophilic conditions in the presence of oxygen (0.02, 0.1, 0.5, 2% tested). Furthermore, no growth is obtained with hydrogen as electron donor and KNO_2, $Na_2S_2O_3$, Na_2SO_3, $FeCl_3$, $Na_2Se_4O_4$, $MnCl_2$, or fumarate as possible electron acceptors (0.01 and 0.1% tested). In addition, no growth is observed with nitrate in combination with 0.5% S^0, 0.1% Na_2SO_3, or 0.05% $Na_2S_2O_3$. *Ammonifex degensii* is also unable to grow under a 300 kPa N_2/CO_2 atmosphere (80:20, v/v) with 0.1 % KNO_3 as electron acceptor in the presence of benzoate, meat extract, peptone, yeast extract (each 0.01 %, final concentration), ethanol, 2-propanol, acetate, propionate, succinate, formamide, formaldehyde, fumarate, DL-malate, L(+)-lactate, butyrate, D(+)-glucose, maltose, glycogen, or Casamino acids (each 0.05 %, final concentration).

Growth of *Ammonifex degensii* was inhibited by addition of ampicillin, chloramphenicol, gentamicin, hygromycin, kanamycin, neomycin, penicillin, phosphomycin, rifampin, and vancomycin (each 10 μg/ml; final conccentration).

Ammonifex degensii has been isolated only once from sample KC4, obtained during a land expediton of R/V Sonne cruise in Indonesia (Huber et al., 1991, 1992). The sample was taken anaerobically within the Kawah Candradimuka crater (Dieng Plateau, Java) from the flux of a strongly gassed, grayish water hole with a diameter of about 6 m. Sample KC4 consisted of a mixture of water and grayish and black sediment; the original temperature was 80 °C, the original pH was 7.5 (Huber et al., 1996).

Enrichment and isolation procedures

For the enrichment and isolation of *Ammonifex degensii*, the anaerobic technique of Balch and Wolfe (1976) is suitable. Modified medium 1 supplemented with 0.1% sodium nitrate (Balch et al., 1979), a pH of 7.5, a gas phase of 300 kPa H_2/CO_2 (80:20, v/v), and an incubation temperature of 75 °C is recommended (Huber et al., 1996). Single colonies of *Ammonifex degensii* can be obtained by plating under anaerobic conditions using a stainless-steel anaerobic jar (Balch et al., 1979) and plates and solidified by addition of 2% agar. White, round colonies about 1 mm in diameter will be formed after anaerobic incubation for about 14 d at 75 °C under a H_2/CO_2 atmosphere (200 kPa; 80:20, v/v). Packed cells obtained from large-scale fermentations exhibited an olive green appearance.

Maintenance procedures

Ammonifex degensii can be stored long-term in liquid nitrogen at −140 °C in the presence of 5% dimethylsulfoxide.

Differentiation of the genus *Ammonifex* from other genera

From other closely related taxa, *Ammonifex degensii* can be phylogenetically differentiated by comparison of the 16S rDNA sequences. Furthermore, *Ammonifex degensii* is unique by its metabolic flexibility i.e., the anaerobic growth under autotrophic conditions with hydrogen as the electron donor and nitrate, sulfate, or sulfur as the electron acceptor.

Further reading

Huber, R. and K.O. Stetter. 2000. Discovery of hyperthermophilic micro-organisms. *In* Adams and Kelly (Editors), Methods in Enzymology. Academic Press, San Diego, CA, pp. 11–24.

List of species of the genus *Ammonifex*

1. **Ammonifex degensii** Huber and Stetter 1996, 836[VP] (Effective publication: Huber, Rossnagel, Woese, Rachel, Langworthy and Stetter 1996, 47.)

de.gen'si.i. N.L. gen. n. *degensii* of Degens honoring Egon T. Degens.

Description is the same as for the genus.
DNA G+C content (mol%): 53 (T_m) and 55.5 (HPLC).
Type strain: KC4, DSM 10501.
GenBank accession number (16S rRNA gene): U34975.

Genus III. **Caldanaerobacter** Fardeau, Bonilla-Salinas, L'Haridon, Jeanthon, Verhé, Cayol, Patel, Garcia and Ollivier 2004, 471[VP]

MARIE-LAURE FARDEAU, BERNARD OLLIVIER AND JEAN-LUC CAYOL

Cald.an.ae.ro.bac'ter. L. adj. *caldus* hot; Gr. pref. *an* not; Gr. n. *aer* air; N.L. masc. n. *bacter* equivalent of Gr. neut. dim. n. *bakterion* rod, staff; N.L. masc. n. *Caldanaerobacter* a rod that grows in the absence of air at high temperatures.

Cells are **straight rods** that stain **Gram-positive or negative**. Endospores may be observed. **Strictly anaerobic heterotrophs**. Acetate and **L-alanine** are major end products of glucose fermentation, with approximately 1 mol of L-alanine being produced per mol of glucose. **Thermophilic** member of the family *Thermoanaerobacteraceae*.

DNA G+C content (mol%): 33–41.

Type species: **Caldanaerobacter subterraneus** (Fardeau, Magot, Patel, Thomas, Garcia and Ollivier 2000) Fardeau, Bonilla-Salinas, L'Haridon, Jeanthon, Verhé, Cayol, Patel and Garcia 2004, 471[VP] (*Thermoanaerobacter subterraneus* Fardeau, Magot, Patel, Thomas, Garcia and Ollivier 2000, 2145).

Further descriptive information

Members of *Caldanaerobacter* were originally classified in the genera *Thermoanaerobacter* (e.g., *Thermoanaerobacter subterraneus*, *Thermoanaerobacter yonseiensis* and *Thermoanaerobacter tengcongensis*; Fardeau et al., 2000; Kim et al., 2001; Xue et al.,

2001) or *Carboxydibrachium* (e.g., *Caldanaerobacter pacificum*; Sokolova et al., 2001). On the basis of phylogenetic, phenotypic, and metabolic characteristics, all of these species were reclassified within *Caldanaerobacter* as different subspecies of *Caldanaerobacter subterraneus*, the only representative of this genus so far characterized. A remarkable metabolic feature of these organisms is their ability to ferment glucose to L-alanine in nearly equal molar amounts. Most strains grow at 80 °C and are, therefore, considered extreme thermophiles. They all require yeast extract for growth, and several subspecies oxidize CO to CO_2 and H_2. Strains may be isolated from hot terrestrial and subterrestrial ecosystems as well as deep-sea hydrothermal vents.

Enrichment and isolation procedure

These strict anaerobes are cultivated using Hungate techniques (Hungate, 1969). Enrichments of *Caldanaerobacter* can be obtained on media containing yeast extract and sugars as energy sources (Fardeau et al., 2000; Kim et al., 2001; Xue et al., 2001). CO may also be utilized (Sokolova et al., 2001). They may require a seawater-based medium, depending on their source of isolation (Sokolova et al., 2001).

One typical *Caldanaerobacter* medium is as follows (per liter of distilled water): 1 g of NH_4Cl, 0.3 g of K_2HPO_4, 0.3 g of KH_2PO_4, 0.1 g of $CaCl_2 \cdot 2H_2O$, 0.2 g of KCl, 0.5 g of $MgCl_2 \cdot 6H_2O$, 2.0 g of yeast extract, 2.0 g of bio-Trypticase, 10.0 g of NaCl, 3.6 g of glucose, 0.5 g of cysteine-HCl, and 10 ml of trace element solution (Balch et al., 1979). The NaCl concentration can be increased up to 18 g/l for recovering marine isolates, and CO may replace sugars as the energy source (Sokolova et al., 2001). The pH is adjusted to 7.0 with 10 M KOH, and the medium is autoclaved for 45 min at 110 °C. Prior to inoculation, $Na_2S \cdot 9H_2O$ and $NaHCO_3$ are added from sterile stock solutions.

Maintenance procedure

Anaerobic cultures can be maintained for at least one year in basal medium plus 20% glycerol at –80 °C or plus 50% glycerol at –20 °C.

Differentiation of the genus *Caldanaerobacter* from other genera

Caldanaerobacter may be distinguished from *Thermoanaerobacter* by differences in the sequences of their 16S rRNA genes (mean sequence similarity of 92–95%), differences in the mol% G+C contents of their DNA, and the ability to produce large amounts of L-alanine during the glucose fermentation.

Taxonomics comments

These subspecies of *Caldanaerobacter subterraneus* (*Caldanaerobacter subterraneus* subsp. *subterraneus*, *Caldanaerobacter subterraneus* subsp. *tengcongensis*, *Caldanaerobacter subterraneus* subsp. *yonseiensis*, and *Caldanaerobacter subterraneus* subsp. *pacificus*) constitute a distinct phylogenetic lineage within the family *Thermoanaerobacteraceae* (Figure 248). For instance, the 16S rRNA gene of *Caldanaerobacter subterraneus* subsp. *subterraneus* possesses 97.2, 97.7, and 98.2% sequence similarity with the genes of *Caldanaerobacter subterraneus* subsp. *pacificus*, *Caldanaerobacter subterraneus* subsp. *tengcongensis*, and *Caldanaerobacter subterraneus* subsp. *yonseiensis*, respectively. Similarly, DNA–DNA similarity values between these subspecies all exceed 68%. Despite these molecular similarities, the phenotypes of these subspecies are significantly different, especially the mol% G+C contents of their DNA and their ability to oxidize CO (Table 246).

Differentiation of the species of the genus *Caldanaerobacter*

Only one species, *Caldanaerobacter subterraneus*, with four subspecies is currently known.

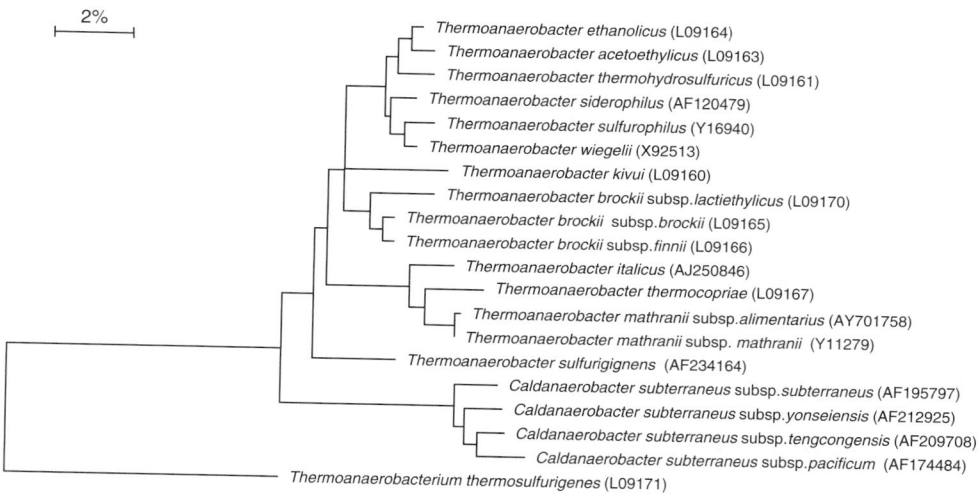

2%

Thermoanaerobacter ethanolicus (L09164)
Thermoanaerobacter acetoethylicus (L09163)
Thermoanaerobacter thermohydrosulfuricus (L09161)
Thermoanaerobacter siderophilus (AF120479)
Thermoanaerobacter sulfurophilus (Y16940)
Thermoanaerobacter wiegelii (X92513)
Thermoanaerobacter kivui (L09160)
Thermoanaerobacter brockii subsp. *lactiethylicus* (L09170)
Thermoanaerobacter brockii subsp. *brockii* (L09165)
Thermoanaerobacter brockii subsp. *finnii* (L09166)
Thermoanaerobacter italicus (AJ250846)
Thermoanaerobacter thermocopriae (L09167)
Thermoanaerobacter mathranii subsp. *alimentarius* (AY701758)
Thermoanaerobacter mathranii subsp. *mathranii* (Y11279)
Thermoanaerobacter sulfurigignens (AF234164)
Caldanaerobacter subterraneus subsp. *subterraneus* (AF195797)
Caldanaerobacter subterraneus subsp. *yonseiensis* (AF212925)
Caldanaerobacter subterraneus subsp. *tengcongensis* (AF209708)
Caldanaerobacter subterraneus subsp. *pacificum* (AF174484)
Thermoanaerobacterium thermosulfurigenes (L09171)

FIGURE 248. Phylogeny of *Caldanaerobacter subterraneus* and *Thermoanaerobacter* species. The phylogenetic tree of the 16S rRNA genes was constructed by the neighbor-joining method using the type species of *Thermoanaerobacterium* as outgroup. Bar indicates 2% inferred substitutions.

TABLE 246. Major discriminating characteristics of subspecies of *Caldanaerobacter subterraneus*[a]

| Characteristic | *Caldanaerobacter subterraneus* subsp. | | | |
	subterraneus[b]	*tengcongensis*[c]	*yonseiensis*[d]	*pacificus*[e]
Source	Oil well	Hot spring	Geothermal water	Hydrothermal vent
Temperature range (°C)	40–75	50–80	50–85	50–80
Optimum temperature (°C)	65	75	75	70
NaCl range (%)	0–3	0–2.5	0–4	ND
Optimum NaCl (%)	0	0.2	0	2–2.5
DNA G+C content (mol%)	41	33	37	33
CO oxidation	–	+	+	+
Diagnostic fermentation products[f]				
Ethanol	–	+	+	–
Lactate	+	–	+	–

[a]Symbols and abbreviations: +, present; –, absent; ND, not determined.
[b]Data from Fardeau et al. (2000).
[c]Data from Xue et al. (2001).
[d]Data from Kim et al. (2001).
[e]Data from Sokolova et al. (2001).
[f]In addition to L-alanine, acetate, H_2, and CO_2, which are produced by all subspecies from glucose.

List of species of the genus *Caldanaerobacter*

1. **Caldanaerobacter subterraneus** (Fardeau, Magot, Patel, Thomas, Garcia and Ollivier 2000) Fardeau, Bonilla-Salinas, L'Haridon, Jeanthon, Verhé, Cayol, Patel and Garcia 2004, 471[VP] (*Thermoanaerobacter subterraneus* Fardeau, Magot, Patel, Thomas, Garcia and Ollivier 2000, 2145.)

sub.ter.ra′ne.us. L. masc. adj. *subterraneus* underground, subterranean, describing its site of isolation.

Same description as that given for the genus. L-alanine, acetate, lactate, H_2, and CO_2 are produced during glucose fermentation.

DNA G+C content (mol%): 38.4–41 (HPLC).

Type strain: ATCC BAA-225, DSM 13054, CNCM I-2383, SEBR 7858.

GenBank accession number (16S rRNA gene): AF195797, AF174484, AF209708, and AF212925.

1a. **Caldanaerobacter subterraneus subsp. subterraneus** (Fardeau, Magot, Patel, Thomas, Garcia and Ollivier 2000) Fardeau, Bonilla-Salinas, L'Haridon, Jeanthon, Verhé, Cayol, Patel and Garcia 2004, 471[VP] (*Thermoanaerobacter subterraneus* Fardeau, Magot, Patel, Thomas, Garcia and Ollivier 2000, 2145.)

Rods (0.5–0.7 × 2–8 μm) that occur singly or in pairs and possess laterally inserted flagella. Spores are not observed, but cultures exposed to 120 °C for 45 min retain viability, indicating the presence of heat-resistant forms. Electron microscopic examination reveals a Gram-positive type cell wall. Round colonies (3 mm in diameter) develop on Phytagel plates or in roll-tubes after 3 d of incubation at 70 °C. Obligately anaerobic chemo-organotroph. Thermophilic. Optimum temperature for growth is 65–75 °C; temperature range is 40–80 °C. Optimum pH is 7.0–7.5; the pH range is 5.7–9.2. Halotolerant; grows in the presence of up to 3% NaCl. Either yeast extract or bio-Trypticase is required for growth on carbohydrates, and both compounds together greatly enhance growth. Yeast extract cannot be replaced by vitamins. Ferments cellobiose, D-fructose, D-galactose, D-glucose, DL-lactose, DL-maltose, D-mannose, melibiose, D-ribose, starch, D-xylose, glycerol, mannitol, pyruvate, and

xylan. L-alanine, acetate, lactate, H_2, and CO_2 are produced during glucose fermentation. Elemental sulfur, thiosulfate, and sulfite, but not sulfate, are used as electron acceptors. The type strain was isolated from oilfield water located in southwest France.

DNA G+C content (mol%): 41 (HPLC).

Type strain: ATCC BAA-225, CNCM I-2383, DSM 13054, SEBR 7858.

GenBank accession number (16S rRNA gene): AF195797.

1b. **Caldanaerobacter subterraneus subsp. pacificus** (Sokolova, González, Kostrikina, Chernyh, Tourova, Kato, Bonch-Osmolovskaya and Robb 2001) Fardeau, Bonilla-Salinas, L'Haridon, Jeanthon, Verhé, Cayol, Patel and Garcia 2004, 54[VP] (*Carboxydobrachium pacificum* Sokolova, González, Kostrikina, Chernyh, Tourova, Kato, Bonch-Osmolovskaya and Robb 2001, 147.)

pa.ci′fi.cus. L. masc. adj. *pacificus* peaceful; pertaining to the Pacific Ocean, the eastern part of which is the source of the type strain.

Cells are long, thin, straight, nonmotile rods, about 0.3 × 4–10 μm. Occur singly or form chains of three to five cells. Cells sometimes branch. They have a Gram-positive type of cell wall that is covered with an S-layer. On solid medium, round, white, translucent colonies are produced. Obligate anaerobe. Extreme thermophile. Grows within the temperature range of 50–80 °C; optimum growth is at 70 °C. The pH growth range is 5.8–7.6; the optimum is 6.8–7.2. Grows chemolithotrophically on CO. Growth and CO consumption are not inhibited by penicillin, but ampicillin, streptomycin, kanamycin, and neomycin inhibit growth and CO utilization. Requires seawater-based medium. The optimum concentration of sea salts is 20.5–25.5 g/l. This species utilizes CO as the sole energy source with formation of H_2 and CO_2 according to the equation: $CO + H_2O \rightarrow CO_2 + H_2$. Requires yeast extract, peptone, or acetate as a carbon source. Cells grow organotrophically on peptone, yeast extract, starch, cellobiose, glucose, galactose, fructose, and pyruvate. The products of organotrophic growth are acetate, CO_2, and H_2. An H_2/CO_2 gas mixture does not support growth. Reduces thiosulfate to sulfide. This species was isolated from a submarine hydrothermal vent in the Okinawa Trough.

DNA G+C content (mol%): 33 (T_m).

Type strain: ATCC BAA-271, JM, DSM 12653.

GenBank accession number (16S rRNA gene): AF174484.

1c. **Caldanaerobacter subterraneus subsp. tengcongensis** (Xue, Xu, Liu, Ma and Zhou 2001) Fardeau, Bonilla-Salinas, L'Haridon, Jeanthon, Verhé, Cayol, Patel and Garcia 2004, 472^{VP} (*Thermoanaerobacter tengcongensis* Xue, Xu, Liu, Ma and Zhou 2001, 1340)

teng.con.gen'sis. N.L. masc. adj. *tengcongensis* pertaining to Tengcong, China.

Cells are rods, 0.5–0.6 × 1–10 μm, that stain Gram-negative. They occur singly, in pairs, or in chains. No spores are observed. Heterotrophic growth requires yeast extract, which cannot be replaced by Tryptone or Casamino acids. Utilizes glucose, galactose, maltose, cellobiose, mannose, fructose, lactose, maltose, and starch. The major fermentation products on glucose are L-alanine and acetate. Oxidizes CO. Hydrogen and sulfite inhibit growth. Thiosulfate and sulfur stimulate growth. Chloromycetin, polymyxin B, streptomycin sulfate, and tetracycline HCl inhibit growth. This bacterium has been isolated from a hot spring in Tengcong in China.

DNA G+C content (mol%): 33 (T_m).

Type strain: Chinese Collection of Microorganisms AS 1.2430, DSM 15242, JCM 11007, MB4.

GenBank accession number (16S rRNA gene): AF209708.

1d. **Caldanaerobacter subterraneus subsp. yonseiensis** (Kim, Grote, Lee, Antranikian and Pyun 2001) Fardeau, Bonilla-Salinas, L'Haridon, Jeanthon, Verhé, Cayol, Patel and Garcia 2004, 472^{VP} (*Thermoanaerobacter yonseiensis* Kim, Grote, Lee, Antranikian and Pyun 2001, 1546)

yon.sei.en'sis. N.L. masc. adj. *yonseiensis* pertaining to Yonsei University, Seoul, Korea, in recognition of its support of research into extreme thermophiles and their thermostable enzymes.

Cells are motile, straight rods, 0.4–0.8 μm × 1.0–3.0 μm. Gram-positive. Under suboptimal growth conditions, cells occur in long chains and form terminal spores. Optimal conditions for growth are 75°C (range 50–85°C) and pH 6.5 (range 4.5–9.0). The doubling time under optimal conditions on medium containing starch, yeast extract, and thiosulfate is 60 min. Cells grow chemoorganotrophically under strictly anaerobic conditions on glucose, fructose, lactose, maltose, D-xylose, D-galactose, mannose, sucrose, cellobiose, starch, pullulan, yeast extract, tryptone, and peptone. Cells require yeast extract, elemental sulfur, L-cysteine, sodium nitrate, or sodium thiosulfate for growth. Oxidizes CO. Major fermentation products on glucose are L-alanine, lactate, acetate, ethanol, CO_2, and H_2S. Propionate, butyrate, isovalerate, 2-propanol, and 1-pentanol are formed in small amounts. Growth on glucose is inhibited by tetracycline, chloramphenicol, penicillin G, neomycin, vancomycin, kanamycin, and rifampin. Respiratory lipoquinones are absent. Approximately 1 mol of L-alanine is produced per mol of glucose fermented. This subspecies was isolated from mud samples taken from a hot stream at Sileri in the Dieng volcanic area located on the island of Java, Indonesia.

DNA G+C content (mol%): 37 (HPLC).

Type strain: DSM 13777, KB-1^P, KFCC 11116^P.

GenBank accession number (16S rRNA gene): AF212925.

Genus IV. **Carboxydothermus** Svetlichny, Sokolova, Gerhardt, Ringpfeil, Kostrikina and Zavarzin 1991b, 580^{VP} (Effective publication: Svetlichny, Sokolova, Gerhardt, Ringpfeil, Kostrikina and Zavarzin 1991a, 258; emend. Slobodkin, Sokolova, Lysenko and Wiegel 2006, 2350.)

JUERGEN WIEGEL

Car.bo.xy.do.ther'mus. N.L. n. *carboxydum* carbon monoxide; Gr. adj. *thermos* hot; N.L. masc. n. *Carboxydothermus* utilizing carbon monoxide while living in hot places.

Cells are straight to slightly curved rods, 0.3–0.5 × 1.3–2.7 μm. **Gram-stain-positive with a Gram-positive cell-wall structure.** Metabolism is **obligately anaerobic, chemolithotrophic or chemo- or organoheterotrophic.** Characteristic feature of the genus is the **utilization of CO.**

The type species is obligately carboxydotrophic, forming H_2 and CO_2 from CO. The other species can also oxidize organic substances such as glycerol, mainly to acetate. While additional electron acceptors are not required, some that are used include Fe(III), sulfite, thiosulfate, elemental sulfur, nitrate, fumarate, and 9,10-anthraquinone-2,6-disulfonate, Growth occurs in the slightly acidic to slightly alkaline pH range. All described species are **thermophilic.** Sulfate is not reduced.

DNA G+C content (mol%): 39–41.

Type species: **Carboxydothermus hydrogenoformans** Svetlichny, Sokolova, Gerhardt, Ringpfeil, Kostrikina and Zavarzin 1991b, 580^{VP} (Effective publication: Svetlichny, Sokolova, Gerhardt, Ringpfeil, Kostrikina and Zavarzin 1991a, 258.).

Further descriptive information

Obligately anaerobic. The optimal temperature for growth of both species is 65–70°C. Both species grow above 70°C; thus, they are regarded as extremely thermophilic anaerobes. The pH^{25C} range for growth is from slightly acidic to slightly alkaline. Both species can grow chemolithoautotrophically with CO but with different electron acceptors. *Carboxydothermus hydrogenoformans* is an obligate CO-utilizer. It uses protons as the electron acceptor in the reaction: CO + H_2O → CO_2 + H_2. *Carboxydothermus ferrireducens* requires electron acceptors other than protons to oxidize CO. It can also grow chemo- or organoheterotrophically. For organoheterotrophic growth, the medium is supplemented with glycerol (30 mM) as an electron donor and a carbon source and fumarate (20 mM) and/or ferrihydrite [90 mM; a poorly crystalline Fe(III) oxide/hydroxide, freshly prepared] as electron acceptors. Other electron acceptors include Fe(III) citrate (20 mM), Fe(III) EDTA (15 mM), thiosulfate (20 mM), fumarate (10 mM), 9,10-anthraquinone-2,6-disulfonate (20 mM), and, with lactate as electron donor, precipitated or sublimated elemental sulfur (2 g/l), sulfite (10 mM), and nitrate (10 mM). Electron donors may include CO (100% of the headspace gas), H_2 (in the presence of CO_2 in a 80:20 ratio), formate (20 mM), lactate (20 mM), or glycerol (30 mM). Growth is poor or absent on hexoses. Both species are inhibited by 100 μg/ml chloramphenicol. The sensitivity to other antibiotics differs between the two species. Under optimal growth conditions, the doubling times are around 2–3 h.

Enrichment and isolation procedures

Habitats are terrestrial geothermally heated features. *Carboxydothermus hydrogenoformans* was isolated from geothermally heated swamp mud of the Kinashir Island, Kamchatka, Far-East Russia (Svetlichny et al., 1991a). *Carboxydothermus ferrireducens* was isolated from a hot spring in Yellowstone National Park, USA (Slobodkin et al., 1997b).

The enrichment and isolation of *Carboxydothermus* is based on either of the two distinguishing characteristics, the ability to oxidize CO to H_2 and CO_2 or to couple CO oxidation with ferrihydrite reduction to magnetite. For routine cultivation, both species are grown at 60–70 °C in a bicarbonate-buffered mineral medium at neutral pH plus 0.2 g/l yeast extract (Slobodkin et al., 1997b; Svetlichny et al., 1991a). For *Carboxydothermus hydrogenoformans*, the headspace gas is 100% CO. For *Carboxydothermus ferrireducens*, the headspace gas is N_2 and the medium is supplemented with 30 mM glycerol, 20 mM fumarate and/or 90 mM ferrihydrite (freshly prepared). Single colonies of *Carboxydothermus hydrogenoformans* have been obtained using roll-flasks with 5% (w/v) agar, whereas *Carboxydothermus ferrireducens* has been isolated from 2.2% agar-shake roll tubes (Ljungdahl and Wiegel, 1986).

Maintenance procedures

Carboxydothermus ferrireducens cultures can be kept for several weeks at room temperature when grown in Fe(III)-containing mineral media. Cultures supplemented with 40–50% glycerol (by vol., final concentration) can be stored for years at –20 °C, which is ideal for frequent withdrawals with a syringe since the stock remains liquid, and at –80 °C for even longer storage. Collections store the culture after lyophilization in sealed glass ampoules.

Taxonomic and phylogenetic comments

The genus *Carboxydothermus* belongs to the family "*Thermoanaerobacteraceae*" of the order "*Thermoanaerobacterales*". Like some of the other genera of this group, sporulation has never been observed. The phylogenetic position of *Carboxydothermus hydrogenoformans* has been confirmed by the recently published whole-genome sequence (Wu et al., 2005).

Differentiation of the genus *Carboxydothermus* from other carboxydothermophilic bacteria

Several anaerobic, thermophilic, carboxydotrophic, hydrogenogenic bacteria have been described. They belong to several phylogenetically diverse taxa. They all grow lithotrophically on CO, performing the metabolic reaction $CO + H_2O \rightarrow CO_2 + H_2$ ($\Delta G^0 = -20$ kJ/mol). The first bacterium to be isolated was *Carboxydothermus hydrogenoformans* (Svetlichny et al., 1991a), followed

by *Caldanaerobacter subterraneus* subsp. *pacificus* (Fardeau et al., 2004; Sokolova et al., 2001), *Carboxydocella thermautotrophica* (Sokolova et al., 2002), *Thermosinus carboxydivorans* (Sokolova et al., 2004a) and *Thermincola carboxydiphila* (Sokolova et al., 2005), as well as others that will be published in the near future (T.G. Sokolova, personal communication). Furthermore, a hyperthermophilic carboxydotrophic hydrogenogenic archaeon belonging to the genus *Thermococcus* was isolated recently (Sokolova et al., 2004b). So far, all these taxa are neutrophiles (pH range 6.5–7.8); only *Thermincola carboxydiphila*, isolated from Lake Baikal, is a moderate alkaliphile. The main difference between bacterial taxa with very similar metabolism and *Carboxydothermus hydrogenoformans* is their different phylogenetic positions based on 16S rRNA sequence analysis. In addition, the other species of *Carboxydothermus*, *Carboxydothermus ferrireducens*, is not a carboxydogenic hydrogenogenic chemolithoautotroph, but a chemoheterotrophic Fe(III)-reducing bacterium. Other taxa and strains also exhibit very similar physiology, i.e., the ability to reduce Fe(III) using H_2 as an electron donor (Gavrilov et al., 2003; Slobodkin et al., 1999b; Slobodkin and Wiegel, 1997; Sokolova et al., 2007), and CO utilization has not been tested for most of them so it is premature at this time to give a key for differentiation of this genus from other taxa; this will only be possible when further species have been isolated and described.

Differentiation of the species of the genus *Carboxydothermus*

The differences between the two species of *Carboxydothermus* may partly reflect the isolation conditions, since only single characterized strains exist for each species. Physiologically similar strains have been isolated, but they have not been characterized and are not available in culture collections. The DNA–DNA hybridization between the two type strains is 53%, which supports their classification into two species despite their 98.4% 16S rRNA gene sequence similarity and some similar metabolic properties. Based on the properties of the type strains, the two species utilize CO differently (Henstra and Stams, 2004; Pusheva and Sokolova, 1995). Whereas *Carboxydothermus hydrogenoformans* is a chemolithoautotrophic CO-utilizer (in the presence of 0.05–0.5%, w/v, yeast extract), *Carboxydothermus ferrireducens* was isolated as a chemoheterotrophic (dissimilatory) Fe(III) reducer. *Carboxydothermus hydrogenoformans* is capable of Fe(III) reduction with H_2 as an electron donor and ferrihydrite as an electron acceptor. It also oxidizes CO to H_2 and CO_2, but it cannot grow chemolithotrophically on these products. In contrast, *Carboxydothermus ferrireducens* grows on CO, utilizing an external electron acceptor such as Fe(III), but without producing hydrogen or acetate. After 7 d of cultivation, the cell yield is 2.5–3.0×10^7 cells/ml, having converted 26–28 mM ferrihydrite to Fe(II) as a black magnetic precipitate and consumed 14 mM equivalents of CO. In contrast, *Carboxydothermus hydrogenoformans* does not grow with CO in the presence of ferrihydrite, Fe(III) citrate or Fe(III) EDTA. However, with H_2, ferrihydrite is reduced to 20–22 mM Fe(II) after 7 d, with a maximal cell yield of 3.5–4.0×10^7 cells/ml. Acetate does not accumulate, indicating that this is not a homoacetogenic fermentation.

List of species of the genus *Carboxydothermus*

1. **Carboxydothermus hydrogenoformans** Svetlichny, Sokolova, Gerhardt, Ringpfeil, Kostrikina and Zavarzin 1991b, 580^{VP} (Effective publication: Svetlichny, Sokolova, Gerhardt, Ringpfeil, Kostrikina and Zavarzin 1991a, 258.)

hy.dro.gen.o.for'mans. N.L. n. *hydrogenum* hydrogen; L. part. adj. *formans* forming, N.L. part. adj. *hydrogenoformans* hydrogen-forming.

This description is based mainly on Svetlichny et al. (1991a) and Henstra and Stams (2004). Cells form round, white, translucent colonies during growth on agar plates under 100% CO. They are straight to slightly curved rods, $0.4–0.5 \times 1.3–2.4\,\mu m$. Spores are not observed. Cells occur singly, in pairs, in short chains, or in palisade-like aggregates. Cells from the early exponential growth phase are motile and contain 1–2 laterally inserted flagella. An S-layer is present. Cells stain Gram-positive. During late exponential growth phase, lysis occurs and L-shaped cells form.

Growth temperature range is 40–78 °C, with an optimum around 70–72 °C (no growth at 37 or 80 °C). The pH range for growth is 6.4–7.7, with an optimum of pH 6.8–7.0 (no growth at pH 6.0 or 8.0).

Obligately anaerobic carboxydotroph, i.e., obligately anaerobic chemolithoautotroph with no growth in the absence of CO using peptone, yeast extract, or other organic substrates.

Sulfate and sulfur are not reduced.

Isolated from a hot swamp in Kinashir Island (Kamchatka, Far-East Russia).

Penicillin, streptomycin, and chloramphenicol prevent growth at $100\,\mu g/ml$.

DNA G+C content (mol%): 39 (T_m).

Type strain: Z-2901, DSM 6008, ATCC BAA-161.

GenBank accession number (16S rRNA gene): AF244579.

2. **Carboxydothermus ferrireducens** (Slobodkin, Reysenbach, Strutz, Dreier and Wiegel 1997b) Slobodkin, Sokolova, Lysenko and Wiegel 2006, 2350[VP] (*Thermoterrabacterium ferrireducens* Slobodkin, Reysenbach, Strutz, Dreier and Wiegel 1997b, 546)

fer.ri.re.du'cens. L. n. *ferrum* iron; L. part. adj. *reducens* converting to a different state; N.L. part. adj. *ferrireducens* reducing ferric iron.

The description is based on Slobodkin et al. (1997b), with additional data from Gavrilov et al. (2003), Henstra & Stams (2004), and Slobodkin et al. (2006).

Cells are straight to slightly curved rods, $0.3–0.4 \times 1.6–2.7\,\mu m$, with rounded ends. They occur in liquid culture as single cells or in pairs. Flagellation is of the retarded peritrichous type leading mainly to a tumbling motility. Cells stain Gram-positive. Spores are not observed. Obligately anaerobic metabolism. Cells do not require complex media compo-

nents for growth. Growth occurs at 50–74 °C, with optimum growth at 65 °C, and at a pH[65C] range of 5.5–7.6, with optimum growth at 6.0–6.2. No growth is observed at or below 48 °C, at or above 76 °C, at or below a pH[65C] of 5.3, or at or above a pH[65C] of 7.8. Grows in 0.0–1.0% (w/v) NaCl, but not 2% NaCl.

In the presence of CO_2, strain JW/AS-Y7[T] grows with glycerol as the sole organic carbon source and forms the incompletely oxidized product acetate while reducing ferrihydrite, Fe(III) citrate, or Fe(III) EDTA. H_2 is not formed. The ratio of acetate formed per glycerol utilized is greater than one and, based on the stoichiometries observed, suggests the use of an acetogenic pathway (not yet elucidated) according to the hypothetical equations $4\,C_3H_8O_3 + 2\,HCO_3^- \rightarrow 7\,C_2H_3O_2^- + 5\,H^+ + 4\,H_2O$ ($\Delta G^0 = -151.5\,kJ/mol$ glycerol) and, in the presence of Fe(III) ions, $8\,C_3H_8O_3 + 2\,HCO_3^- + 8\,Fe^{3+} \rightarrow 13\,C_2H_3O_2^- + 19\,H^+ + 4\,H_2O + 8\,Fe^{2+}$ ($\Delta G^0 = -252.6\,kJ/mol$ glycerol).

The following substrates are also utilized: lactate, 1,2-propanediol, glycerate, pyruvate, yeast extract, and peptone. In addition, glucose, fructose, and mannose support weak growth with acetate as the only organic product. Lithoautotrophic growth is possible with H_2 and ferrihydrite as the electron acceptor. CO_2 or CO can serve as the only carbon source. Carbon monoxide is oxidized with ferrihydrite, fumarate, or 9,10-anthraquinone-2,6-disulfonate as electron acceptors without production of H_2. No growth occurs with acetate, methanol, ethanol, *n*-propanol, *i*-propanol, *n*-butanol, propionate, acetone, ethyleneglycol, 1,3-propanediol, fumarate, succinate, phenol, benzoate, 9,10-anthraquinone-2,6-disulfonate, starch, olive oil, elemental sulfur, sucrose, galactose, xylose, cellobiose, or arabinose, with or without Fe(III). Reduces Fe(III) to Fe(II), 9,10-anthraquinone-2,6-disulfonate to 9,10-anthrahydroquinone-2,6-disulfonate, fumarate (20 mM) to succinate, thiosulfate (20 mM) to elemental sulfur or sulfide, and, with 20 mM lactate as carbon source and electron donor [Henstra and Stams (2004) in contrast to Slobodkin et al. (1997b)], 10 mM sulfite to sulfide, 2 g/l elemental sulfur to sulfide, and 10 mM nitrate to nitrite and ammonium. Does not reduce MnO_2. Chloramphenicol, erythromycin, and rifampin prevent growth at $100\,\mu g/ml$, whereas $100\,\mu g/ml$ ampicillin, streptomycin, and tetracycline have no effect.

Isolated from a freshwater hot spring in the Calcite Springs area in Yellowstone National Park, Wyoming, USA.

DNA G+C content (mol%): 41 (HPLC).

Type strain: JW/AS-Y7, DSM 11255, VKM B-2392.

GenBank accession number (16S rRNA gene): U76363.

Genus V. **Gelria** Plugge, Balk, Zoetendal and Stams 2002, 406[VP]

CAROLINE M. PLUGGE

Gel.ri'a. N.L. fem. n. *Gelria* Gelre or Gelderland, one of the 12 provinces in The Netherlands, in which Wageningen is located.

Cells are **rod-shaped** $0.5 \times 3–20\,\mu m$ in glucose-containing liquid cultures. **Cells vary in length dependent on growth substrate.** Cells are short rods ($<2\,\mu m$) when grown in co-culture with a methanogenic partner on glutamate. **Nonmotile** and **Gram-stain-positive.** Formation of **terminal spores, $0.5 \times 0.5\,\mu m$, during** **late exponential phase. Strictly anaerobic chemo-organotroph. Moderately thermophilic** between 37 and 60 °C with an optimum at 50–55 °C. **Neutrophilic;** pH range 5.5–8. **Saccharolytic growth** in pure culture. **Hydrogen formed can be transferred to a syntrophic methanogenic partner,** not to sulfate, sulfite, thiosulfate, nitrate, or fumarate. Isolated from granular sludge from an upflow anaerobic sludge bed reactor.

DNA G+C content (mol%): 33.8 (T_m).

Type species: **Gelria glutamica** Plugge, Balk, Zoetendal and Stams 2002, 406^{VP}.

Enrichment and isolation procedures

The type species of the genus *Gelria, Gelria glutamica,* has been isolated from methanogenic granular sludge originating from an upflow anaerobic sludge bed reactor. The bicarbonate-buffered mineral salts medium used for enrichment and cultivation is as described under genus *Syntrophobacter* (see McInerney et al., 2004). In addition, 0.5 g yeast extract per liter is added as a source of unknown growth factors.

For primary enrichments, the basal medium with 20 mM sodium glutamate and sodium sulfide as the reductant is used. This medium is also inoculated or pregrown with a pure culture of a methanogen, to allow degradation of glutamate via interspecies electron transfer. The type strain can only convert glutamate in the presence of a methanogenic partner. A pure culture of *Gelria glutamica* was obtained by serial dilution in basal medium containing agar with pyruvate as the carbon and energy source. Defined syntrophic co-cultures of *Gelria glutam-ica* are constructed by growing the pure culture in glutamate basal medium that has also been inoculated with a pure culture of *Methanothermobacter thermautotrophicus* Z-245.

Maintenance procedures

For short-term maintenance, members of this genus can be stored in their original growth medium at 4 °C in the dark. For long-term storage, a glucose-grown pure culture can be freeze-dried. Recovery of a syntrophic co-culture converting glutamate from a freeze-dried culture is achieved by growing the pure culture in glutamate-containing basal medium that has also been inoculated with a pure culture of *Methanothermobacter thermautotrophicus* Z-245. The addition of 5 mM pyruvate to this co-culture significantly shortens the lag phase.

Taxonomic comments

Phylogenetically, the type strain *Gelria glutamica* is only distantly related to *Moorella glycerini* and *Moorella thermoacetica* with similarities of the 16S rRNA gene sequence between 90–92%. The 16S rRNA gene sequence of *Gelria glutamica* has additional loops in the V1, V7, and V9 helices. These additional loops are only present in the 16S rRNA gene sequence and are not transcribed to the 16S rRNA.

List of species of the genus *Gelria*

1. **Gelria glutamica** Plugge, Balk, Zoetendal and Stams 2002, 406^{VP}

glu.ta'mi.ca. N.L. n. *acidum glutamicum* glutamic acid; N.L. fem. adj. *glutamica* referring to glutamic acid, on which the bacterium grows.

Colonies appear white and round at the surface and are 0.7–1.0 mm in diameter when grown on pyruvate-containing soft agar (0.7–0.8%) media at 55 °C. Subsurface colonies are lenticular. Cells are 0.5 by 0.5–6 μm, variations depending on the growth substrate. In pure culture the strain can grow on pyruvate, lactate, glycerol, glucose, rhamnose, and galactose. In syntrophic association with a hydrogenotrophic methanogen, the organism can utilize glutamate, 2-oxoglutarate, proline, Casamino acids, and a variety of sugars. Glutamate and proline are degraded to propionate, H_2, NH_4^+, and HCO_3^-. Propionate formation from glutamate occurs via direct oxidation through the intermediate methylmalonyl-CoA (Plugge et al., 2001). Sugars are converted to acetate, propionate, HCO_3^-, and H_2 as main products. Growth occurs between 37–60 °C with an optimum at 50–55 °C and pH 5.5–8 (optimum 7).

DNA G+C content (mol%): 33.8 (T_m).

Type strain: TGO, ATCC BAA-262, DSM 14054.

GenBank accession number (16S rRNA gene): AF321086.

Genus VI. **Moorella** Collins, Lawson, Willems, Cordoba, Fernández-Garayzábal, Garcia, Cai, Hippe and Farrow 1994, 822^{VP}

JUERGEN WIEGEL

Moo.rel'la. N.L. fem. n. *Moorella* in honor of W.E.C. (Ed) Moore, an American bacteriologist, who worked with anaerobes.

In early exponential growth phase, cells stain Gram-positive. However, some species stain negative during the late exponential and stationary growth phases. **Straight rods with a tendency to polymorphism** under stress conditions such as high glucose or high acetate concentrations. **Physiology is obligately anaerobic, thermophilic, and chemolithoautotrophic and/or heterotrophic;** produces acetate as sole or main fermentation product from sugars, C_1 carbon sources, and other substrates. Produces nearly 3 moles of acetate per mole of glucose consumed, which is sometimes called "homoacetogenic" fermentation. While growing on substrates other than hexoses, CO, or CO_2/H_2, *Moorella* species can produce various products. May use nitrate, nitrite or fumarate as electron acceptors. Forms various aromatic compounds via decarboxylation of arylic acids, which are used as CO_2 donors under CO_2-limited conditions. The cell wall contains LL-diaminopimelate (DAP).

DNA G+C content (mol%): 53–55.

Type species: **Moorella thermoacetica** (Fontaine, Peterson, McCoy, Johnson and Ritter 1942) Collins, Lawson, Willems, Cordoba, Fernández-Garayzábal, Garcia, Cai, Hippe and Farrow 1994, 824^{VP} (*Clostridium thermoaceticum* Fontaine, Peterson, McCoy, Johnson and Ritter 1942, 705).

Further descriptive information

Physiology. Currently, the genus *Moorella* contains four species, all of which use the Wood–Ljungdahl pathway of autotrophic acetyl-CoA biosynthesis, with tetrahydrofolate (THF) as the characteristic cofactor for the fixation of CO_2 or CO into acetate. The pathway has been elucidated mainly in *Moorella thermoacetica*, but has also been demonstrated in other species through analysis of key enzyme activities and cofactors (for reviews see Drake et al., 2008; Ragsdale, 2008). Glucose is first metabolized by enzymes of the Embden–Meyerhof–Parnas pathway to pyruvate, which is then converted via the ferredoxin-dependent pyruvate oxidoreductase, phosphotransacetylase, and acetate kinase to 2 acetate, 2 CO_2, 8 reducing equivalents, and 2 ATP via substrate-level phosphorylation. These 2 CO_2 and 8 reducing equivalents are then converted via the two branches of the Wood–Ljungdahl pathway (the methyl and carbonyl branches) to the third acetate, making it a homoacetogenic fermentation. For that reason, these bacteria have historically been called "homoacetogenic bacteria". However, more recently this term has been replaced with "acetogenic" upon the realization that *Moorella* and other similar bacteria produce a range of products on other, less common substrates, including succinate from fumarate and aromatic compounds from aromatic acids (Daniel et al., 1990; Drake, 1994; Schink, 1994; Seifritz et al., 2002). In the methyl branch, one of the CO_2 molecules is reduced to methyl-THF. The methyl group is then transferred to a corrinoid protein. The other CO_2 is activated via the carbonyl branch and combined with the methyl moiety to form acetyl-CoA by the enzyme acetyl-CoA synthase. Because of its ability to oxidize CO, this enzyme has also been called carbon monoxide dehydrogenase. Due to electron-transport phosphorylation (ETP) during the reduction of formyl-THF to methyl-THF in the methyl branch, this pathway is the most energy-efficient fermentation of glucose, yielding at least 6 moles of ATP per mole of fermented glucose: 2 ATP from the Embden–Meyerhof pathway, 3 ATP from the conversion of acetyl-CoA to acetate, and 1 ATP derived from ETP. This conclusion is in agreement with $Y_{glucose}$ values (Ljungdahl, 1994).

Hexose-utilizing *Moorella* species (as well as other acetogens) contain an anaerobic electron-transport chain that uses H_2 or CO as electron donors and methylene-THF as a physiological electron acceptor and generates a proton motive force for ATP synthesis (Diekert and Thauer, 1978; Daniel et al., (1990) and references therein). The electron-transport chain contains cytochrome b_{554}, cytochrome b_{559} (Das et al., 2005), menaquinone (Das et al., 1987), a flavoprotein, a 4Fe-4S- and an 8Fe-8S-ferredoxin (Elliott and Ljungdahl, 1982; Yang et al., 1977), and rubredoxin (Yang et al., 1980). However, a definite role for rubredoxin is still not clear. Electron-transport also drives the uptake of amino acids (Hugenholtz and Ljungdahl, 1989, 1990a). The electron-transport chain used during CO-dependent growth has also been partly elucidated (Hugenholtz et al., 1987). Lastly, *Moorella thermoautotrophica* contains an F_0F_1 ATPase that is not functionally compatible with the enzyme in *Escherichia coli* and, therefore, appears to be different (Das et al., 1997; Das and Ljungdahl, 1993).

The first step in the methyl branch is catalyzed by formate dehydrogenase, which is one of the most oxygen-sensitive enzymes known (Das, 2003; Drake, 1994; Ljungdahl, 1994; Ragsdale, 2008; Wood and Ljungdahl, 1991; Yamamoto et al., 1983). This enzyme was the first tungsten-dependent enzyme discovered in biology (Andreesen and Ljungdahl, 1973). More recently, tungsten has been shown to replace molybdenum in many pterin-containing enzymes of deep-sea hyperthermophiles (Kletzin and Adams, 1996). All of the remaining enzymes of this important pathway have been purified and characterized from *Moorella* species (mainly *Moorella thermoacetica*). For an extended discussion of their distribution among acetate-forming bacteria and the bioenergetics and biochemistry of acetogenesis, the reader is referred to recent reviews (Das and Ljungdahl, 1993; Drake, 1994; Drake et al., 2008; Hugenholtz and Ljungdahl, 1990b; Ragsdale, 2008; Wood and Ljungdahl, 1991).

O_2 metabolism in *Moorella* has not been fully elucidated (Das et al., 2005). Recent studies with acetogenic bacteria from the termite hindgut (Boga and Brune, 2003) and *Moorella thermoautotrophica* (J. Wiegel, unpublished results) suggested that *Moorella* can reduce O_2 even though it will not grow under microaerophilic conditions.

Under CO_2-limiting conditions, carboxylated aromatic compounds can be used as a source of required CO_2 (Hsu et al., 1990). Two interesting, although unnamed, isolates reveal novel metabolic properties for this taxa. On the basis of their 16S rRNA gene sequences, both isolates are closely related to *Moorella thermoacetica*. Karita et al. (2003) isolated the cellulolytic strain F21, which may make it possible to efficiently produce acetate directly from cellulose. For biotechnological applications, such as production of the road de-icer calcium-magnesium acetate (CMA), this strain makes it possible to avoid the costly saccharification procedures. Inokuma et al. (2007) reported on the enzymology of the *Moorella* strain H4C22-1, which produces some ethanol during growth on H_2/CO_2. Yields are typically 120 mM acetate and 5.2 mM ethanol. The level of *aldh* mRNA, which encodes an aldehyde dehydrogenase, was threefold higher during growth on H_2/CO_2 than on fructose. In contrast, the levels of *adhA, B,* and *C,* which encode the primary and secondary alcohol dehydrogenases, decreased. Thus, it appears likely that *Moorella thermoacetica* and other *Moorella* species may use alternative fermentation pathways under certain conditions. In the presence of viologen dyes, methanol is formed from CO instead of acetate (White et al., 1987). *Moorella* species are apparently able to synthesize branched-chain amino acids via the common pathway or via the reductive carboxylation/transamination of the corresponding fatty acid pathway (Allison et al., 1984; Wiegel et al., 1981; Wiegel and Schlegel, 1977).

All described *Moorella* species produce heat-stable spores, although the extent differs among the species and strains. Sporulation is especially common following chemolithoautotrophic growth on methanol or glycerol with *Moorella glycerini*. For spores, the decimal reduction times or D_{10} times are around 15 min at 120 °C for the type species and type strain following growth on glucose in rich media. D_{10} times have not been determined for spores formed following chemolithoautotrophic growth (Fontaine et al., 1942). The D_{10} time at 121 °C for spores of *Moorella mulderi* (Balk et al., 2003) is 31 min. For spores from cells of *Moorella thermoautotrophica* and *Moorella thermoacetica* strain JW/DB-4 (=ATCC 39073), grown chemolithoautotrophically at 60 °C, D_{10} times at 121 °C are 30–80 min (J. Wiegel, unpublished results) and 111 min, respectively (Byrer et al., 2000).

Full activation of spore suspensions requires incubations of more than 10 min and up to 1 h at 100 °C. In contrast, activation of spores for many other glycolytic thermophiles and mesophiles typically requires 1–3 min at 100 or 80 °C, respectively. Moreover, spores of strains of *Moorella thermoacetica*, *Moorella thermoautotrophica*, and *Moorella mulderi* were not significantly deactivated at incubations for several hours at 100 °C (Balk et al., 2003; Byrer et al., 2000). The spore population is apparently comprised of subpopulations that differ in their heat-resistance, since the courses for spore activation and inactivation at 121 °C are bi- or even multi-phasic (Byrer et al., 2000).

The type strain of *Moorella thermoautotrophica* fixes N_2. It is presently not known whether or not the other species also fix N_2 (Bogdahn and Kleiner, 1986).

Phylogeny. The genus *Moorella* is classified within the family *Thermoanaerobacteraceae*, the first of four families of the order *Thermoanaerobacteriales* of the class *Clostridia* within the phylum *Firmicutes* (see Ludwig et al., 2009). In previously published 16S rRNA gene-based phylogenetic trees, the closest relatives to *Moorella* species have been species of the spore-forming, sulfate-reducing bacterium *Desulfotomaculum*, asporogenic and syntrophic fatty-acid-degrading thermophiles and mesophiles, and several thermophilic glycolytic or saccharolytic thermophiles such as members of *Thermoanaerobacter* and *Thermoanaerobacterium*. However, extensive analysis of the phylogeny of *Firmicutes* by Ludwig and others (this volume) classifies *Moorella* with nonacetogenic heterotrophic (e.g., *Thermoanaerobacter*) and chemolithoautotrophic bacteria (e.g., *Carboxydothermus*) within the family *Thermoanaerobacteraceae* fam. nov. However, this assignment is not strongly supported and future studies may indicate the need for reassignment. Similarly, classification of the genus *Thermoanaerobacterium*, which is assigned to a family *incertae sedis* within the *Thermanaerobacteriales*, is awaiting further clarification. *Desulfotomaculum* and the syntrophic organisms appear to be outside this order and unrelated to *Moorella*. Except for members of the genus *Thermacetogenium* and the species *Thermoanaerobacter kivui*, which are also classified within the *Thermoanaerobacteraceae*, other physiologically similar bacteria exhibiting homoacetogenic fermentation and possessing the Wood–Ljungdahl pathway are classified in different orders. These include: the mesophiles *Clostridium formicaceticum*, *Clostridium aceticum*, *Clostridium ljungdahlii* and *Sporomusa* species; the psychrophilic *Acetobacterium* species; and the thermophile *Caloramator fervidus* (Drake et al., 2008; Ragsdale, 2008; Rainey et al., 2006; Tanner and Woese, 1994).

Industrial applications. *Moorella thermoacetica* and *Moorella thermoautotrophicum* have been used commercially to produce the environmentally friendly road deicer CMA from dolime and hydrolyzed corn starch. CMA, which is sold as Crytech CMA40, can be produced from adapted wild-type strains (Ljungdahl et al., 1986; Wiegel et al., 1991) as well as mutants, e.g., C5-2 (=ATCC 49707; U.S. Patents 4506012 and 4513084) and ATCC 39289, which are both improved acetate-producing mutants derived from ATCC 35608[T], and strain 99-78-22 (=ATCC 31490; U.S. Patent 4371619) (Schwartz and Keller, 1982a, 1982b, 1983). Wiegel et al. (1991) compared two laboratory strains of *Moorella thermoacetica* and two isolates of *Moorella thermoautotrophica* with respect to properties important for the production of CMA. *Moorella thermoacetica* lab-strain LJD and lab-strain Wood both

tolerated 250 mM $CaCl_2$, whereas *Moorella thermoautotrophica* strains JW 701/3 [T] and JW 701/5 only tolerated 25 mM $CaCl_2$. All other properties varied as much between the species as they did within each species. Using a novel felt-pad rotating fermenter and lab-strain LJD-EMS concentrations of 2 M acetate/acetic acid (equivalent to 120 g/l acetic acid) were produced and production rates of up to 4 g/l/h CMA were achieved with cell recycling (Wiegel et al., 1991). Using continuous cultures without cell recycling and dolime as neutralizing reagent, the mutant JW/YS 701/5 P-Na 20 produced over 250 g/l CMA at pH 6.9. In continuous culture with cell recycling, the rates of CMA production were up to 4 g/l/h and the ratio of acetate formed per metabolized glucose moiety was about 2.8. Each *Moorella* species has different advantages and disadvantages (Wiegel et al., 1991).

The improvement of acetate production from 30 mM lactose by growing *Moorella thermoautotrophica* in co-culture with *Methanothermobacter thermautotrophicus* and *Clostridium thermolacticum* has been suggested (Talabardon et al., 2000). Through efficient *in situ* H_2 scavenging within the consortium, acetate formation was improved at the expense of reduced by-products like ethanol. The use of this thermophilic anaerobic consortium opens new opportunities for the efficient valorization of lactose, a major waste product from the cheese industry, and production of CMA (Collet et al., 2003).

The production of acetate from cellulose by *Clostridium thermocellum–Moorella thermoautotrophica* co-cultures was investigated by Freier and Wiegel (1981). Acetate yields of 2.8 mol per mol of cellobiose equivalents fermented were obtained (Wiegel, 1988; Wiegel et al., 1986; J. Wiegel, unpublished results; Simankova and Nozhevnikova, 1989). Supernatants of cultures producing CMA, as well as the CMA itself, have negligible ecological impact on plants and animals. Culture supernatants are not toxic to mice.

Habitats and isolation. Known habitats are horse manure (which was the source of the type strain of the type species), emu droppings (R. Tanner, personal communication), sewage sludge, mesobiotic freshwater sediments (Wiegel et al., 1981), canned food samples (Carlier and Bedora-Faure, 2006), and dahlia tubers (Drent and Gottschal, 1991), but species have been isolated mainly from hot springs, including those in Japan (Karita et al., 2003), Kamchatka (Far-eastern Russia) (J. Wiegel, unpublished results), Wyoming (USA) (Rainey et al., 1993), and New Zealand (J. Wiegel, unpublished results). Since more than 70% of the methane formed in the environment is formed via acetate, it has become evident that acetogenesis, along with methanogenesis, is one of the most important processes in the environment. Thus, these bacteria are no longer considered a curiosity. Instead, they are an ecological and industrially important taxon. When *Moorella thermoautotrophica* and *Moorella thermoacetica* are isolated under autotrophic conditions, inhibitors for methanogens such as bromoethanesulfonate (3–10 μM) should be included. Otherwise, *Methanothermobacter thermautotrophicus*-like strains will out-compete and overgrow the acetogens in the enrichments and on agar plates, presumably due to the higher affinity of methanogens for H_2. Mesobiotic sediments from freshwater streams in Germany contain up to about 100 viable cells of *Moorella*-like H_2- or methanol-utilizing and acetate-forming thermophiles. A proportion (depending on the sample) of those cells are in the vegetative state, as estimated by comparing MPNs before and after 80 °C heat treatment of the initial sample (Braun et al., 1979; J. Wiegel, unpublished results). Because laboratory cultures of this species did not grow at less than 37 °C and the bulk temperature in those sediments is never over 30 °C, it is unclear whether or not *Moorella* is growing in the mesobiotic environment or just remaining viable by maintenance metabolism. J. Wiegel (unpublished results) has shown that the methanogenic counterpart *Methanothermobacter thermautotrophicus* can grow in autoclaved (121 °C, 40 min) natural sediments at temperatures as low as 22 °C and can produce methane at temperatures at least as low as 16 °C (J. Wiegel, unpublished results). To date, comparable experiments have not been performed with *Moorella*. In addition to the possibility of *Moorella thermoautotrophica*-like strains growing at lower temperatures in natural settings, there is also the possibility that growth occurs mainly in microniches with temporarily elevated temperatures due to release of heat from biodegradative processes. *Moorella thermoacetica* has been found growing commensally with *Thermicanus aegyptius* in an Egyptian soil, an ecological situation that may occur much more frequently than presently realized (Drake et al., 2008; Gössner et al., 1999).

Although nickel is a required metal ion for several enzymes in the acetogens, due to the presence of a high-affinity transport system for nickel observed in *Moorella thermoacetica*, the necessity of adding nickel salts is not usually observed during isolations or for growth in media containing complex nutrients such as yeast extract (Lundie et al., 1988).

Taxonomic comments

The original species names were *Clostridium thermoaceticum* (sic) and *Clostridium thermoautotrophicum* (sic). The presently acknowledged proper latinized form would require that the connecting vowel "o" be dropped; and the names should be "*Moorella thermacetica*" and "*Moorella thermautotrophica*", as used intermittently (Trüper, 1999). However, the Notification List (Int. J. Syst. Bacteriol., 1995, 45: 199–200) uses the names with the "o" and thus, according to the new Rule 61, the valid species names are *Moorella thermoacetica* and *Moorella thermoautotrophica* (see J.P. Euzéby "Grammatical or orthographical changes prohibited by Rule 61" at http://www.bacterio.cict.fr/corrections2.html).

Differentiation of the genus *Moorella* from other thermophiles and mesophilic acetogens

In addition to their separate position in 16S rRNA gene-based phylogenetic trees, *Moorella* species differ from other glycolytic and chemolithoautotrophic mesophilic acetogens by being thermophilic. They differ from most other thermophiles by the use of the Wood–Ljungdahl pathway, producing 3 acetate per mol fermented glucose minus the glucose and acetyl-CoA used for biosynthesis. Fermentations typically yield ratios of about 2.85 acetate molecules per glucose. *Moorella* species have cell walls containing LL-DAP, whereas the clostridia *sensu stricto* and *Thermoanaerobacter* and related species have *m*-DAP-direct type cell walls. The relatively high G+C values of 53–55 mol% further separate *Moorella* from *Clostridium sensu stricto*, which possess values below 35 mol%, and other thermophilic acetogens carrying out homoacetogenic fermentation. The latter species are *Thermoanaerobacter kivui*, which possesses an *m*-DAP-direct type of cell wall; *Caloramator fervidus*, which has a much lower mol% G+C content; and *Thermacetogenium phaeum*, which in contrast to all *Moorella* strains can syntrophically oxidize acetate in co-culture with a hydrogenotrophic methanogen. In addition, several phylogenetically unidentified thermophilic acetogenic isolates have been described. For a detailed overview on the various acetogenic bacteria, see Drake et al. (2008).

TABLE 247. Properties that differentiate species of the genus *Moorella*[a]

Property	*M. thermoacetica*[b]	*M. thermoautotrophica*[b]	*M. glycerini*[c]	*M. mulderi*[d]
Origin	Horse manure	Mesobiotic and thermobiotic sediment and soil	Hot spring, Yellowstone National Park (USA)	Anaerobic digester (The Netherlands)
Cell size (μm)	0.4 × 2.8	0.8–1 × 3–6[e]	0.4–0.6 × 3–6.5	0.4–0.6 × 2–8
Temperature range (optimum)[e]	47–65 (56–60)[e]	42–66 (60–62)[e]	43–65 (58)	40–70 (65)
pH^{25C} range (optimum)[e]	5.7–7.65[f] (6.6–6.8)	4.8–7.3 (6.5–6.6 on glucose; 5.5 on glycerate)	5.9–7.8 (6.3–6.5)	5.5–85 (7.0)
Doubling time (glucose) (h)[e]	4–8	5	4	ND (on methanol >20 h)
CaCl$_2$ tolerance (mM)	250	25	ND	ND
MgCl$_2$ tolerance (mM)	250	250–350[e]	ND	ND
DNA G+C content (mol%)	53–55 (T_m)	53–55 (T_m)	54–55 (HPLC)	53–54 (T_m)
Electron donor/carbon source:				
H$_2$/CO$_2$	+	+	–	+
Methanol	+	+	–	+
Glycerol	–	–	–	+
Glycerate	(+)	–	+	–
Lactate	–	+	ND	ND
Nitrate as electron acceptor	+	+	+	–

[a]For all species: spore formation positive, terminal in swollen mother cell (drumstick shape) and non-elongated sporulating cells; Gram-stain is positive; from glucose and fructose, 2.2–2.9 mol acetate is formed per mol hexose, CO$_2$ and H$_2$ are also formed; thiosulfate used as electron acceptor. +, Positive; (+), weakly positive; –, negative; ND, no data.
[b]Wiegel et al. (Wiegel et al., 1981, 1991).
[c]Slobodkin et al. (1997a).
[d]Balk et al. (2003).
[e]Strain-dependent and using glucose as substrate.
[f]For starting growth; final pH of culture can be as low as 4.1.

Differentiation of the species of the genus *Moorella*

Differentiation of the species is largely based on 16S rRNA gene analysis. The type species *Moorella thermoacetica* and *Moorella thermoautotrophica* are very similar in their physiological capabilities. Many of the physiological differences, such as differences in chemolithoautotrophic growth, methanol utilization, growth rates, sporulation, and cell sizes (Wiegel, 1982), which justified the formation of novel species at the time of their publications now to appear to be minor and within the range of variation often observed between strains. Moreover, besides the difference in capability for chemolithoautotrophic growth, originally the two species were differentiated on the ability to use pyruvate. At the time, it was believed the *Moorella thermoautotrophica* was unable to use pyruvate. However, this conclusion was probably due to mistakes in media preparation, because the pyruvate was sterilized by autoclaving instead of filtration (J. Wiegel, unpublished results). DNA–DNA hybridization between strains of *Moorella thermoautotrophica* (type strain JW 701/3[T] and strain KIVU) and *Moorella thermoacetica* DSM 521[T] is between 48 and 52%, which supports the creation of two species (Wiegel et al., 1981). However, more recent hybridization results have yielded higher values (R. Tanner, personal communication). The phylogenetic position of strains *Moorella thermoacetica* JW/DB-4 and JW/DB-2, based on 16S rRNA gene sequences, is intermediary between *Moorella thermoacetica* and *Moorella thermoautotrophica*. This relationship is also reflected in the physiological properties and the extent of sporulation, which is only slightly less for the new strains than for *Moorella thermoautotrophica* (Byrer et al., 2000; Wiegel, 1982). Thus, the extent of differences, including in the 16S rRNA genes, between *Moorella thermoacetica* and *Moorella thermoautotrophica* is borderline for their separation into two different species. Possibly, when more strains related to *Moorella thermoacetica* have been isolated and characterized, a more informed decision can be made as to whether or not *Moorella thermoautotrophica* should remain classified as a separate species or a subspecies of *Moorella thermoacetica*. It is also quite possible that if strains of the two species were subjected to the same growth conditions for extended periods of time, they would become physiologically indistinguishable. However, glyoxylate-dependent growth in medium containing nitrite, as observed with *Moorella thermoacetica*, has not been observed for *Moorella thermoautotrophica* (Seifritz et al., 2003, 2002).

Moorella mulderi and *Moorella glycerini* differ from *Moorella thermoacetica* and *Moorella thermoautotrophica* mainly in their inability to use nitrate as an electron acceptor (Table 247). *Moorella mulderi* differs from *Moorella glycerini* in its inability to use glycerol as a carbon source. *Moorella glycerini* differs from *Moorella thermoautotrophica* and *Moorella thermoacetica* in that it reduces thiosulfate only to elemental sulfur and not to H_2S. It also differs from all three of the other species in that it does not grow chemolithoautotrophically.

The pH optima of these bacteria depend strongly on the growth substrate. Differences in pH optima between *Moorella mulderi*, *Moorella thermoautotrophica*, and *Moorella thermoacetica* are small when the same growth substrates are used. The originally published pH optimum of around 5.7 for *Moorella thermoautotrophica* was determined during chemolithoautotrophic growth or with glycerate as a substrate. Glycerate is convenient because it is converted in a 1:1 ratio to acetate and thus does not significantly change the pH of the culture, making it easier to determine pH profiles. Growth on methanol, H_2/CO_2, and glycerate does not require the use of all the enzymes of the Embden–Meyerhof–Parnas pathway and has more neutral or slight alkaline pH optima than when only the Wood–Ljungdahl pathway is used. The pH optima for *Moorella thermoautotrophica* and *Moorella thermoacetica* during growth on glucose are around 6.6–6.8 (Wiegel et al., 1981, 1991). For *Moorella mulderi*, it is pH 7.0 when the pH is determined at 25 °C using temperature-corrected standards (M. Balk, personal communication).

Additional isolates

Besides the industrially interesting derivatives of the type species, several isolates have been described that possess 16S rRNA gene sequences that are closely related to that of the type species *M thermoacetica*. These include: two strains with highly thermoresistant spores, *Moorella thermoacetica* JW/BD-2 and JW/BD-4 (Byrer et al., 2000); the cellulolytic strain F21 (Karita et al., 2003); and the ethanol-producing strain H4C22-1 (Sakai et al., 2004).

Wiegel et al. (1981) described 15 strains of *Moorella thermoautotrophica*, eight of which were partially characterized in the original publication. These strains are all methylotrophic, able to grow on H_2/CO_2, and able to grow heterotrophically on glycerate, glucose, fructose, and galactose. However, taxonomic assignments were based solely on physiological properties, which are probably not a reliable method for this group of bacteria. Sa (1985) isolated an additional methylotrophic strain. Drent and Gottschal (1991) isolated strain I1 following growth on inulin, a polymer of fructose, from Dahlia tubers. This strain contains one or more inulinases that function optimally at 60 °C and neutral pH. Enzyme activity was found bound to the cells, as well as free in the medium. The strain differs from the type strain of *Moorella thermoautotrophica* with respect to fermentation products, substrate spectrum, and optimum temperature for growth. Berestovskaya et al. (1987) isolated strain Z-99, which released about 500 μmol H_2 per h per mg protein during the lag phase in glucose-containing medium. During growth on methanol at 55 °C, hydrogenase activity was 50 mmol/min/mg protein. In the late exponential growth phase, growth yields were 4.5 g of biomass per mole methanol utilized.

List of species of the genus *Moorella*

1. **Moorella thermoacetica** (Fontaine, Peterson, McCoy, Johnson and Ritter 1942) Collins, Lawson, Willems, Cordoba, Fernández-Garayzábal, Garcia, Cai, Hippe and Farrow 1994, 824[VP] (*Clostridium thermoaceticum* Fontaine, Peterson, McCoy, Johnson and Ritter 1942, 705)

ther.mo.a.ce'ti.ca. Gr. adj. *thermos* hot; L. neut. adj. *aceticum* pertaining to vinegar (acetic acid); N.L. fem. adj. *thermoacetica* producing acetic acid at elevated temperatures.

This description is based mainly on the original reports of Fontaine et al. (1942), Ljungdahl et al. (1985), and Wiegel et al. (1991). The type strain was isolated from horse manure at elevated temperatures in medium containing nitrate. There are significant variations in the properties of cultures of the type strain from different laboratories (termed lab-strains LJD, Wood, and Drake, respectively), although they all were derived from the same culture, i.e., the one isolated by Fontaine et al. (1942). The differences are clearly due to selection during extensive subculturing under different growth conditions.

Cells are straight or slightly bent rods of 0.4 × 2.8 μm, found singly or in pairs, but rarely in chains when grown heterotrophically on peptone-yeast extract-glucose (PYG) or yeast extract-methanol medium. Cells are slightly smaller when grown chemolithoautotrophically. They stain Gram-positive in the exponential growth phase, but may stain Gram-negative in the stationary growth phase. Although flagella were described in the original report of Fontaine et al. (1942), cells are usually nonmotile; if motility does occur, only tumbling motion is seen during growth on methanol or on agar cultures with low nutrient concentrations.

Spores are rarely observed in PYG medium, especially in lab-strain LJD (=ATCC 39079), which was maintained in the laboratory of L.G. Ljungdahl (University of Georgia, Athens, USA) in rich complex medium continuously sparged with CO_2. In contrast, in methanol-grown cultures of lab-strain Wood (obtained by the author from the laboratory of the late H.G. Wood), up to 10% of the cells sporulate. Differences between the two lab-strains are even more pronounced with respect to chemolithoautotrophic growth, e.g., lab-strain LJD is difficult to adapt to chemolithoautotropic growth and is much slower than lab strain Wood. Spores of the type strain survive for 15 min at 120 °C. *Moorella thermoacetica* strain JW/DB-4 (=ATCC BB-48), which was isolated in the author's laboratory as a contaminant in autoclaved and uninoculated chemolithoautotrophic medium, produced spores after chemolithoautotrophic growth at 60 °C with D_{10} times at 121 °C of about 2 h (Byrer et al., 2000).

The temperature ranges for strains LJD and Wood growing on complex medium with glucose as main carbon source are 51–65 °C (optimum 60 °C) and 47–65 °C (optimum 56–58 °C), respectively. The corresponding pH ranges are 5.7–7.65 (optimum pH 6.8) and 5.7–6.8 (optimum pH 6.6), respectively. However, the pH optimum for growth on methanol and glycerate is about pH 6.0. The original description listed growth between 45 and 65 °C, with an optimum between 55 and 60 °C. During growth with glucose, the pH can reach values down to pH 4.1 in the stationary phase; however, no growth occurs upon transfer into new media at this pH. $CaCl_2$ and $MgCl_2$ tolerances for both strains are 250 mM. The minimal concentration of yeast extract required for growth is 0.005% (w/v). It can be replaced by nicotinic acid, nickel, selenium, and tungsten (Lundie and Drake, 1984). Cultures in complex media need to be stirred or agitated by sparging with CO_2. Otherwise, cells flocculate and settle at the bottom of the culture vessel and the growth yield is reduced. To ensure growth, it is recommended that fresh agar shake tubes are prepared. Strain LJD is assumed to be sensitive to reduced water activity conditions such as found in old medium containing 2% (w/v) agar. Glucose, fructose, and trehalose can serve as sole carbon sources.

Moorella thermoacetica was described for more than 40 years as a heterotrophic bacterium. However, *Moorella thermoacetica* can grow on C_1 compounds, including methanol/CO_2, chemolithoautotrophically on carbon monoxide (with or without H_2), and with H_2/CO_2 (Kerby and Zeikus, 1983; Savage et al., 1987; Wiegel, 1982). Under CO_2 limitation, the carboxyl group of hydroxyaromatic acids can serve as a CO_2 source (Hsu et al., 1990). Other recently described acetogenic substrates are oxalate and glyoxylate. Nitrate and nitrite can serve as electron acceptors and reduce the yield of acetate. In an undefined medium containing 0.1% yeast extract, 5 mM nitrate served as an electron acceptor with glyoxylate as substrate, but not with glucose.

DNA G+C content (mol%): 54 (T_m).

Type strain: ATCC 35608, DSM 521, JCM 9319.

GenBank accession number (16S rRNA gene): AY656675.

Additional remarks: the genomic sequence from lab-strain LJD (=ATCC 39073) is available (http://genome.ornl.gov/microbial/).

2. **Moorella glycerini** Slobodkin, Reysenbach, Mayer and Wiegel 1997a, 973[VP]

gly.ce.ri'ni. Gr. adj. *glykeros* sweet; N.L. adj. *glycerini* of glycerol, referring to utilization of glycerol as a substrate.

During all growth phases, cells stain Gram-positive and are rods of 0.4–0.6 × 3.0–6.5 μm, exhibiting tumbling motility due to retarded peritrichous flagellation. Cells usually occur singly. Round endospores with a diameter of 1.0–1.5 μm occur readily in the late exponential growth phase in terminal swollen sporangia. Cells are covered with tetragonal type S-layer proteins containing subunits of center-to-center distances of 10 nm. Heterotrophic growth occurs with glycerol as sole carbon source, although yeast extract stimulates growth rate and yield. Glucose, fructose, xylose, mannose, galactose, lactate, glycerate, and pyruvate serve as carbon and energy sources with acetate as the sole fermentation product. Traces of lactate and branched chain fatty acids occur at elevated yeast extract concentrations. Formate (with or without fumarate), succinate, and benzoate do not support growth; neither do short-chain alcohols (C_1 to C_5). However, fumarate is reduced to succinate in the presence of a metabolizable carbon source such as pyruvate. Nitrate, sulfate, sulfite, amorphous Fe(III), and MnO_2 are not reduced. Ampicillin, chloramphenicol, erythromycin, rifampin, and tetracycline (all at 100 μg/ml) prevent growth.

Isolated from hot springs at the Calcite Spring site in Yellowstone National Park (WY, USA) using mineral medium supplemented with yeast extract and glycerol as the main carbon and energy sources and incubation at 60 °C. Enrichments included autoclaving the culture for 10 min at 121 °C to obtain the thermophilic, endospore-forming culture (for spore resistance, see genus description above).

DNA G+C content (mol%): 54.5 ± 0.4 (HPLC).

Type strain: JW/AS-Y6, ATCC 700316, DSM 11254.

GenBank accession number (16S rRNA gene): U82327.

3. **Moorella mulderi** Balk, Weijma, Friedrich and Stams 2005, 1[VP] (Effective publication: Balk, Weijma, Friedrich and Stams 2003, 319.)

mul.de′ri. N.L. gen. n. *mulderi* of Mulder, named in honor of Eppe G. Mulder, head of the Microbiology Department at Wageningen, The Netherlands, 1956–1981.

Spore-forming, straight rods of 0.4–0.6 μm × 2–8 μm, which stain Gram-positive and grow as single cells or in pairs. Growth on glucose is observed in the temperature range 40–70 °C, with an optimum at 65 °C, and in the pH range 5.5–8.5, with an optimum about 7.0, under obligately anaerobic conditions. NaCl concentrations of 0–4.5% (w/v) NaCl support growth, with an optimum around 1% NaCl. Grows with methanol, H_2/CO_2 (80%/20%, v/v), formate, lactate, pyruvate, glucose, fructose, cellobiose, and pectin. Thiosulfate is reduced to sulfide. Comparison of 16S rRNA genes reveals that strain TMS^T is most closely related to *Moorella glycerini* (96% sequence similarity), *Moorella thermoacetica* (92%), and *Moorella thermoautotrophica* (92%).

Isolated from a thermophilic bioreactor operated at 65 °C with methanol as the energy source. Methanol was used as the carbon source for the enrichment.

DNA G+C content (mol%): 53 (T_m).

Type strain: TMS, ATCC BAA-608, DSM 14980.

GenBank accession number (16S rRNA gene): AF487538.

4. **Moorella thermoautotrophica** (Wiegel, Braun and Gottschalk 1982) Collins, Lawson, Willems, Cordoba, Fernández-Garayzábal, Garcia, Cai, Hippe and Farrow 1994, 824^{VP} (*Clostridium thermoautotrophicum* Wiegel, Braun and Gottschalk 1982, 384)

therm.au.to.tro′phi.ca. Gr. adj. *thermos* hot; Gr. pron. *autos* self; Gr. n. *trophos* a feeder; N.L. fem. adj. *thermautotrophica* indicating that the organism grows at elevated temperatures with CO_2 as carbon source and H_2 as energy source and consequently is chemolithoautotrophic.

The description is based mainly on the reports of Wiegel et al. (1981, 1991) and Clark et al. (1982) and the type strain JW-YS701/3 and isolate JW-YS701/5, both from hot springs in Yellowstone National Park.

Straight or slightly bent rods of 0.8–1.0 × 3.0–6.0 μm (when grown heterotrophically on PYG medium or yeast extract-methanol) or slightly smaller when grown chemolithoauto-trophically, found singly or in pairs, rarely in chains. Cells stain Gram-positive in exponential growth phase, but tend to stain Gram-negative from the beginning of the late exponential growth phase. Cells are slightly motile and have 3–8 peritrichous flagella that sometimes have unusual multiple (up to 8) basal plates. Flagella are 140–150 nm thick and 8–15 μm long. The cell surface is covered with an S-layer with a tetragonal array comprised of subunits 11.2 nm in diameter. The round to slightly oval spores are found in sporangia of 2–3 μm and have a diameter of 1–2.5 μm when released. In PYG medium, the extent of sporulation is about 5–10%. In chemolithoautotrophic medium or following growth on methanol, it is over 90%. In agar shake roll tubes and on agar plates (2–2.2% w/v agar), colonies reach diameters of about 3 mm after about 7–12 d and are circular, flat to convex, smooth, and tannish white in color. No growth is obtained with fucose, cellulose, dulcitol, erythritol, xylitol, ethanol, propanol, inositol, glycerol, glycol, citrate, succinate, malate, or glutamate. Casamino acids and egg yolk cannot replace yeast extract and do not allow for growth in the presence of trace amounts of yeast extract. In the presence of 2 mg/l nicotinic acid, the type strain can be adapted to grow in mineral media without yeast extract (Savage and Drake, 1986). The type strain utilizes glucose, fructose, arabinose, trehalose, and rhamnose and weakly utilizes ribose. The type and other strains grow well on 100% CO after several days lag phase, on mixtures of CO/H_2, and on CO/CO_2 with or without H_2. Growth occurs at pH^{25C} 4.8–7.3, although the pH^{25C} in stationary cultures reaches values around 4.0.

Moorella thermoautotrophica strains can be isolated from nearly every sediment with a bulk pH between 4.0 and 9.0 in a temperate climate, where the sediment/soil freezes in winter, and from geothermally heated features, manure piles, and artificially heated soils. Of the *Moorella* species, this one is nearly ubiquitous.

DNA G+C content (mol%): 54.5 ± 0.4 (HPLC).

Type strain: JW 701/3, ATCC 33924, DSM 1974.

GenBank accession number (16S rRNA gene): L09168, X58353, X77849.

Genus VII. **Thermacetogenium** Hattori, Kamagata, Hanada and Shoun 2000, 1608^{VP}

YOICHI KAMAGATA

Therm.a.ce.to.ge′ni.um. Gr. adj. *thermos* hot; L. n. *acetum* vinegar; Gr. v. suff. *genium* producing; N.L. neut. n. *Thermacetogenium* thermophilic vinegar producer.

Cells are rod-shaped. Strictly anaerobic and thermophilic. Chemoautotrophic and chemo-organotrophic. Gram reaction is negative, but has a Gram-stain-positive cell-wall structure. Round terminal endospores are sometimes formed. Colonies are disk-shaped. **Able to oxidize acetate in co-culture with hydrogenotrophic micro-organisms.** Acetate can also be utilized by sulfate reduction in pure culture. Possesses **menaquinone-7. Grows acetogenically on several alcohols, methoxylated aromatics, organic acids, amino acids, and H_2/CO_2.** Sugars are not utilized. Additional supplements are not required.

DNA G+C content (mol%): 53.5.

Type species: **Thermacetogenium phaeum** Hattori, Kamagata, Hanada and Shoun 2000, 1608^{VP}.

Further descriptive information

On the basis of the phylogenetic analysis, the representative strain *Thermacetogenium phaeum* PB^T was found to be a member of the *Bacillus–Clostridium* subphylum of the Gram-stain-positive bacteria, and the closest relative is *Thermoterrabacterium ferrireducens* (87.4% 16S rRNA gene similarity), since reclassified as *Carboxydothermus ferrireducens*. It is a strictly anaerobic, thermophilic, syntrophic, acetate-oxidizing bacterium.

Acetate is one of the most important intermediates for methanogenesis in anaerobic mineralization of organic materials. Of a number of methanogens described previously, only the genera *Methanosarcina* and *Methanosaeta* are known to produce methane from acetate. Methanogenesis from acetate by these organisms is catalyzed by an aceticlastic reaction in which the methyl group of acetate is reduced to methane (Ferry, 1992). In contrast to these methanogens, some syntrophic proton-reducing organisms are found to oxidize acetate to form methane in association with hydrogenotrophic methanogens (Schnurer et al., 1996; Zinder and Koch, 1984). Metabolically, this syntrophic association consists of two reactions which were originally proposed by Barker (1936). In the first reaction, acetate is oxidized to form H_2 and CO_2 ($CH_3COO^- + 4 H_2O = 4 H_2 + 2 HCO_3^- + H^+$) which are, in the second reaction, converted to methane ($4 H_2 + HCO_3^- + H^+ = CH_4 + 3 H_2O$). As with other anaerobic, syntrophic, fatty acid oxidations, the former reaction is highly endergonic under the standard conditions ($\Delta G^0 = +104$ kJ per mol) unless it couples with the latter reaction ($\Delta G^0 = -135$ kJ per mol). Thus, syntrophic acetate degradation is possible only when syntrophic micro-organisms and H_2-consuming micro-organisms cooperate ($CH_3COO^- + H_2O = CH_4 + HCO_3^-$: $\Delta G^0 = -31$ kJ per mol). As a matter of fact, the H_2 partial pressure is kept below 40–80 Pa in the *Thermacetogenium* and *Methanothermobacter* co-culture (Hattori et al., 2001; Luo et al., 2002).

Two syntrophic acetate oxidizers in association with hydrogenotrophic methanogens were reported before *Thermacetogenium* was isolated. The first description was strain AOR, which was a thermophilic acetate oxidizer isolated from a methanogenic reactor in co-culture with *Methanobacterium* species (Zinder and Koch, 1984). The isolate was later found to be a homoacetogen which forms acetate from H_2/CO_2 in pure culture, whereas acetate oxidation, i.e., reverse reaction of acetogenesis, occurs in co-culture with the methanogen (Lee and Zinder, 1988). The second acetate syntroph was described as *Clostridium ultunense* which mesophilically oxidizes acetate in the presence of hydrogenotrophic methanogens (Schnurer et al., 1996). Although *Clostridium ultunense* is phylogenetically and chemotaxonomically well characterized, the phylogenetic position of the thermophilic strain AOR is not known. Moreover, the isolate was not deposited to any culture collections and is no longer available.

The genus *Thermacetogenium* is the thirdly described microorganism that is able to oxidize acetate in co-culture with methanogens. Since strain AOR is no longer available, the genus *Thermacetogenium* is the only thermophilic acetate-oxidizing syntrophic organism that is phylogenetically and chemotaxonomically characterized.

The type strain of *Thermacetogenium* was isolated from sludge in a thermophilic methane-fermenting reactor which had been treating kraft-pulp wastewater. The ecological significance and abundance is, however, unknown. In moderately thermophilic environments (around 55 °C), *Methanosarcina* and *Methanosaeta* usually predominate as acetate degraders and play a crucial role in the mineralization of acetate. Besides those methanogens, some thermophilic sulfate-reducing bacteria such as *Desulfotomaculum thermoacetoxidans* (Min and Zinder, 1990) are known to be able to oxidize acetate in the presence of sulfate. Whether *Thermacetogenium* contributes to acetate oxidation to a great extent in methanogenic environments remains unclear, but evidence indicates that the organism is very versatile.

First, like strain AOR (Lee and Zinder, 1988), *Thermacetogenium* is a homoacetogen which forms acetate from H_2/CO_2 in pure culture. Acetogenesis is the reverse reaction of acetate oxidation that occurs in co-culture with the methanogen. Secondly, the organism is capable of utilizing a variety of substrates including pyruvate, methanol, ethanol, 1-propanol, 1-butanol, 2,3-butanediol, 3,4,5-trimethoxybenzoate, syringate, vanillate, glycine, cysteine, and formate (Hattori et al., 2000). Acetate is the sole or major fermentation product. In utilization of 3,4,5-trimethoxybenzoate, syringate, and vanillate, the methoxy group of each compound is converted to form acetate with a corresponding demethoxylated skeleton. Thirdly, *Thermacetogenium* is able to reduce sulfate and thiosulfate with acetate as the electron donor. Menaquinone-7 should function as the electron carrier in sulfate and thiosulfate reduction, though genes and enzymes involved in sulfate and thiosulfate reduction have not been studied. The *Desulfotomaculum* group is distantly related to *Thermacetogenium* on 16S rRNA gene basis, but it is likely that sulfate-reducing abilities are widely distributed among the cluster including *Thermacetogenium* and *Desulfotomaculum*. From these traits, it can be concluded that *Thermacetogenium* might thrive as an acetogen that could use a number of alcohols and methoxylated compounds. If sulfate is available, it could grow as sulfate reducer. The oxidation of acetate coupled with an H_2-consuming methanogen may be only one of the alternatives to survive in anaerobic environments.

Regarding the syntrophic acetate-oxidizing reaction, the traits of the partner methanogen may be crucial. *Thermacetogenium phaeum* strain PB was grown on acetate with two different strains of *Methanothermobacter thermautotrophicus* (Hattori et al., 2001). *Thermacetogenium* grew well in co-culture with the strain TM that was isolated from a thermophilic methane reactor from which the *Thermacetogenium* strain was also isolated. By contrast, the organism grew much more slowly when co-cultured with strain ΔH, a well-known representative of *Methanothermobacter thermautotrophicus*. The only recognizable difference between the two strains was that strain ΔH is not capable of utilizing formate, whereas strain TM is. In addition, *Thermacetogenium* was found to produce a trace amount of formate as well as H_2. Both *Thermacetogenium* and *Methanothermobacter thermautotrophicus* strain TM possess formate dehydrogenase, whereas *Methanothermobacter thermautotrophicus* strain ΔH does not. These results strongly indicate that both H_2 and formate are involved in interspecies electron transfer in the syntrophic acetate oxidation by the organism.

The acetate oxidation in co-culture and acetate formation in pure culture are both driven by the carbon monoxide dehydrogenase (CODH)/acetyl-CoA pathway. The major enzymes (CODH, formyl-tetrahydrofolate synthase, methylene-tetrahydrofolate dehydrogenase, formate dehydrogenase, and hydrogenase) related to the pathway were detected both in cells grown syntrophically on acetate and in cells grown in pure culture on pyruvate (Hattori et al., 2005). Cells grown syntrophically on acetate could immediately convert H_2/CO_2 to form acetate in the presence of a methanogenesis inhibitor, bromoethane sulfonate. However, cells grown on methanol or pyruvate in pure culture cannot immediately convert acetate by mixing resting cells of *Methanothermobacter thermautotrophicus* grown on H_2/CO_2 or *Methanothermobacter thermautotrophicus* cells selectively obtained from a syntrophically acetate-oxidizing co-culture after selective lysis of *Thermacetogenium* cells. Such new co-cultures recombined from separate pure cultures could start acetate oxidation and methane formation after a significant lag phase of about a month (Hattori et al., 2005). These results suggest that pure culture cells of *Thermacetogenium* may need time to rearrange necessary proteins for syntrophic acetate oxidation.

Enrichment and isolation procedures

The organism may be present in high temperature (around 55 °C) anaerobic environments where methanogenesis occurs. For the primary enrichment of a syntrophic acetate-oxidizing co-culture, a sample taken from a thermophilic anaerobic reactor is serially diluted and inoculated into the basal medium (see *Maintenance procedures*) containing 40–80 mM sodium acetate. Yeast extract, peptones, or other organic nutrients should be avoided since those could allow irrelevant heterotrophs to outgrow. After months of incubation at 55 °C, the microorganisms may grow. However, *Thermacetogenium* could grow together with aceticlastic methanogens such as *Methanosaeta* or *Methanosarcina*, or those methanogens could outcompete it. This may be very much dependent on the population size of this organism. If it is predominant in the original samples, the culture receiving high dilution may consist primarily of *Methanothermobacter*-like methanogens and rod-shaped organisms. It is a good indication if the culture does not contain *Methanosarcina*- and *Methanosaeta*-like methanogens but instead contains hydrogenotrophic methanogens such as *Methanothermobacter* and produces methane stoichiometrically from acetate. Careful microscopic examination would easily differentiate the *Methanothermobacter* type of methanogens from *Methanosarcina* and *Methanosaeta*. Once an acetate-oxidizing, methane-forming enrichment culture which only contains hydrogenotrophic methanogens has been identified, it is very likely that an H_2-forming species would predominate this culture together with an H_2-consuming methanogen. For highly purifying the two-membered co-culture, repeated transfer to fresh acetate medium with serial dilutions would be desired.

Attempts to isolate *Thermacetogenium* from "pure co-culture" with hydrogenotrophic methanogens have so far been unsuccessful, since they do not form visible colonies consisting of the two species of microbes. Therefore, when isolating *Thermacetogenium*, it would be better to inoculate the highly enriched co-culture on agar medium containing simple organic substrates such as pyruvate, methanol, ethanol, 1-propanol, 1-butanol, 2,3-butanediol, or methoxylated aromatics (such as syringate, vanillate, and 3,4,5-trimethoxybenzoate) since *Thermacetogenium* could grow in pure culture on some of those substrates. Bromoethane sulfonate (BES, 20 mM) should be added to the medium to inhibit the concomitant growth of methanogens.

The only culture representative of this genus, *Thermacetogenium phaeum*, forms colonies on 2% agar medium containing 80 mM pyruvate as the sole carbon and energy source (Hattori et al., 2000). After 2 months of incubation, the colonies that grew up to 2 mm in diameter were picked with a sterile Pasteur pipette and subcultured in liquid medium containing 80 mM pyruvate and 20 mM BES. Purity of the strain was checked by inoculating it into medium containing 0.1% yeast extract and 1% Bacto Tryptone (Difco) and by microscopy. The "partner" hydrogenotrophic methanogen, *Methanothermobacter* was also isolated from the enrichment culture by using the same procedure, except that pyruvate was replaced with H_2/CO_2 (80:20, v/v, 151 kPa) and BES was omitted.

One of the hardest parts of cultivation is reconstruction of the co-culture from pure cultures of *Thermacetogenium* and *Methanothermobacter thermautotrophicus* to verify that the co-culture is able to oxidize acetate and form methane. In the case of *Thermacetogenium phaeum*, to prove that it is a syntrophic acetate-oxidizing bacterium, it was co-inoculated along with the *Methanothermobacter* strain into the basal medium containing 80 mM sodium acetate and incubated at 55 °C. As the control experiments, strains PB[T] and TM were separately inoculated into the basal medium under the same conditions. Growth was observed only in the mixed culture after 40 d of incubation. Acetate was converted to methane. The co-culture was able to retain methanogenic activity from acetate after further transfer into fresh medium.

Maintenance procedures

Because of the strict anaerobic nature of this organism, all manipulations must be done using anaerobic techniques. The composition of the basal medium is based on DSM 334 medium.

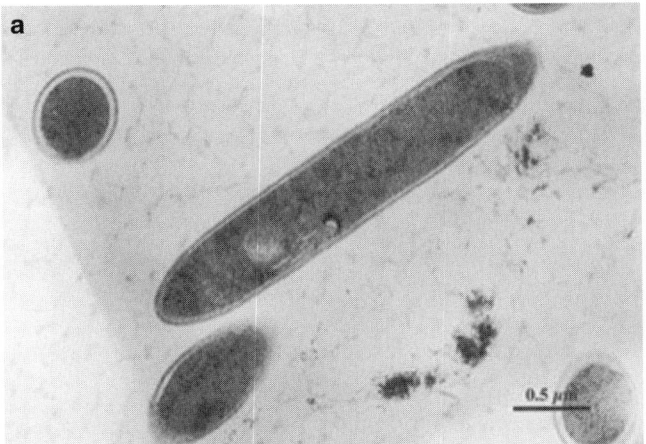
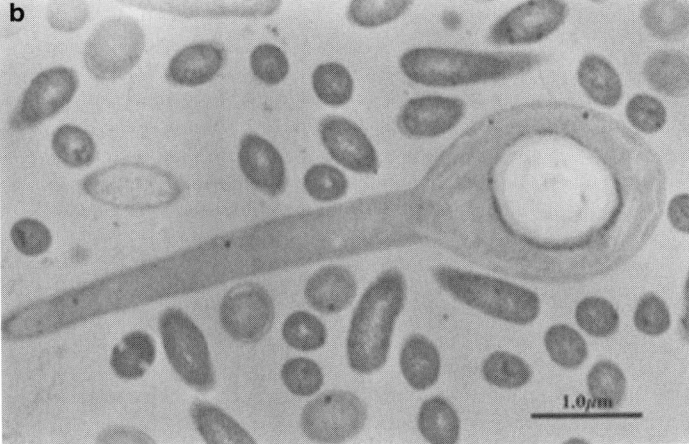

FIGURE 249. Transmission electron micrographs of pure culture of *Thermacetogenium phaeum* (a) and syntrophic coculture showing terminal endospore (b).

The basal medium contains (per liter): NH$_4$Cl, 1.0 g; KH$_2$PO$_4$, 0.3 g; NaCl, 0.6 g; MgCl$_2$·2H$_2$O, 0.1 g; CaCl$_2$·2H$_2$O, 0.08 g; KHCO$_3$, 3.5 g; sodium resazurin, 0.001 g; 10 ml of vitamin solution, 5 ml of trace element solution. The vitamin solution was replaced with that of DSM 141 medium. The composition is as follows (per liter): biotin, 2.0 mg; folic acid, 2.0 mg; pyridoxine.HCl, 10.0 mg; thiamine.HCl, 5.0 mg; riboflavin, 5.0 mg; nicotinic acid, 5.0 mg; DL-calcium pantothenate, 5.0 mg; vitamin B$_9$, 0.1 mg; p-aminobenzoate, 5.0 mg; lipoic acid, 5.0 mg. The trace element solution is based on DSM medium 318 (per liter): nitrilotriacetic acid (NTA), 12.8 g; FeCl$_3$·6H$_2$O, 1.35 g; MnCl$_2$·4H$_2$O, 0.1 g; CoCl$_2$·6H$_2$O, 0.024 g; CaCl$_2$·2H$_2$O, 0.1 g; ZnCl$_2$, 0.1 g; CuCl$_2$·2H$_2$O, 0.025 g; H$_3$BO$_3$, 0.01 g; Na$_2$MoO$_4$·4H$_2$O, 0.024 g; NaCl, 1.0 g; NiCl$_2$·6H$_2$O, 0.12 g; Na$_2$SeO$_3$·5H$_2$O, 4.0 mg; Na$_2$WO$_4$·2H$_2$O, 4.0 mg. The trace element solution is adjusted to pH 6.5 with 1 M KOH. For growth of the syntrophic acetate-oxidizing co-culture, the medium contains 80 mM sodium acetate. For pure culture, 80 mM sodium pyruvate is used as the substrate. The medium is anaerobically dispensed into serum vials or bottles under a N$_2$/CO$_2$ (80:20, v/v) atmosphere. The vials or bottles are sealed with butyl rubber stoppers fitted with caps. The medium is autoclaved for 20 min at 121 °C. The pH of the autoclaved medium is approximately 6.9–7.1. Prior to inoculation, the medium is reduced with sterile stock solutions of Na$_2$S and cysteine/HCl (final concentration: 0.3 g/l each). Cultures are incubated at 55 °C in the dark without shaking.

Once the two-membered co-culture (*Thermacetogenium* and *Methanothermobacter*) is established, it can be maintained stably. Cultures after stationary phase can be left at ambient temperature for at least several months. However, pure culture of *Thermacetogenium* is sometimes very difficult to maintain in this manner since it sometimes lyses very quickly after stationary phase. Pure cultures and co-cultures can be stored in liquid nitrogen or at –80 °C in medium plus 15–20% glycerol.

Differentiation of the genus *Thermacetogenium* from other genera

Thermacetogenium is phylogenetically distant from any known bacterium. *Thermoterrabacterium ferrireducens*, since reclassified as *Carboxydothermus ferrireducens*, is related (Slobodkin et al., 1997b), but it is an iron-reducing micro-organism, whereas *Thermacetogenium* is not able to reduce iron. The most outstanding feature that differentiates *Thermacetogenium* from other organisms is that it oxidizes acetate syntrophically with H$_2$-consuming methanogens. The other acetate-oxidizing syntroph previously known and well-characterized is *Clostridium ultunense*, but it is mesophilic whereas *Thermacetogenium* is thermophilic.

List of species of the genus *Thermacetogenium*

1. **Thermacetogenium phaeum** Hattori, Kamagata, Hanada and Shoun 2000, 1608^{VP}

phae.um. N.L. neut. adj. *phaeum* brown, referring to the color of the colonies.

Straight or slightly curved rod-shaped cells (Figure 249a), 0.4–0.7 μm wide and 2.0–12.6 μm long. It occurs singly, in pairs, or in chains. Round terminal endospores are observed (Figure 249b). Colonies are disk-shaped, 1.0–3.0 mm in diameter, smooth, and brownish. Methanol, ethanol, *n*-propanol, *n*-butanol, 2,3-butanediol, ethanolamine, pyruvate, 3,4,5-trimethoxybenzoate, syringate, vanillate, glycine, cysteine, formate, and H$_2$/CO$_2$ are utilized in pure culture. Acetate is the primary fermentation product. Sugars not utilized. Organic nutrients such as yeast extract are not required for growth. Acetate is oxidized to CO$_2$ in co-culture with H$_2$-utilizing methanogens. Acetate is also oxidized to CO$_2$ in pure culture with reduction of sulfate or thiosulfate as electron acceptor. Sulfite, nitrate, nitrite, Fe(III), and fumarate are not utilized when acetate is used as electron donor. Temperature range for growth is 40–65 °C, with an optimum at 58 °C. Growth is observed at a pH range from 5.9 to 8.4, and the optimum is 6.8. Growth occurs in the presence of NaCl concentrations at 0.05–4.5 % (w/v). The major quinone is menaquinone-7.

DNA G+C content (mol%): 53.5 (HPLC).

Type strain: PB, ATCC BAA-254, DSM12270.

GenBank accession number (16S rRNA gene): AB020336.

Genus VIII. **Thermanaeromonas** Mori, Hanada, Maruyama and Marumo 2002, 1679^{VP}

KOJI MORI AND SATOSHI HANADA

Ther.man.aer.o.mo'nas. Gr. n. *thermos* hot; Gr. pref. *an* not; Gr. n. *aer* air; Gr. n. *monas* a unit, monad; N.L. fem. n. *Thermanaeromonas* a thermophilic, anaerobic monad.

Straight, rod-shaped cells, 0.6 μm in diameter and 2–6 μm in length. Gram stain reaction is negative. Nonmotile. **Terminal endospore formation. Strictly anaerobic. Thermophilic.** Growth occurs between 55 and 73 °C, with an optimum at 70 °C. The pH range for growth is 5.5–8.5, with an optimum at pH 6.5. Growth does not occur above 1% NaCl concentration. **Chemoorganoheterotrophic.** Various organic compounds used for growth. Growth by fermentation or anaerobic respiration. **Thiosulfate, nitrate, and nitrite used as electron acceptor. Menaquinone 7** is the major quinone. Major cellular fatty acids are **C$_{15:0 \text{ iso}}$ and** C$_{17:0 \text{ iso}}$. Isolated from a geothermal aquifer at a depth of 550 m in Toyoha Mines (Hokkaido, Japan).

DNA G+C content (mol%): 49.6.

Type species: **Thermanaeromonas toyohensis** Mori, Hanada, Maruyama and Marumo 2002, 1679^{VP}.

Further descriptive information

Transmission electron microscopic observation showed that *Thermanaeromonas toyohensis* has a Gram-positive cell-wall structure, although the Gram-stain is negative (Figure 250).

FIGURE 250. Cell morphology of *Thermanaeromonas toyohensis*. (a) Phase-contrast micrograph. (b, c) Ultrathin sections of *Thermanaeromonas toyohensis*. Spore is formed at the end of *Thermanaeromonas toyohensis* cell (Reproduced with permission from Mori et al. (2002). Int. J. Syst. Evol. Microbiol. *52*: 1675–1680.)

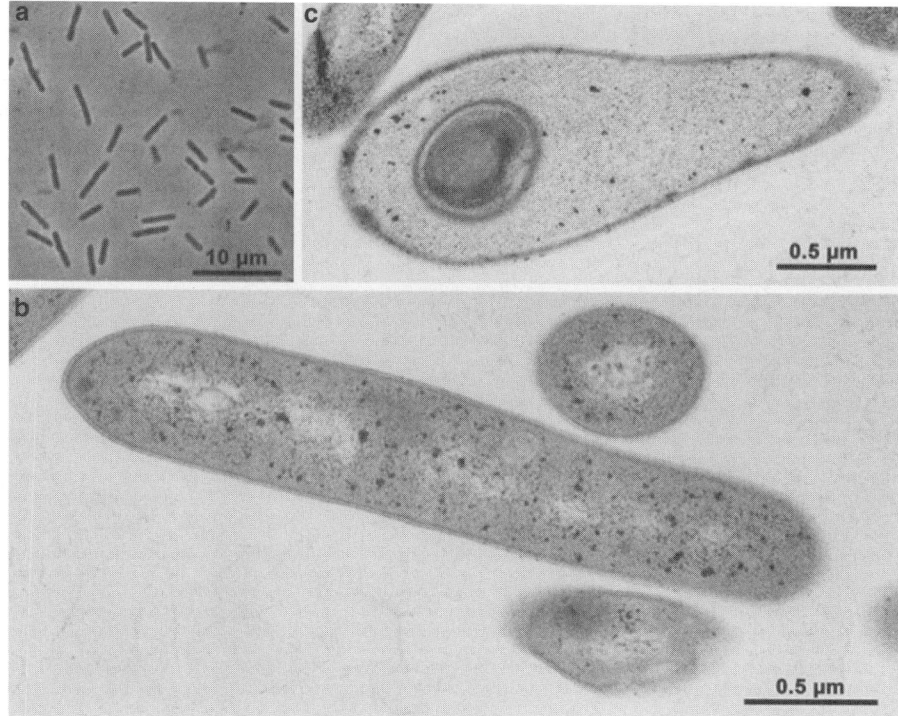

Thermanaeromonas toyohensis also contains menaquinone 8 as a minor quinone in addition to a major quinone, menaquinone 7. $C_{16:0}$ is also detected as the minor cellular fatty acid (8% of the total fatty acids) along with major components ($C_{15:0\ iso}$ and $C_{17:0\ iso}$). The cellular polyamines in *Thermanaeromonas toyohensis* are mainly two quaternary pentaamines, N^4-bis(aminopropyl) spermidine and N^4-bis(aminopropyl)norspermidine (Hosoya et al., 2004).

White colonies of *Thermanaeromonas toyohensis*, showing approximately 1 mm in diameter, take 3 d to develop in a solidified medium containing 0.05% (w/v) glucose, 0.05% yeast extract, and 0.08% thiosulfate. *Thermanaeromonas toyohensis* grows chemoorganoheterotorophically under anaerobic conditions by fermentation or anaerobic respiration. Substrates for anaerobic respiration are arabinose, cellobiose, fructose, glucose, inositol, maltose, mannose, sucrose, trehalose, xylose, yeast extract, formate, lactate, and pyruvate. With the exception of formate, these substrates are also used for fermentative growth. Under anaerobic respiratory conditions, *Thermanaeromonas toyohensis* uses thiosulfate, nitrate, or nitrite as an electron acceptor. The following electron acceptors do not support growth: sulfate, sulfite, elemental sulfur, iron (III) citrate, and fumarate. The cell yield and growth rate under conditions favoring anaerobic respiration with thiosulfate as an electron acceptor are clearly higher than those resulting from fermentative growth.

Only one strain of *Thermanaeromonas toyohensis* has been isolated. The isolate was obtained from a geothermal aquifer that emanated from a crack at 550 m below sea level at the Toyoha Mines in Hokkaido, Japan. The Toyoha Mine is a Ag/Pb/Zn/ Cu polymetallic vein-type deposit and the mineralization of the deposit occurred from 3–0.5 million years ago. The collected hot water was 71 °C and pH 5.8. It welled out from the wall at a rate of 11/min.

Enrichment and isolation procedures

Enrichment and isolation can be performed in the presence of glucose, yeast extract, and thiosulfate under N_2/CO_2 (80/20, v/v) conditions. The medium also contains (per liter): KH_2PO_4, 0.75 g; K_2HPO_4, 0.78 g; NH_4Cl, 0.53 g; $NaHCO_3$, 5.0 g; Na_3EDTA, 0.041 g; $FeSO_4·7H_2O$, 0.011 g; $MgSO_4·7H_2O$, 0.25 g; $CaCl_2·2H_2O$, 0.029 g; NaCl, 0.23 g; trace-element solution DSM 334 (DSMZ, 1993), 10 ml; vitamin solution DSM 141 (DSMZ, 1993), 10 ml. Incubation at approximately 70 °C is recommended. After enrichment, colonies form in this medium solidified with 0.6% (w/v) gellan gum (Wako Pure Chemicals). After 3 d of incubation under N_2/CO_2 (80/20), white colonies can be observed in pour plates.

Maintenance procedures

Liquid cultures retain viability for several months at room temperature. Long-term preservation in liquid nitrogen is possible in the presence of 5% (v/v) dimethylsulfoxide or 10% (v/v) glycerol under anaerobic conditions.

Differentiation of the genus *Thermanaeromonas* from other genera

Thermanaeromonas toyohensis can be clearly differentiated from other genera within the family *Thermoanaerobacteraceae* by comparison of 16S rRNA gene sequences (Figure 251). *Therman-*

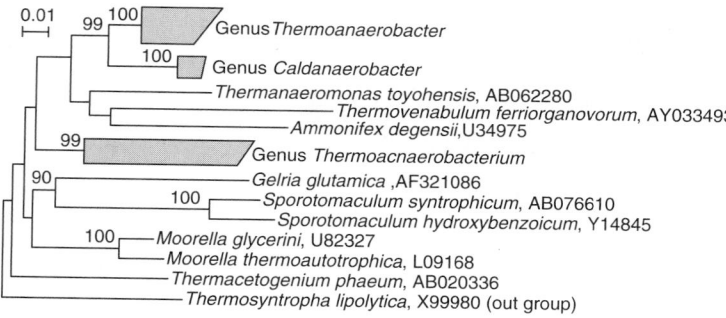

FIGURE 251. Phylogenetic relationships based on 16S rRNA gene sequences in the family *Thermoanaerobacteraceae*, inferred from the neighbor-joining method after alignment using the ARB program. Bootstrap probabilities are indicated at branching points, and scale bar shows substitutions per the compared nucleotides.

TABLE 248. Comparison of properties among *Thermanaeromonas* and related genera[a]

Character	*Thermanaeromonas*	*Ammonifex*	*Caldanaerobacter*	*Thermoanaerobacter*	*Thermovenabulum*
Morphology	Rods	Rods	Rods (sometimes branching)	Rods	Rods (sometimes branching)
Size (μm)	0.6 × 2–6	0.6 × 2–8.5	0.3 × 4–10	0.4–0.8 × 1–10	0.5 × 1.5–7
Habitat	Thermal aquifer	Hot spring	Submarine, thermal vent, and oilfield	Hot spring, oil well, and lake	Hot spring
Temperature range for growth (°C)	55–73	57–77	50–85	35–85	45–76
DNA G+C content (mol%)	49.6	54	33–41	29–41	36
Spore formation	+	−	ND	+[b]	+
Chemolithotrophic growth	−	+	±	−	−
Electron acceptor:					
SO$_4^{2-}$	−	+	−	−[b]	−
SO$_3^{2-}$	−	−	−[b]	±	−
S$_2$O$_3^{2-}$	+	−	±	−[b]	+
S^0	−	+	−[b]	−[b]	+
Fe(III)	−	−	ND	ND	+
Reduction of:					
NO$_3^-$	+	+	ND	−[b]	+
NO$_2^-$	+	−	ND	−[b]	−

[a]Symbols: +, >85% positive; −, 0–15% positive; ND, not data.
[b]With the exception of some species.

aeromonas toyohensis is phylogenetically distant from the other genera, with low sequence similarities of less than 90%. The phylogenetic relatives are the genera *Ammonifex*, *Caldanaerobacter*, *Thermoanaerobacter*, and *Thermovenabulum*. While all species in the related genera are thermophilic anaerobes like *Thermanaeromonas toyohensis*, there are obvious differences in the genotypic and phenotypic characteristics between the genus *Thermanaeromonas* and the related genera (see Table 248). The genomic DNA G+C content of *Thermanaeromonas toyohensis* is 49.6 mol%, and the value is higher than those in the related genera, except the genus *Ammonifex*. The G+C content of *Ammonifex degensii* is high enough to distinguish it from *Thermanaeromonas toyohensis* (Huber et al., 1996). *Thermanaeromonas toyohensis* is not autotrophic, whereas all species in the genus *Thermoanaerobacter* are autotrophic. However, *Ammonifex degensii* can grow autotrophically on hydrogen and carbon dioxide in the presence of nitrate, sulfate, or sulfur as an electron acceptor (Huber et al., 1996), and three of four subspecies of *Caldanaerobacter subterraneus* show autotrophic growth on CO as a sole energy source (Fardeau et al., 2004; Sokolova et al., 2001). Use of electron acceptors is a significant trait for differentiation of *Thermanaero-*

monas toyohensis from the related organisms. In the presence of lactate as a sole substrate, *Thermanaeromonas toyohensis* uses thiosulfate, nitrate, and nitrite as an electron acceptor, but cannot use sulfate, sulfite, or elemental sulfur. *Thermovenabulum ferriorganovorum* can reduce Fe(III) in the medium with molecular hydrogen and organic substrates (Zavarzina et al., 2002).

Taxonomic comments

Sequence analysis of the 16S rRNA gene indicates that *Thermanaeromonas toyohensis* belongs to the family *Thermoanaerobacteraceae* in the phylum *Firmicutes*. The closet relatives are the genera *Themoanaerobacter* and *Caldanaerobacter*, although their sequence similarities to *Thermanaeromonas toyohensis* are very low (88–89%) and the genus is clearly distant from the related organisms in the phylogenetic tree (Figure 251).

The genus *Thermanaeromonas* contains only one species, *Thermanaeromonas toyohensis*, and this species is represented by only a single strain (strain ToBE). However, a very closely related strain, designated BA-5, has been isolated from a geothermal aquifer in a gold mine by A.S. Bonin, and its 16S rRNA gene sequence has been deposited in GenBank database under accession number AY695836 (sequence similarity is 93%). In addition, one environmental clone sequence (accession no. DQ097672) recovered from an oilfield in China is shown in the DDBJ and is thought to be a phylogenetic relative to *Thermanaeromonas*

toyohensis (sequence similarity is 93%) (Nazina et al., 2005). These isolate and clone sequences were both found in subterranean hot environments similar to that where *Thermanaeromonas toyohensis* is found.

List of species of the genus *Thermanaeromonas*

1. **Thermanaeromonas toyohensis** Mori, Hanada, Maruyama and Marumo 2002, 1679[VP]

    to.yo.hen'sis. N.L. adj. *toyohensis* from Toyoha, referring to its isolation from the Toyoha Mines.

    The description is as given for the genus.

Isolated from a geothermal aquifer at a depth of 550 m in Toyoha Mines (Hokkaido, Japan).

DNA G+C content (mol%): 49.6 (HPLC).

Type strain: ToBE, DSM 14490, JCM 11376, NBRC 101528.

GenBank accession number (16S rRNA gene): AB062280.

References

Ahring, B.K., K. Jensen, P. Nielsen, A.B. Bjerre and A.S. Schmidt. 1996. Pretreatment of wheat straw and conversion of xylose and xylan to ethanol by thermophilic anaerobic bacteria. Biores. Technol. *58*: 107–113.

Ahring, B.K., D. Licht, A.S. Schmidt, P. Sommer and A.B. Thomsen. 1999. Production of ethanol from wet oxidized wheat straw by *Thermoanaerobacter mathranii*. Biores. Technol. *68*: 3–9.

Alister, A., E. Herlitz, R. Borquez and M. Roeckel. 1990. Production of thermostable β-galactosidase with *Thermoanaerobacter ethanolicus*. Ann. N.Y. Acad. Sci. *613*: 605–609.

Allison, M.J., A.L. Baetz and J. Wiegel. 1984. Alternative pathways for biosynthesis of leucine and other amino acids in *Bacteroides ruminicola* and *Bacteroides fragilis*. Appl. Environ. Microbiol. *48*: 1111–1117.

Andreesen, J.R. and L.G. Ljungdahl. 1973. Formate dehydrogenase of *Clostridium thermoaceticum*: incorporation of selenium-75, and the effects of selenite, molybdate, and tungstate on the enzyme. J. Bacteriol. *116*: 867–873.

Antranikian, G., P. Zablowski and G. Gottschalk. 1987. Conditions for the overproduction and excretion of thermostable α-amylase and pullulanase from *Clostridium thermohydrosulfuricum* DSM 567. Appl. Microbiol. Biotechnol. *27*: 75–81.

Antranikian, G. 1989. The formation of extracellular, thermoactive amylase and pullulanase in batch culture by *Thermoanaerobacter finnii*. Appl. Biochem. Biotechnol. *20–21*: 267–279.

Arni, R.K., L. Watanabe, M. Fontes, D.S. Burdette and J.G. Zeikus. 1996. Crystallization of the secondary alcohol dehydrogenase from *Thermoanaerobacter ethanolicus* 39E. Prot. Pep.Lett. *3*: 423–426.

Balch, W.E. and R.S. Wolfe. 1976. New approach to the cultivation of methanogenic bacteria: 2-mercaptoethanesulfonic acid (HS-CoM)-dependent growth of *Methanobacterium ruminantium* in a pressurized atmosphere. Appl. Environ. Microbiol. *32*: 781–791.

Balch, W.E., G.E. Fox, L.J. Magrum, C.R. Woese and R.S. Wolfe. 1979. Methanogens: reevaluation of a unique biological group. Microbiol. Rev. *43*: 260–296.

Balk, M., J. Weijma, M.W. Friedrich and A.J.M. Stams. 2003. Methanol utilization by a novel thermophilic homoacetogenic bacterium, *Moorella mulderi* sp. nov., isolated from a bioreactor. Arch. Microbiol. *179*: 315–320.

Balk, M., J. Weijma, M.W. Friedrich and A.J.M. Stams. 2005. *In* Validation of the publication of new names and new combinations previously effectively published outside the IJSEM. List no. 101. Int. J. Syst. Evol. Microbiol. *55*: 1–2.

Barker, H.A. 1936. On the biochemistry of a unique biological group. Microbiol. Rev. *43*: 260–296.

Ben-Bassat, A. and J.G. Zeikus. 1981. *Thermobacteroides acetoethylicus* gen. nov. and sp. nov., a new chemoorganotrophic, anaerobic, thermophilic bacterium. Arch. Microbiol. *128*: 365–370.

Berestovskaya, Y.Y., V.R. Kryokov, I.V. Bodnar and M.A. Pusheva. 1987. Culturing of homoacetogenic bacterium *Clostridium thermoautotrophicum*. Mikrobiology (USSR) *56*: 506–511.

Bock, K., J. Schuster-Kolbe, E. Altman, G. Allmaier, B. Stahl, R. Christian, U.B. Sleytr and P. Messner. 1994. Primary structure of the *O*-glycosidically linked glycan chain of the crystalline surface layer glycoprotein of *Thermoanaerobacter thermohydrosulfuricus* L111-69. Galactosyl tyrosine as a novel linkage unit. J. Biol. Chem. *269*: 7137–7144.

Boga, H.I. and A. Brune. 2003. Hydrogen-dependent oxygen reduction by homoacetogenic bacteria isolated from termite guts. Appl. Environ. Microbiol. *69*: 779–786.

Bogdahn, M. and D. Kleiner. 1986. N_2 fixation and NH_4^+ assimilation in the thermophilic anaerobes *Clostridium thermosaccharolyticum* and *Clostridium thermoautotrophicum*. Arch. Microbiol. *144*: 102–104.

Bogin, O., M. Peretz and Y. Burstein. 1997. *Thermoanaerobacter brockii* alcohol dehydrogenase: characterization of the active site metal and its ligand amino acids. Protein Sci. *6*: 450–458.

Bogin, O., M. Peretz and Y. Burstein. 1998a. Probing structural elements of thermal stability in bacterial oligomeric alcohol dehydrogenases. I. Construction and characterization of chimeras consisting of secondary ADHs from *Thermoanaerobacter brockii* and *Clostridium beijerinckii*. Lett. Peptide Sci. *5*: 399–408.

Bogin, O., M. Peretz, Y. Hacham, Y. Korkhin, F. Frolow, A.J. Kalb and Y. Burstein. 1998b. Enhanced thermal stability of *Clostridium beijerinckii* alcohol dehydrogenase after strategic substitution of amino acid residues with prolines from the homologous thermophilic *Thermoanaerobacter brockii* alcohol dehydrogenase. Protein Sci. *7*: 1156–1163.

Bonch-Osmolovskaya, E.A., M.L. Miroshnichenko, N.A. Chernykh, N.A. Kostrikina, E.V. Pikuta and F.A. Rainey. 1997. Reduction of elemental sulfur by moderately thermophilic organotrophic bacteria and the description of *Thermoanaerobacter sulfurophilus* sp. nov. Microbiology (En. transl. from Mikrobiologiya) *66*: 483–489.

Bonch-Osmolovskaya, E.A., M.L. Miroshnichenko, N.A. Chernykh, N.A. Kostrikina, E.V. Pikuta and F.A. Rainey. 1998. *In* Validation of the publication of new names and new combinations previously effectively published outside the IJSB. List no. 66. Int. J. Syst. Bacteriol. *48*: 631–632.

Braun, M., S. Schoberth and G. Gottschalk. 1979. Enumeration of bacteria forming acetate from H_2 and CO_2 in anaerobic habitats. Arch. Microbiol. *120*: 201–204.

Breves, R., K. Bronnenmeier, N. Wild, F. Lottspeich, W.L. Staudenbauer and J. Hofemeister. 1997. Genes encoding two different beta-glucosidases of *Thermoanaerobacter brockii* are clustered in a common operon. Appl. Environ. Microbiol. *63*: 3902–3910.

Bryant, F.O., J. Wiegel and L.G. Ljungdahl. 1988. Purification and properties of primary and secondary alcohol dehydrogenases from *Thermoanaerobacter ethanolicus*. Appl. Environ. Microbiol. *54*: 460–465.

Bryant, F.O., J. Wiegel and L.G. Ljungdahl. 1992. Comparisons of alcohol dehydrogenases from wild-type and mutant strain, JW200 Fe4, of *Thermoanaerobacter ethanolicus*. Appl. Microbiol. Biotechnol. *37*: 490–495.

Burdette, D. and J.G. Zeikus. 1994. Purification of acetaldehyde dehydrogenase and alcohol dehydrogenases from *Thermoanaerobacter ethanolicus* 39E and characterization of the secondary-alcohol dehydrogenase (2° Adh) as a bifunctional alcohol dehydrogenase–acetyl-CoA reductive thioesterase. Biochem. J. *302*: 163–170.

Burdette, D.S., F. Secundo, R.S. Phillips, J. Dong, R.A. Scott and J.G. Zeikus. 1997. Biophysical and mutagenic analysis of *Thermoanaerobacter ethanolicus* secondary-alcohol dehydrogenase activity and specificity. Biochem. J. *326*: 717–724.

Burdette, D.S., V.V. Tchernajencko and J.G. Zeikus. 2000. Effect of thermal and chemical denaturants on *Thermoanaerobacter ethanolicus* secondary-alcohol dehydrogenase stability and activity. Enzyme Microb. Technol. *27*: 11–18.

Burdette, D.S., S.H. Jung, G.J. Shen, R.I. Hollingsworth and J.G. Zeikus. 2002. Physiological function of alcohol dehydrogenases and long-chain (C-30) fatty acids in alcohol tolerance of *Thermoanaerobacter ethanolicus*. Appl. Environ. Microbiol. *68*: 1914–1918.

Byrer, D.E., F.A. Rainey and J. Wiegel. 2000. Novel strains of *Moorella thermoacetica* form unusually heat-resistant spores. Arch. Microbiol. *174*: 334–339.

Canganella, F. and J. Wiegel. 1993. The potential of thermophilic clostridia in biotechnology. *In* Woods (Editor), The Clostridia and Biotechnology. Butterworths, Stoneham, MA, pp. 391–429.

Carlier, J.P. and M. Bedora-Faure. 2006. Phenotypic and genotypic characterization of some *Moorella* sp. strains isolated from canned foods. Syst. Appl. Microbiol. *29*: 581–588.

Carlier, J.P., I. Bonne and M. Bedora-Faure. 2006. Isolation from canned foods of a novel *Thermoanaerobacter* species phylogenetically related to *Thermoanaerobacter mathranii* (Larsen 1997): emendation of the species description and proposal of *Thermoanaerobacter mathranii* subsp. *alimentarius* subsp. nov. Anaerobe *12*: 153–159.

Carlier, J.P., I. Bonne and M. Bedora-Faure. 2007. List of changes in taxonomic opinion no. 5. Notification of changes in taxonomic opinion previously published outside the IJSEM. Int. J. Syst. Evol. Microbiol. *57*: 4–5.

Carreira, L.H., J. Wiegel and L.G. Ljungdahl. 1983. Production of ethanol from biopolymers by anaerobic, thermophilic, and extreme thermophilic bacteria. I. Regulation of carbohydrate utilization in mutants of *Thermoanaerobacter ethanolicus*. Biotech. Bioeng. Symp. *13*: 183–191.

Cayol, J.-L., B. Ollivier, B.K.C. Patel, G. Ravot, M. Magot, E. Ageron, P.A.D. Grimont and J.-L. Garcia. 1995. Description of *Thermoanaerobacter brockii* subsp. *lactiethylicus* subsp. nov., isolated from a deep subsurface French oil-well, a proposal to reclassify *Thermoanaerobacter finii* as *Thermoanaerobacter brockii* subsp. *finnii* comb. nov., and an emended description of *Thermoanaerobacter brockii*. Int. J. Syst. Bacteriol. *45*: 783–789.

Chaen, H., T. Nishimoto, T. Nakada, S. Fukuda, M. Kurimoto and Y. Tsujisaka. 2001. Enzymatic synthesis of novel oligosaccharides from L-sorbose, maltose, and sucrose using kojibiose phosphorylase. J. Biosci. Bioeng. *92*: 173–176.

Chinn, M.S., S.E. Nokes and H.J. Strobel. 2006. Screening of thermophilic anaerobic bacteria for solid substrate cultivation on lignocellulosic substrates. Biotechnol. Prog. *22*: 53–59.

Christian, R., P. Messner, C. Weiner, U.B. Sleytr and G. Schulz. 1988. Structure of a glycan from the surface-layer glycoprotein of *Clostridium thermohydrosulfuricum* strain L111-69. Carbohydr. Res. *176*: 160–163.

Clark, J.E., S.W. Ragsdale, L.G. Ljungdahl and J. Wiegel. 1982. Levels of enzymes involved in the synthesis of acetate from CO_2 in *Clostridium thermoautotrophicum*. J. Bacteriol. *151*: 507–509.

Collet, C., J.P. Schwitzguebel and P. Peringer. 2003. Improvement of acetate production from lactose by growing *Clostridium thermolacticum* in mixed batch culture. J. Appl. Microbiol. *95*: 824–831.

Collins, M.D., P.A. Lawson, A. Willems, J.J. Cordoba, J. Fernández-Garayzábal, P. Garcia, J. Cai, H. Hippe and J.A. Farrow. 1994. The phylogeny of the genus *Clostridium*: proposal of five new genera and eleven new species combinations. Int. J. Syst. Bacteriol. *44*: 812–826.

Cook, G.M., P.H. Janssen and H.W. Morgan. 1991. Endospore formation by *Thermoanaerobium brockii* HTD4. Syst. Appl. Microbiol. *14*: 240–244.

Cook, G.M. and H.W. Morgan. 1994. Hyperbolic growth of *Thermoanaerobacter thermohydrosulfuricus* (*Clostridium thermohydrosulfuricum*) increases ethanol-production in pH-controlled batch culture. Appl. Microbiol. Biotechnol. *41*: 84–89.

Cook, G.M., F.A. Rainey, B.K.C. Patel and H.W. Morgan. 1996. Characterization of a new obligately anaerobic thermophile, *Thermoanaerobacter wiegelii* sp. nov. Int. J. Syst. Bacteriol. *46*: 123–127.

Cook, G.M. 2000. The intracellular pH of the thermophilic bacterium *Thermoanaerobacter wiegelii* during growth and production of fermentation acids. Extremophiles *4*: 279–284.

Crowther, R.A. and U.B. Sleytr. 1977. An analysis of the fine structure of the surface layers from two strains of clostridia, including correction for distorted images. J. Ultrastruct. Res. *58*: 41–49.

Daniel, S.L., T. Hsu, S.I. Dean and H.L. Drake. 1990. Characterization of the H_2- and CO-dependent chemolithotrophic potentials of the acetogens *Clostridium thermoaceticum* and *Acetogenium kivui*. J. Bacteriol. *172*: 4464–4471.

Das, A., J. Huegenholz, H. van Halbeek and L.G. Lujungdahl. 1987. Structure and function of a menaquinone involved in electron transport in membranes of *Clostridium thermoautotrophicum* and *Clostridium thermoaceticum*. J. Bacteriol. *171*: 5823–5829.

Das, A. and L.G. Ljungdahl. 1993. F_0 and F_1 Parts of ATP synthases from *Clostridium thermoautotrophicum* and *Escherichia coli* are not functionally compatible. FEMS Lett. *317*: 17–21.

Das, A., D.M. Ivey and L.G. Ljungdahl. 1997. Purification and reconstitution into proteoliposomes of the F_1F_0 ATP synthase from the obligately anaerobic gram-positive bacterium *Clostridium thermoautotrophicum*. J. Bacteriol. *179*: 1714–1720.

Das, A., and L.G. Ljungdahl. 2003. Electron-transport system in acetogens. *In* Ljungdahl, Barton, Ferry and Johnson (Editors), Biochemistry and Physiology of Anaerobic Bacteria. Springer-Verlag, New York, pp. 191–204.

Das, A., R. Silaghi-Dumitrescu, L.G. Ljungdahl and D.M. Kurtz, Jr. 2005. Cytochrome *bd* oxidase, oxidative stress, and dioxygen tolerance of the strictly anaerobic bacterium *Moorella thermoacetica*. J. Bacteriol. *187*: 2020–2029.

Diekert, G.B. and R.K. Thauer. 1978. Carbon monoxide oxidation by *Clostridium thermoaceticum* and *Clostridium formicoaceticum*. J. Bacteriol. *136*: 597–606.

Drake, H.L. 1994. Acetogenesis, acetogenic bacteria and the acetyl Co-A "Wood Ljungdahl" pathway: past and current perspectives. *In* Drake (Editor), Acetogenesis. Chapman & Hall, New York, pp. 273–302.

Drake, H.L., A.S. Gössner and S.L. Daniel. 2008. Old acetogens, new light. Ann. N. Y. Acad. Sci. *1125*: 100–128.

Drent, W.J. and J.C. Gottschal. 1991. Fermentation of inulin by a new strain of *Clostridium thermoautotrophicum* isolated from dahlia tubers. FEMS Microbiol. Lett. *78*: 285–291.

DSMZ. 1993. Catalog of Strains. Gesellschaft fur Biotechnologische Forschung, Braunschweig.

Elliott, J.I. and L.G. Ljungdahl. 1982. Isolation and characterization of an Fe_8-S_8 ferredoxin (ferredoxin-II) from *Clostridium thermoaceticum*. J. Bacteriol. *151*: 328–333.

Erbeznik, M., K.A. Dawson and H.J. Strobel. 1998a. Cloning and characterization of transcription of the *xylAB* operon in *Thermoanaerobacter ethanolicus*. J. Bacteriol. *180*: 1103–1109.

Erbeznik, M., H.J. Strobel, K.A. Dawson and C.R. Jones. 1998b. The D-xylose-binding protein, XylF, from *Thermoanaerobacter ethanolicus* 39E: cloning, molecular analysis, and expression of the structural gene. J. Bacteriol. *180*: 3570–3577.

Erbeznik, M., H.J. Strobel and K.A. Dawson. 2000. Organization and sequence of histidine biosynthesis genes *hisH, -A, -F,* and *-IE* in *Thermoanaerobacter ethanolicus*. Curr. Microbiol. *40*: 140–142.

Erbeznik, M., S.E. Hudson, A.B. Herrman and H.J. Strobel. 2004. Molecular analysis of the *xylFGH* operon, coding for xylose ABC transport, in *Thermoanaerobacter ethanolicus*. Curr. Microbiol. *48*: 295–299.

Fardeau, M.-L., J.-L. Cayol, M. Magot and B. Ollivier. 1994. Hydrogen oxidation abilities in the presence of thiosulfate as electron acceptor within the genus *Thermoanaerobacter*. Curr. Microbiol. *29*: 269–272.

Fardeau, M.-L., C. Faudon, J.-L. Cayol, M. Magot, B.K.C. Patel and B. Ollivier. 1996. Effect of thiosulphate as electron acceptor on glucose and xylose oxidation by *Thermoanaerobacter finnii* and a *Thermoanaerobacter* sp. isolated from oil field water. Res. Microbiol. *147*: 159–165.

Fardeau, M.-L., B.K.C. Patel, M. Magot and B. Ollivier. 1997. Utilization of serine, leucine, isoleucine, and valine by *Thermoanaerobacter brockii* in the presence of thiosulfate or *Methanobacterium* sp. as electron acceptors. Anaerobe *3*: 405–410.

Fardeau, M.-L., M. Magot, B.K.C. Patel, P. Thomas, J.-L. Garcia and B. Ollivier. 2000. *Thermoanaerobacter subterraneus* sp. nov., a novel thermophile isolated from oilfield water. Int. J. Syst. Evol. Microbiol. *50*: 2141–2149.

Fardeau, M.-L., M. Bonilla-Salinas, S. L'Haridon, C. Jeanthon, F. Verhé, J.-L. Cayol, B.K.C. Patel, J.-L. Garcia and B. Ollivier. 2004. Isolation from oil reservoirs of novel thermophilic anaerobes phylogenetically related to *Thermoanaerobacter subterraneus*: reassignment of *T. subterraneus, Thermoanaerobacter yonseiensis, Thermoanaerobacter tengcongensis* and *Carboxydibrachium pacificum* to *Caldanaerobacter subterraneus* gen. nov., sp. nov., comb. nov. as four novel subspecies. Int. J. Syst. Evol. Microbiol. *54*: 467–474.

Faudon, C., M.-L. Fardeau, J. Heim, B.K.C. Patel, M. Magot and B. Ollivier. 1995. Peptide and amino acid oxidation in the presence of thiosulfate by members of the genus *Thermoanaerobacter*. Curr. Microbiol. *31*: 152–157.

Ferry, J.G. 1992. Methane from acetate. J. Bacteriol. *174*: 5489–5495.

Fokina, N.A. and G.A. Velikodvorskaia. 1997. Cloning and expression of the gene for thermostable beta-galactosidase from *Thermoanaerobacter ethanolicus* in *Escherichia coli*: purification and properties of the product. Mol. Gen. Mikrobiol. Virusol. *2*: 34–36.

Fontaine, F.E., W.H. Peterson, E. McCoy, M.J. Johnson and G.J. Ritter. 1942. A new type of glucose fermentation by *Clostridium thermoaceticum* n sp. J. Bacteriol. *43*: 701–715.

Freier, D. 1981. Interaction of thermophilic anaerobic bacteria in cellulolytic mixed cultures. Diploma thesis. University of Göttingen, Germany.

Gavrilov, S.N., E.A. Bonch-Osmolovskaya and A.I. Slobodkin. 2003. Physiology of organotrophic and lithotrophic growth of the thermophilic iron-reducing bacteria *Thermoterrabacterium ferrireducens* and *Thermoanaerobacter siderophilus*. Microbiology (En. transl. from Mikrobiologiya) *72*: 132–137.

Germain, P., F. Toukourou and L. Donaduzzi. 1986. Ethanol production by anaerobic thermophilic bacteria: regulation of lactate dehydrogenase activity in *Clostridium thermohydrosulfuricum*. Appl. Microbiol. Biotechnol. *24*: 300–305.

Gössner, A.S., R. Devereux, N. Ohnemüller, G. Acker, E. Stackebrandt and H.L. Drake. 1999. *Thermicanus aegyptius* gen. nov., sp. nov., isolated from oxic soil, a fermentative microaerophile that grows commensally with the thermophilic acetogen *Moorella thermoacetica*. Appl. Environ. Microbiol. *65*: 5124–5133.

Grassia, G.S., K.M. Mclean, P. Glenat, J. Bauld and A.J. Sheehy. 1996. A systematic survey for thermophilic fermentative bacteria and archaea in high temperature petroleum reservoirs. FEMS Microbiol. Lett. *21*: 47–58.

Hattori, S., Y. Kamagata, S. Hanada and H. Shoun. 2000. *Thermacetogenium phaeum* gen. nov., sp. nov., a strictly anaerobic, thermophilic, syntrophic acetate-oxidizing bacterium. Int. J. Syst. Evol. Microbiol. *50*: 1601–1609.

Hattori, S., H.W. Luo, H. Shoun and Y. Kamagata. 2001. Involvement of formate as an interspecies electron carrier in a syntrophic acetate-oxidizing anaerobic microorganism in coculture with methanogens. J. Biosci. Bioeng. *91*: 294–298.

Hattori, S., A.S. Galushko, Y. Kamagata and B. Schink. 2005. Operation of the CO dehydrogenase/acetyl coenzyme A pathway in both acetate oxidation and acetate formation by the syntrophically acetate-oxidizing bacterium *Thermacetogenium phaeum*. J. Bacteriol. *187*: 3471–3476.

Heiss, C., M. Laivenieks, J.G. Zeikus and R.S. Phillips. 2001. Mutation of cysteine-295 to alanine in secondary alcohol dehydrogenase from *Thermoanaerobacter ethanolicus* affects the enantioselectivity and substrate specificity of ketone reductions. Bioorg. Med. Chem. *9*: 1659–1666.

Heitmann, T., E. Wenzig and M. A. 1996. A kinetic model of growth and product formation of the anaerobic microorganism *Thermoanaerobacter thermohydrosulfuricus*. J. Biotechnol. *50*: 213–223.

Henstra, A.M. and A.J.M. Stams. 2004. Novel physiological features of *Carboxydothermus hydrogenoformans* and *Thermoterrabacterium ferrireducens*. Appl. Environ. Microbiol. *70*: 7236–7240.

Hild, H.M., D.C. Stuckey and D.J. Leak. 2003. Effect of nutrient limitation on product formation during continuous fermentation of xylose with *Thermoanaerobacter ethanolicus* JW200 Fe(7). Appl. Microbiol. Biotechnol. *60*: 679–686.

Hollaus, F. and U. Sleytr. 1972. On the taxonomy and fine structure of some hyperthermophilic saccharolytic clostridia. Arch. Mikrobiol. *86*: 129–146.

Hollaus, F. and H. Klaushofer. 1973. Identification of hyperthermophilic obligate anaerobic bacteria from extraction juices of beet sugar factories. Int. Sugar J. *75*: 237–241, 271–275.

Holt, P.J., R.E. Williams, K.N. Jordan, C.R. Lowe and N.C. Bruce. 2000. Cloning, sequencing and expression in *Escherichia coli* of the primary alcohol dehydrogenase gene from *Thermoanaerobacter ethanolicus* JW200. FEMS Microbiol. Lett. *190*: 57–62.

Hosoya, R., K. Hamana, M. Niitsu and T. Itoh. 2004. Polyamine analysis for chemotaxonomy of thermophilic eubacteria: polyamine distribution profiles within the orders *Aquificales, Thermotogales, Thermodesulfobacteriales, Thermales, Thermoanaerobacteriales, Clostridiales* and *Bacillales*. J. Gen. Appl. Microbiol. *50*: 271–287.

Hsu, T., S.L. Daniel, M.F. Lux and H.L. Drake. 1990. Biotransformations of carboxylated aromatic compounds by the acetogen *Clostridium thermoaceticum*: generation of growth-supportive CO_2 equivalents under CO_2-limited conditions. J. Bacteriol. *172*: 212–217.

Huber, G., R. Huber, B.E. Jones, G. Lauerer, A. Neuner, A. Segerer, K.O. Stetter and E.T. Degens. 1991. Hyperthermophilic archaea and bacteria occurring within Indonesian hydrothermal areas. Syst. Appl. Microbiol. *14*: 397–404.

Huber, G., R. Huber, B.E. Jones, G. Lauerer, A. Neuner, A. Segerer, K.O. Stetter and E.T. Degens. 1992. Hyperthermophilic archaea and eubacteria occurring within Indonesian hydrothermal areas *In* Degens, Wong and Zen (Editors), The sea off mount Tambora Mitt. Geol.-Paläont. Inst. Univ. Hamburg, Hamburg, pp. 161–172.

Huber, R., P. Rossnagel, C.R. Woese, R. Rachel, T.A. Langworthy and K.O. Stetter. 1996. Formation of ammonium from nitrate during chemolithoautotrophic growth of the extremely thermophilic bacterium *Ammonifex degensii* gen. nov. sp. nov. Syst. Appl. Microbiol. *19*: 40–49.

Huber, R. and K.O. Stetter. 1996. *In* Validation of the publication of new names and new combinations previously effectively published outside the IJSB. List no. 58. Int. J. Syst. Bacteriol. *46*: 836–837.

Hugenholtz, J., D.M. Ivey and L.G. Ljungdahl. 1987. Carbon monoxide-driven electron transport in *Clostridium thermoautotrophicum* membranes. J. Bacteriol. *169*: 5845–5847.

Hugenholtz, J. and L.G. Ljungdahl. 1989. Electron transport and electrochemical proton gradient in membrane vesicles of *Clostridium thermoautotrophicum*. J. Bacteriol. *171*: 2873–2875.

Hugenholtz, J. and L.G. Ljungdahl. 1990a. Amino acid transport in membrane vesicles of *Clostridium thermoautotrophicum*. FEMS Microbiol. Lett. *57*: 117–121.

Hugenholtz, J. and L.G. Ljungdahl. 1990b. Metabolism and energy generation in homoacetogenic clostridia. FEMS Microbiol. Rev. *7*: 383–389.

Hungate, R.E. 1969. A roll tube method for cultivation of strict anaerobes. *In* Norris and Ribbons (Editors), Methods in Microbiology, Vol. 3B. Academic Press, London and New York, pp. 117–132.

Hyun, H.H. and J.G. Zeikus. 1985. Regulation and genetic enhancement of beta-amylase production in *Clostridium thermosulfurogenes*. J. Bacteriol. *164*: 1146–1152.

Inokuma, K., Y. Nakashimada, T. Akahoshi and N. Nishio. 2007. Characterization of enzymes involved in the ethanol production of *Moorella* sp. HUC22-1. Arch Microbiol. *188*: 37–45.

Jahn-Schmid, B., M. Graninger, M. Glozik, S. Kupcu, C. Ebner, F.M. Unger, U.B. Sleytr and P. Messner. 1996. Immunoreactivity of allergen (Bet v 1) conjugated to crystalline bacterial cell surface layers (S-layers). Immunotechnology *2*: 103–113.

Jin and Toda. 1987. Abstract. Presented at the Annual Meeting of the Society of Fermentation Technology, Japan.

Jin, F.X., K. Yamasato and K. Toda. 1988. *Clostridium thermocopriae* sp. nov., a cellulolytic thermophile from animal feces, compost, soil, and a hot spring in Japan. Int. J. Syst. Bacteriol. *38*: 279–281.

Jones, C.R., M. Ray and H.J. Strobel. 2002a. Cloning and transcriptional analysis of the *Thermoanaerobacter ethanolicus* strain 39E maltose ABC transport system. Extremophiles *6*: 291–299.

Jones, C.R., M. Ray and H.J. Strobel. 2002b. Transcriptional analysis of the xylose ABC transport operons in the thermophilic anaerobe *Thermoanaerobacter ethanolicus*. Curr. Microbiol. *45*: 54–62.

Jorgensen, F., O.C. Hansen and P. Stougaard. 2004. Enzymatic conversion of D-galactose to D-tagatose: heterologous expression and characterisation of a thermostable L-arabinose isomerase from *Thermoanaerobacter mathranii*. Appl. Microbiol. Biotechnol. *64*: 816–822.

Jorgensen, S.T., M. Tangney, R.L. Starnes, K. Amemiya and P.L. Jorgensen. 1997. Cloning and nucleotide sequence of a thermostable cyclodextrin glycosyltransferase gene from *Thermoanaerobacter* sp. ATCC 53627 and its expression in *Escherichia coli*. Biotechnol. Lett. *19*: 1027–1031.

Jung, S., J.G. Zeikus and R.I. Hollingsworth. 1994. A new family of very long chain alpha,omega-dicarboxylic acids is a major structural fatty acyl component of the membrane lipids of *Thermoanaerobacter ethanolicus* 39E. J. Lipid Res. *35*: 1057–1065.

Karita, S., K. Nakayama, M. Goto, K. Sakka, W.J. Kim and S. Ogawa. 2003. A novel cellulolytic, anaerobic, and thermophilic bacterium, *Moorella* sp. strain F21. Biosci Biotechnol Biochem *67*: 183–185.

Karnholz, A., K. Kusel, A. Gössner, A. Schramm and H.L. Drake. 2002. Tolerance and metabolic response of acetogenic bacteria toward oxygen. Appl. Environ. Microbiol. *68*: 1005–1009.

Kerby, R. and J.G. Zeikus. 1983. Growth of *Clostridium thermoaceticum* on H₂/CO₂ or CO as energy source. Curr. Microbiol. *8*: 27–30.

Kevbrina, M.V., A.M. Ryabokon and M.A. Pusheva. 1996. Acetate formation from CO-containing gas mixtures by free and immobilized cells of the thermophilic homoacetogenic bacterium *Thermoanaerobacter kivui*. Microbiology *65*: 656–660.

Kim, B.C., R. Grote, D.W. Lee, G. Antranikian and Y.R. Pyun. 2001. *Thermoanaerobacter yonseiensis* sp. nov., a novel extremely thermophilic,

xylose-utilizing bacterium that grows at up to 85 °C. Int. J. Syst. Evol. Microbiol. *51*: 1539–1548.

Kim, T.J., B.C. Kim and H.S. Lee. 1997. Production of cyclodextrin using raw corn starch without a pretreatment. Enzyme Microb. Technol. *20*: 506–509.

Klaushofer, H. and E. Parkkinen. 1965. Zur Frage der Bedeutung aerober und anaerober thermophiler Sporenbildner als Infektionsursache in Rübenzucker-Fabriken. I. *Clostridium thermohydrosulfuricum* eine neue Art eines saccharoseabbauenden, thermophilen, schwefelwasserstoffbildendem clostridiums. Z. Zukernidustr. (Boehman) *90*: 445–449.

Kleifeld, O., L. Rulisek, O. Bogin, A. Frenkel, Z. Havlas, Y. Burstein and I. Sagi. 2004. Higher metal-ligand coordination in the catalytic site of cobalt-substituted *Thermoanaerobacter brockii* alcohol dehydrogenase lowers the barrier for enzyme catalysis. Biochemistry *43*: 7151–7161.

Kletzin, A. and M.W. Adams. 1996. Tungsten in biological systems. FEMS Microbiol. Rev. *18*: 5–63.

Klingeberg, M., H. Hippe and G. Antranikian. 1990. Production of novel pullulanases at high concentrations by two newly isolated thermophilic clostridia. FEMS Microbiol. Lett. *57*: 145–152.

Klinke, H.B., A.B. Thomsen and B.K. Ahring. 2001. Potential inhibitors from wet oxidation of wheat straw and their effect on ethanol production by *Thermoanaerobacter mathranii*. Appl. Microbiol. Biotechnol. *57*: 631–638.

Korkhin, Y., A.J. Kalb, M. Peretz, O. Bogin, Y. Burstein and F. Frolow. 1998. NADP-dependent bacterial alcohol dehydrogenases: crystal structure, cofactor-binding and cofactor specificity of the ADHs of *Clostridium beijerinckii* and *Thermoanaerobacter brockii*. J. Mol. Biol. *278*: 967–981.

Korkhin, Y., A.J. Kalb, M. Peretz, O. Bogin, Y. Burstein and F. Frolow. 1999. Oligomeric integrity-the structural key to thermal stability in bacterial alcohol dehydrogenases. Protein Sci. *8*: 1241–1249.

Kozianowski, G., F. Canganella, F.A. Rainey, H. Hippe and G. Antranikian. 1997. Purification and characterization of thermostable pectate-lyases from a newly isolated thermophilic bacterium, *Thermoanaerobacter italicus* sp. nov. Extremophiles *1*: 171–182.

Kozianowski, G., F. Canganella, F.A. Rainey, H. Hippe and G. Antranikian. 1998. *In* Validation of the publication of new names and new combinations previously effectively published outside the IJSB. List no. 67. Int. J. Syst. Bacteriol. *48*: 1083–1084.

Kupcu, S., C. Mader and M. Sára. 1995. The crystalline cell-surface layer from *Thermoanaerobacter thermohydrosulfuricus* L111-69 as an immobilization matrix-influence of the morphological properties and the pore-size of the matrix on the loss of activity of covalently bound enzymes. Biotechnol. Appl. Biochem. *21*: 275–286.

Kupcu, S., U.B. Sleytr and M. Sara. 1996. Two-dimensional paracrystalline glycoprotein S-layers as a novel matrix for the immobilization of human IgG and their use as microparticles in immunoassays. J. Immunol. Methods *196*: 73–84.

Kuriki, T. and T. Imanaka. 1999. The concept of the alpha-amylase family: structural similarity and common catalytic mechanism. J. Biosci. Bioeng. *87*: 557–565.

Lacis, L.S. and H.G. Lawford. 1985. *Thermoanaerobacter ethanolicus* in a comparison of the growth efficiencies of thermophilic and mesophilic anaerobes. J. Bacteriol. *163*: 1275–1278.

Lacis, L.S. and H.G. Lawford. 1992. Strain selection in carbon-limited chemostats affects reproducibility of *Thermoanaerobacter ethanolicus* fermentations. Appl. Environ. Microbiol. *58*: 761–764.

Lamed, R. and J.G. Zeikus. 1980. Ethanol production by thermophilic bacteria: relationship between fermentation product yields of and catabolic enzyme activities in *Clostridium thermocellum* and *Thermoanaerobium brockii*. J. Bacteriol. *144*: 569–578.

Langworthy, T.A., G. Holzer, J.G. Zeikus and T.G. Tornabene. 1983. Iso-branched and anteiso-branched glycerol diethers of the thermophilic anaerobe *Thermodesulfotobacterium commune*. Syst. Appl. Microbiol. *4*: 1–17.

Larsen, L., P. Nielsen and B.K. Ahring. 1997. *Thermoanaerobacter mathranii* sp. nov., an ethanol-producing, extremely thermophilic anaerobic bacterium from a hot spring in Iceland. Arch. Microbiol. *168*: 114–119.

Larsen, L., P. Nielsen and B.K. Ahring. 1998. *In* Validation of the publication of new names and new combinations previously effectively published outside the IJSB. List no. 64. Int. J. Syst. Bacteriol. *48*: 327–328.

Lee, M.J. and S.H. Zinder. 1988. Isolation and characterization of a thermophilic bacterium which oxidizes acetate in syntrophic association with a methanogen and which grows acetogenically on H_2/CO_2. Appl. Environ. Microbiol. *54*: 124–129.

Lee, Y.E., M.K. Jain, C. Lee, S.E. Lowe and J.G. Zeikus. 1993. Taxonomic distinction of saccharolytic anaerobes: description of *Thermoanaerobacterium xylanolyticum* gen. nov., sp. nov., and *Thermoanaerobacterium saccharolyticum* gen. nov., sp. nov.; reclassification of *Thermoanaerobium brockii*, *Clostridium thermosulfurogenes*, and *Clostridium thermohydrosulfuricum* E100-69 as *Thermoanaerobacter brockii* comb. nov., *Thermoanaerobacterium thermosulfurogenes* comb. nov., and *Thermoanaerobacter thermohydrosulfuricus* comb. nov., respectively; and transfer of *Clostridium thermohydrosulfuricum* 39E to *Thermoanaerobacter ethanolicus*. Int. J. Syst. Bacteriol. *43*: 41–51.

Lee, Y.J., M. Dashti, A. Prange, F.A. Rainey, M. Rohde, W.B. Whitman and J. Wiegel. 2007a. *Thermoanaerobacter sulfurigignens* sp. nov., an anaerobic thermophilic bacterium that reduces 1 M thiosulfate to elemental sulfur and tolerates 90 mM sulfite. Int. J. Syst. Evol. Microbiol. *57*: 1429–1434.

Lee, Y.J., A. Prange, H. Lichtenberg, M. Rohde, W.B. Whitman and J. Wiegel. 2007b. In situ speciation of sulfur globules produced from thiosulfate reduction by *Thermoanaerobacter sulfurigignens* and *Thermoanaerobacterium thermosulfurigenes*. J. Bacteriol. *189*: 7525–7529.

Leigh, J.A., F. Mayer and R.S. Wolfe. 1981. *Acetogenium kivui*, a new thermophilic hydrogen-oxidizing, acetogenic bacterium. Arch. Microbiol. *129*: 275–280.

Leu, J.Y., C.P. McGovern-Traa, A.J. Porter, W.J. Harris and W.A. Hamilton. 1998. Identification and Phylogenetic analysis of thermophilic sulfate-reducing bacteria in oil field samples by 16S rDNA gene cloning and sequencing. Anaerobe *4*: 165–174.

Li, C., J. Heatwole, S. Soelaiman and M. Shoham. 1999. Crystal structure of a thermophilic alcohol dehydrogenase substrate complex suggests determinants of substrate specificity and thermostability. Proteins *37*: 619–627.

Lin, F.P. and K.L. Leu. 2002. Cloning, expression, and characterization of thermostable region of amylopullulanase gene from *Thermoanaerobacter ethanolicus* 39E. Appl. Biochem. Biotechnol. *97*: 33–44.

Lind, D.L., R.M. Daniel, D.A. Cowan and H.W. Morgan. 1989. β-galactosidase from a strain of the anaerobic thermophile, *Thermoanaerobacter*. Enzyme Microb. Technol. *11*: 180–186.

Ljungdahl, L.G. and J. Wiegel. 1981. Anaerobic thermophilic culture. U.S. Patent 4,292,407.

Ljungdahl, L.G., L.H. Carreira, R. Garrison, N. Rabek and J. Wiegel. 1985. Comparison of three thermophilic homoacetogenic bacteria for Ca-Mg acetate production. Biotech. Bioeng. Symp. *15*: 207–223.

Ljungdahl, L.G., L.H. Carreira, R.J. Garrison, N. Rabek and J. Wiegel. 1986. CMA manufacture II: Improved bacterial strain for acetate production: Final report. FHWA/RD-86/117.

Ljungdahl, L.G. and J. Wiegel. 1986. Anaerobic fermentations. *In* Demain and Solomon (Editors), Manual of Industrial Microbiology and Biotechnology. American Society for Microbiology, pp. 84–96.

Ljungdahl, L.G. 1994. The acetyl Co-A pathway and the chemiosmotic gneration of ATP during acetogenesis. *In* Drake (Editor), Acetogenesis. Chapman & Hall, New York, pp. 63–87.

Londesborough, J. 1986. Phosphorylation of proteins in *Clostridium thermohydrosulfuricum*. J. Bacteriol. *165*: 595–601.

Lovitt, R.W., R. Longin and J.G. Zeikus. 1984. Ethanol production by thermophilic bacteria: physiological comparison of solvent effects on parent and alcohol-tolerant strains of *Clostridium thermohydrosulfuricum*. Appl. Environ. Microbiol. *48*: 171–177.

Ludwig, W., K.-H. Schleifer and W.B. Whitman. 2009. Revised road map to the phylum *Firmicutes*. *In* De Vos, Garrity, Jones, Krieg, Ludwig, Rainey, Schleifer and Whitman (Editors), Bergey's Manual of Systematic Bacteriology, 2nd edn, Vol. 3, The *Firmicutes*, Springer, New York, pp. 1–14.

Lundie, L.L., Jr., H.C. Yang, J.K. Heinonen, S.I. Dean and H.L. Drake. 1988. Energy-dependent, high-affinity transport of nickel by the acetogen *Clostridium thermoaceticum*. J. Bacteriol. *170*: 5705–5708.

Lundie, L.L.J. and H.L. Drake. 1984. Development of a minimally defined medium for the acetogen *Clostridium thermoaceticum*. J. Bacteriol. *159*: 700–703.

Luo, H.W., H. Zhang, T. Suzuki, S. Hattori and Y. Kamagata. 2002. Differential expression of methanogenesis genes of *Methanothermobacter thermoautotrophicus* (formerly *Methanobacterium thermoautotrophicum*) in pure culture and in cocultures with fatty acid-oxidizing syntrophs. Appl. Environ. Microbiol. *68*: 1173–1179.

Lupas, A., H. Engelhardt, J. Peters, U. Santarius, S. Volker and W. Baumeister. 1994. Domain structure of the *Acetogenium kivui* surface layer revealed by electron crystallography and sequence analysis. J. Bacteriol. *176*: 1224–1233.

Mai, V., J. Wiegel and W.W. Lorenz. 2000. Cloning, sequencing, and characterization of the bifunctional xylosidase-arabinosidase from the anaerobic thermophile *Thermoanaerobacter ethanolicus*. Gene *247*: 137–143.

Martin, M.T., M. Alcalde, F.J. Plou, L. Dijkhuizen and A. Ballesteros. 2001. Synthesis of malto-oligosaccharides via the acceptor reaction catalyzed by cyclodextrin glycosyltransferases. Biocatalysis Biotransform. *191*: 21–35.

Martin, M.T., M.A. Cruces, M. Alcalde, F.J. Plou, M. Bernabe and A. Ballesteros. 2004. Synthesis of maltooligosyl fructofuranosides catalyzed by immobilized cyclodextrin glucosyltransferase using starch as donor. Tetrahedron *60*: 529–534.

Maruta, K., H. Watanabe, T. Nishimoto, M. Kubota, H. Chaen, S. Fukuda, M. Kurimoto and Y. Tsujisaka. 2006. Acceptor specificity of trehalose phosphorylase from *Thermoanaerobacter brockii*: production of novel nonreducing trisaccharide, 6-*O*-α-D-galactopyranosyl trehalose. J. Biosci. Bioeng. *101*: 385–390.

Mathupala, S.P., S.E. Lowe, S.M. Podkovyrov and J.G. Zeikus. 1993. Sequencing of the amylopullulanase (*apu*) gene of *Thermoanaerobacter ethanolicus* 39E, and identification of the active site by site-directed mutagenesis. J. Biol. Chem. *268*: 16332–16344.

Mathupala, S.P., J.H. Park and J.G. Zeikus. 1994. Evidence for α-1,6 and α-1,4-glucosidic bond-cleavage in highly branched glycogen by amylopullulanase from *Thermoanaerobacter ethanolicus*. Biotechnol. Lett. *16*: 1311–1316.

Mayer, M.A.G., K. Bronnenmeier, W.H. Schwarz, C. Schertler and W.L. Straudenbauer. 1995. Isolation and properties of acetate kinase-negative and phosphotransacetylase-negative mutants of *Thermoanaerobacter thermohydrosulfuricus*. Microbiology *141*: 2891–2896.

McInerney, M.J., A.J.M. Stams and D.R. Boone. 2004. The genus *Syntrophobacter*. *In* Brenner, Krieg, Staley and Garrity (Editors), Bergey's Manual of Systematic Bacteriology, Vol. 2C, The *Alpha-*, *Beta-*, *Delta-*, and *Epsilonproteobacteria*. Springer, New York, pp. 1021–1027.

McMahon, M. and P. Mulcahy. 2002. Bioaffinity purification of NADP(+)-dependent dehydrogenases: studies with alcohol dehydrogenase from *Thermoanaerobacter brockii*. Biotechnol. Bioeng. *77*: 517–527.

Melasniemi, H. 1987. Characterization of alpha-amylase and pullulanase activities of *Clostridium thermohydrosulfuricum*. Evidence for a novel thermostable amylase. Biochem. J. *246*: 193–197.

Melasniemi, H. and M. Paloheimo. 1989. Cloning and expression of the *Clostridium thermohydrosulfuricum* alpha-amylase-pullulanase gene in *Escherichia coli*. J. Gen. Microbiol. *135*: 1755–1762.

Messner, P., R. Christian, C. Neuninger and G. Schulz. 1995. Similarity of "core" structures in two different glycans of tyrosine-linked eubacterial S-layer glycoproteins. J. Bacteriol. *177*: 2188–2193.

Min, H. and S.H. Zinder. 1990. Isolation and characterization of a thermophilic sulfate-reducing bacterium *Desulfotomaculum thermoacetoxidans* sp. nov. Arch. Microbiol. *153*: 399–404.

Miroliaei, M. and M. Nemat-Gorgani. 2002. Effect of organic solvents on stability and activity of two related alcohol dehydrogenases: a comparative study. Int. J. Biochem. Cell Biol. *34*: 169–175.

Mitchell, R.W., B. Hahnhagerdal, J.D. Ferchak and E.K. Pye. 1982. Characterization of β-1,4-glucosidase activity in *Thermoanaerobacter ethanolicus*. Biotechnol. Bioeng. *12*: 461–467.

Mori, K., S. Hanada, A. Maruyama and K. Marumo. 2002. *Thermanaeromonas toyohensis* gen. nov., sp. nov., a novel thermophilic anaerobe isolated from a subterranean vein in the Toyoha Mines. Int. J. Syst. Evol. Microbiol. *52*: 1675–1680.

Mori, Y. 1990. Characterization of a symbiotic co-culture of *Clostridium thermohydrosulfuricum* YM3 and *Clostridium thermocellum* YM4. Appl. Environ. Microbiol. *56*: 37–42.

Mori, Y. 1995. Nutritional interdependence between *Thermoanaerobacter thermohydrosulfuricus* and *Clostridium thermocellum*. Arch. Microbiol. *164*: 152–154.

Musa, M.M., K.I. Ziegelmann-Fjeld, C. Vieille, J.G. Zeikus and R.S. Phillips. 2007a. Xerogel-encapsulated W110A secondary alcohol dehydrogenase from *Thermoanaerobacter ethanolicus* performs asymmetric reduction of hydrophobic ketones in organic solvents. Angew. Chem. Int. Ed. Engl. *46*: 3091–3094.

Musa, M.M., K.I. Ziegelmann-Fjeld, C. Vieille, J.G. Zeikus and R.S. Phillips. 2007b. Asymmetric reduction and oxidation of aromatic ketones and alcohols using W110A secondary alcohol dehydrogenase from *Thermoanaerobacter ethanolicus*. J. Org. Chem. *72*: 30–34.

Nakano, H., T. Kiso, K. Okamoto, T. Tomita, M.B. Manan and S. Kitahata. 2003. Synthesis of glycosyl glycerol by cyclodextrin glucanotransferases. J. Biosci. Bioeng. *95*: 583–588.

Nazina, T.N., D. Sokolova, N.M. Shestakova, A.A. Grigor'ian, E.M. Mikhailova, T.L. Babich, A.M. Lysenko, T.P. Turova, A.B. Poltaraus, T. Feng, F. Ni and S.S. Beliaev. 2005. The phylogenetic diversity of aerobic organotrophic bacteria from the Dagan high-temperature oil field. Mikrobiologiia *74*: 401–409.

Ng, T.K., A. Ben-Bassat and J.G. Zeikus. 1981. Ethanol production by thermophilic bacteria: fermentation of cellulosic substrates by cocultures of *Clostridium thermocellum* and *Clostridium thermohydrosulfuricum*. Appl. Environ. Microbiol. *41*: 1337–1343.

Ng, T.K. and J.G. Zeikus. 1982. Differential metabolism of cellobiose and glucose by *Clostridium thermocellum* and *Clostridium thermohydrosulfuricum*. J. Bacteriol. *150*: 1391–1399.

Okada, H., E. Fukushi, S. Onodera, T. Nishimoto, J. Kawabata, M. Kikuchi and N. Shiomi. 2003. Synthesis and structural analysis of five novel oligosaccharides prepared by glucosyltransfer from β-D-glucose 1-phosphate to isokestose and nystose using *Thermoanaerobacter brockii* kojibiose phosphorylase. Carbohydr. Res. *338*: 879–885.

Olofsson, L., I.A. Nicholls and S. Wikman. 2005. TBADH activity in water-miscible organic solvents: correlations between enzyme performance, enantioselectivity and protein structure through spectroscopic studies. Org. Biomol. Chem. *3*: 750–755.

Onyenwoke, R.U., J.A. Brill, K. Farahi and J. Wiegel. 2004. Sporulation genes in members of the low G+C Gram-type-positive phylogenetic branch (*Firmicutes*). Arch. Microbiol. *182*: 182–192.

Onyenwoke, R.U., V.V. Kevbrin, A.M. Lysenko and J. Wiegel. 2007. *Thermoanaerobacter pseudethanolicus* sp. nov., a thermophilic heterotrophic anaerobe from Yellowstone National Park. Int. J. Syst. Evol. Microbiol. *57*: 2191–2193.

Orphan, V.J., L.T. Taylor, D. Hafenbradl and E.F. Delong. 2000. Culture-dependent and culture-independent characterization of microbial assemblages associated with high-temperature petroleum reservoirs. Appl. Environ. Microbiol. *66*: 700–711.

Parkkinen, E. 1986. Conversion of starch into ethanol by *Clostridium thermohydrosulfuricum*. Appl. Microbiol. Biotechnol. *25*: 213–219.

Patel, B.K.C., J.A. Hudson, H.W. Morgan and R.M. Daniel. 1988. Nondiauxic fermentation of acid hydrolyzed pine by a strain of *Clostridium thermohydrosulfuricum* isolated from a New Zealand hot spring. FEMS Microbiol. Lett. *56*: 285–288.

Peng, H., B. Fu, Z. Mao and W. Shao. 2006. Electrotransformation of *Thermoanaerobacter ethanolicus* JW200. Biotechnol. Lett. *28*: 1913–1917.

Peretz, M. and Y. Burstein. 1989. Amino acid sequence of alcohol dehydrogenase from the thermophilic bacterium *Thermoanaerobium brockii*. Biochemistry *28*: 6549–6555.

Peretz, M., O. Bogin, S. Tel-Or, A. Cohen, G. Li, J.S. Chen and Y. Burstein. 1997a. Molecular cloning, nucleotide sequencing, and expression of genes encoding alcohol dehydrogenases from the thermophile *Thermoanaerobacter brockii* and the mesophile *Clostridium beijerinckii*. Anaerobe *3*: 259–270.

Peretz, M., L.M. Weiner and Y. Burstein. 1997b. Cysteine reactivity in *Thermoanaerobacter brockii* alcohol dehydrogenase. Protein Sci. *6*: 1074–1083.

Peteranderl, R., F. Canganella, A. Holzenburg and J. Wiegel. 1993. Induction and Regeneration of Autoplasts from *Clostridium thermohydrosulfuricum* JW102 and *Thermoanaerobacter ethanolicus* JW200. Appl. Environ. Microbiol. *59*: 3498–3501.

Pham, V.T. and R.S. Phillips. 1990. Effects of substrate structure and temperature on the stereospecificity of secondary alcohol dehydrogenase from *Thermoanaerobacter ethanolicus*. J. Am. Chem. Soc. *112*: 3629–3632.

Phillips, R.S. 1996. Temperature modulation of the stereochemistry of enzymatic catalysis: prospects for exploitation. Trends Biotechnol. *14*: 13–16.

Phillips, R.S. 2002a. How does active site water affect enzymatic stereorecognition? J. Mol. Catal. B: Enzymatic *19*: 103–107.

Phillips, R.S. 2002b. Tailoring the substrate specificity of secondary alcohol dehydrogenase. Can. J. Chem. *80*: 680–685.

Plugge, C.M., J.M. van Leeuwen, T. Hummelen, M. Balk and A.J.M. Stams. 2001. Elucidation of the pathways of catabolic glutamate conversion in three thermophilic anaerobic bacteria. Arch. Microbiol. *176*: 29–36.

Plugge, C.M., M. Balk, E.G. Zoetendal and A.J. Stams. 2002. *Gelria glutamica* gen. nov., sp. nov., a thermophilic, obligately syntrophic, glutamate-degrading anaerobe. Int. J. Syst. Evol. Microbiol. *52*: 401–407.

Podkovyrov, S.M., D. Burdette and J.G. Zeikus. 1993. Analysis of the catalytic center of cyclomaltodextrinase from *Thermoanaerobacter ethanolicus* 39E. FEBS Lett. *317*: 259–262.

Pusheva, M.A. and T.G. Sokolova. 1995. Distribution of CO-dehydrogenase activity in the anaerobic thermophilic carboxydotrophic bacterium *Carboxydothermus hydrogenoformans* grown at the expense of CO or pyruvate. Microbiology (En. transl. from Mikrobiologiya) *64*: 491–495.

Rabinkov, A., T. Miron, D. Mirelman, M. Wilchek, S. Glozman, E. Yavin and L. Weiner. 2000. S-Allylmercaptoglutathione: the reaction product of allicin with glutathione possesses SH-modifying and antioxidant properties. Biochim. Biophys. Acta. *1499*: 144–153.

Ragsdale, S.W. 2008. Enzymology of the wood-Ljungdahl pathway of acetogenesis. Ann. N. Y. Acad. Sci. *1125*: 129–136.

Rainey, F.A. and E. Stackebrandt. 1993. Transfer of the type species of the genus *Thermobacteroides* to the genus *Thermoanaerobacter* as *Thermoanaerobacter acetoethylicus* (Ben-Bassat and Zeikus 1981) comb. nov., description of *Coprothermobacter* gen. nov., and reclassification of *Thermobacteroides proteolyticus* as *Coprothermobacter proteolyticus* (Ollivier *et al.* 1985) comb. nov. Int. J. Syst. Bacteriol. *43*: 857–859.

Rainey, F.A., N.L. Ward, H.W. Morgan, R. Toalster and E. Stackebrandt. 1993. Phylogenetic analysis of anaerobic thermophilic bacteria: aid for their reclassification. J. Bacteriol. *175*: 4772–4779.

Rainey, F.A., R.Tanner and J. Wiegel. 2006. Family *Clostridiaceae*. *In* The Prokaryotes: A Handbook on the Biology of Bacteria: *Bacteria: Firmicutes, Cyanobacteria*, 3rd edn, release 3.20 Vol. 4. Springer-Verlag, New York, pp. 654–678.

Riessen, S. and G. Antranikian. 2001. Isolation of *Thermoanaerobacter keratinophilus* sp. nov., a novel thermophilic, anaerobic bacterium with keratinolytic activity. Extremophiles *5*: 399–408.

Roh, Y., S.V. Liu, G.S. Li, H.S. Huang, T.J. Phelps and J.Z. Zhou. 2002. Isolation and characterization of metal-reducing *Thermoanaerobacter* strains from deep subsurface environments of the Piceance Basin, Colorado. Appl. Environ. Microbiol. *68*: 6013–6020.

Ryabokon, A.M., M.A. Pusheva, E.N. Detkova and E.I. Rainina. 1995. Reduction of CO_2 to acetate by immobilized cells of the thermophilic homoacetogenic bacterium *Thermoanaerobacter kivui*. Microbiology *64*: 657–661.

Sa, I. 1985. Isolation of *Clostridium thermoautotrophicum*, an anaerobic methylotrophic spore-forming organism. Mikrobiology (USSR) *54*: 416–420.

Sakai, S., Y. Nakashimada, H. Yoshimoto, S. Watanabe, H. Okada and N. Nishio. 2004. Ethanol production from H_2 and CO_2 by a newly isolated thermophilic bacterium, *Moorella* sp. HUC22-1. Biotechnol. Lett. *26*: 1607–1612.

Sára, M., S. Küpcü and U.B. Sleytr. 1989. Localization of the carbohydrate residue of the S-layer of the glycoprotein from *Clostridium thermohydrosulfuricum* L111-69. Arch. Microbiol. *151*: 416–420.

Sára, M. and U.B. Sleytr. 1989. Use of regularly structured bacterial cell envelope layers as matrix for the immobilization of macromolecules. Appl. Microbiol. Biotechnol. *30*: 184–189.

Sára, M. and U.B. Sletyr. 1996. Biotechnology and biomimetic with crystalline bacterial cell surface layers (S-layers). Micron *27*: 141–156.

Savage, M.D. and H.L. Drake. 1986. Adaptation of the acetogen *Clostridium thermoautotrophicum* to minimal medium. J. Bacteriol. *165*: 315–318.

Savage, M.D., Z.R. Wu, S.L. Daniel, L.L. Lundie and H.L. Drake. 1987. Carbon monoxide-dependent chemolithotrophic growth of *Clostridium thermoautotrophicum*. Appl. Environ. Microbiol. *53*: 1902–1906.

Schink, B. and J.G. Zeikus. 1980. Microbial methanol formation: a major end product of pectin metabolism. Curr. Microbiol. *4*: 387–389.

Schink, B. 1994. Diversity, ecology, and isolation of acetogenic bacteria. *In* Drake (Editor), Acetogenesis. Chapman & Hall, New York, pp. 197–235.

Schmid, U., H. Giesel, S.M. Schoberth and H. Sahm. 1986. *Thermoanaerobacter finnii* spec.nov., a new ethanologenic sporogenous bacterium. Syst. Appl. Microbiol. *8*: 80–85.

Schmid, U., K.L. Schimz and H. Sahm. 1989. Determination of intracellular pyridine nucleotide levels by bioluminescence using anaerobic bacteria as a model. Anal. Biochem. *180*: 17–23.

Schnurer, A., B. Schink and B.H. Svensson. 1996. *Clostridium ultunense* sp. nov., a mesophilic bacterium oxidizing acetate in syntrophic association with a hydrogenotrophic methanogenic bacterium. Int. J. Syst. Bacteriol. *46*: 1145–1152.

Schwartz, R.D. and F.A. Keller. 1982a. Acetic acid production by *Clostridium thermoaceticum* in pH-controlled batch fermentations at acidic pH. Appl. Environ. Microbiol. *43*: 1385–1392.

Schwartz, R.D. and F.A. Keller. 1982b. Isolation of a strain of *Clostridium thermoaceticum* capable of growth and acetic acid production at pH 4.5. Appl. Environ. Microbiol. *43*: 117–123.

Schwartz, R.D. and F.A.J. Keller. 1983. Acetic acid by fermentation. U.S. Patent 4,371,619.

Seebach, D., M.F. Züger, F. Giovannini, B. Sonnletner and A. Fiechter. 1984. Preparative microbial reduction of beta-oxoesters with *Thermoanaerobium brockii*. Angew. Chem. Int. Ed. Engl. *23*: 151–152.

Seifritz, C., J.M. Frostl, H.L. Drake and S.L. Daniel. 2002. Influence of nitrate on oxalate- and glyoxylate-dependent growth and acetogenesis by *Moorella thermoacetica*. Arch. Microbiol. *178*: 457–464.

Seifritz, C., H.L. Drake and S.L. Daniel. 2003. Nitrite as an energy-conserving electron sink for the acetogenic bacterium *Moorella thermoacetica*. Curr. Microbiol. *46*: 329–333.

Sha, D., M. Davis and T. Mamone. 1997. Purification and characterization of a thermostable DNA polymerase from *Thermoanaerobacter thermohydrosulfuricus*. FASEB J. *11*: A1134–A1134.

Shao, W. and J. Wiegel. 1992. Purification and characterization of a thermostable β-xylosidase from *Thermoanaerobacter ethanolicus*. J. Bacteriol. *174*: 5848–5853.

Simankova, M.V. and A.N. Nozhevnikova. 1989. Thermophilic fermentation of cellulose by a combined culture of *Clostridium thermocellum* and *Clostridium thermoautotrophicum*. Mikrobiology (USSR) *58*: 723–728.

Sleytr, U.B. and A.M. Glauert. 1973. Evidence for an empty core in a bacterial flagellum. Nature *241*: 542–543.

Sleytr, U.B., P. Messner, D. Pum and M. Sara. 1999. Crystalline bacterial cell surface layers (S layers): from supramolecular cell structure to biomimetics and nanotechnology. Angew. Chem. Int. Ed. Engl. *38*: 1035–1054.

Slobodkin, A., A.L. Reysenbach, F. Mayer and J. Wiegel. 1997a. Isolation and characterization of the homoacetogenic thermophilic bacterium *Moorella glycerini* sp. nov. Int. J. Syst. Bacteriol. *47*: 969–974.

Slobodkin, A., A.L. Reysenbach, N. Strutz, M. Dreier and J. Wiegel. 1997b. *Thermoterrabacterium ferrireducens* gen. nov., sp. nov., a thermophilic anaerobic dissimilatory Fe(III)-reducing bacterium from a continental hot spring. Int. J. Syst. Bacteriol. *47*: 541–547.

Slobodkin, A.I. and J. Wiegel. 1997. Fe(III) as an electron acceptor for H_2 oxidation in thermophilic anaerobic enrichment cultures from geothermal areas. Extremophiles *1*: 106–109.

Slobodkin, A.I., C. Jeanthon, S. L'Haridon, T. Nazina, M. Mirosh-nichenko and E. Bonch-Osmolovskaya. 1999a. Dissimilatory reduction of Fe(III) by thermophilic bacteria and archaea in deep subsurface petroleum reservoirs of Western Siberia. Curr. Microbiol. *39*: 99–102.

Slobodkin, A.I., T.P. Tourova, B.B. Kuznetsov, N.A. Kostrikina, N.A. Chernyh and E.A. Bonch-Osmolovskaya. 1999b. *Thermoanaerobacter siderophilus* sp. nov., a novel dissimilatory Fe(III)-reducing, anaerobic, thermophilic bacterium. Int. J. Syst. Bacteriol. *49*: 1471–1478.

Slobodkin, A.I., T.G. Sokolova, A.M. Lysenko and J. Wiegel. 2006. Reclassification of *Thermoterrabacterium ferrireducens* as *Carboxydothermus ferrireducens* comb. nov., and emended description of the genus *Carboxydothermus*. Int. J. Syst. Evol. Microbiol. *56*: 2349–2351.

Sokolova, T., J. Hanel, R.U. Onyenwoke, A.L. Reysenbach, A. Banta, R. Geyer, J.M. Gonzalez, W.B. Whitman and J. Wiegel. 2007. Novel chemolithotrophic, thermophilic, anaerobic bacteria *Thermolithobacter ferrireducens* gen. nov., sp. nov. and *Thermolithobacter carboxydivorans* sp. nov. Extremophiles *11*: 145–157.

Sokolova, T.G., J.M. Gonzalez, N.A. Kostrikina, N.A. Chernyh, T.P. Tourova, C. Kato, E.A. Bonch-Osmolovskaya and F.T. Robb. 2001. *Carboxydobrachium pacificum* gen. nov., sp. nov., a new anaerobic, thermophilic, CO-utilizing marine bacterium from Okinawa Trough. Int. J. Syst. Evol. Microbiol. *51*: 141–149.

Sokolova, T.G., N.A. Kostrikina, N.A. Chernyh, T.P. Tourova, T.V. Kolganova and E. Bonch-Osmolovskaya. 2002. *Carboxydocella thermautotrophica* gen. nov., sp. nov., a novel anaerobic, CO-utilizing thermophile from a Kamchatkan hot spring. Int. J. Syst. Evol. Microbiol. *52*: 1961–1967.

Sokolova, T.G., J.M. Gonzalez, N.A. Kostrikina, N.A. Chernyh, T.V. Slepova, E.A. Bonch-Osmolovskaya and F.T. Robb. 2004a. *Thermosinus carboxydivorans* gen. nov., sp. nov., a new anaerobic, thermophilic, carbon-monoxide-oxidizing, hydrogenogenic bacterium from a hot pool of Yellowstone National Park. Int. J. Syst. Evol. Microbiol. *54*: 2353–2359.

Sokolova, T.G., C. Jeanthon, N.A. Kostrikina, N.A. Chernyh, A.V. Lebedinsky, E. Stackebrandt and E.A. Bonch-Osmolovskaya. 2004b. The first evidence of anaerobic CO oxidation coupled with H_2 production by a hyperthermophilic archaeon isolated from a deep-sea hydrothermal vent. Extremophiles *8*: 317–323.

Sokolova, T.G., N.A. Kostrikina, N.A. Chernyh, T.V. Kolganova, T.P. Tourova and E.A. Bonch-Osmolovskaya. 2005. *Thermincola carboxydiphila* gen. nov., sp. nov., a novel anaerobic, carboxydotrophic, hydrogenogenic bacterium from a hot spring of the Lake Baikal area. Int. J. Syst. Evol. Microbiol. *55*: 2069–2073.

Sommer, P., T. Georgieva and B.K. Ahring. 2004. Potential for using thermophilic anaerobic bacteria for bioethanol production from hemicellulose. Biochem. Soc. Trans. *32*: 283–289.

Sonne-Hansen, J., I.M. Mathrani and B.K. Ahring. 1993. Xylanolytic anaerobic thermophiles from Icelandic hot-springs. Appl. Microbiol. Biotechnol. *38*: 537–541.

Subbotina, I.V., N.A. Chernyh, T.G. Sokolova, I.V. Kublanov, E.A. Bonch-Osmolovskaya and A.V. Lebedinsky. 2003. Oligonucleotide probes for the detection of representatives of the genus *Thermoanaerobacter*. Microbiology (En. transl. from Mikrobiologiya) *72*: 331–339.

Svensson, B. 1994. Protein engineering in the alpha-amylase family: catalytic mechanism, substrate specificity, and stability. Plant Mol. Biol. *25*: 141–157.

Svetlichny, V.A., T.G. Sokolova, M. Gerhardt, M. Ringpfeil, N.A. Kostrikina and G.A. Zavarzin. 1991a. *Carboxydothermus hydrogenoformans* gen. nov., sp. nov., a CO-utilizing thermophilic anaerobic bacterium from hydrothermal environments of Kunashir Island. Syst. Appl. Microbiol. *14*: 254–260.

Svetlichny, V.A., T.G. Sokolova, M. Gerhardt, M. Ringpfeil, N.A. Kostrikina and G.A. Zavarzin 1991b. *In* Validation of the publication of new names and new combinations previously effectively published outside the IJSB. List no. 39. Int. J. Syst. Bacteriol. *1*: 580–581.

Szewzyk, U., R. Szewzyk and T.A. Stenstrom. 1994. Thermophilic, anaerobic bacteria isolated from a deep borehole in granite in Sweden. Proc Natl Acad Sci. U S A *91*: 1810–1813.

Talabardon, M., J.P. Schwitzguebel, P. Peringer and S.T. Yang. 2000. Acetic acid production from lactose by an anaerobic thermophilic coculture immobilized in a fibrous-bed bioreactor. Biotechnol. Prog. *16*: 1008–1017.

Tanner, R.S. and C.R. Woese. 1994. A phylogenetic assessment of the acetogens. *In* Drake (Editor), Acetogenesis. Chapman & Hall, New York, pp. 254–269.

Thauer, R.K., K. Jungermann and K. Decker. 1977. Energy conservation in chemotropic anaerobic bacteria. Bacteriol. Rev. *41*: 100–180.

Thomas, T.M. and R.K. Scopes. 1998. The effects of temperature on the kinetics and stability of mesophilic and thermophilic 3-phosphoglycerate kinases. Biochem. J. *330*: 1087–1095.

Todd, M.J., S. Walke, G. Lorimer, K. Truscott and R.K. Scopes. 1995. The single-ring *Thermoanaerobacter brockii* chaperonin 60 (Tbr-EL7) dimerizes to Tbr-EL14.Tbr-ES7 under protein folding conditions. Biochemistry *34*: 14932–14941.

Torry-Smith, M., P. Sommer and B.K. Ahring. 2003. Purification of bioethanol effluent in an UASB reactor system with simultaneous biogas formation. Biotechnol. Bioeng. *84*: 7–12.

Tripp, A.E., D.S. Burdette, J.G. Zeikus and R.S. Phillips. 1998. Mutation of serine-39 to threonine in thermostable secondary alcohol dehydrogenase from *Thermoanaerobacter ethanolicus* changes enantiospecificity. J. Am. Chem. Soc. *120*: 5137–5141.

Trüper, H.G. 1999. How to name a prokaryote? Etymological considerations, proposals and practical advice in prokaryote nomenclature. FEMS Microbiol. Rev. *23*: 231–249.

Truscott, K.N., P.B. Hoj and R.K. Scopes. 1994. Purification and characterization of chaperonin 60 and chaperonin 10 from the anaerobic thermophile *Thermoanaerobacter brockii*. Eur. J. Biochem. *222*: 277–284.

Truscott, K.N. and R.K. Scopes. 1998. Sequence analysis and heterologous expression of the groE genes from *Thermoanaerobacter* sp. Rt8. G4. Gene *217*: 15–23.

Tsiroulnikov, K., H. Rezai, E. Bonch-Osmolovskaya, P. Nedkov, A. Gousterova, V. Cueff, A. Godfroy, G. Barbier, F. Metro, J.M. Chobert, P. Clayette, D. Dormont, J. Grosclaude and T. Haertle. 2004. Hydrolysis of the amyloid prion protein and nonpathogenic meat and bone meal by anaerobic thermophilic prokaryotes and streptomyces subspecies. J. Agric. Food. Chem. *52*: 6353–6360.

Turunen, M., E. Parkkinen, J. Londesborough and M. Korhola. 1987. Distinct forms of lactate dehydrogenase purified from ethanol- and lactate-producing cells of *Clostridium thermohydrosulfuricum*. J. Gen. Microbiol. *133*: 2865–2873.

Tyurin, M.V., L.R. Lynd and J. Wiegel. 2006. Gene transfer systems for obligately anaerobic thermophilic bacteria. *In* Oren and Rainey (Editors), Methods in Microbiology. Academic Press/Elsevier, pp. 309–330.

Vieille, C. and G.J. Zeikus. 2001. Hyperthermophilic enzymes: sources, uses, and molecular mechanisms for thermostability. Microbiol. Mol. Biol. Rev. *65*: 1–43.

Wagner, I.D. and J. Wiegel. 2008. Diversity of thermophilic anaerobes. Ann. N. Y. Acad. Sci. *1125*: 1–43.

Wagner, I.D., W. Zhao, C.L. Zhang, C.S. Romanek, M. Rohde and J. Wiegel. 2008. *Thermoanaerobacter uzonensis* sp. nov., an anaerobic, thermophilic bacterium isolated from a hot spring within the Uzon Caldera, Kamchatka, Far East Russia. Int. J. Syst. Evol. Microbiol. *58*: 2565–2573.

Weber, V., S. Weigert, M. Sara, U.B. Sleytr and D. Falkenhagen. 2001. Development of affinity microparticles for extracorporeal blood purification based on crystalline bacterial cell surface proteins. Ther. Apher. *5*: 433–438.

White, H., H. Lebertz, I. Thanos and H. Simon. 1987. *Clostridium-Thermoaceticum* forms methanol from carbon-monoxide in the presence of viologen dyes. FEMS Microbiol. Lett. *43*: 173–176.

Wiegel, J. and H.G. Schlegel. 1977. Leucine biosynthesis: effect of branched-chain amino acids and threonine on alpha-isopropylmalate synthase activity from aerobic and anaerobic microorganisms. Biochem. Syst. Ecol. *5*: 169–176.

Wiegel, J., L.G. Ljungdahl and J.R. Rawson. 1979. Isolation from soil and properties of the extreme thermophile *Clostridium thermohydrosulfuricum*. J. Bacteriol. *139*: 800–810.

Wiegel, J., M. Braun and G. Gottschalk. 1981. *Clostridium thermoautotrophicum* species novum, a thermophile producing acetate from molecular hydrogen and carbon dioxide. Curr. Microbiol. *5*: 255–260.

Wiegel, J. and L.G. Ljungdahl. 1981. *Thermoanaerobacter ethanolicus* gen. nov., spec. nov, a new, extreme thermophilic, anaerobic bacterium. Arch. Microbiol. *128*: 343–348.

Wiegel, J. 1982. *Clostridium thermoautotrophicum*: growth and sporulation in media containing C-1 compounds as substrate. Presented at the Annu. Meet. Am. Soc. Microbiol., Atlanta, Abstr. I107.

Wiegel, J. and L.G. Ljungdahl. 1982. *In* Validation of the publication of new names and new combinations previously effectively published outside the IJSB. List no. 9. Int. J. Syst. Bacteriol. *32*: 384–385.

Wiegel, J., L.H. Carreira, C.P. Mothershed and J. Puls. 1983. Production of ethanol from bio-polymers by anaerobic, thermophilic, and extreme thermophilic bacteria. 2. *Thermoanaerobacter ethanolicus* JW200 and its mutants in batch cultures and resting cell experiments. Biotechnol. Bioeng. *13*: 193–205.

Wiegel, J., Ch.P. Mothershed and J. Puls. 1986. Fermentation of steam-exploded wood hemicellulose by thermophilic anaerobic bacteria. *In* Barry, Houghton, Llewellyn and O'Rear (Editors), Biodeterioration. C.A.B. International, Slough, pp. 524–529.

Wiegel, J. 1988. *Clostridium thermocellum*, a potentially important industrial micro-organism. SIM News *38*: 5–9.

Wiegel, J. 1990. Temperature spans for growth: hypothesis and discussion. FEMS Microbiol. Rev. *75*: 155–169.

Wiegel, J., L.H. Carreira, R. Garrison, N. Rabek and L.G. Ljungdahl. 1991. Calcium magnesium acetate (CMA) manufacture from glucose by fermentation with thermophilic homoacetogenic bacteria. *In* Wise, Levendis and Metghalchi (Editors), Calcium Magnesium Acetate. Elsevier Science, Amsterdam and New York, pp. 359–418.

Wiegel, J. 1998a. Anaerobic alkalithermophiles, a novel group of extremophiles. Extremophiles *2*: 257–267.

Wiegel, J. 1998b. Lateral gene exchange, an evolutionary mechanism for extending the upper or lower temperature limits for growth of a microorganism? A hypothesis. *In* Wiegel and Adams (Editors), Thermophiles, the Molecular Key to the Evolution and Origon of Life? Taylor & Francis, London, pp. 175–185.

Wiegel, J., L.H. Carreira, Ch.P. Mothershed, L. G. Ljungdahl, J. Puls. 1986. Formation of ethanol and acetate from biomass using thermophilic and extreme thermophilic anaerobic bacteria. *In* Industrial Wood Energy Forum 83. Forest Product Research Society, Madison *II*: 482–489.

Wimmer, B., F. Lottspeich, J. Ritter and K. Bronnenmeier. 1997. A novel type of thermostable alpha-D-glucosidase from *Thermoanaerobacter thermohydrosulfuricus* exhibiting maltodextrinohydrolase activity. Biochem. J. *328*: 581–586.

Wood, H.G. and L.G. Ljungdahl. 1991. Autotrophic character of the acetogenic bacteria. *In* Shively and Barton (Editors), Variations in Autotrophic Life. Academic Press, San Diego, pp. 201–250.

Wu, M., Q. Ren, A.S. Durkin, S.C. Daugherty, L.M. Brinkac, R.J. Dodson, R. Madupu, S.A. Sullivan, J.F. Kolonay, D.H. Haft, W.C. Nelson, L.J. Tallon, K.M. Jones, L.E. Ulrich, J.M. Gonzalez, I.B. Zhulin, F.T. Robb and J.A. Eisen. 2005. Life in hot carbon monoxide: the complete genome sequence of *Carboxydothermus hydrogenoformans* Z-2901. PLoS Genet *1*: 0564–0574.

Wynter, C., B.K. Patel, P. Bain, J. de Jersey, S. Hamilton and P.A. Inkerman. 1996. A novel thermostable dextranase from a *Thermoanaerobacter* species cultured from the geothermal waters of the Great Artesian Basin of Australia. FEMS Microbiol. Lett. *140*: 271–276.

Wynter, C.V.A., M. Chang, B. Patel, J. De Jersey, S. Hamilton and P.A. Inkerman. 1997. Isolation and characterization of a thermostable dextranase. Enzyme Microb. Technol. *20*: 242–247.

Xue, Y., Y. Xu, Y. Liu, Y. Ma and P. Zhou. 2001. *Thermoanaerobacter tengcongensis* sp. nov., a novel anaerobic, saccharolytic, thermophilic bacterium isolated from a hot spring in Tengcong, China. Int. J. Syst. Evol. Microbiol. *51*: 1335–1341.

Yamamoto, I., T. Saiki, S.M. Liu and L.G. Ljungdahl. 1983. Purification and properties of NADP-dependent formate dehydrogenase from *Clostridium thermoaceticum*, a tungsten-selenium-iron protein. J. Biol. Chem. *258*: 1826–1832.

Yamamoto, K., R. Murakami and Y. Takamura. 1998. Isoprenoid quinone, cellular fatty acid composition and diaminopimelic acid isomers of newly classified thermophilic anaerobic Gram-positive bacteria. FEMS Microbiol. Lett. *161*: 351–358.

Yamamoto, T., H. Yamashita, K. Mukai, H. Watanabe, M. Kubota, H. Chaen and S. Fukuda. 2006. Construction and characterization of chimeric enzymes of kojibiose phosphorylase and trehalose phosphorylase from *Thermoanaerobacter brockii*. Carbohydr. Res. *341*: 2350–2359.

Yang, H., A. Jonsson, E. Wehtje, P. Adlercreutz and B. Mattiasson. 1997. The enantiomeric purity of alcohols formed by enzymatic reduction of ketones can be improved by optimisation of the temperature and by using a high co-substrate concentration. Biochim. Biophys. Acta. *1336*: 51–58.

Yang, S.S., L.G. Ljungdahl and J. Legall. 1977. 4-Iron, 4-sulfide ferredoxin with high thermostability from *Clostridium thermoaceticum*. J. Bacteriol. *130*: 1084–1090.

Yang, S.S., L.G. Ljungdahl, D.V. Dervartanian and G.D. Watt. 1980. Isolation and characterization of two rubredoxins from *Clostridium thermoaceticum*. Biochim. Biophys. Acta *590*: 24–33.

Zamost, B.L., H.K. Nielsen and R.L. Starnes. 1991. Thermostable enzymes for industrial applications. J. Ind. Microbiol. *8*: 71–81.

Zavarzina, D.G., T.P. Tourova, B.B. Kuznetsov, E.A. Bonch-Osmolovskaya and A.I. Slobodkin. 2002. *Thermovenabulum ferriorganovorum* gen. nov., sp. nov., a novel thermophilic, anaerobic, endospore-forming bacterium. Int. J. Syst. Evol. Microbiol. *52*: 1737–1743.

Zeikus, J.G., P.W. Hegge and M.A. Anderson. 1979. *Thermoanaerobium brockii* gen. nov and sp. nov., a new chemoorganotrophic, caldoactive, anaerobic bacterium. Arch. Microbiol. *122*: 41–48.

Zeikus, J.G., A. Ben-Bassat and P.W. Hegge. 1980. Microbiology of methanogenesis in thermal, volcanic environments. J. Bacteriol. *143*: 432–440.

Zeikus, J.G., A. Ben-Bassat, T.K. Ng and R.J. Lamed. 1981. Thermophilic ethanol fermentations. Basic Life Sci. *18*: 441–461.

Zeikus, J.G., C. Lee, Y.E. Lee and B.C. Saha. 1991. Thermostable saccharidases-new sources, uses, and biodesigns. ACS Symp. Series *460*: 36–51.

Zheng, C.S. and R.S. Phillips. 1992. Effect of coenzyme analogs on enantioselectivity of alcohol-dehydrogenase. J. Chem. Soc. Perkin Transact. *1*: 1083–1084.

Zheng, C.S., V.T. Pham and R.S. Phillips. 1994. Asymmetric reduction of aliphatic and cyclic-ketones with secondary alcohol-dehydrogenase from *Thermoanaerobacter ethanolicus* – effects of substrate structure and temperature. Catal. Today *22*: 607–620.

Zhou, J., S. Liu, B. Xia, C. Zhang, A.V. Palumbo and T.J. Phelps. 2001. Molecular characterization and diversity of thermophilic iron-reducing enrichment cultures from deep subsurface environments. J. Appl. Microbiol. *90*: 96–105.

Ziegelmann-Fjeld, K.I., M.M. Musa, R.S. Phillips, J.G. Zeikus and C. Vieille. 2007. A *Thermoanaerobacter ethanolicus* secondary alcohol dehydrogenase mutant derivative highly active and stereoselective on phenylacetone and benzylacetone. Protein Eng. Des. Sel. *20*: 47–55.

Zinder, S.H. and M. Koch. 1984. Non-aceticlastic methanogenesis from acetate: acetate oxidation by a thermophilic syntrophic coculture. Arch. Microbiol. *138*: 263–272.

Family II. **Thermodesulfobiaceae** Mori, Kim, Kakegawa and Hanada 2004, 1[VP] (Effective publication: Mori, Kim, Kakegawa and Hanada 2003, 288.)

KOJI MORI AND SATOSHI HANADA

Ther.mo.de.sul.fo.bi.a'ce.ae. N.L. neut. n. *Thermodesulfobium* type genus of the family; suff. *-aceae* denoting a family; N.L. fem. pl. n. *Thermodesulfobiaceae* the family of *Thermodesulfobium*.

Rod-shaped cells that **stain Gram-negative. Nonsporeforming. Moderately thermophilic.** Chemoautotrophic and strictly anaerobic. Growth occurs by **anaerobic respiration with sulfate or nitrate** as electron acceptors. Represents a distinct phylogenetic lineage based on 16S rRNA gene sequence analyses. The family contains two genera, *Thermodesulfobium* and *Coprothermobacter*.

DNA G+C content (mol%): 35.1–43.0.

Type genus: **Thermodesulfobium** Mori, Kim, Kakegawa and Hanada 2004, 1[VP] (Effective publication: Mori, Kim, Kakegawa and Hanada 2003, 288.).

Genus I. **Thermodesulfobium** Mori, Kim, Kakegawa and Hanada 2004, 1[VP] (Effective publication: Mori, Kim, Kakegawa and Hanada 2003, 288.)

KOJI MORI AND SATOSHI HANADA

Ther.mo.de.sul.fo'bi.um. Gr. adj. *thermos* hot, L. pref. *de* from, L. neut. n. *sulfur* sulfur, Gr. masc. n. *bios* life, N.L. neut. n. *Thermodesulfobium* a thermophilic organism that reduces a sulfur compound.

Strictly anaerobic, nonmotile rods about $0.5\mu m \times 2$–$4\mu m$. **Stains Gram-negative.** Nonsporeforming. **Moderately thermophilic.** Growth temperature range is 37–65 °C, with an optimum of 50–55 °C. **The pH range for growth is 4.0–6.5.** Growth does not occur above a NaCl concentration of 1% (w/v). **Chemoautotrophic.** Growth occurs on H_2/CO_2 or formate by **anaerobic respiration with sulfate, thiosulfate, nitrate, or nitrite** as electron acceptors. The major cellular fatty acid is $C_{16:0}$. $MK-7(H_2)$ and MK-7 are the major quinones.

DNA G+C content (mol%): 35.1.

Type species: **Thermodesulfobium narugense** Mori, Kim, Kakegawa and Hanada 2004, 1[VP] (Effective publication: Mori, Kim, Kakegawa and Hanada 2003, 288.).

Further descriptive information

This genus contains only one species, *Thermodesulfobium narugense*, which was isolated from a microbial mat in a Japanese hot spring. The species is a moderate thermophile and is able to tolerate temperatures up to 65 °C.

Rod-shaped, Gram-stain-negative, nonsporeforming, and nonmotile. The presence of a typical Gram-negative cell wall with an outer membrane is visualized by transmission electron microscopy (Figure 252). This result is confirmed by the polymyxin B-LPS test (Wiegel and Quandt, 1982) where the fibrous structure and blebs typical of lipopolysaccharides are observed on the surface of polymyxin-B-treated cells.

Thermodesulfobium narugense is a strictly anaerobic bacterium able to grow under an H_2/CO_2 atmosphere. Anaerobic growth is coupled to sulfate reduction. At the optimal growth temperature of 55 °C and pH of 5.5, the doubling time is 14 h. When H_2 and CO_2 are provided as energy and carbon sources, thiosulfate, nitrate, or nitrite also serve as electron acceptors. Sulfite, elemental sulfur, Fe(III), citrate, fumarate, dimethyl sulfoxide, and O_2 do not serve as electron acceptors. In the presence of sulfate, the bacterium also uses formate instead of H_2 as an electron donor, but the growth is clearly slower. The following organic compounds are not utilized as substrates: glucose, ace-tate, lactate, pyruvate, malate, propionate, butyrate, fumarate, succinate, citrate, ethanol, propanol, and methanol.

The major quinones are $MK-7(H_2)$ and MK-7 (approx. 90% of total). Trace amounts of MK-8 and $MK-7(H_4)$ are also found. Palmitic acid $(C_{16:0})$ accounts for approximately 46% of the total cellular fatty acids. In addition, the following fatty acids are present as minor components: $C_{19:0\,\omega9c\,cyclo}$, $C_{18:0}$, $C_{18:1\,\omega9c}$, $C_{20:0}$, $C_{14:0}$, $C_{12:0\,3OH}$, and $C_{12:0}$.

Enrichment and isolation procedures

Thermodesulfobium narugense is occasionally found in a microbial mat formed in a hot spring at 45–60 °C and pH 6–7. For enrichment of the sulfate-reducing bacterium, a piece of the mat is inoculated into an enrichment medium composed of the following salts and solutions (per liter): KH_2PO_4, 0.75 g; K_2HPO_4, 0.78 g; NH_4Cl, 0.53 g; Na_3EDTA, 0.041 g; $FeSO_4 \cdot 7H_2O$, 0.011 g; $MgSO_4 \cdot 7H_2O$, 0.25 g; $CaCl_2 \cdot 2H_2O$, 0.029 g; NaCl, 0.23 g; Na_2SO_4, 2.8 g; sodium acetate, 0.16 g; trace element solution DSM 334, 10 ml; and vitamin solution DSM 141, 10 ml. The medium is autoclaved under a N_2/CO_2 (4:1, v/v) atmosphere in vials with butyl rubber stoppers and aluminum seals. Prior to inoculation, the medium is reduced with a sterile stock solution of cysteine HCl (final concentration of 0.25 g/l), and the gas phase is replaced by H_2/CO_2 (4:1, v/v; 152 kPa). The pH of the medium is adjusted to 6.0. As an indicator for sulfide production, an $FeSO_4$ solution (final concentration of 0.2 g/l) is added to the medium. After 1 week of incubation at 55 °C, pronounced sulfide production is observed. The enrichment culture is transferred several times to new medium of the same composition. For isolation of a pure culture, the same medium is solidified with 2% (w/v) agar and incubated under identical anaerobic conditions for 2–3 weeks.

Maintenance procedures

Thermodesulfobium narugense is maintained in the enrichment medium under a H_2/CO_2 (4:1, v/v) atmosphere at 55 °C. The culture should be transferred every 2 weeks. After growth, the

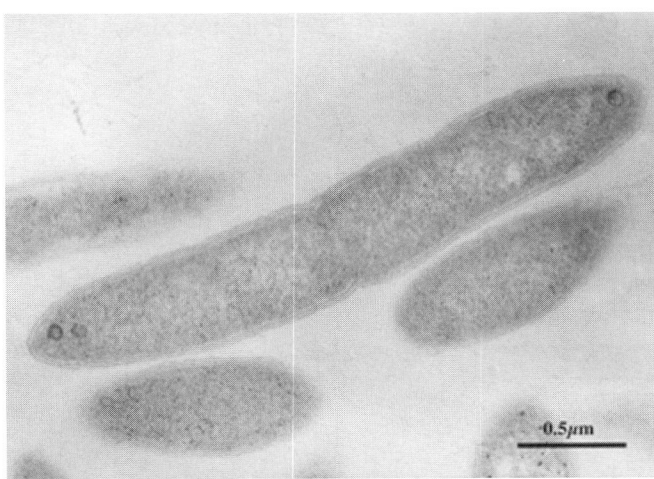

FIGURE 252. Ultrathin section of *Thermodesulfobium narugense* showing a Gram-negative type of cell wall with an outer membrane.

culture can also be stored at room temperature for several weeks. For long-term storage, it can be preserved in liquid nitrogen (−196 °C) under strictly anaerobic conditions with 5% dimethylsulfoxide or 10% glycerol as the cryoprotectant. Liquid drying is also successful with a protective medium composed of 0.1 M potassium phosphate buffer (pH 7.0), 3% sodium glutamate, 1.5% ribitol, and 0.05% cysteine hydrochloride monohydrate.

Differentiation of the genus *Thermodesulfobium* from other genera

Phylogenetic analysis based on the 16S rRNA gene sequence (Figure 253) reveals that *Thermodesulfobium narugense* is only distantly related to other known sulfate-reducing bacteria, such as *Thermodesulfobacterium* spp., *Thermodesulfovibrio* spp., *Desulfobacter* spp. (within the class *Deltaproteobacteria*), and *Desulfotomaculum* spp. (within the phylum *Firmicutes*). With sequence similarities less than 79%, *Thermodesulfobium narugense* is clearly distinguished from these other sulfate-reducing bacteria. Sequence analyses of the *dsrAB* (alpha and beta subunits of dissimilatory sulfite reductase) and *apsA* (alpha subunit of adenosine-5′-phosphosulfate reductase) genes also support the phylogenetically solitary position of *Thermodesulfobium narugense* (Mori et al., 2003).

In spite of their phylogenetic differences, *Thermodesulfobium narugense* shares a number of phenotypic similarities with *Thermodesulfobacterium* spp. and *Thermodesulovibrio* spp., including

cell walls that stain Gram-stain-negative, thermophily, absence of spores, and low G+C content of the DNA (Table 249). However, *Thermodesulfobium narugense* differs in some ways from these genera (Henry et al., 1994; Jeanthon et al., 2002; Sonne-Hansen and Ahring, 1999). The optimum growth temperature of all species in the genera *Thermodesulfovibrio* and *Thermodesulfobacterium* is above 65 °C, while the temperature maximum of *Thermodesulfobium narugense* is 55 °C. With the exception of *Thermodesulfobacterium hydrogeniphilum*, *Thermodesulfovibrio* and *Thermodesulfobacterium* spp. are chemoheterotrophs and can also grow fermentatively. *Thermodesulfobium narugense* grows chemoautotrophically and can not ferment. *Thermodesulfovibrio* and *Thermodesulfobacterium* spp. do not use nitrate or nitrite, but *Thermodesulfobium narugense* does. *Thermodesulfovibrio* and *Thermodesulfobacterium* spp. prefer neutral or basic pH for growth, while *Thermodesulfobium narugense* prefers slightly acidic conditions. Moderate acidophily is rare among sulfate-reducing organisms whether they are thermophilic or mesophilic.

Taxonomic comments

The circumscription of the genus *Thermodesulfobium* is based on the features of a single isolate. Further isolation and characterization of related strains is necessary to elucidate the phenotypic diversity of the genus.

The taxonomic affiliation of the genus *Thermodesulfobium* is still unsettled. In the Bergey's Taxonomic Outline, release 5, May (http://www.bergeysoutline.com), this genus was classified in the order *Thermoanaerobacteriales* in the class *Clostridia* of the phylum *Firmicutes*. However, this taxonomic assignment is probably inaccurate (Ludwig et al., 2009; Mori et al., 2003). Phylogenetic analyses of the 16S rRNA gene clearly demonstrates that *Thermodesulfobium narugense* is only distantly related to members of other phyla including the *Firmicutes* and the candidate divisions of environmental clones (Figure 253). Phylogenetic trees constructed by neighbor-joining, maximum-likelihood, and maximum-parsimony methods all indicate that the closest relative of *Thermodesulfobium narugense* is the environmental clone OPB46 retrieved from a Yellowstone hot spring and classified within the candidate division OP9 (Hugenholtz et al., 1998). However, their sequence similarity is merely 81%, and the bootstrap value of the branching node is quite low. The low bootstrap values suggest that the affiliation of *Thermodesulfobium narugense* with OP9 is unreliable and that they are not in the same clade. These results support the conclusion that *Thermodesulfobium narugense* is sufficiently isolated from previously described bacteria to justify creation of a new phylum even though the analysis is based upon a single isolate.

List of species of the genus *Thermodesulfobium*

1. **Thermodesulfobium narugense** Mori, Kim, Kakegawa and Hanada 2004, 1[VP] (Effective publication: Mori, Kim, Kakegawa and Hanada 2003, 288.)

   na.ru.gen'se. N.L. neut. adj. *narugense* from Narugo, the hot spring in Japan where the species was isolated.

   The description is the same as given for the genus. Isolated from a microbial mat in Narugo hot spring, Miyagi, Japan (the temperature and pH of the hot water were 58 °C

and 6.9, respectively). The mat was composed mainly of a moderately thermophilic sulfur-oxidizing bacterium, *Thiomonas thermosulfata*, which belongs to the class *Betaproteobacteria*.

DNA G+C content (mol%): 35.1 (HPLC).

Type strain: Na82, DSM 14796, JCM 11510, NBRC 100082.

GenBank accession number (16S rRNA gene): AB077817.

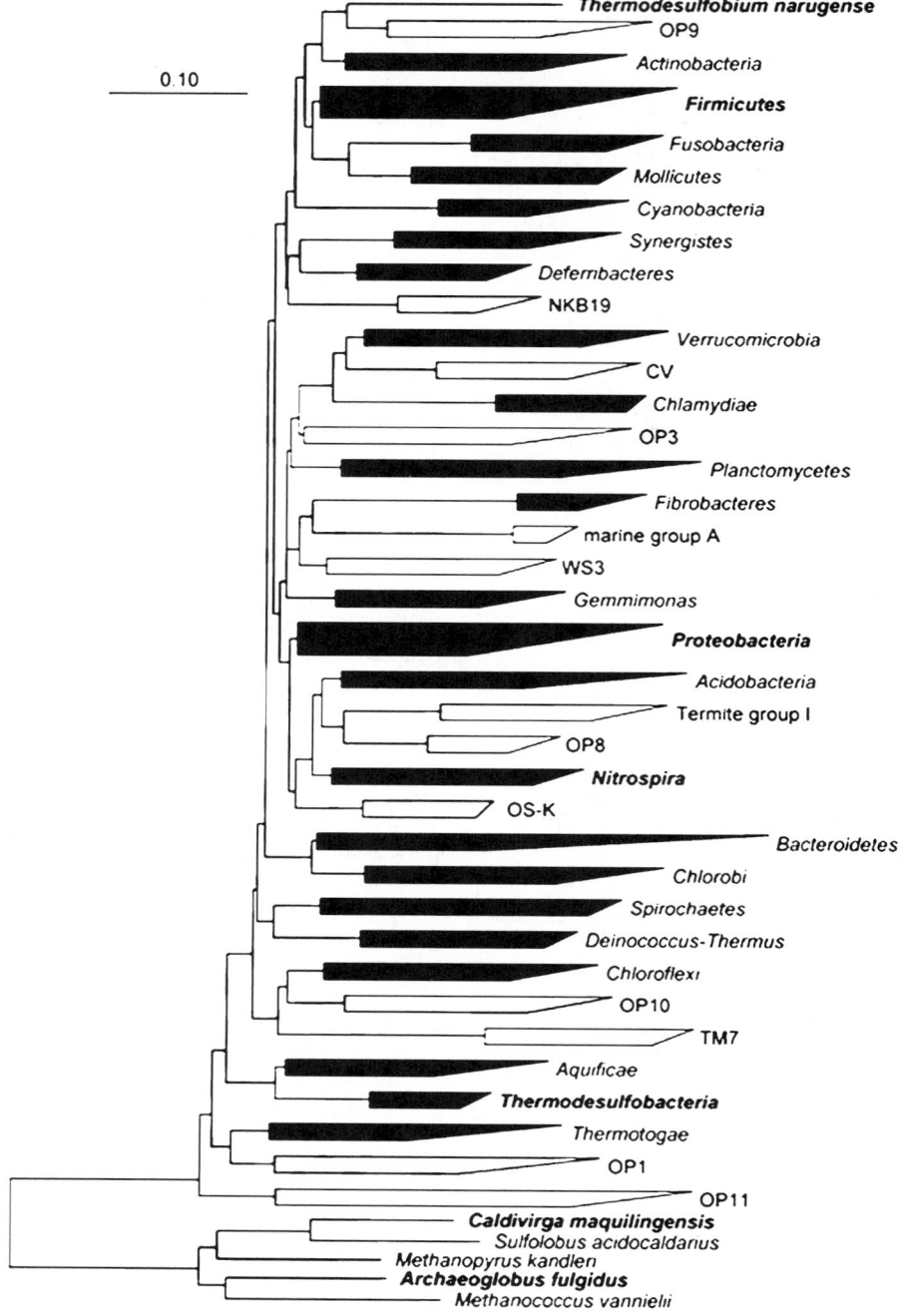

FIGURE 253. Phylogenetic tree of the 16S rRNA gene of *Thermodesulfobium narugense* with representatives of other bacterial phyla. The tree was constructed with the ARB program based upon the tree reported by Hugenholz (2002). The scale bar shows substitutions per compared positions. Phyla or other taxa containing sulfate-reducing organisms are indicated with bold font.

TABLE 249. Characteristics differentiating *Thermodesulfobium* from other sulfate-reducing micro-organisms[a]

Characteristic	*Thermodesulfobium (Firmicutes)*	*Archaeoglobus (Euryarchaeota)*	*Caldivirga (Crenarchaeota)*	*Desulfotomaculum* and relatives *(Firmicutes)*	*Desulfovibrio* and relatives *(Deltaproteobacteria)*	*Thermodesulfobacterium (Thermodesulfobacteria)*	*Thermodesulfovibrio (Nitrospirae)*
Optimum growth temperature (°C)	50–55	82–85	85	20–68	10–38[b]	70–75	65
Chemoautotrophic growth	+	d	–	d	d	–	–
Reduction of nitrate	+	–	+	–	d	–	d
Spore formation	–	–	–	+	–	–	–
Genomic G+C content (mol%)	35	41–46	28	38–57	34—69	28–40	30, 38

[a]Symbols: +, >85% positive; d, different strains give different reactions (16–84% positive); –, 0–15% positive.

[b]Except for *Thermodesulforhabdus norvegica*, *Desulfacinum infernum*, and *Desulfacinum hydrothermale* (optimum temperature for growth of each species is 60 °C).

Genus II. **Coprothermobacter** Rainey and Stackebrandt 1993a, 857[VP]

BERNARD OLLIVIER AND JEAN-LOUIS GARCIA

Co.pro.ther′mo.bac′ter. Gr. fem. n. *kopros* manure; Gr. adj. *thermos* warm; Gr. masc. dim. n. *bakterion* rod; N.L. masc. n. *Coprothermobacter* a thermophilic rod from manure.

Rod-shaped cells, 0.5 × 1.0–6.0 μm; occur singly or in pairs in young cultures; pleomorphic in old cultures. Colonies are white, circular (diameter 1–2 mm), convex, smooth with entire edges. Gram-stain-negative. Nonsporeforming. **Nonmotile. Strictly anaerobic. Thermophilic**. Optimum temperature, 55–70 °C. Neutrophilic. Chemo-organotrophic. Proteolytic using gelatin and peptones. Sugars are used poorly unless yeast extract or rumen fluid is added. Trypticase peptone stimulates glucose utilization. The principal fermentation end products from sugar or protein fermentation are acetic acid, H_2, and CO_2. Thiosulfate, but not sulfate, is used as electron acceptor for growth. Isolated from thermophilic digestors. The type species, *Coprothermobacter proteolyticus*, represents a deeply branched taxon of the domain *Bacteria*.

DNA G+C content (mol%): 43.

Type species: **Coprothermobacter proteolyticus** (Ollivier, Mah, Ferguson, Boone, Garcia and Robinson 1985) Rainey and Stackebrandt 1993a, 857[VP] (*Thermobacteroides proteolyticus* Ollivier, Mah, Ferguson, Boone, Garcia and Robinson 1985, 427).

Further descriptive information

Electron micrographs of thin sections of the two species of the genus *Coprothermobacter*, *Coprothermobacter proteolyticus* and *Coprothermobacter platensis*, reveal a thin inner wall layer and a heavy outer wall. Both species grow poorly on proteins or sugars in mineral medium in the absence of organic compounds (yeast extract), indicating that cofactor(s) present in yeast extract are required for good growth. In the case of *Coprothermobacter proteolyticus*, yeast extract cannot be replaced by trypticase peptones, coenzyme M, or sodium acetate; trypticase peptone stimulates the fermentation of glucose, but Casamino acids do not; synthesis of cell material is not always proportional to the amount of carbon and energy source utilized, particularly with maltose and mannose (Ollivier et al., 1985). *Coprothermobacter platensis* (Etchebehere et al., 1998) differs from *Coprothermobacter proteolyticus* in optimum temperature for growth, antibiotic susceptibility, and sugar utilization (Table 250). Gelatin fermentation by *Coprothermobacter proteolyticus* leads to the formation of ammonium, acetate, H_2, and CO_2 along with smaller amounts of isobutyrate and isovalerate, but also trace amounts of propionate (Kersters et al., 1993; Ollivier et al., 1985). Interestingly, co-cultures of *Coprothermobacter proteolyticus* and a hydrogenotrophic methanogen (*Methanobacterium* species) show an increase in propionate and isobutyrate production (Ollivier et al., 1986) suggesting that some peptides or amino acids are only oxidized via interspecies hydrogen transfer by this microorganism. Both species reduce thiosulfate, but not sulfate, to sulfide with a concomitant increase in growth and glucose utilization (Etchebehere et al., 1998). Degradation of gelatin by *Coprothermobacter proteolyticus* is not influenced by ammonium chloride concentrations up to 6 g/l; growth is inhibited by Na^+ (0.60 M), neomycin (0.15 g/l), and penicillin G (20 U/

TABLE 250. Characteristics differentiating *Coprothermobacter proteolyticus* and *Coprothermobacter platensis*

Characteristic	*C. proteolyticus*[a]	*C. platensis*[b]
Dimensions (µm)	0.5 × 1–6	0.5 × 1.5–2.0
Optimum temperature (°C)	63	55
Growth on:		
Fructose	+	+
Sucrose	+	+
Melibiose	−	ND[c]
Xylose	+	−
Production of ethanol from sugar metabolism	−	−
Resistance to:		
Penicillin G (20 U/ml)	−	+
Polymixin B (20 mg/l)	+	−
C$_{15:0 iso}$ as the major fatty acid	+	ND

[a]Data from Ollivier et al. (1985) and Kersters et al. (1994).
[b]Data from Etchebehere et al. (1998).
[c]ND, Not determined.

ml); vancomycin, polymyxin B, sodium azide, and kanamycin are not effective inhibitors (Kersters et al., 1994). Vancomycin (2.5 mg/l), neomycin (0.15 g/l), and polymyxin (20 mg/l) inhibit growth of *Coprothermobacter platensis* (Etchebehere et al., 1998). The major polyamines synthesized by *Coprothermobacter proteolyticus* are putrescine, spermidine, and spermine (Hamana et al., 1996). *Coprothermobacter proteolyticus* possesses a thermostable protease with optimal temperature of 85 °C and optimal pH of 9.5. The protease retains about 90% of its activity at pH 10.0 and appears quite specific as compared to enzymes from other thermophilic or hyperthermophilic proteolytic micro-organisms (Klingeberg et al., 1991).

Enrichment and isolation procedures

Coprothermobacter species can be enriched from anaerobic thermophilic digesters treating proteinaceous wastes such as tannery wastes and food wastes using media and procedures similar to those described by Ollivier et al. (1986) with gelatin as the energy source and Na$_2$S and cysteine as the reductive agents. Similarly, the use of peptone-yeast medium (Holdeman et al., 1977) is recommended to enrich proteolytic bacteria in general and *Coprothermobacter proteolyticus* in particular as described by Kersters et al. (1994). At least three subcultures in the same growth conditions, at temperature from 55 °C up to 70 °C are needed before isolation.

After several transfers, the enrichment cultures are serially diluted using the method of Hungate (1969) with roll tubes containing the basal medium, gelatin as the energy source, and purified agar at a concentration of 2%. For isolation, agar medium can be also poured in plates within an anaerobic chamber. In order to detect selectively proteolytic colonies during the first steps of isolation, casein can be used as substrate. Colonies, surrounded with large clearing zones are picked and restreaked on gelatin agar plates or roll tubes. At least two colonies are picked, and the process of serial dilution in roll tubes is repeated in order to purify the cultures.

Maintenance procedures

Stock cultures can be maintained on the medium described by Ollivier et al. (1986) and transferred at least monthly. Liquid cultures retain viability after several weeks of storage at room temperature. As a precaution, cultures also have to be refrigerated. Because of easy death/lysis of cells when cultures are further incubated at high temperature, it is recommended to stock cultures before the end of the exponential growth phase.

Taxonomic and phylogenetic considerations

The genus *Coprothermobacter* contains two species, *Coprothermobacter proteolyticus* and *Coprothermobacter platensis* (Etchebehere et al., 1998). The type species, *Coprothermobacter proteolyticus*, was originally placed in the genus *Thermobacteroides*, mainly because of phenotypic similarities to the type species of that genus (*Thermobacteroides acetoethylicus*) (Ben-Bassat and Zeikus, 1981). However, phylogenetic analysis of 16S rRNA sequences later led to the reclassification of *Thermobacteroides acetoethylicus* as *Thermoanaerobacter acetoethylicus* (Rainey and Stackebrandt, 1993b, 1993a). *Coprothermobacter proteolyticus*, being phylogenetically only distantly related to any species known at that time, was placed in the new genus *Coprothermobacter* as the type species. According to the 16S rRNA phylogenetic analysis presented in the roadmap to this volume (Figure 7), the genus *Coprothermobacter* is a member of the family *Thermodesulfobiaceae*, order *Thermoanaerobacterales*, class *Clostridia* in the phylum *Firmicutes*.

Differentiation of the genus *Coprothermobacter* from other genera

Characteristics by which *Coprothermobacter* may be differentiated from phenotypically similar genera and species are shown in Table 251.

List of species of the genus *Coprothermobacter*

1. **Coprothermobacter proteolyticus** (Ollivier, Mah, Ferguson, Boone, Garcia and Robinson 1985) Rainey and Stackebrandt 1993a, 857[VP] (*Thermobacteroides proteolyticus* Ollivier, Mah, Ferguson, Boone, Garcia and Robinson 1985, 427)

pro.te.o.ly′ti.cus. Gr. adj. *protos* first; Gr. adj. *lytikos* loosening, dissolving; N.L. masc. adj. *proteolyticus* proteolytic.

Cells are rod-shaped, 0.5 × 1–6 µm, occurring singly or in pairs in young cultures (Figure 254); pleomorphic in old cultures. No lysis is observed in the stationary phase. Colonies in roll tubes are 1–2 mm in diameter after 3–4 d; they are opaque, whitish, crenated or entire, circular, and slightly convex. Gram-negative cell-wall structure (Figure 255). Nonmotile. Nonsporeforming.

Obligately anaerobic. Ferments peptone, gelatin, casein, trypticase peptone, and bovine serum albumin in the presence of 0.1% yeast extract. Grows on D-glucose,

TABLE 251. Differential characteristics of the genus *Coprothermobacter* and other phylogenetically related and phenotypically similar genera

Characteristic	*Coprothermobacter*	*Aquifex*	*Fervidobacterium*	*Thermoanaerobacter*	*Thermoanaerobacterium*	*Thermosipho*	*Thermotoga*
Morphology	Rods	Rods	Rods with sheath, spheroids	Rods	Rods	Rods with sheath, chains	Rods with sheath
Existence of species growing at 90 °C	−	+	−	−	−	−	+
Existence of species forming spores	−	−	−	+	+	−	−
End product from sugar metabolism:							
Lactate	−	−[a]	+	+	+	+	+
Ethanol	−	−	+	+	+	−	+/−
Oxygen as electron acceptor for growth	−	+[b]	−	−	−	−	−
Reduction of thiosulfate to sulfide	+	ND[c,d]	+	+	−[e]	+	+

[a]*Aquifex pyrophilus* is chemolithoautotrophic.
[b]*Aquifex pyrophilus* uses hydrogen under microaerobic conditions.
[c]ND, Not determined.
[d]Sulfide produced from sulfur reduction.
[e]Sulfur produced from thiosulfate reduction.

D-fructose, maltose, sucrose, D-mannose, and starch when 1% yeast extract is added. Does not grow on D-arabinose, melibiose, and cellulose. Trypticase peptone stimulates glucose utilization. The major fermentation products from gelatin and glucose in the presence of yeast extract are acetic acid, H_2, and CO_2. Ammonium along with smaller quantities of propionic, isobutyric, and isovaleric acids are produced during gelatin fermentation. Thiosulfate and elemental sulfur, but not sulfate, are used as electron acceptors for growth.

Optimal temperature for growth is 63–70 °C (range, 35–70 °C); optimal pH is 6.8–7.5 (range, pH 5.0–9.4). Growth is inhibited by neomycin (0.15 g/l) and penicillin G (20 U/ml). Sensitive to vancomycin (5 mg/l). Concentration of 0.7 M Na⁺ is inhibitory. The major cellular fatty acids are $C_{15:0 iso}$, $C_{16:0}$, $C1_{4:0 iso 3OH}$, $C_{17:0 iso}$ and a fatty acid with an equivalent chain-length of 15.360.

Two strains were isolated from anaerobic thermophilic digesters. Strain BT (=ATCC 35245 = OCM 4 = DSM 5265 = LMG 11567) was isolated from a digestor that was fermenting tannery wastes and cattle manure, whereas strain I8 (=LMG 14268) was isolated from a biokitchen waste digestor.

DNA G+C content (mol%): 43–45 (ultracentrifugation).
Type strain: BT, ATCC 35245, OCM 4, DSM 5265, LMG 11567.
GenBank accession number (16S rRNA gene): X69335.

2. **Coprothermobacter platensis** Etchebehere, Pavan, Zorzópulos, Soubes and Muxi 1998, 1302[VP]

pla.ten'sis. L. masc. adj. *platensis* pertaining to Rio de la Plata, a river between Uruguay and Argentina, the region where the strain was isolated.

Cells are straight rods, 0.5 × 1.5–2 µm, that occur singly or in pairs in young cultures. Lysis is observed in the

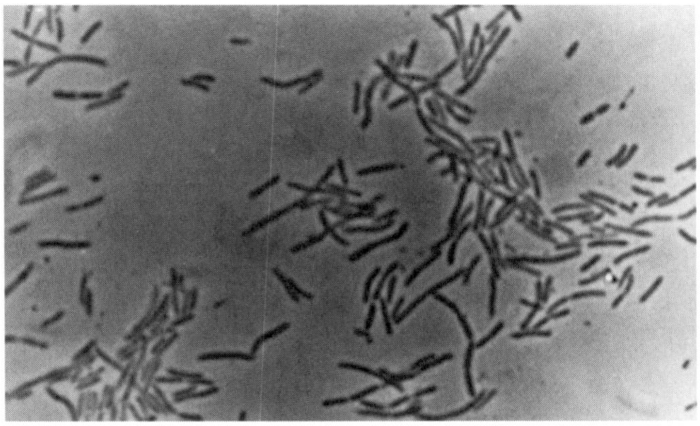

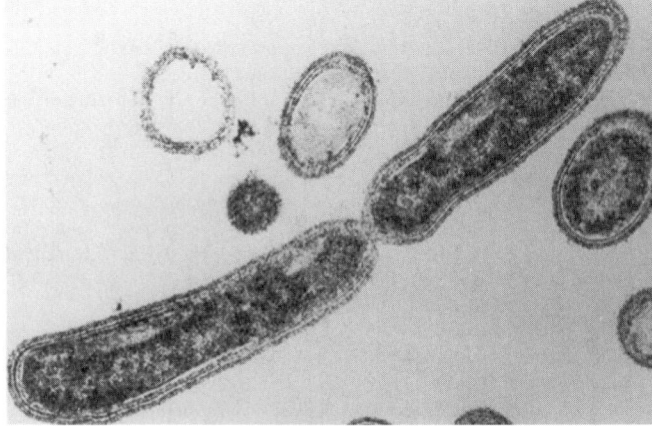

FIGURE 254. Phase-contrast photomicrograph of *Coprothermobacter proteolyticus.* Bar =10 µm.

FIGURE 255. Ultrathin section of *Coprothermobacter proteolyticus* showing a cell-wall structure typical of Gram-negative bacteria. Bar = 0.5 µm.

stationary phase. Colonies in PY agar plates are circular, 1 mm in diameter, with an entire border, transparent to whitish. Stains Gram-negative, but the cell wall under the electron microscope is atypical with a dense inner layer and a less dense outer layer. Nonmotile and nonspore-forming.

Obligately anaerobic. Proteolytic. Ferments gelatin, casein, bovine albumin, peptone, and yeast extract. Glucose, fructose, sucrose, maltose, and starch are poorly fermented. Fermentation products from glucose are acetate, H_2, and CO_2. The major fermentation products from gelatin are acetate, propionate, H_2, and CO_2. Growth on glucose is stimulated by thiosulfate, which is reduced to sulfide. Sulfate and nitrate are not reduced. Moderately thermophilic;

optimum temperature 55 °C (range 35–65 °C). Optimum pH 7.0 (range 4.3–8.3). Yeast extract is required. Growth is inhibited by vancomycin (2.5 mg/l), neomycin (0.15 g/l), and polymyxin B (20 mg/l). Resistant to penicillin G (20 U/ ml), kanamycin (600 ng/ml), and sodium azide (0.5 g/l). NaCl, O.4 M or higher, is inhibitory. Phylogenetically closely related to *Coprothermobacter proteolyticus* according to the 16S rDNA sequence analysis. Both are included in one of the earlier branches of the domain *Bacteria*. Isolated from a mesophilic upflow anaerobic sludge blanket reactor of a baker's yeast factory.

DNA G+C content (mol%): 43 (HPLC).

Type strain: 3R, DSM 11748.

GenBank accession number (16S rRNA gene): Y08935.

References

Ben-Bassat, A. and J.G. Zeikus. 1981. *Thermobacteroides acetoethylicus* gen. nov. and sp. nov., a new chemoorganotrophic, anaerobic, thermophilic bacterium. Arch. Microbiol. *128*: 365–370.

Etchebehere, C., M.E. Pavan, J. Zorzopulos, M. Soubes and L. Muxi. 1998. *Coprothermobacter platensis* sp. nov., a new anaerobic proteolytic thermophilic bacterium isolated from an anaerobic mesophilic sludge. Int. J. Syst. Bacteriol. *48*: 1297–1304.

Hamana, K., H. Hamana, M. Niitsu and K. Samejima. 1996. Polyamines of thermophilic Gram-positive anaerobes belonging to the genera *Caldicellulosiruptor, Caloramator, Clostridium, Coprothermobacter, Moorella, Thermoanaerobacter* and *Thermoanaerobacterium*. Microbios *85*: 213–222.

Henry, E.A., R. Devereux, J.S. Maki, C.C. Gilmour, C.R. Woese, L. Mandelco, R. Schauder, C.C. Remsen and R. Mitchell. 1994. Characterization of a new thermophilic sulfate-reducing bacterium, *Thermodesulfovibrio yellowstonii*, gen. nov. and sp. nov.: its phylogenetic relationship to *Thermodesulfobacterium commune* and their origins deep within the bacterial domain. Arch. Microbiol. *161*: 62–69.

Holdeman, L.V., E.P. Cato and W.E.C. Moore. 1977. Culture methods: Use of pre-reduced media. *In* Holdeman, Cato and Moore (Editors), Anaerobic Laboratory Manual, 4th edn. Anaerobe Laboratory, Virginia Polytechnic Institute and State University, Blacksburg, pp. 117–149.

Hugenholtz, P., C. Pitulle, K.L. Hershberger and N.R. Pace. 1998. Novel division level bacterial diversity in a Yellowstone hot spring. J. Bacteriol. *180*: 366–376.

Hugenholtz, P. 2002. Exploring prokaryotic diversity in the genomic era. Genome Biol. *3*: reviews0003.1-0003.8.

Hungate, R.E. 1969. A roll tube method for cultivation of strict anaerobes. *In* Norris and Ribbons (Editors), Methods in Microbiology, Vol. 3B. Academic Press, London and New York, pp. 117–132.

Jeanthon, C., S. L'Haridon, V. Cueff, A. Banta, A.L. Reysenbach and D. Prieur. 2002. *Thermodesulfobacterium hydrogeniphilum* sp. nov., a thermophilic, chemolithoautotrophic, sulfate-reducing bacterium isolated from a deep-sea hydrothermal vent at Guaymas Basin, and emendation of the genus *Thermodesulfobacterium*. Int. J. Syst. Evol. Microbiol. *52*: 765–772.

Kersters, I., F. Houwen and W. Verstraete. 1993. Thermophilic, anaerobic degradation of gelatin by *Thermobacteroides proteolyticus*. Biotechnol. Lett. *15*: 931–936.

Kersters, I., G.M. Maestrojuan, U. Torck, M. Vancanneyt, K. Kersters and W. Verstraete. 1994. Isolation of *Coprothermobacter proteolyticus*

from an anaerobic digest and further characterization of the species. Syst. Appl. Microbiol. *17*: 289–295.

Klingeberg, M., F. Hashwa and G. Antranikian. 1991. Properties of extremely thermostable proteases from anaerobic hyperthermophilic bacteria. Appl. Microbiol. Biotechnol. *34*: 715–719.

Ludwig, W., K.-H. Schleifer and W.B. Whitman. 2009. Revised road map to the phylum *Firmicutes*. *In* De Vos, Garrity, Jones, Krieg, Ludwig, Rainey, Schleifer and Whitman (Editors), Bergey's Manual of Systematic Bacteriology, 2nd edn, Vol. 3, The *Firmicutes*, Springer, New York, pp. 1–14.

Mori, K., H. Kim, T. Kakegawa and S. Hanada. 2003. A novel lineage of sulfate-reducing microorganisms: *Thermodesulfobiaceae* fam. nov., *Thermodesulfobium narugense*, gen. nov., sp. nov., a new thermophilic isolate from a hot spring. Extremophiles *7*: 283–290.

Mori, K., H. Kim, T. Kakegawa and S. Hanada. 2004. *In* Validation of the publication of new names and new combinations previously effectively published outside the IJSEM. List no. 95. Int. J. Syst. Evol. Microbiol. *54*: 1–2.

Ollivier, B., N. Smiti, R.A. Mah and J.-L. Garcia. 1986. Thermophilic methanogenesis from gelatin by a mixed defined bacterial culture. Appl. Microbiol. Biotechnol. *24*: 79–83.

Ollivier, B.M., R.A. Mah, T.J. Ferguson, D.R. Boone, J.L. Garcia and R. Robinson. 1985. Emendation of the genus *Thermobacteroides*: *Thermobacteroides proteolyticus* sp. nov., a proteolytic acetogen from a methanogenic enrichment. Int. J. Syst. Bacteriol. *35*: 425–428.

Rainey, F.A. and E. Stackebrandt. 1993a. Transfer of the type species of the genus *Thermobacteroides* to the genus *Thermoanaerobacter* as *Thermoanaerobacter acetoethylicus* (Ben-Bassat and Zeikus 1981) comb. nov., description of *Coprothermobacter* gen. nov., and reclassification of *Thermobacteroides proteolyticus* as *Coprothermobacter proteolyticus* (Ollivier et al. 1985) comb. nov. Int. J. Syst. Bacteriol. *43*: 857–859.

Rainey, F.A. and E. Stackebrandt. 1993b. Phylogenetic analysis of the bacterial genus *Thermobacteroides* indicates an ancient origin of *Thermobacteroides proteolyticus*. Lett. Appl. Microbiol. *16*: 282–286.

Sonne-Hansen, J. and B.K. Ahring. 1999. *Thermodesulfobacterium hveragerdense* sp. nov., and *Thermodesulfovibrio islandicus* sp. nov., two thermophilic sulfate reducing bacteria isolated from a Icelandic hot spring. Syst. Appl. Microbiol. *22*: 559–564.

Wiegel, J. and L. Quandt. 1982. Determination of the gram type using the reaction between polymyxin B and lipopolysaccharides of the outer cell wall of whole bacteria. J. Gen. Microbiol. *128*: 2261–2270.

Family III. Incertae Sedis

Members of this monophyletic cluster were previously assigned to the family "*Thermoanaerobacteriaceae*" by Garrity et al. (2005). Because the current outline proposes that *Thermoanaerobacter* and not *Thermoanaerobacterium* is the type of the, now, *Thermoanaerobacteraceae* fam. nov., the taxonomic status of these genera is ambiguous. Although classified within a single family *incertae sedis*, the evidence for the relationship between *Thermoanaerobacterium* and the other members of this group is not strong. Thus, subsequent analyses may suggest reorganization.

Genus I. **Caldicellulosiruptor** Rainey, Donnison, Janssen, Saul, Rodrigo, Bergquisy, Daniel, Stackebrandt and Morgan 1995, 197[VP] (Effective publication: Rainey, Donnison, Janssen, Saul, Rodrigo, Bergquist, Daniel, Stackebrandt and Morgan 1994, 264.) emend. Onyenwoke, Lee, Dabrowski, Ahring and Wiegel 2006, 1394

FRED A. RAINEY

Cal.di.cel.lu.lo.si.rup'tor. L. adj. *caldus* hot; N.L. n. *cellulosum* cellulose; L. masc. n. *ruptor* breaker; N.L. masc. n. *Caldicellulosiruptor* cellulose-breaker living under hot conditions.

Cells are straight **rods. Gram-stain-negative.** Endospores not observed. Growth not obtained after heat treatment at 115 °C for 20 min. Growth **strictly anaerobic**, occurring over the temperature range of 45–82 °C with optima in the range of 65–75 °C. No growth detected below 45 °C or above 82 °C. Monosaccharides, disaccharides and polysaccharides serve as fermentable substrates. **Cellulose is hydrolyzed by most strains.**

DNA G+C content (mol%): 35–37.5 (T_m, HPLC).

Type species: **Caldicellulosiruptor saccharolyticus** Rainey, Donnison, Janssen, Saul, Rodrigo, Bergquist, Daniel, Stackebrandt and Morgan 1995, 197[VP] (Effective publication: Rainey, Donnison, Janssen, Saul, Rodrigo, Bergquist, Daniel, Stackebrandt and Morgan 1994, 265.).

Further descriptive information

The genus *Caldicellulosiruptor* contains five species: *Caldicellulosiruptor saccharolyticus* (Rainey et al., 1994) *Caldicellulosiruptor acetigenus* (Onyenwoke et al., 2006), *Caldicellulosiruptor kristjanssonii* (Bredholt et al., 1999), *Caldicellulosiruptor lactoaceticus* (Mladenovska et al., 1995) and *Caldicellulosiruptor owensensis* (Huang et al., 1998). All strains are rod-shaped, varying in length from 1.5 to 9.4 μm and in diameter from 0.4 to 1.0 μm. Cells have rounded ends and occur singly, in pairs or in short chains. In the case of *Caldicellulosiruptor acetigenus* chains of up to eight cells have been observed (Nielsen et al., 1993). During the exponential-phase, *Caldicellulosiruptor owensensis* cultures contain small coccoid cells (Huang et al., 1998). Although this genus falls within the radiation of the low G+C Gram-positive taxa (*Firmicutes*), all strains tested stain Gram-negative. A Gram-positive cell-wall structure based on electron microscopy studies has been shown in *Caldicellulosiruptor owensensis* (Huang et al., 1998) and *Caldicellulosiruptor acetigenus* (Nielsen et al., 1993; Onyenwoke et al., 2006). Flagella have been observed in two of the five species of this genus; *Caldicellulosiruptor kristjanssonii* is nonmotile, but peritrichous flagella are observed in exponential-phase cultures and two subterminal flagella were seen in old cultures (Bredholt et al., 1999). Although motility has not been observed, young cells of *Caldicellulosiruptor owensensis* have lophotrichous flagella (Huang et al., 1998). One of the main phenotypic characteristics differentiating the cellulose-degrading species of *Caldicellulosiruptor* from cellulolytic species of the genus *Clostridium* is the absence of endospores in species of *Caldicellulosiruptor*. Endospores or similar heat resistant structures have not been observed in any species of the genus *Caldicellulosiruptor* and in some cases this has been confirmed by heat treatments such as boiling for 5 min. *Caldicellulosiruptor kristjanssonii* (Bredholt et al., 1999) and *Caldicellulosiruptor acetigenus* (Nielsen et al., 1993), or heating culture to 115 °C for 20 min., *Caldicellulosiruptor saccharolyticus* (Rainey et al., 1994). Sphaeroplasts have been observed in *Caldicellulosiruptor acetigenus* when it was grown on minimal salts media and xylose (Nielsen et al., 1993). Colony descriptions have been provided for all species of the genus *Caldicellulosiruptor*, but growth on different media and under different conditions makes these characteristics of little comparative value. Colonies of *Caldicellulosiruptor saccharolyticus* grown on 2/1 cellobiose agar are 2–5 mm in diameter, cream-colored, and umbonate (Rainey et al., 1994). *Caldicellulosiruptor acetigenus* on xylan containing agar forms colonies that are off white, milky-colored and have clearing zones around them of up to 2 cm in diameter (Nielsen et al., 1993); after 7 d incubation in agar roll tubes surface colonies of *Caldicellulosiruptor lactoaceticus* are circular, white, opalescent, 1 mm in diameter with an entire edge (Mladenovska et al., 1995). *Caldicellulosiruptor owensensis* forms circular, convex, opaque, yellowish colonies with smooth edges that are up to 2 mm in diameter (Huang et al., 1998). Colonies of *Caldicellulosiruptor kristjanssonii* in roll tubes containing Avicel (microcrystalline cellulose) are 0.5–1.0 mm in diameter, flat with fringed edges and after 9–14 d incubation show 2–4 mm zones of clearing (Bredholt et al., 1999).

The genus *Caldicellulosiruptor* can be differentiated from the thermophilic, cellulolytic *Clostridium* species (*Clostridium stercorarium*, *Clostridium thermocellum* and *Clostridium thermocopriae*) and related strains based on maximum growth temperature. In the numerical taxonomy study of Rainey et al. (1993a) growth at 75 °C and at 80 °C was determined for 51 strains, 43 of which did not form endospores and clustered separately from the eight spore-forming strains (including the type strains of the species *Clostridium stercorarium*, *Clostridium thermocellum* and *Clostridium thermocopriae*). The strains falling in clusters A–D are considered to be *Caldicellulosiruptor* strains (including the strains Tp8T 6331[T], COMP. B1, Wai35. B1, RI2. B1, Z-1203, Rt8. B7, Rt8. B15, and Ok9. B1) and all grow at 75 °C and ~70% of them at 80 °C, in contrast to the endospore forming strains, 60% of which grow at 75 °C and none at 80 °C (Rainey et al., 1993a). Growth of the described species of the genus *Caldicellulosiruptor* is in the temperature range of 45–82 °C with optima in the range of 65–78 °C (see species descriptions below). *Caldicellulosiruptor kristjanssonii* has both the highest temperature maximum for growth at 82 °C and optimum for growth at 78 °C (Bredholt et al., 1999). *Caldicellulosiruptor* species are neutrophiles growing optimally at pH ~7.0. The pH range for growth of the species of this genus is 5.2–9.0 (see species descriptions below). *Caldicellulosiruptor owenensis* is described as an alkalitolerant species growing up to pH 9.0 (Huang et al., 1998).

All species of the genus are obligately anaerobic chemoorganotrophs with fermentative metabolism. Monosaccharides, disaccharides, sugar alcohols, and polysaccharides serve as fermentable substrates. The species of the genus can be differentiated based on substrate utilization patterns (Table 252). *Caldicellulosiruptor lactoaceticus* uses a limited number of substrates including Avicel, starch and xylan (Table 252, Mladenovska et al., 1995). Growth of the species *Caldicellulosiruptor saccharolyticus*, *Caldicellulosiruptor lactoaceticus* and *Caldicellulosiruptor kristjanssonii* is supported by Avicel, starch and xylan (Bredholt et al., 1999; Mladenovska et al., 1995; Rainey et al., 1994). *Caldicellulosiruptor owensensis* also grows on starch, xylan and cellulose but the type of cellulose tested is not provided, therefore the ability of this species to degrade crystalline cellulose is undetermined (Huang et al., 1998). *Caldicellulosiruptor acetigenus* is the only species of the genus *Caldicellulosiruptor* that does not degrade cellulose either filter paper or Avicel, but does display cellulase enzyme activity when grown on carboxymethylcellulose (Nielsen et al., 1993; Onyenwoke et al., 2006). The end products of fermentation of sugars for the species of the genus include acetate, ethanol, CO_2 and H_2. Lactate is produced as the main product of fermentation by *Caldicellulosiruptor lactoaceticus* and in smaller amounts by *Caldicellulosiruptor saccharolyticus*, *Caldicellulosiruptor owensensis* and *Caldicellulosiruptor kristjanssonii*. *Caldicellulosiruptor acetigenus* does not produce lactate as an end product of fermentation, but does produce trace amounts of iso-butyrate (Nielsen et al., 1993; Onyenwoke et al., 2006).

The G+C content is in the range of 35.0–37.5 mol% for the type strains of the species of the genus *Caldicellulosiruptor*. The values for the species *Caldicellulosiruptor acetigenus*, *Caldicellulosiruptor lactoaceticus* and *Caldicellulosiruptor kristjanssonii* were determined by the HPLC method while those of *Caldicellulosiruptor saccharolyticus* and *Caldicellulosiruptor owenensis* were determined by the thermal denaturation and buoyant gradient density methods respectively.

Enrichment, isolation and growth conditions

Four of the five species of the genus *Caldicellulosiruptor* have been isolated from thermal springs or material associated with thermal springs, the exception being *Caldicellulosiruptor owensensis* which was isolated from a freshwater pond with a temperature of 32 °C in the dry lake bed of Owens Lake, California, USA. Species of this genus can be isolated using anaerobic techniques with xylan or cellulose as substrates in enrichment cultures. The roll tube method (Hungate, 1969) can be used for the isolation of single colonies that give zones of clearing in the xylan- or cellulose-containing agar. *Caldicellulosiruptor saccharolyticus* strain Tp8T 6331[T] was isolated from the decomposed end of a *Pinus radiata* plank in the downstream flow of a thermal spring TP10, Taupo, New Zealand (Sissons et al., 1987). Thermal spring TP10 had a temperature of 78 °C and the downstream flow at which the pine wood plank was located (TP8) had a temperature of 48 °C (Sissons et al., 1987). Strain Tp8T 6331[T] was isolated from the homogenized wood sample inoculated into basal salts medium containing 0.5% MN 300 cellulose and subsequent serial dilution with cellobiose (Sissons et al., 1987). A cellulose-clearing colony was selected and further purified on cellobiose agar and liquid cultures containing cellobiose and Sigmacell 50 (Sissons et al., 1987). *Caldicellulosiruptor acetigenus* was originally isolated as a xylanolytic, non-spore-forming bacterium, and described as *Thermoanaerobium acetigenum* (Nielsen et al., 1993) before being transferred to the genus *Caldicellulosiruptor* (Onyenwoke et al., 2006). The type strain XB6[T] of *Caldicellulosiruptor acetigenus* was isolated from sediment and biomat collected from a thermal spring with

TABLE 252. Substrate utilizations differentiating species of the genus *Caldicellulosiruptor*[a]

Substrate	*C. saccharolyticus*[b]	*C. acetigenus*[c]	*C. kristjanssonii*[d]	*C. lactoaceticus*[e]	*C. owensensis*[f]
Arabinose	+	+	−	−	+
Fructose	+	+	+	−	+
Glucose	+	+	+	−	+
Galactose	+	+	+	−	+
Ribose	−	ND	−	−	+
Sucrose	+	+	+	−	+
Raffinose	−	+	−	−	+
Trehalose	+	+	+	−	−
Inositol	−	ND	ND	ND	+
Mannitol	−	ND	−	−	+
Cellulose	+	−	+	+	+
Xylan	+	+	+	+	+

[a]ND, Not determined.
[b]Data from Rainey et al. (1994).
[c]Data from Nielsen et al. (1993) and Onyenwoke et al. (2006).
[d]Data from Bredholt et al. (1999).
[e]Data from Mladenovska et al. (1995).
[f]Data from Huang et al. (1998).

a temperature of 70 °C and a pH of ~8.5 in the Hverðagerdi-Hengill geothermal area, Iceland (Nielsen et al., 1993). The enrichment cultures were established at 68 °C in mineral medium at pH 7.0 containing 4 g/l of beech wood xylan and 10% (v/v) of sediment as an inoculum (Nielsen et al., 1993). Enrichments showing solublization of xylan, and over pressure, were transferred to fresh medium several times before pure cultures were isolated on xylan containing media solidified with Gelrite using the roll tube technique (Nielsen et al., 1993). The isolation of *Caldicellulosiruptor kristjanssonii* followed the same techniques used in the isolation of *Caldicellulosiruptor acetigenus*, except that xylan was replaced with Avicel. The inoculum was a biomat from a hot spring at pH 8.7 and the enrichment was established at 78 °C (Bredholt et al., 1999). Avicel was also used as the substrate in the enrichment culture from which *Caldicellulosiruptor lactoaceticus* was isolated (Mladenovska et al., 1995). The isolation source of *Caldicellulosiruptor owensensis* indicated that species of this genus can be isolated from non-thermal sources. *Caldicellulosiruptor owensensis* was isolated from sediment samples (pH 9.0 and 32 °C) collected from a small freshwater pond in the dry bed of Owens Lake (Huang et al., 1998). The enrichment temperature of 75 °C selected for thermophilic organisms and xylan added to the carbonate buffered culture medium selected for species that were xylanolytic (Huang et al., 1998).

Taxonomic comments

It is the lack of spore formation and ability to grow at >75 °C that differentiated these thermophilic cellulolytic bacteria from the spore-forming, thermophilic, cellulolytic species of the genus *Clostridium*. The numerical taxonomy study of Rainey et al. (1993a) involved the determination of 92 characteristics for 51 strains, 43 that did not form endospores and eight spore-forming strains (including the type strains of the species *Clostridium stercorarium*, *Clostridium thermocellum* and *Clostridium thermocopriae*). At the 72% similarity level in a S_{SM} unweighted pair-group method with averages (UPGMA) analysis, five clusters (designated A–E) were formed (Rainey et al., 1993a). The thermophilic, cellulolytic *Clostridium* species all fell into cluster E, while the non-spore-forming strains that grew at 75 °C or above fell into clusters A–D. This study demonstrated that at the phenotypic level the non-spore-forming, cellulolytic, thermophiles could be differentiated from *Clostridium* species. Phylogenetic analysis based on partial 16S rRNA gene sequences of 16 of the strains from clusters A through D grouped them separately from the cellulolytic thermophilic species of the genus *Clostridium* (Rainey et al., 1993a). Comparison of seven full 16S rRNA gene sequences of strains from clusters A through D with other thermophilic anaerobic bacteria demonstrated that they warranted novel genus status (Rainey et al., 1993b). The genus *Caldicellulosiruptor* was described for the strain Tp8T 6331 with the species *Caldicellulosiruptor saccharolyticus* designated the type species (Rainey et al., 1994). Since then four additional species have been added to the genus based on 16S rRNA gene sequence comparisons, phenotypic differences and DNA–DNA hybridization studies (Bredholt et al., 1999; Huang et al., 1998;

Mladenovska et al., 1995; Nielsen et al., 1993; Onyenwoke et al., 2006). Figure 256 shows the phylogenetic relationships of the type strains of the species of *Caldicellulosiruptor*, as well as a number of undescribed strains for which full 16S rRNA gene sequences are available. The 16S rRNA gene sequence similarities between the described species of *Caldicellulosiruptor* are in the range of 95.5–99.6, with the most closely related species being *Caldicellulosiruptor kristjanssonii* and *Caldicellulosiruptor acetigenus* (clone 1). These two species share 99.6% 16S rRNA gene sequence similarity, but it should be noted that the type strain of the species *Caldicellulosiruptor acetigenus* contains two copies of the 16S rRNA gene which share only 98.87% similarity (Onyenwoke et al., 2006). This considerably large difference between the two gene copies results in the 16S rRNA gene copy designated clone 2 (AY772477) being more similar (99.43%) to the type strain of *Caldicellulosiruptor lactoaceticus* than to the other 16S rRNA gene copy designated clone 1 (AY772476) (98.78% similarity). DNA–DNA hybridization studies demonstrated the true species status of *Caldicellulosiruptor acetigenus* when compared to *Caldicellulosiruptor kristjanssonii*, *Caldicellulosiruptor lactoaceticus* and *Caldicellulosiruptor saccharolyticus*, with reassociation values of 53.1, 50.9 and 34.3 respectively (Onyenwoke et al., 2006). The inclusion of the additional strains of the genus for which full 16S rRNA gene sequences are available (Figure 256) clearly demonstrates the species status of some of these strains including those not validly named, "*Anerocellum thermophilum*" Z-1203 (Svetlichnyi et al., 1990) and "*Thermoanaerobacter cellulolyticus*" NA10 (Taya et al., 1988, 1985).

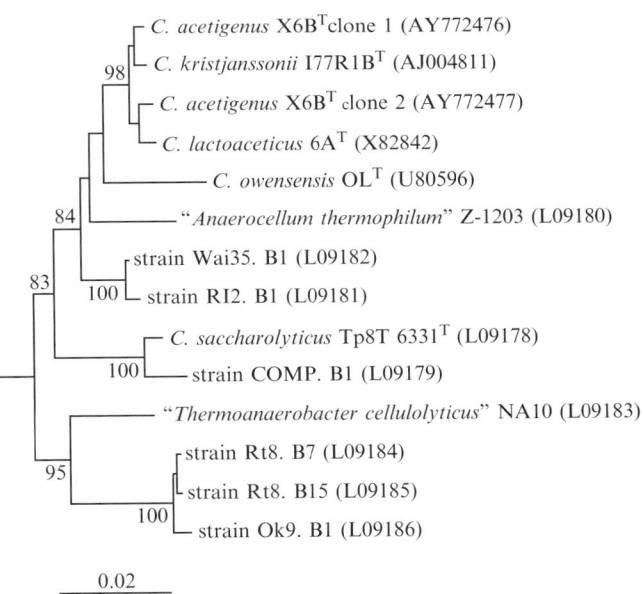

FIGURE 256. 16S rRNA based phylogeny of the species and strains of the genus *Caldicellulosiruptor*. The tree was constructed from distance matrices using the neighbor-joining method. The scale bar represents 2 nucleotide substitutions per 100 nucleotides.

List of species of the genus *Caldicellulosiruptor*

1. **Caldicellulosiruptor saccharolyticus** Rainey, Donnison, Janssen, Saul, Rodrigo, Bergquisy, Daniel, Stackebrandt and Morgan 1995, 197[VP] (Effective publication: Rainey, Donnison, Janssen, Saul, Rodrigo, Bergquist, Daniel, Stackebrandt and Morgan 1994, 265.)

sac.cha.ro.ly'ti.cus. Gr. n. *sakchar* sugar; Gr. adj. *lutikos* able to loosen; N.L. masc. adj. *saccharolyticus* breaking up polysaccharides.

Cells are straight rods 0.4–0.6 μm in diameter and 3.0–4.0 μm in length, occurring both singly and in pairs. Endospores are not formed. On 2/1 cellobiose agar colonies are 2–5 mm in diameter, cream-colored and umbonate. The optimum temperature for growth is 70 °C in the range of 45–80 °C. The pH range is 5.5–8.0, with an optimum at pH 7.0. The main fermentation products on 2/1 cellobiose medium are acetate and lactate with trace amounts of ethanol. Hydrogen sulfide is not produced. Acid is produced from arabinose, amorphous cellulose, Avicel, cellobiose, fructose, galactose, glucose, glycogen, gum guar, gum locust bean, lactose, laminarin, lichenin, maltose, mannose, pullulan, pectin, rhamnose, Sigmacell 20, Sigmacell 50, Sigmacell 100, starch, sucrose, xylan, and xylose. Acid is not produced from amygdalin, erythritol, glycerol, inositol, inulin, mannitol, melibiose, melezitose, raffinose, ribose, sorbitol, or sorbose. Culture supernatants hydrolyse carboxymethylcelluose, gum guar, gum locust bean, laminarin, lichenin, pullulan, starch, and xylan, but not arabinoglactan, chitin, glycogen, gum karaya, inulin, or mannan. Growth is inhibited by penicillin, neomycin and polymyxin B at 10ug/ml and by novobiocin at 100 μg/ml. Rifampin is not inhibitory at 1000 μg/ml. *Caldicellulosiruptor saccharolyticus* was isolated from pinewood in the flow of a geothermal spring, Taupo, New Zealand.

DNA G+C content (mol%): 37.5 (T_m).

Type strain: Tp8T 6331, ATCC 43494, DSM 8903.

GenBank accession number (16S rRNA gene): L09178.

2. **Caldicellulosiruptor acetigenus** (Nielsen, Mathrani and Ahring 1993) Onyenwoke, Lee, Dabrowski, Ahring and Wiegel 2006, 1394[VP] (*Thermoanaerobium acetigenum* Nielsen, Mathrani and Ahring 1993, 464)

a.ce.ti'ge.nus. L. n. *acetum* vinegar; L. v. *genere gignere* to produce; N.L. masc. adj. *acetigenus* vinegar- or acetic acid-producing.

Cells are rod-shaped, 0.7–1.0 μm in diameter and 3.6–5.9 μm in length, occurring singly, in pairs or in chains of up to eight cells. *Caldicellulosiruptor acetigenus* strains are Gram-stain-negative, but have a Gram-positive cell-wall structure. On solidified xylan-containing medium, off-white, milky-colored colonies are observed. It is a strictly anaerobic chemoorganoheterotroph.

The optimum temperature for growth is 65–68 °C in the range of 50–78 °C. The pH range is 5.2–8.6, with an optimum at pH 7.0. Doubling time under optimal conditions is approximately 4h. Growth is supported by arabinose, cellobiose, fructose, D-galactose, D-glucose, lactose, maltose, mannose, raffinose, soluble starch, sucrose, trehalose, D-xylose, and xylan. Growth and CMC-cellulase activity is observed when grown on carboxymethylcellulose (Hercules CMC, 7LT, or 7M) in the presence of traces of yeast extract, but not

with filter paper or crystalline cellulose (Avicel). The end products of growth on glucose or D-xylose include acetate, CO_2, H_2, ethanol and traces of isobutyric acid (but not lactate). Isolated from a combined biomat and sediment sample from a slightly alkaline hot spring at Hverðagerdi, Iceland.

DNA G+C content (mol%): 35.7 (HPLC).

Type strain: X6B, ATCC BAA-1149, DSM 7040.

GenBank accession number (16S rRNA gene): AY772476, AY772477.

3. **Caldicellulosiruptor kristjanssonii** Bredholt, Sonne-Hansen, Nielsen, Mathrani and Ahring. 1999, 995[VP]

kris.tjans.son'ii. N.L. gen. n. *kristjanssonii* named after Jakob K. Kristjansson.

Cells are rod-shaped with rounded ends, 0.7–1.0 μm in diameter and 2.8–9.4 μm in length, occurring singly, in pairs or in short chains, and are Gram-stain-negative. Endospores are not found. Motility is not observed, but two subterminal flagella are present. When grown in roll-tubes on mineral medium plus Avicel, colonies are flat, cream in color, with a fringed edge. After 9–14d, clearing zones around colonies are 2–4mm. The optimum temperature for growth is 78 °C with a range of 45–82 °C. The pH range is 5.8–8.0, with an optimum at pH 7.0. Obligately anaerobic chemoorganotroph. Growth is supported by Avicel, cellobiose, dextrin, D-fructose, D-galactose, D-glucose, lactose, maltose, mannose, pectin, salicin, soluble starch, sucrose, trehalose, xylan, and xylose. It does not utilize D-ribose, L-arabinose, esculin, glycerol, inulin, lactic acid, mannitol, pyruvate, raffinose, L-rhamnose, D-ribose, sorbitol, casein peptone, or yeast extract. Growth is inhibited by air, chloramphenicol, neomycin, penicillin, streptomycin, tetracycline, and vancomycin. The major end products of growth are acetic acid, H_2 and CO_2, with smaller amounts of lactic acid and ethanol and trace amounts of formic acid. Hydrogen sulfide is not produced from sulfate, thiosulfate, or casein peptone. *Caldicellulosiruptor kristjanssonii* was isolated from microbial mats and cellulosic materials, e.g., wood or straw, in Icelandic thermal and slightly alkaline springs.

DNA G+C content (mol%): 35.0 (HPLC).

Type strain: I77R1B, ATCC 700853, DSM 12137.

GenBank accession number (16S rRNA gene): AJ004811.

4. **Caldicellulosiruptor lactoaceticus** Mladenovska, Mathrani and Ahring 1997, 1274[VP] (Effective publication: Mladenovska, Mathrani and Ahring 1995, 229.)

lac.to.acet.i'cus. N.L. adj. *acidum lacticum* lactic acid; L. neut. n. *acetum* vinegar; N.L. neut. adj. *lactoaceticum*, referring to production of lactic acids as major fermentation products.

Cells are Gram-stain-negative, rod-shaped with rounded ends, 0.7 μm in diameter and 1.5–3.5 μm in length, occurring singly, in pairs or in short chains. Endospores are not formed. Cells are non-motile and flagella are not observed. Colonies are 1mm in diameter, round, white, opalescent and circular with an entire edge after incubation in agar roll tubes for approximately 1 week. The optimum temperature for growth is 68 °C with a range of 50–78 °C. The pH range is >5.6–<9.0, with an optimum at pH 7.0. Cells are obligately anaerobic and chemoorganoheterotrophic. Growth occurs with Avicel, cellobiose, lactose, maltose, pectin, soluble starch, xylan, and xylose. No growth occurs on acetate, arabinose, arbutin, casein peptone, dextran, ethanol, fructose, galactose, glucose, lactate, mannitol, mannose, methanol, pyruvate, D-ribose, raffinose, rhamnose, sorbitol, sucrose, trehalose, yeast extract, or $H_2 + CO_2$. Fermentation products are lactate and acetate with small amounts of ethanol, CO_2, and H_2 are produced. Sulfate, thiosulfate, and sulfite are not reduced. Nitrate does not stimulate growth. No growth occurs in the presence of ampicillin, chloramphenicol, kanamycin, neomycin, streptomycin sulfate, penicillin V, tetracycline, or vancomycin (at 10 μg/ml). Cells are resistant to D-cycloserine (100 μg/ml). *Caldicellulosiruptor lactoaceticus* was isolated from microbial mats and cellulosic materials, e.g., wood or straw, in slightly alkaline Icelandic thermal springs.

DNA G+C content (mol%): 35.2 (HPLC).

Type strain: 6A, DSM 9545.

GenBank accession number (16S rRNA gene): X82842.

5. **Caldicellulosiruptor owensensis** Huang, Patel, Mah and Baresi 1998, 95^VP

o.wen.sen'sis. N.L. adj. *owensensis* from Owens Lake, CA, USA.

Cells are rod-shaped, 0.5–0.8 μm in diameter and 2.0–5.0 μm in length, occurring singly, in pairs or in chains and arenonmotile, but have lophotrichous flagella. Gram-stain-negative.

Endospores are not formed. Colonies are <2 mm in diameter, circular with smooth edges, convex, opaque and yellowish.

The optimum temperature for growth is 75 °C with a range of 50–80 °C. The pH range is 5.5–9.0, with an optimum at pH 7.5. Cells are alkali tolerant. Yeast extract or vitamin solutions are not required for growth. Growth is inhibited by penicillin G, streptomycin, chloramphenicol, and ampicillin at 100 μg/ml. Cells are resistant to D-cycloserine (100 μg/ml), erythromycin (200 μg ml/l) and tetracycline (100 μg/ml). Growth is strictly anaerobic. Chemoorganotrophic. Growth occurs with arabinose, cellobiose, cellulose, dextrin, fructose, galactose, glucose, glycogen, inositol, lactose, mannitol, maltose, mannose, pectin, raffinose, rhamnose, ribose, starch, sucrose, tagatose, xylan, xylose, and yeast extract. It does not grow on acetate, amygdalin, arbutin, erythritol, glycerol, lactate, melezitose, melibiose, methanol, pyruvate, sorbitol, trehalose, trypticase peptone, or $H_2 + CO_2$. The end products from glucose fermentation are acetate, lactate, ethanol, H_2, and CO_2. Cells do not reduce nitrate, sulfate, sulfite or thiosulfate. *Caldicellulosiruptor owensensis* was isolated from the Owens Lake in California, USA.

DNA G+C content (mol%): 36.6 (Bd).

Type strain: OL, ATCC 700167, DSM 13100.

GenBank accession number (16S rRNA gene): U80596.

Genus II. **Thermoanaerobacterium** Lee, Jain, Lee, Lowe and Zeikus 1993a, 48^VP

Rob U. Onyenwoke and Juergen Wiegel

Ther.mo.an.ae.ro.bac'te.ri.um. Gr. n. *thermos* hot, Gr. pref. *an* not, Gr. n. *aer* air, Gr. n. *bakterion* a small rod, N.L. neut. n. *Thermoanaerobacterium* a rod which grows in the absence of air at high temperatures.

Extreme thermophiles with growth temperature optima between 55 and 70 °C. Cells have **Gram-positive cell-wall structure**, but many strains are **Gram-stain-negative.** Cells are rod-shaped and motile by peritrichously inserted flagella, except for *Thermoanaerobacterium aciditolerans*, which has a single, polar flagellum. **Endospores are present in some species.** Hexagonal S-layer typically present. However, it may be difficult to visualize because of extracellular hydrolytic enzymes attached to the cells. **Obligate anaerobes** and catalase-negative.

The temperature growth range for the members of the genus is extremely broad (35–75 °C; for discussion of this topic, see Wiegel (1990, 1998). The pH range for growth is equally broad, 3.2–8.5. The lowest pH optimum is 5.2 for *Thermoanaerobacterium aotearoense*, which is the lowest pH optimum for a validly published, anaerobic, thermophilic bacterium. Similarly, *Thermoanaerobacterium aciditolerans* has the lowest minimum pH for growth, 3.2. The highest pH optimum is 7.8–8.0 for *Thermoanaerobacterium thermosaccharolyticum*. No true alkalithermophilic species have been isolated so far (Table 253). **Chemo-organotrophs** and yeast extract stimulates growth for most species. Fermentation end products from hexoses (unless noted otherwise below) are: acetic acid, ethanol, lactic acid, H_2, and CO_2 in various stoichiometries. In some instances, butyrate and butanol are formed (e.g., *Thermoanaerobacterium thermosaccharolyticum*). All strains with the exception of *Thermoanaerobacterium thermosaccharolyticum* contain menaquinone-7 as the major component of their isoprenoid quinone system. Members of the genus include some well-characterized species, such as *Thermoanaero-*

bacterium thermosaccharolyticum (basonym *Clostridium*) or the "swelling can food spoilers" (Collins et al., 1994; Dotzauer et al., 2002; McClung, 1935; Prevot, 1938). A comparison of the properties of the various *Thermoanaerobacterium* species is given in Table 253.

DNA G+C content (mol%): 29–46.

Type species: **Thermoanaerobacterium thermosulfurigenes** (Schink and Zeikus 1983a) Lee, Jain, Lee, Lowe and Zeikus 1993a, 48^VP (*Clostridium thermosulfurogenes* Schink and Zeikus 1983a, 1156).

Further descriptive information

Sporulation. Sporulation has only been observed in some of the species, and it is not regarded as a reliable taxonomic marker because a mutation in one of the many sequentially operating gene products can prevent a species from sporulating. Microorganisms that possess all or most of the sporulation genes but fail to sporulate are termed "asporulating" (Onyenwoke et al., 2004). Common sporulation genes are present in some of the asporulating *Thermoanaerobacterium* species (Brill and Wiegel, 1997; Onyenwoke et al., 2004; unpublished results). The inactivation kinetics of *Thermoanaerobacterium thermosaccharolyticum* endospores during pressure-assisted thermal processing have been studied to better understand the endospore inactivation (Ahn et al., 2007).

Thiosulfate metabolism. The reduction of thiosulfate to elemental sulfur instead of H_2S was once considered a feature

TABLE 253. Comparison of properties among *Thermoananerobacterium* species[a,b]

Characteristics	*T. aciditolerans*	*T. aotearoense*	*T. polysaccharolyticum*	*T. saccharolyticum*	*T. thermosaccharolyticum*	*T. thermosulfurigenes*	*T. xylanolyticum*	*T. zeae*
Spores	+	+	–	–	+	+	+	–
Gram stain	NA	–	v	–	–	–	–	v
Optimum temp. (°C)	55	60–63	65–68	60	55–62	60	60	65–70
pH optimum	5.7	5.2	6.8–7.0	6.0	7.8	5.5–6.5	6.0	ND
DNA G+C content (mol%)	34	34.5–35	46	36	29–32	32.6	36.1	42
Products of glucose fermentation:								
Butyric acid	–	–	–	–	+	–	–	–
Formic acid	–	–	+	–	–	–	–	+
Lactic acid	+	+	+	+	+	+	–	–
Succinic acid	–	–	–	–	+	–	–	–
Pectin fermented	ND	+	–	–	+	+	–	–
Lactose fermented	+	+	+	+	+	–	+	+
Thiosulfate reduction to:								
H_2S	–	–	+	–	+	–	–	–
S^0	+	+	–	+	–	+	+	–

[a]Symbols and abbreviations: +, 90% or more of strains are positive; –, 10% or less of strains are positive; NA, not available although electron microscopy indicates a Gram-positive cell-wall structure; v, variable; ND, not determined.

[b]All species are motile and produce acetic acid, ethanol, H_2, and CO_2 when fermenting glucose.

that distinguished this genus from *Thermoanaerobacter* (Lee et al., 1993a). With the reclassification *Clostridium thermosaccharolyticum*, which produces sulfide from thiosulfate, to *Thermoanaerobacterium* (Collins et al., 1994) and the isolation of *Thermoanaerobacter sulfurigignens*, which only produces sulfur (Lee et al., 2007a, 2007b), this generalization proved unreliable. Sulfur granules in both *Thermoanaerobacter sulfurigignens* and *Thermoanaerobacterium thermosulfurigenes* consist of linear sulfur chains with organic residues at the ends (Lee et al., 2007). Meyer et al. (1990) suggest that a rubredoxin might have a role in cellular sulfur metabolism in *Thermoanaerobacterium thermosaccharolyticum*.

Cell structure. Most *Thermoanaerobacterium* species contain an S-layer. The *Thermoanaerobacterium thermosaccharolyticum* S-layer is unusual because it lacks a covalent linkage between the S-layer glycan, the peptidoglycan, and the S-layer protein (Schaffer et al., 2000, and literature cited therein). The molecular mass of the monomeric subunit of the major S-layer protein from strain E207-71 was determined to be ~75 kDa (Allmaier et al., 1995). Altman et al. (1990) determined the chemical properties from the S layer units from *Clostridium thermosaccharolyticum* D120-70 and Cejka and Baumeister (1987) determined the three-dimensional structure.

Cytoskeletal elements of *Thermoanaerobacterium thermosaccharolyticum* where observed by immunoelectron microscopy and proposed to play roles in maintaining cell shape as well as forming membrane vesicles for release of starch-hydrolyzing enzymes (Mayer et al., 1998; Mayer, 2006, and literature cited therein).

Enzymes and industrial applications. The thermostable enzymes produced by *Thermoanaerobacterium* species have the potential for important industrial applications. A few examples are given below.

The xylose isomerase can be used as a glucose isomerase for fructose production. The enzyme from *Thermoanaerobacterium saccharolyticum* (Lee et al., 1993d) and *Thermoanaerobacterium thermosulfurigenes* 4B is highly thermostable (Lee et al., 1990a; Lee and Zeikus, 1991) and has been heterologously expressed in *Escherichia coli* and *Bacillus* (Lee et al., 1990b). The crystal structure was determined (Huber et al., 1996; Kim et al., 2001; Lloyd et al., 1994), which allowed further optimization of this industrial enzyme (Sriprapundh et al., 2000). The role of the aromatic amino acids for the catalysis and substrate specificity (Meng et al., 1993b) as well as thermal stabilization has been elucidated (Meng et al., 1993a). In contrast to the commercial enzymes, the xylose isomerase from *Thermoanaerobacterium saccharolyticum* JW/SL-YS489 exhibits an acidic pH optimum. Thus, it is more suitable for industrial applications (Liu et al., 1996c) because it is no longer necessary to change the pH after saccharification. Comparison with the xylose isomerase from strain 4B suggests that the lower pH optimum results from substitutions that are outside of the catalytic site. Divalent metal cations such as Co^{2+} are not only important for the catalysis but also for enzyme stability (Epting et al., 2005). Recombinant xylose isomerases have also been used for saccharidase synthesis (Lee et al., 1990a, 1993d) and single-step processes for sweetener production (Lee and Zeikus, 1991). The gene for the D-xylose ketol-isomerase of *Thermoanaerobacterium thermosulfurigenes* has also been proposed as a selectable genetic marker within transgenic plant cells (Kim et al., 2001; Meng et al., 1993b, 1993a).

Since the solubility of starch increases with temperature, *Thermoanaerobacterium* species (including strains of *Thermoanaerobacterium thermosulfurigenes*, *Thermoanaerobacterium thermosaccharolyticum*, and *Thermoanaerobacterium saccharolyticum*) are of considerable interest for production of thermostable and thermoactive amylolytic enzymes. This includes glucoamylase (Specka et al., 1991), pullulanase, glucoamylase, and α-glucosidase from *Thermoanaerobacterium thermosaccharolyticum* (Feng et al., 2002; Ganghofner et al., 1998); and pullanase (Burchhardt et al., 1991) and amylopullulanase (Saha et al., 1990) from *Thermoanaerobacterium thermosulfurigenes* EM1. Some genes have also been heterologously expressed, including the α-amylase (Haeckel and Bahl, 1989), glycosyl hydrolases (Matuschek et al., 1994) and associated ABC transporter (Matuschek et al., 1997), and pullanase (Matuschek et al., 1996) from *Thermoanaerobacterium thermosulfurigenes* EM1. From *Thermoanaerobacterium thermosaccharolyticum*, the endo-acting amylopullanase has also been heterologously expressed (Ramesh et al., 1994). The crystal structures of the glucoamylase from *Thermoanaerobacterium thermosaccharolyticum* (Aleshin et al., 2003) and the cyclodextrin glycosyltransferase of *Thermoanaerobacterium thermosulfurigenes* EM1 have been resolved at 2.3 Å (Knegtel et al., 1996; Wind et al., 1998, and literature cited therein; Leemhuis et al., 2003). The structural information was then used to engineer the glycosyltransferase for specific changes in product formation and pH optima. Hyun and Zeikus (1985) described the enhanced formation of the *Thermoanaerobacterium thermosulfurigenes* amylase during ethanol production. For *Thermoanaerobacterium thermosulfurigenes* strain EM1, the saccharolytic enzymes are excreted via membrane vesicles that are shed during cell wall renewal (Antranikian et al., 1987). The C-termini of the glucosyl hydrolases contain repeated sequences similar to repeated units in the S-layer proteins, suggesting that the enzymes intercalate into the S-layer and thus remain loosely associated with the cell envelope (Matuschek et al., 1996). Presently, it is not clear whether these two mechanisms for excreting hydrolytic enzymes onto the cell surface and localizing extracellular enzymes are general strategies among related thermophilic anaerobes.

The kinetic properties and structure of the β-galactosidase from *Thermoanaerobacterium thermosulfurigenes* EM1 has been reported (Burchhardt and Bahl, 1991; Zolotarev et al., 2003). Xylan and hemicellulose are hydrolyzed and the resulting xylose is fermented by nearly all *Thermoanaerobacterium* species. A number of hemicellulolytic enzymes have been studied in detail. *Thermoanaerobacterium saccharolyticum* JW/SL-YS485 produces high-molecular-weight xylanases which act synergistically with other hydrolytic enzymes (Liu et al., 1996a; Shao et al., 1995a), three xylosidases (Lorenz and Wiegel, 1997; Shao and Wiegel, 1995), glucuronidase (Shao et al., 1995b), and two acetylesterases (Shao and Wiegel, 1995), one of which has cephalosporin C deacetylase activity (Lorenz and Wiegel, 1997). The high-molecular-weight xylanase, similar to the glycosyl hydrolases from *Thermoanaerobacterium sulfurigenes* EM1 (Matuschek et al., 1996), possesses C-terminal repeated modules similar to the modules from the S-layer protein, suggesting that these enzymes also intercalate into the S-layer. This might explain why the enzyme can only be effectively isolated from whole cells and not from the culture supernatant. It also suggests that the enzymes and the reaction products remain associated the cells, a situation similar to that of the cellulosome of another thermophilic anaerobe *Clostridium thermocellum*. The hemicellulolytic system of *Thermoanaerobacterium saccharolyticum* B6A-RI[T] has also been studied in detail and it differs from the one in strain JW/SL-YS485 (Hespell, 1992; Lee et al., 1993b, 1993c; Weimer, 1985). The enzymes studied include the recombinant endoxylanase (Lee et al., 1993c) and β-D-xylosidase (Lee and Zeikus, 1993; Yang et al., 2004). Armand et al. (1996) studied in detail the stereochemistry of the xylanases, and Vocadlo et al. (2002a, 2002b) showed that xylosidase is the acid/base catalyst facilitating the reverse protonation of glutamine residue 160. The three-dimensional structure of the β-D-xylosidase from *Thermoanaerobacterium saccharolyticum* B6A-RI[T], a family 39 glucuronidase, has been determined by Yang et al. (2004). Bronnenmeier et al. (1995) and Haldrup et al. (1998) described the second characterized thermophilic glucuronidase from another strain, *Thermoanaerobacterium thermosaccharolyticum* HG-8. While the sugar transport systems in *Thermoanaerobacterium* have not received much attention, the binding protein-dependent maltose transport system in *Thermoanaerobacterium thermosulfurigenes* EM1 has been characterized (Sahm et al., 1996).

A new enzyme-coupled assay for the quantification of D-xylose was developed from the enzymes of *Thermoanaerobacterium saccharolyticum* JW/SL-YS 485 (Wagschal et al., 2005). The assay uses readily available enzymes and allows for the kinetic evaluation of hemicellulolytic enzymes leading to xylose formation from xylo-oligosaccharide substrates.

Mannases have been rarely determined in this group of bacteria. *Thermoanaerobacterium polysaccharolyticum* produces a highly active multidomain enzyme exhibiting both mannanase and endoglucanase activity (Cann et al., 1999). Furthermore, its thermostable α-galactosidase is useful for pre-treating food ingredients that can elicit gastrointestinal disturbances if not modified by this enzyme (King et al., 2002).

Pectin is degraded at high rates by *Thermoanaerobacterium thermosulfurigenes*, which has very active pectinases, a high ethanol/lactate ratio during growth on pentoses, and polygalacturonase hydrolase activity. A pectin methylesterase leads to the formation of methanol, which is a typical product for anaerobic pectin degradation but otherwise a rare fermentation endproduct (Schink and Zeikus, 1980).

Thermoanaerobacterium thermosaccharolyticum has been implicated in hydrogen gas production during the thermophilic conversion of food waste and an artificial garbage slurry in a continuously stirred tank reactor (Shin and Youn, 2005; Ueno et al., 2006). It has also been used for production of 1,2-propanediol (1,2-PD; Sanchezriera et al., 1987; Cameron et al., 1998; Altaras et al., 2001; and literature cited therein). Co-cultures of *Thermoanaerobacterium thermosaccharolyticum* and *Clostridium thermocellum* have been tested for their potential to produce ethanol from hydrolyzed cellulose (Hyun and Zeikus, 1985; Lynd, 1996; Vancanneyt et al., 1987; Venkateswaran and Demain, 1986). The ethanol tolerance and ethanol yields of *Thermoanaerobacterium thermosaccharolyticum* have been increased through adaption (Saddler and Chan, 1984), using cell recycling (Mistry and Cooney, 1989) during continuous culture fermentation (Baskaran et al., 1995a), and genetic manipulation (Klapatch et al., 1994; Simpson and Cowan, 1997), including the deletion of the lactate dehydrogenase gene (Desai et al., 2004). Lynd et al. (2001) showed that high concentrations of ethanol do not limit growth of *Thermoanaerobacterium thermosaccharolyticum* HG-8 at elevated xylose concentrations. Instead, salt accumulation resulting from the addition of base to control pH inhibits growth.

Since its growth rate is rapid and its substrate range is extensive at slightly acidic pH values (pH$_{opt}^{25C}$ 5.5), *Thermoanaerobacterium aotearoense* may be advantageous for the biodegradation of influent anaerobic granular sludge rich in carbohydrate (Hernon et al., 2006). *Thermoanaerobacterium* species—like many other thermophilic *Firmicutes*—contain heat-stable ferredoxins (e.g., *Thermoanaerobacterium thermosaccharolyticum*; Wilder et al., 1963; Tanaka et al., 1973). *Thermoanaerobacterium thermosulfurigenes* EM1 synthesizes four heat-shock proteins at elevated levels between 10 and 15 min. after a temperature shift-up from 50 °C to 62 °C (Narberhaus et al., 1994).

Genetics and molecular manipulations. *Thermoanaerobacterium saccharolyticum* may be a useful host for expression of recombinant genes from other thermophiles, and vectors have been constructed for this purpose (Mai and Wiegel, 2001; Tyurin et al., 2006). The system is based on a procedure for regenerating autoplasts of *Thermoanaerobacter* developed by Peterandel et al. (1993). It has been adapted for other thermophiles with a Gram-positive cell-wall structure (Tyurin et al., 2006, and literature cited therein), although less successfully for *Thermoanaerobacter ethanolicus* JW200^T (Wiegel and Mai, unpublished data). Plasmids from *Thermoanaerobacterium thermosaccharolyticum* DSM 571^T have been mapped and characterized (Delver et al., 1996; Belogurova et al., 2002, and literature cited therein). Klapatch et al. (1996b) described a procedure for the electrotransformation of *Thermoanaerobacterium thermosaccharolyticum* HG-8. During electrotransformation of *Thermoanaerobacterium saccharolyticum* JW/SL-YS 485 (as well as of *Clostridium thermocellum* JW20), spontaneously arising current oscillations of about 24 MHz occur at field strengths of 10 kV/cm that have a large beneficial effect on transformation efficiency. The maximum tested field strength that did not repress the current oscillation (25 kV/cm) resulted in the highest transformation efficiency (Tyurin et al., 2005, 2006). Klapatch et al. (1996a) reported that a specific restriction endonuclease activity in cell extracts of *Thermoanaerobacterium thermosaccharolyticum* HG-8 could interfere with the genetic manipulations. The utility of a CAK1-derived phagemid, a proposed gene transfer system for the clostridia, was used to express recombinant *Thermoanaerobacterium polysaccharolyticum manA* in *Escherichia coli* (Li and Blaschek, 2002).

The β-galactosidase gene from *Thermoanaerobacterium thermosulfurigenes* EM1 has been used as a reporter for characterization of promoters required for solventogenesis and acidogenesis in *Clostridium acetobutylicum* (Feustel et al., 2004). Mutagenesis of *Thermoanaerobacterium thermosulfurigenes* EM1 illustrates the importance of surface layer homology domains in the attachment of proteins to bacterial cell walls (May et al., 2006).

The unambiguous description of plasmids is rare among the thermophilic anaerobic *Firmicutes*. Belogurova et al. (2002, 1991) characterized a plasmid from *Thermoanaerobacterium thermosaccharolyticum*, which is, so far, the only one described from *Thermoananerobacterium*.

Habitats. Known habitats include geothermal and man-made thermobiotic environments, such as slightly alkaline, neutral and slightly acidic hot springs, high temperature petroleum reservoirs, organic waste piles, industrial streams at elevated temperatures containing carbohydrates (Wiegel, unpublished results), and thermal volcanic algal-bacterial mats as well as some mesobiotic environments, such as various soils (including agricultural fields), sediments of slightly alkaline and acid springs, tartrate infusion of grape residues, pond and river sediments, and fruit juice waste products.

Enrichment and isolation procedures

Isolation procedures for *Thermoanaerobacterium* include enrichments using medium containing yeast extract and glucose, with subsequent isolation on agar plates or agar shake-roll tubes at 60 °C. A more specific isolation scheme for *Thermoanaerobacterium* employs an enrichment at 60 °C and a medium containing 0.1% (w/v) yeast extract, 0.5% (w/v) xylose, and 15–25 mM thiosulfate. Subsequent plating on this same medium solidified with 2% (w/v) agar yields a high percentage of colonies containing extracellular sulfur globules and cells of *Thermoanaerobacterium* containing intracellular sulfur-globules (Wiegel and Liu, unpublished results).

Maintenance procedures

The most reliable method for preservation is storage in liquid nitrogen. Cultures may also be freeze-dried for long-term storage. Some species have also been stored for up to 20 years in pre-reduced medium containing 40–45% (v/v) glycerol as cryoprotectant at −20 °C (unpublished observation). Because the suspension remains liquid, aliquots are easily transferred into subcultures via sterile syringes. For longer term storage, the suspension is kept at −80 °C. Working stock cultures can be maintained by transfer into the appropriate liquid medium at regular intervals of 1–2 weeks, although most species can be maintained for 3–6 months at room temperature after they have reached the late-exponential or early stationary growth phase. Sporulating strains can be maintained on agar slants after growth on low concentrations (0.10% w/v or less) of slowly utilizable substrates or after harvesting the spores. The spores are resuspended in mineral medium or 20 mM (Na, K, H)$_3$-phosphate buffer, pH^{25C} around 7.0, and stored at 4 °C or room temperature. Spores can even be maintained under aerobic conditions for years at 4–20 °C (Wiegel, unpublished results).

Taxonomic comments

The genus *Thermoanaerobacterium* is currently classified within Family III *Incertae Sedis* of the order *Thermoanaerobacterales* ord. nov. (Ludwig et al., 2009). Other genera within this group include *Caldicellulosiruptor*, *Thermosediminibacter*, and *Thermovenabulum*. However, the evidence for an affiliation of *Thermoanaerobacterium* with these genera is not strong, and a reclassification is warranted in the future.

Hamana et al. (2001) and Hosoya et al. (2004) have described the use of polyamines to differentiate the "*Thermoanaerobacteriales*" from thermophiles in the orders *Clostridiales* and *Bacillales*.

Lastly, *Thermoanaerobacterium polysaccharolyticum* and *Thermoanaerobacterium zeae* (Cann et al., 2001) have recently been reassigned to a novel genus *Caldanaerobius*, the type species of which is *Caldanaerobius fijiensis* (Lee et al., 2008). This reclassification is based upon the low sequence similarity of their 16S rRNA genes, less than 90% similarity to the genes from *Thermoanaerobacterium*, *Caldanaerobacter*, and *Thermoanaerobacter* (Figure 257). It is supported

FIGURE 257. Neighbor-joining tree showing the estimated phylogenetic relationships for the type strains of each species within the genus *Thermoanaerobacterium* based on 16S rRNA gene sequence data with Jukes–Cantor correction for synonymous changes. The 16S rRNA gene data used represent *Bacillus subtilis* subsp. *subtilis* DSM 10[T] nucleotide positions 45–1423. Numbers at nodes indicate bootstrap support values for 1000 replicates. Bar indicates 0.05 nucleotide substitutions per site. The superscript "T" denotes the sequence is from the type strain for the species. GenBank accession numbers are given after the strain designations. *Thermoanaerobacterium polysaccharolyticum* and *Thermoanaerobacterium zeae* are indicated by their newly assigned names, i.e., *Caldanaerobius polysaccharolyticus* and *Caldanaerobius zeae*, respectively (Lee et al., 2008).

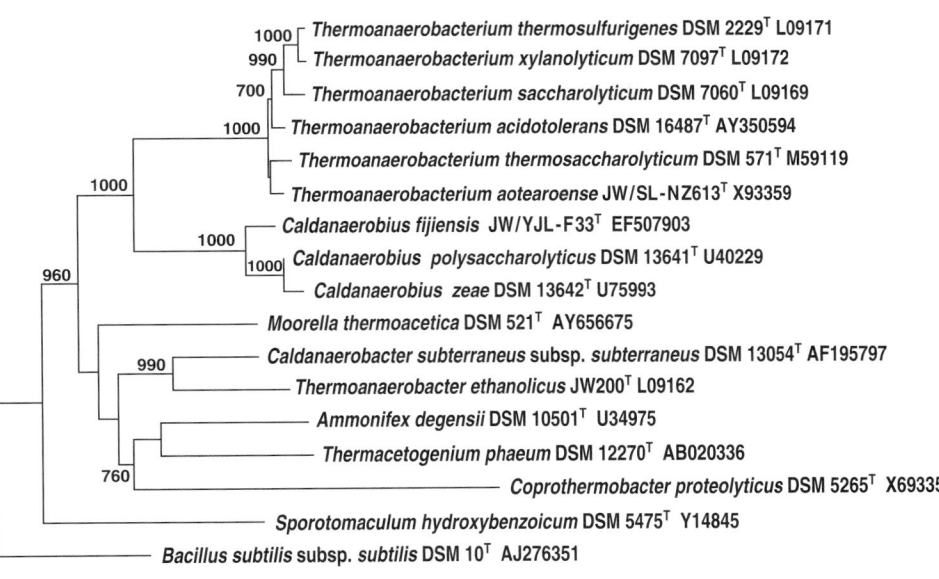

by the higher mol% G+C content of their DNA (42–46 mol%) compared to 29–36 mol% for the remaining species of *Thermoanaerobacterium* as well as the formation of formate, indicating the presence of the pyruvate–formate lyase in these species.

Further reading

Antranikian, G. 1990. Physiology and enzymology of thermophilic anaerobic bacteria degrading starch. FEMS Microbiol. Rev. *75*: 201–208.

Canganella, F. and J. Wiegel. 1993. The potential of thermophilic clostridia in biotechnology. *In* Woods (Editor), The Clostridia and Biotechnology, Butterworths, Stoneham, MA, pp. 391–429.

Lowe, S.E., M.K. Jain and J.G. Zeikus. 1993. Biology, ecology, and biotechnological applications of anaerobic-bacteria adapted to environmental stresses in temperature, pH, salinity, or substrates. Microbiol. Rev. *57*: 451–509.

Mai, V. and J. Wiegel. 1999. Recombinant DNA applications in thermophiles. *In* Demain and Davis (Editors) and Hershberger (Section Editor), ASM Manual of Industrial Microbiology and Biotechnology, 2nd edn. ASM Press, Washington, DC, pp. 511–519.

Rainey, F.A., N.L. Ward, H.W. Morgan, R. Toalster and E. Stackebrandt. 1993. Phylogenetic analysis of anaerobic thermophilic bacteria – aid for their reclassification. J. Bacteriol. *175*: 4772–4779.

Wagner, I.D. and J.Wiegel. 2008. Diversity of thermophilic anaerobes. Ann. N. Y. Acad. Sci. *1125*: 1–43

Wiegel, J. 1990. Temperature spans for growth: A hypothesis and discussion. FEMS Microbiol. Rev. *75*: 155–170.

Wiegel, J. 1992. The obligately anaerobic thermophilic bacteria. *In* Kristjansson (Editor), Thermophilic Bacteria, CRC Press, Boca Raton, FL, pp.105–184.

Wiegel, J. 1998. Lateral gene exchange, an evolutionary mechanism for extending the upper or lower temperature limits for growth of a micro-organism? A hypothesis. *In* Wiegel and Adams (Editors), Thermophiles, the molecular key to the evolution and origin of life? Taylor and Francis, London, pp. 175–185.

List of species in the genus *Thermoanaerobacterium*

1. **Thermoanaerobacterium thermosulfurigenes** (Schink and Zeikus 1983a) Lee, Jain, Lee, Lowe and Zeikus 1993a, 48[VP] (*Clostridium thermosulfurogenes* Schink and Zeikus 1983a, 1156)

ther.mo.sul.fur.i′ge.nes. Gr. adj. *thermos* hot; L.n. *sulfur* brimstone; Gr. suff. *genes* born from; N.L. neut. adj. *thermosulfurigenes* releasing sulfur in heat.

Cells are straight rods (0.5 × >2 μm), which vary in length from 2 μm (single cells) to > 20 μm (filamentous chains). Motile with peritrichous flagella. Stain Gram-negative, but electron micrographs reveal a double-layered cell wall without the presence of an outer membranous layer (Gram-positive cell-wall structure). Other distinctive features include numerous internal membranes that appear vesicular and large and cytoplasmic granules that are electron-dense. The components of the cell envelope of strain EM1 and the

purified S-layer protein have been characterized (Brechtel et al., 1999). In this strain, the S-layer homology domains do not attach to the peptidoglycan (Brechtel and Bahl, 1999). Significant differences in the cell envelope structure were observed when the cells were grown in continuous culture under glucose limitation as compared to starch limitation (Antranikian et al., 1987; Mayer et al., 1998). This phenomenon is likely due to an increase in production of α-amylase and pullulanase during starch limitation (Antranikian et al., 1987; Mayer and Gottschalk, 2004). Agar-embedded colonies are fluffy, 0.5–1.5 mm in diameter, and without pigment. Terminal, spherical, refractile spores with distinctly swollen sporangia are observed when grown with xylose or pectin but never with glucose as the energy and carbon source. Free spores are rare. Catalase-negative. Cytochromes are not present.

Yeast extract and tryptone enhance growth. The optimum temperature is near 60 °C. The minimum and maximum growth temperatures are above 35 °C and below 75 °C, respectively. The pH optimum for growth on glucose is 5.5–6.5. No growth is found below 4.0 or above 7.6. Fermentable carbohydrates include pectin, starch, polygalacturonic acid, amygdalin, esculin, salicin, D-xylose, galactose, glucose, inositol, mannitol, melibiose, rhamnose, trehalose, mannose, cellobiose, maltose, L-arabinose, and sucrose. No growth is found on H_2/CO_2, cellulose, arabinogalactan, galacturonate, citrate, pyruvate, lactate, tartrate, lactose, melezitose, raffinose, D-ribose, sorbitol, methanol, or glycerol. Doubling times on glucose or pectin are approximately 2 h. Sulfate and nitrate are not reduced. Sulfur droplets are localized in the cytoplasm (Liu et al., 1996b). Addition of sulfite inhibits growth and does not lead to the formation of elemental sulfur. Thiosulfate or other inorganic sulfur compounds are not required for growth. Gelatin is liquefied, but indole, acetylmethylcarbinol, and hydrogen sulfide are not produced. Glucose (0.5% w/v) is fermented according to the simplified fermentation balance (Schink and Zeikus, 1983b): 1 glucose → 0.9 ethanol + 0.6 acetic acid + 0.5 lactic acid + 1.5 CO_2 + 1.2 H_2. During pectin fermentation, the acetate/ethanol ratio is 8.6, and the acetate/lactate ratio is 6.8. Less lactate is produced during growth on polygalacturonic acid, and the acetate/lactate ratio is 11.5. *Thermoanaerobacterium thermosulfurigenes* is the main pectin utilizer in hot springs and the major source of methanol. Small amounts of 2-propanol, but not butyrate, are also produced (Schink and Zeikus, 1980).

Cell growth is inhibited by antibiotics (penicillin, streptomycin, cycloserine, tetracycline, or chloramphenicol; 100 µg/ml), sodium azide (500 µg/ml), sodium chloride (2% w/v), O_2 (20.3 kPa), or sulfite. Hydrogen gas does not cause inhibition. The type strain was isolated from an algal-bacterial community of a hot spring at 55–65 °C at Octopus Spring, Yellowstone National Park, Wyoming, USA. The enrichment contained pectin as an energy source (Schink and Zeikus, 1983a). Another strain, EM1 or DSM 3896, has also been obtained from fruit juice waste products in Germany (Madi et al., 1987).

DNA G+C content (mol%): 32.6 (T_m).

Type strain: 4BT, ATCC 33743, DSM 2229.

GenBank accession number (16S rRNA gene): L09171.

2. **Thermoanaerobacterium aciditolerans** Kublanov, Prokofeva, Kostrikina, Kolganova, Tourova, Wiegel and Bonch-Osmolovskaya 2007, 263[VP]

a.ci.di.tol'er.ans. N.L. neut. n. *acidum* an acid; L. pres. part. *tolerans* tolerating; N.L. part. adj. *aciditolerans* acid-tolerating.

Description based on the type strain. Cells are rods, 3–12 × 0.4 µm, and typically sluggishly motile by a single polar flagellum. Cells have a Gram-positive cell-wall structure and form endospores. Moderate acidophile growing at a pH range from 3.2–7.1, with a pH optimum of 5.7 at 55 °C. No growth is observed at or below pH 2.8 and at or above pH 7.5. All pH values reported here are measured at 20 °C. At their optimum pH, strain 761–119[T] grows over a temperature range from 37–68 °C, with an optimum of 55 °C. No

growth is observed within 5 d at or below 35 °C and at or above 70 °C. The doubling time under optimal conditions is 1 h. Grows by fermentation of glucose, maltose, fructose, sucrose, lactose, xylose, ribose, arabinose, galactose, yeast extract, sorbitol, starch, xylan, gelatin, and albumin. Glycerol, ethanol, pyruvate, and citrate are not utilized. Glucose is fermented to acetate, ethanol, and lactate at a ratio of 14.5:9:1. H_2 and CO_2 are also formed. In contrast, *Thermoanaerobacterium aotearoense* forms these same products at ratio of 1:1:0.05. Sulfite is reduced to sulfide. Thiosulfate is converted to sulfur granules, which probably also occurs at low pH due to chemical disproportionation. Growth occurs in the presence of 0–3% (w/v) NaCl. No growth at 4% NaCl. The type strain was isolated from a hydrothermal vent at pH 3.8 and 48 °C in the Orange Field of the Uzon Caldera on Kamchatka in far-eastern Russia (54° 30.237′ N, 160° 00.038′ E). The type strain was isolated under anaerobic conditions on medium containing 2.5% yeast extract and 0.2% sucrose at pH 4.0 and 60 °C.

DNA G+C content (mol%): 34 ±0.5 (T_m).

Type strain: 761-119, DSM 16487, VKM B-2363.

GenBank accession number (16S rRNA gene): AY350594.

3. **Thermoanaerobacterium aotearoense** Liu, Rainey, Morgan, Mayer and Wiegel 1996b, 395[VP]

ao.te.a′ro.en.se. Maori n. *ao* cloud; Maori adj. *tea* white; Maori adj. *roa* long; N.L. adj. *aotearoense* long white cloud, referring to the native Maori name for New Zealand, Aotearoa, Land of the Long White Cloud.

Description is mainly based on the type strain, although several strains have been isolated with less than 0.5% evolutionary distance within their 16S rRNA gene sequences. Rods are 0.7–1.0 × 2.1–14.3 µm, typically peritrichous, and exhibit tumbling motility. Cells stain Gram-negative but have a Gram-positive cell-wall structure covered with hexagonal S-layer lattices. Beginning in the late-exponential growth phase, oval and terminal endospores, 1.4–2.1 × 2.8–2.9 µm, are produced in slightly swollen sporangia by 5–15% of cells. Thiosulfate is converted to intracellular sulfur granules, which are each surrounded by a membrane. Granules are also found in low pH media, but these probably form due to chemical disproportionation. The temperature range and optimum for growth at pH 5.2 is 35–66 °C and 60–63 °C, respectively. The pH range and optimum for growth at 60 °C is pH 3.8–6.8 and pH 5.2, respectively. pH values were measured at 25 °C. Substrates utilized in the presence of 0.055% (w/v) yeast extract are L-arabinose, galactose, cellobiose, glucose, mannose, fructose, lactose, pectin, sucrose, xylose, maltose, starch, rhamnose, pectin, xylan, N-acetylglucosamine, and salacin. Some strains also utilize ribose and raffinose. The fermentation of glucose yields acetate, ethanol, lactate, CO_2, and H_2 in the approximate ratio of 1:1:0.5:2:2. The fermentation of xylose yields the same products in the ratio of 1:1:0.1:2:2. No dissimilatory sulfate reduction with glucose, acetate, and lactate as electron donors.

Thermoanaerobacterium aotearoense has only been isolated from various hot springs in New Zealand (North Island). The type strain was isolated from a small pool in the Weimangu Thermal Valley at the Warbick Terrace.

DNA G+C content (mol%): 34.5–35 (HPLC).

Type strain: JW/SL-NZ613, DSM 10170.
GenBank accession number (16S rRNA gene): X93359.

4. **Thermoanaerobacterium polysaccharolyticum** Cann, Stroot, Mackie, White and Mackie 2001, 299^{VP}

poly.sac.cha.ro.ly'ti.cum. Gr. adj. *polus* many; Gr. n. *sakchar* sugar; Gr. adj. *lutikos* dissolving; N.L. neut. adj. *polysaccharolyticum* dissolving many sugars.

Description based on the type strain. Cells are straight rods occurring singly or in pairs. Tumbling motility by peritrichously inserted flagella. Gram-stain variable, but has a Gram-positive ultrastructure as determined by electron microscopy. Spores not observed. A surface layer-like protein has been observed; however the array structure has not been determined. Cells are catalase-negative. Major fatty acids are $C_{15:0\ iso}$ (70.6%), $C_{17:0\ iso}$ (19.2%), and straight chain $C_{16:0}$ (10.1%).

An obligate anaerobe that tolerates exposure to air, it will only grow in sealed tubes after reduction of the oxidized medium. It only grows in the anaerobic region of a stab culture. Yeast extract is not required for growth. A wide variety of complex and simple carbohydrates are fermented including melibiose, raffinose, arabinose, galactose, lactose, maltose, mannose, rhamnose, sucrose, trehalose, xylose, cellobiose, and melezitose, but not cellulose, starch, xylan, cracked corn, pectin, or sorbose. It also does not use mannitol, malate, fumarate, citrate, glycerol, or H_2/CO_2. The major end products of glucose fermentation are ethanol and carbon dioxide, with H_2, acetate, formate, and lactate formed in lesser amounts. Glucose and xylose are used simultaneously when supplied in equal proportions. Doubling times at 68 °C with glucose, raffinose, and melibiose as the carbon sources are 2.1, 3.4, and 5.5 h, respectively. Nitrate, sulfate, and sulfur are not reduced. Indole is not produced. The optimum temperature for growth is 65–68 °C at pH 6.8 on glucose. The maximum growth temperature is 70 °C. No growth is observed below 45 °C. At 65 °C, the pH range and optimum for growth are 5.0–8.0 and 6.8–7.0, respectively. Strain KMTHCJ was isolated from an organic waste pile at a canning factory in Hoopeston, Illinois, USA, in enrichments containing raffinose as the sole carbon source at 60 °C.

Thermoanaerobacterium polysaccharolyticum has recently been reassigned to the genus *Caldanaerobius* as *Caldanaerobius polysaccharolyticus* (Lee et al., 2008).

DNA G+C content (mol%): 46 (T_m).
Type strain: KMTHCJ, ATCC BAA-17, DSM 13641.
GenBank accession number (16S rRNA gene): U40229.

5. **Thermoanaerobacterium saccharolyticum** Lee, Jain, Lee, Lowe and Zeikus 1993a, 49^{VP}

sac.cha.ro.ly'ti.cum. Gr. n. *sakchar* sugar; Gr. adj. *lutikos* dissolving; N.L. neut. adj. *saccharolyticum* sugar-dissolving.

Description is mainly based upon the type strain B6A-RIT and strain JW/SL-YS 485 (Wiegel and Liu, unpublished result). Cells are rods, 0.8–1.0 × 3.0–30 μm, occurring in chains of varying lengths. Longer cells are formed during nutrient limitation or stationary phase. Cells are motile with peritrichous flagella and stain Gram-negative. Catalase-negative. The cell wall of strain B6A-RI^T contains three electron-dense layers that are 5 nm thick and alternating with electron-transparent layers of similar thickness. Colonies

on agar plates are soft, tan, circular, and convex with hollow centers ("donut"-shaped). Colony diameters range from 0.5–4.0 mm after 4 d at 55 °C. Although spores have not been observed for strain B6A-RI^T, they have been found at a very low frequency for strain JW/SL-YS 485. Cells of the type strain contain a 1.5 Mb plasmid.

Growth is observed with xylan and starch but not cellulose, pectin, ribose, melibiose, melezitose, xylitol, or sorbitol. Other complex and simple carbohydrates fermented include maltose, lactose, sucrose, cellobiose, glucose, xylose, galactose, mannose, fructose, trehalose, rhamnose, raffinose, and mannitol. Landuyt et al. (1995) have also demonstrated growth on paraffin oil. No growth occurs in the absence of a fermentable carbohydrate. The major fermentation products from either glucose or xylan are acetic acid and ethanol in approximately equal amounts. L-rhamnose fermentation yields equimolar amounts of 1,2-propandiol and a mixture of ethanol, acetic acid, lactic acid, H_2, and CO_2.

The pH range for growth is from 5.0 to less than 7.5, and the optimum is about 6.0. The temperature range for growth is 45–70 °C, with an optimum of 60 °C. A culture of the type strain can survive 85 °C for 15 min but not 90 °C for 5 min. Growth is inhibited by antibiotics (penicillin G [200 μg/ml], chloramphenicol, and neomycin [100 μg/ml]), and O_2 (0.2 atm. [or 20 kPa]). Growth occurs in the presence of up to 2% (w/v) NaCl. Strains were isolated from geothermal sites in the Thermopolis areas of Yellowstone National Park, Wyoming, USA, the Steamboat area, Nevada, USA, and New Zealand. The type strain was isolated from sediments of the Frying Pan thermal acid spring in Yellowstone National Park using xylan as the carbon source (Lee et al., 1993a). Strains JW/SL-YS 485 (DSM 8691) and JW/SL-YS 489 (DSM 8685) were isolated from Weimangu Thermal Valley (North Island; NZ) using thiosulfate and xylose containing medium at pH 5.0 (Wiegel and Liu, unpublished results). At this time it is not clear whether the type strain B6A-RI and the strain B6A originally isolated by Paul Weimer (1985) are identical or represent two different strains.

DNA G+C content (mol%): 36 (T_m).
Type strain: B6A-RI, ATCC 49915, DSM 7060.
GenBank accession number (16S rRNA gene): L09169.

6. **Thermoanaerobacterium thermosaccharolyticum** (McClung 1935) Collins, Lawson, Willems, Cordoba, Fernandez-Garayzabal, Garcia, Cai, Hippe and Farrow 1994, 824^{VP} (*Clostridium thermosaccharolyticum* McClung 1935, 200)

ther.mo.sac.cha.ro.ly'ti.cum. Gr. adj. *thermos* hot; Gr. n. *sakchar* sugar; Gr. adj. *lutikos* dissolving; N.L. neut. adj. *thermosaccharolyticum* referring to thermophily and sugar-dissolving.

This description is based on that of McClung (1935) under the subjective synonym *Therminosporus thermosaccharolyticus* Hollaus and Sleytr 1972, and on study of the type strain LMG 2811^T. In many industrially oriented and technical papers, this species is often referred to as the "can-swelling (micro)organism" and named *Clostridium thermosaccharolyticum* McClung 1935. Also *Clostridium tartarivorum* Hollaus and Klaushofer 1973 is now regarded as a tartaric-acid-metabolizing strain of *Thermoanaerobacterium thermosaccharolyticum* (Matteuzzi et al., 1978).

Colonies on blood agar plates are non-hemolytic and small (<0.5 mm), circular or slightly irregular, grayish white in color, smooth and shiny appearance with a low convex, but a mottled internal structure. Colonies on pea-infusion agar are 2–4 mm, granular with indistinct feathered edges, and grayish-white with an often slightly raised center. Cells in PYG (peptone-yeast extract-glucose) broth culture are Gram-stain-negative but possess a Gram-positive cell-wall structure containing *meso*-diaminopimelic acid and an S-layer with a rectangular (tetragonal) matrix and subunits spaced at 9.5 nm. Cells are typically motile by peritrichous flagella and catalase-negative. Cells are rods, 0.4–0.7×2.4–$16\,\mu m$, occurring singly or in pairs but never in chains. Spores are round or oval, 1.3–$1.5\,\mu m$, and located in terminally distended cells, sometimes with elongation of the mother cell (Pheil and Ordal, 1967; Campbell and Ordal, 1968; Hsu and Ordal, 1970; unpublished results). Sporulation is enhanced by carbon sources yielding a low growth rate (e.g., α- or β-methylglucoside, cellobiose, galactose, salicin, and starch; Hsu and Ordal, 1969b, 1969a, 1970). Sporulation is enhanced by the addition of L-xylose or L-arabinose to a basal medium, and the pH optimum for sporulation is 5.0–5.5 (Pheil and Ordal, 1967). Cells become shorter and thicker and do not sporulate (at least less than 1 in 10^8 cells) in medium containing glucose.

No growth occurs in the absence of a fermentable carbohydrate. Fermentation end products during growth on PYG broth include acetic, butyric, lactic and succinic acids, ethanol, and H_2 (Nikitina et al., 1993). Hill et al. (1993) investigated end product regulation during the fermentation of xylose under nutrient limitations using continuous and batch cultures. Under some conditions, butanol was a major end product. The occurrence of ethanol, butyric acid, and butanol as well as succinic acid depends on the strain. In the type strain, the growth conditions also have a large affect. Elongated, sporulating cells form ethanol at an ethanol/glucose ratio above 1.2 (Hsu and Ordal (1970); Landuyt and Hsu, 1985; Freier-Schroder et al., 1989; unpublished results).

Growth optimization has been reported, including effects of antibiotics (Baskaran et al., 1995b; Mosolova et al., 1991). The optimum temperature for growth is 55–62 °C, with some growth at 37 °C and poor if any growth at 30 °C. Growth also occurs up to 69 °C but not at 70 °C (unpublished results). The pH range and optimum for growth are 6.5–8.5 and 7.8, respectively (Liu et al., 1996b; Mosolova et al., 1991). Dextrin, pectin (Hollaus and Sleytr, 1972), arabinose, fructose, galactose, glucose, mannose, xylose, cellobiose, lactose, maltose, sucrose, trehalose, glycogen, starch, xylan, and salicin are fermented but not rhamnose, pyruvate, inulin, or mannitol. Notably, cells originally cultured on pyruvate will ferment pyruvate, while cells originally cultured on glucose are unable to ferment it (Lee and Ordal, 1967). Diauxic growth occurs on glucose and xylose (Aduseopoku and Mitchell, 1988). When present together, the type strain can use glucose as the carbon source and pyruvate solely as an energy source to form ATP. Pectin utilization involves an intracellular oligogalacturonate hydrolase (van Rijssel et al., 1993). Nitrate is reduced. Thiosulfate and sulfite are reduced to H_2S. Most other species of *Thermoanaerobacterium* form S^0. Contains nitrogenase (Bogdahn and Kleiner, 1986; Clarke, 1949). Major fatty acids are C_{14} and C_{16}, but $C_{15\,iso}$ and C_{17} are also present. This is the only species in the genus *Ther-*

moananerobacterium for which a plasmid has been characterized (Belogurova et al., 2002, and literature cited therein).

The original soil isolate was obtained by McClung (1935). *Thermoanaerobacterium thermosaccharolyticum* is ubiquitous and can be isolated from many mesobiotic and thermobiotic environments, including slightly acidic (pH >4) to slightly alkaline (pH <9.5) hot springs and various manures (unpublished results) and pond sediments (van Rijssel and Hansen, 1989). "*Clostridium tartarivorum*", now classified as *Thermoanaerobacterium thermosaccharolyticum* (Matteuzzi et al., 1978), was isolated from tartrate infusion of grape residue (Mercer and Vaughn, 1951). It has also been isolated once from human gingival crevices, but it is not regarded as pathogenic (DeCampos et al., 1981).

In addition to the type, other available strains are HG-6 (ATCC 31925), HG-8 (ATCC 31960; McClung, 1935; Hollaus and Klaushofer, 1973; Cameron and Cooney, 1986), T9-1 (DSM 572, ATCC 27384, *Clostridium tartarivorum*), DSM 573 (Matteuzzi et al., 1978; Mercer and Vaughn, 1951), DSM 869 (ATCC 25773; Prévot, 1938), DSM 7416 (van Rijssel and Hansen, 1989), and FH1 (Hoster et al., 2001). Other strains that have been characterized include strain E207.71 and D120-70 (Altman et al., 1996).

DNA G+C content (mol%): 29–32 (HPLC).

Type strain: ATCC 7956, DSM 571, HAMBI 2225, LMG 2811.

GenBank accession number (16S rRNA gene): M59119.

7. **Thermoanaerobacterium xylanolyticum** Lee, Jain, Lee, Lowe and Zeikus 1993a, 48^VP

xy.lan.o.ly′ti.cum. Gr. n. *xylanosum* xylan; Gr. adj. *lutikos* dissolving; N.L. adj. *xylanolyticum* xylan-dissolving.

Description is based on the type strain. Cells stain Gram-negative but have a Gram-positive cell-wall structure; motile, short rods (0.8–1.0×2.0–$7.0\,\mu m$) that occur singly or in pairs. Unlike *Thermoanaerobacterium saccharolyticum*, this species does not develop elongated cells during nutrient limitation or stationary phase, and the cytoplasm is much less granular. Terminal, spherical spores are formed, and free spores are rarely seen. Surface colonies on agar are circular, cloudy to white in color, 2–5 mm in diameter, with a rough surface texture and with smooth edges.

Growth on xylan and starch but not cellulose, ribose, melibiose, melezitose, xylitol, pectin, or lactate. Other fermented carbohydrates include maltose, lactose, sucrose, cellobiose, glucose, xylose, galactose, mannose, fructose, rhamnose, and mannitol. Major fermentation products from either glucose or xylan are acetic acid and ethanol in approximately equal amounts but not lactic acid. The pH range and optimum for growth are 5.0–7.5 and 6.0, respectively. The temperature range and optimum for growth are 45–70 °C and 60 °C, respectively. Growth is inhibited by penicillin G (200 μg/ml); neomycin sulfate, ampicillin, streptomycin sulfate, rifampin, polymyxin B, erythromycin, tetracycline, or acridine orange (all at 100 μg/ml); or by NaCl (1% w/v). Strain LX-11^T was isolated from sediments of the acidic thermal spring Frying Pan in Yellowstone National Park, Wyoming, USA, using xylan as the carbon source (Lee et al., 1993a).

DNA G+C content (mol%): 36.1 (T_m).

Type strain: LX-11, ATCC 49914, DSM 7097.

GenBank accession number (16S rRNA gene): L09172.

8. **Thermoanaerobacterium zeae** Cann, Stroot, Mackie, White and Mackie 2001, 300[VP]

ze.ae. L. gen. n. *zeae* of corn (*Zea mays*), describing the use of corn as a substrate for growth.

Cells are Gram-stain-variable, straight rods that occur singly or in pairs, and motile by means of peritrichous flagella. Spore formation has never been observed. Catalase-negative. Nitrate, sulfate, and sulfur are not reduced. Indole is not produced. Ferments a wide variety of complex and simple carbohydrates, including cracked corn, starch, xylan, glucose, melibiose, raffinose, arabinose, galactose, lactose, maltose, mannose, rhamnose, salicin, sucrose, trehalose, xylose, cellobiose, melizitose, and pyruvate. Cellulose, pectin, sorbose, mannitol, malate, fumarate, citrate, glycerol, and H_2/CO_2 do not support growth. The optimum temperature for growth on glucose at pH 6.8 is 65–70 °C. The maximum temperature for growth is 72 °C, and growth does not occur below 37 °C. The pH range for growth is 3.9–7.9. Under optimal conditions with glucose, the doubling time is 1.8 h. The major fermentation products of glucose are ethanol and carbon dioxide. Small amounts of H_2, acetate, and formate are formed. The quantities of formate exceed that of acetate. Growth occurs in the presence of 300 mM NaCl, 450 mM KCl, and 150 mM NH_4Cl. Sulfur granules and H_2S are not formed from thiosulfate; sulfate and elemental sulfur are not reduced. Growth is completely inhibited by tetracycline, rifampin, kanamycin, and erythromycin (50 μg/ml for all).

Strain mel2[T] was isolated from an organic waste pile from a canning factory in Hoopeston, Illinois, USA, following enrichment at 60 °C with melibiose as the sole carbon source.

DNA G+C content (mol%): 42 (T_m).
Type strain: mel2, ATCC BAA-16, DSM 13642.
GenBank accession number (16S rRNA gene): U75993.
Taxonomic note: Thermoanaerobacterium zeae has recently been reassigned to the genus *Caldanaerobius* as *Caldanaerobius zeae* (Lee et al., 2008).

Genus III. **Thermosediminibacter** Lee, Wagner, Brice, Kevbrin, Mills, Romanek and Wiegel 2006, 925[VP] (Effective publication: Lee, Wagner, Brice, Kevbrin, Mills, Romanek and Wiegel 2005, 381.)

JUERGEN WIEGEL

Ther.mo.se.di.mi.ni.bac′ter. Gr. adj. *thermos* hot: L. neut. n. *sediment -inis* sediment; N.L. masc. n. *bacter* (from Gr. neut. n. *baktron*) a rod or staff; N.L. masc. n. *Thermosediminibacter* thermophilic rod isolated from sediment.

Gram-positive cell-wall structure, but stains Gram-negative. Slightly curved rods, occasionally branched or **swollen with the tendency to form autoplasts** in the early stationary growth phase (Figure 258). Physiology is obligately **anaerobic, thermophilic, and heterotrophic**, requiring yeast extract for growth. Cell wall is of the *meso*-DAP direct type.

DNA G+C content (mol%): 45–50.

Type species: **Thermosediminibacter oceani** Lee, Wagner, Brice, Kevbrin, Mills, Romanek and Wiegel 2006, 925[VP] (Effective publication: Lee, Wagner, Brice, Kevbrin, Mills, Romanek and Wiegel 2005, 382.).

Further descriptive information

The genus *Thermosediminibacter* is to date represented by two species. The most closely related validly published genus is the thermophile *Thermovenabulum* (Zarvarzina et al., 2002) whose 16S rRNA gene sequence possesses an evolutionary distance of 0.06–0.12, depending upon the mask employed (Figure 259). Phylogenetic trees place this genus within the radiation of the *Thermoanaerobacterales* ord. nov. (Ludwig et al., 2009). Known habitats are ocean sediments from the Peru Margin.

Enrichment and habitat

The strains of *Thermosediminibacter oceani* that have been isolated are from the Trujillo Basin on the Peru continental shelf or the outer shelf edge of the Peruvian high productivity upwelling system at depths of 252–426 m and temperatures of 9–12 °C. The ages of the sediments used for isolation are estimated to be about 50,000 years. Strains were isolated with glucose, mannose, xylose, or pyruvate supplemented media and possessed 16S rRNA sequence similarity of approxiimately 99%.

The second species, *Thermosediminibacter litoriperuensis*, was isolated from sediments from the Peru Margin, which is the lower slope of the Peru Trench, at a depth of 5086 m below sea level and from 10 m depth of sediment at a temperature of 2 °C. This sediment contained volcanic ash, and the age is estimated to be 10,000–15,000 years (Lee et al., 2005). Further details on the sediments can be found in D'Hondt et al. (2003). Other habitats are presently unknown. The selective utilization of Mn(IV) ions as an electron acceptor and the tolerance for 4–6% NaCl may be a hint that these species are indeed marine bacteria. However, the presence of volcanic glass particles in the sediment suggests the possibility that the bacteria could be from terrestrial geothermal features blown into the ocean by volcanic eruptions and/or wind (D'Hondt et al., 2003).

Maintenance procedures

The best method of preservation appears to be in liquid nitrogen or as a freeze-dried culture. Short term storage in 30–50% pre-reduced glycerol at –20 or –80 °C, respectively, has been successful for more than three years. Cultures grown to the late-exponential phase can also be stored at 12–25 °C for 4 months without losing viability.

Differentiation of the genus *Thermosediminibacter* from related taxa

Besides its distinctive 16S rRNA gene sequence (Figure 259) and a difference of more than 10 mol% in the G+C content of its DNA, most phenotypic properties are very similar to those of the heterotrophs within the *Thermoanaerobacterales*. None of the *Thermosediminibacter* isolates can reduce sulfate or Fe(III). Formation of heat-resistant spores has not been observed. Similar to other members of the *Thermoanaerobacterales*, asymmetrical cell division, i.e., formation of coccoid-shaped cells intermittently in cell chains, occurs under various growth conditions (Wiegel and Ljungdahl, 1981). Similar to some other genera, such as *Anaerobranca* within

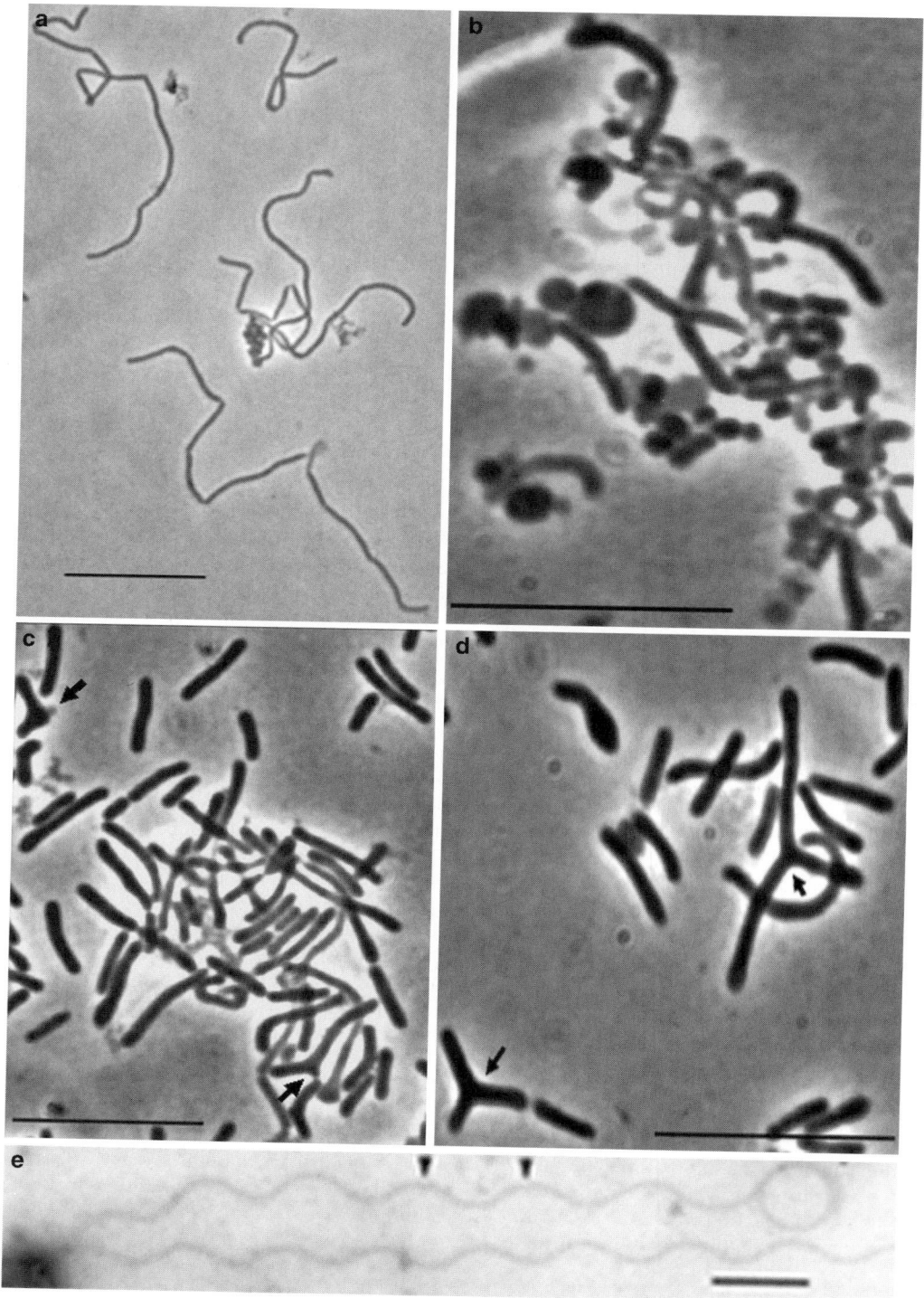

FIGURE 258. Phase-contrast micrographs of *Thermosediminibacter oceani* JW/IW-1228[T] and *Thermosediminibacter lito-riperuensis* JW/YJL-1230-7/2[T]. (a) *Thermosediminibacter oceani* cells from mid-exponential growth phase and (b) the late-exponential growth phase exhibiting partly swollen cells and L-form-like cells. (c) *Thermosediminibacter litoriperu-ensis* cells from the mid-exponential growth phase and (d) late-exponential growth phase showing primary branches. Arrows indicate branched cells. (e) Transmission electron micrograph of *Thermosediminibacter litoriperuensis*. Arrow-heads indicate periodicity. Bars = 10 μm (a, b, c, d) and 1 μm (e).

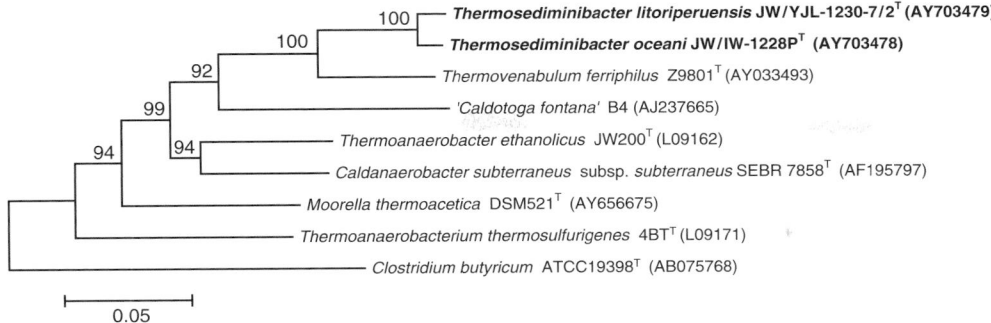

FIGURE 259. Phylogenetic dendrogram based on 16S rRNA gene sequences showing the positions of *Thermosediminibacter oceani* JW/IW-1228P and *Thermosediminibacter litoriperuensis* JW/YJL-1230-7/2 among members of the family *Thermoanaerobacteraceae*. The tree was constructed using the neighbor-joining method with Jukes–Cantor distance corrections. Numbers at the nodes represent the bootstrap values (1000 replicates); values above 90% were considered significant. Scale bar = 5 nucleotide substitutions per 100 nucleotides.

Family XIV *Incertae Sedis* in the *Clostridiales* (Engle et al., 1995; Ludwig et al., 2009), truly branched cells are observed infrequently but not with all strains under all growth conditions (Figure 258).

Further reading

Kevbrin, V.V., C.S. Romanek and J. Wiegel. 2004. Alkalithermophiles: A double challenge from extreme environments. Section VI: Extremophiles and Biodiversity, Vol. 6, Origins: Genesis, Evolution and the Biodiversity of Life. *In* Seckbach (Editor), Cellular Origin, Life in Extreme Habitats and Astrobiology (COLE). Kluwer Academic Publishers, Dordrecht, The Netherlands, pp. 395–412

Wiegel, J. 1998. Anaerobic alkali-thermophiles, a novel group of extremophiles. Extremophiles *2*: 257–267.

Wiegel, J. and V. Kevbrin. 2004. Diversity of aerobic and anaerobic alkalithermophiles. Biochem. Soc. Trans. *32*: 193–198.

List of species of the genus *Thermosediminibacter*

1. **Thermosediminibacter oceani** Lee, Wagner, Brice, Kevbrin, Mills, Romanek and Wiegel 2006, 925[VP] (Effective publication: Lee, Wagner, Brice, Kevbrin, Mills, Romanek and Wiegel 2005, 382.)

   o.ce.a′ni. L. masc. n. *oceanus* ocean; L. gen. masc. n. *oceani* of an ocean, referring to the source of isolation.

   Strains grow in agar-shake roll tubes as irregular shaped, white colonies. Vegetative cells grown in liquid cultures are straight, sometimes highly elongated rods, 0.2–0.7 µm × 1.5–16 µm. Cells stain Gram-negative at all growth phases. Without agitation, cells have the tendency to elongate, form chains or/and aggregates, and flocculate. In the late-exponential or stationary growth phase, the elongated cells exhibit swollen ends and bulging sections, and the cytoplasm becomes granular and heterogeneous, eventually forming autoplasts (L-shaped cells). In chains of 5–10 cells, coccoid cells are infrequently interspersed. Sporulation has never been observed under any growth conditions. Cells possess 1–4 peritrichous flagella. Tumbling, but not swimming motility is observed. The major fatty acids are $C_{15:0\ iso}$, $C_{16:1\ \omega9c}$, $C_{16:0}$, and $C_{18:1\ \omega9c}$, with a small amount of the polyunsaturated $C_{18:2\ \omega6}$ (Lee et al., 2005).

   The temperature range for growth is 52–76 °C, with an optimum at around 68 °C. The pH^{25C} range for growth is 6.3–9.3 with an optimum at 7.5 (Wiegel, 1998). The salinity range for growth is 0–6% (w/v) with an optimum at 1–2% NaCl added to the mineral media.

   Yeast extract is required for growth. In the presence of 0.02% yeast extract, Casamino acids (0.2% w/v), fructose, glucose, mannose, sucrose, and xylose (20 mM) serve as carbon and energy sources. Weak growth is observed on Difco Beef extract, tryptone (0.2% w/v), lactate, pyruvate, methanol, inositol, manitol, sorbitol, cellobiose, maltose, raffinose, and trehalose (20 mM). No growth found on H_2/CO_2 (80:20, v/v) in the presence of 0.02% yeast extract and in the presence or absence of Fe(III). In the presence of 0.3% yeast extract as sole carbon and energy source, thiosulfate, elemental sulfur, and MnO_2 serve as electron acceptors, but sulfate and Fe(III) do not, even though the addition of ferric citrate enhances growth.

   Several strains (available from the author) have been isolated (JW/IW-1227G, JW/IW-1227M, JW/IW-1227X, JW/IW-1228T), although only from samples of two different boreholes of the Trujillo Basin and the Peru continental shelf.

   DNA G+C content (mol%): 50 (HPLC).

   Type strain: JW/IW-1228P, ATCC BAA-1034, DSM 16646.

   GenBank accession number (16S rRNA gene): AY703478.

2. **Thermosediminibacter litoriperuensis** Lee, Wagner, Brice, Kevbrin, Mills, Romanek and Wiegel 2006, 925[VP] (Effective publication: Lee, Wagner, Brice, Kevbrin, Mills, Romanek and Wiegel 2005, 382.)

   li.to.ri.pe.ru.en′sis. L. neut. n. *litus -oris* the seashore, seaside, beach, coast; N.L. masc. adj. *peruensis* pertaining to Peru; N.L. masc. adj. *litoriperuensis* of a Peruvian coast, referring to its origin from the coast of Peru.

   Colonies in agar-shake roll tubes are irregularly shaped, 0.1–1.5 mm in diameter. Cells in liquid cultures are straight rods, 0.3–0.5 µm × 2.0–10.0 µm (or less elongated than *Thermosediminibacter oceani*). Frequently forms chains. Strain JW/YJL-1230-7/2[T] also produces swollen ends, but infrequently forms autoplasts. Gram-stain reaction is negative at all growth phases. Cells possess 2–4 peritrichous flagella with a wavelength periodicity of ~1–1.3 μm (Figure 258). Fewer than 5% of cells are branched. Spores have not been observed either microscopically or by resistance to heat treatment.

The temperature range for growth is 43–76 °C with an optimum at around 64 °C. The pH^{25C} range for growth is from 5–9.5 with an optimum at 7.9–8.4. The salinity range for growth is from 0–4.5% (w/v).

Substrates utilized include tryptone (0.2%), acetate, lactate, inositol, mannitol, xylitol, fructose, galactose, glucose, mannose, raffinose, sucrose, and xylose (20 mM). In the presence of 0.3% yeast extract as sole carbon and energy source, thiosulfate, elemental sulfur, and MnO_2 serve as electron acceptors but sulfate and Fe(III) do not. However, the addition of ferric citrate to the medium enhances growth. The main fermentation end products in glucose-containing medium supplemented with yeast extract are acetate and small amounts of propionate, isobutyrate, and isovalerate.

Strains (available from the author) JW/YJL-1230-7/1 through/3 and JW/YJL-1230-8/1 through/3 were all isolated from sediments of the Peru Margin.

DNA G+C content (mol%): 45–46 (HPLC).

Type strain: JW/YJL-1230-7/2, ATCC BAA-1035, DSM 16647.

GenBank accession number (16S rRNA gene): AY703479.

Genus IV. **Thermovenabulum** Zavarzina, Tourova, Kuznetsov, Bonch-Osmolovskaya and Slobodkin 2002, 1741^{VP}

ALEXANDER SLOBODKIN

Ther.mo.ve.na′bu.lum. Gr. adj. *thermos* hot; L. neut. n. *venabulum* a hunting spear; N.L. neut. n. *Thermovenabulum* hot hunting spear-shaped cell, referring to the branched cell morphology.

Straight, **sometimes branched rods** 0.5–0.6 μm in diameter and 1.5–7.0 μm in length. **Gram-positive cell-wall structure. Forms protrusions of cell envelope.** Cells occur singly or in short chains and exhibit tumbling motility due to peritrichous flagellation. **Form** round refractile **endospores** in terminally swollen sporangia. **Anaerobic. Thermophilic**; the temperature range for growth is 45–76 °C. Neutrophilic. Growth occurs at NaCl concentrations from 0–3.5% (w/v). **Grows organotrophically on a number of proteinaceous substrates and carbohydrates. Utilizes molecular hydrogen in the presence of Fe(III) and an organic electron donor.** Reduces amorphous Fe(III) oxide, Fe(III) citrate, 9,10-anthraquinone-2,6-disulfonate, fumarate, nitrate, sulfite, thiosulfate, elemental sulfur, and MnO_2. None of the electron acceptors utilized, except fumarate, stimulates growth.

DNA G+C content (mol%): 36.

Type species: **Thermovenabulum ferriorganovorum** Zavarzina, Tourova, Kuznetsov, Bonch-Osmolovskaya and Slobodkin 2002, 1741^{VP}.

Further descriptive information

The genus *Thermovenabulum* is currently represented by a single species, *Thermovenabulum ferriorganovorum*, which was isolated from hydrothermal source in Kamchatka, Russia.

About 2% of vegetative cells of *Thermovenabulum* may be branched, having Y-like shape morphology. Some cells have protrusions of cell envelope and cytoplasm, 0.1–0.3 μm in length. In the late-exponential phase of growth, up to 15% of the cells may contain heat-resistant endospores.

Members of the genus *Thermovenabulum* utilize complex proteinaceous substrates such as yeast extract, beef extract, Casamino acids, and peptone in the presence as well as in the absence of an external electron acceptor. They reduce amorphous Fe(III) oxide, Fe(III) citrate, 9,10-anthraquinone-2,6-disulfonate, fumarate, nitrate, sulfite, thiosulfate, elemental sulfur, and MnO_2, but none of these electron acceptors, except fumarate, stimulates growth. Fe(III) stimulates growth only in the medium with H_2 and a low concentration of yeast extract (0.2 g/l). In the presence of Fe(III) and high concentrations of proteinaceous substrate, *Thermovenabulum* consumes molecular hydrogen. *Thermovenabulum ferriorganovorum* also grows fermentatively on a number of sugars but not on glucose. The fermentation products from melibiose are ethanol, acetate, H_2, and CO_2.

Enrichment and isolation procedures

The type and the sole species of the genus *Thermovenabulum* has been isolated from a terrestrial hydrothermal source with Fe(III) deposits located in the Uzon caldera on the Kamchatka peninsula (Zavarzina et al., 2002). There are no reports that members of this genus occur in other environments. Employment of the anaerobic media with amorphous Fe(III) oxide, molecular hydrogen, and proteinaceous substrates and incubation in the temperature range of 60–70 °C may favor the enrichment of *Thermovenabulum* spp. In this medium, formation of a black magnetic precipitate is a helpful indicator for microbial growth. The type species of *Thermovenabulum* forms colonies on 1.5% (w/v) agar in the presence of various electron acceptors.

Maintenance procedures

Members of the genus *Thermovenabulum* may be maintained on the medium of Zavarzina et al. (2002) with yeast extract, molecular hydrogen, and amorphous Fe(III) oxide. The highest growth yields could be obtained in liquid medium of the same mineral composition lacking H_2 and Fe(III), pre-reduced with $Na_2S\cdot9H_2O$ (0.5 g/l), and supplied with beef extract and fumarate. Vacuum-drying of the cultures performed at the DSMZ result in good recovery. Liquid cultures may be stored at 4 °C for 6–7 months without loss of viability.

List of species of the genus *Thermovenabulum*

1. **Thermovenabulum ferriorganovorum** Zavarzina, Tourova, Kuznetsov, Bonch-Osmolovskaya and Slobodkin 2002, 1741[VP]

fer.ri.or.ga.no.vo′rum. L. n. *ferrum* iron; N.L. n. *organum* organic compound; L. v. *voro* to eat, consume; N.L. neut. adj. *ferriorganovorum* using iron and organic compounds.

Cells are straight, sometimes branched rods, 0.5–0.6 μm in diameter and 1.5–7.0 μm in length, forming round, refractile, heat-resistant endospores in terminally swollen sporangia. Cells occur singly or in short chains and exhibit slight tumbling motility due to peritrichous flagellation. The temperature range for growth is 45–76 °C, with an optimum at 63–65 °C. The pH range for growth is 4.8–8.2, with an optimum at 6.7–6.9. Growth occurs in NaCl concentrations of 0–3.5% (w/v). Anaerobic. Substrates utilized include peptone, yeast extract, beef extract, Casamino acids, starch, pyruvate, melibiose, sucrose, fructose, maltose, xylose, and ribose. Utilizes molecular hydrogen in the presence of Fe(III) and an organic electron donor. No growth occurs with formate, acetate, propionate, lactate, methanol, ethanol, glycerol, glucose, mannose, galactose, arabinose, cellobiose, or glycogen. The fermentation products from melibiose are ethanol, acetate, H_2, and CO_2. Reduces amorphous Fe(III) oxide, Fe(III) citrate, 9,10-anthraquinone-2,6-disulfonate, fumarate, nitrate, sulfite, thiosulfate, elemental sulfur, and MnO_2. Nitrate is reduced to ammonium. Sulfite, thiosulfate, and elemental sulfur are reduced to hydrogen sulfide. None of the electron acceptors utilized, except fumarate, stimulates growth. Does not reduce sulfate and is incapable of growth with O_2. Growth is inhibited by chloramphenicol, neomycin, polymyxin B, kanamycin, and streptomycin but not by penicillin.

DNA G+C content (mol%): 36 (T_m).
Type strain: Z-9801, DSM 14006, UNIQEM 210.
GenBank accession number (16S rRNA gene): AY033493.

References

Aduseopoku, J. and W.J. Mitchell. 1988. Diauxic growth of *Clostridium thermosaccharolyticum* on glucose and xylose. FEMS Microbiol. Lett. *50*: 45–49.

Ahn, J., V.M. Balasubramaniam and A.E. Yousef. 2007. Inactivation kinetics of selected aerobic and anaerobic bacterial spores by pressure-assisted thermal processing. Int. J. Food. Microbiol. *113*: 321–329.

Aleshin, A.E., P.H. Feng, R.B. Honzatko and P.J. Reilly. 2003. Crystal structure and evolution of a prokaryotic glucoamylase. J. Mol. Biol. *327*: 61–73.

Allmaier, G., C. Schaffer, P. Messner, U. Rapp and F.J. Mayer-Posner. 1995. Accurate determination of the molecular weight of the major surface layer protein isolated from *Clostridium thermosaccharolyticum* by time-of-flight mass spectrometry. J. Bacteriol. *177*: 1402–1404.

Altaras, N.E., M.R. Etzel and D.C. Cameron. 2001. Conversion of sugars to 1,2-propanediol by *Thermoanaerobacterium thermosaccharolyticum* HG-8. Biotechnol. Prog. *17*: 52–56.

Altman, E., J.R. Brisson, P. Messner and U.B. Sleytr. 1990. Chemical characterization of the regularly arranged surface layer glycoprotein of *Clostridium thermosaccharolyticum* D120-70. Eur. J. Biochem. *188*: 73–82.

Altman, E., C. Schaffer, J.R. Brisson and P. Messner. 1996. Isolation and characterization of an amino sugar-rich glycopeptide from the surface layer glycoprotein of *Thermoanaerobacterium thermosaccharolyticum* E207-71. Carbohydr. Res. *295*: 245–253.

Antranikian, G., C. Herzberg, F. Mayer and G. Gottschalk. 1987. Changes in the cell-envelope structure of *Clostridium* sp strain-EM1 during massive production of alpha-amylase and pullulanase. FEMS Microbiol. Lett. *41*: 193–197.

Armand, S., C. Vieille, C. Gey, A. Heyraud, J.G. Zeikus and B. Henrissat. 1996. Stereochemical course and reaction products of the action of beta-xylosidase from *Thermoanaerobacterium saccharolyticum* strain B6A-RI. Eur. J. Biochem. *236*: 706–713.

Baskaran, S., H.J. Ahn and L.R. Lynd. 1995a. Investigation of the ethanol tolerance of *Clostridium thermosaccharolyticum* in continuous culture. Biotechnol. Prog. *11*: 276–281.

Baskaran, S., D.A.L. Hogsett and L.R. Lynd. 1995b. Optimization of a chemically defined, minimal medium for *Clostridium thermosaccharolyticum*. Appl. Biochem. Biotechnol. *51–52*: 399–411.

Belogurova, N.G., T.P. Mosolova, S.V. Kalyuzhnyy and S.D. Varfolomeyev. 1991. Kinetics of growth and metabolism of *Clostridium thermosaccharolyticum* culture. Isolation and characteristics of its plasmids. Appl. Biochem. Biotechnol. *27*: 1–8.

Belogurova, N.G., E.E. Davydova, M.V. Sobennikova, S.D. Varfolomeev, E.P. Del'ver and A.A. Belogurov. 2002. Replication protein RepN encoded by the RC plasmid of thermophilic bacterium *Thermoanaerobacterium saccharolyticum*: mutational analysis and deletion mapping of domains responsible for its lethal effect. Mol. Biol. (Mosk) *36*: 106–113.

Bogdahn, M. and D. Kleiner. 1986. N_2 fixation and NH_4^+ assimilation in the thermophilic anaerobes *Clostridium thermosaccharolyticum* and *Clostridium thermoautotrophicum*. Arch. Microbiol. *144*: 102–104.

Brechtel, E. and H. Bahl. 1999. In *Thermoanaerobacterium thermosulfurigenes* EM1 S-layer homology domains do not attach to peptidoglycan. J. Bacteriol. *181*: 5017–5023.

Brechtel, E., M. Matuschek, A. Hellberg, E.M. Egelseer, R. Schmid and H. Bahl. 1999. Cell wall of *Thermoanaerobacterium thermosulfurigenes* EM1: isolation of its components and attachment of the xylanase XynA. Arch. Microbiol. *171*: 159–165.

Bredholt, S., J. Sonne-Hansen, P. Nielsen, I.M. Mathrani and B.K. Ahring. 1999. *Caldicellulosiruptor kristjanssonii* sp. nov., a cellulolytic extremely thermophilic, anaerobic bacterium. Int. J. Syst. Bacteriol. *49*: 991–996.

Brill, J.A. and J. Wiegel. 1997. Differentiation between sporeforming and asporogenous bacteria by a PCR and Southern hybridization based method. J. Microbiol. Methods *31*: 29–36.

Bronnenmeier, K., H. Meissner, S. Stocker and W.L. Staudenbauer. 1995. Alpha-D-glucuronidases from the xylanolytic thermophiles *Clostridium stercorarium* and *Thermoanaerobacterium saccharolyticum*. Microbiology *141*: 2033–2040.

Burchhardt, G. and H. Bahl. 1991. Cloning and analysis of the beta-galactosidase-encoding gene from *Clostridium thermosulfurogenes* EM1. Gene *106*: 13–19.

Burchhardt, G., A. Wienecke and H. Bahl. 1991. Isolation of the pullulanase gene from *Clostridium thermosulfurogenes* (DSM 3896) and its expression in *Escherichia coli*. Curr. Microbiol. *22*: 91–95.

Cameron, D.C. and C.L. Cooney. 1986. A novel fermentation: the production of R(−)-1,2-propanediol and acetol by *Clostridium thermosaccharolyticum*. Biotechnology *4*: 651–654.

Cameron, D.C., N.E. Altaras, M.L. Hoffman and A.J. Shaw. 1998. Metabolic engineering of propanediol pathways. Biotechnol. Prog. *14*: 116–125.

Campbell, M.F. and Z.J. Ordal. 1968. Inhibition of sporulation of *Clostridium thermosaccharolyticum*. Appl. Microbiol. *16*: 1949–1951.

Cann, I.K., S. Kocherginskaya, M.R. King, B.A. White and R.I. Mackie. 1999. Molecular cloning, sequencing, and expression of a novel multidomain mannanase gene from *Thermoanaerobacterium polysaccharolyticum*. J. Bacteriol. *181*: 1643–1651.

Cann, I.K.O., P.G. Stroot, K.R. Mackie, B.A. White and R.I. Mackie. 2001. Characterization of two novel saccharolytic, anaerobic thermophiles, *Thermoanaerobacterium polysaccharolyticum* sp. nov. and *Thermoanaerobacterium zeae* sp. nov., and emendation of the genus *Thermoanaerobacterium*. Int. J. Syst. Evol. Microbiol. *51*: 293–302.

Cejka, Z. and W. Baumeister. 1987. 3-Dimensional structure of the surface protein of *Clostridium thermosaccharolyticum*. FEMS Microbiol. Lett. *44*: 13–18.

Clarke, F.M. 1949. Some nitrogen requirements of *Clostridium thermosaccharolyticum*. J. Bacteriol. *57*: 465–471.

Collins, M.D., P.A. Lawson, A. Willems, J.J. Cordoba, J. Fernández-Garayzábal, P. Garcia, J. Cai, H. Hippe and J.A. Farrow. 1994. The phylogeny of the genus *Clostridium*: proposal of five new genera and eleven new species combinations. Int. J. Syst. Bacteriol *44*: 812–826.

D'Hondt, S.L., B.B. Jorgensen, D.J. Miller, I.W. Aiello, B. Bekins, R. Blake, B.A. Cragg, H. Cypionka, G.R. Dickens, T. Ferdelman, K.H. Ford, G.L. Gettemy, G. Guèèrin, K.-U. Hinrichs, N. Holm, C.H. House, F. Inagaki, P. Meister, R.M. Mitterer, T.H. Naehr, S. Niitsuma, R.J. Parkes, A. Schippers, C.G. Skilbeck, D.C. Smith, A.J. Spivack, A. Teske and J. Wiegel. 2003. Volume 201 initial reports, doi:10.2973/odp.proc.ir.201.2003. Ocean Drilling Program, College Station, TX.

DeCampos, S.L., E.P.T. DeOliveira, H.G. Higashi, M.R. Lorenzetti, A.V. Diniz and D. Zappa. 1981. Ocorrencia de microorganismos do genero *Clostridium* em material colhido de bolsa periodontal e de sulco gingival de pacientes com e sem doenca pericontal avancada, habitants de zonas rural e urgana. Rev. Assoc. Paul. Cirurg Dent. *35*: 422–426.

Delver, E.P., N.G. Belogurova, E.E. Tupikova, S.D. Varfolomeyev and A.A. Belogurov. 1996. Characterization, sequence and mode of replication of plasmid pNB2 from the thermophilic bacterium *Clostridium thermosaccharolyticum*. Mol. Gen. Genet. *253*: 166–172.

Desai, S.G., M.L. Guerinot and L.R. Lynd. 2004. Cloning of L-lactate dehydrogenase and elimination of lactic acid production via gene knockout in *Thermoanaerobacterium saccharolyticum* JW/SL-YS485. Appl. Microbiol. Biotechnol. *65*: 600–605.

Dotzauer, C., M.A. Ehrmann and R.F. Vogel. 2002. Occurrence and detection of *Thermoanaerobacterium* and *Thermoanaerobacter* in canned food. Food Technol. Biotechnol. *40*: 21–26.

Engle, M., Y.H. Li, C. Woese and J. Wiegel. 1995. Isolation and characterization of a novel alkalitolerant thermophile, *Anaerobranca horikoshii* gen. nov., sp. nov. Int. J. Syst. Bacteriol. *45*: 454–461.

Feng, P.H., S. Berensmeier, K. Buchholz and P.J. Reilly. 2002. Production, purification, and characterization of *Thermoanaerobacterium thermosaccharolyticum* glucoamylase. Starch-Starke *54*: 328–337.

Feustel, L., S. Nakotte and P. Durre. 2004. Characterization and development of two reporter gene systems for *Clostridium acetobutylicum*. Appl. Environ. Microbiol. *70*: 798–803.

Freier-Schroder, D., J. Wiegel and G. Gottschalk. 1989. Butanol formation by *Clostridium thermosaccharolyticum* at neutral pH. Biotechnol. Lett. *11*: 831–836.

Ganghofner, D., J. Kellermann, W.L. Staudenbauer and K. Bronnenmeier. 1998. Purification and properties of an amylopullanase, a glucoamylase, and an alpha-glucosidase in the amylolytic enzyme system of *Thermoanaerobacterium thermosaccharolyticum*. Biosci. Biotechnol. Biochem. *62*: 302–308.

Garrity, G.M., J.A. Bell and T. Lilburn. 2005. The Revised Road Map to the Manual. *In* Brenner, Krieg, Staley and Garrity (Editors), Bergey's Manual of Systematic Bacteriology, 2nd edn, Vol. 2, The *Proteobacteria*, Part A, Introductory Essays. Springer, New York, pp. 159–220.

Haeckel, K. and H. Bahl. 1989. Cloning and expression of the thermostable alpha-amylase gene from *Clostridium thermosulfurogenes* (DSM 3896) in *Escherichia coli*. FEMS Microbiol. Lett. *51*: 333–337.

Haldrup, A., S.G. Petersen and F.T. Okkels. 1998. The xylose isomerase gene from *Thermoanaerobacterium thermosulfurogenes* allows effective selection of transgenic plant cells using D-xylose as the selection agent. Plant Mol. Biol. *37*: 287–296.

Hamana, K., M. Niitsu, K. Samejima and T. Itoh. 2001. Polyamines of the thermophilic eubacteria belonging to the genera *Thermosipho*, *Thermaerobacter* and *Caldicellulosiruptor*. Microbios *104*: 177–185.

Hernon, F., C. Forbes and E. Colleran. 2006. Identification of mesophilic and thermophilic fermentative species in anaerobic granular sludge. Water Sci. Technol. *54*: 19–24.

Hespell, R.B. 1992. Fermentation of xylans by *Butyrivibrio fibrisolvens* and *Thermoanaerobacter* strain B6A: utilization of uronic acids and xylanolytic activities. Curr. Microbiol. *25*: 189–195.

Hill, P.W., T.R. Klapatch and L.R. Lynd. 1993. Bioenergetics and end product regulation of *Clostridium thermosaccharolyticum* in response to nutrient limitation. Biotechnol. Bioeng. *42*: 873–883.

Hollaus, F. and U. Sleytr. 1972. On the taxonomy and fine structure of some hyperthermophilic saccharolytic clostridia. Arch. Mikrobiol. *86*: 129–146.

Hollaus, F. and H. Klaushofer. 1973. Identification of hyperthermophilic obligate anaerobic bacteria from extraction juices of beet sugar factories. Int. Sugar J. *75*: 237–241, 271–275.

Hosoya, R., K. Hamana, M. Niitsu and T. Itoh. 2004. Polyamine analysis for chemotaxonomy of thermophilic eubacteria: polyamine distribution profiles within the orders *Aquificales*, *Thermotogales*, *Thermodesulfobacteriales*, *Thermales*, *Thermoanaerobacteriales*, *Clostridiales* and *Bacillales*. J. Gen. Appl. Microbiol. *50*: 271–287.

Hoster, F., R. Daniel and G. Gottschalk. 2001. Isolation of a new *Thermoanaerobacterium thermosaccharolyticum* strain (FH1) producing a thermostable dextranase. J. Gen. Appl. Microbiol. *47*: 187–192.

Hsu, E.J. and Z.J. Ordal. 1969a. Sporulation of *Clostridium thermosaccharolyticum* under conditions of restricted growth. J. Bacteriol. *97*: 1511–1512.

Hsu, E.J. and Z.J. Ordal. 1969b. Sporulation of *Clostridium thermosaccharolyticum*. Appl. Microbiol. *18*: 958–960.

Hsu, E.J. and Z.J. Ordal. 1970. Comparative metabolism of vegetative and sporulating cultures of *Clostridium thermosaccharolyticum*. J. Bacteriol. *102*: 369–376.

Huang, C.Y., B.K. Patel, R.A. Mah and L. Baresi. 1998. *Caldicellulosiruptor owensensis* sp. nov., an anaerobic, extremely thermophilic, xylanolytic bacterium. Int. J. Syst. Bacteriol. *48*: 91–97.

Huber, R.E., N.J. Roth and H. Bahl. 1996. Quaternary structure, Mg^{2+} interactions, and some kinetic properties of the beta-galactosidase from *Thermoanaerobacterium thermosulfurigenes* EM1. J. Protein Chem. *15*: 621–629.

Hungate, R.E. 1969. A roll tube method for cultivation of strict anaerobes. *In* Norris and Ribbons (Editors), Methods in Microbiology, Vol. 3B. Academic Press, London and New York, pp. 117–132.

Hyun, H.H. and J.G. Zeikus. 1985. Simultaneous and enhanced production of thermostable amylases and ethanol from starch by cocultures of *Clostridium thermosulfurogenes* and *Clostridium thermohydrosulfuricum*. Appl. Environ. Microbiol. *49*: 1174–1181.

Kim, Y.S., Y.J. Im, S.H. Rho, D. Sriprapundh, C. Vieille, S.W. Suh, J.G. Zeikus and S.H. Eom. 2001. Crystallization and preliminary X-ray studies of Trp138Phe/Val185Thr xylose isomerases from *Thermotoga neapolitana* and *Thermoanaerobacterium thermosulfurigenes*. Acta Crystallogr. Sec. D-Biol. Crystallogr. *57*: 1686–1688.

King, M.R., B.A. White, H.P. Blaschek, B.M. Chassy, R.I. Mackie and I.K. Cann. 2002. Purification of a thermostable alpha-galactosidase from *Thermoanaerobacterium polysaccharolyticum*. J. Agric. Food Chem. *50*: 5676–5682.

Klapatch, T.R., D.A.L. Hogsett, S. Baskaran, S. Pal and L.R. Lynd. 1994. Organism development and characterization for ethanol production using thermophilic bacteria. Appl. Biochem. Biotechnol. *45–46*: 209–223.

Klapatch, T.R., A.L. Demain and L.R. Lynd. 1996a. Restriction endonuclease activity in *Clostridium thermocellum* and *Clostridium thermosaccharolyticum*. Appl. Microbiol. Biotechnol. *45*: 127–131.

Klapatch, T.R., M.L. Guerinot and L.R. Lynd. 1996b. Electrotransformation of *Clostridium thermosaccharolyticum*. J. Ind. Microbiol. *16*: 342–347.

Knegtel, R.M., R.D. Wind, H.J. Rozeboom, K.H. Kalk, R.M. Buitelaar, L. Dijkhuizen and B.W. Dijkstra. 1996. Crystal structure at 2.3 Å resolution and revised nucleotide sequence of the thermostable cyclodextrin glycosyltransferase from *Thermoanaerobacterium thermosulfurigenes* EM1. J. Mol. Biol. *256*: 611–622.

Kublanov, I.V., M.I. Prokofeva, N.A. Kostrikina, T.V. Kolganova, T.P. Tourova, J. Wiegel and E.A. Bonch-Osmolovskaya. 2007. *Thermoanaerobacterium aciditolerans* sp. nov., a moderate thermoacidophile from a Kamchatka hot spring. Int. J. Syst. Evol. Microbiol. *57*: 260–264.

Landuyt, S.M. and E.J. Hsu. 1985. Solvent fermentation proceeds acid fermentation in elongated cells of *Clostridium thermosaccharolyticum*. *In* Dring, Ellar and Gould (Editors), Fundamental and Applied Aspects of Bacterial Spores. Academic Press, London.

Landuyt, S.M., E.J. Hsu, B. Wang and S. Tsay. 1995. Conversion of paraffin oil to alcohols by *Clostridium thermosaccharolyticum*. Appl. Environ. Microbiol. *61*: 1153–1155.

Lee, C., B.C. Saha and J.G. Zeikus. 1990a. Characterization of *Thermoanaerobacter* glucose-isomerase in relation to saccharidase synthesis and development of single-step processes for sweetener production. Appl. Environ. Microbiol. *56*: 2895–2901.

Lee, C. and J.G. Zeikus. 1991. Purification and characterization of thermostable glucose isomerase from *Clostridium thermosulfurogenes* and *Thermoanaerobacter* strain B6A. Biochem. J. *273*: 565–571.

Lee, C.K. and Z.J. Ordal. 1967. Regulatory effect of pyruvate on glucose metabolism of *Clostridium thermosaccharolyticum*. J. Bacteriol. *94*: 530–536.

Lee, C.Y., L. Bhatnagar, B.C. Saha, Y.E. Lee, M. Takagi, T. Imanaka, M. Bagdasarian and J.G. Zeikus. 1990b. Cloning and expression of the *Clostridium thermosulfurogenes* glucose isomerase gene in *Escherichia coli* and *Bacillus subtilis*. Appl. Environ. Microbiol. *56*: 2638–2643.

Lee, Y.E., M.K. Jain, C. Lee, S.E. Lowe and J.G. Zeikus. 1993a. Taxonomic distinction of saccharolytic anaerobes: description of *Thermoanaerobacterium xylanolyticum* gen. nov., sp. nov., and *Thermoanaerobacterium saccharolyticum* gen. nov., sp. nov.; reclassification of *Thermoanaerobium brockii*, *Clostridium thermosulfurogenes*, and *Clostridium thermohydrosulfuricum* E100-69 as *Thermoanaerobacter brockii* comb. nov., *Thermoanaerobacterium thermosulfurogenes* comb. nov., and *Thermoanaerobacter thermohydrosulfuricus* comb. nov., respectively; and transfer of *Clostridium thermohydrosulfuricum* 39E to *Thermoanaerobacter ethanolicus*. Int. J. Syst. Bacteriol. *43*: 41–51.

Lee, Y.E., S.E. Lowe and J.G. Zeikus. 1993b. Gene cloning, sequencing, and biochemical characterization of endoxylanase from *Thermoanaerobacterium saccharolyticum* B6A-RI. Appl. Environ. Microbiol. *59*: 3134–3137.

Lee, Y.E., S.E. Lowe and J.G. Zeikus. 1993c. Regulation and characterization of xylanolytic enzymes of *Thermoanaerobacterium saccharolyticum* B6A-RI. Appl. Environ. Microbiol. *59*: 763–771.

Lee, Y.E., M.V. Ramesh and J.G. Zeikus. 1993d. Cloning, sequencing and biochemical characterization of xylose isomerase from *Thermoanaerobacterium saccharolyticum* strain B6A-RI. J. Gen. Microbiol. *139*: 1227–1234.

Lee, Y.E. and J.G. Zeikus. 1993. Genetic organization, sequence and biochemical-characterization of recombinant beta xylosidase from *Thermoanaerobacterium saccharolyticum* strain B6A-RI. J. Gen. Microbiol. *139*: 1235–1243.

Lee, Y.J., I.D. Wagner, M.E. Brice, V.V. Kevbrin, G.L. MIlls, C.S. Romanek and J. Wiegel. 2005. *Thermosediminibacter oceani* gen. nov., sp. nov. and *Thermosediminibacter litoriperuensis* sp. nov., new anaerobic thermophilic bacteria isolated from Peru Margin. Extremophiles *9*: 375–383.

Lee, Y.J., I.D. Wagner, M.E.Brice, V.V. Kevbrin and J.Wiegel. 2006. *In* Validation of the publication of new names and new combinations previously published outside the IJSEM. List no. 109. Int. J. Syst. Evol. Microbiol. *56*: 925–927.

Lee, Y.J., A. Prange, H. Lichtenberg, M. Rohde, M. Dashti and J. Wiegel. 2007. In situ analysis of sulfur species in sulfur globules produced from thiosulfate by *Thermoanaerobacter sulfurigignens* and *Thermoanaerobacterium thermosulfurigenes*. J. Bacteriol. *189*: 7525–7529.

Lee, Y.J., R.I. Mackie, I.K. Cann and J. Wiegel. 2008. Description of *Caldanaerobius fijiensis* gen. nov., sp. nov., an inulin-degrading, ethanol-producing, thermophilic bacterium from a Fijian hot spring sediment, and reclassification of *Thermoanaerobacterium polysaccharolyticum* and *Thermoanaerobacterium zeae* as *Caldanaerobius polysaccharolyticus* comb. nov. and *Caldanaerobius zeae* comb. nov. Int. J. Syst. Evol. Microbiol. *58*: 666–670.

Leemhuis, H., B.W. Dijkstra and L. Dijkhuizen. 2003. *Thermoanaerobacterium thermosulfurigenes* cyclodextrin glycosyltransferase: mechanism and kinetics of inhibition by acarbose and cyclodextrins. Eur. J. Biochem. *270*: 155–162.

Li, Y. and H.P. Blaschek. 2002. Molecular characterization and utilization of the CAK1 filamentous viruslike particle derived from *Clostridium beijerinckii*. J. Ind. Microbiol. Biotechnol. *28*: 118–126.

Liu, S.Y., F.C. Gherardini, M. Matuschek, H. Bahl and J. Wiegel. 1996a. Cloning, sequencing, and expression of the gene encoding a large S-layer-associated endoxylanase from *Thermoanaerobacterium* sp. strain JW/SL-YS 485 in *Escherichia coli*. J. Bacteriol. *178*: 1539–1547.

Liu, S.Y., F.A. Rainey, H.W. Morgan, F. Mayer and J. Wiegel. 1996b. *Thermoanaerobacterium aotearoense* sp. nov., a slightly acidophilic, anaerobic thermophile isolated from various hot springs in New Zealand, and emendation of the genus *Thermoanaerobacterium*. Int. J. Syst. Bacteriol. *46*: 388–396.

Liu, S.Y., J. Wiegel and F.C. Gherardini. 1996c. Purification and cloning of a thermostable xylose (glucose) isomerase with an acidic pH optimum from *Thermoanaerobacterium* sp. strain JW/SL-YS 489. J. Bacteriol. *178*: 5938–5945.

Lloyd, L.F., O.S. Gallay, J. Akins and J.G. Zeikus. 1994. Crystallization and preliminary X-ray diffraction studies of xylose isomerase from *Thermoanaerobacterium thermosulfurigenes* strain 4b. J. Mol. Biol. *240*: 504–506.

Lorenz, W.W. and J. Wiegel. 1997. Isolation, analysis, and expression of two genes from *Thermoanaerobacterium* sp. strain JW/SL YS485: a beta-xylosidase and a novel acetyl xylan esterase with cephalosporin C deacetylase activity. J. Bacteriol. *179*: 5436–5441.

Ludwig, W., K.-H. Schleifer and W.B. Whitman. 2009. Revised road map to the phylum *Firmicutes*. *In* De Vos, Garrity, Jones, Krieg, Ludwig, Rainey, Schleifer and Whitman (Editors), Bergey's Manual of Systematic Bacteriology, 2nd edn, Vol. 3, The *Firmicutes*, Springer, New York, pp. 1–14.

Lynd, L.R. 1996. Overview and evaluation of fuel ethanol from cellulosic biomass: technology, economics, the environment, and policy. Annu. Rev. Energy Environ. *21*: 403–465.

Lynd, L.R., S. Baskaran and S. Casten. 2001. Salt accumulation resulting from base added for pH control, and not ethanol, limits growth of *Thermoanaerobacterium thermosaccharolyticum* HG-8 at elevated feed xylose concentrations in continuous culture. Biotechnol. Prog. *17*: 118–125.

Madi, E., G. Antranikian, K. Ohmiya and G. Gottschalk. 1987. Thermostable amylolytic enzymes from a new *Clostridium* isolate. Appl. Environ. Microbiol. *53*: 1661–1667.

Mai, V. and J. Wiegel. 2001. Advances in development of a genetic system for *Thermoanaerobacterium* spp.: expression of genes encoding hydrolytic enzymes, development of a second shuttle vector, and integration of genes into the chromosome. Appl. Environ. Microbiol. *66*: 4817–4821.

Matteuzzi, D., F. Hollaus and B. Biavati. 1978. Proposal of neotype for *Clostridium thermohydrosulfuricum* and merging of *Clostridium tartarivorum* with *Clostridium thermosaccharolyticum*. Int. J. Syst. Bacteriol. *28*: 528–531.

Matuschek, M., G. Burchhardt, K. Sahm and H. Bahl. 1994. Pullulanase of *Thermoanaerobacterium thermosulfurigenes* EM1 (*Clostridium thermosulfurogenes*): molecular analysis of the gene, composite structure of the enzyme, and a common model for its attachment to the cell surface. J. Bacteriol. *176*: 3295–3302.

Matuschek, M., K. Sahm, A. Zibat and H. Bahl. 1996. Characterization of genes from *Thermoanaerobacterium thermosulfurigenes* EM1 that encode two glycosyl hydrolases with conserved S-layer-like domains. Mol. Gen. Genet. *252*: 493–496.

Matuschek, M., K. Sahm and H. Bahl. 1997. Molecular characterization of genes encoding a novel ABC transporter in *Thermoanaerobacterium thermosulfurigenes* EM1. Curr. Microbiol. *35*: 237–239.

May, A., T. Pusztahelyi, N. Hoffmann, R.J. Fischer and H. Bahl. 2006. Mutagenesis of conserved charged amino acids in SLH domains of *Thermoanaerobacterium thermosulfurigenes* EM1 affects attachment to cell wall sacculi. Arch. Microbiol. *185*: 263–269.

Mayer, F., B. Vogt and C. Poc. 1998. Immunoelectron microscopic studies indicate the existence of a cell shape preserving cytoskeleton in prokaryotes. Naturwissenschaften *85*: 278–282.

Mayer, F. and G. Gottschalk. 2004. The bacterial cytoskeleton and its putative role in membrane vesicle formation observed in a Gram-positive bacterium producing starch-degrading enzymes. J. Mol. Microbiol. Biotechnol. *6*: 127–132.

Mayer, F. 2006. Cytoskeletal elements in bacteria *Mycoplasma pneumoniae*, *Thermoanaerobacterium* sp., and *Escherichia coli* as revealed by electron microscopy. J. Mol. Microbiol. Biotechnol. *11*: 228–243.

McClung, L.S. 1935. Studies on anaerobic bacteria; IV. Taxonomy of cultures of a thermophilic species causing 'swells' of canned foods. J. Bacteriol. *29*: 189–203.

Meng, M., M. Bagdasarian and J.G. Zeikus. 1993a. Thermal stabilization of xylose isomerase from *Thermoanaerobacterium thermosulfurigenes*. Biotechnology *11*: 1157–1161.

Meng, M., M. Bagdasarian and J.G. Zeikus. 1993b. The role active-site aromatic and polar residues in catalysis and substrate discrimination by xylose isomerase. Proc. Natl. Acad. Sci. U.S.A. *90*: 8459–8463.

Mercer, W.A. and R.H. Vaughn. 1951. The characteristics of some thermophilic, tartrate fermenting anaerobes. J. Bacteriol. *62*: 27–37.

Meyer, J., J. Gagnon, L.C. Sieker, A. Vandorsselaer and J.M. Moulis. 1990. Rubredoxin from *Clostridium thermosaccharolyticum*: amino acid sequence, mass spectrometric and preliminary crystallographic data. Biochem. J. *271*: 839–841.

Mistry, F.R. and C.L. Cooney. 1989. Production of ethanol by *Clostridium thermosaccharolyticum*.1. Effect of cell recycle and environmental parameters. Biotechnol. Bioeng. *34*: 1295–1304.

Mladenovska, Z., I.M. Mathrani and B.K. Ahring. 1995. Isolation and characterization of *Caldicellulosiruptor lactoaceticus* sp. nov., an extremely thermophilic, cellulolytic, anaerobic bacterium. Arch. Microbiol. *163*: 223–230.

Mladenovska, Z., I.M. Mathrani and B.K. Ahring. 1997. *Caldicellulosiruptor lactoaceticus* sp. nov. In Validation of the publication of new names

and new combinations previously effectively published outside the IJSB. List No. 63. Int. J. Syst. Bacteriol. *47*: 1274.

Mosolova, T.P., S.V. Kalyuzhnyi, N.G. Belogurov and S.D. Varfolomeev. 1991. Effects of antibiotics, temperature, and pH on the growth and metabolism of *Clostridium thermosaccharolyticum*. Microbiology (En. transl. from Mikrobiologiya) *60*: 340–345.

Narberhaus, F., A. Pich and H. Bahl. 1994. Synthesis of hea shock proteins in *Thermoanaerobacterium* thermasulfurigenes EM1 (*Clostridium thermosulfurogenes* EM1). Curr. Microbiol. *29*: 13–18.

Nielsen, P., I.M. Mathrani and B.K. Ahring. 1993. *Thermoanaerobium acetigenum* spec. nov., a new anaerobic, extremely thermophilic, xylanolytic non-spore-forming bacterium isolated from an Icelandic hot spring. Arch. Microbiol. *159*: 460–464.

Nikitina, O.A., S.S. Zatsepin, S.V. Kalyuzhnyi, E.I. Rainina, S.D. Varfolomeev, A.L. Zubov and V.I. Lozinskii. 1993. Hydrogen production by thermophilic anaerobic bacteria *Clostridium thermosaccharolyticum*, immobilized in polyvinyl alcohol cryogel. Microbiology (En. transl. from Mikrobiologiya) *62*: 296–301.

Onyenwoke, R.U., J.A. Brill, K. Farahi and J. Wiegel. 2004. Sporulation genes in members of the low G+C Gram-type-positive phylogenetic branch (*Firmicutes*). Arch. Microbiol. *182*: 182–192.

Onyenwoke, R.U., Y.J. Lee, S. Dabrowski, B.K. Ahring and J. Wiegel. 2006. Reclassification of *Thermoanaerobium acetigenum* as *Caldicellulosiruptor acetigenus* comb. nov. and emendation of the genus description. Int. J. Syst. Evol. Microbiol. *56*: 1391–1395.

Peteranderl, R., F. Canganella, A. Holzenburg and J. Wiegel. 1993. Induction and Regeneration of autoplasts from *Clostridium thermohydrosulfuricum* JW102 and *Thermoanaerobacter ethanolicus* JW200. Appl Environ Microbiol. *59*: 3498–3501.

Pheil, C.G. and Z.J. Ordal. 1967. Sporulation of thermophilic anaerobes. Appl. Microbiol. *15*: 893–898.

Prevot, A.R. 1938. Etudes de systematique bacterienne. IV. Critique de la conception actuelle du genre *Clostridium*. Ann. Intst. Pasteur (Paris) *61*: 72–91.

Rainey, F.A., P.H. Janssen, R.M. Daniel, H.W. Morgan and E. Stackebrandt. 1993a. A biphasic approach to the determination of the phenotypic and genotypic diversity of some anaerobic, cellulolytic, thermophilic, rod-shaped bacteria. Antonie van Leeuwenhoek *64*: 341–355.

Rainey, F.A., N.L. Ward, H.W. Morgan, R. Toalster and E. Stackebrandt. 1993b. Phylogenetic analysis of anaerobic thermophilic bacteria: aid for their reclassification. J. Bacteriol *175*: 4772–4779.

Rainey, F.A., A.M. Donnison, P.H. Janssen, D. Saul, A. Rodrigo, P.L. Bergquist, R.M. Daniel, E. Stackebrandt and H.W. Morgan. 1994. Description of *Caldicellulosiruptor saccharolyticus* gen. nov., sp. nov: an obligately anaerobic, extremely thermophilic, cellulolytic bacterium. FEMS Microbiol. Lett. *120*: 263–266.

Rainey, F.A., A.M. Donnison, P.H. Janssen, D. Saul, A. Rodrigo, P.L. Bergquist, R.M. Daniel, E. Stackebrandt and H.W. Morgan. 1995. *In* Validation of the publication of new names and new combinations previously effectively published outside the IJSB. List no. 52. Int. J. Syst. Bacteriol. *45*: 197–198.

Ramesh, M.V., S.M. Podkovyrov, S.E. Lowe and J.G. Zeikus. 1994. Cloning and sequencing of the *Thermoanaerobacterium saccharolyticum* B6A-RI *apu* gene and purification and characterization of the amylopullulanase from *Escherichia coli*. Appl. Environ. Microbiol. *60*: 94–101.

Saddler, J.N. and M.K.H. Chan. 1984. Conversion of pretreated lignocellulosic substrates to ethanol by *Clostridium thermocellum* in mono-culture and co-culture with *Clostridium thermosaccharolyticum* and *Clostridium thermohydrosulphuricum*. Can. J. Microbiol. *30*: 212–220.

Saha, B.C., R. Lamed, C.Y. Lee, S.P. Mathupala and J.G. Zeikus. 1990. Characterization of an endo-acting amylopullulanase from *Thermoanaerobacter* strain B6A. Appl. Environ. Microbiol. *56*: 881–886.

Sahm, K., M. Matuschek, H. Muller, W.J. Mitchell and H. Bahl. 1996. Molecular analysis of the *amy* gene locus of *Thermoanaerobacterium thermosulfurigenes* EM1 encoding starch-degrading enzymes and a binding protein-dependent maltose transport system. J. Bacteriol. *178*: 1039–1046.

Sanchezriera, F., D.C. Cameron and C.L. Cooney. 1987. Influence of environmental factors in the production of R(–)-1,2-propanediol by *Clostridium thermosaccharolyticum*. Biotechnol. Lett. *9*: 449–454.

Schaffer, C., K. Dietrich, B. Unger, A. Scheberl, F.A. Rainey, H. Kahlig and P. Messner. 2000. A novel type of carbohydrate-protein linkage region in the tyrosine-bound S-layer glycan of *Thermoanaerobacterium thermosaccharolyticum* D120-70. Eur. J. Biochem. *267*: 5482–5492.

Schink, B. and J.G. Zeikus. 1980. Microbial methanol formation: a major end product of pectin metabolism. Curr. Microbiol. *4*: 387–389.

Schink, B. and J.G. Zeikus. 1983a. *Clostridium thermosulfurogenes* sp. nov., a new thermophile that produces elemental sulfur from thiosulfate. J. Gen. Microbiol. *129*: 1149–1158.

Schink, B. and J.G. Zeikus. 1983b. In Validation of the publication of new names and new combinations previously effectively published outside the IJSB. List no. 12. Int. J. Syst. Bacteriol. *33*: 896–897.

Shao, W., S. DeBlois and J. Wiegel. 1995a. A high-molecular weight, cell-associated xylanase isolated from exponentially growing *Thermoanaerobacterium* sp. strain JW/SL-YS 485. Appl. Environ. Microbiol. *61*: 937–940.

Shao, W., S.K.C. Obi, J. Puls and J. Wiegel. 1995b. Purification and characterization of the alpha-glucuronidase from *Thermoanaerobacterium* sp. strain JW/SL-YS 485, an important enzyme for the utilization of substituted xylans. Appl. Environ. Microbiol. *61*: 1077–1081.

Shao, W. and J. Wiegel. 1995. Purification and characterization of two thermostable acetly xylan esterases from *Thermoanaerobacterium* sp. strain JW/SL-YS 485. Appl. Environ. Microbiol. *61*: 729–733.

Shin, H.S. and J.H. Youn. 2005. Conversion of food waste into hydrogen by thermophilic acidogenesis. Biodegradation *16*: 33–44.

Simpson, H.D. and D.A. Cowan. 1997. Controlling the enantioselectivity of sec-alcohol dehydrogenase from *Thermoanaerobacterium* sp. Ket4B1. Protein Pept. Lett. *4*: 25–31.

Sissons, C.H., K.R. Sharrock, R.M. Daniel and H.W. Morgan. 1987. Isolation of cellulolytic anaerobic extreme thermophiles from New Zealand thermal sites. Appl. Environ. Microbiol. *53*: 832–838.

Specka, U., F. Mayer and G. Antranikian. 1991. Purification and properties of a thermoactive glucoamylase from *Clostridium thermosaccharolyticum*. Appl. Environ. Microbiol. *57*: 2317–2323.

Sriprapundh, D., C. Vieille and J.G. Zeikus. 2000. Molecular determinants of xylose isomerase thermal stability and activity: analysis of thermozymes by site-directed mutagenesis. Protein Eng. *13*: 259–265.

Svetlichnyi, V.A., T.P. Svetlichnaya, N.A. Chernykh and G.A. Zavarzin. 1990. *Anaerocellum thermophilum* gen. nov. sp. nov.: an extremely thermophilic cellulolytic eubacterium isolated from hot springs in the Valley of Geysers. Microbiology (En. transl. from Mikrobiologiya) *59*: 598–604.

Tanaka, M., M. Haniu, K.T. Yasunobu, R.H. Himes and J.M. Akagi. 1973. The primary structure of the *Clostridium thermosaccharolyticum* ferredoxin, a heat-stable ferredoxin. J. Biol. Chem. *248*: 5215–5217.

Taya, M., H. Hinoki and T. Kobayashi. 1985. Tungsten requirement of an extremely thermophilic, cellulolytic anaerobe (strain NA10). Agric. Biol. Chem. *52*: 2513–2515.

Taya, M., H. Hinok, Y. Suzuki, T. Yagi and T. Kobayashi. 1988. Isolation and characterization of an extremely thermophilic, cellulolytic, anaerobic bacterium. Appl. Microbiol. Biotechnol. *29*: 474–479.

Tyurin, M.V., C.R. Sullivan and L.R. Lynd. 2005. Role of spontaneous current oscillations during high-efficiency electrotransformation of thermophilic anaerobes. Appl. Environ. Microbiol. *71*: 8069–8076.

Tyurin, M.V., L.R. Lynd and J. Wiegel. 2006. Gene transfer systems for obligately anaerobic thermophilic bacteria. *In* Oren and Rainey (Editors), Methods in Microbiology. Academic Press/Elsevier, pp. 309–330.

Ueno, Y., D. Sasaki, H. Fukui, S. Haruta, M. Ishii and Y. Igarashi. 2006. Changes in bacterial community during fermentative hydrogen and acid production from organic waste by thermophilic anaerobic microflora. J. Appl. Microbiol. *101*: 331–343.

van Rijssel, M. and T.A. Hansen. 1989. Fermentation of pectin by a newly isolated *Clostridium thermosaccharolyticum* strain. FEMS Microbiol. Lett. *61*: 41–46.

van Rijssel, M., M.P. Smidt, G. Vankouwen and T.A. Hansen. 1993. Involvement of an intracellular oligogalacturonate hydrolase in metabolism of pectin by *Clostridium thermosaccharolyticum*. Appl. Environ. Microbiol. *59*: 837–842.

Vancanneyt, M., P. Devos and J. Deley. 1987. Ethanol production from glucose by *Clostridium thermosaccharolyticum* strains: effect of pH and temperature. Biotechnol. Lett. *9*: 567–572.

Venkateswaran, S. and A.L. Demain. 1986. The *Clostridium thermocellum*-*Clostridium thermosaccharolyticum* ethanol production process: nutritional studies and scale down. Chem. Eng. Comm. *45*: 53–60.

Vocadlo, D.J., J. Wicki, K. Rupitz and S.G. Withers. 2002a. A case for reverse protonation: Identification of Glu160 as an acid/base catalyst in *Thermoanaerobacterium saccharolyticum* ss-xylosidase and detailed kinetic analysis of a site-directed mutant. Biochemistry *41*: 9736–9746.

Vocadlo, D.J., J. Wicki, K. Rupitz and S.G. Withers. 2002b. Mechanism of *Thermoanaerobacterium saccharolyticum* beta-xylosidase: kinetic studies. Biochemistry *41*: 9727–9735.

Wagschal, K., D. Franqui-Espiet, C.C. Lee, G.H. Robertson and D.W. Wong. 2005. Enzyme-coupled assay for beta-xylosidase hydrolysis of natural substrates. Appl. Environ. Microbiol. *71*: 5318–5323.

Weimer, P.J. 1985. Thermophilic anaerobic fermentation of hemicellulose and hemicellulose-derived aldose sugars by *Thermoanaerobacter* strain B6A. Arch. Microbiol. *143*: 130–136.

Wiegel, J. and L.G. Ljungdahl. 1981. *Thermoanaerobacter ethanolicus* gen. nov., spec. nov, a new, extreme thermophilic, anaerobic bacterium. Arch. Microbiol. *128*: 343–348.

Wiegel, J. 1990. Temperature spans for growth: hypothesis and discussion. FEMS Microbiol. Rev. *75*: 155–169.

Wiegel, J. 1998. Lateral gene exchange, an evolutionary mechanism for extending the upper or lower temperature limits for growth of a microorganism? A hypothesis. *In* Wiegel and Adams (Editors), Thermophiles, the Molecular Key to the Evolution and Origin of Life? Taylor & Francis, London, pp. 175–185.

Wilder, M., J.M. Akagi and R.C. Valentine. 1963. Ferredoxin of *Clostridium thermosaccharolyticum*. J. Bacteriol. *86*: 861–865.

Wind, R.D., J.C.M. Uitdehaag, R.M. Buitelaar, B.W. Dijkstra and L. Dijkhuizen. 1998. Engineering of cyclodextrin product specificity and pH optima of the thermostable cyclodextrin glycosyltransferase from *Thermoanaerobacterium thermosulfurigenes* EM1. J. Biol. Chem. *273*: 5771–5779.

Yang, J.K., H.J. Yoon, H.J. Ahn, B.I. Lee, J.D. Pedelacq, E.C. Liong, J. Berendzen, M. Laivenieks, C. Vieille, J.G. Zeikus, D.J. Vocadlo, S.G. Withers and S.W. Suh. 2004. Crystal structure of β-D-xylosidase from *Thermoanaerobacterium saccharolyticum*, a family 39 glycoside hydrolase. J. Mol. Biol. *335*: 155–165.

Zavarzina, D.G., T.P. Tourova, B.B. Kuznetsov, E.A. Bonch-Osmolovskaya and A.I. Slobodkin. 2002. *Thermovenabulum ferriorganovorum* gen. nov., sp. nov., a novel thermophilic, anaerobic, endospore-forming bacterium. Int. J. Syst. Evol. Microbiol. *52*: 1737–1743.

Zolotarev, Y.A., A.K. Dadayan, Y.A. Borisov, E.M. Dorokhova, V.S. Kozik, N.N. Vtyurin, E.V. Bocharov, R.N. Ziganshin, N.A. Lunina, S.V. Kostrov, T.V. Ovchinnikova and N.F. Myasoedov. 2003. The effect of three-dimensional structure on the solid state isotope exchange of hydrogen in polypeptides with spillover hydrogen. Bioorg. Chem. *31*: 453–463.

Family IV. **Incertae Sedis**

This genus is not closely related to members of any previously described family. While some rRNA analyzes suggests an affiliation with the *Clostridiales*, other analyses and phenotypic properties suggest a relationship to *Thermoanaerobacteraceae* fam. nov. and *Thermoanaerobacterales* fam. nov. Family III *Incertae Sedis*. In view of these ambiguities, this genus is assigned to its own family *incertae sedis* (Figure 6).

Genus I. **Mahella** Bonilla-Salinas, Fardeau, Thomas, Cayol, Patel and Ollivier 2004, 2172[VP]

MONICA BONILLA-SALINAS, BERNARD OLLIVIER, BHARAT K.C. PATEL AND JEAN-LUC CAYOL

Mah.el′la. L. dim. ending -*ella*; N.L. fem. n. *Mahella* named in honor of the American microbiologist Robert A. Mah for his important contribution to the taxonomy of anaerobes.

Cells are **straight rods**, about $3–20\,\mu m \times 0.5\,\mu m$, occurring singly or in pairs. The rods are motile by **peritrichous flagella** and stain **Gram-positive**. **Spores** are formed. **Obligately anaerobic**. Moderately **thermophilic**; the optimum growth temperature is $50\,°C$. **Halotolerant** and grows in the presence of up to 4% NaCl with an optimum of 0.1%. The pH optimum is 7.5. **Heterotroph**. Glucose is fermented to lactate, formate, H_2, and CO_2. Yeast extract is not required for growth. Elemental sulfur, sulfate, thiosulfate, sulfite, nitrate, and nitrite are not used as electron acceptors.

DNA G+C content (mol%): 55.5.

Type species: **Mahella australensis** Bonilla-Salinas, Fardeau, Thomas, Cayol, Patel and Ollivier 2004, 2172[VP].

Further descriptive information

Forms white, round, smooth colonies (1–2 mm in diameter) after 7 d in Hungate tubes with medium solidified with 2% Noble agar. A Gram-positive cell wall structure is visible by electron microscopy of ultrathin sections. Cells form terminal endospores with swollen sporangia (Figure 260).

Cells grow anaerobically with the following carbon and energy sources: arabinose, cellobiose, fructose, galactose, glucose, mannose, sucrose, xylose, and yeast extract. Glucose is fermented into (mol per mol of glucose consumed): lactate, 1.3; formate, 1.5; acetate, 0.2; ethanol, 0.5; H_2, and CO_2. Acetate together with H_2 and CO_2 are the products of pyruvate fermentation. Yeast extract is not required but increases growth. The following compounds are not fermented: ethanol, methanol, 1-propanol, glycerol, 1,2-propanediol, olive oil, starch, benzoate, formate, succinate, lactate, fumarate, acetate, propionate, peptone, and Casamino acids. The doubling time under optimal conditions is 11 h.

Cells are resistant to penicillin and ampicillin, which do not inhibit growth at concentrations of $200\,\mu g/ml$. Cells are sensitive to chloramphenicol at a concentration of $50\,\mu g/ml$.

The organism was isolated from an oil-well water sample collected from the Riverslea oilfield in the Bowen-Surat Basin of Queensland in eastern Australia.

Enrichment and isolation procedure

Cells are strictly anaerobic and their cultivation requires the Hungate technique (Hungate, 1969). The initial enrichment medium contains complex organic substrates (e.g., yeast extract and bioTrypticase as energy sources) and (per liter of distilled water): 0.3 g of K_2HPO_4, 0.3 g of KH_2PO_4, 0.2 g of $MgCl_2 \cdot 6H_2O$, 0.1 g of $CaCl_2 \cdot 2H_2O$, 0.5 g of cysteine hydrochloride, 1 mg of resazurin, 0.1 g of KCl, 1 g of NaCl, 1 g of NH_4Cl, and 10 ml of trace mineral solution (Balch et al., 1979). The pH is adjusted to 7.0 with 10 M KOH. Vessels are autoclaved for 45 min at $110\,°C$. Prior to inoculation, $Na_2S \cdot 9H_2O$ and $NaHCO_3$ are added from sterile stock solutions. The type strain was enriched from 2 ml of oil well water inoculated into 20 ml of enrichment medium and incubated at $50\,°C$. Growth was observed after 3–4 d. The basal medium used for characterization was similar to the enrichment medium supplemented with 20 mM glucose.

TABLE 254. Major discriminating characteristics between *Mahella australiensis* and its closest relatives *Thermoanaerobacterium thermosulfurogenes* and *Thermoanaerobacterium aotearoense*[a]

Characteristic	*M. australiensis*[b]	*T. thermosulfurogenes*[b]	*T. aotearoense*[c]
Temperature range (°C)	30–60	35–75	35–66
Optimum temperature (°C)	50	60	60–63
DNA G + C content (mol%)	55.5	32.6	34.5–35.0
Reduction of thiosulfate	–	+	+
End products of glucose fermentation	**L, F**, E, A, H_2, CO_2	**E, A**, L, H_2, CO_2	**E, A**, L, H_2, CO_2

[a]Symbols and abbreviations: +, present; –, absent; A, acetate; E, ethanol; F, formate; L, lactate. Major end products of metabolism are indicated in bold type.
[b]Data from Bonilla-Salinas et al. (2004).
[c]Data from Schink and Zeikus (1983).
[d]Data from Liu et al. (1996).

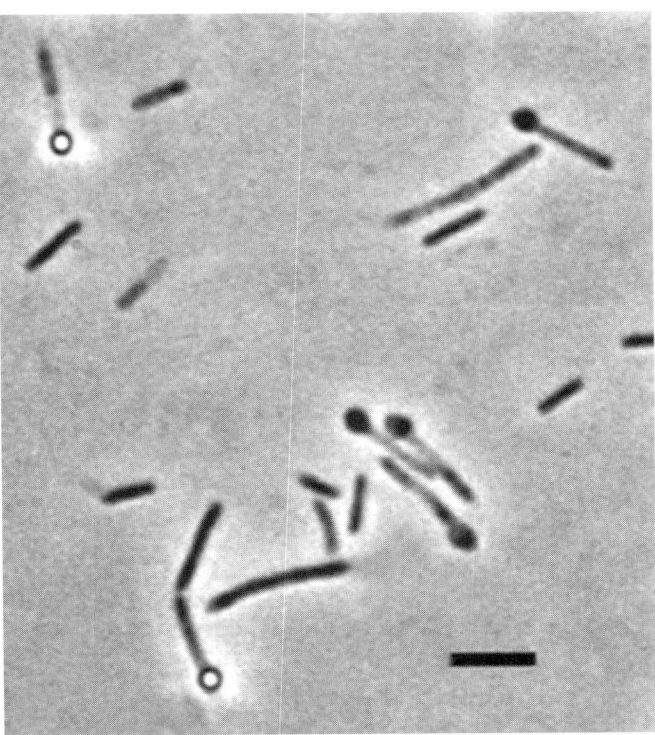

FIGURE 260. Phase-contrast micrograph of *Mahella australiensis* showing terminal spores and swollen sporangia. Bar = 10 μm. (Reprinted with permission from Bonilla-Salinas et al., 2004. Int. J. Syst. Evol. Microbiol. *54*: 2169–2173.)

Maintenance procedure

For a long-term storage of at least one year, anaerobic cultures can be maintained at –80 °C in enrichment medium plus 20% glycerol.

Differentiation of the genus *Mahella* from other genera

On the basis of 16S rRNA gene sequence analyses, *Mahella* is related to members of the family *Thermoanaerobacteraceae*. This family includes thermophiles of the genera *Thermoanaerobacter* and *Thermoanaerobacterium* that use thiosulfate as a terminal electron acceptor, reducing it to sulfide or elemental sulfur, respectively. The 16S rRNA genes of *Thermoanaerobacterium thermosulfurigenes* and *Thermoanaerobacterium aotearoense* possess the highest sequence similarity to *Mahella australiensis*, 85.7 and 85.5%, respectively (Figure 261). However, unlike both species, *Mahella australiensis* is unable to reduce thiosulfate and does not grow at temperatures above 60 °C. In addition, the DNA G+C content of *Mahella australiensis* is much higher than that of the *Thermoanaerobacterium* species (Table 254). Differences are also apparent in the fermentation products. *Mahella australiensis* is also clearly distinguished by molecular and phenotypic criteria from other anaerobic, spore-forming bacteria belonging to the order *Clostridiales*.

Taxonomics comments

By 16S rRNA sequence analyses, *Mahella* is related to cluster VI of the order *Clostridiales* (Figure 261), but the assignment of *Mahella* to a family is not clear. Despite its close relationships with members of the family *Thermoanaerobacteraceae*, there are significant metabolic differences (e.g., its inability to use thiosulfate as terminal electron acceptor and grow above 60 °C) that indicate that *Mahella* should be classified in a novel family. This conclusion is supported by detailed phylogenetic analyses of the *Clostridiales* and suggests that the family *Thermoanaerobacteraceae* is itself paraphyletic and represents multiple lineages (Ludwig et al., 2009).

Differentiation of species of the genus *Mahella*

Only one species, *Mahella australiensis*, is currently known.

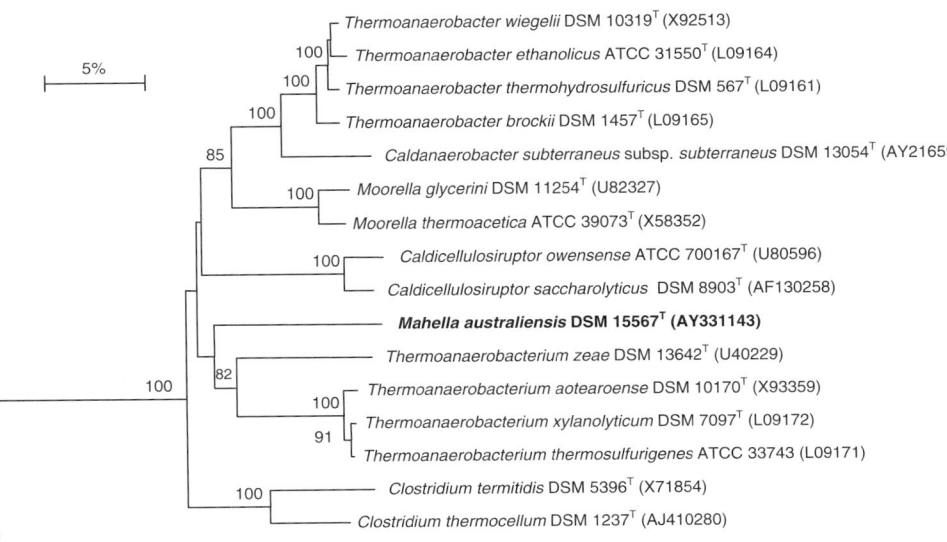

FIGURE 261. Neighbor-joining phylogenetic tree of *Mahella australiensis* and representatives of the *Thermoanaerobacterales* and other members of the *Clostridiales* based upon the 16S rRNA gene. *Escherichia coli* is the outgroup. Bootstrap values from 100 replicates are shown at the nodes. Only values greater than 80 are reported. The bar represents 5 inferred substitutions per 100 positions.

List of species of the genus *Mahella*

1. **Mahella australiensis** Bonilla-Salinas, Fardeau, Thomas, Cayol, Patel and Ollivier 2004, 2172[VP]

    aus.tra.li.en′sis. N.L. fem. adj. *australiensis* related to Australia.

Description is the same as for the genus.
DNA G+C content (mol%): 55.5 (HPLC).
Type strain: 50-1 BON, CIP 107919, DSM 15567.
GenBank accession number (16S rRNA gene): AY331143.

References

Balch, W.E., G.E. Fox, L.J. Magrum, C.R. Woese and R.S. Wolfe. 1979. Methanogens: reevaluation of a unique biological group. Microbiol. Rev. *43*: 260-296.

Bonilla Salinas, M., M.L. Fardeau, P. Thomas, J.L. Cayol, B.K.C. Patel and O. B. 2004. *Mahella australiensis* gen. nov., sp. nov., a moderately thermophilic anaerobic bacterium isolated from an Australian oil well. Int. J. Syst. Evol. Microbiol. *54*: 2169-2173.

Hungate, R.E. 1969. A roll tube method for cultivation of strict anaerobes. *In* Norris and Ribbons (Editors), Methods in Microbiology, Vol. 3B. Academic Press, London and New York, pp. 117-132.

Liu, S.Y., F.A. Rainey, H.W. Morgan, F. Mayer and J. Wiegel. 1996. *Thermoanaerobacterium aotearoense* sp. nov., a slightly acidophilic, anaerobic thermophile isolated from various hot springs in New Zealand, and emendation of the genus *Thermoanaerobacterium*. Int. J. Syst. Bacteriol. *46*: 388-396.

Ludwig, W., K.-H. Schleifer and W.B. Whitman. 2009. Revised road map to the phylum *Firmicutes*. *In* De Vos, Garrity, Jones, Krieg, Ludwig, Rainey, Schleifer and Whitman (Editors), Bergey's Manual of Systematic Bacteriology, 2nd edn, Vol. 3, The *Firmicutes*, Springer, New York, pp. 1–14.

Schink, B. and J.G. Zeikus. 1983. *Clostridium thermosulfurogenes* sp. nov., a new thermophile that produces elemental sulfur from thiosulfate. J. Gen. Microbiol. *129*: 1149-1158.

Class III. **Erysipelotrichia** class. nov.

WOLFGANG LUDWIG, KARL-HEINZ SCHLEIFER AND WILLIAM B. WHITMAN

E.ry.si.pe.lo.tri′chia. N.L. fem. n. *Erysipelothrix* type genus of the type order of the class; suff. -*ia* ending proposed by Gibbons and Murray and by Stackebrandt et al. to denote a class; N.L. neut. pl. n. *Erysipelotrichia* the *Erysipelothrix* class.

The class *Erysipelotrichia* is circumscribed for this volume on the basis of the phylogenetic analyses of the 16S rRNA sequences and includes only the order *Erysipelotrichales* and the family *Erysipelotrichaceae*. The description of the class is the same as the order.

Type order: **Erysipelotrichales** ord. nov.

Order I. **Erysipelotrichales** ord. nov.

WOLFGANG LUDWIG, KARL-HEINZ SCHLEIFER AND WILLIAM B. WHITMAN

E.ry.si.pe.lo.tri′cha.les. N.L. fem. n. *Erysipelothrix* type genus of the order; suff. -*ales* ending to denote an order; N.L. fem. pl. n. *Erysipelotrichales* the *Erysipelothrix* order.

The order *Erysipelotrichales* is circumscribed for this volume on the basis of the phylogenetic analyses of the 16S rRNA sequences and includes only the family *Erysipelotrichaceae* and its eight genera. It is composed of nonmotile, slender rods with a Gram-positive cell-wall structure. Endospores are not produced.

Aerobic to facultatively anaerobic. Chemo-organotrophic, metabolism respiratory and weakly fermentative. Cytochromes and isoprenoid quinones absent. Otherwise, the description of the order is the same as for the family.

Type genus: **Erysipelothrix** Rosenbach 1909, 367[AL].

Reference

Rosenbach, F.J. 1909. Experimentelle, Morphologische und klinische Studien über krankheitserregende Mikroorganismen des Scheinerotlaufs, des Crysipeloids und der Mõusesepticämie. Z. Hyg. Infektionskr. *63*: 343–371.

Family I. **Erysipelotrichaceae** Verbarg, Rheims, Emus, Frühling, Kroppenstedt, Stackebrandt and Schumann 2004, 223^{VP}

ERKO STACKEBRANDT

E.ry.si.pe.lo.thri.cha′ce.ae. N.L. fem. n. *Erysipelothrix* type genus of the family; -*aceae* ending to denote a family; N.L. fem. pl. n. *Erysipelotrichaceae* the *Erysipelothrix* family.

Straight, or slightly curved, slender rods; some strains with a tendency to form long filaments. Nonmotile. Endospores are not produced. Aerobic to **facultatively anaerobic. Chemo-organotrophic, metabolism respiratory and weakly fermentative.** Acid but no gas produced from glucose and other carbohydrates. **Cytochromes and isoprenoid quinones absent.** Members of some genera contain **peptidoglycan belonging to the B-cross-linking type**, having L-alanine in position three of the peptide subunit and an interpeptide bridge consisting of the peptide Gly→L-Lys→L-Lys. Predominant fatty acids are $C_{16:0}$, $C_{18:1\ cis9}$, and $C_{18:0}$. Some strains pathogenic for mammals and birds. Phylogenetically a member of the phylum *Firmicutes* (Figure 4).

DNA G+C content (mol%): 36–40 (HPLC, T_m, Bd).

Type genus: **Erysipelothrix** Rosenbach 1909, 367^{AL}.

Further descriptive information

The hallmark of *Erysipelothrix* is the presence of a cell wall of peptidoglycan group B in which the peptide bridge is formed between amino acids at positions 2 and 4 of adjacent peptide side chains and not, as in the vast majority of bacteria, between amino acids at positions 3 and 4. To link the two carboxylic groups of amino acids at positions 2 and 4, the interpeptide bridge of B-group must contain at least one diamino acid residue. The B-type occurs within the family *Microbacteriaceae*, phylum *Actinobacteria*, order *Actinomycetales*, and in some members of the class *Clostridia*, e.g., *Erysipelothrix, Holdemania, Acetobacterium, Clostridium barkeri*, and *Eubacterium limosum* (Schleifer and Kandler, 1972; Schubert and Fiedler, 2001; Willems et al., 1997).

In comparison to the original description, the family *Erysipelotrichaceae* was extended to include four additional genera. Besides one newly described genus *Allobaculum* (Greetham et al., 2004), three of them were transferred from other families. The family is now organized in eight genera: *Erysipelothrix, Allobaculum, Bulleidia, Catenibacterium* (formerly *Lachnospiraceae*), *Coprobacillus* (formerly *Clostridia*), *Holdemania, Solobacterium*, and *Turicibacter* (formerly *incertae sedis* among *Bacillales*).

Furthermore, a number of type species validly published as members of genera not assigned to the *Erysipelotrichaceae* are members of this family. *Clostridium catenaformis, Clostridium cocleatum, Clostridium innocuum, Clostridium ramosum*, and *Clostridium spiroforme* as well as *Eubacterium biforme, Eubacterium cylindroides, Eubacterium dolichum*, and *Eubacterium tortuosum* apparently are misclassified in *Clostridiales* and consequently should be transferred to *Erysipelotrichaceae*. Similarly, phylogenetically *Lactobacillus catenaformis* and *Lactobacillus vitulinus*, as well as *Streptococcus pleomorphus*, do not belong to the *Lactobacillales* but are members of the *Erysipelotrichaceae*.

Genus I. **Erysipelothrix** Rosenbach 1909, 367^{AL}

ERKO STACKEBRANDT

E.ry.si.pe′lo.thrix. Gr. neut. n. *erysipelas* erysipelas; Gr. fem. n. *thrix* hair; N.L. fem. n. *Erysipelothrix* erysipelas thread.

Straight, or slightly curved, slender rods with a tendency to form long filaments. Rods usually 0.2–0.4 μm to 0.8–2.5 μm or 0.5 μm in diameter and 1.5–3.0 μm in length with rounded ends; occur singly in short chains, in pairs at an angle to give V-forms, or in groups with no particular arrangement. Filaments may be 60 μm or more in length. **Nonmotile.** Capsules not formed. **Do not form spores.** Gram-positive. Not acid-fast. **Facultatively anaerobic.** Colonies small, usually transparent and nonpigmented. Narrow zones of α-hemolysis may occur on blood agar. No β-hemolysis. Optimum temperature 30–37 °C; growth occurs between 5 and 42 °C. Do not survive heating at 60 °C for 15 min. Catalase-negative. Oxidase-negative. Fermentative activity weak. **Acid but no gas produced from glucose and certain other carbohydrates.** Organic growth factors are required. **Cell wall contains a group B peptidoglycan based on lysine. Mycolic acids not present. The fatty acids are primarily of the straight chain mono-unsaturated ($C_{18:1\ cis9}$) and straight chain saturated series ($C_{16:0}$, $C_{18:0}$).** Widely distributed in nature. May be parasitic on mammals, birds and fish; some strains pathogenic for mammals and birds.

DNA G+C content (mol%): 36–40 (T_m, Bd).

Type species: **Erysipelothrix rhusiopathiae** (Migula 1900) Buchanan 1918, 55^{AL}.

Further descriptive information*

The species of clinical interest is *Erysipelothrix rhusiopathiae*. The great variation in serological, biochemical, chemical, and genomic properties of the species was noted (Erler, 1972; Feist, 1972; Flossmann and Erler, 1972; Takahashi et al., 1992; White and Mirikitani, 1976) which led to the description of *Erysipelothrix tonsillarum* (also named *Erysipelothrix tonsillae* in the older literature (Takahashi et al., 1989) for serotype 7 strains that were frequently isolated from the tonsils of apparently healthy pigs (Takahashi et al., 1987a). Takahashi et al. (2000) reported the evidence that some of these strains may actually be canine pathogens. The species *Erysipelothrix inopinata*, as yet represented by a single strain, was isolated from sterile-filtered vegetable broth (Verbarg et al., 2004); serotype 13 strains of *Erysipelothrix rhusiopathiae* may be included in this species (see *Molecular differentiation of strains and species*, below).

---

*Based on the description given in the first edition by Jones (1986a).

As summarized by Jones (1986a), strains of *Erysipelothrix rhusiopathiae* exist in a rough and smooth form each characterized by closely associated morphological and colonial features (a property not mentioned in the species description of *Erysipelothrix tonsillarum* and *Erysipelothrix inopinata*). In the smooth form (S-form) the cells appear as small, straight, or slightly curved Gram-positive rods. Colonies (24–48 h) are very small, 0.3–1.5 mm in diameter, convex, circular, transparent with smooth glistening surface, and entire. In older cultures the colonies become slightly larger and the center becomes opaque. In the rough form (R-form), long filaments often more than 60 μm in length predominate. R-form colonies are larger, flatter, and more opaque; the surface is matt, uneven and the edge irregular. R-form colonies may resemble miniature anthrax colonies (Wilson and Miles, 1975). Both forms may be decolorized easily when Gram-stained. Both rods and filaments may appear Gram-negative, but contain Gram-positive granules that give them a beaded appearance. In smears of blood and tissue taken from acute forms of infection, especially in cases of septicemia, the organisms are of the S-form; in chronic infections with arthritis and endocarditis, the R-form is isolated frequently, and S-forms are usually present along with R-forms (Ewald, 1981). The cell morphology varies to some extent with the growth medium and conditions of incubation. It has been reported that a pH of 7.6–8.2 favors the S-form while at pH values below 7.0 the R-form predominates (Wilson and Miles, 1975). Grieco and Sheldon (1970) claimed that S-forms grow better at 33 °C, and at 37 °C the R-form is favored. The distinction between S-form and R-form colonies is not always sharp; intermediate forms may exist (Barber, 1939; Ewald, 1981). S-form colonies dissociate to give rise to intermediate and R-form colonies, while R-forms may also give rise to S-form colonies.

Erysipelothrix rhusiopathiae and *Erysipelothrix tonsillarum* strains produce a characteristic "pipe cleaner" type of growth in gelatin stab cultures incubated at 22 °C (*Erysipelothrix inopinata* was not tested). At first (24 h), growth is faint and hazy and confined to an area just below the surface. After a few days the growth appears as a column extending to the bottom of the tube. The column of growth is composed of fine lateral outgrowths, which in S-form organisms extend only 2 or 3 mm from the stab but may reach the sides of the tube in R-form organisms. Gelatin is not liquefied when incubation is at 22, 30, or 37 °C.

Members are microaerophilic. Especially on first isolation, *Erysipelothrix rhusiopathiae* grows in a band just below the surface of a soft agar culture. Whether this is due to a preference for CO_2 or reduced oxygen tension is not resolved. Laboratory cultures grow well aerobically and anaerobically.

Growth is favored by a slightly alkaline pH. The optimum pH for growth is between 7.2 and 7.6; pH limits of growth have been reported as 6.8–8.2 (Karlson and Merchant, 1941) and 6.7–9.2 (Sneath et al., 1951). *Erysipelothrix inopinata* grows under aerobic and anaerobic conditions in BHI and Columbia Blood, preferably at pH 8.

Growth of *Erysipelothrix rhusiopathiae* in nutrient agar is improved by the addition of glucose (0.2–0.5% w/v) or serum (5–10% v/v). The exact growth requirements of either type strain have not been determined; for growth of *Erysipelothrix rhusiopathiae* several amino acids, riboflavin, and small amounts of oleic acid are required (Hunter, 1942.). Ewald (1981) noted that tryptophan enhanced growth.

The majority of *Erysipelothrix rhusiopathiae* strains produce H_2S, but the results can vary with the medium used. The use of lead acetate paper with cultures in a liquid or solid medium does not always detect H_2S production. Tests are best carried out in triple-sugar iron agar (Wood, 1965). Notably, an occasional old laboratory or vaccine strain does not produce detectable H_2S in triple-sugar iron agar (R.L. Wood, personal communication to D. Jones).

Most known strains produce hyaluronidase and it has been speculated that virulence is correlated with hyaluronidase production; good hyaluronidase producers usually belong to serovar 1 (Ewald, 1957, 1981). Neuraminidase is produced in differing amounts by all strains, and the level of neuraminidase activity appears to correlate with virulence (Krasemann and Müller, 1975; Müller and Krasemann, 1976; Müller and Seidler, 1975; Nikolov and Abrashev, 1976).

Erysipelothrix rhusiopathiae strains may be identified serologically. Heat and acid-stable, type-specific, and heat-labile species antigens occur. This accounts for differences observed between antisera prepared with boiled and unboiled strains. Most strains agglutinate with unabsorbed sera prepared against unboiled antigens. The type-specific antigens are polysaccharide complexes. Several serovars have been detected and different serotyping schemes proposed (Kucsera, 1973). Dedié (1949) recognized two serovars, A and B, and proposed that all those strains which showed no reaction with A- or B-type specific antiserum be designated a third group, "N". As new or supposedly new serovars were detected within group N, they were designated by consecutive letters of the alphabet. Unfortunately, new serovars were not always compared with strains belonging to different serovars and the serological methods used were different. To overcome this problem, Kucsera (1973) recommended the use of a uniform serological method (in particular, the double agar-gel diffusion precipitation test with autoclaved antigens and type-specific antisera) and a uniform system for designating the serovar using Arabic numbers.

Although capital letters of the alphabet are still used by some workers to designate the two main serovars (A and B), the numbered system introduced by Kucsera (1973) is now preferred by most investigators interested in the serotyping of *Erysipelothrix rhusiopathiae*. Under this system, Dedié's original serovars A and B are designated 1 and 2, respectively. All the other serovars recognized to date have been found within Dedié's group N (Kucsera, 1973; Norrung, 1979; Wood et al., 1978). Of the 23 serovars, 1 and 2 are the most common; the other 21 are relatively rare. Strains with serotype 3, 7 (including the type strains), 10, 14, 20, 22, and 23 are now classified in *Erysipelothrix tonsillarum*. Cultures of serovar 2 agglutinate chicken red blood cells, which lyse when complement is added to the complex (Dinter et al., 1976).

L-Forms of *Erysipelothrix rhusiopathiae* have been described (Ewald, 1981). Bacteriophages active on *Erysipelothrix rhusiopathiae* strains have been isolated, though a functional phage typing system has never been developed. Lysogeny has been reported (Ewald, 1981).

In vitro, most strains of *Erysipelothrix rhusiopathiae* (and probably *Erysipelothrix tonsillarum*) are resistant to sulfonamides, colistin, gentamicin, kanamycin, neomycin, novobiocin, and polymyxin and are sensitive to penicillin, streptomycin, chloramphenicol, tetracycline, and other antibiotics (Fuzi, 1963; Sneath et al., 1951; Wood, 1965).

The organisms grow in the presence of phenol (0.2% w/v), potassium tellurite (0.05% w/v), sodium azide (0.1% w/v), thallous acetate (0.02% w/v), 2,3,5-triphenyltetrazolium chloride (0.2% w/v), and crystal violet (0.001% w/v) (Ewald, 1981; Sneath et al., 1951).

On blood agar, α-hemolysis may be so intense that after 48 h incubation, a slight clearing may be seen. Care must be taken in observing the plates because true β-hemolysis never occurs.

Physiological properties. For *Erysipelothrix rhusiopathiae*, acid production from carbohydrates is usually poor or inconsistent when grown in 1% (w/v) peptone water. Most workers recommend the addition of sterile horse serum (5–10% v/v) to the basal medium (Seeliger, 1974; White and Shuman, 1961; Wood, 1970). It is not always convenient to use serum: good results are achieved by testing for acid production in nutrient broth (Oxoid) with the test carbohydrate (0.5–1% w/v) and phenol red as indicator.

The methyl red test is usually reported as negative, but Wilkinson and Jones (1977) reported all strains examined gave a weak positive reaction when incubated at 35 °C for 7 d in BM broth. Acetoin (Voges–Proskauer test) was not produced in this medium. BM broth consists of peptone (Oxoid), 10.0 g; Lab Lemco (Oxoid), 8.0 g; yeast extract (Oxoid), 4.0 g; glucose, 5.0 g; Tween 80, 1 ml; K_2HPO_4, 5.0 g; $MgSO_4 \cdot 7H_2O$, 0.2 g; $MnSO_4 \cdot 5H_2O$, 0.05 g; and distilled water, 1000 ml. Fermentation of glucose results in the production of mainly L(+)-lactic acid with smaller amounts of acetic acid, formic acid, ethyl alcohol, and carbon dioxide. Glucose metabolism is via the Embden–Meyerhof–Parnas pathway, although a small amount of glucose is dissimilated by the hexose monophosphate pathway (Robertson and McCullough, 1968). The evidence available indicates that the tricarboxylic acid cycle is relatively unimportant in *Erysipelothrix rhusiopathiae* (Robertson and McCullough, 1968). Exogenous citrate is not utilized.

Utilization of substrates and enzyme activities of type strains of *Erysipelothrix* species and some additional *Erysipelothrix rhusiopathiae* strains were determined using Biolog GP and API STREPT microplate plate panels, respectively (Verbarg et al., 2004). Most substrates revealed identical reactions by type strains of the species (see species description), but sufficient differences are observed which allow their discrimination (Table 255).

Habitat. *Erysipelothrix rhusiopathiae* has a very wide distribution in nature. It is parasitic on mammals, birds, and fish. Most frequently it is found in pigs where it is the causative agent of the economically important disease, swine erysipelas. Infection probably occurs by the oral route through the ingestion of contaminated material. It has been isolated, not in association with infection, from the surfaces of fish, shellfish, fish slime, and fish boxes (Ewald, 1981; Grieco and Sheldon, 1970; Woodbine, 1950).

Besides from swine erysipelas, *Erysipelothrix rhusiopathiae* has been isolated from bovine tonsils (Hassanein et al., 2003), causing endocarditis in dogs (Takahasi et al., 2000), polyarthritis in lambs and calves, septicemia in turkeys and ducks, septicemia and urticaria in dolphins, and cutaneous lesions in humans (Takahashi et al., 1987b). Natural infections, with epizootics, also occur in other domestic animals (sheep, lambs, cows, horses, dogs, and mice) and birds (turkeys, chickens, geese, and pheasants) and also in wild and zoo animals. Strains of this species have been isolated from the slime on the bodies of a variety of different fish, and from cephalopods and crustaceans,

TABLE 255. Phenotypic properties that differentiate type strains of *Erysipelothrix* species as determined by API 32 STREPT and Biolog GP microplate panels[a]

Characteristic	1. *E. rhusiopathiae*	2. *E. inopinata*	3. *E. tonsillarum*
API STREPT			
β-Glucosidase	–[b]	+	+
Alkaline phosphatase	–	–	+
Ribose (acid)	–	w	+
Lactose (acid)	+	–	–
Trehalose (acid)	–	+	–
N-Acetyl-β-glucosamidase	+(v)	+	+
β-Mannosidase	–	w	–
Biolog GP Microplate panel			
Utilization of:			
L-Arabinose	+	–	w
N-Acetyl-D-mannosamine	+	–	+
Arbutin	–	+	–
Cellobiose	–	+	–
D-Fructose	+	–	+
D-Galactose	+	–	+
Gentiobiose	–	+	–
Glycerol	–	+	–
α-D-Lactose	+(v)	–	–
D-Mannose	+(v)	–	–
3-Methyl glucose	–(v)	–	–
D-Psicose	+	–	+
D-Ribose	–(v)	w	+
Salicin	–	+	–
D-Trehalose	–	+	–
Xylose	–	–	w

[a]Symbols: +, positive; –, negative; w, weak; v, variable.
[b]The reactions of three additional strains are indicated in parenthesis (Verbarg et al., 2004).

including oysters and lobsters (Fidalgo et al., 2000), from salted or pickled bacon after several weeks, and from ham 3 months after smoking. Viable organisms have been recovered from a buried carcass after 9 months. Heat and direct sunlight diminish the viability of *Erysipelothrix rhusiopathiae*. A low temperature, alkaline conditions, and organic matter favor its survival (Ewald, 1981; Grieco and Sheldon, 1970; Woodbine, 1950).

Pathogenic strains of *Erysipelothrix rhusiopathiae* have been isolated from the feces of apparently healthy swine, and asymptomatic swine commonly harbor this organism in their tonsils and other lymphoid tissues (Wood, 1974a). Some of these isolates are included in *Erysipelothrix tonsillarum* (Takahashi et al., 1987a). Contamination of soil and water occurs not only from the feces and urine of sick animals but also from the activities of asymptomatic carrier pigs (Wood, 1974b). The shedding of the organisms by asymptomatic pigs into the soil of pigpens is probably the reason that *Erysipelothrix rhusiopathiae* may be isolated from farms on which no cases of swine erysipelas have occurred for many years (Wood and Packer, 1972). Isolates of the genus have also been recovered from the tonsils of healthy cattle in Japan (Hassanein et al., 2001). Contaminated surface water, rodents, wild birds, and insects may be responsible for carrying the organisms to farms and fish factories. *Erysipelothrix* spp. were isolated from the lung and intestine of a diseased blue penguin (Boerner et al., 2004), which is the first reported case of erysipelas infection in a captive aquatic bird. In man it causes a skin infection known as erysipeloid. The majority of reported infections result from direct handling of contaminated organic matter such as swine carcasses, fish, and poultry. The infections are largely limited to veterinarians, butchers, and fish handlers. Generally, infection is confined to the skin of the hands and lower arms where the organisms gain entry through cuts and abrasions. Only rarely does the infection become systemic, causing arthritis and endocarditis. Most fatal cases have been shown to occur among excessive users of alcohol (Ewald, 1981). The occurrence of *Erysipelothrix rhusiopathiae* in men has been summarized by Stackebrandt et al. (2005). For a non-sporeforming organism, *Erysipelothrix rhusiopathiae* is remarkably persistent in the environment. Organisms remain viable in drinking water for up to 5 d and for up to 15 d in sewage. However, it does not survive heating at 60 °C for 15 min and does not grow in 10% (w/v) NaCl.

Erysipelothrix inopinata was isolated in the course of the validation of production processes in aseptic manufacturing of pharmaceuticals when a vegetable-based growth medium was tested for its dilution performance (Verbarg et al., 2004).

Enrichment and isolation procedures

A number of procedures have been devised for the isolation of *Erysipelothrix rhusiopathiae*. These methods may also refer to *Erysipelothrix tonsillarum*, classified as *Erysipelothrix rhusiopathiae* in the past. Most procedures are based on the ability of the organism to grow in the presence of various substances which are bacteriocidal or bacteriostatic for other organisms (Ewald, 1981; Wood, 1965).

Erysipelothrix rhusiopathiae can be isolated easily from the blood of infected animals. A small quantity of blood is placed in a tube of semisolid nutrient agar supplemented with 0.2% (w/v) glucose or horse serum (5–10%, v/v), incubated at 35 °C for 24–48 h. The layer of growth, which develops below the surface, is then plated onto a suitable solid medium (Blood Agar Base No. 2 [Difco] plus 0.2%, w/v, glucose), incubated at 35 °C for up to 48 h, and examined for colonies.

Successful isolation of *Erysipelothrix rhusiopathiae* from pig or human skin requires the removal of a small piece of the entire thickness of the dermis because the organisms are located in the deeper parts of the skin. *Erysipelothrix rhusiopathiae* is rarely the only bacterium present in skin samples or in pieces of other tissues (spleen, kidney, liver, lung, and tonsils). Isolation is best achieved by placing the skin or small pieces of tissue in 10 ml of modified ESB medium (Wood, 1965). [ESB medium is Nutrient Broth No. 2 (Oxoid), 25.0 g and distilled water, 1000 ml. After sterilization at 121 °C for 15 min, 50 ml of horse serum, 400 mg of kanamycin, 50 mg of neomycin, and 25 mg of vancomycin are added aseptically. This medium may be stored at 4 °C for not more than 2 weeks.]

After overnight incubation at 35 °C, about 5 ml of the liquid portion is placed in a sterile tube and centrifuged for 20 min at approximately 1400 × *g*. The supernatant is discarded, the sediment resuspended in 1–2 ml physiological buffer (0.85% w/v) and a portion plated on MBA medium (Harrington and Hulse, 1971). [MBA medium is Heart Infusion Agar (Difco), 40.0 g; sodium azide, 0.4 g; distilled water, 1000 ml. After sterilization at 121 °C for 15 min, 20 ml of horse blood and 50 ml of horse serum are added aseptically.] After incubation at 35 °C for 24–48 h, the plate is examined for colonies.

Isolation from feces or contaminated soil may be achieved in much the same way. Samples (approx. 100 g) are placed in a sterile blender containing 220 ml of sterile 0.1 M phosphate buffer. After mixing for 10 min, the whole is transferred to a sterile centrifuge bottle and centrifuged at a low speed for 10 min. The bottle is shaken slightly to resuspend the top 5–10 ml of the sediment and the cloudy supernatant decanted into a sterile 1-l screw-capped flask containing 200 ml of double-strength ESB medium (but the volume of serum is not doubled, i.e., it remains 5%, v/v). After mixing, the flask is incubated at 35 °C for 48 h. Samples are then plated on Packer's agar with 5% (v/v) horse serum (Packer, 1943), incubated at 35 °C for 24–48 h, and the plates then examined for colonies. Packer's agar is recommended for grossly contaminated specimens such as soil or feces because it is more selective for *Erysipelothrix rhusiopathiae* than is MBA medium (R.L. Wood, personal communication to D. Jones).

A fluorescent antibody technique may be used to detect *Erysipelothrix rhusiopathiae* in tissues (Dacres and Groth, 1959; Seidler et al., 1971) and in enrichment broth cultures (Harrington et al., 1974).

Sakuma et al. (1973) used whole-body autobacteriography to study localization of *Erysipelothrix rhusiopathiae* in the whole body of mice. Bacterial localization as demonstrated by this technique was very similar to that observed when infected mice were investigated by conventional bacteriological techniques. The inoculated organisms eventually localized in organs such as the spleen, liver, and subcutaneous and intramuscular tissue. They then multiplied in these tissues and finally distributed widely as in a bacteremia.

Erysipelothrix inopinata was enriched in the course of preparation of a vegetable CSB medium (peptone vegetable 20.0 g; D(+)-glucose, 2.5 g; K$_2$HPO$_4$, 2.5 g; water, 1000 ml). The water used for dilution was heated to 80 °C for 1 h and allowed to cool down to room temperature. The dehydrated medium was then added to the water and the solution was filtered through a membrane filter (pore width, 0.2 μm). Following incubation of a medium sample at room temperature for 3 d, the medium became turbid. Microscopic analysis and plating in medium TSA (tryptic soy agar: casein peptone, 15 g; soy peptone, 5.0 g; NaCl, 5.0 g; agar, 15.0 g, water, 1000 ml; pH 7.3) and TSS (TSA + 5% sheep blood) indicated the presence of a single contaminant, MF-EP02^T (DSM 15511^T), the type strain of *Erysipelothrix inopinata* (Verbarg et al., 2004).

Enrichment of *Erysipelothrix tonsillarum* strains, originally described for serotype 7 of *Erysipelothrix rhusiopathiae*, is identical to that of strains of this species (Takahashi et al., 1987a).

Maintenance procedures

The organisms may be preserved for several months by stab inoculation into screw-capped tubes of nutrient agar (pH 7.4). After overnight growth at 30 °C, the tubes are tightly closed and kept at room or refrigerator temperature in the dark. Longer-term preservation (over 5 years) may be achieved by freezing on glass beads at 70 °C (Feltham et al., 1978). The organisms can also be preserved by freeze drying or storage in liquid nitrogen.

Differentiation of the genus *Erysipelothrix* from other genera

Holdemania filiformis, the closest phylogenetic neighbor of *Erysipelothrix* (Figure 262), differs from members of *Erysipelothrix* in strictly anaerobic growth, composition of fatty acids (presence of $C_{16:1\ cis9}$, $C_{10:1}$, $C_{18:1\ cis9}$ dimethyl acetal, and $C_{16:0}$ dimethyl acetal) and in the presence of an interpeptide bridge consisting of L-Asp→L-Lys instead of L-Lys→L-Lys in the B-group peptidoglycan.

Traditionally and still widely used in clinical laboratories, new isolates of *Erysipelothrix rhusiopathiae* will be identified by examination of cell morphology, growth characteristics in nutrient gelatin, maximum growth temperature, catalase and oxidase tests, production of acid from carbohydrates in a suitable medium, and hydrogen sulfide production (Jones, 1986a). Serological identification and the mouse protection test by specific hyperimmune anti-*Erysipelothrix* serum may be necessary for certain identification of *Erysipelothrix rhusiopathiae*, but they are not normally required. Table 256 lists the features most

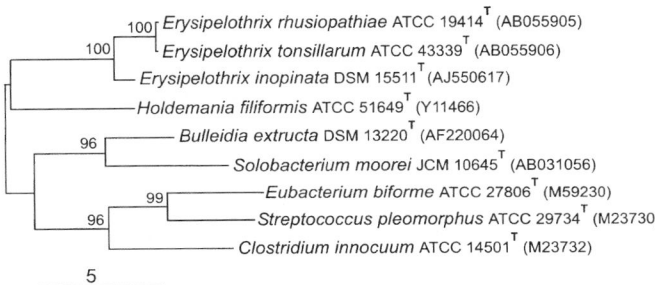

FIGURE 262. 16S rRNA gene sequence dendrogram (DeSoete, 1983) displaying the phylogenetic position *Erysipelothrix* strains among some phylogenetic neighbors. Numbers indicate the percentage of bootstrap (>95%) samplings (Felsenstein, 1993), derived from 1000 resamplings. Bar indicates 5% sequence divergence. Numbers in parentheses are accession numbers of 16S rRNA gene sequences.

TABLE 256. Useful characters in differentiating the genera *Erysipelothrix*, *Brochothrix*, *Kurthia*, *Listeria*, and *Corynebacterium*[a]

Characteristic	Erysipelothrix	Brochothrix	Kurthia	Listeria	Corynebacterium
Smooth form morphology	+	+	−	+	+
Rough form morphology	+	+	+	−	−
Catalase	−	+	+	+	+
Strictly aerobic	−	−	+	−	−
Acid from glucose	+	+	−	+	+[b]
Growth above 30 °C	+	−	+	+	+
Menaquinones	−	MK-7	MK-7	MK-7	MK-9(H_2), MK-8(H_2)
Diagnostic feature of peptidoglycan	B-type, L-Lys	A-type, *meso*-A_2pm	A-type, L-Lys	A-type, *meso*-A_2pm	A-type, *meso*-A_2pm
DNA G+C content (mol%)	36–40	35.6–36.1	36.7–37.9	36–38	51–65

[a]Data from Collins and Cummins (1986), Reyn (1986), and Seeliger and Jones (1986).
[b]Some organisms are aerobic.

useful in differentiating strains of *Erysipelothrix* from some other Gram-positive, non-sporeforming, rod-shaped bacteria with which they may be possibly confused.

Brochothrix thermosphacta also differs from *Erysipelothrix rhusiopathiae* in producing acid from far more carbohydrates, producing acetoin (Voges–Proskauer-positive), not producing H_2S, growing well on the selective medium of Gardner (1966), not exhibiting a "pipe cleaner" type of growth in nutrient gelatin, and containing cytochromes. *Listeria* may be further distinguished from *Erysipelothrix rhusiopathiae* by growing well on the usual nutrient media; producing colonies with a marked blue-green sheen on Tryptose Agar (Difco); exhibiting motility; possessing a more marked saccharolytic activity; hydrolysis of esculin; greater tolerance of NaCl (*Listeria monocytogenes* grows in nutrient broth plus 8.5% w/v NaCl, *Erysipelothrix* does not); producing acetoin; sensitivity to neomycin (Fuzi, 1963); hydrolysis of Tweens 20 and 80; not producing H_2S. Strains of *Listeria* and *Erysipelothrix* are serologically distinct; *Listeria* contains menaquinones and cytochromes. The mouse protection test is considered by some clinical workers, especially veterinarians, to be the best method of identifying new isolates of *Erysipelothrix rhusiopathiae* with complete certainty. However, the test will detect only strains virulent for mice.

Mouse protection test (Jones, 1986a) Commercial horse anti-erysipelothrix serum is satisfactory for the test. It can be obtained from manufacturers specializing in veterinary products. Three to six mice are injected subcutaneously with 0.1 ml of a 24-h broth culture of the suspected strain. The loose skin of the flank is recommended. At the same time, 0.3 ml of antiserum is injected into the opposite flank. If the organism is *Erysipelothrix rhusiopathiae*, the mice that received culture and antiserum should survive; those receiving culture alone should die within 6 d. It is recommended that a control test using a known virulent strain of *Erysipelothrix rhusiopathiae* be set up in parallel as a check on the antiserum.

Chemotaxonomic analyses

To circumscribe novel isolates of *Erysipelothrix*, chemical properties of taxonomic relevance for members of the *Clostridium* subphylum should be analyzed. Cells for isoprenoid quinones analysis can be obtained from biomass grown under aerobic and anaerobic conditions on TSB agar + 5% sheep blood. Menaquinone may be extracted by chloroform-methanol (2:1, v/v) from lyophilized cells, purified by preparative TLC on silica gel and analyzed by the HPLC method (Collins et al., 1977; Groth et al., 1996). None of the *Erysipelothrix* strains contain significant amount of isoprenoid quinones, i.e., menaquinones, no matter under which conditions cells were grown for the preparation of menaquinones. Only traces of MK-7 were detected in cells of *Erysipelothrix tonsillarum* DSM 14972[T] following aerobic cultivation. The lack of significant amounts of menaquinones in strains of *Erysipelothrix rhusiopathiae* confirms the results of Collins and Jones (1981).

The composition of fatty acids reveals high similarity among all *Erysipelothrix* species. The pattern is dominated by $C_{18:1\ cis9}$ (>30%), $C_{16:0}$ (>24%), and $C_{18:0}$ (>10%) fatty acid methyl esters; quantitative values are given in the species description. This pattern differs from that of the phylogenetic neighbor *Holdemania filiformis* ATCC 51649[T] which contains higher amounts of $C_{18:1\ cis9}$ (50%), additional minor components, and significant amounts of dimethyl acetals [$C_{18:1\ cis9}$ (12%) and $C_{16:0}$ (4%)] (Willems et al., 1997).

The method of analysis of the peptidoglycan structure is described by Schleifer and Kandler (1972), modified as described by Willems et al. (1997). The two-dimensional thin-layer chromatogram of the partial acid hydrolysate of *Erysipelothrix* strains (Schleifer and Kandler, 1972; Verbarg et al., 2004) revealed the presence of lysine, glutamic acid, glycine, serine, alanine, muramic acid, and glucosamine, and fragments including D-Glu→Gly, L-Ser→D-Glu, and L-Lys→L-Lys, while aspartic acid or fragments containing aspartic acid were missing. The quantitative amino acid composition of *Erysipelothrix inopinata* was Ala:Gly:Ser:Glu:Lys = 1.7:0.7:0.9:1.0:1.5. It was deduced that the peptidoglycan type is B1δ with the interpeptide bridge being Gly→L-Lys→L-Lys, identical to that reported for *Erysipelothrix rhusiopathiae* (Schubert and Fiedler, 2001).

The range of mol% G+C values of DNA base composition is rapidly determined by HPLC (Mesbah et al., 1989). The HPLC values of 36–40 mol% support those obtained with older methods (T_m, Bd) (Takahashi et al., 1992).

Molecular differentiation of strains and species

Today, molecular analysis by sequencing parts or the almost entire 16S rRNA gene is the most reliable and fastest way to determine the phylogenetic position. Another rapid identification method is the PCR amplification of taxon-specific 16S rRNA gene fragments (Fidalgo and Riley, 2004; Makino et al., 1994; Takeshi et al., 1999), e.g., the forward primer 5′-TGATGC-CATAGAAACTGGTA-3′ and the reverse primer 5′-CTGTATC-CGCCATAACTA-3′) (Makino et al., 1994). *In silico* analysis of the 16S rRNA gene sequence of *Erysipelothrix inopinata* DSM 15511[T] for target sites for a primer pair, indicates that these primers would also identify strains of this species. Shimoji et al. (1998) describe the use of a genus-specific primer pair suitable to amplify a 937-bp long fragment of the chromosome and used this system to identify enriched *Erysipelothrix rhusiopathiae* strains in specimens from swine with arthritis. This method clearly identifies *Erysipelothrix* strains and separates them from genera listed above and from those with which they were confused in the past, at least in certain stages of growth, e.g., streptococci, *Gemella*, and lactobacilli (Jones, 1986a).

The most commonly used methods for determining intra- and interspecific relationships of the closely related species *Erysipelothrix rhusiopathiae* and *Erysipelothrix tonsillarum* were serotyping and DNA–DNA hybridization (Imada et al., 2004; Opriessnig et al., 2004; Takahashi et al., 1987a, 1992, 2000). Based on the latter method, Takahashi et al. (1992) determined serovars 1, 2, 4, 5, 6, 8, 9, 11, 12, 15, 16, 17, 19, and 21 to exhibit more than 73% hybridization with the type strain of *Erysipelothrix rhusiopathiae* but less than 24% hybridization with the type strain of *Erysipelothrix tonsillarum*. This species embraced serovars 3, 7, 10, 14, 20, and 22, as well as 23 strains with higher than 66% hybridization but less than 27% hybridization with the type strain of *Erysipelothrix rhusiopathiae*. The fact that strains representing serovars 13 and 18 exhibited low levels of hybridization (16–47%) with both type strains indicated the presence of two additional genomic species. Strains of serovar 13 may be considered strains of *Erysipelothrix inopinata* inasmuch as the partial 16S rRNA gene sequence of one representative of serovar 13, Pécs 56, is 99.9% similar with that of the type strain DSM 15511[T]. Two species of the genus and two genomospecies

(serovar 13 probably representing *Erysipelothrix inopinata*) can be distinguished by a set of four specific 16S rRNA gene primers (Takeshi et al., 1999). Other methods of high discriminatory potential are pulsed-field gel electrophoresis (PFGE) (Okatani et al., 2001; Opriessnig et al., 2004), restriction fragment length polymorphism (RFLP) (Ahrne et al., 1995), random amplified polymorphic DNA analysis (RAPD) (Imada et al., 2004; Okatani et al., 2000), comparison of protein patterns by sodium dodecyl sulfate-polyacrylamide gel electrophoresis (Bernath et al., 1997, 2001; Tamura et al., 1993), multi-locus enzyme electrophoresis (MLEE) (Chooromoney et al., 1994), 16S rRNA gene sequence analyses, and riboprinting (Verbarg et al., 2004). Except for the latter two methods, which were applied to all three species, all other methods were only applied to *Erysipelothrix rhusiopathiae* and *Erysipelothrix tonsillarum*. Results of serotyping and MLEE were not exclusively in accord with the taxonomic separation of the two species. Ribotyping and 16S rRNA gene sequence analysis demonstrated the close phylogenetic distance between these two species, while *Erysipelothrix inopinata* was significantly different.

Taxonomic comments

As summarized by Jones (1986a), Reboli and Farrar (1991), and Stackebrandt et al. (2005), the close relationship between the genera *Erysipelothrix* and *Listeria* was once discussed (Barber, 1939), but subsequent results of numerical phenetic studies (Davis and Newton, 1969; Feresu and Jones, 1988; Jones, 1975; Stuart and Pease, 1972; Wilkinson and Jones, 1977), chemotaxonomic studies (Schleifer and Kandler, 1972; Tadayon and Carroll, 1971; Verbarg et al., 2004), and DNA hybridization studies (Stuart and Welshimer, 1974) did not support this conclusion. Enzyme, numerical phenetic and DNA-base ratio studies reveal a closer relationship of *Erysipelothrix* to the family *Lactobacillaceae* and to streptococci (Jones, 1986a) than to *Corynebacteriaceae* (Flossmann and Erler, 1972; Jones, 1975; White and Mirikitani, 1976), but *Erysipelothrix* was classified among the regular non-sporeforming Gram-positive rods, containing *Corynebacteriaceae* (Jones, 1986a). In another study, the closest similarity of *Erysipelothrix* was to the genus *Gemella* (Wilkinson and Jones, 1977).

16S rRNA gene sequence analyses indicated relationship of *Erysipelothrix* species to the *Firmicutes* (Kiuchi et al., 2000; Verbarg et al., 2004), belonging to the family *Erysipelotrichaceae* (*Clostridium* cluster XVI according to Collins et al., 1994; Verbarg et al., 2004), a lineage that also embraces *Holdemania filiformis* (Willems et al., 1997). Less closely related are *Bulleidia extructa* (Downes et al., 2000), *Solobacterium moorei* (Kageyama and Benno, 2000a), and non-authentic members of *Eubacterium*, *Streptococcus*, and *Clostridium* (Figure 262). The 16S rRNA gene sequences of the *Erysipelothrix rhusiopathiae* strains ATCC 19414[T] (DSM 555[T]), DSM 5056, and three strains from the Czech National Collection of Type Cultures (CNCTC 6291, CNCTC 6291, and CNCTC 6327) were identical and highly similar to the sequence of *Erysipelothrix tonsillarum* ATCC 4339[T] (99.8% similarity). The gene sequence of *Erysipelothrix inopinata* DSM 15511[T] is less related (96.4% similarity) to those of the type strains of another two species. *Erysipelothrix rhusiopathiae* strains serotype 13 (AB019249) and serotype 18 (AB019250) (Takeshi et al., 1999), covering the 3′ half of the molecule (about 790 nucleotides) only share 97.5 and 97.8% similarity, respectively, with the corresponding fragment of the gene of DSM 15511^T. DNA–DNA reassociation values obtained for these strains range only between 18 and 36%, which confirm differences at the physiological level, hence their separate species status.

Further reading

Imada, Y. Takase, A. Kikuma R., Iwamaru, Y. Akachi, S. and Y. Hayakawa. 2004. Serotyping of 800 strains of *Erysipelothrix* isolated from pigs affected with erysipelas and discrimination of attenuated live vaccine strain by genotyping. J. Clin. Microbiol. *42*: 2121–2126.

List of species of the genus *Erysipelothrix*

1. **Erysipelothrix rhusiopathiae** (Migula 1900) Buchanan 1918, 55^{AL} Epit. spec. cons. Opin. 32, Jud. Comm. 1970, 9 (*Bacterium rhusiopathiae* Migula 1900, 431)

rhu.si.o.pa'thi.ae. Gr. adj. *rhusios* red; Gr. n. *pathos* disease; N.L. gen. n. *rhusiopathiae* of red disease.

In addition to those features given in the generic description, acid is usually produced in a suitable medium from glucose, galactose, fructose, lactose, maltose, dextrin, and N-acetylglucosamine. Weak or delayed acid production from mannose and sometimes sucrose. Acid is not produced usually from glycerol, erythritol, arabinose, xylose, adonitol, β-methylxyloside, sorbose, rhamnose, dulcitol, inositol, mannitol, sorbitol, n-methyl-D-mannoside, α-methyl-D-glucoside, amygdalin, arbutin, esculin, salicin, cellobiose, melibiose, trehalose, inulin, melezitose, raffinose, amidone, glycogen, xylitol, β-gentobiose, D-turanose, D-lyxose, D-tagatose, D-fudose, or L-fucose. Weak acid or no change in litmus milk. Methyl red reaction negative or very weakly positive. Acetoin (Voges–Proskauer test) not produced. Indole not produced. Exogenous citrate not utilized. Hydrogen sulfide produced. Nitrates not reduced. Ammonia not produced from peptone. According to API 32 STREPT substrate panel, positive for glycyl-tryptophan arylamidase, and pyroglutamic acid arylamidase; negative for oxidase, aminopeptidase, hydrolysis of starch, gelatin, DNA, casein, and urease; acid from mannitol, sorbitol, raffinose, sucrose, L-arabinose, D-arabitol, cyclodextrin, glycogen, pullulan, maltose, melibiose, melezitose, tagatose, β-glucuronidase; production of acetoin; hydrolysis of hippurate.

As determined with Biolog GP the following substrates are utilized: adenosine, uridine, methylpyruvate, N-acetylglucosamine and α-D-glucose. All strains are negative in β-methyl-D-glucoside, D-tagatose, lactamide, alaninamide, D-arabitol, lactulose, α-methyl-D-mannoside, D-lactic acid methyl ester, D-alanine, β-cyclodextrin, maltose, palatinose, turanose, L-lactic acid, L-alanine, dextrin, maltotriose, xylitol, D-malic acid, L-asparagine, glycogen, D-mannitol, D-raffinose, L-malic acid, inulin, L-fucose, L-rhamnose, acetic acid, L-glutamic acid, adenosine-5′-monophosphate, mannan, D-melezitose, α-hydroxy-butyric acid, mono-methyl succinate, glycyl-L-glutamic acid, thymidine-5′- monophosphate, Tween 40, D-galacturonic acid, D-melibiose, β-hydroxy-butyric acid, propionic acid, L-pyroglutamic acid, uridine-5′-monophosphate, Tween 60, α-methyl D-galactoside, sedoheptulosan, γ-hydroxy-butyric acid, pyruvic acid, L-serine, fructose 6-phosphate, D-gluconic acid, β-methyl D-galactoside, D-sorbitol, p-hydroxy-phenyl acetic acid, succinamic acid, putrescine, glucose-1-phosphate, stachyose, α-ketoglutaric acid, succinic acid, 2,3-butanediol, glucose 6-phosphate, amygdalin, m-inositol, α-methyl D-glucoside, sucrose, α-keto valeric acid, N-acetyl L-glutamic acid, and D-L-α-glycerol phosphate. Gelatin not liquefied, but most strains exhibit characteristic pipe cleaner growth in gelatin stab cultures incubated at 22 °C. Urea, sodium hippurate, esculin, starch, cellulose, casein, and Tweens 20, 40, and 60 not hydrolyzed. Tyrosine and xanthine not degraded. Neuraminidase but not sulfatase, deoxyribonuclease, and ribonuclease is produced. Most strains produce hyaluronidase. Some strains produce phosphatase and lecithinase.

On potassium tellurite (0.05%, w/v) agar, colonies grayish after 24 h, becoming black at 2–3 d. Grow in presence of, but do not reduce, 2,3,5-triphenyltetrazolium chloride (0.2% w/v). Most strains do not grow in 6.5% (w/v) NaCl; none grows in 10% (w/v) NaCl. Do not grow on the medium of Gardner (1966) or on MRS medium (De Man et al., 1960).

Twenty-two serovars can be distinguished on the basis of heat-stable somatic antigens.

Causes swine erysipelas. Sometimes pathogenic to man causing erysipeloid. Mice and pigeons are very susceptible; septicemia is produced. Rabbits are less susceptible and guinea pigs are more resistant.

Major fatty acids of the type strain are $C_{18:1 \; cis9}$ (39.3%), $C_{16:0}$ (31.7%), and $C_{18:0}$ (10.1%); others in smaller amounts (>1% to <5%) are $C_{14:0}$ (1.6%), $C_{16:1 \; cis8}$ (1.3%), $C_{17:0}$ (1.0%), $C_{18:2}$ (6.5%), and $C_{18:1 \; cis11}$ (1.9%).

DNA G+C content (mol%): 36–40 (T_m, Bd); type strain, 36 (Bd).

Type strain: ATCC 19414, CCUG 221, CIP 105957, DSM 5055, NCTC 8163.

GenBank accession no. (16S rRNA gene): AB055905.

2. **Erysipelothrix inopinata** Verbarg, Rheims, Emus, Frühling, Kroppenstedt, Stackebrandt and Schumann 2004, 224^{VP}

in.o.pi.na'ta. L. adj. *inopinatus -a* unexpected.

In addition to those features given in the generic description, cells are approximately 0.5 μm width and 1.5–3.0 μm long. Surface colonies on BHI (Difco) after 2 d of incubation are punctiform to approximately 1.5 mm in diameter, creamy white, undulate, convex, translucent, and soft. Growth occurs under aerobic and anaerobic conditions in BHI and Columbia Blood, preferably at pH 8. Growth occurs at 20 and 40 °C, but not at 45 °C. The optimal temperature for growth is 25–30 °C. According to API 32 STREPT substrate panel, positive for acid from glucose, glycyl-tryptophan arylamidase, and pyroglutamic acid arylamidase; negative for oxidase, aminopeptidase, hydrolysis of starch, gelatin, DNA and casein, urease, acid from mannitol, sorbitol, raffinose, sucrose, L-arabinose, D-arabitol, cyclodextrin, glycogen, pullulan, maltose, melibiose, melezitose, tagatose, β-glucuronidase, production of acetoin, and hydrolysis of hippurate.

As determined with Biolog GP, the following substrates are utilized by the type strain: adenosine, uridine, methylpyruvate, N-acetylglucosamine and α-D-glucose. Negative in β-methyl-D-glucoside, D-tagatose, lactamide, alaninamide, D-arabitol, lactulose, α-methyl-D-mannoside, D-lactic acid methyl ester, D-alanine, β-cyclodextrin, maltose, palatinose, turanose, L-lactic acid, L-alanine, dextrin, maltotriose, xylitol, D-malic acid, L-asparagine, glycogen, D-mannitol, D-raffinose, L-malic acid, inulin, L-fucose, L-rhamnose, acetic acid, L-glutamic acid, adenosine-5′-monophosphate, mannan, D-melezitose, α-hydroxy-butyric acid, mono-methyl succinate, glycyl-L-glutamic acid, thymidine-5′-monophosphate, Tweens 40 and 60, D-galacturonic acid, D-melibiose, β-hydroxy-butyric acid, propionic acid, L-pyroglutamic acid, uridine-5′-monophosphate, α-methyl-D-galactoside, sedoheptulosan, γ-hydroxy-butyric acid, pyruvic acid, L-serine, fructose 6-phosphate, D-gluconic acid, β-methyl D-galactoside, D-sorbitol, p-hydroxy-phenyl acetic acid, succinamic acid, putrescine, glucose-1-phosphate, stachyose, α-keto glutaric acid, succinic acid, 2,3-butanediol, glucose 6-phosphate, amygdalin, m-inositol, α-methyl D-glucoside, sucrose, α-keto valeric acid, N-acetyl L-glutamic acid, and D-L-α-glycerol phosphate. Major fatty acids (>5%) are $C_{18:1\ cis9}$ (33.1%), $C_{16:0}$ (34.2%), and $C_{18:0}$ (12.9%); smaller amounts (>1% to <5%) are $C_{10:0}$ (3.0%), $C_{14:0}$ (3.0), $C_{16:1\ cis8}$ (1.2%), $C_{17:0}$ (1.2%), $C_{18:2}$ (2.8%), and $C_{18:1\ cis12}$ (2.5%). Isolated from vegetable broth used for preparation of growth media, Germany.

DNA G+C content (mol%): 37.5 (HPLC).

Type strain: CIP 107935, DSM 15511[T], strain MF-EP02.

GenBank accession number (16S rRNA gene): AJ550617.

3. **Erysipelothrix tonsillarum** Takahashi, Fujisawa, Benno, Tamura, Sawada, Suzuki, Muramatsu and Mitsuoka 1987a, 168[VP]

ton.sil.la′rum. L. gen. n. *tonsillarum* of the tonsils.

In addition to characters listed in the generic description, cells are about 0.3 μm wide by 1.0–1.5 μm long. Surface colonies on BI agar after 2 d of incubation are punctiform to approximately 1 mm in diameter, circular, entire, convex, colorless, transparent, and soft. Acid from dextrin maltose, galactose, levulose, lactose, and glucose. Weak acid from sucrose and mannose. Glycerol, salicin, inositol, dulcitol, raffinose, mannitol, rhamnose, trehalose, xylose, sorbitol, and arabinose not fermented. H_2S produced. Gelatin not liquefied. "Test-tube brush" growth in gelatin. Esculin not hydrolyzed. Litmus milk not acidified. According to API 32 STREPT substrate panel, the type strain is positive for acid production from glucose, glycyl-tryptophan arylamidase, and pyroglutamic acid arylamidase; negative for oxidase, aminopeptidase, hydrolysis of starch, gelatin, DNA, and casein, urease, acid from mannitol, sorbitol, raffinose, sucrose, L-arabinose, D-arabitol, cyclodextrin, glycogen, pullulan, maltose, melibiose, melezitose, tagatose, β-glucuronidase, production of acetoin and hydrolysis of hippurate. As determined with Biolog GP, the following substrates are utilized by the type strain: adenosine, uridine, methylpyruvate, N-acetylglucosamine and α-D-glucose. All strains are negative in β-methyl-D-glucoside, D-tagatose, lactamide, alaninamide, D-arabitol, lactulose, α-methyl-D-mannoside, D-lactic acid methyl ester, D-alanine, β-cyclodextrin, maltose, palatinose, turanose, L-lactic acid, L-alanine, dextrin, maltotriose, xylitol, D-malic acid, L-asparagine, glycogen, D-mannitol, D-raffinose, L-malic acid, inulin, L-fucose, L-rhamnose, acetic acid, L-glutamic acid, adenosine-5′-monophosphate, mannan, D-melezitose, α-hydroxy-butyric acid, mono-methyl succinate, glycyl-L-glutamic acid, thymidine-5′-monophosphate, Tween 40, D-galacturonic acid, D-melibiose, β-hydroxy-butyric acid, propionic acid, L-pyroglutamic acid, uridine-5′-monophosphate, Tween 60, α-methyl-D-galactoside, sedoheptulosan, γ-hydroxy-butyric acid, pyruvic acid, L-serine, fructose 6-phosphate, D-gluconic acid, β-methyl D-galactoside, D-sorbitol, p-hydroxy-phenyl acetic acid, succinamic acid, putrescine, glucose-1-phosphate, stachyose, α-keto-glutaric acid, succinic acid, 2,3-butanediol, glucose 6-phosphate, amygdalin, m-inositol, α-methyl D-glucoside, sucrose, α-keto-valeric acid, N-acetyl L-glutamic acid, and D-L-α-glycerol phosphate. Susceptible to ampicillin and erythromycin. Major fatty acids are $C_{18:1\ cis9}$ (32.5%), $C_{16:0}$ (28.2%), and $C_{18:0}$ (19.6%); others in smaller amounts (>1% to <5%) are $C_{14:0}$ (2.7%), $C_{17:0}$ (1.4%), $C_{18:2}$ (5.0%), and $C_{18:1\ cis12}$ (5.8%). Type strain isolated from tonsils of apparently healthy pigs, Japan.

DNA G+C content (mol%): 37.9 ± 2.0 (T_m), 37.4 (HPLC).

Type strain: ATCC 43339[T], CCUG 31352, CIP 105960, DSM 14972, JCM 8533, strain T-305.

GenBank accession number (16S rRNA gene): AB055906.

Genus II. **Allobaculum** Greetham, Gibson, Giffard, Hippe, Merkhoffer, Steiner, Falsen and Collins 2006, 1459[VP] (Effective publication: Greetham, Gibson, Giffard, Hippe, Merkhoffer, Steiner, Falsen and Collins 2004, 306.)

THE EDITORIAL BOARD

Al.lo.bac.u.lum. Gr. pref. *allo* the other; L. neut. n. *baculum* small rod; N.L. neut. n. *Allobaculum* the other small rod.

Nonsporeforming rods in pairs or chains. **Gram-stain-positive.** Nonmotile. **Strictly anaerobic** and catalase-negative. Produce lactic and butyric acids and small amounts of ethanol from glucose. **Hydrolyze esculin** but not gelatin, starch, or urea. Lecithinase- and lipase-negative. **Cannot produce indole or reduce nitrate to nitrate.** Have a **type A wall** containing *meso*-diaminopimelic acid.

DNA G+C content (mol%): 37.9.

Type species: **Allobaculum stercoricanis** Greetham, Gibson, Giffard, Hippe, Merkhoffer, Steiner, Falsen and Collins 2006, 1459[VP] (Effective publication: Greetham, Gibson, Giffard, Hippe, Merkhoffer, Steiner, Falsen and Collins 2004, 306.).

Further descriptive information

16S rRNA gene treeing analysis and sequence divergence reveals that the genus forms a new branch at the periphery of

Clostridium rRNA cluster XVI and related taxa. Nevertheless, sequence divergence (13–15%) with members of this cluster indicates only a distant relationship to these species.

The only species in the genus was isolated from dog feces on reinforced clostridial agar. Its detection in all fecal samples and high mean abundance of 1.5% suggests that occurrence is widespread in the canine gut.

Enrichment and isolation methods

A 10% (w/v) anaerobic slurry of feces collected immediately after defecation was prepared in pre-reduced 0.1 M phosphate buffered saline (pH 7) and homogenized in an atmosphere of 10% H_2, 10% CO_2, and 80% N_2. Serial tenfold dilutions prepared in half-strength peptone water and cysteine-HCl (0.5 g/l) were plated and incubated (37 °C, 48 h) on reinforced clostridial agar (novobiocin 8 mg/l and colistin 8 mg/l).

Differentiation of the genus *Allobaculum* from other genera

Allobaculum can be distinguished from *Eubacterium* by lower DNA G+C content (37 versus 45–55mol%), wall type (type A wall based on *meso*-diaminopimelic acid vs. type B murein), and phylogenetically (approx. 20% 16S rRNA sequence divergence). Also, *Allobaculum* is phylogenetically distant from *Catenibacterium*, *Coprobacillus*, *Erysipelothrix*, *Holdemania*, and *Solobacterium* (greater then 13% sequence divergence). It can be differentiated on the basis of acid end products (lactic and butyric acids from *Allobaculum*; acetic, butyric, isobutyric, and lactic acids from *Catenibacterium*; acetic and lactic acids from *Coprobacillus*; lactic acid from *Erysipelothrix*; and acetic, lactic, and butyric acids from *Solobacterium*) and on the basis of cell-wall composition (type A wall containing *meso*-diaminopimelic acid for *Allobaculum*; *Erysipelothrix*, and *Holdemania* both type B murein based on L-lysine).

List of species of the genus *Allobaculum*

1. **Allobaculum stercoricanis** Greetham, Gibson, Giffard, Hippe, Merkhoffer, Steiner, Falsen and Collins 2006, 1459[VP] (Effective publication: Greetham, Gibson, Giffard, Hippe, Merkhoffer, Steiner, Falsen and Collins 2004, 306.)

   ster.co.ri.ca′nis. L. n. *stercus, oris* feces; L. genit. n. *canis* of the dog; L. gen. n. *stercoricanis* from dog feces.

   Colonies (up to 3 mm in diameter) are irregular, translucent, grayish in color and have erose margins. Cells are rod-shaped (0.75–0.9 × 1.2–2.0 µm), nonhemolytic, and grow at 30–40 °C but not at 20 °C or 45 °C; optimum ~37–40 °C.

   Produces acid from glucose (pH of PY-1% glucose medium decreases after 6 d from 6.8 to 5.7). Acid is produced from cellobiose, fructose, galactose, maltose, salicin, and sucrose but not from amygdalin, arabinose, glycogen, inositol, lactose, mannitol, mannose, melezitose, melibiose, raffinose, rhamnose, ribose, sorbitol, starch, trehalose, or xylose. API tests detect activity for acid phosphatase, alkaline phosphatase, arginine arylamidase, esterase C-4, ester lipase C8, and phosphoamidase but not for alanine arylamidase, alanine phenylalanine, proline arylamidase, arginine dihydrolase, α-arabinosidase, trypsin, chymotrypsin, α-fucosidase, α-glucosidase, β-glucosidase, β-glucuronidase, α-galactosidase, β-galactosidase-6-phosphate, glutamic acid decarboxylase, glutamyl glutamic acid arylamidase, glycyl tryptophan arylamidase, glycine arylamidase, histidine arylamidase, leucine arylamidase, leucyl glycine arylamidase, lipase C14, α-mannosidase, β-mannosidase, phenyl alanine arylamidase, proline arylamidase, pyroglutamic acid arylamidase, serine arylamidase, valine arylamidase, urease, or tyrosine arylamidase, and (depending on the test kit) may or may not detect activity for *N*-acetyl-β-glucosaminidase and β-galactosidase.

   For all other characteristics, refer to the genus description.
   DNA G+C content (mol%): 37.9 (HPLC).
   Type strain: DSM 13633, CCUG 45212.
   GenBank accession number (16S rRNA gene): AJ417075.

Genus III. **Bulleidia** Downes, Olsvik, Hiom, Spratt, Cheeseman, Olsen, Weightman and Wade 2000, 982[VP]

WILLIAM G. WADE AND JULIA DOWNES

Bull′eid′ia. L.n., named to honor Arthur Bulleid, a distinguished British oral microbiologist.

Straight or slightly curved, short rods which occur singly or in pairs. Rods 0.5 µm in diameter and 0.8–2.5 µm in length. **Gram-stain-positive** but intensity of staining variable. **Nonsporeforming and nonmotile. Obligately anaerobic.** Optimal growth temperature 30–37 °C. Colonies are 0.7–0.9 mm in diameter, circular, entire, low convex, gray-white and opaque with a glossy appearance after 7 d incubation on blood agar.

Saccharolytic. Principal end products of glucose fermentation are acetic and lactic acids with trace amounts of succinic acid produced. **Growth in broth media is poor** but stimulated by the addition of 0.5% Tween 80 in the presence of fermentable carbohydrates. Catalase-negative. Nitrate not reduced. Indole and H_2S not produced. Urea not hydrolyzed. Gelatin not liquefied. No growth in 20% bile. **Arginine is hydrolyzed**.

DNA G+C content (mol%): 38 (HPLC).
Type species: **Bulleidia extructa** Downes, Olsvik, Hiom, Spratt, Cheeseman, Olsen, Weightman and Wade 2000, 982[VP].

Further descriptive information

Found in the oral cavity of man where it has been isolated from periodontal pockets and dentoalveolar abscesses.

Enrichment and isolation procedures

Strains of *Bulleidia* species can be isolated on complex media appropriate for culture of anaerobes including blood agar, enriched brain heart infusion agar (Holdeman et al., 1977), and fastidious anaerobe agar (FAA, LabM, Bury, UK).

TABLE 257. Descriptive and differential characteristics of *Bulleidia extructa* and related species[a]

Characteristics	*Bulleidia extructa*	*Erysipelothrix rhusiopathiae*	*Holdemania filiformis*	*Solobacterium moorei*
Growth in air + CO_2	–	+	–	–
Esculin hydrolysis	–	–	+	d
Arginine hydrolysis	+	+	–	+
Growth in 20% bile	–	+	w	–
H_2S production	–	+	w	–
Indole production	–	–	–	–
Nitrate reduction	–	–	–	–
Urea hydrolysis	–	–	–	–
Vancomycin (5 ug disc)	S	R	R	S
Acid produced from:				
Fructose	–	+	v	+
Glucose	+	+	v	+
Maltose	+	+	–	+
Sucrose	–	d	v	v
Metabolic end products[b]	a, l, (s)	a, l	a, l, (s)	a, l, (s)
Enzyme profile[c]	2000 0120 00	2103 4173/7 05	0500 0041 20	650/10 0120 00/1
DNA G+C content (mol%)	38	36–40	38	37–39
Colony morphology	Convex, opaque	Convex, transparent to opaque	Convex, translucent	Umbonate, translucent
Source	Human oral	Animal, environmental, human infection	Human fecal	Human oral and fecal

[a]Symbols: +, 90% or more of strains are positive; –, 90% or more of strains are negative; d, 11–89% of strains are positive; w, weak reaction; v, test unreliable due to poor reproducibility; S, sensitive; R, resistant.

[b]a, Acetate; l, lactate; s, succinate.

[c]Enzyme profile generated by Rapid ID32A anaerobe identification kit (bioMérieux).

Maintenance procedures

Lyophilization of cultures in the early stationary phase of growth is recommended for long-term storage of most strains. Strains can also be stored at –70 °C in brain heart infusion broth supplemented with 10% glycerol.

Procedures and methods for characterization tests

The general methods described for the characterization of anaerobes in the VPI Anaerobe Laboratory Manual (Holdeman et al., 1977) and the Wadsworth-KTL Anaerobic Laboratory Manual (Jousimies-Somer et al., 2002) are suitable for the study of members of this genus. The Rapid ID32A anaerobe identification kit (bioMérieux) is useful for the generation of enzyme profiles that distinguish members of this genus from related taxa.

Differentiation from closely related taxa

Descriptive characteristics of *Bulleidia extructa* and characteristics that differentiate *Bulleidia extructa* from closely related species are shown in Table 257. The description of *Bulleidia extructa* is based on six strains isolated from the oral cavity. The characteristics of *Solobacterium moorei* listed in Table 257 are based on the study of three strains isolated from human feces (Kageyama and Benno, 2000a) and eight strains isolated from the oral cavity (Downes et al., 2001). Demonstration of production of acid from sugars is dependent on good growth in the test broth; false negative reactions occur. Perhaps because of this, Kageyama and Benno (2000a) reported that the three fecal strains did not produce acid from sucrose, although we have found their strains and our oral strains to be positive for this test. Moderate amounts of acetic and lactic acids are produced by *Solobacterium moorei* strains, grown to a turbidity of 3 to 4+ (on a negative to 4+ scale), in peptone-yeast extract-glucose broth supplemented with 0.5 % Tween 80. Kageyama and Benno (2000a) reported the production of moderate amounts of acetic, lactic, and butyric acids when the fecal strains were grown in peptone-yeast extract-Fildes-glucose (PYFG) broth (Kaneuchi et al., 1976).

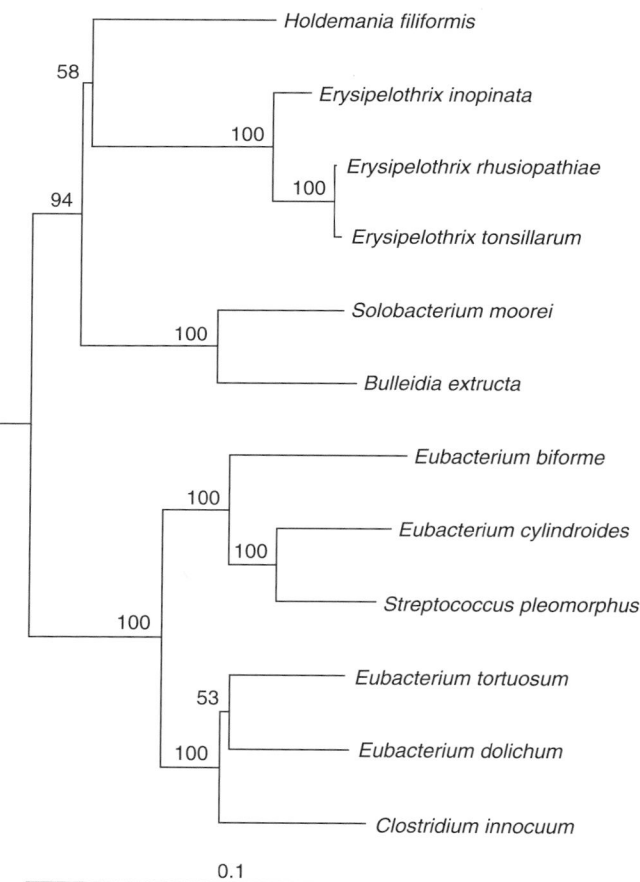

FIGURE 263. Phylogenetic tree based on 16S rRNA gene sequence comparisons over 1355 aligned bases showing relationships between *Bulleidia extructa* and related species. Tree was constructed using the neighbor-joining method following distance analysis of aligned sequences. Numbers represent bootstrap values for each branch based on data for 100 trees.

This discrepancy in reported metabolic end products could in part be due to differences in culture media or composition of the anaerobic atmosphere as both these factors can influence the end products formed.

The characteristics of *Holdemania filiformis* are based on the description of three strains isolated from human feces (Willems et al., 1997) supplemented with data from additional tests (enzyme profile, vancomycin sensitivity, H_2S production, urea hydrolysis, and growth in 20% bile) on the type strain. As with the other species listed in Table 257, growth of *Holdemania filiformis* in broth media is stimulated by both fermentable carbohydrates and Tween 80. Production of acid from fermentable carbohydrates is also unreliable and in part dependent on good growth in the medium. The characteristics for *Erysipelothrix rhusiopathiae* are as described by Jones (1986a) augmented by enzyme profile and bile sensitivity tests performed on the type strain.

Taxonomic comments

Phylogenetic analysis reveals that the genus *Bulleidia* belongs to the family *Erysipelotrichaceae*, class *Mollicutes*, and phylum *Firmicutes*; the order is uncertain. It is also related to the genus *Holdemania* (Figure 263).

Solobacterium moorei (Kageyama and Benno, 2000a) is a closely related species. (Figure 263). Since the type strains of *Solobacterium moorei* and *Bulleidia extructa* share 93.2% similarity in their 16S rRNA gene sequences and many phenotypic characteristics, *Solobacterium moorei* appears to represents a novel species within the genus *Bulleidia*. The generic name *Bulleidia* would have precedence by prior publication under Rule 27 of the Bacteriological Code.

Acknowledgements

The work reported in this chapter was funded, in part, by Wellcome Trust grant ref. 058950.

List of species of the genus *Bulleidia*

1. **Bulleidia extructa** Downes, Olsvik, Hiom, Spratt, Cheeseman, Olsen, Weightman and Wade 2000, 982[VP]

   ex.truc'ta. L. adj. *extructa* slow, referring to the slow growth of the organism.

   Description is based on six strains isolated from the oral cavity. Cells are obligately anaerobic, nonsporeforming, nonmotile, Gram-stain-positive short bacilli ($0.5 \times 0.8–2.5\,\mu m$) occurring singly and in pairs aligned side by side. After 7d incubation on FAA plates, colonies are 0.7–0.9mm in diameter, circular, entire, low convex, gray to off-white, opaque, and glossy.

   Growth is poor in pre-reduced aerobically sterilized media supplemented with 0.5% Tween 80 with arabinose, cellobiose, lactose, mannitol, mannose, melezitose, melibiose, raffinose, rhamnose, salicin, sorbitol, sucrose, trehalose, or xylose. Growth is stimulated by fructose, glucose, and maltose in the presence of 0.5% Tween 80. Glucose and maltose are fermented; fructose is not fermented. Moderate amounts of acetate and lactate and trace amounts of succinate are produced as the end products of glucose metabolism in PYG. Arginine is hydrolyzed. Catalase and indole are not produced; nitrate is not reduced. Esculin and urea are not hydrolyzed. There is no growth in 20% bile. Gelatin is not liquefied and H_2S is not produced.

   Isolated from human periodontal pockets and dentoalveolar infections.

   DNA G+C content (mol%): 38 (HPLC).
   Type strain: W1219, ATCC BAA−170, DSM 13220.
   GenBank accession number (16S rRNA gene): AF220064.

Genus IV. **Catenibacterium** Kageyama and Benno 2000e, 1598[VP]

FRED A. RAINEY

Ca.te.ni.bac.te'ri.um. L. fem. n. *catena* chain; Gr. dim. n. *bakterion* a small rod; N.L. neut. n. *Catenibacterium* chain rodlet.

Cells occur in tangled chains. Gram-stain-positive and obligately anaerobic. Spores are absent. Fermentation products of glucose are acetic, lactic, butyric and isobutyric acids. Cell wall contains an A1c-type peptidoglycan with an (L-Ala)−D-Glu−*m*-Dpm peptide subunit. The genus *Catenibacterium* is a member of the *Clostridium* subphylum of the Gram-positive bacteria and exhibits a close phylogenetic association with *Lactobacillus catenaformis* and *Lactobacillus vitulinus*.

DNA G+C content (mol%): 36–39 (HPLC).

Type species: **Catenibacterium mitsuokai** Kageyama and Benno 2000e, 1598[VP].

List of species of the genus *Catenibacterium*

1. **Catenibacterium mitsuokai** Kageyama and Benno 2000e, 1598[VP]

   mit.su.o'kai. N.L. gen. n. *mitsuokai* of Mitsuoka, named after K. Mitsuoka, a Japanese microbiologist.

   This description is based on a study of six strains isolated from human feces. Cells are $0.4\,\mu m \times 1.2\,\mu m$ long and occur in tangled chains. Nonmotile. Gram-stain-positive and obligately anaerobic. Spores are absent. Can be cultivated in 2d at 37°C on EG agar in an anaerobic jar with 100% CO_2. Cells produce acid from glucose, mannose, galactose, fructose, sucrose, maltose, cellobiose, lactose and salicin but not from arabinose, xylose, rhamnose, ribose, trehalose, raffinose, melezitose, starch, glycogen, mannitol, sorbitol, inositol, erythritol, esculin, or amygdalin. Hydrolysis of starch is positive and that of esculin is negative. Gas formation, indole production, nitrate reduction, gelatin liquefaction and H_2S production are all negative. Cell wall contains A1γ-type peptidoglycan with an (L-Ala)–D-Glu–m-Dpm peptide subunit. The type strain was isolated from human feces.

DNA G+C content (mol%): 36–39 (HPLC).
Type strain: RCA14-39, CIP 106738, JCM 10609.
GenBank accession number (16S rRNA gene): AB030224.

Genus V. **Coprobacillus** Kageyama and Benno 2000d, 949[VP] (Effective publication: Kageyama and Benno 2000c, 27.)

THE EDITORIAL BOARD

Co.pro.ba.cil′lus. L. n. *copro* feces; L. dim. n. *bacillus* a small rod; N.L. n. *Coprobacillus* rodlet isolated from feces.

Gram-positive short rods that occur in chains. Spores are not formed. Nonmotile. **Obligate anaerobe** that produces acetate and lactate from glucose.

DNA G+C content (mol%): 32–34 (HPLC).

Type species: **Coprobacillus cateniformis** corrig. Kageyama and Benno 2000d, 949[VP] (Effective publication: *Coprobacillus catenaformis* [sic] Kageyama and Benno 2000c, 28.).

Further descriptive information

This description is based upon the initial report of Kageyama and Benno (2000c), which describes three closely related strains of the only species *Coprobacillus catenaformis*. These strains occur as short rods, 0.4 × 1.2–1.7 μm, in chains of variable length. All strains produce acid from glucose, mannose, fructose, sucrose, maltose, cellobiose, lactose, trehalose, and salicin. No acid is produced from arabinose, xylose, rhamnose, ribose, raffinose, melezitose, starch, glycogen, mannitol, sorbitol, inositol, erythritol, esculin, and amygdalin. The strains do not form gas, produce indole or H_2S, reduce nitrate, or hydrolyze gelatin. During growth on complex medium, moderate amounts of acetate and lactate and minor amounts of butyrate, valerate, and isobutyrate are produced. Growth is inhibited by 20% bile.

All strains were isolated at 37 °C from human feces on EG agar that contained (per liter): 5% horse blood, 3 g beef extract, 5 g yeast extract, 10 g peptone, 1.5 g glucose, 0.5 g L-cysteine-HCl, 0.2 g L-cystine, 4 g Na_2HPO_4, 0.5 g soluble starch, 0.5 g Tween 80, 0.5 g silicone, and 15 g agar, pH 7.7, in an anaerobic incubation jar under an atmosphere of 100% CO_2. Subsequently, strains may be cultured in PYFG broth comprised of 10 g trypticase, 10 g yeast extract, 0.5 g cystine, and 40 ml of salt solution (per liter: 0.2 g $CaCl_2$, 0.2 g $MgSO_4$, 1 g K_2HPO_4, 1 g KH_2PO_4, 10 g $NaHCO_3$, and 2 g NaCl) and a pH of 7.6.

Coprobacillus is classified within the *Erysipelotrichaceae* on the basis of its 16S rRNA gene sequence (Ludwig et al., 2009). Similarly, *Clostridium ramosum* and *Clostridium spiroforme*, whose 16S rRNA genes possess about 90% sequence similarity that of *Coprobacillus*, should also be reclassified to this family.

List of species of the genus *Coprobacillus*

1. **Coprobacillus cateniformis** corrig. Kageyama and Benno 2000d, 949[VP] (Effective publication: *Coprobacillus catenaformis* [sic] Kageyama and Benno 2000c, 28.)

ca.te.ni.for′mis. L. n. *catena* chain, L. adj. *formis* shaped; N.L. adj. *cateniformis* chain-like.

The description of the species is the same as the genus.

DNA G+C content (mol%): 32–34 (HPLC).
Type strain: JCM 10604.
GenBank accession number (16S rRNA gene): AB030218.

Genus VI. **Holdemania** Willems, Moore, Weiss and Collins 1997, 1203[VP]

ANNE WILLEMS

Hold.eman′ia. L. n. named in honor of Lillian V. Holdeman Moore, a contemporary American microbiologist, for her outstanding contribution to anaerobic bacteriology.

Gram-stain-positive, strictly anaerobic, nonsporeforming rods. Cells occur in pairs and short chains. Catalase-negative. **Fermentative metabolism.** Growth in broth media is enhanced by the addition of Tween 80. **Major end products from glucose fermentation are acetic and lactic acid. Succinic acid is also produced but in smaller quantities. The cell-wall peptidoglycan is of type B1** with an L-Asp–L-Lys interpeptide bridge. **Major cellular fatty acids are nonhydroxylated, monounsaturated or saturated.** The only known representatives are three strains that were isolated from human feces.

DNA G+C content (mol%): 38.

Type species: **Holdemania filiformis** Willems, Moore, Weiss and Collins 1997, 1203[VP].

Further descriptive information

The genus currently has only one species, therefore the following information also describes the species *Holdemania filiformis*. It is based on the study of three strains isolated from the feces of healthy humans (Willems et al., 1997).

Phylogenetic treatment. According to the 16S rRNA phylogenetic analysis presented in the roadmap to this volume (Figure 4), the genus *Holdemania* is a member of the family *Erysipelotrichaceae*, order *Erysipelotrichales*, class *Erysipelotrichia* in the phylum *Firmicutes*. It is closely related to subphylum XVI of Collins et al. (1994). Its closest relatives at 90% or less sequence similarity are *Erysipelothrix rhusiopathiae*, the causal agent of

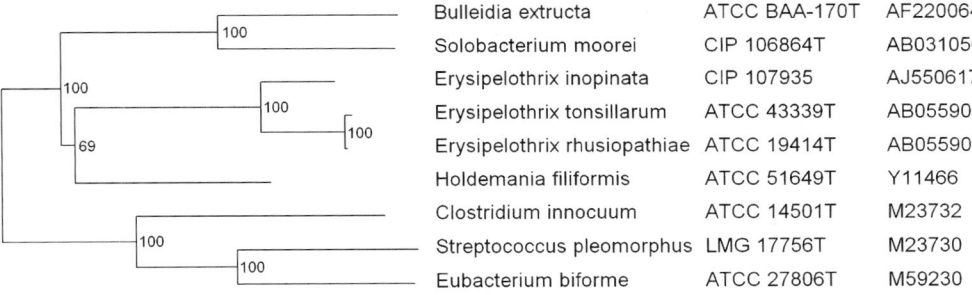

Bulleidia extructa	ATCC BAA-170T	AF220064
Solobacterium moorei	CIP 106864T	AB031056
Erysipelothrix inopinata	CIP 107935	AJ550617
Erysipelothrix tonsillarum	ATCC 43339T	AB055906
Erysipelothrix rhusiopathiae	ATCC 19414T	AB055905
Holdemania filiformis	ATCC 51649T	Y11466
Clostridium innocuum	ATCC 14501T	M23732
Streptococcus pleomorphus	LMG 17756T	M23730
Eubacterium biforme	ATCC 27806T	M59230

FIGURE 264. Neighbor-joining tree on the basis of 16S rDNA sequences showing the phylogenetic position of the genus *Holdemania*. Distances were calculated using the Kimura 2-correction. Bootstrap values, expressed as a percentage of 1000 replicates, are shown at branching points.

erysipelas in swine (Willems et al., 1997), *Erysipelothrix tonsillarum* from the tonsils of healthy pigs (Takahashi et al., 1987a), *Erysipelothrix inopinata* from vegetable peptone broth (Verbarg et al., 2004), *Bulleidia extrucata* from oral infections (Downes et al., 2000), and *Solobacterium moorei* (Kageyama and Benno, 2000a). A phylogentic tree is shown in Figure 264.

Cell morphology. Cells are short to long, Gram-stain-positive rods that decolorize easily. Occur in pairs and short chains. Longer cells may have central to terminal swellings, although no spores or heat resistance have been observed.

Cell-wall composition. The cell-wall peptidoglycan is of type B1δ→ (Schleifer and Kandler, 1972) with an L-Asp→L-Lys interpeptide bridge (Willems et al., 1997). This is similar to that of *Erysipelothrix rusiopathiae* and *Erysipelothrix inopinata* which have type B1δL→ with a Gly→L-Lys→L-Lys interpeptide bridge (Schubert and Fiedler, 2001; Verbarg et al., 2004).

Colonial and cultural characteristics. Anaerobic growth in brain heart infusion agar streak tubes produces circular, low convex colonies with an entire margin, translucent with a granular appearance, 1.0 mm in diameter. No growth on the surface of solid media incubated in air enriched to 10% CO_2. Cultures in peptone-yeast extract-Tween (PYT)-glucose broth are turbid with a smooth, white sediment (Willems et al., 1997).

Nutrition and growth conditions. Good growth in PYT broth medium with esculin, fructose, glucose, salicin, or sucrose, even though the pH is just 5.8–6.2. Weak growth in PYT broth without carbohydrates or with amygdalin, arabinose, cellobiose, erythritol, glycogen, inositol, mannitol, mannose, melezitose, melibiose, raffinose, rhamnose, ribose, sorbitol, starch, trehalose, or xylose. The pH in these media is 6.2–6.5. Moderate growth in peptone-yeast extract broth with lactose or maltose.

Metabolism and metabolic pathways. *Holdemania* has a strictly anaerobic metabolism, is saccharolytic, and is not proteolytic. Fermentation end products in PYT-glucose broth acidified to pH 2.0 are (in milliequivalents per ml of culture): acetic acid (0.8), lactic acid (0.9), and succinic acid (0.2); formic acid is sometimes detected. Catalase and indole are not produced; esculin is hydrolyzed, but gelatin and meat are not digested and milk is unchanged. Nitrate is not reduced and arginine is not deaminated.

Chemotaxonomic characteristics. The cellular fatty acid composition of *Holdemania filiformis* cells grown in PYT-glucose broth comprises the following components at more than 4% (based on four analyses of two isolates; Willems et al., 1997):

$C_{18:1 \omega 9c}$ fatty acid methyl ester (50% ± 11.5% standard deviation), $C_{18:1 \omega 9c}$ dimethyl acetal (12% ± 4.5%), $C_{16:1 \omega 7c}$ fatty acid methyl ester (6% ± 3.4%), $C_{10:1}$ fatty acid methyl ester (5% ± 4.0%), $C_{16:0}$ fatty acid methyl ester (4% ± 3.2%), and $C_{16:0}$ dimethyl acetal (4% ± 0.6%). Hydroxylated or branched-chain fatty acids were not detected.

Antibiotic sensitivity. Resistant to vancomycin by the disk diffusion method using 5 µg disks (Downes et al., 2000).

Ecology. *Holdemania filiformis* is isolated from the feces of healthy people. The significance or abundance of these organisms in the intestinal microflora remains unknown.

Enrichment and isolation procedures

No specific isolation procedures have been described for *Holdemania*. Up to now, three strains (J1-31B-1, J1-37, and S4B-1) have been described. They were isolated from the feces of healthy humans by using routine isolation and cultivation procedures for anaerobic, saccharolytic bacteria (Holdeman et al., 1977).

Maintenance procedures

Strains can be maintained in prereduced, anaerobically sterilized PYT-glucose broth (Holdeman et al., 1977). For long-term preservation, strains can also be lyophilized.

Differentiation of the genus *Holdemania* from other taxa

The nearest phylogenetic neighbor of *Holdemania* are members of the genus *Erysipelothrix*. *Holdemania* is differs from *Erysipelothrix* in the following properties: the amino acid sequence of the interpeptide bridge of its peptidoglycan, the absence of hydroxylated and branched chain fatty acids and the presence of dimethylacetal components in the cell membranes, and the absence of arginine dihydrolase and the presence of esculin hydrolase (Takahashi et al., 1987a; Verbarg et al., 2004; Willems et al., 1997). *Holdemania* can be differentiated from the genus *Bulleidia* by its ability to hydrolyze esculin but not arginine and its resistance to vancomycin (5 µg disk) (Downes et al., 2000). Fermentation products from glucose (acetic and lactic acids) permit differentiation from the genus *Solobacterium* (acetic, lactic, and butyric acids) (Kageyama and Benno, 2000a).

Taxonomic comments

The three human fecal isolates that comprise *Holdemania filiformis* were part of the Moore collection of the Virginia Polytechnic Institute and State University (Blacksburg, VA, USA). As Gram-stain-positive, nonsporeforming, strictly anaerobic, catalase-negative rods that produce mainly acetic and lactic acid and some succinic acid from PYT broth, they were provisionally designated *Eubacterium* group S14. Their phylogenetic affiliation was determined from the 16S rDNA sequence of two strains (J1-31B-1 and S4B-1) and they were found to form a relatively isolated lineage among *Clostridium* groups XVI, XVII, and XVIII (Collins et al., 1994; Willems et al., 1997). Many of the taxa in these groups have been isolated from human feces. The closest relatives, although still quite remote at 90% or less sequence similarity, are *Erysipelothrix*, *Bulleidia*, and *Solobacterium*. In view of this absence of close relatives and its phenotypic distinctness, group S14 was designated a new genus with one species, *Holdemania filiformis* (Willems et al., 1997). According to the 16S rRNA phylogenetic analysis presented in the roadmap to this volume (Figure 4), the genus *Holdemania* is a member of the family *Erysipelotrichaceae*, order *Erysipelotrichales*, class *Erysipelotrichia* in the phylum *Firmicutes*.

Acknowledgements

A.W. is grateful to the Fund for Scientific Research – Flanders, for a postdoctoral research fellowship.

List of species of the genus *Holdemania*

1. **Holdemania filiformis** Willems, Moore, Weiss and Collins 1997, 1203[VP]

   fi.li.for′mis. L. n. *filum* thread; L. n. *forma* shape; N.L. adj. *filiformis* filiform, thread-shaped.

   Morphological and cellular characteristics are as described above for the genus. Isolated from the feces of healthy humans.

   DNA G+C content (mol%): 38 (T_m).

   Type strain: VPI J1-31B-1, ATCC 51649, DSM 12042.

   GenBank accession number (16S rRNA gene): Y11466.

Genus VII. **Solobacterium** Kageyama and Benno 2000b, 1415[VP] (Effective publication: Kageyama and Benno 2000a, 226.)

AKIKO KAGEYAMA AND YOSHIMI BENNO

So.lo.bac.te′ri.um. L. adj. *solus* sole; Gr. dim n. *bakterion* a small rod; N.L. neut. n. *Solobacterium* sole bacterium.

Cells are 0.2 μm × 0.4–0.7 μm and **occur singly. Gram-positive. Obligatorily anaerobic.** Colonies appear after cultivation for 2d at 37 °C on EG agar in an anaerobic jar containing 100% CO_2. **Spores and flagella are absent. Fermentation products of glucose are mainly acetic, lactic, and butyric acids; pyruvic acid is produced in a small amount.** Strains are isolated from humans. Comparative 16S rDNA sequence analysis revealed that strains are members of the *Clostridiales* and that 16S rDNA sequence is most similar to that of members of *Clostridium* cluster XVI (Collins et al., 1994).

DNA G+C content (mol%): 37–39.

Type species: **Solobacterium moorei** Kageyama and Benno 2000b, 1415[VP].

Further descriptive information

The genus *Eubacterium* contains all Gram-positive, anaerobic, nonsporeforming, rod-shaped bacteria, which do not belong to the genera *Propionibacterium*, *Lactobacillus*, or *Bifidobacterium*. Differentiation among these genera is based on fermentation products from glucose: propionic acid as a major end product for *Propionibacterium*, lactic acid as a major end product for *Lactobacillus*, acetic acid (and to a lesser extent) lactic acid as major end products for *Bifidobacterium*, butyric, acetic, or formic acid as a major product for *Eubacterium*, and mainly acetic, lactic, butyric acids, and a small amount of pyruvic acid for *Solobacterium*. On this basis, *Solobacterium* isolates would be identified as members of the genus *Eubacterium*. Over the years, the genus *Eubacterium* has been used as a holding place for a large number of phenotypically diverse species, some of which have been transferred to new genera. From these reasons, the new genus *Solobacterium* was proposed.

Phylogenetic analyses of the 16S rDNA sequence revealed that *Solobacterium* is most closely related to the *Clostridia*, specifically to members of *Clostridium* cluster XVI (Collins et al., 1994) (Figure 265). In an analysis including all taxa from cluster XVI and related clusters, *Solobacterium moorei* forms a distinct lineage, that is most closely aligned with *Holdemania filiformis* (87% sequence similarity; Willems et al., 1997), and *Erysipelothrix rhusiopathiae* (86% sequence similarity). Bootstrap analysis (value 100%) showed this association to be statistically significant.

Enrichment and isolation procedures

The strains were isolated from humans. Fecal samples were diluted and then an aliquot (0.05 ml) of the diluted sample was spread on the EG agar plates. Bacterial strains were cultivated for 2d at 37 °C on EG agar medium [premixed EG agar (Eiken Chemical Co., Tokyo, Japan) supplemented with 5% horse blood containing 3 g of beef extract, 5 g of yeast extract, 10 g of peptone, 1.5 g of glucose, 0.5 g of L-cysteine HCl, 0.2 g of L-cystine, 4 g of Na_2HPO_4, 0.5 g of soluble starch, 0.5 g of Tween 80, 0.5 g of silicone, and 15 g of agar in 1000 ml, pH 7.7] in an anaerobic jar containing 100% CO_2.

Maintenance procedures

Stock cultures were prepared by growing isolates on EG agar plates for 4–5 d, as described above. Cultures remain viable for up to 1 week. For longer term preservation, cultures can be frozen at −80 °C in EG medium containing 10% glycerol, or by lyophilization using 10% skim milk as the cryoprotectant.

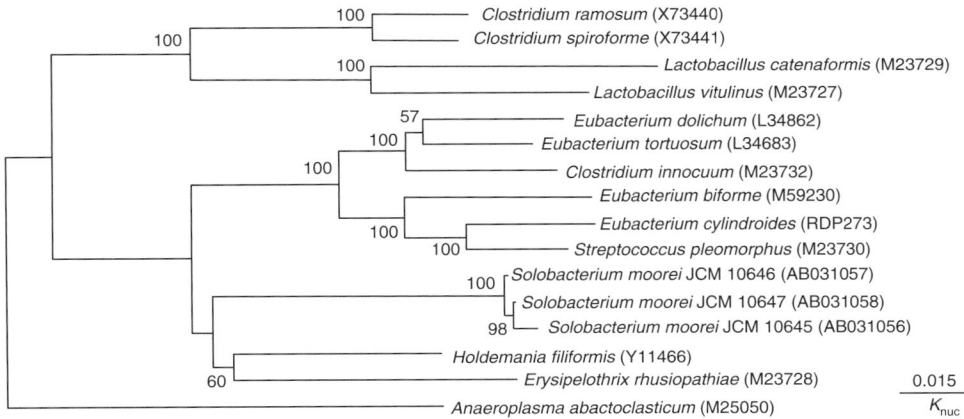

FIGURE 265. Phylogenetic tree derived from 16S rDNA sequences showing the position of *Solobacterium*. The tree was created using the neighbor-joining method and K_{nuc} values. The numbers on the tree indicate bootstrap values for the branch points with greater than 50% bootstrap support.

TABLE 258. Characteristics differentiating *Solobacterium* and related genera[a]

Characteristic	*Solobacterium*	*Holdemania*[b]	*Erysipelothrix*[c]	*Eubacterium sensu stricto*[d]	*Propionibacterium*[e]	*Lactobacillus*[f]	*Bifidobacterium*[g]
Relationship to O_2	Obligate anaerobe	Strict anaerobe	Facultative anaerobe	Obligate anaerobe	Obligate anaerobe	Obligate anaerobe	Obligate anaerobe
DNA G+C content (mol%)	38	38	38–40	45–50	59–66	32–53	57–64
Fermentation products:							
Acetic acid	+	+	−	+	−	−	+[*]
Butyric acid	+	−	−	+	−	−	−
Formic acid	−	−	−	+	−	−	(+)
Lactic acid	+	+	+	+	−	+	+[h]
Propionic acid	−	−	−	−	+	−	−
Pyruvic acid	(+)	−	−	−	−	−	−
Succinic acid	−	(+)	−	−	−	−	−
H_2	−	−	−	+	−	−	−

[a]Symbols: +, >85% positive; d, different strains give different reactions (16–84% positive); −, 0–15% positive; w, weak reaction; (), some strains can ferment this product; ND, not determined.
[b]Data from Willems et al. (1997).
[c]Data from Jones (1986a).
[d]Data from Willems and Collins (1996).
[e]Data from Jones (1986d).
[f]Data from Jones (1986c).
[g]Data from Jones (1986b).
[h]*Bifidobacterium* ferments large amounts of acetic acid better than lactic acid.

Differentiation of the genus *Solobacterium* from other genera

Solobacterium is a Gram-positive, anaerobic, nonsporeforming, rod-shaped bacterium. Other genera that are Gram-positive, anaerobic, nonsporeforming, rod-shaped bacteria, include *Propionibacterium*, *Lactobacillus*, *Bifidobacterium*, and *Eubacterium*. Differentiation among these genera is based on fermentation products from glucose: propionic acid as a major end product for *Propionibacterium*, lactic acid as a major end product for *Lactobacillus*, acetic acid (and to a lesser extent lactic acid) as major end products for *Bifidobacterium*, and butyric, acetic, or formic acid as a major product for *Eubacterium*. Because of its broad

definition, the genus *Eubacterium*, has over the years, acted as a depository for a large number of phenotypically diverse species (Andreesen, 1992). *Solobacterium* produces mainly acetic, lactic, and butyric acids, and a small amount of pyruvic acid from glucose. Using this criterion, *Solobacterium* resembles *Eubacterium* species. The genus *Holdemania* is also a Gram-positive, anaerobic, nonsporeforming, rod-shaped bacterium resembling *Eubacterium*, but it has been proposed as a separate genus.

Table 258 lists the features most useful in differentiating the genus *Solobacterium* from other phenotypically and genetically related Gram-positive, nonsporeforming, anaerobic, rod-shaped bacteria.

List of species of the genus *Solobacterium*

1. **Solobacterium moorei** Kageyama and Benno 2000b, 1415[VP] (Effective publication: Kageyama and Benno 2000a, 226.)

moo'rei. N.L. gen. n. *moorei* of Moore, named in honor of W.E.C. (Ed) Moore, an American microbiologist.

In addition to the features given in the genus description, cells produce acid from ribose, glucose, galactose, fructose, and maltose but not from arabinose, xylose, rhamnose, mannose, sucrose, cellobiose, lactose, trehalose, raffinose, melezitose, starch, glycogen, mannitol, sorbitol, inositol, erythritol, esculin, salicin, or amygdalin. The hydrolysis of starch is negative and that of esculin is positive. Gas formation, indole production, nitrate reduction, gelatin liquefaction, and H₂S production are negative. Its habitat is the human intestinal tract.

DNA G+C content (mol%): 37–39 (HPLC).
Type strain: JCM 10645.

Genus VIII. **Turicibacter** Bosshard, Zbinden and Altwegg 2002, 1266[VP]

PHILIPP P. BOSSHARD

Tu.ri.ci.bac'ter. L. neut. n. *Turicum* the L. name of Zürich; N.L. masc. n. *bacter* equivalent of a Gr. neut. dim. n. *bakterion* a small rod; N.L. masc. n. *Turicibacter* a rod-shaped organism from Zürich, Switzerland, where the bacterium was first isolated.

Cells are irregularly shaped, nonbranching **rods**, 0.5–2.0 × 0.7–7.0 µm in size, chain-forming (up to 30 µm) Figure 266. Cells stain **Gram-stain-positive. Anaerobic, nonsporeforming**. Chemoorganotrophic, fermentative metabolism. Catalase-negative and oxidase-negative. **Lactate** is the main fermentation product.

DNA G+C content (mol%): 36.9.

Type species: **Turicibacter sanguinis** Bosshard, Zbinden and Altwegg 2002, 1266[VP].

Further descriptive information

Phylogenetically, the genus *Turicibacter* is distantly related to other taxa of the *Firmicutes*. Similarities of the 16S rRNA gene sequence with those of other genera are less than 88% (Bosshard et al., 2002). According to the 16S rRNA phylogenetic analysis presented in the roadmap to this volume, the genus *Turicibacter* is a deep branch of the family *Erysipelotrichaceae*, order *Erysipelotrichales*, class *Erysipelotrichia* in the phylum *Firmicutes*. However, there are reasons to be cautious about this assignment. Placements of deep branches in phylogenetic trees are frequently problematic and subject to change as additional sequences or improved methods become available. Moreover, the other *Erysipelotrichaceae* are aerobic or facultatively anaerobic, whereas *Turicibacter* is strictly anaerobic. Therefore, this classification warrants further investigation.

Three strains of *Turicibacter sanguinis* have been isolated from blood cultures; the type strain MOL361[T] was isolated in Zurich (Switzerland) from a febrile 35-year-old man with acute appendicitis, one strain was isolated in Gothenburg (Sweden) from a febrile 79-year-old woman, and one strain was isolated in Edmonton, Alberta (Canada). With 16S rRNA gene sequence libraries, closely related sequences of uncultured bacteria have been isolated from human gut, hot compost, raw milk, dairy waste, and the gut of beetles.

Isolation and maintenance procedures

Turicibacter sanguinis may be maintained lyophilized and stored at −70 °C. The bacteria can not be recultivated when preserved at −70 °C for 1 d in skim milk, Protector tubes (TSC) or glycerol (10, 40, or 70%).

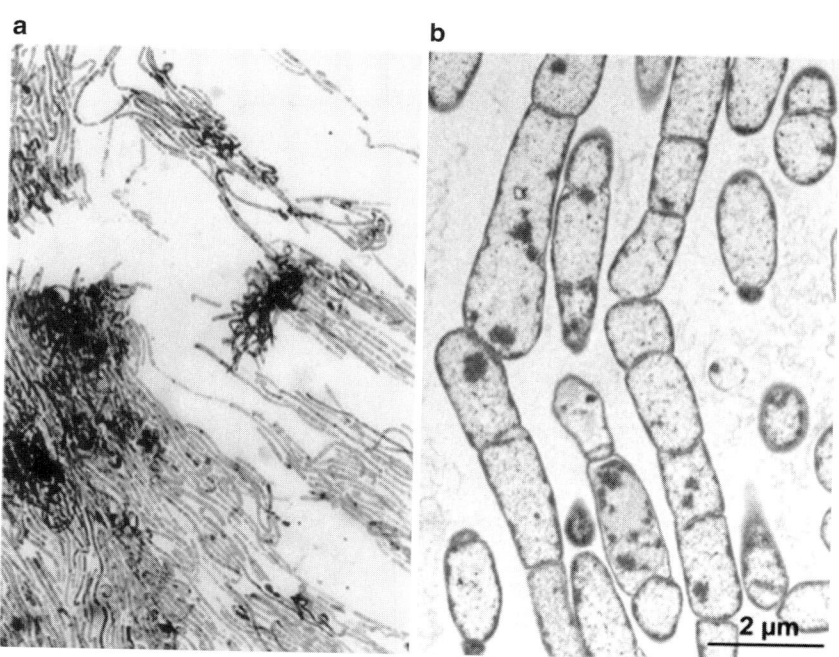

FIGURE 266. Gram-stain (a) and transmission electron micrograph (b) of *Turicibacter sanguinis* isolate MOL361[T] from a 3 d culture on chocolate agar. (Reprinted with permission from Bosshard et al. 2002. Int. J. Syst. Evol. Microbiol. *52*:1263–1266.)

List of species of the genus *Turicibacter*

1. **Turicibacter sanguinis** Bosshard, Zbinden and Altwegg 2002, 1266[VP]

san′gui.nis. L. masc. gen. n. *sanguinis* of blood, indicating that the bacterium was isolated from a blood culture.

The species description is identical to that of the genus with the following additions.

Colonies are nonhemolytic, grayish white with a convex elevation, and irregular in form with spreading, undulating margins. Strictly anaerobic. Growth is observed at 25–46 °C, with an optimum at 37 °C. Optimum pH is 7.5; no growth occurs at pH ≤6.5 or pH ≥8.0. Grows on sheep blood, chocolate and Bru-

cella agar, and in chopped meat carbohydrate and thioglycollate. Maltose and 5-keto-gluconate are the only carbohydrates utilized. Activity for α-glucosidase and α- and β-galactosidase is detected. Nitrates are not reduced. Indole is not produced. Lactate and minimal amounts of acetate are the only fermentation products. Susceptible to penicillin, vancomycin, and kanamycin; resistant to colistin. The major cellular fatty acids are $C_{16:0}$ (37 %), $C_{18:0}$ (15.5 %), and $C_{18:1\,\omega9c}$ (14.5 %).

DNA G+C content (mol%): 36.9 (HPLC).

Type strain: MOL361, DSM 14220, NCCB 100008.

GenBank accession number (16S rDNA gene): AF349724.

References

Ahrne, S., I.M. Stenstrom, N.E. Jensen, B. Pettersson, M. Uhlen and G. Molin. 1995. Classification of *Erysipelothrix* strains on the basis of restriction fragment length polymorphisms. Int. J. Syst. Bacteriol. *45*: 382–385.

Andreesen, J.R. 1992. The genus *Eubacterium*. *In* Balows, Trüper, Dworkin, Harder and Schleifer (Editors), The Prokaryotes. A Handbook on the Biology of Bacteria: Ecophysiology, Isolation, Identification, Applications, 2nd edn. Springer-Verlag, New York, pp. 1914–1924.

Barber, M. 1939. A comparative study of *Listeria* and *Erysipelothrix*. J. Pathol. Bacteriol. *48*: 11–23.

Bernath, S., G. Kucsera, I. Kadar, G. Horvath and G. Morovjan. 1997. Comparison of the protein patterns of *Erysipelothrix rhusiopathiae* strains by SDS-PAGE and autoradiography. Acta Vet. Hung. *45*: 417–425.

Bernath, S., L. Nemet, K. Toth and G. Morovjan. 2001. Computerized comparison of the protein compositions of *Erysipelothrix rhusiopathiae* and *Erysipelothrix tonsillarum* strains. J. Vet. Med. B Infect. Dis. Vet. Public Health *48*: 73–79.

Boerner, L., K.R. Nevis, L.S. Hinckley, E.S. Weber and S. Frasca, Jr. 2004. *Erysipelothrix* septicemia in a little blue penguin (*Eudyptula minor*). J. Vet. Diagn. Invest. *16*: 145–149.

Bosshard, P.P., R. Zbinden and M. Altwegg. 2002. *Turicibacter sanguinis* gen. nov., sp. nov., a novel anaerobic, Gram-positive bacterium. Int. J. Syst. Evol. Microbiol. *52*: 1263–1266.

Buchanan, R.E. 1918. Studies in the nomenclature and classification of the bacteria. J. Bacteriol. *3*: 27–61.

Chooromoney, K.N., D.J. Hampson, G.J. Eamens and M.J. Turner. 1994. Analysis of *Erysipelothrix rhusiopathiae* and *Erysipelothrix tonsillarum* by multilocus enzyme electrophoresis. J. Clin. Microbiol. *32*: 371–376.

Collins, M.D., T. Pirouz, M. Goodfellow and D.E. Minnikin. 1977. Distribution of menaquinones in actinomycetes and corynebacteria. J. Gen. Microbiol. *100*: 221–230.

Collins, M.D. and D. Jones. 1981. Bacterial isoprenoid quinones. Microbiol. Rev. *45*: 316–354.

Collins, M.D. and C.S. Cummins. 1986. Genus *Corynebacterium* Lehman and Neumann 1896, 350[AL]. *In* Sneath, Mair, Sharpe and Holt (Editors), Bergey's Manual of Systematic Bacteriology. The Williams and Wilkins Co., Baltimore, pp. 1266–1276.

Collins, M.D., P.A. Lawson, A. Willems, J.J. Cordoba, J. Fernández-Garayzábal, P. Garcia, J. Cai, H. Hippe and J.A.E. Farrow. 1994. The phylogeny of the genus *Clostridium*: proposal of five new genera and eleven new species combination. Int. J. Syst. Bacteriol. *44*: 812–826.

Dacres, W.G. and A.H. Groth, Jr. 1959. Identification of *Erysipelothrix insidiosa* with fluorescent antibody. J. Bacteriol. *78*: 298–299.

Davis, G.H. and K.G. Newton. 1969. Numerical taxonomy of some named coryneform bacteria. J. Gen. Microbiol. *56*: 195–214.

De Man, J.C., M. Rogosa and M.E. Sharpe. 1960. A medium for the cultivation of lactobacilli. J. Appl. Bacteriol. *23*: 130–135.

Dedié, K. 1949. Die säurelöslichen antigene von *Erysipelothrix rhusiopathiae*. Monatshefte Veterinärmed. *4*: 7–10.

DeSoete, G. 1983. A least square algorithm for fitting additive trees to proximity data. Psychometrika *48*: 621–626.

Dinter, Z., H. Diderholm and G. Rockborn. 1976. Complement-dependent hemolysis following hemagglutination by *Erysipelothrix rhusiopathiae*. Zentbl. Bakteriol. Orig. A *236*: 533–553.

Downes, J., B. Olsvik, S.J. Hiom, D.A. Spratt, S.L. Cheeseman, I. Olsen, A.J. Weightman and W.G. Wade. 2000. *Bulleidia extructa* gen. nov., sp. nov., isolated from the oval cavity. Int. J. Syst. Evol. Microbiol. *50*: 979–983.

Downes, J., M.A. Munson, D.A. Spratt, E. Kononen, E. Tarkka, H. Jousimies-Somer and W.G. Wade. 2001. Characterisation of *Eubacterium*-like strains isolated from oral infections. J. Med. Microbiol. *50*: 947–951.

Erler, W. 1972. Serological, chemical and immunochemical studies on *Erysipelothrix* bacteria. X. Differentiation of *Erysipelothrix* bacteria according to chemical characteristics. Arch. Exp. Veterinärmed. *26*: 809–816.

Ewald, F.W. 1957. Das hyaluronidase-bildungsvermögen von rotlaufbakterien. Monatsh. Tierheilk. *9*: 333–341.

Ewald, F.W. 1981. The genus *Erysipelothrix*. *In* Starr, Stolp, Trüper, Balows and Schlegel (Editors), The Prokaryotes: A Handbook on Habitats, Isolation, and Identification of Bacteria. Springer-Verlag, New York, pp. 1688–1700.

Feist, H. 1972. Serological, chemical and immunochemical studies of *Erysipelothrix* bacteria. XII. Murein in *Erysipelothrix* bacteria. Arch. Exp. Veterinärmed. *26*: 825–834.

Felsenstein, D. 1993. PHYLIP (Phylogeny Inference Package) 3.57 edn. Department of Genetics, University of Washington, Seattle.

Feltham, R.K., A.K. Power, P.A. Pell and P.A. Sneath. 1978. A simple method for storage of bacteria at −76 °C. J. Appl. Bacteriol. *44*: 313–316.

Feresu, S.B. and D. Jones. 1988. Taxonomic studies on *Brochothrix*, *Erysipelothrix*, *Listeria* and atypical lactobacilli. J. Gen. Microbiol. *134*: 1165–1183.

Fidalgo, S.G., Q. Wang and T.V. Riley. 2000. Comparison of methods for detection of *Erysipelothrix* spp. and their distribution in some Australasian seafoods. Appl. Environ. Microbiol. *66*: 2066–2070.

Fidalgo, S.G. and T.V. Riley. 2004. Detection of *Erysipelothrix rhusiopathiae* in clinical and environmental samples. Methods Mol. Biol. *268*: 199–205.

Flossmann, K.D. and W. Erler. 1972. Serologische, chemische und immunchemische untersuchungen an rothufbakterien. XI. Isolierung und charakterisierung von desoxyribonukleinsäuren aus rotlufbakterien. Arch. Exp. Vetinärmed. *26*: 817–824.

Fuzi, M. 1963. A neomycin sensitivity test for the rapid differentiation of *Listeria monocytogenes* and *Erysipelothrix rhusiopathiae*. J. Pathol. Bacteriol. *85*: 524–525.

Gardner, G.A. 1966. A selective medium for the enumeration of *Microbacterium thermosphactum* in meat and meat products. J. Appl. Bacteriol. *29*: 455–460.

Greetham, H.L., G.R. Gibson, C. Giffard, H. Hippe, B. Merkhoffer, U. Steiner, E. Falsen and M.D. Collins. 2004. *Allobaculum stercoricanis* gen. nov., sp. nov., isolated from canine faeces. Anaerobe *10*: 301–307.

Greetham, H.L., G.R. Gibson, C. Giffard, H. Hippe, B. Merkhoffer, U. Steiner, E. Falsen and M.D.Collins. 2006. *In* Validation of publication of new names and new combinations previously effectively published outside the IJSEM. List No. 110.. Int. J. Syst. Evol. Microbiol. *56*: 1459–1460.

Grieco, M.H. and C. Sheldon. 1970. *Erysipelothrix rhusiopathiae*. Ann. N. Y. Acad. Sci. *174*: 523–532.

Groth, I., P. Schumann, N. Weiss, K. Martin and F.A. Rainey. 1996. *Agrococcus jenensis* gen. nov., sp. nov., a new genus of actinomycetes with diaminobutyric acid in the cell wall. Int. J. Syst. Bacteriol. *46*: 234–239.

Harrington, R., Jr. and D.C. Hulse. 1971. Comparison of two plating media for the isolation of *Erysipelothrix rhusiopathiae* from enrichment broth culture. Appl. Microbiol. *22*: 141–142.

Harrington, R., Jr., R.L. Wood and D.C. Hulse. 1974. Comparison of a fluorescent antibody technique and cultural method for the detection of *Erysipelothrix rhusiopathiae* in primary broth cultures. Am. J. Vet. Res. *35*: 461–462.

Hassanein, R., T. Sawada, Y. Kataoka, K. Itoh and Y. Suzuki. 2001. Serovars of *Erysipelothrix* species isolated from the tonsils of healthy cattle in Japan. Vet. Microbiol. *82*: 97–100.

Hassanein, R., T. Sawada, Y. Kataoka, A. Gadallah and Y. Suzuki. 2003. Molecular identification of *Erysipelothrix* isolates from the tonsils of healthy cattle by PCR. Vet. Microbiol. *95*: 239–245.

Holdeman, L.V., E.P. Cato and W.E.C. Moore (Editors). 1977. Anaerobe Laboratory Manual, 4th edn. Anaerobe Laboratory, Virginia Polytechnic Institute and State University, Blacksburg, VA.

Hunter, S.H. 1942. Some growth requirements of *Erysipelothrix* and *Listeria*. J. Bacteriol. *43*: 629–640.

Imada, Y., A. Takase, R. Kikuma, Y. Iwamaru, S. Akachi and Y. Hayakawa. 2004. Serotyping of 800 strains of *Erysipelothrix* isolated from pigs affected with erysipelas and discrimination of attenuated live vaccine strain by genotyping. J. Clin. Microbiol. *42*: 2121–2126.

Jones, D. 1975. A numerical taxonomic study of coryneform and related bacteria. J. Gen Microbiol. *87*: 52–96.

Jones, D. 1986a. Genus *Erysipelothrix* Rosenbach 1909, 367[AL]. *In* Sneath, Mair, Sharpe and Holt (Editors), Bergey's Manual of Systematic Bacteriology, Vol. 2. The Williams & Wilkins Co., Baltimore, pp. 1245–1249.

Jones, D. 1986b. Genus *Bifidobacterium* Orla-Jensen 1924. *In* Sneath, Mair, Sharpe and Holt (Editors), Bergey's Manual of Systematic Bacteriology, Vol. 2. The Williams & Wilkins Co., Baltimore, pp. 1418–1434.

Jones, D. 1986c. Genus *Lactobacillus* Beijerinck 1901. *In* Sneath, Mair, Sharpe and Holt (Editors), Bergey's Manual of Systematic Bacteriology, Vol. 2. The Williams & Wilkins Co., Baltimore, pp. 1209–1234.

Jones, D. 1986d. Genus *Propionibacterium* Orla-Jensen 1909. *In* Sneath, Mair, Sharpe and Holt (Editors), Bergey's Manual of Systematic Bacteriology, Vol. 2. The Williams & Wilkins Co., Baltimore, pp. 1346–1353.

Jousimies-Somer, H.R., P. Summanen, D.M. Citron, E.J. Baron, H.M. Wexler and S.M. Finegold. 2002. Wadsworth-KTL anaerobic bacteriology manual. Star Publishing Company, Belmont, CA.

Judicial Commission. 1970. Opinion 32, Epit. spec. cons. Int. J. Syst. Bacteriol. *20*: 9.

Kageyama, A. and Y. Benno. 2000a. Phylogenic and phenotypic characterization of some *Eubacterium*-like isolates from human feces: description of *Solobacterium moorei* gen. nov., sp. nov. Microbiol. Immunol. *44*: 223–227.

Kageyama, A. and Y. Benno. 2000b. *In* Validation of publication of new names and new combinations previously effectively published outside the IJSEM. List no. 75. Int. J. Syst. Evol. Microbiol. *50*: 1415–1417.

Kageyama, A. and Y. Benno. 2000c. *Coprobacillus catenaformis* gen. nov., sp. nov., a new genus and species isolated from human feces. Microbiol. Immunol. *44*: 23–28.

Kageyama, A. and Y. Benno. 2000d. *In* Validation of publication of new names and new combinations previously effectively published outside the IJSEM. List no. 74. Int. J. Syst. Evol. Microbiol. *50*: 949–950.

Kageyama, A. and Y. Benno. 2000e. *Catenibacterium mitsuokai* gen. nov., sp. nov., a Gram-positive anaerobic bacterium isolated from human faeces. Int. J. Syst. Evol. Microbiol. *50*: 1595–1599.

Kaneuchi, C., K. Watanabe, A. Terada, Y. Benno and T. Mitsuoka. 1976. Taxonomic study of *Bacteroides clostridiiformis* subsp. *clostridiiformis* (Burri and Ankersmit) Holdeman and Moore and of related organisms: proposal of *Clostridium clostridiiformis* (Burri and Ankersmit) comb. nov. and *Clostridium symbiosum* (Stevens) comb. nov. Int. J. Syst. Bacteriol. *26*: 195–204; erratum p. 341.

Karlson, A.G. and I.A. Merchant. 1941. The cultural and biochemical properties of *Erysipelothrix rhusiopathiae*. Am. J. Vet. Res. *2*: 5–10.

Kiuchi, A., M. Hara, H.S. Pham, K. Takikawa and K. Tabuchi. 2000. Phylogenetic analysis of the *Erysipelothrix rhusiopathiae* and *Erysipelothrix tonsillarum* based upon 16S rRNA. DNA Seq. *11*: 257–260.

Krasemann, C. and H.E. Müller. 1975. The virulence of *Erysipelothrix rhusiopathiae* strains and their neuraminidase production (author's transl.). Zentbl. Bakteriol. Orig. A *231*: 206–213.

Kucsera, G. 1973. Proposal for standardization of designations used for serotypes of *Erysipelothrix rhusiopathiae* (Migula) Buchanan. Int. J. Syst. Bacteriol. *23*: 184–188.

Ludwig, W., K.-H. Schleifer and W.B. Whitman. 2009. Revised road map to the phylum *Firmicutes*. *In* De Vos, Garrity, Jones, Krieg, Ludwig, Rainey, Schleifer and Whitman (Editors), Bergey's Manual of Systematic Bacteriology, 2nd edn, Vol. 3, The *Firmicutes*, Springer, New York, pp. 1–14.

Makino, S., Y. Okada, T. Maruyama, K. Ishikawa, T. Takahashi, M. Nakamura, T. Ezaki and H. Morita. 1994. Direct and rapid detection of *Erysipelothrix rhusiopathiae* DNA in animals by PCR. J. Clin. Microbiol. *32*: 1526–1531.

Mesbah, M., U. Premachandran and W.B. Whitman. 1989. Precise measurement of the G + C content of deoxyribonucleic acid by high-performance liquid chromatography. Int. J. Syst. Bacteriol. *39*: 159–167.

Migula, W. 1900. System der Bakterien. Handbuch der Morphologie, Entwicklungsgeschichte und Systematik der bacterien, Vol. 2; p. 583. G. Fischer Verlag Jena.

Müller, H.E. and D. Seidler. 1975. Neuraminidase neutralizing antibodies in pigs with chronic *Erysipelothrix rhusiopathiae* infection (author's transl). Zentbl. Bakteriol. Orig. A *230*: 51–58.

Müller, H.E. and C. Krasemann. 1976. Immunity against *Erysipelothrix rhusiopathiae* infection by means of active immunization using homol"ogous neuraminidase (author's transl). Z. Immunitätsforsch. Exp. Klin. Immunol. *151*: 237–241.

Nikolov, P. and I. Abrashev. 1976. Comparative studies of the neuraminidase activity of virulent strains and avirulent variants of *Erysipelothrix insidiosa*. Acta Microbiol. Virol. Immunol. (Sofiia) *3*: 28–31.

Norrung, V. 1979. Two new serotypes of *Erysipelothrix rhusiopathiae*. Nord. Vet. Med. *31*: 462–465.

Okatani, A.T., H. Hayashidani, T. Takahashi, T. Taniguchi, M. Ogawa and K. Kaneko. 2000. Randomly amplified polymorphic DNA analysis of *Erysipelothrix* spp. J. Clin. Microbiol. *38*: 4332–4336.

Okatani, A.T., T. Uto, T. Taniguchi, T. Horisaka, T. Horikita, K. Kaneko and H. Hayashidani. 2001. Pulsed-field gel electrophoresis in diffe-rentiation of *Erysipelothrix* species strains. J. Clin. Microbiol. *39*: 4032–4036.

Opriessnig, T., L.J. Hoffman, D.L. Harris, S.B. Gaul and P.G. Halbur. 2004. *Erysipelothrix rhusiopathiae*: genetic characterization of midwest US isolates and live commercial vaccines using pulsed-field gel electrophoresis. J. Vet. Diagn. Invest. *16*: 101–107.

Packer, R.A. 1943. The use of sodium azide and crystal violet in a slective medium for *Erysipelothrix rhusiopathiae* and streptococci. J. Bacteriol. *46*: 343–349.

Reboli, A.C. and W.E. Farrar. 1991. The genus *Erysipelothrix*. *In* Balows, Trüper, Harder and Schleifer (Editors), The Prokaryotes. A Handbook on the Biology of Bacteria: Ecophysiology, Isolation, Identification, Applications. Springer-Verlag, New York, pp. 1629–1642.

Reyn, A. 1986. Genus *Gemella* Berger 1960, 253[AL]. *In* Sneath, Mair, Sharpe and Holt (Editors), Bergey's Manual of Systematic Bacteriology. The Williams & Wilkins Co., Baltimore, pp. 1081–1082.

Robertson, D.C. and W.G. McCullough. 1968. Glucose catabolism of *Erysipelothrix rhusiopathiae*. J. Bacteriol. *95*: 2112–2116.

Rosenbach, F.J. 1909. Experimentelle, Morphologische und klinische Studien über krankheitserregende Mikroorganismen des Scheinerotlaufs, des Erysipeloids und der Mäusesepticämie. Z. Hyg. Infektionskr. *63*: 343–371.

Sakuma, S., M. Sakuma, A. Okaniwa and Y. Sato. 1973. Detection of *Erysipelothrix insidiosa* in mice by whole body autobacteriography. Natl. Inst. Anim. Health Q (Tokyo) *13*: 54–58.

Schleifer, K.H. and O. Kandler. 1972. Peptidoglycan types of bacterial cell walls and their taxonomic implications. Bacteriol. Rev. *36*: 407–477.

Schubert, K. and F. Fiedler. 2001. Structural investigations on the cell surface of *Erysipelothrix rhusiopathiae*. Syst. Appl. Microbiol. *24*: 26–30.

Seeliger, H.P.R. 1974. Genus *Erysipelothrix*. *In* Buchanan and Gibbons (Editors), Bergey's Manual of Determinative Bacteriology, 8th edn. The Williams & Wilkins Co., Baltimore, p. 597.

Seeliger, H.P.R. and D. Jones. 1986. Genus *Listeria*. *In* Sneath, Mair, Sharpe and Holt (Editors), Bergey's Manual of Systematic Bacteriology, Vol. 2. The Williams & Wilkins Co., pp. 1235–1245.

Seidler, D., G. Trautwein and K.H. Bohm. 1971. Identification of *Erysipelothrix insidiosa* with fluorescent antibodies. Zentbl. Veterinärmed. B *18*: 280–292.

Shimoji, Y., Y. Mori, K. Hyakutake, T. Sekizaki and Y. Yokomizo. 1998. Use of an enrichment broth cultivation-PCR combination assay for rapid diagnosis of swine erysipelas. J. Clin. Microbiol. *36*: 86–89.

Sneath, P.H., J.D. Abbott and A.C. Cunliffe. 1951. The bacteriology of erysipeloid. Br. Med. J. *2*: 1063–10666.

Stackebrandt, E. 2005. The genus *Erysipelothrix*. *In* Dworkin, Falkow, Rosenberg, Schleifer and Stackebrandt (Editors), The Prokaryotes, 3rd (electronic release 3.1) edn. Springer-Verlag, New York.

Stuart, M.R. and P.E. Pease. 1972. A numerical study on the relationships of *Listeria* and *Erysipelothrix*. J. Gen. Microbiol. *73*: 551–565.

Stuart, S.E. and H.J. Welshimer. 1974. Taxonomic re-examination of *Listeria pirie* and transfer of *Listeria grayi* and *Listeria murrayi* to a new genus, *Murraya*. Int. J. Syst. Bacteriol. *24*: 177–185.

Tadayon, R.A. and K.K. Carroll. 1971. Effect of growth conditions on the fatty acid composition of *Listeria monocytogenes* and comparison with the fatty acids of *Erysipelothrix* and *Corynebacterium*. Lipids *6*: 820–825.

Takahashi, T., T. Fujisawa, Y. Benno, Y. Tamura, T. Sawada, S. Suzuki, M. Muramatsu and T. Mitsuoka. 1987a. *Erysipelothrix tonsillarum* sp. nov. isolated from tonsils of apparently healthy pigs. Int. J. Syst. Bacteriol. *37*: 166–168.

Takahashi, T., N. Hirayama, T. Sawada, Y. Tamura and M. Muramatsu. 1987b. Correlation between adherence of *Erysipelothrix rhusiopathiae* strains of serovar 1a to tissue culture cells originated from porcine kidney and their pathogenicity in mice and swine. Vet. Microbiol. *13*: 57–64.

Takahashi, T., Y. Tamura, T. Sawada, S. Suzuki, M. Muramatsu, T. Fujisawa, Y. Benno and T. Mitsuoka. 1989. Enzymatic profiles of *Erysipelothrix rhusiopathiae* and *Erysipelothrix tonsillae*. Res. Vet. Sci. *47*: 275–276.

Takahashi, T., T. Fujisawa, Y. Tamura, S. Suzuki, M. Muramatsu, T. Sawada, Y. Benno and T. Mitsuoka. 1992. DNA relatedness among *Erysipelothrix rhusiopathiae* strains representing all twenty-three serovars and *Erysipelothrix tonsillarum*. Int. J. Syst. Bacteriol. *42*: 469–473.

Takahasi, T., T. Fujisawa, K. Yamamoto, M. Kijima and T. Takahashi. 2000. Taxonomic evidence that serovar 7 of *Erysipelothrix* strains isolated from dogs with endocarditis are *Erysipelothrix tonsillarum*. J. Vet. Med. B Infect. Dis. Vet. Public Health *47*: 311–313.

Takeshi, K., S. Makino, T. Ikeda, N. Takada, A. Nakashiro, K. Nakanishi, K. Oguma, Y. Katoh, H. Sunagawa and T. Ohyama. 1999. Direct and rapid detection by PCR of *Erysipelothrix* sp. DNAs prepared from bacterial strains and animal tissues. J. Clin. Microbiol. *37*: 4093–4098.

Tamura, Y., T. Takahashi, K. Zarkasie, M. Nakamura and H. Yoshimura. 1993. Differentiation of *Erysipelothrix rhusiopathiae* and *Erysipelothrix tonsillarum* by sodium dodecyl sulfate-polyacrylamide gel electrophoresis of cell proteins. Int. J. Syst. Bacteriol. *43*: 111–114.

Verbarg, S., H. Rheims, S. Emus, A. Frühling, R.M. Kroppenstedt, E. Stackebrandt and P. Schumann. 2004. *Erysipelothrix inopinata* sp. nov., isolated in the course of sterile filtration of vegetable peptone broth, and description of *Erysipelotrichaceae* fam. nov. Int. J. Syst. Evol. Microbiol. *54*: 221–225.

White, T.G. and R.D. Shuman. 1961. Fermentation reactions of *Erysipelothrix rhusiopathiae*. J. Bacteriol. *82*: 595–599.

White, T.G. and F.K. Mirikitani. 1976. Some biological and physical chemical properties of *Erysipelothrix rhusiopathiae*. Cornell Vet. *66*: 152–163.

Wilkinson, B.J. and D. Jones. 1977. A numerical taxonomic survey of *Listeria* and related bacteria. J. Gen Microbiol. *98*: 399–421.

Willems, A. and M.D. Collins. 1996. Phylogenetic relationships of the genera *Acetobacterium* and *Eubacterium* sensu stricto and reclassification of *Eubacterium alactolyticum* as *Pseudoramibacter alactolyticus* gen. nov., comb. nov. Int. J. Syst. Bacteriol. *46*: 1083–1087.

Willems, A., W.E.C. Moore, N. Weiss and M.D. Collins. 1997. Phenotypic and phylogenetic characterization of some *Eubacterium*-like isolates containing a novel type B wall murein from human feces: description of *Holdemania filiformis* gen. nov., sp. nov. Int. J. Syst. Bacteriol. *47*: 1201–1204.

Wilson, G.S. and A.A. Miles (Editors). 1975. Topley and Wilson's Principles of Bacteriology and Immunity, 6th edn, vol. 1. Arnold, London.

Wood, R.L. 1965. A selective liquid medium utilizing antibiotics for isolation of *Erysipelothrix insidiosa*. Am. J. Vet. Res. *26*: 1303–1308.

Wood, R.L. 1970. *Erysipelothrix*. *In* Blair, Lennette and Truant (Editors), Manual of Clinical Microbiology. American Society for Microbiology, Bethesda, MD, pp. 101–105.

Wood, R.L. and R.A. Packer. 1972. Isolation of *Erysipelothrix rhusiopathiae* from soil and manure of swine-raising premises. Am. J. Vet. Res. *33*: 1611–1620.

Wood, R.L. 1974a. Isolation of pathogenic *Erysipelothrix rhusiopathiae* from feces of apparently healthy swine. Am. J. Vet. Res. *35*: 41–43.

Wood, R.L. 1974b. *Erysipelothrix* infection. *In* Hubbert, McCullough and Schnurrenberger (Editors), Diseases Transmitted from Animals to Man. Charles C. Thomas, Springfield, IL, pp. 271–281.

Wood, R.L., D.R. Haubrich and R. Harrington, Jr. 1978. Isolation of previously unreported serotypes of *Erysipelothrix rhusiopathiae* from swine. Am. J. Vet. Res. *39*: 1958–1961.

Woodbine, M. 1950. *Erysipelothrix rhusiopathiae*. Bacteriology and chemotherapy. Bacteriol. Rev. *14*: 161–178.

Appendix 1.   Validly published names, conserved and rejected names, and taxonomic opinions cited in the *International Journal of Systematic and Evolutionary Microbiology* since publication of Volume 2 of the Second Edition of the *Systematics**

JEAN P. EUZÉBY

New phyla

Names above the rank of class are not covered by the Rules of the *Bacteriological Code* (1990 Revision), and the names of phyla are not to be regarded as having been validly published. These names are listed for completeness.

Lentisphaerae Cho et al. 2004 – Valid publication: Validation List no. 98 – Effective publication: J.C. Cho et al. (2004)

Proteobacteria Garrity et al. 2005 – Valid publication: Validation List no. 106 – Effective publication: Garrity et al. (2005i)

New classes

Alphaproteobacteria Garrity et al. 2006, 1[VP] – Valid publication: Validation List no. 107 – Effective publication: Garrity et al. (2005xv)

Anaerolineae Yamada et al. 2006, 1336[VP]

Betaproteobacteria Garrity et al. 2006, 1[VP] – Valid publication: Validation List no. 107 – Effective publication: Garrity et al. (2005xxii)

Caldilineae Yamada et al. 2006, 1339[VP]

Deltaproteobacteria Kuever et al. 2006, 1[VP] – Valid publication: Validation List no. 107 – Effective publication: Kuever et al. (2005a)

Epsilonproteobacteria Garrity et al. 2006, 2[VP] – Valid publication: Validation List no. 107 – Effective publication: Garrity et al. (2005xxxviii)

Gammaproteobacteria Garrity et al. 2005, 2236[VP] – Valid publication: Validation List no. 106 – Effective publication: Garrity et al. (2005ii)

Ktedonobacteria corrig. Cavaletti et al. 2007, 433[VP] – Valid publication: Validation List no. 114 – Effective publication: Cavaletti et al. (2006b)

Thermolithobacteria Sokolova et al. 2007, 1372[VP] – Valid publication: Validation List no. 116 – Effective publication: Sokolova et al. (2007)

Opitutae Choo et al. 2007, 535[VP]

New orders

Acidithiobacillales Garrity et al. 2005, 2235[VP] – Valid publication: Validation List no. 106 – Effective publication: Garrity et al. (2005iii)

Aeromonadales Martin-Carnahan and Joseph 2005, 2235[VP] – Valid publication: Validation List no. 106 – Effective publication: Martin-Carnahan and Joseph (2005)

Alteromonadales Bowman and McMeekin 2005, 2235[VP] – Valid publication: Validation List no. 106 – Effective publication: Bowman and McMeekin (2005)

Anaerolineales Yamada et al. 2006, 1338[VP]

Bdellovibrionales Garrity et al. 2006, 1[VP] – Valid publication: Validation List no. 107 – Effective publication: Garrity et al. (2005xxxvi)

Burkholderiales Garrity et al. 2006, 1[VP] – Valid publication: Validation List no. 107 – Effective publication: Garrity et al. (2005xxiii)

Caldilineales Yamada et al. 2006, 1339[VP]

Campylobacterales Garrity et al. 2006, 1[VP] – Valid publication: Validation List no. 107 – Effective publication: Garrity et al. (2005xxxixi)

Cardiobacteriales Garrity et al. 2005, 2235[VP] – Valid publication: Validation List no. 106 – Effective publication: Garrity et al. (2005vi)

Chromatiales Imhoff 2005, 2235[VP] – Valid publication: Validation List no. 106 – Effective publication: Imhoff (2005)

Desulfarculales corrig. Kuever et al. 2006, 1[VP] – Valid publication: Validation List no. 107 – Effective publication: Kuever et al. (2005o)

Desulfobacterales Kuever et al. 2006, 2[VP] – Valid publication: Validation List no. 107 – Effective publication: Kuever et al. (2005j)

Desulfovibrionales Kuever et al. 2006, 2[VP] – Valid publication: Validation List no. 107 – Effective publication: Kuever et al. (2005d)

Desulfurellales Kuever et al. 2006, 2[VP] – Valid publication: Validation List no. 107 – Effective publication: Kuever et al. (2005b)

Desulfuromonadales corrig. Kuever et al. 2006, 2[VP] – Valid publication: Validation List no. 107 – Effective publication: Kuever et al. (2005r)

Hydrogenophilales Garrity et al. 2006, 2[VP] – Valid publication: Validation List no. 107 – Effective publication: Garrity et al. (2005xxvi)

Kordiimonadales K.K. Kwon et al. 2005, 2036[VP]

Ktedonobacterales corrig. Cavaletti et al. 2007, 433[VP] – Valid publication: Validation List no. 114 – Effective publication: Cavaletti et al. (2006b)

Legionellales Garrity et al. 2005, 2236[VP] – Valid publication: Validation List no. 106 – Effective publication: Garrity et al. (2005ix)

Lentisphaerales Cho et al. 2004, 1005[VP] – Valid publication: Validation List no. 98 – Effective publication: J.C. Cho et al. (2004)

Methylococcales Bowman 2005, 2236[VP] – Valid publication: Validation List no. 106 – Effective publication: Bowman (2005b)

Methylophilales Garrity et al. 2006, 2[VP] – Valid publication: Validation List no. 107 – Effective publication: Garrity et al. (2005xxviii)

---

*Citations for the original authorities for basonyms, synonyms, and emendations do not appear in the bibliography unless cited elsewhere in this book. References for Validation Lists and Lists of Changes in Taxonomic Opinion are provided at the end of the chapter.

Natranaerobiales Mesbah et al. 2007, 2511[VP]

Nautiliales Miroshnichenko et al. 2004, 43[VP]

Neisseriales Tønjum 2006, 2[VP] – Valid publication: Validation List no. 107 – Effective publication: Tønjum (2005)

Nitrosomonadales Garrity et al. 2006, 2[VP] – Valid publication: Validation List no. 107 – Effective publication: Garrity et al. (2005xxx)

Oceanospirillales Garrity et al. 2005, 2236[VP] – Valid publication: Validation List no. 106 – Effective publication: Garrity et al. (2005xi)

Opitutales Choo et al. 2007, 536[VP]

Pasteurellales Garrity et al. 2005, 2236[VP] – Valid publication: Validation List no. 106 – Effective publication: Garrity et al. (2005xiv)

Puniceicoccales Choo et al. 2007, 536[VP]

Rhizobiales Kuykendall 2006, 3[VP] – Valid publication: Validation List no. 107 – Effective publication: Kuykendall (2005)

According to Rule 51b(1), the order name *Rhizobiales* Kuykendall 2006 is illegitimate.

Rhodobacterales Garrity et al. 2006, 3[VP] – Valid publication: Validation List no. 107 – Effective publication: Garrity et al. (2005xvi)

Rhodocyclales Garrity et al. 2006, 3[VP] – Valid publication: Validation List no. 107 – Effective publication: Garrity et al. (2005xxxii)

Sphingomonadales Yabuuchi and Kosako 2006, 3[VP] – Valid publication: Validation List no. 107 – Effective publication: Yabuuchi and Kosako (2005)

Syntrophobacterales Kuever et al. 2006, 3[VP] – Valid publication: Validation List no. 107 – Effective publication: Kuever et al. (2005t)

Thermolithobacterales Sokolova et al. 2007, 1372[VP] – Valid publication: Validation List no. 116 – Effective publication: Sokolova et al. (2007)

Thiotrichales Garrity et al. 2005, 2236[VP] – Valid publication: Validation List no. 106 – Effective publication: Garrity et al. (2005vii)

Victivallales Cho et al. 2004, 1005[VP] – Valid publication: Validation List no. 98 – Effective publication: J.C. Cho et al. (2004)

Xanthomonadales Saddler and Bradbury 2005, 2236[VP] – Valid publication: Validation List no. 106 – Effective publication: Saddler and Bradbury (2005a)

New suborders

Catenulisporineae Cavaletti et al. 2006a, 1751[VP]

Cystobacterineae Reichenbach 2007, 894[VP] – Valid publication: Validation List no. 115 – Effective publication: Reichenbach (2005a)

Nannocystineae Reichenbach 2007, 894[VP] – Valid publication: Validation List no. 115 – Effective publication: Reichenbach (2005a)

Sorangiineae corrig. Reichenbach 2007, 894[VP] – Valid publication: Validation List no. 115 – Effective publication: Reichenbach (2005a)

New families

Acidithiobacillaceae Garrity et al. 2005, 2235[VP] – Valid publication: Validation List no. 106 – Effective publication: Garrity et al. (2005iv)

Actinospicaceae Cavaletti et al. 2006a, 1751[VP]

Alcanivoracaceae corrig. Golyshin et al. 2005, 2235[VP] – Valid publication: Validation List no. 106 – Effective publication: Golyshin et al. (2005)

Anaerolineaceae Yamada et al. 2006, 1338[VP]

Bacteriovoracaceae Davidov and Jurkevitch 2004, 1450[VP]

Bdellovibrionaceae Garrity et al. 2006, 1[VP] – Valid publication: Validation List no. 107 – Effective publication: Garrity et al. (2005xxxvii)

According to Rule 51b(1), the family name *Bdellovibrionaceae* Garrity et al. 2006 is illegitimate.

Beijerinckiaceae Garrity et al. 2006, 1[VP] – Valid publication: Validation List no. 107 – Effective publication: Garrity et al. (2005xviii)

Bradyrhizobiaceae Garrity et al. 2006, 1[VP] – Valid publication: Validation List no. 107 – Effective publication: Garrity et al. (2005xix)

According to Rule 51b(1), the family name *Bradyrhizobiaceae* Garrity et al. 2006 is illegitimate.

Burkholderiaceae Garrity et al. 2006, 1[VP] – Valid publication: Validation List no. 107 – Effective publication: Garrity et al. (2005xxiv)

Caldilineaceae Yamada et al. 2006, 1339[VP]

Catenulisporaceae Busti et al. 2006, 1745[VP]

Colwelliaceae Ivanova et al. 2004d, 1785[VP]

Conexibacteraceae Stackebrandt 2005, 547[VP] – Valid publication: Validation List no. 102 – Effective publication: Stackebrandt (2004)

Coxiellaceae Garrity et al. 2005, 2235[VP] – Valid publication: Validation List no. 106 – Effective publication: Garrity et al. (2005x)

Desulfarculaceae Kuever et al. 2006, 1[VP] – Valid publication: Validation List no. 107 – Effective publication: Kuever et al. (2005p)

Desulfobacteraceae Kuever et al. 2006, 2[VP] – Valid publication: Validation List no. 107 – Effective publication: Kuever et al. (2005k)

Desulfobulbaceae Kuever et al. 2006, 2[VP] – Valid publication: Validation List no. 107 – Effective publication: Kuever et al. (2005n)

Desulfohalobiaceae Kuever et al. 2006, 2[VP] – Valid publication: Validation List no. 107 – Effective publication: Kuever et al. (2005g)

Desulfomicrobiaceae Kuever et al. 2006, 2[VP] – Valid publication: Validation List no. 107 – Effective publication: Kuever et al. (2005f)

Desulfonatronaceae corrig. Kuever et al. 2006, 2[VP] – Valid publication: Validation List no. 107 – Effective publication: Kuever et al. (2005i)

Desulfovibrionaceae Kuever et al. 2006, 2[VP] – Valid publication: Validation List no. 107 – Effective publication: Kuever et al. (2005e)

Desulfurellaceae Kuever et al. 2006, 2[VP] – Valid publication: Validation List no. 107 – Effective publication: Kuever et al. (2005c)

Desulfurobacteriaceae L'Haridon et al. 2006, 2850[VP]

Desulfuromonadaceae corrig. Kuever et al. 2006, 2[VP] – Valid publication: Validation List no. 107 – Effective publication: Kuever et al. (2005s)

Erysipelotrichaceae Verbarg et al. 2004, 223[VP]

Erythrobacteraceae K.B. Lee et al. 2005, 1916[VP]

Ferrimonadaceae Ivanova et al. 2004d, 1785[VP]

Francisellaceae Sjöstedt 2005, 2236[VP] – Valid publication: Validation List no. 106 – Effective publication: Sjöstedt (2005)

Geobacteraceae Garrity et al. 2006, 2[VP] – Valid publication: Validation List no. 107 – Effective publication: Garrity et al. (2005xxxv)

According to Rule 24b(2), *Geobacteraceae* Holmes et al. 2004a has priority.

Geobacteraceae Holmes et al. 2004a, 1597[VP]

Geodermatophilaceae Normand 2006, 2277[VP]

Hahellaceae Garrity et al. 2005, 2236[VP] – Valid publication: Validation List no. 106 – Effective publication: Garrity et al. (2005xiii)

Halothiobacillaceae Kelly and Wood 2005, 2236[VP] – Valid publication: Validation List no. 106 – Effective publication: Kelly and Wood (2005a)

Helicobacteraceae Garrity et al. 2006, 2[VP] – Valid publication: Validation List no. 107 – Effective publication: Garrity et al. (2005xl)

Holosporaceae Görtz and Schmidt 2006, 2[VP] – Valid publication: Validation List no. 107 – Effective publication: Görtz and Schmidt (2005)

Hydrogenophilaceae Garrity et al. 2006, 2[VP] – Valid publication: Validation List no. 107 – Effective publication: Garrity et al. (2005xxvii)

Hyphomonadaceae K.B. Lee et al. 2005, 1915[VP]

Idiomarinaceae Ivanova et al. 2004d, 1784[VP]

Kofleriaceae Reichenbach 2007, 894[VP] – Valid publication: Validation List no. 115 – Effective publication: Reichenbach (2005l)

Ktedonobacteraceae corrig. Cavaletti et al. 2007, 433[VP] – Valid publication: Validation List no. 114 – Effective publication: Cavaletti et al. (2006b)

Litoricolaceae H. Kim et al. 2007a, 1797[VP]

Methermicoccaceae Cheng et al. 2007, 2968[VP]

Methylobacteriaceae Garrity et al. 2006, 2[VP] – Valid publication: Validation List no. 107 – Effective publication: Garrity et al. (2005xx)

Methylocystaceae Bowman 2006, 2[VP] – Valid publication: Validation List no. 107 – Effective publication: Bowman (2005a)

Methylophilaceae Garrity et al. 2006, 2[VP] – Valid publication: Validation List no. 107 – Effective publication: Garrity et al. (2005xxix)

Moritellaceae Ivanova et al. 2004d, 1784[VP]

Nakamurellaceae Tao et al. 2004, 999[VP] – Illegitimate synonym: *Microsphaeraceae* Rainey et al. 1997

Nannocystaceae Reichenbach 2006, 2[VP] – Valid publication: Validation List no. 107 – Effective publication: Reichenbach (2005j)

Natranaerobiaceae Mesbah et al. 2007, 2511[VP]

Nautiliaceae Miroshnichenko et al. 2004, 44[VP]

Nitrosomonadaceae Garrity et al. 2006, 2[VP] – Valid publication: Validation List no. 107 – Effective publication: Garrity et al. (2005xxxi)

Nitrospinaceae Garrity et al. 2006, 3[VP] – Valid publication: Validation List no. 107 – Effective publication: Garrity et al. (2005xxxiv)

Oceanospirillaceae Garrity et al. 2005, 2236[VP] – Valid publication: Validation List no. 106 – Effective publication: Garrity et al. (2005xii)

Opitutaceae Choo et al. 2007, 536[VP]

Oxalobacteraceae Garrity et al. 2006, 3[VP] – Valid publication: Validation List no. 107 – Effective publication: Garrity et al. (2005xxv)

Patulibacteraceae Y. Takahashi et al. 2006, 405[VP]

Phyllobacteriaceae Mergaert and Swings 2006, 3[VP] – Valid publication: Validation List no. 107 – Effective publication: Mergaert and Swings (2005)

Piscirickettsiaceae Fryer and Lannan 2005, 2236[VP] – Valid publication: Validation List no. 106 – Effective publication: Fryer and Lannan (2005)

Pseudoalteromonadaceae Ivanova et al. 2004d, 1784[VP]

Psychromonadaceae Ivanova et al. 2004d, 1785[VP]

Puniceicoccaceae Choo et al. 2007, 536[VP]

Rhodobacteraceae Garrity et al. 2006, 3[VP] – Valid publication: Validation List no. 107 – Effective publication: Garrity et al. (2005xvii)

According to Rule 51b(1), the family name *Rhodobacteraceae* Garrity et al. 2006 is illegitimate.

Rhodobiaceae Garrity et al. 2006, 3[VP] – Valid publication: Validation List no. 107 – Effective publication: Garrity et al. (2005xxi)

Rhodocyclaceae Garrity et al. 2006, 3[VP] – Valid publication: Validation List no. 107 – Effective publication: Garrity et al. (2005xxxiii)

Segniliparaceae Butler et al. 2005, 1621[VP]

Shewanellaceae Ivanova et al. 2004d, 1784[VP]

Solirubrobacteraceae Stackebrandt 2005, 548[VP] – Valid publication: Validation List no. 102 – Effective publication: Stackebrandt (2004)

Syntrophaceae Kuever et al. 2006, 3[VP] – Valid publication: Validation List no. 107 – Effective publication: Kuever et al. (2005v)

Syntrophobacteraceae Kuever et al. 2006, 3[VP] – Valid publication: Validation List no. 107 – Effective publication: Kuever et al. (2005u)

Thermithiobacillaceae Garrity et al. 2005, 2236[VP] – Valid publication: Validation List no. 106 – Effective publication: Garrity et al. (2005v)

Thermoactinomycetaceae Matsuo et al. 2006, 2840[VP]

Thermodesulfobiaceae Mori et al. 2004, 1[VP] – Valid publication: Validation List no. 95 – Effective publication: Mori et al. (2003)

Thermoleophilaceae Stackebrandt 2005, 548^{VP} – Valid publication: Validation List no. 102 – Effective publication: Stackebrandt (2004)

Thermolithobacteraceae Sokolova et al. 2007, 1372^{VP} – Valid publication: Validation List no. 116 – Effective publication: Sokolova et al. (2007)

Thiotrichaceae Garrity et al. 2005, 2236^{VP} – Valid publication: Validation List no. 106 – Effective publication: Garrity et al. (2005viii)

Trueperaceae Rainey et al. 2005 (complete authorship reads Rainey, da Costa and Albuquerque), 1744^{VP} – Valid publication: Validation List no. 105 – Effective publication: Albuquerque et al. (2005)

Xanthobacteraceae K.B. Lee et al. 2005, 1916^{VP}

Xanthomonadaceae Saddler and Bradbury 2005, 2236^{VP} – Valid publication: Validation List no. 106 – Effective publication: Saddler and Bradbury (2005b)

Yaniaceae W.J. Li et al. 2005e, 1936^{VP}

The name *Yaniaceae* W.J. Li et al. 2005e is illegitimate because the type genus *Yania* W.J. Li et al. 2004a, 529^{VP} is illegitimate (see below *Yania* W.J. Li et al. 2004a, 529^{VP}).

New genera

Acaricomes Pukall et al. 2006, 467^{VP}

Acetanaerobacterium Chen and Dong 2004, 2261^{VP}

Acidicaldus Johnson et al. 2006, 1459^{VP} – Valid publication: Validation List no. 110 – Effective publication: Johnson et al. (2006)

Actinocatenispora Thawai et al. 2006, 1792^{VP}

Actinospica Cavaletti et al. 2006a, 1751^{VP}

Actinotalea Yi et al. 2007b, 155^{VP}

Adhaeribacter Rickard et al. 2005, 827^{VP}

Advenella Coenye et al. 2005, 254^{VP}

Aeriscardovia Simpson et al. 2004, 405^{VP}

Aestuariibacter Yi et al. 2004b, 573^{VP}

Aestuariimicrobium Jung et al. 2007b, 2117^{VP}

Agarivorans Kurahashi and Yokota 2004, 695^{VP}

Aggregatibacter Nørskov-Lauritsen and Kilian 2006, 2143^{VP}

Akkermansia Derrien et al. 2004, 1474^{VP}

Algibacter Nedashkovskaya et al. 2004d, 1260^{VP}

Algicola Ivanova et al. 2004d, 1784^{VP}

Aliivibrio Urbanczyk et al. 2007, 2827^{VP}

Alkalibacillus Jeon et al. 2005b, 1894^{VP}

Alkalibacter Garnova et al. 2005, 983^{VP} – Valid publication: Validation List no. 103 – Effective publication: Garnova et al. (2004)

Alkaliflexus Zhilina et al. 2005, 1395^{VP} – Valid publication: Validation List no. 104 – Effective publication: Zhilina et al. (2004)

Alkalimonas Ma et al. 2007, 433^{VP} – Valid publication: Validation List no. 114 – Effective publication: Y. Ma et al. (2004)

Allobaculum Greetham et al. 2006, 1459^{VP} – Valid publication: Validation List no. 110 – Effective publication: Greetham et al. (2004b)

Alloscardovia Huys et al. 2007, 1445^{VP}

Altererythrobacter K.K. Kwon et al. 2007, 2210^{VP}

Aminiphilus Díaz et al. 2007, 1917^{VP}

Anaerofustis Finegold et al. 2004, 1005^{VP} – Valid publication: Validation List no. 98 – Effective publication: Finegold et al. (2004a)

Anaerosporobacter Jeong et al. 2007, 1786^{VP}

Anaerotruncus Lawson et al. 2004b, 415^{VP}

Anaerovirgula Pikuta et al. 2006b, 2628^{VP}

Anderseniella Brettar et al. 2007, 2403^{VP}

Andreprevotia Weon et al. 2007f, 1574^{VP}

Aquicella Santos et al. 2004, 1^{VP} – Valid publication: Validation List no. 95 – Effective publication: Santos et al. (2003)

Aquiflexum Brettar et al. 2004, 2339^{VP}

Aquimarina Nedashkovskaya et al. 2005h, 227^{VP}

Aquimonas Saha et al. 2005a, 1493^{VP}

Aquincola Lechner et al. 2007, 1300^{VP}

Aquisalimonas Márquez et al. 2007, 1140^{VP}

Aquitalea H.T. Lau et al. 2006, 870^{VP}

Arcicella Nikitin et al. 2004, 683^{VP}

Arenimonas S.W. Kwon et al. 2007c, 956^{VP}

Arsenicicoccus Collins et al. 2004d, 607^{VP}

Aspromonas Jin et al. 2007, 1879^{VP}

Atopococcus Collins et al. 2005, 1695^{VP}

Atopostipes Cotta et al. 2004, 1425^{VP} – Valid publication: Validation List no. 99 – Effective publication: Cotta et al. (2004)

Aureispira Hosoya et al. 2006, 2933^{VP}

Avibacterium Blackall et al. 2005, 359^{VP}

Azohydromonas Xie and Yokota 2005f, 2422^{VP}

Balneimonas corrig. Takeda et al. 2004, 631^{VP} – Valid publication: Validation List no. 97 – Effective publication: Takeda et al. (2004)

Balneola Urios et al. 2006, 1886^{VP}

Barnesiella Sakamoto et al. 2007b, 344^{VP}

Belliella Brettar et al. 2004, 69^{VP}

Bellilinea Yamada et al. 2007, 2302^{VP}

Belnapia Reddy et al. 2006, 54^{VP}

Bergeriella Xie and Yokota 2005, 1395^{VP} – Valid publication: Validation List no. 104 – Effective publication: Xie and Yokota (2005a)

Bibersteinia Blackall et al. 2007, 672^{VP}

Bizionia Nedashkovskaya et al. 2005g, 377^{VP}

Blastopirellula Schlesner et al. 2004, 1578^{VP}

Bowmanella Jean et al. 2006a, 2465^{VP}

Brooklawnia Rainey et al. 2006 (authorship reads Rainey, da Costa and Moe), 1981^{VP} – Valid publication: H.S. Bae et al. (2006a)

Bryantella Wolin et al. 2004, 1^{VP} – Valid publication: Validation List no. 95 – Effective publication: Wolin et al. (2003)

The name *Bryantella* Wolin et al. 2004 is illegitimate because it is a later homonym of *Bryantella* Chickering, 1946 (Animalia, Arthropoda, Arachnida, Araneae, Salticidae, Dendryphantinae,

Dendryphantini) and a later homonym of *Bryantella* Britton 1957 (Animalia, Arthropoda, Scarabaeoidea, Scarabaeidae, Melolonthinae). See Principle 2.

Byssovorax Reichenbach 2006, 2362[VP]

Caenispirillum J.H. Yoon et al. 2007xvi, 1219[VP]

Caldalkalibacillus Xue et al. 2006, 1220[VP]

Caldanaerobacter Fardeau et al. 2004, 471[VP]

Castellaniella Kämpfer et al. 2006d, 818[VP]

Catellibacterium Tanaka et al. 2004, 958[VP]

Catellicoccus Lawson et al. 2006, 431[VP]

Catenulispora Busti et al. 2006, 1745[VP]

Cerasibacillus Nakamura et al. 2004, 1067[VP]

Cerasicoccus J. Yoon et al. 2007c, 2070[VP]

Chimaereicella Tiago et al. 2006, 925[VP] – Valid publication: Validation List no. 109 – Effective publication: Tiago et al. (2006a)

Chitinibacter Chern et al. 2004, 1390[VP]

Chitinilyticum S. C. Chang et al. 2007, 2858[VP]

Chitinimonas S.C. Chang et al. 2004, 1005[VP] – Valid publication: Validation List no. 98 – Effective publication: S.C. Chang et al. (2004)

Citreicella Sorokin et al. 2006, 1[VP] – Valid publication: Validation List no. 107 – Effective publication: Sorokin et al. (2005a)

Citreimonas Choi and Cho 2006, 2801[VP]

Cloacibacterium Allen et al. 2006, 1314[VP]

Cohnella Kämpfer et al. 2006g, 784[VP]

Collimonas De Boer et al. 2004, 862[VP]

Conchiformibius corrig. Xie and Yokota 2005, 1395[VP] – Valid publication: Validation List no. 104 – Effective publication: Xie and Yokota (2005a)

Coraliomargarita J. Yoon et al. 2007f, 962[VP]

Corallococcus Reichenbach 2007, 893[VP] – Valid publication: Validation List no. 115 – Effective publication: Clavel et al. (2005)

Costertonia K.K. Kwon et al. 2006c, 1352[VP]

Crabtreella Xie and Yokota 2006a, 623[VP]

According to Rules 51b(1) and 51b (2), *Crabtreella* could be regarded as an illegitimate name because the type species of the genus, *Crabtreella saccharophila* Xie and Yokota 2006a, is a later homotypic synonym of *Shinella zoogloeoides* D.S. An et al. 2006a.

Cryptanaerobacter Juteau et al. 2005, 248[VP]

Curvibacter Ding and Yokota 2004, 2228[VP]

Deefgea Stackebrandt et al. 2007, 643[VP]

Defluviicoccus corrig. Maszenan et al. 2005a, 2109[VP]

Demequina Yi et al. 2007b, 154[VP]

Denitratisoma Fahrbach et al. 2006, 1549[VP]

Desulfarculus Kuever et al. 2006, 1[VP] – Valid publication: Validation List no. 107 – Effective publication: Kuever et al. (2005q)

Desulfatibacillum Cravo-Laureau et al. 2004a, 81[VP]

Desulfatiferula Cravo-Laureau et al. 2007, 2701[VP]

Desulfitibacter M.B. Nielsen et al. 2006, 2835[VP]

Desulfoglaeba Davidova et al. 2006, 2741[VP]

Desulfopila D. Suzuki et al. 2007a, 524[VP]

Desulfothermus Kuever et al. 2006, 2[VP] – Valid publication: Validation List no. 107 – Effective publication: Kuever et al. (2005h)

Desulfovermiculus Belyakova et al. 2007, 1371[VP] – Valid publication: Validation List no. 116 – Effective publication: Belyakova et al. (2006)

Desulfovirgula Kaksonen et al. 2007a, 101[VP]

Desulfurispora Kaksonen et al. 2007b, 1092[VP]

Dethiosulfatibacter Takii et al. 2007, 2324[VP]

Dickeya Samson et al. 2005, 1423[VP]

Dinoroseobacter Biebl et al. 2005a, 1095[VP]

Dokdonella J.H. Yoon et al. 2006g, 149[VP]

Dokdonia J.H. Yoon et al. 2005h, 2326[VP]

Donghaeana J.H. Yoon et al. 2006b, 190[VP]

Donghicola J.H. Yoon et al. 2007xiii, 75[VP]

Dyella Xie and Yokota 2005c, 756[VP]

Echinicola Nedashkovskaya et al. 2006b, 955[VP]

Ectothiorhodosinus Gorlenko et al. 2007, 1371[VP] – Valid publication: Validation List no. 116 – Effective publication: Gorlenko et al. (2004a)

Effluviibacter Suresh et al. 2006, 1706[VP]

Elizabethkingia K.K. Kim et al. 2005b, 1291[VP]

Emticicia Saha and Chakrabarti 2006a, 993[VP]

Endozoicomonas Kurahashi and Yokota 2007, 1371[VP] – Valid publication: Validation List no. 116 – Effective publication: Kurahashi and Yokota (2007)

Epilithonimonas O'Sullivan et al. 2006, 177[VP]

Ethanoligenens Xing et al. 2006, 758[VP]

Fabibacter S.C. Lau et al. 2006b, 1062[VP]

Fangia K.W. Lau et al. 2007, 2668[VP]

Fastidiosipila Falsen et al. 2005, 856[VP]

Flagellimonas S.S.Bae et al. 2007, 1051[VP]

Flaviramulus Einen and Øvreås 2006, 2460[VP]

Flavisolibacter M.H. Yoon and Im 2007, 1837[VP]

Fluviicola O'Sullivan et al. 2005, 2193[VP]

Formosa Ivanova et al. 2004a, 707[VP]

Frondicola L. Zhang et al. 2007b, 1181[VP]

The name *Frondicola* L. Zhang et al. 2007 is illegitimate because it is a later homonym of a fungal genus name *Frondicola* Hyde 1992.

Fulvivirga Nedashkovskaya et al. 2007d, 1048[VP]

Gaetbulibacter Jung et al. 2005a, 1848[VP]

Gaetbulimicrobium J.H. Yoon et al. 2006a, 118[VP]

Galbibacter Khan et al. 2007c, 971[VP]

Geoalkalibacter Zavarzina et al. 2007, 894[VP] – Valid publication: Validation List no. 115 – Effective publication: Zavarzina et al. (2006)

Geopsychrobacter Holmes et al. 2005, 547[VP] – Valid publication: Validation List no. 102 – Effective publication: Holmes et al. (2004b)

Geosporobacter Klouche et al. 2007, 1760[VP]

Geothermobacter Kashefi et al. 2005, 547[VP] – Valid publication: Validation List no. 102 – Effective publication: Kashefi et al. (2003)

Giesbergeria Grabovich et al. 2006, 571[VP]

Gillisia Van Trappen et al. 2004e, 446[VP]

Gilvibacter Khan et al. 2007b, 268[VP]

Goodfellowia Labeda and Kroppenstedt 2006, 1205[VP]

> The name *Goodfellowia* Labeda and Kroppenstedt 2006 is illegitimate because it is a later homonym of *Goodfellowia* Hartert 1903 (Animalia, Chordata, Aves, Passeriformes, Sturnidae). See Principle 2.

Gracilibacter Y.J. Lee et al. 2006, 2092[VP]

Gramella Nedashkovskaya et al. 2005f, 393[VP]

Granulibacter Greenberg et al. 2006, 2614[VP]

Granulicoccus Maszenan et al. 2007, 733[VP]

Guggenheimella Wyss et al. 2005, 669[VP]

Gulbenkiania Vaz-Moreira et al. 2007b, 1110[VP]

Gulosibacter Manaia et al. 2004, 786[VP]

Haematobacter Helsel et al. 2007, 1371[VP] – Valid publication: Validation List no. 116 – Effective publication: Helsel et al. (2007)

Haladaptatus Savage et al. 2007, 23[VP]

Halalkalibacillus Echigo et al. 2007, 1083[VP]

Halalkalicoccus Xue et al. 2005, 2504[VP]

Halolactibacillus Ishikawa et al. 2005, 2435[VP]

Halopiger Gutiérrez et al. 2007, 1404[VP]

Haloplanus Elevi Bardavid et al. 2007, 782[VP]

Haloquadratum Burns et al. 2007, 391[VP]

Halospina Sorokin et al. 2006a, 386[VP]

Halostagnicola Castillo et al. 2006b, 1521[VP]

Halotalea Ntougias et al. 2007b, 1981[VP]

Halovivax Castillo et al. 2006a, 767[VP]

Herminiimonas Fernandes et al. 2005, 2236[VP] – Valid publication: Validation List no. 106 – Effective publication: Fernandes et al. (2005)

Hespellia Whitehead et al. 2004, 244[VP]

Hoeflea Peix et al. 2005, 1165[VP]

Hongiella Yi and Chun 2004a, 160[VP]

Howardella Cook et al. 2007, 2943[VP]

Humicoccus J.H. Yoon et al. 2007v, 59[VP]

Humihabitans A. Kageyama et al. 2007c, 2165[VP]

Hyalangium Reichenbach 2007, 894[VP] – Valid publication: Validation List no. 115 – Effective publication: Reichenbach (2005e)

Hydrocarboniphaga Palleroni et al. 2004, 1207[VP]

Hydrogenimonas Takai et al. 2004b, 30[VP]

Hydrogenivirga S. Nakagawa et al. 2004a, 2083[VP]

Hylemonella Spring et al. 2004, 104[VP]

Ignatzschineria Tóth et al. 2007 (illegitimate synonym *Schineria* Tóth et al. 2001), 180[VP]

Ignisphaera Niederberger et al. 2006, 970[VP]

Insolitispirillum J.H. Yoon et al. 2007xxvi, 2834[VP]

Isopticola Stackebrandt et al. 2004b, 686[VP]

Jahnella corrig Reichenbach 2007, 894[VP] – Valid publication: Validation List no. 115 – Effective publication: Reichenbach (2005h)

Jiangella L. Song et al. 2005, 883[VP]

Jonquetella Jumas-Bilak et al. 2007a, 2747[VP]

Kaistella M.K. Kim et al. 2004, 2323[VP]

Kaistia Im et al. 2005, 983[VP] – Valid publication: Validation List no. 103 – Effective publication: Im et al. (2004c)

Kangiella J.H. Yoon et al. 2004j, 1832[VP]

Kofleria Reichenbach 2007, 894[VP] – Valid publication: Validation List no. 115 – Effective publication: Reichenbach (2005m)

Kordia Sohn et al. 2004b, 678[VP]

Kordiimonas K.K. Kwon et al. 2005, 2036[VP]

Krasilnikovia Ara and Kudo 2007, 2449[VP] – Valid publication: Validation List no. 118 – Effective publication: Ara and Kudo (2007a)

Kribbia Jung et al. 2006a, 2430[VP]

Krokinobacter Khan et al. 2006a, 326[VP]

Ktedonobacter corrig. Cavaletti et al. 2007, 433[VP] – Valid publication: Validation List no. 114 – Effective publication: Cavaletti et al. (2006b)

Labedella S.D. Lee 2007d, 2500[VP]

Labrenzia Biebl et al. 2007, 1105[VP]

Laceyella J.H. Yoon et al. 2005n, 398[VP]

Lacinutrix Bowman and Nichols 2005, 1482[VP]

Lactonifactor Clavel et al. 2007, 894[VP] – Valid publication: Validation List no. 115 – Effective publication: Clavel et al. (2007)

Lactovum Matthies et al. 2005, 547[VP] – Valid publication: Validation List no. 102 – Effective publication: Matthies et al. (2004)

Lapillicoccus S.D. Lee and Lee 2007, 2796[VP]

Larkinella Vancanneyt et al. 2006b, 239[VP]

Leadbetterella Weon et al. 2005, 2299[VP]

Lebetimonas Takai et al. 2005, 188[VP]

Leeia J.M. Lim et al. 2007a, 1205[VP]

Leeuwenhoekiella Nedashkovskaya et al. 2005m, 1035[VP]

Lentisphaera Cho et al. 2004, 1005[VP] – Valid publication: Validation List no. 98 – Effective publication: J.C. Cho et al. (2004)

Leptolinea Yamada et al. 2006, 1339[VP]

Levilinea Yamada et al. 2006, 1338[VP]

Lishizhenia K.W. Lau et al. 2006b, 2321[VP]

Litoricola H. Kim et al. 2007a, 1796[VP]

Loktanella Van Trappen et al. 2004a, 1266[VP]

Longilinea Yamada et al. 2007, 2303[VP]

Luedemannella Ara and Kudo 2007, 1372[VP] – Valid publication: Validation List no. 116 – Effective publication: Ara and Kudo (2007d)

Luteibacter Johansen et al. 2005, 2289[VP]

Lutibacter Choi and Cho 2006, 773[VP]

Lutimonas S.J. Yang et al. 2007, 1682[VP]

Lysinibacillus Ahmed et al. 2007d, 1121[VP]

Mahella Bonilla Salinas et al. 2004b, 2172[VP]

Malikia Spring et al. 2005b, 627[VP]

Maribacter Nedashkovskaya et al. 2004b, 1021[VP]

Maribius Choi et al. 2007, 274[VP]

Marinicola J.H. Yoon et al. 2005c, 860[VP]

Mariniflexile Nedashkovskaya et al. 2006a, 1636[VP]

Marinimicrobium J.M. Lim et al. 2006c, 656[VP]

Marinovum Martens et al. 2006, 1302[VP]

Maritimibacter K. Lee et al. 2007, 1655[VP]

Marixanthomonas Romanenko et al. 2007a, 459[VP]

Martelella Rivas et al. 2005d, 957[VP]

Mechercharimyces Matsuo et al. 2006, 2840[VP]

Meganema Thomsen et al. 2006, 1867[VP]

Metascardovia Okamoto et al. 2007, 2449[VP] – Valid publication: Validation List no. 118 – Effective publication: Okamoto et al. (2007)

Methanomethylovorans Lomans et al. 2004, 307[VP] – Valid publication: Validation List no. 96 – Effective publication: Lomans et al. (1999)

Methermicoccus Cheng et al. 2007, 2968[VP]

Methylibium Nakatsu et al. 2006, 988[VP]

Methylohalobius Heyer et al. 2005, 1824[VP]

Methylohalomonas Sorokin et al. 2007d, 2766[VP]

Methylonatrum Sorokin et al. 2007d, 2768[VP]

Methylosoma Rahalkar et al. 2007, 1078[VP]

Methylotenera Kalyuzhnaya et al. 2006a, 2822[VP]

Methylothermus Tsubota et al. 2005, 1883[VP]

Methylothermus is illegitimate because the type species of the genus, *Methylothermus thermalis*, is not validly published.

Methyloversatilis Kalyuzhnaya et al. 2006b, 2521[VP]

Microcella Tiago et al. 2005, 1743[VP] – Valid publication: Validation List no. 105 – Effective publication: Tiago et al. (2005b)

Millisia Soddell et al. 2006a, 742[VP]

Mitsuaria Amakata et al. 2005, 1930[VP]

Modicisalibacter Ben Ali Gam et al. 2007, 2311[VP]

Moryella Carlier et al. 2007, 726[VP]

Mucilaginibacter Pankratov et al. 2007, 2352[VP]

Mucispirillum Robertson et al. 2005, 1203[VP]

Myceligenerans Cui et al. 2004, 1292[VP]

Nakamurella Tao et al. 2004, 999[VP] – Illegitimate synonym: *Microsphaera* Yoshimi et al. 1996

Natranaerobius Mesbah et al. 2007, 2511[VP]

Natronocella Sorokin et al. 2007, 1372[VP] – Valid publication: Validation List no. 116 – Effective publication: Sorokin et al. (2007e)

Natronolimnobius Itoh et al. 2005, 1744[VP] – Valid publication: Validation List no. 105 – Effective publication: Itoh et al. (2005)

Naxibacter P. Xu et al. 2005, 1151[VP]

Neoasaia Yukphan et al. 2006, 499[VP] – Valid publication: Validation List no. 108 – Effective publication: Yukphan et al. (2005c)

Neptuniibacter Arahal et al. 2007, 1004[VP]

Nereida Pujalte et al. 2005c, 634[VP]

Nesiotobacter Donachie et al. 2006, 564[VP]

Niabella B.Y. Kim et al. 2007, 540[VP]

Niastella Weon et al. 2006f, 1779[VP]

Nicoletella Kuhnert et al. 2005, 547[VP] – Valid publication: Validation List no. 102 – Effective publication: Kuhnert et al. (2004)

Nitratifractor S. Nakagawa et al. 2005c, 931[VP]

Nitratireductor Labbé et al. 2004, 272[VP]

Nitratiruptor S. Nakagawa et al. 2005c, 931[VP]

Nitrincola Dimitriu et al. 2005, 2276[VP]

Nonlabens S.C. Lau et al. 2005c, 2281[VP]

Novispirillum xxvi, 2832[VP]

Oceanibulbus Wagner-Döbler et al. 2004, 1183[VP]

Oceanicola J.C. Cho and Giovannoni 2004b, 1133[VP]

Olivibacter Ntougias et al. 2007a, 402[VP]

Olleya Mancuso Nichols et al. 2005, 1560[VP]

Oribacterium Carlier et al. 2004, 1614[VP]

Ornithinibacillus Mayr et al. 2006, 1386[VP]

Oryzihumus A. Kageyama et al. 2005a, 2557[VP]

Oscillibacter Iino et al. 2007, 1844[VP]

Ottowia Spring et al. 2004, 103[VP]

Owenweeksia K.W. Lau et al. 2005, 1055[VP]

Palleronia Martínez-Checa et al. 2005, 2528[VP]

Paludibacter Ueki et al. 2006a, 43[VP]

Parabacteroides Sakamoto and Benno 2006, 1602[VP]

Paraferrimonas Khan and Harayama 2007, 1496[VP]

Parapedobacter M.K. Kim et al. 2007b, 1339[VP]

Parasporobacterium Lomans et al. 2004, 307[VP] – Valid publication: Validation List no. 96 – Effective publication: Lomans et al. (2001)

Parvibaculum Schleheck et al. 2004, 1496[VP]

Parvimonas Tindall and Euzéby 2006, 2712[VP]

Patulibacter Y. Takahashi et al. 2006, 405[VP]

Paucibacter Rapala et al. 2005, 1566[VP]

Paucisalibacillus Nunes et al. 2006, 1842[VP]

Pelagibaca J.C. Cho and Giovannoni 2006, 857[VP]

Pelagibacillus Y.G. Kim et al. 2007b, 1558[VP]

Pelagicoccus J. Yoon et al. 2007e, 1381[VP]

Pelomonas Xie and Yokota 2005f, 2424[VP]

Pelosinus Shelobolina et al. 2007, 133[VP]

Peredibacter Davidov and Jurkevitch 2004, 1451[VP]

Perexilibacter J. Yoon et al. 2007a, 965[VP]

Persicitalea J. Yoon et al. 2007b, 1016[VP]

Persicivirga O'Sullivan et al. 2006, 177[VP]

Petrimonas Grabowski et al. 2005, 1118[VP]

Petrobacter Bonilla Salinas et al. 2004a, 647[VP]

Phaeobacter Martens et al. 2006, 1301[VP]

Phycicoccus S.D. Lee 2006f, 2371[VP]

Pibocella Nedashkovskaya et al. 2005b, 179[VP]

Pilibacter Higashiguchi et al. 2006, 18[VP]

Piscibacillus Tanasupawat et al. 2007a, 1415[VP]

Planifilum Hatayama et al. 2005b, 2104[VP]

Pleomorphomonas Xie and Yokota 2005d, 1236[VP]

Polymorphospora Tamura et al. 2006, 1961[VP]

Pontibacillus J.M. Lim et al. 2005b, 168[VP]

Pontibacter Nedashkovskaya et al. 2005l, 2585[VP]

Prolixibacter Holmes et al. 2007, 705[VP]

Propionicicella H.S. Bae et al. 2006, 2026[VP] – Valid publication: Validation List no. 111 – Effective publication: H.S. Bae et al. (2006b)

Proteiniphilum Chen and Dong 2005, 2259[VP]

Pseudidiomarina Jean et al. 2006b, 904[VP]

Pseudochrobactrum Kämpfer et al. 2006h, 1825[VP]

Pseudoclavibacter Manaia et al. 2004, 787[VP]

Pseudolabrys Kämpfer et al. 2006j, 2470[VP]

Pseudoruegeria J.H. Yoon et al. 2007xxviii, 546[VP]

Pseudosphingobacterium Vaz-Moreira et al. 2007c, 1537[VP]

Pseudovibrio Shieh et al. 2004, 2311[VP]

Pullulanibacillus Hatayama et al. 2006, 2549[VP]

Puniceicoccus Choo et al. 2007, 536[VP]

Pusillimonas Stolz et al. 2005, 1080[VP]

Pyxidicoccus corrig. Reichenbach 2007, 894[VP] – Valid publication: Validation List no. 115 – Effective publication: Reichenbach (2005c)

Quadrisphaera Maszenan et al. 2005b, 1774[VP]

Quatrionicoccus Tindall and Euzéby 2006, 2712[VP]

Rapidithrix Srisukchayakul et al. 2007, 2277[VP]

Reichenbachiella Nedashkovskaya et al. 2005l, 2587[VP]

Reinekea Romanenko et al. 2004b, 672[VP]

Rhodonellum Schmidt et al. 2006, 2891[VP]

Rhodopirellula Schlesner et al. 2004, 1577[VP]

Rhodovarius Kämpfer et al. 2004, 1909[VP] – Valid publication: Validation List no. 100 – Effective publication: Kämpfer et al. (2004b)

Robiginitalea J.C. Cho and Giovannoni 2004a, 1104[VP]

Robiginitomaculum K. Lee et al. 2007b, 2598[VP]

Roseibacterium T. Suzuki et al. 2006, 420[VP]

Roseicyclus Rathgeber et al. 2005, 1602[VP]

Roseisalinus Labrenz et al. 2005, 45[VP]

Roseivirga Nedashkovskaya et al. 2005c, 232[VP]

Ruania Q. Gu et al. 2007a, 811[VP]

Rubellimicrobium Denner et al. 2006, 1360[VP]

Rubritalea Scheuermayer et al. 2006, 2123[VP]

Saccharibacter Jojima et al. 2004, 2266[VP]

Saccharophagus Ekborg et al. 2005, 1548[VP]

Salicola Maturrano et al. 2006, 1689[VP]

Salimicrobium J.H. Yoon et al. 2007xxi, 2409[VP]

Salinibacillus Ren and Zhou 2005, 952[VP]

Salinimonas Jeon et al. 2005c, 241[VP]

Salinispora Maldonado et al. 2005, 1763[VP]

Salipiger Martínez-Cánovas et al. 2004d, 1739[VP]

Salirhabdus Albuquerque et al. 2007, 1569[VP]

Salsuginibacillus Carrasco et al. 2007b, 2384[VP]

Sandarakinorhabdus Gich and Overmann 2006, 852[VP]

Sandarakinotalea Khan et al. 2006b, 960[VP]

Schlesneria Kulichevskaya et al. 2007, 2685[VP]

Sedimenticola Narasingarao and Häggblom 2006, 2507[VP] – Valid publication: Validation List no. 112 – Effective publication: Narasingarao and Häggblom (2006)

Sediminibacter Khan et al. 2007b, 267[VP]

Sediminicola Khan et al. 2006c, 843[VP]

Sediminitomix Khan et al. 2007d, 1692[VP]

Segetibacter D.S. An et al. 2007a, 1829[VP]

Segniliparus Butler et al. 2005, 1621[VP]

Seinonella J.H. Yoon et al. 2005n, 399[VP]

Sejongia Yi et al. 2005b, 414[VP]

Serinicoccus Yi et al. 2004c, 1587[VP]

Shimazuella D.J. Park et al. 2007, 2663[VP]

Shimia Choi and Cho 2006, 1872[VP]

Shinella D.S. An et al. 2006a, 446[VP]

Silanimonas E.M. Lee et al. 2005, 387[VP]

Silvimonas H.C. Yang et al. 2005, 2331[VP]

Simplicispira Grabovich et al. 2006, 575[VP]

Sinococcus W.J. Li et al. 2006d, 1191[VP]

The name *Sinococcus* W.J. Li et al. 2006d is illegitimate because it is a later homonym of an insect genus name *Sinococcus* Wu and Zheng 2000.

Smaragdicoccus Adachi et al. 2007, 300[VP]

Sneathiella Jordan et al. 2007, 119[VP]

Solimonas M.K. Kim et al. 2007a, 2593[VP]

Sphaerisporangium corrig. Ara and Kudo 2007, 2449[VP] – Valid publication: Validation List no. 118 – Effective publication: Ara and Kudo (2007b)

Sphingosinicella Maruyama et al. 2006, 88[VP]

Sporacetigenium S. Chen et al. 2006, 723[VP]

Sporotalea Boga et al. 2007, 894[VP] – Valid publication: Validation List no. 115 – Effective publication: Boga et al. (2007)

Stackebrandtia Labeda and Kroppenstedt 2005, 1690[VP]

Stanierella Nedashkovskaya et al. 2005h, 227[VP]

Stenothermobacter S.C. Lau et al. 2006a, 183[VP]

Subdoligranulum Holmstrøm et al. 2004, 1909[VP] – Valid publication: Validation List no. 100 – Effective publication: Holmstrøm et al. (2004)

Subsaxibacter Bowman and Nichols 2005, 1481[VP]

Subsaximicrobium Bowman and Nichols 2005, 1480[VP]

Sulfuricurvum Kodama and Watanabe 2004, 2299[VP]

Sulfurivirga Takai et al. 2006a, 1927[VP]

Sulfurovum Inagaki et al. 2004, 1480[VP]

Swaminathania Loganathan and Nair 2004, 1189[VP]

Tamlana S.D. Lee 2007a, 766[VP]

Telmatospirillum Sizova et al. 2007, 1372[VP] – Valid publication: Validation List no. 116 – Effective publication: Sizova et al. (2007)

Tenuibacillus Ren and Zhou 2005, 98[VP]

Tepidanaerobacter Sekiguchi et al. 2006, 1627[VP]

Tepidicella França et al. 2006b, 910[VP]

Tepidimicrobium Slobodkin et al. 2006b, 371[VP]

Terribacillus S.Y. An et al. 2007a, 54[VP]

Terriglobus Eichorst et al. 2007, 1933[VP] – Valid publication: Validation List no. 117 – Effective publication: Eichorst et al. (2007)

Terrimonas Xie and Yokota 2006c, 1120[VP]

Tetrathiobacter Ghosh et al. 2005, 1785[VP]

Thalassobacillus García et al. 2005, 1793[VP]

Thalassobacter Macián et al. 2005a, 109[VP]

Thalassobius Arahal et al. 2005, 2374[VP]

Thalassococcus O.O. Lee et al. 2007, 1923[VP]

Thalassolituus Yakimov et al. 2004, 145[VP]

Thermincola Sokolova et al. 2005, 2072[VP]

Thermodesulfatator Moussard et al. 2004, 231[VP]

Thermodesulfobium Mori et al. 2004, 1[VP] – Valid publication: Validation List no. 95 – Effective publication: Mori et al. (2003)

Thermoflavimicrobium J.H. Yoon et al. 2005n, 399[VP]

Thermogymnomonas Itoh et al. 2007, 2560[VP]

Thermolithobacter Sokolova et al. 2007, 1372[VP] – Valid publication: Validation List no. 116 – Effective publication: Sokolova et al. (2007)

Thermosediminibacter Lee et al. 2006, 925[VP] – Valid publication: Validation List no. 109 – Effective publication: Y.J. Lee et al. (2005)

Thermosinus Sokolova et al. 2004, 2357[VP]

Thermovirga Dahle and Birkeland 2006, 1544[VP]

Thiobacter Hirayama et al. 2005, 471[VP]

Thioclava Sorokin et al. 2005b, 1074[VP]

Thiohalomonas Sorokin et al. 2007c, 1587[VP]

Thiohalophilus Sorokin et al. 2007, 1933[VP] – Valid publication: Validation List no. 117 – Effective publication: Sorokin et al. (2007b)

Thioreductor S. Nakagawa et al. 2005a, 603[VP]

Thiovirga Ito et al. 2005, 1063[VP]

Thorsellia Kämpfer et al. 2006f, 337[VP]

Truepera da Costa et al. 2005 (complete authorship reads da Costa, Rainey and Albuquerque), 1744[VP] – Valid publication:

Validation List no. 105 – Effective publication: Albuquerque et al. (2005)

Tuberibacillus Hatayama et al. 2006, 2549[VP]

Turneriella Levett et al. 2005, 1499[VP]

Ulvibacter Nedashkovskaya et al. 2004c, 121[VP]

Umezawaea Labeda and Kroppenstedt 2007, 2761[VP]

Undibacterium Kämpfer et al. 2007d, 1513[VP]

Uruburuella Vela et al. 2005, 645[VP]

Viridibacillus Albert et al. 2007, 2734[VP]

Volucribacter Christensen et al. 2004, 817[VP]

Vulcanibacillus L'Haridon et al. 2006, 1050[VP]

Wautersia Vaneechoutte et al. 2004, 322[VP]

Wautersiella Kämpfer et al. 2006a, 2328[VP]

Wenxinia Ying et al. 2007b, 1714[VP]

Winogradskyella Nedashkovskaya et al. 2005a, 51[VP]

Woodsholea Abraham et al. 2004, 1232[VP]

Xylanibacter Ueki et al. 2006b, 2220[VP]

Xylanibacterium Rivas et al. 2004e, 560[VP]

Xylanimicrobium Stackebrandt and Schumann 2004, 1385[VP]

Yangia Dai et al. 2006, 531[VP]

Yania W.J. Li et al. 2004a, 529[VP]

The name *Yania* W.J. Li et al. 2004a is illegitimate because it is a later homonym of *Yania* Roewer 1919 (Opiliones, Arachnida, Arthropoda, Animalia) and a later homonym of *Yania* Huang 1997 (Hesperiidae, Papilionoidea, Heteroneura, Glossata, Lepidoptera, Insecta, Arthropoda, Animalia). See Principle 2.

Yeosuana K.K. Kwon et al. 2006a, 731[VP]

Yonghaparkia J.H. Yoon et al. 2006k, 2418[VP]

Zeaxanthinibacter Asker et al. 2007a, 841[VP]

Zhihengliuella Y.Q. Zhang et al. 2007a, 1019[VP]

Zhouia Z.P Liu et al. 2006a, 2826[VP]

Zimmermannella Y.C. Lin et al. 2004, 1674[VP]

Strict application of the Rule 51b(1) indicates that the genus name *Zimmermannella* Y.C. Lin et al. 2004 is illegitimate.

Zobellella Y.T. Lin and Shieh 2006, 1214[VP]

Zunongwangia corrig. Qin et al. 2007, 1372[VP] – Valid publication: Validation List no. 116 – Effective publication: Qin et al. (2007)

New species

Acaricomes phytoseiuli Pukall et al. 2006, 468[VP]

Acetanaerobacterium elongatum Chen and Dong 2004, 2261[VP]

Acetobacter ghanensis Cleenwerck et al. 2007, 1650[VP]

Acetobacter nitrogenifigens Dutta and Gachhui 2006, 1902[VP]

Acetobacter oeni Silva et al. 2006, 22[VP]

Acetobacter senegalensis Ndoye et al. 2007, 1580[VP]

Acidaminococcus intestini Jumas-Bilak et al. 2007b, 2318[VP]

Acidianus sulfidivorans Plumb et al. 2007, 1422[VP]

Acidicaldus organivorans corrig. Johnson et al. 2006, 1459[VP] – Valid publication: Validation List no. 110 – Effective publication: Johnson et al. (2006)

Actinoalloteichus hymeniacidonis H. Zhang et al. 2006, 2311[VP]

Actinoalloteichus spitiensis Singla et al. 2005, 2563[VP]

Actinobaculum massiliense corrig. Greub and Raoult 2006, 2025[VP] – Valid publication: Validation List no. 111 – Effective publication: Greub and Raoult (2002)

Actinocatenispora sera Matsumoto et al. 2007, 2653[VP]

Actinocatenispora thailandica Thawai et al. 2006, 1793[VP]

Actinocorallia aurea Tamura et al. 2007a, 2054[VP]

Actinocorallia cavernae S.D. Lee 2006c, 1087[VP]

Actinomadura alba Y.X. Wang et al. 2007, 1738[VP]

Actinomadura hallensis S.D. Lee and Jeong 2006, 262[VP]

Actinomadura mexicana Quintana et al. 2004, 307[VP] – Valid publication: Validation List no. 96 – Effective publication: Quintana et al. (2003)

Actinomadura meyerae corrig. Quintana et al. 2004, 307[VP] – Valid publication: Validation List no. 96 – Effective publication: Quintana et al. (2003)

Actinomadura napierensis Cook et al. 2005, 705[VP]

Actinomadura rudentiformis Le Roes and Meyers 2007, 48[VP]

Actinomyces dentalis Hall et al. 2005, 430[VP]

Actinomyces hongkongensis Woo et al. 2004, 307[VP] – Valid publication: Validation List no. 96 – Effective publication: Woo et al. (2003)

Actinomyces ruminicola D. An et al. 2006b, 2045[VP]

Actinoplanes couchii Kämpfer et al. 2007a, 722[VP]

Actinoplanes liguriensis Wink et al. 2006, 2128[VP]

Actinoplanes teichomyceticus Wink et al. 2006, 2129[VP]

Actinospica acidiphila Cavaletti et al. 2006a, 1752[VP]

Actinospica robiniae Cavaletti et al. 2006a, 1752[VP]

Adhaeribacter aquaticus Rickard et al. 2005, 827[VP]

Advenella incenata Coenye et al. 2005, 255[VP]

Aeriscardovia aeriphila Simpson et al. 2004, 405[VP]

Aerococcus suis Vela et al. 2007b, 1293[VP]

Aeromicrobium alkaliterrae J.H. Yoon et al. 2005q, 2174[VP]

Aeromicrobium panaciterrae Y.S. Cui et al. 2007a, 690[VP]

Aeromicrobium tamlense S.D. Lee and Kim 2007, 339[VP]

Aeromonas bivalvium Miñana-Galbis et al. 2007, 585[VP]

Aeromonas molluscorum Miñana-Galbis et al. 2004, 2077[VP]

Aeromonas sharmana Saha and Chakrabarti 2006b, 1907[VP]

Aeromonas simiae Harf-Monteil et al. 2004, 484[VP]

Aeropyrum camini S. Nakagawa et al. 2004b, 333[VP]

Aestuariibacter halophilus Yi et al. 2004b, 575[VP]

Aestuariibacter salexigens Yi et al. 2004b, 573[VP]

Aestuariimicrobium kwangyangense Jung et al. 2007b, 2117[VP]

Agarivorans albus Kurahashi and Yokota 2004, 696[VP]

Agrococcus casei Bora et al. 2007, 95[VP]

Agrococcus lahaulensis Mayilraj et al. 2006f, 1809[VP]

Agromyces allii Jung et al. 2007c, 591[VP]

Agromyces humatus Jurado et al. 2005c, 874[VP]

Agromyces italicus Jurado et al. 2005c, 874[VP]

Agromyces lapidis Jurado et al. 2005c, 874[VP]

Agromyces neolithicus Jurado et al. 2005a, 157[VP]

Agromyces salentinus Jurado et al. 2005a, 156[VP]

Agromyces subbeticus Jurado et al. 2005b, 1900[VP]

Agromyces ulmi Rivas et al. 2004c, 1989[VP]

Akkermansia muciniphila Derrien et al. 2004, 1474[VP]

Alcaligenes aquatilis Van Trappen et al. 2005a, 2573[VP]

Alcanivorax balearicus Rivas et al. 2007, 1334[VP]

Alcanivorax dieselolei C. Liu and Shao 2005, 1184[VP]

Algibacter lectus Nedashkovskaya et al. 2004d, 1260[VP]

Algibacter mikhailovii Nedashkovskaya et al. 2007e, 2148[VP]

Algoriphagus antarcticus Van Trappen et al. 2004f, 1972[VP]

Algoriphagus aquimarinus Nedashkovskaya et al. 2004h, 1762[VP]

Algoriphagus chordae Nedashkovskaya et al. 2004h, 1762[VP]

Algoriphagus locisalis J.H. Yoon et al. 2005e, 1638[VP]

Algoriphagus terrigena J.H. Yoon et al. 2006o, 779[VP]

Algoriphagus vanfongensis Nedashkovskaya et al. 2007b, 1990[VP]

Algoriphagus winogradskyi Nedashkovskaya et al. 2004h, 1763[VP]

Algoriphagus yeomjeoni J.H. Yoon et al. 2005b, 869[VP]

Alicyclobacillus contaminans Goto et al. 2007, 1281[VP]

Alicyclobacillus fastidiosus Goto et al. 2007, 1281[VP]

Alicyclobacillus kakegawensis Goto et al. 2007, 1283[VP]

Alicyclobacillus macrosporangiidus Goto et al. 2007, 1283[VP]

Alicyclobacillus sacchari Goto et al. 2007, 1283[VP]

Alicyclobacillus shizuokensis Goto et al. 2007, 1283[VP]

Alicyclobacillus tolerans Karavaiko et al. 2005, 946[VP]

Alicyclobacillus vulcanalis Simbahan et al. 2004, 1706[VP]

Alistipes onderdonkii Y. Song et al. 2006, 1988[VP]

Alistipes shahii Y. Song et al. 2006, 1989[VP]

Alkalibacillus filiformis Romano et al. 2005d, 2397[VP]

Alkalibacillus salilacus Jeon et al. 2005a, 1895[VP]

Alkalibacillus silvisoli Usami et al. 2007, 773[VP]

Alkalibacter saccharofermentans Garnova et al. 2005, 983[VP] – Valid publication: Validation List no. 103 – Effective publication: Garnova et al. (2004)

Alkalibacterium iburiense K. Nakajima et al. 2005, 1529[VP]

Alkalibacterium psychrotolerans Yumoto et al. 2004b, 2382[VP]

Alkaliflexus imshenetskii Zhilina et al. 2005, 1395[VP] – Valid publication: Validation List no. 104 – Effective publication: Zhilina et al. (2004)

Alkalilimnicola ehrlichii Hoeft et al. 2007, 510[VP]

Alkalimonas amylolytica Ma et al. 2007, 433[VP] – Valid publication: Validation List no. 114 – Effective publication: Y. Ma et al. (2004)

Alkalimonas collagenimarina Kurata et al. 2007, 1552[VP]

Alkalimonas delamerensis Ma et al. 2007, 433[VP] – Valid publication: Validation List no. 114 – Effective publication: Y. Ma et al. (2004)

Allobaculum stercoricanis Greetham et al. 2006, 1459[VP] – Valid publication: Validation List no. 110 – Effective publication: Greetham et al. (2004b)

Alloscardovia omnicolens Huys et al. 2007, 1445^VP

Altererythrobacter epoxidivorans K.K. Kwon et al. 2007, 2210^VP

Alteromonas addita Ivanova et al. 2005b, 1067^VP

Alteromonas hispanica Martínez-Checa et al. 2005, 2389^VP

Alteromonas litorea J.H. Yoon et al. 2004m, 1200^VP

Alteromonas simiduii Chiu et al. 2007, 1215^VP

Alteromonas stellipolaris Van Trappen et al. 2004b, 1160^VP

Alteromonas tagae Chiu et al. 2007, 1215^VP

Aminiphilus circumscriptus Díaz et al. 2007, 1917^VP

Aminobacter ciceronei McDonald et al. 2005, 1830^VP

Aminobacter lissarensis McDonald et al. 2005, 1831^VP

Amphibacillus sediminis S.Y. An et al. 2007e, 2491^VP

Amycolatopsis australiensis Tan et al. 2006, 2299^VP

Amycolatopsis benzoatilytica Majumdar et al. 2006, 202^VP

Amycolatopsis decaplanina Wink et al. 2004, 237^VP

Amycolatopsis echigonensis Ding et al. 2007, 1750^VP

Amycolatopsis halotolerans S.D. Lee 2006a, 552^VP

Amycolatopsis jejuensis S.D. Lee 2006a, 552^VP

Amycolatopsis minnesotensis S.D. Lee et al. 2006, 268^VP

Amycolatopsis nigrescens Groth et al. 2007, 517^VP

Amycolatopsis niigatensis Ding et al. 2007, 1750^VP

Amycolatopsis palatopharyngis Huang et al. 2004b, 361^VP

Amycolatopsis plumensis Saintpierre-Bonaccio et al. 2005, 2060^VP

Amycolatopsis regifaucium Tan et al. 2007, 2566^VP

Amycolatopsis rifamycinica Bala et al. 2004, 1148^VP

Amycolatopsis saalfeldensis Carlsohn et al. 2007a, 1644^VP

Amycolatopsis taiwanensis Tseng et al. 2006, 1814^VP

Anaerobranca californiensis Gorlenko et al. 2004b, 742^VP

Anaerofustis stercorihominis Finegold et al. 2004, 1005^VP – Valid publication: Validation List no. 98 – Effective publication: Finegold et al. (2004a)

Anaerolinea thermolimosa Yamada et al. 2006, 1338^VP

Anaerosporobacter mobilis Jeong et al. 2007, 1786^VP

Anaerotruncus colihominis Lawson et al. 2004b, 415^VP

Anaerovirgula multivorans Pikuta et al. 2006b, 2628^VP

Ancylobacter polymorphus Xin et al. 2006, 1187^VP

Ancylobacter rudongensis Xin et al. 2004, 386^VP

Ancylobacter vacuolatus Xin et al. 2006, 1186^VP

Anderseniella baltica Brettar et al. 2007, 2404^VP

Andreprevotia chitinilytica Weon et al. 2007f, 1574^VP

Aneurinibacillus danicus Goto et al. 2004, 425^VP

Aneurinibacillus terranovensis Allan et al. 2005, 1048^VP

Angulomicrobium amanitiforme Fritz et al. 2004, 656^VP

Anoxybacillus amylolyticus Poli et al. 2006, 1459^VP – Valid publication: Validation List no. 110 – Effective publication: Poli et al. (2006)

Anoxybacillus ayderensis Dulger et al. 2004, 1502^VP

Anoxybacillus contaminans De Clerck et al. 2004b, 943^VP

Anoxybacillus kamchatkensis Kevbrin et al. 2006, 925^VP – Valid publication: Validation List no. 109 – Effective publication: Kevbrin et al. (2005)

Anoxybacillus kestanbolensis Dulger et al. 2004, 1503^VP

Anoxybacillus voinovskiensis Yumoto et al. 2004a, 1241^VP

Aquicella lusitana Santos et al. 2004, 1^VP – Valid publication: Validation List no. 95 – Effective publication: Santos et al. (2003)

Aquicella siphonis Santos et al. 2004, 1^VP – Valid publication: Validation List no. 95 – Effective publication: Santos et al. (2003)

Aquiflexum balticum Brettar et al. 2004, 2339^VP

Aquimarina intermedia Nedashkovskaya et al. 2006f, 2039^VP

Aquimarina muelleri Nedashkovskaya et al. 2005h, 227^VP

Aquimonas voraii Saha et al. 2005a, 1493^VP

Aquincola tertiaricarbonis Lechner et al. 2007, 1301^VP

Aquisalimonas asiatica Márquez et al. 2007, 1141^VP

Aquitalea magnusonii H.T. Lau et al. 2006, 870^VP

Arcanobacterium bialowiezense Lehnen et al. 2006, 864^VP

Arcanobacterium bonasi Lehnen et al. 2006, 864^VP

Arcicella aquatica Nikitin et al. 2004, 684^VP

Arcobacter cibarius Houf et al. 2005, 716^VP

Arcobacter halophilus Donachie et al. 2005, 1275^VP

Arenibacter certesii Nedashkovskaya et al. 2004a, 1174^VP

Arenibacter echinorum Nedashkovskaya et al. 2007c, 2657^VP

Arenibacter palladensis Nedashkovskaya et al. 2006g, 159^VP

Arenimonas donghaensis S.W. Kwon et al. 2007c, 956^VP

Arenimonas malthae C.C. Young et al. 2007d, 2792^VP

Arsenicicoccus bolidensis Collins et al. 2004d, 608^VP

Arthrobacter ardleyensis M. Chen et al. 2005, 2235^VP – Valid publication: Validation List no. 106 – Effective publication: M. Chen et al. (2005)

Arthrobacter arilaitensis Irlinger et al. 2005, 459^VP

Arthrobacter bergerei Irlinger et al. 2005, 460^VP

Arthrobacter castelli Heyrman et al. 2005c, 1461^VP

Arthrobacter gangotriensis Gupta et al. 2004, 2376^VP

Arthrobacter kerguelensis Gupta et al. 2004, 2376^VP

Arthrobacter monumenti Heyrman et al. 2005c, 1461^VP

Arthrobacter nitroguajacolicus Kotoučková et al. 2004, 776^VP

Arthrobacter parietis Heyrman et al. 2005c, 1462^VP

Arthrobacter pigmenti Heyrman et al. 2005c, 1462^VP

Arthrobacter psychrophenolicus Margesin et al. 2004, 2070^VP

Arthrobacter russicus Y. Li et al. 2004b, 834^VP

Arthrobacter scleromae Huang et al. 2005, 1743^VP – Valid publication: Validation List no. 105 – Effective publication: Huang et al. (2005b)

Arthrobacter stackebrandtii Tvrzová et al. 2005b, 807^VP

Arthrobacter tecti Heyrman et al. 2005c, 1462^VP

Arthrobacter tumbae Heyrman et al. 2005c, 1463^VP

Asaia krungthepensis Yukphan et al. 2004a, 315^VP

Asanoa iriomotensis Tamura and Sakane 2005, 726^VP

Aspromonas composti Jin et al. 2007, 1879^{VP}

Asticcacaulis benevestitus Vasilyeva et al. 2006, 2087^{VP}

Asticcacaulis taihuensis Z.P. Liu et al. 2005b, 1241^{VP}

Atopococcus tabaci Collins et al. 2005, 1695^{VP}

Atopostipes suicloacalis corrig. Cotta et al. 2004, 1425^{VP} – Valid publication: Validation List no. 99 – Effective publication: Cotta et al. (2004)

Aurantimonas altamirensis Jurado et al. 2006, 2585^{VP}

Aurantimonas ureilytica Weon et al. 2007g, 1719^{VP}

Aureispira marina Hosoya et al. 2006, 2933^{VP}

Aureispira maritima Hosoya et al. 2007, 1950^{VP}

Avibacterium endocarditidis Bisgaard et al. 2007, 1732^{VP}

Azohydromonas australica Xie and Yokota 2005f, 2423^{VP}

Azonexus caeni Quan et al. 2006, 1046^{VP}

Azorhizobium doebereinerae Moreira et al. 2006, 1459^{VP} – Valid publication: Validation List no. 110 – Effective publication: Moreira et al. (2006)

Azospira restricta H.S. Bae et al. 2007, 1525^{VP}

Azospirillum canadense Mehnaz et al. 2007a, 623^{VP}

Azospirillum melinis Peng et al. 2006, 1268^{VP}

Azospirillum oryzae Xie and Yokota 2005e, 1437^{VP}

Azospirillum zeae Mehnaz et al. 2007b, 2808^{VP}

Bacillus acidiceler Peak et al. 2007, 2035^{VP}

Bacillus acidicola Albert et al. 2005, 2129^{VP}

Bacillus aerius Shivaji et al. 2006, 1471^{VP}

Bacillus aerophilus Shivaji et al. 2006, 1471^{VP}

Bacillus akibai Nogi et al. 2005b, 2314^{VP}

Bacillus algicola Ivanova et al. 2004, 1425^{VP} – Valid publication: Validation List no. 99 – Effective publication: Ivanova et al. (2004b)

Bacillus altitudinis Shivaji et al. 2006, 1472^{VP}

Bacillus alveayuensis S.S. Bae et al. 2005, 1214^{VP}

Bacillus arenosi Heyrman et al. 2005b, 114^{VP}

Bacillus arsenicus Shivaji et al. 2005d, 1126^{VP}

Bacillus arvi Heyrman et al. 2005b, 115^{VP}

Bacillus asahii Yumoto et al. 2004c, 1999^{VP}

Bacillus axarquiensis Ruiz-García et al. 2005, 1282^{VP}

Bacillus bataviensis Heyrman et al. 2004a, 55^{VP}

Bacillus bogoriensis Vargas et al. 2005, 901^{VP}

Bacillus boroniphilus Ahmed et al. 2007, 893^{VP} – Valid publication: Validation List no. 115 – Effective publication: Ahmed et al. (2007a)

Bacillus cellulosilyticus Nogi et al. 2005b, 2314^{VP}

Bacillus chagannorensis Carrasco et al. 2007a, 2087^{VP}

Bacillus cibi J.H. Yoon et al. 2005p, 735^{VP}

Bacillus decisifrondis L. Zhang et al. 2007a, 977^{VP}

Bacillus drentensis Heyrman et al. 2004a, 56^{VP}

Bacillus farraginis Scheldeman et al. 2004b, 1361^{VP}

Bacillus foraminis Tiago et al. 2006c, 2573^{VP}

Bacillus fordii Scheldeman et al. 2004b, 1362^{VP}

Bacillus fortis Scheldeman et al. 2004b, 1362^{VP}

Bacillus galactosidilyticus Heyndrickx et al. 2004, 619^{VP}

Bacillus gelatini De Clerck et al. 2004b, 944^{VP}

Bacillus ginsengihumi Ten et al. 2007, 1371^{VP} – Valid publication: Validation List no. 116 – Effective publication: Ten et al. (2006c)

Bacillus hemicellulosilyticus Nogi et al. 2005b, 2312^{VP}

Bacillus herbersteinensis Wieser et al. 2005, 2122^{VP}

Bacillus humi Heyrman et al. 2005b, 115^{VP}

Bacillus hwajinpoensis J.H. Yoon et al. 2004e, 807^{VP}

Bacillus idriensis Ko et al. 2006, 2543^{VP}

Bacillus indicus Suresh et al. 2004a, 1373^{VP}

Bacillus infantis Ko et al. 2006, 2543^{VP}

Bacillus koreensis J.M. Lim et al. 2006b, 61^{VP}

Bacillus kribbensis J.M. Lim et al. 2007b, 2914^{VP}

Bacillus lehensis Ghosh et al. 2007, 241^{VP}

Bacillus litoralis J.H. Yoon and Oh 2005b, 1947^{VP}

Bacillus macauensis T. Zhang et al. 2006, 351^{VP}

Bacillus macyae Santini et al. 2004, 2243^{VP}

Bacillus malacitensis Ruiz-García et al. 2005, 1283^{VP}

Bacillus mannanilyticus Nogi et al. 2005b, 2314^{VP}

Bacillus massiliensis Glazunova et al. 2006b, 1487^{VP}

Bacillus muralis Heyrman et al. 2005a, 128^{VP}

Bacillus murimartini Borchert et al. 2007, 2892^{VP}

Bacillus niabensis S.W. Kwon et al. 2007d, 1910^{VP}

Bacillus novalis Heyrman et al. 2004a, 52^{VP}

Bacillus odysseyi La Duc et al. 2004, 200^{VP}

Bacillus okhensis Nowlan et al. 2006, 1076^{VP}

Bacillus oshimensis Yumoto et al. 2005a, 910^{VP}

Bacillus panaciterrae Ten et al. 2006b, 2864^{VP}

Bacillus patagoniensis Olivera et al. 2005, 446^{VP}

Bacillus plakortidis Borchert et al. 2007, 2892^{VP}

Bacillus pocheonensis Ten et al. 2007, 2535^{VP}

Bacillus qingdaonensis Q.F. Wang et al. 2007a, 1146^{VP}

Bacillus ruris Heyndrickx et al. 2005, 2552^{VP}

Bacillus safensis Satomi et al. 2006a, 1739^{VP}

Bacillus salarius J.M. Lim et al. 2006e, 376^{VP}

Bacillus saliphilus Romano et al. 2005c, 162^{VP}

Bacillus selenatarsenatis S. Yamamura et al. 2007, 1063^{VP}

Bacillus seohaeanensis J.C. Lee et al. 2006, 1896^{VP}

Bacillus shackletonii Logan et al. 2004b, 375^{VP}

Bacillus soli Heyrman et al. 2004a, 55^{VP}

Bacillus stratosphericus Shivaji et al. 2006, 1471^{VP}

Bacillus taeanensis J.M. Lim et al. 2006a, 2905^{VP}

Bacillus tequilensis Gatson et al. 2006, 1481^{VP}

Bacillus thioparans corrig. Pérez-Ibarra et al. 2007, 1933^{VP} – Valid publication: Validation List no. 117 – Effective publication: Pérez-Ibarra et al. (2007)

Bacillus velezensis Ruiz-García et al. 2005, 194[VP]

Bacillus vietnamensis Noguchi et al. 2004, 2119[VP]

Bacillus vireti Heyrman et al. 2004a, 54[VP]

Bacillus wakoensis Nogi et al. 2005b, 2312[VP]

Bacteriovorax litoralis Baer et al. 2004, 1015[VP]

Bacteriovorax marinus Baer et al. 2004, 1015[VP]

Bacteroides barnesiae Lan et al. 2006, 2857[VP]

Bacteroides cellulosilyticus Robert et al. 2007, 1519[VP]

Bacteroides coprocola Kitahara et al. 2005, 2146[VP]

Bacteroides coprophilus Hayashi et al. 2007a, 1325[VP]

Bacteroides coprosuis Whitehead et al. 2005, 2517[VP]

Bacteroides dorei Bakir et al. 2006c, 1642[VP]

Bacteroides finegoldii Bakir et al. 2006b, 934[VP]

Bacteroides gallinarum Lan et al. 2006, 2857[VP]

Bacteroides goldsteinii Song et al. 2006, 499[VP] – Valid publication: Validation List no. 108 – Effective publication: Y. Song et al. (2005)

Bacteroides intestinalis Bakir et al. 2006a, 153[VP]

Bacteroides massiliensis Fenner et al. 2005, 1336[VP]

Bacteroides nordii Song et al. 2005, 983[VP] – Valid publication: Validation List no. 103 – Effective publication: Y.L. Song et al. (2004)

Bacteroides plebeius Kitahara et al. 2005, 2146[VP]

Bacteroides salanitronis Lan et al. 2006, 2857[VP]

Bacteroides salyersiae corrig. Song et al. 2005, 983[VP] – Valid publication: Validation List no. 103 – Effective publication: Y.L. Song et al. (2004)

Balneimonas flocculans corrig. Takeda et al. 2004, 631[VP] – Valid publication: Validation List no. 97 – Effective publication: Takeda et al. (2004)

Balneola vulgaris Urios et al. 2006, 1886[VP]

Barnesiella viscericola Sakamoto et al. 2007b, 345[VP]

Bartonella chomelii Maillard et al. 2004, 219[VP]

Belliella baltica Brettar et al. 2004, 69[VP]

Bellilinea caldifistulae Yamada et al. 2007, 2302[VP]

Belnapia moabensis Reddy et al. 2006, 56[VP]

Bifidobacterium psychraerophilum Simpson et al. 2004, 404[VP]

Bizionia algoritergicola Bowman and Nichols 2005, 1483[VP]

Bizionia gelidisalsuginis Bowman and Nichols 2005, 1482[VP]

Bizionia myxarmorum Bowman and Nichols 2005, 1483[VP]

Bizionia paragorgiae Nedashkovskaya et al. 2005g, 377[VP]

Bizionia saleffrena Bowman and Nichols 2005, 1482[VP]

Blastococcus jejuensis S.D. Lee 2006g, 2395[VP]

Blastococcus saxobsidens Urzì et al. 2004, 258[VP]

Borrelia spielmanii Richter et al. 2006, 880[VP]

Borrelia turcica Güner et al. 2004, 1651[VP]

Bowmanella denitrificans Jean et al. 2006a, 2465[VP]

Brachybacterium phenoliresistens J.H. Chou et al. 2007c, 2677[VP]

Brachybacterium zhongshanense G. Zhang et al. 2007, 2523[VP]

Bradyrhizobium betae Rivas et al. 2004f, 1274[VP]

Bradyrhizobium canariense Vinuesa et al. 2005, 573[VP]

Brevibacill\us ginsengisoli Baek et al. 2006, 2667[VP]

Brevibacillus levickii Allan et al. 2005, 1048[VP]

Brevibacillus limnophilus Goto et al. 2004, 426[VP]

Brevibacterium antiquum Gavrish et al. 2005, 1[VP] – Valid publication: Validation List no. 101 – Effective publication: Gavrish et al. (2004)

Brevibacterium aurantiacum Gavrish et al. 2005, 1[VP] – Valid publication: Validation List no. 101 – Effective publication: Gavrish et al. (2004)

Brevibacterium celere Ivanova et al. 2004c, 2110[VP]

Brevibacterium permense Gavrish et al. 2005, 1[VP] – Valid publication: Validation List no. 101 – Effective publication: Gavrish et al. (2004)

Brevibacterium picturae Heyrman et al. 2004b, 1540[VP]

Brevibacterium samyangense S.D. Lee 2006e, 1891[VP]

Brevibacterium sanguinis Wauters et al. 2004, 1425[VP] – Valid publication: Validation List no. 99 – Effective publication: Wauters et al. (2004)

Brevundimonas aveniformis Ryu et al. 2007a, 1564[VP]

Brevundimonas kwangchunensis J.H. Yoon et al. 2006e, 616[VP]

Brevundimonas lenta J.H. Yoon et al. 2007ix, 2239[VP]

Brevundimonas mediterranea Fritz et al. 2005, 484[VP]

Brevundimonas nasdae Y. Li et al. 2004a, 824[VP]

Brevundimonas terrae J.H. Yoon et al. 2006c, 2918[VP]

Brooklawnia cerclae Rainey et al. 2006 (authorship reads Rainey, da Costa and Moe), 1981[VP] – Valid publication: H.S. Bae et al. (2006a)

Brucella ceti Foster et al. 2007, 2691[VP]

Brucella pinnipedialis Foster et al. 2007, 2691[VP]

Bryantella formatexigens Wolin et al. 2004, 1[VP] – Valid publication: Validation List no. 95 – Effective publication: Wolin et al. (2003)

The name *Bryantella formatexigens* Wolin et al. 2004 is illegitimate because the genus name is illegitimate (see above *Bryantella* Wolin et al. 2004, 1[VP]).

Burkholderia bryophila Vandamme et al. 2007, 2233[VP]

Burkholderia dolosa Vermis et al. 2004, 691[VP]

Burkholderia endofungorum Partida-Martinez et al. 2007, 2589[VP]

Burkholderia ferrariae Valverde et al. 2006a, 2424[VP]

Burkholderia ginsengisoli H.B. Kim et al. 2006, 2531[VP]

Burkholderia megapolitana Vandamme et al. 2007, 2232[VP]

Burkholderia mimosarum W.M. Chen et al. 2006, 1851[VP]

Burkholderia nodosa W.M. Chen et al. 2007, 1058[VP]

Burkholderia oklahomensis Glass et al. 2006, 2175[VP]

Burkholderia phenoliruptrix Coenye et al. 2005, 547[VP] – Valid publication: Validation List no. 102 – Effective publication: Coenye et al. (2004a)

Burkholderia phytofirmans Sessitsch et al. 2005, 1190[VP]

Burkholderia rhizoxinica Partida-Martinez et al. 2007, 2587[VP]

Burkholderia silvatlantica Perin et al. 2006, 1935[VP]

Burkholderia soli Yoo et al. 2007a, 123[VP]

Burkholderia terrae H.C. Yang et al. 2006b, 456[VP]

Burkholderia tropica Reis et al. 2004, 2161[VP]

Burkholderia unamae Caballero-Mellado et al. 2004, 1170[VP]

Burkholderia xenovorans Goris et al. 2004, 1680[VP]

Caenispirillum bisanense J.H. Yoon et al. 2007xvi, 1219[VP]

Caldalkalibacillus thermarum Xue et al. 2006, 1220[VP]

Caldimonas taiwanensis W.M. Chen et al. 2005, 1743[VP] – Valid publication: Validation List no. 105 – Effective publication: W.M. Chen et al. (2005)

Caminibacter mediatlanticus Voordeckers et al. 2005, 777[VP]

Caminibacter profundus Miroshnichenko et al. 2004, 44[VP]

Campylobacter canadensis Inglis et al. 2007, 2642[VP]

Campylobacter insulaenigrae Foster et al. 2004, 2371[VP]

Carboxydocella sporoproducens Slepova et al. 2006, 799[VP]

Cardiobacterium valvarum Han et al. 2004, 1425[VP] – Valid publication: Validation List no. 99 – Effective publication: Han et al. (2004)

Carnobacterium pleistocenium Pikuta et al. 2005, 477[VP]

Castellaniella denitrificans Kämpfer et al. 2006d, 818[VP]

Catellatospora bangladeshensis Ara and Kudo 2006, 399[VP]

Catellatospora chokoriensis Ara and Kudo 2006, 397[VP]

Catellatospora coxensis Ara and Kudo 2006, 398[VP]

Catellibacterium nectariphilum Tanaka et al. 2004, 958[VP]

Catellicoccus marimammalium Lawson et al. 2006, 431[VP]

Catenulispora acidiphila Busti et al. 2006, 1745[VP]

Catenulispora rubra Tamura et al. 2007b, 2273[VP]

Cellulomonas bogoriensis B.E. Jones et al. 2005, 1713[VP]

Cellulomonas composti M.S. Kang et al. 2007, 1259[VP]

Cellulomonas denverensis Brown et al. 2005, 1395[VP] – Valid publication: Validation List no. 104 – Effective publication: J.M. Brown et al. (2005)

Cellulomonas terrae D.S. An et al. 2005b, 1708[VP]

Cellulomonas xylanilytica Rivas et al. 2004b, 535[VP]

Cellulophaga pacifica Nedashkovskaya et al. 2004e, 611[VP]

Cellulosimicrobium funkei J.M. Brown et al. 2006, 804[VP]

Cellulosimicrobium terreum J.H. Yoon et al. 2007xxv, 2496[VP]

Cerasibacillus quisquiliarum Nakamura et al. 2004, 1067[VP]

Cerasicoccus arenae J. Yoon et al. 2007c, 2070[VP]

Chimaereicella alkaliphila Tiago et al. 2006, 925[VP] – Valid publication: Validation List no. 109 – Effective publication: Tiago et al. (2006a)

Chimaereicella boritolerans Ahmed et al. 2007c, 991[VP]

Chitinibacter tainanensis Chern et al. 2004, 1390[VP]

Chitinilyticum aquatile S.C. Chang et al. 2007, 2858[VP]

Chitinimonas koreensis B.Y. Kim et al. 2006d, 1762[VP]

Chitinimonas taiwanensis S.C. Chang et al. 2004, 1005[VP] – Valid publication: Validation List no. 98 – Effective publication: S.C. Chang et al. (2004)

Chitinophaga ginsengisegetis H.G. Lee et al. 2007, 1398[VP]

Chitinophaga ginsengisoli H.G. Lee et al. 2007, 1399[VP]

Chitinophaga skermanii Kämpfer et al. 2006k, 2226[VP]

Chitinophaga terrae M.K. Kim and Jung 2007, 1723[VP]

Chondromyces robustus Reichenbach 2007, 893[VP] – Valid publication: Validation List no. 115 – Effective publication: Reichenbach (2005g)

Chromobacterium subtsugae Martin et al. 2007, 997[VP]

Chromohalobacter japonicus Sánchez-Porro et al. 2007, 2265[VP]

Chromohalobacter nigrandesensis Prado et al. 2006, 650[VP]

Chromohalobacter salarius Aguilera et al. 2007, 1240[VP]

Chromohalobacter sarecensis Quillaguamán et al. 2004a, 1925[VP]

Chryseobacterium caeni Quan et al. 2007, 143[VP]

Chryseobacterium daecheongense K.K. Kim et al. 2005a, 136[VP]

Chryseobacterium daeguense J.H. Yoon et al. 2007xv, 1358[VP]

Chryseobacterium flavum Zhou et al. 2007a, 1767[VP]

Chryseobacterium formosense C.C. Young et al. 2005, 426[VP]

Chryseobacterium haifense Hantsis-Zacharov and Halpern 2007, 2347[VP]

Chryseobacterium hispanicum Gallego et al. 2006b, 1591[VP]

Chryseobacterium hominis Vaneechoutte et al. 2007, 2625[VP]

Chryseobacterium luteum Behrendt et al. 2007b, 1883[VP]

Chryseobacterium miricola Li et al. 2004, 307[VP] – Valid publication: Validation List no. 96 – Effective publication: Y. Li et al. (2003)

Chryseobacterium piscium de Beer et al. 2006, 1321[VP]

Chryseobacterium shigense Shimomura et al. 2005, 1905[VP]

Chryseobacterium soldanellicola M.S. Park et al. 2006, 436[VP]

Chryseobacterium taeanense M.S. Park et al. 2006, 436[VP]

Chryseobacterium taichungense F.T. Shen et al. 2005, 1302[VP]

Chryseobacterium taiwanense Tai et al. 2006, 1774[VP]

Chryseobacterium vrystaatense de Beer et al. 2005, 2151[VP]

Chryseobacterium wanjuense Weon et al. 2006c, 1502[VP]

Citreicella thiooxidans Sorokin et al. 2006, 1[VP] – Valid publication: Validation List no. 107 – Effective publication: Sorokin et al. (2005a)

Citreimonas salinaria Choi and Cho 2006, 2801[VP]

Citricoccus alkalitolerans W.J. Li et al. 2005c, 88[VP]

Cloacibacterium normanense Allen et al. 2006, 1314[VP]

Clostridium aciditolerans Y.J. Lee et al. 2007b, 314[VP]

Clostridium aestuarii S. Kim et al. 2007, 1317[VP]

Clostridium aldenense Warren et al. 2007, 893[VP] – Valid publication: Validation List no. 115 – Effective publication: Warren et al. (2006)

Clostridium alkalicellulosi corrig. Zhilina et al. 2006, 925[VP] – Valid publication: Validation List no. 109 – Effective publication: Zhilina et al. (2005a)

Clostridium asparagiforme Mohan et al. 2007, 1933[VP] – Valid publication: Validation List no. 117 – Effective publication: Mohan et al. (2006)

Clostridium bartlettii Song et al. 2004, 1425[VP] – Valid publication: Validation List no. 99 – Effective publication: Y.L. Song et al. (2004b)

Clostridium carboxidivorans Liou et al. 2005, 2089[VP]

Clostridium citroniae Warren et al. 2007, 893[VP] – Valid publication: Validation List no. 115 – Effective publication: Warren et al. (2006)

Clostridium drakei Liou et al. 2005, 2089[VP]

Clostridium ganghwense S. Kim et al. 2006, 693[VP]

Clostridium glycyrrhizinilyticum Sakuma et al. 2006, 2507[VP] – Valid publication: Validation List no. 112 – Effective publication: Sakuma et al. (2006)

Clostridium jejuense Jeong et al. 2004, 1467[VP]

Clostridium lundense Cirne et al. 2006, 627[VP]

Clostridium nitrophenolicum Suresh et al. 2007b, 1889[VP]

Clostridium saccharogumia Clavel et al. 2007, 893[VP] – Valid publication: Validation List no. 115 – Effective publication: Clavel et al. (2007)

Clostridium schirmacherense Alam et al. 2006, 719[VP]

Clostridium straminisolvens S. Kato et al. 2004, 2046[VP]

Cohnella hongkongensis Kämpfer et al. 2006g, 784[VP]

Cohnella laeviribosi E.A. Cho et al. 2007, 2905[VP]

Cohnella thermotolerans Kämpfer et al. 2006g, 784[VP]

Collimonas fungivorans De Boer et al. 2004, 863[VP]

Colwellia aestuarii Jung et al. 2006b, 37[VP]

Colwellia piezophila Nogi et al. 2004, 1630[VP]

Comamonas badia Tago and Yokota 2005, 983[VP] – Valid publication: Validation List no. 103 – Effective publication: Tago and Yokota (2004)

Comamonas odontotermitis J.H. Chou et al. 2007d, 889[VP]

Conchiformibius kuhniae corrig. Xie and Yokota 2005, 1395[VP] – Valid publication: Validation List no. 104 – Effective publication: Xie and Yokota (2005a)

Coraliomargarita akajimensis J. Yoon et al. 2007f, 962[VP]

Corallococcus macrosporus Reichenbach 2007, 893[VP] – Valid publication: Validation List no. 115 – Effective publication: Reichenbach (2005b)

Corynebacterium caspium Collins et al. 2004b, 926[VP]

Corynebacterium ciconiae Fernández-Garayzábal et al. 2004, 2194[VP]

Corynebacterium halotolerans H.H. Chen et al. 2004, 781[VP]

Corynebacterium hansenii Renaud et al. 2007, 1115[VP]

Corynebacterium nigricans Shukla et al. 2004, 1[VP] – Valid publication: Validation List no. 95 – Effective publication: Shukla et al. (2003)

Corynebacterium resistens Otsuka et al. 2005, 2235[VP] – Valid publication: Validation List no. 106 – Effective publication: Otsuka et al. (2005)

Corynebacterium tuberculostearicum Feurer et al. 2004, 1059[VP]

Corynebacterium tuscaniense corrig. Riegel et al. 2006, 2025[VP] – Valid publication: Validation List no. 111 – Effective publication: Riegel et al. (2006)

Corynebacterium ureicelerivorans Yassin 2007, 1202[VP]

Costertonia aggregata K.K. Kwon et al. 2006c, 1352[VP]

Crabtreella saccharophila Xie and Yokota 2006a, 623[VP]

Crabtreella saccharophila Xie and Yokota 2006a is a later homotypic synonym of *Shinella zoogloeoides* D.S. An et al. 2006a

Cryobacterium psychrotolerans D.C. Zhang et al. 2007, 868[VP]

Cryptanaerobacter phenolicus Juteau et al. 2005, 249[VP]

Cupriavidus laharis Sato et al. 2006, 977[VP]

Cupriavidus pinatubonensis Sato et al. 2006, 977[VP]

Curtobacterium ammoniigenes Aizawa et al. 2007, 1451[VP]

Curvibacter gracilis Ding and Yokota 2004, 2228[VP]

Cyclobacterium amurskyense Nedashkovskaya et al. 2005e, 2392[VP]

Cyclobacterium lianum Ying et al. 2006, 2929[VP]

Cystobacter armeniaca Reichenbach 2007, 893[VP] – Valid publication: Validation List no. 115 – Effective publication: Reichenbach (2005d)

Cystobacter badius Reichenbach 2007, 893[VP] – Valid publication: Validation List no. 115 – Effective publication: Reichenbach (2005d)

Cystobacter miniatus Reichenbach 2007, 893[VP] – Valid publication: Validation List no. 115 – Effective publication: Reichenbach (2005d)

Cystobacter velatus Reichenbach 2007, 893[VP] – Valid publication: Validation List no. 115 – Effective publication: Reichenbach (2005d)

Dechloromonas denitrificans Horn et al. 2005, 1262[VP]

Dechloromonas hortensis Wolterink et al. 2005, 2067[VP]

Deefgea rivuli Stackebrandt et al. 2007, 643[VP]

Defluviicoccus vanus corrig. Maszenan et al. 2005a, 2109[VP]

Deinococcus apachensis Rainey and da Costa 2005, 2235[VP] – Valid publication: Validation List no. 106 – Effective publication: Rainey and da Costa (2005)

Deinococcus cellulosilyticus Weon et al. 2007e, 1687[VP]

Deinococcus deserti de Groot et al. 2005, 2445[VP]

Deinococcus ficus W.A. Lai et al. 2006, 790[VP]

Deinococcus frigens Hirsch et al. 2006, 925[VP] – Valid publication: Validation List no. 109 – Effective publication: Hirsch et al. (2004a)

Deinococcus hohokamensis Rainey and da Costa 2005, 2235[VP] – Valid publication: Validation List no. 106 – Effective publication: Rainey and da Costa (2005)

Deinococcus hopiensis Rainey and da Costa 2005, 2235[VP] – Valid publication: Validation List no. 106 – Effective publication: Rainey and da Costa (2005)

Deinococcus indicus Suresh et al. 2004b, 459[VP]

Deinococcus maricopensis Rainey and da Costa 2005, 2235[VP] – Valid publication: Validation List no. 106 – Effective publication: Rainey and da Costa (2005)

Deinococcus marmoris Hirsch et al. 2006, 925[VP] – Valid publication: Validation List no. 109 – Effective publication: Hirsch et al. (2004a)

Deinococcus mumbaiensis Shashidhar and Bandekar 2006, 925[VP] – Valid publication: Validation List no. 109 – Effective publication: Shashidhar and Bandekar (2006)

Deinococcus navajonensis Rainey and da Costa 2005, 2235^VP – Valid publication: Validation List no. 106 – Effective publication: Rainey and da Costa (2005)

Deinococcus papagonensis Rainey and da Costa 2005, 2235^VP – Valid publication: Validation List no. 106 – Effective publication: Rainey and da Costa (2005)

Deinococcus peraridilitoris Rainey et al. 2007, 1410^VP

Deinococcus pimensis Rainey and da Costa 2005, 2235^VP – Valid publication: Validation List no. 106 – Effective publication: Rainey and da Costa (2005)

Deinococcus saxicola Hirsch et al. 2006, 925^VP – Valid publication: Validation List no. 109 – Effective publication: Hirsch et al. (2004a)

Deinococcus sonorensis Rainey and da Costa 2005, 2236^VP – Valid publication: Validation List no. 106 – Effective publication: Rainey and da Costa (2005)

Deinococcus yavapaiensis Rainey and da Costa 2005, 2236^VP – Valid publication: Validation List no. 106 – Effective publication: Rainey and da Costa (2005)

Deinococcus yunweiensis Y.Q. Zhang et al. 2007b, 373^VP

Demequina aestuarii Yi et al. 2007b, 154^VP

Denitratisoma oestradiolicum Fahrbach et al. 2006, 1550^VP

Dermacoccus abyssi Pathom-aree et al. 2006b, 1235^VP

Dermacoccus barathri Pathom-aree et al. 2006c, 2306^VP

Dermacoccus profundi Pathom-aree et al. 2006c, 2306^VP

Desulfatibacillum aliphaticivorans Cravo-Laureau et al. 2004a, 81^VP

Desulfatibacillum alkenivorans Cravo-Laureau et al. 2004b, 1641^VP

Desulfatiferula olefinivorans Cravo-Laureau et al. 2007, 2701^VP

Desulfitibacter alkalitolerans M.B. Nielsen et al. 2006, 2835^VP

Desulfobacterium niacini Kuever et al. 2006, 2^VP – Valid publication: Validation List no. 107 – Effective publication: Kuever et al. (2005l)

Desulfobacterium vacuolatum Kuever et al. 2006, 2^VP – Valid publication: Validation List no. 107 – Effective publication: Kuever et al. (2005l)

Desulfobulbus japonicus D. Suzuki et al. 2007b, 853^VP

Desulfofaba fastidiosa Abildgaard et al. 2004, 398^VP

Desulfoglaeba alkanexedens Davidova et al. 2006, 2741^VP

Desulfohalobium utahense Jakobsen et al. 2006, 2068^VP

Desulfomicrobium thermophilum Thevenieau et al. 2007, 2449^VP – Valid publication: Validation List no. 118 – Effective publication: Thevenieau et al. (2007)

Desulfonatronum cooperativum Zhilina et al. 2005b, 1004^VP

Desulfopila aestuarii D. Suzuki et al. 2007a, 524^VP

Desulfosarcina ovata Kuever et al. 2006, 2^VP – Valid publication: Validation List no. 107 – Effective publication: Kuever et al. (2005m)

Desulfosporosinus lacus Ramamoorthy et al. 2006, 2734^VP

Desulfothermus naphthae Kuever et al. 2006, 2^VP – Valid publication: Validation List no. 107 – Effective publication: Kuever et al. (2005h)

Desulfothermus okinawensis Nunoura et al. 2007b, 2363^VP

Desulfotignum toluenicum Ommedal and Torsvik 2007, 2868^VP

Desulfotomaculum arcticum Vandieken et al. 2006b, 689^VP

Desulfotomaculum carboxydivorans Parshina et al. 2005, 2164^VP

Desulfotomaculum thermosubterraneum Kaksonen et al. 2006, 2606^VP

Desulfovermiculus halophilus Belyakova et al. 2007, 1371^VP – Valid publication: Validation List no. 116 – Effective publication: Belyakova et al. (2006)

Desulfovibrio alaskensis Feio et al. 2004, 1751^VP

Desulfovibrio alkalitolerans Abildgaard et al. 2006, 1023^VP

Desulfovibrio bastinii Magot et al. 2004, 1696^VP

Desulfovibrio bizertensis Haouari et al. 2006, 2912^VP

Desulfovibrio ferrireducens Vandieken et al. 2006a, 684^VP

Desulfovibrio frigidus Vandieken et al. 2006a, 684^VP

Desulfovibrio gracilis Magot et al. 2004, 1696^VP

Desulfovibrio marinus Ben Dhia Thabet et al. 2007, 2169^VP

Desulfovibrio putealis Basso et al. 2005, 104^VP

Desulfovirgula thermocuniculi Kaksonen et al. 2007a, 101^VP

Desulfurispora thermophila Kaksonen et al. 2007b, 1092^VP

Desulfurobacterium atlanticum L'Haridon et al. 2006, 2850^VP

Desulfurobacterium pacificum L'Haridon et al. 2006, 2850^VP

Desulfurococcus fermentans Perevalova et al. 2005, 998^VP

Desulfuromonas svalbardensis Vandieken et al. 2006c, 1138^VP

Desulfuromusa ferrireducens Vandieken et al. 2006c, 1138^VP

Dethiosulfatibacter aminovorans Takii et al. 2007, 2324^VP

Devosia insulae J.H. Yoon et al. 2007xvii, 1313^VP

Devosia limi Vanparys et al. 2005b, 1998^VP

Devosia soli Yoo et al. 2006, 2691^VP

Devosia subaequoris S.D. Lee 2007c, 2214^VP

Dialister micraerophilus Jumas-Bilak et al. 2005, 2476^VP

Dialister propionicifaciens Jumas-Bilak et al. 2005, 2477^VP

Dickeya dadantii Samson et al. 2005, 1424^VP

Dickeya dianthicola Samson et al. 2005, 1424^VP

Dickeya dieffenbachiae Samson et al. 2005, 1425^VP

Dickeya zeae Samson et al. 2005, 1425^VP

Dietzia cinnamea Yassin et al. 2006, 644^VP

Dietzia kunjamensis Mayilraj et al. 2006e, 1670^VP

Dinoroseobacter shibae Biebl et al. 2005a, 1095^VP

Dokdonella fugitiva Cunha et al. 2006, 1459^VP – Valid publication: Validation List no. 110 – Effective publication: Cunha et al. (2006)

Dokdonella koreensis J.H. Yoon et al. 2006g, 149^VP

Dokdonia donghaensis J.H. Yoon et al. 2005h, 2326^VP

Donghaeana dokdonensis J.H. Yoon et al. 2006b, 190^VP

Donghicola eburneus J.H. Yoon et al. 2007xiii, 75^VP

Duganella violaceinigra W.J. Li et al. 2004f, 1813^VP

Dyadobacter beijingensis Dong et al. 2007, 864^VP

Dyadobacter crusticola Reddy and Garcia-Pichel 2005, 1298^{VP}

Dyadobacter ginsengisoli Q.M. Liu et al. 2006, 1942^{VP}

Dyadobacter hamtensis Chaturvedi et al. 2005, 2114^{VP}

Dyadobacter koreensis Baik et al. 2007a, 1230^{VP}

Dyella japonica Xie and Yokota 2005c, 756^{VP}

Dyella koreensis D.S. An et al. 2005a, 1628^{VP}

Dyella yeojuensis B.Y. Kim et al. 2006c, 2081^{VP}

Echinicola pacifica Nedashkovskaya et al. 2006b, 955^{VP}

Echinicola vietnamensis Nedashkovskaya et al. 2007a, 763^{VP}

Ectothiorhodosinus mongolicus corrig. Gorlenko et al. 2007, 1371^{VP} – Valid publication: Validation List no. 116 – Effective publication: Gorlenko et al. (2004a)

Effluviibacter roseus Suresh et al. 2006, 1706^{VP}

Eggerthella hongkongensis Lau et al. 2006, 2025^{VP} – Valid publication: Validation List no. 111 – Effective publication: S.K.P. Lau et al. (2004)

Eggerthella sinensis Lau et al. 2006, 2025^{VP} – Valid publication: Validation List no. 111 – Effective publication: S.K.P. Lau et al. (2004)

Emticicia oligotrophica Saha and Chakrabarti 2006a, 993^{VP}

Endozoicomonas elysicola Kurahashi and Yokota 2007, 1371^{VP} – Valid publication: Validation List no. 116 – Effective publication: Kurahashi and Yokota (2007)

Enterobacter helveticus Stephan et al. 2007, 825^{VP}

Enterobacter ludwigii Hoffmann et al. 2005, 2236^{VP} – Valid publication: Validation List no. 106 – Effective publication: Hoffmann et al. (2005b)

Enterobacter radicincitans Kämpfer et al. 2005, 1396^{VP} – Valid publication: Validation List no. 104 – Effective publication: Kämpfer et al. (2005b)

Enterobacter turicensis Stephan et al. 2007, 824^{VP}

Enterococcus aquimarinus Švec et al. 2005a, 2186^{VP}

Enterococcus caccae Carvalho et al. 2006, 1507^{VP}

Enterococcus camelliae Sukontasing et al. 2007, 2153^{VP}

Enterococcus canintestini Naser et al. 2005, 2181^{VP}

Enterococcus devriesei Švec et al. 2005b, 2482^{VP}

Enterococcus hermanniensis Koort et al. 2004, 1826^{VP}

Enterococcus italicus Fortina et al. 2004, 1720^{VP}

Enterococcus saccharominimus Vancanneyt et al. 2004c, 2178^{VP}

Enterococcus silesiacus Švec et al. 2006, 580^{VP}

Enterococcus termitis Švec et al. 2006, 580^{VP}

Enterovibrio coralii Thompson et al. 2005, 916^{VP}

Epilithonimonas tenax O'Sullivan et al. 2006, 177^{VP}

Erwinia papayae Gardan et al. 2004, 112^{VP}

Erwinia tasmaniensis Geider et al. 2006, 2942^{VP}

Erwinia toletana Rojas et al. 2004, 2220^{VP}

Erysipelothrix inopinata Verbarg et al. 2004, 224^{VP}

Erythrobacter aquimaris J.H. Yoon et al. 2004b, 1984^{VP}

Erythrobacter gaetbuli J.H. Yoon et al. 2005w, 74^{VP}

Erythrobacter luteolus J.H. Yoon et al. 2005a, 1169^{VP}

Erythrobacter seohaensis J.H. Yoon et al. 2005w, 74^{VP}

Erythrobacter vulgaris Ivanova et al. 2006, 499^{VP} – Valid publication: Validation List no. 108 – Effective publication: Ivanova et al. (2005a)

Ethanoligenens harbinense Xing et al. 2006, 759^{VP}

Exiguobacterium aestuarii I.G. Kim et al. 2005, 888^{VP}

Exiguobacterium artemiae López-Cortés et al. 2006, 1459^{VP} – Valid publication: Validation List no. 110 – Effective publication: López-Cortés et al. (2006b)

Exiguobacterium indicum Chaturvedi and Shivaji 2006, 2769^{VP}

Exiguobacterium marinum I.G. Kim et al. 2005, 888^{VP}

Exiguobacterium mexicanum López-Cortés et al. 2006, 1459^{VP} – Valid publication: Validation List no. 110 – Effective publication: López-Cortés et al. (2006b)

Exiguobacterium oxidotolerans Yumoto et al. 2004d, 2016^{VP}

Exiguobacterium profundum Crapart et al. 2007, 289^{VP}

Exiguobacterium sibiricum Rodrigues et al. 2006, 2507^{VP} – Valid publication: Validation List no. 112 – Effective publication: Rodrigues et al. (2006)

Fabibacter halotolerans S.C. Lau et al. 2006b, 1062^{VP}

Fangia hongkongensis K.W. Lau et al. 2007, 2669^{VP}

Fastidiosipila sanguinis Falsen et al. 2005, 856^{VP}

Ferrimonas futtsuensis T. Nakagawa et al. 2006, 2644^{VP}

Ferrimonas kyonanensis T. Nakagawa et al. 2006, 2644^{VP}

Ferrimonas marina Katsuta et al. 2005, 1854^{VP}

Ferrimonas senticii Campbell et al. 2007, 2672^{VP}

Fervidobacterium changbaicum Cai et al. 2007, 2335^{VP}

Flagellimonas eckloniae S.S. Bae et al. 2007, 1051^{VP}

Flammeovirga kamogawensis Hosoya and Yokota 2007b, 1328^{VP}

Flammeovirga yaeyamensis M. Takahashi et al. 2006, 2099^{VP}

Flaviramulus basaltis Einen and Øvreås 2006, 2460^{VP}

Flavisolibacter ginsengisoli M.H. Yoon and Im 2007, 1838^{VP}

Flavisolibacter ginsengiterrae M.H. Yoon and Im 2007, 1837^{VP}

Flavobacterium antarcticum Yi et al. 2005a, 640^{VP}

Flavobacterium aquidurense Cousin et al. 2007, 247^{VP}

Flavobacterium ceti Vela et al. 2007a, 2607^{VP}

Flavobacterium croceum M. Park et al. 2006, 2445^{VP}

Flavobacterium cucumis Weon et al. 2007j, 1596^{VP}

Flavobacterium daejeonense B.Y. Kim et al. 2006b, 1647^{VP}

Flavobacterium defluvii M. Park et al. 2007, 235^{VP}

Flavobacterium degerlachei Van Trappen et al. 2004d, 89^{VP}

Flavobacterium denitrificans Horn et al. 2005, 1263^{VP}

Flavobacterium filum Ryu et al. 2007b, 2028^{VP}

Flavobacterium frigidimaris Nogi et al. 2005, 1743^{VP} – Valid publication: Validation List no. 105 – Effective publication: Nogi et al. (2005a)

Flavobacterium frigoris Van Trappen et al. 2004d, 90^{VP}

Flavobacterium fryxellicola Van Trappen et al. 2005b, 771^{VP}

Flavobacterium glaciei D.C. Zhang et al. 2006a, 2924^{VP}

Flavobacterium granuli Aslam et al. 2005a, 750^{VP}

Flavobacterium hercynium Cousin et al. 2007, 248^{VP}

Flavobacterium indicum Saha and Chakrabarti 2006c, 2620^{VP}

Flavobacterium micromati Van Trappen et al. 2004d, 89^{VP}

Flavobacterium psychrolimnae Van Trappen et al. 2005b, 771^{VP}

Flavobacterium saliperosum Z.W. Wang et al. 2006, 441^{VP}

Flavobacterium segetis Yi and Chun 2006b, 1243^{VP}

Flavobacterium soli J.H. Yoon et al. 2006i, 999^{VP}

Flavobacterium suncheonense B.Y. Kim et al. 2006b, 1648^{VP}

Flavobacterium terrae Weon et al. 2007j, 1596^{VP}

Flavobacterium terrigena J.H. Yoon et al. 2007viii, 949^{VP}

Flavobacterium weaverense Yi and Chun 2006b, 1242^{VP}

Flectobacillus lacus Hwang and Cho 2006, 1200^{VP}

Fluviicola taffensis O'Sullivan et al. 2005, 2193^{VP}

Formosa agariphila Nedashkovskaya et al. 2006d, 166^{VP}

Formosa algae Ivanova et al. 2004a, 707^{VP}

Frondicola australicus L. Zhang et al. 2007b, 1181^{VP}

The name *Frondicola australicus* L. Zhang et al. 2007b is illegitimate because it is placed in an illegitimate genus (see above *Frondicola* L. Zhang et al. 2007b, 1181^{VP}).

Fulvivirga kasyanovii Nedashkovskaya et al. 2007d, 1048^{VP}

Fusobacterium canifelinum Conrads et al. 2004, 1909^{VP} – Valid publication: Validation List no. 100 – Effective publication: Conrads et al. (2004)

Gaetbulibacter saemankumensis Jung et al. 2005a, 1848^{VP}

Gaetbulimicrobium brevivitae J.H. Yoon et al. 2006a, 118^{VP}

Galbibacter mesophilus Khan et al. 2007c, 972^{VP}

Gelidibacter gilvus Bowman and Nichols 2005, 1483^{VP}

Gelidibacter salicanalis Bowman and Nichols 2005, 1484^{VP}

Geoalkalibacter ferrihydriticus Zavarzina et al. 2007, 894^{VP} – Valid publication: Validation List no. 115 – Effective publication: Zavarzina et al. (2006)

Geobacillus debilis Banat et al. 2004, 2199^{VP}

Geobacillus gargensis Nazina et al. 2004, 2023^{VP}

Geobacillus jurassicus Nazina et al. 2005, 983^{VP} – Valid publication: Validation List no. 103 – Effective publication: Nazina et al. (2005)

Geobacillus lituanicus Kuisiene et al. 2004, 1993^{VP}

Geobacillus tepidamans Schäffer et al. 2004, 2366^{VP}

Geobacter argillaceus Shelobolina et al. 2007, 133^{VP}

Geobacter bemidjiensis Nevin et al. 2005, 1671^{VP}

Geobacter pickeringii Shelobolina et al. 2007, 133^{VP}

Geobacter psychrophilus Nevin et al. 2005, 1672^{VP}

Geopsychrobacter electrodiphilus Holmes et al. 2005, 547^{VP} – Valid publication: Validation List no. 102 – Effective publication: Holmes et al. (2004b)

Georgenia ruanii W.J. Li et al. 2007, 1426^{VP}

Geosporobacter subterraneus Klouche et al. 2007, 1760^{VP}

Geothermobacter ehrlichii Kashefi et al. 2005, 547^{VP} – Valid publication: Validation List no. 102 – Effective publication: Kashefi et al. (2003)

Giesbergeria kuznetsovii Grabovich et al. 2006, 573^{VP}

Giesbergeria voronezhensis Grabovich et al. 2006, 573^{VP}

Gillisia hiemivivida Bowman and Nichols 2005, 1484^{VP}

Gillisia illustrilutea Bowman and Nichols 2005, 1484^{VP}

Gillisia limnaea Van Trappen et al. 2004e, 447^{VP}

Gillisia mitskevichiae Nedashkovskaya et al. 2005d, 322^{VP}

Gillisia myxillae O.O. Lee et al. 2006a, 1797^{VP}

Gillisia sandarakina Bowman and Nichols 2005, 1484^{VP}

Gilvibacter sediminis Khan et al. 2007b, 268^{VP}

Glaciecola agarilytica Yong et al. 2007, 953^{VP}

Glaciecola chathamensis Matsuyama et al. 2006, 2886^{VP}

Glaciecola nitratireducens Baik et al. 2006a, 2187^{VP}

Glaciecola polaris Van Trappen et al. 2004c, 1769^{VP}

Glaciecola psychrophila D.C. Zhang et al. 2006b, 2868^{VP}

Gluconacetobacter kombuchae Dutta and Gachhui 2007, 356^{VP}

Gluconacetobacter nataicola Lisdiyanti et al. 2006, 2109^{VP}

Gluconacetobacter rhaeticus Dellaglio et al. 2005a, 2369^{VP}

Gluconacetobacter saccharivorans Lisdiyanti et al. 2006, 2108^{VP}

Gluconacetobacter swingsii Dellaglio et al. 2005a, 2368^{VP}

Gluconobacter thailandicus Tanasupawat et al. 2005, 983^{VP} – Valid publication: Validation List no. 103 – Effective publication: Tanasupawat et al. (2004)

Glycomyces algeriensis Labeda and Kroppenstedt 2004, 2345^{VP}

Glycomyces arizonensis Labeda and Kroppenstedt 2004, 2345^{VP}

Glycomyces lechevalierae Labeda and Kroppenstedt 2004, 2346^{VP}

Glycomyces sambucus Q. Gu et al. 2007b, 1996^{VP}

Gordonia araii A. Kageyama et al. 2006a, 1820^{VP}

Gordonia defluvii Soddell et al. 2006b, 2267^{VP}

Gordonia effusa A. Kageyama et al. 2006a, 1820^{VP}

Gordonia malaquae Yassin et al. 2007b, 1067^{VP}

Gordonia otitidis K. Iida et al. 2005, 1874^{VP}

Gordonia shandongensis Luo et al. 2007, 607^{VP}

Gordonia soli F.T. Shen et al. 2006, 2599^{VP}

Gracilibacillus boraciitolerans Ahmed et al. 2007b, 800^{VP}

Gracilibacillus orientalis Carrasco et al. 2006, 602^{VP}

Gracilibacter thermotolerans Y.J. Lee et al. 2006, 2092^{VP}

Gramella echinicola Nedashkovskaya et al. 2005f, 393^{VP}

Gramella portivictoriae S.C. Lau et al. 2005d, 2499^{VP}

Granulibacter bethesdensis Greenberg et al. 2006, 2615^{VP}

Granulicoccus phenolivorans Maszenan et al. 2007, 733^{VP}

Guggenheimella bovis Wyss et al. 2005, 669^{VP}

Gulbenkiania mobilis Vaz-Moreira et al. 2007b, 1111^{VP}

Gulosibacter molinativorax Manaia et al. 2004, 787^{VP}

Haematobacter missouriensis Helsel et al. 2007, 1371^{VP} – Valid publication: Validation List no. 116 – Effective publication: Helsel et al. (2007)

Hemophilus pittmaniae Nørskov-Lauritsen et al. 2005, 455^{VP}

Hahella ganghwensis Baik et al. 2005, 683^{VP}

Haladaptatus paucihalophilus Savage et al. 2007, 23[VP]

Halalkalibacillus halophilus Echigo et al. 2007, 1083[VP]

Halalkalicoccus jeotgali Roh et al. 2007b, 2297[VP]

Halalkalicoccus tibetensis Xue et al. 2005, 2504[VP]

Haloarcula amylolytica Y. Yang et al. 2007, 105[VP]

Halobacillus aidingensis W.Y. Liu et al. 2005, 1995[VP]

Halobacillus campisalis J.H. Yoon et al. 2007vi, 2024[VP]

Halobacillus dabanensis W.Y. Liu et al. 2005, 1995[VP]

Halobacillus faecis S.Y. An et al. 2007d, 2478[VP]

Halobacillus kuroshimensis Hua et al. 2007, 1246[VP]

Halobacillus locisalis Yoon et al. 2004, 1425[VP] – Valid publication: Validation List no. 99 – Effective publication: J.H. Yoon et al. (2004)

Halobacillus profundi Hua et al. 2007, 1245[VP]

Halobacillus yeomjeoni J.H. Yoon et al. 2005i, 2416[VP]

Halobacterium jilantaiense Y. Yang et al. 2006, 2355[VP]

Halobacterium noricense Gruber et al. 2005, 983[VP] – Valid publication: Validation List no. 103 – Effective publication: Gruber et al. (2004)

Halobiforma lacisalsi X.W. Xu et al. 2005c, 1951[VP]

Halochromatium roseum Anil Kumar et al. 2007c, 2112[VP]

Halococcus hamelinensis Goh et al. 2006, 1328[VP]

Halococcus qingdaonensis Q.F. Wang et al. 2007b, 603[VP]

Halococcus thailandensis Namwong et al. 2007, 2202[VP]

Haloferax larsenii X.W. Xu et al. 2007a, 718[VP]

Haloferax lucentense corrig. Gutierrez et al. 2004, 1[VP] – Valid publication: Validation List no. 95 – Effective publication: Gutierrez et al. (2002)

Haloferax prahovense Enache et al. 2007, 395[VP]

Haloferax sulfurifontis Elshahed et al. 2004, 2278[VP]

Halolactibacillus halophilus Ishikawa et al. 2005, 2437[VP]

Halolactibacillus miurensis Ishikawa et al. 2005, 2437[VP]

Halomonas alkaliphila Romano et al. 2007, 1933[VP] – Valid publication: Validation List no. 117 – Effective publication: Romano et al. (2006a)

Halomonas almeriensis Martínez-Checa et al. 2005, 2010[VP]

Halomonas anticariensis Martínez-Cánovas et al. 2004a, 1330[VP]

Halomonas arcis X.W. Xu et al. 2007c, 1622[VP]

Halomonas avicenniae Soto-Ramírez et al. 2007, 903[VP]

Halomonas axialensis Kaye et al. 2004, 509[VP]

Halomonas boliviensis Quillaguamán et al. 2004b, 724[VP]

Halomonas campaniensis Romano et al. 2005, 2236[VP] – Valid publication: Validation List no. 106 – Effective publication: Romano et al. (2005b)

Halomonas denitrificans K.K. Kim et al. 2007, 680[VP]

Halomonas gomseomensis K.K. Kim et al. 2007, 678[VP]

Halomonas gudaonensis Y.N. Wang et al. 2007b, 914[VP]

Halomonas hydrothermalis Kaye et al. 2004, 509[VP]

Halomonas indalinina Cabrera et al. 2007, 378[VP]

Halomonas janggokensis K.K. Kim et al. 2007, 678[VP]

Halomonas koreensis J.M. Lim et al. 2004a, 2041[VP]

Halomonas kribbensis Jeon et al. 2007, 2196[VP]

Halomonas neptunia Kaye et al. 2004, 508[VP]

Halomonas organivorans García et al. 2004, 1726[VP]

Halomonas saccharevitans X.W. Xu et al. 2007c, 1621[VP]

Halomonas salaria K.K. Kim et al. 2007, 680[VP]

Halomonas shengliensis Y.N. Wang et al. 2007a, 1224[VP]

Halomonas subterranea X.W. Xu et al. 2007c, 1622[VP]

Halomonas sulfidaeris Kaye et al. 2004, 508[VP]

Halomonas taeanensis J.C. Lee et al. 2005, 2031[VP]

Halomonas ventosae Martínez-Cánovas et al. 2004c, 735[VP]

Halopiger xanaduensis Gutiérrez et al. 2007, 1404[VP]

Haloplanus natans Elevi Bardavid et al. 2007, 783[VP]

Haloquadratum walsbyi Burns et al. 2007, 391[VP]

Halorubrum aidingense Cui et al. 2006b, 1633[VP]

Halorubrum alkaliphilum Feng et al. 2005, 151[VP]

Halorubrum arcis X.W. Xu et al. 2007b, 1071[VP]

Halorubrum ejinorense Castillo et al. 2007b, 2540[VP]

Halorubrum ezzemoulense Kharroub et al. 2006b, 1585[VP]

Halorubrum lipolyticum Cui et al. 2006b, 1632[VP]

Halorubrum litoreum H.L. Cui et al. 2007b, 2206[VP]

Halorubrum orientale Castillo et al. 2006c, 2562[VP]

Halorubrum terrestre Ventosa et al. 2004, 391[VP]

Halorubrum tibetense Fan et al. 2004, 1215[VP]

Halorubrum xinjiangense Feng et al. 2004, 1791[VP]

Halospina denitrificans Sorokin et al. 2006a, 386[VP]

Halostagnicola larsenii Castillo et al. 2006b, 1521[VP]

Halotalea alkalilenta Ntougias et al. 2007b, 1981[VP]

Haloterrigena hispanica Romano et al. 2007, 1501[VP]

Haloterrigena limicola Cui et al. 2006c, 1839[VP]

Haloterrigena longa Cui et al. 2006c, 1838[VP]

Haloterrigena saccharevitans X.W. Xu et al. 2005a, 2541[VP]

Halovibrio denitrificans Sorokin et al. 2006a, 386[VP]

According to Rule 28b(2), *Halovibrio denitrificans* Sorokin et al. 2006a is not validly published (see the Request for an Opinion by Sorokin and Tindall 2006)

Halovivax asiaticus Castillo et al. 2006a, 769[VP]

Halovivax ruber Castillo et al. 2007a, 1026[VP]

Helcococcus sueciensis Collins et al. 2004a, 1559[VP]

Helicobacter anseris Fox et al. 2006, 2025[VP] – Valid publication: Validation List no. 111 – Effective publication: Fox et al. (2006)

Helicobacter brantae Fox et al. 2006, 2025[VP] – Valid publication: Validation List no. 111 – Effective publication: Fox et al. (2006)

Helicobacter cetorum Harper et al. 2006, 2025[VP] – Valid publication: Validation List no. 111 – Effective publication: Harper et al. (2002)

Helicobacter cynogastricus Van den Bulck et al. 2006, 1563[VP]

Helicobacter equorum Moyaert et al. 2007, 217^{VP}

Helicobacter marmotae Fox et al. 2006, 2025^{VP} – Valid publication: Validation List no. 111 – Effective publication: Fox et al. (2002)

Helicobacter mastomyrinus Shen et al. 2006, 2025^{VP} – Valid publication: Validation List no. 111 – Effective publication: Z. Shen et al. (2005)

Herbaspirillum chlorophenolicum Im et al. 2004a, 854^{VP}

Herbaspirillum hiltneri Rothballer et al. 2006, 1347^{VP}

Herbaspirillum putei Ding and Yokota 2004, 2227^{VP}

Herbaspirillum rhizosphaerae Jung et al. 2007d, 2287^{VP}

Herminiimonas aquatilis Kämpfer et al. 2006, 1459^{VP} – Valid publication: Validation List no. 110 – Effective publication: Kämpfer et al. (2006b)

Herminiimonas arsenicoxydans Muller et al. 2006, 1768^{VP}

Herminiimonas fonticola Fernandes et al. 2005, 2236^{VP} – Valid publication: Validation List no. 106 – Effective publication: Fernandes et al. (2005)

Herminiimonas saxobsidens Lang et al. 2007b, 2621^{VP}

Hespellia porcina Whitehead et al. 2004, 244^{VP}

Hespellia stercorisuis Whitehead et al. 2004, 244^{VP}

Hoeflea alexandrii Palacios et al. 2006, 1994^{VP}

Hoeflea marina Peix et al. 2005, 1165^{VP}

Hoeflea phototrophica Biebl et al. 2006, 825^{VP}

Hongiella halophila Yi and Chun 2004a, 160^{VP}

Hongiella mannitolivorans Yi and Chun 2004a, 160^{VP}

Hongiella marincola J.H. Yoon et al. 2004o, 1848^{VP}

Hongiella ornithinivorans Yi and Chun 2004a, 160^{VP}

Howardella ureilytica Cook et al. 2007, 2943^{VP}

Humicoccus flavidus J.H. Yoon et al. 2007v, 59^{VP}

Humihabitans oryzae A. Kageyama et al. 2007c, 2165^{VP}

Hyalangium minutum Reichenbach 2007, 894^{VP} – Valid publication: Validation List no. 115 – Effective publication: Reichenbach (2005e)

Hydrocarboniphaga effusa Palleroni et al. 2004, 1207^{VP}

Hydrogenimonas thermophila Takai et al. 2004b, 31^{VP}

Hydrogenivirga caldilitoris S. Nakagawa et al. 2004a, 2083^{VP}

Hydrogenophaga atypica Kämpfer et al. 2005c, 343^{VP}

Hydrogenophaga caeni Chung et al. 2007, 1129^{VP}

Hydrogenophaga defluvii Kämpfer et al. 2005c, 343^{VP}

Hymenobacter chitinivorans Buczolits et al. 2006, 2077^{VP}

Hymenobacter gelipurpurascens Buczolits et al. 2006, 2077^{VP}

Hymenobacter norwichensis Buczolits et al. 2006, 2077^{VP}

Hymenobacter ocellatus Buczolits et al. 2006, 2076^{VP}

Hymenobacter rigui Baik et al. 2006b, 2191^{VP}

Hymenobacter xinjiangensis Q. Zhang et al. 2007, 1754^{VP}

Idiomarina fontislapidosi Martínez-Cánovas et al. 2004b, 1795^{VP}

Idiomarina homiensis S.W. Kwon et al. 2006, 2232^{VP}

Idiomarina ramblicola Martínez-Cánovas et al. 2004b, 1796^{VP}

Idiomarina salinarum J.H. Yoon et al. 2007i, 2504^{VP}

Idiomarina seosinensis Choi and Cho 2005, 382^{VP}

Ignicoccus hospitaliz Paper et al. 2007, 807^{VP}

Ignisphaera aggregans Niederberger et al. 2006, 970^{VP}

Isoptericola dokdonensis J.H. Yoon et al. 2006r, 2896^{VP}

Isoptericola halotolerans Y.Q. Zhang et al. 2005c, 1869^{VP}

Isoptericola hypogeus Groth et al. 2005, 1718^{VP}

Janibacter anophelis Kämpfer et al. 2006i, 391^{VP}

Janibacter corallicola Kageyama et al. 2007, 2449^{VP} – Valid publication: Validation List no. 118 – Effective publication: A. Kageyama et al. (2007d)

Janibacter melonis J.H. Yoon et al. 2004i, 1979^{VP}

Jannaschia cystaugens Adachi et al. 2004, 1691^{VP}

Jannaschia donghaensis J.H. Yoon et al. 2007xx, 2134^{VP}

Jannaschia rubra Macián et al. 2005b, 651^{VP}

Jannaschia seosinensis Choi et al. 2006, 48^{VP}

Jeotgalicoccus pinnipedialis Hoyles et al. 2004, 747^{VP}

Jiangella gansuensis L. Song et al. 2005, 883^{VP}

Jonesia quinghaiensis Schumann et al. 2004, 2183^{VP}

Jonquetella anthropi Jumas-Bilak et al. 2007a, 2747^{VP}

Kaistella koreensis M.K. Kim et al. 2004, 2323^{VP}

Kaistia adipata Im et al. 2005, 983^{VP} – Valid publication: Validation List no. 103 – Effective publication: Im et al. (2004c)

Kaistia granuli H.W. Lee et al. 2007, 2283^{VP}

Kangiella aquimarina J.H. Yoon et al. 2004j, 1833^{VP}

Kangiella koreensis J.H. Yoon et al. 2004j, 1833^{VP}

Kineococcus marinus S.D. Lee 2006d, 1282^{VP}

Kingella potus Lawson et al. 2005, 1743^{VP} – Valid publication: Validation List no. 105 – Effective publication: Lawson et al. (2005d)

Kitasatospora arboriphila Groth et al. 2004, 2125^{VP}

Kitasatospora gansuensis Groth et al. 2004, 2127^{VP}

Kitasatospora nipponensis Groth et al. 2004, 2127^{VP}

Kitasatospora paranensis Groth et al. 2004, 2128^{VP}

Kitasatospora sampliensis Mayilraj et al. 2006a, 521^{VP}

Kitasatospora terrestris Groth et al. 2004, 2128^{VP}

Kitasatospora viridis Z. Liu et al. 2005a, 709^{VP}

Klebsiella singaporensis X. Li et al. 2004, 2135^{VP}

Klebsiella variicola Rosenblueth et al. 2004, 631^{VP} – Valid publication: Validation List no. 97 – Effective publication: Rosenblueth et al. (2004)

Knoellia aerolata Weon et al. 2007d, 2863^{VP}

Kocuria aegyptia W.J. Li et al. 2006c, 735^{VP}

Kocuria carniphila Tvrzová et al. 2005a, 140^{VP}

Kocuria himachalensis Mayilraj et al. 2006c, 1974^{VP}

Kocuria marina S.B. Kim et al. 2004a, 1619^{VP}

Kordia algicida Sohn et al. 2004b, 678^{VP}

Kordiimonas gwangyangensis K.K. Kwon et al. 2005, 2036^{VP}

Krasilnikovia cinnamomea corrig. Ara and Kudo 2007, 2449^{VP} – Valid publication: Validation List no. 118 – Effective publication: Ara and Kudo (2007a)

Kribbella alba Li et al. 2006, 1459^{VP} – Valid publication: Validation List no. 110 – Effective publication: W.J. Li et al. (2006b)

Kribbella aluminosa Carlsohn et al. 2007b, 1946^{VP}

Kribbella antibiotica Li et al. 2004, 1425^{VP} – Valid publication: Validation List no. 99 – Effective publication: W.J. Li et al. (2004e)

Kribbella jejuensis J. Song et al. 2004, 1347^{VP}

Kribbella karoonensis Kirby et al. 2006, 1100^{VP}

Kribbella lupini Trujillo et al. 2006a, 410^{VP}

Kribbella solani J. Song et al. 2004, 1347^{VP}

Kribbella swartbergensis Kirby et al. 2006, 1100^{VP}

Kribbella yunnanensis Li et al. 2006, 1459^{VP} – Valid publication: Validation List no. 110 – Effective publication: W.J. Li et al. (2006b)

Kribbia dieselivorans Jung et al. 2006a, 2430^{VP}

Krokinobacter diaphorus Khan et al. 2006a, 327^{VP}

Krokinobacter eikastus Khan et al. 2006a, 327^{VP}

Krokinobacter genikus Khan et al. 2006a, 326^{VP}

Ktedonobacter racemifer corrig. Cavaletti et al. 2007, 433^{VP} – Valid publication: Validation List no. 114 – Effective publication: Cavaletti et al. (2006b)

Labedella gwakjiensis S.D. Lee 2007d, 2501^{VP}

Labrenzia alexandrii Biebl et al. 2007, 1105^{VP}

Labrys methylaminiphilus Miller et al. 2005, 1252^{VP}

Labrys miyagiensis Islam et al. 2007, 556^{VP}

Labrys neptuniae Y.J. Chou et al. 2007, 579^{VP}

Labrys okinawensis Islam et al. 2007, 556^{VP}

Lacinutrix copepodicola Bowman and Nichols 2005, 1482^{VP}

Lactobacillus acidifarinae Vancanneyt et al. 2005b, 619^{VP}

Lactobacillus amylotrophicus Naser et al. 2006d, 2526^{VP}

Lactobacillus antri Roos et al. 2005, 80^{VP}

Lactobacillus apodemi Osawa et al. 2006, 1695^{VP}

Lactobacillus camelliae Tanasupawat et al. 2007, 1371^{VP} – Valid publication: Validation List no. 116 – Effective publication: Tanasupawat et al. (2007b)

Lactobacillus composti Endo and Okada 2007a, 872^{VP}

Lactobacillus concavus Tong and Dong 2005, 2201^{VP}

Lactobacillus crustorum Scheirlinck et al. 2007b, 1466^{VP}

Lactobacillus farraginis Endo and Okada 2007b, 711^{VP}

Lactobacillus gastricus Roos et al. 2005, 80^{VP}

Lactobacillus ghanensis D.S. Nielsen et al. 2007, 1471^{VP}

Lactobacillus hammesii Valcheva et al. 2005, 766^{VP}

Lactobacillus harbinensis Miyamoto et al. 2006, 2^{VP} – Valid publication: Validation List no. 107 – Effective publication: Miyamoto et al. (2005)

Lactobacillus hayakitensis Morita et al. 2007, 2838^{VP}

Lactobacillus kalixensis Roos et al. 2005, 81^{VP}

Lactobacillus namurensis Scheirlinck et al. 2007a, 226^{VP}

Lactobacillus nantensis Valcheva et al. 2006, 589^{VP}

Lactobacillus oligofermentans Koort et al. 2005, 2236^{VP} – Valid publication: Validation List no. 106 – Effective publication: Koort et al. (2005)

Lactobacillus parabrevis Vancanneyt et al. 2006a, 1556^{VP}

Lactobacillus paracollinoides K. Suzuki et al. 2004, 116^{VP}

Lactobacillus parafarraginis Endo and Okada 2007b, 711^{VP}

Lactobacillus rennini Chenoll et al. 2006, 451^{VP}

Lactobacillus rossiae corrig. Corsetti et al. 2005, 39^{VP}

Lactobacillus saerimneri Pedersen and Roos 2004, 1367^{VP}

Lactobacillus satsumensis Endo and Okada 2005, 85^{VP}

Lactobacillus secaliphilus Ehrmann et al. 2007, 748^{VP}

Lactobacillus siliginis Aslam et al. 2006, 2212^{VP}

Lactobacillus sobrius Konstantinov et al. 2006, 31^{VP}

Lactobacillus spicheri Meroth et al. 2004, 631^{VP} – Valid publication: Validation List no. 97 – Effective publication: Meroth et al. (2004)

Lactobacillus suntoryeus Cachat and Priest 2005, 33^{VP}

Lactobacillus thailandensis Tanasupawat et al. 2007, 1371^{VP} – Valid publication: Validation List no. 116 – Effective publication: Tanasupawat et al. (2007b)

Lactobacillus ultunensis Roos et al. 2005, 81^{VP}

Lactobacillus vini Rodas et al. 2006, 516^{VP}

Lactobacillus zymae Vancanneyt et al. 2005b, 619^{VP}

Lactonifactor longoviformis Clavel et al. 2007, 894^{VP} – Valid publication: Validation List no. 115 – Effective publication: Clavel et al. (2007)

Lactovum miscens Matthies et al. 2005, 547^{VP} – Valid publication: Validation List no. 102 – Effective publication: Matthies et al. (2004)

Lapillicoccus jejuensis S.D. Lee and Lee 2007, 2797^{VP}

Larkinella insperata Vancanneyt et al. 2006b, 239^{VP}

Leadbetterella byssophila Weon et al. 2005, 2299^{VP}

Lebetimonas acidiphila Takai et al. 2005, 188^{VP}

Lechevalieria fradiae J. Zhang et al. 2007, 834^{VP}

Lechevalieria xinjiangensis W. Wang et al. 2007, 2821^{VP}

Leeia oryzae J.M. Lim et al. 2007a, 1207^{VP}

Leeuwenhoekiella aequorea Nedashkovskaya et al. 2005m, 1036^{VP}

Leeuwenhoekiella blandensis Pinhassi et al. 2006, 1492^{VP}

Legionella drancourtii La Scola et al. 2004, 703^{VP}

Legionella impletisoli Kuroki et al. 2007, 1933^{VP} – Valid publication: Validation List no. 117 – Effective publication: Kuroki et al. (2007)

Legionella yabuuchiae Kuroki et al. 2007, 1933^{VP} – Valid publication: Validation List no. 117 – Effective publication: Kuroki et al. (2007)

Leifsonia ginsengi F. Qiu et al. 2007, 407^{VP}

Lentibacillus halodurans Yuan et al. 2007, 487^{VP}

Lentibacillus halophilus Tanasupawat et al. 2006, 1862^{VP}

Lentibacillus juripiscarius Namwong et al. 2005, 319^{VP}

Lentibacillus kapialis Pakdeeto et al. 2007, 367^{VP}

Lentibacillus lacisalsi J.M. Lim et al. 2005c, 1807^{VP}

Lentibacillus salarius Jeon et al. 2005a, 1342^{VP}

Lentisphaera araneosa Cho et al. 2004, 1005^VP – Valid publication: Validation List no. 98 – Effective publication: J.C. Cho et al. (2004)

Lentzea kentuckyensis Labeda et al. 2007, 1782^VP

Leptolinea tardivitalis Yamada et al. 2006, 1339^VP

Leptospira broomii Levett et al. 2006, 673^VP

Leptotrichia goodfellowii Eribe et al. 2004, 589^VP

Leptotrichia hofstadii Eribe et al. 2004, 589^VP

Leptotrichia shahii Eribe et al. 2004, 589^VP

Leptotrichia wadei Eribe et al. 2004, 591^VP

Leucobacter albus Y.C. Lin et al. 2004, 1675^VP

Leucobacter alluvii Morais et al. 2006, 2507^VP – Valid publication: Validation List no. 112 – Effective publication: Morais et al. (2006)

Leucobacter aridicollis Morais et al. 2005, 547^VP – Valid publication: Validation List no. 102 – Effective publication: Morais et al. (2004)

Leucobacter chromiireducens Morais et al. 2005, 547^VP – Valid publication: Validation List no. 102 – Effective publication: Morais et al. (2004)

Leucobacter iarius Somvanshi et al. 2007, 685^VP

Leucobacter luti Morais et al. 2006, 2507^VP – Valid publication: Validation List no. 112 – Effective publication: Morais et al. (2006)

Leuconostoc durionis Leisner et al. 2005, 1269^VP

Leuconostoc holzapfelii De Bruyne et al. 2007, 2957^VP

Leuconostoc pseudoficulneum Chambel et al. 2006, 1379^VP

Levilinea saccharolytica Yamada et al. 2006, 1338^VP

Lewinella agarilytica S.D. Lee 2007e, 2817^VP

Lewinella lutea Khan et al. 2007a, 2950^VP

Lewinella marina Khan et al. 2007a, 2950^VP

Lishizhenia caseinilytica K.W. Lau et al. 2006b, 2321^VP

Litoricola lipolytica H. Kim et al. 2007a, 1796^VP

Loktanella agnita Ivanova et al. 2005e, 2206^VP

Loktanella atrilutea Hosoya and Yokota 2007d, 1968^VP

Loktanella fryxellensis Van Trappen et al. 2004a, 1268^VP

Loktanella hongkongensis S.C. Lau et al. 2004, 2283^VP

Loktanella koreensis Weon et al. 2006e, 2200^VP

Loktanella maricola J.H. Yoon et al. 2007xii, 1801^VP

Loktanella rosea Ivanova et al. 2005e, 2206^VP

Loktanella salsilacus Van Trappen et al. 2004a, 1267^VP

Loktanella vestfoldensis Van Trappen et al. 2004a, 1268^VP

Longilinea arvoryzae Yamada et al. 2007, 2304^VP

Luedemannella flava Ara and Kudo 2007, 1372^VP – Valid publication: Validation List no. 116 – Effective publication: Ara and Kudo (2007d)

Luedemannella helvata Ara and Kudo 2007, 1372^VP – Valid publication: Validation List no. 116 – Effective publication: Ara and Kudo (2007d)

Luteibacter rhizovicinus Johansen et al. 2005, 2289^VP

Luteimonas composti C.C. Young et al. 2007c, 742^VP

Lutibacter litoralis Choi and Cho 2006, 775^VP

Lutimonas vermicola S.J. Yang et al. 2007, 1682^VP

Lysinibacillus boronitolerans Ahmed et al. 2007d, 1121^VP

Lysobacter concretionis H.S. Bae et al. 2005, 1160^VP

Lysobacter daejeonensis Weon et al. 2006a, 950^VP

Lysobacter defluvii Yassin et al. 2007a, 1134^VP

Lysobacter koreensis J.W. Lee et al. 2006, 234^VP

Lysobacter niabensis Weon et al. 2007c, 549^VP

Lysobacter niastensis Weon et al. 2007c, 549^VP

Lysobacter yangpyeongensis Weon et al. 2006a, 950^VP

Mahella australiensis Bonilla Salinas et al. 2004b, 2172^VP

Malikia granosa Spring et al. 2005b, 627^VP

Maribacter aquivivus Nedashkovskaya et al. 2004b, 1022^VP

Maribacter dokdonensis J.H. Yoon et al. 2005j, 2054^VP

Maribacter orientalis Nedashkovskaya et al. 2004b, 1022^VP

Maribacter polysiphoniae Nedashkovskaya et al. 2007f, 2841^VP

Maribacter sedimenticola Nedashkovskaya et al. 2004b, 1021^VP

Maribacter ulvicola Nedashkovskaya et al. 2004b, 1022^VP

Maribius pelagius Choi et al. 2007, 274^VP

Maribius salinus Choi et al. 2007, 274^VP

Marichromatium bheemlicum Anil Kumar et al. 2007b, 1264^VP

Marichromatium indicum Arunasri et al. 2005, 678^VP

Marinibacillus campisalis J.H. Yoon et al. 2004g, 1320^VP

Marinibacillus campisalis J.H. Yoon et al. 2004g is illegitimate because it is placed in an illegitimate genus.

Marinicola seohaensis J.H. Yoon et al. 2005c, 862^VP

Mariniflexile gromovii Nedashkovskaya et al. 2006a, 1636^VP

Marinilactibacillus piezotolerans Toffin et al. 2005, 349^VP

Marinimicrobium agarilyticum J.M. Lim et al. 2006c, 656^VP

Marinimicrobium koreense J.M. Lim et al. 2006c, 656^VP

Marinitoga hydrogenitolerans Postec et al. 2005, 1220^VP

Marinitoga okinawensis Nunoura et al. 2007a, 470^VP

Marinobacter algicola Green et al. 2006, 526^VP

Marinobacter bryozoorum Romanenko et al. 2005a, 145^VP

Marinobacter daepoensis J.H. Yoon et al. 2004l, 1802^VP

Marinobacter flavimaris J.H. Yoon et al. 2004l, 1802^VP

Marinobacter gudaonensis J. Gu et al. 2007a, 253^VP

Marinobacter koreensis B.Y. Kim et al. 2006f, 2654^VP

Marinobacter maritimus Shivaji et al. 2005b, 1455^VP

Marinobacter salicampi J.H. Yoon et al. 2007xxvii, 2104^VP

Marinobacter salsuginis Antunes et al. 2007, 1038^VP

Marinobacter sediminum Romanenko et al. 2005a, 147^VP

Marinobacter segnicrescens Guo et al. 2007, 1973^VP

Marinobacter vinifirmus Liebgott et al. 2006, 2514^VP

Marinobacterium halophilum H.W. Chang et al. 2007, 79^VP

Marinobacterium litorale H. Kim et al. 2007b, 1661^VP

Marinococcus halotolerans W.J. Li et al. 2005d, 1803^VP

Marinomonas aquimarina corrig. Macián et al. 2005, 983^{VP} – Valid publication: Validation List no. 103 – Effective publication: Macián et al. (2005c)

Marinomonas dokdonensis J.H. Yoon et al. 2005f, 2305^{VP}

Marinomonas ostreistagni K.W. Lau et al. 2006a, 2274^{VP}

Marinomonas polaris Gupta et al. 2006, 363^{VP}

Marinomonas pontica Ivanova et al. 2005c, 278^{VP}

Marinomonas ushuaiensis Prabagaran et al. 2005, 311^{VP}

Marinospirillum insulare Satomi et al. 2004, 166^{VP}

Maritimibacter alkaliphilus K. Lee et al. 2007, 1656^{VP}

Marixanthomonas ophiurae Romanenko et al. 2007a, 459^{VP}

Marmoricola aequoreus S.D. Lee 2007b, 1392^{VP}

Martelella mediterranea Rivas et al. 2005d, 958^{VP}

Massilia albidiflava Y.Q. Zhang et al. 2006, 462^{VP}

Massilia aurea Gallego et al. 2006e, 2451^{VP}

Massilia dura Y.Q. Zhang et al. 2006, 461^{VP}

Massilia lutea Y.Q. Zhang et al. 2006, 463^{VP}

Massilia plicata Y.Q. Zhang et al. 2006, 462^{VP}

Mechercharimyces asporophorigenens Matsuo et al. 2006, 2840^{VP}

Mechercharimyces mesophilus Matsuo et al. 2006, 2840^{VP}

Meganema perideroedes Thomsen et al. 2006, 1867^{VP}

Megasphaera paucivorans Juvonen and Suihko 2006, 700^{VP}

Megasphaera sueciensis Juvonen and Suihko 2006, 700^{VP}

Meiothermus timidus Pires et al. 2005, 1396^{VP} – Valid publication: Validation List no. 104 – Effective publication: Pires et al. (2005)

Mesonia mobilis Nedashkovskaya et al. 2006e, 2435^{VP}

Mesorhizobium albiziae F.Q. Wang et al. 2007, 1196^{VP}

Mesorhizobium septentrionale Gao et al. 2004, 2010^{VP}

Mesorhizobium temperatum Gao et al. 2004, 2011^{VP}

Mesorhizobium thiogangeticum Ghosh and Roy 2006, 95^{VP}

Metascardovia criceti Okamoto et al. 2007, 2449^{VP} – Valid publication: Validation List no. 118 – Effective publication: Okamoto et al. (2007)

Methanobacterium aarhusense Shlimon et al. 2004, 762^{VP}

Methanobacterium beijingense K. Ma et al. 2005, 327^{VP}

Methanobrevibacter millerae Rea et al. 2007, 454^{VP}

Methanobrevibacter olleyae Rea et al. 2007, 454^{VP}

Methanocalculus chunghsingensis M.C. Lai et al. 2004, 189^{VP}

Methanococcoides alaskense Singh et al. 2005, 2537^{VP}

Methanococcus aeolicus Kendall et al. 2006, 1528^{VP}

Methanofollis formosanus S.Y. Wu et al. 2005, 841^{VP}

Methanomethylovorans hollandica Lomans et al. 2004, 307^{VP} – Valid publication: Validation List no. 96 – Effective publication: Lomans et al. (1999)

Methanomethylovorans thermophila Jiang et al. 2005, 2469^{VP}

Methanosaeta harundinacea K. Ma et al. 2006, 130^{VP}

Methanotorris formicicus Takai et al. 2004c, 1099^{VP}

Methermicoccus shengliensis Cheng et al. 2007, 2968^{VP}

Methylibium aquaticum J. Song and Cho 2007, 2126^{VP}

Methylibium fulvum M.H. Yoon et al. 2007c, 2065^{VP}

Methylibium petroleiphilum Nakatsu et al. 2006, 988^{VP}

Methylobacillus pratensis Doronina et al. 2004, 1455^{VP}

Methylobacter tundripaludum Wartiainen et al. 2006a, 112^{VP}

Methylobacterium adhaesivum Gallego et al. 2006a, 340^{VP}

Methylobacterium aquaticum Gallego et al. 2005a, 285^{VP}

Methylobacterium hispanicum Gallego et al. 2005a, 284^{VP}

Methylobacterium isbiliense Gallego et al. 2005c, 2335^{VP}

Methylobacterium jeotgali Aslam et al. 2007a, 568^{VP}

Methylobacterium nodulans Jourand et al. 2004, 2271^{VP}

Methylobacterium oryzae Madhaiyan et al. 2007, 329^{VP}

Methylobacterium platani Y.S. Kang et al. 2007, 2852^{VP}

Methylobacterium podarium Anesti et al. 2006, 2025^{VP} – Valid publication: Validation List no. 111 – Effective publication: Anesti et al. (2004)

Methylobacterium populi Van Aken et al. 2004, 1194^{VP}

Methylobacterium salsuginis X. Wang et al. 2007, 1701^{VP}

Methylobacterium variabile Gallego et al. 2005b, 1432^{VP}

Methylocella tundrae Dedysh et al. 2004, 155^{VP}

Methylocystis heyeri Dedysh et al. 2007, 478^{VP}

Methylocystis hirsuta Lindner et al. 2007, 1898^{VP}

Methylocystis rosea Wartiainen et al. 2006b, 546^{VP}

Methylohalobius crimeensis Heyer et al. 2005, 1824^{VP}

Methylohalomonas lacus Sorokin et al. 2007d, 2767^{VP}

Methylonatrum kenyense Sorokin et al. 2007d, 2768^{VP}

Methylophaga aminisulfidivorans H.G. Kim et al. 2007, 2099^{VP}

Methylosarcina lacus Kalyuzhnaya et al. 2005, 2349^{VP}

Methylosoma difficile Rahalkar et al. 2007, 1079^{VP}

Methylotenera mobilis Kalyuzhnaya et al. 2006a, 2822^{VP}

Methyloversatilis universalis Kalyuzhnaya et al. 2006b, 2521^{VP}

Microbacterium aoyamense A. Kageyama et al. 2006b, 2115^{VP}

Microbacterium deminutum A. Kageyama et al. 2006b, 2115^{VP}

Microbacterium halotolerans W.J. Li et al. 2005a, 69^{VP}

Microbacterium hydrocarbonoxydans Schippers et al. 2005a, 657^{VP}

Microbacterium indicum Shivaji et al. 2007, 1821^{VP}

Microbacterium koreense J.S. Lee et al. 2006a, 426^{VP}

Microbacterium marinilacus A. Kageyama et al. 2007a, 2358^{VP}

Microbacterium natoriense J. Liu et al. 2005, 664^{VP}

Microbacterium oleivorans Schippers et al. 2005a, 657^{VP}

Microbacterium paludicola H.Y. Park et al. 2006, 536^{VP}

Microbacterium pumilum A. Kageyama et al. 2006b, 2115^{VP}

Microbacterium sediminicola A. Kageyama et al. 2007a, 2357^{VP}

Microbacterium terricola corrig. Kageyama et al. 2007, 1372^{VP} – Valid publication: Validation List no. 116 – Effective publication: A. Kageyama et al. (2007b)

Microbacterium ulmi Rivas et al. 2004d, 516^{VP}

Microbacterium xylanilyticum K.K. Kim et al. 2005c, 2077^{VP}

Microbulbifer celer J.H. Yoon et al. 2007ii, 2366^{VP}

Microbulbifer maritimus J.H. Yoon et al. 2004f, 1114^VP

Microcella alkaliphila Tiago et al. 2006b, 2315^VP

Microcella putealis Tiago et al. 2005, 1743^VP – Valid publication: Validation List no. 105 – Effective publication: Tiago et al. (2005b)

Micrococcus flavus X.Y. Liu et al. 2007, 67^VP

Microlunatus ginsengisoli Y.S. Cui et al. 2007b, 715^VP

Micromonospora auratinigra corrig. Thawai et al. 2004, 1425^VP – Valid publication: Validation List no. 99 – Effective publication: Thawai et al. (2004)

Micromonospora chokoriensis Ara and Kudo 2007, 1372^VP – Valid publication: Validation List no. 116 – Effective publication: Ara and Kudo (2007c)

Micromonospora citrea Kroppenstedt et al. 2005, 1743^VP – Valid publication: Validation List no. 105 – Effective publication: Kroppenstedt et al. (2005)

Micromonospora coriariae Trujillo et al. 2006b, 2384^VP

Micromonospora coxensis Ara and Kudo 2007, 1372^VP – Valid publication: Validation List no. 116 – Effective publication: Ara and Kudo (2007c)

Micromonospora eburnea Thawai et al. 2005b, 420^VP

Micromonospora echinaurantiaca Kroppenstedt et al. 2005, 1743^VP – Valid publication: Validation List no. 105 – Effective publication: Kroppenstedt et al. (2005)

Micromonospora echinofusca Kroppenstedt et al. 2005, 1743^VP – Valid publication: Validation List no. 105 – Effective publication: Kroppenstedt et al. (2005)

Micromonospora endolithica Hirsch et al. 2004, 631^VP – Valid publication: Validation List no. 97 – Effective publication: Hirsch et al. (2004b)

Micromonospora fulviviridis Kroppenstedt et al. 2005, 1743^VP – Valid publication: Validation List no. 105 – Effective publication: Kroppenstedt et al. (2005)

Micromonospora inyonensis Kroppenstedt et al. 2005, 1743^VP – Valid publication: Validation List no. 105 – Effective publication: Kroppenstedt et al. (2005)

Micromonospora lupini Trujillo et al. 2007, 2803^VP

Micromonospora mirobrigensis Trujillo et al. 2005a, 879^VP

Micromonospora peucetia Kroppenstedt et al. 2005, 1743^VP – Valid publication: Validation List no. 105 – Effective publication: Kroppenstedt et al. (2005)

Micromonospora saelicesensis Trujillo et al. 2007, 2801^VP

Micromonospora sagamiensis Kroppenstedt et al. 2005, 1743^VP – Valid publication: Validation List no. 105 – Effective publication: Kroppenstedt et al. (2005)

Micromonospora siamensis Thawai et al. 2006, 2^VP – Valid publication: Validation List no. 107 – Effective publication: Thawai et al. (2005a)

Micromonospora viridifaciens Kroppenstedt et al. 2005, 1744^VP – Valid publication: Validation List no. 105 – Effective publication: Kroppenstedt et al. (2005)

Microtetraspora malaysiensis Nakajima et al. 2004, 1^VP – Valid publication: Validation List no. 95 – Effective publication: Y. Nakajima et al. (2003)

Millisia brevis Soddell et al. 2006a, 742^VP

Mitsuaria chitosanitabida Amakata et al. 2005, 1930^VP

Modestobacter versicolor Reddy et al. 2007, 2018^VP

Modicisalibacter tunisiensis Ben Ali Gam et al. 2007, 2312^VP

Moorella mulderi Balk et al. 2005, 1^VP – Valid publication: Validation List no. 101 – Effective publication: Balk et al. (2003)

Moraxella bovoculi Angelos et al. 2007, 793^VP

Moraxella oblonga Xie and Yokota 2005b, 332^VP

Morganella psychrotolerans Emborg et al. 2006, 2478^VP

Moryella indoligenes Carlier et al. 2007, 728^VP

Mucilaginibacter gracilis Pankratov et al. 2007, 2353^VP

Mucilaginibacter paludis Pankratov et al. 2007, 2352^VP

Mucispirillum schaedleri Robertson et al. 2005, 1203^VP

Muricauda aquimarina J.H. Yoon et al. 2005v, 1019^VP

Muricauda flavescens J.H. Yoon et al. 2005v, 1018^VP

Myceligenerans crystallogenes Groth et al. 2006, 286^VP

Myceligenerans xiligouense Cui et al. 2004, 1292^VP

Mycobacterium arupense Cloud et al. 2006, 1417^VP

Mycobacterium aubagnense Adékambi et al. 2006b, 140^VP

Mycobacterium boenickei Schinsky et al. 2004, 1664^VP

Mycobacterium bolletii Adékambi et al. 2006b, 140^VP

Mycobacterium brisbanense Schinsky et al. 2004, 1665^VP

Mycobacterium canariasense Jiménez et al. 2004, 1733^VP

Mycobacterium chimaera Tortoli et al. 2004, 1283^VP

Mycobacterium colombiense Murcia et al. 2006, 2053^VP

Mycobacterium conceptionense Adékambi et al. 2006, 2025^VP – Valid publication: Validation List no. 111 – Effective publication: Adékambi et al. (2006a)

Mycobacterium cosmeticum Cooksey et al. 2004, 2390^VP

Mycobacterium florentinum Tortoli et al. 2005, 1105^VP

Mycobacterium fluoranthenivorans Hormisch et al. 2006, 1459^VP – Valid publication: Validation List no. 110 – Effective publication: Hormisch et al. (2004)

Mycobacterium houstonense Schinsky et al. 2004, 1664^VP

Mycobacterium kumamotonense Masaki et al. 2007, 433^VP – Valid publication: Validation List no. 114 – Effective publication: Masaki et al. (2006)

Mycobacterium massiliense Adékambi et al. 2006, 2025^VP – Valid publication: Validation List no. 111 – Effective publication: Adékambi et al. (2004)

Mycobacterium monacense Reischl et al. 2006, 2578^VP

Mycobacterium nebraskense Mohamed et al. 2004, 2060^VP

Mycobacterium neworleansense Schinsky et al. 2004, 1665^VP

Mycobacterium parascrofulaceum Turenne et al. 2004a, 1550^VP

Mycobacterium parmense Fanti et al. 2004, 1126^VP

Mycobacterium phocaicum Adékambi et al. 2006b, 140^VP

Mycobacterium pseudoshottsii Rhodes et al. 2005, 1144^VP

Mycobacterium psychrotolerans Trujillo et al. 2004, 1461^VP

Mycobacterium pyrenivorans Derz et al. 2004, 2316^VP

Mycobacterium saskatchewanense Turenne et al. 2004b, 665^VP

Mycobacterium seoulense Mun et al. 2007, 597^VP

Mycoplasma amphoriforme Pitcher et al. 2005, 2592^VP

Mycoplasma iguanae D.R. Brown et al. 2006, 763^VP

Mycoplasma testudineum D.R. Brown et al. 2004, 1529^VP

Myroides pelagicus J. Yoon et al. 2006, 1919^VP

Nannocystis pusilla Reichenbach 2007, 894^VP – Valid publication: Validation List no. 115 – Effective publication: Reichenbach (2005k)

Natranaerobius thermophilus Mesbah et al. 2007, 2511^VP

Natrinema altunense X.W. Xu et al. 2005b, 1312^VP

Natrinema ejinorense Castillo et al. 2006d, 2685^VP

Natronocella acetinitrilica Sorokin et al. 2007, 1372^VP – Valid publication: Validation List no. 116 – Effective publication: Sorokin et al. (2007e)

Natronococcus jeotgali Roh et al. 2007a, 2130^VP

Natronolimnobius baerhuensis Itoh et al. 2005, 1744^VP – Valid publication: Validation List no. 105 – Effective publication: Itoh et al. (2005)

Natronolimnobius innermongolicus Itoh et al. 2005, 1744^VP – Valid publication: Validation List no. 105 – Effective publication: Itoh et al. (2005)

Natronorubrum aibiense Cui et al. 2006a, 1517^VP

Natronorubrum sulfidifaciens H.L. Cui et al. 2007a, 739^VP

Naxibacter alkalitolerans P. Xu et al. 2005b, 1152^VP

Neisseria animaloris Vandamme et al. 2006, 1803^VP

Neisseria bacilliformis Han et al. 2006, 1459^VP – Valid publication: Validation List no. 110 – Effective publication: Han et al. (2006)

Neisseria zoodegmatis Vandamme et al. 2006, 1804^VP

Neoasaia chiangmaiensis Yukphan et al. 2006, 499^VP – Valid publication: Validation List no. 108 – Effective publication: Yukphan et al. (2005c)

Neptuniibacter caesariensis Arahal et al. 2007, 1004^VP

Nereida ignava Pujalte et al. 2005c, 635^VP

Nesiotobacter exalbescens Donachie et al. 2006, 566^VP

Nesterenkonia aethiopica Delgado et al. 2006, 1232^VP

Nesterenkonia halotolerans W.J. Li et al. 2004b, 838^VP

Nesterenkonia jeotgali J.H. Yoon et al. 2006l, 2591^VP

Nesterenkonia lutea W.J. Li et al. 2005b, 465^VP

Nesterenkonia sandarakina W.J. Li et al. 2005b, 464^VP

Nesterenkonia xinjiangensis W.J. Li et al. 2004b, 840^VP

Niabella aurantiaca B.Y. Kim et al. 2007, 540^VP

Niastella koreensis Weon et al. 2006f, 1779^VP

Niastella yeongjuensis Weon et al. 2006f, 1781^VP

Nicoletella semolina Kuhnert et al. 2005, 547^VP – Valid publication: Validation List no. 102 – Effective publication: Kuhnert et al. (2004)

Nitratifractor salsuginis S. Nakagawa et al. 2005c, 931^VP

Nitratireductor aquibiodomus Labbé et al. 2004, 272^VP

Nitratiruptor tergarcus S. Nakagawa et al. 2005c, 931^VP

Nitrincola lacisaponensis Dimitriu et al. 2005, 2277^VP

Nocardia acidivorans Kämpfer et al. 2007b, 1185^VP

Nocardia alba Li et al. 2004, 1425^VP – Valid publication: Validation List no. 99 – Effective publication: W.J. Li et al. (2004c)

Nocardia amamiensis H. Yamamura et al. 2007, 1601^VP

Nocardia anemiae Kageyama et al. 2005, 1396^VP – Valid publication: Validation List no. 104 – Effective publication: A. Kageyama et al. (2005b)

Nocardia aobensis Kageyama et al. 2005, 547^VP – Valid publication: Validation List no. 102 – Effective publication: A. Kageyama et al. (2004c)

Nocardia araoensis A. Kageyama et al. 2004h, 2027^VP

Nocardia arthritidis Kageyama et al. 2005, 1^VP – Valid publication: Validation List no. 101 – Effective publication: A. Kageyama et al. (2004d)

Nocardia asiatica A. Kageyama et al. 2004a, 127^VP

Nocardia concava A. Kageyama et al. 2005c, 2082^VP

Nocardia coubleae Rodríguez-Nava et al. 2007, 1485^VP

Nocardia elegans Yassin and Brenner 2005, 1508^VP

Nocardia exalbida S. Iida et al. 2006, 1195^VP

Nocardia gamkensis le Roes and Meyers 2007, 1^VP – Valid publication: Validation List no. 113 – Effective publication: le Roes and Meyers (2006)

Nocardia harenae J.P. Seo and Lee 2006, 2206^VP

Nocardia higoensis A. Kageyama et al. 2004g, 1930^VP

Nocardia inohanensis A. Kageyama et al. 2004f, 567^VP

Nocardia jejuensis S.D. Lee 2006b, 561^VP

Nocardia jiangxiensis Cui et al. 2005, 1923^VP

Nocardia kruczakiae Conville et al. 2005, 547^VP – Valid publication: Validation List no. 102 – Effective publication: Conville et al. (2004)

Nocardia lijiangensis Xu et al. 2006, 2026^VP – Valid publication: Validation List no. 111 – Effective publication: P. Xu et al. (2006)

Nocardia mexicana Rodríguez-Nava et al. 2006, 925^VP – Valid publication: Validation List no. 109 – Effective publication: Rodríguez-Nava et al. (2004)

Nocardia miyunensis Cui et al. 2005, 1924^VP

Nocardia neocaledoniensis Saintpierre-Bonaccio et al. 2004b, 601^VP

Nocardia niigatensis A. Kageyama et al. 2004f, 568^VP

Nocardia ninae Laurent et al. 2007, 664^VP

Nocardia pigrifrangens L. Wang et al. 2004, 1685^VP

Nocardia pneumoniae A. Kageyama et al. 2004h, 2028^VP

Nocardia polyresistens P. Xu et al. 2005a, 1468^VP

Nocardia shimofusensis A. Kageyama et al. 2004g, 1930^VP

Nocardia sienata corrig. Kageyama et al. 2004, 1005^VP – Valid publication: Validation List no. 98 – Effective publication: A. Kageyama et al. (2004e)

Nocardia speluncae J.P. Seo et al. 2007, 2934^VP

Nocardia takedensis H. Yamamura et al. 2005, 435^VP

Nocardia tenerifensis Kämpfer et al. 2004a, 383[VP]

Nocardia terpenica Hoshino et al. 2007, 1458[VP]

Nocardia testacea corrig. Kageyama et al. 2004, 1005[VP] – Valid publication: Validation List no. 98 – Effective publication: A. Kageyama et al. (2004e)

Nocardia thailandica Kageyama et al. 2005, 547[VP] – Valid publication: Validation List no. 102 – Effective publication: A. Kageyama et al. (2004b)

Nocardia vermiculata Kageyama et al. 2005, 547[VP] – Valid publication: Validation List no. 102 – Effective publication: A. Kageyama et al. (2004b)

Nocardia xishanensis J. Zhang et al. 2004, 2302[VP]

Nocardia yamanashiensis A. Kageyama et al. 2004f, 568[VP]

Nocardioides aestuarii Yi and Chun 2004c, 2152[VP]

Nocardioides alkalitolerans J.H. Yoon et al. 2005l, 813[VP]

Nocardioides aquiterrae J.H. Yoon et al. 2004d, 74[VP]

Nocardioides aromaticivorans Kubota et al. 2005, 984[VP] – Valid publication: Validation List no. 103 – Effective publication: Kubota et al. (2005)

Nocardioides dubius J.H. Yoon et al. 2005t, 2211[VP]

Nocardioides exalbidus Li et al. 2007, 2449[VP] – Valid publication: Validation List no. 118 – Effective publication: B. Li et al. (2007b)

Nocardioides furvisabuli S.D. Lee 2007, 37[VP]

Nocardioides ganghwensis Yi and Chun 2004b, 1298[VP]

Nocardioides insulae J.H. Yoon et al. 2007vii, 138[VP]

Nocardioides kongjuensis J.H. Yoon et al. 2006n, 1786[VP]

Nocardioides kribbensis J.H. Yoon et al. 2005m, 1614[VP]

Nocardioides lentus J.H. Yoon et al. 2006m, 274[VP]

Nocardioides marinisabuli D.W. Lee et al. 2007, 2961[VP]

Nocardioides marinus Choi et al. 2007, 778[VP]

Nocardioides oleivorans Schippers et al. 2005b, 1502[VP]

Nocardioides panacihumi D.S. An et al. 2007b, 2145[VP]

Nocardioides terrigena J.H. Yoon et al. 2007xxiii, 2474[VP]

Nocardiopsis aegyptia Sabry et al. 2004, 455[VP]

Nocardiopsis alkaliphila Hozzein et al. 2004, 250[VP]

Nocardiopsis baichengensis W.J. Li et al. 2006a, 1095[VP]

Nocardiopsis chromatogenes W.J. Li et al. 2006a, 1094[VP]

Nocardiopsis gilva W.J. Li et al. 2006a, 1093[VP]

Nocardiopsis rhodophaea W.J. Li et al. 2006a, 1094[VP]

Nocardiopsis rosea W.J. Li et al. 2006a, 1094[VP]

Nocardiopsis salina W.J. Li et al. 2004d, 1808[VP]

Nonlabens tegetincola S.C. Lau et al. 2005c, 2281[VP]

Nonomuraea bangladeshensis Ara et al. 2007b, 1506[VP]

Nonomuraea coxensis Ara et al. 2007b, 1507[VP]

Nonomuraea kuesteri Kämpfer et al. 2005a, 848[VP]

Nonomuraea maheshkhaliensis Ara et al. 2007, 2449[VP] – Valid publication: Validation List no. 118 – Effective publication: Ara et al. (2007a)

Novosphingobium lentum Tiirola et al. 2005, 585[VP]

Novosphingobium nitrogenifigens Addison et al. 2007, 2470[VP]

Novosphingobium pentaromativorans Sohn et al. 2004a, 1486[VP]

Novosphingobium taihuense Z.P. Liu et al. 2005a, 1231[VP]

Oceanibulbus indolifex Wagner-Döbler et al. 2004, 1183[VP]

Oceanicola batsensis J.C. Cho and Giovannoni 2004b, 1133[VP]

Oceanicola granulosus J.C. Cho and Giovannoni 2004b, 1133[VP]

Oceanicola marinus K.Y. Lin et al. 2007, 1627[VP]

Oceanicola nanhaiensis J. Gu et al. 2007b, 159[VP]

Oceanimonas smirnovii Ivanova et al. 2005, 984[VP] – Valid publication: Validation List no. 103 – Effective publication: Ivanova et al. (2005d)

Oceanisphaera donghaensis S.J. Park et al. 2006, 897[VP]

Oceanithermus desulfurans Mori et al. 2004, 1564[VP]

Oceanobacillus chironomi Raats and Halpern 2007, 258[VP]

Oceanobacillus oncorhynchi Yumoto et al. 2005b, 1523[VP]

Oceanobacillus profundus Y.G. Kim et al. 2007a, 412[VP]

Ochrobactrum cytisi Zurdo-Piñeiro et al. 2007, 787[VP]

Ochrobactrum haematophilum Kämpfer et al. 2007e, 2517[VP]

Ochrobactrum lupini Trujillo et al. 2006, 1459[VP] – Valid publication: Validation List no. 110 – Effective publication: Trujillo et al. (2005b)

Ochrobactrum oryzae Tripathi et al. 2006, 1679[VP]

Ochrobactrum pseudintermedium Teyssier et al. 2007, 1011[VP]

Ochrobactrum pseudogrignonense Kämpfer et al. 2007e, 2515[VP]

Oenococcus kitaharae Endo and Okada 2006, 2347[VP]

Olivibacter sitiensis Ntougias et al. 2007a, 402[VP]

Olleya marilimosa Mancuso Nichols et al. 2005, 1560[VP]

Oribacterium sinus Carlier et al. 2004, 1614[VP]

Ornithinibacillus bavariensis Mayr et al. 2006, 1387[VP]

Ornithinibacillus californiensis Mayr et al. 2006, 1388[VP]

Ornithinimicrobium kibberense Mayilraj et al. 2006d, 1660[VP]

Oryzihumus leptocrescens A. Kageyama et al. 2005a, 2558[VP]

Oscillibacter valericigenes Iino et al. 2007, 1844[VP]

Ottowia thiooxydans Spring et al. 2004, 104[VP]

Owenweeksia hongkongensis K.W. Lau et al. 2005, 1055[VP]

Paenibacillus alkaliterrae J.H. Yoon et al. 2005k, 2342[VP]

Paenibacillus anaericanus Horn et al. 2005, 1263[VP]

Paenibacillus antarcticus Montes et al. 2004, 1523[VP]

Paenibacillus assamensis Saha et al. 2005b, 2579[VP]

Paenibacillus barcinonensis Sánchez et al. 2005, 937[VP]

Paenibacillus barengoltzii Osman et al. 2006, 1514[VP]

Paenibacillus cellulosilyticus Rivas et al. 2006b, 2779[VP]

Paenibacillus cineris Logan et al. 2004a, 1075[VP]

Paenibacillus cookii Logan et al. 2004a, 1075[VP]

Paenibacillus elgii D.S. Kim et al. 2004, 2034[VP]

Paenibacillus favisporus Velázquez et al. 2004, 61[VP]

Paenibacillus fonticola J.H. Chou et al. 2007b, 1348^VP

Paenibacillus gansuensis J.M. Lim et al. 2006d, 2133^VP

Paenibacillus ginsengarvi M.H. Yoon et al. 2007a, 1812^VP

Paenibacillus ginsengisoli Lee et al. 2007, 1372^VP – Valid publication: Validation List no. 116 – Effective publication: M. Lee et al. (2007)

Paenibacillus hodogayensis Takeda et al. 2005, 740^VP

Paenibacillus humicus Vaz-Moreira et al. 2007a, 2270^VP

Paenibacillus lactis Scheldeman et al. 2004a, 889^VP

Paenibacillus massiliensis Roux and Raoult 2004, 1051^VP

Paenibacillus mendelii Šmerda et al. 2005, 2353^VP

Paenibacillus motobuensis K. Iida et al. 2005, 1814^VP

Paenibacillus panacisoli Ten et al. 2006a, 2680^VP

Paenibacillus pasadenensis Osman et al. 2006, 1512^VP

Paenibacillus phyllosphaerae Rivas et al. 2005c, 745^VP

Paenibacillus rhizosphaerae Rivas et al. 2005a, 1307^VP

Paenibacillus sabinae Y. Ma et al. 2007a, 9^VP

Paenibacillus sanguinis Roux and Raoult 2004, 1052^VP

Paenibacillus sepulcri Šmerda et al. 2006, 2343^VP

Paenibacillus soli M.J. Park et al. 2007, 149^VP

Paenibacillus taiwanensis F.L. Lee et al. 2007, 1353^VP

Paenibacillus terrigena Xie and Yokota 2007, 71^VP

Paenibacillus timonensis Roux and Raoult 2004, 1053^VP

Paenibacillus wynnii Rodríguez-Díaz et al. 2005, 2097^VP

Paenibacillus xinjiangensis J.M. Lim et al. 2006f, 2581^VP

Paenibacillus xylanilyticus Rivas et al. 2005b, 406^VP

Paenibacillus zanthoxyli Y. Ma et al. 2007b, 876^VP

Palaeococcus helgesonii Amend et al. 2006, 3^VP – Valid publication: Validation List no. 107 – Effective publication: Amend et al. (2003)

Palleronia marisminoris Martínez-Checa et al. 2005, 2528^VP

Paludibacter propionicigenes Ueki et al. 2006a, 43^VP

Parabacteroides johnsonii Sakamoto et al. 2007a, 295^VP

Paracoccus bengalensis Ghosh et al. 2006, 2507^VP – Valid publication: Validation List no. 112 – Effective publication: Ghosh et al. (2006)

Paracoccus haeundaensis J.H. Lee et al. 2004, 1701^VP

Paracoccus homiensis B.Y. Kim et al. 2006e, 2389^VP

Paracoccus koreensis La et al. 2005, 1659^VP

Paracoccus sulfuroxidans X.Y. Liu et al. 2006, 2695^VP

Paraferrimonas sedimenticola Khan and Harayama 2007, 1497^VP

Parapedobacter koreensis M.K. Kim et al. 2007b, 1339^VP

Parasporobacterium paucivorans Lomans et al. 2004, 307^VP – Valid publication: Validation List no. 96 – Effective publication: Lomans et al. (2001)

Parvibaculum lavamentivorans Schleheck et al. 2004, 1496^VP

Patulibacter minatonensis Y. Takahashi et al. 2006, 405^VP

Paucibacter toxinivorans Rapala et al. 2005, 1567^VP

Paucisalibacillus globulus Nunes et al. 2006, 1842^VP

Pectinatus haikarae Juvonen and Suihko 2006, 701^VP

Pectinatus portalensis Gonzalez et al. 2005, 547^VP – Valid publication: Validation List no. 102 – Effective publication: Gonzalez et al. (2004)

Pediococcus cellicola B. Zhang et al. 2005, 2169^VP

Pediococcus ethanolidurans L. Liu et al. 2006, 2407^VP

Pediococcus siamensis Tanasupawat et al. 2007, 1372^VP – Valid publication: Validation List no. 116 – Effective publication: Tanasupawat et al. (2007b)

Pediococcus stilesii Franz et al. 2006, 332^VP

Pedobacter aquatilis Gallego et al. 2006c, 1854^VP

Pedobacter caeni Vanparys et al. 2005a, 1316^VP

Pedobacter duraquae Muurholm et al. 2007, 2225^VP

Pedobacter ginsengisoli Ten et al. 2006e, 2568^VP

Pedobacter hartonius Muurholm et al. 2007, 2225^VP

Pedobacter himalayensis Shivaji et al. 2005a, 1084^VP

Pedobacter insulae J.H. Yoon et al. 2007xviii, 2002^VP

Pedobacter koreensis Baik et al. 2007b, 2081^VP

Pedobacter lentus J.H. Yoon et al. 2007xix, 2093^VP

Pedobacter metabolipauper Muurholm et al. 2007, 2226^VP

Pedobacter panaciterrae M.H. Yoon et al. 2007b, 385^VP

Pedobacter roseus Hwang et al. 2006, 1834^VP

Pedobacter sandarakinus J.H. Yoon et al. 2006q, 1276^VP

Pedobacter steynii Muurholm et al. 2007, 2226^VP

Pedobacter suwonensis S.W. Kwon et al. 2007a, 481^VP

Pedobacter terrae J.H. Yoon et al. 2007xxii, 2465^VP

Pedobacter terricola J.H. Yoon et al. 2007xix, 2094^VP

Pedobacter westerhofensis Muurholm et al. 2007, 2226^VP

Pelagibaca bermudensis J.C. Cho and Giovannoni 2006, 858^VP

Pelagibacillus goriensis Y.G. Kim et al. 2007b, 1559^VP

Pelagicoccus albus J. Yoon et al. 2007g, 1381^VP

Pelagicoccus croceus J. Yoon et al. 2007e, 2877^VP

Pelagicoccus litoralis J. Yoon et al. 2007g, 1383^VP

Pelagicoccus mobilis J. Yoon et al. 2007g, 1381^VP

Pelobacter seleniigenes Narasingarao and Häggblom 2007, 1941^VP

Pelomonas aquatica Gomila et al. 2007, 2634^VP

Pelomonas puraquae Gomila et al. 2007, 2634^VP

Pelosinus fermentans Shelobolina et al. 2007, 133^VP

Pelotomaculum isophthalicicum corrig. Qiu et al. 2006, 2026^VP – Valid publication: Validation List no. 111 – Effective publication: Y.L. Qiu et al. (2006)

Pelotomaculum propionicicum Imachi et al. 2007, 1491^VP

Pelotomaculum schinkii de Bok et al. 2005, 1702^VP

Pelotomaculum terephthalicicum corrig. Qiu et al. 2006, 2026^VP – Valid publication: Validation List no. 111 – Effective publication: Y.L. Qiu et al. (2006)

Peptostreptococcus stomatis Downes and Wade 2006, 753^VP

Perexilibacter aurantiacus J. Yoon et al. 2007a, 966^VP

Persicitalea jodogahamensis J. Yoon et al. 2007b, 1016^VP

Persicivirga xylanidelens O'Sullivan et al. 2006, 178^{VP}

Petrimonas sulfuriphila Grabowski et al. 2005, 1119^{VP}

Petrobacter succinatimandens Bonilla Salinas et al. 2004a, 648^{VP}

Petrotoga halophila Miranda-Tello et al. 2007, 43^{VP}

Petrotoga mexicana Miranda-Tello et al. 2004, 172^{VP}

Phaeobacter daeponensis J.H. Yoon et al. 2007xi, 860^{VP}

Phaeobacter inhibens Martens et al. 2006, 1301^{VP}

Phenylobacterium falsum Tiago et al. 2005, 1744^{VP} – Valid publication: Validation List no. 105 – Effective publication: Tiago et al. (2005a)

Phenylobacterium koreense Aslam et al. 2005b, 2004^{VP}

Phenylobacterium lituiforme Kanso and Patel 2004, 2144^{VP}

Photobacterium aplysiae H.J. Seo et al. 2005b, 2295^{VP}

Photobacterium frigidiphilum H.J. Seo et al. 2005b, 1664^{VP}

Photobacterium ganghwense Y.D. Park et al. 2006a, 747^{VP}

Photobacterium halotolerans Rivas et al. 2006a, 1070^{VP}

Photobacterium kishitanii Ast et al. 2007, 2077^{VP}

Photobacterium lipolyticum J.H. Yoon et al. 2005u, 338^{VP}

Photobacterium lutimaris Jung et al. 2007a, 335^{VP}

Photobacterium rosenbergii Thompson et al. 2005, 915^{VP}

Phycicoccus jejuensis S.D. Lee 2006f, 2371^{VP}

Phyllobacterium bourgognense Mantelin et al. 2006, 837^{VP}

Phyllobacterium brassicacearum Mantelin et al. 2006, 837^{VP}

Phyllobacterium catacumbae Jurado et al. 2005d, 1489^{VP}

Phyllobacterium ifriqiyense Mantelin et al. 2006, 837^{VP}

Phyllobacterium leguminum Mantelin et al. 2006, 837^{VP}

Phyllobacterium trifolii Valverde et al. 2005, 1988^{VP}

Pibocella ponti Nedashkovskaya et al. 2005b, 179^{VP}

Pigmentiphaga daeguensis J.H. Yoon et al. 2007iv, 1190^{VP}

Pilibacter termitis Higashiguchi et al. 2006, 19^{VP}

Piscibacillus salipiscarius Tanasupawat et al. 2007b, 1416^{VP}

Planifilum fimeticola Hatayama et al. 2005b, 2104^{VP}

Planifilum fulgidum Hatayama et al. 2005b, 2104^{VP}

Planifilum yunnanense Y.X. Zhang et al. 2007, 1852^{VP}

Planococcus columbae Suresh et al. 2007a, 1269^{VP}

Planococcus donghaensis Choi et al. 2007, 2648^{VP}

Planococcus maitriensis Alam et al. 2004, 307^{VP} – Valid publication: Validation List no. 96 – Effective publication: Alam et al. (2003)

Planococcus stackebrandtii Mayilraj et al. 2005, 93^{VP}

Planomicrobium chinense Dai et al. 2005, 701^{VP}

Planotetraspora silvatica Tamura and Sakane 2004, 2055^{VP}

Plantibacter auratus Y.C. Lin and Yokota 2006, 2338^{VP}

Pleomorphomonas koreensis Im et al. 2006b, 1665^{VP}

Pleomorphomonas oryzae Xie and Yokota 2005d, 1236^{VP}

Polaribacter butkevichii Nedashkovskaya et al. 2006, 1459^{VP} – Valid publication: Validation List no. 110 – Effective publication: Nedashkovskaya et al. (2005i)

Polaribacter dokdonensis J.H. Yoon et al. 2006j, 1252^{VP}

Polaromonas aquatica Kämpfer et al. 2006c, 607^{VP}

Polaromonas hydrogenivorans Sizova and Panikov 2007, 619^{VP}

Polaromonas naphthalenivorans Jeon et al. 2004, 96^{VP}

Polymorphospora rubra Tamura et al. 2006, 1963^{VP}

Pontibacillus chungwhensis J.M. Lim et al. 2005b, 169^{VP}

Pontibacillus marinus J.M. Lim et al. 2005a, 1030^{VP}

Pontibacter actiniarum Nedashkovskaya et al. 2005l, 2586^{VP}

Pontibacter akesuensis Zhou et al. 2007b, 324^{VP}

Porphyrobacter dokdonensis J.H. Yoon et al. 2006d, 1082^{VP}

Porphyrobacter donghaensis J.H. Yoon et al. 2004h, 2234^{VP}

Porphyromonas somerae Summanen et al. 2006, 925^{VP} – Valid publication: Validation List no. 109 – Effective publication: Summanen et al. (2005)

Porphyromonas uenonis Finegold et al. 2005, 547^{VP} – Valid publication: Validation List no. 102 – Effective publication: Finegold et al. (2004b)

Prevotella baroniae Downes et al. 2005, 1554^{VP}

Prevotella bergensis Downes et al. 2006, 611^{VP}

Prevotella copri Hayashi et al. 2007b, 943^{VP}

Prevotella maculosa Downes et al. 2007, 2938^{VP}

Prevotella marshii Downes et al. 2005, 1554^{VP}

Prevotella multiformis Sakamoto et al. 2005a, 818^{VP}

Prevotella multisaccharivorax Sakamoto et al. 2005b, 1842^{VP}

Prevotella nanceiensis Alauzet et al. 2007, 2218^{VP}

Prevotella paludivivens Ueki et al. 2007, 1808^{VP}

Prevotella pleuritidis Sakamoto et al. 2007c, 1726^{VP}

Prevotella salivae Sakamoto et al. 2004, 882^{VP}

Prevotella shahii Sakamoto et al. 2004, 881^{VP}

Prevotella stercorea Hayashi et al. 2007b, 943^{VP}

Prevotella timonensis Glazunova et al. 2007, 885^{VP}

Prolixibacter bellariivorans Holmes et al. 2007, 705^{VP}

Promicromonospora pachnodae Cazemier et al. 2004, 1^{VP} – Valid publication: Validation List no. 95 – Effective publication: Cazemier et al. (2003)

Propionicicella superfundia H.S. Bae et al. 2006, 2026^{VP} – Valid publication: Validation List no. 111 – Effective publication: H.S. Bae et al. (2006b)

Propionispora hippei Abou-Zeid et al. 2004, 953^{VP}

Proteiniphilum acetatigenes Chen and Dong 2005, 2261^{VP}

Providencia vermicola Somvanshi et al. 2006, 631^{VP}

Pseudidiomarina sediminum Hu and Li 2007, 2576^{VP}

Pseudidiomarina taiwanensis Jean et al. 2006b, 904^{VP}

Pseudoalteromonas aliena Ivanova et al. 2004g, 1436^{VP}

Pseudoalteromonas byunsanensis Y.D. Park et al. 2005a, 2521^{VP}

Pseudoalteromonas marina Y.D. Nam et al. 2007b, 16^{VP}

Pseudoalteromonas spongiae S.C. Lau et al. 2005a, 1595^{VP}

Pseudochrobactrum asaccharolyticum Kämpfer et al. 2006h, 1826^{VP}

Pseudochrobactrum kiredjianiae Kämpfer et al. 2007f, 759^{VP}

Pseudochrobactrum saccharolyticum Kämpfer et al. 2006h, 1827^VP

Pseudoclavibacter helvolus Manaia et al. 2004, 787^VP

Pseudolabrys taiwanensis Kämpfer et al. 2006j, 2470^VP

Pseudomonas antarctica Reddy et al. 2004, 717^VP

Pseudomonas argentinensis Peix et al. 2005, 1110^VP

Pseudomonas azotifigens Hatayama et al. 2005a, 1542^VP

Pseudomonas borbori Vanparys et al. 2006, 1880^VP

Pseudomonas delhiensis Prakash et al. 2007, 529^VP

Pseudomonas guineae Bozal et al. 2007, 2611^VP

Pseudomonas knackmussii Stolz et al. 2007, 575^VP

Pseudomonas lurida Behrendt et al. 2007a, 983^VP

Pseudomonas lutea Peix et al. 2004, 849^VP

Pseudomonas meridiana Reddy et al. 2004, 717^VP

Pseudomonas mohnii Cámara et al. 2007, 930^VP

Pseudomonas moorei Cámara et al. 2007, 929^VP

Pseudomonas moraviensis Tvrzová et al. 2006, 2661^VP

Pseudomonas otitidis Clark et al. 2006, 713^VP

Pseudomonas pachastrellae Romanenko et al. 2005b, 922^VP

Pseudomonas panacis Y.D. Park et al. 2005b, 1723^VP

Pseudomonas peli Vanparys et al. 2006, 1880^VP

Pseudomonas pohangensis Weon et al. 2006d, 2155^VP

Pseudomonas proteolytica Reddy et al. 2004, 718^VP

Pseudomonas psychrotolerans Hauser et al. 2004, 1636^VP

Pseudomonas reinekei Cámara et al. 2007, 928^VP

Pseudomonas segetis Y.D. Park et al. 2006b, 2595^VP

Pseudomonas simiae Vela et al. 2006, 2674^VP

Pseudomonas vranovensis Tvrzová et al. 2006, 2661^VP

Pseudomonas xanthomarina Romanenko et al. 2005, 2236^VP – Valid publication: Validation List no. 106 – Effective publication: Romanenko et al. (2005c)

Pseudonocardia ammonioxydans Z.P. Liu et al. 2006b, 556^VP

Pseudonocardia antarctica Prabahar et al. 2004, 1005^VP – Valid publication: Validation List no. 98 – Effective publication: Prabahar et al. (2004)

Pseudonocardia benzenivorans Kämpfer and Kroppenstedt 2004, 751^VP

Pseudonocardia chloroethenivorans S.B. Lee et al. 2004, 138^VP

Pseudonocardia dioxanivorans Mahendra and Alvarez-Cohen 2005, 597^VP

Pseudonocardia oroxyli Q. Gu et al. 2006, 2194^VP

Pseudonocardia tetrahydrofuranoxydans Kämpfer et al. 2006e, 1536^VP

Pseudoruegeria aquimaris J.H. Yoon et al. 2007xxviii, 546^VP

Pseudosphingobacterium domesticum Vaz-Moreira et al. 2007c, 1538^VP

Pseudovibrio ascidiaceicola Fukunaga et al. 2006, 346^VP

Pseudovibrio denitrificans Shieh et al. 2004, 2311^VP

Pseudovibrio japonicus Hosoya and Yokota 2007c, 1953^VP

Pseudoxanthomonas daejeonensis D.C. Yang et al. 2005, 791^VP

Pseudoxanthomonas japonensis Thierry et al. 2004, 2254^VP

Pseudoxanthomonas kalamensis Harada et al. 2006, 1105^VP

Pseudoxanthomonas kaohsiungensis J.S. Chang et al. 2005, 984^VP – Valid publication: Validation List no. 103 – Effective publication: J.S. Chang et al. (2005)

Pseudoxanthomonas koreensis D.C. Yang et al. 2005, 791^VP

Pseudoxanthomonas mexicana Thierry et al. 2004, 2254^VP

Pseudoxanthomonas spadix C.C. Young et al. 2007a, 1825^VP

Pseudoxanthomonas suwonensis Weon et al. 2006b, 661^VP

Pseudoxanthomonas yeongjuensis Yoo et al. 2007c, 648^VP

Psychrobacter adeliensis Shivaji et al. 2005, 547^VP – Valid publication: Validation List no. 102 – Effective publication: Shivaji et al. (2004)

Psychrobacter alimentarius J.H. Yoon et al. 2005x, 175^VP

Psychrobacter aquaticus Shivaji et al. 2005c, 759^VP

Psychrobacter aquimaris J.H. Yoon et al. 2005r, 1011^VP

Psychrobacter arcticus Bakermans et al. 2006, 1290^VP

Psychrobacter arenosus Romanenko et al. 2004a, 1744^VP

Psychrobacter celer J.H. Yoon et al. 2005o, 1888^VP

Psychrobacter cibarius Jung et al. 2005b, 580^VP

Psychrobacter cryohalolentis Bakermans et al. 2006, 1289^VP

Psychrobacter maritimus Romanenko et al. 2004a, 1744^VP

Psychrobacter namhaensis J.H. Yoon et al. 2005r, 1012^VP

Psychrobacter nivimaris Heuchert et al. 2004, 1909^VP – Valid publication: Validation List no. 100 – Effective publication: Heuchert et al. (2004)

Psychrobacter salsus Shivaji et al. 2005, 547^VP – Valid publication: Validation List no. 102 – Effective publication: Shivaji et al. (2004)

Psychrobacter vallis Shivaji et al. 2005c, 759^VP

Psychroflexus tropicus Donachie et al. 2004, 937^VP

Psychromonas hadalis Nogi et al. 2007, 1362^VP

Psychromonas ingrahamii Auman et al. 2006, 1006^VP

Psychroserpens mesophilus K.K. Kwon et al. 2006b, 1057^VP

Puniceicoccus vermicola Choo et al. 2007, 536^VP

Pusillimonas noertemannii Stolz et al. 2005, 1080^VP

Pyxidicoccus fallax corrig Reichenbach 2007, 894^VP – Valid publication: Validation List no. 115 – Effective publication: Reichenbach (2005c)

Quadrisphaera granulorum Maszenan et al. 2005b, 1774^VP

Rapidithrix thailandica Srisukchayakul et al. 2007, 2277^VP

Reinekea blandensis Pinhassi et al. 2007, 2374^VP

Reinekea marinisedimentorum Romanenko et al. 2004b, 672^VP

Rheinheimera aquimaris J.H. Yoon et al. 2007xxix, 1389^VP

Rheinheimera chironomi Halpern et al. 2007, 1875^VP

Rheinheimera perlucida Brettar et al. 2006, 2182^VP

Rheinheimera texasensis Merchant et al. 2007, 2379^VP

Rhizobium cellulosilyticum García-Fraile et al. 2007, 846^VP

Rhizobium daejeonense Quan et al. 2005, 2547^VP

Rhizobium lusitanum Valverde et al. 2006b, 2635^{VP}

Rhodanobacter fulvus Im et al. 2005, 547^{VP} – Valid publication: Validation List no. 102 – Effective publication: Im et al. (2004b)

Rhodanobacter ginsengisoli Weon et al. 2007a, 2812^{VP}

Rhodanobacter spathiphylli De Clercq et al. 2006, 1758^{VP}

Rhodanobacter terrae Weon et al. 2007a, 2812^{VP}

Rhodanobacter thiooxydans C.S. Lee et al. 2007, 1777^{VP}

Rhodobacter changlensis Anil Kumar et al. 2007d, 2570^{VP}

Rhodobacter massiliensis Greub and Raoult 2006, 2026^{VP} – Valid publication: Validation List no. 111 – Effective publication: Greub and Raoult (2003)

Rhodobacter vinaykumarii Srinivas et al. 2007d, 1985^{VP}

Rhodobium gokarnense Srinivas et al. 2007b, 934^{VP}

Rhodobium pfennigii Caumette et al. 2007, 1254^{VP}

Rhodoblastus sphagnicola Kulichevskaya et al. 2006, 1401^{VP}

Rhodococcus aetherivorans Goodfellow et al. 2004, 1005^{VP} – Valid publication: Validation List no. 98 – Effective publication: Goodfellow et al. (2004)

Rhodococcus baikonurensis Y. Li et al. 2004b, 833^{VP}

Rhodococcus gordoniae A.L. Jones et al. 2004, 409^{VP}

Rhodococcus imtechensis Ghosh et al. 2006, 1968^{VP}

Rhodococcus kroppenstedtii Mayilraj et al. 2006b, 981^{VP}

Rhodococcus kyotonensis B. Li et al. 2007b, 1957^{VP}

Rhodococcus phenolicus Rehfuss and Urban 2006, 499^{VP} – Valid publication: Validation List no. 108 – Effective publication: Rehfuss and Urban (2005b)

Rhodococcus qingshengii J.L. Xu et al. 2007, 2756^{VP}

Rhodococcus triatomae Yassin 2005, 1578^{VP}

Rhodococcus yunnanensis Y.Q. Zhang et al. 2005b, 1135^{VP}

Rhodonellum psychrophilum Schmidt et al. 2006, 2891^{VP}

Rhodopirellula baltica Schlesner et al. 2004, 1577^{VP}

Rhodovarius lipocyclicus Kämpfer et al. 2004, 1909^{VP} – Valid publication: Validation List no. 100 – Effective publication: Kämpfer et al. (2004b)

Rhodovulum imhoffii Srinivas et al. 2007a, 231^{VP}

Rhodovulum marinum Srinivas et al. 2006, 1655^{VP}

Rhodovulum visakhapatnamense Srinivas et al. 2007c, 1764^{VP}

Rickettsia asiatica Fujita et al. 2006, 2367^{VP}

Rickettsia heilongjiangensis Fournier et al. 2006, 499^{VP} – Valid publication: Validation List no. 108 – Effective publication: Fournier et al. (2003)

Rickettsia tamurae Fournier et al. 2006, 1674^{VP}

Robiginitalea biformata J.C. Cho and Giovannoni 2004a, 1105^{VP}

Robiginitomaculum antarcticum K. Lee et al. 2007b, 2598^{VP}

Roseburia faecis Duncan et al. 2006, 2440^{VP}

Roseburia hominis Duncan et al. 2006, 2440^{VP}

Roseburia inulinivorans Duncan et al. 2006, 2440^{VP}

Roseibacterium elongatum T. Suzuki et al. 2006, 420^{VP}

Roseicyclus mahoneyensis Rathgeber et al. 2005, 1602^{VP}

Roseisalinus antarcticus Labrenz et al. 2005, 46^{VP}

Roseivirga echinicomitans Nedashkovskaya et al. 2005k, 1799^{VP}

Roseivirga ehrenbergii Nedashkovskaya et al. 2005c, 233^{VP}

Roseivirga spongicola S.C. Lau et al. 2006b, 1063^{VP}

Roseomonas aquatica Gallego et al. 2006d, 2293^{VP}

Roseomonas lacus Jiang et al. 2006, 26^{VP}

Roseomonas terrae J.H. Yoon et al. 2007xxiv, 2486^{VP}

Roseospira goensis Kalyan Chakravarthy et al. 2007, 2456^{VP}

Roseospira visakhapatnamensis Kalyan Chakravarthy et al. 2007, 2454^{VP}

Roseovarius crassostreae Boettcher et al. 2005, 1535^{VP}

Roseovarius mucosus Biebl et al. 2005b, 2382^{VP}

Rothia aeria Y. Li et al. 2004b, 833^{VP}

Ruania albidiflava Q. Gu et al. 2007a, 812^{VP}

Rubellimicrobium thermophilum Denner et al. 2006, 1360^{VP}

Rubritalea marina Scheuermayer et al. 2006, 2123^{VP}

Rubritalea spongiae J. Yoon et al. 2007d, 2339^{VP}

Rubritalea squalenifaciens Kasai et al. 2007, 1633^{VP}

Rubritalea tangerina J. Yoon et al. 2007d, 2340^{VP}

Rubrivivax benzoatilyticus Ramana et al. 2006, 2162^{VP}

Rubrobacter taiwanensis M.Y. Chen et al. 2004, 1853^{VP}

Ruegeria mobilis Muramatsu et al. 2007, 1307^{VP}

Ruegeria pelagia K. Lee et al. 2007a, 1817^{VP}

Runella defluvii Lu et al. 2007b, 2602^{VP}

Runella limosa Ryu et al. 2006, 2759^{VP}

Saccharibacter floricola Jojima et al. 2004, 2267^{VP}

Saccharophagus degradans Ekborg et al. 2005, 1548^{VP}

Saccharothrix algeriensis Zitouni et al. 2004, 1380^{VP}

Saccharothrix xinjiangensis Hu et al. 2004, 2093^{VP}

Salegentibacter agarivorans Nedashkovskaya et al. 2006c, 884^{VP}

Salegentibacter catena Ying et al. 2007a, 221^{VP}

Salegentibacter flavus Ivanova et al. 2006a, 585^{VP}

Salegentibacter holothuriorum Nedashkovskaya et al. 2004g, 1109^{VP}

Salegentibacter mishustinae Nedashkovskaya et al. 2005j, 237^{VP}

Salegentibacter salarius J.H. Yoon et al. 2007iii, 2740^{VP}

Salicola marasensis Maturrano et al. 2006, 1690^{VP}

Salicola salis Kharroub et al. 2006a, 2650^{VP}

Salimicrobium luteum J.H. Yoon et al. 2007xxi, 2409^{VP}

Salinibacillus aidingensis Ren and Zhou 2005, 952^{VP}

Salinibacillus kushneri Ren and Zhou 2005, 952^{VP}

Salinicoccus jeotgali Aslam et al. 2007b, 637^{VP}

Salinicoccus kunmingensis Y.G. Chen et al. 2007, 2330^{VP}

Salinicoccus luteus Y.Q. Zhang et al. 2007c, 1903^{VP}

Salinicoccus salsiraiae França et al. 2007, 433^{VP} – Valid publication: Validation List no. 114 – Effective publication: França et al. (2006a)

Salinicoccus siamensis Pakdeeto et al. 2007, 2006^{VP}

Salinimonas chungwhensis Jeon et al. 2005c, 242^{VP}

Salinispora arenicola Maldonado et al. 2005, 1764^{VP}

Salinispora tropica Maldonado et al. 2005, 1764^VP

Salipiger mucosus corrig. Martínez-Cánovas et al. 2004d, 1739^VP

Salirhabdus euzebyi Albuquerque et al. 2007, 1569^VP

Salmonella subterranea Shelobolina et al. 2005, 547^VP – Valid publication: Validation List no. 102 – Effective publication: Shelobolina et al. (2004)

Salsuginibacillus kocurii Carrasco et al. 2007b, 2384^VP

Sandarakinorhabdus limnophila Gich and Overmann 2006, 853^VP

Sandarakinotalea sediminis Khan et al. 2006b, 962^VP

Sanguibacter marinus Huang et al. 2005a, 1756^VP

Schlegelella aquatica Y.J. Chou et al. 2006, 2796^VP

Schlesneria paludicola Kulichevskaya et al. 2007, 2686^VP

Sedimenticola selenatireducens Narasingarao and Häggblom 2006, 2507^VP – Valid publication: Validation List no. 112 – Effective publication: Narasingarao and Häggblom (2006)

Sediminibacter furfurosus Khan et al. 2007b, 267^VP

Sediminicola luteus Khan et al. 2006c, 843^VP

Sediminitomix flava Khan et al. 2007d, 1692^VP

Segetibacter koreensis D.S. An et al. 2007a, 1831^VP

Segniliparus rotundus Butler et al. 2005, 1622^VP

Segniliparus rugosus Butler et al. 2005, 1622^VP

Sejongia antarctica Yi et al. 2005b, 414^VP

Sejongia jeonii Yi et al. 2005b, 414^VP

Sejongia marina K. Lee et al. 2007c, 2919^VP

Serinicoccus marinus Yi et al. 2004c, 1587^VP

Serratia ureilytica Bhadra et al. 2005, 2157^VP

Shewanella abyssi Miyazaki et al. 2006, 1611^VP

Shewanella affinis Ivanova et al. 2004h, 1092^VP

Shewanella algidipiscicola Satomi et al. 2007, 351^VP

Shewanella aquimarina J.H. Yoon et al. 2004n, 2351^VP

Shewanella atlantica Zhao et al. 2007, 2159^VP

Shewanella canadensis Zhao et al. 2007, 2159^VP

Shewanella decoloratinonis M. Xu et al. 2005, 366^VP

Shewanella donghaensis S.H. Yang et al. 2007, 210^VP

Shewanella gaetbuli J.H. Yoon et al. 2004c, 490^VP

Shewanella glacialipiscicola Satomi et al. 2007, 351^VP

Shewanella hafniensis Satomi et al. 2006b, 247^VP

Shewanella halifaxensis Zhao et al. 2006, 209^VP

Shewanella haliotis D. Kim et al. 2007, 2928^VP

Shewanella irciniae O.O. Lee et al. 2006b, 2873^VP

Shewanella kaireitica Miyazaki et al. 2006, 1610^VP

Shewanella loihica Gao et al. 2006, 1914^VP

Shewanella marisflavi J.H. Yoon et al. 2004n, 2351^VP

Shewanella morhuae Satomi et al. 2006b, 247^VP

Shewanella pacifica Ivanova et al. 2004e, 1085^VP

Shewanella piezotolerans Xiao et al. 2007, 64^VP

Shewanella pneumatophori Hirota et al. 2005, 2358^VP

Shewanella profunda Toffin et al. 2004, 1947^VP

Shewanella psychrophila Xiao et al. 2007, 64^VP

Shewanella sediminis Zhao et al. 2005, 1517^VP

Shewanella spongiae S.H. Yang et al. 2006, 2881^VP

Shewanella surugensis Miyazaki et al. 2006, 1612^VP

Shimazuella kribbensis D.J. Park et al. 2007, 2663^VP

Shimia marina Choi and Cho 2006, 1872^VP

Shinella granuli D.S.An et al. 2006a, 446^VP

Shinella zoogloeoides D.S. An et al. 2006a, 447^VP

Silanimonas lenta E.M. Lee et al. 2005, 387^VP

Silvimonas terrae H.C. Yang et al. 2005, 2332^VP

Simplicispira limi Lu et al. 2007a, 33^VP

Sinococcus qinghaiensis W.J. Li et al. 2006d, 1191^VP

   The name *Sinococcus qinghaiensis* W.J. Li et al. 2006d is illegitimate because it is placed in an illegitimate genus (see above *Sinococcus* W.J. Li et al. 2006d, 1191^VP)

Sinorhizobium americanum corrig. Toledo et al. 2004, 1909^VP – Valid publication: Validation List no. 100 – Effective publication: Toledo et al. (2003)

Skermanella aerolata Weon et al. 2007b, 1541^VP

Slackia faecicanis Lawson et al. 2005c, 1245^VP

Smaragdicoccus niigatensis Adachi et al. 2007, 300^VP

Sneathiella chinensis Jordan et al. 2007, 119^VP

Solimonas soli M.K. Kim et al. 2007a, 2593^VP

Solirubrobacter soli M.K. Kim et al. 2007c, 1454^VP

Sphaerisporangium cinnabarinum corrig. Ara and Kudo 2007, 2449^VP – Valid publication: Validation List no. 118 – Effective publication: Ara and Kudo (2007b)

Sphaerisporangium melleum corrig. Ara and Kudo 2007, 2449^VP – Valid publication: Validation List no. 118 – Effective publication: Ara and Kudo (2007b)

Sphaerisporangium rubeum corrig. Ara and Kudo 2007, 2449^VP – Valid publication: Validation List no. 118 – Effective publication: Ara and Kudo (2007b)

Sphingobacterium composti Ten et al. 2007, 1372^VP – Valid publication: Validation List no. 116 – Effective publication: Ten et al. (2006d)

Sphingobacterium composti Yoo et al. 2007b, 1592^VP

Sphingobacterium composti Yoo et al. 2007b is a later homonym of *Sphingobacterium composti* Ten et al. 2007

Sphingobacterium daejeonense K.H. Kim et al. 2006, 2035^VP

Sphingobium aromaticiconvertens Wittich et al. 2007b, 308^VP

Sphingobium francense Pal et al. 2005, 1971^VP

Sphingobium fuliginis Prakash and Lal 2006, 2150^VP

Sphingobium indicum Pal et al. 2005, 1970^VP

Sphingobium japonicum Pal et al. 2005, 1971^VP

Sphingobium olei C.C. Young et al. 2007b, 2615^VP

Sphingomonas abaci Busse et al. 2005, 2568^VP

Sphingomonas azotifigens Xie and Yokota 2006b, 892^VP

Sphingomonas desiccabilis Reddy and Garcia-Pichel 2007, 1032^VP

Sphingomonas dokdonensis J.H. Yoon et al. 2006p, 2167^VP

Sphingomonas fennica Wittich et al. 2007a, 1744^VP

Sphingomonas haloaromaticamans Wittich et al. 2007a, 1744^VP

Sphingomonas jaspsi Asker et al. 2007b, 1440^VP

Sphingomonas kaistensis M.K. Kim et al. 2007d, 1533^VP

Sphingomonas molluscorum Romanenko et al. 2007b, 361^VP

Sphingomonas mucosissima Reddy and Garcia-Pichel 2007, 1031^VP

Sphingomonas oligophenolica Ohta et al. 2004, 2188^VP

Sphingomonas panni Busse et al. 2005, 2568^VP

Sphingomonas phyllosphaerae Rivas et al. 2004a, 2148^VP

Sphingomonas pseudosanguinis Kämpfer et al. 2007c, 1344^VP

Sphingomonas soli D.C. Yang et al. 2006, 706^VP

Sphingomonas yabuuchiae Y. Li et al. 2004a, 824^VP

Sphingomonas yunnanensis Y.Q. Zhang et al. 2005a, 2363^VP

Sphingopyxis baekryungensis J.H. Yoon et al. 2005s, 1226^VP

Sphingopyxis flavimaris J.H. Yoon and Oh 2005a, 371^VP

Sphingosinicella microcystinivorans Maruyama et al. 2006, 88^VP

Sphingosinicella xenopeptidilytica Geueke et al. 2007, 111^VP

Spirochaeta bajacaliforniensis Fracek and Stolz 2004, 631^VP – Valid publication: Validation List no. 97 – Effective publication: Fracek and Stolz (1985)

Spirochaeta coccoides Dröge et al. 2006, 1460^VP – Valid publication: Validation List no. 110 – Effective publication: Dröge et al. (2006)

Spiroplasma atrichopogonis Koerber et al. 2005, 291^VP

Spiroplasma leucomae Oduori et al. 2005, 2449^VP

Spiroplasma penaei Nunan et al. 2005, 2320^VP

Spirosoma rigui Baik et al. 2007c, 2872^VP

Sporacetigenium mesophilum S. Chen et al. 2006, 724^VP

Sporomusa rhizae Gößner et al. 2006, 1460^VP – Valid publication: Validation List no. 110 – Effective publication: Gößner et al. (2006)

Sporosarcina koreensis S.W. Kwon et al. 2007b, 1697^VP

Sporosarcina saromensis S.Y. An et al. 2007c, 1870^VP

Sporosarcina soli S.W. Kwon et al. 2007b, 1697^VP

Sporotalea propionica Boga et al. 2007, 894^VP – Valid publication: Validation List no. 115 – Effective publication: Boga et al. (2007)

Stackebrandtia nassauensis Labeda and Kroppenstedt 2005, 1690^VP

Staphylococcus pettenkoferi Trülzsch et al. 2007, 1547^VP

Staphylococcus pseudintermedius Devriese et al. 2005, 1571^VP

Staphylococcus simiae Pantůček et al. 2005, 1957^VP

Stappia alba Pujalte et al. 2006, 3^VP – Valid publication: Validation List no. 107 – Effective publication: Pujalte et al. (2005a)

Stappia marina B.C. Kim et al. 2006, 78^VP

Starkeya koreensis Im et al. 2006a, 2412^VP

Stenothermobacter spongiae S.C. Lau et al. 2006a, 184^VP

Stenotrophomonas dokdonensis J.H. Yoon et al. 2006f, 1366^VP

Stenotrophomonas humi Heylen et al. 2007, 2060^VP

Stenotrophomonas koreensis H.C. Yang et al. 2006a, 83^VP

Stenotrophomonas terrae Heylen et al. 2007, 2059^VP

Stigmatella hybrida Reichenbach 2007, 894^VP – Valid publication: Validation List no. 115 – Effective publication: Reichenbach (2005f)

Streptacidiphilus jiangxiensis Huang et al. 2005, 1744^VP – Valid publication: Validation List no. 105 – Effective publication: Huang et al. (2004a)

Streptacidiphilus oryzae L. Wang et al. 2006, 1260^VP

Streptococcus castoreus Lawson et al. 2005b, 845^VP

Streptococcus devriesei Collins et al. 2004, 631^VP – Valid publication: Validation List no. 97 – Effective publication: Collins et al. (2004c)

Streptococcus halichoeri Lawson et al. 2004a, 1756^VP

Streptococcus ictaluri Shewmaker et al. 2007, 1606^VP

Streptococcus marimammalium Lawson et al. 2005a, 272^VP

Streptococcus massiliensis Glazunova et al. 2006a, 1130^VP

Streptococcus minor Vancanneyt et al. 2004a, 451^VP

Streptococcus orisuis Takada and Hirasawa 2007, 1274^VP

Streptococcus pseudopneumoniae Arbique et al. 2005, 1^VP – Valid publication: Validation List no. 101 – Effective publication: Arbique et al. (2004)

Streptococcus pseudoporcinus Bekal et al. 2007, 894^VP – Valid publication: Validation List no. 115 – Effective publication: Bekal et al. (2006)

Streptomyces africanus Meyers et al. 2004, 1534^VP

Streptomyces bangladeshensis Al-Bari et al. 2005, 1976^VP

Streptomyces cheonanensis H.J. Kim et al. 2006, 474^VP

Streptomyces drozdowiczii Semêdo et al. 2004, 1327^VP

Streptomyces durmitorensis Savic et al. 2007, 2121^VP

Streptomyces emeiensis Sun et al. 2007, 1637^VP

Streptomyces ferralitis Saintpierre-Bonaccio et al. 2004a, 2063^VP

Streptomyces glauciniger Huang et al. 2004c, 2087^VP

Streptomyces guanduensis C. Xu et al. 2006, 1113^VP

Streptomyces hainanensis Jiang et al. 2007, 2697^VP

Streptomyces hebeiensis P. Xu et al. 2004a, 729^VP

Streptomyces jietaisiensis He et al. 2005, 1941^VP

Streptomyces koyangensis J.Y. Lee et al. 2005, 260^VP

Streptomyces paucisporeus C. Xu et al. 2006, 1113^VP

Streptomyces pharetrae le Roes and Meyers 2005, 2236^VP – Valid publication: Validation List no. 106 – Effective publication: le Roes and Meyers (2005)

Streptomyces radiopugnans Mao et al. 2007, 2581^VP

Streptomyces rubidus C. Xu et al. 2006, 1113^VP

Streptomyces scabrisporus P. Xu et al. 2004b, 580^VP

Streptomyces sodiiphilus W.J. Li et al. 2005f, 1332^VP

Streptomyces synnematoformans Hozzein and Goodfellow 2007, 2012^VP

Streptomyces vietnamensis Zhu et al. 2007, 1773^VP

Streptomyces yanglinensis C. Xu et al. 2006, 1114^VP

Streptomyces yanii Z. Liu et al. 2005b, 1608^VP

Streptomyces yeochonensis S.B. Kim et al. 2004b, 213^VP

Streptosporangium purpuratum L.P. Zhang et al. 2005, 723^VP

Streptosporangium yunnanense L.P. Zhang et al. 2005, 723^VP

Subdoligranulum variabile Holmstrøm et al. 2004, 1909^VP – Valid publication: Validation List no. 100 – Effective publication: Holmstrøm et al. (2004)

Subsaxibacter broadyi Bowman and Nichols 2005, 1481^VP

Subsaximicrobium saxinquilinus Bowman and Nichols 2005, 1481^VP

Subsaximicrobium wynnwilliamsii Bowman and Nichols 2005, 1481^VP

Sulfitobacter delicatus Ivanova et al. 2004f, 478^VP

Sulfitobacter donghicola J.H. Yoon et al. 2007x, 1791^VP

Sulfitobacter dubius Ivanova et al. 2004f, 479^VP

Sulfitobacter litoralis J.R. Park et al. 2007, 694^VP

Sulfitobacter marinus J.H. Yoon et al. 2007xiv, 304^VP

Sulfobacillus sibiricus Melamud et al. 2006, 499^VP – Valid publication: Validation List no. 108 – Effective publication: Melamud et al. (2003)

Sulfobacillus thermotolerans Bogdanova et al. 2006, 1041^VP

Sulfuricurvum kujiense Kodama and Watanabe 2004, 2299^VP

Sulfurihydrogenibium azorense Aguiar et al. 2004, 37^VP

Sulfurihydrogenibium yellowstonense S. Nakagawa et al. 2005b, 2267^VP

Sulfurimonas paralvinellae Takai et al. 2006b, 1731^VP

Sulfurivirga caldicuralii Takai et al. 2006a, 1927^VP

Sulfurospirillum cavolei Kodama et al. 2007, 829^VP

Sulfurovum lithotrophicum Inagaki et al. 2004, 1481^VP

Sutterella stercoricanis Greetham et al. 2004a, 1583^VP

Suttonella ornithocola Foster et al. 2005, 2271^VP

Swaminathania salitolerans Loganathan and Nair 2004, 1189^VP

Syntrophobacter sulfatireducens S. Chen et al. 2005, 1323^VP

Syntrophomonas cellicola C. Wu et al. 2006, 2334^VP

Syntrophomonas curvata C. Zhang et al. 2004, 972^VP

Syntrophomonas erecta C. Zhang et al. 2005, 802^VP

Syntrophomonas palmitatica Hatamoto et al. 2007, 2141^VP

Syntrophomonas zehnderi Sousa et al. 2007, 613^VP

Tamlana crocina S.D. Lee 2007, 767^VP

Telmatospirillum siberiense Sizova et al. 2007, 1372^VP – Valid publication: Validation List no. 116 – Effective publication: Sizova et al. (2007)

Tenacibaculum aestuarii Jung et al. 2006c, 1580^VP

Tenacibaculum litopenaei Sheu et al. 2007, 1150^VP

Tenacibaculum litoreum Choi et al. 2006, 639^VP

Tenacibaculum lutimaris J.H. Yoon et al. 2005g, 797^VP

Tenacibaculum skagerrakense Frette et al. 2004, 523^VP

Tenuibacillus multivorans Ren and Zhou 2005, 98^VP

Tepidanaerobacter syntrophicus Sekiguchi et al. 2006, 1627^VP

Tepidibacter formicigenes Urios et al. 2004a, 442^VP

Tepidicella xavieri França et al. 2006b, 911^VP

Tepidimicrobium ferriphilum Slobodkin et al. 2006b, 371^VP

Tepidimonas taiwanensis T.L. Chen et al. 2006, 1460^VP – Valid publication: Validation List no. 110 – Effective publication: T.L. Chen et al. (2006)

Tepidimonas thermarum Albuquerque et al. 2007, 1^VP – Valid publication: Validation List no. 113 – Effective publication: Albuquerque et al. (2006)

Terrabacter aerolatus Weon et al. 2007i, 2108^VP

Terrabacter terrae Montero-Barrientos et al. 2005, 2493^VP

Terribacillus halophilus S.Y. An et al. 2007a, 54^VP

Terribacillus saccharophilus S.Y. An et al. 2007a, 54^VP

Terriglobus roseus Eichorst et al. 2007, 1933^VP – Valid publication: Validation List no. 117 – Effective publication: Eichorst et al. (2007)

Terrimonas lutea Xie and Yokota 2006c, 1120^VP

Tetragenococcus koreensis M. Lee et al. 2005, 1412^VP

Tetrasphaera jenkinsii McKenzie et al. 2006, 2288^VP

Tetrasphaera remsis Osman et al. 2007, 2752^VP

Tetrasphaera vanveenii McKenzie et al. 2006, 2288^VP

Tetrasphaera veronensis McKenzie et al. 2006, 2288^VP

Tetrathiobacter kashmirensis Ghosh et al. 2005, 1786^VP

Tetrathiobacter mimigardefordensis Wübbeler et al. 2006, 1308^VP

Thalassobacillus devorans García et al. 2005, 1793^VP

Thalassobacter stenotrophicus Macián et al. 2005a, 109^VP

Thalassobius aestuarii Yi and Chun 2007, 894^VP – Valid publication: Validation List no. 115 – Effective publication: Yi and Chun (2006a)

Thalassobius mediterraneus Arahal et al. 2005, 2374^VP

Thalassococcus halodurans O.O. Lee et al. 2007, 1923^VP

Thalassolituus oleivorans Yakimov et al. 2004, 146^VP

Thalassomonas agarivorans Jean et al. 2006c, 1249^VP

Thalassomonas ganghwensis Yi et al. 2004a, 379^VP

Thalassomonas loyana Thompson et al. 2006, 366^VP

Thalassospira profundimaris C. Liu et al. 2007, 318^VP

Thalassospira xiamenensis C. Liu et al. 2007, 318^VP

Thermaerobacter litoralis Tanaka et al. 2006, 1533^VP

Thermincola carboxydiphila Sokolova et al. 2005, 2072^VP

Thermincola ferriacetica Zavarzina et al. 2007, 894^VP – Valid publication: Validation List no. 115 – Effective publication: Zavarzina et al. (2007)

Thermoanaerobacter pseudethanolicus Onyenwoke et al. 2007, 2192^VP

Thermoanaerobacter sulfurigignens Y.J. Lee et al. 2007a, 1433^VP

Thermoanaerobacterium aciditolerans Kublanov et al. 2007, 263^VP

Thermobacillus composti Watanabe et al. 2007, 1476^VP

Thermococcus barossii Duffaud et al. 2005, 548^VP – Valid publication: Validation List no. 102 – Effective publication: Duffaud et al. (1998)

Thermococcus celericrescens Kuwabara et al. 2007, 442^VP

Thermococcus coalescens Kuwabara et al. 2005, 2512[VP]

Thermococcus kodakarensis corrig. Atomi et al. 2005, 984[VP] – Valid publication: Validation List no. 103 – Effective publication: Atomi et al. (2004)

Thermococcus thioreducens Pikuta et al. 2007, 1617[VP]

Thermodesulfatator indicus Moussard et al. 2004, 232[VP]

Thermodesulfobium narugense Mori et al. 2004, 1[VP] – Valid publication: Validation List no. 95 – Effective publication: Mori et al. (2003)

Thermogymnomonas acidicola Itoh et al. 2007, 2561[VP]

Thermolithobacter carboxydivorans Sokolova et al. 2007, 1372[VP] – Valid publication: Validation List no. 116 – Effective publication: Sokolova et al. (2007)

Thermolithobacter ferrireducens Sokolova et al. 2007, 1372[VP] – Valid publication: Validation List no. 116 – Effective publication: Sokolova et al. (2007)

Thermomonas koreensis M.K. Kim et al. 2006, 1618[VP]

Thermosediminibacter litoriperuensis Lee et al. 2006, 925[VP] – Valid publication: Validation List no. 109 – Effective publication: Y.J. Lee et al. (2005)

Thermosediminibacter oceani Lee et al. 2006, 925[VP] – Valid publication: Validation List no. 109 – Effective publication: Y.J. Lee et al. (2005)

Thermosinus carboxydivorans Sokolova et al. 2004, 2358[VP]

Thermosipho atlanticus Urios et al. 2004b, 1956[VP]

Thermovibrio ammonificans Vetriani et al. 2004, 180[VP]

Thermovibrio guaymasensis L'Haridon et al. 2006, 2850[VP]

Thermovirga lienii Dahle and Birkeland 2006, 1544[VP]

Thioalkalimicrobium microaerophilum Sorokin et al. 2007, 894[VP] – Valid publication: Validation List no. 115 – Effective publication: Sorokin et al. (2007a)

Thioalkalivibrio halophilus corrig. Banciu et al. 2005, 548[VP] – Valid publication: Validation List no. 102 – Effective publication: Banciu et al. (2004)

Thioalkalivibrio thiocyanodenitrificans corrig. Sorokin et al. 2005, 1396[VP] – Valid publication: Validation List no. 104 – Effective publication: Sorokin et al. (2004)

Thiobacter subterraneus Hirayama et al. 2005, 471[VP]

Thiocapsa marina Caumette et al. 2004, 1035[VP]

Thioclava pacifica Sorokin et al. 2005, 1074[VP]

Thiohalomonas denitrificans Sorokin et al. 2007c, 1587[VP]

Thiohalomonas nitratireducens Sorokin et al. 2007c, 1588[VP]

Thiohalophilus thiocyanatoxydans corrig. Sorokin et al. 2007, 1933[VP] – Valid publication: Validation List no. 117 – Effective publication: Sorokin et al. (2007b)

Thiomicrospira arctica Knittel et al. 2005, 784[VP]

Thiomicrospira halophila Sorokin et al. 2006b, 2379[VP]

Thiomicrospira psychrophila Knittel et al. 2005, 785[VP]

Thiomicrospira thermophila Takai et al. 2004a, 2331[VP]

Thioreductor micantisoli S. Nakagawa et al. 2005a, 603[VP]

Thiorhodococcus bheemlicus Anil Kumar et al. 2007a, 2460[VP]

Thiorhodococcus kakinadensis Anil Kumar et al. 2007a, 2461[VP]

Thiorhodococcus mannitoliphagus Rabold et al. 2006, 1949[VP]

Thiovirga sulfuroxydans Ito et al. 2005, 1063[VP]

Thorsellia anophelis Kämpfer et al. 2006f, 337[VP]

Trabulsiella odontotermitis J.H. Chou et al. 2007a, 699[VP]

Treponema azotonutricium Graber et al. 2004, 631[VP] – Valid publication: Validation List no. 97 – Effective publication: Graber et al. (2004)

Treponema berlinense Nordhoff et al. 2005, 1678[VP]

Treponema porcinum Nordhoff et al. 2005, 1678[VP]

Treponema primitia Graber et al. 2004, 631[VP] – Valid publication: Validation List no. 97 – Effective publication: Graber et al. (2004)

Treponema putidum Wyss et al. 2004, 1121[VP]

Trichococcus patagoniensis Pikuta et al. 2006a, 2060[VP]

Truepera radiovictrix Albuquerque et al. 2005 (complete authorship reads Albuquerque, da Costa and Rainey), 1744[VP] – Valid publication: Validation List no. 105 – Effective publication: Albuquerque et al. (2005)

Tsukamurella pseudospumae S.W. Nam et al. 2004, 1211[VP]

Tsukamurella spongiae Olson et al. 2007, 1480[VP]

Tuberibacillus calidus Hatayama et al. 2006, 2549[VP]

Ulvibacter antarcticus Choi et al. 2007, 2923[VP]

Ulvibacter litoralis Nedashkovskaya et al. 2004c, 121[VP]

Undibacterium pigrum Kämpfer et al. 2007d, 1514[VP]

Ureibacillus composti Weon et al. 2007h, 2910[VP]

Ureibacillus suwonensis B.Y. Kim et al. 2006a, 665[VP]

Ureibacillus thermophilus Weon et al. 2007h, 2911[VP]

Uruburuella suis Vela et al. 2005, 646[VP]

Vagococcus carniphilus Shewmaker et al. 2004, 1508[VP]

Vagococcus elongatus Lawson et al. 2007, 753[VP]

Variovorax dokdonensis J.H. Yoon et al. 2006h, 813[VP]

Variovorax soli B.Y. Kim et al. 2006g, 2901[VP]

Veillonella denticariosi Byun et al. 2007, 2847[VP]

Veillonella montpellierensis Jumas-Bilak et al. 2004, 1315[VP]

Vibrio comitans Sawabe et al. 2007, 920[VP]

Vibrio crassostreae Faury et al. 2004, 2139[VP]

Vibrio ezurae Sawabe et al. 2005, 1[VP] – Valid publication: Validation List no. 101 – Effective publication: Sawabe et al. (2004a)

Vibrio gallicus Sawabe et al. 2004b, 845[VP]

Vibrio gigantis Le Roux et al. 2005, 2254[VP]

Vibrio hispanicus Gomez-Gil et al. 2004, 263[VP]

Vibrio inusitatus Sawabe et al. 2007, 921[VP]

Vibrio litoralis Y.D. Nam et al. 2007a, 564[VP]

Vibrio neonatus Sawabe et al. 2005, 1[VP] – Valid publication: Validation List no. 101 – Effective publication: Sawabe et al. (2004a)

Vibrio ponticus Macián et al. 2005, 1[VP] – Valid publication: Validation List no. 101 – Effective publication: Macián et al. (2004)

Vibrio rarus Sawabe et al. 2007, 920[VP]

Vibrio rhizosphaerae Ramesh Kumar and Nair 2007, 2245[VP]

Virgibacillus dokdonensis J.H. Yoon et al. 2005d, 1836[VP]

Virgibacillus halophilus S.Y. An et al. 2007b, 1609[VP]

Virgibacillus koreensis J.S. Lee et al. 2006b, 254[VP]

Virgibacillus olivae Quesada et al. 2007, 908[VP]

Volucribacter amazonae Christensen et al. 2004, 817[VP]

Volucribacter psittacicida Christensen et al. 2004, 817[VP]

Vulcanibacillus modesticaldus L'Haridon et al. 2006, 1050[VP]

Wautersia numazuensis corrig. Kageyama et al. 2005, 1396[VP] – Valid publication: Validation List no. 104 – Effective publication: C. Kageyama et al. (2005)

Wautersiella falsenii Kämpfer et al. 2006a, 2328[VP]

Wenxinia marina Ying et al. 2007b, 1714[VP]

Williamsia deligens Yassin and Hupfer 2006, 196[VP]

Williamsia marianensis Pathom-aree et al. 2006a, 1125[VP]

Williamsia maris Stach et al. 2004, 193[VP]

Williamsia serinedens Yassin et al. 2007c, 560[VP]

Winogradskyella epiphytica Nedashkovskaya et al. 2005a, 53[VP]

Winogradskyella eximia Nedashkovskaya et al. 2005a, 54[VP]

Winogradskyella poriferorum S.C. Lau et al. 2005b, 1591[VP]

Winogradskyella thalassocola Nedashkovskaya et al. 2005a, 52[VP]

Woodsholea maritima Abraham et al. 2004, 1233[VP]

Xanthomonas euvesicatoria Jones et al. 2006, 926[VP] – Valid publication: Validation List no. 109 – Effective publication: J.B. Jones et al. (2004)

Xanthomonas fuscans Schaad et al. 2007, 895[VP] – Valid publication: Validation List no. 115 – Effective publication: Schaad et al. (2006)

Xanthomonas perforans Jones et al. 2006, 926[VP] – Valid publication: Validation List no. 109 – Effective publication: J.B. Jones et al. (2004)

Xenorhabdus budapestensis Lengyel et al. 2005, 1396[VP] – Valid publication: Validation List no. 104 – Effective publication: Lengyel et al. (2005)

Xenorhabdus cabanillasii Tailliez et al. 2006, 2815[VP]

Xenorhabdus doucetiae Tailliez et al. 2006, 2815[VP]

Xenorhabdus ehlersii Lengyel et al. 2005, 1396[VP] – Valid publication: Validation List no. 104 – Effective publication: Lengyel et al. (2005)

Xenorhabdus griffiniae Tailliez et al. 2006, 2815[VP]

Xenorhabdus hominickii Tailliez et al. 2006, 2816[VP]

Xenorhabdus innexi Lengyel et al. 2005, 1396[VP] – Valid publication: Validation List no. 104 – Effective publication: Lengyel et al. (2005)

Xenorhabdus koppenhoeferi Tailliez et al. 2006, 2816[VP]

Xenorhabdus kozodoii Tailliez et al. 2006, 2816[VP]

Xenorhabdus mauleonii Tailliez et al. 2006, 2816[VP]

Xenorhabdus miraniensis Tailliez et al. 2006, 2816[VP]

Xenorhabdus romanii Tailliez et al. 2006, 2816[VP]

Xenorhabdus stockiae Tailliez et al. 2006, 2817[VP]

Xenorhabdus szentirmaii Lengyel et al. 2005, 1396[VP] – Valid publication: Validation List no. 104 – Effective publication: Lengyel et al. (2005)

Xylanibacter oryzae Ueki et al. 2006b, 2220[VP]

Xylanibacterium ulmi Rivas et al. 2004e, 560[VP]

Yangia pacifica Dai et al. 2006, 531[VP]

Yania flava W.J. Li et al. 2005e, 1937[VP]

The name *Yania flava* W.J. Li et al. 2005e is illegitimate because the genus name is illegitimate (see above *Yania* W.J. Li et al. 2004a, 529[VP]).

Yania halotolerans W.J. Li et al. 2004a, 530[VP]

The name *Yania halotolerans* W.J. Li et al. 2004a is illegitimate because the genus name is illegitimate (see above *Yania* W.J. Li et al. 2004a, 529[VP]).

Yeosuana aromativorans K.K. Kwon et al. 2006a, 731[VP]

Yersinia aleksiciae Sprague and Neubauer 2005, 834[VP]

Yonghaparkia alkaliphila J.H. Yoon et al. 2006k, 2418[VP]

Zeaxanthinibacter enoshimensis Asker et al. 2007a, 841[VP]

Zhihengliuella halotolerans Y.Q. Zhang et al. 2007a, 1020[VP]

Zhouia amylolytica Z.P. Liu et al. 2006a, 2827[VP]

Zimmermannella alba Y.C. Lin et al. 2004, 1675[VP]

According to Rule 37a(1), *Zimmermannella alba* Y.C. Lin et al. 2004 cannot be maintained in the genus *Zimmermannella*.

Zimmermannella bifida Y.C. Lin et al. 2004, 1675[VP]

According to Rule 37a (1), *Zimmermannella bifida* Y.C. Lin et al. 2004 cannot be maintained in the genus *Zimmermannella*.

Zimmermannella faecalis Y.C. Lin et al. 2004, 1675[VP]

According to Rule 37a (1), *Zimmermannella faecalis* Y.C. Lin et al. 2004 cannot be maintained in the genus *Zimmermannella*.

Zimmermannella helvola Lin et al. 2004, 1674[VP]

Strict application of the Rule 51b(1) indicates that the genus name *Zimmermannella helvola* Y.C. Lin et al. 2004 is illegitimate.

Zobellella denitrificans Y.T. Lin and Shieh 2006, 1214[VP]

Zobellella taiwanensis Y.T. Lin and Shieh 2006, 1214[VP]

Zobellia amurskyensis Nedashkovskaya et al. 2004f, 1647[VP]

Zobellia laminariae Nedashkovskaya et al. 2004f, 1647[VP]

Zobellia russellii Nedashkovskaya et al. 2004f, 1647[VP]

Zoogloea oryzae Xie and Yokota 2006a, 622[VP]

Zunongwangia profunda corrig. Qin et al. 2007, 1372[VP] – Valid publication: Validation List no. 116 – Effective publication: Qin et al. (2007)

New subspecies

Alcaligenes faecalis subsp. *phenolicus* Rehfuss and Urban 2006, 1[VP] – Valid publication: Validation List no. 107 – Effective publication: Rehfuss and Urban (2005a)

Bifidobacterium animalis subsp. *animalis* (Mitsuoka 1969) Scardovi and Trovatelli 1974, 1142[VP] – Reference: Masco et al. (2004)

Caldanaerobacter subterraneus subsp. *pacificus* Fardeau et al. 2004, 472[VP]

Caldanaerobacter subterraneus subsp. *subterraneus* (Fardeau et al. 2000) Fardeau et al. 2004, 471^VP

Enterobacter cloacae subsp. *cloacae* (Jordan 1890) Hormaeche and Edwards 1960, 1395^VP – Valid publication: Validation List no. 104 – Effective publication: Hoffmann et al. (2005a)

Francisella philomiragia subsp. *noatunensis* Mikalsen et al. 2007, 1964^VP

Francisella philomiragia subsp. *philomiragia* (Jensen et al. 1969) Hollis et al. 1990, 1964^VP

Lactobacillus delbrueckii subsp. *indicus* Dellaglio et al. 2005b, 403^VP

Lactobacillus kefiranofaciens subsp. *kefiranofaciens* Fujisawa et al. 1988, 555^VP – Reference: Vancanneyt et al. (2004b)

Lactobacillus plantarum subsp. *argentoratensis* Bringel et al. 2005, 1633^VP

Lactobacillus plantarum subsp. *plantarum* (Orla-Jensen 1919) Bergey et al. 1923, 1633^VP – Reference: Bringel et al. (2005)

Leucobacter chromiireducens subsp. *chromiireducens* Morais et al. 2005, 2774^VP

Leucobacter chromiireducens subsp. *solipictus* Muir and Tan 2007, 2774^VP

Oceanobacillus oncorhynchi subsp. *incaldanensis* Romano et al. 2006b, 808^VP

Oceanobacillus oncorhynchi subsp. *oncorhynchi* Yumoto et al. 2005, 809^VP – Reference: Romano et al. (2006b)

Photorhabdus asymbiotica subsp. *asymbiotica* Fischer-Le Saux et al. 1999, 1309^VP – Reference: Akhurst et al. (2004)

Photorhabdus asymbiotica subsp. *australis* Akhurst et al. 2004, 1309^VP

Photorhabdus luminescens subsp. *kayaii* Hazir et al. 2004, 1005^VP – Valid publication: Validation List no. 98 – Effective publication: Hazir et al. (2004)

Photorhabdus luminescens subsp. *thracensis* Hazir et al. 2004, 1005^VP – Valid publication: Validation List no. 98 – Effective publication: Hazir et al. (2004)

Pseudomonas chlororaphis subsp. *chlororaphis* (Guignard and Sauvageau 1894) Bergey et al. 1930, 1289^VP

Salinivibrio costicola subsp. *alcaliphilus* Romano et al. 2005, 984^VP – Valid publication: Validation List no. 103 – Effective publication: Romano et al. (2005a)

Salmonella enterica subsp. *enterica* (*ex* Kauffmann and Edwards 1952) Le Minor and Popoff 1987, 519^VP – Reference: Judicial Commission of the International Committee on Systematics of Prokaryotes 2005b (Opinion 80)

Streptococcus equi subsp. *ruminatorum* Fernández et al. 2004, 2295^VP

Thermoanaerobacter mathranii subsp. *alimentarius* Carlier et al. 2007, 1^VP – Valid publication: Validation List no. 113 – Effective publication: Carlier et al. (2006)

Thermoanaerobacter mathranii subsp. *mathranii* Larsen et al. 1998, 1^VP – Valid publication: Validation List no. 113 – Effective publication: Carlier et al. (2006)

Xanthomonas alfalfae subsp. *alfalfae* (*ex* Riker et al. 1935) Schaad et al. 2007, 895^VP – Valid publication: Validation List no. 115 – Effective publication: Schaad et al. (2006)

Xanthomonas alfalfae subsp. *citrumelonis* Schaad et al. 2007, 895^VP – Valid publication: Validation List no. 115 – Effective publication: Schaad et al. (2006)

Xanthomonas citri subsp. *citri* (*ex* Hasse 1915) Gabriel et al. 1989, 895^VP – Valid publication: Validation List no. 115 – Effective publication: Schaad et al. (2006)

Xanthomonas fuscans subsp. *aurantifolii* Schaad et al. 2007, 895^VP – Valid publication: Validation List no. 115 – Effective publication: Schaad et al. (2006)

Xanthomonas fuscans subsp. *fuscans* Schaad et al. 2007, 895^VP – Valid publication: Validation List no. 115 – Effective publication: Schaad et al. (2006)

Zymomonas mobilis subsp. *francensis* Coton et al. 2006, 125^VP

New combinations

Acholeplasma pleciae (Tully et al. 1994) Knight Jr. 2004, 1952^VP – Basonym: *Mesoplasma pleciae* Tully et al. 1994

Actinotalea fermentans (Bagnara et al. 1985) Yi et al. 2007b, 155^VP – Basonym: *Cellulomonas fermentans* Bagnara et al. 1985

Aerococcus urinaeequi (Garvie 1988) Felis et al. 2005, 1327^VP – Basonym: *Pediococcus urinaeequi* (*ex* Mees 1934) Garvie 1988

Aggregatibacter actinomycetemcomitans (Klinger 1912) Nørskov-Lauritsen and Kilian 2006, 2143^VP – Basonym: *Actinobacillus actinomycetemcomitans* (Klinger 1912) Topley and Wilson 1929 (Approved Lists 1980)

Aggregatibacter aphrophilus (Khairat 1940) Nørskov-Lauritsen and Kilian 2006, 2143^VP – Basonym: *Hemophilus aphrophilus* Khairat 1940 (Approved Lists 1980)

Aggregatibacter segnis (Kilian 1977) Nørskov-Lauritsen and Kilian 2006, 2144^VP – Basonym: *Hemophilus segnis* Kilian 1977 (Approved Lists 1980)

Agromyces hippuratus (Zgurskaya et al. 1992) Ortiz-Martinez et al. 2004, 1555^VP – Basonym: *Agromyces fucosus* subsp. *hippuratus* Zgurskaya et al. 1992

Algicola bacteriolytica (Sawabe et al. 1998) Ivanova et al. 2004d, 1784^VP – Basonym: *Pseudoalteromonas bacteriolytica* Sawabe et al. 1998

Algicola sagamiensis (Kobayashi et al. 2003) Y.D. Nam et al. 2007b, 17^VP – Basonym: *Pseudoalteromonas sagamiensis* Kobayashi et al. 2003

Algoriphagus alkaliphilus (Tiago et al. 2006) Nedashkovskaya et al. 2007b, 1993^VP – Basonym: *Chimaereicella alkaliphila* Tiago et al. 2006a

Algoriphagus boritolerans (Ahmed et al. 2007c) Nedashkovskaya et al. 2007b, 1993^VP – Basonym: *Chimaereicella boritolerans* Ahmed et al. 2007c

Algoriphagus halophilus (Yi and Chun 2004) Nedashkovskaya et al. 2004h, 1763^VP – Basonym: *Hongiella halophila* Yi and Chun 2004a

Algoriphagus mannitolivorans (Yi and Chun 2004) Nedashkovskaya et al. 2007b, 1993^VP – Basonym: *Hongiella mannitolivorans* Yi and Chun 2004a

Algoriphagus marincola (Yoon et al. 2004) Nedashkovskaya et al. 2007b, 1993^VP – Basonym: *Hongiella marincola* J.H. Yoon et al. 2004o

Algoriphagus ornithinivorans (Yi and Chun 2004) Nedashkovskaya et al. 2007b, 1993^VP – Basonym: *Hongiella ornithinivorans* Yi and Chun 2004a

Alicyclobacillus disulfidooxidans (Dufresne et al. 1996) Karavaiko et al. 2005, 946^VP – Basonym: *Sulfobacillus disulfidooxidans* Dufresne et al. 1996

Aliivibrio fischeri (Beijerinck 1889) Urbanczyk et al. 2007, 2827^VP – Basonym: *Vibrio fischeri* (Beijerinck 1889) Lehmann and Neumann 1896 (Approved Lists 1980)

Aliivibrio logei (Harwood et al. 1980) Urbanczyk et al. 2007, 2827^VP – Basonym: *Photobacterium logei* (*ex* Bang et al. 1978) Harwood et al. 1980

Aliivibrio salmonicida (Egidius et al. 1986) Urbanczyk et al. 2007, 2827^VP – Basonym: *Vibrio salmonicida* Egidius et al. 1986

Aliivibrio wodanis (Lunder et al. 2000) Urbanczyk et al. 2007, 2828^VP – Basonym: *Vibrio wodanis* Lunder et al. 2000

Alkalibacillus haloalkaliphilus (Fritze 1996) Jeon et al. 2005b, 1894^VP – Basonym: *Bacillus haloalkaliphilus* Fritze 1996

Altererythrobacter luteolus (Yoon et al. 2005) K.K. Kwon et al. 2007, 2210^VP – Basonym: *Erythrobacter luteolus* J.H. Yoon et al. 2005a

Alysiella crassa (Schmid 1922) Xie and Yokota 2005, 1395^VP – Basonym: *Simonsiella crassa* Schmid 1922 (Approved Lists 1980) – Valid publication: Validation List no. 104 – Effective publication: Xie and Yokota (2005a)

Amycolatopsis lurida (Lechevalier et al. 1986) Stackebrandt et al. 2004a, 267^VP – Basonym: *Amycolatopsis orientalis* subsp. *lurida* (*ex* Grundy et al. 1957) Lechevalier et al. 1986

Aquimarina brevivitae (Yoon et al. 2006) Nedashkovskaya et al. 2006f, 2040^VP – Basonym: *Gaetbulimicrobium brevivitae* J.H. Yoon et al. 2006a

Aquimarina latercula (Lewin 1969) Nedashkovskaya et al. 2006f, 2040^VP – Basonym: *Cytophaga latercula* Lewin 1969 (Approved Lists 1980)

Avibacterium avium (Hinz and Kunjara 1977) Blackall et al. 2005, 360^VP – Basonym: *Hemophilus avium* Hinz and Kunjara 1977 (Approved Lists 1980)

Avibacterium gallinarum (Hall et al. 1955) Blackall et al. 2005, 359^VP – Basonym: *Pasteurella gallinarum* Hall et al. 1955 (Approved Lists 1980)

Avibacterium paragallinarum (Biberstein and White 1969) Blackall et al. 2005, 360^VP – Basonym: *Hemophilus paragallinarum* Biberstein and White 1969 (Approved Lists 1980)

Avibacterium volantium (Mutters et al. 1985) Blackall et al. 2005, 360^VP – Basonym: *Pasteurella volantium* Mutters et al. 1985

Azohydromonas lata (Palleroni and Palleroni 1978) Xie and Yokota 2005f, 2423^VP – Basonym: *Alcaligenes latus* Palleroni and Palleroni 1978 (Approved Lists 1980)

Bergeriella denitrificans (Berger 1962) Xie and Yokota 2005, 1395^VP – Basonym: *Neisseria denitrificans* Berger 1962 (Approved Lists 1980) – Valid publication: Validation List no. 104 – Effective publication: Xie and Yokota (2005a)

Bibersteinia trehalosi (Sneath and Stevens 1990) Blackall et al. 2007, 673^VP – Basonym: *Pasteurella trehalosi* Sneath and Stevens 1990

Bifidobacterium animalis subsp. *lactis* (Meile et al. 1997) Masco et al. 2004, 1142^VP – Basonym: *Bifidobacterium lactis* Meile et al. 1997

Blastopirellula marina (Schlesner 1987) Schlesner et al. 2004, 1578^VP – Basonym: *Pirellula marina* (Schlesner 1987) Schlesner and Hirsch 1987

Brachyspira intermedia (Stanton et al. 1997) Hampson and La 2006, 1011^VP – Basonym: *Serpulina intermedia* Stanton et al. 1997

Brachyspira murdochii (Stanton et al. 1997) Hampson and La 2006, 1011^VP – Basonym: *Serpulina murdochii* Stanton et al. 1997

Caldanaerobacter subterraneus (Fardeau et al. 2000) Fardeau et al. 2004, 471^VP – Basonym: *Thermoanaerobacter subterraneus* Fardeau et al. 2000

Caldanaerobacter subterraneus subsp. *tengcongensis* (Xue et al. 2001) Fardeau et al. 2004, 472^VP – Basonym: *Thermoanaerobacter tengcongensis* Xue et al. 2001

Caldanaerobacter subterraneus subsp. *yonseiensis* (Kim et al. 2001) Fardeau et al. 2004, 472^VP – Basonym: *Thermoanaerobacter yonseiensis* Kim et al. 2001

Caldicellulosiruptor acetigenus (Nielsen et al. 1994) Onyenwoke et al. 2006, 1394^VP – Basonym: *Thermoanaerobium acetigenum* Nielsen et al. 1994

Carboxydothermus ferrireducens (Slobodkin et al. 1997) Slobodkin et al. 2006a, 2350^VP – Basonym: *Thermoterrabacterium ferrireducens* Slobodkin et al. 1997

Castellaniella defragrans (Foss et al. 1998) Kämpfer et al. 2006d, 818^VP – Basonym: *Alcaligenes defragrans* Foss et al. 1998

Catellatospora methionotrophica (Asano and Kawamoto 1988) Ara and Kudo 2006, 399^VP – Basonym: *Catellatospora citrea* subsp. *methionotrophica* Asano and Kawamoto 1988

Chitinophaga arvensicola (Oyaizu et al. 1983) Kämpfer et al. 2006k, 2226^VP – Basonym: *Cytophaga arvensicola* Oyaizu et al. 1983

Chitinophaga filiformis (Reichenbach 1989) Kämpfer et al. 2006k, 2225^VP – Basonym: *Flexibacter filiformis* (*ex* Solntseva 1940) Reichenbach 1989

Chitinophaga japonensis (Fujita et al. 1997) Kämpfer et al. 2006k, 2225^VP – Basonym: *Flexibacter japonensis* Fujita et al. 1997

Chitinophaga sancti (Lewin 1969) Kämpfer et al. 2006k, 2225^VP – Basonym: *Flexibacter sancti* Lewin 1969 (Approved Lists 1980)

Chromohalobacter beijerinckii (Hof 1935) Peçonek et al. 2006, 1956^VP – Basonym: *Pseudomonas beijerinckii* Hof 1935 (Approved Lists 1980)

Conchiformibius steedae corrig. (Kuhn and Gregory 1979) Xie and Yokota 2005, 1395^VP – Basonym: *Simonsiella steedae* Kuhn and Gregory 1979 (Approved Lists 1980) – Valid publication: Validation List no. 104 – Effective publication: Xie and Yokota (2005a)

Corallococcus coralloides (Thaxter 1892) Reichenbach 2007, 893^VP – Basonym: *Myxococcus coralloides* Thaxter 1892 (Approved Lists 1980) – Valid publication: Validation List no. 115 – Effective publication: Reichenbach (2005b)

Cupriavidus basilensis (Steinle et al. 1999) Vandamme and Coenye 2004, 2287^VP – Basonym: *Ralstonia basilensis* Steinle et al. 1999

Cupriavidus campinensis (Goris et al. 2001) Vandamme and Coenye 2004, 2287^VP – Basonym: *Ralstonia campinensis* Goris et al. 2001

Cupriavidus gilardii (Coenye et al. 1999) Vandamme and Coenye 2004, 2287^VP – Basonym: *Ralstonia gilardii* Coenye et al. 1999

Cupriavidus metallidurans (Goris et al. 2001) Vandamme and Coenye 2004, 2287^{VP} – Basonym: *Ralstonia metallidurans* Goris et al. 2001

Cupriavidus oxalaticus (Sahin et al. 2000) Vandamme and Coenye 2004, 2287^{VP} – Basonym: *Ralstonia oxalatica* (*ex* Khambata and Bhat 1953) Sahin et al. 2000

Cupriavidus pauculus (Vandamme et al. 1999) Vandamme and Coenye 2004, 2288^{VP} – Basonym: *Ralstonia paucula* Vandamme et al. 1999

Cupriavidus respiraculi (Coenye et al. 2003) Vandamme and Coenye 2004, 2288^{VP} – Basonym: *Ralstonia respiraculi* Coenye et al. 2003

Cupriavidus taiwanensis (Chen et al. 2001) Vandamme and Coenye 2004, 2288^{VP} – Basonym: *Ralstonia taiwanensis* Chen et al. 2001

Curvibacter delicatus (Leifson 1962) Ding and Yokota 2004, 2229^{VP} – Basonym: *Aquaspirillum delicatum* (Leifson 1962) Hylemon et al. 1973 (Approved Lists 1980)

Curvibacter lanceolatus (Leifson 1962) Ding and Yokota 2004, 2228^{VP} – Basonym: *Pseudomonas lanceolata* Leifson 1962 (Approved Lists 1980)

Desulfarculus baarsii (Widdel 1981) Kuever et al. 2006, 2^{VP} – Basonym: *Desulfovibrio baarsii* Widdel 1981 – Valid publication: Validation List no. 107 – Effective publication: Kuever et al. (2005q)

Desulfofaba hansenii (Finster et al. 2001) Abildgaard et al. 2004, 398^{VP} – Basonym: *Desulfomusa hansenii* Finster et al. 2001

Desulfosarcina cetonica corrig. (Galushko and Rozanova 1994) Kuever et al. 2006, 2^{VP} – Basonym: *Desulfobacterium cetonicum* Galushko and Rozanova 1994 – Valid publication: Validation List no. 107 – Effective publication: Kuever et al. (2005m)

Desulfovibrio oxamicus (Postgate and Campbell 1966) López-Cortés et al. 2006a, 1498^{VP} – Basonym: *Desulfovibrio vulgaris* subsp. *oxamicus* Postgate and Campbell 1966 (Approved Lists 1980)

Dickeya chrysanthemi (Burkholder et al. 1953) Samson et al. 2005, 1423^{VP} – Basonym: *Erwinia chrysanthemi* Burkholder et al. 1953 (Approved Lists 1980)

Dickeya paradisiaca (Fernandez-Borrero and Lopez-Duque 1970) Samson et al. 2005, 1425^{VP} – Basonym: *Erwinia paradisiaca* Fernandez-Borrero and Lopez-Duque 1970 (Approved Lists 1980)

Elizabethkingia meningoseptica (King 1959) K.K. Kim et al. 2005b, 1291^{VP} – Basonym: *Flavobacterium meningosepticum* King 1959 (Approved Lists 1980)

Elizabethkingia miricola (Li et al. 2004) K.K. Kim et al. 2005b, 1292^{VP} – Basonym: *Chryseobacterium miricola* Y. Li et al. 2004

Enterobacter cloacae subsp. *dissolvens* (Rosen 1922) Hoffmann et al. 2005, 1396^{VP} – Basonym: *Erwinia dissolvens* (Rosen 1922) Burkholder 1948 (Approved Lists 1980) – Valid publication: Validation List no. 104 – Effective publication: Hoffmann et al. (2005a)

Geobacillus pallidus (Scholz et al. 1988) Banat et al. 2004, 2200^{VP} – Basonym: *Bacillus pallidus* Scholz et al. 1988

Geobacillus vulcani (Caccamo et al. 2000) Nazina et al. 2004, 2023^{VP} – Basonym: *Bacillus vulcani* Caccamo et al. 2000

Geobacter thiogenes (De Wever et al. 2001) Nevin et al. 2007, 465^{VP} – Basonym: *Trichlorobacter thiogenes* De Wever et al. 2001

Giesbergeria anulus (Williams and Rittenberg 1957) Grabovich et al. 2006, 575^{VP} – Basonym: *Aquaspirillum anulus* (Williams and Rittenberg 1957) Hylemon et al. 1973 (Approved Lists 1980)

Giesbergeria giesbergeri (Williams and Rittenberg 1957) Grabovich et al. 2006, 575^{VP} – Basonym: *Aquaspirillum giesbergeri* (Williams and Rittenberg 1957) Hylemon et al. 1973 (Approved Lists 1980)

Giesbergeria sinuosa (Williams and Rittenberg 1957) Grabovich et al. 2006, 575^{VP} – Basonym: *Aquaspirillum sinuosum* (Williams and Rittenberg 1957) Hylemon et al. 1973 (Approved Lists 1980)

Goodfellowia coeruleoviolacea (Preobrazhenskaya and Sveshnikova 1974) Labeda and Kroppenstedt 2006, 1206^{VP} – Basonym: *Actinomadura coeruleoviolacea* Preobrazhenskaya and Terekhova 1987

The name *Goodfellowia coeruleoviolacea* (Preobrazhenskaya and Sveshnikova 1974) Labeda and Kroppenstedt 2006 is illegitimate because the genus name is illegitimate (see above *Goodfellowia* Labeda and Kroppenstedt 2006, 1205^{VP}).

Haematobacter massiliensis (Greub and Raoult 2006) Helsel et al. 2007, 1371^{VP} – Basonym: *Rhodobacter massiliensis* Greub and Raoult 2006 – Valid publication: Validation List no. 116 – Effective publication: Helsel et al. (2007)

Herbaspirillum autotrophicum (Aragno and Schlegel 1978) Ding and Yokota 2004, 2228^{VP} – Basonym: *Aquaspirillum autotrophicum* Aragno and Schlegel 1978 (Approved Lists 1980)

Herbaspirillum huttiense (Leifson 1962) Ding and Yokota 2004, 2228^{VP} – Basonym: *Pseudomonas huttiensis* Leifson 1962 (Approved Lists 1980)

Hylemonella gracilis (Canale-Parola et al. 1966) Spring et al. 2004, 104^{VP} – Basonym: *Aquaspirillum gracile* (Canale-Parola et al. 1966) Hylemon et al. 1973 (Approved Lists 1980)

Ignatzschineria larvae (Tóth et al. 2001) Tóth et al. 2007, 180^{VP} – Illegitimate basonym: *Schineria larvae* Tóth et al. 2001

Insolitispirillum peregrinum (Pretorius 1963) J.H. Yoon et al. 2007xxvi, 2834^{VP} – Basonym: *Aquaspirillum peregrinum* (Pretorius 1963) Hylemon et al. 1973 (Approved Lists 1980)

Insolitispirillum peregrinum subsp. *integrum* (Terasaki 1973) J.H. Yoon et al. 2007xxvi, 2834^{VP} – Basonym: *Aquaspirillum peregrinum* subsp. *integrum* (Terasaki 1973) Terasaki 1979 (Approved Lists 1980)

Insolitispirillum peregrinum subsp. *peregrinum* (Pretorius 1963) J.H. Yoon et al. 2007xxvi, 2834^{VP} – Basonym: *Aquaspirillum peregrinum* subsp. *peregrinum* (Pretorius 1963) Hylemon et al. 1973 (Approved Lists 1980)

Isoptericola variabilis (Bakalidou et al. 2002) Stackebrandt et al. 2004b, 687^{VP} – Basonym: *Cellulosimicrobium variabile* Bakalidou et al. 2002

Kluyvera intermedia (Izard et al. 1980) Pavan et al. 2005, 441^{VP} – Basonym: *Enterobacter intermedius* corrig. Izard et al. 1980

Labrenzia aggregata (Uchino et al. 1999) Biebl et al. 2007, 1105^{VP} – Basonym: *Stappia aggregata* (*ex* Ahrens 1968) Uchino et al. 1999

Labrenzia alba (Pujalte et al. 2006) Biebl et al. 2007, 1105^{VP} – Basonym: *Stappia alba* Pujalte et al. 2006

Labrenzia marina (Kim et al. 2006) Biebl et al. 2007, 1105^{VP} – Basonym: *Stappia marina* B.C. Kim et al. 2006

Laceyella putida (Lacey and Cross 1989) J.H. Yoon et al. 2005n, 399^VP – Basonym: *Thermoactinomyces putidus* Lacey and Cross 1989

Laceyella sacchari (Lacey 1971) J.H. Yoon et al. 2005n, 398^VP – Basonym: *Thermoactinomyces sacchari* Lacey 1971 (Approved Lists 1980)

Lactobacillus kefiranofaciens subsp. *kefirgranum* (Takizawa et al. 1994) Vancanneyt et al. 2004b, 555^VP – Basonym: *Lactobacillus kefirgranum* Takizawa et al. 1994

Leeuwenhoekiella marinoflava (Reichenbach 1989) Nedashkovskaya et al. 2005m, 1035^VP – Basonym: *Cytophaga marinoflava* (*ex* Colwell et al. 1966) Reichenbach 1989

Lysinibacillus fusiformis (Priest et al. 1989) Ahmed et al. 2007d, 1122^VP – Basonym: *Bacillus fusiformis* (*ex* Meyer and Gottheil 1901) Priest et al. 1989

Lysinibacillus sphaericus (Meyer and Neide 1904) Ahmed et al. 2007d, 1123^VP – Basonym: *Bacillus sphaericus* Meyer and Neide 1904 (Approved Lists 1980)

Malikia spinosa (Leifson 1962) Spring et al. 2005b, 628^VP – Basonym: *Pseudomonas spinosa* Leifson 1962 (Approved Lists 1980)

Marinovum algicola (Lafay et al. 1995) Martens et al. 2006, 1302^VP – Basonym: *Roseobacter algicola* Lafay et al. 1995

Mycoplasma coccoides (Schilling 1928) Neimark et al. 2005, 1389^VP – Basonym: *Eperythrozoon coccoides* Schilling 1928 (Approved Lists 1980)

According to Rule 28b(2), the new combination *Mycoplasma coccoides* (Schilling 1928) Neimark et al. 2005 is not validly published (see the Request for an Opinion by Neimark et al., 2005).

Mycoplasma ovis (Neitz et al. 1934) Neimark et al. 2004, 369^VP – Basonym: *Eperythrozoon ovis* Neitz et al. 1934 (Approved Lists 1980)

Nakamurella multipartita (Yoshimi et al. 1996) Tao et al. 2004, 999^VP – Illegitimate basonym: *Microsphaera multipartita* Yoshimi et al. 1996

Novispirillum itersonii (Giesberger 1936) J.H. Yoon et al. 2007xxvi, 2833^VP – Basonym: *Aquaspirillum itersonii* (Giesberger 1936) Hylemon et al. 1973 (Approved Lists 1980)

Novispirillum itersonii subsp. *itersonii* (Giesberger 1936) J.H. Yoon et al. 2007xxvi, 2833^VP – Basonym: *Aquaspirillum itersonii* subsp. *itersonii* (Giesberger 1936) Hylemon et al. 1973 (Approved Lists 1980)

Novispirillum itersonii subsp. *nipponicum* (Terasaki 1973) J.H. Yoon et al. 2007xxvi, 2834^VP – Basonym: *Aquaspirillum itersonii* subsp. *nipponicum* (Terasaki 1973) Terasaki 1979 (Approved Lists 1980)

Novosphingobium resinovorum (Delaporte and Daste 1956) Y.W. Lim et al. 2007, 1907^VP – Basonym: *Flavobacterium resinovorum* Delaporte and Daste 1956 (Approved Lists 1980)

Oceanobacillus picturae (Heyrman et al. 2003) J.S. Lee et al. 2006b, 256^VP – Basonym: *Virgibacillus picturae* Heyrman et al. 2003

Paenibacillus chitinolyticus (Kuroshima et al. 1996) J.S. Lee et al. 2004, 932^VP – Basonym: *Bacillus chitinolyticus* Kuroshima et al. 1996

Paenibacillus ehimensis (Kuroshima et al. 1996) J.S. Lee et al. 2004, 931^VP – Basonym: *Bacillus ehimensis* Kuroshima et al. 1996

Parabacteroides distasonis (Eggerth and Gagnon 1933) Sakamoto and Benno 2006, 1602^VP – Basonym: *Bacteroides distasonis* Eggerth and Gagnon 1933 (Approved Lists 1980)

Parabacteroides goldsteinii (Song et al. 2006) Sakamoto and Benno 2006, 1602^VP – Basonym: *Bacteroides goldsteinii* Song et al. 2006

Parabacteroides merdae (Johnson et al. 1986) Sakamoto and Benno 2006, 1604^VP – Basonym: *Bacteroides merdae* Johnson et al. 1986

Parvimonas micra (Prévot 1933) Tindall and Euzéby 2006, 2712^VP – Basonym: *Peptostreptococcus micros* (Prévot 1933) Smith 1957 (Approved Lists 1980)

Pelomonas saccharophila (Doudoroff 1940) Xie and Yokota 2005f, 2424^VP – Basonym: *Pseudomonas saccharophila* Doudoroff 1940 (Approved Lists 1980)

Peredibacter starrii (Seidler et al. 1972) Davidov and Jurkevitch 2004, 1451^VP – Basonym: *Bdellovibrio starrii* Seidler et al. 1972 (Approved Lists 1980)

Phaeobacter gallaeciensis (Ruiz-Ponte et al. 1998) Martens et al. 2006, 1301^VP – Basonym: *Roseobacter gallaeciensis* Ruiz-Ponte et al. 1998

Photobacterium indicum (Johnson and Weisrock 1969) Ivanova et al. 2004d, 1785^VP – Basonym: *Hyphomicrobium indicum* Johnson and Weisrock 1969

Photobacterium indicum (Johnson and Weisrock 1969) Xie and Yokota 2004, 2115^VP – Basonym: *Hyphomicrobium indicum* Johnson and Weisrock 1969 (Approved Lists 1980)

According to Rule 24b(2), *Photobacterium indicum* (Johnson and Weisrock 1969) Ivanova et al. 2004d has priority.

Planomicrobium alkanoclasticum (Engelhardt et al. 2001) Dai et al. 2005, 702^VP – Basonym: *Planococcus alkanoclasticus* Engelhardt et al. 2001

Planomicrobium psychrophilum (Reddy et al. 2002) Dai et al. 2005, 702^VP – Basonym: *Planococcus psychrophilus* Reddy et al. 2002

Pseudomonas chlororaphis subsp. *aurantiaca* (Nakhimovskaya 1948) Peix et al. 2007, 1289^VP – Basonym: *Pseudomonas aurantiaca* Nakhimovskaya 1948 (Approved Lists 1980)

Pseudomonas chlororaphis subsp. *aureofaciens* (Kluyver 1956) Peix et al. 2007, 1289^VP – Basonym: *Pseudomonas aureofaciens* Kluyver 1956 (Approved Lists 1980)

Pullulanibacillus naganoensis (Tomimura et al. 1990) Hatayama et al. 2006, 2550^VP – Basonym: *Bacillus naganoensis* Tomimura et al. 1990

Quatrionicoccus australiensis (Maszenan et al. 2002) Tindall and Euzéby 2006, 2712^VP – Illegitimate basonym: *Quadricoccus australiensis* Maszenan et al. 2002

Ralstonia syzygii (Roberts et al. 1990) Vaneechoutte et al. 2004, 321^VP – Basonym: *Pseudomonas syzygii* Roberts et al. 1990

Reichenbachiella agariperforans (Nedashkovskaya et al. 2003) Nedashkovskaya et al. 2005l, 2587^VP – Illegitimate basonym: *Reichenbachia agariperforans* Nedashkovskaya et al. 2003

Rhizobium larrymoorei (Bouzar and Jones 2001) J.M. Young 2004, 149^VP – Basonym: *Agrobacterium larrymoorei* Bouzar and Jones 2001

Rhodococcus corynebacterioides (Serrano et al. 1972) Yassin and Schaal 2005, 1347^VP – Basonym: *Nocardia corynebacterioides* Serrano et al. 1972 (Approved Lists 1980)

Roseivirga seohaensis (Yoon et al. 2005) S.C. Lau et al. 2006b, 1064^VP – Basonym: *Marinicola seohaensis* J.H. Yoon et al. 2005c

Ruegeria lacuscaerulensis (Petursdottir and Kristjansson 1999) Yi et al. 2007, 818^VP – Basonym: *Silicibacter lacuscaerulensis* Petursdottir and Kristjansson 1999

Ruegeria pomeroyi (González et al. 2003) Yi et al. 2007, 818^VP – Basonym: *Silicibacter pomeroyi* González et al. 2003

Salimicrobium album (Hao et al. 1985) J.H. Yoon et al. 2007xxi, 2409^VP – Basonym: *Marinococcus albus* Hao et al. 1985

Salimicrobium halophilum (Ventosa et al. 1990) J.H. Yoon et al. 2007xxi, 2409^VP – Basonym: *Bacillus halophilus* Ventosa et al. 1990

Salmonella enterica subsp. *arizonae* (Borman 1957) Le Minor and Popoff 1987, 519^VP – Basonym: *Salmonella arizonae* (Borman 1957) Kauffmann 1964 (Approved Lists 1980) – Reference: Judicial Commission of the International Committee on Systematics of Prokaryotes 2005b (Opinion 80)

Salmonella enterica subsp. *bongori* (Le Minor et al. 1985) Le Minor and Popoff 1987, 519^VP – Basonym: *Salmonella choleraesuis* subsp. *bongori* corrig. Le Minor et al. 1985 – Reference: Judicial Commission of the International Committee on Systematics of Prokaryotes 2005b (Opinion 80)

Salmonella enterica subsp. *diarizonae* (Le Minor et al. 1985) Le Minor and Popoff 1987, 519^VP – Basonym: *Salmonella choleraesuis* subsp. *diarizonae* corrig. Le Minor et al. 1985 – Reference: Judicial Commission of the International Committee on Systematics of Prokaryotes 2005b (Opinion 80)

Salmonella enterica subsp. *houtenae* (Le Minor et al. 1985) Le Minor and Popoff 1987, 519^VP – Basonym: *Salmonella choleraesuis* subsp. *houtenae* corrig. Le Minor et al. 1985 – Reference: Judicial Commission of the International Committee on Systematics of Prokaryotes 2005b (Opinion 80)

Salmonella enterica subsp. *indica* (Le Minor et al. 1987) Le Minor and Popoff 1987, 519^VP – Basonym: *Salmonella choleraesuis* subsp. *indica* Le Minor et al. 1987 – Reference: Judicial Commission of the International Committee on Systematics of Prokaryotes 2005b (Opinion 80)

Salmonella enterica subsp. *salamae* (Le Minor et al. 1985) Le Minor and Popoff 1987, 519^VP – Basonym: *Salmonella choleraesuis* subsp. *salamae* corrig. Le Minor et al. 1985 – Reference: Judicial Commission of the International Committee on Systematics of Prokaryotes 2005b (Opinion 80)

Seinonella peptonophila (Nonomura and Ohara 1971) J.H. Yoon et al. 2005n, 400^VP – Basonym: *Thermoactinomyces peptonophilus* Nonomura and Ohara 1971 (Approved Lists 1980)

Simplicispira metamorpha (Terasaki 1961) Grabovich et al. 2006, 575^VP – Basonym: *Aquaspirillum metamorphum* (Terasaki 1961) Hylemon et al. 1973 (Approved Lists 1980)

Simplicispira psychrophila (Terasaki 1973) Grabovich et al. 2006, 575^VP – Basonym: *Aquaspirillum psychrophilum* (Terasaki 1973) Terasaki 1979 (Approved Lists 1980)

Sorangium cellulosum (Brockman 1989) Reichenbach 2007, 894^VP – Basonym: *Polyangium cellulosum* (*ex* Imshenetski and Solnt-seva 1936) Brockman 1989 – Valid publication: Validation List no. 115 – Effective publication: Reichenbach (2005i)

Sphaerisporangium viridialbum corrig. (Nonomura and Ohara 1960) Ara and Kudo 2007, 2449^VP – Basonym: *Streptosporangium viridialbum* Nonomura and Ohara 1960 – Valid publication: Validation List no. 118 – Effective publication: Ara and Kudo (2007b)

Sphingobium chungbukense (Kim et al. 2000) Pal et al. 2005, 1971^VP – Basonym: *Sphingomonas chungbukensis* Kim et al. 2000

Sphingobium cloacae (Fujii et al. 2001) Prakash and Lal 2006, 2151^VP – Basonym: *Sphingomonas cloacae* Fujii et al. 2001

Sphingobium xenophagum (Stolz et al. 2000) Pal et al. 2006, 669^VP – Basonym: *Sphingomonas xenophaga* Stolz et al. 2000

Sphingopyxis taejonensis (Lee et al. 2001) Pal et al. 2006, 670^VP – Basonym: *Sphingomonas taejonensis* Lee et al. 2001

Sporolactobacillus laevolacticus (Andersch et al. 1994) Hatayama et al. 2006, 2550^VP – Basonym: *Bacillus laevolacticus* (*ex* Nakayama and Yanoshi 1967) Andersch et al. 1994

Stanierella latercula (Lewin 1969) Nedashkovskaya et al. 2005h, 228^VP – Basonym: *Cytophaga latercula* Lewin 1969 (Approved Lists 1980)

Sulfitobacter guttiformis (Labrenz et al. 2000) J.H. Yoon et al. 2007x, 1791^VP – Basonym: *Staleya guttiformis* Labrenz et al. 2000

Sulfurimonas denitrificans (Timmer-ten Hoor 1975) Takai et al. 2006b, 1732^VP – Basonym: *Thiomicrospira denitrificans* Timmer-ten Hoor 1975 (Approved Lists 1980)

Syntrophomonas bryantii (Stieb and Schink 1985) C. Wu et al. 2006, 2335 – Basonym: *Clostridium bryantii* Stieb and Schink 1985

According to Rule 27(3), *Syntrophomonas bryantii* (Stieb and Schink 1985) C. Wu et al. 2006 is not validly published because the type strain is not deposited in two different collections in two different countries.

Terrimonas ferruginea (Sickles and Shaw 1934) Xie and Yokota 2006c, 1120^VP – Basonym: *Flavobacterium ferrugineum* Sickles and Shaw 1934 (Approved Lists 1980)

Tetragenococcus solitarius (Collins et al. 1989) Ennahar and Cai 2005, 592^VP – Basonym: *Enterococcus solitarius* Collins et al. 1989

Tetrasphaera duodecadis (Lochhead 1958) Ishikawa and Yokota 2006, 1371^VP – Basonym: *Arthrobacter duodecadis* Lochhead 1958 (Approved Lists 1980)

Thalassobius gelatinovorus (Rüger and Höfle 1992) Arahal et al. 2006, 3^VP – Basonym: *Agrobacterium gelatinovorum* (*ex* Ahrens 1968) Rüger and Höfle 1992 – Valid publication: Validation List no. 107 – Effective publication: Arahal et al. (2006)

Thermoflavimicrobium dichotomicum (Krasil'nikov and Agre 1964) J.H. Yoon et al. 2005n, 399^VP – Basonym: *Thermoactinomyces dichotomicus* corrig. (Krasil'nikov and Agre 1964) Cross and Goodfellow 1973 (Approved Lists 1980)

Thermopolyspora flexuosa (Meyer 1989) Goodfellow et al. 2005, 1982^VP – Basonym: *Actinomadura flexuosa* (*ex* Krasil'nikov and Agre 1964) Meyer 1989

Thiomonas delicata (Katayama-Fujimura et al. 1984) Kelly and Wood 2006, 926^VP – Basonym: *Thiobacillus delicatus* (*ex* Mizoguchi et al. 1976) Katayama-Fujimura et al. 1984 – Valid publication: Validation List no. 109 – Effective publication: Kelly and Wood (2005b)

Turneriella parva (Hovind-Hougen et al. 1982) Levett et al. 2005, 1499^VP – Basonym: *Leptospira parva* Hovind-Hougen et al. 1982

Umezawaea tangerina (Kinoshita et al. 2000) Labeda and Kroppenstedt 2007, 2761^VP – Basonym: *Saccharothrix tangerinus* Kinoshita et al. 2000

Virgibacillus halodenitrificans (Denariaz et al. 1989) J.H. Yoon et al. 2004k, 2166^VP – Basonym: *Bacillus halodenitrificans* Denariaz et al. 1989

Viridibacillus arenosi (Heyrman et al. 2005) Albert et al. 2007, 2735^VP – Basonym: *Bacillus arenosi* Heyrman et al. 2005b

Viridibacillus arvi (Heyrman et al. 2005) Albert et al. 2007, 2735^VP – Basonym: *Bacillus arvi* Heyrman et al. 2005b

Viridibacillus neidei (Nakamura et al. 2002) Albert et al. 2007, 2735^VP – Basonym: *Bacillus neidei* Nakamura et al. 2002

Wautersia basilensis (Steinle et al. 1999) Vaneechoutte et al. 2004, 323^VP – Basonym: *Ralstonia basilensis* Steinle et al. 1999

Wautersia campinensis (Goris et al. 2001) Vaneechoutte et al. 2004, 323^VP – Basonym: *Ralstonia campinensis* Goris et al. 2001

Wautersia eutropha (Davis 1969) Vaneechoutte et al. 2004, 323^VP – Basonym: *Alcaligenes eutrophus* Davis 1969 (Approved Lists 1980)

Wautersia gilardii (Coenye et al. 1999) Vaneechoutte et al. 2004, 324^VP – Basonym: *Ralstonia gilardii* Coenye et al. 1999

Wautersia metallidurans (Goris et al. 2001) Vaneechoutte et al. 2004, 324^VP – Basonym: *Ralstonia metallidurans* Goris et al. 2001

Wautersia oxalatica (Sahin et al. 2000) Vaneechoutte et al. 2004, 324^VP – Basonym: *Ralstonia oxalatica* (*ex* Khambata and Bhat 1953) Sahin et al. 2000

Wautersia paucula (Vandamme et al. 1999) Vaneechoutte et al. 2004, 325^VP – Basonym: *Ralstonia paucula* Vandamme et al. 1999

Wautersia respiraculi (Coenye et al. 2003) Vaneechoutte et al. 2004, 325^VP – Basonym: *Ralstonia respiraculi* Coenye et al. 2003

Wautersia taiwanensis (Chen et al. 2001) Vaneechoutte et al. 2004, 325^VP – Basonym: *Ralstonia taiwanensis* Chen et al. 2001

Xylanimicrobium pachnodae (Cazemier et al. 2004) Stackebrandt and Schumann 2004, 1385^VP – Basonym: *Promicromonospora pachnodae* Cazemier et al. 2004

Revived names

Byssovorax cruenta (*ex* Thaxter 1897) Reichenbach 2006, 2363^VP, sp. nov., nom. rev. – Synonym: "*Myxococcus cruentus*" Thaxter 1897

Corallococcus exiguus (*ex* Kofler 1913) Reichenbach 2007, 893^VP, nom. rev., comb. nov. – Synonym: "*Myxococcus exiguus*" Kofler 1913 – Valid publication: Validation List no. 115 – Effective publication: Reichenbach (2005b)

Cystobacter violaceus (*ex* Kühlwein and Gallwitz 1958) Reichenbach 2007, 894^VP, nom. rev., comb. nov. – Synonyms: "*Polyangium violaceum*" Kühlwein and Gallwitz 1958, "*Archangium violaceum*" Kühlwein and reichenbach 1964 – Valid publication: Validation List no. 115 – Effective publication: Reichenbach (2005d)

Flammeovirga arenaria (*ex* Lewin 1969) M. Takahashi et al. 2006, 2099^VP, nom. rev., comb. nov. – Synonym: "*Microscilla arenaria*" Lewin 1969

Gluconobacter albidus (*ex* Kondo and Ameyama 1958) Yukphan et al. 2005, 983^VP, nom. rev., comb. nov. – Synonym: "*Acetobacter albidus*" Kondo and Ameyama 1958 – Valid publication: Validation List no. 103 – Effective publication: Yukphan et al. (2004b)

Jahnella thaxteri corrig. (*ex* Jahn 1924) Reichenbach 2007, 894^VP, nom. rev., comb. nov. – Synonym: "*Archangium thaxteri*" Jahn 1924 – Valid publication: Validation List no. 115 – Effective publication: Reichenbach (2005h)

Kofleria flava (*ex* Kofler 1913) Reichenbach 2007, 894^VP, nom. rev., comb. nov. – Synonyms: "*Polyangium flavum*" Kofler 1913, "*Archangium flavum*" (Kofler 1913) Jahn 1924 – Valid publication: Validation List no. 115 – Effective publication: Reichenbach (2005m)

Mycobacterium salmoniphilum (*ex* Ross 1960) Whipps et al. 2007, 2529^VP, sp. nov., nom. rev. – Synonym: "*Mycobacterium salmoniphilum*" Ross 1960

Salmonella enterica (*ex* Kauffmann and Edwards 1952) Le Minor and Popoff 1987, 519^VP, sp. nov., nom. rev. – Synonym: "*Salmonella enterica*" Kauffmann and Edwards 1952 – Reference: Judicial Commission of the International Committee on Systematics of Prokaryotes 2005b (Opinion 80)

Sorangium (*ex* Jahn 1924) Reichenbach 2007, 894^VP, gen. nov., nom. rev.- Synonym: "*Sorangium*" Jahn 1924 – Valid publication: Validation List no. 115 – Effective publication: Reichenbach (2005i)

Thermopolyspora (*ex* Krasil'nikov and Agre 1964) Goodfellow et al. 2005, 1980^VP, gen. nov., nom. rev. – Synonym: "*Thermopolyspora*" Krassilnikov and Agre 1964

Xanthomonas alfalfae (*ex* Riker et al. 1935) Schaad et al. 2007, 894^VP, sp. nov., nom. rev. – Synonym: "*Xanthomonas alfalfae*" (*ex* Riker et al. 1935) Dowson 1943 – Valid publication: Validation List no. 115 – Effective publication: Schaad et al. (2006)

Xanthomonas citri subsp. *malvacearum* (*ex* Smith 1901) Schaad et al. 2007, 895^VP, subsp. nov., nom. rev. – Synonym: "*Xanthomonas malvacearum*" (Smith 1901) Dowson 1939 – Valid publication: Validation List no. 115 – Effective publication: Schaad et al. (2006)

Xanthomonas gardneri (*ex* Sutic 1957) Jones et al. 2006, 926^VP, nom. rev., comb. nov. – Synonym: "*Pseudomonas gardneri*" Sutic 1957 – Valid publication: Validation List no. 109 – Effective publication: J.B. Jones et al. (2004)

Conserved specific epithet

enterica in *Salmonella enterica* (*ex* Kauffmann and Edwards 1952) Le Minor and Popoff 1987 – Reference: Judicial Commission of the International Committee on Systematics of Prokaryotes 2005b (Opinion 80)

Rejected names

Pelczaria Poston 1994 – Reference: Judicial Commission of the International Committee on Systematics of Prokaryotes 2005a (Opinion 78)

Pelczaria aurantia Poston 1994 – Reference: Judicial Commission of the International Committee on Systematics of Prokaryotes 2005a (Opinion 78)

Emendations

Acidaminococcus Rogosa 1969 emend. Jumas-Bilak et al. 2007b

Acidianus Segerer et al. 1986 emend. Plumb et al. 2007

Acidomonas methanolica (Uhlig et al. 1986) Urakami et al. 1989 emend. Yamashita et al. 2004

Acidomonas Urakami et al. 1989 emend. Yamashita et al. 2004

Actinobacillus capsulatus Arseculeratne 1962 (Approved Lists 1980) emend. Kuhnert et al. 2007

Actinobacillus rossii Sneath and Stevens 1990 emend. Christensen et al. 2005

Aeromicrobium Miller et al. 1991 emend. J.H. Yoon et al. 2005q

Agromyces fucosus Zgurskaya et al. 1992 emend. Ortiz-Martinez et al. 2004

Alcaligenes faecalis Castellani and Chalmers 1919 (Approved Lists 1980) emend. Rehfuss and Urban 2005a – Publication in the IJSEM: List of Changes in Taxonomic Opinion no. 3

Algibacter Nedashkovskaya et al. 2004d emend. Nedashkovskaya et al. 2007e

Algoriphagus Bowman et al. 2003 emend. Nedashkovskaya et al. 2004h

Algoriphagus Bowman et al. 2003 emend. Nedashkovskaya et al. 2007b

Alicyclobacillus acidocaldarius (Darland and Brock 1971) Wisotzkey et al. 1992 emend. Goto et al. 2006 – Publication in the IJSEM: List of Changes in Taxonomic Opinion no. 5

Alicyclobacillus Wisotzkey et al. 1992 emend. Karavaiko et al. 2005

Alteromonadaceae Ivanova and Mikhailov 2001 emend. Ivanova et al. 2004d

Alteromonas Baumann et al. 1972 (Approved Lists 1980) emend. Van Trappen et al. 2004b

Alteromonas macleodii Baumann et al. 1972 (Approved Lists 1980) emend. Yi et al. 2004

Amphibacillus Niimura et al. 1990 emend. S.Y. An et al. 2007e

Anaerolinea Sekiguchi et al. 2003 emend. Yamada et al. 2006

Aquificales Reysenbach 2002 emend. L'Haridon et al. 2006

Aquimarina Nedashkovskaya et al. 2005h emend. Nedashkovskaya et al. 2006f

Arcanobacterium Collins et al. 1983 emend. Lehnen et al. 2006

Arenibacter Ivanova et al. 2001 emend. Nedashkovskaya et al. 2006g

Arenibacter latericius Ivanova et al. 2001 emend. Nedashkovskaya et al. 2006g

Arthrobacter cumminsii Funke et al. 1997 emend. Funke et al. 1998 – Publication in the IJSEM: List of Changes in Taxonomic Opinion no. 1

Aureispira Hosoya et al. 2006 emend. Hosoya et al. 2007

Azospira Reinhold-Hurek and Hurek 2000 emend. H.S. Bae et al. 2007

Bacillus coagulans Hammer 1915 (Approved Lists 1980) emend. De Clerck et al. 2004a – Publication in the IJSEM: List of Changes in Taxonomic Opinion no. 1

Bacillus mojavensis Roberts et al. 1994 emend. L.T. Wang et al. 2007

Bacillus simplex (*ex* Meyer and Gottheil 1901) Priest et al. 1989 emend. Heyrman et al. 2005a

Bifidobacterium animalis (Mitsuoka 1969) Scardovi and Trovatelli 1974 emend. Masco et al. 2004

Blastococcus aggregatus Ahrens and Moll 1970 (Approved Lists 1980) emend. Urzì et al. 2004

Blastococcus Ahrens and Moll 1970 (Approved Lists 1980) emend. S.D. Lee 2006g

Blastococcus Ahrens and Moll 1970 (Approved Lists 1980) emend. Urzì et al. 2004

Burkholderia pyrrocinia (Imanaka et al. 1965) Vandamme et al. 1997 emend. Storms et al. 2004 – Publication in the IJSEM: List of Changes in Taxonomic Opinion no. 1

Caldicellulosiruptor Rainey et al. 1995 emend. Onyenwoke et al. 2006

Carboxydothermus Svetlichny et al. 1991 emend. Slobodkin et al. 2006a

Cardiobacterium valvarum Han et al. 2004 emend. Han and Falsen 2005 – Publication in the IJSEM: List of Changes in Taxonomic Opinion no. 3

Cellulosimicrobium Schumann et al. 2001 emend. J.M. Brown et al. 2006

Cellulosimicrobium Schumann et al. 2001 emend. J.H. Yoon et al. 2007xxv

Chitinimonas S.C. Chang et al. 2004 emend. B.Y. Kim et al. 2006d

Chitinophaga arvensicola (Oyaizu et al. 1983) Kämpfer et al. 2006k emend. Pankratov et al. 2006

Chitinophaga Sangkhobol and Skerman 1981 emend. Kämpfer et al. 2006k

Chloroflexi Garrity and Holt 2001 emend. Hugenholtz and Stackebrandt 2004 (category not covered by the Rules)

Clostridium bifermentans (Weinberg and Séguin 1918) Bergey et al. 1923 (Approved Lists 1980) emend. Chamkha et al. 2001 – Publication in the IJSEM: List of Changes in Taxonomic Opinion no. 1

Clostridium sardiniense Prévot 1938 (Approved Lists 1980) emend. X. Wang et al. 2005

Corynebacterium aurimucosum Yassin et al. 2002 emend. Daneshvar et al. 2004 – Publication in the IJSEM: List of Changes in Taxonomic Opinion no. 1

Cupriavidus Makkar and Casida 1987 emend. Vandamme and Coenye 2004

Cyclobacterium Raj and Maloy 1990 emend. Ying et al. 2006

Desulfatibacillum Cravo-Laureau et al. 2004a emend. Cravo-Laureau et al. 2004b

Desulfitobacterium hafniense Christiansen and Ahring 1996 emend. Niggemyer et al. 2001 – Publication in the IJSEM: List of Changes in Taxonomic Opinion no. 2

Desulfofaba Knoblauch et al. 1999 emend. Abildgaard et al. 2004

Desulfurobacterium L'Haridon et al. 1998 emend. L'Haridon et al. 2006

Desulfurobacterium L'Haridon et al. 1998 emend. Alain et al. 2003 – Publication in the IJSEM: List of Changes in Taxonomic Opinion no. 2

Desulfurococcus Zillig and Stetter 1983 emend. Perevalova et al. 2005

Devosia Nakagawa et al. 1996 emend. Yoo et al. 2006

Devosia Nakagawa et al. 1996 emend. J.H. Yoon et al. 2007xvii

Dialister (*ex* Bergey et al. 1923) Moore and Moore 1994 emend. Jumas-Bilak et al. 2005

Dialister pneumosintes (Olitsky and Gates 1921) Moore and Moore 1994 emend. Jumas-Bilak et al. 2005

Dyadobacter Chelius and Triplett 2000 emend. Reddy and Garcia-Pichel 2005

Enterobacter asburiae Brenner et al. 1988 emend. Hoffmann et al. 2005a – Publication in the IJSEM: List of Changes in Taxonomic Opinion no. 2

Enterobacter kobei Kosako et al. 1997 emend. Hoffmann et al. 2005a – Publication in the IJSEM: List of Changes in Taxonomic Opinion no. 2

Flammeovirga aprica (Reichenbach 1989) Nakagawa et al. 1997 emend. M. Takahashi et al. 2006

Flammeovirga Nakagawa et al. 1997 emend. M. Takahashi et al. 2006

Flexithrix dorotheae Lewin 1970 (Approved Lists 1980) emend. Hosoya and Yokota 2007a

Flexithrix Lewin 1970 (Approved Lists 1980) emend. Hosoya and Yokota 2007a

Formosa algae Ivanova et al. 2004a emend. Nedashkovskaya et al. 2006d

Formosa Ivanova et al. 2004a emend. Nedashkovskaya et al. 2006d

Georgenia Altenburger et al. 2002 emend. W.J. Li et al. 2007

Glaciecola Bowman et al. 1998 emend. Van Trappen et al. 2004c

Gluconacetobacter hansenii corrig. (Gosselé et al. 1983) Yamada et al. 1998

Glycomyces Labeda et al. 1985 emend. Labeda and Kroppenstedt 2004

Glycomycetaceae Rainey et al. 1997 (complete authorship reads Rainey, Ward-Rainey and Stackebrandt) emend. Labeda and Kroppenstedt 2005

Hahella Lee et al. 2001 emend. Baik et al. 2005

Halobacillus Spring et al. 1996 emend. J.H. Yoon et al. 2007vi

Halobacterium salinarum corrig. (Harrison and Kennedy 1922) Elazari-Volcani 1957 (Approved Lists 1980) emend. Gruber et al. 2004 – Publication in the IJSEM: List of Changes in Taxonomic Opinion no. 2

Halochromatium Imhoff et al. 1998 emend. Anil Kumar et al. 2007c

Halomonadaceae Franzmann et al. 1989 emend. Ben Ali Gam et al. 2007

Halomonadaceae Franzmann et al. 1989 emend. Ntougias et al. 2007b

Halovibrio Fendrich 1989 emend. Sorokin et al. 2006a

Hongiella Yi and Chun 2004 emend. Nedashkovskaya et al. 2004h

Hymenobacter Hirsch et al. 1999 emend. Buczolits et al. 2006

Idiomarinaceae Ivanova et al. 2004d emend. Jean et al. 2006b

Labrys monachus corrig. Vasilyeva and Semenov 1985 emend. Islam et al. 2007

Labrys Vasilyeva and Semenov 1985 emend. Islam et al. 2007

Lactobacillus fermentum Beijerinck 1901 (Approved Lists 1980) emend. Dellaglio et al. 2004

Lactobacillus kefiranofaciens Fujisawa et al. 1988 emend. Vancanneyt et al. 2004b

Lactobacillus salivarius Rogosa et al. 1953 (Approved Lists 1980) emend. Y. Li et al. 2006

Y. Li et al. 2006 propose that the infraspecific division of *Lactobacillus salivarius* Rogosa et al. 1953 (Approved Lists 1980) into the subspecies *Lactobacillus salivarius* subsp. *salicinius* Rogosa et al. 1953 (Approved Lists 1980) and *Lactobacillus salivarius* subsp. *salivarius* Rogosa et al. 1953 (Approved Lists 1980) be discontinued.

Lactobacillus vaccinostercus Kozaki and Okada 1983 emend. Dellaglio et al. 2006

Lampropedia hyalina (Ehrenberg 1832) Schroeter 1886 (Approved Lists 1980) emend. N. Lee et al. 2004

Lampropedia hyalina (Ehrenberg 1832) Schroeter 1886 (Approved Lists 1980) emend. Xie and Yokota 2003 – Publication in the IJSEM: List of Changes in Taxonomic Opinion no. 2

Lampropedia Schroeter 1886 (Approved Lists 1980) emend. N. Lee et al. 2004

Leisingera Schaefer et al. 2002 emend. Martens et al. 2006

Lentibacillus Yoon et al. 2002 emend. Jeon et al. 2005a

Leptospiraceae Hovind-Hougen 1979 (Approved Lists 1980) emend. Levett et al. 2005

Leucobacter chromiireducens Morais et al. 2005 emend. Muir and Tan 2007

Lewinella cohaerens (Lewin 1970) Sly et al. 1998 emend. Khan et al. 2007a

Lewinella nigricans (Lewin 1970) Sly et al. 1998 emend. Khan et al. 2007a

Lewinella persica (Lewin 1970) Sly et al. 1998 emend. Khan et al. 2007a

Lewinella Sly et al. 1998 emend. Khan et al. 2007a

Marinibacillus Yoon et al. 2001 emend. J.H. Yoon et al. 2004g

Massilia timonae La Scola et al. 2000 emend. Lindquist et al. 2003 – Publication in the IJSEM: List of Changes in Taxonomic Opinion no. 1

Mesonia Nedashkovskaya et al. 2003 emend. Nedashkovskaya et al. 2006e

Methanoculleus thermophilus corrig. (Rivard and Smith 1982) Maestrojuán et al. 1990 emend. Spring et al. 2005a

Methanogenium Romesser et al. 1981 emend. Spring et al. 2005a

Methanogenium thermophilicum Rivard and Smith 1982 emend. Zabel et al. 1985 – Publication in the IJSEM: List of Changes in Taxonomic Opinion no. 1

Methanosarcina baltica von Klein et al. 2002 emend. Singh et al. 2005

Methylibium Nakatsu et al. 2006 emend. M.H. Yoon et al. 2007c

Methylocella Dedysh et al. 2000 emend. Dedysh et al. 2004

Methylocystis (*ex* Whittenbury et al. 1970) Bowman et al. 1993 emend. Dedysh et al. 2007

Methylosarcina Wise et al. 2001 emend. Kalyuzhnaya et al. 2005

Methylovorus glucosotrophus Govorukhina and Trotsenko 1991 emend. Doronina et al. 2005

Methylovorus Govorukhina and Trotsenko 1991 emend. Doronina et al. 2005

Methylovorus mays Doronina et al. 2001 emend. Doronina et al. 2005

Microcella Tiago et al. 2005b emend. Tiago et al. 2006b

Mobiluncus curtisii Spiegel and Roberts 1984 emend. Hoyles et al. 2004 – Publication in the IJSEM: List of Changes in Taxonomic Opinion no. 1

Mobiluncus mulieris Spiegel and Roberts 1984 emend. Hoyles et al. 2004 – Publication in the IJSEM: List of Changes in Taxonomic Opinion no. 1

Mobiluncus Spiegel and Roberts 1984 emend. Hoyles et al. 2004 – Publication in the IJSEM: List of Changes in Taxonomic Opinion no. 1

Modestobacter Mevs et al. 2000 emend. Reddy et al. 2007

Modestobacter multiseptatus Mevs et al. 2000 emend. Reddy et al. 2007

Muricauda Bruns et al. 2001 emend. J.H. Yoon et al. 2005v

Natronorubrum Xu et al. 1999 emend. Cui et al. 2006a

Nesterenkonia Stackebrandt et al. 1995 emend. W.J. Li et al. 2005b

Nocardiopsis salina W.J. Li et al. 2004d emend. W.J. Li et al. 2006a

Nonlabens tegetincola S.C. Lau et al. 2005c emend. S.C. Lau et al. 2006b

Oceanimonas corrig. Brown et al. 2001 emend. Ivanova et al. 2005d – Publication in the IJSEM: List of Changes in Taxonomic Opinion no. 2

Oceanithermus Miroshnichenko et al. 2003 emend. Mori et al. 2004

Oceanobacillus Lu et al. 2002 emend. J.S. Lee et al. 2006b

Oceanobacillus Lu et al. 2002 emend. Yumoto et al. 2005b

Oceanobacillus oncorhynchi Yumoto et al. 2005b emend. Romano et al. 2006b

Oenococcus Dicks et al. 1995 emend. Endo and Okada 2006

Paenibacillus larvae (White 1906) Ash et al. 1994 emend. Genersch et al. 2006

Pannonibacter Borsodi et al. 2003 emend. Biebl et al. 2007

Pasteurella aerogenes McAllister and Carter 1974 (Approved Lists 1980) emend. Christensen et al. 2005

Pasteurella mairii corrig. Sneath and Stevens 1990 emend. Christensen et al. 2005

Pasteuria nishizawae Sayre et al. 1992 emend. Noel et al. 2005

Pectinatus Lee et al. 1978 (Approved Lists 1980) emend. Juvonen and Suihko 2006

Pedobacter Steyn et al. 1998 emend. Gallego et al. 2006c

Pedobacter Steyn et al. 1998 emend. Hwang et al. 2006

Pedobacter Steyn et al. 1998 emend. Vanparys et al. 2005a

Pelotomaculum Imachi et al. 2002 emend. de Bok et al. 2005

Pelotomaculum Imachi et al. 2002 emend. Y.L. Qiu et al. 2006 – Publication in the IJSEM: List of Changes in Taxonomic Opinion no. 5

Phaeobacter Martens et al. 2006 emend. J.H. Yoon et al. 2007xi

Phenylobacterium Lingens et al. 1985 emend. Kanso and Patel 2004

Phenylobacterium Lingens et al. 1985 emend. Tiago et al. 2005a – Publication in the IJSEM: List of Changes in Taxonomic Opinion no. 3

Photorhabdus asymbiotica Fischer-Le Saux et al. 1999 emend. Akhurst et al. 2004

Phyllobacterium (*ex* Knösel 1962) Knösel 1984 emend. Jurado et al. 2005d

Phyllobacterium (*ex* Knösel 1962) Knösel 1984 emend. Mantelin et al. 2006

Pigmentiphaga Blümel et al. 2001 emend. J.H. Yoon et al. 2007iv

Pirellula Schlesner and Hirsch 1987 emend. Schlesner et al. 2004

Planococcus maritimus Yoon et al. 2003 emend. Ivanova et al. 2006b – Publication in the IJSEM: List of Changes in Taxonomic Opinion no. 6

Planotetraspora Runmao et al. 1993 emend. Tamura and Sakane 2004

Pontibacillus J.M. Lim et al. 2005b emend. J.M. Lim et al. 2005a

Pseudomonas asplenii (Ark and Tompkins 1946) Savulescu 1947 (Approved Lists 1980) emend. Tvrzová et al. 2006

Pseudomonas chlororaphis (Guignard and Sauvageau 1894) Bergey et al. 1930 (Approved Lists 1980) emend. Peix et al. 2007

Pseudomonas citronellolis Seubert 1960 (Approved Lists 1980) emend. Lang et al. 2007a

Pseudomonas nitroreducens Iizuka and Komagata 1964 (Approved Lists 1980) emend. Lang et al. 2007a

Pseudoxanthomonas broegbernensis Finkmann et al. 2000 emend. Thierry et al. 2004

Pseudoxanthomonas Finkmann et al. 2000 emend. Thierry et al. 2004

Ralstonia insidiosa Coenye et al. 2003 emend. Vaneechoutte et al. 2004

Rheinheimera Brettar et al. 2002 emend. Merchant et al. 2007

Rhodobacter Imhoff et al. 1984 emend. Srinivas et al. 2007d

Roseburia intestinalis Duncan et al. 2002 emend. Duncan et al. 2006

Roseibium denhamense Suzuki et al. 2000 emend. Biebl et al. 2007

Roseibium hamelinense Suzuki et al. 2000 emend. Biebl et al. 2007

Roseibium Suzuki et al. 2000 emend. Biebl et al. 2007

Roseivirga Nedashkovskaya et al. 2005c emend. Nedashkovskaya et al. 2005k

Roseobacter Shiba 1991 emend. Martens et al. 2006

Rothia dentocariosa (Onishi 1949) Georg and Brown 1967 (Approved Lists 1980) emend. Daneshvar et al. 2004 – Publication in the IJSEM: List of Changes in Taxonomic Opinion no. 1

Rubrobacteraceae Rainey et al. 1997 (complete authorship reads Rainey, Ward-Rainey and Stackebrandt) emend. Stackebrandt 2004 – Publication in the IJSEM: List of Changes in Taxonomic Opinion no. 2

Rubrobacteridae Rainey et al. 1997 (complete authorship reads Rainey, Ward-Rainey and Stackebrandt) emend. Stackebrandt 2004 – Publication in the IJSEM: List of Changes in Taxonomic Opinion no. 2

Ruegeria atlantica (Rüger and Höfle 1992) Uchino et al. 1999 emend. Yi et al. 2007

Ruegeria atlantica (Rüger and Höfle 1992) Uchino et al. 1999 emend. Muramatsu et al. 2007

Ruegeria Uchino et al. 1999 emend. Martens et al. 2006

Ruegeria Uchino et al. 1999 emend. Yi et al. 2007

Salegentibacter McCammon and Bowman 2000 emend. Ying et al. 2007a

Schlegelella thermodepolymerans Elbanna et al. 2003 emend. Lütke-Eversloh et al. 2004

Skermanella Sly and Stackebrandt 1999 emend. Weon et al. 2007b

Sphingosinicella Maruyama et al. 2006 emend. Geueke et al. 2007

Sphingosinicella microcystinivorans Maruyama et al. 2006 emend. Geueke et al. 2007

Staphylococcus caprae Devriese et al. 1983 emend. Kawamura et al. 1998 – Publication in the IJSEM: List of Changes in Taxonomic Opinion no. 1

Staphylococcus vitulinus corrig. Webster et al. 1994 emend. Švec et al. 2004

Stappia stellulata (Rüger and Höfle 1992) Uchino et al. 1999 emend. Biebl et al. 2007

Stappia Uchino et al. 1999 emend. Biebl et al. 2007

Streptomyces cinereorectus Terekhova and Preobrazhenskaya 1986 emend. Lanoot et al. 2004 – Publication in the IJSEM: List of Changes in Taxonomic Opinion no. 1

Streptomyces colombiensis Pridham et al. 1958 (Approved Lists 1980) emend. Lanoot et al. 2004 – Publication in the IJSEM: List of Changes in Taxonomic Opinion no. 1

Streptomyces corchorusii Ahmad and Bhuiyan 1958 (Approved Lists 1980) emend. Lanoot et al. 2005

Streptomyces filamentosus Okami and Umezawa 1953 (Approved Lists 1980) emend. Lanoot et al. 2004 – Publication in the IJSEM: List of Changes in Taxonomic Opinion no. 1

Streptomyces flavovirens (Waksman 1923) Waksman and Henrici 1948 (Approved Lists 1980) emend. Lanoot et al. 2005

Streptomyces fradiae (Waksman and Curtis 1916) Waksman and Henrici 1948 (Approved Lists 1980) emend. Lanoot et al. 2004 – Publication in the IJSEM: List of Changes in Taxonomic Opinion no. 1

Streptomyces griseus (Krainsky 1914) Waksman and Henrici 1948 (Approved Lists 1980) emend. Z. Liu et al. 2005b

Streptomyces microflavus (Krainsky 1914) Waksman and Henrici 1948 (Approved Lists 1980) emend. Lanoot et al. 2005

Streptomyces minutiscleroticus (Thirumalachar 1965) Pridham 1970 (Approved Lists 1980) emend. Lanoot et al. 2005

Streptomyces phaeopurpureus Shinobu 1957 (Approved Lists 1980) emend. Lanoot et al. 2004 – Publication in the IJSEM: List of Changes in Taxonomic Opinion no. 1

Streptomyces tricolor (Wollenweber 1920) Waksman 1961 (Approved Lists 1980) emend. Lanoot et al. 2004 – Publication in the IJSEM: List of Changes in Taxonomic Opinion no. 1

Streptomyces vinaceus Jones 1952 (Approved Lists 1980) emend. Lanoot et al. 2004 – Publication in the IJSEM: List of Changes in Taxonomic Opinion no. 1

Sulfitobacter Sorokin 1996 emend. J.H. Yoon et al. 2007x

Sulfurihydrogenibium azorense Aguiar et al. 2004 emend. S. Nakagawa et al. 2005b

Sulfurihydrogenibium subterraneum Takai et al. 2003 emend. S. Nakagawa et al. 2005b

Sulfurihydrogenibium Takai et al. 2003 emend. S. Nakagawa et al. 2005b

Sulfurimonas Inagaki et al. 2003 emend. Takai et al. 2006b

Suttonella Dewhirst et al. 1990 emend. Foster et al. 2005

Syntrophobacter Boone and Bryant 1984 emend. S. Chen et al. 2005

Syntrophomonas McInerney et al. 1982 emend. C. Wu et al. 2006

Tepidimonas Moreira et al. 2000 emend. Albuquerque et al. 2006 – Publication in the IJSEM: List of Changes in Taxonomic Opinion no. 5

Tetrasphaera Maszenan et al. 2000 emend. Ishikawa and Yokota 2006

Thalassobacter Macián et al. 2005a emend. Pujalte et al. 2005b

Thalassomonas Macián et al. 2001 emend. Jean et al. 2006c

Thalassospira López-López et al. 2002 emend. C. Liu et al. 2007

Thermoactinomyces Tsilinsky 1899 (Approved Lists 1980) emend. J.H. Yoon et al. 2005n

Thermoanaerobacter mathranii Larsen et al. 1998 emend. Carlier et al. 2006 – Publication in the IJSEM: List of Changes in Taxonomic Opinion no. 5

Thermoanaerobacter Wiegel and Ljungdahl 1982 emend. Y.J. Lee et al. 2007a

Thermomicrobia Garrity and Holt 2002 emend. Hugenholtz and Stackebrandt 2004

Thioalkalivibrio Sorokin et al. 2001 emend. Banciu et al. 2004 – Publication in the IJSEM: List of Changes in Taxonomic Opinion no. 2

Thiomonas cuprina Moreira and Amils 1997 emend. Kelly et al. 2007

Thiomonas delicata (Katayama-Fujimura et al. 1984) Kelly and Wood 2006 emend. Katayama et al. 2006

Thiomonas Moreira and Amils 1997 emend. Kelly et al. 2007

Tissierella Collins and Shah 1986 emend. J.W. Bae et al. 2004

Trichococcus collinsii Liu et al. 2002 emend. Pikuta et al. 2006a

Xanthomonas citri (*ex* Hasse 1915) Gabriel et al. 1989 emend. Schaad et al. 2006 – Publication in the IJSEM: List of Changes in Taxonomic Opinion no. 6

Yania W.J. Li et al. 2004a emend. W.J. Li et al. 2005e

Zymomonas mobilis subsp. *mobilis* (Lindner 1928) De Ley and Swings 1976 emend. Coton et al. 2006

Zymomonas mobilis subsp. *pomaceae* (Millis 1956) De Ley and Swings 1976 emend. Coton et al. 2006

Synonyms

Aeromonas culicicola Pidiyar et al. 2002 pro synon. *Aeromonas veronii* Hickman-Brenner et al. 1988 – Publication in the IJSEM: List of Changes in Taxonomic Opinion no. 3 – Original publication: Huys et al. (2005)

Bacillus axarquiensis Ruiz-García et al. 2005 pro synon. *Bacillus mojavensis* Roberts et al. 1994 – Publication in the IJSEM: Wang et al. (2007)

Bacillus malacitensis Ruiz-García et al. 2005 pro synon. *Bacillus mojavensis* Roberts et al. 1994 – Publication in the IJSEM: L.T. Wang et al. (2007)

Brevibacterium liquefaciens Okabayashi and Masuo 1960 (Approved Lists 1980) pro synon. *Arthrobacter nicotianae* Giovannozzi-Sermanni 1959 (Approved List 1980) – Publication in the IJSEM: Gelsomino et al. (2004)

Caenibacterium Manaia et al. 2003 pro synon. *Schlegelella* Elbanna et al. 2003 – Publication in the IJSEM: Lütke-Eversloh et al. (2004) [Rule 37a(1)]

Caenibacterium thermophilum Manaia et al. 2003 pro synon. *Schlegelella thermodepolymerans* Elbanna et al. 2003 – Publication in the IJSEM: Lütke-Eversloh et al. (2004)

Chimaereicella Tiago et al. 2006a pro synon. *Algoriphagus* Bowman et al. 2003 – Publication in the IJSEM: Nedashkovskaya et al. (2007b) [Rule 37a(1)]

Clostridium absonum Nakamura et al. 1973 (Approved Lists 1980) pro synon. *Clostridium sardiniense* Prévot 1938 (Approved Lists 1980) – Publication in the IJSEM: X. Wang et al. (2005)

Clostridium hastiforme MacLennan 1939 (Approved Lists 1980) pro synon. *Tissierella praeacuta* (Tissier 1908) Collins and Shah 1986 – Publication in the IJSEM: J.W. Bae et al. (2004)

Corynebacterium mooreparkense Brennan et al. 2001 pro synon. *Corynebacterium variabile* corrig. (Müller 1961) Collins 1987 – Publication in the IJSEM: Gelsomino et al. (2005)

Corynebacterium nigricans Shukla et al. 2004 pro synon. *Corynebacterium aurimucosum* Yassin et al. 2002 – Publication in the IJSEM: List of Changes in Taxonomic Opinion no. 1 – Original publication: Daneshvar et al. (2004)

Desulfitobacterium frappieri Bouchard et al. 1996 pro synon. *Desulfitobacterium hafniense* Christiansen and Ahring 1996 –

Publication in the IJSEM: List of Changes in Taxonomic Opinion no. 2 – Original publication: Niggemyer et al. (2001)

Desulfomusa Finster et al. 2001 pro synon. *Desulfofaba* Knoblauch et al. 1999 – Publication in the IJSEM: Abildgaard et al. (2004) [Rule 37a(1)]

Enterococcus flavescens Pompei et al. 1992 pro synon. *Enterococcus casseliflavus* (*ex* Vaughan et al. 1979) Collins et al. 1984 – Publication in the IJSEM: Naser et al. (2006c)

Enterococcus saccharominimus Vancanneyt et al. 2004c pro synon. *Enterococcus italicus* Fortina et al. 2004 – Publication in the IJSEM: Naser et al. (2006c)

Falcivibrio grandis Hammann et al. 1984 pro synon. *Mobiluncus mulieris* Spiegel and Roberts 1984 – Publication in the IJSEM: List of Changes in Taxonomic Opinion no. 1 – Original publication: Hoyles et al. (2004)

Falcivibrio vaginalis Hammann et al. 1984 pro synon. *Mobiluncus curtisii* Spiegel and Roberts 1984 – Publication in the IJSEM: List of Changes in Taxonomic Opinion no. 1 – Original publication: Hoyles et al. (2004)

Flexibacter aggregans (Lewin 1969) Leadbetter 1974 (Approved Lists 1980) pro synon. *Flexithrix dorotheae* Lewin 1970 (Approved Lists 1980) – Publication in the IJSEM: Hosoya and Yokota (2007a)

Gaetbulimicrobium J.H. Yoon et al. 2006a pro synon. *Aquimarina* Nedashkovskaya et al. 2005h – Publication in the IJSEM: Nedashkovskaya et al. (2006f) [Rule 37a(1)]

Gordonia nitida Yoon et al. 2000 pro synon. *Gordonia alkanivorans* Kummer et al. 1999 – Publication in the IJSEM: Arenskötter et al. (2005)

Hemophilus paraphrophilus Zinnemann et al. 1968 pro synon. *Hemophilus aphrophilus* Khairat 1940 (Approved Lists 1980) – Publication in the IJSEM: Nørskov-Lauritsen and Kilian (2006)

Hongiella Yi and Chun 2004 pro synon. *Algoriphagus* Bowman et al. 2003 – Publication in the IJSEM: Nedashkovskaya et al. (2007b) [Rule 37a(1)]

Jannaschia cystaugens Adachi et al. 2004 synon. *Thalassobacter stenotrophicus* Macián et al. 2005a – Publication in the IJSEM: Pujalte et al. (2005b) [Rules 15 and 17]

Kluyvera cochleae Müller et al. 1996 pro synon. *Kluyvera intermedia* (Izard et al. 1980) Pavan et al. 2005 – Publication in the IJSEM: Pavan et al. (2005)

Lactobacillus arizonensis Swezey et al. 2000 pro synon. *Lactobacillus plantarum* (Orla-Jensen 1919) Bergey et al. 1923 (Approved Lists 1980) – Publication in the IJSEM: Kostinek et al. (2005)

Lactobacillus cellobiosus Rogosa et al. 1953 (Approved Lists 1980) pro synon. *Lactobacillus fermentum* Beijerinck 1901 (Approved Lists 1980) – Publication in the IJSEM: Dellaglio et al. (2004)

Lactobacillus curvatus subsp. *melibiosus* Torriani et al. 1996 pro synon. *Lactobacillus sakei* subsp. *carnosus* corrig. Torriani et al. 1996 – Publication in the IJSEM: Koort et al. (2004)

Lactobacillus cypricasei Lawson et al. 2001 pro synon. *Lactobacillus acidipiscis* Tanasupawat et al. 2000 – Publication in the IJSEM: Naser et al. (2006b)

Lactobacillus durianis Leisner et al. 2002 pro synon. *Lactobacillus vaccinostercus* Kozaki and Okada 1983 – Publication in the IJSEM: Dellaglio et al. (2006)

Lactobacillus ferintoshensis Simpson et al. 2002 pro synon. *Lactobacillus parabuchneri* Farrow et al. 1989 – Publication in the IJSEM: Vancanneyt et al. (2005a)

Lactobacillus suntoryeus Cachat and Priest 2005 pro synon. *Lactobacillus helveticus* (Orla-Jensen 1919) Bergey et al. 1925 (Approved Lists 1980) – Publication in the IJSEM: Naser et al. (2006a)

Lactobacillus thermotolerans Niamsup et al. 2003 pro synon. *Lactobacillus ingluviei* Baele et al. 2003 – Publication in the IJSEM: Felis et al. (2006)

Leuconostoc argentinum Dicks et al. 1993 pro synon. *Leuconostoc lactis* Garvie 1960 (Approved Lists 1980) – Publication in the IJSEM: Vancanneyt et al. (2006c)

Marinicola J.H. Yoon et al. 2005c pro synon. *Roseivirga* Nedashkovskaya et al. 2005c – Publication in the IJSEM: Nedashkovskaya et al. (2005c) and S.C. Lau et al. (2006b) [Rule 37a(1)]

Marinobacter aquaeolei Nguyen et al. 1999 pro synon. *Marinobacter hydrocarbonoclasticus* Gauthier et al. 1992 – Publication in the IJSEM: Márquez and Ventosa (2005)

Methanogenium frittonii Harris et al. 1996 pro synon. *Methanoculleus thermophilus* corrig. (Rivard and Smith 1982) Maestrojuán et al. 1990 – Publication in the IJSEM: Spring et al. (2005a)

Methylobacterium chloromethanicum McDonald et al. 2001 pro synon. *Methylobacterium extorquens* (Urakami and Komagata 1984) Bousfield and Green 1985 – Publication in the IJSEM: List of Changes in Taxonomic Opinion no. 4 – Original publication: Y. Kato et al. (2005)

Methylobacterium dichloromethanicum Doronina et al. 2000 pro synon. *Methylobacterium extorquens* (Urakami and Komagata 1984) Bousfield and Green 1985 – Publication in the IJSEM: List of Changes in Taxonomic Opinion no. 4 – Original publication: Y. Kato et al. (2005)

Methylobacterium lusitanum Doronina et al. 2002 pro synon. *Methylobacterium rhodesianum* Green et al. 1988 – Publication in the IJSEM: List of Changes in Taxonomic Opinion no. 4 – Original publication: Y. Kato et al. (2005)

Novosphingobium subarcticum (Nohynek et al. 1996) pro synon. *Novosphingobium resinovorum* (Delaporte and Daste 1956) Lim et al. 2007 – Publication in the IJSEM: Y.M. Lim et al. (2007)

Pseudomonas chloritidismutans Wolterink et al. 2002 pro synon. *Pseudomonas stutzeri* (Lehmann and Neumann 1896) Sijderius 1946 (Approved Lists 1980) – Publication in the IJSEM: List of Changes in Taxonomic Opinion no. 4 – Original publication: Cladera et al. (2006)

Pseudomonas multiresinivorans Mohn et al. 1999 pro synon. *Pseudomonas nitroreducens* Iizuka and Komagata 1964 (Approved Lists 1980) emend. Lang et al. 2007a – Publication in the IJSEM: Lang et al. (2007a)

Roseomonas fauriae Rihs et al. 1998 pro synon. *Azospirillum brasilense* corrig. Tarrand et al. 1979 (Approved Lists 1980) – Publication in the IJSEM: Helsel et al. (2006)

Salmonella choleraesuis corrig. (Smith 1894) Weldin 1927 (Approved Lists 1980) synon. *Salmonella enterica* (ex Kauffmann and Edwards 1952) Le Minor and Popoff 1987 – Publication in the IJSEM: Tindall et al. (2005)

Salmonella choleraesuis subsp. *choleraesuis* corrig. (Smith 1894) Weldin 1927 synon. *Salmonella enterica* subsp. *enterica* (ex Kauffmann and Edwards 1952) Le Minor and Popoff 1987 – Publication in the IJSEM: Tindall et al. (2005)

Salmonella enteritidis (Gaertner 1888) Castellani and Chalmers 1919 (Approved Lists 1980) synon. *Salmonella enterica* subsp. *enterica* (ex Kauffmann and Edwards 1952) Le Minor and Popoff 1987 – Publication in the IJSEM: Tindall et al. (2005)

Salmonella paratyphi (ex Kayser 1902) Ezaki et al. 2000. synon. *Salmonella enterica* subsp. *enterica* (ex Kauffmann and Edwards 1952) Le Minor and Popoff 1987 – Publication in the IJSEM: Tindall et al. (2005)

Salmonella typhi (Schroeter 1886) Warren and Scott 1930 (Approved Lists 1980) synon. *Salmonella enterica* subsp. *enterica* (ex Kauffmann and Edwards 1952) Le Minor and Popoff 1987 – Publication in the IJSEM: Tindall et al. (2005)

Salmonella typhimurium (Loeffler 1892) Castellani and Chalmers 1919 (Approved Lists 1980) synon. *Salmonella enterica* subsp. *enterica* (ex Kauffmann and Edwards 1952) Le Minor and Popoff 1987 – Publication in the IJSEM: Tindall et al. (2005)

Shewanella affinis pro synon. *Shewanella colwelliana* (Weiner et al. 1988) Coyne et al. 1990 – Publication in the IJSEM: Satomi et al. (2007)

Silicibacter Petursdottir and Kristjansson 1999 pro synon. *Ruegeria* Uchino et al. 1999 – Publication in the IJSEM: Yi et al. (2007) [Rule 37a(1)]

Sphingomonas subarctica pro synon. *Novosphingobium resinovorum* (Delaporte and Daste 1956) Lim et al. 2007 – Publication in the IJSEM: Y.W. Lim et al. (2007)

Staleya Labrenz et al. 2000 pro synon. *Sulfitobacter* Sorokin 1996 – Publication in the IJSEM: J.H. Yoon et al. (2007x) [Rule 37a(1)]

Stanierella Nedashkovskaya et al. 2005h pro synon. *Aquimarina* Nedashkovskaya et al. 2005h – Publication in the IJSEM: Nedashkovskaya et al. (2006f) [Rule 37a(1)]

Staphylococcus pulvereri Zakrzewska-Czerwińska et al. 1995 pro synon. *Staphylococcus vitulinus* corrig. Webster et al. 1994 – Publication in the IJSEM: Švec et al. (2004)

Stenotrophomonas africana Drancourt et al. 1997 pro synon. *Stenotrophomonas maltophilia* (Hugh 1981) Palleroni and Bradbury 1993 – Publication in the IJSEM: Coenye et al. (2004b)

Streptococcus difficilis corrig. Eldar et al. 1995 pro synon. *Streptococcus agalactiae* Lehmann and Neumann 1896 (Approved Lists 1980) – Publication in the IJSEM: Kawamura et al. (2005)

Streptomyces arabicus Shibata et al. 1957 (Approved Lists 1980) pro synon. *Streptomyces vinaceus* Jones 1952 (Approved Lists 1980) – Publication in the IJSEM: List of Changes in Taxonomic Opinion no. 1 – Original publication: Lanoot et al. (2004)

Streptomyces argenteolus Tresner et al. 1961 (Approved Lists 1980) pro synon. *Streptomyces griseus* (Krainsky 1914) Waksman and Henrici 1948 (Approved Lists 1980) – Publication in the IJSEM: Z. Liu et al. (2005b)

Streptomyces caviscabies Goyer et al. 1996 pro synon. *Streptomyces griseus* (Krainsky 1914) Waksman and Henrici 1948 (Approved Lists 1980) – Publication in the IJSEM: Z. Liu et al. (2005b)

Streptomyces chibaensis Suzuki et al. 1958 (Approved Lists 1980) pro synon. *Streptomyces corchorusii* Ahmad and Bhuiyan 1958 (Approved Lists 1980) – Publication in the IJSEM: Lanoot et al. (2005)

Streptomyces chrysomallus Lindenbein 1952 (Approved Lists 1980) pro synon. *Streptomyces anulatus* (Beijerinck 1912) Waksman 1953 (Approved Lists 1980) – Publication in the IJSEM: Lanoot et al. (2005) [Rule 37a(1)]

Streptomyces chrysomallus subsp. *chrysomallus* Lindenbein 1952 (Approved Lists 1980) pro synon. *Streptomyces anulatus* (Beijerinck 1912) Waksman 1953 (Approved Lists 1980) – Publication in the IJSEM: Lanoot et al. (2005)

Streptomyces citreofluorescens (Korenyako et al. 1960) Pridham 1970 (Approved Lists 1980) pro synon. *Streptomyces anulatus* (Beijerinck 1912) Waksman 1953 (Approved Lists 1980) – Publication in the IJSEM: Lanoot et al. (2005)

Streptomyces cochleatus Nakagaito et al. 1993 pro synon. *Streptomyces cinereorectus* Terekhova and Preobrazhenskaya 1986 – Publication in the IJSEM: List of Changes in Taxonomic Opinion no. 1 – Original publication: Lanoot et al. (2004)

Streptomyces distallicus (Locci et al. 1969) Witt and Stackebrandt 1991 pro synon. *Streptomyces colombiensis* Pridham et al. 1958 (Approved Lists 1980) – Publication in the IJSEM: List of Changes in Taxonomic Opinion no. 1 – Original publication: Lanoot et al. (2004)

Streptomyces flaviscleroticus (*ex* Pridham 1970) Goodfellow et al. 1986 pro synon. *Streptomyces minutiscleroticus* (Thirumalachar 1965) Pridham 1970 (Approved Lists 1980) – Publication in the IJSEM: Lanoot et al. (2005)

Streptomyces fluorescens (Krasil'nikov 1958) Pridham 1970 (Approved Lists 1980) pro synon. *Streptomyces anulatus* (Beijerinck 1912) Waksman 1953 (Approved Lists 1980) – Publication in the IJSEM: Lanoot et al. (2005)

Streptomyces griseus subsp. *alpha* (Ciferri 1927) Pridham 1970 (Approved Lists 1980) pro synon. *Streptomyces microflavus* (Krainsky 1914) Waksman and Henrici 1948 (Approved Lists 1980) – Publication in the IJSEM: Lanoot et al. (2005)

Streptomyces griseus subsp. *cretosus* Pridham 1970 (Approved Lists 1980) pro synon. *Streptomyces microflavus* (Krainsky 1914) Waksman and Henrici 1948 (Approved Lists 1980) – Publication in the IJSEM: Lanoot et al. (2005)

Streptomyces lipmanii (Waksman and Curtis 1916) Waksman and Henrici 1948 (Approved Lists 1980) pro synon. *Streptomyces microflavus* (Krainsky 1914) Waksman and Henrici 1948 (Approved Lists 1980) – Publication in the IJSEM: Lanoot et al. (2005)

Streptomyces nigrifaciens Waksman 1961 (Approved Lists 1980) pro synon. *Streptomyces flavovirens* (Waksman 1923) Waksman and Henrici 1948 (Approved Lists 1980) – Publication in the IJSEM: Lanoot et al. (2005)

Streptomyces phaeoviridis Shinobu 1957 (Approved Lists 1980) pro synon. *Streptomyces phaeopurpureus* Shinobu 1957 (Approved Lists 1980) – Publication in the IJSEM: List of Changes in Taxonomic Opinion no. 1 – Original publication: Lanoot et al. (2004)

Streptomyces roseodiastaticus (Duché 1934) Waksman 1953 (Approved Lists 1980) pro synon. *Streptomyces tricolor* (Wollenweber 1920) Waksman 1961 (Approved Lists 1980) – Publication in the IJSEM: List of Changes in Taxonomic Opinion no. 1 – Original publication: Lanoot et al. (2004)

Streptomyces roseoflavus Arai 1951 (Approved Lists 1980) pro synon. *Streptomyces fradiae* (Waksman and Curtis 1916) Waksman and Henrici 1948 (Approved Lists 1980) – Publication in the IJSEM: List of Changes in Taxonomic Opinion no. 1 – Original publication: Lanoot et al. (2004)

Streptomyces roseosporus Falcão de Morais and Dália Maia 1961 (Approved Lists 1980) pro synon. *Streptomyces filamentosus* Okami and Umezawa 1953 (Approved Lists 1980) – Publication in the IJSEM: List of Changes in Taxonomic Opinion no. 1 – Original publication: Lanoot et al. (2004)

Streptomyces setonii (Millard and Burr 1926) Waksman 1953 (Approved Lists 1980) pro synon. *Streptomyces griseus* (Krainsky 1914) Waksman and Henrici 1948 (Approved Lists 1980) – Publication in the IJSEM: Z. Liu et al. (2005b)

Streptomyces willmorei (Erikson 1935) Waksman and Henrici 1948 (Approved Lists 1980) pro synon. *Streptomyces microflavus* (Krainsky 1914) Waksman and Henrici 1948 (Approved Lists 1980) – Publication in the IJSEM: Lanoot et al. (2005)

Syntrophospora Zhao et al. 1990 pro synon. *Syntrophomonas* McInerney et al. 1982 – Publication in the IJSEM: C. Wu et al. (2006) [Rule 37a(1)]

C. Wu et al. 2006 propose to transfer *Syntrophospora bryantii* (Stieb and Schink 1985) Zhao et al. 1990 (the type species of the genus *Syntrophospora* Zhao et al. 1990) to the genus *Syntrophomonas* McInerney et al. 1982 as *Syntrophomonas bryantii* (Stieb and Schink 1985) C. Wu et al. 2006, comb. nov. According to Rule 37a(1), bacteriologists adhering to this proposal should change the name *Syntrophospora* Zhao et al. 1990 to *Syntrophomonas* McInerney et al. 1982. However, *Syntrophomonas bryantii* (Stieb and Schink 1985) C. Wu et al. 2006 is not validly published [see above *Syntrophomonas bryantii* (Stieb and Schink 1985) C. Wu et al. 2006, 2335]

Thermoterrabacterium Slobodkin et al. 1997 pro synon. *Carboxydothermus* Svetlichny et al. 1991 – Publication in the IJSEM: Slobodkin et al. (2006a) [Rule 37a(1)]

Trichlorobacter De Wever et al. 2001 pro synon. *Geobacter* Lovley et al. 1995 – Publication in the IJSEM: Nevin et al. (2007) [Rule 37a(1)]

Wautersia eutropha (Davis 1969) Vaneechoutte et al. 2004 pro synon. *Cupriavidus necator* Makkar and Casida 1987 – Publication in the IJSEM: Vandamme and Coenye (2004)

Wautersia Vaneechoutte et al. 2004 pro synon. *Cupriavidus* Makkar and Casida 1987 – Publication in the IJSEM: Vandamme and Coenye 2004 [Rule 37a(1)]

Weissella kimchii Choi et al. 2002 pro synon. *Weissella cibaria* Björkroth et al. 2002 – Publication in the IJSEM: Ennahar and Cai (2004)

References for Validation Lists and Lists of changes in Taxonomic Opinion cited in this chapter

Validation of publication of new names and new combinations previously effectively published outside the IJSEM – Validation List no. 95. 2004. Int. J. Syst. Evol. Microbiol. *54*: 1–2.

Validation of publication of new names and new combinations previously effectively published outside the IJSEM – Validation List no. 96. 2004. Int. J. Syst. Evol. Microbiol. *54*: 307–308.

Validation of publication of new names and new combinations previously effectively published outside the IJSEM – Validation List no. 97. 2004. Int. J. Syst. Evol. Microbiol. *54*: 631–632.

Validation of publication of new names and new combinations previously effectively published outside the IJSEM – Validation List no. 98. 2004. Int. J. Syst. Evol. Microbiol. *54*: 1005–1006.

Validation of publication of new names and new combinations previously effectively published outside the IJSEM – Validation List no. 99. 2004. Int. J. Syst. Evol. Microbiol. *54*: 1425–1426.

Validation of publication of new names and new combinations previously effectively published outside the IJSEM – Validation List no. 100. 2004. Int. J. Syst. Evol. Microbiol. *54*: 1909–1910.

Validation of publication of new names and new combinations previously effectively published outside the IJSEM – Validation List no. 101. 2005. Int. J. Syst. Evol. Microbiol. *55*: 1–2.

Validation of publication of new names and new combinations previously effectively published outside the IJSEM – Validation List no. 102. 2005. Int. J. Syst. Evol. Microbiol. *55*: 547–549.

Validation of publication of new names and new combinations previously effectively published outside the IJSEM – Validation List no. 103. 2005. Int. J. Syst. Evol. Microbiol. *55*: 983–985.

Validation of publication of new names and new combinations previously effectively published outside the IJSEM – Validation List no. 104. 2005. Int. J. Syst. Evol. Microbiol. *55*: 1395–1397.

Validation of publication of new names and new combinations previously effectively published outside the IJSEM – Validation List no. 105. 2005. Int. J. Syst. Evol. Microbiol. *55*: 1743–1745.

Validation of publication of new names and new combinations previously effectively published outside the IJSEM – Validation List no. 106. 2005. Int. J. Syst. Evol. Microbiol. *55*: 2235–2238.

List of new names and new combinations previously effectively, but not validly, published – Validation List no. 107. 2006. Int. J. Syst. Evol. Microbiol. *56*: 1–6.

List of new names and new combinations previously effectively, but not validly, published – Validation List no. 108. 2006. Int. J. Syst. Evol. Microbiol. *56*: 499–500.

List of new names and new combinations previously effectively, but not validly, published – Validation List no. 109. 2006. Int. J. Syst. Evol. Microbiol. *56*: 925–927.

List of new names and new combinations previously effectively, but not validly, published – Validation List no. 110. 2006. Int. J. Syst. Evol. Microbiol. *56*: 1459–1460.

List of new names and new combinations previously effectively, but not validly, published – Validation List no. 111. 2006. Int. J. Syst. Evol. Microbiol. *56*: 2025–2027.

List of new names and new combinations previously effectively, but not validly, published – Validation List no. 112. 2006. Int. J. Syst. Evol. Microbiol. *56*: 2507–2508.

List of new names and new combinations previously effectively, but not validly, published – Validation List no. 113. 2007. Int. J. Syst. Evol. Microbiol. *57*: 1.

List of new names and new combinations previously effectively, but not validly, published – Validation List no. 114. 2007. Int. J. Syst. Evol. Microbiol. *57*: 433–434.

List of new names and new combinations previously effectively, but not validly, published – Validation List no. 115. 2007. Int. J. Syst. Evol. Microbiol. *57*: 893–897.

List of new names and new combinations previously effectively, but not validly, published – Validation List no. 116. 2007. Int. J. Syst. Evol. Microbiol. *57*: 1371–1373.

List of new names and new combinations previously effectively, but not validly, published – Validation List no. 117. 2007. Int. J. Syst. Evol. Microbiol. *57*: 1933–1934.

List of new names and new combinations previously effectively, but not validly, published – Validation List no. 118. 2007. Int. J. Syst. Evol. Microbiol. *57*: 2449–2450.

Notification of changes in taxonomic opinion previously published outside the IJSEM – List of Changes in Taxonomic Opinion no. 1. 2005. Int. J. Syst. Evol. Microbiol. *55*: 7–8.

Notification of changes in taxonomic opinion previously published outside the IJSEM – List of Changes in Taxonomic Opinion no. 2. 2005. Int. J. Syst. Evol. Microbiol. *55*: 1403–1404.

Notification of changes in taxonomic opinion previously published outside the IJSEM – List of Changes in Taxonomic Opinion no. 3. 2006. Int. J. Syst. Evol. Microbiol. *56*: 11.

Notification of changes in taxonomic opinion previously published outside the IJSEM – List of Changes in Taxonomic Opinion no. 4. 2006. Int. J. Syst. Evol. Microbiol. *56*: 1463.

Notification of changes in taxonomic opinion previously published outside the IJSEM – List of Changes in Taxonomic Opinion no. 5. 2007. Int. J. Syst. Evol. Microbiol. *57*: 4–5.

Notification of changes in taxonomic opinion previously published outside the IJSEM – List of Changes in Taxonomic Opinion no. 6. 2007. Int. J. Syst. Evol. Microbiol. *57*: 1376.

References

Abildgaard, L., N.B. Ramsing and K. Finster. 2004. Characterization of the marine propionate-degrading, sulfate-reducing bacterium *Desulfofaba fastidiosa* sp. nov. and reclassification of *Desulfomusa hansenii* as *Desulfofaba hansenii* comb. nov. Int. J. Syst. Evol. Microbiol. *54*: 393–399.

Abildgaard, L., M.B. Nielsen, K.U. Kjeldsen and K. Ingvorsen. 2006. *Desulfovibrio alkalitolerans* sp. nov., a novel alkalitolerant, sulphate-reducing bacterium isolated from district heating water. Int. J. Syst. Evol. Microbiol. *56*: 1019–1024.

Abou-Zeid, D.M., H. Biebl, C. Spröer and R.J. Müller. 2004. *Propionispora hippei* sp. nov., a novel Gram-negative, spore-forming anaerobe that produces propionic acid. Int. J. Syst. Evol. Microbiol. *54*: 951–954.

Abraham, W.R., C. Strömpl, M. Vancanneyt, A. Bennasar, J. Swings, H. Lünsdorf, J. Smit and E.R. Moore. 2004. *Woodsholea maritima* gen. nov., sp. nov., a marine bacterium with a low diversity of polar lipids. Int. J. Syst. Evol. Microbiol. *54*: 1227–1234.

Adachi, K., A. Katsuta, S. Matsuda, X. Peng, N. Misawa, Y. Shizuri, R.M. Kroppenstedt, A. Yokota and H. Kasai. 2007. *Smaragdicoccus niigatensis* gen. nov., sp. nov., a novel member of the suborder *Corynebacterineae*. Int. J. Syst. Evol. Microbiol. *57*: 297–301.

Adachi, M., T. Kanno, R. Okamoto, A. Shinozaki, K. Fujikawa-Adachi and T. Nishijima. 2004. *Jannaschia cystaugens* sp. nov., an *Alexandrium* (Dinophyceae) cyst formation-promoting bacterium from Hiroshima Bay, Japan. Int. J. Syst. Evol. Microbiol. *54*: 1687–1692.

Addison, S.L., S.M. Foote, N.M. Reid and G. Lloyd-Jones. 2007. *Novosphingobium nitrogenifigens* sp. nov., a polyhydroxyalkanoate-accumulating diazotroph isolated from a New Zealand pulp and paper wastewater. Int. J. Syst. Evol. Microbiol. *57*: 2467–2471.

Adékambi, T., M. Reynaud-Gaubert, G. Greub, M.J. Gevaudan, B. La Scola, D. Raoult and M. Drancourt. 2004. Amoebal coculture of "*Mycobacterium massiliense*" sp. nov. from the sputum of a patient with hemoptoic pneumonia. J. Clin. Microbiol. *42*: 5493–5501.

Adékambi, T., A. Stein, J. Carvajal, D. Raoult and M. Drancourt. 2006a. Description of *Mycobacterium conceptionense* sp. nov., a *Mycobacterium fortuitum* group organism isolated from a posttraumatic osteitis inflammation. J. Clin. Microbiol. *44*: 1268–1273.

Adékambi, T., P. Berger, D. Raoult and M. Drancourt. 2006b. *rpoB* gene sequence-based characterization of emerging non-tuberculous mycobacteria with descriptions of *Mycobacterium bolletii* sp. nov., *Mycobacterium phocaicum* sp. nov. and *Mycobacterium aubagnense* sp. nov. Int. J. Syst. Evol. Microbiol. *56*: 133–143.

Aguiar, P., T.J. Beveridge and A.L. Reysenbach. 2004. *Sulfurihydrogenibium azorense* sp. nov., a thermophilic hydrogen-oxidizing microaerophile from terrestrial hot springs in the Azores. Int. J. Syst. Evol. Microbiol. *54*: 33–39.

Aguilera, M., A. Cabrera, C. Incerti, S. Fuentes, N.J. Russell, A. Ramos-Cormenzana and M. Monteoliva-Sánchez. 2007. *Chromohalobacter salarius* sp. nov., a moderately halophilic bacterium isolated from a solar saltern in Cabo de Gata, Almeria, southern Spain. Int. J. Syst. Evol. Microbiol. *57*: 1238–1242.

Ahmed, I., A. Yokota and T. Fujiwara. 2007a. A novel highly boron tolerant bacterium, *Bacillus boroniphilus* sp. nov., isolated from soil, that requires boron for its growth. Extremophiles. *11*: 217–224.

Ahmed, I., A. Yokota and T. Fujiwara. 2007b. *Gracilibacillus boraciitolerans* sp. nov., a highly boron-tolerant and moderately halotolerant bacterium isolated from soil. Int. J. Syst. Evol. Microbiol. *57*: 796–802.

Ahmed, I., A. Yokota and T. Fujiwara. 2007c. *Chimaereicella boritolerans* sp. nov., a boron-tolerant and alkaliphilic bacterium of the family *Flavobacteriaceae* isolated from soil. Int. J. Syst. Evol. Microbiol. *57*: 986–992.

Ahmed, I., A. Yokota, A. Yamazoe and T. Fujiwara. 2007d. Proposal of *Lysinibacillus boronitolerans* gen. nov. sp. nov. and transfer of *Bacillus fusiformis* to *Lysinibacillus fusiformis* comb. nov. and *Bacillus sphaericus* to *Lysinibacillus sphaericus* comb. nov. Int. J. Syst. Evol. Microbiol. *57*: 1117–1125.

Aizawa, T., N.B. Ve, K. Kimoto, N. Iwabuchi, H. Sumida, I. Hasegawa, S. Sasaki, T. Tamura, T. Kudo, K. Suzuki, M. Nakajima and M. Sunairi. 2007. *Curtobacterium ammoniigenes* sp. nov., an ammonia-producing bacterium isolated from plants inhabiting acidic swamps in actual acid sulfate soil areas of Vietnam. Int. J. Syst. Evol. Microbiol. *57*: 1447–1452.

Akhurst, R.J., N.E. Boemare, P.H. Janssen, M.M. Peel, D.A. Alfredson and C.E. Beard. 2004. Taxonomy of Australian clinical isolates of the genus *Photorhabdus* and proposal of *Photorhabdus asymbiotica* subsp. *asymbiotica* subsp. nov. and *P. asymbiotica* subsp. *australis* subsp. nov. Int. J. Syst. Evol. Microbiol. *54*: 1301–1310.

Alain, K., S. Rolland, P. Crassous, F. Lesongeur, M. Zbinden, C. Le Gall, A. Godfroy, A. Page, S.K. Juniper, M.A. Cambon-Bonavita, F. Duchiron and J. Querellou. 2003. *Desulfurobacterium crinifex* sp. nov., a novel thermophilic, pinkish-streamer forming, chemolithoautotrophic bacterium isolated from a Juan de Fuca Ridge hydrothermal vent and amendment of the genus *Desulfurobacterium*. Extremophiles. *7*: 361–370.

Alam, S.I., L. Singh, S. Dube, G.S.N. Reddy and S. Shivaji. 2003. Psychrophilic *Planococcus maitriensis* sp. nov. from Antarctica. Syst. Appl. Microbiol. *26*: 505–510.

Alam, S.I., A. Dixit, G.S. Reddy, S. Dube, M. Palit, S. Shivaji and L. Singh. 2006. *Clostridium schirmacherense* sp. nov., an obligately anaero-bic, proteolytic, psychrophilic bacterium isolated from lake sediment of Schirmacher Oasis, Antarctica. Int. J. Syst. Evol. Microbiol. *56*: 715–720.

Alauzet, C., F. Mory, J.P. Carlier, H. Marchandin, E. Jumas-Bilak and A. Lozniewski. 2007. *Prevotella nanceiensis* sp. nov., isolated from human clinical samples. Int. J. Syst. Evol. Microbiol. *57*: 2216–2220.

Al-Bari, M.A., M.S. Bhuiyan, M.E. Flores, P. Petrosyan, M. García-Varela and M.A. Islam. 2005. *Streptomyces bangladeshensis* sp. nov., isolated from soil, which produces bis-(2-ethylhexyl)phthalate. Int. J. Syst. Evol. Microbiol. *55*: 1973–1977.

Albert, R.A., J. Archambault, R. Rosselló-Mora, B.J. Tindall and M. Matheny. 2005. *Bacillus acidicola* sp. nov., a novel mesophilic, acidophilic species isolated from acidic *Sphagnum* peat bogs in Wisconsin. Int. J. Syst. Evol. Microbiol. *55*: 2125–2130.

Albert, R.A., J. Archambault, M. Lempa, B. Hurst, C. Richardson, S. Gruenloh, M. Duran, H.L. Worliczek, B.E. Huber, R. Rosselló-Mora, P. Schumann and H.J. Busse. 2007. Proposal of *Viridibacillus* gen. nov. and reclassification of *Bacillus arvi*, *Bacillus arenosi* and *Bacillus neidei* as *Viridibacillus arvi* gen. nov., comb. nov., *Viridibacillus arenosi* comb. nov. and *Viridibacillus neidei* comb. nov. Int. J. Syst. Evol. Microbiol. *57*: 2729–2737.

Albuquerque, L., C. Simões, M.F. Nobre, N.M. Pino, J.R. Battista, M.T. Silva, F.A. Rainey and M.S. da Costa. 2005. *Truepera radiovictrix* gen. nov., sp. nov., a new radiation resistant species and the proposal of *Trueperaceae* fam. nov. FEMS Microbiol Lett. *247*: 161–169.

Albuquerque, L., I. Tiago, A. Veríssimo and M.S. da Costa. 2006. *Tepidimonas thermarum* sp. nov., a new slightly thermophilic betaproteobacterium isolated from the Elisenquelle in Aachen and emended description of the genus *Tepidimonas*. Syst. Appl. Microbiol. *29*: 450–456.

Albuquerque, L., I. Tiago, F.A. Rainey, M. Taborda, M.F. Nobre, A. Veríssimo and M.S. da Costa. 2007. *Salirhabdus euzebyi* gen. nov., sp. nov., a Gram-positive, halotolerant bacterium isolated from a sea salt evaporation pond. Int. J. Syst. Evol. Microbiol. *57*: 1566–1571.

Allan, R.N., L. Lebbe, J. Heyrman, P. De Vos, C.J. Buchanan and N.A. Logan. 2005. *Brevibacillus levickii* sp. nov. and *Aneurinibacillus terranovensis* sp. nov., two novel thermoacidophiles isolated from geothermal soils of northern Victoria Land, Antarctica. Int. J. Syst. Evol. Microbiol. *55*: 1039–1050.

Allen, T.D., P.A. Lawson, M.D. Collins, E. Falsen and R.S. Tanner. 2006. *Cloacibacterium normanense* gen. nov., sp. nov., a novel bacterium in the family *Flavobacteriaceae* isolated from municipal wastewater. Int. J. Syst. Evol. Microbiol. *56*: 1311–1316.

Amakata, D., Y. Matsuo, K. Shimono, J.K. Park, C.S. Yun, H. Matsuda, A. Yokota and M. Kawamukai. 2005. *Mitsuaria chitosanitabida* gen. nov., sp. nov., an aerobic, chitosanase-producing member of the '*Betaproteobacteria*'. Int. J. Syst. Evol. Microbiol. *55*: 1927–1932.

Amend, J.P., D.R. Meyer-Dombard, S.N. Sheth, N. Zolotova and A.C. Amend. 2003. *Palaeococcus helgesonii* sp. nov., a facultatively anaerobic, hyperthermophilic archaeon from a geothermal well on Vulcano Island, Italy. Arch. Microbiol. *179*: 394–401.

An, D.S., W.T. Im, H.C. Yang, D.C. Yang and S.T. Lee. 2005a. *Dyella koreensis* sp. nov., a β-glucosidase-producing bacterium. Int. J. Syst. Evol. Microbiol. *55*: 1625–1628.

An, D.S., W.T. Im, H.C. Yang, M.S. Kang, K.K. Kim, L. Jin, M.K. Kim and S.T. Lee. 2005b. *Cellulomonas terrae* sp. nov., a cellulolytic and xylanolytic bacterium isolated from soil. Int. J. Syst. Evol. Microbiol. *55*: 1705–1709.

An, D.S., W.T. Im, H.C. Yang and S.T. Lee. 2006a. *Shinella granuli* gen. nov., sp. nov., and proposal of the reclassification of *Zoogloea ramigera* ATCC 19623 as *Shinella zoogloeoides* sp. nov. Int. J. Syst. Evol. Microbiol. *56*: 443–448.

An, D., S. Cai and X. Dong. 2006b. *Actinomyces ruminicola* sp. nov., isolated from cattle rumen. Int. J. Syst. Evol. Microbiol. *56*: 2043–2048.

An, D.S., H.G. Lee, W.T. Im, Q.M. Liu and S.T. Lee. 2007a. *Segetibacter koreensis* gen. nov., sp. nov., a novel member of the phylum *Bacteroi-*

detes, isolated from the soil of a ginseng field in South Korea. Int. J. Syst. Evol. Microbiol. *57*: 1828–1833.

An, D.S., W.T. Im, S.T. Lee and M.H. Yoon. 2007b. *Nocardioides panacihumi* sp. nov., isolated from soil of a ginseng field. Int. J. Syst. Evol. Microbiol. *57*: 2143–2146.

An, S.Y., M. Asahara, K. Goto, H. Kasai and A. Yokota. 2007a. *Terribacillus saccharophilus* gen. nov., sp. nov. and *Terribacillus halophilus* sp. nov., spore-forming bacteria isolated from field soil in Japan. Int. J. Syst. Evol. Microbiol. *57*: 51–55.

An, S.Y., M. Asahara, K. Goto, H. Kasai and A. Yokota. 2007b. *Virgibacillus halophilus* sp. nov., spore-forming bacteria isolated from soil in Japan. Int. J. Syst. Evol. Microbiol. *57*: 1607–1611.

An, S.Y., T. Haga, H. Kasai, K. Goto and A. Yokota. 2007c. *Sporosarcina saromensis* sp. nov., an aerobic endospore-forming bacterium. Int. J. Syst. Evol. Microbiol. *57*: 1868–1871.

An, S.Y., K. Kanoh, H. Kasai, K. Goto and A. Yokota. 2007d. *Halobacillus faecis* sp. nov., a spore-forming bacterium isolated from a mangrove area on Ishigaki Island, Japan. Int. J. Syst. Evol. Microbiol. *57*: 2476–2479.

An, S.Y., S. Ishikawa, H. Kasai, K. Goto and A. Yokota. 2007e. *Amphibacillus sediminis* sp. nov., an endospore-forming bacterium isolated from lake sediment in Japan. Int. J. Syst. Evol. Microbiol. *57*: 2489–2492.

Anesti, V., J. Vohra, S. Goonetilleka, I.R. McDonald, B. Sträubler, E. Stackebrandt, D.P. Kelly and A.P. Wood. 2004. Molecular detection and isolation of facultatively methylotrophic bacteria, including *Methylobacterium podarium* sp. nov., from the human foot microflora. Environ. Microbiol. *6*: 820–830.

Angelos, J.A., P.Q. Spinks, L.M. Ball and L.W. George. 2007. *Moraxella bovoculi* sp. nov., isolated from calves with infectious bovine keratoconjunctivitis. Int. J. Syst. Evol. Microbiol. *57*: 789–795.

Anil Kumar, P., T.S.S. Jyothsna, T.N. Srinivas, C. Sasikala, V. Ramana Ch. and J.F. Imhoff. 2007a. Two novel species of marine photo trophic *Gammaproteobacteria: Thiorhodococcus bheemlicus* sp. nov. and *Thiorhodococcus kakinadensis* sp. nov. Int. J. Syst. Evol. Microbiol. *57*: 2458–2461.

Anil Kumar, P., T.S.S. Jyothsna, T.N. Srinivas, C. Sasikala, V. Ramana Ch and J.F. Imhoff. 2007b. *Marichromatium bheemlicum* sp. nov., a non-diazotrophic, photosynthetic gammaproteobacterium from a marine aquaculture pond. Int. J. Syst. Evol. Microbiol. *57*: 1261–1265.

Anil Kumar, P., T.N.R. Srinivas, Ch. Sasikala and Ch. V. Ramana. 2007c. *Halochromatium roseum* sp. nov., a non-motile phototrophic gammaproteobacterium with gas vesicles, and emended description of the genus *Halochromatium*. Int. J. Syst. Evol. Microbiol. 57, 2110–2113.

Anil Kumar, P., T.N. Srinivas, C. Sasikala and V. Ramana Ch. 2007d. *Rhodobacter changlensis* sp. nov., a psychrotolerant, phototrophic alphaproteobacterium from the Himalayas of India. Int. J. Syst. Evol. Microbiol. *57*: 2568–2571.

Antunes, A., L. Franca, F.A. Rainey, R. Huber, M.F. Nobre, K.J. Edwards and M.S. da Costa. 2007. *Marinobacter salsuginis* sp. nov., isolated from the brine-seawater interface of the Shaban Deep, Red Sea. Int. J. Syst. Evol. Microbiol. *57*: 1035–1040.

Ara, I. and T. Kudo. 2006. Three novel species of the genus *Catellatospora, Catellatospora chokoriensis* sp. nov., *Catellatospora coxensis* sp. nov. and *Catellatospora bangladeshensis* sp. nov., and transfer of *Catellatospora citrea* subsp. *methionotrophica* Asano and Kawamoto 1988 to *Catellatospora methionotrophica* sp. nov., comb. nov. Int. J. Syst. Evol. Microbiol. *56*: 393–400.

Ara, I. and T. Kudo. 2007a. *Krasilnikovia* gen. nov., a new member of the family *Micromonosporaceae* and description of *Krasilnikovia cinnamonea* sp. nov. Actinomycetologica *21*: 1–10.

Ara, I. and T. Kudo. 2007b. *Sphaerosporangium* gen. nov., a new member of the family *Streptosporangiaceae*, with descriptions of three new species as *Sphaerosporangium melleum* sp. nov., *Sphaerosporangium rubeum* sp. nov. and *Sphaerosporangium cinnabarinum* sp. nov., and transfer of

Streptosporangium viridialbum Nonomura and Ohara 1960 to *Sphaerosporangium viridialbum* comb. nov. Actinomycetologica *21*: 11–21.

Ara, I. and T. Kudo. 2007c. Two new species of the genus *Micromonospora: Micromonospora chokoriensis* sp. nov. and *Micromonospora coxensis* sp. nov., isolated from sandy soil. J. Gen. Appl. Microbiol. *53*: 29–37.

Ara, I. and T. Kudo. 2007d. *Luedemannella* gen. nov., a new member of the family *Micromonosporaceae* and description of *Luedemannella helvata* sp. nov. and *Luedemannella flava* sp. nov. J. Gen. Appl. Microbiol. *53*: 39–51.

Ara, I., T. Kudo, A. Matsumoto, Y. Takahashi and S. Ōmura. 2007a. *Nonomuraea maheshkhaliensis* sp. nov., a novel actinomycete isolated from mangrove rhizosphere mud. J. Gen. Appl. Microbiol. *53*: 159–166.

Ara, I., T. Kudo, A. Matsumoto, Y. Takahashi and S. Ōmura. 2007b. *Nonomuraea bangladeshensis* sp. nov. and *Nonomuraea coxensis* sp. nov. Int. J. Syst. Evol. Microbiol. *57*: 1504–1509.

Arahal, D.R., M.C. Macián, E. Garay and M.J. Pujalte. 2005. *Thalassobius mediterraneus* gen. nov., sp. nov., and reclassification of *Ruegeria gelatinovorans* as *Thalassobius gelatinovorus* comb. nov. Int. J. Syst. Evol. Microbiol. *55*: 2371–2376.

Arahal, D.R., M.C. Macián, E. Garay and M.J. Pujalte. 2006. *Thalassobius mediterraneus* gen. nov., sp. nov., and reclassification of *Ruegeria gelatinovorans* as *Thalassobius gelatinovorus* comb. nov. Int. J. Syst. Evol. Microbiol. *55*: 2371–2376.

Arahal, D.R., I. Lekunberri, J.M. González, J. Pascual, M.J. Pujalte, C. Pedrós-Alió and J. Pinhassi. 2007. *Neptuniibacter caesariensis* gen. nov., sp. nov., a novel marine genome-sequenced gammaproteobacterium. Int. J. Syst. Evol. Microbiol. *57*: 1000–1006.

Arbique, J.C., C. Poyart, P. Trieu-Cuot, G. Quesne, M. da G.S. Carvalho, A.G. Steigerwalt, R.E. Morey, D. Jackson, R.J. Davidson and R.R. Facklam. 2004. Accuracy of phenotypic and genotypic testing for identification of *Streptococcus pneumoniae* and description of *Streptococcus pseudopneumoniae* sp. nov. J. Clin. Microbiol. *42*: 4686–4696.

Arenskötter, M., A. Linos, P. Schumann, R.M. Kroppenstedt and A. Steinbüchel. 2005. *Gordonia nitida* Yoon *et al.* 2000 is a later synonym of *Gordonia alkanivorans* Kummer *et al.* 1999. Int. J. Syst. Evol. Microbiol. *55*: 695–697.

Arunasri, K., C. Sasikala, C.V. Ramana, J. Süling and J.F. Imhoff. 2005. *Marichromatium indicum* sp. nov., a novel purple sulfur gammaproteobacterium from mangrove soil of Goa, India. Int. J. Syst. Evol. Microbiol. *55*: 673–679.

Asker, D., T. Beppu and K. Ueda. 2007a. *Zeaxanthinibacter enoshimensis* gen. nov., sp. nov., a novel zeaxanthin-producing marine bacterium of the family *Flavobacteriaceae*, isolated from seawater off Enoshima Island, Japan. Int. J. Syst. Evol. Microbiol. *57*: 837–843.

Asker, D., T. Beppu and K. Ueda. 2007b. *Sphingomonas jaspsi* sp. nov., a novel carotenoid-producing bacterium isolated from Misasa, Tottori, Japan. Int. J. Syst. Evol. Microbiol. *57*: 1435–1441.

Aslam, Z., W.T. Im, M.K. Kim and S.T. Lee. 2005a. *Flavobacterium granuli* sp. nov., isolated from granules used in a wastewater treatment plant. Int. J. Syst. Evol. Microbiol. *55*: 747–751.

Aslam, Z., W.T. Im, L.N. Ten and S.T. Lee. 2005b. *Phenylobacterium koreense* sp. nov., isolated from South Korea. Int. J. Syst. Evol. Microbiol. *55*: 2001–2005.

Aslam, Z., W.T. Im, L.N. Ten, M.J. Lee, K.H. Kim and S.T. Lee. 2006. *Lactobacillus siliginis* sp. nov., isolated from wheat sourdough in South Korea. Int. J. Syst. Evol. Microbiol. *56*: 2209–2213.

Aslam, Z., C.S. Lee, K.H. Kim, W.T. Im, L.N. Ten and S.T. Lee. 2007a. *Methylobacterium jeotgali* sp. nov., a non-pigmented, facultatively methylotrophic bacterium isolated from jeotgal, a traditional Korean fermented seafood. Int. J. Syst. Evol. Microbiol. *57*: 566–571.

Aslam, Z., J.H. Lim, W.T. Im, M. Yasir, Y.R. Chung and S.T. Lee. 2007b. *Salinicoccus jeotgali* sp. nov., isolated from jeotgal, a traditional Korean fermented seafood. Int. J. Syst. Evol. Microbiol. *57*: 633–638.

Ast, J.C., I. Cleenwerck, K. Engelbeen, H. Urbanczyk, F.L. Thompson, P. De Vos and P.V. Dunlap. 2007. *Photobacterium kishitanii* sp. nov., a luminous marine bacterium symbiotic with deep-sea fishes. Int. J. Syst. Evol. Microbiol. *57*: 2073–2078.

Atomi, H., T. Fukui, T. Kanai, M. Morikawa and T. Imanaka. 2004. Description of *Thermococcus kadakaraensis* sp. nov., a well studied hyperthermophilic archaeon previously reported as *Pyrococcus* sp. KOD1. Archaea. *1*: 263–267.

Auman, A.J., J.L. Breezee, J.J. Gosink, P. Kämpfer and J.T. Staley. 2006. *Psychromonas ingrahamii* sp. nov., a novel gas vacuolate, psychrophilic bacterium isolated from Arctic polar sea ice. Int. J. Syst. Evol. Microbiol. *56*: 1001–1007.

Bae, H.S., W.T. Im and S.T. Lee. 2005. *Lysobacter concretionis* sp. nov., isolated from anaerobic granules in an upflow anaerobic sludge blanket reactor. Int. J. Syst. Evol. Microbiol. *55*: 1155–1161.

Bae, H.S., W.M. Moe, J. Yan, I. Tiago, M.S. da Costa and F.A. Rainey. 2006a. *Brooklawnia cerclae* gen. nov., sp. nov., a propionate-forming bacterium isolated from chlorosolvent-contaminated groundwater. Int. J. Syst. Evol. Microbiol. *56*: 1977–1983.

Bae, H.S., W.M. Moe, J. Yan, I. Tiago, M.S. da Costa and F.A. Rainey. 2006b. *Propionicicella superfundia* gen. nov., sp. nov., a chlorosolvent-tolerant propionate-forming, facultative anaerobic bacterium isolated from contaminated groundwater. Syst. Appl. Microbiol. *29*: 404–413.

Bae, H.S., B.A. Rash, F.A. Rainey, M.F. Nobre, I. Tiago, M.S. da Costa and W.M. Moe. 2007. Description of *Azospira restricta* sp. nov., a nitrogen-fixing bacterium isolated from groundwater. Int. J. Syst. Evol. Microbiol. *57*: 1521–1526.

Bae, J.W., J.R. Park, Y.H. Chang, S.K. Rhee, B.C. Kim and Y.H. Park. 2004. *Clostridium hastiforme* is a later synonym of *Tissierella praeacuta*. Int. J. Syst. Evol. Microbiol. *54*: 947–949.

Bae, S.S., J.H. Lee and S.J. Kim. 2005. *Bacillus alveayuensis* sp. nov., a thermophilic bacterium isolated from deep-sea sediments of the Ayu Trough. Int. J. Syst. Evol. Microbiol. *55*: 1211–1215.

Bae, S.S., K.K. Kwon, S.H. Yang, H.S. Lee, S.J. Kim and J.H. Lee. 2007. *Flagellimonas eckloniae* gen. nov., sp. nov., a mesophilic marine bacterium of the family *Flavobacteriaceae*, isolated from the rhizosphere of *Ecklonia kurome*. Int. J. Syst. Evol. Microbiol. *57*: 1050–1054.

Baek, S.H., W.T. Im, H.W. Oh, J.S. Lee, H.M. Oh and S.T. Lee. 2006. *Brevibacillus ginsengisoli* sp. nov., a denitrifying bacterium isolated from soil of a ginseng field. Int. J. Syst. Evol. Microbiol. *56*: 2665–2669.

Baer, M.L., J. Ravel, S.A. Piñeiro, D. Guether-Borg and H.N. Williams. 2004. Reclassification of salt-water *Bdellovibrio* sp. as *Bacteriovorax marinus* sp. nov. and *Bacteriovorax litoralis* sp. nov. Int. J. Syst. Evol. Microbiol. *54*: 1011–1016.

Baik, K.S., C.N. Seong, E.M. Kim, H. Yi, K.S. Bae and J. Chun. 2005. *Hahella ganghwensis* sp. nov., isolated from tidal flat sediment. Int. J. Syst. Evol. Microbiol. *55*: 681–684.

Baik, K.S., Y.D. Park, C.N. Seong, E.M. Kim, K.S. Bae and J. Chun. 2006a. *Glaciecola nitratireducens* sp. nov., isolated from seawater. Int. J. Syst. Evol. Microbiol. *56*: 2185–2188.

Baik, K.S., C.N. Seong, E.Y. Moon, Y.D. Park, H. Yi and J. Chun. 2006b. *Hymenobacter rigui* sp. nov., isolated from wetland freshwater. Int. J. Syst. Evol. Microbiol. *56*: 2189–2192.

Baik, K.S., M.S. Kim, E.M. Kim, H.R. Kim and C.N. Seong. 2007a. *Dyadobacter koreensis* sp. nov., isolated from fresh water. Int. J. Syst. Evol. Microbiol. *57*: 1227–1231.

Baik, K.S., Y.D. Park, M.S. Kim, S.C. Park, E.Y. Moon, M.S. Rhee, J.H. Choi and C.N. Seong. 2007b. *Pedobacter koreensis* sp. nov., isolated from fresh water. Int. J. Syst. Evol. Microbiol. *57*: 2079–2083.

Baik, K.S., M.S. Kim, S.C. Park, D.W. Lee, S.D. Lee, J.O. Ka, S.K. Choi and C.N. Seong. 2007c. *Spirosoma rigui* sp. nov., isolated from fresh water. Int. J. Syst. Evol. Microbiol. *57*: 2870–2873.

Bakermans, C., H.L. Ayala-del-Río, M.A. Ponder, T. Vishnivetskaya, D. Gilichinsky, M.F. Thomashow and J.M. Tiedje. 2006. *Psychrobacter cryohalolentis* sp. nov. and *Psychrobacter arcticus* sp. nov., isolated from Siberian permafrost. Int. J. Syst. Evol. Microbiol. *56*: 1285–1291.

Bakir, M.A., M. Kitahara, M. Sakamoto, M. Matsumoto and Y. Benno. 2006a. *Bacteroides intestinalis* sp. nov., isolated from human faeces. Int. J. Syst. Evol. Microbiol. *56*: 151–154.

Bakir, M.A., M. Kitahara, M. Sakamoto, M. Matsumoto and Y. Benno. 2006b. *Bacteroides finegoldii* sp. nov., isolated from human faeces. Int. J. Syst. Evol. Microbiol. *56*: 931–935.

Bakir, M.A., M. Sakamoto, M. Kitahara, M. Matsumoto and Y. Benno. 2006c. *Bacteroides dorei* sp. nov., isolated from human faeces. Int. J. Syst. Evol. Microbiol. *56*: 1639–1643.

Bala, S., R. Khanna, M. Dadhwal, S.R. Prabagaran, S. Shivaji, J. Cullum and R. Lal. 2004. Reclassification of *Amycolatopsis mediterranei* DSM 46095 as *Amycolatopsis rifamycinica* sp. nov. Int. J. Syst. Evol. Microbiol. *54*: 1145–1149.

Balk, M., J. Weijma, M.W. Friedrich and A.J.M. Stams. 2003. Methanol utilization by a novel thermophilic homoacetogenic bacterium, *Moorella mulderi* sp. nov., isolated from a bioreactor. Arch. Microbiol. *179*: 315–320.

Banat, I.M., R. Marchant and T.J. Rahman. 2004. *Geobacillus debilis* sp. nov., a novel obligately thermophilic bacterium isolated from a cool soil environment, and reassignment of *Bacillus pallidus* to *Geobacillus pallidus* comb. nov. Int. J. Syst. Evol. Microbiol. *54*: 2197–2201.

Banciu, H., D.Y. Sorokin, E.A. Galinski, G. Muyzer, R. Kleerebezem and J.G. Kuenen. 2004. *Thialkalivibrio halophilus* sp. nov., a novel obligately chemolithoautotrophic, facultatively alkaliphilic, and extremely salt-tolerant, sulfur-oxidizing bacterium from a hypersaline alkaline lake. Extremophiles. *8*: 325–334.

Bardavid, R.E., L. Mana and A. Oren. 2007. *Haloplanus natans* gen. nov., sp. nov., an extremely halophilic, gas-vacuolate archaeon isolated from Dead Sea-Red Sea water mixtures in experimental outdoor ponds. Int. J. Syst. Evol. Microbiol. *57*: 780–783.

Basso, O., P. Caumette and M. Magot. 2005. *Desulfovibrio putealis* sp. nov., a novel sulfate-reducing bacterium isolated from a deep subsurface aquifer. Int. J. Syst. Evol. Microbiol. *55*: 101–104.

Behrendt, U., A. Ulrich, P. Schumann, J.M. Meyer and C. Spröer. 2007a. *Pseudomonas lurida* sp. nov., a fluorescent species associated with the phyllosphere of grasses. Int. J. Syst. Evol. Microbiol. *57*: 979–985.

Behrendt, U., A. Ulrich, C. Spröer and P. Schumann. 2007b. *Chryseobacterium luteum* sp. nov., associated with the phyllosphere of grasses. Int. J. Syst. Evol. Microbiol. *57*: 1881–1885.

Bekal, S., C. Gaudreau, R.A. Laurence, E. Simoneau and L. Raynal. 2006. *Streptococcus pseudoporcinus* sp. nov., a novel species isolated from the genitourinary tract of women. J. Clin. Microbiol. *44*: 2584–2586.

Belyakova, E.V., E.P. Rozanova, I.A. Borzenkov, T.P. Tourova, M.A. Pusheva, A.M. Lysenko and T.V. Kolganova. 2006. The new facultatively chemolithoautotrophic, moderately halophilic, sulfate-reducing bacterium *Desulfovermiculus halophilus* gen. nov., sp. nov., isolated from an oil field. Mikrobiologiya. *75*: 201–211 (in Russian); English translation: Microbiology. *75*: 161–171.

Ben Ali Gam, Z., S. Abdelkafi, L. Casalot, J.L. Tholozan, R. Oueslati and M. Labat. 2007. *Modicisalibacter tunisiensis* gen. nov., sp. nov., an aerobic, moderately halophilic bacterium isolated from an oilfield-water injection sample, and emended description of the family *Halomonadaceae* Franzmann *et al.* 1989 emend Dobson and Franzmann 1996 emend. Ntougias *et al.* 2007. Int. J. Syst. Evol. Microbiol. *57*: 2307–2313.

Bhadra, B., P. Roy and R. Chakraborty. 2005. *Serratia ureilytica* sp. nov., a novel urea-utilizing species. Int. J. Syst. Evol. Microbiol. *55*: 2155–2158.

Biebl, H., M. Allgaier, B.J. Tindall, M. Koblizek, H. Lünsdorf, R. Pukall and I. Wagner-Döbler. 2005a. *Dinoroseobacter shibae* gen. nov., sp. nov., a new aerobic phototrophic bacterium isolated from dinoflagellates. Int. J. Syst. Evol. Microbiol. *55*: 1089–1096.

Biebl, H., M. Allgaier, H. Lünsdorf, R. Pukall, B.J. Tindall and I. Wagner-Döbler. 2005b. *Roseovarius mucosus* sp. nov., a member of the *Roseobacter* clade with trace amounts of bacteriochlorophyll *a*. Int. J. Syst. Evol. Microbiol. *55*: 2377–2383.

Biebl, H., B.J. Tindall, R. Pukall, H. Lünsdorf, M. Allgaier and I. Wagner-Döbler. 2006. *Hoeflea phototrophica* sp. nov., a novel marine aerobic alphaproteobacterium that forms bacteriochlorophyll *a*. Int. J. Syst. Evol. Microbiol. *56*: 821–826.

Biebl, H., R. Pukall, H. Lünsdorf, S. Schulz, M. Allgaier, B.J. Tindall and I. Wagner-Döbler. 2007. Description of *Labrenzia alexandrii* gen. nov., sp. nov., a novel alphaproteobacterium containing bacteriochlorophyll a, and a proposal for reclassification of *Stappia aggregata* as *Labrenzia aggregata* comb. nov., of *Stappia marina* as *Labrenzia marina* comb. nov. and of *Stappia alba* as *Labrenzia alba* comb. nov., and emended descriptions of the genera *Pannonibacter*, *Stappia* and *Roseibium*, and of the species *Roseibium denhamense* and *Roseibium hamelinense*. Int. J. Syst. Evol. Microbiol. *57*: 1095–1107.

Bisgaard, M., J.P. Christensen, A.M. Bojesen and H. Christensen. 2007. *Avibacterium endocarditidis* sp. nov., isolated from valvular endocarditis in chickens. Int. J. Syst. Evol. Microbiol. *57*: 1729–1734.

Blackall, P.J., H. Christensen, T. Beckenham, L.L. Blackall and M. Bisgaard. 2005. Reclassification of *Pasteurella gallinarum*, [*Haemophilus*] *paragallinarum*, *Pasteurella avium* and *Pasteurella volantium* as *Avibacterium gallinarum* gen. nov., comb. nov., *Avibacterium paragallinarum* comb. nov., *Avibacterium avium* comb. nov. and *Avibacterium volantium* comb. nov. Int. J. Syst. Evol. Microbiol. *55*: 353–362.

Blackall, P.J., A.M. Bojesen, H. Christensen and M. Bisgaard. 2007. Reclassification of [*Pasteurella*] *trehalosi* as *Bibersteinia trehalosi* gen. nov., comb. nov. Int. J. Syst. Evol. Microbiol. *57*: 666–674.

Boettcher, K.J., K.K. Geaghan, A.P. Maloy and B.J. Barber. 2005. *Roseovarius crassostreae* sp. nov., a member of the *Roseobacter* clade and the apparent cause of juvenile oyster disease (JOD) in cultured Eastern oysters. Int. J. Syst. Evol. Microbiol. *55*: 1531–1537.

Boga, H.I., R. Ji, W. Ludwig and A. Brune. 2007. *Sporotalea propionica* gen. nov. sp. nov., a hydrogen-oxidizing, oxygen-reducing, propionigenic firmicute from the intestinal tract of a soil-feeding termite. Arch. Microbiol. *187*: 15–27.

Bogdanova, T.I., I.A. Tsaplina, T.F. Kondrat'eva, V.I. Duda, N.E. Suzina, V.S. Melamud, T.P. Tourova and G.I. Karavaiko. 2006. *Sulfobacillus thermotolerans* sp. nov., a thermotolerant, chemolithotrophic bacterium. Int. J. Syst. Evol. Microbiol. *56*: 1039–1042.

Bonilla Salinas, M., M.L. Fardeau, J.L. Cayol, L. Casalot, B.K. Patel, P. Thomas, J.L. García and B. Ollivier. 2004a. *Petrobacter succinatimandens* gen. nov., sp. nov., a moderately thermophilic, nitrate-reducing bacterium isolated from an Australian oil well. Int. J. Syst. Evol. Microbiol. *54*: 645–649.

Bonilla Salinas, M., M.L. Fardeau, P. Thomas, J.L. Cayol, B.K. Patel and B. Ollivier. 2004b. *Mahella australiensis* gen. nov., sp. nov., a moderately thermophilic anaerobic bacterium isolated from an Australian oil well. Int. J. Syst. Evol. Microbiol. *54*: 2169–2173.

Bora, N., M. Vancanneyt, R. Gelsomino, J. Swings, N. Brennan, T.M. Cogan, S. Larpin, N. Desmasures, F.E. Lechner, R.M. Kroppenstedt, A.C. Ward and M. Goodfellow. 2007. *Agrococcus casei* sp. nov., isolated from the surfaces of smear-ripened cheeses. Int. J. Syst. Evol. Microbiol. *57*: 92–97.

Borchert, M.S., P. Nielsen, I. Graeber, I. Kaesler, U. Szewzyk, T. Pape, G. Antranikian and T. Schäfer. 2007. *Bacillus plakortidis* sp. nov. and *Bacillus murimartini* sp. nov., novel alkalitolerant members of rRNA group 6. Int. J. Syst. Evol. Microbiol. *57*: 2888–2893.

Bouchek-Mechiche, K., L. Gardan, D. Andrivon and P. Normand. 2006. *Streptomyces turgidiscabies* and *Streptomyces reticuliscabiei*: one genomic species, two pathogenic groups. Int. J. Syst. Evol. Microbiol. *56*: 2771–2776.

Bowman, J.P. 2005a. Family V. *Methylocystaceae* fam. nov. *In* Brenner, Krieg, Staley and Garrity (Editors), Bergey's Manual of Systematic Bacteriology, 2nd Edition, Volume 2 (The *Proteobacteria*), Part C (The *Alpha-*, *Beta-*, *Delta-*, and *Epsilonproteobacteria*), Springer, New York, pp. 411–413.

Bowman, J.P. 2005b. Order VII. *Methylococcales* ord. nov. *In* Brenner, Krieg, Staley and Garrity (Editors), Bergey's Manual of Systematic Bacteriology, 2nd Edition, Volume 2 (The *Proteobacteria*), Part B (The *Gammaproteobacteria*), Springer, New York, pp. 248–252.

Bowman, J.P. and T.A. McMeekin. 2005. Order X. *Alteromonadales* ord. nov. *In* Brenner, Krieg, Staley and Garrity (Editors), Bergey's Manual of Systematic Bacteriology, 2nd Edition, Volume 2 (The *Proteobacteria*), Part B (The *Gammaproteobacteria*), Springer, New York, p. 443.

Bowman, J.P. and D.S. Nichols. 2005. Novel members of the family *Flavobacteriaceae* from Antarctic maritime habitats including *Subsaximicrobium wynnwilliamsii* gen. nov., sp. nov., *Subsaximicrobium saxinquilinus* sp. nov., *Subsaxibacter broadyi* gen. nov., sp. nov., *Lacinutrix copepodicola* gen. nov., sp. nov., and novel species of the genera *Bizionia*, *Gelidibacter* and *Gillisia*. Int. J. Syst. Evol. Microbiol. *55*: 1471–1486.

Bozal, N., M.J. Montes and E. Mercadé. 2007. *Pseudomonas guineae* sp. nov., a novel psychrotolerant bacterium from an Antarctic environment. Int. J. Syst. Evol. Microbiol. *57*: 2609–2612.

Brettar, I., R. Christen and M.G. Höfle. 2004. *Belliella baltica* gen. nov., sp. nov., a novel marine bacterium of the *Cytophaga–Flavobacterium–Bacteroides* group isolated from surface water of the central Baltic Sea. Int. J. Syst. Evol. Microbiol. *54*: 65–70.

Brettar, I., R. Christen and M.G. Höfle. 2004. *Aquiflexum balticum* gen. nov., sp. nov., a novel marine bacterium of the *Cytophaga–Flavobacterium–Bacteroides* group isolated from surface water of the central Baltic Sea. Int. J. Syst. Evol. Microbiol. *54*: 2335–2341.

Brettar, I., R. Christen and M.G. Höfle. 2006. *Rheinheimera perlucida* sp. nov., a marine bacterium of the *Gammaproteobacteria* isolated from surface water of the central Baltic Sea. Int. J. Syst. Evol. Microbiol. *56*: 2177–2183.

Brettar, I., R. Christen, J. Bötel, H. Lünsdorf and M.G. Höfle. 2007. *Anderseniella baltica* gen. nov., sp. nov., a novel marine bacterium of the *Alphaproteobacteria* isolated from sediment in the central Baltic Sea. Int. J. Syst. Evol. Microbiol. *57*: 2399–2405.

Bringel, F., A. Castioni, D.K. Olukoya, G.E. Felis, S. Torriani and F. Dellaglio. 2005. *Lactobacillus plantarum* subsp. *argentoratensis* subsp. nov., isolated from vegetable matrices. Int. J. Syst. Evol. Microbiol. *55*: 1629–1634.

Brown, D.R., J.L. Merritt, E.R. Jacobson, P.A. Klein, J.G. Tully and M.B. Brown. 2004. *Mycoplasma testudineum* sp. nov., from a desert tortoise (*Gopherus agassizii*) with upper respiratory tract disease. Int. J. Syst. Evol. Microbiol. *54*: 1527–1529.

Brown, D.R., D.L. Demcovitz, D.R. Plourde, S.M. Potter, M.E. Hunt, R.D. Jones and D.S. Rotstein. 2006. *Mycoplasma iguanae* sp. nov., from a green iguana (*Iguana iguana*) with vertebral disease. Int. J. Syst. Evol. Microbiol. *56*: 761–764.

Brown, J.M., R.P. Frazier, R.E. Morey, A.G. Steigerwalt, G.J. Pellegrini, M.I. Daneshvar, D.G. Hollis and M.M. McNeil. 2005. Phenotypic and genetic characterization of clinical isolates of CDC coryneform group A-3: proposal of a new species of *Cellulomonas*, *Cellulomonas denverensis* sp. nov. J. Clin. Microbiol. *43*: 1732–1737.

Brown, J.M., A.G. Steigerwalt, R.E. Morey, M.I. Daneshvar, L.J. Romero and M.M. McNeil. 2006. Characterization of clinical isolates previously identified as *Oerskovia turbata*: proposal of *Cellulosimicrobium funkei* sp. nov. and emended description of the genus *Cellulosimicrobium*. Int. J. Syst. Evol. Microbiol. *56*: 801–804.

Buczolits, S., E.B. Denner, P. Kämpfer and H.J. Busse. 2006. Proposal of *Hymenobacter norwichensis* sp. nov., classification of 'Taxeobacter ocellatus', 'Taxeobacter gelupurpurascens' and 'Taxeobacter chitinovorans' as *Hymenobacter ocellatus* sp. nov., *Hymenobacter gelipurpurascens* sp. nov. and *Hymenobacter chitinivorans* sp. nov., respectively, and emended description of the genus *Hymenobacter* Hirsch *et al.* 1999. Int. J. Syst. Evol. Microbiol. *56*: 2071–2078.

Burns, D.G., P.H. Janssen, T. Itoh, M. Kamekura, Z. Li, G. Jensen, F. Rodríguez-Valera, H. Bolhuis and M.L. Dyall-Smith. 2007. *Haloquadratum walsbyi* gen. nov., sp. nov., the square haloarchaeon of Walsby, isolated from saltern crystallizers in Australia and Spain. Int. J. Syst. Evol. Microbiol. *57*: 387–392.

Busse, H.J., E. Hauser and P. Kämpfer. 2005. Description of two novel species, *Sphingomonas abaci* sp. nov. and *Sphingomonas panni* sp. nov. Int. J. Syst. Evol. Microbiol. *55*: 2565–2569.

Busti, E., L. Cavaletti, P. Monciardini, P. Schumann, M. Rohde, M. Sosio and S. Donadio. 2006. *Catenulispora acidiphila* gen. nov., sp. nov., a novel, mycelium-forming actinomycete, and proposal of *Catenulisporaceae* fam. nov. Int. J. Syst. Evol. Microbiol. *56*: 1741–1746.

Butler, W.R., M.M. Floyd, J.M. Brown, S.R. Toney, M.I. Daneshvar, R.C. Cooksey, J. Carr, A.G. Steigerwalt and N. Charles. 2005. Novel mycolic acid-containing bacteria in the family *Segniliparaceae* fam. nov., including the genus *Segniliparus* gen. nov., with descriptions of *Segniliparus rotundus* sp. nov. and *Segniliparus rugosus* sp. nov. Int. J. Syst. Evol. Microbiol. *55*: 1615–1624.

Byun, R., J.P. Carlier, N.A. Jacques, H. Marchandin and N. Hunter. 2007. *Veillonella denticariosi* sp. nov., isolated from human carious dentine. Int. J. Syst. Evol. Microbiol. *57*: 2844–2848.

Caballero-Mellado, J., L. Martínez-Aguilar, G. Paredes-Valdez and P.E. Santos. 2004. *Burkholderia unamae* sp. nov., an N2-fixing rhizospheric and endophytic species. Int. J. Syst. Evol. Microbiol. *54*: 1165–1172.

Cabrera, A., M. Aguilera, S. Fuentes, C. Incerti, N.J. Russell, A. Ramos-Cormenzana and M. Monteoliva-Sánchez. 2007. *Halomonas indalinina* sp. nov., a moderately halophilic bacterium isolated from a solar saltern in Cabo de Gata, Almeria, southern Spain. Int. J. Syst. Evol. Microbiol. *57*: 376–380.

Cachat, E. and F.G. Priest. 2005. *Lactobacillus suntoryeus* sp. nov., isolated from malt whisky distilleries. Int. J. Syst. Evol. Microbiol. *55*: 31–34.

Cai, J., Y. Wang, D. Liu, Y. Zeng, Y. Xue, Y. Ma and Y. Feng. 2007. *Fervidobacterium changbaicum* sp. nov., a novel thermophilic anaerobic bacterium isolated from a hot spring of the Changbai Mountains, China. Int. J. Syst. Evol. Microbiol. *57*: 2333–2336.

Cámara, B., C. Strömpl, S. Verbarg, C. Spröer, D.H. Pieper and B.J. Tindall. 2007. *Pseudomonas reinekei* sp. nov., *Pseudomonas moorei* sp. nov. and *Pseudomonas mohnii* sp. nov., novel species capable of degrading chlorosalicylates or isopimaric acid. Int. J. Syst. Evol. Microbiol. *57*: 923–931.

Campbell, S., R.M. Harada and Q.X. Li. 2007. *Ferrimonas senticii* sp. nov., a novel gammaproteobacterium isolated from the mucus of a puffer fish caught in Kaneohe Bay, Hawai'i. Int. J. Syst. Evol. Microbiol. *57*: 2670–2673.

Carlier, J.P., G. K'Ouas, I. Bonne, A. Lozniewski and F. Mory. 2004. *Oribacterium sinus* gen. nov., sp. nov., within the family 'Lachnospiraceae' (phylum *Firmicutes*). Int. J. Syst. Evol. Microbiol. *54*: 1611–1615.

Carlier, J.-P., I. Bonne and M. Bedora-Faure. 2006. Isolation from canned foods of a novel *Thermoanaerobacter* species phylogenetically related to *Thermoanaerobacter mathranii* (Larsen 1997): emendation of the species description and proposal of *Thermoanaerobacter mathranii* subsp. *Alimentarius* subsp. nov. Anaerobe *12*: 153–159.

Carlier, J.P., G. K'Ouas and X.Y. Han. 2007. *Moryella indoligenes* gen. nov., sp. nov., an anaerobic bacterium isolated from clinical specimens. Int. J. Syst. Evol. Microbiol. *57*: 725–729.

Carlsohn, M.R., I. Groth, G.Y. Tan, B. Schütze, H.P. Saluz, T. Munder, J. Yang, J. Wink and M. Goodfellow. 2007a. *Amycolatopsis saalfeldensis* sp. nov., a novel actinomycete isolated from a medieval alum slate mine. Int. J. Syst. Evol. Microbiol. *57*: 1640–1646.

Carlsohn, M.R., I. Groth, C. Spröer, B. Schütze, H.P. Saluz, T. Munder and E. Stackebrandt. 2007b. *Kribbella aluminosa* sp. nov., isolated from a medieval alum slate mine. Int. J. Syst. Evol. Microbiol. *57*: 1943–1947.

Carrasco, I.J., M.C. Márquez, X. Yanfen, Y. Ma, D.A. Cowan, B.E. Jones, W.D. Grant and A. Ventosa. 2006. *Gracilibacillus orientalis* sp. nov., a novel moderately halophilic bacterium isolated from a salt lake in Inner Mongolia, China. Int. J. Syst. Evol. Microbiol. *56*: 599–604.

Carrasco, I.J., M.C. Márquez, Y. Xue, Y. Ma, D.A. Cowan, B.E. Jones, W.D. Grant and A. Ventosa. 2007a. *Bacillus chagannorensis* sp. nov., a moderate halophile from a soda lake in Inner Mongolia, China. Int. J. Syst. Evol. Microbiol. *57*: 2084–2088.

Carrasco, I.J., M.C. Márquez, Y. Xue, Y. Ma, D.A. Cowan, B.E. Jones, W.D. Grant and A. Ventosa. 2007b. *Salsuginibacillus kocurii* gen. nov., sp. nov., a moderately halophilic bacterium from soda-lake sediment. Int. J. Syst. Evol. Microbiol. *57*: 2381–2386.

Carvalho M. da G.S, G., P.L. Shewmaker, A.G. Steigerwalt, R.E. Morey, A.J. Sampson, K. Joyce, T.J. Barrett, L.M. Teixeira and R.R. Facklam. 2006. *Enterococcus caccae* sp. nov., isolated from human stools. Int. J. Syst. Evol. Microbiol. *56*: 1505–1508.

Castillo, A.M., M.C. Gutiérrez, M. Kamekura, Y. Ma, D.A. Cowan, B.E. Jones, W.D. Grant and A. Ventosa. 2006a. *Halovivax asiaticus* gen. nov., sp. nov., a novel extremely halophilic archaeon isolated from Inner Mongolia, China. Int. J. Syst. Evol. Microbiol. *56*: 765–770.

Castillo, A.M., M.C. Gutiérrez, M. Kamekura, Y. Xue, Y. Ma, D.A. Cowan, B.E. Jones, W.D. Grant and A. Ventosa. 2006b. *Halostagnicola larsenii* gen. nov., sp. nov., an extremely halophilic archaeon from a saline lake in Inner Mongolia, China. Int. J. Syst. Evol. Microbiol. *56*: 1519–1524.

Castillo, A.M., M.C. Gutiérrez, M. Kamekura, Y. Xue, Y. Ma, D.A. Cowan, B.E. Jones, W.D. Grant and A. Ventosa. 2006c. *Halorubrum orientale* sp. nov., a halophilic archaeon isolated from Lake Ejinor, Inner Mongolia, China. Int. J. Syst. Evol. Microbiol. *56*: 2559–2563.

Castillo, A.M., M.C. Gutiérrez, M. Kamekura, Y. Xue, Y. Ma, D.A. Cowan, B.E. Jones, W.D. Grant and A. Ventosa. 2006d. *Natrinema ejinorense* sp. nov., isolated from a saline lake in Inner Mongolia, China. Int. J. Syst. Evol. Microbiol. *56*: 2683–2687.

Castillo, A.M., M.C. Gutiérrez, M. Kamekura, Y. Xue, Y. Ma, D.A. Cowan, B.E. Jones, W.D. Grant and A. Ventosa. 2007a. *Halovivax ruber* sp. nov., an extremely halophilic archaeon isolated from Lake Xilinhot, Inner Mongolia, China. Int. J. Syst. Evol. Microbiol. *57*: 1024–1027.

Castillo, A.M., M.C. Gutiérrez, M. Kamekura, Y. Xue, Y. Ma, D.A. Cowan, B.E. Jones, W.D. Grant and A. Ventosa. 2007b. *Halorubrum ejinorense* sp. nov., isolated from Lake Ejinor, Inner Mongolia, China. Int. J. Syst. Evol. Microbiol. *57*: 2538–2542.

Caumette, P., R. Guyoneaud, J.F. Imhoff, J. Süling and V. Gorlenko. 2004. *Thiocapsa marina* sp. nov., a novel, okenone-containing, purple sulfur bacterium isolated from brackish coastal and marine environments. Int. J. Syst. Evol. Microbiol. *54*: 1031–1036.

Caumette, P., R. Guyoneaud, R. Duran, C. Cravo-Laureau and R. Matheron. 2007. *Rhodobium pfennigii* sp. nov., a phototrophic purple non-sulfur bacterium with unusual bacteriochlorophyll a antennae, isolated from a brackish microbial mat on Rangiroa atoll, French Polynesia. Int. J. Syst. Evol. Microbiol. *57*: 1250–1255.

Cavaletti, L., P. Monciardini, P. Schumann, M. Rohde, R. Bamonte, E. Busti, M. Sosio and S. Donadio. 2006a. *Actinospica robiniae* gen. nov., sp. nov. and *Actinospica acidiphila* sp. nov.: proposal for *Actinospicaceae* fam. nov. and *Catenulisporineae* subord. nov. in the order *Actinomycetales*. Int. J. Syst. Evol. Microbiol. *56*: 1747–1753.

Cavaletti, L., P. Monciardini, R. Bamonte, P. Schumann, M. Rohde, M. Sosio and S. Donadio. 2006b. New lineage of filamentous, spore-forming, gram-positive bacteria from soil. Appl. Environ. Microbiol. *72*: 4360–4369.

Cazemier, A.E., J.C. Verdoes., F.A.G. Reubsaet, J.H.P. Hackstein, C. van der Drift and H.J.M. Op den Camp. 2003. *Promicromonospora pachnodae* sp. nov., a member of the (hemi)cellulolytic hindgut flora of larvae of the scarab beetle *Pachnoda marginata*. Antonie van Leeuwenhoek. *83*: 135–148.

Chambel, L., I.M. Chelo, L. Zé-Zé, L.G. Pedro, M.A. Santos and R. Tenreiro. 2006. *Leuconostoc pseudoficulneum* sp. nov., isolated from a ripe fig. Int. J. Syst. Evol. Microbiol. *56*: 1375–1381.

Chamkha, M., B.K.C. Bharat, J.L. Garcia and M. Labat. 2001. Isolation of *Clostridium bifermentans* from oil mill wastewaters converting cinnamic acid to 3-phenylpropionic acid and emendation of the species. Anaerobe. *7*: 189–197.

Chang, H.W., Y.D. Nam, H.Y. Kwon, J.R. Park, J.S. Lee, J.H. Yoon, K.G. An and J.W. Bae. 2007. *Marinobacterium halophilum* sp. nov., a marine bacterium isolated from the Yellow Sea. Int. J. Syst. Evol. Microbiol. *57*: 77–80.

Chang, J.S., C.L. Chou, G.H. Lin, S.Y. Sheu and W.M. Chen. 2005. *Pseudoxanthomonas kaohsiungensis*, sp. nov., a novel bacterium isolated from oil-polluted site produces extracellular surface activity. Syst. Appl. Microbiol. *28*: 137–144.

Chang, S.C., J.T. Wang, P. Vandamme, J.H. Hwang, P.S. Chang andW.M. Chen. 2004. *Chitinimonas taiwanensis* gen. nov., sp. nov., a novel chitinolytic bacterium isolated from a freshwater pond for shrimp culture. Syst. Appl. Microbiol. *27*: 43–49.

Chang, S.C., W.M. Chen, J.T. Wang and M.C. Wu. 2007. *Chitinilyticum aquatile* gen. nov., sp. nov., a chitinolytic bacterium isolated from a freshwater pond used for Pacific white shrimp culture. Int. J. Syst. Evol. Microbiol. *57*: 2854–2860.

Chaturvedi, P., G.S. Reddy and S. Shivaji. 2005. *Dyadobacter hamtensis* sp. nov., from Hamta glacier, located in the Himalayas, India. Int. J. Syst. Evol. Microbiol. *55*: 2113–2117.

Chaturvedi, P. and S. Shivaji. 2006. *Exiguobacterium indicum* sp. nov., a psychrophilic bacterium from the Hamta glacier of the Himalayan mountain ranges of India. Int. J. Syst. Evol. Microbiol. *56*: 2765–2770.

Chen, H.H., W.J. Li, S.K. Tang, R.M. Kroppenstedt, E. Stackebrandt, L.H. Xu and C.L. Jiang. 2004. *Corynebacterium halotolerans* sp. nov., isolated from saline soil in the west of China. Int. J. Syst. Evol. Microbiol. *54*: 779–782.

Chen, M., X. Xiao, P. Wang, X. Zeng and F. Wang. 2005. *Arthrobacter ardleyensis* sp. nov., isolated from Antarctic lake sediment and deepsea sediment. Arch. Microbiol. *183*: 301–305.

Chen, M.Y., S.H. Wu, G.H. Lin, C.P. Lu, Y.T. Lin, W.C. Chang and S.S. Tsay. 2004. *Rubrobacter taiwanensis* sp. nov., a novel thermophilic, radiation-resistant species isolated from hot springs. Int. J. Syst. Evol. Microbiol. *54*: 1849–1855.

Chen, S. and X. Dong. 2004. *Acetanaerobacterium elongatum* gen. nov., sp. nov., from paper mill waste water. Int. J. Syst. Evol. Microbiol. *54*: 2257–2262.

Chen, S. and X. Dong. 2005. *Proteiniphilum acetatigenes* gen. nov., sp. nov., from a UASB reactor treating brewery wastewater. Int. J. Syst. Evol. Microbiol. *55*: 2257–2261.

Chen, S., X. Liu and X. Dong. 2005. *Syntrophobacter sulfatireducens* sp. nov., a novel syntrophic, propionate-oxidizing bacterium isolated from UASB reactors. Int. J. Syst. Evol. Microbiol. *55*: 1319–1324.

Chen, S., L. Song and X. Dong. 2006. *Sporacetigenium mesophilum* gen. nov., sp. nov., isolated from an anaerobic digester treating municipal solid waste and sewage. Int. J. Syst. Evol. Microbiol. *56*: 721–725.

Chen, T.L., Y.J. Chou, W.M. Chen, B. Arun and C.C. Young. 2006. *Tepidimonas taiwanensis* sp. nov., a novel alkaline-protease-producing bacterium isolated from a hot spring. Extremophiles. *10*: 35–40.

Chen, W.M., J.S. Chang, C.H. Chiu, S.C. Chang, W.C. Chen and C.M. Jiang. 2005. *Caldimonas taiwanensis* sp. nov., a amylase producing bacterium isolated from a hot spring. Syst. Appl. Microbiol. *28*: 415–420.

Chen, W.M., E.K. James, T. Coenye, J.H. Chou, E. Barrios, S.M. de Faria, G.N. Elliott, S.Y. Sheu, J.I. Sprent and P. Vandamme. 2006. *Burkholderia mimosarum* sp. nov., isolated from root nodules of *Mimosa* spp. from Taiwan and South America. Int. J. Syst. Evol. Microbiol. *56*: 1847–1851.

Chen, W.M., S.M. de Faria, E.K. James, G.N. Elliott, K.Y. Lin, J.H. Chou, S.Y. Sheu, M. Cnockaert, J.I. Sprent and P. Vandamme. 2007. *Burkholderia nodosa* sp. nov., isolated from root nodules of the woody Brazilian legumes *Mimosa bimucronata* and *Mimosa scabrella*. Int. J. Syst. Evol. Microbiol. *57*: 1055–1059.

Chen, Y.G., X.L. Cui, R. Pukall, H.M. Li, Y.L. Yang, L.H. Xu, M.L. Wen, Q. Peng and C.L. Jiang. 2007. *Salinicoccus kunmingensis* sp. nov., a moderately halophilic bacterium isolated from a salt mine in Yunnan, south-west China. Int. J. Syst. Evol. Microbiol. *57*: 2327–2332.

Cheng, L., T.L. Qiu, X.B. Yin, X.L. Wu, G.Q. Hu, Y. Deng and H. Zhang. 2007. *Methermicoccus shengliensis* gen. nov., sp. nov., a thermophilic, methylotrophic methanogen isolated from oil-production water, and proposal of *Methermicoccaceae* fam. nov. Int. J. Syst. Evol. Microbiol. *57*: 2964–2969.

Chenoll, E., M.C. Macián and R. Aznar. 2006. *Lactobacillus rennini* sp. nov., isolated from rennin and associated with cheese spoilage. Int. J. Syst. Evol. Microbiol. *56*: 449–452.

Chern, L.L., E. Stackebrandt, S.F. Lee, F.L. Lee, J.K. Chen and H.M. Fu. 2004. *Chitinibacter tainanensis* gen. nov., sp. nov., a chitin-degrading aerobe from soil in Taiwan. Int. J. Syst. Evol. Microbiol. *54*: 1387–1391.

Chiu, H.H., W.Y. Shieh, S.Y. Lin, C.M. Tseng, P.W. Chiang and I. Wagner-Döbler. 2007. *Alteromonas tagae* sp. nov. and *Alteromonas simiduii* sp. nov., mercury-resistant bacteria isolated from a Taiwanese estuary. Int. J. Syst. Evol. Microbiol. *57*: 1209–1216.

Cho, E.A., J.S. Lee, K.C. Lee, H.C. Jung, J.G. Pan and Y.R. Pyun. 2007. *Cohnella laeviribosi* sp. nov., isolated from a volcanic pond. Int. J. Syst. Evol. Microbiol. *57*: 2902–2907.

Cho, J.C. and S.J. Giovannoni. 2004a. *Robiginitalea biformata* gen. nov., sp. nov., a novel marine bacterium in the family *Flavobacteriaceae* with a higher G+C content. Int. J. Syst. Evol. Microbiol. *54*: 1101–1106.

Cho, J.C. and S.J. Giovannoni. 2004b. *Oceanicola granulosus* gen. nov., sp. nov. and *Oceanicola batsensis* sp. nov., poly-beta-hydroxybutyrate-producing marine bacteria in the order 'Rhodobacterales'. Int. J. Syst. Evol. Microbiol. *54*: 1129–1136.

Cho, J.C. and S.J. Giovannoni. 2006. *Pelagibaca bermudensis* gen. nov., sp. nov., a novel marine bacterium within the *Roseobacter* clade in the order *Rhodobacterales*. Int. J. Syst. Evol. Microbiol. *56*: 855–859.

Cho, J.C., K.L. Vergin, R.M. Morris and S.J. Giovannoni. 2004. *Lentisphaera araneosa* gen. nov., sp. nov., a transparent exopolymer producing marine bacterium, and the description of a novel bacterial phylum, *Lentisphaerae*. Environ. Microbiol. *6*: 611–621.

Choi, D.H. and B.C. Cho. 2005. *Idiomarina seosinensis* sp. nov., isolated from hypersaline water of a solar saltern in Korea. Int. J. Syst. Evol. Microbiol. *55*: 379–383.

Choi, D.H. and B.C. Cho. 2006a. *Lutibacter litoralis* gen. nov., sp. nov., a marine bacterium of the family *Flavobacteriaceae* isolated from tidal flat sediment. Int. J. Syst. Evol. Microbiol. *56*: 771–776.

Choi, D.H. and B.C. Cho. 2006b. *Shimia marina* gen. nov., sp. nov., a novel bacterium of the *Roseobacter* clade isolated from biofilm in a coastal fish farm. Int. J. Syst. Evol. Microbiol. *56*: 1869–1873.

Choi, D.H. and B.C. Cho. 2006c. *Citreimonas salinaria* gen. nov., sp. nov., a member of the *Roseobacter* clade isolated from a solar saltern. Int. J. Syst. Evol. Microbiol. *56*: 2799–2803.

Choi, D.H., Y.G. Kim, C.Y. Hwang, H. Yi, J. Chun and B.C. Cho. 2006a. *Tenacibaculum litoreum* sp. nov., isolated from tidal flat sediment. Int. J. Syst. Evol. Microbiol. *56*: 635–640.

Choi, D.H., H. Yi, J. Chun and B.C. Cho. 2006b. *Jannaschia seosinensis* sp. nov., isolated from hypersaline water of a solar saltern in Korea. Int. J. Syst. Evol. Microbiol. *56*: 45–49.

Choi, D.H., J.C. Cho, B.D. Lanoil, S.J. Giovannoni and B.C. Cho. 2007. *Maribius salinus* gen. nov., sp. nov., isolated from a solar saltern and *Maribius pelagius* sp. nov., cultured from the Sargasso Sea, belonging to the *Roseobacter* clade. Int. J. Syst. Evol. Microbiol. *57*: 270–275.

Choi, J.H., W.T. Im, Q.M. Liu, J.S. Yoo, J.H. Shin, S.K. Rhee and D.H. Roh. 2007. *Planococcus donghaensis* sp. nov., a starch-degrading bacterium isolated from the East Sea, South Korea. Int. J. Syst. Evol. Microbiol. *57*: 2645–2650.

Choi, D.H., H.M. Kim, J.H. Noh and B.C. Cho. 2007. *Nocardioides marinus* sp. nov. Int. J. Syst. Evol. Microbiol. *57*: 775–779.

Choi, T.H., H.K. Lee, K. Lee and J.C. Cho. 2007. *Ulvibacter antarcticus* sp. nov., isolated from Antarctic coastal seawater. Int. J. Syst. Evol. Microbiol. *57*: 2922–2925.

Chou, J.H., W.M. Chen, A.B. Arun and C.C. Young. 2007a. *Trabulsiella odontotermitis* sp. nov., isolated from the gut of the termite *Odontotermes formosanus* Shiraki. Int. J. Syst. Evol. Microbiol. *57*: 696–700.

Chou, J.H., Y.J. Chou, K.Y. Lin, S.Y. Sheu, D.S. Sheu, A.B. Arun, C.C. Young and W.M. Chen. 2007b. *Paenibacillus fonticola* sp. nov., isolated from a warm spring. Int. J. Syst. Evol. Microbiol. *57*: 1346–1350.

Chou, J.H., K.Y. Lin, M.C. Lin, S.Y. Sheu, Y.H. Wei, A.B. Arun, C.C. Young and W.M. Chen. 2007c. *Brachybacterium phenoliresistens* sp. nov., isolated from oil-contaminated coastal sand. Int. J. Syst. Evol. Microbiol. *57*: 2674–2679.

Chou, J.H., S.Y. Sheu, K.Y. Lin, W.M. Chen, A.B. Arun and C.C. Young. 2007d. *Comamonas odontotermitis* sp. nov., isolated from the gut of the termite *Odontotermes formosanus*. Int. J. Syst. Evol. Microbiol. *57*: 887–891.

Chou, Y.J., S.Y. Sheu, D.S. Sheu, J.T. Wang and W.M. Chen. 2006. *Schlegelella aquatica* sp. nov., a novel thermophilic bacterium isolated from a hot spring. Int. J. Syst. Evol. Microbiol. *56*: 2793–2797.

Chou, Y.J., G.N. Elliott, E.K. James, K.Y. Lin, J.H. Chou, S.Y. Sheu, D.S. Sheu, J.I. Sprent and W.M. Chen. 2007. *Labrys neptuniae* sp. nov., isolated from root nodules of the aquatic legume *Neptunia oleracea*. Int. J. Syst. Evol. Microbiol. *57*: 577–581.

Choo, Y.J., K. Lee, J. Song and J.C. Cho. 2007. *Puniceicoccus vermicola* gen. nov., sp. nov., a novel marine bacterium, and description of *Puniceicoccaceae* fam. nov., *Puniceicoccales* ord. nov., *Opitutaceae* fam. nov., *Opitutales* ord. nov. and *Opitutae* classis nov. in the phylum '*Verrucomicrobia*'. Int. J. Syst. Evol. Microbiol. *57*: 532–537.

Christensen, H., M. Bisgaard, B. Aalbaek and J.E. Olsen. 2004. Reclassification of Bisgaard taxon 33, with proposal of *Volucribacter psittacicida* gen. nov., sp. nov. and *Volucribacter amazonae* sp. nov. as new members of the *Pasteurellaceae*. Int. J. Syst. Evol. Microbiol. *54*: 813–818.

Christensen, H., P. Kuhnert, M. Bisgaard, R. Mutters, F. Dziva and J.E. Olsen. 2005. Emended description of porcine [*Pasteurella*] *aerogenes*, [*Pasteurella*] *mairii* and [*Actinobacillus*] *rossii*. Int. J. Syst. Evol. Microbiol. *55*: 209–223.

Chung, B.S., S.H. Ryu, M. Park, Y. Jeon, Y.R. Chung and C.O. Jeon. 2007. *Hydrogenophaga caeni* sp. nov., isolated from activated sludge. Int. J. Syst. Evol. Microbiol. *57*: 1126–1130.

Cirne, D.G., O.D. Delgado, S. Marichamy and B. Mattiasson. 2006. *Clostridium lundense* sp. nov., a novel anaerobic lipolytic bacterium isolated from bovine rumen. Int. J. Syst. Evol. Microbiol. *56*: 625–628.

Cladera, A.M., E. García-Valdés and J. Lalucat. 2006. Genotype versus phenotype in the circumscription of bacterial species: the case of *Pseudomonas stutzeri* and *Pseudomonas chloritidismutans*. Arch. Microbiol. *184*: 353–361.

Clark, L.L., J.J. Dajcs, C.H. McLean, J.G. Bartell and D.W. Stroman. 2006. *Pseudomonas otitidis* sp. nov., isolated from patients with otic infections. Int. J. Syst. Evol. Microbiol. *56*: 709–714.

Clavel, T., R. Lippman, F. Gavini, J. Doré and M. Blaut. 2007. *Clostridium saccharogumia* sp. nov. and *Lactonifactor longoviformis* gen. nov., sp. nov., two novel human faecal bacteria involved in the conversion of the dietary phytoestrogen secoisolariciresinol diglucoside. Syst. Appl. Microbiol. *30*: 16–26.

Cleenwerck, I., N. Camu, K. Engelbeen, T. De Winter, K. Vandemeulebroecke, P. De Vos and L. De Vuyst. 2007. *Acetobacter ghanensis* sp. nov., a novel acetic acid bacterium isolated from traditional heap fermentations of Ghanaian cocoa beans. Int. J. Syst. Evol. Microbiol. *57*: 1647–1652.

Cloud, J.L., J.J. Meyer, J.I. Pounder, K.C. Jost, Jr., A. Sweeney, K.C. Carroll and G.L. Woods. 2006. *Mycobacterium arupense* sp. nov., a nonchromogenic bacterium isolated from clinical specimens. Int. J. Syst. Evol. Microbiol. *56*: 1413–1418.

Coenye, T., D. Henry, D.P. Speert and P. Vandamme. 2004a. *Burkholderia phenoliruptrix* sp. nov., to accommodate the 2,4,5-trichlorophenoxyacetic acid and halophenol-degrading strain AC1100. Syst. Appl. Microbiol. *27*: 623–627.

Coenye, T., E. Vanlaere, E. Falsen and P. Vandamme. 2004b. *Stenotrophomonas africana* Drancourt *et al.* 1997 is a later synonym of *Stenotrophomonas maltophilia* (Hugh 1981) Palleroni and Bradbury 1993. Int. J. Syst. Evol. Microbiol. *54*: 1235–1237.

Coenye, T., E. Vanlaere, E. Samyn, E. Falsen, P. Larsson and P. Vandamme. 2005. *Advenella incenata* gen. nov., sp. nov., a novel member of the *Alcaligenaceae*, isolated from various clinical samples. Int. J. Syst. Evol. Microbiol. *55*: 251–256.

Collins, M.D., E. Falsen, K. Brownlee and P.A. Lawson. 2004a. *Helcococcus suecensis* sp. nov., isolated from a human wound. Int. J. Syst. Evol. Microbiol. *54*: 1557–1560.

Collins, M.D., L. Hoyles, G. Foster and E. Falsen. 2004b. *Corynebacterium caspium* sp. nov., from a Caspian seal (*Phoca caspica*). Int. J. Syst. Evol. Microbiol. *54*: 925–928.

Collins, M.D., T. Lundström, C. Welinder-Olsson, I. Hansson, O. Wattle, R.A. Hutson and E. Falsen. 2004c. *Streptococcus devriesei* sp. nov., from equine teeth. Syst. Appl. Microbiol. *27*: 146–150.

Collins, M.D., J. Routh, A. Saraswathy, P.A. Lawson, P. Schumann, C. Welinder-Olsson and E. Falsen. 2004d. *Arsenicicoccus bolidensis* gen. nov., sp. nov., a novel actinomycete isolated from contaminated lake sediment. Int. J. Syst. Evol. Microbiol. *54*: 605–608.

Collins, M.D., A. Wiernik, E. Falsen and P.A. Lawson. 2005. *Atopococcus tabaci* gen. nov., sp. nov., a novel Gram-positive, catalase-negative, coccus-shaped bacterium isolated from tobacco. Int. J. Syst. Evol. Microbiol. *55*: 1693–1696.

Conrads, G., D.M. Citron, R. Mutters, S. Jang and E.J.C. Goldstein. 2004. *Fusobacterium canifelinum* sp. nov., from the oral cavity of cats and dogs. Syst. Appl. Microbiol. *27*: 407–413.

Conville, P.S., J.M. Brown, A.G. Steigerwalt, J.W. Lee, V.L. Anderson, J.T. Fishbain, S.M. Holland and F.G. Witebsky. 2004. *Nocardia kruczakiae* sp. nov., a pathogen in immunocompromised patients and a member of the "*N. nova* complex". J. Clin. Microbiol. *42*: 5139–5145.

Cook, A.E., M. Roes and P.R. Meyers. 2005. *Actinomadura napierensis* sp. nov., isolated from soil in South Africa. Int. J. Syst. Evol. Microbiol. *55*: 703–706.

Cook, A.R., P.W. Riley, H. Murdoch, P.N. Evans and I.R. McDonald. 2007. *Howardella ureilytica* gen. nov., sp. nov., a Gram-positive, coccoid-shaped bacterium from a sheep rumen. Int. J. Syst. Evol. Microbiol. *57*: 2940–2945.

Cooksey, R.C., J.H. de Waard, M.A. Yakrus, I. Rivera, M. Chopite, S.R. Toney, G.P. Morlock and W.R. Butler. 2004. *Mycobacterium cosmeticum* sp. nov., a novel rapidly growing species isolated from a cosmetic infection and from a nail salon. Int. J. Syst. Evol. Microbiol. *54*: 2385–2391.

Corsetti, A., L. Settanni, D. van Sinderen, G.E. Felis, F. Dellaglio and M. Gobbetti. 2005. *Lactobacillus rossii* sp. nov., isolated from wheat sourdough. Int. J. Syst. Evol. Microbiol. *55*: 35–40.

Coton, M., J.M. Laplace, Y. Auffray and E. Coton. 2006. Polyphasic study of *Zymomonas mobilis* strains revealing the existence of a novel subspecies *Z. mobilis* subsp. *francensis* subsp. nov., isolated from French cider. Int. J. Syst. Evol. Microbiol. *56*: 121–125.

Cotta, M.A., T.R. Whitehead, M.D. Collins and P.A. Lawson. 2004. *Atopostipes suicloacale* gen. nov., sp. nov., isolated from an underground swine manure storage pit. Anaerobe. *10*: 191–195.

Cousin, S., O. Päuker and E. Stackebrandt. 2007. *Flavobacterium aquidurense* sp. nov. and *Flavobacterium hercynium* sp. nov., from a hard-water creek. Int. J. Syst. Evol. Microbiol. *57*: 243–249.

Crapart, S., M.L. Fardeau, J.L. Cayol, P. Thomas, C. Sery, B. Ollivier and Y. Combet-Blanc. 2007. *Exiguobacterium profundum* sp. nov., a moderately thermophilic, lactic acid-producing bacterium isolated from a deep-sea hydrothermal vent. Int. J. Syst. Evol. Microbiol. *57*: 287–292.

Cravo-Laureau, C., R. Matheron, J.L. Cayol, C. Joulian and A. Hirschler-Réa. 2004a. *Desulfatibacillum aliphaticivorans* gen. nov., sp. nov., an n-alkane- and n-alkene-degrading, sulfate-reducing bacterium. Int. J. Syst. Evol. Microbiol. *54*: 77–83.

Cravo-Laureau, C., R. Matheron, C. Joulian, J.L. Cayol and A. Hirschler-Réa. 2004b. *Desulfatibacillum alkenivorans* sp. nov., a novel n-alkene-degrading, sulfate-reducing bacterium, and emended description of the genus *Desulfatibacillum*. Int. J. Syst. Evol. Microbiol. *54*: 1639–1642.

Cravo-Laureau, C., C. Labat, C. Joulian, R. Matheron and A. Hirschler-Réa. 2007. *Desulfatiferula olefinivorans* gen. nov., sp. nov., a long-chain n-alkene-degrading, sulfate-reducing bacterium. Int. J. Syst. Evol. Microbiol. *57*: 2699–2702.

Cui, H.L., D. Tohty, J. Feng, P.J. Zhou and S.J. Liu. 2006a. *Natronorubrum aibiense* sp. nov., an extremely halophilic archaeon isolated from Aibi salt lake in Xin-Jiang, China, and emended description of the genus *Natronorubrum*. Int. J. Syst. Evol. Microbiol. *56*: 1515–1517.

Cui, H.L., D. Tohty, P.J. Zhou and S.J. Liu. 2006b. *Halorubrum lipolyticum* sp. nov. and *Halorubrum aidingense* sp. nov., isolated from two salt lakes in Xin-Jiang, China. Int. J. Syst. Evol. Microbiol. *56*: 1631–1634.

Cui, H.L., D. Tohty, P.J. Zhou and S.J. Liu. 2006c. *Haloterrigena longa* sp. nov. and *Haloterrigena limicola* sp. nov., extremely halophilic archaea isolated from a salt lake. Int. J. Syst. Evol. Microbiol. *56*: 1837–1840.

Cui, H.L., D. Tohty, H.C. Liu, S.J. Liu, A. Oren and P.J. Zhou. 2007a. *Natronorubrum sulfidifaciens* sp. nov., an extremely haloalkaliphilic archaeon isolated from Aiding salt lake in Xin-Jiang, China. Int. J. Syst. Evol. Microbiol. *57*: 738–740.

Cui, H.L., Z.Y. Lin, Y. Dong, P.J. Zhou and S.J. Liu. 2007b. *Halorubrum litoreum* sp. nov., an extremely halophilic archaeon from a solar salt-ern. Int. J. Syst. Evol. Microbiol. *57*: 2204–2206.

Cui, Q., L. Wang, Y. Huang, Z. Liu and M. Goodfellow. 2005. *Nocardia jiangxiensis* sp. nov. and *Nocardia miyunensis* sp. nov., isolated from acidic soils. Int. J. Syst. Evol. Microbiol. *55*: 1921–1925.

Cui, X., P. Schumann, E. Stackebrandt, R.M. Kroppenstedt, R. Pukall, L. Xu, M. Rohde and C. Jiang. 2004. *Myceligenerans xiligouense* gen. nov., sp. nov., a novel hyphae-forming member of the family *Promicromonosporaceae*. Int. J. Syst. Evol. Microbiol. *54*: 1287–1293.

Cui, Y.S., W.T. Im, C.R. Yin, J.S. Lee, K.C. Lee and S.T. Lee. 2007a. *Aeromicrobium panaciterrae* sp. nov., isolated from soil of a ginseng field in South Korea. Int. J. Syst. Evol. Microbiol. *57*: 687–691.

Cui, Y.S., W.T. Im, C.R. Yin, D.C. Yang and S.T. Lee. 2007b. *Microlunatus ginsengisoli* sp. nov., isolated from soil of a ginseng field. Int. J. Syst. Evol. Microbiol. *57*: 713–716.

Cunha, S., I. Tiago, A.L. Pires, M.S. da Costa and A. Veríssimo. 2006. *Dokdonella fugitiva* sp. nov., a *Gammaproteobacterium* isolated from potting soil. Syst. Appl. Microbiol. *29*: 191–196.

Dahle, H. and N.K. Birkeland. 2006. *Thermovirga lienii* gen. nov., sp. nov., a novel moderately thermophilic, anaerobic, amino-acid-degrading bacterium isolated from a North Sea oil well. Int. J. Syst. Evol. Microbiol. *56*: 1539–1545.

Dai, X., Y.N. Wang, B.J. Wang, S.J. Liu and Y.G. Zhou. 2005. *Planomicrobium chinense* sp. nov., isolated from coastal sediment, and transfer of *Planococcus psychrophilus* and *Planococcus alkanoclasticus* to *Planomicrobium* as *Planomicrobium psychrophilum* comb. nov. and *Planomicrobium alkanoclasticum* comb. nov. Int. J. Syst. Evol. Microbiol. *55*: 699–702.

Dai, X., B.J. Wang, Q.X. Yang, N.Z. Jiao and S.J. Liu. 2006. *Yangia pacifica* gen. nov., sp. nov., a novel member of the *Roseobacter* clade from coastal sediment of the East China Sea. Int. J. Syst. Evol. Microbiol. *56*: 529–533.

Daneshvar, M.I., D.G. Hollis, R.S. Weyant, J.G. Jordan, J.P. Macgregor, R.E. Morey, A.M. Whitney, D.J. Brenner, A.G. Steigerwalt, L.O. Helsel, P.M. Raney, J.B. Patel J.B., P.N. Levett and J.M. Brown. 2004. Identification of some charcoal-black-pigmented CDC fermentative coryneform group 4 isolates as *Rothia dentocariosa* and some as *Corynebacterium aurimucosum*: proposal of *Rothia dentocariosa* emend. Georg and Brown 1967, *Corynebacterium aurimucosum* emend. Yassin *et al.* 2002, and *Corynebacterium nigricans* Shukla *et al.* 2003 pro synon. *Corynebacterium aurimucosum*. J. Clin. Microbiol. *42*: 4189–4198.

Davidov, Y. and E. Jurkevitch. 2004. Diversity and evolution of *Bdellovibrio*-and-like organisms (BALOs), reclassification of *Bacteriovorax starrii* as *Peredibacter starrii* gen. nov., comb. nov., and description of the *Bacteriovorax–Peredibacter* clade as *Bacteriovoracaceae* fam. nov. Int. J. Syst. Evol. Microbiol. *54*: 1439–1452.

Davidova, I.A., K.E. Duncan, O.K. Choi and J.M. Suflita. 2006. *Desulfoglaeba alkanexedens* gen. nov., sp. nov., an n-alkane-degrading, sulfate-reducing bacterium. Int. J. Syst. Evol. Microbiol. *56*: 2737–2742.

de Beer, H., C.J. Hugo, P.J. Jooste, A. Willems, M. Vancanneyt, T. Coenye and P.A. Vandamme. 2005. *Chryseobacterium vrystaatense* sp. nov., isolated from raw chicken in a chicken-processing plant. Int. J. Syst. Evol. Microbiol. *55*: 2149–2153.

de Beer, H., C.J. Hugo, P.J. Jooste, M. Vancanneyt, T. Coenye and P. Vandamme. 2006. *Chryseobacterium piscium* sp. nov., isolated from fish of the South Atlantic Ocean off South Africa. Int. J. Syst. Evol. Microbiol. *56*: 1317–1322.

de Boer, W., J.H. Leveau, G.A. Kowalchuk, P.J. Klein Gunnewiek, E.C. Abeln, M.J. Figge, K. Sjollema, J.D. Janse and J.A. van Veen. 2004. *Collimonas fungivorans* gen. nov., sp. nov., a chitinolytic soil bacterium with the ability to grow on living fungal hyphae. Int. J. Syst. Evol. Microbiol. *54*: 857–864.

de Bok, F.A., H.J. Harmsen, C.M. Plugge, M.C. de Vries, A.D. Akkermans, W.M. de Vos and A.J. Stams. 2005. The first true obligately syntrophic propionate-oxidizing bacterium, *Pelotomaculum schinkii* sp. nov., co-cultured with *Methanospirillum hungatei*, and emended description of the genus *Pelotomaculum*. Int. J. Syst. Evol. Microbiol. *55*: 1697–1703.

De Bruyne, K., U. Schillinger, L. Caroline, B. Boehringer, I. Cleenwerck, M. Vancanneyt, L. De Vuyst, C.M. Franz and P. Vandamme. 2007. *Leuconostoc holzapfelii* sp. nov., isolated from Ethiopian coffee fermentation and assessment of sequence analysis of housekeeping genes for delineation of *Leuconostoc* species. Int. J. Syst. Evol. Microbiol. *57*: 2952–2959.

De Clerck, E., M. Rodriguez-Diaz, G. Forsyth, L. Lebbe, N.A. Logan and P. De Vos. 2004a. Polyphasic characterization of *Bacillus coagulans* strains, illustrating heterogeneity within this species, and emended description of the species. Syst. Appl. Microbiol. *27*: 50–60.

De Clerck, E., M. Rodríguez-Díaz, T. Vanhoutte, J. Heyrman, N.A. Logan and P. De Vos. 2004b. *Anoxybacillus contaminans* sp. nov. and *Bacillus gelatini* sp. nov., isolated from contaminated gelatin batches. Int. J. Syst. Evol. Microbiol. *54*: 941–946.

De Clercq, D., S. Van Trappen, I. Cleenwerck, A. Ceustermans, J. Swings, J. Coosemans and J. Ryckeboer. 2006. *Rhodanobacter spathiphylli* sp. nov., a gammaproteobacterium isolated from the roots of *Spathiphyllum* plants grown in a compost-amended potting mix. Int. J. Syst. Evol. Microbiol. *56*: 1755–1759.

de Groot, A., V. Chapon, P. Servant, R. Christen, M.F. Saux, S. Sommer and T. Heulin. 2005. *Deinococcus deserti* sp. nov., a gamma-radiation-tolerant bacterium isolated from the Sahara Desert. Int. J. Syst. Evol. Microbiol. *55*: 2441–2446.

Dedysh, S.N., Y.Y. Berestovskaya, L.V. Vasylieva, S.E. Belova, V.N. Khmelenina, N.E. Suzina, Y.A. Trotsenko, W. Liesack and G.A. Zavarzin. 2004. *Methylocella tundrae* sp. nov., a novel methanotrophic bacterium from acidic tundra peatlands. Int. J. Syst. Evol. Microbiol. *54*: 151–156.

Dedysh, S.N., S.E. Belova, P.L. Bodelier, K.V. Smirnova, V.N. Khmelenina, A. Chidthaisong, Y.A. Trotsenko, W. Liesack and P.F. Dunfield. 2007. *Methylocystis heyeri* sp. nov., a novel type II methanotrophic bacterium possessing 'signature' fatty acids of type I methanotrophs. Int. J. Syst. Evol. Microbiol. *57*: 472–479.

Delgado, O., J. Quillaguamán, S. Bakhtiar, B. Mattiasson, A. Gessesse and R. Hatti-Kaul. 2006. *Nesterenkonia aethiopica* sp. nov., an alkaliphilic, moderate halophile isolated from an Ethiopian soda lake. Int. J. Syst. Evol. Microbiol. *56*: 1229–1232.

Dellaglio, F., S. Torriani and G.E. Felis. 2004. Reclassification of *Lactobacillus cellobiosus* Rogosa *et al.* 1953 as a later synonym of *Lactobacillus fermentum* Beijerinck 1901. Int. J. Syst. Evol. Microbiol. *54*: 809–812.

Dellaglio, F., I. Cleenwerck, G.E. Felis, K. Engelbeen, D. Janssens and M. Marzotto. 2005a. Description of *Gluconacetobacter swingsii* sp. nov. and *Gluconacetobacter rhaeticus* sp. nov., isolated from Italian apple fruit. Int. J. Syst. Evol. Microbiol. *55*: 2365–2370.

Dellaglio, F., G.E. Felis, A. Castioni, S. Torriani and J.E. Germond. 2005b. *Lactobacillus delbrueckii* subsp. *indicus* subsp. nov., isolated from Indian dairy products. Int. J. Syst. Evol. Microbiol. *55*: 401–404.

Dellaglio, F., M. Vancanneyt, A. Endo, P. Vandamme, G.E. Felis, A. Castioni, J. Fujimoto, K. Watanabe and S. Okada. 2006. *Lactobacillus durianis* Leisner *et al.* 2002 is a later heterotypic synonym of *Lactobacillus vaccinostercus* Kozaki and Okada 1983. Int. J. Syst. Evol. Microbiol. *56*: 1721–1724.

Denner, E.B., M. Kolari, D. Hoornstra, I. Tsitko, P. Kämpfer, H.J. Busse and M. Salkinoja-Salonen. 2006. *Rubellimicrobium thermophilum* gen. nov., sp. nov., a red-pigmented, moderately thermophilic bacterium isolated from coloured slime deposits in paper machines. Int. J. Syst. Evol. Microbiol. *56*: 1355–1362.

Derrien, M., E.E. Vaughan, C.M. Plugge and W.M. de Vos. 2004. *Akkermansia muciniphila* gen. nov., sp. nov., a human intestinal mucin-degrading bacterium. Int. J. Syst. Evol. Microbiol. *54*: 1469–1476.

Derz, K., U. Klinner, I. Schuphan, E. Stackebrandt and R.M. Kroppenstedt. 2004. *Mycobacterium pyrenivorans* sp. nov., a novel polycyclic-aromatic-hydrocarbon-degrading species. Int. J. Syst. Evol. Microbiol. *54*: 2313–2317.

Devriese, L.A., M. Vancanneyt, M. Baele, M. Vaneechoutte, E. De Graef, C. Snauwaert, I. Cleenwerck, P. Dawyndt, J. Swings, A. Decostere and F. Haesebrouck. 2005. *Staphylococcus pseudintermedius* sp. nov., a coagulase-positive species from animals. Int. J. Syst. Evol. Microbiol. *55*: 1569–1573.

Dhia Thabet, O.B., M.L. Fardeau, C. Suarez-Nuñez, M. Hamdi, P. Thomas, B. Ollivier and D. Alazard. 2007. *Desulfovibrio marinus* sp. nov., a moderately halophilic sulfate-reducing bacterium isolated from marine sediments in Tunisia. Int. J. Syst. Evol. Microbiol. *57*: 2167–2170.

Díaz, C., S. Baena, M.L. Fardeau and B.K. Patel. 2007. *Aminiphilus circumscriptus* gen. nov., sp. nov., an anaerobic amino-acid-degrading bacterium from an upflow anaerobic sludge reactor. Int. J. Syst. Evol. Microbiol. *57*: 1914–1918.

Dimitriu, P.A., S.K. Shukla, J. Conradt, M.C. Márquez, A. Ventosa, A. Maglia, B.M. Peyton, H.C. Pinkart and M.R. Mormile. 2005. *Nitrincola lacisaponensis* gen. nov., sp. nov., a novel alkaliphilic bacterium isolated from an alkaline, saline lake. Int. J. Syst. Evol. Microbiol. *55*: 2273–2278.

Ding, L. and A. Yokota. 2004. Proposals of *Curvibacter gracilis* gen. nov., sp. nov. and *Herbaspirillum putei* sp. nov. for bacterial strains isolated from well water and reclassification of [*Pseudomonas*] *huttiensis*, [*Pseudomonas*] *lanceolata*, [*Aquaspirillum*] *delicatum* and [*Aquaspirillum*] *autotrophicum* as *Herbaspirillum huttiense* comb. nov., *Curvibacter lanceolatus* comb. nov., *Curvibacter delicatus* comb. nov. and *Herbaspirillum autotrophicum* comb. nov. Int. J. Syst. Evol. Microbiol. *54*: 2223–2230.

Ding, L., T. Hirose and A. Yokota. 2007. *Amycolatopsis echigonensis* sp. nov. and *Amycolatopsis niigatensis* sp. nov., novel actinomycetes isolated from a filtration substrate. Int. J. Syst. Evol. Microbiol. *57*: 1747–1751.

Donachie, S.P., J.P. Bowman and M. Alam. 2004. *Psychroflexus tropicus* sp. nov., an obligately halophilic *Cytophaga–Flavobacterium–Bacteroides* group bacterium from an Hawaiian hypersaline lake. Int. J. Syst. Evol. Microbiol. *54*: 935–940.

Donachie, S.P., J.P. Bowman, S.L. On and M. Alam. 2005. *Arcobacter halophilus* sp. nov., the first obligate halophile in the genus *Arcobacter*. Int. J. Syst. Evol. Microbiol. *55*: 1271–1277.

Donachie, S.P., J.P. Bowman and M. Alam. 2006. *Nesiotobacter exalbescens* gen. nov., sp. nov., a moderately thermophilic alphaproteobacterium from an Hawaiian hypersaline lake. Int. J. Syst. Evol. Microbiol. *56*: 563–567.

Dong, Z., X. Guo, X. Zhang, F. Qiu, L. Sun, H. Gong and F. Zhang. 2007. *Dyadobacter beijingensis* sp. nov., isolated from the rhizosphere of turf grasses in China. Int. J. Syst. Evol. Microbiol. *57*: 862–865.

Doronina, N.V., Y.A. Trotsenko, T.V. Kolganova, T.P. Tourova and M.S. Salkinoja-Salonen. 2004. *Methylobacillus pratensis* sp. nov., a novel non-pigmented, aerobic, obligately methylotrophic bacterium isolated from meadow grass. Int. J. Syst. Evol. Microbiol. *54*: 1453–1457.

Doronina, N.V., E.G. Ivanova and Y.A. Trotsenko. 2005. Phylogenetic position and emended description of the genus *Methylovorus*. Int. J. Syst. Evol. Microbiol. *55*: 903–906.

Downes, J., I. Sutcliffe, A.C. Tanner and W.G. Wade. 2005. *Prevotella marshii* sp. nov. and *Prevotella baroniae* sp. nov., isolated from the human oral cavity. Int. J. Syst. Evol. Microbiol. *55*: 1551–1555.

Downes, J., I.C. Sutcliffe, T. Hofstad and W.G. Wade. 2006. *Prevotella bergensis* sp. nov., isolated from human infections. Int. J. Syst. Evol. Microbiol. *56*: 609–612.

Downes, J. and W.G. Wade. 2006. *Peptostreptococcus stomatis* sp. nov., isolated from the human oral cavity. Int. J. Syst. Evol. Microbiol. *56*: 751–754.

Downes, J., I.C. Sutcliffe, V. Booth and W.G. Wade. 2007. *Prevotella maculosa* sp. nov., isolated from the human oral cavity. Int. J. Syst. Evol. Microbiol. *57*: 2936–2939.

Dröge, S., J. Fröhlich, R. Radek and H. König. 2006. *Spirochaeta coccoides* sp. nov., a novel coccoid spirochete from the hindgut of the termite *Neotermes castaneus*. Appl. Environ. Microbiol. *72*: 392–397.

Duffaud, G.D., O.B. D'Hennezel, A.S. Peek, A.L. Reysenbach and R.M. Kelly. 1998. Isolation and characterization of *Thermococcus barossii*, sp. nov., a hyperthermophilic archaeon isolated from a hydrothermal vent flange formation. Syst. Appl. Microbiol. *21*: 40–49.

Dulger, S., Z. Demirbag and A.O. Belduz. 2004. *Anoxybacillus ayderensis* sp. nov. and *Anoxybacillus kestanbolensis* sp. nov. Int. J. Syst. Evol. Microbiol. *54*: 1499–1503.

Duncan, S.H., R.I. Aminov, K.P. Scott, P. Louis, T.B. Stanton and H.J. Flint. 2006. Proposal of *Roseburia faecis* sp. nov., *Roseburia hominis* sp. nov. and *Roseburia inulinivorans* sp. nov., based on isolates from human faeces. Int. J. Syst. Evol. Microbiol. *56*: 2437–2441.

Dutta, D. and R. Gachhui. 2006. Novel nitrogen-fixing *Acetobacter nitrogenifigens* sp. nov., isolated from Kombucha tea. Int. J. Syst. Evol. Microbiol. *56*: 1899–1903.

Dutta, D. and R. Gachhui. 2007. Nitrogen-fixing and cellulose-producing *Gluconacetobacter kombuchae* sp. nov., isolated from Kombucha tea. Int. J. Syst. Evol. Microbiol. *57*: 353–357.

Echigo, A., T. Fukushima, T. Mizuki, M. Kamekura and R. Usami. 2007. *Halalkalibacillus halophilus* gen. nov., sp. nov., a novel moderately halophilic and alkaliphilic bacterium isolated from a non-saline soil sample in Japan. Int. J. Syst. Evol. Microbiol. *57*: 1081–1085.

Ehrmann, M.A., M. Brandt, P. Stolz, R.F. Vogel and M. Korakli. 2007. *Lactobacillus secaliphilus* sp. nov., isolated from type II sourdough fermentation. Int. J. Syst. Evol. Microbiol. *57*: 745–750.

Eichorst, S.A., J.A. Breznak and T.M. Schmidt. 2007. Isolation and characterization of soil bacteria that define *Terriglobus* gen. nov., in the phylum *Acidobacteria*. Appl. Environ. Microbiol. *73*: 2708–2717.

Einen, J. and L. Øvreås. 2006. *Flaviramulus basaltis* gen. nov., sp. nov., a novel member of the family *Flavobacteriaceae* isolated from seafloor basalt. Int. J. Syst. Evol. Microbiol. *56*: 2455–2461.

Ekborg, N.A., J.M. González, M.B. Howard, L.E. Taylor, S.W. Hutcheson and R.M. Weiner. 2005. *Saccharophagus degradans* gen. nov., sp. nov., a versatile marine degrader of complex polysaccharides. Int. J. Syst. Evol. Microbiol. *55*: 1545–1549.

Elshahed, M.S., K.N. Savage, A. Oren, M.C. Gutiérrez, A. Ventosa and L.R. Krumholz. 2004. *Haloferax sulfurifontis* sp. nov., a halophilic archaeon isolated from a sulfide- and sulfur-rich spring. Int. J. Syst. Evol. Microbiol. *54*: 2275–2279.

Emborg, J., P. Dalgaard and P. Ahrens. 2006. *Morganella psychrotolerans* sp. nov., a histamine-producing bacterium isolated from various seafoods. Int. J. Syst. Evol. Microbiol. *56*: 2473–2479.

Enache, M., T. Itoh, M. Kamekura, G. Teodosiu and L. Dumitru. 2007. *Haloferax prahovense* sp. nov., an extremely halophilic archaeon isolated from a Romanian salt lake. Int. J. Syst. Evol. Microbiol. *57*: 393–397.

Endo, A. and S. Okada. 2005. *Lactobacillus satsumensis* sp. nov., isolated from mashes of shochu, a traditional Japanese distilled spirit made from fermented rice and other starchy materials. Int. J. Syst. Evol. Microbiol. *55*: 83–85.

Endo, A. and S. Okada. 2006. *Oenococcus kitaharae* sp. nov., a non-acidophilic and non-malolactic-fermenting oenococcus isolated from a composting distilled shochu residue. Int. J. Syst. Evol. Microbiol. *56*: 2345–2348.

Endo, A. and S. Okada. 2007a. *Lactobacillus composti* sp. nov., a lactic acid bacterium isolated from a compost of distilled shochu residue. Int. J. Syst. Evol. Microbiol. *57*: 870–872.

Endo, A. and S. Okada. 2007b. *Lactobacillus farraginis* sp. nov. and *Lactobacillus parafarraginis* sp. nov., heterofermentative lactobacilli isolated from a compost of distilled shochu residue. Int. J. Syst. Evol. Microbiol. *57*: 708–712.

Ennahar, S. and Y. Cai. 2004. Genetic evidence that *Weissella kimchii* Choi *et al.* 2002 is a later heterotypic synonym of *Weissella cibaria* Björkroth *et al.* 2002. Int. J. Syst. Evol. Microbiol. *54*: 463–465.

Ennahar, S. and Y. Cai. 2005. Biochemical and genetic evidence for the transfer of *Enterococcus solitarius* Collins *et al.* 1989 to the genus *Tetragenococcus* as *Tetragenococcus* solitarius comb. nov. Int. J. Syst. Evol. Microbiol. *55*: 589–592.

Eribe, E.R., B.J. Paster, D.A. Caugant, F.E. Dewhirst, V.K. Stromberg, G.H. Lacy and I. Olsen. 2004. Genetic diversity of *Leptotrichia* and description of *Leptotrichia goodfellowii* sp. nov., *Leptotrichia hofstadii* sp. nov., *Leptotrichia shahii* sp. nov. and *Leptotrichia wadei* sp. nov. Int. J. Syst. Evol. Microbiol. *54*: 583–592.

Fahrbach, M., J. Kuever, R. Meinke, P. Kämpfer and J. Hollender. 2006. *Denitratisoma oestradiolicum* gen. nov., sp. nov., a 17beta-oestradiol-degrading, denitrifying betaproteobacterium. Int.J. Syst. Evol. Microbiol. *56*: 1547–1552.

Falsen, E., M.D. Collins, C. Welinder-Olsson, Y. Song, S.M. Finegold and P.A. Lawson. 2005. *Fastidiosipila sanguinis* gen. nov., sp. nov., a new Gram-positive, coccus-shaped organism from human blood. Int. J. Syst. Evol. Microbiol. *55*: 853–858.

Fan, H., Y. Xue, Y. Ma, A. Ventosa and W.D. Grant. 2004. *Halorubrum tibetense* sp. nov., a novel haloalkaliphilic archaeon from Lake Zabuye in Tibet, China. Int. J. Syst. Evol. Microbiol. *54*: 1213–1216.

Fanti, F., E. Tortoli, L. Hall, G.D. Roberts, R.M. Kroppenstedt, I. Dodi, S. Conti, L. Polonelli and C. Chezzi. 2004. *Mycobacterium parmense* sp. nov. Int. J. Syst. Evol. Microbiol. *54*: 1123–1127.

Fardeau, M.L., M. Bonilla Salinas, S. L'Haridon, C. Jeanthon, F. Verhé, J.L. Cayol, B.K. Patel, J.L. García and B. Ollivier. 2004. Isolation from oil reservoirs of novel thermophilic anaerobes phylogenetically related to *Thermoanaerobacter subterraneus*: reassignment of *T. subterraneus*, *Thermoanaerobacter yonseiensis*, *Thermoanaerobacter tengcongensis* and *Carboxydibrachium pacificum* to *Caldanaerobacter subterraneus* gen. nov., sp. nov., comb. nov. as four novel subspecies. Int. J. Syst. Evol. Microbiol. *54*: 467–474.

Faury, N., D. Saulnier, F.L. Thompson, M. Gay, J. Swings and F. Le Roux. 2004. *Vibrio crassostreae* sp. nov., isolated from the haemolymph of oysters (*Crassostrea gigas*). Int. J. Syst. Evol. Microbiol. *54*: 2137–2140.

Feio, M.J., V. Zinkevich, I.B. Beech, E. Llobet-Brossa, P. Eaton, J. Schmitt and J. Guezennec. 2004. *Desulfovibrio alaskensis* sp. nov., a sulphate-reducing bacterium from a soured oil reservoir. Int. J. Syst. Evol. Microbiol. *54*: 1747–1752.

Felis, G.E., S. Torriani and F. Dellaglio. 2005. Reclassification of *Pediococcus urinaeequi* (ex Mees 1934) Garvie 1988 as *Aerococcus urinaeequi* comb. nov. Int. J. Syst. Evol. Microbiol. *55*: 1325–1327.

Felis, G.E., M. Vancanneyt, C. Snauwaert, J. Swings, S. Torriani, A. Castioni and F. Dellaglio. 2006. Reclassification of *Lactobacillus thermotolerans* Niamsup *et al.* 2003 as a later synonym of *Lactobacillus ingluviei* Baele *et al.* 2003. Int. J. Syst. Evol. Microbiol. *56*: 793–795.

Feng, J., P.J. Zhou and S.J. Liu. 2004. *Halorubrum xinjiangense* sp. nov., a novel halophile isolated from saline lakes in China. Int. J. Syst. Evol. Microbiol. *54*: 1789–1791.

Feng, J., P. Zhou, Y.G. Zhou, S.J. Liu and K. Warren-Rhodes. 2005. *Halorubrum alkaliphilum* sp. nov., a novel haloalkaliphile isolated from a soda lake in Xinjiang, China. Int. J. Syst. Evol. Microbiol. *55*: 149–152.

Fenner, L., V. Roux, M.N. Mallet and D. Raoult. 2005. *Bacteroides massiliensis* sp. nov., isolated from blood culture of a newborn. Int. J. Syst. Evol. Microbiol. *55*: 1335–1337.

Fernandes, C., F.A. Rainey, M.F. Nobre, I. Pinhal, F. Folhas and M.S. da Costa. 2005. *Herminiimonas fonticola* gen. nov., sp. nov., a Betaproteobacterium isolated from a source of bottled mineral water. Syst. Appl. Microbiol. *28*: 596–603.

Fernandez, E., V. Blume, P. Garrido, M.D. Collins, A. Mateos, L. Domínguez and J.F. Fernández-Garayzábal. 2004. *Streptococcus equi* subsp. *ruminatorum* subsp. nov., isolated from mastitis in small ruminants. Int. J. Syst. Evol. Microbiol. *54*: 2291–2296.

Fernández-Garayzábal, J.F., A.I. Vela, R. Egido, R.A. Hutson, M.P. Lanzarot, M. Fernández-García and M.D. Collins. 2004. *Corynebacterium ciconiae* sp. nov., isolated from the trachea of black storks (*Ciconia nigra*). Int. J. Syst. Evol. Microbiol. *54*: 2191–2195.

Feurer, C., D. Clermont, F. Bimet, A. Candréa, M. Jackson, P. Glaser, C. Bizet and C. Dauga. 2004. Taxonomic characterization of nine strains isolated from clinical and environmental specimens, and proposal of *Corynebacterium tuberculostearicum* sp. nov. Int. J. Syst. Evol. Microbiol. *54*: 1055–1061.

Finegold, S.M., P.A. Lawson., M.L. Vaisanen., D.R. Molitoris, Y. Song, C. Liu and M.D. Collins. 2004a. *Anaerofustis stercorihominis* gen. nov., sp. nov., from human feces. Anaerobe. *10*: 41–45.

Finegold, S.M., M.L. Vaisanen, M. Rautio, E. Eerola, P. Summanen, D. Molitoris, Y. Song, C. Liu and H. Jousimies-Somer. 2004b. *Porphyromonas uenonis* sp. nov., a pathogen for humans distinct from *P. asaccharolytica* and *P. endodontalis*. J. Clin. Microbiol. *42*: 5298–5301.

Fortina, M.G., G. Ricci, D. Mora and P.L. Manachini. 2004. Molecular analysis of artisanal Italian cheeses reveals *Enterococcus italicus* sp. nov. Int. J. Syst. Evol. Microbiol. *54*: 1717–1721.

Foster, G., B. Holmes, A.G. Steigerwalt, P.A. Lawson, P. Thorne, D.E. Byrer, H.M. Ross, J. Xerry, P.M. Thompson and M.D. Collins. 2004. *Campylobacter insulaenigrae* sp. nov., isolated from marine mammals. Int. J. Syst. Evol. Microbiol. *54*: 2369–2373.

Foster, G., H. Malnick, P.A. Lawson, J. Kirkwood, S.K. Macgregor and M.D. Collins. 2005. *Suttonella ornithocola* sp. nov., from birds of the tit families, and emended description of the genus *Suttonella*. Int. J. Syst. Evol. Microbiol. *55*: 2269–2272.

Foster, G., B.S. Osterman, J. Godfroid, I. Jacques and A. Cloeckaert. 2007. *Brucella ceti* sp. nov. and *Brucella pinnipedialis* sp. nov. for *Brucella* strains with cetaceans and seals as their preferred hosts. Int. J. Syst. Evol. Microbiol. *57*: 2688–2693.

Fournier, P.E., J.S. Dumler, G. Greub, J. Zhang, Y. Wu and D. Raoult. 2003. Gene sequence-based criteria for identification of new rickettsia isolates and description of *Rickettsia heilongjiangensis* sp. nov. J. Clin. Microbiol. *41*: 5456–5465.

Fournier, P.E., N. Takada, H. Fujita and D. Raoult. 2006. *Rickettsia tamurae* sp. nov., isolated from *Amblyomma testudinarium* ticks. Int. J. Syst. Evol. Microbiol. *56*: 1673–1675.

Fox, J.G., Z. Shen, S. Xu, Y. Feng, C.A. Dangler, F.E. Dewhirst, B.J. Paster and J.M. Cullen. 2002. *Helicobacter marmotae* sp. nov. isolated from livers of woodchucks and intestines of cats. J. Clin. Microbiol. *40*: 2513–2519.

Fox, J.G., N.S. Taylor, S. Howe, M. Tidd, S. Xu, B.J. Paster and F.E. Dewhirst. 2006. *Helicobacter anseris* sp. nov. and *Helicobacter brantae* sp. nov., isolated from feces of resident Canada geese in the greater Boston area. Appl. Environ. Microbiol. *72*: 4633–4637.

Fracek, S.P. Jr. and J.F. Stolz. 1985. *Spirochaeta bajacaliforniensis* sp. n. from a microbial mat community at Laguna Figueroa, Baja California Norte, Mexico. Arch. Microbiol. *142*: 317–325.

França, L., F.A. Rainey, M.F. Nobre and M.S. da Costa. 2006a. *Salinicoccus salsiraiae* sp. nov.: a new moderately halophilic gram-positive bacterium isolated from salted skate. Extremophiles. *10*: 531–536.

França, L., F.A. Rainey, M.F. Nobre and M.S. da Costa. 2006b. *Tepidicella xavieri* gen. nov., sp. nov., a betaproteobacterium isolated from a hot spring runoff. Int. J. Syst. Evol. Microbiol. *56*: 907–912.

Franz, C.M., M. Vancanneyt, K. Vandemeulebroecke, M. De Wachter, I. Cleenwerck, B. Hoste, U. Schillinger, W.H. Holzapfel and J. Swings. 2006. *Pediococcus stilesii* sp. nov., isolated from maize grains. Int. J. Syst. Evol. Microbiol. *56*: 329–333.

Frette, L., N.O. Jørgensen, H. Irming and N. Kroer. 2004. *Tenacibaculum skagerrakense* sp. nov., a marine bacterium isolated from the pelagic zone in Skagerrak, Denmark. Int. J. Syst. Evol. Microbiol. *54*: 519–524.

Fritz, I., C. Strömpl and W.R. Abraham. 2004. Phylogenetic relationships of the genera *Stella, Labrys* and *Angulomicrobium* within the 'Alphaproteobacteria' and description of *Angulomicrobium amanitiforme* sp. nov. Int. J. Syst. Evol. Microbiol. *54*: 651–657.

Fritz, I., C. Strömpl, D.I. Nikitin, A.M. Lysenko and W.R. Abraham. 2005. *Brevundimonas mediterranea* sp. nov., a non-stalked species from the Mediterranean Sea. Int. J. Syst. Evol. Microbiol. *55*: 479–486.

Fryer, J.L. and C.N. Lannan. 2005. Family II. *Piscirickettsiaceae* fam. nov. *In* Brenner, Krieg, Staley and Garrity (Editors), Bergey's Manual of Systematic Bacteriology, 2nd Edition, Volume 2 (The *Proteobacteria*), Part B (The *Gammaproteobacteria*), Springer, New York, p. 180.

Fujita, H., P.E. Fournier, N. Takada, T. Saito and D. Raoult. 2006. *Rickettsia asiatica* sp. nov., isolated in Japan. Int. J. Syst. Evol. Microbiol. *56*: 2365–2368.

Fukunaga, Y., M. Kurahashi, K. Tanaka, K. Yanagi, A. Yokota and S. Harayama. 2006. *Pseudovibrio ascidiaceicola* sp. nov., isolated from ascidians (sea squirts). Int. J. Syst. Evol. Microbiol. *56*: 343–347.

Funke, G., M. Pagano-Niederer, B. Sjöden and E. Falsen. 1998. Characteristics of *Arthrobacter cumminsii*, the most frequently encountered *Arthrobacter* species in human clinical specimens. J. Clin. Microbiol. *36*: 1539–1543.

Gallego, V., M.T. García and A. Ventosa. 2005a. *Methylobacterium hispanicum* sp. nov. and *Methylobacterium aquaticum* sp. nov., isolated from drinking water. Int. J. Syst. Evol. Microbiol. *55*: 281–287.

Gallego, V., M.T. García and A. Ventosa. 2005b. *Methylobacterium variabile* sp. nov., a methylotrophic bacterium isolated from an aquatic environment. Int. J. Syst. Evol. Microbiol. *55*: 1429–1433.

Gallego, V., M.T. García and A. Ventosa. 2005c. *Methylobacterium isbiliense* sp. nov., isolated from the drinking water system of Sevilla, Spain. Int. J. Syst. Evol. Microbiol. *55*: 2333–2337.

Gallego, V., M.T. García and A. Ventosa. 2006a. *Methylobacterium adhaesivum* sp. nov., a methylotrophic bacterium isolated from drinking water. Int. J. Syst. Evol. Microbiol. *56*: 339–342.

Gallego, V., M.T. García and A. Ventosa. 2006b. *Chryseobacterium hispanicum* sp. nov., isolated from the drinking water distribution system of Sevilla, Spain. Int. J. Syst. Evol. Microbiol. *56*: 1589–1592.

Gallego, V., M.T. García and A. Ventosa. 2006c. *Pedobacter aquatilis* sp. nov., isolated from drinking water, and emended description of the genus *Pedobacter*. Int. J. Syst. Evol. Microbiol. *56*: 1853–1858.

Gallego, V., C. Sánchez-Porro, M.T. García and A. Ventosa. 2006d. *Roseomonas aquatica* sp. nov., isolated from drinking water. Int. J. Syst. Evol. Microbiol. *56*: 2291–2295.

Gallego, V., C. Sánchez-Porro, M.T. García and A. Ventosa. 2006e. *Massilia aurea* sp. nov., isolated from drinking water. Int. J. Syst. Evol. Microbiol. *56*: 2449–2453.

Gao, H., A. Obraztova, N. Stewart, R. Popa, J.K. Fredrickson, J.M. Tiedje, K.H. Nealson and J. Zhou. 2006. *Shewanella loihica* sp. nov., isolated from iron-rich microbial mats in the Pacific Ocean. Int. J. Syst. Evol. Microbiol. *56*: 1911–1916.

Gao, J.L., S.L. Turner, F.L. Kan, E.T. Wang, Z.Y. Tan, Y.H. Qiu, J. Gu, Z. Terefework, J.P. Young, K. Lindström and W.X. Chen. 2004. *Mesorhizobium septentrionale* sp. nov. and *Mesorhizobium temperatum* sp. nov., isolated from *Astragalus adsurgens* growing in the northern regions of China. Int. J. Syst. Evol. Microbiol. *54*: 2003–2012.

García, M.T., E. Mellado, J.C. Ostos and A. Ventosa. 2004. *Halomonas organivorans* sp. nov., a moderate halophile able to degrade aromatic compounds. Int. J. Syst. Evol. Microbiol. *54*: 1723–1728.

García, M.T., V. Gallego, A. Ventosa and E. Mellado. 2005. *Thalassobacillus devorans* gen. nov., sp. nov., a moderately halophilic, phenol-degrading, Gram-positive bacterium. Int. J. Syst. Evol. Microbiol. *55*: 1789–1795.

García-Fraile, P., R. Rivas, A. Willems, A. Peix, M. Martens, E. Martínez-Molina, P.F. Mateos and E. Velázquez. 2007. *Rhizobium cellulosilyticum* sp. nov., isolated from sawdust of *Populus alba*. Int. J. Syst. Evol. Microbiol. *57*: 844–848.

Gardan, L., R. Christen, W. Achouak and P. Prior. 2004. *Erwinia papayae* sp. nov., a pathogen of papaya (*Carica papaya*). Int. J. Syst. Evol. Microbiol. *54*: 107–113.

Garnova, E.S., T.N. Zhilina, T.P. Tourova, N.A. Kostrikina and G.A. Zavarzin. 2004. Anaerobic, alkaliphilic, saccharolytic bacterium *Alkalibacter saccharofermentans* gen. nov., sp. nov. from a soda lake in the Transbaikal region of Russia. Extremophiles. *8*: 309–316.

Garrity, G.M., J.A. Bell and T. Lilburn. 2005i. Phylum XIV. *Proteobacteria* phyl. nov. *In* Brenner, Krieg, Staley and Garrity (Editors), Bergey's Manual of Systematic Bacteriology, 2nd Edition, Volume 2 (The *Proteobacteria*), Part B (The *Gammaproteobacteria*), Springer, New York, p. 1.

Garrity, G.M., J.A. Bell and T. Lilburn. 2005ii. Class III. *Gammaproteobacteria* class. nov. *In* Brenner, Krieg, Staley and Garrity (Editors), Bergey's Manual of Systematic Bacteriology, 2nd Edition, Volume 2 (The *Proteobacteria*), Part B (The *Gammaproteobacteria*), Springer, New York, p. 1.

Garrity, G.M., J.A. Bell and T. Lilburn. 2005iii. Order II. *Acidithiobacillales* ord. nov. *In* Brenner, Krieg, Staley and Garrity (Editors), Bergey's Manual of Systematic Bacteriology, 2nd Edition, Volume 2 (The *Proteobacteria*), Part B (The *Gammaproteobacteria*), Springer, New York, p. 60

Garrity, G.M., J.A. Bell and T. Lilburn. 2005iv. Family I. *Acidithiobacillaceae* fam. nov. *In* Brenner, Krieg, Staley and Garrity (Editors), Bergey's Manual of Systematic Bacteriology, 2nd Edition, Volume 2 (The *Proteobacteria*), Part B (The *Gammaproteobacteria*), Springer, New York, p. 60.

Garrity, G.M., J.A. Bell and T. Lilburn. 2005v.: Family II. *Thermithiobacillaceae* fam. nov. *In* Brenner, Krieg, Staley and Garrity eds, Bergey's Manual of Systematic Bacteriology, 2nd Edition, Volume 2 (The *Proteobacteria*), Part B (The *Gammaproteobacteria*), Springer, New York, p. 62.

Garrity, G.M., J.A. Bell and T. Lilburn. 2005vi. Order IV. *Cardiobacteriales* ord. nov. *In* Brenner, Krieg, Staley and Garrity (Editors), Bergey's Manual of Systematic Bacteriology, 2nd Edition, Volume 2 (The *Proteobacteria*), Part B (The *Gammaproteobacteria*), Springer, New York, p. 123.

Garrity, G.M., J.A. Bell and T. Lilburn. 2005vii. Order V. *Thiotrichales* ord. nov. *In* Brenner, Krieg, Staley and Garrity (Editors), Bergey's Manual of Systematic Bacteriology, 2nd Edition, Volume 2 (The *Proteobacteria*), Part B (The *Gammaproteobacteria*), Springer, New York, p. 131.

Garrity, G.M., J.A. Bell and T. Lilburn. 2005viii. Family I. *Thiotrichaceae* fam. nov. *In* Brenner, Krieg, Staley and Garrity (Editors), Bergey's Manual of Systematic Bacteriology, 2nd Edition, Volume 2 (The *Proteobacteria*), Part B (The *Gammaproteobacteria*), Springer, New York, p. 131.

Garrity, G.M., J.A. Bell and T. Lilburn. 2005ix. Order VI. *Legionellales* ord. nov. *In* Brenner, Krieg, Staley and Garrity (Editors), Bergey's Manual of Systematic Bacteriology, 2nd Edition, Volume 2 (The *Proteobacteria*), Part B (The *Gammaproteobacteria*), Springer, New York, p. 210.

Garrity, G.M., J.A. Bell and T. Lilburn. 2005x. Family II. *Coxiellaceae* fam. nov. *In* Brenner, Krieg, Staley and Garrity (Editors), Bergey's Manual of Systematic Bacteriology, 2nd Edition, Volume 2 (The *Proteobacteria*), Part B (The *Gammaproteobacteria*), Springer, New York, p. 237.

Garrity, G.M., J.A. Bell and T. Lilburn. 2005xi. Order VIII. *Oceanospirillales* ord. nov. *In* Brenner, Krieg, Staley and Garrity (Editors), Bergey's Manual of Systematic Bacteriology, 2nd Edition, Volume 2 (The *Proteobacteria*), Part B (The *Gammaproteobacteria*), Springer, New York, p. 270.

Garrity, G.M., J.A. Bell and T. Lilburn. 2005xii. Family I. *Oceanospirillaceae* fam. nov. *In* Brenner, Krieg, Staley and Garrity (Editors), Bergey's Manual of Systematic Bacteriology, 2nd Edition, Volume 2 (The *Proteobacteria*), Part B (The *Gammaproteobacteria*), Springer, New York, p. 271.

Garrity, G.M., J.A. Bell and T. Lilburn. 2005xiii. Family III. *Hahellaceae* fam. nov. *In* Brenner, Krieg, Staley and Garrity (Editors), Bergey's Manual of Systematic Bacteriology, 2nd Edition, Volume 2 (The *Proteobacteria*), Part B (The *Gammaproteobacteria*), Springer, New York, p. 299.

Garrity, G.M., J.A. Bell and T. Lilburn. 2005xiv. Order XIV. *Pasteurellales* ord. nov. *In* Brenner, Krieg, Staley and Garrity (Editors), Bergey's Manual of Systematic Bacteriology, 2nd Edition, Volume 2 (The *Proteobacteria*), Part B (The *Gammaproteobacteria*), Springer, New York, p. 850.

Garrity, G.M., J.A. Bell and T. Lilburn. 2005xv. Class I. *Alphaproteobacteria* class. nov. *In* Brenner, Krieg, Staley and Garrity (Editors), Bergey's Manual of Systematic Bacteriology, 2nd Edition, Volume 2 (The *Proteobacteria*), Part C (The *Alpha-, Beta-, Delta-,* and *Epsilonproteobacteria*), Springer, New York, p. 1.

Garrity, G.M., J.A. Bell and T. Lilburn. 2005xvi. Order III. *Rhodobacterales* ord. nov. *In* Brenner, Krieg, Staley and Garrity (Editors), Bergey's Manual of Systematic Bacteriology, 2nd Edition, Volume 2 (The *Proteobacteria*), Part C (The *Alpha-, Beta-, Delta-,* and *Epsilonproteobacteria*), Springer, New York, p. 161.

Garrity, G.M., J.A. Bell and T. Lilburn. 2005xvii. Family I. *Rhodobacteraceae* fam. nov. *In* Brenner, Krieg, Staley and Garrity (Editors), Bergey's Manual of Systematic Bacteriology, 2nd Edition, Volume 2 (The *Proteobacteria*), Part C (The *Alpha-, Beta-, Delta-,* and *Epsilonproteobacteria*), Springer, New York, p. 161.

Garrity, G.M., J.A. Bell and T. Lilburn. 2005xviii. Family VI. *Beijerinckiaceae* fam. nov. *In* Brenner, Krieg, Staley and Garrity (Editors), Bergey's Manual of Systematic Bacteriology, 2nd Edition, Volume 2 (The *Proteobacteria*), Part C (The *Alpha-, Beta-, Delta-,* and *Epsilonproteobacteria*), Springer, New York, p. 422.

Garrity, G.M., J.A. Bell and T. Lilburn. 2005xix. Family VII. *Bradyrhizobiaceae* fam. nov. *In* Brenner, Krieg, Staley and Garrity (Editors), Bergey's Manual of Systematic Bacteriology, 2nd Edition, Volume 2 (The *Proteobacteria*), Part C (The *Alpha-, Beta-, Delta-,* and *Epsilonproteobacteria*), Springer, New York, p. 438.

Garrity, G.M., J.A. Bell and T. Lilburn. 2005xx. Family IX. *Methylobacteriaceae* fam. nov. *In* Brenner, Krieg, Staley and Garrity (Editors), Bergey's Manual of Systematic Bacteriology, 2nd Edition, Volume 2 (The *Proteobacteria*), Part C (The *Alpha-, Beta-, Delta-,* and *Epsilonproteobacteria*), Springer, New York, p. 567.

Garrity, G.M., J.A. Bell and T. Lilburn. 2005xxi. Family X. *Rhodobiaceae* fam. nov. *In* Brenner, Krieg, Staley and Garrity (Editors), Bergey's Manual of Systematic Bacteriology, 2nd Edition, Volume 2 (The *Proteobacteria*), Part C (The *Alpha-, Beta-, Delta-,* and *Epsilonproteobacteria*), Springer, New York, p. 571.

Garrity, G.M., J.A. Bell and T. Lilburn. 2005xxii. Class II. *Betaproteobacteria* class. nov. *In* Brenner, Krieg, Staley and Garrity (Editors), Bergey's Manual of Systematic Bacteriology, 2nd Edition, Volume 2 (The *Proteobacteria*), Part C (The *Alpha-, Beta-, Delta-,* and *Epsilonproteobacteria*), Springer, New York, p. 575.

Garrity, G.M., J.A. Bell and T. Lilburn. 2005xxiii. Order I. *Burkholderiales* ord. nov. *In* Brenner, Krieg, Staley and Garrity (Editors), Bergey's Manual of Systematic Bacteriology, 2nd Edition, Volume 2 (The *Proteobacteria*), Part C (The *Alpha-, Beta-, Delta-,* and *Epsilonproteobacteria*), Springer, New York, p. 575.

Garrity, G.M., J.A. Bell and T. Lilburn. 2005xxiv. Family I. *Burkholderiaceae* fam. nov. *In* Brenner, Krieg, Staley and Garrity (Editors), Bergey's Manual of Systematic Bacteriology, 2nd Edition, Volume 2 (The *Proteobacteria*), Part C (The *Alpha-, Beta-, Delta-,* and *Epsilonproteobacteria*), Springer, New York, p. 575.

Garrity, G.M., J.A. Bell and T. Lilburn. 2005xxv. Family II. *Oxalobacteraceae* fam. nov. *In* Brenner, Krieg, Staley and Garrity (Editors), Bergey's Manual of Systematic Bacteriology, 2nd Edition, Volume 2 (The *Proteobacteria*), Part C (The *Alpha-, Beta-, Delta-,* and *Epsilonproteobacteria*), Springer, New York, p. 623.

Garrity, G.M., J.A. Bell and T. Lilburn. 2005xxvi. Order II. *Hydrogenophilales* ord. nov. *In* Brenner, Krieg, Staley and Garrity (Editors), Bergey's Manual of Systematic Bacteriology, 2nd Edition, Volume 2 (The *Proteobacteria*), Part C (The *Alpha-, Beta-, Delta-,* and *Epsilonproteobacteria*), Springer, New York, p. 763.

Garrity, G.M., J.A. Bell and T. Lilburn. 2005xxvii. Family I. *Hydrogenophilaceae* fam. nov. *In* Brenner, Krieg, Staley and Garrity (Editors), Bergey's Manual of Systematic Bacteriology, 2nd Edition, Volume 2 (The *Proteobacteria*), Part C (The *Alpha-, Beta-, Delta-,* and *Epsilonproteobacteria*), Springer, New York, p. 763.

Garrity, G.M., J.A. Bell and T. Lilburn. 2005xxviii. Order III. *Methylophilales* ord. nov. *In* Brenner, Krieg, Staley and Garrity (Editors), Bergey's Manual of Systematic Bacteriology, 2nd Edition, Volume 2 (The *Proteobacteria*), Part C (The *Alpha-, Beta-, Delta-,* and *Epsilonproteobacteria*), Springer, New York, p. 770.

Garrity, G.M., J.A. Bell and T. Lilburn. 2005xxix. Family I. *Methylophilaceae* fam. nov. *In* Brenner, Krieg, Staley and Garrity (Editors), Bergey's Manual of Systematic Bacteriology, 2nd Edition, Volume 2 (The *Proteobacteria*), Part C (The *Alpha-, Beta-, Delta-,* and *Epsilonproteobacteria*), Springer, New York, p. 770.

Garrity, G.M., J.A. Bell and T. Lilburn. 2005xxx. Order V. *Nitrosomonadales* ord. nov. *In* Brenner, Krieg, Staley and Garrity (Editors), Bergey's Manual of Systematic Bacteriology, 2nd Edition, Volume 2 (The *Proteobacteria*), Part C (The *Alpha-, Beta-, Delta-,* and *Epsilonproteobacteria*), Springer, New York, p. 863.

Garrity, G.M., J.A. Bell and T. Lilburn. 2005xxxi. Family I. *Nitrosomonadaceae* fam. nov. *In* Brenner, Krieg, Staley and Garrity (Editors), Bergey's Manual of Systematic Bacteriology, 2nd Edition, Volume 2 (The *Proteobacteria*), Part C (The *Alpha-, Beta-, Delta-,* and *Epsilonproteobacteria*), Springer, New York, p. 864.

Garrity, G.M., J.A. Bell and T. Lilburn. 2005xxxii. Order VI. *Rhodocyclales* ord. nov. *In* Brenner, Krieg, Staley and Garrity (Editors), Bergey's Manual of Systematic Bacteriology, 2nd Edition, Volume 2 (The *Proteobacteria*), Part C (The *Alpha-, Beta-, Delta-,* and *Epsilonproteobacteria*), Springer, New York, p. 887.

Garrity, G.M., J.A. Bell and T. Lilburn. 2005xxxiii. Family I. *Rhodocyclaceae* fam. nov. *In* Brenner, Krieg, Staley and Garrity (Editors), Bergey's Manual of Systematic Bacteriology, 2nd Edition, Volume 2 (The *Proteobacteria*), Part C (The *Alpha-, Beta-, Delta-,* and *Epsilonproteobacteria*), Springer, New York, p. 887.

Garrity, G.M., J.A. Bell and T. Lilburn. 2005xxxiv. Family III. *Nitrospinaceae* fam. nov. *In* Brenner, Krieg, Staley and Garrity (Editors), Bergey's Manual of Systematic Bacteriology, 2nd Edition, Volume 2 (The *Proteobacteria*), Part C (The *Alpha-, Beta-, Delta-,* and *Epsilonproteobacteria*), Springer, New York, p. 999.

Garrity, G.M., J.A. Bell and T. Lilburn. 2005xxxv. Family II. *Geobacteraceae* fam. nov. *In* Brenner, Krieg, Staley and Garrity (Editors), Bergey's Manual of Systematic Bacteriology, 2nd Edition, Volume 2 (The *Proteobacteria*), Part C (The *Alpha-, Beta-, Delta-,* and *Epsilonproteobacteria*), Springer, New York, p. 1017.

Garrity, G.M., J.A. Bell and T. Lilburn. 2005xxxvi. Order VII. *Bdellovibrionales* ord. nov. *In* Brenner, Krieg, Staley and Garrity (Editors), Bergey's Manual of Systematic Bacteriology, 2nd Edition, Volume 2 (The *Proteobacteria*), Part C (The *Alpha-, Beta-, Delta-,* and *Epsilonproteobacteria*), Springer, New York, p. 1040.

Garrity, G.M., J.A. Bell and T. Lilburn. 2005xxxvii. Family I. *Bdellovibrionaceae* fam. nov. *In* Brenner, Krieg, Staley and Garrity (Editors), Bergey's Manual of Systematic Bacteriology, 2nd Edition, Volume 2 (The *Proteobacteria*), Part C (The *Alpha-, Beta-, Delta-,* and *Epsilonproteobacteria*), Springer, New York, pp. 1040–1041.

Garrity, G.M., J.A. Bell and T. Lilburn. 2005xxxviii. Class V. *Epsilonproteobacteria* class. nov. *In* Brenner, Krieg, Staley and Garrity (Editors), Bergey's Manual of Systematic Bacteriology, 2nd Edition, Volume 2 (The *Proteobacteria*), Part C (The *Alpha-, Beta-, Delta-,* and *Epsilonproteobacteria*), Springer, New York, p. 1145.

Garrity, G.M., J.A. Bell and T. Lilburn. 2005xxxix. Order I. *Campylobacterales* ord. nov. *In* Brenner, Krieg, Staley and Garrity (Editors), Bergey's Manual of Systematic Bacteriology, 2nd Edition, Volume 2 (The *Proteobacteria*), Part C (The *Alpha-, Beta-, Delta-,* and *Epsilonproteobacteria*), Springer, New York, p. 1145.

Garrity, G.M., J.A. Bell and T. Lilburn. 2005xl. Family II. *Helicobacteraceae* fam. nov. *In* Brenner, Krieg, Staley and Garrity (Editors), Bergey's Manual of Systematic Bacteriology, 2nd Edition, Volume 2 (The *Proteobacteria*), Part C (The *Alpha-, Beta-, Delta-,* and *Epsilonproteobacteria*), Springer, New York, p. 1168.

Gatson, J.W., B.F. Benz, C. Chandrasekaran, M. Satomi, K. Venkateswaran and M.E. Hart. 2006. *Bacillus tequilensis* sp. nov., isolated from a 2000-year-old Mexican shaft-tomb, is closely related to *Bacillus subtilis*. Int. J. Syst. Evol. Microbiol. *56*: 1475–1484.

Gavrish, E.Y., V.I. Krauzova, N.V. Potekhina, S.G. Karasev, E.G. Plotnikova, O.V. Altyntseva, L.A.Korosteleva and L.I. Evtushenko. 2004. Three new species of brevibacteria, *Brevibacterium antiquum* sp. nov., *Brevibacterium aurantiacum* sp. nov., and *Brevibacterium permense* sp. nov. Mikrobiologiia *73*: 218–225 (in Russian); English translation: Microbiology. *73*: 176–183.

Geider, K., G. Auling, Z. Du, V. Jakovljevic, S. Jock and B. Völksch. 2006. *Erwinia tasmaniensis* sp. nov., a non-phytopathogenic bacterium from apple and pear trees. Int. J. Syst. Evol. Microbiol. *56*: 2937–2943.

Gelsomino, R., M. Vancanneyt and J. Swings. 2004. Reclassification of *Brevibacterium liquefaciens* Okabayashi and Masuo 1960 as *Arthrobacter nicotianae* Giovannozzi-Sermanni 1959. Int. J. Syst. Evol. Microbiol. *54*: 615–616.

Gelsomino, R., M. Vancanneyt, C. Snauwaert, K. Vandemeulebroecke, B. Hoste, T.M. Cogan and J. Swings. 2005. *Corynebacterium mooreparkense*, a later heterotypic synonym of *Corynebacterium variabile*. Int. J. Syst. Evol. Microbiol. *55*: 1129–1131.

Genersch, E., E. Forsgren, J. Pentikäinen, A. Ashiralieva, S. Rauch, J. Kilwinski and I. Fries. 2006. Reclassification of *Paenibacillus larvae* subsp. *pulvifaciens* and *Paenibacillus larvae* subsp. *larvae* as *Paenibacillus larvae* without subspecies differentiation. Int. J. Syst. Evol. Microbiol. *56*: 501–511.

Geueke, B., H.J. Busse, T. Fleischmann, P. Kämpfer and H.P. Kohler. 2007. Description of *Sphingosinicella xenopeptidilytica* sp. nov., a beta-peptide-degrading species, and emended descriptions of the genus *Sphingosinicella* and the species *Sphingosinicella microcystinivorans*. Int. J. Syst. Evol. Microbiol. *57*: 107–113.

Ghosh, A., D. Paul, D. Prakash, S. Mayilraj and R.K. Jain. 2006. *Rhodococcus imtechensis* sp. nov., a nitrophenol-degrading actinomycete. Int. J. Syst. Evol. Microbiol. *56*: 1965–1969.

Ghosh, A., M. Bhardwaj, T. Satyanarayana, M. Khurana, S. Mayilraj and R.K. Jain. 2007. *Bacillus lehensis* sp. nov., an alkalitolerant bacterium isolated from soil. Int. J. Syst. Evol. Microbiol. *57*: 238–242.

Ghosh, W. and P. Roy. 2006. *Mesorhizobium thiogangeticum* sp. nov., a novel sulfur-oxidizing chemolithoautotroph from rhizosphere soil of an Indian tropical leguminous plant. Int. J. Syst. Evol. Microbiol. *56*: 91–97.

Ghosh, W., A. Bagchi, S. Mandal, B. Dam and P. Roy. 2005. *Tetrathiobacter kashmirensis* gen. nov., sp. nov., a novel mesophilic, neutrophilic, tetrathionate-oxidizing, facultatively chemolithotrophic betaproteobacterium isolated from soil from a temperate orchard in Jammu and Kashmir, India. Int. J. Syst. Evol. Microbiol. *55*: 1779–1787.

Ghosh, W., S. Mandal and P. Roy. 2006. *Paracoccus bengalensis* sp. nov., a novel sulfur-oxidizing chemolithoautotroph from the rhizospheric soil of an Indian tropical leguminous plant. Syst. Appl. Microbiol. *29*: 396–403.

Gich, F. and J. Overmann. 2006. *Sandarakinorhabdus limnophila* gen. nov., sp. nov., a novel bacteriochlorophyll *a*-containing, obligately aerobic bacterium isolated from freshwater lakes. Int. J. Syst. Evol. Microbiol. *56*: 847–854.

Glass, M.B., A.G. Steigerwalt, J.G. Jordan, P.P. Wilkins and J.E. Gee. 2006. *Burkholderia oklahomensis* sp. nov., a *Burkholderia pseudomallei*-like species formerly known as the Oklahoma strain of *Pseudomonas pseudomallei*. Int. J. Syst. Evol. Microbiol. *56*: 2171–2176.

Glazunova, O.O., D. Raoult and V. Roux. 2006a. *Streptococcus massiliensis* sp. nov., isolated from a patient blood culture. Int. J. Syst. Evol. Microbiol. *56*: 1127–1131.

Glazunova, O.O., D. Raoult and V. Roux. 2006b. *Bacillus massiliensis* sp. nov., isolated from cerebrospinal fluid. Int. J. Syst. Evol. Microbiol. *56*: 1485–1488.

Glazunova, O.O., T. Launay, D. Raoult and V. Roux. 2007. *Prevotella timonensis* sp. nov., isolated from a human breast abscess. Int. J. Syst. Evol. Microbiol. *57*: 883–886.

Goh, F., S. Leuko, M.A. Allen, J.P. Bowman, M. Kamekura, B.A. Neilan and B.P. Burns. 2006. *Halococcus hamelinensis* sp. nov., a novel halophilic archaeon isolated from stromatolites in Shark Bay, Australia. Int. J. Syst. Evol. Microbiol. *56*: 1323–1329.

Golyshin, P.N., S. Harayama, K.N. Timmis and M.M. Yakimov. 2005. Family II. *Alcanivoraceae* fam. nov. *In* Brenner, Krieg, Staley and Garrity (Editors), Bergey's Manual of Systematic Bacteriology, 2nd Edition, Volume 2 (The *Proteobacteria*), Part B (The *Gammaproteobacteria*), Springer, New York, p. 295.

Gomez-Gil, B., F.L. Thompson, C.C. Thompson, A. García-Gasca, A. Roque and J. Swings. 2004. *Vibrio hispanicus* sp. nov., isolated from Artemia sp. and sea water in Spain. Int. J. Syst. Evol. Microbiol. *54*: 261–265.

Gomila, M., B. Bowien, E. Falsen, E.R. Moore and J. Lalucat. 2007. Description of *Pelomonas aquatica* sp. nov. and *Pelomonas puraquae* sp. nov., isolated from industrial and haemodialysis water. Int. J. Syst. Evol. Microbiol. *57*: 2629–2635.

González, J.M., V. Jurado, L. Laiz, J. Zimmermann, B. Hermosin and C. Saiz-Jimenez. 2004. *Pectinatus portalensis* nov. sp., a relatively fast-growing, coccoidal, novel *Pectinatus* species isolated from a wastewater treatment plant. Antonie van Leeuwenhoek. *86*: 241–248.

Goodfellow, M., A.L. Jones, L.A. Maldonado and J. Salanitro. 2004. *Rhodococcus aetherivorans* sp. nov., a new species that contains methyl t-butyl ether-degrading actinomycetes. Syst. Appl. Microbiol. *27*: 61–65.

Goodfellow, M., L.A. Maldonado and E.T. Quintana. 2005. Reclassification of *Nonomuraea flexuosa* (Meyer 1989) Zhang *et al.* 1998 as *Thermopolyspora flexuosa* gen. nov., comb. nov., nom. rev. Int. J. Syst. Evol. Microbiol. *55*: 1979–1983.

Goris, J., P. De Vos, J. Caballero-Mellado, J. Park, E. Falsen, J.F. Quensen, 3rd, J.M. Tiedje and P. Vandamme. 2004. Classification of the biphenyl- and polychlorinated biphenyl-degrading strain LB400ᵀ and relatives as *Burkholderia xenovorans* sp. nov. Int. J. Syst. Evol. Microbiol. *54*: 1677–1681.

Gorlenko, V.M., I.A. Bryantseva, E.E. Panteleeva, T.P. Tourova, T.V. Kolganova, Z.K. Makhneva and A.A. Moskalenko. 2004a. *Ectothiorhodosinus mongolicum* gen. nov., sp. nov., a new purple bacterium from a soda lake in Mongolia. Mikrobiologiia. 73: 80–88 (in Russian); English translation: Microbiology 73: 66–73.

Gorlenko, V., A. Tsapin, Z. Namsaraev, T. Teal, T. Tourova, D. Engler, R. Mielke and K. Nealson. 2004b. *Anaerobranca californiensis* sp. nov., an anaerobic, alkalithermophilic, fermentative bacterium isolated from a hot spring on Mono Lake. Int. J. Syst. Evol. Microbiol. *54*: 739–743.

Görtz, H.D. and H.J. Schmidt. 2005. Family III. *Holosporaceae* fam. nov. *In* Brenner, Krieg, Staley and Garrity (Editors), Bergey's Manual of Systematic Bacteriology, 2nd Edition, Volume 2 (The *Proteobacteria*), Part C (The *Alpha-, Beta-, Delta-,* and *Epsilonproteobacteria*), Springer, New York, pp. 146–149.

Gößner, A.S., K. Küsel, D. Schulz, S. Trenz, G. Acker, C.R. Lovell and H.L. Drake. 2006. Trophic interaction of the aerotolerant anaerobe *Clostridium intestinale* and the acetogen *Sporomusa rhizae* sp. nov. isolated from roots of the black needlerush *Juncus roemerianus*. Microbiology. *152*: 1209–1219.

Goto, K., R. Fujita, Y. Kato, M. Asahara and A. Yokota. 2004. Reclassification of *Brevibacillus brevis* strains NCIMB 13288 and DSM 6472 (=NRRL NRS-887) as *Aneurinibacillus danicus* sp. nov. and *Brevibacillus limnophilus* sp. nov. Int. J. Syst. Evol. Microbiol. *54*: 419–427.

Goto, K., K. Mochida, Y. Kato, M. Asahara, C. Ozawa, H. Kasai and A. Yokota. 2006. Diversity of *Alicyclobacillus* isolated from fruit juices and their raw materials, and emended description of *Alicyclobacillus acidocaldarius*. Microbiol. Cult. Coll. *22*: 1–14.

Goto, K., K. Mochida, Y. Kato, M. Asahara, R. Fujita, S.Y. An, H. Kasai and A. Yokota. 2007. Proposal of six species of moderately thermophilic, acidophilic, endospore-forming bacteria: *Alicyclobacillus contaminans* sp. nov., *Alicyclobacillus fastidiosus* sp. nov., *Alicyclobacillus kakegawensis* sp. nov., *Alicyclobacillus macrosporangiidus* sp. nov., *Alicyclobacillus sacchari* sp. nov. and *Alicyclobacillus shizuokensis* sp. nov. Int. J. Syst. Evol. Microbiol. *57*: 1276–1285.

Graber, J.R., J.R. Leadbetter and J.A. Breznak. 2004. Description of *Treponema azotonutricium* sp. nov. and *Treponema primitia* sp. nov., the first spirochetes isolated from termite guts. Appl. Environ. Microbiol. *70*: 1315–1320.

Grabovich, M., E. Gavrish, J. Kuever, A.M. Lysenko, D. Podkopaeva and G. Dubinina. 2006. Proposal of *Giesbergeria voronezhensis* gen. nov., sp. nov. and *G. kuznetsovii* sp. nov. and reclassification of [*Aquaspirillum*] *anulus*, [*A.*] *sinuosum* and [*A.*] *giesbergeri* as *Giesbergeria anulus* comb. nov., *G. sinuosa* comb. nov. and *G. giesbergeri* comb. nov., and [*Aquaspirillum*] *metamorphum* and [*A.*] *psychrophilum* as *Simplicispira metamorpha* gen. nov., comb. nov. and *S. psychrophila* comb. nov. Int. J. Syst. Evol. Microbiol. *56*: 569–576.

Grabowski, A., B.J. Tindall, V. Bardin, D. Blanchet and C. Jeanthon. 2005. *Petrimonas sulfuriphila* gen. nov., sp. nov., a mesophilic fermentative bacterium isolated from a biodegraded oil reservoir. Int. J. Syst. Evol. Microbiol. *55*: 1113–1121.

Green, D.H., J.P. Bowman, E.A. Smith, T. Gutiérrez and C.J. Bolch. 2006. *Marinobacter algicola* sp. nov., isolated from laboratory cultures of paralytic shellfish toxin-producing dinoflagellates. Int. J. Syst. Evol. Microbiol. *56*: 523–527.

Greenberg, D.E., S.F. Porcella, F. Stock, A. Wong, P.S. Conville, P.R. Murray, S.M. Holland and A.M. Zelazny. 2006. *Granulibacter bethesdensis* gen. nov., sp. nov., a distinctive pathogenic acetic acid bacterium in the family *Acetobacteraceae*. Int. J. Syst. Evol. Microbiol. *56*: 2609–2616.

Greetham, H.L., M.D. Collins, G.R. Gibson, C. Giffard, E. Falsen and P.A. Lawson. 2004a. *Sutterella stercoricanis* sp. nov., isolated from canine faeces. Int. J. Syst. Evol. Microbiol. *54*: 1581–1584.

Greetham, H.L., G.R. Gibson, C. Giffard, H. Hippe, B. Merkhoffer, U. Steiner, E. Falsen and M.D. Collins. 2004b. *Allobaculum stercoricanis* gen. nov., sp. nov., isolated from canine feces. Anaerobe. *10*: 301–307.

Greub, G. and D. Raoult. 2002. "*Actinobaculum massiliae*," a new species causing chronic urinary tract infection. J. Clin. Microbiol. *40*: 3938–3941.

Greub, G. and D. Raoult. 2003. *Rhodobacter massiliensis* sp. nov., a new amoebae-resistant species isolated from the nose of a patient. Res. Microbiol. *154*: 631–635.

Groth, I., C. Rodríguez, B. Schütze, P. Schmitz, E. Leistner and M. Goodfellow. 2004. Five novel *Kitasatospora* species from soil: *Kitasatospora arboriphila* sp. nov., *K. gansuensis* sp. nov., *K. nipponensis* sp. nov., *K. paranensis* sp. nov. and *K. terrestris* sp. nov. Int. J. Syst. Evol. Microbiol. *54*: 2121–2129.

Groth, I., P. Schumann, B. Schütze, J.M. González, L. Laiz, C. Saiz-Jimenez and E. Stackebrandt. 2005. *Isoptericola hypogeus* sp. nov., isolated from the Roman catacomb of Domitilla. Int. J. Syst. Evol. Microbiol. *55*: 1715–1719.

Groth, I., P. Schumann, B. Schütze, J.M. González, L. Laiz, M.L. Suihko and E. Stackebrandt. 2006. *Myceligenerans crystallogenes* sp. nov., isolated from Roman catacombs. Int. J. Syst. Evol. Microbiol. *56*: 283–287.

Groth, I., G.Y. Tan, J.M. González, L. Laiz, M.R. Carlsohn, B. Schütze, J. Wink and M. Goodfellow. 2007. *Amycolatopsis nigrescens* sp. nov., an actinomycete isolated from a Roman catacomb. Int. J. Syst. Evol. Microbiol. *57*: 513–519.

Gruber, C., A. Legat, M. Pfaffenhuemer, C. Radax, G. Weidler, H.J. Busse and H. Stan-Lotter. 2004. *Halobacterium noricense* sp. nov., an archaeal isolate from a bore core of an alpine Permian salt deposit, classification of *Halobacterium* sp. NRC-1 as a strain of *H. salinarum* and emended description of *H. salinarum*. Extremophiles. *8*: 431–439.

Gu, J., H. Cai, S.L. Yu, R. Qu, B. Yin, Y.F. Guo, J.Y. Zhao and X.L. Wu. 2007a. *Marinobacter gudaonensis* sp. nov., isolated from an oil-polluted

saline soil in a Chinese oilfield. Int. J. Syst. Evol. Microbiol. *57*: 250–254.

Gu, J., B. Guo, Y.N. Wang, S.L. Yu, R. Inamori, R. Qu, Y.G. Ye and X.L. Wu. 2007b. *Oceanicola nanhaiensis* sp. nov., isolated from sediments of the South China Sea. Int. J. Syst. Evol. Microbiol. *57*: 157–160.

Gu, Q., H. Luo, W. Zheng, Z. Liu and Y. Huang. 2006. *Pseudonocardia oroxyli* sp. nov., a novel actinomycete isolated from surface-sterilized *Oroxylum indicum* root. Int. J. Syst. Evol. Microbiol. *56*: 2193–2197.

Gu, Q., M. Pa ciak, H. Luo, A. Gamian, Z. Liu and Y. Huang. 2007a. *Ruania albidiflava* gen. nov., sp. nov., a novel member of the suborder *Micrococcineae*. Int. J. Syst. Evol. Microbiol. *57*: 809–814.

Gu, Q., W. Zheng and Y. Huang. 2007b. *Glycomyces sambucus* sp. nov., an endophytic actinomycete isolated from the stem of *Sambucus adnata* Wall. Int. J. Syst. Evol. Microbiol. *57*: 1995–1998.

Güner, E.S., M. Watanabe, N. Hashimoto, T. Kadosaka, Y. Kawamura, T. Ezaki, H. Kawabata, Y. Imai, K. Kaneda and T. Masuzawa. 2004. *Borrelia turcica* sp. nov., isolated from the hard tick *Hyalomma aegyptium* in Turkey. Int. J. Syst. Evol. Microbiol. *54*: 1649–1652.

Guo, B., J. Gu, Y.G. Ye, Y.Q. Tang, K. Kida and X.L. Wu. 2007. *Marinobacter segnicrescens* sp. nov., a moderate halophile isolated from benthic sediment of the South China Sea. Int. J. Syst. Evol. Microbiol. *57*: 1970–1974.

Gupta, P., G.S. Reddy, D. Delille and S. Shivaji. 2004. *Arthrobacter gangotriensis* sp. nov. and *Arthrobacter kerguelensis* sp. nov. from Antarctica. Int. J. Syst. Evol. Microbiol. *54*: 2375–2378.

Gupta, P., P. Chaturvedi, S. Pradhan, D. Delille and S. Shivaji. 2006. *Marinomonas polaris* sp. nov., a psychrohalotolerant strain isolated from coastal sea water off the subantarctic Kerguelen islands. Int. J. Syst. Evol. Microbiol. *56*: 361–364.

Gutierrez, M.C., M. Kamekura., M.L. Holmes, M.L. Dyall-Smith and A. Ventosa. 2002. Taxonomic characterization of *Haloferax* sp. "*H. alicantei*" strain Aa 2.2: description of *Haloferax lucentensis* sp. nov. Extremophiles. *6*: 479–483.

Gutiérrez, M.C., A.M. Castillo, M. Kamekura, Y. Xue, Y. Ma, D.A. Cowan, B.E. Jones, W.D. Grant and A. Ventosa. 2007. *Halopiger xanaduensis* gen. nov., sp. nov., an extremely halophilic archaeon isolated from saline Lake Shangmatala in Inner Mongolia, China. Int. J. Syst. Evol. Microbiol. *57*: 1402–1407.

Hall, V., M.D. Collins, P.A. Lawson, E. Falsen and B.I. Duerden. 2005. *Actinomyces dentalis* sp. nov., from a human dental abscess. Int. J. Syst. Evol. Microbiol. *55*: 427–431.

Halpern, M., Y. Senderovich and S. Snir. 2007. *Rheinheimera chironomi* sp. nov., isolated from a chironomid (Diptera; Chironomidae) egg mass. Int. J. Syst. Evol. Microbiol. *57*: 1872–1875.

Hampson, D.J. and T. La. 2006. Reclassification of *Serpulina intermedia* and *Serpulina murdochii* in the genus *Brachyspira* as *Brachyspira intermedia* comb. nov. and *Brachyspira murdochii* comb. nov. Int. J. Syst. Evol. Microbiol. *56*: 1009–1012.

Han, X.Y. and E. Falsen. 2005. Characterization of oral strains of *Cardiobacterium valvarum* and emended description of the organism. J. Clin. Microbiol. *43*: 2370–2374.

Han, X.Y., M.C. Meltzer, J.T. Woods and V. Fainstein. 2004. Endocarditis with ruptured cerebral aneurysm caused by *Cardiobacterium valvarum* sp. nov. J. Clin. Microbiol. *42*: 1590–1595.

Han, X.Y., T. Hong and E. Falsen. 2006. *Neisseria bacilliformis* sp. nov. isolated from human infections. J. Clin. Microbiol. *44*: 474–479.

Hantsis-Zacharov, E. and M. Halpern. 2007. *Chryseobacterium haifense* sp. nov., a psychrotolerant bacterium isolated from raw milk. Int. J. Syst. Evol. Microbiol. *57*: 2344–2348.

Haouari, O., M.L. Fardeau, L. Casalot, J.L. Tholozan, M. Hamdi and B. Ollivier. 2006. Isolation of sulfate-reducing bacteria from Tunisian marine sediments and description of *Desulfovibrio bizertensis* sp. nov. Int. J. Syst. Evol. Microbiol. *56*: 2909–2913.

Harada, R.M., S. Campbell and Q.X. Li. 2006. *Pseudoxanthomonas kalamensis* sp. nov., a novel gammaproteobacterium isolated from Johnston

Atoll, North Pacific Ocean. Int. J. Syst. Evol. Microbiol. *56*: 1103–1107.

Harf-Monteil, C., A.L. Flèche, P. Riegel, G. Prévost, D. Bermond, P.A. Grimont and H. Monteil. 2004. *Aeromonas simiae* sp. nov., isolated from monkey faeces. Int. J. Syst. Evol. Microbiol. *54*: 481–485.

Harper, C.G., Y. Feng, S. Xu, N.S. Taylor, M. Kinsel, F.E. Dewhirst, B.J. Paster, M. Greenwell, G. Levine, A. Rogers and J.G. Fox. 2002. *Helicobacter cetorum* sp. nov., a urease-positive *Helicobacter* species isolated from dolphins and whales. J. Clin. Microbiol. *40*: 4536–4543.

Hatamoto, M., H. Imachi, S. Fukayo, A. Ohashi and H. Harada. 2007. *Syntrophomonas palmitatica* sp. nov., an anaerobic, syntrophic, long-chain fatty-acid-oxidizing bacterium isolated from methanogenic sludge. Int. J. Syst. Evol. Microbiol. *57*: 2137–2142.

Hatayama, K., S. Kawai, H. Shoun, Y. Ueda and A. Nakamura. 2005a. *Pseudomonas azotifigens* sp. nov., a novel nitrogen-fixing bacterium isolated from a compost pile. Int. J. Syst. Evol. Microbiol. *55*: 1539–1544.

Hatayama, K., H. Shoun, Y. Ueda and A. Nakamura. 2005b. *Planifilum fimeticola* gen. nov., sp. nov. and *Planifilum fulgidum* sp. nov., novel members of the family 'Thermoactinomycetaceae' isolated from compost. Int. J. Syst. Evol. Microbiol. *55*: 2101–2104.

Hatayama, K., H. Shoun, Y. Ueda and A. Nakamura. 2006. *Tuberibacillus calidus* gen. nov., sp. nov., isolated from a compost pile and reclassification of *Bacillus naganoensis* Tomimura *et al.* 1990 as *Pullulanibacillus naganoensis* gen. nov., comb. nov. and *Bacillus laevolacticus* Andersch *et al.* 1994 as *Sporolactobacillus laevolacticus* comb. nov. Int. J. Syst. Evol. Microbiol. *56*: 2545–2551.

Hauser, E., P. Kämpfer and H.J. Busse. 2004. *Pseudomonas psychrotolerans* sp. nov. Int. J. Syst. Evol. Microbiol. *54*: 1633–1637.

Hayashi, H., K. Shibata, M.A. Bakir, M. Sakamoto, S. Tomita and Y. Benno. 2007a. *Bacteroides coprophilus* sp. nov., isolated from human faeces. Int. J. Syst. Evol. Microbiol. *57*: 1323–1326.

Hayashi, H., K. Shibata, M. Sakamoto, S. Tomita and Y. Benno. 2007b. *Prevotella copri* sp. nov. and *Prevotella stercorea* sp. nov., isolated from human faeces. Int. J. Syst. Evol. Microbiol. *57*: 941–946.

Hazir, S., E. Stackebrandt, E. Lang, P. Schumann, R.U. Ehlers and N. Keskin. 2004. Two new subspecies of *Photorhabdus luminescens*, isolated from *Heterorhabditis bacteriophora* (Nematoda: Heterorhabditidae): *Photorhabdus luminescens* subsp. *kayaii* subsp. nov. and *Photorhabdus luminescens* subsp. *thracensis* subsp. nov. Syst. Appl. Microbiol. *27*: 36–42.

He, L., W. Li, Y. Huang, L. Wang, Z. Liu, B. Lanoot, M. Vancanneyt and J. Swings. 2005. *Streptomyces jietaisiensis* sp. nov., isolated from soil in northern China. Int. J. Syst. Evol. Microbiol. *55*: 1939–1944.

Helsel, L.O., D.G. Hollis, A.G. Steigerwalt and P.N. Levett. 2006. Reclassification of *Roseomonas fauriae* Rihs *et al.* 1998 as a later heterotypic synonym of *Azospirillum brasilense* Tarrand *et al.* 1979. Int. J. Syst. Evol. Microbiol. *56*: 2753–2755.

Helsel, L.O, D. Hollis, A.G. Steigerwalt, R.E. Morey, J. Jordan, T. Aye, J. Radosevic, D. Jannat-Khah, D. Thiry, D.R. Lonsway, J.B. Patel, M.I. Daneshvar and P.N. Levett. 2007. Identification of "*Haematobacter*", a new genus of aerobic Gram-negative rods isolated from clinical specimens, and reclassification of *Rhodobacter massiliensis* as "*Haematobacter massiliensis* comb. nov.". J. Clin. Microbiol. *45*: 1238–1243.

Heuchert, A., F.O. Glöckner, R. Amann and U. Fischer. 2004. *Psychrobacter nivimaris* sp. nov., a heterotrophic bacterium attached to organic particles isolated from the south Atlantic (Antarctica). Syst. Appl. Microbiol. *27*: 399–406.

Heyer, J., U. Berger, M. Hardt and P.F. Dunfield. 2005. *Methylohalobius crimeensis* gen. nov., sp. nov., a moderately halophilic, methanotrophic bacterium isolated from hypersaline lakes of Crimea. Int. J. Syst. Evol. Microbiol. *55*: 1817–1826.

Heylen, K., B. Vanparys, F. Peirsegaele, L. Lebbe and P. De Vos. 2007. *Stenotrophomonas terrae* sp. nov. and *Stenotrophomonas humi* sp. nov., two nitrate-reducing bacteria isolated from soil. Int. J. Syst. Evol. Microbiol. *57*: 2056–2061.

Heyndrickx, M., N.A. Logan, L. Lebbe, M. Rodríguez-Díaz, G. Forsyth, J. Goris, P. Scheldeman and P. De Vos. 2004. *Bacillus galactosidilyticus* sp. nov., an alkali-tolerant beta-galactosidase producer. Int. J. Syst. Evol. Microbiol. *54*: 617–621.

Heyndrickx, M., P. Scheldeman, G. Forsyth, L. Lebbe, M. Rodríguez-Díaz, N.A. Logan and P. De Vos. 2005. *Bacillus ruris* sp. nov., from dairy farms. Int. J. Syst. Evol. Microbiol. *55*: 2551–2554.

Heyrman, J., B. Vanparys, N.A. Logan, A. Balcaen, M. Rodríguez-Díaz, A. Felske and P. De Vos. 2004a. *Bacillus novalis* sp. nov., *Bacillus vireti* sp. nov., *Bacillus soli* sp. nov., *Bacillus bataviensis* sp. nov. and *Bacillus drentensis* sp. nov., from the Drentse A grasslands. Int. J. Syst. Evol. Microbiol. *54*: 47–57.

Heyrman, J., J. Verbeeren, P. Schumann, J. Devos, J. Swings and P. De Vos. 2004b. *Brevibacterium picturae* sp. nov., isolated from a damaged mural painting at the Saint-Catherine chapel (Castle Herberstein, Austria). Int. J. Syst. Evol. Microbiol. *54*: 1537–1541.

Heyrman, J., N.A. Logan, M. Rodríguez-Díaz, P. Scheldeman, L. Lebbe, J. Swings, M. Heyndrickx and P. De Vos. 2005a. Study of mural painting isolates, leading to the transfer of 'Bacillus maroccanus' and 'Bacillus carotarum' to *Bacillus simplex*, emended description of *Bacillus simplex*, re-examination of the strains previously attributed to 'Bacillus macroides' and description of *Bacillus muralis* sp. nov. Int. J. Syst. Evol. Microbiol. *55*: 119–131.

Heyrman, J., M. Rodríguez-Díaz, J. Devos, A. Felske, N.A. Logan and P. De Vos. 2005b. *Bacillus arenosi* sp. nov., *Bacillus arvi* sp. nov. and *Bacillus humi* sp. nov., isolated from soil. Int. J. Syst. Evol. Microbiol. *55*: 111–117.

Heyrman, J., J. Verbeeren, P. Schumann, J. Swings and P. De Vos. 2005c. Six novel *Arthrobacter* species isolated from deteriorated mural paintings. Int. J. Syst. Evol. Microbiol. *55*: 1457–1464.

Higashiguchi, D.T., C. Husseneder, J.K. Grace and J.M. Berestecky. 2006. *Pilibacter termitis* gen. nov., sp. nov., a lactic acid bacterium from the hindgut of the Formosan subterranean termite (*Coptotermes formosanus*). Int. J. Syst. Evol. Microbiol. *56*: 15–20.

Hirayama, H., K. Takai, F. Inagaki, K.H. Nealson and K. Horikoshi. 2005. *Thiobacter subterraneus* gen. nov., sp. nov., an obligately chemolithoautotrophic, thermophilic, sulfur-oxidizing bacterium from a subsurface hot aquifer. Int. J. Syst. Evol. Microbiol. *55*: 467–472.

Hirota, K., Y. Nodasaka, Y. Orikasa, H. Okuyama and I. Yumoto. 2005. *Shewanella pneumatophori* sp. nov., an eicosapentaenoic acid-producing marine bacterium isolated from the intestines of Pacific mackerel (*Pneumatophorus japonicus*). Int. J. Syst. Evol. Microbiol. *55*: 2355–2359.

Hirsch, P., C.A. Gallikowski, J. Siebert, K. Peissl, R. Kroppenstedt, P. Schumann, E. Stackebrandt and R. Anderson. 2004a. *Deinococcus frigens* sp. nov., *Deinococcus saxicola* sp. nov., and *Deinococcus marmoris* sp. nov., low temperature and draught-tolerating, UV-resistant bacteria from continental Antarctica. Syst. Appl. Microbiol. *27*: 636–645.

Hirsch, P., U. Mevs, R.M. Kroppenstedt, P. Schumann and E. Stackebrandt. 2004b. Cryptoendolithic actinomycetes from Antarctic sandstone rock samples: *Micromonospora endolithica* sp. nov. and two isolates related to *Micromonospora coerulea* Jensen 1932. Syst. Appl. Microbiol. *27*: 166–174.

Hoeft, S.E., J.S. Blum, J.F. Stolz, F.R. Tabita, B. Witte, G.M. King, J.M. Santini and R.S. Oremland. 2007. *Alkalilimnicola ehrlichii* sp. nov., a novel, arsenite-oxidizing haloalkaliphilic gammaproteobacterium capable of chemoautotrophic or heterotrophic growth with nitrate or oxygen as the electron acceptor. Int. J. Syst. Evol. Microbiol. *57*: 504–512.

Hoffmann, H., S. Stindl, W. Ludwig, A. Stumpf, A. Mehlen, J. Heesemann, D. Monget, K.H. Schleifer and A. Roggenkamp. 2005a. Reassignment of *Enterobacter dissolvens* to *Enterobacter cloacae* as *E. cloacae* subspecies *dissolvens* comb. nov. and emended description of *Enterobacter asburiae* and *Enterobacter kobei*. Syst. Appl. Microbiol. *28*: 196–205.

Hoffmann, H., S. Stindl, A. Stumpf, A. Mehlen, D. Monget, J. Heesemann, K.H. Schleifer and A. Roggenkamp. 2005b. Description of *Enterobacter ludwigii* sp. nov., a novel *Enterobacter* species of clinical relevance. Syst. Appl. Microbiol. *28*: 206–212.

Holmes, D.E., K.P. Nevin and D.R. Lovley. 2004a. Comparison of 16S rRNA, *nifD, recA, gyrB, rpoB* and *fusA* genes within the family *Geobacteraceae* fam. nov. Int. J. Syst. Evol. Microbiol. *54*: 1591–1599.

Holmes, D.E., J.S. Nicoll, D.R. Bond and D.R. Lovley. 2004b. Potential role of a novel psychrotolerant member of the family *Geobacteraceae, Geopsychrobacter electrodiphilus* gen. nov., sp. nov., in electricity production by a marine sediment fuel cell. Appl. Environ. Microbiol. *70*: 6023–6030.

Holmes, D.E., K.P. Nevin, T.L. Woodard, A.D. Peacock and D.R. Lovley. 2007. *Prolixibacter bellariivorans* gen. nov., sp. nov., a sugar-fermenting, psychrotolerant anaerobe of the phylum *Bacteroidetes*, isolated from a marine-sediment fuel cell. Int. J. Syst. Evol. Microbiol. *57*: 701–707.

Holmstrøm, K., M.D. Collins, T. Møller, E. Falsen and P.A. Lawson. 2004. *Subdoligranulum variabile* gen. nov., sp. nov. from human feces. Anaerobe. *10*: 197–203.

Hormisch, D., I. Brost, G.W. Kohring, F. Giffhorn, R.M. Kroppenstedt, E. Stackebrandt, P. Färber and W.H. Holzapfel. 2004. *Mycobacterium fluoranthenivorans* sp. nov., a fluoranthene and aflatoxin B₁ degrading bacterium from contaminated soil of a former coal gas plant. Syst. Appl. Microbiol. *27*: 653–660.

Horn, M.A., J. Ihssen, C. Matthies, A. Schramm, G. Acker and H.L. Drake. 2005. *Dechloromonas denitrificans* sp. nov., *Flavobacterium denitrificans* sp. nov., *Paenibacillus anaericanus* sp. nov. and *Paenibacillus terrae* strain MH72, N2O-producing bacteria isolated from the gut of the earthworm *Aporrectodea caliginosa.* Int. J. Syst. Evol. Microbiol. *55*: 1255–1265.

Hoshino, Y., K. Watanabe, S. Iida, S. Suzuki, T. Kudo, T. Kogure, K. Yazawa, J. Ishikawa, R.M. Kroppenstedt and Y. Mikami. 2007. *Nocardia terpenica* sp. nov., isolated from Japanese patients with nocardiosis. Int. J. Syst. Evol. Microbiol. *57*: 1456–1460.

Hosoya, S. and A. Yokota. 2007a. Reclassification of *Flexibacter aggregans* (Lewin 1969) Leadbetter 1974 as a later heterotypic synonym of *Flexithrix dorotheae* Lewin 1970. Int. J. Syst. Evol. Microbiol. *57*: 1086–1088.

Hosoya, S. and A. Yokota. 2007b. *Flammeovirga kamogawensis* sp. nov., isolated from coastal seawater in Japan. Int. J. Syst. Evol. Microbiol. *57*: 1327–1330.

Hosoya, S. and A. Yokota. 2007c. *Pseudovibrio japonicus* sp. nov., isolated from coastal seawater in Japan. Int. J. Syst. Evol. Microbiol. *57*: 1952–1955.

Hosoya, S. and A. Yokota. 2007d. *Loktanella atrilutea* sp. nov., isolated from seawater in Japan. Int. J. Syst. Evol. Microbiol. *57*: 1966–1969.

Hosoya, S., V. Arunpairojana, C. Suwannachart, A. Kanjana-Opas and A. Yokota. 2006. *Aureispira marina* gen. nov., sp. nov., a gliding, arachidonic acid-containing bacterium isolated from the southern coastline of Thailand. Int. J. Syst. Evol. Microbiol. *56*: 2931–2935.

Hosoya, S., V. Arunpairojana, C. Suwannachart, A. Kanjana-Opas and A. Yokota. 2007. *Aureispira maritima* sp. nov., isolated from marine barnacle debris. Int. J. Syst. Evol. Microbiol. *57*: 1948–1951.

Houf, K., S.L. On, T. Coenye, J. Mast, J. Van Hoof and P. Vandamme. 2005. *Arcobacter cibarius* sp. nov., isolated from broiler carcasses. Int. J. Syst. Evol. Microbiol. *55*: 713–717.

Hoyles, L., M.D. Collins, E. Falsen, N. Nikolaitchouk and A.L. McCartney. 2004. Transfer of members of the genus *Falcivibrio* to the genus *Mobiluncus*, and emended descriptions of the genus *Mobiluncus*. Syst. Appl. Microbiol. *27*: 72–83.

Hoyles, L., M.D. Collins, G. Foster, E. Falsen and P. Schumann. 2004. *Jeotgalicoccus pinnipedialis* sp. nov., from a southern elephant seal (*Mirounga leonina*). Int. J. Syst. Evol. Microbiol. *54*: 745–748.

Hozzein, W.N. and M. Goodfellow. 2007. *Streptomyces synnematoformans* sp. nov., a novel actinomycete isolated from a sand dune soil in Egypt. Int. J. Syst. Evol. Microbiol. *57*: 2009–2013.

Hozzein, W.N., W.J. Li, M.I. Ali, O. Hammouda, A.S. Mousa, L.H. Xu and C.L. Jiang. 2004. *Nocardiopsis alkaliphila* sp. nov., a novel alkaliphilic actinomycete isolated from desert soil in Egypt. Int. J. Syst. Evol. Microbiol. *54*: 247–252.

Hu, Y.T., P.J. Zhou, Y.G. Zhou, Z.H. Liu and S.J. Liu. 2004. *Saccharothrix xinjiangensis* sp. nov., a pyrene-degrading actinomycete isolated from Tianchi Lake, Xinjiang, China. Int. J. Syst. Evol. Microbiol. *54*: 2091–2094.

Hu, Z.Y. and Y. Li. 2007. *Pseudidiomarina sediminum* sp. nov., a marine bacterium isolated from coastal sediments of Luoyuan Bay in China. Int. J. Syst. Evol. Microbiol. *57*: 2572–2577.

Hua, N.P., A. Kanekiyo, K. Fujikura, H. Yasuda and T. Naganuma. 2007. *Halobacillus profundi* sp. nov. and *Halobacillus kuroshimensis* sp. nov., moderately halophilic bacteria isolated from a deep-sea methane cold seep. Int. J. Syst. Evol. Microbiol. *57*: 1243–1249.

Huang, Y., Q. Cui, L. Wang, C. Rodriguez, E. Quintana, M. Goodfellow and Z. Liu. 2004a. *Streptacidiphilus jiangxiensis* sp. nov., a novel actinomycete isolated from acidic rhizosphere soil in China. Antonie van Leeuwenhoek. *86*: 159–165.

Huang, Y., M. Pa ciak, Z. Liu, Q. Xie and A. Gamian. 2004b. *Amycolatopsis palatopharyngis* sp. nov., a potentially pathogenic actinomycete isolated from a human clinical source. Int. J. Syst. Evol. Microbiol. *54*: 359–363.

Huang, Y., W. Li, L. Wang, B. Lanoot, M. Vancanneyt, C. Rodríguez, Z. Liu, J. Swings and M. Goodfellow. 2004c. *Streptomyces glauciniger* sp. nov., a novel mesophilic streptomycete isolated from soil in south China. Int. J. Syst. Evol. Microbiol. *54*: 2085–2089.

Huang, Y., X. Dai, L. He, Y.N. Wang, B.J. Wang, Z. Liu and S.J. Liu. 2005a. *Sanguibacter marinus* sp. nov., isolated from coastal sediment. Int. J. Syst. Evol. Microbiol. *55*: 1755–1758.

Huang, Y., N. Zhao, L. He, L. Wang, Z. Liu, M. You and F. Guan. 2005b. *Arthrobacter scleromae* sp. nov. isolated from human clinical specimens. J. Clin. Microbiol. *43*: 1451–1455.

Hugenholtz, P. and E. Stackebrandt. 2004. Reclassification of *Sphaerobacter thermophilus* from the subclass *Sphaerobacteridae* in the phylum *Actinobacteria* to the class *Thermomicrobia* (emended description) in the phylum *Chloroflexi* (emended description). Int. J. Syst. Evol. Microbiol. *54*: 2049–2051.

Huys, G., M. Cnockaert and J. Swings. 2005. *Aeromonas culicicola* Pidiyar *et al.* 2002 is a later subjective synonym of *Aeromonas veronii* Hickman-Brenner *et al.* 1987. Syst. Appl. Microbiol. *28*: 604–609.

Huys, G., M. Vancanneyt, K. D'Haene, E. Falsen, G. Wauters and P. Vandamme. 2007. *Alloscardovia omnicolens* gen. nov., sp. nov., from human clinical samples. Int. J. Syst. Evol. Microbiol. *57*: 1442–1446.

Hwang, C.Y. and B.C. Cho. 2006. *Flectobacillus lacus* sp. nov., isolated from a highly eutrophic pond in Korea. Int. J. Syst. Evol. Microbiol. *56*: 1197–1201.

Hwang, C.Y., D.H. Choi and B.C. Cho. 2006. *Pedobacter roseus* sp. nov., isolated from a hypertrophic pond, and emended description of the genus *Pedobacter*. Int. J. Syst. Evol. Microbiol. *56*: 1831–1836.

Iida, K., Y. Ueda, Y. Kawamura, T. Ezaki, A. Takade, S. Yoshida and K. Amako. 2005. *Paenibacillus motobuensis* sp. nov., isolated from a composting machine utilizing soil from Motobu-town, Okinawa, Japan. Int. J. Syst. Evol. Microbiol. *55*: 1811–1816.

Iida, S., H. Taniguchi, A. Kageyama, K. Yazawa, H. Chibana, S. Murata, F. Nomura, R.M. Kroppenstedt and Y. Mikami. 2005. *Gordonia otitidis* sp. nov., isolated from a patient with external otitis. Int. J. Syst. Evol. Microbiol. *55*: 1871–1876.

Iida, S., A. Kageyama, K. Yazawa, N. Uchiyama, T. Toyohara, N. Chohnabayashi, S. Suzuki, F. Nomura, R.M. Kroppenstedt and Y. Mikami. 2006. *Nocardia exalbida* sp. nov., isolated from Japanese patients with nocardiosis. Int. J. Syst. Evol. Microbiol. *56*: 1193–1196.

Iino, T., K. Mori, K. Tanaka, K. Suzuki and S. Harayama. 2007. *Oscillibacter valericigenes* gen. nov., sp. nov., a valerate-producing anaerobic bacterium isolated from the alimentary canal of a Japanese corbicula clam. Int. J. Syst. Evol. Microbiol. *57*: 1840–1845.

L

Im, W.T., H.S. Bae, A. Yokota and S.T. Lee. 2004a. *Herbaspirillum chlorophenolicum* sp. nov., a 4-chlorophenol-degrading bacterium. Int. J. Syst. Evol. Microbiol. *54*: 851–855.

Im, W.T., S.T. Lee and A. Yokota. 2004b. *Rhodanobacter fulvus* sp. nov., a α-galactosidase-producing gammaproteobacterium. J. Gen. Appl. Microbiol. *50*: 143–147.

Im, W.T., A. Yokota, M.K. Kim and S.T. Lee. 2004c. *Kaistia adipata* gen. nov., sp. nov., a novel α-proteobacterium. J. Gen. Appl. Microbiol. *50*: 249–254.

Im, W.T., Z. Aslam, M. Lee, L.N. Ten, D.C. Yang and S.T. Lee. 2006a. *Starkeya koreensis* sp. nov., isolated from rice straw. Int. J. Syst. Evol. Microbiol. *56*: 2409–2414.

Im, W.T., S.H. Kim, M.K. Kim, L.N. Ten and S.T. Lee. 2006b. *Pleomorphomonas koreensis* sp. nov., a nitrogen-fixing species in the order *Rhizobiales*. Int. J. Syst. Evol. Microbiol. *56*: 1663–1666.

Imachi, H., S. Sakai, A. Ohashi, H. Harada, S. Hanada, Y. Kamagata and Y. Sekiguchi. 2007. *Pelotomaculum propionicicum* sp. nov., an anaerobic, mesophilic, obligately syntrophic, propionate-oxidizing bacterium. Int. J. Syst. Evol. Microbiol. *57*: 1487–1492.

Imhoff, J.F. 2005. Order I. *Chromatiales* ord. nov. *In* Brenner, Krieg, Staley and Garrity (Editors), Bergey's Manual of Systematic Bacteriology, 2nd Edition, Volume 2 (The *Proteobacteria*), Part B (The *Gammaproteobacteria*), Springer, New York, pp. 1–3.

Inagaki, F., K. Takai, K.H. Nealson and K. Horikoshi. 2004. *Sulfurovum lithotrophicum* gen. nov., sp. nov., a novel sulfur-oxidizing chemolithoautotroph within the ε-Proteobacteria isolated from Okinawa Trough hydrothermal sediments. Int. J. Syst. Evol. Microbiol. *54*: 1477–1482.

Inglis, G.D., B.M. Hoar, D.P. Whiteside and D.W. Morck. 2007. *Campylobacter canadensis* sp. nov., from captive whooping cranes in Canada. Int. J. Syst. Evol. Microbiol. *57*: 2636–2644.

Irlinger, F., F. Bimet, J. Delettre, M. Lefèvre and P.A. Grimont. 2005. *Arthrobacter bergerei* sp. nov. and *Arthrobacter arilaitensis* sp. nov., novel coryneform species isolated from the surfaces of cheeses. Int. J. Syst. Evol. Microbiol. *55*: 457–462.

Ishikawa, M., K. Nakajima, Y. Itamiya, S. Furukawa, Y. Yamamoto and K. Yamasato. 2005. *Halolactibacillus halophilus* gen. nov., sp. nov. and *Halolactibacillus miurensis* sp. nov., halophilic and alkaliphilic marine lactic acid bacteria constituting a phylogenetic lineage in *Bacillus* rRNA group 1. Int. J. Syst. Evol. Microbiol. *55*: 2427–2439.

Ishikawa, T. and A. Yokota. 2006. Reclassification of *Arthrobacter duodecadis* Lochhead 1958 as *Tetrasphaera duodecadis* comb. nov. and emended description of the genus *Tetrasphaera*. Int. J. Syst. Evol. Microbiol. *56*: 1369–1373.

Islam, M.S., H. Kawasaki, Y. Nakagawa, T. Hattori and T. Seki. 2007. *Labrys okinawensis* sp. nov. and *Labrys miyagiensis* sp. nov., budding bacteria isolated from rhizosphere habitats in Japan, and emended descriptions of the genus *Labrys* and *Labrys monachus*. Int. J. Syst. Evol. Microbiol. *57*: 552–557.

Ito, T., K. Sugita, I. Yumoto, Y. Nodasaka and S. Okabe. 2005. *Thiovirga sulfuroxydans* gen. nov., sp. nov., a chemolithoautotrophic sulfur-oxidizing bacterium isolated from a microaerobic waste-water biofilm. Int. J. Syst. Evol. Microbiol. *55*: 1059–1064.

Itoh, T., T. Yamaguchi, P. Zhou and T. Takashina. 2005. *Natronolimnobius baerhuensis* gen. nov., sp. nov. and *Natronolimnobius innermongolicus* sp. nov., novel haloalkaliphilic archaea isolated from soda lakes in Inner Mongolia, China. Extremophiles. *9*: 111–116.

Itoh, T., N. Yoshikawa and T. Takashina. 2007. *Thermogymnomonas acidicola* gen. nov., sp. nov., a novel thermoacidophilic, cell wall-less archaeon in the order *Thermoplasmatales*, isolated from a solfataric soil in Hakone, Japan. Int. J. Syst. Evol. Microbiol. *57*: 2557–2561.

Ivanova, E.P., Y.V. Alexeeva, S. Flavier, J.P. Wright, N.V. Zhukova, N.M. Gorshkova, V.V. Mikhailov, D.V. Nicolau and R. Christen. 2004a. *Formosa algae* gen. nov., sp. nov., a novel member of the family *Flavobacteriaceae*. Int. J. Syst. Evol. Microbiol. *54*: 705–711.

Ivanova, E.P., Y.A. Alexeeva, N.V. Zhukova, N.M. Gorshkova, V. Buljan, D.V. Nicolau, V.V. Mikhailov and R. Christen. 2004b. *Bacillus algicola*

sp. nov., a novel filamentous organism isolated from brown alga *Fucus evanescens*. Syst. Appl. Microbiol. *27*: 301–307.

Ivanova, E.P., R. Christen, Y.V. Alexeeva, N.V. Zhukova, N.M. Gorshkova, A.M. Lysenko, V.V. Mikhailov and D.V. Nicolau. 2004c. *Brevibacterium celere* sp. nov., isolated from degraded thallus of a brown alga. Int. J. Syst. Evol. Microbiol. *54*: 2107–2111.

Ivanova, E.P., S. Flavier and R. Christen. 2004d. Phylogenetic relationships among marine *Alteromonas*-like proteobacteria: emended description of the family *Alteromonadaceae* and proposal of *Pseudoalteromonadaceae* fam. nov., *Colwelliaceae* fam. nov., *Shewanellaceae* fam. nov., *Moritellaceae* fam. nov., *Ferrimonadaceae* fam. nov., *Idiomarinaceae* fam. nov. and *Psychromonadaceae* fam. nov. Int. J. Syst. Evol. Microbiol. *54*: 1773–1788.

Ivanova, E.P., N.M. Gorshkova, J.P. Bowman, A.M. Lysenko, N.V. Zhukova, A.F. Sergeev, V.V. Mikhailov and D.V. Nicolau. 2004e. *Shewanella pacifica* sp. nov., a polyunsaturated fatty acid-producing bacterium isolated from sea water. Int. J. Syst. Evol. Microbiol. *54*: 1083–1087.

Ivanova, E.P., N.M. Gorshkova, T. Sawabe, N.V. Zhukova, K. Hayashi, V.V. Kurilenko, Y. Alexeeva, V. Buljan, D.V. Nicolau, V.V. Mikhailov and R. Christen. 2004f. *Sulfitobacter delicatus* sp. nov. and *Sulfitobacter dubius* sp. nov., respectively from a starfish (*Stellaster equestris*) and sea grass (*Zostera marina*). Int. J. Syst. Evol. Microbiol. *54*: 475–480.

Ivanova, E.P., N.M. Gorshkova, N.V. Zhukova, A.M. Lysenko, E.A. Zelepuga, N.G. Prokof'eva, V.V. Mikhailov, D.V. Nicolau and R. Christen. 2004g. Characterization of *Pseudoalteromonas distincta*-like sea-water isolates and description of *Pseudoalteromonas aliena* sp. nov. Int. J. Syst. Evol. Microbiol. *54*: 1431–1437.

Ivanova, E.P., O.I. Nedashkovskaya, T. Sawabe, N.V. Zhukova, G.M. Frolova, D.V. Nicolau, V.V. Mikhailov and J.P. Bowman. 2004h. *Shewanella affinis* sp. nov., isolated from marine invertebrates. Int. J. Syst. Evol. Microbiol. *54*: 1089–1093.

Ivanova, E.P., J.P. Bowman, A.M. Lysenko, N.V. Zhukova, N.M. Gorshkova, T.A. Kuznetsova, N.I. Kalinovskaya, L.S. Shevchenko and V.V. Mikhailov. 2005a. *Erythrobacter vulgaris* sp. nov., a novel organism isolated from the marine invertebrates. Syst. Appl. Microbiol. *28*: 123–130.

Ivanova, E.P., J.P. Bowman, A.M. Lysenko, N.V. Zhukova, N.M. Gorshkova, A.F. Sergeev and V.V. Mikhailov. 2005b. *Alteromonas addita* sp. nov. Int. J. Syst. Evol. Microbiol. *55*: 1065–1068.

Ivanova, E.P., O.M. Onyshchenko, R. Christen, A.M. Lysenko, N.V. Zhukova, L.S. Shevchenko and E.A. Kiprianova. 2005c. *Marinomonas pontica* sp. nov., isolated from the Black Sea. Int. J. Syst. Evol. Microbiol. *55*: 275–279.

Ivanova, E.P., O.M. Onyshchenko, R. Christen, N.V. Zhukova, A.M. Lysenko, L.S. Shevchenko, V. Buljan, B. Hambly and E.A. Kiprianova. 2005d. *Oceanimonas smirnovii* sp. nov., a novel organism isolated from the Black Sea. Syst. Appl. Microbiol. *28*: 131–136.

Ivanova, E.P., N.V. Zhukova, A.M. Lysenko, N.M. Gorshkova, A.F. Sergeev, V.V. Mikhailov and J.P. Bowman. 2005e. *Loktanella agnita* sp. nov. and *Loktanella rosea* sp. nov., from the north-west Pacific Ocean. Int. J. Syst. Evol. Microbiol. *55*: 2203–2207.

Ivanova, E.P., J.P. Bowman, R. Christen, N.V. Zhukova, A.M. Lysenko, N.M. Gorshkova, N. Mitik-Dineva, A.F. Sergeev and V.V. Mikhailov. 2006a. *Salegentibacter flavus* sp. nov. Int. J. Syst. Evol. Microbiol. *56*: 583–586.

Ivanova, E.P., J.P. Wright, A.M. Lysenko, N.V. Zhukova, Y.V. Alexeeva, V. Buljan, N.I. Kalinovskaya, D.V. Nicolau, R. Christen and V.V. Mikhailov. 2006b. Characterization of unusual alkaliphilic Grampositive bacteria isolated from degraded brown alga thalluses. Mikrobiol. Z. *68*: 10–20.

Jakobsen, T.F., K.U. Kjeldsen and K. Ingvorsen. 2006. *Desulfohalobium utahense* sp. nov., a moderately halophilic, sulfate-reducing bacterium isolated from Great Salt Lake. Int. J. Syst. Evol. Microbiol. *56*: 2063–2069.

Jean, W.D., J.S. Chen, Y.T. Lin and W.Y. Shieh. 2006a. *Bowmanella denitrificans* gen. nov., sp. nov., a denitrifying bacterium isolated from seawater from An-Ping Harbour, Taiwan. Int. J. Syst. Evol. Microbiol. 56: 2463–2467.

Jean, W.D., W.Y. Shieh and H.H. Chiu. 2006b. *Pseudidiomarina taiwanensis* gen. nov., sp. nov., a marine bacterium isolated from shallow coastal water of An-Ping Harbour, Taiwan, and emended description of the family *Idiomarinaceae*. Int. J. Syst. Evol. Microbiol. 56: 899–905.

Jean, W.D., W.Y. Shieh and T.Y. Liu. 2006c. *Thalassomonas agarivorans* sp. nov., a marine agarolytic bacterium isolated from shallow coastal water of An-Ping Harbour, Taiwan, and emended description of the genus *Thalassomonas*. Int. J. Syst. Evol. Microbiol. 56: 1245–1250.

Jeon, C.O., W. Park, W.C. Ghiorse and E.L. Madsen. 2004. *Polaromonas naphthalenivorans* sp. nov., a naphthalene-degrading bacterium from naphthalene-contaminated sediment. Int. J. Syst. Evol. Microbiol. 54: 93–97.

Jeon, C.O., J.M. Lim, J.C. Lee, G.S. Lee, J.M. Lee, L.H. Xu, C.L. Jiang and C.J. Kim. 2005a. *Lentibacillus salarius* sp. nov., isolated from saline sediment in China, and emended description of the genus *Lentibacillus*. Int. J. Syst. Evol. Microbiol. 55: 1339–1343.

Jeon, C.O., J.M. Lim, J.M. Lee, L.H. Xu, C.L. Jiang and C.J. Kim. 2005b. Reclassification of *Bacillus haloalkaliphilus* Fritze 1996 as *Alkalibacillus haloalkaliphilus* gen. nov., comb. nov. and the description of *Alkalibacillus salilacus* sp. nov., a novel halophilic bacterium isolated from a salt lake in China. Int. J. Syst. Evol. Microbiol. 55: 1891–1896.

Jeon, C.O., J.M. Lim, D.J. Park and C.J. Kim. 2005c. *Salinimonas chungwhensis* gen. nov., sp. nov., a moderately halophilic bacterium from a solar saltern in Korea. Int. J. Syst. Evol. Microbiol. 55: 239–243.

Jeon, C.O., J.M. Lim, J.R. Lee, G.S. Lee, D.J. Park, J.C. Lee, H.W. Oh and C.J. Kim. 2007. *Halomonas kribbensis* sp. nov., a novel moderately halophilic bacterium isolated from a solar saltern in Korea. Int. J. Syst. Evol. Microbiol. 57: 2194–2198.

Jeong, H., H. Yi, Y. Sekiguchi, M. Muramatsu, Y. Kamagata and J. Chun. 2004. *Clostridium jejuense* sp. nov., isolated from soil. Int. J. Syst. Evol. Microbiol. 54: 1465–1468.

Jeong, H., Y.W. Lim, H. Yi, Y. Sekiguchi, Y. Kamagata and J. Chun. 2007. *Anaerosporobacter mobilis* gen. nov., sp. nov., isolated from forest soil. Int. J. Syst. Evol. Microbiol. 57: 1784–1787.

Jiang, B., S.N. Parshina, W. van Doesburg, B.P. Lomans and A.J. Stams. 2005. *Methanomethylovorans thermophila* sp. nov., a thermophilic, methylotrophic methanogen from an anaerobic reactor fed with methanol. Int. J. Syst. Evol. Microbiol. 55: 2465–2470.

Jiang, C.Y., X. Dai, B.J. Wang, Y.G. Zhou and S.J. Liu. 2006. *Roseomonas lacus* sp. nov., isolated from freshwater lake sediment. Int. J. Syst. Evol. Microbiol. 56: 25–28.

Jiang, Y., S.K. Tang, J. Wiese, L.H. Xu, J.F. Imhoff and C.L. Jiang. 2007. *Streptomyces hainanensis* sp. nov., a novel member of the genus *Streptomyces*. Int. J. Syst. Evol. Microbiol. 57: 2694–2698.

Jimenez, M.S., M.I. Campos-Herrero, D. García, M. Luquin, L. Herrera and M.J. García. 2004. *Mycobacterium canariasense* sp. nov. Int. J. Syst. Evol. Microbiol. 54: 1729–1734.

Jin, L., K.K. Kim, W.T. Im, H.C. Yang and S.T. Lee. 2007. *Aspromonas composti* gen. nov., sp. nov., a novel member of the family *Xanthomonadaceae*. Int. J. Syst. Evol. Microbiol. 57: 1876–1880.

Johansen, J.E., S.J. Binnerup, N. Kroer and L. Mølbak. 2005. *Luteibacter rhizovicinus* gen. nov., sp. nov., a yellow-pigmented gammaproteobacterium isolated from the rhizosphere of barley (*Hordeum vulgare* L.). Int. J. Syst. Evol. Microbiol. 55: 2285–2291.

Johnson, D.B., B. Stallwood, S. Kimura and K.B. Hallberg. 2006. Isolation and characterization of *Acidicaldus organivorus*, gen. nov., sp. nov.: a novel sulfur-oxidizing, ferric iron-reducing thermo-acidophilic heterotrophic *Proteobacterium*. Arch. Microbiol. 185: 212–221.

Jojima, Y., Y. Mihara, S. Suzuki, K. Yokozeki, S. Yamanaka and R. Fudou. 2004. *Saccharibacter floricola* gen. nov., sp. nov., a novel osmophilic acetic acid bacterium isolated from pollen. Int. J. Syst. Evol. Microbiol. 54: 2263–2267.

Jones, A.L., J.M. Brown, V. Mishra, J.D. Perry, A.G. Steigerwalt and M. Goodfellow. 2004. *Rhodococcus gordoniae* sp. nov., an actinomycete isolated from clinical material and phenol-contaminated soil. Int. J. Syst. Evol. Microbiol. 54: 407–411.

Jones, B.E., W.D. Grant, A.W. Duckworth, P. Schumann, N. Weiss and E. Stackebrandt. 2005. *Cellulomonas bogoriensis* sp. nov., an alkaliphilic cellulomonad. Int. J. Syst. Evol. Microbiol. 55: 1711–1714.

Jones, J.B., G.H. Lacy, H. Bouzar, R.E. Stall and N.W. Schaad. 2004. Reclassification of the xanthomonads associated with bacterial spot disease of tomato and pepper. Syst. Appl. Microbiol. 27: 755–762.

Jordan, E.M., F.L. Thompson, X.H. Zhang, Y. Li, M. Vancanneyt, R.M. Kroppenstedt, F.G. Priest and B. Austin. 2007. *Sneathiella chinensis* gen. nov., sp. nov., a novel marine alphaproteobacterium isolated from coastal sediment in Qingdao, China. Int. J. Syst. Evol. Microbiol. 57: 114–121.

Jourand, P., E. Giraud, G. Bena, A. Sy, A. Willems, M. Gillis, B. Dreyfus and P. de Lajudie. 2004. *Methylobacterium nodulans* sp. nov., for a group of aerobic, facultatively methylotrophic, legume root-nodule-forming and nitrogen-fixing bacteria. Int. J. Syst. Evol. Microbiol. 54: 2269–2273.

Judicial Commission of the International Committee on Systematics of Prokaryotes. 2005a. Rejection of the genus name *Pelczaria* with the species *Pelczaria aurantia* Poston 1994. Opinion 78. Int. J. Syst. Evol. Microbiol. 55: 515.

Judicial Commission of the International Committee on Systematics of Prokaryotes. 2005b. The type species of the genus *Salmonella* Lignieres 1900 is *Salmonella enterica* (ex Kauffmann and Edwards 1952) Le Minor and Popoff 1987, with the type strain LT2ᵀ, and conservation of the epithet *enterica* in *Salmonella enterica* over all earlier epithets that may be applied to this species. Opinion 80. Int. J. Syst. Evol. Microbiol. 55: 519–520.

Jumas-Bilak, E., J.P. Carlier, H. Jean-Pierre, C. Teyssier, B. Gay, J. Campos and H. Marchandin. 2004. *Veillonella montpellierensis* sp. nov., a novel, anaerobic, Gram-negative coccus isolated from human clinical samples. Int. J. Syst. Evol. Microbiol. 54: 1311–1316.

Jumas-Bilak, E., H. Jean-Pierre, J.P. Carlier, C. Teyssier, K. Bernard, B. Gay, J. Campos, F. Morio and H. Marchandin. 2005. *Dialister micraerophilus* sp. nov. and *Dialister propionicifaciens* sp. nov., isolated from human clinical samples. Int. J. Syst. Evol. Microbiol. 55: 2471–2478.

Jumas-Bilak, E., J.P. Carlier, H. Jean-Pierre, D. Citron, K. Bernard, A. Damay, B. Gay, C. Teyssier, J. Campos and H. Marchandin. 2007a. *Jonquetella anthropi* gen. nov., sp. nov., the first member of the candidate phylum 'Synergistetes' isolated from man. Int. J. Syst. Evol. Microbiol. 57: 2743–2748.

Jumas-Bilak, E., J.P. Carlier, H. Jean-Pierre, F. Mory, C. Teyssier, B. Gay, J. Campos and H. Marchandin. 2007b. *Acidaminococcus intestini* sp. nov., isolated from human clinical samples. Int. J. Syst. Evol. Microbiol. 57: 2314–2319.

Jung, S.Y., S.J. Kang, M.H. Lee, S.Y. Lee, T.K. Oh and J.H. Yoon. 2005a. *Gaetbulibacter saemankumensis* gen. nov., sp. nov., a novel member of the family *Flavobacteriaceae* isolated from a tidal flat sediment in Korea. Int. J. Syst. Evol. Microbiol. 55: 1845–1849.

Jung, S.Y., M.H. Lee, T.K. Oh, Y.H. Park and J.H. Yoon. 2005b. *Psychrobacter cibarius* sp. nov., isolated from jeotgal, a traditional Korean fermented seafood. Int. J. Syst. Evol. Microbiol. 55: 577–582.

Jung, S.Y., H.S. Kim, J.J. Song, S.G. Lee, T.K. Oh and J.H. Yoon. 2006a. *Kribbia dieselivorans* gen. nov., sp. nov., a novel member of the family *Intrasporangiaceae*. Int. J. Syst. Evol. Microbiol. 56: 2427–2432.

Jung, S.Y., T.K. Oh and J.H. Yoon. 2006b. *Colwellia aestuarii* sp. nov., isolated from a tidal flat sediment in Korea. Int. J. Syst. Evol. Microbiol. 56: 33–37.

Jung, S.Y., T.K. Oh and J.H. Yoon. 2006c. *Tenacibaculum aestuarii* sp. nov., isolated from a tidal flat sediment in Korea. Int. J. Syst. Evol. Microbiol. 56: 1577–1581.

Jung, S.Y., Y.T. Jung, T.K. Oh and J.H. Yoon. 2007a. *Photobacterium lutimaris* sp. nov., isolated from a tidal flat sediment in Korea. Int. J. Syst. Evol. Microbiol. 57: 332–336.

Jung, S.Y., H.S. Kim, J.J. Song, S.G. Lee, T.K. Oh and J.H. Yoon. 2007b. *Aestuariimicrobium kwangyangense* gen. nov., sp. nov., an LL-diaminopimelic acid-containing bacterium isolated from tidal flat sediment. Int. J. Syst. Evol. Microbiol. 57: 2114–2118.

Jung, S.Y., S.Y. Lee, T.K. Oh and J.H. Yoon. 2007c. *Agromyces allii* sp. nov., isolated from the rhizosphere of *Allium victorialis* var. *platyphyllum*. Int. J. Syst. Evol. Microbiol. *57*: 588–593.

Jung, S.Y., M.H. Lee, T.K. Oh and J.H. Yoon. 2007d. *Herbaspirillum rhizosphaerae* sp. nov., isolated from rhizosphere soil of *Allium victorialis* var. *platyphyllum*. Int. J. Syst. Evol. Microbiol. *57*: 2284–2288.

Jurado, V., I. Groth, J.M. González, L. Laiz and C. Saiz-Jimenez. 2005a. *Agromyces salentinus* sp. nov. and *Agromyces neolithicus* sp. nov. Int. J. Syst. Evol. Microbiol. *55*: 153–157.

Jurado, V., I. Groth, J.M. González, L. Laiz and C. Saiz-Jimenez. 2005b. *Agromyces subbeticus* sp. nov., isolated from a cave in southern Spain. Int. J. Syst. Evol. Microbiol. *55*: 1897–1901.

Jurado, V., I. Groth, J.M. González, L. Laiz, B. Schuetze and C. Saiz-Jimenez. 2005c. *Agromyces italicus* sp. nov., *Agromyces humatus* sp. nov. and *Agromyces lapidis* sp. nov., isolated from Roman catacombs. Int. J. Syst. Evol. Microbiol. *55*: 871–875.

Jurado, V., L. Laiz, J.M. González, M. Hernandez-Marine, M. Valens and C. Saiz-Jimenez. 2005d. *Phyllobacterium catacumbae* sp. nov., a member of the order '*Rhizobiales*' isolated from Roman catacombs. Int. J. Syst. Evol. Microbiol. *55*: 1487–1490.

Jurado, V., J.M. González, L. Laiz and C. Saiz-Jimenez. 2006. *Aurantimonas altamirensis* sp. nov., a member of the order *Rhizobiales* isolated from Altamira Cave. Int. J. Syst. Evol. Microbiol. *56*: 2583–2585.

Juteau, P., V. Côté, M.F. Duckett, R. Beaudet, F. Lépine, R. Villemur and J.G. Bisaillon. 2005. *Cryptanaerobacter phenolicus* gen. nov., sp. nov., an anaerobe that transforms phenol into benzoate via 4-hydroxybenzoate. Int. J. Syst. Evol. Microbiol. *55*: 245–250.

Juvonen, R. and M.L. Suihko. 2006. *Megasphaera paucivorans* sp. nov., *Megasphaera sueciensis* sp. nov. and *Pectinatus haikarae* sp. nov., isolated from brewery samples, and emended description of the genus *Pectinatus*. Int. J. Syst. Evol. Microbiol. *56*: 695–702.

Kageyama, A., N. Poonwan, K. Yazawa, Y. Mikami and K. Nishimura. 2004a. *Nocardia asiatica* sp. nov., isolated from patients with nocardiosis in Japan and clinical specimens from Thailand. Int. J. Syst. Evol. Microbiol. *54*: 125–130.

Kageyama, A., N. Poonwan, K. Yazawa, S. Suzuki, R. Kroppenstedt and Y. Mikami. 2004b. *Nocardia vermiculata* sp. nov. and *Nocardia thailandica* sp. nov. isolated from clinical specimens. Actinomycetologica. *18*: 27–33.

Kageyama, A., S. Suzuki, K. Yazawa, K. Nishimura, R.M. Kroppenstedt and Y. Mikami. 2004c. *Nocardia aobensis* sp. nov., isolated from patients in Japan. Microbiol. Immunol. *48*: 817–822.

Kageyama, A., K. Torikoe, M. Iwamoto, J.I. Masuyama, Y. Shibuya, H. Okazaki, K. Yazawa, S. Minota, R.M. Kroppenstedt and Y. Mikami. 2004d. *Nocardia arthritidis* sp. nov., a new pathogen isolated from a patient with rheumatoid arthritis in Japan. J. Clin. Microbiol. *42*: 2366–2371.

Kageyama, A., K. Yazawa, K. Nishimura and Y. Mikami. 2004e. *Nocardia testaceus* sp. nov. and *Nocardia senatus* sp. nov., isolated from patients in Japan. Microbiol. Immunol. *48*: 271–276.

Kageyama, A., K. Yazawa, K. Nishimura and Y. Mikami. 2004f. *Nocardia inohanensis* sp. nov., *Nocardia yamanashiensis* sp. nov. and *Nocardia niigatensis* sp. nov., isolated from clinical specimens. Int. J. Syst. Evol. Microbiol. *54*: 563–569.

Kageyama, A., K. Yazawa, A. Mukai, M. Kinoshita, N. Takata, K. Nishimura, R.M. Kroppenstedt and Y. Mikami. 2004g. *Nocardia shimofusensis* sp. nov., isolated from soil, and *Nocardia higoensis* sp. nov., isolated from a patient with lung nocardiosis in Japan. Int. J. Syst. Evol. Microbiol. *54*: 1927–1931.

Kageyama, A., K. Yazawa, A. Mukai, T. Kohara, K. Nishimura, R.M. Kroppenstedt and Y. Mikami. 2004h. *Nocardia araoensis* sp. nov. and *Nocardia pneumoniae* sp. nov., isolated from patients in Japan. Int. J. Syst. Evol. Microbiol. *54*: 2025–2029.

Kageyama, A., Y. Takahashi, T. Seki, H. Tomoda and S. mura. 2005a. *Oryzihumus leptocrescens* gen. nov., sp. nov. Int. J. Syst. Evol. Microbiol. *55*: 2555–2559.

Kageyama, A., K. Yazawa, K. Nishimura and Y. Mikami. 2005b. *Nocardia anaemiae* sp. nov. isolated from an immunocompromised patient and the first isolation report of *Nocardia vinacea* from humans. Jpn. J. Med. Mycol. *46*: 21–26.

Kageyama, A., K. Yazawa, H. Taniguchi, H. Chibana, K. Nishimura, R.M. Kroppenstedt and Y. Mikami. 2005c. *Nocardia concava* sp. nov., isolated from Japanese patients. Int. J. Syst. Evol. Microbiol. *55*: 2081–2083.

Kageyama, A., S. Iida, K. Yazawa, T. Kudo, S. Suzuki, T. Koga, H. Saito, H. Inagawa, A. Wada, R.M. Kroppenstedt and Y. Mikami. 2006a. *Gordonia araii* sp. nov. and *Gordonia effusa* sp. nov., isolated from patients in Japan. Int. J. Syst. Evol. Microbiol. *56*: 1817–1821.

Kageyama, A., Y. Takahashi and S. Ōmura. 2006b. *Microbacterium deminutum* sp. nov., *Microbacterium pumilum* sp. nov. and *Microbacterium aoyamense* sp. nov. Int. J. Syst. Evol. Microbiol. *56*: 2113–2117.

Kageyama, A., Y. Takahashi, Y. Matsuo, H. Kasai, Y. Shizuri and S. Ōmura. 2007a. *Microbacterium sediminicola* sp. nov. and *Microbacterium marinilacus* sp. nov., isolated from marine environments. Int. J. Syst. Evol. Microbiol. *57*: 2355–2359.

Kageyama, A., Y. Takahashi and S. Ōmura. 2007b. *Microbacterium terricolae* sp. nov., isolated from soil in Japan. J. Gen. Appl. Microbiol. *53*: 1–5.

Kageyama, A., Y. Takahashi and S. Ōmura. 2007c. *Humihabitans oryzae* gen. nov., sp. nov. Int. J. Syst. Evol. Microbiol. *57*: 2163–2166.

Kageyama, A., Y. Takahashi, M. Yasumoto-Hirose, H. Kasai, Y. Shizuri and S. Ōmura. 2007d. *Janibacter corallicola* sp. nov., isolated from coral in Palau. J. Gen. Appl. Microbiol. *53*: 185–189.

Kageyama, C., T. Ohta, K. Hiraoka, M. Suzuki, T. Okamoto and K. Ohishi. 2005. Chlorinated aliphatic hydrocarbon-induced degradation of trichloroethylene in *Wautersia numadzuensis* sp. nov. Arch. Microbiol. *183*: 56–65.

Kaksonen, A.H., S. Spring, P. Schumann, R.M. Kroppenstedt and J.A. Puhakka. 2006. *Desulfotomaculum thermosubterraneum* sp. nov., a thermophilic sulfate-reducer isolated from an underground mine located in a geothermally active area. Int. J. Syst. Evol. Microbiol. *56*: 2603–2608.

Kaksonen, A.H., S. Spring, P. Schumann, R.M. Kroppenstedt and J.A. Puhakka. 2007a. *Desulfovirgula thermocuniculi* gen. nov., sp. nov., a thermophilic sulfate-reducer isolated from a geothermal underground mine in Japan. Int. J. Syst. Evol. Microbiol. *57*: 98–102.

Kaksonen, A.H., S. Spring, P. Schumann, R.M. Kroppenstedt and J.A. Puhakka. 2007b. *Desulfurispora thermophila* gen. nov., sp. nov., a thermophilic, spore-forming sulfate-reducer isolated from a sulfidogenic fluidized-bed reactor. Int. J. Syst. Evol. Microbiol. *57*: 1089–1094.

Kalyan Chakravarthy, S., T.N. Srinivas, P. Anil Kumar, C. Sasikala and V. Ramana Ch. 2007. *Roseospira visakhapatnamensis* sp. nov. and *Roseospira goensis* sp. nov. Int. J. Syst. Evol. Microbiol. *57*: 2453–2457.

Kalyuzhnaya, M.G., S.M. Stolyar, A.J. Auman, J.C. Lara, M.E. Lidstrom and L. Chistoserdova. 2005. *Methylosarcina lacus* sp. nov., a methanotroph from Lake Washington, Seattle, USA, and emended description of the genus *Methylosarcina*. Int. J. Syst. Evol. Microbiol. *55*: 2345–2350.

Kalyuzhnaya, M.G., S. Bowerman, J.C. Lara, M.E. Lidstrom and L. Chistoserdova. 2006a. *Methylotenera mobilis* gen. nov., sp. nov., an obligately methylamine-utilizing bacterium within the family *Methylophilaceae*. Int. J. Syst. Evol. Microbiol. *56*: 2819–2823.

Kalyuzhnaya, M.G., P. De Marco, S. Bowerman, C.C. Pacheco, J.C. Lara, M.E. Lidstrom and L. Chistoserdova. 2006b. *Methyloversatilis universalis* gen. nov., sp. nov., a novel taxon within the *Betaproteobacteria* represented by three methylotrophic isolates. Int. J. Syst. Evol. Microbiol. *56*: 2517–2522.

Kämpfer, P. and R.M. Kroppenstedt. 2004. *Pseudonocardia benzenivorans* sp. nov. Int. J. Syst. Evol. Microbiol. *54*: 749–751.

Kämpfer, P., S. Buczolits, U. Jäckel, I. Grün-Wollny and H.J. Busse. 2004a. *Nocardia tenerifensis* sp. nov. Int. J. Syst. Evol. Microbiol. *54*: 381–383.

Kämpfer, P., H.J. Busse, R. Rosselló-Mora, E. Kjellin and E. Falsen. 2004b. *Rhodovarius lipocyclicus* gen. nov. sp. nov., a new genus of the α-1 subclass of the *Proteobacteria*. Syst. Appl. Microbiol. *27*: 511–516.

Kämpfer, P., R.M. Kroppenstedt and I. Grün-Wollny. 2005a. *Nonomuraea kuesteri* sp. nov. Int. J. Syst. Evol. Microbiol. *55*: 847–851.

Kämpfer, P., S. Ruppel and R. Remus. 2005b. *Enterobacter radicincitans* sp. nov., a plant growth promoting species of the family *Enterobacteriaceae*. Syst. Appl. Microbiol. *28*: 213–221.

Kämpfer, P., R. Schulze, U. Jäckel, K.A. Malik, R. Amann and S. Spring. 2005c. *Hydrogenophaga defluvii* sp. nov. and *Hydrogenophaga atypica* sp. nov., isolated from activated sludge. Int. J. Syst. Evol. Microbiol. *55*: 341–344.

Kämpfer, P., V. Avesani, M. Janssens, J. Charlier, T. De Baere and M. Vaneechoutte. 2006a. Description of *Wautersiella falsenii* gen. nov., sp. nov., to accommodate clinical isolates phenotypically resembling members of the genera *Chryseobacterium* and *Empedobacter*. Int. J. Syst. Evol. Microbiol. *56*: 2323–2329.

Kämpfer, P., H.J. Busse and E. Falsen. 2006b. *Herminiimonas aquatilis* sp. nov., a new species from drinking water. Syst. Appl. Microbiol. *29*: 287–291.

Kämpfer, P., H.J. Busse and E. Falsen. 2006c. *Polaromonas aquatica* sp. nov., isolated from tap water. Int. J. Syst. Evol. Microbiol. *56*: 605–608.

Kämpfer, P., K. Denger, A.M. Cook, S.T. Lee, U. Jäckel, E.B. Denner and H.J. Busse. 2006d. *Castellaniella* gen. nov., to accommodate the phylogenetic lineage of *Alcaligenes defragrans*, and proposal of *Castellaniella defragrans* gen. nov., comb. nov. and *Castellaniella denitrificans* sp. nov. Int. J. Syst. Evol. Microbiol. *56*: 815–819.

Kämpfer, P., U. Kohlweyer, B. Thiemer and J.R. Andreesen. 2006e. *Pseudonocardia tetrahydrofuranoxydans* sp. nov. Int. J. Syst. Evol. Microbiol. *56*: 1535–1538.

Kämpfer, P., J.M. Lindh, O. Terenius, S. Haghdoost, E. Falsen, H.J. Busse and I. Faye. 2006f. *Thorsellia anophelis* gen. nov., sp. nov., a new member of the *Gammaproteobacteria*. Int. J. Syst. Evol. Microbiol. *56*: 335–338.

Kämpfer, P., R. Rosselló-Mora, E. Falsen, H.J. Busse and B.J. Tindall. 2006g. *Cohnella thermotolerans* gen. nov., sp. nov., and classification of 'Paenibacillus hongkongensis' as *Cohnella hongkongensis* sp. nov. Int. J. Syst. Evol. Microbiol. *56*: 781–786.

Kämpfer, P., R. Rosselló-Mora, H.C. Scholz, C. Welinder-Olsson, E. Falsen and H.J. Busse. 2006h. Description of *Pseudochrobactrum* gen. nov., with the two species *Pseudochrobactrum asaccharolyticum* sp. nov. and *Pseudochrobactrum saccharolyticum* sp. nov. Int. J. Syst. Evol. Microbiol. *56*: 1823–1829.

Kämpfer, P., O. Terenius, J.M. Lindh and I. Faye. 2006i. *Janibacter anophelis* sp. nov., isolated from the midgut of *Anopheles arabiensis*. Int. J. Syst. Evol. Microbiol. *56*: 389–392.

Kämpfer, P., C.C. Young, A.B. Arun, F.T. Shen, U. Jäckel, R. Rosselló-Mora, W.A. Lai and P.D. Rekha. 2006j. *Pseudolabrys taiwanensis* gen. nov., sp. nov., an alphaproteobacterium isolated from soil. Int. J. Syst. Evol. Microbiol. *56*: 2469–2472.

Kämpfer, P., C.C. Young, K.R. Sridhar, A.B. Arun, W.A. Lai, F.T. Shen and P.D. Rekha. 2006k. Transfer of [*Flexibacter*] *sancti*, [*Flexibacter*] *filiformis*, [*Flexibacter*] *japonensis* and [*Cytophaga*] *arvensicola* to the genus *Chitinophaga* and description of *Chitinophaga skermanii* sp. nov. Int. J. Syst. Evol. Microbiol. *56*: 2223–2228.

Kämpfer, P., B. Huber, K. Thummes, I. Grün-Wollny and H.J. Busse. 2007a. *Actinoplanes couchii* sp. nov. Int. J. Syst. Evol. Microbiol. *57*: 721–724.

Kämpfer, P., B. Huber, S. Buczolits, K. Thummes, I. Grün-Wollny and H.J. Busse. 2007b. *Nocardia acidivorans* sp. nov., isolated from soil of the island of Stromboli. Int. J. Syst. Evol. Microbiol. *57*: 1183–1187.

Kämpfer, P., U. Meurer, M. Esser, T. Hirsch and H.J. Busse. 2007c. *Sphingomonas pseudosanguinis* sp. nov., isolated from the water reservoir of an air humidifier. Int. J. Syst. Evol. Microbiol. *57*: 1342–1345.

Kämpfer, P., R. Rosselló-Mora, M. Hermansson, F. Persson, B. Huber, E. Falsen and H.J. Busse. 2007d. *Undibacterium pigrum* gen. nov., sp. nov., isolated from drinking water. Int. J. Syst. Evol. Microbiol. *57*: 1510–1515.

Kämpfer, P., H.C. Scholz, B. Huber, E. Falsen and H.J. Busse. 2007e. *Ochrobactrum haematophilum* sp. nov. and *Ochrobactrum pseudogrignonense* sp. nov., isolated from human clinical specimens. Int. J. Syst. Evol. Microbiol. *57*: 2513–2518.

Kämpfer, P., H. Scholz, B. Huber, K. Thummes, H.J. Busse, E.W. Maas and E. Falsen. 2007f. Description of *Pseudochrobactrum kiredjianiae* sp. nov. Int. J. Syst. Evol. Microbiol. *57*: 755–760.

Kang, M.S., W.T. Im, H.M. Jung, M.K. Kim, M. Goodfellow, K.K. Kim, H.C. Yang, D.S. An and S.T. Lee. 2007. *Cellulomonas composti* sp. nov., a cellulolytic bacterium isolated from cattle farm compost. Int. J. Syst. Evol. Microbiol. *57*: 1256–1260.

Kang, Y.S., J. Kim, H.D. Shin, Y.D. Nam, J.W. Bae, C.O. Jeon and W. Park. 2007. *Methylobacterium platani* sp. nov., isolated from a leaf of the tree *Platanus orientalis*. Int. J. Syst. Evol. Microbiol. *57*: 2849–2853.

Kanso, S. and B.K. Patel. 2004. *Phenylobacterium lituiforme* sp. nov., a moderately thermophilic bacterium from a subsurface aquifer, and emended description of the genus *Phenylobacterium*. Int. J. Syst. Evol. Microbiol. *54*: 2141–2146.

Karavaiko, G.I., T.I. Bogdanova, T.P. Tourova, T.F. Kondrat'eva, I.A. Tsaplina, M.A. Egorova, E.N. Krasil'nikova and L.M. Zakharchuk. 2005. Reclassification of 'Sulfobacillus thermosulfidooxidans subsp. *thermotolerans*' strain K1 as *Alicyclobacillus tolerans* sp. nov. and *Sulfobacillus disulfidooxidans* Dufresne *et al.* 1996 as *Alicyclobacillus disulfidooxidans* comb. nov., and emended description of the genus *Alicyclobacillus*. Int. J. Syst. Evol. Microbiol. *55*: 941–947.

Kasai, H., A. Katsuta, H. Sekiguchi, S. Matsuda, K. Adachi, K. Shindo, J. Yoon, A. Yokota and Y. Shizuri. 2007. *Rubritalea squalenifaciens* sp. nov., a squalene-producing marine bacterium belonging to subdivision 1 of the phylum 'Verrucomicrobia'. Int. J. Syst. Evol. Microbiol. *57*: 1630–1634.

Kashefi, K., D.E. Holmes, J.A. Baross and D.R. Lovley. 2003. Thermophily in the *Geobacteraceae*: *Geothermobacter ehrlichii* gen. nov., sp. nov., a novel thermophilic member of the *Geobacteraceae* from the "Bag City" hydrothermal vent. Appl. Environ. Microbiol. *69*: 2985–2993.

Katayama, Y., Y. Uchino, A.P. Wood and D.P. Kelly. 2006. Confirmation of *Thiomonas delicata* (formerly *Thiobacillus delicatus*) as a distinct species of the genus *Thiomonas* Moreira and Amils 1997 with comments on some species currently assigned to the genus. Int. J. Syst. Evol. Microbiol. *56*: 2553–2557.

Kato, S., S. Haruta, Z.J. Cui, M. Ishii, A. Yokota and Y. Igarashi. 2004. *Clostridium straminisolvens* sp. nov., a moderately thermophilic, aerotolerant and cellulolytic bacterium isolated from a cellulose-degrading bacterial community. Int. J. Syst. Evol. Microbiol. *54*: 2043–2047.

Kato, Y., M. Asahara, D. Arai, K. Goto and A. Yokota. 2005. Reclassification of *Methylobacterium chloromethanicum* and *Methylobacterium dichloromethanicum* as later subjective synonyms of *Methylobacterium extorquens* and of *Methylobacterium lusitanum* as a later subjective synonym of *Methylobacterium rhodesianum*. J. Gen. Appl. Microbiol. *51*: 287–299.

Katsuta, A., K. Adachi, S. Matsuda, Y. Shizuri and H. Kasai. 2005. *Ferrimonas marina* sp. nov. Int. J. Syst. Evol. Microbiol. *55*: 1851–1855.

Kawamura, Y., X.G. Hou, F. Sultana, K. Hirose, M. Miyake, S.E. Shu and T. Ezaki. 1998. Distribution of *Staphylococcus* species among human clinical specimens and emended description of *Staphylococcus caprae*. J. Clin. Microbiol. *36*: 2038–2042.

Kawamura, Y., Y. Itoh, N. Mishima, K. Ohkusu, H. Kasai and T. Ezaki. 2005. High genetic similarity of *Streptococcus agalactiae* and *Streptococcus difficilis*: *S. difficilis* Eldar *et al.* 1995 is a later synonym of *S. agalactiae* Lehmann and Neumann 1896 (Approved Lists 1980). Int. J. Syst. Evol. Microbiol. *55*: 961–965.

Kaye, J.Z., M.C. Márquez, A. Ventosa and J.A. Baross. 2004. *Halomonas neptunia* sp. nov., *Halomonas sulfidaeris* sp. nov., *Halomonas axialensis* sp. nov. and *Halomonas hydrothermalis* sp. nov.: halophilic bacteria isolated from deep-sea hydrothermal-vent environments. Int. J. Syst. Evol. Microbiol. *54*: 499–511.

Kelly, D.P. and A.P. Wood. 2005a. Family III. *Halothiobacillaceae* fam. nov. Kelly and Wood 2003. *In* Brenner, Krieg, Staley and Garrity (Editors),

Bergey's Manual of Systematic Bacteriology, 2nd Edition, Volume 2 (The *Proteobacteria*), Part B (The *Gammaproteobacteria*), Springer, New York, p. 58.

Kelly, D.P. and A.P. Wood. 2005b. Genus *Incertae Sedis* XVIII. *Thiomonas* Moreira and Amils 1997, 527^VP. *In* Brenner, Krieg, Staley and Garrity (Editors), Bergey's Manual of Systematic Bacteriology, 2nd Edition, Volume 2 (The *Proteobacteria*), Part C (The *Alpha-, Beta-, Delta-,* and *Epsilonproteobacteria*), Springer, New York, pp. 757–759.

Kelly, D.P., Y. Uchino, H. Huber, R. Amils and A.P. Wood. 2007. Reassessment of the phylogenetic relationships of *Thiomonas cuprina*. Int. J. Syst. Evol. Microbiol. *57*: 2720–2724.

Kendall, M.M., Y. Liu, M. Sieprawska-Lupa, K.O. Stetter, W.B. Whitman and D.R. Boone. 2006. *Methanococcus aeolicus* sp. nov., a mesophilic, methanogenic archaeon from shallow and deep marine sediments. Int. J. Syst. Evol. Microbiol. *56*: 1525–1529.

Kevbrin, V.V., K. Zengler, A.M. Lysenko and J. Wiegel. 2005. *Anoxybacillus kamchatkensis* sp. nov., a novel thermophilic facultative aerobic bacterium with a broad pH optimum from the Geyser valley, Kamchatka. Extremophiles. *9*: 391–398.

Khan, S.T. and S. Harayama. 2007. *Paraferrimonas sedimenticola* gen. nov., sp. nov., a marine bacterium of the family *Ferrimonadaceae*. Int. J. Syst. Evol. Microbiol. *57*: 1493–1498.

Khan, S.T., Y. Nakagawa and S. Harayama. 2006a. *Krokinobacter* gen. nov., with three novel species, in the family *Flavobacteriaceae*. Int. J. Syst. Evol. Microbiol. *56*: 323–328.

Khan, S.T., Y. Nakagawa and S. Harayama. 2006b. *Sandarakinotalea sediminis* gen. nov., sp. nov., a novel member of the family *Flavobacteriaceae*. Int. J. Syst. Evol. Microbiol. *56*: 959–963.

Khan, S.T., Y. Nakagawa and S. Harayama. 2006c. *Sediminicola luteus* gen. nov., sp. nov., a novel member of the family *Flavobacteriaceae*. Int. J. Syst. Evol. Microbiol. *56*: 841–845.

Khan, S.T., Y. Fukunaga, Y. Nakagawa and S. Harayama. 2007a. Emended descriptions of the genus *Lewinella* and of *Lewinella cohaerens, Lewinella nigricans* and *Lewinella persica*, and description of *Lewinella lutea* sp. nov. and *Lewinella marina* sp. nov. Int. J. Syst. Evol. Microbiol. *57*: 2946–2951.

Khan, S.T., Y. Nakagawa and S. Harayama. 2007b. *Sediminibacter furfurosus* gen. nov., sp. nov. and *Gilvibacter sediminis* gen. nov., sp. nov., novel members of the family *Flavobacteriaceae*. Int. J. Syst. Evol. Microbiol. *57*: 265–269.

Khan, S.T., Y. Nakagawa and S. Harayama. 2007c. *Galbibacter mesophilus* gen. nov., sp. nov., a novel member of the family *Flavobacteriaceae*. Int. J. Syst. Evol. Microbiol. *57*: 969–973.

Khan, S.T., Y. Nakagawa and S. Harayama. 2007d. *Sediminitomix flava* gen. nov., sp. nov., of the phylum *Bacteroidetes*, isolated from marine sediment. Int. J. Syst. Evol. Microbiol. *57*: 1689–1693.

Kharroub, K., M. Aguilera, T. Quesada, J.A. Morillo, A. Ramos-Cormenzana, A. Boulharouf and M. Monteoliva-Sánchez. 2006a. *Salicola salis* sp. nov., an extremely halophilic bacterium isolated from Ezzemoul sabkha in Algeria. Int. J. Syst. Evol. Microbiol. *56*: 2647–2652.

Kharroub, K., T. Quesada, R. Ferrer, S. Fuentes, M. Aguilera, A. Boulahrouf, A. Ramos-Cormenzana and M. Monteoliva-Sánchez. 2006b. *Halorubrum ezzemoulense* sp. nov., a halophilic archaeon isolated from Ezzemoul sabkha, Algeria. Int. J. Syst. Evol. Microbiol. *56*: 1583–1588.

Kim, B.C., J.R. Park, J.W. Bae, S.K. Rhee, K.H. Kim, J.W. Oh and Y.H. Park. 2006. *Stappia marina* sp. nov., a marine bacterium isolated from the Yellow Sea. Int. J. Syst. Evol. Microbiol. *56*: 75–79.

Kim, B.Y., S.Y. Lee, H.Y. Weon, S.W. Kwon, S.J. Go, Y.K. Park, P. Schumann and D. Fritze. 2006a. *Ureibacillus suwonensis* sp. nov., isolated from cotton waste composts. Int. J. Syst. Evol. Microbiol. *56*: 663–666.

Kim, B.Y., H.Y. Weon, S. Cousin, S.H. Yoo, S.W. Kwon, S.J. Go and E. Stackebrandt. 2006b. *Flavobacterium daejeonense* sp. nov. and *Flavobacterium suncheonense* sp. nov., isolated from greenhouse soils in Korea. Int. J. Syst. Evol. Microbiol. *56*: 1645–1649.

Kim, B.Y., H.Y. Weon, K.H. Lee, S.J. Seok, S.W. Kwon, S.J. Go and E. Stackebrandt. 2006c. *Dyella yeojuensis* sp. nov., isolated from greenhouse soil in Korea. Int. J. Syst. Evol. Microbiol. *56*: 2079–2082.

Kim, B.Y., H.Y. Weon, S.H. Yoo, W.M. Chen, S.W. Kwon, S.J. Go and E. Stackebrandt. 2006d. *Chitinimonas koreensis* sp. nov., isolated from greenhouse soil in Korea. Int. J. Syst. Evol. Microbiol. *56*: 1761–1764.

Kim, B.Y., H.Y. Weon, S.H. Yoo, S.W. Kwon, Y.H. Cho, E. Stackebrandt and S.J. Go. 2006e. *Paracoccus homiensis* sp. nov., isolated from a sea-sand sample. Int. J. Syst. Evol. Microbiol. *56*: 2387–2390.

Kim, B.Y., H.Y. Weon, S.H. Yoo, J.S. Kim, S.W. Kwon, E. Stackebrandt and S.J. Go. 2006f. *Marinobacter koreensis* sp. nov., isolated from sea sand in Korea. Int. J. Syst. Evol. Microbiol. *56*: 2653–2656.

Kim, B.Y., H.Y. Weon, S.H. Yoo, S.Y. Lee, S.W. Kwon, S.J. Go and E. Stackebrandt. 2006g. *Variovorax soli* sp. nov., isolated from greenhouse soil. Int. J. Syst. Evol. Microbiol. *56*: 2899–2901.

Kim, B.Y., H.Y. Weon, S.H. Yoo, S.B. Hong, S.W. Kwon, E. Stackebrandt and S.J. Go. 2007. *Niabella aurantiaca* gen. nov., sp. nov., isolated from a greenhouse soil in Korea. Int. J. Syst. Evol. Microbiol. *57*: 538–541.

Kim, D., K.S. Baik, M.S. Kim, B.M. Jung, T.S. Shin, G.H. Chung, M.S. Rhee and C.N. Seong. 2007. *Shewanella haliotis* sp. nov., isolated from the gut microflora of abalone, *Haliotis discus hannai*. Int. J. Syst. Evol. Microbiol. *57*: 2926–2931.

Kim, D.S., C.Y. Bae, J.J. Jeon, S.J. Chun, H.W. Oh, S.G. Hong, K.S. Baek, E.Y. Moon and K.S. Bae. 2004. *Paenibacillus elgii* sp. nov., with broad antimicrobial activity. Int. J. Syst. Evol. Microbiol. *54*: 2031–2035.

Kim, H., Y.J. Choo and J.C. Cho. 2007a. *Litoricolaceae* fam. nov., to include *Litoricola lipolytica* gen. nov., sp. nov., a marine bacterium belonging to the order *Oceanospirillales*. Int. J. Syst. Evol. Microbiol. *57*: 1793–1798.

Kim, H., Y.J. Choo, J. Song, J.S. Lee, K.C. Lee and J.C. Cho. 2007b. *Marinobacterium litorale* sp. nov. in the order *Oceanospirillales*. Int. J. Syst. Evol. Microbiol. *57*: 1659–1662.

Kim, H.B., M.J. Park, H.C. Yang, D.S. An, H.Z. Jin and D.C. Yang. 2006. *Burkholderia ginsengisoli* sp. nov., a beta-glucosidase-producing bacterium isolated from soil of a ginseng field. Int. J. Syst. Evol. Microbiol. *56*: 2529–2533.

Kim, H.G., N.V. Doronina, Y.A. Trotsenko and S.W. Kim. 2007. *Methylophaga aminisulfidivorans* sp. nov., a restricted facultatively methylotrophic marine bacterium. Int. J. Syst. Evol. Microbiol. *57*: 2096–2101.

Kim, H.J., S.C. Lee and B.K. Hwang. 2006. *Streptomyces cheonanensis* sp. nov., a novel streptomycete with antifungal activity. Int. J. Syst. Evol. Microbiol. *56*: 471–475.

Kim, I.G., M.H. Lee, S.Y. Jung, J.J. Song, T.K. Oh and J.H. Yoon. 2005. *Exiguobacterium aestuarii* sp. nov. and *Exiguobacterium marinum* sp. nov., isolated from a tidal flat of the Yellow Sea in Korea. Int. J. Syst. Evol. Microbiol. *55*: 885–889.

Kim, K.H., L.N. Ten, Q.M. Liu, W.T. Im and S.T. Lee. 2006. *Sphingobacterium daejeonense* sp. nov., isolated from a compost sample. Int. J. Syst. Evol. Microbiol. *56*: 2031–2036.

Kim, K.K., H.S. Bae, P. Schumann and S.T. Lee. 2005a. *Chryseobacterium daecheongense* sp. nov., isolated from freshwater lake sediment. Int. J. Syst. Evol. Microbiol. *55*: 133–138.

Kim, K.K., M.K. Kim, J.H. Lim, H.Y. Park and S.T. Lee. 2005b. Transfer of *Chryseobacterium meningosepticum* and *Chryseobacterium miricola* to *Elizabethkingia* gen. nov. as *Elizabethkingia meningoseptica* comb. nov. and *Elizabethkingia miricola* comb. nov. Int. J. Syst. Evol. Microbiol. *55*: 1287–1293.

Kim, K.K., H.Y. Park, W. Park, I.S. Kim and S.T. Lee. 2005c. *Microbacterium xylanilyticum* sp. nov., a xylan-degrading bacterium isolated from a biofilm. Int. J. Syst. Evol. Microbiol. *55*: 2075–2079.

Kim, K.K., L. Jin, H.C. Yang and S.T. Lee. 2007. *Halomonas gomseomensis* sp. nov., *Halomonas janggokensis* sp. nov., *Halomonas salaria* sp. nov. and *Halomonas denitrificans* sp. nov., moderately halophilic bacteria isolated from saline water. Int. J. Syst. Evol. Microbiol. *57*: 675–681.

Kim, M.K. and H.Y. Jung. 2007. *Chitinophaga terrae* sp. nov., isolated from soil. Int. J. Syst. Evol. Microbiol. *57*: 1721–1724.

Kim, M.K., W.T. Im, Y.K. Shin, J.H. Lim, S.H. Kim, B.C. Lee, M.Y. Park, K.Y. Lee and S.T. Lee. 2004. *Kaistella koreensis* gen. nov., sp. nov., a novel member of the *Chryseobacterium–Bergeyella–Riemerella* branch. Int. J. Syst. Evol. Microbiol. *54*: 2319–2324.

Kim, M.K., W.T. Im, J.G. In, S.H. Kim and D.C. Yang. 2006. *Thermomonas koreensis* sp. nov., a mesophilic bacterium isolated from a ginseng field. Int. J. Syst. Evol. Microbiol. *56*: 1615–1619.

Kim, M.K., Y.J. Kim, D.H. Cho, T.H. Yi, N.K. Soung and D.C. Yang. 2007a. *Solimonas soli* gen. nov., sp. nov., isolated from soil of a ginseng field. Int. J. Syst. Evol. Microbiol. *57*: 2591–2594.

Kim, M.K., J.R. Na, D.H. Cho, N.K. Soung and D.C. Yang. 2007b. *Parapedobacter koreensis* gen. nov., sp. nov. Int. J. Syst. Evol. Microbiol. *57*: 1336–1341.

Kim, M.K., J.R. Na, T.H. Lee, W.T. Im, N.K. Soung and D.C. Yang. 2007c. *Solirubrobacter soli* sp. nov., isolated from soil of a ginseng field. Int. J. Syst. Evol. Microbiol. *57*: 1453–1455.

Kim, M.K., K. Schubert, W.T. Im, K.H. Kim, S.T. Lee and J. Overmann. 2007d. *Sphingomonas kaistensis* sp. nov., a novel alphaproteobacterium containing *pufLM* genes. Int. J. Syst. Evol. Microbiol. *57*: 1527–1534.

Kim, S., H. Jeong, S. Kim and J. Chun. 2006. *Clostridium ganghwense* sp. nov., isolated from tidal flat sediment. Int. J. Syst. Evol. Microbiol. *56*: 691–693.

Kim, S., H. Jeong and J. Chun. 2007. *Clostridium aestuarii* sp. nov., from tidal flat sediment. Int. J. Syst. Evol. Microbiol. *57*: 1315–1317.

Kim, S.B., O.I. Nedashkovskaya, V.V. Mikhailov, S.K. Han, K.O. Kim, M.S. Rhee and K.S. Bae. 2004a. *Kocuria marina* sp. nov., a novel actinobacterium isolated from marine sediment. Int. J. Syst. Evol. Microbiol. *54*: 1617–1620.

Kim, S.B., C.N. Seong, S.J. Jeon, K.S. Bae and M. Goodfellow. 2004b. Taxonomic study of neutrotolerant acidophilic actinomycetes isolated from soil and description of *Streptomyces yeochonensis* sp. nov. Int. J. Syst. Evol. Microbiol. *54*: 211–214.

Kim, Y.G., D.H. Choi, S. Hyun and B.C. Cho. 2007a. *Oceanobacillus profundus* sp. nov., isolated from a deep-sea sediment core. Int. J. Syst. Evol. Microbiol. *57*: 409–413.

Kim, Y.G., C.Y. Hwang, K.W. Yoo, H.T. Moon, J.H. Yoon and B.C. Cho. 2007b. *Pelagibacillus goriensis* gen. nov., sp. nov., a moderately halotolerant bacterium isolated from coastal water off the east coast of Korea. Int. J. Syst. Evol. Microbiol. *57*: 1554–1560.

Kirby, B.M., M. Le Roes and P.R. Meyers. 2006. *Kribbella karoonensis* sp. nov. and *Kribbella swartbergensis* sp. nov., isolated from soil from the Western Cape, South Africa. Int. J. Syst. Evol. Microbiol. *56*: 1097–1101.

Kitahara, M., M. Sakamoto, M. Ike, S. Sakata and Y. Benno. 2005. *Bacteroides plebeius* sp. nov. and *Bacteroides coprocola* sp. nov., isolated from human faeces. Int. J. Syst. Evol. Microbiol. *55*: 2143–2147.

Klouche, N., M.L. Fardeau, J.F. Lascourrèges, J.L. Cayol, H. Hacene, P. Thomas and M. Magot. 2007. *Geosporobacter subterraneus* gen. nov., sp. nov., a spore-forming bacterium isolated from a deep subsurface aquifer. Int. J. Syst. Evol. Microbiol. *57*: 1757–1761.

Knight, T.F., Jr. 2004. Reclassification of *Mesoplasma pleciae* as *Acholeplasma pleciae* comb. nov. on the basis of 16S rRNA and gyrB gene sequence data. Int. J. Syst. Evol. Microbiol. *54*: 1951–1952.

Knittel, K., J. Kuever, A. Meyerdierks, R. Meinke, R. Amann and T. Brinkhoff. 2005. *Thiomicrospira arctica* sp. nov. and *Thiomicrospira psychrophila* sp. nov., psychrophilic, obligately chemolithoautotrophic, sulfur-oxidizing bacteria isolated from marine Arctic sediments. Int. J. Syst. Evol. Microbiol. *55*: 781–786.

Ko, K.S., W.S. Oh, M.Y. Lee, J.H. Lee, H. Lee, K.R. Peck, N.Y. Lee and J.H. Song. 2006. *Bacillus infantis* sp. nov. and *Bacillus idriensis* sp. nov., isolated from a patient with neonatal sepsis. Int. J. Syst. Evol. Microbiol. *56*: 2541–2544.

Kodama, Y. and K. Watanabe. 2004. *Sulfuricurvum kujiense* gen. nov., sp. nov., a facultatively anaerobic, chemolithoautotrophic, sulfur-oxidizing bacterium isolated from an underground crude-oil storage cavity. Int. J. Syst. Evol. Microbiol. *54*: 2297–2300.

Kodama, Y., T. Ha le and K. Watanabe. 2007. *Sulfurospirillum cavolei* sp. nov., a facultatively anaerobic sulfur-reducing bacterium isolated from an underground crude oil storage cavity. Int. J. Syst. Evol. Microbiol. *57*: 827–831.

Koerber, R.T., G.E. Gasparich, M.F. Frana and W.L. Grogan, Jr. 2005. *Spiroplasma atrichopogonis* sp. nov., from a ceratopogonid biting midge. Int. J. Syst. Evol. Microbiol. *55*: 289–292.

Konstantinov, S.R., E. Poznanski, S. Fuentes, A.D. Akkermans, H. Smidt and W.M. de Vos. 2006. *Lactobacillus sobrius* sp. nov., abundant in the intestine of weaning piglets. Int. J. Syst. Evol. Microbiol. *56*: 29–32.

Koort, J., P. Vandamme, U. Schillinger, W. Holzapfel and J. Björkroth. 2004. *Lactobacillus curvatus* subsp. *melibiosus* is a later synonym of *Lactobacillus sakei* subsp. *carnosus*. Int. J. Syst. Evol. Microbiol. *54*: 1621–1626.

Koort, J., T. Coenye, P. Vandamme, A. Sukura and J. Björkroth. 2004. *Enterococcus hermanniensis* sp. nov., from modified-atmosphere-packaged broiler meat and canine tonsils. Int. J. Syst. Evol. Microbiol. *54*: 1823–1827.

Koort, J., A. Murros, T. Coenye., S. Eerola, P. Vandamme, A. Sukura and J. Björkroth. 2005. *Lactobacillus oligofermentans* sp. nov., associated with spoilage of modified-atmosphere-packaged poultry products. Appl. Environ. Microbiol. *71*: 4400–4406.

Kostinek, M., R. Pukall, A.P. Rooney, U. Schillinger, C. Hertel, W.H. Holzapfel and C.M. Franz. 2005. *Lactobacillus arizonensis* is a later heterotypic synonym of *Lactobacillus plantarum*. Int. J. Syst. Evol. Microbiol. *55*: 2485–2489.

Kotou ková, L., P. Schumann, E. Durnová, C. Spröer, I. Sedlá ek, J. Ne a, Z. Zdráhal and M. N mec. 2004. *Arthrobacter nitroguajacolicus* sp. nov., a novel 4-nitroguaiacol-degrading actinobacterium. Int. J. Syst. Evol. Microbiol. *54*: 773–777.

Kroppenstedt, R.M., S. Mayilraj, J.M. Wink, W. Kallow, P. Schumann, C. Secondini and E. Stackebrandt. 2005. Eight new species of the genus *Micromonospora*, *Micromonospora citrea* sp. nov., *Micromonospora echinaurantiaca* sp. nov., *Micromonospora echinofusca* sp. nov. *Micromonospora fulviviridis* sp. nov., *Micromonospora inyonensis* sp. nov., *Micromonospora peucetia* sp. nov., *Micromonospora sagamiensis* sp. nov., and *Micromonospora viridifaciens* sp. nov. Syst. Appl. Microbiol. *28*: 328–339.

Kublanov, I.V., M.I. Prokofeva, N.A. Kostrikina, T.V. Kolganova, T.P. Tourova, J. Wiegel and E.A. Bonch-Osmolovskaya. 2007. *Thermoanaerobacterium aciditolerans* sp. nov., a moderate thermoacidophile from a Kamchatka hot spring. Int. J. Syst. Evol. Microbiol. *57*: 260–264.

Kubota, M., K. Kawahara, K. Sekiya, T. Uchida, Y. Hattori, H. Futamata and A. Hiraishi. 2005. *Nocardioides aromaticivorans* sp. nov., a dibenzofuran-degrading bacterium isolated from dioxin-polluted environments. Syst. Appl. Microbiol. *28*: 165–174.

Kuever, J., F.A. Rainey and F. Widdel. 2005a. Class IV. *Deltaproteobacteria* class. nov. *In* Brenner, Krieg, Staley and Garrity (Editors), Bergey's Manual of Systematic Bacteriology, 2nd Edition, Volume 2 (The *Proteobacteria*), Part C (The *Alpha-, Beta-, Delta-,* and *Epsilonproteobacteria*), Springer, New York, p. 922.

Kuever, J., F.A. Rainey and F. Widdel. 2005b. Order I. *Desulfurellales* ord. nov. *In* Brenner, Krieg, Staley and Garrity (Editors), Bergey's Manual of Systematic Bacteriology, 2nd Edition, Volume 2 (The *Proteobacteria*), Part C (The *Alpha-, Beta-, Delta-,* and *Epsilonproteobacteria*), Springer, New York, p. 922.

Kuever, J., F.A. Rainey and F. Widdel. 2005c. Family I. *Desulfurellaceae* fam. nov. *In* Brenner, Krieg, Staley and Garrity (Editors), Bergey's Manual of Systematic Bacteriology, 2nd Edition, Volume 2 (The *Proteobacteria*), Part C (The *Alpha-, Beta-, Delta-,* and *Epsilonproteobacteria*), Springer, New York, p. 923.

Kuever, J., F.A. Rainey and F. Widdel. 2005d. Order II. *Desulfovibrionales* ord. nov. *In* Brenner, Krieg, Staley and Garrity (Editors), Bergey's Manual of Systematic Bacteriology, 2nd Edition, Volume 2 (The *Proteobacteria*), Part C (The *Alpha-, Beta-, Delta-,* and *Epsilonproteobacteria*), Springer, New York, pp. 925–926.

Kuever, J., F.A. Rainey and F. Widdel. 2005e. Family I. *Desulfovibrionaceae* fam. nov. *In* Brenner, Krieg, Staley and Garrity (Editors), Bergey's Manual of Systematic Bacteriology, 2nd Edition, Volume 2 (The *Proteobacteria*), Part C (The *Alpha-, Beta-, Delta-*, and *Epsilonproteobacteria*), Springer, New York, p. 926.

Kuever, J., F.A. Rainey and F. Widdel. 2005f. Family II. *Desulfomicrobiaceae* fam. nov. *In* Brenner, Krieg, Staley and Garrity (Editors), Bergey's Manual of Systematic Bacteriology, 2nd Edition, Volume 2 (The *Proteobacteria*), Part C (The *Alpha-, Beta-, Delta-*, and *Epsilonproteobacteria*), Springer, New York, p. 944.

Kuever, J., F.A. Rainey and F. Widdel. 2005g. Family III. *Desulfohalobiaceae* fam. nov. *In* Brenner, Krieg, Staley and Garrity (Editors), Bergey's Manual of Systematic Bacteriology, 2nd Edition, Volume 2 (The *Proteobacteria*), Part C (The *Alpha-, Beta-, Delta-*, and *Epsilonproteobacteria*), Springer, New York, pp. 948–949.

Kuever, J., F.A. Rainey and F. Widdel. 2005h. Genus III. *Desulfothermus* gen. nov. *In* Brenner, Krieg, Staley and Garrity (Editors), Bergey's Manual of Systematic Bacteriology, 2nd Edition, Volume 2 (The *Proteobacteria*), Part C (The *Alpha-, Beta-, Delta-*, and *Epsilonproteobacteria*), Springer, New York, pp. 955–956.

Kuever, J., F.A. Rainey and F. Widdel. 2005i. Family IV. *Desulfonatronumaceae* fam. nov. *In* Brenner, Krieg, Staley and Garrity (Editors), Bergey's Manual of Systematic Bacteriology, 2nd Edition, Volume 2 (The *Proteobacteria*), Part C (The *Alpha-, Beta-, Delta-*, and *Epsilonproteobacteria*), Springer, New York, p. 956.

Kuever, J., F.A. Rainey and F. Widdel. 2005j. Order III. *Desulfobacterales* ord. nov. *In* Brenner, Krieg, Staley and Garrity (Editors), Bergey's Manual of Systematic Bacteriology, 2nd Edition, Volume 2 (The *Proteobacteria*), Part C (The *Alpha-, Beta-, Delta-*, and *Epsilonproteobacteria*), Springer, New York, p. 959.

Kuever, J., F.A. Rainey and F. Widdel. 2005k. Family I. *Desulfobacteraceae* fam. nov. *In* Brenner, Krieg, Staley and Garrity (Editors), Bergey's Manual of Systematic Bacteriology, 2nd Edition, Volume 2 (The *Proteobacteria*), Part C (The *Alpha-, Beta-, Delta-*, and *Epsilonproteobacteria*), Springer, New York, pp. 959–960.

Kuever, J., F.A. Rainey and F. Widdel. 2005l. Genus II. *Desulfobacterium* Bak and Widdel 1988, 136^{VP} (Effective publication: Bak and Widdel 1986, 175). *In* Brenner, Krieg, Staley and Garrity (Editors), Bergey's Manual of Systematic Bacteriology, 2nd Edition, Volume 2 (The *Proteobacteria*), Part C (The *Alpha-, Beta-, Delta-*, and *Epsilonproteobacteria*), Springer, New York, pp. 965–967.

Kuever, J., F.A. Rainey and F. Widdel. 2005m. Genus X. *Desulfosarcina* Widdel 1981, 382^{VP} (Effective publication: Widdel 1980, 382). *In* Brenner, Krieg, Staley and Garrity (Editors), Bergey's Manual of Systematic Bacteriology, 2nd Edition, Volume 2 (The *Proteobacteria*), Part C (The *Alpha-, Beta-, Delta-*, and *Epsilonproteobacteria*), Springer, New York, pp. 981–984.

Kuever, J., F.A. Rainey and F. Widdel. 2005n. Family II. *Desulfobulbaceae* fam. nov. *In* Brenner, Krieg, Staley and Garrity (Editors), Bergey's Manual of Systematic Bacteriology, 2nd Edition, Volume 2 (The *Proteobacteria*), Part C (The *Alpha-, Beta-, Delta-*, and *Epsilonproteobacteria*), Springer, New York, p. 988.

Kuever, J., F.A. Rainey and F. Widdel. 2005o. Order IV. *Desulfarcales* ord. nov. *In* Brenner, Krieg, Staley and Garrity (Editors), Bergey's Manual of Systematic Bacteriology, 2nd Edition, Volume 2 (The *Proteobacteria*), Part C (The *Alpha-, Beta-, Delta-*, and *Epsilonproteobacteria*), Springer, New York, p. 1003.

Kuever, J., F.A. Rainey and F. Widdel. 2005p. Family I. *Desulfarculaceae* fam. nov. *In* Brenner, Krieg, Staley and Garrity (Editors), Bergey's Manual of Systematic Bacteriology, 2nd Edition, Volume 2 (The *Proteobacteria*), Part C (The *Alpha-, Beta-, Delta-*, and *Epsilonproteobacteria*), Springer, New York, p. 1003.

Kuever, J., F.A. Rainey and F. Widdel. 2005q. Genus I. *Desulfarculus* gen. nov. *In* Brenner, Krieg, Staley and Garrity (Editors), Bergey's Manual of Systematic Bacteriology, 2nd Edition, Volume 2 (The *Proteobacteria*),

Part C (The *Alpha-, Beta-, Delta-*, and *Epsilonproteobacteria*), Springer, New York, pp. 1004–1005.

Kuever, J., F.A. Rainey and F. Widdel. 2005r. Order V. *Desulfuromonales* ord. nov. *In* Brenner, Krieg, Staley and Garrity (Editors), Bergey's Manual of Systematic Bacteriology, 2nd Edition, Volume 2 (The *Proteobacteria*), Part C (The *Alpha-, Beta-, Delta-*, and *Epsilonproteobacteria*), Springer, New York, pp. 1005–1006.

Kuever, J., F.A. Rainey and F. Widdel. 2005s. Family I. *Desulfuromonaceae* fam. nov. *In* Brenner, Krieg, Staley and Garrity (Editors), Bergey's Manual of Systematic Bacteriology, 2nd Edition, Volume 2 (The *Proteobacteria*), Part C (The *Alpha-, Beta-, Delta-*, and *Epsilonproteobacteria*), Springer, New York, p. 1006.

Kuever, J., F.A. Rainey and F. Widdel. 2005t. Order VI. *Syntrophobacterales* ord. nov. *In* Brenner, Krieg, Staley and Garrity (Editors), Bergey's Manual of Systematic Bacteriology, 2nd Edition, Volume 2 (The *Proteobacteria*), Part C (The *Alpha-, Beta-, Delta-*, and *Epsilonproteobacteria*), Springer, New York, p. 1021.

Kuever, J., F.A. Rainey and F. Widdel. 2005u. Family I. *Syntrophobacteraceae* fam. nov. *In* Brenner, Krieg, Staley and Garrity (Editors), Bergey's Manual of Systematic Bacteriology, 2nd Edition, Volume 2 (The *Proteobacteria*), Part C (The *Alpha-, Beta-, Delta-*, and *Epsilonproteobacteria*), Springer, New York, p. 1021.

Kuever, J., F.A. Rainey and F. Widdel. 2005v. Family II. *Syntrophaceae* fam. nov. *In* Brenner, Krieg, Staley and Garrity (Editors), Bergey's Manual of Systematic Bacteriology, 2nd Edition, Volume 2 (The *Proteobacteria*), Part C (The *Alpha-, Beta-, Delta-*, and *Epsilonproteobacteria*), Springer, New York, p. 1033.

Kuhnert, P., B. Korczak, E. Falsen, R. Straub, A. Hoops, P. Boerlin, J. Frey and R. Mutters. 2004. *Nicoletella semolina* gen. nov., sp. nov., a new member of *Pasteurellaceae* isolated from horses with airway disease. J. Clin. Microbiol. *42*: 5542–5548.

Kuhnert, P., B.M. Korczak, H. Christensen and M. Bisgaard. 2007. Emended description of *Actinobacillus capsulatus* Arseculeratne 1962, 38^{AL}. Int. J. Syst. Evol. Microbiol. *57*: 625–632.

Kuisiene, N., J. Raugalas and D. Chitavichius. 2004. *Geobacillus lituanicus* sp. nov. Int. J. Syst. Evol. Microbiol. *54*: 1991–1995.

Kulichevskaya, I.S., V.S. Guzev, V.M. Gorlenko, W. Liesack and S.N. Dedysh. 2006. *Rhodoblastus sphagnicola* sp. nov., a novel acidophilic purple non-sulfur bacterium from *Sphagnum* peat bog. Int. J. Syst. Evol. Microbiol. *56*: 1397–1402.

Kulichevskaya, I.S., A.O. Ivanova, S.E. Belova, O.I. Baulina, P.L. Bodelier, W.I. Rijpstra, J.S. Sinninghe Damsté, G.A. Zavarzin and S.N. Dedysh. 2007. *Schlesneria paludicola* gen. nov., sp. nov., the first acidophilic member of the order *Planctomycetales*, from *Sphagnum*-dominated boreal wetlands. Int. J. Syst. Evol. Microbiol. *57*: 2680–2687.

Kumar, N.R. and S. Nair. 2007. *Vibrio rhizosphaerae* sp. nov., a red-pigmented bacterium that antagonizes phytopathogenic bacteria. Int. J. Syst. Evol. Microbiol. *57*: 2241–2246.

Kurahashi, M. and A. Yokota. 2004. *Agarivorans albus* gen. nov., sp. nov., a γ-proteobacterium isolated from marine animals. Int. J. Syst. Evol. Microbiol. *54*: 693–697.

Kurahashi, M. and A. Yokota. 2007. *Endozoicomonas elysicola* gen. nov., sp. nov., a γ-proteobacterium isolated from the sea slug *Elysia ornata*. Syst. Appl. Microbiol. *30*: 202–206.

Kurata, A., M. Miyazaki, T. Kobayashi, Y. Nogi and K. Horikoshi. 2007. *Alkalimonas collagenimarina* sp. nov., a psychrotolerant, obligate alkaliphile isolated from deep-sea sediment. Int. J. Syst. Evol. Microbiol. *57*: 1549–1553.

Kuroki, H., H Miyamoto, K. Fukuda, H. Iihara, Y. Kawamura, M. Ogawa, Y. Wang, T. Ezaki and H. Taniguchi. 2007. *Legionella impletisoli* sp. nov. and *Legionella yabuuchiae* sp. nov., isolated from soils contaminated with industrial wastes in Japan. Syst. Appl. Microbiol. *30*: 273–279.

Kuwabara, T., M. Minaba, Y. Iwayama, I. Inouye, M. Nakashima, K. Marumo, A. Maruyama, A. Sugai, T. Itoh, J. Ishibashi, T. Urabe and M. Kamekura. 2005. *Thermococcus coalescens* sp. nov., a cell-fusing hyperthermophilic archaeon from Suiyo Seamount. Int. J. Syst. Evol. Microbiol. 55: 2507–2514.

Kuwabara, T., M. Minaba, N. Ogi and M. Kamekura. 2007. *Thermococcus celericrescens* sp. nov., a fast-growing and cell-fusing hyperthermophilic archaeon from a deep-sea hydrothermal vent. Int. J. Syst. Evol. Microbiol. 57: 437–443.

Kuykendall, L.D. 2005. Order VI. *Rhizobiales* ord. nov. *In* Brenner, Krieg, Staley and Garrity (Editors), Bergey's Manual of Systematic Bacteriology, 2nd Edition, Volume 2 (The *Proteobacteria*), Part C (The *Alpha-, Beta-, Delta-,* and *Epsilonproteobacteria*), Springer, New York, p. 324.

Kwon, K.K., H.S. Lee, S.H. Yang and S.J. Kim. 2005. *Kordiimonas gwangyangensis* gen. nov., sp. nov., a marine bacterium isolated from marine sediments that forms a distinct phyletic lineage (*Kordiimonadales* ord. nov.) in the 'Alphaproteobacteria'. Int. J. Syst. Evol. Microbiol. 55: 2033–2037.

Kwon, K.K., H.S. Lee, H.B. Jung, J.H. Kang and S.J. Kim. 2006a. *Yeosuana aromativorans* gen. nov., sp. nov., a mesophilic marine bacterium belonging to the family *Flavobacteriaceae*, isolated from estuarine sediment of the South Sea, Korea. Int. J. Syst. Evol. Microbiol. 56: 727–732.

Kwon, K.K., S.J. Lee, J.H. Park, T.Y. Ahn and H.K. Lee. 2006b. *Psychroserpens mesophilus* sp. nov., a mesophilic marine bacterium belonging to the family *Flavobacteriaceae* isolated from a young biofilm. Int. J. Syst. Evol. Microbiol. 56: 1055–1058.

Kwon, K.K., Y.K. Lee and H.K. Lee. 2006c. *Costertonia aggregata* gen. nov., sp. nov., a mesophilic marine bacterium of the family *Flavobacteriaceae*, isolated from a mature biofilm. Int. J. Syst. Evol. Microbiol. 56: 1349–1353.

Kwon, K.K., J.H. Woo, S.H. Yang, J.H. Kang, S.G. Kang, S.J. Kim, T. Sato and C. Kato. 2007. *Altererythrobacter epoxidivorans* gen. nov., sp. nov., an epoxide hydrolase-active, mesophilic marine bacterium isolated from cold-seep sediment, and reclassification of *Erythrobacter luteolus* Yoon *et al.* 2005 as *Altererythrobacter luteolus* comb. nov. Int. J. Syst. Evol. Microbiol. 57: 2207–2211.

Kwon, S.W., B.Y. Kim, H.Y. Weon, Y.K. Baek, B.S. Koo and S.J. Go. 2006. *Idiomarina homiensis* sp. nov., isolated from seashore sand in Korea. Int. J. Syst. Evol. Microbiol. 56: 2229–2233.

Kwon, S.W., B.Y. Kim, K.H. Lee, K.Y. Jang, S.J. Seok, J.S. Kwon, W.G. Kim and H.Y. Weon. 2007a. *Pedobacter suwonensis* sp. nov., isolated from the rhizosphere of Chinese cabbage (*Brassica campestris*). Int. J. Syst. Evol. Microbiol. 57: 480–484.

Kwon, S.W., B.Y. Kim, J. Song, H.Y. Weon, P. Schumann, B.J. Tindall, E. Stackebrandt and D. Fritze. 2007b. *Sporosarcina koreensis* sp. nov. and *Sporosarcina soli* sp. nov., isolated from soil in Korea. Int. J. Syst. Evol. Microbiol. 57: 1694–1698.

Kwon, S.W., B.Y. Kim, H.Y. Weon, Y.K. Baek and S.J. Go. 2007c. *Arenimonas donghaensis* gen. nov., sp. nov., isolated from seashore sand. Int. J. Syst. Evol. Microbiol. 57: 954–958.

Kwon, S.W., S.Y. Lee, B.Y. Kim, H.Y. Weon, J.B. Kim, S.J. Go and G.B. Lee. 2007d. *Bacillus niabensis* sp. nov., isolated from cotton-waste composts for mushroom cultivation. Int. J. Syst. Evol. Microbiol. 57: 1909–1913.

La Duc, M.T., M. Satomi and K. Venkateswaran. 2004. *Bacillus odysseyi* sp. nov., a round-spore-forming bacillus isolated from the Mars Odyssey spacecraft. Int. J. Syst. Evol. Microbiol. 54: 195–201.

La Scola, B., R.J. Birtles, G. Greub, T.J. Harrison, R.M. Ratcliff and D. Raoult. 2004. *Legionella drancourtii* sp. nov., a strictly intracellular amoebal pathogen. Int. J. Syst. Evol. Microbiol. 54: 699–703.

La, H.J., W.T. Im, L.N. Ten, M.S. Kang, D.Y. Shin and S.T. Lee. 2005. *Paracoccus koreensis* sp. nov., isolated from anaerobic granules in an upflow anaerobic sludge blanket (UASB) reactor. Int. J. Syst. Evol. Microbiol. 55: 1657–1660.

Labbe, N., S. Parent and R. Villemur. 2004. *Nitratireductor aquibiodomus* gen. nov., sp. nov., a novel α-proteobacterium from the marine denitrification system of the Montreal Biodome (Canada). Int. J. Syst. Evol. Microbiol. 54: 269–273.

Labeda, D.P. and R.M. Kroppenstedt. 2004. Emended description of the genus *Glycomyces* and description of *Glycomyces algeriensis* sp. nov., *Glycomyces arizonensis* sp. nov. and *Glycomyces lechevalierae* sp. nov. Int. J. Syst. Evol. Microbiol. 54: 2343–2346.

Labeda, D.P. and R.M. Kroppenstedt. 2005. *Stackebrandtia nassauensis* gen. nov., sp. nov. and emended description of the family *Glycomycetaceae*. Int. J. Syst. Evol. Microbiol. 55: 1687–1691.

Labeda, D.P. and R.M. Kroppenstedt. 2006. *Goodfellowia* gen. nov., a new genus of the *Pseudonocardineae* related to *Actinoalloteichus*, containing *Goodfellowia coeruleoviolacea* gen. nov., comb. nov. Int. J. Syst. Evol. Microbiol. 56: 1203–1207.

Labeda, D.P. and R.M. Kroppenstedt. 2007. Proposal of *Umezawaea* gen. nov., a new genus of the *Actinosynnemataceae* related to *Saccharothrix*, and transfer of *Saccharothrix tangerinus* Kinoshita *et al.* 2000 as *Umezawaea tangerina* gen. nov., comb. nov. Int. J. Syst. Evol. Microbiol. 57: 2758–2761.

Labeda, D.P., J.M. Donahue, S.F. Sells and R.M. Kroppenstedt. 2007. *Lentzea kentuckyensis* sp. nov., of equine origin. Int. J. Syst. Evol. Microbiol. 57: 1780–1783.

Labrenz, M., P.A. Lawson, B.J. Tindall, M.D. Collins and P. Hirsch. 2005. *Roseisalinus antarcticus* gen. nov., sp. nov., a novel aerobic bacteriochlorophyll α-producing α-proteobacterium isolated from hypersaline Ekho Lake, Antarctica. Int. J. Syst. Evol. Microbiol. 55: 41–47.

Lai, M.C., C.C. Lin, P.H. Yu, Y.F. Huang and S.C. Chen. 2004. *Methanocalculus chunghsingensis* sp. nov., isolated from an estuary and a marine fishpond in Taiwan. Int. J. Syst. Evol. Microbiol. 54: 183–189.

Lai, W.A., P. Kämpfer, A.B. Arun, F.T. Shen, B. Huber, P.D. Rekha and C.C. Young. 2006. *Deinococcus ficus* sp. nov., isolated from the rhizosphere of *Ficus religiosa* L. Int. J. Syst. Evol. Microbiol. 56: 787–791.

Lan, P.T., M. Sakamoto, S. Sakata and Y. Benno. 2006. *Bacteroides barnesiae* sp. nov., *Bacteroides salanitronis* sp. nov. and *Bacteroides gallinarum* sp. nov., isolated from chicken caecum. Int. J. Syst. Evol. Microbiol. 56: 2853–2859.

Lang, E., B. Griese, C. Spröer, P. Schumann, M. Steffen and S. Verbarg. 2007a. Characterization of 'Pseudomonas azelaica' DSM 9128, leading to emended descriptions of *Pseudomonas citronellolis* Seubert 1960 (Approved Lists 1980) and *Pseudomonas nitroreducens* Iizuka and Komagata 1964 (Approved Lists 1980), including *Pseudomonas multiresinivorans* as its later heterotypic synonym. Int. J. Syst. Evol. Microbiol. 57: 878–882.

Lang, E., J. Swiderski, E. Stackebrandt, P. Schumann, C. Spröer and N. Sahin. 2007b. *Herminiimonas saxobsidens* sp. nov., isolated from a lichen-colonized rock. Int. J. Syst. Evol. Microbiol. 57: 2618–2622.

Lanoot, B., M. Vancanneyt, P. Dawyndt, M. Cnockaert, J. Zhang, Y. Huang, Z. Liu and J. Swings. 2004. BOX-PCR fingerprinting as a powerful tool to reveal synonymous names in the genus *Streptomyces*. Emended descriptions are proposed for the species *Streptomyces cinereorectus, S. fradiae, S. tricolor, S. colombiensis, S. filamentosus, S. vinaceus* and *S. phaeopurpureus*. Syst. Appl. Microbiol. 27: 84–92.

Lanoot, B., M. Vancanneyt, A. Van Schoor, Z. Liu and J. Swings. 2005. Reclassification of *Streptomyces nigrifaciens* as a later synonym of *Streptomyces flavovirens*; *Streptomyces citreofluorescens, Streptomyces chrysomallus* subsp. *chrysomallus* and *Streptomyces fluorescens* as later synonyms of *Streptomyces anulatus*; *Streptomyces chibaensis* as a later synonym of *Streptomyces corchorusii*; *Streptomyces flaviscleroticus* as a later synonym of *Streptomyces minutiscleroticus*; and *Streptomyces lipmanii, Streptomyces griseus* subsp. *alpha, Streptomyces griseus* subsp. *cretosus* and *Streptomyces willmorei* as later synonyms of *Streptomyces microflavus*. Int. J. Syst. Evol. Microbiol. 55: 729–731.

Lau, H.T., J. Faryna and E.W. Triplett. 2006. *Aquitalea magnusonii* gen. nov., sp. nov., a novel Gram-negative bacterium isolated from a humic lake. Int. J. Syst. Evol. Microbiol. 56: 867–871.

Lau, K.W., C.Y. Ng, J. Ren, S.C. Lau, P.Y. Qian, P.K. Wong, T.C. Lau and M. Wu. 2005. *Owenweeksia hongkongensis* gen. nov., sp. nov., a novel marine bacterium of the phylum 'Bacteroidetes'. Int. J. Syst. Evol. Microbiol. 55: 1051–1057.

Lau, K.W., J. Ren, N.L. Wai, S.C. Lau, P.Y. Qian, P.K. Wong and M. Wu. 2006a. *Marinomonas ostreistagni* sp. nov., isolated from a pearl-oyster culture pond in Sanya, Hainan Province, China. Int. J. Syst. Evol. Microbiol. *56*: 2271–2275.

Lau, K.W., J. Ren, N.L. Wai, P.Y. Qian, P.K. Wong and M. Wu. 2006b. *Lishizhenia caseinilytica* gen. nov., sp. nov., a marine bacterium of the phylum *Bacteroidetes*. Int. J. Syst. Evol. Microbiol. *56*: 2317–2322.

Lau, K.W., J. Ren, M.C. Fung, P.C. Woo, K.Y. Yuen, K.K. Chan, P.Y. Qian, P.K. Wong and M. Wu. 2007. *Fangia hongkongensis* gen. nov., sp. nov., a novel gammaproteobacterium of the order *Thiotrichales* isolated from coastal seawater of Hong Kong. Int. J. Syst. Evol. Microbiol. *57*: 2665–2669.

Lau, S.C., M.M. Tsoi, X. Li, I. Plakhotnikova, M. Wu, P.K. Wong and P.Y. Qian. 2004. *Loktanella hongkongensis* sp. nov., a novel member of the α-Proteobacteria originating from marine biofilms in Hong Kong waters. Int. J. Syst. Evol. Microbiol. *54*: 2281–2284.

Lau, S.C., M.M. Tsoi, X. Li, S. Dobretsov, Y. Plakhotnikova, P.K. Wong and P.Y. Qian. 2005a. *Pseudoalteromonas spongiae* sp. nov., a novel member of the γ-Proteobacteria isolated from the sponge *Mycale adhaerens* in Hong Kong waters. Int. J. Syst. Evol. Microbiol. *55*: 1593–1596.

Lau, S.C., M.M. Tsoi, X. Li, I. Plakhotnikova, S. Dobretsov, K.W. Lau, M. Wu, P.K. Wong, J.R. Pawlik and P.Y. Qian. 2005b. *Winogradskyella poriferorum* sp. nov., a novel member of the family *Flavobacteriaceae* isolated from a sponge in the Bahamas. Int. J. Syst. Evol. Microbiol. *55*: 1589–1592.

Lau, S.C., M.M. Tsoi, X. Li, I. Plakhotnikova, S. Dobretsov, P.K. Wong, J.R. Pawlik and P.Y. Qian. 2005c. *Nonlabens tegetincola* gen. nov., sp. nov., a novel member of the family *Flavobacteriaceae* isolated from a microbial mat in a subtropical estuary. Int. J. Syst. Evol. Microbiol. *55*: 2279–2283.

Lau, S.C., M.M. Tsoi, X. Li, I. Plakhotnikova, S. Dobretsov, P.K. Wong and P.Y. Qian. 2005d. *Gramella portivictoriae* sp. nov., a novel member of the family *Flavobacteriaceae* isolated from marine sediment. Int. J. Syst. Evol. Microbiol. *55*: 2497–2500.

Lau, S.C., M.M. Tsoi, X. Li, I. Plakhotnikova, S. Dobretsov, M. Wu, P.K. Wong, J.R. Pawlik and P.Y. Qian. 2006a. *Stenothermobacter spongiae* gen. nov., sp. nov., a novel member of the family *Flavobacteriaceae* isolated from a marine sponge in the Bahamas, and emended description of *Nonlabens tegetincola*. Int. J. Syst. Evol. Microbiol. *56*: 181–185.

Lau, S.C., M.M. Tsoi, X. Li, I. Plakhotnikova, S. Dobretsov, M. Wu, P.K. Wong, J.R. Pawlik and P.Y. Qian. 2006b. Description of *Fabibacter halotolerans* gen. nov., sp. nov. and *Roseivirga spongicola* sp. nov., and reclassification of [*Marinicola*] *seohaensis* as *Roseivirga seohaensis* comb. nov. Int. J. Syst. Evol. Microbiol. *56*: 1059–1065.

Lau, S.K.P., P.C.Y. Woo, G.K.S. Woo, A.M.Y. Fung, M.K.M. Wong, K. Chan, D.M.W. Tam and K. Yuen. 2004. *Eggerthella hongkongensis* sp. nov. and *Eggerthella sinensis* sp. nov., two novel *Eggerthella* species, account for half of the cases of *Eggerthella* bacteremia. Diagn. Microbiol. Infect. Dis. *49*: 255–263.

Laurent, F., V. Rodríguez-Nava, L. Noussair, A. Couble, M.H. Nicolas-Chanoine and P. Boiron. 2007. *Nocardia ninae* sp. nov., isolated from a bronchial aspirate. Int. J. Syst. Evol. Microbiol. *57*: 661–665.

Lawson, P.A., G. Foster, E. Falsen, N. Davison and M.D. Collins. 2004a. *Streptococcus halichoeri* sp. nov., isolated from grey seals (*Halichoerus grypus*). Int. J. Syst. Evol. Microbiol. *54*: 1753–1756.

Lawson, P.A., Y. Song, C. Liu, D.R. Molitoris, M.L. Väisänen, M.D. Collins and S.M. Finegold. 2004b. *Anaerotruncus colihominis* gen. nov., sp. nov., from human faeces. Int. J. Syst. Evol. Microbiol. *54*: 413–417.

Lawson, P.A., G. Foster, E. Falsen and M.D. Collins. 2005a. *Streptococcus marimammalium* sp. nov., isolated from seals. Int. J. Syst. Evol. Microbiol. *55*: 271–274.

Lawson, P.A., G. Foster, E. Falsen, S.J. Markopoulos and M.D. Collins. 2005b. *Streptococcus castoreus* sp. nov., isolated from a beaver (*Castor fiber*). Int. J. Syst. Evol. Microbiol. *55*: 843–846.

Lawson, P.A., H.L. Greetham, G.R. Gibson, C. Giffard, E. Falsen and M.D. Collins. 2005c. *Slackia faecicanis* sp. nov., isolated from canine faeces. Int. J. Syst. Evol. Microbiol. *55*: 1243–1246.

Lawson, P.A., H. Malnick, M.D. Collins, J.J. Shah, M.A. Chattaway, R. Bendall and J.W. Hartley. 2005d. Description of *Kingella potus* sp. nov., an organism isolated from a wound caused by an animal bite. J. Clin. Microbiol. *43*: 3526–3529.

Lawson, P.A., M.D. Collins, E. Falsen and G. Foster. 2006. *Catellicoccus marimammalium* gen. nov., sp. nov., a novel Gram-positive, catalase-negative, coccus-shaped bacterium from porpoise and grey seal. Int. J. Syst. Evol. Microbiol. *56*: 429–432.

Lawson, P.A., E. Falsen, M.A. Cotta and T.R. Whitehead. 2007. *Vagococcus elongatus* sp. nov., isolated from a swine-manure storage pit. Int. J. Syst. Evol. Microbiol. *57*: 751–754.

le Roes, M. and P.R. Meyers. 2005. *Streptomyces pharetrae* sp. nov., isolated from soil from the semi-arid Karoo region. Syst. Appl. Microbiol. *28*: 488–493.

le Roes, M. and P.R. Meyers. 2006. *Nocardia gamkensis* sp. nov. Antonie Van Leeuwenhoek. *90*: 291–298.

le Roes, M. and P.R. Meyers. 2007. *Actinomadura rudentiformis* sp. nov., isolated from soil. Int. J. Syst. Evol. Microbiol. *57*: 45–50.

Le Roux, F., A. Goubet, F.L. Thompson, N. Faury, M. Gay, J. Swings and D. Saulnier. 2005. *Vibrio gigantis* sp. nov., isolated from the haemolymph of cultured oysters (*Crassostrea gigas*). Int. J. Syst. Evol. Microbiol. *55*: 2251–2255.

Lechner, U., D. Brodkorb, R. Geyer, G. Hause, C. Härtig, G. Auling, F. Fayolle-Guichard, P. Piveteau, R.H. Müller and T. Rohwerder. 2007. *Aquincola tertiaricarbonis* gen. nov., sp. nov., a tertiary butyl moiety-degrading bacterium. Int. J. Syst. Evol. Microbiol. *57*: 1295–1303.

Lee, C.S., K.K. Kim, Z. Aslam and S.T. Lee. 2007. *Rhodanobacter thiooxydans* sp. nov., isolated from a biofilm on sulfur particles used in an autotrophic denitrification process. Int. J. Syst. Evol. Microbiol. *57*: 1775–1779.

Lee, D.W., C.G. Hyun and S.D. Lee. 2007. *Nocardioides marinisabuli* sp. nov., a novel actinobacterium isolated from beach sand. Int. J. Syst. Evol. Microbiol. *57*: 2960–2963.

Lee, E.M., C.O. Jeon, I. Choi, K.S. Chang and C.J. Kim. 2005. *Silanimonas lenta* gen. nov., sp. nov., a slightly thermophilic and alkaliphilic gammaproteobacterium isolated from a hot spring. Int. J. Syst. Evol. Microbiol. *55*: 385–389.

Lee, F.L., H.P. Kuo, C.J. Tai, A. Yokota and C.C. Lo. 2007. *Paenibacillus taiwanensis* sp. nov., isolated from soil in Taiwan. Int. J. Syst. Evol. Microbiol. *57*: 1351–1354.

Lee, H.G., D.S. An, W.T. Im, Q.M. Liu, J.R. Na, D.H. Cho, C.W. Jin, S.T. Lee and D.C. Yang. 2007. *Chitinophaga ginsengisegetis* sp. nov. and *Chitinophaga ginsengisoli* sp. nov., isolated from soil of a ginseng field in South Korea. Int. J. Syst. Evol. Microbiol. *57*: 1396–1401.

Lee, H.W., H.S. Yu, Q.M. Liu, H.M. Jung, D.S. An, W.T. Im, F.X. Jin and S.T. Lee. 2007. *Kaistia granuli* sp. nov., isolated from anaerobic granules in an upflow anaerobic sludge blanket reactor. Int. J. Syst. Evol. Microbiol. *57*: 2280–2283.

Lee, J.C., C.O. Jeon, J.M. Lim, S.M. Lee, J.M. Lee, S.M. Song, D.J. Park, W.J. Li and C.J. Kim. 2005. *Halomonas taeanensis* sp. nov., a novel moderately halophilic bacterium isolated from a solar saltern in Korea. Int. J. Syst. Evol. Microbiol. *55*: 2027–2032.

Lee, J.C., J.M. Lim, D.J. Park, C.O. Jeon, W.J. Li and C.J. Kim. 2006. *Bacillus seohaeanensis* sp. nov., a halotolerant bacterium that contains L-lysine in its cell wall. Int. J. Syst. Evol. Microbiol. *56*: 1893–1898.

Lee, J.H., Y.S. Kim, T.J. Choi, W.J. Lee and Y.T. Kim. 2004. *Paracoccus haeundaensis* sp. nov., a Gram-negative, halophilic, astaxanthin-producing bacterium. Int. J. Syst. Evol. Microbiol. *54*: 1699–1702.

Lee, J.S., Y.R. Pyun and K.S. Bae. 2004. Transfer of *Bacillus ehimensis* and *Bacillus chitinolyticus* to the genus *Paenibacillus* with emended descriptions of *Paenibacillus ehimensis* comb. nov. and *Paenibacillus chitinolyticus* comb. nov. Int. J. Syst. Evol. Microbiol. *54*: 929–933.

Lee, J.S., K.C. Lee and Y.H. Park. 2006a. *Microbacterium koreense* sp. nov., from sea water in the South Sea of Korea. Int. J. Syst. Evol. Microbiol. *56*: 423–427.

Lee, J.S., J.M. Lim, K.C. Lee, J.C. Lee, Y.H. Park and C.J. Kim. 2006b. *Virgibacillus koreensis* sp. nov., a novel bacterium from a salt field, and transfer of *Virgibacillus picturae* to the genus *Oceanobacillus* as *Oceanobacillus picturae* comb. nov. with emended descriptions. Int. J. Syst. Evol. Microbiol. *56*: 251–257.

Lee, J.W., W.T. Im, M.K. Kim and D.C. Yang. 2006. *Lysobacter koreensis* sp. nov., isolated from a ginseng field. Int. J. Syst. Evol. Microbiol. *56*: 231–235.

Lee, J.Y., J.Y. Lee, H.W. Jung and B.K. Hwang. 2005. *Streptomyces koyangensis* sp. nov., a novel actinomycete that produces 4-phenyl-3-butenoic acid. Int. J. Syst. Evol. Microbiol. *55*: 257–262.

Lee, K., Y.J. Choo, S.J. Giovannoni and J.C. Cho. 2007. *Maritimibacter alkaliphilus* gen. nov., sp. nov., a genome-sequenced marine bacterium of the *Roseobacter* clade in the order *Rhodobacterales*. Int. J. Syst. Evol. Microbiol. *57*: 1653–1658.

Lee, K., Y.J. Choo, S.J. Giovannoni and J.C. Cho. 2007a. *Ruegeria pelagia* sp. nov., isolated from the Sargasso Sea, Atlantic Ocean. Int. J. Syst. Evol. Microbiol. *57*: 1815–1818.

Lee, K., H.K. Lee, T.H. Choi and J.C. Cho. 2007b. *Robiginitomaculum antarcticum* gen. nov., sp. nov., a member of the family *Hyphomonadaceae*, from Antarctic seawater. Int. J. Syst. Evol. Microbiol. *57*: 2595–2599.

Lee, K., H.K. Lee, T.H. Choi and J.C. Cho. 2007c. *Sejongia marina* sp. nov., isolated from Antarctic seawater. Int. J. Syst. Evol. Microbiol. *57*: 2917–2921.

Lee, K.B., C.T. Liu, Y. Anzai, H. Kim, T. Aono and H. Oyaizu. 2005. The hierarchical system of the '*Alphaproteobacteria*': description of *Hyphomonadaceae* fam. nov., *Xanthobacteraceae* fam. nov. and *Erythrobacteraceae* fam. nov. Int. J. Syst. Evol. Microbiol. *55*: 1907–1919.

Lee, M., M.K. Kim, M. Vancanneyt, J. Swings, S.H. Kim, M.S. Kang and S.T. Lee. 2005. *Tetragenococcus koreensis* sp. nov., a novel rhamnolipid-producing bacterium. Int. J. Syst. Evol. Microbiol. *55*: 1409–1413.

Lee, M., L.N. Ten, S.-H. Baek, W.-T. Im, Z. Aslam and S.-T. Lee. 2007. *Paenibacillus ginsengisoli* sp. nov., a novel bacterium isolated from soil of a ginseng field in Pocheon province, South Korea. Antonie van Leeuwenhoek. *91*: 127–135.

Lee, N., C.M. Cellamare, C. Bastianutti, R. Rosselló-Mora, P. Kämpfer, W. Ludwig, K.H. Schleifer and L. Stante. 2004. Emended description of the species *Lampropedia hyalina*. Int. J. Syst. Evol. Microbiol. *54*: 1709–1715.

Lee, O.O., S.C. Lau, M.M. Tsoi, X. Li, I. Plakhotnikova, S. Dobretsov, M.C. Wu, P.K. Wong and P.Y. Qian. 2006a. *Gillisia myxillae* sp. nov., a novel member of the family *Flavobacteriaceae*, isolated from the marine sponge *Myxilla incrustans*. Int. J. Syst. Evol. Microbiol. *56*: 1795–1799.

Lee, O.O., S.C. Lau, M.M. Tsoi, X. Li, I. Plakhotnikova, S. Dobretsov, M.C. Wu, P.K. Wong, M. Weinbauer and P.Y. Qian. 2006b. *Shewanella irciniae* sp. nov., a novel member of the family *Shewanellaceae*, isolated from the marine sponge Ircinia dendroides in the Bay of Villefranche, Mediterranean Sea. Int. J. Syst. Evol. Microbiol. *56*: 2871–2877.

Lee, O.O., M.M. Tsoi, X. Li, P.K. Wong and P.Y. Qian. 2007. *Thalassococcus halodurans* gen. nov., sp. nov., a novel halotolerant member of the *Roseobacter* clade isolated from the marine sponge *Halichondria panicea* at Friday Harbor, USA. Int. J. Syst. Evol. Microbiol. *57*: 1919–1924.

Lee, S.B., S.E. Strand, H.D. Stensel and R.P. Herwig. 2004. *Pseudonocardia chloroethenivorans* sp. nov., a chloroethene-degrading actinomycete. Int. J. Syst. Evol. Microbiol. *54*: 131–139.

Lee, S.D. 2006a. *Amycolatopsis jejuensis* sp. nov. and *Amycolatopsis halotolerans* sp. nov., novel actinomycetes isolated from a natural cave. Int. J. Syst. Evol. Microbiol. *56*: 549–553.

Lee, S.D. 2006b. *Nocardia jejuensis* sp. nov., a novel actinomycete isolated from a natural cave on Jeju Island, Republic of Korea. Int. J. Syst. Evol. Microbiol. *56*: 559–562.

Lee, S.D. 2006c. *Actinocorallia cavernae* sp. nov., isolated from a natural cave in Jeju, Korea. Int. J. Syst. Evol. Microbiol. *56*: 1085–1088.

Lee, S.D. 2006d. *Kineococcus marinus* sp. nov., isolated from marine sediment of the coast of Jeju, Korea. Int. J. Syst. Evol. Microbiol. *56*: 1279–1283.

Lee, S.D. 2006e. *Brevibacterium samyangense* sp. nov., an actinomycete isolated from a beach sediment. Int. J. Syst. Evol. Microbiol. *56*: 1889–1892.

Lee, S.D. 2006f. *Phycicoccus jejuensis* gen. nov., sp. nov., an actinomycete isolated from seaweed. Int. J. Syst. Evol. Microbiol. *56*: 2369–2373.

Lee, S.D. 2006g. *Blastococcus jejuensis* sp. nov., an actinomycete from beach sediment, and emended description of the genus *Blastococcus* Ahrens and Moll 1970. Int. J. Syst. Evol. Microbiol. *56*: 2391–2396.

Lee, S.D. 2007a. *Tamlana crocina* gen. nov., sp. nov., a marine bacterium of the family *Flavobacteriaceae*, isolated from beach sediment in Korea. Int. J. Syst. Evol. Microbiol. *57*: 764–769.

Lee, S.D. 2007b. *Marmoricola aequoreus* sp. nov., a novel actinobacterium isolated from marine sediment. Int. J. Syst. Evol. Microbiol. *57*: 1391–1395.

Lee, S.D. 2007c. *Devosia subaequoris* sp. nov., isolated from beach sediment. Int. J. Syst. Evol. Microbiol. *57*: 2212–2215.

Lee, S.D. 2007d. *Labedella gwakjiensis* gen. nov., sp. nov., a novel actinomycete of the family *Microbacteriaceae*. Int. J. Syst. Evol. Microbiol. *57*: 2498–2502.

Lee, S.D. 2007e. *Lewinella agarilytica* sp. nov., a novel marine bacterium of the phylum *Bacteroidetes*, isolated from beach sediment. Int. J. Syst. Evol. Microbiol. *57*: 2814–2818.

Lee, S.D. 2007f. *Nocardioides furvisabuli* sp. nov., isolated from black sand. Int. J. Syst. Evol. Microbiol. *57*: 35–39.

Lee, S.D. and H.S. Jeong. 2006. *Actinomadura hallensis* sp. nov., a novel actinomycete isolated from Mt. Halla in Korea. Int. J. Syst. Evol. Microbiol. *56*: 259–264.

Lee, S.D. and S.J. Kim. 2007. *Aeromicrobium tamlense* sp. nov., isolated from dried seaweed. Int. J. Syst. Evol. Microbiol. *57*: 337–341.

Lee, S.D. and D.W. Lee. 2007. *Lapillicoccus jejuensis* gen. nov., sp. nov., a novel actinobacterium of the family *Intrasporangiaceae*, isolated from stone. Int. J. Syst. Evol. Microbiol. *57*: 2794–2798.

Lee, S.D., L.L. Kinkel and D.A. Samac. 2006. *Amycolatopsis minnesotensis* sp. nov., isolated from a prairie soil. Int. J. Syst. Evol. Microbiol. *56*: 265–269.

Lee, Y.J., I.D. Wagner, M.E. Brice, V.V. Kevbrin, G.L. Mills, C.S. Romanek and J. Wiegel. 2005. *Thermosediminibacter oceani* gen. nov., sp. nov. and *Thermosediminibacter litoriperuensis* sp. nov., new anaerobic thermophilic bacteria isolated from Peru Margin. Extremophiles *9*: 375–383.

Lee, Y.J., C.S. Romanek, G.L. Mills, R.C. Davis, W.B. Whitman and J. Wiegel. 2006. *Gracilibacter thermotolerans* gen. nov., sp. nov., an anaerobic, thermotolerant bacterium from a constructed wetland receiving acid sulfate water. Int. J. Syst. Evol. Microbiol. *56*: 2089–2093.

Lee, Y.J., M. Dashti, A. Prange, F.A. Rainey, M. Rohde, W.B. Whitman and J. Wiegel. 2007a. *Thermoanaerobacter sulfurigignens* sp. nov., an anaerobic thermophilic bacterium that reduces 1 M thiosulfate to elemental sulfur and tolerates 90 mM sulfite. Int. J. Syst. Evol. Microbiol. *57*: 1429–1434.

Lee, Y.J., C.S. Romanek and J. Wiegel. 2007b. *Clostridium aciditolerans* sp. nov., an acid-tolerant spore-forming anaerobic bacterium from constructed wetland sediment. Int. J. Syst. Evol. Microbiol. *57*: 311–315.

Lehnen, A., H.J. Busse, K. Frolich, M. Krasinska, P. Kämpfer and S. Speck. 2006. *Arcanobacterium bialowiezense* sp. nov. and *Arcanobacterium bonasi* sp. nov., isolated from the prepuce of European bison bulls (*Bison bonasus*) suffering from balanoposthitis, and emended description of the genus *Arcanobacterium* Collins *et al.* 1983. Int. J. Syst. Evol. Microbiol. *56*: 861–866.

Leisner, J.J., M. Vancanneyt, R. Van der Meulen, K. Lefebvre, K. Engelbeen, B. Hoste, B.G. Laursen, L. Bay, G. Rusul, L. De Vuyst and J. Swings. 2005. *Leuconostoc durionis* sp. nov., a heterofermenter with no detectable gas production from glucose. Int. J. Syst. Evol. Microbiol. *55*: 1267–1270.

Lengyel, K., E. Lang, A. Fodor, E. Szállás, P. Schumann and E. Stackebrandt. 2005. Description of four novel species of *Xenorhabdus*, family *Enterobacteriaceae*: *Xenorhabdus budapestensis* sp. nov., *Xenorhabdus ehlersii* sp. nov., *Xenorhabdus innexi* sp. nov., and *Xenorhabdus szentir-maii* sp. nov. Syst. Appl. Microbiol. *28*: 115–122.

Levett, P.N., R.E. Morey, R. Galloway, A.G. Steigerwalt and W.A. Ellis. 2005. Reclassification of *Leptospira parva* Hovind-Hougen *et al.* 1982 as *Turneriella parva* gen. nov., comb. nov. Int. J. Syst. Evol. Microbiol. *55*: 1497–1499.

Levett, P.N., R.E. Morey, R.L. Galloway and A.G. Steigerwalt. 2006. *Leptospira broomii* sp. nov., isolated from humans with leptospirosis. Int. J. Syst. Evol. Microbiol. *56*: 671–673.

L'Haridon, S., M.L. Miroshnichenko, N.A. Kostrikina, B.J. Tindall, S. Spring, P. Schumann, E. Stackebrandt, E.A. Bonch-Osmolovskaya and C. Jeanthon. 2006. *Vulcanibacillus modesticaldus* gen. nov., sp. nov., a strictly anaerobic, nitrate-reducing bacterium from deep-sea hydrothermal vents. Int. J. Syst. Evol. Microbiol. *56*: 1047–1053.

L'Haridon, S., A.L. Reysenbach, B.J. Tindall, P. Schönheit, A. Banta, U. Johnsen, P. Schumann, A. Gambacorta, E. Stackebrandt and C. Jeanthon. 2006. *Desulfurobacterium atlanticum* sp. nov., *Desulfurobacterium pacificum* sp. nov. and *Thermovibrio guaymasensis* sp. nov., three thermophilic members of the *Desulfurobacteriaceae* fam. nov., a deep branching lineage within the *Bacteria*. Int. J. Syst. Evol. Microbiol. *56*: 2843–2852.

Li, B., K. Furihata, L.X. Ding and A. Yokota. 2007a. *Rhodococcus kyotonensis* sp. nov., a novel actinomycete isolated from soil. Int. J. Syst. Evol. Microbiol. *57*: 1956–1959.

Li, B., C.H. Xie and A. Yokota. 2007b. *Nocardioides exalbidus* sp. nov., a novel actinomycete isolated from lichen in Izu-Oshima Island, Japan. Actinomycetologica *21*: 22–26.

Li, W.J., H.H. Chen, P. Xu, Y.Q. Zhang, P. Schumann, S.K. Tang, L.H. Xu and C.L. Jiang. 2004a. *Yania halotolerans* gen. nov., sp. nov., a novel member of the suborder *Micrococcineae* from saline soil in China. Int. J. Syst. Evol. Microbiol. *54*: 525–531.

Li, W.J., H.H. Chen, Y.Q. Zhang, P. Schumann, E. Stackebrandt, L.H. Xu and C.L. Jiang. 2004b. *Nesterenkonia halotolerans* sp. nov. and *Nesterenkonia xinjiangensis* sp. nov., actinobacteria from saline soils in the west of China. Int. J. Syst. Evol. Microbiol. *54*: 837–841.

Li, W.J., Y. Jiang, R.M. Kroppenstedt, L.H. Xu and C.L. Jiang. 2004c. *Nocardia alba* sp.nov., a novel actinomycete strain isolated from soil in China. Syst. Appl. Microbiol. *27*: 308–312.

Li, W.J., D.J. Park, S.K. Tang, D. Wang, J.C. Lee, L.H. Xu, C.J. Kim and C.L. Jiang. 2004d. *Nocardiopsis salina* sp. nov., a novel halophilic actinomycete isolated from saline soil in China. Int. J. Syst. Evol. Microbiol. *54*: 1805–1809.

Li, W.J., D. Wang, Y.Q. Zhang, P. Schumann, E. Stackebrandt, L.H. Xu and C.L. Jiang. 2004e. *Kribbella antibiotica* sp. nov., a novel nocardioform actinomycete strain isolated from soil in Yunnan, China. Syst. Appl. Microbiol. *27*: 160–165.

Li, W.J., Y.Q. Zhang, D.J. Park, C.T. Li, L.H. Xu, C.J. Kim and C.L. Jiang. 2004f. *Duganella violaceinigra* sp. nov., a novel mesophilic bacterium isolated from forest soil. Int. J. Syst. Evol. Microbiol. *54*: 1811–1814.

Li, W.J., H.H. Chen, C.J. Kim, D.J. Park, S.K. Tang, J.C. Lee, L.H. Xu and C.L. Jiang. 2005a. *Microbacterium halotolerans* sp. nov., isolated from a saline soil in the west of China. Int. J. Syst. Evol. Microbiol. *55*: 67–70.

Li, W.J., H.H. Chen, C.J. Kim, Y.Q. Zhang, D.J. Park, J.C. Lee, L.H. Xu and C.L. Jiang. 2005b. *Nesterenkonia sandarakina* sp. nov. and *Nesterenkonia lutea* sp. nov., novel actinobacteria, and emended description of the genus *Nesterenkonia*. Int. J. Syst. Evol. Microbiol. *55*: 463–466.

Li, W.J., H.H. Chen, Y.Q. Zhang, C.J. Kim, D.J. Park, J.C. Lee, L.H. Xu and C.L. Jiang. 2005c. *Citricoccus alkalitolerans* sp. nov., a novel actinobacterium isolated from a desert soil in Egypt. Int. J. Syst. Evol. Microbiol. *55*: 87–90.

Li, W.J., P. Schumann, Y.Q. Zhang, G.Z. Chen, X.P. Tian, L.H. Xu, E. Stackebrandt and C.L. Jiang. 2005d. *Marinococcus halotolerans* sp. nov., isolated from Qinghai, north-west China. Int. J. Syst. Evol. Microbiol. *55*: 1801–1804.

Li, W.J., P. Schumann, Y.Q. Zhang, P. Xu, G.Z. Chen, L.H. Xu, E. Stackebrandt and C.L. Jiang. 2005e. Proposal of *Yaniaceae* fam. nov. and

Yania flava sp. nov. and emended description of the genus *Yania*. Int. J. Syst. Evol. Microbiol. *55*: 1933–1938.

Li, W.J., Y.G. Zhang, Y.Q. Zhang, S.K. Tang, P. Xu, L.H. Xu and C.L. Jiang. 2005f. *Streptomyces sodiiphilus* sp. nov., a novel alkaliphilic actinomycete. Int. J. Syst. Evol. Microbiol. *55*: 1329–1333.

Li, W.J., R.M. Kroppenstedt, D. Wang, S.K. Tang, J.C. Lee, D.J. Park, C.J. Kim, L.H. Xu and C.L. Jiang. 2006a. Five novel species of the genus *Nocardiopsis* isolated from hypersaline soils and emended description of *Nocardiopsis salina* Li *et al.* 2004. Int. J. Syst. Evol. Microbiol. *56*: 1089–1096.

Li, W.J., D. Wang, Y.Q. Zhang, L.H. Xu and C.L. Jiang. 2006b. *Kribbella yunnanensis* sp. nov., *Kribbella alba* sp. nov., two novel species of genus *Kribbella* isolated from soils in Yunnan, China. Syst. Appl. Microbiol. *29*: 29–35.

Li, W.J., Y.Q. Zhang, P. Schumann, H.H. Chen, W.N. Hozzein, X.P. Tian, L.H. Xu and C.L. Jiang. 2006c. *Kocuria aegyptia* sp. nov., a novel actinobacterium isolated from a saline, alkaline desert soil in Egypt. Int. J. Syst. Evol. Microbiol. *56*: 733–737.

Li, W.J., Y.Q. Zhang, P. Schumann, X.P. Tian, Y.Q. Zhang, L.H. Xu and C.L. Jiang. 2006d. *Sinococcus qinghaiensis* gen. nov., sp. nov., a novel member of the order *Bacillales* from a saline soil in China. Int. J. Syst. Evol. Microbiol. *56*: 1189–1192.

Li, W.J., P. Xu, P. Schumann, Y.Q. Zhang, R. Pukall, L.H. Xu, E. Stackebrandt and C.L. Jiang. 2007. *Georgenia ruanii* sp. nov., a novel actinobacterium isolated from forest soil in Yunnan (China), and emended description of the genus *Georgenia*. Int. J. Syst. Evol. Microbiol. *57*: 1424–1428.

Li, X., D. Zhang, F. Chen, J. Ma, Y. Dong and L. Zhang. 2004. *Klebsiella singaporensis* sp. nov., a novel isomaltulose-producing bacterium. Int. J. Syst. Evol. Microbiol. *54*: 2131–2136.

Li, Y., Y. Kawamura, N. Fujiwara, T. Naka, H. Liu, X. Huang, K. Kobayashi and T. Ezaki. 2003. *Chryseobacterium miricola* sp. nov., a novel species isolated from condensation water of space station Mir. Syst. Appl. Microbiol. *26*: 523–528.

Li, Y., Y. Kawamura, N. Fujiwara, T. Naka, H. Liu, X. Huang, K. Kobayashi and T. Ezaki. 2004a. *Sphingomonas yabuuchiae* sp. nov. and *Brevundimonas nasdae* sp. nov., isolated from the Russian space laboratory Mir. Int. J. Syst. Evol. Microbiol. *54*: 819–825.

Li, Y., Y. Kawamura, N. Fujiwara, T. Naka, H. Liu, X. Huang, K. Kobayashi and T. Ezaki. 2004b. *Rothia aeria* sp. nov., *Rhodococcus baikonurensis* sp. nov. and *Arthrobacter russicus* sp. nov., isolated from air in the Russian space laboratory Mir. Int. J. Syst. Evol. Microbiol. *54*: 827–835.

Li, Y., E. Raftis, C. Canchaya, G.F. Fitzgerald, D. van Sinderen and P.W. O'Toole. 2006. Polyphasic analysis indicates that *Lactobacillus salivarius* subsp. *salivarius* and *Lactobacillus salivarius* subsp. *salicinius* do not merit separate subspecies status. Int. J. Syst. Evol. Microbiol. *56*: 2397–2403.

Liebgott, P.P., L. Casalot, S. Paillard, J. Lorquin and M. Labat. 2006. *Marinobacter vinifirmus* sp. nov., a moderately halophilic bacterium isolated from a wine-barrel-decalcification wastewater. Int. J. Syst. Evol. Microbiol. *56*: 2511–2516.

Lim, J.M., J.H. Yoon, J.C. Lee, C.O. Jeon, D.J. Park, C. Sung and C.J. Kim. 2004. *Halomonas koreensis* sp. nov., a novel moderately halophilic bacterium isolated from a solar saltern in Korea. Int. J. Syst. Evol. Microbiol. *54*: 2037–2042.

Lim, J.M., C.O. Jeon, D.J. Park, H.R. Kim, B.J. Yoon and C.J. Kim. 2005a. *Pontibacillus marinus* sp. nov., a moderately halophilic bacterium from a solar saltern, and emended description of the genus *Pontibacillus*. Int. J. Syst. Evol. Microbiol. *55*: 1027–1031.

Lim, J.M., C.O. Jeon, S.M. Song and C.J. Kim. 2005b. *Pontibacillus chungwhensis* gen. nov., sp. nov., a moderately halophilic Gram-positive bacterium from a solar saltern in Korea. Int. J. Syst. Evol. Microbiol. *55*: 165–170.

Lim, J.M., C.O. Jeon, S.M. Song, J.C. Lee, Y.J. Ju, L.H. Xu, C.L. Jiang and C.J. Kim. 2005c. *Lentibacillus lacisalsi* sp. nov., a moderately halophilic bacterium isolated from a saline lake in China. Int. J. Syst. Evol. Microbiol. *55*: 1805–1809.

Lim, J.M., C.O. Jeon and C.J. Kim. 2006a. *Bacillus taeanensis* sp. nov., a halophilic Gram-positive bacterium from a solar saltern in Korea. Int. J. Syst. Evol. Microbiol. *56*: 2903–2908.

Lim, J.M., C.O. Jeon, J.C. Lee, Y.J. Ju, D.J. Park and C.J. Kim. 2006b. *Bacillus koreensis* sp. nov., a spore-forming bacterium, isolated from the rhizosphere of willow roots in Korea. Int. J. Syst. Evol. Microbiol. *56*: 59–63.

Lim, J.M., C.O. Jeon, J.C. Lee, S.M. Song, K.Y. Kim and C.J. Kim. 2006c. *Marinimicrobium koreense* gen. nov., sp. nov. and *Marinimicrobium agarilyticum* sp. nov., novel moderately halotolerant bacteria isolated from tidal flat sediment in Korea. Int. J. Syst. Evol. Microbiol. *56*: 653–657.

Lim, J.M., C.O. Jeon, J.C. Lee, L.H. Xu, C.L. Jiang and C.J. Kim. 2006d. *Paenibacillus gansuensis* sp. nov., isolated from desert soil of Gansu Province in China. Int. J. Syst. Evol. Microbiol. *56*: 2131–2134.

Lim, J.M., C.O. Jeon, S.M. Lee, J.C. Lee, L.H. Xu, C.L. Jiang and C.J. Kim. 2006e. *Bacillus salarius* sp. nov., a halophilic, spore-forming bacterium isolated from a salt lake in China. Int. J. Syst. Evol. Microbiol. *56*: 373–377.

Lim, J.M., C.O. Jeon, D.J. Park, L.H. Xu, C.L. Jiang and C.J. Kim. 2006f. *Paenibacillus xinjiangensis* sp. nov., isolated from Xinjiang province in China. Int. J. Syst. Evol. Microbiol. *56*: 2579–2582.

Lim, J.M., C.O. Jeon, G.S. Lee, D.J. Park, U.G. Kang, C.Y. Park and C.J. Kim. 2007a. *Leeia oryzae* gen. nov., sp. nov., isolated from a rice field in Korea. Int. J. Syst. Evol. Microbiol. *57*: 1204–1208.

Lim, J.M., C.O. Jeon, J.R. Lee, D.J. Park and C.J. Kim. 2007b. *Bacillus kribbensis* sp. nov., isolated from a soil sample in Jeju, Korea. Int. J. Syst. Evol. Microbiol. *57*: 2912–2916.

Lim, Y.W., E.Y. Moon and J. Chun. 2007c. Reclassification of *Flavobacterium resinovorum* Delaporte and Daste 1956 as *Novosphingobium resinovorum* comb. nov., with *Novosphingobium subarcticum* (Nohynek *et al.* 1996) Takeuchi *et al.* 2001 as a later heterotypic synonym. Int. J. Syst. Evol. Microbiol. *57*: 1906–1908.

Lin, K.Y., S.Y. Sheu, P.S. Chang, J.C. Cho and W.M. Chen. 2007. *Oceanicola marinus* sp. nov., a marine alphaproteobacterium isolated from seawater collected off Taiwan. Int. J. Syst. Evol. Microbiol. *57*: 1625–1629.

Lin, Y.C. and A. Yokota. 2006. *Plantibacter auratus* sp. nov., in the family *Microbacteriaceae*. Int. J. Syst. Evol. Microbiol. *56*: 2337–2339.

Lin, Y.C., K. Uemori, D.A. de Briel, V. Arunpairojana and A. Yokota. 2004. *Zimmermannella helvola* gen. nov., sp. nov., *Zimmermannella alba* sp. nov., *Zimmermannella bifida* sp. nov., *Zimmermannella faecalis* sp. nov. and *Leucobacter albus* sp. nov., novel members of the family *Microbacteriaceae*. Int. J. Syst. Evol. Microbiol. *54*: 1669–1676.

Lin, Y.T. and W.Y. Shieh. 2006. *Zobellella denitrificans* gen. nov., sp. nov. and *Zobellella taiwanensis* sp. nov., denitrifying bacteria capable of fermentative metabolism. Int. J. Syst. Evol. Microbiol. *56*: 1209–1215.

Lindner, A.S., A. Pacheco, H.C. Aldrich, A. Costello Staniec, I. Uz and D.J. Hodson. 2007. *Methylocystis hirsuta* sp. nov., a novel methanotroph isolated from a groundwater aquifer. Int. J. Syst. Evol. Microbiol. *57*: 1891–1900.

Lindquist, D., D. Murrill, W.P. Burran, G. Winans, J.M. Janda and W. Probert. 2003. Characteristics of *Massilia timonae* and *Massilia timonae*-like isolates from human patients, with an emended description of the species. J. Clin. Microbiol. *41*: 192–196.

Liou, J.S., D.L. Balkwill, G.R. Drake and R.S. Tanner. 2005. *Clostridium carboxidivorans* sp. nov., a solvent-producing clostridium isolated from an agricultural settling lagoon, and reclassification of the acetogen *Clostridium scatologenes* strain SL1 as *Clostridium drakei* sp. nov. Int. J. Syst. Evol. Microbiol. *55*: 2085–2091.

Lisdiyanti, P., R.R. Navarro, T. Uchimura and K. Komagata. 2006. Reclassification of *Gluconacetobacter hansenii* strains and proposals of *Gluconacetobacter saccharivorans* sp. nov. and *Gluconacetobacter nataicola* sp. nov. Int. J. Syst. Evol. Microbiol. *56*: 2101–2111.

Liu, C. and Z. Shao. 2005. *Alcanivorax dieselolei* sp. nov., a novel alkane-degrading bacterium isolated from sea water and deep-sea sediment. Int. J. Syst. Evol. Microbiol. *55*: 1181–1186.

Liu, C., Y. Wu, L. Li, Y. Ma and Z. Shao. 2007. *Thalassospira xiamenensis* sp. nov. and *Thalassospira profundimaris* sp. nov. Int. J. Syst. Evol. Microbiol. *57*: 316–320.

Liu, J., T. Nakayama, H. Hemmi, Y. Asano, N. Tsuruoka, K. Shimomura, M. Nishijima and T. Nishino. 2005. *Microbacterium natoriense* sp. nov., a novel D-aminoacylase-producing bacterium isolated from soil in Natori, Japan. Int. J. Syst. Evol. Microbiol. *55*: 661–665.

Liu, L., B. Zhang, H. Tong and X. Dong. 2006. *Pediococcus ethanolidurans* sp. nov., isolated from the walls of a distilled-spirit-fermenting cellar. Int. J. Syst. Evol. Microbiol. *56*: 2405–2408.

Liu, Q.M., W.T. Im, M. Lee, D.C. Yang and S.T. Lee. 2006. *Dyadobacter ginsengisoli* sp. nov., isolated from soil of a ginseng field. Int. J. Syst. Evol. Microbiol. *56*: 1939–1944.

Liu, W.Y., J. Zeng, L. Wang, Y.T. Dou and S.S. Yang. 2005. *Halobacillus dabanensis* sp. nov. and *Halobacillus aidingensis* sp. nov., isolated from salt lakes in Xinjiang, China. Int. J. Syst. Evol. Microbiol. *55*: 1991–1996.

Liu, X.Y., B.J. Wang, C.Y. Jiang and S.J. Liu. 2006. *Paracoccus sulfuroxidans* sp. nov., a sulfur oxidizer from activated sludge. Int. J. Syst. Evol. Microbiol. *56*: 2693–2695.

Liu, X.Y., B.J. Wang, C.Y. Jiang and S.J. Liu. 2007. *Micrococcus flavus* sp. nov., isolated from activated sludge in a bioreactor. Int. J. Syst. Evol. Microbiol. *57*: 66–69.

Liu, Z., C. Rodríguez, L. Wang, Q. Cui, Y. Huang, E.T. Quintana and M. Goodfellow. 2005a. *Kitasatospora viridis* sp. nov., a novel actinomycete from soil. Int. J. Syst. Evol. Microbiol. *55*: 707–711.

Liu, Z., Y. Shi, Y. Zhang, Z. Zhou, Z. Lu, W. Li, Y. Huang, C. Rodríguez and M. Goodfellow. 2005b. Classification of *Streptomyces griseus* (Krainsky 1914) Waksman and Henrici 1948 and related species and the transfer of '*Microstreptospora cinerea*' to the genus *Streptomyces* as *Streptomyces yanii* sp. nov. Int. J. Syst. Evol. Microbiol. *55*: 1605–1610.

Liu, Z.P., B.J. Wang, Y.H. Liu and S.J. Liu. 2005a. *Novosphingobium taihuense* sp. nov., a novel aromatic-compound-degrading bacterium isolated from Taihu Lake, China. Int. J. Syst. Evol. Microbiol. *55*: 1229–1232.

Liu, Z.P., B.J. Wang, S.J. Liu and Y.H. Liu. 2005b. *Asticcacaulis taihuensis* sp. nov., a novel stalked bacterium isolated from Taihu Lake, China. Int. J. Syst. Evol. Microbiol. *55*: 1239–1242.

Liu, Z.P., B.J. Wang, X. Dai, X.Y. Liu and S.J. Liu. 2006a. *Zhouia amylolytica* gen. nov., sp. nov., a novel member of the family *Flavobacteriaceae* isolated from sediment of the South China Sea. Int. J. Syst. Evol. Microbiol. *56*: 2825–2829.

Liu, Z.P., J.F. Wu, Z.H. Liu and S.J. Liu. 2006b. *Pseudonocardia ammonioxydans* sp. nov., isolated from coastal sediment. Int. J. Syst. Evol. Microbiol. *56*: 555–558.

Logan, N.A., E. De Clerck, L. Lebbe, A. Verhelst, J. Goris, G. Forsyth, M. Rodríguez-Díaz, M. Heyndrickx and P. De Vos. 2004a. *Paenibacillus cineris* sp. nov. and *Paenibacillus cookii* sp. nov., from Antarctic volcanic soils and a gelatin-processing plant. Int. J. Syst. Evol. Microbiol. *54*: 1071–1076.

Logan, N.A., L. Lebbe, A. Verhelst, J. Goris, G. Forsyth, M. Rodríguez-Díaz, M. Heyndrickx and P. De Vos. 2004b. *Bacillus shackletonii* sp. nov., from volcanic soil on Candlemas Island, South Sandwich archipelago. Int. J. Syst. Evol. Microbiol. *54*: 373–376.

Loganathan, P. and S. Nair. 2004. *Swaminathania salitolerans* gen. nov., sp. nov., a salt-tolerant, nitrogen-fixing and phosphate-solubilizing bacterium from wild rice (*Porteresia coarctata* Tateoka). Int. J. Syst. Evol. Microbiol. *54*: 1185–1190.

Lomans, B.P., R. Maas, R. Luderer, H.J.M. Op Den Camp, A. Pol, C. Van Der Drift and G.D. Vogels. 1999. Isolation and characterization of *Methanomethylovorans hollandica* gen. nov., sp. nov., isolated from freshwater sediment, a methylotrophic methanogen able to grow on dimethyl sulfide and methanethiol. Appl. Environ. Microbiol. *65*: 3641–3650.

Lomans, B.P., P. Leijdekkers, J.J. Wesselink, P. Bakkes, A. Pol, C. van der Drift and H.J.M. op den Camp. 2001. Obligate sulfide-dependent degradation of methoxylated aromatic compounds and formation of methanethiol and dimethyl sulfide by a freshwater sediment isolate, *Parasporobacterium paucivorans* gen. nov., sp. nov. Appl. Environ. Microbiol. *67*: 4017–4023.

López-Cortés, A., M.L. Fardeau, G. Fauque, C. Joulian and B. Ollivier. 2006a. Reclassification of the sulfate- and nitrate-reducing bacterium *Desulfovibrio vulgaris* subsp. *oxamicus* as *Desulfovibrio oxamicus* sp. nov., comb. nov. Int. J. Syst. Evol. Microbiol. *56*: 1495–1499.

López-Cortés, A., P. Schumann, R. Pukall and E. Stackebrandt. 2006b. *Exiguobacterium mexicanum* sp. nov. and *Exiguobacterium artemiae* sp. nov., isolated from the brine shrimp *Artemia franciscana*. Syst. Appl. Microbiol. *29*: 183–190.

Lu, S., S.H. Ryu, B.S. Chung, Y.R. Chung, W. Park and C.O. Jeon. 2007a. *Simplicispira limi* sp. nov., isolated from activated sludge. Int. J. Syst. Evol. Microbiol. *57*: 31–34.

Lu, S., J.R. Lee, S.H. Ryu, B.S. Chung, W.S. Choe and C.O. Jeon. 2007b. *Runella defluvii* sp. nov., isolated from a domestic wastewater treatment plant. Int. J. Syst. Evol. Microbiol. *57*: 2600–2603.

Luo, H., Q. Gu, J. Xie, C. Hu, Z. Liu and Y. Huang. 2007. *Gordonia shandongensis* sp. nov., isolated from soil in China. Int. J. Syst. Evol. Microbiol. *57*: 605–608.

Lütke-Eversloh, T., K. Elbanna, M.C. Cnockaert, J. Mergaert, J. Swings, C.M. Manaia and A. Steinbüchel. 2004. *Caenibacterium thermophilum* is a later synonym of *Schlegelella thermodepolymerans*. Int. J. Syst. Evol. Microbiol. *54*: 1933–1935.

Ma, K., X. Liu and X. Dong. 2005. *Methanobacterium beijingense* sp. nov., a novel methanogen isolated from anaerobic digesters. Int. J. Syst. Evol. Microbiol. *55*: 325–329.

Ma, K., X. Liu and X. Dong. 2006. *Methanosaeta harundinacea* sp. nov., a novel acetate-scavenging methanogen isolated from a UASB reactor. Int. J. Syst. Evol. Microbiol. *56*: 127–131.

Ma, Y., Y. Xue, W.D. Grant, N.C. Collins, A.W. Duckworth, R.P. van Steenbergen and B.E. Jones. 2004. *Alkalimonas amylolytica* gen. nov., sp. nov., and *Alkalimonas delamerensis* gen. nov., sp. nov., novel alkaliphilic bacteria from soda lakes in China and East Africa. Extremophiles. *8*: 193–200.

Ma, Y., Z. Xia, X. Liu and S. Chen. 2007a. *Paenibacillus sabinae* sp. nov., a nitrogen-fixing species isolated from the rhizosphere soils of shrubs. Int. J. Syst. Evol. Microbiol. *57*: 6–11.

Ma, Y., J. Zhang and S. Chen. 2007b. *Paenibacillus zanthoxyli* sp. nov., a novel nitrogen-fixing species isolated from the rhizosphere of *Zanthoxylum simulans*. Int. J. Syst. Evol. Microbiol. *57*: 873–877.

Macián, M.C., E. Garay, P.A.D. Grimont and M.J. Pujalte. 2004. *Vibrio ponticus* sp. nov., a neighbour of *V. fluvialis*-*V. furnissii* clade, isolated from gilthead sea bream, mussels and seawater. Syst. Appl. Microbiol. *27*: 535–540.

Macián, M.C., D.R. Arahal, E. Garay, W. Ludwig, K.H. Schleifer and M.J. Pujalte. 2005a. *Thalassobacter stenotrophicus* gen. nov., sp. nov., a novel marine α-proteobacterium isolated from Mediterranean sea water. Int. J. Syst. Evol. Microbiol. *55*: 105–110.

Macián, M.C., D.R. Arahal, E. Garay, W. Ludwig, K.H. Schleifer and M.J. Pujalte. 2005b. *Jannaschia rubra* sp. nov., a red-pigmented bacterium isolated from sea water. Int. J. Syst. Evol. Microbiol. *55*: 649–653.

Macián, M.C., D.R. Arahal, E. Garay and M.J. Pujalte. 2005c. *Marinomonas aquamarina* sp. nov., isolated from oysters and seawater. Syst. Appl. Microbiol. *28*: 145–150.

Madhaiyan, M., B.Y. Kim, S. Poonguzhali, S.W. Kwon, M.H. Song, J.H. Ryu, S.J. Go, B.S. Koo and T.M. Sa. 2007. *Methylobacterium oryzae* sp. nov., an aerobic, pink-pigmented, facultatively methylotrophic, 1-aminocyclopropane-1-carboxylate deaminase-producing bacterium isolated from rice. Int. J. Syst. Evol. Microbiol. *57*: 326–331.

Magot, M., O. Basso, C. Tardy-Jacquenod and P. Caumette. 2004. *Desulfovibrio bastinii* sp. nov. and *Desulfovibrio gracilis* sp. nov., moderately halophilic, sulfate-reducing bacteria isolated from deep subsurface oilfield water. Int. J. Syst. Evol. Microbiol. *54*: 1693–1697.

Mahendra, S. and L. Alvarez-Cohen. 2005. *Pseudonocardia dioxanivorans* sp. nov., a novel actinomycete that grows on 1,4-dioxane. Int. J. Syst. Evol. Microbiol. *55*: 593–598.

Maillard, R., P. Riegel, F. Barrat, C. Bouillin, D. Thibault, C. Gandoin, L. Halos, C. Demanche, A. Alliot, J. Guillot, Y. Piémont, H.J. Boulouis and M. Vayssier-Taussat. 2004. *Bartonella chomelii* sp. nov., isolated from French domestic cattle (*Bos taurus*). Int. J. Syst. Evol. Microbiol. *54*: 215–220.

Majumdar, S., S.R. Prabhagaran, S. Shivaji and R. Lal. 2006. Reclassification of *Amycolatopsis orientalis* DSM 43387 as *Amycolatopsis benzoatilytica* sp. nov. Int. J. Syst. Evol. Microbiol. *56*: 199–204.

Maldonado, L.A., W. Fenical, P.R. Jensen, C.A. Kauffman, T.J. Mincer, A.C. Ward, A.T. Bull and M. Goodfellow. 2005. *Salinispora arenicola* gen. nov., sp. nov. and *Salinispora tropica* sp. nov., obligate marine actinomycetes belonging to the family *Micromonosporaceae*. Int. J. Syst. Evol. Microbiol. *55*: 1759–1766.

Manaia, C.M., B. Nogales, N. Weiss and O.C. Nunes. 2004. *Gulosibacter molinativorax* gen. nov., sp. nov., a molinate-degrading bacterium, and classification of 'Brevibacterium helvolum' DSM 20419 as *Pseudoclavibacter helvolus* gen. nov., sp. nov. Int. J. Syst. Evol. Microbiol. *54*: 783–789.

Mantelin, S., M.F. Saux, F. Zakhia, G. Bena, S. Bonneau, H. Jeder, P. de Lajudie and J.C. Cleyet-Marel. 2006. Emended description of the genus *Phyllobacterium* and description of four novel species associated with plant roots: *Phyllobacterium bourgognense* sp. nov., *Phyllobacterium ifriqiyense* sp. nov., *Phyllobacterium leguminum* sp. nov. and *Phyllobacterium brassicacearum* sp. nov. Int. J. Syst. Evol. Microbiol. *56*: 827–839.

Mao, J., Q. Tang, Z. Zhang, W. Wang, D. Wei, Y. Huang, Z. Liu, Y. Shi and M. Goodfellow. 2007. *Streptomyces radiopugnans* sp. nov., a radiation-resistant actinomycete isolated from radiation-polluted soil in China. Int. J. Syst. Evol. Microbiol. *57*: 2578–2582.

Margesin, R., P. Schumann, C. Spröer and A.M. Gounot. 2004. *Arthrobacter psychrophenolicus* sp. nov., isolated from an alpine ice cave. Int. J. Syst. Evol. Microbiol. *54*: 2067–2072.

Márquez, M.C. and A. Ventosa. 2005. *Marinobacter hydrocarbonoclasticus* Gauthier *et al.* 1992 and *Marinobacter aquaeolei* Nguyen *et al.* 1999 are heterotypic synonyms. Int. J. Syst. Evol. Microbiol. *55*: 1349–1351.

Márquez, M.C., I.J. Carrasco, Y. Xue, Y. Ma, D.A. Cowan, B.E. Jones, W.D. Grant and A. Ventosa. 2007. *Aquisalimonas asiatica* gen. nov., sp. nov., a moderately halophilic bacterium isolated from an alkaline, saline lake in Inner Mongolia, China. Int. J. Syst. Evol. Microbiol. *57*: 1137–1142.

Martens, T., T. Heidorn, R. Pukall, M. Simon, B.J. Tindall and T. Brinkhoff. 2006. Reclassification of *Roseobacter gallaeciensis* Ruiz-Ponte *et al.* 1998 as *Phaeobacter gallaeciensis* gen. nov., comb. nov., description of *Phaeobacter inhibens* sp. nov., reclassification of *Ruegeria algicola* (Lafay *et al.* 1995) Uchino *et al.* 1999 as *Marinovum algicola* gen. nov., comb. nov., and emended descriptions of the genera *Roseobacter*, *Ruegeria* and *Leisingera*. Int. J. Syst. Evol. Microbiol. *56*: 1293–1304.

Martin, P.A., D. Gundersen-Rindal, M. Blackburn and J. Buyer. 2007. *Chromobacterium subtsugae* sp. nov., a betaproteobacterium toxic to Colorado potato beetle and other insect pests. Int. J. Syst. Evol. Microbiol. *57*: 993–999.

Martin-Carnahan, A. and S.W. Joseph. 2005. Order XII. *Aeromonadales* ord. nov. *In* Brenner, Krieg, Staley and Garrity (Editors), Bergey's Manual of Systematic Bacteriology, 2nd Edition, Volume 2 (The *Proteobacteria*), Part B (The *Gammaproteobacteria*), Springer, New York, p. 556.

Martínez-Cánovas, M.J., V. Béjar, F. Martínez-Checa and E. Quesada. 2004a. *Halomonas anticariensis* sp. nov., from Fuente de Piedra, a saline-wetland wildfowl reserve in Malaga, southern Spain. Int. J. Syst. Evol. Microbiol. *54*: 1329–1332.

Martínez-Cánovas, M.J., V. Béjar, F. Martínez-Checa, R. Páez and E. Quesada. 2004b. *Idiomarina fontislapidosi* sp. nov. and *Idiomarina ramblicola* sp. nov., isolated from inland hypersaline habitats in Spain. Int. J. Syst. Evol. Microbiol. *54*: 1793–1797.

Martínez-Cánovas, M.J., E. Quesada, I. Llamas and V. Béjar. 2004c. *Halomonas ventosae* sp. nov., a moderately halophilic, denitrifying, exopolysaccharide-producing bacterium. Int. J. Syst. Evol. Microbiol. *54*: 733–737.

Martínez-Cánovas, M.J., E. Quesada, F. Martínez-Checa, A. del Moral and V. Béjar. 2004d. *Salipiger mucescens* gen. nov., sp. nov., a moderately halophilic, exopolysaccharide-producing bacterium isolated from hypersaline soil, belonging to the α-*Proteobacteria*. Int. J. Syst. Evol. Microbiol. *54*: 1735–1740.

Martínez-Checa, F., V. Béjar, M.J. Martínez-Cánovas, I. Llamas and E. Quesada. 2005. *Halomonas almeriensis* sp. nov., a moderately halophilic, exopolysaccharide-producing bacterium from Cabo de Gata, Almeria, south-east Spain. Int. J. Syst. Evol. Microbiol. *55*: 2007–2011.

Martínez-Checa, F., V. Béjar, I. Llamas, A. Del Moral and E. Quesada. 2005. *Alteromonas hispanica* sp. nov., a polyunsaturated-fatty-acid-producing, halophilic bacterium isolated from Fuente de Piedra, southern Spain. Int. J. Syst. Evol. Microbiol. *55*: 2385–2390.

Martínez-Checa, F., E. Quesada, M.J. Martínez-Cánovas, I. Llamas and V. Béjar. 2005. *Palleronia marisminoris* gen. nov., sp. nov., a moderately halophilic, exopolysaccharide-producing bacterium belonging to the '*Alphaproteobacteria*', isolated from a saline soil. Int. J. Syst. Evol. Microbiol. *55*: 2525–2530.

Maruyama, T., H.D. Park, K. Ozawa, Y. Tanaka, T. Sumino, K. Hamana, A. Hiraishi and K. Kato. 2006. *Sphingosinicella microcystinivorans* gen. nov., sp. nov., a microcystin-degrading bacterium. Int. J. Syst. Evol. Microbiol. *56*: 85–89.

Masaki, T., K. Ohkusu, H. Hata, N. Fujiwara, H. Iihara, M. Yamada-Noda, P.H. Nhung, M. Hayashi, Y. Asano, Y. Kawamura and T. Ezaki. 2006. *Mycobacterium kumamotonense* sp. nov. recovered from clinical specimen and the first isolation report of *Mycobacterium arupense* in Japan: novel slowly growing, nonchromogenic clinical isolates related to *Mycobacterium terrae* complex. Microbiol. Immunol. *50*: 889–897.

Masco, L., M. Ventura, R. Zink, G. Huys and J. Swings. 2004. Polyphasic taxonomic analysis of *Bifidobacterium animalis* and *Bifidobacterium lactis* reveals relatedness at the subspecies level: reclassification of *Bifidobacterium animalis* as *Bifidobacterium animalis* subsp. *animalis* subsp. nov. and *Bifidobacterium lactis* as *Bifidobacterium animalis* subsp. *lactis* subsp. nov. Int. J. Syst. Evol. Microbiol. *54*: 1137–1143.

Maszenan, A.M., R.J. Seviour, B.K. Patel, P.H. Janssen and J. Wanner. 2005a. *Defluviicoccus vanus* gen. nov., sp. nov., a novel Gram-negative coccus/coccobacillus in the '*Alphaproteobacteria*' from activated sludge. Int. J. Syst. Evol. Microbiol. *55*: 2105–2111.

Maszenan, A.M., J.H. Tay, P. Schumann, H.L. Jiang and S.T. Tay. 2005b. *Quadrisphaera granulorum* gen. nov., sp. nov., a Gram-positive polyphosphate-accumulating coccus in tetrads or aggregates isolated from aerobic granules. Int. J. Syst. Evol. Microbiol. *55*: 1771–1777.

Maszenan, A.M., H.L. Jiang, J.H. Tay, P. Schumann, R.M. Kroppenstedt and S.T. Tay. 2007. *Granulicoccus phenolivorans* gen. nov., sp. nov., a Gram-positive, phenol-degrading coccus isolated from phenol-degrading aerobic granules. Int. J. Syst. Evol. Microbiol. *57*: 730–737.

Matsumoto, A., Y. Takahashi, M. Fukumoto and S. mura. 2007. *Actinocatenispora sera* sp. nov., isolated by long-term culturing. Int. J. Syst. Evol. Microbiol. *57*: 2651–2654.

Matsuo, Y., A. Katsuta, S. Matsuda, Y. Shizuri, A. Yokota and H. Kasai. 2006. *Mechercharimyces mesophilus* gen. nov., sp. nov. and *Mechercharimyces asporophorigenens* sp. nov., antitumour substance-producing marine bacteria, and description of *Thermoactinomycetaceae* fam. nov. Int. J. Syst. Evol. Microbiol. *56*: 2837–2842.

Matsuyama, H., T. Hirabayashi, H. Kasahara, H. Minami, T. Hoshino and I. Yumoto. 2006. *Glaciecola chathamensis* sp. nov., a novel marine polysaccharide-producing bacterium. Int. J. Syst. Evol. Microbiol. *56*: 2883–2886.

Matthies, C., A. Gößner, G. Acker, A. Schramm and H.L. Drake. 2004. *Lactovum miscens* gen. nov., sp. nov., an aerotolerant, psychrotolerant, mixed-fermentative anaerobe from acidic forest soil. Res. Microbiol. *155*: 847–854.

Maturrano, L., M. Valens-Vadell, R. Rosselló-Mora and J. Antón. 2006. *Salicola marasensis* gen. nov., sp. nov., an extremely halophilic bacterium isolated from the Maras solar salterns in Peru. Int. J. Syst. Evol. Microbiol. *56*: 1685–1691.

Mayilraj, S., G.S. Prasad, K. Suresh, H.S. Saini, S. Shivaji and T. Chakrabarti. 2005. *Planococcus stackebrandtii* sp. nov., isolated from a cold desert of the Himalayas, India. Int. J. Syst. Evol. Microbiol. *55*: 91–94.

Mayilraj, S., S. Krishnamurthi, P. Saha and H.S. Saini. 2006a. *Kitasatospora sampliensis* sp. nov., a novel actinobacterium isolated from soil of a sugar-cane field in India. Int. J. Syst. Evol. Microbiol. *56*: 519–522.

Mayilraj, S., S. Krishnamurthi, P. Saha and H.S. Saini. 2006b. *Rhodococcus kroppenstedtii* sp. nov., a novel actinobacterium isolated from a cold desert of the Himalayas, India. Int. J. Syst. Evol. Microbiol. *56*: 979–982.

Mayilraj, S., R.M. Kroppenstedt, K. Suresh and H.S. Saini. 2006c. *Kocuria himachalensis* sp. nov., an actinobacterium isolated from the Indian Himalayas. Int. J. Syst. Evol. Microbiol. *56*: 1971–1975.

Mayilraj, S., P. Saha, K. Suresh and H.S. Saini. 2006d. *Ornithinimicrobium kibberense* sp. nov., isolated from the Indian Himalayas. Int. J. Syst. Evol. Microbiol. *56*: 1657–1661.

Mayilraj, S., K. Suresh, R.M. Kroppenstedt and H.S. Saini. 2006e. *Dietzia kunjamensis* sp. nov., isolated from the Indian Himalayas. Int. J. Syst. Evol. Microbiol. *56*: 1667–1671.

Mayilraj, S., K. Suresh, P. Schumann, R.M. Kroppenstedt and H.S. Saini. 2006f. *Agrococcus lahaulensis* sp. nov., isolated from a cold desert of the Indian Himalayas. Int. J. Syst. Evol. Microbiol. *56*: 1807–1810.

Mayr, R., H.J. Busse, H.L. Worliczek, M. Ehling-Schulz and S. Scherer. 2006. *Ornithinibacillus* gen. nov., with the species *Ornithinibacillus bavariensis* sp. nov. and *Ornithinibacillus californiensis* sp. nov. Int. J. Syst. Evol. Microbiol. *56*: 1383–1389.

McDonald, I.R., P. Kämpfer, E. Topp, K.L. Warner, M.J. Cox, T.L. Hancock, L.G. Miller, M.J. Larkin, V. Ducrocq, C. Coulter, D.B. Harper, J.C. Murrell and R.S. Oremland. 2005. *Aminobacter ciceronei* sp. nov. and *Aminobacter lissarensis* sp. nov., isolated from various terrestrial environments. Int. J. Syst. Evol. Microbiol. *55*: 1827–1832.

McKenzie, C.M., E.M. Seviour, P. Schumann, A.M. Maszenan, J.R. Liu, R.I. Webb, P. Monis, C.P. Saint, U. Steiner and R.J. Seviour. 2006. Isolates of '*Candidatus* Nostocoida limicola' Blackall *et al.* 2000 should be described as three novel species of the genus *Tetrasphaera*, as *Tetrasphaera jenkinsii* sp. nov., *Tetrasphaera vanveenii* sp. nov. and *Tetrasphaera veronensis* sp. nov. Int. J. Syst. Evol. Microbiol. *56*: 2279–2290.

Mehnaz, S., B. Weselowski and G. Lazarovits. 2007a. *Azospirillum canadense* sp. nov., a nitrogen-fixing bacterium isolated from corn rhizosphere. Int. J. Syst. Evol. Microbiol. *57*: 620–624.

Mehnaz, S., B. Weselowski and G. Lazarovits. 2007b. *Azospirillum zeae* sp. nov., a diazotrophic bacterium isolated from rhizosphere soil of *Zea mays*. Int. J. Syst. Evol. Microbiol. *57*: 2805–2809.

Melamud, V.S., T.A. Pivovarova, T.P. Tourova, T.V. Kolganova, G.A. Osipov, A.M. Lysenko, T.F. Kondrat'eva and G.I. Karavaiko. 2003. *Sulfobacillus sibiricus* sp. nov., a new moderately thermophilic bacterium. Microbiology *72*: 605–612.

Merchant, M.M., A.K. Welsh and R.J. McLean. 2007. *Rheinheimera texasensis* sp. nov., a halointolerant freshwater oligotroph. Int. J. Syst. Evol. Microbiol. *57*: 2376–2380.

Mergaert, J. and J. Swings. 2005. Family IV. *Phyllobacteriaceae* fam. nov. *In* Brenner, Krieg, Staley and Garrity (Editors), Bergey's Manual of Systematic Bacteriology, 2nd Edition, Volume 2 (The *Proteobacteria*), Part C (The *Alpha-*, *Beta-*, *Delta-*, and *Epsilonproteobacteria*), Springer, New York, p. 393.

Meroth, C.B., W.P. Hammes and C. Hertel. 2004. Characterisation of the microbiota of rice sourdoughs and description of *Lactobacillus spicheri* sp. nov. Syst. Appl. Microbiol. *27*: 151–159.

Mesbah, N.M., D.B. Hedrick, A.D. Peacock, M. Rohde and J. Wiegel. 2007. *Natranaerobius thermophilus* gen. nov., sp. nov., a halophilic, alkalithermophilic bacterium from soda lakes of the Wadi An Natrun, Egypt, and proposal of *Natranaerobiaceae* fam. nov. and *Natranaerobiales* ord. nov. Int. J. Syst. Evol. Microbiol. *57*: 2507–2512.

Meyers, P.R., C.M. Goodwin, J.A. Bennett, B.L. Aken, C.E. Price and J.M. van Rooyen. 2004. *Streptomyces africanus* sp. nov., a novel streptomycete with blue aerial mycelium. Int. J. Syst. Evol. Microbiol. 54: 1531–1535.

Mikalsen, J., A.B. Olsen, T. Tengs and D.J. Colquhoun. 2007. *Francisella philomiragia* subsp. *noatunensis* subsp. nov., isolated from farmed Atlantic cod (*Gadus morhua* L.). Int. J. Syst. Evol. Microbiol. 57: 1960–1965.

Miller, J.A., M.G. Kalyuzhnaya, E. Noyes, J.C. Lara, M.E. Lidstrom and L. Chistoserdova. 2005. *Labrys methylaminiphilus* sp. nov., a novel facultatively methylotrophic bacterium from a freshwater lake sediment. Int. J. Syst. Evol. Microbiol. 55: 1247–1253.

Miñana-Galbis, D., M. Farfán, M.C. Fusté and J.G. Lorén. 2004. *Aeromonas molluscorum* sp. nov., isolated from bivalve molluscs. Int. J. Syst. Evol. Microbiol. 54: 2073–2078.

Miñana-Galbis, D., M. Farfán, M.C. Fusté and J.G. Lorén. 2007. *Aeromonas bivalvium* sp. nov., isolated from bivalve molluscs. Int. J. Syst. Evol. Microbiol. 57: 582–587.

Miranda-Tello, E., M.L. Fardeau, P. Thomas, F. Ramirez, L. Casalot, J.L. Cayol, J.L. García and B. Ollivier. 2004. *Petrotoga mexicana* sp. nov., a novel thermophilic, anaerobic and xylanolytic bacterium isolated from an oil-producing well in the Gulf of Mexico. Int. J. Syst. Evol. Microbiol. 54: 169–174.

Miranda-Tello, E., M.L. Fardeau, C. Joulian, M. Magot, P. Thomas, J.L. Tholozan and B. Ollivier. 2007. *Petrotoga halophila* sp. nov., a thermophilic, moderately halophilic, fermentative bacterium isolated from an offshore oil well in Congo. Int. J. Syst. Evol. Microbiol. 57: 40–44.

Miroshnichenko, M.L., S. L'Haridon, P. Schumann, S. Spring, E.A. Bonch-Osmolovskaya, C. Jeanthon and E. Stackebrandt. 2004. *Caminibacter profundus* sp. nov., a novel thermophile of *Nautiliales* ord. nov. within the class '*Epsilonproteobacteria*', isolated from a deep-sea hydrothermal vent. Int. J. Syst. Evol. Microbiol. 54: 41–45.

Miyamoto, M., Y. Seto, D.H. Hao, T. Teshima, Y.B. Sun., T. Kabuki, L.B. Yao and H. Nakajima. 2005. *Lactobacillus harbinensis* sp. nov., consisted of strains isolated from traditional fermented vegetables 'Suan cai' in Harbin, Northeastern China and *Lactobacillus perolens* DSM 12745. Syst. Appl. Microbiol. 28: 688–694.

Miyazaki, M., Y. Nogi, R. Usami and K. Horikoshi. 2006. *Shewanella surugensis* sp. nov., *Shewanella kaireitica* sp. nov. and *Shewanella abyssi* sp. nov., isolated from deep-sea sediments of Suruga Bay, Japan. Int. J. Syst. Evol. Microbiol. 56: 1607–1613.

Mohamed, A.M., P.C. Iwen, S. Tarantolo and S.H. Hinrichs. 2004. *Mycobacterium nebraskense* sp. nov., a novel slowly growing scotochromogenic species. Int. J. Syst. Evol. Microbiol. 54: 2057–2060.

Mohan, R., P. Namsolleck, P.A. Lawson, M. Osterhoff, M.D. Collins, C.A. Alpert and M. Blaut. 2006. *Clostridium asparagiforme* sp. nov., isolated from a human faecal sample. Syst. Appl. Microbiol. 29: 292–299.

Montero-Barrientos, M., R. Rivas, E. Velázquez, E. Monte and M.G. Roig. 2005. *Terrabacter terrae* sp. nov., a novel actinomycete isolated from soil in Spain. Int. J. Syst. Evol. Microbiol. 55: 2491–2495.

Montes, M.J., E. Mercadé, N. Bozal and J. Guinea. 2004. *Paenibacillus antarcticus* sp. nov., a novel psychrotolerant organism from the Antarctic environment. Int. J. Syst. Evol. Microbiol. 54: 1521–1526.

Morais, P.V., R. Francisco, R. Branco, A.P. Chung and M.S. da Costa. 2004. *Leucobacter chromiireducens* sp. nov., and *Leucobacter aridicollis* sp. nov., two new species isolated from a chromium contaminated environment. Syst. Appl. Microbiol. 27: 646–652.

Morais, P.V., C. Paulo, R. Francisco, R. Branco, A.P. Chung and M.S. da Costa. 2006. *Leucobacter luti* sp. nov., and *Leucobacter alluvii* sp. nov., two new species of the genus *Leucobacter* isolated under chromium stress. Syst. Appl. Microbiol. 29: 414–421.

Moreira, F.M.S., L. Cruz, S.M. Faria, T. Marsh, E. Martínez-Romero, F.O. Pedrosa, R.M. Pitard and J.P.W. Young. 2006. *Azorhizobium doebereinerae* sp. nov. Microsymbiont of *Sesbania virgata* Caz. Pers. Syst. Appl. Microbiol. 29: 197–206.

Mori, K., H. Kim, T. Kakegawa and S. Hanada. 2003. A novel lineage of sulfate-reducing microorganisms: *Thermodesulfobiaceae* fam. nov., *Thermodesulfobium narugense*, gen. nov., sp. nov., a new thermophilic isolate from a hot spring. Extremophiles 7: 283–290.

Mori, K., T. Kakegawa, Y. Higashi, K. Nakamura, A. Maruyama and S. Hanada. 2004. *Oceanithermus desulfurans* sp. nov., a novel thermophilic, sulfur-reducing bacterium isolated from a sulfide chimney in Suiyo Seamount. Int. J. Syst. Evol. Microbiol. 54: 1561–1566.

Morita, H., C. Shiratori, M. Murakami, H. Takami, Y. Kato, A. Endo, F. Nakajima, M. Takagi, H. Akita, S. Okada and T. Masaoka. 2007. *Lactobacillus hayakitensis* sp. nov., isolated from intestines of healthy thoroughbreds. Int. J. Syst. Evol. Microbiol. 57: 2836–2839.

Moussard, H., S. L'Haridon, B.J. Tindall, A. Banta, P. Schumann, E. Stackebrandt, A.L. Reysenbach and C. Jeanthon. 2004. *Thermodesulfatator indicus* gen. nov., sp. nov., a novel thermophilic chemolithoautotrophic sulfate-reducing bacterium isolated from the Central Indian Ridge. Int. J. Syst. Evol. Microbiol. 54: 227–233.

Moyaert, H., A. Decostere, P. Vandamme, L. Debruyne, J. Mast, M. Baele, L. Ceelen, R. Ducatelle and F. Haesebrouck. 2007. *Helicobacter equorum* sp. nov., a urease-negative *Helicobacter* species isolated from horse faeces. Int. J. Syst. Evol. Microbiol. 57: 213–218.

Muir, R.E. and M.W. Tan. 2007. *Leucobacter chromiireducens* subsp. *solipictus* subsp. nov., a pigmented bacterium isolated from the nematode *Caenorhabditis elegans*, and emended description of *L. chromiireducens*. Int. J. Syst. Evol. Microbiol. 57: 2770–2776.

Müller, D., D.D. Simeonova, P. Riegel, S. Mangenot, S. Koechler, D. Lièvremont, P.N. Bertin and M.C. Lett. 2006. *Herminiimonas arsenicoxydans* sp. nov., a metalloresistant bacterium. Int. J. Syst. Evol. Microbiol. 56: 1765–1769.

Mun, H.S., H.J. Kim, E.J. Oh, H. Kim, G.H. Bai, H.K. Yu, Y.G. Park, C.Y. Cha, Y.H. Kook and B.J. Kim. 2007. *Mycobacterium seoulense* sp. nov., a slowly growing scotochromogenic species. Int. J. Syst. Evol. Microbiol. 57: 594–599.

Muramatsu, Y., Y. Uchino, H. Kasai, K. Suzuki and Y. Nakagawa. 2007. *Ruegeria mobilis* sp. nov., a member of the *Alphaproteobacteria* isolated in Japan and Palau. Int. J. Syst. Evol. Microbiol. 57: 1304–1309.

Murcia, M.I., E. Tortoli, M.C. Menendez, E. Palenque and M.J. García. 2006. *Mycobacterium colombiense* sp. nov., a novel member of the *Mycobacterium avium* complex and description of MAC-X as a new ITS genetic variant. Int. J. Syst. Evol. Microbiol. 56: 2049–2054.

Muurholm, S., S. Cousin, O. Päuker, E. Brambilla and E. Stackebrandt. 2007. *Pedobacter duraquae* sp. nov., *Pedobacter westerhofensis* sp. nov., *Pedobacter metabolipauper* sp. nov., *Pedobacter hartonius* sp. nov. and *Pedobacter steynii* sp. nov., isolated from a hard-water rivulet. Int. J. Syst. Evol. Microbiol. 57: 2221–2227.

Nakagawa, S., S. Nakamura, F. Inagaki, K. Takai, N. Shirai and Y. Sako. 2004a. *Hydrogenivirga caldilitoris* gen. nov., sp. nov., a novel extremely thermophilic, hydrogen- and sulfur-oxidizing bacterium from a coastal hydrothermal field. Int. J. Syst. Evol. Microbiol. 54: 2079–2084.

Nakagawa, S., K. Takai, K. Horikoshi and Y. Sako. 2004b. *Aeropyrum camini* sp. nov., a strictly aerobic, hyperthermophilic archaeon from a deep-sea hydrothermal vent chimney. Int. J. Syst. Evol. Microbiol. 54: 329–335.

Nakagawa, S., F. Inagaki, K. Takai, K. Horikoshi and Y. Sako. 2005a. *Thioreductor micantisoli* gen. nov., sp. nov., a novel mesophilic, sulfur-reducing chemolithoautotroph within the ε-*Proteobacteria* isolated from hydrothermal sediments in the Mid-Okinawa Trough. Int. J. Syst. Evol. Microbiol. 55: 599–605.

Nakagawa, S., Z. Shtaih, A. Banta, T.J. Beveridge, Y. Sako and A.L. Reysenbach. 2005b. *Sulfurihydrogenibium yellowstonense* sp. nov., an extremely thermophilic, facultatively heterotrophic, sulfur-oxidizing bacterium from Yellowstone National Park, and emended descriptions of the genus *Sulfurihydrogenibium*, *Sulfurihydrogenibium subterraneum* and *Sulfurihydrogenibium azorense*. Int. J. Syst. Evol. Microbiol. 55: 2263–2268.

Nakagawa, S., K. Takai, F. Inagaki, K. Horikoshi and Y. Sako. 2005c. *Nitratiruptor tergarcus* gen. nov., sp. nov. and *Nitratifractor salsuginis* gen. nov., sp. nov., nitrate-reducing chemolithoautotrophs of the ε-*Proteobacteria* isolated from a deep-sea hydrothermal system in the Mid-Okinawa Trough. Int. J. Syst. Evol. Microbiol. 55: 925–933.

Nakagawa, T., T. Iino, K. Suzuki and S. Harayama. 2006. *Ferrimonas futtsuensis* sp. nov. and *Ferrimonas kyonanensis* sp. nov., selenate-reducing bacteria belonging to the *Gammaproteobacteria* isolated from Tokyo Bay. Int. J. Syst. Evol. Microbiol. *56*: 2639–2645.

Nakajima, K., K. Hirota, Y. Nodasaka and I. Yumoto. 2005. *Alkalibacterium iburiense* sp. nov., an obligate alkaliphile that reduces an indigo dye. Int. J. Syst. Evol. Microbiol. *55*: 1525–1530.

Nakajima, Y., C.C. Ho and T. Kudo. 2003. *Microtetraspora malaysiensis* sp. nov., isolated from Malaysian primary dipterocarp forest soil. J. Gen. Appl. Microbiol. *49*: 181–189.

Nakamura, K., S. Haruta, S. Ueno, M. Ishii, A. Yokota and Y. Igarashi. 2004. *Cerasibacillus quisquiliarum* gen. nov., sp. nov., isolated from a semi-continuous decomposing system of kitchen refuse. Int. J. Syst. Evol. Microbiol. *54*: 1063–1069.

Nakatsu, C.H., K. Hristova, S. Hanada, X.Y. Meng, J.R. Hanson, K.M. Scow and Y. Kamagata. 2006. *Methylibium petroleiphilum* gen. nov., sp. nov., a novel methyl tert-butyl ether-degrading methylotroph of the *Betaproteobacteria*. Int. J. Syst. Evol. Microbiol. *56*: 983–989.

Nam, S.W., W. Kim, J. Chun and M. Goodfellow. 2004. *Tsukamurella pseudospumae* sp. nov., a novel actinomycete isolated from activated sludge foam. Int. J. Syst. Evol. Microbiol. *54*: 1209–1212.

Nam, Y.D., H.W. Chang, J.R. Park, H.Y. Kwon, Z.X. Quan, Y.H. Park, B.C. Kim and J.W. Bae. 2007a. *Vibrio litoralis* sp. nov., isolated from a Yellow Sea tidal flat in Korea. Int. J. Syst. Evol. Microbiol. *57*: 562–565.

Nam, Y.D., H.W. Chang, J.R. Park, H.Y. Kwon, Z.X. Quan, Y.H. Park, J.S. Lee, J.H. Yoon and J.W. Bae. 2007b. *Pseudoalteromonas marina* sp. nov., a marine bacterium isolated from tidal flats of the Yellow Sea, and reclassification of *Pseudoalteromonas sagamiensis* as *Algicola sagamiensis* comb. nov. Int. J. Syst. Evol. Microbiol. *57*: 12–18.

Namwong, S., S. Tanasupawat, T. Smitinont, W. Visessanguan, T. Kudo and T. Itoh. 2005. Isolation of *Lentibacillus salicampi* strains and *Lentibacillus juripiscarius* sp. nov. from fish sauce in Thailand. Int. J. Syst. Evol. Microbiol. *55*: 315–320.

Namwong, S., S. Tanasupawat, W. Visessanguan, T. Kudo and T. Itoh. 2007. *Halococcus thailandensis* sp. nov., from fish sauce in Thailand. Int. J. Syst. Evol. Microbiol. *57*: 2199–2203.

Nandasena, K.G., G.W. O'Hara, R.P. Tiwari, A. Willems and J.G. Howieson. 2007. *Mesorhizobium ciceri* biovar biserrulae, a novel biovar nodulating the pasture legume *Biserrula pelecinus* L. Int. J. Syst. Evol. Microbiol. *57*: 1041–1045.

Narasingarao, P. and M.M. Häggblom. 2006. *Sedimenticola selenatireducens*, gen. nov., sp. nov., an anaerobic selenate-respiring bacterium isolated from estuarine sediment. Syst. Appl. Microbiol. *29*: 382–388.

Narasingarao, P. and M.M. Häggblom. 2007. *Pelobacter seleniigenes* sp. nov., a selenate-respiring bacterium. Int. J. Syst. Evol. Microbiol. *57*: 1937–1942.

Naser, S.M., M. Vancanneyt, E. De Graef, L.A. Devriese, C. Snauwaert, K. Lefebvre, B. Hoste, P. Švec, A. Decostere, F. Haesebrouck and J. Swings. 2005. *Enterococcus canintestini* sp. nov., from faecal samples of healthy dogs. Int. J. Syst. Evol. Microbiol. *55*: 2177–2182.

Naser, S.M., K.E. Hagen, M. Vancanneyt, I. Cleenwerck, J. Swings and T.A. Tompkins. 2006a. *Lactobacillus suntoryeus* Cachat and Priest 2005 is a later synonym of *Lactobacillus helveticus* (Orla-Jensen 1919) Bergey *et al.* 1925 (Approved Lists 1980). Int. J. Syst. Evol. Microbiol. *56*: 355–360.

Naser, S.M., M. Vancanneyt, B. Hoste, C. Snauwaert and J. Swings. 2006b. *Lactobacillus cypricasei* Lawson *et al.* 2001 is a later heterotypic synonym of *Lactobacillus acidipiscis* Tanasupawat *et al.* 2000. Int. J. Syst. Evol. Microbiol. *56*: 1681–1683.

Naser, S.M., M. Vancanneyt, B. Hoste, C. Snauwaert, K. Vandemeulebroecke and J. Swings. 2006c. Reclassification of *Enterococcus flavescens* Pompei *et al.* 1992 as a later synonym of *Enterococcus casseliflavus* (ex Vaughan *et al.* 1979) Collins *et al.* 1984 and *Enterococcus saccharominimus* Vancanneyt *et al.* 2004 as a later synonym of *Enterococcus italicus* Fortina *et al.* 2004. Int. J. Syst. Evol. Microbiol. *56*: 413–416.

Naser, S.M., M. Vancanneyt, C. Snauwaert, G. Vrancken, B. Hoste, L. De Vuyst and J. Swings. 2006d. Reclassification of *Lactobacillus amylophi-lus* LMG 11400 and NRRL B-4435 as *Lactobacillus amylotrophicus* sp. nov. Int. J. Syst. Evol. Microbiol. *56*: 2523–2527.

Nazina, T.N., E.V. Lebedeva, A.B. Poltaraus, T.P. Tourova, A.A. Grigoryan, D. Sokolova, A.M. Lysenko and G.A. Osipov. 2004. *Geobacillus gargensis* sp. nov., a novel thermophile from a hot spring, and the reclassification of *Bacillus vulcani* as *Geobacillus vulcani* comb. nov. Int. J. Syst. Evol. Microbiol. *54*: 2019–2024.

Nazina, T.N., D.S. Sokolova, A.A. Grigoryan, N.M. Shestakova, E.M. Mikhailova, A.B. Poltaraus, T.P. Tourova, A.M. Lysenko, G.A. Osipov and S.S. Belyaev. 2005. *Geobacillus jurassicus* sp. nov., a new thermophilic bacterium isolated from a high-temperature petroleum reservoir, and the validation of the *Geobacillus* species. Syst. Appl. Microbiol. *28*: 43–53.

Ndoye, B., I. Cleenwerck, K. Engelbeen, R. Dubois-Dauphin, A.T. Guiro, S. Van Trappen, A. Willems and P. Thonart. 2007. *Acetobacter senegalensis* sp. nov., a thermotolerant acetic acid bacterium isolated in Senegal (sub-Saharan Africa) from mango fruit (*Mangifera indica* L.). Int. J. Syst. Evol. Microbiol. *57*: 1576–1581.

Nedashkovskaya, O.I., S.B. Kim, S.K. Han, A.M. Lysenko, V.V. Mikhailov and K.S. Bae. 2004a. *Arenibacter certesii* sp. nov., a novel marine bacterium isolated from the green alga *Ulva fenestrata*. Int. J. Syst. Evol. Microbiol. *54*: 1173–1176.

Nedashkovskaya, O.I., S.B. Kim, S.K. Han, A.M. Lysenko, M. Rohde, M.S. Rhee, G.M. Frolova, E. Falsen, V.V. Mikhailov and K.S. Bae. 2004b. *Maribacter* gen. nov., a new member of the family *Flavobacteriaceae*, isolated from marine habitats, containing the species *Maribacter sedimenticola* sp. nov., *Maribacter aquivivus* sp. nov., *Maribacter orientalis* sp. nov. and *Maribacter ulvicola* sp. nov. Int. J. Syst. Evol. Microbiol. *54*: 1017–1023.

Nedashkovskaya, O.I., S.B. Kim, S.K. Han, M.S. Rhee, A.M. Lysenko, E. Falsen, G.M. Frolova, V.V. Mikhailov and K.S. Bae. 2004c. *Ulvibacter litoralis* gen. nov., sp. nov., a novel member of the family *Flavobacteriaceae* isolated from the green alga *Ulva fenestrata*. Int. J. Syst. Evol. Microbiol. *54*: 119–123.

Nedashkovskaya, O.I., S.B. Kim, S.K. Han, M.S. Rhee, A.M. Lysenko, M. Rohde, N.V. Zhukova, G.M. Frolova, V.V. Mikhailov and K.S. Bae. 2004d. *Algibacter lectus* gen. nov., sp. nov., a novel member of the family *Flavobacteriaceae* isolated from green algae. Int. J. Syst. Evol. Microbiol. *54*: 1257–1261.

Nedashkovskaya, O.I., M. Suzuki, A.M. Lysenko, C. Snauwaert, M. Vancanneyt, J. Swings, M.V. Vysotskii and V.V. Mikhailov. 2004e. *Cellulophaga pacifica* sp. nov. Int. J. Syst. Evol. Microbiol. *54*: 609–613.

Nedashkovskaya, O.I., M. Suzuki, M. Vancanneyt, I. Cleenwerck, A.M. Lysenko, V.V. Mikhailov and J. Swings. 2004f. *Zobellia amurskyensis* sp. nov., *Zobellia laminariae* sp. nov. and *Zobellia russellii* sp. nov., novel marine bacteria of the family *Flavobacteriaceae*. Int. J. Syst. Evol. Microbiol. *54*: 1643–1648.

Nedashkovskaya, O.I., M. Suzuki, M. Vancanneyt, I. Cleenwerck, N.V. Zhukova, M.V. Vysotskii, V.V. Mikhailov and J. Swings. 2004g. *Salegentibacter holothuriorum* sp. nov., isolated from the edible holothurian *Apostichopus japonicus*. Int. J. Syst. Evol. Microbiol. *54*: 1107–1110.

Nedashkovskaya, O.I., M. Vancanneyt, S. Van Trappen, K. Vandemeulebroecke, A.M. Lysenko, M. Rohde, E. Falsen, G.M. Frolova, V.V. Mikhailov and J. Swings. 2004h. Description of *Algoriphagus aquimarinus* sp. nov., *Algoriphagus chordae* sp. nov. and *Algoriphagus winogradskyi* sp. nov., from sea water and algae, transfer of *Hongiella halophila* Yi and Chun 2004 to the genus *Algoriphagus* as *Algoriphagus halophilus* comb. nov. and emended descriptions of the genera *Algoriphagus* Bowman *et al.* 2003 and *Hongiella* Yi and Chun 2004. Int. J. Syst. Evol. Microbiol. *54*: 1757–1764.

Nedashkovskaya, O.I., S.B. Kim, S.K. Han, C. Snauwaert, M. Vancanneyt, J. Swings, K.O. Kim, A.M. Lysenko, M. Rohde, G.M. Frolova, V.V. Mikhailov and K.S. Bae. 2005a. *Winogradskyella thalassocola* gen. nov., sp. nov., *Winogradskyella epiphytica* sp. nov. and *Winogradskyella eximia* sp. nov., marine bacteria of the family *Flavobacteriaceae*. Int. J. Syst. Evol. Microbiol. *55*: 49–55.

Nedashkovskaya, O.I., S.B. Kim, K.H. Lee, K.S. Bae, G.M. Frolova, V.V. Mikhailov and I.S. Kim. 2005b. *Pibocella ponti* gen. nov., sp. nov., a novel marine bacterium of the family *Flavobacteriaceae* isolated from the green alga *Acrosiphonia sonderi*. Int. J. Syst. Evol. Microbiol. *55*: 177–181.

Nedashkovskaya, O.I., S.B. Kim, D.H. Lee, A.M. Lysenko, L.S. Shevchenko, G.M. Frolova, V.V. Mikhailov, K.H. Lee and K.S. Bae. 2005c. *Roseivirga ehrenbergii* gen. nov., sp. nov., a novel marine bacterium of the phylum 'Bacteroidetes', isolated from the green alga *Ulva fenestrata*. Int. J. Syst. Evol. Microbiol. *55*: 231–234.

Nedashkovskaya, O.I., S.B. Kim, K.H. Lee, V.V. Mikhailov and K.S. Bae. 2005d. *Gillisia mitskevichiae* sp. nov., a novel bacterium of the family *Flavobacteriaceae*, isolated from sea water. Int. J. Syst. Evol. Microbiol. *55*: 321–323.

Nedashkovskaya, O.I., S.B. Kim, M.S. Lee, M.S. Park, K.H. Lee, A.M. Lysenko, H.W. Oh, V.V. Mikhailov and K.S. Bae. 2005e. *Cyclobacterium amurskyense* sp. nov., a novel marine bacterium isolated from sea water. Int. J. Syst. Evol. Microbiol. *55*: 2391–2394.

Nedashkovskaya, O.I., S.B. Kim, A.M. Lysenko, G.M. Frolova, V.V. Mikhailov, K.S. Bae, D.H. Lee and I.S. Kim. 2005f. *Gramella echinicola* gen. nov., sp. nov., a novel halophilic bacterium of the family *Flavobacteriaceae* isolated from the sea urchin *Strongylocentrotus intermedius*. Int. J. Syst. Evol. Microbiol. *55*: 391–394.

Nedashkovskaya, O.I., S.B. Kim, A.M. Lysenko, G.M. Frolova, V.V. Mikhailov and K.S. Bae. 2005g. *Bizionia paragorgiae* gen. nov., sp. nov., a novel member of the family *Flavobacteriaceae* isolated from the soft coral *Paragorgia arborea*. Int. J. Syst. Evol. Microbiol. *55*: 375–378.

Nedashkovskaya, O.I., S.B. Kim, A.M. Lysenko, G.M. Frolova, V.V. Mikhailov, K.H. Lee and K.S. Bae. 2005h. Description of *Aquimarina muelleri* gen. nov., sp. nov., and proposal of the reclassification of [*Cytophaga*] *latercula* Lewin 1969 as *Stanierella latercula* gen. nov., comb. nov. Int. J. Syst. Evol. Microbiol. *55*: 225–229.

Nedashkovskaya, O.I., S.B. Kim., A.M. Lysenko, N.I. Kalinovskaya, V.V. Mikhailov, I.S. Kim and K.S. Bae. 2005i. *Polaribacter butkevichii* sp. nov., a novel marine mesophilic bacterium of the family *Flavobacteriaceae*. Curr. Microbiol. *51*: 408–412.

Nedashkovskaya, O.I., S.B. Kim, A.M. Lysenko, V.V. Mikhailov, K.S. Bae and I.S. Kim. 2005j. *Salegentibacter mishustinae* sp. nov., isolated from the sea urchin *Strongylocentrotus intermedius*. Int. J. Syst. Evol. Microbiol. *55*: 235–238.

Nedashkovskaya, O.I., S.B. Kim, A.M. Lysenko, M.S. Park, V.V. Mikhailov, K.S. Bae and H.Y. Park. 2005k. *Roseivirga echinicomitans* sp. nov., a novel marine bacterium isolated from the sea urchin *Strongylocentrotus intermedius*, and emended description of the genus *Roseivirga*. Int. J. Syst. Evol. Microbiol. *55*: 1797–1800.

Nedashkovskaya, O.I., S.B. Kim, M. Suzuki, L.S. Shevchenko, M.S. Lee, K.H. Lee, M.S. Park, G.M. Frolova, H.W. Oh, K.S. Bae, H.Y. Park and V.V. Mikhailov. 2005l. *Pontibacter actiniarum* gen. nov., sp. nov., a novel member of the phylum 'Bacteroidetes', and proposal of *Reichenbachiella* gen. nov. as a replacement for the illegitimate prokaryotic generic name *Reichenbachia* Nedashkovskaya *et al.* 2003. Int. J. Syst. Evol. Microbiol. *55*: 2583–2588.

Nedashkovskaya, O.I., M. Vancanneyt, P. Dawyndt, K. Engelbeen, K. Vandemeulebroecke, I. Cleenwerck, B. Hoste, J. Mergaert, T.L. Tan, G.M. Frolova, V.V. Mikhailov and J. Swings. 2005m. Reclassification of [*Cytophaga*] *marinoflava* Reichenbach 1989 as *Leeuwenhoekiella marinoflava* gen. nov., comb. nov. and description of *Leeuwenhoekiella aequorea* sp. nov. Int. J. Syst. Evol. Microbiol. *55*: 1033–1038.

Nedashkovskaya, O.I., S.B. Kim, J. Kwak, V.V. Mikhailov and K.S. Bae. 2006a. *Mariniflexile gromovii* gen. nov., sp. nov., a gliding bacterium isolated from the sea urchin *Strongylocentrotus intermedius*. Int. J. Syst. Evol. Microbiol. *56*: 1635–1638.

Nedashkovskaya, O.I., S.B. Kim, M. Vancanneyt, A.M. Lysenko, D.S. Shin, M.S. Park, K.H. Lee, W.J. Jung, N.I. Kalinovskaya, V.V. Mikhailov, K.S. Bae and J. Swings. 2006b. *Echinicola pacifica* gen. nov., sp. nov., a novel flexibacterium isolated from the sea urchin *Strongylocentrotus intermedius*. Int. J. Syst. Evol. Microbiol. *56*: 953–958.

Nedashkovskaya, O.I., S.B. Kim, M. Vancanneyt, D.S. Shin, A.M. Lysenko, L.S. Shevchenko, V.B. Krasokhin, V.V. Mikhailov, J. Swings and K.S. Bae. 2006c. *Salegentibacter agarivorans* sp. nov., a novel marine bacterium of the family *Flavobacteriaceae* isolated from the sponge *Artemisina* sp. Int. J. Syst. Evol. Microbiol. *56*: 883–887.

Nedashkovskaya, O.I., S.B. Kim, M. Vancanneyt, C. Snauwaert, A.M. Lysenko, M. Rohde, G.M. Frolova, N.V. Zhukova, V.V. Mikhailov, K.S. Bae, H.W. Oh and J. Swings. 2006d. *Formosa agariphila* sp. nov., a budding bacterium of the family *Flavobacteriaceae* isolated from marine environments, and emended description of the genus *Formosa*. Int. J. Syst. Evol. Microbiol. *56*: 161–167.

Nedashkovskaya, O.I., S.B. Kim, N.V. Zhukova, J. Kwak, V.V. Mikhailov and K.S. Bae. 2006e. *Mesonia mobilis* sp. nov., isolated from seawater, and emended description of the genus *Mesonia*. Int. J. Syst. Evol. Microbiol. *56*: 2433–2436.

Nedashkovskaya, O.I., M. Vancanneyt, L. Christiaens, N.I. Kalinovskaya, V.V. Mikhailov and J. Swings. 2006f. *Aquimarina intermedia* sp. nov., reclassification of *Stanierella latercula* (Lewin 1969) as *Aquimarina latercula* comb. nov. and *Gaetbulimicrobium brevivitae* Yoon *et al.* 2006 as *Aquimarina brevivitae* comb. nov. and emended description of the genus *Aquimarina*. Int. J. Syst. Evol. Microbiol. *56*: 2037–2041.

Nedashkovskaya, O.I., M. Vancanneyt, I. Cleenwerck, C. Snauwaert, S.B. Kim, A.M. Lysenko, L.S. Shevchenko, K.H. Lee, M.S. Park, G.M. Frolova, V.V. Mikhailov, K.S. Bae and J. Swings. 2006g. *Arenibacter palladensis* sp. nov., a novel marine bacterium isolated from the green alga *Ulva fenestrata*, and emended description of the genus *Arenibacter*. Int. J. Syst. Evol. Microbiol. *56*: 155–160.

Nedashkovskaya, O.I., S.B. Kim, B. Hoste, D.S. Shin, I.A. Beleneva, M. Vancanneyt and V.V. Mikhailov. 2007a. *Echinicola vietnamensis* sp. nov., a member of the phylum *Bacteroidetes* isolated from seawater. Int. J. Syst. Evol. Microbiol. *57*: 761–763.

Nedashkovskaya, O.I., S.B. Kim, K.K. Kwon, D.S. Shin, X. Luo, S.J. Kim and V.V. Mikhailov. 2007b. Proposal of *Algoriphagus vanfongensis* sp. nov., transfer of members of the genera *Hongiella* Yi and Chun 2004 emend. Nedashkovskaya *et al.* 2004 and *Chimaereicella* Tiago *et al.* 2006 to the genus *Algoriphagus*, and emended description of the genus *Algoriphagus* Bowman *et al.* 2003 emend. Nedashkovskaya *et al.* 2004. Int. J. Syst. Evol. Microbiol. *57*: 1988–1994.

Nedashkovskaya, O.I., S.B. Kim, A.M. Lysenko, K.H. Lee, K.S. Bae and V.V. Mikhailov. 2007c. *Arenibacter echinorum* sp. nov., isolated from the sea urchin *Strongylocentrotus intermedius*. Int. J. Syst. Evol. Microbiol. *57*: 2655–2659.

Nedashkovskaya, O.I., S.B. Kim, D.S. Shin, I.A. Beleneva and V.V. Mikhailov. 2007d. *Fulvivirga kasyanovii* gen. nov., sp. nov., a novel member of the phylum *Bacteroidetes* isolated from seawater in a mussel farm. Int. J. Syst. Evol. Microbiol. *57*: 1046–1049.

Nedashkovskaya, O.I., M. Vancanneyt, S.B. Kim, B. Hoste and K.S. Bae. 2007e. *Algibacter mikhailovii* sp. nov., a novel marine bacterium of the family *Flavobacteriaceae*, and emended description of the genus *Algibacter*. Int. J. Syst. Evol. Microbiol. *57*: 2147–2150.

Nedashkovskaya, O.I., M. Vancanneyt, P. De Vos, S.B. Kim, M.S. Lee and V.V. Mikhailov. 2007f. *Maribacter polysiphoniae* sp. nov., isolated from a red alga. Int. J. Syst. Evol. Microbiol. *57*: 2840–2843.

Neimark, H., B. Hoff and M. Ganter. 2004. *Mycoplasma ovis* comb. nov. (formerly *Eperythrozoon ovis*), an epierythrocytic agent of haemolytic anaemia in sheep and goats. Int. J. Syst. Evol. Microbiol. *54*: 365–371.

Neimark, H., W. Peters, B.L. Robinson and L.B. Stewart. 2005. Phylogenetic analysis and description of *Eperythrozoon coccoides*, proposal to transfer to the genus *Mycoplasma* as *Mycoplasma coccoides* comb. nov. and Request for an Opinion. Int. J. Syst. Evol. Microbiol. *55*: 1385–1391.

Nevin, K.P., D.E. Holmes, T.L. Woodard, E.S. Hinlein, D.W. Ostendorf and D.R. Lovley. 2005. *Geobacter bemidjiensis* sp. nov. and *Geobacter psychrophilus* sp. nov., two novel Fe(III)-reducing subsurface isolates. Int. J. Syst. Evol. Microbiol. *55*: 1667–1674.

Nevin, K.P., D.E. Holmes, T.L. Woodard, S.F. Covalla and D.R. Lovley. 2007. Reclassification of *Trichlorobacter thiogenes* as *Geobacter thiogenes* comb. nov. Int. J. Syst. Evol. Microbiol. *57*: 463–466.

Nichols, C.M., J.P. Bowman and J. Guezennec. 2005. *Olleya marilimosa* gen. nov., sp. nov., an exopolysaccharide-producing marine bacterium from the family *Flavobacteriaceae*, isolated from the Southern Ocean. Int. J. Syst. Evol. Microbiol. *55*: 1557–1561.

Niederberger, T.D., D.K. Götz, I.R. McDonald, R.S. Ronimus and H.W. Morgan. 2006. *Ignisphaera aggregans* gen. nov., sp. nov., a novel hyperthermophilic crenarchaeote isolated from hot springs in Rotorua and Tokaanu, New Zealand. Int. J. Syst. Evol. Microbiol. *56*: 965–971.

Nielsen, D.S., U. Schillinger, C.M. Franz, J. Bresciani, W. Amoa-Awua, W.H. Holzapfel and M. Jakobsen. 2007. *Lactobacillus ghanensis* sp. nov., a motile lactic acid bacterium isolated from Ghanaian cocoa fermentations. Int. J. Syst. Evol. Microbiol. *57*: 1468–1472.

Nielsen, M.B., K.U. Kjeldsen and K. Ingvorsen. 2006. *Desulfitibacter alkalitolerans* gen. nov., sp. nov., an anaerobic, alkalitolerant, sulfite-reducing bacterium isolated from a district heating plant. Int. J. Syst. Evol. Microbiol. *56*: 2831–2836.

Niggemyer, A., S. Spring, E. Stackebrandt and R.F. Rosenzweig. 2001. Isolation and characterization of a novel AsV-reducing bacterium: implications for arsenic mobilization and the genus *Desulfitobacterium*. Appl. Environ. Microbiol. *67*: 5568–5580.

Nikitin, D.I., C. Strömpl, M.S. Oranskaya and W.R. Abraham. 2004. Phylogeny of the ring-forming bacterium *Arcicella aquatica* gen. nov., sp. nov. (ex Nikitin *et al.* 1994), from a freshwater neuston biofilm. Int. J. Syst. Evol. Microbiol. *54*: 681–684.

Noel, G.R., N. Atibalentja and L.L. Domier. 2005. Emended description of *Pasteuria nishizawae*. Int. J. Syst. Evol. Microbiol. *55*: 1681–1685.

Nogi, Y., S. Hosoya, C. Kato and K. Horikoshi. 2004. *Colwellia piezophila* sp. nov., a novel piezophilic species from deep-sea sediments of the Japan Trench. Int. J. Syst. Evol. Microbiol. *54*: 1627–1631.

Nogi, Y., K. Soda and T. Oikawa. 2005a. *Flavobacterium frigidimaris* sp. nov., isolated from Antarctic seawater. Syst. Appl. Microbiol. *28*: 310–315.

Nogi, Y., H. Takami and K. Horikoshi. 2005b. Characterization of alkaliphilic *Bacillus* strains used in industry: proposal of five novel species. Int. J. Syst. Evol. Microbiol. *55*: 2309–2315.

Nogi, Y., S. Hosoya, C. Kato and K. Horikoshi. 2007. *Psychromonas hadalis* sp. nov., a novel piezophilic bacterium isolated from the bottom of the Japan Trench. Int. J. Syst. Evol. Microbiol. *57*: 1360–1364.

Noguchi, H., M. Uchino, O. Shida, K. Takano, L.K. Nakamura and K. Komagata. 2004. *Bacillus vietnamensis* sp. nov., a moderately halotolerant, aerobic, endospore-forming bacterium isolated from Vietnamese fish sauce. Int. J. Syst. Evol. Microbiol. *54*: 2117–2120.

Nordhoff, M., D. Taras, M. Macha, K. Tedin, H.J. Busse and L.H. Wieler. 2005. *Treponema berlinense* sp. nov. and *Treponema porcinum* sp. nov., novel spirochaetes isolated from porcine faeces. Int. J. Syst. Evol. Microbiol. *55*: 1675–1680.

Normand, P. 2006. *Geodermatophilaceae* fam. nov., a formal description. Int. J. Syst. Evol. Microbiol. *56*: 2277–2278.

Nørskov-Lauritsen, N., B. Bruun and M. Kilian. 2005. Multilocus sequence phylogenetic study of the genus *Haemophilus* with description of *Haemophilus pittmaniae* sp. nov. Int. J. Syst. Evol. Microbiol. *55*: 449–456.

Nørskov-Lauritsen, N. and M. Kilian. 2006. Reclassification of *Actinobacillus actinomycetemcomitans*, *Haemophilus aphrophilus*, *Haemophilus paraphrophilus* and *Haemophilus segnis* as *Aggregatibacter actinomycetemcomitans* gen. nov., comb. nov., *Aggregatibacter aphrophilus* comb. nov. and *Aggregatibacter segnis* comb. nov., and emended description of *Aggregatibacter aphrophilus* to include V factor-dependent and V factor-independent isolates. Int. J. Syst. Evol. Microbiol. *56*: 2135–2146.

Nowlan, B., M.S. Dodia, S.P. Singh and B.K. Patel. 2006. *Bacillus okhensis* sp. nov., a halotolerant and alkalitolerant bacterium from an Indian saltpan. Int. J. Syst. Evol. Microbiol. *56*: 1073–1077.

Ntougias, S., C. Fasseas and G.I. Zervakis. 2007a. *Olivibacter sitiensis* gen. nov., sp. nov., isolated from alkaline olive-oil mill wastes in the region of Sitia, Crete. Int. J. Syst. Evol. Microbiol. *57*: 398–404.

Ntougias, S., G.I. Zervakis and C. Fasseas. 2007b. *Halotalea alkalilenta* gen. nov., sp. nov., a novel osmotolerant and alkalitolerant bacterium from alkaline olive mill wastes, and emended description of the family *Halomonadaceae* Franzmann *et al.* 1989, emend. Dobson and Franzmann 1996. Int. J. Syst. Evol. Microbiol. *57*: 1975–1983.

Nunan, L.M., D.V. Lightner, M.A. Oduori and G.E. Gasparich. 2005. *Spiroplasma penaei* sp. nov., associated with mortalities in *Penaeus vannamei*, Pacific white shrimp. Int. J. Syst. Evol. Microbiol. *55*: 2317–2322.

Nunes, I., I. Tiago, A.L. Pires, M.S. da Costa and A. Veríssimo. 2006. *Paucisalibacillus globulus* gen. nov., sp. nov., a Gram-positive bacterium isolated from potting soil. Int. J. Syst. Evol. Microbiol. *56*: 1841–1845.

Nunoura, T., H. Oida, M. Miyazaki, Y. Suzuki, K. Takai and K. Horikoshi. 2007a. *Marinitoga okinawensis* sp. nov., a novel thermophilic and anaerobic heterotroph isolated from a deep-sea hydrothermal field, Southern Okinawa Trough. Int. J. Syst. Evol. Microbiol. *57*: 467–471.

Nunoura, T., H. Oida, M. Miyazaki, Y. Suzuki, K. Takai and K. Horikoshi. 2007b. *Desulfothermus okinawensis* sp. nov., a thermophilic and heterotrophic sulfate-reducing bacterium isolated from a deep-sea hydrothermal field. Int. J. Syst. Evol. Microbiol. *57*: 2360–2364.

Oduori, M.A., J.J. Lipa and G.E. Gasparich. 2005. *Spiroplasma leucomae* sp. nov., isolated in Poland from white satin moth (*Leucoma salicis* L.) larvae. Int. J. Syst. Evol. Microbiol. *55*: 2447–2450.

Ohta, H., R. Hattori, Y. Ushiba, H. Mitsui, M. Ito, H. Watanabe, A. Tonosaki and T. Hattori. 2004. *Sphingomonas oligophenolica* sp. nov., a halo- and organo-sensitive oligotrophic bacterium from paddy soil that degrades phenolic acids at low concentrations. Int. J. Syst. Evol. Microbiol. *54*: 2185–2190.

Okamoto, M., Y. Benno, K.P. Leung and N. Maeda. 2007. *Metascardovia criceti* gen. nov., sp. nov., from hamster dental plaque. Microbiol. Immunol. *51*: 747–754.

Olivera, N., F. Siñeriz and J.D. Breccia. 2005. *Bacillus patagoniensis* sp. nov., a novel alkalitolerant bacterium from the rhizosphere of *Atriplex lampa* in Patagonia, Argentina. Int. J. Syst. Evol. Microbiol. *55*: 443–447.

Olson, J.B., D.K. Harmody, A.K. Bej and P.J. McCarthy. 2007. *Tsukamurella spongiae* sp. nov., a novel actinomycete isolated from a deep-water marine sponge. Int. J. Syst. Evol. Microbiol. *57*: 1478–1481.

Ommedal, H. and T. Torsvik. 2007. *Desulfotignum toluenicum* sp. nov., a novel toluene-degrading, sulphate-reducing bacterium isolated from an oil-reservoir model column. Int. J. Syst. Evol. Microbiol. *57*: 2865–2869.

Onyenwoke, R.U., Y.J. Lee, S. Dabrowski, B.K. Ahring and J. Wiegel. 2006. Reclassification of *Thermoanaerobium acetigenum* as *Caldicellulosiruptor acetigenus* comb. nov. and emendation of the genus description. Int. J. Syst. Evol. Microbiol. *56*: 1391–1395.

Onyenwoke, R.U., V.V. Kevbrin, A.M. Lysenko and J. Wiegel. 2007. *Thermoanaerobacter pseudethanolicus* sp. nov., a thermophilic heterotrophic anaerobe from Yellowstone National Park. Int. J. Syst. Evol. Microbiol. *57*: 2191–2193.

Ortiz-Martinez, A., J.M. González, L.I. Evtushenko, V. Jurado, L. Laiz, I. Groth and C. Saiz-Jimenez. 2004. Reclassification of *Agromyces fucosus* subsp. *hippuratus* as *Agromyces hippuratus* sp. nov., comb. nov. and emended description of *Agromyces fucosus*. Int. J. Syst. Evol. Microbiol. *54*: 1553–1556.

Osawa, R., T. Fujisawa and R. Pukall. 2006. *Lactobacillus apodemi* sp. nov., a tannase-producing species isolated from wild mouse faeces. Int. J. Syst. Evol. Microbiol. *56*: 1693–1696.

Osman, S., M. Satomi and K. Venkateswaran. 2006. *Paenibacillus pasadenensis* sp. nov. and *Paenibacillus barengoltzii* sp. nov., isolated from a spacecraft assembly facility. Int. J. Syst. Evol. Microbiol. *56*: 1509–1514.

Osman, S., C. Moissl, N. Hosoya, A. Briegel, S. Mayilraj, M. Satomi and K. Venkateswaran. 2007. *Tetrasphaera remsis* sp. nov., isolated from the Regenerative Enclosed Life Support Module Simulator (REMS) air system. Int. J. Syst. Evol. Microbiol. *57*: 2749–2753.

O'Sullivan, L.A., J. Rinna, G. Humphreys, A.J. Weightman and J.C. Fry. 2005. *Fluviicola taffensis* gen. nov., sp. nov., a novel freshwater bacterium of the family *Cryomorphaceae* in the phylum 'Bacteroidetes'. Int. J. Syst. Evol. Microbiol. *55*: 2189–2194.

O'Sullivan, L.A., J. Rinna, G. Humphreys, A.J. Weightman and J.C. Fry. 2006. Culturable phylogenetic diversity of the phylum 'Bacteroidetes' from river epilithon and coastal water and description of novel members of the family *Flavobacteriaceae*: *Epilithonimonas tenax* gen. nov., sp. nov. and *Persicivirga xylanidelens* gen. nov., sp. nov. Int. J. Syst. Evol. Microbiol. *56*: 169–180.

Otsuka, Y., Y. Kawamura, T. Koyama, H. Iihara, K. Ohkusu and T. Ezaki. 2005. *Corynebacterium resistens* sp. nov., a new multidrug-resistant coryneform bacterium isolated from human infections. J. Clin. Microbiol. *43*: 3713–3717.

Pakdeeto, A., S. Tanasupawat, C. Thawai, S. Moonmangmee, T. Kudo and T. Itoh. 2007. *Lentibacillus kapialis* sp. nov., from fermented shrimp paste in Thailand. Int. J. Syst. Evol. Microbiol. *57*: 364–369.

Pakdeeto, A., S. Tanasupawat, C. Thawai, S. Moonmangmee, T. Kudo and T. Itoh. 2007. *Salinicoccus siamensis* sp. nov., isolated from fermented shrimp paste in Thailand. Int. J. Syst. Evol. Microbiol. *57*: 2004–2008.

Pal, R., S. Bala, M. Dadhwal, M. Kumar, G. Dhingra, O. Prakash, S.R. Prabagaran, S. Shivaji, J. Cullum, C. Holliger and R. Lal. 2005. Hexachlorocyclohexane-degrading bacterial strains *Sphingomonas paucimobilis* B90A, UT26 and Sp+, having similar lin genes, represent three distinct species, *Sphingobium indicum* sp. nov., *Sphingobium japonicum* sp. nov. and *Sphingobium francense* sp. nov., and reclassification of [*Sphingomonas*] *chungbukensis* as *Sphingobium chungbukense* comb. nov. Int. J. Syst. Evol. Microbiol. *55*: 1965–1972.

Pal, R., V.K. Bhasin and R. Lal. 2006. Proposal to reclassify [*Sphingomonas*] *xenophaga* Stolz *et al.* 2000 and [*Sphingomonas*] *taejonensis* Lee *et al.* 2001 as *Sphingobium xenophagum* comb. nov. and *Sphingopyxis taejonensis* comb. nov., respectively. Int. J. Syst. Evol. Microbiol. *56*: 667–670.

Palacios, L., D.R. Arahal, B. Reguera and I. Marín. 2006. *Hoeflea alexandrii* sp. nov., isolated from the toxic dinoflagellate *Alexandrium minutum* AL1V. Int. J. Syst. Evol. Microbiol. *56*: 1991–1995.

Palleroni, N.J., A.M. Port, H.K. Chang and G.J. Zylstra. 2004. *Hydrocarboniphaga effusa* gen. nov., sp. nov., a novel member of the γ-*Proteobacteria* active in alkane and aromatic hydrocarbon degradation. Int. J. Syst. Evol. Microbiol. *54*: 1203–1207.

Pankratov, T.A., I.S. Kulichevskaya, W. Liesack and S.N. Dedysh. 2006. Isolation of aerobic, gliding, xylanolytic and laminarinolytic bacteria from acidic *Sphagnum* peatlands and emended description of *Chitinophaga arvensicola* Kämpfer *et al.* 2006. Int. J. Syst. Evol. Microbiol. *56*: 2761–2764.

Pankratov, T.A., B.J. Tindall, W. Liesack and S.N. Dedysh. 2007. *Mucilaginibacter paludis* gen. nov., sp. nov. and *Mucilaginibacter gracilis* sp. nov., pectin-, xylan- and laminarin-degrading members of the family *Sphingobacteriaceae* from acidic *Sphagnum* peat bog. Int. J. Syst. Evol. Microbiol. *57*: 2349–2354.

Pantůcek, R., I. Sedlácek, P. Petráš, D. Koukalová, P. Švec, V. Štětina, M. Vancanneyt, L. Chrastinová, J. Vokurková, V. Růžičková, J. Doškař, J. Swings and V. Hájek. 2005. *Staphylococcus simiae* sp. nov., isolated from South American squirrel monkeys. Int. J. Syst. Evol. Microbiol. *55*: 1953–1958.

Paper, W., U. Jahn, M.J. Hohn, M. Kronner, D.J. Näther, T. Burghardt, R. Rachel, K.O. Stetter and H. Huber. 2007. *Ignicoccus hospitalis* sp. nov., the host of 'Nanoarchaeum equitans'. Int. J. Syst. Evol. Microbiol. *57*: 803–808.

Park, D.J., S.G. Dastager, J.C. Lee, S.H. Yeo, J.H. Yoon and C.J. Kim. 2007. *Shimazuella kribbensis* gen. nov., sp. nov., a mesophilic representative of the family *Thermoactinomycetaceae*. Int. J. Syst. Evol. Microbiol. *57*: 2660–2664.

Park, H.Y., K.K. Kim, L. Jin and S.T. Lee. 2006. *Microbacterium paludicola* sp. nov., a novel xylanolytic bacterium isolated from swamp forest. Int. J. Syst. Evol. Microbiol. *56*: 535–539.

Park, J.R., J.W. Bae, Y.D. Nam, H.W. Chang, H.Y. Kwon, Z.X. Quan and Y.H. Park. 2007. *Sulfitobacter litoralis* sp. nov., a marine bacterium isolated from the East Sea, Korea. Int. J. Syst. Evol. Microbiol. *57*: 692–695.

Park, M., S. Lu, S.H. Ryu, B.S. Chung, W. Park, C.J. Kim and C.O. Jeon. 2006. *Flavobacterium croceum* sp. nov., isolated from activated sludge. Int. J. Syst. Evol. Microbiol. *56*: 2443–2447.

Park, M., S.H. Ryu, T.H. Vu, H.S. Ro, P.Y. Yun and C.O. Jeon. 2007. *Flavobacterium defluvii* sp. nov., isolated from activated sludge. Int. J. Syst. Evol. Microbiol. *57*: 233–237.

Park, M.J., H.B. Kim, D.S. An, H.C. Yang, S.T. Oh, H.J. Chung and D.C. Yang. 2007. *Paenibacillus soli* sp. nov., a xylanolytic bacterium isolated from soil. Int. J. Syst. Evol. Microbiol. *57*: 146–150.

Park, M.S., S.R. Jung, K.H. Lee, M.S. Lee, J.O. Do, S.B. Kim and K.S. Bae. 2006. *Chryseobacterium soldanellicola* sp. nov. and *Chryseobacterium taeanense* sp. nov., isolated from roots of sand-dune plants. Int. J. Syst. Evol. Microbiol. *56*: 433–438.

Park, S.J., C.H. Kang, Y.D. Nam, J.W. Bae, Y.H. Park, Z.X. Quan, D.S. Moon, H.J. Kim, D.H. Roh and S.K. Rhee. 2006. *Oceanisphaera donghaensis* sp. nov., a halophilic bacterium from the East Sea, Korea. Int. J. Syst. Evol. Microbiol. *56*: 895–898.

Park, Y.D., K.S. Baik, H. Yi, K.S. Bae and J. Chun. 2005a. *Pseudoalteromonas byunsanensis* sp. nov., isolated from tidal flat sediment in Korea. Int. J. Syst. Evol. Microbiol. *55*: 2519–2523.

Park, Y.D., H.B. Lee, H. Yi, Y. Kim, K.S. Bae, J.E. Choi, H.S. Jung and J. Chun. 2005b. *Pseudomonas panacis* sp. nov., isolated from the surface of rusty roots of Korean ginseng. Int. J. Syst. Evol. Microbiol. *55*: 1721–1724.

Park, Y.D., K.S. Baik, C.N. Seong, K.S. Bae, S. Kim and J. Chun. 2006a. *Photobacterium ganghwense* sp. nov., a halophilic bacterium isolated from sea water. Int. J. Syst. Evol. Microbiol. *56*: 745–749.

Park, Y.D., H. Yi, K.S. Baik, C.N. Seong, K.S. Bae, E.Y. Moon and J. Chun. 2006b. *Pseudomonas segetis* sp. nov., isolated from soil. Int. J. Syst. Evol. Microbiol. *56*: 2593–2595.

Parshina, S.N., J. Sipma, Y. Nakashimada, A.M. Henstra, H. Smidt, A.M. Lysenko, P.N. Lens, G. Lettinga and A.J. Stams. 2005. *Desulfotomaculum carboxydivorans* sp. nov., a novel sulfate-reducing bacterium capable of growth at 100% CO. Int. J. Syst. Evol. Microbiol. *55*: 2159–2165.

Partida-Martinez, L.P., I. Groth, I. Schmitt, W. Richter, M. Roth and C. Hertweck. 2007. *Burkholderia rhizoxinica* sp. nov. and *Burkholderia endofungorum* sp. nov., bacterial endosymbionts of the plant-pathogenic fungus *Rhizopus microsporus*. Int. J. Syst. Evol. Microbiol. *57*: 2583–2590.

Pathom-aree, W., Y. Nogi, I.C. Sutcliffe, A.C. Ward, K. Horikoshi, A.T. Bull and M. Goodfellow. 2006a. *Williamsia marianensis* sp. nov., a novel actinomycete isolated from the Mariana Trench. Int. J. Syst. Evol. Microbiol. *56*: 1123–1126.

Pathom-aree, W., Y. Nogi, I.C. Sutcliffe, A.C. Ward, K. Horikoshi, A.T. Bull and M. Goodfellow. 2006b. *Dermacoccus abyssi* sp. nov., a piezotolerant actinomycete isolated from the Mariana Trench. Int. J. Syst. Evol. Microbiol. *56*: 1233–1237.

Pathom-aree, W., Y. Nogi, A.C. Ward, K. Horikoshi, A.T. Bull and M. Goodfellow. 2006c. *Dermacoccus barathri* sp. nov. and *Dermacoccus profundi* sp. nov., novel actinomycetes isolated from deep-sea mud of the Mariana Trench. Int. J. Syst. Evol. Microbiol. *56*: 2303–2307.

Pavan, M.E., R.J. Franco, J.M. Rodríguez, P. Gadaleta, S.L. Abbott, J.M. Janda and J. Zorzópulos. 2005. Phylogenetic relationships of the genus *Kluyvera*: transfer of *Enterobacter intermedius* Izard *et al.* 1980 to the genus *Kluyvera* as *Kluyvera intermedia* comb. nov. and reclassification of *Kluyvera cochleae* as a later synonym of *K. intermedia*. Int. J. Syst. Evol. Microbiol. *55*: 437–442.

Peak, K.K., K.E. Duncan, W. Veguilla, V.A. Luna, D.S. King, L. Heller, L. Heberlein-Larson, F. Reeves, A.C. Cannons, P. Amuso and J. Cattani.

2007. *Bacillus acidiceler* sp. nov., isolated from a forensic specimen, containing *Bacillus anthracis* pX02 genes. Int. J. Syst. Evol. Microbiol. *57*: 2031–2036.

Peçonek, J., C. Gruber, V. Gallego, A. Ventosa, H.J. Busse, P. Kämpfer, C. Radax and H. Stan-Lotter. 2006. Reclassification of *Pseudomonas beijerinckii* Hof 1935 as *Chromohalobacter beijerinckii* comb. nov., and emended description of the species. Int. J. Syst. Evol. Microbiol. *56*: 1953–1957.

Pedersen, C. and S. Roos. 2004. *Lactobacillus saerimneri* sp. nov., isolated from pig faeces. Int. J. Syst. Evol. Microbiol. *54*: 1365–1368.

Peix, A., R. Rivas, I. Santa-Regina, P.F. Mateos, E. Martínez-Molina, C. Rodríguez-Barrueco and E. Velázquez. 2004. *Pseudomonas lutea* sp. nov., a novel phosphate-solubilizing bacterium isolated from the rhizosphere of grasses. Int. J. Syst. Evol. Microbiol. *54*: 847–850.

Peix, A., O. Berge, R. Rivas, A. Abril and E. Velázquez. 2005. *Pseudomonas argentinensis* sp. nov., a novel yellow pigment-producing bacterial species, isolated from rhizospheric soil in Cordoba, Argentina. Int. J. Syst. Evol. Microbiol. *55*: 1107–1112.

Peix, A., R. Rivas, M.E. Trujillo, M. Vancanneyt, E. Velázquez and A. Willems. 2005. Reclassification of *Agrobacterium ferrugineum* LMG 128 as *Hoeflea marina* gen. nov., sp. nov. Int. J. Syst. Evol. Microbiol. *55*: 1163–1166.

Peix, A., A. Valverde, R. Rivas, J.M. Igual, M.H. Ramírez-Bahena, P.F. Mateos, I. Santa-Regina, C. Rodríguez-Barrueco, E. Martínez-Molina and E. Velázquez. 2007. Reclassification of *Pseudomonas aurantiaca* as a synonym of *Pseudomonas chlororaphis* and proposal of three subspecies, *P. chlororaphis* subsp. *chlororaphis* subsp. nov., *P. chlororaphis* subsp. *aureofaciens* subsp. nov., comb. nov. and *P. chlororaphis* subsp. *aurantiaca* subsp. nov., comb. nov. Int. J. Syst. Evol. Microbiol. *57*: 1286–1290.

Peng, G., H. Wang, G. Zhang, W. Hou, Y. Liu, E.T. Wang and Z. Tan. 2006. *Azospirillum melinis* sp. nov., a group of diazotrophs isolated from tropical molasses grass. Int. J. Syst. Evol. Microbiol. *56*: 1263–1271.

Perevalova, A.A., V.A. Svetlichny, I.V. Kublanov, N.A. Chernyh, N.A. Kostrikina, T.P. Tourova, B.B. Kuznetsov and E.A. Bonch-Osmolovskaya. 2005. *Desulfurococcus fermentans* sp. nov., a novel hyperthermophilic archaeon from a Kamchatka hot spring, and emended description of the genus *Desulfurococcus*. Int. J. Syst. Evol. Microbiol. *55*: 995–999.

Pérez-Ibarra, B.M., M.E. Flores and M. García-Varela. 2007. Isolation and characterization of *Bacillus thioparus* sp. nov., chemolithoautotrophic, thiosulfate-oxidizing bacterium. FEMS Microbiol. Lett. *271*: 289–296.

Perin, L., L. Martínez-Aguilar, G. Paredes-Valdez, J.I. Baldani, P. Estrada-de Los Santos, V.M. Reis and J. Caballero-Mellado. 2006. *Burkholderia silvatlantica* sp. nov., a diazotrophic bacterium associated with sugar cane and maize. Int. J. Syst. Evol. Microbiol. *56*: 1931–1937.

Pikuta, E.V., D. Marsic, A. Bej, J. Tang, P. Krader and R.B. Hoover. 2005. *Carnobacterium pleistocenium* sp. nov., a novel psychrotolerant, facultative anaerobe isolated from permafrost of the Fox Tunnel in Alaska. Int. J. Syst. Evol. Microbiol. *55*: 473–478.

Pikuta, E.V., R.B. Hoover, A.K. Bej, D. Marsic, W.B. Whitman, P.E. Krader and J. Tang. 2006a. *Trichococcus patagoniensis* sp. nov., a facultative anaerobe that grows at −5°C, isolated from penguin guano in Chilean Patagonia. Int. J. Syst. Evol. Microbiol. *56*: 2055–2062.

Pikuta, E.V., T. Itoh, P. Krader, J. Tang, W.B. Whitman and R.B. Hoover. 2006b. *Anaerovirgula multivorans* gen. nov., sp. nov., a novel spore-forming, alkaliphilic anaerobe isolated from Owens Lake, California, USA. Int. J. Syst. Evol. Microbiol. *56*: 2623–2629.

Pikuta, E.V., D. Marsic, T. Itoh, A.K. Bej, J. Tang, W.B. Whitman, J.D. Ng, O.K. Garriott and R.B. Hoover. 2007. *Thermococcus thioreducens* sp. nov., a novel hyperthermophilic, obligately sulfur-reducing archaeon from a deep-sea hydrothermal vent. Int. J. Syst. Evol. Microbiol. *57*: 1612–1618.

Ping, X., Y. Takahashi, A. Seino, Y. Iwai and S. Ōmura. 2004. *Streptomyces scabrisporus* sp. nov. Int. J. Syst. Evol. Microbiol. *54*: 577–581.

Pinhassi, J., J.P. Bowman, O.I. Nedashkovskaya, I. Lekunberri, L. Gomez-Consarnau and C. Pedrós-Alió. 2006. *Leeuwenhoekiella blandensis* sp. nov., a genome-sequenced marine member of the family *Flavobacteriaceae*. Int. J. Syst. Evol. Microbiol. *56*: 1489–1493.

Pinhassi, J., M.J. Pujalte, M.C. Macián, I. Lekunberri, J.M. González, C. Pedrós-Alió and D.R. Arahal. 2007. *Reinekea blandensis* sp. nov., a marine, genome-sequenced gammaproteobacterium. Int. J. Syst. Evol. Microbiol. *57*: 2370–2375.

Pires, A.L., L. Albuquerque, I. Tiago, M.F. Nobre, N. Empadinhas, A. Veríssimo and M.S. da Costa. 2005. *Meiothermus timidus* sp. nov., a new slightly thermophilic yellow-pigmented species. FEMS Microbiol. Lett. *245*: 39–45.

Pitcher, D.G., D. Windsor, H. Windsor, J.M. Bradbury, C. Yavari, J.S. Jensen, C. Ling and D. Webster. 2005. *Mycoplasma amphoriforme* sp. nov., isolated from a patient with chronic bronchopneumonia. Int. J. Syst. Evol. Microbiol. *55*: 2589–2594.

Plumb, J.J., C.M. Haddad, J.A. Gibson and P.D. Franzmann. 2007. *Acidianus sulfidivorans* sp. nov., an extremely acidophilic, thermophilic archaeon isolated from a solfatara on Lihir Island, Papua New Guinea, and emendation of the genus description. Int. J. Syst. Evol. Microbiol. *57*: 1418–1423.

Poli, A., E. Esposito, L. Lama, P. Orlando, G. Nicolaus, F. De Appolonia, A. Gambacorta and B. Nicolaus. 2006. *Anoxybacillus amylolyticus* sp. nov., a thermophilic amylase producing bacterium isolated from Mount Rittmann (*Antarctica*). Syst. Appl. Microbiol. *29*: 300–307.

Postec, A., C. Le Breton, M.L. Fardeau, F. Lesongeur, P. Pignet, J. Querellou, B. Ollivier and A. Godfroy. 2005. *Marinitoga hydrogenitolerans* sp. nov., a novel member of the order *Thermotogales* isolated from a black smoker chimney on the Mid-Atlantic Ridge. Int. J. Syst. Evol. Microbiol. *55*: 1217–1221.

Prabagaran, S.R., K. Suresh, R. Manorama, D. Delille and S. Shivaji. 2005. *Marinomonas ushuaiensis* sp. nov., isolated from coastal sea water in Ushuaia, Argentina, sub-Antarctica. Int. J. Syst. Evol. Microbiol. *55*: 309–313.

Prabahar, V., S. Dube, G.S.N. Reddy and S. Shivaji. 2004. *Pseudonocardia antarctica* sp. nov. an *Actinomycetes* from McMurdo Dry Valleys, Antarctica. Syst. Appl. Microbiol. *27*: 66–71.

Prado, B., C. Lizama, M. Aguilera, A. Ramos-Cormenzana, S. Fuentes, V. Campos and M. Monteoliva-Sánchez. 2006. *Chromohalobacter nigrandesensis* sp. nov., a moderately halophilic, Gram-negative bacterium isolated from Lake Tebenquiche on the Atacama Saltern, Chile. Int. J. Syst. Evol. Microbiol. *56*: 647–651.

Prakash, O. and R. Lal. 2006. Description of *Sphingobium fuliginis* sp. nov., a phenanthrene-degrading bacterium from a fly ash dumping site, and reclassification of *Sphingomonas cloacae* as *Sphingobium cloacae* comb. nov. Int. J. Syst. Evol. Microbiol. *56*: 2147–2152.

Prakash, O., K. Kumari and R. Lal. 2007. *Pseudomonas delhiensis* sp. nov., from a fly ash dumping site of a thermal power plant. Int. J. Syst. Evol. Microbiol. *57*: 527–531.

Pujalte, M.J., M.C. Macián, D.R. Arahal and E. Garay. 2005a. *Stappia alba* sp. nov., isolated from Mediterranean oysters. Syst. Appl. Microbiol. *28*: 672–678.

Pujalte, M.J., M.C. Macián, D.R. Arahal and E. Garay. 2005b. *Thalassobacter stenotrophicus* Macián *et al.* 2005 is a later synonym of *Jannaschia cystaugens* Adachi *et al.* 2004, with emended description of the genus *Thalassobacter*. Int. J. Syst. Evol. Microbiol. *55*: 1959–1963.

Pujalte, M.J., M.C. Macián, D.R. Arahal, W. Ludwig, K.H. Schleifer and E. Garay. 2005c. *Nereida ignava* gen. nov., sp. nov., a novel aerobic marine α-proteobacterium that is closely related to uncultured *Prionitis* (alga) gall symbionts. Int. J. Syst. Evol. Microbiol. *55*: 631–636.

Pukall, R., P. Schumann, C. Schütte, R. Gols and M. Dicke. 2006. *Acaricomes phytoseiuli* gen. nov., sp. nov., isolated from the predatory mite *Phytoseiulus persimilis*. Int. J. Syst. Evol. Microbiol. *56*: 465–469.

Qin, Q.-L., D.-L. Zhao, J. Wang, X.-L. Chen, H.-Y. Dang, T.-G. Li, Y.-Z. Zhang and P.-J. Gao. 2007. *Wangia profunda* gen. nov., sp. nov., a novel marine bacterium of the family *Flavobacteriaceae* isolated from southern Okinawa Trough deep-sea sediment. FEMS Microbiol. Lett. *271*: 53–58.

Qiu, F., Y. Huang, L. Sun, X. Zhang, Z. Liu and W. Song. 2007. *Leifsonia ginsengi* sp. nov., isolated from ginseng root. Int. J. Syst. Evol. Microbiol. *57*: 405–408.

Qiu, Y.L., Y. Sekiguchi, S. Hanada, H. Imachi, I.C. Tseng, S.S. Cheng, A. Ohashi, H. Harada and Y. Kamagata. 2006. *Pelotomaculum terephthalicum* sp. nov. and *Pelotomaculum isophthalicum* sp. nov.: two anaerobic bacteria that degrade phthalate isomers in syntrophic association with hydrogenotrophic methanogens. Arch. Microbiol. *185*: 172–182.

Quan, Z.X., H.S. Bae, J.H. Baek, W.F. Chen, W.T. Im and S.T. Lee. 2005. *Rhizobium daejeonense* sp. nov. isolated from a cyanide treatment bioreactor. Int. J. Syst. Evol. Microbiol. *55*: 2543–2549.

Quan, Z.X., W.T. Im and S.T. Lee. 2006. *Azonexus caeni* sp. nov., a denitrifying bacterium isolated from sludge of a wastewater treatment plant. Int. J. Syst. Evol. Microbiol. *56*: 1043–1046.

Quan, Z.X., K.K. Kim, M.K. Kim, L. Jin and S.T. Lee. 2007. *Chryseobacterium caeni* sp. nov., isolated from bioreactor sludge. Int. J. Syst. Evol. Microbiol. *57*: 141–145.

Quesada, T., M. Aguilera, J.A. Morillo, A. Ramos-Cormenzana and M. Monteoliva-Sánchez. 2007. *Virgibacillus olivae* sp. nov., isolated from waste wash-water from processing of Spanish-style green olives. Int. J. Syst. Evol. Microbiol. *57*: 906–910.

Quillaguamán, J., O. Delgado, B. Mattiasson and R. Hatti-Kaul. 2004a. *Chromohalobacter sarecensis* sp. nov., a psychrotolerant moderate halophile isolated from the saline Andean region of Bolivia. Int. J. Syst. Evol. Microbiol. *54*: 1921–1926.

Quillaguamán, J., R. Hatti-Kaul, B. Mattiasson, M.T. Alvarez and O. Delgado. 2004b. *Halomonas boliviensis* sp. nov., an alkalitolerant, moderate halophile isolated from soil around a Bolivian hypersaline lake. Int. J. Syst. Evol. Microbiol. *54*: 721–725.

Quintana, E.T., M.E. Trujillo and M. Goodfellow. 2003. *Actinomadura mexicana* sp. nov. and *Actinomadura meyerii* sp. nov., two novel soil sporoactinomycetes. Syst. Appl. Microbiol. *26*: 511–517.

Raats, D. and M. Halpern. 2007. *Oceanobacillus chironomi* sp. nov., a halotolerant and facultatively alkaliphilic species isolated from a chironomid egg mass. Int. J. Syst. Evol. Microbiol. *57*: 255–259.

Rabold, S., V.M. Gorlenko and J.F. Imhoff. 2006. *Thiorhodococcus mannitoliphagus* sp. nov., a purple sulfur bacterium from the White Sea. Int. J. Syst. Evol. Microbiol. *56*: 1945–1951.

Rahalkar, M., I. Bussmann and B. Schink. 2007. *Methylosoma difficile* gen. nov., sp. nov., a novel methanotroph enriched by gradient cultivation from littoral sediment of Lake Constance. Int. J. Syst. Evol. Microbiol. *57*: 1073–1080.

Rainey, F.A., K. Ray, M. Ferreira, B.Z. Gatz, M.F. Nobre., D. Bagaley, B.A. Rash, M.J. Park, A.M. Earl, N.C. Shank, A.M. Small, M.C. Henk, J.R. Battista, P. Kämpfer and M.S. da Costa S. 2005. Extensive diversity of ionizing-radiation-resistant bacteria recovered from Sonoran Desert soil and description of nine new species of the genus *Deinococcus* obtained from a single soil sample. Appl. Environ. Microbiol. *71*: 5225–5235.

Rainey, F.A., M. Ferreira, M.F. Nobre, K. Ray, D. Bagaley, A.M. Earl, J.R. Battista, B. Gómez-Silva, C.P. McKay and M.S. da Costa. 2007. *Deinococcus peraridilitoris* sp. nov., isolated from a coastal desert. Int. J. Syst. Evol. Microbiol. *57*: 1408–1412.

Ramamoorthy, S., H. Sass, H. Langner, P. Schumann, R.M. Kroppenstedt, S. Spring, J. Overmann and R.F. Rosenzweig. 2006. *Desulfosporosinus lacus* sp. nov., a sulfate-reducing bacterium isolated from pristine freshwater lake sediments. Int. J. Syst. Evol. Microbiol. *56*: 2729–2736.

Ramana Ch, V., C. Sasikala, K. Arunasri, P. Anil Kumar, T.N. Srinivas, S. Shivaji, P. Gupta, J. Süling and J.F. Imhoff. 2006. *Rubrivivax benzoatilyticus* sp. nov., an aromatic, hydrocarbon-degrading purple betaproteobacterium. Int. J. Syst. Evol. Microbiol. *56*: 2157–2164.

Rapala, J., K.A. Berg, C. Lyra, R.M. Niemi, W. Manz, S. Suomalainen, L. Paulin and K. Lahti. 2005. *Paucibacter toxinivorans* gen. nov., sp. nov.,

a bacterium that degrades cyclic cyanobacterial hepatotoxins microcystins and nodularin. Int. J. Syst. Evol. Microbiol. *55*: 1563–1568.

Rathgeber, C., N. Yurkova, E. Stackebrandt, P. Schumann, J.T. Beatty and V. Yurkov. 2005. *Roseicyclus mahoneyensis* gen. nov., sp. nov., an aerobic phototrophic bacterium isolated from a meromictic lake. Int. J. Syst. Evol. Microbiol. *55*: 1597–1603.

Rea, S., J.P. Bowman, S. Popovski, C. Pimm and A.D. Wright. 2007. *Methanobrevibacter millerae* sp. nov. and *Methanobrevibacter olleyae* sp. nov., methanogens from the ovine and bovine rumen that can utilize formate for growth. Int. J. Syst. Evol. Microbiol. *57*: 450–456.

Reddy, G.S. and F. García-Pichel. 2005. *Dyadobacter crusticola* sp. nov., from biological soil crusts in the Colorado Plateau, USA, and an emended description of the genus *Dyadobacter* Chelius and Triplett 2000. Int. J. Syst. Evol. Microbiol. *55*: 1295–1299.

Reddy, G.S. and F. García-Pichel. 2007. *Sphingomonas mucosissima* sp. nov. and *Sphingomonas desiccabilis* sp. nov., from biological soil crusts in the Colorado Plateau, USA. Int. J. Syst. Evol. Microbiol. *57*: 1028–1034.

Reddy, G.S., G.I. Matsumoto, P. Schumann, E. Stackebrandt and S. Shivaji. 2004. Psychrophilic pseudomonads from Antarctica: *Pseudomonas antarctica* sp. nov., *Pseudomonas meridiana* sp. nov. and *Pseudomonas proteolytica* sp. nov. Int. J. Syst. Evol. Microbiol. *54*: 713–719.

Reddy, G.S., M. Nagy and F. García-Pichel. 2006. *Belnapia moabensis* gen. nov., sp. nov., an alphaproteobacterium from biological soil crusts in the Colorado Plateau, USA. Int. J. Syst. Evol. Microbiol. *56*: 51–58.

Reddy, G.S., R.M. Potrafka and F. García-Pichel. 2007. *Modestobacter versicolor* sp. nov., an actinobacterium from biological soil crusts that produces melanins under oligotrophy, with emended descriptions of the genus *Modestobacter* and *Modestobacter multiseptatus* Mevs *et al.* 2000. Int. J. Syst. Evol. Microbiol. *57*: 2014–2020.

Rehfuss, M. and J. Urban. 2005a. *Alcaligenes faecalis* subsp. *phenolicus* subsp. nov. a phenol-degrading, denitrifying bacterium isolated from a graywater bioprocessor. Syst. Appl. Microbiol. *28*: 421–429.

Rehfuss, M. and J. Urban. 2005b. *Rhodococcus phenolicus* sp. nov., a novel bioprocessor isolated actinomycete with the ability to degrade chlorobenzene, dichlorobenzene and phenol as sole carbon sources. Syst. Appl. Microbiol. *28*: 695–701.

Reichenbach, H. 2005a. Order VIII. *Myxococcales* Tchan, Pochon and Prévot 1948, 398[AL]. *In* Brenner, Krieg, Staley and Garrity (Editors), Bergey's Manual of Systematic Bacteriology, 2nd Edition, Volume 2 (*The Proteobacteria*), Part C (*The Alpha-, Beta-, Delta-,* and *Epsilonproteobacteria*), Springer, New York, pp. 1059–1072.

Reichenbach, H. 2005b. Genus II. *Corallococcus* gen. nov. (*Chondrococcus* Jahn 1924, 85). *In* Brenner, Krieg, Staley and Garrity (Editors), Bergey's Manual of Systematic Bacteriology, 2nd Edition, Volume 2 (*The Proteobacteria*), Part C (*The Alpha-, Beta-, Delta-,* and *Epsilonproteobacteria*), Springer, New York, pp. 1079–1082.

Reichenbach, H. 2005c. Genus III. *Pyxicoccus* gen. nov. *In* Brenner, Krieg, Staley and Garrity (Editors), Bergey's Manual of Systematic Bacteriology, 2nd Edition, Volume 2 (*The Proteobacteria*), Part C (*The Alpha-, Beta-, Delta-,* and *Epsilonproteobacteria*), Springer, New York, pp. 1083–1084.

Reichenbach, H. 2005d. Genus I. *Cystobacter* Schroeter 1886, 170[AL]. *In* Brenner, Krieg, Staley and Garrity (Editors), Bergey's Manual of Systematic Bacteriology, 2nd Edition, Volume 2 (*The Proteobacteria*), Part C (*The Alpha-, Beta-, Delta-,* and *Epsilonproteobacteria*), Springer, New York, pp. 1086–1096.

Reichenbach, H. 2005e. Genus III. *Hyalangium* gen. nov. *In* Brenner, Krieg, Staley and Garrity (Editors), Bergey's Manual of Systematic Bacteriology, 2nd Edition, Volume 2 (*The Proteobacteria*), Part C (*The Alpha-, Beta-, Delta-,* and *Epsilonproteobacteria*), Springer, New York, pp. 1099–1101.

Reichenbach, H. 2005f. Genus V. *Stigmatella* Berkeley and Curtis in Berkeley 1857, 313. *In* Brenner, Krieg, Staley and Garrity (Editors), Bergey's Manual of Systematic Bacteriology, 2nd Edition, Volume 2 (*The Proteobacteria*), Part C (*The Alpha-, Beta-, Delta-,* and *Epsilonproteobacteria*), Springer, New York, pp. 1104–1108.

Reichenbach, H. 2005g. Genus III. *Chondromyces* Berkeley and Curtis in Berkeley 1874, 64^AL. *In* Brenner, Krieg, Staley and Garrity (Editors), Bergey's Manual of Systematic Bacteriology, 2nd Edition, Volume 2 (*The Proteobacteria*), Part C (*The Alpha-, Beta-, Delta-, and Epsilonproteobacteria*), Springer, New York, pp. pp. 1121–1129.

Reichenbach, H. 2005h. Genus V. *Jahnia* gen. nov. *In* Brenner, Krieg, Staley and Garrity (Editors), Bergey's Manual of Systematic Bacteriology, 2nd Edition, Volume 2 (*The Proteobacteria*), Part C (*The Alpha-, Beta-, Delta-, and Epsilonproteobacteria*), Springer, New York, pp. 1130–1131.

Reichenbach, H. 2005i. Genus VI. *Sorangium* (Jahn 1924) gen. nov., nom. rev. *In* Brenner, Krieg, Staley and Garrity (Editors), Bergey's Manual of Systematic Bacteriology, 2nd Edition, Volume 2 (*The Proteobacteria*), Part C (*The Alpha-, Beta-, Delta-, and Epsilonproteobacteria*), Springer, New York, pp. 1132–1136.

Reichenbach, H. 2005j. Family IV. *Nannocystaceae* fam. nov. *In* Brenner, Krieg, Staley and Garrity (Editors), Bergey's Manual of Systematic Bacteriology, 2nd Edition, Volume 2 (The *Proteobacteria*), Part C (The *Alpha-, Beta-, Delta-, and Epsilonproteobacteria*), Springer, New York, pp. 1136–1137.

Reichenbach, H. 2005k. Genus I. *Nannocystis* Reichenbalch 1970, 137^AL. *In* Brenner, Krieg, Staley and Garrity (Editors), Bergey's Manual of Systematic Bacteriology, 2nd Edition, Volume 2 (*The Proteobacteria*), Part C (*The Alpha-, Beta-, Delta-, and Epsilonproteobacteria*), Springer, New York, pp. 1137–1142.

Reichenbach, H. 2005l. Family V. *Kofleriaceae* fam. nov. *In* Brenner, Krieg, Staley and Garrity (Editors), Bergey's Manual of Systematic Bacteriology, 2nd Edition, Volume 2 (*The Proteobacteria*), Part C (*The Alpha-, Beta-, Delta-, and Epsilonproteobacteria*), Springer, New York, p. 1143.

Reichenbach, H. 2005m. Genus I. *Kofleria* gen. nov. *In* Brenner, Krieg, Staley and Garrity (Editors), Bergey's Manual of Systematic Bacteriology, 2nd Edition, Volume 2 (*The Proteobacteria*), Part C (*The Alpha-, Beta-, Delta-, and Epsilonproteobacteria*), Springer, New York, pp. 1143–1144.

Reichenbach, H., E. Lang, P. Schumann and C. Spröer. 2006. *Byssovorax cruenta* gen. nov., sp. nov., nom. rev., a cellulose-degrading myxobacterium: rediscovery of '*Myxococcus cruentus*' Thaxter 1897. Int. J. Syst. Evol. Microbiol. *56*: 2357–2363.

Reis, V.M., P. Estrada-de los Santos, S. Tenorio-Salgado, J. Vogel, M. Stoffels, P. Guyon, P. Mavingui, V.L. Baldani, M. Schmid, J.I. Baldani, J. Balandreau, A. Hartmann and J. Caballero-Mellado. 2004. *Burkholderia tropica* sp. nov., a novel nitrogen-fixing, plant-associated bacterium. Int. J. Syst. Evol. Microbiol. *54*: 2155–2162.

Reischl, U., H. Melzl, R.M. Kroppenstedt, T. Miethke, L. Naumann, A. Mariottini, G. Mazzarelli and E. Tortoli. 2006. *Mycobacterium monacense* sp. nov. Int. J. Syst. Evol. Microbiol. *56*: 2575–2578.

Ren, P.G. and P.J. Zhou. 2005. *Tenuibacillus multivorans* gen. nov., sp. nov., a moderately halophilic bacterium isolated from saline soil in Xin-Jiang, China. Int. J. Syst. Evol. Microbiol. *55*: 95–99.

Ren, P.G. and P.J. Zhou. 2005. *Salinibacillus aidingensis* gen. nov., sp. nov. and *Salinibacillus kushneri* sp. nov., moderately halophilic bacteria isolated from a neutral saline lake in Xin-Jiang, China. Int. J. Syst. Evol. Microbiol. *55*: 949–953.

Renaud, F.N., A.L. Coustumier, N. Wilhem, D. Aubel, P. Riegel, C. Bollet and J. Freney. 2007. *Corynebacterium hansenii* sp. nov., an alpha-glucosidase-negative bacterium related to *Corynebacterium xerosis*. Int. J. Syst. Evol. Microbiol. *57*: 1113–1116.

Rhodes, M.W., H. Kator, A. McNabb, C. Deshayes, J.M. Reyrat, B.A. Brown-Elliott, R. Wallace, Jr., K.A. Trott, J.M. Parker, B. Lifland, G. Osterhout, I. Kaattari, K. Reece, W. Vogelbein and C.A. Ottinger. 2005. *Mycobacterium pseudoshottsii* sp. nov., a slowly growing chromogenic species isolated from Chesapeake Bay striped bass (*Morone saxatilis*). Int. J. Syst. Evol. Microbiol. *55*: 1139–1147.

Richter, D., D. Postic, N. Sertour, I. Livey, F.R. Matuschka and G. Baranton. 2006. Delineation of *Borrelia burgdorferi sensu lato* species by multilocus sequence analysis and confirmation of the delineation of *Borrelia spielmanii* sp. nov. Int. J. Syst. Evol. Microbiol. *56*: 873–881.

Rickard, A.H., A.T. Stead, G.A. O'May, S. Lindsay, M. Banner, P.S. Handley and P. Gilbert. 2005. *Adhaeribacter aquaticus* gen. nov., sp. nov., a Gram-negative isolate from a potable water biofilm. Int. J. Syst. Evol. Microbiol. *55*: 821–829.

Riegel, P., R. Creti, R. Mattei, A. Nieri and C. Von Hunolstein. 2006. Isolation of *Corynebacterium tuscaniae* sp. nov. from blood cultures of a patient with endocarditis. J. Clin. Microbiol. *44*: 307–312.

Rivas, R., A. Abril, M.E. Trujillo and E. Velázquez. 2004a. *Sphingomonas phyllosphaerae* sp. nov., from the phyllosphere of Acacia caven in Argentina. Int. J. Syst. Evol. Microbiol. *54*: 2147–2150.

Rivas, R., M.E. Trujillo, P.F. Mateos, E. Martínez-Molina and E. Velázquez. 2004b. *Cellulomonas xylanilytica* sp. nov., a cellulolytic and xylanolytic bacterium isolated from a decayed elm tree. Int. J. Syst. Evol. Microbiol. *54*: 533–536.

Rivas, R., M.E. Trujillo, P.F. Mateos, E. Martínez-Molina and E. Velázquez. 2004c. *Agromyces ulmi* sp. nov., a xylanolytic bacterium isolated from *Ulmus nigra* in Spain. Int. J. Syst. Evol. Microbiol. *54*: 1987–1990.

Rivas, R., M.E. Trujillo, M. Sánchez, P.F. Mateos, E. Martínez-Molina and E. Velázquez. 2004d. *Microbacterium ulmi* sp. nov., a xylanolytic, phosphate-solubilizing bacterium isolated from sawdust of *Ulmus nigra*. Int. J. Syst. Evol. Microbiol. *54*: 513–517.

Rivas, R., M.E. Trujillo, P. Schumann, R.M. Kroppenstedt, M. Sánchez, P.F. Mateos, E. Martínez-Molina and E. Velázquez. 2004e. *Xylanibacterium ulmi* gen. nov., sp. nov., a novel xylanolytic member of the family *Promicromonosporaceae*. Int. J. Syst. Evol. Microbiol. *54*: 557–561.

Rivas, R., A. Willems, J.L. Palomo, P. García-Benavides, P.F. Mateos, E. Martínez-Molina, M. Gillis and E. Velázquez. 2004f. *Bradyrhizobium betae* sp. nov., isolated from roots of *Beta vulgaris* affected by tumour-like deformations. Int. J. Syst. Evol. Microbiol. *54*: 1271–1275.

Rivas, R., C. Gutiérrez, A. Abril, P.F. Mateos, E. Martínez-Molina, A. Ventosa and E. Velázquez. 2005a. *Paenibacillus rhizosphaerae* sp. nov., isolated from the rhizosphere of *Cicer arietinum*. Int. J. Syst. Evol. Microbiol. *55*: 1305–1309.

Rivas, R., P.F. Mateos, E. Martínez-Molina and E. Velázquez. 2005b. *Paenibacillus xylanilyticus* sp. nov., an airborne xylanolytic bacterium. Int. J. Syst. Evol. Microbiol. *55*: 405–408.

Rivas, R., P.F. Mateos, E. Martínez-Molina and E. Velázquez. 2005c. *Paenibacillus phyllosphaerae* sp. nov., a xylanolytic bacterium isolated from the phyllosphere of *Phoenix dactylifera*. Int. J. Syst. Evol. Microbiol. *55*: 743–746.

Rivas, R., S. Sánchez-Márquez, P.F. Mateos, E. Martínez-Molina and E. Velázquez. 2005d. *Martelella mediterranea* gen. nov., sp. nov., a novel α-proteobacterium isolated from a subterranean saline lake. Int. J. Syst. Evol. Microbiol. *55*: 955–959.

Rivas, R., P. García-Fraile, P.F. Mateos, E. Martínez-Molina and E. Velázquez. 2006a. *Photobacterium halotolerans* sp. nov., isolated from Lake Martel in Spain. Int. J. Syst. Evol. Microbiol. *56*: 1067–1071.

Rivas, R., P. García-Fraile, P.F. Mateos, E. Martínez-Molina and E. Velázquez. 2006b. *Paenibacillus cellulosilyticus* sp. nov., a cellulolytic and xylanolytic bacterium isolated from the bract phyllosphere of Phoenix dactylifera. Int. J. Syst. Evol. Microbiol. *56*: 2777–2781.

Rivas, R., P. García-Fraile, A. Peix, P.F. Mateos, E. Martínez-Molina and E. Velázquez. 2007. *Alcanivorax balearicus* sp. nov., isolated from Lake Martel. Int. J. Syst. Evol. Microbiol. *57*: 1331–1335.

Robert, C., C. Chassard, P.A. Lawson and A. Bernalier-Donadille. 2007. *Bacteroides cellulosilyticus* sp. nov., a cellulolytic bacterium from the human gut microbial community. Int. J. Syst. Evol. Microbiol. *57*: 1516–1520.

Robertson, B.R., J.L. O'Rourke, B.A. Neilan, P. Vandamme, S.L. On, J.G. Fox and A. Lee. 2005. *Mucispirillum schaedleri* gen. nov., sp. nov., a spiral-shaped bacterium colonizing the mucus layer of the gastrointestinal tract of laboratory rodents. Int. J. Syst. Evol. Microbiol. *55*: 1199–1204.

Rodas, A.M., E. Chenoll, M.C. Macián, S. Ferrer, I. Pardo and R. Aznar. 2006. *Lactobacillus vini* sp. nov., a wine lactic acid bacterium homofermentative for pentoses. Int. J. Syst. Evol. Microbiol. *56*: 513–517.

Rodrigues, D.F., J. Goris, T. Vishnivetskaya, D.Gilichinsky, M.F. Thomashow and J.M. Tiedje. 2006. Characterization of *Exiguobacterium* isolates from the Siberian permafrost. Description of *Exiguobacterium sibiricum* sp. nov. Extremophiles *10*: 285–294.

Rodríguez-Díaz, M., L. Lebbe, B. Rodelas, J. Heyrman, P. De Vos and N.A. Logan. 2005. *Paenibacillus wynnii* sp. nov., a novel species harbouring the *nifH* gene, isolated from Alexander Island, Antarctica. Int. J. Syst. Evol. Microbiol. *55*: 2093–2099.

Rodríguez-Nava, V., A. Couble, C. Molinard, H. Sandoval, P. Boiron and F. Laurent. 2004. *Nocardia mexicana* sp. nov., a new pathogen isolated from human mycetomas. J. Clin. Microbiol. *42*: 4530–4535.

Rodríguez-Nava, V., Z.U. Khan, G. Potter, R.M. Kroppenstedt, P. Boiron and F. Laurent. 2007. *Nocardia coubleae* sp. nov., isolated from oil-contaminated Kuwaiti soil. Int. J. Syst. Evol. Microbiol. *57*: 1482–1486.

Roh, S.W., Y.D. Nam, H.W. Chang, Y. Sung, K.H. Kim, H.J. Lee, H.M. Oh and J.W. Bae. 2007a. *Natronococcus jeotgali* sp. nov., a halophilic archaeon isolated from shrimp jeotgal, a traditional fermented seafood from Korea. Int. J. Syst. Evol. Microbiol. *57*: 2129–2131.

Roh, S.W., Y.D. Nam, H.W. Chang, Y. Sung, K.H. Kim, H.M. Oh and J.W. Bae. 2007b. *Halalkalicoccus jeotgali* sp. nov., a halophilic archaeon from shrimp jeotgal, a traditional Korean fermented seafood. Int. J. Syst. Evol. Microbiol. *57*: 2296–2298.

Rojas, A.M., J.E. de Los Rios, M. Fischer-Le Saux, P. Jimenez, P. Reche, S. Bonneau, L. Sutra, F. Mathieu-Daudé and M. McClelland. 2004. *Erwinia toletana* sp. nov., associated with *Pseudomonas savastanoi*-induced tree knots. Int. J. Syst. Evol. Microbiol. *54*: 2217–2222.

Romanenko, L.A., A.M. Lysenko, M. Rohde, V.V. Mikhailov and E. Stackebrandt. 2004a. *Psychrobacter maritimus* sp. nov. and *Psychrobacter arenosus* sp. nov., isolated from coastal sea ice and sediments of the Sea of Japan. Int. J. Syst. Evol. Microbiol. *54*: 1741–1745.

Romanenko, L.A., P. Schumann, M. Rohde, V.V. Mikhailov and E. Stackebrandt. 2004b. *Reinekea marinisedimentorum* gen. nov., sp. nov., a novel gammaproteobacterium from marine coastal sediments. Int. J. Syst. Evol. Microbiol. *54*: 669–673.

Romanenko, L.A., P. Schumann, M. Rohde, N.V. Zhukova, V.V. Mikhailov and E. Stackebrandt. 2005a. *Marinobacter bryozoorum* sp. nov. and *Marinobacter sediminum* sp. nov., novel bacteria from the marine environment. Int. J. Syst. Evol. Microbiol. *55*: 143–148.

Romanenko, L.A., M. Uchino, E. Falsen, G.M. Frolova, N.V. Zhukova and V.V. Mikhailov. 2005b. *Pseudomonas pachastrellae* sp. nov., isolated from a marine sponge. Int. J. Syst. Evol. Microbiol. *55*: 919–924.

Romanenko, L.A., M. Uchino, E. Falsen, A.M. Lysenko, N.V. Zhukova and V.V. Mikhailov. 2005c. *Pseudomonas xanthomarina* sp. nov., a novel bacterium isolated from marine ascidian. J. Gen. Appl. Microbiol. *51*: 65–71.

Romanenko, L.A., M. Uchino, G.M. Frolova and V.V. Mikhailov. 2007a. *Marixanthomonas ophiurae* gen. nov., sp. nov., a marine bacterium of the family *Flavobacteriaceae* isolated from a deep-sea brittle star. Int. J. Syst. Evol. Microbiol. *57*: 457–462.

Romanenko, L.A., M. Uchino, G.M. Frolova, N. Tanaka, N.I. Kalinovskaya, N. Latyshev and V.V. Mikhailov. 2007b. *Sphingomonas molluscorum* sp. nov., a novel marine isolate with antimicrobial activity. Int. J. Syst. Evol. Microbiol. *57*: 358–363.

Romano, I., A. Gambacorta, L. Lama, B. Nicolaus and A. Giordano. 2005a. *Salinivibrio costicola* subsp. *alcaliphilus* subsp. nov., a haloalkaliphilic aerobe from Campania Region (Italy). Syst. Appl. Microbiol. *28*: 34–42.

Romano, I., A. Giordano, L. Lama, B. Nicolaus and A. Gambacorta. 2005b. *Halomonas campaniensis* sp. nov., a haloalkaliphilic bacterium isolated from a mineral pool of Campania Region, Italy. Syst. Appl. Microbiol. *28*: 610–618.

Romano, I., L. Lama, B. Nicolaus, A. Gambacorta and A. Giordano. 2005c. *Bacillus saliphilus* sp. nov., isolated from a mineral pool in Campania, Italy. Int. J. Syst. Evol. Microbiol. *55*: 159–163.

Romano, I., L. Lama, B. Nicolaus, A. Gambacorta and A. Giordano. 2005d. *Alkalibacillus filiformis* sp. nov., isolated from a mineral pool in Campania, Italy. Int. J. Syst. Evol. Microbiol. *55*: 2395–2399.

Romano, I., L. Lama, B. Nicolaus, A. Poli, A. Gambacorta and A. Giordano. 2006a. *Halomonas alkaliphila* sp. nov., a novel halotolerant alkaliphilic bacterium isolated from a salt pool in Campania (Italy). J. Gen. Appl. Microbiol. *52*: 339–348.

Romano, I., L. Lama, B. Nicolaus, A. Poli, A. Gambacorta and A. Giordano. 2006b. *Oceanobacillus oncorhynchi* subsp. *incaldanensis* subsp. nov., an alkalitolerant halophile isolated from an algal mat collected from a sulfurous spring in Campania (Italy), and emended description of *Oceanobacillus oncorhynchi*. Int. J. Syst. Evol. Microbiol. *56*: 805–810.

Romano, I., A. Poli, I. Finore, F.J. Huertas, A. Gambacorta, S. Pelliccione, G. Nicolaus, L. Lama and B. Nicolaus. 2007. *Haloterrigena hispanica* sp. nov., an extremely halophilic archaeon from Fuente de Piedra, southern Spain. Int. J. Syst. Evol. Microbiol. *57*: 1499–1503.

Roos, S., L. Engstrand and H. Jonsson. 2005. *Lactobacillus gastricus* sp. nov., *Lactobacillus antri* sp. nov., *Lactobacillus kalixensis* sp. nov. and *Lactobacillus ultunensis* sp. nov., isolated from human stomach mucosa. Int. J. Syst. Evol. Microbiol. *55*: 77–82.

Rosenblueth, M., L. Martínez, J. Silva and E. Martínez-Romero. 2004. *Klebsiella variicola*, a novel species with clinical and plant-associated isolates. Syst. Appl. Microbiol. *27*: 27–35.

Rothballer, M., M. Schmid, I. Klein, A. Gattinger, S. Grundmann and A. Hartmann. 2006. *Herbaspirillum hiltneri* sp. nov., isolated from surface-sterilized wheat roots. Int. J. Syst. Evol. Microbiol. *56*: 1341–1348.

Roux, V. and D. Raoult. 2004. *Paenibacillus massiliensis* sp. nov., *Paenibacillus sanguinis* sp. nov. and *Paenibacillus timonensis* sp. nov., isolated from blood cultures. Int. J. Syst. Evol. Microbiol. *54*: 1049–1054.

Ruiz-García, C., V. Béjar, F. Martínez-Checa, I. Llamas and E. Quesada. 2005. *Bacillus velezensis* sp. nov., a surfactant-producing bacterium isolated from the river Velez in Malaga, southern Spain. Int. J. Syst. Evol. Microbiol. *55*: 191–195.

Ruiz-García, C., E. Quesada, F. Martínez-Checa, I. Llamas, M.C. Urdaci and V. Béjar. 2005. *Bacillus axarquiensis* sp. nov. and *Bacillus malacitensis* sp. nov., isolated from river-mouth sediments in southern Spain. Int. J. Syst. Evol. Microbiol. *55*: 1279–1285.

Ryu, S.H., T.T. Nguyen, W. Park, C.J. Kim and C.O. Jeon. 2006. *Runella limosa* sp. nov., isolated from activated sludge. Int. J. Syst. Evol. Microbiol. *56*: 2757–2760.

Ryu, S.H., M. Park, J.R. Lee, P.Y. Yun and C.O. Jeon. 2007a. *Brevundimonas aveniformis* sp. nov., a stalked species isolated from activated sludge. Int. J. Syst. Evol. Microbiol. *57*: 1561–1565.

Ryu, S.H., M. Park, Y. Jeon, J.R. Lee, W. Park and C.O. Jeon. 2007b. *Flavobacterium filum* sp. nov., isolated from a wastewater treatment plant in Korea. Int. J. Syst. Evol. Microbiol. *57*: 2026–2030.

Sabry, S.A., N.B. Ghanem, G.A. Abu-Ella, P. Schumann, E. Stackebrandt and R.M. Kroppenstedt. 2004. *Nocardiopsis aegyptia* sp. nov., isolated from marine sediment. Int. J. Syst. Evol. Microbiol. *54*: 453–456.

Saddler, G.S. and J.F. Bradbury. 2005a. Order III. *Xanthomonadales* ord. nov. *In* Brenner, Krieg, Staley and Garrity (Editors), Bergey's Manual of Systematic Bacteriology, 2nd Edition, Volume 2 (The *Proteobacteria*), Part B (The *Gammaproteobacteria*), Springer, New York, p. 63.

Saddler, G.S. and J.F. Bradbury. 2005b. Family I. *Xanthomonadaceae* fam. nov. *In* Brenner, Krieg, Staley and Garrity (Editors), Bergey's Manual of Systematic Bacteriology, 2nd Edition, Volume 2 (The *Proteobacteria*), Part B (The *Gammaproteobacteria*), Springer, New York, p. 63.

Saha, P. and T. Chakrabarti. 2006a. *Emticicia oligotrophica* gen. nov., sp. nov., a new member of the family 'Flexibacteraceae', phylum *Bacteroidetes*. Int. J. Syst. Evol. Microbiol. *56*: 991–995.

Saha, P. and T. Chakrabarti. 2006b. *Aeromonas sharmana* sp. nov., isolated from a warm spring. Int. J. Syst. Evol. Microbiol. *56*: 1905–1909.

Saha, P. and T. Chakrabarti. 2006c. *Flavobacterium indicum* sp. nov., isolated from warm spring water in Assam, India. Int. J. Syst. Evol. Microbiol. *56*: 2617–2621.

Saha, P., S. Krishnamurthi, S. Mayilraj, G.S. Prasad, T.C. Bora and T. Chakrabarti. 2005a. *Aquimonas voraii* gen. nov., sp. nov., a novel gammaproteobacterium isolated from a warm spring of Assam, India. Int. J. Syst. Evol. Microbiol. *55*: 1491–1495.

Saha, P., A.K. Mondal, S. Mayilraj, S. Krishnamurthi, A. Bhattacharya and T. Chakrabarti. 2005b. *Paenibacillus assamensis* sp. nov., a novel bacterium isolated from a warm spring in Assam, India. Int. J. Syst. Evol. Microbiol. *55*: 2577–2581.

Saintpierre-Bonaccio, D., H. Amir, R. Pineau, S. Lemriss and M. Goodfellow. 2004a. *Streptomyces ferralitis* sp. nov., a novel streptomycete isolated from a New-Caledonian ultramafic soil. Int. J. Syst. Evol. Microbiol. *54*: 2061–2065.

Saintpierre-Bonaccio, D., L.A. Maldonado, H. Amir, R. Pineau and M. Goodfellow. 2004b. *Nocardia neocaledoniensis* sp. nov., a novel actinomycete isolated from a New-Caledonian brown hypermagnesian ultramafic soil. Int. J. Syst. Evol. Microbiol. *54*: 599–603.

Saintpierre-Bonaccio, D., H. Amir, R. Pineau, G.Y. Tan and M. Goodfellow. 2005. *Amycolatopsis plumensis* sp. nov., a novel bioactive actinomycete isolated from a New-Caledonian brown hypermagnesian ultramafic soil. Int. J. Syst. Evol. Microbiol. *55*: 2057–2061.

Sakamoto, M. and Y. Benno. 2006. Reclassification of *Bacteroides distasonis*, *Bacteroides goldsteinii* and *Bacteroides merdae* as *Parabacteroides distasonis* gen. nov., comb. nov., *Parabacteroides goldsteinii* comb. nov. and *Parabacteroides merdae* comb. nov. Int. J. Syst. Evol. Microbiol. *56*: 1599–1605.

Sakamoto, M., M. Suzuki, Y. Huang, M. Umeda, I. Ishikawa and Y. Benno. 2004. *Prevotella shahii* sp. nov. and *Prevotella salivae* sp. nov., isolated from the human oral cavity. Int. J. Syst. Evol. Microbiol. *54*: 877–883.

Sakamoto, M., Y. Huang, M. Umeda, I. Ishikawa and Y. Benno. 2005a. *Prevotella multiformis* sp. nov., isolated from human subgingival plaque. Int. J. Syst. Evol. Microbiol. *55*: 815–819.

Sakamoto, M., M. Umeda, I. Ishikawa and Y. Benno. 2005b. *Prevotella multisaccharivorax* sp. nov., isolated from human subgingival plaque. Int. J. Syst. Evol. Microbiol. *55*: 1839–1843.

Sakamoto, M., M. Kitahara and Y. Benno. 2007a. *Parabacteroides johnsonii* sp. nov., isolated from human faeces. Int. J. Syst. Evol. Microbiol. *57*: 293–296.

Sakamoto, M., P.T. Lan and Y. Benno. 2007b. *Barnesiella viscericola* gen. nov., sp. nov., a novel member of the family *Porphyromonadaceae* isolated from chicken caecum. Int. J. Syst. Evol. Microbiol. *57*: 342–346.

Sakamoto, M., K. Ohkusu, T. Masaki, H. Kako, T. Ezaki and Y. Benno. 2007c. *Prevotella pleuritidis* sp. nov., isolated from pleural fluid. Int. J. Syst. Evol. Microbiol. *57*: 1725–1728.

Sakuma, K., M. Kitahara, R. Kibe, M. Sakamoto and Y. Benno. 2006. *Clostridium glycyrrhizinilyticum* sp. nov., a glycyrrhizin-hydrolysing bacterium isolated from human faeces. Microbiol. Immunol. *50*: 481–485.

Samson, R., J.B. Legendre, R. Christen, M. Fischer-Le Saux, W. Achouak and L. Gardan. 2005. Transfer of *Pectobacterium chrysanthemi* (Burkholder *et al.* 1953) Brenner *et al.* 1973 and *Brenneria paradisiaca* to the genus *Dickeya* gen. nov. as *Dickeya chrysanthemi* comb. nov. and *Dickeya paradisiaca* comb. nov. and delineation of four novel species, *Dickeya dadantii* sp. nov., *Dickeya dianthicola* sp. nov., *Dickeya dieffenbachiae* sp. nov. and *Dickeya zeae* sp. nov. Int. J. Syst. Evol. Microbiol. *55*: 1415–1427.

Sánchez, M.M., D. Fritze, A. Blanco, C. Spröer, B.J. Tindall, P. Schumann, R.M. Kroppenstedt, P. Diaz and F.I. Pastor. 2005. *Paenibacillus barcinonensis* sp. nov., a xylanase-producing bacterium isolated from a rice field in the Ebro River delta. Int. J. Syst. Evol. Microbiol. *55*: 935–939.

Sánchez-Porro, C., H. Tokunaga, M. Tokunaga and A. Ventosa. 2007. *Chromohalobacter japonicus* sp. nov., a moderately halophilic bacterium isolated from a Japanese salty food. Int. J. Syst. Evol. Microbiol. *57*: 2262–2266.

Santini, J.M., I.C. Streimann and R.N. vanden Hoven. 2004. *Bacillus macyae* sp. nov., an arsenate-respiring bacterium isolated from an Australian gold mine. Int. J. Syst. Evol. Microbiol. *54*: 2241–2244.

Santos, P., I. Pinhal, F.A. Rainey, N. Empadinhas, J. Costa, B. Fields, R. Benson, A. Veríssimo and M.S. da Costa. 2003. Gamma-*Proteobacteria* *Aquicella lusitana* gen. nov., sp. nov., and *Aquicella siphonis* sp. nov. infect protozoa and require activated charcoal for growth in laboratory media. Appl. Environ. Microbiol. *69*: 6533–6540.

Sato, Y., H. Nishihara, M. Yoshida, M. Watanabe, J.D. Rondal, R.N. Concepcion and H. Ohta. 2006. *Cupriavidus pinatubonensis* sp. nov. and *Cupriavidus laharis* sp. nov., novel hydrogen-oxidizing, facultatively chemolithotrophic bacteria isolated from volcanic mudflow deposits from Mt. Pinatubo in the Philippines. Int. J. Syst. Evol. Microbiol. *56*: 973–978.

Satomi, M., B. Kimura, M. Hayashi, M. Okuzumi and T. Fujii. 2004. *Marinospirillum insulare* sp. nov., a novel halophilic helical bacterium isolated from kusaya gravy. Int. J. Syst. Evol. Microbiol. *54*: 163–167.

Satomi, M., M.T. La Duc and K. Venkateswaran. 2006a. *Bacillus safensis* sp. nov., isolated from spacecraft and assembly-facility surfaces. Int. J. Syst. Evol. Microbiol. *56*: 1735–1740.

Satomi, M., B.F. Vogel, L. Gram and K. Venkateswaran. 2006b. *Shewanella hafniensis* sp. nov. and *Shewanella morhuae* sp. nov., isolated from marine fish of the Baltic Sea. Int. J. Syst. Evol. Microbiol. *56*: 243–249.

Satomi, M., B.F. Vogel, K. Venkateswaran and L. Gram. 2007. Description of *Shewanella glacialipiscicola* sp. nov. and *Shewanella algidipiscicola* sp. nov., isolated from marine fish of the Danish Baltic Sea, and proposal that *Shewanella affinis* is a later heterotypic synonym of *Shewanella colwelliana*. Int. J. Syst. Evol. Microbiol. *57*: 347–352.

Savage, K.N., L.R. Krumholz, A. Oren and M.S. Elshahed. 2007. *Haladaptatus paucihalophilus* gen. nov., sp. nov., a halophilic archaeon isolated from a low-salt, sulfide-rich spring. Int. J. Syst. Evol. Microbiol. *57*: 19–24.

Savic, M., I. Bratic and B. Vasiljevic. 2007. *Streptomyces durmitorensis* sp. nov., a producer of an FK506-like immunosuppressant. Int. J. Syst. Evol. Microbiol. *57*: 2119–2124.

Sawabe, T., K. Hayashi., J. Moriwaki, Y. Fukui, F.L. Thompson, J. Swings and R. Christen. 2004a. *Vibrio neonatus* sp. nov. and *Vibrio ezurae* sp. nov. isolated from the gut of Japanese abalones. Syst. Appl. Microbiol. *27*: 527–534.

Sawabe, T., K. Hayashi, J. Moriwaki, F.L. Thompson, J. Swings, P. Potin, R. Christen and Y. Ezura. 2004b. *Vibrio gallicus* sp. nov., isolated from the gut of the French abalone *Haliotis tuberculata*. Int. J. Syst. Evol. Microbiol. *54*: 843–846.

Sawabe, T., Y. Fujimura, K. Niwa and H. Aono. 2007. *Vibrio comitans* sp. nov., *Vibrio rarus* sp. nov. and *Vibrio inusitatus* sp. nov., from the gut of the abalones *Haliotis discus discus*, *H. gigantea*, *H. madaka* and *H. rufescens*. Int. J. Syst. Evol. Microbiol. *57*: 916–922.

Schaad, N.W., E. Postnikova, G. Lacy, A. Sechler, I. Agarkova, P.E. Stromberg, V.K. Stromberg and A.K. Vidaver. 2006. Emended classification of xanthomonad pathogens on citrus. Syst. Appl. Microbiol. *29*: 690–695.

Schäffer, C., W.L. Franck, A. Scheberl, P. Kosma, T.R. McDermott and P. Messner. 2004. Classification of isolates from locations in Austria and Yellowstone National Park as *Geobacillus tepidamans* sp. nov. Int. J. Syst. Evol. Microbiol. *54*: 2361–2368.

Scheirlinck, I., R. Van der Meulen, A. Van Schoor, I. Cleenwerck, G. Huys, P. Vandamme, L. De Vuyst and M. Vancanneyt. 2007a. *Lactobacillus namurensis* sp. nov., isolated from a traditional Belgian sourdough. Int. J. Syst. Evol. Microbiol. *57*: 223–227.

Scheirlinck, I., R. Van der Meulen, A. Van Schoor, G. Huys, P. Vandamme, L. De Vuyst and M. Vancanneyt. 2007b. *Lactobacillus crustorum* sp. nov., isolated from two traditional Belgian wheat sourdoughs. Int. J. Syst. Evol. Microbiol. *57*: 1461–1467.

Scheldeman, P., K. Goossens, M. Rodríguez-Díaz, A. Pil, J. Goris, L. Herman, P. De Vos, N.A. Logan and M. Heyndrickx. 2004a. *Paenibacillus lactis* sp. nov., isolated from raw and heat-treated milk. Int. J. Syst. Evol. Microbiol. *54*: 885–891.

Scheldeman, P., M. Rodríguez-Díaz, J. Goris, A. Pil, E. De Clerck, L. Herman, P. De Vos, N.A. Logan and M. Heyndrickx. 2004b. *Bacillus farraginis* sp. nov., *Bacillus fortis* sp. nov. and *Bacillus fordii* sp. nov., isolated at dairy farms. Int. J. Syst. Evol. Microbiol. *54*: 1355–1364.

Scheuermayer, M., T.A. Gulder, G. Bringmann and U. Hentschel. 2006. *Rubritalea marina* gen. nov., sp. nov., a marine representative of the phylum '*Verrucomicrobia*', isolated from a sponge (*Porifera*). Int. J. Syst. Evol. Microbiol. *56*: 2119–2124.

Schinsky, M.F., R.E. Morey, A.G. Steigerwalt, M.P. Douglas, R.W. Wilson, M.M. Floyd, W.R. Butler, M.I. Daneshvar, B.A. Brown-Elliott, R.J. Wallace, Jr., M.M. McNeil, D.J. Brenner and J.M. Brown. 2004. Taxonomic variation in the *Mycobacterium fortuitum* third biovariant complex: description of *Mycobacterium boenickei* sp. nov., *Mycobacterium houstonense* sp. nov., *Mycobacterium neworleansense* sp. nov. and *Mycobacterium brisbanense* sp. nov. and recognition of *Mycobacterium porcinum* from human clinical isolates. Int. J. Syst. Evol. Microbiol. *54*: 1653–1667.

Schippers, A., K. Bosecker, C. Spröer and P. Schumann. 2005a. *Microbacterium oleivorans* sp. nov. and *Microbacterium hydrocarbonoxydans* sp. nov., novel crude-oil-degrading Gram-positive bacteria. Int. J. Syst. Evol. Microbiol. *55*: 655–660.

Schippers, A., P. Schumann and C. Spröer. 2005b. *Nocardioides oleivorans* sp. nov., a novel crude-oil-degrading bacterium. Int. J. Syst. Evol. Microbiol. *55*: 1501–1504.

Schleheck, D., B.J. Tindall, R. Rosselló-Mora and A.M. Cook. 2004. *Parvibaculum lavamentivorans* gen. nov., sp. nov., a novel heterotroph that initiates catabolism of linear alkylbenzenesulfonate. Int. J. Syst. Evol. Microbiol. *54*: 1489–1497.

Schlesner, H., C. Rensmann, B.J. Tindall, D. Gade, R. Rabus, S. Pfeiffer and P. Hirsch. 2004. Taxonomic heterogeneity within the *Planctomycetales* as derived by DNA-DNA hybridization, description of *Rhodopirellula baltica* gen. nov., sp. nov., transfer of *Pirellula marina* to the genus *Blastopirellula* gen. nov. as *Blastopirellula marina* comb. nov. and emended description of the genus *Pirellula*. Int. J. Syst. Evol. Microbiol. *54*: 1567–1580.

Schmidt, M., A. Priemé and P. Stougaard. 2006. *Rhodonellum psychrophilum* gen. nov., sp. nov., a novel psychrophilic and alkaliphilic bacterium of the phylum *Bacteroidetes* isolated from Greenland. Int. J. Syst. Evol. Microbiol. *56*: 2887–2892.

Schumann, P., X. Cui, E. Stackebrandt, R.M. Kroppenstedt, L. Xu and C. Jiang. 2004. *Jonesia quinghaiensis* sp. nov., a new member of the suborder *Micrococcineae*. Int. J. Syst. Evol. Microbiol. *54*: 2181–2184.

Sekiguchi, Y., H. Imachi, A. Susilorukmi, M. Muramatsu, A. Ohashi, H. Harada, S. Hanada and Y. Kamagata. 2006. *Tepidanaerobacter syntrophicus* gen. nov., sp. nov., an anaerobic, moderately thermophilic, syntrophic alcohol- and lactate-degrading bacterium isolated from thermophilic digested sludges. Int. J. Syst. Evol. Microbiol. *56*: 1621–1629.

Semêdo, L.T., R.C. Gomes, A.A. Linhares, G.F. Duarte, R.P. Nascimento, A.S. Rosado, M. Margis-Pinheiro, R. Margis, K.R. Silva, C.S. Alviano, G.P. Manfio, R.M. Soares, L.F. Linhares and R.R. Coelho. 2004. *Streptomyces drozdowiczii* sp. nov., a novel cellulolytic streptomycete from soil in Brazil. Int. J. Syst. Evol. Microbiol. *54*: 1323–1328.

Seo, H.J., S.S. Bae, J.H. Lee and S.J. Kim. 2005a. *Photobacterium frigidiphilum* sp. nov., a psychrophilic, lipolytic bacterium isolated from deep-sea sediments of Edison Seamount. Int. J. Syst. Evol. Microbiol. *55*: 1661–1666.

Seo, H.J., S.S. Bae, S.H. Yang, J.H. Lee and S.J. Kim. 2005b. *Photobacterium aplysiae* sp. nov., a lipolytic marine bacterium isolated from eggs of the sea hare *Aplysia kurodai*. Int. J. Syst. Evol. Microbiol. *55*: 2293–2296.

Seo, J.P. and S.D. Lee. 2006. *Nocardia harenae* sp. nov., an actinomycete isolated from beach sand. Int. J. Syst. Evol. Microbiol. *56*: 2203–2207.

Seo, J.P., Y.W. Yun and S.D. Lee. 2007. *Nocardia speluncae* sp. nov., isolated from a cave. Int. J. Syst. Evol. Microbiol. *57*: 2932–2935.

Sessitsch, A., T. Coenye, A.V. Sturz, P. Vandamme, E.A. Barka, J.F. Salles, J.D. Van Elsas, D. Faure, B. Reiter, B.R. Glick, G. Wang-Pruski and J. Nowak. 2005. *Burkholderia phytofirmans* sp. nov., a novel plant-associated bacterium with plant-beneficial properties. Int. J. Syst. Evol. Microbiol. *55*: 1187–1192.

Shashidhar, R. and J.R. Bandekar. 2006. *Deinococcus mumbaiensis* sp. nov., a radiation-resistant pleomorphic bacterium isolated from Mumbai, India. FEMS Microbiol. Lett. *254*: 275–280.

Shelobolina, E.S., S.A. Sullivan, K.R. O'neill, K.P. Nevin and D.R. Lovley. 2004. Isolation, characterization, and U(VI)-reducing potential of a facultatively anaerobic, acid-resistant bacterium from low-pH, nitrate- and U(VI)-contaminated subsurface sediment and description of *Salmonella subterranea* sp. nov. Appl. Environ. Microbiol. *70*: 2959–2965.

Shelobolina, E.S., K.P. Nevin, J.D. Blakeney-Hayward, C.V. Johnsen, T.W. Plaia, P. Krader, T. Woodard, D.E. Holmes, C.G. Vanpraagh and D.R. Lovley. 2007. *Geobacter pickeringii* sp. nov., *Geobacter argillaceus* sp. nov. and *Pelosinus fermentans* gen. nov., sp. nov., isolated from subsurface kaolin lenses. Int. J. Syst. Evol. Microbiol. *57*: 126–135.

Shen, F.T., P. Kämpfer, C.C. Young, W.A. Lai and A.B. Arun. 2005. *Chryseobacterium taichungense* sp. nov., isolated from contaminated soil. Int. J. Syst. Evol. Microbiol. *55*: 1301–1304.

Shen, F.T., M. Goodfellow, A.L. Jones, Y.P. Chen, A.B. Arun, W.A. Lai, P.D. Rekha and C.C. Young. 2006. *Gordonia soli* sp. nov., a novel actinomycete isolated from soil. Int. J. Syst. Evol. Microbiol. *56*: 2597–2601.

Shen, Z., S. Xu, F.E. Dewhirst, B.J. Paster, J.A. Pena, I.M. Modlin, M. Kidd and J.G. Fox. 2005. A novel enterohepatic *Helicobacter* species '*Helicobacter mastomyrinus*' isolated from the liver and intestine of rodents. Helicobacter *10*: 59–70.

Sheu, S.Y., K.Y. Lin, J.H. Chou, P.S. Chang, A.B. Arun, C.C. Young and W.M. Chen. 2007. *Tenacibaculum litopenaei* sp. nov., isolated from a shrimp mariculture pond. Int. J. Syst. Evol. Microbiol. *57*: 1148–1153.

Shewmaker, P.L., A.G. Steigerwalt, R.E. Morey, M. da G.S Carvalho, J.A. Elliott, K. Joyce, T.J. Barrett, L.M. Teixeira and R.R. Facklam. 2004. *Vagococcus carniphilus* sp. nov., isolated from ground beef. Int. J. Syst. Evol. Microbiol. *54*: 1505–1510.

Shewmaker, P.L., A.C. Camus, T. Bailiff, A.G. Steigerwalt, R.E. Morey and M. da G. Carvalho. 2007. *Streptococcus ictaluri* sp. nov., isolated from Channel Catfish *Ictalurus punctatus* broodstock. Int. J. Syst. Evol. Microbiol. *57*: 1603–1606.

Shieh, W.Y., Y.T. Lin and W.D. Jean. 2004. *Pseudovibrio denitrificans* gen. nov., sp. nov., a marine, facultatively anaerobic, fermentative bacterium capable of denitrification. Int. J. Syst. Evol. Microbiol. *54*: 2307–2312.

Shimomura, K., S. Kaji and A. Hiraishi. 2005. *Chryseobacterium shigense* sp. nov., a yellow-pigmented, aerobic bacterium isolated from a lactic acid beverage. Int. J. Syst. Evol. Microbiol. *55*: 1903–1906.

Shivaji, S., G.S.N. Reddy, P.U.M. Raghavan, N.B. Sarita and D. Delille. 2004. *Psychrobacter salsus* sp. nov. and *Psychrobacter adeliensis* sp. nov. isolated from fast ice from Adelie Land, Antarctica. Syst. Appl. Microbiol. *27*: 628–635.

Shivaji, S., P. Chaturvedi, G.S. Reddy and K. Suresh. 2005a. *Pedobacter himalayensis* sp. nov., from the Hamta glacier located in the Himalayan mountain ranges of India. Int. J. Syst. Evol. Microbiol. *55*: 1083–1088.

Shivaji, S., P. Gupta, P. Chaturvedi, K. Suresh and D. Delille. 2005b. *Marinobacter maritimus* sp. nov., a psychrotolerant strain isolated from sea water off the subantarctic Kerguelen islands. Int. J. Syst. Evol. Microbiol. *55*: 1453–1456.

Shivaji, S., G.S. Reddy, K. Suresh, P. Gupta, S. Chintalapati, P. Schumann, E. Stackebrandt and G.I. Matsumoto. 2005c. *Psychrobacter vallis* sp. nov. and *Psychrobacter aquaticus* sp. nov., from Antarctica. Int. J. Syst. Evol. Microbiol. *55*: 757–762.

Shivaji, S., K. Suresh, P. Chaturvedi, S. Dube and S. Sengupta. 2005d. *Bacillus arsenicus* sp. nov., an arsenic-resistant bacterium isolated from a siderite concretion in West Bengal, India. Int. J. Syst. Evol. Microbiol. *55*: 1123–1127.

Shivaji, S., P. Chaturvedi, K. Suresh, G.S. Reddy, C.B. Dutt, M. Wainwright, J.V. Narlikar and P.M. Bhargava. 2006. *Bacillus aerius* sp. nov.,

Bacillus aerophilus sp. nov., *Bacillus stratosphericus* sp. nov. and *Bacillus altitudinis* sp. nov., isolated from cryogenic tubes used for collecting air samples from high altitudes. Int. J. Syst. Evol. Microbiol. *56*: 1465–1473.

Shivaji, S., B. Bhadra, R.S. Rao, P. Chaturvedi, P.K. Pindi and C. Raghukumar. 2007. *Microbacterium indicum* sp. nov., isolated from a deep-sea sediment sample from the Chagos Trench, Indian Ocean. Int. J. Syst. Evol. Microbiol. *57*: 1819–1822.

Shlimon, A.G., M.W. Friedrich, H. Niemann, N.B. Ramsing and K. Finster. 2004. *Methanobacterium aarhusense* sp. nov., a novel methanogen isolated from a marine sediment (Aarhus Bay, Denmark). Int. J. Syst. Evol. Microbiol. *54*: 759–763.

Shukla, S.K., K.A. Bernard, M. Harney, D. N. Frank and K.D. Reed. 2003. *Corynebacterium nigricans* sp. nov.: proposed name for a black-pigmented *Corynebacterium* species recovered from the human female urogenital tract. J. Clin. Microbiol. *41*: 4353–4358.

Silva, L.R., I. Cleenwerck, R. Rivas, J. Swings, M.E. Trujillo, A. Willems and E. Velázquez. 2006. *Acetobacter oeni* sp. nov., isolated from spoiled red wine. Int. J. Syst. Evol. Microbiol. *56*: 21–24.

Simbahan, J., R. Drijber and P. Blum. 2004. *Alicyclobacillus vulcanalis* sp. nov., a thermophilic, acidophilic bacterium isolated from Coso Hot Springs, California, USA. Int. J. Syst. Evol. Microbiol. *54*: 1703–1707.

Simpson, P.J., R.P. Ross, G.F. Fitzgerald and C. Stanton. 2004. *Bifidobacterium psychraerophilum* sp. nov. and *Aeriscardovia aeriphila* gen. nov., sp. nov., isolated from a porcine caecum. Int. J. Syst. Evol. Microbiol. *54*: 401–406.

Singh, N., M.M. Kendall, Y. Liu and D.R. Boone. 2005. Isolation and characterization of methylotrophic methanogens from anoxic marine sediments in Skan Bay, Alaska: description of *Methanococcoides alaskense* sp. nov., and emended description of *Methanosarcina baltica*. Int. J. Syst. Evol. Microbiol. *55*: 2531–2538.

Singla, A.K., S. Mayilraj, T. Kudo, S. Krishnamurthi, G.S. Prasad and R.M. Vohra. 2005. *Actinoalloteichus spitiensis* sp. nov., a novel actinobacterium isolated from a cold desert of the Indian Himalayas. Int. J. Syst. Evol. Microbiol. *55*: 2561–2564.

Sizova, M. and N. Panikov. 2007. *Polaromonas hydrogenivorans* sp. nov., a psychrotolerant hydrogen-oxidizing bacterium from Alaskan soil. Int. J. Syst. Evol. Microbiol. *57*: 616–619.

Sizova, M.V., N.S. Panikov, E.M. Spiridonova, N.V. Slobodova and T.P. Tourova. 2007. Novel facultative anaerobic acidotolerant *Telmatospirillum siberiense* gen. nov. sp. nov. isolated from mesotrophic fen. Syst. Appl. Microbiol. *30*: 213–220.

Sjöstedt, A.B. 2005. Family III. *Francisellaceae* fam. nov. *In* Brenner, Krieg, Staley and Garrity (Editors), Bergey's Manual of Systematic Bacteriology, 2nd Edition, Volume 2 (The *Proteobacteria*), Part B (The *Gammaproteobacteria*), Springer, New York, pp. 199–200.

Slepova, T.V., T.G. Sokolova, A.M. Lysenko, T.P. Tourova, T.V. Kolganova, O.V. Kamzolkina, G.A. Karpov and E.A. Bonch-Osmolovskaya. 2006. *Carboxydocella sporoproducens* sp. nov., a novel anaerobic CO-utilizing/H₂-producing thermophilic bacterium from a Kamchatka hot spring. Int. J. Syst. Evol. Microbiol. *56*: 797–800.

Slobodkin, A.I., T.G. Sokolova, A.M. Lysenko and J. Wiegel. 2006a. Reclassification of *Thermoterrabacterium ferrireducens* as *Carboxydothermus ferrireducens* comb. nov., and emended description of the genus *Carboxydothermus*. Int. J. Syst. Evol. Microbiol. *56*: 2349–2351.

Slobodkin, A.I., T.P. Tourova, N.A. Kostrikina, A.M. Lysenko, K.E. German, E.A. Bonch-Osmolovskaya and N.K. Birkeland. 2006b. *Tepidimicrobium ferriphilum* gen. nov., sp. nov., a novel moderately thermophilic, Fe(III)-reducing bacterium of the order *Clostridiales*. Int. J. Syst. Evol. Microbiol. *56*: 369–372.

Šmerda, J., I. Sedláček, Z. Páčová, E. Durnová, A. Smíšková and L. Havel. 2005. *Paenibacillus mendelii* sp. nov., from surface-sterilized seeds of *Pisum sativum* L. Int. J. Syst. Evol. Microbiol. *55*: 2351–2354.

Šmerda, J., I. Sedláček, Z. Páčová, E. Krejčí and L. Havel. 2006. *Paenibacillus sepulcri* sp. nov., isolated from biodeteriorated mural paintings in the Servilia tomb. Int. J. Syst. Evol. Microbiol. *56*: 2341–2344.

Soddell, J.A., F.M. Stainsby, K.L. Eales, R.M. Kroppenstedt, R.J. Seviour and M. Goodfellow. 2006a. *Millisia brevis* gen. nov., sp. nov., an actinomycete isolated from activated sludge foam. Int. J. Syst. Evol. Microbiol. *56*: 739–744.

Soddell, J.A., F.M. Stainsby, K.L. Eales, R.J. Seviour and M. Goodfellow. 2006b. *Gordonia defluvii* sp. nov., an actinomycete isolated from activated sludge foam. Int. J. Syst. Evol. Microbiol. *56*: 2265–2269.

Sohn, J.H., K.K. Kwon, J.H. Kang, H.B. Jung and S.J. Kim. 2004a. *Novosphingobium pentaromativorans* sp. nov., a high-molecular-mass polycyclic aromatic hydrocarbon-degrading bacterium isolated from estuarine sediment. Int. J. Syst. Evol. Microbiol. *54*: 1483–1487.

Sohn, J.H., J.H. Lee, H. Yi, J. Chun, K.S. Bae, T.Y. Ahn and S.J. Kim. 2004b. *Kordia algicida* gen. nov., sp. nov., an algicidal bacterium isolated from red tide. Int. J. Syst. Evol. Microbiol. *54*: 675–680.

Sokolova, T.G., J.M. González, N.A. Kostrikina, N.A. Chernyh, T.V. Slepova, E.A. Bonch-Osmolovskaya and F.T. Robb. 2004. *Thermosinus carboxydivorans* gen. nov., sp. nov., a new anaerobic, thermophilic, carbon-monoxide-oxidizing, hydrogenogenic bacterium from a hot pool of Yellowstone National Park. Int. J. Syst. Evol. Microbiol. *54*: 2353–2359.

Sokolova, T.G., N.A. Kostrikina, N.A. Chernyh, T.V. Kolganova, T.P. Tourova and E.A. Bonch-Osmolovskaya. 2005. *Thermincola carboxydiphila* gen. nov., sp. nov., a novel anaerobic, carboxydotrophic, hydrogenogenic bacterium from a hot spring of the Lake Baikal area. Int. J. Syst. Evol. Microbiol. *55*: 2069–2073.

Sokolova, T., J. Hanel, R.U. Onyenwoke, A.-L. Reysenbach, A. Banta, R. Geyer, J.M. González, W.B. Whitman and J. Wiegel. 2007. Novel chemolithotrophic, thermophilic, anaerobic bacteria *Thermolithobacter ferrireducens* gen. nov., sp. nov. and *Thermolithobacter carboxydivorans* sp. nov. Extremophiles *11*: 145–157.

Somvanshi, V.S., E. Lang, B. Sträubler, C. Spröer, P. Schumann, S. Ganguly, A.K. Saxena and E. Stackebrandt. 2006. *Providencia vermicola* sp. nov., isolated from infective juveniles of the entomopathogenic nematode *Steinernema thermophilum*. Int. J. Syst. Evol. Microbiol. *56*: 629–633.

Somvanshi, V.S., E. Lang, P. Schumann, R. Pukall, R.M. Kroppenstedt, S. Ganguly and E. Stackebrandt. 2007. *Leucobacter iarius* sp. nov., in the family *Microbacteriaceae*. Int. J. Syst. Evol. Microbiol. *57*: 682–686.

Song, J. and J.C. Cho. 2007. *Methylibium aquaticum* sp. nov., a betaproteobacterium isolated from a eutrophic freshwater pond. Int. J. Syst. Evol. Microbiol. *57*: 2125–2128.

Song, J., B.Y. Kim, S.B. Hong, H.S. Cho, K. Sohn, J. Chun and J.W. Suh. 2004. *Kribbella solani* sp. nov. and *Kribbella jejuensis* sp. nov., isolated from potato tuber and soil in Jeju, Korea. Int. J. Syst. Evol. Microbiol. *54*: 1345–1348.

Song, L., W.J. Li, Q.L. Wang, G.Z. Chen, Y.S. Zhang and L.H. Xu. 2005. *Jiangella gansuensis* gen. nov., sp. nov., a novel actinomycete from a desert soil in north-west China. Int. J. Syst. Evol. Microbiol. *55*: 881–884.

Song, Y., C. Liu, J. Lee, M. Bolašos, M.L. Vaisanen and S.M. Finegold. 2005. "*Bacteroides goldsteinii* sp. nov." isolated from clinical specimens of human intestinal origin. J. Clin. Microbiol. *43*: 4522–4527.

Song, Y., E. Könönen, M. Rautio, C. Liu, A. Bryk, E. Eerola and S.M. Finegold. 2006. *Alistipes onderdonkii* sp. nov. and *Alistipes shahii* sp. nov., of human origin. Int. J. Syst. Evol. Microbiol. *56*: 1985–1990.

Song, Y.L., C.X. Liu, M. McTeague and S.M. Finegold. 2004a. "*Bacteroides nordii*" sp. nov. and "*Bacteroides salyersae*" sp. nov. isolated from clinical specimens of human intestinal origin. J. Clin. Microbiol. *42*: 5565–5570.

Song, Y.L., C.X. Liu, M. McTeague., P. Summanen and S.M. Finegold. 2004b. *Clostridium bartlettii* sp. nov., isolated from human faeces. Anaerobe *10*: 179–184.

Sorokin, D.Y. and B.J. Tindall. 2006. The status of the genus name *Halovibrio* Fendrich 1989 and the identity of the strains *Pseudomonas halophila* DSM 3050 and *Halomonas variabilis* DSM 3051. Request for an Opinion. Int. J. Syst. Evol. Microbiol. *56*: 487–489.

Sorokin, D.Y., T.P. Tourova, A.N. Antipov, G. Muyzer and J.G. Kuenen. 2004. Anaerobic growth of the haloalkaliphilic denitrifying sulfur-oxidizing bacterium *Thialkalivibrio thiocyanodenitrificans* sp. nov. with thiocyanate. Microbiology *150*: 2435–2442.

Sorokin, D.Y., T.P. Tourova and G. Muyzer. 2005a. *Citreicella thiooxidans* gen. nov., sp. nov., a novel lithoheterotrophic sulfur-oxidizing bacterium from the Black Sea. Syst. Appl. Microbiol. *28*: 679–687.

Sorokin, D.Y., T.P. Tourova, E.M. Spiridonova, F.A. Rainey and G. Muyzer. 2005b. *Thioclava pacifica* gen. nov., sp. nov., a novel facultatively autotrophic, marine, sulfur-oxidizing bacterium from a near-shore sulfidic hydrothermal area. Int. J. Syst. Evol. Microbiol. *55*: 1069–1075.

Sorokin, D.Y., T.P. Tourova, E.A. Galinski, C. Belloch and B.J. Tindall. 2006a. Extremely halophilic denitrifying bacteria from hypersaline inland lakes, *Halovibrio denitrificans* sp. nov. and *Halospina denitrificans* gen. nov., sp. nov., and evidence that the genus name *Halovibrio* Fendrich 1989 with the type species *Halovibrio variabilis* should be associated with DSM 3050. Int. J. Syst. Evol. Microbiol. *56*: 379–388.

Sorokin, D.Y., T.P. Tourova, T.V. Kolganova, E.M. Spiridonova, I.A. Berg and G. Muyzer. 2006b. *Thiomicrospira halophila* sp. nov., a moderately halophilic, obligately chemolithoautotrophic, sulfur-oxidizing bacterium from hypersaline lakes. Int. J. Syst. Evol. Microbiol. *56*: 2375–2380.

Sorokin, D.Y., M. Foti, H.C. Pinkart and G. Muyzer. 2007a. Sulfur-oxidizing bacteria in Soap Lake (Washington State), a meromictic, haloalkaline lake with an unprecedented high sulfide content. Appl. Environ. Microbiol. *73*: 451–455.

Sorokin, D.Y., T.P. Tourova, E.Y. Bezsoudnova, A. Pol and G. Muyzer. 2007b. Denitrification in a binary culture and thiocyanate metabolism in *Thiohalophilus thiocyanoxidans* gen. nov. sp. nov. - a moderately halophilic chemolithoautotrophic sulfur-oxidizing *Gammaproteobacterium* from hypersaline lakes. Arch. Microbiol. *187*: 441–450.

Sorokin, D.Y., T.P. Tourova, G. Braker and G. Muyzer. 2007c. *Thiohalomonas denitrificans* gen. nov., sp. nov. and *Thiohalomonas nitratireducens* sp. nov., novel obligately chemolithoautotrophic, moderately halophilic, thiodenitrifying *Gammaproteobacteria* from hypersaline habitats. Int. J. Syst. Evol. Microbiol. *57*: 1582–1589.

Sorokin, D.Y., Y.A. Trotsenko, N.V. Doronina, T.P. Tourova, E.A. Galinski, T.V. Kolganova and G. Muyzer. 2007d. *Methylohalomonas lacus* gen. nov., sp. nov. and *Methylonatrum kenyense* gen. nov., sp. nov., methylotrophic gammaproteobacteria from hypersaline lakes. Int. J. Syst. Evol. Microbiol. *57*: 2762–2769.

Sorokin, D.Y., S. van Pelt, T.P. Tourova, S. Takaichi and G. Muyzer. 2007e. Acetonitrile degradation under haloalkaline conditions by *Natronocella acetinitrilica* gen. nov., sp. nov. Microbiology *153*: 1157–1164.

Soto-Ramírez, N., C. Sánchez-Porro, S. Rosas, W. González, M. Quiñones, A. Ventosa and R. Montalvo-Rodríguez. 2007. *Halomonas avicenniae* sp. nov., isolated from the salty leaves of the black mangrove *Avicennia germinans* in Puerto Rico. Int. J. Syst. Evol. Microbiol. *57*: 900–905.

Sousa, D.Z., H. Smidt, M.M. Alves and A.J. Stams. 2007. *Syntrophomonas zehnderi* sp. nov., an anaerobe that degrades long-chain fatty acids in co-culture with *Methanobacterium formicicum*. Int. J. Syst. Evol. Microbiol. *57*: 609–615.

Sprague, L.D. and H. Neubauer. 2005. *Yersinia aleksiciae* sp. nov. Int. J. Syst. Evol. Microbiol. *55*: 831–835.

Spring, S., U. Jäckel, M. Wagner and P. Kämpfer. 2004. *Ottowia thiooxydans* gen. nov., sp. nov., a novel facultatively anaerobic, N_2O-producing bacterium isolated from activated sludge, and transfer of *Aquaspirillum gracile* to *Hylemonella gracilis* gen. nov., comb. nov. Int. J. Syst. Evol. Microbiol. *54*: 99–106.

Spring, S., P. Schumann and C. Spröer. 2005a. *Methanogenium frittonii* Harris *et al.* 1996 is a later synonym of *Methanoculleus thermophilus* (Rivard and Smith 1982) Maestrojuan *et al.* 1990. Int. J. Syst. Evol. Microbiol. *55*: 1097–1099.

Spring, S., M. Wagner, P. Schumann and P. Kämpfer. 2005b. *Malikia granosa* gen. nov., sp. nov., a novel polyhydroxyalkanoate- and polyphosphate-accumulating bacterium isolated from activated sludge, and reclassification of *Pseudomonas spinosa* as *Malikia spinosa* comb. nov. Int. J. Syst. Evol. Microbiol. *55*: 621–629.

Srinivas, T.N., P.A. Kumar, C. Sasikala, Ch V. Ramana, J. Süling and J.F. Imhoff. 2006. *Rhodovulum marinum* sp. nov., a novel phototrophic purple non-sulfur alphaproteobacterium from marine tides of Visakhapatnam, India. Int. J. Syst. Evol. Microbiol. *56*: 1651–1656.

Srinivas, T.N., P.A. Kumar, C. Sasikala and Ch V. Ramana. 2007a. *Rhodovulum imhoffii* sp. nov. Int. J. Syst. Evol. Microbiol. *57*: 228–232.

Srinivas, T.N., P.A. Kumar, C. Sasikala, Ch V. Ramana and J.F. Imhoff. 2007b. *Rhodobium gokarnense* sp. nov., a novel phototrophic alphaproteobacterium from a saltern. Int. J. Syst. Evol. Microbiol. *57*: 932–935.

Srinivas, T.N., P.A. Kumar, C. Sasikala, Ch V. Ramana and J.F. Imhoff. 2007c. *Rhodovulum visakhapatnamense* sp. nov. Int. J. Syst. Evol. Microbiol. *57*: 1762–1764.

Srinivas, T.N., P.A. Kumar, C. Sasikala, Ch V. Ramana and J.F. Imhoff. 2007d. *Rhodobacter vinaykumarii* sp. nov., a marine phototrophic alphaproteobacterium from tidal waters, and emended description of the genus *Rhodobacter*. Int. J. Syst. Evol. Microbiol. *57*: 1984–1987.

Srisukchayakul, P., C. Suwanachart, Y. Sangnoi, A. Kanjana-Opas, S. Hosoya, A. Yokota and V. Arunpairojana. 2007. *Rapidithrix thailandica* gen. nov., sp. nov., a marine gliding bacterium isolated from samples collected from the Andaman sea, along the southern coastline of Thailand. Int. J. Syst. Evol. Microbiol. *57*: 2275–2279.

Stach, J.E., L.A. Maldonado, A.C. Ward, A.T. Bull and M. Goodfellow. 2004. *Williamsia maris* sp. nov., a novel actinomycete isolated from the Sea of Japan. Int. J. Syst. Evol. Microbiol. *54*: 191–194.

Stackebrandt, E. 2004. Will we ever understand? The undescribable diversity of the prokaryotes. Acta Microbiol. Immunol. Hung. *51*: 449–462.

Stackebrandt, E. and P. Schumann. 2004. Reclassification of *Promicromonospora pachnodae* Cazemier *et al.* 2004 as *Xylanimicrobium pachnodae* gen. nov., comb. nov. Int. J. Syst. Evol. Microbiol. *54*: 1383–1386.

Stackebrandt, E., R.M. Kroppenstedt, J. Wink and P. Schumann. 2004a. Reclassification of *Amycolatopsis orientalis* subsp. *lurida* Lechevalier *et al.* 1986 as *Amycolatopsis lurida* sp. nov., comb. nov. Int. J. Syst. Evol. Microbiol. *54*: 267–268.

Stackebrandt, E., P. Schumann and X.L. Cui. 2004b. Reclassification of *Cellulosimicrobium variabile* Bakalidou *et al.* 2002 as *Isoptericola variabilis* gen. nov., comb. nov. Int. J. Syst. Evol. Microbiol. *54*: 685–688.

Stackebrandt, E., E. Lang, S. Cousin, O. Päuker, E. Brambilla, R. Kroppenstedt and H. Lünsdorf. 2007. *Deefgea rivuli* gen. nov., sp. nov., a member of the class *Betaproteobacteria*. Int. J. Syst. Evol. Microbiol. *57*: 639–645.

Stephan, R., S. Van Trappen, I. Cleenwerck, M. Vancanneyt, P. De Vos and A. Lehner. 2007. *Enterobacter turicensis* sp. nov. and *Enterobacter helveticus* sp. nov., isolated from fruit powder. Int. J. Syst. Evol. Microbiol. *57*: 820–826.

Stolz, A., S. Bürger, A. Kuhm, P. Kämpfer and H.J. Busse. 2005. *Pusillimonas noertemannii* gen. nov., sp. nov., a new member of the family *Alcaligenaceae* that degrades substituted salicylates. Int. J. Syst. Evol. Microbiol. *55*: 1077–1081.

Stolz, A., H.J. Busse and P. Kämpfer. 2007. *Pseudomonas knackmussii* sp. nov. Int. J. Syst. Evol. Microbiol. *57*: 572–576.

Storms, V., N. Van Den Vreken, T. Coenye, E. Mahenthiralingam, J.J. Lipuma, M. Gillis and P. Vandamme. 2004. Polyphasic characterisation of *Burkholderia cepacia*-like isolates leading to the emended description of *Burkholderia pyrrocinia*. Syst. Appl. Microbiol. *27*: 517–526.

Sukontasing, S., S. Tanasupawat, S. Moonmangmee, J.S. Lee and K. Suzuki. 2007. *Enterococcus camelliae* sp. nov., isolated from fermented tea leaves in Thailand. Int. J. Syst. Evol. Microbiol. *57*: 2151–2154.

Summanen, P.H., B. Durmaz, M.L. Väisänen, C. Liu, D. Molitoris, E. Eerola, I.M. Helander and S.M. Finegold. 2005. *Porphyromonas somerae* sp. nov., a pathogen isolated from humans and distinct from *Porphyromonas levii*. J. Clin. Microbiol. *43*: 4455–4459.

Sun, W., Y. Huang, Y.Q. Zhang and Z.H. Liu. 2007. *Streptomyces emeiensis* sp. nov., a novel streptomycete from soil in China. Int. J. Syst. Evol. Microbiol. *57*: 1635–1639.

Suresh, K., S.R. Prabagaran, S. Sengupta and S. Shivaji. 2004a. *Bacillus indicus* sp. nov., an arsenic-resistant bacterium isolated from an aquifer in West Bengal, India. Int. J. Syst. Evol. Microbiol. *54*: 1369–1375.

Suresh, K., G.S. Reddy, S. Sengupta and S. Shivaji. 2004b. *Deinococcus indicus* sp. nov., an arsenic-resistant bacterium from an aquifer in West Bengal, India. Int. J. Syst. Evol. Microbiol. *54*: 457–461.

Suresh, K., S. Mayilraj and T. Chakrabarti. 2006. *Effluviibacter roseus* gen. nov., sp. nov., isolated from muddy water, belonging to the family "*Flexibacteraceae*". Int. J. Syst. Evol. Microbiol. *56*: 1703–1707.

Suresh, K., S. Mayilraj, A. Bhattacharya and T. Chakrabarti. 2007a. *Planococcus columbae* sp. nov., isolated from pigeon faeces. Int. J. Syst. Evol. Microbiol. *57*: 1266–1271.

Suresh, K., D. Prakash, N. Rastogi and R.K. Jain. 2007b. *Clostridium nitrophenolicum* sp. nov., a novel anaerobic p-nitrophenol-degrading bacterium, isolated from a subsurface soil sample. Int. J. Syst. Evol. Microbiol. *57*: 1886–1890.

Suzuki, D., A. Ueki, A. Amaishi and K. Ueki. 2007a. *Desulfopila aestuarii* gen. nov., sp. nov., a Gram-negative, rod-like, sulfate-reducing bacterium isolated from an estuarine sediment in Japan. Int. J. Syst. Evol. Microbiol. *57*: 520–526.

Suzuki, D., A. Ueki, A. Amaishi and K. Ueki. 2007b. *Desulfobulbus japonicus* sp. nov., a novel Gram-negative propionate-oxidizing, sulfate-reducing bacterium isolated from an estuarine sediment in Japan. Int. J. Syst. Evol. Microbiol. *57*: 849–855.

Suzuki, K., W. Funahashi, M. Koyanagi and H. Yamashita. 2004. *Lactobacillus paracollinoides* sp. nov., isolated from brewery environments. Int. J. Syst. Evol. Microbiol. *54*: 115–117.

Suzuki, T., Y. Mori and Y. Nishimura. 2006. *Roseibacterium elongatum* gen. nov., sp. nov., an aerobic, bacteriochlorophyll-containing bacterium isolated from the west coast of Australia. Int. J. Syst. Evol. Microbiol. *56*: 417–421.

Švec, P., M. Vancanneyt, I. Sedláček, K. Engelbeen, V. Štětina, J. Swings and P. Petráš. 2004. Reclassification of *Staphylococcus pulvereri* Zakrzewska-Czerwińska *et al.* 1995 as a later synonym of *Staphylococcus vitulinus* Webster *et al.* 1994. Int. J. Syst. Evol. Microbiol. *54*: 2213–2215.

Švec, P., M. Vancanneyt, L.A. Devriese, S.M. Naser, C. Snauwaert, K. Lefebvre, B. Hoste and J. Swings. 2005a. *Enterococcus aquimarinus* sp. nov., isolated from sea water. Int. J. Syst. Evol. Microbiol. *55*: 2183–2187.

Švec, P., M. Vancanneyt, J. Koort, S.M. Naser, B. Hoste, E. Vihavainen, P. Vandamme, J. Swings and J. Björkroth. 2005b. *Enterococcus devriesei* sp. nov., associated with animal sources. Int. J. Syst. Evol. Microbiol. *55*: 2479–2484.

Švec, P., M. Vancanneyt, I. Sedláček, S.M. Naser, C. Snauwaert, K. Lefebvre, B. Hoste and J. Swings. 2006. *Enterococcus silesiacus* sp. nov. and *Enterococcus termitis* sp. nov. Int. J. Syst. Evol. Microbiol. *56*: 577–581.

Tago, Y. and A. Yokota. 2004. *Comamonas badia* sp. nov., a floc-forming bacterium isolated from activated sludge. J. Gen. Appl. Microbiol. *50*: 243–248.

Tai, C.J., H.P. Kuo, F.L. Lee, H.K. Chen, A. Yokota and C.C. Lo. 2006. *Chryseobacterium taiwanense* sp. nov., isolated from soil in Taiwan. Int. J. Syst. Evol. Microbiol. *56*: 1771–1776.

Tailliez, P., S. Pagès, N. Ginibre and N. Boemare. 2006. New insight into diversity in the genus *Xenorhabdus*, including the description of ten novel species. Int. J. Syst. Evol. Microbiol. *56*: 2805–2818.

Takada, K. and M. Hirasawa. 2007. *Streptococcus orisuis* sp. nov., isolated from the pig oral cavity. Int. J. Syst. Evol. Microbiol. *57*: 1272–1275.

Takahashi, M., K. Suzuki and Y. Nakagawa. 2006. Emendation of the genus *Flammeovirga* and *Flammeovirga aprica* with the proposal of *Flammeovirga arenaria* nom. rev., comb. nov. and *Flammeovirga yaeyamensis* sp. nov. Int. J. Syst. Evol. Microbiol. *56*: 2095–2100.

Takahashi, Y., A. Matsumoto, K. Morisaki and S. Ōmura. 2006. *Patulibacter minatonensis* gen. nov., sp. nov., a novel actinobacterium isolated using an agar medium supplemented with superoxide dismutase, and proposal of *Patulibacteraceae* fam. nov. Int. J. Syst. Evol. Microbiol. *56*: 401–406.

Takai, K., H. Hirayama, T. Nakagawa, Y. Suzuki, K.H. Nealson and K. Horikoshi. 2004a. *Thiomicrospira thermophila* sp. nov., a novel microaer-obic, thermotolerant, sulfur-oxidizing chemolithomixotroph isolated from a deep-sea hydrothermal fumarole in the TOTO caldera, Mariana Arc, Western Pacific. Int. J. Syst. Evol. Microbiol. *54*: 2325–2333.

Takai, K., K.H. Nealson and K. Horikoshi. 2004b. *Hydrogenimonas thermophila* gen. nov., sp. nov., a novel thermophilic, hydrogen-oxidizing chemolithoautotroph within the ε-*Proteobacteria*, isolated from a black smoker in a Central Indian Ridge hydrothermal field. Int. J. Syst. Evol. Microbiol. *54*: 25–32.

Takai, K., K.H. Nealson and K. Horikoshi. 2004c. *Methanotorris formicicus* sp. nov., a novel extremely thermophilic, methane-producing archaeon isolated from a black smoker chimney in the Central Indian Ridge. Int. J. Syst. Evol. Microbiol. *54*: 1095–1100.

Takai, K., H. Hirayama, T. Nakagawa, Y. Suzuki, K.H. Nealson and K. Horikoshi. 2005. *Lebetimonas acidiphila* gen. nov., sp. nov., a novel thermophilic, acidophilic, hydrogen-oxidizing chemolithoautotroph within the '*Epsilonproteobacteria*', isolated from a deep-sea hydrothermal fumarole in the Mariana Arc. Int. J. Syst. Evol. Microbiol. *55*: 183–189.

Takai, K., M. Miyazaki, T. Nunoura, H. Hirayama, H. Oida, Y. Furushima, H. Yamamoto and K. Horikoshi. 2006a. *Sulfurivirga caldicuralii* gen. nov., sp. nov., a novel microaerobic, thermophilic, thiosulfate-oxidizing chemolithoautotroph, isolated from a shallow marine hydrothermal system occurring in a coral reef, Japan. Int. J. Syst. Evol. Microbiol. *56*: 1921–1929.

Takai, K., M. Suzuki, S. Nakagawa, M. Miyazaki, Y. Suzuki, F. Inagaki and K. Horikoshi. 2006b. *Sulfurimonas paralvinellae* sp. nov., a novel mesophilic, hydrogen- and sulfur-oxidizing chemolithoautotroph within the *Epsilonproteobacteria* isolated from a deep-sea hydrothermal vent polychaete nest, reclassification of *Thiomicrospira denitrificans* as *Sulfurimonas denitrificans* comb. nov. and emended description of the genus *Sulfurimonas*. Int. J. Syst. Evol. Microbiol. *56*: 1725–1733.

Takeda, M., I. Suzuki and J.I. Koizumi. 2004. *Balneomonas flocculans* gen. nov., sp. nov., a new cellulose-producing member of the α-2 subclass of *Proteobacteria*. Syst. Appl. Microbiol. *27*: 139–145.

Takeda, M., I. Suzuki and J. Koizumi. 2005. *Paenibacillus hodogayensis* sp. nov., capable of degrading the polysaccharide produced by *Sphaerotilus natans*. Int. J. Syst. Evol. Microbiol. *55*: 737–741.

Takii, S., S. Hanada, H. Tamaki, Y. Ueno, Y. Sekiguchi, A. Ibe and K. Matsuura. 2007. *Dethiosulfatibacter aminovorans* gen. nov., sp. nov., a novel thiosulfate-reducing bacterium isolated from coastal marine sediment via sulfate-reducing enrichment with Casamino acids. Int. J. Syst. Evol. Microbiol. *57*: 2320–2326.

Tamura, T. and T. Sakane. 2004. *Planotetraspora silvatica* sp. nov. and emended description of the genus *Planotetraspora*. Int. J. Syst. Evol. Microbiol. *54*: 2053–2056.

Tamura, T. and T. Sakane. 2005. *Asanoa iriomotensis* sp. nov., isolated from mangrove soil. Int. J. Syst. Evol. Microbiol. *55*: 725–727.

Tamura, T., K. Hatano and K. Suzuki. 2006. A new genus of the family *Micromonosporaceae*, *Polymorphospora* gen. nov., with description of *Polymorphospora rubra* sp. nov. Int. J. Syst. Evol. Microbiol. *56*: 1959–1964.

Tamura, T., K. Hatano and K. Suzuki. 2007a. Classification of '*Sarraceniospora aurea*' Furihata *et al.* 1989 as *Actinocorallia aurea* sp. nov. Int. J. Syst. Evol. Microbiol. *57*: 2052–2055.

Tamura, T., Y. Ishida, T. Sakane and K. Suzuki. 2007b. *Catenulispora rubra* sp. nov., an acidophilic actinomycete isolated from forest soil. Int. J. Syst. Evol. Microbiol. *57*: 2272–2274.

Tan, G.Y., S. Robinson, E. Lacey and M. Goodfellow. 2006. *Amycolatopsis australiensis* sp. nov., an actinomycete isolated from arid soils. Int. J. Syst. Evol. Microbiol. *56*: 2297–2301.

Tan, G.Y., S. Robinson, E. Lacey, R. Brown, W. Kim and M. Goodfellow. 2007. *Amycolatopsis regifaucium* sp. nov., a novel actinomycete that produces kigamicins. Int. J. Syst. Evol. Microbiol. *57*: 2562–2567.

Tanaka, R., S. Kawaichi, H. Nishimura and Y. Sako. 2006. *Thermaerobacter litoralis* sp. nov., a strictly aerobic and thermophilic bacterium isolated from a coastal hydrothermal field. Int. J. Syst. Evol. Microbiol. *56*: 1531–1534.

Tanaka, Y., S. Hanada, A. Manome, T. Tsuchida, R. Kurane, K. Nakamura and Y. Kamagata. 2004. *Catellibacterium nectariphilum* gen. nov., sp. nov., which requires a diffusible compound from a strain related to the genus *Sphingomonas* for vigorous growth. Int. J. Syst. Evol. Microbiol. *54*: 955–959.

Tanasupawat, S., C. Thawai, P. Yukphan, D. Moonmangmee, T. Itoh, O. Adachi and Y. Yamada. 2004. *Gluconobacter thailandicus* sp. nov., an acetic acid bacterium in the α-*Proteobacteria*. J. Gen. Appl. Microbiol. *50*: 159–167.

Tanasupawat, S., A. Pakdeeto, S. Namwong, C. Thawai, T. Kudo and T. Itoh. 2006. *Lentibacillus halophilus* sp. nov., from fish sauce in Thailand. Int. J. Syst. Evol. Microbiol. *56*: 1859–1863.

Tanasupawat, S., S. Namwong, T. Kudo and T. Itoh. 2007a. *Piscibacillus salipiscarius* gen. nov., sp. nov., a moderately halophilic bacterium from fermented fish (pla-ra) in Thailand. Int. J. Syst. Evol. Microbiol. *57*: 1413–1417.

Tanasupawat, S., A. Pakdeeto, C. Thawai, P. Yukphan and S. Okada. 2007b. Identification of lactic acid bacteria from fermented tea leaves (miang) in Thailand and proposals of *Lactobacillus thailandensis* sp. nov., *Lactobacillus camelliae* sp. nov., and *Pediococcus siamensis* sp. nov. J. Gen. Appl. Microbiol. *53*: 7–15.

Tao, T.S., Y.Y. Yue, W.X. Chen and W.F. Chen. 2004. Proposal of *Nakamurella* gen. nov. as a substitute for the bacterial genus *Microsphaera* Yoshimi *et al.* 1996 and *Nakamurellaceae* fam. nov. as a substitute for the illegitimate bacterial family *Microsphaeraceae* Rainey *et al.* 1997. Int. J. Syst. Evol. Microbiol. *54*: 999–1000.

Ten, L.N., S.H. Baek, W.T. Im, M. Lee, H.W. Oh and S.T. Lee. 2006a. *Paenibacillus panacisoli* sp. nov., a xylanolytic bacterium isolated from soil in a ginseng field in South Korea. Int. J. Syst. Evol. Microbiol. *56*: 2677–2681.

Ten, L.N., S.H. Baek, W.T. Im, Q.M. Liu, Z. Aslam and S.T. Lee. 2006b. *Bacillus panaciterrae* sp. nov., isolated from soil of a ginseng field. Int. J. Syst. Evol. Microbiol. *56*: 2861–2866.

Ten, L.N., W.-T. Im, S.-H. Baek, J.-S. Lee, H.-M. Oh and S.-T. Lee. 2006c. *Bacillus ginsengihumi* sp. nov., a novel species isolated from soil of a ginseng field in pocheon Province, South Korea. J. Microbiol. Biotechnol. *16*: 1554–1560.

Ten, L.N., Q.-M. Liu, W.-T. Im, Z. Aslam and S.-T. Lee. 2006d. *Sphingobacterium composti* sp. nov., a novel DNase-producing bacterium isolated from compost. J. Microbiol. Biotechnol. *16*: 1728–1733.

Ten, L.N., Q.M. Liu, W.T. Im, M. Lee, D.C. Yang and S.T. Lee. 2006e. *Pedobacter ginsengisoli* sp. nov., a DNase-producing bacterium isolated from soil of a ginseng field in South Korea. Int. J. Syst. Evol. Microbiol. *56*: 2565–2570.

Ten, L.N., S.H. Baek, W.T. Im, L.L. Larina, J.S. Lee, H.M. Oh and S.T. Lee. 2007. *Bacillus pocheonensis* sp. nov., a moderately halotolerant, aerobic bacterium isolated from soil of a ginseng field. Int. J. Syst. Evol. Microbiol. *57*: 2532–2537.

Teyssier, C., H. Marchandin, H. Jean-Pierre, A. Masnou, G. Dusart and E. Jumas-Bilak. 2007. *Ochrobactrum pseudintermedium* sp. nov., a novel member of the family *Brucellaceae*, isolated from human clinical samples. Int. J. Syst. Evol. Microbiol. *57*: 1007–1013.

Thawai, C., S. Tanasupawat, T. Itoh, K. Suwanborirux and T. Kudo. 2004. *Micromonospora aurantionigra* sp. nov., isolated from a peat swamp forest in Thailand. Actinomycetologica *18*: 8–14.

Thawai, C., S. Tanasupawat, T. Itoh, K. Suwanborirux and T. Kudo. 2005a. *Micromonospora siamensis* sp. nov., isolated from Thai peat swamp forest. J. Gen. Appl. Microbiol. *51*: 229–234.

Thawai, C., S. Tanasupawat, T. Itoh, K. Suwanborirux, K. Suzuki and T. Kudo. 2005b. *Micromonospora eburnea* sp. nov., isolated from a Thai peat swamp forest. Int. J. Syst. Evol. Microbiol. *55*: 417–422.

Thawai, C., S. Tanasupawat, T. Itoh and T. Kudo. 2006. *Actinocatenispora thailandica* gen. nov., sp. nov., a new member of the family *Micromonosporaceae*. Int. J. Syst. Evol. Microbiol. *56*: 1789–1794.

Thevenieau, F., M.L. Fardeau, B. Ollivier, C. Joulian and S. Baena. 2007. *Desulfomicrobium thermophilum* sp. nov., a novel thermophilic sulphate-reducing bacterium isolated from a terrestrial hot spring in Colombia. Extremophiles *11*: 295–303.

Thierry, S., H. Macarie, T. Iizuka, W. Geißdörfer, E.A. Assih, M. Spanevello, F. Verhé, P. Thomas, R. Fudou, O. Monroy, M. Labat and A.S. Ouattara. 2004. *Pseudoxanthomonas mexicana* sp. nov. and *Pseudoxanthomonas japonensis* sp. nov., isolated from diverse environments, and emended descriptions of the genus *Pseudoxanthomonas* Finkmann *et al.* 2000 and of its type species. Int. J. Syst. Evol. Microbiol. *54*: 2245–2255.

Thompson, F.L., C.C. Thompson, S. Naser, B. Hoste, K. Vandemeulebroecke, C. Munn, D. Bourne and J. Swings. 2005. *Photobacterium rosenbergii* sp. nov. and *Enterovibrio coralii* sp. nov., vibrios associated with coral bleaching. Int. J. Syst. Evol. Microbiol. *55*: 913–917.

Thompson, F.L., Y. Barash, T. Sawabe, G. Sharon, J. Swings and E. Rosenberg. 2006. *Thalassomonas loyana* sp. nov., a causative agent of the white plague-like disease of corals on the Eilat coral reef. Int. J. Syst. Evol. Microbiol. *56*: 365–368.

Thomsen, T.R., L.L. Blackall, M.A. de Muro, J.L. Nielsen and P.H. Nielsen. 2006. *Meganema perideroedes* gen. nov., sp. nov., a filamentous alphaproteobacterium from activated sludge. Int. J. Syst. Evol. Microbiol. *56*: 1865–1868.

Tiago, I., V. Mendes, C. Pires, P.V. Morais and A. Veríssimo. 2005a. *Phenylobacterium falsum* sp. nov., an *Alphaproteobacterium* isolated from a nonsaline alkaline groundwater, and emended description of the genus *Phenylobacterium*. Syst. Appl. Microbiol. *28*: 295–302.

Tiago, I., C. Pires, V. Mendes, P.V. Morais, M. da Costa and A. Veríssimo. 2005b. *Microcella putealis* gen. nov., sp. nov., a Gram-positive alkaliphilic bacterium isolated from a nonsaline alkaline groundwater. Syst. Appl. Microbiol. *28*: 479–487.

Tiago, I., V. Mendes, C. Pires, P.V. Morais and A. Veríssimo. 2006a. *Chimaereicella alkaliphila* gen. nov., sp. nov., a Gram-negative alkaliphilic bacterium isolated from a nonsaline alkaline groundwater. Syst. Appl. Microbiol. *29*: 100–108.

Tiago, I., P.V. Morais, M.S. da Costa and A. Veríssimo. 2006b. *Microcella alkaliphila* sp. nov., a novel member of the family *Microbacteriaceae* isolated from a non-saline alkaline groundwater, and emended description of the genus *Microcella*. Int. J. Syst. Evol. Microbiol. *56*: 2313–2316.

Tiago, I., C. Pires, V. Mendes, P.V. Morais, M.S. da Costa and A. Veríssimo. 2006c. *Bacillus foraminis* sp. nov., isolated from a non-saline alkaline groundwater. Int. J. Syst. Evol. Microbiol. *56*: 2571–2574.

Tiirola, M.A., H.J. Busse, P. Kämpfer and M.K. Männistö. 2005. *Novosphingobium lentum* sp. nov., a psychrotolerant bacterium from a polychlorophenol bioremediation process. Int. J. Syst. Evol. Microbiol. *55*: 583–588.

Tindall, B.J. and J.P. Euzéby. 2006. Proposal of *Parvimonas* gen. nov. and *Quatrionicoccus* gen. nov. as replacements for the illegitimate, prokaryotic, generic names *Micromonas* Murdoch and Shah 2000 and *Quadricoccus* Maszenan *et al.* 2002, respectively. Int. J. Syst. Evol. Microbiol. *56*: 2711–2713.

Tindall, B.J., P.A. Grimont, G.M. Garrity and J.P. Euzéby. 2005. Nomenclature and taxonomy of the genus *Salmonella*. Int. J. Syst. Evol. Microbiol. *55*: 521–524.

Toffin, L., A. Bidault, P. Pignet, B.J. Tindall, A. Slobodkin, C. Kato and D. Prieur. 2004. *Shewanella profunda* sp. nov., isolated from deep marine sediment of the Nankai Trough. Int. J. Syst. Evol. Microbiol. *54*: 1943–1949.

Toffin, L., K. Zink, C. Kato, P. Pignet, A. Bidault, N. Bienvenu, J.L. Birrien and D. Prieur. 2005. *Marinilactibacillus piezotolerans* sp. nov., a novel marine lactic acid bacterium isolated from deep sub-seafloor sediment of the Nankai Trough. Int. J. Syst. Evol. Microbiol. *55*: 345–351.

Toledo, I., L. Lloret and E. Martínez-Romero. 2003. *Sinorhizobium americanus* sp. nov., a new sinorhizobium species nodulating native *Acacia* spp. in Mexico. Syst. Appl. Microbiol. *26*: 54–64.

Tong, H. and X. Dong. 2005. *Lactobacillus concavus* sp. nov., isolated from the walls of a distilled spirit fermenting cellar in China. Int. J. Syst. Evol. Microbiol. *55*: 2199–2202.

Tønjum, T. 2005. Order IV. *Neisseriales* ord. nov. *In* Brenner, Krieg, Staley and Garrity (Editors), Bergey's Manual of Systematic Bacteriology, 2nd Edition, Volume 2 (The *Proteobacteria*), Part C (The *Alpha-, Beta-, Delta-,* and *Epsilonproteobacteria*), Springer, New York, p. 774.

Tortoli, E., L. Rindi, M.J. García, P. Chiaradonna, R. Dei, C. Garzelli, R.M. Kroppenstedt, N. Lari, R. Mattei, A. Mariottini, G. Mazzarelli, M.I. Murcia, A. Nanetti, P. Piccoli and C. Scarparo. 2004. Proposal to elevate the genetic variant MAC-A, included in the *Mycobacterium avium* complex, to species rank as *Mycobacterium chimaera* sp. nov. Int. J. Syst. Evol. Microbiol. *54*: 1277–1285.

Tortoli, E., L. Rindi, K.S. Goh, M.L. Katila, A. Mariottini, R. Mattei, G. Mazzarelli, S. Suomalainen, P. Torkko and N. Rastogi. 2005. *Mycobacterium florentinum* sp. nov., isolated from humans. Int. J. Syst. Evol. Microbiol. *55*: 1101–1106.

Tóth, E.M., A.K. Borsodi, J.P. Euzéby, B.J. Tindall and K. Márialigeti. 2007. Proposal to replace the illegitimate genus name *Schineria* Tóth *et al.* 2001 with the genus name *Ignatzschineria* gen. nov. and to replace the illegitimate combination *Schineria larvae* Tóth *et al.* 2001 with *Ignatzschineria larvae* comb. nov. Int. J. Syst. Evol. Microbiol. *57*: 179–180.

Tripathi, A.K., S.C. Verma, S.P. Chowdhury, M. Lebuhn, A. Gattinger and M. Schloter. 2006. *Ochrobactrum oryzae* sp. nov., an endophytic bacterial species isolated from deep-water rice in India. Int. J. Syst. Evol. Microbiol. *56*: 1677–1680.

Trujillo, M.E., E. Velázquez, R.M. Kroppenstedt, P. Schumann, R. Rivas, P.F. Mateos and E. Martínez-Molina. 2004. *Mycobacterium psychrotolerans* sp. nov., isolated from pond water near a uranium mine. Int. J. Syst. Evol. Microbiol. *54*: 1459–1463.

Trujillo, M.E., C. Fernández-Molinero, E. Velázquez, R.M. Kroppenstedt, P. Schumann, P.F. Mateos and E. Martínez-Molina. 2005a. *Micromonospora mirobrigensis* sp. nov. Int. J. Syst. Evol. Microbiol. *55*: 877–880.

Trujillo, M.E., A. Willems, A. Abril, A.M. Planchuelo, R. Rivas, D. Ludeña, P.F. Mateos, E. Martínez-Molina and E. Velázquez. 2005b. Nodulation of *Lupinus albus* by strains of *Ochrobactrum lupini* sp. nov. Appl. Environ. Microbiol. *71*: 1318–1327.

Trujillo, M.E., R.M. Kroppenstedt, P. Schumann and E. Martínez-Molina. 2006a. *Kribbella lupini* sp. nov., isolated from the roots of *Lupinus angustifolius*. Int. J. Syst. Evol. Microbiol. *56*: 407–411.

Trujillo, M.E., R.M. Kroppenstedt, P. Schumann, L. Carro and E. Martínez-Molina. 2006b. *Micromonospora coriariae* sp. nov., isolated from root nodules of *Coriaria myrtifolia*. Int. J. Syst. Evol. Microbiol. *56*: 2381–2385.

Trujillo, M.E., R.M. Kroppenstedt, C. Fernández-Molinero, P. Schumann and E. Martínez-Molina. 2007. *Micromonospora lupini* sp. nov. and *Micromonospora saelicesensis* sp. nov., isolated from root nodules of *Lupinus angustifolius*. Int. J. Syst. Evol. Microbiol. *57*: 2799–2804.

Trülzsch, K., B. Grabein, P. Schumann, A. Mellmann, U. Antonenka, J. Heesemann and K. Becker. 2007. *Staphylococcus pettenkoferi* sp. nov., a novel coagulase-negative staphylococcal species isolated from human clinical specimens. Int. J. Syst. Evol. Microbiol. *57*: 1543–1548.

Tseng, M., S.F. Yang, W.J. Li and C.L. Jiang. 2006. *Amycolatopsis taiwanensis* sp. nov., from soil. Int. J. Syst. Evol. Microbiol. *56*: 1811–1815.

Tsubota, J., B. Eshinimaev, V.N. Khmelenina and Y.A. Trotsenko. 2005. *Methylothermus thermalis* gen. nov., sp. nov., a novel moderately thermophilic obligate methanotroph from a hot spring in Japan. Int. J. Syst. Evol. Microbiol. *55*: 1877–1884.

Turenne, C.Y., V.J. Cook, T.V. Burdz, R.J. Pauls, L. Thibert, J.N. Wolfe and A. Kabani. 2004. *Mycobacterium parascrofulaceum* sp. nov., novel slowly growing, scotochromogenic clinical isolates related to *Mycobacterium simiae*. Int. J. Syst. Evol. Microbiol. *54*: 1543–1551.

Turenne, C.Y., L. Thibert, K. Williams, T.V. Burdz, V.J. Cook, J.N. Wolfe, D.W. Cockcroft and A. Kabani. 2004. *Mycobacterium saskatchewanense* sp. nov., a novel slowly growing scotochromogenic species from human clinical isolates related to *Mycobacterium interjectum* and Accuprobe-positive for *Mycobacterium avium* complex. Int. J. Syst. Evol. Microbiol. *54*: 659–667.

Tvrzová, L., P. Schumann, I. Sedláček, Z. Páčová, C. Spröer, S. Verbarg and R.M. Kroppenstedt. 2005a. Reclassification of strain CCM 132, previously classified as *Kocuria varians*, as *Kocuria carniphila* sp. nov. Int. J. Syst. Evol. Microbiol. *55*: 139–142.

Tvrzová, L., P. Schumann, C. Spröer, I. Sedláček, S. Verbarg, R.M. Kroppenstedt and Z. Páčová. 2005b. Polyphasic taxonomic study of strain CCM 2783 resulting in the description of *Arthrobacter stackebrandtii* sp. nov. Int. J. Syst. Evol. Microbiol. *55*: 805–808.

Tvrzová, L., P. Schumann, C. Spröer, I. Sedláček, Z. Páčová, O. Šedo, Z. Zdráhal, M. Steffen and E. Lang. 2006. *Pseudomonas moraviensis* sp. nov. and *Pseudomonas vranovensis* sp. nov., soil bacteria isolated on nitroaromatic compounds, and emended description of *Pseudomonas asplenii*. Int. J. Syst. Evol. Microbiol. *56*: 2657–2663.

Ueki, A., H. Akasaka, D. Suzuki and K. Ueki. 2006a. *Paludibacter propionicigenes* gen. nov., sp. nov., a novel strictly anaerobic, Gram-negative, propionate-producing bacterium isolated from plant residue in irrigated rice-field soil in Japan. Int. J. Syst. Evol. Microbiol. *56*: 39–44.

Ueki, A., H. Akasaka, D. Suzuki, S. Hattori and K. Ueki. 2006b. *Xylanibacter oryzae* gen. nov., sp. nov., a novel strictly anaerobic, Gram-negative, xylanolytic bacterium isolated from rice-plant residue in flooded rice-field soil in Japan. Int. J. Syst. Evol. Microbiol. *56*: 2215–2221.

Ueki, A., H. Akasaka, A. Satoh, D. Suzuki and K. Ueki. 2007. *Prevotella paludivivens* sp. nov., a novel strictly anaerobic, Gram-negative, hemicellulose-decomposing bacterium isolated from plant residue and rice roots in irrigated rice-field soil. Int. J. Syst. Evol. Microbiol. *57*: 1803–1809.

Urbanczyk, H., J.C. Ast, M.J. Higgins, J. Carson and P.V. Dunlap. 2007. Reclassification of *Vibrio fischeri, Vibrio logei, Vibrio salmonicida* and *Vibrio wodanis* as *Aliivibrio fischeri* gen. nov., comb. nov., *Aliivibrio logei* comb. nov., *Aliivibrio salmonicida* comb. nov. and *Aliivibrio wodanis* comb. nov. Int. J. Syst. Evol. Microbiol. *57*: 2823–2829.

Urios, L., V. Cueff, P. Pignet and G. Barbier. 2004a. *Tepidibacter formicigenes* sp. nov., a novel spore-forming bacterium isolated from a Mid-Atlantic Ridge hydrothermal vent. Int. J. Syst. Evol. Microbiol. *54*: 439–443.

Urios, L., V. Cueff-Gauchard, P. Pignet, A. Postec, M.L. Fardeau, B. Ollivier and G. Barbier. 2004b. *Thermosipho atlanticus* sp. nov., a novel member of the *Thermotogales* isolated from a Mid-Atlantic Ridge hydrothermal vent. Int. J. Syst. Evol. Microbiol. *54*: 1953–1957.

Urios, L., H. Agogué, F. Lesongeur, E. Stackebrandt and P. Lebaron. 2006. *Balneola vulgaris* gen. nov., sp. nov., a member of the phylum *Bacteroidetes* from the north-western Mediterranean Sea. Int. J. Syst. Evol. Microbiol. *56*: 1883–1887.

Urzì, C., P. Salamone, P. Schumann, M. Rohde and E. Stackebrandt. 2004. *Blastococcus saxobsidens* sp. nov., and emended descriptions of the genus *Blastococcus* Ahrens and Moll 1970 and *Blastococcus aggregatus* Ahrens and Moll 1970. Int. J. Syst. Evol. Microbiol. *54*: 253–259.

Usami, R., A. Echigo, T. Fukushima, T. Mizuki, Y. Yoshida and M. Kamekura. 2007. *Alkalibacillus silvisoli* sp. nov., an alkaliphilic moderate halophile isolated from non-saline forest soil in Japan. Int. J. Syst. Evol. Microbiol. *57*: 770–774.

Valcheva, R., M. Korakli, B. Onno, H. Prévost, I. Ivanova, M.A. Ehrmann, X. Dousset, M.G. Gänzle and R.F. Vogel. 2005. *Lactobacillus hammesii* sp. nov., isolated from French sourdough. Int. J. Syst. Evol. Microbiol. *55*: 763–767.

Valcheva, R., M.F. Ferchichi, M. Korakli, I. Ivanova, M.G. Gänzle, R.F. Vogel, H. Prévost, B. Onno and X. Dousset. 2006. *Lactobacillus nantensis* sp. nov., isolated from French wheat sourdough. Int. J. Syst. Evol. Microbiol. *56*: 587–591.

Valverde, A., E. Velázquez, F. Fernández-Santos, N. Vizcaíno, R. Rivas, P.F. Mateos, E. Martínez-Molina, J.M. Igual and A. Willems. 2005. *Phyllobacterium trifolii* sp. nov., nodulating *Trifolium* and *Lupinus* in Spanish soils. Int. J. Syst. Evol. Microbiol. *55*: 1985–1989.

Valverde, A., P. Delvasto, A. Peix, E. Velázquez, I. Santa-Regina, A. Ballester, C. Rodríguez-Barrueco, C. García-Balboa and J.M. Igual. 2006a.

Burkholderia ferrariae sp. nov., isolated from an iron ore in Brazil. Int. J. Syst. Evol. Microbiol. 56: 2421–2425.

Valverde, A., J.M. Igual, A. Peix, E. Cervantes and E. Velázquez. 2006b. *Rhizobium lusitanum* sp. nov. a bacterium that nodulates *Phaseolus vulgaris*. Int. J. Syst. Evol. Microbiol. 56: 2631–2637.

Van Aken, B., C.M. Peres, S.L. Doty, J.M. Yoon and J.L. Schnoor. 2004. *Methylobacterium populi* sp. nov., a novel aerobic, pink-pigmented, facultatively methylotrophic, methane-utilizing bacterium isolated from poplar trees (*Populus deltoides* x *nigra* DN34). Int. J. Syst. Evol. Microbiol. 54: 1191–1196.

Van den Bulck, K., A. Decostere, M. Baele, P. Vandamme, J. Mast, R. Ducatelle and F. Haesebrouck. 2006. *Helicobacter cynogastricus* sp. nov., isolated from the canine gastric mucosa. Int. J. Syst. Evol. Microbiol. 56: 1559–1564.

Van Trappen, S., J. Mergaert and J. Swings. 2004a. *Loktanella salsilacus* gen. nov., sp. nov., *Loktanella fryxellensis* sp. nov. and *Loktanella vestfoldensis* sp. nov., new members of the *Rhodobacter* group, isolated from microbial mats in Antarctic lakes. Int. J. Syst. Evol. Microbiol. 54: 1263–1269.

Van Trappen, S., T.L. Tan, J. Yang, J. Mergaert and J. Swings. 2004b. *Alteromonas stellipolaris* sp. nov., a novel, budding, prosthecate bacterium from Antarctic seas, and emended description of the genus *Alteromonas*. Int. J. Syst. Evol. Microbiol. 54: 1157–1163.

Van Trappen, S., T.L. Tan, J. Yang, J. Mergaert and J. Swings. 2004c. *Glaciecola polaris* sp. nov., a novel budding and prosthecate bacterium from the Arctic Ocean, and emended description of the genus *Glaciecola*. Int. J. Syst. Evol. Microbiol. 54: 1765–1771.

Van Trappen, S., I. Vandecandelaere, J. Mergaert and J. Swings. 2004d. *Flavobacterium degerlachei* sp. nov., *Flavobacterium frigoris* sp. nov. and *Flavobacterium micromati* sp. nov., novel psychrophilic bacteria isolated from microbial mats in Antarctic lakes. Int. J. Syst. Evol. Microbiol. 54: 85–92.

Van Trappen, S., I. Vandecandelaere, J. Mergaert and J. Swings. 2004e. *Gillisia limnaea* gen. nov., sp. nov., a new member of the family *Flavobacteriaceae* isolated from a microbial mat in Lake Fryxell, Antarctica. Int. J. Syst. Evol. Microbiol. 54: 445–448.

Van Trappen, S., I. Vandecandelaere, J. Mergaert and J. Swings. 2004f. *Algoriphagus antarcticus* sp. nov., a novel psychrophile from microbial mats in Antarctic lakes. Int. J. Syst. Evol. Microbiol. 54: 1969–1973.

Van Trappen, S., T.L. Tan, E. Samyn and P. Vandamme. 2005a. *Alcaligenes aquatilis* sp. nov., a novel bacterium from sediments of the Weser Estuary, Germany, and a salt marsh on Shem Creek in Charleston Harbor, USA. Int. J. Syst. Evol. Microbiol. 55: 2571–2575.

Van Trappen, S., I. Vandecandelaere, J. Mergaert and J. Swings. 2005b. *Flavobacterium fryxellicola* sp. nov. and *Flavobacterium psychrolimnae* sp. nov., novel psychrophilic bacteria isolated from microbial mats in Antarctic lakes. Int. J. Syst. Evol. Microbiol. 55: 769–772.

Vancanneyt, M., L.A. Devriese, E.M. De Graef, M. Baele, K. Lefebvre, C. Snauwaert, P. Vandamme, J. Swings and F. Haesebrouck. 2004a. *Streptococcus minor* sp. nov., from faecal samples and tonsils of domestic animals. Int. J. Syst. Evol. Microbiol. 54: 449–452.

Vancanneyt, M., J. Mengaud, I. Cleenwerck, K. Vanhonacker, B. Hoste, P. Dawyndt, M.C. Degivry, D. Ringuet, D. Janssens and J. Swings. 2004b. Reclassification of *Lactobacillus kefirgranum* Takizawa *et al.* 1994 as *Lactobacillus kefiranofaciens* subsp. *kefirgranum* subsp. nov. and emended description of L. *kefiranofaciens* Fujisawa *et al.* 1988. Int. J. Syst. Evol. Microbiol. 54: 551–556.

Vancanneyt, M., M. Zamfir, L.A. Devriese, K. Lefebvre, K. Engelbeen, K. Vandemeulebroecke, M. Amar, L. De Vuyst, F. Haesebrouck and J. Swings. 2004c. *Enterococcus saccharominimus* sp. nov., from dairy products. Int. J. Syst. Evol. Microbiol. 54: 2175–2179.

Vancanneyt, M., K. Engelbeen, M. De Wachter, K. Vandemeulebroecke, I. Cleenwerck and J. Swings. 2005a. Reclassification of *Lactobacillus ferintoshensis* as a later heterotypic synonym of *Lactobacillus parabuchneri*. Int. J. Syst. Evol. Microbiol. 55: 2195–2198.

Vancanneyt, M., P. Neysens, M. De Wachter, K. Engelbeen, C. Snauwaert, I. Cleenwerck, R. Van der Meulen, B. Hoste, E. Tsakalidou, L.

De Vuyst and J. Swings. 2005b. *Lactobacillus acidifarinae* sp. nov. and *Lactobacillus zymae* sp. nov., from wheat sourdoughs. Int. J. Syst. Evol. Microbiol. 55: 615–620.

Vancanneyt, M., S.M. Naser, K. Engelbeen, M. De Wachter, R. Van der Meulen, I. Cleenwerck, B. Hoste, L. De Vuyst and J. Swings. 2006a. Reclassification of *Lactobacillus brevis* strains LMG 11494 and LMG 11984 as *Lactobacillus parabrevis* sp. nov. Int. J. Syst. Evol. Microbiol. 56: 1553–1557.

Vancanneyt, M., O.I. Nedashkovskaya, C. Snauwaert, S. Mortier, K. Vandemeulebroecke, B. Hoste, P. Dawyndt, G.M. Frolova, D. Janssens and J. Swings. 2006b. *Larkinella insperata* gen. nov., sp. nov., a bacterium of the phylum '*Bacteroidetes*' isolated from water of a steam generator. Int. J. Syst. Evol. Microbiol. 56: 237–241.

Vancanneyt, M., M. Zamfir, M. De Wachter, I. Cleenwerck, B. Hoste, F. Rossi, F. Dellaglio, L. De Vuyst and J. Swings. 2006c. Reclassification of *Leuconostoc argentinum* as a later synonym of *Leuconostoc lactis*. Int. J. Syst. Evol. Microbiol. 56: 213–216.

Vandamme, P. and T. Coenye. 2004. Taxonomy of the genus *Cupriavidus*: a tale of lost and found. Int. J. Syst. Evol. Microbiol. 54: 2285–2289.

Vandamme, P., B. Holmes, H. Bercovier and T. Coenye. 2006. Classification of Centers for Disease Control Group Eugonic Fermenter (EF)-4a and EF-4b as *Neisseria animaloris* sp. nov. and *Neisseria zoodegmatis* sp. nov., respectively. Int. J. Syst. Evol. Microbiol. 56: 1801–1805.

Vandamme, P., K. Opelt, N. Knochel, C. Berg, S. Schönmann, E. De Brandt, L. Eberl, E. Falsen and G. Berg. 2007. *Burkholderia bryophila* sp. nov. and *Burkholderia megapolitana* sp. nov., moss-associated species with antifungal and plant-growth-promoting properties. Int. J. Syst. Evol. Microbiol. 57: 2228–2235.

Vandieken, V., C. Knoblauch and B.B. Jørgensen. 2006a. *Desulfovibrio frigidus* sp. nov. and *Desulfovibrio ferrireducens* sp. nov., psychrotolerant bacteria isolated from Arctic fjord sediments (Svalbard) with the ability to reduce Fe(III). Int. J. Syst. Evol. Microbiol. 56: 681–685.

Vandieken, V., C. Knoblauch and B.B. Jørgensen. 2006b. *Desulfotomaculum arcticum* sp. nov., a novel spore-forming, moderately thermophilic, sulfate-reducing bacterium isolated from a permanently cold fjord sediment of Svalbard. Int. J. Syst. Evol. Microbiol. 56: 687–690.

Vandieken, V., M. Mussmann, H. Niemann and B.B. Jørgensen. 2006c. *Desulfuromonas svalbardensis* sp. nov. and *Desulfuromusa ferrireducens* sp. nov., psychrophilic, Fe(III)-reducing bacteria isolated from Arctic sediments, Svalbard. Int. J. Syst. Evol. Microbiol. 56: 1133–1139.

Vaneechoutte, M., P. Kämpfer, T. De Baere, E. Falsen and G. Verschraegen. 2004. *Wautersia* gen. nov., a novel genus accommodating the phylogenetic lineage including *Ralstonia eutropha* and related species, and proposal of *Ralstonia* [*Pseudomonas*] *syzygii* (Roberts *et al.* 1990) comb. nov. Int. J. Syst. Evol. Microbiol. 54: 317–327.

Vaneechoutte, M., P. Kämpfer, T. De Baere, V. Avesani, M. Janssens and G. Wauters. 2007. *Chryseobacterium hominis* sp. nov., to accommodate clinical isolates biochemically similar to CDC groups II-h and II-c. Int. J. Syst. Evol. Microbiol. 57: 2623–2628.

Vanparys, B., K. Heylen, L. Lebbe and P. De Vos. 2005a. *Pedobacter caeni* sp. nov., a novel species isolated from a nitrifying inoculum. Int. J. Syst. Evol. Microbiol. 55: 1315–1318.

Vanparys, B., K. Heylen, L. Lebbe and P. De Vos. 2005b. *Devosia limi* sp. nov., isolated from a nitrifying inoculum. Int. J. Syst. Evol. Microbiol. 55: 1997–2000.

Vanparys, B., K. Heylen, L. Lebbe and P. De Vos. 2006. *Pseudomonas peli* sp. nov. and *Pseudomonas borbori* sp. nov., isolated from a nitrifying inoculum. Int. J. Syst. Evol. Microbiol. 56: 1875–1881.

Vargas, V.A., O.D. Delgado, R. Hatti-Kaul and B. Mattiasson. 2005. *Bacillus bogoriensis* sp. nov., a novel alkaliphilic, halotolerant bacterium isolated from a Kenyan soda lake. Int. J. Syst. Evol. Microbiol. 55: 899–902.

Vasilyeva, L.V., M.V. Omelchenko, Y.Y. Berestovskaya, A.M. Lysenko, W.R. Abraham, S.N. Dedysh and G.A. Zavarzin. 2006. *Asticcacaulis benevestitus* sp. nov., a psychrotolerant, dimorphic, prosthecate bacterium from tundra wetland soil. Int. J. Syst. Evol. Microbiol. 56: 2083–2088.

Vaz-Moreira, I., C. Faria, M.F. Nobre, P. Schumann, O.C. Nunes and C.M. Manaia. 2007a. *Paenibacillus humicus* sp. nov., isolated from poultry litter compost. Int. J. Syst. Evol. Microbiol. *57*: 2267–2271.

Vaz-Moreira, I., M.F. Nobre, O.C. Nunes and C.M. Manaia. 2007b. *Gulbenkiania mobilis* gen. nov., sp. nov., isolated from treated municipal wastewater. Int. J. Syst. Evol. Microbiol. *57*: 1108–1112.

Vaz-Moreira, I., M.F. Nobre, O.C. Nunes and C.M. Manaia. 2007c. *Pseudosphingobacterium domesticum* gen. nov., sp. nov., isolated from home-made compost. Int. J. Syst. Evol. Microbiol. *57*: 1535–1538.

Vela, A.I., M.D. Collins, P.A. Lawson, N. García, L. Domínguez and J.F. Fernández-Garayzábal. 2005. *Uruburuella suis* gen. nov., sp. nov., isolated from clinical specimens of pigs. Int. J. Syst. Evol. Microbiol. *55*: 643–647.

Vela, A.I., M.C. Gutiérrez, E. Falsen, E. Rollán, I. Simarro, P. García, L. Domínguez, A. Ventosa and J.F. Fernández-Garayzábal. 2006. *Pseudomonas simiae* sp. nov., isolated from clinical specimens from monkeys (*Callithrix geoffroyi*). Int. J. Syst. Evol. Microbiol. *56*: 2671–2676.

Vela, A.I., A. Fernandez, C. Sánchez-Porro, E. Sierra, M. Mendez, M. Arbelo, A. Ventosa, L. Domínguez and J.F. Fernández-Garayzábal. 2007a. *Flavobacterium ceti* sp. nov., isolated from beaked whales (*Ziphius cavirostris*). Int. J. Syst. Evol. Microbiol. *57*: 2604–2608.

Vela, A.I., N. García, M.V. Latre, A. Casamayor, C. Sánchez-Porro, V. Briones, A. Ventosa, L. Domínguez and J.F. Fernández-Garayzábal. 2007b. *Aerococcus suis* sp. nov., isolated from clinical specimens from swine. Int. J. Syst. Evol. Microbiol. *57*: 1291–1294.

Velázquez, E., T. de Miguel, M. Poza, R. Rivas, R. Rosselló-Mora and T.G. Villa. 2004. *Paenibacillus favisporus* sp. nov., a xylanolytic bacterium isolated from cow faeces. Int. J. Syst. Evol. Microbiol. *54*: 59–64.

Ventosa, A., M.C. Gutiérrez, M. Kamekura, I.S. Zvyagintseva and A. Oren. 2004. Taxonomic study of *Halorubrum distributum* and proposal of *Halorubrum terrestre* sp. nov. Int. J. Syst. Evol. Microbiol. *54*: 389–392.

Verbarg, S., H. Rheims, S. Emus, A. Frühling, R.M. Kroppenstedt, E. Stackebrandt and P. Schumann. 2004. *Erysipelothrix inopinata* sp. nov., isolated in the course of sterile filtration of vegetable peptone broth, and description of *Erysipelotrichaceae* fam. nov. Int. J. Syst. Evol. Microbiol. *54*: 221–225.

Vermis, K., T. Coenye, J.J. LiPuma, E. Mahenthiralingam, H.J. Nelis and P. Vandamme. 2004. Proposal to accommodate *Burkholderia cepacia* genomovar VI as *Burkholderia dolosa* sp. nov. Int. J. Syst. Evol. Microbiol. *54*: 689–691.

Vetriani, C., M.D. Speck, S.V. Ellor, R.A. Lutz and V. Starovoytov. 2004. *Thermovibrio ammonificans* sp. nov., a thermophilic, chemolithotrophic, nitrate-ammonifying bacterium from deep-sea hydrothermal vents. Int. J. Syst. Evol. Microbiol. *54*: 175–181.

Vinuesa, P., M. León-Barrios, C. Silva, A. Willems, A. Jarabo-Lorenzo, R. Pérez-Galdona, D. Werner and E. Martínez-Romero. 2005. *Bradyrhizobium canariense* sp. nov., an acid-tolerant endosymbiont that nodulates endemic genistoid legumes (Papilionoideae: Genisteae) from the Canary Islands, along with *Bradyrhizobium japonicum* bv. *genistearum*, *Bradyrhizobium* genospecies alpha and *Bradyrhizobium* genospecies beta. Int. J. Syst. Evol. Microbiol. *55*: 569–575.

Voordeckers, J.W., V. Starovoytov and C. Vetriani. 2005. *Caminibacter mediatlanticus* sp. nov., a thermophilic, chemolithoautotrophic, nitrate-ammonifying bacterium isolated from a deep-sea hydrothermal vent on the Mid-Atlantic Ridge. Int. J. Syst. Evol. Microbiol. *55*: 773–779.

Wagner-Döbler, I., H. Rheims, A. Felske, A. El-Ghezal, D. Flade-Schröder, H. Laatsch, S. Lang, R. Pukall and B.J. Tindall. 2004. *Oceanibulbus indolifex* gen. nov., sp. nov., a North Sea alphaproteobacterium that produces bioactive metabolites. Int. J. Syst. Evol. Microbiol. *54*: 1177–1184.

Wang, F.Q., E.T. Wang, J. Liu, Q. Chen, X.H. Sui, W.F. Chen and W.X. Chen. 2007. *Mesorhizobium albiziae* sp. nov., a novel bacterium that nodulates *Albizia kalkora* in a subtropical region of China. Int. J. Syst. Evol. Microbiol. *57*: 1192–1199.

Wang, L., Y. Zhang, Y. Huang, L.A. Maldonado, Z. Liu and M. Goodfellow. 2004. *Nocardia pigrifrangens* sp. nov., a novel actinomycete isolated from a contaminated agar plate. Int. J. Syst. Evol. Microbiol. *54*: 1683–1686.

Wang, L., Y. Huang, Z. Liu, M. Goodfellow and C. Rodríguez. 2006. *Streptacidiphilus oryzae* sp. nov., an actinomycete isolated from rice-field soil in Thailand. Int. J. Syst. Evol. Microbiol. *56*: 1257–1261.

Wang, L.T., F.L. Lee, C.J. Tai, A. Yokota and H.P. Kuo. 2007. Reclassification of *Bacillus axarquiensis* Ruiz-García *et al.* 2005 and *Bacillus malacitensis* Ruiz-García *et al.* 2005 as later heterotypic synonyms of *Bacillus mojavensis* Roberts *et al.* 1994. Int. J. Syst. Evol. Microbiol. *57*: 1663–1667.

Wang, Q.F., W. Li, Y.L. Liu, H.H. Cao, Z. Li and G.Q. Guo. 2007a. *Bacillus qingdaonensis* sp. nov., a moderately haloalkaliphilic bacterium isolated from a crude sea-salt sample collected near Qingdao in eastern China. Int. J. Syst. Evol. Microbiol. *57*: 1143–1147.

Wang, Q.F., W. Li, H. Yang, Y.L. Liu, H.H. Cao, M. Dornmayr-Pfaffenhuemer, H. Stan-Lotter and G.Q. Guo. 2007b. *Halococcus qingdaonensis* sp. nov., a halophilic archaeon isolated from a crude sea-salt sample. Int. J. Syst. Evol. Microbiol. *57*: 600–604.

Wang, W., Z. Zhang, Q. Tang, J. Mao, D. Wei, Y. Huang, Z. Liu, Y. Shi and M. Goodfellow. 2007. *Lechevalieria xinjiangensis* sp. nov., a novel actinomycete isolated from radiation-polluted soil in China. Int. J. Syst. Evol. Microbiol. *57*: 2819–2822.

Wang, X., T. Maegawa, T. Karasawa, E. Ozaki and S. Nakamura. 2005. *Clostridium sardiniense* Prevot 1938 and *Clostridium absonum* Nakamura *et al.* 1973 are heterotypic synonyms: evidence from phylogenetic analyses of phospholipase C and 16S rRNA sequences, and DNA relatedness. Int. J. Syst. Evol. Microbiol. *55*: 1193–1197.

Wang, X., F. Sahr, T. Xue and B. Sun. 2007. *Methylobacterium salsuginis* sp. nov., isolated from seawater. Int. J. Syst. Evol. Microbiol. *57*: 1699–1703.

Wang, Y.N., H. Cai, C.Q. Chi, A.H. Lu, X.G. Lin, Z.F. Jiang and X.L. Wu. 2007a. *Halomonas shengliensis* sp. nov., a moderately halophilic, denitrifying, crude-oil-utilizing bacterium. Int. J. Syst. Evol. Microbiol. *57*: 1222–1226.

Wang, Y.N., H. Cai, S.L. Yu, Z.Y. Wang, J. Liu and X.L. Wu. 2007b. *Halomonas gudaonensis* sp. nov., isolated from a saline soil contaminated by crude oil. Int. J. Syst. Evol. Microbiol. *57*: 911–915.

Wang, Y.X., X.Y. Zhi, H.H. Chen, Y.Q. Zhang, S.K. Tang, C.L. Jiang, L.H. Xu and W.J. Li. 2007. *Actinomadura alba* sp. nov., isolated from soil in Yunnan, China. Int. J. Syst. Evol. Microbiol. *57*: 1735–1739.

Wang, Z.W., Y.H. Liu, X. Dai, B.J. Wang, C.Y. Jiang and S.J. Liu. 2006. *Flavobacterium saliperosum* sp. nov., isolated from freshwater lake sediment. Int. J. Syst. Evol. Microbiol. *56*: 439–442.

Warren, Y.A., K.L. Tyrrell, D.M. Citron and E.J.C. Goldstein. 2006. *Clostridium aldenense* sp. nov. and *Clostridium citroniae* sp. nov. isolated from human clinical infections. J. Clin. Microbiol. *44*: 2416–2422.

Wartiainen, I., A.G. Hestnes, I.R. McDonald and M.M. Svenning. 2006a. *Methylobacter tundripaludum* sp. nov., a methane-oxidizing bacterium from Arctic wetland soil on the Svalbard islands, Norway (78°N). Int. J. Syst. Evol. Microbiol. *56*: 109–113.

Wartiainen, I., A.G. Hestnes, I.R. McDonald and M.M. Svenning. 2006b. *Methylocystis rosea* sp. nov., a novel methanotrophic bacterium from Arctic wetland soil, Svalbard, Norway (78°N). Int. J. Syst. Evol. Microbiol. *56*: 541–547.

Watanabe, K., N. Nagao, S. Yamamoto, T. Toda and N. Kurosawa. 2007. *Thermobacillus composti* sp. nov., a moderately thermophilic bacterium isolated from a composting reactor. Int. J. Syst. Evol. Microbiol. *57*: 1473–1477.

Wauters, G., G. Haase, V. Avesani, J. Charlier, M. Janssens, J. Van Broeck and M. Delmée. 2004. Identification of a novel *Brevibacterium* species isolated from humans and description of *Brevibacterium sanguinis* sp. nov. J. Clin. Microbiol. *42*: 2829–2832.

Weon, H.Y., B.Y. Kim, S.W. Kwon, I.C. Park, I.B. Cha, B.J. Tindall, E. Stackebrandt, H.G. Trüper and S.J. Go. 2005. *Leadbetterella byssophila* gen. nov., sp. nov., isolated from cotton-waste composts for the cultivation of oyster mushroom. Int. J. Syst. Evol. Microbiol. *55*: 2297–2302.

Weon, H.Y., B.Y. Kim, Y.K. Baek, S.H. Yoo, S.W. Kwon, E. Stackebrandt and S.J. Go. 2006a. Two novel species, *Lysobacter daejeonensis* sp. nov. and *Lysobacter yangpyeongensis* sp. nov., isolated from Korean greenhouse soils. Int. J. Syst. Evol. Microbiol. *56*: 947–951.

Weon, H.Y., B.Y. Kim, J.S. Kim, S.Y. Lee, Y.H. Cho, S.J. Go, S.B. Hong, W.T. Im and S.W. Kwon. 2006b. *Pseudoxanthomonas suwonensis* sp. nov., isolated from cotton waste composts. Int. J. Syst. Evol. Microbiol. *56*: 659–662.

Weon, H.Y., B.Y. Kim, S.H. Yoo, S.W. Kwon, Y.H. Cho, S.J. Go and E. Stackebrandt. 2006c. *Chryseobacterium wanjuense* sp. nov., isolated from greenhouse soil in Korea. Int. J. Syst. Evol. Microbiol. *56*: 1501–1504.

Weon, H.Y., B.Y. Kim, S.H. Yoo, Y.K. Baek, S.Y. Lee, S.W. Kwon, S.J. Go and E. Stackebrandt. 2006d. *Pseudomonas pohangensis* sp. nov., isolated from seashore sand in Korea. Int. J. Syst. Evol. Microbiol. *56*: 2153–2156.

Weon, H.Y., B.Y. Kim, S.H. Yoo, J.S. Kim, S.W. Kwon, S.J. Go and E. Stackebrandt. 2006e. *Loktanella koreensis* sp. nov., isolated from sea sand in Korea. Int. J. Syst. Evol. Microbiol. *56*: 2199–2202.

Weon, H.Y., B.Y. Kim, S.H. Yoo, S.Y. Lee, S.W. Kwon, S.J. Go and E. Stackebrandt. 2006f. *Niastella koreensis* gen. nov., sp. nov. and *Niastella yeongjuensis* sp. nov., novel members of the phylum *Bacteroidetes*, isolated from soil cultivated with Korean ginseng. Int. J. Syst. Evol. Microbiol. *56*: 1777–1782.

Weon, H.Y., B.Y. Kim, S.B. Hong, Y.A. Jeon, S.W. Kwon, S.J. Go and B.S. Koo. 2007a. *Rhodanobacter ginsengisoli* sp. nov. and *Rhodanobacter terrae* sp. nov., isolated from soil cultivated with Korean ginseng. Int. J. Syst. Evol. Microbiol. *57*: 2810–2813.

Weon, H.Y., B.Y. Kim, S.B. Hong, J.H. Joa, S.S. Nam, K.H. Lee and S.W. Kwon. 2007b. *Skermanella aerolata* sp. nov., isolated from air, and emended description of the genus *Skermanella*. Int. J. Syst. Evol. Microbiol. *57*: 1539–1542.

Weon, H.Y., B.Y. Kim, M.K. Kim, S.H. Yoo, S.W. Kwon, S.J. Go and E. Stackebrandt. 2007c. *Lysobacter niabensis* sp. nov. and *Lysobacter niastensis* sp. nov., isolated from greenhouse soils in Korea. Int. J. Syst. Evol. Microbiol. *57*: 548–551.

Weon, H.Y., B.Y. Kim, P. Schumann, R.M. Kroppenstedt, H.J. Noh, C.W. Park and S.W. Kwon. 2007d. *Knoellia aerolata* sp. nov., isolated from an air sample in Korea. Int. J. Syst. Evol. Microbiol. *57*: 2861–2864.

Weon, H.Y., B.Y. Kim, P. Schumann, J.A. Son, J. Jang, S.J. Go and S.W. Kwon. 2007e. *Deinococcus cellulosilyticus* sp. nov., isolated from air. Int. J. Syst. Evol. Microbiol. *57*: 1685–1688.

Weon, H.Y., B.Y. Kim, S.H. Yoo, J.H. Joa, S.W. Kwon and W.G. Kim. 2007f. *Andreprevotia chitinilytica* gen. nov., sp. nov., isolated from forest soil from Halla Mountain, Jeju Island, Korea. Int. J. Syst. Evol. Microbiol. *57*: 1572–1575.

Weon, H.Y., B.Y. Kim, S.H. Yoo, J.H. Joa, K.H. Lee, Y.S. Zhang, S.W. Kwon and B.S. Koo. 2007g. *Aurantimonas ureilytica* sp. nov., isolated from an air sample. Int. J. Syst. Evol. Microbiol. *57*: 1717–1720.

Weon, H.Y., S.Y. Lee, B.Y. Kim, H.J. Noh, P. Schumann, J.S. Kim and S.W. Kwon. 2007h. *Ureibacillus composti* sp. nov. and *Ureibacillus thermophilus* sp. nov., isolated from livestock-manure composts. Int. J. Syst. Evol. Microbiol. *57*: 2908–2911.

Weon, H.Y., P. Schumann, R.M. Kroppenstedt, B.Y. Kim, J. Song, S.W. Kwon, S.J. Go and E. Stackebrandt. 2007i. *Terrabacter aerolatus* sp. nov., isolated from an air sample. Int. J. Syst. Evol. Microbiol. *57*: 2106–2109.

Weon, H.Y., M.H. Song, J.A. Son, B.Y. Kim, S.W. Kwon, S.J. Go and E. Stackebrandt. 2007j. *Flavobacterium terrae* sp. nov. and *Flavobacterium cucumis* sp. nov., isolated from greenhouse soil. Int. J. Syst. Evol. Microbiol. *57*: 1594–1598.

Whipps, C.M., W.R. Butler, F. Pourahmad, V.G. Watral and M.L. Kent. 2007. Molecular systematics support the revival of *Mycobacterium salmoniphilum* (ex Ross 1960) sp. nov., nom. rev., a species closely related to *Mycobacterium chelonae*. Int. J. Syst. Evol. Microbiol. *57*: 2525–2531.

Whitehead, T.R., M.A. Cotta, M.D. Collins and P.A. Lawson. 2004. *Hespellia stercorisuis* gen. nov., sp. nov. and *Hespellia porcina* sp. nov., isolated from swine manure storage pits. Int. J. Syst. Evol. Microbiol. *54*: 241–245.

Whitehead, T.R., M.A. Cotta, M.D. Collins, E. Falsen and P.A. Lawson. 2005. *Bacteroides coprosuis* sp. nov., isolated from swine-manure storage pits. Int. J. Syst. Evol. Microbiol. *55*: 2515–2518.

Wieser, M., H. Worliczek, P. Kämpfer and H.J. Busse. 2005. *Bacillus herbersteinensis* sp. nov. Int. J. Syst. Evol. Microbiol. *55*: 2119–2123.

Wink, J., J. Gandhi, R.M. Kroppenstedt, G. Seibert, B. Sträubler, P. Schumann and E. Stackebrandt. 2004. *Amycolatopsis decaplanina* sp. nov., a novel member of the genus with unusual morphology. Int. J. Syst. Evol. Microbiol. *54*: 235–239.

Wink, J.M., R.M. Kroppenstedt, P. Schumann, G. Seibert and E. Stackebrandt. 2006. *Actinoplanes liguriensis* sp. nov. and *Actinoplanes teichomyceticus* sp. nov. Int. J. Syst. Evol. Microbiol. *56*: 2125–2130.

Wittich, R.M., H.J. Busse, P. Kämpfer, A.J. Macedo, M. Tiirola, M. Wieser and W.R. Abraham. 2007a. *Sphingomonas fennica* sp. nov. and *Sphingomonas haloaromaticamans* sp. nov., outliers of the genus *Sphingomonas*. Int. J. Syst. Evol. Microbiol. *57*: 1740–1746.

Wittich, R.M., H.J. Busse, P. Kämpfer, M. Tiirola, M. Wieser, A.J. Macedo and W.R. Abraham. 2007b. *Sphingobium aromaticiconvertens* sp. nov., a xenobiotic-compound-degrading bacterium from polluted river sediment. Int. J. Syst. Evol. Microbiol. *57*: 306–310.

Wolin, M.J., T.L. Miller, M.D. Collins and P.A. Lawson. 2003. Formate-dependent growth and homoacetogenic fermentation by a bacterium from human feces: description of *Bryantella formatexigens* gen. nov., sp. nov. Appl. Environ. Microbiol. *69*: 6321–6326.

Wolterink, A., S. Kim, M. Muusse, I.S. Kim, P.J. Roholl, C.G. van Ginkel, A.J. Stams and S.W. Kengen. 2005. *Dechloromonas hortensis* sp. nov. and strain ASK-1, two novel (per)chlorate-reducing bacteria, and taxonomic description of strain GR-1. Int. J. Syst. Evol. Microbiol. *55*: 2063–2068.

Woo, P.C.Y., A.M.Y. Fung, S.K.P. Lau, J.L.L.Teng, B.H.L.Wong, M.K.M. Wong, E. Hon, G.W.K. Tang and K.Y. Yuen. 2003. *Actinomyces hongkongensis* sp. nov. A novel *Actinomyces* species isolated from a patient with pelvic actinomycosis. Syst. Appl. Microbiol. *26*: 518–522.

Wu, C., X. Liu and X. Dong. 2006. *Syntrophomonas cellicola* sp. nov., a spore-forming syntrophic bacterium isolated from a distilled-spirit-fermenting cellar, and assignment of *Syntrophospora bryantii* to *Syntrophomonas bryantii* comb. nov. Int. J. Syst. Evol. Microbiol. *56*: 2331–2335.

Wu, S.Y., S.C. Chen and M.C. Lai. 2005. *Methanofollis formosanus* sp. nov., isolated from a fish pond. Int. J. Syst. Evol. Microbiol. *55*: 837–842.

Wübbeler, J.H., T. Lütke-Eversloh, S. Van Trappen, P. Vandamme and A. Steinbüchel. 2006. *Tetrathiobacter mimigardefordensis* sp. nov., isolated from compost, a betaproteobacterium capable of utilizing the organic disulfide 3,3′-dithiodipropionic acid. Int. J. Syst. Evol. Microbiol. *56*: 1305–1310.

Wyss, C., A. Moter, B.K. Choi, F.E. Dewhirst, Y. Xue, P. Schüpbach, U.B. Göbel, B.J. Paster and B. Guggenheim. 2004. *Treponema putidum* sp. nov., a medium-sized proteolytic spirochaete isolated from lesions of human periodontitis and acute necrotizing ulcerative gingivitis. Int. J. Syst. Evol. Microbiol. *54*: 1117–1122.

Wyss, C., F.E. Dewhirst, B.J. Paster, T. Thurnheer and A. Luginbühl. 2005. *Guggenheimella bovis* gen. nov., sp. nov., isolated from lesions of bovine dermatitis digitalis. Int. J. Syst. Evol. Microbiol. *55*: 667–671.

Xiao, X., P. Wang, X. Zeng, D.H. Bartlett and F. Wang. 2007. *Shewanella psychrophila* sp. nov. and *Shewanella piezotolerans* sp. nov., isolated from west Pacific deep-sea sediment. Int. J. Syst. Evol. Microbiol. *57*: 60–65.

Xie, C.H. and A. Yokota. 2004. Transfer of *Hyphomicrobium indicum* to the genus *Photobacterium* as *Photobacterium indicum* comb. nov. Int. J. Syst. Evol. Microbiol. *54*: 2113–2116.

Xie, C.H. and A. Yokota. 2005a. Phylogenetic analysis of *Alysiella* and related genera of *Neisseriaceae*: proposal of *Alysiella crassa* comb. nov., *Conchiformibium steedae* gen. nov., comb. nov., *Conchiformibium kuhniae* sp. nov. and *Bergeriella denitrificans* gen. nov., comb. nov. J. Gen. Appl. Microbiol. *51*: 1–10.

Xie, C.H. and A. Yokota. 2005b. Transfer of the misnamed [*Alysiella*] sp. IAM 14971 (=ATCC 29468) to the genus *Moraxella* as *Moraxella oblonga* sp. nov. Int. J. Syst. Evol. Microbiol. 55: 331–334.

Xie, C.H. and A. Yokota. 2005c. *Dyella japonica* gen. nov., sp. nov., a γ-proteobacterium isolated from soil. Int. J. Syst. Evol. Microbiol. 55: 753–756.

Xie, C.H. and A. Yokota. 2005d. *Pleomorphomonas oryzae* gen. nov., sp. nov., a nitrogen-fixing bacterium isolated from paddy soil of *Oryza sativa*. Int. J. Syst. Evol. Microbiol. 55: 1233–1237.

Xie, C.H. and A. Yokota. 2005e. *Azospirillum oryzae* sp. nov., a nitrogen-fixing bacterium isolated from the roots of the rice plant *Oryza sativa*. Int. J. Syst. Evol. Microbiol. 55: 1435–1438.

Xie, C.H. and A. Yokota. 2005f. Reclassification of *Alcaligenes latus* strains IAM 12599^T and IAM 12664 and *Pseudomonas saccharophila* as *Azohydromonas lata* gen. nov., comb. nov., *Azohydromonas australica* sp. nov. and *Pelomonas saccharophila* gen. nov., comb. nov., respectively. Int. J. Syst. Evol. Microbiol. 55: 2419–2425.

Xie, C.H. and A. Yokota. 2006a. *Zoogloea oryzae* sp. nov., a nitrogen-fixing bacterium isolated from rice paddy soil, and reclassification of the strain ATCC 19623 as *Crabtreella saccharophila* gen. nov., sp. nov. Int. J. Syst. Evol. Microbiol. 56: 619–624.

Xie, C.H. and A. Yokota. 2006b. *Sphingomonas azotifigens* sp. nov., a nitrogen-fixing bacterium isolated from the roots of *Oryza sativa*. Int. J. Syst. Evol. Microbiol. 56: 889–893.

Xie, C.H. and A. Yokota. 2006c. Reclassification of [*Flavobacterium*] *ferrugineum* as *Terrimonas ferruginea* gen. nov., comb. nov., and description of *Terrimonas lutea* sp. nov., isolated from soil. Int. J. Syst. Evol. Microbiol. 56: 1117–1121.

Xie, C.H. and A. Yokota. 2007. *Paenibacillus terrigena* sp. nov., isolated from soil. Int. J. Syst. Evol. Microbiol. 57: 70–72.

Xin, Y.H., Y.G. Zhou, H.L. Zhou and W.X. Chen. 2004. *Ancylobacter rudongensis* sp. nov., isolated from roots of *Spartina anglica*. Int. J. Syst. Evol. Microbiol. 54: 385–388.

Xin, Y.H., Y.G. Zhou and W.X. Chen. 2006. *Ancylobacter polymorphus* sp. nov. and *Ancylobacter vacuolatus* sp. nov. Int. J. Syst. Evol. Microbiol. 56: 1185–1188.

Xing, D., N. Ren, Q. Li, M. Lin, A. Wang and L. Zhao. 2006. *Ethanoligenens harbinense* gen. nov., sp. nov., isolated from molasses wastewater. Int. J. Syst. Evol. Microbiol. 56: 755–760.

Xu, C., L. Wang, Q. Cui, Y. Huang, Z. Liu, G. Zheng and M. Goodfellow. 2006. Neutrotolerant acidophilic *Streptomyces* species isolated from acidic soils in China: *Streptomyces guanduensis* sp. nov., *Streptomyces paucisporeus* sp. nov., *Streptomyces rubidus* sp. nov. and *Streptomyces yanglinensis* sp. nov. Int. J. Syst. Evol. Microbiol. 56: 1109–1115.

Xu, J.L., J. He, Z.C. Wang, K. Wang, W.J. Li, S.K. Tang and S.P. Li. 2007. *Rhodococcus qingshengii* sp. nov., a carbendazim-degrading bacterium. Int. J. Syst. Evol. Microbiol. 57: 2754–2757.

Xu, M., J. Guo, Y. Cen, X. Zhong, W. Cao and G. Sun. 2005. *Shewanella decolorationis* sp. nov., a dye-decolorizing bacterium isolated from activated sludge of a waste-water treatment plant. Int. J. Syst. Evol. Microbiol. 55: 363–368.

Xu, P., W.J. Li, W.L. Wu, D. Wang, L.H. Xu and C.L. Jiang. 2004a. *Streptomyces hebeiensis* sp. nov. Int. J. Syst. Evol. Microbiol. 54: 727–731.

Xu, P., Y. Takahashi, A. Seino, Y. Iwai and S. Omura. 2004b *Streptomyces scabrisporus* sp. nov. Int. J. Syst. Evol. Microbiol. 54: 577–581.

Xu, P., W.J. Li, S.K. Tang, Y. Jiang, H.H. Chen, L.H. Xu and C.L. Jiang. 2005a. *Nocardia polyresistens* sp. nov. Int. J. Syst. Evol. Microbiol. 55: 1465–1470.

Xu, P., W.J. Li, S.K. Tang, Y.Q. Zhang, G.Z. Chen, H.H. Chen, L.H. Xu and C.L. Jiang. 2005b. *Naxibacter alkalitolerans* gen. nov., sp. nov., a novel member of the family 'Oxalobacteraceae' isolated from China. Int. J. Syst. Evol. Microbiol. 55: 1149–1153.

Xu, P., W.J. Li, S.K. Tang, Y. Jiang, H.Y. Gao, L.H. Xu and C.L. Jiang. 2006. *Nocardia lijiangensis* sp. nov., a novel actinomycete strain isolated from soil in China. Syst. Appl. Microbiol. 29: 308–314.

Xu, X.W., S.J. Liu, D. Tohty, A. Oren, M. Wu and P.J. Zhou. 2005a. *Haloterrigena saccharevitans* sp. nov., an extremely halophilic archaeon from Xin-Jiang, China. Int. J. Syst. Evol. Microbiol. 55: 2539–2542.

Xu, X.W., P.G. Ren, S.J. Liu, M. Wu and P.J. Zhou. 2005b. *Natrinema altunense* sp. nov., an extremely halophilic archaeon isolated from a salt lake in Altun Mountain in Xinjiang, China. Int. J. Syst. Evol. Microbiol. 55: 1311–1314.

Xu, X.W., M. Wu, P.J. Zhou and S.J. Liu. 2005c. *Halobiforma lacisalsi* sp. nov., isolated from a salt lake in China. Int. J. Syst. Evol. Microbiol. 55: 1949–1952.

Xu, X.W., Y.H. Wu, C.S. Wang, A. Oren, P.J. Zhou and M. Wu. 2007a. *Haloferax larsenii* sp. nov., an extremely halophilic archaeon from a solar saltern. Int. J. Syst. Evol. Microbiol. 57: 717–720.

Xu, X.W., Y.H. Wu, H.B. Zhang and M. Wu. 2007b. *Halorubrum arcis* sp. nov., an extremely halophilic archaeon isolated from a saline lake on the Qinghai-Tibet Plateau, China. Int. J. Syst. Evol. Microbiol. 57: 1069–1072.

Xu, X.W., Y.H. Wu, Z. Zhou, C.S. Wang, Y.G. Zhou, H.B. Zhang, Y. Wang and M. Wu. 2007c. *Halomonas saccharevitans* sp. nov., *Halomonas arcis* sp. nov. and *Halomonas subterranea* sp. nov., halophilic bacteria isolated from hypersaline environments of China. Int. J. Syst. Evol. Microbiol. 57: 1619–1624.

Xue, Y., H. Fan, A. Ventosa, W.D. Grant, B.E. Jones, D.A. Cowan and Y. Ma. 2005. *Halalkalicoccus tibetensis* gen. nov., sp. nov., representing a novel genus of haloalkaliphilic archaea. Int. J. Syst. Evol. Microbiol. 55: 2501–2505.

Xue, Y., X. Zhang, C. Zhou, Y. Zhao, D.A. Cowan, S. Heaphy, W.D. Grant, B.E. Jones, A. Ventosa and Y. Ma. 2006. *Caldalkalibacillus thermarum* gen. nov., sp. nov., a novel alkalithermophilic bacterium from a hot spring in China. Int. J. Syst. Evol. Microbiol. 56: 1217–1221.

Yabuuchi, E. and Y. Kosako. 2005. Order IV. *Sphingomonadales* ord. nov. *In* Brenner, Krieg, Staley and Garrity (Editors), Bergey's Manual of Systematic Bacteriology, 2nd Edition, Volume 2 (The *Proteobacteria*), Part C (The *Alpha-*, *Beta-*, *Delta-*, and *Epsilonproteobacteria*), Springer, New York, pp. 230–233.

Yakimov, M.M., L. Giuliano, R. Denaro, E. Crisafi, T.N. Chernikova, W.R. Abraham, H. Luensdorf, K.N. Timmis and P.N. Golyshin. 2004. *Thalassolituus oleivorans* gen. nov., sp. nov., a novel marine bacterium that obligately utilizes hydrocarbons. Int. J. Syst. Evol. Microbiol. 54: 141–148.

Yamada, T., Y. Sekiguchi, S. Hanada, H. Imachi, A. Ohashi, H. Harada and Y. Kamagata. 2006. *Anaerolinea thermolimosa* sp. nov., *Levilinea saccharolytica* gen. nov., sp. nov. and *Leptolinea tardivitalis* gen. nov., sp. nov., novel filamentous anaerobes, and description of the new classes *Anaerolineae* classis nov. and *Caldilineae* classis nov. in the bacterial phylum *Chloroflexi*. Int. J. Syst. Evol. Microbiol. 56: 1331–1340.

Yamada, T., H. Imachi, A. Ohashi, H. Harada, S. Hanada, Y. Kamagata and Y. Sekiguchi. 2007. *Bellilinea caldifistulae* gen. nov., sp. nov. and *Longilinea arvoryzae* gen. nov., sp. nov., strictly anaerobic, filamentous bacteria of the phylum *Chloroflexi* isolated from methanogenic propionate-degrading consortia. Int. J. Syst. Evol. Microbiol. 57: 2299–2306.

Yamamura, H., M. Hayakawa, Y. Nakagawa, T. Tamura, T. Kohno, F. Komatsu and Y. Iimura. 2005. *Nocardia takedensis* sp. nov., isolated from moat sediment and scumming activated sludge. Int. J. Syst. Evol. Microbiol. 55: 433–436.

Yamamura, H., T. Tamura, Y. Sakiyama and S. Harayama. 2007. *Nocardia amamiensis* sp. nov., isolated from a sugar-cane field in Japan. Int. J. Syst. Evol. Microbiol. 57: 1599–1602.

Yamamura, S., M. Yamashita, N. Fujimoto, M. Kuroda, M. Kashiwa, K. Sei, M. Fujita and M. Ike. 2007. *Bacillus selenatarsenatis* sp. nov., a selenate- and arsenate-reducing bacterium isolated from the effluent drain of a glass-manufacturing plant. Int. J. Syst. Evol. Microbiol. 57: 1060–1064.

APPENDIX 1

Yamashita, S., T. Uchimura and K. Komagata. 2004. Emendation of the genus *Acidomonas* Urakami, Tamaoka, Suzuki and Komagata 1989. Int. J. Syst. Evol. Microbiol. *54*: 865–870.

Yang, D.C., W.T. Im, M.K. Kim and S.T. Lee. 2005. *Pseudoxanthomonas koreensis* sp. nov. and *Pseudoxanthomonas daejeonensis* sp. nov. Int. J. Syst. Evol. Microbiol. *55*: 787–791.

Yang, D.C., W.T. Im, M.K. Kim, H. Ohta and S.T. Lee. 2006. *Sphingomonas soli* sp. nov., a beta-glucosidase-producing bacterium in the family *Sphingomonadaceae* in the alpha-4 subgroup of the *Proteobacteria*. Int. J. Syst. Evol. Microbiol. *56*: 703–707.

Yang, H.C., W.T. Im, D.S. An, W.S. Park, I.S. Kim and S.T. Lee. 2005. *Silvimonas terrae* gen. nov., sp. nov., a novel chitin-degrading facultative anaerobe belonging to the '*Betaproteobacteria*'. Int. J. Syst. Evol. Microbiol. *55*: 2329–2332.

Yang, H.C., W.T. Im, M.S. Kang, D.Y. Shin and S.T. Lee. 2006a. *Stenotrophomonas koreensis* sp. nov., isolated from compost in South Korea. Int. J. Syst. Evol. Microbiol. *56*: 81–84.

Yang, H.C., W.T. Im, K.K. Kim, D.S. An and S.T. Lee. 2006b. *Burkholderia terrae* sp. nov., isolated from a forest soil. Int. J. Syst. Evol. Microbiol. *56*: 453–457.

Yang, S.H., K.K. Kwon, H.S. Lee and S.J. Kim. 2006. *Shewanella spongiae* sp. nov., isolated from a marine sponge. Int. J. Syst. Evol. Microbiol. *56*: 2879–2882.

Yang, S.H., J.H. Lee, J.S. Ryu, C. Kato and S.J. Kim. 2007. *Shewanella donghaensis* sp. nov., a psychrophilic, piezosensitive bacterium producing high levels of polyunsaturated fatty acid, isolated from deepsea sediments. Int. J. Syst. Evol. Microbiol. *57*: 208–212.

Yang, S.J., Y.J. Choo and J.C. Cho. 2007. *Lutimonas vermicola* gen. nov., sp. nov., a member of the family *Flavobacteriaceae* isolated from the marine polychaete *Periserrula leucophryna*. Int. J. Syst. Evol. Microbiol. *57*: 1679–1684.

Yang, Y., H.L. Cui, P.J. Zhou and S.J. Liu. 2006. *Halobacterium jilantaiense* sp. nov., a halophilic archaeon isolated from a saline lake in Inner Mongolia, China. Int. J. Syst. Evol. Microbiol. *56*: 2353–2355.

Yang, Y., H.L. Cui, P.J. Zhou and S.J. Liu. 2007. *Haloarcula amylolytica* sp. nov., an extremely halophilic archaeon isolated from Aibi salt lake in Xin-Jiang, China. Int. J. Syst. Evol. Microbiol. *57*: 103–106.

Yassin, A.F. 2005. *Rhodococcus triatomae* sp. nov., isolated from a bloodsucking bug. Int. J. Syst. Evol. Microbiol. *55*: 1575–1579.

Yassin, A.F. 2007. *Corynebacterium ureicelerivorans* sp. nov., a lipophilic bacterium isolated from blood culture. Int. J. Syst. Evol. Microbiol. *57*: 1200–1203.

Yassin, A.F. and S. Brenner. 2005. *Nocardia elegans* sp. nov., a member of the *Nocardia vaccinii* clade isolated from sputum. Int. J. Syst. Evol. Microbiol. *55*: 1505–1509.

Yassin, A.F. and H. Hupfer. 2006. *Williamsia deligens* sp. nov., isolated from human blood. Int. J. Syst. Evol. Microbiol. *56*: 193–197.

Yassin, A.F. and K.P. Schaal. 2005. Reclassification of *Nocardia corynebacterioides* Serrano *et al.* 1972 (Approved Lists 1980) as *Rhodococcus corynebacterioides* comb. nov. Int. J. Syst. Evol. Microbiol. *55*: 1345–1348.

Yassin, A.F., H. Hupfer and K.P. Schaal. 2006. *Dietzia cinnamea* sp. nov., a novel species isolated from a perianal swab of a patient with a bone marrow transplant. Int. J. Syst. Evol. Microbiol. *56*: 641–645.

Yassin, A.F., W.M. Chen, H. Hupfer, C. Siering, R.M. Kroppenstedt, A.B. Arun, W.A. Lai, F.T. Shen, P.D. Rekha and C.C. Young. 2007a. *Lysobacter defluvii* sp. nov., isolated from municipal solid waste. Int. J. Syst. Evol. Microbiol. *57*: 1131–1136.

Yassin, A.F., F.T. Shen, H. Hupfer, A.B. Arun, W.A. Lai, P.D. Rekha and C.C. Young. 2007b. *Gordonia malaquae* sp. nov., isolated from sludge of a wastewater treatment plant. Int. J. Syst. Evol. Microbiol. *57*: 1065–1068.

Yassin, A.F., C.C. Young, W.A. Lai, H. Hupfer, A.B. Arun, F.T. Shen, P.D. Rekha and M.J. Ho. 2007c. *Williamsia serinedens* sp. nov., isolated from an oil-contaminated soil. Int. J. Syst. Evol. Microbiol. *57*: 558–561.

Yi, H. and J. Chun. 2004a. *Hongiella mannitolivorans* gen. nov., sp. nov., *Hongiella halophila* sp. nov. and *Hongiella ornithinivorans* sp. nov., isolated from tidal flat sediment. Int. J. Syst. Evol. Microbiol. *54*: 157–162.

Yi, H. and J. Chun. 2004b. *Nocardioides ganghwensis* sp. nov., isolated from tidal flat sediment. Int. J. Syst. Evol. Microbiol. *54*: 1295–1299.

Yi, H. and J. Chun. 2004c. *Nocardioides aestuarii* sp. nov., isolated from tidal flat sediment. Int. J. Syst. Evol. Microbiol. *54*: 2151–2154.

Yi, H. and J. Chun. 2006a. *Thalassobius aestuarii* sp. nov., isolated from tidal flat sediment. J. Microbiol. *44*: 171–176.

Yi, H. and J. Chun. 2006b. *Flavobacterium weaverense* sp. nov. and *Flavobacterium segetis* sp. nov., novel psychrophiles isolated from the Antarctic. Int. J. Syst. Evol. Microbiol. *56*: 1239–1244.

Yi, H., K.S. Bae and J. Chun. 2004a. *Thalassomonas ganghwensis* sp. nov., isolated from tidal flat sediment. Int. J. Syst. Evol. Microbiol. *54*: 377–380.

Yi, H., K.S. Bae and J. Chun. 2004b. *Aestuariibacter salexigens* gen. nov., sp. nov. and *Aestuariibacter halophilus* sp. nov., isolated from tidal flat sediment, and emended description of *Alteromonas macleodii*. Int. J. Syst. Evol. Microbiol. *54*: 571–576.

Yi, H., P. Schumann, K. Sohn and J. Chun. 2004c. *Serinicoccus marinus* gen. nov., sp. nov., a novel actinomycete with L-ornithine and L-serine in the peptidoglycan. Int. J. Syst. Evol. Microbiol. *54*: 1585–1589.

Yi, H., H.M. Oh, J.H. Lee, S.J. Kim and J. Chun. 2005a. *Flavobacterium antarcticum* sp. nov., a novel psychrotolerant bacterium isolated from the Antarctic. Int. J. Syst. Evol. Microbiol. *55*: 637–641.

Yi, H., H.I. Yoon and J. Chun. 2005b. *Sejongia antarctica* gen. nov., sp. nov. and *Sejongia jeonii* sp. nov., isolated from the Antarctic. Int. J. Syst. Evol. Microbiol. *55*: 409–416.

Yi, H., Y.W. Lim and J. Chun. 2007a. Taxonomic evaluation of the genera *Ruegeria* and *Silicibacter*: a proposal to transfer the genus *Silicibacter* Petursdottir and Kristjansson 1999 to the genus *Ruegeria* Uchino *et al.* 1999. Int. J. Syst. Evol. Microbiol. *57*: 815–819.

Yi, H., P. Schumann and J. Chun. 2007b. *Demequina aestuarii* gen. nov., sp. nov., a novel actinomycete of the suborder *Micrococcineae*, and reclassification of *Cellulomonas fermentans* Bagnara *et al.* 1985 as *Actinotalea fermentans* gen. nov., comb. nov. Int. J. Syst. Evol. Microbiol. *57*: 151–156.

Ying, J.Y., B.J. Wang, S.S. Yang and S.J. Liu. 2006. *Cyclobacterium lianum* sp. nov., a marine bacterium isolated from sediment of an oilfield in the South China Sea, and emended description of the genus *Cyclobacterium*. Int. J. Syst. Evol. Microbiol. *56*: 2927–2930.

Ying, J.Y., Z.P. Liu, B.J. Wang, X. Dai, S.S. Yang and S.J. Liu. 2007a. *Salegentibacter catena* sp. nov., isolated from sediment of the South China Sea, and emended description of the genus *Salegentibacter*. Int. J. Syst. Evol. Microbiol. *57*: 219–222.

Ying, J.Y., B.J. Wang, X. Dai, S.S. Yang, S.J. Liu and Z.P. Liu. 2007b. *Wenxinia marina* gen. nov., sp. nov., a novel member of the *Roseobacter* clade isolated from oilfield sediments of the South China Sea. Int. J. Syst. Evol. Microbiol. *57*: 1711–1716.

Yong, J.J., S.J. Park, H.J. Kim and S.K. Rhee. 2007. *Glaciecola agarilytica* sp. nov., an agar-digesting marine bacterium from the East Sea, Korea. Int. J. Syst. Evol. Microbiol. *57*: 951–953.

Yoo, S.H., H.Y. Weon, B.Y. Kim, S.B. Hong, S.W. Kwon, Y.H. Cho, S.J. Go and E. Stackebrandt. 2006. *Devosia soli* sp. nov., isolated from greenhouse soil in Korea. Int. J. Syst. Evol. Microbiol. *56*: 2689–2692.

Yoo, S.H., B.Y. Kim, H.Y. Weon, S.W. Kwon, S.J. Go and E. Stackebrandt. 2007a. *Burkholderia soli* sp. nov., isolated from soil cultivated with Korean ginseng. Int. J. Syst. Evol. Microbiol. *57*: 122–125.

Yoo, S.H., H.Y. Weon, H.B. Jang, B.Y. Kim, S.W. Kwon, S.J. Go and E. Stackebrandt. 2007b. *Sphingobacterium composti* sp. nov., isolated from cotton-waste composts. Int. J. Syst. Evol. Microbiol. *57*: 1590–1593.

Yoo, S.H., H.Y. Weon, B.Y. Kim, J.H. Kim, Y.K. Baek, S.W. Kwon, S.J. Go and E. Stackebrandt. 2007c. *Pseudoxanthomonas yeongjuensis* sp. nov., isolated from soil cultivated with Korean ginseng. Int. J. Syst. Evol. Microbiol. *57*: 646–649.

Yoon, J., S. Maneerat, F. Kawai and A. Yokota. 2006. *Myroides pelagicus* sp. nov., isolated from seawater in Thailand. Int. J. Syst. Evol. Microbiol. *56*: 1917–1920.

Yoon, J., S. Ishikawa, H. Kasai and A. Yokota. 2007a. *Perexilibacter aurantiacus* gen. nov., sp. nov., a novel member of the family '*Flammeovirgaceae*' isolated from sediment. Int. J. Syst. Evol. Microbiol. *57*: 964–968.

Yoon, J., S. Ishikawa, H. Kasai and A. Yokota. 2007b. *Persicitalea jodogahamensis* gen. nov., sp. nov., a marine bacterium of the family '*Flexibacteraceae*', isolated from seawater in Japan. Int. J. Syst. Evol. Microbiol. *57*: 1014–1017.

Yoon, J., Y. Matsuo, S. Matsuda, K. Adachi, H. Kasai and A. Yokota. 2007c. *Cerasicoccus arenae* gen. nov., sp. nov., a carotenoid-producing marine representative of the family *Puniceicoccaceae* within the phylum '*Verrucomicrobia*', isolated from marine sand. Int. J. Syst. Evol. Microbiol. *57*: 2067–2072.

Yoon, J., Y. Matsuo, S. Matsuda, K. Adachi, H. Kasai and A. Yokota. 2007d. *Rubritalea spongiae* sp. nov. and *Rubritalea tangerina* sp. nov., two carotenoid- and squalene-producing marine bacteria of the family *Verrucomicrobiaceae* within the phylum '*Verrucomicrobia*', isolated from marine animals. Int. J. Syst. Evol. Microbiol. *57*: 2337–2343.

Yoon, J., N. Oku, S. Matsuda, H. Kasai and A. Yokota. 2007e. *Pelagicoccus croceus* sp. nov., a novel marine member of the family *Puniceicoccaceae* within the phylum '*Verrucomicrobia*' isolated from seagrass. Int. J. Syst. Evol. Microbiol. *57*: 2874–2880.

Yoon, J., M. Yasumoto-Hirose, A. Katsuta, H. Sekiguchi, S. Matsuda, H. Kasai and A. Yokota. 2007f. *Coraliomargarita akajimensis* gen. nov., sp. nov., a novel member of the phylum '*Verrucomicrobia*' isolated from seawater in Japan. Int. J. Syst. Evol. Microbiol. *57*: 959–963.

Yoon, J., M. Yasumoto-Hirose, Y. Matsuo, M. Nozawa, S. Matsuda, H. Kasai and A. Yokota. 2007g. *Pelagicoccus mobilis* gen. nov., sp. nov., *Pelagicoccus albus* sp. nov. and *Pelagicoccus litoralis* sp. nov., three novel members of subdivision 4 within the phylum '*Verrucomicrobia*', isolated from seawater by in situ cultivation. Int. J. Syst. Evol. Microbiol. *57*: 1377–1385.

Yoon, J.H., K.H. Kang, T.K. Oh and Y.H. Park. 2004a. *Halobacillus locisalis* sp. nov., a halophilic bacterium isolated from a marine solar saltern of the Yellow Sea in Korea. Extremophiles *8*: 23–28.

Yoon, J.H., K.H. Kang, T.K. Oh and Y.H. Park. 2004b. *Erythrobacter aquimaris* sp. nov., isolated from sea water of a tidal flat of the Yellow Sea in Korea. Int. J. Syst. Evol. Microbiol. *54*: 1981–1985.

Yoon, J.H., K.H. Kang, T.K. Oh and Y.H. Park. 2004c. *Shewanella gaetbuli* sp. nov., a slight halophile isolated from a tidal flat in Korea. Int. J. Syst. Evol. Microbiol. *54*: 487–491.

Yoon, J.H., I.G. Kim, K.H. Kang, T.K. Oh and Y.H. Park. 2004d. *Nocardioides aquiterrae* sp. nov., isolated from groundwater in Korea. Int. J. Syst. Evol. Microbiol. *54*: 71–75.

Yoon, J.H., I.G. Kim, K.H. Kang, T.K. Oh and Y.H. Park. 2004e. *Bacillus hwajinpoensis* sp. nov. and an unnamed *Bacillus* genomospecies, novel members of *Bacillus* rRNA group 6 isolated from sea water of the East Sea and the Yellow Sea in Korea. Int. J. Syst. Evol. Microbiol. *54*: 803–808.

Yoon, J.H., I.G. Kim, T.K. Oh and Y.H. Park. 2004f. *Microbulbifer maritimus* sp. nov., isolated from an intertidal sediment from the Yellow Sea, Korea. Int. J. Syst. Evol. Microbiol. *54*: 1111–1116.

Yoon, J.H., I.G. Kim, P. Schumann, T.K. Oh and Y.H. Park. 2004g. *Marinibacillus campisalis* sp. nov., a moderate halophile isolated from a marine solar saltern in Korea, with emended description of the genus *Marinibacillus*. Int. J. Syst. Evol. Microbiol. *54*: 1317–1321.

Yoon, J.H., M.H. Lee and T.K. Oh. 2004h. *Porphyrobacter donghaensis* sp. nov., isolated from sea water of the East Sea in Korea. Int. J. Syst. Evol. Microbiol. *54*: 2231–2235.

Yoon, J.H., H.B. Lee, S.H. Yeo and J.E. Choi. 2004i. *Janibacter melonis* sp. nov., isolated from abnormally spoiled oriental melon in Korea. Int. J. Syst. Evol. Microbiol. *54*: 1975–1980.

Yoon, J.H., T.K. Oh and Y.H. Park. 2004j. *Kangiella koreensis* gen. nov., sp. nov. and *Kangiella aquimarina* sp. nov., isolated from a tidal flat of the Yellow Sea in Korea. Int. J. Syst. Evol. Microbiol. *54*: 1829–1835.

Yoon, J.H., T.K. Oh and Y.H. Park. 2004k. Transfer of *Bacillus halodenitrificans* Denariaz et al. 1989 to the genus *Virgibacillus* as *Virgibacillus halodenitrificans* comb. nov. Int. J. Syst. Evol. Microbiol. *54*: 2163–2167.

Yoon, J.H., S.H. Yeo, I.G. Kim and T.K. Oh. 2004l. *Marinobacter flavimaris* sp. nov. and *Marinobacter daepoensis* sp. nov., slightly halophilic organisms isolated from sea water of the Yellow Sea in Korea. Int. J. Syst. Evol. Microbiol. *54*: 1799–1803.

Yoon, J.H., S.H. Yeo, T.K. Oh and Y.H. Park. 2004m. *Alteromonas litorea* sp. nov., a slightly halophilic bacterium isolated from an intertidal sediment of the Yellow Sea in Korea. Int. J. Syst. Evol. Microbiol. *54*: 1197–1201.

Yoon, J.H., S.H. Yeo, I.G. Kim and T.K. Oh. 2004n. *Shewanella marisflavi* sp. nov. and *Shewanella aquimarina* sp. nov., slightly halophilic organisms isolated from sea water of the Yellow Sea in Korea. Int. J. Syst. Evol. Microbiol. *54*: 2347–2352.

Yoon, J.H., S.H. Yeo and T.K. Oh. 2004o. *Hongiella marincola* sp. nov., isolated from sea water of the East Sea in Korea. Int. J. Syst. Evol. Microbiol. *54*: 1845–1848.

Yoon, J.H. and T.K. Oh. 2005a. *Sphingopyxis flavimaris* sp. nov., isolated from sea water of the Yellow Sea in Korea. Int. J. Syst. Evol. Microbiol. *55*: 369–373.

Yoon, J.H. and T.K. Oh. 2005b. *Bacillus litoralis* sp. nov., isolated from a tidal flat of the Yellow Sea in Korea. Int. J. Syst. Evol. Microbiol. *55*: 1945–1948.

Yoon, J.H., K.H. Kang, S.H. Yeo and T.K. Oh. 2005a. *Erythrobacter luteolus* sp. nov., isolated from a tidal flat of the Yellow Sea in Korea. Int. J. Syst. Evol. Microbiol. *55*: 1167–1170.

Yoon, J.H., S.J. Kang, S.Y. Jung, C.H. Lee and T.K. Oh. 2005b. *Algoriphagus yeomjeoni* sp. nov., isolated from a marine solar saltern in the Yellow Sea, Korea. Int. J. Syst. Evol. Microbiol. *55*: 865–870.

Yoon, J.H., S.J. Kang, C.H. Lee and T.K. Oh. 2005c. *Marinicola seohaensis* gen. nov., sp. nov., isolated from sea water of the Yellow Sea, Korea. Int. J. Syst. Evol. Microbiol. *55*: 859–863.

Yoon, J.H., S.J. Kang, S.Y. Lee, M.H. Lee and T.K. Oh. 2005d. *Virgibacillus dokdonensis* sp. nov., isolated from a Korean island, Dokdo, located at the edge of the East Sea in Korea. Int. J. Syst. Evol. Microbiol. *55*: 1833–1837.

Yoon, J.H., S.J. Kang and T.K. Oh. 2005e. *Algoriphagus locisalis* sp. nov., isolated from a marine solar saltern. Int. J. Syst. Evol. Microbiol. *55*: 1635–1639.

Yoon, J.H., S.J. Kang and T.K. Oh. 2005f. *Marinomonas dokdonensis* sp. nov., isolated from sea water. Int. J. Syst. Evol. Microbiol. *55*: 2303–2307.

Yoon, J.H., S.J. Kang and T.K. Oh. 2005g. *Tenacibaculum lutimaris* sp. nov., isolated from a tidal flat in the Yellow Sea, Korea. Int. J. Syst. Evol. Microbiol. *55*: 793–798.

Yoon, J.H., S.J. Kang, C.H. Lee and T.K. Oh. 2005h. *Dokdonia donghaensis* gen. nov., sp. nov., isolated from sea water. Int. J. Syst. Evol. Microbiol. *55*: 2323–2328.

Yoon, J.H., S.J. Kang, C.H. Lee, H.W. Oh and T.K. Oh. 2005i. *Halobacillus yeomjeoni* sp. nov., isolated from a marine solar saltern in Korea. Int. J. Syst. Evol. Microbiol. *55*: 2413–2417.

Yoon, J.H., S.J. Kang, S.Y. Lee, C.H. Lee and T.K. Oh. 2005j. *Maribacter dokdonensis* sp. nov., isolated from sea water off a Korean island, Dokdo. Int. J. Syst. Evol. Microbiol. *55*: 2051–2055.

Yoon, J.H., S.J. Kang, S.H. Yeo and T.K. Oh. 2005k. *Paenibacillus alkaliterrae* sp. nov., isolated from an alkaline soil in Korea. Int. J. Syst. Evol. Microbiol. *55*: 2339–2344.

Yoon, J.H., I.G. Kim, M.H. Lee, C.H. Lee and T.K. Oh. 2005l. *Nocardioides alkalitolerans* sp. nov., isolated from an alkaline serpentinite soil in Korea. Int. J. Syst. Evol. Microbiol. *55*: 809–814.

Yoon, J.H., I.G. Kim, M.H. Lee and T.K. Oh. 2005m. *Nocardioides kribbensis* sp. nov., isolated from an alkaline soil. Int. J. Syst. Evol. Microbiol. *55*: 1611–1614.

Yoon, J.H., I.G. Kim, Y.K. Shin and Y.H. Park. 2005n. Proposal of the genus *Thermoactinomyces sensu stricto* and three new genera, *Laceyella*, *Thermoflavimicrobium* and *Seinonella*, on the basis of phenotypic, phylogenetic and chemotaxonomic analyses. Int. J. Syst. Evol. Microbiol. *55*: 395–400.

Yoon, J.H., C.H. Lee, S.J. Kang and T.K. Oh. 2005o. *Psychrobacter celer* sp. nov., isolated from sea water of the South Sea in Korea. Int. J. Syst. Evol. Microbiol. *55*: 1885–1890.

Yoon, J.H., C.H. Lee and T.K. Oh. 2005p. *Bacillus cibi* sp. nov., isolated from jeotgal, a traditional Korean fermented seafood. Int. J. Syst. Evol. Microbiol. *55*: 733–736.

Yoon, J.H., C.H. Lee and T.K. Oh. 2005q. *Aeromicrobium alkaliterrae* sp. nov., isolated from an alkaline soil, and emended description of the genus *Aeromicrobium*. Int. J. Syst. Evol. Microbiol. 55: 2171–2175.

Yoon, J.H., C.H. Lee, S.H. Yeo and T.K. Oh. 2005r. *Psychrobacter aquimaris* sp. nov. and *Psychrobacter namhaensis* sp. nov., isolated from sea water of the South Sea in Korea. Int. J. Syst. Evol. Microbiol. 55: 1007–1013.

Yoon, J.H., C.H. Lee, S.H. Yeo and T.K. Oh. 2005s. *Sphingopyxis baekryungensis* sp. nov., an orange-pigmented bacterium isolated from sea water of the Yellow Sea in Korea. Int. J. Syst. Evol. Microbiol. 55: 1223–1227.

Yoon, J.H., C.H. Lee and T.K. Oh. 2005t. *Nocardioides dubius* sp. nov., isolated from an alkaline soil. Int. J. Syst. Evol. Microbiol. 55: 2209–2212.

Yoon, J.H., J.K. Lee, Y.O. Kim and T.K. Oh. 2005u. *Photobacterium lipolyticum* sp. nov., a bacterium with lipolytic activity isolated from the Yellow Sea in Korea. Int. J. Syst. Evol. Microbiol. 55: 335–339.

Yoon, J.H., M.H. Lee, T.K. Oh and Y.H. Park. 2005v. *Muricauda flavescens* sp. nov. and *Muricauda aquimarina* sp. nov., isolated from a salt lake near Hwajinpo Beach of the East Sea in Korea, and emended description of the genus *Muricauda*. Int. J. Syst. Evol. Microbiol. 55: 1015–1019.

Yoon, J.H., T.K. Oh and Y.H. Park. 2005w. *Erythrobacter seohaensis* sp. nov. and *Erythrobacter gaetbuli* sp. nov., isolated from a tidal flat of the Yellow Sea in Korea. Int. J. Syst. Evol. Microbiol. 55: 71–75.

Yoon, J.H., S.H. Yeo, T.K. Oh and Y.H. Park. 2005x. *Psychrobacter alimentarius* sp. nov., isolated from squid jeotgal, a traditional Korean fermented seafood. Int. J. Syst. Evol. Microbiol. 55: 171–176.

Yoon, J.H., S.J. Kang, S.Y. Jung, H.W. Oh and T.K. Oh. 2006a. *Gaetbulimicrobium brevivitae* gen. nov., sp. nov., a novel member of the family *Flavobacteriaceae* isolated from a tidal flat of the Yellow Sea in Korea. Int. J. Syst. Evol. Microbiol. 56: 115–119.

Yoon, J.H., S.J. Kang, C.H. Lee and T.K. Oh. 2006b. *Donghaeana dokdonensis* gen. nov., sp. nov., isolated from sea water. Int. J. Syst. Evol. Microbiol. 56: 187–191.

Yoon, J.H., S.J. Kang, J.S. Lee and T.K. Oh. 2006c. *Brevundimonas terrae* sp. nov., isolated from an alkaline soil in Korea. Int. J. Syst. Evol. Microbiol. 56: 2915–2919.

Yoon, J.H., S.J. Kang, M.H. Lee, H.W. Oh and T.K. Oh. 2006d. *Porphyrobacter dokdonensis* sp. nov., isolated from sea water. Int. J. Syst. Evol. Microbiol. 56: 1079–1083.

Yoon, J.H., S.J. Kang, H.W. Oh, J.S. Lee and T.K. Oh. 2006e. *Brevundimonas kwangchunensis* sp. nov., isolated from an alkaline soil in Korea. Int. J. Syst. Evol. Microbiol. 56: 613–617.

Yoon, J.H., S.J. Kang, H.W. Oh and T.K. Oh. 2006f. *Stenotrophomonas dokdonensis* sp. nov., isolated from soil. Int. J. Syst. Evol. Microbiol. 56: 1363–1367.

Yoon, J.H., S.J. Kang and T.K. Oh. 2006g. *Dokdonella koreensis* gen. nov., sp. nov., isolated from soil. Int. J. Syst. Evol. Microbiol. 56: 145–150.

Yoon, J.H., S.J. Kang and T.K. Oh. 2006h. *Variovorax dokdonensis* sp. nov., isolated from soil. Int. J. Syst. Evol. Microbiol. 56: 811–814.

Yoon, J.H., S.J. Kang and T.K. Oh. 2006i. *Flavobacterium soli* sp. nov., isolated from soil. Int. J. Syst. Evol. Microbiol. 56: 997–1000.

Yoon, J.H., S.J. Kang and T.K. Oh. 2006j. *Polaribacter dokdonensis* sp. nov., isolated from seawater. Int. J. Syst. Evol. Microbiol. 56: 1251–1255.

Yoon, J.H., S.J. Kang, P. Schumann and T.K. Oh. 2006k. *Yonghaparkia alkaliphila* gen. nov., sp. nov., a novel member of the family *Microbacteriaceae* isolated from an alkaline soil. Int. J. Syst. Evol. Microbiol. 56: 2415–2420.

Yoon, J.H., S.Y. Jung, W. Kim, S.W. Nam and T.K. Oh. 2006l. *Nesterenkonia jeotgali* sp. nov., isolated from jeotgal, a traditional Korean fermented seafood. Int. J. Syst. Evol. Microbiol. 56: 2587–2592.

Yoon, J.H., C.H. Lee and T.K. Oh. 2006m. *Nocardioides lentus* sp. nov., isolated from an alkaline soil. Int. J. Syst. Evol. Microbiol. 56: 271–275.

Yoon, J.H., J.K. Lee, S.Y. Jung, J.A. Kim, H.K. Kim and T.K. Oh. 2006n. *Nocardioides kongjuensis* sp. nov., an *N*-acylhomoserine lactone-degrading bacterium. Int. J. Syst. Evol. Microbiol. 56: 1783–1787.

Yoon, J.H., M.H. Lee, S.J. Kang and T.K. Oh. 2006o. *Algoriphagus terrigena* sp. nov., isolated from soil. Int. J. Syst. Evol. Microbiol. 56: 777–780.

Yoon, J.H., M.H. Lee, S.J. Kang, S.Y. Lee and T.K. Oh. 2006p. *Sphingomonas dokdonensis* sp. nov., isolated from soil. Int. J. Syst. Evol. Microbiol. 56: 2165–2169.

Yoon, J.H., M.H. Lee, S.J. Kang, S.Y. Park and T.K. Oh. 2006q. *Pedobacter sandarakinus* sp. nov., isolated from soil. Int. J. Syst. Evol. Microbiol. 56: 1273–1277.

Yoon, J.H., P. Schumann, S.J. Kang, S.Y. Jung and T.K. Oh. 2006r. *Isoptericola dokdonensis* sp. nov., isolated from soil. Int. J. Syst. Evol. Microbiol. 56: 2893–2897.

Yoon, J.H., S.Y. Jung, Y.T. Jung and T.K. Oh. 2007i. *Idiomarina salinarum* sp. nov., isolated from a marine solar saltern in Korea. Int. J. Syst. Evol. Microbiol. 57: 2503–2506.

Yoon, J.H., S.Y. Jung, S.J. Kang and T.K. Oh. 2007ii. *Microbulbifer celer* sp. nov., isolated from a marine solar saltern of the Yellow Sea in Korea. Int. J. Syst. Evol. Microbiol. 57: 2365–2369.

Yoon, J.H., S.Y. Jung, S.J. Kang, Y.T. Jung and T.K. Oh. 2007iii. *Salegentibacter salarius* sp. nov., isolated from a marine solar saltern. Int. J. Syst. Evol. Microbiol. 57: 2738–2742.

Yoon, J.H., S.J. Kang, W. Kim and T.K. Oh. 2007iv. *Pigmentiphaga daeguensis* sp. nov., isolated from wastewater of a dye works, and emended description of the genus *Pigmentiphaga*. Int. J. Syst. Evol. Microbiol. 57: 1188–1191.

Yoon, J.H., S.J. Kang, S.Y. Jung and T.K. Oh. 2007v. *Humicoccus flavidus* gen. nov., sp. nov., isolated from soil. Int. J. Syst. Evol. Microbiol. 57: 56–59.

Yoon, J.H., S.J. Kang, Y.T. Jung and T.K. Oh. 2007vi. *Halobacillus campisalis* sp. nov., containing *meso*-diaminopimelic acid in the cell-wall peptidoglycan, and emended description of the genus *Halobacillus*. Int. J. Syst. Evol. Microbiol. 57: 2021–2025.

Yoon, J.H., S.J. Kang, C.H. Lee and T.K. Oh. 2007vii. *Nocardioides insulae* sp. nov., isolated from soil. Int. J. Syst. Evol. Microbiol. 57: 136–140.

Yoon, J.H., S.J. Kang, J.S. Lee and T.K. Oh. 2007viii. *Flavobacterium terrigena* sp. nov., isolated from soil. Int. J. Syst. Evol. Microbiol. 57: 947–950.

Yoon, J.H., S.J. Kang, J.S. Lee, H.W. Oh and T.K. Oh. 2007ix. *Brevundimonas lenta* sp. nov., isolated from soil. Int. J. Syst. Evol. Microbiol. 57: 2236–2240.

Yoon, J.H., S.J. Kang, M.H. Lee and T.K. Oh. 2007x. Description of *Sulfitobacter donghicola* sp. nov., isolated from seawater of the East Sea in Korea, transfer of *Staleya guttiformis* Labrenz *et al.* 2000 to the genus *Sulfitobacter* as *Sulfitobacter guttiformis* comb. nov. and emended description of the genus *Sulfitobacter*. Int. J. Syst. Evol. Microbiol. 57: 1788–1792.

Yoon, J.H., S.J. Kang, S.Y. Lee and T.K. Oh. 2007xi. *Phaeobacter daeponensis* sp. nov., isolated from a tidal flat of the Yellow Sea in Korea. Int. J. Syst. Evol. Microbiol. 57: 856–861.

Yoon, J.H., S.J. Kang, S.Y. Lee and T.K. Oh. 2007xii. *Loktanella maricola* sp. nov., isolated from seawater of the East Sea in Korea. Int. J. Syst. Evol. Microbiol. 57: 1799–1802.

Yoon, J.H., S.J. Kang and T.K. Oh. 2007xiii. *Donghicola eburneus* gen. nov., sp. nov., isolated from seawater of the East Sea in Korea. Int. J. Syst. Evol. Microbiol. 57: 73–76.

Yoon, J.H., S.J. Kang and T.K. Oh. 2007xiv. *Sulfitobacter marinus* sp. nov., isolated from seawater of the East Sea in Korea. Int. J. Syst. Evol. Microbiol. 57: 302–305.

Yoon, J.H., S.J. Kang and T.K. Oh. 2007xv. *Chryseobacterium daeguense* sp. nov., isolated from wastewater of a textile dye works. Int. J. Syst. Evol. Microbiol. 57: 1355–1359.

Yoon, J.H., S.J. Kang, S. Park and T.K. Oh. 2007xvi. *Caenispirillum bisanense* gen. nov., sp. nov., isolated from sludge of a dye works. Int. J. Syst. Evol. Microbiol. 57: 1217–1221.

Yoon, J.H., S.J. Kang, S. Park and T.K. Oh. 2007xvii. *Devosia insulae* sp. nov., isolated from soil, and emended description of the genus *Devosia*. Int. J. Syst. Evol. Microbiol. 57: 1310–1314.

Yoon, J.H., S.J. Kang, H.W. Oh and T.K. Oh. 2007xviii. *Pedobacter insulae* sp. nov., isolated from soil. Int. J. Syst. Evol. Microbiol. *57*: 1999–2003.

Yoon, J.H., S.J. Kang, S. Park and T.K. Oh. 2007xix. *Pedobacter lentus* sp. nov. and *Pedobacter terricola* sp. nov., isolated from soil. Int. J. Syst. Evol. Microbiol. *57*: 2089–2095.

Yoon, J.H., S.J. Kang, S. Park and T.K. Oh. 2007xx. *Jannaschia donghaensis* sp. nov., isolated from seawater of the East Sea, Korea. Int. J. Syst. Evol. Microbiol. *57*: 2132–2136.

Yoon, J.H., S.J. Kang and T.K. Oh. 2007xxi. Reclassification of *Marinococcus albus* Hao *et al.* 1985 as *Salimicrobium album* gen. nov., comb. nov. and *Bacillus halophilus* Ventosa *et al.* 1990 as *Salimicrobium halophilum* comb. nov., and description of *Salimicrobium luteum* sp. nov. Int. J. Syst. Evol. Microbiol. *57*: 2406–2411.

Yoon, J.H., S.J. Kang and T.K. Oh. 2007xxii. *Pedobacter terrae* sp. nov., isolated from soil. Int. J. Syst. Evol. Microbiol. *57*: 2462–2466.

Yoon, J.H., S.J. Kang, S.Y. Lee and T.K. Oh. 2007xxiii. *Nocardioides terrigena* sp. nov., isolated from soil. Int. J. Syst. Evol. Microbiol. *57*: 2472–2475.

Yoon, J.H., S.J. Kang, H.W. Oh and T.K. Oh. 2007xxiv. *Roseomonas terrae* sp. nov. Int. J. Syst. Evol. Microbiol. *57*: 2485–2488.

Yoon, J.H., S.J. Kang, P. Schumann and T.K. Oh. 2007xxv. *Cellulosimicrobium terreum* sp. nov., isolated from soil. Int. J. Syst. Evol. Microbiol. *57*: 2493–2497.

Yoon, J.H., S.J. Kang, S. Park, S.Y. Lee and T.K. Oh. 2007xxvi. Reclassification of *Aquaspirillum itersonii* and *Aquaspirillum peregrinum* as *Novispirillum itersonii* gen. nov., comb. nov. and *Insolitispirillum peregrinum* gen. nov., comb. nov. Int. J. Syst. Evol. Microbiol. *57*: 2830–2835.

Yoon, J.H., M.H. Lee, S.J. Kang and T.K. Oh. 2007xxvii. *Marinobacter salicampi* sp. nov., isolated from a marine solar saltern in Korea. Int. J. Syst. Evol. Microbiol. *57*: 2102–2105.

Yoon, J.H., S.Y. Lee, S.J. Kang, C.H. Lee and T.K. Oh. 2007xxviii. *Pseudoruegeria aquimaris* gen. nov., sp. nov., isolated from seawater of the East Sea in Korea. Int. J. Syst. Evol. Microbiol. *57*: 542–547.

Yoon, J.H., S.E. Park, S.J. Kang and T.K. Oh. 2007xxix. *Rheinheimera aquimaris* sp. nov., isolated from seawater of the East Sea in Korea. Int. J. Syst. Evol. Microbiol. *57*: 1386–1390.

Yoon, M.H. and W.T. Im. 2007. *Flavisolibacter ginsengiterrae* gen. nov., sp. nov. and *Flavisolibacter ginsengisoli* sp. nov., isolated from ginseng cultivating soil. Int. J. Syst. Evol. Microbiol. *57*: 1834–1839.

Yoon, M.H., L.N. Ten and W.T. Im. 2007a. *Paenibacillus ginsengarvi* sp. nov., isolated from soil from ginseng cultivation. Int. J. Syst. Evol. Microbiol. *57*: 1810–1814.

Yoon, M.H., L.N. Ten, W.T. Im and S.T. Lee. 2007b. *Pedobacter panaciterrae* sp. nov., isolated from soil in South Korea. Int. J. Syst. Evol. Microbiol. *57*: 381–386.

Yoon, M.H., L.N. Ten, W.T. Im and S.T. Lee. 2007c. *Methylibium fulvum* sp. nov., a member of the *Betaproteobacteria* isolated from ginseng field soil, and emended description of the genus *Methylibium*. Int. J. Syst. Evol. Microbiol. *57*: 2062–2066.

Young, C.C., P. Kämpfer, F.T. Shen, W.A. Lai and A.B. Arun. 2005. *Chryseobacterium formosense* sp. nov., isolated from the rhizosphere of *Lactuca sativa* L. (garden lettuce). Int. J. Syst. Evol. Microbiol. *55*: 423–426.

Young, C.C., M.J. Ho, A.B. Arun, W.M. Chen, W.A. Lai, F.T. Shen, P.D. Rekha and A.F. Yassin. 2007a. *Pseudoxanthomonas spadix* sp. nov., isolated from oil-contaminated soil. Int. J. Syst. Evol. Microbiol. *57*: 1823–1827.

Young, C.C., M.J. Ho, A.B. Arun, W.M. Chen, W.A. Lai, F.T. Shen, P.D. Rekha and A.F. Yassin. 2007b. *Sphingobium olei* sp. nov., isolated from oil-contaminated soil. Int. J. Syst. Evol. Microbiol. *57*: 2613–2617.

Young, C.C., P. Kämpfer, W.M. Chen, W.S. Yen, A.B. Arun, W.A. Lai, F.T. Shen, P.D. Rekha, K.Y. Lin and J.H. Chou. 2007c. *Luteimonas composti* sp. nov., a moderately thermophilic bacterium isolated from food waste. Int. J. Syst. Evol. Microbiol. *57*: 741–744.

Young, C.C., P. Kämpfer, M.J. Ho, H.J. Busse, B.E. Huber, A.B. Arun, F.T. Shen, W.A. Lai and P.D. Rekha. 2007d. *Arenimonas malthae* sp. nov., a

gammaproteobacterium isolated from an oil-contaminated site. Int. J. Syst. Evol. Microbiol. *57*: 2790–2793.

Young, J.M. 2004. Renaming of *Agrobacterium larrymoorei* Bouzar and Jones 2001 as *Rhizobium larrymoorei* (Bouzar and Jones 2001) comb. nov. Int. J. Syst. Evol. Microbiol. *54*: 149.

Yuan, S., P. Ren, J. Liu, Y. Xue, Y. Ma and P. Zhou. 2007. *Lentibacillus halodurans* sp. nov., a moderately halophilic bacterium isolated from a salt lake in Xin-Jiang, China. Int. J. Syst. Evol. Microbiol. *57*: 485–488.

Yukphan, P., W. Potacharoen, S. Tanasupawat, M. Tanticharoen and Y. Yamada. 2004a. *Asaia krungthepensis* sp. nov., an acetic acid bacterium in the α-*Proteobacteria*. Int. J. Syst. Evol. Microbiol. *54*: 313–316.

Yukphan, P., M. Takahashi, W. Potacharoen, S. Tanasupawat, Y. Nakagawa, M. Tanticharoen and Y. Yamada. 2004b. *Gluconobacter albidus ex* Kondo and Ameyama 1958 sp. nov., nom. rev., an acetic acid bacterium in the α-*Proteobacteria*. J. Gen. Appl. Microbiol. *50*: 235–242.

Yukphan, P., T. Malimas, W. Potacharoen, S. Tanasupawat, M. Tanticharoen and Y. Yamada. 2005c. *Neoasaia chiangmaiensis* gen. nov., sp. nov., a novel osmotolerant acetic acid bacterium in the α-*Proteobacteria*. J. Gen. Appl. Microbiol. *51*: 301–311.

Yumoto, I., K. Hirota, T. Kawahara, Y. Nodasaka, H. Okuyama, H. Matsuyama, Y. Yokota, K. Nakajima and T. Hoshino. 2004a. *Anoxybacillus voinovskiensis* sp. nov., a moderately thermophilic bacterium from a hot spring in Kamchatka. Int. J. Syst. Evol. Microbiol. *54*: 1239–1242.

Yumoto, I., K. Hirota, Y. Nodasaka, Y. Yokota, T. Hoshino and K. Nakajima. 2004b. *Alkalibacterium psychrotolerans* sp. nov., a psychrotolerant obligate alkaliphile that reduces an indigo dye. Int. J. Syst. Evol. Microbiol. *54*: 2379–2383.

Yumoto, I., K. Hirota, S. Yamaga, Y. Nodasaka, T. Kawasaki, H. Matsuyama and K. Nakajima. 2004c. *Bacillus asahii* sp. nov., a novel bacterium isolated from soil with the ability to deodorize the bad smell generated from short-chain fatty acids. Int. J. Syst. Evol. Microbiol. *54*: 1997–2001.

Yumoto, I., M. Hishinuma-Narisawa, K. Hirota, T. Shingyo, F. Takebe, Y. Nodasaka, H. Matsuyama and I. Hara. 2004d. *Exiguobacterium oxidotolerans* sp. nov., a novel alkaliphile exhibiting high catalase activity. Int. J. Syst. Evol. Microbiol. *54*: 2013–2017.

Yumoto, I., K. Hirota, T. Goto, Y. Nodasaka and K. Nakajima. 2005a. *Bacillus oshimensis* sp. nov., a moderately halophilic, non-motile alkaliphile. Int. J. Syst. Evol. Microbiol. *55*: 907–911.

Yumoto, I., K. Hirota, Y. Nodasaka and K. Nakajima. 2005b. *Oceanobacillus oncorhynchi* sp. nov., a halotolerant obligate alkaliphile isolated from the skin of a rainbow trout (*Oncorhynchus mykiss*), and emended description of the genus *Oceanobacillus*. Int. J. Syst. Evol. Microbiol. *55*: 1521–1524.

Zabel, H.P., H. König and J. Winter. 1985. Emended description of *Methanogenium thermophilicum*, Rivard and Smith, and assignment of new isolates to this species. Syst. Appl. Microbiol. *6*: 72–78.

Zavarzina, D.G., T.V. Kolganova, E.S. Boulygina, N.A. Kostrikina, T.P. Tourova and G.A. Zavarzin. 2006. *Geoalkalibacter ferrihydriticus* gen. nov. sp. nov., the first alkaliphilic representative of the family *Geobacteraceae*, isolated from a soda lake. Mikrobiologiia. *75*: 775–785 (in Russian); English translation: Microbiology. *75*: 673–682.

Zavarzina, D.G., T.G. Sokolova, T.P. Tourova, N.A. Chernyh, N.A. Kostrikina and E.A. Bonch-Osmolovskaya. 2007. *Thermincola ferriacetica* sp. nov., a new anaerobic, thermophilic, facultatively chemolithoautotrophic bacterium capable of dissimilatory Fe(III) reduction. Extremophiles *11*: 1–7.

Zhang, B., H. Tong and X. Dong. 2005. *Pediococcus cellicola* sp. nov., a novel lactic acid coccus isolated from a distilled-spirit-fermenting cellar. Int. J. Syst. Evol. Microbiol. *55*: 2167–2170.

Zhang, C., X. Liu and X. Dong. 2004. *Syntrophomonas curvata* sp. nov., an anaerobe that degrades fatty acids in co-culture with methanogens. Int. J. Syst. Evol. Microbiol. *54*: 969–973.

Zhang, C., X. Liu and X. Dong. 2005. *Syntrophomonas erecta* sp. nov., a novel anaerobe that syntrophically degrades short-chain fatty acids. Int. J. Syst. Evol. Microbiol. *55*: 799–803.

Zhang, D.C., H.X. Wang, H.C. Liu, X.Z. Dong and P.J. Zhou. 2006a. *Flavobacterium glaciei* sp. nov., a novel psychrophilic bacterium isolated from the China No.1 glacier. Int. J. Syst. Evol. Microbiol. *56*: 2921–2925.

Zhang, D.C., Y. Yu, B. Chen, H.X. Wang, H.C. Liu, X.Z. Dong and P.J. Zhou. 2006b. *Glaciecola psychrophila* sp. nov., a novel psychrophilic bacterium isolated from the Arctic. Int. J. Syst. Evol. Microbiol. *56*: 2867–2869.

Zhang, D.C., H.X. Wang, H.L. Cui, Y. Yang, H.C. Liu, X.Z. Dong and P.J. Zhou. 2007. *Cryobacterium psychrotolerans* sp. nov., a novel psychrotolerant bacterium isolated from the China No. 1 glacier. Int. J. Syst. Evol. Microbiol. *57*: 866–869.

Zhang, G., G. Zeng, X. Cai, S. Deng, H. Luo and G. Sun. 2007. *Brachybacterium zhongshanense* sp. nov., a cellulose-decomposing bacterium from sediment along the Qijiang River, Zhongshan City, China. Int. J. Syst. Evol. Microbiol. *57*: 2519–2524.

Zhang, H., W. Zheng, J. Huang, H. Luo, Y. Jin, W. Zhang, Z. Liu and Y. Huang. 2006. *Actinoalloteichus hymeniacidonis* sp. nov., an actinomycete isolated from the marine sponge *Hymeniacidon perleve*. Int. J. Syst. Evol. Microbiol. *56*: 2309–2312.

Zhang, J., Z. Liu and M. Goodfellow. 2004. *Nocardia xishanensis* sp. nov., a novel actinomycete isolated from soil. Int. J. Syst. Evol. Microbiol. *54*: 2301–2305.

Zhang, J., Q. Xie, Z. Liu and M. Goodfellow. 2007. *Lechevalieria fradiae* sp. nov., a novel actinomycete isolated from soil in China. Int. J. Syst. Evol. Microbiol. *57*: 832–836.

Zhang, L., Z. Xu and B.K. Patel. 2007a. *Bacillus decisifrondis* sp. nov., isolated from soil underlying decaying leaf foliage. Int. J. Syst. Evol. Microbiol. *57*: 974–978.

Zhang, L., Z. Xu and B.K. Patel. 2007b. *Frondicola australicus* gen. nov., sp. nov., isolated from decaying leaf litter from a pine forest. Int. J. Syst. Evol. Microbiol. *57*: 1177–1182.

Zhang, L.P., C.L. Jiang and W.X. Chen. 2005. *Streptosporangium yunnanense* sp. nov. and *Streptosporangium purpuratum* sp. nov., from soil in China. Int. J. Syst. Evol. Microbiol. *55*: 719–724.

Zhang, Q., C. Liu, Y. Tang, G. Zhou, P. Shen, C. Fang and A. Yokota. 2007. *Hymenobacter xinjiangensis* sp. nov., a radiation-resistant bacterium isolated from the desert of Xinjiang, China. Int. J. Syst. Evol. Microbiol. *57*: 1752–1756.

Zhang, T., X. Fan, S. Hanada, Y. Kamagata and H.H. Fang. 2006. *Bacillus macauensis* sp. nov., a long-chain bacterium isolated from a drinking water supply. Int. J. Syst. Evol. Microbiol. *56*: 349–353.

Zhang, Y.Q., Y.G. Chen, W.J. Li, X.P. Tian, L.H. Xu and C.L. Jiang. 2005a. *Sphingomonas yunnanensis* sp. nov., a novel gram-negative bacterium from a contaminated plate. Int. J. Syst. Evol. Microbiol. *55*: 2361–2364.

Zhang, Y.Q., W.J. Li, R.M. Kroppenstedt, C.J. Kim, G.Z. Chen, D.J. Park, L.H. Xu and C.L. Jiang. 2005b. *Rhodococcus yunnanensis* sp. nov., a mesophilic actinobacterium isolated from forest soil. Int. J. Syst. Evol. Microbiol. *55*: 1133–1137.

Zhang, Y.Q., P. Schumann, W.J. Li, G.Z. Chen, X.P. Tian, E. Stackebrandt, L.H. Xu and C.L. Jiang. 2005c. *Isoptericola halotolerans* sp. nov., a novel actinobacterium isolated from saline soil from Qinghai Province, north-west China. Int. J. Syst. Evol. Microbiol. *55*: 1867–1870.

Zhang, Y.Q., W.J. Li, K.Y. Zhang, X.P. Tian, Y. Jiang, L.H. Xu, C.L. Jiang and R. Lai. 2006. *Massilia dura* sp. nov., *Massilia albidiflava* sp. nov., *Massilia plicata* sp. nov. and *Massilia lutea* sp. nov., isolated from soils in China. Int. J. Syst. Evol. Microbiol. *56*: 459–463.

Zhang, Y.Q., P. Schumann, L.Y. Yu, H.Y. Liu, Y.Q. Zhang, L.H. Xu, E. Stackebrandt, C.L. Jiang and W.J. Li. 2007a. *Zhihengliuella halotolerans* gen. nov., sp. nov., a novel member of the family *Micrococcaceae*. Int. J. Syst. Evol. Microbiol. *57*: 1018–1023.

Zhang, Y.Q., C.H. Sun, W.J. Li, L.Y. Yu, J.Q. Zhou, Y.Q. Zhang, L.H. Xu and C.L. Jiang. 2007b. *Deinococcus yunweiensis* sp. nov., a gamma- and UV-radiation-resistant bacterium from China. Int. J. Syst. Evol. Microbiol. *57*: 370–375.

Zhang, Y.Q., L.Y. Yu, H.Y. Liu, Y.Q. Zhang, L.H. Xu and W.J. Li. 2007c. *Salinicoccus luteus* sp. nov., isolated from a desert soil. Int. J. Syst. Evol. Microbiol. *57*: 1901–1905.

Zhang, Y.X., C. Dong and S. Biao. 2007. *Planifilum yunnanense* sp. nov., a thermophilic thermoactinomycete isolated from a hot spring. Int. J. Syst. Evol. Microbiol. *57*: 1851–1854.

Zhao, J.S., D. Manno, C. Beaulieu, L. Paquet and J. Hawari. 2005. *Shewanella sediminis* sp. nov., a novel Na+-requiring and hexahydro-1,3,5-trinitro-1,3,5-triazine-degrading bacterium from marine sediment. Int. J. Syst. Evol. Microbiol. *55*: 1511–1520.

Zhao, J.S., D. Manno, C. Leggiadro, D. O'Neil and J. Hawari. 2006. *Shewanella halifaxensis* sp. nov., a novel obligately respiratory and denitrifying psychrophile. Int. J. Syst. Evol. Microbiol. *56*: 205–212.

Zhao, J.S., D. Manno, S. Thiboutot, G. Ampleman and J. Hawari. 2007. *Shewanella canadensis* sp. nov. and *Shewanella atlantica* sp. nov., manganese dioxide- and hexahydro-1,3,5-trinitro-1,3,5-triazine-reducing, psychrophilic marine bacteria. Int. J. Syst. Evol. Microbiol. *57*: 2155–2162.

Zhilina, T.N., R. Appel, C. Probian, E. Llobet Brossa, J. Harder, F. Widdel and G.A. Zavarzin. 2004. *Alkaliflexus imshenetskii* gen. nov. sp. nov., a new alkaliphilic gliding carbohydrate-fermenting bacterium with propionate formation from a soda lake. Arch. Microbiol. *182*: 244–253.

Zhilina, T.N., V.V. Kevbrin, T.P. Tourova, A.M. Lysenko, N.A. Kostrikina and G.A. Zavarzin. 2005. *Clostridium alkalicellum* sp. nov., an obligately alkaliphilic cellulolytic bacterium from a soda lake in the Baikal region. Mikrobiologiia. *74*: 642–653 (in Russian); English translation: Microbiology. *74*: 557–566.

Zhilina, T.N., D.G. Zavarzina, J. Kuever, A.M. Lysenko and G.A. Zavarzin. 2005. *Desulfonatronum cooperativum* sp. nov., a novel hydrogenotrophic, alkaliphilic, sulfate-reducing bacterium, from a syntrophic culture growing on acetate. Int. J. Syst. Evol. Microbiol. *55*: 1001–1006.

Zhou, Y., X. Wang, H. Liu, K.Y. Zhang, Y.Q. Zhang, R. Lai and W.J. Li. 2007. *Pontibacter akesuensis* sp. nov., isolated from a desert soil in China. Int. J. Syst. Evol. Microbiol. *57*: 321–325.

Zhou, Y., J. Dong, X. Wang, X. Huang, K.Y. Zhang, Y.Q. Zhang, Y.F. Guo, R. Lai and W.J. Li. 2007. *Chryseobacterium flavum* sp. nov., isolated from polluted soil. Int. J. Syst. Evol. Microbiol. *57*: 1765–1769.

Zhu, H.H., J. Guo, Q. Yao, S.Z. Yang, M.R. Deng, T.B. Phuong le, V.T. Hanh and M.J. Ryan. 2007. *Streptomyces vietnamensis* sp. nov., a streptomycete with violet blue diffusible pigment isolated from soil in Vietnam. Int. J. Syst. Evol. Microbiol. *57*: 1770–1774.

Zitouni, A., L. Lamari, H. Boudjella, B. Badji, N. Sabaou, A. Gaouar, F. Mathieu, A. Lebrihi and D.P. Labeda. 2004. *Saccharothrix algeriensis* sp. nov., isolated from Saharan soil. Int. J. Syst. Evol. Microbiol. *54*: 1377–1381.

Zurdo-Piñeiro, J.L., R. Rivas, M.E. Trujillo, N. Vizcaíno, J.A. Carrasco, M. Chamber, A. Palomares, P.F. Mateos, E. Martínez-Molina and E. Velázquez. 2007. *Ochrobactrum cytisi* sp. nov., isolated from nodules of *Cytisus scoparius* in Spain. Int. J. Syst. Evol. Microbiol. *57*: 784–788.

Index of scientific names of *Archaea* and *Bacteria*

Key to the fonts and symbols used in this index:

Nomenclature
    Lower case, Roman   Genera, species, and subspecies of bacteria. Every bacterial name mentioned in the text is listed in the index. Specific epithets are listed individually and also under the genus.*

    CAPITALS, ROMAN: Names of taxa higher than genus (tribes, families, orders, classes, divisions, kingdoms).

Pagination
    Roman: Pages on which taxa are mentioned.

    Boldface: Indicates page on which the description of a taxon is given.†

---

* Infrasubspecific names, such as serovars, biovars, and pathovars, are not listed in the index.
† A description may not necessarily be given in the *Manual* for a taxon that is considered as *incertae sedis* or that is listed in an addendum or note added in proof; however, the page on which the complete citation of such a taxon is given is indicated in boldface type.

Index of scientific names of *Archaea* and *Bacteria*

ISBN-13: 978-0-387-95041-9
ISBN-10: 0-387-95041-9

EAN

9 780387 950419 >